JANAF Thermochemical Tables
Third Edition
Part I, Al–Co

Journal of

Physical and Chemical Reference Data

David R. Lide, Jr., Editor

The Journal of Physical and Chemical Reference Data (ISSN 0047-2689) is published quarterly by the American Chemical Society (1155 16th St., N. W., Washington, DC 20036-9976) and the American Institute of Physics (335 E. 45th St., New York, NY 10017-3483) for the National Bureau of Standards. Second-class postage paid at Washington, DC and additional mailing offices. POSTMASTER: Send address changes to Membership and Subscription Services, P. O. Box 3337, Columbus, Ohio 43210.

The objective of the Journal is to provide critically evaluated physical and chemical property data, fully documented as to the original sources and the criteria used for evaluation. Critical reviews of measurement techniques, whose aim is to assess the accuracy of available data in a given technical area, are also included. The Journal is not intended as a publication outlet for original experimental measurements such as are normally reported in the primary research literature, nor for review articles of a descriptive or primarily theoretical nature.

Supplements to the Journal are published at irregular intervals and are not included in subscriptions to the Journal. They contain compilations which are too lengthy for a journal format.

The Editor welcomes appropriate manuscripts for consideration by the Editorial Board. Potential contributors who are interested in preparing a compilation are invited to submit an outline of the nature and scope of the proposed compilation, with criteria for evaluation of the data and other pertinent factors, to:

David R. Lide, Jr., Editor
J. Phys. Chem. Ref. Data
National Bureau of Standards
Gaithersburg, MD 20899

One source of contributions to the Journal is The National Standard Reference Data System (NSRDS), which was established in 1963 as a means of coordinating on a national scale the production and dissemination of critically evaluated reference data in the physical sciences. Under the Standard Reference Data Act (Public Law 90-396) the National Bureau of Standards of the U.S. Department of Commerce has the primary responsibility in the Federal Government for providing reliable scientific and technical reference data. The Office of Standard Reference Data of NBS coordinates a complex of data evaluation centers, located in university, industrial, and other Government laboratories as well as within the National Bureau of Standards, which are engaged in the compilation and critical evaluation of numerical data on physical and chemical properties retrieved from the world scientific literature. The participants in this NBS-sponsored program, together with similar groups under private or other Government support which are pursuing the same ends, comprise the National Standard Reference Data System.

The primary focus of the NSRDS is on well-defined physical and chemical properties of well-characterized materials or systems. An effort is made to assess the accuracy of data reported in the primary research literature and to prepare compilations of critically evaluated data which will serve as reliable and convenient reference sources for the scientific and technical community.

Information for Contributors

Manuscripts submitted for publication must be prepared in accordance with *Instructions for Preparation of Manuscripts for the Journal of Physical and Chemical Reference Data,* available on request from the Editor.

New and renewal subscriptions should be sent with payment to the Office of the Co troller at the American Chemical Society, 1155 Sixteenth Street, N.W., Washington, D 20036-9976. **Address changes**, with at least six weeks advance notice, should be se to Manager, Membership and Subscription Services, American Chemical Society, P. Box 3337, Columbus, OH 43210. Changes of address must include both old and ne addresses and ZIP codes and, if possible, the address label from the mailing wrapper o recent issue. Claims for missing numbers will not be allowed: if loss was due to failure the change-of-address notice to be received in the time specified; if claim is dated (North America: more than 90 days beyond issue date, (b) all other foreign: more than o year beyond issue date.

Members of AIP member and affiliate societies requesting member subscription rat should direct subscriptions, renewals, and address changes to American Institute Physics, Dept. S/F, 335 E. 45th St., NY 10017-3483.

Subscription Prices (1986) *(not including supplements)*	U.S.A.	Foreign (surface mail)	Optional air freight Europe Mideast N. Africa	Asia and Oceania
Members (of ACS, AIP, or affiliated society)	$ 55.00	$ 65.00	$ 80.00	$ 90.00
Regular rate	$220.00	$230.00	$245.00	$255.00

Rates above do not apply to nonmember subscribers in Japan, who must enter subscrip tion orders with Maruzen Company Ltd., 3-10 Nihonbashi 2-chome, Chuo-ku Tokyo 103, Japan. Tel: (03) 272-7211.

Back numbers are available at a cost of $65 per single copy and $250 per volume.

Orders for reprints, supplements, and back numbers should be addressed to th American Chemical Society, 1155 Sixteenth Street, N. W., Washington, DC 20036-997 Prices for reprints and supplements are listed at the end of this issue.

Microfilm subscriptions of the *Journal of Physical and Chemical Reference Data* a available on 16 mm and 35 mm. This journal also appears in Sec. I of *Current Phys Microform* (CPM) along with 26 other journals published by the American Institute Physics and its member societies. A *Microfilm Catalog* is available on request.

Journal of
Physical and Chemical Reference Data

Volume 14, 1985
Supplement No. 1

JANAF Thermochemical Tables
Third Edition
Part I, Al–Co

M. W. Chase, Jr., C. A. Davies, J. R. Downey, Jr.,
D. J. Frurip, R. A. McDonald, and A. N. Syverud

Thermal Group, 1707 Building,
Dow Chemical U.S.A.,
Midland, Michigan 48674, U.S.A.

Published by the **American Chemical Society**
and the **American Institute of Physics** for
the **National Bureau of Standards**

Library of Congress Catalog Card Number 85-71387

International Standard Book Number
0-88318-473-7

American Institute of Physics, Inc.
335 East 45th Street
New York, New York 10017

Printed in the United States of America

Foreword

The goal of this journal is to provide reliable reference data on physical and chemical properties to the technical community. Of the almost 300 titles we have published in the last 14 years, the four supplements to the *JANAF Thermochemical Tables*, which appeared in 1974, 1975, 1978, and 1982, have probably been the most widely distributed and used. The reasons for the long-standing popularity of the *JANAF Thermochemical Tables* are apparent: an important topic that affects many scientific and engineering areas; a highly professional and critical approach to the selection of the data; and a concise, easy-to-use format. It is difficult to find another reference source of physical data that has achieved the JANAF level of success in understanding users' needs and satisfying them.

The *Journal of Physical and Chemical Reference Data* is therefore very pleased to publish this Third Edition of the *JANAF Thermochemical Tables*. It combines the four supplements mentioned above with the revised and updated material from the Second Edition (which appeared in 1971) and a number of previously unpublished tables. Thus the tables for thermodynamic properties of about 1800 substances have been brought together into a single publication. In addition, all tables have been expressed in SI units and the notation has been made consistent with current international recommendations.

The appearance of the Third Edition comes a short time after the death of Charles W. Beckett, who was for many years an active member of the JANAF Thermochemical Working Group. Charles Beckett made a major contribution to shaping the initial plans for the JANAF Tables some 25 years ago, and his tireless efforts helped assure the continuing high standards of the project. The Editor believes he speaks for the entire community of JANAF users in recognizing our debt of gratitude.

David R. Lide, Jr., Editor
Journal of Physical and Chemical Reference Data

Preface to the First Edition

Beginning in the mid-1950s, when elements other than the conventional carbon, hydrogen, oxygen, nitrogen, chlorine, and fluorine came into consideration as rocket propellant ingredients, formidable difficulties were encountered in conducting rigorous theoretical performance calculations for these new propellants. The first major problem was calculational technique. The second was the lack of accurate thermodynamic data.

By the end of 1959, the calculational technique problem had been substantially resolved by applying the method of minimization of free energy to large, high-speed digital computers. At this point the calculations became as accurate as the thermodynamic data upon which they were based. However, serious gaps were present in the available data. For propellant ingredients, only the standard heat of formation is required to conduct a performance calculation. However, this must be known to a high degree of accuracy. For combustion products, the enthalpy and entropy must be known, as a function of temperature, in addition to the standard heat of formation.

In order to resolve the problem, a substantial experimental thermodynamic research program was initiated under the sponsorship and technical direction of Project PRINCIPIA of the Advanced Research Projects Agency. Simultaneously, a project was initiated to critically evaluate and compile consistent tables of thermodynamic properties of propellant combustion products for use by the aerospace industry. This project, known as the "JANAF Thermochemical Tables," was undertaken by the Dow Chemical Company. Since the objective of the project was to have one single source of "best available data" prepared for use by the entire industry, the JANAF Thermochemical Panel undertook the task of furnishing a critical review of the Tables prior to their publication and distribution. This approach was designed to ensure that the Tables be of the highest possible quality.

Washington, D.C.
July 1964

C. V. Mock
Advanced Research Projects Agency

Preface to the Second Edition

It is appropriate to call attention to some of the reasons for the phenomenal success of the *JANAF Thermochemical Tables* in achieving, first, the initial limited objective of providing the standard data for the chemical rocket propulsion industry, and later, upon publication, worldwide recognition as thermodynamic reference data of the highest quality and timeliness.

First, and most obvious, there was the selection and continued support of a highly competent evaluation team, themselves engaged in a broad spectrum of thermodynamic research. The personnel of the Thermal Research Laboratory of the Dow Chemical Company, under the direction of Dr. D. R. Stull and Dr. H. Prophet, have filled this role to a degree of excellence not likely to be exceeded anywhere. Moreover, the group has heroically remained productive in spite of many battles to retain continuing support, and the actual sharp reduction of funding over the past two years to a less-than-viable level.

A second important factor is the unusual approach to format, evaluation, and distribution of the Tables, as it has been followed since their inception. The primary distribution is in frequently issued loose-leaf supplements. Each previously issued table may thus be revised as often as necessary to take account of improved data. Each loose-leaf table is accompanied on its reverse side by a complete explanation of the selection of the key data, together with all references.

The third vital distinction of these Tables has been the existence of a continuing cognizant working group composed of technological users of data, thermodynamicists, and government sponsors of both research and development. Independent prepublication review of the Tables has been an important contribution of some of the members of this group; but its annual technical meetings have resulted in even more far-reaching benefits. Together, the users and generators of data have been able to establish realistic priorities for the species to be included in the Tables; at the same time the course of experimental research has been guided by the demonstration of absence or inadequacy of needed data. Although the working group no longer enjoys official recognition, the members and participants have enthusiastically volunteered to continue meeting in this important work.

Special words of appreciation are due to Dr. Charles W. Beckett, who has lined up all the technical presentations at the last six annual meetings of the working group; and to Mr. Curtis C. Selph who has served with wisdom and insight as the Air Force project monitor for the *JANAF Thermochemical Tables* contracts.

Arlington, Virginia
October 1970

Joseph F. Masi
Air Force Office of Scientific Research

Preface to the Second Edition

[illegible]

Preface to the Third Edition

The United States Air Force (USAF) for more than 20 years has supported, and continues to support, the preparation of the *JANAF Thermochemical Tables*. The tables, funded through Air Force Office of Scientific Research (AFOSR), were intended originally for calculating performance of thermochemical reactors, such as rocket engines. The computation of such performance figures as thrust and exhaust temperature for a rocket require data such as those in these tables.

The Department of Energy (DOE) [and predecessor agencies Energy Research and Development Administration (ERDA), Office of Coal Research, and Bureau of Mines] became interested in the *JANAF Thermochemical Tables* because the tables had become *the* benchmark for thermochemical data for performance calculations. However, reactors and reagents of interest to DOE are different from those of the USAF.

The Department of Energy needs performance calculations for several types of reactors that are of interest in fossil fuel research. These reactors include air pollution control equipment, automotive internal combustion engines, coal gasifiers and furnaces, fuel cells, liquefaction reactors and their catalyst structures, and magnetohydrodynamic generators. For example, researchers might wish to calculate the first-law efficiency of an electrical power plant, or the exhaust concentration of a pollutant such as sulfur dioxide or mercury.

The calculations needed for DOE interests have as their basis the same mathematics and physical chemistry as required for the performance calculations of USAF interest, but DOE needs tables for more and different chemicals. For example, a rocket fuel probably would not be formulated to include silicon or sulfur, but all known coals contain both these elements. Silicon and sulfur are two distinguishing elements accounting for many peculiarities of coal utilization chemistry (which include the chemistries of all the polysulfides, thiosulfates, alums, etc., and of the many glasses and mixed silicates).

For more than six years (1975–1982), Dow Chemical has worked on separate but complementary contracts to satisfy the thermochemical table needs of USAF and DOE. This Third Edition of the *JANAF Thermochemical Tables* presents the results of these efforts, along with reformatted versions of previous tables, to provide an extensive set of tables of wide utility that are consistent with each other and with the requirements of thermodynamic theory.

There remain open questions about calculation methods for equilibrium problems. I mention this because the foreword for the First Edition endorses "free-energy-minimization" as if it solved all problems of numerical instability. I would remove this apparent endorsement as confusing, inappropriate, and uninformative for users of these tables. For example, the paper of White, Johnson, and Dantzig [J. Chem. Phys. **28**, 751 (1958)] does not address numerical stability in any way. Removing this endorsement would not diminish Dow's JANAF work.

I want to thank several people. All the Dow personnel did outstanding work, and several fellow federal employees helped with orchestrating the various contracts to nourish this finished product. In AFOSR, Joseph Masi, Robert Sperlein, and Leonard Caveny were always encouraging. Within the Bureau of Mines, then in ERDA, and finally in DOE, I was helped and encouraged by Daniel Bienstock, Irving Wender, Jim Hendrie, Alex Mills, Kermit Woodcock, and Mike Hogan.

Pittsburgh, Pennsylvania
April 1984

Francis E. Spencer, Jr.
Pittsburgh Energy Technology Center
Department of Energy

JANAF Thermochemical Tables
Third Edition

M. W. Chase, Jr.,[a)] C. A. Davies, J. R. Downey, Jr., D. J. Frurip, R. A. McDonald, and A. N. Syverud

Thermal Group, 1707 Building, Dow Chemical U.S.A., Midland, Michigan 48674, U.S.A.

Recommended temperature-dependent values are provided for chemical thermodynamic properties of inorganic substances and for organic substances containing only one or two carbon atoms. These tables cover the thermodynamic properties over a wide temperature range with single-phase and multiphase tables for the crystal, liquid, and ideal gas state. The properties tabulated are heat capacity, entropy, Gibbs energy function, enthalpy, enthalpy of formation, Gibbs energy of formation, and the logarithm of the equilibrium constant for formation of each compound from the elements in their standard reference states. All values are given in SI units and are for a standard-state pressure of 100 000 Pa (1 bar). Each tabulation is accompanied by a critical evaluation of the literature upon which the thermochemical table is based. Literature references are given. This volume is a new collective edition of five previous publications. In it all tabulations have been rewritten in a consistent style. Many, but not all, tabulations have been revised as a result of a re-evaluation of the data.

Key words: critically evaluated data; enthalpy; enthalpy of formation; entropy; equilibrium constant of formation; Gibbs energy function; Gibbs energy of formation; heat capacity; thermochemical tables.

[a)] Address correspondence to Malcolm W. Chase, Jr., Ph.D., Room A158, Chemistry Building, National Bureau of Standards, Gaithersburg, MD 20899.

Contents

The Table of Contents is a cumulative listing of Parts I and II.
Part I contains pp. I–X and 1–926 (Tables for Al–Co), and
Part II contains pp. 927–1856 (Tables for Cr–Zr).

1. Introduction

The Third Edition to the *JANAF Thermochemical Tables* has been assembled under the sponsorship of the U.S. Department of Energy. It contains tables of recommended temperature-dependent values for the standard enthalpy (heat) of formation, Gibbs (free) energy of formation, the logarithm of the equilibrium constant of formation, the heat capacity, entropy, enthalpy, and Gibbs energy function for 46 elements and many of their compounds. This publication is a current collection of all tables issued through 31 December 1984 under contracts with the U.S. Department of Energy and the U.S. Air Force Office of Scientific Research. It supersedes the Second Edition[1] and the four supplemental updates published in the *Journal of Physical and Chemical Reference Data*.[2–5]

Numerous new and revised thermochemical tables are included in this collection. However, this Third Edition is primarily a rewriting and a recalculation of all the tables; no attempt has been made to reanalyze the data for all tables. The rewritten tables adhere more closely to the current IUPAC recommendations on symbols and notation. The recalculated tables are all based on the current IUPAC and CODATA recommendations for relative molecular masses and fundamental constants. As a result, a comparison of a table in this Third Edition and its previously published form (i.e., same revision date) will reveal differences; however, these result from the adjustments mentioned above rather than from a reanalysis of the data.

This procedure ensures that all thermal functions will be calculated using the same auxiliary data. Following a similar procedure for the formation properties has introduced slight inconsistencies in those tables for which the formation properties were derived from the previous thermal functions.

This Third Edition is the third major U.S. compendium on thermochemical properties to be published in SI units with a standard-state pressure of 1 bar (0.1 MPa). The other two publications originated with the U.S. Bureau of Standards[6] and the U.S. Geological Survey.[7]

2. History of the JANAF Thermochemical Tables

Between 1955 and 1958, severe difficulties were encountered by individuals attempting to conduct rigorous performance calculations for propellant systems that gave multiphase combustion products characterized by complex chemical and thermal equilibria. Several of these individuals approached the Armed Services requesting that a group be assembled to assess the validity of the calculation methods and thermochemical data which were being employed at that time.

In January 1958, the Armed Services jointly instructed the Solid Propellant Information Agency to organize the Joint Army–Navy–Air Force Ad Hoc Panel on Performance Calculation Methods and Thermodynamic Data. This panel, which consisted of 38 representatives of military facilities, defense contractors, and research organizations, terminated its operations in June 1959, with a recommendation that future activities under its purview could more appropriately be handled by a smaller working group. An additional recommendation included in the panel's final report was that this working group initiate the establishment of a thermochemical data compilation, evaluation, and dissemination program utilizing the available personnel and facilities of the Dow Chemical Company.

On 1 September 1959, the Joint Army–Navy–Air Force Thermochemical Panel was formed under the sponsorship of the Bureau of Naval Weapons, Department of the Navy; Office, Chief of Ordinance, Department of the Army; Air Research and Development Command, Department of the Air Force; and the Advanced Research Projects Agency, Department of Defense. The panel operated in accordance with the Rules of Operations of Solid Propellant Panels as adopted by representatives of the above offices on 1 September 1959.

The JANAF Thermochemical Panel Membership, which consisted originally of approximately 15 individuals with special experience in the technological subject under panel purview, met at the Pentagon on 16 November 1959; at which time, plans for the thermochemical compilation project were reviewed and the project formally initiated.

Urgency required that as large a set of consistent tables be assembled as quickly as possible. Computer programs had to be developed, and at first it was not possible to adequately assess the input information for every table. When the original data were evaluated, the table was printed on white paper; otherwise the table was printed on gray paper. By the end of 1960, the first set was ready for distribution to some 1000 qualified recipients. For a number of years, at the end of each quarter a supplement was issued which contained additional tables and revised tables. Some of the gray tables were revised to white tables.

The distribution of quarterly supplements continued through 31 December 1967, at which time the distribution shifted to semiannual supplements. Simultaneously, the tables were prepared under the auspices of the Interagency Chemical Rocket Propulsion Group, Working Group on Thermochemistry. On 31 December 1970, the working group became the Ad Hoc Working Group on Thermochemistry. By 30 June 1971, the tables project was sponsored by the U. S. Air Force Office of Scientific Research (AFOSR) and prepared under the advice and assistance of a thermochemistry group, referred to as reviewers. In late 1976, the U.S. Energy Research and Development Administration (now reorganized to be the U.S. Department of Energy) joined in sponsoring the tables. The distribution reverted to a quarterly schedule with an expanded list of reviewers. AFOSR and DOE have jointly funded this activity until late 1982. Since that time, AFOSR has been the sole sponsor.

In summary, since 1967, the *JANAF Thermochemical Tables* have been sponsored by the U.S. Air Force Office of Scientific Research and the U.S. Department of Energy under the following contracts:

Air Force Contracts	Department of Energy Contracts
F04611-67-C-0009	DE AC22-76ET10637
F04611-70-C-0028	DE AC22-81PC41514
F44620-70-C-0104	
F44620-75-C-0048	
F49620-82-C-0016	

In addition, there was an interagency transfer of funds, from AFOSR to DOE, in the period 1976–1981.

In 1970 the output of the JANAF project was incorporated into the National Standard Reference Data System, a national program on evaluated physical and chemical data which is coordinated by the National Bureau of Standards. Since that time the *JANAF Thermochemical Tables* have been disseminated as NSRDS publications.

2.1. Members of the JANAF Thermochemical Panel (1959–1961) and the JANAF Thermochemical Working Group (1961–1964)

T. O. Dobbins, Advanced Research Projects Agency (Past Chairman)
W. H. Jones, Institute for Defense Analyses; Aerospace Corporation (Past Chairman)
W. G. May, Institute for Defense Analyses; Esso Research and Engineering Company (Past Chairman)
C. W. Beckett, National Bureau of Standards (Vice Chairman)
G. W. Avery, Chemical Propulsion Information Agency (Past Secretary)
B. K. Farris, Chemical Propulsion Information Agency (Past Secretary)
T. L. Reedy, Chemical Propulsion Information Agency (Secretary)
B. J. Alley, Army Materiel Command
P. W. Bender, University of Wisconsin
W. A. Bernett, Bureau of Naval Weapons
B. Brown, Hercules Powder Company
C. L. Funk, Jet Propulsion Laboratory
J. S. Gordon, Thiokol Chemical Corporation
L. J. Gordon, Aerojet-General Corporation
S. A. Greene, North American Aviation
C. B. Henderson, Atlantic Research Corporation
D. L. Hildenbrand, Philco Corporation
J. L. Margrave, University of Wisconsin
J. P. McCullough, U.S. Bureau of Mines
P. L. Nichols, Jr., Jet Propulsion Laboratory
C. C. Selph, Air Force Systems Command
D. R. Stull, Dow Chemical Company

The Thermochemical Working Group of the Interagency Chemical Rocket Propulsion Group (ICRPG) has consisted of the following at various times during its existence (1964–1969).

Members
- J. F. Masi, Air Force Office of Scientific Research (Chairman)
- C. W. Beckett, National Bureau of Standards (Vice Chairman)
- T. O. Dobbins, Advanced Research Projects Agency
- S. Gordon, National Aeronautics and Space Administration
- R. Jackel, Office of Naval Research
- J. Murrin, Naval Ordnance Systems Command
- R. Odom, Army Materiel Command
- C. C. Selph, Air Force Rocket Propulsion Laboratory
- D. Squire, Army Research Office

Participants
- G. S. Bahn, Marquardt Corporation
- B. Brown, Hercules, Inc.
- H. F. Calcote, AeroChem Research Laboratories
- J. P. Coughlin, Aerojet-General Corporation
- J. S. Gordon, Atlantic Research Corporation
- D. R. Douslin, U.S. Bureau of Mines
- M. Farber, Space Sciences, Inc.
- R. M. Fristrom, Applied Physics Laboratory
- C. B. Henderson, Atlantic Research Corporation
- D. L. Hildenbrand, Douglas Aircraft Company
- J. L. Margrave, Rice University
- W. G. May, Esso Research and Engineering Company
- W. Mitchell, Thiokol Chemical Corporation
- C. F. Robillard, Jet Propulsion Laboratory
- L. Schieler, Aerospace Corporation
- C. H. Shomate, Naval Weapons Center
- D. R. Stull, Dow Chemical Company
- M. Zimmer, Naval Ordnance Station

Secretariat (Chemical Propulsion Information Agency)
- T. Gilliland
- M. McCormack
- T. L. Reedy

In the period 1969–1984, various scientists donated their time to critique the loose-leaf supplements to the JANAF Thermochemical Tables prior to their distribution. Their efforts certainly increased the quality of the tables. As of January 1982 the reviewers of the JANAF Thermochemical Tables were:

DOE Reviewers
- Francis E. Spencer, Jr., DOE/PETC Combustion Technical Division
- Joseph W. Martin, DOE/METC, Coal Projects Management Division
- Thomas C. Ehlert, Marquette University
- R. Howald, Montana State University
- Fred C. Fehsenfeld, NOAA Environmental Research Laboratory
- Gerd Rosenblatt, Los Alamos National Laboratory
- Martin Steinberg, University of California, Santa Barbara
- C. B. Alcock, University of Toronto
- J. Brian Pedley, University of Sussex
- Malcolm H. Rand, AERE Harwell

AFOSR Reviewers
- David White, University of Pennsylvania
- W. L. Worrell, University of Pennsylvania
- Stanley Abramowitz, U.S. National Bureau of Standards
- David A. Ditmars, U.S. National Bureau of Standards

H. M. Rosenstock, U.S. National Bureau of Standards
Leonard Caveny, AFOSR/NA, Bolling AFB
John L. Haas, Jr., U.S. Geological Survey
John S. Gordon, TRW Energy Systems
Joseph F. Masi, AFOSR (retired)
Sanford Gordon, NASA Lewis Research Center
Edgar F. Westrum, Jr., University of Michigan
William D. Good, DOE/BETC
R. H. Hauge, Rice University
John L. Margrave, Rice University
Milton Farber, Space Sciences, Inc.
Curtis C. Selph, AFRPL/LKDH
Donald L. Hildenbrand, SRI International
Leo Brewer, University of California
James P. Coughlin, Aerojet Tactical Systems Co.
N. A. Gokcen, U.S. Bureau of Mines, AMRC

2.2. Project Personnel

The Tables began under the direction of Daniel R. Stull in 1959. Major contributions in the first two years were made by Thomas E. Dergazarian, Samuel Levine, and Louis A. DuPlessis. In 1969 Harold Prophet succeeded Daniel R. Stull as project director and continued in that position until his untimely death in late 1972. Malcolm W. Chase has been the project director since late 1972.

The following professional personnel of the Dow Chemical Company have been involved in the preparation of the JANAF Thermochemical Tables:

Project Director, D. R. Stull	1959–1969
Project Director, H. Prophet	1969–1972
Project Director, M. W. Chase	1972–1984
P. A. Andreozzi	1980–83
J. Chao	1961–69
J. L. Curnutt	1969–79
T. E. Dergazarian	1959–63
C. A. Davies	1980–84
J. R. Downey, Jr.	1976–84
L. A. DuPlessis	1960–61
D. J. Frurip	1983–84
S. T. Hadden	1962
A. T. Hu	1965–70
B. H. Justice	1964
G. C. Karris	1965–68
S. Levine	1960–61
R. A. McDonald	1971–84
F. L. Oetting	1961
R. S. Orehotsky	1963–64
R. V. Petrella	1961–62
E. W. Phillips	1963–66
J. A. Rizos	1962–63
G. C. Sinke	1962,65,69,70
A. C. Swanson	1962–64
A. N. Syverud	1963–84
H. K. Unger	1963–64
E. A. Valenzuela	1975–81
L. C. Walker	1971–73
D. U. Webb	1967–68
S. K. Wollert	1963–65

We also wish to mention the valuable assistance in the typing of these Tables of Norma Dumont, 1959–64; Viola E. Harrington, 1964–66; Carol S. Scheffler, 1966–67; Wildene B. Harris, 1967–68; Mary J. Walter, 1968–71; and for the time period of 1971–78 of Joan Weldon, Cheri Snow, Barbara Boman, Sheila Knoerr, Patricia Grochowski, and Lisa Ittner.Since 1978 Betty Clark, Jan Crouch, and Rhoda Toth have typed the tables.Isabel Carr provided valuable services to the group in abstracting, searching and ordering documents, and proofreading the Tables for the Second Edition. For the period since 1978 Dana Donley, Debbie Simpson, Marge Sheets, Barbara Meier, Diana Scribner, and Anne Schmidt have provided similar assistance. Milton D. Marks has been very instrumental in all computer aspects of this project for the past few years.

3. Notation and Terminology

The symbols and terminology for physicochemical quantities and units are those recommended by IUPAC through its Physical Chemistry Division.[8,9] For the thermodynamic notation needed but not specified by these two sources, the recommendations of the Bulletin of Chemical Thermodynamics[10] are used. Similarly, for spectroscopic nomenclature, the common practice of Moore[11] and Herzberg[12–15] is followed.

3.1. Definition of the Standard State

This description of the standard state is an abbreviated version of that given by the IUPAC Physical Chemistry Division.[9]

Absolute values of some thermodynamic quantities are unknown. Only changes in values caused by changes in parameters such as temperature and pressure can be determined. It is therefore important to define a base line for substances, to which the effect of such variations may be referred. The standard state is such a base line. The properties of these standard states are indicated by use of the symbol °.

For a pure substance the concept of standard state applies to the substance in a well-defined state of aggregation at a well-defined but arbitrarily chosen standard pressure.

Historically, the defined pressure for the standard state, i.e., the standard-state pressure, has been one standard atmosphere (101 325 Pa) and most existing data use this pressure. With the growing use of SI units, continued use of the atmosphere is inconvenient. IUPAC has recommended that the thermodynamic data should be reported for a defined standard-state pressure of 100 000 Pa. The standard-state pressure in general is symbolized as $p°$. Previously in all JANAF thermochemical publications, $p°$ was taken as 1 atm. In the current set of JANAF Thermochemical Tables $p°$ is taken as 100 000 Pa (1 bar). It should be understood that the present change in the standard-state pressure carries no implication for "standard pressures" used in other contexts, e.g., the convention that "normal boiling points" refer to a pressure of 101 325 Pa (1 atm).

The change of the customary value of $p°$ from 1 atm (101 325 Pa) to 1 bar (100 000 Pa) results in small alterations in standard values of some thermodynamic quantities for all substances. For condensed phases the magnitude of these alterations is nearly always negligible in comparison with the uncertainty of current data. However, the alterations in values of entropy and the Gibbs energy function for a pure gaseous substance and of the Gibbs energy of reaction for a reaction involving gaseous species may not be negligible. This will be discussed in detail later.

Hence, in all the tables appearing in this publication the standard state for a pure gaseous substance is that of the substance as an (hypothetical) ideal gas at 1 bar; the standard state for a pure liquid substance is that of the pure liquid under a pressure of 1 bar; the standard state for a pure solid substance is that of the pure crystalline substance under a pressure of 1 bar.

Since an enthalpy measurement is made as a difference between an initial and a final state, the reference temperature for enthalpy must be arbitrarily chosen. It is taken to be 298.15 K in all tables. Refer to the discussion of types of thermochemical tables for further details.

3.2. Symbols

States of Aggregation

The symbols and their corresponding descriptions for the states of aggregation are exactly those given by IUPAC.[9] They are rewritten here for the convenience of the reader.

Single letters are used to denote the three basic states of aggregation, gas, liquid, and solid, while combinations of letters are used for more subtle descriptions of states.

g	gas or a vapor
ℓ	liquid
s	solid
cd	condensed phase (i.e., either the solid or the liquid state)
fl	fluid (i.e., either the gaseous or the liquid state)
cr	crystalline solid; where polymorphism occurs, it may be necessary to augment the symbol cr with a descriptor for the crystal modification under discussion; the preferred descriptors are Roman numerals, with textual definition of the crystallographic significance of the numerals used
am	amorphous solid
vit	vitreous substance (a glass)
mon	monomeric form
pol	polymeric form (in many cases the monomeric or polymeric character of the entity will be clear from the context without the symbol and the symbol should be used only in cases where ambiguity might result)
sln	solution; in many contexts it will be clear whether a liquid solution or a solid solution is meant, but where this is unclear it must be made clear by supplemental notation
aq	solution in which water is the solvent (an aqueous solution); in the past this symbol has sometimes been used to denote an infinitely dilute aqueous solution, but infinite dilution should hencefoward be denoted by the extra symbol ∞.
sat	saturated; the state of equilibrium between phases, whether of a pure substance or a system of more than one component.

Processes

The symbols and their corresponding descriptions for 11 processes are exactly those given by IUPAC.[9] The remaining six processes are denoted by the symbols recommended by the Bulletin of Chemical Thermodynamics.[10] The use of special symbols to denote a process are as follows:

vap	vaporization (evaporation) of a liquid
sub	sublimation (evaporation) of a solid
fus	melting (fusion) of a solid
trs	transition of one solid phase to another
mix	the mixing of fluids
sol	the process of solution (dissolution)
r	chemical reaction in general
c	combustion reaction; also can denote critical properties
f	a reaction in which a compound is formed from its elements (formation)
at	a process in which a substance is separated into its constituent gaseous atoms (atomization)
dcm	a process in which a substance decomposes (not given by IUPAC)
dso	a process in which a substance dissociates (not given by IUPAC)
dim	a process in which a substance dimerizes (not given by IUPAC)
dil	the dilution of a solution
gt	glass transition (not given by IUPAC)
hyd	hydrolysis (not given by IUPAC)
ion	ionization (not given by IUPAC)

Fundamental Constants

c	speed of light in vacuum
h	Planck constant
N_A	Avogadro constant
e	elementary charge
F	Faraday constant
R	gas constant
T_0	absolute temperature of "ice point"
k	Boltzmann constant
c_2	second radiation constant

Units

cal	calorie
eV	electron volt
g-atom	a mole of atoms
J	joule
K	kelvin
kcal	kilocalorie
kJ	kilojoule
mol	mole
L	liter

Mathematical Quantities

ln	logarithm to the base e
log	logarithm to the base 10

Spectroscopic Quantities

The spectroscopic symbols, use, and description follow the well-established practices of Moore[11] and Herzberg.[12–15] These sources provide more extensive discussion as to the meaning and use of the various spectroscopic quantities.

The measured quantities often are reported in the literature for a specific isotopic molecule. The thermochemical tables in this publication normally use spectroscopic information for a natural isotopic abundance "molecule." Care should be exercised in noting to which molecule the spectroscopic data pertains.

Å	Angstrom
AP	appearance potential
B_0, B_e	rotational constant
D	centrifugal distortion constant
D_0^0	dissociation energy is defined as the energy of the ground-state products relative to the lowest existing level of the molecule; $D_0^0 = \Delta_{at} H°(0\ K)$.
EA	electron affinity
g_i	quantum (or statistical) weight of the *i*th electronic state; quantity is the product of the spin multiplicity and state degeneracy of the electronic level under consideration
I_A, I_B, I_C	principal moments of inertia of a molecule
IP	ionization potential
k	Hooke's law force constant
n	number of potential maxima in an internal rotation; also refers to principal quantum number in describing energy levels of atoms
N	number of atoms in molecule
r_e	internuclear distance (for the "equilibrium" structure of the molecule)
V_0	potential barrier to internal rotation
α_e	first-order rotation–vibration interaction constant
ϵ_i	*i*th electronic energy level
σ	symmetry number
μ	reduced mass of the molecule
ν	observed vibrational fundamental
ω_e	vibrational fundamental for infinitesimal amplitude
$\omega_e x_e$, $\omega_e y_e$	vibrational anharmonicity constants

Thermodynamic and Other Quantities

A_r	relative atomic mass (atomic weight)
C_p	molar heat capacity at constant pressure
C_s	molar heat capacity at saturation
ΔU	change in internal energy
ΔG	change in Gibbs energy, $G = H - TS = U + PV - TS$
ΔH	change in enthalpy, $H = U + PV$
K	equilibrium constant
M_r	relative molecular mass (molecular weight)
p	pressure
T	temperature, in kelvin
t	temperature, in degrees Celsius (°C)
V	volume
Δ	indicates the increment in a given property for a given process or reaction, taken as the value for the final state (or sum for the products) less that for the initial state (or sum for reactants).
$\Delta_f H°$	represents the standard enthalpy of formation, which is the increment in enthalpy associated with the reaction of forming the given compound from its elements in their reference states, with each substance in its thermodynamic standard state at the given temperature.

Experimental Method / Techniques

dsc	differential scanning calorimetry
dta	differential thermal analysis
emf	electromotive force
esr	electron spin resonance
glc	gas–liquid chromatography
gpc	gel permeation chromatography
ir	infrared spectroscopy
nmr	nuclear magnetic resonance
tga	thermogravimetric analysis
LLE	liquid–liquid equilibria
VLE	vapor–liquid equilibria
VP/xxxx	vapor pressure/experimental technique
/ebul	= ebulliometric
/Knud	= Knudsen effusion
/stat	= static
/Kems	= Knudsen effusion–mass spectrometry
/Kete	= Knudsen effusion–torsion effusion
/tran	= transpiration
/Lang	= Langmuir free evaporation
/Lams	= Langmuir-mass spectrometry

3.3. Relative Atomic Masses and Natural Isotopic Composition of the Elements

The Table of Atomic Weights of the Elements 1981[16] is reprinted below. Added to this list are the relative atomic masses for the electron (e^-) and deuterium (D). The relative atomic mass for the electron is given by Cohen and Taylor.[17]

The 1981 report on atomic weights[18] includes a complete review of the natural isotopic composition of the elements and also tabulates the relative atomic masses for selected radioisotopes. This information is required for the conversion of spectroscopic data from that corresponding to specific isotopes to the naturally occurring isotopic abundance.

As in the IUPAC report, values in parentheses are used for radioactive elements whose relative masses cannot be quoted exactly without knowledge of the origin of the elements; the value given is the atomic mass number of the isotope of that element of longest known half life. Alternatively, the relative atomic mass of the isotope with the longest known half life or that of a commonly used isotope could be listed, since the relative atomic masses of the individual isotopes are accurately known.

Table of relative atomic masses of the elements, 1981
[Pure Appl. Chem. **55**, 1101 (1983), alphabetical order, based on the relative atomic mass, $A_r(^{12}C) = 12$]

Name	Symbol	Atomic number	Relative atomic mass
Actinium	Ac	89	227.027 8
Aluminum	Al	13	26.981 54
Americium	Am	95	(243)
Antimony (Stibium)	Sb	51	121.75
Argon	Ar	18	39.948
Arsenic	As	33	74.921 6
Astatine	At	85	(210)
Barium	Ba	56	137.33
Berkelium	Bk	97	(247)
Beryllium	Be	4	9.012 18
Bismuth	Bi	83	208.980 4
Boron	B	5	10.81
Bromine	Br	35	79.904
Cadmium	Cd	48	112.41
Caesium	Cs	55	132.905 4
Calcium	Ca	20	40.08
Californium	Cf	98	(251)
Carbon	C	6	12.011
Cerium	Ce	58	140.12
Chlorine	Cl	17	35.453
Chromium	Cr	24	51.996
Cobalt	Co	27	58.933 2
Copper	Cu	29	63.546
Curium	Cm	96	(247)
Deuterium	D	1	2.014 102
Dysprosium	Dy	66	162.50
Einsteinium	Es	99	(252)
Electron	e^-		0.000 548 58
Erbium	Er	68	167.26
Europium	Eu	63	151.96
Fermium	Fm	100	(257)
Fluorine	F	9	18.998 403
Francium	Fr	87	(223)
Gadolinium	Gd	64	157.25
Gallium	Ga	31	69.72
Germanium	Ge	32	72.59
Gold	Au	79	196.966 5
Hafnium	Hf	72	178.49
Helium	He	2	4.002 60
Holmium	Ho	67	164.930 4
Hydrogen	H	1	1.007 94
Indium	In	49	114.82
Iodine	I	53	126.904 5
Iridium	Ir	77	192.22
Iron	Fe	26	55.847
Krypton	Kr	36	83.80
Lanthanum	La	57	138.905 5
Lawrencium	Lr	103	(260)
Lead	Pb	82	207.2
Lithium	Li	3	6.941
Lutetium	Lu	71	174.967
Magnesium	Mg	12	24.305
Manganese	Mn	25	54.938 0
Medelevium	Md	101	(258)
Mercury	Hg	80	200.59
Molybdenum	Mo	42	95.94
Neodymium	Nd	60	144.24
Neon	Ne	10	20.179
Neptunium	Np	93	237.048 2
Nickel	Ni	28	58.69
Niobium	Nb	41	92.906 4
Nitrogen	N	7	14.006 7
Nobelium	No	102	(259)
Osmium	Os	76	190.2
Oxygen	O	8	15.999 4
Palladium	Pd	46	106.42
Phosphorus	P	15	30.973 76
Platinum	Pt	78	195.08
Plutonium	Pu	94	(244)
Polonium	Po	84	(209)
Potassium (Kalium)	K	19	39.098 3
Praseodymium	Pr	59	140.907 7
Promethium	Pm	61	(145)
Protactinium	Pa	91	231.035 9
Radium	Ra	88	226.025 4
Radon	Rn	86	(222)
Rhenium	Re	75	186.207
Rhodium	Rh	45	102.905 5
Rubidium	Rb	37	85.467 8
Ruthenium	Ru	44	101.07
Samarium	Sm	62	150.36
Scandium	Sc	21	44.955 9
Selenium	Se	34	78.96
Silicon	Si	14	28.085 5
Silver	Ag	47	107.868 2
Sodium (Natrium)	Na	11	22.989 77
Strontium	Sr	38	87.62
Sulfur	S	16	32.06
Tantalum	Ta	73	180.947 9
Technetium	Tc	43	(98)
Tellurium	Te	52	127.60
Terbium	Tb	65	158.925 4
Thallium	Tl	81	204.383
Thorium	Th	90	232.038 1
Thulium	Tm	69	168.934 2
Tin	Sn	50	118.69
Titanium	Ti	22	47.88
Tungsten (Wolfram)	W	74	183.85
(Unnilhexium)	(Unh)	106	(263)
(Unnilpentium)	(Unp)	105	(262)
(Unnilquadium)	(Unq)	104	(261)
Uranium	U	92	238.028 9
Vanadium	V	23	50.941 5
Xenon	Xe	54	131.29
Ytterbium	Yb	70	173.04
Yttrium	Y	39	88.905 9
Zinc	Zn	30	65.38
Zirconium	Zr	40	91.22

3.4. Fundamental Constants and Conversion Factors

The numerical values of the fundamental constants used in our calculations are those obtained by Cohen and Taylor[17] based on a least-squares adjustment of data existing in 1973. These values have been adopted by CODATA.[17]

Fundamental constants

Quantity	Symbol	Value[a]
Speed of light in vacuum	c	299 792 458(1.2) m s^{-1}
Planck constant	h	6.626 176(36)$\times 10^{-34}$ J s
	$h/2\pi$	1.054 588 7(57)$\times 10^{-34}$ J s
Avogadro constant	N_A	6.022 045(31)$\times 10^{23}$ mol^{-1}
Elementary charge	e	1.602 189 2(46)$\times 10^{-19}$ C
Faraday constant, $N_A e$	F	9.648 456(27)$\times 10^{4}$ C mol^{-1}
Absolute temperature of "ice point," 0 °C	T_0	273.150 0 K
Molar gas constant	R	8.314 41(26) J mol^{-1} K^{-1}
Boltzmann constant, RN_A^{-1}	k	1.380 662(44)$\times 10^{-23}$ J K^{-1}
Second radiation constant, hck^{-1}	c_2	1.438 786(45)$\times 10^{-2}$ m K

[a] Numbers in parentheses are the one standard deviation uncertainties in the last digits of the quoted value.

Conversion factors for energy-related units. 1973 CODATA constants

	kJ mol^{-1}	kcal mol^{-1}	eV (per molecule)	cm^{-1}	MHz
1 kJ mol^{-1}	1	0.239 005 7	0.010 364 35	83.593 47	2.506 069$\times 10^{6}$
1 kcal mol^{-1}	4.184	1	0.043 364 45	349.755 1	1.048 539$\times 10^{7}$
1 eV (per molecule)	96.484 6	23.060 36	1	8065.479	2.417 970$\times 10$
1 MHz	3.990 313$\times 10^{-7}$	9.537 077$\times 10^{-8}$	4.135 701$\times 10^{-9}$	3.335 641$\times 10^{-5}$	1
1 cm^{-1}	0.011 962 66	2.859 144$\times 10^{-3}$	1.239 852$\times 10^{-4}$	1	2.997 924$\times 10^{4}$
1 K (degree)	8.314 41$\times 10^{-3}$	1.987 19$\times 10^{-3}$	8.617 349$\times 10^{-5}$	0.695 030 4[a]	2.083 649$\times 10^{4}$
1 hartree	2625.500	627.509 5	27.211 606	2.194 746$\times 10^{5}$	6.579 684$\times 10^{9}$
1 rydberg	1312.750	313.754 7	13.605 803	1.097 373$\times 10^{5}$	3.289 842$\times 10^{9}$

[a] This value is inverse of the second radiation constant c_2; $(hc/k)^{-1} = c_2^{-1} = 0.695\ 030\ 4$.

3.5. Temperature Scale

Since the JANAF Thermochemical Tables deal with properties that vary with temperature, the temperature scale is of significance. The International Temperature Scales are devised to approximate, as closely as possible, the absolute thermodynamic (kelvin) temperature suggested by W. Thomson (Lord Kelvin) in 1854. The International Temperature Scale is established by the Comité International des Poids et Mesures (International Committee of Weights and Measures).

There have been three international temperature scales: The International Temperature Scale of 1927,[19] The International Temperature Scale of 1948 which was renamed the International Practical Temperature Scale of 1948 (IPTS-48) in 1960,[20,21] and The International Practical Temperature Scale of 1968 (IPTS-68).[22]

Our concern in the preparation of the JANAF Thermochemical Tables is to adjust any experimental measurements to the thermodynamic scale.In practice, this means converting to the IPTS-68 scale since it is the currently accepted closest approximation to the thermodynamic scale.

To facilitate temperature adjustments, the article by Corruccini[20] is used for the 1927 to 1948 conversions and the article by Douglas[23] for the 1948 to 1968 conversion. Guildner and Edsinger[24] included a plot showing the deviation of IPTS-68 from the thermodynamic temperatures from 273.16 to 730 K.

Ideal-gas tables of thermodynamic properties derived from statistical mechanics are based on the thermodynamic temperatures (as well as on the values of the physical constants used) and are hence independent of any practical temperature scale. The enthalpy of formation, Gibbs energy of formation, and logarithm of the equilibrium constant might depend on temperature-adjusted data.

For condensed phase tables, the temperature scale conversion is often not made, since the correction would be small compared to the uncertainty in the data.

4. Reference States and Conversions

There are three types of tables included in this publication: single-phase, multiphase, and reference state tables. The tables are clearly labeled as to the type at the top of each page and by the state designations associated with the chemical formula. The reference state can involve a single phase or multiphases. The major difference between a multiphase table and a single-phase table is the existence of a discontinuity in each function (or in its first derivative) at a first-order transition temperature.

4.1. Reference State

A related set of thermodynamic property tables requires that there be a reference table of the thermodynamic properties for each element to which all other forms of that element or any compound involving that element may be referred. If the temperature range of interest (0–6000 K herein) can be represented by a single phase for that material (say hydrogen) the tabulated values in the reference table will be smooth and regular. If the temperature range of interest includes more than one phase (say magnesium), there will be a solid phase (from 0 to 923 K, the melting point), a liquid phase (923 to 1378 K, where the vapor pressure reaches 1 bar), and a gas phase (1378 to 6000 K), and the tabulated values in the reference table will be discontinuous at these temperatures or phase boundaries. Practical usage dictates that in so far as possible the phase most stable at 1 bar pressure be selected. This practice is followed in these tables, and does lead to discontinuities in the thermodynamic functions. Attention is called to these discontinuities in the reference state tables by insertion of a comment as to the type of transition at the temperature of the phase transition.

Since these tables deal with ideal gases, the pressure is really the fugacity. Calculating a boiling point as the temperature at which $\Delta_{vap} G^\circ = 0$ for the process $Mg(\ell) \rightarrow Mg(g)$ corresponds to an (ideal) fugacity of 1 bar. This will differ from the real (observed) boiling point at 1 bar which is also different from the real (observed) boiling point at 1 atm. Each of these "three boiling points" corresponds to a different $\Delta_{vap} G^\circ$ value.

With the use of f and °, as in $\Delta_f H^\circ$, the implication is that both the compound in question and its constituent elements are in standard states and that the elements, moreover, are in their reference states; for any given temperature the reference states of the elements will normally be those that are stable at the chosen standard-state pressure and at that temperature. A resulting feature of tabulations of $\Delta_f H^\circ$ and $\Delta_f G^\circ$ as functions of temperature for compounds is that discontinuous changes are sometimes to be seen; these correspond to changes in the stable reference states of the elements, as phase-transition temperatures are passed. Thus, values of $\Delta_f H^\circ$ (SO_2Cl_2,g) would show discontinuous changes at three temperatures corresponding to the transitions S(cr,I)→S(cr,II), S(cr,II)→S(ℓ), and S(ℓ)→1/2 S_2(g), where I refers to rhombic and II to monoclinic crystal forms.

4.2. Single-Phase and Multiphase Tables

A multiphase table contains the data for two or more phases of a pure substance. Each phase is limited to the temperature range in which it is more stable (i.e., has a lower $\Delta_f G^\circ$) than the other phases in the table. There are discontinuities in the properties C_p°, S°, and H° at the temperatures where each phase transforms ($\Delta G^\circ = 0$) to the next phase. Alternatively, the properties of a pure substance can be given in a series of tables where each table is devoted to a single phase. Single-phase tables do not have discontinuities in the properties C_p°, S°, and H° at transition temperatures. Moreover, the properties are extrapolated beyond the transition temperatures into regions where that phase is less stable than its other phases. These characteristics of continuity and extrapolation of properties lead directly to the advantages and problems of single-phase tables. The enthalpy of formation and Gibbs energy of formation may have discontinuities, even in single-phase tables, corresponding to transitions in the reference states.

Single-phase tables are more convenient than multiphase tables in certain computer applications. Interpolation and approximation (curve fitting) apply directly to the continuous properties in a single-phase table. Multiphase tables must first be divided into segments using the stored values of the transition temperatures; then interpolation and approximation must be limited to the temperatures in the pertinent segment.

Single-phase tables are essential in uses that involve unstable phases. This includes phases that persist in a metastable region or those that are unstable at all temperatures. The real world abounds in such cases. Single-phase tables are often applied to solid–solid or solid–liquid equilibrium in mixtures, where an unstable phase is "stabilized" in the presence of other components. Analysis of such equilibria often requires liquid or solid properties that are many hundreds of degrees into the metastable region of each pure phase. Single-phase tables are designed to supply these properties, but their preparation is not trivial.

The JANAF Thermochemical Tables often include estimated glass transitions in the liquid tables. The glass temperature is usually assumed to be ~0.7 times the melting temperature. The heat capacity of the glass is usually assumed to be equal to that of the stable crystalline phase, since existing glass data show a decrease in C_p° to values near those of the crystal. Such tables are, in fact, not single-phase tables but two-phase glass–liquid tables.

The glass transition is of practical concern when thermochemical tables are used near the glass temperature of any of the pure liquids. Although the ratio of the glass temperature to the melting temperature can vary from perhaps 0.35 to >0.8, it often is ~0.7. Temperatures this low are not uncommon when liquid tables are used in the analysis of binary or multicomponent phase diagrams.

It is important to ask whether glass transitions should be included in liquid tables intended for such use. Estimated glass transitions are included for two reasons: the transition is usually observed in supercooled liquids and its inclusion avoids an entropy paradox. The first reason presumes that glass transitions do occur but cannot be observed in liquids that readily crystallize. The more compelling reason is the avoidance of the paradox in which the entropy of the supercooled liquid (a disordered state) becomes *less* than that of the stable crystalline state.

The choice remains as to what temperature and what

phase should be used as the reference point for enthalpy in the extrapolated single-phase tables. Since 298.15 K was adopted as the reference temperature for multiphase and reference state tables, there is strong incentive for retention of that temperature in single-phase tables. The choice has been made to have the reference at the same arbitrary point in each phase. Thus, the JANAF Thermochemical Tables use a reference temperature of 298.15 K in the table for the high-temperature phase even when that phase is not stable at 298.15 K.

Enthalpies at all temperatures depend on the arbitrary extrapolations needed to relate the observed data to the reference temperature. Each single-phase table has a different reference $\Delta_f H^\circ$ dependent on an arbitrary extrapolation of enthalpy.

For the reference temperature of 298.15 K, the enthalpy at 0 K is always negative. This means that the function $-(G^\circ - H^\circ)/T$ approaches positive infinity at 0 K and is therefore difficult to interpolate near 0 K.

4.3. Conversion to SI Units and the Standard-State Pressure

In all previous JANAF Thermochemical Tables, the standard-state pressure was one atmosphere (101 325 Pa) and the unit of energy was the thermochemical calorie (4.184 J). For this publication, the standard-state pressure is changed to one bar (100 000 Pa) and the energy unit to the joule. The values from previous JANAF tabulations have been converted as described below. This information is provided not only to make clear the correspondence between this publication and previous JANAF Thermochemical Tables but also to assist the reader in making comparisons with other tables. This information is the same as that provided in "The NBS Tables of Chemical Thermodynamic Properties."[6]

Conversion of Values in Calories to Joules

For the energy units the defined conversion factor of 4.184 J cal^{-1} is used.

Conversion for Change in Pressure

The following expressions define the effect of pressure on the properties of all substances.

$$(\partial H/\partial p)_T = V - T(\partial V/\partial T)_p = V(1 - \alpha T),$$

$$(\partial C_p/\partial p)_T = -T(\partial^2 V/\partial T^2),$$

$$(\partial S/\partial p)_T = -V\alpha = -(\partial V/\partial T)_p,$$

$$(\partial G/\partial p)_T = V,$$

$$\alpha \equiv (1/V)(\partial V/\partial T)_p.$$

For the small pressure change from 1 atm to 1 bar, the pressure coefficients may be taken as constants for condensed phases. There are no condensed phase cases in this publication for which the pressure coefficients are large enough to affect the tabulated values. Typical values of α for solids are 10^{-5} to 2×10^{-4} K^{-1} and $V < 10^{-3}$ m^3 mol^{-1}. For ideal gases, $V = RT/p$ and $\alpha = R/pV = 1/T$. The pressure coefficients for gases are large enough to affect high-accuracy data. In practice the only properties tabulated here that are affected are $S^\circ(T)$ and $-(G^\circ - H^\circ[298.15\ K])/T$ for all gases, and $\Delta_f G^\circ(T)$ and $\log K_f$ for any substance whenever gases are involved.

The conversion equations used for the change in standard-state pressure are listed below together with some that apply to thermal functions. Superscript ° and * denote values at 1 bar and 1 atm, respectively. These equations apply to the small change in pressure from 1 atm to 1 bar.

For all substances, exactly for ideal gases and within the limit of experimental inaccuracy for all other phases:

$$\Delta_f H^\circ(T) - \Delta_f H^*(T) = 0,$$

$$[H^\circ(T) - H^\circ(T_1)] = [H^*(T) - H^*(T_1)] = 0,$$

$$C_p^\circ(T) - C_p^*(T) = 0,$$

where T_1 is a reference temperature.

For condensed phases, within the limit of experimental inaccuracy:

$$S^\circ(T) - S^*(T) = 0,$$

$$G^\circ(T) - G^*(T) = 0.$$

For substances that are gaseous:

$$S^\circ(T) - S^*(T) = R \ln p^*/p^\circ$$

$$= R \ln \frac{101\ 325}{100\ 000} = 0.109\ 442\ \mathrm{J\ K^{-1}\ mol^{-1}}.$$

For substances that are gaseous at T but may or may not be gaseous at T_1:

$$-[G^\circ(T) - H^\circ(T_1)]/T + [G^*(T) - H^*(T_1)]/T$$

$$= R \ln p^*/p^\circ \text{ as for } S.$$

$$[G^\circ(T) - G^\circ(T_1)] - [G^*(T) - G^*(T_1)]$$

$$= -R(T - T_j)\ln p^*/p^\circ,$$

where $T > T_j \geqslant T_1$ and T_j is the temperature at which the function begins to apply to the ideal gas, usually the boiling point at 1 bar.

For each substance with gases in its formation reaction:

$$\Delta_f G^\circ(T) - \Delta_f G^*(T)$$

$$= -\delta RT \ln 1.013\ 25$$

$$= -0.032\ 63\ \delta\ \mathrm{kJ\ mol^{-1}} \text{ at } 298.15\ \mathrm{K},$$

where δ is the net increase in the number of moles of gas in the formation reaction.

Conversions for Entropies of Gases

To convert from:	$\frac{S^\circ(g,0.1\ MPa)}{J\ K^{-1}\ mol^{-1}}$	$\frac{S^*(g,1\ atm)}{cal\ K^{-1}\ mol^{-1}}$	$\frac{S^\dagger(g,1\ atm)}{J\ K^{-1}\ mol^{-1}}$
To:	Use		
$\frac{S^\circ(g,0.1\ MPa)}{J\ K^{-1}\ mol^{-1}}$	$=S^\circ$	$4.184(S^* + 0.026\ 157)$	$S^\dagger + 0.109\ 442$
$\frac{S^\dagger(g,1\ atm)}{J\ K^{-1}\ mol^{-1}}$	$=S^\circ - 0.109\ 442$	$4.184\ S^*$	$S^\dagger$
$\frac{S(g,0.1\ MPa)}{cal\ K^{-1}\ mol^{-1}}$	$=S^\circ/4.184$	$S^* + 0.026\ 157$	$(S^\dagger + 0.109\ 442)/4.184$
$\frac{S^*(g,1\ atm)}{cal\ K^{-1}\ mol^{-1}}$	$=(S^\circ - 0.109\ 442)/4.184$	S^*	$S^\dagger/4.184$

S°—entropy for a gas at 0.1 MPa in $J\ K^{-1}\ mol^{-1}$.
S^*—entropy for a gas at 1 atm in $cal\ K^{-1}\ mol^{-1}$.
$S^\dagger$—entropy for a gas at 1 atm in $J\ K^{-1}\ mol^{-1}$.

4.4. Boiling Point and the Standard-State Pressure

In these tables, the defined pressure for the standard state is 1 bar (100 000 Pa). As stated by IUPAC,[9] the present recommended change in the standard-state pressure carries no implication for "standard pressures" used in other contexts, e.g., the convention that "normal boiling points" refer to a pressure of 1 atm (101 325 Pa). Instances may occur where it is desirable to adopt yet other values for p° and so all authors must clearly state the value adopted.

In these tables, the temperature at which $\Delta_r G^\circ = 0$ for the process liquid to gas, refers to a fugacity of 1 bar. Typically the real (i.e., observed) boiling point (corresponding to a nonideal gas) is slightly higher. In many cases the experimental uncertainties in measuring a boiling point (at 1 bar) and the uncertainties in the calculated value due to uncertainties in the thermal functions make this difference meaningless. Such a situation exists whether the standard-state pressure is 1 bar or 1 atm.

The phrase "boiling point" must be used carefully for other reasons. Care must be exercised such that the boiling process is indeed defined. Two examples which illustrate this point are MgO and LiF. Magnesium oxide does not vaporize congruently (so that the total vapor pressure depends on the ambient oxygen potential), whereas LiF vaporizes congruently but monomer, dimer, and trimer species exist in the vapor.

5. Evaluation of Thermodynamic Data

Interconsistency

The basic aim of these thermodynamic property tables is to provide a related and consistent set of enthalpies of formation and Gibbs energies of formation. This allows the prediction of the enthalpy and Gibbs energy changes and the equilibrium constants for any reaction among the constituents of the tables.

The enthalpy of formation and Gibbs energy of formation are related to each other through the entropy of formation of the substance. Self-consistency can be assured by use of this constraint to derive any one property from selected values of the other two. The entropy, which is an absolute quantity in thermochemical calculations, is suitable for one of the selected properties. The enthalpy of formation is convenient for the other, since enthalpy of reaction is the most commonly measured link between different substances.

Enthalpy changes are relative quantities, so it is important to relate each substance to an interconsistent base. This is done by referring each enthalpy of formation to the chemical elements in their standard reference states. By convention, the enthalpy of formation of an element in its standard reference state is zero at all temperatures.

The reference state table may involve a single phase or several phases and examples of both kinds are found in this compilation. For elements which are solid at room temperature, the reference state is normally the stable solid state(s) up to the melting point, the liquid up to the boiling point at 1 bar, and thereafter the gas phase which may be monatomic or diatomic. For other elements, the choices for the reference states are:

(1) the ideal monatomic gas for the inert gases and the electron gas—He, Ne, Ar, Kr, Xe, Rn, and e^-. The classical statistical mechanical values are assumed for $H^\circ(T) - H^\circ(0)$ for the electron.

(2) the ideal diatomic gas for the gases H_2, D_2, N_2, O_2, F_2, and Cl_2.

(3) crystalline white phosphorus and the ideal diatomic gas, $1/2\ P_2(g)$. Use of white phosphorus is a change from our previous publication in which red phosphorus was the reference state. Although red phosphorus is the more stable form, it is less well characterized and its properties are less reproducible.

(4) the ideal diatomic gas combined with the appropriate condensed phases for Br_2, I_2, and S; for sulfur, the gaseous portion of the reference state is $1/2\ S_2(g)$.

(5) graphite (Acheson-spectroscopic grade) for carbon at all temperatures.

These choices are arbitrary and vary in different compilations. Enthalpies or Gibbs energies of formation taken from different sources should be converted to common reference states before use.

The establishment of reference states is the first step in

any scheme for interconsistency. The next step, ideally, would be to solve simultaneously for all $\Delta_f H°$, $\Delta_f G°$, and $S°$ using all the suitable data for $\Delta_r H°$, $\Delta_r G°$, $\Delta_r S°$, and $S°$. This would provide values for enthalpies, Gibbs energies, and entropies which, when combined, would yield the minimum overall deviations from the measured values. Simultaneous solution is limited in practice to small groups of interrelated substances.

We have used this simultaneous approach in a limited way for treating groups of interrelated fluorides. Dr. John L. Haas, Jr. (U.S. Geological Survey) has treated, simultaneously, the data for some geologically important systems as a function of temperature and pressure.[25] Dr. J. Brian Pedley (University of Sussex) has treated $\Delta_r H°$ data for several systems, the most pertinent involving all the diatomic oxides.[26] Dr. Vivian Parker and Dr. David Garvin (NBS) have refined the latter method and applied it to some species that were adopted by CODATA.[27] JANAF policy is to use the results of these treatments.

Since simultaneous solution has not been attempted for all substances it is necessary to fix, simultaneously when possible, certain key values for common substances. These are then used in the usual sequential way to help fix other substances. JANAF policy is to adopt the key values recommended by the CODATA Task Group for Key Values in Chemical Thermodynamics.[28] Exceptions to this policy only occur when additional recent data are available.

Because of the many revisions of these tables, perfect interconsistency is not always attained. However, any inconsistency of the tables is a prime concern and is a cause for revision if the effects are of the same order of magnitude as the stated uncertainties.

An approximate type of consistency relates to trends in properties within families in the periodic table. Expected trends can aid both in estimation of missing data and in evaluation of observed data, for example, in a series from fluoride to iodide or from polyhalide to monohalide.

General Evaluation Techniques

Heat Capacities and Enthalpy Difference

The evaluation of reported values is a judgmental process, but the analysis should be as objective as possible and therefore certain ground rules must be established. The first step involves the examination of a plot of all information so that a general idea of the agreement can be visually obtained. If certain data sets differ from the majority, they are examined for possible causes of the difference. Calibration data, sample purity, and experimental scatter are checked. If the source of error can be located, the data are given appropriately less weight. Otherwise all data of equal reliability are considered equal, even if they disagree. Such data are then smoothed by a weighted least-squares curve fit. Enthalpy data are smoothed similarly and heat capacities are derived directly from the differentiated polynomial. The smoothing of enthalpy data often requires constraining the fit so that it passes through zero at the reference temperature. Additional constraints are often used to fix the heat capacity at 298.15 K or to join enthalpy data smoothly with low-temperature heat capacities.

Transitions

The evaluation of solid-state transitions involves first the recognition of the type of transition, which may not always be obvious. A first-order transition such as fusion involves a discontinuous change of enthalpy and entropy at the transition point, whereas second-order transitions involve only discontinuities in heat capacity. Because of impurities and other factors, first-order transitions often do not occur sharply at one temperature; instead, they spread a little on either side and are sometimes difficult to distinguish from λ-type second-order transitions.

Many enthalpies of fusion and enthalpies of solid-state transitions are obtained from enthalpy measurements. Here the enthalpies are given directly by the difference in enthalpy at the transition temperature. Normally, the enthalpy above and below can be smoothly extended to the temperature of transition, although the transition temperature may not be clearly obtainable from the enthalpy data.

The enthalpies of the transitions involving vaporization and sublimation are evaluated from equilibrum data by the second- and third-law methods. When reliable calorimetric measurements are available, the adopted enthalpy of transition is usually based on them.

Equilibrium

Perhaps the most significant facts to be established in evaluating equilibrium data are whether equilibrium was attained and whether the process was properly described. For example, in a sublimation process the most accurate measurements are valueless if the vapor phase is not uniquely defined. If polymerization or breakdown might have occurred, it is important to establish the exact reaction first. Because mass spectroscopy has shown that vapor phases are often extremely complex, it is almost a requirement that the vapor composition be established by a mass spectrometer before equilibrium data can be accepted. Some methods of measurement are less accurate than others and should be given less weight. For example, equilibria determined in a mass spectrometer may have significant uncertainty because of the approximations involved in obtaining absolute pressures. On the other hand, measurements of the electromotive force of reversible cells can have great accuracy. Mass spectra provide identification of the vapor species while a very precise emf measurement tells nothing as to the reaction which is occurring. Each method has inherent limitations; these may be more critical in some uses than in other uses. Clues to the existence of these limitations often appear in the second- and third-law analyses.

The Second-Law Method

Starting from the equation $\Delta G° = -RT \ln K$, by differentiation with respect to T and substitution of $d(\Delta G°)/dT = -\Delta S°$, we obtain $\Delta H° = RT^2 d(\ln K)/dT$, the well-known van't Hoff equation. By substituting $dT = -T^2 d(1/T)$, one obtains $\Delta H° = -R\, d(\ln K)/d(1/T)$; thus, the slope of a $\ln K$ versus $1/T$ plot is $-\Delta H°/R$. If $\Delta H°$ is constant, then the slope is constant and the plot is a straight line. Since the variation of $\Delta H°$ with temperature is often quite small, it is customary to assume a straight-line relationship. This method of obtaining heats of reaction from equilibrium measurements is known as a "second-law" calculation. For greatest accuracy the equilibrium measurements should extend over a wide range of temperature, and in this case $\Delta H°$ is usually not constant.

Curvature corrections can be applied by assuming a specific form for the variation of $\Delta H°$ with temperature. In this manner we can calculate $\Delta_r H°(298.15\ K)$ from $\Delta_r H°(T)$. The effects of such corrections may be of significance. It should be noted that the second law cannot be applied to a single observation, but the third-law method, which is described below, can be so used. The second-law method also can be applied when only relative values of the equilibrium constant are available, for example, from mass-spectroscopic intensity measurements.

The Third-Law Method

The "third-law" method is based on a knowledge of the absolute entropy of the reactants and products. It allows the calculation of a reaction enthalpy from each data point when the change in the Gibbs energy function for the reaction is known. The Gibbs energy function used here is defined as

$$\mathrm{gef}(T) = [G°(T) - H°(298.15\ K)]/T$$

and is easily calculated from the relation

$$S°(T) = -[G°(T) - H°(T)]/T$$

$$= [H°(T) - H°(298.15\ K)]/T$$

$$- [G°(T) - H°(298.15\ K)]/T,$$

thus,

$$\mathrm{gef}(T) = -S°(T) + [H°(T) - H°(298.15\ K)]/T.$$

From the definition we can write for the change in a reaction

$$\Delta G°/T = \mathrm{gef}(T) + \Delta H°(298.15\ K)/T = -R \ln K,$$

thus,

$$\Delta H°(298.15\ K)/T = -R \ln K - \Delta \mathrm{gef}(T),$$

where Δ signifies Σ products $- \Sigma$ reactants.

In the JANAF Thermochemical Tables, Gibbs energy functions are based on an enthalpy reference temperature of 298.15 K and thus yield enthalpy changes at 298.15 K regardless of the temperature of the reaction. It should be noted that the Gibbs energy function in a single-phase table is based on the enthalpy of that phase at 298.15 K even though the phase may not be stable at 298.15 K. This differs from the usual convention of combining all condensed phases into one multiphase table such that the enthalpy of reaction refers to the phase stable at the reference temperature. In the JANAF tables the functions always refer to the designated phase for 298.15 K regardless of its stability. For example, if the vapor pressure over liquid copper is analyzed using Cu(ℓ) Gibbs energy functions, the result is the enthalpy of vaporization of the liquid at 298.15 K. To calculate the enthalpy of sublimation of Cu(cr) it is necessary to add the enthalpy of fusion at 298.15 K, which is the difference in the enthalpy of formation of Cu(ℓ) and Cu(cr) at 298.15 K. It should also be noted that Gibbs energy functions are always negative, thus the negative of the function is usually tabulated and the proper sign must be remembered when using these functions.

The analysis of data by the third-law is generally considered superior to the second-law analysis. It is definitely superior if the Gibbs energy functions are accurately known because each equilibrium point yields one value for the enthalpy change. These values will often reveal trends that indicate nonequilibrium or erroneous values in the equilibrium constants of a set of data. If the data are good, the second-law value should agree. When the Gibbs energy functions are estimated, the third-law values derived from them must be handled carefully. Values that are constant and that agree with the second-law enthalpy indicate that the Gibbs energy functions and equilibrium data are mutually consistent. A drift with temperature of the third-law values indicates errors in either the data or the functions. The magnitude of such drifts is often given in the table write-up and represents the entropy change required to bring the second- and third-law values into agreement. If the drift is within the uncertainty of the experimental entropy, then the data are acceptable. Drifts that are much larger than reasonable entropy errors generally indicate failure to attain equilibrium or improperly estimated Gibbs energy functions.

We have found a combination of the two methods to be so valuable that a third-law calculation is always accompanied by a second-law treatment of the same data. The calculation is done in ascending temperature order to give $\Delta_r H°(298.15\ K)$ and its deviation from the mean along with the calculated log K and its deviation from the least-square line. Third-law analysis enables bad points to be seen as deviations from the general trend of the differences; thus the method can be used even when the Gibbs energy functions are estimated. If these bad points are located on the ends of the data set, the second-law line often fits them quite well and by itself does not arouse suspicion. But the dropping of such points can bring widely discordant $\Delta H°$ values into agreement with each other and often with the third-law values.

In our analysis of equilibrium data, tabulations for both the second- and third-law results are normally given. In addition, we often list values for the drift of δS, as defined by

$$-\,\mathrm{drift} = \delta S = \Delta_r S°(\text{2nd law, 298.15 K})$$

$$- \Delta_r S°(\text{3rd law, 298.15 K}).$$

δS, previously called a drift (in $\Delta_r H°$), gives a direct comparison of the deviation of the calculated second-law entropies from our adopted functions.The current policy is to use δS rather than drift. $\Delta_r S$ (2nd law, 298.15 K) is not calculated, however, via the usual second-law method or the Σ method.[29,31] It is instead derived from the slope of the assumed linear temperature dependence of the deviations from the mean of the third-law enthalpies of reaction. Our experience indicates that this method agrees closely, but not exactly, with the Σ method.

6. Construction of the Tables

The JANAF Thermochemical Tables consist of thermal functions and formation functions, both of which are temperature dependent. The thermal functions consist of heat capacity, enthalpy increments, entropy, and Gibbs energy function. The normal relationships hold between the thermal functions:

$$H°(T) = \int_0^T C_p°(T)\mathrm{d}T,$$

$$S°(T) = \int_0^T C_p°(T)/T\,\mathrm{d}T,$$

$$[G°(T) - H°(298.15\text{ K})]/T$$
$$= -S°(T) + [H°(T) - H°(298.15\text{ K})]/T.$$

The formation functions consist of enthalpy of formation, Gibbs energy of formation, and the logarithm of the equilibrium constant of formation. Enthalpies of formation at temperatures other than 298.15 K require a knowledge of the enthalpies of the reference elements:

$$\Delta_f H°(T) = \Delta_f H°(298.15\text{ K})$$
$$+ [H°(T) - H°(298.15\text{ K})]_{\text{compound}}$$
$$- \sum [H°(T) - H°(298.15\text{ K})]_{\text{elements}}.$$

The Gibbs energy of formation is readily calculated from the enthalpy of formation when the entropies of the elements are known. Thus

$$\Delta_f G°(T) = \Delta_f H°(T)$$
$$- T\{S°(T)_{\text{compound}} - \sum S°(T)_{\text{elements}}\}.$$

The logarithm of the equilibrium constant of formation is then found from the relation $\Delta_f G°(T) = -RT \ln K_f$.

Solid Phase Thermal Functions

The solid phase thermal functions are derived from recommended heat capacity and/or enthalpy increment values. Although the JANAF Thermochemical Tables place more emphasis on high temperatures ($T > 298.15$ K), it is desirable to have data in the temperature range of 0 to 298.15 K. There is no basic difference in evaluating low-temperature data, but many more heat capacity points are needed to describe fully the large curvature in the temperature dependence of the heat capacity. In order to evaluate enthalpy increment it is necessary to start with the temperature $T = 0$ and integrate up to 298.15 K; then the value at 298.15 K is subtracted from the intermediate values, giving negative values for the enthalpy below 298.15 K.

The tables extend to well above the normal melting point to provide data in a metastable region which in this case is a superheated region. Explanations are inserted in the tabulations to indicate the end of the phase stability and any solid-state transitions.

Liquid Phase Thermal Functions

The construction of the liquid single-phase thermal functions is identical with that used for the solid phase; however, the required data at 298.15 K are not usually readily available. The data are obtained by calculating a preliminary table using the chosen heat capacities with zero values for $\Delta_f H°(298.15\text{ K})$ and $S°(298.15\text{ K})$. The correct starting values are then determined by comparing the values from the tables of crystal and liquid, using the following equations:

$$\Delta_{\text{fus}} H° = \Delta_f H°(T_{\text{fus}}, \ell) - \Delta_f H°(T_{\text{fus}}, \text{cr}),$$

$$\Delta_{\text{fus}} S° = \Delta_{\text{fus}} H°/T_{\text{fus}} = S°(T_{\text{fus}}, \ell) - S°(T_{\text{fus}}, \text{cr}).$$

The correct values at 298.15 K may be obtained from the above relations. A typical liquid table is extrapolated both below the melting point (if the melting point is above 298.15 K) and above the boiling point to facilitate interpolation and to provide data in the metastable regions, both in the supercooled and superheated regions. A glass transition may be included, in which case the tabulation becomes a multiphase table.

Gas Phase Thermal Functions

The gas phase thermal functions are generated using statistical mechanical relationships. Minimal data required for the various types of molecules are summarized below. Some of the equations used are given in Sec. 6.1 on Calculational Methods. The relative molecular mass is required for all molecules. Minimum information required is:

Monatomic species:
: Low-lying atomic energy levels and their statistical weights.

Diatomic species:
: Spectroscopic parameters such as the vibrational–rotational constants, symmetry number, and low-lying electronic energy levels.

Linear polyatomic species:
: Rotational constant, symmetry number, vibrational frequencies, and low-lying electronic levels.

Nonlinear polyatomic species:
: Principal moments of inertia, symmetry number, vibrational frequencies and low-lying electronic levels.

Multiphase Tables

Tables which show only values for the stable phases at 1 bar pressure are multiphase tables. Multiphase tables can always be recognized by the presence of solid lines, indicating phase transitions, on the table. They are prepared in a manner similar to tables for condensed phases. The functions are evaluated in the same manner as for a solid up to the first transition point; then the enthalpy and entropy of transition are added and the integration is continued using the heat capacities of the next phase. At each transition, the above process is repeated.

6.1. Calculational Methods

The calculation of the temperature-dependent values of the thermal function is performed using a variety of procedures. Some of these procedures were developed internally while others were developed at other locations. A general description of the traditional equations used in the JANAF project follows. Numerous higher order calculational schemes are used whenever the data are sufficient. These procedures are detailed on the appropriate tables.

In gas phase calculations, all thermodynamic variables are calculated in dimensionless form, e.g., C_p/R, to at least seven significant figures. This is an arbitrary choice since most data do not warrant this accuracy. This has been done to permit the calculation of the crossover temperatures and enthalpies between single-phase tables accurately and to permit conversion to the desired units without loss of accuracy. The printed tables, however, will only contain three decimal places. Condensed phase tables do not depend directly on R (the gas constant) and they are not calculated in dimensionless form, but are calculated to five decimal places for units conversion.

Condensed Phase Species

Thermodynamic data for condensed states are derived from either measured or estimated information. The thermochemical table is obtained normally by the appropriate integration of heat capacity data. At the appropriate temperatures, transition enthalpies are added to the enthalpy total, while the quotient of the transition enthalpy divided by the thermodynamic temperature is added to the entropy total.

Gaseous Species

Calculation of the contributions of rotation and translation involves the use of quantum statistics, but to obtain a numerical solution the quantum statistics are usually replaced by classical statistics at temperatures above about 10 K; below 10 K this classical approximation no longer holds. For this reason the equations presented here fail in the vicinity of 0 K. In agreement with the third-law concept, C_p° and S° are zero at 0 K. For a reference element, log K_f is zero at 0 K, while for compounds the absolute values of the Gibbs energy function and log K_f become infinite at 0 K, for the choice of the enthalpy reference temperature of 298.15 K.

In addition, classical statistics can sometimes lead to erroneous results at high temperatures. For example, the classical approximation could give abnormally high C_p° values for a diatomic molecule having a shallow potential energy curve.

The functions calculated for gases are obtained, in general, from molecular constants that have been adjusted to the natural isotopic abundance. Normally, the approximation of adjusting the molecular parameters by a weighting according to a weighted natural isotopic abundance can be made. The proper correction (which is used for all reference states) is based on a weighted isotopic average of the tabulations.

An electronic term of unit statistical weight lying at 30 000 cm^{-1} and above contributes a negligible amount to the thermodynamic functions at temperatures of 6000 K and below. On the other hand, a number of such terms cannot be neglected. In these cases, the number of terms and their values have been summed and their contribution included.

Some fundamental vibrational frequencies have been estimated by analogy with related molecules. Occasionally, two or more fundamentals are estimated at the same value. This is not to be confused with a true degeneracy which is indicated by placing the degeneracy value in parentheses following the frequency. However, when the information has been taken from another compilation, the degeneracies indicated by the compiler have been retained.

The method used in generating the thermal functions depends not only on the quality/quantity of information available but also on the specifics of the nature of the information (e.g., split ground state, shallow potential wells, etc.). The gas phase tables are generated using the following relationships unless stated otherwise. These relationships are given in Mayer and Mayer[30] and Pitzer and Brewer.[31] Use of a more sophisticated approach will always be noted and details given on the appropriate tables. The more common alternative approaches are given by McBride and Gordon[32] and McDowell.[33]

I. Ideal Monatomic Gas

(a) Translation

$$C_p^\circ/R = 5/2,$$

$$[H^\circ(T) - H^\circ(0\ \text{K})]/RT = 5/2,$$

$$S^\circ/R = 3/2 \ln M_r + 5/2 \ln T + 5/2 + \ln \frac{k}{p^\circ}\left(\frac{2\pi k}{Nh^2}\right)^{3/2},$$

$$-[G^\circ(T) - H^\circ(0\ \text{K})]/RT = 3/2 \ln M_r + 5/2 \ln T + \ln \frac{k}{p^\circ}\left(\frac{2\pi k}{Nh^2}\right)^{3/2}.$$

(b) Electronic

$$C_p^\circ/R = T^2 \frac{\mathrm{d}^2 \ln Q}{\mathrm{d}T^2} + 2T \frac{\mathrm{d} \ln Q}{\mathrm{d}T},$$

$$[H^\circ(T) - H^\circ(0\ \text{K})]/RT = T \frac{\mathrm{d} \ln Q}{\mathrm{d}T},$$

$$S^\circ/R = T \frac{\mathrm{d} \ln Q}{\mathrm{d}T} + \ln Q,$$

$$-[G^\circ(T) - H^\circ(0\ \text{K})]/RT = \ln Q,$$

where

$$Q = \sum_i g_i e^{-c_2\epsilon_i/kT},$$

where ϵ_i is in cm^{-1} and the electronic levels to be included are discussed on the appropriate gas phase table.

II. Ideal Diatomic Gas

(a) Translation

Same as in I(a).

(b) Rotation

$$C_p^\circ/R = 1 + (c_2B/T)^2/45,$$

$$[H^\circ(T) - H^\circ(0\ \text{K})]/RT = 1 - (c_2B/T)/3 - (c_2B/T)^2/45.$$

$$S^\circ/R = 1 - \ln(c_2B\sigma/T) - (c_2B/T)/90.$$

$$-[G^\circ(T) - H^\circ(0\ \text{K})]/RT = -\ln(c_2B\sigma/T) + (c_2B/T)/3 + (c_2B/T)^2/90,$$

where

$$B = (B_e - \alpha_e/2);$$

in cm^{-1} when spectroscopic constants are available, or

$$B = h/8\pi^2 cI = 2.799\ 320 \times 10^{-39}/I;$$

in cm^{-1} when calculated from molecular model, I in g cm^{-2}.

(c) Vibration

$$C^\circ_p/R = u^2e^{-u}/(1-e^{-u})^2,$$
$$[H^\circ(T)-H^\circ(0\ \mathrm{K})]/RT = ue^{-u}/(1-e^{-u}),$$
$$S^\circ/R = ue^{-u}/(1-e^{-u}) - \ln(1-e^{-u}),$$
$$-[G^\circ(T)-H^\circ(0\ \mathrm{K})]/RT = -\ln(1-e^{-u}),$$

where

$$u = c_2(\omega_e - 2\omega_e x_e)/T;$$

when spectroscopic constants (in cm^{-1}) are available, or

$$u = c_2\omega/T;$$

where ω (in cm^{-1}) is the fundamental frequency of a harmonic oscillator.

(d) Corrections for anharmonic vibrations

$$C^\circ_p/R = \frac{16}{u} - \frac{\delta u^2 e^u}{(e^u-1)^2} + \frac{u^2e^u(2\delta e^u - 4Xu - 8x)}{(e^u-1)^3} + \frac{12Xu^3e^2u}{(e^u-1)^4},$$

$$[H^\circ(T)-H^\circ(0\ \mathrm{K})]/RT = \frac{\beta\gamma}{u} + \frac{u(\delta e^u - 2X)}{(e^u-1)^2} + \frac{4Xu^2e^u}{(e^u-1)^3},$$

$$S^\circ/R = \frac{16\gamma}{u} + \frac{\delta}{(e^u-1)} + \frac{\delta u e^u}{(e^u-1)^2} + \frac{4Xu^2e^u}{(e^u-1)^3},$$

$$-[G^\circ(T)-H^\circ(0\ \mathrm{K})]/RT = \frac{\beta\gamma}{u} + \frac{\delta}{(e^u-1)} + \frac{2Xu}{(e^u-1)^2},$$

where $u = (\omega_e - 2\omega_e x_e)(c_2/T)$, $X = \omega_e x_e/\omega_e$, $\delta = \alpha_e/B_e$, and $\gamma = B_e/\omega_e$; ω_e, $\omega_e x_e$, B_e and α_e are in cm^{-1}.

(e) Electronic

Same as I(b) when the ith-state vibrational partition function, Q^i_v, and the ith-state rotational partition function, Q^i_r, are equal to the respective ground-state partition functions. In this case the partition function

$$Q = Q_t Q_v Q_{r_i} Q^i_e,$$

otherwise all the thermodynamic functions are derived from

$$Q = Q_t \sum_i Q^i_e Q^i_v Q^i_r,$$

where Q_t is the translational partition function and

$$Q^i_e = g_i \exp(-c_2\epsilon_i/T).$$

III. Linear Polyatomic Molecule

(a) Translation

Same as in I(a).

(b) Rotation

Same as in II(b).

(c) Vibration

Same as in II(c) for $3N-5$ vibrational degrees of freedom.

(d) Corrections for anharmonic vibrations.

This contribution is normally neglected.

(e) Electronic

Same as II(e) where levels and quantum weights are known.

IV. Nonlinear Polyatomic Molecule (Rigid Rotator, Harmonic Oscillator)

(a) Translation

Same as in I(a).

(b) Rotation

$$C^\circ_p/R = 3/2,$$
$$[H^\circ(T)-H^\circ(0\ \mathrm{K})]/RT = 3/2,$$
$$S^\circ/R = 3/2 + (1/2)\ln(\pi\sigma/I_A I_B I_C),$$
$$-[G^\circ(T)-H^\circ(0\ \mathrm{K})]/RT = (1/2)\ln(\pi\sigma/I_A I_B I_C).$$

(c) Vibration

Same as II(c) for $3N-6$ vibrational degree of freedom.

(d) Corrections for anharmonic vibrations

This contribution is normally neglected.

(e) Electronic

Same as II(e) where levels and quantum weight are known.

6.2. Dates

At the bottom of the tabulation side of each table, there are one or two dates. These dates refer to the approximate time at which the evaluation for these tables was completed. The data for the most recent and the preceding evaluation are given. These dates will be at the bottom of the page at the right-hand side and left-hand side, respectively. Previously, all revision dates had been listed. Current practice will limit the dates to two. For the gas phase tables, the standard-state pressure is given in parenthesis after the date(s). If the dates are identical, then the difference in this current table and its predecessor lies in the changes due to the standard-state pressure. Contrary to previous practice, the latest issue date of each table is not given in either of the two indices in this publication.

7. Acknowledgments

The authors gratefully acknowledge the financial support of the U.S. Department of Energy (Contract No. DE AC22-81PC41514) for the publication of this third edition. The encouragement and continued cooperation by our Technical Project Officer, Mr. F. E. Spencer, Jr., is also greatly appreciated. The preparation of the individual JANAF Thermochemical Tables has been supported by the DOE (Mr. F. E. Spencer, Jr.) and the AFOSR (Dr. L. H. Caveny). The authors sincerely appreciate the support of those two organizations.A special note of thanks is extended to Dr. Joseph F. Masi (AFOSR, retired) who fought for the support and continuation of this project during some very difficult funding periods in the late 1960s and early 1970s.

The JANAF Thermochemical Tables have benefitted substantially over the past 25 years from the many scientists scattered throughout the world who have provided us with unpublished data, data in advance of publication, clarifications of fine points of experimentation and interpretation, and free and helpful discussions on a wide range of scientific matters. Our heartfelt thanks are long overdue to all involved. Although it is not possible to list all those who assisted, we particularly acknowledge Prof. Leo Brewer (Berkeley), Dr. David Garvin (NBS), Dr. Donald L. Hildenbrand (SRI), Dr. David R. Lide, Jr. (OSRD), Dr. John L. Haas (USGS), and Dr. Malcolm H. Rand (AERE). Lastly, those who reviewed this manuscript certainly contributed significantly to the quality of this publication.

The preparation of this publication requires the help and cooperation of many individuals. Extensive assistance with the computer application in all aspects of the project (especially assembling the indices and the tabulation printouts) was received from Dr. Milton D. Marks. The initial conversion of the tables into the new style and format was coordinated by Miss Barbara K. Meier. The long tedious task of typing all these tables for the third edition on a word processor was handled admirably by Mrs. Betty Clark and Mrs. Jan Crouch. Their patience, hard work, and talents are very much appreciated and will be long remembered.

Finally, we would like to thank the management of the Dow Chemical Company for their long-standing encouragement and support of our activities on the JANAF Thermochemical Tables Project.

8. References

[1]D. R. Stull and H. Prophet, *JANAF Thermochemical Tables, 2nd Edition*, NSRDS-NBS-37 (1971).

[2]M. W. Chase, J. L. Curnutt, A. T. Hu, H. Prophet, A. N. Syverud, and L. C. Walker, "JANAF Thermochemical Tables, 1974 Supplement," J. Phys. Chem. Ref. Data **3**, 311 (1974).

[3]M. W. Chase, J. L. Curnutt, H. Prophet, R. A. McDonald, and A. N. Syverud, "JANAF Thermochemical Tables, 1975 Supplement," J. Phys. Chem. Ref. Data **4**, 1 (1975).

[4]M. W. Chase, J. L. Curnutt, R. A. McDonald, and A. N. Syverud, "JANAF Thermochemical Tables, 1978 Supplement," J. Phys. Chem. Ref. Data **7**, 793 (1978).

[5]M. W. Chase, J. L. Curnutt, J. R. Downey, Jr., R. A. McDonald, A. N. Syverud, and E. A. Valenzuela, "JANAF Thermochemical Tables, 1982 Supplement," J. Phys. Chem. Ref. Data **11**, 695 (1982).

[6]D. D. Wagman, W. H. Evans, V. B. Parker, R. H. Schumm, I. Halow, S. M. Bailey, K. L. Churney, and R. L. Nuttall, *The NBS tables of chemical thermodynamic properties*, J. Phys. Chem. Ref. Data **11**, Suppl. 2 (1982).

[7]R. A. Robie, B. S. Hemingway, and J. R. Fisher, *Thermodynamic Properties of Minerals and Related Substances at 298.15 K and 1 Bar (10^5 Pascals) Pressure and at Higher Temperatures,* U.S. Geol. Surv. Bull. 1452 (1978).

[8]*Manual of Symbols and Terminology for Physicochemical Quantities and Units: 1979 edition,* edited by D. H. Whiffen (Pergamon, Oxford, 1979).

[9]"Notation of States and Processes, Significance of the Word Standard in Chemical Thermodynamics, and Remarks on Commonly Tabulated Forms of Thermodynamic Functions," edited by J. D. Cox, Pure Appl. Chem. **54**, 1239 (1982), App. IV.

[10]Bulletin of Chemical Thermodynamics, edited by R. D. Freeman, Oklahoma State University, Stillwater, Oklahoma, extant.

[11]C. E. Moore, *Atomic Energy Levels*, U.S. Natl. Bur. Stand., NSRDS-NBS-35, Vols. I–III (1970).

[12]G. Herzberg, *Molecular Spectra and Molecular Structure. I. Spectra of Diatomic Molecules* (Van Nostrand, New York, 1950).

[13]G. Herzberg, *Molecular Spectra and Molecular Structure. II. Infrared and Raman Spectra of Polyatomic Molecules* (Van Nostrand, New York, 1945).

[14]G. Herzberg, *Molecular Spectra and Molecular Structure. III. Electronic Spectra of Polyatomic Molecules* (Van Nostrand, New York, 1966).

[15]K. P. Huber and G. Herzberg, *Molecular Spectra and Molecular Structure. IV. Constants of Diatomic Molecules* (Van Nostrand Reinhold, New York, 1979).

[16]"Atomic Weights of the Elements—1981," edited by N. E. Holden and R. L. Martin, Pure Appl. Chem. **55**, 1101 (1983).

[17]E. R. Cohen and B. N. Taylor, "The 1973 Least-Squares Adjustment of the Fundamental Constants," J. Phys. Chem. Ref. Data **2**, 663 (1973).

[18]"Isotopic Compositions of the Elements—1981," edited by N. E. Holden, R. L. Martin, and I. L. Barnes, Pure Appl. Chem. **55**, 1119 (1983).

[19]G. K. Burgess, "The International Temperature Scale," RP22, J. Res. Natl. Bur. Stand. (U.S.) **1**, 635 (1928). Describes the International Temperature Scale of 1927.

[20]R. J. Corruccini, "Differences Between the International Temperature Scales at 1948 and 1927," RP 2014, J. Res. Natl. Bur. Stand. (U.S.) **43**, 133 (1949).

[21]H. F. Stimson, "International Practical Temperature Scale of 1948. Text Revision of 1960," J. Res. Natl. Bur. Stand. (U.S.) Sect. A **65**, 139 (1961).

[22]C. R. Barber, "The International Practical Temperature Scale of 1968, Adopted by the Comité International des Poids et Mesures," Metrologia **5**, 35 (1969).

[23]T. B. Douglas, "Conversion of Existing Calorimetrically Determined Thermodynamic Properties to the Basis of the International Practical Temperature Scale of 1968," J. Res. Natl. Bur. Stand. (U.S.) Sect. A **73**, 451 (1969). (Formulas for converting H, Cp, S, and G from one temperature scale to another. Differences between the two scales are tabulated from 90 to 10 000 K.)

[24]L. A. Guildner and R. E. Edsinger, "Deviation of International Practical Temperatures from Thermodynamic Temperatures in the Temperature Range from 273.16 K to 730 K," J. Res. Natl. Bur. Stand. (U.S.) Sect. A **80**, 703 (1976).

[25]J. L. Haas, Jr., G. R. Robinson, Jr., and B. S. Hemingway, "Thermodynamic Tabulations for Selected Phases in the system $CaO \cdot Al_2O_3 \cdot SiO_2 \cdot H_2O$ at 101.325 kPa (1 atm) between 273.15 and 1800 K," J. Phys. Chem. Ref. Data **10**, 575 (1981).

[26]J. B. Pedley and E. M. Marshall, "Thermochemical Data for Gaseous Monoxides," J. Phys. Chem. Ref. Data **12**, 967 (1983).

[27]D. Garvin, V. B. Parker, D. D. Wagman, and W. H. Evans, "A Combined Least Sums and Least Squares Approach to the Solution of Thermodynamic Data Networks," *The Proceedings of the Fifth Biennial International CODATA Conference* (Pergamon, New York, 1977).

[28]"CODATA Recommended Key Values for Thermodynamics—1977," edited by J. D. Cox, J. Chem. Thermodyn. **10**, 903 (1978).

[29]D. Cubicciotti, "A New Σ-plot Treatment of Equilibrium Data and its Application to the Vaporization of Bismuth Chloride," J. Phys. Chem. **70**, 2410 (1966).

[30]J. E. Mayer and M. G. Mayer, *Statistical Mechanics*, 7th printing (Wiley, New York, 1957).

[31]G. N. Lewis and M. Randall. *Thermodynamics*, 2nd ed., revised by K. S. Pitzer and L. Brewer (McGraw–Hill, New York, 1961).

[32]B. J. McBride and S. Gordon, FORTRAN IV Program for Calculation of Thermodynamic Data, NASA TN D-4097, 1967, 129 pp.

[33]R. E. McDowell, Quantum Chemistry Program Exchange, Indiana University, QCPE #269, 1974.

9. Indices to Tables

There are two indices provided with this publication. Both indices are alphabetical, one by chemical name and one by chemical formula. In both cases the state of the compound is given in parentheses after the chemical formula. The tables and discussions are arranged in the same order as given in the chemical formula index.

9.1. Description of the Chemical Name Index to the JANAF Thermochemical Tables

The nomenclature for the chemical species included in this publication does not fully correspond to the current Chemical Abstracts practice. The attempt has been made to use names as easily understandable as possible while not deviating far from the Chemical Abstracts system. Cited are two examples where the Chemical Abstracts nomenclature is followed, $AlBr_3$(g) and BBr_3(g) are listed as aluminum bromide and tribromoborane, respectively. For historical reasons, we would have preferred to name BBr_3(g) as boron bromide. However, Chemical Abstracts refers to most boron halides as borane derivatives. On the other hand, whereas Al_2Br_6 is called di-μ-bromotetrabromoaluminum in Chemical Abstracts, this publication uses aluminum bromide.

Other general comments on the nomenclature are:

(1) The carbon and silicon neutral species follow the Chemical Abstracts organic-type nomenclature in which the name reflects the type of bonding.

(2) The "-ium" suffix that denotes positive ions is not used. CH^+(g) is named "methylidyne, ion" whereas Chemical Abstracts uses "methyliumylidene."

(3) Salts are not named as an "acid, metal salt of." For example, $BaSO_4$ is barium sulfate, not sulfuric acid, barium salt.

(4) Many mineral names used.

(5) The sometimes confusing nomenclature for bridged structures in dimers is not used. Normally the monomer name is used.

In adopting this modified Chemical Abstracts nomenclature, there are numerous species for which the name will be significantly different from earlier JANAF publications. It is impractical to list all name changes, but in the more severe cases, a comment is made on the tables as to the previously used name.

9.2. Description of the Chemical Formula Index to the JANAF Thermochemical Tables

The tables included in this publication are ordered as given by the Chemical Formula Index. The modified Hill indexing system [J. Am. Chem. Soc. **22**, 478 (1900)] for chemical compounds as used by Chemical Abstracts is employed here to order the tables.

The Hill empirical formula is obtained for a particular compound by placing the elements in their alphabetical order unless C is present, in which case C is placed first, H second, and the remaining elements in alphabetical order. Substances with the same Hill formula are ordered alphabetically by their name. However, tables representing different states of the same species are in the order crystal, liquid, crystal–liquid, and ideal gas.

As examples, BeCl(g) would appear as written, whereas NaCl(g) would be filed under ClNa(g). To expand on the ordering of carbon compounds, the C always comes first, followed immediately by H if hydrogen is present, and then followed by any other elments present in their normal alphabetical order. The number of atoms of the element indexed first also influences the order of indexing the compound; e.g., all formulas with C (one carbon atom only) come before those with C_2 (two carbon atoms), followed by C compounds, and so on. Thus: $CHBF_4O$ (formyl fluoride, compound with BF_3), $CHCl_3$ (chloroform), CH_2F_3P (trifluormethylphosphine), CH_3NO_3 (methyl nitrate), CH_4Cl_2Si (dichloromethylsilane), $CH_{10}N_4O_3$ (hydrazine carbonate), CO (carbon monoxide), CO_2 (carbon dioxide), CO_3Zn (zinc carbonate), CZr (zirconium carbide), C_2Ba (barium acetylide), C_2BrClF_2 (bromochlorodifluoroethylene), C_2Ca (calcium carbide), and so on. See any formula index of Chemical Abstracts for further details and examples.

The electron gas, e^- (ref st), is inserted in the filing order after the deuterium species. Nonstoichiometric compounds are inserted just prior to or just after the corresponding stoichiometric order.

9.3. Chemical Name Index

CHEMICAL NAME INDEX - CONTINUED

CHEMICAL NAME INDEX - CONTINUED

CHEMICAL NAME INDEX - CONTINUED

CHEMICAL NAME INDEX - CONTINUED

CHEMICAL NAME INDEX - CONTINUED

CHEMICAL NAME INDEX - CONTINUED

CHEMICAL NAME INDEX - CONTINUED

CHEMICAL NAME INDEX - CONTINUED

CHEMICAL NAME INDEX - CONTINUED

CHEMICAL NAME INDEX - CONTINUED

CHEMICAL NAME INDEX - CONTINUED

CHEMICAL NAME INDEX - CONTINUED

CHEMICAL NAME INDEX - CONTINUED

CHEMICAL NAME INDEX - CONTINUED

CHEMICAL NAME INDEX - CONTINUED

CHEMICAL NAME INDEX - CONTINUED

CHEMICAL NAME INDEX - CONTINUED

CHEMICAL NAME INDEX - CONTINUED

9.4. Chemical Formula Index

CHEMICAL FORMULA INDEX - CONTINUED

CHEMICAL FORMULA INDEX - CONTINUED

CHEMICAL FORMULA INDEX - CONTINUED

CHEMICAL FORMULA INDEX - CONTINUED

CHEMICAL FORMULA INDEX - CONTINUED

CHEMICAL FORMULA INDEX - CONTINUED

CHEMICAL FORMULA INDEX - CONTINUED

CHEMICAL FORMULA INDEX - CONTINUED

CHEMICAL FORMULA INDEX - CONTINUED

CHEMICAL FORMULA INDEX - CONTINUED

CHEMICAL FORMULA INDEX - CONTINUED

CHEMICAL FORMULA INDEX - CONTINUED

CHEMICAL FORMULA INDEX - CONTINUED

CHEMICAL FORMULA INDEX - CONTINUED

CHEMICAL FORMULA INDEX - CONTINUED

CHEMICAL FORMULA INDEX - CONTINUED

CHEMICAL FORMULA INDEX - CONTINUED

CHEMICAL FORMULA INDEX - CONTINUED

CHEMICAL FORMULA INDEX - CONTINUED

ALUMINUM (Al) REFERENCE STATE A_r = 26.98154 Aluminum (Al) Al_1(ref)

0 to 933.45	K	crystal
933.45 to 2790.812	K	liquid
above 2790.812	K	ideal monatomic gas

Refer to the individual tables for details.

Enthalpy Reference Temperature = T_r = 298.15 K Standard State Pressure = p° = 0.1 MPa

	J K⁻¹mol⁻¹			kJ mol⁻¹			
T/K	C_p°	S°	$-[G°-H°(T_r)]/T$	$H°-H°(T_r)$	$\Delta_fH°$	$\Delta_fG°$	Log K_f
0	0.	0.	INFINITE	-4.539	0.	0.	0.
100	12.997	6.987	47.543	-4.056	0.	0.	0.
200	21.338	19.144	30.413	-2.254	0.	0.	0.
298.15	24.209	28.275	28.275	0.	0.	0.	0.
300	24.247	28.425	28.276	0.045	0.	0.	0.
400	25.784	35.630	29.248	2.553	0.	0.	0.
500	26.842	41.501	31.129	5.186	0.	0.	0.
600	27.886	46.485	33.283	7.921	0.	0.	0.
700	29.100	50.872	35.488	10.769	0.	0.	0.
800	30.562	54.850	37.663	13.749	0.	0.	0.
900	32.308	58.548	39.780	16.890	0.	0.	0.
933.450	32.959	59.738	40.474	17.982	CRYSTAL <--> LIQUID TRANSITION		
933.450	31.751	71.213	40.474	28.693			
1000	31.751	73.400	42.594	30.806	0.	0.	0.
1100	31.751	76.426	45.534	33.981	0.	0.	0.
1200	31.751	79.189	48.225	37.156	0.	0.	0.
1300	31.751	81.730	50.706	40.331	0.	0.	0.
1400	31.751	84.083	53.007	43.506	0.	0.	0.
1500	31.751	86.273	55.153	46.681	0.	0.	0.
1600	31.751	88.323	57.162	49.856	0.	0.	0.
1700	31.751	90.247	59.052	53.031	0.	0.	0.
1800	31.751	92.062	60.836	56.207	0.	0.	0.
1900	31.751	93.779	62.525	59.382	0.	0.	0.
2000	31.751	95.408	64.129	62.557	0.	0.	0.
2100	31.751	96.957	65.656	65.732	0.	0.	0.
2200	31.751	98.434	67.112	68.907	0.	0.	0.
2300	31.751	99.845	68.505	72.082	0.	0.	0.
2400	31.751	101.196	69.839	75.257	0.	0.	0.
2500	31.751	102.493	71.120	78.432	0.	0.	0.
2600	31.751	103.738	72.350	81.607	0.	0.	0.
2700	31.751	104.936	73.535	84.782	0.	0.	0.
2790.812	31.751	105.986	74.574	87.665	LIQUID <--> IDEAL GAS FUGACITY = 1 bar		
2790.812	20.795	211.333	74.574	381.667			
2800	20.795	211.401	75.023	381.858	0.	0.	0.
2900	20.796	212.131	79.738	383.938	0.	0.	0.
3000	20.798	212.836	84.163	386.017	0.	0.	0.
3100	20.800	213.518	88.325	388.097	0.	0.	0.
3200	20.804	214.178	92.248	390.178	0.	0.	0.
3300	20.808	214.818	95.952	392.258	0.	0.	0.
3400	20.815	215.440	99.458	394.339	0.	0.	0.
3500	20.823	216.043	102.780	396.421	0.	0.	0.
3600	20.833	216.630	105.934	398.504	0.	0.	0.
3700	20.846	217.201	108.934	400.588	0.	0.	0.
3800	20.862	217.757	111.790	402.673	0.	0.	0.
3900	20.881	218.299	114.515	404.760	0.	0.	0.
4000	20.904	218.828	117.116	406.849	0.	0.	0.
4100	20.932	219.345	119.603	408.941	0.	0.	0.
4200	20.964	219.849	121.984	411.036	0.	0.	0.
4300	21.002	220.343	124.265	413.134	0.	0.	0.
4400	21.046	220.826	126.455	415.236	0.	0.	0.
4500	21.088	221.299	128.557	417.341	0.	0.	0.
4600	21.143	221.763	130.578	419.452	0.	0.	0.
4700	21.206	222.219	132.523	421.570	0.	0.	0.
4800	21.276	222.666	134.396	423.694	0.	0.	0.
4900	21.352	223.105	136.202	425.824	0.	0.	0.
5000	21.439	223.537	137.945	427.964	0.	0.	0.
5100	21.535	223.963	139.627	430.112	0.	0.	0.
5200	21.641	224.382	141.253	432.271	0.	0.	0.
5300	21.757	224.795	142.825	434.441	0.	0.	0.
5400	21.884	225.203	144.347	436.623	0.	0.	0.
5500	22.021	225.606	145.821	438.818	0.	0.	0.
5600	22.170	226.004	147.249	441.027	0.	0.	0.
5700	22.330	226.398	148.634	443.252	0.	0.	0.
5800	22.496	226.787	149.978	445.491	0.	0.	0.
5900	22.680	227.173	151.284	447.749	0.	0.	0.
6000	22.836	227.552	152.551	450.005	0.	0.	0.

PREVIOUS: June 1979 (1 atm) CURRENT: June 1983 (1 bar)

Aluminum (Al) Al_1(ref)

ALUMINUM (Al) CRYSTAL A_r = 26.98154

$S°(298.15\ K) = 28.275 \pm 0.08\ J\ K^{-1}\ mol^{-1}$

$T_{fus} = 933.45 \pm 0.01\ K$

$\Delta_f H°(0\ K) = 0\ kJ\ mol^{-1}$

$\Delta_f H°(298.15\ K) = 0\ kJ\ mol^{-1}$

$\Delta_{fus} H° = 10.711 \pm 0.21\ kJ\ mol^{-1}$

Enthalpy of Formation

Zero by definition.

Heat Capacity and Entropy

The adopted thermal functions for Al(cr) are derived from the studies of Giauque and Meads (1), Berg (2), Downie and Martin (3), Takahashi (4, 5), McDonald (6), Brooks (7), Pochapsky (8) and Ditmars et al. (9). The mathematical and graphical treatment of these eight studies yields a continuous and smooth heat capacity curve. In contrast, CODATA (9) has recommended an S°(298.15 K) value which is 0.075 $J\ K^{-1}\ mol^{-1}$ greater than ours. Their result was based on an analysis of the low temperature heat capacity data of Giauque and Meads (1) and Berg (2). At the time of their analysis, the studies of Downie and Martin (3) and Takahashi (4, 5) had not been published. The three more recent studies (3, 4, 5, 8) suggest a lower heat capacity in the vicinity of 298.15 K, thus the lower entropy in comparison with CODATA (10).

Roberts (11) surveyed the superconductive properties of the elements and recommended a critical temperature of 1.175±0.002 K for Al(cr). Since this temperature is so low, the effects of superconductivity on the thermodynamic functions are not considered. The entropy contribution due to superconductivity will be less than 0.002 $J\ K^{-1}\ mol^{-1}$. The data of Giauque and Meads (1) and Downie and Martin (3) agree at temperatures up to 150 K but drift apart by 0.2 $J\ K^{-1}\ mol^{-1}$ at 200 K and 0.17 $J\ K^{-1}\ mol^{-1}$ at 300 K, with the Downie and Martin study being lower. The Takahashi (4, 5) study is even lower at ~298 K. The high temperature heat capacity values are derived from the enthalpy study of Ditmars et al. (9). Their curve is intermediate between those derived from previous studies (4, 5, 6, 7, 8) and implies a flatter C_p curve near the melting point (in comparison to previous interpretations). Numerous other heat capacity and enthalpy studies are available but were omitted in this analysis. A detailed discussion of the Group IIIA metals (B, Al, and Ga) is in preparation by the JANAF staff.

Fusion Data

See the liquid table for details.

Sublimation Data

There are no sublimation studies for Al(cr). The adopted value of $\Delta_{sub}H°(298.15\ K)$ is the value recommended by CODATA (10). See the ideal gas table for details.

References

1. W. F. Giauque and P. F. Meads, J. Amer. Chem. Soc. 63, 1897 (1941).
2. W. T. Berg, Phys. Rev. 167, 583 (1968).
3. D. B. Downie and J. F. Martin, J. Chem. Thermodynamics 12, 779 (1980).
4. Y. Takahashi, Pure & Appl. Chem. 47, 323 (1976).
5. Y. Takahashi, T. Azumi, H. Kadokura, Y. Sekino, H. Yokakawa, and M. Kamimoto, Therm. Anal., [Proc. Int. Conf.], 5th, 152 (1977).
6. R. A. McDonald, J. Chem. Eng. Data 12, 115 (1967).
7. C. R. Brooks and R. E. Bingham, J. Phys. Chem. Solids 29, 1553 (1968).
8. T. E. Pochapsky, Acta Met. 1, 747 (1963).
9. D. A. Ditmars, C. A. Plint, and R. C. Shukla, Int. J. Thermophys., submitted for publication, 1985.
10. J. D. Cox, chairman, ICSU – CODATA Task Group on Key Values for Thermodynamics, J. Chem. Thermodynamics 10, 903 (1978).
11. B. W. Roberts, J. Phys. Chem. Ref. Data 5, 581 (1976); U. S. Nat. Bur. Stand., Tech. Note 983, 1978.

Aluminum (Al) $Al_1(cr)$

Enthalpy Reference Temperature = T_r = 298.15 K — Standard State Pressure = p° = 0.1 MPa

	$J\ K^{-1}mol^{-1}$			$kJ\ mol^{-1}$			
T/K	$C_p°$	S°	$-[G°-H°(T_r)]/T$	$H°-H°(T_r)$	$\Delta_f H°$	$\Delta_f G°$	Log K_f
0	0.	0.	INFINITE	-4.539	0.	0.	0.
100	12.997	6.987	47.543	-4.056	0.	0.	0.
200	21.338	19.144	30.413	-2.254	0.	0.	0.
250	23.084	24.108	28.668	-1.140	0.	0.	0.
298.15	24.209	28.275	28.275	0.	0.	0.	0.
300	24.247	28.425	28.276	0.045	0.	0.	0.
350	25.113	32.231	28.574	1.280	0.	0.	0.
400	25.784	35.630	29.248	2.553	0.	0.	0.
450	26.335	38.699	30.130	3.856	0.	0.	0.
500	26.842	41.501	31.129	5.186	0.	0.	0.
600	27.886	46.485	33.283	7.921	0.	0.	0.
700	29.100	50.872	35.488	10.769	0.	0.	0.
800	30.562	54.850	37.663	13.749	0.	0.	0.
900	32.308	58.548	39.780	16.890	0.	0.	0.
933.450	32.959	59.738	40.474	17.982	------ CRYSTAL <--> LIQUID -----		
1000	34.358	62.055	41.834	20.221	-10.585	0.760	-0.040
1100	36.722	65.438	43.827	23.772	-10.209	1.878	-0.089
1200	39.408	68.746	45.766	27.576	-9.580	2.951	-0.128

PREVIOUS: June 1979 CURRENT: June 1983

Aluminum (Al) $Al_1(cr)$

$S°(298.15\ K) = [39.549]\ J\ K^{-1}\ mol^{-1}$

$T_{fus} = 933.45 \pm 0.01\ K$

$\Delta_f H°(298.15\ K) = [10.563]\ kJ\ mol^{-1}$

$\Delta_{fus} H° = 10.711 \pm 0.21\ kJ\ mol^{-1}$

Enthalpy of Formation

The enthalpy of formation of liquid aluminum is calculated from that of the crystal by adding $\Delta_{fus}H°$ and the difference in enthalpy, H°(933.45 K)-H°(298.15 K), between the crystal and liquid.

Heat Capacity and Entropy

The adopted heat capacity values are derived from the enthalpy study of McDonald (1). The enthalpy data, ten experimental points in the liquid region (941-1647 K), were measured in crucibles of BN and TiB_2 sealed in a platinum-rhodium capsule. This containment procedure reduced the problem with the reactivity of liquid aluminum. This data suggests a constant liquid heat capacity of 7.589 cal $K^{-1}\ mol^{-1}$ for this 700 K range.

The liquid enthalpy data of Awbery and Griffiths (2), four points in the liquid region (937-1036 K), agree within 1% with our adopted enthalpy values. The enthalpy studies of Wust et al. (3) and Umino (4) are lower by as much as 4% and 12%, respectively. The heat capacity values of Schmidt et al. (5), 17 values in the range 933-1300 K, are not constant but lie 6-11% lower than our adopted values.

Fusion Data

There are numerous enthalpy of fusion studies reported in the literature. Twelve studies (1, 2, 3, 5, 6, 7, 8, 9, 10, 11, 12, and 13) reported values in the range 2.46 to 2.66 kcal mol^{-1}, with a rather isolated value at 2.15 kcal mol^{-1} (4). We adopt a value 2.560 ± 0.05 kcal mol^{-1} based on the enthalpy data of McDonald (1). The chosen value has eliminated uncertainty due to reaction of liquid aluminum and its vessel by use of BN and TiB_2 containers.

The adopted melting temperature, T_{fus} = 933.45 K, is a secondary standard on IPTS-68 (14, 15).

Vaporization Data

The vaporization studies are summarized on the Al(g) table. The boiling point (fugacity of one bar), T_{vap} = 2790.812 K, is calculated as the temperature for which $\Delta_r G° = 0$ for Al(ℓ) = Al(g). The difference in the enthalpy of formation of Al(ℓ) and Al(g) at T_{vap} is the enthalpy of vaporization, $\Delta_{vap}H°$ = 294.001 kJ mol^{-1}.

References

1. R. A. McDonald, J. Chem. Eng. Data 12, 115 (1967).
2. J. H. Awbery and E. Griffiths, Proc. Phys. Soc. (London) 38, 378 (1926).
3. F. Wust, A. Meuthen, and R. Durer, Forsch. Arb. Ver. deut. Ing., 204 (1918).
4. S. Umino, Sci. Reps. Tohoku Imp. Univ., Ser. 1, 15, 297 (1926).
5. U. Schmidt, O. Vollmer, and R. Kohlhaas, Z. Naturforsch. 25, 1258 (1970).
6. R. Vogel, Z. Anorg. Chem. 75, 41 (1912).
7. J. H. Awbery, Phil. Mag. 26, 776 (1938).
8. O. Kubaschewski, Z. Elektrochem. 54, 275 (1950).
9. F. E. Wittig, Z. Metallk. 43, 158 (1952).
10. W. Oelsen, O. Oelsen, and D. Thiel, Z. Metallk. 46, 555 (1955).
11. W. Oelsen, K. H. Rieskamp, and Ol Oelsen, Arch. Eisenhuttenw. 26 253 (1955).
12. E. Schurmann and H. Trager, Arch. Eisenhuttenw. 32, 397 (1961).
13. D. M. Speros and R. L. Woodhouse, J. Phys. Chem. 67, 2164 (1963).
14. The International Practical Temperature Scale of 1968, Metrologia 5, 35 (1969).
15. G. T. Furukawa, J. L. Riddle et al., U.S. Nat. Bur. Stand., NBS Spec. Pub. 260-77, 1982.

Enthalpy Reference Temperature = T_r = 298.15 K — Standard State Pressure = p° = 0.1 MPa

T/K	$C_p°$ (J K⁻¹mol⁻¹)	S° (J K⁻¹mol⁻¹)	$-[G°-H°(T_r)]/T$ (J K⁻¹mol⁻¹)	$H°-H°(T_r)$ (kJ mol⁻¹)	$\Delta_f H°$ (kJ mol⁻¹)	$\Delta_f G°$ (kJ mol⁻¹)	Log K_f
0							
100							
200							
250							
298.15	24.209	39.549	39.549	0.	10.562	7.201	-1.262
300	24.247	39.699	39.550	0.045	10.562	7.180	-1.250
350	25.113	43.505	39.848	1.280	10.562	6.617	-0.987
400	25.784	46.904	40.522	2.553	10.562	6.053	-0.790
450	26.335	49.973	41.404	3.856	10.562	5.489	-0.637
500	26.842	52.774	42.403	5.186	10.562	4.926	-0.515
600	27.886	57.759	44.557	7.921	10.562	3.798	-0.331
700	29.100	62.146	46.762	10.769	10.562	2.671	-0.199
720.000	29.371	62.969	47.201	11.353	GLASS <--> LIQUID TRANSITION		
720.000	31.751	62.969	47.201	11.353			
800	31.751	66.315	48.948	13.893	10.706	1.535	-0.100
900	31.751	70.054	51.089	17.068	10.740	0.384	-0.022
933.450	31.751	71.213	51.790	18.131	------ CRYSTAL <--> LIQUID -----		
1000	31.751	73.400	53.156	20.244	0.	0.	0.
1100	31.751	76.426	55.136	23.419	0.	0.	0.
1200	31.751	79.189	57.027	26.594	0.	0.	0.
1300	31.751	81.730	58.831	29.769	0.	0.	0.
1400	31.751	84.083	60.552	32.944	0.	0.	0.
1500	31.751	86.273	62.194	36.119	0.	0.	0.
1600	31.751	88.323	63.764	39.294	0.	0.	0.
1700	31.751	90.247	65.266	42.469	0.	0.	0.
1800	31.751	92.062	66.704	45.644	0.	0.	0.
1900	31.751	93.779	68.085	48.819	0.	0.	0.
2000	31.751	95.408	69.410	51.994	0.	0.	0.
2100	31.751	96.957	70.686	55.169	0.	0.	0.
2200	31.751	98.434	71.914	58.344	0.	0.	0.
2300	31.751	99.845	73.098	61.519	0.	0.	0.
2400	31.751	101.196	74.240	64.694	0.	0.	0.
2500	31.751	102.493	75.345	67.870	0.	0.	0.
2600	31.751	103.738	76.413	71.045	0.	0.	0.
2700	31.751	104.936	77.447	74.220	0.	0.	0.
2790.812	31.751	105.986	78.359	77.103	------- FUGACITY = 1 bar -------		
2800	31.751	106.091	78.450	77.395	-293.901	0.968	-0.018
2900	31.751	107.205	79.422	80.570	-292.806	11.479	-0.207
3000	31.751	108.281	80.366	83.745	-291.710	21.953	-0.382

PREVIOUS: June 1979 CURRENT: June 1983

ALUMINUM (Al) CRYSTAL-LIQUID A_r = 26.98154

0 to 933.45 K crystal
above 933.45 K liquid

Refer to the individual tables for details.

Enthalpy Reference Temperature = T_r = 298.15 K Standard State Pressure = p° = 0.1 MPa

	J K⁻¹mol⁻¹			kJ mol⁻¹			
T/K	C_p°	S°	$-[G^\circ-H^\circ(T_r)]/T$	$H^\circ-H^\circ(T_r)$	Δ_fH°	Δ_fG°	Log K_f
0	0.	0.	INFINITE	-4.539	0.	0.	0.
100	12.997	6.987	47.543	-4.056	0.	0.	0.
200	21.338	19.144	30.413	-2.254	0.	0.	0.
250	23.084	24.108	28.668	-1.140	0.	0.	0.
298.15	24.209	28.275	28.275	0.	0.	0.	0.
300	24.247	28.425	28.276	0.045	0.	0.	0.
350	25.113	32.231	28.574	1.280	0.	0.	0.
400	25.784	35.630	29.248	2.553	0.	0.	0.
450	26.335	38.699	30.130	3.856	0.	0.	0.
500	26.842	41.501	31.129	5.186	0.	0.	0.
600	27.886	46.485	33.283	7.921	0.	0.	0.
700	29.100	50.872	35.488	10.769	0.	0.	0.
800	30.562	54.850	37.663	13.749	0.	0.	0.
900	32.308	58.548	39.780	16.890	0.	0.	0.
933.450	32.959	59.738	40.474	17.982	CRYSTAL <--> LIQUID		
933.450	31.751	71.213	40.474	28.693	TRANSITION		
1000	31.751	73.400	42.594	30.806	0.	0.	0.
1100	31.751	76.426	45.534	33.981	0.	0.	0.
1200	31.751	79.189	48.225	37.156	0.	0.	0.
1300	31.751	81.730	50.706	40.331	0.	0.	0.
1400	31.751	84.083	53.007	43.506	0.	0.	0.
1500	31.751	86.273	55.153	46.681	0.	0.	0.
1600	31.751	88.323	57.162	49.856	0.	0.	0.
1700	31.751	90.247	59.052	53.031	0.	0.	0.
1800	31.751	92.062	60.836	56.207	0.	0.	0.
1900	31.751	93.779	62.525	59.382	0.	0.	0.
2000	31.751	95.408	64.129	62.557	0.	0.	0.
2100	31.751	96.957	65.656	65.732	0.	0.	0.
2200	31.751	98.434	67.112	68.907	0.	0.	0.
2300	31.751	99.845	68.505	72.082	0.	0.	0.
2400	31.751	101.196	69.839	75.257	0.	0.	0.
2500	31.751	102.493	71.120	78.432	0.	0.	0.
2600	31.751	103.738	72.350	81.607	0.	0.	0.
2700	31.751	104.936	73.535	84.782	0.	0.	0.
2790.812	31.751	105.986	74.574	87.665	------- FUGACITY = 1 bar -------		
2800	31.751	106.091	74.677	87.957	-293.901	0.968	-0.018
2900	31.751	107.205	75.780	91.132	-292.805	11.479	-0.207
3000	31.751	108.281	76.846	94.307	-291.710	21.953	-0.382

PREVIOUS: CURRENT: June 1983

ALUMINUM (Al) IDEAL GAS A_r = 26.98154

IP(Al, g) = 48278.37 ± 0.02 cm^{-1}
S°(298.15 K) = 164.553 ± 0.03 $J\ K^{-1}\ mol^{-1}$

$\Delta_f H°$(0 K) = 327.3 ± 4.2 kJ mol^{-1}
$\Delta_f H°$(298.15 K) = 329.7 ± 4.2 kJ mol^{-1}

Electronic Levels and Quantum Weights

State	ε_i, cm^{-1}	g_i
$^2P°_{1/2}$	0	2
$^2P°_{3/2}$	112.061	4
$^2S_{1/2}$	25347.756	2
.	.	.
.	.	.
$^2D_{5/2}$	48182.	6
IP	48278.37	

Enthalpy of Formation

The enthalply of formation is derived from the vaporization measurements of Brewer and Searcy (1383-1468 K, 1), Porter et al. (932 K, 2), Potter and Hildenbrand (1400-1611 K, 3), Rao and Motzfeld (1556 K, 4), and Johnson et al. (1327 K, 5). The adopted value $\Delta_f H°$(Al, g, 298.15 K) = 78.8 ± 1.0 kcal mol^{-1} is that adopted by CODATA (6). The difference in S°(Al, cr, 298.15 K) between our adopted value and that of CODATA (6) would create a change of only 0.02 $kcal^{-1}$ mol^{-1} in $\Delta_f H°$(Al, g, 298.15 K). Other vaporization studies have been omitted in this analysis: Farkas (1476 K, 7), Baur and Brunner (1734-2237 K, 8), and Priselkov et al. (1273-1473 K, 9).

Heat Capacity and Entropy

The electronic energy levels are given in the compilation by Martin and Zalubas (10). Although we list only the ground state, the two lowest excited states, the highest observed excited state (below the ionization limit), and the ionization potential for Al(g), all levels below the ionization limit listed by Martin and Zalubas (10) are used in the calculation. The 105 observed levels below the ionization limit are too numerous to list completely. However, they include contributions from observed levels to n = 35. The contributions are from excitations of the 3p electron in the ground state configuration, $1s^2 2s^2 2p^6 3s^2 3p$. There is a contribution at ~29000 cm^{-1} due to the excitation of a 3s electron, giving rise to a $1s^2 2s^2 2p^6 3s 3p^2$ configuration. Other contributions of this latter type lie above the ionization limit.

The electronic level at 112.061 cm^{-1} has a significant effect on the thermodynamic values at low temperatures. The higher electronic states have an effect on the values above 3000 K. The Gibbs energy function values up to 4000 K are essentially independent of the estimated missing levels (for n < 35), the cut-off procedure, and the inclusion of levels for n > 3.

The thermal functions at 298.15 K agree with recent CODATA recommendations (6) except for two minor differences. First, the entropy differs by 0.1094 $J\ K^{-1}\ mol^{-1}$ because this table uses a standard state pressure of 1 bar, whereas the CODATA recommendations are based on 1 atm. Second, an entropy difference of 0.004 $J\ K^{-1}\ mol^{-1}$ arises due to the use of slightly different values for the fundamental constants.

References

1. L. Brewer and A. W. Searcy, J. Amer. Chem. Soc. 73, 5308 (1951).
2. R. E. Porter, P. Schissel, and M. G. Inghram, J. Chem. Phys. 23, 339 (1955).
3. N. D. Potter and D. L. Hildenbrand, Philco Corporation, Newport Beach, California; personal communication, Jan. 11, 1966.
4. D. B. Rao and Motzfeldt, Acta. Chem. Scand. 24, 2796 (1970).
5. R. G. Johnson, D. G. Hudson, W. C. Caldwell, F. H. Spedding, and W. R. Savage, J. Chem. Phys. 25, 917 (1958).
6. J. D. Cox, chairman, ICSU-CODATA Task Group on Key Values for Thermodynamics, J. Chem. Thermodynamics 10, 903 (1978).
7. L. Farkas, Z. Phys. 7, 735 (1931).
8. E. Baur and R. Brunner, Helv. Chim. Acta 17, 958 (1934).
9. Yu. A. Priselkov, Yu. A. Sapozhorikov, and A. V. Tseplyseva, Izv. Akad. Nauk., SSR, Otd. Tech. Nauk., No. 1, 106 (1959).
10. W. C. Martin and R. Zalubas, J. Phys. Chem. Ref. Data 8, 817 (1979).

Aluminum (Al) $Al_1(g)$

Enthalpy Reference Temperature = T_r = 298.15 K — Standard State Pressure = p° = 0.1 MPa

	J $K^{-1}mol^{-1}$			kJ mol^{-1}			
T/K	$C°_p$	S°	$-[G°-H°(T_r)]/T$	$H°-H°(T_r)$	$\Delta_f H°$	$\Delta_f G°$	Log K_f
0	0.	0.	INFINITE	-6.919	327.320	327.320	INFINITE
100	25.192	139.619	184.197	-4.458	329.297	316.034	-165.079
200	22.133	155.883	166.528	-2.129	329.824	302.476	-78.999
250	21.650	160.764	164.907	-1.036	329.804	295.639	-61.770
298.15	21.390	164.553	164.553	0.	329.699	289.068	-50.643
300	21.383	164.686	164.554	0.040	329.694	288.816	-50.287
350	21.221	167.969	164.813	1.104	329.524	282.016	-42.089
400	21.117	170.795	165.388	2.163	329.309	275.243	-35.943
450	21.046	173.278	166.130	3.217	329.060	268.499	-31.167
500	20.995	175.492	166.957	4.268	328.781	261.785	-27.348
600	20.930	179.314	168.708	6.363	328.141	248.444	-21.629
700	20.891	182.537	170.460	8.454	327.385	235.219	-17.552
800	20.866	185.325	172.147	10.542	326.492	222.112	-14.502
900	20.849	187.781	173.751	12.628	325.436	209.126	-12.137
1000	20.836	189.977	175.266	14.712	313.605	197.027	-10.292
1100	20.827	191.963	176.695	16.795	312.513	185.422	-8.805
1200	20.821	193.775	178.044	18.877	311.420	173.917	-7.570
1300	20.816	195.441	179.319	20.959	310.327	162.503	-6.529
1400	20.811	196.984	180.526	23.041	309.233	151.172	-5.640
1500	20.808	198.419	181.672	25.122	308.139	139.921	-4.872
1600	20.805	199.762	182.761	27.202	307.045	128.742	-4.203
1700	20.803	201.023	183.798	29.283	305.950	117.631	-3.614
1800	20.801	202.212	184.789	31.363	304.856	106.585	-3.093
1900	20.800	203.337	185.736	33.443	303.760	95.600	-2.628
2000	20.798	204.404	186.643	35.523	302.665	84.673	-2.211
2100	20.797	205.419	187.513	37.603	301.570	73.800	-1.836
2200	20.796	206.386	188.349	39.682	300.475	62.979	-1.495
2300	20.795	207.311	189.153	41.762	299.379	52.209	-1.186
2400	20.795	208.196	189.928	43.841	298.284	41.486	-0.903
2500	20.794	209.044	190.676	45.921	297.188	30.808	-0.644
2600	20.794	209.860	191.398	48.000	296.092	20.175	-0.405
2700	20.794	210.645	192.097	50.080	294.997	9.583	-0.185
2790.812	20.795	211.333	192.712	51.968	------- FUGACITY = 1 bar -------		
2800	20.795	211.401	192.773	52.159	0.	0.	0.
2900	20.796	212.131	193.428	54.239	0.	0.	0.
3000	20.798	212.836	194.063	56.318	0.	0.	0.
3100	20.800	213.518	194.680	58.398	0.	0.	0.
3200	20.804	214.178	195.279	60.478	0.	0.	0.
3300	20.808	214.818	195.861	62.559	0.	0.	0.
3400	20.815	215.440	196.428	64.640	0.	0.	0.
3500	20.823	216.043	196.980	66.722	0.	0.	0.
3600	20.833	216.630	197.518	68.805	0.	0.	0.
3700	20.846	217.201	198.042	70.889	0.	0.	0.
3800	20.862	217.757	198.553	72.974	0.	0.	0.
3900	20.881	218.299	199.053	75.061	0.	0.	0.
4000	20.904	218.828	199.541	77.150	0.	0.	0.
4100	20.932	219.345	200.017	79.242	0.	0.	0.
4200	20.964	219.849	200.484	81.337	0.	0.	0.
4300	21.002	220.343	200.940	83.435	0.	0.	0.
4400	21.046	220.826	201.386	85.537	0.	0.	0.
4500	21.058	221.299	201.823	87.642	0.	0.	0.
4600	21.143	221.763	202.252	89.753	0.	0.	0.
4700	21.206	222.219	202.672	91.870	0.	0.	0.
4800	21.276	222.666	203.084	93.995	0.	0.	0.
4900	21.352	223.105	203.488	96.125	0.	0.	0.
5000	21.439	223.537	203.885	98.264	0.	0.	0.
5100	21.535	223.963	204.274	100.413	0.	0.	0.
5200	21.641	224.382	204.657	102.572	0.	0.	0.
5300	21.757	224.795	205.033	104.742	0.	0.	0.
5400	21.884	225.203	205.403	106.924	0.	0.	0.
5500	22.021	225.606	205.766	109.119	0.	0.	0.
5600	22.170	226.004	206.124	111.328	0.	0.	0.
5700	22.330	226.398	206.476	113.553	0.	0.	0.
5800	22.496	226.787	206.823	115.792	0.	0.	0.
5900	22.680	227.173	207.165	118.050	0.	0.	0.
6000	22.836	227.552	207.501	120.306	0.	0.	0.

PREVIOUS: June 1979 (1 atm) CURRENT: June 1983 (1 bar)

Aluminum (Al) $Al_1(g)$

ALUMINUM, ION (Al^+) IDEAL GAS M_r = 26.98099

IP(Al^+, g) = 151862.7 ± 0.4 cm^{-1}
S°(298.15 K) = 149.950 ± 0.04 J K^{-1} mol^{-1}

$\Delta_f H°$(0 K) = 904.86 ± 4.2 kJ mol^{-1}
$\Delta_f H°$(298.15 K) = [912.713] kJ mol^{-1}

Electronic Levels and Quantum Weights

State	ε_i, cm^{-1}	g_i
$^1S_0$	0	1
$^3P_0$	37393.03	1
$^3P_1$	37453.91	3
$^3P_2$	37577.79	5
$^1P_1$	59852.02	3

Enthalpy of Formation

$\Delta_f H°(Al^+$, g, 0 K) is calculated from $\Delta_f H°$(Al, g, 0 K) (1) using the spectroscopic value of IP(Al) = 48278.37 ± 0.02 cm^{-1} (577.538 ± 0.0002 kJ mol^{-1}) from Martin and Zalubas (2). The ionization limit is converted from cm^{-1} to kJ mol^{-1} using the factor 1 cm^{-1} = 0.01196266 kJ cm^{-1}, which is derived from the 1973 CODATA fundamental constants (3). Rosenstock et al. (4) and Levin and Lias (5) have summarized additional ionization and appearance potential data.

$\Delta_f H°(Al^+$, g, 298.15 K) is calculated from $\Delta_f H°$(Al, g, 0 K) by using IP(Al) with JANAF (1) enthalpies, H°(0 K) - H°(298.15 K), for Al(g), Al^+(g), and e^-(g). $\Delta_f H°(Al + Al^+ + e^-$, 298.15 K) differs from a room temperature threshold energy due to inclusion of these enthalpies and to threshold effects discussed by Rosenstock et al. (4). $\Delta_f H°$(298.15 K) should be changed by -6.197 kJ mol^{-1} if it is to be used in the ion convention that excludes the enthalpy of the electron.

Heat Capacity and Entropy

The information on electronic energy levels and quantum weights, given by Martin and Zalubas (2), is incomplete because many theoretically predicted levels have not been observed. Our calculations indicate that any reasonable method of filling in these missing levels and cutting off the summation in the partition function (6) has no effect on the thermodynamic functions to 6000 K. This is a result of the high energy of all levels other than the ground state and the first four excited states; the next excited state is 85481.35 cm^{-1} above the ground state. Since inclusion of these excited states has no effect on the thermodynamic functions (to 6000 K), we list only the ground state and the first four excited states. The reported uncertainty in S°(298.15 K) is due to uncertainties in the relative ionic mass and the fundamental constants. Extension of these calculations above 6000 K may require consideration of the excited states and use of different fill and cutoff procedures (6).

References

1. JANAF Thermochemical Tables: Al(g), 6-30-83; e^-(g), 3-31-82.
2. W. C. Martin and R. Zalubas, J. Phys. Chem. Ref. Data 8, 817 (1979).
3. E. R. Cohen and B. N. Taylor, J. Phys. Chem. Ref. Data 2, 663 (1973).
4. H. M. Rosenstock, K. Draxl et al., J. Phys. Chem. Ref. Data 6, Supp. 1 (1977).
5. R. D. Levin and S. G. Lias, U. S. Nat. Bur. Stand., NSRDS-NBS-71, 1982.
6. J. R. Downey, Jr., The Dow Chemical Company, Report AFOSR-TR-78-0960, Contract No. F44620-75-1-0048, 1978.

Aluminum, Ion (Al^+) Al_1^+(g)

Enthalpy Reference Temperature = T_r = 298.15 K — Standard State Pressure = p° = 0.1 MPa

T/K	$C_p°$ (J $K^{-1}mol^{-1}$)	S° (J $K^{-1}mol^{-1}$)	-[G°-H°(T_r)]/T (J $K^{-1}mol^{-1}$)	H°-H°(T_r) (kJ mol^{-1})	$\Delta_f H°$ (kJ mol^{-1})	$\Delta_f G°$ (kJ mol^{-1})	Log K_f
0	0.	0.	INFINITE	-6.197	904.858		
100	20.786	127.243	168.431	-4.119			
200	20.786	141.651	151.852	-2.040			
250	20.786	146.289	150.292	-1.001			
298.15	20.786	149.950	149.950	0.	912.713	870.181	-152.452
300	20.786	150.079	149.951	0.038	912.745	869.917	-151.466
350	20.786	153.283	150.204	1.078	913.589	862.712	-128.753
400	20.786	156.059	150.766	2.117	914.394	855.388	-111.702
450	20.786	158.507	151.493	3.156	915.169	847.965	-98.429
500	20.786	160.697	152.306	4.196	915.919	840.458	-87.802
600	20.786	164.487	154.029	6.274	917.340	825.230	-71.843
700	20.786	167.691	155.758	8.353	918.650	809.774	-60.426
800	20.786	170.466	157.427	10.431	919.826	794.138	-51.852
900	20.786	172.915	159.014	12.510	920.843	778.364	-45.175
1000	20.786	175.105	160.516	14.589	911.084	763.246	-39.868
1100	20.786	177.086	161.934	16.667	912.066	748.415	-35.539
1200	20.786	178.894	163.273	18.746	913.049	733.494	-31.928
1300	20.786	180.558	164.539	20.824	914.031	718.492	-28.869
1400	20.786	182.099	165.739	22.903	915.013	703.413	-26.245
1500	20.786	183.533	166.878	24.982	915.995	688.265	-23.968
1600	20.786	184.874	167.961	27.060	916.977	673.051	-21.973
1700	20.786	186.134	168.994	29.139	917.959	657.775	-20.211
1800	20.786	187.322	169.979	31.217	918.941	642.442	-18.643
1900	20.786	188.446	170.922	33.296	919.924	627.054	-17.239
2000	20.786	189.512	171.825	35.375	920.906	611.615	-15.974
2100	20.786	190.527	172.692	37.453	921.888	596.126	-14.828
2200	20.786	191.493	173.524	39.532	922.870	580.590	-13.785
2300	20.786	192.417	174.326	41.610	923.852	565.010	-12.832
2400	20.786	193.302	175.098	43.689	924.834	549.387	-11.957
2500	20.786	194.151	175.844	45.768	925.816	533.723	-11.152
2600	20.786	194.966	176.563	47.846	926.799	518.020	-10.407
2700	20.786	195.750	177.260	49.925	927.781	502.280	-9.717
2800	20.786	196.506	177.934	52.004	634.862	487.470	-9.094
2900	20.786	197.236	178.587	54.082	636.939	482.169	-8.685
3000	20.786	197.940	179.220	56.161	639.017	476.797	-8.302
3100	20.787	198.622	179.835	58.239	641.094	471.355	-7.942
3200	20.787	199.282	180.433	60.318	643.172	465.847	-7.604
3300	20.788	199.922	181.013	62.397	645.248	460.273	-7.286
3400	20.788	200.542	181.579	64.476	647.325	454.637	-6.985
3500	20.790	201.145	182.129	66.555	649.400	448.939	-6.700
3600	20.791	201.730	182.666	68.634	651.475	443.183	-6.430
3700	20.793	202.300	183.189	70.713	653.549	437.368	-6.175
3800	20.796	202.855	183.699	72.792	655.622	431.498	-5.931
3900	20.800	203.395	184.197	74.872	657.693	425.573	-5.700
4000	20.805	203.922	184.684	76.952	659.763	419.594	-5.479
4100	20.811	204.435	185.159	79.033	661.830	413.565	-5.269
4200	20.818	204.937	185.624	81.115	663.896	407.484	-5.068
4300	20.828	205.427	186.079	83.197	665.958	401.355	-4.875
4400	20.839	205.906	186.524	85.280	668.018	395.178	-4.691
4500	20.853	206.374	186.960	87.365	670.077	388.953	-4.515
4600	20.869	206.833	187.387	89.451	672.130	382.684	-4.346
4700	20.888	207.282	187.806	91.539	674.179	376.369	-4.183
4800	20.910	207.722	188.216	93.628	676.223	370.011	-4.027
4900	20.935	208.153	188.618	95.721	678.264	363.610	-3.876
5000	20.965	208.577	189.013	97.816	680.298	357.168	-3.731
5100	20.999	208.992	189.401	99.914	682.326	350.686	-3.592
5200	21.037	209.400	189.782	102.015	684.347	344.163	-3.457
5300	21.079	209.801	190.156	104.121	686.362	337.602	-3.327
5400	21.127	210.196	190.523	106.232	688.369	331.003	-3.202
5500	21.181	210.584	190.884	108.347	690.368	324.366	-3.081
5600	21.240	210.966	191.240	110.468	692.358	317.694	-2.963
5700	21.305	211.342	191.589	112.595	694.339	310.986	-2.850
5800	21.376	211.714	191.933	114.729	696.313	304.243	-2.740
5900	21.453	212.080	192.271	116.870	698.274	297.466	-2.634
6000	21.538	212.441	192.604	119.020	700.247	290.654	-2.530

PREVIOUS: June 1979 (1 atm) CURRENT: June 1983 (1 bar)

Aluminum, Ion (Al^+) Al_1^+(g)

ALUMINUM, ION (Al^-) IDEAL GAS $M_r = 26.98209$

Aluminum, Ion (Al^-) $Al_1^-(g)$

EA(Al, g) = 0.441 ± 0.010 eV
S°(298.15 K) = 168.135 ± 0.08 J K^{-1} mol^{-1}

$\Delta_f H°$(0 K) = 284.77 ± 5.2 kJ mol^{-1}
$\Delta_f H°$(298.15 K) = [280.790] kJ mol^{-1}

Electonic Levels and Quantum Weights

State	ε_i, cm^{-1}	g_i
$^3P_0$	0.0	1
$^3P_0$	26.0	3
$^3P_2$	76.0	5
$^1D_2$	2678	5

Enthalpy of Formation

$\Delta_f H°(Al^-$, g, 0 K) is calculated from $\Delta_f H°$(Al, g, 0 K) (1) using the adopted election affinity of EA(Al) = 0.441 ± 0.010 eV (42.550 ± 0.965 kJ mol^{-1}). This value, recommended by Hotop and Lineberger (2), is based on a laser photodetachment electron spectrometry study (3). Additional information on $Al^-(g)$ may be obtained in the critical discussions of Hotop and Lineberger (2, 4), Rosenstock et al. (5) and Massey (6).

$\Delta_f H°(Al^-$, g, 298.15 K) is obtained from $\Delta_f H°$(Al, g, 0 K) by using EA(Al) with JANAF (1) enthalpies, H°(0 K)-H°(298.15 K), for $Al^-(g)$, Al(g), and $e^-(g)$. $\Delta_r H°(Al^- \rightarrow Al + e^-$, 298.15 K) differs from a room-temperature threshold energy due to inclusion of these enthalpies and to threshold effects discussed by Rosenstock et al. (5). $\Delta_f H°$(298.15 K) should be changed by +6.197 kJ mol^{-1} if it is to be used in the ion convention that excludes the enthalpy of the electron.

Heat Capacity and Entropy

The ground state electronic configuration for $Al^-(g)$ is given by Hotop and Lineberger (2, 4), Rosenstock et al. (5), and Massey (6). The fine-structure separation has been calculated via an isoelectronic extrapolation of ratios of fine structure separations (4) and is that recommended by Hotop and Lineberger (2).

A comparison of the isoelectronic sequence - $Al^-(g)$, Si(g), $P^+(g)$, and $S^{++}(g)$ - would suggest that two states, $^1D_2$ and $^1S_0$, may exist at low wave numbers. Feigirle et al. (3), using laser photodetachment electron spectrometry, observed the bound state, $^1D_2$, at 0.332 eV. The $^1S_0$ state would correspond to a metastable state.

References

1. JANAF Thermochemical Tables: Al(g), 6-30-83; $e^-(g)$, 3-31-82.
2. H. Hotop and W. C. Lineberger, J. Phys. Chem. Ref. Data, to be published, 1985.
3. C. S. Feigerle, R. R. Corderman and W. C. Lineberger, J. Chem. Phys. 74, 1513 (1981).
4. H. Hotop and W. C. Lineberger, J. Phys. Chem. Ref. Data 4, 539 (1975).
5. H. M. Rosenstock, K. Draxl et al., J. Phys. Chem. Ref. Data 6, Supp. 1 (1977).
6. H. S. W. Massey, "Negative Ions", 3rd ed., Cambridge University Press, Cambridge, 1976.

Enthalpy Reference Temperature = T_r = 298.15 K; Standard State Pressure = p° = 0.1 MPa

	J $K^{-1}mol^{-1}$			kJ mol^{-1}			
T/K	$C_p°$	S°	$-[G°-H°(T_r)]/T$	$H°-H°(T_r)$	$\Delta_f H°$	$\Delta_f G°$	Log K_f
0	0.	0.	INFINITE	-6.756	284.770		
100	22.297	144.748	186.955	-4.221			
200	21.171	159.730	170.057	-2.065			
250	21.031	164.437	168.480	-1.011			
298.15	20.960	168.135	168.135	0.	280.790	245.345	-42.983
300	20.958	168.265	168.135	0.039	280.745	245.125	-42.680
350	20.921	171.492	168.390	1.086	279.518	239.285	-35.711
400	20.913	174.285	168.957	2.131	278.251	233.624	-30.508
450	20.933	176.749	169.688	3.177	276.954	228.123	-26.480
500	20.983	178.957	170.506	4.225	275.633	222.768	-23.272
600	21.163	182.797	172.244	6.332	272.926	212.448	-18.495
700	21.421	186.078	173.992	8.460	270.129	202.588	-15.117
800	21.715	188.957	175.686	10.617	267.226	193.136	-12.610
900	22.006	191.532	177.306	12.803	264.192	184.055	-10.682
1000	22.270	193.864	178.847	15.017	250.412	176.081	-9.198
1100	22.493	195.998	180.311	17.256	247.397	168.794	-8.015
1200	22.670	197.963	181.701	19.514	244.402	161.780	-7.042
1300	22.804	199.783	183.023	21.788	241.422	155.016	-6.229
1400	22.897	201.477	184.281	24.074	238.454	148.481	-5.540
1500	22.955	203.058	185.481	26.367	235.493	142.157	-4.950
1600	22.985	204.541	186.626	28.664	232.537	136.031	-4.441
1700	22.991	205.935	187.721	30.963	229.582	130.090	-3.997
1800	22.978	207.249	188.770	33.261	226.627	124.323	-3.608
1900	22.950	208.490	189.776	35.558	223.670	118.720	-3.264
2000	22.912	209.666	190.741	37.851	220.709	113.273	-2.958
2100	22.865	210.783	191.669	40.140	217.744	107.974	-2.686
2200	22.812	211.846	192.562	42.424	214.775	102.816	-2.441
2300	22.756	212.859	193.423	44.702	211.799	97.794	-2.221
2400	22.697	213.826	194.253	46.975	208.818	92.901	-2.022
2500	22.636	214.751	195.054	49.241	205.831	88.133	-1.841
2600	22.576	215.638	195.829	51.502	202.838	83.484	-1.677
2700	22.515	216.489	196.579	53.757	199.839	78.950	-1.527
2800	22.455	217.306	197.304	56.005	-97.067	75.496	-1.408
2900	22.397	218.093	198.008	58.248	-98.983	81.692	-1.471
3000	22.340	218.852	198.690	60.485	-100.904	87.955	-1.531
3100	22.285	219.583	199.352	62.716	-102.831	94.282	-1.589
3200	22.231	220.290	199.996	64.942	-104.764	100.672	-1.643
3300	22.180	220.973	200.621	67.162	-106.703	107.121	-1.696
3400	22.130	221.635	201.229	69.378	-108.647	113.630	-1.746
3500	22.082	222.275	201.822	71.588	-110.597	120.196	-1.794
3600	22.036	222.897	202.398	73.794	-112.552	126.818	-1.840
3700	21.992	223.500	202.961	75.996	-114.514	133.494	-1.885
3800	21.950	224.086	203.509	78.193	-116.480	140.223	-1.928
3900	21.910	224.656	204.044	80.386	-118.453	147.004	-1.969
4000	21.871	225.210	204.566	82.575	-120.432	153.836	-2.009
4100	21.834	225.749	205.076	84.760	-122.417	160.717	-2.048
4200	21.799	226.275	205.575	86.942	-124.409	167.647	-2.085
4300	21.765	226.788	206.062	89.120	-126.407	174.624	-2.121
4400	21.733	227.288	206.539	91.295	-128.413	181.648	-2.156
4500	21.702	227.776	207.005	93.466	-130.424	188.717	-2.191
4600	21.672	228.252	207.462	95.635	-132.446	195.832	-2.224
4700	21.644	228.718	207.909	97.801	-134.476	202.990	-2.256
4800	21.617	229.174	208.348	99.964	-136.515	210.192	-2.287
4900	21.591	229.619	208.777	102.124	-138.564	217.436	-2.318
5000	21.566	230.055	209.198	104.282	-140.624	224.722	-2.348
5100	21.543	230.482	209.612	106.438	-142.696	232.050	-2.377
5200	21.520	230.900	210.017	108.591	-144.780	239.418	-2.405
5300	21.498	231.310	210.415	110.742	-146.878	246.826	-2.433
5400	21.477	231.711	210.806	112.890	-148.990	254.275	-2.460
5500	21.457	232.105	211.189	115.037	-151.117	261.762	-2.486
5600	21.438	232.492	211.566	117.182	-153.260	269.288	-2.512
5700	21.419	232.871	211.937	119.325	-155.421	276.853	-2.537
5800	21.401	233.243	212.301	121.466	-157.597	284.455	-2.562
5900	21.384	233.609	212.659	123.605	-159.795	292.096	-2.586
6000	21.368	233.968	213.011	125.743	-161.991	299.771	-2.610

PREVIOUS: June 1979 (1 atm) CURRENT: June 1983 (1 bar)

ALUMINUM BORATE ($AlBO_2$) IDEAL GAS M_r = 69.79034

$S°(298.15\ K) = [269.6 \pm 10]\ J\ K^{-1}\ mol^{-1}$

$\Delta_f H°(0\ K) = -539.0 \pm 17\ kJ\ mol^{-1}$
$\Delta_f H°(298.15\ K) = -541.4 \pm 17\ kJ\ mol^{-1}$

Vibrational Frequencies and Degeneracies

ν, cm^{-1}	ν, cm^{-1}
[2000](1)	[600](1)
[1000](1)	[350](1)
[600](1)	[800](1)

Ground State Quantum Weight: [1] $\sigma = 1$
Point Group: $[C_s]$
Bond Distances: Al-O = [1.66] Å O-B = [1.36] Å B=O = [1.20] Å
Bond Angles: Al-O-B = [145]° O-B=O = [180]°
Product of the Moments of Inertia: $I_A I_B I_C = [6.527732] \times 10^{-115}\ g^3\ cm^6$

Enthalpy of Formation

The enthalpy of formation, $\Delta_f H°(AlBO_2$, g, 298.15 K) = -129.4 ± 4 kcal mol^{-1} is calculated from the enthalpy of reaction, $\Delta_r H°(1500\ K) = -14.7 \pm 3.2$ kcal mol^{-1} for $1/2\ Al_2O(g) + 1/2\ B_2O_3(g) = AlBO_2(g)$ with all JANAF auxiliary data. The value of $\Delta_r H°(1500\ K)$ was obtained from the mass spectrometric determination of ion intensity ratios by Buchler et al. (1).

Heat Capacity and Entropy

The bent molecular structure, the bond distances O-B and B=O, and the bond angle O-B-O are assumed to be the same as those in $LiBO_2(g)$. The bond distance Al-O and the bond angle Al-O-(BO) are estimated to be the same as those in $Al_2O(g)$. The principal moments of inertia are $I_A = 0.7404 \times 10^{-39}$, $I_B = 29.3249 \times 10^{-39}$, and $I_C = 30.0653 \times 10^{-39}\ g\ cm^2$.

The vibrational frequencies were estimated by comparison with those in $LiBO_2(g)$, $Al_2O(g)$, $B_2O_3(g)$, and AlCl(g).

Reference

1. A. Buchler, P. E. Blackburn, and J. L. Stauffer, ARPA Order No. 315-67, Arthur D. Little, Inc., Cambridge, MA.

Aluminum Borate ($AlBO_2$) $Al_1B_1O_2(g)$

Enthalpy Reference Temperature = T_r = 298.15 K Standard State Pressure = p° = 0.1 MPa

	J $K^{-1}mol^{-1}$			kJ mol^{-1}			
T/K	$C_p°$	S°	$-[G°-H°(T_r)]/T$	$H°-H°(T_r)$	$\Delta_f H°$	$\Delta_f G°$	Log K_f
0	0.	0.	INFINITE	-12.008	-538.981	-538.981	INFINITE
100	34.879	223.687	310.212	-8.653	-539.038	-543.346	283.815
200	43.736	250.340	274.120	-4.756	-540.187	-547.246	142.926
250	48.632	260.633	270.413	-2.445	-540.819	-548.937	114.694
298.15	52.833	269.567	269.567	0.	-541.410	-550.447	96.436
300	52.982	269.894	269.568	0.098	-541.432	-550.503	95.851
350	56.680	278.347	270.227	2.842	-542.026	-551.967	82.377
400	59.785	286.124	271.734	5.756	-542.616	-553.347	72.260
450	62.400	293.321	273.738	8.812	-543.207	-554.653	64.382
500	64.622	300.014	276.035	11.989	-543.805	-555.893	58.074
600	68.173	312.125	281.062	18.638	-545.033	-558.196	48.595
700	70.861	322.845	286.280	25.596	-546.332	-560.288	41.809
800	72.936	332.449	291.462	32.790	-547.739	-562.188	36.707
900	74.562	341.137	296.506	40.168	-549.289	-563.902	32.728
1000	75.853	349.062	301.371	47.691	-561.605	-564.675	29.496
1100	76.890	356.342	306.042	55.330	-563.180	-564.906	26.825
1200	77.731	363.070	310.518	63.062	-564.761	-564.993	24.594
1300	78.420	369.320	314.804	70.871	-566.352	-564.948	22.700
1400	78.991	375.153	318.908	78.743	-567.958	-564.780	21.072
1500	79.468	380.620	322.842	86.666	-569.585	-564.497	19.658
1600	79.870	385.761	326.615	94.634	-571.236	-564.104	18.416
1700	80.211	390.614	330.239	102.638	-572.915	-563.606	17.318
1800	80.503	395.207	333.722	110.674	-574.624	-563.010	16.338
1900	80.754	399.567	337.073	118.737	-576.367	-562.317	15.459
2000	80.972	403.715	340.303	126.824	-578.146	-561.532	14.666
2100	81.162	407.670	343.417	134.931	-579.962	-560.657	13.946
2200	81.328	411.449	346.424	143.056	-581.818	-559.694	13.289
2300	81.475	415.068	349.331	151.196	-583.714	-558.646	12.687
2400	81.605	418.538	352.142	159.350	-635.911	-556.446	12.111
2500	81.721	421.872	354.865	167.517	-637.970	-553.093	11.556
2600	81.824	425.079	357.505	175.694	-640.039	-549.657	11.043
2700	81.916	428.169	360.065	183.881	-642.119	-546.141	10.566
2800	81.999	431.150	362.551	192.077	-938.112	-541.580	10.103
2900	82.074	434.028	364.966	200.280	-939.120	-527.401	9.500
3000	82.142	436.812	367.315	208.491	-940.141	-513.186	8.935
3100	82.204	439.506	369.600	216.709	-941.174	-498.937	8.407
3200	82.260	442.117	371.826	224.932	-942.220	-484.655	7.911
3300	82.312	444.649	373.995	233.161	-943.278	-470.340	7.445
3400	82.359	447.107	376.109	241.394	-944.348	-455.992	7.005
3500	82.402	449.495	378.172	249.632	-945.431	-441.614	6.591
3600	82.442	451.817	380.185	257.875	-946.525	-427.203	6.199
3700	82.479	454.077	382.152	266.121	-947.632	-412.763	5.827
3800	82.513	456.277	384.074	274.370	-948.752	-398.291	5.475
3900	82.544	458.420	385.953	282.623	-949.883	-383.790	5.140
4000	82.573	460.511	387.791	290.879	-951.027	-369.261	4.822
4100	82.600	462.550	389.589	299.138	-952.184	-354.702	4.519
4200	82.625	464.541	391.350	307.399	-1433.203	-333.091	4.143
4300	82.649	466.485	393.075	315.663	-1433.298	-306.897	3.728
4400	82.671	468.385	394.765	323.929	-1433.410	-280.701	3.332
4500	82.691	470.243	396.422	332.197	-1433.536	-254.502	2.954
4600	82.710	472.061	398.047	340.467	-1433.682	-228.300	2.592
4700	82.728	473.840	399.640	348.739	-1433.846	-202.094	2.246
4800	82.745	475.582	401.204	357.012	-1434.031	-175.884	1.914
4900	82.761	477.288	402.740	365.288	-1434.237	-149.670	1.596
5000	82.776	478.960	404.248	373.565	-1434.466	-123.452	1.290
5100	82.790	480.600	405.729	381.843	-1434.720	-97.229	0.996
5200	82.804	482.208	407.184	390.123	-1435.002	-71.002	0.713
5300	82.816	483.785	408.614	398.404	-1435.312	-44.768	0.441
5400	82.828	485.333	410.021	406.686	-1435.653	-18.529	0.179
5500	82.840	486.853	411.404	414.969	-1436.027	7.717	-0.073
5600	82.850	488.346	412.765	423.254	-1436.437	33.972	-0.317
5700	82.860	489.812	414.104	431.539	-1436.885	60.233	-0.552
5800	82.870	491.253	415.421	439.826	-1437.370	86.502	-0.779
5900	82.879	492.670	416.719	448.113	-1437.900	112.779	-0.998
6000	82.888	494.063	417.996	456.402	-1438.455	139.065	-1.211

PREVIOUS: June 1966 (1 atm) CURRENT: June 1966 (1 bar)

Aluminum Borate ($AlBO_2$) $Al_1B_1O_2(g)$

ALUMINUM BROMIDE (AlBr) IDEAL GAS M_r = 106.88554

$S°(298.15\ K) = 239.62 \pm 0.21\ J\ K^{-1}\ mol^{-1}$

$\Delta_f H°(0\ K) = 23.13 \pm 12.55\ kJ\ mol^{-1}$
$\Delta_f H°(298.15\ K) = 15.90 \pm 12.55\ kJ\ mol^{-1}$

Electronic Levels and Quantum Weights

State	ε_i, cm^{-1}	g_i
$X^1\Sigma^+$	0	1
	23663.	2
$a^3\Pi$	23795.3	2
	[23920]	[2]
$A^1\Pi$	35837.8	2
$[b^3\Sigma^+]$	[43000]	[3]

$\omega_e = 377.61\ cm^{-1}$ $\omega_e x_e = 1.3239\ cm^{-1}$ $\sigma = 1$
$B_e = 0.15871\ cm^{-1}$ $\alpha_e = 0.0008583\ cm^{-1}$ $r_e = 2.294803$ Å

Enthalpy of Formation

Ram (1), Gaydon (2), Singh (3), Rosen (4) and Barrow (5) reference possible spectroscopic values for the dissociation energy of AlBr(g). The $A^1\Pi$ state yielded three values for the dissociation energy: 106.9 kcal mol^{-1} by a short extrapolation, a minimum of 104.4 kcal mol^{-1} as indicated by the highest observed vibrational level, and a maximum of 105.7 kcal mol^{-1} as determined by predissociation which sets in at v' = 4. Ram (1) has observed rotational predissociation in the AlBr band system and reports the dissociation energy at 105.85±0.46 kcal mol^{-1}. Gaydon (2) reports a value of 86.5 kcal mol^{-1} based on a ground state linear Birge-Sponer extrapolation (v, 0-11). However, substituting the values tabulated above for ω_e and $x_e\omega_e$ into the equation $D° = \omega_e^2/4X_e\omega_e - 0.5\omega_e$, a value of 76.5 kcal mol^{-1} is obtained for the dissociation energy. A value of 105 kcal mol^{-1} is suggested by Barrow (5) based on wavelengths for maximum photoionization.

Semenkovich (6) measured the vapor pressure of the reaction Al(ℓ) + NaBr(ℓ) → Na(g) + AlBr(g) at 1175 K. By the 3rd Law method $\Delta_r H°(298.15\ K)$ is equal to 108.134 kcal mol^{-1}. Using JANAF auxiliary data (7), $\Delta_f H°$(AlBr, g, 298.15 K) is equal to 3.829 kcal mol^{-1} and the dissociation energy is 100.87 kcal mol^{-1}.

Gross (8) studied the equilibrium involved in the reaction 2 Al(ℓ) + $AlBr_3$(g) → 3 AlBr(g) in the range 550 - 1300°C. The reported $\Delta_r H°(298.15\ K)$ is 106.27 kcal mol^{-1}. Using auxiliary JANAF data (7), $\Delta_f H°(298.15\ K)$ is 4.43 kcal mol^{-1} and the dissociation energy is 100.26 kcal mol^{-1}.

It should be noted that the highest observed vibrational level in the $A^1\Pi$ state lies approximately 4 kcal mol^{-1} above the thermochemical dissociation energy of Semenkovich (6) and Gross (8). This suggests a potential maximum in the $A^1\Pi$ state as mentioned by Barrow (5). A similar maximum is suggested for AlF(g) and AlCl(g) (7, 9, 5). In addition, Wyse and Gordy (10), from nuclear quadrupole coupling, deduce that the ionic character of the AlBr bond is >8%. Thus the value of D_0^0 = 76.5 kcal mol^{-1} as obtained by a linear Birge-Sponer extrapolation is undoubtedly too low (2). This ground state extrapolation value was adjusted to 99.5 kcal mol^{-1} and 102.2 kcal mol^{-1} using two ionic character correction factors as suggested by Hildenbrand (9). The former value is more suitable for metal fluorides while the latter is for the more general case. A comparison of D_0^0 values (7, 2, 4) for the group IIIA gaseous monohalides suggests a value of 96-103 kcal mol^{-1}, which is consistent with the value obtained by Semenkovich (6) and Gross (8).

Due to the high ionic character in the AlBr bond and the possible unreliability of extrapolations of excited states which contain a potential maximum, the spectroscopic data is suspect as to the determination of $\Delta_f H°(298.15\ K)$. The chosen value for $\Delta_f H°(298.15\ K)$, 15.90 kJ mol^{-1}, is a rounded value from the vapor pressure work of Semenkovich (6).

Heat Capacity and Entropy

The spectroscopic constants for $Al^{79}Br$ and $Al^{81}Br$ were taken from Wyse and Gordy (10) and averaged according to the natural isotopic abundance of bromine. The electronic states are taken from Rosen (4) with the exception of the upper $^3\Pi$ state and the $b^3\Sigma^+$ state. The upper $^3\Pi$ state is estimated to be approximately in a linear relationship with the two lower $^3\Pi$ state. The $b^3\Sigma^+$ state is assigned by analogy with AlF and AlCl (7). Lakshminarayana and Haranath (11) also reported values for two of the $a^3\Pi$ states which are in excellent agreement with Rosen's values (4))

References

1. R. S. Ram, Spectroscopy Lett. 9, 435 (1976); Opt. Pura. Y. Appli (Spain) 6, 38 (1973).
2. A. G. Gaydon, "Dissociation Energies and Spectra of Diatomic Molecules" Chapman and Hall, Ltd., London, 1968.
3. J. Singh, K. P. R. Nair and D. K. Rai, J. Mol. Struct. 6, 328 (1970).
4. B. Rosen (ed.), "Donnes Spectroscopiques Relatives Aux Molecules Diatomiques," Pergamon Press, New York, 1970.
5. R. F. Barrow, Trans. Faraday Soc. 56, 952 (1960); Nature 189, 480 (1961).
6. S. A. Semenkovich, Zh. Prikl. Khim. 30, 933 (1957).
7. JANAF Thermochemical Tables: Al(ℓ), Al(g), AlCl(g), AlF(g), and AlI(g), 9-30-79; NaBr(ℓ), 9-30-64; Na(g), 6-30-62; Br(g), 6-30-74.
8. P. Gross, Fulmer Research Institute, Ltd., personal communication to H. Prophet, July, 1966.
9. D. L. Hildenbrand, "Advances in High Temperature Chemistry," Vol. 1, L. Eyring (ed) pp. 193 - 217, Academic Press, New York, 1967; D. L. Hildenbrand and E. Murad, J. Chem. Phys. 51, 807 (1969).
10. F. C. Wyse and W. Gordy, J. Chem. Phys. 56, 2130 (1974).
11. A. Lakshminarayana and P. B. V. Haranath, Curr. Sci. 39, 228 (1970).

Aluminum Bromide (AlBr) $Al_1Br_1(g)$

Enthalpy Reference Temperature = T_r = 298.15 K Standard State Pressure = p° = 0.1 MPa

T/K	$C_p°$ (J K^{-1}mol^{-1})	S° (J K^{-1}mol^{-1})	$-[G°-H°(T_r)]/T$ (J K^{-1}mol^{-1})	$H°-H°(T_r)$ (kJ mol^{-1})	$\Delta_f H°$ (kJ mol^{-1})	$\Delta_f G°$ (kJ mol^{-1})	Log K_f
0	0.	0.	INFINITE	-9.568	23.125	23.125	INFINITE
100	30.226	203.571	269.949	-6.638	24.175	7.209	-3.766
200	33.855	225.727	242.827	-3.420	23.141	-9.434	2.464
250	34.918	233.404	240.199	-1.699	22.340	-17.488	3.654
298.15	35.600	239.616	239.616	0.	15.899	-24.422	4.279
300	35.622	239.836	239.617	0.066	15.850	-24.672	4.296
350	36.102	245.366	240.052	1.860	0.084	-30.552	4.560
400	36.443	250.210	241.026	3.674	-0.290	-34.903	4.558
450	36.695	254.518	242.290	5.503	-0.685	-39.207	4.551
500	36.887	258.394	243.710	7.342	-1.100	-43.465	4.541
600	37.162	265.146	246.736	11.046	-1.992	-51.855	4.514
700	37.351	270.889	249.787	14.772	-2.982	-60.089	4.484
800	37.492	275.886	252.743	18.514	-4.097	-68.173	4.451
900	37.605	280.309	255.565	22.269	-5.365	-76.108	4.417
1000	37.700	284.276	258.241	26.035	-17.402	-83.132	4.342
1100	37.783	287.873	260.774	29.809	-18.695	-89.643	4.257
1200	37.858	291.164	263.172	33.591	-19.983	-96.036	4.180
1300	37.927	294.197	265.443	37.380	-21.269	-102.321	4.111
1400	37.992	297.010	267.599	41.176	-22.553	-108.508	4.048
1500	38.054	299.634	269.648	44.979	-23.835	-114.602	3.991
1600	38.114	302.091	271.600	48.787	-25.117	-120.612	3.938
1700	38.171	304.404	273.462	52.601	-26.398	-126.541	3.888
1800	38.228	306.587	275.242	56.421	-27.680	-132.395	3.842
1900	38.283	308.656	276.947	60.247	-28.964	-138.177	3.799
2000	38.337	310.621	278.582	64.078	-30.251	-143.892	3.758
2100	38.391	312.492	280.152	67.914	-31.541	-149.542	3.720
2200	38.445	314.280	281.663	71.756	-32.837	-155.131	3.683
2300	38.499	315.990	283.119	75.603	-34.137	-160.661	3.649
2400	38.553	317.629	284.523	79.456	-35.443	-166.134	3.616
2500	38.609	319.204	285.879	83.314	-36.755	-171.552	3.584
2600	38.666	320.720	287.190	87.178	-38.073	-176.918	3.554
2700	38.725	322.180	288.459	91.047	-39.397	-182.233	3.526
2800	38.788	323.590	289.689	94.923	-334.626	-186.532	3.480
2900	38.853	324.952	290.881	98.805	-334.862	-181.238	3.264
3000	38.924	326.270	292.039	102.694	-335.100	-175.937	3.063
3100	38.999	327.548	293.164	106.590	-335.340	-170.628	2.875
3200	39.081	328.787	294.258	110.494	-335.579	-165.311	2.698
3300	39.170	329.991	295.323	114.406	-335.815	-159.986	2.532
3400	39.267	331.162	296.359	118.328	-336.048	-154.654	2.376
3500	39.372	332.302	297.370	122.260	-336.274	-149.316	2.228
3600	39.487	333.412	298.356	126.203	-336.492	-143.971	2.089
3700	39.612	334.496	299.318	130.158	-336.699	-138.620	1.957
3800	39.749	335.554	300.258	134.126	-336.894	-133.265	1.832
3900	39.896	336.589	301.176	138.108	-337.074	-127.903	1.713
4000	40.056	337.601	302.074	142.105	-337.238	-122.538	1.600
4100	40.228	338.592	302.953	146.120	-337.384	-117.169	1.493
4200	40.413	339.563	303.813	150.152	-337.509	-111.796	1.390
4300	40.612	340.517	304.656	154.203	-337.612	-106.421	1.293
4400	40.823	341.453	305.481	158.274	-337.692	-101.043	1.200
4500	41.048	342.373	306.291	162.368	-337.744	-95.664	1.110
4600	41.286	343.277	307.085	166.484	-337.772	-90.284	1.025
4700	41.537	344.168	307.865	170.625	-337.772	-84.904	0.944
4800	41.801	345.045	308.630	174.792	-337.744	-79.524	0.865
4900	42.078	345.910	309.382	178.986	-337.685	-74.145	0.790
5000	42.368	346.763	310.121	183.208	-337.596	-68.768	0.718
5100	42.669	347.605	310.848	187.460	-337.477	-63.392	0.649
5200	42.982	348.436	311.563	191.743	-337.326	-58.019	0.583
5300	43.305	349.258	312.266	196.057	-337.144	-52.650	0.519
5400	43.639	350.071	312.959	200.404	-336.930	-47.284	0.457
5500	43.983	350.875	313.641	204.785	-336.684	-41.922	0.398
5600	44.336	351.670	314.313	209.201	-336.407	-36.566	0.341
5700	44.697	352.458	314.975	213.652	-336.099	-31.214	0.286
5800	45.066	353.239	315.628	218.141	-335.757	-25.868	0.233
5900	45.441	354.012	316.272	222.666	-335.388	-20.528	0.182
6000	45.822	354.779	316.908	227.229	-334.967	-15.197	0.132

PREVIOUS: September 1979 (1 atm) CURRENT: September 1979 (1 bar)

Aluminum Bromide (AlBr) $Al_1Br_1(g)$

ALUMINUM BROMIDE ($AlBr_3$) CRYSTAL M_r = 266.69354

$S°(298.15 K) = 180.22 \pm 1.05 J K^{-1} mol^{-1}$
$T_{fus} = 370.6 \pm 0.2 K$

$\Delta_f H°(0 K) = -491.90 \pm 0.84 kJ mol^{-1}$
$\Delta_f H°(298.15 K) = -511.28 \pm 0.84 kJ mol^{-1}$
$\Delta_{fus} H° = 11.25 \pm 0.84 kJ mol^{-1}$

Enthalpy of Formation

The value chosen for the enthalpy of formation is the weighted average of the direct combination value reported by Gross et al. (1) and the solution calorimetry value given by Efimov, Kislova and Medvedev (2).

Gross, Hayman, and Stuart (1) measured calorimetrically the standard enthalpy of formation of $AlBr_3$(cr)at 25°C by reacting Al with liquid bromine in a glass combustion vessel. They reported $\Delta_f H°(AlBr_3$, cr, 298.15 K) = -122.16±0.10 kcal mol^{-1}.

Efimov, Kislova, and Medvedev (2) measured the heat of solution of $AlBr_3$(cr) in an aqueous HCl solution containing KCl(cr) and the heat of solution of $AlCl_3$(cr) in an aqueous HCl solution containing KBr(cr). From these heats of solution measurements and using the heats of formation of $AlCl_3$(cr)(3), KCl(cr)(4, 5) and KBr(cr)(4, 6) as auxiliary data, they obtained $\Delta_f H°(AlBr_3$, cr, 298.15 K) = -122.5±0.5 kcal mol^{-1}.

Anthony et al. (10) measured the heat of solution of $AlBr_3$(cr) and $AlCl_3$(cr) in an aqueous alkaline solution containing dissolved sodium halides. They reported $\Delta_f H°(AlBr_3$, cr, 298.15 K) = -118.3±0.6 kcal mol^{-1}. Eley and Watts (11), in studying heats of solution of $AlBr_3$ and its complexes with pyridine, trimethylamine, and triethylamine in HCl, reported $\Delta_f H°(AlBr_3$,cr, 298.15 K) = -125.6 kcal mol^{-1}. Klemm and Tanke (12) measured the heat of solution of $AlBr_3$(cr) at 0°C in aqueous HCl. Using auxiliary data, including estimated heat capacities, Klemm and Tanke (11) reported $\Delta_f H°(AlBr_3$, cr, 298.15 K) = -121 kcal mol^{-1}. Gross, Hayman, and Stuart (1) have re-examined the data of Klemm and Tanke (12) and have calculated $\Delta_f H°(AlBr_3$, cr, 298.15 K)= -122.4 kcal mol^{-1}. This latter value was based on the heat of solution measurements (13), $\Delta_f H°(AlCl_3$, cr, 298.15 K) = -168.65 kcal mol^{-1}, and on the enthalpies of formation difference of chloride and bromide ions at the proper concentration.

A number of other studies on the enthalpies of formation of $AlBr_3$(cr) are based on solution calorimetry. The earliest measurements in water (7, 8, 9) are probably not reliable because of hydrolysis of the aluminum halide, but estimates of the enthalpy of formation of $AlBr_3$(cr) based on measurements in aqueous acid or base are also in disagreement.

Heat Capacity and Entropy

The heat capacities, 11.6 - 308 K, have been measured by Webb, Justice, and Prophet (13). Above 308 K they are derived (13) from the enthalpy data measured by Fischer (14). The entropy of the crystal is based on a value of S°(12 K) = 4.021 J K^{-1} mol^{-1}, as discussed in Webb et al. (13).

Fusion Data

Fischer (14) determined the melting temperature (T_{fus}) for $AlBr_3$(cr). The heat of melting ($\Delta_{fus}H°$) is calculated from the enthalpies of $AlBr_3$(cr) and $AlBr_3$(ℓ) at the melting temperature. The enthalpy of the crystal at T_{fus} was determined from the curve used for the heat capacity calculations. The enthalpy of the liquid at T_{fus} was determined from a linear least squares fit to the enthalpy data for $AlBr_3$(ℓ) reported by Fischer (14). See Webb et al. (13) for details.

Sublimation Data

The heat of sublimation is calculated from the enthalpies of formation of the dimer and the crystal. For the sublimation process, 2 $AlBr_3$(cr) = Al_2Br_6(g), $\Delta_{sub}H°(298.15 K) = +85.4 \pm 4.2 kJ mol^{-1}$. Also refer to the Al_2Br_6(g) table for additional comments.

References

1. P. Gross, C. Hayman, and M. C. Stuart, Fulmer Research Institute, Ltd., Sci. Rep. No. 9, Contract F61052-70-C0021, July 23, 1971.
2. M. E. Efimov, G. N. Kislova and V. A. Medvedev, J. Chem. Thermodynamics 12, 1149 (1980).
3. JANAF Thermochemical Tables: $AlCl_3$(cr), 9-30-79.
4. ICSU-CODATA Task Group on Key Values for Thermodynamics, J. Chem. Thermodynamics 10, 903 (1978).
5. V. B. Parker, NSRDS-NBS 2, 1965; U. S. Nat. Bur. Stand. Tech. Note 270-8, 1981.
6. M. E. Efimov, G. N. Klevaichuck, V. A. Medvedev, M. V. Kilday, J. Res. Natl. Bur. Stand. 84, 273 (1979).
7. M. P. E. Berthelot, Ann. Chim. Phys. 15, 185 (1878).
8. G. Gustavson, J. Russ. Phys. Chem. Soc. 51, 96 (1906)
9. V. A. Plotnikov, S. I. Jacubson, Zh. Fiz. Khim. 12, 113 (1938).
10. M. E. Anthony, A. Finch and P. J. Gardner, J. Chem. Soc. Dalton Trans. 1973, 659 (1973).
11. D. D. Eley and H. Watts, J. Chem. Soc. (London) 1954, 1319 (1954).
12. W. Klemm and E. Tanke, Z. anorg. allgem. Chem. 200, 343 (1931).
13. D. U. Webb, B. H. Justice, and H. Prophet, J. Chem. Thermodynamics 1, 227 (1969).
14. W. Fischer, Z. anorg. allgem. Chem. 200, 332 (1931).

Aluminum Bromide ($AlBr_3$) Al_1Br_3(cr)

Enthalpy Reference Temperature = T_r = 298.15 K Standard State Pressure = p° = 0.1 MPa

	J $K^{-1}mol^{-1}$			kJ mol^{-1}			
T/K	$C_p°$	S°	$-[G°-H°(T_r)]/T$	$H°-H°(T_r)$	$\Delta_f H°$	$\Delta_f G°$	Log K_f
0	0.	0.	INFINITE	-21.915	-491.896	-491.896	INFINITE
100	70.819	86.415	261.717	-17.530	-492.185	-492.051	257.021
200	89.947	142.206	189.044	-9.368	-493.176	-491.562	128.383
298.15	100.578	180.216	180.216	0.	-511.285	-488.515	85.586
300	100.816	180.839	180.218	0.186	-511.353	-488.374	85.033
370.600	113.078	203.294	182.510	7.703	------ CRYSTAL <--> LIQUID -----		
400	117.780	211.773	184.330	10.977	-554.791	-471.592	61.584
500	123.846	238.919	192.612	23.154	-550.784	-451.247	47.141
600	124.265	261.541	202.271	35.562	-546.690	-431.725	37.585
700	124.348	280.703	212.142	47.993	-542.715	-412.880	30.810
800	124.432	297.313	221.773	60.432	-538.885	-394.596	25.764
900	124.516	311.974	230.997	72.879	-535.225	-376.781	21.868
1000	124.600	325.097	239.762	85.335	-542.346	-358.604	18.732
1100	124.683	336.977	248.068	97.799	-538.731	-340.405	16.164
1200	124.767	347.829	255.936	110.272	-535.122	-322.535	14.040
1300	124.851	357.819	263.394	122.753	-531.517	-304.965	12.254
1400	124.934	367.075	270.473	135.242	-527.917	-287.674	10.733
1500	125.018	375.697	277.204	147.739	-524.323	-270.639	9.424

PREVIOUS: June 1972 CURRENT: September 1979

Aluminum Bromide ($AlBr_3$) Al_1Br_3(cr)

ALUMINUM BROMIDE ($AlBr_3$) LIQUID M_r = 266.69354

$S°(298.15\ K) = [206.481]\ J\ K^{-1}\ mol^{-1}$

$T_{fus} = 370.6 \pm 0.2\ K$

$\Delta_f H°(298.15\ K) = [-501.381]\ kJ\ mol^{-1}$

$\Delta_{fus} H° = 11.25 \pm 0.84\ kJ\ mol^{-1}$

Enthalpy of Formation

$\Delta_f H°(AlBr_3$, ℓ, 298.15 K) is calculated from $\Delta_f H°(AlBr_3$, cr, 298.15 K) by adding the enthalpy of fusion, $\Delta_{fus} H°$, and the difference in enthalpy, H°(370.6 K)−H°(298.15 K), between the crystal and liquid.

Heat Capacity and Entropy

Fischer (1) has measured the enthalpy of $AlBr_3$(ℓ) from 374 K to 406 K. The heat capacity is derived from a linear least squares fit to his data. The resulting constant value for the heat capacity is adopted for temperatures above 400 K.

S°($AlBr_3$, ℓ, 298.15 K) is calculated in a manner analogous to that used for the enthalpy of formation.

Fusion Data

See the crystal table for details.

Vaporization Data

See Al_2Br_6(g) table for details.

Reference

1. W. Fischer, Z. anorg. allgem. Chem. 200, 332 (1931).

Aluminum Bromide ($AlBr_3$) Al_1Br_3(l)

Enthalpy Reference Temperature = T_r = 298.15 K — Standard State Pressure = p° = 0.1 MPa

T/K	$C_p°$ (J K⁻¹mol⁻¹)	S° (J K⁻¹mol⁻¹)	−[G°−H°(T_r)]/T (J K⁻¹mol⁻¹)	H°−H°(T_r) (kJ mol⁻¹)	$\Delta_f H°$ (kJ mol⁻¹)	$\Delta_f G°$ (kJ mol⁻¹)	Log K_f
0							
100							
200							
298.15	124.965	206.481	206.481	0.	-501.381	-486.442	85.223
300	124.965	207.254	206.483	0.231	-501.404	-486.350	84.681
370.600	124.965	233.664	209.234	9.054	------ CRYSTAL <--> LIQUID -----		
400	124.965	243.204	211.385	12.728	-543.136	-472.510	61.703
500	124.965	271.089	220.641	25.224	-538.810	-455.357	47.571
600	124.965	293.873	231.005	37.721	-534.628	-439.062	38.224
700	124.965	313.136	241.397	50.217	-530.587	-423.455	31.599
800	124.965	329.823	251.431	62.714	-526.700	-408.418	26.667
900	124.965	344.542	260.975	75.210	-522.990	-393.858	22.859
1000	124.965	357.708	270.001	87.707	-530.070	-378.940	19.794
1100	124.965	369.618	278.525	100.203	-526.424	-364.003	17.285
1200	124.965	380.492	286.576	112.699	-522.790	-349.398	15.209
1300	124.965	390.494	294.190	125.196	-519.170	-335.096	13.464
1400	124.965	399.755	301.403	137.692	-515.562	-321.072	11.979
1500	124.965	408.377	308.251	150.189	-511.969	-307.305	10.701

PREVIOUS: June 1972 CURRENT: September 1979

Aluminum Bromide ($AlBr_3$) Al_1Br_3(l)

ALUMINUM BROMIDE ($AlBr_3$)   CRYSTAL-LIQUID M_r = 266.69354

0 to 370.6 K   crystal
above   370.6 K   liquid

Refer to the individual tables for details.

Aluminum Bromide ($AlBr_3$) Al_1Br_3(cr,l)

Enthalpy Reference Temperature = T_r = 298.15 K   Standard State Pressure = $p°$ = 0.1 MPa

T/K	$C_p°$ (J K⁻¹mol⁻¹)	$S°$ (J K⁻¹mol⁻¹)	$-[G°-H°(T_r)]/T$ (J K⁻¹mol⁻¹)	$H°-H°(T_r)$ (kJ mol⁻¹)	$\Delta_fH°$ (kJ mol⁻¹)	$\Delta_fG°$ (kJ mol⁻¹)	Log K_f
0	0.	0.	INFINITE	-21.915	-491.896	-491.896	INFINITE
100	70.819	86.415	261.717	-17.530	-492.185	-492.051	257.021
200	89.947	142.206	189.044	-9.368	-493.176	-491.562	128.383
298.15	100.578	180.216	180.216	0.	-511.285	-488.515	85.586
300	100.816	180.839	180.218	0.186	-511.353	-488.374	85.033
370.600	113.078	203.294	182.510	7.703	CRYSTAL <--> LIQUID		
370.600	124.965	233.664	182.510	18.958	TRANSITION		
400	124.965	243.204	186.625	22.632	-543.136	-472.510	61.703
500	124.965	271.089	200.833	35.128	-538.810	-455.357	47.571
600	124.965	293.873	214.498	47.625	-534.628	-439.061	38.224
700	124.965	313.136	227.249	60.121	-530.587	-423.455	31.599
800	124.965	329.823	239.051	72.618	-526.700	-408.418	26.667
900	124.965	344.541	249.970	85.114	-522.990	-393.857	22.859
1000	124.965	357.708	260.097	97.610	-530.070	-378 939	19.794
1100	124.965	369.618	269.521	110.107	-526.424	-364.003	17.285
1200	124.965	380.492	278.322	122.603	-522.790	-349.398	15.209
1300	124.965	390.494	286.571	135.100	-519.170	-335.095	13.464
1400	124.965	399.755	294.329	147.596	-515.562	-321.071	11.979
1500	124.965	408.377	301.648	160.093	-511.969	-307.305	10.701

PREVIOUS: CURRENT: September 1979

Aluminum Bromide ($AlBr_3$) Al_1Br_3(cr,l)

ALUMINUM BROMIDE ($AlBr_3$) IDEAL GAS M_r = 266.69354

$\Delta_{at}H°(0\ K) = 1068.2 \pm 2.1\ kJ\ mol^{-1}$
$S°(298.15\ K) = 349.44 \pm 1.26\ J\ K^{-1}\ mol^{-1}$

$\Delta_f H°(0\ K) = -387.16 \pm 1.7\ kJ\ mol^{-1}$
$\Delta_f H°(298.15\ K) = -410.45 \pm 1.7\ kJ\ mol^{-1}$

Vibrational Frequencies and Degeneracies

ν, cm^{-1}	ν, cm^{-1}
228(1)	508.4(2)
155(1)	93.0(2)

Ground State Quantum Weight: [1] σ = 6
Point Group: D_{3h}
Bond Distance: Al-Br = [2.27 ± 0.10] Å
Bond Angle: Br-Al-Br = 120°
Product of the Moments of Inertia: $I_A I_B I_C = [2.157410 \times 10^{-111}]\ g^3\ cm^6$

Enthalpy of Formation

Fischer, Rahlfs, and Benze (1) have determined the equilibrium constants for the reaction $Al_2Br_6(g)$=2 $AlBr_3(g)$ from vapor density measurements. The heats of reaction are calculated from these data using both 2nd law and 3rd law methods. The results of the calculations are as follows, where the $\Delta_f H°(298.15\ K)$ is based on the 3rd law value of $\Delta_r H°(298.15\ K)$, for the reaction as written above, and $\Delta_f H°(Al_2Br_6,\ g,\ 298.15\ K) = -224.0\ kcal\ mol^{-1}$.

T/K	Data Points	$\Delta_r H°(298.15\ K)$, kcal mol^{-1} Al_2Br_6 2nd law	3rd law	Drift cal K^{-1} mol^{-1}	$\Delta_f H°(298.15\ K)$ kcal mol^{-1}
613-846	4	29.6±1.0	27.78±0.6	-2.6±1.3	-98.11
605-854	7	29.4±0.5	27.73±0.5	-2.2±0.7	-98.14

The adopted value for the enthalpy of formation is $\Delta_f H°(AlBr_3,\ g,\ 298.15\ K) = -98.1\pm0.4\ kcal\ mol^{-1}(-410.45\pm1.7\ kj\ mol^{-1})$.

Heat Capacity and Entropy

Beattie and Horder (2) studied the gas phase Raman spectra of monomeric $AlBr_3$. The results are consistent with a planar structure of D_{3h} symmetry with ν_1=228 cm^{-1} (polarized) and ν_4=93 cm^{-1}. The weak antisymmetric stretching mode was tentatively assigned as ν_3=360 cm^{-1}.

Pong et al. (3), in examining the infrared spectrum of the monomeric $AlBr_3$ isolated in solid argon, reported ν_2=155.0 cm^{-1}, ν_3=510 cm^{-1} and ν_4=92 cm^{-1} for $AlBr_3$ isolated in xenon.

The adopted values for ν_1 and ν_4 are taken from Beattie and Horder (2) while ν_2 and ν_3 are taken from Pong et al. (3). The Al-Br bond distance was estimated by Heise and Weiland (5) and is consistent with the trend of experimentally known bond distances in the gaseous molecules AlF_3, $AlCl_3$, and AlI_3(6). The principal moments of inertia are $I_A = I_B = 102.5575 \times 10^{-39}$, and I_C = 205.1151 x 10^{-39} g cm^2.

References

1. W. Fischer, O. Rahlfs and R. B. Benze, J. anorg. allgem. Chem. 205, 1 (1932).
2. I. R. Beattie and J. R. Horder, J. Chem. Soc. (London) 1969A, 2655 (1969).
3. P. G. S. Pong, A. E. Shirk and J. S. Shirk, J. Chem. Phys. 70, 525 (1979).
4. P. A. Perov, S. V. Nedyak and M. A. Mal'tsev, Vest. Moskov. Univ. Khim. 29, 2901 (1974).
5. M. Heise and K. Weiland, Helv. Chim. Acta. 34, 2182 (1951).
6. JANAF Thermochemical Tables: $AlF_3(g)$, $AlCl_3(g)$, and $AlI_3(g)$, 9-30-79.

Aluminum Bromide ($AlBr_3$) $Al_1Br_3(g)$

Enthalpy Reference Temperature = T_r = 298.15 K   Standard State Pressure = p° = 0.1 MPa

T/K	$C_p°$ (J $K^{-1}mol^{-1}$)	S°	$-[G°-H°(T_r)]/T$	$H°-H°(T_r)$ (kJ mol^{-1})	$\Delta_f H°$	$\Delta_f G°$	Log K_f
0	0.	0.	INFINITE	-18.015	-387.162	-387.162	INFINITE
100	57.421	276.467	412.068	-13.560	-387.381	-406.252	212.204
200	69.507	320.454	356.177	-7.144	-390.119	-424.154	110.778
250	73.019	336.366	350.669	-3.576	-391.887	-432.465	90.359
298.15	75.376	349.441	349.441	0.	-410.450	-438.135	76.759
300	75.451	349.907	349.442	0.140	-410.566	-438.307	76.316
350	77.155	361.673	350.367	3.957	-456.958	-440.380	65.723
400	78.376	372.060	352.443	7.847	-457.086	-438.002	57.197
450	79.271	381.346	355.147	11.789	-457.209	-435.609	50.564
500	79.943	389.734	358.193	15.770	-457.333	-433.203	45.256
600	80.861	404.398	364.707	23.814	-457.604	-428.352	37.291
700	81.439	416.909	371.293	31.931	-457.942	-423.451	31.598
800	81.825	427.810	377.691	40.096	-458.387	-418.495	27.325
900	82.094	437.464	383.806	48.293	-458.977	-413.475	23.997
1000	82.289	446.124	389.612	56.512	-470.334	-407.620	21.292
1100	82.434	453.975	395.112	64.749	-470.948	-401.319	19.057
1200	82.546	461.152	400.321	72.998	-471.561	-394.962	17.192
1300	82.633	467.763	405.258	81.257	-472.178	-388.554	15.612
1400	82.703	473.889	409.944	89.524	-472.800	-382.098	14.256
1500	82.759	479.597	414.399	97.797	-473.431	-375.597	13.079
1600	82.805	484.940	418.643	106.075	-474.071	-369.054	12.048
1700	82.844	489.961	422.692	114.358	-474.725	-362.470	11.137
1800	82.876	494.697	426.562	122.644	-475.395	-355.848	10.326
1900	82.903	499.179	430.267	130.933	-476.084	-349.188	9.600
2000	82.927	503.432	433.820	139.225	-476.796	-342.490	8.945
2100	82.947	507.479	437.232	147.518	-477.534	-335.757	8.352
2200	82.964	511.338	440.513	155.814	-478.299	-328.988	7.811
2300	82.979	515.026	443.673	164.111	-479.095	-322.184	7.317
2400	82.993	518.558	446.720	172.410	-479.923	-315.344	6.863
2500	83.005	521.946	449.662	180.709	-480.783	-308.469	6.445
2600	83.015	525.202	452.505	189.010	-481.677	-301.559	6.058
2700	83.024	528.335	455.256	197.312	-482.603	-294.613	5.700
2800	83.033	531.354	457.920	205.615	-777.462	-286.666	5.348
2900	83.040	534.268	460.503	213.919	-777.353	-269.139	4.848
3000	83.047	537.084	463.009	222.223	-777.272	-251.616	4.381
3100	83.053	539.807	465.443	230.528	-777.214	-234.095	3.944
3200	83.059	542.444	467.808	238.834	-777.177	-216.576	3.535
3300	83.064	545.000	470.109	247.140	-777.157	-199.057	3.151
3400	83.069	547.479	472.348	255.447	-777.150	-181.540	2.789
3500	83.073	549.887	474.529	263.754	-777.154	-164.022	2.448
3600	83.077	552.228	476.655	272.061	-777.163	-146.503	2.126
3700	83.080	554.504	478.728	280.369	-777.174	-128.985	1.821
3800	83.084	556.720	480.752	288.677	-777.184	-111.466	1.532
3900	83.087	558.878	482.728	296.986	-777.188	-93.947	1.258
4000	83.089	560.981	484.658	305.295	-777.185	-76.428	0.998
4100	83.092	563.033	486.544	313.604	-777.171	-58.910	0.751
4200	83.095	565.035	488.389	321.913	-777.144	-41.391	0.515
4300	83.097	566.991	490.195	330.223	-777.102	-23.874	0.290
4400	83.099	568.901	491.962	338.532	-777.042	-6.358	0.075
4500	83.101	570.769	493.693	346.842	-776.960	11.157	-0.130
4600	83.103	572.595	495.388	355.153	-776.860	28.670	-0.326
4700	83.105	574.382	497.050	363.463	-776.738	46.180	-0.513
4800	83.106	576.132	498.679	371.773	-776.595	63.687	-0.693
4900	83.108	577.846	500.277	380.084	-776.428	81.191	-0.866
5000	83.109	579.525	501.846	388.395	-776.239	98.692	-1.031
5100	83.110	581.170	503.385	396.706	-776.028	116.189	-1.190
5200	83.112	582.784	504.896	405.017	-775.795	133.681	-1.343
5300	83.113	584.367	506.381	413.328	-775.540	151.169	-1.490
5400	83.114	585.921	507.840	421.640	-775.265	168.651	-1.631
5500	83.115	587.446	509.273	429.951	-774.970	186.128	-1.768
5600	83.116	588.944	510.682	438.263	-774.656	203.600	-1.899
5700	83.117	590.415	512.068	446.574	-774.325	221.066	-2.026
5800	83.118	591.860	513.432	454.886	-773.975	238.525	-2.148
5900	83.119	593.281	514.773	463.198	-773.613	255.979	-2.266
6000	83.120	594.678	516.093	471.510	-773.215	273.424	-2.380

PREVIOUS: September 1979 (1 atm)   CURRENT: September 1979 (1 bar)

Aluminum Bromide ($AlBr_3$) $Al_1Br_3(g)$

ALUMINUM CHLORIDE (AlCl) IDEAL GAS M_r = 62.43454

$S°(298.15\ K) = 227.957 \pm 0.08\ J\ K^{-1}\ mol^{-1}$

$\Delta_f H°(0\ K) = -51.66 \pm 6.3\ kJ\ mol^{-1}$

$\Delta_f H°(298.15\ K) = -51.46 \pm 6.3\ kJ\ mol^{-1}$

Electronic Levels and Quantum Weights

State	ε_i, cm^{-1}	g_i
$X\ ^1\Sigma^+$	0	1
	24541.65	2
$a\ ^3\pi$	24615.31	2
	24680.00	2
$A\ ^1\pi$	38237.70	2
$b\ ^3\Sigma^+$	43590.00	3

$\omega_e = 480.2903\ cm^{-1}$ $\omega_e x_e = 2.064\ cm^{-1}$ $\sigma = 1$

$B_e = 0.24255\ cm^{-1}$ $\alpha_e = 0.001598\ cm^{-1}$ $r_e = 2.130113$ Å

Enthalpy of Formation

Equilibrium constants as a function of temperature for the reaction 2 Al(ℓ) + $AlCl_3$(g) → 3 AlCl(g) have been reported by many investigators. The heats of reaction are calculated from these data using second and third law methods. Unfortunately, most of the data are very discordant and of no help in defining $\Delta_f H°$(AlCl, g, 298.15 K). The results of Rao and Dadape (1) cover a temperature range of 300 K, show good compatibility with the thermodynamic functions, are near an average of all the data, and are judged most acceptable. The third-law heat of reaction derived from their data is adopted and combined with the heat of formation of $AlCl_3$(g) (12) to give $\Delta_f H°$(AlCl, g, 298.15 K) = -12.30±1.5 kcal mol^{-1} (-51.46±6.3 kJ mol^{-1}).

Source	T/K	Data Points	$\Delta_r H°$(298.15 K), kcal mol^{-1} 2nd law	3rd law	Drift cal K^{-1} mol^{-1}	$\Delta_f H°$(298.15 K) kcal mol^{-1}
Rao and Dadape (1)	1125-1425	11	97.25±1.6	97.71±0.8	0.4±1.3	-12.30
Frish et al. (2)	931-1034	9	94.58±1.0	101.64±0.5	7.2±1.0	-10.99
Gross et al. (3)	1225-1278[a]	7	--	104.60±2.8	--	-10.00
Ginsberg et al. (4)	1423-1500	8	86.34±1.8	99.66±0.8	9.1±1.2	-11.65
Semenkovich (5)	1223-1473	6	84.19±0.4	93.42±1.3	6.9±0.3	-13.73
Heimgartner (6)	978-1383	13	55.52±2.6	94.90±9.5	33.0±2.2	-13.24
Weiss (7)	1273-1473	3	88.29±3.7	81.26±1.2	-5.1±2.7	-17.78
Pinchuk et al. (8)	1173-1373	3	80.20±7.4	89.92±1.9	7.6±5.8	-14.90
Mitani et al. (9)	1273-1473[b]	2	81.29±0.0	94.18±2.7	9.4±0.0	-13.48
Tanabe et al. (10)	1173-1573[b]	2	88.23±0.0	89.49±0.5	0.9±0.0	-15.04
Kikuch et al. (11)	1273-1523[b]	2	73.61±0.0	93.60±5.1	14.3±0.0	-13.67

a. Temperature range too small for significant second law calculation.

b. Individual points not reported; calculation based on reported equation at extremes of temperature range cited.

Gross et al. (3) and Pinchuk et al. (8) have studied the reaction of sodium and potassium chloride with liquid aluminum to form aluminum monochloride. These studies lead to $\Delta_f H°$(298.15 K) values for AlCl(g) in the range of -12.69±1.0 to -13.60±1.5 kcal mol^{-1}.

The adopted enthalpy of formation yields a dissociation energy of 498.57 kJ mol^{-1} for AlCl(g). Barrow (13) and Ram (14) reference possible spectroscopic values for the dissociation energy. A short extrapolation of vibrational levels yields D° = 519.2 kJ mol^{-1} but Barrow (13) speculates that a potential maximum in the $A^1\pi$ state of AlF and AlCl causes extrapolation of levels to give erroneous heats of dissociation. Ram et al. has observed rotational predissociation in the AlCl bond system and reports the dissociation energy at 507.5±0.97 kJ mol^{-1}. This yields $\Delta_f H°$(AlCl, g, 298.15 K) = -60.42 kJ mol^{-1} in reasonable agreement with the selected value.

Heat Capacity and Entropy

The spectrum of AlCl has been observed and analyzed by a number of investigators (14, 15, 16, 17). The spectroscopic constants for $Al^{35}Cl$ and $Al^{37}Cl$ are taken from the microwave study of Wyse and Gordy (17) and averaged according to the natural isotopic abundance of chlorine. The electronic levels are those reported by Sharma (18); the value for the $b^3\Sigma^+$ state is somewhat uncertain. The assignment of electronic states is supported by theoretical calculations (19).

References

1. D. B. Rao and V. V. Dadape, J. Phys. Chem. 70, 1349 (1966).
2. M. A. Frisch, M. A. Greenbaum, and M. Farber, J. Phys. Chem. 69, 3001 (1965).
3. P. Gross, C. S. Campbell, P. J. C. Kent, and D. L. Levi, Discuss. Faraday Soc. 4, 206 (1948).
4. H. Ginsberg and V. Sparwald, Aluminum 41, 219 (1965).
5. S. A. Semenkovich, Zh. Prikl Khim. 33, 1281 (1960).
6. R. Heimgartner, Schweiz, Arch. Angew. Wiss. Tech. 18, 241 (1952).
7. P. Weiss, Z. Erzbergau Metallhuettenw. 3, 241 (1950).
8. Y. M. Pinchuk and A. I. Belyaev, Izv. Vysshikh Uchebn. Zavedenii, Tsvetn. Met. 7, 71 (1964).
9. H. Mitani and H. Nagai, Nippon Kinzoku Gakkaishi 31, 1296 (1967); CA 68, 90375g.
10. T. Kukuchi, T. Kurosawa and T. Yagihashi, Trans. Japan Inst. Metals 5, 122-7 (1964); CA 62, 4957h.
11. I. Tanabe, H. Konno, Y. Sawada, and T. Takahashi, Denki Kagaku 32, 285 (1964); CA 62, 4993f.
12. JANAF Thermochemical Tables: Al(ℓ), $AlCl_3$(g), 9-30-79; NaCl(cr), NaCl(ℓ), 9-30-64; Na(g), 6-30-62; KCl(cr), 3-31-66; K(g), 6-30-62.
13. R. F. Barrow, Trans. Faraday Soc. 56, 952 (1960).
14. R. S. Ram, S. B. Rai, D. K. Bai and K. N. Upadhya, Indian J. Phys. 53B, 56 (1979); ibid., J. Chem. Phys. Phys.-Chim. Biol. 76, 560 (1979).
15. D. R. Lide, J. Chem. Phys. 42, 1013 (1965).
16. A. K. Chaudry and K. N. Upadhya, Indian J. Phys. 42, 544 (1968).
17. F. C. Wyse and W. Gordy, J. Chem. Phys. 56, 2130 (1972).
18. D. Sharma, Astrophys. J. 113, 210 (1951).
19. P. Carsky and J. Málek, Collec. Czech. Chem. Commun. 42, 2758 (1977).

Aluminum Chloride (AlCl) $Al_1Cl_1(g)$

Enthalpy Reference Temperature = T_r = 298.15 K — Standard State Pressure = p° = 0.1 MPa

T/K	$C_p°$ (J $K^{-1}mol^{-1}$)	S°	$-[G°-H°(T_r)]/T$	$H°-H°(T_r)$ (kJ mol^{-1})	$\Delta_f H°$	$\Delta_f G°$	Log K_f
0	0.	0.	INFINITE	-9.323	-51.656	-51.656	INFINITE
100	29.527	193.170	257.245	-6.407	-50.681	-59.849	31.262
200	32.545	214.524	231.068	-3.309	-50.903	-68.982	18.016
250	33.787	221.927	228.523	-1.649	-51.166	-73.473	15.351
298.15	34.660	227.957	227.957	0.	-51.463	-77.743	13.620
300	34.689	228.171	227.957	0.064	-51.475	-77.906	13.565
350	35.338	233.570	228.382	1.816	-51.818	-82.284	12.280
400	35.814	238.321	229.334	3.595	-52.188	-86.611	11.310
450	36.170	242.561	230.572	5.395	-52.579	-90.891	10.550
500	36.445	246.387	231.965	7.211	-52.990	-95.126	9.938
600	36.835	253.068	234.942	10.876	-53.876	-103.472	9.008
700	37.099	258.768	237.948	14.574	-54.862	-111.661	8.332
800	37.291	263.735	240.868	18.294	-55.973	-119.701	7.816
900	37.439	268.136	243.658	22.030	-57.237	-127.593	7.405
1000	37.559	272.087	246.306	25.780	-69.271	-134.575	7.029
1100	37.660	275.671	248.815	29.541	-70.560	-141.042	6.698
1200	37.749	278.952	251.192	33.312	-71.846	-147.393	6.416
1300	37.829	281.977	253.445	37.091	-73.128	-153.637	6.173
1400	37.901	284.783	255.585	40.878	-74.407	-159.782	5.962
1500	37.969	287.400	257.619	44.671	-75.684	-165.835	5.775
1600	38.033	289.853	259.558	48.471	-76.959	-171.804	5.609
1700	38.095	292.160	261.409	52.278	-78.232	-177.693	5.460
1800	38.153	294.340	263.178	56.090	-79.503	-183.507	5.325
1900	38.211	296.404	264.873	59.908	-80.774	-189.250	5.203
2000	38.266	298.365	266.499	63.732	-82.044	-194.926	5.091
2100	38.321	300.234	268.061	67.562	-83.314	-200.539	4.988
2200	38.375	302.018	269.565	71.396	-84.584	-206.092	4.893
2300	38.428	303.725	271.013	75.236	-85.856	-211.586	4.805
2400	38.482	305.361	272.410	79.082	-87.130	-217.025	4.723
2500	38.536	306.933	273.760	82.933	-88.407	-222.411	4.647
2600	38.590	308.446	275.065	86.789	-89.687	-227.746	4.575
2700	38.647	309.903	276.329	90.651	-90.970	-233.032	4.508
2800	38.705	311.310	277.553	94.519	-386.158	-237.302	4.427
2900	38.766	312.669	278.741	98.392	-386.353	-231.983	4.178
3000	38.830	313.984	279.894	102.272	-386.552	-226.656	3.946
3100	38.898	315.258	281.014	106.158	-386.754	-221.323	3.729
3200	38.971	316.495	282.103	110.052	-386.959	-215.983	3.526
3300	39.050	317.695	283.164	113.953	-387.165	-210.637	3.334
3400	39.135	318.862	284.197	117.862	-387.372	-205.284	3.154
3500	39.228	319.998	285.203	121.780	-387.579	-199.926	2.984
3600	39.328	321.104	286.185	125.708	-387.783	-194.561	2.823
3700	39.437	322.183	287.144	129.646	-387.984	-189.191	2.671
3800	39.556	323.237	288.080	133.595	-388.179	-183.816	2.527
3900	39.684	324.266	288.995	137.557	-388.367	-178.436	2.390
4000	39.823	325.272	289.889	141.533	-388.546	-173.050	2.260
4100	39.973	326.257	290.764	145.522	-388.713	-167.661	2.136
4200	40.134	327.222	291.621	149.528	-388.868	-162.268	2.018
4300	40.308	328.169	292.460	153.550	-389.006	-156.871	1.906
4400	40.493	329.098	293.282	157.589	-389.128	-151.471	1.798
4500	40.690	330.010	294.088	161.649	-389.229	-146.068	1.696
4600	40.900	330.906	294.878	165.728	-389.311	-140.664	1.597
4700	41.122	331.788	295.654	169.829	-389.370	-135.258	1.503
4800	41.357	332.656	296.416	173.953	-389.405	-129.851	1.413
4900	41.603	333.512	297.165	178.101	-389.414	-124.443	1.327
5000	41.862	334.355	297.900	182.274	-389.397	-119.036	1.244
5100	42.132	335.186	298.623	186.473	-389.352	-113.629	1.164
5200	42.414	336.007	299.334	190.701	-389.278	-108.224	1.087
5300	42.706	336.818	300.034	194.957	-389.174	-102.820	1.013
5400	43.010	337.619	300.722	199.242	-389.040	-97.418	0.942
5500	43.323	338.411	301.400	203.559	-388.876	-92.019	0.874
5600	43.645	339.195	302.068	207.907	-388.680	-86.623	0.808
5700	43.977	339.970	302.726	212.288	-388.454	-81.231	0.744
5800	44.317	340.738	303.375	216.703	-388.193	-75.844	0.683
5900	44.665	341.498	304.015	221.152	-387.903	-70.461	0.624
6000	45.019	342.252	304.646	225.636	-387.560	-65.086	0.567

PREVIOUS: September 1979 (1 atm) CURRENT: September 1979 (1 bar)

Aluminum Chloride (AlCl) $Al_1Cl_1(g)$

ALUMINUM CHLORIDE, ION ($AlCl^+$) IDEAL GAS M_r = 62.43399

$S°(298.15\ K) = [232.19 \pm 1.7]\ J\ K^{-1}\ mol^{-1}$

$\Delta_f H°\ (0\ K) = 855.7 \pm 40\ kJ\ mol^{-1}$
$\Delta_f H°(298.15\ K) = [861.904]\ kJ\ mol^{-1}$

Electronic Levels and Quantum Weights

State	ε_i, cm^{-1}	g_i
$[^2\Sigma]$	0	[2]
$[^2\Pi]$	[10000]	[4]

$\omega_e = [570]\ cm^{-1}$ $\omega_e x_e = [2]\ cm^{-1}$ $\sigma = 1$
$B_e = [0.2618]\ cm^{-1}$ $\alpha_e = [0.0015]\ cm^{-1}$ $r_e = [2.05]$ Å

Enthalpy of Formation

We adopt $\Delta_f H°(0\ K) = 204.5 \pm 10\ kcal\ mol^{-1}$ based on that of AlCl(g) (1) combined with the ionization potential IP(AlCl) = 9.4±0.4 eV (216.8±9 kcal mol^{-1}). IP(AlCl) is assumed to be equal to the electron-impact appearance potential of $AlCl^+$ measured as 9.4±0.4 (2), 9.5±0.3 (3) and 9.3±0.6 eV (4). Theoretical values include IP = 8.83 eV from an extended Huckel calcualtion (5) and from ab initio calculations (6, 7) either 8.4 eV (direct value) or 9.8 eV derived from D°($AlCl^+$). Electron-impact formation of $AlCl^+$ from $AlCl_3$ (8) apparently yields fragment ions with excess energy; i.e., we derive the upper limit IP≤11.7 eV.

For the dissociation process $AlCl^+(X^2\Sigma) \rightarrow Al^+(^1S) + Cl(^2P)$, our adopted $\Delta_f H°$ yields $D°(AlCl^+) = 39 \pm 10\ kcal\ mol^{-1}$. This compares with ~31 kcal mol^{-1} derived from the theoretical potential energy curve (6, 7) which, however, has a maximum of ~8 kcal mol^{-1} arising from an avoided crossing with another $^2\Sigma$ state.

Heat Capacity and Entropy

Bonding in Group III monohalides and their ions was characterized by Berkowitz and Dehmer (7) from photoelectron spectra and theoretical calculations (6, 7). They concluded that ionization removes an electron from an antibonding orbital centered mainly on the metal. This shortens the bond length and enhances the ionic nature of the bond. A decrease in bond length is observed for BF^+ (9), predicted for AlF^+ (1) from Rydberg states of AlF, and predicted for $AlCl^+$ from theoretical calculations (6, 7).

For $AlCl^+$, we adopt r_e = 2.05 Å based mainly on the theoretical prediction (6, 7). This may be compared with 2.13 Å for AlCl and 2.06 Å for $AlCl_3$ (1). Lacking data for Rydberg states of AlCl, we estimate ω_e from $k(AlCl^+)/k(MgCl) = 1.7$ transferred from the similar isoelectronic pair AlF^+ and MgF (1). B_e is calculated from r_e, while α_e and $\omega_e x_e$ are estimated by comparison with AlF^+, AlF and AlCl (1).

The electronic ground state and an approximate value for the $^2\Pi$ level are based on theoretical calculations (6, 7). These suggest that $^2\Pi$ has only a shallow minimum (r_e>3.6 Å) or possibly is repulsive. They also indicate the presence of a $^2\Sigma$ state having a double minimum above 30000 cm^{-1}. We include only the $^2\Pi$ state. If it is repulsive, our thermodynamic functions are upper-limit values. The bias would be quite small even at 3000 K.

References

1. JANAF Thermochemical Tables: AlCl(g), $AlCl_3$(g) 6-30-70; AlF^+(g) 12-31-76; AlF(g), MgF(g) 6-30-76.
2. D. L. Hildenbrand and L. P. Theard, J. Chem. Phys. 50, 5350 (1969).
3. D. L. Hildenbrand, J. Chem. Phys. 52, 5751 (1970).
4. N. D. Potter, AD-715567, Appendix 6 (1970).
5. J. W. Hastie and J. L. Margrave, J. Phys. Chem. 73, 1105 (1969).
6. M. Attermeyer, G. Das and A. C. Wahl, Argonne National Laboratory, unpublished calculations cited in ref. (7).
7. J. Berkowitz and J. L. Dehmer, J. Chem. Phys. 57, 3194 (1972).
8. R. F. Porter and E. F. Zeller, J. Chem. Phys. 33, 858 (1960).
9. R. B. Caton and A. E. Douglas, Can. J. Phys. 48, 432 (1970).

Aluminum Chloride, Ion ($AlCl^+$) $Al_1Cl_1^+(g)$

Enthalpy Reference Temperature = T_r = 298.15 K Standard State Pressure = p° = 0.1 MPa

T/K	$C_p°$	S°	$-[G°-H°(T_r)]/T$	$H°-H°(T_r)$	$\Delta_f H°$	$\Delta_f G°$	Log K_f
	J $K^{-1}mol^{-1}$			kJ mol^{-1}			
0	0.	0.	INFINITE	-9.154	855.682		
100	29.269	198.245	260.674	-6.243			
200	31.566	219.136	235.214	-3.216			
250	32.825	226.319	232.738	-1.605			
298.15	33.794	232.187	232.187	0.	861.904	828.108	-145.081
300	33.826	232.396	232.188	0.063	861.929	827.899	-144.150
350	34.588	237.671	232.603	1.774	862.585	822.175	-122.703
400	35.165	242.329	233.533	3.518	863.220	816.358	-106.605
450	35.606	246.497	234.746	5.288	863.838	810.463	-94.076
500	35.948	250.267	236.112	7.077	864.439	804.500	-84.046
600	36.436	256.867	239.037	10.698	865.587	792.404	-68.985
700	36.762	262.510	241.997	14.359	866.643	780.121	-58.213
800	36.993	267.435	244.875	18.047	867.580	767.696	-50.125
900	37.166	271.802	247.629	21.756	868.365	755.161	-43.828
1000	37.302	275.725	250.246	25.479	858.384	743.309	-38.826
1100	37.415	279.286	252.727	29.215	859.148	731.764	-34.749
1200	37.516	282.546	255.078	32.962	859.917	720.150	-31.347
1300	37.612	285.553	257.308	36.718	860.691	708.471	-28.467
1400	37.711	288.344	259.426	40.485	861.470	696.733	-25.995
1500	37.818	290.949	261.442	44.261	862.254	684.939	-23.852
1600	37.939	293.394	263.363	48.049	863.046	673.092	-21.974
1700	38.077	295.698	265.198	51.849	863.846	661.195	-20.316
1800	38.235	297.878	266.954	55.665	864.656	649.251	-18.841
1900	38.414	299.950	268.636	59.497	865.478	637.262	-17.520
2000	38.615	301.926	270.252	63.348	866.314	625.229	-16.329
2100	38.837	303.815	271.805	67.221	867.166	613.154	-15.251
2200	39.079	305.627	273.302	71.116	868.034	601.038	-14.270
2300	39.338	307.370	274.745	75.037	868.922	588.882	-13.374
2400	39.613	309.050	276.140	78.984	869.828	576.687	-12.551
2500	39.901	310.673	277.489	82.960	870.755	564.454	-11.794
2600	40.199	312.244	278.796	86.965	871.702	552.183	-11.093
2700	40.503	313.766	280.063	91.000	872.671	539.875	-10.445
2800	40.811	315.245	281.293	95.066	579.760	528.499	-9.859
2900	41.120	316.682	282.489	99.162	581.866	526.631	-9.486
3000	41.428	318.082	283.652	103.290	583.994	524.691	-9.136
3100	41.732	319.445	284.785	107.448	586.142	522.679	-8.807
3200	42.029	320.775	285.889	111.636	588.310	520.597	-8.498
3300	42.319	322.072	286.965	115.853	590.499	518.447	-8.206
3400	42.600	323.340	288.017	120.099	592.707	516.230	-7.931
3500	42.870	324.579	289.044	124.373	594.935	513.949	-7.670
3600	43.127	325.790	290.048	128.673	597.181	511.603	-7.423
3700	43.373	326.975	291.030	132.998	599.446	509.195	-7.189
3800	43.604	328.135	291.991	137.347	601.729	506.725	-6.965
3900	43.823	329.270	292.932	141.718	604.029	504.195	-6.753
4000	44.027	330.383	293.855	146.111	606.346	501.606	-6.550
4100	44.217	331.472	294.759	150.523	608.680	498.958	-6.357
4200	44.393	332.540	295.646	154.954	611.030	496.254	-6.172
4300	44.555	333.586	296.516	159.401	613.395	493.493	-5.995
4400	44.703	334.612	297.370	163.864	615.775	490.677	-5.825
4500	44.838	335.618	298.209	168.341	618.171	487.807	-5.662
4600	44.961	336.605	299.033	172.831	620.578	484.884	-5.506
4700	45.071	337.573	299.843	177.333	622.999	481.907	-5.356
4800	45.169	338.523	300.639	181.845	625.430	478.880	-5.211
4900	45.256	339.456	301.422	186.367	627.874	475.801	-5.072
5000	45.333	340.371	302.191	190.896	630.326	472.673	-4.938
5100	45.399	341.269	302.949	195.433	632.786	469.496	-4.809
5200	45.456	342.151	303.694	199.976	635.255	466.270	-4.684
5300	45.504	343.017	304.428	204.524	637.729	462.996	-4.563
5400	45.543	343.868	305.151	209.076	640.208	459.676	-4.446
5500	45.575	344.704	305.862	213.632	642.690	456.310	-4.334
5600	45.600	345.526	306.563	218.191	645.175	452.899	-4.224
5700	45.618	346.333	307.254	222.752	647.660	449.444	-4.119
5800	45.630	347.127	307.934	227.314	650.148	445.944	-4.016
5900	45.636	347.907	308.605	231.878	652.630	442.402	-3.917
6000	45.637	348.674	309.267	236.441	655.131	438.816	-3.820

PREVIOUS: June 1976 (1 atm) CURRENT: June 1976 (1 bar)

ALUMINUM CHLORIDE FLUORIDE (AlClF) IDEAL GAS M_r = 81.432943

$S°(298.15\ K)$ = [282.9 ± 4] $J\ K^{-1}\ mol^{-1}$

$\Delta_f H°(0\ K)$ = [-488.1 ± 63] $kJ\ mol^{-1}$
$\Delta_f H°(298.15\ K)$ = [-489.5 ± 63] $kJ\ mol^{-1}$

Electronic Levels and Quantum Weights

ε_i, cm^{-1}	g_i
[0]	2
[20000]	2
[25000]	2

Vibrational Frequencies and Degeneracies

ν, cm^{-1}
[750](1)
[200](1)
[550](1)

Point Group: C_s σ = 1
Bond Distances: Al-F = [1.65] Å Al-Cl = [2.10] Å
Bond Angle: Cl-Al-F = [120]°
Product of the Moments of Inertia: $I_A I_B I_C$ = [1.440598 x 10^{-114}] $g^3\ cm^6$

Enthalpy of Formation

We tentatively adopt $\Delta_f H°(298.15\ K)$ = -117±15 kcal mol^{-1} and $\Delta_{at} H°(0\ K)$ = 241±15 kcal mol^{-1} based on our average bond energies for $AlF_2(g)$ and $AlCl_2(g)$ (1). Mass-spectrometric data (2, 3) yield $\Delta_f H°(298.15\ K)$ = -142.6 and $\Delta_{at} H°(0\ K)$ = 267 kcal mol^{-1}, corresponding to an enhancement of 26 kcal mol^{-1} in the stability of AlClF(g). The same study (2, 3) indicated stability enhancements of 22 (or 13) kcal mol^{-1} for AlF_2, 19 kcal mol^{-1} for $AlClF_2$, and 22 kcal mol^{-1} for $AlCl_2F$. See the appropriate tables (1) for further details.

Farber et al. (2, 3) used molecular flow effusion of vapors of AlF_3 and $AlCl_3$ over Al(ℓ) in Al_2O_3(cr) to study the reaction AlCl(g) + AlF(g) = AlClF(g) + Al(g). They calculated Kp (5 points, 1338-1537 K) directly from mass-spectrometric ion intensities. Using JANAF auxiliary data (1), we derive [$\Delta_r S°$(2nd law) - $\Delta_r S°$(3rd law)] = 0.8±18 cal $K^{-1}\ mol^{-1}$, $\Delta_r H°(298.15\ K)$ = 12±25 (2nd law) or 11.2±5 (3rd law) and $\Delta_f H°(298.15\ K)$ = -142.6±5 kcal mol^{-1}. These results are internally consistent but, as in the case of AlF_2(2, 3), we suspect that common problems may cause a negative bias in $\Delta_f H°$. The ionizing energy is ambiguous: 20 eV (3) or AP+3 eV (2). The appearance potential AP($AlClF^+$) = 11±1 eV is biased by ~2 eV, raising several questions about interpretation of the data (see $AlCl_2^+$, 1). It is conceivable that $AlClF^+$ arises from a process other than direct ionization of AlClF.

We conclude from these speculations that the experimental data (2, 3) are insufficient to establish the enhanced stability of AlClF. Mass-spectrometric data indicate very little enhanced stability in BClF (4), in agreement with average bond energies summarized on the table for $AlCl_2$ (1).

Heat Capacity and Entropy

The electronic ground state, excited levels, bond distances and bond angle are assumed to be the same as in $AlCl_2$ and AlF_2 (1). Vibrational frequencies are estimated by comparison with $AlCl_2$ and AlF_2 (1). The principal moments of inertia are I_A = 2.3419 x 10^{-39}, I_B =23.6618 x 10^{-39} and I_C 26.0037 x 10^{-39} g cm^2.

References

1. JANAF Thermochemical Tables: $AlCl_2(g)$, $AlCl_2^+(g)$, $AlF_2(g)$, $AlClF_2(g)$, $AlCl_2F(g)$ 6-30-76; AlF(g) 12-31-75; AlCl(g) 6-30-70; Al(g) 12-31-65; Cl(g) 6-30-72; F(g) 9-30-65.
2. M. Farber and S. P. Harris, High Temp. Sci. 3, 231 (1971).
3. M. Farber, R. D. Srivastava and O. M. Uy, Space Sciences, Inc., AFRPL-TR-70-97, July, 1970.
4. R. D. Srivastava and M. Farber, Trans. Faraday Soc. 67, 2298 (1971).

Aluminum Chloride Fluoride (AlClF) $Al_1Cl_1F_1(g)$

Enthalpy Reference Temperature = T_r = 298.15 K Standard State Pressure = p° = 0.1 MPa

T/K	$C_p°$	S°	$-[G°-H°(T_r)]/T$	$H°-H°(T_r)$	$\Delta_f H°$	$\Delta_f G°$	Log K_f
	J $K^{-1}mol^{-1}$			kJ mol^{-1}			
0	0.	0.	INFINITE	-12.133	-488.119	-488.119	INFINITE
100	37.818	236.245	322.865	-8.662	-488.041	-492.998	257.516
200	43.971	264.376	287.217	-4.568	-488.732	-497.717	129.990
250	46.668	274.486	283.688	-2.300	-489.139	-499.916	104.452
298.15	48.806	282.895	282.895	0.	-489.528	-501.957	87.941
300	48.879	283.198	282.896	0.090	-489.543	-502.034	87.412
350	50.619	290.869	283.498	2.580	-489.942	-504.084	75.230
400	51.969	297.720	284.855	5.146	-490.340	-506.077	66.087
450	53.019	303.904	286.634	7.772	-490.739	-508.020	58.969
500	53.842	309.534	288.647	10.444	-491.143	-509.919	53.271
600	55.016	319.463	292.977	15.891	-491.984	-513.596	44.712
700	55.785	328.005	297.386	21.434	-492.901	-517.126	38.588
800	56.312	335.491	301.690	27.040	-493.930	-520.518	33.986
900	56.686	342.146	305.823	32.691	-495.107	-523.773	30.399
1000	56.961	348.134	309.759	38.374	-507.053	-526.127	27.482
1100	57.168	353.573	313.499	44.081	-508.256	-527.976	25.072
1200	57.328	358.554	317.049	49.806	-509.460	-529.716	23.058
1300	57.454	363.148	320.421	55.546	-510.666	-531.355	21.350
1400	57.554	367.410	323.627	61.296	-511.877	-532.901	19.883
1500	57.636	371.383	326.679	67.056	-513.093	-534.361	18.608
1600	57.703	375.105	329.591	72.823	-514.316	-535.739	17.490
1700	57.759	378.605	332.372	78.596	-515.544	-537.040	16.501
1800	57.806	381.908	335.033	84.374	-516.780	-538.269	15.620
1900	57.846	385.034	337.583	90.157	-518.022	-539.429	14.830
2000	57.881	388.002	340.031	95.943	-519.270	-540.523	14.117
2100	57.911	390.827	342.383	101.733	-520.523	-541.555	13.470
2200	57.938	393.522	344.647	107.526	-521.781	-542.527	12.881
2300	57.963	396.098	346.828	113.321	-523.042	-543.442	12.342
2400	57.985	398.565	348.933	119.118	-524.306	-544.302	11.846
2500	58.007	400.933	350.966	124.918	-525.571	-545.109	11.389
2600	58.028	403.208	352.932	130.719	-526.836	-545.865	10.967
2700	58.049	405.399	354.834	136.523	-528.101	-546.573	10.574
2800	58.070	407.510	356.678	142.329	-823.264	-546.267	10.191
2900	58.093	409.548	358.466	148.137	-823.427	-536.371	9.661
3000	58.117	411.518	360.202	153.948	-823.586	-526.470	9.167
3100	58.143	413.424	361.888	159.761	-823.739	-516.563	8.704
3200	58.172	415.271	363.528	165.576	-823.884	-506.652	8.270
3300	58.204	417.061	365.123	171.395	-824.021	-496.736	7.863
3400	58.238	418.799	366.676	177.217	-824.147	-486.817	7.479
3500	58.276	420.488	368.190	183.043	-824.262	-476.894	7.117
3600	58.318	422.130	369.666	188.873	-824.364	-466.967	6.776
3700	58.364	423.729	371.105	194.707	-824.451	-457.038	6.452
3800	58.414	425.286	372.511	200.546	-824.522	-447.108	6.146
3900	58.468	426.804	373.883	206.390	-824.576	-437.175	5.855
4000	58.526	428.285	375.225	212.239	-824.612	-427.241	5.579
4100	58.589	429.731	376.537	218.095	-824.627	-417.306	5.317
4200	58.655	431.143	377.820	223.957	-824.622	-407.372	5.066
4300	58.727	432.524	379.076	229.826	-824.594	-397.437	4.828
4400	58.802	433.875	380.307	235.703	-824.544	-387.504	4.600
4500	58.881	435.198	381.512	241.587	-824.468	-377.572	4.383
4600	58.965	436.493	382.693	247.479	-824.369	-367.642	4.175
4700	59.052	437.762	383.851	253.380	-824.246	-357.715	3.976
4800	59.143	439.006	384.987	259.290	-824.097	-347.790	3.785
4900	59.237	440.226	386.102	265.209	-823.922	-337.869	3.602
5000	59.335	441.424	387.197	271.137	-823.723	-327.951	3.426
5100	59.436	442.600	388.272	277.076	-823.498	-318.038	3.257
5200	59.539	443.755	389.327	283.024	-823.248	-308.130	3.095
5300	59.645	444.890	390.365	288.984	-822.974	-298.226	2.939
5400	59.754	446.006	391.385	294.954	-822.675	-288.328	2.789
5500	59.865	447.104	392.388	300.935	-822.354	-278.435	2.644
5600	59.977	448.183	393.375	306.927	-822.010	-268.549	2.505
5700	60.091	449.246	394.346	312.930	-821.644	-258.669	2.370
5800	60.207	450.292	395.302	318.945	-821.255	-248.796	2.241
5900	60.323	451.322	396.242	324.971	-820.849	-238.929	2.115
6000	60.441	452.337	397.169	331.010	-820.403	-229.072	1.994

PREVIOUS: June 1976 (1 atm) CURRENT: June 1976 (1 bar)

Aluminum Chloride Fluoride (AlClF) $Al_1Cl_1F_1(g)$

ALUMINUM CHLORIDE FLUORIDE, ION ($AlClF^+$) IDEAL GAS M_r = 81.432394

$S°(298.15 K) = [259.72 \pm 63]\ J\ K^{-1}\ mol^{-1}$

$\Delta_f H°(0 K) = 271 \pm 126\ kJ\ mol^{-1}$

$\Delta_f H°(298.15 K) = [276.144]\ kJ\ mol^{-1}$

Vibrational Frequencies and Degeneracies

ν, cm^{-1}
[850](1)
[170](2)
[500](1)

Ground State Quantum Weight: [1] $\sigma = 1$

Point Group: [$C_{\infty v}$]

Bond Distances: Al-F = [1.60] Å Al-Cl = [2.05] Å

Bond Angle: F-Al-Cl = [180]°

Rotational Constant: $B_0 = [0.095960]\ cm^{-1}$

Enthalpy of Formation

We adopt $\Delta_f H°(298.15 K) = 66 \pm 30\ kcal\ mol^{-1}$ based on $\Delta_f H°$ of AlClF (1) combined with the ionization potential IP(AlClF) = $182 \pm 23\ kcal\ mol^{-1}$ (7.9±1 eV). IP is estimated by comparison with $IP(AlCl_2)$ = 7.8±0.9 eV and $IP(AlF_2)$ = 8.1±0.9 eV (1). Farber and Harris (2) reported an appearance potential of $AP(AlClF^+)$ = 11±1 eV which appears to be seriously biased. This is discussed on the tables for $AlCl_2^+(g)$ and AlClF(g).

Heat Capacity and Entropy

We assume the electronic ground state to be linear $^1\Sigma^+$ and neglect excited states (1). Bond distances are estimated to be the same as in $AlCl_2^+$ and AlF_2^+ (1). Vibrational frequencies are estimated by comparison with AlClF, $AlCl_2^+$ and AlF_2^+ (1).

References

1. JANAF Thermochemical Tables: AlClF(g), $AlCl_2^+(g)$, $AlF_2^+(g)$ 6-30-76.
2. M. Farber and S. P. Harris, High Temp. Sci. 3, 231 (1971).

Aluminum Chloride Fluoride, Ion ($AlClF^+$) $Al_1Cl_1F_1^+(g)$

Enthalpy Reference Temperature = T_r = 298.15 K   Standard State Pressure = p° = 0.1 MPa

T/K	$C_p°$ (J K⁻¹mol⁻¹)	S° (J K⁻¹mol⁻¹)	-[G°-H°(T_r)]/T (J K⁻¹mol⁻¹)	H°-H°(T_r) (kJ mol⁻¹)	$\Delta_f H°$ (kJ mol⁻¹)	$\Delta_f G°$ (kJ mol⁻¹)	Log K_f
0	0.	0.	INFINITE	-12.635	270.854		
100	39.763	209.503	302.848	-9.334			
200	47.610	239.743	264.378	-4.927			
250	50.330	250.669	260.574	-2.476			
298.15	52.450	259.722	259.722	0.	276.144	264.370	-46.316
300	52.522	260.047	259.723	0.097	276.174	264.297	-46.018
350	54.264	268.279	260.369	2.768	276.996	262.252	-39.139
400	55.638	275.618	261.825	5.517	277.820	260.090	-33.964
450	56.723	282.237	263.731	8.327	278.645	257.824	-29.927
500	57.585	288.259	265.887	11.186	279.466	255.466	-26.688
600	58.835	298.877	270.525	17.011	281.082	250.513	-21.809
700	59.668	308.013	275.243	22.939	282.629	245.295	-18.304
800	60.245	316.020	279.850	28.936	284.070	239.862	-15.661
900	60.658	323.141	284.272	34.983	285.366	234.256	-13.596
1000	60.963	329.549	288.484	41.064	275.898	229.276	-11.976
1100	61.194	335.370	292.486	47.173	277.175	224.551	-10.663
1200	61.372	340.703	296.285	53.301	278.453	219.711	-9.564
1300	61.513	345.621	299.893	59.446	279.730	214.764	-8.629
1400	61.626	350.184	303.325	65.603	281.005	209.719	-7.825
1500	61.718	354.439	306.592	71.770	282.275	204.583	-7.124
1600	61.794	358.425	309.708	77.946	283.540	199.362	-6.508
1700	61.857	362.173	312.685	84.129	284.799	194.062	-5.963
1800	61.910	365.710	315.534	90.317	286.052	188.689	-5.476
1900	61.955	369.059	318.264	96.511	287.300	183.245	-5.038
2000	61.994	372.237	320.883	102.708	288.542	177.737	-4.642
2100	62.027	375.263	323.401	108.909	289.778	172.166	-4.282
2200	62.056	378.149	325.825	115.113	291.011	166.537	-3.954
2300	62.082	380.908	328.160	121.320	292.240	160.851	-3.653
2400	62.104	383.551	330.414	127.530	293.467	155.112	-3.376
2500	62.124	386.087	332.590	133.741	294.692	149.322	-3.120
2600	62.141	388.523	334.695	139.954	295.917	143.483	-2.883
2700	62.157	390.869	336.732	146.169	297.142	137.597	-2.662
2800	62.171	393.130	338.706	152.386	4.468	132.633	-2.474
2900	62.184	395.312	340.621	158.603	6.793	137.170	-2.471
3000	62.195	397.420	342.479	164.822	9.121	141.626	-2.466
3100	62.205	399.459	344.284	171.042	11.454	146.004	-2.460
3200	62.215	401.435	346.040	177.263	13.793	150.307	-2.454
3300	62.223	403.349	347.748	183.485	16.138	154.537	-2.446
3400	62.231	405.207	349.410	189.708	18.491	158.696	-2.438
3500	62.238	407.011	351.030	195.931	20.852	162.785	-2.429
3600	62.245	408.764	352.610	202.156	23.223	166.807	-2.420
3700	62.251	410.470	354.151	208.380	25.605	170.762	-2.411
3800	62.256	412.130	355.655	214.606	27.999	174.653	-2.401
3900	62.261	413.747	357.124	220.832	30.406	178.481	-2.390
4000	62.266	415.323	358.559	227.058	32.826	182.248	-2.380
4100	62.270	416.861	359.962	233.285	35.260	185.953	-2.369
4200	62.275	418.362	361.335	239.512	37.709	189.599	-2.358
4300	62.278	419.827	362.678	245.740	40.173	193.186	-2.347
4400	62.282	421.259	363.993	251.968	42.654	196.716	-2.335
4500	62.285	422.659	365.282	258.196	45.153	200.189	-2.324
4600	62.288	424.028	366.544	264.425	47.667	203.607	-2.312
4700	62.291	425.367	367.781	270.654	50.197	206.969	-2.300
4800	62.294	426.679	368.995	276.883	52.744	210.278	-2.288
4900	62.297	427.963	370.185	283.113	55.308	213.534	-2.276
5000	62.299	429.222	371.353	289.342	57.887	216.736	-2.264
5100	62.301	430.455	372.500	295.572	60.482	219.888	-2.252
5200	62.304	431.665	373.626	301.803	63.092	222.988	-2.240
5300	62.306	432.852	374.733	308.033	65.716	226.038	-2.228
5400	62.307	434.017	375.820	314.264	68.354	229.038	-2.216
5500	62.309	435.160	376.888	320.495	71.004	231.989	-2.203
5600	62.311	436.283	377.939	326.726	73.666	234.892	-2.191
5700	62.313	437.386	378.972	332.957	76.338	237.747	-2.179
5800	62.314	438.469	379.989	339.188	79.022	240.556	-2.166
5900	62.316	439.535	380.989	345.420	81.711	243.318	-2.154
6000	62.317	440.582	381.973	351.651	84.429	246.032	-2.142

PREVIOUS: June 1976 (1 atm)   CURRENT: June 1976 (1 bar)

Aluminum Chloride Fluoride, Ion ($AlClF^+$) $Al_1Cl_1F_1^+(g)$

ALUMINUM CHLORIDE FLUORIDE ($AlClF_2$) IDEAL GAS M_r = 100.431346

$S°(298.15\ K)$ = [297.8 ± 4] J K^{-1} mol^{-1}

$\Delta_f H°(0\ K)$ = -995.98 ± 6.3 kJ mol^{-1}
$\Delta_f H°(298.15\ K)$ = -999.14 ± 6.3 kJ mol^{-1}

Vibrational Frequencies and Degeneracies

ν, cm^{-1}	ν, cm^{-1}
[800](1)	[910](1)
[520](1)	[210](1)
[240](1)	[260](1)

Ground State Quantum Weight: [1] σ = 2
Point Group: [C_{2v}]
Bond Distances: Al-F = [1.63] Å Al-Cl = [2.06] Å
Bond Angles: Cl-Al-F = [120]° F-Al-F = [120]°
Product of the Moments of Inertia: $I_A I_B I_C$ = [1.281169] x 10^{-113}] g^3 cm^6

Enthalpy of Formation

We adopt $\Delta_f H°(298.15\ K)$ = -238.8±1.5 kcal mol^{-1} and $\Delta_{at}H°$ = 380.8±3 kcal mol^{-1} based on $\Delta_r H°(298.15\ K)$ = 0.5±1 kcal mol^{-1} for 1/3 $AlCl_3(g)$ + 2/3 $AlF_3(g)$ = $AlClF_2(g)$. Krause and Douglas (1) obtained this result by an entrainment method which measured enhanced volatility of AlF_3 in the presence of $AlCl_3$. Corrections were made for all the possible dimers which can form in this system. The entrainment data establish that aluminum fluoride-chloride exchange reactions are almost thermoneutral.

Our adopted $\Delta_f H°$ is compared below with an approximate mass-spectrometric Kp for reaciton of AlF_2 with $AlCl_2$ (2). The discrepancy in $\Delta_f H°(AlClF_2)$ is -19 kcal mol^{-1} if we use $\Delta_f H°$ = -180 (AlF_2) and -71 ($AlCl_2$) derived from the mass-spectrometric data (2, 3). The discrepancy disappears if we use $\Delta_f H°$ = -166 (AlF_2) and -67 ($AlCl_2$) derived from average bond energies in AlF_3 and $AlCl_3$ (4). This favors the latter values of $\Delta_f H°$ for AlF_2 and $AlCl_2$ (4), although the discrepancy might arise from the observed ion intensity of $AlClF_2^+$. The signal from $AlClF_2^+$ was very weak at 40 eV (5); this high ionizing energy might bias Kp and cause considerable fragmentation. We can eliminate AlF_2 and $AlCl_2$ by taking the difference between the mass-spectrometric reactions for $AlClF_2$ and $AlCl_2F$ (4). This yields [$\Delta_f H°(AlClF_2)$-$\Delta_f H°(AlCl_2F)$ = -47 kcal mol^{-1} which agrees with -49.8 kcal mol^{-1} from entrainment data (1). Thus, we tentatively ascribe the discrepancy to AlF_2 (4) and $AlCl_2$.

Source	Method	Reaction	T/K	$\Delta_r H°(T)$ (kcal mol^{-1})	$\Delta_r H°(298.15\ K)$ (kcal mol^{-1})	$\Delta_f H°(298.15\ K)$ (kcal mol^{-1})
(1)	Entrainment	1/3 $AlCl_3(g)$ + 2/3 $AlF_3(g)$ = $AlClF_2(g)$	1196-1257	0.5±1	0.5±1	-238.8
(2)	Mass Spec.	$AlF_2(g)$ + $AlCl_2(g)$ = $AlCl(g)$ + $AlClF_2(g)$	1491	--	-19.2	-240[a] or -258[b]

[a] Assuming $\Delta_f H°(298.15\ K)$ = -166 (AlF_2) and -67 ($AlCl_2$) kcal mol^{-1}
[b] Assuming $\Delta_f H°(298.15\ K)$ = -180 (AlF_2) and -71 ($AlCl_2$) kcal mol^{-1}

Heat Capacity and Entropy

We adopt a C_{2v} structure with bond angles and bond distances assumed equal to those in AlF_3 and $AlCl_3$ (4). We assume a singlet electronic ground state and neglect excited states. Vibrational frequencies are estimated by comparison of AlF_3 and $AlCl_3$ with the series BF_3, $BClF_2$, BCl_2F and BCl_3 (6, 7). The principal moments of inertia are I_A = 12.5730 x 10^{-39}, I_B = 26.2481 x 10^{-39}, and I_C = 38.8212 x 10^{-39} g cm^2.

References

1. R. F. Krause and T. B. Douglas, J. Phys. Chem. 72, 3444 (1968).
2. M. Farber and S. P. Harris, High Temp. Sci. 3, 231 (1971).
3. O. M. Uy, R. D. Srivastava and M. Farber, High Temp. Sci. 4, 227 (1972).
4. JANAF Thermochemical Tables: $AlF_2(g)$, $AlCl_2(g)$, $AlCl_2F(g)$ 6-30-76; $AlF_3(g)$, $AlCl_3(g)$ 6-30-70.
5. M. Farber, Space Sciences, Inc., Monrovia, Calif., personal communication, August 30, 1976.
6. D. F. Wolfe and G. L. Humphrey, J. Mol. Struct. 3, 293 (1969).
7. L. P. Lindeman and M. K. Wilson, J. Chem. Phys. 24, 242 (1956).

Aluminum Chloride Fluoride ($AlClF_2$) $Al_1Cl_1F_2(g)$

Enthalpy Reference Temperature = T_r = 298.15 K Standard State Pressure = p° = 0.1 MPa

T/K	$C_p°$ (J $K^{-1}mol^{-1}$)	S° (J $K^{-1}mol^{-1}$)	-[G°-H°(T_r)]/T (J $K^{-1}mol^{-1}$)	H°-H°(T_r) (kJ mol^{-1})	$\Delta_f H°$ (kJ mol^{-1})	$\Delta_f G°$ (kJ mol^{-1})	Log K_f
0	0.	0.	INFINITE	-14.795	-995.980	-995.980	INFINITE
100	43.865	238.385	350.065	-11.168	-997.199	-993.852	519.135
200	57.230	273.348	303.584	-6.047	-998.328	-990.042	258.572
250	61.827	286.629	298.897	-3.067	-998.774	-987.917	206.414
298.15	65.447	297.839	297.839	0.	-999.139	-985.793	172.707
300	65.572	298.245	297.841	0.121	-999.152	-985.710	171.627
350	68.588	308.588	298.651	3.478	-999.479	-983.443	146.771
400	70.996	317.910	300.485	6.970	-999.766	-981.132	128.123
450	72.917	326.387	302.899	10.569	-1000.025	-978.787	113.615
500	74.454	334.152	305.642	14.255	-1000.265	-976.414	102.005
600	76.700	347.939	311.571	21.821	-1000.724	-971.601	84.585
700	78.209	359.883	317.639	29.571	-1001.209	-966.709	72.137
800	79.259	370.399	323.590	37.447	-1001.773	-961.744	62.795
900	80.014	379.780	329.322	45.413	-1002.462	-956.700	55.525
1000	80.573	388.241	334.797	53.444	-1013.906	-950.811	49.665
1100	80.997	395.941	340.011	61.523	-1014.596	-944.469	44.849
1200	81.325	403.003	344.970	69.640	-1015.281	-938.063	40.833
1300	81.584	409.524	349.688	77.786	-1015.966	-931.600	37.432
1400	81.792	415.577	354.181	85.955	-1016.653	-925.085	34.515
1500	81.962	421.227	358.465	94.143	-1017.346	-918.520	31.986
1600	82.101	426.521	362.554	102.346	-1018.047	-911.909	29.771
1700	82.218	431.502	366.465	110.562	-1018.755	-905.254	27.815
1800	82.316	436.204	370.210	118.789	-1019.472	-898.556	26.075
1900	82.399	440.657	373.802	127.025	-1020.197	-891.819	24.518
2000	82.471	444.885	377.251	135.269	-1020.928	-885.044	23.115
2100	82.532	448.911	380.568	143.519	-1021.664	-878.231	21.845
2200	82.586	452.751	383.763	151.775	-1022.403	-871.384	20.689
2300	82.633	456.423	386.843	160.036	-1023.143	-864.503	19.633
2400	82.674	459.941	389.816	168.301	-1023.880	-857.590	18.665
2500	82.711	463.317	392.689	176.570	-1024.613	-850.646	17.773
2600	82.743	466.561	395.468	184.843	-1025.338	-843.673	16.950
2700	82.772	469.685	398.159	193.119	-1026.054	-836.672	16.186
2800	82.798	472.695	400.768	201.398	-1320.659	-828.677	15.459
2900	82.821	475.601	403.298	209.678	-1320.252	-811.113	14.610
3000	82.842	478.409	405.756	217.962	-1319.829	-793.564	13.817
3100	82.861	481.126	408.143	226.247	-1319.386	-776.029	13.076
3200	82.879	483.757	410.465	234.534	-1318.924	-758.509	12.381
3300	82.894	486.308	412.725	242.823	-1318.438	-741.003	11.729
3400	82.909	488.783	414.926	251.113	-1317.929	-723.513	11.115
3500	82.922	491.186	417.071	259.404	-1317.394	-706.038	10.537
3600	82.934	493.522	419.162	267.697	-1316.831	-688.578	9.991
3700	82.945	495.795	421.203	275.991	-1316.241	-671.135	9.475
3800	82.955	498.007	423.195	284.286	-1315.621	-653.708	8.986
3900	82.965	500.162	425.141	292.582	-1314.972	-636.298	8.522
4000	82.974	502.262	427.043	300.879	-1314.291	-618.904	8.082
4100	82.982	504.311	428.902	309.177	-1313.579	-601.528	7.664
4200	82.990	506.311	430.722	317.475	-1312.834	-584.170	7.265
4300	82.997	508.264	432.502	325.775	-1312.057	-566.830	6.886
4400	83.003	510.172	434.246	334.075	-1311.248	-549.509	6.523
4500	83.009	512.038	435.954	342.375	-1310.403	-532.206	6.178
4600	83.015	513.862	437.628	350.677	-1309.528	-514.922	5.847
4700	83.021	515.647	439.269	358.978	-1308.621	-497.658	5.531
4800	83.026	517.395	440.879	367.281	-1307.682	-480.413	5.228
4900	83.030	519.107	442.458	375.583	-1306.710	-463.188	4.938
5000	83.035	520.785	444.007	383.887	-1305.709	-445.984	4.659
5100	83.039	522.429	445.529	392.190	-1304.678	-428.800	4.392
5200	83.043	524.042	447.023	400.495	-1303.619	-411.636	4.135
5300	83.047	525.624	448.492	408.799	-1302.531	-394.493	3.888
5400	83.050	527.176	449.934	417.104	-1301.418	-377.370	3.650
5500	83.054	528.700	451.353	425.409	-1300.279	-360.269	3.422
5600	83.057	530.196	452.747	433.715	-1299.117	-343.188	3.201
5700	83.060	531.666	454.119	442.021	-1297.932	-326.129	2.989
5800	83.063	533.111	455.469	450.327	-1296.724	-309.090	2.784
5900	83.066	534.531	456.797	458.633	-1295.500	-292.072	2.586
6000	83.068	535.927	458.104	466.940	-1294.236	-275.078	2.395

PREVIOUS: June 1976 (1 atm) CURRENT: June 1976 (1 bar)

Aluminum Chloride Fluoride ($AlClF_2$) $Al_1Cl_1F_2(g)$

ALUMINUM CHLORIDE OXIDE (OAlCl) CRYSTAL M_r = 78.43394

$S°$(298.15 K) = [54.4 ± 4] J K^{-1} mol^{-1}
T_{fus} = Unknown
T_{dcm} = [500] K

$\Delta_f H°$(0 K) = Unknown
$\Delta_f H°$(298.15 K) = -793.3 ± 1.3 kJ mol^{-1}
$\Delta_{fus} H°$ = Unknown

Enthalpy of Formation

$\Delta_f H°$(AlOCl, cr, 298.15 K) is calculated from $\Delta_r H°$(298.15 K) = 21.05 kcal mol^{-1} for the reaction AlOCl(cr) + Cl_2(g) = $AlCl_3$(cr) + 1/2 O_2(g) reported by Schafer et al. (1). This reaction is the overall reaction of the following four separate reactions:

Reaction	$\Delta_r H°$(298.15 K), kcal mol^{-1}
AlOCl(cr) + 2(HCl·299.5 H_2O) = $AlCl_3$·600 H_2O(ℓ)	-46.9
$AlCl_3$·600 H_2O(ℓ) = $AlCl_3$(cr) + 600 H_2O(ℓ)	+79.3
H_2O(ℓ) = H_2(g) + 1/2 O_2(g)	+68.32
H_2(g) + Cl_2(g) + 599 H_2O(ℓ) = 2(HCl·299.5 H_2O)	-79.67
AlOCl(cr) + Cl_2(g) = $AlCl_3$(cr) + 1/2 O_2(g)	+21.05

Heat Capacity and Entropy

The heat capacities for AlOCl(cr) are estimated based on the assumption that ΔC_p = 0 for the reaction $AlCl_3$(cr) + Al_2O_3(cr) = 3 AlOCl(cr). $S°$(298.15 K) = 13 cal K^{-1} mol^{-1} was estimated by Schafer et al. (1).

Decomposition Data

T_{dcm} is calculated as the temperature at which the Gibbs energy change of the reaction 6 AlOCl(cr) + 3 Al_2O_3(cr) + Al_2Cl_6(g) is zero.

References

1. H. Schafer, F. E. Wittig and W. Wilborn, Z. anorg. allgem. Chem. 297, 48 (1958).

Aluminum Chloride Oxide (OAlCl) $Al_1Cl_1O_1$(cr)

Enthalpy Reference Temperature = T_r = 298.15 K; Standard State Pressure = p° = 0.1 MPa

T/K	$C_p°$ (J $K^{-1}mol^{-1}$)	S° (J $K^{-1}mol^{-1}$)	-[G°-H°(T_r)]/T (J $K^{-1}mol^{-1}$)	H°-H°(T_r) (kJ mol^{-1})	$\Delta_f H°$ (kJ mol^{-1})	$\Delta_f G°$ (kJ mol^{-1})	Log K_f
0							
100							
200							
298.15	56.902	54.392	54.392	0.	-793.286	-737.235	129.161
300	57.112	54.745	54.393	0.105	-793.284	-736.888	128.304
400	64.308	72.215	56.726	6.196	-792.923	-718.130	93.778
500	69.203	87.118	61.351	12.883	-792.183	-699.511	73.077
600	72.634	100.052	66.748	19.983	-791.215	-681.065	59.292
700	74.977	111.431	72.335	27.367	-790.141	-662.791	49.458
800	76.860	121.570	77.867	34.963	-789.045	-644.673	42.093
900	78.168	130.701	83.239	42.716	-787.995	-626.691	36.372
1000	79.245	138.995	88.406	50.589	-797.638	-608.060	31.762
1100	80.040	146.586	93.355	58.554	-796.477	-589.158	27.977
1200	80.709	153.580	98.086	66.592	-795.269	-570.363	24.827
1300	81.243	160.062	102.607	74.691	-794.023	-551.672	22.166
1400	81.672	166.099	106.929	82.837	-792.749	-533.077	19.889
1500	81.961	171.744	111.064	91.019	-791.459	-514.574	17.919
1600	82.216	177.041	115.024	99.228	-790.158	-496.157	16.198
1700	82.435	182.033	118.820	107.461	-788.850	-477.822	14.682
1800	82.620	186.750	122.464	115.714	-787.539	-459.564	13.336
1900	82.770	191.221	125.966	123.984	-786.228	-441.379	12.134
2000	82.885	195.470	129.336	132.267	-784.920	-423.263	11.055
2100	82.958	199.515	132.582	140.560	-783.619	-405.212	10.079
2200	83.011	203.376	135.713	148.858	-782.330	-387.223	9.194
2300	83.036	207.067	138.736	157.161	-781.055	-369.292	8.387
2400	83.052	210.601	141.657	165.465	-779.797	-351.417	7.648
2500	83.052	213.991	144.483	173.770	-778.557	-333.593	6.970
2600	83.052	217.249	147.220	182.075	-777.335	-315.819	6.345
2700	83.052	220.383	149.872	190.381	-776.134	-298.091	5.767
2800	83.052	223.403	152.444	198.686	-1068.853	-279.439	5.213
2900	83.052	226.318	154.942	206.991	-1066.595	-251.286	4.526
3000	83.052	229.133	157.368	215.296	-1064.357	-223.210	3.886

PREVIOUS: CURRENT: March 1964

Aluminum Chloride Oxide (OAlCl) $Al_1Cl_1O_1$(cr)

ALUMINUM CHLORIDE OXIDE (OAlCl) IDEAL GAS M_r = 78.43394

$S°(298.15\ K) = [248.9 \pm 20]\ J\ K^{-1}\ mol^{-1}$

$\Delta_f H°(0\ K) = [-346.0 \pm 20]\ kJ\ mol^{-1}$

$\Delta_f H°(298.15\ K) = [-348.1 \pm 20]\ kJ\ mol^{-1}$

Vibrational Frequencies and Degeneracies

ν, cm^{-1}
[450](1)
[350](2)
[900](1)

Ground State Quantum Weight: [1]

Point Group: $C_{\infty v}$ $\sigma = 1$

Bond Distance: O-Al = [1.62] Å Al-Cl = [2.14] Å

Bond Angle: O-Al-Cl = 180°

Rotational Constant: B_o = [0.097704] cm^{-1}

Enthalpy of Formation

The Gibbs energy change of the reaction 1/3 $Al_2O_3(\ell)$ + 1/3 $AlCl_3(g)$ = AlOCl(g), $\Delta_r G°(2400\ K) = 28 \pm 3\ kcal\ mol^{-1}$, has been determined by Greenbaum et al. (1). Incorporating appropriate auxiliary data, the enthalpy of reaction is derived to be 82.9 kcal mol^{-1}, yielding $\Delta_f H°$(AlOCl, g, 298.15 K) = $-82.2 \pm 5\ kcal\ mol^{-1}$.

Heat Capacity and Entropy

Point group, ground state quantum weight and vibrational frequencies are obtained from Hilsenrath et al. (2). The bond distances of r(O-Al) and r(Al-Cl) are assumed to be the same as those in AlO(g) and AlCl(g) molecules, respectively. The three principal moments of inertia are: $I_A = 0$ and $I_B = I_C = 2.86467 \times 10^{-38} g\ cm^2$.

References

1. M. A. Greenbaum et al., Marimont Corporation, California, personal communication, August 26, 1964.
2. J. Hilsenrath, W. H. Evans and H. W. Wooley, U. S. Nat. Bur. Stand. Report 6484. July 1, 1959.

Aluminum Chloride Oxide (OAlCl) $Al_1Cl_1O_1(g)$

Enthalpy Reference Temperature = T_r = 298.15 K Standard State Pressure = p° = 0.1 MPa

T/K	$C_p°$ (J K⁻¹mol⁻¹)	S° (J K⁻¹mol⁻¹)	$-[G°-H°(T_r)]/T$ (J K⁻¹mol⁻¹)	$H°-H°(T_r)$ (kJ mol⁻¹)	$\Delta_f H°$ (kJ mol⁻¹)	$\Delta_f G°$ (kJ mol⁻¹)	Log K_f
0	0.	0.	INFINITE	-11.407	-346.045	-346.045	INFINITE
100	32.420	204.221	288.567	-8.435	-346.464	-348.072	181.814
200	43.404	230.239	253.322	-4.617	-347.422	-349.296	91.227
250	47.248	240.357	249.740	-2.346	-347.803	-349.719	73.070
298.15	50.068	248.931	248.931	0.	-348.109	-350.059	61.329
300	50.162	249.241	248.932	0.093	-348.120	-350.071	60.953
350	52.394	257.148	249.551	2.659	-348.386	-350.375	52.291
400	54.118	264.262	250.953	5.324	-348.617	-350.643	45.789
450	55.464	270.717	252.796	8.065	-348.826	-350.884	40.730
500	56.527	276.618	254.887	10.865	-349.023	-351.102	36.679
600	58.059	287.069	259.403	16.600	-349.420	-351.481	30.599
700	59.076	296.101	264.015	22.460	-349.871	-351.789	26.251
800	59.780	304.038	268.532	28.405	-350.425	-352.027	22.985
900	60.284	311.110	272.877	34.410	-351.124	-352.187	20.440
1000	60.656	317.482	277.024	40.458	-362.591	-351.500	18.360
1100	60.938	323.276	280.969	46.538	-363.316	-350.356	16.637
1200	61.155	328.588	284.719	52.643	-364.041	-349.146	15.198
1300	61.327	333.491	288.285	58.767	-364.769	-347.875	13.978
1400	61.465	338.041	291.678	64.907	-365.502	-346.548	12.930
1500	61.577	342.285	294.912	71.060	-366.241	-345.169	12.020
1600	61.670	346.262	297.999	77.222	-366.987	-343.740	11.222
1700	61.747	350.003	300.949	83.393	-367.741	-342.263	10.516
1800	61.811	353.535	303.773	89.571	-368.505	-340.743	9.888
1900	61.867	356.878	306.481	95.755	-369.279	-339.180	9.325
2000	61.914	360.053	309.081	101.944	-370.065	-337.575	8.817
2100	61.955	363.075	311.581	108.137	-370.864	-335.931	8.356
2200	61.990	365.958	313.987	114.335	-371.676	-334.248	7.936
2300	62.021	368.714	316.307	120.535	-372.503	-332.529	7.552
2400	62.048	371.354	318.546	126.739	-373.346	-330.773	7.199
2500	62.072	373.887	320.709	132.945	-374.204	-328.981	6.874
2600	62.094	376.322	322.802	139.153	-375.080	-327.155	6.573
2700	62.113	378.666	324.828	145.363	-375.974	-325.295	6.293
2800	62.130	380.925	326.791	151.576	-670.786	-322.433	6.015
2900	62.145	383.106	328.696	157.789	-670.620	-309.995	5.584
3000	62.159	385.213	330.545	164.005	-670.471	-297.563	5.181
3100	62.172	387.251	332.341	170.221	-670.341	-285.134	4.804
3200	62.183	389.225	334.088	176.439	-670.229	-272.710	4.452
3300	62.193	391.139	335.788	182.658	-670.133	-260.289	4.120
3400	62.203	392.996	337.444	188.878	-670.053	-247.871	3.808
3500	62.212	394.799	339.057	195.098	-669.989	-235.455	3.514
3600	62.220	396.552	340.629	201.320	-669.939	-223.040	3.236
3700	62.227	398.256	342.164	207.542	-669.903	-210.627	2.974
3800	62.234	399.916	343.662	213.765	-669.879	-198.215	2.725
3900	62.240	401.533	345.125	219.989	-669.866	-185.802	2.489
4000	62.246	403.109	346.555	226.213	-669.863	-173.391	2.264
4100	62.251	404.646	347.953	232.438	-669.870	-160.979	2.051
4200	62.256	406.146	349.321	238.663	-669.885	-148.567	1.848
4300	62.261	407.611	350.660	244.889	-669.907	-136.154	1.654
4400	62.265	409.042	351.970	251.116	-669.935	-123.741	1.469
4500	62.269	410.441	353.254	257.342	-669.967	-111.328	1.292
4600	62.273	411.810	354.512	263.569	-670.006	-98.913	1.123
4700	62.277	413.149	355.746	269.797	-670.050	-86.497	0.961
4800	62.280	414.461	356.955	276.025	-670.099	-74.081	0.806
4900	62.283	415.745	358.142	282.253	-670.151	-61.663	0.657
5000	62.286	417.003	359.307	288.481	-670.209	-49.244	0.514
5100	62.289	418.237	360.450	294.710	-670.272	-36.824	0.377
5200	62.292	419.446	361.573	300.939	-670.341	-24.404	0.245
5300	62.294	420.633	362.676	307.168	-670.415	-11.981	0.118
5400	62.296	421.797	363.760	313.398	-670.497	0.443	-0.004
5500	62.299	422.940	364.826	319.628	-670.586	12.869	-0.122
5600	62.301	424.063	365.874	325.858	-670.685	25.296	-0.236
5700	62.303	425.165	366.904	332.088	-670.793	37.725	-0.346
5800	62.305	426.249	367.918	338.318	-670.910	50.157	-0.452
5900	62.306	427.314	368.916	344.549	-671.043	62.589	-0.554
6000	62.308	428.361	369.898	350.779	-671.167	75.023	-0.653

PREVIOUS: September 1964 (1 atm) CURRENT: September 1964 (1 bar)

Aluminum Chloride Oxide (OAlCl) $Al_1Cl_1O_1(g)$

ALUMINUM CHLORIDE ($AlCl_2$) IDEAL GAS $M_r = 97.88754$

$S°(298.15\ K) = [289.42 \pm 2.4]\ J\ K^{-1}\ mol^{-1}$

$\Delta_fH(0\ K) = -279.4 \pm 20\ kJ\ mol^{-1}$
$\Delta_fH°(298.15\ K) = -280.3 \pm 20\ kJ\ mol^{-1}$

Electronic Levels and Quantum Weights

State	ε_i, cm^{-1}	g_i
$^2A_1$	[0]	2
$^2B_1$	[20000]	2
$^2B_2$	[25000]	2

Vibrational Frequencies and Degeneracies

ν, cm^{-1}
[430](1)
[160](1)
[570](1)

Point Group: $[C_{2v}]$ $\sigma = 2$
Bond Distance: Al-Cl = [2.10] Å
Bond Angle: Cl-Al-Cl = [120]°
Product of the Moments of Inertia: $I_AI_BI_C = [5.925281 \times 10^{-114}]\ g^3\ cm^6$

Enthalpy of Formation

We adopt $\Delta_fH°(298.15\ K) = -67\pm5\ kcal\ mol^{-1}$ and $\Delta_{at}H° = 201.4\pm6\ kcal\ mol^{-1}$; i.e., the average bond energy $\Delta_{at}H°/2 = 101\ kcal\ mol^{-1}$ is taken equal to that of $AlCl_3$ (1). $\Delta_fH°$ is consistent with experimental results of -66 ± 3 (2) and $\geq -71\pm5\ kcal\ mol^{-1}$ (3, 4), but there may be significant bias in these values.

Chai et al. (2) derived $\Delta_fH°(298.15\ K) = -66\pm3\ kcal\ mol^{-1}$ from study of reactions of Al(c,ℓ) with the vapor of $AlCl_3$ in an argon carrier gas. Weight-loss data (800-1000 K) for Al and $AlCl_3$ were analyzed assuming five vapor species (AlCl, $AlCl_2$, $AlCl_3$, Al_2Cl_4, Al_2Cl_6) in four simultaneous equilibrium reactions. Experimental data were combined with auxiliary data, including $S°(AlCl_2)$ and Kp for two reactions, to derive $\Delta_fH°(AlCl_2)$ and Kp for another reaction involving Al_2Cl_4. Despite minor changes in $S°(AlCl_2)$, $\Delta_fH° = -66$ is till appropriate for the $AlCl_2$ equilibrium (2). We suspect, however, that this equilibrium is strongly dependent on the other equilibria, particularly that involving $Al_2Cl_4(g)$.

Farber and Harris (3, 4) reported Kp (5 points, 1150-1430 K) for the reaction $AlCl(g) + AlCl_3(g) = 2\ AlCl_2(g)$. We obtain $[\Delta_rS°(2nd\ law)-\Delta_rS°(3rd\ law)] = -35\pm10\ cal\ K^{-1}mol^{-1}$, $\Delta_rH°(298.15\ K) = -36\pm13$(2nd law) or 10.5 ± 9(3rd law) and $\Delta_fH°(298.15\ K) \geq -70.7\pm5\ kcal\ mol^{-1}$. We consider $\Delta_fH°$ to be a negative limit due to possible bias arising from the ionizing energy. Energy of the ionizing electrons was variously reported as 20 eV (4) or as 3 eV above the AP (3). If 20 eV was used, we expect $AlCl_2^+$ to be enhanced due to fragmentation of $AlCl_3$. The alternative ionizing energy leads to another problem, since the authors (3, 4) used literature values for the AP of $AlCl^+$ but their own value for $AlCl_2^+$. Their value has a bias of ~3 eV which is discussed in detail on the table for $AlCl_2^+(g)$ (1). We suspect that the mass spectrometer had a biased energy scale during the measurments (3, 4). This would induce negative bias in the ionizing energy used for $AlCl^+$ and $AlCl_3^+$, since the AP's were not based on the experimental energy scale. Again we expect $AlCl_2^+$ to be enhanced relative to $AlCl^+$ and $AlCl_3^+$. The large discrepancy in $\Delta_rS°$ is symptomatic of a temperature-dependent error; we expect that bias in Kp is greatest at low T.

Average bond energies (BE) suggest that MX_n (M=Al, B and X=F, Cl) does not show the progression to multiple bonding of NO_3, NO_2, and NO. Although $[BE(MX)-BE(MX_3)] = \sim 20\ kcal\ mol^{-1}$, only the mass-spectrometric data of Farber et al. (3-6) yield $BE(MX_2) > BE(MX_3)$. Other data (1, 2, 7) and theoretical calculations (8) suggest that $BE(MX_2) \leq BE(MX_3)$.

Source	Method	Species	Average Bond Energy ($\Delta_{at}H°/n$), kcal mol^{-1}				
			$AlCl_n$	BCl_n	AlF_n	BF_n	NO_n
(1) JANAF	Review	MX	118.4	127.1	159.3	179.4	150.0
(1) JANAF	Review	MX_3	100.8	104.5	140.2	152.5	90.3
(1) JANAF	Review	MX_2	101	--	140	--	111.0
(3, 4) Farber et al.	Mass spec.	MX_2	102.6	--	146.5, 151	--	--
(5, 6) Srivastava et al.	Mass spec.	MX_2	--	104.0	147.0	155.0	--
(7) Dibeler et al	Photoionization	MX_2	--	102.6	--	145.6	--
(2, 1) Others	Various	MX_2	100.2(2)	103.5(1)	132.4(1)	--	--

Heat Capacity and Entropy

The electronic ground state and excited levels are assumed to be the same as in AlF_2 (1). Vibrational frequencies are calculated from the estimated force constants $f_r = 2.2$ and $f_\alpha/r^2 = 0.13$ mdyn/Å. Frequencies and force constants are estimated from simultaneous consideration of $AlCl_2^+$, $AlCl_2$, $AlCl_2^-$ and the analogous boron species. We assume that frequency changes are qualitatively similar to those of NO_2^+, NO_2, and NO_2^-. Thus, in going from $AlCl_2^+$ to $AlCl_2$, we expect a large decrease in ν_3, an increase in ν_2 and a small increase in ν_1. By analogy with BF_2 (8), the odd electron of $AlCl_2$ should occupy an anti-bonding orbital which increases the bond length but decreases the bond angle and stretching force constant. We assume a bond length 0.05 Å longer than in $AlCl_2^+$ (1) and a bond angle equal to that in BF_2 (8). The principal moments of inertia are $I_A = 3.5781 \times 10^{-39}$, $I_B = 38.9438 \times 10^{-39}$ and $I_C = 42.5220 \times 10^{-39}\ g\ cm^2$.

References

1. JANAF Thermochemical Tables: $AlCl_2^+(g)$, $AlCl_2^-(g)$ and $AlF_2(g)$, 6-30-76; AlF(g), 12-31-75; $AlCl_3(g)$, AlCl(g) and $AlF_3(g)$, 6-30-70; Al(g), 12-31-65; F(g), 9-30-65; $BCl_2(g)$, 6-30-72; $BCl_3(g)$ and BCl(g), 12-31-64; $BF_3(g)$, 6-30-69.
2. B. J. Chai, H. C. Ko, M. A. Greenbaum and M. Farber, J. Phys. Chem. 71, 3331 (1967).
3. M. Farber and S. P. Harris, High Temp. Sci. 3, 231 (1971).
4. M. Farber, R. D. Srivastava and O. M. Uy, Space Science, Inc., AFRPL-TR-70-97, July, 1970.
5. R. D. Srivastava and M. Farber, Trans. Faraday Soc. 67, 2298 (1971).
6. O. M. Uy, R. D. Srivastava and M. Farber, High Temp. Sci. 4, 227 (1972).
7. V. H. Dibeler and J. A. Walker, Inorg. Chem. 8, 50 (1969); 7, 1742 (1968).
8. C. Thomsom and D. A. Brochie, Theoret. Chim. Acta 32, 101 (1973); Mol. Phys. 28, 301 (1974).

Aluminum Chloride ($AlCl_2$) $Al_1Cl_2(g)$

Enthalpy Reference Temperature = T_r = 298.15 K Standard State Pressure = p° = 0.1 MPa

T/K	$C°_p$ (J K^{-1}mol^{-1})	S°	$-[G°-H°(T_r)]/T$	$H°-H°(T_r)$ (kJ mol^{-1})	$\Delta_fH°$	$\Delta_fG°$	Log K_f
0	0.	0.	INFINITE	-12.764	-279.371	-279.371	INFINITE
100	39.512	239.866	331.991	-9.212	-279.216	-283.603	148.139
200	47.065	269.680	294.017	-4.867	-279.710	-287.825	75.172
250	49.753	280.487	290.260	-2.443	-280.019	-289.818	60.554
298.15	51.635	289.420	289.420	0.	-280.328	-291.677	51.101
300	51.696	289.739	289.421	0.096	-280.340	-291.748	50.798
350	53.093	297.819	290.056	2.717	-280.672	-293.623	43.821
400	54.109	304.979	291.482	5.399	-281.015	-295.450	38.582
450	54.863	311.397	293.345	8.124	-281.369	-297.233	34.502
500	55.434	317.209	295.445	10.882	-281.736	-298.976	31.234
600	56.219	327.391	299.945	16.468	-282.518	-302.351	26.322
700	56.716	336.097	304.502	22.116	-283.389	-305.590	22.803
800	57.050	343.693	308.937	27.805	-284.380	-308.695	20.156
900	57.284	350.427	313.180	33.523	-285.524	-311.668	18.089
1000	57.453	356.472	317.212	39.260	-297.439	-313.743	16.388
1100	57.580	361.954	321.034	45.012	-298.612	-315.317	14.973
1200	57.677	366.968	324.656	50.775	-299.786	-316.784	13.789
1300	57.754	371.588	328.091	56.547	-300.961	-318.153	12.784
1400	57.814	375.870	331.352	62.325	-302.140	-319.431	11.918
1500	57.864	379.861	334.455	68.109	-303.322	-320.625	11.165
1600	57.904	383.597	337.411	73.898	-304.508	-321.740	10.504
1700	57.938	387.108	340.232	79.690	-305.700	-322.780	9.918
1800	57.966	390.421	342.929	85.485	-306.897	-323.751	9.395
1900	57.990	393.555	345.512	91.283	-308.102	-324.654	8.925
2000	58.011	396.530	347.989	97.083	-309.314	-325.494	8.501
2100	58.030	399.361	350.368	102.885	-310.535	-326.273	8.116
2200	58.046	402.061	352.657	108.689	-311.767	-326.993	7.764
2300	58.062	404.642	354.862	114.494	-313.011	-327.658	7.441
2400	58.076	407.113	356.988	120.301	-314.268	-328.268	7.145
2500	58.091	409.484	359.040	126.110	-315.540	-328.825	6.870
2600	58.105	411.763	361.025	131.919	-316.827	-329.331	6.616
2700	58.121	413.956	362.945	137.731	-318.131	-329.787	6.380
2800	58.137	416.070	364.805	143.544	-613.353	-329.227	6.142
2900	58.155	418.111	366.608	149.358	-613.596	-319.075	5.747
3000	58.175	420.082	368.358	155.175	-613.857	-308.915	5.379
3100	58.198	421.990	370.057	160.993	-614.135	-298.746	5.034
3200	58.223	423.838	371.709	166.814	-614.430	-288.567	4.710
3300	58.252	425.631	373.316	172.638	-614.741	-278.379	4.406
3400	58.284	427.370	374.880	178.465	-615.066	-268.182	4.120
3500	58.319	429.060	376.404	184.295	-615.403	-257.974	3.850
3600	58.359	430.703	377.890	190.129	-615.751	-247.757	3.595
3700	58.403	432.303	379.339	195.967	-616.107	-237.530	3.353
3800	58.450	433.861	380.753	201.810	-616.469	-227.294	3.124
3900	58.503	435.380	382.135	207.657	-616.833	-217.047	2.907
4000	58.559	436.862	383.484	213.510	-617.199	-206.792	2.700
4100	58.620	438.309	384.804	219.369	-617.563	-196.527	2.504
4200	58.685	439.722	386.095	225.234	-617.922	-186.254	2.316
4300	58.755	441.104	387.358	231.106	-618.273	-175.972	2.138
4400	58.829	442.455	388.595	236.986	-618.616	-165.682	1.967
4500	58.907	443.778	389.807	242.872	-618.944	-155.384	1.804
4600	58.990	445.074	390.994	248.767	-619.260	-145.079	1.647
4700	59.076	446.343	392.158	254.670	-619.559	-134.768	1.498
4800	59.166	447.588	393.300	260.582	-619.841	-124.450	1.354
4900	59.259	448.809	394.421	266.504	-620.103	-114.127	1.217
5000	59.356	450.007	395.520	272.434	-620.345	-103.798	1.084
5100	59.456	451.184	396.600	278.375	-620.564	-93.465	0.957
5200	59.559	452.339	397.661	284.326	-620.762	-83.128	0.835
5300	59.664	453.475	398.703	290.287	-620.935	-72.787	0.717
5400	59.772	454.591	399.728	296.259	-621.086	-62.443	0.604
5500	59.882	455.689	400.736	302.241	-621.212	-52.096	0.495
5600	59.994	456.769	401.727	308.235	-621.314	-41.748	0.389
5700	60.107	457.831	402.702	314.240	-621.393	-31.398	0.288
5800	60.222	458.878	403.661	320.257	-621.445	-21.047	0.190
5900	60.338	459.908	404.606	326.285	-621.477	-10.695	0.095
6000	60.455	460.923	405.536	332.324	-621.465	-0.345	0.003

PREVIOUS: June 1976 (1 atm) CURRENT: June 1976 (1 bar)

Aluminum Chloride ($AlCl_2$) $Al_1Cl_2(g)$

ALUMINUM CHLORIDE, ION ($AlCl_2^+$) IDEAL GAS M_r = 97.88699

$\Delta_fH°(0\ K) = 475 \pm 75\ kJ\ mol^{-1}$

$S°(298.15\ K) = [268.26 \pm 4.2]\ J\ K^{-1}\ mol^{-1}$ $\Delta_fH°(298.15\ K) = [481.160]\ kJ\ mol^{-1}$

Vibrational Frequencies and Degeneracies

ν, cm^{-1}
[390](1)
[120](2)
[690](1)

Ground State Quantum Weight: [1] $\sigma = 2$

Point Group: $[D_{\infty h}]$

Bond Distance: Al-Cl = [2.05] Å

Bond Angle: Cl-Al-Cl = [180]°

Rotational Constant: $B_o = [0.056573]\ cm^{-1}$

Enthalpy of Formation

We adopt $\Delta_fH°(298.15\ K) = 115\pm18\ kcal\ mol^{-1}$ and $IP(AlCl_2) = 7.8\pm0.9$ eV. Electron-impact studies gave 13.4±0.5 eV (1) and 13.4±0.7 eV (2) for the appearance potential (AP) of $AlCl_2^+$ from $AlCl_3$. The analogous process for onset of BCl_2^+ from BCl_3 has been studied by both photoionization (3) and electron impact (2). Comparison indicates that BCl_2^+ from electron impact (9) carried an excess energy (E*) of ~0.9 eV. We estimate the excess energy for $AlCl_2^+$ as 1.2±0.6 eV. Thus, we take AP-E* = $\Delta_rH°$ = 12.2±0.8 eV (281.3±18 $kcal\ mol^{-1}$) for the reaction $AlCl_3(g) + e^-(g) = AlCl_2^+(g) + Cl(g) + 2\ e^-(g)$. With JANAF auxiliary data (4) this yields $\Delta_fH°(AlCl_2^+) = 113.5\pm18$, $\Delta_fH°(298.15\ K) = 115\pm18$ and $IP(AlCl_2) = 180\pm20\ kcal\ mol^{-1}$ (7.8±0.9 eV).

Electron impact studies (5-7) of $AlCl_2$ and BCl_2 gave approximate appearance potentials $AP(MCl_2^+)$ = 12±1 eV. This is ~4 eV greater than our adiabatic IP values. Such a difference could arise from several factors including bias in the ionizing-energy scale, bias in threshold detection, excess energy in MCl_2^+ and misassignment of the ionization process. The studies (5-7) make no mention of a calibrant gas for the energy scale. Reactive radicals (e.g. $AlCl_2$ from a reducing environment) may induce a bias in the energy scale. Consistent with this hypothesis, the values reported (6, 7) for $AP(BCl^+/BCl)$ and $AP(BCl_2^+/BCl_3)$ are ~2 and 3-4 eV higher than other electron-impact data (2, 8). $AP(MCl_2^+/MCl_2)$ was obtained (5-7) by a vanishing current (or initial break) method. We expect this to have a positive bias which increases as the pressure of MCl_2 decreases (9). Electron impact on MCl_2 should yield a vertical AP corresponding to a nonlinear (excited) configuration of MCl_2^+. Excitation energies of >1 eV are expected if MCl_2^+ is similar to BF_2^+ and AlF_2^+ (4). Alternatively, $AP(MCl_2^+)$ = 12±1 eV (5-7) is a magnitude which may not exclude processes such as fragmentation or ion-molecular reactions of MCl_3. $AP(MCl_2^+)$ almost overlaps with the onset of MCl_2^+ from MCl_3; i.e., 12.30 eV (3, photoionization) or 13.0-13.2 eV (2, 8 electron impact) from BCl_3 and 13.4±0.5 eV (1, 2, electron impact) from $AlCl_3$. On the other hand, the coincidence of $AP(BCl^+) = AP(BCl_2^+)$ = 12±1 eV (6, 7) suggests that the two ions might be related, possibly via $BCl^+ + BCl_3 = BCl_2^+$. The authors (5, 6) did not measure $AP(AlCl^+)$, but the analogous ion-molecular reaction is also energetically feasible. In summary, there are three plausible factors which may contribute to positive bias in the observed $AP(AlCl_2^+)$ and some conceivable alternative interpretations of the data (5-7). This precludes use of $AP(AlCl_2^+)$ in calculating $\Delta_fH°$ of $AlCl_2^+$.

Hastie and Margrave (10) used an extended Hucket method to calculate $IP(AlCl_2)$ = 6.65 eV. This result should approximate the vertical IP which we expect at ~9 eV.

Heat Capacity and Entropy

We assume the electronic ground state to be linear $^1\Sigma^+$ and neglect excited states by analogy with $MgCl_2$ (4, 11) and other tratomic species having sixteen valence electrons (12). We estimate the bond distance to be almost the same as in $AlCl_3$ and 0.05 Å shorter than in $AlCl_2$ (4). Likewise, we transfer the stretching force constant from $AlCl_3$ and the ratio $f_r/f_\alpha = 19\ Å^{-2}$ from $MgCl_2$ (11). Thus, we calculate vibrational frequencies from the estimated force constants $f_r = 2.9$, $f_{rr} = 0.2$ and $f_\alpha/r^2 = 0.036$ mdyn/Å. The resulting values have been rounded upward.

References

1. R. F. Porter and E. F. Zeller, J. Chem. Phys. 33, 858 (1960).
2. N. D. Potter, Rept. AD-715567, 160 pp. (1970).
3. V. H. Dibeler and J. A. Walker, Inorg. Chem. 8, 50 (1969).
4. JANAF Thermochemical Tables: $AlCl_2(g)$ and $AlF_2^+(g)$, 6-30-76; Cl(g), 6-30-72; $AlCl_3(g)$, 6-30-70; $MgCl_2(g)$, 12-31-69.
5. M. Farber and S. P. Harris, High Temp. Sci. 3, 231 (1971).
6. M. Farber, R. D. Srivastava and O. M. Uy, Space Sciences, Inc., AFRPL-TR-70-97, July 1970.
7. R. D. Srivastava and M. Farber, Trans. Faraday Soc. 67, 2298 (1971).
8. J. L. Franklin, J. G. Dilland, et al., NSRDS-NBS 26, 1969.
9. R. W. Kiser, "Introduction to Mass Spectrometry and Its Application", Prentice-Hall, Inc., Englewood Cliffs, N. J., pp. 166-173, 1965.
10. J. W. Hastie and J. L. Margrave, J. Phys. Chem. 73, 1105 (1969).
11. M. L. Lesiecki and J. W. Nibler, J. Chem. Phys. 64, 871 (1976).
12. J. W. Rabalais, J. M. McDonald, V. Scherr and S. P. McGlynn, Chem. Rev. 71, 73 (1971).

Aluminum Chloride, Ion ($AlCl_2^+$) $Al_1Cl_2^+(g)$

Enthalpy Reference Temperature = T_r = 298.15 K Standard State Pressure = p° = 0.1 MPa

T/K	$C_p°$ (J K⁻¹mol⁻¹)	S° (J K⁻¹mol⁻¹)	$-[G°-H°(T_r)]/T$ (J K⁻¹mol⁻¹)	$H°-H°(T_r)$ (kJ mol⁻¹)	$\Delta_fH°$ (kJ mol⁻¹)	$\Delta_fG°$ (kJ mol⁻¹)	Log K_f
0	0.	0.	INFINITE	-13.481	475.202		
100	43.152	214.688	314.010	-9.932			
200	50.669	247.132	273.173	-5.208			
250	53.206	258.723	269.158	-2.609			
298.15	55.063	268.261	268.261	0.	481.160	469.865	-82.318
300	55.124	268.602	268.262	0.102	481.193	469.794	-81.798
350	56.563	277.213	268.939	2.896	482.072	467.825	-69.819
400	57.646	284.840	270.459	5.752	482.944	465.730	-60.818
450	58.470	291.679	272.443	8.656	483.807	463.526	-53.805
500	59.106	297.874	274.681	11.596	484.662	461.227	-48.184
600	60.000	308.736	279.477	17.555	486.332	456.382	-39.732
700	60.579	318.031	284.337	23.586	487.922	451.264	-33.674
800	60.971	326.147	289.067	29.665	489.399	445.924	-29.116
900	61.249	333.346	293.594	35.776	490.728	440.408	-25.561
1000	61.452	339.810	297.898	41.912	481.290	435.514	-22.749
1100	61.604	345.674	301.979	48.065	482.596	430.873	-20.460
1200	61.721	351.040	305.847	54.232	483.905	426.113	-18.548
1300	61.813	355.984	309.516	60.408	485.213	421.244	-16.926
1400	61.887	360.568	313.001	66.594	486.520	416.275	-15.531
1500	61.946	364.840	316.316	72.785	487.824	411.211	-14.320
1600	61.996	368.839	319.475	78.983	489.125	406.062	-13.257
1700	62.036	372.599	322.490	85.184	490.422	400.830	-12.316
1800	62.071	376.146	325.374	91.390	491.713	395.523	-11.478
1900	62.100	379.503	328.135	97.598	492.998	390.144	-10.726
2000	62.125	382.688	330.784	103.809	494.275	384.698	-10.047
2100	62.146	385.720	333.328	110.023	495.544	379.187	-9.432
2200	62.165	388.612	335.776	116.239	496.802	373.617	-8.871
2300	62.181	391.375	338.134	122.456	498.049	367.990	-8.357
2400	62.196	394.022	340.408	128.675	499.282	362.309	-7.885
2500	62.208	396.561	342.603	134.895	500.501	356.576	-7.450
2600	62.219	399.001	344.726	141.116	501.704	350.795	-7.048
2700	62.230	401.350	346.780	147.339	502.890	344.969	-6.674
2800	62.238	403.613	348.769	153.562	210.157	340.066	-6.344
2900	62.247	405.797	350.698	159.786	212.402	344.666	-6.208
3000	62.254	407.908	352.570	166.012	214.629	349.189	-6.080
3100	62.260	409.949	354.389	172.237	216.836	353.638	-5.959
3200	62.266	411.926	356.156	178.464	219.025	358.016	-5.844
3300	62.272	413.842	357.875	184.690	221.196	362.326	-5.735
3400	62.277	415.701	359.549	190.918	223.350	366.570	-5.632
3500	62.281	417.506	361.179	197.146	225.489	370.751	-5.533
3600	62.286	419.261	362.768	203.374	227.615	374.871	-5.439
3700	62.289	420.967	364.318	209.603	229.728	378.933	-5.350
3800	62.293	422.629	365.831	215.832	231.832	382.937	-5.264
3900	62.296	424.247	367.308	222.062	233.927	386.886	-5.182
4000	62.299	425.824	368.751	228.291	236.017	390.782	-5.103
4100	62.302	427.362	370.162	234.521	238.103	394.625	-5.028
4200	62.305	428.864	371.542	240.752	240.188	398.417	-4.955
4300	62.307	430.330	372.892	246.982	242.273	402.160	-4.885
4400	62.309	431.762	374.214	253.213	244.361	405.854	-4.818
4500	62.312	433.163	375.508	259.444	246.456	409.501	-4.753
4600	62.314	434.532	376.777	265.676	248.555	413.101	-4.691
4700	62.315	435.872	378.020	271.907	250.662	416.655	-4.631
4800	62.317	437.184	379.239	278.139	252.778	420.165	-4.572
4900	62.319	438.469	380.434	284.370	254.906	423.630	-4.516
5000	62.320	439.728	381.608	290.602	257.044	427.051	-4.461
5100	62.322	440.962	382.759	296.834	259.194	430.430	-4.409
5200	62.323	442.173	383.890	303.067	261.357	433.767	-4.357
5300	62.325	443.360	385.001	309.299	263.533	437.061	-4.307
5400	62.326	444.525	386.093	315.532	265.723	440.315	-4.259
5500	62.327	445.668	387.166	321.764	267.925	443.528	-4.212
5600	62.328	446.791	388.220	327.997	270.140	446.701	-4.167
5700	62.329	447.895	389.258	334.230	272.368	449.834	-4.122
5800	62.330	448.979	390.278	340.463	274.611	452.928	-4.079
5900	62.331	450.044	391.282	346.696	276.862	455.983	-4.037
6000	62.332	451.092	392.270	352.929	279.147	458.997	-3.996

PREVIOUS: June 1976 (1 atm) CURRENT: June 1976 (1 bar)

Aluminum Chloride, Ion ($AlCl_2^+$) $Al_1Cl_2^+(g)$

ALUMINUM CHLORIDE, ION ($AlCl_2^-$) IDEAL GAS M_r = 97.88809

$S°(298.15\ K) = [285.8 \pm 4]\ J\ K^{-1}\ mol^{-1}$

$\Delta_f H°(0\ K) = [-474 \pm 105]\ kJ\ mol^{-1}$
$\Delta_f H°(298.15\ K) = [-481.160]\ kJ\ mol^{-1}$

Electronic Levels and Quantum Weights

State	ϵ_i, cm^{-1}	g_i
$^1A_1$	[0]	1
$^3B_1$	[23000]	3
$^1B_1$	[30000]	1

Vibrational Frequencies and Degeneracies

ν, cm^{-1}
[430](1)
[160](1)
[460](1)

Point Group: [C_{2v}] σ = 2
Bond Distance: Al-Cl = [2.15] Å
Bond Angle: Cl-Al-Cl = [105]°
Product of the Moments of Inertia: $I_A I_B I_C = [7.583526 \times 10^{-114}]\ g^3\ cm^6$

Enthalpy of Formation

We adopt $\Delta_f H°(298.15\ K) = -115 \pm 25\ kcal\ mol^{-1}$ based on an estimated electron affinity of $EA(AlCl_2) = 46 \pm 23\ kcal\ mol^{-1}$ (2.0±1 eV) and a fluoride-ion affinity of IA(AlCl) = 46±25 kcal mol^{-1}. JANAF auxiliary data (1) are used in converting from one quantity to another. If AlX_2 radicals were analogous to X atoms (2, 3), then $EA(AlCl_2)$ should be approximately equal to $EA(AlF_2)$ = 0.2±0.4 eV (1). This analogy may not be justified, since the pertinent orbitals are quite different. An electron added to AlX_2 presumably occupies a molecular orbital centered mainly on the metal opposite the two halogens. We assume that the orbital energy is characteristic of the metal Al and is perturbed relatively little by substitution of Cl for F. This implies that $EA(AlCl_2) \simeq EA(AlF_2)$.

The energy for the self-ionization process, $2\ AlF_2 = AlF_2^+ + AlF_2^-$, is $\Delta_r H° = [IP(AlF_2)-EA(AlF_2)]$ = 5.9±1 eV (1). Assuming $AlCl_2$ is similar, we estimate $EA(AlCl_2)$ = 1.9 eV. The energy difference between the reactions $AlF_2^- = AlF + F^-$ and $AlF_2 = AlF + F$ is $\Delta_r H° = [IA(AlF)-D_0^0\ (AlF_2)] = -29\ kcal\ mol^{-1}$. Assuming $AlCl_2^-$ is analogous, we estimate $EA(AlCl_2)$ = 2.3 eV. We adopt the intermediate value $EA(AlCl_2)$ = 2.0±1 eV but emphasize that all estimates depend on AlF_2^- (1).

Heat Capacity and Entropy

Electronic levels and quantum weights are assumed equal to those of isoelectronic $SiCl_2$ (1). Vibrational frequencies are calculated from the estimated force constants $f_r = 1.73$, $f_{rr} = 0.07$ and $f_\alpha/r^2 = 0.156$ mdyn/Å. Frequencies and force constants are estimated from simultaneous consideration of $AlCl_2$, $AlCl_2^+$, $AlCl_2^-$ and the analogous boron species. We assume that frequency changes are qualitatively similar in the two series $AlCl_2^+$, $AlCl_2$, $AlCl_2^-$ and NO_2^+, NO_2, NO_2^- (1) which have the same number of valence electrons. Furthermore, we expect isoelectronic $SiCl_2$ (1) to provide upper-limit frequencies for $AlCl_2^-$. Thus, in going from $AlCl_2$ to $AlCl_2^-$, we assume little change in ν_1 and ν_2 but a significant decrease in ν_3.

By analogy with BF_2 (4, 5), we expect that the odd electron in $AlCl_2$ occupies a molecular orbital centered mainly on the metal opposite the two chlorides. This orbital should be antibonding (4) in the sense that addition of an electron increases the bond length but decreases the bond angle and stretching force constant. We assume that the bond length is 0.05 Å longer and the bond angle is 15° smaller than in $AlCl_2$ (1). The principal moments of inertia are: $I_A = 5.5597 \times 10^{-39}$, $I_B = 34.2573 \times 10^{-39}$ and $I_C = 39.8170 \times 10^{-39}\ g\ cm^2$.

References

1. JANAF Thermochemical Tables: $AlCl_2(g)$, $AlCl_2^+(g)$, $AlF_2(g)$, $AlF_2^+(g)$ and $AlF_2^-(g)$, 6-30-76; $NO_2^-(g)$, 6-30-72; $SiCl_2(g)$, 12-31-70; AlCl(g), 6-30-70; $Cl^-(g)$, 6-30-65.
2. J. L. Franklin and P. W. Harland, Ann. Rev. Phys. Chem. 25, 485 (1974).
3. H. Hotop and W. C. Lineberger, J. Phys. Chem. Ref. Data 4, 539 (1975).
4. C. Thomson and D. A. Brotchie, Theoret. Chim. Acta 32, 101 (1973).
5. W. Nelson and W. Gordy, J. Chem. Phys. 51, 4710 (1969).

Aluminum Chloride, Ion ($AlCl_2^-$) $Al_1Cl_2^-(g)$

Enthalpy Reference Temperature = T_r = 298.15 K Standard State Pressure = p° = 0.1 MPa

T/K	$C_p°$	S°	$-[G°-H°(T_r)]/T$	$H°-H°(T_r)$	$\Delta_f H°$	$\Delta_f G°$	Log K_f
	J K⁻¹mol⁻¹			kJ mol⁻¹			
0	0.	0.	INFINITE	-12.969	-474.211		
100	39.846	235.193	329.316	-9.412			
200	48.255	265.583	290.476	-4.979			
250	50.901	276.654	286.636	-2.496			
298.15	52.650	285.778	285.778	0.	-481.160	-485.169	85.000
300	52.706	286.104	285.779	0.097	-481.209	-485.193	84.480
350	53.953	294.328	286.426	2.766	-482.533	-485.753	72.495
400	54.836	301.593	287.877	5.487	-483.876	-486.122	63.481
450	55.478	308.091	289.768	8.245	-485.236	-486.321	56.451
500	55.957	313.962	291.899	11.032	-486.614	-486.368	50.810
600	56.606	324.227	296.456	16.663	-489.429	-486.055	42.315
700	57.013	332.985	301.064	22.345	-492.345	-485.264	36.211
800	57.283	340.617	305.541	28.061	-495.388	-484.047	31.605
900	57.471	347.376	309.821	33.799	-498.590	-482.439	28.000
1000	57.607	353.438	313.885	39.553	-512.566	-479.704	25.057
1100	57.708	358.934	317.734	45.319	-515.804	-476.261	22.616
1200	57.786	363.958	321.380	51.094	-519.045	-472.524	20.568
1300	57.847	368.586	324.836	56.876	-522.289	-468.515	18.825
1400	57.895	372.875	328.116	62.663	-525.537	-464.257	17.322
1500	57.934	376.871	331.234	68.455	-528.790	-459.767	16.010
1600	57.966	380.611	334.205	74.250	-532.049	-455.059	14.856
1700	57.993	384.126	337.039	80.048	-535.313	-450.147	13.831
1800	58.015	387.441	339.748	85.848	-538.584	-445.043	12.915
1900	58.034	390.578	342.341	91.651	-541.862	-439.757	12.090
2000	58.051	393.556	344.828	97.455	-545.149	-434.298	11.343
2100	58.065	396.388	347.217	103.261	-548.445	-428.674	10.663
2200	58.078	399.090	349.514	109.068	-551.752	-422.893	10.041
2300	58.090	401.672	351.726	114.876	-555.072	-416.962	9.469
2400	58.101	404.144	353.859	120.686	-558.405	-410.886	8.943
2500	58.112	406.516	355.918	126.496	-561.753	-404.670	8.455
2600	58.124	408.796	357.908	132.308	-565.116	-398.321	8.002
2700	58.136	410.990	359.834	138.121	-568.497	-391.841	7.581
2800	58.150	413.104	361.699	143.935	-865.796	-384.269	7.169
2900	58.166	415.145	363.507	149.751	-868.117	-367.030	6.611
3000	58.184	417.117	365.261	155.569	-870.456	-349.712	6.089
3100	58.205	419.025	366.965	161.388	-872.812	-332.315	5.599
3200	58.230	420.874	368.621	167.210	-875.185	-314.842	5.139
3300	58.259	422.666	370.231	173.034	-877.574	-297.294	4.706
3400	58.293	424.406	371.799	178.862	-879.976	-279.674	4.297
3500	58.332	426.096	373.326	184.693	-882.391	-261.982	3.910
3600	58.376	427.740	374.815	190.528	-884.816	-244.222	3.544
3700	58.427	429.340	376.267	196.368	-887.248	-226.394	3.196
3800	58.484	430.899	377.685	202.214	-889.686	-208.501	2.866
3900	58.547	432.419	379.069	208.065	-892.125	-190.543	2.552
4000	58.618	433.902	380.421	213.924	-894.564	-172.523	2.253
4100	58.696	435.350	381.743	219.789	-897.000	-154.441	1.968
4200	58.782	436.766	383.036	225.663	-899.429	-136.301	1.695
4300	58.875	438.150	384.302	231.546	-901.848	-118.103	1.435
4400	58.975	439.505	385.541	237.438	-904.256	-99.848	1.185
4500	59.083	440.831	386.755	243.341	-906.647	-81.539	0.946
4600	59.199	442.131	387.945	249.255	-909.022	-63.177	0.717
4700	59.322	443.405	389.112	255.181	-911.378	-44.764	0.497
4800	59.452	444.656	390.256	261.120	-913.712	-26.301	0.286
4900	59.589	445.883	391.378	267.072	-916.021	-7.789	0.083
5000	59.733	447.088	392.481	273.038	-918.306	10.769	-0.113
5100	59.884	448.273	393.563	279.018	-920.564	29.373	-0.301
5200	60.041	449.437	394.626	285.015	-922.795	48.021	-0.482
5300	60.204	450.582	395.671	291.027	-924.996	66.712	-0.657
5400	60.372	451.709	396.699	297.056	-927.168	85.444	-0.827
5500	60.545	452.818	397.709	303.101	-929.310	104.216	-0.990
5600	60.724	453.911	398.703	309.165	-931.421	123.026	-1.148
5700	60.906	454.987	399.681	315.246	-933.502	141.874	-1.300
5800	61.093	456.048	400.644	321.346	-935.549	160.758	-1.448
5900	61.283	457.094	401.592	327.465	-937.569	179.677	-1.591
6000	61.476	458.126	402.525	333.603	-939.537	198.628	-1.729

PREVIOUS: June 1976 (1 atm) CURRENT: June 1976 (1 bar)

Aluminum Chloride, Ion ($AlCl_2^-$) $Al_1Cl_2^-(g)$

ALUMINUM CHLORIDE FLUORIDE ($AlCl_2F$) IDEAL GAS M_r = 116.885943

S°(298.15 K) = [311.4 ± 4] J K^{-1} mol^{-1}

$\Delta_fH°$(0 K) = -788.4 ± 6 kJ mol^{-1}
$\Delta_fH°$(298.15 K) = -790.8 ± 6 kJ mol^{-1}

Vibrational Frequencies and Degeneracies

ν, cm^{-1}	ν, cm^{-1}
[840](1)	[640](1)
[430](1)	[190](1)
[160](1)	[220](1)

Ground State Quantum Weight: [1] σ = 2
Point Group: C_{2v}
Bond Distances: Al-F = [1.63] Å Al-Cl = [2.06] Å
Bond Angles: Cl-Al-F = [120]° Cl-Al-Cl = [120]°
Product of the Moments of Inertia: $I_AI_BI_C$ = [3.841449 x 10^{-113}] g^3 cm^6

Enthalpy of Formation

We adopt $\Delta_fH°$(298.15 K) = -189.0±1.5 kcal mol^{-1} and $\Delta_aH°$ = 341.4±3 kcal mol^{-1} based on $\Delta_rH°$(298.15 K) = 0.5±1 kcal mol^{-1} for 2/3 $AlCl_3$(g) + 1/3 AlF_3(g) = $AlCl_2F$(g). Krause and Douglas (1) obtained this result by an entrainment method which measured enhanced volatility of AlF_3 in the presence of $AlCl_3$. Corrections were made for all the possible dimers which can form in this system. The entrainment data establish that aluminum fluoride-chloride exchange reactions are almost thermoneutral.

Our adopted $\Delta_fH°$ is compared below with an approximate mass-spectrometric Kp for reaction of AlF_2 with $AlCl_2$ (2). The discrepancy in $\Delta_fH°(AlCl_2F)$ is -22 kcal mol^{-1} if we use $\Delta_fH°$ = -180 (AlF_2) and -71 ($AlCl_2$) derived from the mass-spectrometric data (2, 3). The discrepancy almost disappears if we use $\Delta_fH°$ = -166 (AlF_2) and -67 ($AlCl_2$) derived from average bond energies in AlF_3 and $AlCl_3$ (4). This favors the latter values of $\Delta_fH°$ for AlF_2 and $AlCl_2$ (4), although the discrepancy might arise from the observed ion intensity of $AlCl_2F^+$. The signal from $AlCl_2F^+$ was very weak at 40 eV (5); this high ionizing energy might bias Kp and cause considerable fragmentation. We can eliminate AlF_2 and $AlCl_2$ by taking the difference between the mass-spectrometric reactions for $AlClF_2$ (4) and $AlCl_2F$. This yields [$\Delta_fH°(AlClF_2)-\Delta_fH°(AlCl_2F)$] = -47 kcal mol^{-1} which agrees with -49.8 kcal mol^{-1} from entrainment data (1). Thus, we tentatively ascribe the discrepancy to AlF_2 (4) and $AlCl_2$.

Source	Method	Reaction	T/K	$\Delta_rH°$(T) kcal mol^{-1}	$\Delta_rH°$(298.15 K) kcal mol^{-1}	$\Delta_fH°$(298.15 K) kcal mol^{-1}
(1)	Entrainment	2/3 $AlCl_3$(g) + 1/3 AlF_3(g) = $AlCl_2F$(g)	1196-1257	0.5±1	0.5±1	-189.0
(2)	Mass Spec.	AlF_2(g) + $AlCl_2$(g) = AlF(g) + $AlCl_2F$(g)	1491	--	-23.4	-194[a] or -211[b]

[a] Assuming $\Delta_fH°$(298.15 K) = -166 (AlF_2) and -67 ($AlCl_2$) kcal mol^{-1}.
[b] Assuming $\Delta_fH°$(298.15 K) = -180 (AlF_2) and -71 ($AlCl_2$) kcal mol^{-1}.

Heat Capacity and Entropy

We adopt a C_{2v} structure with bond angles and bond distances assumed equal to those in AlF_3 and $AlCl_3$ (4). We assume a singlet electronic ground state and neglect excited states. Vibrational frequencies are estimated by comparison of AlF_3 and $AlCl_3$ with the series BF_3, $BClF_2$, BCl_2F and BCl_3 (6, 7). The principal moments of inertia are: I_A = 18.3596 x 10^{-39}, I_B = 37.4744 x 10^{-39} and I_C = 55.8339 x 10^{-39} g cm^2.

References

1. R. F. Krause and T. B. Douglas, J. Phys. Chem. 72, 3444 (1968).
2. M. Farber and S. P. Harris, High Temp. Sci. 3, 231 (1971).
3. O. M. Uy, R. D. Srivastava and M. Farber, High Temp. Sci. 4, 227 (1972).
4. JANAF Thermochemical Tables: AlF_2(g), $AlCl_2$(g), and $AlCl_2$(g), 6-30-76; AlF_3(g) and $AlCl_3$(g), 6-30-70.
5. M. Farber, Space Sciences, Inc., Monrovia, Calif., personal communication, August 30, 1976.
6. D. F. Wolfe and G. L. Humphrey, J. Mol. Struct. 3, 293 (1969).
7. L. P. Lindeman and M. K. Wilson, J. Chem. Phys. 24, 242 (1956).

Aluminum Chloride Fluoride ($AlCl_2F$) $Al_1Cl_2F_1$(g)

Enthalpy Reference Temperature = T_r = 298.15 K Standard State Pressure = p° = 0.1 MPa

T/K	$C_p°$ (J $K^{-1}mol^{-1}$)	S° (J $K^{-1}mol^{-1}$)	-[G°-H°(T_r)]/T (J $K^{-1}mol^{-1}$)	H°-H°(T_r) (kJ mol^{-1})	$\Delta_fH°$ (kJ mol^{-1})	$\Delta_fG°$ (kJ mol^{-1})	Log K_f
0	0.	0.	INFINITE	-15.747	-788.390	-788.390	INFINITE
100	47.886	247.731	366.958	-11.923	-789.415	-786.070	410.601
200	61.104	285.446	317.475	-6.406	-790.202	-782.405	204.343
250	65.494	299.572	312.518	-3.236	-790.517	-780.418	163.059
298.15	68.804	311.403	311.403	0.	-790.776	-778.449	136.381
300	68.916	311.829	311.404	0.127	-790.785	-778.372	135.527
350	71.562	322.660	312.253	3.642	-791.019	-776.284	115.854
400	73.607	332.355	314.171	7.274	-791.227	-774.165	101.096
450	75.197	341.120	316.686	10.995	-791.418	-772.021	89.614
500	76.446	349.110	319.535	14.788	-791.600	-769.856	80.426
600	78.234	363.218	325.671	22.528	-791.963	-765.473	66.640
700	79.411	375.372	331.923	30.414	-792.373	-761.026	56.788
800	80.220	386.032	338.034	38.398	-792.874	-756.516	49.395
900	80.797	395.515	343.904	46.451	-793.510	-751.934	43.641
1000	81.221	404.051	349.499	54.553	-804.905	-746.513	38.994
1100	81.541	411.808	354.816	62.691	-805.551	-740.643	35.170
1200	81.788	418.914	359.866	70.858	-806.194	-734.714	31.981
1300	81.983	425.469	364.663	79.047	-806.837	-728.731	29.281
1400	82.139	431.550	369.226	87.254	-807.483	-722.698	26.964
1500	82.265	437.222	373.572	95.474	-808.134	-716.620	24.955
1600	82.370	442.534	377.718	103.706	-808.791	-710.497	23.195
1700	82.457	447.531	381.679	111.947	-809.456	-704.334	21.642
1800	82.530	452.246	385.470	120.197	-810.129	-698.131	20.259
1900	82.592	456.710	389.103	128.453	-810.811	-691.890	19.021
2000	82.645	460.948	392.590	136.715	-811.502	-685.613	17.906
2100	82.691	464.981	395.942	144.982	-812.203	-679.302	16.897
2200	82.731	468.829	399.168	153.253	-812.912	-672.956	15.978
2300	82.765	472.507	402.278	161.528	-813.631	-666.579	15.138
2400	82.796	476.030	405.278	169.806	-814.358	-660.170	14.368
2500	82.823	479.411	408.176	178.087	-815.094	-653.730	13.659
2600	82.847	482.659	410.979	186.370	-815.839	-647.261	13.004
2700	82.869	485.787	413.692	194.656	-816.591	-640.762	12.396
2800	82.888	488.801	416.321	202.944	-1111.252	-633.269	11.814
2900	82.905	491.710	418.870	211.234	-1110.923	-616.204	11.099
3000	82.921	494.520	421.345	219.525	-1110.600	-599.150	10.432
3100	82.935	497.240	423.750	227.818	-1110.282	-582.107	9.808
3200	82.948	499.873	426.088	236.112	-1109.967	-565.074	9.224
3300	82.959	502.426	428.363	244.407	-1109.654	-548.051	8.675
3400	82.970	504.902	430.578	252.704	-1109.341	-531.037	8.158
3500	82.980	507.308	432.736	261.001	-1109.027	-514.033	7.672
3600	82.989	509.645	434.840	269.299	-1108.709	-497.037	7.212
3700	82.997	511.919	436.892	277.599	-1108.386	-480.051	6.777
3800	83.005	514.133	438.896	285.899	-1108.056	-463.074	6.365
3900	83.012	516.289	440.853	294.200	-1107.715	-446.105	5.975
4000	83.018	518.391	442.765	302.501	-1107.364	-429.145	5.604
4100	83.024	520.441	444.635	310.803	-1106.998	-412.194	5.251
4200	83.030	522.441	446.464	319.106	-1106.617	-395.252	4.916
4300	83.035	524.395	448.253	327.409	-1106.218	-378.319	4.596
4400	83.040	526.304	450.006	335.713	-1105.801	-361.396	4.290
4500	83.044	528.170	451.722	344.017	-1105.359	-344.483	3.999
4600	83.049	529.996	453.404	352.322	-1104.898	-327.580	3.720
4700	83.053	531.782	455.053	360.627	-1104.413	-310.687	3.453
4800	83.057	533.530	456.669	368.932	-1103.904	-293.804	3.197
4900	83.060	535.243	458.255	377.238	-1103.368	-276.933	2.952
5000	83.063	536.921	459.812	385.544	-1102.807	-260.073	2.717
5100	83.066	538.566	461.340	393.851	-1102.220	-243.223	2.491
5200	83.069	540.179	462.841	402.158	-1101.607	-226.386	2.274
5300	83.072	541.761	464.315	410.465	-1100.967	-209.561	2.065
5400	83.075	543.314	465.764	418.772	-1100.302	-192.748	1.864
5500	83.077	544.838	467.188	427.080	-1099.610	-175.948	1.671
5600	83.080	546.335	468.588	435.388	-1098.894	-159.161	1.485
5700	83.082	547.806	469.965	443.696	-1098.153	-142.387	1.305
5800	83.084	549.251	471.319	452.004	-1097.386	-125.626	1.131
5900	83.086	550.671	472.652	460.312	-1096.599	-108.878	0.964
6000	83.088	552.068	473.964	468.621	-1095.768	-92.146	0.802

PREVIOUS: June 1976 (1 atm) CURRENT: June 1976 (1 bar)

Aluminum Chloride Fluoride ($AlCl_2F$) $Al_1Cl_2F_1$(g)

ALUMINUM CHLORIDE ($AlCl_3$) CRYSTAL M_r = 133.34054

$S°(298.15\ K) = 109.29 \pm 0.42\ J\ K^{-1}\ mol^{-1}$

$T_{fus} = 465.7 \pm 0.2\ K$

$\Delta_f H°(0\ K) = -704.30 \pm 0.84\ kJ\ mol^{-1}$

$\Delta_f H°(298.15\ K) = -705.63 \pm 0.84\ kJ\ mol^{-1}$

$\Delta_{fus} H° = 35.35 \pm 0.84\ kJ\ mol^{-1}$

Enthalpy of Formation

The adopted enthalpy of formation, $\Delta_f H°(AlCl_3$, cr, 298.15 K) = −168.65±0.20 kcal mol^{-1} (−705.632±0.84 kJ mol^{-1}) is a weighted average of the values of Gross and Hayman (1) and of Coughlin (3). NBS (4) recommended a value of −704.2 kJ mol^{-1} based on data available prior to 1968.

Gross and Hayman (1) measured the enthalpy of direct combination of "spectrographically pure" aluminum and carefully purified chlorine and derived $\Delta_f H°$(298.15 K) = −168.80±0.16 kcal mol^{-1}. Siemonsen (2) also reported the enthalpy of direct combination and is in agreement at −167.5 kcal mol^{-1} within his relatively large uncertainty interval of ±1.1 kcal mol^{-1}. Coughlin (3) determined the enthalpy of solution of 99.99+% pure Al in 4.360 m HCl and also the enthalpy of solution of anhydrous $AlCl_3$ in the same solvent. Enthalpies of mixing were measured to close a thermochemical cycle at 303.15 K (adjusted to 1981 relative atomic masses).

	Reaction	$\Delta_r H°$(303.15 K), cal mol^{-1}
(1)	$Al(cr) + 3H^+(sln) = Al^{+++}(sln) + 3/2\ H_2(g)$	−127057 ± 120
(2)	$3(HCl\cdot 12.731\ H_2O)(\ell) = 3H^+(sln) + 3\ Cl^-(sln) + 38.193\ H_2O(sln)$	0 ± 10
(3)	$AlCl_3(cr) = Al^{+++}(sln) + 3\ Cl^-(sln)$	−72504 ± 50
(4)	$38.193\ H_2O(\ell) = 38.193\ H_2O(sln)$	−3050 ± 20
(5)	$Al(cr) + 3\ (HCl\cdot 12.731\ H_2O)(\ell) = AlCl_3(cr) + 38.193\ H_2O(\ell) + 3/2\ H_2(g)$	−51503 ± 140

With auxiliary data from JANAF (12) and CODATA (5), the enthalpy of reaction (5) at 298.15 K is −51910 ± 140 cal mol^{-1} $AlCl_3$. Interpolation of data for $\Delta_f H°$(HCl, aq, 298.15 K) (4) yields $\Delta_f H°(AlCl_3$, cr, 298.15 K) = −168.32 kcal mol^{-1}. The overall uncertainty is estimated as 300 cal mol^{-1}. Semenenko et al. (6) also determined the enthalpy of solution of Al in (HCl, 11.13 H_2O) and of $AlCl_3$in (HCl, 11.26 H_2O). Their results yield a more negative value, $\Delta_f H°(AlCl_3$, cr, 298.15 K) = −170.1±0.4 kcal mol^{-1}. This discrepancy is due both to differences in the solution enthalpies measured for Al and $AlCl_3$. When adjusted to the conditions of Coughlin's experiments, Semenenko's enthalpy of reaction (1) is calculated to be −128.20 kcal mol^{-1} which deviates more significantly than the values calculated for other authors (7). Semenenko's enthalpy of solution of $AlCl_3$, recalculated to solution in pure water at 20°C is −77.46 kcal mol^{-1}, a more positive value than those found for other investigators (7). The enthalpy of solution of $AlCl_3$ measured by Krivtsov et al. (8) and recalculated to these same conditions is −78.66 kcal mol^{-1}.

Heat Capacity and Entropy

Justice (9) reported low temperature heat capacity data (13 to 310 K) and presented smoothed thermodynamic functions, including S°(298.15 K) = 109.29 J K^{-1} mol^{-1} of which 0.360 J K mol^{-1} is extrapolation below 13 K. Enthalpies relative to 273.15 K were measured by Fischer (10) from 298.15 K to 504 K and enthalpies relative to 298.15 K were measured by McDonald (11) from 310 K to 493 K. The enthalpy data for the crystal can be fit within about 2% by integration of $C_p°(cr) = 64.94 + 0.088\ T$ in J mol^{-1} K^{-1}, which is an extrapolation of $C_p°$ data reported by Justice (9).

Fusion Data

See the liquid table for details.

Sublimation Data

The heat of sublimation is calculated from the enthalpies of formation of the dimer and the crystal for the sublimation process, 2 $AlCl_3(cr) = Al_2Cl_6(g)$, $\Delta_{sub}H°$(298.15 K) = 115.52±2.3 kJ mol^{-1}. Also refer to the $Al_2Cl_6(g)$ table for additional comments.

References

1. P. Gross and C. Hayman, Trans. Faraday Soc. 66, 30 (1970).
2. H. Siemonsen, Z. Elektrochem. 55, 327 (1951).
3. J. P. Coughlin, J. Amer. Chem. Soc. 78, 5479 (1956); J. Phys. Chem. 62, 419 (1958).
4. U.S. Nat. Bur. Stand. Tech. Note 270-3, 1968.
5. ICSU-CODATA Task Group on Key Values for Thermodynamics, J. Chem. Thermodynamics 10, 903 (1978); V. B. Parker, U.S. Nat. Bur. Stand., NSRDS-NBS 2, 1965.
6. K. N. Semenenko, A. P. Savchenkova, T. S. Il'ina and V. N. Surov, Russ. J. Inorg. Chem. 16, 1561 (1971).
7. JANAF Thermochemical Tables, 2nd ed., NSRDS-NBS 37, 1971.
8. K. N. Krivtsov. V. Y. Rosolovskii and G. N. Sherokova, Russ. J. Inorg. Chem. 16, 1402 (1971).
9. B. H. Justice, J. Chem. Eng. Data 14, 4 (1969).
10. W. Fischer, Z. anorg. allgem. chem. 200, 332 (1931).
11. R. A. McDonald, The Dow Chemical Co., Midland, Michigan, unpublished data, 1960; CPIA Publ. No. 44U, 213 (1964).
12. JANAF Thermochemical Tables: Al(cr), 6-30-79; H_2(g), 3-31-77.

Aluminum Chloride ($AlCl_3$) $Al_1Cl_3(cr)$

Enthalpy Reference Temperature = T_r = 298.15 K — Standard State Pressure = p° = 0.1 MPa

T/K	$C_p°$ (J $K^{-1}mol^{-1}$)	S° (J $K^{-1}mol^{-1}$)	−[G°−H°(T_r)]/T (J $K^{-1}mol^{-1}$)	H°−H°(T_r) (kJ mol^{-1})	$\Delta_f H°$ (kJ mol^{-1})	$\Delta_f G°$ (kJ mol^{-1})	Log K_f
0	0.	0.	INFINITE	-16.979	-704.300	-704.300	INFINITE
100	47.112	30.798	180.962	-15.016	-707.189	-681.219	355.832
200	79.203	75.145	117.298	-8.431	-706.962	-655.173	171.114
298.15	91.128	109.286	109.286	0.	-705.632	-630.018	110.377
300	91.295	109.850	109.288	0.169	-705.602	-629.549	109.614
400	100.081	137.318	112.974	9.738	-703.746	-604.464	78.935
465.700	105.854	152.965	117.529	16.503	------ CRYSTAL <--> LIQUID -----		
500	108.868	160.594	120.224	20.185	-701.288	-579.914	60.583
600	117.654	181.219	128.701	31.511	-698.146	-555.922	48.397
700	126.440	200.016	137.565	43.716	-694.297	-532.513	39.737
800	135.227	217.473	146.474	56.799	-689.744	-509.703	33.280
900	144.013	233.908	155.284	70.761	-684.503	-487.506	28.294
1000	152.800	249.536	163.934	85.602	-689.183	-465.168	24.298
1100	161.586	264.511	172.401	101.321	-682.264	-443.094	21.041
1200	170.372	278.948	180.682	117.919	-674.484	-421.689	18.356
1300	179.159	292.932	188.781	135.396	-665.841	-400.967	16.111
1400	187.945	306.531	196.709	153.751	-656.333	-380.943	14.213
1500	196.732	319.797	204.474	172.985	-645.961	-361.629	12.593

PREVIOUS: June 1970 CURRENT: September 1979

Aluminum Chloride ($AlCl_3$) $Al_1Cl_3(cr)$

ALUMINUM CHLORIDE ($AlCl_3$) LIQUID M_r = 133.34054

$S°(298.15\ K) = [172.908]\ J\ K^{-1}\ mol^{-1}$ $\Delta_f H°(298.15\ K) = [-674.805]\ kJ\ mol^{-1}$

$T_{fus} = 465.7 \pm 0.2\ K$ $\Delta_{fus}H° = 35.35 \pm 0.84\ kJ\ mol^{-1}$

Enthalpy of Formation

$\Delta_f H°(AlCl_3, \ell, 298.15\ K)$ is calculated from $\Delta_f H°(AlCl_3, cr, 298.15\ K)$ by adding the enthalpy of fusion, $\Delta_{fus}H°$, and the difference in enthalpy, H°(465.7 K)-H°(298.15 K), between the crystal and the liquid.

Heat Capacity and Entropy

Fischer (1) and McDonald (2) measured enthalpy data for the crystal and liquid. In each case, the liquid data extended over a short range of temperature, 473-504 K and 475-493 K, respectively. Their data can be represented within 2% by the heat capacity equation derived for the crystal [see $AlCl_3$(cr) table], a heat of fusion of 35.35 kJ mol^{-1}, and a liquid heat capacity of 125.5 J K^{-1} mol^{-1}. Fischer's data are high, while McDonald's points lie below the selected values.

S°(298.15 K) is calculated in a manner similiar to that used for the enthalpy of formation.

Fusion Data

Kendall et al. (3) observed $AlCl_3$ to melt "sharply" at 463.4 K. Foster (4) noted that $AlCl_3$ began melting and melted completely in the range 465.7±0.2 K. Johnson et al. (5), Nisel'son et al. (6), and Ostrikova et al. (7) reported melting temperatures of 465.9±0.3 K, 467.2 K, and 466 K, respectively. Smits and Meyering (8), Treadwell and Terebesi (9) and Viola et al. (10) derived triple point temperatures of 465.8 K, 466.5 K, and 466.86 K, respectively, from intersection of vapor pressure curves for crystal and liquid. The value of Foster, T_{fus} =465.7±0.2 K, is adopted as most accurate. The vapor pressure (monomer and dimer contributions) at the triple point is 2.3 atm. $\Delta_{fus}H°$ is based on enthalpies measured by Fischer (1) and McDonald (2).

References

1. W. Fischer, Z. anorg. allgem. Chem. 200, 332 (1931).
2. R. A. McDonald, The Dow Chemical Co., Midland, Michigan, unpublished measurements, 1960; CPIA Publ. No. 44U, 213 (1964).
3. J. Kendall, E. D. Crittenden, and H. K. Miller, J. Amer. Chem. Soc. 45, 963 (1923).
4. L. M. Foster, J. Amer. Chem. Soc. 72, 1902 (1950).
5. J. W. Johnson, D. Cubicciotti, and W. J. Silva, High Temp. Sci. 3, 523 (1971).
6. L. A. Nisel'son, A. I. Pustinl'nik, O. R. Gavrilov, and V. A. Rodin, Zh. Neorg. Khim. 10, 2339 (1965).
7. N. V. Ostrikova, Z. A. Miroshnik, V. D. Kuleshova, N. P. Dudkina, and A. I. Ilyushchenko, Zh. Neorg. Khim. 14, 2229 (1969).
8. A. Smits and J. L. Meyering, Z. Physik Chem. B41, 98 (1938).
9. W. D. Treadwell and L. Terebesi, Helv. Chim. Acta 15, 1053 (1932).
10. J. T. Viola, D. W. Seegmiller, A. A. Fannin and L. A. King, J. Chem. Eng. Data 22, 367 (1977).

Aluminum Chloride ($AlCl_3$) $Al_1Cl_3(l)$

Enthalpy Reference Temperature = T_r = 298.15 K Standard State Pressure = p° = 0.1 MPa

	J K⁻¹mol⁻¹			kJ mol⁻¹			
T/K	$C_p°$	S°	$-[G°-H°(T_r)]/T$	$H°-H°(T_r)$	$\Delta_f H°$	$\Delta_f G°$	Log K_f
0							
100							
200							
298.15	125.520	172.908	172.908	0.	-674.805	-618.161	108.299
300	125.520	173.684	172.910	0.232	-674.712	-617.809	107.570
400	125.520	209.794	177.833	12.784	-669.873	-599.581	78.297
465.700	125.520	228.883	183.723	21.031	------ CRYSTAL <--> LIQUID -----		
500	125.520	237.803	187.130	25.336	-665.311	-582.541	60.858
600	125.520	260.688	197.541	37.888	-660.942	-566.400	49.309
700	125.520	280.037	207.979	50.440	-656.746	-550.977	41.114
800	125.520	296.798	218.057	62.992	-652.724	-536.143	35.007
900	125.520	311.582	227.644	75.544	-648.894	-521.803	30.285
1000	125.520	324.807	236.710	88.096	-655.862	-507.118	26.489
1100	125.520	336.770	245.272	100.648	-652.110	-492.425	23.383
1200	125.520	347.692	253.358	113.200	-648.376	-478.074	20.810
1300	125.520	357.739	261.006	125.752	-644.658	-464.032	18.645
1400	125.520	367.041	268.252	138.304	-640.953	-450.277	16.800
1500	125.520	375.701	275.130	150.856	-637.263	-436.786	15.210

PREVIOUS: June 1970 CURRENT: September 1979

Aluminum Chloride ($AlCl_3$) $Al_1Cl_3(l)$

ALUMINUM CHLORIDE ($AlCl_3$) CRYSTAL-LIQUID

0 to 465.7 K crystal
above 465.7 K liquid

Refer to the individual tables for details.

M_r = 133.34054 Aluminum Chloride ($AlCl_3$) Al_1Cl_3(cr,l)

Enthalpy Reference Temperature = T_r = 298.15 K Standard State Pressure = p° = 0.1 MPa

T/K	C_p° (J K⁻¹mol⁻¹)	S° (J K⁻¹mol⁻¹)	−[G°−H°(T_r)]/T (J K⁻¹mol⁻¹)	H°−H°(T_r) (kJ mol⁻¹)	Δ_fH° (kJ mol⁻¹)	Δ_fG° (kJ mol⁻¹)	Log K_f
0	0.	0.	INFINITE	-16.979	-704.300	-704.300	INFINITE
100	47.112	30.798	180.962	-15.016	-707.189	-681.219	355.832
200	79.203	75.145	117.298	-8.431	-706.962	-655.173	171.114
298.15	91.128	109.286	109.286	0.	-705.632	-630.018	110.377
300	91.295	109.850	109.288	0.169	-705.602	-629.549	109.614
400	100.081	137.318	112.974	9.738	-703.746	-604.464	78.935
465.700	105.854	152.965	117.529	16.503	CRYSTAL <--> LIQUID		
465.700	125.520	228.883	117.529	51.857	TRANSITION		
500	125.520	237.803	125.478	56.163	-665.311	-582.541	60.858
600	125.520	260.688	146.164	68.715	-660.942	-566.400	49.309
700	125.520	280.037	163.942	81.267	-656.746	-550.977	41.114
800	125.520	296.798	179.525	93.819	-652.724	-536.144	35.007
900	125.520	311.582	193.393	106.371	-648.894	-521.803	30.285
1000	125.520	324.807	205.884	118.923	-655.862	-507.118	26.489
1100	125.520	336.770	217.248	131.475	-652.110	-492.426	23.383
1200	125.520	347.692	227.670	144.027	-648.376	-478.074	20.810
1300	125.520	357.739	237.294	156.579	-644.658	-464.033	18.645
1400	125.520	367.041	246.233	169.131	-640.953	-450.278	16.800
1500	125.520	375.701	254.579	181.683	-637.263	-436.787	15.210

PREVIOUS: CURRENT: September 1979

Aluminum Chloride ($AlCl_3$) Al_1Cl_3(cr,l)

$\Delta_{at}H°$ = 1269.0 ± 3.4 kJ mol^{-1}

$S°$(298.15 K) = 314.49 ± 2.9 J K^{-1} mol^{-1}

$\Delta_f H°$(0 K) = -582.85 ± 2.9 kJ mol^{-1}

$\Delta_f H°$(298.15 K) = -584.59 ± 2.9 kJ mol^{-1}

Vibrational Frequencies and Degeneracies

ν, cm^{-1}	ν, cm^{-1}	ν, cm^{-1}	ν, cm^{-1}
371(1)	183(1)	610(2)	146(2)

Ground State Quantum Weight: [1]   σ = 6

Point Group: D_{3h}

Bond Distance: Al-Cl = 2.06 Å

Bond Angle: Cl-Al-Cl = 120°

Product of the Moments of Inertia: $I_A I_B I_C$ = 1.052528 x 10^{-112} g^3 cm^6

Enthalpy of Formation

The enthalpy of formation is calculated from the value adopted for Al_2Cl_6(g) and the heat of dissociation of Al_2Cl_6→$2AlCl_3$ derived from equilibrium P-V-T data on aluminum chloride vapor as listed below.

The four investigations are in reasonable agreement. Giving consideration also to equilibrium data on the reaction of $AlCl_3$(g) and Al(ℓ) to form AlCl(g) [see AlCl(g) table], there is selected as a best value $\Delta_r H°$(298.15 K) = 30.25 kcal mol^{-1} Al_2Cl_6, which yields $\Delta_f H°$($AlCl_3$, g, 298.15 K) = -139.72 kcal mol^{-1} (-584.59 ± 2.9 kJ mol^{-1}). NBS (15) recommended a value of -139.4 kcal mol^{-1} (-583.250 kJ mol^{-1}) for the enthalpy of formation of $AlCl_3$(g) based on data available prior to 1968.

Source	T/K	Data Points	$\Delta_r H°$ (298.15 K), kcal mol^{-1} Al_2Cl_6 2nd law	3rd law	Drift cal K^{-1} mol^{-1}
Smits and Meijering (1)	669-816	6	29.45±0.7	30.21±0.2	1.0±0.9
Fischer et al. (2)	647-869	17[a]	31.04±0.3	30.42±0.2	-0.8±0.4
Vrieland and Stull (3)	669-825	19[b]	30.27±0.3	30.50±0.2	0.3±0.4
Polyachenok et al. (4)	650-1100	2[c]	30.68±0.0	30.03±0.5	-0.7±0.0

[a]Twenty-three points reported, points at 938, 881, 944, 734, 805 and 605 K rejected by statistical test.

[b]Twenty-six points reported, points at 627.9, 643.1, 671.3, 638.4, 671.6, 722.5 and 686.9 K rejected by statistical test.

[c]Individual points not reported, an equation for the equilibrium constant was used to calculate two points at the extremes of temperature.

Heat Capacity and Entropy

The structure of monomeric $AlCl_3$ has been the subject of considerable debate due to conflicting assignments made on the basis of the electron diffraction data of Zasorin and Rambidi (5) and the matrix-isolation IR study by Lesiecki and Shirk (6). Electron diffraction of the vapor at 800 K yielded an Al-Cl bond length of 2.06±0.01 Å and a Cl-Al-Cl angle of 118±1.5°, which allowing for the possibility of "shrinkage" was taken to indicate a planar equilibrium D_{3h} geometry. The argon matrix spectra obtained by Lesiecki and Shirk (6) showed four infrared-active fundamentals including the symmetric stretch, implying a pyramidal C_{3v} structure; their normal coordinate analysis suggested a Cl-Al-Cl angle near 112°. Hargittai and Hargittai (7) argued, however, that the electron diffraction results for $AlCl_3$ are not compatible with a pyramidal model having an inversion frequency of 183 cm^{-1} and a barrier height equal to 2-3 vibrational quanta. Beattie et al. (8) were unable to reproduce the IR frequencies observed by Lesiecki and Shirk (6) in their own matrix isolation studies and it was later concluded that the earlier work by Lesiecki and Shirk was based on incorrect band assignments (9). The D_{3h} structure is supported by the original electron diffraction work as well as theoretical calculations (10, 11) and correlations (9) of the frequencies of $AlCl_3$, $AlCl_2Br$, $AlClBr_2$ and $AlBr_3$.

For aluminum chloride vapor at high temperatures (monomer predominant), Klemperer (12) observed the band at 610 cm^{-1} in the IR spectrum and Beattie and Horder (13) reported Raman bands at 371 cm^{-1} and 146 cm^{-1}. The remaining fundamental is that reported by Lesiecki and Shirk (6, 9); a second experimental observation of ν_2=174 cm^{-1} was reported by Perov et al. (14) in a xenon matrix along with ν_1=350 cm^{-1}, ν_4=142 cm^{-1}. The principal moments of inertia are I_A= I_B=37.4744 x 10^{-39} g cm^2 and I_C=74.9488 x 10^{-39} g cm^2.

References

1. A. Smits and J. L. Meijering, Z. Phys. Chem. B41, 98 (1938).
2. W. Fischer, O. Rahlfs and B. Benze, Z. Anorg. Chem 205, 1 (1932).
3. E. Vrieland and D. R. Stull, J. Chem. Eng. Data 12, 532 (1967).
4. L. D. Polyachenok, G. P. Dudchik and O. G. Polyachenok, Russ. J. Phys. Chem. 50, 227 (1976).
5. E. Z. Zasorin and N. G. Rambidi, Zh. Strukt. Khim. 8, 391 (1967).
6. M. L. Lesiecki and J. S. Shirk, J. Chem. Phys. 56, 4171 (1972).
7. I. Hargittai and M. Hargittai, J. Chem. Phys. 60, 2563 (1974).
8. I. R. Beattie, H. E. Blayden and J. S. Ogden, J. Chem. Phys. 64, 909 (1976); J. Chem. Soc. Dalton Trans. 1976, 666.
9. J. S. Shirk and A. E. Shirk, J. Chem. Phys. 64, 910 (1976); R. G. S. Pong et al., ibid. 70, 525 (1979).
10. A. L. Companion, J. Chem. Phys. 57, 1807 (1972).
11. S. P. So and W. G. Richards, Chem. Phys. Lett. 32, 231 (1975).
12. W. Klemperer, J. Chem. Phys. 24, 353 (1956).
13. I. R. Beattie and J. R. Horder, J. Chem. Soc. A1969, 2655 (1969).
14. P. A. Perov, S. V. Nedyak and M. A. Mal'tsev, Vest. Moskov. Univer. Khim. 29, 2901 (1974).
15. U.S. Nat. Bur. Stand. Tech. Note 270-3, 1968.

Enthalpy Reference Temperature = T_r = 298.15 K   Standard State Pressure = p° = 0.1 MPa

T/K	$C_p°$ (J $K^{-1}$$mol^{-1}$)	S° (J $K^{-1}$$mol^{-1}$)	-[G°-H°(T_r)]/T (J $K^{-1}$$mol^{-1}$)	H°-H°(T_r) (kJ mol^{-1})	$\Delta_f H°$ (kJ mol^{-1})	$\Delta_f G°$ (kJ mol^{-1})	Log K_f
0	0.	0.	INFINITE	-16.575	-582.852	-582.852	INFINITE
100	51.072	247.272	372.957	-12.568	-583.698	-579.375	302.635
200	64.468	287.225	320.862	-6.728	-584.215	-574.842	150.133
250	68.798	302.100	315.662	-3.390	-584.420	-572.475	119.612
298.15	71.876	314.494	314.494	0.	-584.588	-570.158	99.889
300	71.976	314.939	314.496	0.133	-584.594	-570.069	99.258
350	74.300	326.218	315.381	3.793	-584.748	-567.635	84.715
400	76.014	336.257	317.375	7.553	-584.888	-565.181	73.805
450	77.300	345.288	319.983	11.387	-585.021	-562.710	65.318
500	78.281	353.485	322.930	15.278	-585.152	-560.223	58.526
600	79.643	367.888	329.255	23.179	-585.435	-555.212	48.335
700	80.514	380.235	335.677	31.190	-585.779	-550.149	41.053
800	81.100	391.027	341.936	39.273	-586.227	-545.029	35.587
900	81.513	400.604	347.932	47.405	-586.816	-539.846	31.332
1000	81.813	409.209	353.637	55.572	-598.170	-533.828	27.884
1100	82.038	417.017	359.049	63.765	-598.777	-527.364	25.042
1200	82.211	424.163	364.182	71.978	-599.382	-520.846	22.672
1300	82.346	430.749	369.053	80.206	-599.987	-514.276	20.664
1400	82.454	436.856	373.680	88.446	-600.595	-507.660	18.941
1500	82.542	442.548	378.084	96.696	-601.206	-501.001	17.446
1600	82.614	447.877	382.281	104.954	-601.823	-494.300	16.137
1700	82.674	452.888	386.289	113.218	-602.447	-487.561	14.981
1800	82.724	457.615	390.121	121.488	-603.078	-480.785	13.952
1900	82.767	462.088	393.792	129.763	-603.720	-473.974	13.030
2000	82.803	466.335	397.314	138.041	-604.372	-467.128	12.200
2100	82.835	470.376	400.698	146.323	-605.038	-460.249	11.448
2200	82.862	474.230	403.953	154.608	-605.719	-453.339	10.764
2300	82.886	477.914	407.089	162.896	-606.418	-446.397	10.138
2400	82.907	481.442	410.114	171.185	-607.137	-439.424	9.564
2500	82.925	484.826	413.036	179.477	-607.877	-432.421	9.035
2600	82.942	488.079	415.860	187.770	-608.642	-425.388	8.546
2700	82.956	491.210	418.593	196.065	-609.432	-418.324	8.093
2800	82.970	494.227	421.241	204.362	-904.150	-410.264	7.654
2900	82.981	497.139	423.808	212.659	-903.900	-392.629	7.072
3000	82.992	499.952	426.299	220.958	-903.677	-375.003	6.529
3100	83.002	502.673	428.719	229.258	-903.483	-357.384	6.022
3200	83.010	505.309	431.072	237.558	-903.317	-339.771	5.546
3300	83.018	507.863	433.360	245.860	-903.177	-322.162	5.099
3400	83.026	510.342	435.588	254.162	-903.061	-304.557	4.679
3500	83.032	512.748	437.759	262.465	-902.969	-286.956	4.283
3600	83.038	515.088	439.874	270.768	-902.896	-269.356	3.908
3700	83.044	517.363	441.938	279.072	-902.841	-251.759	3.554
3800	83.049	519.578	443.952	287.377	-902.800	-234.163	3.219
3900	83.054	521.735	445.919	295.682	-902.770	-216.567	2.901
4000	83.058	523.838	447.841	303.988	-902.748	-198.973	2.598
4100	83.063	525.889	449.719	312.294	-902.730	-181.378	2.311
4200	83.066	527.890	451.557	320.600	-902.712	-163.785	2.037
4300	83.070	529.845	453.355	328.907	-902.692	-146.191	1.776
4400	83.073	531.755	455.115	337.214	-902.666	-128.599	1.527
4500	83.076	533.622	456.839	345.522	-902.629	-111.007	1.289
4600	83.079	535.448	458.528	353.829	-902.581	-93.415	1.061
4700	83.082	537.234	460.184	362.138	-902.519	-75.826	0.843
4800	83.085	538.984	461.807	370.446	-902.439	-58.237	0.634
4900	83.087	540.697	463.400	378.754	-902.340	-40.651	0.433
5000	83.089	542.375	464.963	387.063	-902.220	-23.067	0.241
5100	83.091	544.021	466.497	395.372	-902.077	-5.485	0.056
5200	83.093	545.634	468.003	403.682	-901.910	12.093	-0.121
5300	83.095	547.217	469.483	411.991	-901.719	29.669	-0.292
5400	83.097	548.770	470.937	420.301	-901.501	47.240	-0.457
5500	83.099	550.295	472.366	428.610	-901.257	64.808	-0.615
5600	83.100	551.792	473.771	436.920	-900.987	82.370	-0.768
5700	83.102	553.263	475.153	445.230	-900.690	99.927	-0.916
5800	83.103	554.709	476.512	453.541	-900.363	117.479	-1.058
5900	83.105	556.129	477.849	461.851	-900.014	135.025	-1.195
6000	83.106	557.526	479.166	470.162	-899.616	152.562	-1.328

PREVIOUS: September 1979 (1 atm)   CURRENT: September 1979 (1 bar)

POTASSIUM TETRACHLOROALUMINATE ($KAlCl_4$) CRYSTAL M_r = 207.89184

$S°(298.15\ K) = [196.6 \pm 8]\ J\ K^{-1}\ mol^{-1}$

T_{fus} = 529 K

$\Delta_f H°(0\ K)$ = Unknown

$\Delta_f H°(298.15\ K) = -1196.6 \pm 10\ kJ\ mol^{-1}$

$\Delta_{fus} H°$ = Unknown

Enthalpy of Formation

The enthalpy of reaction of aluminum chloride and potassium chloride was measured by Baud (1). From $\Delta_r H° = -26.38\ kcal\ mol^{-1}$ for the reaction: $Al_2Cl_6(cr) + 2\ KCl(cr) = 2\ KAlCl_4(cr)$, the value of $\Delta_f H°(KAlCl_4, cr, 298.15\ K)$ is calculated to be -286 kcal mol^{-1}.

Heat Capacity and Entropy

Heat capacities are estimated by comparison with those for $KMgCl_3(cr)$ reported by Kelley (5). $S°(298.15\ K) = 47 \pm 2\ cal\ K^{-1}\ mol^{-1}$ is estimated according to the methods suggested by Kubaschewski and Evans (2) and Kelley (3).

Fusion Data

T_{fus} was reported by Fischer and Simon (4).

References

1. E. Baud, Ann. chim. phys. 1, 8 (1904).
2. O. Kubaschewski and E. L. Evans, 'Metallurgical Thermochemistry', Pergamon Press, 1958.
3. K. K. Kelley, U. S. Bur. Mines, personal communication, June, 1960.
4. W. Fischer and A.-L. Simon, Z. anorg. allgem. Chem. 306, 1 (1960).
5. K. K. Kelley, U. S. Bur. Mines Bull. 584, 1960.

Potassium Tetrachloroaluminate ($KAlCl_4$) $Al_1Cl_4K_1(cr)$

Enthalpy Reference Temperature = T_r = 298.15 K Standard State Pressure = p° = 0.1 MPa

T/K	$C_p°$ (J K⁻¹mol⁻¹)	S° (J K⁻¹mol⁻¹)	$-[G°-H°(T_r)]/T$ (J K⁻¹mol⁻¹)	$H°-H°(T_r)$ (kJ mol⁻¹)	$\Delta_f H°$ (kJ mol⁻¹)	$\Delta_f G°$ (kJ mol⁻¹)	Log K_f
0							
100							
200							
298.15	156.482	196.648	196.648	0.	-1196.624	-1094.521	191.755
300	156.649	197.616	196.651	0.290	-1196.559	-1093.887	190.463
400	165.477	243.891	202.901	16.396	-1195.380	-1059.743	138.388
500	174.305	281.763	214.994	33.385	-1191.274	-1026.295	107.216
600	183.176	314.346	228.897	51.269	-1186.429	-993.742	86.513
700	190.790	343.177	243.203	69.982	-1180.903	-962.056	71.789
800	196.648	369.064	257.345	89.375	-1174.867	-931.199	60.801
900	199.995	392.431	271.078	109.218	-1168.589	-901.117	52.299
1000	202.087	413.622	284.290	129.332	-1172.876	-870.976	45.495

PREVIOUS: June 1962 CURRENT: June 1963

Potassium Tetrachloroaluminate ($KAlCl_4$) $Al_1Cl_4K_1(cr)$

SODIUM TETRACHLOROALUMINATE ($NaAlCl_4$) CRYSTAL M_r = 191.78331

$\Delta_f H°(0\ K)$ = Unknown
$\Delta_f H°(298.15\ K) = -1142.2 \pm 4\ kJ\ mol^{-1}$
$\Delta_{fus}H°$ = Unknown

$S°(298.15\ K) = [188.3 \pm 8]\ J\ K^{-1}\ mol^{-1}$
T_{fus} = 424 K

Enthalpy of Formation

$\Delta_f H°(NaAlCl_4$, cr, 298.15 K) is calculated from $\Delta_r H° = -11.9\ kcal\ mol^{-1}$ for the reaction $Al_2Cl_6(cr) + 2\ NaCl(cr) = 2\ NaAlCl_4(cr)$, as measured by Baud (1).

Heat Capacity and Entropy

Heat capacities are estimated by comparison with those of $KAlCl_4$(cr). $S°(298.15\ K) = 45 \pm 2\ cal\ K^{-1}\ mol^{-1}$ is estimated according to the methods suggested by Kubaschewski and Evans (2) and Kelley (3).

Fusion Data

T_{fus} was reported by Fischer and Simon (4).

References

1. E. Baud, Ann. chim. phys. 1, 8 (1904).
2. O. Kubaschewski and E. L. Evans, 'Metallurgical Thermochemistry', Pergamon Press, 1958.
3. K. K. Kelley, U. S Bur. Mines, personal communication, June, 1960.
4. W. Fischer and A.-L. Simon, Z. anorg. allgem. Chem. 306, 1 (1960).

Sodium Tetrachloroaluminate ($NaAlCl_4$) $Al_1Cl_4Na_1$(cr)

Enthalpy Reference Temperature = T_r = 298.15 K Standard State Pressure = p° = 0.1 MPa

	J K⁻¹mol⁻¹			kJ mol⁻¹			
T/K	$C_p°$	S°	$-[G°-H°(T_r)]/T$	$H°-H°(T_r)$	$\Delta_f H°$	$\Delta_f G°$	Log K_f
0							
100							
200							
298.15	154.975	188.280	188.280	0.	-1142.232	-996.454	174.575
300	154.808	189.238	188.283	0.287	-1142.174	-995.549	173.341
400	164.808	235.161	194.477	16.274	-1138.854	-947.163	123.687
500	174.054	272.947	206.498	33.224	-1135.045	-899.670	93.988
600	182.422	305.445	220.342	51.062	-1130.680	-852.996	74.260
700	189.117	334.098	234.587	69.658	-1125.827	-807.093	60.226
800	193.510	359.645	248.652	88.795	-1120.680	-761.909	49.748
900	197.276	382.673	262.285	108.349	-1115.361	-717.381	41.636
1000	199.158	403.572	275.385	128.187	-1120.602	-672.688	35.138

PREVIOUS: CURRENT: June 1963

Sodium Tetrachloroaluminate ($NaAlCl_4$) $Al_1Cl_4Na_1$(cr)

POTASSIUM HEXACHLOROALUMINATE (K_3AlCl_6) CRYSTAL M_r = 356.99444

$\Delta_f H°(0\ K)$ = Unknown
$S°(298.15\ K) = [376.6 \pm 8]\ J\ K^{-1}\ mol^{-1}$ $\Delta_f H°(298.15\ K) = -2092 \pm 4\ kJ\ mol^{-1}$
$T_{fus} = [800 \pm 20]\ K$ $\Delta_{fus}H°$ = Unknown

Enthalpy of Formation

$\Delta_f H°(K_3AlCl_6$, cr, 298.15 K) is recalculated from the same quantity given by Rossini et al. (1). The latter value was derived from the enthalpy change for the reaction $Al_2Cl_6(cr) + 6\ KCl(cr) = 2\ K_3AlCl_6(cr)$, as measured by Baud (2).

Heat Capacity and Entropy

Heat capacities are estimated by comparison with the corresponding values of the sum of the constituent halides for K_3AlCl_6(cr). In a similar manner, S°(298.15 K) is estimated to be $90 \pm 2\ cal\ K^{-1}\ mol^{-1}$.

Fusion Data

T_{fus} is estimated by comparison with that for K_3AlF_6(cr).

References

1. U. S. Nat. Bur. Stand. Circ. 500, 1952.
2. E. Baud, Ann. chem. phys. 1, 8 (1904).

Potassium Hexachloroaluminate (K_3AlCl_6) $Al_1Cl_6K_3$(cr)

Enthalpy Reference Temperature = T_r = 298.15 K Standard State Pressure = p° = 0.1 MPa

	J K⁻¹mol⁻¹			kJ mol⁻¹			
T/K	$C_p°$	S°	$-[G°-H°(T_r)]/T$	$H°-H°(T_r)$	$\Delta_f H°$	$\Delta_f G°$	Log K_f
0							
100							
200							
298.15	248.906	376.560	376.560	0.	-2092.000	-1938.463	339.611
300	249.283	378.101	376.565	0.461	-2091.936	-1937.511	337.351
400	259.157	451.152	386.458	25.878	-2095.873	-1885.176	246.179
500	269.659	510.122	405.474	52.324	-2092.096	-1832.919	191.484
600	279.491	560.178	427.189	79.794	-2087.377	-1781.512	155.094
700	287.859	603.896	449.375	108.165	-2081.856	-1730.961	129.166
800	295.809	642.863	471.168	137.355	-2075.676	-1681.247	109.774
900	302.880	678.124	492.235	167.300	-2068.982	-1632.341	94.739
1000	308.779	710.343	512.458	197.885	-2072.560	-1583.440	82.710
1100	314.532	740.048	531.815	229.056	-2301.871	-1521.060	72.229
1200	319.658	767.640	550.331	260.771	-2290.853	-1450.560	63.141
1300	324.103	793.405	568.049	292.963	-2279.389	-1380.998	55.489
1400	328.026	817.571	585.018	325.574	-2267.536	-1312.333	48.964
1500	331.320	840.316	601.287	358.544	-2255.353	-1244.528	43.338
1600	334.302	861.796	616.904	391.827	-2242.885	-1177.544	38.443
1700	336.969	882.145	631.913	425.394	-2230.167	-1111.349	34.148
1800	339.322	901.473	646.356	459.211	-2217.233	-1045.908	30.351
1900	341.362	919.876	660.272	493.248	-2204.120	-981.191	26.975
2000	343.088	937.430	673.694	527.473	-2190.865	-917.169	23.954

PREVIOUS: June 1962 CURRENT: September 1963

Potassium Hexachloroaluminate (K_3AlCl_6) $Al_1Cl_6K_3$(cr)

SODIUM HEXACHLOROALUMINATE (Na_3AlCl_6) CRYSTAL M_r = 308.66885

$S°(298.15\ K) = [347.3 \pm 8]\ J\ K^{-1}\ mol^{-1}$

$T_{fus} = [780 \pm 20]\ K$

$\Delta_f H°(0\ K)$ = Unknown

$\Delta_f H°(298.15\ K) = -1979 \pm 4\ kJ\ mol^{-1}$

$\Delta_{fus} H°$ = Unknown

Enthalpy of Formation

$\Delta_f H°(Na_3AlCl_6$, cr, 298.15 K) is recalculated from the same quantity given by Rossini et al. (1). The latter value was derived from the enthalpy change for the reaction $Al_2Cl_6(cr) + 6\ NaCl(cr) = 2\ Na_3AlCl_6(cr)$, as measured by Baud (2).

Heat Capacity and Entropy

Heat capacities are estimated by comparison with the corresponding values of the sum of the constituent halides for Na_3AlF_6(cr). In a similar manner, S°(298.15 K) is estimated to be $83 \pm 2\ cal\ K^{-1}\ mol^{-1}$.

Fusion Data

T_{fus} is estimated by comparison with that for Na_3AlF_6(cr).

References

1. U. S. Nat. Bur. Stand. Circ. 500, 1952.
2. E. Baud, Ann. chim. phys. 1, 8 (1904).

Sodium Hexachloroaluminate (Na_3AlCl_6) $Al_1Cl_6Na_3$(cr)

Enthalpy Reference Temperature = T_r = 298.15 K Standard State Pressure = p° = 0.1 MPa

T/K	$C_p°$ (J K⁻¹mol⁻¹)	S° (J K⁻¹mol⁻¹)	$-[G°-H°(T_r)]/T$ (J K⁻¹mol⁻¹)	$H°-H°(T_r)$ (kJ mol⁻¹)	$\Delta_f H°$ (kJ mol⁻¹)	$\Delta_f G°$ (kJ mol⁻¹)	Log K_f
0							
100							
200							
298.15	244.136	347.272	347.272	0.	-1979.032	-1828.584	320.360
300	244.513	348.783	347.277	0.452	-1978.970	-1827.651	318.222
400	254.387	420.484	356.986	25.399	-1983.806	-1777.071	232.062
500	264.010	478.294	375.646	51.324	-1980.531	-1725.747	180.288
600	273.006	527.237	396.933	78.182	-1976.354	-1675.168	145.836
700	281.165	569.941	418.662	105.896	-1971.360	-1625.355	121.286
800	288.696	607.991	439.992	134.399	-1965.663	-1576.307	102.922
900	294.972	642.363	460.599	163.588	-1959.438	-1528.007	88.683
1000	300.620	673.738	480.367	193.372	-1963.442	-1479.661	77.290
1100	305.720	702.633	499.277	223.692	-1956.272	-1431.627	67.982
1200	310.453	729.439	517.353	254.503	-2239.053	-1376.941	59.937
1300	314.741	754.459	534.640	285.765	-2228.519	-1305.523	52.457
1400	318.821	777.935	551.189	317.445	-2217.595	-1234.930	46.076
1500	322.796	800.069	567.050	349.529	-2206.295	-1165.132	40.574
1600	326.352	821.018	582.274	381.990	-2194.644	-1096.100	35.784
1700	329.385	840.895	596.907	414.778	-2182.693	-1027.805	31.581
1800	332.210	859.803	610.992	447.860	-2170.477	-960.222	27.865
1900	334.825	877.835	624.565	481.213	-2158.018	-893.324	24.559
2000	337.230	895.072	637.663	514.818	-2145.341	-827.088	21.601

PREVIOUS: CURRENT: September 1963

Sodium Hexachloroaluminate (Na_3AlCl_6) $Al_1Cl_6Na_3$(cr)

ALUMINUM FLUORIDE (AlF) IDEAL GAS $M_r = 45.979943$

$S°(298.15\ K) = 215.16 \pm 0.04\ J\ K^{-1}\ mol^{-1}$

$\Delta_fH°(0\ K) = -265.624 \pm 3.35\ kJ\ mol^{-1}$

$\Delta_fH°(298.15\ K) = -265.684 \pm 3.35\ kJ\ mol^{-1}$

Electronic Levels and Molecular Constants ($\sigma = 1$)

State	ε_i, cm^{-1}	g_i	ω_e, cm^{-1}	ω_ex_e, cm^{-1}	B_e, cm^{-1}	α_e, cm^{-1}	r_e, Å
$X^1\Sigma^+$	0.0	1	802.26	4.77	0.5525	0.00495	1.6544
$a^3\Pi$	27254.	6	830.30	4.60	0.55718	0.00468	1.6474
$A^1\Pi$	43949.7	2	803.94	5.99	0.55640	0.00534	1.6485
$b^3\Sigma^+$	44804.5	3	786.37	7.64	0.56280	0.00651	1.6391
$B^1\Sigma^+$	54282.5	1	866.60	7.45	0.57968	0.00560	1.6151
$c^3\Sigma^+$	55023.4	3	938.22	4.81	0.58992	0.00458	1.6010
$C^1\Sigma^+$	57755.9	1	938.22	5.09	0.58992	0.00458	1.6010

Enthalpy of Formation

We adopt $\Delta_fH°(298.15\ K) = -63.5 \pm 0.8\ kcal\ mol^{-1}$ ($-265.684 \pm 3.35\ kJ\ mol^{-1}$) and $D_0^{\circ} = 160.0 \pm 1.5\ kcal\ mol^{-1}$ ($669.440 \pm 6.3\ kJ\ mol^{-1}$) based on our analysis of numerous equilibrium data (1-12). We only tabulate our analysis of three studies; the other analyses are essentially those which were reported earlier (22). However, the previously calculated D_0° values will differ due to a more current heat of formation for Al(g). Greatest weight is given to torsion effusion (2) and transport (5, 12) data for the Al-AlF_3 system. Hildenbrand et al. (2) found that orifice area affected the torsion-effusion pressures of AlF from reaction A. They derived equilibrium pressures from a semi-empirical correlation of data for the four cells listed below. Mass-spectrometric studies (9) of the Al-AlF system near 950 K indicated that the vapor consists of AlF with a small amount of AlF . The adopted $D_0^{\circ} = 669.440 \pm 6.3\ kJ\ mol^{-1}$ is consistent with $D_0^{\circ} > 665.6\ kJ\ mol^{-1}$ derived from the highest observed level in $A^1\Pi$ (14). The potential energy curve of this excited state may have a maximum (14, 15).

Source	Method	Reaction[a]	T/K	Data Points	δS cal K^{-1} mol^{-1}	$\Delta_rH°$(298.15 K), kcal mol 2nd Law	3rd Law	$\Delta_fH°$(298.15 K) kcal mol^{-1}	D_0°
Hildenbrand (1963)	torsion eff.[b]								
Cell 7	P=1.30Pt	A	882-931	8	-2.7±2.1	54.1±2.0	56.60±0.2	-63.73±0.8	160.3
Cell 9	P=1.72Pt	A	856-932	8	0.4±1.5	57.3±1.4	56.91±0.2	-63.42±0.8	160.0
Cell 15X	P=2.07Pt	A	866-931	8	1.0±0.6	57.8±0.5	56.85±0.1	-63.48±0.8	160.1
Cell 16X	P=3.13Pt	A	867-929	7[c]	0.5±1.4	57.2±1.3	56.79±0.1	-63.54±0.8	160.1
Semenkovitch (1960)	transport	B	1170-1373	5	-0.2±0.2	55.3±0.3	55.56±0.1	-63.40±0.8	160.0
Hildenbrand (1981)	torsion eff.[b]								
Cell G7S	P=1.12Pt	A	863-930	13	-1.0±0.2	58.25±0.21	57.37±0.05	-62.96±0.8	160.1
Cell G5Z	P=1.31Pt	A	839-928	6	-2.4±0.2	57.61±0.16	57.22±0.04	-63.11±0.8	159.7
Cell G6Z	P=1.65Pt	A	835-929	13	-0.4±0.1	57.68±0.07	57.25±0.04	-63.08±0.8	159.7
Cell G7L	P=2.10Pt	A	840-930	13	-1.0±0.2	58.20±0.19	57.31±0.08	-63.02±0.8	159.6

[a]Reactions: A) 1/3 AlF_3(cr) + 2/3 Al(cr) = AlF(g); B) 1/3 AlF_3(cr) + 2/3 Al(ℓ) = AlF(g);
[b]P = equilibrium pressure and Pt = torsion pressure. [c]One point rejected.

Heat Capacity and Entropy

Electronic levels and molecular constants are based on the detailed analysis of electronic spectra published by Barrow, Kapp and Malmberg (16) and a follow-up to this investigation published by Kapp and coworkers (20). Constnats for the ground state are confirmed by microwave spectra (11, 17, 18). The triplet-singlet separation, $a^3\Pi$-$X^1\Sigma^+$, calculated by Barrow et al. (16) is in excellent agreement with the results from the low resolution chemiluminiscence work of Rosenwaks et al. (21). We omit electronic states between 60000 and 68000 cm^{-1} including nine observed levels and one predicted level (16); these would have negligible effects on the thermodynamic functions. The analysis of Barrow et al. (16) is supported by theoretical calculations (20, 21).

References

1. P. E. Blackburn (A. D. Little, Inc., Cambridge, Mass.), AD-618523, 65 pp. (1965).
2. D. L. Hildenbrand et al., Ford Motor Co., Aeronutronic Div., Rept. No. U-2055, Contract No. 61-0905-c, March 15, 1963.
3. W. P. Witt and R. F. Barrow, Trans. Faraday Soc. 55, 730 (1959).
4. K. Ono, T. Matsushima and T. Ito, Light Metals 14, 363(1964).
5. S. A. Semenkovich, Trudy Vsesoyuz. Alyumin.-Magnievyi Inst. 1960, No. 44, 113 (1960).
6. A. Yu. Baimakov, Trudy Leningrad. Politekh. Inst., Elektromet. Tsvetnylkh Metal. 1957, No. 188, 156 (1957).
7. H. C. Ko, M. A. Greenbaum, J. A. Blauer and M. Farber, J. Phys. Chem. 69, 2311 (1965).
8. D. L. Hildenbrand and L. P. Theard, J. Chem. Phys. 42, 3230 (1965).
9. R. F. Porter, J. Chem. Phys. 33, 951 (1960).
10. P. Gross, C. Hayman and D. L. Levi, Trans. Faraday Soc. 50, 477 (1954).
11. D. R. Lide, J. Chem. Phys. 42, 1013 (1965).
12. D. L. Hildenbrand; Stanford Research Institute, personal communication, May 8, 1981 (based on 1976 work by D. L. Hildenbrand and W. Laughlin).
13. JANAF Thermochemical Tables: AlF_3(cr, g) 9-30-79; Al(cr), 6-30-79.
14. R. F. Barrow, J. W. C. Johns and F. J. Smith, Trans. Faraday Soc. 52, 913 (1956).
15. E. Murad, D. L. Hildenbrand and R. P. Main, J. Chem. Phys. 45, 263 (1966).
16. R. F. Barrow, I. Kopp and C. Malmberg, Phys. Scripta 10, 86 (1974).
17. F. C. Wyse, W. Gordy and E. F. Pearson, J. Chem. Phys. 52, 3887 (1970).
18. J. Hoeft, F. J. Lovas, E. Tiemann and T. Torring, Z. Naturforsch. A25, 1029 (1970).
19. S. P. So and W. G. Richards, J. Phys. B 7, 1973 (1974).
20. I. Kopp, B. Lindgren, and C. Malmberg, Phys. Scripta 14, 170 (1976).
21. S. Rosenwaks, R. E. Steele and H. P. Broida, Chem. Phys. Lett. 38, 121 (1976).
22. M. W. Chase, J. L. Curnutt, R. A. McDonald and A. N. Syverud, J. Phys. Chem. Ref. Data 7, 792 (1978).

Aluminum Fluoride (AlF) $Al_1F_1(g)$

Enthalpy Reference Temperature = T_r = 298.15 K Standard State Pressure = p° = 0.1 MPa

T/K	C_p° (J $K^{-1}mol^{-1}$)	S° (J $K^{-1}mol^{-1}$)	$-[G°-H°(T_r)]/T$ (J $K^{-1}mol^{-1}$)	$H°-H°(T_r)$ (kJ mol^{-1})	$\Delta_fH°$ (kJ mol^{-1})	$\Delta_fG°$ (kJ mol^{-1})	Log K_f
0	0.	0.	INFINITE	-8.891	-265.623	-265.623	INFINITE
100	29.121	182.450	242.280	-5.983	-264.653	-273.680	142.956
200	30.034	202.816	218.015	-3.040	-264.976	-282.645	73.819
298.15	31.932	215.157	215.157	0.	-265.684	-291.172	51.012
300	31.967	215.354	215.157	0.059	-265.699	-291.330	50.725
400	33.602	224.787	216.432	3.342	-266.534	-299.750	39.143
500	34.734	232.415	218.890	6.763	-267.429	-307.952	32.171
600	35.502	238.820	221.692	10.277	-268.387	-315.967	27.507
700	36.037	244.335	224.542	13.855	-269.431	-323.816	24.163
800	36.423	249.174	227.325	17.479	-270.593	-331.508	21.645
900	36.712	253.481	229.996	21.137	-271.904	-339.045	19.678
1000	36.937	257.362	232.542	24.819	-283.981	-345.667	18.056
1100	37.117	260.891	234.961	28.522	-285.313	-351.771	16.704
1200	37.265	264.127	237.259	32.242	-286.642	-357.754	15.573
1300	37.390	267.115	239.442	35.975	-287.969	-363.627	14.611
1400	37.499	269.890	241.519	39.719	-289.295	-369.397	13.782
1500	37.594	272.480	243.498	43.474	-290.620	-375.072	13.061
1600	37.679	274.909	245.386	47.238	-291.946	-380.659	12.427
1700	37.757	277.196	247.190	51.010	-293.272	-386.163	11.865
1800	37.828	279.356	248.918	54.789	-294.598	-391.589	11.364
1900	37.894	281.403	250.574	58.575	-295.922	-396.941	10.913
2000	37.957	283.348	252.165	62.368	-297.245	-402.224	10.505
2100	38.016	285.202	253.694	66.166	-298.565	-407.440	10.135
2200	38.073	286.972	255.167	69.971	-299.881	-412.594	9.796
2300	38.127	288.665	256.587	73.781	-301.190	-417.688	9.486
2400	38.180	290.289	257.957	77.596	-302.491	-422.725	9.200
2500	38.232	291.849	259.282	81.417	-303.783	-427.708	8.936
2600	38.282	293.349	260.564	85.242	-305.063	-432.640	8.692
2700	38.332	294.795	261.805	89.073	-306.331	-437.522	8.464
2800	38.382	296.190	263.008	92.909	-601.485	-441.391	8.234
2900	38.432	297.538	264.176	96.750	-601.627	-435.670	7.847
3000	38.483	298.841	265.310	100.595	-601.751	-429.945	7.486
3100	38.535	300.104	266.412	104.446	-601.858	-424.217	7.148
3200	38.589	301.328	267.484	108.303	-601.945	-418.485	6.831
3300	38.645	302.517	268.527	112.164	-602.011	-412.751	6.533
3400	38.704	303.671	269.544	116.032	-602.057	-407.015	6.253
3500	38.767	304.794	270.535	119.905	-602.081	-401.278	5.989
3600	38.833	305.887	271.502	123.785	-602.083	-395.541	5.739
3700	38.904	306.952	272.446	127.672	-602.063	-389.804	5.503
3800	38.979	307.990	273.368	131.566	-602.019	-384.067	5.279
3900	39.061	309.004	274.269	135.468	-601.953	-378.333	5.067
4000	39.148	309.994	275.149	139.378	-601.863	-372.600	4.866
4100	39.242	310.962	276.011	143.298	-601.749	-366.870	4.674
4200	39.344	311.909	276.855	147.227	-601.612	-361.142	4.491
4300	39.453	312.836	277.681	151.167	-601.451	-355.419	4.317
4400	39.570	313.744	278.490	155.118	-601.267	-349.699	4.151
4500	39.695	314.635	279.283	159.081	-601.056	-343.984	3.993
4600	39.829	315.509	280.061	163.057	-600.824	-338.274	3.841
4700	39.972	316.367	280.825	167.047	-600.569	-332.569	3.696
4800	40.125	317.210	281.574	171.052	-600.290	-326.870	3.557
4900	40.286	318.039	282.310	175.072	-599.987	-321.177	3.424
5000	40.458	318.854	283.032	179.110	-599.663	-315.490	3.296
5100	40.638	319.657	283.743	183.164	-599.316	-309.810	3.173
5200	40.829	320.448	284.441	187.238	-598.947	-304.137	3.055
5300	41.029	321.228	285.128	191.330	-598.556	-298.471	2.942
5400	41.239	321.997	285.803	195.444	-598.145	-292.813	2.832
5500	41.458	322.755	286.468	199.578	-597.712	-287.163	2.727
5600	41.687	323.504	287.123	203.736	-597.260	-281.521	2.626
5700	41.925	324.244	287.768	207.916	-596.788	-275.887	2.528
5800	42.171	324.976	288.403	212.121	-596.294	-270.262	2.434
5900	42.427	325.699	289.029	216.351	-595.784	-264.644	2.343
6000	42.691	326.414	289.646	220.607	-595.234	-259.039	2.255

PREVIOUS: September 1979 (1 atm) CURRENT: September 1979 (1 bar)

Aluminum Fluoride (AlF) $Al_1F_1(g)$

ALUMINUM FLUORIDE, ION (AlF^+) IDEAL GAS M_r = 45.979394

$S°(298.15 K) = [220.07 \pm 0.9] J K^{-1} mol^{-1}$

$\Delta_fH (0 K) = 686.0 \pm 24 kJ mol^{-1}$
$\Delta_fH°(298.15 K) = [692.034] kJ mol^{-1}$

Electronic Levels and Quantum Weights

State	ε_i, cm^{-1}	g_i
$[^2\Sigma^+]$	0	[2]
$[^2\Pi]$	[5000]	[4]
$[^2\Sigma^+]$	[20000]	[2]
$[^2\Pi]$	[33000]	[4]

$\omega_e = [925] cm^{-1}$ $\omega_ex_e = [5.5] cm^{-1}$ $\sigma = 1$
$B_e = [0.587] cm^{-1}$ $\alpha_e = [0.005] cm^{-1}$ $r_e = [1.605]$ Å

Enthalpy of Formation

We adopt $\Delta_fH° = 164 \pm 6 kcal mol^{-1}$ based on IP(AlF) = 9.86 ± 0.25 eV (227.5 ± 6 kcal mol^{-1}). Barrow, Kopp and Malmberg (1) used the observed data for BF to predict that, to a good approximation, IP(AlF) = T°(AlF, 3d) + IP(Al, 3d). This gave IP(AlF) = 79335 cm^{-1} (9.84 eV) which the authors (1) rounded to 80000 cm^{-1} (9.92 eV). We adopt an intermediate value corresponding to $\Delta_fH°$ = 164 kcal mol^{-1}. This yields $D_0^0 = 70 \pm 6 kcal mol^{-1}$ for $AlF^+(g) \rightarrow Al^+(g) + F(g)$.

IP(AlF) = 9.86 ± 0.25 eV is consistent with electron impact data which gave the following values for the appearance potential of AlF^+ from AlF: 8.9 ± 0.6 (2), 9.0 ± 1 (3), 9.2 (4), 9.5 ± 0.5 (5), 9 to 10 (6), 9.7 ± 0.3 (7), 9.7 ± 0.5 (8), 9.9 ± 0.3 (8) and 10.1 ± 0.3 eV (8).

Heat Capacity and Entropy

We assume the ground-state configuration to be the same as observed for BF^+ (9) and the isoelectronic molecules AlO and AlF (10). Theoretical calculations for $AlCl^+$ (11) predict the same ground state; they also suggest that the $^2\Pi$ excited state is either repulsive or has a shallow potential minimum at a much longer bond length. We assume the $^2\Pi$ state in AlF^+ to be nonrepulsive and estimate excited state levels equal to those in AlO (10. Comparison with $AlCl^+$ suggests that $^2\Pi$ should be in the range 5000-15000 cm $^{-1}$. Our thermodynamic functions would be upper-limit values if the $^2\Pi$ state were repulsive.

Bonding in Group III monohalides and their ions was characterized by Berkowitz and Dehmer (11) from photoelectron spectra and theoretical calculations. They concluded that ionization removes an electron from an antibonding orbital centered mainly on the metal. This shortens the bond length and enhances the ionic nature of the bond. A decrease in bond length is observed for BF^+ (9) and predicted by theroretical calculations for $AlCl^+$ (11, 10). By analogy with BF^+ and BF (9), we expect AlF^+ to have vibrational and rotational constants similar to those of the Rydberg excited states of AlF (1, 10). Our adopted constants are approximate avarages from the Rydberg states. The adopted bond length is 0.05 Å shorter than in ground state AlF; a similar difference is observed for BF^+.

References

1. R. F. Barrow, I. Kopp and C. Malmberg, Phys. Scripta 10, 86 (1974).
2. D. L. Hildenbrand and L. P. Theard, J. Chem. Phys. 42, 3230 (1965).
3. R. D. Srivastava and M. Farber, J. Phys. Chem. 75, 1760 (1971).
4. D. L. Hildenbrand and E. Murad, J. Chem. Phys. 44, 1524 (1966).
5. E. Murad, D. L. Hildenbrand and R. P. Main, J. Chem. Phys. 45, 263 (1966).
6. R. F. Porter, J. Chem. Phys. 33, 951 (1960).
7. T. C. Ehlert and J. L. Margrave, J. Amer. Chem. Soc. 86, 3901 (1964).
8. T. C. Ehlert, G. D. Blue, J. W. Green and J. L. Margrave, J. Chem. Phys. 41, 2250 (1964).
9. R. B. Caton and A. E. Douglas, Can. J. Phys. 48, 432 (1970).
10. JANAF Thermochemical Tables: AlO(g), 6-30-75; AlF(g) and MgF(g), 12-31-75; $AlCl^+$(g), 6-30-76.
11. J. Berkowitz and J. L. Dehmer, J. Chem. Phys. 57, 3194 (1972).

Aluminum Fluoride, Ion (AlF^+) $Al_1F_1^+(g)$

Enthalpy Reference Temperature = T_r = 298.15 K Standard State Pressure = p° = 0.1 MPa

	J K⁻¹mol⁻¹			kJ mol⁻¹			
T/K	$C_p°$	S°	$-[G°-H°(T_r)]/T$	$H°-H°(T_r)$	$\Delta_fH°$	$\Delta_fG°$	Log K_f
0	0.	0.	INFINITE	-8.813	685.976		
100	29.110	187.705	246.750	-5.905			
200	29.620	207.970	222.861	-2.978			
250	30.336	214.653	220.573	-1.480			
298.15	31.155	220.065	220.065	0.	692.034	658.828	-115.424
300	31.187	220.258	220.066	0.058	692.056	658.621	-114.676
350	32.031	225.129	220.448	1.638	692.646	653.002	-97.455
400	32.797	229.458	221.309	3.259	693.219	647.299	-84.529
450	33.463	233.360	222.435	4.916	693.778	641.526	-74.466
500	34.030	236.916	223.708	6.604	694.326	635.690	-66.410
600	34.924	243.204	226.447	10.054	695.383	623.863	-54.312
700	35.599	248.640	229.238	13.582	696.365	611.864	-45.658
800	36.167	253.431	231.968	17.170	697.247	599.731	-39.158
900	36.705	257.722	234.595	20.814	698.001	587.495	-34.097
1000	37.260	261.618	237.106	24.512	688.017	575.942	-30.084
1100	37.850	265.196	239.499	28.267	688.816	564.696	-26.815
1200	38.476	268.516	241.780	32.083	689.663	553.376	-24.088
1300	39.127	271.622	243.958	35.963	690.562	541.982	-21.777
1400	39.785	274.545	246.039	39.909	691.516	530.518	-19.794
1500	40.434	277.313	248.033	43.920	692.525	518.983	-18.073
1600	41.055	279.942	249.945	47.995	693.589	507.379	-16.564
1700	41.637	282.449	251.784	52.130	694.705	495.707	-15.231
1800	42.169	284.844	253.555	56.320	695.869	483.968	-14.044
1900	42.644	287.137	255.262	60.561	697.078	472.163	-12.981
2000	43.060	289.335	256.912	64.847	698.326	460.293	-12.022
2100	43.417	291.445	258.506	69.171	699.611	448.360	-11.152
2200	43.715	293.472	260.050	73.529	700.927	436.365	-10.361
2300	43.958	295.420	261.545	77.913	702.270	424.310	-9.636
2400	44.150	297.296	262.996	82.318	703.638	412.195	-8.971
2500	44.297	299.101	264.405	86.741	705.027	400.023	-8.358
2600	44.402	300.840	265.773	91.176	706.434	387.795	-7.791
2700	44.471	302.518	267.103	95.620	707.859	375.513	-7.265
2800	44.509	304.136	268.397	100.070	415.397	364.146	-6.793
2900	44.521	305.698	269.656	104.521	417.945	362.270	-6.525
3000	44.511	307.207	270.883	108.973	420.505	360.307	-6.273
3100	44.483	308.666	272.078	113.423	423.076	358.258	-6.037
3200	44.441	310.078	273.244	117.869	425.658	356.126	-5.813
3300	44.387	311.445	274.381	122.311	428.249	353.913	-5.602
3400	44.324	312.769	275.490	126.746	430.850	351.621	-5.402
3500	44.255	314.053	276.574	131.175	433.460	349.253	-5.212
3600	44.181	315.298	277.632	135.597	436.078	346.810	-5.032
3700	44.105	316.508	278.667	140.011	438.705	344.294	-4.861
3800	44.028	317.683	279.678	144.418	441.340	341.707	-4.697
3900	43.951	318.826	280.667	148.817	443.982	339.051	-4.541
4000	43.875	319.937	281.635	153.208	446.632	336.327	-4.392
4100	43.801	321.020	282.583	157.592	449.288	333.536	-4.249
4200	43.729	322.074	283.511	161.969	451.951	330.681	-4.113
4300	43.661	323.103	284.419	166.338	454.620	327.762	-3.982
4400	43.596	324.106	285.310	170.701	457.295	324.781	-3.856
4500	43.535	325.085	286.183	175.057	459.978	321.739	-3.735
4600	43.478	326.041	287.039	179.408	462.663	318.637	-3.618
4700	43.425	326.975	287.879	183.753	465.352	315.477	-3.506
4800	43.376	327.889	288.703	188.093	468.044	312.260	-3.398
4900	43.332	328.783	289.512	192.429	470.741	308.986	-3.294
5000	43.292	329.658	290.306	196.760	473.438	305.658	-3.193
5100	43.256	330.515	291.086	201.087	476.136	302.276	-3.096
5200	43.225	331.355	291.852	205.411	478.834	298.840	-3.002
5300	43.197	332.178	292.606	209.732	481.532	295.353	-2.911
5400	43.174	332.985	293.346	214.051	484.227	291.815	-2.823
5500	43.155	333.777	294.074	218.367	486.920	288.227	-2.737
5600	43.139	334.554	294.790	222.682	489.609	284.590	-2.655
5700	43.128	335.318	295.494	226.995	492.292	280.905	-2.574
5800	43.119	336.068	296.187	231.307	494.972	277.173	-2.496
5900	43.115	336.805	296.869	235.619	497.642	273.395	-2.420
6000	43.113	337.529	297.541	239.930	500.326	269.570	-2.347

PREVIOUS: June 1976 (1 atm) CURRENT: June 1976 (1 bar)

Aluminum Fluoride, Ion (AlF^+) $Al_1F_1^+(g)$

ALUMINUM FLUORIDE OXIDE (OAlF) IDEAL GAS $M_r = 61.979343$

$\Delta_f H°(0\ K) = -579 \pm 16$ or 28 kJ mol^{-1}

$S°(298.15\ K) = [237.3 \pm 4$ or $16]$ J K^{-1} mol^{-1} $\Delta_f H°(298.15\ K) = -582 \pm 16$ or 28 kJ mol^{-1}

Vibrational Frequencies and Degeneracies

ν, cm^{-1}	ν, cm^{-1}	ν, cm^{-1}
[675](1)	386 (2)	1022(1)

Ground State Quantum Weight: [1] σ = 1

Point Group: $[C_{\infty v}]$

Bond Distances: O-Al = [1.61] Å Al-F = [1.63] Å

Bond Angle: O-Al-F = [180]°

Rotational Constant: $B_o = [0.184214]$ cm^{-1}

Enthalpy of Formation

We adopt $\Delta_f H°(298.15\ K) = -139\pm4$ kcal/mol and $\Delta_{at}H° = 293.2\pm4$ kcal mol^{-1}, derived from effusion data for reaction A (1). This reaction is analyzed along with limited equilibrium data for three other reactions. The latter are in satisfactory agreement with reaction A. Dissociation energies calculated from our adopted values (5) are 13.9 kcal mol^{-1} larger than those of the corresponding diatomic molecules; i.e., $\Delta_{at}H°$ (FAl-O) = 133.9±4, $\Delta_{at}H°$(OAl-F) = 173.2±4, or $\Delta_r H°$ = 13.9±5 kcal mol^{-1} for the reaction OAlF(g) + Al(g) = AlF(g) + AlO(g). These values suggest that O=Al-F has enhanced stability, consistent with the observation of this reaction (4) by mass-spectrometric sampling of flames containing aluminum species. We assign $\Delta_f H°$ the alternative uncertainty of ±7 kcal mol^{-1} in case OAlF is nonlinear. Farber and Srivastava (4) emphasized that $\Delta_f H°$ is much too negative to be compatible with the alternative arrangement Al-O-F.

Source	Method	Reaction[a]	T/K	Data Points	δS cal K^{-1}mol^{-1}	$\Delta_r H°$(298.15 K), kcal mol^{-1} 2nd law	3rd law	$\Delta_f H°$(298.15 K) kcal mol^{-1}
(1)	Reaction effus.	A	2203-2228[b]	8	-22±26	43±25	91.0±1.6	-138.9±3
(2, 3)	Mass spec.[c]	B	1540-1923	8	--	107.1±3.2	--	-143.6±7
(4)	Flame mass spec.	C	2250	1	--	--	-10.2	-135.3±6
	Flame mass spec.	D	2250	1	--	--	-11.7	-133.0±6

[a] Reactions: A) 1/3 AlF_3(g) + 1/3 $Al_2O_3(\alpha)$ = OAlF(g) C) AlF(g) + AlO(g) = OAlF(g) + Al(g)
B) AlF(g) + Al_2O(g) = OAlF(g) + 2 Al(g) D) AlF(g) + H_2O(g) = OAlF(g) + H_2(g)

[b] T converted to IPTS-68 assuming published values to be IPTS-48.
[c] Analysis based on relative ion intensities corrected (3) as follows: I(OAlF) = 66 at 1773 K and I(AlF) = 120 at 1923 K. Values of K_p were not measured, thus precluding third law analysis.

Heat Capacity and Entropy

Snelson (6) observed infrared spectra of OAlF, OBF, OBCl and OBBr isolated in inert gas matrices. He proposed linear structures for OBCl and OBBr based on agreement of observed and calculated shifts in isotopic frequencies. Isotopic shifts for OBF were consistent with linearity but were less conclusive. Lacking isotopic data, Snelson (6) assumed OAlF to be linear.

For triatomics having 16 valence electrons (either BAB or BAC types), a linear ground state is predicted by Walsh's correlation (7) and related semiempirical calculations (8). These correlations derive mainly from covalent triatomics in which the central atom is carbon or some less electropositive element. Ionic triatomics can behave differently, e.g., the alkaine earth difluorides change from linear (BeF_2, 5) to bent configurations (CaF_2, SrF_2 and BaF_2, 9). OAlF is isoelectronic with and intermediate between MgF_2 and SiO_2. Electric deflection data indicated SiO_2 to be linear (10). Data for MgF_2 (5) are contradictory, indicating either a linear or slightly bent (~158°) structure. This comparison with SiO_2 and MgF_2 favors linearity of OAlF but does not rule out a slightly bent structure.

We adopt the linear structure O=Al-F with an O-Al distance slightly shorter than in AlO (5) and an Al-F distance equal to that in AlF_3 (5). We assume the ground-state configuration to be $^1\Sigma^+$ and neglect excited states which should be relatively unimportant (8). Vibrational frequencies are those assigned by Snelson (6) from a comparison of derived force constants for OAlF and OBX molecules. Values of ν_2 and ν_3 from IR spectra in an argon matrix. Snelson (6) used a stretching force constant transferred from AlF_3 in order to estimate ν_1 = 675 cm^{-1}. Although a weak absorption was observed at 687 cm^{-1}, this band was not assigned (6) to ν_1 because it was not conclusively associated with ν_2 and ν_3.

We estimate the uncertainty to be ~1 gibbs/mol in the entropy and Gibbs-energy function of linear OAlF. If OAlF were nonlinear with an angle of 160°, S° would change by +3.8 (298 K) and +2.1 (2000 K) gibbs/mol, while the Gibbs-energy function would change by +2.9 gibbs/mol at 2000 K.

References

1. M. Farber and H. L. Petersen, Trans. Faraday Soc. 59, 836 (1963).
2. R. D. Srivastava and M. Farber, J. Phys. Chem. 75, 1760 (1971).
3. M. Farber, Space Sciences, Inc., Monrovia, Ca., personal communication, December 5, 1975.
4. M. Farber and R. D. Srivastava, submitted for publication in Combustion and Flame.
5. JANAF Thermochemical Tables: AlF(g), 12-31-75; AlO(g) and MgF_2(g), 6-30-75; BeF_2(g) and AlF_3(g), 6-30-70.
6. A. Snelson, High Temp. Sci. 5, 77 (1973); 4, 141, 318 (1972).
7. A. D. Walsh, J. Chem. Soc. 1953, 2266 (1953).
8. J. W. Rabalais, J. M. McDonald, V. Scherr and S. P. McGlynn, Chem. Rev. 71, 73 (1971).
9. V. Calder, D. E. Mann, K. S. Seshadri, M. Allavena and D. White, J. Chem. Phys. 51, 2093 (1969).
10. M. Kaufman, J. Muenter and W. Klemperer, J. Chem. Phys. 47, 3365 (1967).

Aluminum Fluoride Oxide (OAlF) $Al_1F_1O_1(g)$

Enthalpy Reference Temperature = T_r = 298.15 K Standard State Pressure = p° = 0.1 MPa

T/K	$C_p°$ (J K^{-1}mol^{-1})	S°	$-[G°-H°(T_r)]/T$	$H°-H°(T_r)$ (kJ mol^{-1})	$\Delta_f H°$	$\Delta_f G°$	Log K_f
0	0.	0.	INFINITE	-10.785	-579.067	-579.067	INFINITE
100	31.151	195.688	274.080	-7.839	-579.511	-581.198	303.586
200	40.013	219.989	241.397	-4.281	-580.675	-582.431	152.115
250	43.792	229.337	238.071	-2.183	-581.171	-582.810	121.772
298.15	46.806	237.316	237.316	0.	-581.576	-583.088	102.155
300	46.910	237.606	237.317	0.087	-581.590	-583.098	101.526
350	49.445	245.034	237.898	2.498	-581.947	-583.320	87.056
400	51.493	251.776	239.218	5.023	-582.257	-583.495	76.197
450	53.145	257.939	240.960	7.640	-582.535	-583.632	67.746
500	54.482	263.610	242.946	10.332	-582.793	-583.740	60.983
600	56.463	273.731	247.254	15.886	-583.291	-583.883	50.832
700	57.813	282.542	251.679	21.604	-583.824	-583.940	43.574
800	58.763	290.328	256.033	27.435	-584.446	-583.916	38.126
900	59.452	297.291	260.237	33.348	-585.205	-583.806	33.883
1000	59.964	303.582	264.262	39.320	-596.724	-582.842	30.445
1100	60.355	309.317	268.101	45.337	-597.497	-581.416	27.609
1200	60.658	314.582	271.758	51.388	-598.268	-579.920	25.243
1300	60.899	319.447	275.242	57.466	-599.041	-578.360	23.239
1400	61.092	323.967	278.563	63.566	-599.818	-576.740	21.518
1500	61.250	328.188	281.732	69.684	-600.602	-575.064	20.026
1600	61.380	332.145	284.760	75.815	-601.393	-573.336	18.718
1700	61.489	335.870	287.658	81.959	-602.193	-571.558	17.562
1800	61.581	339.387	290.435	88.113	-603.003	-569.733	16.533
1900	61.659	342.719	293.100	94.275	-603.821	-567.862	15.612
2000	61.726	345.883	295.661	100.444	-604.649	-565.948	14.781
2100	61.783	348.896	298.125	106.619	-605.485	-563.993	14.029
2200	61.834	351.771	300.498	112.800	-606.328	-561.997	13.344
2300	61.878	354.521	302.788	118.986	-607.177	-559.963	12.717
2400	61.916	357.155	304.999	125.176	-608.030	-557.892	12.142
2500	61.950	359.683	307.136	131.369	-608.886	-555.785	11.613
2600	61.981	362.114	309.204	137.566	-609.744	-553.644	11.123
2700	62.008	364.454	311.207	143.765	-610.602	-551.470	10.669
2800	62.032	366.709	313.149	149.967	-905.358	-548.297	10.229
2900	62.054	368.886	315.034	156.172	-905.115	-535.549	9.646
3000	62.074	370.990	316.864	162.378	-904.868	-522.810	9.103
3100	62.092	373.026	318.643	168.586	-904.614	-510.078	8.595
3200	62.108	374.998	320.374	174.796	-904.354	-497.356	8.118
3300	62.123	376.909	322.058	181.008	-904.087	-484.641	7.671
3400	62.136	378.764	323.699	187.221	-903.811	-471.935	7.250
3500	62.149	380.565	325.298	193.435	-903.526	-459.237	6.854
3600	62.160	382.316	326.858	199.651	-903.232	-446.547	6.479
3700	62.171	384.019	328.380	205.867	-902.929	-433.865	6.125
3800	62.180	385.677	329.866	212.085	-902.616	-421.192	5.790
3900	62.189	387.293	331.318	218.303	-902.294	-408.527	5.472
4000	62.198	388.867	332.737	224.523	-901.963	-395.870	5.170
4100	62.205	390.403	334.125	230.743	-901.623	-383.222	4.882
4200	62.213	391.902	335.482	236.964	-901.275	-370.583	4.609
4300	62.219	393.366	336.812	243.185	-900.919	-357.951	4.348
4400	62.225	394.797	338.113	249.407	-900.556	-345.328	4.100
4500	62.231	396.195	339.389	255.630	-900.185	-332.714	3.862
4600	62.237	397.563	340.638	261.854	-899.811	-320.108	3.635
4700	62.242	398.902	341.864	268.078	-899.433	-307.509	3.418
4800	62.247	400.212	343.066	274.302	-899.052	-294.919	3.209
4900	62.251	401.496	344.245	280.527	-898.669	-282.337	3.010
5000	62.255	402.753	345.403	286.752	-898.287	-269.763	2.818
5100	62.259	403.986	346.539	292.978	-897.906	-257.196	2.634
5200	62.263	405.195	347.656	299.204	-897.528	-244.637	2.457
5300	62.267	406.381	348.753	305.430	-897.155	-232.085	2.287
5400	62.270	407.545	349.831	311.657	-896.789	-219.540	2.124
5500	62.273	408.688	350.891	317.884	-896.432	-207.002	1.966
5600	62.276	409.810	351.933	324.112	-896.085	-194.469	1.814
5700	62.279	410.912	352.958	330.340	-895.750	-181.944	1.667
5800	62.282	411.995	353.966	336.568	-895.426	-169.424	1.526
5900	62.284	413.060	354.959	342.796	-895.122	-156.910	1.389
6000	62.287	414.107	355.936	349.024	-894.813	-144.402	1.257

PREVIOUS: December 1975 (1 atm) CURRENT: December 1975 (1 bar)

Aluminum Fluoride Oxide (OAlF) $Al_1F_1O_1(g)$

ALUMINUM FLUORIDE (AlF_2) IDEAL GAS M_r = 64.978346

$S°(298.15\ K) = [264.21 \pm 2.1]\ J\ K^{-1}\ mol^{-1}$

$\Delta_f H°(0\ K) = -693 \pm 40\ kJ\ mol^{-1}$

$\Delta_f H°(298.15\ K) = -695 \pm 40\ kJ\ mol^{-1}$

Electronic Levels and Quantum Weights

State	ε_i, cm^{-1}	g_i
$^2A_1$	[0]	2
$^2B_1$	[20000]	2
$^2B_2$	[25000]	2

Vibrational Frequencies and Degeneracies

ν, cm^{-1}
[730](1)
[280](1)
[880](1)

Point Group: [C_{2v}] $\sigma = 2$

Bond Distance: Al-F = [1.65] Å

Bond Angle: F-Al-F = [120]°

Product of the Moments of Inertia: $I_A I_B I_C = [3.369562 \times 10^{-115}]\ g^3\ cm^6$

Enthalpy of Formation

We tentatively adopt $\Delta_f H°(298.15\ K) = -166 \pm 10\ kcal\ mol^{-1}$ which corresponds to an average bond energy, $\Delta_{at}H°/2$ = 140 kcal mol^{-1}, similar to that in $AlF_3(g)$ (1). $\Delta_f H°$ is a compromise between discrepant results (-180 and -151 kcal mol^{-1}) from two mass-spectrometric equilibria analyzed below. In our opinion there is possible cause to suspect Kp for either reaction. The large entropy discrepancy (δS) for reaction B suggests a temperature-dependent bias in the mass-spectrometric Kp (5). If both AlF_2 and AlF_3 were determined from the intensity of AlF_2^+ at different ionizing energies, then we expect $\Delta_f H°$ (5) to be a positive limit. Reaction A was studied twice by molecular flow effusion of $AlF_3(g)$ over the system Al(ℓ)-Al_2O_3(cr). The earlier study (3, 4) yields $\Delta_f H°(298.15\ K)$ = -188 or -179 kcal mol^{-1}, depending on the ionizing energy. Unfortunately, the ionizing energy in the first case is ambiguous. We conclude that $\Delta_f H°$ is quite dependent on ionizing energy. In addition, Kp and $\Delta_f H°$ may be affected by bias in the ionizing energy scale or related problems. These are discussed in more detail on tables for AlF_2^+, $AlCl_2^+$ and $AlCl_2$ (1). We believe that $\Delta_f H°$ (3, 4) may be a negative limit.

The later study (2) of reaction A yields $\Delta_f H°$ = -180 kcal mol^{-1}. Internal evidence (2) gives us no reason to suspect this value. The possibility of negative bias is suggested, however, by inference from average bond energies (see $AlCl_2$, 1) and by using $\Delta_f H°$ = -180 to derive $\Delta_f H°$ of other species in related mass-spectrometric studies. Effects of alternative values of $\Delta_f H°(AlF_2)$ are discussed on the tables for Al_2^-, AlF_4^-, AlF_2O, AlClF, $AlClF_2$, and $AlCl_2F$ (1). Only for $AlClF_2$ and $AlCl_2F$ do we have reliable, independent values of $\Delta_f H°$. These are consistent with $\Delta_f H°(AlF_2)$ = -166 instead of -180. Although an alternative explanation is conceivable (see $AlClF_2$, 1) we choose $\Delta_f H°(AlF_2)$ = -166 kcal mol^{-1}.

Source	Reaction[a]	Ionizing Energy, eV	Data Points	T/K	δS cal K^{-1} mol^{-1}	$\Delta_r H°$(298.15 K), kcal mol^{-1} 2nd law	3rd law	$\Delta_f H°$(298.15 K) kcal mol^{-1}	$\Delta_{at}H°/2$
(2)	A	AP+(? or 3)	7	1453-1675	4.0±1.7	31.0±2.6	24.6±0.9	-180.4	147.0
(3, 4)	A	A.+~2	1	1538	--	--	25.8	-179.2	146.5
(3, 4)	A	20(or AP+3)	5	1408-1633	3.6±0.6	22.5±0.9	17.0±0.8	-188.0	150.9
(5)	B	?	4	1243-1301	-76±10	-47±13	50.2±3.7	<-151.2	>132.4

[a] Reactions: A) 2 AlF(g) = Al(g) + AlF_2(g) B) AlF_3(g) + AlF(g) = 2 AlF_2(g)

Heat Capacity and Entropy

The electronic ground state is that predicted for BF_2 by theoretical calculations (6); it is consistent with semiempirical calculations (7) for BF_2, AlF_2, BCl_2 and $AlCl_2$. We estimate excited states of BF_2 at ~30000 ($^2B_1$) and ~40000 ($^2B_2$) by assuming that approximate theoretical calculations (6) have a bias similar to analogous calculations (8) of NO_2 (9). Dissociative electron-attachment data (10) yield ~43000 cm^{-1} for $BF_2(^2B_2)$ confirming the estimation method. We use $BF_2(^2B_1)$ to estimate a crude correlation-energy correction for $AlF_2(^2B_1)$ (7). Dissociative electron-attachment data (11) for $AlF_2(^2B_2)$ are approximate; they yield ~20000±10000 ($\Delta_f H°$ = -166) or ~25000±10000 cm^{-1} ($\Delta_f H°$ = -180).

Vibrational frequencies are calculated from the estimated force constants f_r = 4.25 and f_α/r^2 = 0.25 mdyn/Å. Frequencies and force constants are estimated from simultaneous consideration of AlF_2^+, AlF_2, AlF_2^- (1) and the analogous boron species. We assume that frequency changes are qualitatively similar to those of NO_2^+, NO_2 and NO_2^-. In going from AlF_2^+ to AlF_2, we expect a large decrease in ν_3, an increase in ν_2 and some increase in ν_1. By analogy with BF_2 (6), the odd electron in AlF_2 should occupy an antibonding orbital which increases the bond length but decreases the bond angle and stretching force constant. We assume a bond length 0.05 Å longer than AlF_2^+ (1) and a bond angle equal to that in BF_2 (6). The principal moments of inertia are: $I_A = 1.7832 \times 10^{-39}$, $I_B = 12.8835 \times 10^{-39}$ and $I_C = 14.6667 \times 10^{-39}\ g\ cm^2$.

References
1. JANAF Thermochemical Tables: $AlCl_2$(g), AlF_2^+(g), AlF_2^-(g), AlF_4^-(g), AlF_2O(g), AlClF(g), $AlClF_2$(g), $AlCl_2F$(g), 6-30-76; AlF(g), 12-31-75; AlF_3(g), 6-30-70; Al(g), 12-31-65; F(g), 9-30-65.
2. O. M. Uy, R. D. Srivastava and M. Farber, High Temp. Sci. 4, 227 (1972).
3. M. Farber and S. P. Harris, High Temp. Sci. 3, 231 (1971).
4. M. Farber, R. D. Srivastava and O. M. Uy, Space Sciences, Inc., AFRPL-TR-70-97, July 1970.
5. T. C. Ehlert and J. L. Margrave, J. Amer. Chem. Soc. 86, 3901 (1964); T. C. Ehlert, Marquette University, personal communication, Aug. 13, 1964.
6. C. Thomson and D. A. Brotchie, Theoret. Chim. Acta 32, 101 (1973); Mol. Phys. 28, 301 (1974).
7. J. W. Hastie and J. L. Margrave, J. Phys. Chem. 73, 1105 (1969).
8. R. A. Gangi and L. Burnelle, J. Chem. Phys. 55, 843 (1971); 49, 561 (1968).
9. G. D. Gillispie, A. U. Khan, A. C. Wahl, R. P. Hosteny and M. Krauss, J. Chem. Phys. 63, 3425 (1975).
10. P. W. Harland and J. L. Franklin, J. Chem. Phys. 61, 1621 (1974).
11. F. Petty, J. L. Wang, R. P. Steiger, P. W. Harland, J. L. Franklin and J. L. Margrave, High Temp. Sci. 5, 25 (1973).

Aluminum Fluoride (AlF_2) $Al_1F_2(g)$

Enthalpy Reference Temperature = T_r = 298.15 K Standard State Pressure = p° = 0.1 MPa

T/K	$C_p°$ (J $K^{-1}mol^{-1}$)	S°	-[G°-H°(T_r)]/T	H°-H°(T_r) (kJ mol^{-1})	$\Delta_f H°$	$\Delta_f G°$	Log K_f
0	0.	0.	INFINITE	-11.505	-692.684	-692.684	INFINITE
100	35.777	220.448	301.626	-8.118	-692.688	-696.997	364.074
200	41.060	246.892	268.255	-4.273	-693.574	-700.993	183.081
250	43.614	256.331	264.952	-2.155	-694.073	-702.790	146.840
298.15	45.857	264.209	264.209	0.	-694.544	-704.426	123.413
300	45.937	264.493	264.210	0.085	-694.562	-704.487	122.662
350	47.921	271.728	264.777	2.433	-695.038	-706.104	105.380
400	49.558	278.237	266.059	4.871	-695.503	-707.653	92.410
450	50.888	284.154	267.746	7.384	-695.961	-709.144	82.315
500	51.967	289.573	269.662	9.956	-696.417	-710.584	74.234
600	53.559	299.199	273.803	15.237	-697.344	-713.331	62.101
700	54.637	307.541	278.040	20.651	-698.330	-715.919	53.423
800	55.391	314.889	282.196	26.154	-699.417	-718.359	46.904
900	55.935	321.446	286.199	31.722	-700.644	-720.655	41.826
1000	56.338	327.361	290.025	37.337	-712.635	-722.045	37.716
1100	56.644	332.746	293.667	42.986	-713.880	-722.925	34.329
1200	56.882	337.685	297.132	48.663	-715.124	-723.693	31.502
1300	57.070	342.246	300.429	54.361	-716.371	-724.356	29.105
1400	57.220	346.481	303.569	60.076	-717.622	-724.924	27.047
1500	57.343	350.433	306.563	65.804	-718.879	-725.401	25.261
1600	57.444	354.137	309.422	71.544	-720.143	-725.795	23.695
1700	57.529	357.622	312.156	77.292	-721.415	-726.109	22.311
1800	57.600	360.912	314.774	83.049	-722.693	-726.348	21.078
1900	57.661	364.028	317.285	88.812	-723.977	-726.517	19.973
2000	57.713	366.987	319.697	94.581	-725.264	-726.617	18.977
2100	57.759	369.804	322.016	100.355	-726.553	-726.653	18.075
2200	57.799	372.492	324.250	106.132	-727.840	-726.628	17.252
2300	57.835	375.062	326.404	111.914	-729.121	-726.544	16.500
2400	57.868	377.524	328.483	117.699	-730.394	-726.404	15.810
2500	57.898	379.887	330.492	123.488	-731.656	-726.212	15.173
2600	57.927	382.159	332.436	129.279	-732.902	-725.970	14.585
2700	57.955	384.345	334.318	135.073	-734.129	-725.680	14.039
2800	57.983	386.454	336.143	140.870	-1029.236	-724.377	13.513
2900	58.012	388.489	337.913	146.670	-1029.321	-713.487	12.851
3000	58.041	390.456	339.632	152.472	-1029.380	-702.596	12.233
3100	58.072	392.360	341.302	158.278	-1029.408	-691.702	11.655
3200	58.105	394.204	342.927	164.087	-1029.406	-680.808	11.113
3300	58.141	395.992	344.508	169.899	-1029.370	-669.915	10.604
3400	58.179	397.729	346.048	175.715	-1029.299	-659.024	10.125
3500	58.221	399.416	347.549	181.535	-1029.193	-648.135	9.673
3600	58.266	401.057	349.012	187.359	-1029.050	-637.249	9.246
3700	58.314	402.654	350.441	193.188	-1028.869	-626.368	8.843
3800	58.366	404.209	351.835	199.022	-1028.651	-615.493	8.461
3900	58.423	405.726	353.198	204.862	-1028.395	-604.623	8.098
4000	58.483	407.206	354.529	210.707	-1028.102	-593.761	7.754
4100	58.548	408.651	355.832	216.559	-1027.770	-582.907	7.426
4200	58.617	410.063	357.106	222.417	-1027.401	-572.060	7.115
4300	58.689	411.443	358.354	228.282	-1026.996	-561.224	6.818
4400	58.766	412.793	359.576	234.155	-1026.554	-550.397	6.534
4500	58.847	414.114	360.773	240.036	-1026.074	-539.581	6.263
4600	58.932	415.409	361.947	245.924	-1025.562	-528.775	6.004
4700	59.021	416.677	363.098	251.822	-1025.016	-517.981	5.757
4800	59.113	417.921	364.227	257.729	-1024.437	-507.199	5.519
4900	59.209	419.141	365.335	263.645	-1023.826	-496.430	5.292
5000	59.308	420.338	366.424	269.571	-1023.186	-485.673	5.074
5100	59.409	421.513	367.492	275.506	-1022.517	-474.930	4.864
5200	59.514	422.668	368.542	281.453	-1021.821	-464.199	4.663
5300	59.621	423.802	369.574	287.409	-1021.099	-453.482	4.469
5400	59.730	424.918	370.589	293.377	-1020.353	-442.780	4.283
5500	59.842	426.015	371.587	299.355	-1019.584	-432.091	4.104
5600	59.955	427.094	372.568	305.345	-1018.794	-421.416	3.931
5700	60.070	428.156	373.534	311.347	-1017.985	-410.757	3.764
5800	60.186	429.202	374.485	317.359	-1017.155	-400.111	3.603
5900	60.303	430.232	375.421	323.384	-1016.312	-389.479	3.448
6000	60.421	431.246	376.343	329.420	-1015.432	-378.864	3.298

PREVIOUS: June 1976 (1 atm) CURRENT: June 1976 (1 bar)

Aluminum Fluoride (AlF_2) $Al_1F_2(g)$

ALUMINUM FLUORIDE, ION (AlF_2^+) IDEAL GAS M_r = 64.977797

$\Delta_fH°(0\ K)$ = 87 ± 63 kJ mol^{-1}

S°(298.15 K) = [239.3 ± 4] J K^{-1} mol^{-1} $\Delta_fH°(298.15\ K)$ = [92.048] kJ mol^{-1}

Vibrational Frequencies and Degeneracies

ν, cm^{-1}

[670](1)

[240](2)

[1000](1)

Ground State Quantum Weight: [1] σ = 2

Point Group: [$D_{\infty h}$]

Bond Distance: Al-F = [1.60] Å

Bond Angle: F-AL-F = [180]°

Rotational Constant: B_0 = [0.173304] cm^{-1}

Enthalpy of Formation

We adopt $\Delta_fH°(298.15\ K)$ = 22±15 kcal mol^{-1} and IP(AlF_2) = 8.1±0.9 eV. Electron-impact studies (1, 2) gave 15.2±0.3 eV for the appearance potential (AP) of AlF_2^+ from AlF_3. The analogous process for onset of BF_2^+ from BF_3 has been studied by both photoionization (3) and electron impact (4). Comparison indicates that BF_2^+ from electron impact (4) carried an excess energy (E*) of ~0.4 to ~0.9 eV. We estimate the excess energy for AlF_2^+ as 1.0±0.5 eV. Thus, we take AP-E* = $\Delta_rH°$ = 14.2±0.6 eV (327.5±14 kcal mol^{-1}) for the reaction $AlF_3(g) + e^-(g) = AlF_2^+(g) + F(g) + 2\ e^-(g)$. With JANAF auxiliary data (5) this yields $\Delta_fH°(AlF_2^+)$ = 21±15, $\Delta_fH°(298.15\ K)$ = 22±15 and IP(AlF_2) = 187±20 kcal mol^{-1} (8.1±0.9 eV).

Electron-impact studies of AlF_2 gave approximate appearance potentials AP(AlF_2^+) = 9±1 (1), 10±1 (6) and 11±1 eV (7). We dismiss the last value (7) due to possible bias analogous to that of $AlCl_2^+(g)$ (5). Electron impact on AlF_2 should yield a vertical AP corresponding to a nonlinear (excited) configuration of AlF_2^+. Theoretical calculations for the linear $^2B_1$ state of BF_2 (8) suggest an excitation energy of ~1.1 eV at a bond angle of 120°. Extended Huckel calculations for AlF_2 (9) suggest ~1.4 eV at 130°. Thus, we combine E* = 1.2±0.5 eV with the observed AP values (1, 6) to get IP(AlF_2) = 7.8±1.2 and 8.8±1.2 eV. These are consistent with the adopted value.

The extended Huckel calculation (9) gave IP(AlF_2) = 7.8 eV at a bond angle of 130°. This result should approximate the vertical IP which we expect at ~9.3 eV.

Heat Capacity and Entropy

We assume the electronic ground state to be $^1\Sigma^+$ and neglect excited states. We expect the ground state to be linear by analogy with other triatomic species having sixteen valence electrons (10). Although conflicting data (5) suggest that isoelectronic MgF_2 is either linear or slightly nonlinear, recent Raman data (11) favor a linear structure. We estimate the bond distance to be slightly shorter than that in AlF_3 and 0.05 Å shorter than that estimated for AlF_2 (5). We transfer the stretching force constant from AlF_3 and the ratio f_r/f_α = ~13 Å^{-2} from MgF_2. This ratio is a compromise between very different values of ν_2 observed for MgF_2 in the gas (5) and matrix (11) phases. Thus, we calculate vibratonal frequencies for AlF_2^+ from the force constants f_r = 4.9, f_{rr} = 0.2 and f_α/r^2 = 0.14 mdyn/Å. The resulting values have been rounded downward.

References

1. T. C. Ehlert and J. L. Margrave, J. Amer. Chem. Soc. 86, 390 (1964).
2. T. C. Ehlert and G. D. Blue, J. W. Green and J. L. Margrave, J. Chem. Phys. 41, 2250 (1964).
3. V. H. Dibeler and S. K. Liston, Inorg. Chem. 7, 1742 (1968).
4. J. L. Franklin, J. G. Dillard et al., NSRDS-NBS 26, 1969.
5. JANAF Thermochemical Tables: AlF(g) and $AlCl_2^+$(g), 6-30-76; AlFO(g), 12-31-75; MgF_2(g), 6-30-75; AlF_3(g), 6-30-70; F(g), 9-30-65.
6. O. M. Uy, R. D. Srivastava and M. Farber, High Temp. Sci. 4, 227 (1972).
7. M. Farber and S. P. Harris, High Temp. Sci. 3, 231 (1971).
8. C. Thomson and D. A. Brotchie, Theoret. Chim. Acta 32, 101 (1973).
9. J. W. Hastie and J. L. Margrave, J. Phys. Chem. 73, 1105 (1969).
10. J. W. Rabalais, J. M. McDonald, V. Scherr and S. P. McGlynn, Chem. Rev. 71, 73 (1971).
11. M. L. Lesiecki and J. W. Nibler, J. Chem. Phys. 64, 871 (1976).

Aluminum Fluoride, Ion (AlF_2^+) $Al_1F_2^+(g)$

Enthalpy Reference Temperature = T_r = 298.15 K Standard State Pressure = p° = 0.1 MPa

T/K	$C_p°$ (J K^{-1}mol^{-1})	S° (J K^{-1}mol^{-1})	-[G°-H°(T_r)]/T (J K^{-1}mol^{-1})	H°-H°(T_r) (kJ mol^{-1})	$\Delta_fH°$ (kJ mol^{-1})	$\Delta_fG°$ (kJ mol^{-1})	Log K_f
0	0.	0.	INFINITE	-11.730	87.485		
100	35.844	193.010	279.338	-8.633			
200	44.056	220.679	243.650	-4.594			
250	46.927	230.826	240.097	-2.318			
298.15	49.281	239.299	239.299	0.	92.048	83.338	-14.601
300	49.364	239.604	239.300	0.091	92.075	83.284	-14.501
350	51.404	247.371	239.908	2.612	92.811	81.761	-12.202
400	53.086	254.349	241.285	5.226	93.561	80.131	-10.464
450	54.463	260.684	243.094	7.916	94.319	78.407	-9.101
500	55.589	266.482	245.147	10.668	95.083	76.598	-8.002
600	57.271	276.776	249.582	16.316	96.601	72.758	-6.334
700	58.428	285.697	254.119	22.105	98.069	68.668	-5.124
800	59.245	293.555	258.567	27.991	99.443	64.372	-4.203
900	59.839	300.569	262.851	33.946	100.682	59.912	-3.477
1000	60.282	306.898	266.944	39.953	91.162	56.083	-2.929
1100	60.619	312.660	270.842	45.999	92.392	52.516	-2.494
1200	60.882	317.946	274.550	52.075	93.625	48.836	-2.126
1300	61.091	322.828	278.079	58.174	94.858	45.054	-1.810
1400	61.258	327.361	281.439	64.292	96.089	41.176	-1.536
1500	61.395	331.593	284.643	70.424	97.315	37.211	-1.296
1600	61.508	335.559	287.703	76.570	98.535	33.165	-1.083
1700	61.603	339.291	290.629	82.725	99.749	29.042	-0.892
1800	61.682	342.814	293.431	88.890	100.957	24.848	-0.721
1900	61.750	346.151	296.118	95.062	102.160	20.586	-0.566
2000	61.808	349.320	298.700	101.240	103.361	16.262	-0.425
2100	61.858	352.337	301.183	107.423	104.561	11.878	-0.295
2200	61.902	355.215	303.574	113.611	105.763	7.436	-0.177
2300	61.940	357.968	305.879	119.803	106.970	2.940	-0.067
2400	61.974	360.605	308.105	125.999	108.186	-1.609	0.035
2500	62.004	363.135	310.256	132.198	109.414	-6.209	0.130
2600	62.030	365.568	312.337	138.400	110.657	-10.859	0.218
2700	62.054	367.909	314.352	144.604	111.918	-15.556	0.301
2800	62.075	370.166	316.305	150.810	-180.700	-19.333	0.361
2900	62.094	372.345	318.200	157.019	-178.298	-13.613	0.245
3000	62.111	374.450	320.041	163.229	-175.870	-7.975	0.139
3100	62.127	376.487	321.829	169.441	-173.414	-2.419	0.041
3200	62.141	378.460	323.568	175.654	-170.928	3.057	-0.050
3300	62.154	380.372	325.260	181.869	-168.412	8.456	-0.134
3400	62.165	382.228	326.909	188.085	-165.862	13.777	-0.212
3500	62.176	384.030	328.515	194.302	-163.280	19.023	-0.284
3600	62.186	385.782	330.082	200.520	-160.665	24.195	-0.351
3700	62.195	387.486	331.610	206.739	-158.016	29.293	-0.414
3800	62.204	389.144	333.102	212.959	-155.333	34.320	-0.472
3900	62.211	390.760	334.560	219.180	-152.617	39.276	-0.526
4000	62.219	392.335	335.985	225.401	-149.869	44.161	-0.577
4100	62.225	393.872	337.378	231.623	-147.088	48.977	-0.624
4200	62.231	395.371	338.741	237.846	-144.276	53.726	-0.668
4300	62.237	396.836	340.075	244.070	-141.433	58.406	-0.709
4400	62.243	398.267	341.382	250.294	-138.562	63.020	-0.748
4500	62.248	399.665	342.661	256.518	-135.660	67.569	-0.784
4600	62.252	401.034	343.915	262.743	-132.733	72.053	-0.818
4700	62.257	402.372	345.145	268.969	-129.780	76.473	-0.850
4800	62.261	403.683	346.351	275.195	-126.804	80.830	-0.880
4900	62.265	404.967	347.534	281.421	-123.804	85.125	-0.907
5000	62.269	406.225	348.695	287.648	-120.785	89.358	-0.934
5100	62.272	407.458	349.836	293.875	-117.746	93.531	-0.958
5200	62.275	408.667	350.955	300.102	-114.690	97.644	-0.981
5300	62.278	409.854	352.056	306.330	-111.618	101.698	-1.002
5400	62.281	411.018	353.137	312.558	-108.533	105.694	-1.022
5500	62.284	412.161	354.200	318.786	-105.436	109.632	-1.041
5600	62.287	413.283	355.245	325.015	-102.328	113.515	-1.059
5700	62.289	414.385	356.272	331.243	-99.213	117.341	-1.075
5800	62.292	415.469	357.284	337.472	-96.088	121.112	-1.091
5900	62.294	416.534	358.279	343.702	-92.962	124.831	-1.105
6000	62.296	417.581	359.259	349.931	-89.810	128.494	-1.119

PREVIOUS: June 1976 (1 atm) CURRENT: June 1976 (1 bar)

Aluminum Fluoride, Ion (AlF_2^+) $Al_1F_2^+(g)$

ALUMINUM FLUORIDE, ION (AlF_2^-) IDEAL GAS M_r = 64.978895

$\Delta_fH°(0\ K) = -900 \pm 84\ kJ\ mol^{-1}$

$S°(298.15\ K) = [260.05 \pm 3.3]\ J\ K^{-1}\ mol^{-1}$ $\Delta_fH°(298.15\ K) = [-907.928]\ kJ\ mol^{-1}$

Electronic Levels and Quantum Weights

State	ε_i, cm^{-1}	g_i
$^1A_1$	[0]	1
$^3B_1$	[26000]	3
$^1B_1$	[44000]	1

Vibrational Frequencies and Degeneracies

ν, cm^{-1}
[730](1)
[280](1)
[780](1)

Point Group: [C_{2v}] σ = 2
Bond Distance: Al-F = [1.70] Å
Bond Angle: F-Al-F = [105]°
Product of the Moments of Inertia: $I_AI_BI_C = [4.600099 \times 10^{-115}]\ g^3\ cm^6$

Enthalpy of Formation

We adopt $\Delta_fH°(298.15\ K) = -217\pm20\ kcal\ mol^{-1}$ which corresponds to an electron affinity of $EA(AlF_2) = 50\pm8\ kcal\ mol^{-1}$ (2.2±0.4 eV) and a fluoride-ion affinity of IA(AlF) = 92±20 kcal mol^{-1}. $\Delta_fH°$ is based on Kp data for the reaction $AlF_2(g) = F(g) + AlF_2^-(g)$ reported by Srivastava et al. (1). The authors measured Kp with a molecular-flow-effusion method using a mass spectrometer operating in both positive- and negative-ion modes. They give three points (1705-1900 K) obtained from study of the vapor species over the system AlF_3(cr)-KF(cr)-Al(ℓ). Our analysis gives $[\Delta_rS°(2nd\ law)-\Delta\ S°(3rd\ law)] = -0.2\pm4.8\ cal\ K^{-1}\ mol^{-1}$ and $\Delta_rH°(298.15\ K) = 28.4\pm8.8$(2nd law) or 28.8±1(3rd law) kcal mol^{-1}. Reducing the 3rd law $\Delta_rH°$ to absolute zero and combining with EA(F) = 78.38 kcal mol^{-1} (2), we derive $EA(AlF_2) = 49.7\pm8\ kcal\ mol^{-1}$. The uncertainity of ±8 kcal mol^{-1} is our estimate of a reasonable bound for error in $\Delta_rH°$.

Depending on the value used for $\Delta_fH°(AlF_2)$, the above reaction yields $\Delta_fH°(AlF_2^-, g, 298.15\ K) = -231$ (3, 2), -217 (2) or -202 (4, 2) kcal mol^{-1}. In contrast, we derive $\Delta_fH°(AlF_2^-, g, 298.15\ K) = -181\pm25\ kcal\ mol^{-1}$, independent of $\Delta_fH°(AlF_2)$, from $\Delta_rH°(0\ K) = AP - E^* = 127$ kcal (5) for the reaction $AlF_3(g) + e^-(g) = AlF_2^-(g) + F(g)$. Depending on $\Delta_fH°(AlF_2)$, this yields $EA(AlF_2) = 0\pm25$ (3, 2), 14±25 (2) or 29±25 (4, 2) kcal mol^{-1}. Petty et al. (5) measured the appearance potential (AP) and excess translational energy of AlF_2^- in the dissociative electron-attachment reaction. E^*, the excess vibrational-translational energy of the products at threshold, was estimated (5) from the measured translational energy via an approximate empirical correlation. We estimate ±25 kcal mol^{-1} as an approximate bound for error in E^* (6). The resulting $\Delta_fH°(AlF_2^-) = -181\pm25\ kcal^{-1}\ mol^{-1}$ is inconsistent with the two most likely values (-231 and -217) derived from AlF_2. Bias might exist in either experiment (5, 2). It is conceivable that AlF_2^- is formed (5) in an excited electronic state. Thus far, electronic excitation has been observed (7) in only one negative molecular ion, AsF_4^-. If it occurs for AlF_2^-, then $\Delta_fH° = -181\pm25$ refers to an excited electronic state. The electronic energy of this state would be ~13000±9000 cm^{-1}, based on the two most likely values of $\Delta_fH°(AlF_2^-)$. Such values are not unreasonable, so we cannot rule out electronic excitation (5). We conclude that additional data are needed to confirm $\Delta_fH°(AlF_2^-)$.

Heat Capacity and Entropy

Electronic levels and quantum weights are assumed equal to those of isoelectronic SiF_2 (2, 8); however, see the above comment on excited levels. Vibrational frequencies are calculated from the estimated force constants $f_r = 3.7$, $f_{rr} = 0.1$ and $f_\alpha/r^2 = 0.29$ mdyn/A. Frequencies and force constants are estimated from simultaneous consideration of AlF_2^+, AlF_2, AlF_2^- and the analogous boron species. We assume that frequency changes are qualitatively similar in the two series AlF_2^+, AlF_2, AlF_2^- and NO_2^+, NO_2, NO_2^-(2) which have the same number of valence electrons. Furthermore, we expect isoelectronic SiF_2 (2) to provide upper-limit frequencies for AlF_2^-. Thus, in going from AlF_2 to AlF_2^-, we assume little change in ν_1 and ν_2 but a significant decrease in ν_3.

By analogy with BF_2 (8, 9), we expect that the odd electron in AlF_2 occupies a molecular orbital centered mainly on the metal opposite the two fluorines. This orbital should be antibonding (8) in the sense that addition of an electron increases the bond length but decreases the bond angle and stretching force constant. We assume that the bond length is 0.05 A longer and the bond angle is 15° smaller than in AlF_2 (2). The principal moments of inertia are $I_A = 2.8061 \times 10^{-39}$, $I_B = 11.4773 \times 10^{-39}$ and $I_C = 14.2834 \times 10^{-39}\ g\ cm^2$.

References

1. R. D. Srivastava, O. M. Uy and M. Farber, J. Chem. Soc., Faraday Trans. I 70, 1033 (1974); AD-731303, 50 pp. (1971).
2. JANAF Thermochemical Tables: AlF_2(g) and SiF_2(g), 6-30-76; NO_2^-(g), 6-30-72; F^-(g), 12-31-71; F(g), 9-30-65.
3. O. M. Uy, R. D. Srivastava and M. Farber, High Temp. Sci. 4, 227 (1972).
4. T. C. Ehlert and J. L. Margrave, J. Amer. Chem. Soc. 86, 3901 (1964); T. C. Ehlert, Marquette University, personal communication, Aug. 13, 1964.
5. F. Petty, J. L. Wang, R. P. Steiger, P. W. Harland, J. L. Franklin and J. L. Margrave, High Temp. Sci. 5, 25 (1973).
6. J. L. Franklin, Science 193, 725 (1976).
7. J. L. Franklin, Rice University, Houston, Texas, personal communications, Sept., 1976.
8. C. Thomson and D. A. Brotchie, Theoret. Chim. Acta 32, 101 (1973).
9. W. Nelson and W. Gordy, J. Chem. Phys. 51, 4710 (1969).

Aluminum Fluoride, Ion (AlF_2^-) $Al_1F_2^-$(g)

Enthalpy Reference Temperature = T_r = 298.15 K Standard State Pressure = p° = 0.1 MPa

T/K	$C_p°$	S°	$-[G°-H°(T_r)]/T$	$H°-H°(T_r)$	$\Delta_fH°$	$\Delta_fG°$	Log K_f
	J $K^{-1}mol^{-1}$			kJ mol^{-1}			
0	0.	0.	INFINITE	-11.573	-899.939		
100	35.786	215.980	297.844	-8.186			
200	41.429	242.516	264.145	-4.326			
250	44.174	252.060	260.799	-2.185			
298.15	46.511	260.046	260.046	0.	-907.928	-910.314	159.483
300	46.594	260.334	260.047	0.086	-907.983	-910.329	158.502
350	48.597	267.672	260.622	2.468	-909.465	-910.602	135.900
400	50.209	274.271	261.923	4.939	-910.936	-910.664	118.921
450	51.494	280.262	263.633	7.483	-912.402	-910.542	105.693
500	52.520	285.742	265.573	10.084	-913.868	-910.257	95.094
600	54.009	295.459	269.765	15.416	-916.824	-909.258	79.158
700	55.003	303.864	274.050	20.870	-919.848	-907.759	67.738
800	55.690	311.256	278.248	26.407	-922.980	-905.820	59.144
900	56.182	317.845	282.288	32.001	-926.259	-903.480	52.437
1000	56.545	323.785	286.146	37.639	-940.306	-900.006	47.011
1100	56.820	329.187	289.817	43.308	-943.610	-895.816	42.539
1200	57.032	334.141	293.307	49.001	-946.917	-891.325	38.798
1300	57.199	338.713	296.626	54.712	-950.228	-886.558	35.622
1400	57.333	342.957	299.786	60.439	-953.545	-881.536	32.891
1500	57.442	346.916	302.797	66.178	-956.871	-876.276	30.515
1600	57.532	350.626	305.672	71.927	-960.204	-870.794	28.429
1700	57.607	354.116	308.420	77.684	-963.546	-865.104	26.581
1800	57.670	357.411	311.051	83.448	-966.896	-859.217	24.934
1900	57.724	360.530	313.574	89.218	-970.251	-853.143	23.455
2000	57.770	363.493	315.996	94.993	-973.611	-846.893	22.119
2100	57.809	366.312	318.326	100.772	-976.973	-840.474	20.906
2200	57.844	369.002	320.568	106.554	-980.334	-833.895	19.799
2300	57.874	371.574	322.731	112.340	-983.690	-827.164	18.785
2400	57.901	374.038	324.817	118.129	-987.038	-820.287	17.853
2500	57.925	376.402	326.834	123.920	-990.375	-813.270	16.992
2600	57.947	378.674	328.784	129.714	-993.697	-806.120	16.195
2700	57.967	380.862	330.673	135.510	-997.001	-798.843	15.455
2800	57.986	382.970	332.503	141.307	-1294.186	-790.476	14.747
2900	58.005	385.005	334.279	147.107	-1296.350	-772.448	13.913
3000	58.023	386.972	336.003	152.908	-1298.488	-754.346	13.134
3100	58.041	388.875	337.678	158.711	-1300.598	-736.173	12.404
3200	58.060	390.718	339.306	164.517	-1302.678	-717.932	11.719
3300	58.079	392.505	340.892	170.323	-1304.726	-699.627	11.074
3400	58.101	394.239	342.435	176.132	-1306.741	-681.261	10.466
3500	58.124	395.923	343.939	181.944	-1308.722	-662.835	9.892
3600	58.149	397.561	345.406	187.757	-1310.668	-644.353	9.349
3700	58.177	399.155	346.838	193.574	-1312.579	-625.818	8.835
3800	58.209	400.707	348.235	199.393	-1314.455	-607.232	8.347
3900	58.243	402.219	349.600	205.215	-1316.294	-588.597	7.883
4000	58.281	403.694	350.934	211.042	-1318.098	-569.915	7.442
4100	58.324	405.134	352.238	216.872	-1319.866	-551.189	7.022
4200	58.370	406.540	353.514	222.706	-1321.600	-532.419	6.622
4300	58.422	407.914	354.764	228.546	-1323.298	-513.609	6.239
4400	58.477	409.258	355.987	234.391	-1324.963	-494.760	5.874
4500	58.538	410.572	357.185	240.242	-1326.592	-475.873	5.524
4600	58.604	411.860	358.360	246.099	-1328.190	-456.951	5.189
4700	58.675	413.121	359.512	251.963	-1329.756	-437.994	4.868
4800	58.751	414.357	360.641	257.834	-1331.292	-419.004	4.560
4900	58.832	415.569	361.750	263.713	-1332.796	-399.982	4.264
5000	58.919	416.759	362.838	269.601	-1334.273	-380.930	3.980
5100	59.011	417.926	363.907	275.497	-1335.722	-361.849	3.706
5200	59.108	419.073	364.957	281.403	-1337.145	-342.739	3.443
5300	59.210	420.200	365.989	287.319	-1338.542	-323.603	3.189
5400	59.317	421.308	367.003	293.245	-1339.916	-304.440	2.945
5500	59.429	422.397	368.000	299.182	-1341.267	-285.252	2.709
5600	59.545	423.469	368.981	305.131	-1342.597	-266.040	2.482
5700	59.666	424.524	369.946	311.092	-1343.907	-246.804	2.262
5800	59.792	425.563	370.896	317.064	-1345.195	-227.546	2.049
5900	59.921	426.586	371.832	323.050	-1346.470	-208.265	1.844
6000	60.055	427.594	372.753	329.049	-1347.706	-188.965	1.645

PREVIOUS: June 1976 (1 atm) CURRENT: June 1976 (1 bar)

Aluminum Fluoride, Ion (AlF_2^-) $Al_1F_2^-$(g)

ALUMINUM FLUORIDE OXIDE ($OAlF_2$) IDEAL GAS M_r = 80.977746

$\Delta_f H°(0\ K) = -1105.4 \pm 30\ kJ\ mol^{-1}$

$S°(298.15\ K) = [292.7 \pm 13]\ J\ K^{-1}\ mol^{-1}$ $\Delta_f H°(298.15\ K) = -1108.8 \pm 30\ kJ\ mol^{-1}$

Electronic Levels and Quantum Weights

State	ε_i, cm^{-1}	g_i
$[^2B_2]$	0	[2]
$[^2B_1]$	[10000]	[2]
$[^2B]$	[20000]	[2]

Vibrational Frequencies and Degeneracies

ν, cm^{-1}	ν, cm^{-1}
[900](1)	[900](1)
[640](1)	[240](1)
[260](1)	[270](1)

Point Group: $[C_{2v}]$ σ = [2]

Bond Distances: Al-F = [1.63] Å Al-O = [1.72] Å

Bond Angles: F-Al-F = [120]° F-Al-O = [120]°

Product of the Moments of Inertia: $I_A I_B I_C = [3.719682 \times 10^{-114}]\ g^3\ cm^6$

Enthalpy of Formation

We adopt $\Delta_f H°(298.15\ K) = -265\pm7\ kcal\ mol^{-1}$ and $\Delta_{at}H° = 437\pm8\ kcal\ mol^{-1}$ based on JANAF auxiliary data (1) combined with $\Delta_r H°(298.15\ K) = 15.5\pm5\ kcal\ mol^{-1}$ for the reaction $AlF(g) + AlFO(g) = Al(g) + AlF_2O(g)$. $\Delta_r H°$ is from our 3rd law analysis of Kp data reported by Uy et al. (2). They calculated Kp directly from ion intensities observed at 1453-1675 K in a mass-spectrometric study of vapor species over the system $AlF_3(g)$-$Al(\ell)$-$Al_2O_3(cr)$. 2nd law analysis gives $\Delta_r H°(298.15\ K) = 20.8\pm0.8\ kcal\ mol^{-1}$ and the difference $[\Delta_r S°(2nd\ law) - \Delta_r S°(3rd\ law)] = 3.4\pm0.5\ cal\ K^{-1}\ mol^{-1}$. Uy et al. (2) assigned $\Delta_r H°$(2nd law) an overall uncertainty (including contributions from T) of ±5 $kcal\ mol^{-1}$.

The adopted $\Delta_{at}H° = 437\ kcal\ mol^{-1}$ implies that AlF_2O is surprisingly stable. Average bond energies from AlF_3 (140 $kcal\ mol^{-1}$) and AlO, AlO_2 and Al_2O (~120 $kcal\ mol^{-1}$) yield an estimate of $\Delta_a H°(AlF_2O) = 400\ kcal\ mol^{-1}$. If we attribute all of the difference to the Al-O bond, then the AlF_2O data imply an Al-O bond energy of 157 $kcal\ mol^{-1}$. This enhancement of 37 $kcal\ mol^{-1}$ contrasts with the near equality of Al-O bond energies in the oxides (1) and hydroxides (8). An analogous treatment of AlFO(g) (1) yields an Al-O bond energy of 153 $kcal\ mol^{-1}$, similar to that in AlF_2O. Existing data for BF_2O and BFO imply bond energies of ~166 and ~200 $kcal\ mol^{-1}$ compared with values of ~130 (B-O) and ~190 (B=O) from the oxides. There are obvious differences between the Al and B systems which we are unable to rationalize via existing knowledge of electronic structure. We feel that confirmation of the oxyfluoride data is desirable.

Heat Capacity and Entropy

Molecular parameters are all estimated by analogy with BF_2O and BF_2O^+. Two emission bands due to BF_2O or BF_2O^+ were analyzed by Mathews (3). Vibrational analysis of the band with $\nu_\infty = 17171\ cm^{-1}$ gave the three symmetrical stretching modes of the lower state. Rotational analysis of the band with $\nu_o = 22391\ cm^{-1}$ gave structural parameters including the bond length B-O = 1.40±0.05 Å. This value indicates a single bond which is longer than those of 1.36, 1.263 and 1.20 Å observed for the various bond types in B_2O_3, BO_2 and BO. It is not known (3) whether the two emission bonds have a common lower state. Zahradnik and Carsky (4) used semiempirical CNDO calculations to predict the electronic transitions of BF_2O and BF_2O^+. They favored assignment of the 22391 cm^{-1} transition to BF_2O^+ and the other transition to either BF_2O^+ or BF_2O.

Due to the lengthened B-O bond and the theoretical results, we tentatively attribute both observed bands to BF_2O^+. We estimate the missing vibrational frequencies of BF_2O^+ and all those of BF_2O and BF_2O^- by comparison with HBF_2, BF_3 and $ClBF_2$. We assume that addition of an electron causes a large increase in ν_1 and ν_5 but a smaller increase in ν_2 and ν_6.

Vibrational frequencies of AlF_2O are estimated from those of BF_2O by comparison of AlF_3, AlFO, AlO and Al_2O (1) with the analogous boron species. Electronic levels are crude estimates based on calculations or data for the 24-valence-electron species BF_2O (4), BF_3^+ (5, 6) and NO_3 (5, 7). Bond angles are assumed to be 120°. Bond distances are assumed equal to those of AlF_3 and Al_2O (1). The principal moments of inertia are: $I_A = 12.0265 \times 10^{-39}$, $I_B = 12.5730 \times 10^{-39}$ and $I_C = 24.5995 \times 10^{-39}\ g\ cm^2$.

References

1. JANAF Thermochemical Tables: AlFO(g) and AlO_2(g), 12-31-75; AlF(g), AlO(g) and Al_2O(g), 6-30-75; AlF_3(g), 6-30-70; Al(g), 12-31-65.
2. O. M. Uy, R. D. Srivastava and M. Farber, High Temp. Sci. 4, 227 (1972); AD-731303, 50 pp. (1971).
3. C. W. Mathews and K. K. Innes, J. Mol. Spectrosc. 15, 199 (1965); ibid. 19, 203 (1966).
4. R. Zahradnik and P. Carsky, Theoret. Chim. Acta 27, 121 (1972).
5. T. E. H. Walker and J. A. Horsley, Mol. Phys. 21, 939 (1971).
6. P. T. Bassett and D. R. Lloyd, J. Chem. Soc. A1971, 1551 (1971).
7. G. Herzberg, "Electronic Spectra of Polyatomic Molecules", D. Van Nostrand Co., Inc., New York, 1966.
8. H. A. Skinner, Farday Symposia of The Chemical Society, No. 8, 167 (1973).

Aluminum Fluoride Oxide ($OAlF_2$) $Al_1F_2O_1(g)$

Enthalpy Reference Temperature = T_r = 298.15 K Standard State Pressure = p° = 0.1 MPa

T/K	$C_p°$ (J $K^{-1}mol^{-1}$)	S° (J $K^{-1}mol^{-1}$)	$-[G°-H°(T_r)]/T$ (J $K^{-1}mol^{-1}$)	$H°-H°(T_r)$ (kJ mol^{-1})	$\Delta_f H°$ (kJ mol^{-1})	$\Delta_f G°$ (kJ mol^{-1})	Log K_f
0	0.	0.	INFINITE	-14.325	-1105.379	-1105.379	INFINITE
100	42.267	235.438	343.053	-10.761	-1106.659	-1103.801	576.567
200	55.050	269.080	298.254	-5.835	-1107.918	-1100.427	287.402
250	59.640	281.871	293.729	-2.964	-1108.393	-1098.497	229.518
298.15	63.390	292.705	292.705	0.	-1108.760	-1096.556	192.112
300	63.521	293.098	292.707	0.117	-1108.773	-1096.480	190.914
350	66.737	303.139	293.492	3.377	-1109.076	-1094.407	163.331
400	69.357	312.228	295.275	6.781	-1109.322	-1092.294	142.639
450	71.477	320.524	297.626	10.304	-1109.528	-1090.152	126.542
500	73.192	328.147	300.303	13.922	-1109.709	-1087.990	113.661
600	75.724	341.730	306.103	21.376	-1110.043	-1083.614	94.337
700	77.441	353.540	312.055	29.040	-1110.406	-1079.182	80.530
800	78.644	363.964	317.905	36.848	-1110.857	-1074.691	70.170
900	79.513	373.280	323.549	44.758	-1111.445	-1070.137	62.109
1000	80.158	381.693	328.950	52.743	-1122.796	-1064.748	55.617
1100	80.650	389.357	334.098	60.784	-1123.404	-1058.914	50.284
1200	81.035	396.391	339.000	68.869	-1124.015	-1053.024	45.837
1300	81.344	402.890	343.668	76.989	-1124.631	-1047.084	42.072
1400	81.600	408.928	348.116	85.136	-1125.256	-1041.095	38.844
1500	81.819	414.566	352.360	93.308	-1125.891	-1035.062	36.044
1600	82.013	419.852	356.415	101.499	-1126.537	-1028.985	33.593
1700	82.190	424.830	360.295	109.710	-1127.193	-1022.868	31.429
1800	82.358	429.532	364.012	117.937	-1127.858	-1016.712	29.504
1900	82.521	433.990	367.579	126.181	-1128.531	-1010.519	27.781
2000	82.683	438.226	371.006	134.441	-1129.208	-1004.291	26.229
2100	82.845	442.265	374.304	142.718	-1129.886	-998.028	24.825
2200	83.008	446.122	377.481	151.010	-1130.562	-991.733	23.547
2300	83.175	449.816	380.546	159.319	-1131.232	-985.408	22.379
2400	83.344	453.359	383.507	167.645	-1131.891	-979.054	21.309
2500	83.516	456.765	386.370	175.988	-1132.535	-972.672	20.323
2600	83.690	460.044	389.141	184.349	-1133.160	-966.265	19.413
2700	83.866	463.206	391.826	192.726	-1133.762	-959.834	18.569
2800	84.043	466.259	394.430	201.122	-1428.239	-952.414	17.768
2900	84.220	469.211	396.958	209.535	-1427.690	-935.430	16.849
3000	84.398	472.069	399.414	217.966	-1427.109	-918.466	15.992
3100	84.574	474.840	401.803	226.414	-1426.493	-901.521	15.191
3200	84.749	477.527	404.127	234.881	-1425.839	-884.597	14.440
3300	84.922	480.138	406.391	243.364	-1425.148	-867.694	13.734
3400	85.092	482.676	408.598	251.865	-1424.416	-850.812	13.071
3500	85.260	485.145	410.750	260.383	-1423.644	-833.952	12.446
3600	85.424	487.549	412.850	268.917	-1422.831	-817.115	11.856
3700	85.585	489.892	414.900	277.467	-1421.976	-800.302	11.298
3800	85.742	492.176	416.904	286.034	-1421.080	-783.511	10.770
3900	85.894	494.405	418.863	294.615	-1420.143	-766.745	10.269
4000	86.042	496.582	420.779	303.212	-1419.165	-750.004	9.794
4100	86.185	498.708	422.654	311.824	-1418.148	-733.287	9.342
4200	86.324	500.787	424.489	320.449	-1417.092	-716.596	8.912
4300	86.458	502.820	426.287	329.088	-1416.000	-699.931	8.502
4400	86.587	504.809	428.049	337.741	-1414.871	-683.291	8.112
4500	86.711	506.756	429.777	346.406	-1413.706	-666.677	7.739
4600	86.830	508.663	431.471	355.083	-1412.511	-650.090	7.382
4700	86.944	510.532	433.133	363.771	-1411.285	-633.529	7.041
4800	87.053	512.363	434.765	372.471	-1410.031	-616.994	6.714
4900	87.157	514.159	436.367	381.182	-1408.749	-600.485	6.401
5000	87.256	515.921	437.941	389.903	-1407.445	-584.003	6.101
5100	87.350	517.650	439.487	398.633	-1406.119	-567.547	5.813
5200	87.440	519.347	441.006	407.372	-1404.773	-551.118	5.536
5300	87.524	521.013	442.500	416.121	-1403.411	-534.714	5.270
5400	87.604	522.650	443.969	424.877	-1402.035	-518.337	5.014
5500	87.679	524.258	445.414	433.641	-1400.648	-501.985	4.767
5600	87.750	525.839	446.837	442.413	-1399.252	-485.658	4.530
5700	87.817	527.393	448.236	451.191	-1397.850	-469.357	4.301
5800	87.879	528.920	449.614	459.976	-1396.442	-453.080	4.080
5900	87.937	530.423	450.971	468.767	-1395.036	-436.827	3.867
6000	87.990	531.902	452.308	477.563	-1393.610	-420.600	3.662

PREVIOUS: June 1976 (1 atm) CURRENT: June 1976 (1 bar)

Aluminum Fluoride Oxide ($OAlF_2$) $Al_1F_2O_1(g)$

ALUMINUM FLUORIDE OXIDE, ION ($OAlF_2^-$)   IDEAL GAS M_r = 80.978295

$\Delta_f H°(0\ K)$ = -1300 ± 105 kJ mol^{-1}

S°(298.15 K) = [284.9 ± 13] J K^{-1} mol^{-1} $\Delta_f H°(298.15\ K)$ = [-1309.592] kJ mol^{-1}

Vibrational Frequencies and Degeneracies

ν, cm^{-1}	ν, cm^{-1}
[950](1)	[900](1)
[650](1)	[270](1)
[260](1)	[300](1)

Ground State Quantum Weight: [1]   σ = [2]

Point Group: [C_{2v}]

Bond Distances: Al-F = [1.63] Å   Al-O = [1.66] Å

Bond Angles: F-Al-F = [120]°   F-Al-O = [120]°

Product of the Moments of Inertia: $I_A I_B I_C$ = [3.468404 x 10^{-114}] g^3 cm^6

Enthalpy of Formation

We adopt $\Delta_f H°(298.15\ K)$ = -313±25 kcal mol^{-1} which corresponds to EA(AlF_2O) = 46±25 kcal mol^{-1} (2.0±1.1 eV) and the fluoride-ion affinity IA(AlFO) = 112 kcal mol^{-1}. $\Delta_f H°$ is derived from Kp for reaction A (see below) observed by Faraber et al. (1) with a mass spectrometer operating in both positive- and negative-ion modes. Farber et al. (1) used a molecular-flow-effusion technique to study vapor species in the systems KF(g)-AlF_3(g)-Al_2O_3(cr), KF(g)-AlF_3(g)-Al_2O_3(cr), and KCl(g)-AlF_3(g)-Al_2O_3(cr). They considered the results to be limiting values; we presume this is due to very low intensities of AlF_2O^-.

The three reactions reported by Farber et al. are analyzed below using JANAF auxiliary data (2). We give inequalities consistent with the authors' limiting values. Reactions B and C yield less stringent limits are are not used. Reaction A involves AlF_2^- which depends on our possibly biased value (2) of $\Delta_f H°(AlF_2)$. Use of the alternative $\Delta_f H°$ (2) based on mass spectrometry (1, 3) reduces the discrepancy between reactions A and C form 35 to 21 kcal mol^{-1}. This may mot be significant if B and C yield only limiting values. The alternative result from A would yield EA(AlF_2O) = 61 and IA(AlFO) = 127 kcal mol^{-1}.

Reaction	Data Points	T/K	$\Delta_r H°$(298.15 K) kcal mol^{-1}	$\Delta_f H°$(298.15 K) kcal mol^{-1}
A) AlFO(g) + AlF_2^-(g) = AlF_2O^-(g) + AlF(g)	1	1793	≥-21.1	≥-313.6[a] or ≥-327.8[a]
B) AlF_2O(g) + F^-(g) = Alf_2O^-(g) + F(g)	1	1753	≥0.1	≥-344.8
C) AlF_2O(g) + Cl^-(g) = AlF_2O(g) + Cl(g)	2	1653-1793	≥0.8	≥-349.1

[a]Assuming $\Delta_f H°(AlF_2^-$, g, 298.15 K) = -217 or -231.2 kcal mol^{-1} depending on choice of $\Delta_f H°(AlF_2)$ (2).

Heat Capacity and Entropy

All molecular parameters are estimated via data for BF_2O^+ (or BF_2O) as discussed on the table for AlF_2O(g) (2). We assume a C_{2v} structure with equal bond angles. Formation of the negative ion is assumed to decrease the Al-O bond distance by 0.06 Å but leave the Al-F distance unchanged. By analogy with AlF_3 (2) we assume a singlet electronic ground state and neglect excited states. Formation of the negative ion is assumed to cause a significant increase in ν_1, ν_5, and ν_6. The principal moments of inertia are I_A = 11.4723 x 10^{-39}, I_B = 12.5732 x 10^{-39} and I_C = 24.0455 x 10^{-39} g cm^2.

References

1. M. Farber, R. D. Srivastava and O. M. Uy, AD-731303, 50 pp. (1971).
2. JANAF Thermochemical Tables: AlF_2^-(g), AlF_2(g) and AlF_2O(g), 6-30-76; AlFO(g), 12-31-75; AlF(g), 6-30-75; AlF_3(g), 6-30-70; Cl(g), 6-30-72; Cl^-(g), 6-30-65; F^-(g), 12-31-71; F(g), 9-30-65.
3. O. M. Uy, T. D. Srivastava and M. Farber, High Temp. Sci. 4, 227 (1972).

Aluminum Fluoride Oxide, Ion ($OAlF_2^-$) $Al_1F_2O_1^-$(g)

Enthalpy Reference Temperature = T_r = 298.15 K   Standard State Pressure = p° = 0.1 MPa

T/K	$C_p°$ (J K^{-1}mol^{-1})	S° (J K^{-1}mol^{-1})	-[G°-H°(T_r)]/T (J K^{-1}mol^{-1})	H°-H°(T_r) (kJ mol^{-1})	$\Delta_f H°$ (kJ mol^{-1})	$\Delta_f G°$ (kJ mol^{-1})	Log K_f
0	0.	0.	INFINITE	-14.077	-1299.765		
100	41.036	228.775	334.358	-10.558			
200	54.051	261.603	290.340	-5.747			
250	58.755	274.184	285.880	-2.924			
298.15	62.592	284.870	284.870	0.	-1309.592	-1288.797	225.792
300	62.726	285.258	284.871	0.116	-1309.644	-1288.668	224.377
350	66.024	295.183	285.648	3.337	-1311.024	-1285.062	191.785
400	68.721	304.181	287.411	6.708	-1312.344	-1281.262	167.316
450	70.912	312.407	289.738	10.201	-1313.619	-1277.300	148.265
500	72.691	319.973	292.388	13.793	-1314.866	-1273.197	133.010
600	75.329	333.475	298.138	21.202	-1317.323	-1264.632	110.096
700	77.127	345.231	304.044	28.831	-1319.800	-1255.656	93.698
800	78.390	355.617	309.854	36.610	-1322.358	-1246.319	81.376
900	79.304	364.905	315.464	44.497	-1325.048	-1236.654	71.774
1000	79.983	373.298	320.835	52.463	-1338.497	-1225.921	64.036
1100	80.500	380.946	325.957	60.488	-1341.199	-1214.532	57.673
1200	80.902	387.969	330.836	68.559	-1343.903	-1202.898	52.361
1300	81.220	394.457	335.484	76.666	-1346.610	-1191.038	47.856
1400	81.475	400.486	339.914	84.801	-1349.326	-1178.969	43.988
1500	81.684	406.115	344.142	92.959	-1352.053	-1166.706	40.628
1600	81.856	411.392	348.182	101.137	-1354.791	-1154.260	37.683
1700	81.999	416.359	352.047	109.330	-1357.543	-1141.642	35.078
1800	82.120	421.049	355.752	117.536	-1360.309	-1128.863	32.759
1900	82.223	425.492	359.306	125.753	-1363.087	-1115.929	30.679
2000	82.311	429.712	362.722	133.980	-1365.876	-1102.848	28.803
2100	82.387	433.730	366.009	142.215	-1368.674	-1089.628	27.103
2200	82.454	437.564	369.175	150.457	-1371.479	-1076.274	25.554
2300	82.511	441.231	372.228	158.705	-1374.288	-1062.793	24.137
2400	82.562	444.743	375.177	166.959	-1377.098	-1049.190	22.835
2500	82.608	448.115	378.028	175.218	-1379.905	-1035.469	21.635
2600	82.648	451.355	380.786	183.480	-1382.707	-1021.637	20.525
2700	82.683	454.475	383.458	191.747	-1385.499	-1007.696	19.495
2800	82.715	457.483	386.048	200.017	-1682.180	-992.685	18.519
2900	82.744	460.386	388.562	208.290	-1683.849	-968.030	17.436
3000	82.770	463.192	391.003	216.566	-1685.502	-943.319	16.425
3100	82.794	465.906	393.376	224.844	-1687.134	-918.552	15.477
3200	82.815	468.535	395.684	233.124	-1688.746	-893.734	14.589
3300	82.835	471.084	397.930	241.407	-1690.334	-868.865	13.753
3400	82.852	473.557	400.118	249.691	-1691.897	-843.948	12.966
3500	82.869	475.959	402.251	257.977	-1693.435	-818.986	12.223
3600	82.884	478.293	404.331	266.265	-1694.947	-793.980	11.520
3700	82.898	480.564	406.361	274.554	-1696.432	-768.933	10.855
3800	82.910	482.775	408.343	282.844	-1697.891	-743.846	10.225
3900	82.922	484.929	410.279	291.136	-1699.322	-718.720	9.626
4000	82.933	487.029	412.171	299.429	-1700.727	-693.559	9.057
4100	82.943	489.077	414.022	307.723	-1702.106	-668.363	8.515
4200	82.953	491.075	415.833	316.017	-1703.460	-643.133	7.999
4300	82.961	493.027	417.606	324.313	-1704.789	-617.872	7.506
4400	82.970	494.935	419.342	332.610	-1706.096	-592.579	7.035
4500	82.977	496.799	421.042	340.907	-1707.377	-567.258	6.585
4600	82.984	498.623	422.709	349.205	-1708.639	-541.908	6.154
4700	82.991	500.408	424.343	357.504	-1709.882	-516.531	5.741
4800	82.997	502.155	425.946	365.803	-1711.107	-491.127	5.345
4900	83.003	503.867	427.519	374.103	-1712.314	-465.698	4.964
5000	83.009	505.544	429.063	382.404	-1713.508	-440.245	4.599
5100	83.014	507.188	430.579	390.705	-1714.690	-414.768	4.248
5200	83.019	508.800	432.068	399.007	-1715.861	-389.268	3.910
5300	83.024	510.381	433.530	407.309	-1717.023	-363.745	3.585
5400	83.028	511.933	434.968	415.611	-1718.180	-338.201	3.271
5500	83.032	513.456	436.381	423.914	-1719.332	-312.636	2.969
5600	83.036	514.953	437.771	432.218	-1720.483	-287.049	2.677
5700	83.040	516.422	439.138	440.522	-1721.634	-261.442	2.396
5800	83.043	517.867	440.483	448.826	-1722.785	-235.814	2.124
5900	83.047	519.286	441.806	457.130	-1723.945	-210.167	1.861
6000	83.050	520.682	443.109	465.435	-1725.089	-184.502	1.606

PREVIOUS: June 1976 (1 atm)   CURRENT: June 1976 (1 bar)

Aluminum Fluoride Oxide, Ion ($OAlF_2^-$) $Al_1F_2O_1^-$(g)

ALUMINUM FLUORIDE (AlF_3) CRYSTAL M_r = 83.976749

$S°(298.15\ K) = 66.48 \pm 0.42\ J\ K^{-1}\ mol^{-1}$
$T_{trs} = 728 \pm 1\ K$
$T_{fus} = [2523 \pm 100]\ K$

$\Delta_fH°(0\ K) = -1504.3 \pm 1.3\ kJ\ mol^{-1}$
$\Delta_fH°(298.15\ K) = -1510.4 \pm 1.3\ kJ\ mol^{-1}$
$\Delta_{trs}H° = 0.5632 \pm 0.042\ kJ\ mol^{-1}$
$\Delta_{fus}H° = [98.3 \pm 12]\ kJ\ mol^{-1}$

Enthalpy of Formation

The adopted value is that recommended by CODATA (1). It is based on separate calorimetric studies of the combustion in fluorine of aluminum (2) and aluminum-Teflon mixtures (3). More extensive auxiliary data (3) for Teflon bring the aluminum-Teflon study into exact agreement with the aluminum study (2). Calorimetric data relating AlF_3 to PbF_2 (2, 4) tend to confirm this value, but PbF_2 is probably less certain than is AlF_3.

Values for the $\Delta_fH°(298.15\ K)$ derived indirectly from equilibrium studies (5) and emf of solid galvanic cells (6, 7, 8, 9) are tabulated below. The original data are reanalyzed with the use of the most recent auxiliary data (4).

Source	Method	Reaction	T/K	$\Delta_rH°(298.15\ K)$, kcal mol^{-1} 2nd law	3rd law	Drift cal K^{-1} mol^{-1}	$\Delta_fH°(298.15\ K)$* kcal mol^{-1}
(5)	equilibria	A	849-1312	33.3±0.20	33.11±0.2	-0.2±0.2	-358.6
(6)	emf	B	873	--	-43.01	--	-360.0
(7)	emf	B	873	--	-42.73	--	-360.3
(8)	emf	B	720-860	-40.26±0.01	-41.47±0.2	-1.5±0.01	-361.6
(9)	emf	B	748-903	-42.1±0.01	-40.58±0.2	-1.8±0.02	-362.5

*Derived from the third law heat of reaction and is per mole of $AlF_3(cr)$.

A. $2/3\ AlF_3(cr) + H_2O(g) = 1/3\ Al_2O_3(\alpha,cr) + 2\ HF(g)$ B. $AlF_3(cr) + 1.5\ Mg(cr) = Al(cr) + 1.5\ MgF_2(cr)$

With the exception of the equilibrium study of Mashovets and Yudin (5), the results are in reasonable agreement with the directly measured calorimetric value. It has been suggested (2) that interaction of the condensed phases (activities≠1) caused the equilibria to deviate from the standard-state reaction assumed above. An alternative may be that the reaction with water vapor produced Al_2O_3 in a metastable form; such forms require ignition temperatures greater than 1400 K for complete conversion to α-Al_2O_3 (10).

Heat Capacity and Entropy

$C_p°$ and $S°(298.15\ K)$ are taken from Douglas and Ditmars (11), whose tables are based on heat capacity measurements from 54 to 298 K (12) and relative enthalpy measurements from 323 to 1173° (11). The enthalpy data confirm earlier data (401-1401 K) of O'Brien and Kelley (13) rather than the adjusted values proposed by Frank (14). Douglas and Ditmars (11) gave a detailed comparison with other data. The entropy is derived from $C_p°$ using the extrapolation $S°(50\ K) = 2.22 \pm 0.25\ J\ K^{-1}\ mol^{-1}$.

Transition Data

T_{trs} and $\Delta_{trs}H°$ for this crystalline first order transition are from Douglas and Ditmars (11). The values were obtained from extensive enthalpy data for the transition region assuming the presence of a very small amount of impurity in solid solution.

Sublimation Data

T_{sub} = 1549 K is the temperature at which the calculated total pressure (monomer plus dimer) reaches one atm; the mole fraction of dimer is calculated as 0.109 at T_{sub}. Sublimation temperatures of 1533 and 1566 K were obtained in two sublimation-pressure studies (15, 16) at the Techische Hochschule, Breslau. The selected enthalpies of sublimation to $AlF_3(g)$ and $Al_2F_6(g)$ reproduce closely the entrainment data of Krause and Douglas (17). Selection of these values of $\Delta_{sub}H°$ and comparisons with other vapor-pressure data are discussed on the gas-phase tables.

Fusion Data

See the liquid table for details.

References

1. ICSU-CODATA Task Group on Key Values for Thermodynamics, J. Chem. Thermodynamics 10, 903 (1978).
2. E. Rudzitis, H. M. Feder and W. N. Hubbard, Inorg. Chem. 6, 1716 (1967).
3. E. S. Domalski and G. T. Armstrong, J. Res. Nat. Bur. Stand. 71A, 105 (1967); 69A, 137 (1965).
4. JANAF Thermochemical Tables: $PbF_2(cr)$, 12-31-73; $H_2O(g)$, 3-31-79; $Al_2O_3(\alpha, cr)$, 12-31-79; HF(g), 6-30-77; Mg(cr), 9-30-62; and $MgF_2(cr)$, 6-30-75.
5. V. P. Mashovets and B. F. Yudin, Izv. Vyssh. Ucheb. Zaved., Tsvet. Met. 5, 95 (1962).
6. R. J. Heus and J. J. Egan, Z. Phys. Chem. (Frankfurt) 49, 38 (1966).
7. H. Tanaka, A. Yamaguchi and J. Moriyana, Nippon Kinz. Gakkaishi 35, 1161 (1971).
8. T. N. Rezukhina, T. F. Sisoeva, L. I. Holokhonova and E. G. Ippolitov, J. Chem. Thermodynamics 6, 883 (1974).
9. W. F. Skelton and J. W. Patterson, J. Less Common Metals 31, 47 (1973).
10. T. Yokokawa and O. J. Kleppa, J. Phys. Chem. 68, 3246 (1964).
11. T. B. Douglas and D. A. Ditmars, J. Res. Nat. Bur. Stand. 71A, 185 (1967).
12. E. G. King, J. Amer. Chem. Soc. 79, 2056 (1957).
13. C. J. O'Brien and K. K. Kelley, J. Amer. Chem. Soc. 79, 5616 (1957).
14. W. B. Frank, J. Phys. Chem. 65, 2081 (1961).
15. O. Ruff and L. LeBoucher, Z. anorg. allg. Chem. 219, 376 (1934).
16. W. Olbrich, Dissertation, Technische Hochschule, Breslau, 1928.
17. R. F. Krause and T. B. Douglas, J. Phys. Chem. 72, 475 (1968).

Aluminum Fluoride (AlF_3) $Al_1F_3(cr)$

Enthalpy Reference Temperature = T_r = 298.15 K   Standard State Pressure = p° = 0.1 MPa

T/K	$C_p°$ (J $K^{-1}mol^{-1}$)	S°	$-[G°-H°(T_r)]/T$	$H°-H°(T_r)$ (kJ mol^{-1})	$\Delta_fH°$	$\Delta_fG°$	Log K_f
0	0.	0.	INFINITE	-11.625	-1504.272	-1504.272	INFINITE
100	24.932	12.058	119.836	-10.778	-1508.269	-1483.221	774.755
200	56.877	39.991	72.814	-6.565	-1510.252	-1457.225	380.588
298.15	75.132	66.484	66.484	0.	-1510.424	-1431.123	250.727
300	75.400	66.949	66.485	0.139	-1510.416	-1430.631	249.095
400	86.287	90.293	69.594	8.280	-1509.613	-1404.138	183.362
500	92.274	110.245	75.781	17.232	-1508.342	-1377.911	143.949
600	97.311	127.494	82.993	26.700	-1506.819	-1351.963	117.699
700	105.357	143.032	90.474	36.791	-1504.904	-1326.297	98.970
728.000	108.692	147.228	92.577	39.786	ALPHA <--> BETA		
728.000	97.579	148.001	92.577	40.349	TRANSITION		
800	98.516	157.248	97.987	47.409	-1502.681	-1300.966	84.944
900	99.717	168.922	105.231	57.322	-1501.390	-1275.830	74.047
1000	100.830	179.487	112.137	67.350	-1510.813	-1250.072	65.297
1100	101.893	189.147	118.705	77.486	-1509.431	-1224.064	58.126
1200	102.918	198.057	124.951	87.727	-1507.984	-1198.185	52.156
1300	103.918	206.334	130.897	98.069	-1506.471	-1172.429	47.109
1400	104.897	214.071	136.564	108.510	-1504.892	-1146.793	42.787
1500	105.864	221.341	141.976	119.048	-1503.244	-1121.271	39.046
1600	106.822	228.204	147.153	129.682	-1501.528	-1095.862	35.776
1700	107.771	234.709	152.114	140.412	-1499.741	-1070.562	32.894
1800	108.717	240.896	156.876	151.236	-1497.882	-1045.369	30.336
1900	109.663	246.799	161.454	162.155	-1495.946	-1020.283	28.050
2000	110.604	252.448	165.864	173.169	-1493.929	-995.299	25.995
2100	111.545	257.867	170.117	184.276	-1491.827	-970.419	24.138
2200	112.487	263.078	174.225	195.478	-1489.635	-945.641	22.452
2300	113.428	268.099	178.198	206.773	-1487.347	-920.965	20.916
2400	114.370	272.946	182.045	218.163	-1484.957	-896.391	19.509
2500	115.311	277.634	185.775	229.647	-1482.460	-871.918	18.218
2523.000	115.529	278.691	186.618	232.302	------- BETA <--> LIQUID -------		
2600	116.252	282.175	189.396	241.225	-1479.850	-847.547	17.027
2700	117.194	286.580	192.914	252.898	-1477.122	-823.279	15.927
2800	118.135	290.859	196.336	264.664	-1768.173	-798.147	14.890
2900	119.077	295.021	199.668	276.525	-1764.101	-763.575	13.753
3000	120.018	299.074	202.914	288.480	-1759.898	-729.145	12.696

PREVIOUS: June 1970   CURRENT: September 1979

Aluminum Fluoride (AlF_3) $Al_1F_3(cr)$

ALUMINUM FLUORIDE (AlF_3) LIQUID, p ~ 4340 bars M_r = 83.976749

$S°(298.15\ K) = [95.379]\ J\ K^{-1}\ mol^{-1}$ $\Delta Hf_{298.15} = [-1422.760]\ kJ\ mol^{-1}$

$T_{fus} = [2523 \pm 100]\ K$ $\Delta_{fus}H° = 98.3 \pm 12\ kJ\ mol^{-1}$

Enthalpy of Formation

$\Delta_f H°(AlF_3, \ell, 298.15\ K)$ is calculated from $\Delta_f H°(AlF_3, cr, 298.15\ K)$ by adding the enthalpy of fusion, $\Delta_{fus}H°$, and the difference in enthalpy, H°(2523 K)-H°(298.15 K), between the crystal and liquid.

Heat Capacity and Entropy

There are no heat capacity or enthalpy measurements covering the liquid region of AlF_3. The pure solid does not melt at ordinary pressures. This $AlF_3(\ell)$ table, therefore, is a non-standard state table for P = ~4400 atm. By analogy (1) with $AlCl_3(\ell)$, $AlBr_3(\ell)$, and $AlI_3(\ell)$, we estimate $Cp(\ell) = 30\ cal\ K^{-1}\ mol^{-1}$; Holm (2, 3) also assumed the same value. We assume a glass transition at 1850 K, below which we adopt the heat capacity data of $AlF_3(cr)$; the transition at 728 K for $AlF_3(cr)$ is not included.

The entropy is calculated in a manner analogous to that used for the enthalpy of formation.

Fusion Data

The pure solid sublimes before melting at ordinary pressures. From studies on CaF_2-AlF_3 and NaCl-AlF_3 binary systems, Millet et al. (4) estimated a triple point temperature for AlF_3 of 2523 K at 30000 atm. Our adopted values lead to a pressure for $AlF_3(\ell) = AlF_3(g)$ of ~2200 atm. The calculated dimer pressure at this temperature is also ~2200 atm.

Holm (3), attempted to relate his drop-calorimetric data on the melting of a Li_3AlF_6-AlF_3 system to the enthalpy of fusion of AlF_3. Based on the assumption that the enthalpy of mixing of $Li_3AlF_6(\ell)$ with $AlF_3(\ell)$ to form the eutectic mixture is small and positive, he obtained $\Delta_{fus}H° = 11.4\ kcal\ mol^{-1}$ for AlF_3 at 1298 K. This result was challenged by Hong and Kleppa (4, 5) who measured the enthalpy of mixing of $AlF_3(cr)$ with a series of liquid alkali fluorides and derived $\Delta_{fus}H° = 26.5 \pm 2\ kcal\ mol^{-1}$ at this temperature. These results suggest that Holm's assumed enthalpy of mixing is in error, both as to sign and magnitude. We adopt the $\Delta_{fus}H°$ value of Hong and Kleppa (5, 6) but adjust the value to the adopted triple point of 2523 K (from 1298 K).

References

1. JANAF Thermochemical Tables: $AlBr_3(\ell)$, $AlCl_3(\ell)$, $AlI_3(\ell)$, and $AlF_3(cr)$, 9-30-79.
2. K. Grjotheim, J. L. Holm, and B. J. Welch, Acta Chem. Scand. 25, 2370 (1971).
3. J. L. Holm, High Temp. Sci. 6, 16 (1974).
4. J. P. Millet, H. Pham and M. Rolin, Rev. Int. Hautes Temp. Refract. 11, 277 (1974).
5. K. C. Hong and O. J. Kleppa, High Temp. Sci. 8, 299 (1976).
6. K. C. Hong and O. J. Kleppa, J. Phys. Chem. 82, 176 (1978).

Aluminum Fluoride (AlF_3) $Al_1F_3(l)$

Enthalpy Reference Temperature = T_r = 298.15 K Standard State Pressure = p° = 0.1 MPa

T/K	$C_p°$ (J K⁻¹mol⁻¹)	S° (J K⁻¹mol⁻¹)	$-[G°-H°(T_r)]/T$ (J K⁻¹mol⁻¹)	$H°-H°(T_r)$ (kJ mol⁻¹)	$\Delta_f H°$ (kJ mol⁻¹)	$\Delta_f G°$ (kJ mol⁻¹)	Log K_f
0							
100							
200							
298.15	97.475	95.379	95.379	0.	-1422.760	-1352.074	236.878
300	97.479	95.982	95.381	0.180	-1422.711	-1351.636	235.341
400	97.763	124.065	99.208	9.943	-1420.286	-1328.320	173.461
500	98.014	145.908	106.444	19.732	-1418.178	-1305.578	136.393
600	98.224	163.797	114.556	29.544	-1416.311	-1283.237	111.716
700	98.391	178.951	122.701	39.375	-1414.655	-1261.191	94.111
800	98.516	192.086	130.571	49.212	-1413.214	-1239.369	80.922
900	99.717	203.759	138.066	59.124	-1411.923	-1217.717	70.674
1000	100.830	214.324	145.172	69.152	-1421.346	-1195.442	62.444
1100	101.893	223.984	151.904	79.289	-1419.964	-1172.918	55.697
1200	102.918	232.894	158.286	89.529	-1418.518	-1150.523	50.081
1300	103.918	241.172	164.348	99.871	-1417.005	-1128.251	45.334
1400	104.897	248.909	170.114	110.312	-1415.425	-1106.098	41.269
1500	105.864	256.179	175.612	120.850	-1413.777	-1084.060	37.750
1600	106.822	263.042	180.864	131.485	-1412.061	-1062.135	34.675
1700	107.771	269.546	185.891	142.214	-1410.275	-1040.319	31.965
1800	108.717	275.733	190.712	153.039	-1408.415	-1018.610	29.559
1850.000	109.190	278.718	193.050	158.486	GLASS <--> LIQUID TRANSITION		
1850.000	125.520	278.718	193.050	158.486			
1900	125.520	282.066	195.349	164.762	-1405.674	-997.017	27.410
2000	125.520	288.504	199.847	177.314	-1402.119	-975.601	25.480
2100	125.520	294.628	204.216	189.866	-1398.573	-954.363	23.738
2200	125.520	300.467	208.459	202.418	-1395.030	-933.293	22.159
2300	125.520	306.047	212.582	214.970	-1391.485	-912.384	20.721
2400	125.520	311.389	216.588	227.522	-1387.933	-891.629	19.406
2500	125.520	316.513	220.483	240.074	-1384.368	-871.023	18.199
2523.000	125.520	317.662	221.364	242.961	------- BETA <--> LIQUID -------		
2600	125.520	321.436	224.272	252.626	-1380.785	-850.560	17.088
2700	125.520	326.173	227.959	265.178	-1377.177	-830.235	16.062
2800	125.520	330.738	231.549	277.730	-1667.443	-809.077	15.094
2900	125.520	335.143	235.045	290.282	-1662.679	-778.505	14.022
3000	125.520	339.398	238.453	302.834	-1657.878	-748.098	13.026
3100	125.520	343.514	241.776	315.386	-1653.038	-717.852	12.096
3200	125.520	347.499	245.018	327.938	-1648.155	-687.762	11.227
3300	125.520	351.361	248.182	340.490	-1643.228	-657.826	10.413
3400	125.520	355.108	251.272	353.042	-1638.253	-628.040	9.649
3500	125.520	358.747	254.291	365.594	-1633.231	-598.401	8.931
3600	125.520	362.283	257.242	378.146	-1628.159	-568.905	8.255
3700	125.520	365.722	260.128	390.698	-1623.038	-539.552	7.617
3800	125.520	369.069	262.951	403.250	-1617.867	-510.338	7.015
3900	125.520	372.330	265.714	415.802	-1612.647	-481.260	6.446
4000	125.520	375.508	268.419	428.354	-1607.378	-452.317	5.907

PREVIOUS: CURRENT: September 1979

Aluminum Fluoride (AlF_3) $Al_1F_3(l)$

0 to 2523 K crystal
above 2523 K liquid*, p = ~4340 bars

*The liquid does not exist at 1 bar.

Refer to the individual tables for details.

Enthalpy Reference Temperature = T_r = 298.15 K   Standard State Pressure = p° = 0.1 MPa

T/K	C_p° (J K⁻¹mol⁻¹)	S° (J K⁻¹mol⁻¹)	$-[G^\circ-H^\circ(T_r)]/T$ (J K⁻¹mol⁻¹)	$H^\circ-H^\circ(T_r)$ (kJ mol⁻¹)	Δ_fH° (kJ mol⁻¹)	Δ_fG° (kJ mol⁻¹)	Log K_f
0	0.	0.	INFINITE	-11.625	-1504.272	-1504.272	INFINITE
100	24.932	12.058	119.836	-10.778	-1508.269	-1483.221	774.755
200	56.877	39.991	72.814	-6.565	-1510.252	-1457.225	380.588
298.15	75.132	66.484	66.484	0.	-1510.424	-1431.123	250.727
300	75.400	66.949	66.485	0.139	-1510.416	-1430.631	249.095
400	86.287	90.293	69.594	8.280	-1509.613	-1404.138	183.362
500	92.274	110.245	75.781	17.232	-1508.342	-1377.911	143.949
600	97.311	127.494	82.993	26.700	-1506.819	-1351.963	117.699
700	105.357	143.032	90.474	36.791	-1504.904	-1326.297	98.970
728.000	108.692	147.228	92.577	39.786	ALPHA <--> BETA TRANSITION		
728.000	97.579	148.001	92.577	40.349			
800	98.516	157.248	97.987	47.409	-1502.681	-1300.966	84.944
900	99.717	168.922	105.231	57.322	-1501.390	-1275.830	74.047
1000	100.830	179.487	112.137	67.350	-1510.813	-1250.072	65.297
1100	101.893	189.147	118.705	77.486	-1509.431	-1224.064	58.126
1200	102.918	198.057	124.951	87.727	-1507.984	-1198.185	52.156
1300	103.918	206.334	130.897	98.069	-1506.471	-1172.429	47.109
1400	104.897	214.071	136.564	108.510	-1504.892	-1146.793	42.787
1500	105.864	221.341	141.976	119.048	-1503.244	-1121.271	39.046
1600	106.822	228.204	147.153	129.682	-1501.528	-1095.862	35.776
1700	107.771	234.709	152.114	140.412	-1499.741	-1070.562	32.894
1800	108.717	240.896	156.876	151.236	-1497.882	-1045.369	30.336
1900	109.663	246.799	161.454	162.155	-1495.946	-1020.283	28.050
2000	110.604	252.448	165.864	173.169	-1493.929	-995.299	25.995
2100	111.545	257.867	170.117	184.276	-1491.827	-970.419	24.138
2200	112.487	263.078	174.225	195.478	-1489.635	-945.641	22.452
2300	113.428	268.099	178.198	206.773	-1487.347	-920.965	20.916
2400	114.368	272.946	182.045	218.163	-1484.957	-896.391	19.509
2500	115.311	277.634	185.775	229.647	-1482.460	-871.918	18.218
2523.000	115.529	278.691	186.618	232.302	BETA <--> LIQUID TRANSITION		
2523.000	125.520	317.662	186.618	330.626			
2600	125.520	321.436	190.555	340.291	-1380.785	-850.560	17.088
2700	125.520	326.173	195.490	352.843	-1377.177	-830.235	16.062
2800	125.520	330.738	200.240	365.395	-1667.443	-809.076	15.094
2900	125.520	335.142	204.816	377.947	-1662.679	-778.504	14.022
3000	125.520	339.398	209.231	390.499	-1657.878	-748.097	13.026
3100	125.520	343.513	213.497	403.051	-1653.038	-717.851	12.096
3200	125.520	347.499	217.623	415.603	-1648.155	-687.761	11.227
3300	125.520	351.361	221.617	428.155	-1643.228	-657.825	10.413
3400	125.520	355.108	225.489	440.707	-1638.253	-628.039	9.649
3500	125.520	358.747	229.244	453.259	-1633.231	-598.400	8.931
3600	125.520	362.283	232.891	465.811	-1628.159	-568.904	8.255
3700	125.520	365.722	236.435	478.363	-1623.038	-539.551	7.617
3800	125.520	369.069	239.881	490.915	-1617.867	-510.337	7.015
3900	125.520	372.330	243.236	503.467	-1612.647	-481.259	6.446
4000	125.520	375.508	246.503	516.019	-1607.378	-452.316	5.907

PREVIOUS: CURRENT: September 1979

ALUMINUM FLUORIDE (AlF_3) IDEAL GAS M_r = 83.976749

$S°(298.15 K) = 276.69 \pm 0.8 J K^{-1} mol^{-1}$

$\Delta_f H°(0 K) = -1205.60 \pm 2.5 kJ mol^{-1}$
$\Delta_f H°(298.15 K) = -1209.32 \pm 2.5 kJ mol^{-1}$

Vibrational Frequencies and Degeneracies

ν, cm^{-1}	ν, cm^{-1}
[650](1)	[935](2)
297 (1)	263 (2)

Ground State Quantum Weight: [1] $\sigma = 6$
Point Group: D_{3h}
Bond Distance: Al-F = 1.63 ± 0.01 Å
Bond Angle: F-Al-F = 120°
Product of the Moments of Inertia: $I_A I_B I_C = 3.975102 \times 10^{-114} g^3 cm^6$

Enthalpy of Formation

The enthalpy of formation is calculated from that of the crystal by use of $\Delta_{sub}H°(298.15 K) = 71.965 \pm 0.5$ kcal/mol. Sublimation studies by mass-spectrometric and entrainment techniques suggest the presence of a small amount of dimer in the saturated vapor. These studies are reviewed on the table for Al_2F_6(g). The heats of sublimation for the monomer and dimer are selected by adopting the JANAF entropies for crystal, monomer and dimer, the mass spectrometric value of 2 $\Delta_{sub}H°$(monomer) – $\Delta_{sub}H°$(dimer) = -49 kcal mol^{-1} at 1000 K, and the entrainment data of Krause and Douglas (1). The resulting calculated pressures deviate by <0.5% from the entrainment data (1), by <10% from the mass-spectrometric equation (3) and by <3% from the mass-effusion equation (4).

Further comparisons with experimental data are given below. The adopted values favor the later (11) of two sublimation-point studies (11, 12); both show considerable scatter and similar deviations from the calculated dP/dT. Torsion-effusion data suggest the possibility of more serious discrepancy. Although data of Witt and Barrow (9) are consistent with the JANAF pressures, Hildenbrand et al. (7, 8, 19) noted a trend toward increasing pressure with decreasing orifice area. The most recent study by Hildenbrand and Laughlin (19) yielded equilibrium pressures that are reasonably consistent with the adopted values, unlike the earlier study (7). The 1981 communication (19) yielded a heat of sublimation value which is within 0.25 kcal/mol of our adopted value. In contrast, the JANAF predictions agree closely with mass-effusion data of Ko et al. (4), who reported no significant effect from a twelvefold variation in orifice area.

Comparison of Calculated and Observed Total Pressures Over AlF_3(cr)

Source	Method	T/K	Mole Fraction Dimer Calc. Range	P_{calc}/P_{obs} Range	Difference[a] in dP/dT
(1)	Entrainment	1194-1258	0.019-0.029	1.004-0.996	Very small (-)
(2)	Entrainment	1108-1273	0.010-0.032	0.73 -2.22	Uncertain (?)
(3)	Mass Spec[b]	983-1165	0.003-0.016	1.09 -0.91	Small (-)
(4)	Mass Effusion	1027-1184	0.005-0.018	0.994-0.972	Very small (-)
(5)	Mass Effusion	889-1006	0.001-0.004	1.42 -1.66	Uncertain (+)
(6)	Mass Effusion	980-1123	0.003-0.012	1.64 -0.88	Large (-)
(19)	Torsion Effusion	902-1098	0.001-0.010	0.88	Small
(8)	Torsion Effusion				
	Cell Gr-7	992-1101	0.004-0.010	0.89 -1.01	Small (+)
	Cell Gr-9	938-1080	0.002-0.008	0.94 -1.34	Medium (+)
(9)	Torsion Effusion	955-1063	0.003-0.007	1.11 -1.07	Small (-)
(10)	Dynamic B.P.	1335-1453	0.044-0.076	0.84 -1.64	Large (+)
(11)	Dynamic B.P.	1367-1524	0.052-0.100	1.33 -0.79	Large (-)
(12)	Dynamic B.P.	1371-1567	0.053-0.116	2.70 -1.28	Large (-)

[a]Magnitude and sign of the difference, calculated minus observed, in dP/dT
[b]Absolute pressures of monomer were obtained by integration of the ion intensity during total sublimation.

Heat Capacity and Entropy

The molecular structure is that derived from electron-diffraction data by Akishin et al. (13). Observed vibrational frequencies are those reported by Buchler et al. (14) from infrared spectra of the vapor at 1000-1200 K. These frequencies are confirmed by infrared spectra of AlF_3 isolated in matrices of neon, argon and krypton (15, 16). Drake and Rosenblatt (17) have estimated the Raman-active fundamentals based on a correlation of vibrational frequencies and internuclear distances for the Group IIIA trihalides. They estimated $\nu_1 = 685 cm^{-1}$ which would decrease S°(298.15 K) by 0.05 cal $K^{-1} mol^{-1}$ from our adopted value. We adopt $\nu_1 = 650 \pm 50 cm^{-1}$ as derived from three other methods: (a) calculation of the force constant k_1 from ν_3 and ν_4; (b) comparison of $k_1(XY_3)/k(XY)$ for X = B, Al and Y = F, Cl, and Br; and (c) extrapolation of $\nu_1(BY_3)/\nu_1(AlY_3)$ to Y=F from Y=Cl, Br and I, using recent gas-phase Raman data (18) for AlY_3. The principal moments of inertia are: $I_A = I_B = 12.5730 \times 10^{-39}$ and $I_C = 25.1460 \times 10^{-39} g cm^2$.

References

References are listed on the Al_2F_6(g) table.

Aluminum Fluoride (AlF_3) Al_1F_3(g)

Enthalpy Reference Temperature = T_r = 298.15 K — Standard State Pressure = p° = 0.1 MPa

T/K	$C_p°$ (J K⁻¹mol⁻¹)	S° (J K⁻¹mol⁻¹)	$-[G°-H°(T_r)]/T$ (J K⁻¹mol⁻¹)	$H°-H°(T_r)$ (kJ mol⁻¹)	$\Delta_f H°$ (kJ mol⁻¹)	$\Delta_f G°$ (kJ mol⁻¹)	Log K_f
0	0.	0.	INFINITE	-14.049	-1205.595	-1205.595	INFINITE
100	41.148	220.719	325.981	-10.526	-1206.916	-1202.734	628.244
200	53.850	253.541	282.132	-5.718	-1208.303	-1197.987	312.882
250	58.444	266.064	277.695	-2.908	-1208.861	-1195.342	249.753
298.15	62.239	276.691	276.691	0.	-1209.322	-1192.695	208.955
300	62.373	277.076	276.692	0.115	-1209.339	-1192.592	207.649
350	65.670	286.947	277.464	3.319	-1209.754	-1189.767	177.563
400	68.387	295.899	279.218	6.673	-1210.119	-1186.887	154.991
450	70.606	304.086	281.533	10.149	-1210.446	-1183.963	137.431
500	72.415	311.622	284.170	13.726	-1210.746	-1181.004	123.379
600	75.107	325.079	289.894	21.111	-1211.307	-1175.002	102.293
700	76.949	336.804	295.776	28.720	-1211.873	-1168.907	87.225
800	78.245	347.169	301.565	36.483	-1212.505	-1162.727	75.918
900	79.185	356.442	307.156	44.357	-1213.254	-1156.461	67.119
1000	79.884	364.822	312.511	52.312	-1224.749	-1149.344	60.036
1100	80.416	372.462	317.619	60.328	-1225.488	-1141.768	54.218
1200	80.830	379.478	322.485	68.391	-1226.218	-1134.124	49.367
1300	81.158	385.961	327.122	76.491	-1226.947	-1126.420	45.260
1400	81.421	391.986	331.542	84.621	-1227.679	-1118.660	41.738
1500	81.636	397.611	335.761	92.774	-1228.416	-1110.847	38.683
1600	81.814	402.885	339.794	100.947	-1229.162	-1102.985	36.009
1700	81.962	407.850	343.652	109.136	-1229.916	-1095.076	33.648
1800	82.087	412.538	347.350	117.338	-1230.678	-1087.122	31.547
1900	82.193	416.979	350.899	125.552	-1231.447	-1079.126	29.667
2000	82.284	421.198	354.309	133.776	-1232.220	-1071.089	27.974
2100	82.362	425.214	357.591	142.009	-1232.993	-1063.014	26.441
2200	82.431	429.047	360.752	150.248	-1233.763	-1054.901	25.047
2300	82.491	432.713	363.802	158.495	-1234.524	-1046.754	23.773
2400	82.543	436.225	366.747	166.746	-1235.272	-1038.574	22.604
2500	82.590	439.595	369.594	175.003	-1236.002	-1030.363	21.528
2600	82.631	442.835	372.349	183.264	-1236.710	-1022.124	20.535
2700	82.668	445.955	375.018	191.529	-1237.389	-1013.857	19.614
2800	82.701	448.962	377.605	199.798	-1531.938	-1004.599	18.741
2900	82.731	451.864	380.116	208.069	-1531.455	-985.773	17.756
3000	82.758	454.669	382.555	216.344	-1530.932	-966.966	16.836
3100	82.782	457.383	384.925	224.621	-1530.367	-948.176	15.977
3200	82.804	460.012	387.231	232.900	-1529.756	-929.405	15.171
3300	82.824	462.560	389.475	241.182	-1529.099	-910.654	14.414
3400	82.843	465.033	391.661	249.465	-1528.393	-891.924	13.703
3500	82.860	467.435	393.792	257.750	-1527.638	-873.215	13.032
3600	82.875	469.769	395.870	266.037	-1526.832	-854.528	12.399
3700	82.889	472.040	397.898	274.325	-1525.974	-835.865	11.800
3800	82.903	474.251	399.879	282.615	-1525.066	-817.226	11.234
3900	82.915	476.404	401.813	290.906	-1524.107	-798.611	10.696
4000	82.926	478.504	403.704	299.198	-1523.097	-780.021	10.186
4100	82.936	480.552	405.554	307.491	-1522.038	-761.457	9.701
4200	82.946	482.550	407.363	315.785	-1520.931	-742.919	9.240
4300	82.955	484.502	409.135	324.080	-1519.776	-724.409	8.800
4400	82.964	486.409	410.869	332.376	-1518.576	-705.926	8.380
4500	82.972	488.274	412.569	340.673	-1517.328	-687.471	7.980
4600	82.979	490.098	414.234	348.970	-1516.040	-669.044	7.597
4700	82.986	491.882	415.868	357.268	-1514.710	-650.645	7.231
4800	82.992	493.629	417.470	365.567	-1513.341	-632.275	6.881
4900	82.999	495.341	419.041	373.867	-1511.935	-613.934	6.545
5000	83.004	497.018	420.584	382.167	-1510.493	-595.622	6.222
5100	83.010	498.661	422.099	390.468	-1509.018	-577.339	5.913
5200	83.015	500.273	423.587	398.769	-1507.513	-559.085	5.616
5300	83.020	501.855	425.049	407.071	-1505.978	-540.861	5.331
5400	83.024	503.406	426.486	415.373	-1504.417	-522.666	5.056
5500	83.028	504.930	427.898	423.676	-1502.831	-504.500	4.791
5600	83.033	506.426	429.287	431.979	-1501.223	-486.363	4.537
5700	83.036	507.896	430.653	440.282	-1499.595	-468.255	4.291
5800	83.040	509.340	431.997	448.586	-1497.947	-450.176	4.054
5900	83.044	510.759	433.320	456.890	-1496.285	-432.125	3.826
6000	83.047	512.155	434.623	465.195	-1494.588	-414.105	3.605

PREVIOUS: September 1979 (1 atm) CURRENT: September 1979 (1 bar)

Aluminum Fluoride (AlF_3) Al_1F_3(g)

TETRAFLUOROALUMINATE, ION (AlF_4^-) IDEAL GAS M_r = 102.975701

$S°(298.15\ K) = 294.75 \pm 2.1\ J\ K^{-1}\ mol^{-1}$

$\Delta_f H°(0\ K) = -1980 \pm 105\ kJ\ mol^{-1}$
$\Delta_f H°(298.15\ K) = [-1991.584]\ kJ\ mol^{-1}$

Vibrational Frequencies and Degeneracies

ν, cm^{-1}	ν, cm^{-1}
622(1)	760(3)
210(2)	322(3)

Ground State Quantum Weight: [1] σ = 12
Point Group: T_d
Bond Distance: Al-F = 1.69 Å
Bond Angle: F-Al-F = 109.4712°
Product of the Moments of Inertia: $I_A I_B I_C = 1.387252 \times 10^{-113}\ g^3\ cm^6$

Enthalpy of Formation

We adopt $\Delta_f H°(298.15\ K) = -476 \pm 25\ kcal\ mol^{-1}$ which corresponds to the fluoride-ion affinity $IA(AlF_3) = 125 \pm 25\ kcal\ mol^{-1}$ for $AlF_4^-(g) = AlF_3(g) + F^-(g)$. $\Delta_f H°$ is a compromise based mainly on Kp data (7 points, 1130-1485 K) for the reaction $2\ AlF_2(g) + AlF_2^-(g) = 2\ AlF(g) + AlF_4^-(g)$. Kp data were measured by Srivastava et al. (1) with a mass spectrometer operating in both positive- and negative-ion modes. The authors used a molecular-flow-effusion technique to react vapors from AlF_3 and K_3AlF_6 or KF with Al(ℓ). Analysis with JANAF auxiliary data (2) yields $\Delta_r H°(298.15\ K)(3rd\ law) = -38.2 \pm 2.7$ and $\Delta_r H°(298.15\ K)(2nd\ law) = -33.9 \pm 6.3\ kcal\ mol^{-1}$ with $[\Delta_r S°(2nd\ law) - \Delta_r S°(3rd\ law)] = -3.3 \pm 4.8\ cal\ K^{-1}\ mol^{-1}$. The 3rd law $\Delta_r H°$ yields $\Delta_f H°(AlF_4^-, g, 298.15\ K) = -460.4 \pm 14\ kcal\ mol^{-1}$ or $-502.4 \pm 14\ kcal\ mol^{-1}$, depending on the choice of $\Delta_f H°(AlF_2, g, 298.15\ K) = -166 \pm 10$ (2) or $-180 \pm 5\ kcal\ mol^{-1}$ (2, 3).

The difference of 14 kcal mol^{-1} in $\Delta_f H°(AlF_2)$ causes a three-fold change (42 kcal mol^{-1}) in $\Delta_f H°(AlF_4^-)$ and increases the fluoride-ion affinity $IA(AlF_3)$ from 109±25 to 151±15 kcal mol^{-1}. The latter value seems rather large in comparison to related species (2, 1): ≤112 or ≤127(OAlF), 92 or 106 (AlF), ≤101 (BF_3), ≤92 (OBF) and 98 or 103 (BF).

By combining published data for $NaAlF_4(g)$ with a coulomb-energy calculation, Holm (4) estimated $IA(AlF_3) = 131\ kcal\ mol^{-1}$. This result has an unknown bias due to limitations of the calculational model and auxiliary data. Refined lattice energy calculations (5) yield halide-ion affinities IA = 87±7 ($AlCl_3$) and ($AlBr_3$) kcal mol^{-1}. $\Delta_f H°(BF_4^-) = -421\ kcal\ mol^{-1}$ (6) derived from lattice energy yields $IA(BF_3) = 90$ compared to ≤101 kcal mol^{-1} from mass spectrometry. Comparison of stretching force constants (2, 7, 8) with average bond energies in AlX_3 and AlX_4^- suggests a slight, but not definitive, preference for $IA(AlF_3)$ = 109 kcal mol^{-1}. As a compromise, we adopt $IA(AlF_3) = 125 \pm 25$ and $\Delta_f H°(AlF_4^-, g, 298.15\ K) = -476 \pm 25\ kcal\ mol^{-1}$.

Heat Capacity and Entropy

Vibrational frequencies are from Raman spectra of AlF_4^- observed in molten salt mixtures of AlF_3 with NaF or KF (8). Similar spectra were observed in mixtures of AlF_3 with LiF (9). The spectra (8) are consistent with T_d symmetry except for some asymmetry in ν_3 which was attributed to influence of the metal ion. By analogy with SiF_4(2), we assume a singlet electronic ground state and neglect excited states. The bond distance is that derived from high-temperature electron diffraction (10) of $NaAlF_4(g)$. Within experimental uncertainty, the AlF_4 grouping was found to be tetrahedral (2, 10). The principal moments of inertia are $I_A = I_B = I_C = 24.0280 \times 10^{-39}\ g\ cm^2$.

References

1. R. D. Srivastava, O. M. Uy and M. Farber, J. Chem. Soc., Faraday Trans. I, 70, 1033 (1974); AD-731303, 50 pp. (1971).
2. JANAF Thermochemical Tables: $AlF_2(g)$, $AlF_2^-(g)$, $AlF_2O^-(g)$, and $SiF_4(g)$, 6-30-76; AlF(g), 6-30-76; AlF(g), 6-30-75; $NaAlF_4(g)$, 6-30-70.
3. O. M. Uy, R. D. Srivastava and M. Farber, High Temp. Sci. 4, 227 (1972).
4. J. L. Holm, Acta Chem. Scand. 27, 1410 (1973).
5. R. C. Gearhart, J. D. Beck and R. H. Wood, Inorg. Chem. 14, 2413 (1975).
6. F. Kutek, Sb. Vys. Sk. Chem.-Technol. Praze, Anorg. Chem. Technol. B14, 77 (1972); Chem. Abstr. 79, 35721 (1973).
7. V. G. Solomonik and K. S. Krasnov, Zh. Priklad. Spektrosk. 21, 360 (1974).
8. B. Gilbert, G. Mamantov and G. M. Begun, Inorg. Nucl. Chem. Lett. 10, 1123 (1974); J. Chem. Phys. 62, 950 (1975).
9. E. Rytter and S. K. Ratkje, Acta Chem. Scand. A29, 565 (1975).
10. V. P. Spiridonov and E. V. Erokhin, Dokl. Akad. Nauk 180 (1), 161 (1968); Proc. Acad. Sci. USSR (English Transl.) 180, 335 (1968).

Tetrafluoroaluminate, Ion (AlF_4^-) $Al_1F_4^-(g)$

Enthalpy Reference Temperature = T_r = 298.15 K; Standard State Pressure = p° = 0.1 MPa

	J $K^{-1}mol^{-1}$			kJ mol^{-1}			
T/K	$C°_p$	S°	$-[G°-H°(T_r)]/T$	$H°-H°(T_r)$	$\Delta_f H°$	$\Delta_f G°$	Log K_f
0	0.	0.	INFINITE	-16.981	-1980.178		
100	46.882	225.025	357.855	-13.283			
200	68.417	264.662	301.864	-7.440			
250	76.195	280.796	296.068	-3.818			
298.15	82.157	294.747	294.747	0.	-1991.584	-1943.855	340.555
300	82.359	295.256	294.749	0.152	-1991.631	-1943.558	338.404
350	87.157	308.328	295.770	4.395	-1992.840	-1935.449	288.850
400	90.872	320.219	298.094	8.850	-1993.959	-1927.173	251.663
450	93.760	331.096	301.166	13.469	-1995.016	-1918.761	222.724
500	96.025	341.097	304.666	18.216	-1996.036	-1910.233	199.561
600	99.261	358.912	312.259	27.992	-1998.020	-1892.886	164.790
700	101.387	374.383	320.053	38.031	-2000.010	-1875.206	139.930
800	102.844	388.022	327.714	48.247	-2002.073	-1857.238	121.265
900	103.881	400.199	335.103	58.586	-2004.262	-1839.003	106.733
1000	104.643	411.185	342.171	69.014	-2017.208	-1819.757	95.054
1100	105.218	421.187	348.907	79.508	-2019.407	-1799.905	85.470
1200	105.661	430.362	355.318	90.053	-2021.607	-1779.853	77.475
1300	106.010	438.834	361.420	100.637	-2023.815	-1759.618	70.702
1400	106.289	446.700	367.234	111.253	-2026.035	-1739.212	64.891
1500	106.516	454.042	372.779	121.893	-2028.269	-1718.646	59.849
1600	106.702	460.922	378.075	132.555	-2030.519	-1697.932	55.432
1700	106.858	467.396	383.141	143.233	-2032.785	-1677.075	51.530
1800	106.989	473.507	387.993	153.925	-2035.066	-1656.086	48.058
1900	107.100	479.295	392.648	164.630	-2037.359	-1634.969	44.948
2000	107.195	484.791	397.119	175.345	-2039.660	-1613.731	42.146
2100	107.277	490.023	401.419	186.069	-2041.964	-1592.378	39.608
2200	107.348	495.015	405.561	196.800	-2044.265	-1570.915	37.298
2300	107.411	499.789	409.555	207.538	-2046.557	-1549.347	35.187
2400	107.465	504.361	413.410	218.282	-2048.834	-1527.681	33.249
2500	107.514	508.749	417.137	229.031	-2051.088	-1505.920	31.464
2600	107.557	512.967	420.742	239.784	-2053.312	-1484.069	29.815
2700	107.595	517.027	424.233	250.542	-2055.501	-1462.33	28.287
2800	107.629	520.940	427.618	261.303	-2351.550	-1439.150	26.848
2900	107.660	524.718	430.901	272.068	-2352.555	-1406.546	25.335
3000	107.688	528.368	434.090	282.835	-2353.508	-1373.909	23.922
3100	107.713	531.899	437.188	293.605	-2354.406	-1341.241	22.600
3200	107.736	535.320	440.201	304.378	-2355.244	-1308.544	21.360
3300	107.757	538.635	443.134	315.152	-2356.020	-1275.823	20.195
3400	107.776	541.852	445.991	325.929	-2356.732	-1243.079	19.098
3500	107.794	544.977	448.774	336.708	-2357.377	-1210.314	18.063
3600	107.810	548.014	451.489	347.488	-2357.955	-1177.533	17.086
3700	107.824	550.968	454.138	358.269	-2358.465	-1144.736	16.161
3800	107.838	553.843	456.724	369.053	-2358.907	-1111.927	15.285
3900	107.851	556.645	459.250	379.837	-2359.282	-1079.106	14.453
4000	107.862	559.375	461.720	390.623	-2359.588	-1046.277	13.663
4100	107.873	562.039	464.134	401.409	-2359.828	-1013.441	12.911
4200	107.883	564.638	466.496	412.197	-2360.003	-980.600	12.196
4300	107.892	567.177	468.808	422.986	-2360.114	-947.756	11.513
4400	107.901	569.658	471.072	433.776	-2360.163	-914.910	10.861
4500	107.909	572.082	473.290	444.566	-2360.148	-882.063	10.239
4600	107.917	574.454	475.464	455.358	-2360.077	-849.218	9.643
4700	107.924	576.775	477.594	466.150	-2359.950	-816.374	9.073
4800	107.931	579.047	479.684	476.942	-2359.768	-783.534	8.527
4900	107.937	581.273	481.735	487.736	-2359.533	-750.698	8.003
5000	107.943	583.454	483.748	498.530	-2359.249	-717.868	7.500
5100	107.949	585.591	485.724	509.324	-2358.918	-685.043	7.016
5200	107.954	587.688	487.665	520.120	-2358.543	-652.225	6.552
5300	107.959	589.744	489.571	530.915	-2358.126	-619.415	6.105
5400	107.964	591.762	491.445	541.711	-2357.669	-586.614	5.674
5500	107.968	593.743	493.287	552.508	-2357.175	-553.822	5.260
5600	107.972	595.688	495.098	563.305	-2356.647	-521.037	4.860
5700	107.976	597.600	496.880	574.102	-2356.087	-488.265	4.474
5800	107.980	599.477	498.633	584.900	-2355.495	-455.501	4.102
5900	107.984	601.323	500.358	595.698	-2354.880	-422.747	3.743
6000	107.987	603.138	502.055	606.497	-2354.218	-390.007	3.395

PREVIOUS: June 1976 (1 atm) CURRENT: June 1976 (1 bar)

Tetrafluoroaluminate, Ion (AlF_4^-) $Al_1F_4^-(g)$

LITHIUM TETRAFLUOROALUMINATE ($LiAlF_4$) IDEAL GAS M_r = 109.916152

$S°(298.15 K) = [326.5 \pm 8]\ J\ K^{-1}\ mol^{-1}$

$\Delta_fH°(0 K) = -1846 \pm 12\ kJ\ mol^{-1}$

$\Delta_fH°(298.15 K) = -1854 \pm 12\ kJ\ mol^{-1}$

Vibrational Frequencies and Degeneracies

ν, cm^{-1}	ν, cm^{-1}	ν, cm^{-1}	ν, cm^{-1}	ν, cm^{-1}	ν, cm^{-1}
817(1)	560 (1)	220 (1)	911 (1)	[157](1)	450(1)
608(1)	[361](1)	[269](1)	316 (1)	644(1)	270(1)

Ground State Quantum Weight: [1] σ = [12]

Point Group: [C_{2v}]

Bond Distances: $Al\text{-}F_{br} = Al\text{-}F_{ter} = [1.69 \pm 0.02]$ Å $Li\text{-}F_{br} = [1.68]$ Å

Bond Angles: $F_{br}\text{-}Al\text{-}F_{br} = [110]°$ $F_{ter}\text{-}Al\text{-}F_{ter} = [109.47]°$ $F_{br}\text{-}Li\text{-}F_{br} = [110.98]°$

Product of the Moments of Inertia: $I_AI_BI_C = 1.887733\times10^{-113}\ g^3\ cm^6$

Enthalpy of Formation

Mass spectrometric studies (1, 2, 3) indicate that the vapor above the $LiF\text{-}AlF_3$ system is composed mainly of $LiAlF_4$, $(LiAlF_4)_2$, and Li_3AlF_5 with $LiAlF_4$ being the major component, 90% according to (1) or 77% according to (3). For $(LiAlF_4)_2(g) = 2\ LiAlF_4(g)$, ln Kp = -48500/RT + 33.5/R at 946 K (3). This is in agreement with the equilibrium constant for the dimer-monomer equilibrium of the sodium analog (4, 5) which has been determined over a wider temperature range. As a first approximation, we ignore any gaseous species other than $LiAlF_4(g)$ and $(LiAlF_4)2$ (g) and adjust the total pressure measurements over the $LiF\text{-}AlF_3$ system to the partial pressure of $LiAlF_4(g)$ by assuming that ln Kp = -43900/RT + 28.50/R for the $(NaAlF_4)_2(g) = 2\ NaAlF_4(g)$ equilibrium (5) also represents the equilibrium of the lithium analog. Rao (6) has made a similar assumption.

$\Delta_fH°$(298.15 K) results from our 3rd law analysis of pressure measurements and equilibrium constants of the following:

[1] $Li_2F_2(g) + AlF_3(g) = LiF(g) + LiAlF_4(g)$ [3] $1/3\ Li_3AlF_6(cr) = 2/3\ AlF_3(cr) + LiAlF_4(g)$ [5] $LiF(\ell) + AlF_3(cr) = LiAlF_4(g)$

[2] $Li_3AlF_6(cr) = 2\ LiF(cr) + LiAlF_4(g)$ [4] $LiAlF_4(g) = LiF(g) + AlF_3(g)$

Source	Method	Reaction	T, K	$\Delta_rH°$(298.15 K), kcal mol^{-1} 2nd law	3rd law	Drift cal K mol^{-1}	$\Delta_fH°(298.15 K)^a$ kcal mol^{-1}
Porter and Zeller (7)	Mass Spec	1	1000-1100	- 0.86	-6.036	- 4.9	-438.95
Hildenbrand et al. (8)	Torsion-Effusion	2	987-1058	101.93	68.86	-32.3	-444.94
		3(cell 7)	899-980	71.10	66.97	- 4.4	-443.27
		3(cell 9)	873-976	67.28	67.10	- 0.2	-443.13
		3(cell 15x)	854-980	67.53	67.23	- 0.3	-443.00
Rao (6)	Effusion	3	872-979	68.59	67.27	- 1.4	-442.97
Shol'ts and Sidorov (3)	Mass Spec	3	946		68.28		-441.95
Sidorov et al. (9)	Mass Spec	4	946		72.02		-442.51
Dewing (10)	b	5	1293		59.155		-444.93

a. From third law heat of reaction plus auxiliary data (11).

b. Dewing has determined activities of LiF and AlF_3 in $LiF\text{-}AlF_3$ melts and combined them with the vapor pressure measurement over the melts of Kuxmann and Tillessen (12) and Chin (13) to obtain an equilibrium constant for Reaction 5.

We give more weight to the results which lead to reasonable agreement in 2nd and 3rd law enthalpies of reaction and adopt $\Delta_fH°(LiAlF_4$, g, 298.15 K) = -443±3 kcal mol^{-1}.

Heat Capacity and Entropy

We adopt the structural and vibrational frequency assignments given by Huglen et al. (14). While the C_{2v} structure has not been conclusively established, it is strongly supported by the interpolation of the infra-red spectroscopic studies (14), by an electron diffraction study of $NaAlF_4$ (15), and by ab initio molecular orbital calculations (16). The C_{2v} structure consists of an irregular AlF_4 tetrahedron combined with Li to form a planar ring, AlF_2Li, through coordination with two fluorines. Thus aluminum forms two types of bonds, $Al\text{-}F_{bridge}$ and $Al\text{-}F_{terminal}$, considered by Huglen et al. (14) to be the same length. A C_s structure and two C_{3v} structures have also been considered.

The AlF_4 tetrahedral angles and the $Al\text{-}F_{br}$ and $Al\text{-}F_{ter}$ bond distances are those from an electron diffraction study of $NaAlF_4$ (15). These are assumed to be the same for $LiAlF_4$. The $Li\text{-}F_{br}$ bond length is estimated (17). The $F_{br}\text{-}Al\text{-}F_{br}$ and the $F_{ter}\text{-}Al\text{-}F_{ter}$ angles are from the sodium analog (17); the $F_{br}\text{-}Li\text{-}F_{br}$ angle results from the adopted dimensions. The slightly different dimensions for the $LiAlF_4$ C_{2v} structures used by Cyvin (17) lead to a S°(298.15 K) value 0.12 cal K mol^{-1} less than the adopted values. The dimensions derived by Curtiss (16) in his ab initio study are quite different, but S°(298.15 K) is only 0.20 cal K mol^{-1} greater.

Nine vibrational frequencies were assigned by Huglen et al. (14) from infra-red matrix-isolation studies; the remaining three were calculated by force field analysis (14).

The principal moments of inertia are $I_A = 24.1060 \times 10^{-39}$, $I_B = 27.9447 \times 10^{-39}$ and $I_C = 28.0231 \times 10^{-39}\ g\ cm^2$.

References

1. A. Buchler and J. B. Berkowitz-Mattuck in "Advances in High Temperature Chemistry," Vol. 1, L. Eyring, Ed., Academic Press, New York, 1967, pp. 129-132.
2. E. N. Kolosov, V. B. Shol'ts and L. N. Sidorov, Vestn. Mosk. Univ., Khim. 13, 49 (1972).
3. V. B. Shol'ts and L. N. Sidorov, Vestn. Mosk. Univ., Khim. 13, 371 (1972).
4. L. N. Sidorov and E. N. Kolosov, Zh. Fiz. Khim 42, 2617 (1968); Russ J. Phys. Chem. 42, 1382 (1968).
5. K. Grjotheim, H. Kvande and K. Motzveldt, Light Met., Proc. Sess. AIME Annu. Meet. 1975, 125 (1975).
6. D. B. Rao, High Temp. Sci. 2, 381 (1970).
7. R. F. Porter and E. E. Zeller, J. Chem. Phys. 33, 858 (1960).
8. D. L. Hildenbrand, L. P. Theard, W. F. Hall, F. Ju, F. S. LaViola and N. D. Potter, Aeronutronic Publ. No. U-2055, Mar. 1963.
9. L. N. Sidorov, E. N. Kolosov, V. A. Davydov, and V. B. Shol'ts, Vestn. Mosk. Univ., Khim. 14, 35 (1973)
10. E. W. Dewing, Metall. Trans. B. 11B, 245 (1980).
11. JANAF Thermochemical Tables: LiF(cr), LiF(ℓ), LiF(g), and $Li_2F_2(g)$, 12-31-68; $AlF_3(cr)$, $AlF_3(g)$, $Li_3AlF_6(cr)$, 6-30-70.
12. U. Kuxmann and U. Tillessen, Z. Erzbergbau Metallhuettenw. 20, 147 (1967).
13. D. A. Chin, unpublished research, Aluminum Company of Canada, Ltd., Arvida, Quebec, 1964.
14. R. Huglen, S. J. Cyvin and H. A. Oye, Proc. Electrochem. Soc. 78-1, 352 (1978).
15. V. B. Spiridonov and E. V. Erokin, Zh. Neorg. Khim. 14, 636 (1969).
16. L. A. Curtiss, Chem. Phys. Lett. 68, 225 (1979).
17. S. J. Cyvin, B. N. Cyvin and A. Snelson, J. Phys. Chem. 75, 2609 (1971).

Lithium Tetrafluoroaluminate ($LiAlF_4$) $Al_1F_4Li_1(g)$

Enthalpy Reference Temperature = T_r = 298.15 K Standard State Pressure = p° = 0.1 MPa

T/K	$C_p°$ (J $K^{-1}mol^{-1}$)	S°	$-[G°-H°(T_r)]/T$	$H°-H°(T_r)$ (kJ mol^{-1})	$\Delta_fH°$	$\Delta_fG°$	Log K_f
0	0.	0.	INFINITE	-19.620	-1846.320	-1846.320	INFINITE
100	52.024	244.477	401.931	-15.745	-1849.239	-1838.179	960.167
200	81.507	290.240	335.102	-8.972	-1851.967	-1825.960	476.892
250	91.971	309.600	328.099	-4.625	-1852.866	-1819.349	380.132
298.15	99.790	326.497	326.497	0.	-1853.512	-1812.832	317.601
300	100.052	327.116	326.499	0.185	-1853.533	-1812.579	315.598
350	106.249	343.025	327.742	5.349	-1854.045	-1805.712	269.488
400	111.011	357.537	330.573	10.786	-1854.480	-1798.777	234.896
450	114.701	370.835	334.318	16.433	-1854.894	-1791.789	207.985
500	117.591	383.075	338.590	22.243	-1858.320	-1784.448	186.420
600	121.719	404.907	347.868	34.223	-1859.005	-1769.606	154.058
700	124.434	423.888	357.402	46.540	-1859.564	-1754.661	130.934
800	126.299	440.633	366.780	59.082	-1860.118	-1739.638	113.587
900	127.627	455.589	375.832	71.782	-1860.759	-1724.541	100.090
1000	128.603	469.089	384.493	84.596	-1872.126	-1708.605	89.248
1100	129.340	481.383	392.751	97.495	-1872.723	-1692.224	80.357
1200	129.909	492.662	400.614	110.458	-1873.303	-1675.789	72.945
1300	130.357	503.079	408.100	123.472	-1873.874	-1659.306	66.672
1400	130.716	512.753	415.234	136.527	-1874.442	-1642.780	61.293
1500	131.008	521.782	422.039	149.613	-1875.010	-1626.213	56.630
1600	131.248	530.244	428.540	162.726	-1875.578	-1609.608	52.548
1700	131.448	538.207	434.760	175.862	-2021.386	-1585.791	48.725
1800	131.616	545.726	440.717	189.015	-2021.208	-1560.173	45.275
1900	131.759	552.846	446.433	202.184	-2021.040	-1534.565	42.188
2000	131.881	559.607	451.924	215.366	-2020.878	-1508.965	39.410
2100	131.987	566.044	457.207	228.560	-2020.721	-1483.373	36.897
2200	132.079	572.187	462.294	241.763	-2020.561	-1457.789	34.612
2300	132.159	578.060	467.201	254.975	-2020.396	-1432.212	32.527
2400	132.229	583.686	471.938	268.194	-2020.220	-1406.642	30.615
2500	132.292	589.085	476.517	281.421	-2020.027	-1381.081	28.856
2600	132.347	594.275	480.947	294.653	-2019.812	-1355.527	27.233
2700	132.396	599.270	485.237	307.890	-2019.572	-1329.982	25.730
2800	132.441	604.086	489.396	321.132	-2313.202	-1303.479	24.317
2900	132.480	608.734	493.432	334.378	-2311.802	-1267.442	22.829
3000	132.516	613.226	497.350	347.628	-2310.365	-1231.454	21.442
3100	132.549	617.572	501.159	360.881	-2308.890	-1195.515	20.144
3200	132.578	621.781	504.863	374.137	-2307.373	-1159.623	18.929
3300	132.605	625.861	508.468	387.396	-2305.814	-1123.780	17.788
3400	132.630	629.820	511.979	400.658	-2304.212	-1087.985	16.715
3500	132.652	633.665	515.401	413.922	-2302.568	-1052.237	15.704
3600	132.673	637.402	518.738	427.188	-2300.881	-1016.537	14.750
3700	132.692	641.037	521.995	440.457	-2299.153	-980.884	13.848
3800	132.709	644.576	525.174	453.727	-2297.385	-945.280	12.994
3900	132.725	648.024	528.280	466.998	-2295.578	-909.721	12.184
4000	132.740	651.384	531.316	480.272	-2293.734	-874.209	11.416
4100	132.754	654.662	534.285	493.547	-2291.857	-838.745	10.686
4200	132.767	657.861	537.189	506.823	-2289.948	-803.326	9.991
4300	132.779	660.985	540.032	520.100	-2288.011	-767.953	9.329
4400	132.791	664.038	542.816	533.378	-2286.027	-732.627	8.697
4500	132.801	667.022	545.543	546.658	-2284.030	-697.345	8.095
4600	132.811	669.941	548.215	559.939	-2282.018	-662.108	7.518
4700	132.820	672.798	550.836	573.220	-2279.989	-626.914	6.967
4800	132.829	675.594	553.406	586.503	-2277.947	-591.764	6.440
4900	132.837	678.333	555.928	599.786	-2275.895	-556.656	5.934
5000	132.845	681.017	558.403	613.070	-2273.839	-521.591	5.449
5100	132.852	683.647	560.833	626.355	-2271.776	-486.567	4.983
5200	132.859	686.227	563.219	639.640	-2269.721	-451.582	4.536
5300	132.865	688.758	565.564	652.927	-2267.673	-416.638	4.106
5400	132.871	691.242	567.869	666.214	-2265.636	-381.732	3.693
5500	132.877	693.680	570.134	679.501	-2263.614	-346.864	3.294
5600	132.882	696.074	572.362	692.789	-2261.612	-312.031	2.911
5700	132.887	698.426	574.553	706.077	-2259.634	-277.236	2.541
5800	132.892	700.737	576.709	719.366	-2257.681	-242.474	2.184
5900	132.897	703.009	578.830	732.656	-2255.763	-207.745	1.839
6000	132.901	705.243	580.918	745.946	-2253.859	-173.051	1.507

PREVIOUS: December 1979 (1 atm) CURRENT: December 1979 (1 bar)

Lithium Tetrafluoroaluminate ($LiAlF_4$) $Al_1F_4Li_1(g)$

SODIUM TETRAFLUOROALUMINATE ($NaAlF_4$) IDEAL GAS Mr = 125.964922

$S°(298.15 K) = 344.9 \pm 8 J K^{-1}mol^{-1}$

$\Delta_fH°(0 K) = -1834 \pm 12 kJ mol^{-1}$

$\Delta_fH°(298.15 K) = -1841 \pm 12 kJ mol^{-1}$

Vibrational Frequencies and Degeneracies

ν, cm^{-1}	ν, cm^{-1}	ν, cm^{-1}	ν, cm^{-1}
808(1)	[291](1)	893(1)	669(1)
613(1)	[180](1)	304(1)	339(1)
378(1)	[269](1)	[102](1)	[200](1)

Ground State Quantum Weight: [1] σ = [2]

Point Group: $[C_{2v}]$

Bond Distances: $Al-F_{br} = Al-F_{ter} = 1.69 \pm 0.02$ Å $Na-F_{br} = 2.11 \pm 0.02$ Å

Bond Angles: $F_{br}-Al-F_{br} = 110°$ $F_{ter}-Al-F_{ter}$ 109.47° $F_{br}-Na-F_{br} = 82°$

Product of the Moments of Inertia: $I_AI_BI_C = 4.775246 \times 10^{-113} g^3 cm^6$

Enthalpy of Formation

$\Delta_fH°(298.15 K)$ is based on our 3rd law analysis of pressure measurements of the NaF-AlF_3 system for the equilibria:

[1] $Na_3AlF_6(cr, \beta)=2 NaF(cr)+NaAlF_4(g)$ [3] $0.2 Na_5Al_3F_{14}(cr)+0.4 AlF_3(cr)=NaAlF_4(g)$

[2] $Na_3AlF_6(\ell)=2 NaF(\ell) + NaAlF_4(g)$ [4] $0.5 Na_5Al_3F_{14}(cr)=0.5 Na_3AlF_6(cr, \beta)+NaAlF4(g)$

The mass-spectrometric measurements of Sidorov et al. (1) showed that the most important species over solid phases in the NaF-AlF_3 system are $NaAlF_4(g)$ and the dimer $[NaAlF_4]_2(g)$. We ignore any other gaseous species in our data treatment. The enthalpy and entropy for the $[NaAlF_4]_2(g) = 2 NaAlF_4(g)$ dissociation have been determined.

Investigator	T/K	$\Delta_{dso}H°$, cal mol^{-1}	$\Delta_{dso}S°$, cal K^{-1} mol^{-1}
Sidorov and Kolosov (2)	910	42600	27.14
Shol'ts and Sidorov (3)	910	45000	33.6
Grjotheim, Kvande, and Motzfeldt (4)	1190-1420	43900	28.50

A plot of ln Kp vs 1/T shows Sidorov and Kolosov (2) and Grjotheim et al. (4) to be in excellent agreement. We use ln Kp = -43900/RT + 28.50/R to calculate the mole fraction $NaAlF_4$ monomer in the gas in order to correct the total pressure over the NaF-AlF_3 system to the partial pressure of monomer for the 3rd law analysis. The use of the total pressure rather than partial pressure leads to a heat of formation which is more negative by less than one kcal mol^{-1}.

Source	Method	Reaction	T/K	$\Delta_rH°(298.15 K)$ 2nd law	kcal mol^{-1} 3rd law	Drift cal $K^{-1}mol^{-1}$	$\Delta_fH°(298.15 K)$ kcal mol^{-1}
Sidorov and Kolosov (2)*	Mass Spec	1	900-1050	63.27	62.27	-1.5	-449.16
Shol'ts and Sidorov (3)	Mass Spec	1	1100		67.76		-443.68
Kolosov et al. (5)	Mass Spec	1	910		65.78		-445.66
Grjotheim et al. (6)	Effusion	3	840-960	65.14	65.49	0.4	-440.03
Shol'ts and Sidorov (3)	Mass Spec	3	910		65.98		-439.54
Kolosov et al. (5)	Mass Spec	3	910		66.90		-439.43
Grjotheim et al. (6)	Effusion	4	840-960	66.41	70.21	4.3	-439.35
Kolosov et al. (5)	Mass Spec	4	910		71.62		-437.94

*Partial pressures by the investigator.

The total vapor pressure over molten cryolite, reaction [2], measured by effusion by Grjotheim et al. (4), corrected to monomer partial pressure, leads to $\Delta_fH°(NaAlF_4, g, 298.15 K) = -447.70$ kcal mol^{-1}. Unit activity of NaF in molten cryolite was used in this calculation so that -447.70 kcal mol^{-1} is a lower limit. There is no evident explanation for the more negative grouping of heats of formation from reactions [1] and [2] compared with the grouping from reactions [3] and [4].

In view of the difficulties and uncertainties in working with the NaF-AlF_3 system, we adopt $\Delta_fH°(NaAlF_4, g, 298.15 K) = -440 \pm 3$ kcal mol^{-1}.

Heat Capacity and Entropy

The adopted structure, C_{2v}, consists of an irregular AlF_4 tetrahedron combined with Na to form the planar ring, AlF_2Na, through coordination with two fluorines. The Al atom forms two types of bonds, $Al-F_{bridge}$ and $Al-F_{terminal}$, which are of the samelength (8). Other structures have been considered, a C_s symmetry in which the ring is not planar and two C_{3v} structures where the Na is coordinated with one or three fluorines, but interpretation of the infra-red spectra supports the C_{2v} model (9).

The bond distances and bond angles are from the electron diffraction study of Spiridonov and Erokin (8). Seven vibrational frequencies were assigned from infrared matrx-isolation studies by Huglen et al. (9); the remaining five were calculated by force field analysis (9). The principal moments of inertia are $I_A = 24.1060 \times 10^{-39}$, $I_B = 44.4685 \times 10^{-39}$, and $I_C = 44.5470 \times 10^{-39}$ g cm^2.

References

1. L. N. Sidorov, E. V. Erokin, P. A. Akishin, and E. N. Lolosov, Dokl. Akad. Nauk SSSR 173, 370 (1967).
2. L. N. Sidorov and E. M. Kolosov, Zh. Fiz. Khim. 42, 2617 (1968); Russ. J. Phys. Chem. 42, 1382 (1968).
3. V. B. Shol'ts and L. N. Sidorov, Vesta, Mosk. Univ., Khim. 13, 371 (1972).
4. K. Grjotheim, H. Kvande, and K. Motzveldt, Light Met., Proc. Sess. AIME Annu. Meet. 1975, 125 (1975).
5. E. N. Kolosov, V. B. Shol'ts, and L. N. Sidorov, Zh. Fiz. Khim. 48, 2199 (1974); Russ. J. Phys. Chem. 48, 1303 (1974).
6. K. Grjotheim, K. Motzfeldt, and D. B. Rao, Symposium at 100th AIME Meeting, New York, March 3, 1971.
7. JANAF Thermochemical Tables: $Na_3AlF_6(cr, \beta)$, $Na_3AlF_6(\ell)$, $Na_5Al_3F_{14}(cr)$, 9-30-79; NaF(cr), 12-31-68; $AlF_3(cr)$, 6-30-70.
8. V. P. Spridonov and E. V. Erokin, Zh. Neorg. Khim. 14, 636 (1969); Russ. J. Inorg. Chem. 14, 332 (1969).
9. R. Huglen, S. J. Cyvin, and H. A. Oye, Proc. Electrochem. Soc. 78-1, 352 (1978).

Sodium Tetrafluoroaluminate ($NaAlF_4$) $Al_1F_4Na_1(g)$

Enthalpy Reference Temperature = T_r = 298.15 K Standard State Pressure = p° = 0.1 MPa

T/K	$C_p°$ (J $K^{-1}mol^{-1}$)	S° (J $K^{-1}mol^{-1}$)	$-[G°-H°(T_r)]/T$ (J $K^{-1}mol^{-1}$)	$H°-H°(T_r)$ (kJ mol^{-1})	$\Delta_fH°$ (kJ mol^{-1})	$\Delta_fG°$ (kJ mol^{-1})	Log K_f
0	0.	0.	INFINITE	-21.223	-1833.547	-1833.547	INFINITE
100	59.284	255.634	425.562	-16.993	-1836.954	-1825.364	953.472
200	87.953	306.516	353.949	-9.487	-1839.564	-1812.638	473.412
250	97.151	327.177	346.571	-4.849	-1840.369	-1805.808	377.303
298.15	103.954	344.896	344.896	0.	-1840.960	-1799.096	315.194
300	104.182	345.539	344.898	0.193	-1840.980	-1798.836	313.205
350	109.570	362.022	346.187	5.542	-1841.496	-1791.771	267.407
400	113.718	376.936	349.113	11.129	-1844.612	-1784.428	233.022
450	116.938	390.523	352.971	16.899	-1845.043	-1776.878	206.255
500	119.464	402.980	357.358	22.811	-1845.395	-1769.285	184.836
600	123.080	425.105	366.852	34.952	-1845.954	-1754.008	152.700
700	125.464	444.269	376.573	47.387	-1846.421	-1738.646	129.739
800	127.102	461.135	386.111	60.020	-1846.897	-1723.218	112.515
900	128.270	476.177	395.298	72.791	-1847.462	-1707.725	99.114
1000	129.129	489.738	404.075	85.663	-1858.773	-1691.402	88.350
1100	129.778	502.077	412.432	98.610	-1859.349	-1674.637	79.522
1200	130.279	513.392	420.380	111.614	-1956.723	-1655.378	72.057
1300	130.674	523.836	427.942	124.663	-1956.466	-1630.277	65.505
1400	130.990	533.532	435.142	137.746	-1956.218	-1605.195	59.891
1500	131.247	542.578	442.006	150.859	-1955.980	-1580.130	55.025
1600	131.459	551.056	448.559	163.994	-1955.756	-1555.081	50.768
1700	131.635	559.031	454.826	177.149	-1955.545	-1530.045	47.013
1800	131.783	566.559	460.826	190.320	-1955.348	-1505.021	43.675
1900	131.909	573.688	466.580	203.505	-1955.161	-1480.009	40.688
2000	132.017	580.457	472.106	216.701	-1954.982	-1455.005	38.001
2100	132.110	586.900	477.420	229.908	-1954.806	-1430.011	35.570
2200	132.191	593.048	482.537	243.123	-1954.627	-1405.025	33.360
2300	132.262	598.926	487.471	256.346	-1954.440	-1380.047	31.342
2400	132.324	604.556	492.233	269.575	-1954.240	-1355.077	29.492
2500	132.379	609.959	496.835	282.810	-1954.019	-1330.117	27.791
2600	132.428	615.152	501.286	296.051	-1953.773	-1305.165	26.221
2700	132.471	620.150	505.596	309.296	-1953.496	-1280.224	24.767
2800	132.510	624.969	509.774	322.545	-2247.085	-1254.326	23.400
2900	132.545	629.619	513.827	335.798	-2245.638	-1218.895	21.955
3000	132.577	634.113	517.762	349.054	-2244.148	-1183.516	20.607
3100	132.605	638.461	521.586	362.313	-2242.612	-1148.187	19.347
3200	132.631	642.672	525.304	375.575	-2241.029	-1112.908	18.166
3300	132.655	646.753	528.923	388.839	-2239.397	-1077.679	17.058
3400	132.677	650.714	532.447	402.106	-2237.716	-1042.501	16.016
3500	132.697	654.560	535.882	415.374	-2235.986	-1007.372	15.034
3600	132.715	658.298	539.230	428.645	-2234.208	-972.294	14.108
3700	132.732	661.935	542.498	441.917	-2232.384	-937.266	13.232
3800	132.747	665.475	545.688	455.191	-2230.515	-902.288	12.403
3900	132.761	668.923	548.803	468.467	-2228.604	-867.359	11.617
4000	132.775	672.285	551.849	481.744	-2226.653	-832.480	10.871
4100	132.787	675.563	554.826	495.022	-2224.667	-797.651	10.162
4200	132.798	678.763	557.739	508.301	-2222.647	-762.869	9.488
4300	132.809	681.888	560.590	521.581	-2220.601	-728.138	8.845
4400	132.819	684.941	563.382	534.863	-2218.489	-693.456	8.232
4500	132.828	687.926	566.116	548.145	-2216.375	-658.821	7.647
4600	132.837	690.846	568.796	561.428	-2214.254	-624.232	7.088
4700	132.845	693.703	571.424	574.712	-2212.120	-589.689	6.554
4800	132.853	696.500	574.000	587.997	-2209.979	-555.192	6.042
4900	132.860	699.239	576.528	601.283	-2207.834	-520.740	5.551
5000	132.867	701.923	579.010	614.569	-2205.686	-486.332	5.081
5100	132.873	704.555	581.445	627.856	-2203.550	-451.966	4.629
5200	132.879	707.135	583.838	641.144	-2201.428	-417.641	4.195
5300	132.885	709.666	586.188	654.432	-2199.323	-383.358	3.778
5400	132.890	712.150	588.498	667.721	-2197.242	-349.114	3.377
5500	132.895	714.588	590.768	681.010	-2195.189	-314.909	2.991
5600	132.900	716.983	593.001	694.300	-2193.169	-280.740	2.619
5700	132.904	719.335	595.197	707.590	-2191.187	-246.608	2.260
5800	132.909	721.647	597.357	720.881	-2189.246	-212.510	1.914
5900	132.913	723.919	599.483	734.172	-2187.354	-178.443	1.580
6000	132.917	726.153	601.575	747.463	-2185.335	-144.431	1.257

PREVIOUS: December 1979 (1 atm) CURRENT: December 1979 (1 bar)

Sodium Tetrafluoroaluminate ($NaAlF_4$) $Al_1F_4Na_1(g)$

POTASSIUM HEXAFLUOROALUMINATE (K_3AlF_6) CRYSTAL M_r = 258.266858

$S°(298.15\ K) = [284.5 \pm 4]\ J\ K^{-1}\ mol^{-1}$

$T_{fus} = 1293\ K$

$\Delta_f H°(0\ K)$ = Unknown

$\Delta_f H°(298.15\ K) = [-3326 \pm 21]\ kJ\ mol^{-1}$

$\Delta_{fus} H°$ = Unknown

Enthalpy of Formation

$\Delta_f H°(K_3AlF_6$, cr, 298.15 K) is calculated from $\Delta_r H° = -30.2\ kcal\ mol^{-1}$ for the reaction, $AlF_3(cr) + 3\ KF(cr) = K_3AlF_6(cr)$, which was estimated by comparison with the values of $\Delta_r H°$ of the similar reactions for $Na_3AlF_6(cr)$, $Na_3AlCl_6(cr)$ and $K_3AlCl_6(cr)$.

Heat Capacity and Entropy

Heat capacities are estimated by the addition of a correction term $3(M_{KF}-M_{NaF})$ to the corresponding values for $Na_3AlF_6(cr)$ where M denotes heat capacity. In a similar manner, S°(298.15 K) is calculated to be 68 ± 1 cal $K^{-1}\ mol^{-1}$.

Fusion Data

T_{fus} was determined from Hall and Insley (1).

Reference

1. F. P. Hall and H. Insley, J. Am. Ceram. Soc. 21, 113 (1938).

Potassium Hexafluoraluminate (K_3AlF_6) $Al_1F_6K_3(cr)$

Enthalpy Reference Temperature = T_r = 298.15 K; Standard State Pressure = p° = 0.1 MPa

T/K	$C_p°$	S°	$-[G°-H°(T_r)]/T$	$H°-H°(T_r)$	$\Delta_f H°$	$\Delta_f G°$	Log K_f
	J $K^{-1}mol^{-1}$			kJ mol^{-1}			
0							
100							
200							
298.15	222.798	284.512	284.512	0.	-3326.280	-3163.447	554.223
300	223.384	285.892	284.516	0.413	-3326.250	-3162.437	550.629
400	241.082	352.662	293.500	23.665	-3331.598	-3106.741	405.699
500	255.308	408.026	311.022	48.502	-3328.814	-3050.825	318.718
600	267.399	455.652	331.248	74.642	-3324.948	-2995.574	260.788
700	278.905	497.740	352.082	101.960	-3320.120	-2941.047	219.464
800	290.077	535.740	372.702	130.431	-3314.390	-2887.274	188.520
900	298.738	570.425	392.772	159.888	-3307.984	-2834.264	164.496
1000	305.432	602.254	412.151	190.103	-3311.793	-2781.230	145.277
1100	311.290	631.646	430.786	220.946	-3541.341	-2714.695	128.910
1200	316.310	658.953	448.676	252.333	-3530.603	-2640.016	114.917
1300	320.390	684.437	465.842	284.173	-3519.481	-2566.249	103.113
1400	323.842	708.311	482.318	316.390	-3508.049	-2493.351	93.028
1500	326.509	730.746	498.139	348.910	-3496.374	-2421.279	84.316
1600	328.862	751.895	513.345	381.681	-3484.506	-2349.992	76.719
1700	330.954	771.898	527.971	414.676	-3472.470	-2279.452	70.030
1800	332.628	790.864	542.053	447.858	-3460.298	-2209.625	64.122
1900	333.883	808.883	555.627	481.187	-3448.026	-2140.478	58.846
2000	334.720	826.033	568.722	514.621	-3435.694	-2071.978	54.115

PREVIOUS: September 1963 CURRENT: December 1963

Potassium Hexafluoraluminate (K_3AlF_6) $Al_1F_6K_3(cr)$

LITHIUM HEXAFLUOROALUMINATE (Li_3AlF_6) CRYSTAL (BETA-GAMMA-DELTA-EPSILON) M_r =161.794958

$S°(298.15\ K) = 187.89 \pm 0.21\ J\ K^{-1}\ mol^{-1}$
$T_{trs1}\ (\beta \to \gamma) = 748\ K$
$T_{trs2}\ (\gamma \to \delta) = 848\ K$
$T_{trs3}\ (\delta \to \epsilon) = 978\ K$
$T_{fus} = 1058 \pm 3\ K$

$\Delta_fH°(0\ K) = -3370.9 \pm 4.6\ kJ\ mol^{-1}$
$\Delta_fH°(298.15\ K) = -3383.6 \pm 4.6\ kJ\ mol^{-1}$
$\Delta_{trs1}H° = [2.092]\ kJ\ mol^{-1}$
$\Delta_{trs2}H° = [1.255]\ kJ\ mol^{-1}$
$\Delta_{trs3}H° = [0.418]\ kJ\ mol^{-1}$
$\Delta_{fus}H° = 86.19 \pm 4.2\ kJ\ mol^{-1}$

Enthalpy of Formation

$\Delta_fH°$ is derived from $\Delta_rH°(298.15\ K) = -5.4\pm0.5\ kcal\ mol^{-1}$ for the reaction 3 LiF(cr) + AlF_3(cr) → Li_3AlF_6(α, β mixture). Greene et al. (1) obtained this result by heating the fluorides under standardized conditions in a small electrical furnace contained in their calorimeter at 25°C. $\Delta_rH°$ includes a small endothermic effect observed during heating of the reaction product. The authors suggest that the enthalpy of conversion between the α- and β-forms is small and thus their result should apply to both forms.

Heat Capacity and Entropy

Furukawa et al. (2) measured $C_p°$ (15-370 K) using a sample prepared by stoichiometric proportions of LiF and AlF_3 in graphite. X-ray diffraction and petrographic examination of separate portions of the sample indicated a single phase identified as β-Li_3AlF_6. The authors tabulated values of $C_p°$ and S° based on their data and the extrapolation S°(15 K) = 0.042 cal K^{-1} mol^{-1}. These values are adopted.

Douglas and Neufer (3) reported relative enthalpy data (323-973 K) for part of the same sample used in measurement of $C_p°$ (2). Their study was made prior to discovery of the five crystalline phases, and their drop-calorimetric data reveal only the transition near 748 K. Reinterpretation of the data above 748 K is complicated by the proximity of the furnace temperatures to the transition temperatures and by the tendency of high-temperature forms to revert to both α- and β-forms under different conditions of cooling. Bjorge and and Jenssen (4) and Rolin et al. (5) also reported crystalline enthalpy data, but these are relatively imprecise and include similar uncertainties in the final state after the drop. These difficulties preclude the derivation of accurate heat capacities for the high-temperature forms.

$C_p°$ for the β-form is adopted from (2, 3), since the two methods are in good agreement. The curve is extrapolated linearly above 748 K to obtain $C_p°$ for the high-temperature forms. Obvious differences in $C_p°$ are not apparent in the observed data for the different forms, so this should be an adequate approximation.

Transition Data

Existence of five polymorphic forms was shown by high-temperature X-ray diffraction (1, 6, 7) and differential thermal analysis (6, 7). Both α- and β-forms persisted at room temperature, but the α-form appeared only on quenching of high-temperature forms (1). The α-form transformed to β near 490 K (1, 6), but the reverse transformation was not observed. Single-crystal X-ray diffraction showed the α-form to be orthorhombic (8).

The adopted values of T_{trs} are the lowest temperatures at which the high-temperature form was observed growing at the expense of the low-temperature form (1). DTA data (6, 7) were higher by roughly 30 K for T_{trs1} and 25 K for T_{trs2}. Combination of the observed enthalpies (3) with the adopted $C_p°$ yields $\Delta_{trs1}H° = 0.76$ and $\Delta_{trs2}H° = 0.03$ or 0.15 kcal mol^{-1}; we prefer to adopt $\Delta_{trs}H°$ values (0.5 and 0.3 kcal mol^{-1}) which are more consistent with the areas of peaks in the DTA curve (7). Evidence suggests that $\Delta_{trs3}H°$ is small, so we estimate 0.1 kcal mol^{-1}.

Fusion Data

See the liquid table for details.

References

1. P. D. Greene, P. Gross and C. Hayman, Trans. Faraday Soc. 64, 633 (1968).
2. G. T. Furukawa, W. G. Saba and J. C. Ford, U. S. Nat. Bur. Stand. Report 10074, p. 47, July 1, 1969.
3. T. B. Douglas and J. E. Neufer, U. S. Nat. Bur. Stand. Report 8186, p. 68, January 1, 1964.
4. B. Bjorge and B. Jenssen, Acta Chem. Scand. 22, 1347 (1968).
5. M. Rolin, H. Latreille and H. Pham, Bull. Soc. Chim. Fr. 1969 (7), 2271.
6. G. Garton and B. M. Wanklyn, J. Inorg. Nucl. Chem. 27, 2466 (1965).
7. J. L. Holm, Acta Chem. Scand. 20, 1167 (1966).
8. J. H. Burns, A. C. Tennissen and G. D. Brunton, Acta Cryst. B24, 225 (1968).

Lithium Hexafluoroaluminate (Li_3AlF_6) $Al_1F_6Li_3$(cr)

Enthalpy Reference Temperature = T_r = 298.15 K Standard State Pressure = p° = 0.1 MPa

	J K⁻¹mol⁻¹			kJ mol⁻¹			
T/K	$C_p°$	S°	$-[G°-H°(T_r)]/T$	$H°-H°(T_r)$	$\Delta_fH°$	$\Delta_fG°$	Log K_f
0	0.	0.	INFINITE	-32.146	-3370.866	-3370.866	INFINITE
100	70.982	38.225	333.657	-29.543	-3378.953	-3328.763	1738.767
200	156.222	115.960	205.121	-17.832	-3383.356	-3276.431	855.716
298.15	202.506	187.891	187.891	0.	-3383.601	-3223.790	564.795
300	203.133	189.145	187.895	0.375	-3383.581	-3222.799	561.139
400	229.869	251.549	196.222	22.131	-3381.793	-3169.443	413.887
500	248.111	304.904	212.753	46.075	-3388.377	-3115.708	325.496
600	262.295	351.440	232.072	71.621	-3384.937	-3061.477	266.525
700	274.052	392.779	252.131	98.453	-3380.383	-3007.913	224.453
748.000	279.073	411.132	261.751	111.737	BETA <--> GAMMA TRANSITION		
748.000	279.073	413.929	261.751	113.829			
800	284.512	432.865	272.262	128.482	-3372.851	-2955.211	192.956
848.000	289.533	449.589	281.830	142.259	GAMMA <--> DELTA TRANSITION		
848.000	289.533	451.069	281.830	143.515			
900	294.972	468.460	292.114	158.712	-3365.399	-2903.447	168.512
978.000	303.131	493.311	307.178	182.038	DELTA <--> EPSILON TRANSITION		
978.000	303.131	493.739	307.178	182.456			
1000	305.432	500.508	311.357	189.150	-3368.604	-2851.735	148.959
1058.000	311.499	517.898	322.206	207.041	EPSILON <--> LIQUID		
1100	315.892	530.109	329.912	220.217	-3360.516	-2800.432	132.981
1200	326.352	558.044	347.770	252.329	-3351.444	-2749.908	119.700
1300	336.812	584.579	364.974	285.487	-3341.386	-2700.181	108.495
1400	347.272	609.922	381.571	319.691	-3330.327	-2651.266	98.920
1500	357.732	634.238	397.611	354.941	-3318.256	-2603.177	90.651
1600	368.192	657.660	413.136	391.238	-3305.169	-2555.926	83.442

PREVIOUS: December 1963 CURRENT: June 1970

Lithium Hexafluoroaluminate (Li_3AlF_6) $Al_1F_6Li_3$(cr)

LITHIUM HEXAFLUOROALUMINATE (Li_3AlF_6) LIQUID M_r = 161.794958

$S°(298.15\ K) = [245.847]\ J\ K^{-1}\ mol^{-1}$ $\Delta_f H°(298.15\ K) = [-3317.640]\ kJ\ mol^{-1}$

$T_{fus} = 1058 \pm 3\ K$ $\Delta_{fus}H° = 86.190 \pm 4.2\ kJ\ mol^{-1}$

Enthalpy of Formation

$\Delta_f H°(Li_3AlF_6, \ell, 298.15\ K)$ is calculated from $\Delta_f H°(Li_3AlF_6, \ell, 298.15\ K)$ by adding the enthalpy of fusion, $\Delta_{fus}H°$, and the difference in enthalpy, H°(1058 K)−H°(298.15 K), between the crystal and liquid.

Heat Capacity and Entropy

Relative enthalpy data were measured by Bjorge and Jenssen (1) from 1065 to 1088 K and by Rolin et al. (2) from about 1075 to 1145 K. The data are consistent with liquid heat capacities in the range from roughly 82 to 92 cal K^{-1} mol^{-1}. We adopt the mean value of 86 cal K^{-1} mol^{-1} (359.824 J K^{-1} mol^{-1}) and extrapolate this to higher and lower temperatures. A glass transition is assumed at 750 K and $C_p°$ at lower temperatures is taken to be the same as that of β-Li_3AlF_6.

S°(298.15 K) is calculated in a manner analogous to that used for the enthalpy of formation.

Fusion Data

T_{fus} was observed at 780°C (3), 782°C (2, 4), 783°C (5, 8) and 785°C (1). The last and highest value is adopted for this table. Combination of the observed liquid enthalpies with the adopted crystal enthalpies yields values for $\Delta_{fus}H°$ of 19.7 kcal mol^{-1} (1) (authors gave 20.2±1.0) and 21.5±0.3 kcal mol^{-1} (2) (authors gave 21.5). We adopt the mean value of 20.6 kcal mol^{-1} (86.190 kJ mol^{-1}) which is confirmed by values of 20.7 (8), 20.8 (4), 22.2 (4), 20.1 (4) and 22.0 (2) derived from crystal-liquid phase data for binary systems. The agreement among values for $\Delta_{fus}H°$ seemingly excludes the possibility of large discrepancies in the enthalpies due to formation of the α-phase instead of β during drop-calorimetry from high temperatures.

Vaporization Data

Li_3AlF_6 vaporizes mainly to $LiAlF_4(g)$ with small contributions from species such as $Li_2AlF_5(g)$, $Li_2F_2(g)$ and LiF(g) (6, 7).

References

1. B. Bjorge and B. Jenssen, Acta Chem. Scand. 22, 1347 (1968).
2. R. Rolin, H. Latreille and H. Pham, Bull. Soc. Chim. Fr. 1969 (7), 2271.
3. J. L. Holm, Acta Chem. Scand. 20, 1167 (1967).
4. M. Malinovsky, Chem. Zvesti 21, 783 (1967); ibid., p. 794.
5. G. Garton and B. Wanklyn, J. Inorg. Nucl. Chem. 27, 2466 (1965).
6. A. Buchler and J. Berkowitz-Mattuck, AD 653163, pp 41-3, June, 1967.
7. R. F. Porter and E. E. Zeller, J. Chem. Phys. 33, 858 (1960).
8. T. Matsushima, Denki Kagaku Oyobi Kogyo Butsuri Kagaku (Electrochemistry and Industrial Physical Chemistry) 37, 778 (1969).

Lithium Hexafluoroaluminate (Li_3AlF_6) $Al_1F_6Li_3(l)$

Enthalpy Reference Temperature = T_r = 298.15 K Standard State Pressure = p° = 0.1 MPa

T/K	$C_p°$	S°	−[G°−H°(T_r)]/T	H°−H°(T_r)	$\Delta_f H°$	$\Delta_f G°$	Log K_f
	J K⁻¹mol⁻¹			kJ mol⁻¹			
0							
100							
200							
298.15	202.506	245.847	245.847	0.	-3317.640	-3175.109	556.266
300	203.133	247.101	245.851	0.375	-3317.620	-3174.224	552.682
400	229.869	309.505	254.178	22.131	-3315.832	-3126.664	408.301
500	248.111	362.860	270.709	46.075	-3322.416	-3078.725	321.632
600	262.295	409.396	290.028	71.621	-3318.976	-3030.290	263.810
700	274.052	450.736	310.087	98.454	-3314.421	-2982.521	222.558
700.000	359.824	450.736	310.087	98.454	GLASS <--> LIQUID TRANSITION		
700.000	359.824	450.736	310.087	98.454			
800	359.824	498.783	330.738	134.436	-3300.936	-2936.031	191.703
900	359.824	541.165	351.811	170.419	-3287.731	-2891.213	167.802
1000	359.824	579.076	372.675	206.401	-3285.392	-2847.092	148.717
1058.000	359.824	599.363	384.551	227.271	------ EPSILON <--> LIQUID -----		
1100	359.824	613.371	393.022	242.383	-3272.388	-2803.892	133.146
1200	359.824	644.680	412.708	278.366	-3259.446	-2761.873	120.221
1300	359.824	673.481	431.675	314.348	-3246.563	-2720.931	109.328
1400	359.824	700.147	449.911	350.331	-3233.726	-2680.980	100.029
1500	359.824	724.972	467.430	386.313	-3220.923	-2641.945	92.001
1600	359.824	748.194	484.260	422.295	-3208.150	-2603.763	85.004
1700	359.824	770.009	500.433	458.278	-3631.111	-2544.851	78.194
1800	359.824	790.576	515.987	494.260	-3616.124	-2481.386	72.008
1900	359.824	810.030	530.955	530.243	-3601.178	-2418.753	66.496
2000	359.824	828.487	545.374	566.225	-3586.265	-2356.906	61.556
2100	359.824	846.043	559.277	602.207	-3571.378	-2295.804	57.105
2200	359.824	862.782	572.696	638.190	-3556.510	-2235.411	53.075
2300	359.824	878.777	585.658	674.172	-3541.653	-2175.693	49.412
2400	359.824	894.091	598.193	710.155	-3526.799	-2116.622	46.067
2500	359.824	908.779	610.324	746.137	-3511.942	-2058.170	43.003
2600	359.824	922.892	622.077	782.119	-3497.076	-2000.313	40.187
2700	359.824	936.472	633.471	818.102	-3482.194	-1943.028	37.590
2800	359.824	949.558	644.528	854.084	-3761.194	-1885.326	35.171
2900	359.824	962.184	655.265	890.067	-3745.173	-1818.614	32.757
3000	359.824	974.383	665.700	926.049	-3729.127	-1752.454	30.513

PREVIOUS: CURRENT: June 1970

Lithium Hexafluoroaluminate (Li_3AlF_6) $Al_1F_6Li_3(l)$

LITHIUM HEXAFLUOROALUMINATE (Li_3AlF_6) BETA-GAMMA-DELTA-EPSILON-LIQUID M_r = 161.794958 Lithium Hexafluoroaluminate (Li_3AlF_6) $Al_1F_6Li_3$(cr,l)

0	to	748 K	crystal, beta
748	to	848 K	crystal, gamma
848	to	978 K	crystal, delta
978	to	1058 K	crystal, epsilon
above		1058 K	liquid

Refer to the individual tables for details.

Enthalpy Reference Temperature = T_r = 298.15 K Standard State Pressure = $p°$ = 0.1 MPa

T/K	$C_p°$ (J K⁻¹mol⁻¹)	$S°$ (J K⁻¹mol⁻¹)	$-[G°-H°(T_r)]/T$ (J K⁻¹mol⁻¹)	$H°-H°(T_r)$ (kJ mol⁻¹)	$\Delta_fH°$ (kJ mol⁻¹)	$\Delta_fG°$ (kJ mol⁻¹)	Log K_f
0	0.	0.	INFINITE	-32.146	-3370.866	-3370.866	INFINITE
100	70.982	38.225	333.657	-29.543	-3378.953	-3328.763	1738.767
200	156.222	115.960	205.121	-17.832	-3383.356	-3276.431	855.716
298.15	202.506	187.891	187.891	0.	-3383.601	-3223.790	564.795
300	203.133	189.145	187.895	0.375	-3383.581	-3222.799	561.139
400	229.869	251.549	196.222	22.131	-3381.793	-3169.443	413.887
500	248.111	304.904	212.753	46.075	-3388.377	-3115.708	325.496
600	262.295	351.440	232.072	71.621	-3384.937	-3061.477	266.525
700	274.052	392.779	252.131	98.453	-3380.383	-3007.913	224.453
748.000	279.073	411.132	261.751	111.737	BETA <--> GAMMA	TRANSITION	
748.000	279.073	413.929	261.751	113.829			
800	284.512	432.865	272.262	128.482	-3372.851	-2955.211	192.956
848.000	289.533	449.589	281.830	142.259	GAMMA <--> DELTA	TRANSITION	
848.000	289.533	451.069	281.830	143.515			
900	294.972	468.460	292.114	158.712	-3365.399	-2903.447	168.512
978.000	303.131	493.311	307.178	182.038	DELTA <--> EPSILON	TRANSITION	
978.000	303.131	493.739	307.178	182.456			
1000	305.432	500.508	311.357	189.150	-3368.604	-2851.735	148.959
1058.000	311.499	517.898	322.206	207.041	EPSILON <--> LIQUID	TRANSITION	
1058.000	359.824	599.363	322.206	293.232			
1100	359.824	613.371	333.058	308.344	-3272.388	-2803.892	133.146
1200	359.824	644.680	357.741	344.327	-3259.446	-2761.873	120.221
1300	359.824	673.481	380.935	380.309	-3246.563	-2720.931	109.328
1400	359.824	700.147	402.796	416.292	-3233.726	-2680.980	100.029
1500	359.824	724.972	423.456	452.274	-3220.923	-2641.945	92.001
1600	359.824	748.195	443.034	488.256	-3208.150	-2603.763	85.004
1700	359.824	770.009	461.633	524.239	-3631.111	-2544.851	78.194
1800	359.824	790.576	479.342	560.221	-3616.125	-2481.386	72.008
1900	359.824	810.030	496.239	596.204	-3601.178	-2418.753	66.496
2000	359.824	828.487	512.394	632.186	-3586.265	-2356.906	61.556
2100	359.824	846.043	527.867	668.168	-3571.378	-2295.805	57.105
2200	359.824	862.782	542.713	704.151	-3556.510	-2235.412	53.075
2300	359.824	878.777	556.980	740.133	-3541.653	-2175.694	49.412
2400	359.824	894.091	570.709	776.116	-3526.799	-2116.623	46.067
2500	359.824	908.779	583.940	812.098	-3511.942	-2058.171	43.003
2600	359.824	922.892	596.707	848.080	-3497.076	-2000.313	40.187
2700	359.824	936.472	609.041	884.063	-3482.194	-1943.028	37.590
2800	359.824	949.558	620.970	920.045	-3761.194	-1885.327	35.171
2900	359.824	962.184	632.520	956.028	-3745.173	-1818.614	32.757
3000	359.824	974.383	643.713	992.010	-3729.127	-1752.455	30.513

PREVIOUS: CURRENT: June 1970

Lithium Hexafluoroaluminate (Li_3AlF_6) $Al_1F_6Li_3$(cr,l)

$S°(298.15\ K) = 238.47 \pm 1.7\ J\ K^{-1}\ mol^{-1}$

$T_{trs}(\alpha \rightarrow \beta) = 836 \pm 2\ K$

$T_{fus} = 1285 \pm 3\ K$

$\Delta_f H°(0\ K) = -3304.65 \pm 3.3\ kJ\ mol^{-1}$

$\Delta_f H°(298.15\ K) = -3316.91 \pm 3.3\ kJ\ mol^{-1}$

$\Delta_{trs} H°\ (\alpha \rightarrow \beta) = 9.456 \pm 1.7\ kJ\ mol^{-1}$

$\Delta_{fus} H° = 110.039 \pm 5.0\ kJ\ mol^{-1}$

Enthalpy of Formation

The heats of solution of Al(cr), NaCl(cr), and Na_3AlF_6(cr) in 4.36 m HCl acid solution were determined at 303.15 K by Coughlin (1). The enthalpy change for the reaction Al(cr) + 2.96 NaCl(cr) + 2.96(HF·5.716 H_2O) + 3.617 H_2O(ℓ) = $Na_{2.96}AlF_{5.96}$(cr) + 2.96(HCl·12.731 H_2O) + 3/2 H_2(g) was derived as $\Delta_r H°$(298.15 K) = -153.22±0.14 kcal mol . Using auxiliary $\Delta_f H°$(298.15 K) data from (2, 3), we obtain $\Delta_f H°$(($Na_{2.96}AlF_{5.96}$, cr, 298.15 K) = -787.25±0.34 kcal mol^{-1}. Assuming $\Delta_r H°$(298.15 K) = 0 for the reaction $Na_{2.96}AlF_{5.96}$ + 0.04 NaF = Na_3AlF_6, we obtain $\Delta_f H°$(Na_3AlF_6, cr, 298.15 K) = -792.76±0.4 kcal mol^{-1}).

Gross et al. (4) measured the heat of the reaction 3/2 PbF_2(cr) + Al(cr) + 3 NaF(cr) = 3/2 Pb(cr) + Na_3AlF_6(cr) directly in a bomb calorimeter. The average heat of the reaction was found to be $\Delta_r H°$(298.15 K) = -138.3 kcal mol^{-1}. Using auxiliary data (2), we obtain $\Delta_f H°$(Na_3AlF_6, cr, 298.15 K) = -794.40±1.5 kcal mol^{-1}. Gross et al. (4) also measured the heat of reaction for 3/2 PbF_2(cr) + Al(cr) = 3/2 Pb(cr) + AlF_3(cr) as $\Delta_r H°$(AlF_3, cr, 298.15 K) = -360.99±0.31 kcal mol^{-1}), we obtain $\Delta_f H°$(Na_3AlF_6, cr, 298.15 K) = -794.19±1.0 kcal mol^{-1}.

Baud (6) reported $\Delta_r H°$ = -40.7 kcal mol^{-1} for the reaction 2 AlF_3(cr) + 6 NaF(cr) = 2 Na_3AlF_6(cr). Using auxiliary data from 2, as above, we obtain $\Delta_f H°$(Na_3AlF_6, cr, 298.15 K) = -792.8±2.0 kcal mol^{-1}.

Grjotheim et al. (7) measured the enthalpy of mixing of 3/4 NaF(cr) + 1/4 Na_3AlF_6(cr) as $\Delta_r H°$(910 K) = -3.8±0.8 kcal mol^{-1}. When this result is combined with JANAF auxiliary data for the enthalpies and $\Delta_f H°$(298.15 K) values from (2) we obtain $\Delta_f H°$(Na_3AlF_6, cr, 298.15 K) = -792.8±2.0 kcal mol^{-1}.

The amounts of sodium over mixtures of cryolite and aluminum have been measured by Ono et al. (8), Dewing (9), and by Stokes and Frank (10). The data have been corrected for the formation of sodium dimer by using JANAF functions for Na_2(g) and Na(g) to determine the equilibrium constants for the reaction Na_2(g) = 2 Na(g). Our evaluation of this data is presented below. We were unable to evaluate the data of Stokes and Frank (10) due to lack of appropriate data on the activities of Na_3AlF_6(ℓ). Also shown in the table are the emf work of Dewing (11), and the equilibrium study of Mashovets and Yudin (12).

Our adopted value for the heat of formation of cryolite, $\Delta_f H°$(Na_3AlF_6, cr, 298.15 K) = -792.76±0.8 kcal mol^{-1}(-3316.91 kJ mol^{-1}) is based on the study by Coughlin (1) and discussions with Dewing (23).

Source	Method	Reaction[a]	T/K	Data	δS cal K^{-1} mol^{-1}	$\Delta_r H°$(298.15 K), kcal mol^{-1} 2nd Law	3rd Law	$\Delta_f H°$(298.15 K) kcal mol^{-1}
Mashovets (12) (1962)	transpiration	A	1101-1239	5	1.32±1.3	48.11±0.8	46.57±0.2	-792.78±1.0
Mashovets (12) (1962)	transpiration	B	1274	1	-	-	60.54	-792.81±1.0
Ono (8) (1965)	DTA	C	1063-1153	eqn	4.98	115.86	110.38±2.0	-793.22±3.0
Dewing (8) (1969)	transpiration	C	949-1089	11	-3.71±1.0	104.21±1.0	108.03±0.50	-795.57±2.0
Dewing (9) (1969)	transpiration	D	908-913	2	-12.0	99.3	110.20±0.1	-793.40±1.5
Dewing (11) (1970)	emf	E	839-943	9	1.02±0.2	-18.20±0.2	-19.115±0.09	-793.52±0.6

[a]Reactions:

A 2/3 Na_3AlF_6(cr) + H_2O(g) = 1/3 Al_2O_3(cr) + 2 HF(g) + 2 NaF(cr)

B 2/3 Na_3AlF_6(cr) + H_2O(g) = 1/3 Al_2O_3(cr) + 2 HF(g) + 2 NaF(ℓ)

C 6 NaF(cr) + Al(ℓ) = 3 Na(g) + Na_3AlF_6(cr)

D 6 NaF(cr) + Al(cr) = 3 Na(g) + Na_3AlF_6(cr)

E 3 NaF(cr) + AlF_3(cr) = Na_3AlF_6(cr)

Heat Capacity and Entropy

The low temperature heat capacities, 53.66-296.00 K, were measured by King (13). The values of $C°_p$ below 53.66 K are extrapolated using the $C°_p$ equation (combination of Debye and Einstein functions) given by King. S°(298.15 K) is derived based on the low temperature heat capacities S°(50 K) = 3.953 cal K^{-1} mol^{-1} and H°(50;0) = 0.140 kcal mol^{-1}.

High temperature enthalpies have been determined by drop calorimetry by O'Brien and Kelley (14) (401.3-809.7 K), and by Albright (15) (349.99-812.70 K). The enthalpies reported by Albright are 7% higher than those reported by O'Brien and Kelley. Enthalpy data from these two calorimeters have been compared previously (see MoO_3(cr) table) indicating that the CRT calorimeter (15) results were 2% higher than the Bureau of Mines calorimeter (14). Frank (16) commented that the enthalpies reported by O'Brien and Kelley were incorrect due to an apparent error in temperature measurement but Douglas and Ditmars (17) have questioned this correction (see NaF(cr) table for details). Our analysis indicates that the enthalpies of O'Brien and Kelley join smoothly with the low temperature $C°_p$'s of King. This is not true of Albright's enthalpy data. We adopt heat capacities for the α-phase derived from the original enthalpy data reported by O'Brien and Kelley (14).

The high temperature enthalpies have also been measured by the following investigators: July (17) (284.5-373.3 K), Baud (19) (289-328 K), Roth and Bertram (20) (369-802 K), Krestovnikov and Karetnikov (21) (288-1173 K), and by Lyshenko (22) (293-838 K). Their data are not used for evaluation due to inconsistencies among the reported values.

Transition Data and Fusion Data

See the β-crystal table for details.

References

References are listed on the liquid table.

Cryolite, Alpha (Na_3AlF_6) $Al_1F_6Na_3$(cr)

Enthalpy Reference Temperature = T_r = 298.15 K Standard State Pressure = p° = 0.1 MPa

T/K	$C°_p$ (J $K^{-1}mol^{-1}$)	S°	-[G°-H°(T_r)]/T	H°-H°(T_r) (kJ mol^{-1})	$\Delta_f H°$	$\Delta_f G°$	Log K_f
0	0.	0.	INFINITE	-38.095	-3304.649	-3304.649	INFINITE
100	100.039	62.287	402.572	-34.028	-3313.806	-3261.080	1703.414
200	179.033	159.302	257.165	-19.573	-3317.306	-3206.528	837.459
298.15	215.727	238.467	238.467	0.	-3316.908	-3152.168	552.247
300	216.212	239.803	238.471	0.400	-3316.883	-3151.146	548.663
400	236.379	305.007	247.223	23.114	-3323.200	-3095.511	404.232
500	248.634	358.934	264.326	47.304	-3321.044	-3038.816	317.463
600	262.793	405.519	284.058	72.876	-3317.675	-2982.666	259.664
700	276.822	447.083	304.429	99.858	-3313.053	-2927.179	218.429
800	290.780	484.959	324.661	128.238	-3307.209	-2872.439	187.551
836.000	295.809	497.868	331.843	138.797	-------- ALPHA <--> BETA	-------	
900	304.691	520.012	344.443	158.013	-3300.200	-2818.502	163.582
1000	318.582	552.834	363.657	189.176	-3302.683	-2764.645	144.410
1100	332.461	583.849	382.277	221.729	-3293.191	-2711.290	128.748
1200	346.310	613.371	400.315	255.668	-3572.796	-2651.533	115.418

PREVIOUS: December 1968 CURRENT: December 1979

Cryolite, Alpha (Na_3AlF_6) $Al_1F_6Na_3$(cr)

CRYOLITE, BETA (β-Na_3AlF_6) CRYSTAL (BETA) M_r = 209.941268

$S°(298.15\ K)$ = [284.881] $J\ K^{-1}\ mol^{-1}$

T_{trs} ($\alpha \to \beta$) = 836 ± 2 K

T_{fus} ($\beta \to \ell$) = 1285 ± 3 K

$\Delta_f H°(298.15\ K)$ = [-3288.781] $kJ\ mol^{-1}$

$\Delta_{trs}H°$ ($\alpha \to \beta$) = 9.456 ± 1.7 $kJ\ mol^{-1}$

$\Delta_{fus}H°$ ($\beta \to \ell$) = 110.039 ± 5.0 $kJ\ mol^{-1}$

Enthalpy of Formation

$\Delta_f H°(Na_3AlF_6$, cr, β, 298.15 K) is calculated from $\Delta_f H°(Na_3AlF_6$, cr, α, 298.15 K) by adding the enthalpy of transition, $\Delta_{trs}H°$, and the difference in enthalpy, H°(836 K)-H°(298.15 K), between the α- and β-phases.

Heat Capacity and Entropy

High-temperature enthalpies of the β-phase have been determined by drop calorimetry by O'Brien and Kelley (1) (846.7-1263.5 K), Albright (2) (824.21-1267.17 K), and by Holm and Gronvold (3) (1070-1175 K). The data of Holm and Gronvold is in very good agreement with the data of Albright but the data of O'Brien and Kelley are systematically lower by 2%. We have already discussed the discrepancy between the data of O'Brien and Kelley and that of Albright (see the discussion on the Na_3AlF_6(cr, α) table). Our experience with the enthalpies reported by Holm, which are primarily aimed at determining heats of fusion, is that the enthalpies show a positive bias (see the $Na_5Al_3F_{14}$(cr) table) of 2-4%. For these reasons, we adopt the enthalpies from the work of O'Brien and Kelley (1). Heat capacities derived from these enthalpies have been extended smoothly below 836 K and above 1285 K.

S°(298.15 K) is calculated in a manner analogous to that used for the enthalpy of formation.

Transition Data

The α-phase is the low temperature monoclinic modification of cryolite (sodium hexafluoroaluminate), space group $P2_1/n$ (4, 10). The transition to the cubic high-temperature β-phase (space group Fm3m) occurs at 836±2 K. Reported values for $T_{trs}(\alpha \to \beta)$ are 823 K (2), 832 K (6), 833 K (7), 833.5 K (6), 834 K (8), 834.5 K (9), 837 K (9), 838 K (10, 11, 12), and 845 K (1). The value of $\Delta H_{trs}H°(\alpha \to \beta)$ is calculated as the difference between the adopted enthalpies for the α- and β-phases at 836 K. Reported values of $\Delta H_{trs}H°(\alpha \to \beta)$ are (in $kJ\ mol^{-1}$) = 5.19 (12), 9.04 (1), 9.37±1.7 (9), 9.87 (9), and 9.96 (2). The cause of the discrepancies among the reported values for T_{trs}, $\Delta_{trs}H°$ and the high temperature enthalpies may be due to differences in the sample composition. The samples used have normally been hand-picked naturally occurring crystals.

Fusion Data

Our adopted value for the melting point of cryolite is 1285±3 K. Reported values include (all in K): 1273 (12), 1279 (2), 1281 (7), 1282 (13, 14), 1284 (3, 15), 1285 (10), 1293 (8), and 1300 (1). The value of $\Delta_{fus}H°$ is calculated as the difference between the adopted enthalpies of the β-crystal and liquid at 1285 K. Other reported values of $\Delta_{fus}H°$ are (in $kJ\ mol^{-1}$): 69.62 (12), 106.7±10.5 (15), 111.3±0.8 (16), 111.75 (6), 113.4±2.1 (3), 115.65 (1), and 116.77 (2).

References

1. C. J. O'Brien and K. K. Kelley, J. Amer. Chem. Soc. 74, 5616 (1957).
2. D. M. Albright, Ph.D. Thesis, Carnegie Institute of Technology, 1956.
3. B. J. Holm and F. Gronvold, Acta. Chem. Scand. 27, 2043 (1973).
4. J. D. H. Donnay and H. M. Ondik, "Crystal Data, Determinative Tables, Third Edition," Vol. II, 1973.
5. JANAF Thermochemical Tables: Na_3AlF_6(cr, α), Na_3AlF_6(ℓ), $Na_5Al_3F_{14}$(cr), 12-31-79.
6. W. B. Frank, J. Phys. Chem. 65, 2081 (1961).
7. J. Brynestad, K. Grjotheim, and S. Urnes, Metallurgia Ital. 52, 495 (1960).
8. F. C. Kracek, "Handbook of Physical Constants," Geological Society of America, Special Paper No. 36, 140 (1942).
9. A. J. Majumdar and R. Roy, J. Inorg. Nucl. Chem. 27, 1961 (1965).
10. G. J. Landon and A. R. Ubbelohde, Proc. Roy. Soc. (London) 240A, 160 (1957).
11. V. S. Lyashenko, Metallurg. 10, 85 (1935).
12. W. A. Roth and W. Bertram, Z. Elektrochem. 35, 297 (1929).
13. M. Rolin, Bull. Soc. Chim. France 1960, 671 (1960).
14. N. W. F. Phillips, R. H. Singleton, and E. A. Hollingshead, J. Electro Chem. Soc. 102, 690 (1955).
15. B. J. Holm and J. L. Holm, Thermochemica Acta 5, 273 (1973).
16. M. Rolin and M. Bernard, Bull. Soc. Chim. France, 1962, 423 (1962).

Cryolite, Beta (Na_3AlF_6) $Al_1F_6Na_3$(cr)

Enthalpy Reference Temperature = T_r = 298.15 K — Standard State Pressure = p° = 0.1 MPa

T/K	$C_p°$ (J K⁻¹mol⁻¹)	S° (J K⁻¹mol⁻¹)	-[G°-H°(T_r)]/T (J K⁻¹mol⁻¹)	H°-H°(T_r) (kJ mol⁻¹)	$\Delta_f H°$ (kJ mol⁻¹)	$\Delta_f G°$ (kJ mol⁻¹)	Log K_f
0							
100							
200							
298.15	194.723	284.881	284.881	0.	-3288.781	-3137.880	549.744
300	194.744	286.086	284.885	0.360	-3288.796	-3136.943	546.190
400	200.916	342.763	292.572	20.077	-3298.111	-3085.524	402.928
500	212.464	388.778	307.340	40.719	-3299.502	-3032.196	316.771
600	226.103	428.705	324.306	62.639	-3299.786	-2978.688	259.318
700	240.747	464.652	341.827	85.977	-3298.807	-2925.231	218.283
800	255.977	497.791	359.276	110.812	-3296.509	-2872.005	187.523
836.000	261.500	509.179	363.999	121.370	-------- ALPHA <--> BETA		-------
900	271.458	528.831	376.408	137.181	-3292.905	-2819.145	163.619
1000	287.148	558.243	393.133	165.109	-3298.624	-2765.994	144.481
1100	302.963	586.352	409.430	194.615	-3292.178	-2713.032	128.831
1200	318.863	613.395	425.307	225.706	-3574.631	-2653.397	115.499
1285.000	332.461	635.677	438.489	253.387	------- BETA <--> LIQUID		-------
1300	334.846	639.549	440.786	258.392	-3562.664	-2577.103	103.549
1400	350.870	664.950	455.895	292.678	-3549.161	-2501.787	93.343
1500	366.895	689.704	470.661	328.566	-3534.115	-2427.490	84.533

PREVIOUS: CURRENT: December 1979

Cryolite, Beta (Na_3AlF_6) $Al_1F_6Na_3$(cr)

CRYOLITE (Na_3AlF_6) LIQUID M_r = 209.941268

$S°(298.15\ K)$ = [286.058] J K^{-1} mol^{-1}

T_{fus} (β → ℓ) = 1285 ± 3 K

$\Delta_f H°(298.15\ K)$ = [-3246.472] kJ mol^{-1}

$\Delta_{fus}H°$ (β → ℓ) = 110.039 ± 5.0 kJ mol^{-1}

Enthalpy of Formation

$\Delta_f H°(Na_3AlF_6$, ℓ, 298.15 K) is calculated from $\Delta_f H°(Na_3AlF_6$, cr, β, 298.15 K) by adding $\Delta_{fus}H°$ and the difference in enthalpy, H°(1285 K)-H°(298.15 K), between the β-crystal and liquid (1).

Hong and Kleppa (2), using a twin micro calorimeter, have determined the enthalpy of mixing for the reaction 3 NaF(ℓ) + AlF_3(cr) = $NaAlF_6$(ℓ) as $\Delta_r H°$(1298 K) = -10.595±0.05 kcal mol^{-1}. When this result is combined with JANAF auxiliary data and heats of formation from 3, we obtain $\Delta_f H°(Na_3AlF_6$, ℓ, 298.15 K) = -775.1±1.0 kcal mol^{-1}.

Yudin and Mashovets (5) studied the equilibrium reaction between water vapor and cryolite (reaction B below). Results of our analysis of these studies are given below. The average of the three experimental values, $\Delta_f H°(Na_3AlF_6$, ℓ, 298.15 K) = -776.2±1.5 kcal mol^{-1}, is in good agreement with our adopted value.

Source	Reaction[a]	T/K	Data Points	δS cal K^{-1} mol^{-1}	$\Delta_r H°$(298.15 K), kcal mol^{-1} 2nd law	3rd law	$\Delta_f H°(Na_3AlF_6$, ℓ, 298.15 K) kcal mol^{-1}
Yudin (3) (1963)	A	1300	1	-	-	24.30±2.0	-777.76±2.0
Mashovets (2) (1962)	B	1286-1315	3	-6.15±0.1	41.10±0.1	49.12±0.2	-775.67±2.0

[a]Reactions: A $Na_3AlF_6(\ell) = 3\ NaF(\ell) + AlF_3(cr)$

B $2/3\ Na_3AlF_6(\ell) + H_2O(g) = 1/3\ Al_2O_3(cr) + 2\ HF(g) + 2\ NaF(\ell)$

Heat Capacity and Entropy

The high temperature enthalpies, 1285.5 - 1370.5 K, were measured by O'Brien and Kelley (6). Using their data and smoothing, we derive a constant $C_p°$ = 94.53 cal K^{-1} mol^{-1} (395.514 J K^{-1} mol^{-1}) for Na_3AlF_6(ℓ). A hypothetical glass transition is assumed at 790 K. The $C_p°$ values above 1370.5 K are obtained by linear extrapolation. S°(298.15 K) is calculated in a manner analogous to that used for the enthalpy of formation.

Albright (7) meaured the high temperature enthalpies of Na_3AlF_6(ℓ) in the temperature range 1298.1 - 1311.8 K. His data are systematically higher (5%) than the values determined by O'Brien and Kelley. Holm and Gronvold (8) also measured enthalpies of Na_3AlF_6(ℓ) by drop calorimetry (1290.3 - 1333.1 K). Their data are intermediate between that of Albright and that of O'Brien and Kelley. For reasons discussed previously (see the Na_3AlF_6(cr, β) table), we believe both sets (7, 8) are biased and have not used the data in this evaluation.

Fusion Data

See the β-crystal table for details.

References-Na_3AlF_6(ℓ)

1. JANAF Thermochemical Tables: Na_3AlF_6(cr, β), 12-1-79.
2. K. C. Hong and O. J. Kleppa, J. Phys. Chem. 82, 176 (1978).
3. J. L. Holm and Gronvold, Acta. Chem. Scand. 26, 1733 (1972).
4. B. F. Yudin and V. P. Mashovets, J. Appl. Chem. 36, 1192 (1963).
5. V. P. Mashovets and B. F. Yudin, Izvest. Vysshikh Ucheb. Zavedenil Tsvetnaya met. 5, 95 (1962).
6. C. J. O'Brien and K. K. Kelley, J. Amer. Chem. Soc. 79, 5616 (1957).
7. D. M. Albright, Ph.D. Thesis, Carnegie Institute of Technology, 1956.
8. B. J. Holm and F. Gronvold, Acta. Chem. Scand. 27, 2043 (1973).

References-Na_3AlF_6-(cr, α)

1. J. P. Coughlin, J. Amer. Chem. Soc. 80, 1802 (1948).
2. U. S. Nat. Bur. Stand. IR 75-968, 1976.
3. U. S. Nat. Bur. Stand. Tech. Note 270-3, 1968.
4. P. Gross, C. Haymen, and D. I. Levi, Met. Soc. Conf. 8, 903 (1961); see Chem. Abstr. 62, 8419f (1962).
5. JANAF Thermochemical Tables: PbF_2(cr, α), 12-31-73; Na_3AlF_6(cr, β), Na_3AlF_6(ℓ), 12-31-79.
6. E. Baud, Ann. Chim. Phys. 1, 8 (1904).
7. K. Grjotheim, K. Motzfeldt, and D. B. Rao, in "Light Metals 1971. Proceedings of Symposia at 100th AIME Annual Meeting," (T. G. Edgeworth, Ed.). p. 223, AIME, New York 1971; as quoted in J. L. Holm, High Temp. Sci. 6, 16 (1974).
8. K. Ono, T. Matsushima, and T. Ito, Nippon Kinzoku Gakkaishi 29, 501 (1965).
9. E. W. Dewing, Met. Trans. 1, 1691 (1970).
10. J. J. Stokes, Jr., and W. B. Frank, "Extractive Metallurgy of Aluminum," Vol. II, Interscience Publishers, 1963, p.3.
11. E. W. Dewing, Met. Trans. 1, 2211 (1970).
12. V. P. Mashovets and B. F. Yudin, Izvest. Ucheb. Zavedenii Tsvetnaya Met. 5, 95 (1962); see Chem. Abstr. 5874h.
13. E. G. King, J. Am. Chem. Soc. 79, 2056 (1957).
14. C. J. O'Brien and K. K. Kelley, J. Am. Chem. Soc. 79, 5616 (1957).
15. D. M. Albright, Ph.D. Thesis, Carnegie Institute of Technology, 1956.
16. W. B. Frank, J. Phys. Chem. 65, 2081 (1961).
17. T. B. Douglas and D. A. Ditmars, Nat. Bur. Stand. Report 9389, 1966.
18. J. Joly, Proc. Roy. Soc. (London) 41, 250 (1886).
19. M. E. Baud, J. Phys. Radium Ser. 4, 2, 569 (1903).
20. W. A. Roth and Bertram, A. Elektrochem. 35, 297 (1929).
21. A. N. Krestovnikov and G. A. Karetnikov, Legkie Metal. 3, 29 (1934).
22. V. S. Lyashenko, Metallurg. 10, 85 (1935).
23. E. W. Dewing, Aluminum Company of Canada, Ltd. personal communication, October, 1980.

Cryolite (Na_3AlF_6) $Al_1F_6Na_3$(l)

Enthalpy Reference Temperature = T_r = 298.15 K   Standard State Pressure = p° = 0.1 MPa

T/K	$C_p°$ (J $K^{-1}mol^{-1}$)	S° (J $K^{-1}mol^{-1}$)	$-[G°-H°(T_r)]/T$ (J $K^{-1}mol^{-1}$)	$H°-H°(T_r)$ (kJ mol^{-1})	$\Delta_f H°$ (kJ mol^{-1})	$\Delta_f G°$ (kJ mol^{-1})	Log K_f
0							
100							
200							
298.15	215.727	286.058	286.058	0.	-3246.472	-3095.921	542.393
300	216.212	287.394	286.062	0.400	-3246.447	-3094.987	538.885
400	236.379	352.597	294.813	23.114	-3252.764	-3044.111	397.520
500	248.634	406.524	311.917	47.304	-3250.608	-2992.175	312.590
600	262.793	453.110	331.649	72.876	-3247.240	-2940.784	256.018
700	276.822	494.674	352.020	99.858	-3242.617	-2890.057	215.659
790.000	289.366	528.901	370.246	125.337	GLASS <--> LIQUID		
790.000	395.514	528.901	370.246	125.337	TRANSITION		
800	395.514	533.876	372.260	129.292	-3235.719	-2840.082	185.438
836.000	395.514	551.285	379.597	143.531	ALPHA <--> BETA		
900	395.514	580.461	392.856	168.844	-3218.933	-2791.638	162.022
1000	395.514	622.132	413.737	208.395	-3213.029	-2744.289	143.347
1100	395.514	659.829	434.423	247.946	-3196.537	-2698.214	128.128
1200	395.514	694.243	454.661	287.498	-3470.530	-2646.313	115.191
1285.000	395.514	721.311	471.415	321.116	BETA <--> LIQUID		
1300	395.514	725.901	474.324	327.049	-3451.697	-2578.393	103.601
1400	395.514	755.211	493.354	366.600	-3432.929	-2511.920	93.721
1500	395.514	782.499	511.731	406.152	-3414.220	-2446.787	85.205
1600	395.514	808.025	529.460	445.703	-3395.566	-2382.900	77.794
1700	395.514	832.003	546.559	485.254	-3376.962	-2320.178	71.290
1800	395.514	854.610	563.051	524.806	-3358.402	-2258.550	65.541
1900	395.514	875.994	578.964	564.357	-3339.880	-2197.953	60.426
2000	395.514	896.281	594.327	603.909	-3321.387	-2138.329	55.847
2100	395.514	915.578	609.169	643.460	-3302.915	-2079.631	51.728
2200	395.514	933.978	623.518	683.011	-3284.455	-2021.812	48.004
2300	395.514	951.559	637.401	722.563	-3265.996	-1964.832	44.623
2400	395.514	968.392	650.844	762.114	-3247.531	-1908.656	41.541
2500	395.514	984.537	663.871	801.665	-3229.049	-1853.249	38.722
2600	395.514	1000.050	676.505	841.217	-3210.544	-1798.582	36.134
2700	395.514	1014.977	688.766	880.768	-3192.008	-1744.628	33.752
2800	395.514	1029.360	700.675	920.319	-3467.335	-1690.393	31.535
2900	395.514	1043.240	712.250	959.871	-3447.624	-1627.279	29.310
3000	395.514	1056.648	723.507	999.422	-3427.867	-1564.845	27.246

PREVIOUS: December 1968   CURRENT: December 1979

Cryolite (Na_3AlF_6) $Al_1F_6Na_3$(l)

CRYOLITE (Na_3AlF_6) ALPHA-BETA-LIQUID M_r = 209.941268 Cryolite (Na_3AlF_6) $Al_1F_6Na_3$(cr,l)

0 to 836 K crystal, alpha
836 to 1285 K crystal, beta
above 1285 K liquid

Refer to the individual tables for details.

Enthalpy Reference Temperature = T_r = 298.15 K — Standard State Pressure = $p°$ = 0.1 MPa

T/K	$C_p°$ (J K⁻¹mol⁻¹)	$S°$ (J K⁻¹mol⁻¹)	$-[G°-H°(T_r)]/T$ (J K⁻¹mol⁻¹)	$H°-H°(T_r)$ (kJ mol⁻¹)	$\Delta_fH°$ (kJ mol⁻¹)	$\Delta_fG°$ (kJ mol⁻¹)	Log K_f
0	0.	0.	INFINITE	-38.095	-3304.649	-3304.649	INFINITE
100	100.039	62.287	402.572	-34.028	-3313.806	-3261.080	1703.414
200	179.033	159.302	257.165	-19.573	-3317.306	-3206.528	837.459
298.15	215.727	238.467	238.467	0.	-3316.908	-3152.168	552.247
300	216.212	239.803	238.471	0.400	-3316.883	-3151.146	548.663
400	236.379	305.007	247.223	23.114	-3323.200	-3095.511	404.232
500	248.634	358.934	264.326	47.304	-3321.044	-3038.816	317.463
600	262.793	405.519	284.058	72.876	-3317.675	-2982.666	259.664
700	276.822	447.083	304.429	99.858	-3313.053	-2927.179	218.429
800	290.780	484.959	324.661	128.238	-3307.209	-2872.439	187.551
836.000	295.809	497.868	331.843	138.797	ALPHA <--> BETA	TRANSITION	
836.000	261.500	509.179	331.843	148.253			
900	271.458	528.831	345.157	165.307	-3292.905	-2819.145	163.619
1000	287.148	558.243	365.007	193.236	-3298.624	-2765.994	144.481
1100	302.963	586.352	383.860	222.741	-3292.178	-2713.032	128.831
1200	318.863	613.395	401.868	253.832	-3574.631	-2653.397	115.499
1285.000	332.461	635.677	416.601	281.513	BETA <--> LIQUID	TRANSITION	
1285.000	395.514	721.311	416.601	391.552			
1300	395.514	725.901	420.143	397.485	-3451.697	-2578.392	103.601
1400	395.514	755.211	443.043	437.036	-3432.929	-2511.920	93.721
1500	395.514	782.499	464.774	476.588	-3414.220	-2446.787	85.205
1600	395.514	808.025	485.438	516.139	-3395.566	-2382.900	77.794
1700	395.514	832.003	505.126	555.690	-3376.962	-2320.178	71.290
1800	395.514	854.610	523.920	595.242	-3358.402	-2258.550	65.541
1900	395.514	875.994	541.892	634.793	-3339.880	-2197.953	60.426
2000	395.514	896.281	559.109	674.345	-3321.387	-2138.329	55.847
2100	395.514	915.578	575.628	713.896	-3302.915	-2079.631	51.728
2200	395.514	933.978	591.502	753.447	-3284.454	-2021.812	48.004
2300	395.514	951.559	606.777	792.999	-3265.996	-1964.832	44.623
2400	395.514	968.392	621.496	832.550	-3247.531	-1908.656	41.541
2500	395.514	984.537	635.697	872.101	-3229.049	-1853.249	38.722
2600	395.514	1000.050	649.414	911.653	-3210.544	-1798.582	36.134
2700	395.514	1014.977	662.679	951.204	-3192.008	-1744.628	33.752
2800	395.514	1029.360	675.519	990.755	-3467.335	-1690.393	31.535
2900	395.514	1043.240	687.961	1030.307	-3447.624	-1627.278	29.310
3000	395.514	1056.648	700.029	1069.858	-3427.867	-1564.845	27.246

PREVIOUS: CURRENT: December 1979

Cryolite (Na_3AlF_6) $Al_1F_6Na_3$(cr,l)

ALUMINUM HYDRIDE (AlH) IDEAL GAS M_r = 27.98948

$S°(298.15\ K) = 187.86 \pm 0.5\ J\ K^{-1}\ mol^{-1}$

$\Delta_f H°(0\ K) = 259.5 \pm 20\ kJ\ mol^{-1}$
$\Delta_f H°(298.15\ K) = 259.4 \pm 20\ kJ\ mol^{-1}$

Electronic Level and Quantum Weight

State	ε_i, cm^{-1}	g_i
$^1\Sigma^+$	0	1

$\omega_e = 1682.56\ cm^{-1}$ $\omega_e x_e = 29.09\ cm^{-1}$ $\sigma = 1$
$B_e = 6.3907\ cm^{-1}$ $\alpha_e = 0.1858\ cm^{-1}$ $r_e = 1.6477\ Å$

Enthalpy of Formation

Gaydon (1) extrapolated the available $X^1\Sigma$ vibrational levels, also considered predissociation in the $A^1\pi$ state and concluded that the best value of $D_0°$ was 2.9 ± 0.2 ev. The linear Birge-Sponer extrapolation gave a value of 3.05 ev while the predissociation limit was 3.07 ev. The value recommended by Gaydon (1) is adopted here and leads to a value of $\Delta_f H°$(AlH, g, 298.15 K) = 62.16 ± 5.0 kcal mol^{-1}, using JANAF auxiliary data.

Heat Capacity and Entropy

Zeeman and Ritter (2) examined the UV band system of AlH under high resolution and reported the molecular constants adopted above.

References

1. A. G. Gaydon, 'Dissociation Energies', 2nd ed., Chapman and Hall, Ltd., London, 1953.
2. P. B. Zeeman and E. G. Ritter, Can. J. Phys. 32, 555 (1954).

Aluminum Hydride (AlH) $Al_1H_1(g)$

Enthalpy Reference Temperature = T_r = 298.15 K Standard State Pressure = p° = 0.1 MPa

T/K	$C_p°$ (J $K^{-1}mol^{-1}$)	S° (J $K^{-1}mol^{-1}$)	$-[G°-H°(T_r)]/T$ (J $K^{-1}mol^{-1}$)	$H°-H°(T_r)$ (kJ mol^{-1})	$\Delta_f H°$ (kJ mol^{-1})	$\Delta_f G°$ (kJ mol^{-1})	Log K_f
0	0.	0.	INFINITE	-8.669	259.512	259.512	INFINITE
100	29.124	156.000	213.828	-5.783	260.415	250.550	-130.874
200	29.154	176.194	190.540	-2.869	260.180	240.711	-62.867
250	29.220	182.705	188.346	-1.410	259.827	235.882	-49.285
298.15	29.373	187.864	187.864	0.	259.408	231.308	-40.524
300	29.381	188.045	187.864	0.054	259.391	231.133	-40.244
350	29.658	192.594	188.223	1.530	258.907	226.462	-33.798
400	30.041	196.578	189.023	3.022	258.397	221.861	-28.972
450	30.498	200.142	190.064	4.535	257.877	217.325	-25.226
500	30.996	203.381	191.236	6.072	257.354	212.848	-22.236
600	32.008	209.122	193.751	9.223	256.304	204.045	-17.764
700	32.942	214.128	196.312	12.471	255.236	195.419	-14.582
800	33.754	218.581	198.822	15.807	254.115	186.949	-12.207
900	34.441	222.597	201.244	19.218	252.897	178.626	-10.367
1000	35.018	226.257	203.565	22.692	240.954	171.204	-8.943
1100	35.505	229.618	205.783	26.218	239.786	164.286	-7.801
1200	35.917	232.725	207.900	29.790	238.643	157.473	-6.855
1300	36.271	235.615	209.922	33.400	237.518	150.755	-6.057
1400	36.577	238.314	211.855	37.043	236.403	144.123	-5.377
1500	36.844	240.847	213.704	40.714	235.296	137.570	-4.791
1600	37.081	243.232	215.476	44.410	234.191	131.091	-4.280
1700	37.293	245.487	217.176	48.129	233.089	124.681	-3.831
1800	37.484	247.624	218.808	51.868	231.985	118.336	-3.434
1900	37.658	249.655	220.379	55.626	230.881	112.052	-3.081
2000	37.818	251.591	221.891	59.399	229.775	105.826	-2.764
2100	37.966	253.440	223.350	63.189	228.667	99.656	-2.479
2200	38.104	255.209	224.758	66.992	227.556	93.539	-2.221
2300	38.234	256.906	226.119	70.809	226.442	87.472	-1.987
2400	38.356	258.536	227.436	74.639	225.326	81.454	-1.773
2500	38.473	260.104	228.712	78.480	224.207	75.482	-1.577
2600	38.584	261.615	229.948	82.333	223.086	69.555	-1.397
2700	38.691	263.073	231.148	86.197	221.963	63.671	-1.232
2800	38.793	264.482	232.314	90.071	-73.064	58.797	-1.097
2900	38.893	265.845	233.447	93.955	-73.096	63.507	-1.144
3000	38.989	267.165	234.549	97.850	-73.130	68.218	-1.188
3100	39.082	268.445	235.622	101.753	-73.166	72.930	-1.229
3200	39.174	269.688	236.667	105.666	-73.205	77.644	-1.267
3300	39.263	270.894	237.686	109.588	-73.245	82.358	-1.304
3400	39.350	272.068	238.680	113.518	-73.288	87.074	-1.338
3500	39.435	273.210	239.650	117.458	-73.333	91.791	-1.370
3600	39.520	274.322	240.598	121.406	-73.380	96.510	-1.400
3700	39.602	275.406	241.524	125.362	-73.430	101.230	-1.429
3800	39.684	276.463	242.430	129.326	-73.483	105.951	-1.456
3900	39.764	277.495	243.316	133.298	-73.540	110.674	-1.482
4000	39.844	278.503	244.183	137.279	-73.600	115.398	-1.507
4100	39.922	279.487	245.032	141.267	-73.664	120.124	-1.530
4200	40.000	280.450	245.864	145.263	-73.732	124.851	-1.553
4300	40.077	281.393	246.679	149.267	-73.805	129.580	-1.574
4400	40.154	282.315	247.479	153.279	-73.883	134.311	-1.594
4500	40.229	283.218	248.263	157.298	-73.965	139.043	-1.614
4600	40.305	284.103	249.032	161.324	-74.055	143.778	-1.633
4700	40.379	284.971	249.788	165.359	-74.152	148.514	-1.651
4800	40.454	285.821	250.530	169.400	-74.256	153.253	-1.668
4900	40.527	286.656	251.259	173.449	-74.367	157.994	-1.684
5000	40.601	287.476	251.975	177.506	-74.488	162.737	-1.700
5100	40.674	288.281	252.679	181.570	-74.618	167.483	-1.715
5200	40.746	289.071	253.371	185.641	-74.758	172.231	-1.730
5300	40.819	289.848	254.052	189.719	-74.909	176.982	-1.744
5400	40.891	290.612	254.722	193.804	-75.071	181.737	-1.758
5500	40.963	291.363	255.381	197.897	-75.246	186.494	-1.771
5600	41.034	292.101	256.030	201.997	-75.433	191.254	-1.784
5700	41.106	292.828	256.670	206.104	-75.634	196.018	-1.796
5800	41.177	293.544	257.299	210.218	-75.846	200.786	-1.808
5900	41.247	294.248	257.920	214.339	-76.076	205.557	-1.820
6000	41.318	294.942	258.531	218.467	-76.300	210.331	-1.831

PREVIOUS: June 1963 (1 atm) CURRENT: June 1963 (1 bar)

Aluminum Hydride (AlH) $Al_1H_1(g)$

ALUMINUM OXIDE HYDRIDE (OAlH) IDEAL GAS M_r = 43.98888

$S°(298.15\ K) = 215.3 \pm 2.4\ JK^{-1}mol^{-1}$

$\Delta_f H°(0\ K) = [37.6 \pm 84]\ kJ\ mol^{-1}$
$\Delta_f H°(298.15\ K) = [33.5 \pm 84]\ kJ\ mol^{-1}$

Vibrational Frequencies and Degeneracies

ν, cm^{-1}
[900] (1)
[1000] (2)
[1600] (1)

Ground State Quantum Weight: [1] $\sigma = 1$
Point Group: [$C_{\infty v}$]
Bond Distances: Al-O = [1.62] Å Al-H = [1.64] Å
Bond Angle: H-Al-O = [180]°
$B_0 = [0.538347]\ cm^{-1}$

Enthalpy of Formation

The enthalpy of formation is estimated by summing the constituent bond energies. The Al=O was taken as that in AlO(g) and the Al-H bond was taken as that in AlH(g); this gave a heat of atomization of 182 kcal mol^{-1}.

Heat Capacity and Entropy

Vibrational frequencies, bond lengths and angles and point group are taken from NBS(1).

References

1. U. S. Nat. Bur. Stand. Report 6297, Jan. 1, 1959.

Aluminum Hydride Oxide (OAlH) $Al_1H_1O_1(g)$

Enthalpy Reference Temperature = T_r = 298.15 K   Standard State Pressure = p° = 0.1 MPa

	J K⁻¹mol⁻¹			kJ mol⁻¹			
T/K	$C_p°$	S°	$-[G°-H°(T_r)]/T$	$H°-H°(T_r)$	$\Delta_f H°$	$\Delta_f G°$	Log K_f
0	0.	0.	INFINITE	-9.018	37.569	37.569	INFINITE
100	29.106	182.067	243.166	-6.110	37.041	33.235	-17.360
200	30.298	202.437	218.263	-3.165	35.382	30.013	-7.839
250	32.196	209.387	215.811	-1.606	34.400	28.784	-6.014
298.15	34.569	215.256	215.256	0.	33.472	27.787	-4.868
300	34.666	215.470	215.257	0.064	33.437	27.752	-4.832
350	37.323	221.013	215.688	1.864	32.539	26.876	-4.011
400	39.919	226.168	216.680	3.795	31.722	26.124	-3.411
450	42.325	231.011	218.006	5.852	30.987	25.469	-2.956
500	44.491	235.584	219.537	8.024	30.327	24.892	-2.600
600	48.094	244.029	222.927	12.661	29.184	23.916	-2.082
700	50.851	251.659	226.496	17.614	28.194	23.118	-1.725
800	52.953	258.593	230.081	22.809	27.263	22.457	-1.466
900	54.567	264.927	233.606	28.189	26.312	21.912	-1.272
1000	55.822	270.744	237.033	33.711	14.685	22.238	-1.162
1100	56.810	276.112	240.345	39.344	13.870	23.033	-1.094
1200	57.598	281.091	243.536	45.066	13.103	23.901	-1.040
1300	58.235	285.727	246.605	50.859	12.369	24.830	-0.998
1400	58.755	290.062	249.556	56.709	11.655	25.816	-0.963
1500	59.186	294.131	252.393	62.607	10.953	26.852	-0.935
1600	59.545	297.963	255.123	68.544	10.256	27.934	-0.912
1700	59.847	301.582	257.750	74.514	9.558	29.061	-0.893
1800	60.105	305.010	260.281	80.512	8.856	30.228	-0.877
1900	60.325	308.266	262.722	86.534	8.147	31.434	-0.864
2000	60.514	311.365	265.077	92.576	7.428	32.679	-0.853
2100	60.679	314.322	267.352	98.636	6.697	33.959	-0.845
2200	60.823	317.148	269.552	104.711	5.954	35.275	-0.838
2300	60.949	319.854	271.681	110.800	5.197	36.625	-0.832
2400	61.061	322.451	273.742	116.900	4.425	38.007	-0.827
2500	61.160	324.945	275.741	123.011	3.639	39.423	-0.824
2600	61.248	327.346	277.680	129.132	2.837	40.870	-0.821
2700	61.327	329.659	279.562	135.261	2.020	42.348	-0.819
2800	61.397	331.891	281.392	141.397	-292.713	44.825	-0.836
2900	61.461	334.046	283.170	147.540	-292.466	56.875	-1.024
3000	61.519	336.131	284.901	153.689	-292.233	68.918	-1.200
3100	61.571	338.149	286.586	159.844	-292.017	80.953	-1.364
3200	61.619	340.104	288.229	166.003	-291.815	92.980	-1.518
3300	61.662	342.001	289.829	172.167	-291.629	105.002	-1.662
3400	61.702	343.843	291.391	178.335	-291.458	117.019	-1.798
3500	61.738	345.632	292.915	184.507	-291.302	129.030	-1.926
3600	61.772	347.371	294.404	190.683	-291.161	141.038	-2.046
3700	61.803	349.064	295.859	196.862	-291.036	153.041	-2.161
3800	61.831	350.713	297.280	203.043	-290.926	165.042	-2.269
3900	61.858	352.319	298.671	209.228	-290.831	177.041	-2.371
4000	61.882	353.886	300.032	215.415	-290.752	189.036	-2.469
4100	61.905	355.414	301.364	221.604	-290.689	201.030	-2.561
4200	61.926	356.906	302.669	227.796	-290.643	213.022	-2.649
4300	61.945	358.363	303.947	233.989	-290.613	225.014	-2.733
4400	61.964	359.788	305.200	240.185	-290.600	237.005	-2.814
4500	61.981	361.181	306.429	246.382	-290.603	248.996	-2.890
4600	61.997	362.543	307.634	252.581	-290.626	260.987	-2.964
4700	62.012	363.876	308.817	258.781	-290.668	272.979	-3.034
4800	62.026	365.182	309.977	264.983	-290.729	284.972	-3.101
4900	62.040	366.461	311.117	271.187	-290.810	296.967	-3.166
5000	62.052	367.715	312.236	277.391	-290.913	308.964	-3.228
5100	62.064	368.944	313.336	283.597	-291.038	320.962	-3.287
5200	62.075	370.149	314.417	289.804	-291.186	332.963	-3.345
5300	62.086	371.331	315.480	296.012	-291.359	344.968	-3.400
5400	62.096	372.492	316.525	302.221	-291.557	356.976	-3.453
5500	62.105	373.632	317.553	308.431	-291.781	368.987	-3.504
5600	62.114	374.751	318.565	314.642	-292.033	381.004	-3.554
5700	62.122	375.850	319.560	320.854	-292.314	393.025	-3.602
5800	62.130	376.931	320.540	327.067	-292.621	405.051	-3.648
5900	62.138	377.993	321.505	333.280	-292.962	417.082	-3.693
6000	62.145	379.037	322.455	339.494	-293.314	429.118	-3.736

PREVIOUS: March 1964 (1 atm)   CURRENT: March 1964 (1 bar)

Aluminum Hydride Oxide (OAlH) $Al_1H_1O_1(g)$

ALUMINUM HYDROXIDE (AlOH) IDEAL GAS $M_r = 43.98888$

$S°(298.15\ K) = [216.4 \pm 2]\ J\ K^{-1}\ mol^{-1}$

$\Delta_f H°(0\ K) = [-175.6 \pm 13]\ kJ\ mol^{-1}$

$\Delta_f H°(298.15\ K) = [-179.9 \pm 13]\ kJ\ mol^{-1}$

Vibrational Frequencies and Degeneracies

ν, cm^{-1}
[1000](1)
[1300](2)
[3600](1)

Ground State Quantum Weight: [1] $\sigma = 1$

Point Group: $[C_{\infty v}]$

Bond Distances: Al–O = [1.87] Å O–H = [0.96] Å

Bond Angle: Al–O–H = [180]°

Rotational Constant: $B_0 = [0.425629]\ cm^{-1}$

Enthalpy of Formation

From a mass-spectrometric investigation of the equilibrium 1/3 Al_2O_3(cr) + 1/3 Al(g) + 1/2 D_2(g) = AlOD(g), Farber et al. (1) report second law heats of reaction of 56.7 + 0.6 kcal mol^{-1} at 1980 K, 57.1 ± 2.8 kcal mol^{-1} at 2060 K and 62.7 ± 1.4 kcal mol^{-1} at 2000 K. Using JANAF auxiliary data, assuming AlOD to be AlOH, we obtain $\Delta_f H°$(AlOD, g, 298.15 K) = -44.6 ± 1, -44.0 ± 3.2 and -38.6 ± 1.8 kcal mol^{-1}. A second equilibrium in the same investigation was Al_2O(g) + D_2O(g) = 2 AlOD(g). for which second law heats of reaction of 6.3 ± 2.1 kcal mol^{-1} at 2060 K and 5.7 ± 0.8 kcal mol^{-1} at 2000 K were reported. Farber et al. (1) also measured the heat of reaction 1/3 Al_2O_3(cr) + 4/3 Al(g) = Al_2O(g) at the same temperatures as -10.7 ± 2.8 kcal mol^{-1} at 2060° and -10.3 ± 10.2 kcal mol^{-1} at 2000 K. Combining these reactions and eliminating Al_2O(g) yields $\Delta_r H°$(2060 K) = -4.4 ± 5.0 kcal mol^{-1} and $\Delta H°$(2000 K) = -5 ± 11 kcal mol^{-1} for the reaction 1/3 Al_2O_3(cr) + 4/3 Al(g) + 2 AlOD(g).

Using auxiliary JANAF values for Al_2O_3(cr) and Al(g), $\Delta_f H°$(D_2O, g, 298.15 K) = -59.56 kcal mol^{-1} (2) and functions for D_2O(g) from Friedman and Haar (3), we obtain $\Delta_f H°$(AlOD, g, 298.15 K) = -43.9 ± 6 kcal mol^{-1} and -43.4 ± 3 kcal mol^{-1}. From all the above results we adopt a value of $\Delta_f H°$(AlOD, g, 298.15 K) = -44 ± 3 kcal mol^{-1}, assuming that AlOH is less stable than AlOD by 1 kcal mol^{-1} as is the case for H_2O, HDO and D_2O, we obtain $\Delta_f H°$(AlOH, g, 298.15 K) = -43 ± 3 kcal mol^{-1}.

Farber et al. (1) also performed weight loss experiments under molecular flow conditions by allowing hydrogen to flow over Al_2O_3(cr). The products assumed were AlOH(g), Al(g), Al_2O(g), H(g) and H_2O(g) and the equilibrium amounts of H(g), Al(g) and Al_2O(g) were assumed to be those predicted by the free energies from the JANAF tables. A 3rd law analysis of the equilibrium constants given by Farber for the reaction 2 H_2(g) + Al_2O_3(cr) = 2 AlOH(g) + H_2O(g) yields $\Delta_r H°$(298.15 K) = 238 ± 12 kcal mol^{-1} and $\Delta_f H°$(AlOH, g, 298.15 K) = -52 ± 6 kcal mol^{-1}. However, the assumed equilibria involving Al_2O(g) and Al(g) may be significantly in error and thus the $\Delta_f H°$(298.15 K) should be considered as simply confirming the magnitude determined mass-spectrometrically.

Heat Capacity and Entropy

The vibrational frequencies are estimated by using the AlO(g) fundamental as approximately the Al–O stretch; the O–H stretch is that in H_2O(g); and the bending frequency is approximately the OH bend in CD_3OH(g) reported by Shimanouchi (4).

The AlO bond length was taken from NBS (5), and the OH bond length is assumed to be that in H_2O(g). The bond angle is assumed to be 180° using the prediction of Walsh (6) for "HAB" molecules with less than 11 valence electrons. Confirmation of this prediction is scanty for molecules having A more electronegative than B; however, Kuczkowski et al. (7) confirmed the linearity of KOH and CsOH which have B valence electrons.

References

1. M. Farber, M. A. Frisch, G. Grenier and H. C. Ko, Space Sciences, Inc., Final Report, USAF contract F04611-67-C-0010, AFRPL-TR-67-244, November 1967.
2. U. S. Nat. Bur. Stand. Tech. Note 270-1, 1965.
3. A. S. Friedman and L. Haar, J. Chem. Phys. 22, 2051 (1954).
4. T. Shimanouchi, NSRDS-NBS 6, 1967.
5. U. S. Nat. Bur. Stand. Report 6297, Jan. 1, 1959.
6. A. D. Walsh, J. Chem. Soc. 2288 (1953).
7. R. L. Kuczkowski, D. R. Lide, Jr., and L. C. Krishner, J. Chem. Phys. 44, 3131 (1966).

Aluminum Hydroxide (AlOH) $Al_1H_1O_1(g)$

Enthalpy Reference Temperature = T_r = 298.15 K Standard State Pressure = p° = 0.1 MPa

T/K	$C_p°$ (J K^{-1}mol^{-1})	S° (J K^{-1}mol^{-1})	$-[G°-H°(T_r)]/T$ (J K^{-1}mol^{-1})	$H°-H°(T_r)$ (kJ mol^{-1})	$\Delta_f H°$ (kJ mol^{-1})	$\Delta_f G°$ (kJ mol^{-1})	Log K_f
0	0.	0.	INFINITE	-8.830	-175.627	-175.627	INFINITE
100	29.101	184.020	243.238	-5.922	-176.155	-180.156	94.104
200	29.550	204.257	219.257	-3.000	-177.837	-183.570	47.944
250	30.503	210.942	216.947	-1.501	-178.879	-184.884	38.629
298.15	31.918	216.430	216.430	0.	-179.912	-185.947	32.577
300	31.979	216.628	216.431	0.059	-179.952	-185.984	32.383
350	33.772	221.689	216.826	1.702	-181.006	-186.906	27.894
400	35.672	226.323	217.728	3.438	-182.019	-187.679	24.508
450	37.532	230.633	218.925	5.269	-182.981	-188.328	21.861
500	39.273	234.678	220.300	7.189	-183.892	-188.873	19.731
600	42.302	242.116	223.327	11.273	-185.588	-189.708	16.516
700	44.763	248.828	226.499	15.631	-187.174	-190.268	14.198
800	46.782	254.941	229.678	20.211	-188.719	-190.604	12.445
900	48.471	260.552	232.801	24.976	-190.285	-190.747	11.071
1000	49.912	265.735	235.838	29.897	-202.512	-189.951	9.922
1100	51.158	270.552	238.778	34.952	-203.907	-188.626	8.957
1200	52.245	275.051	241.615	40.123	-205.224	-187.179	8.148
1300	53.200	279.271	244.351	45.397	-206.477	-185.624	7.458
1400	54.041	283.245	246.989	50.759	-207.679	-183.974	6.864
1500	54.784	287.000	249.532	56.201	-208.836	-182.241	6.346
1600	55.442	290.557	251.986	61.713	-209.959	-180.431	5.890
1700	56.026	293.936	254.355	67.287	-211.052	-178.552	5.486
1800	56.546	297.153	256.644	72.916	-212.123	-176.609	5.125
1900	57.010	300.223	258.857	78.595	-213.176	-174.607	4.800
2000	57.424	303.158	261.000	84.317	-214.215	-172.550	4.507
2100	57.796	305.969	263.075	90.078	-215.245	-170.441	4.239
2200	58.130	308.665	265.086	95.875	-216.267	-168.284	3.996
2300	58.430	311.256	267.037	101.703	-217.284	-166.080	3.772
2400	58.702	313.749	268.932	107.560	-218.300	-163.832	3.566
2500	58.948	316.150	270.773	113.442	-219.314	-161.542	3.375
2600	59.171	318.466	272.563	119.348	-220.330	-159.211	3.199
2700	59.373	320.703	274.305	125.276	-221.349	-156.841	3.034
2800	59.558	322.866	276.001	131.223	-516.272	-153.465	2.863
2900	59.727	324.959	277.653	137.187	-516.203	-140.509	2.531
3000	59.882	326.987	279.264	143.167	-516.139	-127.555	2.221
3100	60.023	328.952	280.835	149.163	-516.081	-114.603	1.931
3200	60.154	330.860	282.369	155.172	-516.030	-101.653	1.659
3300	60.274	332.713	283.867	161.193	-515.987	-88.705	1.404
3400	60.385	334.514	285.330	167.226	-515.951	-75.757	1.164
3500	60.487	336.266	286.760	173.270	-515.923	-62.811	0.937
3600	60.582	337.971	288.159	179.323	-515.905	-49.865	0.724
3700	60.670	339.632	289.528	185.386	-515.895	-36.920	0.521
3800	60.752	341.251	290.868	191.457	-515.896	-23.974	0.330
3900	60.828	342.831	292.180	197.536	-515.907	-11.029	0.148
4000	60.899	344.371	293.466	203.623	-515.928	1.917	-0.025
4100	60.965	345.876	294.726	209.716	-515.962	14.863	-0.189
4200	61.027	347.346	295.961	215.815	-516.007	27.811	-0.346
4300	61.085	348.783	297.173	221.921	-516.065	40.760	-0.495
4400	61.139	350.188	298.362	228.032	-516.137	53.710	-0.638
4500	61.190	351.562	299.529	234.149	-516.220	66.661	-0.774
4600	61.238	352.908	300.675	240.270	-516.320	79.615	-0.904
4700	61.283	354.225	301.800	246.396	-516.437	92.572	-1.029
4800	61.325	355.516	302.906	252.527	-516.570	105.531	-1.148
4900	61.365	356.781	303.993	258.661	-516.719	118.493	-1.263
5000	61.403	358.021	305.061	264.800	-516.888	131.458	-1.373
5100	61.439	359.237	306.111	270.942	-517.077	144.427	-1.479
5200	61.472	360.430	307.144	277.087	-517.287	157.399	-1.581
5300	61.504	361.602	308.161	283.236	-517.519	170.377	-1.679
5400	61.535	362.752	309.161	289.388	-517.774	183.358	-1.774
5500	61.563	363.881	310.146	295.543	-518.053	196.344	-1.865
5600	61.591	364.990	311.115	301.701	-518.358	209.336	-1.953
5700	61.616	366.081	312.070	307.861	-518.690	222.334	-2.037
5800	61.641	367.153	313.011	314.024	-519.048	235.337	-2.119
5900	61.664	368.207	313.937	320.189	-519.437	248.346	-2.199
6000	61.687	369.243	314.850	326.357	-519.835	261.361	-2.275

PREVIOUS: December 1967 (1 atm) CURRENT: December 1967 (1 bar)

Aluminum Hydroxide (AlOH) $Al_1H_1O_1(g)$

ALUMINUM HYDROXIDE, ION ($AlOH^+$) IDEAL GAS M_r = 43.98833

$\Delta_f H°(0\ K)$ = [541.9 ± 109] kJ mol^{-1}

S°(298.15 K) = [228.7 ± 2] J K^{-1} mol^{-1} $\Delta_f H°(298.15\ K)$ = [543.920] kJ mol^{-1}

Vibrational Frequencies and Degeneracies

ν, cm^{-1}

[900](1)

[1100](2)

[3600](1)

Ground State Quantum Weight: [4] σ = 1

Point Group: [$C_{\infty v}$]

Bond Distances: Al-O = [1.90] A O-H = [0.96] Å

Bond Angle: Al-O-H = [180]°

Rotational Constant: B_0 = [0.412959] cm^{-1}

Enthalpy of Formation

The enthalpy of formation is obtained from $\Delta_f H°$(AlOH, g, 298.15 K) and its ionization potential, which is estimated to be 7.5 ± 1 eV (173 ± 23 kcal mol^{-1}). The ionization potential is estimated from a reported value of 7.5 eV for the appearance potential of $AlOD^+$, presumably from the parent AlOD, by Farber et al. (1).

Heat Capacity and Entropy

According to the correlation diagram given by Walsh (2) for "HAB" molecules, a compound with 9 valence electrons would have the three outermost electrons is an unfilled Π orbital. Thus, it would be a $^2\Pi$ state since the molecule ought to be linear if it has less than 11 valence electrons. The outermost orbital is a bonding orbital and thus the loss of an electron from it would weaken the bonding. The vibrational frequencies are estimated from those for AlOH(g) with somewhat lower frequencies due to the weaker bonding. Similarly, the Al-O bond length is slightly increased over that in AlOH(g).

References

1. M. Farber, M. A. Frisch, G. Grenier, and H. C. Ko, Space Sciences, Inc., Final Report, USAF contract F04611-67-C-0010, AFRPL-TR-67-144, November 1967.

2. A. D. Walsh, J. Chem. Soc. 2288 (1953).

Aluminum Hydroxide, Ion ($AlOH^+$) $Al_1H_1O_1^+(g)$

Enthalpy Reference Temperature = T_r = 298.15 K Standard State Pressure = p° = 0.1 MPa

T/K	$C_p°$	S° (J K^{-1}mol^{-1})	$-[G°-H°(T_r)]/T$ (J K^{-1}mol^{-1})	$H°-H°(T_r)$ (kJ mol^{-1})	$\Delta_f H°$ (kJ mol^{-1})	$\Delta_f G°$ (kJ mol^{-1})	Log K_f
0	0.	0.	INFINITE	-8.947	541.890		
100	29.104	195.798	256.187	-6.039			
200	30.021	216.119	231.632	-3.103			
250	31.562	222.970	229.234	-1.566			
298.15	33.535	228.693	228.693	0.	543.920	527.974	-92.499
300	33.616	228.901	228.694	0.062	543.922	527.875	-91.911
350	35.837	234.250	229.111	1.798	544.000	525.195	-78.381
400	37.986	239.177	230.066	3.645	544.136	522.500	-68.232
450	39.947	243.766	231.336	5.594	544.333	519.784	-60.335
500	41.684	248.067	232.796	7.636	544.582	517.044	-54.015
600	44.529	255.930	236.009	11.953	545.198	511.480	-44.528
700	46.724	262.966	239.367	16.520	545.900	505.806	-37.744
800	48.476	269.324	242.720	21.283	546.616	500.029	-32.649
900	49.928	275.119	246.003	26.205	547.286	494.165	-28.681
1000	51.167	280.446	249.184	31.261	537.272	488.990	-25.542
1100	52.245	285.374	252.253	36.433	538.074	484.124	-22.989
1200	53.191	289.961	255.207	41.706	538.936	479.182	-20.858
1300	54.029	294.253	258.047	47.068	539.850	474.166	-19.052
1400	54.771	298.284	260.778	52.508	540.805	469.077	-17.501
1500	55.432	302.086	263.407	58.019	541.795	463.920	-16.155
1600	56.020	305.683	265.938	63.592	542.813	458.695	-14.975
1700	56.544	309.095	268.377	69.221	543.852	453.406	-13.931
1800	57.013	312.341	270.730	74.899	544.909	448.055	-13.002
1900	57.432	315.434	273.002	80.622	545.979	442.645	-12.169
2000	57.809	318.390	275.198	86.384	547.059	437.178	-11.418
2100	58.147	321.219	277.323	92.182	548.145	431.658	-10.737
2200	58.451	323.931	279.380	98.013	549.235	426.085	-10.117
2300	58.726	326.536	281.374	103.872	550.327	420.463	-9.549
2400	58.975	329.040	283.308	109.757	551.419	414.793	-9.028
2500	59.200	331.452	285.186	115.666	552.509	409.078	-8.547
2600	59.405	333.778	287.010	121.596	553.596	403.319	-8.103
2700	59.591	336.024	288.784	127.546	554.678	397.519	-7.690
2800	59.761	338.194	290.511	133.514	261.855	392.647	-7.325
2900	59.917	340.294	292.191	139.498	264.022	397.279	-7.156
3000	60.059	342.328	293.829	145.497	266.183	401.838	-6.997
3100	60.190	344.299	295.425	151.509	268.337	406.324	-6.847
3200	60.311	346.212	296.982	157.534	270.482	410.741	-6.705
3300	60.422	348.070	298.503	163.571	272.620	415.091	-6.570
3400	60.524	349.875	299.987	169.618	274.749	419.376	-6.443
3500	60.619	351.631	301.438	175.676	276.868	423.599	-6.322
3600	60.707	353.340	302.856	181.742	278.978	427.761	-6.207
3700	60.788	355.004	304.243	187.817	281.078	431.865	-6.097
3800	60.864	356.626	305.600	193.899	283.168	435.913	-5.992
3900	60.934	358.208	306.929	199.989	285.247	439.905	-5.892
4000	61.000	359.752	308.230	206.086	287.314	443.844	-5.796
4100	61.061	361.259	309.505	212.189	289.369	447.732	-5.704
4200	61.119	362.731	310.755	218.298	291.412	451.570	-5.616
4300	61.173	364.170	311.981	224.413	293.441	455.359	-5.532
4400	61.223	365.576	313.183	230.533	295.457	459.101	-5.450
4500	61.270	366.953	314.362	236.657	297.460	462.798	-5.372
4600	61.315	368.300	315.520	242.787	299.446	466.450	-5.297
4700	61.357	369.619	316.657	248.920	301.416	470.059	-5.224
4800	61.396	370.911	317.774	255.058	303.369	473.627	-5.154
4900	61.433	372.178	318.872	261.199	305.305	477.154	-5.087
5000	61.468	373.419	319.950	267.345	307.221	480.642	-5.021
5100	61.501	374.637	321.011	273.493	309.117	484.091	-4.958
5200	61.533	375.831	322.053	279.645	310.992	487.503	-4.897
5300	61.563	377.004	323.079	285.800	312.845	490.880	-4.838
5400	61.591	378.155	324.088	291.957	314.674	494.222	-4.781
5500	61.617	379.285	325.082	298.118	316.479	497.530	-4.725
5600	61.643	380.395	326.060	304.281	318.258	500.806	-4.671
5700	61.667	381.487	327.023	310.446	320.010	504.050	-4.619
5800	61.690	382.559	327.971	316.614	321.736	507.264	-4.568
5900	61.711	383.614	328.905	322.784	323.430	510.447	-4.519
6000	61.732	384.652	329.825	328.956	325.115	513.601	-4.471

PREVIOUS: December 1967 (1 atm) CURRENT: December 1967 (1 bar)

Aluminum Hydroxide, Ion ($AlOH^+$) $Al_1H_1O_1^+(g)$

ALUMINUM HYDROXIDE, ION ($AlOH^-$) IDEAL GAS M_r = 43.98943

$\Delta_fH°$(0 K) = [-221 ± 95] kJ mol^{-1}

S°(298.15 K) = [239.8 ± 2] J K^{-1} mol^{-1} $\Delta_fH°$(298.15 K) = [-230.120] kJ mol^{-1}

Vibrational Frequencies and Degeneracies

ν, cm^{-1}

[900](1)
[1100](1)
[3600](1)

Ground State Quantum Weight: [2] σ = 1

Point Group: [C_s]

Bond Distances: Al–O = [1.90] Å O-H = [0.96] Å

Bond Angle: Al-O-H = [120]°

Product of the Moments of Inertia: $I_AI_BI_C$ = [4.487473 x 10^{-117}] g^3 cm^6

Enthalpy of Formation

The enthalpy of formation is obtained from $\Delta_fH°$(AlOH, g, 298.15 K) and its electron affinity, which is estimated to be 0.5 ± 0.9 eV (12 ± 20 kcal mol^{-1}). The electron affinity is estimated to be small, or even negative, since the additional electron must go into an antibonding orbital, according to the correlation diagram of Walsh (1).

Heat Capacity and Entropy

Since the molecule now has 11 valence electrons, the Walsh correlation diagram predicts a bent molecule. The extra electron is antibonding thus making the bonds weaker than in AlOH(g). The vibrational frequencies and bond lengths are taken to be the same as in $AlOH^+$(g), which is also less strongly bound than AlOH(g). The bond angle is arbitrarily chosen as 120°. The principal moments of inertia are I_A = 0.1049x 10^{-39} g cm^2, I_B = 6.4867x 10^{-39} g cm^2 and I_C = 6.5917 x 10^{-39} g cm^2.

Reference

1. A. D. Walsh, J. Chem. Soc. 2288 (1953).

Aluminum Hydroxide, Ion ($AlOH^-$) $Al_1H_1O_1^-$(g)

Enthalpy Reference Temperature = T_r = 298.15 K Standard State Pressure = p° = 0.1 MPa

T/K	$C_p°$ (J K^{-1}mol^{-1})	S° (J K^{-1}mol^{-1})	-[G°-H°(T_r)]/T (J K^{-1}mol^{-1})	H°-H°(T_r) (kJ mol^{-1})	$\Delta_fH°$ (kJ mol^{-1})	$\Delta_fG°$ (kJ mol^{-1})	Log K_f
0	0.	0.	INFINITE	-10.123	-220.931		
100	33.261	202.633	270.605	-6.797			
200	33.987	225.808	243.059	-3.450			
250	35.123	233.506	240.403	-1.724			
298.15	36.521	239.810	239.810	0.	-230.120	-236.871	41.499
300	36.578	240.036	239.810	0.068	-230.190	-236.912	41.250
350	38.106	245.789	240.261	1.935	-232.059	-237.885	35.502
400	39.556	250.973	241.281	3.877	-233.906	-238.591	31.157
450	40.861	255.709	242.625	5.888	-235.726	-239.067	27.750
500	42.009	260.075	244.155	7.960	-237.524	-239.341	25.004
600	43.887	267.908	247.476	12.259	-241.084	-239.370	20.839
700	45.359	274.788	250.896	16.724	-244.641	-238.804	17.820
800	46.572	280.926	254.273	21.322	-248.247	-237.725	15.522
900	47.619	286.473	257.548	26.033	-251.946	-236.188	13.708
1000	48.548	291.539	260.697	30.842	-266.364	-233.474	12.195
1100	49.385	296.206	263.716	35.740	-269.994	-230.008	10.922
1200	50.141	300.536	266.606	40.717	-273.584	-226.214	9.847
1300	50.826	304.577	269.373	45.766	-277.141	-222.123	8.925
1400	51.446	308.367	272.025	50.880	-280.669	-217.758	8.125
1500	52.005	311.936	274.567	56.053	-284.175	-213.142	7.422
1600	52.508	315.309	277.009	61.279	-287.661	-208.292	6.800
1700	52.962	318.506	279.357	66.553	-291.134	-203.225	6.244
1800	53.371	321.545	281.617	71.870	-294.595	-197.955	5.744
1900	53.740	324.440	283.795	77.226	-298.049	-192.491	5.292
2000	54.072	327.206	285.897	82.616	-301.498	-186.846	4.880
2100	54.372	329.851	287.928	88.039	-304.945	-181.029	4.503
2200	54.644	332.387	289.891	93.490	-308.391	-175.047	4.156
2300	54.889	334.821	291.792	98.967	-311.839	-168.909	3.836
2400	55.112	337.162	293.634	104.467	-315.289	-162.621	3.539
2500	55.315	339.416	295.421	109.989	-318.744	-156.189	3.263
2600	55.500	341.589	297.155	115.529	-322.204	-149.618	3.006
2700	55.668	343.687	298.840	121.088	-325.670	-142.915	2.765
2800	55.822	345.714	300.478	126.663	-623.043	-135.115	2.521
2900	55.963	347.676	302.072	132.252	-625.428	-117.647	2.119
3000	56.092	349.575	303.624	137.855	-627.820	-100.096	1.743
3100	56.211	351.416	305.136	143.470	-630.221	-82.466	1.390
3200	56.321	353.203	306.610	149.097	-632.631	-64.758	1.057
3300	56.422	354.938	308.048	154.734	-635.050	-46.975	0.744
3400	56.515	356.623	309.452	160.381	-637.479	-29.118	0.447
3500	56.602	358.263	310.824	166.037	-639.918	-11.189	0.167
3600	56.682	359.858	312.164	171.701	-642.367	6.809	-0.099
3700	56.756	361.413	313.474	177.373	-644.827	24.875	-0.351
3800	56.826	362.927	314.755	183.052	-647.298	43.009	-0.591
3900	56.890	364.404	316.010	188.738	-649.781	61.208	-0.820
4000	56.950	365.845	317.238	194.430	-652.276	79.469	-1.038
4100	57.007	367.252	318.440	200.128	-654.783	97.794	-1.246
4200	57.059	368.626	319.619	205.831	-657.303	116.180	-1.445
4300	57.109	369.970	320.774	211.540	-659.837	134.627	-1.635
4400	57.155	371.283	321.907	217.253	-662.385	153.131	-1.818
4500	57.198	372.568	323.019	222.971	-664.946	171.694	-1.993
4600	57.239	373.826	324.110	228.692	-667.525	190.315	-2.161
4700	57.278	375.057	325.181	234.418	-670.120	208.992	-2.323
4800	57.314	376.263	326.232	240.148	-672.732	227.724	-2.478
4900	57.348	377.445	327.266	245.881	-675.362	246.510	-2.628
5000	57.380	378.604	328.281	251.617	-678.011	265.351	-2.772
5100	57.411	379.741	329.279	257.357	-680.682	284.245	-2.911
5200	57.440	380.856	330.260	263.100	-683.373	303.191	-3.046
5300	57.467	381.950	331.225	268.845	-686.087	322.190	-3.175
5400	57.493	383.025	332.174	274.593	-688.824	341.239	-3.301
5500	57.517	384.080	333.108	280.343	-691.587	360.339	-3.422
5600	57.541	385.117	334.028	286.096	-694.375	379.491	-3.540
5700	57.563	386.135	334.933	291.851	-697.191	398.692	-3.654
5800	57.584	387.136	335.825	297.609	-700.032	417.943	-3.764
5900	57.604	388.121	336.703	303.368	-702.906	437.242	-3.871
6000	57.623	389.089	337.568	309.129	-705.789	456.589	-3.975

PREVIOUS: December 1967 (1 atm) CURRENT: December 1967 (1 bar)

Aluminum Hydroxide, Ion ($AlOH^-$) $Al_1H_1O_1^-$(g)

ALUMINUM HYDROXIDE OXIDE (OAlOH) IDEAL GAS M_r = 59.98828

Aluminum Hydroxide Oxide (OAlOH) $Al_1H_1O_2(g)$

$\Delta_fH°(0\ K) = [-455 \pm 63]\ kJ\ mol^{-1}$

$S°(298.15\ K) = 254.4 \pm 6\ J\ K^{-1}\ mol^{-1}$ $\Delta_fH°(298.15\ K) = [-460 \pm 63]\ kJ\ mol^{-1}$

Vibrational Frequencies and Degeneracies

ν, cm^{-1}	ν, cm^{-1}
[3400](1)	[700](1)
[1200](1)	[500](1)
[1100](1)	[400](1)

Ground State Quantum Weight: 1 σ = 1

Point Group: $[C_s]$

Bond Distances: Al=O = [1.62] Å Al-O = [1.87] Å O-H = [0.94] Å

Bond Angles: O-Al-O = [180]° Al-O-H = [105]°

Product of the Moments of Inertia: $I_AI_BI_C$ = $[37.497324 \times 10^{-117}]\ gm^3\ cm^6$

Enthalpy of Formation

The hydroxyl group can often be considered as intermediate between F and Cl as is illustrated by the following examples of bond energies from these tables. D(F-H) = 136 kcal mol^{-1}, D(Cl-H) = 103 kcal mol^{-1} and D(HO-H) = 119 kcal mol^{-1}; also, D(F-BO) = 163 kcal mol^{-1}, D(Cl-BO) = 105 kcal mol^{-1} and D(HO-BO) = 143 kcal mol^{-1}. Thus from D(F-AlO) = 181 kcal mol^{-1} and D(Cl-AlO) = 134 kcal mol^{-1}, we estimate D(HO-AlO) = 158 ± 20 kcal mol^{-1}. This yields $\Delta_fH°$(HOAlO, g, 298.15 K) = -127 ± 20 kcal mol^{-1}.

Consider also D(FB-O) = 175 kcal mol^{-1} and D(ClB-O) = 169 kcal mol^{-1}; and D(FAl-O) = 138 kcal mol^{-1} and D(ClAl-O) = 132 kcal mol^{-1}; it is evident that the substituent F or Cl has little effect on the B-O and Al-O bonds. Thus, we assume that D(HOAl-O) = 135 ± 10 kcal mol^{-1} which leads to $\Delta_fH°$(HOAlO, g, 298.15 K) = -119 ± 10 kcal mol^{-1}.

Farber et al. (1) found no evidence for the ion $DOAlO^+$ when reacting $D_2O(g) + Al_2O_3(cr)$. Form estimates of the pressure needed to ensure detection of DOAlO and a knowledge of the D_2O pressure, they calculated the most negative value of $\Delta_fH°$(HOAlO, g, 298.15 K) = -104 ± 10 kcal mol^{-1}. The absence of $DOAlO^+$ may be caused by fragmentation by the ionizing electrons and thus the value quoted masy not be a real limit.

We adopt $\Delta_fH°$(HOAlO, g, 298.15 K) = -110 ± 15 kcal mol^{-1} which encompasses all the values.

Heat Capacity and Entropy

Vibrational frequencies, bond distances, bond angles, and the product of the moments of inertia are taken from NBS (2). The principal moments of inertia are: $I_A = 0.1301 \times 10^{-39}\ g\ cm^2$, $I_B = 16.9121 \times 10^{-39}\ g\ cm^2$, and $I_C = 17.0422 \times 10^{-39}\ g\ cm^2$.

References

1. M. Farber, M. A. Frisch, G. Grenier, and H. C. Ko, "Investigation of the Thermodynamic Properties of Rocket Combustion Products", AFRPL-TR-67-244, Nov., 1967.
2. U. S. Nat. Bur. Stand. Report 6297, Jan. 1, 1959.

Enthalpy Reference Temperature = T_r = 298.15 K Standard State Pressure = p° = 0.1 MPa

T/K	$C_p°$ (J $K^{-1}mol^{-1}$)	S° (J $K^{-1}mol^{-1}$)	$-[G°-H°(T_r)]/T$ (J $K^{-1}mol^{-1}$)	$H°-H°(T_r)$ (kJ mol^{-1})	$\Delta_fH°$ (kJ mol^{-1})	$\Delta_fG°$ (kJ mol^{-1})	Log K_f
0	0.	0.	INFINITE	-11.722	-454.505	-454.505	INFINITE
100	34.495	209.798	293.555	-8.376	-456.047	-453.961	237.125
200	42.414	235.968	258.738	-4.554	-458.285	-451.012	117.792
250	46.534	245.881	255.197	-2.329	-459.330	-449.071	93.828
298.15	50.137	254.391	254.391	0.	-460.240	-447.011	78.314
300	50.267	254.702	254.392	0.093	-460.273	-446.929	77.817
350	53.561	262.704	255.017	2.690	-461.111	-444.637	66.358
400	56.415	270.047	256.443	5.442	-461.856	-442.231	57.750
450	58.859	276.837	258.336	8.325	-462.524	-439.738	51.043
500	60.945	283.149	260.506	11.322	-463.130	-437.173	45.671
600	64.267	294.569	265.252	17.590	-464.220	-431.877	37.598
700	66.781	304.673	270.176	24.148	-465.234	-426.406	31.819
800	68.767	313.725	275.063	30.929	-466.247	-420.790	27.475
900	70.398	321.921	279.821	37.890	-467.320	-415.045	24.089
1000	71.774	329.412	284.411	45.000	-479.089	-408.414	21.333
1100	72.954	336.309	288.820	52.238	-480.054	-401.300	19.056
1200	73.978	342.702	293.047	59.586	-480.970	-394.099	17.155
1300	74.871	348.659	297.098	67.029	-481.845	-386.825	15.543
1400	75.653	354.237	300.983	74.556	-482.688	-379.483	14.159
1500	76.341	359.481	304.709	82.157	-483.508	-372.083	12.957
1600	76.948	364.427	308.289	89.822	-484.311	-364.629	11.904
1700	77.484	369.109	311.730	97.544	-485.102	-357.124	10.973
1800	77.959	373.551	315.042	105.317	-485.888	-349.574	10.144
1900	78.382	377.778	318.234	113.134	-486.671	-341.979	9.402
2000	78.758	381.808	321.312	120.991	-487.456	-334.343	8.732
2100	79.094	385.659	324.285	128.884	-488.247	-326.668	8.125
2200	79.396	389.345	327.160	136.809	-489.045	-318.955	7.573
2300	79.667	392.881	329.941	144.762	-489.853	-311.206	7.068
2400	79.911	396.277	332.634	152.742	-490.672	-303.421	6.604
2500	80.131	399.543	335.246	160.744	-491.505	-295.602	6.176
2600	80.330	402.690	337.780	168.767	-492.352	-287.749	5.781
2700	80.512	405.725	340.240	176.809	-493.214	-279.863	5.414
2800	80.676	408.656	342.632	184.869	-787.993	-270.977	5.055
2900	80.827	411.490	344.957	192.944	-787.792	-252.517	4.548
3000	80.964	414.232	347.221	201.034	-787.607	-234.062	4.075
3100	81.090	416.889	349.426	209.137	-787.440	-215.613	3.633
3200	81.206	419.466	351.575	217.251	-787.290	-197.170	3.218
3300	81.313	421.966	353.670	225.377	-787.157	-178.731	2.829
3400	81.411	424.395	355.715	233.514	-787.042	-160.295	2.463
3500	81.501	426.756	357.711	241.659	-786.945	-141.863	2.117
3600	81.585	429.053	359.661	249.814	-786.865	-123.433	1.791
3700	81.663	431.290	361.567	257.976	-786.803	-105.006	1.482
3800	81.735	433.469	363.430	266.146	-786.759	-86.579	1.190
3900	81.802	435.593	365.253	274.323	-786.732	-68.153	0.913
4000	81.865	437.665	367.038	282.506	-786.725	-49.729	0.649
4100	81.923	439.687	368.785	290.696	-786.736	-31.304	0.399
4200	81.978	441.662	370.497	298.891	-786.767	-12.878	0.160
4300	82.029	443.591	372.175	807.091	-786.817	5.549	-0.067
4400	82.077	445.477	373.819	315.297	-786.888	23.976	-0.285
4500	82.121	447.322	375.432	323.507	-786.976	42.405	-0.492
4600	82.164	449.128	377.015	331.721	-787.089	60.837	-0.691
4700	82.203	450.895	378.568	339.939	-787.224	79.272	-0.881
4800	82.240	452.626	380.093	348.161	-787.383	97.710	-1.063
4900	82.276	454.322	381.590	356.387	-787.565	116.151	-1.238
5000	82.309	455.985	383.062	364.616	-787.774	134.597	-1.406
5100	82.340	457.615	384.508	372.849	-788.010	153.047	-1.568
5200	82.370	459.214	385.929	381.084	-788.274	171.500	-1.723
5300	82.398	460.784	387.327	389.323	-788.567	189.961	-1.872
5400	82.424	462.324	388.701	397.564	-788.893	208.426	-2.016
5500	82.449	463.837	390.054	405.808	-789.250	226.898	-2.155
5600	82.473	465.323	391.384	414.054	-789.642	245.378	-2.289
5700	82.496	466.783	392.694	422.302	-790.071	263.864	-2.418
5800	82.517	468.217	393.984	430.553	-790.534	282.358	-2.543
5900	82.538	469.628	395.254	438.806	-791.040	300.859	-2.664
6000	82.557	471.016	396.506	447.060	-791.565	319.369	-2.780

PREVIOUS: December 1968 (1 atm) CURRENT: December 1968 (1 bar)

Aluminum Hydroxide Oxide (OAlOH) $Al_1H_1O_2(g)$

LITHIUM TETRAHYDROALUMINATE ($LiAlH_4$) CRYSTAL M_r = 37.95414

$\Delta_f H°(0\ K)$ = Unknown

S°(298.15 K) = [87.86 ± 0.4] J K^{-1} mol^{-1} $\Delta_f H°(298.15\ K) = -117.15 \pm 8.4\ kJ\ mol^{-1}$

Enthalpy of Formation

The heats of solution of $LiAlH_4$(cr) in HCl(aq) were measured calorimetrically by Davis et al. (1), Fasolino et al. (2), and Smith and Bass (3). The results are given in the following table.

Source		Reaction	$\Delta_r H°(298.15\ K)$ kcal mol^{-1}
Davis et al. (1)	(1)	$LiAlH_4$(cr) + 10 HCl(aq, 50 H_2O) → [LiCl + $AlCl_3$ + 6 HCl](aq, 500 H_2O) + 4 H_2(g)	-165.87
	(2)	[$AlCl_3$ + LiCl + 6 HCl](aq, 500 H_2O) + 41020 H_2O(ℓ) → [$AlCl_3$ + LiCl + 6 HCl](aq, 41520 H_2O)	-4.43
	(3)	[$AlCl_3$ + LiCl + 6 HCl](aq, 41520 H_2O) → [$AlCl_3$ + LiCl + 6 HCl](aq, 5190 H_2O)	0.00(assumed)
	(1)	+ (2) + (3)	-170.30
Fasolino et al. (2)	(4)	Al + 60 HCl(aq, 14 H_2O) → [$AlCl_3$ + 57 HCl](aq, 840 H_2O) + 3/2 H_2(g)	-128.14
	(5)	Li + 60 HCl(aq, 14 H_2O) → [LiCl + 59 HCl](aq, 840 H_2O) + 1/2 H_2(g)	-67.05
	(6)	$LiAlH_4$(cr) + 60 HCl(aq, 14 H_2O) → [LiCl + $AlCl_3$ + 56 HCl](aq, 840 H_2O) + 4 H_2(g)	-170.52
	(4)	+ (5) - (6)	-24.67
Smith and Bass (3)	(7)	$LiAlH_4$(cr) + 39.715 HCl(aq, 11.624 H_2O) → [LiCl + $AlCl_3$ + 35.715 HCl](aq, 12.926 H_2O) + 4 H_2(g)	-164.6
	(8)	LiCl(cr) + [$AlCl_3$ + 35.715 HCl](aq, 12.926 H_2O) → [LiCl + $AlCl_3$ + 35.715 HCl](aq, 12.926 H_2O)	-6.75
	(7)	- (8)	-157.85

Incorporating appropriate thermal data, the values for $\Delta_f H°(LiAlH_4$, cr, 298.15 K) were derived as -28.51 (recalculated value, by Smith and Bass (3)), -24.67±1.31 and -28.4±1.5 kcal mol^{-1}, respectively. The adopted value, -28.0±2.0 kcal mol^{-1} (-117.152±8.4 kJ mol^{-1}), is the weighted average of the above three values.

Heat Capacity and Entropy

Bonnetot et al (4) measured $C_p°$ in the region 15 to 308 K. Our adopted thermal functions at 300 K and below are based on the smoothed values and integration by those authors using S°(15 K) = 0.10 J K^{-1} mol^{-1}. Above 300 K we have used $C_p°$ values estimated by comparison with those for $LiBH_4$(cr), $LiBO_2$(cr) and $LiAlO_2$(cr) (5).

Decomposition Data

T_{dcm}(410±10 K) was reported by Finholt et al. (6). The decomposition products are aluminum, hydrogen, and lithium hydride. This reaction is apparently kinetically controlled since the free energy for the decomposition is negative even at room temperature.

References

1. W. D. Davis, L. S. Mason and G. Stegeman, J. Am. Chem. Soc. 71, 2775 (1949).
2. L. G. Fasolino et al., "Heat of Formation of Lithium Aluminum Hydride," Special Report, National Research Corporation, April 26, 1963.
3. M. B. Smith and G. E. Bass, Jr., J. Chem. Eng. Data 8, 342 (1963).
4. P. Bonnetot, P. Claudy, M. Diot and J. M. Letoffe, J. Chem. Thermodynamics 11, 1197 (1979).
5. JANAF Thermochemical Tables: $LiBH_4$(cr) 12-31-64; $LiBO_2$(cr) 6-30-71; $LiAlO_2$(cr) 12-31-79.
6. A. E. Finholt, A. C. Bond, Jr., and H. I. Schlesinger, J. Am. Chem. Soc. 69, 1199 (1947).

Lithium Tetrahydroaluminate ($LiAlH_4$) $Al_1H_4Li_1$(cr)

Enthalpy Reference Temperature = T_r = 298.15 K; Standard State Pressure = p° = 0.1 MPa

	J $K^{-1}mol^{-1}$			kJ mol^{-1}			
T/K	$C_p°$	S°	$-[G°-H°(T_r)]/T$	$H°-H°(T_r)$	$\Delta_f H°$	$\Delta_f G°$	Log K_f
0							
100							
200							
298.15	86.400	87.864	87.864	0.	-117.152	-48.323	8.466
300	86.944	88.400	87.866	0.160	-117.189	-47.895	8.339
400	93.722	114.323	91.354	9.187	-119.082	-24.508	3.200
500	101.253	136.080	98.176	18.952	-123.729	-0.360	0.038
600	106.901	155.059	106.109	29.370	-124.888	24.431	-2.127
700	111.294	171.886	114.327	40.292	-125.614	49.379	-3.685
800	114.223	186.947	122.480	51.574	-126.114	74.414	-4.859
900	116.200	200.519	130.409	63.099	-126.571	99.507	-5.775
1000	117.780	212.847	138.046	74.801	-137.678	125.414	-6.551
1100	118.847	224.124	145.366	86.634	-137.979	151.738	-7.205
1200	119.746	234.505	152.368	98.565	-138.256	178.088	-7.752
1300	120.416	244.116	159.060	110.573	-138.536	204.461	-8.215
1400	121.043	253.063	165.459	122.646	-138.832	230.857	-8.613
1500	121.615	261.434	171.581	134.779	-139.148	257.274	-8.959
1600	122.173	269.301	177.445	146.969	-139.485	283.713	-9.262
1700	122.717	276.724	183.069	159.213	-285.080	317.350	-9.751
1800	123.247	283.753	188.469	171.512	-284.705	352.776	-10.237
1900	123.763	290.431	193.661	183.862	-284.357	388.182	-10.672
2000	124.265	296.792	198.660	196.264	-284.034	423.570	-11.063

PREVIOUS: June 1963 CURRENT: March 1964

Lithium Tetrahydroaluminate ($LiAlH_4$) $Al_1H_4Li_1$(cr)

ALUMINUM IODIDE (AlI) IDEAL GAS M_r = 153.88604

$\Delta_f H°(0\ K) = 69.34 \pm 4.18\ kJ\ mol^{-1}$

$S°(298.15\ K) = 247.830 \pm 0.21\ J\ K^{-1}\ mol^{-1}$

$\Delta_f H°(298.15\ K) = 67.95 \pm 4.18\ kJ\ mol^{-1}$

Electronic Levels and Quantum Weights

State	ε_i, cm^{-1}	g_i
$X^1\Sigma^+$	0	1
	21899.6	2
$a^3\pi$	22097.9	2
	[22290.]	[2]
$A^1\pi$	31487.	2
$[b^3\Sigma^+]$	[43000.]	[3]

$\omega_e = 316.16\ cm^{-1}$ $\omega_e x_e = 0.960\ cm^{-1}$ $\sigma = 1$

$B_e = 0.11770\ cm^{-1}$ $\alpha_e = 0.0005586\ cm^{-1}$ $r_e = 2.5371$

Enthalpy of Formation

Gaydon (1), Rosen (2) and Huber and Herzberg (3) reference the analysis by Barrow (4) for values of the dissociation energy of AlI(g). Barrow (4) reports $D_0^0 = 364.0\ kJ\ mol^{-1}$ from an analysis of fluctuation bonds in the AlI spectrum. Martin and Barrow (5) report a value of $365.97 \pm 1.92\ kJ\ mol^{-1}$ for the dissociation energy of AlI based on previous observations of the $A^1\pi$-$X^1\Sigma^+$ continuum and calculated potentials for these states. Our adopted dissociation energy is the average of these values, $365.01 \pm 0.21\ kJ\ mol^{-1}$, and yields $\Delta_f H°(AlI, g, 298.15\ K) = 67.95 \pm 4.18\ kJ\ mol^{-1}$.

Heat Capacity and Entropy

Molecular and spectroscopic constants are based on the analysis of electronic spectra published by Martin and Barrow (5). The constants are in excellent agreement with those from earlier studies (6, 7). The electronic states are taken from Rosen (2) with the exception of the upper $^3\pi$ state and the $b^3\Sigma^+$ state. The upper $^3\pi$ state is estimated to be approximately in a linear relationship with the two lower $^3\pi$ states. The $b^3\Sigma^+$ state is assigned by analogy with AlF, AlCl and AlBr (8).

References

1. A. G. Gaydon, "Dissociation Energies and Spectra of Diatomic Molecules", Chapman and Hall, Ltd., London, 1968.
2. B. Rosen (ed.), "Donnes Spectroscopiques Relatives Aux Molecules Diatomiques", Pergamon Press., New York, 1970.
3. K. P. Huber and G. Herzberg, "Molecular Spectra and Molecular Structure IV. Constants of Diatomic Molecules", Van Nostrand Reinnold Company, New York, 1979.
4. R. F. Barrow, Trans. Faraday Soc. 56, 952 (1960).
5. E. Martin and R. F. Barrow, Phys. Scripta 17, 501 (1978).
6. F. C. Wyse and W. Gordy, J. Chem. Phys. 56, 2130 (1974).
7. A. Lakshminarayana and P. B. V. Haranath, Curr. Sci. India 39, 344 (1970).
8. JANAF Thermochemical Tables: AlF(g), AlCl(g), and AlBr(g), 9-30-79.

Aluminum Iodide (AlI) $Al_1I_1(g)$

Enthalpy Reference Temperature = T_r = 298.15 K Standard State Pressure = p° = 0.1 MPa

T/K	$C_p°$ (J K⁻¹mol⁻¹)	S°	$-[G°-H°(T_r)]/T$	$H°-H°(T_r)$ (kJ mol⁻¹)	$\Delta_f H°$	$\Delta_f G°$	Log K_f
0	0.	0.	INFINITE	-9.749	69.337	69.337	INFINITE
100	31.012	210.863	278.841	-6.798	70.280	52.942	-27.654
200	34.703	233.666	251.098	-3.486	69.323	35.915	-9.380
250	35.586	241.513	248.422	-1.727	68.655	27.638	-5.775
298.15	36.127	247.830	247.830	0.	67.948	19.802	-3.469
300	36.144	248.054	247.831	0.067	67.920	19.503	-3.396
350	36.516	253.655	248.273	1.884	67.096	11.497	-1.716
400	36.778	258.549	249.258	3.717	58.242	3.887	-0.508
450	36.971	262.893	250.536	5.561	56.766	-2.819	0.327
500	37.119	266.796	251.970	7.413	35.208	-7.460	0.779
600	37.333	273.584	255.023	11.136	34.318	-15.912	1.385
700	37.484	279.351	258.097	14.877	33.328	-24.207	1.806
800	37.601	284.364	261.074	18.632	32.212	-32.351	2.112
900	37.697	288.798	263.913	22.397	30.941	-40.347	2.342
1000	37.780	292.774	266.604	26.171	18.899	-47.432	2.478
1100	37.856	296.379	269.149	29.953	17.598	-54.002	2.564
1200	37.925	299.676	271.558	33.742	16.296	-60.454	2.632
1300	37.990	302.714	273.839	37.538	14.988	-66.797	2.684
1400	38.053	305.532	276.003	41.340	13.672	-73.039	2.725
1500	38.113	308.159	278.060	45.148	12.342	-79.186	2.758
1600	38.171	310.621	280.019	48.962	10.992	-85.244	2.783
1700	38.228	312.937	281.888	52.782	9.619	-91.217	2.803
1800	38.284	315.123	283.674	56.608	8.219	-97.108	2.818
1900	38.340	317.195	285.385	60.439	6.790	-102.921	2.829
2000	38.396	319.163	287.025	64.276	5.332	-108.658	2.838
2100	38.451	321.037	288.600	68.118	3.845	-114.321	2.844
2200	38.507	322.827	290.115	71.966	2.333	-119.912	2.847
2300	38.564	324.540	291.575	75.820	0.799	-125.435	2.849
2400	38.623	326.183	292.983	79.679	-0.752	-130.890	2.849
2500	38.685	327.761	294.343	83.544	-2.314	-136.280	2.847
2600	38.750	329.279	295.658	87.416	-3.880	-141.608	2.845
2700	38.819	330.743	296.930	91.295	-5.445	-146.875	2.841
2800	38.893	332.156	298.163	95.180	-300.904	-151.117	2.819
2900	38.974	333.522	299.359	99.073	-301.352	-145.760	2.625
3000	39.063	334.845	300.520	102.975	-301.782	-140.387	2.444
3100	39.159	336.127	301.648	106.886	-302.188	-135.000	2.275
3200	39.266	337.372	302.745	110.807	-302.565	-129.601	2.116
3300	39.383	338.582	303.813	114.740	-302.910	-124.191	1.966
3400	39.512	339.760	304.853	118.684	-303.218	-118.770	1.825
3500	39.653	340.907	305.867	122.643	-303.488	-113.341	1.692
3600	39.808	342.027	306.856	126.616	-303.717	-107.905	1.566
3700	39.977	343.120	307.821	130.605	-303.902	-102.463	1.447
3800	40.160	344.188	308.764	134.611	-304.041	-97.017	1.334
3900	40.359	345.234	309.686	138.637	-304.134	-91.568	1.226
4000	40.572	346.258	310.587	142.684	-304.178	-86.117	1.125
4100	40.802	347.263	311.470	146.752	-304.173	-80.665	1.028
4200	41.047	348.249	312.334	150.844	-304.118	-75.214	0.935
4300	41.307	349.218	313.180	154.962	-304.012	-69.765	0.847
4400	41.583	350.171	314.010	159.106	-303.856	-64.319	0.764
4500	41.875	351.108	314.824	163.279	-303.646	-58.878	0.683
4600	42.180	352.032	315.623	167.482	-303.387	-53.441	0.607
4700	42.500	352.943	316.407	171.716	-303.076	-48.011	0.534
4800	42.834	353.841	317.178	175.982	-302.714	-42.588	0.463
4900	43.181	354.728	317.935	180.283	-302.301	-37.173	0.396
5000	43.540	355.604	318.680	184.619	-301.838	-31.767	0.332
5100	43.910	356.469	319.412	188.991	-301.325	-26.370	0.270
5200	44.291	357.326	320.133	193.401	-300.764	-20.984	0.211
5300	44.681	358.173	320.843	197.850	-300.154	-15.610	0.154
5400	45.080	359.012	321.542	202.338	-299.497	-10.247	0.099
5500	45.486	359.843	322.231	206.866	-298.794	-4.897	0.047
5600	45.899	360.666	322.910	211.435	-298.045	0.440	-0.004
5700	46.318	361.482	323.579	216.046	-297.253	5.763	-0.053
5800	46.740	362.291	324.240	220.699	-296.416	11.072	-0.100
5900	47.167	363.094	324.892	225.394	-295.540	16.366	-0.145
6000	47.595	363.890	325.535	230.132	-294.602	21.642	-0.188

PREVIOUS: September 1979 (1 atm) CURRENT: September 1979 (1 bar)

Aluminum Iodide (AlI) $Al_1I_1(g)$

ALUMINUM IODIDE (AlI_3) CRYSTAL M_r = 407.69504

$\Delta_f H°$(0 K) = Unknown

$S°$(298.15 K) = [196.10] J K^{-1} mol^{-1} $\Delta_f H°$(298.15 K) = -302.9±2.0 kJ mol^{-1}

T_{fus} = 464.15 ± 0.2 K $\Delta_{fus}H°$ = 15.9 ± 1.3 kJ mol^{-1}

Enthalpy of Formation

Literature values for the enthalpy of formation of AlI_3(cr) are based on solution calorimetry measurements and differ considerably in magnitude. The heat of solution of the crystal in H_2O at 298.15 K was measured by Dear and Eley (1) and reported as -91.4±1.5 kcal mol^{-1}. Adopting CODATA's (2) values for the enthalpies of formation of the aqueous ions, this yields $\Delta_f H°$(AlI_3, cr, 298.15 K) =-78.0±2.0 kcal mol^{-1}. This estimate is undoubtedly high because of hydrolysis of the salt.

More reliable measurements have been attempted in aqueous acid and base. Klemm and Tanke (3) measured the heat of solution of AlI_3(cr) at 273 K in aqueous HCl. Using auxiliary data including estimated heat capacities, they reported $\Delta_f H°$(AlI_3, cr, 298.15 K)= -70.6±0.7 kcal mol^{-1}. Efimov, Kislova and Medvedev (4) measured the heat of solution of AlI_3(cr) in an aqueous HCl solution containing KCl(cr) and the heat of solution of $AlCl_3$(cr) in an aqueous HCl solution containing KI(cr). From these heats of solution measurements and using the heats of formation of $AlCl_3$(5), KCl(6, 7) and KI (6, 8) as auxiliary data, they obtained $\Delta_f H°$(AlI_3, cr, 298.15 K) = -72.4±0.5 kcal mol^{-1}. In a similar study, Anthoney et al. (9) measured the heat of solution of AlI_3(cr) and $AlCl_3$(cr) in an aqueous alkaline solution containing dissolved sodium halides. They reported $\Delta_f H°$(AlI_3, cr, 298.15 K = -67.0±0.6 kcal mol^{-1}.

Corbett and Gregory (10) investigated the equilibrium AlI_3(cr) + 3 HCl(g) + $AlCl_3$(cr) + 3 HI(g) and reported $\Delta_r G°$(298.15 K) = -8.7±0.2 kcal mol^{-1}. Using auxiliary data, the derived heat of reaction is -10.7±0.2 kcal mol^{-1} and the heat of formation of AlI_3, $\Delta_f H°$(AlI_3, cr, 298.15 K) is evaluated to be -73.6±0.8 kcal mol^{-1}. The value reported by Efimov et al. (4) is chosen as most reliable and is adopted for the heat of formation.

Heat Capacity and Entropy

The heat capacities, 298.15-464 K, were taken from Kelley (11), whose values were derived from the enthalpy data measured by Fischer (12). Above 464 K the $C_p°$ values are obtained by graphical extrapolation. The entropy, $S°$(AlI_3, cr, 298.15 K), is calculated from $S°$(AlI_3, ℓ, 298.15 K) by subtracting $\Delta_{fus}S°$ and the difference between $S°$(464.15 K)-$S°$(298.15 K) for crystal and liquid.

Fusion Data

The temperature and enthalpy of melting were measured by Fischer (12).

Sublimation Data

The heat of sublimation, $\Delta_{sub}H°$(298.15 K) = 116.3 kJ mol^{-1} for the process 2 AlI_3(cr) = 2 Al_2I_6(g), is calculated from vapor pressure data reported by Fischer, Rahlfs and Benze (13) by both the 2nd and 3rd law methods. See Al_2I_6(g) table for details.

References

1. D. J. A. Dear and D. D. Eley, J. Chem. Soc. 1954, 4684 (1954).
2. ISCU CODATA-Task Group on Key Values for Thermodynamics, J. Chem. Thermodynamics 10, 903 (1978)
3. W. Klemm and H. Jacobi, Z. Anorg. Chem. 207, 186 (1932).
4. M. E. Efimov, G. N. Kislova and V. A. Medvedev, J. Chem. Thermodynamics 12, 1149 (1980).
5. L. V. Gurvich, I. V. Veits, et al., "Thermodynamic Properties of Individual Substances", 3rd ed., Volume III, Nauka, Moscow, 1980.
6. V. B. Parker, NSRDS-NBS 6, 1965.
7. M. E. Efimov, G. N. Klevarchuk, V. A. Medvedev and M. V. Kilday, J. Res. Nat. Bur. Stand. 84, 273 (1979).
8. M. E. Anthoney, A. Finch and P. J. Gardener, J. Chem. Soc., Dalton Trans. II 1973, 659 (1973).
9. J. D. Corbett and N. W. Gregory, J. Amer. Chem. Soc. 76, 1446 (1954).
10. K. K. Kelley, U.S. Bur. Mines Bull. 584, 1960.
11. W. Fischer, Z. anorg. allgem. Chem. 200, 332 (1931).
12. W. Fischer, O. Rahlfs and B. Benze, Z. anorg. allgem. Chem., 205, 1 (1932).

Aluminum Iodide (AlI_3) Al_1I_3(cr)

Enthalpy Reference Temperature = T_r = 298.15 K Standard State Pressure = p° = 0.1 MPa

	J $K^{-1}mol^{-1}$			kJ mol^{-1}			
T/K	$C_p°$	$S°$	$-[G°-H°(T_r)]/T$	$H°-H°(T_r)$	$\Delta_f H°$	$\Delta_f G°$	Log K_f
0							
100							
200							
298.15	98.910	196.104	196.104	0.	-302.922	-301.018	52.737
300	99.077	196.716	196.106	0.183	-302.934	-301.006	52.410
400	108.533	226.513	200.105	10.563	-327.520	-299.436	39.102
464.150	114.614	243.098	204.919	17.721	------ CRYSTAL <--> LIQUID -----		
500	116.901	251.742	207.972	21.885	-391.126	-286.306	29.910
600	121.986	273.543	217.126	33.850	-387.527	-265.674	23.129
700	125.063	292.599	226.576	46.216	-383.660	-245.669	18.332
800	126.918	309.430	235.902	58.823	-379.703	-226.226	14.771
900	127.901	324.444	244.921	71.571	-375.781	-207.277	12.030
1000	128.449	337.949	253.560	84.390	-382.581	-187.999	9.820
1100	128.875	350.213	261.798	97.257	-378.611	-168.733	8.012
1200	129.160	361.440	269.640	110.160	-374.632	-149.829	6.522
1300	129.319	371.784	277.105	123.084	-370.667	-131.256	5.274
1400	129.501	381.374	284.214	136.024	-366.732	-112.988	4.216
1500	129.704	390.316	290.993	148.984	-362.838	-95.000	3.308

PREVIOUS: June 1964 CURRENT: September 1979

Aluminum Iodide (AlI_3) Al_1I_3(cr)

ALUMINUM IODIDE (AlI_3) LIQUID M_r = 407.69504

$S°(298.15 K) = [223.648] J K^{-1} mol^{-1}$ $\Delta_f H°(298.15 K) = [-289.444] kJ mol^{-1}$

$T_{fus} = 464.15 \pm 0.2 K$ $\Delta_{fus}H° = 15.9 \pm 1.3 kJ mol^{-1}$

Enthalpy of Formation

$\Delta_f H°(AlI_3, \ell, 298.15 K)$ is calculated from $\Delta_f H°(AlI_3, cr, 298.15 K)$ by adding the enthalpy of fusion, $\Delta_{fus}H°$, and the difference in enthalpy, H°(464.15 K)−H°(298.15 K), between the crystal and liquid.

Heat Capacity and Entropy

The heat capacities, 464-500 K, were taken from Kelley (1), whose values were derived from the enthalpy data measured by Fischer (2). The $C_p°$ values above 500 K were assumed to be the same as that at 500 K. S°(298.15 K) is estimated so that the values of $\Delta_{vap}H°$ calculated from vapor pressure data by the 2nd and 3rd law methods are in reasonable agreement. See the $Al_2I_6(g)$ table for details.

Fusion Data

The temperature and enthalpy of melting were reported by Fischer (2).

References

1. K. K. Kelley, U. S. Bur. Mines Bull. 584, 1960.
2. W. Fischer, Z. anorg. allgem. Chem. 200 , 332 (1931).

Aluminum Iodide (AlI_3) $Al_1I_3(l)$

Enthalpy Reference Temperature = T_r = 298.15 K Standard State Pressure = p° = 0.1 MPa

T/K	$C_p°$ (J K⁻¹mol⁻¹)	S° (J K⁻¹mol⁻¹)	−[G°−H°(T_r)]/T (J K⁻¹mol⁻¹)	H°−H°(T_r) (kJ mol⁻¹)	$\Delta_f H°$ (kJ mol⁻¹)	$\Delta_f G°$ (kJ mol⁻¹)	Log K_f
0							
100							
200							
298.15	121.336	223.648	223.648	0.	-289.444	-295.752	51.815
300	121.336	224.398	223.650	0.224	-289.415	-295.792	51.502
400	121.336	259.305	228.409	12.358	-312.247	-297.280	38.821
464.150	121.336	277.352	233.957	20.142	------ CRYSTAL <--> LIQUID -----		
500	121.336	286.380	237.397	24.492	-375.041	-287.541	30.039
600	121.336	308.502	247.460	36.625	-371.274	-270.397	23.540
700	121.336	327.206	257.551	48.759	-367.639	-253.873	18.944
800	121.336	343.408	267.293	60.892	-364.155	-237.860	15.531
900	121.336	357.700	276.559	73.026	-360.848	-222.274	12.900
1000	121.336	370.484	285.324	85.160	-368.333	-206.286	10.775
1100	121.336	382.048	293.600	97.293	-365.097	-190.238	9.034
1200	121.336	392.606	301.417	109.427	-361.887	-174.483	7.595
1300	121.336	402.318	308.810	121.560	-358.713	-158.995	6.389
1400	121.336	411.310	315.814	133.694	-355.585	-143.750	5.363
1500	121.336	419.681	322.463	145.828	-352.517	-128.726	4.483
1600	121.336	427.512	328.786	157.961	-349.524	-113.905	3.719
1700	121.336	434.868	334.812	170.095	-346.619	-99.268	3.050
1800	121.336	441.803	340.565	182.228	-343.812	-84.800	2.461
1900	121.336	448.364	346.068	194.362	-341.109	-70.485	1.938
2000	121.336	454.587	351.339	206.496	-338.511	-56.309	1.471

PREVIOUS: June 1964 CURRENT: September 1979

Aluminum Iodide (AlI_3) $Al_1I_3(l)$

ALUMINUM IODIDE (AlI_3) CRYSTAL-LIQUID M_r = 407.69504

0 to 464.15 K crystal
above 464.15 K liquid

Refer to the individual tables for details.

Aluminum Iodide (AlI_3) Al_1I_3(cr,l)

Enthalpy Reference Temperature = T_r = 298.15 K Standard State Pressure = p° = 0.1 MPa

T/K	C_p° (J K⁻¹mol⁻¹)	S° (J K⁻¹mol⁻¹)	$-[G^\circ-H^\circ(T_r)]/T$ (J K⁻¹mol⁻¹)	$H^\circ-H^\circ(T_r)$ (kJ mol⁻¹)	Δ_fH° (kJ mol⁻¹)	Δ_fG° (kJ mol⁻¹)	Log K_f
0							
100							
200							
298.15	98.910	196.104	196.104	0.	-302.922	-301.018	52.737
300	99.077	196.716	196.106	0.183	-302.934	-301.006	52.410
400	108.533	226.513	200.105	10.563	-327.520	-299.436	39.102
464.150	114.614	243.098	204.919	17.721	CRYSTAL <--> LIQUID		
464.150	121.336	277.352	204.919	33.620	TRANSITION		
500	121.336	286.380	210.440	37.970	-375.041	-287.540	30.039
600	121.336	308.502	224.996	50.103	-371.274	-270.396	23.540
700	121.336	327.206	238.296	62.237	-367.639	-253.873	18.944
800	121.336	343.408	250.445	74.371	-364.155	-237.860	15.531
900	121.336	357.699	261.584	86.504	-360.848	-222.274	12.900
1000	121.336	370.483	271.846	98.638	-368.332	-206.285	10.775
1100	121.336	382.048	281.347	110.771	-365.097	-190.237	9.034
1200	121.336	392.606	290.185	122.905	-361.887	-174.483	7.595
1300	121.336	402.318	298.442	135.039	-358.712	-158.995	6.388
1400	121.336	411.310	306.187	147.172	-355.584	-143.750	5.363
1500	121.336	419.681	313.477	159.306	-352.517	-128.726	4.483

PREVIOUS: CURRENT: September 1979

Aluminum Iodide (AlI_3) Al_1I_3(cr,l)

ALUMINUM IODIDE (AlI_3) IDEAL GAS M_r = 407.69504

$S°(298.15 K) = 373.62 ± 08 J K^{-1} mol^{-1}$

$\Delta_f H°(0 K) = -187.9 ± 8.4 kJ mol^{-1}$
$\Delta_f H°(298.15 K) = -193.3 ± 8.4 kJ mol^{-1}$

Vibrational Frequencies and Degeneracies

ν, cm^{-1}	ν, cm^{-1}
156 (1)	[450](2)
[140](1)	64 (2)

Ground State Quantum Weight: [1] $\sigma = 6$
Point Group: D_{3h}
Bond Distance: Al-I = 2.449 ± 0.013 Å
Bond Angle: I-Al-I = 120 ± 4°
Product of the Moments of Inertia: $I_A I_B I_C = 1.362810 \times 10^{-110}\ g^3\ cm^6$

Enthalpy of Formation

Fischer, Rahlfs and Benze (1) have determined the equilibrium constants for the reaction $Al_2I_6(g)$=2 $AlI_3(g)$ from vapor density measurements. The heats of reaction are calculated from these data using both 2nd and 3rd Law methods. The results are summarized below:

T/K	Data Points	$\Delta_r H°(298.15 K)$, kcal mol^{-1} Al_2I_6 2nd law	3rd law	Drift cal K^{-1} mol^{-1}	$\Delta_f H°(298.15 K)$ kcal mol^{-1}
656-819	11[a]	24.39±0.5	24.56±0.2	0.2±0.6	-46.2

[a]Points at 614.15, 644.15, 843.15 rejected due to statistical test.

The adopted value, $\Delta_f H°(AlI_3, g, 298.15 K)$= -46.2±2.0 kcal mol^{-1} (-193.3±8.4 kJ)mol^{-1}, is based on the 3rd law enthalpy of reaction and $\Delta_f H°(Al_2I_6, g, 298.15 K) = -489.5±8.4 kJ mol^{-1}$.

Heat Capacity and Entropy

The molecular structure, bond distances and angles were taken from the electron diffraction work reported by Shen (2). These values are in good agreement with those determined by Akishin, Rambidi and Zasorin (3). Beattie and Horder (4) studied the gas phase Raman spectra of monomeric AlI_3 and reported ν_1=156 cm^{-1} and ν_4=64 cm^{-1}. The other vibrational frequencies were adjusted not only for consistency with the frequencies chosen for $AlF_3(g)$, $AlCl_3(g)$ and $AlBr_3(g)$ (5) but also for agreement between the 2nd and 3rd Law $\Delta_r H°(298.15 K)$ in the data set tabulated above. The principal moments of inertia are $I_A = 189.5840 \times 10^{-39}$, $I_B = 189.5841 \times 10^{-39}$, and $I_C = 379.1681 \times 10^{-39}\ g\ cm^2$

References

1. W. Fischer, O. Rahlfs and B. Benze, Z. anorg. allgem. Chem. 205, 1 (1932).
2. W. Shen, Diss. Abstr. Int. B34, 3735 (1974)
3. P. A. Akishin, N. G. Rambidi, and E. Z. Zasorin, Kristallografiya 4, 167 (1959).
4. I. R. Beattie and J. R. Horder, J. Chem. Soc. (London) A1969, 2655 (1969).
5. JANAF Thermochemical Tables: $AlF_3(g)$, $AlCl_3(g)$ and $AlBr_3(g)$, 9-30-79.

Aluminum Iodide (AlI_3) $Al_1I_3(g)$

Enthalpy Reference Temperature = T_r = 298.15 K Standard State Pressure = p° = 0.1 MPa

T/K	$C_p°$ (J $K^{-1}mol^{-1}$)	S°	$-[G°-H°(T_r)]/T$	$H°-H°(T_r)$ (kJ mol^{-1})	$\Delta_f H°$	$\Delta_f G°$	Log K_f
0	0.	0.	INFINITE	-18.930	-187.894	-187.894	INFINITE
100	61.385	297.474	438.424	-14.095	-188.118	-208.018	108.658
200	72.150	343.761	380.536	-7.355	-190.578	-227.014	59.290
250	75.138	360.205	374.875	-3.668	-191.946	-235.966	49.302
298.15	77.081	373.616	373.616	0.	-193.301	-244.323	42.804
300	77.142	374.093	373.617	0.143	-193.354	-244.639	42.595
350	78.515	386.094	374.562	4.036	-194.914	-253.067	37.768
400	79.482	396.645	376.676	7.987	-220.475	-260.444	34.010
450	80.184	406.049	379.427	11.980	-223.836	-265.238	30.788
500	80.706	414.526	382.520	16.003	-287.387	-263.960	27.576
600	81.414	429.308	389.122	24.112	-287.645	-259.251	22.570
700	81.856	441.894	395.784	32.277	-287.978	-254.494	18.991
800	82.149	452.844	402.247	40.478	-288.427	-249.681	16.302
900	82.353	462.533	408.417	48.704	-289.027	-244.803	14.208
1000	82.501	471.217	414.271	56.947	-300.403	-239.089	12.489
1100	82.611	479.086	419.811	65.203	-301.045	-232.927	11.061
1200	82.695	486.278	425.054	73.468	-301.703	-226.706	9.868
1300	82.760	492.900	430.022	81.741	-302.389	-220.428	8.857
1400	82.813	499.035	434.735	90.020	-303.116	-214.097	7.988
1500	82.855	504.750	439.214	98.303	-303.899	-207.711	7.233
1600	82.890	510.098	443.479	106.590	-304.752	-201.271	6.571
1700	82.919	515.124	447.547	114.881	-305.690	-194.775	5.985
1800	82.943	519.865	451.435	123.174	-306.724	-188.222	5.462
1900	82.963	524.350	455.155	131.469	-307.859	-181.608	4.993
2000	82.981	528.606	458.722	139.767	-309.097	-174.931	4.569
2100	82.996	532.655	462.147	148.065	-310.435	-168.191	4.184
2200	83.009	536.516	465.441	156.366	-311.866	-161.384	3.832
2300	83.021	540.206	468.612	164.667	-313.377	-154.510	3.509
2400	83.031	543.740	471.669	172.970	-314.955	-147.570	3.212
2500	83.040	547.129	474.620	181.273	-316.583	-140.562	2.937
2600	83.047	550.386	477.472	189.578	-318.243	-133.488	2.682
2700	83.054	553.521	480.231	197.883	-319.918	-126.350	2.444
2800	83.061	556.541	482.902	206.189	-615.492	-118.183	2.205
2900	83.066	559.456	485.492	214.495	-616.052	-100.412	1.809
3000	83.071	562.272	488.005	222.802	-616.580	-82.622	1.439
3100	83.076	564.996	490.445	231.109	-617.063	-64.815	1.092
3200	83.080	567.634	492.816	239.417	-617.490	-46.994	0.767
3300	83.084	570.190	495.122	247.725	-617.852	-29.160	0.462
3400	83.087	572.671	497.367	256.034	-618.142	-11.317	0.174
3500	83.091	575.079	499.553	264.343	-618.353	6.535	-0.098
3600	83.094	577.420	501.683	272.652	-618.483	24.390	-0.354
3700	83.096	579.697	503.761	280.961	-618.528	42.249	-0.596
3800	83.099	581.913	505.789	289.271	-618.486	60.108	-0.826
3900	83.101	584.071	507.769	297.581	-618.356	77.964	-1.044
4000	83.103	586.175	509.703	305.891	-618.139	95.815	-1.251
4100	83.105	588.227	511.593	314.202	-617.836	113.661	-1.448
4200	83.107	590.230	513.441	322.512	-617.448	131.498	-1.635
4300	83.109	592.186	515.250	330.823	-616.977	149.324	-1.814
4400	83.110	594.096	517.020	339.134	-616.426	167.139	-1.984
4500	83.112	595.964	518.754	347.445	-615.794	184.940	-2.147
4600	83.113	597.791	520.452	355.756	-615.090	202.726	-2.302
4700	83.114	599.578	522.117	364.068	-614.314	220.497	-2.451
4800	83.116	601.328	523.749	372.379	-613.469	238.250	-2.593
4900	83.117	603.042	525.350	380.691	-612.558	255.985	-2.729
5000	83.118	604.721	526.921	389.003	-611.587	273.700	-2.859
5100	83.119	606.367	528.462	397.315	-610.557	291.396	-2.985
5200	83.120	607.981	529.976	405.626	-609.473	309.071	-3.105
5300	83.121	609.564	531.463	413.939	-608.337	326.724	-3.220
5400	83.122	611.118	532.924	422.251	-607.154	344.356	-3.331
5500	83.122	612.643	534.359	430.563	-605.926	361.965	-3.438
5600	83.123	614.141	535.770	438.875	-604.658	379.552	-3.540
5700	83.124	615.612	537.158	447.187	-603.352	397.116	-3.639
5800	83.125	617.058	538.524	455.500	-602.009	414.656	-3.734
5900	83.125	618.479	539.867	463.812	-600.637	432.173	-3.826
6000	83.126	619.876	541.189	472.125	-599.215	449.664	-3.915

PREVIOUS: September 1979 (1 atm) CURRENT: September 1979 (1 bar)

Aluminum Iodide (AlI_3) $Al_1I_3(g)$

LITHIUM ALUMINUM OXIDE ($LiAlO_2$) CRYSTAL M_r = 65.92134

$S°(298.15\ K) = 53.31 \pm 1.2\ J\ K^{-1}\ mol^{-1}$

$T_{fus} = 1973 \pm 15\ K$

$\Delta_f H°(0\ K) = -1180.52 \pm 2.1\ kJ\ mol^{-1}$

$\Delta_f H°(298.15\ K) = -1188.67 \pm 2.1\ kJ\ mol^{-1}$

$\Delta_{fus} H° = [87.864 \pm 21]\ kJ\ mol^{-1}$

Enthalpy of Formation

The adopted heat of formation is based on the solution calorimetric studies of Coughlin (1). From measurements of the heat of solution of $LiAlO_2$(cr) in aqueous HCl and of four auxiliary reactions, Coughlin reported $\Delta_r H°(303.15\ K) = -89.12 \pm 0.14$ kcal mol^{-1} for Al(cr) + LiCl(cr) + 14.731 H_2O(ℓ) + $LiAlO_2$(cr) + HCl·12.731 H_2O(aq) + 3/2 H_2(g). This value is corrected to $\Delta_r H°(298.15\ K) = -88.9 \pm 0.14$ kcal mol^{-1} using heat capacities of the materials. This value, when combined with auxiliary heats of formation (2, 3) yields our adopted value of $\Delta_f H°$($LiAlO_2$, cr, 298.15 K) = -284.1 ± 0.5 kcal mol^{-1} (−1188.674 kJ mol^{-1}).

Gross et al. (4) measured the heats of solution of $LiAlO_2$(cr) and of Al_2O_3·3 H_2O(cr, gibbsite)+ Li_2O(cr) in aqueous HF. From these and the heat of solution of three moles of H_2O in HF(aq) they report $\Delta_r H°(298.15\ K) = -11.3$ kcal mol^{-1} for Li_2O(cr) + Al_2O_3·3 H_2O(cr, gibbsite) + 2 $LiAlO_2$(cr) + 3 H_2O(ℓ). Combining this $\Delta_r H°$ with auxiliary heats of formation (2, 5) yields $\Delta_f H°$($LiAlO_2$, cr, 298.15 K) = -283.8 ± 0.7 kcal mol^{-1}, in excellent agreement with the adopted value.

Equilibrium studies of vapor pressure over $LiAlO_2$(cr) by the torsion effusion technique support the adopted value of $\Delta_f H°$(298.15 K). Potter et al. (6) and Popkov and Semenov (7) both studied the equilibrium $LiAlO_2$(cr) + Li(g) + x/4 O_2(g) + (1−x)/2 O(g) + 1/2 Al_2O_3(cr, α) in the temperature range from 1500 to 1820 K where 1−x ≡ 0.13. Neither author reports the raw data but they do report 2nd and 3rd law values for $\Delta_r H°$(298.15 K). Based on these reported values and auxiliary $\Delta_f H°$(298.15 K) data (2) we calculate 2nd and 3rd law values for $\Delta_f H°$($LiAlO_2$, cr, 298.15 K) of −283.9 and −282.8 kcal mol^{-1}, respectively, from the data of Potter et al. (6). 2nd and 3rd law values for $\Delta_f H°$(298.15 K) calculated from the results of Popkov and Semenov (7) are −283.8 and −277.9 kcal mol^{-1}, respectively. The expected uncertainty in these determinations is of the order of 2 to 5 kcal mol^{-1}.

Heat Capacity and Entropy

The adopted values of $C_p°$ below 300 K are taken from Douglas and Beckett (8) based on the low temperature $C_p°$ measurements of King (53–296 K, 9). At higher temperatures the adopted $C_p°$ data are based on the drop calorimetric enthalpy data of Christensen et al. (399–1795 K, 10). No transitions were detected in either study.

Phase Data

The data in this table are for the γ or tetragonal phase of $LiAlO_2$. The crystal structure is $P4_12_12$ (11, 12). A hexagonal (α) phase may be prepared but converts to the γ phase on heating. Byker et al. (13) recently reviewed the phase data and constructed a phase diagram for the Al_2O·Li_2O system. The reader is referred to their paper for details but the following comments may be informative. The α to γ phase transformation occurs above 1173 K but may occur at lower temperature. The transition has frequently been reported as irreversible although Lejus and Collongues (14) report that the γ to α transformation does occur if aided by prolonged grinding. Measurements of the heat of the α–γ transition are hampered by its slow nature but apparently imply that it is slightly exothermic near 1200 K. Based on the above data and the known density change between α and γ phases, Byker et al. constructed a phase diagram which indicates that the α phase is the thermodynamically stable phase below ~550 K at 1 atm pressure. This seems to support the observation of Lejus and Collongues. The uncertainties are such that this temperature may be off by several hundred Kelvin. Thus, while it is clear that the γ phase is the thermodynamically stable phase above 1200 K, its stability at low temperature may be controlled by kinetic factors.

Fusion Data

See the liquid table for details.

References

1. J. P. Coughlin, J. Amer. Chem. Soc. 79, 2397 (1957).
2. JANAF Thermochemical Tables: LiCl(cr), 6-30-62; H_2O(ℓ), 3-31-79; Li_2O(cr), 3-31-64; O(g), 3-31-77; Li(g), 6-30-62.
3. V. B. Parker, D. D. Wagman and D. Garvin, "Selected Thermochemical Data Compatible with CODATA Recommendations", NBSIR 75-968, 1976; avail NTIS, PB-250845.
4. P. Gross, J. Christie and C. Hayman, Fulmer Research Institute Report R163/SR/6, June 1970.
5. J. L. Haas, Jr., G. R. Robinson, Jr. and B. S. Hemingway, J. Phys. Chem. Ref. Data 10, 575 (1981). $\Delta_f H°$(cr, gibbsite, 298.15 K) = -618.2 ± 0.3 kcal mol^{-1}.
6. N. D. Potter, M. H. Boyer, F. Ju, D. L. Hildenbrand and E. Murad, AFOSR 70-2311TR, 1970; avail. NTIS AD-715567.
7. O. S. Popkov and G. A. Semenov, Zh. Fiz. Khim. 45, 476 (1971).
8. T. B. Douglas and C. W. Beckett, U.S. Nat. Bur. Stand. Report 6484, 1959.
9. E. G. King, J. Amer. Chem. Soc. 77, 3189 (1955).
10. A. U. Christensen, K. C. Conway and K. K. Kelley, U. S. Bur. Mines RI 5565, 1960.
11. M. Marezio, Acta Cryst. 19, 396 (1965).
12. F. Bertaut, A. Delapalme, G. Bassi, A. Durif-Varambon and J. C. Joubert, Bull. Soc. France. Mineral. Crist. 88, 103 (1965).
13. H. J. Byker, I. Eliezer, N. Eliezer and R. A. Howald, J. Phys. Chem. 83, 2349 (1979).
14. A. M. Lejus and R. Collongues, Compt. Rend. 254, 2005 (1962).

Lithium Aluminum Oxide ($LiAlO_2$) $Al_1Li_1O_2$(cr)

Enthalpy Reference Temperature = T_r = 298.15 K Standard State Pressure = p° = 0.1 MPa

T/K	$C_p°$	S°	$-[G°-H°(T_r)]/T$	$H°-H°(T_r)$	$\Delta_f H°$	$\Delta_f G°$	Log K_f
	J $K^{-1}mol^{-1}$			kJ mol^{-1}			
0	0.	0.	INFINITE	-9.693	-1180.523	-1180.523	INFINITE
100	17.966	8.046	99.257	-9.121	-1183.834	-1165.875	608.991
200	47.999	30.154	58.942	-5.758	-1187.025	-1146.565	299.452
298.15	67.777	53.313	53.313	0.	-1188.674	-1126.303	197.324
300	68.128	53.733	53.314	0.126	-1188.693	-1125.916	196.039
400	81.546	75.421	56.180	7.697	-1189.202	-1104.891	144.284
500	88.408	94.425	61.974	16.226	-1192.298	-1083.492	113.192
600	92.688	110.947	68.790	25.294	-1192.108	-1061.744	92.433
700	95.751	125.475	75.871	34.723	-1191.706	-1040.047	77.609
800	98.165	138.424	82.896	44.423	-1191.220	-1018.414	66.496
900	100.203	150.107	89.725	54.343	-1190.737	-996.843	57.855
1000	102.006	160.759	96.303	64.455	-1200.889	-974.563	50.906
1100	103.654	170.559	102.614	74.740	-1200.170	-951.964	45.205
1200	105.198	179.645	108.660	85.183	-1199.326	-929.436	40.457
1300	106.671	188.124	114.450	95.777	-1198.363	-906.983	36.443
1400	108.089	196.082	119.999	106.515	-1197.279	-884.609	33.005
1500	109.466	203.586	125.324	117.393	-1196.075	-862.317	30.029
1600	110.813	210.694	130.439	128.408	-1194.752	-840.109	27.427
1700	112.140	217.452	135.361	139.556	-1338.549	-810.812	24.913
1800	113.449	223.899	140.102	150.835	-1336.240	-779.835	22.630
1900	114.746	230.068	144.676	162.245	-1333.827	-748.988	20.591
1973.000	115.679	234.411	147.916	170.656	------ CRYSTAL <--> LIQUID -----		
2000	116.022	235.986	149.094	173.784	-1331.309	-718.272	18.759
2100	117.294	241.678	153.368	185.449	-1328.691	-687.684	17.105
2200	118.558	247.163	157.508	197.242	-1325.971	-657.223	15.604

PREVIOUS: June 1972 CURRENT: December 1979

Lithium Aluminum Oxide ($LiAlO_2$) $Al_1Li_1O_2$(cr)

LITHIUM ALUMINUM OXIDE ($LiAlO_2$) LIQUID M_r = 65.92134

$S°(298.15\ K) = [93.711]\ J\ K^{-1}\ mol^{-1}$ $\Delta_f H°(298.15\ K) = [-1107.739]\ kJ\ mol^{-1}$

$T_{fus} = 1973 \pm 15\ K$ $\Delta_{fus}H° = [87.864 \pm 21]\ kJ\ mol^{-1}$

Enthalpy of Formation

$\Delta_f H°(LiAlO_2$, ℓ, 298.15 K) is calculated from $\Delta_f H°(LiAlO_2$, cr, 298.15 K) by adding the enthalpy of fusion, $\Delta_{fus}H°$, and the difference in enthalpy, H°(1973 K)-H°(298.15 K), between the crystal and liquid.

Heat Capacity and Entropy

The heat capacity of $LiAlO_2$(ℓ) is estimated to be 30 cal $K^{-1}\ mol^{-1}$ (125.52 J $K^{-1}\ mol^{-1}$) above a hypothetical glass transition temperature of 1450 K. Below this temperature the heat capacity is taken to be the same as the crystal. The adopted $C°_p$ = 30 cal $K^{-1}\ mol^{-1}$ is based on an estimated value of $C°_p$ = 7.5 cal K^{-1} g-atom^{-1} recommended by Kubaschewski and Alcock (2).

S°(298.15 K) is calculated in a manner analogous to that used for the enthalpy formation.

Fusion Data

The adopted melting point, T_{fus} = 1973±15 K, is based on the phase study of Strickler and Roy (3) and is suppported by the value 1975 K reported by Prophet (4). The older study of Ballo and Dittler (5) reported T_{fus} to lie between 2173 and 2273 K and Hummel et al. (6) reported T_{fus} = 1883±15 K. Hummel et al. also stated that their samples were only partially fused after considerable time at 1973 K. Prophet also reported that his samples slowly decomposed on fusion and that successive melting points approached that of aluminum oxide at 2308 K. After fusing seven times x-ray analysis indicated the sample was 95% α-aluminum oxide and 5% $LiAl_5O_8$ with Li_2O being lost by vaporization. The adopted $\Delta_{fus}H°$ is estimated by assuming the entropy of melting to be approximately the same on a gram-atom basis as that of $BeAl_2O_4$ (1).

References

1. JANAF Thermochemical Tables: $BeAl_2O_4$(ℓ), 12-31-79.
2. O. Kubaschewski and C. B. Alcock, "Metallurgical Thermochemistry", 5th ed., Pergamon, New York, 1979.
3. D. W. Strickler and R. Roy, J. Am. Ceram. Soc. 44, 225 (1961).
4. H. Prophet, Dow Chemical Company, personal communication, March 9, 1961.
5. R. Ballo and E. Dittler, Z. Anorg. Chem. 76, 39 (1912).
6. F. A. Hummel, B. S. R. Sastry and D. Wotring, J. Am. Ceram. Soc. 41, 88 (1958).

Lithium Aluminum Oxide ($LiAlO_2$) $Al_1Li_1O_2$(l)

Enthalpy Reference Temperature = T_r = 298.15 K — Standard State Pressure = p° = 0.1 MPa

T/K	$C°_p$	S°	$-[G°-H°(T_r)]/T$	$H°-H°(T_r)$	$\Delta_f H°$	$\Delta_f G°$	Log K_f
	J K⁻¹mol⁻¹			kJ mol⁻¹			
0							
100							
200							
298.15	67.777	93.711	93.711	0.	-1107.739	-1057.412	185.254
300	68.128	94.131	93.712	0.126	-1107.758	-1057.100	184.057
400	81.546	115.820	96.578	7.697	-1108.267	-1040.115	135.825
500	88.408	134.823	102.372	16.226	-1111.362	-1022.756	106.847
600	92.688	151.345	109.189	25.294	-1111.172	-1005.048	87.497
700	95.751	165.874	116.270	34.723	-1110.771	-987.390	73.680
800	98.165	178.822	123.294	44.423	-1110.285	-969.797	63.321
900	100.203	190.505	130.124	54.343	-1109.801	-952.266	55.268
1000	102.006	201.158	136.702	64.455	-1119.953	-934.026	48.789
1100	103.654	210.958	143.013	74.740	-1119.234	-915.467	43.472
1200	105.198	220.044	149.058	85.183	-1118.391	-896.979	39.044
1300	106.671	228.523	154.848	95.777	-1117.427	-878.566	35.301
1400	108.089	236.480	160.398	106.515	-1116.343	-860.232	32.096
1450.000	108.784	240.285	163.087	111.937	______ GLASS <--> LIQUID	TRANSITION	______
1450.000	125.520	240.285	163.087	111.937			
1500	125.520	244.541	165.732	118.213	-1114.319	-841.994	29.321
1600	125.520	252.642	170.913	130.765	-1111.459	-823.932	26.899
1700	125.520	260.251	175.947	143.317	-1253.851	-798.873	24.546
1800	125.520	267.426	180.832	155.869	-1250.271	-772.213	22.409
1900	125.520	274.212	185.570	168.421	-1246.715	-745.751	20.502
1973.000	125.520	278.945	188.937	177.584	------ CRYSTAL <--> LIQUID -----		
2000	125.520	280.651	190.164	180.973	-1243.184	-719.476	18.791
2100	125.520	286.775	194.620	193.525	-1239.679	-693.377	17.247
2200	125.520	292.614	198.943	206.077	-1236.201	-667.443	15.847
2300	125.520	298.194	203.137	218.629	-1232.750	-641.669	14.573
2400	125.520	303.536	207.210	231.181	-1229.327	-616.043	13.408
2500	125.520	308.660	211.166	243.733	-1225.934	-590.560	12.339
2600	125.520	313.583	215.011	256.285	-1222.571	-565.211	11.355
2700	125.520	318.320	218.750	268.837	-1219.240	-539.991	10.447
2800	125.520	322.885	222.388	281.389	-1509.843	-513.926	9.587
2900	125.520	327.289	225.930	293.941	-1505.483	-478.435	8.618
3000	125.520	331.545	229.380	306.493	-1501.159	-443.093	7.715
3100	125.520	335.660	232.743	319.045	-1496.871	-407.895	6.873
3200	125.520	339.645	236.021	331.597	-1492.620	-372.835	6.086
3300	125.520	343.508	239.220	344.149	-1488.408	-337.907	5.349
3400	125.520	347.255	242.343	356.701	-1484.235	-303.106	4.657
3500	125.520	350.894	245.393	369.253	-1480.103	-268.428	4.006
3600	125.520	354.430	248.373	381.805	-1476.013	-233.866	3.393
3700	125.520	357.869	251.286	394.357	-1471.966	-199.418	2.815
3800	125.520	361.216	254.135	406.909	-1467.963	-165.078	2.269
3900	125.520	364.477	256.922	419.461	-1464.005	-130.843	1.752
4000	125.520	367.654	259.651	432.013	-1460.094	-96.710	1.263

PREVIOUS: June 1972 CURRENT: December 1979

Lithium Aluminum Oxide ($LiAlO_2$) $Al_1Li_1O_2$(l)

LITHIUM ALUMINUM OXIDE ($LiAlO_2$) CRYSTAL-LIQUID M_r = 65.92134

0 to 1973 K		crystal
above	1973 K	liquid

Refer to the individual tables for details.

Lithium Aluminum Oxide ($LiAlO_2$) $Al_1Li_1O_2(cr,l)$

Enthalpy Reference Temperature = T_r = 298.15 K Standard State Pressure = $p°$ = 0.1 MPa

T/K	$C_p°$ (J K⁻¹mol⁻¹)	$S°$ (J K⁻¹mol⁻¹)	$-[G°-H°(T_r)]/T$ (J K⁻¹mol⁻¹)	$H°-H°(T_r)$ (kJ mol⁻¹)	$\Delta_f H°$ (kJ mol⁻¹)	$\Delta_f G°$ (kJ mol⁻¹)	Log K_f
0	0.	0.	INFINITE	-9.693	-1180.523	-1180.523	INFINITE
100	17.966	8.046	99.257	-9.121	-1183.834	-1165.875	608.991
200	47.999	30.154	58.942	-5.758	-1187.025	-1146.565	299.452
298.15	67.777	53.313	53.313	0.	-1188.674	-1126.303	197.324
300	68.128	53.733	53.314	0.126	-1188.693	-1125.916	196.039
400	81.546	75.421	56.180	7.697	-1189.202	-1104.891	144.284
500	88.408	94.425	61.974	16.226	-1192.298	-1083.492	113.192
600	92.688	110.947	68.790	25.294	-1192.108	-1061.744	92.433
700	95.751	125.475	75.871	34.723	-1191.706	-1040.047	77.609
800	98.165	138.424	82.896	44.423	-1191.220	-1018.414	66.496
900	100.203	150.107	89.725	54.343	-1190.737	-996.843	57.855
1000	102.006	160.759	96.303	64.455	-1200.889	-974.563	50.906
1100	103.654	170.559	102.614	74.740	-1200.170	-951.964	45.205
1200	105.198	179.645	108.660	85.183	-1199.326	-929.436	40.457
1300	106.671	188.124	114.450	95.777	-1198.363	-906.983	36.443
1400	108.089	196.082	119.999	106.515	-1197.279	-884.609	33.005
1500	109.466	203.586	125.324	117.393	-1196.075	-862.317	30.029
1600	110.813	210.694	130.439	128.408	-1194.752	-840.109	27.427
1700	112.140	217.452	135.361	139.556	-1338.549	-810.812	24.913
1800	113.449	223.899	140.102	150.835	-1336.240	-779.835	22.630
1900	114.746	230.068	144.676	162.245	-1333.827	-748.988	20.591
1973.000	115.679	234.411	147.916	170.656	CRYSTAL <--> LIQUID TRANSITION		
1973.000	125.520	278.945	147.916	258.520			
2000	125.520	280.651	149.696	261.909	-1243.184	-719.476	18.791
2100	125.520	286.775	156.079	274.461	-1239.679	-693.377	17.247
2200	125.520	292.614	162.154	287.013	-1236.201	-667.444	15.847
2300	125.520	298.194	167.948	299.565	-1232.750	-641.669	14.573
2400	125.520	303.536	173.487	312.117	-1229.327	-616.044	13.408
2500	125.520	308.660	178.792	324.669	-1225.934	-590.560	12.339
2600	125.520	313.583	183.882	337.221	-1222.571	-565.211	11.355
2700	125.520	318.320	188.774	349.773	-1219.240	-539.991	10.447
2800	125.520	322.885	193.483	362.325	-1509.843	-513.926	9.587
2900	125.520	327.289	198.021	374.877	-1505.483	-478.436	8.618
3000	125.520	331.545	202.402	387.429	-1501.159	-443.094	7.715
3100	125.520	335.660	206.634	399.981	-1496.871	-407.895	6.873
3200	125.520	339.646	210.729	412.533	-1492.621	-372.836	6.086
3300	125.520	343.508	214.694	425.085	-1488.408	-337.908	5.349
3400	125.520	347.255	218.538	437.637	-1484.235	-303.106	4.657
3500	125.520	350.894	222.268	450.189	-1480.104	-268.428	4.006
3600	125.520	354.430	225.891	462.741	-1476.013	-233.866	3.393
3700	125.520	357.869	229.411	475.293	-1471.966	-199.419	2.815
3800	125.520	361.216	232.836	487.845	-1467.963	-165.078	2.269
3900	125.520	364.477	236.170	500.397	-1464.005	-130.843	1.752
4000	125.520	367.654	239.417	512.949	-1460.094	-96.710	1.263

PREVIOUS: CURRENT: December 1979

Lithium Aluminum Oxide ($LiAlO_2$) $Al_1Li_1O_2(cr,l)$

ALUMINUM NITRIDE (AlN) CRYSTAL M_r = 40.98824

$S°(298.15\ K) = 20.14 \pm 0.21\ J\ K^{-1}\ mol^{-1}$

$\Delta_f H°(0\ K) = -312.98 \pm 2.5\ kJ\ mol^{-1}$

$\Delta_f H°(298.15\ K) = -317.98 \pm 2.5\ kJ\ mol^{-1}$

Enthalpy of Formation

The adopted enthalpy of formation is the average value of the calorimetric determinations of Neugebauer and Margrave (1) and Mah et al. (2). Neugebauer and Margrave obtained $\Delta_f H°(298.15\ K) = -76.47\pm0.2\ kcal\ mol^{-1}$ by direct nitridation. Mah et al. obtained $\Delta_f H°(298.15\ K) = -75.6\pm0.4\ kcal\ mol^{-1}$ via oxygen combustion calorimetry. The older direct nitridation study of Neumann et al. (3) yields $\Delta_f H°(298.15\ K) = -57.4\pm0.8\ kcal\ mol^{-1}$; the source of bias in this study is not apparent. Fichter and Jenny (4) also made measurements using oxygen combustion calorimetry; our recalculation of their data using the modern accepted value for the heat of formation of Al_2O_3(cr, α) (5) yields $\Delta_f H°(298.15\ K) = -71.5\pm1.0\ kcal\ mol^{-1}$.

Equilibrium data supports the calorimetric determinations but is less certain due largely to corrections for a condensation coefficient significantly different from unity. We have adopted a condensation coefficient (fraction of incident molecules which condense on the vaporizing surface) of $\alpha = 2 \times 10^{-3}$ from Hildenbrand and Hall (6) and used that value to convert measured pressures to equilibrium pressures, $P_e = P_m/\alpha$. This is a large correction, equivalent in some cases to 50 kcal mol^{-1} in the $\Delta_r H°(298.15\ K)$ values. The data below corresponds to the dissociation of 2 AlN(cr) = 2 Al(g) + N_2(g).

Source	Method	Data Point	T/K	$\Delta_r H°(298.15\ K)$ kcal mol^{-1} 2nd law	3rd law	Drift cal K^{-1} mol^{-1}	$\Delta_f H°(298.15\ K)$ kcal mol^{-1}
Hildenbrand and Hall[a] (6)	torsion-effusion	29[b]	1776-1972	309.3±6.0	307.2±1.5	-1.1±3.2	-74.8±1.2
Hoch and White[a] (7)	Langmuir	13[b]	1660-1948	342.6±18.5	307.6±7.6	-19.3±10.2	-75.0±3.9
Dreger et al[a] (8)	Knudsen-effusion	11[c]	1450-1870	391.5±17.6	308.9±12.0	-48.2±10.2	-75.6±6.1
Bolgar et al[a] (9)	Langmuir	10	1718-1829	388.5±9.7	315.3±4.2	-41.5±5.5	-78.8±2.3
Linevsky (10)	atomic resonance	7	1641-1707	290.1±18.8	299.7±1.8	5.8±11.2	-71.0±1.4

a. Raw pressure data corrected using $\alpha = 2 \times 10^{-3}$
b. Two points rejected by a statistical test.
c. One point rejected by a statistical test.

Studies not re-evaluated above include those of Blank (11) and Wu (12). Blank's (11) Knudsen and Langmuir effusion studies yield a third law value of $\Delta_r H°(298.15\ K) = -306.4\ kcal\ mol^{-1}$ ($\Delta_f H°$(AlN, cr, 298.15 K) = -74.4 kcal mol^{-1}) when corrected for condensation coefficient according to the author. We have not reanalyzed this data since not all of the measured points from the graph are given in tabular form. Wu's Knudsen effusion study (12) leads to a value of $\Delta_f H°$(AlN, cr, 298.15 K) = -62.2±1.1 kcal mol^{-1} as reported by the author based on a second-law analysis. The data are presented graphically only. Wu also suggests that AlN decomposes incongruently with loss of N_2 until a composition of Al_2N is reached.

Heat Capacity and Entropy

The adopted heat capacity data is derived by using orthogonal polynomials to smoothly merge the heat capacity-enthalpy data of Mah et al. (2). These authors reported heat capacity data in the 53-296 K range and enthalpy data (via drop calorimetry) in the 399-1799 K range. The entropy is based on $S°(50\ K) = 0.0692\ cal\ K^{-1}\ mol^{-1}$. The $C_p°$ data above 1799 K are a smooth extrapolation of the merged data and will approach a constant $C_p° = 12.521\ cal\ K^{-1}\ mol^{-1}$ at very high temperature. Demidenko et al. (13) measured heat capacity over the range 55 to 330 K and their values for $S°(298.15\ K) = 4.804\ cal\ K^{-1}\ mol^{-1}$ and $H°(298.15\ K) - H°(0\ K) = 925\ cal\ K^{-1}\ mol^{-1}$ are in excellent agreement with the adopted values; their complete data are not available. Mezaki et al. (14) measured enthalpies via drop calorimetry over the range 478 to 1113 K; their values are ~1.5% lower than the adopted values.

References

1. C. A. Neugebauer and J. L. Margrave, Z. anorg. allgem. Chem. 290, 82 (1957).
2. A. D. Mah, E. G. King, W. W. Weller and A. U. Christensen, U.S. Bur. Mines RI 5716, 1961.
3. B. Neumann, C. Kroger and H. Haebler, Z. anorg. allgem. Chem. 204, 81 (1932).
4. F. Fichter and E. Jenny, Helv. Chim. Acta 5, 498 (1922).
5. JANAF Thermochemical Tables: Al_2O_3(cr, α), 12-31-79; Al(g), 6-30-79; N_2(g), 3-31-77.
6. D. L. Hildenbrand and W. F. Hall, J. Phys. Chem. 67, 888 (1963).
7. M. Hoch and D. White, AD-142616, 1956.
8. L. H. Dreger, V. V. Dadape et al., J. Phys. Chem. 66, 1556 (1962).
9. A. S. Bolgar, S. P. Gordienko et al., Khim. Fiz. Nitridov 151 (1968).
10. M. J. Linevsky, AD-670626 1968.
11. B. A. H. Blank, UCRL-16018, 1965.
12. C. H. Wu, Ber. Kernforschiengsanlage Juelich JUEL 739-PC, 1971.
13. A. F. Demidenko, V. I. Koshchenko, L. D. Sabanova and Yu. M. Gran, Zhur. Fiz. Khim. 49, 1585 (1975).
14. R. Mezaki, E. W. Tilleux, T. F. Jambois and J. L. Margrave, Symp. Thermophys. Properties, Papers, 3rd, Lafayette, Ind, 138-45 (1965).

Aluminum Nitride (AlN) Al_1N_1(cr)

Enthalpy Reference Temperature = T_r = 298.15 K; Standard State Pressure = p° = 0.1 MPa

T/K	$C_p°$ (J K^{-1}mol^{-1})	S°	-[G°-H°(T$_r$)]/T	H°-H°(T$_r$) (kJ mol^{-1})	$\Delta_f H°$	$\Delta_f G°$	Log K_f
0	0.	0.	INFINITE	-3.871	-312.980	-312.980	INFINITE
100	5.678	2.164	39.274	-3.711	-314.756	-306.283	159.986
200	19.332	10.267	22.581	-2.463	-316.764	-296.990	77.566
298.15	30.097	20.142	20.142	0.	-317.984	-286.995	50.280
300	30.254	20.329	20.143	0.056	-318.000	-286.803	49.937
400	36.692	29.987	21.416	3.428	-318.594	-276.301	36.081
500	40.799	38.647	24.013	7.317	-318.808	-265.697	27.757
600	43.538	46.341	27.106	11.541	-318.811	-255.072	22.206
700	45.434	53.201	30.353	15.994	-318.727	-244.455	18.241
800	46.791	59.361	33.601	20.608	-318.648	-233.850	15.269
900	47.792	64.932	36.777	25.339	-318.647	-223.252	12.957
1000	48.550	70.008	39.850	30.158	-329.363	-211.887	11.068
1100	49.136	74.664	42.807	35.043	-329.302	-200.142	9.504
1200	49.598	78.960	45.643	39.981	-329.214	-188.404	8.201
1300	49.970	82.945	48.361	44.960	-329.107	-176.674	7.099
1400	50.272	86.660	50.965	49.972	-328.986	-164.953	6.154
1500	50.521	90.137	53.462	55.012	-328.856	-153.240	5.336
1600	50.728	93.404	55.857	60.075	-328.717	-141.537	4.621
1700	50.903	96.485	58.157	65.157	-328.573	-129.843	3.990
1800	51.052	99.399	60.368	70.255	-328.425	-118.157	3.429
1900	51.180	102.162	62.496	75.366	-328.273	-106.480	2.927
2000	51.290	104.790	64.545	80.490	-328.119	-94.810	2.476
2100	51.385	107.295	66.522	85.624	-327.963	-83.149	2.068
2200	51.469	109.688	68.430	90.767	-327.805	-71.495	1.697
2300	51.543	111.977	70.274	95.917	-327.646	-59.848	1.359
2400	51.609	114.172	72.058	101.075	-327.486	-48.208	1.049
2500	51.666	116.280	73.785	106.239	-327.325	-36.574	0.764
2600	51.718	118.308	75.458	111.408	-327.165	-24.948	0.501
2700	51.765	120.260	77.082	116.582	-327.003	-13.327	0.258
2800	51.807	122.144	78.658	121.761	-620.743	-0.744	0.014
2900	51.845	123.962	80.189	126.943	-619.486	21.376	-0.385
3000	51.878	125.721	81.677	132.130	-618.229	43.454	-0.757

PREVIOUS: December 1962 CURRENT: December 1979

Aluminum Nitride (AlN) Al_1N_1(cr)

ALUMINUM NITRIDE (AlN) IDEAL GAS $M_r = 40.98824$

$S°(298.15\ K) = 228.6 \pm 0.8\ J\ K^{-1}\ mol^{-1}$

$\Delta_f H°(0\ K) = 523 \pm 38\ kJ\ mol^{-1}$
$\Delta_f H°(298.15\ K) = 523 \pm 38\ kJ\ mol^{-1}$

Electronic Levels and Quantum Weights

State	ε_i, cm^{-1}	g_i
$^3\Pi$	0	6
$^3\Pi$	19727.4	6

$\omega_e = 746.9\ cm^{-1}$ $\omega_e x_e = [5.59\ cm^{-1}]$ $\sigma = 1$
$B_e = 0.5730\ cm^{-1}$ $\alpha_e = 0.0056\ cm^{-1}$ $r_e = 1.65\ Å$

Enthalpy of Formation

The enthalpy of formation is calculated from the adopted dissociation energy using auxiliary JANAF data (1). Simmons and McDonald (2) observed an apparent predissociation from the J=47 rotational level during their emission spectroscopic study of AlN. They have shown that it is possible to assign this predissociation as being from the excited $^3\Pi$ state of AlN(g) to ground state Al(g) + N(g). If this assignment is correct, we calculate $D_0^0 = 21{,}000\ cm^{-1}$ using their value of $B_o = 0.5811\ cm^{-1}$ for the excited $^3\Pi$ state rotational constant. A linear Birge-Sponer extrapolation using the adopted molecular constant data yields $D_0^0 = 25{,}000\ cm^{-1}$. Considering the nature of the uncertainties in these calculations, we adopt an average value of $D_0^0 = 23{,}000 \pm 3{,}000\ cm^{-1}$ or $D_0^0 = 66 \pm 9\ kcal\ mol^{-1}$. With auxiliary JANAF data (1) this yields an adopted value of $\Delta_f H°(298.15\ K) = 125 \pm 9\ kcal\ mol^{-1}$ ($523.000 \pm 38\ kJ\ mol^{-1}$).

Heat Capacity and Entropy

The adopted molecular constant data are taken from the emission spectroscopy work of Simmons and McDonald (2) except for $\omega_e x_e$ which we estimate from the Pekeris relationship $\omega_e x_e = B_e(1 + \omega_e \alpha_e/6B_e^2)^2$ (3). Although the energy difference between the two $^3\Pi$ states is well established, it is not certain that the lowest $^3\Pi$ state is the ground state since it has only been studied in emission. We assume it to be the ground state by analogy with the BN molecule where the $^3\Pi$ state has been shown theoretically and experimentally (via absorption spectroscopy) to be the ground state (4). For BN, SCF calculations (5) indicate $^3\Sigma$, $^1\Sigma$ and $^1\Pi$ states lying between 3400 and 4800 cm^{-1}. If these states lie at the same energy for AlN, they would have no effect on the thermochemical properties below 500 K but would increase the entropy at 6000 K by 5.0 $J\ K^{-1}\ mol^{-1}$.

References

1. JANAF Thermochemical Tables: Al(g), 6-30-79; N(g), 3-31-77.
2. J. D. Simmons and J. K. McDonald, J. Mol. Spectrosc. 41, 584 (1972).
3. C. L. Pekeris, Phys. Rev. 45, 98 (1934).
4. K. P. Huber and G. Herzberg "Molecular Spectra and Molecular Structure IV. Constants of Diatomic Molecules", Van Nostrand-Reinhold, New York, 1979.
5. G. Verhaegen, W. G. Richards and C. M. Moser, J. Chem. Phys., 46, 160 (1967).

Aluminum Nitride (AlN) $Al_1N_1(g)$

Enthalpy Reference Temperature = T_r = 298.15 K Standard State Pressure = p° = 0.1 MPa

	J K⁻¹mol⁻¹			kJ mol⁻¹			
T/K	$C_p°$	S°	$-[G°-H°(T_r)]/T$	$H°-H°(T_r)$	$\Delta_f H°$	$\Delta_f G°$	Log K_f
0	0.	0.	INFINITE	-8.942	522.933	522.933	INFINITE
100	29.134	195.617	255.951	-6.033	523.906	513.034	-267.981
200	30.316	216.062	231.448	-3.077	523.605	502.220	-131.166
250	31.378	222.940	229.080	-1.535	523.306	496.908	-103.823
298.15	32.367	228.553	228.553	0.	523.000	491.851	-86.170
300	32.403	228.753	228.553	0.060	522.988	491.658	-85.605
350	33.288	233.817	228.951	1.703	522.668	486.462	-72.600
400	34.019	238.311	229.846	3.386	522.348	481.311	-62.853
450	34.612	242.353	231.015	5.102	522.028	476.201	-55.276
500	35.093	246.026	232.335	6.845	521.704	471.126	-49.218
600	35.808	252.491	235.170	10.393	521.025	461.074	-40.140
700	36.304	258.051	238.051	14.000	520.263	451.141	-33.665
800	36.663	262.923	240.862	17.649	519.377	441.325	-28.816
900	36.934	267.258	243.558	21.330	518.328	431.630	-25.051
1000	37.147	271.160	246.127	25.034	506.497	422.821	-22.086
1100	37.320	274.709	248.566	28.758	505.396	414.507	-19.683
1200	37.465	277.963	250.882	32.497	504.286	406.293	-17.685
1300	37.589	280.967	253.082	36.250	503.167	398.172	-15.999
1400	37.699	283.756	255.175	40.014	502.040	390.138	-14.556
1500	37.797	286.361	257.168	43.789	500.905	382.184	-13.309
1600	37.886	288.803	259.070	47.573	499.765	374.307	-12.220
1700	37.969	291.102	260.887	51.366	498.620	366.501	-11.261
1800	38.046	293.275	262.627	55.167	497.471	358.762	-10.411
1900	38.119	295.334	264.294	58.975	496.320	351.087	-9.652
2000	38.189	297.291	265.896	62.791	495.166	343.473	-8.971
2100	38.257	299.156	267.435	66.613	494.011	335.917	-8.355
2200	38.322	300.937	268.918	70.442	492.855	328.416	-7.798
2300	38.387	302.642	270.347	74.278	491.698	320.967	-7.289
2400	38.450	304.277	271.727	78.119	490.542	313.569	-6.825
2500	38.514	305.848	273.061	81.968	489.387	306.219	-6.398
2600	38.577	307.360	274.351	85.822	488.234	298.915	-6.005
2700	38.641	308.817	275.601	89.683	487.082	291.655	-5.642
2800	38.706	310.223	276.812	93.550	192.031	285.406	-5.324
2900	38.772	311.583	277.988	97.424	191.979	288.742	-5.201
3000	38.840	312.898	279.130	101.305	191.930	292.079	-5.086
3100	38.910	314.173	280.240	105.192	191.884	295.418	-4.978
3200	38.981	315.409	281.320	109.087	191.842	298.759	-4.877
3300	39.056	316.610	282.371	112.989	191.804	302.101	-4.782
3400	39.133	317.777	283.395	116.898	191.770	305.444	-4.693
3500	39.212	318.913	284.394	120.815	191.741	308.787	-4.608
3600	39.294	320.018	285.368	124.741	191.716	312.132	-4.529
3700	39.379	321.096	286.319	128.674	191.696	315.477	-4.454
3800	39.467	322.148	287.248	132.616	191.681	318.823	-4.383
3900	39.558	323.174	288.157	136.568	191.671	322.169	-4.315
4000	39.652	324.177	289.045	140.528	191.665	325.515	-4.251
4100	39.748	325.157	289.913	144.498	191.665	328.861	-4.190
4200	39.847	326.116	290.764	148.478	191.670	332.207	-4.132
4300	39.949	327.055	291.597	152.468	191.679	335.553	-4.076
4400	40.053	327.974	292.413	156.468	191.693	338.899	-4.023
4500	40.159	328.876	293.214	160.478	191.713	342.244	-3.973
4600	40.268	329.759	293.999	164.500	191.735	345.589	-3.924
4700	40.378	330.627	294.769	168.532	191.760	348.934	-3.878
4800	40.490	331.478	295.525	172.575	191.788	352.277	-3.834
4900	40.604	332.314	296.267	176.630	191.819	355.621	-3.791
5000	40.720	333.135	296.996	180.696	191.851	358.963	-3.750
5100	40.836	333.943	297.713	184.774	191.884	362.305	-3.711
5200	40.954	334.737	298.417	188.864	191.917	365.646	-3.673
5300	41.073	335.518	299.110	192.965	191.948	368.987	-3.637
5400	41.192	336.287	299.791	197.078	191.978	372.327	-3.602
5500	41.313	337.044	300.462	201.203	192.005	375.667	-3.568
5600	41.433	337.790	301.122	205.341	192.028	379.006	-3.535
5700	41.554	338.524	301.771	209.490	192.046	382.344	-3.504
5800	41.675	339.248	302.411	213.651	192.060	385.682	-3.473
5900	41.795	339.961	303.042	217.825	192.064	389.021	-3.444
6000	41.916	340.665	303.663	222.010	192.082	392.357	-3.416

PREVIOUS: December 1979 (1 atm) CURRENT: December 1979 (1 bar)

Aluminum Nitride (AlN) $Al_1N_1(g)$

SODIUM ALUMINUM OXIDE ($NaAlO_2$) CRYSTAL M_r = 81.97011

$S°(298.15\ K) = 70.40 \pm 0.04\ J\ K^{-1}\ mol^{-1}$

$T_{trs} = 740\ K$

$T_{fus} = > 1923\ K$

$\Delta_f H°(0\ K) = -1137.32 \pm 0.71\ kJ\ mol^{-1}$

$\Delta_f H°(298.15\ K) = -1133.19 \pm 0.71\ kJ\ mol^{-1}$

$\Delta_{trs} H° = 1.297\ kJ\ mol^{-1}$

$\Delta_{fus} H°$ = Unknown

Enthalpy of Formation

$\Delta_f H°(298.15\ K)$ was reported by Coughlin (1). The value was calculated from the measured heat of solution of $NaAlO_2$(cr) at 303.15 K in 4.360m hydrochloric acid solution.

Heat Capacity and Entropy

The low temperature heat capacities, below 350 K, were measured by Westrum (2). Heat capacities above 350 K, 399.4 - 1698.6 K, were measured by Christensen et al. (3). These two sets of $C_p°$ data were plotted graphically and joined smoothly at 350 K. Low temperature heat capacities 51 K - 298.15 K, were also determined by King (4) and agree very well with those reported by Westrum (2). S°(298.15 K) was reported by Westrum (2) based on $S°(10\ K) = 0.01\ J\ K^{-1}\ mol^{-1}$.

Transition Data

T_{trs} and $\Delta_{trs}H°$ were obtained from Christensen et al. (3).

Fusion Data

T_{trs} is taken from Kammermeyer et al. (5).

References

1. J. P. Coughlin, J. Am. Chem. Soc. 79, 2397 (1957).
2. E. F. Westrum, University of Michigan, Ann Arbor, Michigan, personal communication, May 19, 1960.
3. A. U. Christensen, K. C. Conway and K. K. Kelley, U. S. Bur. Mines RI 5565, 1960.
4. E. G. King, J. Am. Chem. Soc. 77, 3189 (1955).
5. K. Kammermeyer and A. B. Peck, J. Am. Ceram. Soc. 16, 363 (1933).

Sodium Aluminum Oxide ($NaAlO_2$) $Al_1Na_1O_2$(cr)

Enthalpy Reference Temperature = T_r = 298.15 K Standard State Pressure = p° = 0.1 MPa

T/K	$C_p°$	S°	$-[G°-H°(T_r)]/T$	$H°-H°(T_r)$	$\Delta_f H°$	$\Delta_f G°$	Log K_f
	J K⁻¹mol⁻¹			kJ mol⁻¹			
0	0.	0.	INFINITE	-23.794	-1137.319	-1137.319	INFINITE
100	28.259	15.004	123.662	-10.866	-1129.119	-1110.209	579.914
200	56.907	44.279	76.663	-6.477	-1131.899	-1090.089	284.702
298.15	73.638	70.400	70.400	0.	-1133.195	-1069.248	187.328
300	73.889	70.856	70.401	0.136	-1133.209	-1068.851	186.104
400	83.412	93.498	73.424	8.030	-1136.417	-1047.095	136.737
500	89.563	112.798	79.417	16.690	-1136.549	-1024.739	107.054
600	94.270	129.554	86.408	25.888	-1136.263	-1002.398	87.267
700	98.291	144.393	93.651	35.519	-1135.686	-980.130	73.138
740.000	99.788	149.896	96.544	39.481		ALPHA <--> BETA	
740.000	97.696	151.649	96.544	40.778		TRANSITION	
800	98.751	159.306	100.967	46.671	-1133.761	-958.058	62.555
900	100.525	171.040	108.112	56.635	-1133.231	-936.128	54.331
1000	102.299	181.723	114.946	66.776	-1143.354	-913.493	47.716
1100	104.073	191.556	121.470	77.095	-1142.628	-890.541	42.288
1200	105.847	200.688	127.695	87.591	-1238.567	-865.220	37.662
1300	107.621	209.230	133.642	98.264	-1236.730	-834.182	33.518
1400	109.395	217.270	139.331	109.115	-1234.746	-803.290	29.971
1500	111.169	224.878	144.783	120.143	-1232.613	-772.545	26.902
1600	112.943	232.110	150.017	131.349	-1230.328	-741.948	24.222
1700	114.751	239.011	155.050	142.734	-1227.889	-711.498	21.862
1800	116.491	245.620	159.900	154.297	-1225.296	-681.196	19.768
1900	118.233	251.965	164.579	166.034	-1222.553	-651.042	17.898
2000	119.858	258.072	169.102	177.939	-1219.665	-621.037	16.220
2100	121.366	263.957	173.480	190.002	-1216.644	-591.179	14.705
2200	122.759	269.635	177.722	202.209	-1213.503	-561.468	13.331
2300	124.034	275.121	181.838	214.549	-1210.252	-531.903	12.080
2400	125.194	280.424	185.836	227.012	-1206.905	-502.481	10.936
2500	126.237	285.557	189.723	239.584	-1203.473	-473.200	9.887
2600	127.163	290.526	193.505	252.255	-1199.968	-444.058	8.921
2700	127.973	295.341	197.188	265.013	-1196.403	-415.052	8.030
2800	128.666	300.008	200.777	277.846	-1486.690	-385.211	7.186
2900	129.243	304.533	204.277	290.742	-1481.946	-345.957	6.231
3000	129.704	308.923	207.693	303.691	-1477.178	-306.866	5.343

PREVIOUS: December 1960 CURRENT: March 1963

Sodium Aluminum Oxide ($NaAlO_2$) $Al_1Na_1O_2$(cr)

ALUMINUM OXIDE (AlO) IDEAL GAS M_r = 42.98094

$S°(298.15\ K) = 218.38 \pm 0.08\ J\ K^{-1}\ mol^{-1}$

$\Delta_f H°(0\ K) = 67.0 \pm 8\ kJ\ mol^{-1}$

$\Delta_f H°(298.15\ K) = 66.9 \pm 8\ kJ\ mol^{-1}$

Electronic States and Molecular Constants (σ = 1)

State	ε_i, cm^{-1}	g_i	ω_e, cm^{-1}	$\omega_e x_e$, cm^{-1}	B_e, cm^{-1}	α_e, cm^{-1}	r_e, Å	Source
$X^2\Sigma^+$	0.0	2	979.23	6.97	0.64136	0.0058	1.6178	(1)
$A^2\Pi_i$	5282.	4	728.5	4.15	0.5365	0.0050	1.769	(2, 1)
$B^2\Sigma^+$	20635.2	2	870.05	3.52	0.60408	0.00447	1.6670	(1)
$C^2\Pi_r$	33055.	4	856.	6.	[0.60]	[0.004]	[1.67]	(1, 3-4)
$^4\Sigma^+$	[30200]	4	[820]	[5.0]	[0.565]	[0.004]	[1.724]	(3-4)
$^4\Delta$	[31600]	8	[820]	[5.0]	[0.565]	[0.004]	[1.724]	(3-4)
$^4\Sigma^-$	[33000]	4	[820]	[5.0]	[0.565]	[0.004]	[1.724]	(3-4)
$^2\Delta$	[34700]	4	[820]	[5.0]	[0.565]	[0.004]	[1.724]	(3-4)
$^2\Sigma^-$	[34900]	2	[820]	[5.0]	[0.565]	[0.004]	[1.724]	(3-4)
$D^2\Sigma^+$	40187	2	817.5	4.8	0.5652	0.0046	1.7234	(1-2)

Enthalpy of Formation

We adopt $\Delta_f H°(298.15\ K) = 16.0 \pm 2\ kcal\ mol^{-1}$ ($D_0^0 = 121.193 \pm 2\ kcal\ mol^{-1}$) based mainly on $D_0^0 \geq 121.0 \pm 0.6$ kcal mol-1 (5) from laser-induced fluorescence of AlO and on $D_0^0 = 122.0 \pm 2.5$ and 121.6 kcal mol^{-1} (6) from recent mass spectrometry of the evaporation of Al_2O_3. The mass-spectrometric values of D_0^0 (6) become 121.5±3 and ~120.8±3 kcal mol^{-1} when we convert to $\Delta_f H°(298.15\ K)$ = 78.8±1.0 kcal mol^{-1} for Al(g) and to a temperature scale based on T_{fus} = 2327 K for α-Al_2O_3 instead of 2318 K.

The adopted $\Delta_f H°$ was selected by Pedley and Marshall (7) from a network analysis of thermochemical data on D° of AlO, BO, BaO, SiO and TiO. They considered over two dozen studies on AlO plus many studies on related oxides. We concur in their selection.

Our previous JANAF Thermochemical Table (8) summarized older mass-spectrometric values of D° ranging from 113.5 to 119.9 kcal mol^{-1}. We now give less weight to the lower values for two reasons: they are less consistent with the network of data for other oxides and they all may have a negative bias arising from ionization cross sections. These lower values converge toward the adopted value if they are recalculated using alternative approximations for ionization cross sections (9-11) of reactants and calibrants. Ho and Burns (6) and Drowart (11) gave some examples of similar recalculations. We conclude that errors in cross sections provide a plausible, though unproved, explanation of much of the discrepancy between the mass-spectrometric and fluorescence values of D_0^0. Errors in the interpretation of fluorescence data (5) appear to be less likely, though not impossible. Plausibility of the fluorescence interpretation rests in part on the spectroscopic limit $D_0^0 \leq 122.1 \pm 0.6$ kcal mol^{-1} derived by Drowart (11) from reinterpretation of absorption edges observed by Tyte and previously interpreted to give $D_0^0 \leq 119.9$ kcal mol^{-1} (12). Additional results on D° have been summarized elsewhere (5). Gurvich et al. (13) selected $D_0^0 = 121.176 \pm 1.2$ kcal mol^{-1}.

Heat Capacity and Entropy

Electronic levels (ν_{oo}) and vibration-rotation constants of the observed states are from Huber and Herzberg (1) and recent studies (2). Schamps (3) made extensive predictions of the unobserved levels of AlO and MgO. We adopt these estimates since additional data for MgO (4) and the F state of AlO confirm that they are reasonably accurate. Estimated molecular constants are derived from isoconfigurational levels whenever possible. We omit levels above 40187 cm^{-1} since they will have a negligible effect. The thermodynamic functions are calculated using first-order anharmonic corrections to Q_r^i and Q_v^i in the partition function $Q = Q_t \sum_i Q_r^i Q_v^i g_i \exp(-c_2 \varepsilon_i / T)$. Functions calculated by Gurvich et al. (13) are in close agreement; their values of the Gibbs energy function differ by ≤0.01 at T≤5000 K and by only 0.026 cal $K^{-1}\ mol^{-1}$ at 6000 K.

References

1. K. P. Huber and G. Herzberg, "Constants of Diatomic Molecules," Van Nostrand Reinhold Co., New York, 1979.
2. M. Singh and M. D. Saksena, Can. J. Phys. 59, 955 (1981); Proc. Indian Acad. Sci., Sect. A 77, 139 (1973).
3. J. Schamps, Chem. Phys. 2, 352 (1973); J. Chem. Phys. 56, 573 (1972).
4. JANAF Thermochemical Tables: AlO(g), 12-31-79; MgO(g), 12-31-74.
5. L. Pasternack and P. J. Dagdigian, J. Chem. Phys. 67, 3854 (1977); P. J. Dagdigian et al., ibid. 62, 1824 (1975).
6. P. Ho and R. P. Burns, High Temp. Sci. 12, 31 (1980); C. M. Fu and R. P. Burns, ibid. 8, 353 (1976).
7. J. B. Pedley and E. M. Marshall, personal communication from J. B. Pedley, Univ. of Sussex, Falmer, Brighton, UK, Sept. 30, 1981.
8. JANAF Thermochemical Tables, J. Phys. Chem. Ref. Data 7, 827 (1978).
9. J. B. Mann, in "Recent Developments in Mass Spectrometry," edited by K. Ogata and T. Hayakawa, University Park Press, Baltimore, 1970, p. 814.
10. R. J. Ackermann and E. G. Rauh, J. Chem. Phys. 60, 2266 (1974).
11. J. Drowart, Faraday Symp. of The Chem. Soc., No. 8, 165 (1973); personal communication, Free Univ. of Brussels, Brussels, Belgium, Sept. 9, 1976.
12. J. K. McDonald and K. K. Innes, J. Mol. Spectrosc. 32, 501 (1969).
13. L. V. Gurvich, I. V. Veits et al., "Thermodynamic Properties of Individual Substances," 3rd ed., Volume III, Nauka, Moscow, 1981.

Aluminum Oxide (AlO) $Al_1O_1(g)$

Enthalpy Reference Temperature = T_r = 298.15 K; Standard State Pressure = p° = 0.1 MPa

T/K	$C_p°$ (J $K^{-1}mol^{-1}$)	S° (J $K^{-1}mol^{-1}$)	$-[G°-H°(T_r)]/T$ (J $K^{-1}mol^{-1}$)	$H°-H°(T_r)$ (kJ mol^{-1})	$\Delta_f H°$ (kJ mol^{-1})	$\Delta_f G°$ (kJ mol^{-1})	Log K_f
0	0.	0.	INFINITE	-8.788	67.037	67.037	INFINITE
100	29.108	186.129	244.933	-5.880	68.009	58.760	-30.693
200	29.504	206.370	221.161	-2.958	67.673	49.577	-12.948
298.15	30.883	218.386	218.386	0.	66.944	40.845	-7.156
300	30.913	218.577	218.387	0.057	66.929	40.683	-7.084
400	32.489	227.691	219.619	3.229	66.107	32.057	-4.186
500	33.750	235.083	221.995	6.544	65.260	23.642	-2.470
600	34.685	241.323	224.710	9.968	64.369	15.401	-1.341
700	35.399	246.726	227.478	13.474	63.400	7.315	-0.546
800	36.001	251.492	230.187	17.044	62.321	-0.625	0.041
900	36.583	255.766	232.796	20.673	61.106	-8.422	0.489
1000	37.205	259.652	235.290	24.362	49.149	-15.315	0.800
1100	37.898	263.230	237.670	28.116	47.973	-21.704	1.031
1200	38.669	266.560	239.940	31.944	46.852	-27.988	1.218
1300	39.506	269.688	242.109	35.852	45.793	-34.182	1.373
1400	40.386	272.648	244.186	39.847	44.806	-40.296	1.503
1500	41.283	275.465	246.178	43.930	43.893	-46.343	1.614
1600	42.167	278.157	248.093	48.103	43.057	-52.331	1.708
1700	43.015	280.739	249.938	52.362	42.296	-58.269	1.790
1800	43.805	283.221	251.719	56.704	41.605	-64.164	1.862
1900	44.523	285.609	253.440	61.121	40.977	-70.023	1.925
2000	45.158	287.909	255.106	65.606	40.405	-75.850	1.981
2100	45.706	290.126	256.721	70.150	39.881	-81.649	2.031
2200	46.165	292.263	258.288	74.744	39.396	-87.425	2.076
2300	46.538	294.324	259.811	79.380	38.942	-93.179	2.116
2400	46.830	296.311	261.290	84.049	38.509	-98.914	2.153
2500	47.047	298.227	262.730	88.743	38.091	-104.632	2.186
2600	47.196	300.075	264.131	93.456	37.681	-110.333	2.217
2700	47.286	301.858	265.495	98.180	37.272	-116.018	2.244
2800	47.325	303.579	266.825	102.911	-257.042	-120.719	2.252
2900	47.321	305.240	268.121	107.644	-256.368	-115.863	2.087
3000	47.279	306.843	269.385	112.374	-255.706	-111.029	1.933
3100	47.209	308.393	270.619	117.099	-255.059	-106.217	1.790
3200	47.115	309.890	271.823	121.815	-254.430	-101.426	1.656
3300	47.003	311.338	272.998	126.521	-253.820	-96.654	1.530
3400	46.878	312.739	274.147	131.215	-253.231	-91.901	1.412
3500	46.744	314.096	275.269	135.896	-252.663	-87.164	1.301
3600	46.605	315.411	276.366	140.564	-252.118	-82.443	1.196
3700	46.464	316.686	277.438	145.217	-251.596	-77.737	1.097
3800	46.324	317.923	278.487	149.857	-251.096	-73.045	1.004
3900	46.187	319.125	279.514	154.482	-250.619	-68.365	0.916
4000	46.055	320.293	280.519	159.094	-250.163	-63.698	0.832
4100	45.930	321.428	281.503	163.694	-249.730	-59.042	0.752
4200	45.813	322.534	282.467	168.281	-249.318	-54.396	0.677
4300	45.704	323.610	283.411	172.857	-248.928	-49.760	0.604
4400	45.606	324.660	284.337	177.422	-248.557	-45.133	0.536
4500	45.518	325.684	285.244	181.978	-248.205	-40.514	0.470
4600	45.441	326.683	286.134	186.526	-247.874	-35.902	0.408
4700	45.376	327.660	287.008	191.067	-247.561	-31.297	0.348
4800	45.322	328.615	287.864	195.601	-247.268	-26.699	0.291
4900	45.280	329.549	288.706	200.131	-246.993	-22.106	0.236
5000	45.250	330.463	289.532	204.658	-246.736	-17.520	0.183
5100	45.232	331.359	290.343	209.182	-246.498	-12.937	0.133
5200	45.226	332.237	291.140	213.705	-246.278	-8.360	0.084
5300	45.231	333.099	291.924	218.227	-246.077	-3.787	0.037
5400	45.248	333.944	292.694	222.751	-245.894	0.783	-0.008
5500	45.276	334.775	293.452	227.277	-245.730	5.349	-0.051
5600	45.315	335.591	294.197	231.807	-245.585	9.913	-0.092
5700	45.365	336.394	294.930	236.341	-245.461	14.474	-0.133
5800	45.425	337.183	295.652	240.880	-245.354	19.034	-0.171
5900	45.494	337.960	296.363	245.426	-245.270	23.591	-0.209
6000	45.574	338.725	297.062	249.980	-245.186	28.145	-0.245

PREVIOUS: December 1979 (1 atm) CURRENT: December 1979 (1 bar)

Aluminum Oxide (AlO) $Al_1O_1(g)$

ALUMINUM OXIDE, ION (AlO^+) IDEAL GAS GFW = 42.980391

$S°(298.15\ K) = [231.0 \pm 13]\ J\ K^{-1}mol^{-1}$

$\Delta_f H°(0\ K) = 986.6 \pm 20\ kJ\ mol^{-1}$
$\Delta_H H°(298.15\ K) = [992.993]\ kJ\ mol^{-1}$

Electronic States and Molecular Constants (σ = 1)

State	ε_i, cm^{-1}	g_i	ω_e, cm^{-1}	$\omega_e x_e$, cm^{-1}	B_e, cm^{-1}	α_e, cm^{-1}	r_e, Å
$^3\Pi$	[0]	6	[710]	[4]	[0.5066]	[0.004]	[1.82]
$^1\Pi$	[2000]	2	[710]	[4]	[0.5066]	[0.004]	[1.82]
$^1\Sigma$	[300]	1	[820]	[5]	[0.5806]	[0.005]	[1.70]
$^3\Sigma$	[12000]	3	[870]	[5]	[0.5806]	[0.005]	[1.70]
$^1\Sigma$	[20000]	1	[870]	[5]	[0.5806]	[0.005]	[1.70]

Enthalpy of Formation

$\Delta_f H°$ is calculated from that of AlO (1) using Hildenbrand's appearance potential (2) of 9.53±0.15 eV (219.8±3.5 kcal mol^{-1}), assuming that it is identical with the ionization potential for AlO(g) → AlO^+(g) + e (g). Other reported values for the appearance potential range from 9 to 10.3±1 eV, including several values of 9.5 (3-5). These values are consistent with, but less precise than, the adopted value. $\Delta_r H°(0\ K)$ yields D_0^0 = 39.5±5 kcal/mol for AlO^+ → Al^+ + O; this is larger than the value of ~26 kcal mol^{-1} predicted by Schamps (6) from his SCF calculation of IP(AlO) = 10.08 eV. MNDO calculations (7) differ by only 3.5 kcal mol^{-1} from our [$\Delta_f H°(AlO^+$, 298.15 K) − $\Delta_f H°$(AlO, 298.15 K) even though the MNDO values of $\Delta_f H°$(298.15 K) deviate by -14 and -18 kcal mol^{-1}, respectively.

$\Delta_f H°(AlO^+$, g, 298.15 K) is obtained from $\Delta_f H°$(AlO) by using IP(AlO) with JANAF (1) enthalpies, H°(0 K) −H°(298.15 K), for AlO(g), AlO^+(g) and e^-(g). $\Delta_r H°$ (AlO → AlO^+ + e^-, 298.15 K) differs from a room-temperature threshold energy due to inclusion of these enthalpies and to threshold effects discussed by Rosenstock et al. (5). $\Delta_f H°$(298.15 K) should be changed by -1.481 kcal mol^{-1} if it is to be used in the ion convention that excludes the enthalpy of the electron.

Heat Capacity and Entropy

No spectroscopic data have been observed for AlO^+ but Schamps (6) recently predicted electronic levels and molecular constants from variational calculations with semiempirical estimates of correlation energy differences. Similar calculations were reasonably accurate for MgO (6). The predictions for AlO^+ yield almost equal energies for $^3\Pi$ and $^1\Sigma^+$; thus they do not distinquish which is the ground state. AlO^+ is isoelectronic with MgO and AlN. MgO has a $^1\Sigma^+$ ground state with the low-lying $^3\Pi$ level at 2600 cm^{-1} (8). Triplet-triplet bands are observed for AlN (9, 10) and it is quite likely that the lower $^3\Pi$ level is the ground state (10). We conclude for AlO^+ that $^3\Pi$ is very low lying and probably the ground state; it will dominate the electronic partition function.

The adopted electronic levels are minor modifications of the predictions of Schamps (6). Molecular constants of the $^3\Pi$ state are estimated to be intermediate between those of MgO (1) and AlN (9). The r_e adopted for $^3\Pi$ is ~0.1 Å longer than that of Al_2O (1). Values of r_e adopted for AlO^+ are also longer than those of their Π and Σ counterparts in AlO (1). Values of ω_e are based on force-constant comparisons. Other constants are based on comparisons with the observed and calculated values for MgO (1, 6).

The low-lying electronic levels cause an entropy uncertainty of perhaps 3 cal/(mol K) at 298 K but <1 cal/(mol K) at T>2000 K. Uncertainty in the enthalpy is large (perhaps 4 kcal mol^{-1}) at high temperature. The thermodynamic functions are calculated using first-order anharmonic corrections to Q_r^i and Q_v^i in the partition function $Q=Q_t\Sigma_i Q_r^i Q_v^i g_i \exp(-c_2\varepsilon_i/T)$.

References

1. JANAF Thermochemical Tables: AlO(g), Al_2O(g) 12-31-79; MgO(g) 12-31-74; Al^+(g) 6-30-79; e^-(g) 3-31-77.
2. D. L. Hildenbrand, Chem. Phys. Lett. 20, 127 (1973).
3. R. P. Burns et al., J. Chem. Phys. 44, 3307 (1966); 32, 1366 (1960); High Temp. Sci. 12, 31 (1980); 8, 353 (1976).
4. M. Farber, R. D. Srivastava and O. M. Uy, J. Chem. Soc., Faraday Trans. I 68, 249 (1972); J. Chem. Phys. 56, 5312 (1972).
5. H. M. Rosenstock, K. Draxl et al., J. Phys. Chem. Ref. Data 6, Supp. 1 (1977).
6. J. Schamps, Chem. Phys. 2, 352 (1973); J. Chem. Phys. 56, 573 (1972).
7. L. P. Davis et al., J. Comput. Chem. 2, 433 (1981).
8. T. Ikeda, N. B. Wong, D. O. Harris and R. W. Field, J. Mol. Spectrosc. 68, 452 (1977).
9. J. D. Simmons and J. K. McDonald, J. Mol. Spectrosc. 41, 584 (1972).
10. T. M. Dunn, Univ. Michigan, Ann Arbor, personal communication, Aug., 1975.

Aluminum Oxide, Ion (AlO^+) $Al_1O_1^+(g)$

Enthalpy Reference Temperature = T_r = 298.15 K Standard State Pressure = p° = 0.1 MPa

T/K	$C_p°$ (J K⁻¹mol⁻¹)	S° (J K⁻¹mol⁻¹)	−[G°−H°(T_r)]/T (J K⁻¹mol⁻¹)	H°−H°(T_r) (kJ mol⁻¹)	$\Delta_f H°$ (kJ mol⁻¹)	$\Delta_f G°$ (kJ mol⁻¹)	Log K_f
0	0.	0.	INFINITE	-9.090	986.587		
100	29.445	197.312	259.050	-6.174			
200	31.118	218.161	233.943	-3.156			
298.15	33.135	230.975	230.975	0.	992.993	956.886	-167.642
300	33.168	231.180	230.975	0.061	993.021	956.661	-166.570
400	34.685	240.942	232.296	3.458	994.503	944.317	-123.315
500	35.770	248.805	234.837	6.984	995.945	931.603	-97.324
600	36.583	255.401	237.729	10.603	997.327	918.604	-79.972
700	37.206	261.090	240.669	14.294	998.622	905.379	-67.560
800	37.686	266.091	243.541	18.040	999.797	891.977	-58.240
900	38.052	270.551	246.298	21.828	1000.820	878.437	-50.983
1000	38.330	274.576	248.928	25.647	991.071	865.552	-45.212
1100	38.539	278.239	251.429	29.491	992.064	852.952	-40.503
1200	38.696	281.600	253.805	33.354	993.056	840.261	-36.576
1300	38.813	284.702	256.064	37.229	994.043	827.488	-33.249
1400	38.900	287.581	258.214	41.115	995.026	814.639	-30.395
1500	38.966	290.268	260.262	45.009	996.003	801.720	-27.918
1600	39.017	292.784	262.217	48.908	996.972	788.737	-25.750
1700	39.057	295.151	264.085	52.812	997.933	775.692	-23.834
1800	39.091	297.384	265.874	56.719	998.886	762.592	-22.130
1900	39.120	299.499	267.588	60.630	999.831	749.439	-20.604
2000	39.147	301.506	269.234	64.543	1000.766	736.236	-19.229
2100	39.174	303.417	270.817	68.459	1001.693	722.986	-17.983
2200	39.202	305.240	272.341	72.378	1002.611	709.693	-16.850
2300	39.231	306.983	273.809	76.300	1003.521	696.358	-15.815
2400	39.262	308.653	275.226	80.224	1004.423	682.984	-14.865
2500	39.295	310.256	276.596	84.152	1005.317	669.572	-13.990
2600	39.331	311.798	277.920	88.083	1006.203	656.125	-13.182
2700	39.369	313.283	279.203	92.018	1007.083	642.644	-12.433
2800	39.410	314.716	280.446	95.957	714.056	630.098	-11.755
2900	39.453	316.100	281.651	99.900	716.019	627.064	-11.295
3000	39.499	317.438	282.822	103.848	717.977	623.964	-10.864
3100	39.546	318.734	283.960	107.800	719.930	620.798	-10.460
3200	39.595	319.990	285.066	111.757	721.879	617.569	-10.081
3300	39.647	321.209	286.143	115.719	723.824	614.279	-9.723
3400	39.699	322.394	287.192	119.686	725.764	610.930	-9.386
3500	39.753	323.545	288.214	123.659	727.702	607.525	-9.067
3600	39.808	324.666	289.211	127.637	729.636	604.064	-8.765
3700	39.863	325.757	290.184	131.621	731.567	600.549	-8.478
3800	39.919	326.821	291.134	135.610	733.495	596.982	-8.206
3900	39.976	327.859	292.063	139.604	735.421	593.365	-7.947
4000	40.034	328.872	292.970	143.605	737.343	589.698	-7.701
4100	40.091	329.861	293.858	147.611	739.262	585.983	-7.466
4200	40.149	330.828	294.727	151.623	741.177	582.221	-7.241
4300	40.206	331.773	295.577	155.641	743.088	578.414	-7.026
4400	40.264	332.698	296.411	159.664	744.995	574.562	-6.821
4500	40.321	333.604	297.227	163.694	746.900	570.667	-6.624
4600	40.378	334.490	298.028	167.729	748.796	566.730	-6.435
4700	40.435	335.359	298.813	171.769	750.687	562.751	-6.254
4800	40.491	336.211	299.583	175.816	752.570	558.733	-6.080
4900	40.547	337.047	300.339	179.868	754.447	554.675	-5.913
5000	40.602	337.866	301.081	183.925	756.313	550.579	-5.752
5100	40.657	338.671	301.811	187.988	758.168	546.447	-5.597
5200	40.711	339.461	302.527	192.056	760.013	542.277	-5.447
5300	40.765	340.237	303.231	196.130	761.844	538.072	-5.303
5400	40.818	340.999	303.924	200.209	763.660	533.833	-5.164
5500	40.871	341.749	304.605	204.294	765.461	529.560	-5.029
5600	40.922	342.486	305.274	208.384	767.244	525.255	-4.899
5700	40.974	343.211	305.934	212.478	769.008	520.918	-4.774
5800	41.024	343.924	306.583	216.578	770.755	516.550	-4.652
5900	41.074	344.625	307.221	220.683	772 476	512.152	-4.534
6000	41.123	345.316	307.851	224.793	774.195	507.724	-4.420

PREVIOUS: December 1979 (1 atm) CURRENT: December 1979 (1 bar)

Aluminum Oxide, Ion (AlO^+) $Al_1O_1^+(g)$

$S°(298.15\ K) = [212.81 \pm 1.7]\ J\ K^{-1}\ mol^{-1}$

$\Delta_fH°(0\ K) = -263.2 \pm 17\ kJ\ mol^{-1}$
$\Delta_fH°(298.15\ K) = [-269.425]\ kJ\ mol^{-1}$

Electronic Levels and Quantum Weights

State	ε_i, cm^{-1}	g_i
$[^1\Sigma^+]$	0	[1]
$[^3\Pi]$	[27000]	[6]

$\omega_e = [900]\ cm^{-1}$ $\omega_ex_e = [5.4]\ cm^{-1}$ $\sigma = 1$
$B_e = [0.640]\ cm^{-1}$ $\alpha_e = [0.006]\ cm^{-1}$ $r_e = [1.62]$ Å

Enthalpy of Formation

We adopt $\Delta_fH°(0\ K) = -62.9\pm4\ kcal\ mol^{-1}$ based on equilibrium data for $AlO(g) + Cl^-(g) = Cl(g) + AlO^-(g)$ obtained using effusion mass spectrometry by Srivastava et al. (1, 2). Our analysis of the data is summarized below. For the process $AlO^-(g) \rightarrow Al(g) + O^-(g)$ we calculate a dissociation energy $D_0^O = 166.4\pm4\ kcal\ mol^{-1}$; this is comparable to $D_0^O(AlF) = 159.3\pm1.5\ kcal\ mol^{-1}$ (3) and much larger than $D_0^O(AlO) = 121.2\pm2\ kcal\ mol^{-1}$ (3). The corresponding electron affinity, EA(AlO) = 78.9±4 $kcal\ mol^{-1}$ (3.42 eV), is 4.4 $kcal\ mol^{-1}$ less than EA(Cl) = 83.3 $kcal\ mol^{-1}$ (3). Gaines and Page (4) used a semi-empirical method to predict EA(AlO) = 60 and EA(BO) = 49 $kcal\ mol^{-1}$, corresponding to a difference of 11 $kcal\ mol^{-1}$. Although this difference is comparable with the experimental difference of 9 $kcal\ mol^{-1}$, the values predicted for EA(AlO) and EA(BO) are ~20 $kcal\ mol^{-1}$ lower than the experimental results (1, 2). MNDO calculations (5) predicted $\Delta_fH°(298.15\ K) = -55.9$ and $[\Delta_fH°(AlO^-, 298.15\ K) - \Delta_fH°(AlO, 298.15\ K)] = 54\ kcal\ mol^{-1}$, but these values have a large uncertainty. Gurvich et al. (6) selected EA(AlO) = 79.8±2.4 $kcal\ mol^{-1}$ based on the equilibrium data (1).

The value of $\Delta_rH°$ ($AlO^- \rightarrow AlO + e^-$, 298.15 K) calculated from our tables will differ from a room temperature threshold energy due to inclusion of the enthalpies, H°(0 K)−H°(298.15 K), of $AlO^-(g)$, AlO(g) and $e^-(g)$ and to the threshold effects discussed by Rosenstock et al. (7).

Source	T/K	Data Points	δS cal K^{-1} mol^{-1}	$\Delta_rH°$(298.15 K), kcal mol^{-1} 2nd law	3rd law	$\Delta_fH°$(298.15 K) kcal mol^{-1}
(1, 2) Srivastava (1972)	2082-2222	5	-5.8±3.8	-8.1±8.3	4.48±1.1	-64.39

Heat Capacity and Entropy

All of the molecular constants are estimated by comparison with AlF and AlO (3). We estimate the electronic states and levels from the low-lying states of isoelectronic AlF. The bond distance is taken equal to that in AlO and slightly shorter than that in AlF. B_e is calculated from r_e. We derive $\omega_e = 900\pm100\ cm^{-1}$ from k = 4.8 mdyn/Å estimated from k(AlF) = 4.23 and k(AlO) = 5.68. Similar comparisons of ω_ex_e/ω_e and α_e/B_e are used to estimated ω_ex_e and α_e.

Gurvich et al. (6) calculated $C_p°$, S° and H° using somewhat larger estimates for ω_e and B_e. Their values of S° differ from ours by -0.21 (298.15 K), -0.56 (2000 K) and -0.63 (4000 K) cal K^{-1} mol^{-1}.

References

1. R. D. Srivastava, O. M. Uy and M. Farber, J. Chem. Soc., Faraday Trans. II 68, 1388 (1972); Trans. Faraday Soc. 67, 2941 (1971).
2. M. Farber, R. D. Srivastava and O. M. Uy (Space Sci. Inc., Monrovia, Calif.), AD-731303, 50 pp. (1971).
3. JANAF Thermochemical Tables: AlF(g), 12-31-75; AlO(g), 12-31-79; Cl^-(g), 6-30-65.
4. A. F. Gaines and F. M. Page, Trans. Faraday Soc. 62, 3086 (1966).
5. L. P. Davis et al., J. Comput. Chem. 2, 433 (1981).
6. L. V. Gurvich, I. V. Veits et al., "Thermodynamic Properties of Individual Substances," 3rd ed., Vol. III, Nauka, Moscow, 1981.
7. H. M. Rosenstock, K. Draxl et al., J. Phys. Chem. Ref. Data 6, Suppl. 1 (1977).

Enthalpy Reference Temperature = T_r = 298.15 K   Standard State Pressure = p° = 0.1 MPa

T/K	$C_p°$ (J $K^{-1}mol^{-1}$)	S°	$-[G°-H°(T_r)]/T$	$H°-H°(T_r)$ (kJ mol^{-1})	$\Delta_fH°$	$\Delta_fG°$	Log K_f
0	0.	0.	INFINITE	-8.826	-263.172		
100	29.112	180.387	239.570	-5.918			
200	29.690	200.668	215.615	-2.989			
250	30.453	207.371	213.318	-1.487			
298.15	31.303	212.807	212.807	0.	-269.424	-287.606	50.387
300	31.335	213.001	212.808	0.058	-269.477	-287.718	50.096
350	32.194	217.897	213.193	1.646	-270.901	-290.646	43.377
400	32.961	222.247	214.057	3.276	-272.332	-293.369	38.310
450	33.622	226.168	215.189	4.941	-273.768	-295.913	34.349
500	34.179	229.740	216.468	6.636	-275.212	-298.296	31.163
600	35.041	236.053	219.220	10.100	-278.142	-302.639	26.347
700	35.654	241.503	222.023	13.636	-281.159	-306.484	22.870
800	36.102	246.295	224.763	17.225	-284.298	-309.890	20.234
900	36.439	250.567	227.397	20.853	-287.593	-312.893	18.160
1000	36.699	254.420	229.910	24.510	-301.660	-314.759	16.441
1100	36.907	257.928	232.300	28.191	-304.988	-315.908	15.001
1200	37.077	261.147	234.572	31.890	-308.317	-316.754	13.788
1300	37.219	264.121	236.732	35.605	-311.647	-317.322	12.750
1400	37.341	266.883	238.788	39.334	-314.979	-317.633	11.851
1500	37.446	269.463	240.748	43.073	-318.314	-317.706	11.064
1600	37.539	271.883	242.619	46.822	-321.652	-317.557	10.367
1700	37.623	274.161	244.408	50.581	-324.993	-317.199	9.746
1800	37.699	276.314	246.122	54.347	-328.339	-316.644	9.189
1900	37.769	278.354	247.765	58.120	-331.688	-315.903	8.685
2000	37.835	280.293	249.343	61.900	-335.043	-314.985	8.227
2100	37.896	282.141	250.861	65.687	-338.403	-313.900	7.808
2200	37.954	283.905	252.323	69.479	-341.768	-312.654	7.423
2300	38.009	285.593	253.734	73.278	-345.139	-311.255	7.069
2400	38.062	287.212	255.095	77.081	-348.516	-309.710	6.741
2500	38.113	288.767	256.411	80.890	-351.898	-308.023	6.436
2600	38.164	290.263	257.684	84.704	-355.286	-306.202	6.152
2700	38.213	291.704	258.918	88.523	-358.679	-304.249	5.886
2800	38.262	293.095	260.114	92.346	-655.979	-301.203	5.619
2900	38.312	294.438	261.274	96.175	-658.287	-288.492	5.196
3000	38.362	295.738	262.402	100.009	-660.601	-275.701	4.800
3100	38.414	296.997	263.497	103.848	-662.918	-262.833	4.429
3200	38.467	298.217	264.563	107.692	-665.240	-249.890	4.079
3300	38.523	299.402	265.601	111.541	-667.565	-236.874	3.749
3400	38.582	300.552	266.612	115.396	-669.893	-223.788	3.438
3500	38.645	301.672	267.598	119.258	-672.224	-210.634	3.144
3600	38.712	302.761	268.560	123.126	-674.557	-197.412	2.864
3700	38.784	303.823	269.499	127.000	-676.892	-184.126	2.599
3800	38.861	304.858	270.416	130.882	-679.228	-170.777	2.347
3900	38.944	305.869	271.312	134.773	-681.565	-157.366	2.108
4000	39.035	306.856	272.188	138.672	-683.902	-143.895	1.879
4100	39.132	307.821	273.045	142.580	-686.238	-130.366	1.661
4200	39.237	308.765	273.885	146.498	-688.573	-116.780	1.452
4300	39.350	309.690	274.707	150.427	-690.908	-103.138	1.253
4400	39.471	310.596	275.512	154.368	-693.241	-89.442	1.062
4500	39.602	311.484	276.302	158.322	-695.569	-75.694	0.879
4600	39.741	312.356	277.076	162.289	-697.897	-61.893	0.703
4700	39.890	313.212	277.836	166.271	-700.223	-48.042	0.534
4800	40.048	314.054	278.581	170.267	-702.546	-34.141	0.372
4900	40.216	314.881	279.314	174.281	-704.866	-20.192	0.215
5000	40.393	315.696	280.033	178.311	-707.184	-6.195	0.065
5100	40.580	316.497	280.741	182.359	-709.500	7.848	-0.080
5200	40.776	317.287	281.436	186.427	-711.814	21.935	-0.220
5300	40.982	318.066	282.120	190.515	-714.126	36.069	-0.355
5400	41.197	318.834	282.792	194.624	-716.437	50.245	-0.486
5500	41.420	319.592	283.455	198.755	-718.747	64.463	-0.612
5600	41.652	320.340	284.107	202.908	-721.057	78.725	-0.734
5700	41.893	321.080	284.749	207.085	-723.368	93.027	-0.852
5800	42.141	321.810	285.381	211.287	-725.677	107.370	-0.967
5900	42.397	322.533	286.005	215.514	-727.991	121.752	-1.078
6000	42.660	323.248	286.620	219.767	-730.286	136.172	-1.185

PREVIOUS: December 1979 (1 atm)   CURRENT: December 1979 (1 bar)

ALUMINUM OXIDE (AlO_2) IDEAL GAS M_r = 58.98034

$\Delta_{at}H°(0\ K)$ = 216.52 ± 8 kJ mol^{-1} $\Delta_f H°(0\ K)$ -85.01 ± 32 kJ mol^{-1}

S°(298.15 K) = [251.83 ± 8] J K^{-1} mol^{-1} $\Delta_f H°(298.15\ K)$ = -86.19 ± 32 kJ mol^{-1}

Electronic Levels and Quantum Weights		Vibrational Frequencies and Degeneracies
ε_i, cm^{-1}	g_i	ν, cm^{-1}
0	[4]	[700](1)
[15000]	[4]	[200](2)
[20000]	[2]	[930](1)

Point Group: [$D_{\infty h}$] σ = 2
Bond Distance: Al-O = [1.69] Å
Bond Angle: O-Al-O = [180]°
Rotational Constant: B_0 = [0.184455] cm^{-1}

Enthalpy of Formation

We adopt $\Delta_f H°(298.15\ K)$ = -20.6±8 kcal mol^{-1} and $\Delta_{at}H°(0\ K)$ = 216.52 kcal mol^{-1} derived from mass-spectrometric data measured by Ho and Burns (1) during laser evaporation of single crystals of Al_2O_3. $\Delta_f H°$ and $\Delta_{at}H°$ are consistent with limiting values from studies in two other laboratories that were unable to detect AlO_2^+ (2-3). AlO_2 appears to be much less stable than predicted by earlier data (4-5); those differ from the adopted $\Delta_f H°$ by -19 to -23 kcal mol^{-1}. Our analyses of the data are given below.

AlO_2 is the least abundant of the five Al-containing species identified in the vapor over Al_2O_3. Ho and Burns (1), unlike Farber et al. (4-5), identified AlO_2 as the source of the mass=59 ion by accurate mass data combined with intensities and appearance potentials of AlO_2^+ and conceivable impurity ions. Their partial pressure (1) should be free of large error from mass=59 ions other than AlO_2^+. Error does arise from uncertainties in the evaporation coefficients, ionization cross sections and total flux of species (1). An extreme example of this error is the formation of α-Al_2O_3, which is biased by -9 kcal mol^{-1} if calculated from the reported (1) pressures of Al and O. Our assigned uncertainty includes contributions from this error and that of the Gibbs-energy function.

Semiempirical calculations predicted $\Delta_f H°(298.15\ K)$ = -14.6 and -24.3 kcal mol^{-1} by the MNDO half-electron (6) and unrestricted Hartree-Fock (7) methods. These values agree with our adopted $\Delta_f H°$ but may be too uncertain to exclude the more negative value. Recent summaries (1, 8) cite many other AlO_2 papers, including emission spectra and kinetic data that may not involve AlO_2. New studies (9) reassign much of the emission to an aluminum hydrate. We discount the bond-dissociation energies estimated from kinetics since formation of AlO_2 was inferred only from disappearance of AlO. Gurvich et al. (10) selected $\Delta_{at}H°$ = 215.1 ± 7.2 kcal mol^{-1} based on different Gibbs-energy functions.

Source	Method	Reaction[a]	T/K	δS[b] cal K^{-1} mol^{-1}	$\Delta_r H°(298.15)$, kcal mol^{-1} 2nd Law	3rd Law	$\Delta_f H°(298.15\ K)$ kcal mol^{-1}	$\Delta_{at}H°(0\ K)$
(1)	Mass spec.	A	2221-2327[c]	-26±5	-279±12	-218.5±2.7	-20.6±8	216.5
(2)	Eff. mass spec.	B	2600	--	--	>25.1[d]	>-21.7±8	<216.6
(3)	Mass spec.	B	2450	--	--	>24.9[d]	>-18.9±8	<214.8
(4)	Eff. mass spec.	B	1943-2093	-1±1	2.5±1.4	3.6±0.2	-43.2±8	239.1
(4)	Flow eff. mass spec	B	1663-1983	-9±6	-14±12	3.2±3.8	-43.6±8	239.5
(5)	Flame mass spec.	C	2250	--	--	2.5	-39.3±12	235.2

[a]Reactions: A) Al(g) + 2 O(g) = AlO_2(g); B) AlO(g) = AlO_2(g) + Al(g); C) AlO(g) + H_2O(g) = AlO_2(g) + H_2(g).
[b]δS = $\Delta_r S°$(2nd law) - $\Delta_r S°$(3rd law). [c]Temperature scale adjusted.
[d]AlO_2 not detected; detection limits estimated as $P(AlO_2) \le P(Al_2O_2)/10$ in (3) and $P(AlO_2) \le P(Al_2O_2$, 2300 K) in (2).

Heat Capacity and Entropy

We adopt a linear OAlO structure and estimate the bond length by correcting the predicted value (6) for OAlO using the bias (6, 11) in predicted values for AlO and Al_2O. Electronic states are estimated from BO_2 and vibrational frequencies from several related molecules. The adopted frequencies correspond to k_1=4.2, k_{12}=0.5 and (k_δ/ℓ^2)=0.086 mdyn/Å. The Al-O stretching constant k_1 may be compared to 3.6 for Al_2O (11), 4.0 for AlOH (12), 5.5 for AlO (11) and 6.6 for FAlO (12). We assume that $k_1/(k_\delta/\ell^2)$=49 compared to 31 in BO_2 and 39 in FBO.

Linear OAlO is the most stable structure according to MNDO (6) and unrestricted Hartree-Fock (7) calculations. The latter predicts that linear AlOO and cyclic AlO_2 are less stable by 22 and 38 kcal mol^{-1} (7). These predictions are not decisive but they are the best available. Spectral assignments are conflicting. IR bands in gas-plus-O_2 matrices have been assigned to each of three structures. A band at 523 cm^{-1} in neon was tentatively assigned (13) to OAlO, but comparison with GaO_2 (14) suggests that it is more likely to be ν_2 of cyclic AlO_2. Assignments for ν_2 in argon include 496 (15) and 717 (16) cm^{-1}. Other assignments for cyclic AlO_2 include ν_1=1096 (15, Ar), 1177 (16, Ar), 1116 (17, Ar) and 1096 (17, N_2) and ν_3=525 cm^{-1} (16, Ar). Assignments to AlOO include 1336 and 1091 cm^{-1} (18, N_2). Gurvich et al. (10) calculated functions for cyclic AlO_2 using 1096, 496 and 550 cm^{-1}; their values of the Gibbs energy function differ from ours by 1.4 and 0.2 J K^{-1} mol^{-1} at 298 and 2000 K.

References

1. P. Ho and R. P. Burns, High Temp. Sci. 12, 31 (1980).
2. A. D. Chervonnyi, V. A. Piven', O. E. Kashireninov and G. B. Manelis, High Temp. Sci. 9, 99 (1977).
3. R. C. Paule, High Temp. Sci. 8, 257 (1976).
4. M. Farber, R. D. Srivastava and O. M. Uy, J. Chem. Soc., Faraday Trans. I, 68, 249 (1972); J. Chem. Phys. 55, 4142 (1971).
5. M. Farber, R. D. Srivastava et al., Faraday Symp. of the Chem. Soc., No. 8, 121 (1973); Combust. Flame 27, 99 (1976).
6. L. P. Dvis et al., J. Comput. Chem. 2, 433 (1981).
7. A. G. Turner, USAF Academy, Colorado, personal communication, April, 1982.
8. M. Farber and R. D. Srivastava, High Temp. Sci. 11, 1 (1979).
9. S. B. Oblath and J. L. Gole, Combust. Flame 37, 293 (1980); J. Chem. Phys. 70, 581 (1979); J. Phys. Chem. 85, 2651 (1981).
10. L. V. Gurvich, I. V. Veits et al., "Thermodynamic Properties of Individual Substances", 3rd ed., Vol. III, Nauka, Moscow, 1981.
11. JANAF Thermochemical Tables: AlO(g) and Al_2O(g), 12-31-79.
12. R. H. Hauge, J. W. Kauffman and J. L. Margrave, J. Am. Chem. Soc. 102, 6005 (1980); H. Schnoeckel, J. Mol. Struct. 50, 267 (1978).
13. A. Snelson, IIT Research Inst., AFRPL-TR-70-113, Contract F04611-69-C-0093, 1970.
14. M. J. Zehe, D. A. Lynch, B. J. Kelsall and K. D. Carlson, J. Phys. Chem. 83, 656 (1979).
15. S. B. Osin, L. V. Serebrennikov, V. F. Shevel'kova and A. A. Mal'tsev, Vestn. Mosk. Univ., Ser. 2: Khim. 19(2), 229 (1978).
16. P. A. Finn, D. M. Gruen, and D. L. Page, Adv. Chem. Ser. 158, 30 (1976).
17. D. A. Lynch, Case Western Reserve Univ., Cleveland, NASA-CR-133100, 1970.
18. S. M. Sonchik, Ph.D. Thesis, Case Western Reserve Univ., 1980.

Aluminum Oxide (AlO_2) $Al_1O_2(g)$

Enthalpy Reference Temperature = T_r = 298.15 K Standard State Pressure = p° = 0.1 MPa

T/K	$C_p°$ (J K^{-1}mol^{-1})	S°	-[G°-H°(T_r)]/T	H°-H°(T_r) (kJ mol^{-1})	$\Delta_f H°$	$\Delta_f G°$	Log K_f
0	0.	0.	INFINITE	-12.042	-85.010	-85.010	INFINITE
100	37.838	204.213	292.685	-8.847	-85.203	-87.595	45.755
200	44.987	232.915	256.245	-4.666	-85.735	-89.792	23.451
250	47.638	243.244	252.641	-2.349	-85.990	-90.776	18.967
298.15	49.892	251.832	251.832	0.	-86.190	-91.679	16.062
300	49.972	252.140	251.833	0.092	-86.197	-91.713	15.969
350	51.950	259.997	252.448	2.642	-86.359	-92.619	13.823
400	53.583	267.044	253.840	5.282	-86.487	-93.504	12.210
450	54.914	273.435	255.668	7.995	-86.594	-94.375	10.955
500	55.998	279.279	257.741	10.769	-86.692	-95.234	9.949
600	57.605	289.640	262.216	16.454	-86.901	-96.924	8.438
700	58.700	298.607	266.789	22.273	-87.185	-98.573	7.356
800	59.469	306.499	271.269	28.184	-87.592	-100.174	6.541
900	60.025	313.537	275.582	34.160	-88.162	-101.714	5.903
1000	60.438	319.883	279.700	40.184	-99.515	-102.421	5.350
1100	60.753	325.659	283.619	46.244	-100.139	-102.681	4.876
1200	60.997	330.956	287.346	52.332	-100.775	-102.885	4.478
1300	61.190	335.847	290.891	58.442	-101.424	-103.034	4.140
1400	61.346	340.387	294.267	64.569	-102.085	-103.133	3.848
1500	61.473	344.624	297.484	70.710	-102.760	-103.185	3.593
1600	61.578	348.595	300.556	76.863	-103.450	-103.191	3.369
1700	61.667	352.331	303.493	83.025	-104.154	-103.153	3.170
1800	61.744	355.858	306.305	89.196	-104.875	-103.074	2.991
1900	61.812	359.198	309.001	95.374	-105.611	-102.953	2.830
2000	61.873	362.370	311.591	101.558	-106.365	-102.794	2.685
2100	61.930	365.390	314.082	107.748	-107.135	-102.597	2.552
2200	61.984	368.273	316.480	113.944	-107.923	-102.362	2.430
2300	62.038	371.029	318.792	120.145	-108.727	-102.091	2.319
2400	62.092	373.670	321.024	126.351	-109.549	-101.786	2.215
2500	62.148	376.206	323.181	132.563	-110.387	-101.445	2.120
2600	62.207	378.645	325.268	138.781	-111.240	-101.070	2.031
2700	62.269	380.994	327.288	145.005	-112.109	-100.663	1.947
2800	62.335	383.260	329.247	151.235	-406.892	-99.255	1.852
2900	62.405	385.448	331.148	157.472	-406.693	-88.272	1.590
3000	62.480	387.565	332.993	163.716	-406.505	-77.295	1.346
3100	62.560	389.615	334.787	169.968	-406.329	-66.323	1.118
3200	62.644	391.603	336.531	176.228	-406.162	-55.359	0.904
3300	62.734	393.532	338.229	182.497	-406.005	-44.399	0.703
3400	62.828	395.406	339.884	188.775	-405.856	-33.443	0.514
3500	62.927	397.228	341.496	195.063	-405.714	-22.492	0.336
3600	63.030	399.003	343.069	201.361	-405.578	-11.545	0.168
3700	63.137	400.731	344.604	207.669	-405.448	-0.601	0.008
3800	63.247	402.416	346.104	213.988	-405.322	10.340	-0.142
3900	63.361	404.061	347.569	220.319	-405.201	21.277	-0.285
4000	63.478	405.666	349.001	226.661	-405.084	32.210	-0.421
4100	63.596	407.235	350.402	233.014	-404.971	43.141	-0.550
4200	63.717	408.769	351.774	239.380	-404.861	54.069	-0.672
4300	63.840	410.270	353.117	245.758	-404.755	64.995	-0.790
4400	63.963	411.739	354.432	252.148	-404.653	75.918	-0.901
4500	64.088	413.178	355.722	258.551	-404.553	86.838	-1.008
4600	64.212	414.588	356.986	264.966	-404.460	97.757	-1.110
4700	64.337	415.970	358.227	271.393	-404.372	108.674	-1.208
4800	64.461	417.326	359.444	277.833	-404.291	119.589	-1.301
4900	64.585	418.656	360.639	284.285	-404.217	130.503	-1.391
5000	64.707	419.962	361.812	290.750	-404.153	141.415	-1.477
5100	64.828	421.245	362.965	297.227	-404.099	152.326	-1.560
5200	64.947	422.505	364.098	303.715	-404.058	163.235	-1.640
5300	65.065	423.743	365.212	310.216	-404.030	174.145	-1.716
5400	65.180	424.960	366.307	316.728	-404.018	185.053	-1.790
5500	65.293	426.157	367.384	323.252	-404.024	195.961	-1.861
5600	65.403	427.335	368.444	329.787	-404.049	206.871	-1.930
5700	65.511	428.493	369.488	336.332	-404.098	217.781	-1.996
5800	65.615	429.634	370.515	342.889	-404.167	228.692	-2.060
5900	65.717	430.756	371.526	349.455	-404.267	239.602	-2.121
6000	65.816	431.861	372.523	356.032	-404.373	250.515	-2.181

PREVIOUS: December 1979 (1 atm) CURRENT: December 1979 (1 bar)

Aluminum Oxide (AlO_2) $Al_1O_2(g)$

ALUMINUM OXIDE, ION (AlO_2^-) IDEAL GAS M_r = 58.98089

$\Delta_{at}H°(AlO_2^- + Al + 2O + e^-) = 311.2 \pm 12$ kJ mol^{-1}
$S°(298.15\ K) = [235.09 \pm 6.3]$ J K^{-1} mol^{-1}

$\Delta_fH°(0\ K) = -481 \pm 50$ kJ mol^{-1}
$\Delta_fH°(298.15\ K) = [-489.210]$ kJ mol^{-1}

Vibrational Frequencies and Degeneracies

ν, cm^{-1}
[780](1)
[270](2)
[1120](1)

Ground State Quantum Weight: [1] $\sigma = 2$
Point Group: $[D_{\infty h}]$
Bond Distance: Al–O = [1.68] Å
Bond Angle: O–Al–O = [180]°
Rotational Constant: B_0 = [0.186657] cm^{-1}

Enthalpy of Formation

We adopt the rounded value $\Delta_fH°(0\ K) = -115\pm12$ kcal mol^{-1}, which corresponds to the electron affinity EA = 94.68 kcal mol^{-1} for AlO_2(1). This is based on equilibrium data (2-3) that yield [EA(AlO_2) – EA(Cl)] = 11.6 kcal mol^{-1} and EA(AlO_2) = 94.9 kcal mol^{-1}. Data for the reaction $AlO_2(g) + Cl^-(g) = AlO_2^-(g) + Cl(g)$ at 2080-2222 K were obtained by Srivastava, Uy and Farber (2-3) using effusion mass spectrometry. From their data we derive the values $\Delta_rH°(298.15\ K) = -11.7\pm0.9$ (3rd law) and -8.7 ± 8.6 (2nd law) kcal mol^{-1}plus $[\Delta_rS°(2nd\ law) - \Delta_rS°(3rd\ law)] = 1.4\pm4.0$ cal mol^{-1} K). The third-law $\Delta_rH°$ yields $\Delta_fH°(AlO_2^-, g, 0\ K) = -115.2$ and EA(AlO_2) = 94.9 kcal mol^{-1}. Gurvich et al. (9) selected $\Delta_fH°(0\ K) = -115.7\pm14.3$ and EA = 96.8±12 kcal mol^{-1} based on (2).

It is conceivable that this electron affinity is biased. New data show that similar studies of Farber et al. on $\Delta_{at}H°(AlO_2, g, 0\ K)$ may be biased by >20 kcal mol^{-1} (1). Two calculations (4-5) provide a test for bias in EA = $\Delta_rH°(AlO_2^- + AlO_2 + e^-)$ = 94.9 kcal mol^{-1}. Vertical ionization potentials of 85.3 and 78.4 kcal mol^{-1} were calculated for AlO_2^- by Gutsev and Boldyrev (4) using two different bases in the non-empirical discrete-variation Xα method. MNDO calculations of Davis et al. (5) gave 76.0 kcal mol^{-1} instead of our adopted difference $[\Delta_fH°(AlO_2) - \Delta_fH°(AlO_2^-)]$ = 96.3 kcal mol^{-1} at 298.15 K. Although both calculations favor smaller values of EA, they may be too uncertain to yield a definite conclusion. New data are desirable for both AlO_2^- and AlO^- (1).

The value of $\Delta_rH°(AlO_2^- + AlO_2 + e^-$, 298.15 K) calculated from our tables will differ from a room temperature threshold energy due to inclusion of JANAF (1) enthalpies, H°(0 K)–H°(298.15 K) of $AlO_2^-(g)$, $AlO_2(g)$ and $e^-(g)$ and to threshold effects discussed by Rosenstock et al. (6).

Heat Capacity and Entropy

We adopt a linear, symmetric structure and estimate the bond length to be 0.01 Å shorter than in AlO_2 (1). MNDO calculations (5) support this linear structure and also predict that the bond length of AlO_2^- is 0.007 Å shorter than in AlO_2. AlO_2^- is isoelectronic with SiO_2 and MgF_2, which are also believed to be linear.

We estimate the vibrational frequencies of AlO_2^-, together with those of AlO_2, from comparisons with SiO_2, OAlF, MgF_2, CO_2, OBF, BeF_2, BO_2^-, and BO_2. Our adopted values correspond to k_1=5.6, k_{12}=0.2 and (k_δ/ℓ^2)=0.16 mdyn/Å. The Al–O stretching constant k_1 may be compared to 5.5 in AlO (1), 6.6 in OAlF and OAlCl (7) and 4.2 assumed in AlO_2 (1). We assume that $k_1/(k_\delta/\ell^2)$=35. We take the ground state to be $^1\Sigma^+$ by analogy to other triatomics with 16 valence electrons (8).

Gurvich et al. (9) calculated functions that differ by -1.3 cal K^{-1} mol^{-1} in entropy and Gibbs energy function.

References

1. JANAF Thermochemical Tables: $AlO_2(g)$, AlO(g), and AlO (g), 12-31-79; e (g), 3-31-77.
2. R. D. Srivastava, O. M. Uy and M. Farber, J. Chem. Soc., Faraday Trans. II, 68, 1388 (1972).
3. M. Farber, R. D. Srivastava and O. M. Uy, Space Sci. Inc., Monrovia, Calif., AD-731303, 50 pp. (1971).
4. G. L. Gutsev and A. I. Boldyrev, Russ. J. Inorg. Chem. 26, 1375 and 1266 (1981); Chem. Phys. 56, 277 (1981).
5. L. P. Davis et al., J. Comput. Chem. 2, 433 (1981).
6. H. M. Rosenstock, K. Draxl et al., J. Phys. Chem. Ref. Data 6, Supp. 1 (1977).
7. H. Schnoeckel, J. Mol. Struct. 50, 267 (1978).
8. J. W. Rabalais, J. M. McDonald, V. Scherr and S. P. McGlynn, Chem. Rev. 71, 73 (1971).
9. L. V. Gurvich, I. V. Veits et al., "Thermodynamic Properties of Individual Substances," 3rd ed., Vol. III, Nauka, Moscow, 1981.

Aluminum Oxide, Ion (AlO_2^-) $Al_1O_2^-(g)$

Enthalpy Reference Temperature = T_r = 298.15 K Standard State Pressure = p° = 0.1 MPa

	J K^{-1}mol^{-1}			kJ mol^{-1}			
T/K	$C_p°$	S°	$-[G°-H°(T_r)]/T$	$H°-H°(T_r)$	$\Delta_fH°$	$\Delta_fG°$	Log K_f
0	0.	0.	INFINITE	-11.368	-481.158		
100	34.491	190.471	273.699	-8.323			
200	42.492	217.143	239.288	-4.429			
250	45.226	226.927	235.864	-2.234			
298.15	47.528	235.094	235.094	0.	-489.210	-483.454	84.699
300	47.611	235.388	235.095	0.088	-489.260	-483.418	84.171
350	49.683	242.887	235.682	2.522	-490.577	-482.339	71.985
400	51.455	249.640	237.012	5.051	-491.854	-481.075	62.822
450	52.950	255.789	238.761	7.662	-493.103	-479.652	55.677
500	54.203	261.435	240.750	10.342	-494.334	-478.092	49.946
600	56.131	271.498	245.057	15.865	-496.785	-474.613	41.319
700	57.493	280.259	249.474	21.550	-499.280	-470.722	35.126
800	58.475	288.004	253.815	27.351	-501.876	-466.466	30.457
900	59.198	294.935	258.006	33.236	-504.615	-461.877	26.807
1000	59.743	301.201	262.017	39.184	-518.123	-456.214	23.830
1100	60.161	306.916	265.843	45.180	-520.890	-449.889	21.363
1200	60.489	312.165	269.487	51.214	-523.659	-443.312	19.297
1300	60.750	317.018	272.959	57.276	-526.434	-436.504	17.539
1400	60.960	321.528	276.269	63.362	-529.215	-429.482	16.024
1500	61.133	325.739	279.428	69.467	-532.005	-422.261	14.704
1600	61.275	329.690	282.447	75.587	-534.805	-414.853	13.544
1700	61.395	333.408	285.337	81.721	-537.617	-407.270	12.514
1800	61.496	336.920	288.106	87.866	-540.442	-399.522	11.594
1900	61.582	340.248	290.763	94.020	-543.281	-391.615	10.766
2000	61.656	343.408	293.317	100.182	-546.135	-383.559	10.018
2100	61.719	346.418	295.775	106.350	-549.005	-375.360	9.337
2200	61.775	349.290	298.143	112.525	-551.893	-367.023	8.714
2300	61.824	352.038	300.427	118.705	-554.797	-358.555	8.143
2400	61.867	354.670	302.632	124.890	-557.719	-349.961	7.617
2500	61.904	357.196	304.765	131.078	-560.659	-341.244	7.130
2600	61.938	359.625	306.828	137.271	-563.617	-332.409	6.678
2700	61.968	361.963	308.827	143.466	-566.592	-323.459	6.258
2800	61.995	364.217	310.765	149.664	-863.486	-313.432	5.847
2900	62.020	366.393	312.646	155.865	-865.401	-293.754	5.291
3000	62.042	368.496	314.473	162.068	-867.334	-274.009	4.771
3100	62.061	370.530	316.249	168.273	-869.283	-254.199	4.283
3200	62.079	372.501	317.976	174.480	-871.248	-234.327	3.825
3300	62.096	374.412	319.657	180.689	-873.230	-214.392	3.394
3400	62.111	376.266	321.295	186.899	-875.227	-194.397	2.987
3500	62.125	378.066	322.892	193.111	-877.239	-174.344	2.602
3600	62.137	379.816	324.449	199.324	-879.267	-154.232	2.238
3700	62.149	381.519	325.968	205.539	-881.309	-134.065	1.893
3800	62.160	383.177	327.452	211.754	-883.366	-113.841	1.565
3900	62.170	384.791	328.902	217.971	-885.437	-93.563	1.253
4000	62.179	386.366	330.319	224.188	-887.523	-73.232	0.956
4100	62.188	387.901	331.704	230.406	-889.624	-52.849	0.673
4200	62.196	389.400	333.060	236.626	-891.739	-32.414	0.403
4300	62.203	390.863	334.388	242.845	-893.869	-11.929	0.145
4400	62.210	392.293	335.687	249.066	-896.016	8.606	-0.102
4500	62.217	393.692	336.961	255.287	-898.175	29.189	-0.339
4600	62.223	395.059	338.209	261.509	-900.354	49.821	-0.566
4700	62.228	396.397	339.433	267.732	-902.550	70.501	-0.784
4800	62.234	397.707	340.633	273.955	-904.764	91.227	-0.993
4900	62.239	398.991	341.811	280.179	-906.998	112.001	-1.194
5000	62.243	400.248	342.968	286.403	-909.252	132.820	-1.388
5100	62.248	401.481	344.103	292.627	-911.529	153.684	-1.574
5200	62.252	402.690	345.218	298.852	-913.830	174.592	-1.754
5300	62.256	403.875	346.314	305.078	-916.156	195.546	-1.927
5400	62.260	405.039	347.390	311.303	-918.509	216.543	-2.095
5500	62.263	406.182	348.449	317.530	-920.891	237.584	-2.256
5600	62.266	407.303	349.490	323.756	-923.304	258.670	-2.413
5700	62.270	408.406	350.514	329.983	-925.750	279.798	-2.564
5800	62.273	409.489	351.521	336.210	-928.227	300.970	-2.711
5900	62.276	410.553	352.513	342.437	-930.744	322.183	-2.852
6000	62.278	411.600	353.489	348.665	-933.279	343.439	-2.990

PREVIOUS: December 1979 (1 atm) CURRENT: December 1979 (1 bar)

Aluminum Oxide, Ion (AlO_2^-) $Al_1O_2^-(g)$

ALUMINUM SULFIDE (AlS) IDEAL GAS M_r = 59.04154

$\Delta_f H°(0\ K)$ = 238.4 ± 8.5 kJ mol^{-1}

S°(298.15 K) = 230.160 ± 0.4 J K^{-1} mol^{-1} $\Delta_f H°(298.15\ K)$ = 238.5 ± 8.5 kJ mol^{-1}

Electronic States Molecular Constants (σ = 1)

State	ε_i, cm^{-1}	g_i	ω_e, cm^{-1}	$\omega_e x_e$, cm^{-1}	B_e, cm^{-1}	α_e, cm^{-1}	r_e, Å
X $^2\Sigma^+$	0	2	617.1	3.33	0.2799	0.0018	2.029
A $^2\Pi_i$	[4500]	4	[459]	[2]	[0.234]	[0.001]	
B $^2\Sigma^+$	23381.2	2	510.9	1.45	0.2461	0.0012	2.164
C $^2\Pi_{r1/2}$	29942.5	2	545	6	[0.2600]	[0.0028]	3.090
C$^2\Pi_{r3/2}$	30059.7	2	559	14	[0.2539]	[0.0044]	3.127
D $^2\Sigma^+$	35714.9	2	430	14	0.2402	0.0036	2.190

Enthalpy of Formation

The adopted value of $\Delta_f H°(298.15\ K)$ is an average of 3rd law values from the equilibrium studies of Ficalora et al. (1) and Uy and Drowart (2). Ficalora et al. (1) used mass spectrometric ion intensity data to study the equilibrium Al(g) + 1/2 S_2(g) = AlS(g) and they reported 2nd and 3rd law values for $\Delta_f H°(298.15\ K)$ of -37.3±3.0 kcal mol^{-1} and -35.4±2.0 kcal mol^{-1}, respectively. Uy and Drowart (2) used mass spectrometric ion intensity data to study two different equilibria and report an average value of D° = 88.4±1.9 kcal mol^{-1} via 3rd law techniques. Using auxiliary JANAF data (3) we calculate 3rd law values of $\Delta_f H°(298.15\ K)$ = 58.8±2.0 kcal mol^{-1} and $\Delta_f H°(298.15\ K)$ = 55.2±1.9 kcal mol^{-1} from the data of (1) and (2), respectively. A complete reanalysis is impossible since neither set of authors report their raw data. Reanalysis is desirable since the free energy functions in this table differ from those in the older table (4) used by both sets of authors, due to inclusion of excited electronic states. A crude calculation shows that these differences are large enough to affect third-law results by as much as ±0.2 kcal mol^{-1} which is within the uncertainty of the reported values. The adopted value of $\Delta_f H°(298.15\ K)$, coupled with auxiliary JANAF data (3), yields an adopted D_0^0 = 86.9±2.0 kcal mol^{-1}.

Heat Capacity and Entropy

With the exception of the two $^2\Pi$ states the adopted molecular constant data are from Huber and Herzberg (5) based on earlier absorption and emission spectroscopic studies by several authors. We have adopted a modified nomenclature system consistent with states for AlO (3), but different from that of Huber and Herzberg for AlS (5). This is desirable to accommodate the predicted low lying $^2\Pi_i$ state which is unobserved for AlS but has been observed for AlO. The energy of this $A^2\Pi$ state for AlS is predicted from relative energies of the two $^2\Pi$ states for AlO (3) and is consistent with INDO/S calculations (6) if the bias in these calculations for AlO is directly transferable to AlS. Molecular constant data for the $A^2\Pi$ state is predicted assuming the same relative differences between this state and the ground state as in AlO.

We adopt Lavendy's (7) molecular constant data for the $C^2\Pi$ state, except we use his experimental values of B_0 = 0.2572 cm^{-1} and B_0 = 0.2495 cm^{-1} for the 1/2 and 3/2 components in conjunction with the Pekeris relationship (8) to estimate values of B_e and α_e. Lavendy's emission spectroscopy measurements indicate that this state is highly perturbed and this is reflected in the unusually large differences between molecular constant data for the 1/2 and 3/2 components. The nature of the perturbing state was not identified. We treat the $C^2\Pi$ state as two separate electronic states separated by the spin-orbit coupling constant of 117.2 cm^{-1}. This approximation introduces little uncertainty since the entire $C^2\Pi$ state makes an entropy contribution of less than 0.1 cal mol^{-1} K^{-1} at 6000 K.

The largest uncertainty in the table is due to uncertainty in the estimated $A^2\Pi$ state. This state makes no contribution to the thermochemical functions below 500 K so its uncertainty is not reflected in that given for the S°(298.15 K) value. This state makes the following entropy contributions in cal K^{-1} mol^{-1}; 0.05 at 1000 K, 0.9 at 2000 K, 1.8 at 3000 K, 2.2 at 4000 K, 2.4 at 5000 K and 2.5 at 6000 K.

References

1. P. J. Ficalora, J. W. Hastie, and J. L. Margrave, J. Phys. Chem. 72, 1660 (1968).
2. O. M. Uy and J. Drowart, Trans. Faraday Soc. 1293 (1971).
3. JANAF Thermochemical Tables: S_2(g) and S(g), 9-30-77; Al(g), 6-30-79; AlO(g), 6-30-75.
4. JANAF Thermochemical Tables, 2nd ed., NSRDS-NBS 37, 1971.
5. K. P. Huber and G. Herzberg, "Molecular Spectra and Molecular Structure; Constants of Diatomic Molecules," Van Nostrand-Reinhold, New York, 1979.
6. P. Carsky and J. Malek, Collect. Czech. Chem. Commun. 42, 2758 (1977).
7. H. Lavendy, J. Phys. B: Atom. Molec. Phys. 13, 1151 (1980).
8. C. L. Pekeris, Phys. Rev. 45, 98 (1934).

Aluminum Sulfide (AlS) $Al_1S_1(g)$

Enthalpy Reference Temperature = T_r = 298.15 K Standard State Pressure = p° = 0.1 MPa

T/K	$C_p°$ (J K^{-1}mol^{-1})	S° (J K^{-1}mol^{-1})	-[G°-H°(T_r)]/T (J K^{-1}mol^{-1})	H°-H°(T_r) (kJ mol^{-1})	$\Delta_f H°$ (kJ mol^{-1})	$\Delta_f G°$ (kJ mol^{-1})	Log K_f
0	0.	0.	INFINITE	-9.087	238.352	238.352	INFINITE
100	29.206	196.985	258.754	-6.177	240.089	222.341	-116.139
200	31.167	217.716	233.589	-3.175	239.647	204.661	-53.452
298.15	33.398	230.601	230.601	0.	238.488	187.722	-32.888
300	33.432	230.807	230.601	0.062	238.463	187.407	-32.630
400	34.871	240.640	231.932	3.483	234.779	170.692	-22.290
500	35.749	248.523	234.488	7.017	231.752	155.008	-16.194
600	36.348	255.096	237.390	10.624	229.039	139.919	-12.181
700	36.861	260.738	240.332	14.284	226.505	125.267	-9.348
800	37.417	265.695	243.199	17.997	224.026	110.974	-7.246
900	38.085	270.140	245.949	21.771	168.402	98.073	-5.692
1000	38.884	274.193	248.574	25.619	156.478	91.088	-4.758
1100	39.797	277.941	251.075	29.552	155.365	84.604	-4.018
1200	40.783	281.445	253.462	33.581	154.331	78.218	-3.405
1300	41.794	284.750	255.742	37.709	153.382	71.914	-2.890
1400	42.782	287.883	257.927	41.939	152.519	65.680	-2.451
1500	43.708	290.867	260.025	46.264	151.735	59.505	-2.072
1600	44.543	293.715	262.042	50.677	151.025	53.380	-1.743
1700	45.266	296.438	263.986	55.169	150.379	47.297	-1.453
1800	45.870	299.043	265.862	59.726	149.785	41.251	-1.197
1900	46.353	301.536	267.674	64.339	149.233	35.236	-0.969
2000	46.720	303.924	269.427	68.993	148.711	29.250	-0.764
2100	46.979	306.210	271.125	73.679	148.209	23.289	-0.579
2200	47.144	308.400	272.770	78.386	147.719	17.352	-0.412
2300	47.225	310.497	274.365	83.105	147.231	11.437	-0.260
2400	47.235	312.508	275.912	87.828	146.738	5.544	-0.121
2500	47.188	314.435	277.415	92.550	146.236	-0.329	0.007
2600	47.094	316.284	278.875	97.264	145.720	-6.181	0.124
2700	46.963	318.059	280.293	101.968	145.186	-12.014	0.232
2800	46.804	319.764	281.673	106.656	-149.270	-16.858	0.314
2900	46.624	321.404	283.015	111.328	-148.751	-12.138	0.219
3000	46.431	322.981	284.321	115.980	-148.257	-7.436	0.129
3100	46.229	324.500	285.593	120.614	-147.787	-2.750	0.046
3200	46.024	325.965	286.831	125.226	-147.342	1.922	-0.031
3300	45.818	327.378	288.039	129.818	-146.922	6.580	-0.104
3400	45.614	328.742	289.216	134.390	-146.528	11.225	-0.172
3500	45.415	330.062	290.364	138.941	-146.159	15.860	-0.237
3600	45.223	331.338	291.485	143.473	-145.814	20.484	-0.297
3700	45.039	332.575	292.579	147.986	-145.494	25.099	-0.354
3800	44.864	333.774	293.647	152.481	-145.197	29.705	-0.408
3900	44.698	334.937	294.691	156.959	-144.923	34.304	-0.459
4000	44.543	336.067	295.711	161.421	-144.672	38.897	-0.508
4100	44.399	337.165	296.709	165.868	-144.443	43.483	-0.554
4200	44.265	338.233	297.685	170.301	-144.236	48.064	-0.598
4300	44.142	339.273	298.640	174.721	-144.050	52.640	-0.639
4400	44.031	340.287	299.575	179.130	-143.885	57.213	-0.679
4500	43.930	341.275	300.491	183.528	-143.739	61.781	-0.717
4600	43.840	342.239	301.388	187.916	-143.615	66.347	-0.753
4700	43.761	343.181	302.267	192.296	-143.511	70.910	-0.788
4800	43.692	344.102	303.129	196.669	-143.427	75.471	-0.821
4900	43.633	345.002	303.975	201.035	-143.363	80.031	-0.853
5000	43.584	345.883	304.804	205.396	-143.319	84.590	-0.884
5100	43.545	346.746	305.618	209.752	-143.296	89.148	-0.913
5200	43.515	347.591	306.417	214.105	-143.293	93.705	-0.941
5300	43.495	348.420	307.202	218.456	-143.311	98.263	-0.968
5400	43.484	349.233	307.973	222.804	-143.350	102.821	-0.995
5500	43.481	350.031	308.730	227.153	-143.409	107.381	-1.020
5600	43.487	350.814	309.475	231.501	-143.490	111.941	-1.044
5700	43.501	351.584	310.207	235.850	-143.593	116.503	-1.068
5800	43.524	352.341	310.927	240.201	-143.715	121.067	-1.090
5900	43.554	353.085	311.635	244.555	-143.862	125.634	-1.112
6000	43.592	353.817	312.332	248.912	-144.010	130.201	-1.133

PREVIOUS: December 1979 (1 atm) CURRENT: December 1979 (1 bar)

Aluminum Sulfide (AlS) $Al_1S_1(g)$

ALUMINUM, DIATOMIC (Al_2) IDEAL GAS M_r = 53.96308

$S°(298.15\ K) = 233.5 \pm 0.3\ J\ K^{-1}\ mol^{-1}$

$\Delta_f H°(0\ K) = 486.3 \pm 3.5\ kJ\ mol^{-1}$

$\Delta_f H°(298.15\ K) = 487.0 \pm 3.5\ kJ\ mol^{-1}$

Electronic Levels and Quantum Weights

State	ε_i, cm^{-1}	g_i
$^3\Sigma_g^-$	0	3
$^5\Sigma_u^-$	[1000.]	[5]
$1\ ^3\Sigma_u^-$	[10000]	[3]
$2\ ^3\Sigma_u^-$	17269.36	3

$\omega_e = 350.01\ cm^{-1}$ $\omega_e x_e = 2.022\ cm^{-1}$ $\omega_e y_e = -0.0105\ cm^{-1}$ $\sigma = 2$

$B_e = 0.2054\ cm^{-1}$ $\alpha_e = 0.0012\ cm^{-1}$ $D_e = 3.07\times10^{-7}\ cm^{-1}$ $r_e = 2.4668$ Å

Enthalpy of Formation

From a mass-spectrometric investigation of the equilibria

$$AlAu(g) + Al(g) = Al_2(g) + Au(g)$$
$$2\ AlAu(g) = Al_2(g) + Au_2(g)$$

at 1921 - 2032 K and a 2nd law and 3rd law treatment of the data, Sterns and Kohl (1) derived the dissociation energy, D_0^0, of 1.55±0.15 eV for Al_2. Using current auxiliary data for $Al_2(g)$ and Al(g) (7), this value is corrected to 1.56 eV. As discussed by Stearns and Kohl (1), it is likely that all of the previous mass spectrometic determinations (2, 3, 4) resulted in slightly high values for D_0^0 (<1.96±0.4 eV, 1.78±0.19 eV and 1.80 eV, respectively) because of interferences due to the measurement of Al^+ ion intensities from extraneous sources. The dissociation energy of Al_2 has also been evaluated by a linear Birge-Sponer extrapolation (5) as 1.87±0.2 eV and a graphical Birge-Sponer extrapolation (6) as 1.6±0.2 eV. An ab initio calculation (LCAO-STO-MO-SCF-CI) by Leleyter and Joyes (9) yielded a dissociation energy of 1.03 eV. Based on the Stearns and Kohl result and auxiliary JANAF data (7), we adopt $\Delta_f H°(298.15\ K)$ = 116.4±3.5 kcal/mol for $Al_2(g)$.

Heat Capacity and Entropy

The bond spectrum of Al_2 was first reported by Zeeman (6) who wrongly attributed the emission to the AlC molecule. The vibrational constants are those reported by him. Values for the rotational constants are taken from Ginter et al. (5) who also assigned the upper and lower $^3\Sigma$ state configurations and energies. The electronic energies for the $^5\Sigma$ state and $1\ ^3\Sigma$ state have been estimated by analogy with B_2. An ab initio SCF calcualtion of the ground state potential curve for Al_2 (8) is reported but without inclusion of configuration interaction the calcualted separations between states are not reliable.

References

1. C. A. Stearns and F. J. Kohl, High Temp. Sci. 5, 113 (1973).
2. W. A. Chupka, J. Berkowitz, C. F. Giese and M. G. Inghram, J. Phys. Chem. 62, 611 (1958).
3. O. M. Uy and J. Drowart, Trans. Faraday Soc. 67, 1293 (1971).
4. G. D. Blue and K. A. Gingerich, Sixteenth Annual Conference on Mass Spectroscopy and Allied Topics ASTM E-14, No. 129, Pittsburg, Pa. 1968; values as quoted in reference (1) above.
5. D. S. Ginter, M. L. Ginter and K. K. Innes, Astrophys. J. 139, 365 (1964).
6. P. B. Zeeman, Can. J. Phys. 32, 9 (1954).
7. JANAF Thermochemical Tables: Al(g), 6-30-79.
8. N. H. Sabelli, R. Benedek and T. L. Gilbert, Phys. Rev. A 20, 677 (1979).
9. M. Leleyter and P. Joyes, J. Phys. B 13, 2165 (1980).

Aluminum (Al_2) $Al_2(g)$

Enthalpy Reference Temperature = T_r = 298.15 K Standard State Pressure = p° = 0.1 MPa

T/K	$C_p°$ (J K⁻¹mol⁻¹)	S° (J K⁻¹mol⁻¹)	-[G°-H°(T_r)]/T (J K⁻¹mol⁻¹)	H°-H°(T_r) (kJ mol⁻¹)	$\Delta_f H°$ (kJ mol⁻¹)	$\Delta_f G°$ (kJ mol⁻¹)	Log K_f
0	0.	0.	INFINITE	-9.816	486.280	486.280	INFINITE
100	30.569	196.386	265.163	-6.878	488.251	470.010	-245.508
200	34.832	218.934	236.936	-3.600	487.925	451.796	-117.997
298.15	38.448	233.531	233.531	0.	487.018	434.251	-76.079
300	38.511	233.769	233.532	0.071	486.999	433.923	-75.553
400	41.208	245.250	235.079	4.068	485.980	416.384	-54.374
500	42.539	254.607	238.079	8.264	484.910	399.107	-41.694
600	42.884	262.401	241.502	12.539	483.715	382.056	-33.261
700	42.718	269.002	244.971	16.821	482.301	365.221	-27.253
800	42.350	274.682	248.339	21.075	480.594	348.608	-22.762
900	41.941	279.646	251.547	25.289	478.526	332.230	-19.282
1000	41.559	284.045	254.581	29.464	454.869	317.624	-16.591
1100	41.230	287.990	257.442	33.603	452.658	304.006	-14.436
1200	40.961	291.566	260.139	37.712	450.417	290.591	-12.649
1300	40.749	294.836	262.684	41.797	448.152	277.363	-11.145
1400	40.587	297.849	265.090	45.863	445.868	264.311	-9.862
1500	40.470	300.645	267.368	49.916	443.571	251.423	-8.755
1600	40.392	303.255	269.530	53.959	441.263	238.688	-7.792
1700	40.348	305.702	271.587	57.995	438.950	226.098	-6.947
1800	40.334	308.008	273.547	62.029	436.634	213.645	-6.200
1900	40.345	310.188	275.418	66.063	434.317	201.319	-5.535
2000	40.380	312.259	277.209	70.099	432.003	189.116	-4.939
2100	40.435	314.230	278.925	74.140	429.694	177.029	-4.403
2200	40.507	316.113	280.573	78.187	427.391	165.051	-3.919
2300	40.595	317.915	282.158	82.242	425.096	153.178	-3.479
2400	40.697	319.645	283.684	86.306	422.810	141.404	-3.078
2500	40.810	321.309	285.156	90.382	420.535	129.726	-2.710
2600	40.934	322.912	286.578	94.469	418.272	118.138	-2.373
2700	41.066	324.459	287.952	98.569	416.022	106.638	-2.063
2800	41.207	325.955	289.283	102.682	-174.017	97.155	-1.812
2900	41.355	327.404	290.572	106.810	-174.048	106.841	-1.924
3000	41.509	328.808	291.824	110.953	-174.064	116.527	-2.029
3100	41.668	330.172	293.039	115.112	-174.065	126.213	-2.127
3200	41.831	331.497	294.220	119.287	-174.050	135.899	-2.218
3300	41.999	332.787	295.369	123.479	-174.020	145.585	-2.304
3400	42.169	334.043	296.488	127.687	-173.974	155.269	-2.385
3500	42.343	335.268	297.579	131.913	-173.912	164.952	-2.462
3600	42.519	336.464	298.643	136.156	-173.834	174.633	-2.534
3700	42.696	337.631	299.681	140.416	-173.741	184.311	-2.602
3800	42.876	338.772	300.694	144.695	-173.634	193.987	-2.667
3900	43.056	339.888	301.685	148.992	-173.511	203.659	-2.728
4000	43.238	340.980	302.654	153.306	-173.375	213.329	-2.786
4100	43.420	342.050	303.602	157.639	-173.225	222.995	-2.841
4200	43.604	343.099	304.530	161.990	-173.064	232.657	-2.894
4300	43.787	344.127	305.439	166.360	-172.890	242.314	-2.944
4400	43.970	345.136	306.329	170.748	-172.707	251.968	-2.991
4500	44.155	346.126	307.203	175.154	-172.510	261.617	-3.037
4600	44.338	347.098	308.060	179.579	-172.308	271.263	-3.080
4700	44.522	348.054	308.900	184.022	-172.100	280.903	-3.122
4800	44.706	348.993	309.726	188.483	-171.886	290.539	-3.162
4900	44.889	349.917	310.537	192.963	-171.668	300.171	-3.200
5000	45.073	350.826	311.333	197.461	-171.449	309.798	-3.236
5100	45.255	351.720	312.117	201.977	-171.229	319.421	-3.272
5200	45.438	352.600	312.887	206.512	-171.012	329.039	-3.305
5300	45.620	353.468	313.644	211.065	-170.799	338.653	-3.338
5400	45.802	354.322	314.390	215.636	-170.592	348.264	-3.369
5500	45.984	355.164	315.123	220.225	-170.393	357.870	-3.399
5600	46.165	355.994	315.846	224.833	-170.204	367.473	-3.428
5700	46.346	356.813	316.557	229.458	-170.028	377.073	-3.455
5800	46.526	357.621	317.258	234.102	-169.862	386.669	-3.482
5900	46.707	358.418	317.949	238.764	-169.717	396.264	-3.508
6000	46.887	359.204	318.630	243.443	-169.548	405.851	-3.533

PREVIOUS: June 1979 (1 atm) CURRENT: June 1979 (1 bar)

Aluminum (Al_2) $Al_2(g)$

BERYLLIUM ALUMINUM OXIDE ($BeAl_2O_4$) CRYSTAL M_r = 126.97286

$S°$(298.15 K) = 66.29 ± 0.13 J K^{-1} mol^{-1}
T_{fus} = 2146 ± 10 K

$\Delta_f H°$(0 K) = -2285.5 ± 3.8 kJ mol^{-1}
$\Delta_f H°$(298.15 K) = -2300.78 ± 3.8 kJ mol^{-1}
$\Delta_{fus} H°$ = 170.008 ± 3.8 kJ mol^{-1}

Enthalpy of Formation

The adopted value of $\Delta_f H°$(298.15 K) is calculated from $\Delta_r H°$(968 K) = -4.0±0.3 kcal mol^{-1} for BeO(cr, α) + Al_2O_3(cr, α) → $BeAl_2O_4$(cr) using current values of $\Delta_f H°$ for the reactants (1). The value of $\Delta_r H°$(968 K) was determined by Holm and Kleppa (2) from the difference in heats of solution of the three components in an oxide melt.

For comparison, a second-law treatment of the equilibrium data of Young (3) yields $\Delta_r H°$(298.15 K) = -2.3 kcal mol^{-1} for the same reaction. This is obtained from a suitable linear combination of the three equilibria studied by Young and leads to $\Delta_f H°$(298.15 K) = -548.2 kcal mol^{-1} in good agreement with the adopted value.

Heat Capacity and Entropy

Heat capacity and derived properties at and below 298.15 K are taken from the calorimetric data (16 - 380 K) of Furukawa and Saba (4). The entropy is based on $S°$(16 K) = 0.004 cal K^{-1} mol^{-1}. Above 298.15 K the adopted $C_p°$ data are based on Ishihara and West's (5) merging of enthalpy data of Ditmar and Douglas (6) (323 - 1173 K) and Ishihara and West (5) (1182 - 2137) with the $C_p°$ data of Furukawa and Saba (4). Values above 2137 K are our extrapolation. Ishihara and West report a sharp rise in enthalpy data above 2030 K, apparently associated with premelting phenomena. Data in this region was ignored by Ishihara and West during their treatment.

Fusion Data

See the liquid table for details.

References

1. JANAF Thermochemical Tables: BeO(cr, α), 6-30-75; Al_2O_3(cr, α), 12-31-79.
2. J. L. Holm and O. J. Kleppa, Acta Chem. Scand. 20, 2568 (1966).
3. W. A. Young, J. Phys. Chem. 64, 1003 (1960).
4. G. T. Furukawa and W. G. Saba, J. Res. Nat. Bur. Stand. 69A, 13 (1965).
5. S. Ishihara and E. D. West, J. Res. Nat. Bur. Stand. 80A, 65 (1976).
6. D. A. Ditmars and T. B. Douglas, J. Res. Nat. Bur. Stand. 71A, 89 (1967).

Beryllium Aluminum Oxide ($BeAl_2O_4$) $Al_2Be_1O_4$(cr)

Enthalpy Reference Temperature = T_r = 298.15 K — Standard State Pressure = $p°$ = 0.1 MPa

T/K	$C_p°$	$S°$	$-[G°-H°(T_r)]/T$	$H°-H°(T_r)$	$\Delta_f H°$	$\Delta_f G°$	Log K_f
	J $K^{-1}mol^{-1}$			kJ mol^{-1}			
0	0.	0.	INFINITE	-13.088	-2285.491	-2285.491	INFINITE
100	16.365	5.523	132.214	-12.669	-2291.889	-2256.332	1178.587
200	66.430	31.982	74.785	-8.560	-2297.784	-2218.294	579.359
298.15	105.380	66.291	66.291	0.	-2300.782	-2178.542	381.672
300	105.940	66.945	66.293	0.195	-2300.815	-2177.784	379.186
400	130.280	101.047	70.777	12.108	-2301.703	-2136.594	279.011
500	145.250	131.854	79.971	25.942	-2301.355	-2095.338	218.899
600	154.970	159.252	90.946	40.983	-2300.371	-2054.220	178.836
700	161.750	183.668	102.480	56.832	-2299.118	-2013.292	150.234
800	166.830	205.612	114.024	73.271	-2297.811	-1972.550	128.794
900	170.870	225.503	125.322	90.163	-2296.605	-1931.966	112.128
1000	174.200	243.683	136.263	107.421	-2316.784	-1889.987	98.723
1100	177.030	260.423	146.799	124.986	-2315.359	-1847.375	87.725
1200	179.480	275.934	156.922	142.814	-2313.836	-1804.898	78.565
1300	181.670	290.388	166.639	160.873	-2312.239	-1762.550	70.820
1400	183.750	303.927	175.967	179.144	-2310.577	-1720.329	64.186
1500	185.910	316.677	184.926	197.625	-2308.849	-1678.228	58.441
1600	188.410	328.752	193.541	216.337	-2321.567	-1635.720	53.401
1700	191.560	340.263	201.836	235.326	-2319.211	-1592.925	48.945
1800	195.790	351.326	209.835	254.683	-2316.556	-1550.278	44.988
1900	201.570	362.058	217.566	274.535	-2313.473	-1507.788	41.452
2000	209.490	372.588	225.054	295.068	-2309.778	-1465.476	38.274
2100	220.190	383.055	232.329	316.525	-2305.227	-1423.369	35.404
2146.000	226.250	387.890	235.612	326.790	------ CRYSTAL <--> LIQUID -----		
2200	233.098	393.597	239.419	339.191	-2299.533	-1381.503	32.801
2300	248.548	404.290	246.354	363.252	-2292.510	-1339.927	30.431
2400	266.455	415.238	253.162	388.982	-2283.885	-1298.690	28.265

PREVIOUS: June 1972 CURRENT: December 1979

Beryllium Aluminum Oxide ($BeAl_2O_4$) $Al_2Be_1O_4$(cr)

BERYLLIUM ALUMINUM OXIDE ($BeAl_2O_4$) LIQUID M_r = 126.97286

S°(298.15 K) = [126.406] J K^{-1} mol^{-1} $\Delta_fH°$(298.15 K) = [-2163.842] kJ mol^{-1}

T_{fus} = 2146 ± 10 K $\Delta_{fus}H°$ = 170.008 ± 3.8 kJ mol^{-1}

Enthalpy of Formation

$\Delta_fH°(BeAl_2O_4$, ℓ, 298.15 K) is calculated from $\Delta_fH°(BeAl_2O_4$, cr, 298.5 K) by adding the enthalpy of fusion, $\Delta_{fus}H°$, and the difference in enthalpy, H°(2146 K)-H°(298.15 K), between the crystal and liquid.

Heat Capacity and Entropy

Ishihara and West (2) reported enthalpy data from 2177 to 2355 K and reported a quadratic enthalpy equation for the liquid. This equation yields a temperature dependence for $C_p°$ which is unreasonably large. Therefore we have fit their enthalpy data to a linear equation which fits their measured points with an average deviation of 0.15 kcal mol^{-1} and a maximum deviation of 0.35 kcal mol^{-1}. Based on the scatter in the data we believe higher fits are unjustified and adopt the resulting constant value of $C_p°$ = 58.893 cal K^{-1} mol^{-1} (246.408 J K^{-1} mol^{-1}) above the assumed glass transition at 1450 K. Below T_{gls}, $C_p°$ is taken to be the same as the crystal.

S°(298.15 K) is calculated in a manner analagous to that used for enthalpy of formation.

Fusion Data

The adopted melting point is taken from the phase study of Lang et al. (3) after adjustment to IPTS-68. We calculate $\Delta_{fus}H°$ from the difference between H°(2146;298.15) for the crystal (1) and liquid using our adopted linear enthalpy equation for the liquid.

References

1. JANAF Thermochemical Tables: $BeAl_2O_4$(cr), 12-31-79.
2. S. Ishihara and E. D. West, J. Res. Nat. Bur. Stand. 80A, 65 (1976).
3. S. M. Lang, C. L. Fillmore and L. H. Maxwell, J. Res. Nat. Bur. Stand. 48, 298 (1952).

Beryllium Aluminum Oxide ($BeAl_2O_4$) $Al_2Be_1O_4$(l)

Enthalpy Reference Temperature = T_r = 298.15 K   Standard State Pressure = p° = 0.1 MPa

T/K	$C_p°$ (J $K^{-1}mol^{-1}$)	S° (J $K^{-1}mol^{-1}$)	$-[G°-H°(T_r)]/T$ (J $K^{-1}mol^{-1}$)	$H°-H°(T_r)$ (kJ mol^{-1})	$\Delta_fH°$ (kJ mol^{-1})	$\Delta_fG°$ (kJ mol^{-1})	Log K_f
0							
100							
200							
298.15	105.380	126.406	126.406	0.	-2163.842	-2059.526	360.820
300	105.940	127.060	126.408	0.195	-2163.876	-2058.879	358.483
400	130.280	161.162	130.892	12.108	-2164.764	-2023.701	264.268
500	145.250	191.969	140.086	25.942	-2164.416	-1988.456	207.733
600	154.970	219.367	151.061	40.983	-2163.431	-1953.350	170.054
700	161.750	243.783	162.595	56.832	-2162.179	-1918.434	143.155
800	166.830	265.727	174.139	73.271	-2160.872	-1883.703	122.993
900	170.870	285.618	185.438	90.163	-2159.665	-1849.131	107.321
1000	174.200	303.799	196.378	107.421	-2179.845	-1813.163	94.710
1100	177.030	320.538	206.914	124.986	-2178.419	-1776.562	84.362
1200	179.480	336.049	217.037	142.814	-2176.897	-1740.096	75.744
1300	181.670	350.503	226.754	160.873	-2175.299	-1703.761	68.458
1400	183.750	364.042	236.082	179.144	-2173.638	-1667.550	62.217
1450.000	184.800	370.508	240.606	188.358	GLASS <--> LIQUID TRANSITION		
1450.000	246.409	370.508	240.606	188.358			
1500	246.409	378.862	245.076	200.679	-2168.857	-1631.514	56.814
1600	246.409	394.765	253.940	225.320	-2175.646	-1595.419	52.085
1700	246.409	409.703	262.668	249.960	-2167.637	-1559.400	47.915
1800	246.409	423.788	271.231	274.601	-2159.697	-1523.851	44.221
1900	246.409	437.110	279.615	299.242	-2151.827	-1488.741	40.928
2000	246.409	449.750	287.808	323.883	-2144.023	-1454.044	37.976
2100	246.409	461.772	295.808	348.524	-2136.288	-1419.736	35.314
2146.000	246.409	467.111	299.423	359.859	------ CRYSTAL <--> LIQUID -----		
2200	246.409	473.235	303.614	373.165	-2128.619	-1385.793	32.903
2300	246.409	484.188	311.229	397.806	-2121.017	-1352.200	30.709
2400	246.409	494.675	318.656	422.447	-2113.480	-1318.935	28.706
2500	246.409	504.734	325.899	447.088	-2106.009	-1285.983	26.869
2600	246.409	514.399	332.964	471.729	-2098.603	-1253.328	25.180
2700	246.409	523.698	339.858	496.370	-2091.260	-1220.956	23.621
2800	246.409	532.660	346.584	521.011	-2962.738	-1180.698	22.026
2900	246.409	541.306	353.151	545.652	-2952.264	-1117.240	20.124
3000	246.409	549.660	359.562	570.293	-2941.834	-1054.141	18.354
3100	246.409	557.740	365.826	594.934	-2931.448	-991.388	16.705
3200	246.409	565.563	371.946	619.575	-2921.107	-928.972	15.164
3300	246.409	573.145	377.929	644.216	-2910.809	-866.877	13.722
3400	246.409	580.501	383.779	668.857	-2900.558	-805.093	12.369
3500	246.409	587.644	389.502	693.498	-2890.353	-743.611	11.098
3600	246.409	594.586	395.103	718.139	-2880.194	-682.419	9.902
3700	246.409	601.337	400.586	742.779	-2870.085	-621.510	8.774
3800	246.409	607.909	405.956	767.420	-2860.026	-560.871	7.710
3900	246.409	614.309	411.216	792.061	-2850.018	-500.497	6.703
4000	246.409	620.548	416.372	816.702	-2840.064	-440.380	5.751

PREVIOUS: June 1972   CURRENT: December 1979

Beryllium Aluminum Oxide ($BeAl_2O_4$) $Al_2Be_1O_4$(l)

BERYLLIUM ALUMINUM OXIDE ($BeAl_2O_4$) CRYSTAL - LIQUID

0 to 2146 K crystal

avove 2146 K liquid

Refer to the individual tables for details.

M_r = 126.97286 Beryllium Aluminum Oxide ($BeAl_2O_4$) $Al_2Be_1O_4$(cr,l)

Enthalpy Reference Temperature = T_r = 298.15 K — Standard State Pressure = p° = 0.1 MPa

	J K⁻¹mol⁻¹			kJ mol⁻¹			
T/K	C_p°	S°	$-[G^\circ-H^\circ(T_r)]/T$	$H^\circ-H^\circ(T_r)$	Δ_fH°	Δ_fG°	Log K_f
0	0.	0.	INFINITE	-13.088	-2285.491	-2285.491	INFINITE
100	16.365	5.523	132.214	-12.669	-2291.889	-2256.332	1178.587
200	66.430	31.982	74.785	-8.560	-2297.784	-2218.294	579.359
298.15	105.380	66.291	66.291	0.	-2300.782	-2178.542	381.672
300	105.940	66.945	66.293	0.195	-2300.815	-2177.784	379.186
400	130.280	101.047	70.777	12.108	-2301.703	-2136.594	279.011
500	145.250	131.854	79.971	25.942	-2301.355	-2095.338	218.899
600	154.970	159.252	90.946	40.983	-2300.371	-2054.220	178.836
700	161.750	183.668	102.480	56.832	-2299.118	-2013.292	150.234
800	166.830	205.612	114.024	73.271	-2297.811	-1972.550	128.794
900	170.870	225.503	125.322	90.163	-2296.605	-1931.966	112.128
1000	174.200	243.683	136.263	107.421	-2316.784	-1889.987	98.723
1100	177.030	260.423	146.799	124.986	-2315.359	-1847.375	87.725
1200	179.480	275.934	156.922	142.814	-2313.836	-1804.898	78.565
1300	181.670	290.388	166.639	160.873	-2312.239	-1762.550	70.820
1400	183.750	303.927	175.967	179.144	-2310.577	-1720.329	64.186
1500	185.910	316.677	184.926	197.625	-2308.849	-1678.228	58.441
1600	188.410	328.752	193.541	216.337	-2321.567	-1635.720	53.401
1700	191.560	340.263	201.836	235.326	-2319.211	-1592.925	48.945
1800	195.790	351.326	209.835	254.683	-2316.556	-1550.278	44.988
1900	201.570	362.058	217.566	274.535	-2313.473	-1507.788	41.452
2000	209.490	372.588	225.054	295.068	-2309.778	-1465.476	38.274
2100	220.190	383.055	232.329	316.525	-2305.227	-1423.369	35.404
2146.000	226.250	387.890	235.612	326.790	CRYSTAL <--> LIQUID		
2146.000	246.409	467.111	235.612	496.798	TRANSITION		
2200	246.409	473.235	241.369	510.105	-2128.619	-1385.793	32.903
2300	246.409	484.188	251.690	534.745	-2121.017	-1352.200	30.709
2400	246.409	494.675	261.598	559.386	-2113.480	-1318.935	28.706
2500	246.409	504.734	271.123	584.027	-2106.009	-1285.983	26.869
2600	246.409	514.399	280.295	608.668	-2098.603	-1253.328	25.180
2700	246.409	523.698	289.139	633.309	-2091.260	-1220.957	23.621
2800	246.409	532.660	297.677	657.950	-2962.738	-1180.698	22.026
2900	246.409	541.306	305.930	682.591	-2952.264	-1117.240	20.124
3000	246.409	549.660	313.916	707.232	-2941.834	-1054.141	18.354
3100	246.409	557.740	321.652	731.873	-2931.448	-991.388	16.705
3200	246.409	565.563	329.152	756.514	-2921.107	-928.972	15.164
3300	246.409	573.145	336.432	781.155	-2910.810	-866.878	13.722
3400	246.409	580.501	343.503	805.796	-2900.558	-805.093	12.369
3500	246.409	587.644	350.377	830.437	-2890.353	-743.611	11.098
3600	246.409	594.586	357.064	855.078	-2880.194	-682.419	9.902
3700	246.409	601.337	363.575	879.719	-2870.085	-621.510	8.774
3800	246.409	607.909	369.919	904.360	-2860.026	-560.871	7.710
3900	246.409	614.309	376.104	929.001	-2850.018	-500.497	6.703
4000	246.409	620.548	382.137	953.642	-2840.064	-440.380	5.751

PREVIOUS: CURRENT: December 1979

Beryllium Aluminum Oxide ($BeAl_2O_4$) $Al_2Be_1O_4$(cr,l)

ALUMINUM BROMIDE $((AlBr_3)_2)$ IDEAL GAS M_r = 533.38708

$\Delta_f H°(0\ K)$ = -892.9 ± 3.4 kJ mol^{-1}

$S°(298.15\ K)$ = 547.2 ± 4.2 J K^{-1} mol^{-1} $\Delta_f H°(298.15\ K)$ = -937.2 ± 3.4 kJ mol^{-1}

Vibrational Frequencies and Degeneracies

ν, cm^{-1}	ν, cm^{-1}	ν, cm^{-1}	ν, cm^{-1}	ν, cm^{-1}	ν, cm^{-1}
409 (1)	59 (1)	76 (1)	[8] (1)	346 (1)	376 (1)
203 (1)	[30](1)	500 (1)	489 (1)	90 (1)	199 (1)
139 (1)	[250](1)	[110](1)	114 (1)	[67](1)	112 (1)

Ground State Quantum Weight: [1] σ = 4

Point Group: D_{2h}

Bond Distances: Al-Br = 2.222 ± 0.005 Å Al-Br(bridge) = 2.414 ± 0.007 Å

Bond Angles: Br-Al-Br = 122.8 ± 3.3° Br(bridge)-Al-Br(bridge) = 92.3 ± 0.9°

Product of the Moments of Inertia: $I_A I_B I_C = 8.865729 \times 10^{-110}\ g^3\ cm^6$

Enthalpy of Formation

Vapor pressures over $AlBr_3$(cr) and $AlBr_3$(ℓ) have been measured by Dunne and Gregory (1), Fischer, Rahlfs and Benze (2) and Sulzmann (3). The vapor pressures were corrected for vapor non-ideality by means of the equation ΔG°/T=-RlnP-BP/T. The Bertholot equation of state and critical constants T_c=763 K and P_c=28.5 atm reported by Johnson, Silva and Cubicciotti (4) were used to calculate B. The corrected vapor pressures were used to calculate $\Delta_r H°(298.15\ K)$ by both 2nd and 3rd law methods. The results of the calculations are as follows with reaction (A) corresponding to 2 $AlBr_3$(cr)=Al_2Br_6(g) and reaction (B) corresponding to 2 $AlBr_3$(ℓ)=Al_2Br_6(g).

Source	Reaction	Method	Data Points	T/K	$\Delta_r H°(298.15\ K)$, kcal mol^{-1} 2nd law	3rd law	Drift cal K^{-1} mol^{-1}	$\Delta_f H°(298.15\ K)$ kcal mol^{-1}
(3)	A	Manometric	13	302-334	21.2±0.2	20.54±0.1	-2.2±0.5	-223.86
(1)	A	Effusion	3	273-310	19.6±0.7	20.26±0.1	2.2±2.4	-224.14
(2)	B	Mass Loss	18[a]	413-522	15.6±0.0	15.62±0.0	0.1±0.1	-223.98
(2)	B	Transport	4	396-470	16.0±0.1	15.61±0.1	-0.9±0.3	-223.99

[a]Point at 436.2 K rejected due to statistical test.

Heat Capacity and Entropy

Structural assignments for the dimer of aluminum tribromide are based on the electron diffraction data reported by Shen (5). These values are in good agreement with earlier diffraction studies by Palmer and Elliott (6) and Akishin et al. (7). The principal moments of inertia are $I_A = 282.4212 \times 10^{-39}$, $I_B = 502.7859 \times 10^{-39}$ and $I_C = 624.3585 \times 10^{-39}$ g cm^2. Beattie, Gilson and Ozin (8) measured the condensed phase IR and Raman bands for aluminum tribromide while Beattie and Horder (9) measured the gas phase Raman spectra. Perov et al (10) investigated the IR spectra of monomeric and dimeric aluminum tribromide in a xenon matrix at 20-30 K. It has been established that the condensed phases of $AlBr_3$ are dimeric (11. 12), thus spectroscopic studies of the crystal and liquid pertain to the structure of Al_2Br_6(g). However, gas phase values are adopted in preference to condensed phase frequencies whenever possible. Five frequencies (ν_5, ν_6, ν_9, ν_{10}, ν_{15}) were adjusted not only for consistency with the frequencies chosen for Al_2F_6(g) and Al_2Cl_6(g) but also for agreement between 2nd and 3rd law $\Delta_r H°(298.15\ K)$ values in the second data set tabulated above.

References

1. T. G. Dunne and N. W. Gregory, J. Amer. Chem. Soc. 80, 1526 (1958).
2. W. Fischer, O. Rahlfs and B. Benze, Z. Anorg. Chem. 205, 1 (1932).
3. K. G. P. Sulzmann, J. Electrochem. Soc. 121, 1239 (1974).
4. J. W. Johnson, W. J. Silva, D. Cubicciotti, J. Phys. Chem. 72, 1669 (1968).
5. Q. Shen, Diss. Abstr. Int. B34, 3735 (1974).
6. K. J. Palmer and N. Elliott, J. Amer. Chem. Soc. 60, 1852 (1938).
7. P. A. Akishin, N. G. Rambidi and E. Z. Zasorin, Kristallografiya 4, 167 (1959).
8. I. R. Beattie, T. Gilson and G. A. Ozin, J. Chem. Soc. (London) 1968A, 2655 (1969).
9. I. R. Beattie and J. R. Horder, J. Chem. Soc. (London) 1969A, 2655 (1969).
10. P. A. Perov, S. V. Nedyak and A. A. Mal'tsev, Vest. Mosk. Univ. Khim. 29, 201 (1974).
11. R. A. Renes and C. H. MacGillavry, Rec. Trav. Chim. 64, 275 (1945).
12. K. N. Smenenko and T. N. Naumova, Russ. J. Inorg. Chem. 9, 718 (1964).

Aluminum Bromide $((AlBr_3)_2)$ Al_2Br_6(g)

Enthalpy Reference Temperature = T_r = 298.15 K Standard State Pressure = p° = 0.1 MPa

T/K	$C_p°$ (J $K^{-1}mol^{-1}$)	S° (J $K^{-1}mol^{-1}$)	$-[G°-H°(T_r)]/T$ (J $K^{-1}mol^{-1}$)	$H°-H°(T_r)$ (kJ mol^{-1})	$\Delta_f H°$ (kJ mol^{-1})	$\Delta_f G°$ (kJ mol^{-1})	Log K_f
0	0.	0.	INFINITE	-38.309	-892.918	-892.918	INFINITE
100	122.051	386.605	686.066	-29.946	-893.902	-915.011	477.953
200	154.379	482.768	562.199	-15.886	-898.150	-934.593	244.091
250	162.423	518.147	549.952	-7.951	-900.889	-943.399	197.112
298.15	167.514	547.220	547.220	0.	-937.216	-947.368	165.975
300	167.672	548.257	547.223	0.310	-937.415	-947.430	164.962
350	171.215	574.387	549.279	8.788	-1029.359	-944.065	140.894
400	173.690	597.420	553.885	17.414	-1028.768	-931.920	121.696
450	175.474	617.986	559.886	26.145	-1028.167	-919.850	106.773
500	176.796	636.546	566.639	34.954	-1027.568	-907.847	94.842
600	178.581	668.951	581.068	52.730	-1026.422	-884.012	76.960
700	179.692	696.569	595.644	70.648	-1025.414	-860.359	64.201
800	180.428	720.615	609.795	88.656	-1024.625	-836.837	54.640
900	180.939	741.898	623.313	106.726	-1024.129	-813.397	47.208
1000	181.309	760.982	636.142	124.839	-1045.169	-788.473	41.186
1100	181.584	778.276	648.290	142.985	-1044.723	-762.825	36.224
1200	181.794	794.085	659.790	161.154	-1044.280	-737.217	32.090
1300	181.959	808.643	670.688	179.342	-1043.844	-711.647	28.594
1400	182.090	822.133	681.029	197.545	-1043.419	-686.109	25.599
1500	182.195	834.699	690.860	215.759	-1043.012	-660.601	23.004
1600	182.282	846.461	700.221	233.983	-1042.625	-635.120	20.735
1700	182.354	857.514	709.152	252.215	-1042.266	-609.662	18.733
1800	182.415	867.939	717.687	270.453	-1041.940	-584.225	16.954
1900	182.466	877.803	725.857	288.698	-1041.652	-558.804	15.363
2000	182.510	887.163	733.690	306.946	-1041.410	-533.397	13.931
2100	182.548	896.069	741.212	325.199	-1041.219	-508.001	12.636
2200	182.580	904.562	748.445	343.456	-1041.085	-482.613	11.459
2300	182.609	912.678	755.411	361.715	-1041.012	-457.230	10.384
2400	182.634	920.451	762.127	379.977	-1041.002	-431.848	9.399
2500	182.656	927.906	768.610	398.242	-1041.059	-406.466	8.493
2600	182.676	935.071	774.875	416.509	-1041.182	-381.081	7.656
2700	182.693	941.965	780.937	434.777	-1041.370	-355.688	6.881
2800	182.709	948.610	786.807	453.047	-1629.423	-328.353	6.126
2900	182.723	955.022	792.498	471.319	-1627.541	-281.919	5.078
3000	182.736	961.216	798.019	489.592	-1625.713	-235.551	4.101
3100	182.747	967.208	803.381	507.866	-1623.934	-189.241	3.189
3200	182.758	973.011	808.591	526.141	-1622.196	-142.989	2.334
3300	182.767	978.634	813.659	544.417	-1620.492	-96.789	1.532
3400	182.776	984.091	818.592	562.694	-1618.815	-50.642	0.778
3500	182.784	989.389	823.397	580.972	-1617.158	-4.544	0.068
3600	182.791	994.538	828.080	599.251	-1615.512	41.508	-0.602
3700	182.798	999.547	832.647	617.531	-1613.871	87.514	-1.235
3800	182.804	1004.422	837.103	635.811	-1612.226	133.475	-1.835
3900	182.810	1009.170	841.454	654.091	-1610.572	179.393	-2.403
4000	182.815	1013.799	845.705	672.373	-1608.903	225.268	-2.942
4100	182.820	1018.313	849.861	690.654	-1607.211	271.101	-3.454
4200	182.824	1022.718	853.924	708.936	-1605.494	316.893	-3.941
4300	182.829	1027.020	857.900	727.219	-1603.744	362.644	-4.405
4400	182.833	1031.224	861.791	745.502	-1601.961	408.353	-4.848
4500	182.836	1035.332	865.602	763.786	-1600.134	454.021	-5.270
4600	182.840	1039.351	869.336	782.069	-1598.271	499.649	-5.674
4700	182.843	1043.283	872.995	800.354	-1596.364	545.235	-6.060
4800	182.846	1047.133	876.583	818.638	-1594.414	590.781	-6.429
4900	182.849	1050.903	880.102	836.923	-1592.417	636.285	-6.783
5000	182.852	1054.597	883.555	855.208	-1590.376	681.748	-7.122
5100	182.854	1058.218	886.945	873.493	-1588.290	727.171	-7.448
5200	182.857	1061.769	890.273	891.779	-1586.161	772.550	-7.760
5300	182.859	1065.252	893.541	910.064	-1583.988	817.890	-8.061
5400	182.861	1068.670	896.753	928.350	-1581.774	863.188	-8.350
5500	182.863	1072.025	899.909	946.637	-1579.520	908.444	-8.628
5600	182.865	1075.320	903.012	964.923	-1577.229	953.659	-8.895
5700	182.867	1078.557	906.064	983.210	-1574.904	998.833	-9.153
5800	182.868	1081.737	909.065	1001.496	-1572.541	1043.964	-9.402
5900	182.870	1084.863	912.018	1019.783	-1570.153	1089.056	-9.642
6000	182.872	1087.937	914.925	1038.070	-1567.695	1134.103	-9.873

PREVIOUS: September 1979 (1 atm) CURRENT: September 1979 (1 bar)

Aluminum Bromide $((AlBr_3)_2)$ Al_2Br_6(g)

ALUMINUM CHLORIDE $((AlCl_3)_2)$ IDEAL GAS M_r = 266.68108

$\Delta_f H°(0\ K) = -1293.02 \pm 3.35$ kJ mol^{-1}

$S°(298.15\ K) = 475.0 \pm 4.2$ J K^{-1} mol^{-1} $\Delta_f H°(298.15\ K) = -1295.74 \pm 3.35$ kJ mol^{-1}

Vibrational Frequencies (all singly degenerate), cm^{-1}

A_g	A_g	A_u	B_{1g}	B_{1g}	B_{1u}	B_{2g}	B_{2u}	B_{3g}	B_{3u}	B_{3u}
501	217	[60]	289	[10]	625	610	420	[161]	484	144
336	99		166		175	116	135		320	

Ground State Quantum Weight: [1] σ = 4
Point Group: D_{2h}
Bond Distances: Al-Cl = 2.065±0.002Å, Al-Cl(bridge) = 2.252±0.004 Å
Bond Angles: Cl-Al-Cl=123.4±1.6°, Cl(bridge)-Al-Cl(bridge) = 91.0±0.5°
Product of the Moments of Inertia: $I_A I_B I_C = 5.687094 \times 10^{-111}$ g^3 cm^6

Enthalpy of Formation

The enthalpy of formation is calculated from the value for $AlCl_3$(cr) and the heat of sublimation as derived from vapor pressure data. Vapor pressures were corrected for vapor non-ideality by means of the equation $\Delta G°/T = -R\ln P - BP/T$. The Berthelot equation of state and critical constants T_c = 620 K (1) and P_c = 24.3 atm were used to calculate B. P_c is an extrapolation of the data given by Denisova and Baskova (2). The corrected vapor pressures were used to derive $\Delta_{sub}H°(298.15\ K)$ by both 2nd law and 3rd law methods. The first nine data sets show good agreement between 2nd and 3rd law values and an average of $\Delta_{sub}H°(298.15\ K)$ = 27.61±0.55 kcal mol^{-1} Al_2Cl_6 is adopted. The remaining data sets show serious discrepancies between 2nd and 3rd law values, indicating temperature dependent errors in measurement, and are given no weight.

Source	T/K	Data Points	$\Delta_{sub}H°(298.15\ K)$, kcal mol^{-1} 2nd Law	3rd Law	Drift cal K^{-1} mol^{-1}
Viola et al. (3)	389.9-466.3	44(4)	27.48±0.0	27.62±0.0	0.3±0.1
Viola et al. (3)	467.7-510.0	67(4)	26.82±0.0	27.65±0.1	1.7±0.1
Smits and Meijering (4) liq.	444.7-464.6	7(2)	27.44±0.1	27.59±0.0	0.3±0.3
Smits and Meijering (4)	462.4-481.8	10	27.07±0.1	27.61±0.0	1.1±0.2
Treadwell and Terebesi (5) liq.	388.7-461.7	11	28.31±0.2	27.55±0.1	-1.8±0.3
Treadwell and Terebesi (5)	466.5-476.2	3	27.39±0.1	27.60±0.0	0.5±0.2
Fischer et al. (6) liq.	392.8-428.3	9	27.65±0.4	27.61±0.1	-0.1±1.1
Fischer et al. (6)	403.8-450.4	12(2)	27.70±0.1	27.60±0.0	-0.2±0.3
Nisel'son et al. (7) liq.	471.2-506.5	18(1)	27.28±0.1	27.64±0.0	0.7±0.1
Friedel and Crafts (8)	441.0-455.9	5	29.80±1.6	27.80±0.1	-4.5±3.5
Friedel and Crafts (8) liq.	477.4-486.2	3	29.92±1.4	27.76±0.0	-4.5±3.0
Maier (9)	392.8-454.4	8(1)	25.87±0.7	27.38±0.2	3.5±1.7
Dunne and Gregory (10)	294.2-322.2	2	29.97±0.0	27.79±0.3	-7.1±0.0
Denisova and Baskova (2) liq.	467.8-511.6	8	26.31±0.2	27.56±0.1	2.5±0.3

First number indicates number of data points used in calculation; number in parentheses indicates number of data points rejected.

Heat Capacity and Entropy

Structural assignments for the dimer of aluminum trichloride are based on the electron diffraction data reported by Shen (11). These values are in good agreement with earlier electron diffraction studies by Palmer and Elliott (12) and Akishin et al (13) and a theoretical investigation by Curtiss (14). The principal moments of inertia are: $I_A = 108.2257 \times 10^{-39}$, $I_B = 206.7252 \times 10^{-39}$ and $I_C = 254.1948 \times 10^{-39}$ g cm^2. Klemperer (15) observed three frequencies in the infrared spectrum of Al_2Cl_6 vapor, Beattie and Horder (16) and Maroni et al (17) reported gas phase Raman spectra, and Gerding and Smit (18) and Pershina and Raskin (19) measured liquid phase Raman frequencies. These earlier vibrational assignments were supplemented by matrix isolated/IR and Raman studies reported by Beattie et al (20) and Perov et al (21) and the detailed vibrational analysis given by Tranquille and Fouassier (22). Normal coordinate calculations have been reported by several investigators (23, 24, 17, 25) but their basis on incomplete data makes many assignments doubtful. The new vibrational assignment incorporates the additional frequencies observed by Tranquille and Fouassier and differs with earlier calculations primarily in reassignment of the frequencies at 289 cm^{-1} and 161 cm^{-1}. The estimated fundamentals are based on comparisons with Al_2F_6, Al_2Br_6, and Al_2I_6 and are adjusted such that the entropy is consistent for the 2nd and 3rd law calculations above.

References

1. J. W. Johnson, D. Cubicciotti, and W. J. Silva, High Temp. Sci. 3, 523 (1971).
2. N. D. Denisova and A. P. Baskova, Zh. Fiz. Khim. 43, 2353 (1969).
3. J. T. Viola, D. W. Seegmiller, A. A. Fannin, and L. A. King, J. Chem. Eng. Data 22, 367 (1977).
4. A. Smits and J. L. Meijering, Z. Phys. Chem. B41, 98 (1938).
5. W. D. Treadwell and L. Terebesi, Helv. Chim. Acta 15, 1053 (1932).
6. W. Fischer, O. Rahlfs, and B. Benze, Z. Anorg. Chem. 205, 1 (1932).
7. L. A. Nisel'son, A. I. Pustil'nik, O. R. Gavrilov, and V. A. Rodin, Zh. Neorg. Khim. 10, 2339 (1965).
8. C. Friedel and J. M. Crafts, Compt. Rend. 106, 1764 (1888).
9. C. G. Maier, U.S. Bur. Mines Tech. Paper 360, (1929).
10. T. G. Dunne and N. W. Gregory, J. Amer. Chem. Soc. 80, 1530 (1958).
11. Q. Shen, Diss. Abstr. Int. B34, 3735 (1974).
12. K. J. Palmer and N. Elliott, J. Amer. Chem. Soc. 60, 1852 (1938).
13. P. A. Akishin, N. G. Rambidi and E. Z. Zasorin, Kristallografiya 4, 186 (1959).
14. L. A. Curtiss, Int. J. Quantum Chem. 14, 709 (1978).
15. W. Klemperer, J. Chem. Phys. 24, 353 (1956).
16. I. R. Beattie and J. R. Horder, J. Chem. Soc. 1969A, 2655.
17. V. A. Maroni, D. M. Green, R. L. McBeth, and E. J. Cairns, Spectrochim Acta 26A, 418 (1970).
18. H. Gerding and E. Smit, Z. Phys. Chem. B50, 171 (1941).
19. E. V. Pershina and Sh. Sh. Raskin, Opt. Spect. 13, 272 (1962).
20. I. R. Beattie, H. E. Bloyden, S. M. Hall, S. N. Jenny, and J. S. Ogden, J. Chem. Soc., Dalton Trans. 1976, 666.
21. P. A. Perov, S. V. Nedyak, A. A. Mal'tsen, Vest. Mask. Univ. Khim. 29, 201 (1974).
22. M. Tranquille and M. Fouassier, J. Chem. Soc., Faraday Trans. 76, 26 (1980).
23. T. Onishi and T. Shimanouchi, Spectrochim Acta, 20, 325 (1964).
24. R. H. Miller, PhD Thesis, Ohio University, August 21, 1965.
25. S. J. Cyvin and O. Torset, Revue de Chim. Miner. 9, 179 (1972).

Aluminum Chloride $((AlCl_3)_2)$ Al_2Cl_6(g)

Enthalpy Reference Temperature = T_r = 298.15 K Standard State Pressure = p° = 0.1 MPa

T/K	$C_p°$ (J K^{-1}mol^{-1})	S°	$-[G°-H°(T_r)]/T$	$H°-H°(T_r)$ (kJ mol^{-1})	$\Delta_f H°$	$\Delta_f G°$	Log K_f
0	0.	0.	INFINITE	-33.898	-1293.019	-1293.019	INFINITE
100	101.982	330.759	602.707	-27.195	-1296.020	-1270.996	663.900
200	140.807	415.028	489.094	-14.813	-1296.355	-1245.724	325.350
250	151.631	447.687	477.628	-7.485	-1296.111	-1233.091	257.640
298.15	158.858	475.050	475.050	0.	-1295.743	-1220.985	213.912
300	159.089	476.033	475.053	0.294	-1295.727	-1220.522	212.512
350	164.322	500.973	477.011	8.387	-1295.260	-1208.023	180.288
400	168.079	523.174	481.419	16.702	-1294.746	-1195.596	156.129
450	170.844	543.138	487.186	25.178	-1294.204	-1183.234	137.346
500	172.925	561.251	493.701	33.775	-1293.652	-1170.934	122.327
600	175.775	593.052	507.683	51.221	-1292.573	-1146.493	99.811
700	177.576	620.293	521.871	68.895	-1291.609	-1122.224	83.741
800	178.779	644.089	535.692	86.717	-1290.848	-1098.082	71.697
900	179.621	665.197	548.931	104.639	-1290.369	-1074.018	62.334
1000	180.232	684.155	561.522	122.633	-1311.416	-1048.469	54.766
1100	180.688	701.356	573.464	140.680	-1310.969	-1022.196	48.540
1200	181.038	717.093	584.787	158.768	-1310.518	-995.964	43.353
1300	181.312	731.595	595.529	176.886	-1310.067	-969.770	38.966
1400	181.530	745.040	605.734	195.028	-1309.620	-943.610	35.207
1500	181.707	757.571	615.444	213.190	-1309.181	-917.482	31.950
1600	181.852	769.302	624.697	231.368	-1308.752	-891.382	29.101
1700	181.972	780.331	633.531	249.560	-1308.336	-865.310	26.588
1800	182.074	790.735	641.978	267.762	-1307.937	-839.261	24.355
1900	182.159	800.582	650.069	285.974	-1307.557	-813.234	22.357
2000	182.233	809.927	657.830	304.194	-1307.200	-787.226	20.560
2100	182.296	818.820	665.286	322.420	-1306.869	-761.236	18.935
2200	182.351	827.302	672.459	340.653	-1306.569	-735.260	17.457
2300	182.399	835.408	679.369	358.890	-1306.304	-709.298	16.109
2400	182.441	843.172	686.034	377.132	-1306.078	-683.346	14.873
2500	182.478	850.621	692.469	395.378	-1305.897	-657.403	13.736
2600	182.511	857.778	698.691	413.628	-1305.763	-631.466	12.686
2700	182.541	864.667	704.711	431.880	-1305.681	-605.533	11.715
2800	182.567	871.306	710.543	450.136	-1893.454	-577.667	10.777
2900	182.591	877.713	716.198	468.393	-1891.290	-530.713	9.559
3000	182.612	883.903	721.685	486.654	-1889.183	-483.832	8.424
3100	182.631	889.891	727.015	504.916	-1887.132	-437.021	7.364
3200	182.649	895.690	732.196	523.180	-1885.136	-390.276	6.371
3300	182.665	901.311	737.236	541.445	-1883.193	-343.591	5.439
3400	182.679	906.764	742.142	559.713	-1881.300	-296.965	4.562
3500	182.693	912.059	746.922	577.981	-1879.451	-250.394	3.737
3600	182.705	917.206	751.581	596.251	-1877.644	-203.875	2.958
3700	182.716	922.212	756.125	614.522	-1875.870	-157.406	2.222
3800	182.727	927.085	760.560	632.794	-1874.126	-110.985	1.526
3900	182.736	931.832	764.891	651.068	-1872.403	-64.609	0.865
4000	182.745	936.458	769.123	669.342	-1870.696	-18.277	0.239
4100	182.753	940.971	773.260	687.617	-1868.996	28.013	-0.357
4200	182.761	945.375	777.305	705.892	-1867.299	74.261	-0.924
4300	182.768	949.675	781.264	724.169	-1865.595	120.468	-1.463
4400	182.775	953.877	785.140	742.446	-1863.880	166.636	-1.978
4500	182.781	957.985	788.935	760.724	-1862.143	212.764	-2.470
4600	182.787	962.002	792.654	779.002	-1860.385	258.854	-2.939
4700	182.792	965.933	796.299	797.281	-1858.597	304.905	-3.389
4800	182.798	969.782	799.873	815.561	-1856.776	350.918	-3.819
4900	182.802	973.551	803.379	833.841	-1854.914	396.893	-4.231
5000	182.807	977.244	806.820	852.121	-1853.011	442.828	-4.626
5100	182.811	980.864	810.197	870.402	-1851.063	488.726	-5.006
5200	182.815	984.414	813.513	888.683	-1849.067	534.584	-5.370
5300	182.819	987.896	816.771	906.965	-1847.020	580.404	-5.720
5400	182.823	991.314	819.972	925.247	-1844.922	626.185	-6.057
5500	182.826	994.668	823.118	943.530	-1842.771	671.927	-6.381
5600	182.829	997.963	826.211	961.812	-1840.567	717.629	-6.694
5700	182.832	1001.199	829.252	980.095	-1838.311	763.291	-6.995
5800	182.835	1004.379	832.244	998.379	-1835.995	808.913	-7.285
5900	182.838	1007.504	835.188	1016.662	-1833.633	854.494	-7.565
6000	182.841	1010.577	838.086	1034.946	-1831.175	900.030	-7.835

PREVIOUS: September 1979 (1 atm) CURRENT: September 1979 (1 bar)

Aluminum Chloride $((AlCl_3)_2)$ Al_2Cl_6(g)

ALUMINUM POTASSIUM CHLORIDE ($K_3Al_2Cl_9$) CRYSTAL M_r = 490.33498

$\Delta_fH°$(0 K) = Unknown

S°(298.15 K) = [468.608] J K^{-1} mol^{-1} $\Delta_fH°$(298.15 K) = -2860 ± 12 kJ mol^{-1}

T_{fus} = Unknown

Enthalpy of Formation

$\Delta_fH°$($K_3Al_2Cl_9$, cr, 298.15 K) is calculated from the heat of solution, $\Delta_rH°$ = -112.07 kcal mol^{-1}, of $K_3Al_2Cl_9$(cr) in water (one mole of solute in 36 l. of water), as determined by Baud (1).

Heat Capacity and Entropy

Heat capacities are estimated based on the assumption that the property of the mixed chloride ($2AlCl_3+3KCl$) is the sum of the corresponding values for its component chlorides, i.e. AlCl(cr) and KCl(cr). In a similar manner, S°(298.15 K) is estimated to be 112 cal K^{-1} mol^{-1}.

Reference

1. E. Baud, Ann. chim. Phys. 1, 8 (1904).

Potassium Aluminum Chloride ($K_3Al_2Cl_9$) $Al_2Cl_9K_3$(cr)

Enthalpy Reference Temperature = T_r = 298.15 K Standard State Pressure = p° = 0.1 MPa

	J K⁻¹mol⁻¹			kJ mol⁻¹			
T/K	$C_p°$	S°	-[G°-H°(T_r)]/T	H°-H°(T_r)	$\Delta_fH°$	$\Delta_fG°$	Log K_f
0							
100							
200							
298.15	337.230	468.608	468.608	0.	-2860.182	-2625.893	460.046
300	337.858	470.696	468.614	0.624	-2860.094	-2624.440	456.956
400	353.130	570.010	482.052	35.183	-2862.602	-2545.238	332.374
500	367.272	650.359	507.925	71.217	-2857.227	-2466.497	257.673
600	379.740	718.443	537.478	108.579	-2850.799	-2388.938	207.976
700	390.807	777.827	567.660	147.116	-2843.468	-2312.529	172.563
800	400.660	830.668	597.294	186.700	-2835.425	-2237.222	146.076
900	407.940	878.265	625.912	227.117	-2826.979	-2162.952	125.534
1000	416.726	921.687	653.349	268.338	-2839.443	-2088.115	109.072
1100	427.019	961.876	679.592	310.513	-3066.550	-1999.618	94.954
1200	438.818	999.527	704.700	353.792	-3052.785	-1903.224	82.845
1300	452.123	1035.166	728.761	398.327	-3037.812	-1808.025	72.647
1400	466.934	1069.205	751.871	444.267	-3021.478	-1714.029	63.951
1500	483.252	1101.968	774.125	491.764	-3003.629	-1621.254	56.457

PREVIOUS: June 1962 CURRENT: September 1964

Potassium Aluminum Chloride ($K_3Al_2Cl_9$) $Al_2Cl_9K_3$(cr)

ALUMINUM FLUORIDE $(AlF_3)_2$ IDEAL GAS M_r = 167.953498

$S°(298.15\ K) = 387.3 \pm 12\ J\ K^{-1}\ mol^{-1}$

$\Delta_f H°(0\ K) = -2624.08 \pm 16\ kJ\ mol^{-1}$

$\Delta_f H°(298.15\ K) = -2633.62 \pm 16\ kJ\ mol^{-1}$

Vibrational Frequencies, ν, cm^{-1} (all single degenerate)

[830]	[200]	[210]	[30]	600	805
[600]	[90]	980	[970]	[230]	575
[370]	[630]	340	[260]	[310]	300

Ground State Quantum Weight: [1] σ = 4

Point Group: $[D_{2h}]$

Bond Distances: Al-F = [1.63] Å Al-F(bridge) = [1.80] Å

Bond Angles: F-Al-F = [120°] F(bridge)-Al-F(bridge) = [90]°

Product of the Moments of Inertia: $I_A I_B I_C = [2.669982 \times 10^{-112}]\ g^3\ cm^6$

Enthalpy of Formation

The existence of small amounts of dimer in the saturated vapor was inferred from the intensities of AlF_2^+, AlF_3^+ and $Al_2F_5^+$ observed by mass spectrometry (1, 2, 3). Buchler et al. (1) concluded that the precursor of the first two ions was mainly AlF_3, while the dimer was the precursor of $Al_2F_5^+$. Intensities of $Al_2F_5^+$ relative to those of AlF_2^+ were observed as 0.0057 (1), 0.007 (2) and 0.025 (3) at ionizing energies of 60, 90 and 100 V for vapor at 1025, 1100 and 1025 K, respectively.

The temperature dependence of ion intensities was used to derive 2nd law values of $\Delta_{sub}H°$ for monomer and dimer (1, 2) and a value of $\Delta_{dim}H°$ based on an estimate for the $\Delta_{dim}S°$ (3). The absolute pressure of the monomer was obtained (2) from the relative ion intensity assuming a relative ionization cross section equal to two and neglecting other effects. Krause and Douglas (4) reported values for $\Delta_{sub}H°$(monomer), $\Delta_{dim}H°$, and $\Delta_{dim}S°$; these were derived by combination of $\Delta_{sub}H°$(dimer) (1) with precise sublimation pressures from entrainment data near 1200 K, where the dimer is more abundant. After reduction to 1000 K with JANAF functions, these results may be summarized as follows:

Source	Method	$\Delta_r H°$, kcal mol^{-1}*			$\Delta_r S°$, cal K^{-1} mol^{-1}*		
		c → m	2c → d	2m → d	c → m	2c → d	2m → d
(4)	Entrainment	68.1 ± 0.4	[86]	-50 ± 3	[44.04]	---	-35 ± 3
(1)	Ion Intensities	67.3 ± 3	85.8 ± 3	-48.8 ± 3	---	---	---
(2)	Ion Intensities	70.5	92.1	-49.0	46.4	55.1	-37.7
(3)	Ion Intensities	---	---	-48.0 ± 4	---	---	[-32 ± 3]
JANAF	Adopted	68.4	87.8	[-49]	[44.37]	[52.83]	[-35.9]

*Reactions are abbreviated by c = crystal, d = dimer gas and m = monomer gas. Brackets enclose those values assumed in deriving the other values.

JANAF values are given in the last line. These are based on JANAF entropies, on $\Delta_{dim}H° = -49\ kcal\ mol^{-1}$ for which there is close agreement, and on the entrainment pressures of Krause and Douglas (4). Reduction of the dimer value to 298.15 K gives $\Delta_{sub}H° = 92.55 \pm 4$ and $\Delta_f H° = -629.45 \pm 4\ kcal\ mol^{-1}$. The table for AlF_3(g) gives a detailed comparison of observed and predicted total pressure along with predictions of the mole fraction of dimer in the saturated vapor. The latter are reasonably consistent with the relative ion intensities.

Heat Capacity and Entropy

The molecular structure is assumed to be similar to that of Al_2Cl_6. Bond distances and bond angles are estimated by comparison with AlF_3 (5), $AlCl_3$ (5) and Al_2Cl_6 (5). The principal moments of inertia are $I_A = 35.3676 \times 10^{-39}$, $I_B = 79.7440 \times 10^{-39}$, $I_C = 94.6684 \times 10^{-39}\ g\ cm^2$.

Six observed vibrational frequencies are from infrared spectra of the dimer isolated in rare-gas matrices (6). Unobserved fundamentals (2 in infrared, 9 in Raman and 1 inactive) are estimated such that the entropy is consistent with the result of Krause and Douglas (4) as given above. The estimated fundamentals are based on comparisons with Al_2Cl_6, Al_2Br_6 and Al_2I_6. Assignments used in the comparisons are minor modifications of those given by Beattie et al. (7).

References - Al_2F_6(g)

1. A. Buchler, E. P Marram and J. L. Stauffer, J. Phys. Chem. 71, 4139 (1967).
2. E. V. Erokhin, N. A. Zhegul'skaya, L. N. Sidorov and P. kA. kAkishin, Izv. Akad. Nauk SSSR, Neorg. Mater. 3 (5), 873 (1967).
3. R. F. Porter and E. E. Zeller, J. Chem. Phys. 33, 858 (1960).
4. R. F. Krause and T. B. Douglas, J. Phys. Chem. 72, 475 (1968).
5. JANAF Thermochemical Tables: AlF_3(g), $AlCl_3$(g), Al_2Cl_6(g), Al_2Br_6(g), and Al_2I_6(g), 9-30-79.
6. A. Snelson, J. Phys. Chem. 71, 3202 (1967).
7. I. R. Bettie and J. R. Horder, J. Chem. Soc. A1969, 2655; I. R. Beattie, T. Gilson and G. A. Ozin, ibid., A1968, 813.

References - AlF_3(g)

1. R. F. Krause and T. B. Douglas, J. Phys. Chem. 72, 475 (1968).
2. I. I. Naryshkin, Zh. Fiz. Khim. 13, 528 (1939).
3. E. V. Erokhin, N. A. Zhegul'skaya, L. N. Sidorov and P. A. Akishin, Izv. Akad. Nauk SSSR, Neorg. Mater. 3 (5), 873 (1967).
4. H. C. Ko, M. A. Greenbaum, J. A. Blauer and M. Farber, J. Phys. Chem. 69, 2311 (1965).
5. P. E. Blackburn, OAR Final Report, A. D. Little, Inc., May 31, 1965.
6. A. M. Evseev, G. V. Pozharskaya, An. N. Nesmeyanov and Ya. I. Gerasimov, Zh. Neorg. Khim. 4, 2196 (1959).
7. D. L. Hildenbrand, Aeronutronics, Inc., personal communication, July 30, 1963; D. L. Hildenbrand et al., Aeronutronic Rept. No. U-2055, Contract NOw 61-0905-c, March 15, 1963.
8. D. L. Hildenbrand and L. P. Theard, AD 258410, 24 pp (1961).
9. W. P. Witt and R. F. Barrow, Trans. Faraday Soc. 55, 730 (1959).
10. M. M. Vetyukov, M. L. Blyushtein and V. P. Poddymov, Izv. Vyssh. Ucheb. Zaved., Tsvet. Met. 2, (6), 126 (1959).
11. O. Ruff and L. Le Boucher, Z. anorg. Allgem. Chem. 219, 376 (1934).
12. W. Olbrich, Dissertation, Technische Hochschule, Breslau, 1928.
13. P. A. Akishin, N. G. Rambidi and E. Z. Zasorin, Kristallografiya 4, 186 (1959); English translation, p. 167.
14. A. Buchler, E. P. Marram and J. L. Stauffer, J. Phys. Chem. 71, 4139 (1967).
15. A. Snelson, J. Phys. Chem. 71, 3202 (1967).
16. I. R. Beattie and J. R. Horder, J. Chem. Soc. A1969, 2655.
17. M. C. Drake and G. M. Rosenblatt, Proc. - Electrochem. Soc. 78-1, 234 (1978).
18. JANAF Thermochemical Tables: $AlCl_3$(g), $AlBr_3$(g), and AlI_3(g), 9-30-79.
19. D. L. Hildenbrand, SRI International, personal communication, May 8, 1981.

Aluminum Fluoride $((AlF_3)_2)$ Al_2F_6(g)

Enthalpy Reference Temperature = T_r = 298.15 K Standard State Pressure = p° = 0.1 MPa

T/K	$C_p°$ (J K⁻¹mol⁻¹)	S° (J K⁻¹mol⁻¹)	$-[G°-H°(T_r)]/T$ (J K⁻¹mol⁻¹)	$H°-H°(T_r)$ (kJ mol⁻¹)	$\Delta_f H°$ (kJ mol⁻¹)	$\Delta_f G°$ (kJ mol⁻¹)	Log K_f
0	0.	0.	INFINITE	-26.015	-2624.080	-2624.080	INFINITE
100	70.831	277.577	487.750	-21.017	-2628.771	-2604.020	1360.201
200	108.448	339.120	398.728	-11.922	-2632.066	-2577.841	673.263
250	122.124	364.842	389.423	-6.145	-2633.025	-2564.166	535.753
298.15	132.744	387.294	387.294	0.	-2633.619	-2550.845	446.897
300	133.107	388.117	387.297	0.246	-2633.636	-2550.331	444.052
350	141.826	409.316	388.951	7.128	-2633.991	-2536.416	378.539
400	148.717	428.722	392.727	14.398	-2634.158	-2522.464	329.400
450	154.175	446.566	397.731	21.976	-2634.188	-2508.499	291.179
500	158.525	463.044	403.449	29.797	-2634.122	-2494.536	260.602
600	164.854	492.546	415.899	45.988	-2633.822	-2466.645	214.741
700	169.093	518.297	428.727	62.699	-2633.461	-2438.810	181.986
800	172.039	541.080	441.375	79.764	-2633.186	-2411.024	157.424
900	174.155	561.472	453.606	97.079	-2633.115	-2383.260	138.321
1000	175.720	579.906	465.329	114.577	-2654.519	-2353.970	122.959
1100	176.906	596.712	476.521	132.211	-2654.394	-2323.921	110.354
1200	177.826	612.146	487.188	149.949	-2654.244	-2293.884	99.850
1300	178.551	626.410	497.356	167.770	-2654.082	-2263.861	90.963
1400	179.134	639.664	507.053	185.655	-2653.918	-2233.850	83.346
1500	179.608	652.039	516.311	203.593	-2653.762	-2203.851	76.745
1600	179.999	663.644	525.160	221.574	-2653.618	-2173.861	70.969
1700	180.325	674.566	533.631	239.590	-2653.487	-2143.880	65.873
1800	180.600	684.881	541.750	257.637	-2653.370	-2113.907	61.344
1900	180.833	694.652	549.542	275.709	-2653.264	-2083.940	57.292
2000	181.033	703.933	557.032	293.802	-2653.164	-2053.978	53.644
2100	181.206	712.770	564.239	311.915	-2653.063	-2024.022	50.345
2200	181.356	721.203	571.184	330.043	-2652.953	-1994.070	47.345
2300	181.487	729.268	577.883	348.185	-2652.826	-1964.123	44.607
2400	181.602	736.994	584.353	366.340	-2652.671	-1934.183	42.096
2500	181.704	744.410	590.608	384.505	-2652.480	-1904.249	39.787
2600	181.795	751.538	596.661	402.680	-2652.242	-1874.325	37.656
2700	181.876	758.401	602.525	420.864	-2651.947	-1844.410	35.682
2800	181.948	765.016	608.211	439.055	-3239.391	-1812.572	33.814
2900	182.013	771.402	613.729	457.253	-3236.769	-1761.660	31.731
3000	182.072	777.574	619.088	475.458	-3234.068	-1710.841	29.788
3100	182.126	783.545	624.297	493.668	-3231.281	-1660.112	27.973
3200	182.174	789.328	629.365	511.883	-3228.404	-1609.475	26.272
3300	182.218	794.934	634.297	530.102	-3225.433	-1558.929	24.676
3400	182.258	800.375	639.102	548.326	-3222.365	-1508.475	23.175
3500	182.295	805.659	643.786	566.554	-3219.196	-1458.112	21.761
3600	182.329	810.794	648.354	584.785	-3215.926	-1407.841	20.427
3700	182.360	815.791	652.812	603.019	-3212.553	-1357.663	19.167
3800	182.389	820.654	657.165	621.257	-3209.078	-1307.578	17.974
3900	182.416	825.392	661.418	639.497	-3205.501	-1257.584	16.843
4000	182.440	830.011	665.576	657.740	-3201.824	-1207.683	15.771
4100	182.463	834.516	669.642	675.985	-3198.047	-1157.877	14.752
4200	182.485	838.913	673.620	694.233	-3194.173	-1108.163	13.782
4300	182.504	843.207	677.514	712.482	-3190.204	-1058.544	12.859
4400	182.523	847.403	681.327	730.733	-3186.143	-1009.017	11.979
4500	182.540	851.505	685.064	748.987	-3181.989	-959.583	11.139
4600	182.556	855.518	688.726	767.241	-3177.752	-910.243	10.336
4700	182.571	859.444	692.317	785.498	-3173.433	-860.996	9.569
4800	182.586	863.288	695.839	803.756	-3169.036	-811.841	8.835
4900	182.599	867.053	699.294	822.015	-3164.562	-762.779	8.131
5000	182.612	870.742	702.687	840.275	-3160.019	-713.809	7.457
5100	182.623	874.358	706.017	858.537	-3155.409	-664.931	6.810
5200	182.635	877.904	709.289	876.800	-3150.737	-616.143	6.189
5300	182.645	881.383	712.503	895.064	-3146.007	-567.446	5.593
5400	182.655	884.797	715.662	913.329	-3141.224	-518.839	5.019
5500	182.664	888.149	718.768	931.595	-3136.392	-470.321	4.467
5600	182.673	891.440	721.822	949.862	-3131.515	-421.890	3.935
5700	182.682	894.674	724.826	968.130	-3126.599	-373.549	3.423
5800	182.690	897.851	727.782	986.398	-3121.640	-325.294	2.930
5900	182.697	900.974	730.691	1004.668	-3116.657	-277.122	2.453
6000	182.705	904.045	733.555	1022.938	-3111.601	-229.042	1.994

PREVIOUS: September 1979 (1 atm) CURRENT: September 1979 (1 bar)

Aluminum Fluoride $((AlF_3)_2)$ Al_2F_6(g)

ALUMINUM IODIDE $((AlI_3)_2)$ IDEAL GAS M_r = 815.39008

$S°(298.15\ K) = [597.38 \pm 1]\ J\ K^{-1}\ mol^{-1}$

$\Delta_f H°(0\ K) = -481.80 \pm 8.4\ kJ\ mol^{-1}$
$\Delta_f H°(298.15\ K) = -489.53 \pm 8.4\ kJ\ mol^{-1}$

Vibrational Frequencies and Degeneracies

ν, cm^{-1}	ν, cm^{-1}	ν, cm^{-1}	ν, cm^{-1}	ν, cm^{-1}	ν, cm^{-1}
339 (1)	42 (1)	54 (1)	[7] (1)	296 (1)	316 (1)
145 (1)	[20](1)	416 (1)	408 (1)	65 (1)	141 (1)
93 (1)	[220](1)	[100](1)	82 (1)	[50](1)	75 (1)

Ground State Quantum Weight: [1] $\sigma = 4$
Point Group: D_{2h}
Bond Distances: Al-I = 2.449 ± 0.013 Å Al-I (bridge) = 2.634 ± 0.03 Å
Bond Angles: I-Al-I = 115.0 ± 7.4° I (bridge)-Al-I (bridge) = 99.6 ± 4.5°
Product of the Moments of Inertia: $I_A I_B I_C = 5.884591 \times 10^{-109}\ g^3\ cm^6$

Enthalpy of Formation

The vapor pressure over $AlBr_3(\ell)$ has been measured by Fischer, Rahlfs and Benze (1). The vapor consists of both $Al_2I_6(g)$ and $AlI_3(g)$. Equilibrium constants for the reaction $Al_2I_6(g)=2\ AlI_3(g)$ have also been determined by Fischer et al. (1) in the temperature range 614-843 K. From this data equilibrium constants at other temperatures were estimated and the partial pressures of $Al_2I_6(g)$ over $AlI_3(\ell)$ were calculated from the total pressure using these estimated values. The heats of reaction are calculated from the vaporization data using 2nd and 3rd law methods. The results of the calculations are as follows, where the $\Delta_f H°(298.15\ K)$ value is based on the 3rd law value of $\Delta_r H°(298.15\ K)$ and $\Delta_f H°(AlI_3, \ell, 298.15\ K) = -69.2\ kcal\ mol^{-1}$.

Method	T/K	Data Points	$\Delta_r H°(298.15\ K)$, $kcal\ mol^{-1}$ 2nd law	3rd law	Drift $cal\ K^{-1}\ mol^{-1}$	$\Delta_f H°(298.15\ K)$ $kcal\ mol^{-1}$
Mass Loss	513.15-645.35	19[a]	21.50±0.1	21.34±0.0	-0.3±0.1	-117.06
Transport	513.95-578.15	7[b]	20.80±0.1	21.51±0.1	1.3±0.3	-116.89

[a]One point at 493.95 rejected due to statistical test.
[b]One point at 482.95 rejected due to statistical test.

The adopted enthalpy of formation is $\Delta_f H°(Al_2I_6, g, 298.15\ K) = -117.0 \pm 2.0\ kcal\ mol^{-1}$ $(-489.528 \pm 8.4\ kJ\ mol^{-1})$

Heat Capacity and Entropy

The molecular structure, bond distances and angles were obtained from the electron diffraction work of Shen (2). These values are in good agreement with the earlier electron diffraction work by Akishin, Rambidi and Zasorin (3) and Palmer and Elliot (4). The principal moments of inertia are $I_A = 530.1962 \times 10^{-39}$, $I_B = 963.2338 \times 10^{-39}$, and $I_C = 1152.2532 \times 10^{-39}\ g\ cm^2$. Beattie, Gilson and Ozin (5) measured the condensed phase IR and Raman bonds for Al_2I_6 and Beattie and Horder (6) measured the gas phase Raman spectra. Five frequencies (ν_5, ν_6, ν_9, ν_{10}, ν_{15}) were adjusted not only for consistency with the frequencies chosen for $Al_2F_6(g)$, $Al_2Cl_6(g)$ and $Al_2Br_6(g)$ (7), but also for agreement between 2nd and 3rd law $\Delta_r H°(298.15\ K)$ values in the data sets tabulated above.

References

1. W. Fischer, O. Rahlfs and B. Benze, A. Anorg. Allgem. Chem. 205, 1 (1932).
2. Q. Shen, Diss. Abstr. Int. B34, 3735 (1974).
3. P. A. Akishin, N. G. Rambidi and E. Z. Zasorin, Kristallografiya 4, 167 (1959).
4. K. J. Palmer and N. Elliot, J. Amer. Chem. Soc. 60, 1852 (1938).
5. I. R. Beattie, T. Gilson and G. A. Ozin, J. Chem. Soc. (London) A1968, 813.
6. I. R. Beattie and J. R. Horder, J. Chem. Soc. (London) A1969, 2655.
7. JANAF Thermochemical Tables: $Al_2F_6(g)$, $Al_2Cl_6(g)$ and $Al_2Br_6(g)$, 9-30-79.

Aluminum Iodide $((AlI_3)_2)$ $Al_2I_6(g)$

Enthalpy Reference Temperature = T_r = 298.15 K Standard State Pressure = p° = 0.1 MPa

T/K	$C_p°$ ($J\ K^{-1}mol^{-1}$)	S°	$-[G°-H°(T_r)]/T$	$H°-H°(T_r)$ ($kJ\ mol^{-1}$)	$\Delta_f H°$	$\Delta_f G°$	Log K_f
0	0.	0.	INFINITE	-40.941	-481.796	-481.796	INFINITE
100	133.463	427.756	742.263	-31.451	-482.424	-505.504	264.048
200	161.741	530.588	612.847	-16.452	-485.826	-527.310	137.719
250	168.093	567.419	600.192	-8.193	-487.677	-537.470	112.298
298.15	171.953	597.380	597.380	0.	-489.528	-546.893	95.813
300	172.071	598.444	597.384	0.318	-489.602	-547.249	95.285
350	174.684	625.179	599.489	8.991	-491.834	-556.688	83.081
400	176.475	648.629	604.196	17.773	-542.078	-564.151	73.671
450	177.749	669.492	610.313	26.630	-547.927	-566.559	65.764
500	178.685	688.270	617.186	35.542	-674.164	-556.918	58.181
600	179.935	720.969	631.838	53.479	-672.960	-533.585	46.453
700	180.707	748.768	646.606	71.514	-671.923	-510.440	38.090
800	181.215	772.934	660.920	89.611	-671.124	-487.429	31.826
900	181.567	794.299	674.575	107.751	-670.637	-464.500	26.959
1000	181.820	813.443	687.522	125.921	-691.704	-440.085	22.988
1100	182.009	830.782	699.769	144.113	-691.308	-414.943	19.704
1200	182.153	846.625	711.357	162.322	-690.948	-389.835	16.969
1300	182.265	861.210	722.330	180.543	-690.644	-364.755	14.656
1400	182.354	874.720	732.739	198.774	-690.424	-339.696	12.674
1500	182.426	887.304	742.628	217.013	-690.317	-314.648	10.957
1600	182.485	899.079	752.043	235.259	-690.353	-289.603	9.455
1700	182.534	910.144	761.021	253.510	-690.559	-264.551	8.129
1800	182.576	920.579	769.598	271.765	-690.957	-239.481	6.950
1900	182.611	930.451	777.806	290.025	-691.558	-214.384	5.894
2000	182.640	939.818	785.675	308.287	-692.367	-189.249	4.943
2100	182.666	948.730	793.229	326.553	-693.376	-164.069	4.081
2200	182.688	957.228	800.492	344.820	-694.570	-138.838	3.296
2300	182.708	965.349	807.484	363.090	-695.926	-113.547	2.579
2400	182.725	973.126	814.225	381.362	-697.414	-88.195	1.920
2500	182.740	980.585	820.731	399.635	-699.003	-62.778	1.312
2600	182.753	987.753	827.018	417.910	-700.658	-37.296	0.749
2700	182.765	994.650	833.100	436.186	-702.342	-11.750	0.227
2800	182.776	1001.297	838.989	454.463	-1291.825	15.793	-0.295
2900	182.785	1007.711	844.697	472.741	-1291.280	62.484	-1.125
3000	182.794	1013.908	850.235	491.020	-1290.670	109.156	-1.901
3100	182.802	1019.902	855.612	509.299	-1289.971	155.805	-2.625
3200	182.809	1025.706	860.837	527.580	-1289.160	202.430	-3.304
3300	182.815	1031.331	865.919	545.861	-1288.219	249.028	-3.942
3400	182.821	1036.789	870.864	564.143	-1287.134	295.596	-4.541
3500	182.826	1042.088	875.681	582.425	-1285.893	342.129	-5.106
3600	182.831	1047.239	880.375	600.708	-1284.488	388.624	-5.639
3700	182.836	1052.248	884.953	618.992	-1282.913	435.079	-6.142
3800	182.840	1057.124	889.420	637.276	-1281.164	481.488	-6.619
3900	182.844	1061.874	893.781	655.560	-1279.241	527.849	-7.070
4000	182.848	1066.503	898.042	673.844	-1277.143	574.158	-7.498
4100	182.851	1071.018	902.206	692.129	-1274.873	620.412	-7.904
4200	182.854	1075.424	906.278	710.415	-1272.432	666.610	-8.291
4300	182.857	1079.727	910.262	728.700	-1269.826	712.747	-8.658
4400	182.860	1083.931	914.161	746.986	-1267.060	758.822	-9.008
4500	182.862	1088.040	917.980	765.272	-1264.133	804.831	-9.342
4600	182.865	1092.059	921.720	783.558	-1261.060	850.774	-9.661
4700	182.867	1095.992	925.387	801.845	-1257.844	896.650	-9.965
4800	182.869	1099.842	928.981	820.132	-1254.491	942.456	-10.256
4900	182.871	1103.613	932.507	838.419	-1251.006	988.189	-10.534
5000	182.873	1107.307	935.966	856.706	-1247.399	1033.850	-10.801
5100	182.874	1110.928	939.361	874.993	-1243.676	1079.439	-11.056
5200	182.876	1114.480	942.695	893.281	-1239.844	1124.953	-11.300
5300	182.878	1117.963	945.969	911.568	-1235.909	1170.391	-11.535
5400	182.879	1121.381	949.186	929.856	-1231.879	1215.756	-11.760
5500	182.880	1124.737	952.347	948.144	-1227.760	1261.045	-11.976
5600	182.882	1128.032	955.455	966.432	-1223.560	1306.259	-12.184
5700	182.883	1131.269	958.511	984.721	-1219.284	1351.396	-12.384
5800	182.884	1134.450	961.517	1003.009	-1214.935	1396.457	-12.576
5900	182.885	1137.576	964.475	1021.297	-1210.528	1441.443	-12.762
6000	182.886	1140.650	967.386	1039.586	-1206.021	1486.350	-12.940

PREVIOUS: September 1979 (1 atm) CURRENT: September 1979 (1 bar)

Aluminum Iodide $((AlI_3)_2)$ $Al_2I_6(g)$

MAGNESIUM ALUMINUM OXIDE ($MgAl_2O_4$) CRYSTAL M_r = 142.26568

$S°(298.15\ K) = 88.7 \pm 4.2\ J\ K^{-1}\ mol^{-1}$

$T_{fus} = 2048 \pm 20\ K$

$\Delta_f H°(0\ K) = -2283.07 \pm 1.3\ kJ\ mol^{-1}$

$\Delta_f H°(298.15\ K) = -2299.11 \pm 1.3\ kJ\ mol^{-1}$

$\Delta_{fus} H° = [192.464 \pm 21]\ kJ\ mol^{-1}$

Enthalpy of Formation

From differences in heats of solution of MgO(cr), Al_2O_3(cr, α) and $MgAl_2O_4$(cr) in a melt of $2PbO \cdot B_2O_3$ at 970 K, Charlu et al. (1), derived $\Delta_r H°(970\ K) = -5.38 \pm 0.18\ kcal\ mol^{-1}$ for Al_2O_3(cr, α) + MgO(cr) → $MgAl_2O_4$(cr). When combined with auxiliary JANAF data (2), this yields our adopted heat of formation. This value is within 3 kcal mol^{-1} of earlier values determined via a similar method (3, 4). The direct measurement for formation from the elements, $\Delta_f H°(298.15\ K) = -567.1 \pm 0.3\ kcal\ mol^{-1}$, by Zaionchkovskii and Rubalskaya (5) appears to be widely discordant with other values.

Our 2nd and 3rd law analysis of all the equilibrium data is given below. Heats of formation based on the 3rd law analysis support the adopted heat of formation even though agreement between 2nd and 3rd law values is only fair. The two studies by Grojtheim et al. involving reaction 1 require significant activity corrections since the Al(ℓ) is not pure as indicated but is alloyed with Mg (6) or Si (8). Activities were obtained assuming the AlMg(ℓ) solution is ideal as stated by the authors (6) and supported independently (13). Activities used to correct the other study (8) are given by the authors in their paper.

Source	Reaction	T/K	Data Points	$\Delta_r H°$(298.15 kcal mol^{-1}) 2nd law	3rd law	Drift J $K^{-1} mol^{-1}$	$\Delta_f H°$(298.15 K) kcal mol^{-1}
Grjotheim et al. (6)	1	1143-1414	9	140.2±1.7	126.0±2.3	-10.8±1.3	-549.5±2.4
Grjotheim et al. (7)	2	1324-1452	7	307.8±6.6	306.6±1.0	-0.8±4.8	-553.1±1.1
Grjotheim et al. (8)	1	1216-1423	7	133.5±0.7	127.0±0.8	-5.0±0.5	-548.5±1.0
Rao and Dadape (9)	3	1025-1192	7	25.9±3.4	25.3±0.7	-0.5±2.5	-552.6±4.6
Taylor and Schmalzreid (10)	4	803	1	---	-0.4	---	-544.6
Rein and Chipman (11)	4	1873	1	---	-2.2	---	-546.4
Rosen and Muan (12)	4	1673	1	---	-4.0	---	-548.2

Reactions: (1) 4 MgO(cr) + 2 Al(ℓ) → $MgAl_2O_4$(cr) + 3 Mg(g) (3) 4 MgO(cr) + 3 AlCl(g) → $MgAl_2O_4$(cr) + $AlCl_3$(g) + 3 Mg(g)

(2) 8 MgO(cr) + Al_4C_3(cr) → 2 $MgAl_2O_4$(cr) + 3 C(ref) 6 Mg(g) (4) MgO(cr) + Al_2O_3(cr, α) → $MgAl_2O_4$(cr)

Heat Capacity and Entropy

The adopted low temperature $C_p°$ to 300 K is taken from the analysis of Douglas and Beckett (14) and based on the measurements of King (15) (54-296 K). The high temperature heat capacities are calculated from the equation given by Landa and Naumova (16) based on their enthalpy measurements (1500-2200 K) as well as those of Bonnickson (17) (421-1805 K). Enthalpies in the overlapping regions of these two studies agree within ±1%.

Integration of the low temperature heat capacity data leads to $S°(50\ K) = 0.299\ cal\ K^{-1}\ mol^{-1}$ and $S°(298.15\ K) = 19.268\ cal\ K^{-1}\ mol^{-1}$ (14). We assume that $S°(0\ K) = 1.93\ cal\ K^{-1}\ mol^{-1}$ to correct for the ~0.15 mole fraction of inverted cations present in synthetic samples of $MgAl_2O_4$; the method of calculation is given by Navrostsky and Kleppa (18). This configurational entropy term cannot be determined from integration of the $C_p°$ data since the disorder remains at zero Kelvin.

Fusion Data

See the liquid table for details.

Phase Data

This table refers to $MgAl_2O_4$ (spinel) with 15% inversion. Navrotsky and Kleppa (18) have discussed the thermodynamics of spinels in detail and the 15% inversion is typical of most synthetic spinels or of natural spinel after heating above ~1300 K. Natural spinel which has been annealed through geologic time may have the completely normal (ordered) structure. If so its entropy will be 1.93 cal $K^{-1}\ mol^{-1}$ less than in this table and its heat of formation will be ~0.9 kcal mol^{-1} more exothermic than given here based on annealing-heat of solution experiments on natural spinel (18). These measurements found the order-disorder transition to occur in the range 973-1300 K. There have been recent reports of second-order phase transitions in the 723-873 K range detected by x-ray (19) and thermal expansion studies (20) but these are not obvious in the enthalpy data given above.

References

1. T. V. Charlu, R. C. Newton and O. J. Kleppa, Geochim. Cosmochim. Acta 39, 1487 (1975).
2. JANAF Thermochemical Tables: Al_2O_3(cr, α), 12-31-79; MgO(cr), 12-31-74; Al(ℓ), 6-30-79; Mg(g), 9-30-72; Al_4C_3(cr), 9-30-65; C(ref st), 3-31-78; AlCl(g), 9-30-79; $AlCl_3$(g), 9-30-79.
3. A. Navrotsky and O. J. Kleppa, J. Inorg. Nucl. Chem. 30, 479 (1968); A. Navrotsky and O. J. Kleppa, Inorg. Chem. 5, 192 (1966).
4. J. A. Shearer and O. J. Kleppa, J. Inorg. Nucl. Chem. 35, 1073 (1973).
5. Ya. A. Zaionchkovskii and E. V. Rubalskaya, Izv. Akad. Nauk SSSR, Neorgan. Materialy 1, 1376 (1965).
6. K. Grjotheim, O. Herstad and J. M. Toguri, Can. J. Chem. 39, 443 (1961).
7. K. Grjotheim, O. Herstad and K. S. Johannessen, Z. anorg. allgem. Chem. 328, 267 (1964).
8. K. Grjotheim, O. Herstad, R. Skarbo and J. M. Toguri, Can. J. Chem. 41, 739 (1963).
9. D. B. Rao and V. V. Dadape, J. Phys. Chem. 71, 537 (1967).
10. R. W. Taylor and H. Schmalzried, J. Phys. Chem. 68, 2444 (1964).
11. R. H. Rein and J. Chipman, Trans. Metall. Soc. AIME 233, 415 (1965).
12. E. Rosen and A. Muan, J. Amer. Ceramic Soc. 49, 107 (1966).
13. Y. J. Bhatt and S. P. Garg, Metall. Trans. B 7B, 271 (1976).
14. T. B. Douglas and C. W. Beckett, U. S. Nat. Bur. Stand. Report 6484, 1959.
15. E. G. King, J. Phys. Chem. 59, 218 (1955).
16. Ya. A. Landa and I. A. Naumova, Ogneupory 9 (1979).
17. K. R. Bonnickson, J. Phys. Chem. 59, 220 (1955).
18. A. Naurotsky and O. J. Kleppa, J. Inorg. Nucl. Chem. 29, 2701 (1967).
19. R. K. Mishra and G. Thomas, Acta Crystallogr., Sect. A A33, 678 (1977).
20. I. Suzuki and M. Kumazawa, Phys. and Chem. Miner. (Germany) 5, 279 (1980).

Magnesium Aluminum Oxide ($MgAl_2O_4$) $Al_2Mg_1O_4$(cr)

Enthalpy Reference Temperature = T_r = 298.15 K Standard State Pressure = p° = 0.1 MPa

T/K	$C_p°$ (J $K^{-1} mol^{-1}$)	S°	$-[G°-H°(T_r)]/T$	$H°-H°(T_r)$ (kJ mol^{-1})	$\Delta_f H°$	$\Delta_f G°$	Log K_f
0	0.	8.075	INFINITE	-15.411	-2283.075	-2283.075	INFINITE
100	22.669	16.757	164.497	-14.774	-2289.857	-2254.524	1177.642
200	78.069	49.819	98.221	-9.681	-2296.197	-2216.480	578.885
298.15	116.198	88.692	88.692	0.	-2299.108	-2176.621	381.335
300	116.533	89.412	88.695	0.215	-2299.137	-2175.861	378.851
400	137.988	126.244	93.571	13.069	-2299.797	-2134.626	278.754
500	149.829	158.409	103.398	27.506	-2299.410	-2093.365	218.692
600	157.887	186.474	114.956	42.911	-2298.562	-2052.230	178.663
700	164.155	211.299	126.980	59.023	-2297.527	-2011.255	150.082
800	169.473	233.573	138.935	75.710	-2296.463	-1970.433	128.656
900	174.238	253.814	150.592	92.899	-2295.485	-1929.739	111.999
1000	178.657	272.403	161.856	110.546	-2324.450	-1886.914	98.562
1100	182.853	289.629	172.698	128.623	-2323.172	-1843.219	87.527
1200	186.895	305.713	183.120	147.112	-2321.562	-1799.655	78.337
1300	190.828	320.829	193.137	165.999	-2319.622	-1756.239	70.567
1400	194.686	335.112	202.773	185.275	-2444.763	-1709.818	63.794
1500	198.485	348.674	212.051	204.934	-2440.816	-1657.458	57.718
1600	202.238	361.604	220.997	224.970	-2436.542	-1605.372	52.410
1700	205.962	373.976	229.635	245.381	-2431.944	-1553.563	47.735
1800	209.652	385.853	237.986	266.162	-2427.024	-1502.034	43.588
1900	213.325	397.287	246.071	287.311	-2421.783	-1450.787	39.885
2000	216.982	408.322	253.909	308.826	-2416.221	-1399.824	36.560
2100	220.622	418.997	261.518	330.706	-2410.341	-1349.148	33.558
2200	224.254	429.344	268.912	352.950	-2404.143	-1298.758	30.836
2300	227.877	439.393	276.107	375.557	-2397.627	-1248.658	28.358
2400	231.488	449.167	283.115	398.525	-2390.796	-1198.849	26.092
2408.000	231.777	449.938	283.668	400.378	------ CRYSTAL <--> LIQUID -----		
2500	235.095	458.690	289.949	421.854	-2383.648	-1149.330	24.014
2600	238.697	467.981	296.618	445.544	-2376.185	-1100.104	22.101
2700	242.291	477.057	303.133	469.594	-2368.406	-1051.170	20.336
2800	245.885	485.934	309.504	494.002	-2948.114	-1000.595	18.666

PREVIOUS: June 1972 CURRENT: December 1979

Magnesium Aluminum Oxide ($MgAl_2O_4$) $Al_2Mg_1O_4$(cr)

MAGNESIUM ALUMINUM OXIDE ($MgAl_2O_4$) LIQUID M_r = 142.26568

S°(298.15 K) = [168.473] J K^{-1} mol^{-1}

T_{fus} = 2408 ± 20 K

$\Delta_fH°$(298.15 K) = [-2106.527] kJ mol^{-1}

$\Delta_{fus}H°$ = [192.464 ± 21] kJ mol^{-1}

Enthalpy of Formation

The enthalpy of formation of the liquid is calculated from that of the crystal by adding the enthalpy of fusion, $\Delta_{fus}H°$, and the difference in enthalpy, H°(2408 K)-H°(298.15 K), between the crystal and liquid.

Heat Capacity and Entropy

The adopted $C_p°$ for the liquid is estimated to be 52.5 cal K^{-1} mol^{-1} (219.66 J K^{-1} mol^{-1}) above an assumed glass transition at 1750 K. Below T_{gls}, the heat capacity is taken to be that of the crystal (1). The heat capacity is estimated by assuming $C_p°$=7.5 cal K^{-1} g-atom^{-1} as suggested by Kubaschewski and Alcock (2).

S°(298.15 K) is calculated in manner similar to that used for the enthalpy of formation.

Fusion Data

Rankin and Merwin (3) studied the binary system alumina + magnesia and reported T_{fus}=2408±20 K for $MgAl_2O_4$(cr). Wilde and Rees (4) confirm this value in a later paper. The adopted heat of melting is 46±5 kcal mol^{-1}. This is estimated assuming that the entropy of melting per g-atom (2.7 cal K^{-1}) is the same as that of Be_2O_4(cr)(1).

References

1. JANAF Thermochemical Tables: $MgAl_2O_4$(cr), 12-31-79; Be_2O_4(ℓ), 12-31-79.
2. O. Kubaschewski and C. B. Alcock, "Metallurgical Thermochemistry", 5th ed., Pergamon, Oxford, 1979.
3. G. A. Rankin and H. E. Merwin, J. Amer. Chem. Soc. 38, 568 (1916).
4. W. T. Wilde and W. J. Rees, Trans. Brit. Ceramic Soc. 42, 123 (1943).

Magnesium Aluminum Oxide ($MgAl_2O_4$) $Al_2Mg_1O_4$(l)

Enthalpy Reference Temperature = T_r = 298.15 K — Standard State Pressure = p° = 0.1 MPa

T/K	$C_p°$ (J K⁻¹mol⁻¹)	S° (J K⁻¹mol⁻¹)	-[G°-H°(T_r)]/T (J K⁻¹mol⁻¹)	H°-H°(T_r) (kJ mol⁻¹)	$\Delta_fH°$ (kJ mol⁻¹)	$\Delta_fG°$ (kJ mol⁻¹)	Log K_f
0							
100							
200							
298.15	116.198	168.473	168.473	0.	-2106.527	-2007.826	351.763
300	116.533	169.193	168.475	0.215	-2106.556	-2007.214	349.487
400	137.988	206.024	173.351	13.069	-2107.216	-1973.957	257.772
500	149.829	238.190	183.178	27.506	-2106.829	-1940.674	202.741
600	157.887	266.255	194.736	42.911	-2105.981	-1907.517	166.064
700	164.155	291.079	206.760	59.023	-2104.945	-1874.521	139.878
800	169.473	313.354	218.716	75.710	-2103.882	-1841.676	120.249
900	174.238	333.594	230.373	92.899	-2102.904	-1808.961	104.989
1000	178.657	352.183	241.637	110.546	-2131.868	-1774.113	92.670
1100	182.853	369.409	252.479	128.623	-2130.591	-1738.396	82.550
1200	186.895	385.494	262.900	147.112	-2128.981	-1702.811	74.121
1300	190.828	400.609	272.918	165.999	-2127.041	-1667.373	66.996
1400	194.686	414.893	282.553	185.275	-2252.182	-1628.930	60.776
1500	198.485	428.454	291.832	204.934	-2248.234	-1584.548	55.179
1600	202.238	441.384	300.778	224.970	-2243.961	-1540.439	50.290
1700	205.962	453.757	309.415	245.381	-2239.363	-1496.608	45.985
1750.000	207.811	459.754	313.625	255.725	GLASS <--> LIQUID		
1750.000	219.660	459.754	313.625	255.725	TRANSITION		
1800	219.660	465.942	317.771	266.708	-2233.896	-1453.066	42.167
1900	219.660	477.818	325.885	288.674	-2227.838	-1409.852	38.760
2000	219.660	489.085	333.765	310.640	-2221.826	-1366.955	35.701
2100	219.660	499.803	341.419	332.606	-2215.860	-1324.359	32.942
2200	219.660	510.021	348.852	354.572	-2209.940	-1282.044	30.440
2300	219.660	519.785	356.073	376.538	-2204.065	-1239.999	28.161
2400	219.660	529.134	363.091	398.504	-2198.236	-1198.209	26.078
2408.000	219.660	529.865	363.644	400.261	------ CRYSTAL <--> LIQUID -----		
2500	219.660	538.101	369.913	420.470	-2192.452	-1156.661	24.167
2600	219.660	546.716	376.549	442.436	-2186.712	-1115.342	22.408
2700	219.660	555.006	383.006	464.402	-2181.016	-1074.243	20.782
2800	219.660	562.995	389.292	486.368	-2763.167	-1031.419	19.241
2900	219.660	570.703	395.415	508.334	-2755.368	-969.710	17.466
3000	219.660	578.150	401.383	530.300	-2747.614	-908.268	15.814
3100	219.660	585.352	407.202	552.266	-2739.904	-847.082	14.273
3200	219.660	592.326	412.879	574.232	-2732.239	-786.147	12.833
3300	219.660	599.086	418.420	596.198	-2724.618	-725.450	11.483
3400	219.660	605.643	423.830	618.164	-2717.043	-664.982	10.216
3500	219.660	612.010	429.116	640.130	-2709.515	-604.738	9.025
3600	219.660	618.199	434.283	662.096	-2702.035	-544.707	7.903
3700	219.660	624.217	439.335	684.062	-2694.603	-484.885	6.845
3800	219.660	630.075	444.278	706.028	-2687.223	-425.260	5.846
3900	219.660	635.781	449.116	727.994	-2679.895	-365.830	4.900
4000	219.660	641.342	453.852	749.960	-2672.622	-306.588	4.004

PREVIOUS: June 1972 CURRENT: December 1979

Magnesium Aluminum Oxide ($MgAl_2O_4$) $Al_2Mg_1O_4$(l)

MAGNESIUM ALUMINUM OXIDE ($MgAl_2O_4$) CRYSTAL–LIQUID M_r = 142.26568 Magnesium Aluminum Oxide ($MgAl_2O_4$) $Al_2Mg_1O_4$(cr,l)

0 to 2408 K crystal

above 2408 K liquid

Refer to the individual tables for details.

Enthalpy Reference Temperature = T_r = 298.15 K Standard State Pressure = p° = 0.1 MPa

T/K	C_p° (J K⁻¹mol⁻¹)	S° (J K⁻¹mol⁻¹)	$-[G°-H°(T_r)]/T$ (J K⁻¹mol⁻¹)	$H°-H°(T_r)$ (kJ mol⁻¹)	$\Delta_f H°$ (kJ mol⁻¹)	$\Delta_f G°$ (kJ mol⁻¹)	Log K_f
0	0.	8.075	INFINITE	-15.411	-2283.075	-2283.075	INFINITE
100	22.669	16.757	164.497	-14.774	-2289.857	-2254.524	1177.642
200	78.069	49.819	98.221	-9.681	-2296.197	-2216.480	578.885
298.15	116.198	88.692	88.692	0.	-2299.108	-2176.621	381.335
300	116.533	89.412	88.695	0.215	-2299.137	-2175.861	378.851
400	137.988	126.244	93.571	13.069	-2299.797	-2134.626	278.754
500	149.829	158.409	103.398	27.506	-2299.410	-2093.365	218.692
600	157.887	186.474	114.956	42.911	-2298.562	-2052.230	178.663
700	164.155	211.299	126.980	59.023	-2297.527	-2011.255	150.082
800	169.473	233.573	138.935	75.710	-2296.463	-1970.433	128.656
900	174.238	253.814	150.592	92.899	-2295.485	-1929.739	111.999
1000	178.657	272.403	161.856	110.546	-2324.450	-1886.914	98.562
1100	182.853	289.629	172.698	128.623	-2323.172	-1843.219	87.527
1200	186.895	305.713	183.120	147.112	-2321.562	-1799.655	78.337
1300	190.828	320.829	193.137	165.999	-2319.622	-1756.239	70.567
1400	194.686	335.112	202.773	185.275	-2444.763	-1709.818	63.794
1500	198.485	348.674	212.051	204.934	-2440.816	-1657.458	57.718
1600	202.238	361.604	220.997	224.970	-2436.542	-1605.372	52.410
1700	205.962	373.976	229.635	245.381	-2431.944	-1553.563	47.735
1800	209.652	385.853	237.986	266.162	-2427.024	-1502.034	43.588
1900	213.325	397.287	246.071	287.311	-2421.783	-1450.787	39.885
2000	216.982	408.322	253.909	308.826	-2416.221	-1399.824	36.560
2100	220.622	418.997	261.518	330.706	-2410.341	-1349.148	33.558
2200	224.254	429.344	268.912	352.950	-2404.143	-1298.758	30.836
2300	227.877	439.393	276.107	375.557	-2397.627	-1248.658	28.358
2400	231.488	449.167	283.115	398.525	-2390.796	-1198.849	26.092
2408.000	231.777	449.938	283.668	400.378	CRYSTAL <--> LIQUID		
2408.000	219.660	529.865	283.668	592.842	TRANSITION		
2500	219.660	538.101	292.881	613.051	-2192.452	-1156.661	24.167
2600	219.660	546.716	302.479	635.017	-2186.712	-1115.343	22.408
2700	219.660	555.006	311.679	656.983	-2181.016	-1074.244	20.782
2800	219.660	562.995	320.513	678.949	-2763.167	-1031.420	19.241
2900	219.660	570.703	329.008	700.915	-2755.369	-969.711	17.466
3000	219.660	578.150	337.189	722.881	-2747.614	-908.268	15.814
3100	219.660	585.352	345.079	744.847	-2739.904	-847.083	14.273
3200	219.660	592.326	352.697	766.813	-2732.239	-786.148	12.833
3300	219.660	599.086	360.062	788.779	-2724.618	-725.451	11.483
3400	219.660	605.643	367.189	810.745	-2717.043	-664.983	10.216
3500	219.660	612.011	374.093	832.711	-2709.515	-604.739	9.025
3600	219.660	618.199	380.788	854.677	-2702.035	-544.708	7.904
3700	219.660	624.217	387.286	876.643	-2694.603	-484.885	6.845
3800	219.660	630.075	393.599	898.609	-2687.223	-425.260	5.846
3900	219.660	635.781	399.736	920.575	-2679.895	-365.830	4.900
4000	219.660	641.342	405.707	942.541	-2672.622	-306.589	4.004

PREVIOUS: CURRENT: December 1979

Magnesium Aluminum Oxide ($MgAl_2O_4$) $Al_2Mg_1O_4$(cr,l)

ALUMINUM OXIDE (Al_2O) IDEAL GAS M_r = 69.96248

$\Delta_{at}H° = 250 \pm 4$ kJ mol^{-1}
S°(298.15 K) = 252.33 ± 3.3 J K^{-1} mol^{-1}

$\Delta_f H°$(0 K) = −144.5 ± 17 kJ mol^{-1}
$\Delta_f H°$(298.15 K) = −145.2 ± 17 kJ mol^{-1}

Vibrational Frequencies and Degeneracies

ν, cm^{-1}	ν, cm^{-1}	ν, cm^{-1}
471 (1)	[160](2)	994 (1)

Electronic Levels and Quantum Weights

State	ε_i, cm^{-1}	g_i	State	ε_i, cm^{-1}	g_i
X	0	1	B	34331	[3]
A	23286	[1]	C	36233	[1]

Bond Distance: Al–O = 1.73 Å
Bond Angle: Al–O–Al = 180° ± 10°
Rotational Constant: B_0 = 0.104378 cm^{-1}

σ = 2
Point Group: $D_{\infty h}$

Enthalpy of Formation

We adopt the heat of atomization $\Delta_{at}H°$ = 250±4 kcal mol^{-1} and $\Delta_f H°$(298.15 K) = −34.7±4 kcal mol^{-1}. Six reactions analyzed below yield D° values ranging from 247.4 to 252.7 kcal mol^{-1} (1–10). These values show no obvious dependence on the reaction and eight of the eleven fall in the range 250±2 kcal mol^{-1}. The overall uncertainty is about twice this range due to possible error in the data and the thermodynamic functions. The mass-spectrometric methods are uncertain due to use of several different approximations for ionization cross sections; some of these must be significantly in error. Other methods have their own uncertainties. We omit below two studies that were recently revised by Ho and Burns (1) and others, such as (11), that we believe to be seriously biased.

Source	Method	Reaction[a]	T/K	δS cal K^{-1} mol^{-1}	$\Delta_r H°$(298.15 K), kcal mol^{-1} 2nd law	3rd law	$\Delta_f H°$(298.15 K) kcal mol^{-1}	$\Delta_{at}H°$(0 K)
(1) Ho (1980)	Mass spec.	A	2221–2327[b]	4.6±3.2	−240.6±7.3	−251.0±0.6	−33.9±4	249.2
(1) Fu (1976)	Mass spec.	A	2056–2327[b]	0.4±1.3	−251.4±3.0	−252.4±0.8	−35.2±4	250.5
(2) Smoes (1976)	Mass spec.	B	1198–1461	−0.2±0.3	18.8±0.4	19.1±0.2	−34.8±3	250.1
(3) Hildenbrand (1973)	Mass spec.	C	2104–2256	−0.5±1.4	−10.0±3.1	−9.0±0.4	−36.6±5	251.9
(4) Farber (1972)	Mass spec.	D	1943–2093	0.0±5.5	−129±11	−128.8±1.3	−34.0±5	249.3
(5) Thompson (1973)	Mass spec.	E	1438–1580	−0.2±1.9	92.3±2.8	92.7±0.6	−37.4±4	252.7
(6) Rao (1970)	Knudsen eff.	E	1556	--	--	97.4	−32.7±5	248.0
(7) Kulifeev (1969)	Knudsen eff.	E	1443–1576	19±5	123±7	93.6±2.0	−36.5±5	251.8
(8) Herstad (1966)	K.E. + P.C.[c]	E	1585–2129	0.0±1.0	96.5±2.0	96.4±1.1	−33.7±4	249.0
(9) DeMaria (1968)	Mass spec.	F	1313–1511	−1.0±0.5	−5.1±0.7	−3.7±0.3	−32.1±5	247.4
(10) Brewer (1951)	Volatilization	F	1466–1725	5.1±2.7	−0.5±4.3	−8.6±1.4	−37.0±5	252.3

[a]Reactions: A) 2 Al(g) + O(g) → Al_2O(g); B) 1/3 Al_2O_3(cr) + 1/3 Al(ℓ) + Al(g) = Al_2O(g); D) Al(g) + AlO(g) = Al_2O(g); E) 1/3 Al_2O_3(cr) + 4/3 Al(ℓ) = Al_2O(g); F) 1/3 Al_2O_3(cr) + 4/3 Al(g) = Al_2O(g).
[b]Temperature scale adjusted. [c]Knudsen effusion plus pressure compensation.

Heat Capacity and Entropy

We adopt a linear AlOAl structure with a low bending frequency. The long-standing controversy over these properties has been greatly resolved by Douglas (12) who studied electronic absorption and emission spectra of Al_2O in Kr and Ar matrices. Vibrational progressions in the spectra (12) gave 471±15, ~482 and ~452 cm^{-1} for ν_1 of the X, A and C states, respectively, plus ~133 and ~170 cm^{-1} for ν_2 of the A and C states. This supplements previous evidence of linearity, i.e. absence of ν_1 in IR spectra (13–18) absence of deflection by an inhomogenous electric field (19) and theoretical predictions (20). IR spectra (13–18) also failed to show ν_2, which has been estimated as 102 or 120±30 cm^{-1} from SCF calculations (20) or force constants (17). The evidence suggests that the X state has $\nu_2 \leq 170$ cm^{-1}, implying that the equilibrium bond angle (21) is much closer to 180° than the uncorrected angle of 141° derived from electron diffraction. Likewise, the angle of 140°–150° (15) derived from the isotopic shift of ν_3 becomes nearly 180° (12) if corrected for anharmonicity.

We conclude that the bond angle is 180°±10° and adopt the bond distance from electron diffraction (21). Frequencies ν_1 and ν_3 are from (12) and (13–18), while ν_2 = 160 cm^{-1} is chosen to reproduce the entropy derived from equilibria (see δS in the table above). Electronic levels are from (12); states A and C probably are bent. The band found (18) at ~24400 cm^{-1} is from Al_2(12). Values of Gibbs energy function from Gurvich et al. (22) are 1.0 cal K^{-1} mol^{-1} larger than ours at T>1400 K.

References

1. P. Ho and R. P. Burns, High Temp. Sci. 12, 31 (1980); C. M. Fu and R. P. Burns, ibid., 8, 353 (1976).
2. S. Smoes and J. Drowart, personal communication from J. Drowart, Free Univ. Brussels, Brussels, Belgium, Sept. 9, 1976.
3. D. L. Hildenbrand, Chem. Phys. Lett. 20, 127 (1973).
4. M. Farber, R. D. Srivastava and O. M. Uy, J. Chem. Soc., Faraday Trans. I 68, 249 (1972).
5. K. R. Thompson, High Temp. Sci. 5, 62 (1973).
6. D. B. Rao and K. Motzfeldt, Acta Chem. Scand. 24, 2796 (1970).
7. V. K. Kulifeev and G. A. Ukhlinov, Izv. Vyssh. Ucheb. Zaved., Tsvet. Met. 12(2), 72 (1969).
8. O. Herstad and K. Motzfeldt, Rev. Int. Hautes Temp. Refract. 3, 291 (1966).
9. G. A. DeMaria, K. A. Gingerich and V. Piacente, J. Chem. Phys. 49, 4705 (1968).
10. L. Brewer and A. W. Searcy, J. Amer. Chem. Soc. 73, 5308 (1951).
11. A. D. Chervonnyi, V. A. Piven', O. E. Kashireninov and G. B. Manelis, High Temp. Sci. 9, 99 (1977).
12. M. A. Douglas, Thesis, Rice Univ., Houston, personal communication, April 15, 1982.
13. D. A. Lynch, M. J. Zehe, and K. D. Carlson, J. Phys. Chem. 78, 236 (1974); 75, 1963 (1971).
14. C. P. Marino and D. White, J. Phys. Chem. 78, 2929 (1973).
15. M. J. Linevsky, D. White and D. E. Mann, J. Chem. Phys. 41, 542 (1964).
16. A. A. Mal'tsev and V. F. Shevelkov, Teplofiz. Vysok. Temp., Akad. Nauk SSSR 2, 650 (1964).
17. A. Snelson, J. Phys. Chem. 74, 2574 (1970).
18. P. A. Finn, D. M. Gruen and D. L. Page, Adv. Chem. Ser. 158, 30 (1976).
19. A. Buchler, J. L. Stauffer, W. Klemperer and L. Wharton, J. Chem. Phys. 39, 2299 (1963).
20. E. L. Wagner, Theor. Chim. Acta 32, 295 (1974).
21. A. A. Ivanov, S. M. Tomalchev et al., J. Struct. Chem. 14, 854 (1973); High Temp. Sci. 5, 385 (1973).
22. L. V. Gurvich, I. V. Veits et al., "Thermodynamic Properties of Individual Substances," 3rd ed., Vol. III, Nauka, Moscow, 1981.

Aluminum Oxide (Al_2O) Al_2O_1(g)

Enthalpy Reference Temperature = T_r = 298.15 K; Standard State Pressure = p° = 0.1 MPa

T/K	$C_p°$ (J K^{-1}mol^{-1})	S° (J K^{-1}mol^{-1})	−[G°−H°(T_r)]/T (J K^{-1}mol^{-1})	H°−H°(T_r) (kJ mol^{-1})	$\Delta_f H°$ (kJ mol^{-1})	$\Delta_f G°$ (kJ mol^{-1})	Log K_f
0	0.	0.	INFINITE	-12.710	-144.474	-144.474	INFINITE
100	40.424	201.813	295.494	-9.368	-143.552	-153.671	80.269
200	47.795	232.403	256.967	-4.913	-144.156	-163.630	42.736
250	50.162	243.332	253.179	-2.462	-144.662	-168.442	35.194
298.15	52.032	252.332	252.332	0.	-145.185	-172.975	30.304
300	52.098	252.654	252.333	0.096	-145.205	-173.147	30.148
350	53.696	260.809	252.973	2.742	-145.767	-177.760	26.529
400	55.010	268.068	254.415	5.461	-146.342	-182.291	23.805
450	56.087	274.611	256.301	8.239	-146.929	-186.750	21.677
500	56.971	280.568	258.435	11.067	-147.532	-191.142	19.968
600	58.297	291.080	263.023	16.834	-148.815	-199.746	17.389
700	59.213	300.139	267.693	22.713	-150.259	-208.123	15.530
800	59.863	308.091	272.256	28.668	-151.933	-216.278	14.121
900	60.337	315.171	276.638	34.679	-153.907	-224.206	13.013
1000	60.691	321.547	280.815	40.732	-177.417	-230.375	12.034
1100	60.961	327.345	284.786	46.815	-179.438	-235.573	11.186
1200	61.172	332.658	288.557	52.922	-181.456	-240.587	10.472
1300	61.339	337.562	292.140	59.048	-183.472	-245.433	9.862
1400	61.473	342.112	295.549	65.188	-185.488	-250.124	9.332
1500	61.583	346.357	298.796	71.341	-187.505	-254.670	8.868
1600	61.674	350.335	301.895	77.504	-189.526	-259.082	8.458
1700	61.750	354.076	304.855	83.676	-191.551	-263.368	8.092
1800	61.814	357.608	307.689	89.854	-193.581	-267.534	7.764
1900	61.869	360.951	310.405	96.038	-195.616	-271.587	7.466
2000	61.915	364.126	313.012	102.227	-197.658	-275.532	7.196
2100	61.956	367.148	315.519	108.421	-199.708	-279.376	6.949
2200	61.991	370.031	317.931	114.618	-201.764	-283.121	6.722
2300	62.022	372.787	320.257	120.819	-203.829	-286.773	6.513
2400	62.050	375.427	322.501	127.023	-205.902	-290.335	6.319
2500	62.075	377.961	324.669	133.229	-207.984	-293.810	6.139
2600	62.097	380.396	326.766	139.438	-210.073	-297.202	5.971
2700	62.118	382.740	328.796	145.649	-212.171	-300.513	5.814
2800	62.138	384.999	330.763	151.861	-802.079	-301.811	5.630
2900	62.156	387.180	332.671	158.076	-802.003	-283.946	5.114
3000	62.174	389.288	334.523	164.293	-801.934	-266.083	4.633
3100	62.192	391.327	336.323	170.511	-801.873	-248.222	4.183
3200	62.210	393.301	338.073	176.731	-801.820	-230.363	3.760
3300	62.229	395.216	339.776	182.953	-801.775	-212.506	3.364
3400	62.248	397.074	341.434	189.177	-801.737	-194.650	2.990
3500	62.269	398.879	343.049	195.403	-801.707	-176.795	2.639
3600	62.291	400.633	344.625	201.631	-801.684	-158.941	2.306
3700	62.315	402.340	346.162	207.861	-801.669	-141.087	1.992
3800	62.341	404.002	347.662	214.094	-801.661	-123.233	1.694
3900	62.369	405.622	349.127	220.329	-801.661	-105.380	1.411
4000	62.399	407.202	350.560	226.567	-801.669	-87.526	1.143
4100	62.433	408.743	351.960	232.809	-801.685	-69.673	0.888
4200	62.470	410.248	353.330	239.054	-801.710	-51.818	0.644
4300	62.509	411.718	354.671	245.303	-801.744	-33.963	0.413
4400	62.553	413.156	355.984	251.556	-801.788	-16.108	0.191
4500	62.600	414.562	357.270	257.814	-801.839	1.749	-0.020
4600	62.651	415.938	358.530	264.076	-801.904	19.607	-0.223
4700	62.705	417.286	359.766	270.344	-801.983	37.467	-0.416
4800	62.764	418.607	360.978	276.617	-802.075	55.328	-0.602
4900	62.827	419.902	362.168	282.897	-802.180	73.192	-0.780
5000	62.895	421.172	363.335	289.183	-802.303	91.058	-0.951
5100	62.967	422.418	364.481	295.476	-802.445	108.927	-1.116
5200	63.044	423.641	365.607	301.777	-802.606	126.798	-1.274
5300	63.125	424.843	366.714	308.085	-802.789	144.673	-1.426
5400	63.211	426.024	367.801	314.402	-802.995	162.551	-1.572
5500	63.301	427.184	368.870	320.727	-803.227	180.434	-1.714
5600	63.396	428.326	369.922	327.062	-803.487	198.321	-1.850
5700	63.496	429.449	370.956	333.407	-803.776	216.213	-1.981
5800	63.601	430.554	371.974	339.761	-804.092	234.110	-2.108
5900	63.710	431.642	372.977	346.127	-804.448	252.012	-2.231
6000	63.823	432.714	373.963	352.503	-804.796	269.918	-2.350

PREVIOUS: December 1979 (1 atm) CURRENT: December 1979 (1 bar)

Aluminum Oxide (Al_2O) Al_2O_1(g)

ALUMINUM OXIDE, ION (Al_2O^+) IDEAL GAS M_r = 69.961931

$\Delta_f H°(0\ K)$ = 646.7 ± 25 kJ mol^{-1}

$S°(298.15\ K)$ = [260.6 ± 8] J K^{-1} mol^{-1} $\Delta_f H°(298.15\ K)$ = [652.436] kJ mol^{-1}

Vibrational Frequencies and Degeneracies

ν, cm^{-1}
[450] (1)
[140] (2)
[900] (1)

Ground State Quantum Weight: [2] σ = 2

Point Group: [$D_{\infty h}$]

Bond Distance: Al-O = [1.73] Å

Bond Angle: Al-O-Al = [180]°

Rotational Constant: B_0 = [0.104379] cm^{-1}

Enthalpy of Formation

$\Delta_f H°(Al_2O^+$, g, 0 K) is calculated from that of Al_2O (1) using Hildenbrand's appearance potential (2) of 8.20±0.15 eV (189.1±3.5 kcal mol^{-1}). We assume that the appearance potential is identical with the ionization potential for $Al_2O(g) \rightarrow Al_2O^+(g) + e^-(g)$. Other values reported for the appearance potential include 7.7±0.5 (3, 4), 7.9±0.3 (5), 8.0±0.5 (6), 8.5±0.5 (7) and 8.5±0.2 eV (8). These values are consistent with, but less precise than, the adopted value.

$\Delta_f H°(298.15\ K)$ is obtained from $\Delta_f H°(Al_2O)$ by using IP(Al_2O) with JANAF (1) enthalpies, H°(0 K)-H°(298.15 K), for $Al_2O(g)$, $Al_2O^+(g)$ and e^- (g). $\Delta_r H°(Al_2O \rightarrow Al_2O^+ + e^-$, 298.15 K) differs from a room-temperature threshold energy due to inclusion of these enthalpies and to threshold effects discussed by Rosenstock et al. (9). $\Delta_f H°(298.15\ K)$ should be changed by -1.481 kcal mol^{-1} if it is to be used in the ion convention that excludes the enthalpy of the electron. MNDO calculations (10) differ by only -9.9 kcal mol^{-1} from our value of [$\Delta_f H°(Al_2O^+$, 298.15 K) - $\Delta_f H°(Al_2O$, 298.15 K)] even though the MNDO values of $\Delta_f H°(298.15\ K)$ differ by -47 and -37 kcal mol^{-1}.

Heat Capacity and Entropy

Al_2O^+ is assumed to be linear as predicted by MNDO calculations (10) and by the correlation of Walsh (11). We assume that the bond distance is the same as that in Al_2O (1) since the change in bond length on ionization was predicted to be only 0.004 Å (10). Vibrational frequencies are estimated to be somewhat lower than those of Al_2O since the ion has one less bonding electron. The electronic ground state is doublet due to the odd number of electrons.

References

1. JANAF Thermochemical Tables: $Al_2O(g)$, 12-31-79; $e^-(g)$, 3-31-77.
2. D. L. Hildenbrand, Chem. Phys. Lett. 20, 127 (1973).
3. K. R. Thompson, High Temp. Sci. 5, 62 (1973).
4. R. P. Burns et al., J. Chem. Phys. 44, 3307 (1966); 32, 1366 (1960).
5. G. DeMaria, K. A. Gingerich and V. Piacente, J. Chem. Phys. 49, 4705 (1968).
6. S. Smoes, J. Drowart and C. E. Meyers, J. Chem. Thermodynamics 8, 225 (1976).
7. P. E. Blackburn, A. Buchler and J. L. Stauffer, J. Phys. Chem. 70, 2469 (1966).
8. P. Ho and R. P. Burns, High Temp. Sci. 12, 31 (1980).
9. H. M. Rosenstock, K. Draxl et al., J. Phys. Chem. Ref. Data 6, Suppl. 1 (1977).
10. L. P. Davis et al., J. Comput. Chem. 2, 433 (1981).
11. A. D. Walsh, J. Chem. Soc. 1953, 2266 (1953).

Aluminum Oxide, Ion (Al_2O^+) $Al_2O_1^+(g)$

Enthalpy Reference Temperature = T_r = 298.15 K Standard State Pressure = p° = 0.1 MPa

T/K	$C_p°$ (J $K^{-1}mol^{-1}$)	S° (J $K^{-1}mol^{-1}$)	-[G°-H°(T_r)]/T (J $K^{-1}mol^{-1}$)	H°-H°(T_r) (kJ mol^{-1})	$\Delta_f H°$ (kJ mol^{-1})	$\Delta_f G°$ (kJ mol^{-1})	Log K_f
0	0.	0.	INFINITE	-12.981	646.679		
100	41.629	209.128	304.597	-9.547			
200	48.639	240.389	265.373	-4.997			
250	51.019	251.507	261.521	-2.503			
298.15	52.897	260.660	260.660	0.	652.436	615.909	-107.905
300	52.962	260.987	260.661	0.098	652.456	615.682	-107.200
350	54.537	269.274	261.312	2.787	652.976	609.512	-90.965
400	55.804	276.642	262.776	5.546	653.481	603.268	-78.779
450	56.821	283.276	264.691	8.363	653.972	596.961	-69.293
500	57.641	289.306	266.856	11.225	654.444	590.602	-61.700
600	58.846	299.930	271.507	17.054	655.300	577.750	-50.298
700	59.661	309.066	276.235	22.982	655.984	564.769	-42.144
800	60.231	317.072	280.850	28.978	656.429	551.704	-36.023
900	60.642	324.191	285.277	35.023	656.567	538.602	-31.260
1000	60.947	330.597	289.494	41.103	635.164	527.023	-27.529
1100	61.179	336.417	293.499	47.210	635.245	516.205	-24.513
1200	61.358	341.748	297.301	53.337	635.326	505.379	-21.999
1300	61.500	346.665	300.912	59.480	635.406	494.547	-19.871
1400	61.615	351.227	304.345	65.636	635.484	483.709	-18.047
1500	61.708	355.482	307.614	71.802	635.558	472.865	-16.467
1600	61.784	359.467	310.731	77.977	635.628	462.017	-15.083
1700	61.848	363.214	313.709	84.159	635.692	451.164	-13.863
1800	61.902	366.751	316.559	90.346	635.750	440.308	-12.777
1900	61.948	370.099	319.289	96.539	635.801	429.448	-11.806
2000	61.987	373.278	321.910	102.736	635.845	418.586	-10.932
2100	62.021	376.303	324.429	108.936	635.882	407.723	-10.142
2200	62.051	379.189	326.853	115.140	635.910	396.857	-9.423
2300	62.077	381.948	329.189	121.346	635.929	385.991	-8.766
2400	62.099	384.590	331.442	127.555	635.940	375.123	-8.164
2500	62.119	387.126	333.619	133.766	635.942	364.256	-7.611
2600	62.137	389.562	335.724	139.979	635.935	353.389	-7.100
2700	62.153	391.908	337.762	146.193	635.919	342.522	-6.626
2800	62.167	394.168	339.736	152.409	48.093	333.592	-6.223
2900	62.180	396.350	341.651	158.627	50.251	343.749	-6.192
3000	62.192	398.458	343.510	164.845	52.401	353.834	-6.161
3100	62.202	400.498	345.315	171.065	54.541	363.846	-6.131
3200	62.212	402.473	347.071	177.286	56.673	373.789	-6.101
3300	62.220	404.387	348.779	183.507	58.797	383.666	-6.073
3400	62.228	406.245	350.442	189.730	60.912	393.479	-6.045
3500	62.236	408.049	352.062	195.953	63.018	403.230	-6.018
3600	62.242	409.802	353.642	202.177	65.115	412.921	-5.991
3700	62.248	411.508	355.183	208.401	67.204	422.553	-5.965
3800	62.254	413.168	356.687	214.626	69.282	432.129	-5.940
3900	62.259	414.785	358.156	220.852	71.351	441.651	-5.915
4000	62.264	416.361	359.592	227.078	73.410	451.120	-5.891
4100	62.269	417.899	360.995	233.305	75.458	460.537	-5.867
4200	62.273	419.399	362.368	239.532	77.493	469.905	-5.844
4300	62.277	420.865	363.711	245.760	79.516	479.224	-5.821
4400	62.280	422.296	365.026	251.987	81.525	488.496	-5.799
4500	62.284	423.696	366.315	258.216	83.524	497.722	-5.777
4600	62.287	425.065	367.577	264.444	85.503	506.905	-5.756
4700	62.290	426.405	368.815	270.673	87.465	516.045	-5.735
4800	62.293	427.716	370.028	276.902	89.407	525.143	-5.715
4900	62.295	429.000	371.218	283.132	91.330	534.201	-5.695
5000	62.298	430.259	372.387	289.361	93.229	543.220	-5.675
5100	62.300	431.493	373.534	295.591	95.103	552.201	-5.656
5200	62.302	432.702	374.660	301.821	96.950	561.145	-5.637
5300	62.304	433.889	375.766	308.052	98.768	570.055	-5.618
5400	62.306	435.054	376.853	314.282	100.554	578.930	-5.600
5500	62.308	436.197	377.922	320.513	102.306	587.772	-5.582
5600	62.310	437.320	378.973	326.744	104.021	596.584	-5.565
5700	62.312	438.423	380.006	332.975	105.696	605.365	-5.548
5800	62.313	439.506	381.023	339.206	107.335	614.116	-5.531
5900	62.315	440.572	382.023	345.438	108.924	622.839	-5.514
6000	62.316	441.619	383.008	351.669	110.510	631.533	-5.498

PREVIOUS: December 1979 (1 atm) CURRENT: December 1979 (1 bar)

ALUMINUM OXIDE ((AlO)$_2$) IDEAL GAS M_r = 85.96188

$\Delta_{at}H°$ = 1540 ± 32 kJ mol^{-1}
S°(298.15 K) = [280.98 ± 12] J K^{-1} mol^{-1}

$\Delta_f H°$(0 K) = -391.3 ± 32 kJ mol^{-1}
$\Delta_f H°$(298.15 K) = -394.6 ± 32 kJ mol^{-1}

Vibrational Frequencies and Degeneracies

ν, cm^{-1}	ν, cm^{-1}	ν, cm^{-1}
[600](1)	[350](1)	496 (1)
[400](1)	[200](1)	[650](1)

Ground State Quantum Weight: [1] σ = [4]
Bond Distance: Al-O = [1.80] Å Point Group: [D_{2h}]
Bond Angles: O-Al-O = [90]° Al-O-Al = [90]°
Product of the Moments of Inertia: $I_A I_B I_C$ = [2.889675 x 10^{-114}] g^3 cm^6

Enthalpy of Formation

We adopt $\Delta_f H°$(298.15 K) = -94.3±8 and $\Delta_{at}H°$ = 368±8 kcal mol^{-1} based on the mass-spectrometric studies summarized below. Data of Chervonnyi (2) and Drowart (4) are analyzed using the isomolecular reaction B instead of the atomization reaction A, which is much affected by probable bias in calibrations and relative ionization cross sections. Reaction A would give discrepancies of -16 to +5 kcal mol^{-1} in $\Delta_{at}H°$. Use of reaction B emphasizes the consistency among the studies. Only the result of Farber et al. (3) appears to be discrepant. JANAF auxiliary data (5) have been used to calculate the $\Delta_f H°$ values; their uncertainties include a large contribution from possible bias in the Gibbs-energy function. Gurvich et al. (6) selected $\Delta_f H°$(298.15 K) = -96.96 and $\Delta_{at}H°$ = 370.5±12 kcal mol^{-1} based on somewhat different functions and without benefit of the data of Ho (1).

Source	Method	Reaction[a]	T/K	δS cal K^{-1} mol^{-1}	Δ H°(298.15 K), kcal mol^{-1} 2nd law	3rd law	$\Delta_f H°$(298.15 K) kcal mol^{-1}	$\Delta_{at}H°$
(1) Ho (1980)	Mass spec.	A	2221-2327[c]	-31±1	-442±3	-371.3±2.8	-94.6±8	368.2
(1) Fu (1976)	Mass spec.	A	2327[c]	--	--	-368.4	-91.7±8	365.4[d]
(2) Chervonnyi (1977)	Eff. mass spec.[e]	B[b]	2300-2600	-0.3±3	1±6	2.1±1.0	-95.4±10	369.0
(3) Farber (1972)	Eff. mass spec.	B	1943-2093	-7.1±1	-22±2	-7.3±0.8	104.8±10	378.5
(4) Drowart (1960)	Eff. mass spec.[e]	B[b]	2464-2594	31±23	79±57	1.3±5.4	-96.2±13	369.9[d]
(4) Drowart (1960)	Eff. mass spec.[f]	B[b]	2281-2466	-32±17	-72±39	4.1±6.9	-93.4±13	367.1[d]

[a]Reactions: A) 2 Al(g) + 2 O(g) = Al_2O_2(g); B) Al_2O(g) + AlO(g) = Al(g) + Al_2O_2(g).
[b]Reaction B used instead of A to minimize bias from calibrations and ionization cross sections.
[c]Temperature scale adjusted to T_{fus} = 2327 K instead of 2318 K for α-Al_2O_3.
[d]Ho and Burn's recalculations (1) ofthe results of Fu (1) and Drowart (4) yield $\Delta_{at}H°$ = 367.0 (1) and ~364.6 (4) kcal mol^{-1} when converted to our new Al_2O_2 functions.
[e]Tungsten cell.
[f]Molybdenum cell.

Heat Capacity and Entropy

We adopt a square-planar structure of D_{2h} symmetry. The Al-O bond distance is taken as 1.8 Å, i.e., 0.1 Å longer than the predictions of ab initio (7) and MNDO (8) calculations and 0.07 Å longer than observed in Al_2O (5). Similar predictions for Al_2O (7, 9) have a bias of -0.09 to -0.12 Å. The principal moments of inertia are I_A = 8.6080 x 10^{-39}, I_B = 14.5167 x 10^{-39} and I_C = 23.1247 x 10^{-39} g cm^2.

Vibrational frequencies are estimated to be lower than values from MNDO calculations (8) but higher than Al_2F_2 values from ionic model calculations (10). Matrix IR spectra gave a band near 496 cm^{-1} that was assigned to D_{2h} Al_2O_2 in three studies (11) but to cyclic AlO_2 in another (12). A band at 686 cm^{-1}, once assigned to Al_2O_2 (13), was reassigned to $Al(O_2)_2$ by Finn (11) who also attributed bands at 490 and ~1215 cm^{-1} to a C_{2v} form of Al_2O_2.

Evidence is contradictory on the most stable geometry. The square-planar form is 15 kcal mol^{-1} more stable than linear AlOAlO according to ab initio calculations (7) but 27 kcal mol^{-1} less stable according to MNDO calculations (8-9). Also possible is rhombic D_{2h} AlO_2Al with an O-O bond similar to that predicted by ab initio methods for LiO_2Li (14). This geometry was not considered in the ab initio study (7) and showed no minimum in the MNDO study (8). Better calculations are needed to resolve the question. Thermochemical and spectral evidence is inconclusive (15). Uncertainties in the structure and vibrational frequencies cause an uncertainty of ≤3 cal K^{-1} mol^{-1} in S° and -(G°-H°)/T at 2000 K. Estimated values for the rhombic peroxide (6) are ~1 cal K^{-1} mol^{-1} lower at 2000 K but this difference could be plausibly reversed by use of lower frequencies.

References

1. P. Ho and R. P. Burns, High Temp. Sci. 12, 31 (1980); C. M. Fu and R. P. Burns, ibid., 8, 353 (1976).
2. A. D. Chervonnyi et al., High Temp. Sci. 9, 99 (1977).
3. M. Farber, R. D. Srivastava and O. M. Uy, J. Chem. Soc., Faraday Trans. I 68, 249 (1972).
4. J. Drowart, G. DeMaria, R. P. Burns and M. G. Inghram, J. Chem. Phys. 32, 1366 (1960).
5. JANAF Thermochemical Tables: Al(g), 6-30-79; AlO(g), Al_2O(g), and AlO_2(g), 12-31-79; O(g), 3-31-77.
6. L. V. Gurvich, I. V. Veits et al., "Thermodynamic Properties of Individual Substances," 3rd ed., Vol. III, Nauka, Moscow, 1981.
7. R. L. DeKock and M. R. Barbachyn, J. Inorg. Nucl. Chem. 43, 2645 (1981).
8. A. G. Turner, personal communication, USAF Academy, Colorado, April, 1982.
9. L. P. Davis et al., J. Comput. Chem. 2, 433 (1981).
10. V. G. Solomonik, Russ. J. Phys. Chem. 53, 313 (1979).
11. C. P. Marino and D. White, J. Phys. Chem. 77, 2929 (1973); P. A. Finn, D. M. Gruen and D. L. Page, Adv. Chem. Ser. 158, 30 (1976); S. M. Sonchik, Ph.D. Thesis, Case Western Reserve Univ., 1980.
12. S. B. Osin et al., Vestn. Mosk. Univ., Ser. 2: Khim 19 (2), 229 (1978).
13. D. A. Lynch, Case Western Reserve Univ., Rept. NASA-CR-133100 (N73-26102), 138 pp., 1972.
14. J. H. Yates and R. M. Pitzer, J. Chem. Phys. 66, 3592 (1977).
15. A. N. Syverud, J. Inorg. Nucl. Chem. 38, 2163 (1976).

Aluminum Oxide ((AlO)$_2$) Al_2O_2(g)

Enthalpy Reference Temperature = T_r = 298.15 K Standard State Pressure = p° = 0.1 MPa

T/K	$C_p°$ (J K^{-1}mol^{-1})	S°	-[G°-H°(T_r)]/T	H°-H°(T_r) (kJ mol^{-1})	$\Delta_f H°$	$\Delta_f G°$	Log K_f
0	0.	0.	INFINITE	-14.518	-391.307	-391.307	INFINITE
100	40.385	223.077	333.078	-11.000	-391.661	-395.241	206.453
200	56.664	256.193	286.829	-6.127	-393.303	-398.187	103.996
250	62.784	269.527	282.062	-3.134	-393.995	-399.325	83.434
298.15	67.164	280.980	280.980	0.	-394.551	-400.300	70.131
300	67.307	281.395	280.981	0.124	-394.571	-400.336	69.705
350	70.614	292.032	281.813	3.577	-395.065	-401.257	59.884
400	73.052	301.629	283.700	7.171	-395.511	-402.110	52.510
450	74.877	310.343	286.184	10.872	-395.935	-402.910	46.769
500	76.268	318.308	289.004	14.652	-396.355	-403.662	42.170
600	78.197	332.397	295.093	22.383	-397.255	-405.041	35.262
700	79.429	344.551	301.310	30.268	-398.319	-406.257	30.315
800	80.259	355.215	307.395	38.255	-399.630	-407.305	26.594
900	80.842	364.703	313.246	46.312	-401.261	-408.170	23.690
1000	81.266	373.244	318.825	54.418	-424.448	-407.314	21.276
1100	81.583	381.005	324.131	62.562	-426.164	-405.518	19.256
1200	81.827	388.114	329.171	70.733	-427.892	-403.565	17.567
1300	82.018	394.672	333.960	78.925	-429.632	-401.467	16.131
1400	82.171	400.756	338.517	87.135	-431.386	-399.235	14.896
1500	82.295	406.430	342.857	95.359	-433.154	-396.877	13.820
1600	82.396	411.744	346.998	103.593	-434.937	-394.400	12.876
1700	82.481	416.742	350.955	111.837	-436.735	-391.811	12.039
1800	82.552	421.459	354.742	120.089	-438.549	-389.117	11.292
1900	82.612	425.924	358.372	128.347	-440.380	-386.321	10.621
2000	82.663	430.162	361.857	136.611	-442.229	-383.428	10.014
2100	82.708	434.197	365.206	144.880	-444.096	-380.442	9.463
2200	82.746	438.045	368.430	153.152	-445.982	-377.366	8.960
2300	82.780	441.724	371.538	161.429	-447.886	-374.205	8.498
2400	82.810	445.248	374.536	169.708	-449.810	-370.961	8.074
2500	82.836	448.629	377.432	177.991	-451.752	-367.635	7.681
2600	82.859	451.878	380.234	186.275	-453.714	-364.232	7.318
2700	82.879	455.006	382.945	194.562	-455.694	-360.752	6.979
2800	82.898	458.020	385.573	202.851	-1045.495	-355.264	6.628
2900	82.915	460.929	388.122	211.142	-1045.321	-330.617	5.955
3000	82.930	463.741	390.596	219.434	-1045.166	-305.974	5.327
3100	82.943	466.460	392.999	227.728	-1045.027	-281.336	4.740
3200	82.955	469.094	395.336	236.023	-1044.907	-256.703	4.190
3300	82.967	471.646	397.610	244.319	-1044.803	-232.074	3.673
3400	82.977	474.123	399.825	252.616	-1044.715	-207.447	3.187
3500	82.986	476.529	401.982	260.914	-1044.645	-182.823	2.728
3600	82.995	478.867	404.085	269.213	-1044.590	-158.200	2.295
3700	83.003	481.141	406.137	277.513	-1044.552	-133.579	1.886
3800	83.010	483.354	408.140	285.814	-1044.531	-108.957	1.498
3900	83.017	485.511	410.097	294.115	-1044.526	-84.337	1.130
4000	83.023	487.613	412.008	302.417	-1044.538	-59.717	0.780
4100	83.029	489.663	413.877	310.720	-1044.567	-35.096	0.447
4200	83.034	491.664	415.706	319.023	-1044.615	-10.474	0.130
4300	83.039	493.618	417.495	327.327	-1044.681	14.149	-0.172
4400	83.044	495.527	419.247	335.631	-1044.767	38.774	-0.460
4500	83.049	497.393	420.963	343.935	-1044.869	63.400	-0.736
4600	83.053	499.218	422.644	352.240	-1044.998	88.029	-1.000
4700	83.056	501.004	424.293	360.546	-1045.150	112.662	-1.252
4800	83.060	502.753	425.909	368.852	-1045.327	137.298	-1.494
4900	83.063	504.466	427.495	377.158	-1045.529	161.939	-1.726
5000	83.067	506.144	429.051	385.464	-1045.763	186.583	-1.949
5100	83.070	507.789	430.579	393.771	-1046.028	211.233	-2.163
5200	83.073	509.402	432.079	402.078	-1046.326	235.887	-2.370
5300	83.075	510.984	433.553	410.386	-1046.661	260.549	-2.568
5400	83.078	512.537	435.001	418.693	-1047.036	285.217	-2.759
5500	83.080	514.062	436.425	427.001	-1047.453	309.891	-2.943
5600	83.082	515.559	437.825	435.309	-1047.915	334.575	-3.121
5700	83.085	517.029	439.202	443.618	-1048.425	359.266	-3.292
5800	83.087	518.474	440.556	451.926	-1048.981	383.968	-3.458
5900	83.088	519.895	441.889	460.235	-1049.597	408.677	-3.618
6000	83.090	521.291	443.200	468.544	-1050.227	433.396	-3.773

PREVIOUS: December 1979 (1 atm) CURRENT: December 1979 (1 bar)

Aluminum Oxide ((AlO)$_2$) Al_2O_2(g)

ALUMINUM OXIDE, ION ($Al_2O_2^+$) IDEAL GAS M_r = 85.961331

$\Delta_fH°(0\ K)$ = 564.0 ± 60 kJ mol^{-1}

S°(298.15 K) = [289.7 ± 16] J K^{-1} mol^{-1} $\Delta_fH°(298.15\ K)$ = [567.414] kJ mol^{-1}

Vibrational Frequencies and Degeneracies

ν, cm^{-1}	ν, cm^{-1}	ν, cm^{-1}
[560](1)	[329](1)	[460](1)
[370](1)	[180](1)	[600](1)

Ground State Quantum Weight: [2] σ = [4]

Point Group: [D_{2h}]

Bond Distance: Al-O = [1.80] Å

Bond Angles: O-Al-O = [90]° Al-O-Al = [90]°

Product of the Moments of Inertia: [2.889607 x 10^{-114}] g^3 cm^6

Enthalpy of Formation

$\Delta_fH°(0\ K)$ is calculated from that of Al_2O_2 (1) using the appearance potential 9.9±0.5 eV (2-3). We assume that the appearance potential is equal to the ionization potential for $Al_2O_2(g) \rightarrow Al_2O_2(g) + e^-(g)$. Other reported appearance potentials include 10±1 (4-5) and 10 eV (6).

$\Delta_fH°(298.15\ K)$ is obtained from $\Delta_fH°(Al_2O_2,\ 0\ K)$ by using $IP(Al_2O_2)$ with JANAF (1) enthalpies H°(0 K)-H°(298.15 K) for $Al_2O_2(g)$, $Al_2O_2^+$ (g) and e (g). $\Delta_rH°(Al_2O_2 \rightarrow Al_2O_2^+ + e^-$, 298.15 K) differs from a room-temperature threshold energy due to inclusion of these enthalpies and to threshold effects discussed by Rosenstock et al. (7). $\Delta_fH°(298.15\ K)$ should be changed by -1.481 kcal mol^{-1} if it is to be used in the convention that excludes the enthalpy of the electron. MNDO calculations (8) predicted $\Delta_fH°(Al_2O_2^+$, 298.15 K) = 78.7 and [$\Delta_fH°(Al_2O_2^+$, 298.15 K) - $\Delta_fH°(Al_2O_2$, 298.15 K)] = 186.7 kcal mol^{-1} for the linear AlOAlO ion and neutral molecule; our adopted values are larger by 43 and 57 kcal mol^{-1}, respectively.
1

Heat Capacity and Entropy

The structure of $Al_2O_2^+$ is probably even more uncertain than that of Al_2O_2 (1). MNDO calculations (8) predict an AlOAlO ion that is essentially linear, but a similar prediction for the neutral molecule is contradicted by ab initio calculations (1). Pending resolution of this discrepancy, we arbitrarily adopt a square-planar structure like that adopted for Al_2O_2 (1). We assume the same bond distance as in Al_2O_2 since the changes on ionization are predicted to be small (8). Vibrational frequencies are estimated to be slightly lower than those of Al_2O_2(1). We assume the ground electronic state to be doublet, due to the odd number of electrons, and neglect excited states. The principal moments of inertia are: I_A = 8.6079 x 10^{-39}, I_B = 14.5167 x 10^{-39} and I_C = 23.1246 x 10^{-39} g cm^2.

References

1. JANAF Thermochemical Tables: Al_2O_2(g), 12-31-79; Al(g), 6-30-79; e^-(g), 3-31-77.
2. J. Drowart, G. DeMaria, R. P. Burns and M. G. Inghram, J. Chem. Phys. 32, 1366 (1960).
3. P. Ho and R. P. Burns, High Temp. Sci. 12, 31 (1980); C. M. Fu and R. P. Burns, ibid., 8, 353 (1976).
4. M. Farber, R. D. Srivastava and O. M. Uy, J. Chem. Soc., Faraday Trans. I 68, 249 (1972).
5. R. C. Paule, High Temp. Sci. 8, 257 (1976).
6. A. D. Chervonnyi, V. A. Piven', O. E. Kashireninov and G. B. Manelis, High Temp. Sci. 9, 99 (1977).
7. H. M. Rosenstock, K. Draxl et al., J. Phys. Chem. Ref. Data 6, Supp. 1 (1977).
8. L. P. Davis et al., J. Comput. Chem. 2, 433 (1981).

Aluminum Oxide, Ion ($Al_2O_2^+$) $Al_2O_2^+$(g)

Enthalpy Reference Temperature = T_r = 298.15 K Standard State Pressure = p° = 0.1 MPa

T/K	$C_p°$ (J $K^{-1}mol^{-1}$)	S° (J $K^{-1}mol^{-1}$)	$-[G°-H°(T_r)]/T$ (J $K^{-1}mol^{-1}$)	$H°-H°(T_r)$ (kJ mol^{-1})	$\Delta_fH°$ (kJ mol^{-1})	$\Delta_fG°$ (kJ mol^{-1})	Log K_f
0	0.	0.	INFINITE	-14.975	564.004		
100	41.878	229.595	343.628	-11.403			
200	58.897	264.099	295.729	-6.326			
250	64.816	277.913	290.814	-3.225			
298.15	68.923	289.700	289.700	0.	567.414	552.810	-96.850
300	69.056	290.127	289.701	0.128	567.436	552.719	-96.237
350	72.091	301.013	290.554	3.660	568.061	550.218	-82.115
400	74.295	310.790	292.483	7.323	568.722	547.624	-71.512
450	75.927	319.640	295.017	11.080	569.395	544.946	-63.256
500	77.162	327.706	297.888	14.909	570.062	542.194	-56.643
600	78.860	341.937	304.076	22.717	571.318	536.500	-46.706
700	79.936	354.180	310.379	30.661	572.391	530.609	-39.594
800	80.657	364.904	316.539	38.692	573.203	524.581	-34.252
900	81.162	374.435	322.452	46.785	573.687	518.470	-30.091
1000	81.529	383.006	328.086	54.920	552.608	513.846	-26.841
1100	81.803	390.790	333.438	63.087	552.994	509.951	-24.216
1200	82.013	397.918	338.519	71.279	553.365	506.021	-22.027
1300	82.178	404.489	343.344	79.489	553.720	502.061	-20.173
1400	82.309	410.584	347.932	87.713	554.060	498.075	-18.583
1500	82.415	416.266	352.300	95.950	554.384	494.064	-17.205
1600	82.503	421.588	356.466	104.196	554.691	490.033	-15.998
1700	82.575	426.592	360.445	112.450	554.982	485.983	-14.932
1800	82.636	431.314	364.253	120.710	555.255	481.916	-13.985
1900	82.688	435.783	367.901	128.977	555.510	477.834	-13.137
2000	82.732	440.026	371.402	137.248	555.747	473.740	-12.373
2100	82.770	444.063	374.766	145.523	555.965	469.634	-11.682
2200	82.803	447.914	378.005	153.801	556.164	465.519	-11.053
2300	82.832	451.596	381.125	162.083	556.344	461.394	-10.479
2400	82.857	455.122	384.135	170.368	556.504	457.262	-9.952
2500	82.880	458.504	387.043	178.655	556.644	453.124	-9.468
2600	82.900	461.755	389.854	186.944	556.766	448.981	-9.020
2700	82.917	464.884	392.575	195.234	556.868	444.834	-8.606
2800	82.933	467.900	395.212	203.527	-30.851	442.619	-8.257
2900	82.947	470.811	397.769	211.821	-28.595	459.488	-8.276
3000	82.960	473.623	400.251	220.116	-26.358	476.280	-8.293
3100	82.972	476.343	402.662	228.413	-24.138	492.999	-8.307
3200	82.982	478.978	405.006	236.711	-21.936	509.645	-8.319
3300	82.992	481.531	407.286	245.010	-19.750	526.223	-8.329
3400	83.001	484.009	409.506	253.309	-17.582	542.735	-8.338
3500	83.009	486.415	411.670	261.610	-15.431	559.183	-8.345
3600	83.016	488.754	413.779	269.911	-13.295	575.571	-8.351
3700	83.023	491.028	415.836	278.213	-11.177	591.899	-8.356
3800	83.029	493.243	417.844	286.516	-9.075	608.171	-8.360
3900	83.035	495.399	419.805	294.819	-6.989	624.387	-8.363
4000	83.041	497.502	421.721	303.123	-4.921	640.549	-8.365
4100	83.046	499.552	423.595	311.427	-2.870	656.661	-8.366
4200	83.050	501.554	425.427	319.732	-0.837	672.722	-8.367
4300	83.054	503.508	427.220	328.037	1.177	688.736	-8.366
4400	83.059	505.417	428.976	336.343	3.170	704.702	-8.366
4500	83.062	507.284	430.695	344.649	5.148	720.623	-8.365
4600	83.066	509.110	432.380	352.955	7.100	736.500	-8.363
4700	83.069	510.896	434.032	361.262	9.028	752.336	-8.361
4800	83.072	512.645	435.651	369.569	10.931	768.131	-8.359
4900	83.075	514.358	437.240	377.876	12.808	783.887	-8.356
5000	83.078	516.036	438.799	386.184	14.654	799.605	-8.353
5100	83.080	517.681	440.330	394.492	16.469	815.286	-8.350
5200	83.083	519.295	441.833	402.800	18.250	830.931	-8.347
5300	83.085	520.877	443.310	411.108	19.994	846.543	-8.343
5400	83.087	522.430	444.761	419.417	21.699	862.122	-8.339
5500	83.089	523.955	446.187	427.726	23.362	877.670	-8.335
5600	83.091	525.452	447.589	436.035	24.980	893.189	-8.331
5700	83.093	526.923	448.968	444.344	26.549	908.678	-8.327
5800	83.095	528.368	450.324	452.653	28.072	924.141	-8.323
5900	83.096	529.788	451.659	460.963	29.536	939.576	-8.318
6000	83.098	531.185	452.973	469.273	30.986	954.985	-8.314

PREVIOUS: December 1979 (1 atm) CURRENT: December 1979 (1 bar)

Aluminum Oxide, Ion ($Al_2O_2^+$) $Al_2O_2^+$(g)

ALUMINUM OXIDE, ALPHA (α-Al_2O_3, CORUNDUM) CRYSTAL (ALPHA) M_r = 101.96128

$\Delta_fH°(0\ K) = -1663.6 \pm 1.2$ kJ mol^{-1}

$S°(298.15\ K) = 50.950 \pm 0.08$ J K^{-1} mol^{-1} $\Delta_fH°(298.15\ K) = -1675.7 \pm 1.2$ kJ mol^{-1}

$T_{fus} = 2327 \pm 6$ K $\Delta_{fus}H° = 111.1 \pm 4$ kJ mol^{-1}

Enthalpy of Formation

The adopted $\Delta_fH°$ is from calorimetric heats of combustion measured by Mah (1) and Holley and Huber (2). Early measurements of the heat of combustion of Al were seriously biased (3), but the sources of bias were minimized in later studies. After conversion to the present atomic weight of Al these studies yield $\Delta_fH°$ values of -400.5 ± 0.25 (1), -400.5 ± 0.3 (2), -399.2 ± 0.3 (3), and -402 ± 2 or -400.6 ± 1.4 (4) kcal mol^{1}.

$\Delta_fH°$ is the value adopted by CODATA (5). This direct calorimetric result has been criticized by various authors who proposed less negative $\Delta_fH°$ values, for example -393 (6) and -397.1 (7) kcal mol^{1} based on emf and mineralogical data, respectively. Plausible explanations have been suggested (8, 9) for these different values. Moreover, the adopted $\Delta_fH°$ is confirmed by several direct and indirect measurements.

New direct confirmation includes $\Delta_fH° = -400.2\pm0.4$ and $\Delta_fH° < -399.9$ kcal mol^{-1} obtained, respectively, from emf measurements (10) of $\Delta_fG°$(1030-1250 K) and from heat of combustion (11). Indirect confirmation comes from data on reactions that link α-Al_2O_3 to other aluminum compounds. Kp data (12, 13) at high temperature yield $\Delta_fH° = -401.9\pm1.5$ kcal mol^{-1} from $AlCl_3$(g) and -405.2 ± 1.5 from AlF_3(cr), although the latter value may be biased if aluminum oxyfluorides (14) were formed. $\Delta_fH°$ is confirmed to within 1.0 kcal mol^{-1} by equilibria and calorimetry for independent reaction cycles involving each of the minerals mullite (11), kaolinite (9) and gibbsite (15). Data on many related minerals have been carefully evaluated by Haas et al. (16); they found no inconsistency large enough to require a change in $\Delta_fH°$ of Al_2O_3.

Heat Capacity and Entropy

$C_p°$ is from the equations of Ditmars et al. (17, 19) for NBS Standard Reference Material 720 (α-Al_2O_3). These equations are based on NBS data: $C_p°$ of Chang (18) (8.6 to 371 K) and enthalpy data of Ditmars and Douglas (19) (323 to 1173 K) and Ditmars et al. (17) (1173 to 2257 K). We increase $C_p°$ slightly (by <0.2%) above 1700 K to avoid a rapidly rising extrapolation above T_{fus}. $C_p°$ from the equation (17, 19) has an upward inflection point near 2050 K that is not necessarily justified by the data (17).

$S°(298.15\ K) = 12.177\pm0.02$ cal K^{-1} mol^{-1} is obtained from $C_p°$ based on $S°(9\ K) = 0.00053$ cal K^{-1} mol^{-1}. CODATA (5) and Gurvich et al. (20) selected the slightly smaller values $S° = 12.17\pm0.02$ cal K^{-1} mol^{-1} and $H°(298.15\ K)-H°(0\ K)= 2.394\pm0.005$ kcal mol^{-1}. We choose instead to be consistent with newly certified values for SRM 720.

Fusion Data and Sublimation Data

See the liquid table for details.

References

1. A. D. Mah, J. Phys. Chem. 61, 1572 (1957).
2. C. E. Holley and E. J. Huber, J. Amer. Chem. Soc. 73, 5577 (1951).
3. P. E. Snyder and H. Seltz, J. Amer. Chem. Soc. 67, 683 (1945).
4. A. Schneider and G. Gattow, Z. Anorg. Allgem. Chem. 277, 40 (1954).
5. ICSU-CODATA Task Group on Key Values for Thermodynamics, J. Chem. Thermodynamics 8, 603 (1976).
6. D. Ghosh and D. A. R. Kay, J. Electrochem. Soc. 124, 1836 (1977); D. A. R. Kay, Rev. Int. Hautes Temp. Refract. 16, 21 (1979).
7. H. C. Helgeson, J. M. Delany, H. W. Nesbitt and D. K. Bird, Amer. J. Sci. 278A, 1 (1978).
8. H. Byker and R. A. Howald, J. Electrochem. Soc. 125, 889 (1978).
9. J. J. Hemley, J. W. Montoya, J. W. Marinenko and R. W. Luce, Econ. Geol. 75, 210 (1980).
10. A. Sterten, S. Haugen and K. Hamberg, Electrochim. Acta 21, 589 (1976).
11. I. D. Zenkov, Russ. J. Phys. Chem. 55, 1698 (1981).
12. W. Fischer and R. Gewehr, Z. Anorg. Allgem. Chem. 209, 17 (1932).
13. V. P. Mashovets and B. F. Yudin, Izv. Vyssh. Ucheb. Zaved., Tsvet. Met. 5 (4), 95 (1962).
14. B. Siegel, Inorg. Chim. Acta, Rev. 2, 137 (1968).
15. B. S. Hemingway, R. A. Robie and J. A. Kittrick, Geochim. Cosmochim. Acta 42, 1533 (1978).
16. J. L. Haas, Jr., G. R. Robinson, Jr., and B. S. Hemingway, J. Phys. Chem. Ref. Data 10, 575 (1981).
17. D. A. Ditmars, S. Ishihara, S. S. Chang and G. Bernstein, J. Res. Nat. Bur. Stand. 87, 159 (1982).
18. S. S. Chang, Proc. Symp. Thermophys. Prop. 7, 83 (1977).
19. D. A. Ditmars and T. B. Douglas, J. Res. Nat. Bur. Stand. 75A, 401 (1971).
20. L. V. Gurvich, I. V. Veits et al., "Thermodynamic Properties of Individual Substances," 3rd ed., Vol. III, Nauka, Moscow, 1981.

Aluminum Oxide, Alpha (Al_2O_3) Al_2O_3(cr)

Enthalpy Reference Temperature = T_r = 298.15 K Standard State Pressure = p° = 0.1 MPa

	J K^{-1}mol^{-1}			kJ mol^{-1}			
T/K	$C_p°$	S°	$-[G°-H°(T_r)]/T$	$H°-H°(T_r)$	$\Delta_fH°$	$\Delta_fG°$	Log K_f
0	0.	0.	INFINITE	-10.020	-1663.608	-1663.608	INFINITE
100	12.855	4.295	101.230	-9.693	-1668.606	-1641.642	857.506
200	51.120	24.880	57.381	-6.500	-1673.383	-1612.656	421.183
298.15	79.015	50.950	50.950	0.	-1675.692	-1582.275	277.208
300	79.416	51.440	50.951	0.147	-1675.717	-1581.696	275.398
400	96.086	76.779	54.293	8.995	-1676.342	-1550.226	202.439
500	106.131	99.388	61.098	19.145	-1676.045	-1518.718	158.659
600	112.545	119.345	69.177	30.101	-1675.300	-1487.319	129.483
700	116.926	137.041	77.632	41.586	-1674.391	-1456.059	108.652
800	120.135	152.873	86.065	53.447	-1673.498	-1424.931	93.038
900	122.662	167.174	94.296	65.591	-1672.744	-1393.908	80.900
1000	124.771	180.210	102.245	77.965	-1693.394	-1361.437	71.114
1100	126.608	192.189	109.884	90.535	-1692.437	-1328.286	63.075
1200	128.252	203.277	117.211	103.280	-1691.366	-1295.228	56.380
1300	129.737	213.602	124.233	116.180	-1690.190	-1262.264	50.718
1400	131.081	223.267	130.965	129.222	-1688.918	-1229.393	45.869
1500	132.290	232.353	137.425	142.392	-1687.561	-1196.617	41.670
1600	133.361	240.925	143.628	155.675	-1686.128	-1163.934	37.999
1700	134.306	249.039	149.592	169.060	-1684.632	-1131.342	34.762
1800	135.143	256.740	155.333	182.533	-1683.082	-1098.841	31.888
1900	135.896	264.067	160.864	196.085	-1681.489	-1066.426	29.318
2000	136.608	271.056	166.201	209.710	-1679.858	-1034.096	27.008
2100	137.319	277.738	171.354	223.407	-1678.190	-1001.849	24.920
2200	138.030	284.143	176.336	237.174	-1676.485	-969.681	23.023
2300	138.741	290.294	181.158	251.013	-1674.743	-937.593	21.293
2327.000	138.934	291.914	182.434	254.761	------- ALPHA <--> LIQUID ------		
2400	139.453	296.214	185.829	264.922	-1672.963	-905.582	19.709
2500	140.206	301.922	190.360	278.905	-1671.142	-873.645	18.254
2600	140.959	307.435	194.757	292.963	-1669.279	-841.781	16.912
2700	141.754	312.770	199.030	307.098	-1667.369	-809.990	15.670
2800	142.591	317.940	203.185	321.315	-2253.212	-776.335	14.483
2900	143.511	322.960	207.229	335.620	-2249.002	-723.665	13.035
3000	144.474	327.841	211.168	350.019	-2244.729	-671.139	11.686

PREVIOUS: June 1975 CURRENT: December 1979

Aluminum Oxide, Alpha (Al_2O_3) Al_2O_3(cr)

ALUMINUM OXIDE, DELTA (δ-Al_2O_3) CRYSTAL (DELTA) M_r = 101.96128

$S°(298.15 K) = [50.6 \pm 4] J K^{-1} mol^{-1}$ $\Delta_fH°(0 K) = [-1653.9 \pm 4] kJ mol^{-1}$

$T_{fus} = [2308] K$ $\Delta_fH°(298.15 K) = -1666.5 \pm 4 kJ mol^{-1}$

$\Delta_{fus}H° = [93.3] kJ mol^{-1}$

Enthalpy of Formation

$\Delta_fH°$ is calculated from that of α-Al_2O_3 (1) using $\Delta_rH°(978 K) = -2.7\pm0.4 kcal mol^{-1}$ for the irrreversible process $\delta\rightarrow\alpha$. This yields $\Delta_rH°(298.15 K) = -2.2 kcal mol^{-1}$ based on our adopted functions. Yokokawa and Kleppa (2) determined $\Delta_rH°$ for the two crystalline forms from the difference in their heats of solution in an oxide melt at 978 K. DTA studies by Gani and McPherson (3) gave $\Delta_rH° = -2.8\pm0.5 kcal mol^{-1}$. Assuming T = ~1400 K, we derive $\Delta_rH°(298.15 K) = -1.9 kcal mol^{-1}$. This confirms the calometric result. Both samples of δ-Al_2O_3 had been obtained by rapid quenching from high temperature, one (2) from combustion of Al_4C_3 and the other (3) by a plasma method.

Heat Capacity and Entropy

$C_p°$ is assumed to be 3% larger than that of α-Al_2O_3 (1) by comparison with the observed values for κ-Al_2O_3 (1) and the adopted values for γ-Al_2O_3 (1). S° is selected such that $\Delta_rG°(\kappa\rightarrow\delta)<0$ below T_{fus} and $\Delta_rG°(\delta\rightarrow\alpha)<0$ at T<2400 K.

Marchidan et al. (4) measured enthalpy data (573-1177 K) for a sample designated only as TA-600 alumina. They attributed a transition at ~993 K (720°C) to the irreversible process $\eta\rightarrow\delta$. This temperature corresponds roughly to processes designated as either (γ or η)$\rightarrow\theta$ or $\gamma\rightarrow\delta$ by Lippens and Steggerda (5). DTA data of Alevra et al. (6) showed an endotherm near 750°C subsequent to the dehydration of hydroxide samples; however, TGA data indicated that the resulting alumina still retained considerable water at this temperature. The enthalpy data (4, 1003-1177 K) are 4.7±0.3% larger than those of α-Al_2O_3 (1) and presumably correspond to an alumina containing a significant amount of water. The δ-Al_2O_3 used in $\Delta_fH°$ studies was presumably essentially anhydrous. Thus, we omit the enthalpy data (4) on the presumption that they are larger than those of our standard state which is anhydrous δ-Al_2O_3.

Phase Data

The stable crystalline form is corundum (α-alumina). Delta-alumina is one of several structurally related, metastable forms. These occur in "active alumina" and are much studied (5) due to their importance in adsorbents and catalysts. Lippens and Steggerda (5) summarized the classifications of metastable forms, their crystal stuctures and conditions of formaiton by dehydration. Mixtures of the closely related δ- and θ-aluminas are also formed by rapid cooling of droplets of molten alumina following their passage through a flame (7). There is some controversy (5) over the X-ray data and unit-cell dimensions of δ-Al_2O_3 formed by the two methods. Recent data (8) for θ-Al_2O_3 (5) suggest a similar controversy. Delta-alumina belongs to the "nearly anhydrous", high-temperature classification (5) based on the temperature of 600° to 900°C at which it forms during dehydration. The irreversible transitions $\delta\rightarrow\theta(\rightarrow\alpha)\rightarrow\alpha$ are thermally activated at tempertures of 800°-1050°C for the first step (5, 6] and ~1200°C for the final step (5).

Fusion Data

The hypothetical melting point of metastable δ-phase is calculated as the temperature at which $\Delta_rG°(\delta\rightarrow\ell)=0$. $\Delta_{fus}H°$ is the corresponding difference in $\Delta_fH°$.

References

1. JANAF Thermochemical Tables: Al_2O_3(cr, α), Al_2O_3(cr, κ), and Al_2O_3(cr, γ), 12-31-79.
2. T. Yokokawa and O. J. Kleppa, J. Phys. Chem. 68, 3246 (1964); J. Amer. Chem. Soc. 68, 3246 (1964).
3. M. S. J. Gani and R. McPherson, Thermochim. Acta 7, 251 (1973).
4. D. I. Marchidan, L. Pandele and A. Nicolescu, Rev. Roum. Chim. 17, 1493 (1972).
5. B. C. Lippens and J. J. Steggerda in "Physical and Chemical Aspects of Adsorbents and Catalysts," B. G. Linsen, Ed., Academic Press, New York, 1970, pp. 171-211.
6. V. Alevra, D. Ciomirtan and M. Ionescu, Rev. Roum. Chim. 17, 1379 (1972).
7. M. Plummer, J. Appl. Chem. 8, 35 (1958).
8. G. Yamaguchi, I. Yasui and W.-C. Chiu, Bull. Chem. Soc. Japan 43, 2487 (1970).

Aluminum Oxide, Delta (Al_2O_3) Al_2O_3(cr)

Enthalpy Reference Temperature = T_r = 298.15 K Standard State Pressure = p° = 0.1 MPa

	J K⁻¹mol⁻¹			kJ mol⁻¹			
T/K	$C_p°$	S°	$-[G°-H°(T_r)]/T$	$H°-H°(T_r)$	$\Delta_fH°$	$\Delta_fG°$	Log K_f
0							
100							
200							
298.15	81.385	50.626	50.626	0.	-1666.487	-1572.974	275.579
300	81.799	51.131	50.628	0.151	-1666.507	-1572.394	273.778
400	98.968	77.231	54.070	9.264	-1666.867	-1540.932	201.225
500	109.315	100.518	61.079	19.719	-1666.266	-1509.504	157.697
600	115.922	121.073	69.400	31.004	-1665.192	-1478.248	128.693
700	120.434	139.300	78.109	42.834	-1663.938	-1447.188	107.991
800	123.739	155.608	86.795	55.050	-1662.689	-1416.311	92.476
900	126.342	170.338	95.273	67.558	-1661.571	-1385.582	80.417
1000	128.514	183.764	103.460	80.304	-1681.850	-1353.447	70.697
1100	130.406	196.103	111.329	93.251	-1680.516	-1320.670	62.713
1200	132.100	207.524	118.875	106.378	-1679.063	-1288.021	56.066
1300	133.630	218.159	126.108	119.666	-1677.500	-1255.496	50.446
1400	135.013	228.113	133.042	133.099	-1675.837	-1223.096	45.634
1500	136.258	237.471	139.696	146.664	-1674.084	-1190.819	41.468
1600	137.362	246.301	146.085	160.346	-1672.253	-1158.660	37.826
1700	138.336	254.658	152.228	174.132	-1670.355	-1126.618	34.617
1800	139.197	262.590	158.141	188.009	-1668.402	-1094.690	31.767
1900	139.973	270.137	163.838	201.968	-1666.402	-1062.872	29.220
2000	140.706	277.336	169.335	216.002	-1664.362	-1031.160	26.931
2100	141.438	284.218	174.643	230.109	-1662.283	-999.550	24.862
2200	142.171	290.815	179.774	244.289	-1660.165	-968.040	22.984
2300	142.904	297.151	184.741	258.543	-1658.008	-936.629	21.272
2400	143.636	303.248	189.552	272.870	-1655.810	-905.312	19.704
2500	144.412	309.128	194.219	287.272	-1653.570	-874.087	18.263
2600	145.188	314.806	198.748	301.752	-1651.285	-842.953	16.935
2700	146.007	320.301	203.149	316.311	-1648.952	-811.907	15.707
2800	146.868	325.626	207.428	330.954	-2234.367	-779.013	14.533
2900	147.817	330.797	211.594	345.688	-2229.729	-727.118	13.097
3000	148.808	335.824	215.651	360.519	-2225.023	-675.384	11.759

PREVIOUS: June 1975 CURRENT: December 1979

Aluminum Oxide, Delta (Al_2O_3) Al_2O_3(cr)

ALUMINUM OXIDE, GAMMA (γ-Al_2O_3) CRYSTAL (GAMMA) M_r = 101.96128

$S°(298.15\ K) = [52.3 \pm 8]\ J\ K^{-1}\ mol^{-1}$
$T_{fus} = [2289]\ K$

$\Delta_fH°(0\ K) = [-1644.3 \pm 6.3]\ kJ\ mol^{-1}$
$\Delta_fH°(298.15\ K) = -1656.9 \pm 6.3\ kJ\ mol^{-1}$
$\Delta_{fus}H° = [78.49]\ kJ\ mol^{-1}$

Enthalpy of Formation

$\Delta_fH°$ is calculated from that of α-Al_2O_3 (1) using $\Delta_rH°(978\ K) = -5.3\ kcal\ mol^{-1}$ for the irreversible process $\gamma\to\alpha$. Yokokawa and Kleppa (2) determined $\Delta_rH°$ for the two crystalline forms from the difference in their heats of solution in an oxide melt at 978 K. This yields $\Delta_rH°(298.15\ K) = -4.5\ kcal\ mol^{-1}$ based on our adopted functions. DTA studies by Gani and McPherson (3) gave $\Delta_rH°(\gamma\to\alpha) = -5.8\pm1.6\ kcal\ mol^{-1}$. Assuming T = ~1400 K, we derive $\Delta_rH°(298.15\ K) = -4.4\ kcal\ mol^{-1}$ which confirms the calorimetric result (2). Yamada et al. (4) used dynamic, adiabatic calorimetry to measure $\Delta_rH°(773\ K) = 12.6\pm1.1\ kcal\ mol^{-1}$ for $Al_2O_3\cdot H_2O$(cr, boehmite) → $Al_2O_3(\gamma) + H_2O(g)$. Enthalpies from (4, 1) convert this to $\Delta_rH°(298.15\ K) = 17.54\pm1.3\ kcal\ mol^{-1}$, yielding $\Delta_fH°(\gamma$-Al_2O_3, 298.15 K) = -398.1 ± 1.5 or $-396.7\pm2\ kcal\ mol^{-1}$ based on new or old values for $\Delta_fH°$(boehmite, 298.15 K)) = -473.4 ± 0.4 (5) or -472.0 (6) $kcal\ mol^{-1}$.

We adopt $\Delta_fH°(\gamma, 298.15\ K) = -396.0\pm1.5\ kcal\ mol^{-1}$ and $\Delta_rH°(\gamma\to\alpha, 298.15\ K) = -4.5\ kcal\ mol^{-1}$. These values derive (2) from the highest ignition temperature prior to the appearance of a more stable phase (κ- or α-Al_2O_3) in the sample. This standard state should involve maximum attainable crystal development and minimum residual water. γ-alumina shows much variability depending on its thermal history. Lower ignition temperatures gave values of $\Delta_rH°(\gamma\to\alpha)$ more negative by as much as 2.5 kcal mol^{-1} (2). Other reported values of $\Delta_rH°(\gamma\to\alpha)$, including -4.5 to -7.5 (3), -7.7 (7), -9.1 (8), -11.0 (9) and -7.8 (10) kcal mol^{-1}, tend to be skewed toward more negative values. These values correspond to less stable samples, presumably with more residual water and less well developed crystal structure. This is consistent with $\Delta_rH°(298.15\ K) = -388\pm2\ kcal\ mol^{-1}$ observed (4) for ρ'-Al_2O_3 which is nearly amorphous.

Heat Capacity and Entropy

$C_p°$ is assumed to be 4.7% larger than that of α-Al_2O_3 (1). Structural and related characteristics (11) suggest that $C_p°(\gamma) > C_p°(\kappa)$ which in turn is ~2.2% larger (1, 4) than $C_p°(\alpha)$. Our estimate is derived from Marchidan et al. (12) who measured enthalpies (1003-1177 K) which are 4.7±0.3% larger than those of α-Al_2O_3. S° is estimated such that $\Delta_rG°(\gamma\to\alpha)<0$ at T<2400 K. This is consistent with observed stability relationships (11, 2).

Marchidan et al. (12) measured enthalpy data (573-1177 K) for a sample designated only as TA-600 alumina. They attributed a transition at ~993 K to the irreversible process $\eta\to\delta$. Comtemporary studies (13) from the same institution lead us to presume that the TA-600 sample retained a significant amount of water even above the transition. The "low-temperature" forms γ and η are difficult to distinquish (11); they retain more water (11, 13) than the "high-temperature" forms. This may explain the unusual enthalpy data attributed (12) to η-Al_2O_3; these deviate from α-Al_2O_3 by +12.4% at 573 to 620 K and then decrease linearly to +3.8% at 991 K. These data do not seem suitable for standard-state γ-Al_2O_3 having minimum water content and maximum crystal development. We adopt instead the constant deviation of +4.7% found at higher temperature. Differential enthalpy analysis (8) gave an enthalpy 2% larger than α-Al_2O_3 at 1290 K, but the error is large (±4% in enthalpy) and no information is given on the sample history. The observed $\Delta_rH(\gamma\to\alpha) = -9.1\ kcal\ mol^{-1}$ (8) suggests the sample differed from our desired standard state.

The adopted entropy may be compared with the value $\Delta S°(\alpha\to\gamma) = 1.56\ cal\ K^{-1}\ mol^{-1}$ estimated by Borer and Gunthard (14) for the "defect" spinel structure of γ-Al_2O_3. This yields S°(γ, 298.15 K)) = 13.7 or 14.3 cal $K^{-1}\ mol^{-1}$ assuming $C_p°(\gamma)/C_p°(\alpha)$ = 1.00 or 1.047, respectively. The authors' assumption (14) concerning the structure of γ-Al_2O_3 (11) is overly simplified.

Phase Data

The stable crystalline form is corundum (α-alumina). Gamma-alumina is one of several structurally related, metastable forms. These occur in "active alumina" and are much studied (11) due to their importance in adsorbents and catalysts. Lippens and Steggerda (11) summarized the classifications of metastable forms, their crystal structures and conditions of formation by dehydration. "Low-temperature" forms γ and η have similar X-ray diffraction patterns. Their nomenclature is confused in the literature. The term γ-Al_2O_3 has been applied to either form and as a generic term for all low temperature forms. γ and η are often poorly crystallized and difficult to distinquish (11). They retain various amounts of water (11, 13) depending on their thermal history. The low-temperature forms are obtained by dehydrating temperatures not exceeding 600°C and change irreversibly to "high-temperature" forms (δ, θ or κ) at 600° to 900°C (11, 13, 2).

Fusion Data

The hypothetical melting point of metastable γ-phase is calculated as the temperature at which $\Delta_rG°(\gamma\to\ell) = 0$. $\Delta_{fus}H°$ is the corresponding difference in $\Delta_fH°$.

References

1. JANAF Thermochemical Tables: Al_2O_3(cr, α), Al_2O_3(cr, δ) and Al_2O_3(cr, κ); 12-31-79; H_2O(g), 3-31-79.
2. T. Yokokawa and O. J. Kleppa, J. Phys. Chem. 68, 3246 (1964).
3. M. S. J. Gani and R. McPherson, Thermochim. Acta 7, 251 (1973).
4. K. Yamada, T. Fukunaga, Y. Takahashi and T. Mukaibo, Denki Kagaku 41, 290 and 287 (1973).
5. J. L. Haas, G. R. Robinson and B. S. Hemingway, J. Phys. Chem. Ref. Data 10, 575 (1981).
6. U. S. Natl. Bur. Std. Tech. Note 270-3, 1968.
7. P. Gross, C. Hayman and R. H. Lewin, Fulmer Res. Inst. Rept. R.163/32/September 1968.
8. D. Phan Xuan, R. Castenet and M. Laffitte, Rev. Int. Hautes Temp. Refract. 11, 285 (1974).
9. V. Kostomaroff and M. Rey, Silicates Ind. 28, 9 (1963).
10. W. A. Roth, Angew. Chem. 49, 198 (1936).
11. B. C. Lippens and J. J. Steggerda in "Physical and Chemical Aspects of Adsorbents and Catalysts," B. G. Linsen, Ed., Academic Press, New York, 1970, pp. 171-211.
12. D. I. Marchidan, L. Pandele and A. Nicolescu, Rev. Roum. Chim. 17, 1493 (1972).
13. V. Alevra, D. Ciomirtan and M. Ionescu, Rev. Roum. Chim. 17, 1379 (1972).
14. W. J. Borer and H. H. Gunthard, Helv. Chim. Acta 53, 1043 (1970).

Aluminum Oxide, Gamma (Al_2O_3) Al_2O_3(cr)

Enthalpy Reference Temperature = T_r = 298.15 K   Standard State Pressure = p° = 0.1 MPa

T/K	$C_p°$ (J K⁻¹mol⁻¹)	S° (J K⁻¹mol⁻¹)	$-[G°-H°(T_r)]/T$ (J K⁻¹mol⁻¹)	$H°-H°(T_r)$ (kJ mol⁻¹)	$\Delta_fH°$ (kJ mol⁻¹)	$\Delta_fG°$ (kJ mol⁻¹)	Log K_f
0							
100							
200							
298.15	82.729	52.300	52.300	0.	-1656.864	-1563.850	273.980
300	83.149	52.813	52.302	0.153	-1656.882	-1563.273	272.190
400	100.602	79.343	55.800	9.417	-1657.091	-1532.001	200.059
500	111.120	103.015	62.925	20.045	-1656.317	-1500.804	156.788
600	117.835	123.909	71.384	31.515	-1655.057	-1469.815	127.959
700	122.422	142.438	80.236	43.541	-1653.608	-1439.054	107.384
800	125.782	159.014	89.066	55.959	-1652.158	-1408.504	91.966
900	128.427	173.987	97.683	68.674	-1650.833	-1378.128	79.985
1000	130.635	187.635	106.006	81.629	-1670.901	-1346.370	70.327
1100	132.558	200.178	114.005	94.790	-1669.354	-1313.990	62.396
1200	134.280	211.787	121.675	108.134	-1667.684	-1281.757	55.793
1300	135.835	222.597	129.028	121.641	-1665.902	-1249.668	50.212
1400	137.241	232.716	136.076	135.296	-1664.017	-1217.720	45.434
1500	138.507	242.229	142.839	149.084	-1662.040	-1185.911	41.297
1600	139.629	251.204	149.334	162.992	-1659.983	-1154.236	37.682
1700	140.619	259.700	155.579	177.006	-1657.858	-1122.691	34.496
1800	141.495	267.762	161.589	191.112	-1655.675	-1091.274	31.668
1900	142.283	275.434	167.381	205.301	-1653.445	-1059.979	29.141
2000	143.028	282.751	172.968	219.567	-1651.174	-1028.802	26.870
2100	143.773	289.747	178.363	233.907	-1648.862	-997.740	24.817
2200	144.518	296.453	183.580	248.321	-1646.510	-966.788	22.954
2300	145.262	302.893	188.628	262.810	-1644.117	-935.946	21.256
2400	146.007	309.091	193.519	277.374	-1641.683	-905.209	19.701
2500	146.796	315.068	198.262	292.014	-1639.206	-874.573	18.273
2600	147.584	320.840	202.866	306.732	-1636.681	-844.037	16.957
2700	148.416	326.426	207.340	321.532	-1634.108	-813.599	15.740
2800	149.292	331.839	211.690	336.417	-2219.282	-781.322	14.576
2900	150.256	337.094	215.924	351.394	-2214.400	-730.053	13.150
3000	151.264	342.205	220.049	366.469	-2209.450	-678.952	11.822

PREVIOUS: June 1975   CURRENT: December 1979

Aluminum Oxide, Gamma (Al_2O_3) Al_2O_3(cr)

ALUMINUM OXIDE, KAPPA (κ-Al_2O_3) CRYSTAL (KAPPA) M_r = 101.96128

$S°(298.15\ K) = [53.6 \pm 4]\ J\ K^{-1}\ mol^{-1}$

$T_{fus} = [2312]\ K$

$\Delta_f H°(0\ K) = [-1650.0 \pm 4]\ kJ\ mol^{-1}$

$\Delta_f H°(298.15\ K) = -1662.3 \pm 4\ kJ\ mol^{-1}$

$\Delta_{fus} H° = [91.2]\ kJ\ mol^{-1}$

Enthalpy of Formation

$\Delta_f H°$ is calculated from that of α-Al_2O_3 (1) using $\Delta_r H°(978\ K) = -3.6\ kcal\ mol^{-1}$ for the irreversible process $\kappa \to \alpha$. This yields $\Delta_r H°(298.15\ K) = -3.2\ kcal\ mol^{-1}$ based on our adopted functions. Yokokawa and Kleppa (2) determined $\Delta_r H°$ for the two crystalline forms from the difference in their heats of solution in an oxide melt at 978 K. Yamada et al. (3) derived $\Delta_f H°(298.15\ K) = -397 \pm 1\ kcal\ mol^{-1}$ from the same data.

Heat Capacity and Entropy

$C°_p$ (380 to 1110 K) was measured by Takahashi et al. (4) using a dynamic, adiabatic calorimeter with an open-type container for the sample. Reproducibility of the data was reported to be within ±1.5% up to 1100 K. We represent the observed data by a $C°_p$ curve which is 2.2% larger than that of α-Al_2O_3 (1). Deviations from the adopted curve exceed 1.0% for only 3 of the 39 points (4): i.e., -1.6% (400 K), +1.4% (993.8 K) and +1.3% (1070.8 K). Takahashi et al. (4) prepared their κ-Al_2O_3 by dehydration of tohdite ($Al_2O_3 \cdot 1/5\ H_2O$) at 840°C under a vacuum of 10^{-5} torr.

S° is selected arbitrarily so that $\Delta_r G°(\kappa \to \alpha) = 0$ near 2400 K. This is consistent with the view that κ-phase is unstable with respect to α-phase at all temperatures below T_{fus}.

Phase Data

The stable crystalline form is corundum (α-alumina). Kappa-alumina is one of several structurally related, metastable forms. These occur in "active alumina" and are much studied (5) due to their importance in adsorbents and catalysts. Lippens and Steggerda (5) summarized the classifications of metastable forms, their crystal structures and conditions of formation by dehydration. Okumiya et al. (6) recently published an extensive study on the formation and structural relations of κ-Al_2O_3 and its precursors κ'-Al_2O_3 and tohdite. Kappa-alumina belongs to the "nearly anhydrous", high-temperature classification (5) based on its formation temperture of 900° to 1000°C (somewhat lower in vacuo). The irreversible transition $\kappa \to \alpha$ is thermally activated at ~1200°C (2, 5). Thermochemical studies cited in this table presume that the samples of κ-Al_2O_3 were essentially anhydrous.

Fusion Data

The hypothetical melting point of metastable κ-phase is calculated as the temperature at which $\Delta_r G°(\kappa \to \ell) = 0$. $\Delta_{fus} H°$ is the corresponding difference in $\Delta_f H°$.

References

1. JANAF Thermochemical Tables: Al_2O_3(cr, α), 12-31-79.
2. T. Yokokawa and O. J. Kleppa, J. Phys. Chem. 68, 3246 (1964).
3. K. Yamada, T. Fukunaga, Y. Takahashi and T. Mukaibo, Denki Kagaku 41, 290 (1973).
4. Y. Takahashi, K. Yamada, T. Fukunaga and T. Mukaibo, Denki Kagaku 41, 287 (1973).
5. B. C. Lippens and J. J. Steggerda in "Physical and Chemical Aspects of Adsorbents and Catalysts", B. G. Linsen, Ed., Academic Press, New York, 1970, pp. 171-211.
6. M. Okumiya, G. Yamaguchi, O. Yamada and S. Ono, Bull. Chem. Soc. Japan 44, 418 and 1567 (1971).

Aluminum Oxide, Kappa (Al_2O_3) Al_2O_3(cr)

Enthalpy Reference Temperature = T_r = 298.15 K — Standard State Pressure = p° = 0.1 MPa

	J K⁻¹mol⁻¹			kJ mol⁻¹			
T/K	$C°_p$	S°	$-[G°-H°(T_r)]/T$	$H°-H°(T_r)$	$\Delta_f H°$	$\Delta_f G°$	Log K_f
0							
100							
200							
298.15	80.753	53.555	53.555	0.	-1662.303	-1569.663	274.998
300	81.164	54.056	53.557	0.150	-1662.325	-1569.088	273.203
400	98.199	79.953	56.972	9.192	-1662.755	-1537.909	200.830
500	108.466	103.059	63.926	19.566	-1662.235	-1506.744	157.408
600	115.021	123.455	72.183	30.763	-1661.249	-1475.734	128.474
700	119.498	141.540	80.824	42.501	-1660.087	-1444.905	107.820
800	122.778	157.721	89.443	54.622	-1658.933	-1414.245	92.341
900	125.361	172.337	97.855	67.034	-1657.912	-1383.722	80.309
1000	127.516	185.659	105.979	79.680	-1678.290	-1351.782	70.610
1100	129.393	197.902	113.786	92.527	-1677.056	-1319.189	62.643
1200	131.074	209.234	121.274	105.552	-1675.705	-1286.715	56.009
1300	132.592	219.786	128.451	118.736	-1674.245	-1254.358	50.401
1400	133.964	229.663	135.331	132.065	-1672.687	-1222.116	45.598
1500	135.200	238.949	141.933	145.525	-1671.039	-1189.990	41.439
1600	136.295	247.710	148.272	159.100	-1669.314	-1157.976	37.804
1700	137.261	256.003	154.368	172.779	-1667.524	-1126.072	34.600
1800	138.116	263.873	160.235	186.549	-1665.678	-1094.275	31.755
1900	138.886	271.361	165.888	200.399	-1663.787	-1062.582	29.212
2000	139.613	278.504	171.342	214.324	-1661.856	-1030.989	26.927
2100	140.340	285.333	176.608	228.322	-1659.886	-999.494	24.861
2200	141.067	291.878	181.700	242.392	-1657.878	-968.093	22.985
2300	141.794	298.165	186.628	256.535	-1655.832	-936.785	21.275
2400	142.521	304.215	191.402	270.751	-1653.746	-905.567	19.709
2500	143.290	310.049	196.032	285.041	-1651.617	-874.437	18.270
2600	144.060	315.683	200.526	299.408	-1649.445	-843.392	16.944
2700	144.873	321.135	204.893	313.855	-1647.225	-812.432	15.717
2800	145.728	326.419	209.139	328.384	-2232.754	-779.619	14.544
2900	146.668	331.549	213.272	343.003	-2228.230	-727.802	13.109
3000	147.652	336.538	217.298	357.719	-2223.639	-676.141	11.773

PREVIOUS: June 1975 CURRENT: December 1979

Aluminum Oxide, Kappa (Al_2O_3) Al_2O_3(cr)

ALUMINUM OXIDE (Al_2O_3) LIQUID M_r = 101.96128

$S°(298.15 K)$ = [67.298] J K^{-1} mol^{-1} $\Delta_fH°(298.15 K)$ = [-1620.567] kJ mol^{-1}

T_{fus} = 2327 ± 6 K $\Delta_{fus}H°$ = 111.1 ± 4.2 kJ mol^{-1}

Enthalpy of Formation

$\Delta_fH°$ is calculated from that of the crystal by adding the enthalpy of fusion and the difference in enthalpy, H°(2327 K)-H°(298.15 K), between the crystal and liquid.

Heat Capacity and Entropy

Enthalpy data for the liquid include those of Shpil'rain et al. (1, 2, 2326 to 3104 K), Sheindlin et al. (3, 2350 to2800 K), Kantor et al. (4, 5) (2337 to 2480 K), and approximate values of West and Ishihara (6) (2345 to 2495 K). The latest study (1, 2) is the most extensive. It satisfactorily resolves the discrepancy in $C_p°$ which was reported earlier to be 47.7 (3) or 34.6 (4) cal K^{-1} mol^{-1}. The new data (1, 2) yield 46.0 cal K^{-1} mol^{-1}, assuming $C_p°$ is independent of temperature. Although Shpil'rain et al. (1, 2) derived a $C_p°$ which decreases linearly with increasing temperature, this does not significantly improve the fit of the enthalpy data. The apparent value of the temperature coefficient of $C_p°$ is very data dependent; it changes sign on omission of the point at 2326 K (2323 K, IPTS-48).

We adopt $C_p°$ = 46.0 cal K^{-1} mol^{-1}. Compared to our adopted functions, the enthalpy data deviate by <±1.0% (1, 2), -0.9 to -0.1% (3, equation), +2.4 to +0.2% (4, 5) and +2 to +5% (6). We assume a glass transition at 1350 K, below which $C_p°$ is taken to be the same as for Al_2O_3(cr, α). The entropy is calculated in a manner analogous to that used for the enthalpy of formation.

Fusion Data

T_{fus} = 2054±6°C was recommended by Schneider (7) as the result of a cooperative measurement of the melting point by nine groups in seven countries. We adopt this value. It is confirmed by several recent studies (e.g. 8-10); however, Nelson et al. (11) suggested the possibility that T_{fus} may be somewhat different in an oxygen atmosphere than in inert gases or vacuum.

$\Delta_{fus}H°$ is the difference at T_{fus} between the adopted enthalpy fits of Al_2O_3(ℓ) and Al_2O_3(α). Reported calorimetric values of $\Delta_{fus}H°$ include 25.7±1.3 (1, 2), 25.9 (3), and 28.3±0.6 (4, 5) kcal mol^{-1}.

Sublimation and Vaporization Data

Vaporization of Al_2O_3(α) and Al_2O_3(ℓ) has been studied by several methods (12-22). Additional references are cited by Farber et al. (14) and Burns (16). Vaporization is sometimes considered to be congruent but the vapor composition is complex (14, 16). Atomic species predominate below T_{fus}, but molecular species (e.g., AlO and Al_2O) become increasingly important at higher temperatures. For the crystal the average vaporization coefficient is α < 0.3 (12, 14-16). The rate of evaporation increases discontinuously during melting by radiative heating; this has been attributed to a change in α (16) or to a change in absorbance and emittance (19). Other work (11) suggests that the radiative properties change on melting. Container materials such as Mo or W cause changes in the partial pressures of O, Al and the molecular species observed in effusion studies (14, 16, 21); these effects can be predicted with a model developed by Banon et al. (20). Partial pressures reported (14, 22) for all observed vapor species yield an overall O/Al ratio of ~1.1 instead of the value 1.5 required for congruent vaporization; however, these data do not yield a definite conclusion about congruency due to their relatively large uncertainties.

References

1. E. E. Shpil'rain, D. N. Kagan and L. S. Barkhatov, High Temp.-High Pressures 4, 605 (1972).
2. L. S. Barkhatov, D. N. Kagan, A. F. Tsytsarkin, E. E. Shpil'rain and K. A. Yakimovich, High Temp. 11, 1063 (1973).
3. A. E. Sheindlin, V. Ya. Chekhovskoi and V. A. Petrov, Inzh-Fiz. Zh., Akad. Nauk Belorussk. SSR 7 (5), 63 (1964); High Temp.-High Pressures 2, 1 (1970).
4. P. B. Kantor, L. S. Lazareva, V. V. Kandyba and E. N. Fomichev, Ukr. Fiz. Zh. 7, 205 (1962).
5. E. N. Fomichev, V. P. Bondarenko and V. V. Kandyba, High Temp.-High Pressures 5, 1 (1973).
6. E. D. West and S. Ishihara, U. S. Nat. Bur. Stand. Report 9028, 71, Jan., 1966.
7. S. J. Schneider, Pure Appl. Chem. 21, 115 (1970).
8. E. N. Fomichev, P. B. Kantor and V. V. Kandyba, Heat Transfer-Sov. Res. 5 (3), 176 (1973).
9. E. E. Shpil'rain, K. A. Yakimovich and A. F. Tsytsarkin, High Temp.-High Pressures 5, 191 (1973).
10. T. Sata and T. Takahashi, Yogyo Kyokai Shi 79, 70 (1971).
11. L. S. Nelson, N. L. Richardson, K. Keil and S. R. Skaggs, High Temp. Sci. 5, 138 (1973).
12. M. Peleg and C. B. Alcock, High Temp. Sci. 6, 52 (1974); Trans. Brit. Ceram. Soc. 66, 217 (1967); Bull. Soc. Fr. Ceram., No. 72, 25 (1966).
13. E. M. Akulenok, Yu. K. Danileiko et al., Kristallografiya 18, 1044 (1973).
14. M. Farber, R. D. Srivastava and O. M. Uy, J. Chem. Soc., Faraday Trans. I 68, 249 (1972); Proc. Symp. Thermophys. Prop., 5th, 483 (1970).
15. T. Sasamoto and T. Sata, Kogyo Kagaku Zasshi 74, 832 (1971).
16. R. P. Burns et al., J. Chem. Phys. 44, 3307 (1966); 40, 2739 (1964); 32, 1366 (1960).
17. W. Kroenert and A. Boehm., Glas-Email-Keramo-Tech. 23, 319 (1972).
18. H. von Wartenberg, Z. Anorg. Allgem. Chem. 269, 76 (1952); 270, 328 (1952).
19. R. C. Paule, High Temp. Sci. 8, 257 (1976).
20. S. Banon, C. Chatillon and M. Allibert, manuscript, from C. Chatillon, CNRS, personal communication, July, 1981.
21. Y. Ikeda, M. Tamaki and G. Matsumoto, Shitsuryo Bunseki 26, 159 (1978).
22. P. Ho and R. P. Burns, High Temp. Sci. 12, 31 (1980).

Aluminum Oxide (Al_2O_3) Al_2O_3(l)

Enthalpy Reference Temperature = T_r = 298.15 K   Standard State Pressure = p° = 0.1 MPa

T/K	$C_p°$ (J $K^{-1}mol^{-1}$)	S° (J $K^{-1}mol^{-1}$)	$-[G°-H°(T_r)]/T$ (J $K^{-1}mol^{-1}$)	$H°-H°(T_r)$ (kJ mol^{-1})	$\Delta_fH°$ (kJ mol^{-1})	$\Delta_fG°$ (kJ mol^{-1})	Log K_f
0							
100							
200							
298.15	79.015	67.298	67.298	0.	-1620.567	-1532.025	268.404
300	79.416	67.788	67.299	0.147	-1620.592	-1531.475	266.654
400	96.086	93.127	70.641	8.995	-1621.217	-1501.641	196.094
500	106.131	115.736	77.446	19.145	-1620.920	-1471.768	153.755
600	112.545	135.693	85.525	30.101	-1620.175	-1442.003	125.538
700	116.926	153.389	93.980	41.586	-1619.266	-1412.378	105.393
800	120.135	169.222	102.413	53.447	-1618.373	-1382.885	90.293
900	122.662	183.522	110.644	65.591	-1617.619	-1353.496	78.555
1000	124.771	196.558	118.593	77.965	-1638.269	-1322.660	69.089
1100	126.608	208.537	126.233	90.535	-1637.312	-1291.144	61.311
1200	128.252	219.625	133.559	103.280	-1636.241	-1259.721	54.834
1300	129.737	229.951	140.581	116.180	-1635.065	-1228.392	49.357
1350.000	130.424	234.860	143.983	122.684	GLASS <--> LIQUID TRANSITION		
1350.000	192.464	234.860	143.983	122.684			
1400	192.464	241.859	147.354	132.308	-1630.708	-1197.213	44.669
1500	192.464	255.138	154.102	151.554	-1623.274	-1166.509	40.621
1600	192.464	267.559	160.809	170.800	-1615.878	-1136.299	37.096
1700	192.464	279.227	167.435	190.047	-1608.520	-1106.551	34.000
1800	192.464	290.228	173.954	209.293	-1601.197	-1077.235	31.261
1900	192.464	300.634	180.350	228.540	-1593.910	-1048.325	28.820
2000	192.464	310.507	186.613	247.786	-1586.658	-1019.797	26.634
2100	192.464	319.897	192.739	267.032	-1579.440	-991.632	24.665
2200	192.464	328.850	198.724	286.279	-1572.256	-963.809	22.884
2300	192.464	337.406	204.569	305.525	-1565.106	-936.313	21.264
2327.000	192.464	339.652	206.123	310.722	------- ALPHA <--> LIQUID ------		
2400	192.464	345.597	210.275	324.772	-1557.989	-909.127	19.787
2500	192.464	353.454	215.846	344.018	-1550.905	-882.237	18.433
2600	192.464	361.002	221.285	363.264	-1543.853	-855.629	17.190
2700	192.464	368.266	226.595	382.511	-1536.832	-829.292	16.044
2800	192.464	375.265	231.781	401.757	-2117.645	-801.279	14.948
2900	192.464	382.019	236.845	421.004	-2108.494	-754.429	13.589
3000	192.464	388.544	241.794	440.250	-2099.372	-707.892	12.325
3100	192.464	394.855	246.630	459.496	-2090.279	-661.658	11.149
3200	192.464	400.965	251.358	478.743	-2081.214	-615.720	10.051
3300	192.464	406.888	255.982	497.989	-2072.175	-570.063	9.023
3400	192.464	412.633	260.505	517.236	-2063.163	-524.679	8.061
3500	192.464	418.212	264.932	536.482	-2054.176	-479.561	7.157
3600	192.464	423.634	269.265	555.728	-2045.214	-434.699	6.307
3700	192.464	428.908	273.509	574.975	-2036.276	-390.086	5.507
3800	192.464	434.040	277.666	594.221	-2027.363	-345.712	4.752
3900	192.464	439.040	281.740	613.468	-2018.474	-301.574	4.039
4000	192.464	443.912	285.734	632.714	-2009.610	-257.664	3.365

PREVIOUS: June 1975   CURRENT: December 1979

Aluminum Oxide (Al_2O_3) Al_2O_3(l)

ALUMINUM OXIDE (Al_2O_3) ALPHA-LIQUID M_r = 101.96128 Aluminum Oxide (Al_2O_3) Al_2O_3(cr,l)

0 to 2327 K		crystal, alpha(corundum)
above	2327 K	liquid

Refer to the individual tables for details.

Enthalpy Reference Temperature = T_r = 298.15 K Standard State Pressure = p° = 0.1 MPa

	J K⁻¹mol⁻¹			kJ mol⁻¹			
T/K	C_p°	S°	$-[G^\circ-H^\circ(T_r)]/T$	$H^\circ-H^\circ(T_r)$	$\Delta_f H^\circ$	$\Delta_f G^\circ$	Log K_f
0	0.	0.	INFINITE	-10.020	-1663.608	-1663.608	INFINITE
100	12.855	4.295	101.230	-9.693	-1668.606	-1641.642	857.506
200	51.120	24.880	57.381	-6.500	-1673.383	-1612.656	421.183
298.15	79.015	50.950	50.950	0.	-1675.692	-1582.275	277.208
300	79.416	51.440	50.951	0.147	-1675.717	-1581.696	275.398
400	96.086	76.779	54.293	8.995	-1676.342	-1550.226	202.439
500	106.131	99.388	61.098	19.145	-1676.045	-1518.718	158.659
600	112.545	119.345	69.177	30.101	-1675.300	-1487.319	129.483
700	116.926	137.041	77.632	41.586	-1674.391	-1456.059	108.652
800	120.135	152.873	86.065	53.447	-1673.498	-1424.931	93.038
900	122.662	167.174	94.296	65.591	-1672.744	-1393.908	80.900
1000	124.771	180.210	102.245	77.965	-1693.394	-1361.437	71.114
1100	126.608	192.189	109.884	90.535	-1692.437	-1328.286	63.075
1200	128.252	203.277	117.211	103.280	-1691.366	-1295.228	56.380
1300	129.737	213.602	124.233	116.180	-1690.190	-1262.264	50.718
1400	131.081	223.267	130.965	129.222	-1688.918	-1229.393	45.869
1500	132.290	232.353	137.425	142.392	-1687.561	-1196.617	41.670
1600	133.361	240.925	143.628	155.675	-1686.128	-1163.934	37.999
1700	134.306	249.039	149.592	169.060	-1684.632	-1131.342	34.762
1800	135.143	256.740	155.333	182.533	-1683.082	-1098.841	31.888
1900	135.896	264.067	160.864	196.085	-1681.489	-1066.426	29.318
2000	136.608	271.056	166.201	209.710	-1679.858	-1034.096	27.008
2100	137.319	277.738	171.354	223.407	-1678.190	-1001.849	24.920
2200	138.030	284.143	176.336	237.174	-1676.485	-969.681	23.023
2300	138.741	290.294	181.158	251.013	-1674.743	-937.593	21.293
2327.000	138.934	291.914	182.434	254.761	ALPHA <--> LIQUID		
2327.000	192.464	339.652	182.434	365.847	TRANSITION		
2400	192.464	345.597	187.307	379.896	-1557.989	-909.127	19.787
2500	192.464	353.454	193.796	399.143	-1550.905	-882.237	18.433
2600	192.464	361.002	200.083	418.389	-1543.853	-855.629	17.190
2700	192.464	368.266	206.179	437.636	-1536.832	-829.292	16.044
2800	192.464	375.265	212.093	456.882	-2117.645	-801.279	14.948
2900	192.464	382.019	217.837	476.128	-2108.494	-754.429	13.589
3000	192.464	388.544	223.419	495.375	-2099.372	-707.892	12.325
3100	192.464	394.855	228.848	514.621	-2090.279	-661.658	11.149
3200	192.464	400.965	234.132	533.868	-2081.214	-615.719	10.051
3300	192.464	406.888	239.277	553.114	-2072.175	-570.063	9.023
3400	192.464	412.633	244.292	572.360	-2063.162	-524.679	8.061
3500	192.464	418.212	249.182	591.607	-2054.176	-479.561	7.157
3600	192.464	423.634	253.953	610.853	-2045.213	-434.699	6.307
3700	192.464	428.908	258.610	630.100	-2036.276	-390.086	5.507
3800	192.464	434.040	263.160	649.346	-2027.363	-345.712	4.752
3900	192.464	439.040	267.606	668.592	-2018.474	-301.574	4.039
4000	192.464	443.912	271.953	687.839	-2009.609	-257.664	3.365

PREVIOUS: CURRENT: December 1979

Aluminum Oxide (Al_2O_3) Al_2O_3(cr,l)

ALUMINUM SILICATE, ANDALUSITE (Al_2SiO_5) CRYSTAL M_r = 162.04558

$\Delta_fH°(0\ K) = -2575.16 \pm 2.1\ kJ\ mol^{-1}$

$S°(298.15\ K) = 93.22 \pm 0.4\ J\ K^{-1}\ mol^{-1}$ $\Delta_fH°(298.15\ K) = -2592.07 \pm 2.1\ kJ\ mol^{-1}$

Enthalpy of Formation

The heat of the reaction $Al_2O_3(cr, \alpha) + SiO_2(quartz) \rightarrow Al_2SiO_5(andalusite)$ has been determined at 968 K by Holm and Kleppa (1,2) to be -1.99 ± 0.17 kcal mol^{-1}. This value was obtained from heat of solution measurements in a high temperature oxide melt calorimeter. Using the JANAF functions, this reduces to $\Delta_rH°(298.15\ K) = -1.42\ kcal\ mol^{-1}$ which is used to calculate the adopted value for the enthalpy of formation.

The above value is in serious disagreement with the solution data of Neumann (3); this work, however, has recently been questioned by several workers. Flood et al. (4) suggested that $\Delta_rH°$ was between 0 and -10 kcal mol^{-1}. Pankratz et al. (5) conclude that $\Delta_rH°$ should lie between 0 and -6 kcal mol^{-1}. Waldbaum (6) calculated $\Delta_rH° = -1.108\ kcal\ mol^{-1}$ from high temperature - high pressure equilibria. These latter values are all in good agreement with the adopted calorimetric value.

Heat Capacity and Entropy

The low temperature heat capacity has been measured from 54.9 to 296.2 K by Todd (7). The entropy is based on S°(51 K) = 0.62 cal $K^{-1}\ mol^{-1}$. The high temperature enthalpy has been measured to 1600 K by Pankratz et al. (5). The low and high temperature data were joined smoothly together by means of a Shomate function plot (8). Since all the aluminum and silicon atoms occupy differently coordinated sites, there is no possibility of any residual entropy of mixing in this polymorph.

References

1. J. L. Holm and O. J. Kleppa, J. Phys. Chem. 70, 1690 (1966).
2. J. L. Holm and O. J. Kleppa, Am. Mineralogist 51, 1608 (1966).
3. F. Neumann, Z. anorg. Allgem. Chem. 145, 193 (1925).
4. H. Flood and W. J. Knapp, J. Am. Ceram. soc. 40, 206 (1957).
5. L. B. Pankratz and K. K. Kelley, U. S. Bur. Mines RI 6370, 1964.
6. C. R. Waldbaum, Am. Mineralogist 50, 186 (1965).
7. S. S. Todd, J. Am. Chem. Soc. 72, 4742 (1950).
8. C. H. Shomate, J. Phys. Chem. 58, 368 (1954).

Aluminum Silicate, Andalusite (Al_2SiO_5) $Al_2O_5Si_1(cr)$

Enthalpy Reference Temperature = T_r = 298.15 K Standard State Pressure = p° = 0.1 MPa

	$J\ K^{-1}mol^{-1}$			$kJ\ mol^{-1}$			
T/K	$C_p°$	S°	$-[G°-H°(T_r)]/T$	$H°-H°(T_r)$	$\Delta_fH°$	$\Delta_fG°$	Log K_f
0	0.	0.	INFINITE	-17.093	-2575.159	-2575.159	INFINITE
100	30.323	13.563	174.899	-16.134	-2582.696	-2538.944	1326.209
200	84.605	51.844	103.323	-10.296	-2588.917	-2492.553	650.988
298.15	122.755	93.221	93.221	0.	-2592.072	-2444.482	428.263
300	123.320	93.982	93.223	0.228	-2592.107	-2443.567	425.463
400	149.565	133.277	98.402	13.950	-2592.950	-2393.872	312.608
500	165.368	168.473	108.968	29.752	-2592.338	-2344.152	244.892
600	174.515	199.471	121.524	46.768	-2591.068	-2294.628	199.765
700	181.067	226.884	134.656	64.560	-2589.556	-2245.340	167.549
800	186.121	251.401	147.744	82.926	-2588.003	-2196.273	143.402
900	190.351	273.572	160.512	101.754	-2586.532	-2147.396	124.632
1000	194.041	293.822	172.845	120.977	-2606.407	-2097.155	109.544
1100	197.380	312.475	184.702	140.550	-2604.612	-2046.314	97.171
1200	200.481	329.783	196.080	160.444	-2602.639	-1995.646	86.868
1300	203.414	345.947	206.993	180.640	-2600.491	-1945.149	78.157
1400	206.225	361.125	217.466	201.122	-2598.175	-1894.824	70.697
1500	208.949	375.446	227.525	221.882	-2595.693	-1844.671	64.237
1600	211.602	389.016	237.198	242.910	-2593.049	-1794.688	58.591
1700	214.200	401.923	246.511	264.200	-2640.423	-1744.428	53.600
1800	216.756	414.238	255.489	285.748	-2637.235	-1691.815	49.095
1900	219.283	426.026	264.157	307.550	-2633.851	-1639.382	45.070
2000	221.781	437.337	272.535	329.604	-2630.273	-1587.133	41.452
2100	224.258	448.218	280.644	351.906	-2626.505	-1535.068	38.183
2200	226.714	458.707	288.500	374.455	-2622.546	-1483.186	35.215
2300	229.149	468.839	296.122	397.248	-2618.399	-1431.490	32.510
2400	231.563	478.642	303.524	420.284	-2614.066	-1379.978	30.034
2500	233.957	488.144	310.720	443.560	-2609.546	-1328.650	27.761
2600	236.329	497.366	317.722	467.075	-2604.842	-1277.506	25.665
2700	238.680	506.329	324.542	490.825	-2599.954	-1226.546	23.729
2800	241.011	515.052	331.191	514.810	-3182.686	-1173.836	21.898
2900	243.321	523.550	337.678	539.027	-3175.242	-1102.224	19.853
3000	245.609	531.837	344.013	563.473	-3167.617	-1030.869	17.949

PREVIOUS: September 1966 CURRENT: September 1967

Aluminum Silicate, Andalusite (Al_2SiO_5) $Al_2O_5Si_1(cr)$

ALUMINUM SILICATE, KYANITE (Al_2SiO_5) CRYSTAL M_r = 162.04558

Aluminum Silicate, Kyanite (Al_2SiO_5) $Al_2O_5Si_1(cr)$

$S°(298.15\ K) = 83.815 \pm 0.4\ J\ K^{-1}mol^{-1}$

$\Delta_fH°\ (0\ K) = -2576.37 \pm 2.1\ kJ\ mol^{-1}$

$\Delta_fH°\ (298.15\ K) = -2594.33 \pm 2.1\ kJ\ mol^{-1}$

Enthalpy of Formation

The heat of the reaction $Al_2O_3(cr, \alpha) + SiO_2(quartz) \rightarrow Al_2SiO_5(cr, kyanite)$ has been determined at 968 K by Holm and Kleppa (1,2) to be $-2.37 \pm 0.15\ kcal\ mol^{-1}$. This value was obtained from heat of solution measurements in a high temperature oxide melt calorimeter. Using the JANAF functions, this reduces to $\Delta_rH°(298.15\ K) = -1.96\ kcal\ mol^{-1}$, which is used to calculate the adopted value for the heat of formation of kyanite.

The above value is in serious disagreement with the solution data of Neumann (3). This work, however, has recently been questioned by several workers. Flood and Knapp (4) suggested that $\Delta_rH°$ was between 0 and $-10\ kcal\ mol^{-1}$. Pankratz and Kelley (5) conclude that $\Delta_rH°$ should lie between 0 and -6 kcal/mol. From phase equilibria, Haskell and DeVries (6) obtained $\Delta_rH° = -2 \pm 7\ kcal\ mol^{-1}$. Waldbaum (7) calculated $\Delta_rH° = -0.622\ kcal\ mol^{-1}$ from high temperature - high pressure equilibria. These latter values are all in good agreement with the adopted calorimetric value.

Heat Capacity and Entropy

The low temperature heat capacity has been measured from 54.8 to 296.3 K by Todd (8). The entropy is based on $S°(51\ K) = 0.33\ cal\ K^{-1}\ mol^{-1}$. The high temperature enthalpy has been measured to 1500K by Pankratz and Kelley (5). The low and high temperature data were joined smoothly together by means of a Shomate function plot (9). The high temperature heat capacity has been measured between 100° and 1200°C by Leonidov, Barskii and Khitarov (10). The values reported are in excellent agreement with the adopted functions. Since the aluminum atoms and the silicon atoms occupy different types of sites there is no possibility of any residual entropy of mixing in this polymorph.

References

1. J. L. Holm and O. J. Kleppa, J. Phys. Chem. 70, 1690 (1966).
2. J. L. Holm and O. J. Kleppa, Am. Mineralogist 51, 1608 (1966).
3. F. Neumann, Z. anorg. allgem. chem. 145, 193 (1925).
4. H. Flood and W. J. Knapp, J. Am. Ceram. Soc. 40, 206 (1957).
5. L. B. Pankratz and K. K. Kelly, U. S. Bur. Mines RI 6370, 1964.
6. R. W. Haskell and R. C. Devries, J. Am. Ceram. Soc. 47, 202 (1964).
7. C. R. Waldbaum, Am. Minnalogist 50, 186 (1965).
8. S. S. Todd, J. Am. Chem. Soc. 72, 4742 (1950).
9. C. H., Shomate, J. Phys. Chem. 58, 368 (1954).
10. V. Ya. Leonidov, Yu. P. Barskii, and N. I. Khitarov, Geochemistry, 409 (1964).

Enthalpy Reference Temperature = T_r = 298.15 K　　Standard State Pressure = p° = 0.1 MPa

	J K⁻¹mol⁻¹			kJ mol⁻¹			
T/K	$C_p°$	S°	$-[G°-H°(T_r)]/T$	$H°-H°(T_r)$	$\Delta_fH°$	$\Delta_fG°$	Log K_f
0	0.	0.	INFINITE	-16.045	-2576.370	-2576.370	INFINITE
100	23.695	8.935	162.808	-15.387	-2584.209	-2539.995	1326.757
200	80.977	43.273	93.769	-10.099	-2590.980	-2492.902	651.079
298.15	121.751	83.815	83.815	0.	-2594.331	-2443.937	428.168
300	122.306	84.570	83.817	0.226	-2594.368	-2443.004	425.365
400	148.277	123.509	88.949	13.824	-2595.336	-2392.351	312.409
500	165.498	158.566	99.433	29.567	-2594.783	-2341.644	244.630
600	176.230	189.745	111.939	46.684	-2593.412	-2291.137	199.461
700	183.238	217.463	125.072	64.674	-2591.701	-2240.890	167.217
800	188.280	242.271	138.199	83.258	-2589.930	-2190.896	143.051
900	192.359	264.685	151.027	102.292	-2588.254	-2141.119	124.267
1000	196.213	285.154	163.431	121.723	-2607.920	-2090.000	109.170
1100	199.761	304.023	175.366	141.523	-2605.899	-2038.304	96.791
1200	203.167	321.554	186.826	161.673	-2603.669	-1986.801	86.483
1300	206.100	337.933	197.827	182.138	-2601.253	-1935.492	77.769
1400	208.861	353.308	208.389	202.887	-2598.670	-1884.376	70.307
1500	211.451	367.806	218.538	223.902	-2595.932	-1833.450	63.846
1600	214.129	381.538	228.300	245.181	-2593.037	-1782.711	58.200
1700	216.727	394.598	237.701	266.724	-2640.158	-1731.712	53.209
1800	219.302	407.059	246.766	288.526	-2636.716	-1678.373	48.705
1900	221.853	418.984	255.519	310.584	-2633.077	-1625.229	44.681
2000	224.380	430.428	263.980	332.896	-2629.241	-1572.283	41.064
2100	226.855	441.436	272.170	355.458	-2625.212	-1519.533	37.796
2200	229.321	452.046	280.107	378.267	-2620.994	-1466.979	34.831
2300	231.776	462.294	287.806	401.322	-2616.585	-1414.623	32.127
2400	234.221	472.210	295.284	424.622	-2611.987	-1362.462	29.653
2500	236.655	481.821	302.554	448.165	-2607.200	-1310.496	27.381
2600	239.079	491.150	309.630	471.952	-2602.224	-1258.725	25.288
2700	241.493	500.218	316.521	495.981	-2597.058	-1207.149	23.354
2800	243.896	509.044	323.240	520.251	-3179.505	-1153.832	21.525
2900	246.289	517.644	329.796	544.760	-3171.768	-1081.625	19.482
3000	248.672	526.034	336.198	569.508	-3163.841	-1009.684	17.580

PREVIOUS: September 1966　　CURRENT: September 1967

Aluminum Silicate, Kyanite (Al_2SiO_5) $Al_2O_5Si_1(cr)$

ALUMINUM SILICATE, SILLIMANITE (Al_2SiO_5) CRYSTAL M_r = 162.04558

$\Delta_f H°(0\ K) = -2572.52 \pm 2.1\ kJ\ mol^{-1}$

$S°(298.15\ K) = 96.179 \pm 0.4\ J\ K^{-1}\ mol^{-1}$ $\Delta_f H°(298.15\ K) = -2589.10 \pm 2.1\ kJ\ mol^{-1}$

Enthalpy of Formation

The heat of the reaction Al_2O_3(cr, α) + SiO_2(quartz) → Al_2SiO_5(sillimanite) has been determined at 968 K by Holm and Kleppa (1, 2) to be −1.51 ± 0.15 kcal mol^{-1}. This value was obtained from heat of solution measurements in a high temperature oxide melt calorimeter. Using the JANAF functions, this reduces to $\Delta_r H°(298.15\ K) = -0.71\ kcal\ mol^{-1}$ which is used to calculate the adopted value for the enthalpy of formation.

The above value is in serious disagreement with the solution data of Neumann (3); this work, however, has recently been questioned by several workers. Flood and Knapp (4) suggested that $\Delta_r H°$ was between 0 and −10 kcal mol^{-1}. Pankratz and Kelley (5) conclude that $\Delta_r H°$ should lie between 0 and −6 kcal mol^{-1}. Waldbaum (6) calculated $\Delta_r H° = -0.356\ kcal\ mol^{-1}$ from high temperature - high pressure equilibria. These latter values are all in good agreement with the adopted calorimetric value.

Heat Capacity and Entropy

The low temperature heat capacity has been measured from 54.4 to 296.5 K by Todd (7). The entropy is based on S°(51 K) = 0.73 cal $K^{-1}\ mol^{-1}$. The high temperature enthalpy has been measured to 1500 K by Pankratz and Kelley (5). The low and high temperature data were joined smoothly together by means of a Shomate function plot (8). The crystal structure of sillimanite indicates that there is no mixing of the tetra-coordinated aluminum and silicon atoms (9). Thus there should be no residual entropy at 0 K.

References

1. J. L. Holm and O. J. Kleppa, J. Phys. Chem. 70, 1690 (1966).
2. J. L. Holm and O. J. Kleppa, Am. Mineralogist 51, 1608 (1966).
3. F. Neumann, Z. anorg. allgem. Chem. 145, 193 (1925).
4. H. Flood and W. J. Knapp, J. Am. Ceram. Soc. 40, 206 (1957).
5. L. B. Pankratz and K. K. Kelley, U. S. Bur. Mines RI 6370, 1964.
6. C. R. Waldbaum, Am. Mineralogist 50, 186 (1965).
7. S. S. Todd, J. Am. Chem. Soc. 72, 4742 (1950).
8. C. H. Shomate, J. Phys. Chem. 58, 368 (1954).
9. L. Bragg and G. F. Claringbull, "Crystal Structure of Minerals", Vol. IV, G. Bell and Sons. Ltd., London, 1965.

Aluminum Silicate, Sillimanite (Al_2SiO_5) $Al_2O_5Si_1$(cr)

Enthalpy Reference Temperature = T_r = 298.15 K Standard State Pressure = p° = 0.1 MPa

T/K	$C_p°$ (J K⁻¹mol⁻¹)	S° (J K⁻¹mol⁻¹)	−[G°−H°(T_r)]/T (J K⁻¹mol⁻¹)	H°−H°(T_r) (kJ mol⁻¹)	$\Delta_f H°$ (kJ mol⁻¹)	$\Delta_f G°$ (kJ mol⁻¹)	Log K_f
0	0.	0.	INFINITE	-17.424	-2572.519	-2572.519	INFINITE
100	32.277	15.099	178.603	-16.350	-2579.942	-2536.344	1324.850
200	85.773	54.593	106.303	-10.342	-2585.993	-2490.179	650.368
298.15	122.215	96.179	96.179	0.	-2589.101	-2442.394	427.897
300	122.769	96.937	96.182	0.227	-2589.137	-2441.483	425.100
400	147.478	135.871	101.322	13.820	-2590.110	-2392.070	312.372
500	163.197	170.578	111.773	29.402	-2589.718	-2342.584	244.728
600	173.025	201.248	124.183	46.239	-2588.627	-2293.253	199.645
700	179.757	228.445	137.172	63.891	-2587.254	-2244.131	167.459
800	184.996	252.798	150.130	82.135	-2585.824	-2195.211	143.333
900	189.510	274.853	162.782	100.864	-2584.452	-2146.468	124.578
1000	193.535	295.031	175.012	120.019	-2604.393	-2096.351	109.502
1100	197.158	313.647	186.780	139.554	-2602.637	-2045.629	97.139
1200	200.765	330.958	198.082	159.451	-2600.661	-1995.078	86.843
1300	204.292	347.168	208.933	179.706	-2598.455	-1944.701	78.139
1400	207.577	362.429	219.357	200.300	-2596.027	-1894.501	70.685
1500	210.777	376.860	229.381	221.218	-2593.386	-1844.484	64.231
1600	213.911	390.563	239.030	242.453	-2590.535	-1794.649	58.589
1700	216.999	403.624	248.331	263.999	-2637.654	-1744.552	53.604
1800	220.045	416.114	257.308	285.852	-2634.161	-1692.117	49.104
1900	223.062	428.092	265.983	308.007	-2630.424	-1639.881	45.083
2000	226.053	439.610	274.378	330.463	-2626.444	-1587.849	41.470
2100	229.013	450.711	282.512	353.216	-2622.224	-1536.022	38.206
2200	231.964	461.432	290.403	376.265	-2617.765	-1484.400	35.244
2300	234.906	471.809	298.066	399.609	-2613.068	-1432.989	32.544
2400	237.840	481.868	305.516	423.246	-2608.133	-1381.787	30.074
2500	240.764	491.637	312.766	447.176	-2602.959	-1330.795	27.805
2600	243.680	501.136	319.829	471.399	-2597.547	-1280.014	25.716
2700	246.586	510.387	326.716	495.912	-2591.897	-1229.446	23.785
2800	249.484	519.408	333.438	520.716	-3173.809	-1177.155	21.960
2900	252.373	528.213	340.003	545.809	-3165.489	-1105.994	19.921
3000	255.253	536.817	346.420	571.190	-3156.929	-1035.121	18.023

PREVIOUS: September 1966 CURRENT: September 1967

Aluminum Silicate, Sillimanite (Al_2SiO_5) $Al_2O_5Si_1$(cr)

ALUMINUM SULFIDE (Al_2S_3) CRYSTAL M_r = 150.14308

$\Delta_fH°(0\ K) = -647.35 \pm 3.8\ kJ\ mol^{-1}$

$S°(298.15\ K) = 116.85 \pm 0.21\ J\ K^{-1}\ mol^{-1}$ $\Delta_fH°(298.15\ K) = -651.03 \pm 3.8\ kJ\ mol^{-1}$

Enthalpy of Formation

The adopted value for $\Delta_fH°(298.15\ K)$ is taken from the recent solution calorimetric studies of Ko et al. (1). The discrepancy between this value and other measured values is quite large. Korshunov (2) and Kapustinskii and Golutvin (3) determined values of $\Delta_fH°(298.15\ K) = -121.5\pm0.4\ kcal\ mol^{-1}$ and $-172.9\pm4.3\ kcal\ mol^{-1}$, respectively, by direct combination of the elements. Sabatier (4) measured a value of $-75.0\ kcal\ mol^{-1}$ for the reaction $Al_2S_3(cr) + 3H_2O(\ell) \rightarrow Al_2O_3 \cdot 3H_2O(cr) + H_2S(g)$. The heat of formation of Al_2S_3 based on this heat has been widely quoted (1, 5) as $-140.5\ kcal\ mol^{-1}$ which is an apparent misinterpretation. Sabatier gives a value of $\Delta_fH°(298.15\ K) = -124.4\ kcal\ mol^{-1}$ based on heats of formation for the other materials as accepted at that time. Our reanalysis of Sabatier's data using currently accepted heats of formation (6, 7) yields $\Delta_fH°(298.15\ K) = -148.0\ kcal\ mol^{-1}$.

Heat Capacity and Entropy

The adopted $C_p°$ data are based on the low temperature $C_p°$ measurements (12 - 312 K) of Ho et al. (1) and the high temperature enthalpy measurements (403 -852 K) of Ferrante and McCune (8). We have not reanalyzed these results except to extrapolate them to higher temperature. The entropy is based on $S°(15\ K) = 0.046\ cal\ K^{-1}\ mol^{-1}$.

Phase Data

This table is for the hexagonal phase of Al_2S_3 usually designated as the α phase (8). Other phases known to exist at higher temperatures are the rhombohedral (γ) and face centered cubic phases. Ferrante and McCune (8) observed an extra enthalpy effect at 878.5 K which they confirmed to be partial conversion to the metastable γ phase via x-ray diffraction. They further showed via x-ray techniques that conversion to the γ phase is essentially complete after heating in the 1320 - 1500 K range. They were unable to measure the enthalpy of the γ phase since conversion to the γ phase appears to be slow and rapid quenching in the drop calorimeter leads to mixed phases. The α - γ transition temperature was reported as 1273 K (9) and melting points of 1403 K (10) and 1373 K (9) were reported.

References

1. H. C. Ko, J. M. Stuve and R. R. Brown, U. S. Bur. Mines RI 8203, 1976.
2. I. A. Korshunov, J. Phys. Chem. (USSR) 13, 703 (1939).
3. A. F. Kapustinskii and Yu. M. Golutvin, Izvest. Akad. Nauk SSSR, Otdel. Khim. Nauk, 192-200 (1951).
4. P. Sabatier, Ann. Chim. Phys. 22, 5 (1891).
5. A. K. Mills, "Thermodynamic Data for Inorganic Sulphides, Selenides and Tellurides", Butterworths, London, 1974.
6. JANAF Thermochemical Tables: $H_2O(\ell)$, 3-31-79; $H_2S(g)$, 6-30-77.
7. J. L. Haas, Jr., G. R. Robinson, Jr. and B. S. Hemingway, J. Phys. Chem. Ref. Data 10, 575 (1981). $\Delta_fH°(cr, gibbsite, 298.15\ K) = -618.2\pm0.3\ kcal\ mol^{-1}$.
8. M. J. Ferrante and R. A. McCune, U.S. Bur. Mines RI 8526, 1981.
9. J. Flahaut, Ann. Chim. (Paris) 7 1, 632 (1952).
10. J. Flahaut, Compt. Rend. 232, 2100 (1951).

Aluminum Sulfide (Al_2S_3) $Al_2S_3(cr)$

Enthalpy Reference Temperature = T_r = 298.15 K Standard State Pressure = $p°$ = 0.1 MPa

T/K	$C_p°$ (J K⁻¹mol⁻¹)	$S°$ (J K⁻¹mol⁻¹)	$-[G°-H°(T_r)]/T$ (J K⁻¹mol⁻¹)	$H°-H°(T_r)$ (kJ mol⁻¹)	$\Delta_fH°$ (kJ mol⁻¹)	$\Delta_fG°$ (kJ mol⁻¹)	Log K_f
0	0.	0.	INFINITE	-18.634	-647.350	-647.350	INFINITE
100	48.677	30.970	197.042	-16.607	-648.360	-646.303	337.594
200	87.571	78.115	126.026	-9.582	-649.864	-643.647	168.103
298.15	105.064	116.847	116.847	0.	-651.030	-640.335	112.184
300	105.282	117.497	116.849	0.195	-651.052	-640.269	111.481
400	114.947	149.264	121.114	11.260	-658.794	-636.244	83.085
500	120.365	175.538	129.449	23.045	-664.059	-630.029	65.819
600	124.135	197.832	139.035	35.278	-668.050	-622.827	54.222
700	127.127	217.199	148.848	47.846	-671.218	-615.032	45.894
800	129.700	234.346	158.484	60.690	-673.970	-606.814	39.621
900	132.039	249.760	167.784	73.778	-835.933	-595.021	34.534
1000	134.051	263.777	176.693	87.084	-856.026	-566.794	29.606
1100	135.825	276.639	185.202	100.580	-854.495	-537.943	25.545
1200	137.327	288.523	193.323	114.240	-852.846	-509.238	22.167
1300	138.566	299.566	201.076	128.037	-851.107	-480.674	19.314
1400	139.536	309.872	208.483	141.944	-849.304	-452.247	16.874
1500	140.235	319.524	215.568	155.935	-847.463	-423.949	14.763

PREVIOUS: CURRENT: December 1979

CHIOLITE ($Na_5Al_3F_{14}$) (CRYSTAL) M_r = 461.871112

$S°_{298.15}$ = 483.54 ± 0.8 J K^{-1} mol^{-1}
T_{fus} = 1010 ± 2 K

$\Delta_f H°$(0 K) = -7554.7 ± 16 kJ mol^{-1}
$\Delta_f H°$(298.15) = -7581.4 ± 16 kJ mol^{-1}
$\Delta_{fus} H°$ = 228.9 ± 8 kJ mol^{-1}

Enthalpy of Formation

Our adopted value of the heat of formation, $\Delta_f H°(Na_5Al_3F_{14}$, cr, 298.15 K) = -1812.0±4 kcal mol^{-1} is based on the emf work of Dewing (1). dewing used solid-electrolyte emf cells of the type Al/Na_3AlF_6, $Na_5Al_3F_{14}$/Na /$Na_5Al_3F_{14}$, AlF_3/Al to determine the free energies for the reaction 5 NaF(cr) + 3 AlF_3(cr) = $Na_5Al_3F_{14}$(cr) over the range 844-996 K. Based on Dewing's (1) reported $\Delta_r G°$ values (20 points), we evaluate $\Delta_r H°$(298.15 K) = -36.58±0.36 kcal mol^{-1} and -39.97±0.42 kcal mol^{-1} by the 2nd and 3rd law methods, respectively. From the 3rd law value, we calculate $\Delta_f H°(Na_5Al_3F_{14}$, cr, 298.15 K) = -1811.95±2.1 kcal mol^{-1}. There has been some confusion over the exact reaction to which Dewing's $\Delta_r G°$ values apply. Both Dewing (1) and Stuve and Ferrante (2) applied the reported $\Delta_r G°$ d values to the erroneous reaction 3 Na_3AlF_6(cr) = $Na_5Al_3F_{14}$(cr) + 4 NaF(cr). Our adopted heat of formation is 58 kcal mol^{-1} less stable than their reported values (1, 2).

Holm (3) estimated $\Delta_r H°$(910 K) = -2.7±1.2 kcal mol^{-1} for the reaction 5/8 NaF(cr) + 3/8 AlF_3(cr) = 1/8 $Na_5Al_3F_{14}$(cr). Converting to a molar basis, we obtain $\Delta_r H°$(910 K) = -21.6±10 kcal mol^{-1} for the reaction 5 NaF(cr) + 3 AlF_3(cr) = $Na_5Al_3F_{14}$(cr). Using JANAF enthalpies (4) and auxiliary data from (4, 5), we obtain $\Delta_f H°(Na_5Al_3F_{14}$, cr, 298.15 K) = -1791.7±12 kcal mol^{-1}.

Grjotheim et al. (6) studied the vapor pressures over the solid systems $Na_5Al_3F_{14}/AlF_3$ and $Na_5Al_3F_{14}/Na_3AlF_6$ by the weight-loss effusion method. Their results are in good agreement with the mass-spectrometric observations of Sidorov et al. (7) on similar systems. Results have been sumamrized by Grjotheim et al. (6) who obtained $\Delta_r H°$(910 K) = 60.65±1.5 kcal mol^{-1} for the reaction 1/5 $Na_5Al_3F_{14}$(cr) + 2/5 AlF_3(cr) = $NaAlF_4$(g). Using JANAF enthalpies (4) and $\Delta_f H°$(298.15 K) values from (4, 5), we obtain $\Delta_f H°(Na_5Al_3F_{14}$, cr, 298.15 K) = -1831.0±20 kcal mol^{-1}. Similarly, Grjotheim et al. (6) obtained $\Delta_r H°$(910 K) = 59.74±2.0 kcal mol^{-1} and recalculated $\Delta_r H°$(910 K) = 60.4±2 kcal mol^{-1} from the data of Sidorov et al. (7) for the reaction 1/2 $Na_5Al_3F_{14}$(cr) = 1/2 Na_3AlF_6(cr) + $NaAlF_4$(g). Using the same auxiliary data as above, we obtain $\Delta_f H°(Na_5Al_3F_{14}$, cr, 298.15 K) = -1808.3±10 kcal mol^{-1} from (6) and $\Delta_f H°(Na_5Al_3F_{14}$, cr, 298.15 K) = -1809.6±10 kcal mol^{-1} from (7). These results are in very good agreement with the recalculated value from Dewing (1).

Heat Capacity and Entropy

Low-temperature heat capacities of chiolite were determined by Stuve and Ferrante (2) by adiabatic calorimetry over the range 5.83-315.75 K. A slight bump was noted in the heat capacity curve between 140 and 175 K. S°(298.15 K) = 115.57±0.2 K^{-1} mol^{-1} is based on S°(5 K) = 0.022 cal K^{-1} mol^{-1} obtained by extrapolating the function Cp/T versus T^2 (2). H°(298.15 K)-H°(0 K) = 19.338 kcal mol^{-1}.

High temperature enthalpies were measured by Stuve and Ferrante (2) (402.8-991.5 K) and by Holm (3) (677.0-931.7 K). The data of Holm are consistently higher (2.3±0.3 kcal, corresponding to 3-4%) than the data of Stuve and Ferrante and show more scatter. This positive bias ascribed to the calorimeter of Holm was also evident in a study on cryolite (see the Na_3AlF_6, cr, β table). We adopt heat capacities for chiolite based on the drop calorimetry work of Stuve and ferante (2).

Fusion Data

Chiolite is the mineral name for $Na_5Al_3F_{14}$ (corresponds to 5 NaF + 3 AlF_3). The crystals are tetragonal, space group P_4/mnc (8). Chiolite melts incongruently at 1010 K (1, 3, 6) to a combination of cryolite and AlF_3. The value of $\Delta_{fus}H°$ for this process is calculated as the difference between the enthalpies of the crystal and liquid at 1010 K.

References

1. E. W. Dewing, Met. Trans. 2, 2211 (1970).
2. J. M. Stuve and M. J. Ferrante, U.S. Bur. Mines RI 8442, 1980.
3. J. L. Holm, High Temp. Sci. 6, 16 (1974).
4. JANAF Thermochemical tables: NaF(cr), 12-31-68; AlF_3(cr), 9-30-79.
5. U.S. Nat. Bur. Stand. IR 75-968, 1976.
6. K. Grjotheim, K. Motzfeldt, and D. B. Rao, paper No. A71-22 presented at the Metallurgical Society of AIME conference, TMS Session, New York, 1971.
7. L. N. Sidorov, E. N. Kolosov and V. B. Shcl'ts, Zh. Fiz. Khim. 42, 2620 (1968).
8. J. D. H. Donnay and H. M. Ondik, "Crystal Data, Determinative Tables, Third Edition," Vol. II, U. S. Dept. of Commerce, 1973.

Chiolite ($Na_5Al_3F_{14}$) $Al_3F_{14}Na_5$(cr)

Enthalpy Reference Temperature = T_r = 298.15 K Standard State Pressure = p° = 0.1 MPa

T/K	$C°_p$ (J $K^{-1}mol^{-1}$)	S° (J $K^{-1}mol^{-1}$)	-[G°-H°(T_r)]/T (J $K^{-1}mol^{-1}$)	H°-H°(T_r) (kJ mol^{-1})	$\Delta_f H°$ (kJ mol^{-1})	$\Delta_f G°$ (kJ mol^{-1})	Log K_f
0	0.	0.	INFINITE	-80.910	-7554.692	-7554.692	INFINITE
100	214.723	142.800	861.276	-71.848	-7574.128	-7455.146	3894.169
200	370.912	317.691	522.895	-41.041	-7581.508	-7325.950	1913.342
298.15	454.006	483.545	483.545	0.	-7581.408	-7200.348	1261.471
300	455.094	486.357	483.554	0.841	-7581.368	-7197.984	1253.280
400	499.570	623.823	501.993	48.732	-7591.644	-7069.564	923.190
500	525.092	738.188	538.126	100.031	-7587.309	-6939.508	724.965
600	542.665	835.534	579.783	153.451	-7581.492	-6810.474	592.904
700	556.472	920.245	622.500	208.421	-7574.685	-6682.498	498.654
800	568.606	995.357	664.500	264.686	-7567.175	-6555.548	428.033
900	579.484	1062.962	705.080	322.094	-7559.205	-6429.570	373.163
1000	589.944	1124.591	743.994	380.597	-7582.715	-6302.216	329.194
1010.000	590.781	1130.465	747.792	386.500	------ CRYSTAL <--> LIQUID -----		
1100	600.319	1181.296	781.205	440.100	-7573.318	-6174.615	293.208
1200	610.864	1233.983	816.767	500.660	-8047.057	-6035.700	262.727
1300	621.324	1283.291	850.776	562.269	-8031.751	-5868.701	235.807
1400	631.784	1329.719	883.344	624.925	-8015.548	-5702.916	212.778
1500	642.244	1373.664	914.580	688.626	-7998.439	-5538.319	192.861

PREVIOUS: CURRENT: December 1979

Chiolite ($Na_5Al_3F_{14}$) $Al_3F_{14}Na_5$(cr)

CHIOLITE ($Na_5Al_3F_{14}$) LIQUID M_r = 461.871112

$S°(298.15 K) = [532.124] J K^{-1} mol^{-1}$ $\Delta_f H°(298.15 K) = [-7497.408] kJ mol^{-1}$

$T_{fus} = 1010 \pm 2 K$ $\Delta_{fus}H° = 228.9 \pm 8 kJ mol^{-1}$

Enthalpy of Formation

$\Delta_f H°(Na_5Al_3F_{14}$, ℓ, 298.15 K) is calculated from $\Delta_f H°(Na_5Al_3F_{14}$, cr, 298.15 K) by adding the enthalpy of fusion, $\Delta_{fus}H°$, and the difference in enthalpy, H°(1010 K)-H°(298.15 K), between the crystal and liquid.

Hong and Kleppa (2) measured the enthalpies of mixing for the reaction 5/8 NaF(ℓ) + 3/8 AlF (cr) = 1/8 $Na_5Al_3F_{14}$(ℓ) as $\Delta_r H°$(298.15 K) = -0.591 kcal. Converting to a molar basis, we obtain $\Delta_r H°(1298 K) = -4.725 kcal mol^{-1}$ for the reaction 5 NaF(ℓ) + 3 AlF_3(cr) = $Na_5Al_3F_{14}$(ℓ), using JANAF data (1) for the enthalpies of formation from (1, 3), we obtain $\Delta_f H°(Na_5Al_3F_{14}$, ℓ, 298.15 K) = $-1793.5 \pm 10 kcal mol^{-1}$.

Hong and Kleppa (4) also measured the enthalpy of mixing for the reaction 5/9 Na_3AlF_6(ℓ) + 4/9 AlF_3(cr) = 1/3 $Na_5Al_3F_{14}$(ℓ) as $\Delta_r H°$(1298 K) = 4.49 kcal. Converting to a molar basis, we obtain $\Delta_r H°(1298 K) = 40.40 kcal mol^{-1}$ for the reaction 5 Na_3AlF_6(ℓ) + 4 AlF_3(cr) = 3 $Na_5Al_3F_{14}$(ℓ). Using enthalpies from JANAF (1) and enthalpies of formation from (1, 3), we obtain $\Delta_f H°(Na_5Al_3F_{14}$, ℓ, 298.15 K) = $-1794.1 \pm 10 kcal mol^{-1}$. These results are in good agreement with adopted enthalpy of formation.

Heat Capacity and Entropy

High temperature enthalpies were measured by drop calorimetry by Holm (5, 1046.6-1172.9 K). Our analysis of his experimental data leads to a constant heat capacity of $232.76 cal K^{-1} mol^{-1}$. We have corrected his experimental enthalpies for what we believe to be a positive bias by subtracting $2.3 kcal mol^{-1}$ from his measured data. Our adopted enthalpies can be represented by the equation H°(T)-H°(298.15 K) = $-88,000 + 232.76 T cal mol^{-1}$ (1010-1200 K). A hypothetical glass transition is assumed at 650 K. The $C_p°$ values above 1172.9 K are obtained by linear extrapolation. The entropy, S°(298.15 K), is calculated in a manner analogous to that used for the enthalpy of formation.

Fusion Data

See the crystal table for details.

References

1. JANAF Thermochemical Tables: $Na_5Al_3F_{14}$(cr) and Na_3AlF_6(ℓ), 12-30-79; AlF_3(cr), 9-30-79.
2. K. C. Hong and U. J. Kleppa, J. Phys. Chem. 82, 176 (1978).
3. U. S. Nat. Bur. Stand. IR 75-968, 1976.
4. K. C. Hong and U. J. Kleppa, High Temp. Sci. 8, 299 (1976).
5. J. L. Holm, High Temp. Sci. 6, 16 (1974).

Chiolite ($Na_5Al_3F_{14}$) $Al_3F_{14}Na_5$(l)

Enthalpy Reference Temperature = T_r = 298.15 K Standard State Pressure = p° = 0.1 MPa

	J K⁻¹mol⁻¹			kJ mol⁻¹			
T/K	$C_p°$	S°	$-[G°-H°(T_r)]/T$	$H°-H°(T_r)$	$\Delta_f H°$	$\Delta_f G°$	Log K_f
0							
100							
200							
298.15	454.006	532.124	532.124	0.	-7497.408	-7130.832	1249.292
300	455.094	534.935	532.132	0.841	-7497.367	-7128.557	1241.192
400	499.570	672.402	550.572	48.732	-7507.644	-7004.995	914.759
500	525.092	786.767	586.704	100.031	-7503.309	-6879.797	718.727
600	542.665	884.124	628.363	153.456	-7497.486	-6755.621	588.129
650.000	549.778	927.848	649.737	180.772	GLASS <--> LIQUID		
650.000	973.868	927.848	649.737	180.772	TRANSITION		
700	973.868	1000.020	672.212	229.465	-7469.641	-6633.296	494.983
800	973.868	1130.062	721.496	326.852	-7421.008	-6517.144	425.526
900	973.868	1244.767	773.390	424.239	-7373.059	-6407.048	371.855
1000	973.868	1347.374	825.748	521.626	-7357.686	-6299.969	329.077
1010.000	973.868	1357.064	830.961	531.365	------ CRYSTAL <--> LIQUID -----		
1100	973.868	1440.193	877.455	619.013	-7310.405	-6196.490	294.247
1200	973.868	1524.931	927.932	716.399	-7747.317	-6085.098	264.877
1300	973.868	1602.882	976.893	813.786	-7696.233	-5948.652	239.020
1400	973.868	1675.053	1024.216	911.173	-7645.300	-5816.135	217.003
1500	973.868	1742.243	1069.870	1008.560	-7594.505	-5687.253	198.048
1600	973.868	1805.095	1113.879	1105.947	-7543.838	-5561.756	181.573
1700	973.868	1864.136	1156.293	1203.333	-7493.288	-5439.424	167.133
1800	973.868	1919.800	1197.178	1300.720	-7442.840	-5320.064	154.384
1900	973.868	1972.455	1236.609	1398.107	-7392.479	-5203.508	143.054
2000	973.868	2022.408	1274.661	1495.494	-7342.185	-5089.599	132.927
2100	973.868	2069.923	1311.408	1592.880	-7291.938	-4978.207	123.826
2200	973.868	2115.227	1346.924	1690.267	-7241.715	-4869.206	115.610
2300	973.868	2158.517	1381.276	1787.654	-7191.493	-4762.488	108.160
2400	973.868	2199.965	1414.531	1885.041	-7141.249	-4657.956	101.378
2500	973.868	2239.720	1446.749	1982.428	-7090.962	-4555.519	95.182
2600	973.868	2277.916	1477.987	2079.814	-7040.610	-4455.095	89.504
2700	973.868	2314.670	1508.299	2177.201	-6990.174	-4356.609	84.284
2800	973.868	2350.087	1537.734	2274.588	-7821.342	-4257.091	79.417
2900	973.868	2384.261	1566.339	2371.975	-7767.406	-4130.747	74.403
3000	973.868	2417.277	1594.157	2469.362	-7713.346	-4006.267	69.755

PREVIOUS: CURRENT: December 1979

Chiolite ($Na_5Al_3F_{14}$) $Al_3F_{14}Na_5$(l)

CHIOLITE ($Na_5Al_3F_{14}$) CRYSTAL-LIQUID M_r = 461.871112

0 to 1010 K crystal
above 1010 K liquid

Refer to the individual tables for details.

Adopted enthalpy equations used to prepare this tabulation:

$Na_5Al_3F_{14}$(cr)
$H_T - H_{298} = 17.894 \times 10^5/T - 43,416 + 122.461\ T + 10.158 \times 10^{-3}\ T^2$ cal mol^{-1}(298.15-1010 K)

$Na_5Al_3F_{14}$(ℓ)
$H_T - H_{298} = -88,000 + 232.76\ T$ cal mol^{-1}(1010-1200 K)

Chiolite ($Na_5Al_3F_{14}$) $Al_3F_{14}Na_5$(cr,l)

Enthalpy Reference Temperature = T_r = 298.15 K — Standard State Pressure = p° = 0.1 MPa

T/K	C_p° (J K⁻¹mol⁻¹)	S° (J K⁻¹mol⁻¹)	−[G°−H°(T_r)]/T (J K⁻¹mol⁻¹)	H°−H°(T_r) (kJ mol⁻¹)	$\Delta_f H^\circ$ (kJ mol⁻¹)	$\Delta_f G^\circ$ (kJ mol⁻¹)	Log K_f
0	0.	0.	INFINITE	-80.910	-7554.692	-7554.692	INFINITE
100	214.723	142.800	861.276	-71.848	-7574.128	-7455.146	3894.169
200	370.912	317.691	522.895	-41.041	-7581.508	-7325.950	1913.342
298.15	454.006	483.545	483.545	0.	-7581.408	-7200.348	1261.471
300	455.094	486.357	483.554	0.841	-7581.368	-7197.984	1253.280
400	499.570	623.823	501.993	48.732	-7591.644	-7069.564	923.190
500	525.092	738.188	538.126	100.031	-7587.309	-6939.508	724.965
600	542.665	835.534	579.783	153.451	-7581.492	-6810.474	592.904
700	556.472	920.245	622.500	208.421	-7574.685	-6682.498	498.654
800	568.606	995.357	664.500	264.686	-7567.175	-6555.548	428.033
900	579.484	1062.962	705.080	322.094	-7559.205	-6429.570	373.163
1000	589.944	1124.591	743.994	380.597	-7582.715	-6302.216	329.194
1010.000	590.781	1130.465	747.792	386.500	CRYSTAL <--> LIQUID		
1010.000	973.868	1357.064	747.792	615.365	TRANSITION		
1100	973.868	1440.193	801.090	703.013	-7310.405	-6196.490	294.247
1200	973.868	1524.931	857.931	800.400	-7747.317	-6085.097	264.877
1300	973.868	1602.882	912.277	897.787	-7696.233	-5948.652	239.020
1400	973.868	1675.053	964.215	995.174	-7645.299	-5816.135	217.003
1500	973.868	1742.243	1013.870	1092.560	-7594.504	-5687.253	198.048
1600	973.868	1805.095	1061.378	1189.947	-7543.838	-5561.756	181.573
1700	973.868	1864.136	1106.880	1287.334	-7493.287	-5439.424	167.133
1800	973.868	1919.800	1150.511	1384.721	-7442.840	-5320.064	154.384
1900	973.868	1972.455	1192.398	1482.108	-7392.479	-5203.507	143.054
2000	973.868	2022.408	1232.660	1579.494	-7342.185	-5089.598	132.927
2100	973.868	2069.923	1271.408	1676.881	-7291.937	-4978.206	123.826
2200	973.868	2115.227	1308.742	1774.268	-7241.714	-4869.206	115.610
2300	973.868	2158.517	1344.754	1871.655	-7191.493	-4762.488	108.159
2400	973.868	2199.965	1379.531	1969.042	-7141.249	-4657.956	101.378
2500	973.868	2239.720	1413.149	2066.428	-7090.962	-4555.518	95.182
2600	973.868	2277.916	1445.679	2163.815	-7040.610	-4455.094	89.504
2700	973.868	2314.670	1477.188	2261.202	-6990.174	-4356.609	84.284
2800	973.868	2350.087	1507.734	2358.589	-7821.342	-4257.090	79.417
2900	973.868	2384.261	1537.373	2455.975	-7767.406	-4130.747	74.403
3000	973.868	2417.277	1566.156	2553.362	-7713.346	-4006.267	69.755

PREVIOUS: CURRENT: December 1979

Chiolite ($Na_5Al_3F_{14}$) $Al_3F_{14}Na_5$(cr,l)

$S°(298.15\ K) = 175.55 \pm 0.3\ J\ K^{-1}\ mol^{-1}$

$T_{fus} = 2186 \pm 10\ K$

$\Delta_fH°(0\ K) = -5585.7 \pm 5\ kJ\ mol^{-1}$

$\Delta_fH°(298.15\ K) = -5624.1 \pm 5\ kJ\ mol^{-1}$

$\Delta_{fus}H° = [402]\ kJ\ mol^{-1}$

Enthalpy of Formation

$\Delta_fH°(Al_6BeO_{10}$, cr, 298.15 K) is calculated from $\Delta_rH°(298.15\ K) = 2.71 \pm 0.45\ kcal\ mol^{-1}$ for BeO(cr, α) + 3 Al_2O_3(cr, α) → Al_6BeO_{10}(cr) using current value of $\Delta_fH°$ for the reactants (1). $\Delta_rH°$ is based on the value 2.9 ± 0.45 $kcal\ mol^{-1}$ at 968 K determined by Holm and Kleppa (2) from the difference in heats of solution of the three components in an oxide melt.

For comparison with the above reaction, we derive $\Delta_rH°(298.15\ K) = 0.6\ kcal\ mol^{-1}$ from $\Delta_rG°(1800\ K) = -3.95\ kcal\ mol^{-1}$ based on equilibrium data. Young (3) used water vapor as a carrier gas to determine the transport of BeO, presumably as $Be(OH)_2$, in equilibrium with BeO(cr) and Al_6BeO_{10}(cr). The difference between the two equilibria gives the above reaction. We compare at 1800 K, near the upper limit of the data, where the experimental method is most sensitive. The resulting difference in $\Delta_rH°(298.15\ K)$ is -2.1 kcal/mol, roughly twice the uncertainty claimed for $\Delta_rG°$ (3).

Heat Capacity and Entropy

$C_p°$ and the derived properties below 298.15 K are taken from calorimetric data (15-380 K) of Furukawa and Saba (4). The entropy is based on $S°(15\ K) = 0.011\ cal\ K^{-1}\ mol^{-1}$. The heat capacity is extrapolated above 1200 K by comparison with current values (1) for $BeAl_2O_4$(cr), Al_2O_3(cr, α), and BeO(cr, α).

Holm and Kleppa (2) combined their $\Delta_rH°$ for formation from the oxides with the corresponding $\Delta_rG°$ from Young (3) in order to calculate $\Delta_fS°$. Based on the present functions and Young's result at 1800 K, we calculate an entropy which is 1.2 $cal\ K^{-1}\ mol^{-1}$ larger than the adopted calorimetric value. To explain the difference, Holm and Kleppa (2) proposed Be-Al disorder in the crystal, which might provide extra entropy up to 2.76 $cal\ K^{-1}\ mol^{-1}$. Crystal structure data were recently published (6), but they do not appear to be sufficient to test the hypothesis of disorder. Although we believe that the discrepancy is within the uncertainty of the equilibrium data, further information is desirable.

Fusion Data

See the liquid table for details.

References

1. JANAF Thermochemical Tables: BeO(cr, α), 6-30-71; Al_2O_3(cr, α) and $BeAl_2O_4$(cr), 6-30-77.
2. J. L. Holm and O. J. Kleppa, Acta Chem. Scand. 20, 2568 (1966).
3. W. A. Young, J. Phys. Chem. 64, 1003 (1960).
4. G. T. Furukawa and W. G. Saba, J. Res. Nat. Bur. Stand. 71A, 3 (1967).
5. D. A. Ditmars and T. B. Douglas, J. Res. Nat. Bur. Stand. 71A, 97 (1967).
6. L. A. Harris and H. L. Yakel, J. Amer. Ceram. Soc. 53, 359 (1970).

Enthalpy Reference Temperature = T_r = 298.15 K; Standard State Pressure = p° = 0.1 MPa

T/K	$C_p°$ (J K⁻¹mol⁻¹)	S° (J K⁻¹mol⁻¹)	$-[G°-H°(T_r)]/T$ (J K⁻¹mol⁻¹)	$H°-H°(T_r)$ (kJ mol⁻¹)	$\Delta_fH°$ (kJ mol⁻¹)	$\Delta_fG°$ (kJ mol⁻¹)	Log K_f
0	0.	0.	INFINITE	-34.112	-5585.659	-5585.659	INFINITE
100	47.049	16.606	345.259	-32.865	-5601.878	-5512.643	2879.509
200	172.929	87.939	197.142	-21.840	-5616.797	-5417.092	1414.799
298.15	265.186	175.552	175.552	0.	-5624.133	-5317.254	931.561
300	266.546	177.197	175.557	0.492	-5624.212	-5315.350	925.485
400	324.268	262.481	186.792	30.275	-5626.175	-5211.985	680.616
500	358.527	338.838	209.722	64.558	-5625.087	-5108.521	533.684
600	380.560	406.278	236.976	101.581	-5622.541	-5005.431	435.762
700	396.083	466.167	265.522	140.452	-5619.419	-4902.822	365.853
800	407.827	519.854	294.016	180.671	-5616.267	-4800.669	313.452
900	417.249	568.451	321.852	221.939	-5613.464	-4698.893	272.717
1000	425.178	612.834	348.763	264.071	-5674.818	-4592.838	239.905
1100	431.998	653.685	374.651	306.937	-5671.319	-4484.805	212.966
1200	437.939	691.532	399.500	350.439	-5667.470	-4377.108	190.531
1300	443.295	726.800	423.336	394.503	-5663.316	-4269.745	171.560
1400	448.316	759.835	446.204	439.084	-5658.886	-4162.711	155.313
1500	453.336	790.936	468.159	484.165	-5654.182	-4056.005	141.243
1600	458.566	820.358	489.260	529.756	-5663.723	-3949.095	128.925
1700	464.215	848.326	509.566	575.892	-5657.995	-3842.104	118.053
1800	470.282	875.028	529.133	622.612	-5651.825	-3735.464	108.400
1900	476.976	900.632	548.016	669.969	-5645.155	-3629.179	99.773
2000	484.298	925.281	566.267	718.028	-5637.922	-3523.259	92.018
2100	492.248	949.100	583.933	766.850	-5630.062	-3417.716	85.011
2186.000	499.586	969.001	598.692	809.495	------ CRYSTAL <--> LIQUID -----		
2200	500.825	972.195	601.059	816.498	-5621.511	-3312.561	78.650
2300	510.030	994.658	617.686	867.036	-5612.205	-3207.816	72.852
2400	519.862	1016.570	633.851	918.525	-5602.079	-3103.493	67.546
2500	530.322	1038.002	649.590	971.029	-5591.070	-2999.608	62.673
2600	541.410	1059.015	664.934	1024.610	-5579.112	-2896.182	58.185
2700	553.125	1079.666	679.914	1079.332	-5566.140	-2793.234	54.038
2800	565.468	1100.003	694.554	1135.256	-7606.453	-2678.757	49.973
2900	578.438	1120.071	708.882	1192.446	-7583.620	-2503.171	45.087
3000	592.036	1139.908	722.920	1250.965	-7559.563	-2328.388	40.541

PREVIOUS: CURRENT: June 1972

BERYLLIUM ALUMINUM OXIDE ($BeAl_6O_{10}$) LIQUID M_r = 330.89542

$S°(298.15\ K) = [314.902]\ J\ K^{-1}\ mol^{-1}$ $\Delta_fH°(298.15\ K) = [-5299.313]\ kJ\ mol^{-1}$

$T_{fus} = 2186 \pm 10\ K$ $\Delta_{fus}H° = [402]\ kJ\ mol^{-1}$

Enthalpy of Formation

$\Delta_fH°(Al_6BeO_{10}$, ℓ, 298.15 K) is calculated from $\Delta_fH°(Al_6BeO_{10}$, cr, 298.15 K) by adding the enthalpy of fusion, $\Delta_{fus}H°$, and the difference in enthalpy, H°(2186 K) - H°(298.15 K), between the crystal and liquid.

Heat Capacity and Entropy

$C_p°$ is assumed to be 136 cal $K^{-1}\ mol^{-1}$, corresponding to 8 cal $K^{-1}g\text{-}atom^{-1}$, by analogy with current tables (1) for $BeAl_2O_4$(ℓ) and Al_2O_3(ℓ). Below the assumed glass transition at 1400 K, $C_p°$ is taken to be the same as that of the crystal. The entropy is calculated in a manner analogous to that used for the enthalpy of formation.

Fusion Data

T_{fus}, adjusted to IPTS-68, is from the phase study of Lang et al. (2). $\Delta_{fus}H°$ is estimated as being slightly smaller than the sum of $\Delta_{fus}H°$ for the component oxides (1). Data for $BeAl_2O_4$ are consistent with this approach.

References

1. JANAF Thermochemical Tables: BeO(ℓ), 6-30-71; Al_2O_3(ℓ) and $BeAl_2O_4$(ℓ), 6-30-72.
2. S. M. Lang, C. L. Fillmore, and L. H. Maxwell, J. Res. Nat. Bur. Stand. 48, 298 (1952).

Beryllium Aluminum Oxide ($BeAl_6O_{10}$) $Al_6Be_1O_{10}(l)$

Enthalpy Reference Temperature = T_r = 298.15 K Standard State Pressure = p° = 0.1 MPa

T/K	$C_p°$ (J K⁻¹mol⁻¹)	S° (J K⁻¹mol⁻¹)	-[G°-H°(T_r)]/T (J K⁻¹mol⁻¹)	H°-H°(T_r) (kJ mol⁻¹)	$\Delta_fH°$ (kJ mol⁻¹)	$\Delta_fG°$ (kJ mol⁻¹)	Log K_f
0							
100							
200							
298.15	265.186	314.902	314.902	0.	-5299.313	-5033.981	881.933
300	266.546	316.547	314.907	0.492	-5299.392	-5032.335	876.207
400	324.268	401.830	326.142	30.275	-5301.355	-4942.905	645.477
500	358.527	478.187	349.072	64.558	-5300.266	-4853.375	507.029
600	380.560	545.628	376.326	101.581	-5297.721	-4764.221	414.762
700	396.083	605.517	404.871	140.452	-5294.599	-4675.547	348.894
800	407.827	659.204	433.366	180.671	-5291.447	-4587.329	299.522
900	417.249	707.801	461.202	221.939	-5288.643	-4499.488	261.143
1000	425.178	752.184	488.113	264.071	-5349.998	-4407.368	230.217
1100	431.998	793.034	514.000	306.937	-5346.499	-4313.270	204.820
1200	437.939	830.882	538.850	350.439	-5342.650	-4219.508	183.670
1300	443.295	866.150	562.685	394.503	-5338.496	-4126.079	165.788
1400	448.316	899.187	585.554	439.087	-5334.063	-4032.981	150.472
1400.010	448.316	899.190	585.556	439.091	GLASS <--> LIQUID TRANSITION		
1400.010	569.024	899.190	585.556	439.091			
1500	569.024	938.445	607.786	495.988	-5317.539	-3940.625	137.225
1600	569.024	975.169	629.612	552.890	-5315.768	-3848.838	125.652
1700	569.024	1009.666	650.964	609.793	-5299.274	-3757.661	115.459
1800	569.024	1042.190	671.804	666.695	-5282.921	-3667.452	106.427
1900	569.024	1072.956	692.115	723.598	-5266.707	-3578.146	98.370
2000	569.024	1102.143	711.893	780.500	-5250.630	-3489.690	91.141
2100	569.024	1129.906	731.143	837.402	-5234.690	-3402.036	84.621
2186.000	569.024	1152.745	747.284	886.338	------ CRYSTAL <--> LIQUID -----		
2200	569.024	1156.377	749.874	894.305	-5218.884	-3315.134	78.711
2300	569.024	1181.671	768.102	951.207	-5203.213	-3228.954	73.332
2400	569.024	1205.888	785.843	1008.110	-5187.674	-3143.452	68.415
2500	569.024	1229.117	803.112	1065.012	-5172.267	-3058.593	63.906
2600	569.024	1251.434	819.929	1121.914	-5156.987	-2974.347	59.755
2700	569.024	1272.910	836.311	1178.817	-5141.835	-2890.686	55.924
2800	569.024	1293.604	852.275	1235.719	-7181.169	-2795.555	52.152
2900	569.024	1313.571	867.840	1292.622	-7158.625	-2639.328	47.539
3000	569.024	1332.862	883.021	1349.524	-7136.184	-2483.871	43.248
3100	569.024	1351.520	897.834	1406.426	-7113.842	-2329.160	39.246
3200	569.024	1369.586	912.296	1463.329	-7091.602	-2175.174	35.506
3300	569.024	1387.096	926.420	1520.231	-7069.458	-2021.880	32.004
3400	569.024	1404.083	940.220	1577.134	-7047.413	-1869.252	28.718
3500	569.024	1420.578	953.710	1634.036	-7025.465	-1717.275	25.629
3600	569.024	1436.608	966.902	1690.938	-7003.614	-1565.920	22.721
3700	569.024	1452.198	979.809	1747.841	-6981.861	-1415.175	19.979
3800	569.024	1467.373	992.441	1804.743	-6960.208	-1265.010	17.389
3900	569.024	1482.154	1004.809	1861.646	-6938.653	-1115.416	14.939
4000	569.024	1496.560	1016.923	1918.548	-6917.202	-966.378	12.620

PREVIOUS: CURRENT: June 1972

Beryllium Aluminum Oxide ($BeAl_6O_{10}$) $Al_6Be_1O_{10}(l)$

BERYLLIUM ALUMINUM OXIDE ($BeAl_6O_{10}$) CRYSTAL–LIQUID

0 to 2186 K crystal
above 2186 K liquid

Refer to the individual tables for details.

M_r = 330.89542 Beryllium Aluminum Oxide ($BeAl_6O_{10}$) $Al_6Be_1O_{10}$(cr,l)

Enthalpy Reference Temperature = T_r = 298.15 K Standard State Pressure = p° = 0.1 MPa

T/K	C_p° (J K⁻¹mol⁻¹)	S° (J K⁻¹mol⁻¹)	−[G°−H°(T_r)]/T (J K⁻¹mol⁻¹)	H°−H°(T_r) (kJ mol⁻¹)	Δ_fH° (kJ mol⁻¹)	Δ_fG° (kJ mol⁻¹)	Log K_f
0	0.	0.	INFINITE	-34.112	-5585.659	-5585.659	INFINITE
100	47.049	16.606	345.259	-32.865	-5601.878	-5512.643	2879.509
200	172.929	87.939	197.142	-21.840	-5616.797	-5417.092	1414.799
298.15	265.186	175.552	175.552	0.	-5624.133	-5317.254	931.561
300	266.546	177.197	175.557	0.492	-5624.212	-5315.350	925.485
400	324.268	262.481	186.792	30.275	-5626.175	-5211.985	680.616
500	358.527	338.838	209.722	64.558	-5625.087	-5108.521	533.684
600	380.560	406.278	236.976	101.581	-5622.541	-5005.431	435.762
700	396.083	466.167	265.522	140.452	-5619.419	-4902.822	365.853
800	407.827	519.854	294.016	180.671	-5616.267	-4800.669	313.452
900	417.249	568.451	321.852	221.939	-5613.464	-4698.893	272.717
1000	425.178	612.834	348.763	264.071	-5674.818	-4592.838	239.905
1100	431.998	653.685	374.651	306.937	-5671.319	-4484.805	212.966
1200	437.939	691.532	399.500	350.439	-5667.470	-4377.108	190.531
1300	443.295	726.800	423.336	394.503	-5663.316	-4269.745	171.560
1400	448.316	759.835	446.204	439.084	-5658.886	-4162.711	155.313
1500	453.336	790.936	468.159	484.165	-5654.182	-4056.005	141.243
1600	458.566	820.358	489.260	529.756	-5663.723	-3949.095	128.925
1700	464.215	848.326	509.566	575.892	-5657.995	-3842.104	118.053
1800	470.282	875.028	529.133	622.612	-5651.825	-3735.464	108.400
1900	476.976	900.632	548.016	669.969	-5645.155	-3629.179	99.773
2000	484.298	925.281	566.267	718.028	-5637.922	-3523.259	92.018
2100	492.248	949.100	583.933	766.850	-5630.062	-3417.716	85.011
2186.000	499.586	969.001	598.692	809.495	CRYSTAL <--> LIQUID		
2186.000	569.024	1152.745	598.692	1211.159	TRANSITION		
2200	569.024	1156.378	602.230	1219.126	-5218.884	-3315.136	78.711
2300	569.024	1181.672	626.877	1276.028	-5203.212	-3228.956	73.332
2400	569.024	1205.889	650.502	1332.930	-5187.673	-3143.454	68.415
2500	569.024	1229.118	673.185	1389.833	-5172.266	-3058.595	63.906
2600	569.024	1251.436	694.999	1446.735	-5156.987	-2974.350	59.755
2700	569.024	1272.911	716.008	1503.638	-5141.834	-2890.689	55.924
2800	569.024	1293.605	736.269	1560.540	-7181.169	-2795.558	52.152
2900	569.024	1313.573	755.834	1617.442	-7158.624	-2639.331	47.539
3000	569.024	1332.863	774.748	1674.345	-7136.183	-2483.874	43.248
3100	569.024	1351.522	793.055	1731.247	-7113.842	-2329.163	39.246
3200	569.024	1369.587	810.791	1788.150	-7091.601	-2175.178	35.506
3300	569.024	1387.097	827.991	1845.052	-7069.458	-2021.884	32.004
3400	569.024	1404.084	844.686	1901.954	-7047.412	-1869.255	28.718
3500	569.024	1420.579	860.905	1958.857	-7025.465	-1717.278	25.629
3600	569.024	1436.609	876.676	2015.759	-7003.614	-1565.924	22.721
3700	569.024	1452.199	892.021	2072.662	-6981.861	-1415.179	19.979
3800	569.024	1467.374	906.963	2129.564	-6960.207	-1265.014	17.389
3900	569.024	1482.155	921.523	2186.466	-6938.652	-1115.420	14.939
4000	569.024	1496.561	935.719	2243.369	-6917.201	-966.382	12.620

PREVIOUS: CURRENT: June 1972

Beryllium Aluminum Oxide ($BeAl_6O_{10}$) $Al_6Be_1O_{10}$(cr,l)

ALUMINUM SILICATE, MULLITE ($Al_6Si_2O_{13}$) CRYSTAL M_r = 426.05244

$S°(298.15\ K) = 274.9 \pm 12\ J\ K^{-1}\ mol^{-1}$

$T_{fus} = 2123 \pm 15\ K$

$\Delta_f H°(0\ K) = -6775.17 \pm 6.3\ kJ\ mol^{-1}$

$\Delta_f H°(298.15\ K) = -6819.21 \pm 6.3\ kJ\ mol^{-1}$

Enthalpy of Formation

The heat of the reaction 3 Al_2O_3(cr, α) + 2 SiO_2(quartz) + $Al_6Si_2O_{13}$(cr) has been determined at 968 K by Holm and Kleppa (1, 2) to be +5.44 ± 0.35 kcal mol^{-1}. This value was obtained from heat of solution measurements in a high temperature oxide melt calorimeter. Using the JANAF functions, this reduces to $\Delta_r H°(298.15\ K) = 6.77$ kcal mol^{-1}, which is used to calculate the adopted value for the enthalpy of formation.

Heat Capacity and Entropy

The low temperature heat capacity and high temperature enthalpy of mullite have been measured by Pankratz et al. (3). The low temperature investigations extended from 53.5 K to 296.4 K; the Debye-Einstein equation 3D(182/T) + 15E(350/T) + 10E(502/T) + 15E(897/T) represented the data from 53.5 to 114.6 K within 1.4 percent and gave a smooth extrapolation to zero. From this equation, the entropy at 51 K was found to be 2.17 cal K^{-1} mol^{-1} as opposed to the value 1.96 cal K^{-1} mol^{-1} found by Pankratz et al. (5). However, since it is known that mullite has some of its Al atoms on Si sites, there will be configurational entropy remaining at 0 K. This is borne out by the free energy determinations of Rein and Chipman (4) and Kay and Taylor (5), who obtain $\Delta_r G°(1823\ K) = -5.7 \pm 0.1$ kcal mol^{-1} for the formation from oxide. If the entropies and enthalpies reported by Pankratz et al. are used with the heat of formation, they yield $\Delta_r G°(1823\ K) = +2.9$ kcal mol^{-1}. Holm and Kleppa (1, 2) calcualte that random mixing of tetra-coordinated aluminum and silicon atoms could produce 7 cal K^{-1} mol^{-1} residual entropy. We have added 4.7 cal K^{-1} mol^{-1} in order to match the measured $\Delta_r G°(1823\ K)$.

A review of the enthalpy determination indicated that the values might easily be 2 percent low if as little as 4 percent of kyanite remained unchanged in the sample used for measurement. Since x-ray could not detect this amount, it is not an unreasonable assumption. Accordingly, we have checked the effect of increasing the heat capacity above 298.15 K by 2 percent; this then gives $\Delta_r G°(1823\ K) = -2.10$ kcal mol^{-1}. In order to bring this within reasonable agreement with the observations, a residual configurational entropy of 2 cal K^{-1} mol^{-1} would be needed. Thus the uncertainty in the residual entropy is of the order of 2 cal K^{-1} mol^{-1} due to possible uncertainty in the high temperature heat capacity.

Fusion Data

The congruent melting point of mullite was reported by Aramaki and Roy (6).

References

1. J. L. Holm and O. J. Kleppa, J. Phys. Chem. 70, 1690 (1966).
2. J. L. Holm and O. J. Kleppa, Am. Mineralogist 51, 1608 (1966).
3. L. B. Pankratz, W. W. Weller, and K. K. Kelley, U. S. Bur. Mines RI 6287, 1963.
4. R. H. Rein and J. Chapman, Trans. Met. Soc. AIME 233, 415 (1965).
5. D. A. R. Kay and J. Taylor, Trans. Faraday Soc. 56, 1372 (1960).
6. S. Aramaki and R. Roy, Nature 184, 631 (1959).

Aluminum Silicate, Mullite ($Al_6Si_2O_{13}$) $Al_6O_{13}Si_2$(cr)

Enthalpy Reference Temperature = T_r = 298.15 K Standard State Pressure = p° = 0.1 MPa

	J K⁻¹mol⁻¹			kJ mol⁻¹			
T/K	$C_p°$	S°	$-[G°-H°(T_r)]/T$	$H°-H°(T_r)$	$\Delta_f H°$	$\Delta_f G°$	Log K_f
0	0.	0.	INFINITE	-46.072	-6775.166	-6775.166	INFINITE
100	85.264	60.703	493.090	-43.239	-6794.649	-6683.111	3490.899
200	225.957	164.662	301.776	-27.423	-6810.923	-6564.686	1714.520
298.15	325.641	274.889	274.889	0.	-6819.209	-6441.793	1128.575
300	327.031	276.907	274.895	0.604	-6819.302	-6439.451	1121.208
400	390.744	380.251	288.552	36.680	-6821.830	-6312.326	824.305
500	431.998	472.124	316.267	77.929	-6820.816	-6185.013	646.144
600	459.822	553.470	349.157	122.588	-6817.859	-6058.108	527.405
700	479.486	625.893	383.614	169.595	-6813.987	-5931.782	442.635
800	494.130	690.919	418.033	218.309	-6809.866	-5806.035	379.095
900	504.800	749.756	451.675	268.273	-6806.008	-5680.793	329.705
1000	513.377	803.395	484.205	319.190	-6866.308	-5551.389	289.975
1100	519.705	852.625	515.491	370.848	-6861.822	-5420.111	257.380
1200	525.618	898.102	545.504	423.117	-6857.067	-5289.254	230.235
1300	531.115	940.394	574.273	475.957	-6852.049	-5158.805	207.283
1400	536.198	979.942	601.852	529.326	-6846.783	-5028.752	187.625
1500	540.866	1017.097	628.309	583.183	-6841.293	-4899.084	170.601
1600	544.325	1052.115	653.713	637.443	-6835.652	-4769.787	155.718
1700	547.766	1085.217	678.131	692.047	-6930.264	-4639.951	142.568
1800	551.188	1116.624	701.626	746.995	-6923.960	-4505.410	130.744
1900	554.593	1146.516	724.261	802.285	-6917.466	-4371.222	120.173
2000	557.978	1175.049	746.093	857.913	-6910.783	-4237.380	110.669
2100	561.098	1202.348	767.174	913.866	-6903.926	-4103.878	102.078
2200	564.323	1228.525	787.554	970.136	-6896.898	-3970.703	94.276
2300	567.653	1253.683	807.277	1026.734	-6889.690	-3837.857	87.160
2400	571.089	1277.914	826.385	1083.670	-6882.287	-3705.329	80.644
2500	574.631	1301.299	844.916	1140.956	-6874.678	-3573.111	74.656
2600	578.277	1323.907	862.907	1198.600	-6866.849	-3441.202	69.135
2700	582.030	1345.801	880.388	1256.615	-6858.786	-3309.596	64.028
2800	585.887	1367.038	897.391	1315.010	-8613.882	-3172.488	59.183
2900	589.850	1387.666	913.943	1373.796	-8598.737	-2978.424	53.647
3000	593.919	1407.731	930.070	1432.983	-8583.318	-2784.880	48.489

PREVIOUS: September 1966 CURRENT: September 1967

Aluminum Silicate, Mullite ($Al_6Si_2O_{13}$) $Al_6O_{13}Si_2$(cr)

ARGON (Ar) REFERENCE STATE A_r = 39.948

0 to 6000 K Ideal Monatomic Gas

IP(Ar, g) = 127109.9 ± 0.1 cm^{-1}
S°(298.15 K) = 154.845 ± 0.003 J K^{-1} mol^{-1}

$\Delta_fH°$(0 K) = 0 kJ mol^{-1}
$\Delta_fH°$(298.15 K) = 0 kJ mol^{-1}

Electronic Levels and Quantum Weights

State	ϵ_i, cm^{-1}	g_i
$^1S_0$	0	1

Enthalpy of Formation

Zero by definition.

Heat Capacity and Entropy

Information on the electronic energy levels and quantum weights is taken from Moore (1, 9). All predicted levels have been observed for n=3 and 4 but above that many predicted levels are missing. Our calculations indicate that any reasonable method of filling in these missing levels and cutting off the summation in the partition function (2) has no effect on the thermodynamic properties to 6000 K. This is undoubtedly a result of the high energy of these levels; the first excited level is over 93000 cm^{-1} above the ground state. Therefore, we list the ground state only. Extension to higher temperatures may require consideration of excited states and utilization of different fill and cutoff procedures (2).

The thermodynamic functions at 298.15 K agree exactly with recent CODATA recommendations (3) except for two minor differences. First, the entropy differs by 0.1094 J K^{-1} mol^{-1} because this table uses a standard-state pressure of 1 bar, whereas the CODATA recommendations are based on 1 atm. Second, entropy differences of the order of 0.001-0.004 J K^{-1} mol^{-1} for the rare gases arise due to the use of slightly different values for R; this table uses R = 8.31441 J K^{-1} mol^{-1}. Considering these minor changes, this table agrees within the estimated uncertainty with those by Hultgren et al. (4), McBride et al. (5), Gurvich et al. (8) and Wagman et al. (12). The estimated uncertainty is due to uncertainties in the relative atomic mass and fundamental constants which in this tabulation are based on the 1981 scale (6) and the 1973 values (7), respectively.

Phase Data

The triple point, 83.798 K, and boiling point, 87.294 K, are a defining fixed point and a secondary reference point, respectively of IPTS-68 (10, 11). Hultgren et al. (4) had recommended a triple point of 83.81 K (0.6801 atm) and a boiling point of 87.30 K (1 atm). These values are provided for the convenience of the reader and have not been evaluated by the present authors. Because of these low values, the reference state for argon is chosen to be the ideal gas at all temperatures. This may differ from the choice of other authors.

References

1. C. E. Moore, U. S. Nat. Bur. Stand., NSRDS-NBS-35, Volume I, 1970 [Reprint of NBS Circular 467, Volume I, 1949].
2. J. R. Downey, Jr., The Dow Chemical Co., AFOSR-TR-78-0960, Contract No. F44620-75-1-0048, 1978.
3. J. D. Cox, chairman, ICSU-CODATA Task Group on Key Values for Thermodynamics, J. Chem. Thermodynamics 10, 903 (1978).
4. R. Hultgren and P. D. Desai et al, "Selected Values of the Thermodynamic Properties of the Elements", American Society for Metals, Metals Park, Ohio, 1973.
5. B. J. McBride, S. Heimel, J. G. Ehlers, and S. Gordon, NASA SP-3001, 1973.
6. N. E. Holden and R. L. Martin, Pure Appl. Chem. 55, 1101 (1983).
7. E. R. Cohen and B. N. Taylor, J. Phys. Chem. Ref. Data 2, 663 (1973).
8. L. V. Gurvich, I. V. Veits et al., "Thermodynamic Properties of Individual Substances," 3rd ed., Vol. I, Nauka, Moscow, 1978.
9. C. E. Moore, U. S. Nat. Bur. Stand., NSRDS-NBS-34, 1970.
10. H. Preston-Thomas, Metrologia 12, 7 (1976).
11. L. Crovini, R. E. Bedford and A. Moser, Metrologia 13, 197 (1977).
12. D. D. Wagman, W. H. Evans et al., J. Phys. Chem. Ref. Data 11, Supp. 2 (1982).

Argon (Ar) Ar_1(ref)

Enthalpy Reference Temperature = T_r = 298.15 K   Standard State Pressure = p° = 0.1 MPa

T/K	$C_p°$ (J $K^{-1}mol^{-1}$)	S° (J $K^{-1}mol^{-1}$)	$-[G°-H°(T_r)]/T$ (J $K^{-1}mol^{-1}$)	$H°-H°(T_r)$ (kJ mol^{-1})	$\Delta_fH°$ (kJ mol^{-1})	$\Delta_fG°$ (kJ mol^{-1})	Log K_f
0	0.	0.	INFINITE	-6.197	0.	0.	0.
100	20.786	132.137	173.325	-4.119	0.	0.	0.
200	20.786	146.545	156.746	-2.040	0.	0.	0.
250	20.786	151.183	155.187	-1.001	0.	0.	0.
298.15	20.786	154.845	154.845	0.	0.	0.	0.
300	20.786	154.973	154.845	0.038	0.	0.	0.
350	20.786	158.177	155.098	1.078	0.	0.	0.
400	20.786	160.953	155.660	2.117	0.	0.	0.
450	20.786	163.401	156.387	3.156	0.	0.	0.
500	20.786	165.591	157.200	4.196	0.	0.	0.
600	20.786	169.381	158.924	6.274	0.	0.	0.
700	20.786	172.585	160.653	8.353	0.	0.	0.
800	20.786	175.361	162.321	10.431	0.	0.	0.
900	20.786	177.809	163.909	12.510	0.	0.	0.
1000	20.786	179.999	165.410	14.589	0.	0.	0.
1100	20.786	181.980	166.828	16.667	0.	0.	0.
1200	20.786	183.789	168.167	18.746	0.	0.	0.
1300	20.786	185.453	169.434	20.824	0.	0.	0.
1400	20.786	186.993	170.634	22.903	0.	0.	0.
1500	20.786	188.427	171.773	24.982	0.	0.	0.
1600	20.786	189.769	172.856	27.060	0.	0.	0.
1700	20.786	191.029	173.888	29.139	0.	0.	0.
1800	20.786	192.217	174.874	31.217	0.	0.	0.
1900	20.786	193.341	175.816	33.296	0.	0.	0.
2000	20.786	194.407	176.719	35.375	0.	0.	0.
2100	20.786	195.421	177.586	37.453	0.	0.	0.
2200	20.786	196.388	178.419	39.532	0.	0.	0.
2300	20.786	197.312	179.220	41.610	0.	0.	0.
2400	20.786	198.197	179.993	43.689	0.	0.	0.
2500	20.786	199.045	180.738	45.768	0.	0.	0.
2600	20.786	199.860	181.458	47.846	0.	0.	0.
2700	20.786	200.645	182.154	49.925	0.	0.	0.
2800	20.786	201.401	182.828	52.004	0.	0.	0.
2900	20.786	202.130	183.481	54.082	0.	0.	0.
3000	20.786	202.835	184.115	56.161	0.	0.	0.
3100	20.786	203.516	184.730	58.239	0.	0.	0.
3200	20.786	204.176	185.327	60.318	0.	0.	0.
3300	20.786	204.816	185.908	62.397	0.	0.	0.
3400	20.786	205.436	186.473	64.475	0.	0.	0.
3500	20.786	206.039	187.024	66.554	0.	0.	0.
3600	20.786	206.625	187.560	68.632	0.	0.	0.
3700	20.786	207.194	188.083	70.711	0.	0.	0.
3800	20.786	207.748	188.593	72.790	0.	0.	0.
3900	20.786	208.288	189.091	74.868	0.	0.	0.
4000	20.786	208.815	189.578	76.947	0.	0.	0.
4100	20.786	209.328	190.053	79.025	0.	0.	0.
4200	20.786	209.829	190.518	81.104	0.	0.	0.
4300	20.786	210.318	190.973	83.183	0.	0.	0.
4400	20.786	210.796	191.418	85.261	0.	0.	0.
4500	20.786	211.263	191.854	87.340	0.	0.	0.
4600	20.786	211.720	192.281	89.418	0.	0.	0.
4700	20.786	212.167	192.699	91.497	0.	0.	0.
4800	20.786	212.604	193.109	93.576	0.	0.	0.
4900	20.786	213.033	193.512	95.654	0.	0.	0.
5000	20.786	213.453	193.906	97.733	0.	0.	0.
5100	20.786	213.864	194.294	99.811	0.	0.	0.
5200	20.786	214.268	194.674	101.890	0.	0.	0.
5300	20.786	214.664	195.047	103.969	0.	0.	0.
5400	20.786	215.053	195.414	106.047	0.	0.	0.
5500	20.786	215.434	195.775	108.126	0.	0.	0.
5600	20.786	215.808	196.129	110.204	0.	0.	0.
5700	20.786	216.176	196.478	112.283	0.	0.	0.
5800	20.786	216.538	196.820	114.362	0.	0.	0.
5900	20.786	216.893	197.158	116.440	0.	0.	0.
6000	20.786	217.243	197.489	118.519	0.	0.	0.

PREVIOUS: March 1977 (1 atm)   CURRENT: March 1982 (1 bar)

Argon (Ar) Ar_1(ref)

ARGON, ION (Ar^+) IDEAL GAS M_r = 39.94745

$IP(Ar^+, g)$ = 222848.2 ± 0.1 cm^{-1}

$S°(298.15\ K)$ = 166.404 ± 0.003 J K^{-1} mol^{-1}

$\Delta_f H°(0\ K)$ = 1520.573 ± 0.001 kJ mol^{-1}

$\Delta_f H°(298.15\ K)$ = [1526.778] kJ mol^{-1}

Electronic Levels and Quantum Weights

State	ε_i, cm^{-1}	g_i
$^2P_{3/2}$	0	4
$^2P_{1/2}$	1431.41	2

Enthalpy of Formation

The ionization limit of neutral argon (127109.9±0.1 cm^{-1}) reported by Moore (1) is adopted as $\Delta_f H°(0\ K)$ for $Ar^+(g)$. The ionization limit is converted from cm^{-1} to kJ mol^{-1} using the factor, 1 cm^{-1} = 0.01196266 kJ mol^{-1}, which is derived from the 1973 CODATA fundamental constants (2). The uncertainty in the ionization limit corresponds to an uncertainty of ±0.001 kJ mol^{-1} in the enthalpy of formation. Rosenstock et al. (9) and Levin and Lias (11) have summarized additional ionization potential and appearance potential data. A recent atomic spectroscopic study by Minnhagen (12) yielded an ionization potential within 0.13 cm^{-1} of our adopted value. Gurvich et al. (8) and Wagman et al. (10) adopted the same ionization potential, but the use of slightly different fundamental constants by Wagman et al. (10) resulted in a enthalpy of formation difference of 0.015 kJ mol^{-1}.

$\Delta_f H°(Ar^+, g, 298.15\ K)$ is obtained from $\Delta_f H°(Ar, g, 0\ K)$ by using IP(Ar) with JANAF (3) enthalpies H°(0 K)-H°(298.15 K) for $Ar^+(r)$, Ar(g), and $e^-(r)$. $\Delta_r H°(Ar \rightarrow Ar^+ + e^-, 298.15\ K)$ differs from a room temperature threshold energy due to inclusion of these enthalpies and to threshold effects discussed by Rosenstock et al. (9). $\Delta_f H°(298.15\ K)$ should be changed by -6.197 kJ mol^{-1} if it is to be used in the ion convention that excludes the enthalpy of the electron.

Heat Capacity and Entropy

The information on electronic energy levels and quantum weights given by Moore (4) is incomplete because many theoretically predicted levels have not been observed. Our calculations indicate that any reasonable method of filling in these missing levels and cutting off the summation in the partition function (5) has no effect on the thermodynamic functions to 6000 K. This is a result of the high energy of all levels other than the ground state and the $^2P_{1/2}$ level; the next lowest level is over 108000 cm^{-1} above the ground state. Since inclusion of these upper levels has no effect on the thermodynamic functions (to 6000 K), we list only the ground state and the $^2P_{1/2}$ state, with the energy of the latter state taken from a more recent compilation by Moore (1). The reported uncertainty in S°(298.15 K) is due to uncertainties in the relative ionic mass and fundamental constants. Extension of these calculations above 6000 K may require consideration of the higher excited states and use of different fill and cutoff procedures (5).

The thermodynamic functions reported here agree with those of Green et al. (6), Hilsenrath et al. (7), and Gurvich et al. (8) except for one or two minor changes. First, the entropy differs by 0.1094 J K^{-1} mol^{-1} because this table uses a standard-state pressure of 1 bar, whereas the cited references used a pressure of 1 atm. Second, smaller differences arise from the use of slightly different values for the fundamental constants, the relative ionic mass, and the position of the $^2P_{1/2}$ electronic level.

References

1. C. E. Moore, U. S. Nat. Bur. Stand., NSRDS-NBS-34, 1970.
2. E. R. Cohen and B. N. Taylor, J. Phys. Chem. Ref. Data 2, 663 (1973).
3. JANAF Thermochemical Tables: Ar(r), 3-31-82; e⁻(r), 3-31-82.
4. C. E. Moore, U. S. Nat. Bur. Stand., NSRDS-NBS-35, Volume I, 1970 [Reprint of NBS Circular 467, Volume I, 1949].
5. J. R. Downey, Jr., The Dow Chemical Co., AFOSR-TR-78-0960, Contract No. F44620-75-1-0048, 1978.
6. J. W. Green, D. E. Poland and J. L. Margrave, University of Wisconsin, Report AD-275542, 1961.
7. J. Hilsenrath, C. G. Messina and W. H. Evans, U.S. Nat. Bur. Stand., Report AD-606163, 1964.
8. L. V. Gurvich, I. V. Veits et al., "Thermodynamic Properties of Individual Substances", 3rd ed., Vol. 1, Nauka, Moscow, 1978.
9. H. M. Rosenstock, K. Draxl, B. W. Steiner and J. T. Herron, J. Phys. Chem. Ref. Data 6, Supp. 1 (1977).
10. D. D. Wagman, W. H. Evans et al., J. Phys. Chem. Ref. Data 11, Supp. 2 (1982).
11. R. D. Levin and S. G. Lias, U. S. Nat. Bur. Stand., NSRDS-NBS-71, 1982.
12. L. Minnhagen, J. Opt. Soc. Am. 63, 1185 (1973).

Argon, Ion (Ar^+) $Ar_1^+(g)$

Enthalpy Reference Temperature = T_r = 298.15 K Standard State Pressure = p° = 0.1 MPa

	J $K^{-1}mol^{-1}$			kJ mol^{-1}			
T/K	$C_p°$	S°	$-[G°-H°(T_r)]/T$	$H°-H°(T_r)$	$\Delta_f H°$	$\Delta_f G°$	Log K_f
0	0.	0.	INFINITE	-6.206	1520.573		
100	20.786	143.663	184.937	-4.127			
200	20.801	158.073	168.315	-2.048			
250	20.861	162.720	166.748	-1.007			
298.15	20.984	166.404	166.404	0.	1526.778	1517.077	-265.786
300	20.990	166.533	166.404	0.039	1526.817	1517.017	-264.136
350	21.185	169.783	166.660	1.093	1527.871	1515.300	-226.146
400	21.422	172.627	167.232	2.158	1528.936	1513.432	-197.634
450	21.673	175.165	167.975	3.235	1530.014	1511.429	-175.442
500	21.915	177.461	168.810	4.325	1531.104	1509.306	-157.676
600	22.318	181.494	170.597	6.538	1533.316	1504.740	-130.999
700	22.588	184.956	172.407	8.784	1535.563	1499.800	-111.916
800	22.734	187.983	174.169	11.051	1537.830	1494.536	-97.583
900	22.786	190.665	175.856	13.328	1540.106	1488.988	-86.419
1000	22.773	193.065	177.459	15.606	1542.385	1483.186	-77.474
1100	22.718	195.234	178.978	17.881	1544.660	1477.156	-70.144
1200	22.639	197.207	180.416	20.149	1546.928	1470.919	-64.027
1300	22.546	199.015	181.778	22.409	1549.187	1464.493	-58.844
1400	22.449	200.683	183.070	24.658	1551.437	1457.893	-54.395
1500	22.350	202.228	184.296	26.898	1553.677	1451.134	-50.533
1600	22.254	203.668	185.462	29.129	1555.907	1444.225	-47.149
1700	22.162	205.014	186.573	31.349	1558.128	1437.176	-44.159
1800	22.076	206.278	187.633	33.561	1560.340	1429.998	-41.497
1900	21.995	207.470	188.646	35.765	1562.543	1422.696	-39.113
2000	21.919	208.596	189.616	37.960	1564.739	1415.279	-36.963
2100	21.849	209.664	190.545	40.149	1566.927	1407.752	-35.016
2200	21.785	210.679	191.437	42.330	1569.109	1400.121	-33.243
2300	21.725	211.646	192.295	44.506	1571.284	1392.391	-31.622
2400	21.670	212.569	193.121	46.675	1573.454	1384.567	-30.134
2500	21.619	213.453	193.917	48.840	1575.618	1376.652	-28.764
2600	21.571	214.299	194.684	50.999	1577.778	1368.651	-27.497
2700	21.528	215.113	195.426	53.154	1579.933	1360.567	-26.322
2800	21.487	215.895	196.143	55.305	1582.083	1352.403	-25.229
2900	21.450	216.648	196.837	57.452	1584.230	1344.162	-24.211
3000	21.415	217.375	197.510	59.595	1586.374	1335.847	-23.259
3100	21.383	218.077	198.162	61.735	1588.513	1327.461	-22.368
3200	21.353	218.755	198.795	63.872	1590.650	1319.006	-21.531
3300	21.326	219.412	199.410	66.006	1592.784	1310.484	-20.743
3400	21.300	220.048	200.008	68.137	1594.915	1301.897	-20.001
3500	21.276	220.665	200.589	70.266	1597.044	1293.248	-19.301
3600	21.253	221.264	201.155	72.392	1599.171	1284.538	-18.638
3700	21.232	221.846	201.706	74.516	1601.295	1275.769	-18.011
3800	21.212	222.412	202.244	76.639	1603.417	1266.943	-17.415
3900	21.194	222.963	202.768	78.759	1605.537	1258.060	-16.850
4000	21.176	223.499	203.280	80.877	1607.656	1249.124	-16.312
4100	21.160	224.022	203.779	82.994	1609.773	1240.134	-15.800
4200	21.145	224.532	204.267	85.110	1611.888	1231.093	-15.311
4300	21.130	225.029	204.744	87.223	1614.002	1222.002	-14.844
4400	21.117	225.515	205.211	89.336	1616.114	1212.861	-14.398
4500	21.104	225.989	205.667	91.447	1618.225	1203.672	-13.972
4600	21.092	226.453	206.114	93.556	1620.335	1194.437	-13.563
4700	21.080	226.906	206.552	95.665	1622.443	1185.155	-13.172
4800	21.070	227.350	206.981	97.773	1624.551	1175.829	-12.796
4900	21.059	227.784	207.401	99.879	1626.657	1166.459	-12.435
5000	21.050	228.210	207.813	101.984	1628.763	1157.046	-12.088
5100	21.040	228.626	208.217	104.089	1630.867	1147.590	-11.754
5200	21.032	229.035	208.613	106.193	1632.971	1138.094	-11.432
5300	21.023	229.435	209.002	108.295	1635.074	1128.557	-11.123
5400	21.015	229.828	209.384	110.397	1637.176	1118.980	-10.824
5500	21.008	230.214	209.759	112.498	1639.277	1109.365	-10.536
5600	21.001	230.592	210.128	114.599	1641.377	1099.711	-10.258
5700	20.994	230.964	210.490	116.699	1643.477	1090.020	-9.989
5800	20.987	231.329	210.847	118.798	1645.576	1080.292	-9.729
5900	20.981	231.688	211.197	120.896	1647.675	1070.527	-9.478
6000	20.975	232.040	211.541	122.994	1649.772	1060.728	-9.234

PREVIOUS: March 1977 (1 atm) CURRENT: March 1982 (1 bar)

Argon, Ion (Ar^+) $Ar_1^+(g)$

BORON (B) REFERENCE STATE A_r = 10.81

0 to 2350 K	crystal
2350 to 4137.895 K	liquid
above 4137.895 K	ideal monatomic gas

Refer to the individual tables for details.

Boron (B) B_1(ref)

Enthalpy Reference Temperature = T_r = 298.15 K; Standard State Pressure = p° = 0.1 MPa

T/K	C_p° (J K⁻¹mol⁻¹)	S° (J K⁻¹mol⁻¹)	$-[G^\circ-H^\circ(T_r)]/T$ (J K⁻¹mol⁻¹)	$H^\circ-H^\circ(T_r)$ (kJ mol⁻¹)	$\Delta_f H^\circ$ (kJ mol⁻¹)	$\Delta_f G^\circ$ (kJ mol⁻¹)	Log K_f
0	0.	0.	INFINITE	-1.214	0.	0.	0.
100	1.076	0.308	12.207	-1.190	0.	0.	0.
200	5.998	2.419	6.704	-0.857	0.	0.	0.
298.15	11.315	5.834	5.834	0.	0.	0.	0.
300	11.405	5.904	5.834	0.021	0.	0.	0.
400	15.693	9.794	6.335	1.384	0.	0.	0.
500	18.722	13.644	7.415	3.115	0.	0.	0.
600	20.778	17.251	8.758	5.096	0.	0.	0.
700	22.249	20.570	10.212	7.251	0.	0.	0.
800	23.361	23.617	11.699	9.534	0.	0.	0.
900	24.245	26.421	13.181	11.915	0.	0.	0.
1000	24.978	29.014	14.637	14.378	0.	0.	0.
1100	25.606	31.425	16.055	16.908	0.	0.	0.
1200	26.161	33.677	17.430	19.496	0.	0.	0.
1300	26.663	35.792	18.762	22.138	0.	0.	0.
1400	27.125	37.785	20.051	24.828	0.	0.	0.
1500	27.557	39.671	21.296	27.562	0.	0.	0.
1600	27.966	41.463	22.501	30.338	0.	0.	0.
1700	28.356	43.170	23.667	33.155	0.	0.	0.
1800	28.732	44.801	24.796	36.009	0.	0.	0.
1900	29.097	46.365	25.891	38.901	0.	0.	0.
2000	29.452	47.866	26.952	41.828	0.	0.	0.
2100	29.799	49.312	27.983	44.791	0.	0.	0.
2200	30.140	50.706	28.984	47.788	0.	0.	0.
2300	30.475	52.053	29.958	50.819	0.	0.	0.
2350.000	30.641	52.710	30.435	52.346	BETA <--> LIQUID TRANSITION		
2350.000	31.750	74.075	30.435	102.554			
2400	31.750	74.744	31.351	104.142	0.	0.	0.
2500	31.750	76.040	33.113	107.317	0.	0.	0.
2600	31.750	77.285	34.788	110.492	0.	0.	0.
2700	31.750	78.483	36.384	113.667	0.	0.	0.
2800	31.750	79.638	37.909	116.842	0.	0.	0.
2900	31.750	80.752	39.367	120.017	0.	0.	0.
3000	31.750	81.828	40.764	123.192	0.	0.	0.
3100	31.750	82.869	42.106	126.367	0.	0.	0.
3200	31.750	83.877	43.396	129.542	0.	0.	0.
3300	31.750	84.854	44.637	132.717	0.	0.	0.
3400	31.750	85.802	45.834	135.892	0.	0.	0.
3500	31.750	86.723	46.989	139.067	0.	0.	0.
3600	31.750	87.617	48.105	142.242	0.	0.	0.
3700	31.750	88.487	49.185	145.417	0.	0.	0.
3800	31.750	89.334	50.231	148.592	0.	0.	0.
3900	31.750	90.158	51.244	151.767	0.	0.	0.
4000	31.750	90.962	52.227	154.942	0.	0.	0.
4100	31.750	91.746	53.181	158.117	0.	0.	0.
4139.449	31.750	92.050	53.550	159.369	LIQUID <--> IDEAL GAS FUGACITY = 1 bar		
4139.449	20.860	208.131	53.550	639.878			
4200	20.870	208.434	55.781	641.141	0.	0.	0.
4300	20.887	208.925	59.337	643.229	0.	0.	0.
4400	20.906	209.405	62.742	645.319	0.	0.	0.
4500	20.928	209.875	66.006	647.410	0.	0.	0.
4600	20.952	210.336	69.139	649.504	0.	0.	0.
4700	20.979	210.786	72.148	651.601	0.	0.	0.
4800	21.009	211.228	75.041	653.700	0.	0.	0.
4900	21.042	211.662	77.825	655.803	0.	0.	0.
5000	21.078	212.087	80.506	657.909	0.	0.	0.
5100	21.117	212.505	83.090	660.018	0.	0.	0.
5200	21.160	212.916	85.583	662.132	0.	0.	0.
5300	21.206	213.319	87.989	664.251	0.	0.	0.
5400	21.255	213.716	90.313	666.374	0.	0.	0.
5500	21.308	214.107	92.561	668.502	0.	0.	0.
5600	21.365	214.491	94.735	670.635	0.	0.	0.
5700	21.425	214.870	96.839	672.775	0.	0.	0.
5800	21.489	215.243	98.877	674.921	0.	0.	0.
5900	21.557	215.611	100.853	677.073	0.	0.	0.
6000	21.628	215.974	102.768	679.232	0.	0.	0.

PREVIOUS: March 1979 (1 atm) CURRENT: June 1983 (1 bar)

Boron (B) B_1(ref)

BORON, BETA-RHOMBOHEDRAL (β-B) CRYSTAL A_r = 10.81

$S°(298.15\ K) = 5.83 \pm 0.08\ J\ K^{-1}\ mol^{-1}$
$T_{fus} = 2350 \pm 50\ K$

$\Delta_f H°(0\ K) = 0\ kJ\ mol^{-1}$
$\Delta_f H°(298.15\ K) = 0\ kJ\ mol^{-1}$
$\Delta_{fus} H° = 50.2 \pm 1.7\ kJ\ mol^{-1}$

Enthalpy of Formation

Zero by definition.

Heat Capacity and Entropy

The adopted low temperature heat capacity values (T < 298.15 K) are based on the heat capacity studies of Biler (1, 2), Johnston et al. (3), and Bogdanov et al. (4). The adopted high temperature heat capacities (T > 298.15 K) are derived from the enthalpy studies of McDonald and Stull (5) and Stout et al. (6). The mathematical and graphical treatment of these five studies, providing experimental measurements in the 2-2218 K range, yield a continuous and smooth heat capacity curve. The resulting entropy, S°(298.15 K) = 5.83 ± 0.08 $J\ K^{-1}\ mol^{-1}$, is 0.07 $J\ K^{-1}\ mol^{-1}$ less than the CODATA recommended entropy (7) due to the differing interpretation of the Cp results for T<50 K. In this analysis, the scatter in the experimental data obscures any differences that may arise in the thermal functions due to crystallographic differences in the various crystalline samples.

Biler (1, 2) measured the heat capacity of B(cr, β) in the region 2-15 K. He reported θ_D = 1480 K derived from his extrapolation to T/K = 0 of a graph of $(Cp-\gamma T)/T^3$ vs T^2. For boron, there is no electronic contribution, i.e. γ = 0. Only graphical data was presented. The sample, stated to be the β-rhombohedral modification based on X-ray data, contained 50 ppm Cu whereas the other impurities were reported as <1ppm (via electron micro-probe analysis).

Johnston et al. (3) measured 46 heat capacity values in the range 16.90-303.71 K. This study indicated a shallow maximum near 25 K at which temperature Cp°=0.038 $J\ K^{-1}\ mol^{-1}$. A graph of the results by Biler (1, 2) tends to agree with the observations of Johnston et al. (3) and to support the presence of a Cp anomaly. The purity of Johnston's sample was not reported. Hultgren et al. (8) stated that the sample was of the tetragonal (α) modification. This inference is presumably based on Johnston's reported method of preparation and X-ray data.

Bogdanov et al. (4) measured the heat capacity from 16 to 280 K. Metallic impurity content was stated to be <0.01 %. X-ray analysis was used to infer that the sample was the β-polymorph. Bogdanov et al. also inferred that the sample of Johnston et al. was not of a specific crystalline modification. They surmised that impurities caused the shallow maximum seen by Johnston et al. (3) near 25 K. Bogdanov et al. found no maximum, but their smooth values level out below 40 K in contradiction to theory. Their last five values (20-40 K) vary linearly with T and extrapolate to Cp = 0.0126 $J\ K^{-1}\ mol^{-1}$ at T/K = 0. The authors did not publish their experimental data, so it is difficult to interpret this behavior. It may arise from bias of their calorimeter near its limit of sensitivity or from anomalous behavior of their sample.

McDonald and Stull (5) measured enthalpies in a copper block drop calorimeter, 20 points in the range 282.8 to 1667.8 K. The sample was reported to be the β-rhombohedral form (based on X-ray data) with a metallic impurity of 0.59%. Stout et al. (6), also using a copper block drop calorimeter, measured enthalpies, five data points in the range 1820-2218 K. The sample was reported to have 0.04% metallic impurities and to contain 300 ppm O and 6300 ppm C. No information as to crystalline form was given. Wise et al. (9) measured the enthalpy of two crystalline boron samples (each reportedly containing some of the β-modification) in the temperature range 515-1103 K.

The adopted heat capacity values below 25 K are derived from the θ_D value of Biler (1, 2). From 25 K to 300 K, the derived values are a compromise of the data of Johnston et al. (3) and Bogdanov et al. (4), but excluding in each the effect of the possible anomalous behavior. In the region 25 to 95 K, there is large scatter in each data set, so that large deviations occur, up to 23%. In the region 95-300 K, Johnston's data shows deviations of -2.1 to +1.8% while Bogdanov's data shows derviations of -2.9 to +1.7% from the adopted values. The two sets of data are intertwined, one is not always above the other.

In the enthalpy data, 26 of the 30 points of McDonald and Stull (5) are within +0.6% to -0.5%. Stout's (6) data deviates from -0.7% at 1819.5 K to +0.7% at 2217.7 K. The data of McDonald and Stull and Stout et al. are slightly skewed to each other. The observed enthalpies of McDonald and Stull (5) fall between the enthalpies of the two samples measured by Wise et al. (9).

Other heat capacity and enthalpy studies are available but were excluded from this analysis. A detailed discussion of the Group IIIA metals (B, Al, and Ga) is in preparation by the JANAF staff.

Fusion Data

See the liquid table for details.

Sublimation Data

The adopted value of $\Delta_{sub}H°(298.15\ K)$ is the value recommended by CODATA (7). See the ideal gas table for details.

References

1. N. Biler, Stanford University, Ph. D. Dissertation, 1974.
2. N. Biler, W. A. Phillips, and T. H. Geballe, Proc. Int. Conf. Low Temp. Phys., 9 (1975).
3. H. L. Johnston, H. N. Hersh, and E. C. Kerr, J. Am. Chem. Soc. 73, 1112 (1951).
4. V. I. Bogdanov, Yu. Kh. Vekilov, G. V. Tsagareishvili, and I. M. Zhgenti, Sov. Phys. - Solid State 12, 2701 (1971).
5. R. A. McDonald and D. R. Stull, J. Chem. Eng. Data 7, 84 (1962).
6. N. D. Stout, R. W. Mar, and W. O. J. Boo, High Temp. Sci. 5, 241 (1973).
7. J. D. Cox, chairman, ISCU-CODATA Task Group on Key Values for Thermodynamics, J. Chem. Thermodynamics 10, 903 (1978).
8. R. Hultgren, P. D. Desai et al., "Selected values of the Thermodynamic Properties of the Elements," American Society for Metals, Metals Park, Ohio, 1973.
9. S. S. Wise, J. L. Margrave, and R. L. Altman, J. Phys. Chem. 64, 915 (1960).

Boron, Beta-Rhombohedral (B) $B_1(cr)$

Enthalpy Reference Temperature = T_r = 298.15 K Standard State Pressure = p° = 0.1 MPa

T/K	$C_p°$ (J K⁻¹mol⁻¹)	S° (J K⁻¹mol⁻¹)	$-[G°-H°(T_r)]/T$ (J K⁻¹mol⁻¹)	$H°-H°(T_r)$ (kJ mol⁻¹)	$\Delta_f H°$ (kJ mol⁻¹)	$\Delta_f G°$ (kJ mol⁻¹)	Log K_f
0	0.	0.	INFINITE	-1.214	0.	0.	0.
100	1.076	0.308	12.207	-1.190	0.	0.	0.
200	5.998	2.419	6.704	-0.857	0.	0.	0.
250	8.821	4.063	6.007	-0.486	0.	0.	0.
298.15	11.315	5.834	5.834	0.	0.	0.	0.
300	11.405	5.904	5.834	0.021	0.	0.	0.
350	13.654	7.834	5.981	0.648	0.	0.	0.
400	15.693	9.794	6.335	1.384	0.	0.	0.
450	17.361	11.742	6.828	2.211	0.	0.	0.
500	18.722	13.644	7.415	3.115	0.	0.	0.
600	20.778	17.251	8.758	5.096	0.	0.	0.
700	22.249	20.570	10.212	7.251	0.	0.	0.
800	23.361	23.617	11.699	9.534	0.	0.	0.
900	24.245	26.421	13.181	11.915	0.	0.	0.
1000	24.978	29.014	14.637	14.378	0.	0.	0.
1100	25.606	31.425	16.055	16.908	0.	0.	0.
1200	26.161	33.677	17.430	19.496	0.	0.	0.
1300	26.663	35.792	18.762	22.138	0.	0.	0.
1400	27.125	37.785	20.051	24.828	0.	0.	0.
1500	27.557	39.671	21.296	27.562	0.	0.	0.
1600	27.966	41.463	22.501	30.338	0.	0.	0.
1700	28.356	43.170	23.667	33.155	0.	0.	0.
1800	28.732	44.801	24.796	36.009	0.	0.	0.
1900	29.097	46.365	25.891	38.901	0.	0.	0.
2000	29.452	47.866	26.952	41.828	0.	0.	0.
2100	29.799	49.312	27.983	44.791	0.	0.	0.
2200	30.140	50.706	28.984	47.788	0.	0.	0.
2300	30.475	52.053	29.958	50.819	0.	0.	0.
2350.000	30.641	52.710	30.435	52.346	------- BETA <--> LIQUID -------		
2400	30.806	53.357	30.906	53.883	-50.259	1.069	-0.023
2500	31.133	54.621	31.829	56.980	-50.337	3.209	-0.067
2600	31.457	55.848	32.730	60.109	-50.383	5.352	-0.108
2700	31.778	57.042	33.608	63.271	-50.396	7.496	-0.145
2800	32.096	58.203	34.466	66.465	-50.377	9.640	-0.180

PREVIOUS: March 1979 CURRENT: June 1983

Boron, Beta-Rhombohedral (B) $B_1(cr)$

BORON (B) LIQUID A_r = 10.81

$S°(298.15\ K) = [26.555]\ J\ K^{-1}\ mol^{-1}$

$T_{fus} = 2350 \pm 50\ K$

$\Delta_f H°(298.15\ K) = [48.927]\ kJ\ mol^{-1}$

$\Delta_{fus}H° = 50.2 \pm 1.7\ kJ\ mol^{-1}$

Enthalpy of Formation

The enthalpy of formation of liquid boron is calculated from that of the crystal (β-rhomohedral) by adding $\Delta_{fus}H°$ and the difference in enthalpy, H°(2350 K) - H°(298.15 K), between the crystal and liquid.

Heat Capacity and Entropy

There are no heat capacity or enthalpy studies which adequately define the liquid phase heat capacity values. Tsagareishvili and Tsagareishvili (1) estimated $Cp°(\ell) = 39.04\ J\ K^{-1}\ mol^{-1}$ at T_{fus} by means of quasi-thermodynamic correlations. Using our adopted Cp° value for the crystal at T_{fus}, we recalculate this to be $Cp°(\ell) = 35.98\ J\ K^{-1}\ mol^{-1}$ at T_{fus}. This value appears to be too large. Even though B(cr) and Al(cr) have different crystal structures, the heat capacity values of the crystals at their respective T_{fus} are very similar, 30.64 $J\ K^{-1}\ mol^{-1}$ for B(β-cr) and 32.96 $J\ K^{-1}\ mol^{-1}$ for Al(cr). We estimate that the B(ℓ) heat capacity is 31.75 $J\ K^{-1}\ mol^{-1}$, as is the value for Al(ℓ).

A glass transition is assumed at 1760 K. The entropy at 298.15 K is calculated in a manner analogous to that used for the enthalpy of formation.

Fusion Data

The adopted value for the melting point of β-rhombohedral boron is T_{fus} = 2350±50 K. This value is within 2 K of the value recommended by Charlesworth (2). The value is based primarily on the work of Kimpel and Moss (3). These authors (3) used a disappearing hole technique to measure T_{fus} = 2352±20 K for a 99.5% pure sample. Other studies are by Wisnyi and Pijanowski (4) (2313±30 K), Cueilleron (5) (2273-2348 K), and Mar (6) (2315±20 K).

The enthalpy of melting was measured by Stout et al (7) using drop calorimetry. The enthalpy of melting, $\Delta_{fus}H° = 50.2 \pm 1.7\ kJ\ mol^{-1}$, was the result of three calorimetric determinations involving different samples and containers. It leads to an entropy of fusion, $\Delta_{fus}S° = 21.4\ J\ K^{-1}\ mol^{-1}$, which is more than twice as large as previous estimates.

Vaporization Data

There are no vaporization studies involving liquid boron. The boiling point, T_{vap}(1 bar) = 4137.895 K, is calculated as the temperature for which $\Delta_r G°=0$ for B(ℓ)=B(g). The difference in the enthalpy of formation of B(ℓ) and B(g) at T_{vap} is the enthalpy of vaporization, $\Delta_{vap}H° = 480.344\ kJ\ mol^{-1}$.

References

1. D. Sh. Tsagareishvili and G. V. Tsagareishvili, J. Less Common Met. 67, 541 (1979).
2. J. H. Charlesworth, AFML-TR-70-137, 1970.
3. R. F. Kimpel and R. G. Moss, J. Chem. Eng. Data 13, 231 (1968).
4. L. G. Wisnyi and S. W. Pijanowski, USAEC Rept. No. TID 7530 (Part 1), 46 (1957).
5. J. Cueilleron, Compt. rend. 219, 209 (1944); Ann. Chim. 19, 459 (1944).
6. R. W. Mar, Thermochemical Acta 4, 367 (1972).
7. N. D. Stout, R. W. Mar, and W. O. J. Boo, High Temp. Sci. 5, 241 (1973).

Boron (B) $B_1(l)$

Enthalpy Reference Temperature = T_r = 298.15 K — Standard State Pressure = p° = 0.1 MPa

T/K	$C_p°$ (J K⁻¹mol⁻¹)	S° (J K⁻¹mol⁻¹)	$-[G°-H°(T_r)]/T$ (J K⁻¹mol⁻¹)	$H°-H°(T_r)$ (kJ mol⁻¹)	$\Delta_f H°$ (kJ mol⁻¹)	$\Delta_f G°$ (kJ mol⁻¹)	Log K_f
298.15	11.315	26.555	26.555	0.	48.927	42.749	-7.490
300	11.405	26.625	26.555	0.021	48.927	42.711	-7.437
350	13.654	28.554	26.702	0.648	48.927	41.675	-6.220
400	15.693	30.515	27.056	1.384	48.927	40.639	-5.307
450	17.361	32.463	27.549	2.211	48.927	39.603	-4.597
500	18.722	34.365	28.136	3.115	48.927	38.567	-4.029
600	20.778	37.971	29.479	5.096	48.927	36.495	-3.177
700	22.249	41.291	30.932	7.251	48.927	34.423	-2.569
800	23.361	44.337	32.420	9.534	48.927	32.351	-2.112
900	24.245	47.142	33.902	11.915	48.927	30.279	-1.757
1000	24.978	49.735	35.357	14.378	48.927	28.207	-1.473
1100	25.606	52.146	36.775	16.908	48.927	26.135	-1.241
1200	26.161	54.398	38.151	19.496	48.927	24.062	-1.047
1300	26.663	56.512	39.483	22.138	48.927	21.990	-0.884
1400	27.125	58.505	40.771	24.828	48.927	19.918	-0.743
1500	27.557	60.392	42.017	27.562	48.927	17.846	-0.621
1600	27.966	62.183	43.222	30.338	48.927	15.774	-0.515
1700	28.356	63.891	44.388	33.155	48.927	13.702	-0.421
1750.000	28.546	64.715	44.957	34.577	GLASS <--> LIQUID		
1750.000	31.750	64.715	44.957	34.577	TRANSITION		
1800	31.750	65.610	45.518	36.165	49.083	11.628	-0.337
1900	31.750	67.326	46.621	39.340	49.366	9.539	-0.262
2000	31.750	68.955	47.698	42.515	49.614	7.436	-0.194
2100	31.750	70.504	48.747	45.690	49.826	5.322	-0.132
2200	31.750	71.981	49.770	48.865	50.004	3.199	-0.076
2300	31.750	73.392	50.766	52.040	50.148	1.068	-0.024
2350.000	31.750	74.075	51.255	53.627	------- BETA <--> LIQUID -------		
2400	31.750	74.744	51.737	55.215	0.	0.	0.
2500	31.750	76.040	52.684	58.390	0.	0.	0.
2600	31.750	77.285	53.606	61.565	0.	0.	0.
2700	31.750	78.483	54.506	64.740	0.	0.	0.
2800	31.750	79.638	55.383	67.915	0.	0.	0.
2900	31.750	80.752	56.238	71.090	0.	0.	0.
3000	31.750	81.828	57.074	74.265	0.	0.	0.
3100	31.750	82.869	57.889	77.440	0.	0.	0.
3200	31.750	83.877	58.685	80.615	0.	0.	0.
3300	31.750	84.854	59.464	83.790	0.	0.	0.
3400	31.750	85.802	60.224	86.965	0.	0.	0.
3500	31.750	86.723	60.968	90.140	0.	0.	0.
3600	31.750	87.617	61.696	93.315	0.	0.	0.
3700	31.750	88.487	62.409	96.490	0.	0.	0.
3800	31.750	89.334	63.106	99.665	0.	0.	0.
3900	31.750	90.158	63.789	102.840	0.	0.	0.
4000	31.750	90.962	64.459	106.015	0.	0.	0.
4100	31.750	91.746	65.115	109.190	0.	0.	0.
4139.449	31.750	92.050	65.370	110.442	------- FUGACITY = 1 bar -------		
4200	31.750	92.511	65.758	112.365	-479.849	7.024	-0.087
4300	31.750	93.258	66.389	115.540	-478.762	18.603	-0.226
4400	31.750	93.988	67.008	118.715	-477.677	30.158	-0.358
4500	31.750	94.702	67.615	121.890	-476.593	41.687	-0.484

PREVIOUS: March 1979 CURRENT: June 1983

Boron (B) $B_1(l)$

BORON (B) CRYSTAL-LIQUID A_r = 10.81

0 to 2350 K crystal
above 2350 K liquid

Refer to the individual tables for details.

Enthalpy Reference Temperature = T_r = 298.15 K Standard State Pressure = p° = 0.1 MPa

T/K	C_p° (J K⁻¹mol⁻¹)	S° (J K⁻¹mol⁻¹)	$-[G°-H°(T_r)]/T$ (J K⁻¹mol⁻¹)	$H°-H°(T_r)$ (kJ mol⁻¹)	$\Delta_f H°$ (kJ mol⁻¹)	$\Delta_f G°$ (kJ mol⁻¹)	Log K_f
0	0.	0.	INFINITE	-1.214	0.	0.	0.
100	1.076	0.308	12.207	-1.190	0.	0.	0.
200	5.998	2.419	6.704	-0.857	0.	0.	0.
250	8.821	4.063	6.007	-0.486	0.	0.	0.
298.15	11.315	5.834	5.834	0.	0.	0.	0.
300	11.405	5.904	5.834	0.021	0.	0.	0.
350	13.654	7.834	5.981	0.648	0.	0.	0.
400	15.693	9.794	6.335	1.384	0.	0.	0.
450	17.361	11.742	6.828	2.211	0.	0.	0.
500	18.722	13.644	7.415	3.115	0.	0.	0.
600	20.778	17.251	8.758	5.096	0.	0.	0.
700	22.249	20.570	10.212	7.251	0.	0.	0.
800	23.361	23.617	11.699	9.534	0.	0.	0.
900	24.245	26.421	13.181	11.915	0.	0.	0.
1000	24.978	29.014	14.637	14.378	0.	0.	0.
1100	25.606	31.425	16.055	16.908	0.	0.	0.
1200	26.161	33.677	17.430	19.496	0.	0.	0.
1300	26.663	35.792	18.762	22.138	0.	0.	0.
1400	27.125	37.785	20.051	24.828	0.	0.	0.
1500	27.557	39.671	21.296	27.562	0.	0.	0.
1600	27.966	41.463	22.501	30.338	0.	0.	0.
1700	28.356	43.170	23.667	33.155	0.	0.	0.
1800	28.732	44.801	24.796	36.009	0.	0.	0.
1900	29.097	46.365	25.891	38.901	0.	0.	0.
2000	29.452	47.866	26.952	41.828	0.	0.	0.
2100	29.799	49.312	27.983	44.791	0.	0.	0.
2200	30.140	50.706	28.984	47.788	0.	0.	0.
2300	30.475	52.053	29.958	50.819	0.	0.	0.
2350.000	30.641	52.710	30.435	52.346	BETA <--> LIQUID		
2350.000	31.750	74.075	30.435	102.554	TRANSITION		
2400	31.750	74.744	31.351	104.142	0.	0.	0.
2500	31.750	76.040	33.113	107.317	0.	0.	0.
2600	31.750	77.285	34.788	110.492	0.	0.	0.
2700	31.750	78.483	36.384	113.667	0.	0.	0.
2800	31.750	79.638	37.909	116.842	0.	0.	0.
2900	31.750	80.752	39.367	120.017	0.	0.	0.
3000	31.750	81.828	40.764	123.192	0.	0.	0.
3100	31.750	82.869	42.106	126.367	0.	0.	0.
3200	31.750	83.877	43.396	129.542	0.	0.	0.
3300	31.750	84.854	44.637	132.717	0.	0.	0.
3400	31.750	85.802	45.834	135.892	0.	0.	0.
3500	31.750	86.723	46.989	139.067	0.	0.	0.
3600	31.750	87.617	48.105	142.242	0.	0.	0.
3700	31.750	88.487	49.185	145.417	0.	0.	0.
3800	31.750	89.334	50.231	148.592	0.	0.	0.
3900	31.750	90.158	51.244	151.767	0.	0.	0.
4000	31.750	90.962	52.227	154.942	0.	0.	0.
4100	31.750	91.746	53.181	158.117	0.	0.	0.
4139.449	31.750	92.050	53.550	159.369	------- FUGACITY = 1 bar -------		
4200	31.750	92.511	54.109	161.292	-479.849	7.024	-0.087
4300	31.750	93.258	55.010	164.467	-478.762	18.604	-0.226
4400	31.750	93.988	55.888	167.642	-477.677	30.158	-0.358
4500	31.750	94.702	56.743	170.817	-476.593	41.687	-0.484

PREVIOUS: March 1979 CURRENT: June 1983

BORON (B) IDEAL GAS A_r = 10.81 Boron (B) $B_1(g)$

IP(B, g) = 66928.10 ± 1 cm^{-1}
S°(298.15 K) = 153.435 ± 0.035 $J K^{-1} mol^{-1}$

$\Delta_fH°$(0 K) = 555 ± 12 kJ mol^{-1}
$\Delta_fH°$(298.15 K) = 560 ± 12 kJ mol^{-1}

Electronic Levels and Quantum Weights

State	ε_i, cm^{-1}	g_i
$^2P°_{1/2}$	0.00	2
$^2P°_{3/2}$	15.254	4
$^4P_{1/2}$	28870.0	2
.	.	.
.	.	.
.	.	.
$^2D_{5/2}$	66855.78	6
IP	66928.10	

Enthalpy of Formation

The enthalpy of formation is derived from the measurements of Paule and Margrave (1), Robson and Gilles (2), Hildenbrand and Hall (3), Akishin et al. (4), and Schissel and Trulson (5). $\Delta_fH°$(B, g, 298.15 K) = 560 ± 12 kJ mol^{-1} is the value recommended by CODATA (6). CODATA stated that these measurements were selected from the many sets of data available as being the most self-consistent. The difference in S°(B, cr, 298.15 K) between our adopted value and that of CODATA (6), will create a change of less than 0.2 kJ mol^{-1} in $\Delta_fH°$(B, g, 298.15 K). Not included in this analysis are other vaporization studies by Trulson et al. (1923 - 2249 K, 7), Burns et al. (2403 K, 8), Mar and Bedford (1823 - 2253 K, 9) and Storms and Mueller (1750 - 2050 K, 10). The two latter studies were published after the time of the CODATA analysis. Inclusion of these studies would suggest a somewhat larger value for Δ_fH. The $\Delta_fH°$(B, g, 298.15 K) values reported by Mar and Bedford (9) are 563.6 ± 33.4 kJ mol^{-1}(2nd law) and 561.1 ± 3.3 kJ mol^{-1}(3rd law) from torsion data and 566.1 ± 13.3 kJ mol^{-1} from mass spectrometric data; Storms and Mueller (10) reported 574.9 ± 0.8 kJ mol^{-1} as the 2nd law value derived from a mass spectrometric study.

Heat Capacity and Entropy

The electronic levels are from the recent analysis by Odintzova and Striganov (11). Although only the ground state, first two excited states, and the highest observed state (below the ionization limit) are listed, all reported states below the ionization limit are used in the calculations. The 164 observed levels are too numerous to list completely, but do include contributions from observed levels to n=39.

The level at 15.254 cm^{-1} has a large effect on the heat capacity and entropy below 100 K. The heat capacity effect decreases to zero above 600 K where the 15.254 cm^{-1} level is fully populated. The higher excited states affect the heat capacity values above 3000 K. The Gibbs energy function values up to 6000 K are essentially independent of the cut-off procedure, the inclusion of levels for n>2, and the estimated missing levels (for n<39).

The thermal functions at 298.15 K agree with recent CODATA recommendations (1) except for two minor differences. First, the entropy differs by 0.1094 $J K^{-1} mol^{-1}$ because this table uses a standard state pressure of 1 bar, whereas the CODATA recommendations are based on 1 atm. Second, an entropy difference of 0.001 $J K^{-1} mol^{-1}$ arises due to the use of slightly different values for the fundamental constants.

References

1. R. C. Paule and J. L. Margrave, J. Phys. Chem. 67, 1368 (1963).
2. H. E. Robson and P. W. Gilles, J. Phys. Chem. 68, 983 (1964).
3. D. L. Hildenbrand and W. F. Hall, J. Phys. Chem. 68, 989 (1964).
4. P. A. Akishin, O. J. Nikitin, and G. L. Gorokhov, Dokl. Akad. Nauk SSSR 129, 1075 (1959).
5. P. O. Schissel and O. C. Trulson, J. Phys. Chem. 66, 1492 (1962).
6. J. D. Cox, chairman, ICSU-CODATA Task Group on Key Values for Thermodynamics, J. Chem. Thermodynamics 10, 903 (1978); CODATA Bulletin 17, 1976.
7. O. C. Trulson and H. W. Goldstein, J. Phys. Chem. 69, 2531 (1965).
8. R. P. Burns, A. J. Jason, and M. G. Inghram, J. Chem. Phys. 46, 394 (1967).
9. R. W. Mar and R. G. Bedford, High Temp. Sci. 8, 365 (1976).
10. E. Storms and B. Mueller, J. Phys. Chem. 81, 318 (1977).
11. G. A. Odintzova and A. R. Striganov, J. Phys. Chem. Ref. Data 8, 63 (1979).

Enthalpy Reference Temperature = T_r = 298.15 K　　Standard State Pressure = p° = 0.1 MPa

	J $K^{-1}mol^{-1}$			kJ mol^{-1}			
T/K	$C°_p$	S°	$-[G°-H°(T_r)]/T$	$H°-H°(T_r)$	$\Delta_fH°$	$\Delta_fG°$	Log K_f
0	0.	0.	INFINITE	-6.316	554.898	554.898	INFINITE
100	20.881	130.687	171.936	-4.125	557.065	544.027	-284.171
200	20.809	145.130	155.338	-2.042	558.815	530.273	-138.493
250	20.801	149.772	153.778	-1.001	559.485	523.057	-109.287
298.15	20.796	153.435	153.435	0.	560.000	515.993	-90.400
300	20.796	153.564	153.436	0.038	560.017	515.720	-89.795
350	20.793	156.769	153.689	1.078	560.430	508.302	-75.860
400	20.792	159.546	154.251	2.118	560.734	500.834	-65.402
450	20.790	161.995	154.978	3.157	560.946	493.332	-57.265
500	20.790	164.185	155.791	4.197	561.082	485.812	-50.752
600	20.789	167.975	157.516	6.276	561.180	470.745	-40.982
700	20.788	171.180	159.245	8.355	561.104	455.677	-34.003
800	20.787	173.956	160.914	10.433	560.900	440.628	-28.770
900	20.787	176.404	162.502	12.512	560.597	425.612	-24.702
1000	20.787	178.594	164.003	14.591	560.213	410.633	-21.449
1100	20.787	180.575	165.421	16.669	559.762	395.697	-18.790
1200	20.787	182.384	166.761	18.748	559.252	380.804	-16.576
1300	20.787	184.048	168.027	20.827	558.689	365.955	-14.704
1400	20.786	185.588	169.227	22.905	558.078	351.152	-13.102
1500	20.786	187.023	170.366	24.984	557.422	336.395	-11.714
1600	20.786	188.364	171.450	27.063	556.724	321.682	-10.502
1700	20.786	189.624	172.482	29.141	555.987	307.014	-9.433
1800	20.786	190.812	173.468	31.220	555.211	292.391	-8.485
1900	20.786	191.936	174.411	33.299	554.398	277.812	-7.638
2000	20.786	193.002	175.314	35.377	553.549	263.277	-6.876
2100	20.786	194.017	176.180	37.456	552.665	248.785	-6.188
2200	20.786	194.984	177.013	39.535	551.747	234.335	-5.564
2300	20.786	195.908	177.815	41.613	550.795	219.929	-4.995
2400	20.786	196.792	178.587	43.692	499.550	206.633	-4.497
2500	20.786	197.641	179.333	45.770	498.453	194.451	-4.063
2600	20.787	198.456	180.052	47.849	497.357	182.312	-3.663
2700	20.787	199.240	180.749	49.928	496.261	170.216	-3.293
2800	20.787	199.996	181.423	52.006	495.165	158.160	-2.951
2900	20.788	200.726	182.076	54.085	494.068	146.144	-2.632
3000	20.789	201.431	182.709	56.164	492.972	134.165	-2.336
3100	20.791	202.112	183.324	58.243	491.876	122.223	-2.059
3200	20.793	202.772	183.922	60.322	490.780	110.316	-1.801
3300	20.795	203.412	184.503	62.402	489.685	98.444	-1.558
3400	20.798	204.033	185.068	64.481	488.589	86.604	-1.331
3500	20.803	204.636	185.619	66.561	487.494	74.797	-1.116
3600	20.808	205.222	186.155	68.642	486.400	63.021	-0.914
3700	20.814	205.792	186.678	70.723	485.306	51.276	-0.724
3800	20.822	206.348	187.188	72.805	484.213	39.560	-0.544
3900	20.831	206.889	187.687	74.887	483.120	27.873	-0.373
4000	20.842	207.416	188.173	76.971	482.029	16.214	-0.212
4100	20.855	207.931	188.649	79.056	480.939	4.582	-0.058
4139.449	20.860	208.131	188.834	79.879	------- FUGACITY = 1 bar -------		
4200	20.870	208.434	189.114	81.142	0.	0.	0.
4300	20.887	208.925	189.569	83.230	0.	0.	0.
4400	20.906	209.405	190.015	85.319	0.	0.	0.
4500	20.928	209.875	190.451	87.411	0.	0.	0.
4600	20.952	210.336	190.878	89.505	0.	0.	0.
4700	20.979	210.786	191.297	91.602	0.	0.	0.
4800	21.009	211.228	191.707	93.701	0.	0.	0.
4900	21.042	211.662	192.110	95.803	0.	0.	0.
5000	21.078	212.087	192.506	97.909	0.	0.	0.
5100	21.117	212.505	192.894	100.019	0.	0.	0.
5200	21.160	212.916	193.275	102.133	0.	0.	0.
5300	21.206	213.319	193.649	104.251	0.	0.	0.
5400	21.255	213.716	194.017	106.374	0.	0.	0.
5500	21.308	214.107	194.379	108.502	0.	0.	0.
5600	21.365	214.491	194.735	110.636	0.	0.	0.
5700	21.425	214.870	195.084	112.776	0.	0.	0.
5800	21.489	215.243	195.429	114.921	0.	0.	0.
5900	21.557	215.611	195.768	117.073	0.	0.	0.
6000	21.628	215.974	196.101	119.233	0.	0.	0.

PREVIOUS: March 1979 (1 atm)　　CURRENT: June 1983 (1 bar)

Boron (B) $B_1(g)$

BORON, ION (B^+) IDEAL GAS M_r = 10.80945

$IP(B^+, g)$ = 202887.4 ± 0.8 cm^{-1} $\Delta_f H°(0 K)$ = 1356 ± 12 kJ mol^{-1}

S°(298.15 K) = 138.542 ± 0.04 J K^{-1} mol^{-1} $\Delta_f H°(298.15 K)$ = [1366.717] kJ mol^{-1}

Electronic Levels and Quantum Weights

State	ε_i, cm^{-1}	g_i
$^2S_0$	0	1
$^3P_0$	37333.6	1
$^3P_1$	37340.0	3
$^3P_2$	37356.4	5

Enthalpy of Formation

$\Delta_f H°(B^+, g, 0 K)$ is calculated from $\Delta_f H°(B, g, 0 K)$ (1) using the spectroscopic value of IP(B) = 66928.10 ± 0.1 cm^{-1} (800.638 ± 0.001 kJ mol^{-1}) from Odintzova and Striganov (2). The ionization limit is converted from cm^{-1} to kJ mol^{-1} using the factor, 1 cm^{-1} = 0.01196266 kJ cm^{-1}, which is derived from the 1973 CODATA fundamental constants (3). Rosenstock et al. (4) and Levin and Lias (5) have summarized additional ionization and appearance potential data.

$\Delta_f H°(B^+, g, 298.15 K)$ is calculated from $\Delta_f H°(B, g, 0 K)$ by using IP(B) with JANAF (1) enthalpies, H°(0 K) - H°(298.15 K), for B(g), $B^+(g)$, and $e^-(g)$. $\Delta_f H°(B \rightarrow B^+ + e^-, 298.15 K)$ differs from a room temperature threshold energy due to inclusion of these enthalpies and to threshold effects discussed by Rosenstock et al. (4). $\Delta_f H°(298.15 K)$ should be changed by -6.197 kJ mol^{-1} if it is to be used in the ion convention that excludes the enthalpy of the electron.

Heat Capacity and Entropy

The information on electronic energy levels and quantum weights, given by Moore (6,7), is incomplete because many theoretically predicted levels have not been observed. Our calculations indicate that any reasonable method of filling in these missing levels and cutting off the summation in the partition function (8) has no effect on the thermodynamic functions to 6000 K. This is a result of the high energy of all levels other than the ground state and the first three excited states; the next excited state is 73396.7 cm^{-1} above the ground state. Since inclusion of these excited states has no effect on the thermodynamic functions (to 6000 K), we list only the ground state and the first three excited states. The energies of the ^3P levels are uncertain by 249 cm^{-1}. No intersystem combinations have been observed. The reported uncertainty in S°(298.15 K) is due to uncertainties in the relative ionic mass and the fundamental constants. Extension of these calculations above 6000 K may require consideration of the excited states and use of different fill and cutoff procedures (8)

References

1. JANAF Thermochemical Tables: B(g), 6-30-83; e^-(g), 3-31-82.
2. G. A. Odintzova and A. R. Striganov, J. Phys. Chem. Ref. Data 8, 63-67 (1979)
3. E. R. Cohen and B. N. Taylor, J. Phys. Chem. Ref. Data 2, 663 (1973)
4. H. M. Rosenstock, K. Draxl et al., J. Phys. Chem. Ref. Data 6, Supp. 1 (1977).
5. R. D. Levin and S. G. Lias, U.S. Nat. Bur. Stand., NSRDS-NBS-71, 1982.
6. C. E. Moore, U.S. Nat. Bur. Stand., NSRD-NBS-34, 1970.
7. C. E. Moore, U.S. Nat. Bur. Stand., NSRDS-NBS-35, Volume I, 1970 [Reprint of NBS Circular 467, Volume I, 1949].
8. J. R. Downey, Jr., The Dow Chemical Company, AFOSR-TR-78-0960, Contract No. F44620-75-1-0048, 1978.

Boron, Ion (B^+) $B_1^+(g)$

Enthalpy Reference Temperature = T_r = 298.15 K Standard State Pressure = p° = 0.1 MPa

T/K	$C_p°$	S°	$-[G°-H°(T_r)]/T$	$H°-H°(T_r)$	$\Delta_f H°$	$\Delta_f G°$	Log K_f
	J $K^{-1}mol^{-1}$			kJ mol^{-1}			
0	0.	0.	INFINITE	-6.197	1355.536		
100	20.786	115.835	157.023	-4.119			
200	20.786	130.243	140.444	-2.040			
250	20.786	134.881	138.885	-1.001			
298.15	20.786	138.542	138.542	0.	1366.717	1320.895	-231.415
300	20.786	138.671	138.543	0.038	1366.773	1320.611	-229.939
350	20.786	141.875	138.796	1.078	1368.224	1312.801	-195.925
400	20.786	144.651	139.358	2.117	1369.567	1304.790	-170.388
450	20.786	147.099	140.085	3.156	1370.818	1296.617	-150.507
500	20.786	149.289	140.898	4.196	1371.994	1288.309	-134.589
600	20.786	153.079	142.622	6.274	1374.170	1271.364	-110.682
700	20.786	156.283	144.350	8.353	1376.172	1254.069	-93.580
800	20.786	159.058	146.019	10.431	1378.046	1236.497	-80.735
900	20.786	161.507	147.607	12.510	1379.822	1218.696	-70.731
1000	20.786	163.697	149.108	14.589	1381.517	1200.701	-62.718
1100	20.786	165.678	150.526	16.667	1383.144	1182.540	-56.154
1200	20.786	167.486	151.865	18.746	1384.712	1164.234	-50.678
1300	20.786	169.150	153.131	20.824	1386.228	1145.799	-46.039
1400	20.786	170.691	154.331	22.903	1387.695	1127.249	-42.058
1500	20.786	172.125	155.470	24.982	1389.118	1108.596	-38.605
1600	20.786	173.466	156.553	27.060	1390.499	1089.849	-35.580
1700	20.786	174.726	157.586	29.139	1391.840	1071.017	-32.908
1800	20.786	175.914	158.571	31.217	1393.143	1052.108	-30.531
1900	20.786	177.038	159.514	33.296	1394.408	1033.127	-28.403
2000	20.786	178.104	160.417	35.375	1395.638	1014.080	-26.485
2100	20.786	179.119	161.284	37.453	1396.833	994.972	-24.749
2200	20.786	180.086	162.117	39.532	1397.993	975.809	-23.169
2300	20.786	181.010	162.918	41.610	1399.119	956.593	-21.725
2400	20.786	181.894	163.690	43.689	1349.953	938.398	-20.424
2500	20.786	182.743	164.436	45.768	1350.935	921.230	-19.248
2600	20.786	183.558	165.156	47.846	1351.917	904.023	-18.162
2700	20.786	184.342	165.852	49.925	1352.900	886.777	-17.156
2800	20.786	185.098	166.526	52.004	1353.882	869.495	-16.221
2900	20.786	185.828	167.179	54.082	1354.864	852.179	-15.349
3000	20.786	186.532	167.812	56.161	1355.846	834.828	-14.536
3100	20.787	187.214	168.427	58.239	1356.829	817.444	-13.774
3200	20.787	187.874	169.025	60.318	1357.811	800.029	-13.059
3300	20.788	188.514	169.606	62.397	1358.793	782.583	-12.387
3400	20.789	189.134	170.171	64.476	1359.776	765.108	-11.754
3500	20.790	189.737	170.721	66.555	1360.758	747.603	-11.157
3600	20.792	190.323	171.258	68.634	1361.741	730.070	-10.593
3700	20.794	190.892	171.781	70.713	1362.724	712.511	-10.059
3800	20.797	191.447	172.291	72.792	1363.707	694.924	-9.552
3900	20.801	191.987	172.789	74.872	1364.690	677.312	-9.072
4000	20.806	192.514	173.276	76.953	1365.674	659.674	-8.614
4100	20.812	193.028	173.751	79.034	1366.659	642.012	-8.179
4200	20.820	193.529	174.216	81.115	887.795	631.350	-7.852
4300	20.830	194.019	174.671	83.198	889.868	625.219	-7.595
4400	20.841	194.498	175.116	85.281	891.940	619.041	-7.349
4500	20.856	194.967	175.552	87.366	894.012	612.815	-7.113
4600	20.872	195.425	175.979	89.452	896.083	606.544	-6.888
4700	20.892	195.874	176.398	91.540	898.153	600.227	-6.671
4800	20.915	196.314	176.808	93.631	900.223	593.866	-6.463
4900	20.941	196.746	177.211	95.724	902.292	587.462	-6.262
5000	20.972	197.169	177.606	97.819	904.360	581.016	-6.070
5100	21.006	197.585	177.993	99.918	906.428	574.529	-5.884
5200	21.046	197.993	178.374	102.021	908.495	568.001	-5.706
5300	21.090	198.395	178.748	104.127	910.562	561.433	-5.533
5400	21.139	198.789	179.115	106.239	912.629	554.827	-5.367
5500	21.194	199.178	179.477	108.355	914.696	548.182	-5.206
5600	21.254	199.560	179.832	110.478	916.764	541.499	-5.051
5700	21.321	199.937	180.181	112.606	918.831	534.779	-4.901
5800	21.393	200.308	180.525	114.742	920.900	528.024	-4.755
5900	21.473	200.675	180.864	116.885	922.970	521.232	-4.615
6000	21.558	201.036	181.197	119.037	925.040	514.406	-4.478

PREVIOUS: March 1979 (1 atm) CURRENT: June 1983 (1 bar)

Boron, Ion (B^+) $B_1^+(g)$

BORON, ION (B^-) IDEAL GAS M_r = 10.81055 Boron, Ion (B^-) $B_1^-(g)$

EA(B, g) = 0.277 ± 0.010 eV

S°(298.15 K) = 156.811 ± 0.005 J K^{-1} mol^{-1}

$\Delta_f H°$(0 K) = 528 ± 12 kJ mol^{-1}

$\Delta_f H°$(298.15 K) = [527.033] kJ mol^{-1}

Electonic Levels and Quantum Weights

State	ε_i, cm^{-1}	g_i
$^3P_0$	0.0	1
$^3P_1$	4.0	3
$^3P_2$	9.0	5

Enthalpy of Formation

$\Delta_f H°(B^-$, g, 0 K) is calculated from $\Delta_f H°$(B, g, 0 K) (1) using the adopted election affinity of EA(B) = 0.277 ± 0.010 eV (26.726 ± 0.965 kJ mol^{-1}). This value, recommended by Hotop and Lineberger (2), is based on a laser photodetachment electron spectrometry study (3). Additional information on B^-(g) may be obtained in the critical discussions of Hotop and Lineberger (2, 4), Rosenstock et al. (5) and Massey (6).

$\Delta_f H°(B^-$, g, 298.15 K) is obtained from $\Delta_f H°$(B, g, 0 K) by using EA(B) with JANAF (1) enthalpies, H°(0 K) − H°(298.15 K), for B^-(g), B(g), and e^-(g). $\Delta_r H°(B^- \rightarrow B + e^-$, 298.15 K) differs from a room-temperature threshold energy due to inclusion of these enthalpies and to threshold effects discussed by Rosenstock et al. (5). $\Delta_f H°$(298.15 K) should be changed by +6.197 kJ mol^{-1} if it is to be used in the ion convention that excludes the enthalpy of the electron.

Heat Capacity and Entropy

The ground state electronic configuration for B^-(g) is given by Hotop and Lineberger (2, 4), Rosenstock et al. (5), and Massey (6). The fine-structure separation has been calculated via an isoelectronic extrapolation of ratios of fine structure separations (4) and is that recommended by Hotop and Lineberger (2)

A comparison of the isoelectronic sequence B^-(g), C(g), N^+(g), O^{++}(g) would suggest that a state, $^1D_2$, may exist at low energy, ~5000 cm^{-1}. However, this is most likely a metastable state lying at an energy greater than the electron affinity. As discussed by Hotop and Lineberger (2, 4), no stable excited states have been observed. Thus we assume no stable excited states exist.

References

1. JANAF Thermochemical Tables: B(g), 6-30-83; e^-(g), 3-31-82.
2. H. Hotop and W. C. Lineberger, J. Phys. Chem. Ref. Data, to be published, 1985.
3. C. S. Feigerle, R. R. Corderman and W. C. Lineberger, J. Chem. Phys. 74, 1513 (1981).
4. H. Hotop and W. C. Lineberger, J. Phys. Chem. Ref. Data 4, 539 (1975).
5. H. M. Rosenstock, K. Draxl et al., J. Phys. Chem. Ref. Data 6, Supp. 1 (1977).
6. H. S. W. Massey, "Negative Ions", 3rd ed., Cambridge University Press, Cambridge, 1976.

Enthalpy Reference Temperature = T_r = 298.15 K Standard State Pressure = p° = 0.1 MPa

	J $K^{-1}mol^{-1}$			kJ mol^{-1}			
T/K	$C_p°$	S°	−[G°−H°(T_r)]/T	H°−H°(T_r)	$\Delta_f H°$	$\Delta_f G°$	Log K_f
0	0.	0.	INFINITE	-6.273	528.172		
100	20.804	134.096	175.295	-4.120			
200	20.790	148.511	158.713	-2.040			
250	20.789	153.150	157.153	-1.001			
298.15	20.788	156.811	156.811	0.	527.033	488.274	-85.544
300	20.788	156.940	156.812	0.038	527.012	488.034	-84.974
350	20.787	160.144	157.065	1.078	526.385	481.585	-71.873
400	20.787	162.920	157.627	2.117	525.650	475.234	-62.059
450	20.787	165.368	158.354	3.157	524.822	468.981	-54.438
500	20.787	167.558	159.167	4.196	523.919	462.824	-48.351
600	20.787	171.348	160.891	6.275	521.938	450.788	-39.245
700	20.786	174.553	162.619	8.353	519.783	439.098	-32.766
800	20.786	177.328	164.288	10.432	517.500	427.726	-27.928
900	20.786	179.776	165.876	12.510	515.118	416.647	-24.182
1000	20.786	181.966	167.377	14.589	512.656	405.837	-21.199
1100	20.786	183.948	168.795	16.668	510.126	395.277	-18.770
1200	20.786	185.756	170.134	18.746	507.537	384.950	-16.756
1300	20.786	187.420	171.401	20.825	504.896	374.841	-15.061
1400	20.786	188.960	172.601	22.904	502.206	364.938	-13.616
1500	20.786	190.395	173.740	24.982	499.472	355.228	-12.370
1600	20.786	191.736	174.823	27.061	496.695	345.702	-11.286
1700	20.786	192.996	175.855	29.139	493.879	336.351	-10.335
1800	20.786	194.184	176.841	31.218	491.025	327.167	-9.494
1900	20.786	195.308	177.784	33.297	488.133	318.142	-8.746
2000	20.786	196.374	178.687	35.375	485.206	309.271	-8.077
2100	20.786	197.388	179.553	37.454	482.243	300.547	-7.476
2200	20.786	198.355	180.386	39.532	479.246	291.965	-6.932
2300	20.786	199.279	181.188	41.611	476.215	283.520	-6.439
2400	20.786	200.164	181.960	43.690	422.892	276.276	-6.013
2500	20.786	201.013	182.705	45.768	419.717	270.232	-5.646
2600	20.786	201.828	183.425	47.847	416.542	264.315	-5.310
2700	20.786	202.612	184.121	49.925	413.367	258.521	-5.001
2800	20.786	203.368	184.795	52.004	410.192	252.844	-4.717
2900	20.786	204.098	185.448	54.083	407.017	247.280	-4.454
3000	20.786	204.802	186.082	56.161	403.842	241.826	-4.211
3100	20.786	205.484	186.697	58.240	400.667	236.478	-3.985
3200	20.786	206.144	187.294	60.318	397.492	231.232	-3.774
3300	20.786	206.783	187.875	62.397	394.317	226.086	-3.579
3400	20.786	207.404	188.441	64.476	391.142	221.036	-3.396
3500	20.786	208.007	188.991	66.554	387.967	216.079	-3.225
3600	20.786	208.592	189.527	68.633	384.792	211.213	-3.065
3700	20.786	209.162	190.050	70.711	381.617	206.435	-2.914
3800	20.786	209.716	190.561	72.790	378.442	201.743	-2.773
3900	20.786	210.256	191.059	74.869	375.267	197.134	-2.640
4000	20.786	210.782	191.545	76.947	372.092	192.607	-2.515
4100	20.786	211.295	192.021	79.026	368.917	188.159	-2.397
4200	20.786	211.796	192.486	81.104	-114.108	190.813	-2.373
4300	20.786	212.285	192.940	83.183	-116.195	198.097	-2.406
4400	20.786	212.763	193.386	85.262	-118.285	205.431	-2.439
4500	20.786	213.230	193.821	87.340	-120.377	212.811	-2.470
4600	20.786	213.687	194.248	89.419	-122.471	220.239	-2.501
4700	20.786	214.134	194.667	91.498	-124.567	227.712	-2.531
4800	20.786	214.572	195.077	93.576	-126.666	235.229	-2.560
4900	20.786	215.000	195.479	95.655	-128.769	242.790	-2.588
5000	20.786	215.420	195.874	97.733	-130.875	250.395	-2.616
5100	20.786	215.832	196.261	99.812	-132.985	258.041	-2.643
5200	20.786	216.236	196.641	101.891	-135.099	265.729	-2.669
5300	20.786	216.632	197.015	103.969	-137.217	273.457	-2.695
5400	20.786	217.020	197.382	106.048	-139.340	281.225	-2.720
5500	20.786	217.401	197.742	108.126	-141.468	289.033	-2.745
5600	20.786	217.776	198.097	110.205	-143.602	296.880	-2.769
5700	20.786	218.144	198.445	112.284	-145.741	304.765	-2.793
5800	20.786	218.505	198.788	114.362	-147.887	312.687	-2.816
5900	20.786	218.861	199.125	116.441	-150.039	320.646	-2.839
6000	20.786	219.210	199.457	118.519	-152.198	328.642	-2.861

PREVIOUS: March 1979 (1 atm) CURRENT: June 1983 (1 bar)

Boron, Ion (B^-) $B_1^-(g)$

BERYLLIUM BORATE (BeO_2) IDEAL GAS M_r = 51.82098

$S°(298.15\ K) = 265.4\ J\ K^{-1}\ mol^{-1}$

$\Delta_f H°(0\ K) = -481.8\ kJ\ mol^{-1}$
$\Delta_f H°(298.15\ K) = -482.0\ kJ\ mol^{-1}$

Vibrational Frequencies and Degeneracies

ν, cm^{-1}	ν, cm^{-1}
[2000](1)	[600](1)
[1000](1)	[400](1)
[900](1)	[800](1)

Ground State Quantum Weight: [2] σ = [1]
Point Group: [C_s]
Bond Distances: Be–O = [1.40] Å O–B = [1.36] Å B=O = [1.20] Å
Bond Angles: Be–O–B = [120]° O–B = [180]°
Product of the Moments of Inertia: $I_A I_B I_C = [2.451196 \times 10^{-115}]\ g^3\ cm^6$

Enthalpy of Formation

Since Blackburn and Buchler (1) have postulated that the BO_2^- group has the pseudochloride character in bond strength, the enthalpy of formation of $BeBO_2(g)$ is estimated by adding the difference of the enthalpies of formation of BeCl(g) and AlCl(g) to the enthalpies of formation of $AlBO_2(g)$, using all JANAF values: $\Delta_f H°$(BeCl, g, 298.15 K) = 3.0, $\Delta_f H°$(AlCl, g, 298.15 K) = -11.2 and $\Delta_f H°(AlBO_2$, g, 298.15 K) = -129.4 kcal mol^{-1}.

Heat Capacity and Entropy

All molecular constants are estimated by comparison with those in $LiBO_2(g)$, $AlBO_2(g)$, $B_2O_3(g)$ and BeCl(g). The principal moments of inertia are $I_A = 1.1016 \times 10^{-39}$, $I_B = 14.3762 \times 10^{-39}$ and $I_C = 15.4778 \times 10^{-39}\ g\ cm^2$.

Reference

1. P. E. Blackburn and A. Buchler, J. Phys. Chem. 69, 4250 (1965).

Beryllium Borate ($BeBO_2$) $B_1Be_1O_2(g)$

Enthalpy Reference Temperature = T_r = 298.15 K Standard State Pressure = p° = 0.1 MPa

T/K	$C_p°$ (J K⁻¹mol⁻¹)	S° (J K⁻¹mol⁻¹)	$-[G°-H°(T_r)]/T$ (J K⁻¹mol⁻¹)	$H°-H°(T_r)$ (kJ mol⁻¹)	$\Delta_f H°$ (kJ mol⁻¹)	$\Delta_f G°$ (kJ mol⁻¹)	Log K_f
0	0.	0.	INFINITE	-11.594	-481.761	-481.761	INFINITE
100	34.261	221.501	304.015	-8.251	-481.387	-486.125	253.926
200	41.473	247.191	269.710	-4.504	-481.461	-490.883	128.206
250	46.006	256.932	266.199	-2.317	-481.694	-493.214	103.051
298.15	50.168	265.396	265.396	0.	-481.997	-495.406	86.793
300	50.320	265.707	265.397	0.093	-482.010	-495.489	86.272
350	54.165	273.759	266.024	2.707	-482.377	-497.707	74.279
400	57.492	281.215	267.463	5.501	-482.778	-499.870	65.276
450	60.347	288.155	269.381	8.449	-483.201	-501.982	58.269
500	62.797	294.644	271.586	11.529	-483.642	-504.045	52.657
600	66.738	306.459	276.434	18.015	-484.564	-508.040	44.229
700	69.721	316.981	281.488	24.845	-485.535	-511.876	38.197
800	72.017	326.447	286.527	31.936	-486.560	-515.570	33.663
900	73.810	335.037	291.447	39.231	-487.645	-519.132	30.130
1000	75.228	342.890	296.204	46.686	-488.797	-522.569	27.296
1100	76.363	350.115	300.781	54.267	-490.026	-525.888	24.972
1200	77.282	356.800	305.174	61.951	-491.338	-529.091	23.031
1300	78.033	363.017	309.387	69.718	-492.741	-532.180	21.383
1400	78.655	368.823	313.428	77.554	-494.241	-535.159	19.967
1500	79.173	374.268	317.304	85.446	-495.845	-538.026	18.736
1600	79.609	379.392	321.026	93.385	-512.094	-540.257	17.638
1700	79.978	384.230	324.603	101.365	-513.520	-541.974	16.653
1800	80.295	388.810	328.044	109.379	-514.997	-543.606	15.775
1900	80.567	393.159	331.358	117.423	-516.525	-545.154	14.987
2000	80.802	397.298	334.552	125.491	-518.109	-546.620	14.276
2100	81.007	401.245	337.635	133.582	-519.751	-548.005	13.631
2200	81.187	405.018	340.612	141.692	-521.452	-549.310	13.042
2300	81.346	408.630	343.492	149.819	-523.213	-550.538	12.503
2400	81.486	412.095	346.278	157.961	-575.297	-550.618	11.984
2500	81.611	415.424	348.978	166.116	-577.262	-549.550	11.482
2600	81.722	418.627	351.596	174.282	-579.258	-548.402	11.018
2700	81.822	421.714	354.136	182.460	-581.286	-547.177	10.586
2800	81.911	424.691	356.603	190.646	-874.304	-539.653	10.067
2900	81.992	427.567	359.001	198.842	-875.334	-527.684	9.505
3000	82.066	430.348	361.333	207.045	-876.380	-515.678	8.979
3100	82.132	433.040	363.603	215.255	-877.445	-503.636	8.486
3200	82.193	435.648	365.814	223.471	-878.529	-491.561	8.024
3300	82.248	438.178	367.968	231.693	-879.634	-479.452	7.589
3400	82.299	440.634	370.070	239.920	-880.761	-467.308	7.179
3500	82.346	443.021	372.120	248.153	-881.911	-455.131	6.792
3600	82.388	445.341	374.122	256.389	-883.088	-442.921	6.427
3700	82.428	447.599	376.077	264.630	-884.291	-430.678	6.080
3800	82.464	449.798	377.988	272.875	-885.524	-418.400	5.751
3900	82.498	451.940	379.857	281.123	-886.788	-406.091	5.439
4000	82.530	454.029	381.686	289.374	-888.085	-393.750	5.142
4100	82.559	456.067	383.475	297.629	-889.417	-381.375	4.859
4200	82.586	458.057	385.227	305.886	-1370.635	-361.943	4.501
4300	82.611	460.001	386.944	314.146	-1370.957	-337.922	4.105
4400	82.635	461.900	388.626	322.408	-1371.321	-313.895	3.726
4500	82.657	463.758	390.275	330.673	-1371.730	-289.858	3.365
4600	82.678	465.575	391.892	338.940	-1372.188	-265.812	3.018
4700	82.697	467.353	393.479	347.208	-1372.695	-241.754	2.687
4800	82.715	469.094	395.036	355.479	-1373.255	-217.686	2.369
4900	82.732	470.800	396.565	363.751	-1373.870	-193.605	2.064
5000	82.748	472.471	398.066	372.025	-1374.542	-169.511	1.771
5100	82.764	474.110	399.541	380.301	-1375.272	-145.403	1.489
5200	82.778	475.717	400.991	388.578	-1376.064	-121.281	1.218
5300	82.792	477.294	402.416	396.857	-1376.919	-97.142	0.957
5400	82.804	478.842	403.817	405.136	-1377.840	-72.987	0.706
5500	82.817	480.361	405.195	413.417	-1378.827	-48.815	0.464
5600	82.828	481.854	406.550	421.700	-1379.883	-24.622	0.230
5700	82.839	483.320	407.884	429.983	-1381.010	-0.412	0.004
5800	82.849	484.761	409.197	438.267	-1382.209	23.820	-0.215
5900	82.859	486.177	410.490	446.553	-1383.481	48.071	-0.426
6000	82.868	487.570	411.763	454.839	-1384.829	72.347	-0.630

PREVIOUS: June 1966 (1 atm) CURRENT: June 1966 (1 bar)

Beryllium Borate ($BeBO_2$) $B_1Be_1O_2(g)$

BROMOBORANE (BBr) IDEAL GAS M_r = 90.714

$\Delta_f H°$(0 K) = 239 ± 42 kJ mol^{-1}

S°(298.15 K) = 225.00 J K^{-1} mol^{-1} $\Delta_f H°$(298.15 K) = 234 ± 42 kJ mol^{-1}

Electronic Level and Quantum Weight

State	ε_i, cm^{-1}	g_i
$^1\Sigma^+$	0	1

ω_e = 689.04 cm^{-1} $\omega_e x_e$ = 3.57 cm^{-1} σ = 1

B_e = 0.497 cm^{-1} α_e = 0.0036 cm^{-1} r_e = 1.887 Å

Enthalpy of Formation

Herzberg (1) reported D_0^0 = 4.1 eV which corresponds to a enthalpy of formation of 64 kcal mol^{-1}. Gaydon (2) gives essentially the same value, with a provisor that the value is unreliable due to the ionic nature of the bonding. Barrow (3) made a critical evaluation of the bond dissociation energies of the group III halides and concluded D_0^0(BBr) = 103.5 which gives $\Delta_f H°$(298.15 K) = 56 kcal mol^{-1}. An estimate of the dissociation energy by taking the ratio D_0^0(BF)/[$\Delta_{at}H°$(BF_3, g, 0 K)/3] = 1.177 and applying this to BBr_3 yields D_0^0(BBr) = 103 kcal mol^{-1}; this was the value adopted.

Heat Capacity and Entropy

The vibrational and rotational constants are all taken from NBS (4).

References

1. G. Herzberg, "Spectra of Diatomic Molecules", D. van Nostrand Co., New York, 1950.
2. A. G. Gaydon, "Dissociation Energies and Spectra of Diatomic Molecules", Chapman and Hall, London, 1953.
3. R. F. Barrow, Trans. Faraday Soc. 56, 952 (1960).
4. U. S. Nat. Bur. Stand. Report 7093, 1962.

Bromoborane (BBr) B_1Br_1(g)

Enthalpy Reference Temperature = T_r = 298.15 K Standard State Pressure = p° = 0.1 MPa

T/K	$C_p°$	S°	−[G°−H°(T_r)]/T	H°−H°(T_r)	$\Delta_f H°$	$\Delta_f G°$	Log K_f
	J K^{-1}mol^{-1}			kJ mol^{-1}			
0	0.	0.	INFINITE	-8.997	238.775	238.775	INFINITE
100	29.154	191.806	252.688	-6.088	240.264	223.806	-116.904
200	30.637	212.350	227.930	-3.116	240.453	207.208	-54.117
250	31.784	219.311	225.531	-1.555	240.234	198.918	-41.562
298.15	32.788	224.997	224.997	0.	234.304	191.651	-33.576
300	32.824	225.200	224.998	0.061	234.274	191.386	-33.323
350	33.684	230.327	225.400	1.724	218.985	185.073	-27.621
400	34.373	234.871	226.306	3.426	219.037	180.225	-23.535
450	34.921	238.953	227.488	5.159	219.021	175.373	-20.357
500	35.359	242.656	228.822	6.917	218.950	170.527	-17.815
600	35.999	249.163	231.685	10.487	218.679	160.865	-14.005
700	36.435	254.747	234.590	14.110	218.278	151.260	-11.287
800	36.747	259.633	237.422	17.770	217.779	141.719	-9.253
900	36.981	263.976	240.135	21.456	217.202	132.245	-7.675
1000	37.164	267.882	242.718	25.164	216.560	122.839	-6.416
1100	37.311	271.431	245.169	28.888	215.863	113.500	-5.390
1200	37.434	274.683	247.495	32.626	215.116	104.227	-4.537
1300	37.539	277.684	249.703	36.374	214.322	95.019	-3.818
1400	37.631	280.469	251.803	40.133	213.487	85.872	-3.204
1500	37.714	283.068	253.801	43.900	212.610	76.787	-2.674
1600	37.788	285.505	255.707	47.675	211.694	67.762	-2.212
1700	37.857	287.798	257.528	51.458	210.740	58.796	-1.807
1800	37.921	289.963	259.271	55.247	209.747	49.886	-1.448
1900	37.982	292.015	260.941	59.042	208.717	41.033	-1.128
2000	38.039	293.965	262.543	62.843	207.647	32.235	-0.842
2100	38.094	295.822	264.084	66.650	206.540	23.491	-0.584
2200	38.147	297.596	265.568	70.462	205.393	14.801	-0.351
2300	38.199	299.292	266.997	74.279	204.207	6.165	-0.140
2400	38.249	300.919	268.377	78.101	152.722	-1.351	0.029
2500	38.297	302.482	269.710	81.929	151.379	-7.743	0.162
2600	38.345	303.985	271.000	85.761	150.029	-14.081	0.283
2700	38.392	305.433	272.248	89.598	148.674	-20.368	0.394
2800	38.438	306.830	273.459	93.439	147.312	-26.604	0.496
2900	38.483	308.179	274.633	97.285	145.944	-32.791	0.591
3000	38.528	309.485	275.773	101.136	144.572	-38.931	0.678
3100	38.573	310.749	276.881	104.991	143.196	-45.025	0.759
3200	38.617	311.974	277.958	108.850	141.818	-51.074	0.834
3300	38.660	313.163	279.007	112.714	140.438	-57.081	0.904
3400	38.703	314.318	280.029	116.582	139.059	-63.046	0.969
3500	38.746	315.440	281.025	120.455	137.680	-68.970	1.029
3600	38.789	316.532	281.996	124.332	136.304	-74.854	1.086
3700	38.831	317.596	282.944	128.213	134.931	-80.701	1.139
3800	38.873	318.632	283.869	132.098	133.564	-86.511	1.189
3900	38.915	319.642	284.774	135.987	132.203	-92.284	1.236
4000	38.957	320.628	285.658	139.881	130.850	-98.023	1.280
4100	38.999	321.590	286.522	143.779	129.505	-103.728	1.322
4200	39.040	322.531	287.369	147.681	-351.680	-102.377	1.273
4300	39.082	323.450	288.197	151.587	-351.918	-96.438	1.171
4400	39.123	324.349	289.009	155.497	-352.147	-90.494	1.074
4500	39.164	325.228	289.804	159.411	-352.365	-84.545	0.981
4600	39.205	326.090	290.583	163.330	-352.574	-78.591	0.892
4700	39.246	326.933	291.348	167.252	-352.771	-72.633	0.807
4800	39.287	327.760	292.098	171.179	-352.959	-66.670	0.726
4900	39.328	328.571	292.834	175.110	-353.135	-60.704	0.647
5000	39.369	329.365	293.557	179.045	-353.300	-54.734	0.572
5100	39.409	330.145	294.266	182.984	-353.455	-48.761	0.499
5200	39.450	330.911	294.964	186.927	-353.599	-42.786	0.430
5300	39.491	331.663	295.649	190.874	-353.732	-36.807	0.363
5400	39.531	332.401	296.323	194.825	-353.855	-30.826	0.298
5500	39.571	333.127	296.985	198.780	-353.969	-24.843	0.236
5600	39.612	333.841	297.637	202.739	-354.073	-18.858	0.176
5700	39.652	334.542	298.279	206.702	-354.168	-12.871	0.118
5800	39.693	335.232	298.910	210.669	-354.254	-6.883	0.062
5900	39.733	335.911	299.531	214.641	-354.332	-0.893	0.008
6000	39.773	336.579	300.143	218.616	-354.402	5.098	-0.044

PREVIOUS: December 1964 (1 atm) CURRENT: December 1964 (1 bar)

Bromoborane (BBr) B_1Br_1(g)

BROMOCHLOROBORANE (BBrCl) IDEAL GAS M_r = 126.167

$S°(298.15\ K) = 289.03\ J\ K^{-1}\ mol^{-1}$

$\Delta_f H°(0\ K) = -4.3\ kJ\ mol^{-1}$
$\Delta_f H°(298.15\ K) = -10.5\ kJ\ mol^{-1}$

Vibrational Frequencies and Degeneracies

ν, cm^{-1}
[660](1)
[195](1)
[905](1)

Ground State Quantum Weight: [2] $\sigma = 1$

Point Group: C_s

Bond Distances: B–Cl = [1.73] Å B–Br = [1.87] Å

Bond Angle: Cl–B–Br = [120]°

Product of the Moments of Inertia: $I_A I_B I_C = [2.193041 \times 10^{-114}]\ g^3\ cm^6$

Enthalpy of Formation

The enthalpy of formation is estimated by assuming $\Delta_r H° = 0$ for the reaction $BCl_2(g) + BBr_2(g) = 2\ BBrCl(g)$.

Heat Capacity and Entropy

The vibrational frequencies are estimated by comparison with those for BBr_2 and BCl_2. The bond distances, angles, and quantum weight are assumed the same as in the dihalides. The principal moments of inertia are $I_A = 1.2989 \times 10^{-39}\ g\ cm^2$, $I_B = 40.4453 \times 10^{-39}\ g\ cm^2$ and $I_C = 41.7442 \times 10^{-39}\ g\ cm^2$.

Bromochloroborane (BBrCl) $B_1Br_1Cl_1(g)$

Enthalpy Reference Temperature = T_r = 298.15 K Standard State Pressure = $p°$ = 0.1 MPa

T/K	$C_p°$ (J K⁻¹mol⁻¹)	S° (J K⁻¹mol⁻¹)	$-[G°-H°(T_r)]/T$ (J K⁻¹mol⁻¹)	$H°-H°(T_r)$ (kJ mol⁻¹)	$\Delta_f H°$ (kJ mol⁻¹)	$\Delta_f G°$ (kJ mol⁻¹)	Log K_f
0	0.	0.	INFINITE	-11.889	-4.290	-4.290	INFINITE
100	37.800	243.541	327.661	-8.412	-3.690	-15.870	8.290
200	42.512	271.227	293.183	-4.391	-3.971	-27.994	7.311
250	44.812	280.962	289.793	-2.208	-4.376	-33.957	7.095
298.15	46.846	289.033	289.033	0.	-10.460	-38.950	6.824
300	46.919	289.323	289.034	0.087	-10.496	-39.127	6.813
350	48.726	296.695	289.612	2.479	-25.915	-43.053	6.425
400	50.221	303.303	290.917	4.954	-25.966	-45.498	5.941
450	51.439	309.291	292.631	7.497	-26.060	-47.934	5.564
500	52.429	314.763	294.575	10.094	-26.188	-50.358	5.261
600	53.895	324.461	298.769	15.415	-26.524	-55.162	4.802
700	54.891	332.848	303.052	20.858	-26.942	-59.903	4.470
800	55.589	340.226	307.247	26.384	-27.425	-64.580	4.217
900	56.093	346.804	311.283	31.969	-27.963	-69.192	4.016
1000	56.467	352.735	315.137	37.598	-28.552	-73.742	3.852
1100	56.752	358.130	318.803	43.260	-29.187	-78.231	3.715
1200	56.973	363.078	322.290	48.946	-29.866	-82.660	3.598
1300	57.147	367.646	325.605	54.653	-30.588	-87.030	3.497
1400	57.287	371.886	328.761	60.375	-31.351	-91.343	3.408
1500	57.401	375.843	331.770	66.109	-32.156	-95.601	3.329
1600	57.496	379.550	334.641	71.854	-33.001	-99.803	3.258
1700	57.574	383.038	337.387	77.608	-33.889	-103.951	3.194
1800	57.641	386.331	340.015	83.369	-34.818	-108.046	3.135
1900	57.697	389.449	342.536	89.136	-35.791	-112.088	3.082
2000	57.745	392.410	344.956	94.908	-36.808	-116.077	3.032
2100	57.787	395.228	347.283	100.684	-37.870	-120.015	2.985
2200	57.823	397.917	349.524	106.465	-38.979	-123.901	2.942
2300	57.855	400.488	351.685	112.249	-40.135	-127.735	2.901
2400	57.883	402.951	353.770	118.036	-91.600	-130.449	2.839
2500	57.907	405.315	355.785	123.825	-92.933	-132.041	2.759
2600	57.929	407.586	357.734	129.617	-94.284	-133.579	2.684
2700	57.949	409.773	359.621	135.411	-95.652	-135.064	2.613
2800	57.966	411.881	361.450	141.207	-97.040	-136.498	2.546
2900	57.982	413.915	363.224	147.004	-98.445	-137.883	2.484
3000	57.996	415.881	364.947	152.803	-99.867	-139.219	2.424
3100	58.009	417.783	366.621	158.604	-101.306	-140.507	2.368
3200	58.021	419.625	368.248	164.405	-102.761	-141.748	2.314
3300	58.032	421.410	369.832	170.208	-104.229	-142.943	2.263
3400	58.041	423.143	371.375	176.012	-105.708	-144.095	2.214
3500	58.050	424.826	372.878	181.816	-107.197	-145.202	2.167
3600	58.059	426.461	374.344	187.622	-108.694	-146.267	2.122
3700	58.066	428.052	375.774	193.428	-110.196	-147.290	2.079
3800	58.073	429.601	377.170	199.235	-111.701	-148.272	2.038
3900	58.080	431.109	378.534	205.042	-113.207	-149.215	1.999
4000	58.085	432.580	379.867	210.851	-114.711	-150.119	1.960
4100	58.091	434.014	381.170	216.659	-116.210	-150.985	1.924
4200	58.096	435.414	382.445	222.469	-597.552	-144.792	1.801
4300	58.101	436.781	383.693	228.279	-597.949	-134.007	1.628
4400	58.105	438.117	384.915	234.089	-598.337	-123.213	1.463
4500	58.110	439.423	386.112	239.900	-598.715	-112.410	1.305
4600	58.114	440.700	387.284	245.711	-599.080	-101.599	1.154
4700	58.117	441.950	388.434	251.522	-599.432	-90.781	1.009
4800	58.121	443.173	389.562	257.334	-599.768	-79.955	0.870
4900	58.124	444.372	390.668	263.147	-600.089	-69.122	0.737
5000	58.127	445.546	391.754	268.959	-600.393	-58.283	0.609
5100	58.130	446.697	392.820	274.772	-600.680	-47.438	0.486
5200	58.132	447.826	393.867	280.585	-600.948	-36.587	0.368
5300	58.135	448.933	394.896	286.398	-601.198	-25.732	0.254
5400	58.137	450.020	395.907	292.212	-601.429	-14.872	0.144
5500	58.140	451.087	396.900	298.026	-601.640	-4.007	0.038
5600	58.142	452.134	397.877	303.840	-601.833	6.860	-0.064
5700	58.144	453.163	398.838	309.654	-602.006	17.731	-0.162
5800	58.146	454.175	399.784	315.469	-602.160	28.605	-0.258
5900	58.148	455.169	400.714	321.283	-602.295	39.482	-0.350
6000	58.149	456.146	401.630	327.098	-602.412	50.360	-0.438

PREVIOUS: December 1964 (1 atm) CURRENT: December 1964 (1 bar)

Bromochloroborane (BBrCl) $B_1Br_1Cl_1(g)$

BROMODICHLOROBORANE ($BBrCl_2$) IDEAL GAS M_r = 161.620

$S°$(298.15 K) = 310.40 J K^{-1} mol^{-1}

$\Delta_f H°$(0 K) = -328.7±10 kJ mol^{-1}
$\Delta_f H°$(298.15 K) = -336.8±10 kJ mol^{-1}

Vibrational Frequencies and Degeneracies

ν, cm^{-1}	ν, cm^{-1}
885(1)	959 (1)
406(1)	[208](1)
218(1)	433 (1)

Ground State Quantum Weight: [1] σ = 2
Point Group: C_{2v}
Bond Distances: B-Cl = 1.73 Å B-Br = 1.87 Å
Bond Angles: Cl-B-Cl = [120]° Br-B-Cl = [120]°
Product of the Moments of Inertia: $I_A I_B I_C$ = 9.196849 x 10^{-113} g^3 cm^6

Enthalpy of Formation

The enthalpy of formation is estimated by assuming $\Delta_r H°$ = 0 for the reaction BBr_3(g) + 2 BCl_3(g) = 3 $BBrCl_2$(g).

Heat Capacity and Entropy

Vibrational frequencies are obtained from Lindeman and Wilson (1). The b_1 mode was the only unobserved frequency and was calculated from a normal coordinate treatment. Bond distances were reported by Wentink and Tiensuu (2). Bond angles were estimated by analogy with the other trihalides of boron. The principal moments of inertia are I_A = 26.4297 x 10^{-39} g cm^2, I_B = 47.2365 x 10^{-39} g cm^2 and I_C = 73.6662 x 10^{-39} g cm^2.

References

1. L. P. Lindeman and M. K. Wilson, J. Chem. Phys. 24, 242 (1956).
2. T. Wentink and V. H. Tiensuu, J. Chem. Phys. 28, 82b (1958).

Bromodichloroborane ($BBrCl_2$) $B_1Br_1Cl_2$(g)

Enthalpy Reference Temperature = T_r = 298.15 K Standard State Pressure = p° = 0.1 MPa

	J K^{-1} mol^{-1}			kJ mol^{-1}			
T/K	$C_p°$	S°	-[G°-H°(T_r)]/T	H°-H°(T_r)	$\Delta_f H°$	$\Delta_f G°$	Log K_f
0	0.	0.	INFINITE	-14.547	-328.709	-328.709	INFINITE
100	42.752	252.188	361.654	-10.947	-329.442	-333.037	173.960
200	56.153	286.331	316.048	-5.943	-330.260	-336.307	87.834
250	60.770	299.375	311.440	-3.016	-330.730	-337.767	70.573
298.15	64.398	310.400	310.400	0.	-336.812	-338.417	59.289
300	64.523	310.798	310.401	0.119	-336.847	-338.427	58.925
350	67.577	320.982	311.198	3.424	-352.212	-337.848	50.421
400	70.050	330.173	313.005	6.867	-352.171	-335.799	43.851
450	72.048	338.543	315.384	10.422	-352.141	-333.754	38.741
500	73.665	346.221	318.090	14.066	-352.121	-331.712	34.654
600	76.059	359.878	323.945	21.560	-352.100	-327.633	28.523
700	77.689	371.733	329.943	29.252	-352.103	-323.555	24.144
800	78.834	382.185	335.833	37.082	-352.133	-319.475	20.860
900	79.662	391.521	341.512	45.009	-352.190	-315.390	18.305
1000	80.278	399.948	346.941	53.007	-352.277	-311.297	16.260
1100	80.747	407.622	352.114	61.059	-352.397	-307.193	14.587
1200	81.111	414.664	357.037	69.153	-352.550	-303.078	13.193
1300	81.399	421.168	361.723	77.279	-352.738	-298.947	12.012
1400	81.631	427.209	366.187	85.431	-352.962	-294.802	10.999
1500	81.819	432.848	370.446	93.604	-353.224	-290.638	10.121
1600	81.975	438.134	374.513	101.794	-353.525	-286.456	9.352
1700	82.106	443.107	378.403	109.998	-353.866	-282.254	8.673
1800	82.215	447.804	382.129	118.214	-354.249	-278.031	8.068
1900	82.309	452.251	385.704	126.440	-354.676	-273.785	7.527
2000	82.388	456.475	389.138	134.675	-355.148	-269.516	7.039
2100	82.458	460.497	392.441	142.918	-355.669	-265.222	6.597
2200	82.518	464.334	395.622	151.167	-356.240	-260.901	6.195
2300	82.570	468.003	398.690	159.421	-356.863	-256.554	5.827
2400	82.616	471.518	401.652	167.680	-407.800	-251.110	5.465
2500	82.657	474.892	404.514	175.944	-408.611	-244.564	5.110
2600	82.694	478.134	407.284	184.212	-409.447	-237.986	4.781
2700	82.726	481.256	409.966	192.483	-410.309	-231.375	4.476
2800	82.755	484.265	412.566	200.757	-411.197	-224.732	4.192
2900	82.781	487.170	415.089	209.034	-412.112	-218.056	3.928
3000	82.805	489.976	417.539	217.313	-413.053	-211.349	3.680
3100	82.826	492.692	419.919	225.595	-414.019	-204.609	3.448
3200	82.846	495.322	422.235	233.878	-415.009	-197.839	3.229
3300	82.863	497.871	424.488	242.164	-416.021	-191.036	3.024
3400	82.880	500.345	426.683	250.451	-417.053	-184.203	2.830
3500	82.894	502.748	428.822	258.739	-418.101	-177.340	2.647
3600	82.908	505.083	430.909	267.030	-419.162	-170.446	2.473
3700	82.920	507.355	432.944	275.321	-420.234	-163.522	2.309
3800	82.932	509.567	434.931	283.614	-421.313	-156.570	2.152
3900	82.943	511.721	436.873	291.907	-422.395	-149.589	2.004
4000	82.953	513.821	438.771	300.202	-423.477	-142.580	1.862
4100	82.962	515.869	440.626	308.498	-424.555	-135.544	1.727
4200	82.970	517.869	442.442	316.795	-905.475	-121.458	1.511
4300	82.978	519.821	444.218	325.092	-905.447	-102.791	1.249
4400	82.986	521.729	445.958	333.390	-905.407	-84.125	0.999
4500	82.993	523.594	447.663	341.689	-905.351	-65.460	0.760
4600	82.999	525.418	449.334	349.989	-905.277	-46.796	0.531
4700	83.005	527.203	450.971	358.289	-905.183	-28.135	0.313
4800	83.011	528.951	452.578	366.590	-905.066	-9.476	0.103
4900	83.016	530.662	454.154	374.891	-904.924	9.181	-0.098
5000	83.021	532.340	455.701	383.193	-904.755	27.835	-0.291
5100	83.026	533.984	457.220	391.495	-904.558	46.485	-0.476
5200	83.031	535.596	458.712	399.798	-904.331	65.130	-0.654
5300	83.035	537.178	460.177	408.101	-904.074	83.771	-0.826
5400	83.039	538.730	461.618	416.405	-903.785	102.407	-0.991
5500	83.043	540.253	463.034	424.709	-903.463	121.038	-1.150
5600	83.046	541.750	464.426	433.014	-903.108	139.662	-1.303
5700	83.050	543.220	465.795	441.318	-902.720	158.279	-1.450
5800	83.053	544.664	467.143	449.624	-902.299	176.890	-1.593
5900	83.056	546.084	468.469	457.929	-901.844	195.493	-1.731
6000	83.059	547.480	469.774	466.235	-901.356	214.087	-1.864

PREVIOUS: December 1964 (1 atm) CURRENT: December 1964 (1 bar)

Bromodichloroborane ($BBrCl_2$) $B_1Br_1Cl_2$(g)

BROMOFLUOROBORANE (BBrF) IDEAL GAS M_r = 109.712403

$S°(298.15\ K)$ = [275.77] J K^{-1} mol^{-1}

$\Delta_fH°(0\ K)$ = -233.9 ± 10 kJ mol^{-1}
$\Delta_fH°(298.15\ K)$ = -240.6 ± 10 kJ mol^{-1}

Vibrational Frequencies and Degeneracies

ν, cm^{-1}
[860](1)
[310](1)
[1140](1)

Ground State Quantum Weight: [2] σ = 1
Point Group: C_s
Bond Distances: B-F = [1.3] Å B-Br = [1.87] Å
Bond Angle: Br-B-F = [120]°
Product of the Moments of Inertia: $I_AI_BI_C$ = [4.131563 x 10^{-115}]g^3 cm^6

Enthalpy of Formation

The enthalpy of formation is estimated by assuming $\Delta_rH°$ = 0 for the reaction $BBr_2(g) + BF_2(g) \rightarrow 2\ BBrF(g)$.

Heat Capacity and Entropy

The vibrational frequencies, bond lengths, angles and quantum weight are estimated by comparison with the respective dihalides. The principal moments of inertia are I_A = 0.8401 x 10^{-39} g cm^2, I_B = 21.7603 x 10^{-39} g cm^2 and I_C = 22.6004 x 10^{-39} g cm^2.

Bromofluoroborane (BBrF) $B_1Br_1F_1(g)$

Enthalpy Reference Temperature = T_r = 298.15 K Standard State Pressure = p° = 0.1 MPa

T/K	$C_p°$ (J K⁻¹mol⁻¹)	S° (J K⁻¹mol⁻¹)	-[G°-H°(T_r)]/T (J K⁻¹mol⁻¹)	H°-H°(T_r) (kJ mol⁻¹)	$\Delta_fH°$ (kJ mol⁻¹)	$\Delta_fG°$ (kJ mol⁻¹)	Log K_f
0	0.	0.	INFINITE	-11.207	-233.906	-233.906	INFINITE
100	35.220	233.363	311.742	-7.838	-233.411	-245.506	128.239
200	39.650	259.208	279.627	-4.084	-233.905	-257.455	67.240
250	41.650	268.270	276.476	-2.051	-234.402	-263.290	55.011
298.15	43.546	275.770	275.770	0.	-240.580	-268.140	46.977
300	43.617	276.039	275.770	0.081	-240.619	-268.311	46.717
350	45.462	282.903	276.308	2.308	-256.138	-272.073	40.605
400	47.115	289.084	277.525	4.624	-256.289	-274.339	35.825
450	48.554	294.719	279.127	7.016	-256.478	-276.585	32.105
500	49.787	299.900	280.949	9.476	-256.696	-278.807	29.127
600	51.720	309.158	284.897	14.556	-257.193	-283.184	24.653
700	53.110	317.241	288.953	19.801	-257.748	-287.473	21.451
800	54.123	324.402	292.945	25.166	-258.348	-291.678	19.045
900	54.874	330.822	296.803	30.617	-258.987	-295.807	17.168
1000	55.442	336.635	300.500	36.134	-259.664	-299.862	15.663
1100	55.881	341.940	304.030	41.702	-260.378	-303.848	14.429
1200	56.225	346.818	307.395	47.307	-261.130	-307.766	13.397
1300	56.499	351.330	310.603	52.944	-261.920	-311.621	12.521
1400	56.721	355.525	313.664	58.606	-262.748	-315.413	11.768
1500	56.903	359.445	316.587	64.287	-263.616	-319.145	11.114
1600	57.054	363.122	319.381	69.985	-264.523	-322.818	10.539
1700	57.180	366.585	322.057	75.697	-265.471	-326.432	10.030
1800	57.286	369.856	324.623	81.420	-266.459	-329.990	9.576
1900	57.377	372.956	327.086	87.154	-267.487	-333.492	9.168
2000	57.456	375.901	329.453	92.895	-268.556	-336.938	8.800
2100	57.523	378.706	331.733	98.644	-269.665	-340.330	8.465
2200	57.582	381.384	333.929	104.400	-270.814	-343.668	8.160
2300	57.634	383.944	336.048	110.161	-272.001	-346.953	7.880
2400	57.679	386.398	338.096	115.926	-323.484	-349.117	7.598
2500	57.719	388.754	340.075	121.696	-324.821	-350.158	7.316
2600	57.755	391.018	341.991	127.470	-326.160	-351.145	7.055
2700	57.787	393.198	343.848	133.247	-327.499	-352.080	6.811
2800	57.816	395.300	345.648	139.027	-328.836	-352.966	6.585
2900	57.841	397.330	347.395	144.810	-330.169	-353.804	6.373
3000	57.865	399.291	349.093	150.595	-331.497	-354.597	6.174
3100	57.886	401.189	350.743	156.383	-332.818	-355.345	5.988
3200	57.905	403.027	352.348	162.173	-334.129	-356.051	5.812
3300	57.922	404.809	353.911	167.964	-335.429	-356.716	5.646
3400	57.938	406.538	355.433	173.757	-336.716	-357.341	5.490
3500	57.953	408.218	356.918	179.552	-337.989	-357.930	5.342
3600	57.967	409.851	358.365	185.348	-339.245	-358.481	5.201
3700	57.979	411.439	359.778	191.145	-340.483	-358.999	5.068
3800	57.990	412.986	361.158	196.943	-341.703	-359.483	4.941
3900	58.001	414.492	362.507	202.743	-342.902	-359.935	4.821
4000	58.011	415.961	363.825	208.543	-344.080	-360.357	4.706
4100	58.020	417.393	365.114	214.345	-345.235	-360.749	4.596
4200	58.028	418.791	366.375	220.147	-826.217	-354.090	4.404
4300	58.036	420.157	367.610	225.951	-826.238	-342.848	4.165
4400	58.044	421.491	368.820	231.755	-826.237	-331.607	3.937
4500	58.050	422.796	370.005	237.559	-826.214	-320.365	3.719
4600	58.057	424.072	371.166	243.365	-826.168	-309.125	3.510
4700	58.063	425.320	372.305	249.171	-826.099	-297.885	3.311
4800	58.069	426.543	373.423	254.977	-826.009	-286.648	3.119
4900	58.074	427.740	374.519	260.784	-825.896	-275.412	2.936
5000	58.079	428.914	375.595	266.592	-825.761	-264.179	2.760
5100	58.084	430.064	376.652	272.400	-825.606	-252.949	2.591
5200	58.088	431.192	377.690	278.209	-825.430	-241.722	2.428
5300	58.092	432.298	378.710	284.018	-825.234	-230.498	2.272
5400	58.096	433.384	379.712	289.827	-825.018	-219.279	2.121
5500	58.100	434.450	380.698	295.637	-824.784	-208.064	1.976
5600	58.104	435.497	381.667	301.447	-824.532	-196.853	1.836
5700	58.107	436.525	382.621	307.258	-824.264	-185.647	1.701
5800	58.110	437.536	383.559	313.069	-823.978	-174.446	1.571
5900	58.113	438.529	384.482	318.880	-823.678	-163.249	1.445
6000	58.116	439.506	385.391	324.691	-823.363	-152.058	1.324

PREVIOUS: December 1964 (1 atm) CURRENT: December 1964 (1 bar)

BROMODIFLUOROBORANE ($BBrF_2$) IDEAL GAS M_r = 128.710806

$S°(298.15\ K) = 286.39\ J\ K^{-1}\ mol^{-1}$

$\Delta_f H°(0\ K) = [-811 \pm 10]\ kJ\ mol^{-1}$
$\Delta_f H°(298.15\ K) = [-820 \pm 10]\ kJ\ mol^{-1}$

Vibrational Frequencies and Degeneracies

ν, cm^{-1}	ν, cm^{-1}
1215 (1)	1427 (1)
633 (1)	[346](1)
[330](1)	573 (1)

Ground State Quantum Weight: [1] $\sigma = 2$
Point Group: C_{2v}
Bond Distances: B-F = 1.295 Å B-Br = 1.87 Å
Bond Angles: F-B-F = [120]° Br-B-F = [120]°
Product of the Moments of Inertia: $I_A I_B I_C = 8.471599 \times 10^{-114}\ g^3\ cm^6$

Enthalpy of Formation

The enthalpy of formation is estimated by assuming $\Delta_r H° = 0$ for the reaction 2 $BF_3(g)$ + $BBr_3(g)$ → 3 $BBrF_2(g)$.

Heat Capacity and Entropy

All but two of the vibrational frequencies have been observed and assigned by Lindeman and Wilson (1). The two remaining frequencies were calculated by a normal coordinate treatment. The bond distances were reported by Wentink and Tiensuu (2). The bond angles were estimated by analogy with the other boron trihalides. The principal moments of inertia are $I_A = 7.9360 \times 10^{-39}\ g\ cm^2$, $I_B = 28.9444 \times 10^{-29}\ g\ cm^2$ and $I_C = 36.8805 \times 10^{-39}\ g\ cm^2$.

References

1. L. P. Lindeman and M. K. Wilson, J. Chem. Phys. 24, 242 (1956).
2. T. Wentink and V. H. Tiensuu, J. Chem. Phys. 28, 826 (1958).

Bromodifluoroborane ($BBrF_2$) $B_1Br_1F_2(g)$

Enthalpy Reference Temperature = T_r = 298.15 K Standard State Pressure = p° = 0.1 MPa

	J K⁻¹mol⁻¹			kJ mol⁻¹			
T/K	$C_p°$	S°	$-[G°-H°(T_r)]/T$	$H°-H°(T_r)$	$\Delta_f H°$	$\Delta_f G°$	Log K_f
0	0.	0.	INFINITE	-12.778	-810.549	-810.549	INFINITE
100	36.575	236.647	330.512	-9.387	-811.485	-815.389	425.916
200	47.998	265.547	291.299	-5.150	-812.961	-818.714	213.826
250	52.679	276.777	287.296	-2.630	-813.722	-820.066	171.343
298.15	56.450	286.387	286.387	0.	-820.064	-820.559	143.759
300	56.583	286.737	286.388	0.105	-820.108	-820.562	142.873
350	59.897	295.715	287.090	3.019	-835.735	-819.775	122.345
400	62.745	303.903	288.687	6.087	-835.948	-817.480	106.752
450	65.198	311.439	290.802	9.287	-836.164	-815.159	94.621
500	67.311	318.420	293.219	12.601	-836.376	-812.813	84.914
600	70.693	331.007	298.491	19.509	-836.782	-808.062	70.348
700	73.200	342.102	303.945	26.710	-837.157	-803.245	59.939
800	75.075	352.005	309.344	34.128	-837.508	-798.376	52.129
900	76.495	360.933	314.589	41.710	-837.844	-793.465	46.051
1000	77.589	369.052	319.635	49.417	-838.177	-788.516	41.188
1100	78.444	376.489	324.470	57.220	-838.515	-783.534	37.207
1200	79.122	383.344	329.095	65.099	-838.866	-778.520	33.888
1300	79.667	389.700	333.515	73.040	-839.237	-773.476	31.079
1400	80.111	395.620	337.742	81.029	-839.632	-768.403	28.669
1500	80.477	401.160	341.787	89.059	-840.056	-763.301	26.580
1600	80.781	406.364	345.662	97.123	-840.513	-758.169	24.752
1700	81.038	411.269	349.379	105.214	-841.003	-753.007	23.137
1800	81.255	415.908	352.947	113.329	-841.530	-747.816	21.701
1900	81.440	420.306	356.378	121.464	-842.093	-742.595	20.415
2000	81.600	424.488	359.679	129.616	-842.692	-737.343	19.257
2100	81.739	428.472	362.861	137.783	-843.326	-732.060	18.209
2200	81.860	432.278	365.931	145.963	-843.995	-726.746	17.255
2300	81.966	435.919	368.895	154.155	-844.695	-721.401	16.384
2400	82.060	439.409	371.761	162.356	-895.684	-714.956	15.561
2500	82.143	442.761	374.534	170.566	-896.519	-707.408	14.780
2600	82.217	445.984	377.221	178.785	-897.344	-699.827	14.060
2700	82.283	449.088	379.825	187.010	-898.159	-692.215	13.392
2800	82.342	452.082	382.353	195.241	-898.958	-684.573	12.771
2900	82.395	454.972	384.807	203.478	-899.740	-676.902	12.192
3000	82.444	457.766	387.193	211.720	-900.502	-669.205	11.652
3100	82.487	460.470	389.513	219.966	-901.241	-661.483	11.146
3200	82.527	463.090	391.772	228.217	-901.954	-653.738	10.671
3300	82.564	465.630	393.972	236.472	-902.639	-645.970	10.225
3400	82.597	468.095	396.116	244.730	-903.293	-638.183	9.804
3500	82.627	470.490	398.207	252.991	-903.915	-630.376	9.408
3600	82.655	472.818	400.247	261.255	-904.502	-622.552	9.033
3700	82.681	475.083	402.239	269.522	-905.053	-614.712	8.678
3800	82.705	477.288	404.185	277.791	-905.567	-606.859	8.342
3900	82.727	479.437	406.087	286.063	-906.043	-598.991	8.023
4000	82.747	481.532	407.947	294.336	-906.478	-591.113	7.719
4100	82.766	483.575	409.767	302.612	-906.874	-583.224	7.430
4200	82.784	485.570	411.548	310.890	-1387.078	-568.301	7.068
4300	82.800	487.518	413.293	319.169	-1386.304	-548.816	6.667
4400	82.816	489.422	415.001	327.450	-1385.491	-529.349	6.284
4500	82.830	491.283	416.676	335.732	-1384.638	-509.900	5.919
4600	82.843	493.104	418.318	344.016	-1383.746	-490.471	5.569
4700	82.856	494.885	419.928	352.301	-1382.816	-471.062	5.235
4800	82.868	496.630	421.508	360.587	-1381.847	-451.673	4.915
4900	82.879	498.339	423.058	368.874	-1380.842	-432.305	4.608
5000	82.889	500.013	424.581	377.162	-1379.800	-412.958	4.314
5100	82.899	501.655	426.076	385.452	-1378.722	-393.631	4.032
5200	82.908	503.264	427.545	393.742	-1377.610	-374.327	3.760
5300	82.917	504.844	428.988	402.034	-1376.464	-355.044	3.499
5400	82.925	506.394	430.408	410.326	-1375.285	-335.783	3.248
5500	82.933	507.915	431.803	418.619	-1374.075	-316.544	3.006
5600	82.941	509.410	433.175	426.912	-1372.835	-297.327	2.773
5700	82.948	510.878	434.526	435.207	-1371.566	-278.133	2.549
5800	82.954	512.321	435.855	443.502	-1370.269	-258.961	2.332
5900	82.961	513.739	437.163	451.798	-1368.945	-239.812	2.123
6000	82.967	515.133	438.451	460.094	-1367.596	-220.686	1.921

PREVIOUS: December 1964 (1 atm) CURRENT: December 1964 (1 bar)

Bromodifluoroborane ($BBrF_2$) $B_1Br_1F_2(g)$

BORON BROMIDE OXIDE (OBBr) IDEAL GAS M_r = 106.7134

$S°(298.15\ K) = 248.89\ J\ K^{-1}\ mol^{-1}$

$\Delta_f H°(0\ K) = [-244 \pm 7]\ kJ\ mol^{-1}$
$\Delta_f H°(298.15\ K) = [-251 \pm 7]\ kJ\ mol^{-1}$

Vibrational Frequencies and Degeneracies

ν, cm^{-1}
[550](1)
[370](2)
[1800](1)

Ground State Quantum Weight: [1] $\sigma = 1$
Point Group: $[C_{\infty v}]$
Bond Distances: O-B = [1.20] Å B-Br = [1.87] Å
Bond Angle: O-B-Br = [180]°
Rotation Constant: $B_0 = [0.117436]\ cm^{-1}$

Enthalpy of Formation

The value of $\Delta_f H°$(BOBr, g, 298.15 K) is calculated based on an assumption that D_0^0(BO-Br) = 92.5 kcal mol^{-1} which is estimated by comparison with the following related bond energy values: $D_0^0(BF_2-F)$ = 159, D_0^0(BO-F) = 169, $D_0^0(BCl_2-Cl)$ = 105, D_0^0(BO-Cl) = 110 and $D_0^0(BBr_2-Br)$ = 90.5 kcal mol^{-1}.

Heat Capacity and Entropy

The molecular structure is assumed to be linear. The vibrational frequencies (ν) are estimated from the ν values calculated by the valence force method by Herzberg (1) using estimated force constants, $k_1 = 4.2 \times 10^5$, $k_2 = 16.9 \times 10^5$ dynes cm^{-1} and kg = 0.37×10^{-11} dynes cm $rad.^{-1}$. The B=O and B-F bond distances are assumed to be the same as those in BO(g) and BBr_3(g), respectively.

Reference

1. G. Herzberg, "Infrared and Raman Spectra", D. Van Nostrand Company, Inc., 1945.

Boron Bromide Oxide (OBBr) $B_1Br_1O_1(g)$

Enthalpy Reference Temperature = T_r = 298.15 K Standard State Pressure = p° = 0.1 MPa

	J K⁻¹mol⁻¹			kJ mol⁻¹			
T/K	$C_p°$	S°	$-[G°-H°(T_r)]/T$	$H°-H°(T_r)$	$\Delta_f H°$	$\Delta_f G°$	Log K_f
0	0.	0.	INFINITE	-10.961	-244.191	-244.191	INFINITE
100	31.611	206.321	286.380	-8.006	-244.108	-253.352	132.338
200	41.199	231.278	253.022	-4.349	-244.690	-262.372	68.524
250	44.510	240.848	249.654	-2.201	-245.051	-266.753	55.735
298.15	46.810	248.895	248.895	0.	-251.040	-270.236	47.344
300	46.885	249.185	248.896	0.087	-251.072	-270.355	47.073
350	48.664	256.552	249.474	2.477	-266.371	-272.732	40.703
400	50.078	263.145	250.778	4.947	-266.299	-273.647	35.735
450	51.262	269.114	252.489	7.481	-266.272	-274.567	31.871
500	52.290	274.569	254.428	10.071	-266.282	-275.489	28.780
600	54.011	284.261	258.612	15.389	-266.384	-277.322	24.143
700	55.392	292.694	262.892	20.862	-266.563	-279.132	20.829
800	56.505	300.166	267.093	26.459	-266.794	-280.912	18.342
900	57.401	306.875	271.147	32.155	-267.063	-282.661	16.405
1000	58.127	312.962	275.029	37.933	-267.366	-284.378	14.854
1100	58.716	318.530	278.734	43.776	-267.699	-286.063	13.584
1200	59.198	323.661	282.267	49.673	-268.062	-287.717	12.524
1300	59.596	328.415	285.636	55.613	-268.455	-289.339	11.626
1400	59.927	332.844	288.852	61.590	-268.879	-290.930	10.855
1500	60.204	336.989	291.924	67.597	-269.337	-292.489	10.185
1600	60.438	340.882	294.864	73.629	-269.829	-294.017	9.599
1700	60.637	344.552	297.680	79.683	-270.357	-295.513	9.080
1800	60.808	348.023	300.381	85.756	-270.924	-296.976	8.618
1900	60.955	351.314	302.976	91.844	-271.532	-298.407	8.204
2000	61.082	354.444	305.471	97.946	-272.181	-299.805	7.830
2100	61.193	357.427	307.875	104.060	-272.875	-301.169	7.491
2200	61.291	360.276	310.193	110.184	-273.613	-302.499	7.182
2300	61.377	363.003	312.430	116.317	-274.399	-303.795	6.899
2400	61.453	365.617	314.592	122.459	-325.491	-303.987	6.616
2500	61.521	368.127	316.684	128.608	-326.450	-303.071	6.332
2600	61.582	370.541	318.709	134.763	-327.424	-302.117	6.070
2700	61.636	372.866	320.672	140.924	-328.415	-301.125	5.826
2800	61.685	375.108	322.576	147.090	-329.421	-300.096	5.598
2900	61.729	377.274	324.425	153.261	-330.442	-299.031	5.386
3000	61.769	379.367	326.222	159.436	-331.479	-297.930	5.187
3100	61.805	381.393	327.969	165.614	-332.529	-296.794	5.001
3200	61.838	383.356	329.669	171.797	-333.591	-295.624	4.826
3300	61.868	385.259	331.325	177.982	-334.665	-294.422	4.660
3400	61.896	387.107	332.939	184.170	-335.748	-293.186	4.504
3500	61.922	388.901	334.512	190.361	-336.841	-291.918	4.357
3600	61.945	390.646	336.047	196.554	-337.940	-290.619	4.217
3700	61.966	392.343	337.546	202.750	-339.045	-289.289	4.084
3800	61.986	393.996	339.010	208.948	-340.154	-287.929	3.958
3900	62.005	395.607	340.441	215.147	-341.266	-286.541	3.838
4000	62.022	397.177	341.839	221.349	-342.379	-285.123	3.723
4100	62.038	398.708	343.208	227.552	-343.493	-283.678	3.614
4200	62.053	400.203	344.547	233.756	-824.456	-275.182	3.422
4300	62.067	401.664	345.859	239.962	-824.481	-262.103	3.184
4400	62.079	403.091	347.143	246.169	-824.506	-249.025	2.956
4500	62.092	404.486	348.402	252.378	-824.529	-235.945	2.739
4600	62.103	405.851	349.636	258.588	-824.551	-222.866	2.531
4700	62.113	407.187	350.846	264.799	-824.572	-209.785	2.332
4800	62.123	408.494	352.034	271.010	-824.591	-196.704	2.141
4900	62.133	409.775	353.199	277.223	-824.609	-183.623	1.957
5000	62.142	411.031	354.343	283.437	-824.626	-170.541	1.782
5100	62.150	412.261	355.467	289.652	-824.642	-157.459	1.613
5200	62.158	413.468	356.571	295.867	-824.658	-144.378	1.450
5300	62.165	414.652	357.656	302.083	-824.674	-131.295	1.294
5400	62.172	415.814	358.722	308.300	-824.691	-118.212	1.143
5500	62.179	416.955	359.770	314.517	-824.709	-105.129	0.998
5600	62.185	418.076	360.802	320.736	-824.729	-92.046	0.859
5700	62.191	419.176	361.816	326.955	-824.753	-78.962	0.724
5800	62.197	420.258	362.814	333.174	-824.781	-65.877	0.593
5900	62.202	421.321	363.797	339.394	-824.814	-52.793	0.467
6000	62.207	422.367	364.764	345.614	-824.853	-39.707	0.346

PREVIOUS: March 1965 (1 atm) CURRENT: March 1965 (1 bar)

Boron Bromide Oxide (OBBr) $B_1Br_1O_1(g)$

DIBROMOBORANE (BBr_2) IDEAL GAS M_r = 170.618

$S°(298.15\ K) = [294.66]\ J\ K^{-1}\ mol^{-1}$

$\Delta_f H°(0\ K) = [76.3 \pm 15]\ kJ\ mol^{-1}$

$\Delta_f H°(298.15\ K) = [62.8 \pm 15]\ kJ\ mol^{-1}$

Vibrational Frequencies and Degeneracies

ν, cm^{-1}
[600](1)
[150](1)
[830](1)

Ground State Quantum Weight: [2] $\sigma = 2$

Point Group: [C_{2v}]

Bond Distance: [1.87] Å

Bond Angle: [120]°

Product of the Moments of Inertia: $I_A I_B I_C = [7.270321 \times 10^{-114}]\ g^3\ cm^6$

Enthalpy of Formation

The enthalpy of formation is estimated by analogy with BF_3, specifically the ratio $\Delta_r H°(BX_3 + BX_2 + X)/\Delta_r H°(BX_3 + B + 3X)$ is taken as 0.347.

Heat Capacity and Entropy

The vibrational frequencies were obtained from force constants transferred from the trihalide given by Herzberg (1). The bond length is taken between those of the mono and trihalide and the angle is estimated to be the same as that in the trihalides. The ground state quantum weight is taken as two due to the one unpaired electron.

The principal moments of inertia are $I_A = 1.4699 \times 10^{-39}\ g\ cm^2$, $I_B = 69.5984 \times 10^{-39}\ g\ cm^2$ and $I_C = 71.0683 \times 10^{-39}\ g\ cm^2$.

Reference

1. G. Herzberg, "Infrared and Raman Spectra of Polyatomic Molecules", D. Van Nostrand Inc., 1945.

Dibromoborane (BBr_2) $B_1Br_2(g)$

Enthalpy Reference Temperature = T_r = 298.15 K Standard State Pressure = p° = 0.1 MPa

T/K	$C_p°$ (J K⁻¹mol⁻¹)	S°	$-[G°-H°(T_r)]/T$	$H°-H°(T_r)$ (kJ mol⁻¹)	$\Delta_f H°$	$\Delta_f G°$	Log K_f
0	0.	0.	INFINITE	-12.209	76.274	76.274	INFINITE
100	39.095	247.877	334.353	-8.648	77.019	57.646	-30.111
200	43.694	276.369	298.924	-4.511	75.921	38.615	-10.085
250	46.046	286.375	295.442	-2.267	74.979	29.392	-6.141
298.15	48.056	294.662	294.662	0.	62.760	22.026	-3.859
300	48.127	294.959	294.663	0.089	62.688	21.774	-3.791
350	49.854	302.512	295.255	2.540	31.862	16.646	-2.484
400	51.245	309.264	296.592	5.069	31.825	14.475	-1.890
450	52.357	315.366	298.344	7.660	31.747	12.310	-1.429
500	53.245	320.930	300.329	10.301	31.635	10.156	-1.061
600	54.539	330.760	304.603	15.694	31.328	5.888	-0.513
700	55.403	339.237	308.959	21.194	30.934	1.678	-0.125
800	56.003	346.676	313.218	26.766	30.471	-2.471	0.161
900	56.432	353.299	317.311	32.389	29.948	-6.557	0.381
1000	56.750	359.262	321.213	38.049	29.371	-10.583	0.553
1100	56.990	364.682	324.922	43.737	28.746	-14.549	0.691
1200	57.176	369.649	328.445	49.445	28.074	-18.455	0.803
1300	57.322	374.232	331.793	55.170	27.357	-22.304	0.896
1400	57.440	378.484	334.978	60.909	26.596	-26.096	0.974
1500	57.535	382.451	338.012	66.658	25.792	-29.832	1.039
1600	57.614	386.166	340.907	72.415	24.944	-33.512	1.094
1700	57.680	389.661	343.673	78.180	24.051	-37.139	1.141
1800	57.735	392.960	346.320	83.951	23.114	-40.711	1.181
1900	57.782	396.083	348.858	89.727	22.129	-44.230	1.216
2000	57.822	399.047	351.294	95.507	21.096	-47.697	1.246
2100	57.857	401.869	353.636	101.291	20.014	-51.110	1.271
2200	57.887	404.562	355.890	107.078	18.880	-54.470	1.293
2300	57.914	407.135	358.062	112.868	17.694	-57.778	1.312
2400	57.937	409.601	360.159	118.661	-33.805	-59.964	1.305
2500	57.957	411.966	362.184	124.455	-35.175	-61.026	1.275
2600	57.975	414.240	364.143	130.252	-36.567	-62.033	1.246
2700	57.992	416.428	366.039	136.050	-37.979	-62.985	1.219
2800	58.006	418.537	367.877	141.850	-39.411	-63.886	1.192
2900	58.019	420.573	369.659	147.652	-40.861	-64.734	1.166
3000	58.031	422.540	371.389	153.454	-42.329	-65.532	1.141
3100	58.042	424.443	373.070	159.258	-43.812	-66.281	1.117
3200	58.052	426.286	374.704	165.063	-45.308	-66.982	1.093
3300	58.061	428.073	376.294	170.868	-46.814	-67.636	1.071
3400	58.069	429.806	377.843	176.675	-48.329	-68.245	1.048
3500	58.076	431.489	379.352	182.482	-49.849	-68.808	1.027
3600	58.083	433.126	380.823	188.290	-51.372	-69.328	1.006
3700	58.089	434.717	382.258	194.098	-52.895	-69.806	0.985
3800	58.095	436.266	383.659	199.908	-54.416	-70.243	0.966
3900	58.100	437.775	385.027	205.717	-55.932	-70.640	0.946
4000	58.105	439.246	386.365	211.528	-57.441	-70.998	0.927
4100	58.110	440.681	387.672	217.338	-58.941	-71.318	0.909
4200	58.114	442.082	388.951	223.150	-540.279	-64.578	0.803
4300	58.118	443.449	390.202	228.961	-540.668	-53.247	0.647
4400	58.122	444.785	391.428	234.773	-541.044	-41.907	0.498
4500	58.125	446.091	392.628	240.586	-541.406	-30.559	0.355
4600	58.128	447.369	393.804	246.398	-541.753	-19.203	0.218
4700	58.131	448.619	394.957	252.211	-542.084	-7.840	0.087
4800	58.134	449.843	396.088	258.025	-542.399	3.531	-0.038
4900	58.137	451.042	397.197	263.838	-542.696	14.907	-0.159
5000	58.140	452.216	398.286	269.652	-542.977	26.290	-0.275
5100	58.142	453.368	399.355	275.466	-543.240	37.678	-0.386
5200	58.144	454.497	400.404	281.280	-543.486	49.071	-0.493
5300	58.146	455.604	401.435	287.095	-543.714	60.469	-0.596
5400	58.148	456.691	402.449	292.910	-543.925	71.870	-0.695
5500	58.150	457.758	403.445	298.724	-544.119	83.276	-0.791
5600	58.152	458.806	404.424	304.540	-544.296	94.684	-0.883
5700	58.154	459.835	405.387	310.355	-544.458	106.096	-0.972
5800	58.155	460.847	406.335	316.170	-544.603	117.510	-1.058
5900	58.157	461.841	407.267	321.986	-544.734	128.927	-1.141
6000	58.158	462.818	408.185	327.802	-544.850	140.346	-1.222

PREVIOUS: December 1964 (1 atm) CURRENT: December 1964 (1 bar)

Dibromoborane (BBr_2) $B_1Br_2(g)$

DIBROMOCHLOROBORANE (BBr_2Cl) IDEAL GAS M_r = 206.071

$\Delta_fH°(0\ K)$ = [-257 ± 10] kJ mol^{-1}

$S°(298.15\ K)$ = 321.89 J K^{-1} mol^{-1} $\Delta_fH°(298.15\ K)$ = [-272 ± 10] kJ mol^{-1}

Vibrational Levels and Degeneracies

ν, cm^{-1}	ν, cm^{-1}
925(1)	834(1)
344(1)	195(1)
166(1)	409(1)

Ground State Quantum Weight: [1] σ = 2

Point Group: C_{2v}

Bond Distances: B-Br = 1.87 Å B-Cl = 1.75 Å

Bond Angles: Br-B-Br = [120]° Cl-B-Br = [120]°

Product of the Moments of Inertia: $I_AI_BI_C = 2.555411 \times 10^{-112}\ g^3\ cm^6$

Enthalpy of Formation

The enthalpy of formation is estimated by assuming $\Delta_rH° = 0$ for the reaction 2 $BBr_3(g)$ + $BCl_3(g)$ → 3 $BBr_2Cl(g)$.

Heat Capacity and Entropy

The vibrational frequencies have all been observed and assigned by Lindeman and Wilson (1). The bond lengths were reported by Wentink and Tiensuu (2). Bond angles are estimated by analogy with the other trihalides of boron. The principal moments of inertia are $I_A = 35.0767 \times 10^{-39}\ g\ cm^2$, $I_B = 69.5984 \times 10^{-39}\ g\ cm^2$, and $I_C = 104.6751 \times 10^{-39}\ g\ cm^2$.

References

1. L. P. Lindeman and M. K. Wilson, J. Chem. Phys. 24, 242 (1956).
2. T. Wentink and V. H. Tiensuu, J. Chem. Phys. 28, 826 (1958).

Dibromochloroborane (BBr_2Cl) $B_1Br_2Cl_1(g)$

Enthalpy Reference Temperature = T_r = 298.15 K Standard State Pressure = p° = 0.1 MPa

T/K	$C_p°$ (J $K^{-1}mol^{-1}$)	S° (J $K^{-1}mol^{-1}$)	$-[G°-H°(T_r)]/T$ (J $K^{-1}mol^{-1}$)	$H°-H°(T_r)$ (kJ mol^{-1})	$\Delta_fH°$ (kJ mol^{-1})	$\Delta_fG°$ (kJ mol^{-1})	Log K_f
0	0.	0.	INFINITE	-15.092	-256.738	-256.738	INFINITE
100	45.297	261.227	374.950	-11.372	-257.292	-268.549	140.275
200	58.289	297.067	327.702	-6.127	-258.799	-279.248	72.932
250	62.626	310.557	322.958	-3.100	-259.769	-284.253	59.391
298.15	66.028	321.889	321.889	0.	-271.960	-287.556	50.379
300	66.145	322.298	321.890	0.122	-272.030	-287.652	50.085
350	68.994	332.716	322.707	3.503	-302.786	-288.570	43.067
400	71.286	342.084	324.553	7.012	-302.717	-286.544	37.419
450	73.127	350.591	326.981	10.624	-302.663	-284.525	33.027
500	74.609	358.375	329.736	14.319	-302.619	-282.512	29.514
600	76.790	372.184	335.690	21.896	-302.558	-278.497	24.245
700	78.265	384.139	341.776	29.654	-302.531	-274.490	20.483
800	79.296	394.661	347.742	37.535	-302.535	-270.484	17.661
900	80.040	404.046	353.486	45.503	-302.572	-266.476	15.466
1000	80.591	412.509	358.972	53.536	-302.644	-262.462	13.710
1100	81.010	420.211	364.195	61.617	-302.751	-258.439	12.272
1200	81.336	427.274	369.161	69.735	-302.895	-254.404	11.074
1300	81.593	433.795	373.885	77.882	-303.076	-250.356	10.059
1400	81.799	439.849	378.383	86.052	-303.296	-246.293	9.189
1500	81.967	445.499	382.672	94.241	-303.556	-242.212	8.435
1600	82.106	450.793	386.765	102.445	-303.858	-238.113	7.774
1700	82.222	455.775	390.680	110.661	-304.202	-233.994	7.190
1800	82.319	460.477	394.428	118.888	-304.593	-229.853	6.670
1900	82.402	464.930	398.022	127.125	-305.030	-225.689	6.205
2000	82.473	469.159	401.474	135.369	-305.518	-221.501	5.785
2100	82.535	473.184	404.794	143.619	-306.059	-217.287	5.405
2200	82.588	477.025	407.991	151.875	-306.654	-213.046	5.058
2300	82.635	480.697	411.073	160.136	-307.306	-208.777	4.741
2400	82.676	484.215	414.047	168.402	-358.275	-203.409	4.427
2500	82.712	487.591	416.922	176.671	-359.124	-196.939	4.115
2600	82.745	490.835	419.703	184.944	-360.000	-190.434	3.826
2700	82.773	493.959	422.396	193.220	-360.904	-183.895	3.558
2800	82.799	496.969	425.006	201.499	-361.837	-177.322	3.308
2900	82.822	499.875	427.537	209.780	-362.797	-170.716	3.075
3000	82.843	502.684	429.996	218.063	-363.783	-164.076	2.857
3100	82.862	505.400	432.385	226.349	-364.793	-157.402	2.652
3200	82.879	508.031	434.708	234.636	-365.824	-150.696	2.460
3300	82.895	510.582	436.968	242.924	-366.875	-143.956	2.279
3400	82.909	513.057	439.170	251.215	-367.941	-137.186	2.108
3500	82.923	515.460	441.316	259.506	-369.019	-130.383	1.946
3600	82.935	517.797	443.408	267.799	-370.107	-123.549	1.793
3700	82.946	520.069	445.449	276.093	-371.199	-116.685	1.647
3800	82.956	522.281	447.442	284.388	-372.294	-109.792	1.509
3900	82.966	524.436	449.389	292.684	-373.386	-102.870	1.378
4000	82.974	526.537	451.291	300.981	-374.473	-95.920	1.253
4100	82.982	528.586	453.152	309.279	-375.552	-88.943	1.133
4200	82.990	530.585	454.972	317.578	-856.467	-74.915	0.932
4300	82.997	532.538	456.753	325.877	-856.431	-56.307	0.684
4400	83.004	534.446	458.497	334.177	-856.378	-37.700	0.448
4500	83.010	536.312	460.206	342.478	-856.307	-19.095	0.222
4600	83.016	538.136	461.880	350.779	-856.215	-0.491	0.006
4700	83.021	539.922	463.522	359.081	-856.100	18.111	-0.201
4800	83.026	541.670	465.132	367.383	-855.961	36.710	-0.399
4900	83.031	543.382	466.711	375.686	-855.796	55.305	-0.590
5000	83.035	545.059	468.261	383.989	-855.603	73.897	-0.772
5100	83.039	546.704	469.783	392.293	-855.382	92.485	-0.947
5200	83.043	548.316	471.278	400.597	-855.133	111.068	-1.116
5300	83.047	549.898	472.747	408.902	-854.854	129.647	-1.278
5400	83.051	551.450	474.190	417.207	-854.544	148.219	-1.434
5500	83.054	552.974	475.608	425.512	-854.205	166.786	-1.584
5600	83.057	554.471	477.003	433.818	-853.835	185.346	-1.729
5700	83.060	555.941	478.375	442.123	-853.436	203.899	-1.869
5800	83.063	557.385	479.725	450.430	-853.006	222.445	-2.003
5900	83.066	558.805	481.054	458.736	-852.546	240.983	-2.134
6000	83.068	560.202	482.361	467.043	-852.058	259.514	-2.259

PREVIOUS: December 1964 (1 atm) CURRENT: December 1964 (1 bar)

Dibromochloroborane (BBr_2Cl) $B_1Br_2Cl_1(g)$

$\Delta_fH°(0 K)$ = [-499 ± 10] kJ mol^{-1}

S°(298.15 K) = [309.99] J K^{-1} mol^{-1}

$\Delta_fH°(298.15 K)$ = [-515 ± 10] kJ mol^{-1}

Vibrational Frequencies and Degeneracies

ν, cm^{-1}	ν, cm^{-1}
1310 (1)	869 (1)
[418](1)	[283](1)
[183](1)	[496](1)

Ground State Quantum Weight: [1] σ = 2

Point Group: C_{2v}

Bond Distances: B-Br = 1.87 Å B-Cl = 1.295 Å

Bond Angles: Br-B-Br = [120]° Br-B-F = [120]°

Product of the Moments of Inertia: $I_AI_BI_C$ = 8.725604 x 10^{-113} g^3 cm^6

Enthalpy of Formation

The enthalpy of formation is estimated by assuming $\Delta_rH°$ = 0 for the reaction 2 $BBr_3(g)$ + $BF_3(g)$ → 3 $BBr_2F(g)$.

Heat Capacity and Entropy

Two of the vibrational frequencies were observed and assigned by Lindeman and Wilson (1). The remaining frequencies were calculated by a normal coordinate treatment using force constants obtained by comparison with the other mixed trihalides. The bond distances were reported by Wentink and Tiensuu (2). The bond angles are estimated by analogy with the other trihalides. The principal moments of inertia are I_A = 14.8465 x 10^{-39} g cm^2, I_B = 69.5984 x 10^{-39} g cm^2 and I_C = 84.4449 x 10^{-39} g cm^2.

References

1. L. P. Lindeman and M. K. Wilson, J. Chem. Phys. 24, 242 (1956).
2. T. Wentink and V. H. Tiensuu, J. Chem. Phys. 28, 826 (1958).

Enthalpy Reference Temperature = T_r = 298.15 K Standard State Pressure = p° = 0.1 MPa

T/K	$C_p°$ (J K^{-1}mol^{-1})	S°	-[G°-H°(T_r)]/T	H°-H°(T_r) (kJ mol^{-1})	$\Delta_fH°$	$\Delta_fG°$	Log K_f
0	0.	0.	INFINITE	-14.170	-498.666	-498.666	INFINITE
100	41.579	253.624	359.617	-10.599	-499.366	-510.795	266.812
200	54.385	286.688	315.462	-5.755	-501.220	-521.525	136.208
250	58.848	299.322	311.001	-2.920	-502.323	-526.478	110.001
298.15	62.316	309.994	309.994	0.	-514.632	-529.706	92.802
300	62.436	310.380	309.995	0.115	-514.707	-529.799	92.246
350	65.384	320.233	310.766	3.313	-545.581	-530.619	79.191
400	67.832	329.128	312.515	6.645	-545.629	-528.478	69.012
450	69.875	337.239	314.818	10.090	-545.687	-526.331	61.095
500	71.585	344.692	317.438	13.627	-545.752	-524.177	54.760
600	74.230	357.991	323.115	20.926	-545.891	-519.849	45.257
700	76.122	369.584	328.943	28.448	-546.038	-515.497	38.467
800	77.502	379.843	334.677	36.133	-546.193	-511.124	33.373
900	78.530	389.034	340.215	43.937	-546.362	-506.730	29.410
1000	79.310	397.350	345.519	51.831	-546.550	-502.317	26.238
1100	79.913	404.939	350.581	59.793	-546.760	-497.884	23.643
1200	80.388	411.913	355.406	67.809	-546.998	-493.430	21.478
1300	80.768	418.363	360.003	75.868	-547.266	-488.955	19.646
1400	81.076	424.360	364.389	83.960	-547.568	-484.459	18.075
1500	81.329	429.963	368.576	92.081	-547.906	-479.939	16.713
1600	81.538	435.219	372.578	100.225	-548.282	-475.396	15.520
1700	81.714	440.167	376.410	108.388	-548.699	-470.828	14.467
1800	81.863	444.842	380.083	116.567	-549.158	-466.235	13.530
1900	81.990	449.272	383.609	124.759	-549.662	-461.614	12.691
2000	82.099	453.480	386.998	132.964	-550.211	-456.966	11.935
2100	82.193	457.488	390.260	141.179	-550.806	-452.289	11.250
2200	82.276	461.314	393.404	149.402	-551.448	-447.583	10.627
2300	82.348	464.973	396.436	157.633	-552.137	-442.847	10.057
2400	82.411	468.479	399.366	165.872	-603.132	-437.010	9.511
2500	82.468	471.844	402.198	174.116	-603.991	-430.071	8.986
2600	82.518	475.080	404.939	182.365	-604.861	-423.097	8.500
2700	82.563	478.195	407.595	190.619	-605.740	-416.090	8.050
2800	82.603	481.198	410.171	198.877	-606.628	-409.050	7.631
2900	82.639	484.097	412.670	207.139	-607.520	-401.977	7.240
3000	82.672	486.900	415.098	215.405	-608.416	-394.874	6.875
3100	82.701	489.611	417.458	223.674	-609.311	-387.742	6.533
3200	82.728	492.237	419.754	231.945	-610.203	-380.580	6.212
3300	82.753	494.783	421.989	240.219	-611.089	-373.391	5.910
3400	82.775	497.254	424.167	248.496	-611.966	-366.174	5.626
3500	82.796	499.653	426.289	256.774	-612.830	-358.933	5.357
3600	82.815	501.986	428.360	265.055	-613.680	-351.666	5.103
3700	82.832	504.255	430.381	273.337	-614.512	-344.377	4.862
3800	82.848	506.465	432.354	281.621	-615.323	-337.065	4.633
3900	82.863	508.617	434.282	289.907	-616.111	-329.732	4.416
4000	82.877	510.715	436.167	298.194	-616.875	-322.379	4.210
4100	82.890	512.762	438.010	306.482	-617.611	-315.008	4.013
4200	82.901	514.759	439.814	314.771	-1098.169	-300.594	3.738
4300	82.913	516.710	441.579	323.062	-1097.759	-281.609	3.421
4400	82.923	518.616	443.308	331.354	-1097.319	-262.634	3.118
4500	82.933	520.480	445.003	339.647	-1096.849	-243.669	2.828
4600	82.942	522.303	446.663	347.940	-1096.348	-224.715	2.552
4700	82.950	524.087	448.292	356.235	-1095.814	-205.772	2.287
4800	82.958	525.833	449.889	364.530	-1095.249	-186.841	2.033
4900	82.965	527.544	451.457	372.827	-1094.652	-167.922	1.790
5000	82.973	529.220	452.995	381.123	-1094.022	-149.015	1.557
5100	82.979	530.863	454.506	389.421	-1093.361	-130.122	1.333
5200	82.985	532.474	455.990	397.719	-1092.668	-111.241	1.117
5300	82.991	534.055	457.448	406.018	-1091.944	-92.374	0.910
5400	82.997	535.606	458.881	414.318	-1091.190	-73.522	0.711
5500	83.002	537.129	460.290	422.618	-1090.407	-54.683	0.519
5600	83.007	538.625	461.675	430.918	-1089.594	-35.859	0.334
5700	83.012	540.094	463.038	439.219	-1088.753	-17.050	0.156
5800	83.016	541.538	464.379	447.520	-1087.885	1.744	-0.016
5900	83.021	542.957	465.699	455.822	-1086.991	20.523	-0.182
6000	83.025	544.353	466.998	464.124	-1086.071	39.287	-0.342

PREVIOUS: December 1964 (1 atm) CURRENT: December 1964 (1 bar)

DIBROMOBORANE (BBr_2H) IDEAL GAS M_r = 171.62594

$\Delta_f H°(0\ K)$ = [-87 ± 5] kJ mol^{-1}

S°(298.15 K) = [292.27] J K^{-1} mol^{-1} $\Delta_f H°(298.15\ K)$ = [-105 ± 5] kJ mol^{-1}

Vibrational Frequencies and Degeneracies

ν, cm^{-1}	ν, cm^{-1}
2600 (1)	1044(1)
[532](1)	775(1)
[158](1)	783(1)

Ground State Quantum Weight: 1 σ = 2

Point Group: C_{2v}

Bond Distances: B-H = 1.20 Å B-Br = [1.87] Å

Bond Angles: Br-B-Br = 119.3 ± 2° Br-B-H = 120.3 ± 1°

Product of the Moments of Inertia: $I_A I_B I_C = 1.096494 \times 10^{-113}\ g^3\ cm^6$

Enthalpy of Formation

The value of $\Delta_f H°(BBr_2H$, g, 298.15 K) is derived by assuming the B-H bond strength to be 92 kcal mol^{-1}, the same as that in $BCl_2H(g)$.

Heat Capacity and Entropy

The molecular structure, B-H bond distance and bond angle were obtained from Lynds and Bass (1). The B-Br bond distance is taken from that in $BBr_3(g)$ molecule. The vibrational frequencies adopted and corrected to the average isotopic species were assigned to Brieux de Mandirola and Westerkamp (2) from infrared spectrum. The values in brackets are calculated by the Wilson's FG-method. Infrared spectrum of $BBr_2H(g)$ was also reported by Wason and Porter (3) and three frequencies, ν_1, ν_4 and ν_5, were assigned. The principal moments of inertia are: $I_A = 2.2244 \times 10^{-39}$, $I_B = 69.1057 \times 10^{-39}$ and $I_C = 71.3302 \times 10^{-39}\ g\ cm^2$.

References

1. L. Lynds and C. D. Bass, J. Phys. Chem. 41, 3165 (1964).
2. O. Brieux de Mandirola and J. F. Westerkamp, Spectrochim. Acta 21, 1101 (1965).
3. S. K. Wason and R. F. Porter, J. Phys. Chem. 69, 2461 (1965).

Dibromoborane ($BHBr_2$) $B_1Br_2H_1(g)$

Enthalpy Reference Temperature = T_r = 298.15 K Standard State Pressure = p° = 0.1 MPa

T/K	$C_p°$ (J $K^{-1}mol^{-1}$)	S° (J $K^{-1}mol^{-1}$)	$-[G°-H°(T_r)]/T$ (J $K^{-1}mol^{-1}$)	$H°-H°(T_r)$ (kJ mol^{-1})	$\Delta_f H°$ (kJ mol^{-1})	$\Delta_f G°$ (kJ mol^{-1})	Log K_f
0	0.	0.	INFINITE	-12.624	-87.267	-87.267	INFINITE
100	39.016	243.626	334.406	-9.078	-88.038	-101.949	53.252
200	45.697	272.588	296.897	-4.862	-90.403	-115.011	30.038
250	49.642	283.208	293.122	-2.478	-91.904	-120.994	25.280
298.15	53.246	292.265	292.265	0.	-104.600	-125.138	21.924
300	53.377	292.595	292.266	0.099	-104.689	-125.265	21.811
350	56.661	301.076	292.929	2.852	-135.938	-126.969	18.949
400	59.453	308.830	294.439	5.756	-136.327	-125.661	16.410
450	61.805	315.972	296.440	8.790	-136.693	-124.305	14.429
500	63.795	322.590	298.728	11.931	-137.036	-122.910	12.840
600	66.971	334.516	303.721	18.477	-137.655	-120.025	10.449
700	69.410	345.030	308.886	25.301	-138.194	-117.043	8.734
800	71.355	354.430	314.002	32.342	-138.664	-113.989	7.443
900	72.944	362.929	318.974	39.560	-139.080	-110.879	6.435
1000	74.259	370.685	323.763	46.922	-139.456	-107.725	5.627
1100	75.357	377.815	328.357	54.405	-139.806	-104.535	4.964
1200	76.281	384.413	332.757	61.988	-140.142	-101.313	4.410
1300	77.063	390.550	336.969	69.656	-140.477	-98.064	3.940
1400	77.728	396.286	341.003	77.396	-140.817	-94.789	3.537
1500	78.296	401.669	344.870	85.198	-141.173	-91.489	3.186
1600	78.785	406.738	348.580	93.053	-141.549	-88.165	2.878
1700	79.207	411.527	352.143	100.953	-141.953	-84.816	2.606
1800	79.573	416.065	355.569	108.893	-142.389	-81.443	2.363
1900	79.893	420.376	358.868	116.866	-142.862	-78.044	2.146
2000	80.174	424.482	362.047	124.870	-143.376	-74.619	1.949
2100	80.420	428.399	365.114	132.900	-143.936	-71.168	1.770
2200	80.639	432.146	368.076	140.953	-144.543	-67.689	1.607
2300	80.833	435.735	370.940	149.027	-145.201	-64.181	1.458
2400	81.006	439.179	373.712	157.119	-196.170	-59.574	1.297
2500	81.160	442.489	376.398	165.227	-197.012	-53.865	1.125
2600	81.299	445.674	379.001	173.350	-197.876	-48.123	0.967
2700	81.424	448.745	381.528	181.487	-198.762	-42.346	0.819
2800	81.537	451.708	383.982	189.635	-199.671	-36.537	0.682
2900	81.640	454.571	386.367	197.794	-200.601	-30.694	0.553
3000	81.733	457.341	388.687	205.963	-201.551	-24.819	0.432
3100	81.818	460.022	390.945	214.140	-202.520	-18.912	0.319
3200	81.896	462.621	393.144	222.326	-203.506	-12.974	0.212
3300	81.967	465.142	395.288	230.519	-204.506	-7.004	0.111
3400	82.032	467.590	397.379	238.719	-205.520	-1.004	0.015
3500	82.092	469.969	399.419	246.925	-206.543	5.026	-0.075
3600	82.147	472.282	401.411	255.137	-207.574	11.086	-0.161
3700	82.199	474.534	403.357	263.355	-208.611	17.174	-0.242
3800	82.246	476.727	405.259	271.577	-209.651	23.290	-0.320
3900	82.290	478.864	407.119	279.804	-210.692	29.434	-0.394
4000	82.331	480.947	408.939	288.035	-211.731	35.604	-0.465
4100	82.369	482.981	410.720	296.270	-212.767	41.800	-0.533
4200	82.404	484.966	412.464	304.508	-693.648	55.046	-0.685
4300	82.437	486.906	414.173	312.751	-693.584	72.871	-0.885
4400	82.468	488.801	415.848	320.996	-693.515	90.695	-1.077
4500	82.497	490.655	417.489	329.244	-693.437	108.517	-1.260
4600	82.524	492.468	419.100	337.495	-693.351	126.337	-1.435
4700	82.550	494.243	420.680	345.749	-693.255	144.155	-1.602
4800	82.574	495.982	422.230	354.005	-693.149	161.972	-1.763
4900	82.596	497.684	423.753	362.264	-693.032	179.785	-1.917
5000	82.618	499.353	425.248	370.524	-692.903	197.597	-2.064
5100	82.638	500.989	426.718	378.787	-692.762	215.405	-2.206
5200	82.656	502.594	428.161	387.052	-692.610	233.211	-2.343
5300	82.674	504.169	429.581	395.318	-692.445	251.014	-2.474
5400	82.691	505.714	430.976	403.587	-692.269	268.813	-2.600
5500	82.707	507.232	432.349	411.857	-692.080	286.609	-2.722
5600	82.723	508.722	433.699	420.128	-691.878	304.402	-2.839
5700	82.737	510.187	435.029	428.401	-691.665	322.191	-2.953
5800	82.751	511.626	436.337	436.675	-691.440	339.976	-3.062
5900	82.764	513.040	437.625	444.951	-691.203	357.757	-3.167
6000	82.776	514.432	438.893	453.228	-690.954	375.533	-3.269

PREVIOUS: March 1966 (1 atm) CURRENT: March 1966 (1 bar)

Dibromoborane ($BHBr_2$) $B_1Br_2H_1(g)$

TRIBROMOBORANE (BBr_3) LIQUID M_r = 250.522

$S°(298.15 K) = 228.9 J K^{-1} mol^{-1}$ $\Delta_f H°(298.15 K) = -238.49±0.2 kJ mol^{-1}$

T_{fus} = 228.9 K $\Delta_{fus}H°$ = Unknown

Enthalpy of Formation

The enthalpy of formation of boron tribromide liquid has been determined by Skinner and Smith (1) by hydrolysis to H_3BO_3 and HBr. The adopted enthalpy of formation is calculated from the hydrolysis data (1) using $\Delta_f H°(H_3BO_3$, cr, 298.15 K) from these tables, $\Delta_{sol}H°(H_3BO_3$, cr), from Fasilino (2) with the dilution data of Smisko and Mason (3). The $\Delta_f H°$(HBr, aq) was taken from Evans (4).

Heat Capacity and Entropy

The heat capacity was estimated from that of the gas by Kelley (5). The entropy is back calculated from that of the gas by adjusting the free energy functions to give the best fit of the vapor pressure data.

Fusion Data

The temperature of melting was reported by Boubeau and Keller (6), and Stock and Kuss (7). This value was recently confirmed by the triple point measurement of 227.31 K by Barber, Boynton and Gallagher (8).

Vaporization Data

The boiling point is from the data of Stock and Kuss (7). A 2nd and 3rd law analysis of the vapor pressure data of Stock and Kuss (7) and Barber et al. (8) indicated that the data of Stock and Kuss (7) gave the best agreement as indicated below.

$\Delta_{vap}H°(298.15 K)$, kcal mol^{-1}

Sources	2nd law	3rd law	
		S°(298.15 K) = 54.7	S°(298.15 K) = 54.26
(7)	8.175 ± 0.03	8.222 (-0.02)	8.361 (0.42)
(8)	8.229 ± 0.03	8.203 (-0.44)	8.347 (0.00)

Stock and Kuss (7) had the first eight data points rejected due to a statistical test on the 3rd law values. The values in parentheses are values of the 3rd law drift measured as $\Delta(\Delta H)/\Delta T$. The 2nd law entropy also confirmed the choice of the data of Stock and Kuss (7) and an entropy of 54.7 cal K^{-1} mol^{-1}.

References

1. H. A. Skinner and N. B. Smith, Trans. Faraday Soc. 51, 19 (1955).
2. L. G. Fasilino, National Research Corporation, Special Report NONR-3608(00), June 2, 1964.
3. J. Smisko and L. S. Mason, J. Am. Chem. Soc. 72, 3679 (1950).
4. W. H. Evans, U. S. Nat. Bur. Stand. Report 8504, Appendix III, 1964.
5. K. K. Kelley, U. S. Bur. Mines Bull. 383, 1935.
6. J. Goubeau and H. Keller, Z. anorg. allgem. Chem. 267, 1 (1951).
7. A. Stock and E. Kuss, Ber. 47, 3113 (1914).
8. W. F. Barber, C. F. Boynton and P. E. Gallagher, J. Chem. Eng. Data 9, 137 (1964).

Tribromoborane (BBr_3) $B_1Br_3(l)$

Enthalpy Reference Temperature = T_r = 298.15 K Standard State Pressure = p° = 0.1 MPa

T/K	$C_p°$ (J $K^{-1}mol^{-1}$)	S° (J $K^{-1}mol^{-1}$)	-[G°-H°(T_r)]/T (J $K^{-1}mol^{-1}$)	H°-H°(T_r) (kJ mol^{-1})	$\Delta_f H°$ (kJ mol^{-1})	$\Delta_f G°$ (kJ mol^{-1})	Log K_f
0							
100							
200	128.030	177.745	240.576	-12.566	-224.975	-233.814	61.066
298.15	128.030	228.865	228.865	0.	-238.488	-236.914	41.506
300	128.030	229.657	228.867	0.237	-238.482	-236.904	41.249
400	128.030	266.489	233.889	13.040	-278.762	-227.783	29.745
500	128.030	295.058	243.372	25.843	-273.227	-215.687	22.533
600	128.030	318.400	253.990	38.646	-267.984	-204.675	17.819
700	128.030	338.136	264.638	51.449	-262.944	-194.524	14.516
800	128.030	355.233	274.917	64.252	-258.053	-185.085	12.085
900	128.030	370.312	284.696	77.055	-253.277	-176.252	10.229
1000	128.030	383.802	293.944	89.858	-248.598	-167.946	8.773

PREVIOUS: September 1962 CURRENT: December 1964

Tribromoborane (BBr_3) $B_1Br_3(l)$

TRIBROMOBORANE (BBr_3) IDEAL GAS M_r = 250.522

$S°(298.15\ K) = 324.31\ J\ K^{-1}\ mol^{-1}$

$\Delta_f H°(0\ K) = -181.9 \pm 0.2\ kJ\ mol^{-1}$

$\Delta_f H°(298.15\ K) = -204.2 \pm 0.2\ kJ\ mol^{-1}$

Vibrational Frequencies and Degeneracies

ν, cm^{-1}
278(1)
379(1)
827(2)
150(2)

Ground State Quantum Weight: [1] $\sigma = 6$

Point Group: D_{3h}

Bond Distance: B-Br = 1.87 ± 0.02 Å

Bond Angle: Br-B-Br = 120° ± 6°

Product of the Moments of Inertia: $I_A I_B I_C = 6.742600 \times 10^{-112}\ g^3\ cm^6$

Enthalpy of Formation

$\Delta_f H°(298.15\ K)$ is obtained from the enthalpy of formation of the liquid plus the heat of vaporization at 298.15 K, taken as 8.20 kcal mol^{-1} as the average 2nd and 3rd law value derived from the vapor pressures of Stock and Kuss (1). See $BBr_3(\ell)$ for details.

Heat Capacity and Entropy

Vibrational frequencies were obtained from Wentink and Tiensuu (2) and modified for the natural isotopic abundance (B^{10} 18.83%, B^{11} 81.1%). B-Br bond distance was reported by Wentink and Tiensuu (2). Br-B-Br angle was taken from Sutton (3). The principal moments of inertia are $I_A = I_B = 69.5984 \times 10^{-39}\ g\ cm^2$, $I_C = 139.1968 \times 10^{-39}\ g\ cm^2$.

References

1. A. Stock and E. Kuss, Ber. 47, 3113 (1914).
2. T. Wentink, Jr., and V. H. Tiensuu, J. Chem. Phys. 28, 825 (1958).
3. L. E. Sutton, Ed., "Tables of Interatomic Distances and Configuration in Molecules and Ions", Special Publication No. 11, The Chemical Society, London, 1958.

Tribromoborane (BBr_3) $B_1Br_3(g)$

Enthalpy Reference Temperature = T_r = 298.15 K Standard State Pressure = p° = 0.1 MPa

T/K	$C_p°$	S°	$-[G°-H°(T_r)]/T$	$H°-H°(T_r)$	$\Delta_f H°$	$\Delta_f G°$	Log K_f
	J $K^{-1}mol^{-1}$			kJ mol^{-1}			
0	0.	0.	INFINITE	-15.711	-181.912	-181.912	INFINITE
100	48.309	260.958	379.306	-11.835	-182.250	-200.238	104.593
200	60.557	298.686	330.295	-6.322	-184.421	-217.449	56.792
250	64.597	312.649	325.406	-3.189	-185.884	-225.543	47.125
298.15	67.771	324.307	324.307	0.	-204.179	-231.062	40.481
300	67.880	324.727	324.309	0.125	-204.285	-231.228	40.260
350	70.521	335.397	325.145	3.588	-250.425	-233.188	34.801
400	72.626	344.956	327.034	7.169	-250.324	-230.733	30.131
450	74.301	353.611	329.514	10.843	-250.239	-228.289	26.499
500	75.640	361.511	332.325	14.593	-250.168	-225.854	23.595
600	77.590	375.487	338.384	22.261	-250.060	-221.002	19.240
700	78.897	387.552	344.566	30.090	-249.994	-216.165	16.130
800	79.804	398.150	350.615	38.028	-249.968	-211.335	13.799
900	80.455	407.589	356.431	46.042	-249.981	-206.505	11.985
1000	80.936	416.092	361.979	54.113	-250.034	-201.672	10.534
1100	81.301	423.824	367.255	62.226	-250.126	-196.832	9.347
1200	81.583	430.910	372.268	70.370	-250.258	-191.981	8.357
1300	81.806	437.450	377.034	78.540	-250.430	-187.118	7.518
1400	81.985	443.519	381.569	86.730	-250.644	-182.240	6.799
1500	82.130	449.180	385.889	94.936	-250.901	-177.346	6.176
1600	82.250	454.485	390.013	103.155	-251.202	-172.432	5.629
1700	82.350	459.474	393.953	111.386	-251.549	-167.499	5.147
1800	82.434	464.184	397.725	119.625	-251.946	-162.544	4.717
1900	82.506	468.643	401.342	127.872	-252.393	-157.565	4.332
2000	82.567	472.876	404.813	136.126	-252.895	-152.561	3.984
2100	82.620	476.906	408.151	144.385	-253.454	-147.531	3.670
2200	82.666	480.751	411.364	152.649	-254.073	-142.472	3.383
2300	82.706	484.426	414.462	160.918	-254.753	-137.385	3.120
2400	82.742	487.947	417.451	169.191	-305.756	-131.197	2.855
2500	82.773	491.325	420.339	177.466	-306.640	-123.906	2.589
2600	82.801	494.572	423.132	185.745	-307.556	-116.579	2.342
2700	82.825	497.697	425.836	194.026	-308.503	-109.215	2.113
2800	82.848	500.710	428.457	202.310	-309.480	-101.817	1.899
2900	82.867	503.618	430.998	210.596	-310.484	-94.382	1.700
3000	82.885	506.427	433.466	218.883	-311.515	-86.913	1.513
3100	82.902	509.145	435.864	227.173	-312.568	-79.409	1.338
3200	82.917	511.778	438.195	235.464	-313.640	-71.871	1.173
3300	82.930	514.329	440.464	243.756	-314.729	-64.298	1.018
3400	82.942	516.805	442.673	252.050	-315.829	-56.693	0.871
3500	82.954	519.210	444.826	260.344	-316.938	-49.056	0.732
3600	82.964	521.547	446.924	268.640	-318.051	-41.386	0.600
3700	82.974	523.820	448.972	276.937	-319.164	-33.685	0.476
3800	82.982	526.033	450.971	285.235	-320.273	-25.955	0.357
3900	82.991	528.188	452.923	293.534	-321.376	-18.196	0.244
4000	82.998	530.290	454.831	301.833	-322.468	-10.408	0.136
4100	83.005	532.339	456.697	310.133	-323.546	-2.593	0.033
4200	83.012	534.340	458.522	318.434	-804.457	12.272	-0.153
4300	83.018	536.293	460.308	326.736	-804.412	31.718	-0.385
4400	83.023	538.202	462.057	335.038	-804.347	51.162	-0.607
4500	83.029	540.067	463.769	343.340	-804.260	70.604	-0.820
4600	83.034	541.892	465.448	351.644	-804.150	90.045	-1.022
4700	83.038	543.678	467.094	359.947	-804.014	109.482	-1.217
4800	83.043	545.426	468.707	368.251	-803.853	128.917	-1.403
4900	83.047	547.139	470.291	376.556	-803.664	148.347	-1.581
5000	83.051	548.816	471.844	384.860	-803.448	167.774	-1.753
5100	83.054	550.461	473.370	393.166	-803.203	187.197	-1.917
5200	83.058	552.074	474.868	401.471	-802.931	206.613	-2.075
5300	83.061	553.656	476.340	409.777	-802.630	226.025	-2.228
5400	83.064	555.209	477.786	418.083	-802.301	245.431	-2.374
5500	83.067	556.733	479.207	426.390	-801.943	264.830	-2.515
5600	83.070	558.230	480.605	434.697	-801.559	284.222	-2.651
5700	83.072	559.700	481.980	443.004	-801.147	303.608	-2.782
5800	83.075	561.145	483.332	451.311	-800.708	322.985	-2.909
5900	83.077	562.565	484.663	459.619	-800.244	342.355	-3.031
6000	83.079	563.961	485.973	467.927	-799.755	361.717	-3.149

PREVIOUS: December 1964 (1 atm) CURRENT: December 1964 (1 bar)

Tribromoborane (BBr_3) $B_1Br_3(g)$

CHLOROBORANE (BCl) IDEAL GAS M_r = 46.263

$S°(298.15\ K) = 213.24\ J\ K^{-1}\ mol^{-1}$

$\Delta_f H°(0\ K) = 138.4\ kJ\ mol^{-1}$
$\Delta_f H°(298.15\ K) = 141.4\ kJ\ mol^{-1}$

Electronic Level and Quantum Weight

State	ϵ_i, cm^{-1}	g_i
$^1\Sigma^+$	0	1

$\omega_e = 843.65\ cm^{-1}$ $\omega_e x_e = 5.167\ cm^{-1}$ $\sigma = 1$
$B_e = 0.6914\ cm^{-1}$ $\alpha_e = 0.00657\ cm^{-1}$ $r_e = 1.7157$ Å

Enthalpy of Formation

Dissociation energies of BCl(g) has been reported by Herzberg (1), Gaydon (2), and Barrow (3). The corresponding values of $\Delta_f H°$(BCl, g, 298.15 K) are derived. The results are given as follows:

Source	Dissociation Energy, D_0^0 eV	Dissociation Energy, D_0^0 $kcal\ mol^{-1}$	$\Delta_f H°(298.15\ K)$ $kcal\ mol^{-1}$
(1)	4.2	96.9	62.6
(2)	5.1 ± 0.4	117.6 ± 9.2	41.8 ± 9.2
(3)	--	127	33.8 ± 4

The last value of $\Delta_f H°$(298.15 K) listed in the above table is adopted. Using the bond energy, D_0^0(B-Cl), obtained from BCl_3(g) molecule, $\Delta_f H°$(298.15 K) for BCl(g) is calculated as 37.8 kcal mol^{-1} employing the same ratio of D_0^0(B-X)/ average bond energy of BX_3 as found in BF(g) and BF_3(g), which is within the uncertainty of the value evaluated from dissociation energy.

Heat Capacity and Entropy

The spectroscopic constants are obtained from Herzberg (1) and corrected to the average isotopic species.

References

1. G. Herzberg, "Spectra of Diatomic Molecules", D. Van Nostrand Company, Inc., New York, 1950.
2. A. G. Gaydon, "Dissociation Energies", Chapman and Hall Ltd., London, 1953.
3. R. F. Barrow, Trans. Faraday Soc. 56, 952 (1960).

Chloroborane (BCl) $B_1Cl_1(g)$

Enthalpy Reference Temperature = T_r = 298.15 K Standard State Pressure = p° = 0.1 MPa

T/K	$C_p°$ (J K⁻¹mol⁻¹)	S° (J K⁻¹mol⁻¹)	$-[G°-H°(T_r)]/T$ (J K⁻¹mol⁻¹)	$H°-H°(T_r)$ (kJ mol⁻¹)	$\Delta_f H°$ (kJ mol⁻¹)	$\Delta_f G°$ (kJ mol⁻¹)	Log K_f
0	0.	0.	INFINITE	-8.862	138.362	138.362	INFINITE
100	29.117	180.668	240.206	-5.954	139.790	131.204	-68.534
200	29.874	200.994	216.081	-3.017	140.875	122.156	-31.904
250	30.748	207.751	213.761	-1.502	141.209	117.435	-24.537
298.15	31.659	213.245	213.245	0.	141.419	112.835	-19.768
300	31.693	213.441	213.245	0.059	141.425	112.658	-19.615
350	32.572	218.394	213.635	1.666	141.546	107.852	-16.096
400	33.335	222.794	214.510	3.314	141.583	103.036	-13.455
450	33.976	226.759	215.654	4.997	141.551	98.219	-11.401
500	34.510	230.367	216.948	6.710	141.462	93.408	-9.758
600	35.322	236.735	219.729	10.204	141.159	83.823	-7.297
700	35.892	242.225	222.560	13.766	140.730	74.300	-5.544
800	36.307	247.046	225.325	17.377	140.209	64.844	-4.234
900	36.619	251.342	227.982	21.024	139.614	55.459	-3.219
1000	36.861	255.213	230.514	24.698	138.958	46.143	-2.410
1100	37.054	258.735	232.922	28.394	138.249	36.895	-1.752
1200	37.214	261.967	235.210	32.108	137.493	27.714	-1.206
1300	37.348	264.951	237.384	35.836	136.693	18.598	-0.747
1400	37.464	267.723	239.453	39.577	135.853	9.545	-0.356
1500	37.566	270.311	241.425	43.329	134.975	0.554	-0.019
1600	37.656	272.738	243.307	47.090	134.060	-8.378	0.274
1700	37.739	275.024	245.106	50.860	133.109	-17.251	0.530
1800	37.814	277.183	246.829	54.637	132.124	-26.068	0.756
1900	37.884	279.230	248.481	58.422	131.103	-34.829	0.958
2000	37.950	281.174	250.067	62.214	130.049	-43.535	1.137
2100	38.012	283.028	251.593	66.012	128.960	-52.187	1.298
2200	38.071	284.797	253.063	69.816	127.837	-60.787	1.443
2300	38.128	286.491	254.479	73.626	126.679	-69.335	1.575
2400	38.182	288.115	255.847	77.442	75.227	-76.763	1.671
2500	38.235	289.674	257.169	81.263	73.920	-83.069	1.736
2600	38.286	291.175	258.449	85.089	72.611	-89.323	1.795
2700	38.336	292.621	259.688	88.920	71.297	-95.526	1.848
2800	38.385	294.016	260.889	92.756	69.978	-101.681	1.897
2900	38.433	295.364	262.055	96.597	68.655	-107.788	1.941
3000	38.480	296.668	263.187	100.443	67.327	-113.850	1.982
3100	38.527	297.930	264.287	104.293	65.994	-119.867	2.020
3200	38.572	299.154	265.358	108.148	64.656	-125.841	2.054
3300	38.617	300.342	266.400	112.008	63.313	-131.774	2.086
3400	38.662	301.495	267.415	115.872	61.967	-137.665	2.115
3500	38.706	302.617	268.405	119.740	60.618	-143.517	2.142
3600	38.750	303.708	269.371	123.613	59.266	-149.330	2.167
3700	38.793	304.770	270.313	127.490	57.913	-155.106	2.190
3800	38.837	305.805	271.234	131.372	56.560	-160.845	2.211
3900	38.879	306.814	272.133	135.257	55.208	-166.549	2.231
4000	38.922	307.799	273.012	139.147	53.859	-172.218	2.249
4100	38.964	308.761	273.873	143.042	52.513	-177.853	2.266
4200	39.006	309.700	274.714	146.940	-428.678	-176.431	2.194
4300	39.048	310.619	275.539	150.843	-428.926	-170.422	2.070
4400	39.090	311.517	276.346	154.750	-429.168	-164.408	1.952
4500	39.131	312.396	277.138	158.661	-429.404	-158.388	1.839
4600	39.173	313.256	277.914	162.576	-429.632	-152.363	1.730
4700	39.214	314.099	278.675	166.495	-429.852	-146.333	1.626
4800	39.255	314.925	279.421	170.419	-430.063	-140.298	1.527
4900	39.296	315.735	280.154	174.346	-430.264	-134.259	1.431
5000	39.337	316.529	280.874	178.278	-430.455	-128.217	1.339
5100	39.378	317.309	281.580	182.214	-430.635	-122.170	1.251
5200	39.419	318.074	282.275	186.154	-430.803	-116.120	1.166
5300	39.459	318.825	282.957	190.098	-430.960	-110.067	1.085
5400	39.500	319.563	283.629	194.046	-431.106	-104.011	1.006
5500	39.541	320.288	284.288	197.998	-431.239	-97.952	0.930
5600	39.581	321.001	284.938	201.954	-431.359	-91.892	0.857
5700	39.621	321.702	285.577	205.914	-431.468	-85.829	0.787
5800	39.662	322.391	286.205	209.878	-431.565	-79.764	0.718
5900	39.702	323.070	286.824	213.846	-431.650	-73.698	0.652
6000	39.742	323.737	287.434	217.818	-431.723	-67.630	0.589

PREVIOUS: December 1964 (1 atm) CURRENT: December 1964 (1 bar)

Chloroborane (BCl) $B_1Cl_1(g)$

CHLOROBORANE, ION (BCl^+) IDEAL GAS M_r = 46.26245

$S°(298.15 K) = [219.1 \pm 2]\ J\ K^{-1}\ mol^{-1}$

$\Delta_f H°(0 K) = [1225 \pm 42]\ kJ\ mol^{-1}$
$\Delta_f H°(298.15 K) = [1234 \pm 42]\ kJ\ mol^{-1}$

Electronic Levels and Quantum Weights

State	ε_i, cm^{-1}	g_i
[$^2\Sigma$]	0	[2]
	[28000]	[4]

$\omega_e = [845]\ cm^{-1}$ $\omega_e x_e = [5]\ cm^{-1}$ $\sigma = 1$
$B_e = [0.68]\ cm^{-1}$ $\alpha_e = [0.0056]\ cm^{-1}$ $r_e = [1.73]$ Å

Enthalpy of Formation

Several values of the appearance potential of BCl^+ from BCl_3 have been reported. Osberghaus (1) gave 19.2 eV, Marriott and Craggs (2) reported 18.54 ± 0.07 eV, and Koski et al. (3) obtained two values of 17.2 ± 0.2 Ev and 20 ± 0.2 eV. We make the presumption that the higher values, 19.2 and 20, refer to the reaction $BCl_3(g) + e^-$, $2\ Cl(g) + 2e^-$ and the other values to $BCl_3(g) + e^- \rightarrow BCl^+ + Cl_2(g) + 2\ e^-$. Using the lower value in each case to help eliminate any excess kinetic energy, we obtain $\Delta_f H°(BCl^+$, g, 298.15 K) = 289 and 300 kcal mol^{-1}, respectively. We adopt an average value of $\Delta_f H°(BCl^+$, g, 298.15 K) = 295 ± 10 kcal mol^{-1}, which corresponds to an ionization potential for BCl(g) of 11.3 ± 0.3 eV.

Heat Capacity and Entropy

BCl^+ is isoelectronic with BeCl, and the ground state configuration and electronic levels are estimated by comparison with BeCl. The bond length, frequency and anharmonicity constant are estimated between the value for BCl and BeCl; the rotational constant is calculated from the bond length and atomic masses, and α_e is obtained from the above constants assuming a Morse potential function.

References
1. O. Osberghaus, Z. Phys. 128, 366 (1950).
2. J. Marriott and J. D. Craggs, J. Electron. Contr. 3, 194 (1957).
3. W. S. Koski, J. J. Kaufman and C. F. Pachuki, J. Am. Chem. Soc. 81, 1326 (1959).

Chloroborane, Ion (BCl^+) $B_1Cl_1^+(g)$

Enthalpy Reference Temperature = T_r = 298.15 K Standard State Pressure = p° = 0.1 MPa

	J K⁻¹mol⁻¹			kJ mol⁻¹			
T/K	$C_p°$	S°	$-[G°-H°(T_r)]/T$	$H°-H°(T_r)$	$\Delta_f H°$	$\Delta_f G°$	Log K_f
0	0.	0.	INFINITE	-8.860	1225.027		
100	29.117	186.560	246.084	-5.952			
200	29.867	206.884	221.965	-3.016			
250	30.736	213.639	219.647	-1.502			
298.15	31.643	219.131	219.131	0.	1234.280	1197.686	-209.830
300	31.678	219.326	219.131	0.059	1234.325	1197.459	-208.496
350	32.555	224.277	219.520	1.665	1235.484	1191.222	-177.780
400	33.316	228.675	220.395	3.312	1236.559	1184.825	-154.722
450	33.957	232.637	221.538	4.994	1237.565	1178.297	-136.773
500	34.490	236.243	222.831	6.706	1238.515	1171.660	-122.402
600	35.300	242.608	225.611	10.198	1240.288	1158.120	-100.823
700	35.869	248.094	228.440	13.758	1241.936	1144.294	-85.388
800	36.282	252.912	231.204	17.367	1243.491	1130.238	-73.797
900	36.592	257.205	233.859	21.011	1244.972	1115.991	-64.770
1000	36.832	261.073	236.390	24.683	1246.391	1101.584	-57.541
1100	37.024	264.593	238.796	28.376	1247.758	1087.036	-51.619
1200	37.181	267.821	241.082	32.086	1249.077	1072.367	-46.679
1300	37.313	270.803	243.255	35.811	1250.353	1057.589	-42.494
1400	37.426	273.572	245.323	39.548	1251.588	1042.714	-38.904
1500	37.526	276.158	247.294	43.296	1252.785	1027.752	-35.790
1600	37.614	278.582	249.174	47.053	1253.944	1012.712	-33.062
1700	37.694	280.865	250.972	50.819	1255.068	997.601	-30.653
1800	37.767	283.022	252.693	54.592	1256.156	982.424	-28.509
1900	37.835	285.065	254.343	58.372	1257.210	967.188	-26.590
2000	37.898	287.008	255.929	62.158	1258.229	951.897	-24.861
2100	37.957	288.858	257.453	65.951	1259.213	936.556	-23.296
2200	38.014	290.625	258.921	69.750	1260.163	921.169	-21.871
2300	38.068	292.316	260.336	73.554	1261.078	905.739	-20.570
2400	38.120	293.938	261.703	77.363	1211.698	891.340	-19.400
2500	38.171	295.495	263.024	81.178	1212.464	877.976	-18.344
2600	38.220	296.993	264.302	84.997	1213.226	864.581	-17.370
2700	38.268	298.436	265.539	88.822	1213.984	851.157	-16.467
2800	38.315	299.829	266.739	92.651	1214.738	837.706	-15.628
2900	38.361	301.174	267.903	96.485	1215.486	824.227	-14.846
3000	38.407	302.475	269.034	100.323	1216.229	810.723	-14.116
3100	38.453	303.735	270.133	104.166	1216.967	797.193	-13.433
3200	38.499	304.957	271.203	108.014	1217.700	783.641	-12.792
3300	38.545	306.142	272.244	111.866	1218.429	770.065	-12.189
3400	38.591	307.294	273.258	115.723	1219.154	756.467	-11.622
3500	38.638	308.413	274.246	119.584	1219.876	742.848	-11.086
3600	38.686	309.502	275.210	123.450	1220.597	729.209	-10.581
3700	38.735	310.563	276.152	127.321	1221.316	715.549	-10.102
3800	38.785	311.597	277.071	131.197	1222.037	701.870	-9.648
3900	38.837	312.605	277.969	135.079	1222.759	688.172	-9.217
4000	38.890	313.589	278.847	138.965	1223.484	674.456	-8.807
4100	38.945	314.550	279.706	142.857	1224.214	660.721	-8.418
4200	39.002	315.489	280.547	146.754	745.101	653.993	-8.134
4300	39.062	316.407	281.371	150.657	746.932	651.802	-7.918
4400	39.124	317.306	282.177	154.566	748.770	649.568	-7.711
4500	39.188	318.186	282.968	158.482	750.618	647.293	-7.514
4600	39.255	319.048	283.743	162.404	752.475	644.976	-7.324
4700	39.324	319.893	284.503	166.333	754.343	642.619	-7.142
4800	39.397	320.721	285.249	170.269	756.223	640.222	-6.967
4900	39.472	321.535	285.981	174.213	758.117	637.786	-6.799
5000	39.550	322.333	286.700	178.164	760.024	635.311	-6.637
5100	39.631	323.117	287.406	182.123	761.946	632.798	-6.481
5200	39.715	323.887	288.101	186.090	763.883	630.246	-6.331
5300	39.802	324.644	288.783	190.066	765.837	627.658	-6.186
5400	39.892	325.389	289.454	194.051	767.807	625.032	-6.046
5500	39.985	326.122	290.114	198.044	769.795	622.370	-5.911
5600	40.081	326.843	290.764	202.048	771.800	619.671	-5.780
5700	40.180	327.554	291.403	206.061	773.822	616.937	-5.654
5800	40.282	328.253	292.032	210.084	775.863	614.167	-5.531
5900	40.386	328.943	292.652	214.117	777.922	611.361	-5.413
6000	40.493	329.623	293.262	218.161	780.000	608.520	-5.298

PREVIOUS: June 1968 (1 atm) CURRENT: June 1968 (1 bar)

CHLOROFLUOROBORANE (BClF) IDEAL GAS M_r = 65.261403

Chlorofluoroborane (BClF) $B_1Cl_1F_1(g)$

$S°(298.15\ K) = [264.65]\ J\ K^{-1}\ mol^{-1}$

$\Delta_fH°(0\ K) = [-315 \pm 29]\ kJ\ mol^{-1}$

$\Delta_fH°(298.15\ K) = [-314 \pm 29]\ kJ\ mol^{-1}$

Vibrational Frequencies and Degeneracies

ν, cm^{-1}
[920](1)
[360](1)
[1220](1)

Ground State Quantum Weight: [2] $\sigma = 1$

Point Group: C_s

Bond Distances: B-Cl = [1.73] Å B-F = [1.295] Å

Bond Angle: F-B-Cl = [120]°

Product of the Moments of Inertia: $I_AI_BI_C = [1.811033 \times 10^{-115}]\ g^3\ cm^6$

Enthalpy of Formation

The value of $\Delta_fH°(298.15\ K)$ for BClF(g) is estimated based on an assumption that $\Delta_rH° = 0$ for the reaction $BCl_2(g) + BF_2(g) = 2\ BClF(g)$. The values of $\Delta_fH°(298.15\ K)$ of $BCl_2(g)$ and $BF_2(g)$, used for calculation, are -20 and -130 kcal mol^{-1}, respectively.

Heat Capacity and Entropy

The ground state quantum weight and bond angle are assumed to be the same as those for $BF_2(g)$. The bond distances of B-Cl and B-F are estimated by comparison with those in $BCl_3(g)$ and $BF_3(g)$ molecules, respectively. The vibrational frequencies were estimated from the corresponding values for $BCl_2(g)$ and $BF_2(g)$. The principal moments of inertia are: $I_A = 0.7794 \times 10^{-39}$, $I_B = 14.8586 \times 10^{-39}$ and $I_C = 15.6380 \times 10^{-39}\ g\ cm^2$.

Enthalpy Reference Temperature = T_r = 298.15 K Standard State Pressure = p° = 0.1 MPa

T/K	$C_p°$ (J K⁻¹mol⁻¹)	S°	$-[G°-H°(T_r)]/T$	$H°-H°(T_r)$ (kJ mol⁻¹)	$\Delta_fH°$	$\Delta_fG°$	Log K_f
0	0.	0.	INFINITE	-11.008	-314.591	-314.591	INFINITE
100	34.530	223.216	299.791	-7.657	-314.174	-318.496	166.365
200	38.735	248.470	268.419	-3.990	-313.823	-322.971	84.351
250	40.690	257.324	265.340	-2.004	-313.769	-325.267	67.961
298.15	42.547	264.650	264.650	0.	-313.800	-327.480	57.373
300	42.617	264.913	264.651	0.079	-313.803	-327.564	57.034
350	44.450	271.622	265.176	2.256	-313.906	-329.851	49.228
400	46.125	277.669	266.366	4.521	-314.067	-332.118	43.370
450	47.610	283.189	267.933	6.865	-314.273	-334.363	38.812
500	48.901	288.274	269.716	9.279	-314.509	-336.582	35.163
600	50.964	297.382	273.586	14.277	-315.045	-340.948	29.682
700	52.475	305.357	277.567	19.453	-315.636	-345.219	25.761
800	53.590	312.441	281.492	24.759	-316.267	-349.402	22.814
900	54.425	318.803	285.290	30.162	-316.933	-353.505	20.517
1000	55.061	324.572	288.934	35.638	-317.633	-357.531	18.676
1100	55.554	329.844	292.417	41.169	-318.366	-361.486	17.166
1200	55.942	334.695	295.741	46.745	-319.134	-365.372	15.904
1300	56.253	339.185	298.912	52.355	-319.936	-369.193	14.834
1400	56.505	343.364	301.940	57.994	-320.773	-372.951	13.915
1500	56.712	347.270	304.833	63.655	-321.647	-376.648	13.116
1600	56.884	350.935	307.601	69.335	-322.557	-380.285	12.415
1700	57.028	354.388	310.253	75.031	-323.505	-383.864	11.795
1800	57.150	357.652	312.796	80.740	-324.489	-387.386	11.242
1900	57.254	360.744	315.239	86.460	-325.510	-390.853	10.745
2000	57.344	363.683	317.588	92.190	-326.566	-394.265	10.297
2100	57.421	366.483	319.851	97.929	-327.658	-397.623	9.890
2200	57.489	369.156	322.031	103.674	-328.785	-400.928	9.519
2300	57.548	371.713	324.136	109.426	-329.945	-404.182	9.179
2400	57.600	374.163	326.170	115.183	-381.397	-406.315	8.843
2500	57.646	376.516	328.137	120.946	-382.699	-407.327	8.511
2600	57.687	378.777	330.042	126.713	-384.000	-408.286	8.203
2700	57.724	380.955	331.887	132.483	-385.297	-409.195	7.916
2800	57.757	383.055	333.677	138.257	-386.592	-410.057	7.650
2900	57.787	385.082	335.415	144.034	-387.881	-410.872	7.401
3000	57.813	387.042	337.104	149.815	-389.166	-411.643	7.167
3100	57.838	388.938	338.745	155.597	-390.444	-412.371	6.948
3200	57.860	390.775	340.343	161.382	-391.715	-413.058	6.742
3300	57.880	392.555	341.898	167.169	-392.978	-413.705	6.548
3400	57.898	394.283	343.413	172.958	-394.231	-414.315	6.365
3500	57.915	395.962	344.891	178.749	-395.474	-414.887	6.192
3600	57.931	397.594	346.332	184.541	-396.706	-415.424	6.028
3700	57.945	399.181	347.739	190.335	-397.924	-415.928	5.872
3800	57.958	400.727	349.114	196.130	-399.129	-416.398	5.724
3900	57.970	402.232	350.456	201.926	-400.318	-416.837	5.583
4000	57.982	403.700	351.769	207.724	-401.492	-417.246	5.449
4100	57.992	405.132	353.053	213.523	-402.647	-417.625	5.321
4200	58.002	406.530	354.310	219.322	-883.634	-410.952	5.111
4300	58.011	407.895	355.540	225.123	-883.665	-399.698	4.855
4400	58.019	409.228	356.745	230.924	-883.677	-388.443	4.611
4500	58.027	410.532	357.926	236.727	-883.670	-377.187	4.378
4600	58.035	411.808	359.084	242.530	-883.643	-365.932	4.155
4700	58.042	413.056	360.219	248.334	-888.595	-354.678	3.942
4800	58.048	414.278	361.332	254.138	-883.527	-343.425	3.737
4900	58.054	415.475	362.425	259.943	-883.438	-332.174	3.541
5000	58.060	416.648	363.498	265.749	-883.328	-320.925	3.353
5100	58.066	417.798	364.551	271.555	-883.196	-309.678	3.172
5200	58.071	418.925	365.586	277.362	-883.044	-298.434	2.998
5300	58.076	420.031	366.603	283.170	-882.870	-287.193	2.830
5400	58.080	421.117	367.603	288.977	-882.674	-275.956	2.669
5500	58.084	422.183	368.585	294.786	-882.459	-264.722	2.514
5600	58.089	423.229	369.552	300.594	-882.222	-253.493	2.364
5700	58.092	424.258	370.503	306.403	-881.966	-242.268	2.220
5800	58.096	425.268	371.438	312.213	-881.689	-231.048	2.081
5900	58.100	426.261	372.359	318.023	-881.394	-219.832	1.946
6000	58.103	427.238	373.265	323.833	-881.079	-208.622	1.816

PREVIOUS: December 1964 (1 atm) CURRENT: December 1964 (1 bar)

Chlorofluoroborane (BClF) $B_1Cl_1F_1(g)$

CHLORODIFLUOROBORANE ($BClF_2$) IDEAL GAS M_r = 84.259806

$\Delta_f H$(0 K) = -883.0 ± 20 kJ mol^{-1}

S°(298.15 K) = 274.83 J K^{-1} mol^{-1} $\Delta_f H$°(298.15 K) = -885.3 ± 20 kJ mol^{-1}

Vibrational Frequencies and Degeneracies

ν, cm^{-1}	ν, cm^{-1}
1250(1)	1430(1)
697(1)	366(1)
429(1)	608(1)

Ground State Quantum Weight: 1 σ = 2

Point Group: C_{2v}

Bond Distances: B-Cl = [1.73] Å B-F = [1.295] Å

Bond Angles: F-B-Cl = [120]° F-B-F = [120]°

Product of the Moments of Inertia: $I_A I_B I_C$ = [3.568886 x 10^{-114}] g^3 cm^6

Enthalpy of Formation

The values for BCl_2F and $BClF_2$ of $\Delta_f H$°(298.15 K) = -154 and $\Delta_f H$°(298.15 K) = -211.6 kcal mol^{-1}, respectively, are obtained from the average bond energies of BF_3 and BCl_3. The calculated values are each made less negative by about 0.8 kcal mol^{-1} in order to make them consistent with the equilibrium measurements of Higgins, Leisegang, Raw and Rossow (2) and Gunn and Sanborn (2) for the reaction BF_3(g) + BCl_3(g) + $BClF_2$(g) + BCl_2F(g). Values for the heat of reaction are $\Delta_r H$°(298.15 K) = +1.68 kcal mol^{-1} and $\Delta_r H$°(298.15 K) = 1.1 (uncertainty of -0.5 to +0.8) kcal mol^{-1}, the former calculated from the measured equilibrium constant and the statistical entropy change and the latter obtained from the variation of the equilibrium constant with temperature. Selection of $\Delta_r H$°(298.15 K) = 1.6 kcal mol^{-1} fixes the sum of the enthalpy of formation of $BClF_2$ and BCl_2F as -365.6 kcal mol^{-1}, but it does not determine the individual values. Thus, the selected enthalpies of formation are only as accurate as the average bond energy calculations on which they are based. A reasonable estimate of the uncertainty would be ±5 kcal mol^{-1}.

Heat Capacity and Entropy

The observed frequencies of Lindeman and Wilson (3) were adjusted to the normal isotopic composition of 81.2% ^{10}B and 18.8% ^{11}B. The bond angles are assumed to be 120° and the bond lengths are taken to be the same as in the respective trihalides. The principal moments of inertia are: I_A = 7.9360 x 10^{-39}, I_B = 17.6063 x 10^{-39}, and I_C = 25.5423 x 10^{-39} g cm^2.

References

1. T. H. S. Higgins, E. C. Leisegang, C. J. G. Raw, and A. J. Rossouw, J. Chem. Phys. 23, 1544 (1955).
2. S. R. Gunn and R. H. Sanborn, J. Chem. Phys. 33, 955 (1960).
3. L. P. Lindeman and M. K. Wilson, J. Chem. Phys. 24, 242 (1956).

Chlorodifluoroborane ($BClF_2$) $B_1Cl_1F_2$(g)

Enthalpy Reference Temperature = T_r = 298.15 K Standard State Pressure = p° = 0.1 MPa

T/K	C_p° (J $K^{-1}mol^{-1}$)	S° (J $K^{-1}mol^{-1}$)	-[G°-H°(T_r)]/T (J $K^{-1}mol^{-1}$)	H°-H°(T_r) (kJ mol^{-1})	$\Delta_f H$° (kJ mol^{-1})	$\Delta_f G$° (kJ mol^{-1})	Log K_f
0	0.	0.	INFINITE	-12.320	-883.026	-883.026	INFINITE
100	35.262	227.389	316.986	-8.960	-884.052	-880.272	459.807
200	45.615	254.879	279.545	-4.933	-884.806	-876.171	228.832
250	50.457	265.591	275.705	-2.528	-885.084	-873.980	182.608
298.15	54.466	274.830	274.830	0.	-885.334	-871.818	152.739
300	54.609	275.167	274.831	0.101	-885.344	-871.734	151.782
350	58.172	283.860	275.509	2.923	-885.597	-869.446	129.758
400	61.242	291.833	277.058	5.910	-885.852	-867.121	113.234
450	63.887	299.203	279.115	9.040	-886.105	-864.765	100.379
500	66.163	306.055	281.470	12.292	-886.351	-862.381	90.092
600	69.800	318.456	286.623	19.100	-886.815	-857.542	74.656
700	72.492	329.428	291.970	26.221	-887.236	-852.630	63.624
800	74.504	339.246	297.276	33.576	-887.624	-847.659	55.346
900	76.027	348.113	302.440	41.106	-887.990	-842.641	48.906
1000	77.199	356.187	307.417	48.769	-888.347	-837.583	43.751
1100	78.114	363.589	312.192	56.537	-888.704	-832.490	39.532
1200	78.840	370.418	316.763	64.386	-889.071	-827.363	36.014
1300	79.424	376.753	321.137	72.300	-889.453	-822.206	33.037
1400	79.899	382.657	325.323	80.267	-889.858	-817.018	30.483
1500	80.290	388.183	329.331	88.277	-890.288	-811.800	28.269
1600	80.616	393.375	333.174	96.323	-890.747	-806.553	26.331
1700	80.891	398.271	336.860	104.399	-891.237	-801.276	24.620
1800	81.123	402.902	340.402	112.500	-891.759	-795.969	23.098
1900	81.322	407.293	343.808	120.622	-892.314	-790.632	21.736
2000	81.493	411.469	347.087	128.763	-892.900	-785.266	20.509
2100	81.641	415.449	350.249	136.920	-893.517	-779.869	19.398
2200	81.771	419.250	353.299	145.091	-894.163	-774.442	18.388
2300	81.884	422.887	356.246	153.274	-894.837	-768.985	17.464
2400	81.984	426.374	359.096	161.467	-945.794	-762.430	16.594
2500	82.073	429.723	361.855	169.670	-946.593	-754.773	15.770
2600	82.152	432.943	364.527	177.881	-947.380	-747.084	15.009
2700	82.223	436.045	367.119	186.100	-948.153	-739.366	14.304
2800	82.286	439.036	369.634	194.326	-948.909	-731.619	13.649
2900	82.343	441.925	372.078	202.557	-949.648	-723.846	13.038
3000	82.395	444.717	374.453	210.794	-950.366	-716.047	12.467
3100	82.442	447.420	376.763	219.036	-951.062	-708.225	11.934
3200	82.484	450.038	379.012	227.282	-951.735	-700.381	11.433
3300	82.523	452.577	381.203	235.533	-952.382	-692.516	10.962
3400	82.559	455.041	383.339	243.787	-953.002	-684.632	10.518
3500	82.591	457.434	385.422	252.044	-953.594	-676.730	10.100
3600	82.621	459.762	387.455	260.305	-954.157	-668.811	9.704
3700	82.649	462.026	389.440	268.569	-954.688	-660.878	9.330
3800	82.674	464.230	391.379	276.835	-955.187	-652.931	8.975
3900	82.698	466.378	393.275	285.103	-955.652	-644.970	8.638
4000	82.720	468.472	395.128	293.374	-956.083	-636.998	8.318
4100	82.740	470.515	396.942	301.647	-956.479	-629.016	8.014
4200	82.759	472.509	398.718	309.922	-1436.688	-614.001	7.636
4300	82.776	474.456	400.457	318.199	-1435.923	-594.422	7.221
4400	82.793	476.360	402.160	326.477	-1435.123	-574.861	6.824
4500	82.808	478.220	403.830	334.757	-1434.286	-555.319	6.446
4600	82.822	480.041	405.467	343.039	-1433.413	-535.796	6.084
4700	82.836	481.822	407.073	351.322	-1432.504	-516.292	5.738
4800	82.848	483.566	408.648	359.606	-1431.558	-496.808	5.406
4900	82.860	485.274	410.194	367.891	-1430.576	-477.345	5.089
5000	82.871	486.948	411.713	376.178	-1429.558	-457.901	4.784
5100	82.882	488.590	413.204	384.466	-1428.504	-438.479	4.491
5200	82.892	490.199	414.669	392.754	-1427.415	-419.077	4.210
5300	82.901	491.778	416.109	401.044	-1426.291	-399.696	3.939
5400	82.910	493.328	417.525	409.335	-1425.133	-380.337	3.679
5500	82.919	494.849	418.917	417.626	-1423.941	-361.000	3.428
5600	82.926	496.343	420.287	425.918	-1422.716	-341.685	3.187
5700	82.934	497.811	421.634	434.211	-1421.459	-322.392	2.954
5800	82.941	499.254	422.960	442.505	-1420.171	-303.122	2.730
5900	82.948	500.672	424.265	450.800	-1418.852	-283.873	2.513
6000	82.954	502.066	425.550	459.095	-1417.504	-264.648	2.304

PREVIOUS: December 1964 (1 atm) CURRENT: December 1964 (1 bar)

Chlorodifluoroborane ($BClF_2$) $B_1Cl_1F_2$(g)

BORON CHLORIDE OXIDE (OBCl) IDEAL GAS $M_r = 62.2624$

$S°(298.15\ K) = [237.42]\ J\ K^{-1}\ mol^{-1}$

$\Delta_f H°(0\ K) = [-316.8 \pm 29]\ kJ\ mol^{-1}$
$\Delta_f H°(298.15\ K) = -316.3 \pm 29\ kJ\ mol^{-1}$

Vibrational Frequencies and Degeneracies

ν, cm^{-1}
[690](1)
[400](2)
[1850](1)

Ground State Quantum Weight: [1] $\sigma = 1$

Point Group: $[C_{\infty v}]$

Bond Distances: Cl-B = [1.75] Å B-O = [1.20] Å

Bond Angle: Cl-B-O = [180]°

Rotational Constant: $B_0 = [0.165056]\ cm^{-1}$

Enthalpy of Formation

The vapor-phase equilibria existing in the $B_2O_3 + BCl_3$ system were determined by an effusion method in the temperature range 1234-1389 K by Blauer and Farber (1). Based on the equilibrium constants reported for the reaction $1/3\ B_2O_3(\ell) + 1/3\ BCl_3(g) = BOCl(g)$, the 2nd and 3rd law values of $\Delta_r H°(298.15\ K)$ were evaluated as 57.7 and 56.3 kcal mol^{-1}, respectively. The 3rd law value for $\Delta_r H°(298.15\ K)$ was adopted for the calculation of the enthalpy of formation, $\Delta_f H°(BOCl, g, 298.15\ K)$.

Heat Capacity and Entropy

The molecular structure is assumed to be linear. The vibrational frequencies are estimated from the ν values calculated by the valence force method described by Herzberg (2) using estimated force constants, $k_1 = 5.2 \times 10^5$, $k_2 = 16.7 \times 10^5$ dynes cm^{-1} and $k_3 = 0.40 \times 10^{-11}$ dynes cm rad^{-1}. The B=O and B-Cl bond distances were assumed to be the same as those in BO(g) and $BCl_3(g)$, respectively.

References

1. J. Blauer and M. Farber, Trans. Faraday Soc. 60, 301 (1964).
2. G. Herzberg, "Infrared and Raman Spectra", D. van Nostrand Company, Inc., 1945.

Boron Chloride Oxide (OBCl) $B_1Cl_1O_1(g)$

Enthalpy Reference Temperature = T_r = 298.15 K Standard State Pressure = p° = 0.1 MPa

T/K	$C°_p$ (J K⁻¹mol⁻¹)	S° (J K⁻¹mol⁻¹)	$-[G°-H°(T_r)]/T$ (J K⁻¹mol⁻¹)	$H°-H°(T_r)$ (kJ mol⁻¹)	$\Delta_f H°$ (kJ mol⁻¹)	$\Delta_f G°$ (kJ mol⁻¹)	Log K_f
0	0.	0.	INFINITE	-10.609	-316.774	-316.774	INFINITE
100	30.896	196.590	273.283	-7.669	-316.766	-318.278	166.252
200	39.254	220.565	241.372	-4.161	-316.565	-319.850	83.536
250	42.577	229.698	238.145	-2.112	-316.425	-320.687	67.004
298.15	45.040	237.417	237.417	0.	-316.310	-321.519	56.329
300	45.123	237.696	237.418	0.083	-316.307	-321.551	55.987
350	47.105	244.806	237.975	2.391	-316.224	-322.432	48.120
400	48.708	251.204	239.235	4.788	-316.185	-323.323	42.222
450	50.054	257.021	240.893	7.258	-316.190	-324.215	37.634
500	51.216	262.356	242.776	9.790	-316.229	-325.105	33.963
600	53.146	271.871	246.852	15.012	-316.384	-326.867	28.456
700	54.680	280.184	251.032	20.406	-316.608	-328.597	24.520
800	55.909	287.568	255.146	25.938	-316.878	-330.291	21.566
900	56.898	294.213	259.124	31.580	-317.180	-331.950	19.266
1000	57.696	300.250	262.939	37.311	-317.511	-333.574	17.424
1100	58.344	305.781	266.586	43.114	-317.867	-335.163	15.916
1200	58.875	310.881	270.068	48.976	-318.250	-336.719	14.657
1300	59.313	315.611	273.391	54.886	-318.659	-338.242	13.591
1400	59.677	320.020	276.566	60.836	-319.096	-339.732	12.676
1500	59.982	324.148	279.602	66.819	-319.563	-341.190	11.881
1600	60.239	328.028	282.509	72.831	-320.062	-342.615	11.185
1700	60.459	331.687	285.295	78.866	-320.593	-344.009	10.570
1800	60.647	335.148	287.969	84.921	-321.159	-345.370	10.022
1900	60.809	338.431	290.540	90.994	-321.760	-346.699	9.531
2000	60.949	341.554	293.013	97.083	-322.400	-347.995	9.089
2100	61.072	344.531	295.396	103.184	-323.078	-349.258	8.687
2200	61.180	347.374	297.694	109.296	-323.797	-350.488	8.322
2300	61.275	350.096	299.914	115.419	-324.557	-351.685	7.987
2400	61.359	352.706	302.060	121.551	-375.620	-351.778	7.656
2500	61.434	355.212	304.136	127.691	-376.545	-350.766	7.329
2600	61.500	357.623	306.147	133.837	-377.482	-349.716	7.026
2700	61.560	359.945	308.097	139.991	-378.433	-348.630	6.745
2800	61.614	362.185	309.989	146.149	-379.397	-347.509	6.483
2900	61.663	364.348	311.826	152.313	-380.376	-346.353	6.238
3000	61.707	366.439	313.612	158.482	-381.370	-345.163	6.010
3100	61.747	368.463	315.349	164.655	-382.379	-343.939	5.795
3200	61.784	370.424	317.039	170.831	-383.402	-342.683	5.594
3300	61.817	372.326	318.686	177.011	-384.440	-341.395	5.404
3400	61.848	374.172	320.291	183.194	-385.491	-340.074	5.225
3500	61.876	375.965	321.856	189.381	-386.554	-338.723	5.055
3600	61.902	377.708	323.384	195.570	-387.629	-337.341	4.895
3700	61.925	379.405	324.875	201.761	-388.715	-335.930	4.742
3800	61.947	381.057	326.332	207.955	-389.810	-334.488	4.598
3900	61.968	382.666	327.756	214.150	-390.913	-333.018	4.460
4000	61.987	384.235	329.148	220.348	-392.023	-331.519	4.329
4100	62.004	385.766	330.510	226.548	-393.138	-329.993	4.204
4200	62.021	387.260	331.844	232.749	-874.106	-321.415	3.997
4300	62.036	388.720	333.150	238.952	-874.141	-308.255	3.745
4400	62.050	390.146	334.429	245.156	-874.179	-295.095	3.503
4500	62.064	391.541	335.683	251.362	-874.219	-281.934	3.273
4600	62.076	392.905	336.912	257.569	-874.261	-268.771	3.052
4700	62.088	394.240	338.117	263.777	-874.303	-255.608	2.841
4800	62.099	395.547	339.300	269.986	-874.346	-242.444	2.638
4900	62.109	396.828	340.461	276.197	-874.388	-229.278	2.444
5000	62.119	398.083	341.601	282.408	-874.429	-216.113	2.258
5100	62.128	399.313	342.721	288.620	-874.470	-202.946	2.079
5200	62.137	400.519	343.821	294.834	-874.509	-189.779	1.906
5300	62.145	401.703	344.902	301.048	-874.547	-176.610	1.741
5400	62.153	402.865	345.964	307.263	-874.585	-163.441	1.581
5500	62.160	404.005	347.009	313.478	-874.621	-150.272	1.427
5600	62.167	405.125	348.037	319.695	-874.657	-137.101	1.279
5700	62.174	406.226	349.048	325.912	-874.694	-123.930	1.136
5800	62.180	407.307	350.044	332.129	-874.731	-110.758	0.997
5900	62.186	408.370	351.023	338.348	-874.769	-97.587	0.864
6000	62.191	409.415	351.988	344.566	-874.809	-84.414	0.735

PREVIOUS: December 1965 (1 atm) CURRENT: December 1965 (1 bar)

DICHLOROBORANE (BCl_2) IDEAL GAS M_r = 81.716

$S°(298.15 K) = [272.7 \pm 2.1] J K^{-1} mol^{-1}$

$\Delta_f H°(0 K) = -80.9 \pm 17.6 kJ mol^{-1}$
$\Delta_f H°(298.15 K) = -79.5 \pm 12.6 kJ mol^{-1}$

Electronic Levels and Quantum Weights

ε_i, cm^{-1}	g_i
0	2
[11000]	[2]
28003	1
28153	1
29455	1
29542	1

Vibrational Frequencies and Degeneracies

ν, cm^{-1}
[700] (1)
250 (1)
[725] (1)

Point Group: $[C_{2v}]$ $\sigma = 2$
Bond Distances: B-Cl = [1.73] Å
Bond Angle: Cl-B-Cl = [112]°
Product of the Moments of Inertia: $I_A I_B I_C = 9.065976 \times 10^{-115}\ g^3\ cm^6$

Enthalpy of Formation

Srivastava and Farber (1) have investigated the equilibrium vapors over the boron trichloride-boron system by mass spectrometry. Gaseous BCl_3 was allowed to flow into a boron nitride effusion cell which contained metallic boron. A flow variation study indicated that thermal equilibrium had been achieved during their experiments. Ion intensities for the species $BCl_3(g)$, $BCl_2(g)$, and BCl(g) were measured at electron energies slightly above appearance potentials in the temperature range 1513 to 1813 K. JANAF analysis of equilibrium constants for the reaction $BCl_3(g) + BCl(g) = 2BCl_2(g)$ which are obtained from their ion intenesity data yields a 3rd law $\Delta_r H$(298.15 K) value of 24.56 ± 1.00 kcal mol^{-1} with a drift of -2.1 ± 2.0 cal $K^{-1}mol^{-1}$. The 2nd law $\Delta_r H$ value at 298.15 K is 28.1 kcal mol^{-1}. Using the 3rd law $\Delta_r H$ value in conjunction with JANAF enthalpies of formation (2) for $BCl_3(g)$ and BCl(g) and BCl(g), we calculate $\Delta_f H°(BCl_2$, g, 298.15 K) = -19.0 ± 3.0 kcal mol^{-1}.

The only other reported data on the enthalpy of formation of BCl_2 comes from explosion studies carried out at elevated temperatures by Rusin and Tatevskii (3). The BCl_2 radical was found to be stable at 2250 K in H_2-Cl_2-BCl_3 mixtures. These workers reported $\Delta_f H°(BCl_2$, g, 298.15 K) = -18.4 ± 1.5 kcal mol^{-1} which confirms the results of Srivastava and Farber (1).

We adopt $\Delta_f H°(BCl_2$, g, 298.15 K) = -19.0 ± 3.0 kcal mol^{-1}. This value also leads to a realistic progression in the stepwise bond dissociation energies for BCl(g), $BCl_2(g)$, and $BCl_3(g)$. Using all JANAF data (2), we calculate D_1(B-Cl) = 127.9, D_2(BCl-Cl) = 81.7, and $D_3(BCl_2$-Cl) = 106.2 kcal mol^{-1}. This bond energy scheme, $D_1 > D_2 < D_3$, is consistent with that predicted for the Group III A fluorides by Hildenbrand (4).

Heat Capacity and Entropy

The point group and the Cl-B-Cl bond angle are assumed to be the same as those for BF_2 (2). The B-Cl bond length is estimated to lie between those for BCl and BCl_3 (2). The principal moments of inertia are: $I_A = 1.4577 \times 10^{-39}$, $I_B = 24.2203 \times 10^{-39}$, and $I_C = 25.6780 \times 10^{-39}\ g\ cm^2$.

Dessaux et al. (5, 6) observed a diffuse system of bands in the near ultraviolet which they attributed to the BCl_2 radical. The emission spectrum was obtained as chemiluminescence in the reaction between BCl_3 and atomic hydrogen. They indicated that the frequencies ν_1 = 470, ν_2 = 240, and ν_3 = 990 cm^{-1} were consistent with their spectral data; however, alternate assignments are possible. We prefer the frequencies ν_1 = 700, ν_2 250, and ν_3 = 725 cm^{-1}, since these values are more consistent with those for CCl_2, $SiCl_2$, and $AlCl_2$ (2).

By analogy with the isoelectronic molecule NO_2 (7), the ground state for BCl_2 is assumed to be $^2A_1$. Huckel-molecular-orbital calculations by Hastie and Margrave (8) suggest a doublet excited state ($^2B_1$) near 11,000 cm^{-1} which is included. The upper electronic levels are those observed by Dessaux et al. (6).

References

1. R. D. Srivastava and M. Farber, Trans. Faraday Soc. 67, 2298 (1971).
2. JANAF Thermochemical Tables: $BCl_3(g)$, BCl(g), and B(g), 12-31-64; Cl(g), $BF_2(g)$, and $AlCl_2(g)$, 6-30-72; $CCl_2(g)$, 12-31-68; $SiCl_2(g)$, 12-31-70.
3. A. D. Rusin and V. M. Tatevskii, Teplofiz. Vys. Temp. 7, 62 (1969); CA 70 100335b.
4. D. L. Hildenbrand in "Advances in High Temperature Chemistry," Volume I, Ed. by L. Eyring, Academic Press. New York, 1967.
5. O. Dessaux, P. Goudmand, and G. Pannetier, C. R. Acad. Sci., Paris, Ser. C265, 480 (1967).
6. O. Dessaux, P. Goudmand, and G. Pannetier, Bull. soc. Chim. Fr. 447 (1969).
7. G. Herzberg, "Electronic Spectra of Polyatomic Molecules," D. Van Nostrand Company. Inc., New York, 1966.
8. J. W. Hastie and J. L. margrave, J. Phys. Chem. 73, 1105 (1969).

Dichloroborane (BCl_2) $B_1Cl_2(g)$

Enthalpy Reference Temperature = T_r = 298.15 K Standard State Pressure = p° = 0.1 MPa

T/K	$C_p°$ (J $K^{-1}mol^{-1}$)	S° (J $K^{-1}mol^{-1}$)	$-[G°-H°(T_r)]/T$ (J $K^{-1}mol^{-1}$)	$H°-H°(T_r)$ (kJ mol^{-1})	$\Delta_f H°$ (kJ mol^{-1})	$\Delta_f G°$ (kJ mol^{-1})	Log K_f
0	0.	0.	INFINITE	-11.759	-80.860	-80.860	INFINITE
100	36.437	227.742	311.228	-8.349	-80.386	-84.228	43.996
200	42.282	254.796	276.857	-4.412	-79.820	-88.303	23.062
250	45.066	264.535	273.444	-2.227	-79.625	-90.447	18.898
298.15	47.375	272.677	272.677	0.	-79.496	-92.544	16.213
300	47.455	272.970	272.678	0.088	-79.492	-92.625	16.127
350	49.389	280.436	273.263	2.511	-79.415	-94.821	14.151
400	50.918	287.135	274.585	5.020	-79.392	-97.024	12.670
450	52.122	293.205	276.322	7.597	-79.419	-99.226	11.518
500	53.073	298.747	278.292	10.228	-79.487	-101.424	10.596
600	54.439	308.553	282.540	15.608	-79.720	-105.792	9.210
700	55.341	317.018	286.874	21.100	-80.055	-110.112	8.217
800	55.962	324.450	291.116	26.667	-80.470	-114.378	7.468
900	56.404	331.069	295.194	32.287	-80.953	-118.588	6.883
1000	56.729	337.029	299.085	37.944	-81.494	-122.741	6.411
1100	56.975	342.448	302.784	43.630	-82.088	-126.838	6.023
1200	57.166	347.414	306.299	49.337	-82.732	-130.878	5.697
1300	57.319	351.996	309.641	55.062	-83.421	-134.862	5.419
1400	57.445	356.248	312.820	60.800	-84.154	-138.792	5.178
1500	57.553	360.215	315.849	66.550	-84.930	-142.668	4.968
1600	57.650	363.933	318.739	72.310	-85.745	-146.491	4.782
1700	57.740	367.431	321.501	78.080	-86.601	-150.262	4.617
1800	57.828	370.733	324.146	83.858	-87.494	-153.981	4.468
1900	57.917	373.862	326.681	89.646	-88.426	-157.649	4.334
2000	58.009	376.836	329.115	95.442	-89.395	-161.268	4.212
2100	58.106	379.668	331.455	101.248	-90.400	-164.837	4.100
2200	58.207	382.374	333.708	107.063	-91.442	-168.357	3.997
2300	58.315	384.963	335.881	112.889	-92.521	-171.829	3.902
2400	58.428	387.448	337.978	118.726	-143.896	-174.185	3.791
2500	58.547	389.835	340.005	124.575	-145.127	-175.422	3.665
2600	58.671	392.134	341.966	130.436	-146.363	-176.609	3.548
2700	58.800	394.350	343.865	136.309	-147.605	-177.748	3.439
2800	58.932	396.491	345.707	142.196	-148.852	-178.842	3.336
2900	59.067	398.562	347.494	148.096	-150.105	-179.891	3.240
3000	59.205	400.566	349.230	154.010	-151.365	-180.897	3.150
3100	59.344	402.510	350.917	159.937	-152.629	-181.860	3.064
3200	59.484	404.396	352.559	165.878	-153.899	-182.783	2.984
3300	59.623	406.229	354.158	171.834	-155.172	-183.666	2.907
3400	59.762	408.011	355.716	177.803	-156.448	-184.510	2.835
3500	59.900	409.745	357.235	183.786	-157.726	-185.317	2.766
3600	60.037	411.435	358.717	189.783	-159.003	-186.087	2.700
3700	60.171	413.081	360.164	195.793	-160.278	-186.822	2.637
3800	60.302	414.688	361.578	201.817	-161.548	-187.523	2.578
3900	60.431	416.256	362.960	207.854	-162.812	-188.190	2.521
4000	60.557	417.787	364.312	213.903	-164.066	-188.824	2.466
4100	60.680	419.284	365.634	219.965	-165.310	-189.428	2.413
4200	60.800	420.748	366.929	226.039	-646.390	-182.977	2.276
4300	60.917	422.180	368.197	232.125	-646.518	-171.942	2.089
4400	61.030	423.582	369.440	238.223	-646.629	-160.904	1.910
4500	61.140	424.954	370.659	244.331	-646.723	-149.864	1.740
4600	61.247	426.299	371.854	250.451	-646.796	-138.821	1.576
4700	61.351	427.618	373.026	256.581	-646.848	-127.778	1.420
4800	61.452	428.910	374.177	262.721	-646.878	-116.733	1.270
4900	61.550	430.179	375.307	268.871	-646.882	-105.689	1.127
5000	61.645	431.423	376.417	275.031	-646.862	-94.644	0.989
5100	61.737	432.645	377.507	281.200	-646.814	-83.600	0.856
5200	61.827	433.844	378.579	287.378	-646.739	-72.558	0.729
5300	61.914	435.023	379.633	293.565	-646.635	-61.517	0.606
5400	61.999	436.181	380.670	299.761	-646.503	-50.478	0.488
5500	62.081	437.319	381.689	305.965	-646.341	-39.441	0.375
5600	62.161	438.439	382.693	312.177	-646.149	-28.409	0.265
5700	62.239	439.540	383.680	318.397	-645.927	-17.380	0.159
5800	62.316	440.623	384.653	324.625	-645.675	-6.354	0.057
5900	62.390	441.689	385.611	330.860	-645.393	4.666	-0.041
6000	62.462	442.738	386.554	337.102	-645.082	15.681	-0.137

PREVIOUS: June 1972 (1 atm) CURRENT: June 1972 (1 bar)

Dichloroborane (BCl_2) $B_1Cl_2(g)$

DICHLOROBORANE, ION (BCl_2^+) IDEAL GAS M_r = 81.71545

$S°(298.15\ K) = [257.8 \pm 8.4]\ J\ K^{-1}\ mol^{-1}$

$\Delta_fH°(0\ K) = 665.4 \pm 25\ kJ\ mol^{-1}$
$\Delta_fH°(298.15\ K) = 674.0 \pm 25\ kJ\ mol^{-1}$

Electronic Levels and Quantum Weights

State	ε_i, cm^{-1}	g_i
$^1\Sigma_g^+$	0	1
$^1B_2$	[16000]	1
$^1A_1$	[35000]	1

Vibrational Frequencies and Degeneracies

ν, cm^{-1}
[500] (1)
[150] (2)
[800] (1)

Point Group: $[D_{\infty h}]$ $\sigma = 2$

Bond Distances: B–Cl = [1.74] Å

Bond Angle: Cl–B–Cl = [180]°

Product of the Moments of Inertia: $B_0 = [0.078526]\ cm^{-1}$

Enthalpy of Formation

The appearance potential of the positive ion fragment $BCl_2^+(g)$ from $BCl_3(g)$ was determined mass spectrometrically by Osberghaus (1), Marriott (2), and Koski (3) as 13.2 ± 0.5, 13.01 ± 0.02, and 11.8 ± 0.2 eV, respectively. From the reaction $BCl_3 + e^- = BCl_2^+ + Cl + 2e^-$, the corresponding values of $\Delta_fH°(298.15\ K)$ for $BCl_2^+(g)$ are derived to be 181.9 ± 11.5, 177.5 ± 0.5, and 149.6 ± 4.6 kcal mol^{-1}, using $\Delta_fH°(0\ K)$ = -96.078 and 28.520 kcal mol^{-1} for $BCl_3(g)$ and Cl(g). These electron impact values are superseded by the photoionization results of Dibeler and Walker (7) who report the onset of BCl_2^+ from BCl_3 as 12.30 ± 0.02 eV (283.65 ± 0.05 kcal mol^{-1}). This yields $\Delta_fH°(0\ K)$ = 159.05 ± 0.6 kcal mol^{-1} or $\Delta_fH°(298.15\ K)$ = 161.1 ± 0.6 kcal mol^{-1}, which are adopted.

Heat Capacity and Entropy

The molecular structure is assumed to be linear using the Walsh (4) predictions for 16 valence electron XY_2 molecules. The electronic states, levels and vibrational frequencies are estimated by comparison with those for CO_2 (5). The B-Cl bond distance is estimated using the method of Krasnov (6).

References

1. O. Osberghaus, Z. Phys. 128, 366 (1950).
2. J. Marriott and J. D. Craggs, J. Electron. Contr. 3, 194 (1957).
3. W. S. Koski, J. J. Kaufman, and C. E. Pachucki, J. Amer. Chem. Soc. 81, 1326 (1959).
4. A. D. Walsh, J. Chem. Soc. 2266 (1953).
5. G. Herzberg, "Electronic Spectra of Polyatomic Molecules," D. Van Nostrand Co., Inc., New York, 1966.
6. K. S. Krasnov, Zhur. Strukt. Khim. 1, 209 (1960).
7. V. H. Dibeler and J. A. Walker, Inorg. Chem. 8, 50 (1969).

Dichloroborane, Ion (BCl_2^+) $B_1Cl_2^+(g)$

Enthalpy Reference Temperature = T_r = 298.15 K Standard State Pressure = p° = 0.1 MPa

	J K⁻¹mol⁻¹			kJ mol⁻¹			
T/K	$C_p°$	S°	$-[G°-H°(T_r)]/T$	$H°-H°(T_r)$	$\Delta_fH°$	$\Delta_fG°$	Log K_f
0	0.	0.	INFINITE	-12.848	665.391		
100	40.874	206.815	301.467	-9.465			
200	48.200	237.603	262.511	-4.982			
250	50.884	248.657	258.666	-2.502			
298.15	52.974	257.805	257.805	0.	674.042	659.173	-115.484
300	53.046	258.133	257.806	0.098	674.095	659.081	-114.756
350	54.747	266.443	258.459	2.795	675.485	656.468	-97.972
400	56.076	273.844	259.928	5.567	676.810	653.660	-85.359
450	57.115	280.511	261.851	8.397	678.076	650.690	-75.530
500	57.934	286.573	264.024	11.274	679.294	647.581	-67.652
600	59.110	297.247	268.696	17.131	681.616	641.018	-55.806
700	59.887	306.422	273.445	23.083	683.820	634.076	-47.315
800	60.422	314.455	278.080	29.100	685.933	626.825	-40.927
900	60.803	321.595	282.526	35.163	687.972	619.313	-35.944
1000	61.084	328.017	286.759	41.258	689.946	611.578	-31.946
1100	61.296	333.849	290.779	47.377	691.864	603.648	-28.665
1200	61.460	339.190	294.594	53.515	693.730	595.546	-25.923
1300	61.588	344.115	298.216	59.668	695.548	587.290	-23.598
1400	61.692	348.683	301.660	65.832	697.319	578.895	-21.599
1500	61.776	352.942	304.938	72.006	699.046	570.376	-19.862
1600	61.846	356.931	308.064	78.187	700.730	561.743	-18.339
1700	61.904	360.682	311.050	84.374	702.371	553.006	-16.992
1800	61.955	364.222	313.907	90.567	703.970	544.173	-15.791
1900	61.999	367.573	316.644	96.765	705.528	535.253	-14.715
2000	62.038	370.754	319.271	102.967	707.044	526.252	-13.744
2100	62.075	373.782	321.795	109.173	708.517	517.176	-12.864
2200	62.110	376.671	324.224	115.382	709.947	508.030	-12.062
2300	62.145	379.432	326.565	121.595	711.333	498.821	-11.329
2400	62.180	382.078	328.823	127.811	662.416	490.621	-10.678
2500	62.216	384.617	331.005	134.031	663.634	483.438	-10.101
2600	62.254	387.058	333.114	140.254	664.840	476.207	-9.567
2700	62.295	389.408	335.156	146.482	666.031	468.929	-9.072
2800	62.339	391.674	337.134	152.713	667.207	461.607	-8.611
2900	62.387	393.863	339.053	158.949	668.369	454.244	-8.182
3000	62.437	395.979	340.915	165.191	669.516	446.841	-7.780
3100	62.492	398.027	342.724	171.437	670.649	439.399	-7.404
3200	62.551	400.012	344.484	177.689	671.768	431.922	-7.050
3300	62.613	401.937	346.196	183.947	672.876	424.410	-6.718
3400	62.679	403.808	347.863	190.212	673.974	416.864	-6.404
3500	62.749	405.625	349.487	196.483	675.063	409.286	-6.108
3600	62.822	407.394	351.071	202.762	676.147	401.677	-5.828
3700	62.899	409.117	352.617	209.048	677.226	394.038	-5.563
3800	62.978	410.795	354.126	215.342	678.305	386.369	-5.311
3900	63.061	412.432	355.600	221.644	679.385	378.673	-5.072
4000	63.145	414.030	357.041	227.954	680.469	370.949	-4.844
4100	63.232	415.590	358.450	234.273	681.561	363.197	-4.627
4200	63.321	417.115	359.829	240.601	202.814	362.443	-4.508
4300	63.412	418.606	361.178	246.937	205.015	366.218	-4.449
4400	63.504	420.065	362.500	253.283	207.231	369.941	-4.392
4500	63.597	421.493	363.795	259.638	209.462	373.614	-4.337
4600	63.690	422.891	365.065	266.002	211.712	377.237	-4.284
4700	63.785	424.262	366.310	272.376	213.982	380.811	-4.232
4800	63.879	425.606	367.531	278.759	216.275	384.336	-4.182
4900	63.974	426.924	368.730	285.152	218.591	387.814	-4.134
5000	64.069	428.218	369.907	291.554	220.933	391.243	-4.087
5100	64.163	429.487	371.063	297.966	223.302	394.626	-4.042
5200	64.257	430.734	372.198	304.387	225.699	397.962	-3.998
5300	64.351	431.959	373.314	310.817	228.124	401.251	-3.955
5400	64.443	433.163	374.411	317.257	230.579	404.495	-3.913
5500	64.535	434.346	375.490	323.706	233.065	407.693	-3.872
5600	64.626	435.510	376.552	330.164	235.581	410.845	-3.832
5700	64.715	436.654	377.596	336.631	238.129	413.953	-3.793
5800	64.804	437.781	378.624	343.107	240.707	417.015	-3.756
5900	64.891	438.889	379.636	349.592	243.317	420.032	-3.719
6000	64.977	439.981	380.633	356.085	245.958	423.005	-3.683

PREVIOUS: December 1970 (1 atm) CURRENT: December 1970 (1 bar)

Dichloroborane, Ion (BCl_2^+) $B_1Cl_2^+(g)$

DICHLOROBORANE, ION (BCl_2^-) IDEAL GAS M_r = 81.71655

$\Delta_f H°(0\ K)$ = [-142 ± 105] kJ mol^{-1}

$S°(298.15\ K)$ = [266.3 ± 4.2] J K^{-1} mol^{-1} $\Delta_f H°(298.15\ K)$ = [-146 ± 105] kJ mol^{-1}

Electronic Levels and Quantum Weights

State	ε_i, cm^{-1}	g_i
$^1A_1$	0	1
$^3B_1$	[12000]	3
$^1B_1$	[17000]	1

Vibrational Frequencies and Degeneracies

ν, cm^{-1}
[750] (1)
[250] (1)
[775] (1)

Point Group: [C_{2v}] σ = 2

Bond Distances: B-Cl = [1.716] Å

Bond Angle: Cl-B-Cl = [100]°

Product of the Moments of Inertia: $I_A I_B I_C$ = [8.575781 x 10^{-115}] g^3 cm^6

Enthalpy of Formation

The enthalpy of formation $\Delta_f H°(BCl_2^-$, g, 0 K) = -33.8 ± 25 kcal mol^{-1}, is obtained from an estimated electron affinity for BCl_2 and $\Delta_f H°(BCl_2$, g, 0 K) = -19.3 ± 3.0 kcal mol^{-1} (1). Comparison of the ionization potentials and electron affinities for the isoelectronic molecules NO_2 (1, 2), BF_2 (1), and AlF_2 (1) indicates that in each case the difference between these two quantities (IP-EA) is roughly 7.1 eV. From this relationship, we estimate EA (BCl_2) = 0.63 ± 1.0 eV with IP (BCl_2) = 7.73 eV (1).

Heat Capacity and Entropy

The correlation diagram of Walsh (3) predicts a bent configuration for BCl_2^- (18 valence electrons) with a bond angle considerably reduced from that for BCl_2. Based on the adopted angle for BCl_2 (112°) (1), we estimate the angle in BCl_2^- at 100°. Comparison with angles for CF_2 and SiF_2 (1) indicates that the estimated angle is reasonable. The B-Cl bond length is taken equal to that for BCl (1). The principal moments of inertia are: I_A = 1.8951 x 10^{-39}, I_B = 20.3463 x 10^{-39}, and I_C = 22.2413 x 10^{-39} g cm^2.

The ground state electronic configuration ($^1A_1$) and first excited singlet state ($^1B_1$) for BCl_2^- are estimated by analogy with those for CCl_2 and $SiCl_2$ (1). Also included is a triplet level at 12000 cm^{-1}. The energy separation (5000 cm^{-1}) of the $^3B_1$ and $^1B_1$ levels is estimated by comparison with the corresponding values for SO_2 (4), $SiCl_2$ (1), $GeCl_2$, and $SnCl_2$ (5). The vibrational frequencies are estimated from those for BCl_2, CCl_2, and $SiCl_2$ (1).

References

1. JANAF Thermochemical Tables: BCl_2(g), NO_2^-(g), BF_2^-(g), AlF_2^-(g), 6-30-72; BF_2^+(g), BCl_2^+(g), and $SiCl_2$(g), 12-31-70; CF_2(g), 6-30-70; SiF_2(g) and CCl_2(g), 12-31-68; BCl(g), 12-31-64.
2. J. L. Franklin, J. G. Dillard, H. M. Rosenstock, J. T. Herron, K. Draxl, and F. H. Field, U. S. Nat. Bur. Std. NSRDS-NBS 26, 1969.
3. A. D. Walsh, J. Chem. Soc. 2266 (1953).
4. G. Herzberg, "Electronic Spectra of Polyatomic Molecules," D. Van Nostrand Company, Inc., New York, 1966.
5. J. W. Hastie, R. H. Hauge, and J. L. Margrave, J. Mol. Spectroscopy, 29, 152 (1969).

Dichloroborane, Ion (BCl_2^-) $B_1Cl_2^-(g)$

Enthalpy Reference Temperature = T_r = 298.15 K Standard State Pressure = p° = 0.1 MPa

T/K	$C_p°$ (J $K^{-1}mol^{-1}$)	S° (J $K^{-1}mol^{-1}$)	-[G°-H°(T_r)]/T (J $K^{-1}mol^{-1}$)	H°-H°(T_r) (kJ mol^{-1})	$\Delta_f H°$ (kJ mol^{-1})	$\Delta_f G°$ (kJ mol^{-1})	Log K_f
0	0.	0.	INFINITE	-11.666	-141.514		
100	36.409	221.745	304.298	-8.255			
200	41.747	248.642	270.365	-4.345			
250	44.356	258.240	267.006	-2.191			
298.15	46.614	266.251	266.251	0.	-146.440	-151.317	26.510
300	46.694	266.539	266.252	0.086	-146.476	-151.348	26.352
350	48.651	273.889	266.828	2.472	-147.476	-152.081	22.697
400	50.236	280.493	268.130	4.945	-148.528	-152.668	19.936
450	51.507	286.486	269.842	7.490	-149.627	-153.120	17.774
500	52.524	291.968	271.784	10.092	-150.763	-153.447	16.031
600	54.006	301.684	275.979	15.423	-153.123	-153.764	13.386
700	54.997	310.089	280.265	20.877	-155.575	-153.679	11.468
800	55.684	317.480	284.464	26.413	-158.100	-153.236	10.005
900	56.176	324.068	288.505	32.007	-160.687	-152.473	8.849
1000	56.540	330.007	292.363	37.643	-163.328	-151.420	7.909
1100	56.815	335.409	296.035	43.312	-166.018	-150.099	7.128
1200	57.030	340.362	299.525	49.005	-168.754	-148.531	6.465
1300	57.203	344.934	302.845	54.716	-171.535	-146.734	5.896
1400	57.346	349.179	306.004	60.444	-174.357	-144.720	5.400
1500	57.472	353.140	309.016	66.185	-177.220	-142.504	4.962
1600	57.590	356.853	311.891	71.938	-180.122	-140.095	4.574
1700	57.705	360.347	314.640	77.703	-183.060	-137.503	4.225
1800	57.826	363.649	317.272	83.480	-186.035	-134.738	3.910
1900	57.958	366.779	319.796	89.269	-189.043	-131.806	3.624
2000	58.105	369.756	322.220	95.072	-192.083	-128.715	3.362
2100	58.270	372.594	324.552	100.890	-195.155	-125.471	3.121
2200	58.456	375.309	326.797	106.726	-198.255	-122.080	2.899
2300	58.664	377.912	328.964	112.582	-201.383	-118.548	2.692
2400	58.896	380.414	331.056	118.460	-254.796	-113.810	2.477
2500	59.151	382.823	333.078	124.362	-258.052	-107.869	2.254
2600	59.428	385.149	335.037	130.291	-261.299	-101.797	2.045
2700	59.727	387.397	336.935	136.248	-264.535	-95.601	1.850
2800	60.044	389.575	338.776	142.237	-267.759	-89.285	1.666
2900	60.379	391.688	340.564	148.258	-270.970	-82.855	1.492
3000	60.728	393.740	342.303	154.313	-274.166	-76.313	1.329
3100	61.090	395.737	343.994	160.404	-277.346	-69.666	1.174
3200	61.460	397.683	345.642	166.531	-280.508	-62.916	1.027
3300	61.837	399.580	347.248	172.696	-283.650	-56.068	0.887
3400	62.218	401.432	348.814	178.899	-286.772	-49.124	0.755
3500	62.600	403.241	350.343	185.140	-289.870	-42.090	0.628
3600	62.980	405.009	351.837	191.419	-292.943	-34.967	0.507
3700	63.355	406.740	353.298	197.736	-295.990	-27.759	0.392
3800	63.725	408.435	354.727	204.090	-299.009	-20.469	0.281
3900	64.086	410.095	356.125	210.480	-301.997	-13.100	0.175
4000	64.436	411.722	357.495	216.907	-304.954	-5.654	0.074
4100	64.775	413.317	358.837	223.367	-307.878	1.865	-0.024
4200	65.100	414.882	360.153	229.861	-790.616	16.478	-0.205
4300	65.411	416.417	361.443	236.387	-792.383	35.716	-0.434
4400	65.706	417.924	362.710	242.943	-794.114	54.994	-0.653
4500	65.984	419.404	363.954	249.527	-795.810	74.311	-0.863
4600	66.246	420.857	365.175	256.139	-797.470	93.666	-1.064
4700	66.490	422.285	366.375	262.776	-799.094	113.055	-1.256
4800	66.716	423.687	367.554	269.437	-800.681	132.480	-1.442
4900	66.924	425.065	368.714	276.119	-802.233	151.937	-1.620
5000	67.114	426.419	369.855	282.821	-803.748	171.425	-1.791
5100	67.287	427.749	370.977	289.541	-805.228	190.943	-1.956
5200	67.442	429.058	372.081	296.277	-806.673	210.490	-2.114
5300	67.580	430.344	373.168	303.029	-808.084	230.064	-2.267
5400	67.701	431.608	374.239	309.793	-809.461	249.665	-2.415
5500	67.806	432.851	375.293	316.568	-810.806	269.291	-2.558
5600	67.896	434.074	376.332	323.354	-812.120	288.941	-2.695
5700	67.970	435.276	377.356	330.147	-813.404	308.614	-2.828
5800	68.030	436.459	378.364	336.947	-814.658	328.310	-2.957
5900	68.076	437.622	379.359	343.753	-815.885	348.027	-3.081
6000	68.109	438.767	380.340	350.562	-817.085	367.764	-3.202

PREVIOUS: June 1972 (1 atm) CURRENT: June 1972 (1 bar)

Dichloroborane, Ion (BCl_2^-) $B_1Cl_2^-(g)$

DICHLOROFLUOROBORANE (BCl_2F) IDEAL GAS M_r = 100.714403

$S°(298.15\ K) = 287.7\ J\ K^{-1}\ mol^{-1}$

$\Delta_fH°(0\ K) = -643 \pm 21\ kJ\ mol^{-1}$
$\Delta_fH°(298.15\ K) = -644 \pm 21\ kJ\ mol^{-1}$

Vibrational Frequencies and Degeneracies

ν, cm^{-1}	ν, cm^{-1}
1320 (1)	1000 (1)
[554] (1)	[339] (1)
[266] (1)	528 (1)

Ground State Quantum Weight: 1 $\sigma = 2$
Point Group: C_{2v}
Bond Distances: B-Cl = [1.73] Å B-F = [1.295] Å
Bond Angle: Cl-B-Cl = [120]° F-B-Cl = [120]°
Product of the Moments of Inertia: $I_AI_BI_C = 1.202356 \times 10^{-113}\ g^3\ cm^6$

Enthalpy of Formation

The values for BCl_2F and $BClF_2$ of $\Delta_fH°(298.15\ K) = -154$ and $\Delta_fH°(298.15\ K) = -211.6\ kcal\ mol^{-1}$, respectively, are obtained from the average bond energies of BF_3 and BCl_3. The calculated values were each made less negative by about 0.8 kcal mol^{-1} in order to make them consistent with the equilibrium measurements of Higgins, Leisegang, Raw, and Rossouw (1), and Gunn and Sanborn (2) for the reaction.

$$BF_3(g) + BCl_3(g) + BClF_2(g) + BCl_2F(g).$$

Values for the enthalpy of reaction are $\Delta_rH(298.15\ K) = +1.68\ kcal\ mol^{-1}$ and $\Delta_rH(298.15\ K) = 1.1$ (uncertainty of -0.5 to +0.8) kcal mol^{-1}, the former calculated from the measured equilibrium constant and the statistical entropy change and the latter obtained from the variation of the equilibrium constant with temperature. Selection of $\Delta_rH(298.15\ K) = 1.6\ kcal\ mol^{-1}$ fixes the sum of the enthalpies of formation of $BClF_2$ and BCl_2F as -365.6 kcal mol^{-1}, but it does not determine the individual values. Thus, the selected enthalpies of formation are only as accurate as the average bond energy calculations on which they are based. A reasonable estimate of the uncertainty would be ± 5 kcal mol^{-1}.

Heat Capacity and Entropy

The spectroscopic constants are based on the work of Lindeman and Wilson (3), who observed ν_1, ν_4, and ν_6 and calculated ν_2, ν_3, and ν_5. The frequencies were adjusted to the normal isotopic composition of 81.2% ^{11}B and 18.8% ^{10}B. The bond angles were taken to be 120° while the bond lengths were assumed to be the same as in the respective trihalides. The principal moments of inertia are $I_A = 11.8761 \times 10^{-39}$, $I_B = 26.4297 \times 10^{-39}$, and $I_C = 38.3059 \times 10^{-39}\ g\ cm^2$.

References

1. T. H. S. Higgins, E. C. Leisegang, C. J. G. Raw, and A. J. Rossouw, J. Chem. Phys. 23, 1544 (1955).
2. S. R. Gunn and R. H. Sanborn, J. Chem. Phys. 33, 955 (1960).
3. L. P. Lindeman and M. K. Wison, J. Chem. Phys. 24, 242 (1956).

Dichlorofluoroborane (BCl_2F) $B_1Cl_2F_1(g)$

Enthalpy Reference Temperature = T_r = 298.15 K Standard State Pressure = p° = 0.1 MPa

T/K	$C_p°$ (J K⁻¹mol⁻¹)	S° (J K⁻¹mol⁻¹)	-[G°-H°(T_r)]/T (J K⁻¹mol⁻¹)	H°-H°(T_r) (kJ mol⁻¹)	$\Delta_fH°$ (kJ mol⁻¹)	$\Delta_fG°$ (kJ mol⁻¹)	Log K_f
0	0.	0.	INFINITE	-13.267	-642.796	-642.796	INFINITE
100	37.983	235.579	333.920	-9.834	-643.752	-639.860	334.229
200	50.361	265.837	292.820	-5.397	-644.150	-635.776	166.048
250	55.201	277.612	288.627	-2.754	-644.248	-633.670	132.398
298.15	59.063	287.675	287.675	0.	-644.336	-631.625	110.658
300	59.198	288.041	287.677	0.109	-644.339	-631.546	109.962
350	62.535	297.425	288.410	3.155	-644.434	-629.407	93.934
400	65.335	305.963	290.078	6.354	-644.537	-627.253	81.911
450	67.688	313.799	292.285	9.681	-644.647	-625.086	72.558
500	69.666	321.036	294.803	13.117	-644.759	-622.906	65.075
600	72.738	334.024	300.282	20.245	-644.981	-618.515	53.847
700	74.942	345.412	305.933	27.635	-645.194	-614.087	45.824
800	76.552	355.529	311.512	35.214	-645.402	-609.629	39.805
900	77.751	364.618	316.916	42.932	-645.613	-605.145	35.122
1000	78.662	372.860	322.105	50.755	-645.834	-600.636	31.374
1100	79.368	380.391	327.066	58.658	-646.071	-596.105	28.307
1200	79.923	387.322	331.803	66.623	-646.330	-591.552	25.750
1300	80.367	393.737	336.323	74.638	-646.613	-586.975	23.585
1400	80.726	399.707	340.640	82.694	-646.924	-582.377	21.729
1500	81.022	405.287	344.766	90.782	-647.267	-577.754	20.119
1600	81.267	410.524	348.714	98.896	-647.643	-573.108	18.710
1700	81.472	415.457	352.496	107.034	-648.053	-568.437	17.466
1800	81.646	420.119	356.125	115.190	-648.499	-563.741	16.359
1900	81.795	424.537	359.610	123.362	-648.981	-559.020	15.369
2000	81.922	428.736	362.962	131.548	-649.501	-554.271	14.476
2100	82.033	432.736	366.190	139.746	-650.057	-549.496	13.668
2200	82.129	436.554	369.302	147.954	-650.652	-544.694	12.933
2300	82.213	440.207	372.306	156.171	-651.284	-539.864	12.261
2400	82.288	443.708	375.209	164.396	-702.212	-533.936	11.621
2500	82.353	447.068	378.017	172.629	-702.997	-526.909	11.009
2600	82.412	450.299	380.735	180.867	-703.787	-519.850	10.444
2700	82.464	453.410	383.369	189.111	-704.581	-512.760	9.920
2800	82.511	456.410	385.925	197.360	-705.380	-505.641	9.433
2900	82.554	459.307	388.406	205.613	-706.183	-498.493	8.979
3000	82.592	462.106	390.816	213.870	-706.989	-491.318	8.555
3100	82.626	464.815	393.159	222.131	-707.798	-484.115	8.157
3200	82.658	467.438	395.440	230.395	-708.607	-476.887	7.784
3300	82.687	469.982	397.660	238.663	-709.417	-469.633	7.434
3400	82.713	472.451	399.824	246.933	-710.225	-462.355	7.103
3500	82.737	474.849	401.933	255.205	-711.029	-455.053	6.791
3600	82.759	477.180	403.991	263.480	-711.827	-447.728	6.496
3700	82.779	479.448	406.000	271.757	-712.617	-440.381	6.217
3800	82.798	481.656	407.962	280.036	-713.398	-433.013	5.952
3900	82.815	483.807	409.880	288.316	-714.166	-425.624	5.701
4000	82.832	485.904	411.754	296.599	-714.919	-418.216	5.461
4100	82.846	487.949	413.588	304.883	-715.655	-410.789	5.234
4200	82.860	489.946	415.382	313.168	-1196.221	-396.321	4.929
4300	82.873	491.896	417.139	321.455	-1195.828	-377.280	4.583
4400	82.885	493.801	418.860	329.743	-1195.413	-358.249	4.253
4500	82.897	495.664	420.546	338.032	-1194.975	-339.227	3.938
4600	82.907	497.486	422.199	346.322	-1194.510	-320.216	3.636
4700	82.917	499.269	423.820	354.613	-1194.018	-301.215	3.348
4800	82.927	501.015	425.410	362.905	-1193.497	-282.224	3.071
4900	82.935	502.725	426.970	371.198	-1192.946	-263.245	2.806
5000	82.944	504.401	428.502	379.492	-1192.364	-244.278	2.552
5100	82.951	506.043	430.006	387.787	-1191.750	-225.322	2.308
5200	82.959	507.654	431.484	396.083	-1191.103	-206.378	2.073
5300	82.965	509.234	432.936	404.379	-1190.423	-187.448	1.847
5400	82.972	510.785	434.364	412.676	-1189.709	-168.530	1.630
5500	82.978	512.308	435.767	420.973	-1188.961	-149.626	1.421
5600	82.984	513.803	437.147	429.271	-1188.178	-130.736	1.219
5700	82.990	515.272	438.505	437.570	-1187.361	-111.861	1.025
5800	82.995	516.715	439.841	445.869	-1186.510	-93.000	0.838
5900	83.000	518.134	441.156	454.169	-1185.625	-74.153	0.657
6000	83.005	519.529	442.451	462.469	-1184.707	-55.323	0.482

PREVIOUS: December 1964 (1 atm) CURRENT: December 1964 (1 bar)

Dichlorofluoroborane (BCl_2F) $B_1Cl_2F_1(g)$

DICHLOROBORANE ($BHCl_2$) IDEAL GAS M_r = 82.72394

$S°(298.15 K) = [268.3] J K^{-1} mol^{-1}$

$\Delta_f H°(0 K) = -245 \pm 4 kJ mol^{-1}$
$\Delta_f H°(298.15 K) = -248 \pm 4 kJ mol^{-1}$

Vibrational Frequencies and Degeneracies

ν, cm^{-1}	ν, cm^{-1}
[2610] (1)	1091 (1)
744 (1)	[894] (1)
[287] (1)	786 (1)

Ground State Quantum Weight: 1 $\sigma = 2$
Point Group: C_{2v}
Bond Distances: B-H = 1.13 ± 0.2 Å B-Cl = [1.75] Å
Bond Angles: Cl-B-Cl = 119.7 ± 3° Cl-B-H = 120.15 ± 1.5°
Product of the Moments of Inertia: $I_A I_B I_C = 1.392331 \times 10^{-114} g^3 cm^6$

Enthalpy of Formation

The chemical equilibria for the following two reactions: (A) $BCl_3(g) + H_2(g) = BHCl_2(g) + HCl(g)$, and (B) $6BHCl_2(g) = B_2H_6(g) + 4BCl_3(l, g)$ were investigated by Murib, Horvitz, and Bonecutter (1), Mook, Seiler, and Watson (2) and Lynds and Bass (3). Using the reported equilibrium constants, the corresponding enthalpy changes and the respective $\Delta_f H°(298.15 K)$ values were evaluated. The results obtained are presented as follows.

Source	Reaction	T/K	$\Delta_r H°(298.15 K)$, kcal mol^{-1} 2nd law	3rd law	$\Delta_f H°(298.15 K)$ kcal mol^{-1}
(1)	A	873 - 973	15.57 ± 0.8	16.19	-58.06 ± 0.5
(2)	A	1073 - 1273	13.50 ± 3.0	13.91	-60.3 ± 0.7
(1)	B	273	---	-17.85	-59.6
(3)	B	298	---	-19.09	-59.4

The value of $\Delta_f H°(298.15 K)$ for $BHCl_2(g)$ is selected as -59.3 ± 1 kcal mol^{-1}.

Heat Capacity and Entropy

The molecular structure, B-H bond distance and Cl-B-Cl bond angle were obtained from Lynds and Bass (4). The B-Cl bond distance was assumed to be the same as that in $BCl_3(g)$. The vibrational frequencies adopted were reported by Bass, Lynds, Wolfram, DeWames (5) and corrected to the average isotopic species. It has been pointed out that the complete assignment of the infrared spectra of $HBCl_2(g)$ is quite difficult due to the unavoidable presence of $BCl_3(g)$. The adopted frequencies were obtained using the Green's function and partitioning techniques, taking $HBCl_2$ as an "isotopic" species of BCl_3. An excellent agreement between the calculated and observed spectra reported Lynds and Bass (5) was claimed. Spectra of the mixture of BCl_3, $HBCl_2$ and $DBCl_2$ are quite complex and have been observed by Lynds and Stern (6), Nadean and Oaks (7), Mayers and Putnam (8) and only three bands were correlated with $HBCl_2$. Analysis of infrared data from previous works, according to the conventional Wilson's method of symmetry coordinates, has been undertaken by de Mandirola and Westerkamp (9). Based on four observed frequencies, a set of six frequencies was derived and assigned. The principal moments of inertia are: $I_A = 1.7957 \times 10^{-39}$, $I_B = 26.9625 \times 10^{-39}$ and $I_C = 28.7581 \times 10^{-39} g cm^2$.

References

1. J. H. Murib, D. Horvitz and C. A. Bonecutter, Ind. Eng. Chem. Prod. Res. Devel. 4, 273 (1965).
2. P. H. Mook, G. R. Seiler, and E. R. Watson, "Process Development, Prepilot and Pilot Plant Investigation - High Energy Boron Fuels", ASDTR 62-7-558A, Vol. IV, March, 1962.
3. L. Lynds and C. D. Bass, Inorg. Chem. 3, 1147 (1964).
4. L. Lynds and C. D. Bass, J. Chem. Phys. 40, 1590 (1964).
5. C. D. Bass, L. Lynds, T. Wolfram, and R. E. DeWames, J. Chem. Phys. 40, 3611 (1964).
6. L. Lynds and D. R. Stern, J. Am. Chem. Soc. 81, 5006 (1959).
7. H. G. Nadean and D. M. Oaks, Jr., Anal. Chem. 32, 1481 (1961).
8. H. W. Mayers and R. F. Putnam, Inorg. Chem. 2, 655 (1963).
9. O. B. de Mandirola and J. F. Westerkamp, Spectrochim. Acta 20, 1633 (1964).

Dichloroborane ($BHCl_2$) $B_1Cl_2H_1(g)$

Enthalpy Reference Temperature = T_r = 298.15 K Standard State Pressure = p° = 0.1 MPa

T/K	$C_p°$ (J $K^{-1}mol^{-1}$)	S°	$-[G°-H°(T_r)]/T$	$H°-H°(T_r)$ (kJ mol^{-1})	$\Delta_f H°$	$\Delta_f G°$	Log K_f
0	0.	0.	INFINITE	-11.744	-245.227	-245.227	INFINITE
100	35.653	223.535	307.156	-8.362	-246.280	-244.666	127.800
200	41.988	250.084	272.565	-4.496	-247.132	-242.731	63.395
250	45.882	259.867	269.069	-2.301	-247.624	-241.574	50.474
298.15	49.644	268.272	268.272	0.	-248.111	-240.365	42.111
300	49.784	268.580	268.273	0.092	-248.130	-240.317	41.843
350	53.357	276.528	268.893	2.672	-248.620	-238.976	35.665
400	56.480	283.862	270.311	5.420	-249.087	-237.566	31.023
450	59.157	290.673	272.200	8.313	-249.528	-236.099	27.406
500	61.447	297.027	274.369	11.329	-249.941	-234.584	24.507
600	65.120	308.571	279.127	17.666	-250.682	-231.442	20.149
700	67.933	318.829	284.080	24.325	-251.320	-228.183	17.027
800	70.157	328.051	289.009	31.233	-251.870	-224.839	14.680
900	71.956	336.422	293.820	38.342	-252.351	-221.431	12.852
1000	73.433	344.082	298.468	45.614	-252.780	-217.972	11.386
1100	74.657	351.140	302.940	53.020	-253.173	-214.472	10.184
1200	75.681	357.682	307.233	60.539	-253.544	-210.937	9.182
1300	76.543	363.774	311.350	68.151	-253.906	-207.372	8.332
1400	77.274	369.474	315.301	75.843	-254.268	-203.779	7.603
1500	77.896	374.827	319.092	83.602	-254.638	-200.160	6.970
1600	78.430	379.872	322.735	91.419	-255.022	-196.515	6.416
1700	78.890	384.641	326.238	99.286	-255.427	-192.846	5.925
1800	79.289	389.162	329.609	107.195	-255.857	-189.153	5.489
1900	79.636	393.458	332.857	115.142	-256.316	-185.435	5.098
2000	79.941	397.551	335.990	123.121	-256.806	-181.692	4.745
2100	80.208	401.458	339.016	131.129	-257.332	-177.923	4.426
2200	80.445	405.195	341.939	139.162	-257.897	-174.129	4.134
2300	80.654	408.775	344.768	147.217	-258.502	-170.308	3.868
2400	80.841	412.212	347.507	155.292	-309.410	-165.391	3.600
2500	81.009	415.516	350.162	163.384	-310.182	-159.374	3.330
2600	81.159	418.696	352.737	171.493	-310.969	-153.326	3.080
2700	81.294	421.761	355.237	179.616	-311.774	-147.248	2.849
2800	81.416	424.720	357.666	187.751	-312.597	-141.139	2.633
2900	81.526	427.579	360.028	195.898	-313.440	-135.001	2.432
3000	81.627	430.345	362.326	204.056	-314.303	-128.833	2.243
3100	81.718	433.023	364.563	212.223	-315.188	-122.636	2.066
3200	81.802	435.618	366.744	220.399	-316.094	-116.411	1.900
3300	81.878	438.137	368.869	228.584	-317.021	-110.156	1.744
3400	81.949	440.582	370.942	236.775	-317.967	-103.874	1.596
3500	82.013	442.959	372.966	244.973	-318.932	-97.563	1.456
3600	82.073	445.270	374.943	253.177	-319.914	-91.224	1.324
3700	82.128	447.519	376.874	261.388	-320.911	-84.858	1.198
3800	82.179	449.710	378.762	269.603	-321.921	-78.464	1.079
3900	82.226	451.845	380.609	277.823	-322.943	-72.044	0.965
4000	82.270	453.928	382.416	286.048	-323.974	-65.598	0.857
4100	82.311	455.960	384.185	294.277	-325.011	-59.126	0.753
4200	82.349	457.944	385.917	302.510	-805.902	-45.604	0.567
4300	82.385	459.882	387.615	310.747	-805.857	-27.502	0.334
4400	82.418	461.776	389.279	318.987	-805.813	-9.401	0.112
4500	82.449	463.629	390.911	327.231	-805.769	8.698	-0.101
4600	82.478	465.441	392.511	335.477	-805.720	26.797	-0.304
4700	82.506	467.215	394.082	343.726	-805.667	44.895	-0.499
4800	82.532	468.953	395.624	351.978	-805.606	62.991	-0.685
4900	82.556	470.655	397.138	360.232	-805.537	81.086	-0.864
5000	82.579	472.323	398.625	368.489	-805.456	99.180	-1.036
5100	82.600	473.958	400.086	376.748	-805.364	117.272	-1.201
5200	82.621	475.562	401.522	385.009	-805.258	135.361	-1.360
5300	82.640	477.136	402.934	393.272	-805.138	153.449	-1.512
5400	82.658	478.681	404.322	401.537	-805.002	171.535	-1.659
5500	82.675	480.198	405.688	409.804	-804.849	189.617	-1.801
5600	82.691	481.688	407.032	418.072	-804.679	207.697	-1.937
5700	82.707	483.152	408.355	426.342	-804.491	225.773	-2.069
5800	82.722	484.590	409.657	434.613	-804.283	243.847	-2.196
5900	82.736	486.004	410.939	442.886	-804.056	261.916	-2.319
6000	82.749	487.395	412.202	451.161	-803.809	279.981	-2.437

PREVIOUS: March 1966 (1 atm) CURRENT: March 1966 (1 bar)

Dichloroborane ($BHCl_2$) $B_1Cl_2H_1(g)$

TRICHLOROBORANE (BCl_3) IDEAL GAS M_r = 117.169

S°(298.15 K) = 290.17 J K^{-1} mol^{-1}

$\Delta_f H°$(0 K) = -402.0 ± 2.1 kJ mol^{-1}
$\Delta_f H°$(298.15 K) = -403.0 ± 2.1 kJ mol^{-1}

Vibrational Frequencies and Degeneracies

ν, cm^{-1}	ν, cm^{-1}
471.0 (1)	986.3 (2)
470.6 (1)	243.0 (1)

Ground State Quantum Weight: 1
Point Group: D_{3h} σ = 6
Bond Distance: B-Cl = 1.75 Å
Bond Angle: Cl-B-Cl = 120°
Product of the Moments of Inertia: $I_A I_B I_C = 3.956022 \times 10^{-113}$ g^3 cm^6

Enthalpy of Formation

The enthalpy change of the reaction B(am) + 3/2 Cl_2(g) = BCl_3(g) has been determined in a calorimeter by Johnson, Miller and Prosen (1). From the reported value $\Delta_r H°$(298.15 K) = -97.51 + 0.32 kcal mol^{-1}, the enthalpy of formation for BCl_3(g) was derived as -96.31 ± 0.50 kcal mol^{-1}, using $\Delta_r H°$(298.15 K) = 1.20 ± 0.40 kcal mol^{-1} for the reaction B(cr) = B(am). Some other related enthalpies of reaction reported by previous investigators are presented as follows.

Source	Reaction	$\Delta_r H°$(298.15 K) kcal mol^{-1}	$\Delta_f H°$(298.15 K) kcal mol^{-1}
(2)	B(am) + 3/2 Cl_2(g) = BCl_3(g)	-97.4 ± 0.7	-96.2 ± 0.8
	B(am) + 3/2 Cl_2(g) = BCl_3(l)	-103.3 ± 0.6	-96.5 ± 0.7
(3)	B_2H_6(g) + 6 Cl_2(g) = 2 BCl_3(g) + 6 HCl(g)	-342.9	-100.4 ± 5.0
(4)	BCl_3(l) + (n+3)H_2O(l) = (H_3BO_3+3HCl)·n H_2O (Sln)	-69.2 ± 1.0	-97.2 ± 1.0
(5)	BCl_3(l) + 2403 H_2O(l) = (H_3BO_3+3HCl)·2400 H_2O (Sln)	-68.68 ± 0.06	-96.3 ± 0.2
(6)	BCl_3(l) + 593 H_2O(l) = (H_3BO_3+3HCl)·590 H_2O (Sln)	-69.12 ± 0.11	-95.44 ± 0.39

The enthalpies of formation for BCl_3(l) derived from the enthalpies of hydrolysis reported by other investigatoars were reviewed by Johnson, Miller and Prosen (1).

Heat Capacity and Entropy

The bond distance of B and Cl atoms has been reported to be 1.73 ± 0.02 and 1.76 ± 0.02, 1.75 Å by Levy and Brockway (7), Gregg, Hampson and Jenkins, Jones, and Sutton (8), and Spencer and Lipscomb (9), respectively. The last value is adopted. The values of vibrational frequencies used were reported by Lindeman and Wilson (10). The vibrational frequencies were also determined by Cassie (11), Lassettre and Yost (12), Scruby, Lacher and Park (13). The molecular structure was given by Levy amd Brockway (7). The principal moments of inertia are: $I_A = I_B = 27.0443 \times 10^{-39}$ and $I_C = 54.0887 \times 10^{-39}$ g cm^2.

References

1. W. H. Johnson, R. G. Miller, and E. J. Prosen, J. Res. Nat. Bur. Stand. 62, 213 (1959).
2. G. L. Gal'chenko, B. I. Timofeev and S. M. Skuratov, Russ. J. Inorg. Chem. 5, 1279 (1960). $\Delta_f H°$(298.15 K) were derived from $\Delta_r H°$(298.15 K) using $\Delta_r H°$ = -1.2 ± 0.4 kcal mol^{-1} for B(am) = B(cr) and $\Delta_{vap}H°$(BCl_3, l, 298.15 K) = 5.6 kcal mol^{-1}.
3. J. R. Lacher, R. E. Scruby and J. D. Park, J. Am. Chem. Soc 74, 5292 (1952). The $\Delta_f H°$(298.15 K) value was calculated based on $\Delta_f H°$(298.15 K) = 9.80 kcal mol^{-1} for B_2H_6(g).
4. H. A. Skinner and N. B. Smith, Trans, Faraday Soc. 49, 601 (1953), where n = 4978, average value of seven sets of measurement.
5. S. R. Gunn and L. G. Green, J. Phys. Chem. 64, 61 (1960).
6. L. G. Fasolino, "Heat of Formation of Boron Trichloride", NONR-3608(00), Natl Res. Corp., Massachusettts, July 15, 1964.
7. H. A. Levy and L. O. Brockway, J. Am. Chem. Soc. 59, 2085 (1937).
8. A. H. Gregg, G. C. Hampson, G. I. Jenkins, P. L. F. Jones, and L. E. Sutton, Trans. Faraday Soc. 33, 852 (1937).
9. C. Spencer and W. N. Lipscomb, J. Chem. Phys. 28, 355 (1958).
10. L. P. Lindeman nd M. K. Wilson, J. Chem. Phys. 24, 242 (1956).
11. A. B. D. Cassie, Proc. Roy. Soc. A148, 87 (1935).
12. T. F. Anderson, E. N. Lassettre and D. M. Yost, J. Chem. Phys. 4, 703 (1936).
13. R. E. Scruby, J. R. Lacher, and J. D. Park, J. Chem. Phys. 19, 386 (1951).

Trichloroborane (BCl_3) B_1Cl_3(g)

Enthalpy Reference Temperature = T_r = 298.15 K   Standard State Pressure = p° = 0.1 MPa

T/K	$C_p°$ (J $K^{-1}mol^{-1}$)	S° (J $K^{-1}mol^{-1}$)	-[G°-H°(T_r)]/T (J $K^{-1}mol^{-1}$)	H°-H°(T_r) (kJ mol^{-1})	$\Delta_f H°$ (kJ mol^{-1})	$\Delta_f G°$ (kJ mol^{-1})	Log K_f
0	0.	0.	INFINITE	-14.021	-401.997	-401.997	INFINITE
100	40.687	234.433	339.441	-10.501	-402.868	-397.930	207.857
200	53.832	266.971	295.623	-5.730	-402.987	-392.909	102.617
250	58.598	279.513	291.176	-2.916	-402.972	-390.391	81.568
298.15	62.392	290.169	290.169	0.	-402.961	-387.969	67.971
300	62.524	290.555	290.170	0.116	-402.961	-387.876	67.535
350	65.763	300.445	290.944	3.325	-402.956	-385.362	57.512
400	68.421	309.406	292.700	6.682	-402.962	-382.849	49.995
450	70.595	317.594	295.018	10.159	-402.976	-380.334	44.148
500	72.374	325.127	297.657	13.735	-402.997	-377.817	39.470
600	75.037	338.574	303.383	21.114	-403.047	-372.776	32.453
700	76.872	350.287	309.265	28.715	-403.109	-367.726	27.440
800	78.172	360.641	315.053	36.471	-403.185	-362.667	23.680
900	79.118	369.906	320.642	44.338	-403.281	-357.596	20.754
1000	79.824	378.280	325.994	52.286	-403.399	-352.514	18.413
1100	80.363	385.915	331.099	60.297	-403.544	-347.419	16.498
1200	80.783	392.926	335.963	68.355	-403.717	-342.309	14.900
1300	81.116	399.406	340.597	76.451	-403.922	-337.184	13.548
1400	81.384	405.427	345.016	84.576	-404.159	-332.041	12.389
1500	81.603	411.050	349.233	92.726	-404.430	-326.881	11.383
1600	81.783	416.322	353.263	100.895	-404.736	-321.701	10.502
1700	81.934	421.285	357.119	109.081	-405.079	-316.501	9.725
1800	82.062	425.972	360.816	117.281	-405.460	-311.280	9.033
1900	82.170	430.412	364.363	125.493	-405.881	-306.036	8.414
2000	82.263	434.629	367.771	133.715	-406.343	-300.770	7.855
2100	82.344	438.645	371.052	141.945	-406.848	-295.479	7.350
2200	82.414	442.477	374.212	150.183	-407.398	-290.163	6.889
2300	82.475	446.142	377.260	158.428	-407.995	-284.821	6.468
2400	82.529	449.653	380.204	166.678	-458.902	-278.383	6.059
2500	82.576	453.023	383.050	174.933	-459.679	-270.845	5.659
2600	82.619	456.262	385.804	183.193	-460.477	-263.276	5.289
2700	82.656	459.381	388.471	191.457	-461.298	-255.676	4.946
2800	82.690	462.388	391.058	199.724	-462.144	-248.045	4.627
2900	82.721	465.290	393.568	207.995	-463.016	-240.383	4.330
3000	82.748	468.095	396.005	216.268	-463.914	-232.691	4.052
3100	82.773	470.809	398.375	224.544	-464.839	-224.968	3.791
3200	82.796	473.437	400.680	232.823	-465.789	-217.216	3.546
3300	82.816	475.985	402.923	241.104	-466.764	-209.432	3.315
3400	82.835	478.458	405.109	249.386	-467.762	-201.620	3.098
3500	82.853	480.859	407.239	257.671	-468.781	-193.777	2.892
3600	82.868	483.193	409.317	265.957	-469.818	-185.905	2.697
3700	82.883	485.464	411.344	274.244	-470.871	-178.004	2.513
3800	82.896	487.675	413.324	282.533	-471.935	-170.075	2.338
3900	82.909	489.828	415.258	290.823	-473.008	-162.117	2.171
4000	82.921	491.927	417.149	299.115	-474.086	-154.132	2.013
4100	82.931	493.975	418.997	307.408	-475.164	-146.120	1.862
4200	82.941	495.973	420.807	315.701	-956.089	-131.057	1.630
4300	82.950	497.925	422.577	323.996	-956.071	-111.414	1.353
4400	82.959	499.832	424.312	332.291	-956.044	-91.771	1.089
4500	82.967	501.697	426.011	340.588	-956.005	-72.129	0.837
4600	82.975	503.520	427.676	348.885	-955.950	-52.487	0.596
4700	82.982	505.305	429.309	357.182	-955.878	-32.848	0.365
4800	82.989	507.052	430.910	365.481	-955.783	-13.210	0.144
4900	82.995	508.763	432.482	373.780	-955.665	6.426	-0.069
5000	83.001	510.440	434.024	382.080	-955.521	26.059	-0.272
5100	83.006	512.084	435.539	390.380	-955.348	45.689	-0.468
5200	83.012	513.696	437.026	398.681	-955.145	65.315	-0.656
5300	83.016	515.277	438.488	406.983	-954.909	84.937	-0.837
5400	83.021	516.829	439.924	415.284	-954.641	104.554	-1.011
5500	83.026	518.352	441.336	423.587	-954.337	124.167	-1.179
5600	83.030	519.848	442.725	431.890	-953.998	143.773	-1.341
5700	83.034	521.318	444.091	440.193	-953.623	163.372	-1.497
5800	83.037	522.762	445.435	448.496	-953.210	182.966	-1.648
5900	83.041	524.181	446.758	456.800	-952.760	202.551	-1.793
6000	83.044	525.577	448.060	465.105	-952.273	222.128	-1.934

PREVIOUS: December 1964 (1 atm)   CURRENT: December 1964 (1 bar)

Trichloroborane (BCl_3) B_1Cl_3(g)

FLUOROBORANE (BF) IDEAL GAS M_r = 29.808403

$\Delta_f H°$(0 K) = -119.0 ± 13.8 kJ mol^{-1}

S°(298.15 K) = 200.479 J K^{-1} mol^{-1} $\Delta_f H°$(298.15 K) = -115.9 ± 13.8 kJ mol^{-1}

Electronic Level and Quantum Weight

State	ε_i, cm^{-1}	g_i
$^1\Sigma^+$	0	1

ω_e = 1410.3 cm^{-1} $\omega_e x_e$ = 11.98 cm^{-1} σ = 1

B_e = 1.5286 cm^{-1} α_e = 0.0168 cm^{-1} r_e = 1.265 Å

Enthalpy of Formation

The equilibria of the reaction 2/3 B(cr) + 1/3 BF_3(g) = BF(g) were studied by Blauer, Greenbaum and Farber (1) by means of transpiratio at pressures below 300 μ. The values of log $K_p°$ at 1375 and 1476 K were reported as -3.99 ± 0.02 and -3.30 ± 0.05, respectively. By the 3rd law method the value of $\Delta_r H°$(298.15 K) for the reaction was derived to be 61.38 kcal mol^{-1}, yielding $\Delta_f H°$(BF, g, 298.15 K) = -28.7 ± 2.6 kcal mol^{-1}. The system B + CaF_2 was studied mass spectrometrically in a graphite Knudsen cell by Hildenbrand (6). The equilibria involving BF(g) were observed. Intensity measurements were made at temperatures, 1583-1734 K. From the 3rd law value, $\Delta_r H°$(298.15 K) = 12.2 + 0.3 kcal mol^{-1}, for the reaction BF_3(g) + Ca(g) = BF(g) + CaF_2(g), the value of Δ_fHd(298.15 K) for BF(g) was derived as -26.7 ± 2 kcal mol^{-1}. The dissociation energy of BF(g) molecule has been reported by Herzberg (2), Gaydon (3) and Barrow (4). The corresponding values of $\Delta_f H°$(298.15 K) for BF(g) were derived. The results are compared with those obtained from equilibrium data as follows:

Source	Reaction	$\Delta_r H°$(298.15 K) kcal mol^{-1}	$\Delta_f H°$(298.15 K) kcal mol^{-1}
(1)	2/3 B(cr) + 1/3 BF_3(g) = BF(g)	61.38	-28.7 ± 2.6
(2)	BF(g) = B(g) + F(g)	100.2	+51.5
(3)	BF(g) = B(g) + F(g)	197.0	-45.3 ± 11.5
(4)	BF(g) = B(g) + F(g)	186.0	-34.3
(5)	2/3 B(cr) + 1/3 BF_3 = BF(g)	--	-28 ± 3
(6)	BF_3(g) + Ca(g) = BF(g) + CaF_2(g)	12.2	-26.7 ± 2

The value of $\Delta_f H°$(298.15 K) adopted is the average of the first and last $\Delta_f H°$(298.15 K) values listed in the above table.

Heat Capacity and Entropy

The spectroscopic constants used were obtained from Onaka (7) and corrected to the average isotopic species. Spectroscopic constants for BF(g) were also reported by Herzberg (2).

References

1. J. Blauer, M. A. Greenbaum and M. Farber, J. Phys. Chem. 68, 2332 (1964).
2. G. Herzberg, "Spectra of Diatomic Molecules", D. Van Nostrand Company, New York, 1950.
3. A. G. Gaydon, Dissociation Energies", Chapman and Hall Ltd., London, 1953.
4. R. F. Barrow, Trans. Faraday Soc. 56, 952 (1960).
5. P. Gross, C. Hayman, D. L. Levi and R. H. Lewin, "Stability of Lower Boron Halides", Final Technical Report, August 22, 1963, Fulmer Research Institute Limited, England.
6. D. L. Hildenbrand, "Thermodynamic Properties of Propellant Combustion Products", QLR-65-3, Philco Corporation, California.
7. R. Onaka, J. Chem. Phys. 27, 374 (1957).

Fluoroborane (BF) B_1F_1(g)

Enthalpy Reference Temperature = T_r = 298.15 K Standard State Pressure = p° = 0.1 MPa

T/K	$C_p°$ (J K^{-1}mol^{-1})	S° (J K^{-1}mol^{-1})	-[G°-H°(T_r)]/T (J K^{-1}mol^{-1})	H°-H°(T_r) (kJ mol^{-1})	$\Delta_f H°$ (kJ mol^{-1})	$\Delta_f G°$ (kJ mol^{-1})	Log K_f
0	0.	0.	INFINITE	-8.694	-118.965	-118.965	INFINITE
100	29.108	168.594	226.494	-5.790	-117.538	-125.848	65.736
200	29.154	188.777	203.167	-2.878	-116.423	-134.630	35.162
250	29.302	195.296	200.965	-1.417	-116.085	-139.224	29.089
298.15	29.591	200.479	200.479	0.	-115.897	-143.699	25.176
300	29.605	200.662	200.480	0.055	-115.892	-143.872	25.050
350	30.047	205.258	200.842	1.546	-115.823	-148.542	22.169
400	30.579	209.304	201.652	3.061	-115.858	-153.215	20.008
450	31.149	212.939	202.707	4.604	-115.976	-157.879	18.326
500	31.719	216.250	203.898	6.176	-116.157	-162.525	16.979
600	32.768	222.128	206.459	9.401	-116.649	-171.755	14.953
700	33.646	227.247	209.071	12.724	-117.258	-180.893	13.498
800	34.356	231.788	211.632	16.125	-117.944	-189.937	12.402
900	34.926	235.869	214.102	19.590	-118.688	-198.892	11.543
1000	35.385	239.573	216.467	23.107	-119.479	-207.762	10.852
1100	35.757	242.964	218.724	26.664	-120.311	-216.550	10.283
1200	36.064	246.089	220.876	30.256	-121.181	-225.261	9.805
1300	36.319	248.986	222.928	33.875	-122.088	-233.898	9.398
1400	36.535	251.686	224.887	37.518	-123.030	-242.464	9.046
1500	36.719	254.213	226.759	41.181	-124.006	-250.961	8.739
1600	36.879	256.588	228.549	44.861	-125.017	-259.392	8.468
1700	37.018	258.828	230.265	48.556	-126.061	-267.758	8.227
1800	37.141	260.947	231.911	52.264	-127.137	-276.062	8.011
1900	37.252	262.958	233.493	55.984	-128.245	-284.306	7.816
2000	37.351	264.872	235.014	59.714	-129.383	-292.490	7.639
2100	37.441	266.696	236.480	63.454	-130.549	-300.617	7.477
2200	37.524	268.440	237.893	67.202	-131.743	-308.688	7.329
2300	37.601	270.110	239.258	70.959	-132.961	-316.704	7.193
2400	37.672	271.712	240.577	74.722	-184.463	-323.597	7.043
2500	37.739	273.251	241.854	78.493	-185.804	-329.367	6.882
2600	37.801	274.732	243.090	82.270	-187.134	-335.083	6.732
2700	37.861	276.160	244.288	86.053	-188.449	-340.748	6.592
2800	37.917	277.538	245.451	89.842	-189.749	-346.365	6.462
2900	37.971	278.869	246.581	93.636	-191.032	-351.936	6.339
3000	38.023	280.157	247.679	97.436	-192.298	-357.463	6.224
3100	38.072	281.405	248.747	101.241	-193.545	-362.948	6.116
3200	38.120	282.615	249.786	105.051	-194.774	-368.393	6.013
3300	38.167	283.788	250.799	108.865	-195.982	-373.799	5.917
3400	38.212	284.928	251.786	112.684	-197.170	-379.170	5.825
3500	38.256	286.037	252.749	116.507	-198.338	-384.506	5.738
3600	38.299	287.115	253.689	120.335	-199.484	-389.808	5.656
3700	38.341	288.165	254.606	124.167	-200.610	-395.080	5.578
3800	38.382	289.188	255.503	128.003	-201.714	-400.321	5.503
3900	38.422	290.185	256.379	131.843	-202.797	-405.533	5.432
4000	38.461	291.159	257.237	135.687	-203.859	-410.718	5.363
4100	38.500	292.109	258.076	139.536	-204.900	-415.877	5.298
4200	38.539	293.037	258.897	143.388	-685.770	-413.986	5.149
4300	38.576	293.944	259.702	147.243	-685.683	-407.516	4.950
4400	38.614	294.832	260.490	151.103	-685.577	-401.048	4.761
4500	38.651	295.700	261.263	154.966	-685.454	-394.583	4.580
4600	38.687	296.550	262.021	158.833	-685.313	-388.120	4.407
4700	38.723	297.382	262.764	162.703	-685.156	-381.661	4.242
4800	38.759	298.198	263.494	166.578	-684.984	-375.206	4.083
4900	38.794	298.997	264.210	170.455	-684.796	-368.754	3.931
5000	38.830	299.781	264.914	174.336	-684.594	-362.306	3.785
5100	38.865	300.551	265.605	178.221	-684.378	-355.863	3.645
5200	38.899	301.306	266.285	182.109	-684.149	-349.423	3.510
5300	38.934	302.047	266.952	186.001	-683.908	-342.989	3.380
5400	38.968	302.775	267.609	189.896	-683.656	-336.558	3.256
5500	39.002	303.490	268.255	193.795	-683.393	-330.133	3.135
5600	39.036	304.193	268.890	197.696	-683.120	-323.713	3.019
5700	39.069	304.885	269.516	201.602	-682.837	-317.297	2.908
5800	39.103	305.564	270.132	205.510	-682.547	-310.887	2.800
5900	39.136	306.233	270.738	209.422	-682.248	-304.481	2.696
6000	39.169	306.891	271.335	213.338	-681.943	-298.081	2.595

PREVIOUS: December 1964 (1 atm) CURRENT: December 1964 (1 bar)

Fluoroborane (BF) B_1F_1(g)

BORON FLUORIDE OXIDE (OBF) IDEAL GAS M_r = 45.807803

$S°(298.15\ K) = [224.8]\ J\ K^{-1}\ mol^{-1}$

$\Delta_f H°(0\ K) = -602 \pm 13\ kJ\ mol^{-1}$
$\Delta_f H°(298.15\ K) = -602 \pm 13\ kJ\ mol^{-1}$

Vibrational Frequencies and Degeneracies

ν, cm^{-1}
[1050](1)
[500](2)
[1900](1)

Ground State Quantum Weight: [1] $\sigma = 1$
Point Group: $[C_{\infty v}]$
Bond Distances: B–F = [1.30] Å B–O = [1.20] Å
Bond Angle: O–B–F = [180]°
Rotational Constant: B_0 = [0.309392] cm^{-1}

Enthalpy of Formation

The equilibrium constants for the reaction $B_2O_3(\ell) + BF_3(g) = 3\ BOF(g)$ in the temperature range 1054–1253 K were determined by Farber and Blauer (1) using the so-called Molecular Flow Reaction Method. The value of $\Delta_f H°$ for this reaction was evaluated by both the 2nd and 3rd law methods, using the equilibrium constants corrected for the trimer and correcting $K_p°$ at 1200 K from 9.6×10^{-11} to 9.6×10^{-10}. Hildebrand, Theard and Saul (2) analyzed the vapor above mixture of MgF_2 and B_2O_3 contained in a Knudsen cell with a mass spectrometer and found that in the temperature range 980–1230 K the vapor is composed of $BF_3(g)$, $(BOF)_3(g)$ and BOF(g) species. From the reported temperature dependence of appropriate ion intensities the enthalpy change of the reaction $(BOF)_3(g) = 3\ BOF(g)$ was derived by the 2nd law method as 127.9 kcal mol^{-1} at 1115 K or 129.75 kcal mol^{-1} at 298.15 K. The results obtained are presented as follows, using JANAF free energy functions and auxiliary data.

Source	Reaction	$\Delta_r H°(298.15\ K)$, kcal mol^{-1} 2nd law	3rd law	$\Delta_f H°(298.15\ K)$ kcal mol^{-1}
Farber and Blauer	$B_2O_3(\ell) + BF_3(g) = 3\ BOF(g)$	150.2±16.6	149.0	-140.1±6*
Hildenbrand et al.	$(BOF)_3(g) = 3\ BOF(g)$	129.8 ± 3	--	-145.2 ± 1.5

*Calculation based on the 3rd law value and one point rejected due to statistical test.

The adopted value of $\Delta_f H°(298.15\ K)$ for BOF(g) is the weighted average of the two values listed above.

Heat Capacity and Entropy

The molecular structure is assumed to be linear. The vibrational frequencies (ν) are estimated from the ν values calculated by the valence force method described by Herzberg (3) using estimated force constants, $k_1 = 10 \times 10^5$ dynes cm^{-1} and $k_2 = 0.47 \times 10^{-11}$ dynes cm rad^{-1}. The B=O and B–F bond distances are assumed to be the same as those in BO(g) and $BF_3(g)$ molecules.

References

1. M. Farber and J. Blauer, Trans. Faraday Soc. 58, 2090 (1962).
2. D. L. Hildenbrand, L. P. Theard, and A. M. Saul, J. Chem. Phys. 39, 1973 (1963).
3. G. Herzberg, "Infrared and Raman Spectra", D. Van Nostrand Company, Inc., New York, 1945.

Boron Fluoride Oxide (OBF) $B_1F_1O_1(g)$

Enthalpy Reference Temperature = T_r = 298.15 K Standard State Pressure = $p°$ = 0.1 MPa

T/K	$C_p°$ (J $K^{-1}mol^{-1}$)	S°	$-[G°-H°(T_r)]/T$	$H°-H°(T_r)$ (kJ mol^{-1})	$\Delta_f H°$	$\Delta_f G°$	Log K_f
0	0.	0.	INFINITE	-9.934	-602.462	-602.462	INFINITE
100	29.748	187.280	257.440	-7.016	-602.474	-603.987	315.491
200	35.585	209.533	228.390	-3.771	-602.482	-605.491	158.138
250	38.542	217.799	225.465	-1.916	-602.478	-606.244	126.668
298.15	40.996	224.803	224.803	0.	-602.496	-606.969	106.338
300	41.083	225.057	224.804	0.076	-602.497	-606.996	105.687
350	43.261	231.558	225.312	2.186	-602.547	-607.743	90.701
400	45.151	237.461	226.468	4.397	-602.634	-608.479	79.459
450	46.811	242.877	227.994	6.697	-602.754	-609.203	70.714
500	48.281	247.886	229.736	9.075	-602.899	-609.912	63.717
600	50.752	256.916	233.530	14.032	-603.240	-611.284	53.217
700	52.719	264.893	237.452	19.209	-603.621	-612.594	45.712
800	54.286	272.039	241.336	24.562	-604.024	-613.849	40.080
900	55.539	278.508	245.113	30.056	-604.442	-615.052	35.697
1000	56.547	284.413	248.752	35.662	-604.874	-616.208	32.187
1100	57.362	289.842	252.244	41.359	-605.322	-617.320	29.314
1200	58.028	294.863	255.589	47.129	-605.787	-618.390	26.918
1300	58.575	299.530	258.792	52.960	-606.274	-619.421	24.889
1400	59.030	303.888	261.859	58.841	-606.785	-620.413	23.148
1500	59.410	307.974	264.798	64.764	-607.323	-621.368	21.638
1600	59.731	311.819	267.618	70.721	-607.889	-622.286	20.316
1700	60.004	315.449	270.326	76.708	-608.487	-623.168	19.148
1800	60.238	318.885	272.929	82.721	-609.117	-624.014	18.108
1900	60.439	322.147	275.434	88.755	-609.780	-624.823	17.178
2000	60.614	325.252	277.848	94.808	-610.477	-625.597	16.339
2100	60.766	328.213	280.177	100.877	-611.206	-626.335	15.579
2200	60.900	331.043	282.425	106.960	-611.969	-627.038	14.888
2300	61.017	333.753	284.598	113.056	-612.763	-627.705	14.256
2400	61.121	336.352	286.701	119.163	-663.847	-627.269	13.652
2500	61.214	338.849	288.737	125.280	-664.780	-625.725	13.074
2600	61.297	341.252	290.711	131.406	-665.709	-624.145	12.539
2700	61.371	343.566	292.626	137.539	-666.632	-622.528	12.044
2800	61.438	345.800	294.485	143.680	-667.549	-620.878	11.583
2900	61.498	347.957	296.292	149.827	-668.459	-619.196	11.153
3000	61.553	350.042	298.049	155.979	-669.361	-617.481	10.751
3100	61.603	352.061	299.759	162.137	-670.253	-615.737	10.375
3200	61.648	354.018	301.424	168.300	-671.135	-613.965	10.022
3300	61.689	355.916	303.047	174.466	-672.007	-612.165	9.690
3400	61.727	357.758	304.629	180.637	-672.867	-610.338	9.377
3500	61.762	359.548	306.173	186.812	-673.715	-608.487	9.081
3600	61.794	361.288	307.680	192.989	-674.552	-606.611	8.802
3700	61.823	362.981	309.152	199.170	-675.375	-604.713	8.537
3800	61.850	364.631	310.590	205.354	-676.186	-602.792	8.286
3900	61.876	366.237	311.996	211.540	-676.984	-600.850	8.047
4000	61.899	367.804	313.372	217.729	-677.769	-598.888	7.821
4100	61.921	369.333	314.718	223.920	-678.542	-596.906	7.605
4200	61.941	370.825	316.037	230.113	-1159.151	-587.882	7.311
4300	61.960	372.283	317.328	236.308	-1158.811	-574.284	6.976
4400	61.978	373.708	318.593	242.505	-1158.461	-560.694	6.656
4500	61.994	375.101	319.833	248.704	-1158.101	-547.113	6.351
4600	62.009	376.464	321.050	254.904	-1157.733	-533.540	6.059
4700	62.024	377.797	322.243	261.106	-1157.356	-519.974	5.779
4800	62.038	379.103	323.414	267.309	-1156.972	-506.417	5.511
4900	62.050	380.383	324.564	273.513	-1156.581	-492.867	5.254
5000	62.062	381.636	325.693	279.719	-1156.185	-479.326	5.007
5100	62.074	382.865	326.802	285.925	-1155.784	-465.793	4.771
5200	62.084	384.071	327.891	292.133	-1155.380	-452.268	4.543
5300	62.095	385.254	328.963	298.342	-1154.973	-438.750	4.324
5400	62.104	386.414	330.016	304.552	-1154.565	-425.241	4.113
5500	62.113	387.554	331.052	310.763	-1154.157	-411.739	3.910
5600	62.122	388.673	332.071	316.975	-1153.750	-398.243	3.715
5700	62.130	389.773	333.073	323.188	-1153.345	-384.756	3.526
5800	62.138	390.853	334.060	329.401	-1152.943	-371.275	3.344
5900	62.145	391.916	335.032	335.615	-1152.546	-357.802	3.168
6000	62.152	392.960	335.989	341.830	-1152.155	-344.335	2.998

PREVIOUS: December 1965 (1 atm) CURRENT: December 1965 (1 bar)

Boron Fluoride Oxide (OBF) $B_1F_1O_1(g)$

DIFLUOROBORANE (BF_2) IDEAL GAS M_r = 48.806806

$S°(298.15\ K) = [247.16 \pm 2.1]\ J\ K^{-1}\ mol^{-1}$

$\Delta_fH°(0\ K) = -590.5 \pm 13\ kJ\ mol^{-1}$

$\Delta_fH°(298.15\ K) = -589.9 \pm 13\ kJ\ mol^{-1}$

Electronic Levels and Quantum Weights

State	ε_i, cm^{-1}	g_i
$^2A_1$	0	2
$^2B_1$	[16000]	2

Vibrational Frequencies and Degeneracies

ν, cm^{-1}
[1080](1)
[500](1)
[1213](1)

Point Group: C_{2v} σ = 2

Bond Distance: B-F = [1.30] Å

Bond Angle: F-B-F = 112°

Product of Moments of Inertia: $I_AI_BI_C = [4.36647 \times 10^{-116}]\ g^3\ cm^6$

Enthalpy of Formation

Srivastava and Farber (1), using a gas-inlet boron nitride effusion cell, performed a mass spectrometric study of the equilibrium vapor species over the system $BF_3(g) + B(cr)$. Ion intensities for $BF_3(g)$, BF(g) and $BF_2(g)$ were measured at ionizing electron energies slightly above appearance potentials (~3 eV) in the temperature range 1477 to 1850 K and were used to obtain equilibrium data for the reaction $BF_3(g) + BF(g) = 2\ BF_2(g)$. We have analyzed their results and obtain $\Delta_rH°(298.15\ K) = 17.1 \pm 1.0\ kcal\ mol^{-1}$ from the 3rd law method with a drift of $-7.4 \pm 2.0\ cal\ K^{-1}\ mol^{-1}$ with JANAF heats of formation (2) for $BF_3(g)$ and BF(g).

Although no other experimental determinations of enthalpy of formation of BF_2 have been reported, Margrave (3) combined the appearance potential of BF_2^+ from BF_3 of 16.5 eV with an estimated ionization potential for BF_2 (IP = 9.4 ± 0.1 eV) (4) and calculated $\Delta_fH°(BF_2, g) \leq -124 \pm 9\ kcal\ mol^{-1}$. Based on the JANAF value for $AP(BF_2^+)$ of 15.81 eV (2), we calculate $\Delta_fH°(BF_2, g, 0\ K) = -141.3\ kcal\ mol^{-1}$ by the same method. Margrave (3) also derived $\Delta_fH° = -135\ kcal\ mol^{-1}$ for BF_2 from average bond energy calculations and selected $\Delta_fH°(BF_2, g) = -130 \pm 6\ kcal\ mol^{-1}$ as the "best" value. It is now apparent that this value is several kilocalories too positive.

We note that the adopted enthalpy of formation leads to a progression in the stepwise bond dissociation energies for BF(g), $BF_2(g)$ and $BF_3(g)$ which is consistent with the bond energy scheme predicted by Hildenbrand (5) for the Group III A fluorides, namely $D_1 > D_2 < D_3$. Using all JANAF data (2) we calculate bond dissociation energies of $D_1(B\text{-}F) = 179.4$, $D_2(BF\text{-}F) = 132.2$, and $D_3(BF_2\text{-}F) = 149.3\ kcal\ mol^{-1}$. The adopted value also corresponds to an ionization potential for BF_2 of 9.4 ± 0.2 eV (2) which appears reasonable in comparison with IP's for other Group III A dihalides (6) and is the value estimated for BF_2 by Steele et al. (4).

Heat Capacity and Entropy

Nelson and Gordy (7) were successful in producing trapped BF_2 radicals by gamma irradiation of solid xenon which contained small concentrations of BF_3. From observations on the electron-spin-resonance spectra of these radicals, a bond angle of 112° for BF_2 was derived by these workers to account for the hybridization of the σ-bonding orbitals of the boron atom. We adopt their result and assume the B-F bond length to be the same as that for BHF_2 (2). The three principal moments of inertia are: $I_A = 0.7385 \times 10^{-39}$, $I_B = 7.3289 \times 10^{-39}$, and $I_C = 8.0674 \times 10^{-39}\ g\ cm^2$.

The ground state electronic configuration ($^2A_1$) and doublet excited state ($^2B_1$) are estimated by analogy with those for NO_2 (8). The vibrational frequencies are calculated from the stretching and bending force constants of $k = 4.82 \times 10^5$ and $k\delta/\ell^2 = 0.84 \times 10^5$ dynes/cm by the valence force method. These force constants are estimated from those for CF_2 and SiF_2 (2).

References

1. R. D. Srivastava and M. Farber, Trans. Faraday Soc. 67, 2298 (1971);‡
2. JANAF Thermochemical Tables: $BF_3(g)$, 6-30-69; BF(g), 12-31-64; $BF_2^+(g)$, 12-31-70; $BNF_2(g)$, 12-31-65; $CF_2(g)$, 6-30-70; $SiF_2(g)$, 12-31-68.
3. J. L. Margrave, J. Phys. Chem. 66, 1209 (1962).
4. W. C. Steele, L. D. Nichols, and F. G. A. Stone, J. Amer. Chem. Soc. 84, 1154 (1962).
5. D. L. Hildenbrand in "Advances in High Temperature Chemistry", Volume I, Ed. by L. Eyring, Academic Press, New York, 1967.
6. J. W. Hastie and J. L. Margrave, J. Phys. Chem. 73, 1105 (1969).
7. W. Nelson and W. Gordy, J. Chem. Phys. 51, 4710 (1969).
8. G. Herzberg, "Electronic Spectra of Polyatomic Molecules", D. Van Nostrand Company, Inc., New York, 1966.
9. Thomson and Brotchie, "A Theoretical Discussion of Some Novel Excited States of NH_2 and BF_2", Molec. Phys. 28, 301 (1974).

Difluoroborane (BF_2) $B_1F_2(g)$

Enthalpy Reference Temperature = T_r = 298.15 K Standard State Pressure = p° = 0.1 MPa

	J K⁻¹mol⁻¹			kJ mol⁻¹			
T/K	$C_p°$	S°	$-[G°-H°(T_r)]/T$	$H°-H°(T_r)$	$\Delta_fH°$	$\Delta_fG°$	Log K_f
0	0.	0.	INFINITE	-10.617	-590.522	-590.522	INFINITE
100	33.582	207.676	280.540	-7.286	-590.123	-593.822	310.181
200	36.689	231.795	250.739	-3.789	-589.887	-597.632	156.085
250	38.629	240.188	247.813	-1.906	-589.878	-599.571	125.273
298.15	40.555	247.157	247.157	0.	-589.944	-601.433	105.369
300	40.629	247.408	247.157	0.075	-589.948	-601.504	104.731
350	42.585	253.819	247.659	2.156	-590.083	-603.420	90.056
400	44.409	259.626	248.798	4.331	-590.273	-605.313	79.046
450	46.050	264.953	250.301	6.594	-590.506	-607.179	70.480
500	47.494	269.882	252.015	8.933	-590.768	-609.018	63.624
600	49.827	278.757	255.749	13.805	-591.351	-612.614	53.333
700	51.554	286.575	259.606	18.878	-591.985	-616.109	45.975
800	52.837	293.547	263.421	24.101	-592.654	-619.510	40.450
900	53.801	299.828	267.123	29.435	-593.356	-622.825	36.148
1000	54.538	305.537	270.683	34.854	-594.090	-626.060	32.702
1100	55.111	310.763	274.092	40.337	-594.856	-629.220	29.879
1200	55.563	315.578	277.352	45.872	-595.656	-632.309	27.524
1300	55.925	320.040	280.466	51.447	-596.492	-635.330	25.528
1400	56.219	324.196	283.443	57.055	-597.364	-638.285	23.815
1500	56.461	328.083	286.291	62.689	-598.275	-641.176	22.328
1600	56.662	331.734	289.018	68.345	-599.223	-644.006	21.025
1700	56.831	335.174	291.633	74.020	-600.210	-646.775	19.873
1800	56.975	338.427	294.143	79.711	-601.234	-649.484	18.848
1900	57.100	341.510	296.555	85.415	-602.294	-652.136	17.928
2000	57.209	344.442	298.877	91.130	-603.387	-654.731	17.100
2100	57.306	347.236	301.114	96.856	-604.510	-657.271	16.349
2200	57.393	349.904	303.271	102.591	-605.662	-659.757	15.665
2300	57.474	352.457	305.355	108.334	-606.838	-662.189	15.039
2400	57.550	354.904	307.369	114.086	-658.293	-663.501	14.441
2500	57.622	357.255	309.317	119.844	-659.584	-663.692	13.867
2600	57.693	359.517	311.205	125.610	-660.855	-663.831	13.337
2700	57.762	361.695	313.035	131.383	-662.104	-663.922	12.844
2800	57.832	363.797	314.810	137.163	-663.327	-663.967	12.386
2900	57.903	365.828	316.535	142.949	-664.521	-663.968	11.959
3000	57.975	367.792	318.211	148.743	-665.683	-663.929	11.560
3100	58.048	369.694	319.841	154.544	-666.812	-663.852	11.186
3200	58.124	371.538	321.428	160.353	-667.904	-663.739	10.834
3300	58.202	373.328	322.974	166.169	-668.959	-663.593	10.504
3400	58.283	375.067	324.480	171.993	-669.974	-663.415	10.192
3500	58.365	376.757	325.950	177.826	-670.948	-663.207	9.898
3600	58.450	378.403	327.384	183.667	-671.881	-662.973	9.619
3700	58.538	380.005	328.785	189.516	-672.771	-662.713	9.356
3800	58.627	381.568	330.153	195.374	-673.619	-662.430	9.106
3900	58.718	383.092	331.491	201.241	-674.423	-662.125	8.868
4000	58.811	384.580	332.800	207.118	-675.184	-661.800	8.642
4100	58.904	386.033	334.081	213.004	-675.901	-661.456	8.427
4200	58.999	387.453	335.335	218.899	-1156.425	-654.071	8.135
4300	59.095	388.843	336.563	224.803	-1155.970	-642.116	7.800
4400	59.191	390.203	337.767	230.718	-1155.473	-630.172	7.481
4500	59.288	391.534	338.947	236.642	-1154.938	-618.239	7.176
4600	59.385	392.838	340.104	242.575	-1154.363	-606.319	6.885
4700	59.481	394.116	341.240	248.519	-1153.751	-594.411	6.606
4800	59.577	395.369	342.354	254.472	-1153.101	-582.517	6.339
4900	59.672	396.599	343.449	260.434	-1152.416	-570.637	6.083
5000	59.766	397.805	344.524	266.406	-1151.696	-558.771	5.837
5100	59.858	398.990	345.580	272.387	-1150.943	-546.920	5.602
5200	59.950	400.153	346.619	278.377	-1150.158	-535.084	5.375
5300	60.040	401.296	347.640	284.377	-1149.341	-523.263	5.157
5400	60.128	402.419	348.644	290.385	-1148.495	-511.458	4.947
5500	60.215	403.523	349.632	296.403	-1147.621	-499.670	4.745
5600	60.299	404.609	350.604	302.428	-1146.719	-487.896	4.551
5700	60.381	405.677	351.560	308.462	-1145.792	-476.140	4.363
5800	60.462	406.728	352.503	314.504	-1144.840	-464.401	4.182
5900	60.540	407.762	353.430	320.555	-1143.864	-452.677	4.008
6000	60.615	408.780	354.344	326.612	-1142.867	-440.970	3.839

PREVIOUS: June 1972 (1 atm) CURRENT: June 1972 (1 bar)

Difluoroborane (BF_2) $B_1F_2(g)$

DIFLUOROBORANE, ION (BF_2^+) IDEAL GAS M_r = 48.806257

$S°(298.15\ K) = [225.1 \pm 1]\ J\ K^{-1}\ mol^{-1}$

$\Delta_f H°(0\ K) = 315.8 \pm 2.5\ kJ\ mol^{-1}$
$\Delta_f H°(298.15\ K) = 322.6 \pm 2.5\ kJ\ mol^{-1}$

Electronic Levels and Quantum Weights

State	ε_i, cm^{-1}	g_i
$^1\Sigma_g^+$	0	1
$^1B_2$	[25000]	1

Vibrational Frequencies and Degeneracies

ν, cm^{-1}
[900](1)
[370](2)
[1550](1)

Point Group: $[D_{\infty h}]$ $\sigma = 2$
Bond Distance: B-F = [1.31] Å
Bond Angle: F-B-F = [180]°
Rotational Constant: $B_0 = [0.258528]\ cm^{-1}$

Enthalpy of Formation

The electron impact reaction at low pressure in BF_3(g) has been studied with a mass spectrometer by Osberghaus (1), Law (2) and Marriott (3). From the reported appearance potentials 17.0 ± 0.5, 16.2 ± 0.2, and 16.7 ± 0.05 eV and assumed process $BF_3 + e^- = BF_2^+ + F + 2e^-$, we calculate the corresponding values of $\Delta_f H°(BF_2^+$, g, 298.15 K) as 104.5 ± 11.5, 86.1 ± 4.6, and 85.4 ± 1.2 kcal mol^{-1}, using $\Delta_f H°(0\ K)$ = -170.747 and 18.357 kcal mol^{-1} for BF_3(g) and F(g). These electron impact values are superseded by the photoionization results of Dibeler and Liston (7) who report the onset of BF_2^+ from BF_3 as 15.81 eV (364.6 kcal mol^{-1}). This yields $\Delta_f H°(0\ K)$ = 75.5 ± 0.6 kcal mol^{-1} or $\Delta_f H°(298.15\ K)$ = 77.1 ± 0.6 kcal mol^{-1}, which are adopted.

Heat Capacity and Entropy

The molecular structure is assumed to be linear using the Walsh (4) predictio nfor 16 valence electron XY_2 molecules. The electronic states, levels, and vibrational frequencies are estimated by comparison with those for CO_2 (5). The B-F bond distances is calculated by the method recommended by Krasnov (5).

References

1. O. Osberghaus, Z. Phys. 128, 366 (1950).
2. R. W. Law and J. L. Margrave, J. Chem. Phys. 25, 1086 (1956).
3. J. Marriott and J. D. Craggs, J. Electron Contr. 3, 194 (1957).
4. A. D. Walsh, J. Chem. Soc. 1953, 2266 (1953).
5. G. Herzberg, "Electronic Spectra of Polyatomic Molecules", D. Van Nostrand Co., New York, 1966.
6. K. S. Krasnov, Zhur. Strukt. Khim. 1, 209 (1960).
7. V. H. Dibeler and S. K. Liston, Inorg. Chem. 7, 1742 (1968).

Difluoroborane, Ion (BF_2^+) $B_1F_2^+$(g)

Enthalpy Reference Temperature = T_r = 298.15 K Standard State Pressure = p° = 0.1 MPa

T/K	$C_p°$	S°	$-[G°-H°(T_r)]/T$	$H°-H°(T_r)$	$\Delta_f H°$	$\Delta_f G°$	Log K_f
	J K⁻¹mol⁻¹			kJ mol⁻¹			
0	0.	0.	INFINITE	-10.611	315.817		
100	31.424	184.214	260.803	-7.659			
200	39.162	208.493	229.047	-4.111			
250	42.008	217.548	225.865	-2.079			
298.15	44.291	225.148	225.148	0.	322.586	311.405	-54.557
300	44.371	225.422	225.149	0.082	322.628	311.335	-54.208
350	46.401	232.418	225.697	2.353	323.722	309.366	-46.170
400	48.172	238.732	226.938	4.718	324.761	307.244	-40.122
450	49.727	244.498	228.573	7.166	325.753	304.994	-35.403
500	51.091	249.809	230.435	9.687	326.712	302.636	-31.616
600	53.333	259.332	234.476	14.914	328.562	297.645	-25.912
700	55.047	267.688	238.636	20.336	330.357	292.350	-21.815
800	56.360	275.128	242.741	25.910	332.116	286.800	-18.726
900	57.375	281.827	246.718	31.598	333.848	281.032	-16.311
1000	58.168	287.915	250.538	37.377	335.553	275.071	-14.368
1100	58.794	293.489	254.193	43.226	337.231	268.942	-12.771
1200	59.296	298.627	257.684	49.132	338.880	262.661	-11.433
1300	59.702	303.390	261.019	55.082	340.499	256.243	-10.296
1400	60.035	307.827	264.206	61.070	342.084	249.702	-9.317
1500	60.311	311.979	267.254	67.088	343.636	243.049	-8.464
1600	60.541	315.879	270.172	73.131	345.152	236.294	-7.714
1700	60.736	319.555	272.970	79.195	346.634	229.445	-7.050
1800	60.901	323.031	275.656	85.277	348.080	222.509	-6.457
1900	61.043	326.328	278.236	91.374	349.492	215.494	-5.924
2000	61.165	329.462	280.720	97.485	350.873	208.406	-5.443
2100	61.272	332.449	283.113	103.607	352.224	201.249	-5.006
2200	61.365	335.302	285.421	109.739	353.548	194.029	-4.607
2300	61.446	338.031	287.649	115.879	354.848	186.749	-4.241
2400	61.519	340.648	289.803	122.028	305.868	180.482	-3.928
2500	61.583	343.161	291.888	128.183	307.052	175.233	-3.661
2600	61.640	345.577	293.907	134.344	308.255	169.936	-3.414
2700	61.692	347.905	295.864	140.511	309.479	164.593	-3.184
2800	61.739	350.149	297.763	146.682	310.727	159.204	-2.970
2900	61.782	352.316	299.607	152.858	312.001	153.770	-2.770
3000	61.821	354.411	301.399	159.038	313.303	148.292	-2.582
3100	61.858	356.439	303.142	165.222	314.636	142.770	-2.406
3200	61.893	358.404	304.838	171.410	316.001	137.204	-2.240
3300	61.925	360.309	306.490	177.601	317.400	131.595	-2.083
3400	61.956	362.158	308.100	183.795	318.833	125.943	-1.935
3500	61.986	363.954	309.671	189.992	320.302	120.249	-1.795
3600	62.016	365.701	311.203	196.192	321.808	114.512	-1.662
3700	62.045	367.400	312.699	202.395	323.350	108.732	-1.535
3800	62.074	369.055	314.160	208.601	324.928	102.911	-1.415
3900	62.103	370.668	315.589	214.810	326.544	97.047	-1.300
4000	62.133	372.241	316.985	221.022	328.198	91.142	-1.190
4100	62.163	373.775	318.352	227.237	329.888	85.194	-1.085
4200	62.194	375.274	319.689	233.454	-148.235	86.230	-1.072
4300	62.226	376.738	320.999	239.675	-145.385	91.778	-1.115
4400	62.259	378.169	322.282	245.900	-142.500	97.261	-1.155
4500	62.293	379.568	323.540	252.127	-139.582	102.677	-1.192
4600	62.328	380.938	324.773	258.358	-136.631	108.028	-1.227
4700	62.365	382.278	325.982	264.593	-133.649	113.315	-1.259
4800	62.402	383.592	327.169	270.831	-130.636	118.538	-1.290
4900	62.442	384.879	328.333	277.073	-127.592	123.697	-1.319
5000	62.482	386.141	329.477	283.320	-124.519	128.794	-1.346
5100	62.524	387.379	330.600	289.570	-121.418	133.830	-1.371
5200	62.568	388.593	331.704	295.825	-118.290	138.804	-1.394
5300	62.613	389.785	332.788	302.084	-115.136	143.718	-1.416
5400	62.659	390.956	333.855	308.347	-111.956	148.573	-1.437
5500	62.706	392.106	334.903	314.615	-108.752	153.368	-1.457
5600	62.755	393.237	335.935	320.888	-105.524	158.105	-1.475
5700	62.805	394.348	336.950	327.166	-102.274	162.783	-1.492
5800	62.856	395.440	337.949	333.449	-99.003	167.405	-1.508
5900	62.908	396.515	338.933	339.737	-95.711	171.970	-1.523
6000	62.961	397.573	339.901	346.031	-92.400	176.479	-1.536

PREVIOUS: December 1970 (1 atm) CURRENT: December 1970 (1 bar)

DIFLUOROBORANE, ION (BF_2^-) IDEAL GAS M_r = 48.807355

$\Delta_f H°(0\ K)$ = -796.4 ± 32 kJ mol^{-1}

S°(298.15 K) = [240.7 ± 2] J K^{-1} mol^{-1} $\Delta_f H°(298.15\ K)$ = -802.1 ± 32 kJ mol^{-1}

Electronic Levels and Quantum Weights

State	ε_i, cm^{-1}	g_i
$^1A_1$	0	1
$^3B_1$	[15000]	3
$^1B_1$	[35000]	1

Vibrational Frequencies and Degeneracies

ν, cm^{-1}
[1175](1)
[495](1)
[1188](1)

Point Group: [C_{2v}] σ = 2
Bond Distance: B-F = [1.265] Å
Bond Angle: F-B-F = [100]°
Product of Moments of Inertia: $I_A I_B I_C$ = [3.74972 x 10^{-116}] g^3 cm^6

Enthalpy of Formation

The BF_2^- ion has been detected mass spectrometrically by MacNeil and Thynne (1) and Farber et al. (2). MacNeil and Thynne (1) reported the formation of the negative ion at electron energies of approximately 7.6 eV as a result of electron bombardment of BF_3(g). Assuming the ionization process to be $BF_3(g) + e^- = BF_2^-(g) + F(g)$, we calculate $\Delta_f H°(BF_2^-$, g, 0 K) = -113.8 kcal mol^{-1} from their results with JANAF enthalpies of formation (3) for BF_3(g) and F(g). This value leads to an unrealistic electron affinity (EA) for BF_2 of -1.1 eV. It seems most likely that the BF_2^- ion was formed in the dissociative ionization of BF_3(g) with considerable excess energy.

Farber et al. (2) observed the formation of the BF_2^- ion in a molecular flow effusion-mass spectrometric study of the BF_3(g) + KF(g) + B(cr) system. Equilibrium data for the electron transfer reaction $BF_2^-(g) + F(g) = BF_2(g) + F^-(g)$ were reported for the temperature range 1253 to 1640 K. We have analyzed their results with JANAF functions and obtain $\Delta_r H°$(298.15 K) = -26.3 ± 1.8 kcal mol^{-1} by the 3rd law method with a drift of -1.5 ± 3.1 cal K^{-1} mol^{-1}, the 2nd law $\Delta_r H°$(298.15 K) value is -24.2 kcal mol^{-1}. We adopt the 3rd law value and calculate $\Delta_f H°(BF_2^-$, g, 298.15 K) = -191.7 ± 8.0 kcal mol^{-1} with JANAF auxiliary data. The adopted enthalpy of formation leads to EA (BF_2) = 2.14 ± 0.5 eV.

Heat Capacity and Entropy

The correlation diagram of Walsh (4) predicts a bent configuration for BF_2^- (18 valence electrons) with a bond angle considerably reduced from that for BF_2(112°) (3). These predictions are supported by the experimentally determined bond angles for the isoelectronic molecules CF_2(104.9°) (3), SiF_2(101°) (3) and GeF_2(94°) (5). Thus we adopt a bond angle for BF_2^- of 100° and assume the B-F bond length to be the same as that for BF (3). The principal moments of inertia are: I_A = 0.9240 x 10^{-39}, I_B = 5.9251 x 10^{-39}, and I_C = 6.8491 x 10^{-39} g cm^2.

The electronic states, levels, and quantum weights are estimated by analogy with those for CF_2 (3). The vibrational frequencies are calculated form the stretching and bending force constants of k = 5.16 x 10^5 and $k\delta/\ell^2$ = 0.91 x 10^5 dynes/cn by the valence force method. These force constants are estimated by comparison with those for CF_2, SiF_2, and BF_2 (3).

References

1. K. A. G. MacNeil and J. C. J. Thynne, J. Phys. Chem. 74, 2257 (1970).
2. M. Farber, R. D. Srivastava, and O. M. Uy, Space Sciences, Inc., Final Report Under USAF Contract F04611-70-C-0041, AFRPL-TR-71-94, July, 1971.
3. JANAF Thermochemical Tables: BF_3(g), 6-30-69; F(g), 9-30-65; CF_2(g), 6-30-70; SiF_2(g), 12-31-68; BF(g), 12-31-64; BF_2(g), 6-30-72.
4. A. D. Walsh, J. Chem. Soc. 2266 (1953).
5. J. W. Hastie, R. Hauge, and J. L. Margrave, J. Phys. Chem. 72, 4492 (1968).

Difluoroborane, Ion (BF_2^-) $B_1F_2^-$(g)

Enthalpy Reference Temperature = T_r = 298.15 K Standard State Pressure = p° = 0.1 MPa

T/K	$C_p°$ (J $K^{-1}mol^{-1}$)	S° (J $K^{-1}mol^{-1}$)	-[G°-H°(T_r)]/T (J $K^{-1}mol^{-1}$)	H°-H°(T_r) (kJ mol^{-1})	$\Delta_f H°$ (kJ mol^{-1})	$\Delta_f G°$ (kJ mol^{-1})	Log K_f
0	0.	0.	INFINITE	-10.609	-796.445		
100	33.599	201.283	274.065	-7.278			
200	36.676	225.419	244.310	-3.778			
250	38.519	233.799	241.394	-1.899			
298.15	40.352	240.740	240.740	0.	-802.073	-805.394	141.102
300	40.423	240.990	240.741	0.075	-802.115	-805.414	140.235
350	42.310	247.364	241.240	2.143	-803.302	-805.871	120.270
400	44.094	253.131	242.371	4.304	-804.547	-806.153	105.273
450	45.720	258.420	243.865	6.550	-805.835	-806.277	93.590
500	47.165	263.314	245.568	8.873	-807.153	-806.256	84.229
600	49.525	272.132	249.277	13.713	-809.846	-805.825	70.153
700	51.290	279.905	253.108	18.758	-812.586	-804.938	60.065
800	52.610	286.845	256.899	23.956	-815.359	-803.657	52.473
900	53.607	293.101	260.580	29.269	-818.160	-802.026	46.548
1000	54.372	298.791	264.121	34.670	-820.991	-800.082	41.792
1100	54.968	304.002	267.513	40.138	-823.851	-797.853	37.887
1200	55.439	308.806	270.756	45.659	-826.743	-795.362	34.621
1300	55.817	313.259	273.857	51.223	-829.669	-792.629	31.848
1400	56.124	317.407	276.821	56.820	-832.630	-789.669	29.463
1500	56.378	321.288	279.657	62.446	-835.628	-786.496	27.388
1600	56.591	324.934	282.374	68.095	-838.663	-783.121	25.566
1700	56.774	328.370	284.980	73.763	-841.735	-779.556	23.953
1800	56.933	331.620	287.482	79.449	-844.842	-775.809	22.513
1900	57.075	334.702	289.886	85.149	-847.984	-771.889	21.221
2000	57.207	337.633	292.201	90.863	-851.157	-767.801	20.053
2100	57.333	340.427	294.432	96.590	-854.358	-763.555	18.992
2200	57.457	343.097	296.583	102.330	-857.584	-759.156	18.025
2300	57.583	345.654	298.662	108.082	-860.829	-754.609	17.138
2400	57.713	348.107	300.671	113.847	-914.350	-748.852	16.298
2500	57.850	350.466	302.616	119.625	-917.700	-741.887	15.501
2600	57.995	352.738	304.500	125.417	-921.024	-734.789	14.762
2700	58.151	354.929	306.328	131.224	-924.317	-727.564	14.076
2800	58.319	357.047	308.102	137.048	-927.574	-720.217	13.436
2900	58.498	359.097	309.825	142.888	-930.793	-712.754	12.838
3000	58.689	361.083	311.501	148.748	-933.968	-705.182	12.278
3100	58.893	363.011	313.131	154.627	-937.098	-697.504	11.753
3200	59.108	364.884	314.719	160.526	-940.177	-689.726	11.259
3300	59.334	366.706	316.267	166.448	-943.205	-681.852	10.793
3400	59.570	368.481	317.777	172.394	-946.177	-673.887	10.353
3500	59.816	370.211	319.250	178.363	-949.093	-665.836	9.937
3600	60.070	371.900	320.690	184.357	-951.951	-657.702	9.543
3700	60.331	373.549	322.096	190.377	-954.750	-649.490	9.169
3800	60.597	375.162	323.471	196.423	-957.488	-641.203	8.814
3900	60.868	376.739	324.817	202.497	-960.164	-632.845	8.476
4000	61.142	378.284	326.134	208.597	-962.780	-624.419	8.154
4100	61.418	379.797	327.425	214.725	-965.334	-615.928	7.847
4200	61.694	381.280	328.690	220.881	-1447.676	-600.351	7.466
4300	61.970	382.735	329.930	227.064	-1449.020	-580.161	7.048
4400	62.244	384.163	331.146	233.275	-1450.307	-559.940	6.647
4500	62.515	385.565	332.340	239.513	-1451.535	-539.690	6.265
4600	62.782	386.942	333.512	245.778	-1452.708	-519.414	5.898
4700	63.045	388.295	334.663	252.069	-1453.826	-499.113	5.547
4800	63.302	389.625	335.794	258.386	-1454.891	-478.789	5.210
4900	63.552	390.933	336.906	264.729	-1455.904	-458.443	4.887
5000	63.795	392.219	338.000	271.096	-1456.867	-438.076	4.577
5100	64.031	393.485	339.075	277.488	-1457.782	-417.692	4.278
5200	64.258	394.730	340.134	283.902	-1458.652	-397.289	3.991
5300	64.477	395.956	341.175	290.339	-1459.476	-376.870	3.714
5400	64.687	397.163	342.201	296.797	-1460.259	-356.436	3.448
5500	64.887	398.352	343.211	303.276	-1461.002	-335.988	3.191
5600	65.078	399.523	344.206	309.775	-1461.706	-315.527	2.943
5700	65.260	400.677	345.187	316.292	-1462.374	-295.054	2.704
5800	65.431	401.813	346.153	322.826	-1463.008	-274.569	2.473
5900	65.593	402.933	347.106	329.377	-1463.611	-254.073	2.249
6000	65.745	404.037	348.046	335.944	-1464.183	-233.568	2.033

PREVIOUS: June 1972 (1 atm) CURRENT: June 1972 (1 bar)

Difluoroborane, Ion (BF_2^-) $B_1F_2^-$(g)

DIFLUOROBORANE (BHF_2) IDEAL GAS M_r = 49.814746

$\Delta_fH°(0\ K) = -730.3 \pm 3.3\ kJ\ mol^{-1}$

$S°(298.15\ K) = 244.0\ J\ K^{-1}\ mol^{-1}$ $\Delta_fH°(298.15\ K) = -733.9 \pm 3.3\ kJ\ mol^{-1}$

Vibrational Frequencies and Degeneracies

ν, cm^{-1}	ν, cm^{-1}
2640(1)	1411 (1)
1174(1)	[1158](1)
544(1)	928 (1)

Ground State Quantum Weight: 1 $\sigma = 2$

Point Group: C_{2v}

Bond Distance: B-H = 1.15 Å B-F = 1.30 ± 0.15 Å

Bond Angles: F-B-F = 120 ± 5° F-B-H = 120 ± 2.5°

Product of the Moments of Inertia: $I_AI_BI_C = 7.51811 \times 10^{-116}\ g^3\ cm^6$

Enthalpy of Formation

The equilibrium pressures at 296 and 360 K for the reaction 1/6 $B_2H_6(g)$ + 2/3 $BF_3(g)$ = $HBF_2(g)$ were determined by Porter and Wason (1). By the 3rd law method, the enthalpy change was evaluated to be 3.06 ± 0.33 kcal mol^{-1}. Based on the values of $\Delta_fH°(298.15\ K)$ for $B_2H_6(g)$ and $BF_3(g)$ as 9.8 and -270.1 kcal mol^{-1}, respectively, the enthalpy of formation for $BHF_2(g)$ was derived to be -175.4 ± 0.8 kcal mol^{-1}. The drift in the 3rd law enthalpy of reaction is 3.6 cal K^{-1} mol^{-1}.

Heat Capacity and Entropy

The infrared spectrum of $BHF_2(g)$ has been observed by Coyle, Ritter and Farrar (2), Perec and Becka (3), Lynds (4), and Porter and Wason (1).

The vibrational frequencies, except ν_5, were obtained form Porter and Wason (1) and corrected to the average isotopic species. The value of ν_5 was estimated by comparison with the same value reported by other investigators. The molecular structure, bond distances and angles were obtained from Perec and Becka (3). The principal moments of inertia are: $I_A = 1.0402 \times 10^{-39}$, $I_B = 7.9974 \times 10^{-39}$ and $I_C = 9.0376 \times 10^{-39}\ g\ cm^2$.

The vibrational-rotational spectrum of $BHF_2(g)$ has been studied by Lynds and Bass (_). The values of ν_1, ν_2, ν_3, and ν_6 were assigned as 1164.2, 2620.8, 541.5 and 923.5 cm^{-1}, respectively, which are in fair agreement with the adopted values reported by Porter and Wason (1).

References

1. R. F. Porter and S. K. Wason, J. Phys. Chem. 69, 2208 (1965).
2. T. D. Coyle, J. J. Ritter and T. C. Farrar, Proc. Chem. Soc. 25 (1964).
3. M. Perec and L. N. Becks, J. Chem. Phys. 43, 721 (1965).
4. L. Lynds, J. Chem. Phys. 42, 1124 (1965).

Difluoroborane (BHF_2) $B_1F_2H_1(g)$

Enthalpy Reference Temperature = T_r = 298.15 K Standard State Pressure = p° = 0.1 MPa

T/K	$C°_p$ (J K⁻¹mol⁻¹)	S° (J K⁻¹mol⁻¹)	$-[G°-H°(T_r)]/T$ (J K⁻¹mol⁻¹)	$H°-H°(T_r)$ (kJ mol⁻¹)	$\Delta_fH°$ (kJ mol⁻¹)	$\Delta_fG°$ (kJ mol⁻¹)	Log K_f
0	0.	0.	INFINITE	-10.671	-730.272	-730.272	INFINITE
100	33.464	204.405	277.831	-7.343	-731.374	-729.711	381.162
200	36.674	228.378	247.707	-3.866	-732.506	-727.627	190.037
250	39.376	236.842	244.709	-1.967	-733.179	-726.330	151.758
298.15	42.358	244.029	244.029	0.	-733.874	-724.949	127.008
300	42.476	244.291	244.030	0.078	-733.901	-724.893	126.215
350	45.706	251.082	244.559	2.283	-734.637	-723.334	107.952
400	48.856	257.392	245.773	4.648	-735.366	-721.669	94.240
450	51.806	263.319	247.396	7.165	-736.074	-719.914	83.565
500	54.504	268.919	249.271	9.824	-736.748	-718.082	75.018
600	59.136	279.281	253.424	15.514	-737.977	-714.231	62.179
700	62.868	288.688	257.800	21.621	-739.046	-710.187	52.995
800	65.882	297.286	262.206	28.064	-739.971	-705.999	46.097
900	68.332	305.192	266.549	34.779	-740.780	-701.703	40.726
1000	70.340	312.499	270.783	41.716	-741.497	-697.322	36.424
1100	71.997	319.283	274.888	48.835	-742.147	-692.873	32.902
1200	73.374	325.609	278.854	56.106	-742.750	-688.366	29.964
1300	74.528	331.529	282.681	63.503	-743.325	-683.811	27.476
1400	75.500	337.089	286.370	71.005	-743.884	-679.212	25.342
1500	76.325	342.326	289.928	78.598	-744.441	-674.573	23.491
1600	77.030	347.275	293.359	86.266	-745.003	-669.897	21.870
1700	77.636	351.964	296.670	94.000	-745.577	-665.185	20.439
1800	78.159	356.417	299.866	101.791	-746.168	-660.439	19.165
1900	78.614	360.655	302.955	109.630	-746.779	-655.660	18.025
2000	79.011	364.698	305.942	117.512	-747.411	-650.848	16.998
2100	79.360	368.561	308.832	125.431	-748.064	-646.004	16.068
2200	79.667	372.260	311.632	133.382	-748.738	-641.128	15.222
2300	79.940	375.808	314.346	141.363	-749.432	-636.221	14.449
2400	80.182	379.215	316.978	149.369	-800.403	-630.215	13.716
2500	80.399	382.493	319.534	157.398	-801.209	-623.107	13.019
2600	80.593	385.650	322.016	165.448	-801.995	-615.968	12.375
2700	80.767	388.695	324.430	173.516	-802.760	-608.798	11.778
2800	80.925	391.635	326.778	181.601	-803.503	-601.601	11.223
2900	81.068	394.477	329.063	189.701	-804.221	-594.377	10.706
3000	81.197	397.228	331.290	197.814	-804.912	-587.129	10.223
3100	81.315	399.892	333.460	205.940	-805.576	-579.859	9.771
3200	81.423	402.476	335.577	214.077	-806.211	-572.568	9.346
3300	81.522	404.983	337.642	222.224	-806.816	-565.257	8.947
3400	81.612	407.418	339.659	230.381	-807.391	-557.928	8.572
3500	81.695	409.785	341.629	238.546	-807.935	-550.583	8.217
3600	81.772	412.087	343.554	246.720	-808.447	-543.223	7.882
3700	81.843	414.329	345.437	254.901	-808.928	-535.848	7.565
3800	81.908	416.512	347.279	263.088	-809.378	-528.462	7.264
3900	81.969	418.641	349.081	271.282	-809.798	-521.064	6.979
4000	82.025	420.717	350.846	279.482	-810.186	-513.655	6.708
4100	82.078	422.743	352.575	287.687	-810.545	-506.238	6.450
4200	82.127	424.721	354.269	295.897	-1290.723	-491.787	6.116
4300	82.172	426.654	355.930	304.112	-1289.936	-472.775	5.743
4400	82.215	428.544	357.559	312.332	-1289.123	-453.780	5.387
4500	82.255	430.392	359.157	320.555	-1288.284	-434.804	5.047
4600	82.292	432.200	360.726	328.783	-1287.421	-415.848	4.722
4700	82.328	433.970	362.265	337.014	-1286.534	-396.910	4.411
4800	82.361	435.704	363.777	345.248	-1285.625	-377.991	4.113
4900	82.392	437.403	365.263	353.486	-1284.695	-359.092	3.828
5000	82.421	439.067	366.722	361.726	-1283.744	-340.211	3.554
5100	82.448	440.700	368.157	369.970	-1282.773	-321.351	3.291
5200	82.475	442.301	369.567	378.216	-1281.785	-302.508	3.039
5300	82.499	443.872	370.954	386.465	-1280.778	-283.686	2.796
5400	82.522	445.415	372.319	394.716	-1279.755	-264.882	2.562
5500	82.545	446.929	373.662	402.969	-1278.716	-246.098	2.337
5600	82.565	448.416	374.984	411.225	-1277.663	-227.332	2.120
5700	82.585	449.878	376.285	419.482	-1276.595	-208.586	1.911
5800	82.604	451.314	377.566	427.742	-1275.514	-189.859	1.710
5900	82.622	452.727	378.828	436.003	-1274.420	-171.150	1.515
6000	82.639	454.116	380.071	444.266	-1273.314	-152.460	1.327

PREVIOUS: December 1965 (1 atm) CURRENT: December 1965 (1 bar)

Difluoroborane (BHF_2) $B_1F_2H_1(g)$

DIFLUOROHYDROXYBORANE (BF_2OH) IDEAL GAS M_r = 65.814146

$S°(298.15 K) = [269.17] J K^{-1} mol^{-1}$

$\Delta_fH°(0 K) = [-1077 \pm 20] kJ mol^{-1}$
$\Delta_fH°(298.15 K) = [-1084 \pm 20] kJ mol^{-1}$

Vibrational Frequencies and Degeneracies

ν, cm^{-1}	ν, cm^{-1}
[3680](1)	[888](1)
[1463](1)	[696](1)
[1420](1)	[480](1)
[1250](1)	[480](1)
[1250](1)	

Ground State Quantum Weight: 1 σ = 1

Point Group: C_s

Bond Distances: B-F = [1.30] Å B-O = [1.36] Å O-H = [0.96] Å

Bond Angles: F-B-F = [120]° F-B-O = [120]° B-O-H = [105]°

Product of the Moments of Inertia: $I_AI_BI_C = [1.051510 \times 10^{-114}] g^3 cm^6$

Enthalpy of Formation

The value of enthalpy of formation, $\Delta_fH°(298.15 K)$ was calculated based on an assumption that $\Delta_rH°(298.15 K) = 0$ for the reaction 2/3 $BF_3(g)$ + 1/3 $B(OH)_3(g)$ = $BOHF_2(g)$.

Heat Capacity and Entropy

The vibrational frequencies were obtained from Gordon (1). The B-F and O-H bond distances were assumed to be the same as those in $BF_3(g)$ and $H_2O(g)$ molecules, respectively. The B-O bond distance and the bond angles were estimated from related molecules. The principal moments of inertia are: $I_A = 7.8128 \times 10^{-39}$, $I_B = 8.3349 \times 10^{-39}$ and $I_C = 16.1476 \times 10^{-39} g cm^2$.

Reference

1. J. S. Gordon, Astrosystems International, Inc. (1962).

Difluorohydroxyborane (BF_2OH) $B_1F_2H_1O_1(g)$

Enthalpy Reference Temperature = T_r = 298.15 K Standard State Pressure = p° = 0.1 MPa

T/K	$C_p°$ (J $K^{-1}mol^{-1}$)	S° (J $K^{-1}mol^{-1}$)	$-[G°-H°(T_r)]/T$ (J $K^{-1}mol^{-1}$)	$H°-H°(T_r)$ (kJ mol^{-1})	$\Delta_fH°$ (kJ mol^{-1})	$\Delta_fG°$ (kJ mol^{-1})	Log K_f
0	0.	0.	INFINITE	-11.726	-1076.768	-1076.768	INFINITE
100	34.095	224.715	308.596	-8.388	-1079.313	-1071.015	559.441
200	42.159	250.509	273.610	-4.620	-1081.609	-1061.807	277.316
250	47.167	260.453	270.002	-2.387	-1082.677	-1056.732	220.792
298.15	51.982	269.174	269.174	0.	-1083.656	-1051.646	184.244
300	52.164	269.496	269.175	0.096	-1083.692	-1051.447	183.073
350	56.990	277.902	269.828	2.826	-1084.641	-1045.997	156.106
400	61.514	285.812	271.336	5.790	-1085.519	-1040.415	135.864
450	65.653	293.300	273.364	8.971	-1086.322	-1034.728	120.108
500	69.373	300.413	275.716	12.349	-1087.048	-1028.956	107.494
600	75.605	313.636	280.952	19.610	-1088.285	-1017.217	88.557
700	80.471	325.671	286.494	27.424	-1089.275	-1005.291	75.016
800	84.305	336.676	292.089	35.670	-1090.065	-993.237	64.852
900	87.380	346.790	297.613	44.260	-1090.702	-981.094	56.941
1000	89.894	356.131	303.003	53.127	-1091.219	-968.887	50.609
1100	91.983	364.799	308.232	62.224	-1091.646	-956.632	45.427
1200	93.744	372.881	313.286	71.513	-1092.006	-944.342	41.106
1300	95.244	380.445	318.165	80.964	-1092.317	-932.023	37.449
1400	96.532	387.552	322.870	90.555	-1092.596	-919.683	34.314
1500	97.647	394.251	327.407	100.265	-1092.855	-907.323	31.596
1600	98.617	400.584	331.785	110.079	-1093.105	-894.946	29.217
1700	99.465	406.589	336.010	119.984	-1093.354	-882.553	27.118
1800	100.210	412.296	340.091	129.969	-1093.609	-870.146	25.251
1900	100.868	417.732	344.035	140.024	-1093.874	-857.724	23.580
2000	101.451	422.921	347.851	150.140	-1094.152	-845.288	22.077
2100	101.969	427.883	351.544	160.312	-1094.446	-832.838	20.716
2200	102.432	432.638	355.123	170.532	-1094.755	-820.373	19.478
2300	102.847	437.200	358.593	180.796	-1095.081	-807.893	18.348
2400	103.219	441.585	361.960	191.100	-1145.681	-794.331	17.288
2500	103.554	445.806	365.230	201.439	-1146.114	-779.683	16.291
2600	103.857	449.873	368.408	211.810	-1146.527	-765.017	15.369
2700	104.132	453.798	371.499	222.210	-1146.920	-750.336	14.516
2800	104.382	457.590	374.506	232.635	-1147.290	-735.641	13.724
2900	104.609	461.257	377.434	243.085	-1147.637	-720.933	12.985
3000	104.817	464.807	380.288	253.557	-1147.959	-706.213	12.296
3100	105.007	468.247	383.070	264.048	-1148.255	-691.484	11.651
3200	105.182	471.584	385.784	274.558	-1148.524	-676.745	11.047
3300	105.342	474.823	388.434	285.084	-1148.766	-661.998	10.479
3400	105.490	477.970	391.021	295.626	-1148.979	-647.244	9.944
3500	105.626	481.030	393.549	306.182	-1149.164	-632.485	9.439
3600	105.753	484.007	396.021	316.751	-1149.321	-617.720	8.963
3700	105.870	486.906	398.438	327.332	-1149.449	-602.951	8.512
3800	105.978	489.731	400.803	337.924	-1149.548	-588.180	8.085
3900	106.079	492.485	403.119	348.527	-1149.620	-573.405	7.680
4000	106.173	495.172	405.387	359.140	-1149.663	-558.630	7.295
4100	106.260	497.795	407.609	369.761	-1149.679	-543.854	6.929
4200	106.342	500.356	409.787	380.392	-1629.519	-522.054	6.493
4300	106.419	502.860	411.922	391.030	-1628.395	-495.699	6.022
4400	106.490	505.307	414.017	401.675	-1627.249	-469.371	5.572
4500	106.558	507.701	416.072	412.328	-1626.080	-443.068	5.143
4600	106.621	510.043	418.090	422.987	-1624.890	-416.793	4.733
4700	106.680	512.337	420.071	433.652	-1623.681	-390.542	4.340
4800	106.736	514.584	422.017	444.323	-1622.453	-364.318	3.965
4900	106.789	516.785	423.928	454.999	-1621.208	-338.120	3.604
5000	106.838	518.943	425.807	465.680	-1619.947	-311.947	3.259
5100	106.885	521.059	427.654	476.366	-1618.671	-285.800	2.927
5200	106.929	523.135	429.470	487.057	-1617.381	-259.678	2.608
5300	106.971	525.172	431.257	497.752	-1616.080	-233.580	2.302
5400	107.011	527.172	433.015	508.451	-1614.768	-207.508	2.007
5500	107.049	529.136	434.744	519.154	-1613.447	-181.460	1.723
5600	107.085	531.065	436.447	529.861	-1612.118	-155.435	1.450
5700	107.118	532.961	438.124	540.571	-1610.782	-129.435	1.186
5800	107.151	534.824	439.775	551.285	-1609.440	-103.458	0.932
5900	107.181	536.656	441.402	562.001	-1608.095	-77.505	0.686
6000	107.211	538.458	443.004	572.721	-1606.746	-51.574	0.449

PREVIOUS: March 1965 (1 atm) CURRENT: March 1965 (1 bar)

Difluorohydroxyborane (BF_2OH) $B_1F_2H_1O_1(g)$

BORON FLUORIDE OXIDE (OBF_2) IDEAL GAS M_r = 64.806206

$S°(298.15\ K) = [267.8]\ J\ K^{-1}\ mol^{-1}$

$\Delta_f H°(0\ K) = [-834 \pm 15]\ kJ\ mol^{-1}$

$\Delta_f H°(298.15\ K) = [-837 \pm 15]\ kJ\ mol^{-1}$

Electronic Levels and Quantum Weights

ε_i, cm^{-1}	g_i
0	[2]
17171	[2]
22390	[2]

Vibrational Frequencies and Degeneracies

ν, cm^{-1}	ν, cm^{-1}
1377(1)	[850](1)
856(1)	[1100](1)
491(1)	[500](1)

Point Group: C_{2v} $\sigma = 2$

Bond Distances: B-F = 1.30 ± 0.05 Å B-O = 1.40 ± 0.05 Å

Bond Angles: F-B-O = 117 ± 5° F-B-F = 126 ± 5°

Product of the Moments of Inertia: $I_A I_B I_C = 9.948921 \times 10^{-115}\ g^3\ cm^6$

Enthalpy of Formation

The molecule $BF_2O(g)$ is similar to $BO_2(g)$ in that both have an extra electron. Based on an assumption that $D_0^0(F_2B\text{-}O) = D_0^0(OB\text{-}O) - 10 = 134.3\ kcal\ mol^{-1}$ where the quantity $-10\ kcal\ mol^{-1}$ is estimated as the resonance stabilization energy in $BO_2(g)$, $\Delta_f H°(F_2BO,\ g,\ 298.15\ K)$ is calculated to be $-206\ kcal\ mol^{-1}$.

Assuming $D_0^0(F_2BO\text{-}H) = D_0^0(HO\text{-}H) = 118\ kcal\ mol^{-1}$, the value of $\Delta_f H°(F_2BO,\ g,\ 298.15\ K)$ is evaluated as $-191.6\ kcal\ mol^{-1}$, using $\Delta_f H°(0\ K) = 51.63$ and $-257.35\ kcal\ mol^{-1}$ for H(g) and $F_2BOH(g)$, respectively.

The value of $\Delta_f H°(F_2BO,\ g,\ 298.15\ K)$ is tentatively adopted as $-200 \pm 15\ kcal\ mol^{-1}$

Heat Capacity and Entropy

The 4465-A emission spectrum was observed in a discharge through BF_3 and O_2 by Mathews (1). The emitter of the discharge was shown to be either the planar BOF_2 molecule or molecule-ion BOF_2^+. No experimental evidence has been found to distinguish between them. The molecular structure, bond distances and angles were reported by Mathews (1), based on a rotational analysis of the emission spectrum. These values are adopted here.

Three vibrational frequencies for both $^{11}B^{16}OF_2$ and $^{10}B^{16}OF_2$ molecules (or ions) were deermined by vibrational analysis of the 5800 - Å bands of the emission spectrum by Mathews and Innes (2). These values are corrected to the average isotopic species and adopted. The last three frequencies are estimated from values calculated by the valence-force method, using force constants transferred from $COF_2(g)$.

The ground state quantum weight is taken as 2 because of the extra electron. The other two electronic levels are estimated from the band systems at 5800 and 4465 Å, assuming that two systems have a common lower state, i.e. the ground state.

The principal moments of inertia are: $I_A = 7.4051 \times 10^{-39}$, $I_B = 8.4655 \times 10^{-39}$, and $I_C = 15.8706 \times 10^{-39}\ g\ cm^2$.

References

1. C. W. Mathews, J. Mol. Spectry. 19, 203 (1966).
2. C. W. Mathews and K. K. Innes, J. Mol. Spectry. 15, 199 (1965).

Boron Fluoride Oxide (OBF_2) $B_1F_2O_1(g)$

Enthalpy Reference Temperature = T_r = 298.15 K Standard State Pressure = p° = 0.1 MPa

T/K	$C_p°$ (J K⁻¹mol⁻¹)	S° (J K⁻¹mol⁻¹)	$-[G°-H°(T_r)]/T$ (J K⁻¹mol⁻¹)	$H°-H°(T_r)$ (kJ mol⁻¹)	$\Delta_f H°$ (kJ mol⁻¹)	$\Delta_f G°$ (kJ mol⁻¹)	Log K_f
0	0.	0.	INFINITE	-11.535	-833.955	-833.955	INFINITE
100	33.949	224.266	306.266	-8.200	-835.003	-831.696	434.434
200	41.185	249.703	272.146	-4.489	-836.009	-827.987	216.248
250	45.839	259.392	268.645	-2.313	-836.436	-825.931	172.569
298.15	50.199	267.842	267.842	0.	-836.800	-823.874	144.339
300	50.360	268.153	267.843	0.093	-836.813	-823.794	143.435
350	54.500	276.233	268.472	2.716	-837.144	-821.597	122.617
400	58.160	283.755	269.918	5.535	-837.438	-819.355	106.997
450	61.331	290.793	271.850	8.524	-837.703	-817.079	94.844
500	64.045	297.399	274.078	11.660	-837.939	-814.774	85.119
600	68.325	309.475	278.991	18.290	-838.344	-810.102	70.526
700	71.432	320.254	284.130	25.287	-838.682	-805.368	60.097
800	73.715	329.949	289.262	32.550	-838.979	-800.588	52.273
900	75.422	338.734	294.278	40.010	-839.257	-795.772	46.185
1000	76.721	346.751	299.131	47.620	-839.530	-790.926	41.314
1100	77.729	354.113	303.799	55.345	-839.810	-786.052	37.327
1200	78.523	360.911	308.279	63.159	-840.105	-781.152	34.003
1300	79.158	367.222	312.573	71.044	-840.422	-776.227	31.189
1400	79.674	373.108	316.689	78.987	-840.767	-771.276	28.777
1500	80.097	378.620	320.636	86.976	-841.143	-766.299	26.685
1600	80.449	383.801	324.424	95.004	-841.554	-761.297	24.854
1700	80.744	388.687	328.061	103.064	-842.001	-756.267	23.237
1800	80.994	393.310	331.559	111.151	-842.487	-751.210	21.800
1900	81.209	397.695	334.926	119.261	-843.009	-746.125	20.512
2000	81.394	401.865	338.169	127.392	-843.569	-741.011	19.353
2100	81.557	405.840	341.298	135.540	-844.163	-735.869	18.304
2200	81.700	409.638	344.318	143.702	-844.791	-730.698	17.349
2300	81.829	413.272	347.238	151.879	-845.449	-725.497	16.477
2400	81.945	416.757	350.062	160.068	-896.393	-719.198	15.653
2500	82.053	420.105	352.798	168.268	-897.180	-711.799	14.872
2600	82.153	423.325	355.449	176.478	-897.955	-704.369	14.151
2700	82.249	426.427	358.020	184.698	-898.715	-696.908	13.483
2800	82.341	429.420	360.517	192.928	-899.457	-689.420	12.861
2900	82.431	432.311	362.943	201.166	-900.178	-681.906	12.282
3000	82.520	435.107	365.302	209.414	-900.875	-674.368	11.742
3100	82.610	437.814	367.598	217.671	-901.546	-666.806	11.236
3200	82.701	440.439	369.834	225.936	-902.188	-659.224	10.761
3300	82.793	442.985	372.012	234.211	-902.800	-651.622	10.314
3400	82.888	445.458	374.136	242.495	-903.379	-644.001	9.894
3500	82.985	447.862	376.208	250.788	-903.924	-636.365	9.497
3600	83.085	450.201	378.231	259.092	-904.434	-628.713	9.122
3700	83.189	452.479	380.207	267.405	-904.907	-621.047	8.768
3800	83.296	454.699	382.138	275.730	-905.343	-613.369	8.431
3900	83.407	456.864	384.027	284.065	-905.740	-605.680	8.112
4000	83.521	458.977	385.874	292.411	-906.099	-597.982	7.809
4100	83.639	461.041	387.683	300.769	-906.418	-590.275	7.520
4200	83.760	463.058	389.453	309.139	-1386.548	-575.536	7.158
4300	83.884	465.030	391.188	317.521	-1385.702	-556.237	6.757
4400	84.011	466.960	392.888	325.916	-1384.818	-536.957	6.374
4500	84.141	468.850	394.555	334.324	-1383.898	-517.698	6.009
4600	84.273	470.700	396.191	342.744	-1382.941	-498.460	5.660
4700	84.408	472.514	397.795	351.178	-1381.949	-479.242	5.326
4800	84.544	474.293	399.371	359.626	-1380.923	-460.046	5.006
4900	84.682	476.037	400.917	368.087	-1379.863	-440.872	4.700
5000	84.822	477.750	402.437	376.563	-1378.770	-421.720	4.406
5100	84.962	479.431	403.930	385.052	-1377.646	-402.590	4.123
5200	85.103	481.082	405.398	393.555	-1376.492	-383.483	3.852
5300	85.244	482.704	406.841	402.072	-1375.309	-364.397	3.591
5400	85.385	484.299	408.261	410.604	-1374.099	-345.335	3.340
5500	85.526	485.867	409.658	419.149	-1372.863	-326.296	3.099
5600	85.667	487.409	411.033	427.709	-1371.604	-307.278	2.866
5700	85.806	488.927	412.386	436.283	-1370.321	-288.284	2.642
5800	85.945	490.420	413.718	444.870	-1369.017	-269.312	2.425
5900	86.082	491.891	415.031	453.472	-1367.694	-250.363	2.217
6000	86.218	493.339	416.324	462.087	-1366.354	-231.437	2.015

PREVIOUS: December 1966 (1 atm) CURRENT: December 1966 (1 bar)

Boron Fluoride Oxide (OBF_2) $B_1F_2O_1(g)$

TRIFLUOROBORANE (BF_3) IDEAL GAS M_r = 67.805209

$S°(298.15\ K) = 254.36 \pm 0.04\ J\ K^{-1}\ mol^{-1}$

$\Delta_f H°(0\ K) = -1132.8 \pm 1.7\ kJ\ mol^{-1}$
$\Delta_f H°(298.15\ K) = -1135.6 \pm 1.7\ kJ\ mol^{-1}$

Vibrational Frequencies and Degeneracies

ν, cm^{-1}
888 (1)
696.7(1)
1463.3(2)
480.7(2)

Ground State Quantum Weight: 1 $\sigma = 6$
Point Group: D_{3h}
Bond Distance: B-F = 1.307 ± 0.002 Å
Bond Angle: F-B-F = 120°
Products of the Moments of Inertia: $I_A I_B I_C = 1.056517 \times 10^{-114}\ g^3\ cm^6$

Enthalpy of Formation

The adopted enthalpy of formation was selected by a simultaneous adjustment of several interrelated pieces of data relating to the heat of formation of HF (see HF, g table 12-31-68). The data which were considered relating directly to BF_3 were as follows:

A. $NF_3(g) + B(cr, \beta) \rightarrow BF_3(g) + 0.5\ N_2$ and $\Delta_r H°(298.15\ K) = -239.46 \pm 1.2\ kcal\ mol^{-1}$.

Ludwig and Cooper (1) report the above value after corretion for 0.44 percent impurities in the boron. As much as 5 percent unburned boron was determined by analysis.

B. $1.5\ F_2(g) + B(cr, \beta) \rightarrow BF_3(g)$

Wise et al. (2) originally reported $\Delta_r H°(298.15\ K) = -270.1 \pm 0.24\ kcal\ mol^{-1}$ but later reanalysis (3) of the impurities and due allowance for their effect changed this to $-271.6 \pm 0.9\ kcal\ mol^{-1}$. Johnson et al. (3) using zone refined boron obtained $\Delta_r H°(298.15\ K) = -271.65 \pm 0.22\ kcal\ mol^{-1}$ after correction for 0.12 percent impurities. Domalski and Armstrong (4) obtained $\Delta_r H°(298.15\ K) = -271.03 \pm 0.51\ kcal\ mol^{-1}$ from a combustion in the presence of Teflon, the boron contained 0.32 percent impurities and corrections were included. The above reaction constituted about 0.35 of the total energy measured. The measurements of Gross et al. (5), although on a zone refined sample, did not include a complete purity analysis and their data were not included since probable impurities could introduce serious errors.

C. $3\ HF(50\ H_2O) + B(cr, \beta) + 0.75\ O_2(g) \rightarrow BF_3(g) + 1.5\ H_2O(\ell)$

From the measurements of Gunn (6) on the reaction of $BF_3(g) + 15.67\ HF(3.747\ H_2O) \rightarrow$ [solution] and Good and Mansson (7) for the reaction $B(cr, \beta) + 0.75\ O_2(g) + 18.57\ HF(3.065\ H_2O) \rightarrow$ [solution] + $1.5\ H_2O(\ell)$, where the resulting solutions have the same composition, we obtain $\Delta_f H°(298.15\ K) = -142.77 \pm 0.5\ kcal\ mol^{-1}$. Further details of the simultaneous solution are given by Syverud (8).

Heat Capacity and Entropy

The fundamental frequencies were selected from the measurements of Lindeman and Wilson (9), McKean (10), Nielsen (11), Anderson et al. (12) and Yost et al. (13). The bond length and structure of those reported by Gin et al. (14) which are in agreement with those of Kuchitsu and Konaka (15) from gas phase electron diffraction. These measurements disagree with the earlier determinations of Nielsen (11) and Levy and Brockway (16) but are considerably more precise and are adopted.

The principal moments of inertia are $I_A = I_B = 8.0838 \times 10^{-39}\ g\ cm^2$ and $I_C = 16.1676 \times 10^{-39}\ g\ cm^2$.

References

1. J. R. Ludwig and W. J. Cooper, J. Chem. Eng. Data 8, 76 (1963).
2. S. S. Wise, J. L. Margrave, H. M. Feder and W. N. Hubbard, J. Phys. Chem. 65, 2157 (1961).
3. G. K. Johnson, H. M. Feder and W. N. Hubbard, J. Phys. Chem. 70, 1 (1966).
4. E. S. Domalski and G. T. Armstrong, J. Res. Natl. Bur. Stand. 71A, 195 (1967).
5. P. Gross, C. Hayman, D. L. Levi and M. C. Stuart, Fulmer Research Institute Report R.146/4/23 Nov. 1960.
6. S. R. Gunn, J. Phys. Chem. 69, 1010 (1965).
7. W. D. Good and M. Mansson, J. Phys. Chem. 70, 97 (1966).
8. A. N. Syverud, AFRPL-TR-69-70, January 1969.
9. L. P. Lindeman and M. K. Wilson, J. Chem. Phys. 24, 242 (1956).
10. D. C. McKean, J. Chem. Phys. 24, 1002 (1956).
11. A. H. Nielsen, J. Chem. Phys. 22, 659 (1954).
12. T. F. Anderson, E. N. Lassettre and D. M. Yost, J. Chem. Phys. 4, 703 (1936).
13. D. M. Yost, D. DeVault, T. F. Anderson, and E. N. Lassettre, J. Chem. Phys. 6, 424 (1938).
14. S. G. W. Gin, J. K. Kenney and J. Overend, J. Chem. Phys. 48, 1571 (1968).
15. K. Kuchitsu and S. Konaka, J. Chem. Phys. 45, 4342 (1966).
16. H. Levy and L. O. Brockway, J. Amer. Chem. Soc. 59, 2085 (1937).

Trifluoroborane (BF_3) $B_1F_3(g)$

Enthalpy Reference Temperature = T_r = 298.15 K Standard State Pressure = p° = 0.1 MPa

T/K	$C_p°$ (J K⁻¹mol⁻¹)	S°	$-[G°-H°(T_r)]/T$	$H°-H°(T_r)$ (kJ mol⁻¹)	$\Delta_f H°$	$\Delta_f G°$	Log K_f
0	0.	0.	INFINITE	-11.647	-1132.817	-1132.817	INFINITE
100	34.089	210.208	293.304	-8.310	-1133.864	-1129.299	589.885
200	41.964	235.971	258.706	-4.547	-1134.828	-1124.343	293.648
250	46.471	245.824	255.164	-2.335	-1135.240	-1121.673	234.361
298.15	50.446	254.355	254.355	0.	-1135.621	-1119.025	196.049
300	50.590	254.668	254.356	0.093	-1135.636	-1118.922	194.822
350	54.273	262.749	254.986	2.717	-1136.023	-1116.106	166.570
400	57.539	270.213	256.429	5.514	-1136.407	-1113.235	145.374
450	60.420	277.160	258.351	8.464	-1136.785	-1110.315	128.882
500	62.945	283.660	260.560	11.550	-1137.150	-1107.355	115.685
600	67.071	295.518	265.418	18.060	-1137.831	-1101.330	95.879
700	70.195	306.103	270.488	24.931	-1138.443	-1095.198	81.725
800	72.569	315.638	275.546	32.074	-1138.997	-1088.981	71.103
900	74.389	324.295	280.489	39.426	-1139.508	-1082.698	62.838
1000	75.802	332.209	285.271	46.938	-1139.993	-1076.360	56.223
1100	76.914	339.488	289.873	54.576	-1140.465	-1069.974	50.809
1200	77.800	346.220	294.292	62.314	-1140.935	-1063.545	46.295
1300	78.516	352.476	298.530	70.131	-1141.413	-1057.076	42.474
1400	79.101	358.317	302.594	78.013	-1141.907	-1050.570	39.197
1500	79.584	363.792	306.493	85.948	-1142.422	-1044.028	36.356
1600	79.988	368.941	310.237	93.927	-1142.963	-1037.451	33.869
1700	80.327	373.801	313.834	101.943	-1143.531	-1030.840	31.674
1800	80.616	378.401	317.295	109.991	-1144.127	-1024.194	29.721
1900	80.863	382.766	320.627	118.065	-1144.752	-1017.514	27.973
2000	81.076	386.919	323.838	126.162	-1145.404	-1010.800	26.399
2100	81.261	390.880	326.937	134.279	-1146.080	-1004.054	24.974
2200	81.423	394.664	329.930	142.413	-1146.777	-997.274	23.678
2300	81.564	398.286	332.824	150.563	-1147.491	-990.462	22.494
2400	81.689	401.760	335.625	158.726	-1198.477	-982.551	21.385
2500	81.800	405.097	338.337	166.900	-1199.289	-973.537	20.341
2600	81.899	40[illegible].308	340.967	175.085	-1200.072	-964.492	19.377
2700	81.988	411.400	343.519	183.280	-1200.822	-955.416	18.484
2800	82.067	414.383	345.997	191.483	-1201.536	-946.314	17.654
2900	82.139	417.265	348.405	199.693	-1202.209	-937.187	16.881
3000	82.203	420.050	350.747	207.910	-1202.839	-928.038	16.159
3100	82.262	422.747	353.026	216.133	-1203.422	-918.868	15.483
3200	82.315	425.359	355.246	224.362	-1203.957	-909.680	14.849
3300	82.364	427.893	357.409	232.596	-1204.442	-900.476	14.253
3400	82.408	430.352	359.519	240.835	-1204.875	-891.259	13.693
3500	82.449	432.742	361.577	249.078	-1205.255	-882.029	13.164
3600	82.487	435.065	363.586	257.325	-1205.581	-872.789	12.664
3700	82.521	437.326	365.549	265.575	-1205.852	-863.541	12.191
3800	82.553	439.527	367.466	273.829	-1206.069	-854.286	11.743
3900	82.583	441.671	369.342	282.086	-1206.232	-845.026	11.318
4000	82.610	443.763	371.176	290.346	-1206.341	-835.763	10.914
4100	82.636	445.803	372.972	298.608	-1206.396	-826.498	10.530
4200	82.660	447.794	374.730	306.873	-1686.248	-810.208	10.076
4300	82.682	449.740	376.451	315.140	-1685.110	-789.364	9.589
4400	82.702	451.641	378.139	323.409	-1683.924	-768.546	9.124
4500	82.721	453.500	379.793	331.680	-1682.689	-747.755	8.680
4600	82.739	455.318	381.415	339.953	-1681.408	-726.993	8.255
4700	82.756	457.097	383.006	348.228	-1680.081	-706.259	7.849
4800	82.772	458.840	384.568	356.504	-1678.710	-685.554	7.460
4900	82.787	460.547	386.101	364.782	-1677.297	-664.878	7.088
5000	82.801	462.219	387.607	373.062	-1675.842	-644.231	6.730
5100	82.814	463.859	389.086	381.343	-1674.349	-623.613	6.387
5200	82.827	465.468	390.540	389.625	-1672.817	-603.025	6.057
5300	82.839	467.045	391.968	397.908	-1671.249	-582.467	5.741
5400	82.850	468.594	393.373	406.192	-1669.647	-561.939	5.436
5500	82.860	470.114	394.755	414.478	-1668.011	-541.442	5.142
5600	82.870	471.607	396.114	422.764	-1666.344	-520.973	4.859
5700	82.880	473.074	397.451	431.052	-1664.647	-500.536	4.587
5800	82.889	474.516	398.767	439.340	-1662.921	-480.128	4.324
5900	82.897	475.933	400.063	447.630	-1661.168	-459.749	4.070
6000	82.905	477.326	401.339	455.920	-1659.389	-439.401	3.825

PREVIOUS: June 1969 (1 atm) CURRENT: June 1969 (1 bar)

Trifluoroborane (BF_3) $B_1F_3(g)$

POTASSIUM TETRAFLUOROBORATE (KBF_4) CRYSTAL M_r = 125.901912

$S°(298.15\ K) = 133.89 \pm 4.0\ J\ K^{-1}\ mol^{-1}$
$T_{trs} = 556\ K$
$T_{fus} = 843\ K$

$\Delta_f H°(0\ K)$ = Unknown
$\Delta_f H°(298.15\ K) = -1887 \pm 4\ kJ\ mol^{-1}$
$\Delta_{trs} H° = 14.06 \pm 0.16\ kJ\ mol^{-1}$
$\Delta_{fus} H° = 17.66 \pm 0.16\ kJ\ mol^{-1}$

Enthalpy of Formation

Bills and Cotton (1) measured the enthalpies of several reactions at 25°C which led to the following result:

$H_3BO_3(cr) + 4(HF{\cdot}23\ H_2O) + KNO_3(cr) + KBF_4(cr) + HNO_3{\cdot}92\ H_2O + 3\ H_2O$.

Combining this result with the following enthalpy of formation data,

$\Delta_f H°(H_3BO_3, cr) = -261.47 \pm 0.2\ kcal\ mol^{-1}$ (2)
$\Delta_f H°(HF{\cdot}23\ H_2O) = -76.75 \pm 0.1\ kcal\ mol^{-1}$ (3)
$\Delta_f H°(KNO_3, cr) = -118.22\ kcal\ mol^{-1}$ (4)
$\Delta_f H°(HNO_3{\cdot}92\ H_2O) = -49.44\ kcal\ mol^{-1}$ (5)
$\Delta_f H°(H_2O, \ell) = -68.315\ kcal\ mol^{-1}$

We derive, $\Delta_f H°(KBF_4, cr) = -451.6 \pm 1.2\ kcal\ mol^{-1}$.

Gross, Hayman and Joel (6) recently measured the following enthalpies of reaction:

$NaF(cr) + BF_3(g) \rightarrow NaBF_4(cr)$ $\Delta_r H° = -32.04 \pm 0.12\ kcal\ mol^{-1}$
$[KBF_4 + NaF](cr) + HF(aq) \rightarrow$ Soln A $\Delta_{sol} H° = +8.14 \pm 0.18\ kcal\ mol^{-1}$
$[NaBF_4 + KF](cr) + HF(aq) \rightarrow$ Soln A $\Delta_{sol} H° = -2.94 \pm 0.13\ kcal\ mol^{-1}$

These data combine to give $\Delta H_r = -43.13 \pm 0.22\ kcal\ mol^{-1}$ for the reaction $KF(cr) + BF_3(g) \rightarrow KBF_4(cr)$. Combining this result with the following enthalpies of formation: $\Delta_f H°(KF, cr) = -135.9 \pm 0.1\ kcal\ mol^{-1}$ (7); $\Delta_f H°(BF_3, g) = -271.42 \pm 0.4\ kcal\ mol^{-1}$ (7), we derive $\Delta_f H°(KBF_4, cr) = -450.5 \pm 0.7\ kcal\ mol^{-1}$. An average value, $\Delta_f H°(KBF_4, cr) = -451.0 \pm 1\ kcal\ mol^{-1}$, of these two results is adopted here.

Heat Capacity and Entropy

The heat capacity data are calculated from the high temperature enthalpy data of Dworkin and Bredig (8). S°(298.15 K) is chosen such that a 3rd law analysis of the equilibrium data for the reaction $KBF_4(cr, \ell) = KF(cr, \ell) + BF_3(g)$ reported by de Boer and van Liempt (9) gives the experimentally determined heat of reaction at 25 C (6).

Transition Data

T_{trs} is from the high temperature studies of Dworkin and Bredig (8). $\Delta_{trs}H°$ is calcualted from their high temperature enthalpy data.

Fusion Data

T_{fus} has been reported as 803 K by de Boer and van Liempt (9) and 843 K by Dworkin and Bredig (8). Dworkin and Bredig's value is adopted here. $\Delta_{fus}H°$ is calculated from their high temperature enthalpy data.

Sublimation Data

$\Delta_{sub}H°(298.15\ K)$ is estimated by comparison with data for $LiALF_4(cr)$ reported by Hildenbrand and Theard (10).

References

1. J. L. Bills and F. A. Cotton, J. Phys. Chem. 64, 1477 (1960).
2. JANAF Thermochemical Tables: H_3BO_3(cr), 12-31-64.
3. JANAF Thermochemical Tables: HF(g), 12-31-68 and Ref. 4.
4. V. B. Parker, NSRDS-NBS 2, 1965 and Ref. 5.
5. U. S. Nat. Bur. Stand. Tech. Note 270-3, 1968.
6. P. Gross, C. Hayman and H. A. Joel, Trans. Faraday Soc. 64, 317 (1968).
7. JANAF Thermochemical Tables: KF(cr), 6-30-69.
8. A. S. Dworkin and M. A. Bredig, private communication, Oak Ridge Natl. Lab., Sept. 23, 1969.
9. J. H. de Boer and J. A. M. van Liempt, Rec. Trav. Chim. 46, 124 (1927).
10. D. L. Hildenbrand and L. P. Theard, Aeronutronic, Division of Ford Motor Company, Publication No. U-2174, June 15, 1961.

Potassium Tetrafluoroborate (KBF_4) $B_1F_4K_1$(cr)

Enthalpy Reference Temperature = T_r = 298.15 K Standard State Pressure = p° = 0.1 MPa

T/K	$C_p°$ (J K⁻¹mol⁻¹)	S° (J K⁻¹mol⁻¹)	$-[G°-H°(T_r)]/T$ (J K⁻¹mol⁻¹)	$H°-H°(T_r)$ (kJ mol⁻¹)	$\Delta_f H°$ (kJ mol⁻¹)	$\Delta_f G°$ (kJ mol⁻¹)	Log K_f
0							
100							
200							
298.15	114.483	133.888	133.888	0.	-1886.984	-1784.958	312.717
300	114.780	134.597	133.890	0.212	-1886.963	-1784.325	310.679
400	130.817	169.809	138.583	12.490	-1887.964	-1749.854	228.507
500	146.963	200.734	147.976	26.379	-1885.646	-1715.571	179.224
556.000	156.027	216.808	154.105	34.863		I <--> II	
556.000	140.146	242.093	154.105	48.921		TRANSITION	
600	142.097	252.840	160.956	55.130	-1868.862	-1682.950	146.514
700	146.519	275.076	175.703	69.561	-1866.685	-1652.134	123.284
800	150.938	294.930	189.387	84.434	-1864.291	-1621.644	105.882
843.000	152.838	302.881	194.975	90.965	-------- II	<--> LIQUID	--------
900	155.356	312.963	202.131	99.749	-1861.649	-1591.469	92.366
1000	159.779	329.560	214.054	115.505	-1858.748	-1561.602	81.570
1100	164.197	344.996	225.264	131.704	-1934.537	-1527.429	72.532
1200	168.619	359.472	235.851	148.345	-1930.056	-1490.614	64.885
1300	173.038	373.143	245.891	165.428	-1925.232	-1454.187	58.430
1400	177.456	386.129	255.448	182.953	-1920.057	-1418.145	52.912
1500	181.878	398.523	264.576	200.919	-1914.526	-1382.485	48.142

PREVIOUS: March 1965 CURRENT: December 1969

Potassium Tetrafluoroborate (KBF_4) $B_1F_4K_1$(cr)

POTASSIUM TETRAFLUOROBORATE (KBF_4) LIQUID M_r = 125.901912

$S°(298.15\ K)$ = [150.129] J K^{-1} mol^{-1}

T_{fus} = 843 K

$\Delta_f H°(0\ K)$ = Unknown

$\Delta_f H°(298.15\ K)$ = [-1869.415] kJ mol^{-1}

$\Delta_{fus} H°$ = 17.66 ± 0.16 kJ mol^{-1}

Enthalpy of Formation

$\Delta_f H°(\ell$, 298.15 K) is calculated from $\Delta_f H°(cr)$ by adding $\Delta_{fus} H°$ and the difference in enthalpy, H°(843 K)-H°(298.15 K), between the crystal and liquid.

Heat Capacity and Entropy

The heat capacity data are calculated from the high temperature enthalpy data of Dworkin and Bredig (1). S°(298.15 K) is calculated in a manner analogous to that used for the enthalpy of formation.

Fusion Data

T_{fus} has been reported as 803 K by de Boer and van Liempt (2) and 843 K by Dworkin and Bredig (1). Dworkin and Bredig's value is adopted here. $\Delta_{fus} H°$ is calculated from their high temperature enthalpy data.

References

1. A. S. Dworkin and M. A. Bredig, Oak Ridge Natl. Lab., personal communication, Sept. 23, 1969.
2. J. H. de Boer and J. A. M. van Liempt, Rec. Trav. Chim. 46, 124 (1927).

Potassium Tetrafluoroborate (KBF_4) $B_1F_4K_1(l)$

Enthalpy Reference Temperature = T_r = 298.15 K Standard State Pressure = p° = 0.1 MPa

T/K	$C_p°$	S°	$-[G°-H°(T_r)]/T$	$H°-H°(T_r)$	$\Delta_f H°$	$\Delta_f G°$	Log K_f
	J K⁻¹mol⁻¹			kJ mol⁻¹			
0							
100							
200							
298.15	167.117	150.129	150.129	0.	-1869.415	-1772.232	310.488
300	167.117	151.163	150.133	0.309	-1869.298	-1771.630	308.468
400	167.117	199.240	156.688	17.021	-1865.865	-1739.527	227.159
500	167.117	236.531	169.066	33.733	-1860.724	-1708.547	178.491
600	167.117	267.000	182.926	50.444	-1855.979	-1678.563	146.132
700	167.117	292.761	196.824	67.156	-1851.522	-1649.350	123.076
800	167.117	315.077	210.242	83.868	-1847.289	-1620.759	105.825
843.000	167.117	323.826	215.814	91.054	-------- II	<--> LIQUID	--------
900	167.117	334.760	223.005	100.580	-1843.250	-1592.688	92.437
1000	167.117	352.368	235.077	117.291	-1839.393	-1565.056	81.750
1100	167.117	368.296	246.475	134.003	-1914.670	-1533.192	72.805
1200	167.117	382.837	257.241	150.715	-1910.118	-1498.713	65.237
1300	167.117	396.213	267.424	167.426	-1905.665	-1464.611	58.849
1400	167.117	408.598	277.071	184.138	-1901.303	-1430.848	53.386
1500	167.117	420.128	286.228	200.850	-1897.026	-1397.394	48.662

PREVIOUS: CURRENT: December 1969

Potassium Tetrafluoroborate (KBF_4) $B_1F_4K_1(l)$

POTASSIUM TETRAFLUOROBORATE (KBF_4) CRYSTAL–CRYSTAL–LIQUID

0 to 556 K crystal, I
556 to 843 K crystal, II
above 843 K liquid

Refer to the individual tables for details.

M_r = 125.901912 Potassium Tetrafluoroborate (KBF_4) $B_1F_4K_1$(cr,l)

Enthalpy Reference Temperature = T_r = 298.15 K Standard State Pressure = p° = 0.1 MPa

T/K	C_p° (J K⁻¹mol⁻¹)	S° (J K⁻¹mol⁻¹)	−[G°−H°(T_r)]/T (J K⁻¹mol⁻¹)	H°−H°(T_r) (kJ mol⁻¹)	Δ_fH° (kJ mol⁻¹)	Δ_fG° (kJ mol⁻¹)	Log K_f
0							
100							
200							
298.15	114.483	133.888	133.888	0.	-1886.984	-1784.958	312.717
300	114.780	134.597	133.890	0.212	-1886.963	-1784.325	310.679
400	130.817	169.809	138.583	12.490	-1887.964	-1749.854	228.507
500	146.963	200.734	147.976	26.379	-1885.646	-1715.571	179.224
556.000	156.027	216.808	154.105	34.863		I <--> II	
556.000	140.146	242.093	154.105	48.921		TRANSITION	
600	142.097	252.840	160.956	55.130	-1868.862	-1682.950	146.514
700	146.519	275.076	175.703	69.561	-1866.685	-1652.134	123.284
800	150.938	294.930	189.387	84.434	-1864.291	-1621.644	105.882
843.000	152.838	302.881	194.975	90.965		II <--> LIQUID	
843.000	167.117	323.826	194.975	108.622		TRANSITION	
900	167.117	334.760	203.485	118.147	-1843.251	-1592.688	92.437
1000	167.117	352.368	217.509	134.859	-1839.394	-1565.057	81.750
1100	167.117	368.296	230.504	151.571	-1914.671	-1533.192	72.805
1200	167.117	382.837	242.601	168.283	-1910.118	-1498.714	65.237
1300	167.117	396.213	253.910	184.994	-1905.665	-1464.612	58.849
1400	167.117	408.598	264.522	201.706	-1901.304	-1430.849	53.386
1500	167.117	420.128	274.516	218.418	-1897.027	-1397.394	48.662

PREVIOUS: CURRENT: December 1969

Potassium Tetrafluoroborate (KBF_4) $B_1F_4K_1$(cr,l)

POTASSIUM TETRAFLUOROBORATE (KBF_4) IDEAL GAS M_r = 125.901912

$S°(298.15 K) = [315.4 \pm 3] J K^{-1} mol^{-1}$

$\Delta_f H°(0 K) = [1543.9 \pm 5] kJ mol^{-1}$

$\Delta_f H°(298.15 K) = [-1552.3 \pm 5] kJ mol^{-1}$

Vibrational Frequencies and Degeneracies

ν, cm^{-1}	ν, cm^{-1}	ν, cm^{-1}
[900](1)	[1000](1)	[300](2)
[450](1)	[300](1)	[500](2)
[350](1)	[250](1)	[1400](2)

Ground State Quantum Weight: [1] σ = [3]

Point Group: [C_{3v}]

Bond Distances: B-F = 1.380 Å K-F = 2.752 Å B-K = [2.863] Å

Bond Angles: F-B-F = 109°28' F_{ax}-B-K = [180]°

Product of Moments of Inertia: $I_A I_B I_C = [4.547923 \times 10^{-113}] g^3 cm^6$

Enthalpy of Formation

$\Delta_f H°(g)$ is calculated from $\Delta_f H°(cr)$ and the estimated enthalpy of sublimation.

Heat Capacity and Entropy

The vibrational frequencies are estimated by comparison with data for BF_4^-, BF_3, KF, and K_2F_2. The infrared and Raman spectra of BF_4^- in various phases have been reported by (1, 2, 3, 4). The molecular structure is assumed to be the same as for $LiAlF_4$ as discussed by Porter and Zeller (5). The B-F and K-F bond lengths and F-B-F bond angle are average values of the crystallographic data recently reported by Clark and Lynton (6). The B-K bond length is calcualted from the above structural data according to the assumed molecular model. The principal moments of inertia are $I_A = I_B = 16.0214 \times 10^{-39} g cm^2$ and $I_B = I_C = 53.2791 \times 10^{-39} g cm^2$.

References

1. G. L. Cote and H. W. Thompson, Proc. Royal Soc. (London), 210A, 217 (1951).
2. J. O. Edwards, G. C. Morrison, V. F. Ross, and J. W. Schultz, J. Amer. Chem. Soc. 77, 266 (1955).
3. J. Lecomte, C. Duval, and C. Wadler, Compt. Rend. 249, 1991 (1959).
4. J. A. A. Ketelaar and R. L. Fulton, Z. Elektrochem. 64, 641 (1960).
5. R. F. Porter and E. E. Zeller, J. Chem. Phys. 33, 858 (1960).
6. M. J. R. Clark and H. Lynton, Can. J. Chem. 47, 2579 (1969).

Potassium Tetrafluoroborate (KBF_4) $B_1F_4K_1(g)$

Enthalpy Reference Temperature = T_r = 298.15 K Standard State Pressure = p° = 0.1 MPa

T/K	$C_p°$ (J K⁻¹mol⁻¹)	S° (J K⁻¹mol⁻¹)	$-[G°-H°(T_r)]/T$ (J K⁻¹mol⁻¹)	$H°-H°(T_r)$ (kJ mol⁻¹)	$\Delta_f H°$ (kJ mol⁻¹)	$\Delta_f G°$ (kJ mol⁻¹)	Log K_f
0	0.	0.	INFINITE	-17.598	-1543.916	-1543.916	INFINITE
100	45.326	242.588	382.561	-13.997	-1547.886	-1534.479	801.530
200	72.713	283.084	323.060	-7.995	-1550.669	-1519.843	396.942
250	81.920	300.343	316.820	-4.119	-1551.543	-1512.031	315.922
298.15	88.941	315.394	315.394	0.	-1552.264	-1504.354	263.557
300	89.184	315.944	315.395	0.165	-1552.291	-1504.057	261.880
350	95.129	330.153	316.504	4.777	-1555.374	-1495.865	223.246
400	100.093	343.189	319.036	9.661	-1556.073	-1487.315	194.223
450	104.277	355.227	322.397	14.774	-1556.687	-1478.683	171.641
500	107.817	366.402	326.245	20.078	-1557.226	-1469.986	153.568
600	113.376	386.578	334.657	31.153	-1558.120	-1452.450	126.447
700	117.425	404.375	343.370	42.703	-1558.823	-1434.781	107.065
800	120.418	420.260	352.007	54.603	-1559.403	-1417.020	92.522
900	122.668	434.579	360.399	66.762	-1559.916	-1399.191	81.207
1000	124.391	447.597	368.478	79.119	-1560.414	-1381.306	72.152
1100	125.731	459.518	376.220	91.628	-1639.894	-1358.760	64.522
1200	126.792	470.505	383.625	104.256	-1639.425	-1333.223	58.034
1300	127.643	480.689	390.705	116.979	-1638.961	-1307.725	52.545
1400	128.334	490.174	397.475	129.779	-1638.511	-1282.262	47.842
1500	128.903	499.049	403.954	142.642	-1638.083	-1256.831	43.767
1600	129.376	507.383	410.160	155.557	-1637.683	-1231.427	40.202
1700	129.774	515.239	416.113	168.515	-1637.314	-1206.048	37.057
1800	130.111	522.666	421.828	181.509	-1636.979	-1180.689	34.263
1900	130.399	529.709	427.322	194.535	-1636.677	-1155.348	31.763
2000	130.646	536.404	432.610	207.588	-1636.408	-1130.022	29.513
2100	130.861	542.784	437.706	220.663	-1636.171	-1104.709	27.478
2200	131.048	548.876	442.622	233.759	-1635.962	-1079.406	25.628
2300	131.212	554.705	447.369	246.872	-1635.777	-1054.112	23.940
2400	131.357	560.292	451.959	260.001	-1685.874	-1027.757	22.369
2500	131.485	565.657	456.400	273.143	-1685.808	-1000.337	20.901
2600	131.600	570.816	460.702	286.298	-1685.723	-972.920	19.546
2700	131.702	575.785	464.873	299.463	-1685.618	-945.506	18.292
2800	131.793	580.576	468.920	312.638	-1685.491	-918.097	17.127
2900	131.876	585.203	472.850	325.821	-1685.339	-890.692	16.043
3000	131.950	589.675	476.671	339.012	-1685.162	-863.294	15.031
3100	132.018	594.002	480.386	352.211	-1684.959	-835.902	14.085
3200	132.079	598.195	484.002	365.416	-1684.732	-808.516	13.198
3300	132.135	602.260	487.525	378.626	-1684.472	-781.139	12.364
3400	132.186	606.205	490.958	391.843	-1684.195	-753.769	11.580
3500	132.233	610.038	494.305	405.064	-1683.895	-726.408	10.841
3600	132.277	613.764	497.572	418.289	-1683.580	-699.056	10.143
3700	132.316	617.388	500.762	431.519	-1683.251	-671.712	9.483
3800	132.353	620.917	503.877	444.752	-1682.913	-644.379	8.858
3900	132.387	624.356	506.923	457.989	-1682.570	-617.053	8.264
4000	132.419	627.708	509.901	471.230	-1682.221	-589.737	7.701
4100	132.448	630.978	512.814	484.473	-1681.879	-562.430	7.165
4200	132.475	634.170	515.666	497.719	-2161.400	-528.105	6.568
4300	132.500	637.288	518.458	510.968	-2160.002	-489.235	5.943
4400	132.524	640.334	521.193	524.219	-2158.351	-450.422	5.347
4500	132.546	643.312	523.874	537.473	-2156.940	-411.621	4.778
4600	132.567	646.226	526.502	550.728	-2155.552	-372.852	4.234
4700	132.586	649.077	529.080	563.986	-2154.174	-334.114	3.713
4800	132.604	651.869	531.609	577.246	-2152.719	-295.418	3.215
4900	132.622	654.603	534.091	590.507	-2151.402	-256.738	2.737
5000	132.638	657.283	536.529	603.770	-2150.122	-218.084	2.278
5100	132.653	659.909	538.922	617.034	-2148.883	-179.456	1.838
5200	132.667	662.485	541.274	630.300	-2147.687	-140.851	1.415
5300	132.681	665.013	543.585	643.568	-2146.536	-102.269	1.008
5400	132.693	667.493	545.856	656.837	-2145.433	-63.709	0.616
5500	132.706	669.928	548.090	670.106	-2144.378	-25.168	0.239
5600	132.717	672.319	550.287	683.378	-2143.373	13.354	-0.125
5700	132.728	674.668	552.449	696.650	-2142.074	51.810	-0.475
5800	132.738	676.976	554.576	709.923	-2141.126	90.291	-0.813
5900	132.748	679.246	556.670	723.197	-2140.225	128.757	-1.140
6000	132.757	681.477	558.731	736.473	-2139.370	167.206	-1.456

PREVIOUS: December 1969 (1 atm) CURRENT: December 1969 (1 bar)

Potassium Tetrafluoroborate (KBF_4) $B_1F_4K_1(g)$

BORANE (BH) IDEAL GAS M_r = 11.81794

$S°(298.15\ K) = [171.85]\ J\ K^{-1}\ mol^{-1}$

$\Delta_f H°(0\ K) = 439.5 \pm 8.4\ kJ\ mol^{-1}$
$\Delta_f H°(298.15\ K) = 442.7 \pm 8.4\ kJ\ mol^{-1}$

Electronic Level and Quantum Weight

State	ε_i, cm^{-1}	g_i
$^1\Sigma^+$	0	1

$\omega_e = [2368]\ cm^{-1}$ $\omega_e x_e = [49]\ cm^{-1}$ $\sigma = 1$
$B_e = 12.036\ cm^{-1}$ $\alpha_e = .413\ cm^{-1}$ $r_e = 1.2325\ Å$

Enthalpy of Formation

$\Delta_f H°(298.15\ K)$ was calculated from $D_0^0(BH) = 3.39 \pm 0.04$ e.v. reported by Hurley (1). Hurley stated "by combining an analysis of the spectroscopic data with theoretical calculations, we obtain a greatly improved estimate of $D_0^0(BH)$". This value and the other $D_0^0(BH)$ values which are tabulated below are based on the same spectroscopic data but differ due to different methods of analysis.

$D_0^0(BH)$, eV	Method	Source
3.0 ± 0.4	Birge-Sponer extrapolations	(2)
3.6	Observed predissociation by rotation in BH, A'Π and deduced $D_0^0(BH)$	(3)
<3.51	Pre-dissociation limit. This is only an upper limit due to a maximum in the potential curve of the A'Π state	(4)
3.38 ± 0.04	The potential energy curve of the A'Π state was determined and was found to have a maximum and a minimum in agreement with Herzberg and Mundie	The selected $D_0^0(BH)$ value as reported by Hurley (1).

Heat Capacity and Entropy

The molecular constants from Herzberg (4) were adjusted for the natural abundance of B^{10} and B^{11}. Because the constants ω_e and $\omega_e x_e$ could not be determined directly from the spectrum, Almy and Horsfall (3) obtained estimates for the ground state X $^1\Sigma^+$ from the rotational constants for this state and the relations $D_e = 4\ B_3^3/\omega_e^2$ and $\omega_e x_e/\omega_e = 0.6\ \alpha_e/B_e$.

References

1. A. C. Hurley, Proc. Royal Soc. (London) A261, 237 (1961).
2. A. G. Gaydon, "Dissociation Energies and Spectra of Diatomic Molecules" Chapman and Hall, Ltd., London, (1953).
3. G. M. Almy and R. B. Horsfall, Jr., Phys. Rev. 51, 491 (1937).
4. G. Herzberg and L. G. Mundle, J. Chem. Phys. 8, 263 (1940) and G. Herzberg, "Spectra of Diatomic Molecules", D. Van Nostrand Co., Inc., New York, (1950).

Borane (BH) $B_1H_1(g)$

Enthalpy Reference Temperature = T_r = 298.15 K Standard State Pressure = p° = 0.1 MPa

	J K⁻¹mol⁻¹			kJ mol⁻¹			
T/K	$C_p°$	S°	$-[G°-H°(T_r)]/T$	$H°-H°(T_r)$	$\Delta_f H°$	$\Delta_f G°$	Log K_f
0	0.	0.	INFINITE	-8.639	439.476	439.476	INFINITE
100	29.126	140.014	197.766	-5.775	440.816	431.882	-225.592
200	29.143	160.207	174.516	-2.862	442.050	422.433	-110.328
250	29.156	166.712	172.329	-1.404	442.438	417.481	-87.228
298.15	29.181	171.849	171.849	0.	442.667	412.651	-72.295
300	29.182	172.029	171.849	0.054	442.673	412.465	-71.816
350	29.240	176.532	172.205	1.514	442.782	407.420	-60.804
400	29.347	180.442	172.995	2.979	442.783	402.367	-52.544
450	29.514	183.908	174.019	4.450	442.696	397.319	-46.120
500	29.741	187.029	175.166	5.931	442.543	392.285	-40.982
600	30.340	192.502	177.612	8.934	442.100	382.272	-33.280
700	31.052	197.231	180.084	12.003	441.545	372.344	-27.785
800	31.792	201.426	182.495	15.145	440.928	362.500	-23.669
900	32.503	205.212	184.812	18.360	440.274	352.735	-20.472
1000	33.160	208.671	187.027	21.644	439.594	343.045	-17.919
1100	33.752	211.860	189.142	24.990	438.890	333.424	-15.833
1200	34.280	214.820	191.160	28.392	438.164	323.868	-14.098
1300	34.748	217.583	193.087	31.844	437.414	314.373	-12.632
1400	35.162	220.173	194.930	35.340	436.639	304.938	-11.377
1500	35.530	222.612	196.695	38.875	435.835	295.558	-10.292
1600	35.859	224.916	198.388	42.445	435.003	286.233	-9.345
1700	36.152	227.099	200.013	46.046	434.141	276.961	-8.510
1800	36.417	229.173	201.576	49.674	433.248	267.741	-7.770
1900	36.657	231.148	203.081	53.328	432.324	258.571	-7.109
2000	36.875	233.034	204.531	57.005	431.368	249.451	-6.515
2100	37.075	234.838	205.932	60.703	430.381	240.379	-5.979
2200	37.259	236.567	207.285	64.419	429.361	231.356	-5.493
2300	37.429	238.227	208.595	68.154	428.309	222.379	-5.050
2400	37.587	239.823	209.863	71.905	376.966	214.517	-4.669
2500	37.736	241.361	211.092	75.671	375.772	207.773	-4.341
2600	37.875	242.844	212.285	79.452	374.579	201.077	-4.040
2700	38.007	244.275	213.444	83.246	373.386	194.426	-3.761
2800	38.132	245.660	214.570	87.053	372.193	187.820	-3.504
2900	38.250	247.000	215.665	90.872	371.001	181.256	-3.265
3000	38.364	248.299	216.731	94.703	369.808	174.733	-3.042
3100	38.472	249.559	217.770	98.544	368.615	168.251	-2.835
3200	38.577	250.782	218.783	102.397	367.421	161.806	-2.641
3300	38.677	251.970	219.770	106.260	366.227	155.399	-2.460
3400	38.775	253.126	220.734	110.132	365.033	149.028	-2.290
3500	38.869	254.252	221.676	114.015	363.837	142.693	-2.130
3600	38.961	255.348	222.596	117.906	362.641	136.391	-1.979
3700	39.050	256.417	223.496	121.807	361.445	130.123	-1.837
3800	39.137	257.459	224.376	125.716	360.247	123.887	-1.703
3900	39.222	258.477	225.237	129.634	359.048	117.683	-1.576
4000	39.305	259.471	226.081	133.560	357.848	111.509	-1.456
4100	39.386	260.443	226.907	137.495	356.648	105.366	-1.342
4200	39.467	261.393	227.717	141.437	-124.404	106.276	-1.322
4300	39.545	262.322	228.511	145.388	-124.520	111.769	-1.358
4400	39.623	263.232	229.290	149.346	-124.639	117.266	-1.392
4500	39.699	264.123	230.054	153.313	-124.761	122.765	-1.425
4600	39.774	264.997	230.804	157.286	-124.886	128.267	-1.457
4700	39.848	265.853	231.541	161.267	-125.015	133.771	-1.487
4800	39.922	266.693	232.264	165.256	-125.148	139.279	-1.516
4900	39.994	267.517	232.975	169.252	-125.285	144.789	-1.543
5000	40.066	268.325	233.674	173.255	-125.425	150.303	-1.570
5100	40.137	269.119	234.362	177.265	-125.570	155.818	-1.596
5200	40.207	269.900	235.038	181.282	-125.719	161.337	-1.621
5300	40.277	270.666	235.703	185.306	-125.872	166.859	-1.644
5400	40.346	271.420	236.357	189.337	-126.030	172.384	-1.667
5500	40.415	272.161	237.001	193.375	-126.192	177.911	-1.690
5600	40.483	272.889	237.636	197.420	-126.358	183.442	-1.711
5700	40.551	273.606	238.261	201.472	-126.529	188.976	-1.732
5800	40.618	274.312	238.876	205.530	-126.704	194.513	-1.752
5900	40.685	275.007	239.483	209.596	-126.884	200.052	-1.771
6000	40.752	275.692	240.080	213.667	-127.068	205.595	-1.790

PREVIOUS: December 1964 (1 atm) CURRENT: December 1964 (1 bar)

Borane (BH) $B_1H_1(g)$

BOROН HYDRIDE OXIDE (HBO) IDEAL GAS Mr = 27.81734

$S°(298.15\ K) = [202.6 \pm 0.5]\ J\ K^{-1}\ mol^{-1}$

$\Delta_fH°(0\ K) = -197.7 \pm 3.0\ kcal\ mol^{-1}$
$\Delta_fH°(298.15\ K) = -198.3 \pm 3.0\ kcal\ mol^{-1}$

Ground State Configuration: $[^1\Sigma^+]$ σ = 1
Point Group: $C_{\infty v}$
Bond Distances: H-B = [1.17] Å H-O = [1.19] A
Bond Angle: H-B-O = [180°]
Rotational Constant: B_0 = [1.341144] cm^{-1}

Electronic Levels and Quantum Weights

ε_i, cm^{-1}	g_i	ε_i, cm^{-1}	g_i
0	[1]	[35000]	[6]
[30000]	[3]	[40000]	[3]

Vibrational Frequencies and Degeneracies

ν, cm^{-1}
[2802](1)
757 (2)
1822 (1)

Enthalpy of Formation

We calculate $\Delta_fH°$ from a selected value, $\Delta_fH°(DBO, g) = -47.95\ kcal\ mol^{-1}$, by combining this value with zero-point energies and relative enthalpies, $H°(298.15\ K) - E_0°$, for HBO, D_2, DBO and H_2. All ancillary data are from JANAF (1) unless otherwise indicated. The zero-point energies are estimated as one-half the sum of the vibrational frequencies. The vibrational frequencies for DBO are taken from the matrix-isolation results of Lory and Porter (2). We assume that DBO and HBO have similar structures, and we calculate the relative enthalpy for DBO as $H°(298.15\ K) - E_0° = -2.27\ kcal\ mol^{-1}$. Data for D_2 are taken from NBS (3).

The selected value of $\Delta_fH°$ for DBO is obtained from the results of a mass-spectrometric study (4, 5) of the reaction of D (g) with $B_2O_3(\ell)$. The thermal results which were initially reported (4) for a study of the reactions (A) $D_2(g) + B_2O_3(\ell) = DBO(g) + DOBO(g)$ and (B) $D_2O(g) + B_2O_3(\ell) = 2DOBO(g)$ are unreliable due to erroneous equilibrium constants. 2nd-law enthalpies for these two reactions have been reported in revised form in a later publication (5). Apparently, no attempt was made in this study to perform calibration experiments which would have allowed the ion intensity data to be converted to absolute partial pressures; thus, precluding a 3rd law analysis. We combine their 2nd law values (5) for Δ_rH (1296 K) for reactions (A) and (B) to give Δ_rH (1296 K) = 135±4 kcal mol^{-1} for the process $2D_2(g) + B_2O_3(\ell) = 2DBO(g) + D_2O(g)$. Corrected to 298.15 K this value is 144.1 kcal mol^{-1} which leads to $\Delta_fH°(DBO,g,\ 298.15\ K) = -47.95 \pm 2.3\ kcal\ mol^{-1}$ with $\Delta_fH°(D_2O,g,\ 298.15\ K) = -59.561 \pm 0.02\ kcal\ mol^{-1}$ (3) and $\Delta_fH°(B_2O_3,\ \ell,\ 298.15\ K) = 299.56 \pm 0.6\ kcal\ mol^{-1}$ (1).

Comparison of values for the stretching force constants in HBO, BH_3(6), BO, and B_2O_2 (1) suggests a similarity in the bonds of these molecules. Using $D°(298.15\ K)\ (H-B) = \bar{D}°(298.15\ K)\ (BH_3) = 87.9\ kcal\ mol^{-1}$ and $D°(298.15\ K)\ (B=O) = D°(298.15\ K)\ (BO) + D_0^0(298.15\ K)\ (B_2O_2)\ /2 = 202\ kcal\ mol^{-1}$ (1), we calculate an enthalpy of atomization for HBO of 298.9 kcal mol . This corresponds to $\Delta_fH°(HBO,\ g,\ 298.15\ K) = -45.4\ kcal\ mol^{-1}$ which lends support to the experimental measurements at Farber et al. (4, 5). Pervious JANAF estimates (7) of $\Delta_fH°$ via bond energy calculations led to results which were too positive due to the use of an inaccurate value (168 kcal mol^{-1}) for D_0^0 (B=O). A recent ab-initio LCAO-MO-SCF investigation (8) of HBO yields $\Delta_{at}H$ = 232.2 kcal mol^{-1}; however, this value does not include a contribution from correlation effects (9, 16) and therefore is too low.

Heat Capacity and Entropy

The B=O stretching (ν_3) and bending (ν_2) frequencies for the isotopes $H^{11}BO$ and $H^{10}BO$ have been measured in a low-temperature argon matrix (2). These values are corrected for the natural isotopic abundances of boron. The H-B stretching frequency (ν_1) is calculated from an estimated force constant by the valence force method (10). The stretching force constant K_1 is estimated from the ration $K_1/K_D/\ell_1\ell_2$ = 17.47 which is the value we calculate for DBO from the measured frequencies of Lory and Porter (2). These workers (2) used a slightly different value for K_1 and obtained a value for ν_1 which is roughly 50 cm^{-1} higher than our result. We believe our estimate is probably more nearly correct, since it agrees much better with the H-B stretching frequency (2808 cm^{-1}) observed for BH_3 (6). MO calculations show that the pattern of valence orbitals for HBO is similar to that in HCN (8) and HCP (11, 16). The ordering is substantiated by the photoelectron spectra (12) which have been observed for HCN and HCP. This suggests that the ground and excited electronic states for these isoelectronic molecules are quite similar. We assume that the ground state configuration is $^1\Sigma^+$ by analogy with those for HCN and HCP (1). We also include three triplet levels which are estimated from those observed for HCP (1) and predicted for HCN (13).

There have been no measurements reported for HBO which provide direct experimental information on its structure. However, the millimeter-wave spectra of the related species HBS (14) have been observed. These results show that this molecule is linear and has a B=S bond length which is 0.01 Å less than that for BS (1). We assume a similar decrease in r_e(B=O) for the HBO-BO pair, and we obtain r_e(HB-O) = 1.19 Å with r_e(B=O) = 1.20 Å (1). The H-B bond length is assumed the same as that for HBS (14). Ab-initio calculations (8) of bond lengths for HBO by an optimization procedure predict values only slightly lower (~0.01 Å) than our estimates. We note that bond lengths computed by this procedure are normally slightly lower (~0.005 Å) than the true values as indicated by the results for HCP (16) and HBS (17). By analogy with HBS (14), we assume HBO to be linear. This assumption is in agreement with predictions from the Walsh diagram (15) for HAB molecules with ten valence electrons. In direct evidence available from product rule calculations (2) also tend to support a linear configuration.

References

1. JANAF Thermochemical Tables: H_2(g), 3-31-61; B_2O_3(ℓ), 6-30-71; BO(g), 6-30-68; B_2O_2(g) and BH_3(g), 12-31-64; HCP(g) and HCN(g), 12-31-69; BS(g), 6-30-72.
2. E. R. Lory and R. F. Porter, J. Amer. Chem. Soc. 93, 6301 (1971).
3. U. S. Nat. Bur. Stand. Report 10904, 1972.
4. M. Farber, M. A. Greenbaum, and M. A. Frisch, Rocket Power, Inc., Report No. AFRPL-TR-66-220, September, 1966.
5. M. Farber and M. A. Frisch, Proc. First Intern. Conf. Calorimetry and Thermodynamics, Warsaw, August-September, 1969.
6. A. Kaldor and R. F. Porter, J. Amer. Chem. Soc. 93, 2140 (1971).
7. JANAF Thermochemical Tables, 2nd ed., NSRDS-NBS 37, 1971.
8. C. Thomson and B. J. Wishart, Theoret. Chim. Acta, 35, 267 (1974).
9. C. Thomson, J. Chem. Phys. 58, 216 (1973).
10. G. Herzberg, "Infrared and Raman Spectra of Polyatomic Molecules", D. Van Nostrand Company, Inc., New York, 1945.
11. J. Robert, H. Marsmann, I. Absar, and J. R. VanWazer, J. Amer. Chem. Soc. 93, 3320 (1971).
12. D. C. Frost, S. T. Lee, and C. A. McDowell, Chem. Phys. Letters, 23, 472 (1973).
13. G. M. Schwenzer, S. V. O'Neil, H. F. Schaefer, C. P. Baskin, and C. F. Bender, J. Chem. Phys. 60, 2787 (1974).
14. E. F. Pearson and R. V. McCormick, J. Chem. Phys. 58, 1619 (1973).
15. A. D. Walsh, J. Chem. Soc. 1953, 2288 (1953).
16. C. Thomson, Theoret. Chim. Acta, 35, 237 (1974).
17. C. Thomson, Chem. Phys. Letters, 25, 59 (1974).

Boron Hydride Oxide (HBO) $B_1H_1O_1(g)$

Enthalpy Reference Temperature = T_r = 298.15 K   Standard State Pressure = p° = 0.1 MPa

	J $K^{-1}mol^{-1}$			kJ mol^{-1}			
T/K	$C_p°$	S°	$-[G°-H°(T_r)]/T$	$H°-H°(T_r)$	$\Delta_fH°$	$\Delta_fG°$	Log K_f
0	0.	0.	INFINITE	-9.156	-197.688	-197.688	INFINITE
100	29.137	168.766	231.276	-6.251	-197.759	-200.903	104.941
200	31.249	189.398	205.712	-3.263	-197.906	-204.013	53.283
250	33.279	196.585	203.187	-1.650	-198.092	-205.519	42.941
298.15	35.260	202.617	202.617	0.	-198.322	-206.929	36.253
300	35.334	202.836	202.618	0.065	-198.331	-206.983	36.039
350	37.224	208.427	203.055	1.880	-198.606	-208.403	31.102
400	38.923	213.510	204.049	3.785	-198.913	-209.782	27.395
450	40.455	218.185	205.363	5.770	-199.244	-211.121	24.506
500	41.853	222.521	206.865	7.828	-199.591	-212.422	22.192
600	44.343	230.377	210.142	12.141	-200.304	-214.921	18.711
700	46.511	237.379	213.542	16.686	-201.010	-217.302	16.215
800	48.408	243.716	216.924	21.434	-201.690	-219.582	14.337
900	50.061	249.516	220.228	26.359	-202.336	-221.779	12.872
1000	51.495	254.866	223.427	31.439	-202.952	-223.907	11.696
1100	52.734	259.834	226.514	36.652	-203.543	-225.973	10.731
1200	53.802	264.469	229.486	41.980	-204.117	-227.987	9.924
1300	54.724	268.813	232.346	47.407	-204.683	-229.953	9.240
1400	55.520	272.898	235.098	52.920	-205.249	-231.876	8.651
1500	56.208	276.753	237.748	58.508	-205.820	-233.758	8.140
1600	56.806	280.400	240.300	64.159	-206.404	-235.601	7.692
1700	57.327	283.860	242.762	69.866	-207.006	-237.408	7.295
1800	57.783	287.150	245.137	75.622	-207.629	-239.178	6.941
1900	58.182	290.285	247.431	81.421	-208.278	-240.914	6.623
2000	58.534	293.278	249.650	87.257	-208.956	-242.614	6.336
2100	58.846	296.142	251.796	93.126	-209.665	-244.279	6.076
2200	59.122	298.886	253.874	99.025	-210.407	-245.910	5.839
2300	59.369	301.519	255.889	104.950	-211.183	-247.507	5.621
2400	59.589	304.051	257.843	110.898	-262.256	-248.000	5.398
2500	59.786	306.487	259.741	116.867	-263.184	-247.387	5.169
2600	59.964	308.836	261.584	122.855	-264.119	-246.737	4.957
2700	60.125	311.102	263.376	128.859	-265.060	-246.050	4.760
2800	60.271	313.291	265.120	134.879	-266.008	-245.329	4.577
2900	60.404	315.409	266.818	140.913	-266.965	-244.574	4.405
3000	60.525	317.458	268.472	146.960	-267.931	-243.785	4.245
3100	60.636	319.445	270.084	153.018	-268.906	-242.964	4.094
3200	60.739	321.372	271.657	159.087	-269.890	-242.112	3.952
3300	60.835	323.242	273.192	165.165	-270.883	-241.228	3.818
3400	60.924	325.060	274.691	171.253	-271.886	-240.314	3.692
3500	61.008	326.827	276.155	177.350	-272.899	-239.371	3.572
3600	61.088	328.547	277.587	183.455	-273.921	-238.399	3.459
3700	61.165	330.221	278.987	189.567	-274.953	-237.398	3.351
3800	61.239	331.854	280.357	195.688	-275.994	-236.369	3.249
3900	61.312	333.445	281.698	201.815	-277.044	-235.312	3.152
4000	61.385	334.998	283.011	207.950	-278.103	-234.229	3.059
4100	61.457	336.515	284.297	214.092	-279.171	-233.119	2.970
4200	61.531	337.997	285.558	220.242	-760.096	-224.959	2.798
4300	61.606	339.446	286.795	226.398	-760.092	-212.217	2.578
4400	61.684	340.863	288.008	232.563	-760.098	-199.476	2.368
4500	61.766	342.250	289.198	238.735	-760.113	-186.735	2.168
4600	61.851	343.608	290.366	244.916	-760.136	-173.993	1.976
4700	61.941	344.940	291.513	251.106	-760.168	-161.250	1.792
4800	62.036	346.245	292.640	257.305	-760.208	-148.507	1.616
4900	62.138	347.525	293.747	263.513	-760.256	-135.762	1.447
5000	62.246	348.781	294.835	269.732	-760.311	-123.017	1.285
5100	62.361	350.015	295.905	275.963	-760.372	-110.270	1.129
5200	62.483	351.227	296.957	282.205	-760.441	-97.523	0.980
5300	62.614	352.419	297.992	288.460	-760.515	-84.773	0.835
5400	62.753	353.590	299.011	294.728	-760.595	-72.023	0.697
5500	62.901	354.743	300.014	301.010	-760.679	-59.271	0.563
5600	63.058	355.878	301.001	307.308	-760.768	-46.517	0.434
5700	63.225	356.995	301.974	313.622	-760.861	-33.762	0.309
5800	63.401	358.096	302.932	319.954	-760.957	-21.005	0.189
5900	63.587	359.182	303.876	326.303	-761.056	-8.247	0.073
6000	63.783	360.252	304.807	332.671	-761.158	4.514	-0.039

PREVIOUS: December 1975 (1 atm)   CURRENT: December 1975 (1 bar)

Boron Hydride Oxide (HBO) $B_1H_1O_1(g)$

BORON HYDRIDE OXIDE, ION (HBO^+) IDEAL GAS Mr = 27.81679

$S°(298.15\ K) = [214.57 \pm 6.3]\ J\ K^{-1}\ mol^{-1}$

$\Delta_f H°(0K) = [1182.0 \pm 12]\ kJ\ mol^{-1}$

$\Delta_f H°(298.15\ K) = [1187.5 \pm 12]\ kJ\ mol^{-1}$

Electronic Levels and Quantum Weights

State	ε_i, cm^{-1}	g_i
$X^2\Pi$	0	[4]
$A^2\Sigma^+$	[6000]	[2]
$B^2\Sigma^+$	[40000]	[2]

Vibrational Frequencies and Degeneracies

ν, cm^{-1}
[2559](1)
[801](2)
[1617](1)

Point Group: $[C_{\infty v}]$ $\sigma = 1$

Bond Distances: H-B = [1.19] Å B-O = [1.25] Å

Bond Angle: H-B-O = [180]°

Rotational Constant: $B_0 = [1.228629]\ cm^{-1}$

Enthalpy of Formation

The HBO^+ ion has been detected mass spectrometrically by Sholette and Porter (1) and Farber and Frisch (2). However, no appearance potential data have been reported for the ion. Kroto et al. (3) recently attempted to measure the photelectron spectrum of HBO by passing water vapor over heated boron. The spectrum showed no bands which could be definitely assigned to HBO monomer. Unfortunately, the region of their spectrum (~14-15 eV) where the first photoelectron band of HBO would be expected to lie shows a broad band which also appeared in the spectrum of HBS above 1150°C. Kroto et al. (4) have assigned this band to diborane. We believe the HBO band may well be hidden under this broad band.

We employ Koopmans' theorem (4) to obtain the ionization potential of HBO as 14.3 eV from the one electron orbital energies reported by Thomson and Wishart (5). A comparison of Koopmans' theorem IP's for the related species HBS (6), HCN (5), and HCP (7) with experimental values (8, 9) shows that the theoretical calculations correctly predict these ionization potentials to within about ±0.4 eV. We adopt $\Delta_r H° = 329.7_6 \pm 9\ kcal\ mol^{-1}$ for the ionization process $HBO(g) + e^- = HBO^+(g) + 2e^-$ at 0 K, and we obtain $\Delta_f H°(HBO^+, g, 0\ K) = 282.6 \pm 12.0\ kcal\ mol^{-1}$ when the former value is combined with $\Delta_f H°(HBO, g, 0\ K) = -47.2 \pm 3.0\ kcal\ mol^{-1}$(8). $\Delta_f H°(298.15\ K)$ is $283.8 \pm 12.0\ kcal\ mol^{-1}$.

Heat Capacity and Entropy

Extensive ab initio MO calculations (5) reported for HBO show that the highest occupied orbital is of Π-symmetry and is localized mainly on the oxygen atom. The Walsh diagram (10) for HAB molecules predicts that this orbital is bonding. We would expect therefore that the ion is less strongly bound than the parent molecule. For HCN which has the same orbital ordering as HBO (5), ionization (11) occurs with a 2.2% increase in the H-C bond length and a 4.7% increase in the C≡N bond length. We assume a similar increase in the bond lengths of HBO (8) upon loss of the bonding electron. A linear configuration is assumed by analogy with that for HBS^+ (8). This assumption is supported by predictions from the Walsh diagram (10).

All vibrational frequencies are calculated from estimated force constants by a valence force method (12). The force constants are estimated from those for HBO (8) by comparison with the changes in the changes in the force constants for HCN (8) produced on ionization (9). The electronic states and levels are estimated from those observed for isoelectronic ions HBS^+, and HCP^+ (9). We estimate the uncertainties in our calculated frequencies as $\pm 50\ cm^{-1}$ which introduces an error of only about ±0.1 $cal\ K^{-1}\ mol^{-1}$ in the value of S°(298.15 K). The uncertainty in the electronic levels contribute the majority of the error in the entropy. We predict that the energy separation of the $^2\Pi$-$^2\Sigma$ states is small. Thus, it is possible that the ground state is 2Σ as is the case for the isoelectronic ions CO^+ and N_2^+(13). If the ground state is $^2\Sigma$, then our entropies should be decreased by roughly 1.5 gibbs/mol at all temperatures above 298.15 K. The enthalpy is much more certain than the entropy. Below 2000 K the uncertainty in our relative enthalpies probably does not exceed a few calories. The uncertainty increases at higher temperatures and is about $1.0\ kcal\ mol^{-1}$ at 4000 K. All excited states are expected to be linear as observed for HBS^+ (3).

References

1. W. P. Sholette and R. F. Porter, J. Phys. Chem. 67, 177 (1963).
2. M. Farber and M. A. Frisch, Proc. 1st Intern. Conf. Calorimetry and Thermodynamics, Warsaw, August-September, 1969.
3. H. W. Kroto, R. J. Suffolk, and N. P. C. Westwood, Chem. Phys. Letters, 22, 495 (1973).
4. T. Koopmans, Physica, 1, 104 (1934).
5. C. Thomson and B. J. Wishart, Theoret. Chim. Acta, 35, 267 (1974).
6. C. Thomson, Chem. Phys. Letters, 25, 59 (1974).
7. C. Thomson, Theoret. Chim. Acta, 35, 237 (1974).
8. JANAF Thermochemical Tables: HBS(g), HBO(g), and HBS+(g), 12-31-75; HCN(g), 12-31-69.
9. D. C. Frost, S. T. Lee, and C. A. McDowell, Chem. Phys. Letters, 23, 472 (1973).
10. A. D. Walsh, J. Chem. Soc. 1953, 2288 (1953).
11. S. P. So and W. G. Richards, J. Chem. Soc. Faraday Trans. II 71, 62 (1975).
12. G. Herzberg, "Infrared and Raman Spectra of Polyatomic Molecules", D. Van Nostrand company, Inc., New York, 1962.
13. B. Rosen, ed., "Spectroscopic Data Relative to Diatomic Molecules", Pergamon Press, New York, 1970.

Boron Hydride Oxide, Ion (HBO^+) $B_1H_1O_1^+(g)$

Enthalpy Reference Temperature = T_r = 298.15 K Standard State Pressure = p° = 0.1 MPa

T/K	$C_p°$ (J K⁻¹mol⁻¹)	S° (J K⁻¹mol⁻¹)	$-[G°-H°(T_r)]/T$ (J K⁻¹mol⁻¹)	$H°-H°(T_r)$ (kJ mol⁻¹)	$\Delta_f H°$ (kJ mol⁻¹)	$\Delta_f G°$ (kJ mol⁻¹)	Log K_f
0	0.	0.	INFINITE	-9.090	1182.047		
100	29.122	181.019	242.861	-6.184			
200	30.857	201.543	217.612	-3.214			
250	32.755	208.626	215.126	-1.625			
298.15	34.744	214.565	214.565	0.	1187.545	1169.120	-204.825
300	34.821	214.780	214.566	0.064	1187.573	1169.006	-203.542
350	36.828	220.300	214.997	1.856	1188.314	1165.852	-173.994
400	38.701	225.342	215.979	3.745	1189.031	1162.594	-151.819
450	40.425	230.001	217.282	5.724	1189.732	1159.248	-134.562
500	42.011	234.343	218.773	7.785	1190.428	1155.823	-120.748
600	44.817	242.258	222.041	12.130	1191.826	1148.771	-100.009
700	47.203	249.351	225.444	16.734	1193.258	1141.483	-85.178
800	49.235	255.790	228.841	21.559	1194.733	1133.986	-74.042
900	50.965	261.692	232.168	26.571	1196.252	1126.302	-65.369
1000	52.442	267.140	235.396	31.744	1197.808	1118.447	-58.422
1100	53.711	272.199	238.515	37.053	1199.392	1110.434	-52.730
1200	54.808	276.921	241.520	42.480	1200.995	1102.276	-47.981
1300	55.765	281.346	244.416	48.010	1202.610	1093.984	-43.957
1400	56.606	285.510	247.204	53.629	1204.230	1085.568	-40.503
1500	57.350	289.442	249.890	59.328	1205.848	1077.035	-37.506
1600	58.012	293.164	252.479	65.096	1207.460	1068.395	-34.880
1700	58.605	296.699	254.977	70.928	1209.061	1059.655	-32.559
1800	59.138	300.065	257.389	76.816	1210.648	1050.820	-30.494
1900	59.618	303.275	259.720	82.754	1212.217	1041.898	-28.644
2000	60.050	306.344	261.975	88.738	1213.765	1032.894	-26.976
2100	60.441	309.284	264.159	94.762	1215.291	1023.813	-25.466
2200	60.794	312.104	266.274	100.824	1216.791	1014.660	-24.091
2300	61.112	314.813	268.326	106.920	1218.263	1005.439	-22.834
2400	61.400	317.420	270.318	113.046	1169.447	997.223	-21.704
2500	61.659	319.932	272.253	119.199	1170.782	990.020	-20.685
2600	61.893	322.355	274.133	125.377	1172.116	982.763	-19.744
2700	62.103	324.695	275.963	131.577	1173.449	975.455	-18.871
2800	62.292	326.957	277.744	137.797	1174.779	968.097	-18.060
2900	62.462	329.146	279.479	144.035	1176.105	960.692	-17.304
3000	62.614	331.266	281.170	150.289	1177.425	953.242	-16.597
3100	62.749	333.321	282.819	156.557	1178.739	945.747	-15.936
3200	62.870	335.316	284.429	162.838	1180.046	938.210	-15.315
3300	62.978	337.252	286.000	169.130	1181.345	930.633	-14.731
3400	63.074	339.133	287.535	175.433	1182.635	923.016	-14.180
3500	63.159	340.963	289.036	181.745	1183.916	915.362	-13.661
3600	63.233	342.743	290.503	188.065	1185.187	907.671	-13.170
3700	63.299	344.477	291.939	194.391	1186.448	899.944	-12.705
3800	63.356	346.166	293.344	200.724	1187.698	892.185	-12.264
3900	63.407	347.812	294.719	207.062	1188.938	884.392	-11.845
4000	63.450	349.418	296.067	213.405	1190.165	876.567	-11.447
4100	63.487	350.985	297.387	219.752	1191.381	868.712	-11.068
4200	63.519	352.515	298.681	226.102	712.735	867.852	-10.793
4300	63.546	354.010	299.951	232.456	715.014	871.518	-10.587
4400	63.569	355.472	301.196	238.812	717.278	875.131	-10.389
4500	63.588	356.900	302.418	245.169	719.527	878.693	-10.200
4600	63.603	358.298	303.618	251.529	721.761	882.206	-10.018
4700	63.615	359.666	304.796	257.890	723.979	885.670	-9.843
4800	63.624	361.005	305.953	264.252	726.181	889.087	-9.675
4900	63.630	362.317	307.090	270.614	728.366	892.458	-9.514
5000	63.635	363.603	308.207	276.978	730.534	895.785	-9.358
5100	63.637	364.863	309.306	283.341	732.684	899.069	-9.208
5200	63.638	366.099	310.386	289.705	734.816	902.310	-9.064
5300	63.637	367.311	311.449	296.069	736.929	905.511	-8.924
5400	63.634	368.500	312.495	302.432	739.023	908.672	-8.790
5500	63.631	369.668	313.523	308.796	741.098	911.794	-8.659
5600	63.627	370.815	314.536	315.158	743.153	914.880	-8.534
5700	63.621	371.941	315.534	321.521	745.187	917.928	-8.412
5800	63.615	373.047	316.516	327.883	747.200	920.942	-8.294
5900	63.609	374.135	317.483	334.244	749.191	923.919	-8.180
6000	63.602	375.204	318.436	340.604	751.160	926.864	-8.069

PREVIOUS: December 1975 (1 atm) CURRENT: December 1975 (1 bar)

Boron Hydride Oxide, Ion (HBO^+) $B_1H_1O_1^+(g)$

$\Delta_fH° = [-238.5 \pm 24]\ kJ\ mol^{-1}$

$S°(298.15\ K) = [227.1 \pm 2.0]\ J\ K^{-1}\ mol^{-1}$ $\Delta_fH°(298.15\ K) = [-244.5 \pm 24]\ kJ\ mol^{-1}$

Electronic Levels and Quantum Weights

ϵ_i, cm^{-1}	g_i
0	[2]
[10000]	[2]

Vibrational Frequencies and Degeneracies

ν, cm^{-1}
[2000] (1)
[1000] (1)
[1500] (1)

Ground State Quantum Weight: [2] $\sigma = 1$

Point Group: C_s

Bond Distances: H-B = [1.25] Å B-O = [1.37] Å

Bond Angle: H-B-O = 123°

Product of Moments of Inertia: $I_AI_BI_C = [9.09027 \times 10^{-118}]\ g^3\ cm^6$

Enthalpy of Formation

The identification of the HBO^- radical has been made from observations of its electron spin resonance spectra which was measured (1, 2) in γ-irradiated polycrystalline potassium borohydride. This assignment is confirmed by the fact that hyperfine coupling constants calculated by INDO (3) and UHF (4, 5) methods are in reasonable agreement with the spectral data (2). No experimental measurement of the electron affinity (EA) has been reported: however, recent MO calculations (6) predict that the value is small. We note that these calculations predict incorrectly the relative stabilities of HBO and HBO^- and therefore are of no use in the establishment of the absolute value of EA. This inversion in the predicted stabilities of the radical and parent molecule most likely arises from the neglect of correlation effects.

We estimate $\Delta_fH°$ from a consideration of bond-energy schemes. For the dissociative process $HBO^-(g) = H(g) + B(g) + O^-(g)$, we assume $\Delta_{at}H°(HBO,g,\ 0\ K) = 289.4\ kcal\ mol^{-1}$ (7). This atomization energy gives $\Delta_fH°(HBO^-,\ g,\ 0\ K) = -81.0\ kcal\ mol^{-1}$ which we believe is an upper limit since the unpaired electron in HBO^- is antibonding (8) and consequently, one would expect $\Delta_{at}H°(HBO^-) < \Delta_{at}H°(HBO)$. A lower limit of $\Delta_fH°$ can be obtained from a consideration of the H-B bond dissociation energy. It is very unlikely that $D_0^0(H\text{-}BO^-) < D_0^0(H\text{-}CO)$ which is equal to 14.1 $kcal\ mol^{-1}$. Using this value for $D_0^0(H\text{-}BO^-)$, we obtain $\Delta_fH°(HBO^-,\ g,\ 0\ K) = -33\ kcal\ mol^{-1}$ with $\Delta_fH°(H,g,\ 0\ K) = 52.103\ kcal\ mol^{-1}$ (7) and $\Delta_fH°(BO^-,\ g,\ 0\ K) = -71.0\ kcal\ mol^{-1}$ (9). We adopt an average ($-57\pm24\ kcal\ mol^{-1}$) of the upper and lower limit values. Our adopted $\Delta_fH°$ value corresponds to an $EA(HBO^-)$ of 0.42±1.0 eV.

Heat Capacity and Entropy

The bond lengths are those reported by Thomson (3) who performed an INDO calculation in which the bond lengths were determined by minimization of the total energy. The esr spectrum (1, 2) of HBO^- is consistent with a bent structure. Catton et al. (2) obtained bond angles of 121° and 125° from the p:s ratios deduced from estimated and observed anisotropic components of the hyperfine tensors. A similar calculation (2) for HCO and HCN^- led to bond angles which agreed with independent values to within ±5°. We adopt a bond angle of 123±5°. MO calculations (3) give an angle of 129°. However, the energy minimization procedure with the INDO method generally over-estimates the angle as evidenced by the results for HCO (10) and HCN (2, 3). The principal moments of inertia are: $I_A = 0.1484 \times 10^{-39}$, $I_B = 2.4017 \times 10^{-39}$, and $I_C = 2.5501 \times 10^{-39}\ g\ cm^2$.

The electronic levels and quantum weights are taken from those for HCO (7), which is isoelectronic with HBO^-. The vibrational frequencies arae estimated by comparison with the corresponding values for HBO, HBO^+, HCO and HCO^+ (7).

References

1. M. C. R. Symons and H. W. Wardale, Chem. Commun. 1969, 1483 (1968).
2. R. C. Catton, M. C. R. Symons, and H. W. Wardale, J. Chem. Soc. (A), 1969, 2622 (1966).
3. C. Thomson, Theoret. Chim. Acta, 17, 320 (1970).
4. T. A. Claxton and N. A. Smith, Trans. Faraday Soc. 67, 1859 (1971).
5. T. A. Claxton, Trans. Faraday Soc. 67, 897 (1971).
6. C. Thomson and B. J. Wishart, Theoret. Chim. Acta, 35, 267 (1974).
7. JANAF Thermochemical Tables: HBO(g) and HBO^+(g), 12-31-75; H(g), 6-30-74; B(g), 12-31-64; O^-(g), 6-30-65; HCO(g) and HCO^+(g), 12-31-70.
8. A. D. Walsh, J. Chem. Soc. 1953, 2288 (1953).
9. R. D. Srivastava, O. M. Uy, and M. Farber, Trans. Faraday Soc. 67, 2941 (1971).
10. M. S. Gordon and J. A. Pople, J. Chem. Phys. 49, 4643 (1968).

Enthalpy Reference Temperature = T_r = 298.15 K Standard State Pressure = p° = 0.1 MPa

T/K	$C_p°$ (J K⁻¹mol⁻¹)	S° (J K⁻¹mol⁻¹)	$-[G°-H°(T_r)]/T$ (J K⁻¹mol⁻¹)	$H°-H°(T_r)$ (kJ mol⁻¹)	$\Delta_fH°$ (kJ mol⁻¹)	$\Delta_fG°$ (kJ mol⁻¹)	Log K_f
0	0.	0.	INFINITE	-10.027	-238.827		
100	33.259	190.280	257.292	-6.701			
200	33.602	213.386	230.216	-3.366			
250	34.257	220.947	227.632	-1.671			
298.15	35.199	227.058	227.058	0.	-244.471	-254.111	44.519
300	35.240	227.276	227.058	0.065	-244.519	-254.171	44.255
350	36.456	232.797	227.492	1.857	-245.857	-255.674	38.157
400	37.806	237.752	228.470	3.713	-247.251	-256.982	33.558
450	39.211	242.286	229.756	5.638	-248.682	-258.113	29.961
500	40.609	246.490	231.222	7.634	-250.131	-259.083	27.066
600	43.240	254.131	234.416	11.829	-253.040	-260.600	22.687
700	45.531	260.973	237.729	16.270	-255.928	-261.632	19.523
800	47.454	267.182	241.029	20.923	-258.782	-262.251	17.123
900	49.042	272.866	244.255	25.750	-261.605	-262.515	15.236
1000	50.345	278.103	247.381	30.722	-264.407	-262.466	13.710
1100	51.417	282.953	250.397	35.811	-267.200	-262.136	12.448
1200	52.304	287.466	253.301	40.999	-269.994	-261.553	11.385
1300	53.044	291.683	256.093	46.267	-272.797	-260.736	10.476
1400	53.669	295.637	258.778	51.604	-275.618	-259.702	9.690
1500	54.205	299.359	261.360	56.998	-278.461	-258.467	9.001
1600	54.672	302.873	263.846	62.443	-281.331	-257.040	8.391
1700	55.085	306.200	266.240	67.931	-284.230	-255.433	7.849
1800	55.457	309.359	268.549	73.458	-287.161	-253.655	7.361
1900	55.798	312.367	270.777	79.021	-290.124	-251.713	6.920
2000	56.115	315.237	272.928	84.617	-293.120	-249.614	6.519
2100	56.413	317.982	275.009	90.244	-296.150	-247.364	6.153
2200	56.697	320.613	277.022	95.899	-299.214	-244.969	5.816
2300	56.968	323.139	278.973	101.583	-302.311	-242.435	5.506
2400	57.230	325.569	280.864	107.293	-355.700	-238.696	5.195
2500	57.484	327.911	282.699	113.028	-358.940	-233.754	4.884
2600	57.729	330.170	284.482	118.789	-362.180	-228.682	4.594
2700	57.968	332.353	286.215	124.574	-365.419	-223.486	4.324
2800	58.199	334.466	287.901	130.382	-368.658	-218.170	4.070
2900	58.423	336.512	289.542	136.213	-371.896	-212.739	3.832
3000	58.639	338.496	291.141	142.067	-375.134	-207.195	3.608
3100	58.848	340.422	292.699	147.941	-378.371	-201.544	3.396
3200	59.049	342.294	294.220	153.836	-381.608	-195.788	3.196
3300	59.242	344.114	295.705	159.751	-384.844	-189.931	3.006
3400	59.427	345.885	297.155	165.684	-388.080	-183.976	2.826
3500	59.603	347.610	298.572	171.636	-391.316	-177.926	2.655
3600	59.771	349.292	299.957	177.604	-394.553	-171.783	2.493
3700	59.930	350.932	301.313	183.590	-397.791	-165.551	2.337
3800	60.081	352.532	302.640	189.590	-401.030	-159.230	2.189
3900	60.223	354.095	303.939	195.605	-404.271	-152.825	2.047
4000	60.357	355.621	305.212	201.634	-407.515	-146.337	1.911
4100	60.482	357.113	306.460	207.676	-410.761	-139.767	1.781
4200	60.599	358.572	307.683	213.731	-893.860	-126.094	1.568
4300	60.707	359.999	308.884	219.796	-896.027	-107.788	1.309
4400	60.808	361.396	310.061	225.872	-898.200	-89.432	1.062
4500	60.902	362.763	311.217	231.957	-900.380	-71.026	0.824
4600	60.988	364.103	312.352	238.052	-902.568	-52.572	0.597
4700	61.066	365.415	313.468	244.155	-904.766	-34.070	0.379
4800	61.138	366.702	314.563	250.265	-906.973	-15.521	0.169
4900	61.203	367.963	315.640	256.382	-909.190	3.073	-0.033
5000	61.262	369.200	316.699	262.505	-911.420	21.714	-0.227
5100	61.315	370.414	317.740	268.634	-913.662	40.399	-0.414
5200	61.362	371.605	318.765	274.768	-915.917	59.127	-0.594
5300	61.404	372.774	319.773	280.907	-918.186	77.900	-0.768
5400	61.441	373.922	320.765	287.049	-920.470	96.716	-0.936
5500	61.472	375.050	321.742	293.195	-922.770	115.573	-1.098
5600	61.499	376.158	322.704	299.343	-925.087	134.474	-1.254
5700	61.522	377.247	323.651	305.494	-927.422	153.415	-1.406
5800	61.541	378.317	324.584	311.648	-929.775	172.398	-1.553
5900	61.555	379.369	325.504	317.802	-932.147	191.420	-1.695
6000	61.566	380.404	326.410	323.959	-934.539	210.484	-1.832

PREVIOUS: December 1975 (1 atm) CURRENT: December 1975 (1 bar)

BORIC ACID (HBO_2) CRYSTAL Mr = 43.81674

$S°(298.15\ K) = [48.95 \pm 4.2]\ J\ K\ mol^{-1}$
$T_{fus} = 509 \pm 1\ K$

$\Delta_f H°(0\ K)$ = Unknown
$\Delta_f H°(298.15\ K) = -802.8 \pm 0.84\ kJ\ mol^{-1}$
$\Delta_{fus} H°$ = Unknown

Enthalpy of Formation

The enthalpy change ($\Delta_r H°$) of the reaction HBO_2 (cr, I) + H_2O(l) = H_3BO_3(cr) has been determined Kilday. From the value, $\Delta_r H°(298.15\ K) = -1.28\ kcal\ mol^{-1}$, the enthalpy of formation $\Delta_f H°(298.15\ K)$ for HBO_2 (cr, I) is derived to be -191.87 ± 0.20 kcal mol^{-1} which is adopted. The enthalpies of solution of HBO_2 (cr, I) in H_2O(l) were measured by Sokolova, Skuratov, Shemonaeva and Yuldasheva. Based on the reported value, $\Delta_r H°(298.15\ K) = 1.76 \pm 0.01\ kcal\ mol^{-1}$ for the reaction HBO_2(cr, I) + 501 H_2O(l) = $H_3BO_3 \cdot 500\ H_2O$(Sol), the enthalpy of formation for HBO_2(cr, I) was derived as $-189.46 \pm 0.20\ kcal\ mol^{-1}$, using $\Delta_f H°(H_2O$, l, 298.15 K) = −68.317 and $\Delta_f H°(H_3BO_3 \cdot 500\ H_2O$, Sol, 298.15 K), = $-256.02 \pm 0.20\ kcal\ mol^{-1}$.

Heat Capacity and Entropy

The values of heat capacity for HBO_2(c, I) were estimated by comparison with those of its constituent oxides. S°(298.15 K) was calculated as the sum of the entropy contributions from H^+ and BO_2^- in HBO_2(c) suggested by Kelley and Kubaschewski and Evans.

Fusion Data

Kracek, Morey and Merwin reported the existence of three crystal forms of HBO_2(c), namely HBO_2(cr, I), T_{fus} = 509 ± 1K; HBO_2(cr, II), T_{fus} = 474.1 ± 0.5 K; and HBO_2(cr, III), T_{fus} = 449.2 ± 0.2 K. However, there was no solid phase equilibrium found among these crystals. Sokolova, Skuratov, Shemonaeva and Yuldasheva measured the enthalpies of solution in water of HBO_2(cr, I) and HBO_2(cr, II) and derived the enthalpy of transition as $\Delta_{trs} H°$(II → I, 298.15 K) = 1.29 kcal mol^{-1}.

Sublimation Data

The value of $\Delta_{sub} H°(298.15\ K)$ is calculated as the difference between the $\Delta_f H°(298.15\ K)$ values for HBO_2(g) and HBO_2(c).

References

1. M. V. Kilday and E. J. Prosen, J. Am. Chem. Soc. 82, 5508 (1960).
2. N. D. Sokolova, S. M. Skuratov, A. M. Shemonaeva, and V. M. Yuldasheva, Russ. J. Inorg. Chem. 6, 395 (1961).
3. O. Kubaschewski and E. L. Evans, "Metallurgical Thermochemistry" Pergamon Press, New York (1958).
4. F. C. kracek, G. W. Morey, and H. E. Merwin, Am. J. Sci. 5(35A), 143 (19380.

Boric Acid (HBO_2) $B_1H_1O_2$(cr)

Enthalpy Reference Temperature = T_r = 298.15 K — Standard State Pressure = p° = 0.1 MPa

T/K	$C_p°$ (J K⁻¹mol⁻¹)	S° (J K⁻¹mol⁻¹)	$-[G°-H°(T_r)]/T$ (J K⁻¹mol⁻¹)	$H°-H°(T_r)$ (kJ mol⁻¹)	$\Delta_f H°$ (kJ mol⁻¹)	$\Delta_f G°$ (kJ mol⁻¹)	Log K_f
0							
100							
200							
298.15	54.392	48.953	48.953	0.	-802.784	-486.201	85.180
300	54.810	49.291	48.954	0.101	-803.137	-484.236	84.313
400	61.505	66.015	51.189	5.931	-822.007	-375.080	48.980
500	66.526	80.279	55.615	12.332	-840.448	-261.206	27.288
600	71.546	92.835	60.791	19.227	-858.451	-143.659	12.507
700	78.921	104.435	66.205	26.761	-875.889	-23.139	1.727
800	84.935	115.381	71.675	34.965	-892.782	99.840	-6.519
900	89.642	125.671	77.109	43.706	-909.325	224.915	-13.054
1000	92.885	135.297	82.452	52.845	-925.719	351.820	-18.377
1100	94.056	144.207	87.666	62.194	-942.200	480.371	-22.811
1200	94.977	152.432	92.725	71.648	-958.913	610.431	-26.571
1300	95.646	160.062	97.615	81.181	-975.905	741.897	-29.810
1400	96.065	167.167	102.332	90.769	-993.213	874.682	-32.635
1500	96.232	173.802	106.878	100.386	-1010.864	1008.717	-35.127

PREVIOUS: September 1962 CURRENT: December 1964

Boric Acid (HBO_2) $B_1H_1O_2$(cr)

BORIC ACID (HBO_2) IDEAL GAS M_r = 43.81674

$\Delta_f H°(0\ K) = -557.2 \pm 4.2\ kJ\ mol^{-1}$

S°(298.15 K) = [239.73] J K^{-1} mol^{-1} $\Delta_f H°(298.15\ K) = -560.7 \pm 4.2\ kJ\ mol^{-1}$

Vibrational Frequencies and Degeneracies

ν, cm^{-1}	ν, cm^{-1}
3680(1)	[1250](1)
2030(1)	[600](1)
1420(1)	[700](1)

Ground State Quantum Weight: 1 σ = 1

Point Group: [C_s]

Bond Distances: H-O = [1.0] Å O-B = [1.34] Å B-O = [1.20] Å

Bond Angles: H-O-B = [120]° O-B-O = [180]°

Product of the Moments of Inertia: $I_A I_B I_C = [9.70076 \times 10^{-117}]\ g^3\ cm^6$

Enthalpy of Formation

The equilibrium pressures of the reaction 1/2 $B_2O_3(\ell)$ + 1/2 $H_2O(g)$ = $HBO_2(g)$ have been measured by several investigators. Using the equilibrium data reported, the enthalpies of reaction, $\Delta_r H°$(298.15 K), were evaluated by both the 2nd and 3rd law methods. The corresponding enthalpies of formation for $HBO_2(g)$ were derived based on the $\Delta_r H°$(298.15 K) values obtained by the 3rd law method. The results are presented as follows:

Source	T/K	$\Delta_r H°$(298.15 K), kcal mol^{-1} 2nd law	3rd law	$\Delta_f H°$(298.15 K) kcal mol^{-1}
(1)	1061 - 1451	45.4 ± 0.6	45.16	-133.38 ± 0.50
	1323 - 1451	0.25± 6.0	-5.15	-133.62 ± 0.50
(2)	1250 - 1450	39.7 ± 2.5	---	-138.87 ± 2.50*
(3)	1273	---	44.12	-134.42 ± 1.0
(4)	1070 - 1323	31.0 ± 10.0	44.07	-134.47 ± 3.20
	1143	---	40.9	-137.64 ± 0.70*
	1071 - 1325	44.2 ± 1.8	39.3	-139.3 ± 1.0*
	1148	---	43.2	-135.3 ± 0.4

*The value not used for the calculation of the weighted average of $\Delta_f H°$(298.15 K) for $HBO_2(g)$

Heat Capacity and Entropy

All molecular and spectroscopic constants were taken from White and Mann et al. (2). The principal moments of inertia are: $I_A = 0.1155 \times 10^{-39}$, $I_B = 9.1066 \times 10^{-39}$ and $I_C = 9.2221 \times 10^{-39}\ g\ cm^2$.

References

1. D. J. Meschi, W. A. Chupka and J. Berkowitz, J. Chem. Phys. 33, 530 (1960).
2. D. E. White, P. N. Walsh and A. Sommer, J. Chem. Phys. 32, 488 (1960).
3. S. P. Randall and J. L. Margrave, J. Inorg. Nucl. Chem. 16, 29 (1960).
4. M. Farber et al., "Thermodynamics of Reactions Involving Light Metal Oxides and Propellant Gases", QR-1987-5, May 9 - Aug 9, 1961; QR-1987-7, Nov. 9 - Feb. 9, 1962 and Final Report, May 9, 1960 through June 30, 1962, Rocket Power, Inc., California.

Boric Acid (HBO_2) $B_1H_1O_2(g)$

Enthalpy Reference Temperature = T_r = 298.15 K Standard State Pressure = p° = 0.1 MPa

T/K	$C_p°$ (J $K^{-1}mol^{-1}$)	S°	$-[G°-H°(T_r)]/T$	$H°-H°(T_r)$ (kJ mol^{-1})	$\Delta_f H°$	$\Delta_f G°$	Log K_f
0	0.	0.	INFINITE	-10.687	-557.212	-557.212	INFINITE
100	33.404	200.044	273.638	-7.359	-558.313	-555.919	290.382
200	36.887	224.017	243.422	-3.881	-559.425	-553.107	144.457
250	39.583	232.535	240.414	-1.970	-560.041	-551.456	115.221
298.15	42.232	239.734	239.734	0.	-560.656	-549.748	96.313
300	42.332	239.995	239.735	0.078	-560.680	-549.680	95.708
350	45.007	246.723	240.260	2.262	-561.324	-547.796	81.754
400	47.553	252.901	241.459	4.577	-561.968	-545.819	71.277
450	49.941	258.641	243.053	7.015	-562.605	-543.762	63.118
500	52.157	264.019	244.883	9.568	-563.228	-541.635	56.584
600	56.073	273.885	248.910	14.985	-564.416	-537.203	46.768
700	59.364	282.783	253.124	20.762	-565.518	-532.579	39.742
800	62.136	290.896	257.345	26.841	-566.535	-527.803	34.462
900	64.489	298.354	261.493	33.175	-567.476	-522.904	30.349
1000	66.505	305.256	265.529	39.727	-568.349	-517.905	27.053
1100	68.244	311.678	269.436	46.467	-569.168	-512.820	24.352
1200	69.752	317.682	273.209	53.368	-569.944	-507.663	22.098
1300	71.066	323.318	276.849	60.411	-570.686	-502.443	20.188
1400	72.215	328.628	280.359	67.576	-571.406	-497.166	18.549
1500	73.223	333.645	283.746	74.849	-572.113	-491.839	17.127
1600	74.111	338.400	287.015	82.216	-572.814	-486.464	15.881
1700	74.894	342.917	290.171	89.668	-573.518	-481.046	14.781
1800	75.588	347.218	293.222	97.192	-574.231	-475.586	13.801
1900	76.203	351.321	296.173	104.782	-574.958	-470.086	12.924
2000	76.752	355.244	299.029	112.431	-575.705	-464.547	12.133
2100	77.242	359.001	301.796	120.131	-576.475	-458.970	11.416
2200	77.681	362.605	304.479	127.877	-577.273	-453.356	10.764
2300	78.075	366.067	307.082	135.666	-578.102	-447.705	10.168
2400	78.430	369.397	309.609	143.491	-629.224	-440.949	9.597
2500	78.751	372.605	312.065	151.350	-630.199	-433.084	9.049
2600	79.041	375.700	314.454	159.240	-631.179	-425.180	8.542
2700	79.305	378.688	316.778	167.158	-632.166	-417.238	8.072
2800	79.545	381.576	319.041	175.100	-633.161	-409.259	7.635
2900	79.764	384.372	321.245	183.066	-634.165	-401.246	7.227
3000	79.964	387.079	323.395	191.053	-635.179	-393.196	6.846
3100	80.148	389.704	325.492	199.058	-636.204	-385.113	6.489
3200	80.316	392.252	327.539	207.082	-637.240	-376.997	6.154
3300	80.472	394.725	329.537	215.121	-638.288	-368.849	5.838
3400	80.615	397.130	331.490	223.176	-639.349	-360.668	5.541
3500	80.747	399.469	333.399	231.244	-640.422	-352.456	5.260
3600	80.869	401.745	335.266	239.325	-641.508	-344.213	4.994
3700	80.982	403.962	337.093	247.417	-642.607	-335.940	4.743
3800	81.088	406.123	338.881	255.521	-643.719	-327.636	4.504
3900	81.185	408.231	340.632	263.635	-644.844	-319.303	4.277
4000	81.277	410.288	342.348	271.758	-645.982	-310.942	4.060
4100	81.362	412.296	344.030	279.890	-647.134	-302.551	3.855
4200	81.441	414.257	345.679	288.030	-1128.149	-287.109	3.571
4300	81.516	416.174	347.296	296.178	-1128.242	-267.082	3.244
4400	81.585	418.049	348.883	304.333	-1128.350	-247.055	2.933
4500	81.651	419.883	350.440	312.495	-1128.474	-227.025	2.635
4600	81.712	421.679	351.969	320.663	-1128.615	-206.991	2.350
4700	81.770	423.437	353.471	328.837	-1128.774	-186.954	2.078
4800	81.824	425.159	354.947	337.017	-1128.950	-166.913	1.816
4900	81.876	426.846	356.397	345.202	-1129.146	-146.868	1.566
5000	81.924	428.501	357.823	353.392	-1129.360	-126.819	1.325
5100	81.970	430.124	359.225	361.586	-1129.595	-106.766	1.094
5200	82.013	431.716	360.603	369.786	-1129.850	-86.709	0.871
5300	82.054	433.279	361.960	377.989	-1130.127	-66.645	0.657
5400	82.092	434.813	363.295	386.196	-1130.427	-46.577	0.451
5500	82.129	436.319	364.609	394.407	-1130.750	-26.504	0.252
5600	82.164	437.799	365.903	402.622	-1131.098	-6.422	0.060
5700	82.197	439.254	367.177	410.840	-1131.472	13.664	-0.125
5800	82.229	440.684	368.432	419.061	-1131.872	33.758	-0.304
5900	82.258	442.090	369.668	427.286	-1132.299	53.857	-0.477
6000	82.287	443.473	370.887	435.513	-1132.756	73.966	-0.644

PREVIOUS: December 1964 (1 atm) CURRENT: December 1964 (1 bar)

Boric Acid (HBO_2) $B_1H_1O_2(g)$

BORON HYDRIDE SULFIDE (HBS) IDEAL GAS Mr = 43.87794

$\Delta_f H°(0\ K)$ = [50.8 ± 10.0] kJ mol^{-1}

$\Delta_f H°(298.15\ K)$ = [50.2 ± 10.0] kJ mol^{-1}

S°(298.15 K) = [214.96 ± 0.4] J K^{-1} mol^{-1}

Electronic Levels and Quantum Weights			
ε_i, cm^{-1}	g_i	ε_i, cm^{-1}	g_i
0	[1]	[35000]	[1]
[25000]	[3]	[37000]	[2]
[30000]	[6]	[40000]	[6]
[31000]	[3]		

Vibrational Frequencies and Degeneracies
ν, cm^{-1}
2742(1)
[715](2)
[1156](1)

Point Group: $C_{\infty v}$ σ = 1

Bond Distances: H-B = 1.1692 Å B-S = 1.5994 Å

Bond Angle: H-B-S = 180°

Rotational constant: B_0= 0.643679 cm^{-1}

Enthalpy of Formation

The transient thioborine molecule, HBS, has been identified by Kirk and Timms (1) in the mass spectra of the products fromed by the high temperature (1150-1300°C) reaction of hydrogen sulfide on crystalline boron. The best yields of HBS were found at low pressures (<1 torr) but even under thses conditions the half-life is relatively short (<3 min) (1, 2). No experimental measurement of its enthalpy of formation has been reported.

We estimate $\Delta_f H°$ from a consideration of bond-energy schemes. A comparison of the values for the H-B stretching force constants in HBS (K = 4.021 x 10^5 dynes/cm) and HBO (K = 4.080 x 10^5 dynes/cm, 3) suggests that the H-B bonds in these two molecules are quite similar. Assuming $\Delta_{at}H°$(H-BO, 298.15 K) = 99.5 kcal mol^{-1} (3), we calculate $\Delta_f H°$(298.15 K) = 10.7 kcal mol^{-1} with $\Delta_f H°$(H, g, 298.15 K) = 52.103 kcal mol^{-1} and $\Delta_f H°$(BS, g 298.15 K) = 58.1±4.0 kcal mol^{-1} (3).

A comparison of force constants also suggests that the H-B bonds in BH_3(4) and the B=S bond in BS (3) are similar to those in HBS. Using $\Delta_{at}H°$(H-B, 298.15 K) = $\Delta_{at}H°$(BH_3, g, 298.15 K)/3 = 87.9 kcal mol^{-1} and $\Delta_{at}H°$(B=S, 298.15 K) = $\Delta_{at}H°$(BS, 298.15 K) = 141.0 kcal mol^{-1}, we calculate $\Delta_{at}H°$(298.15 K) = 228.9 kcal mol^{-1} which leads to $\Delta_f H°$(298.15 K) = 22.3 kcal mol^{-1}. All bond dissociation energies are derived from JANAF data (3).

A value for $\Delta_{at}H°$(298.15 K) can also be obtained from the results of MO calculations on HBS. Thomson (5) has shown that dissociation energies obtained from MO calculations when combined with estimates of the molecular extra correlation energy (MECE) give reliable atomization energies. The change in the relativistic energy between the isolated atoms and molecule is assumed small. It is reasonable to assume that the contribution to the atomization energy from correlation effects would be similar for the isoelectronic molecules HBS and HBO. We estimate the MECE for HBO as 56.5 kcal mol^{-1} from the difference in the experimental $\Delta_{at}H°$(3) and the reported Hartree-Fock dissociation energy (6). This value of MECE when added to the computed dissociation energy of 191.7 kcal mol^{-1} (7) gives $\Delta_{at}H°$(298.15 K) = 248.2 kcal mol^{-1} which corresponds to $\Delta_f H°$(HBS, g, 298.15 K) = 3.0 kcal mol^{-1}.

We believe these three estimates of $\Delta_f H°$(10.7, 22.3, 3.0 kcal mol^{-1}) to be equally probable; therefore, we adopt the mean value of 12.0 kcal mol^{-1}. An uncertainty of ±10.0 kcal mol^{-1} is assigned which covers the range of possible values.

Heat Capacity and Entropy

The microwave measurements on HBS by Pearson and McCormick (8) have shown that the molecule is linear in its ground state. They observed the millimeter-wave spectra for eight isotopic species and evaluated structural parameters by the substitution method. We adopt their mean values for the H-B and B=S bond lengths.

Sams and Maki (2) have observed the ν_1 vibrational fundamental (2742 cm^{-1}) in the high-resolution infrared spectra of HBS, (g). Frequencies (ν_1) were recorded for four isotopic species. The value of ν_1(2742 cm^{-1}) is corrected for the natural isotopic abundances of boron: the correction for the sulfur isotopes is negligible. Values of ν_2 and ν_3 are calculated from force constants by a valence force method (9). The B=S stretching force constant (K = 6.9 x 10^5 dynes/cm) is estimated from the value for BS (K = 6.7 x 10^5 dynes/cm) by comparison with the B=O stretching force constants for BO and HBO (3). The bending force constant (K_D = 3.2 x 10^{12} dynes/cm) is taken equal to the value for HBO (3). Sams and Maki (2) also performed a detailed force field calculation on HBS which gave ν_2 = 635 cm^{-1} for $H^{11}B^{32}S$. Results of a similar calculation on HCN led these workers to conclude that their ν_2 values were probably lower (~50 cm^{-1}) than the true values. They suggested a better estimate of ν_2 as 690±30 cm^{-1}. Pearson and McCormick (8) estimated ν_2 = 680 cm^{-1} from the ℓ-doubling constant determined from their microwave measurements. These latter two values support our selection of ν_2 = 715 cm^{-1}. Photoelectron spectrometric results (10, 11) have led to predictions of ν_3 = 1100 cm^{-1} from the changes observed in vibrational frequencies produced upon ionization. We believe our calculated frequencies are uncertain to no more than ±50 cm^{-1} which amounts to an S°(298.15 K) error of ±0.1 cal K^{-1} mol^{-1}.

Evidence is available which suggests that the ground state configuration and excited states of HBS are similar to those for the isoelectronic molecule HCP. Theoretical calculations (7, 12) predict the same ordering of valence orbitals for these two molecules. This orbital order is corroborated by the results (10, 11, and 13) of photoelectron spectroscopy where the similarity (11) in their PE spectra has been noted. In addition, we note that the Zeeman parameters recently reported for HBS and HCP by Pearson et al. (14) are nearly identical. We assume the same ground state configuration ($^1\Sigma^+$) for HBS as that for HCP (3), and we estimate the electronic levels by analogy with those observed for HCP.

References

1. R. W. Kirk and P. L. Timms, Chem. Commun. 3, 18 (1967).
2. R. L. Sams and A. G. maki, J. Mol. Struct. 26, 107 (1975).
3. JANAF Thermochemical Tables: BH_3(g), 12-31-64; BS(g), 6-30-72; HBO(g), 12-31-75; HCP(g), 12-31-69; H(g), 6-30-74; BO(g), 6-30-68.
4. A. Kaldor and R. F. Porter, J. Amer. Chem. soc. 93, 2140 (1971).
5. C. Thomson, J. Chem. Phys. 58, 216 (1973).
6. C. Thomson and B. J. Wishart, Theoret. Chim. Acta, 35, 267 (1974).
7. C. Thomson, Chem. Phys. Letters, 25, 59 (1974).
8. E. F. Pearson and R. V. McCormick, J. Chem. Phys. 58, 1619 (1973).
9. G. herzberg, "Infrared and Raman Spectra of Polyatomic Molecules", D. Van Nostrand Company, Inc., New York, 1962.
10. T. P. Fehler and D. W. Turner, J. Amer. Chem. Soc. 95, 7175 (1973).
11. H. W. Kroto, R. J. Suffolk, and N. P. C. Westwood, Chem. Phys. Letters, 22, 495 (1973).
12. C. Thomson, Theoret. Chim. Acta, 35, 237 (1974).
13. D. C. Frost, S. T. Lee, and C. A. McDonell, Chem. Phys. Letters, 23, 472 (1973).
14. E. F. Pearson, C. L. Norris, and W. H. Flygare, J. Chem. Phys. 60, 1761 (1974).

Boron Hydride Sulfide (HBS) $B_1H_1S_1$(g)

Enthalpy Reference Temperature = T_r = 298.15 K Standard State Pressure = p° = 0.1 MPa

T/K	$C_p°$ (J $K^{-1}mol^{-1}$)	S° (J $K^{-1}mol^{-1}$)	$-[G°-H°(T_r)]/T$ (J $K^{-1}mol^{-1}$)	$H°-H°(T_r)$ (kJ mol^{-1})	$\Delta_f H°$ (kJ mol^{-1})	$\Delta_f G°$ (kJ mol^{-1})	Log K_f
0	0.	0.	INFINITE	-9.287	50.781	50.781	INFINITE
100	29.161	180.556	244.344	-6.379	51.475	39.739	-20.758
200	31.839	201.339	218.167	-3.366	51.167	28.051	-7.326
250	34.328	208.706	215.556	-1.713	50.731	22.320	-4.664
298.15	36.789	214.964	214.964	0.	50.208	16.894	-2.960
300	36.881	215.192	214.965	0.068	50.186	16.688	-2.906
350	39.237	221.057	215.422	1.972	49.573	11.152	-1.664
400	41.313	226.435	216.468	3.987	46.693	5.797	-0.757
450	43.113	231.408	217.855	6.099	45.321	0.758	-0.088
500	44.673	236.033	219.444	8.294	43.880	-4.114	0.430
600	47.240	244.414	222.922	12.895	41.450	-13.478	1.173
700	49.284	251.855	226.534	17.725	39.309	-22.460	1.676
800	50.966	258.549	230.124	22.740	37.354	-31.149	2.034
900	52.379	264.636	233.626	27.909	-17.603	-38.520	2.236
1000	53.577	270.219	237.010	33.209	-18.124	-40.816	2.132
1100	54.598	275.374	240.266	38.619	-18.634	-43.061	2.045
1200	55.470	280.163	243.394	44.123	-19.145	-45.259	1.970
1300	56.219	284.633	246.396	49.709	-19.664	-47.414	1.905
1400	56.863	288.824	249.278	55.364	-20.199	-49.529	1.848
1500	57.419	292.766	252.047	61.078	-20.756	-51.605	1.797
1600	57.901	296.488	254.710	66.845	-21.339	-53.642	1.751
1700	58.320	300.011	257.272	72.657	-21.953	-55.643	1.710
1800	58.686	303.355	259.740	78.507	-22.601	-57.606	1.672
1900	59.008	306.537	262.120	84.392	-23.283	-59.532	1.637
2000	59.291	309.571	264.417	90.308	-24.002	-61.421	1.604
2100	59.541	312.470	266.637	96.249	-24.758	-63.274	1.574
2200	59.763	315.245	268.784	102.215	-25.551	-65.090	1.545
2300	59.961	317.906	270.862	108.201	-26.383	-66.868	1.519
2400	60.139	320.462	272.876	114.206	-77.513	-67.541	1.470
2500	60.299	322.920	274.829	120.228	-78.499	-67.106	1.402
2600	60.444	325.288	276.724	126.266	-79.492	-66.630	1.339
2700	60.576	327.572	278.565	132.317	-80.490	-66.117	1.279
2800	60.697	329.777	280.355	138.380	-81.494	-65.566	1.223
2900	60.810	331.909	282.096	144.456	-82.504	-64.979	1.170
3000	60.915	333.972	283.791	150.542	-83.520	-64.358	1.121
3100	61.016	335.971	285.443	156.639	-84.541	-63.702	1.073
3200	61.112	337.910	287.052	162.745	-85.569	-63.014	1.029
3300	61.207	339.792	288.622	168.861	-86.601	-62.293	0.986
3400	61.301	341.620	290.154	174.987	-87.639	-61.541	0.945
3500	61.396	343.399	291.650	181.121	-88.682	-60.758	0.907
3600	61.493	345.130	293.111	187.266	-89.730	-59.945	0.870
3700	61.594	346.816	294.540	193.420	-90.781	-59.103	0.834
3800	61.701	348.460	295.938	199.585	-91.836	-58.233	0.800
3900	61.814	350.064	297.305	205.761	-92.894	-57.335	0.768
4000	61.935	351.631	298.644	211.948	-93.955	-56.410	0.737
4100	62.066	353.162	299.955	218.148	-95.017	-55.458	0.707
4200	62.206	354.659	301.239	224.361	-575.929	-47.456	0.590
4300	62.359	356.124	302.499	230.590	-575.903	-34.874	0.424
4400	62.524	357.560	303.734	236.834	-575.878	-22.292	0.265
4500	62.703	358.967	304.946	243.095	-575.852	-9.711	0.113
4600	62.897	360.347	306.135	249.375	-575.824	2.870	-0.033
4700	63.106	361.702	307.303	255.675	-575.793	15.450	-0.172
4800	63.331	363.033	308.450	261.996	-575.757	28.029	-0.305
4900	63.574	364.341	309.578	268.342	-575.716	40.607	-0.433
5000	63.833	365.628	310.686	274.712	-575.667	53.185	-0.556
5100	64.111	366.895	311.776	281.109	-575.609	65.762	-0.674
5200	64.407	368.143	312.848	287.535	-575.541	78.337	-0.787
5300	64.721	369.372	313.903	293.991	-575.461	90.911	-0.896
5400	65.054	370.585	314.941	300.479	-575.366	103.483	-1.001
5500	65.405	371.782	315.964	307.002	-575.256	116.053	-1.102
5600	65.774	372.964	316.971	313.561	-575.129	128.621	-1.200
5700	66.162	374.132	317.964	320.158	-574.982	141.187	-1.294
5800	66.567	375.286	318.942	326.794	-574.814	153.750	-1.385
5900	66.989	376.427	319.907	333.471	-574.623	166.310	-1.472
6000	67.428	377.557	320.858	340.192	-574.408	178.866	-1.557

PREVIOUS: December 1975 (1 atm) CURRENT: December 1975 (1 bar)

Boron Hydride Sulfide (HBS) $B_1H_1S_1$(g)

BORON HYDRIDE SULFIDE, ION (HBS^+) IDEAL GAS M_r = 43.87739

$\Delta_f H°(0\ K)$ = [1123.2 ± 11] kJ mol^{-1}

$S°(298.15\ K)$ = [225.4 ± 0.4] J K^{-1} mol^{-1} $\Delta_f H°(298.15\ K)$ = [1129.5 ± 11 kJ mol^{-1}

Electronic Levels and Quantum Weights

State	ε_i, cm^{-1}	g_i
$X^2\Pi_{1/2}$	0	2
$X^2\Pi_{3/2}$	[300]	2
$A^2\Sigma^+$	19599	2
$B^2\Sigma^+$	38069	2

Vibrational Frequencies and Degeneracies

ν, cm^{-1}
[2550](1)
[764](2)
973 (1)

Point Group: $C_{\infty v}$

Bond Distances: H-B = [1.192] Å B-S = [1.680] Å

Bond Angle: H-B-S = 180°

Rotational Constant: B_o = [0.586979] cm^{-1}

Enthalpy of Formation

Kroto et al. (1) and Fehler and Turner (2) have independently measured the 584 Å photoelectron spectra of HBS(g). The reported value of 11.11±0.03 eV (1) and 11.12 eV (2) for the first adiabatic ionization potential are in excellent agreement. We adopt an average (11.115±0.03 eV) of these two IP values, and we assume $\Delta_r H°$ = 256.32±0.7 kcal mol^{-1} refers to the ionization process HBS(g) + e^- + HBS^+(g) + 2 e^- at 0 K. The latter value yields $\Delta_f H°(HBS^+$, g, 0 K) = 268.4±11.0 kcal mol^{-1} when combined with $\Delta_f H°$(HBS, g, 0 K) = 12.1±10.0 kcal mol^{-1} (3). The adopted $\Delta_f H°$ value is placed in brackets to emphasize the uncertainty in the estimated $\Delta_f H°$ value (3) for HBS. $\Delta_f H°(298.15\ K)$ corresponds to 270.0±11.0 kcal mol^{-1}.

Heat Capacity and Entropy

Extensive ab initio MO calculations (4) recently reported for HBS show that the highest occupied orbital is of Π-symmetry and is localized mainly on the sulfur atom. The shape of the first photoelectron band in the HBS spectra (1, 2) suggests that this orbital is bonding. The bonding nature of this Π orbital is also predicted from the Walsh diagram (5) for HAB molecules. As a consequence, we would expect the ion to be less strongly bound than the parent molecule. In the related species HCN (6) which has the same orbital ordering as HBS (4), ionization (7) occurs with a 2.2% increase in the H-C bond length and a 4.7% increase in the C≡N bond length. We assume a similar increase in the bond lengths of HBS (3) upon ionization. If the ion was nonlinear, one would expect to find a vibrational spacing in the photoelectron spectra of HBS which could be associated with the bending frequency of the ion. No frequency as low as 764 cm^{-1} has been observed (1, 2); thus, it is likely that HBS^+ is linear in its ground state. This prediction is supported by the Walsh diagram (5). We assume that the ion is linear.

The progression in the first photoelectron band of the HBS spectra has been reported as 955±40 cm^{-1} (1) and 990 cm^{-1} (2). By comparison with changes observed in vibrational frequencies on ionization of HCN and HCP (3, 8), the vibrational mode excited is most certainly the B≡S stretching frequency, ν_3. We adopt the average (973 cm^{-1}) of these two values. The values of ν_1 and ν_2 are calculated from estimated force constants by a valence force method (9). The two force constants are estimated form those for HCN^+ (8), NCH and NBS (3). We estimate the uncertainties in our frequencies are no greater than ±50 cm^{-1} which corresponds to an error in the value of $S°(298.15\ K)$ of only ±0.12 kcal mol^{-1}.

The electronic states and upper levels are taken from the photoelectron spectrometric study of Kroto et al. (1). Relative term values for the $A^2\Sigma^+$ and $B^2\Sigma^+$ states are calculated as differences in the reported adiabatic ionization potentials (1). Fehler and Turner (2) have estimated the ground state splitting to be ~300 cm^{-1} from the observed peak widths in the first band of the HBS photoelectron spectra. We adopt this result since it compares favorably with the splitting (~278 cm^{-1}) observed for the $^2\Pi$ state of the isoelectronic molecules CS^+ (10). Relative term values for the two $^2\Sigma$ states calculated from the adiabatic ionization potentials of Fehler and Turner (2) agree within about 160 cm^{-1} with the values adopted here. Fehler and Turner (2) have also reported a photoelectron band at 17.9 eV which was not observed by Kroto et al. (1). This band could not be definitely assigned to HBS^+. If this state does exist, it lies above 50000 cm^{-1} and has a negligible effect on the thermodynamic functions of the ion. We prefer not to include it. Kroto et al. (1) have interpreted their photoelectron results in terms of a linear configuration for the three observed states of HBS^+. Thus, we feel our functions are not significantly influenced by the use of the ground state vibrational frequencies and rotational constant for the excited states.

References

1. H. W. Kroto, R. J. Suffolk, and N. P. C. Westwood, Chem. Phys. Letters 22, 495 (1973).
2. T. P. Fehler and D. W. Turner, J. Amer. Chem. Soc. 95, 7175 (1973).
3. JANAF Thermochemical Tables: HBS(g), 12-31-75; HCN(g) and HCP(g), 12-31-69.
4. C. Thomson, Chem. Phys. Letters 25, 59 (1974).
5. A. D. Walsh, J. Chem. Soc. 1953, 2288 (1953).
6. C. Thomson and B. J. Wishart, Theoret. Chim. Acta 35, 267 (1974).
7. S. P. So and W. G. Richards, J. Chem. Soc. Faraday Trans. II, 71, 62 (1975).
8. D. C. Frost, S. T. Lee, and C. A. McDowell, Chem. Phys. Letters 23, 472 (1973).
9. G. Herzberg, "Infrared and Raman Spectra of Polyatomic Molecules", D. Van Nostrand Company, Inc., New York, 1962.
10. D. C. Frost, S. T. Lee, and C. A. McDowell, Chem. Phys. Letters 17, 153 (1972).

Boron Hydride Sulfide, Ion (HBS^+) $B_1H_1S_1^+(g)$

Enthalpy Reference Temperature = T_r = 298.15 K Standard State Pressure = p° = 0.1 MPa

	J $K^{-1}mol^{-1}$			kJ mol^{-1}			
T/K	$C_p°$	S°	$-[G°-H°(T_r)]/T$	$H°-H°(T_r)$	$\Delta_f H°$	$\Delta_f G°$	Log K_f
0	0.	0.	INFINITE	-9.934	1123.199		
100	31.149	187.666	257.455	-6.979			
200	35.146	210.523	228.818	-3.659			
250	37.309	218.594	225.988	-1.848			
298.15	39.455	225.350	225.350	0.	1129.471	1086.806	-190.404
300	39.535	225.594	225.351	0.073	1129.493	1086.541	-189.184
350	41.615	231.848	225.840	2.103	1130.044	1079.338	-161.082
400	43.459	237.528	226.951	4.231	1128.316	1072.148	-140.008
450	45.064	242.741	228.420	6.445	1128.086	1065.132	-123.637
500	46.464	247.563	230.096	8.734	1127.777	1058.156	-110.545
600	48.787	256.248	233.747	13.501	1127.593	1044.255	-90.910
700	50.657	263.914	237.520	18.476	1127.676	1030.362	-76.887
800	52.207	270.783	241.256	23.621	1127.929	1016.444	-66.367
900	53.508	277.009	244.888	28.909	1075.170	1003.568	-58.246
1000	54.607	282.705	248.389	34.316	1076.835	995.523	-52.001
1100	55.538	287.955	251.750	39.825	1078.502	987.311	-46.883
1200	56.329	292.822	254.973	45.419	1080.160	978.948	-42.612
1300	57.004	297.358	258.061	51.087	1081.801	970.447	-38.993
1400	57.581	301.604	261.021	56.817	1083.420	961.820	-35.886
1500	58.076	305.594	263.861	62.600	1085.011	953.079	-33.189
1600	58.503	309.356	266.588	68.430	1086.568	944.232	-30.826
1700	58.873	312.914	269.209	74.299	1088.091	935.289	-28.738
1800	59.195	316.289	271.732	80.203	1089.575	926.258	-26.879
1900	59.477	319.497	274.162	86.137	1091.020	917.145	-25.214
2000	59.725	322.554	276.506	92.097	1092.425	907.957	-23.713
2100	59.943	325.474	278.769	98.081	1093.790	898.700	-22.354
2200	60.137	328.267	280.955	104.085	1095.114	889.379	-21.117
2300	60.309	330.944	283.071	110.107	1096.396	879.999	-19.985
2400	60.464	333.514	285.120	116.146	1047.379	871.632	-18.971
2500	60.603	335.985	287.105	122.200	1048.502	864.286	-18.058
2600	60.729	338.364	289.031	128.266	1049.618	856.895	-17.215
2700	60.844	340.659	290.901	134.345	1050.726	849.462	-16.434
2800	60.949	342.873	292.718	140.435	1051.827	841.988	-15.707
2900	61.046	345.014	294.485	146.535	1052.920	834.474	-15.031
3000	61.136	347.085	296.204	152.644	1054.005	826.923	-14.398
3100	61.220	349.091	297.877	158.762	1055.083	819.336	-13.806
3200	61.299	351.036	299.508	164.887	1056.154	811.714	-13.250
3300	61.373	352.923	301.099	171.021	1057.218	804.059	-12.727
3400	61.444	354.756	302.650	177.162	1058.274	796.371	-12.235
3500	61.512	356.538	304.164	183.310	1059.323	788.653	-11.770
3600	61.577	358.272	305.643	189.464	1060.364	780.905	-11.331
3700	61.641	359.960	307.089	195.625	1061.398	773.128	-10.915
3800	61.702	361.605	308.502	201.792	1062.423	765.323	-10.520
3900	61.762	363.208	309.884	207.966	1063.442	757.491	-10.145
4000	61.821	364.773	311.237	214.145	1064.452	749.633	-9.789
4100	61.879	366.300	312.561	220.330	1065.453	741.750	-9.450
4200	61.937	367.792	313.858	226.521	586.597	740.867	-9.214
4300	61.993	369.250	315.130	232.717	588.670	744.516	-9.044
4400	62.049	370.676	316.376	238.919	590.732	748.116	-8.881
4500	62.105	372.071	317.598	245.127	592.783	751.670	-8.725
4600	62.160	373.437	318.797	251.340	594.823	755.178	-8.575
4700	62.216	374.774	319.974	257.559	596.851	758.643	-8.431
4800	62.270	376.084	321.129	263.783	598.868	762.064	-8.293
4900	62.325	377.369	322.264	270.013	600.873	765.443	-8.160
5000	62.380	378.629	323.379	276.248	602.865	768.781	-8.031
5100	62.434	379.864	324.474	282.489	604.845	772.080	-7.908
5200	62.488	381.077	325.551	288.735	606.813	775.340	-7.788
5300	62.542	382.268	326.610	294.987	608.767	778.562	-7.673
5400	62.596	383.438	327.652	301.244	610.708	781.747	-7.562
5500	62.649	384.587	328.677	307.506	612.636	784.897	-7.454
5600	62.703	385.716	329.685	313.773	614.551	788.012	-7.350
5700	62.756	386.826	330.678	320.046	616.453	791.092	-7.250
5800	62.808	387.918	331.655	326.325	618.341	794.140	-7.152
5900	62.861	388.992	332.618	332.608	620.216	797.155	-7.057
6000	62.913	390.049	333.566	338.897	622.078	800.138	-6.966

PREVIOUS: December 1975 (1 atm) CURRENT: December 1975 (1 bar)

Boron Hydride Sulfide, Ion (HBS^+) $B_1H_1S_1^+(g)$

BORANE (BH_2) IDEAL GAS M_r = 12.82588

$S°(298.15\ K)$ = [180.19] J K^{-1} mol^{-1}

$\Delta_f H°(0\ K)$ = 202 ± 63 kJ mol^{-1}
$\Delta_f H°(298.15\ K)$ = 201 ± 63 kJ mol^{-1}

Vibrational Frequencies and Degeneracies

ν, cm^{-1}	ν, cm^{-1}	ν, cm^{-1}
[2430](1)	[840](2)	[2650](1)

Ground State Quantum Weight: [2] σ = 2
Point Group: [$D_{\infty h}$]
Bond Distance: B-H = [1.20] Å
Bond Angle: H-B-H = [180]°
Rotational Constant: B_o = [5.807235] cm^{-1}

Enthalpy of Formation

$\Delta_f H°(298.15\ K)$ was calculated from appearance potentials reported by Fehlner and Koski (1) and the JANAF Thermochemical Table (2). For example:

	$\Delta_r H°(0\ K)$
$B_2H_6(g) \rightarrow BH_2^+(ion) + BH_2(g) + H_2(g)$	13.4 eV
$BH_2(g) \rightarrow BH_2^+(ion)$	9.8 eV
$B_2H_6(g) \rightarrow 2\ BH_2(g) + H_2(g)$	3.6 eV

The $\Delta_f H°$ data is summarized below.

Source	Method	$\Delta_f H°(298.15\ K)$, kcal mol^{-1}
(2)	Appearance Potentials	48
(3)	Estimate	66
(4)	Estimate	93

Heat Capacity and Entropy

The bond distance B-H = 1.20 Å was estimated from a consideration of BH(g), $BH_3(g)$ and other molecules. The bond angle H-B-H = 180° and the electronic configuration of the ground state [$^2A_2$] were estimated by Jordan and Longuet-Higgins (4). The frequencies were estimated assuming valence forces, Herzberg (5), and using force constants, $k_1 = 3.5 \times 10^5$ dynes cm^{-1} and $k\delta\ l^{-1} = 0.35 \times 10^5$ dynes cm^{-1}, estimated from BH_3, Shepp and Bauer (6).

References

1. T. P. Fehlner and W. S. Koski, J. Am. Chem. Soc. 86, 2733 (1964).
2. JANAF Thermochemical Tables: $B_2H_6(g)$, 12-31-64.
3. C. J. O'Brien and J. R. Perrin, Marquardt Corp., Van Nuys, California and J. Perine, Olin Mathieson Chemical Corp., New York, "Estimation of the Heats of Formation of Gaseous Combustion Product Molecules", October 1959.
4. P. C. H. Jordan and H. C. Longuet-Higgins, Mol. Phys. 5, 121 (1962).
5. G. Herzberg, "Molecular Spectra and Molecular Structure II Infrared and Raman Spectra of Polyatomic Molecules", D. Van Nostrand Co., Inc., New York, pp 172 (1945).
6. A. Shepp and S. H. Bauer, J. Am. Chem. Soc. 76, 265 (1954).

Borane (BH_2) $B_1H_2(g)$

Enthalpy Reference Temperature = T_r = 298.15 K Standard State Pressure = p° = 0.1 MPa

T/K	$C_p°$ (J $K^{-1}mol^{-1}$)	S°	$-[G°-H°(T_r)]/T$	$H°-H°(T_r)$ (kJ mol^{-1})	$\Delta_f H°$	$\Delta_f G°$	Log K_f
0	0.	0.	INFINITE	-9.008	201.505	201.505	INFINITE
100	29.115	146.922	208.138	-6.122	201.368	196.780	-102.787
200	30.550	167.371	183.190	-3.164	201.300	192.192	-50.195
250	32.243	174.363	180.745	-1.595	201.101	189.935	-39.685
298.15	34.026	180.195	180.195	0.	200.832	187.809	-32.903
300	34.094	180.405	180.195	0.063	200.821	187.728	-32.686
350	35.849	185.795	180.617	1.812	200.494	185.571	-27.695
400	37.432	190.687	181.574	3.645	200.134	183.463	-23.958
450	38.855	195.179	182.840	5.553	199.754	181.402	-21.057
500	40.154	199.341	184.284	7.528	199.364	179.384	-18.740
600	42.509	206.874	187.435	11.664	198.589	175.461	-15.275
700	44.645	213.590	190.700	16.023	197.855	171.665	-12.810
800	46.593	219.681	193.948	20.587	197.183	167.970	-10.967
900	48.350	225.272	197.122	25.335	196.576	164.356	-9.539
1000	49.913	230.449	200.199	30.250	196.025	160.806	-8.400
1100	51.287	235.272	203.170	35.312	195.517	157.309	-7.470
1200	52.487	239.787	206.036	40.502	195.040	153.857	-6.697
1300	53.531	244.030	208.797	45.804	194.580	150.444	-6.045
1400	54.439	248.032	211.458	51.203	194.126	147.066	-5.487
1500	55.228	251.815	214.023	56.688	193.668	143.720	-5.005
1600	55.916	255.402	216.498	62.246	193.198	140.406	-4.584
1700	56.517	258.810	218.888	67.868	192.711	137.121	-4.213
1800	57.043	262.056	221.197	73.546	192.201	133.865	-3.885
1900	57.505	265.153	223.429	79.274	191.664	130.639	-3.592
2000	57.913	268.113	225.590	85.046	191.098	127.442	-3.328
2100	58.274	270.947	227.683	90.855	190.500	124.274	-3.091
2200	58.595	273.666	229.712	96.699	189.868	121.135	-2.876
2300	58.880	276.277	231.680	102.573	189.200	118.025	-2.680
2400	59.136	278.788	233.591	108.474	138.236	116.014	-2.525
2500	59.366	281.207	235.447	114.399	137.417	115.105	-2.405
2600	59.572	283.539	237.252	120.347	136.591	114.229	-2.295
2700	59.759	285.791	239.009	126.313	135.758	113.385	-2.194
2800	59.927	287.968	240.718	132.298	134.918	112.571	-2.100
2900	60.081	290.073	242.384	138.298	134.070	111.788	-2.014
3000	60.220	292.112	244.008	144.313	133.213	111.035	-1.933
3100	60.348	294.089	245.592	150.342	132.347	110.309	-1.859
3200	60.464	296.007	247.137	156.383	131.470	109.613	-1.789
3300	60.571	297.869	248.647	162.434	130.584	108.943	-1.724
3400	60.669	299.679	250.121	168.496	129.687	108.301	-1.664
3500	60.760	301.439	251.562	174.568	128.778	107.685	-1.607
3600	60.843	303.152	252.972	180.648	127.859	107.095	-1.554
3700	60.921	304.820	254.351	186.737	126.927	106.531	-1.504
3800	60.992	306.446	255.700	192.832	125.984	105.993	-1.457
3900	61.059	308.031	257.022	198.935	125.028	105.479	-1.413
4000	61.120	309.577	258.316	205.044	124.060	104.990	-1.371
4100	61.178	311.087	259.585	211.159	123.079	104.526	-1.332
4200	61.232	312.562	260.829	217.279	-357.764	111.109	-1.382
4300	61.282	314.004	262.049	223.405	-357.684	122.272	-1.485
4400	61.329	315.413	263.246	229.536	-357.618	133.433	-1.584
4500	61.373	316.792	264.420	235.671	-357.568	144.593	-1.678
4600	61.414	318.141	265.574	241.810	-357.533	155.751	-1.769
4700	61.452	319.462	266.706	247.953	-357.513	166.909	-1.855
4800	61.489	320.756	267.819	254.100	-357.509	178.068	-1.938
4900	61.523	322.025	268.912	260.251	-357.521	189.225	-2.017
5000	61.555	323.268	269.987	266.405	-357.548	200.384	-2.093
5100	61.586	324.487	271.044	272.562	-357.591	211.542	-2.167
5200	61.614	325.683	272.083	278.722	-357.650	222.703	-2.237
5300	61.642	326.857	273.105	284.885	-357.724	233.864	-2.305
5400	61.667	328.010	274.112	291.050	-357.813	245.027	-2.370
5500	61.692	329.141	275.102	297.218	-357.917	256.191	-2.433
5600	61.715	330.253	276.077	303.389	-358.036	267.358	-2.494
5700	61.737	331.346	277.037	309.561	-358.169	278.527	-2.552
5800	61.758	332.420	277.982	315.736	-358.316	289.699	-2.609
5900	61.778	333.476	278.914	321.913	-358.476	300.872	-2.664
6000	61.796	334.514	279.832	328.091	-358.649	312.049	-2.717

PREVIOUS: December 1964 (1 atm) CURRENT: December 1964 (1 bar)

Borane (BH_2) $B_1H_2(g)$

DIHYDROXYBORANE ($B(OH)_2$) IDEAL GAS M_r = 44.82468

$S°(298.15\ K)$ = [243.59] J K^{-1} mol^{-1}

$\Delta_fH°(0\ K)$ = [-470 ± 15] kJ mol^{-1}
$\Delta_fH°(298.15\ K)$ = [-477 ± 15] kJ mol^{-1}

Vibrational Frequencies and Degeneracies

ν, cm^{-1}	ν, cm^{-1}	ν, cm^{-1}
[2500](1)	[750](1)	[1300](1)
[1150](1)	[460](1)	[1000](1)
[1100](1)	[3000](1)	[450](1)

Ground State Quantum Weight: [2] σ = 2
Point Group: [C_{2h}]
Bond Distances: B-O = [1.27] Å O-H = [0.97] Å
Bond Angles: H-O-B = [105]° O-B-O = [180]°
Product of the Moments of Inertia: $I_AI_BI_C$ = [2.62751 x 10^{-116}] g^3 cm^6

Enthalpy of Formation

The value of $\Delta_fH°(B(OH)_2$, g, 298.15 K) was calculated by using the bond energy, D_0^0(B-OH) = 132.7 kcal mol^{-1}, and the enthalpies of formation for B(g) and OH(g) as 132.80 and 9.33 kcal mol^{-1}, respectively.

Heat Capacity and Entropy

The vibrational frequencies were estimated by comparison with those for BO_2(g), $B(OH)_3$(g) and other related molecules. The bond distances and angles were estimated by comparison with those for BO_2(g) and H_2O_2(g) molecules. The molecular configuration is assumed to be trans which is considered as more stable than the cis-form. The principal moments of inertia are: I_A = 0.2939 x 10^{-39}, I_B = 9.3100 x 10^{-39} and I_C = 9.6038 x 10^{-39} g cm^2.

Dihydroxyborane ($B(OH)_2$) $B_1H_2O_2(g)$

Enthalpy Reference Temperature = T_r = 298.15 K Standard State Pressure = p° = 0.1 MPa

	J $K^{-1}mol^{-1}$			kJ mol^{-1}			
T/K	$C_p°$	S°	$-[G°-H°(T_r)]/T$	$H°-H°(T_r)$	$\Delta_fH°$	$\Delta_fG°$	Log K_f
0	0.	0.	INFINITE	-11.772	-470.383	-470.383	INFINITE
100	34.306	198.871	283.171	-8.430	-472.969	-465.422	243.112
200	42.381	224.875	248.034	-4.632	-475.109	-457.021	119.362
250	47.278	234.855	244.419	-2.391	-476.093	-452.384	94.520
298.15	52.015	243.589	243.589	0.	-476.976	-447.736	78.441
300	52.195	243.911	243.590	0.096	-477.008	-447.554	77.926
350	56.883	252.313	244.243	2.825	-477.833	-442.579	66.051
400	61.190	260.195	245.749	5.778	-478.566	-437.492	57.131
450	65.065	267.630	247.771	8.936	-479.213	-432.318	50.182
500	68.519	274.668	250.112	12.278	-479.779	-427.076	44.616
600	74.344	287.695	255.309	19.432	-480.695	-416.445	36.255
700	79.038	299.520	260.793	27.109	-481.365	-405.680	30.272
800	82.894	310.334	266.319	35.212	-481.835	-394.834	25.780
900	86.106	320.289	271.770	43.666	-482.142	-383.939	22.283
1000	88.805	329.504	277.089	52.416	-482.321	-373.017	19.484
1100	91.088	338.079	282.248	61.414	-482.401	-362.081	17.194
1200	93.028	346.090	287.238	70.622	-482.408	-351.143	15.285
1300	94.683	353.603	292.057	80.010	-482.366	-340.205	13.670
1400	96.102	360.673	296.708	89.551	-482.292	-329.272	12.285
1500	97.324	367.346	301.197	99.224	-482.203	-318.346	11.086
1600	98.380	373.662	305.530	109.010	-482.111	-307.425	10.036
1700	99.298	379.654	309.716	118.895	-482.028	-296.509	9.111
1800	100.098	385.353	313.761	128.866	-481.961	-285.599	8.288
1900	100.800	390.784	317.673	138.911	-481.920	-274.691	7.552
2000	101.417	395.971	321.459	149.023	-481.908	-263.785	6.889
2100	101.962	400.932	325.126	159.192	-481.932	-252.878	6.290
2200	102.445	405.687	328.681	169.413	-481.995	-241.969	5.745
2300	102.876	410.250	332.129	179.680	-482.101	-231.057	5.247
2400	103.260	414.637	335.476	189.987	-532.512	-219.070	4.768
2500	103.605	418.860	338.727	200.330	-532.788	-206.005	4.304
2600	103.915	422.929	341.888	210.707	-533.081	-192.928	3.876
2700	104.195	426.856	344.963	221.112	-533.392	-179.839	3.479
2800	104.448	430.650	347.956	231.545	-533.721	-166.738	3.111
2900	104.678	434.320	350.871	242.001	-534.071	-153.626	2.767
3000	104.887	437.872	353.712	252.480	-534.442	-140.500	2.446
3100	105.078	441.314	356.483	262.978	-534.834	-127.362	2.146
3200	105.253	444.653	359.186	273.495	-535.248	-114.212	1.864
3300	105.413	447.894	361.825	284.028	-535.684	-101.048	1.599
3400	105.560	451.044	364.403	294.577	-536.143	-87.870	1.350
3500	105.696	454.105	366.923	305.140	-536.623	-74.679	1.115
3600	105.821	457.085	369.386	315.716	-537.127	-61.473	0.892
3700	105.937	459.986	371.796	326.304	-537.652	-48.254	0.681
3800	106.044	462.812	374.154	336.903	-538.201	-35.019	0.481
3900	106.143	465.568	376.463	347.512	-538.772	-21.769	0.292
4000	106.236	468.257	378.724	358.131	-539.366	-8.506	0.111
4100	106.322	470.881	380.940	368.759	-539.982	4.773	-0.061
4200	106.402	473.444	383.112	379.395	-1020.471	25.092	-0.312
4300	106.477	475.949	385.242	390.039	-1020.046	49.981	-0.607
4400	106.547	478.397	387.331	400.691	-1019.646	74.860	-0.889
4500	106.613	480.792	389.382	411.349	-1019.270	99.731	-1.158
4600	106.674	483.136	391.394	422.013	-1018.920	124.593	-1.415
4700	106.732	485.431	393.371	432.683	-1018.596	149.449	-1.661
4800	106.787	487.679	395.312	443.359	-1018.298	174.298	-1.897
4900	106.838	489.881	397.220	454.041	-1018.027	199.141	-2.123
5000	106.886	492.040	399.095	464.727	-1017.783	223.979	-2.340
5100	106.932	494.157	400.938	475.418	-1017.567	248.812	-2.548
5200	106.975	496.234	402.751	486.113	-1017.378	273.641	-2.749
5300	107.016	498.272	404.534	496.813	-1017.218	298.467	-2.942
5400	107.054	500.273	406.288	507.516	-1017.088	323.290	-3.127
5500	107.091	502.238	408.015	518.223	-1016.987	348.110	-3.306
5600	107.125	504.167	409.715	528.934	-1016.916	372.930	-3.479
5700	107.158	506.064	411.389	539.648	-1016.877	397.748	-3.645
5800	107.189	507.928	413.037	550.366	-1016.869	422.567	-3.806
5900	107.219	509.760	414.661	561.086	-1016.893	447.383	-3.961
6000	107.247	511.563	416.261	571.810	-1016.949	472.202	-4.111

PREVIOUS: December 1964 (1 atm) CURRENT: December 1964 (1 bar)

Dihydroxyborane ($B(OH)_2$) $B_1H_2O_2(g)$

BORANE (BH_3) IDEAL GAS M_r = 13.83382

$\Delta_f H°(0\ K)$ = 110.5 ± 10 kJ mol^{-1}
$S°(298.15\ K)$ = [187.88] J K^{-1} mol^{-1} $\Delta_f H°(298.15\ K)$ = 106.7 ± 10 kJ mol^{-1}

Vibrational Frequencies and Degeneracies

ν, cm^{-1}	ν, cm^{-1}
[2384](1)	[802](1)
[1765](2)	[2976](2)

Ground State Quantum Weight: [1] σ = 6
Point Group: [D_{3h}]
Bond Distance: B-H = [1.16] Å
Bond Angle: H-B-H = [120]°
Product of the Moments of Inertia: $I_A I_B I_C$ = 7.71122 x 10^{-119}] g^3 cm^6

Enthalpy of Formation

$\Delta_f H°(298.15\ K)$ was calculated from appearance potentials by Fehlner and Koski (1). For example:

$B_2H_6(g)$	→ $BH_3^+(ion) + BH_3(g)$	13.1 eV
$BH_3^+(ion)$	+ $BH_3(g)$	-11.4 eV
$B_2H_6(g)$	→ 2 $BH_3(g)$	1.7 eV

The enthalpy of formation data is summarized below.

Reaction: $B_2H_6(g) \rightarrow 2\ BH_3(g)$*

*The auxiliary $\Delta_f H°(B_2H_6$, g, 298.15 K) was from the JANAF table dated Dec. 31, 1964.

Source	Method	$\Delta_r H°(0\ K)$, kcal mol^{-1}	$\Delta_f H°(298.15\ K)$, kcal mol^{-1}*
(2)	Bond energy parameters obtained by linear extrapolation from NH and CH bonds	69.9	41.7
(3)	Apparent equilibrium constants	55 ± 8	33
(3)	A recalculation by JANAF using α values and P = 10^{-7} atm (α = degree of dissociation)	49 ± 8	30
(4)	Correlations of ionization and dissociation energies of hydrides and fluorides in relation to electronic structure	41.5	26.6
(1)	Appearance potentials	39.2	25.5
(5)	From kinetic data for the decomposition of BH_3CO	32-38	22-25
(6)	From calorimetric measurements of analogous compounds	28.4 ± 2	20.1
(7)	From electronegativities	23.9	18

Heat Capacity and Entropy

The molecular configuration, [planar, H-B-H angle - 120°], was estimated by Shepp and Bauer (8) as well as by Jordan and Longuet-Higgins (2). In addition, Shepp and Bauer estimated the bond distance and the vibrational frequencies. Jordan and Longuet-Higgins estimated the electronic configuration [$^1A_1'$]. The principal moments of inertia are: $I_A = I_B = 0.3378 \times 10^{-39}$ and $I_C = 0.6757 \times 10^{-39}$ g cm^2.

References

1. T. P. Fehlner and W. S. Koski, J. Am. Chem. Soc. 86, 2733 (1964).
2. P. C. H. Jordan and H. C. Longuet-Higgins, Mol. Phys. 5, 121 (1962).
3. E. J. Sinke, G. A. Pressley, Jrs. et al., J. Chem. Phys. 41, 2207 (1964).
4. W. C. Price, T. R. Passmore et al., Discussions, Faraday Soc. 35, 201 (1963).
5. M. E. Gerabedian and S. W. Benson, J. Am. Chem. Soc. 86, 176 (1964).
6. R. E. McCoy and S. H. Bauer, J. Am. Chem. Soc. 78, 2061 (1956).
7. L. Pauling, "The Nature of the Chemical Bond", Cornell University Press, Ithica, 3rd ed., (1960).
8. A. Shepp and S. H. Bauer, J. Am. Chem. Soc. 76, 265 (1954).

Borane (BH_3) $B_1H_3(g)$

Enthalpy Reference Temperature = T_r = 298.15 K   Standard State Pressure = p° = 0.1 MPa

T/K	$C_p°$ (J $K^{-1}mol^{-1}$)	S° (J $K^{-1}mol^{-1}$)	$-[G°-H°(T_r)]/T$ (J $K^{-1}mol^{-1}$)	$H°-H°(T_r)$ (kJ mol^{-1})	$\Delta_f H°$ (kJ mol^{-1})	$\Delta_f G°$ (kJ mol^{-1})	Log K_f
0	0.	0.	INFINITE	-10.129	110.478	110.478	INFINITE
100	33.268	150.652	218.682	-6.803	109.281	109.356	-57.122
200	34.135	173.880	191.116	-3.447	108.263	109.795	-28.675
250	35.114	181.598	188.466	-1.717	107.528	110.259	-23.037
298.15	36.221	187.876	187.876	0.	106.692	110.859	-19.422
300	36.266	188.100	187.876	0.067	106.658	110.885	-19.307
350	37.539	193.784	188.323	1.912	105.702	111.665	-16.665
400	38.929	198.886	189.329	3.823	104.692	112.585	-14.702
450	40.434	203.557	190.654	5.806	103.658	113.633	-13.190
500	42.038	207.900	192.164	7.868	102.622	114.797	-11.993
600	45.449	215.863	195.462	12.241	100.620	117.423	-10.223
700	48.945	223.132	198.902	16.961	98.778	120.371	-8.982
800	52.339	229.891	202.358	22.026	97.132	123.571	-8.068
900	55.510	236.241	205.774	27.421	95.683	126.964	-7.369
1000	58.397	242.242	209.123	33.119	94.413	130.510	-6.817
1100	60.980	247.931	212.395	39.090	93.296	134.175	-6.371
1200	63.268	253.337	215.583	45.305	92.305	137.935	-6.004
1300	65.282	258.483	218.687	51.735	91.412	141.775	-5.697
1400	67.049	263.387	221.706	58.353	90.594	145.680	-5.435
1500	68.599	268.067	224.642	65.137	89.832	149.642	-5.211
1600	69.959	272.538	227.497	72.066	89.108	153.653	-5.016
1700	71.153	276.816	230.273	79.123	88.409	157.708	-4.846
1800	72.206	280.914	232.973	86.292	87.722	161.804	-4.695
1900	73.135	284.843	235.601	93.560	87.040	165.939	-4.562
2000	73.959	288.616	238.158	100.916	86.353	170.109	-4.443
2100	74.690	292.242	240.647	108.349	85.655	174.314	-4.336
2200	75.342	295.732	243.072	115.851	84.942	178.552	-4.239
2300	75.925	299.094	245.435	123.415	84.209	182.824	-4.152
2400	76.447	302.337	247.739	131.034	33.193	188.196	-4.096
2500	76.917	305.467	249.986	138.703	32.331	194.673	-4.067
2600	77.341	308.492	252.178	146.416	31.473	201.183	-4.042
2700	77.725	311.419	254.319	154.170	30.615	207.727	-4.019
2800	78.073	314.252	256.409	161.960	29.756	214.302	-3.998
2900	78.389	316.997	258.451	169.783	28.894	220.908	-3.979
3000	78.677	319.659	260.447	177.637	28.026	227.544	-3.962
3100	78.941	322.244	262.399	185.518	27.153	234.209	-3.946
3200	79.183	324.754	264.309	193.424	26.271	240.902	-3.932
3300	79.404	327.194	266.177	201.354	25.380	247.624	-3.920
3400	79.608	329.567	268.007	209.305	24.480	254.372	-3.908
3500	79.796	331.878	269.799	217.275	23.568	261.146	-3.897
3600	79.970	334.128	271.555	225.263	22.644	267.947	-3.888
3700	80.131	336.321	273.276	233.269	21.707	274.774	-3.879
3800	80.280	338.460	274.963	241.289	20.756	281.627	-3.871
3900	80.418	340.547	276.618	249.324	19.791	288.504	-3.864
4000	80.547	342.585	278.242	257.372	18.811	295.407	-3.858
4100	80.667	344.575	279.836	265.433	17.816	302.334	-3.852
4200	80.779	346.521	281.400	273.506	-463.045	316.310	-3.934
4300	80.884	348.423	282.937	281.589	-462.986	334.865	-4.068
4400	80.982	350.283	284.446	289.682	-462.945	353.419	-4.196
4500	81.073	352.104	285.930	297.785	-462.924	371.973	-4.318
4600	81.160	353.887	287.388	305.897	-462.921	390.526	-4.435
4700	81.240	355.633	288.821	314.017	-462.939	409.079	-4.546
4800	81.317	357.345	290.231	322.145	-462.976	427.634	-4.654
4900	81.388	359.022	291.618	330.280	-463.033	446.188	-4.756
5000	81.456	360.667	292.983	338.422	-463.109	464.745	-4.855
5100	81.520	362.281	294.326	346.571	-463.206	483.302	-4.950
5200	81.580	363.864	295.648	354.726	-463.322	501.863	-5.041
5300	81.637	365.419	296.950	362.887	-463.457	520.425	-5.129
5400	81.691	366.945	298.232	371.053	-463.611	538.990	-5.214
5500	81.742	368.445	299.495	379.225	-463.783	557.558	-5.295
5600	81.791	369.918	300.739	387.401	-463.973	576.130	-5.374
5700	81.837	371.366	301.965	395.583	-464.181	594.705	-5.450
5800	81.880	372.790	303.174	403.769	-464.404	613.284	-5.523
5900	81.922	374.190	304.366	411.959	-464.644	631.866	-5.594
6000	81.962	375.567	305.541	420.153	-464.898	650.453	-5.663

PREVIOUS: December 1964 (1 atm)   CURRENT: December 1964 (1 bar)

Borane (BH_3) $B_1H_3(g)$

BORIC ACID (H_3BO_3) CRYSTAL M_r = 61.83202

$S°(298.15\ K) = 88.74 \pm 0.4\ J\ K^{-1}\ mol^{-1}$
$T_{fus} = 444.1 \pm 0.2\ K$
$T_{dcm} = [330\ K]$

$\Delta_f H°(0\ K) = -1080 \pm 0.8\ kJ\ mol^{-1}$
$\Delta_f H°(298.15\ K) = -1094 \pm 0.8\ kJ\ mol^{-1}$
$\Delta_{fus}H°$ = Unknown

Enthalpy of Formation

The enthalpy of formation of $\Delta_f H°(H_3BO_3$, cr, 298.15 K) was taken from Good and Mansson et al. (1). The value was derived based on the following measurements.

	$\Delta_r H°(298.15\ K)$, kcal mol^{-1}
$B(cr) + 0.75\ O_2(g) + 18.674\ HF{\cdot}57.219\ H_2O(\ell) = HBF_4{\cdot}14.674\ HF{\cdot}58.719\ H_2O(\ell)$	-173.4 ± 0.2
$H_3BO_3(cr) + 18.674\ HF{\cdot}55.719\ H_2O(\ell) = HBF_4{\cdot}14.674\ HF{\cdot}58.719\ H_2O(\ell)$	+14.6 ± 0.1

The enthalpy of solution of H_3BO_3(cr) in water have been measured by many investigators. Fasolino (2) obtained the value, $\Delta_r H°(298.15\ K) = 5.45 \pm 0.01\ kcal\ mol^{-1}$ for the reaction $H_3BO_3(cr) + 504\ H_2O(\ell) = H_3BO_3{\cdot}504\ H_2O(sol)$ which leads to $\Delta_f H°(H_3BO_3{\cdot}504\ H_2O$, sol, 298.15 K) = -256.02 kcal mol^{-1}. The $\Delta_r H°$(298.15 K) value is in reasonable agreement with the value reported by Smisko and Mason (3) in which the enthalpies of solution and dilution of H_3BO_3(cr) were given. The enthalpies of solution of H_3BO_3(cr) in water measured by other previous investigators were reviewed by Fasolino (2).

Heat Capacity and Entropy

The low temperature (16-298.15 K) heat capacities were reported by Johnston and Kerr (5). The heat capacities above 300 K were estimated with those of its constituent oxides. S°(298.15 K) was obtained from Johnston and Kerr using S°(17 K) = 0.18 kcal mol^{-1}.

Fusion Data

The value of T_{fus} was taken from Kracek and Morey et al. (6). However, H_3BO_3(cr) is in a metastable state at T_{fus}. Measurements of the decomposition pressure of H_3BO_3(cr) to form HBO_2(cr, II or III) and H_2O(g) have been made by Menzel, Schulz and Deckert (7), Thiel and Stiebeneck (8), Stackelberg, Quatran and Dressel (9), and Bezzi and Hazaki (10). The decomposition pressures of the system $H_3BO_3(cr) = H_2O(g) + HBO_2$(cr, I), at the temperatures of 312 to 352 K, have been reported by West (11).

Decomposition Data

The decomposition temperature (T_{dcm}) is calculated as the temperature at which the Gibbs energy change of the reaction $H_3BO_3(cr) \rightarrow HBO_2(cr) + H_2O(g)$ is zero.

Sublimation Data

The enthalpy of sublimation, $\Delta_{sub}H°$(298.15 K), is derived as the difference between the $\Delta_f H°$(298.15 K) values for H_3BO_3(g) and H_3BO_3(cr).

References

1. W. D. Good, M. Mansson and J. P. McCullough, Tech. Report No. 9 (1963).
2. L. G. Fasolino, "Heat of Formation of B_2O_3(cr) and B_2O_3(am.)", NONR-3608 (1964).
3. J. Smisko and I. S. Mason, J. Am. Chem. Soc. 72, 3679 (1950).
4. H. L. Johnston and E. C. Kerr, J. Am. Chem. Soc. 72, 4733 (1950).
5. F. C. Kracek, G. W. Morey and H. E. Merwin, Am. J. Sci. [5] 35A, 143 (1938).
6. H. Menzel, H. Schulz and H. Deckert, Z. anorg. allgem. Chem. 220, 49 (1934).
7. A. Thiel and H. Siebeneck, Z. anorg. allgem. Chem. 220, 236 (1934).
8. M. V. Stackelberg, F. Quatram and J. Dressel, Z. Elektrochem. 43, 14 (1937).
9. Z. Bezzi, Gass. chim. ital. 65, 766 (1935).
10. H. Tazaki, J. Sci. Hiroshima Univ. Ser. A10, 37, 55, 63, 109, 113 (1939).
11. E. D. West, U.S. Nat. Bur. Stand. ARL Tech. Report 60-276, 1960.

Boric Acid (H_3BO_3) $B_1H_3O_3$(cr)

Enthalpy Reference Temperature = T_r = 298.15 K Standard State Pressure = p° = 0.1 MPa

T/K	$C_p°$ (J $K^{-1}mol^{-1}$)	S° (J $K^{-1}mol^{-1}$)	$-[G°-H°(T_r)]/T$ (J $K^{-1}mol^{-1}$)	$H°-H°(T_r)$ (kJ mol^{-1})	$\Delta_f H°$ (kJ mol^{-1})	$\Delta_f G°$ (kJ mol^{-1})	Log K_f
0	0.	0.	INFINITE	-13.393	-1080.444	-1080.444	INFINITE
100	35.920	28.983	145.340	-11.636	-1087.566	-1049.328	548.113
200	58.743	61.128	95.458	-6.866	-1091.537	-1009.409	263.631
298.15	81.337	88.743	88.743	0.	-1093.990	-968.520	169.681
300	81.755	89.247	88.744	0.151	-1094.022	-967.741	168.499
400	100.207	115.390	92.184	9.282	-1095.069	-925.455	120.852
500	114.642	139.336	99.255	20.041	-1095.014	-883.037	92.250
600	127.194	161.372	107.795	32.146	-1094.023	-840.720	73.191
700	137.915	181.802	116.926	45.413	-1092.200	-798.636	59.595
800	147.277	200.844	126.239	59.684	-1089.646	-756.865	49.418
900	155.017	218.646	135.529	74.806	-1086.476	-715.452	41.524
1000	161.921	235.343	144.684	90.659	-1082.783	-674.421	35.228
1100	167.974	251.066	153.647	107.161	-1078.633	-633.781	30.096
1200	173.218	265.913	162.390	124.227	-1074.097	-593.538	25.836
1300	177.653	279.958	170.898	141.778	-1069.244	-553.686	22.247
1400	181.279	293.261	179.168	159.731	-1064.147	-514.218	19.186
1500	184.096	305.869	187.198	178.006	-1058.879	-475.121	16.545

PREVIOUS: September 1962 CURRENT: December 1964

BORIC ACID (H_3BO_3) IDEAL GAS M_r = 61.83202

$\Delta_f H°(0\ K)$ = Unknown

S°(298.15 K) = [295.237] J K^{-1} mol^{-1} $\Delta_f H°(298.15\ K) = -992.28 \pm 2.5$ kJ mol^{-1}

Vibrational Frequencies and Degeneracies

ν, cm^{-1}	ν, cm^{-1}	ν, cm^{-1}
3250(1)	Rotation	1185(2)
1060(1)	3150(2)	Rotation
861(1)	1440(2)	Rotation
652(1)	544(2)	

Ground State Quantum Weight: 1 σ = 3

Point Group: C_{3h}

Bond Distances: B–O = [1.36] Å O–H = [0.96] Å

Bond Angles: H–O–B = [114]° O–B–O = [120]°

Product of the Moments of Inertia: $I_A I_B I_C = [1.157449 \times 10^{-114}]\ g^3\ cm^6$

Enthalpy of Formation

The vapor pressures of H_3BO_3(cr), 109 – 140°C, were measured by Stackelberg, Quatram and Dressel (1). Using the vapor pressures reported, the enthalpies of sublimation of H_3BO_3(cr) were evaluated to be 24.4 ± 0.6 and 24.22 kcal mol^{-1} by the 2nd and 3rd law methods, respectively. $\Delta_f H°(H_3BO_3$, g, 298.15 K) was calculated from the $\Delta_f H°(H_3BO_3$, cr, 298.15 K) and $\Delta_{sub} H°(H_3BO_3$, cr, 298.15 K). The $\Delta_{sub} H°$(298.15 K) value adopted is the average of the 2nd and 3rd law values mentioned previously.

Heat Capacity and Entropy

The molecular structure and constants were taken from Pistorius (2) based on an x-ray diffraction study on H_3BO_3(cr) by Zachariasen (2). The O-H bond distance was reestimated as 0.96 Å instead of 0.88 Å reported. From the existing infrared and Raman spectra data on H_3BO_3(cr) and its aqueous solution, Pistorius carried out a normal coordinate treatment by means of Wilson's F-G matrix method and calculated the fifteen fundamental frequencies for H_3BO_3(cr). For H_3BO_3(g), twelve reported frequencies were used and corrected to the average isotopic species. The three torsional frequencies, 824(1) and 209(2) cm^{-1}, were not used based on an assumption that there are three internal rotations, i.e. three OH groups, each rotates about its own B-O bond. Employing the model of having three free rotating OH tops, the value of S°(400 K) was evaluated statistically as 75.522 kcal mol^{-1}. The corresponding value obtained by using the 2nd law value of $\Delta_{sub} H°$(400 K), 48.39 ± 1.42 kcal mol^{-1}, for the reaction H_3BO_3(g), is 75.97 ± 1.43 kcal mol^{-1} which is in good agreement with the S°(400 K) value based on a model having three free rotating groups. Thus the assumed model is adopted to evaluate the thermodynamic functions. The reduced moment of inertia for the OH top is $I_r = 1.15321 \times 10^{-40}$ g cm^2. The principal moments of inertia are: $I_A = I_B = 8.3334 \times 10^{-39}$ and $I_C = 16.6669 \times 10^{-39}$ g cm^2. The infrared absorption spectrum of H_3BO_3(g) was observed under conditions where one atmosphere of water vapor with a path length of approximately one meter as background, by Greene, Leroi, Randall et al. (4). The following fundamentals were obtained: 3220 cm^{-1} S., 1490 cm^{-1} S., 1199 cm^{-1} S., 881 cm^{-1} W., and 669 cm^{-1} W.

References

1. M. V. Stackelberg, F. Quatram and J. Dressel, Z. Elecktrochem. 43, 14 (1937).
2. C. W. F. T. Pistorius, J. Chem. Phys. 31, 1454 (1959).
3. W. H. Zachariasen, Acta Cryst. 7, 305 (1954).
4. F. T. Greene, G. E. Leroi, S. P. Randall et al., "Spectra and Thermodynamic Properties of Light-Element Oxides and Hydroxides", Proceedings of the Conference on Propellant Thermodynamics and Handling, Columbus, Ohio, July 1959.

Boric Acid (H_3BO_3) $B_1H_3O_3$(g)

Enthalpy Reference Temperature = T_r = 298.15 K Standard State Pressure = p° = 0.1 MPa

	J $K^{-1}mol^{-1}$			kJ mol^{-1}			
T/K	$C_p°$	S°	$-[G°-H°(T_r)]/T$	$H°-H°(T_r)$	$\Delta_f H°$	$\Delta_f G°$	Log K_f
0	12.472	0.	INFINITE	-15.356	-980.694	-980.694	INFINITE
100	46.202	237.715	345.491	-10.778	-984.995	-967.630	505.438
200	53.861	271.626	300.833	-5.841	-988.799	-948.771	247.794
250	59.585	284.250	296.280	-3.007	-990.617	-938.553	196.100
298.15	65.338	295.237	295.237	0.	-992.277	-928.373	162.647
300	65.558	295.642	295.238	0.121	-992.339	-927.976	161.575
350	71.395	306.189	296.058	3.546	-993.930	-917.121	136.873
400	76.864	316.085	297.949	7.254	-995.384	-906.048	118.318
450	81.857	325.431	300.488	11.224	-996.708	-894.800	103.866
500	86.355	334.293	303.429	15.432	-997.910	-883.411	92.289
600	94.017	350.739	309.966	24.464	-999.992	-860.309	74.897
700	100.271	365.717	316.876	34.188	-1001.711	-836.888	62.449
800	105.498	379.457	323.852	44.484	-1003.133	-813.242	53.099
900	109.953	392.147	330.744	55.262	-1004.306	-789.433	45.818
1000	113.798	403.935	337.481	66.454	-1005.275	-765.505	39.986
1100	117.137	414.942	344.028	78.005	-1006.076	-741.487	35.210
1200	120.047	425.262	350.372	89.868	-1006.743	-717.404	31.228
1300	122.589	434.973	356.510	102.002	-1007.306	-693.269	27.856
1400	124.814	444.141	362.445	114.375	-1007.789	-669.093	24.964
1500	126.764	452.821	368.183	126.956	-1008.217	-644.886	22.457
1600	128.478	461.058	373.733	139.720	-1008.607	-620.651	20.262
1700	129.988	468.893	379.102	152.645	-1008.976	-596.392	18.325
1800	131.321	476.361	384.299	165.711	-1009.338	-572.113	16.602
1900	132.501	483.494	389.334	178.904	-1009.706	-547.812	15.060
2000	133.550	490.317	394.214	192.207	-1010.089	-523.492	13.672
2100	134.484	496.856	398.947	205.610	-1010.495	-499.152	12.416
2200	135.319	503.132	403.541	219.101	-1010.932	-474.792	11.273
2300	136.067	509.164	408.003	232.671	-1011.405	-450.412	10.229
2400	136.740	514.969	412.340	246.312	-1062.178	-424.942	9.249
2500	137.346	520.564	416.557	260.017	-1062.816	-398.377	8.324
2600	137.893	525.962	420.662	273.779	-1063.470	-371.787	7.469
2700	138.390	531.175	424.659	287.594	-1064.142	-345.170	6.678
2800	138.841	536.216	428.554	301.456	-1064.836	-318.529	5.942
2900	139.251	541.096	432.351	315.360	-1065.553	-291.863	5.257
3000	139.626	545.823	436.055	329.305	-1066.296	-265.171	4.617
3100	139.970	550.407	439.670	343.285	-1067.064	-238.454	4.018
3200	140.284	554.856	443.200	357.298	-1067.860	-211.713	3.456
3300	140.574	559.177	446.650	371.341	-1068.683	-184.946	2.927
3400	140.840	563.378	450.021	385.412	-1069.535	-158.153	2.430
3500	141.086	567.464	453.319	399.508	-1070.417	-131.335	1.960
3600	141.313	571.442	456.545	413.628	-1071.328	-104.491	1.516
3700	141.524	575.316	459.703	427.770	-1072.269	-77.622	1.096
3800	141.719	579.093	462.795	441.932	-1073.241	-50.726	0.697
3900	141.901	582.777	465.825	456.114	-1074.243	-23.804	0.319
4000	142.070	586.372	468.794	470.312	-1075.276	3.142	-0.041
4100	142.227	589.882	471.704	484.527	-1076.340	30.116	-0.384
4200	142.374	593.311	474.559	498.757	-1557.285	64.140	-0.798
4300	142.512	596.663	477.360	513.002	-1557.325	102.747	-1.248
4400	142.641	599.940	480.109	527.259	-1557.399	141.353	-1.678
4500	142.762	603.147	482.807	541.530	-1557.507	179.961	-2.089
4600	142.875	606.286	485.458	555.812	-1557.650	218.573	-2.482
4700	142.982	609.360	488.061	570.104	-1557.828	257.189	-2.858
4800	143.082	612.371	490.620	584.408	-1558.042	295.809	-3.219
4900	143.176	615.323	493.135	598.721	-1558.293	334.434	-3.565
5000	143.266	618.216	495.608	613.043	-1558.581	373.064	-3.897
5100	143.350	621.054	498.040	627.374	-1558.907	411.700	-4.217
5200	143.429	623.838	500.432	641.713	-1559.272	450.342	-4.524
5300	143.505	626.571	502.786	656.059	-1559.675	488.993	-4.819
5400	143.576	629.254	505.104	670.413	-1560.119	527.651	-5.104
5500	143.643	631.889	507.385	684.774	-1560.604	566.317	-5.378
5600	143.707	634.478	509.631	699.142	-1561.130	604.994	-5.643
5700	143.768	637.022	511.844	713.516	-1561.698	643.679	-5.899
5800	143.826	639.523	514.024	727.895	-1562.309	682.377	-6.145
5900	143.881	641.982	516.172	742.281	-1562.964	721.081	-6.384
6000	143.934	644.401	518.289	756.672	-1563.664	759.801	-6.615

PREVIOUS: December 1964 (1 atm) CURRENT: December 1964 (1 bar)

Boric Acid (H_3BO_3) $B_1H_3O_3$(g)

POTASSIUM TETRAHYDROBORATE (KBH_4) CRYSTAL M_r = 53.94006

$S°(298.15\ K) = 106.61\ J\ K^{-1}\ mol^{-1}$

$\Delta_f H°(0\ K) = -218.19 \pm 2.30\ kJ\ mol^{-1}$

$\Delta_f H°(298.15\ K) = -226.90 \pm 2.30\ kJ\ mol^{-1}$

Enthalpy of Formation

The enthalpy of reaction of KBH_4(cr) with hydrochloric acid has been measured by solution calorimetry and reported by Johnson, Schumm, Wilson, and Prosen (4).

$KBH_4(cr) + HCl(g) + 3H_2O(l) \rightarrow KCl(cr) + H_3BO_3(cr) + 4H_2(g)$ $\Delta_r H° = -84.62 \pm 0.44\ kcal\ mol^{-1}$

The $\Delta_f H$(298.15 K) values for HCl(g) and H_3BO_3(cr) were obtained from the JANAF Tables. The $\Delta_f H°$(298.15 K) value for KCl(cr) was calculated from $\Delta_f H°(K^+, aq, \infty, 298.15\ K) = -60.32\ kcal\ mol^{-1}$, $\Delta_f H°(Cl^-, aq, \infty, 298.15\ K) = -39.952\ kcal\ mol^{-1}$ and $\Delta_{dil} H° = 4.12\ kcal\ mol^{-1}$ obtained by Evans (4). $\Delta_f H°$(298.15 K) for H_2O(liq) was obtained from Evans (4). Unpublished data of The Dow Chemical Company Thermal laboratory give a value of $\Delta_f H°(298.15\ K) = -55.0 \pm 1.0\ kcal\ mol^{-1}$ by solution calorimetry.

Heat Capacity and Entropy

C_p and S° at temperatures below 298.15 K were reported by Furukawa, Reilly, and Henning (2). The values of C_p in the range from 298.15 to 700 K were obtained from Douglas and Harman (3). These two sets of C_p data were plotted and joined smoothly at 298.15 K. The values of C_p above 700 K are estimated by graphical extrapolation.

Decomposition Data

The temperature of decomposition (into constituent elements) was estimated to be 770 K, according to a report by Douglas and Harman (3).

References

1. W. H. Johnson, R. H. Schumm, I. H. Wilson, and E. J. Prosen, J. Res. Nat. Bur. Stand. 65A, 97 (1961).
2. G. T. Furukawa, M. L. Reilly, and J. M. Henning, J. Res. Nat. Bur. Stand. 68A, 651 (1964).
3. T. B. Douglas and A. W. Harman, J. Res. Nat. Bur. Stand. 60, 117 (1958).
4. W. H. Evans, U. S. Nat. Bur. Stand. Report 8504, Appendix III, 164 (1964).

Potassium Tetrahydroborate (KBH_4) $B_1H_4K_1$(cr)

Enthalpy Reference Temperature = T_r = 298.15 K   Standard State Pressure = p° = 0.1 MPa

T/K	$C_p°$ (J K⁻¹mol⁻¹)	S° (J K⁻¹mol⁻¹)	$-[G°-H°(T_r)]/T$ (J K⁻¹mol⁻¹)	$H°-H°(T_r)$ (kJ mol⁻¹)	$\Delta_f H°$ (kJ mol⁻¹)	$\Delta_f G°$ (kJ mol⁻¹)	Log K_f
0	0.	0.	INFINITE	-16.523	-218.190	-218.190	INFINITE
100	43.388	32.560	176.531	-14.397	-223.820	-203.339	106.213
200	73.580	72.530	114.809	-8.456	-226.193	-181.752	47.469
298.15	96.571	106.608	106.608	0.	-226.898	-159.738	27.986
300	96.734	107.206	106.610	0.179	-226.902	-159.322	27.740
400	100.918	135.797	110.480	10.127	-229.606	-136.305	17.800
500	102.131	158.434	117.887	20.273	-230.144	-112.920	11.797
600	106.033	177.333	126.258	30.645	-230.651	-89.424	7.785
700	115.227	194.306	134.780	41.668	-230.655	-65.874	4.916
800	118.407	209.909	143.213	53.356	-230.132	-42.368	2.766
900	120.625	223.988	151.419	65.312	-229.490	-18.935	1.099
1000	122.340	236.790	159.326	77.464	-228.820	4.424	-0.231
1100	123.386	248.501	166.908	89.752	-307.159	32.327	-1.535
1200	124.265	259.276	174.162	102.136	-305.599	63.120	-2.748
1300	124.919	269.248	181.098	114.596	-304.101	93.786	-3.768
1400	125.520	278.528	187.730	127.118	-302.676	124.339	-4.639
1500	126.060	287.207	194.075	139.697	-301.326	154.792	-5.390
1600	126.566	295.359	200.153	152.329	-300.054	185.158	-6.045
1700	127.037	303.046	205.982	165.009	-298.859	215.447	-6.620
1800	127.473	310.320	211.578	177.735	-297.741	245.667	-7.129
1900	127.873	317.223	216.958	190.503	-296.700	275.828	-7.583
2000	128.240	323.791	222.137	203.309	-295.735	305.936	-7.990

PREVIOUS: March 1963   CURRENT: December 1964

Potassium Tetrahydroborate (KBH_4) $B_1H_4K_1$(cr)

LITHIUM TETRAHYDROBORATE ($LiBH_4$) CRYSTAL M_r = 21.78276

$S°(298.15 K) = 75.82 J K^{-1} mol^{-1}$
T_{fus} = [750] K

$\Delta_f H°(0 K) = -180.391 \pm 0.21 kJ mol^{-1}$
$\Delta_f H°(298.15 K) = -190.464 \pm 0.21 kJ mol^{-1}$
$\Delta_{fus} H°$ = Unknown

Enthalpy of Formation

The enthalpy of formation of $LiBH_4$(cr) was calculated from the solution calorimetry of Davis, Mason and Stegeman (1). The enthalpies of reaction were measured for the following reactions:

$LiBH_4(cr) + 1.25(HCL \cdot 200 H_2O) + 3 H_2O(\ell) \rightarrow (NaCl + H_3BO_3 + 0.25 HCL) \cdot 247 H_2O + 4 H_2(g)$ $\Delta H_1(298.15 K) = -72.14 kcal mol^{-1}$

$LiCl \cdot 60 H_2O + H_3BO_3 \cdot 60 H_2O + 0.25(HCL \cdot 506 H_2O) \rightarrow (LiCl + H_3BO_3 + 0.25 HCl) \cdot 247 H_2O$ $\Delta H_2(298.15 K) = 0.00 kcal mol^{-1}$ (assumed)

$H_3BO_3 \cdot 60 H_2O + \infty H_2O \rightarrow H_3BO_3 \cdot \infty H_2O$ $*\Delta H_3(298.15 K) = 0.200 kcal mol^{-1}$

The summation of the above reactions yields:

$LiBH_4(cr) + 1.25(HCL \cdot 200 H_2O) + 3 H_2O(\ell) + \infty H_2O \rightarrow LiCl \cdot 60 H_2O + H_3BO_3 \cdot \infty H_2O + 0.25(HCL \cdot 508 H_2O) + 4 H_2(g)$ $\Delta H(298.15 K) = -71.94$

The enthalpy of formation of lithium borohydride crystal was calculated to be -45.522 kcal mol^{-1} with the following auxiliary data:

Compound	$\Delta_f H°$(298.15 K)	Source
HCL (508 H_2O)	-39.815	HCl(g) JANAF; dilution data from NBS Rept. 8504, 1 July 1964 Appendix III by W. H. Evans
HCl (200 H_2O)	-39.696	HCl(g) JANAF; dilution data from W. H. Evans, loc. cit.
LiCl (60 H_2O)	-106.023	Li^+ see $NaBH_4$ table; Cl^- and dilution data, W. H. Evans, loc. cit.
$H_3BO_3(\infty H_2O)$	-256.05	JANAF H_3BO_3(cr) and dilution data
		Chem. Soc. 72, 3679 (1950).

*This $\Delta_r H°$(298.15 K) value was obtained from J. Smisko and L. S. Mason loc. cit.

Heat Capacity and Entropy

The low temperature heat capacities, 15.72 - 302.88 K, were obtained from the measurements by Hallett and Johnston (2). The $C°_p$ values above 298.15 K were estimated by comparison with the $C°_p$'s for $NaBH_4$.

The entropy was based on S°(15 K, extrap) = 0.0874 cal $K^{-1} mol^{-1}$ reported by Hallett and Johnston (2).

Fusion Data

T_{fus} was estimated by comparison with the T_{fus} for $NaBH_4$.

References

1. W. D. Davis, L. S. Mason and G. Stegeman, J. Am. Chem. Soc. 71, 2775 (1949).
2. N. C. Hallett and H. L. Johnston, J. Am. Chem. Soc. 75, 1496 (1953).

Lithium Tetrahydroborate ($LiBH_4$) $B_1H_4Li_1$(cr)

Enthalpy Reference Temperature = T_r = 298.15 K Standard State Pressure = p° = 0.1 MPa

T/K	$C°_p$ (J K⁻¹mol⁻¹)	S° (J K⁻¹mol⁻¹)	$-[G°-H°(T_r)]/T$ (J K⁻¹mol⁻¹)	$H°-H°(T_r)$ (kJ mol⁻¹)	$\Delta_f H°$ (kJ mol⁻¹)	$\Delta_f G°$ (kJ mol⁻¹)	Log K_f
0	0.	0.	INFINITE	-12.697	-180.391	-180.391	INFINITE
100	30.019	17.086	132.466	-11.538	-185.749	-166.547	86.995
200	59.575	47.990	82.538	-6.910	-188.682	-146.067	38.149
298.15	82.539	75.815	75.815	0.	-190.464	-124.733	21.853
300	82.989	76.326	75.817	0.153	-190.484	-124.325	21.647
400	91.002	101.384	79.181	8.881	-191.531	-102.115	13.335
500	98.282	122.480	85.783	18.349	-195.573	-79.333	8.288
600	105.144	141.037	93.475	28.537	-196.207	-56.016	4.877
700	109.914	157.625	101.476	39.304	-196.396	-32.631	2.435
800	113.052	172.527	109.442	50.468	-196.317	-9.239	0.603
900	114.349	185.920	117.209	61.840	-196.167	14.136	-0.820
1000	115.353	198.020	124.695	73.325	-196.037	37.496	-1.959
1100	116.392	209.064	131.870	84.913	-195.938	60.844	-2.889
1200	117.319	219.232	138.732	96.600	-195.873	84.185	-3.664
1300	118.019	228.650	145.291	108.367	-195.861	107.522	-4.320
1400	118.684	237.421	151.562	120.203	-195.910	130.861	-4.882
1500	119.316	245.631	157.563	132.103	-196.017	154.205	-5.370
1600	119.913	253.351	163.311	144.065	-196.184	177.558	-5.797
1700	120.477	260.638	168.823	156.084	-341.644	208.100	-6.394
1800	121.007	267.539	174.118	168.159	-341.173	240.424	-6.977
1900	121.502	274.095	179.209	180.285	-340.766	272.723	-7.498
2000	121.964	280.339	184.110	192.458	-340.424	305.003	-7.966

PREVIOUS: March 1963 CURRENT: December 1964

Lithium Tetrahydroborate ($LiBH_4$) $B_1H_4Li_1$(cr)

SODIUM TETRAHYDROBORATE ($NaBH_4$) CRYSTAL M_r = 37.83153

$S°(298.15 K) = 101.39 J K^{-1} mol^{-1}$

T_{trs} = 189.9 K

T_{fus} = 770 K

$\Delta_f H°(0 K) = -183.3 \pm 0.39 kJ mol^{-1}$

$\Delta H°(298.15 K) = -191.84 \pm 0.39 kJ mol^{-1}$

$\Delta_{trs}H° = 0.999 kJ mol^{-1}$

$\Delta_{fus}H°$ = Unknown

Enthalpy of Formation

The enthalpy of formation is calculated from the solution calorimetry of Davis, Mason and Stegeman (1). The enthalpies of reaction and mixing were measured for the following reactions:

$NaBH_4 + 1.25(HCl{\cdot}200H_2O) \rightarrow (NaCl + H_3BO_3 + 0.25HCl){\cdot}247H_2O + 4H_2$ $\Delta_r H = -63.880 kcal mol^{-1}$

$NaCl{\cdot}60H_2O + H_3BO_3{\cdot}60H_2O \rightarrow (NaCl + H_3BO_3){\cdot}120H_2O$ $\Delta_r H = -0.026 kcal mol^{-1}$

$NaCl + (H_3BO_3){\cdot}120H_2O + 0.25(HCl{\cdot}508H_2O) \rightarrow (NaCl + H_3BO_3 + 0.25HCl){\cdot}247H_2O$ $\Delta_r H = +0.083 kcal mol^{-1}$

$N_3BO_3{\cdot}60H_2O + \infty H_2O \rightarrow H_3BO_3{\cdot}\infty H_2O$ $\Delta_r H = +0.068 kcal mol^{-1}$

The summation of the above equations yields:

$NaBH_4(cr) + 1.25(HCl{\cdot}200H_2O) + 3H_2O(l) + \infty H_2O \rightarrow NaCl{\cdot}60H_2O + H_3BO_3{\cdot}\infty H_2O + 0.25(HCl{\cdot}508H_2O) + 4H_2(g)$

$\Delta_r H = -63.869 kcal mol^{-1}$

The $\Delta_f H(298.15 K)$ values for HCl were obtained from JANAF value (5) for HCl(g) and enthalpy of dilution data of Evans (6). The $\Delta_f H(298.15 K)$ for $H_2O(l)$ value was obtained from Evans (6). The $\Delta_f H(298.15 K)$ value for NaCl was obtained from JANAF value (5) of NaCl(cr) together with dilution data from Evans (6). The $\Delta_f H(298.15 K)$ value for H_3BO_3 was obtained from JANAF value (5) for $H_3BO_3(cr)$ together with dilution data of Smisko and Mason (2).

Heat Capacity And Entropy

The low temperature heat capacity (15.76 - 301.26 K) was obtained from Johnston and Hallett (3). In the area of transition (169-193.5 K) the average $\Delta_{trs}H = 635.3 cal mol^{-1}$ and the average $\Delta_{trs}S = 3.35 cal K^{-1} mol^{-1}$ were adopted from Johnston and Hallett (3). Using the present base line these ΔH and ΔS values yield a enthalpy of transition of 0.2388 kcal mol^{-1}. The low temperature heat capacity was fitted with the smoothed enthalpy capacity data (298.15 - 700 K) of Douglas and Harman (4) The heat capacity above 700 K was extrapolated from the plot of both sets of data.

Transition Data

The temperature and heat of transition were obtained from Johnston and Hallett (3).

Fusion Data

The fusion data was obtained from Douglas and Harman (4).

References

1. W. D. Davis, L. S. Mason, and G. Stegeman, J. Amer. Chem. Soc. 71, 2775 (1949).
2. J. Smisko and L. S. Mason, J. Am. Chem. Soc. 72, 3679 (1950).
3. H. L. Johnston and N. C. Hallett, J. Amer. Chem. Soc. 75, 1467 (1953).
4. T. B. douglas and A. W. Harman, J. Res. Nat. Bur. Stand. 60, 117 (1958).
5. JANAF Thermochemical Tables: HCl(g), 9-30-64; NaCl(cr), 9-30-64; H_3BO_3(cr), 12-31-64.
6. W. H. Evans, U.S. Nat. Bur. Stand., Report 8504, Appendix III, 1964).

Sodium Tetrahydroborate ($NaBH_4$) $B_1H_4Na_1(cr)$

Enthalpy Reference Temperature = T_r = 298.15 K Standard State Pressure = p° = 0.1 MPa

	J K⁻¹mol⁻¹			kJ mol⁻¹			
T/K	$C_p°$	S°	$-[G°-H°(T_r)]/T$	$H°-H°(T_r)$	$\Delta_f H°$	$\Delta_f G°$	Log K_f
0	0.	0.	INFINITE	-16.288	-183.530	-183.530	INFINITE
100	40.158	26.309	172.080	-14.577	-189.180	-169.254	88.409
200	71.446	69.772	108.872	-7.820	-190.599	-148.166	38.697
298.15	86.483	101.387	101.387	0.	-191.836	-127.060	22.260
300	86.642	101.922	101.388	0.160	-191.856	-126.658	22.053
400	94.558	127.958	104.885	9.229	-195.583	-104.556	13.654
500	101.797	149.845	111.743	19.051	-196.439	-81.690	8.534
600	108.617	169.013	119.722	29.575	-196.771	-58.700	5.110
700	115.269	186.259	128.014	40.771	-196.558	-35.697	2.664
800	120.248	201.987	136.293	52.555	-195.870	-12.759	0.833
900	124.265	216.397	144.404	64.793	-194.851	10.071	-0.585
1000	127.445	229.654	152.275	77.379	-193.622	32.776	-1.712
1100	130.039	241.923	159.874	90.253	-192.264	55.351	-2.628
1200	131.754	253.314	167.192	103.346	-287.627	80.239	-3.493
1300	132.675	263.902	174.229	116.575	-285.360	110.802	-4.452
1400	133.135	273.754	180.990	129.869	-283.162	141.193	-5.268
1500	133.302	282.945	187.485	143.191	-281.069	171.431	-5.970
1600	133.428	291.553	193.723	156.528	-279.090	201.532	-6.579
1700	133.511	299.644	199.718	169.875	-277.224	231.513	-7.114
1800	133.553	307.276	205.483	183.228	-275.473	261.388	-7.585
1900	133.679	314.500	211.033	196.589	-273.829	291.168	-8.005
2000	133.888	321.362	216.379	209.966	-272.279	320.864	-8.380

PREVIOUS: December 1960 CURRENT: December 1964

Sodium Tetrahydroborate ($NaBH_4$) $B_1H_4Na_1(cr)$

IODOBORANE (BI) IDEAL GAS M_r = 137.7145

$S°(298.15\ K) = [232.75]\ J\ K^{-1}\ mol^{-1}$

$\Delta_f H°(0\ K) = [304 \pm 42]\ kJ\ mol^{-1}$
$\Delta_f H°(298.15\ K) = [305 \pm 42]\ kJ\ mol^{-1}$

Electronic Level and Quantum Weight

State	ε_i, cm^{-1}	g_i
$[^1\Sigma^+]$	0	[1]

$\omega_e = [600]\ cm^{-1}$ $\omega_e x_e = [3.04]\ cm^{-1}$ $\sigma = 1$
$B_e = [0.3912]\ cm^{-1}$ $\alpha_e = [0.0027]\ cm^{-1}$ $r_e = [2.08]$ Å

Enthalpy of Formation

The ratio, $D_0^0(BF) / (\Delta_{at}H°(BF_3,\ g,\ 0\ K)/3) = 1.177$, was applied to BI_3 giving $D_0^0(BI) = 84.4\ kcal\ mol^{-1}$ and $\Delta_f H°(298.15\ K) = 73 \pm 10\ kcal\ mol^{-1}$.

Heat Capacity and Entropy

The bond length was first estimated by comparison with the ratio of the bond lengths in the other boron trihalides and monohalides. The vibrational frequency was then calculated using Guggenheimer's Relation (1) which was checked using BBr_3 and shown to give an excellent fit. The quantity $x_e\mu^{1/2}$, where μ is the reduced mass, has been shown to be constant for similar molecules and its value was transferred from BBr to give x_e and $\omega_e x_e$ for BI. α_e was then calculated using the Morse potential function and B_e was determined from the bond length. The ground state configuration was assumed to be the same as that in BBr.

Reference

1. Guggenheimer, Proc. Phys. Soc. (London) 58, 456 (1946).

Iodoborane (BI) $B_1I_1(g)$

Enthalpy Reference Temperature = T_r = 298.15 K Standard State Pressure = $p°$ = 0.1 MPa

	J K⁻¹mol⁻¹			kJ mol⁻¹			
T/K	$C_p°$	S°	$-[G°-H°(T_r)]/T$	$H°-H°(T_r)$	$\Delta_f H°$	$\Delta_f G°$	Log K_f
0	0.	0.	INFINITE	-9.113	304.132	304.132	INFINITE
100	29.230	199.010	261.032	-6.202	305.494	288.673	-150.787
200	31.317	219.800	235.754	-3.191	305.706	271.726	-70.967
250	32.566	226.926	233.297	-1.593	305.619	263.238	-55.001
298.15	33.558	232.750	232.750	0.	305.432	255.091	-44.691
300	33.592	232.958	232.751	0.062	305.423	254.779	-44.361
350	34.389	238.198	233.163	1.763	305.090	246.362	-36.767
400	35.002	242.832	234.087	3.498	296.677	238.274	-31.115
450	35.475	246.983	235.294	5.260	295.595	231.038	-26.818
500	35.846	250.741	236.654	7.044	274.394	225.825	-23.592
600	36.381	257.327	239.566	10.657	274.149	216.132	-18.816
700	36.742	262.964	242.515	14.314	273.767	206.491	-15.409
800	37.001	267.888	245.386	18.002	273.282	196.912	-12.857
900	37.197	272.258	248.133	21.712	272.716	187.399	-10.876
1000	37.351	276.186	250.746	25.440	272.080	177.953	-9.295
1100	37.478	279.752	253.223	29.182	271.384	168.573	-8.005
1200	37.586	283.017	255.572	32.935	270.633	159.259	-6.932
1300	37.679	286.030	257.800	36.698	269.826	150.011	-6.028
1400	37.763	288.825	259.918	40.471	268.965	140.826	-5.254
1500	37.839	291.433	261.933	44.251	268.048	131.705	-4.586
1600	37.909	293.877	263.854	48.038	267.070	122.647	-4.004
1700	37.975	296.178	265.688	51.832	266.030	113.652	-3.492
1800	38.037	298.350	267.443	55.633	264.926	104.720	-3.039
1900	38.096	300.408	269.124	59.440	263.756	95.851	-2.635
2000	38.153	302.364	270.738	63.252	262.521	87.046	-2.273
2100	38.208	304.227	272.288	67.070	261.222	78.304	-1.948
2200	38.262	306.005	273.781	70.894	259.863	69.625	-1.653
2300	38.314	307.707	275.219	74.723	258.449	61.009	-1.386
2400	38.365	309.339	276.607	78.557	206.724	53.525	-1.165
2500	38.415	310.906	277.948	82.396	205.136	47.174	-0.986
2600	38.465	312.414	279.245	86.240	203.542	40.888	-0.821
2700	38.514	313.866	280.500	90.089	201.948	34.662	-0.671
2800	38.562	315.268	281.717	93.942	200.358	28.495	-0.532
2900	38.609	316.622	282.897	97.801	198.780	22.385	-0.403
3000	38.657	317.932	284.044	101.664	197.216	16.329	-0.284
3100	38.704	319.200	285.157	105.532	195.673	10.325	-0.174
3200	38.750	320.429	286.240	109.405	194.152	4.370	-0.071
3300	38.797	321.623	287.295	113.282	192.658	-1.537	0.024
3400	38.843	322.781	288.321	117.164	191.192	-7.400	0.114
3500	38.888	323.908	289.322	121.051	189.758	-13.220	0.197
3600	38.934	325.004	290.298	124.942	188.355	-18.999	0.276
3700	38.979	326.072	291.251	128.838	186.986	-24.740	0.349
3800	39.025	327.112	292.181	132.738	185.650	-30.444	0.418
3900	39.070	328.126	293.089	136.642	184.349	-36.114	0.484
4000	39.115	329.116	293.978	140.552	183.082	-41.751	0.545
4100	39.160	330.082	294.847	144.465	181.849	-47.356	0.603
4200	39.204	331.026	295.697	148.384	-299.200	-45.908	0.571
4300	39.249	331.949	296.529	152.306	-299.279	-39.876	0.484
4400	39.293	332.852	297.345	156.233	-299.328	-33.843	0.402
4500	39.338	333.736	298.143	160.165	-299.346	-27.809	0.323
4600	39.382	334.601	298.927	164.101	-299.336	-21.775	0.247
4700	39.427	335.448	299.695	168.041	-299.298	-15.741	0.175
4800	39.471	336.279	300.448	171.986	-299.233	-9.708	0.106
4900	39.515	337.093	301.188	175.936	-299.143	-3.678	0.039
5000	39.559	337.892	301.914	179.889	-299.029	2.351	-0.025
5100	39.603	338.676	302.627	183.847	-298.892	8.377	-0.086
5200	39.647	339.445	303.328	187.810	-298.733	14.401	-0.145
5300	39.691	340.201	304.016	191.777	-298.553	20.421	-0.201
5400	39.735	340.943	304.693	195.748	-298.354	26.437	-0.256
5500	39.779	341.672	305.359	199.724	-298.136	32.450	-0.308
5600	39.823	342.390	306.014	203.704	-297.901	38.459	-0.359
5700	39.867	343.095	306.658	207.689	-297.650	44.463	-0.407
5800	39.911	343.789	307.293	211.677	-297.383	50.463	-0.454
5900	39.955	344.471	307.917	215.671	-297.103	56.457	-0.500
6000	39.999	345.143	308.532	219.668	-296.810	62.448	-0.544

PREVIOUS: December 1964 (1 atm) CURRENT: December 1964 (1 bar)

Iodoborane (BI) $B_1I_1(g)$

DIIODOBORANE (BI_2) IDEAL GAS M_r = 264.6190

$\Delta_f H°(0\ K)$ = [245 ± 63] kJ mol^{-1}

$S°(298.15\ K)$ = [309.56] J K^{-1} mol^{-1} $\Delta_f H°(298.15\ K)$ = [243 ± 63] kJ mol^{-1}

Vibrational Frequencies and Degeneracies

ν, cm^{-1}
[550](1)
[110](1)
[770](1)

Ground State Quantum Weight: [2] σ = 2

Point Group: [C_{2v}]

Bond Distance: B-I = [2.06] Å

Bond Angle: I-B-I = [120]°

Product of the Moments of Inertia: $I_A I_B I_C$ = [$3.331449 \times 10^{-113}$] $g^3\ cm^6$

Enthalpy of Formation

The enthalpy of formation was estimated by analogy with BF_3, specifically the ratio $\Delta_r H°(BX_3 + BX_2 + X)$ / $\Delta_r H°(BX_3 + B + 3X)$ was taken as 0.347.

Heat Capacity and Entropy

The vibrational frequencies were obtained from force constants transferred from the trihalide given by Herzberg (1). The bond length was taken between those of the mono and trihalide and the angle was estimated to be the same as that in the trihalides. The ground state quantum weight was taken as two on account of the one unpaired electron.

The principal moments of inertia are $I_A = 1.8266 \times 10^{-39}$ g cm^2, $I_B = 134.1401 \times 10^{-39}$ g cm^2 and $I_C = 135.9667 \times 10^{-39}$ g cm^2.

Reference

1. G. Herzberg, "Infrared and Raman Spectra of Polyatomic Molecules", D. Van Nostrand, Inc., 1945.

Diiodoborane (BI_2) $B_1I_2(g)$

Enthalpy Reference Temperature = T_r = 298.15 K Standard State Pressure = p° = 0.1 MPa

T/K	$C_p°$ (J $K^{-1}mol^{-1}$)	S° (J $K^{-1}mol^{-1}$)	$-[G°-H°(T_r)]/T$ (J $K^{-1}mol^{-1}$)	$H°-H°(T_r)$ (kJ mol^{-1})	$\Delta_f H°$ (kJ mol^{-1})	$\Delta_f G°$ (kJ mol^{-1})	Log K_f
0	0.	0.	INFINITE	-12.522	244.562	244.562	INFINITE
100	40.240	261.642	350.181	-8.854	245.156	225.122	-117.592
200	44.749	290.842	313.919	-4.615	244.129	205.436	-53.654
250	47.122	301.087	310.357	-2.318	243.428	195.842	-40.919
298.15	49.083	309.560	309.560	0.	242.672	186.744	-32.717
300	49.151	309.864	309.561	0.091	242.641	186.397	-32.455
350	50.786	317.568	310.165	2.591	241.702	177.094	-26.430
400	52.076	324.437	311.528	5.164	224.713	168.481	-22.001
450	53.089	330.632	313.312	7.794	222.482	161.585	-18.756
500	53.889	336.268	315.330	10.469	180.091	158.739	-16.583
600	55.037	346.202	319.670	15.920	179.806	154.494	-13.450
700	55.796	354.747	324.085	21.464	179.428	150.304	-11.216
800	56.317	362.234	328.395	27.071	178.973	146.173	-9.544
900	56.689	368.890	332.532	32.722	178.452	142.104	-8.247
1000	56.962	374.877	336.472	38.405	177.871	138.096	-7.213
1100	57.168	380.316	340.214	44.112	177.233	134.149	-6.370
1200	57.328	385.298	343.767	49.838	176.537	130.262	-5.670
1300	57.453	389.892	347.140	55.577	175.779	126.436	-5.080
1400	57.554	394.153	350.348	61.327	174.952	122.671	-4.577
1500	57.635	398.127	353.402	67.087	174.050	118.968	-4.143
1600	57.702	401.849	356.315	72.854	173.064	115.327	-3.765
1700	57.758	405.349	359.098	78.627	171.985	111.751	-3.434
1800	57.805	408.651	361.760	84.405	170.808	108.242	-3.141
1900	57.845	411.778	364.311	90.188	169.529	104.800	-2.881
2000	57.880	414.746	366.759	95.974	168.147	101.429	-2.649
2100	57.909	417.571	369.112	101.764	166.666	98.129	-2.441
2200	57.935	420.265	371.376	107.556	165.091	94.902	-2.253
2300	57.957	422.841	373.558	113.350	163.429	91.748	-2.084
2400	57.977	425.308	375.663	119.147	111.432	89.737	-1.953
2500	57.994	427.675	377.697	124.946	109.552	88.872	-1.857
2600	58.010	429.950	379.663	130.746	107.651	88.082	-1.770
2700	58.024	432.140	381.566	136.548	105.741	87.366	-1.690
2800	58.036	434.250	383.410	142.351	103.833	86.720	-1.618
2900	58.047	436.287	385.199	148.155	101.937	86.143	-1.552
3000	58.057	438.255	386.935	153.960	100.064	85.630	-1.491
3100	58.066	440.159	388.621	159.766	98.222	85.179	-1.435
3200	58.074	442.002	390.261	165.573	96.418	84.787	-1.384
3300	58.082	443.789	391.856	171.381	94.657	84.451	-1.337
3400	58.089	445.523	393.409	177.190	92.946	84.168	-1.293
3500	58.095	447.207	394.922	182.999	91.288	83.934	-1.253
3600	58.101	448.844	396.397	188.809	89.685	83.747	-1.215
3700	58.106	450.436	397.836	194.619	88.140	83.603	-1.180
3800	58.111	451.986	399.241	200.430	86.655	83.501	-1.148
3900	58.116	453.495	400.613	206.241	85.229	83.437	-1.118
4000	58.120	454.967	401.953	212.053	83.863	83.408	-1.089
4100	58.124	456.402	403.264	217.865	82.556	83.413	-1.063
4200	58.127	457.802	404.546	223.678	-398.541	90.473	-1.125
4300	58.131	459.170	405.800	229.490	-398.644	102.117	-1.240
4400	58.134	460.507	407.029	235.304	-398.692	113.764	-1.351
4500	58.137	461.813	408.232	241.117	-398.686	125.411	-1.456
4600	58.139	463.091	409.410	246.931	-398.630	137.057	-1.556
4700	58.142	464.341	410.566	252.745	-398.524	148.701	-1.653
4800	58.144	465.565	411.699	258.559	-398.371	160.343	-1.745
4900	58.147	466.764	412.811	264.374	-398.173	171.981	-1.833
5000	58.149	467.939	413.901	270.189	-397.931	183.614	-1.918
5100	58.151	469.091	414.972	276.004	-397.648	195.242	-2.000
5200	58.153	470.220	416.024	281.819	-397.326	206.864	-2.078
5300	58.155	471.328	417.057	287.634	-396.967	218.480	-2.153
5400	58.156	472.415	418.072	293.450	-396.572	230.088	-2.226
5500	58.158	473.482	419.070	299.266	-396.144	241.689	-2.295
5600	58.159	474.530	420.051	305.081	-395.685	253.282	-2.363
5700	58.161	475.559	421.016	310.897	-395.196	264.867	-2.427
5800	58.162	476.571	421.965	316.714	-394.680	276.442	-2.490
5900	58.164	477.565	422.899	322.530	-394.137	288.009	-2.550
6000	58.165	478.542	423.818	328.346	-393.570	299.566	-2.608

PREVIOUS: December 1964 (1 atm) CURRENT: December 1964 (1 bar)

Diiodoborane (BI_2) $B_1I_2(g)$

TRIIODOBORANE (BI_3) IDEAL GAS M_r = 391.5235

$S°(298.15\ K) = [348.72]\ J\ K^{-1}\ mol^{-1}$

$\Delta_f H°(0\ K) = 75 \pm 50\ kJ\ mol^{-1}$
$\Delta_f H°(298.15\ K) = 71 \pm 50\ kJ\ mol^{-1}$

Vibrational Frequencies and Degeneracies

ν, cm^{-1}
190(1)
339(1)
710(2)
100(2)

Ground State Quantum Weight: [1] $\sigma = 6$
Point Group: D_{3h}
Bond Distance: B-I = [2.03 ± 0.03] Å
Bond Angle: I-B-I = [120]°
Product of the Moments of Inertia: $I_A I_B I_C = [4.420581 \times 10^{-111}]\ g^3\ cm^6$

Enthalpy of Formation

Koski et al. (2) measured the appearance potential of the B^+ ion from BI_3. Assuming the reaction $BI_3 + e^- \rightarrow B^+ + 3I + 2e^-$, one can calculate the enthalpy of atomization of BI_3 as 8.31 ± 0.5 eV. The analogous reaction for BBr_3 can be checked independently and shows the process to be valid. This value leads to $\Delta_f H°(298.15\ K) = 17.0 \pm 12\ kcal\ mol^{-1}$ if the process is assumed to occur at 298.15 K.

Heat Capacity and Entropy

Vibrational frequencies were taken from Wentink and Tiensuu (1), modified according to the natural abundance of boron. The B-I distance was estimated by Koski et al. (2) and the angle was estimated by analogy with the other boron trihalides. The three principal moments of inertia are: $I_A = I_B = 130.2616 \times 10^{-39}$ and $I_C = 260.5232 \times 10^{-39}\ g\ cm^2$.

References

1. T. Wentink and V. H. Tiensuu, J. Chem. Phys. 28, 826 (1958).
2. W. S. Koski, J. J. Kaufman, and C. F. Pochuki, J. Am. Chem. Soc. 81, 1326 (1959).

Triiodoborane (BI_3) $B_1I_3(g)$

Enthalpy Reference Temperature = T_r = 298.15 K Standard State Pressure = p° = 0.1 MPa

T/K	$C_p°$ (J K⁻¹mol⁻¹)	S° (J K⁻¹mol⁻¹)	$-[G°-H°(T_r)]/T$ (J K⁻¹mol⁻¹)	$H°-H°(T_r)$ (kJ mol⁻¹)	$\Delta_f H°$ (kJ mol⁻¹)	$\Delta_f G°$ (kJ mol⁻¹)	Log K_f
0	0.	0.	INFINITE	-16.839	75.300	75.300	INFINITE
100	53.503	281.071	406.863	-12.579	74.961	56.033	-29.269
200	64.157	321.770	355.004	-6.647	73.162	37.779	-9.867
250	67.907	336.504	349.871	-3.342	72.154	29.048	-6.069
298.15	70.787	348.721	348.721	0.	71.128	20.838	-3.651
300	70.884	349.159	348.723	0.131	71.087	20.526	-3.574
350	73.197	360.268	349.595	3.736	69.846	12.193	-1.820
400	74.982	370.163	351.559	7.442	44.578	4.867	-0.636
450	76.365	379.078	354.129	11.227	41.485	0.089	-0.010
500	77.445	387.182	357.035	15.073	-21.816	1.355	-0.142
600	78.984	401.449	363.281	22.901	-21.602	5.967	-0.519
700	79.991	413.705	369.630	30.853	-21.456	10.550	-0.787
800	80.680	424.434	375.824	38.888	-21.372	15.115	-0.987
900	81.169	433.966	381.764	46.982	-21.345	19.674	-1.142
1000	81.527	442.538	387.420	55.118	-21.375	24.233	-1.266
1100	81.798	450.321	392.790	63.284	-21.461	28.797	-1.367
1200	82.006	457.448	397.885	71.475	-21.608	33.372	-1.453
1300	82.170	464.019	402.723	79.684	-21.824	37.962	-1.525
1400	82.301	470.113	407.322	87.908	-22.120	42.572	-1.588
1500	82.407	475.795	411.699	96.144	-22.510	47.206	-1.644
1600	82.495	481.116	415.873	104.389	-23.007	51.869	-1.693
1700	82.568	486.120	419.860	112.642	-23.624	56.567	-1.738
1800	82.629	490.841	423.673	120.902	-24.370	61.305	-1.779
1900	82.681	495.310	427.327	129.168	-25.251	66.088	-1.817
2000	82.726	499.552	430.833	137.438	-26.268	70.921	-1.852
2100	82.765	503.589	434.202	145.713	-27.418	75.808	-1.886
2200	82.798	507.440	437.445	153.991	-28.693	80.753	-1.917
2300	82.827	511.122	440.569	162.272	-30.080	85.759	-1.948
2400	82.853	514.647	443.582	170.556	-81.825	91.895	-2.000
2500	82.875	518.030	446.493	178.843	-83.470	99.167	-2.072
2600	82.896	521.281	449.307	187.131	-85.145	106.506	-2.140
2700	82.914	524.410	452.031	195.422	-86.835	113.909	-2.204
2800	82.930	527.425	454.670	203.714	-88.522	121.375	-2.264
2900	82.944	530.336	457.230	212.007	-90.190	128.901	-2.322
3000	82.957	533.148	459.714	220.303	-91.825	136.484	-2.376
3100	82.969	535.868	462.127	228.599	-93.414	144.121	-2.428
3200	82.980	538.503	464.472	236.896	-94.946	151.807	-2.478
3300	82.989	541.056	466.755	245.195	-96.412	159.541	-2.525
3400	82.998	543.534	468.977	253.494	-97.805	167.319	-2.571
3500	83.007	545.940	471.141	261.794	-99.119	175.136	-2.614
3600	83.014	548.278	473.252	270.095	-100.349	182.989	-2.655
3700	83.021	550.553	475.310	278.397	-101.492	190.876	-2.695
3800	83.027	552.767	477.320	286.700	-102.547	198.792	-2.733
3900	83.033	554.924	479.282	295.003	-103.512	206.735	-2.769
4000	83.039	557.026	481.199	303.306	-104.388	214.701	-2.804
4100	83.044	559.076	483.074	311.610	-105.174	222.688	-2.837
4200	83.048	561.078	484.907	319.915	-585.722	237.718	-2.956
4300	83.053	563.032	486.702	328.220	-585.246	257.318	-3.126
4400	83.057	564.941	488.458	336.526	-584.688	276.906	-3.287
4500	83.061	566.808	490.179	344.831	-584.049	296.481	-3.441
4600	83.064	568.633	491.864	353.138	-583.332	316.040	-3.589
4700	83.068	570.420	493.517	361.444	-582.540	335.583	-3.730
4800	83.071	572.169	495.137	369.751	-581.675	355.109	-3.864
4900	83.074	573.882	496.727	378.058	-580.741	374.616	-3.993
5000	83.077	575.560	498.287	386.366	-579.740	394.102	-4.117
5100	83.079	577.205	499.818	394.674	-578.675	413.569	-4.236
5200	83.082	578.818	501.322	402.982	-577.550	433.014	-4.350
5300	83.084	580.401	502.799	411.290	-576.366	452.436	-4.459
5400	83.086	581.954	504.251	419.599	-575.128	471.836	-4.564
5500	83.088	583.479	505.677	427.907	-573.837	491.213	-4.665
5600	83.090	584.976	507.080	436.216	-572.496	510.566	-4.762
5700	83.092	586.446	508.460	444.525	-571.108	529.894	-4.856
5800	83.094	587.892	509.817	452.835	-569.675	549.197	-4.946
5900	83.096	589.312	511.152	461.144	-568.200	568.475	-5.033
6000	83.097	590.709	512.466	469.454	-566.685	587.728	-5.117

PREVIOUS: December 1964 (1 atm) CURRENT: December 1964 (1 bar)

Triiodoborane (BI_3) $B_1I_3(g)$

POTASSIUM BORATE (KBO_2) CRYSTAL M_r = 81.9071

$S°(298.15\ K) = 79.98 \pm 0.12\ J\ K^{-1}\ mol^{-1}$

$T_{fus} = 1220 \pm 3\ K$

$\Delta_f H°(0\ K) = -990.1 \pm 8\ kJ\ mol^{-1}$

$\Delta_f H°(298.15\ K) = -995.0 \pm 8\ kJ\ mol^{-1}$

$\Delta_{fus} H° = 31.38 \pm 4.2\ kJ\ mol^{-1}$

Enthalpy of Formation

Shartsis and Capps (1) measured the enthalpies of solution in 2N nitric acid of various K_2O-B_2O glass and crystalline mixtures containing 1.1-41.0 mole % of K_2O. When these results are extrapolated to 50 mole %, we obtain $\Delta_{sol}H°$ (25°C) = -11.9 ± 2 kcal mol^{-1} for KBO_2(cr) + HNO_3(27.75 H_2O, aq) + H_2O(ℓ) → KNO_3(27.75 H_2O, aq) + H_3BO_3(27.75 H_2O, aq) which leads to $\Delta_f H°$(KBO_2, cr, 298.15 K) = -237.8 ± 2 kcal mol^{-1}, using the following auxiliary data: $\Delta_f H°$(HNO_3·27.75 H_2O, aq) = -256.37 kcal mol^{-1} (4). The value, $\Delta_f H°$$KBO_2$, cr, 298.15 K) = -237.8 ± 2 kcal mol^{-1}, is adopted in the tabulation. Since the enthalpies of formation of $LiBO_2$(cr) and $NaBO_2$(cr) derived from the enthalpy of solution data are in good agreement with those values (5) adopted, their enthalpy of solution data should be reasonably reliable, although the calorimetric treatment was loose.

Heat Capacity and Entropy

Paukov et al. (6) have measured low temperature heat capacities from 12.11 to 312.22 K in an adiabatic calorimeter. The adopted C_p are derived from their experimental heat capacites by a polynomial curve fitting technique. The derived entropy S°(298.15 K) is 19.116 ± 0.03 cal K^{-1} mol^{-1} based on S°(12.11 K) = 0.0353 cal K^{-1} mol^{-1}. Heat capacities above 310 K are estimated by comparison with those of $NaBO_2$(cr) (5) since both $NaBO_2$ and KBO_2 have same type of crystal structure (7).

Thermal analysis data (8) have been interpreted in terms of a crystal transition at 785°C. Details of these studies are not available and it is not clear how reliable this interpretation is. We note, however, that the reported temperature (785 C) is essentially the same as the eutectic temperature of KBO_2 with $K_2B_4O_7$ (9, 10). We tentatively adopt a single C_p curve without a transition.

Fusion Data

Our analysis of several sets of binary phase data (8, 9, 11-14) yields the enthalpy of fusion, $\Delta_{fus}H°$ = 7.5 ± 1 kcal mol^{-1}. The adopted melting point (947°C) is a weighted average of the observed data (8-15) which vary from 940 to 950°C.

References

1. L. Shartsia and W. Capps, J. Amer. Ceram. Soc. 37, 27 (1954).
2. U. S. Nat. Bur. Stand. Tech. Note 270-3, 1968.
3. The heat of formation of KNO_3(aq) is calculated from JANAF $\Delta_f H°$(KNO_3, ∞H_2O, aq, 298.15 K) = -109.86 kcal mol^{-1} with the enthalpy of dilution ϕ_L given by V. B. Parker, U. S. Nat. Bur. Stand. NSRDS-NBS 2, 1965.
4. This value, -256.37 kcal mol^{-1}, is extrapolated from the enthalpies of formation of H_3BO_3·60 H_2O and H_3BO_3·100 H_2O listed in reference 2.
5. JANAF Thermochemical Tables: $LiBO_2$(cr) and $NaBO_2$(cr), 6-30-71.
6. I. E. Paukov, L. M. Khriplovich and A. P. Popov, Russ. J. Phys. Chem. 44, 309 (1970).
7. R. W. Wyckoff, "Crystal Structure", 2nd edition, Vol. 2, pp. 327, Interscience, New York, 1964.
8. A. G. Bergman and L. N. Mikhalkovich, Russ. J. Inorg. Chem. 15, 1173 (1970); 14, 1668 (1969).
9. A. G. Bergman and L. N. Mikhalkovich, Russ. J. Inorg. Chem. 15, 859 (1970).
10. A. P. Rollet, Compt. Rend. 200, 1763 (1935).
11. A. G. Bergman et al., Russ. J. Inorg. Chem. 15, 1339 (1970); 14 710 (1969); Zh. Neog. Khim. 2, 642, 2641 (1957); Russ. J. Gen. Chem. 25, 1831, 1993 (1955).
12. D. S. Lesnykh, I. G. Eikhenbaum and S. A. Chernyakhovskaya, Russ. J. Inorg. Chem. 15, 420 (1970).
13. H. S. Van Klooster, Z. Anorg. Chem. 69, 122, 135 (1910).
14. I. Ya. Fedoseev, Russ. J. Inorg. Chem. 5, 441 (1960).
15. Z. D. Mamedova, F. R. Samedov and P. F. Rza-Zade, Azerb. Khim. Zh. (3), 82 (1969).

Potassium Borate (KBO_2) $B_1K_1O_2$(cr)

Enthalpy Reference Temperature = T_r = 298.15 K Standard State Pressure = p° = 0.1 MPa

T/K	$C_p°$	S°	$-[G°-H°(T_r)]/T$	$H°-H°(T_r)$	$\Delta_f H°$	$\Delta_f G°$	Log K_f
	J $K^{-1}mol^{-1}$			kJ mol^{-1}			
0	0.	0.	INFINITE	-12.108	-990.083	-990.083	INFINITE
100	35.012	25.125	130.260	-10.513	-993.150	-974.741	509.152
200	54.095	55.873	85.716	-5.968	-994.444	-955.739	249.613
298.15	67.028	79.981	79.981	0.	-994.955	-936.616	164.091
300	67.237	80.397	79.983	0.124	-994.961	-936.254	163.016
400	76.651	101.079	82.739	7.336	-997.561	-916.196	119.643
500	83.973	118.997	88.238	15.379	-997.415	-895.863	93.590
600	89.830	134.842	94.710	24.079	-996.896	-875.596	76.227
700	94.558	149.054	101.476	33.305	-996.076	-855.441	63.834
800	98.491	161.945	108.241	42.963	-995.014	-835.420	54.547
900	101.755	173.739	114.873	52.980	-993.767	-815.544	47.333
1000	104.474	184.607	121.310	63.297	-992.387	-795.814	41.569
1100	106.525	194.666	127.527	73.853	-1069.889	-771.617	36.641
1200	107.864	204.001	133.516	84.582	-1067.376	-744.612	32.412
1220.000	107.989	205.785	134.686	86.741	------ CRYSTAL <--> LIQUID -----		
1300	108.575	212.662	139.275	95.403	-1064.859	-717.817	28.842
1400	109.383	220.738	144.809	106.301	-1062.343	-691.216	25.790
1500	110.190	228.312	150.126	117.280	-1059.820	-664.795	23.150
1600	110.997	235.449	155.237	128.339	-1057.284	-638.542	20.846
1700	111.804	242.203	160.156	139.479	-1054.735	-612.449	18.818
1800	112.611	248.616	164.894	150.700	-1052.169	-586.506	17.020
1900	113.417	254.726	169.462	162.001	-1049.588	-560.706	15.415
2000	114.223	260.564	173.873	173.383	-1046.990	-535.043	13.974

PREVIOUS: CURRENT: June 1971

Potassium Borate (KBO_2) $B_1K_1O_2$(cr)

POTASSIUM BORATE (KBO_2) LIQUID M_r = 81.9071

S°(298.15 K) = [88.455 J K^{-1} mol^{-1} $\Delta_f H°$(298.15 K) = [-980.730] kJ mol^{-1}

T_{fus} = 1220 ± 3 K $\Delta_{fus}H°$ = 31.38 ± 4.2 kJ mol^{-1}

Enthalpy of Formation

$\Delta_f H°$(KBO_2, ℓ, 298.15 K) is calculated from that of the crystal by adding $\Delta_{fus}H°$ and the difference in enthalpy, H°(1220 K)-H°(298.15 K), between the crystal and liquid.

Heat Capacity and Entropy

The constant heat capacity of the liquid is estimated to be 36 J K^{-1} mol^{-1} based on that of $LiBO_2$(ℓ) (1) which was derived from high temperature enthalpy measurements. A glass transition is assumed at 812 K below which the heat capacities are assumed to be the same as the crystal.

S°(KBO_2, ℓ, 298.15 K) is calculated in a manner analogous to that used for the enthalpy of formation.

Fusion Data

See the crystal table for details.

Vaporization Data

The boiling point is calculated as the temperature at which the fugacity is 1 atm for KOB_2(ℓ) + KBO_2(g). The enthalpy of vaporization is the difference in $\Delta_f H°$ at the boiling point between liquid and gas.

Reference

1. JANAF Thermochemical Table: $LiBO_2$(ℓ), 6-30-71.

Potassium Borate (KBO_2) $B_1K_1O_2$(1)

Enthalpy Reference Temperature = T_r = 298.15 K Standard State Pressure = p° = 0.1 MPa

T/K	$C_p°$	S°	-[G°-H°(T_r)]/T	H°-H°(T_r)	$\Delta_f H°$	$\Delta_f G°$	Log K_f
	J $K^{-1}mol^{-1}$			kJ mol^{-1}			
0							
100							
200							
298.15	67.028	88.455	88.455	0.	-980.730	-924.917	162.042
300	67.237	88.871	88.457	0.124	-980.735	-924.571	160.982
400	76.651	109.553	91.213	7.336	-983.335	-905.360	118.228
500	83.973	127.471	96.712	15.379	-983.190	-885.875	92.547
600	89.830	143.316	103.184	24.079	-982.670	-866.455	75.432
700	94.558	157.528	109.950	33.305	-981.850	-847.147	63.215
800	98.491	170.420	116.715	42.964	-980.788	-827.974	54.061
812.000	98.910	171.890	117.520	44.149	GLASS <--> LIQUID TRANSITION		
812.000	146.440	171.890	117.520	44.149			
900	146.440	186.958	123.585	57.035	-975.486	-809.159	46.962
1000	146.440	202.387	130.707	71.679	-969.779	-790.986	41.317
1100	146.440	216.344	137.868	86.323	-1043.193	-768.766	36.506
1200	146.440	229.086	144.946	100.967	-1036.765	-744.103	32.390
1220.000	146.440	231.506	145.346	103.896	CRYSTAL <--> LIQUID		
1300	146.440	240.807	151.876	115.611	-1030.425	-719.972	28.929
1400	146.440	251.660	158.620	130.255	-1024.163	-696.326	25.980
1500	146.440	261.763	165.163	144.899	-1017.974	-673.126	23.440
1600	146.440	271.214	171.499	159.543	-1011.854	-650.336	21.231
1700	146.440	280.092	177.629	174.187	-1005.801	-627.927	19.294
1800	146.440	288.462	183.556	188.831	-999.812	-605.872	17.582
1900	146.440	296.380	189.287	203.475	-993.888	-584.148	16.059
2000	146.440	303.891	194.832	218.119	-988.029	-562.735	14.697

PREVIOUS: CURRENT: June 1971

Potassium Borate (KBO_2) $B_1K_1O_2$(1)

POTASSIUM BORATE (KBO_2) CRYSTAL-LIQUID M_r = 81.9071 Potassium Borate (KBO_2) $B_1K_1O_2$(cr,l)

0 to 1220 K crystal
above 1220 K liquid

Refer to the individual tables for details.

Enthalpy Reference Temperature = T_r = 298.15 K Standard State Pressure = p° = 0.1 MPa

T/K	C_p° (J K⁻¹mol⁻¹)	S° (J K⁻¹mol⁻¹)	$-[G^\circ-H^\circ(T_r)]/T$ (J K⁻¹mol⁻¹)	$H^\circ-H^\circ(T_r)$ (kJ mol⁻¹)	Δ_fH° (kJ mol⁻¹)	Δ_fG° (kJ mol⁻¹)	Log K_f
0	0.	0.	INFINITE	-12.108	-990.083	-990.083	INFINITE
100	35.012	25.125	130.260	-10.513	-993.150	-974.741	509.152
200	54.095	55.873	85.716	-5.968	-994.444	-955.739	249.613
298.15	67.028	79.981	79.981	0.	-994.955	-936.616	164.091
300	67.237	80.397	79.983	0.124	-994.961	-936.254	163.016
400	76.651	101.079	82.739	7.336	-997.561	-916.196	119.643
500	83.973	118.997	88.238	15.379	-997.415	-895.863	93.590
600	89.830	134.842	94.710	24.079	-996.896	-875.596	76.227
700	94.558	149.054	101.476	33.305	-996.076	-855.441	63.834
800	98.491	161.945	108.241	42.963	-995.014	-835.420	54.547
900	101.755	173.739	114.873	52.980	-993.767	-815.544	47.333
1000	104.474	184.607	121.310	63.297	-992.387	-795.814	41.569
1100	106.525	194.666	127.527	73.853	-1069.889	-771.617	36.641
1200	107.864	204.001	133.516	84.582	-1067.376	-744.612	32.412
1220.000	107.989	205.785	134.686	86.741	CRYSTAL <--> LIQUID	TRANSITION	
1220.000	146.440	231.506	134.686	118.121			
1300	146.440	240.807	140.933	129.836	-1030.426	-719.973	28.929
1400	146.440	251.660	148.460	144.480	-1024.164	-696.327	25.980
1500	146.440	261.763	155.680	159.124	-1017.975	-673.127	23.440
1600	146.440	271.214	162.609	173.768	-1011.855	-650.337	21.231
1700	146.440	280.092	169.261	188.412	-1005.802	-627.928	19.294
1800	146.440	288.462	175.653	203.056	-999.813	-605.873	17.582
1900	146.440	296.380	181.801	217.700	-993.889	-584.149	16.059
2000	146.440	303.891	187.719	232.344	-988.030	-562.736	14.697

PREVIOUS: CURRENT: June 1971

Potassium Borate (KBO_2) $B_1K_1O_2$(cr,l)

POTASSIUM BORATE (KBO_2) IDEAL GAS M_r = 81.9071

$S°(298.15\ K) = [297.4 \pm 4]\ J\ K^{-1}\ mol^{-1}$

$\Delta_f H°(0\ K) = -671 \pm 25\ kJ\ mol^{-1}$
$\Delta_f H°(298.15\ K) = -674 \pm 25\ kJ\ mol^{-1}$

Vibrational Frequencies and Degeneracies

ν, cm^{-1}	ν, cm^{-1}
1952	577
1081	265
581	88

Ground State Quantum Weight: 1 σ = 1
Point Group: C_s
Bond Distances: K–O = [2.4] Å O–B = [1.36] Å B–O = [1.20] Å
Bond Angles K–O–B = [90]° O–B–O = [180]°
Product of the Moments of Inertia: $I_A I_B I_C = [5.790543 \times 10^{-114}]\ g^3\ cm^6$

Enthalpy of Formation

Jensen (1) determined the equilibrium constants for the reaction: $K(g) + HBO_2(g) \rightarrow KBO_2(g) + H(g)$ by flame studies. This technique involved several assumptions, the most basic of which is that boron added to $H_2/N_2/O_2$ flames was converted completely to HBO_2. The hydrogen atom concentration was taken from previous studies on such flames. Potassium metaborate was assumed to be formed in the flame. The concentration of K atom was determined by atomic absorption spectrophotometry. JANAF 3rd law analysis of his equilibrium constant equation in the temperature range from 2000 to 2600 K yields $\Delta_r H°(298.15\ K) = 3.66 \pm 0.54\ kcal\ mol^{-1}$ and the drift is 1.26 cal K^{-1} mol^{-1}. (2nd law enthalpy of reaction is $\Delta_r H°(298.15\ K) = 0.78\ kcal\ mol^{-1}$.) Using the 3rd law $\Delta_r H°(298.15\ K)$ and JANAF auxiliary data, we obtain $\Delta_f H°(KBO_2, g, 298.15\ K) = -161.1\ kcal\ mol^{-1}$. The uncertainty in this measurement was about ±6 kcal mol^{-1}.

Heat Capacity and Entropy

Seshadri, Nimon and White (2) observed the infrared spectra of KBO_2 in an argon matrix and made a complete assignment of all six fundamentals which are adopted in the tabulation. The uncertainty of the calculated statistical entropy at 298.15 K is estimated to be ±1 eu which is due to the magnitude of the matrix shifts from the free molecule spectrum, particularly for the lower frequency modes.

The molecular structure, bond distance and bond angle of $KBO_2(g)$ are assumed to be the same as those of $LiBO_2(g)$ and $NaBO_2(g)$ (1), except the bond distance K–O which was estimated to be intermediate between those of potassium fluoride and chloride. The principal moments of inertia are: $I_A = 6.1073 \times 10^{-39}$, $I_B = 27.8891 \times 10^{-39}$ and $I_C = 33.9965 \times 10^{-39}\ g\ cm^2$.

References

1. D. E. Jensen, Trans. Faraday Soc. 65, 2123 (1969).
2. K. S. Seshadri, L. A. Nimon and D. White, J. Mol. Specry. 30, 128 (1969).
3. JANAF Thermochemical Tables: $LiBO_2(g)$ and $NaBO_2(g)$, 6-30-71.

Potassium Borate (KBO_2) $B_1K_1O_2(g)$

Enthalpy Reference Temperature = T_r = 298.15 K Standard State Pressure = p° = 0.1 MPa

T/K	$C_p°$ (J $K^{-1}mol^{-1}$)	S° (J $K^{-1}mol^{-1}$)	$-[G°-H°(T_r)]/T$ (J $K^{-1}mol^{-1}$)	$H°-H°(T_r)$ (kJ mol^{-1})	$\Delta_f H°$ (kJ mol^{-1})	$\Delta_f G°$ (kJ mol^{-1})	Log K_f
0	0.	0.	INFINITE	-14.103	-671.166	-671.166	INFINITE
100	43.615	242.225	345.119	-10.289	-672.013	-675.314	352.748
200	52.329	275.193	302.607	-5.483	-673.045	-678.205	177.129
250	56.018	287.277	298.364	-2.772	-673.537	-679.438	141.961
298.15	59.027	297.408	297.408	0.	-674.042	-680.529	119.226
300	59.133	297.774	297.410	0.109	-674.063	-680.569	118.498
350	61.757	307.092	298.139	3.133	-677.034	-681.511	101.710
400	63.992	315.488	299.792	6.279	-677.706	-682.105	89.074
450	65.921	323.139	301.967	9.528	-678.366	-682.615	79.236
500	67.602	330.174	304.441	12.867	-679.015	-683.052	71.358
600	70.385	342.756	309.803	19.772	-680.291	-683.739	59.525
700	72.568	353.776	315.314	26.924	-681.544	-684.215	51.057
800	74.296	363.584	320.746	34.270	-682.794	-684.511	44.694
900	75.672	372.417	326.004	41.771	-684.063	-684.650	39.736
1000	76.776	380.449	331.053	49.395	-685.376	-684.645	35.762
1100	77.669	387.810	335.883	57.119	-765.710	-679.895	32.286
1200	78.397	394.600	340.497	64.924	-766.122	-672.076	29.255
1300	78.997	400.900	344.904	72.794	-766.554	-664.222	26.689
1400	79.495	406.773	349.116	80.720	-767.011	-656.333	24.488
1500	79.911	412.272	353.145	88.691	-767.496	-648.411	22.580
1600	80.263	417.441	357.003	96.700	-768.011	-640.455	20.909
1700	80.562	422.316	360.703	104.741	-768.559	-632.466	19.433
1800	80.818	426.928	364.256	112.811	-769.146	-624.444	18.121
1900	81.038	431.304	367.670	120.904	-769.773	-616.388	16.946
2000	81.230	435.466	370.957	129.017	-770.443	-608.298	15.887
2100	81.397	439.433	374.124	137.149	-771.161	-600.174	14.929
2200	81.543	443.223	377.179	145.296	-771.930	-592.013	14.056
2300	81.672	446.851	380.130	153.457	-772.752	-583.817	13.259
2400	81.787	450.329	382.983	161.630	-823.891	-574.516	12.504
2500	81.888	453.670	385.744	169.814	-824.908	-564.104	11.786
2600	81.979	456.883	388.419	178.007	-825.956	-553.651	11.123
2700	82.061	459.979	391.012	186.209	-827.039	-543.157	10.508
2800	82.134	462.964	393.529	194.419	-828.159	-532.622	9.936
2900	82.200	465.848	395.973	202.636	-829.320	-522.048	9.403
3000	82.260	468.636	398.349	210.859	-830.526	-511.432	8.905
3100	82.314	471.334	400.660	219.088	-831.780	-500.774	8.438
3200	82.364	473.948	402.910	227.322	-833.085	-490.076	8.000
3300	82.409	476.483	405.101	235.560	-834.436	-479.338	7.587
3400	82.451	478.944	407.237	243.803	-835.852	-468.555	7.198
3500	82.489	481.334	409.320	252.051	-837.326	-457.732	6.831
3600	82.524	483.659	411.353	260.301	-838.868	-446.864	6.484
3700	82.557	485.920	413.338	268.555	-840.479	-435.954	6.155
3800	82.586	488.122	415.277	276.812	-842.165	-424.997	5.842
3900	82.614	490.268	417.172	285.072	-843.928	-413.996	5.545
4000	82.640	492.360	419.026	293.335	-845.768	-402.950	5.262
4100	82.664	494.401	420.840	301.600	-847.696	-391.856	4.992
4200	82.686	496.393	422.615	309.868	-1329.568	-373.690	4.648
4300	82.707	498.339	424.353	318.138	-1330.600	-350.919	4.263
4400	82.726	500.240	426.057	326.409	-1331.457	-328.150	3.896
4500	82.744	502.100	427.726	334.683	-1332.630	-305.335	3.544
4600	82.761	503.919	429.363	342.958	-1333.904	-282.493	3.208
4700	82.777	505.699	430.968	351.235	-1335.260	-259.623	2.885
4800	82.792	507.442	432.543	359.513	-1336.612	-236.736	2.576
4900	82.806	509.149	434.089	367.793	-1338.175	-213.805	2.279
5000	82.819	510.822	435.607	376.075	-1339.846	-190.842	1.994
5100	82.832	512.462	437.098	384.357	-1341.626	-167.844	1.719
5200	82.844	514.071	438.563	392.641	-1343.518	-144.811	1.455
5300	82.855	515.649	440.002	400.926	-1345.523	-121.739	1.200
5400	82.865	517.198	441.418	409.212	-1347.643	-98.630	0.954
5500	82.875	518.718	442.809	417.499	-1349.876	-75.480	0.717
5600	82.885	520.211	444.178	425.787	-1352.225	-52.287	0.488
5700	82.894	521.679	445.525	434.076	-1354.344	-29.101	0.267
5800	82.902	523.120	446.850	442.365	-1356.878	-5.828	0.052
5900	82.910	524.538	448.155	450.656	-1359.522	17.487	-0.155
6000	82.918	525.931	449.440	458.947	-1362.276	40.850	-0.356

PREVIOUS: June 1971 (1 atm) CURRENT: June 1971 (1 bar)

Potassium Borate (KBO_2) $B_1K_1O_2(g)$

LITHIUM BORATE ($LiBO_2$) CRYSTAL M_r = 49.7498

$S°(298.15\ K) = 51.73 \pm 0.2\ J\ K^{-1}\ mol^{-1}$

$T_{fus} = 1117 \pm 1\ K$

$\Delta_f H°(0\ K) = -1013.7 \pm 0.8\ kJ\ mol^{-1}$

$\Delta_f H°(298.15\ K) = -1019.2 \pm 0.8\ kJ\ mol^{-1}$

$\Delta_{fus} H° = 33.81 \pm 0.5\ kJ\ mol^{-1}$

Enthalpy of Formation

Sinke (1) measured calorimetrically the enthalpy of solution of $LiBO_2$(cr) in 0.5 N HNO_3, $\Delta_{sol}H°(25°C) = -10.93 \pm 0.05$ kcal mol^{-1} for $LiBO_2(cr) + HNO_3(aq, 111\ H_2O) \rightarrow H_3BO_3(aq, 111\ H_2O) + LiNO_3(aq, 111\ H_2O)$, which leads to the adopted enthalpy of formation, $\Delta_f H°(LiBO_2$, cr, 298.15 K) = -243.6 ± 0.2 kcal mol^{-1} using the following auxiliary data: $\Delta_f H°(HNO_3 \cdot 111\ H_2O$, aq, 298.15 K) = -49.411 kcal mol^{-1}(2), $\Delta_f H°(H_2O$, ℓ, 298.15 K) = -68.315 kcal mol^{-1} (2), $\Delta_f H°(H_3BO_3 \cdot 111\ H_2O$, aq, 298.15 K) = -256.336 kcal mol^{-1} (2) and $\Delta_f H°(LiNO_3 \cdot 111\ H_2O$, aq, 298.15 K) = -115.944 kcal mol^{-1} (3).

Shartsis and Capps (4) measured enthalpies of solution in 2N nitric acid of various Li_2O-B_2O_3 glass and crystalline mixtures containing 2.5-48.4 mole % of Li_2O. When their results are extrapolated to 50 mole %, we obtain $\Delta_{sol}H°(25°C) = -10.8 \pm 0.5$ kcal mol^{-1} for $LiBO_2(cr) + HNO_3(27.75\ H_2O, aq) + H_2O(ℓ) \rightarrow LiNO_3(27.75\ H_2O, aq) + H_3BO_3(27.75\ H_2O, aq)$. The calculated enthalpy of formation is based on the following data: $\Delta_f H°(HNO_3 \cdot 27.75\ H_2O$, aq, 298.15 K) = -49.433 kcal mol^{-1} (2), $\Delta_f H°(LiNO_3 \cdot 27.75\ H_2O$, aq, 298.15 K) = -115.854 kcal mol^{-1} (5) and $\Delta_f H°(H_3BO_3 \cdot 27.75\ H_2O$, aq, 298.15 K) = -256.37 kcal mol^{-1} (6). The value is $\Delta_f H°(LiBO_2$, cr, 298.15 K) = -243.9 ± 0.5 kcal mol^{-1}, which is in very good agreement with the adopted value.

Heat Capacity and Entropy

Stull et al. (7) measured low temperature Cp data from 15 to 320 K. We use their smoothed Cp values to derive S°(298.15 K) = 12.36 ± 0.05 cal K^{-1} mol^{-1} based on S°(15 K) = 0.017 cal K^{-1} mol^{-1}. McDonald (8) determined high temperature enthalpy data from 288 to 1116 K by drop calorimetry. The low temperature Cp and high temperature enthalpy data are smoothly joined at 298.15 K by a polynomial curve fitting technique. The average deviation of the observed enthalpy data from the adopted values is about 0.3% in the temperature range from 428-1062 K, and the maximum is 0.56% at 922 K. The cp values above T_{fus} are extrapolated from the adopted polynomial function.

Turdakin and Tarasov (9) also measured low temperature heat capacities (55-300 K) in an adiabatic calorimeter. Their values deviate from the adopted Cp by approximately 2%.

Melting curves from binary phase studies (10, 11, 12) have been interpreted in terms of a crystal transition near 800°C, although the most recent paper (13) indicated no transition. Two high-pressure polymorphs have been observed (14) but there is no evidence of their stability at atmospheric pressure. Enthalpy data (8) showed no obvious transition near 800°C; points at 829 and 815 C showed reasonable premelting contributions of about 200 and 300 kcal mol^{-1}, respectively. Although this evidence does not preclude the existence of a transition, definite evidence would be needed to establish such a transition.

Fusion Data

See the liquid table for details.

References

1. G. C. Sinke, Thermal Research Lab., The Dow Chemical Company, Midland, Michigan, personal communication, April, 1961.
2. U. S. Nat. Bur. Stand. Tech. Note 270-3, 1968.
3. This value, ($LiNO_3 \cdot 111\ H_2O$, aq $\Delta_f H°$298.15 K) = -115.944 kcal mol^{-1}, is calculated based on the follwoing auxiliary data:

   a. ϕ_L = 0.171 kcal mol^{-1} for $LiNO_3 \cdot (\infty\ H_2O) \rightarrow LiNO_3 \cdot (111\ H_2O)$

   ϕ_L was obtained from V. B. Parker, U. S. Nat. Bur. Stand., NSRDS-NBS 2, 1965.

   b. $\Delta_f H°(LiNO_3$ [aq, ∞], 298.15 K) = -116.115 kcal mol^{-1}.

   This value is the sum of $\Delta_f H°(Li^+$ [aq, ∞] 298.15 K) = -66.555 kcal mol^{-1} and $\Delta_f H°(NO_3^-$ [aq, ∞], 298.15 K) = 49.56 kcal mol^{-1} (2). The former is derived from $\Delta_f H°$(LiOH [aq, ∞], 298.15 K = -121.525 kcal mol^{-1} and $\Delta_f H°(OH^-$ [aq, ∞] = -54.97 kcal mol^{-1} (2).

5. This value, -115.854 kcal mol^{-1}, is calculated from $\Delta_f H°(LiNO_3$ [aq. ∞], 298.15 K = -116.115 kcal mol^{-1} and ϕ_L = 0.261 kcal mol^{-1} for $LiNO_3 \cdot (\infty\ H_2O) \rightarrow LiNO_3 \cdot (27.75\ H_2O)$ reported by V. B. Parker, U. S. Nat. Bur. Stand. NSRDS-NBS 2, 1965.
6. This value, -256.37 kcal mol^{-1}, is extrapolated from the enthalpies of formation of $H_3BO_3 \cdot 60\ H_2O$ and $H_3BO_3 \cdot 100\ H_2O$ listed in the U. S. Nat. Bur. Stand. Tech. Note 270-3, 1969.
7. D. R. Stull, D. L. Hildenbrand, F. L. Oetting and G. C. Sinke, J. Chem. Eng. Data 15, 52 (1970).
8. R. A. McDonald, Thermal Res. Laboratory, The Dow Chemical Company, personal communication, April, 1961; CPIA Publication No. 44(U), Vol. 1, pp. 213-245, Feb., 1964.
9. V. A. Turdakin and V. V. Tarasov, Russ. J. Phys. Chem. 42, 1483 (1968).
10. I. I. Kitaigorodskii, T. A. Popova and O. K. Botvinkin, J. Phys. Chem., USSR 4, 380 (1933).
11. A. G. Bergman, A. I. Kislova and V. I. Posypaiko, Russ. J. Gen. Chem. 25, 1831, 1993 (1955).
12. A. G. Bergman and D. I. Bondareva, Russ. J. Inorg. Chem. 14, 586, 1339 (1969).
13. A. G. Bergman and D. I. Bondareva, Russ. J. Inorg. Chem. 15, 1339 (1970).
14. C. H. Chang and J. L. Margrave, J. Amer. Chem. Soc. 90, 2020 (1968).

Lithium Borate ($LiBO_2$) $B_1Li_1O_2$(cr)

Enthalpy Reference Temperature = T_r = 298.15 K Standard State Pressure = p° = 0.1 MPa

	J K⁻¹mol⁻¹			kJ mol⁻¹			
T/K	$C_p°$	S°	$-[G°-H°(T_r)]/T$	$H°-H°(T_r)$	$\Delta_f H°$	$\Delta_f G°$	Log K_f
0	0.	0.	INFINITE	-8.975	-1013.678	-1013.678	INFINITE
100	18.795	10.192	93.035	-8.284	-1016.411	-999.334	521.999
200	43.133	31.104	56.731	-5.125	-1018.338	-981.412	256.319
298.15	60.371	51.731	51.731	0.	-1019.222	-963.070	168.726
300	60.618	52.105	51.732	0.112	-1019.231	-962.722	167.625
400	71.145	71.112	54.252	6.744	-1019.534	-943.833	123.252
500	78.605	87.818	59.329	14.244	-1022.756	-924.575	96.590
600	85.057	102.728	65.341	22.432	-1022.692	-904.938	78.782
700	91.098	116.297	71.665	31.242	-1022.217	-885.344	66.065
800	96.939	128.844	78.038	40.645	-1021.330	-865.847	56.534
900	102.675	140.594	84.342	50.627	-1020.027	-846.485	49.129
1000	108.353	151.706	90.528	61.178	-1018.286	-827.292	43.213
1100	113.989	162.298	96.574	72.296	-1016.088	-808.295	38.383
1117.000	114.943	164.053	97.588	74.241	------ CRYSTAL <--> LIQUID -----		
1200	119.604	172.457	102.477	83.975	-1013.422	-789.519	34.367
1300	125.202	182.251	108.239	96.216	-1010.279	-770.984	30.979
1400	130.788	191.734	113.866	109.015	-1006.649	-752.710	28.084
1500	136.369	200.948	119.366	122.373	-1002.524	-734.713	25.585
1600	141.938	209.927	124.747	136.288	-997.901	-717.007	23.408
1700	147.507	218.699	130.016	150.761	-1138.015	-692.429	21 276
1800	153.072	227.288	135.182	165.790	-1131.636	-666.400	19.338
1900	158.636	235.713	140.252	181.375	-1124.764	-640.739	17.615
2000	164.193	243.991	145.233	197.517	-1117.396	-615.452	16.074

PREVIOUS: December 1964 CURRENT: June 1971

Lithium Borate ($LiBO_2$) $B_1Li_1O_2$(cr)

LITHIUM BORATE ($LiBO_2$) LIQUID M_r = 49.7498

$S°(298.15 K) = [65.547] J K^{-1} mol^{-1}$ $\Delta_f H°(298.15 K) = [-1000.259] kJ mol^{-1}$

$T_{fus} = 1117 \pm 1 K$ $\Delta_{fus}H° = 33.81 \pm 0.5 kJ mol^{-1}$

Enthalpy of Formation

$\Delta_f H°(LiBO_2$, ℓ, 298.15 K) is calculated from that of the crystal by adding $\Delta_{fus}H°$ and the difference in enthalpy, H°(1117 K)-H°(298.15 K), between the crystal and liquid.

Heat Capacity and Entropy

McDonald (1) measured high temperature enthalpy data of the liquid from 1118 to 1707 K by drop calorimetry. The adopted heat capacities are derived from his observed data. The average deviation of the observed enthalpy data from the adopted values is about 0.1%. A glass transition is assumed at 745 K below which the Cp is assumed to be the same as that of the crystal.

S°($LiBO_2$, ℓ, 298.15 K) is calculated in a manner analogous to that used for the enthalpy of formation.

Fusion Data

The adopted enthalpy of fusion, $\Delta_{fus}H°(1117 K) = 8.08 \pm 0.12 kcal mol^{-1}$, is calculated from the observed enthalpies (1) by use of the adopted Cp functions of both crystal and liquid.

Petit and Jaeger (2) derived $\Delta_{fus}H°(1109 K) = 7.4 kcal mol^{-1}$ from phase data for the $LiBO_2$-LiF system. Darmois and Zarzycki (3) determined $\Delta_{fus}H°(1113 K) = 8.3 \pm 0.6 kcal mol^{-1}$ from cryoscopic studies involving several secondary components.

Vaporization Data

The boiling point is calculated as the temperature at which the fugacity is 1 atm for $LiBO_2$(ℓ) → $LiBO_2$(g). The enthalpy of vaporization is the difference in $\Delta_f H°$ at the boiling point between liquid and gas.

Prophet (4) determined the boiling point under argon atmosphere as 2050 ± 100 K which is in good agreement with the value calculated.

References

1. R. A. McDonald, Thermal Res. Laboratory, The Dow Chemical Company, personal communication, April, 1961; CPTA Publication No. 44(U), Vol. 1, pp. 213-245, 1964.
2. C. Petit and M. Jaeger, Compt. Rend. 244, 1734 (1957).
3. G. Zarzycki, Compt. Rend. 233, 1110 (1951).
4. H. Prophet, Thermal Res. Laboratory, The Dow Chemical Company, personal communication, April, 1961.

Lithium Borate ($LiBO_2$) $B_1Li_1O_2(1)$

Enthalpy Reference Temperature = T_r = 298.15 K Standard State Pressure = p° = 0.1 MPa

T/K	$C_p°$ (J K⁻¹mol⁻¹)	S° (J K⁻¹mol⁻¹)	$-[G°-H°(T_r)]/T$ (J K⁻¹mol⁻¹)	$H°-H°(T_r)$ (kJ mol⁻¹)	$\Delta_f H°$ (kJ mol⁻¹)	$\Delta_f G°$ (kJ mol⁻¹)	Log K_f
0							
100							
200							
298.15	60.371	65.547	65.547	0.	-1000.259	-948.226	166.125
300	60.618	65.921	65.548	0.112	-1000.268	-947.903	165.045
400	71.145	84.928	68.069	6.744	-1000.570	-930.396	121.497
500	78.605	101.634	73.146	14.244	-1003.793	-912.520	95.330
600	85.057	116.544	79.158	22.432	-1003.729	-894.264	77.853
700	91.098	130.113	85.482	31.242	-1003.254	-876.052	65.372
745.000	93.741	135.871	88.352	35.401	GLASS <--> LIQUID TRANSITION		
745.000	144.310	135.871	88.352	35.401			
800	144.310	146.149	91.977	43.338	-999.673	-858.034	56.024
900	144.310	163.147	98.959	57.769	-993.921	-840.677	48.792
1000	144.310	178.351	106.151	72.200	-988.300	-823.952	43.039
1100	144.310	192.106	113.350	86.631	-982.789	-807.784	38.358
1117.000	144.310	194.319	114.565	89.085	CRYSTAL <--> LIQUID		
1200	144.310	204.662	120.444	101.062	-977.372	-792.115	34.480
1300	144.310	216.213	127.372	115.493	-972.038	-776.894	31.216
1400	144.310	226.908	134.105	129.924	-966.776	-762.080	28.434
1500	144.310	236.864	140.627	144.355	-961.578	-747.641	26.035
1600	144.310	246.178	146.936	158.786	-956.440	-733.546	23.948
1700	144.310	254.927	153.034	173.218	-1096.594	-712.595	21.895
1800	144.310	263.175	158.926	187.649	-1090.814	-690.175	20.028
1900	144.310	270.978	164.620	202.080	-1085.096	-668.073	18.367
2000	144.310	278.380	170.124	216.511	-1079.438	-646.271	16.879
2100	144.310	285.421	175.448	230.942	-1073.842	-624.751	15.540
2200	144.310	292.134	180.601	245.373	-1068.306	-603.495	14.329
2300	144.310	298.549	185.591	259.804	-1062.832	-582.490	13.229
2400	144.310	304.691	190.426	274.235	-1107.678	-560.653	12.202
2500	144.310	310.582	195.115	288.666	-1102.406	-537.969	11.240
2600	144.310	316.242	199.666	303.097	-1097.164	-515.495	10.356
2700	144.310	321.688	204.085	317.528	-1091.954	-493.222	9.542
2800	144.310	326.936	208.379	331.959	-1086.777	-471.141	8.789
2900	144.310	332.000	212.555	346.390	-1081.634	-449.246	8.092
3000	144.310	336.893	216.619	360.821	-1076.526	-427.526	7.444

PREVIOUS: December 1964 CURRENT: June 1971

Lithium Borate ($LiBO_2$) $B_1Li_1O_2(1)$

LITHIUM BORATE ($LiBO_2$) CRYSTAL-LIQUID M_r = 49.7498

0 to 1117 K crystal
above 1117 K liquid

Refer to the individual tables for details.

Lithium Borate ($LiBO_2$) $B_1Li_1O_2(cr,l)$

Enthalpy Reference Temperature = T_r = 298.15 K   Standard State Pressure = $p°$ = 0.1 MPa

T/K	$C_p°$	$S°$	$-[G°-H°(T_r)]/T$	$H°-H°(T_r)$	$\Delta_fH°$	$\Delta_fG°$	Log K_f
	J K⁻¹mol⁻¹			kJ mol⁻¹			
0	0.	0.	INFINITE	-8.975	-1013.678	-1013.678	INFINITE
100	18.795	10.192	93.035	-8.284	-1016.411	-999.334	521.999
200	43.133	31.104	56.731	-5.125	-1018.338	-981.412	256.319
298.15	60.371	51.731	51.731	0.	-1019.222	-963.070	168.726
300	60.618	52.105	51.732	0.112	-1019.231	-962.722	167.625
400	71.145	71.112	54.252	6.744	-1019.534	-943.833	123.252
500	78.605	87.818	59.329	14.244	-1022.756	-924.575	96.590
600	85.057	102.728	65.341	22.432	-1022.692	-904.938	78.782
700	91.098	116.297	71.665	31.242	-1022.217	-885.344	66.065
800	96.939	128.844	78.038	40.645	-1021.330	-865.847	56.534
900	102.675	140.594	84.342	50.627	-1020.027	-846.485	49.129
1000	108.353	151.706	90.528	61.178	-1018.286	-827.292	43.213
1100	113.989	162.298	96.574	72.296	-1016.088	-808.295	38.383
1117.000	114.943	164.053	97.588	74.241	CRYSTAL <--> LIQUID		
1117.000	144.310	194.319	97.588	108.048	TRANSITION		
1200	144.310	204.662	104.641	120.026	-977.372	-792.115	34.480
1300	144.310	216.213	112.785	134.457	-972.038	-776.894	31.216
1400	144.310	226.908	120.559	148.888	-966.776	-762.080	28.434
1500	144.310	236.864	127.985	163.319	-961.578	-747.641	26.035
1600	144.310	246.178	135.084	177.750	-956.440	-733.546	23.948
1700	144.310	254.927	141.879	192.181	-1096.594	-712.595	21.895
1800	144.310	263.175	148.391	206.612	-1090.814	-690.175	20.028
1900	144.310	270.978	154.639	221.043	-1085.096	-668.073	18.367
2000	144.310	278.380	160.643	235.474	-1079.438	-646.271	16.879
2100	144.310	285.421	166.418	249.905	-1073.842	-624.751	15.540
2200	144.310	292.134	171.981	264.336	-1068.306	-603.495	14.329
2300	144.310	298.549	177.346	278.767	-1062.832	-582.490	13.229
2400	144.310	304.691	182.525	293.198	-1107.678	-560.653	12.202
2500	144.310	310.582	187.530	307.629	-1102.406	-537.969	11.240
2600	144.310	316.242	192.372	322.060	-1097.164	-515.495	10.356
2700	144.310	321.688	197.061	336.491	-1091.954	-493.222	9.542
2800	144.310	326.936	201.607	350.923	-1086.777	-471.141	8.789
2900	144.310	332.000	206.016	365.354	-1081.634	-449.246	8.092
3000	144.310	336.893	210.298	379.785	-1076.526	-427.526	7.444

PREVIOUS: CURRENT: June 1971

Lithium Borate ($LiBO_2$) $B_1Li_1O_2(cr,l)$

LITHIUM BORATE ($LiBO_2$) IDEAL GAS M_r = 49.7498

$S°(298.15\ K) = 274.7 \pm 42\ J\ K^{-1}\ mol^{-1}$

$\Delta_f H°(0\ K) = -645.8 \pm 12.6\ kJ\ mol^{-1}$
$\Delta_f H°(298.15\ K) = -646.8 \pm 12.6\ kJ\ mol^{-1}$

Vibrational Frequencies and Degeneracies

ν, cm^{-1}	ν, cm^{-1}
1976(1)	569(1)
1094(1)	471(1)
578(1)	107(1)

Ground State Quantum Weight: 1 σ = 1
Point Group: C_s
Bond Distances: Li-O = 1.82 Å O-B = 1.36 Å B-O = 1.20 Å
Bond Angles: Li-O-B = 90° O-B-O = 180°
Product of the Moments of Inertia: $I_A I_B I_C = 3.915323 \times 10^{-115}\ g^3\ cm^6$

Enthalpy of Formation

Jensen (1) determined the equilibrium constants for reaction (A) $Li(g) + HBO_2(g) \rightarrow LiBO_2(g) + H(g)$ by flame studies. This technique involves several assumptions, the most basic of which is that boron added to $H_2/N_2/O_2$ flames is converted completely to HBO_2. The hydrogen atom concentration was taken from previous studies on such flames. Lithium metaborate was assumed to be formed in the flame. The concentration of Li atoms was determined by atomic absorption spectrophotometry. JANAF 3rd law analysis of his equilibrium constant equation in the temperature range from 2000 to 2600 K yields $\Delta_r H°(298.15\ K) = -2.95 \pm 1.07\ kcal\ mol^{-1}$ and the drift 2.47 cal $K^{-1}\ mol^{-1}$ (the 2nd law enthalpy of reaction being -8.62 kcal mol^{-1}). Using the 3rd law $\Delta_r H°(298.15\ K)$ and JANAF auxiliary data, we obtain $\Delta_f H°(LiBO_2,\ g,\ 298.15\ K) = -150.6\ kcal\ mol^{-1}$. The uncertainty in this measurement is about ±6 kcal mol^{-1}.

Buchler and Berkowitz-Mattuck (2) studied mass spectrometrically the vaporization of lithium metaborate and reported $\Delta_{vap}H°(1160\ K) = 70 \pm 3\ kcal\ mol^{-1}$ for $LiBO_2(\ell) \rightarrow LiBO_2(g)$ and $\Delta_{sub}H°(1060\ K) = 79 \pm 3\ kcal\ mol^{-1}$ for $LiBO_2(cr) \rightarrow LiBO_2(g)$ by a 2nd law method. JANAF reduction to 298.15 K gives $\Delta_{vap}H°(298.15\ K) = 78.1 \pm 3\ kcal\ mol^{-1}$ and $\Delta_{sub}H°(298.15\ K) = 82.4 \pm 3\ kcal\ mol^{-1}$. They also determined the vapor pressure of $LiBO_2$ at 1160 K as 2.0×10^{-6} atm by comparison with the vapor pressure of silver. JANAF 3rd law analysis of this single vapor pressure point gives $\Delta_{vap}H°(298.15\ K) = 85.0\ kcal\ mol^{-1}$ which leads to $\Delta_f H°(LiBO_2,\ g,\ 298.15\ K) = -154.0 \pm 3\ kcal\ mol^{-1}$. The latter is in good agreement with the value adopted in the tabulation. The 2nd law enthalpies of vaporization and sublimation are probably in error by about 6 kcal mol^{-1}.

Hildenbrand et al. (3) measured the total vapor pressure over liquid $LiBO_2$ in the temperature range from 1120 to 1280 K by a torsion-effusion method. Assuming 100% monomeric vapor species, JANAF 3rd law analysis of their vapor pressure data yields the enthalpy of vaporization, $\Delta_{vap}H°(298.15\ K) = 84.47\ kcal\ mol^{-1}$, and a drift of 0.93 ± 1.0 eu, this corresponds to $\Delta_{sub}H°(298.15\ K) = 89.0\ kcal\ mol^{-1}$. (2nd law heat of vaporization is $\Delta_{vap}H°(298.15\ K) = 83.25 \pm 1.20\ kcal\ mol^{-1}$. Using the 3rd law $\Delta_{vap}H°(298.15\ K)$, we obtain $\Delta_f H°(LiBO_2,\ g,\ 298.15\ K) = -154.6\ kcal\ mol^{-1}$. If there are 10% of $LiBO_2$ dimer molecules present in the vapor phase as suggested by Buchler (2), the correction in the 3rd law $\Delta_{vap}H°(298.15\ K)$ is about 0.3 kcal mol^{-1} is about 0.3 kcal mol^{-1} which is much less than the uncertainty assigned to derive $\Delta_{vap}H°(298.15\ K)$. The value, $\Delta_f H°(LiBO,\ g,\ 298.15\ K) = -154.6 \pm 3\ kcal\ mol^{-1}$, is adopted in the tabulation.

Heat Capacity and Entropy

The adopted vibrational frequencies are obtained from Seshadri, Nimon and White (4) who observed the infrared spectra of $LiBO_2$ in an argon matrix and made a complete assignement of all six fundamentals. The uncertainty of the calculated statistical entropy at 298.15 K is estimated to be ±1 cal $K^{-1}\ mol^{-1}$ which is due to the magnitude of the matrix shifts from the free molecule spectrum, particularly for the lower frequency modes.

Buchler and Marram (5) observed only two vibrational frequencies, 1935 and 600 cm^{-1}, in their infrared spectra. These two frequencies are in reasonable agreement with those reported by Seshadri (4).

By electron diffraction studies, Akishin and Spiridonov (6) determined the molecular structure of gaseous lithium metaborate and found the bond lengths and angles which are adopted in the tabulation. We tentatively select the bond angles Li-O-B as 90° from their reported data, i.e, 90-105° because Seshadri et al. (4) treated this bond angle as a parameter in the force constant calculations and found that the best fit to the infrared spectrometric data occurs at approximately 90°. A recent electron diffraction study (7) on lithium metaborate confirms the results given by Akishin (6).

The principal moments of inertia are $I_A = 2.5800 \times 10^{-39}$, $I_B = 11.0962 \times 10^{-39}$ and $I_C = 13.6763 \times 10^{-39}\ g\ cm^2$.

References

1. D. E. Jensen, Trans. Faraday Soc. 65, 2123 (1969).
2. A. Buchler and J. B. Berkowitz-Mattuck, J. Chem. Phys. 39, 286 (1963).
3. D. L. Hildenbrand, W. F. Hall and N. D. Potter, J. Chem. Phys. 39, 296 (1963).
4. K. S. Seshadri, L. A. Nimon and D. White, J. Mol. Spectry. 30, 128 (1969).
5. A. Buchler and E. P. Marram, J. Chem. Phys. 39, 292 (1963).
6. P. A. Akishin and V. P. Spiridonov, Zhur. Strukt. Khim. 2, 63 (1961); 3, 267 (1962).
7. Yu. S. Ezhov, S. M. Tolmachev, V. P. Spiridonov and N. G. Rambidi, Teplofiz Vysokikh Temp. Akad. Nauk SSSR 6, 68 (1968).

Lithium Borate ($LiBO_2$) $B_1Li_1O_2(g)$

Enthalpy Reference Temperature = T_r = 298.15 K Standard State Pressure = p° = 0.1 MPa

	J K⁻¹mol⁻¹			kJ mol⁻¹			
T/K	$C_p°$	S°	$-[G°-H°(T_r)]/T$	$H°-H°(T_r)$	$\Delta_f H°$	$\Delta_f G°$	Log K_f
0	0.	0.	INFINITE	-13.445	-645.772	-645.772	INFINITE
100	40.840	222.599	320.197	-9.760	-645.510	-649.674	339.355
200	49.547	253.423	279.734	-5.262	-646.098	-653.637	170.712
250	53.797	264.948	275.651	-2.676	-646.469	-655.479	136.955
298.15	57.246	274.727	274.727	0.	-646.846	-657.181	115.135
300	57.366	275.082	274.728	0.106	-646.861	-657.245	114.436
350	60.331	284.154	275.438	3.051	-647.283	-658.943	98.342
400	62.816	292.377	277.050	6.131	-647.770	-660.576	86.262
450	64.932	299.901	279.176	9.326	-648.343	-662.143	76.860
500	66.756	306.839	281.600	12.619	-652.005	-663.335	69.298
600	69.739	319.286	286.868	19.451	-653.298	-665.478	57.935
700	72.055	330.217	292.296	26.545	-654.538	-667.410	49.803
800	73.876	339.963	297.656	33.845	-655.754	-669.166	43.692
900	75.323	348.751	302.853	41.308	-656.970	-670.769	38.930
1000	76.481	356.749	307.849	48.900	-658.188	-672.237	35.114
1100	77.416	364.084	312.632	56.597	-659.411	-673.582	31.986
1200	78.179	370.853	317.205	64.378	-660.644	-674.816	29.374
1300	78.806	377.137	321.577	72.228	-661.891	-675.947	27.160
1400	79.327	382.996	325.757	80.135	-663.153	-676.981	25.258
1500	79.762	388.485	329.758	88.090	-664.431	-677.925	23.607
1600	80.130	393.645	333.591	96.085	-665.728	-678.782	22.160
1700	80.443	398.512	337.268	104.115	-812.285	-672.381	20.660
1800	80.711	403.118	340.800	112.173	-812.878	-664.135	19.273
1900	80.941	407.488	344.196	120.255	-813.507	-655.855	18.031
2000	81.142	411.645	347.465	128.360	-814.177	-647.540	16.912
2100	81.316	415.608	350.616	136.483	-814.888	-639.191	15.899
2200	81.469	419.395	353.657	144.622	-815.644	-630.806	14.977
2300	81.604	423.019	356.595	152.776	-816.447	-622.387	14.135
2400	81.724	426.495	359.435	160.943	-867.558	-612.863	13.339
2500	81.830	429.833	362.185	169.121	-868.539	-602.231	12.583
2600	81.925	433.044	364.849	177.308	-869.540	-591.559	11.885
2700	82.011	436.138	367.432	185.505	-870.564	-580.847	11.237
2800	82.087	439.122	369.940	193.710	-871.613	-570.098	10.635
2900	82.156	442.004	372.375	201.922	-872.689	-559.311	10.074
3000	82.219	444.790	374.743	210.141	-873.793	-548.486	9.550
3100	82.276	447.487	377.046	218.366	-874.927	-537.623	9.059
3200	82.328	450.100	379.288	226.596	-876.093	-526.724	8.598
3300	82.375	452.634	381.473	234.832	-877.292	-515.789	8.164
3400	82.419	455.094	383.602	243.071	-878.526	-504.815	7.756
3500	82.459	457.483	385.679	251.315	-879.795	-493.805	7.370
3600	82.496	459.807	387.706	259.563	-881.101	-482.759	7.005
3700	82.529	462.068	389.685	267.814	-882.446	-471.676	6.659
3800	82.561	464.269	391.619	276.069	-883.830	-460.554	6.331
3900	82.590	466.414	393.510	284.326	-885.254	-449.397	6.019
4000	82.617	468.505	395.359	292.587	-886.720	-438.203	5.722
4100	82.642	470.546	397.168	300.850	-888.230	-426.971	5.440
4200	82.665	472.537	398.939	309.115	-1369.633	-408.678	5.083
4300	82.686	474.483	400.673	317.382	-1370.146	-385.792	4.686
4400	82.707	476.384	402.372	325.652	-1370.688	-362.896	4.308
4500	82.726	478.243	404.037	333.924	-1371.291	-339.985	3.946
4600	82.743	480.061	405.670	342.197	-1371.948	-317.061	3.600
4700	82.760	481.841	407.272	350.472	-1372.659	-294.120	3.269
4800	82.776	483.583	408.844	358.749	-1373.425	-271.165	2.951
4900	82.790	485.290	410.387	367.028	-1374.248	-248.192	2.646
5000	82.804	486.963	411.902	375.307	-1375.132	-225.202	2.353
5100	82.817	488.603	413.390	383.588	-1376.072	-202.195	2.071
5200	82.830	490.211	414.851	391.871	-1377.080	-179.169	1.800
5300	82.841	491.789	416.288	400.154	-1378.156	-156.121	1.539
5400	82.852	493.338	417.701	408.439	-1379.302	-133.054	1.287
5500	82.863	494.858	419.090	416.725	-1380.521	-109.964	1.044
5600	82.873	496.351	420.456	425.011	-1381.815	-86.851	0.810
5700	82.882	497.818	421.801	433.299	-1383.187	-63.715	0.584
5800	82.891	499.260	423.124	441.588	-1384.640	-40.553	0.365
5900	82.899	500.677	424.426	449.877	-1386.177	-17.368	0.154
6000	82.907	502.070	425.709	458.168	-1387.802	5.847	-0.051

PREVIOUS: June 1971 (1 atm) CURRENT: June 1971 (1 bar)

Lithium Borate ($LiBO_2$) $B_1Li_1O_2(g)$

BORON NITRIDE (BN) CRYSTAL M_r = 24.8167

$S°(298.15\ K) = 14.795\ J\ K^{-1}\ mol^{-1}$

$T_{dcm} = 2600 \pm 100\ K$

$\Delta_f H°(0\ K) = -247.99 \pm 1.55\ kJ\ mol^{-1}$

$\Delta_f H°(298.15\ K) = -250.91 \pm 1.55\ kJ\ mol^{-1}$

Enthalpy of Formation

The adopted enthalpy of formation was determined by Wise et al. (5) using combustion in fluorine. Other recent calorimetric values are in good agreement as summarized below. Torsion-effusion studies of the decomposition BN(cr) → B(am) + 0.5 N_2(g) by Hildebrand and Hall (7) yield essentially the same $\Delta_f H°$ and indicate that the condensation coefficient $\alpha < 6 \times 10^{-3}$ for N_2(g) on the surface of the sample. Use of $\alpha = 6 \times 10^{-3}$ brings the following decomposition studies into reasonable agreement: Langmuir studies by Dreger et al. (8); mass spectrometric studies by Schissel and Williams (9); and Knudsen effusion studies by Hoch and White (10).

Source	Reaction	$\Delta_r H°$(298.15 K) kcal mol^{-1}	$\Delta_f H°$(298.15 K) kcal mol^{-1}
(1)	BN(cr) + 3/4 O_2(g) → 0.5 B_2O_3(am) + 0.5 N_2	-90.1 ± 0.5	-59.7 ± 0.7
(2)	B(am) + 0.5 N_2(g) → BN(cr)	-60.8 ± 0.3	-59.6 ± 0.8
(3)	BN(cr) + 1.5 F_2(g) → BF_3(g) + 0.5 N_2	-211.2 ± 0.1	
		-271.4	-60.2
(4)	BN(cr) + NF_3(g) → BF_3(g) + N_2	-177.63± 0.3	
	B(cr) + NF_3(g) → BF_3(g) + 0.5 N_2	-237.77± 0.5	-60.1 ± 2
(5)	Combustion in O_2 - H_2O	--	-59.9 ± 1.5
(6)	BN(cr) + 1.5 F_2(g) → BF_3(g) + 0.5 N_2	-211.68± 0.3	-59.97± 0.37

Heat Capacity and Entropy

The low temperature heat capacity was taken from Westrum (11). The intermediate heat capacity (300-1650°C) was obtained from McDonald and Stull (12). The high temperature Cp was taken from Prophet and Stull (13). The heat capacities in the three temperature ranges were plotted and then graphically smoothed. Dworkin et al. (1) also reported that heat capacity (50-300 K) not employed in this table.

The entropy was obtained by integration of the heat capacity data from Westrum (11) based upon $S°(10\ K) = 0.0028\ cal\ K^{-1}\ mol^{-1}$.

Decomposition Data

Observations by the Dow Chemical Company under contract No. AF 33(616)-6194 show decomposition to elements at 1 atm total pressure at 2600 ± 100 K. The decomposition temperature calculated from these tables is 2832 K; presumably this discrepancy is associated with uncertainties in the high temperature enthalpies of BN(cr), B(cr), and B(ℓ) and the enthalpy of melting of B(cr).

References

1. A. S. Dworkin, D. J. Sasmor and E. R. Van Artsdalen, J. Chem. Phys. 22, 837 (1954).
2. G. L. Gal'chenko, A. N. Kornilov, and S. M. Skuratov, Russian J. Inorg. Chem. 5, 1282 (1960).
3. P. Gross, Fulmer Research Institute, Report R. 163/6/15 December 1961, Stoke Poges, England.
4. C. J. Thompson and G. C. Sinke, the Dow Chemical Company, Report ARPA 1-164, January 1961.
5. J. J. Keavney, Stamford Research Laboratories, American Cyanamid Co., Stamford, Conn., personal communication to Wise et al., loc. cit.
6. S. S. Wise, J. L. Margrave, H. M. Feder and W. N. Hubbard, J. Phys. Chem. 70, 7 (1966).
7. D. L. Hildenbrand and W. F. Hall, J. Phys. Chem. 67, 888 (1963).
8. L. H. Dreger, V. V. Dadape, and J. L. Margrave, J. Phys. Chem. 66, 1556 (1962).
9. P. D. Schissel and W. S. Williams, Bull. Am. Phys. Soc. II 4, 139 (1959).
10. M. Hock and D. White, Tech. Research Report MCC-1023-TR-214, Ohio State University Research Foundation, Columbus, Ohio (1956).
11. E. F. Westrum, Jr., Univ. of Michigan, personal communication (1960).
12. R. A. McDonald and D. R. Stull, J. Phys. Chem. 65, 1918 (1961).
13. H. Prophet and D. R. Stull, J. Chem. Eng. Data 8, 78 (1963).

Boron Nitride (BN) B_1N_1(cr)

Enthalpy Reference Temperature = T_r = 298.15 K — Standard State Pressure = p° = 0.1 MPa

T/K	$C_p°$ (J $K^{-1}mol^{-1}$)	S° (J $K^{-1}mol^{-1}$)	$-[G°-H°(T_r)]/T$ (J $K^{-1}mol^{-1}$)	$H°-H°(T_r)$ (kJ mol^{-1})	$\Delta_f H°$ (kJ mol^{-1})	$\Delta_f G°$ (kJ mol^{-1})	Log K_f
0	0.	0.	INFINITE	-2.628	-247.993	-247.993	INFINITE
100	4.950	2.791	27.183	-2.439	-249.280	-241.538	126.166
200	12.414	8.473	16.359	-1.577	-250.206	-233.418	60.963
298.15	19.719	14.795	14.795	0.	-250.914	-225.022	39.423
300	19.849	14.917	14.795	0.037	-250.926	-224.861	39.152
400	26.276	21.539	15.654	2.354	-251.430	-216.091	28.219
500	31.380	27.974	17.480	5.247	-251.737	-207.217	21.648
600	35.229	34.051	19.741	8.586	-251.871	-198.298	17.263
700	38.074	39.701	22.194	12.255	-251.879	-189.367	14.131
800	40.459	44.943	24.714	16.184	-251.787	-180.442	11.782
900	42.593	49.836	27.236	20.339	-251.602	-171.534	9.956
1000	44.350	54.417	29.728	24.689	-251.334	-162.652	8.496
1100	45.815	58.716	32.170	29.201	-251.001	-153.799	7.303
1200	46.861	62.749	34.552	33.836	-250.629	-144.979	6.311
1300	47.698	66.534	36.868	38.566	-250.237	-136.190	5.472
1400	48.283	70.091	39.115	43.367	-249.844	-127.432	4.755
1500	48.702	73.438	41.293	48.218	-249.461	-118.702	4.134
1600	48.911	76.589	43.401	53.099	-249.105	-109.996	3.591
1700	48.939	79.555	45.442	57.992	-248.791	-101.312	3.113
1800	48.953	82.352	47.415	62.887	-248.526	-92.645	2.688
1900	48.953	84.999	49.324	67.782	-248.307	-83.991	2.309
2000	48.953	87.510	51.171	72.677	-248.134	-75.347	1.968
2100	48.953	89.899	52.959	77.573	-248.003	-66.711	1.659
2200	48.953	92.176	54.690	82.468	-247.915	-58.080	1.379
2300	48.953	94.352	56.368	87.363	-247.867	-49.453	1.123
2400	48.953	96.435	57.994	92.258	-298.118	-39.758	0.865
2500	48.953	98.434	59.572	97.154	-298.226	-28.990	0.606
2600	48.953	100.354	61.104	102.049	-298.339	-18.219	0.366
2700	48.953	102.201	62.592	106.944	-298.456	-7.443	0.144
2800	48.953	103.981	64.039	111.840	-298.578	3.338	-0.062
2900	48.953	105.699	65.446	116.735	-298.704	14.123	-0.254
3000	48.953	107.359	66.815	121.630	-298.834	24.912	-0.434
3100	48.953	108.964	68.149	126.525	-298.966	35.706	-0.602
3200	48.953	110.518	69.449	131.421	-299.103	46.504	-0.759
3300	48.953	112.024	70.717	136.316	-299.242	57.306	-0.907
3400	48.953	113.486	71.953	141.211	-299.384	68.113	-1.046
3500	48.953	114.905	73.160	146.107	-299.528	78.923	-1.178

PREVIOUS: March 1964 CURRENT: June 1966

Boron Nitride (BN) B_1N_1(cr)

BORON NITRIDE (BN) IDEAL GAS $M_r = 24.8167$

Boron Nitride (BN) $B_1N_1(g)$

$S°(298.15\ K) = 212.36\ J\ K^{-1}\ mol^{-1}$

$\Delta_f H°(0\ K) = 474 \pm 125\ kJ\ mol^{-1}$

$\Delta_f H°(298.15\ K) = 477 \pm 125\ kJ\ mol^{-1}$

Electronic Levels and Quantum Weights

State	ε_i, cm^{-1}	g_i
$^3\Pi$	0	6
	[8500]	1
	[8900]	2
	27877	6

$\omega_e = 1514.6\ cm^{-1}$ $\omega_e x_e = 12.3\ cm^{-1}$ $\sigma = 1$

$B_e = 1.666\ cm^{-1}$ $\alpha_e = 0.025\ cm^{-1}$ $r_e = 1.281$ Å

Enthalpy of Formation

The enthalpy of formation is based on $D_0^0 = 5.7$ eV which was calculated from ω_e and $\omega_e x_e$ observed by Douglas and Herzberg (1). An uncertainty of ±1.3 eV was assigned from consideration of other information which places the dissociation energy between 4 and 7 eV. The value $D_0^0 = 4.0 \pm 0.5$ eV was selected by Gaydon (2) from his own extrapolation of the spectroscopic data. An upper limit of $D_0^0 \leq 7.0$ eV results from the upper limit for the pressure of BN(cr) + BN(g) obtained from the B^+ ion intensity at 1600°C observed by Akishin and Klodeev (3). It is likely that the BN(g) pressure is even smaller since it is known that BN(cr) + B(s) + 0.5 N_2(g) and Akishin's pressure limit is the same magnitude as that of B(cr) + B(g). Fesenko (4) reports a standard enthalpy of sublimation of 160 kcal mol^{-1} for BN(cr) + BN(g) and also gives pressures for BN(g) at 2500 and 3000 K which are stated to be two orders less than the equilibrium pressure of nitrogen. Fesenko's original article is not yet available to us and it is not clear from the abstract whether the values are based on new data or are merely calculations. The reported pressures yield $D_0^0 = 6.2$ eV but here again the BN(g) pressures are of the same magnitude as the JANAF values for B(s) + B(g).

Some estimated values for D_0^0 (ranging from 5.4 to 6.4 eV) are summarized below along with the experimental evidence.

Source	D_0^0(ev)	$\Delta_{sub}H°$(298.15 K), kcal mol^{-1}	$\Delta_f H°$(298.15 K), kcal mol^{-1}
Gaydon's extrapolation of spectroscopic data	4.0 ± 0.5	213	153
Linear Birge-Sponer extrapolation	5.7	174	114
Fesenko's "BN(g) pressures"	6.2	161	101
Akishin's upper limit for BN(g) pressure	7.0	145	85
Estimate from geometric mean of D° for B_2 and N_2	5.4	180	120
Estimate from D° for series N_2, CN, BN, BeN	5.6	177	117
Estimate $M_2 + N_2 = 2$ MN for M=B from M=C	6.0	166	106
Estimate from arithmetic mean of D° for B_2 and N_2	6.4	158	98

Heat Capacity and Entropy

The molecular constants are based on emission spectra of the $^3\Pi$ - $^3\Pi$ transition otained by Douglas and Herzberg (1). Observation of the same triplet in absorption by Thrush (5) strongly suggests that the ground state of BN is $^3\Pi$ in contrast to $^1\Sigma$ as observed for the isoelectronic molecules BeO and C_2. Electronic levels for the $^3\Pi$ states are taken from Douglas and Herzberg (1) while values for $^1\Sigma$+ and $^1\Pi$ have been estimated from the SCF calculations of Masse and Barlochter (6).

References

1. A. E. Douglas and G. Herzberg, Can. J. Research 18A, 179 (1940).
2. A. G. Gaydon, "Dissocation Energies", second edition, Chapman and Hall Ltd., London, 1953.
3. P. A. Akishin and Yu. S. Klodeev, Zh. Neorgan. Khim. 7, 941 (1962).
4. V. V. Fesenko, Poroshkoivaya Met., Akad. Nauk Ukr. USSR 1, No. 4, 80-5 (1961).
5. B. A. Thrush, Nature 160, 1044 (1960).
6. J. L. Maase and M. Barlocher, Helv. Chim. Acta 47, 314 (1964).

Enthalpy Reference Temperature = T_r = 298.15 K   Standard State Pressure = p° = 0.1 MPa

T/K	$C_p°$ (J $K^{-1}mol^{-1}$)	S° (J $K^{-1}mol^{-1}$)	$-[G°-H°(T_r)]/T$ (J $K^{-1}mol^{-1}$)	$H°-H°(T_r)$ (kJ mol^{-1})	$\Delta_f H°$ (kJ mol^{-1})	$\Delta_f G°$ (kJ mol^{-1})	Log K_f
0	0.	0.	INFINITE	-8.686	473.839	473.839	INFINITE
100	29.107	180.503	238.329	-5.783	475.267	465.238	-243.016
200	29.136	200.682	215.038	-2.871	476.391	454.736	-118.765
250	29.235	207.192	212.842	-1.412	476.751	449.278	-93.872
298.15	29.450	212.358	212.358	0.	476.976	443.965	-77.781
300	29.461	212.541	212.359	0.054	476.983	443.760	-77.265
350	29.820	217.107	212.719	1.536	477.108	438.212	-65.400
400	30.280	221.119	213.523	3.038	477.145	432.652	-56.499
450	30.797	224.715	214.570	4.565	477.111	427.091	-49.575
500	31.333	227.987	215.751	6.118	477.024	421.538	-44.038
600	32.362	233.792	218.286	9.304	476.737	410.465	-35.734
700	33.257	238.850	220.870	12.586	476.343	399.450	-29.807
800	34.001	243.341	223.403	15.950	475.870	388.497	-25.366
900	34.609	247.382	225.847	19.382	475.331	377.607	-21.916
1000	35.105	251.055	228.187	22.868	474.735	366.780	-19.159
1100	35.514	254.421	230.421	26.400	474.088	356.015	-16.906
1200	35.857	257.526	232.552	29.969	473.394	345.312	-15.031
1300	36.150	260.408	234.585	33.570	472.656	334.668	-13.447
1400	36.405	263.096	236.527	37.198	471.878	324.082	-12.092
1500	36.633	265.616	238.383	40.850	471.061	313.554	-10.919
1600	36.841	267.987	240.160	44.523	470.209	303.081	-9.895
1700	37.034	270.226	241.863	48.217	469.324	292.662	-8.992
1800	37.218	272.348	243.498	51.930	468.408	282.296	-8.192
1900	37.394	274.365	245.070	55.661	467.462	271.983	-7.477
2000	37.565	276.288	246.584	59.409	466.488	261.719	-6.835
2100	37.732	278.125	248.042	63.174	465.488	251.505	-6.256
2200	37.895	279.884	249.450	66.955	464.463	241.340	-5.730
2300	38.055	281.572	250.810	70.752	463.413	231.221	-5.251
2400	38.211	283.195	252.126	74.566	412.080	222.217	-4.836
2500	38.364	284.758	253.400	78.395	410.905	214.331	-4.478
2600	38.514	286.265	254.635	82.239	409.741	206.491	-4.148
2700	38.661	287.722	255.834	86.097	408.587	198.695	-3.844
2800	38.803	289.130	256.998	89.971	407.443	190.942	-3.562
2900	38.941	290.494	258.130	93.858	406.309	183.230	-3.300
3000	39.076	291.817	259.230	97.759	405.185	175.557	-3.057
3100	39.205	293.100	260.302	101.673	404.071	167.921	-2.829
3200	39.331	294.347	261.347	105.600	402.967	160.321	-2.617
3300	39.452	295.559	262.365	109.539	401.872	152.755	-2.418
3400	39.569	296.738	263.359	113.490	400.786	145.223	-2.231
3500	39.681	297.887	264.329	117.452	399.708	137.722	-2.055
3600	39.789	299.006	265.277	121.426	398.639	130.252	-1.890
3700	39.893	300.098	266.203	125.410	397.579	122.811	-1.734
3800	39.994	301.163	267.110	129.404	396.526	115.399	-1.586
3900	40.090	302.203	267.996	133.409	395.481	108.015	-1.447
4000	40.183	303.220	268.864	137.422	394.443	100.657	-1.314
4100	40.272	304.213	269.714	141.445	393.412	93.325	-1.189
4200	40.359	305.184	270.547	145.477	-87.461	93.043	-1.157
4300	40.442	306.135	271.364	149.517	-87.391	97.339	-1.182
4400	40.523	307.066	272.165	153.565	-87.317	101.635	-1.207
4500	40.602	307.977	272.950	157.621	-87.238	105.928	-1.230
4600	40.678	308.871	273.722	161.685	-87.156	110.220	-1.252
4700	40.752	309.746	274.479	165.757	-87.070	114.510	-1.273
4800	40.824	310.605	275.222	169.836	-86.982	118.798	-1.293
4900	40.894	311.447	275.953	173.922	-86.892	123.084	-1.312
5000	40.963	312.274	276.671	178.014	-86.800	127.368	-1.331
5100	41.031	313.086	277.377	182.114	-86.706	131.650	-1.348
5200	41.097	313.883	278.072	186.221	-86.612	135.931	-1.365
5300	41.163	314.667	278.755	190.334	-86.517	140.210	-1.382
5400	41.227	315.437	279.427	194.453	-86.422	144.487	-1.398
5500	41.291	316.194	280.089	198.579	-86.327	148.762	-1.413
5600	41.353	316.939	280.740	202.711	-86.233	153.035	-1.427
5700	41.416	317.671	281.382	206.850	-86.141	157.307	-1.442
5800	41.477	318.392	282.014	210.994	-86.051	161.578	-1.455
5900	41.538	319.101	282.636	215.145	-85.963	165.846	-1.468
6000	41.599	319.800	283.250	219.302	-85.878	170.113	-1.481

PREVIOUS: June 1966 (1 atm)   CURRENT: June 1966 (1 bar)

Boron Nitride (BN) $B_1N_1(g)$

SODIUM BORATE ($NaBO_2$) CRYSTAL M_r = 65.79857

$S°(298.15\ K) = 73.538 \pm 0.08\ J\ K^{-1}\ mol^{-1}$

$T_{fus} = 1240 \pm 2\ K$

$\Delta_f H°(0\ K) = -971.0 \pm 2.5\ kJ\ mol^{-1}$

$\Delta_f H°(298.15\ K) = -975.7 \pm 2.1\ kJ\ mol^{-1}$

$\Delta_{fus} H° = 33.47 \pm 2.0\ kJ\ mol^{-1}$

Enthalpy of Formation

Adami and Joe (1) measured enthalpies of solution of B_2O_3(cr), NaCl(cr) and $NaBO_2$(cr) in aqueous HCl solution and derived $\Delta_r H°(298.15\ K) = -12.43 \pm 0.1\ kcal\ mol^{-1}$ for $NaBO_2$(cr) + HCl(12.732 H_2O, aq) → NaCl(cr) + 1/2 B_2O_3(cr) + 1/2 H_2O(ℓ) which leads to $\Delta_f H°(NaBO_2$, cr, 298.15 K) = $-233.2 \pm 0.6\ kcal\ mol^{-1}$ using JANAF auxiliary data (2). This value is adopted.

Grenier and White (3) measured the enthalpy of solution of crystal $Na_2O{\cdot}B_2O_3$ in 2N nitric acid solution at 0°C as $-20.43 \pm 0.36\ kcal\ mol^{-1}$. Since the correction term for $\Delta_{sol}H°$ from 0° to 25°C is generally small, we may assume the enthalpy of solution is the same at 0°C as at 25°C within the uncertainty of ±1 kcal mol^{-1}. Thus we obtain $\Delta_{sol}H°(25°C) = -10.22 \pm 1\ kcal\ mol^{-1}$ for $NaBO_2$(cr) + HNO_3(27.75 H_2O, aq) + H_2O(ℓ) → $NaNO_3$(27.75 H_2O, aq) + H_3BO_3(27.75 H_2O, aq) from which we derived $\Delta_f H°(NaBO_2$, cr, 298.15 K) = $-234.7 \pm 1.5\ kcal\ mol^{-1}$ based on JANAF auxiliary data (4).

Shartsis and Capps (5) measured the heats of solution in 2N nitric acid of various Na_2O-B_2O_3 glass and crystalline mixtures containing 1.0-38.7 mole % of Na_2O. When their data are extrapolated to 50 mole %, we obtain $\Delta_{sol}H°(25°C) = -11.5 \pm 1.5$ kcal mol^{-1} for the same reaction as given before. The derived enthalpy of formation is $-233.4 \pm 1.5\ kcal\ mol^{-1}$ which is in good agreement with the value adopted.

Heat Capacity and Entropy

Grenier and Westrum (6) measured low temperature heat capacities from 5 to 346 K. The adopted Cp are derived from their experimental heat capacities by a polynomial curve fitting technique. The derived entropy, S°(298.15 K), is $17.576 \pm 0.02\ cal\ K^{-1}\ mol^{-1}$ based on $S° = 0.001\ cal\ K^{-1}\ mol^{-1}$ at 5.48 K. Pankratz (7) measured high temperature enthalpy data by drop calorimetry in the temperature range 404-1200.3 K. We adopt the smooth Cp derived by Pankratz. The average deviation of the observed enthalpy data from the adopted values is about 0.1%.

Fusion Data

Our analysis of several sets of binary phase data (8, 9, 10) yields the enthalpy of fusion, $\Delta_{fus}H° = 8.0 \pm 0.5\ kcal\ mol^{-1}$. Petit and Jaeger (11) derived $\Delta_{fus}H°(1239\ K) = 8.0\ kcal\ mol^{-1}$ from their phase data for the $NaBO_2$-NaF system. The adopted melting point (1240 K) is obtained from Pankratz (7). Literature value include 1239 K (8, 9, 10, 11, 12), 1238 K (13) and 1237 K (14).

References

1. L. H. Adami and C. J. Joe, U. S. Bur. Mines RI 7167, 1968.
2. JANAF Thermochemical Tables: B_2O_3(cr), 6-30-71; NaCl(cr), 9-30-64; HCl·12.731 H_2O; U. S. Nat. Bur. Stand. Tech. Note 270-3, 1968.
3. G. Grenier and D. White, J. Chem. Phys. 61, 1681 (1957).
4. U. S. Nat. Bur. Stand. Tech. Note 270-3, 1968; V. B. Parker, NSRDS-NBS 2, 1965.
5. L. Shartsis and W. Capps, J. Amer. Ceram. Soc. 37, 27 (1954).
6. G. Grenier and E. F. Westrum, J. Amer. Chem. Soc. 78, 6226 (1956).
7. L. B. Pankratz, personal communication from E. G. King, Bureau of Mines, Albany Met. Res. Center, Thermodynamics Laboratory, Albany, Oregon, July 30, 1971.
8. A. G. Bergman, et al., Russ. J. Inorg. Chem. 15, 713 (1970); 14, 1339 (1969); 14, 876 (1969); Zh. Neorg. Khim. 2, 2641 (1957).
9. D. S. Lesnykh, I. G. Eikhenbaum and S. A. Chernyakhovskaya, Russ. J. Inorg. Chem. 15, 420 (1970).
10. H. S. Van Klooster, Z. Anorg. Chem. 69, 122, 136 (1910).
11. G. Petit and M. Jaeger, Compt. Rend. 244, 1734 (1957).
12. G. W. Morey and H. E. Merwin, J. Amer. Chem. Soc. 58, 2246 (1936).
13. S. S. Cole, S. R. Scholes and C. R. Amberg, J. Amer. Ceram. Soc. 18, 58 (1935).
14. A. G. Bergman and N. A. Litovskaya, Russ. Inorg. Chem. 14, 876 (1969).

Sodium Borate ($NaBO_2$) $B_1Na_1O_2$(cr)

Enthalpy Reference Temperature = T_r = 298.15 K Standard State Pressure = p° = 0.1 MPa

T/K	$C_p°$ (J K⁻¹mol⁻¹)	S° (J K⁻¹mol⁻¹)	$-[G°-H°(T_r)]/T$ (J K⁻¹mol⁻¹)	$H°-H°(T_r)$ (kJ mol⁻¹)	$\Delta_f H°$ (kJ mol⁻¹)	$\Delta_f G°$ (kJ mol⁻¹)	Log K_f
0	0.	0.	INFINITE	-11.634	-970.998	-970.998	INFINITE
100	32.430	20.000	123.003	-10.300	-973.933	-956.190	499.463
200	53.371	49.752	79.193	-5.888	-975.221	-937.851	244.942
298.15	65.940	73.538	73.538	0.	-975.709	-919.389	161.073
300	66.149	73.947	73.539	0.122	-975.714	-919.039	160.019
400	75.396	94.289	76.250	7.215	-978.576	-899.905	117.516
500	82.718	111.927	81.660	15.133	-978.549	-880.232	91.957
600	88.575	127.543	88.030	23.707	-978.132	-860.601	74.922
700	93.303	141.562	94.693	32.808	-977.394	-841.067	62.761
800	97.236	154.285	101.359	42.340	-976.390	-821.658	53.649
900	100.500	165.932	107.896	52.232	-975.173	-802.387	46.569
1000	103.219	176.667	114.244	62.423	-973.793	-783.262	40.913
1100	105.269	186.607	120.376	72.854	-972.310	-764.280	36.293
1200	106.608	195.831	126.284	83.456	-1067.556	-742.995	32.342
1240.000	106.859	199.331	128.584	87.726	------ CRYSTAL <--> LIQUID -----		
1300	107.320	204.391	131.967	94.151	-1065.165	-716.045	28.771
1400	108.127	212.374	137.429	104.923	-1062.774	-689.279	25.717
1500	108.935	219.861	142.677	115.776	-1060.375	-662.685	23.077
1600	109.742	226.917	147.724	126.710	-1057.963	-636.251	20.771
1700	110.549	233.595	152.580	137.724	-1055.536	-609.968	18.742
1800	111.356	239.936	157.259	148.820	-1053.090	-583.829	16.942
1900	112.162	245.979	161.770	159.996	-1050.624	-557.826	15.336
2000	112.968	251.752	166.126	171.252	-1048.138	-531.953	13.893

PREVIOUS: March 1965 CURRENT: June 1971

Sodium Borate ($NaBO_2$) $B_1Na_1O_2$(cr)

SODIUM BORATE ($NaBO_2$) LIQUID M_r = 65.79857

$S°(298.15\ K)$ = [82.958] J K^{-1} mol^{-1}

T_{fus} = 1240 ± 2 K

$\Delta_f H°(298.15\ K)$ = [-960.019] kJ mol^{-1}

$\Delta_{fus}H°$ = 33.47 ± 2.0 kJ mol^{-1}

Enthalpy of Formation

$\Delta_f H°(NaBO_2, \ell, 298.15\ K)$ is calculated from that of the crystal by adding $\Delta_{fus}H°$ and the difference in enthalpy, H°(1240 K)-H°(298.15 K), between the crystal and liquid.

Heat Capacity and Entropy

The constant heat capacity of the liquid is estimated to be 35 cal K^{-1} mol^{-1} based on that of $LiBO_2(\ell)$, 34.491 cal K^{-1} mol^{-1} (1), which was derived from high temperature enthalpy measurements. A glass transition is assumed at 826 K below which the heat capacities are assumed to be the same as the crystal.

$S°(NaBO_2, \ell, 298.15\ K)$ is calculated in a manner analogous to that used for the enthalpy of formation.

Fusion Data

See the crystal table for details.

Vaporization Data

The boiling point is calculated as the temperature at which the fugacity is 1 bar for the reaction $NaBO_2(\ell) \rightarrow NaBO_2(g)$. The enthalpy of vaporization is the difference in $\Delta_f H°$ at the boiling point between liquid and gas.

Reference

1. JANAF Thermochemical Table: $LiBO_2(\ell)$, 6-30-71.

Sodium Borate ($NaBO_2$) $B_1Na_1O_2(l)$

Enthalpy Reference Temperature = T_r = 298.15 K Standard State Pressure = p° = 0.1 MPa

T/K	$C_p°$	S°	$-[G°-H°(T_r)]/T$	$H°-H°(T_r)$	$\Delta_f H°$	$\Delta_f G°$	Log K_f
	J $K^{-1}mol^{-1}$			kJ mol^{-1}			
0							
100							
200							
298.15	65.940	82.958	82.958	0.	-960.019	-906.507	158.816
300	66.149	83.366	82.959	0.122	-960.024	-906.175	157.779
400	75.396	103.708	85.670	7.215	-962.886	-887.983	115.959
500	82.718	121.346	91.080	15.133	-962.859	-869.251	90.810
600	88.575	136.962	97.450	23.707	-962.442	-850.563	74.048
700	93.303	150.981	104.113	32.808	-961.704	-831.971	62.082
800	97.236	163.706	110.779	42.342	-960.699	-813.504	53.116
826.000	98.129	166.830	112.494	44.881	GLASS <--> LIQUID TRANSITION		
826.000	146.440	166.830	112.494	44.881			
900	146.440	179.395	117.486	55.718	-955.997	-795.328	46.160
1000	146.440	194.824	124.462	70.362	-950.164	-777.790	40.628
1100	146.440	208.781	131.503	85.006	-944.468	-760.829	36.129
1200	146.440	221.523	138.481	99.650	-1035.672	-741.941	32.296
1240.000	146.440	226.325	141.238	105.508	CRYSTAL <--> LIQUID		
1300	146.440	233.244	145.326	114.294	-1029.331	-717.722	28.838
1400	146.440	244.097	151.998	128.938	-1023.069	-693.987	25.893
1500	146.440	254.200	158.479	143.582	-1016.879	-670.698	23.356
1600	146.440	263.651	164.760	158.226	-1010.757	-647.819	21.149
1700	146.440	272.529	170.841	172.870	-1004.700	-625.321	19.214
1800	146.440	280.899	176.725	187.514	-998.706	-603.178	17.504
1900	146.440	288.817	182.418	202.158	-992.772	-581.366	15.983
2000	146.440	296.328	187.927	216.802	-986.898	-559.866	14.622
2100	146.440	303.473	193.261	231.446	-981.083	-538.657	13.398
2200	146.440	310.285	198.426	246.090	-975.327	-517.724	12.292
2300	146.440	316.795	203.432	260.734	-969.629	-497.052	11.288
2400	146.440	323.027	208.287	275.378	-1014.248	-475.558	10.350
2500	146.440	329.005	212.997	290.022	-1008.744	-453.225	9.470

PREVIOUS: March 1965 CURRENT: June 1971

Sodium Borate ($NaBO_2$) $B_1Na_1O_2(l)$

SODIUM BORATE ($NaBO_2$) CRYSTAL–LIQUID M_r = 65.79857

0 to 1240 K crystal
above 1240 K liquid

Refer to the individual tables for details.

Sodium Borate ($NaBO_2$) $B_1Na_1O_2$(cr,l)

Enthalpy Reference Temperature = T_r = 298.15 K Standard State Pressure = p° = 0.1 MPa

T/K	C_p° (J K⁻¹mol⁻¹)	S° (J K⁻¹mol⁻¹)	$-[G°-H°(T_r)]/T$ (J K⁻¹mol⁻¹)	$H°-H°(T_r)$ (kJ mol⁻¹)	$\Delta_fH°$ (kJ mol⁻¹)	$\Delta_fG°$ (kJ mol⁻¹)	Log K_f
0	0.	0.	INFINITE	-11.634	-970.998	-970.998	INFINITE
100	32.430	20.000	123.003	-10.300	-973.933	-956.190	499.463
200	53.371	49.752	79.193	-5.888	-975.221	-937.851	244.942
298.15	65.940	73.538	73.538	0.	-975.709	-919.389	161.073
300	66.149	73.947	73.539	0.122	-975.714	-919 039	160.019
400	75.396	94.289	76.250	7.215	-978.576	-899.905	117.516
500	82.718	111.927	81.660	15.133	-978.549	-880.232	91.957
600	88.575	127.543	88.030	23.707	-978.132	-860.601	74.922
700	93.303	141.562	94.693	32.808	-977.394	-841.067	62.761
800	97.236	154.285	101.359	42.340	-976.390	-821.658	53.649
900	100.500	165.932	107.896	52.232	-975.173	-802.387	46.569
1000	103.219	176.667	114.244	62.423	-973.793	-783.262	40.913
1100	105.269	186.607	120.376	72.854	-972.310	-764.280	36.293
1200	106.608	195.831	126.284	83.456	-1067.556	-742.995	32.342
1240.000	106.859	199.331	128.584	87.726	CRYSTAL <--> LIQUID		
1240.000	146.440	226.325	128.584	121.198	TRANSITION		
1300	146.440	233.244	133.256	129.984	-1029.331	-717.721	28.838
1400	146.440	244.097	140.791	144.628	-1023.069	-693.986	25.893
1500	146.440	254.200	148.019	159.272	-1016.879	-670.697	23.356
1600	146.440	263.651	154.953	173.916	-1010.757	-647.818	21.149
1700	146.440	272.529	161.611	188.560	-1004.700	-625.321	19.214
1800	146.440	280.899	168.008	203.204	-998.705	-603.178	17.504
1900	146.440	288.817	174.160	217.848	-992.772	-581.366	15.983
2000	146.440	296.328	180.082	232.492	-986.898	-559.865	14.622
2100	146.440	303.473	185.789	247.136	-981.083	-538.657	13.398
2200	146.440	310.285	191.294	261.780	-975.326	-517.723	12.292
2300	146.440	316.795	196.610	276.424	-969.628	-497.052	11.288
2400	146.440	323.027	201.749	291.068	-1014.248	-475.557	10.350
2500	146.440	329.005	206.720	305.712	-1008.744	-453.225	9.470

PREVIOUS: CURRENT: June 1971

Sodium Borate ($NaBO_2$) $B_1Na_1O_2$(cr,l)

SODIUM BORATE ($NaBO_2$) IDEAL GAS M_r = 65.79857

$S°(298.15\ K) = 287.2 \pm 4.2\ J\ K^{-1}\ mol^{-1}$

$\Delta_f H°(0\ K) = -645.9 \pm 12.6\ kJ\ mol^{-1}$

$\Delta_f H°(298.15\ K) = -648.5 \pm 12.6\ kJ\ mol^{-1}$

Vibrational Frequencies and Degeneracies

$\nu,\ cm^{-1}$	$\nu,\ cm^{-1}$
1960	573
1087	363
576	108

Ground State Quantum Weight: 1 $\sigma = 1$

Point Group: C_s

Bond Distance: Na–O = 2.14 Å O–B = 1.36 Å B=O = 1.20 Å

Bond Angles: Na–O–B = 90° O–B–O = 180°

Product of the Moments of Inertia: $I_A I_B I_C = 2.407470 \times 10^{-114}\ g^3\ cm^6$

Enthalpy of Formation

Jensen (1) determined equilibrium constants for the reaction: $Na(g) + HBO_2(g) \rightarrow NaBO_2(g) + H(g)$ by flame studies. This technique involved several assumptions, the most basic of which was that boron added to $H_2/N_2/O_2$ flames was converted completely to HBO_2. The hydrogen atom concentration was taken from previous studies on such flames. Sodium metaborate was assumed to be formed in the flame. The concentration of Na atom was determined by atomic absorption spectrophotometry. JANAF 3rd law analysis of his equilibrium constant equation in the temperature range from 2000 to 2600 K yields $\Delta_r H(298.15\ K) = 7.55 \pm 1.08\ kcal\ mol^{-1}$ and the drift 2.50 cal $K^{-1}\ mol^{-1}$. (The 2nd law enthalpy of reaction is $\Delta_r H°(298.15\ K) = 1.81\ kcal\ mol^{-1}$.) Using the 3rd law $\Delta_r H°(298.15\ K)$ and JANAF auxiliary data, we obtain $\Delta_f H°(NaBO_2,\ g,\ 298.15\ K) = -152.8\ kcal\ mol^{-1}$. The uncertainty in this result is about ±6 kcal mol^{-1}.

Buchler and Berkowitz-Mattuck (2) studied mass spectrometrically the sublimation of sodium metaborate and obtained $\Delta_{sub}H°(1070\ K) = 73 \pm 3\ kcal\ mol^{-1}$ for $NaBO_2(cr) \rightarrow NaBO_2(g)$ by a 2nd law method. JANAF reduction to 298.15 K gives $\Delta_{sub}H°(298.15\ K) = 76.6 \pm 3\ kcal\ mol^{-1}$ which leads to $\Delta_f H°(298.15\ K)(NaBO_2,\ g) = -156.6 \pm 3\ kcal\ mol^{-1}$. They also determined the vapor pressure of $NaBO_2$ at 1070 K as 5×10^{-7} atm by comparison with the vapor pressure of silver. JANAF 3rd law analysis of this single vapor pressure point gives $\Delta_{sub}H°(298.15\ K) = 83.37\ kcal\ mol^{-1}$ which leads to $\Delta_f H°(NaBO_2,\ g,\ 298.15\ K) = -149.8\ kcal\ mol^{-1}$.

Cole and Taylor (3) determined the vapor pressures of $NaBO_2(g)$ by a dynamic method with dry N_2 as carrier gas. Their vapor pressure data were taken over a flow-rate range where the apparent vapor pressure increased with decreasing flow rate, and the data were then extrapolated to zero flow rate. The vapor species of the sample was also assumed to be monomeric $NaBO_2(g)$. JANAF 3rd law analyses of their reported vapor pressure data over the liquid $NaBO_2$ in the temperature range 1150°C–1350°C yield $\Delta_{vap}H°(298.15\ K) = 75.37 \pm 3.32\ kcal\ mol^{-1}$ and the drift $-20.6\ cal\ K^{-1}\ mol^{-1}$. (The 2nd law enthalpy of vaporization is $\Delta_{vap}H°(298.15\ K) = 107\ kcal\ mol^{-1}$). Using the 3rd law $\Delta_{vap}H°(298.15\ K)$, we obtain $\Delta_f H°(298.15\ K) = -154.1 \pm 4\ kcal\ mol^{-1}$.

A weighted average, $\Delta_f H°(NaBO_2,\ g,\ 298.15\ K) = -155 \pm 3\ kcal\ mol^{-1}$, is adopted in the tabulation.

Heat Capacity and Entropy

The adopted vibrational frequencies are obtained from Seshadri, Nimon and White (4) who observed the infrared spectra of $NaBO_2$ in an argon matrix and made a complete assignment of all six fundamentals. The uncertainty of the calculated statistical entropy at 298.15 K is estimated to be ± 1 cal $mol^{-1}\ K^{-1}$ which is due to the magnitude of the matrix shifts from the free molecule spectrum, particularly for the lower frequency modes.

Buchler and Marran (5) observed only two vibrational frequencies, 1935 and 600 cm^{-1}, in their gas-phase infrared spectra. These two frequencies are in reasonable agreement with those reported by Seshadri (4).

By electron diffraction studies, Akishin and Spiridonov (6) determined the molecular structure of gaseous sodium metaborate and found the bond length and angles which are adopted in the tabulation. The angle Na–O–B is taken as 90° based on that of $LiBO_2(g)$ (7).

The three principal moments of inertia are $I_A = 5.1904 \times 10^{-39}$, $I_B = 19.0974 \times 10^{-39}$ and $I_C = 24.2878 \times 10^{-39}\ g\ cm^2$.

References

1. D. E. Jensen, Trans. Faraday Soc. 65, 2123 (1969).
2. A. Buchler and J. B. Berkowitz-Mattuck, J. Chem. Phys. 39, 286 (1963).
3. S. S. Cole and N. W. Taylor, J. Am. Ceram. Soc. 18, 82 (1935).
4. K. S. Seshadri, L. A. Nimon and D. White, J. Mol. Spectry. 30, 128 (1969).
5. A. Buchler and P. E. Marram, J. Chem. Phys. 39, 392 (1963).
6. P. A. Akishin and V. P. Spiridonov, Zhur. Strukt. Khim. 2, 63 (1961).
7. JANAF Thermochemical Tables: $LiBO_2(g)$, 6-30-71.

Sodium Borate ($NaBO_2$) $B_1Na_1O_2(g)$

Enthalpy Reference Temperature = T_r = 298.15 K Standard State Pressure = p° = 0.1 MPa

T/K	$C_p°$ (J K⁻¹mol⁻¹)	S° (J K⁻¹mol⁻¹)	$-[G°-H°(T_r)]/T$ (J K⁻¹mol⁻¹)	$H°-H°(T_r)$ (kJ mol⁻¹)	$\Delta_f H°$ (kJ mol⁻¹)	$\Delta_f G°$ (kJ mol⁻¹)	Log K_f
0	0.	0.	INFINITE	-13.703	-645.879	-645.879	INFINITE
100	41.612	233.779	333.819	-10.004	-646.448	-650.083	339.569
200	50.926	265.464	292.357	-5.379	-647.523	-653.295	170.623
250	54.980	277.277	288.189	-2.728	-648.025	-654.680	136.788
298.15	58.233	287.248	287.248	0.	-648.520	-655.918	114.914
300	58.346	287.608	287.249	0.108	-648.540	-655.964	114.213
350	61.142	296.819	287.970	3.097	-649.107	-657.157	98.075
400	63.496	305.141	289.604	6.215	-652.388	-658.058	85.934
450	65.510	312.739	291.759	9.441	-653.069	-658.726	76.463
500	67.255	319.734	294.211	12.761	-653.733	-659.318	68.879
600	70.125	332.261	299.533	19.636	-655.015	-660.314	57.485
700	72.365	343.246	305.009	26.765	-656.247	-661.100	49.332
800	74.131	353.029	310.412	34.094	-657.448	-661.711	43.205
900	75.536	361.844	315.645	41.580	-658.636	-662.172	38.431
1000	76.662	369.863	320.672	49.192	-659.836	-662.501	34.605
1100	77.572	377.214	325.482	56.905	-661.070	-662.708	31.469
1200	78.314	383.996	330.079	64.700	-759.123	-660.360	28.745
1300	78.924	390.290	334.472	72.563	-759.563	-652.112	26.202
1400	79.431	396.158	338.671	80.482	-760.027	-643.830	24.022
1500	79.855	401.653	342.688	88.446	-760.516	-635.513	22.131
1600	80.212	406.818	346.537	96.450	-761.034	-627.163	20.475
1700	80.517	411.690	350.227	104.487	-761.584	-618.779	19.013
1800	80.777	416.300	353.771	112.552	-762.169	-610.363	17.712
1900	81.002	420.674	357.178	120.641	-762.790	-601.912	16.548
2000	81.196	424.834	360.458	128.752	-763.450	-593.428	15.499
2100	81.366	428.799	363.619	136.880	-764.150	-584.910	14.549
2200	81.515	432.588	366.668	145.024	-764.894	-576.357	13.684
2300	81.647	436.215	369.614	153.182	-765.681	-567.770	12.894
2400	81.763	439.692	372.462	161.353	-816.774	-558.079	12.146
2500	81.866	443.032	375.218	169.535	-817.733	-547.280	11.435
2600	81.959	446.244	377.888	177.726	-818.708	-536.443	10.777
2700	82.042	449.339	380.478	185.926	-819.700	-525.567	10.168
2800	82.116	452.324	382.991	194.134	-820.711	-514.655	9.601
2900	82.184	455.207	385.431	202.349	-821.743	-503.707	9.073
3000	82.245	457.994	387.804	210.570	-822.798	-492.722	8.579
3100	82.300	460.692	390.112	218.798	-823.876	-481.701	8.117
3200	82.351	463.306	392.359	227.030	-824.979	-470.646	7.683
3300	82.397	465.840	394.547	235.268	-826.107	-459.556	7.274
3400	82.439	468.301	396.680	243.509	-827.264	-448.431	6.889
3500	82.478	470.691	398.761	251.755	-828.451	-437.272	6.526
3600	82.513	473.015	400.791	260.005	-829.668	-426.078	6.182
3700	82.546	475.276	402.774	268.258	-830.919	-414.851	5.857
3800	82.577	477.478	404.711	276.514	-832.205	-403.588	5.548
3900	82.605	479.623	406.605	284.773	-833.527	-392.291	5.254
4000	82.631	481.715	408.456	293.035	-834.889	-380.960	4.975
4100	82.655	483.756	410.268	301.299	-836.292	-369.595	4.709
4200	82.678	485.748	412.042	309.566	-1317.585	-351.171	4.367
4300	82.699	487.694	413.779	317.835	-1317.991	-328.156	3.986
4400	82.719	489.595	415.480	326.106	-1318.406	-305.135	3.622
4500	82.737	491.454	417.148	334.379	-1318.893	-282.102	3.275
4600	82.754	493.273	418.783	342.653	-1319.443	-259.056	2.942
4700	82.771	495.053	420.387	350.930	-1320.051	-235.998	2.623
4800	82.786	496.796	421.961	359.207	-1320.719	-212.926	2.317
4900	82.800	498.503	423.505	367.487	-1321.451	-189.839	2.024
5000	82.814	500.176	425.022	375.767	-1322.243	-166.737	1.742
5100	82.826	501.816	426.512	384.049	-1323.111	-143.618	1.471
5200	82.838	503.424	427.976	392.333	-1324.054	-120.483	1.210
5300	82.850	505.002	429.414	400.617	-1325.075	-97.327	0.959
5400	82.860	506.551	430.828	408.903	-1326.178	-74.152	0.717
5500	82.870	508.071	432.219	417.189	-1327.366	-50.956	0.484
5600	82.880	509.565	433.587	425.477	-1328.643	-27.736	0.259
5700	82.889	511.032	434.933	433.765	-1330.012	-4.494	0.041
5800	82.898	512.473	436.257	442.054	-1331.478	18.774	-0.169
5900	82.906	513.890	437.561	450.345	-1333.043	42.066	-0.372
6000	82.914	515.284	438.845	458.636	-1334.552	65.369	-0.569

PREVIOUS: June 1971 (1 atm) CURRENT: June 1971 (1 bar)

Sodium Borate ($NaBO_2$) $B_1Na_1O_2(g)$

BOROON OXIDE (BO) IDEAL GAS M_r = 26.8094

Boron Oxide (BO) $B_1O_1(g)$

$S°(298.15\ K) = 203.47 \pm 0.04\ J\ K^{-1}\ mol^{-1}$

$\Delta_fH°(0\ K) = -3.1 \pm 8\ kJ\ mol^{-1}$
$\Delta_fH°(298.15\ K) = 0 \pm 8\ kJ\ mol^{-1}$

Electronic Levels and Quantum Weights

State	ε_i, cm^{-1}	g_i
$X^2\Sigma^+$	0	2
$A^2\Pi$	23836	2
$A^2\Pi$	23959	2
$B'^2\Sigma$	39957	2
$B^2\Sigma^+$	43175	2

$\omega_e = 1895.66\ cm^{-1}$ $\quad \omega_ex_e = 11.90\ cm^{-1}$ $\quad \sigma = 1$
$B_e = 1.800\ cm^{-1}$ $\quad \alpha_e = 0.01676\ cm^{-1}$ $\quad r_e = 1.2049$ Å

Enthalpy of Formation

The enthalpy of formation has recently been determined by several workers whose values are in close agreement. Blackburn, et al. (1) measured the ion currents corresponding to BO^+ and $B_2O_2^+$ as a function of temperature in a mass spectrometer. From a least squares fit of the data they report $\Delta H = 56.6 \pm 1.8\ kcal\ mol^{-1}$ for the reaction $1/2B_2O_2(g) \rightarrow BO(g)$; this yields $\Delta_fH°(BO, g, 298.15\ K) = 2.1 \pm 2.8\ kcal\ mol^{-1}$ using $\Delta_fH°(B_2O_2, g, 298.15\ K) = -109 \pm 2\ kcal\ mol^{-1}$.

De Galan, et al. (2) from flame photometric measurements reports $D_0^0(BO) = 8.3eV(191.4\ kcal\ mol^{-1})$, which corresponds to $\Delta_fH°(BO, g, 298.15\ K) = -0.1\ kcal\ mol^{-1}$.

Coppens, et al. (3) from a mass spectometric study of several isomolecular exchange reactions, conclude that $D_0^0(BO) = 8.29 \pm 0.1\ eV(191.2 \pm 2.3\ kcal\ mol^{-1})$ or $\Delta_fH°(BO, g, 298.15\ K) = 0.1 \pm 2.3\ kcal\ mol^{-1}$.

Farber, et al. (4) investigated the same reaction as Blackburn et al. and reported $\Delta_rH°(1800\ K) = 53.8 \pm 1.9\ kcal\ mol^{-1}$, which yields $\Delta_fH°(BO, g, 298.15\ K) = 0.9 \pm 2.9\ kcal\ mol^{-1}$.

We adopt a median value of $0 \pm 2\ kcal\ mol^{-1}$ which includes all the determinations

Heat Capacity and Entropy

The molecular and vibrational constants for the ground state are from Herzberg (5). The electronic levels are also given by Herzberg with the exception of the level at 39957, the $A_2\Sigma$ level designated B' by Mal'tsev and Kataev (6). The molecular constants were all adjusted to reflect normal isotopic abundancies.

References

1. P. E. Blackburn, A. Buchler, and J. L. Stauffer, J. Phys. Chem. 70, 2469 (1966).
2. L. De Galan, Physica 31, 1286 (1965).
3. P. Coppens, S. Smoes and J. Drowart, Trans. Faraday Soc. 64, 630 (1968).
4. M. Farber, M. A. Frisch, G. Grenier and H. C. Ko, Rocket Power, Inc., AFRPL-TR-67-244, 1967 (AD 830 237).
5. G. Herzberg, 'Spectra of Diatomic Molecules', D. Van Nostrand Co., Inc., New York, 1950.
6. A. A. Mal'tsev and D. I. Kataev, Vestn. Mosk. Univ. Ser. II 22, 2 (1967).

Enthalpy Reference Temperature = T_r = 298.15 K — Standard State Pressure = p° = 0.1 MPa

T/K	$C_p°$ (J K⁻¹mol⁻¹)	S° (J K⁻¹mol⁻¹)	$-[G°-H°(T_r)]/T$ (J K⁻¹mol⁻¹)	$H°-H°(T_r)$ (kJ mol⁻¹)	$\Delta_fH°$ (kJ mol⁻¹)	$\Delta_fG°$ (kJ mol⁻¹)	Log K_f
0	0.	0.	INFINITE	-8.674	-3.118	-3.118	INFINITE
100	29.105	171.662	229.370	-5.771	-1.692	-10.162	5.308
200	29.112	191.838	206.138	-2.860	-0.569	-19.105	4.990
250	29.133	198.336	203.952	-1.404	-0.213	-23.783	4.969
298.15	29.196	203.472	203.472	0.	0.	-28.343	4.966
300	29.200	203.652	203.472	0.054	0.006	-28.519	4.966
350	29.343	208.163	203.828	1.517	0.104	-33.283	4.967
400	29.574	212.096	204.621	2.990	0.094	-38.053	4.969
450	29.882	215.596	205.649	4.476	-0.007	-42.816	4.970
500	30.250	218.763	206.805	5.979	-0.178	-47.564	4.969
600	31.076	224.350	209.276	9.045	-0.673	-56.997	4.962
700	31.910	229.204	211.783	12.194	-1.306	-66.336	4.950
800	32.677	233.516	214.235	15.425	-2.027	-75.578	4.935
900	33.350	237.404	216.597	18.727	-2.809	-84.725	4.917
1000	33.926	240.949	218.858	22.091	-3.638	-93.783	4.899
1100	34.415	244.206	221.016	25.509	-4.505	-102.756	4.879
1200	34.829	247.219	223.075	28.972	-5.405	-111.648	4.860
1300	35.181	250.021	225.042	32.473	-6.337	-120.464	4.840
1400	35.481	252.639	226.920	36.006	-7.300	-129.208	4.821
1500	35.739	255.096	228.718	39.568	-8.294	-137.881	4.801
1600	35.961	257.410	230.439	43.153	-9.318	-146.487	4.782
1700	36.155	259.596	232.091	46.759	-10.375	-155.028	4.763
1800	36.324	261.667	233.677	50.383	-11.463	-163.505	4.745
1900	36.473	263.635	235.202	54.023	-12.584	-171.922	4.726
2000	36.606	265.510	236.671	57.677	-13.739	-180.278	4.708
2100	36.724	267.299	238.087	61.344	-14.928	-188.576	4.691
2200	36.831	269.010	239.454	65.022	-16.151	-196.816	4.673
2300	36.928	270.649	240.775	68.710	-17.409	-205.000	4.656
2400	37.016	272.222	242.053	72.407	-68.961	-212.060	4.615
2500	37.098	273.735	243.290	76.113	-70.368	-217.994	4.555
2600	37.173	275.192	244.489	79.826	-71.778	-223.871	4.498
2700	37.244	276.596	245.653	83.547	-73.190	-229.694	4.444
2800	37.310	277.952	246.782	87.275	-74.607	-235.464	4.393
2900	37.374	279.262	247.880	91.009	-76.026	-241.185	4.344
3000	37.435	280.530	248.947	94.749	-77.449	-246.855	4.298
3100	37.495	281.759	249.986	98.496	-78.876	-252.478	4.254
3200	37.554	282.950	250.997	102.248	-80.305	-258.056	4.212
3300	37.613	284.106	251.983	106.007	-81.737	-263.589	4.172
3400	37.672	285.230	252.945	109.771	-83.172	-269.078	4.134
3500	37.731	286.323	253.883	113.541	-84.609	-274.525	4.097
3600	37.792	287.387	254.799	117.317	-86.047	-279.930	4.062
3700	37.855	288.423	255.693	121.100	-87.487	-285.296	4.028
3800	37.920	289.433	256.568	124.888	-88.927	-290.623	3.995
3900	37.987	290.419	257.424	128.684	-90.368	-295.912	3.963
4000	38.057	291.382	258.261	132.486	-91.809	-301.164	3.933
4100	38.130	292.323	259.080	136.295	-93.249	-306.380	3.903
4200	38.206	293.242	259.882	140.112	-574.537	-304.537	3.787
4300	38.285	294.142	260.669	143.936	-574.887	-298.104	3.621
4400	38.368	295.023	261.439	147.769	-575.237	-291.663	3.462
4500	38.455	295.887	262.195	151.610	-575.586	-285.215	3.311
4600	38.546	296.733	262.937	155.460	-575.935	-278.758	3.165
4700	38.640	297.563	263.665	159.319	-576.284	-272.294	3.026
4800	38.738	298.377	264.380	163.188	-576.632	-265.823	2.893
4900	38.840	299.177	265.082	167.067	-576.980	-259.343	2.765
5000	38.946	299.963	265.771	170.956	-577.327	-252.858	2.642
5100	39.055	300.735	266.449	174.856	-577.674	-246.365	2.523
5200	39.169	301.495	267.116	178.768	-578.020	-239.865	2.409
5300	39.285	302.242	267.772	182.690	-578.368	-233.359	2.300
5400	39.406	302.977	268.417	186.625	-578.715	-226.846	2.194
5500	39.529	303.701	269.052	190.571	-579.064	-220.327	2.092
5600	39.656	304.415	269.677	194.531	-579.414	-213.801	1.994
5700	39.787	305.118	270.293	198.503	-579.766	-207.269	1.899
5800	39.920	305.811	270.899	202.488	-580.120	-200.731	1.808
5900	40.055	306.494	271.497	206.487	-580.477	-194.187	1.719
6000	40.194	307.169	272.086	210.499	-580.838	-187.637	1.634

PREVIOUS: June 1968 (1 atm) CURRENT: June 1968 (1 bar)

Boron Oxide (BO) $B_1O_1(g)$

$\Delta_fH°(0\ K) = -285 \pm 8$ kJ mol^{-1}
$\Delta_fH°(298.15\ K) = -285 \pm 8$ kJ mol^{-1}

$S°(298.15\ K) = 229.81 \pm 0.20$ J K^{-1} mol^{-1}

Electronic Levels and Quantum Weights

State	ε_i, cm^{-1}	g_i
$^2\Pi_g$	0	2
$^2\Pi_g$	149	2
$^2\Pi_u$	18291.6	4
$^2\Sigma_u$	24508.0	2

Vibrational Frequencies and Degeneracies

ν, cm^{-1}
1056 (1)
454 (1)
1321.7 (1)

Point Group: $D_{\infty h}$ $\sigma = 2$
Bond Distance: B–O = 1.263 Å
Bond Angle: O–B–O = 180°
Rotational Constant : B_0 = 0.33026 cm^{-1}

T/K	$C_p°$	S°	−[G°−H°(T_r)]/T	H°−H°(T_r)	$\Delta_fH°$	$\Delta_fG°$	Log K_f
	J K^{-1}mol^{-1}			kJ mol^{-1}			
0	0.	0.	INFINITE	-10.731	-285.346	-285.346	INFINITE
100	33.725	188.764	264.960	-7.620	-285.163	-286.678	149.745
200	38.528	213.526	233.623	-4.019	-284.807	-288.331	75.304
250	41.020	222.394	230.514	-2.030	-284.646	-289.231	60.432
298.15	43.276	229.814	229.814	0.	-284.512	-290.127	50.829
300	43.359	230.082	229.815	0.080	-284.507	-290.162	50.522
350	45.527	236.931	230.351	2.303	-284.388	-291.114	43.446
400	47.495	243.141	231.567	4.630	-284.292	-292.082	38.142
450	49.246	248.838	233.174	7.049	-284.217	-293.060	34.018
500	50.780	254.108	235.007	9.550	-284.160	-294.046	30.719
600	53.263	263.597	238.999	14.759	-284.093	-296.030	25.772
700	55.110	271.953	243.122	20.182	-284.080	-298.021	22.239
800	56.490	279.406	247.200	25.765	-284.116	-300.011	19.589
900	57.533	286.123	251.158	31.469	-284.200	-301.994	17.527
1000	58.333	292.228	254.964	37.264	-284.329	-303.964	15.877
1100	58.957	297.818	258.610	43.129	-284.502	-305.920	14.527
1200	59.451	302.970	262.094	49.051	-284.719	-307.857	13.401
1300	59.847	307.745	265.425	55.016	-284.978	-309.776	12.447
1400	60.169	312.192	268.608	61.018	-285.280	-311.672	11.629
1500	60.435	316.353	271.654	67.048	-285.624	-313.546	10.919
1600	60.655	320.260	274.571	73.103	-286.013	-315.395	10.297
1700	60.841	323.943	277.368	79.178	-286.446	-317.218	9.747
1800	60.998	327.425	280.053	85.270	-286.924	-319.015	9.258
1900	61.133	330.727	282.634	91.377	-287.449	-320.783	8.819
2000	61.250	333.866	285.118	97.496	-288.019	-322.523	8.423
2100	61.352	336.857	287.511	103.627	-288.637	-324.233	8.065
2200	61.443	339.713	289.819	109.766	-289.302	-325.912	7.738
2300	61.523	342.446	292.048	115.915	-290.016	-327.561	7.439
2400	61.597	345.066	294.203	122.071	-341.036	-328.109	7.141
2500	61.664	347.582	296.288	128.234	-341.923	-327.552	6.844
2600	61.728	350.002	298.308	134.404	-342.824	-326.959	6.569
2700	61.788	352.332	300.266	140.579	-343.741	-326.332	6.313
2800	61.847	354.581	302.166	146.761	-344.672	-325.670	6.075
2900	61.904	356.752	304.011	152.949	-345.617	-324.975	5.853
3000	61.961	358.852	305.804	159.142	-346.575	-324.247	5.646
3100	62.019	360.884	307.548	165.341	-347.547	-323.486	5.451
3200	62.078	362.854	309.246	171.546	-348.531	-322.695	5.267
3300	62.138	364.765	310.900	177.757	-349.526	-321.872	5.095
3400	62.201	366.621	312.511	183.974	-350.532	-321.019	4.932
3500	62.265	368.425	314.083	190.197	-351.548	-320.136	4.778
3600	62.333	370.180	315.617	196.427	-352.572	-319.224	4.632
3700	62.403	371.889	317.115	202.663	-353.604	-318.284	4.493
3800	62.475	373.554	318.579	208.907	-354.644	-317.315	4.362
3900	62.551	375.178	320.009	215.159	-355.690	-316.318	4.237
4000	62.629	376.763	321.408	221.418	-356.741	-315.296	4.117
4100	62.711	378.310	322.777	227.685	-357.798	-314.247	4.004
4200	62.795	379.822	324.117	233.960	-838.708	-306.148	3.808
4300	62.882	381.301	325.430	240.244	-838.686	-293.467	3.565
4400	62.971	382.747	326.716	246.536	-838.669	-280.789	3.333
4500	63.062	384.164	327.977	252.838	-838.657	-268.110	3.112
4600	63.156	385.551	329.214	259.149	-838.650	-255.432	2.901
4700	63.252	386.910	330.427	265.469	-838.649	-242.753	2.698
4800	63.350	388.243	331.618	271.799	-838.653	-230.074	2.504
4900	63.449	389.550	332.787	278.139	-838.664	-217.395	2.317
5000	63.549	390.833	333.935	284.489	-838.680	-204.715	2.139
5100	63.651	392.092	335.063	290.849	-838.705	-192.036	1.967
5200	63.754	393.329	336.172	297.219	-838.736	-179.357	1.802
5300	63.857	394.545	337.262	303.600	-838.777	-166.675	1.643
5400	63.961	395.739	338.333	309.991	-838.828	-153.994	1.490
5500	64.065	396.914	339.388	316.392	-838.889	-141.312	1.342
5600	64.169	398.069	340.425	322.804	-838.962	-128.627	1.200
5700	64.273	399.206	341.447	329.226	-839.048	-115.942	1.062
5800	64.377	400.324	342.452	335.658	-839.149	-103.254	0.930
5900	64.480	401.426	343.443	342.101	-839.266	-90.567	0.802
6000	64.582	402.510	344.418	348.554	-839.400	-77.875	0.678

Enthalpy Reference Temperature = T_r = 298.15 K; Standard State Pressure = p° = 0.1 MPa

PREVIOUS: June 1968 (1 atm) CURRENT: June 1968 (1 bar)

Enthalpy of Formation

Greene (1) measured the vapor pressure of B_2O_3 and examined the infrared absorption spectra of B_2O_3, B_2O_2 and BO_2. The isotope shift and oxygen and temperature dependencies of the boric acid fluctuation bands were studied. The enthalpy change, $\Delta_rH°(2100\ K) = 76 \pm 2$ kcal mol^{-1}, for the reaction 1/2 $B_2O_3(\ell)$ + 1/4 $O_2(g)$ = $BO_2(g)$, was obtained spectroscopically. Based on this data and $\Delta_fH°(B_2O_3, \ell, 2100\ K) = -291.05$ kcal mol^{-1}, we derive $\Delta_fH°(BO_2, g, 298.15\ K) = -68.4 \pm 2$ kcal mol^{-1}.

Rusin (2) determined the equilibrium constant of the reaction $HBO_2(g) + OH(g) = BO_2(g) + H_2O(g)$ by simultaneously recording the pressure and spectra at 3000-3200 K of HBO_2 and BO_2, which were produced from the combustion in a bomb of mixtures of H, O, CO and a small amount of B_2H_6. Using the experimental value Kp(3100 K) = 0.9, and JANAF Gibbs energy functions for the reactants and products, we obtain $\Delta_rH°(298.15\ K) = 0.53$ kcal mol^{-1} for that reaction. Employing $\Delta_fH°(298.15\ K) = -134.00$, 9.43 and -57.80 kcal mol^{-1} for $HBO_2(g)$, OH(g) and $H_2O(g)$, respectively, we derive the value $\Delta_fH°(BO_2, g, 298.15\ K) = -66.24$ kcal mol^{-1}.

In a study of gases from flames containing trimethylborate stabilized on flat porous metal burners, Kaskan (3) obtained data consistent with the bands being due to the molecule $BO_2(g)$. The temperature dependence of the equilibrium constants for the reaction $OH(g) + HBO_2(g) = H_2O(g) + BO_2(g)$ was investigated and the data were reported to be consistent with a value $\Delta_rH°(1800\ K) = -16$ kcal mol^{-1}. Based on this $\Delta_rH°(1800\ K)$ value, we calculate $\Delta_fH°(BO_2, g, 298.15\ K) = -83.62$ kcal mol^{-1}. Kaskan (4) has studied the effect of oxygen on the intensity of absorption (5470 Å) in the green or "fluctuation" bands of the gas phase in equilibrium with $B_2O_3(\ell)$. At constant temperature the absorptivity was found to depend on the 1/4 power of the oxygen pressure. This is interpreted as evidence that the absorbing species is $BO_2(g)$. The reaction 1/2 $B_2O_3(\ell)$ + 1/4 $O_2(g)$ = $BO_2(g)$ was found to be 70 kcal mol^{-1} endothermic in the temperature range 1660-1950 K, yielding $\Delta_fH°(BO_2, g, 298.15\ K) = -75.28$ kcal mol^{-1}. Rusin et al. (2) discussed the discrepancy between their results and those of the above two investigations and commented that the results of Kaskan (3, 4) were derived based on the assumption that Beer's law was obeyed over the temperature range 1400-2200 K. Actually, the relative population α of the principal electronic state $X^2\Pi$ of the BO_2 molecule changes greatly over this temperature range. Therefore the expression for the dependence of the optical density on the concentration of BO_2 must be corrected to allow for the variation of α with temperature.

The enthalpy of formation of $BO_2(g)$ at 298.15 K is selected as -68 ± 2 kcal mol^{-1}.

Heat Capacity and Entropy

The electronic levels and quantum weights are obtained from Johns (5). Vibrational frequencies, bond distance and rotational constant were reported by Johns (5), Sommer (6), and Snowden (7). The values adopted are those determined by Snowden (7). Snowden used a (1:1) mixture of KBF_4 and B_2O_3 placed in a 13 mm Vycor tube. The mixture was heated to produce a vapor which passed through an external electrodeless discharge. The intense emission spectrum of the BO_2 molecule was observed. The fluorescence spectrum obtained by Johns was too weak for a full characterization of the electronic ground state.

References

1. F. T. Greene, Diss. Abstr. 21, 1838 (1961); Ph.D. Thesis, University of Wisconsin, 1961.
2. A. D. Rusin and V. M. Tatevskii, Russ. J. Phys. Chem. 37, 376 (1963).
3. W. E. Kaskan and R. C. Millikan, J. Chem. Phys. 32, 1273 (1960).
4. W. E. Kaskan, J. D. Mackenzie, and R. C. Millikan, J. Chem. Phys. 34, 570 (1961).
5. J. W. C. Johns, Can. J. Phys. 39, 1738 (1961).
6. A. Sommer, D. White, M. J. Linevsky, and D. E. Mann, J. Chem. Phys. 38, 87 (1963).
7. B. S. Snowden, Ph.D. Thesis, Vanderbilt University, 1963; Diss. Abstr. 24, 5030 (1964).

BORON OXIDE, ION (BO_2^-) IDEAL GAS M_r = 42.80935

$S°(298.15\ K) = [215.9 \pm 8.4]\ J\ K^{-1}\ mol^{-1}$

$\Delta_f H°(0\ K) = -688 \pm 25\ kJ\ mol^{-1}$
$\Delta_f H°(298.15\ K) = -695 \pm 25\ kJ\ mol^{-1}$

Electronic Levels and Quantum Weights

State	ϵ_i, cm^{-1}	g_i
$^1\Sigma_g^+$	0	1
$^1B_2$	[35000]	1

Vibrational Frequencies and Degeneracies

ν, cm^{-1}
[1150] (1)
[580] (2)
[2200] (1)

Point Group: $[D_{\infty h}]$ $\sigma = 2$
Bond Distance: B-O = [1.25] Å
Bond Angle: O-B-O = [180]°
Rotational Constant: $B_0 = [0.337159]\ cm^{-1}$

Enthalpy of Formation

Jensen (4) has obtained equilibrium constants for the reaction $HBO_2 + e^- + H + BO_2^-$. This involves several assumptions the most basic of which is that boron added to $H_2/N_2/O_2$ flames is converted completely to HBO_2. The free electrons are produced by addition of potassium to the flame, and their concentration is measured directly by a microwave cavity resonance method. The hydrogen atom concentration is taken from previous studies on such flames. The BO_2^- concentration is obtained from the difference between the K^+ concentration and the free electron measurement. The K^+ concentration is measured by an electrostatic probe.

By 2nd and 3rd analysis of the data, 11 points read back from a plot, we obtain $\Delta_r H°(II, 298.15\ K) = 27.5 \pm 4\ kcal\ mol^{-1}$ and $\Delta_r H°\ (III, 298.15\ K) = 20 \pm 2\ kcal\ mol^{-1}$ with a drift of $-3.7 \pm 1.5\ cal\ K^{-1}\ mol^{-1}$. The 3rd law value yields $\Delta_f H°(BO_2^-, 298.15\ K) = -166 \pm 6\ kcal\ mol^{-1}$ where the uncertainty includes possible errors in the functions. This corresponds to an electron affinity of BO_2 equal to $98 \pm 6\ kcal\ mol^{-1}$ (4.25 eV).

Heat Capacity and Entropy

The molecular structure is assumed to be linear using the Walsh (1) prediction for 16 valence electron XY_2 molecules. The electronic states are taken from those for $CO_2(g)$ (2), which is isoelectronic with $BO_2^-(g)$. The electronic level 35000 cm^{-1} and vibrational frequencies are estimated by comparison with the corresponding values for CO_2, CO_2^+, and BO_2, and, the data of Vasco and Srb (3) on the BO_2^- ion in alkali halide lattices. The B-O bond distance in $BO_2^-(g)$ is assumed to be smaller than the B-O bond in $BO_2(g)$, because the $BO_2^-(g)$ molecule has one more bonding electron than the $BO_2(g)$ molecule.

References

1. A. D. Walsh, J. Chem. Soc. 1953, 2266 (1953).
2. G. Herzberg, "Electronic Spectra of Polyatomic Molecules", D. Van Nostrand Co., Inc., New York, 1966.
3. A. Vasco and I. Srb, Czech. J. Phys. 17, 1110 (1967).
4. D. E. Jensen, AeroChem Research Laboratories, Inc., AeroChem TP-187A, 1968.

Boron Oxide, Ion (BO_2^-) $B_1O_2^-(g)$

Enthalpy Reference Temperature = T_r = 298.15 K Standard State Pressure = p° = 0.1 MPa

T/K	$C_p°$	S°	$-[G°-H°(T_r)]/T$	$H°-H°(T_r)$	$\Delta_f H°$	$\Delta_f G°$	Log K_f
	J K⁻¹mol⁻¹			kJ mol⁻¹			
0	0.	0.	INFINITE	-9.629	-688.078		
100	29.376	179.892	247.061	-6.717			
200	33.849	201.391	219.337	-3.589			
250	36.665	209.250	216.552	-1.826			
298.15	39.119	215.922	215.922	0.	-694.544	-689.762	120.843
300	39.207	216.164	215.923	0.072	-694.585	-689.733	120.093
350	41.425	222.378	216.408	2.090	-695.711	-688.835	102.803
400	43.352	228.039	217.514	4.210	-696.860	-687.774	89.814
450	45.040	233.244	218.976	6.421	-698.033	-686.569	79.695
500	46.533	238.069	220.647	8.711	-699.228	-685.231	71.586
600	49.058	246.784	224.293	13.495	-701.663	-682.204	59.391
700	51.105	254.506	228.068	18.507	-704.140	-678.765	50.650
800	52.779	261.443	231.813	23.704	-706.641	-674.970	44.071
900	54.151	267.741	235.461	29.052	-709.158	-670.860	38.936
1000	55.280	273.507	238.981	34.526	-711.687	-666.469	34.813
1100	56.213	278.821	242.365	40.102	-714.229	-661.824	31.427
1200	56.988	283.746	245.610	45.763	-716.784	-656.947	28.596
1300	57.635	288.334	248.722	51.495	-719.355	-651.857	26.192
1400	58.179	292.626	251.707	57.287	-721.945	-646.567	24.124
1500	58.638	296.656	254.570	63.128	-724.558	-641.092	22.325
1600	59.030	300.453	257.320	69.012	-727.196	-635.442	20.745
1700	59.365	304.042	259.964	74.932	-729.863	-629.626	19.346
1800	59.654	307.444	262.508	80.884	-732.561	-623.652	18.098
1900	59.904	310.676	264.959	86.862	-735.292	-617.527	16.977
2000	60.122	313.754	267.322	92.863	-738.059	-611.258	15.964
2100	60.314	316.692	269.604	98.885	-740.864	-604.849	15.045
2200	60.482	319.502	271.809	104.925	-743.707	-598.305	14.206
2300	60.631	322.194	273.941	110.981	-746.592	-591.632	13.436
2400	60.763	324.777	276.006	117.051	-799.777	-583.763	12.705
2500	60.881	327.260	278.007	123.133	-802.823	-574.700	12.008
2600	60.986	329.650	279.947	129.227	-805.879	-565.515	11.361
2700	61.081	331.953	281.831	135.330	-808.947	-556.211	10.761
2800	61.166	334.176	283.661	141.442	-812.026	-546.794	10.201
2900	61.243	336.324	285.440	147.563	-815.117	-537.268	9.677
3000	61.314	338.401	287.171	153.691	-818.219	-527.634	9.187
3100	61.377	340.413	288.856	159.825	-821.334	-517.896	8.726
3200	61.436	342.363	290.498	165.966	-824.461	-508.058	8.293
3300	61.489	344.254	292.099	172.112	-827.599	-498.123	7.885
3400	61.538	346.090	293.660	178.264	-830.749	-488.091	7.499
3500	61.583	347.875	295.183	184.420	-833.910	-477.967	7.133
3600	61.625	349.610	296.671	190.580	-837.083	-467.752	6.787
3700	61.663	351.299	298.125	196.745	-840.266	-457.450	6.458
3800	61.699	352.944	299.546	202.913	-843.460	-447.059	6.145
3900	61.733	354.547	300.936	209.085	-846.664	-436.586	5.847
4000	61.764	356.111	302.296	215.259	-849.878	-426.031	5.563
4100	61.794	357.636	303.627	221.437	-853.103	-415.395	5.292
4200	61.822	359.125	304.931	227.618	-1336.186	-397.656	4.946
4300	61.848	360.580	306.208	233.802	-1338.343	-375.284	4.559
4400	61.873	362.003	307.460	239.988	-1340.511	-352.863	4.189
4500	61.897	363.393	308.688	246.176	-1342.690	-330.392	3.835
4600	61.921	364.754	309.892	252.367	-1344.882	-307.873	3.496
4700	61.943	366.086	311.073	258.560	-1347.087	-285.305	3.171
4800	61.965	367.390	312.233	264.756	-1349.304	-262.690	2.859
4900	61.986	368.668	313.372	270.953	-1351.536	-240.029	2.559
5000	62.008	369.921	314.490	277.153	-1353.781	-217.322	2.270
5100	62.028	371.149	315.589	283.355	-1356.042	-194.570	1.993
5200	62.049	372.353	316.669	289.559	-1358.319	-171.775	1.726
5300	62.070	373.536	317.731	295.765	-1360.613	-148.934	1.468
5400	62.091	374.696	318.775	301.973	-1362.925	-126.051	1.219
5500	62.112	375.836	319.802	308.183	-1365.256	-103.125	0.979
5600	62.133	376.955	320.813	314.395	-1367.607	-80.155	0.748
5700	62.154	378.055	321.808	320.609	-1369.980	-57.144	0.524
5800	62.176	379.136	322.787	326.826	-1372.375	-34.091	0.307
5900	62.198	380.199	323.751	333.045	-1374.795	-10.998	0.097
6000	62.221	381.245	324.700	339.265	-1377.239	12.139	-0.106

PREVIOUS: December 1968 (1 atm) CURRENT: December 1968 (1 bar)

Boron Oxide, Ion (BO_2^-) $B_1O_2^-(g)$

$S°(298.15\ K) = 216.19 \pm 0.42\ J\ K^{-1}\ mol^{-1}$

$\Delta_f H°(0\ K) = 240.0 \pm 17\ kJ\ mol^{-1}$
$\Delta_f H°(298.15\ K) = 243.1 \pm 17.0\ kJ\ mol^{-1}$

Electronic Levels and Quantum Weights

State	ε_i, cm^{-1}	g_i
$X^2\Sigma^+$	0	2
$A^2\Pi$	{15663, 15996}	2
$B^2\Sigma^+$	35849	2
$C^2\Pi$	{38782, 38897}	2

$\omega_e = 1188.14\ cm^{-1}$ $\omega_e x_e = 6.397\ cm^{-1}$ $\sigma = 1$

$B_e = 0.80586\ cm^{-1}$ $\alpha_e = 0.00617\ cm^{-1}$ $r_e = 1.609$ Å

Enthalpy of Formation

Gingerich (1), empolying mass spectrometric investigations of the Au-Ce-CeS-BN-C system from a tungsten Knudsen effusion cell, reports two values of the BS(g) dissociation energy. The reaction enthalpies were evaluated by the 3rd law method. The two values reported are 140.4 ± 6.0 kcal mol^{-1}, based on the dissociation energy of CS(g), and 145.4 ± 6.0 kcal mol^{-1}, based on the dissociation energy of CeS(g).

Uy and Drowart (2) also studied the dissociation energy of BS(g) by two different exchange reactions using mass spectrometric Knudsen cell techniques. Molybdenum cells were used. Based on an exchange between B(g) and YS(g), they reported D_0^0 = 138.8 ± 3.2 kcal mol^{-1} for BS(g). A second determination based on an exchange between BO(g) and YS(g) was reported as D_0^0 = 137.0 ± 3.2 kcal mol^{-1}

Gaydon (3) suggests a value of 118 kcal mol^{-1} for the dissociation energy based on two linear Birge-Sponer extrapolations from the data of Zeeman (4). This data gave values of 154.5 kcal mol^{-1} and 131.4 kcal mol^{-1} for extrapolations from the ground state and $A^2\Pi$ state, respectively. McDonald and Innes (5) extended Zeeman's analysis (4), examined perturbative effects, and calculated $\omega_e y_e$ = -0.004 cm^{-1} for the ground state. This perturbative interpretation was supported by Singh, Tewari, and Mohan (7). Accounting for $\omega_e y_e$ value, the Birge-Sponer extrapolation yields a value D_0^0 = 142.3 kcal mol^{-1} for BS(g). This value was adjusted to 140.0 kcal mol^{-1} using an ionic character correction factor as suggested by Hildenbrand and Murad (6).

The value of 140.0 ± 4.0 kcal mol^{-1} for the dissociation energy of BS(g) is chosen as the best value. It is an average of the four reported thermochemical values and is the adjusted Birge-Sponer value, the adjustment arising from a correction technique which has been shown to be quite dependable. This value leads to a enthalpy of formation $\Delta_f H°(298.15\ K)$ = 58.1 ± 4.0 kcal mol^{-1}.

Heat Capacity and Entropy

The spectroscopic constants for the $X^2\Sigma^+$ state are from Zeeman (4) and are corrected to account for the natural abundances of the elements. The electronic states X, A, and C are reported by Zeeman (4). The $B^2\Sigma^+$ state is reported by McDonald and Innes (5).

References

1. K. A. Gingerich, Chem. Comm. 1970, 580 (1970).
2. O. M. Uy and J. Drowart, High Temp. Sci. 2, 293 (1970).
3. A. G. Gaydon, "Dissociation Energies and Spectra of Diatomic Molecules", Chapman and Hall, Ltd., London, 1968.
4. P. B. Zeeman, Can. J. Phys. 29, 336 (1951).
5. J. K. McDonald and K. K. Innes, J. Mol. Spectrosc. 29, 251 (1969).
6. D. L. Hildenbrand and E. Murad, J. Chem. Phys. 51, 807 (1969).
7. J. Singh, D. P. Tewari, and H. Mohan, Indian J. Pure Appl. Phys. 9, 269 (1971).

Enthalpy Reference Temperature = T_r = 298.15 K Standard State Pressure = p° = 0.1 MPa

T/K	$C_p°$ (J K^{-1}mol^{-1})	S°	$-[G°-H°(T_r)]/T$	$H°-H°(T_r)$ (kJ mol^{-1})	$\Delta_f H°$	$\Delta_f G°$	Log K_f
0	0.	0.	INFINITE	-8.724	239.992	239.992	INFINITE
100	29.106	184.196	242.369	-5.817	242.185	225.049	-117.553
200	29.239	204.389	218.907	-2.903	243.124	207.457	-54.182
250	29.557	210.944	216.682	-1.434	243.203	198.527	-41.480
298.15	30.051	216.190	216.190	0.	243.090	189.930	-33.275
300	30.073	216.376	216.191	0.056	243.083	189.600	-33.012
350	30.706	221.058	216.559	1.575	242.809	180.706	-26.969
400	31.376	225.202	217.385	3.127	240.195	171.948	-22.454
450	32.028	228.936	218.465	4.712	239.027	163.478	-18.976
500	32.631	232.342	219.685	6.329	237.738	155.155	-16.209
600	33.656	238.386	222.310	9.645	235.488	138.854	-12.088
700	34.449	243.636	224.990	13.052	233.393	122.915	-9.172
800	35.057	248.278	227.617	16.529	231.376	107.271	-7.004
900	35.527	252.435	230.147	20.059	176.267	92.958	-5.395
1000	35.895	256.198	232.567	23.631	175.521	83.741	-4.374
1100	36.188	259.634	234.874	27.236	174.724	74.601	-3.542
1200	36.426	262.793	237.071	30.867	173.879	65.536	-2.853
1300	36.622	265.716	239.163	34.519	172.988	56.543	-2.272
1400	36.786	268.437	241.158	38.190	172.051	47.620	-1.777
1500	36.926	270.979	243.062	41.876	171.069	38.766	-1.350
1600	37.046	273.367	244.882	45.575	170.043	29.979	-0.979
1700	37.153	275.616	246.625	49.285	168.974	21.258	-0.653
1800	37.249	277.742	248.295	53.005	167.864	12.601	-0.366
1900	37.338	279.758	249.898	56.734	166.712	4.006	-0.110
2000	37.421	281.676	251.440	60.472	165.521	-4.527	0.118
2100	37.502	283.504	252.923	64.218	164.292	-12.999	0.323
2200	37.583	285.250	254.353	67.973	163.027	-21.412	0.508
2300	37.664	286.922	255.733	71.735	161.727	-29.767	0.676
2400	37.749	288.527	257.066	75.506	110.133	-36.995	0.805
2500	37.838	290.070	258.356	79.285	108.689	-43.096	0.900
2600	37.933	291.556	259.604	83.073	107.246	-49.139	0.987
2700	38.033	292.989	260.815	86.872	105.808	-55.126	1.066
2800	38.142	294.374	261.989	90.680	104.373	-61.060	.39
2900	38.257	295.715	263.129	94.500	102.945	-66.944	1.206
3000	38.381	297.014	264.236	98.332	101.523	-72.777	1.267
3100	38.513	298.274	265.314	102.177	100.109	-78.564	1.324
3200	38.653	299.499	266.363	106.035	98.705	-84.305	1.376
3300	38.801	300.691	267.386	109.908	97.310	-90.003	1.425
3400	38.957	301.852	268.382	113.795	95.927	-95.658	1.470
3500	39.120	302.983	269.355	117.699	94.555	-101.273	1.511
3600	39.289	304.088	270.304	121.620	93.196	-106.849	1.550
3700	39.466	305.167	271.232	125.557	91.850	-112.387	1.587
3800	39.647	306.221	272.139	129.513	90.518	-117.889	1.621
3900	39.834	307.254	273.026	133.487	89.200	-123.356	1.652
4000	40.026	308.265	273.895	137.480	87.897	-128.790	1.682
4100	40.221	309.255	274.745	141.492	86.607	-134.191	1.710
4200	40.419	310.227	275.578	145.524	-394.516	-132.537	1.648
4300	40.620	311.180	276.395	149.576	-394.688	-126.297	1.534
4400	40.822	312.117	277.197	153.648	-394.847	-120.054	1.425
4500	41.026	313.036	277.983	157.741	-394.994	-113.807	1.321
4600	41.230	313.940	278.755	161.853	-395.128	-107.557	1.221
4700	41.435	314.829	279.513	165.987	-395.250	-101.304	1.126
4800	41.639	315.704	280.258	170.140	-395.360	-95.048	1.034
4900	41.842	316.564	280.990	174.314	-395.460	-88.791	0.947
5000	42.043	317.412	281.710	178.509	-395.549	-82.531	0.862
5100	42.243	318.246	282.418	182.723	-395.629	-76.270	0.781
5200	42.441	319.068	283.115	186.957	-395.700	-70.008	0.703
5300	42.637	319.879	283.801	191.211	-395.763	-63.744	0.628
5400	42.830	320.677	284.477	195.484	-395.818	-57.479	0.556
5500	43.019	321.465	285.142	199.777	-395.866	-51.213	0.486
5600	43.206	322.242	285.797	204.088	-395.908	-44.946	0.419
5700	43.390	323.008	286.444	208.418	-395.945	-38.678	0.354
5800	43.569	323.764	287.081	212.766	-395.978	-32.410	0.292
5900	43.746	324.511	287.709	217.132	-396.007	-26.142	0.231
6000	43.918	325.247	288.328	221.515	-396.032	-19.872	0.173

PREVIOUS: June 1972 (1 atm) CURRENT: June 1972 (1 bar)

TITANIUM BORIDE (TiB) CRYSTAL M_r = 58.69

$S°(298.15\ K) = [34.7 \pm 6.3]\ J\ K^{-1}\ mol^{-1}$

T_{dcm} = [2500] K

$\Delta_f H°(0\ K)$ = Unknown

$\Delta_f H°(298.15\ K) = -160.2 \pm 38\ kJ\ mol^{-1}$

Entholpy of Formation

Schissel and Trulson (1) used a mass spectrometer with Knudsen cells to study the vaporization of the titanium-boron system and obtained $\Delta_r G°(2340\ K) = -118.54\ kcal\ mol^{-1}$ for the reaction TiB(c) = Ti(g) + B(g). This corresponds to $\Delta_f G°$(TiB, cr, 2340 K) $= -33.266 \pm 9\ kcal\ mol^{-1}$, where 4 $kcal\ mol^{-1}$ of the uncertainty is due to the enthalpy of sublimation of boron. When this is reduced to 298.15 K one obtains $\Delta_f H°$(TiB, cr, 298.15 K) $= -38.3 \pm 9\ kcal\ mol^{-1}$. This value implies a high stability for TiB. This is also indicated by the phase studies of Nowotny et al. (2).

Heat Capacity and Entropy

The heat capacities fron 298 to 1200 K were estimated from that of TiB_2 by assuming the difference to be the same as that between CrB and CrB_2 as determined by R. Mezaki, E. W. Tilleux, D. W. Barnes and J. L. Margrave (Paper presented at the International Symposium on Nuclear Materials, Vienna, May 1962). A linear extrapolation was assumed above 1200 K.

The heat capacities above 298 K were also estimated from the relationship $C_p(Ti) + C_p(2B) - C_p(TiB_2) = 2/3\ [C_p(Ti) + C_p(B) - C_p(TiB)]$. This estimation agreed with the above to 4% over the temperature range 400 to 1000 K.

S°(TiB, cr, 298.15 K) = 8.3 $cal\ K^{-1}\ mol^{-1}$ was calculated by the method of Latimer, (3). Using JANAF S°(TiB_2, cr, 298.15 K) = 6.8 $cal^{-1}\ K^{-1}\ mol^{-1}$ and Latimer's value of 9.8 $cal\ K^{-1}\ mol^{-1}$ for the entropy contribution of Ti, an entropy contribution of -1.5 $cal\ K^{-1}\ mol^{-1}$ per boron atom was calculated.

Decomposition Data

The phase diagram given by McQuillan and McQuillan (4) was used to estimate the decomposition temperature. This diagram shows the decomposition products are TiB_2 and Ti_2B. However, Nowotny et al. (2) have shown Ti_2B to be nonexistent and so the decomposition products are not defined; probably a eutectic of TiB and TiB_2 is involved.

Reference

1. P. O. Schissel and O. C. Trulson, J. Phys. Chem. 66, 1492 (1962).
2. H. Nowotny, F. Benesovsky, C. Brukl and O. Schob, Monatshefte fur Chem. 92, 403 (1961).
3. W. M. Latimer, J. Am. Chem. Soc. 73, 1480 (1951).
4. A. D. McQuillan and M. K. McQuillan, "The Metallurgy of the Rarer Metals - No. 4 Titanium", Academic Press, New York (1956).

Titanium Boride (TiB) $B_1Ti_1(cr)$

Enthalpy Reference Temperature = T_r = 298.15 K Standard State Pressure = p° = 0.1 MPa

	J K⁻¹mol⁻¹			kJ mol⁻¹			
T/K	$C_p°$	S°	$-[G°-H°(T_r)]/T$	$H°-H°(T_r)$	$\Delta_f H°$	$\Delta_f G°$	Log K_f
0							
100							
200							
298.15	29.673	34.727	34.727	0.	-160.247	-159.691	27.977
300	29.916	34.911	34.728	0.055	-160.260	-159.687	27.804
400	40.334	45.065	36.041	3.610	-160.681	-159.420	20.818
500	45.714	54.690	38.824	7.933	-160.829	-159.085	16.620
600	48.568	63.297	42.200	12.658	-160.911	-158.729	13.819
700	50.124	70.910	45.769	17.599	-161.013	-158.358	11.817
800	50.911	77.657	49.341	22.652	-161.167	-157.968	10.314
900	51.488	83.688	52.829	27.774	-161.419	-157.555	9.144
1000	51.915	89.137	56.191	32.945	-161.831	-157.105	8.206
1100	52.191	94.099	59.415	38.152	-162.469	-156.604	7.436
1200	52.325	98.646	62.498	43.377	-167.327	-155.914	6.787
1300	52.459	102.839	65.442	48.617	-167.710	-154.948	6.226
1400	52.593	106.732	68.254	53.869	-168.206	-153.948	5.744
1500	52.727	110.365	70.941	59.135	-168.824	-152.909	5.325
1600	52.861	113.772	73.513	64.415	-169.576	-151.824	4.957
1700	52.995	116.981	75.976	69.707	-170.472	-150.688	4.630
1800	53.128	120.014	78.339	75.013	-171.524	-149.495	4.338
1900	53.262	122.890	80.609	80.333	-172.744	-148.239	4.075
2000	53.396	125.625	82.792	85.666	-188.831	-146.460	3.825
2100	53.530	128.233	84.894	91.012	-191.171	-144.284	3.589
2200	53.664	130.727	86.921	96.372	-193.532	-141.996	3.371
2300	53.798	133.115	88.878	101.745	-195.913	-139.600	3.170
2400	53.932	135.408	90.770	107.132	-248.574	-136.032	2.961
2500	54.066	137.612	92.599	112.531	-251.073	-131.291	2.743
2600	54.200	139.735	94.372	117.945	-253.558	-126.451	2.540
2700	54.333	141.783	96.090	123.371	-256.030	-121.515	2.351
2800	54.467	143.762	97.757	128.811	-258.489	-116.488	2.173
2900	54.601	145.675	99.377	134.265	-260.934	-111.373	2.006
3000	54.735	147.528	100.951	139.732	-263.366	-106.174	1.849
3100	54.869	149.325	102.483	145.212	-265.785	-100.895	1.700
3200	55.003	151.070	103.974	150.705	-268.190	-95.537	1.559
3300	55.137	152.764	105.427	156.212	-270.582	-90.105	1.426
3400	55.271	154.412	106.844	161.733	-272.960	-84.600	1.300
3500	55.405	156.016	108.226	167.267	-275.325	-79.026	1.179
3600	55.538	157.579	109.575	172.814	-277.677	-73.384	1.065
3700	55.672	159.102	110.893	178.374	-689.113	-59.888	0.845
3800	55.806	160.589	112.181	183.948	-690.204	-42.868	0.589
3900	55.940	162.040	113.441	189.535	-691.336	-25.818	0.346
4000	56.074	163.458	114.674	195.136	-692.509	-8.739	0.114

PREVIOUS: March 1963 CURRENT: June 1965

Titanium Boride (TiB) $B_1Ti_1(cr)$

$S°(298.15\ K) = 202.07 \pm 0.17\ J\ K^{-1}\ mol^{-1}$

$\Delta_f H°(0\ K) = 823.3 \pm 33.5\ kJ\ mol^{-1}$
$\Delta_f H°(298.15\ K) = 829.7 \pm 33.5\ kJ\ mol^{-1}$

Electronic and Molecular Constants (σ = 2)

State	ε_i, cm^{-1}	g_i	ω_e, cm^{-1}	$\omega_e x_e$, cm^{-1}	B_e, cm^{-1}	α_e, cm^{-1}	r_e, Å	Source
$^3\Sigma_g^-$	0.0	3	1061.61	9.536	1.236121	0.01442411	1.589	(1,2,3)
$^5\Sigma_u^-$	[1271.]	[5]	[1215.81]	[9.995]	[1.305]	[0.0113]	[1.546]	(4)
$1^3\Sigma_u^-$	[14829.]	[3]	[1114.83]	[16.318]	[1.275]	[0.0175]	[1.565]	(4)
$2^3\Sigma_u^-$	30573.4	3	946.59	2.652	1.183086	0.01133323	1.625	(1,2,3)

Enthalpy of Formation

The equilibrium vapor pressure of the reaction $B_2(g) \rightarrow B(g) + B(cr)$ was determined by Verhaegen and Drowart (5). Using the value, $p(B_2,g)/p(B,g) = 9 \times 10^{-5}$ at 2330 K where p is partial pressure, the enthalpy of reaction, $\Delta_r H°(298.15\ K)$, was evaluated by the 3rd law method to be -64.5 ± 5.0 kcal mol^{-1}. Using auxiliary data (6), this leads to $\Delta_f H°(298.15\ K) = 198.3 \pm 8.0$ kcal mol^{-1} for $B_2(g)$ or $D_0^0 = 2.97 \pm 0.61$ eV.

Heat Capacity and Entropy

The thermodynamic functions are calculated using first-order anharmonic corrections to Q_r^i and Q_v^i in the partition function $Q = Q_t \Sigma Q_r^i Q_v^i g_i \exp(-c_2 \varepsilon_i/T)$. Molecular and spectroscopic constants for the $^3\Sigma_g^-$ ground state and the $2^3\Sigma_u^-$ state are taken from a study (3) which revises the values of Douglas and Herzberg (1, 2). The molecular constants and electronic levels for the lowest $^5\Sigma_u^-$ state and $1^3\Sigma_u^-$ state are based on the ab initio configuration interaction calculations of Dupuis and Liu (4). These calculations, along with the optical absorption and ESR spectra of B_2 at 4 K reported by Graham and Weltner (7), support a $^3\Sigma_g^-$ ground state rather than $^5\Sigma_u^-$ favored by previous ab initio work (8, 9). All molecular and spectroscopic constants have been corrected to the average isotopic species.

Differences between our calculated thermodynamic functions and those given by Feber and Herrick (10) using direct summation techniques can be attributed to inclusion of the low-lying $^5\Sigma_u^-$ state and $1^3\Sigma_u^-$ state in our calculations. The JANAF adopted entropy is 0.35-1.6 cal K^{-1} mol^{-1} higher than their values in the 500-600 K temperature range. The JANAF adopted heat capacities are also higher in the low temperature range, 0.2-0.9 cal K^{-1} mol^{-1} larger between 298-1000 K. A comparison of the thermodynamic functions calculated using only the ground and upper $2^3\Sigma_u^-$ state shows no major numerical difference with this earlier tabulation (10), however.

References

1. A. E. Douglas and G. Herzberg, Phys. Rev. 57, 752 (1940).
2. A. E. Douglas and G. Herzberg, Can. J. Res. A18, 165 (1940).
3. R. W. Nicholls, P. A. Fraser and W. R. Jarmain, Combus. and Flame 3, 13 (1959).
4. M. DuPuis and B. Liu, J. Chem. Phys. 68, 2902 (1978).
5. G. Verhaegen and J. Drowart, J. Chem. Phys. 37, 1367 (1962).
6. JANAF Thermochemical Tables: B(cr) and B(g), 3-31-79.
7. W. R. M. Graham and W. Weltner, Jr., J. Chem. Phys. 65, 1516 (1976).
8. A. A. Padgett and V. Griffing, J. Chem. Phys. 30, 1286 (1959).
9. C. F. Bender and E. R. Davidson, J. Chem. Phys. 46, 3313 (1967).
10. R. C. Feber and C. C. Herrick, LASL Report LA-3597 (1966).

Enthalpy Reference Temperature = T_r = 298.15 K — Standard State Pressure = p° = 0.1 MPa

	J K^{-1}mol^{-1}			kJ mol^{-1}			
T/K	$C_p°$	S°	$-[G°-H°(T_r)]/T$	$H°-H°(T_r)$	$\Delta_f H°$	$\Delta_f G°$	Log K_f
0	0.	0.	INFINITE	-8.810	823.305	823.305	INFINITE
100	29.111	169.727	228.775	-5.905	826.162	809.251	-422.709
200	29.500	189.962	204.880	-2.984	828.418	791.393	-206.691
250	30.345	196.626	202.585	-1.490	829.169	782.044	-163.399
298.15	31.594	202.072	202.072	0.	829.687	772.918	-135.412
300	31.648	202.268	202.073	0.058	829.704	772.566	-134.516
350	33.187	207.260	202.464	1.679	830.070	763.012	-113.873
400	34.737	211.794	203.351	3.377	830.297	753.415	-98.386
450	36.146	215.969	204.524	5.150	830.415	743.797	-86.338
500	37.335	219.841	205.864	6.988	830.446	734.170	-76.698
600	39.012	226.809	208.788	10.813	830.309	714.924	-62.240
700	39.916	232.899	211.807	14.765	829.950	695.719	-51.915
800	40.310	238.259	214.785	18.779	829.399	676.578	-44.176
900	40.410	243.015	217.663	22.817	828.673	657.517	-38.161
1000	40.355	247.271	220.415	26.856	827.788	638.546	-33.354
1100	40.226	251.111	223.033	30.886	826.758	619.671	-29.426
1200	40.071	254.605	225.521	34.901	825.595	600.895	-26.156
1300	39.914	257.806	227.883	38.900	824.311	582.221	-23.394
1400	39.766	260.758	230.127	42.884	822.915	563.651	-21.030
1500	39.632	263.497	232.262	46.853	821.417	545.183	-18.985
1600	39.516	266.051	234.295	50.811	819.821	526.819	-17.199
1700	39.416	268.444	236.234	54.757	818.135	508.558	-15.626
1800	39.332	270.694	238.086	58.694	816.363	490.398	-14.231
1900	39.263	272.819	239.859	62.624	814.510	472.339	-12.986
2000	39.207	274.832	241.558	66.547	812.578	454.380	-11.867
2100	39.164	276.743	243.188	70.466	810.572	436.519	-10.858
2200	39.132	278.564	244.755	74.381	808.492	418.755	-9.943
2300	39.110	280.303	246.263	78.293	806.343	401.088	-9.109
2400	39.098	281.968	247.716	82.203	703.606	385.653	-8.394
2500	39.095	283.564	249.119	86.112	701.166	372.455	-7.782
2600	39.099	285.097	250.473	90.022	698.725	359.355	-7.220
2700	39.112	286.573	251.783	93.933	696.286	346.349	-6.701
2800	39.131	287.996	253.051	97.845	693.848	333.432	-6.220
2900	39.156	289.369	254.280	101.759	691.412	320.603	-5.775
3000	39.188	290.697	255.472	105.676	688.979	307.858	-5.360
3100	39.224	291.983	256.629	109.597	686.550	295.195	-4.974
3200	39.266	293.229	257.753	113.521	684.124	282.609	-4.613
3300	39.312	294.438	258.847	117.450	681.703	270.099	-4.275
3400	39.362	295.612	259.911	121.384	679.287	257.662	-3.958
3500	39.416	296.754	260.947	125.323	676.876	245.297	-3.661
3600	39.474	297.865	261.957	129.267	674.470	233.000	-3.381
3700	39.534	298.947	262.943	133.217	672.071	220.770	-3.117
3800	39.597	300.002	263.904	137.174	669.677	208.605	-2.867
3900	39.663	301.032	264.843	141.137	667.290	196.502	-2.632
4000	39.731	302.037	265.760	145.107	664.910	184.461	-2.409
4100	39.801	303.019	266.657	149.083	662.536	172.479	-2.197
4200	39.872	303.979	267.534	153.067	-299.529	174.604	-2.172
4300	39.945	304.918	268.393	157.058	-299.713	185.895	-2.258
4400	40.020	305.837	269.233	161.056	-299.894	197.190	-2.341
4500	40.095	306.737	270.057	165.062	-300.072	208.489	-2.420
4600	40.172	307.619	270.864	169.075	-300.247	219.793	-2.496
4700	40.249	308.484	271.655	173.096	-300.418	231.100	-2.568
4800	40.328	309.332	272.431	177.125	-300.588	242.411	-2.638
4900	40.406	310.164	273.193	181.162	-300.757	253.725	-2.705
5000	40.486	310.982	273.940	185.206	-300.924	265.042	-2.769
5100	40.565	311.784	274.675	189.259	-301.091	276.363	-2.831
5200	40.645	312.573	275.396	193.319	-301.258	287.688	-2.890
5300	40.725	313.348	276.105	197.388	-301.426	299.015	-2.947
5400	40.805	314.110	276.801	201.464	-301.596	310.346	-3.002
5500	40.885	314.859	277.487	205.549	-301.768	321.680	-3.055
5600	40.966	315.596	278.160	209.641	-301.942	333.017	-3.106
5700	41.046	316.322	278.824	213.742	-302.121	344.357	-3.156
5800	41.126	317.037	279.476	217.850	-302.303	355.700	-3.203
5900	41.206	317.740	280.119	221.967	-302.491	367.046	-3.250
6000	41.285	318.434	280.752	226.092	-302.685	378.396	-3.294

PREVIOUS: March 1979 (1 atm) CURRENT: March 1979 (1 bar)

BERYLLIUM BORATE ($Be(BO_2)_2$) IDEAL GAS M_r = 94.62978

$\Delta_f H°(0\ K) = -1348 \pm 42\ kJ\ mol^{-1}$

$S°(298.15\ K) = [326.79]\ J\ K^{-1}\ mol^{-1}$ $\Delta_f H°(298.15\ K) = -1351 \pm 42\ kJ\ mol^{-1}$

Vibrational Frequencies and Degeneracies

ν, cm^{-1}	ν, cm^{-1}	ν, cm^{-1}
[2000](1)	[1050](1)	[400](1)
[1600](1)	[650](1)	[350](1)
[1200](1)	[650](1)	[200](1)
[1200](1)	[600](1)	[150](1)
[1050](1)	[500](1)	[120](1)

Ground State Quantum Weight: [1] σ = 2

Point Point: [C_{2h}]

Bond Distance: Be-O = [1.63] Å O-B = [1.34] Å B=O = [1.20] Å

Bond Angle: O=B-O = [180]° O-Be-O = [180]° B-O-Be = [95]°

Product of the Moments of Inertia: $I_A I_B I_C = [5.382534 \times 10^{-113}]\ g^3\ cm^6$

Enthalpy of Formation

The enthalpy of formation was obtained from mass spectrometric studies; the following reactions at 1500°K were reported by Blackburn and Buchler (1).

Reaction	$\Delta_r H°$
1. $B_2O_3(\ell) \rightarrow B_2O_3(g)$	ΔH_1 = 92.226 kcal mol^{-1}
2. 1/3 $Be_3B_2O_6(cr)$ + 2/3 $B_2O_3(\ell) \rightarrow Be(BO_2)_2(g)$	ΔH_2 = 118 ± 2 kcal mol^{-1}
3. $Be_3B_2O_6(cr) \rightarrow$ 3 BeO(cr) + $B_2O_3(g)$	ΔH_3 = 112 ± 1 kcal mol^{-1}
4. $Be_3B_2O_6(cr) \rightarrow$ 2 BeO(cr) + $Be(BO_2)_2(g)$	ΔH_4 = 137 ± 3 kcal mol^{-1}
5. 3 BeO(cr) + $B_2O_3(\ell) \rightarrow Be_3B_2O_6(cr)$	ΔH_5 = -23 kcal mol^{-1}
6. BeO(cr) + $B_2O_3(g) \rightarrow Be(BO_2)_2(g)$	ΔH_6 = 22 kcal mol^{-1}

ΔH_1 was obtained from JANAF values (2) for $B_2O_3(\ell)$ and $B_2O_3(g)$. The values for ΔH_2, ΔH_3, and ΔH_4 were obtained from Blackburn and Buchler (1). ΔH_5 was obtained by taking the average of $\Delta H_5 = -20\ kcal\ mol^{-1}$ in a weight loss experiment, $\Delta H_5 = \Delta H_1 - \Delta H_3 = -19.774\ kcal\ mol^{-1}$, and $\Delta H_5 = 3/2\ (\Delta H_2 - \Delta H_4) = -28.5\ kcal\ mol^{-1}$. The ΔH_6 value was obtained by taking the average of $\Delta H_6 = \Delta H_4 - \Delta H_3 = 25\ kcal\ mol^{-1}$ and $\Delta H_6 = \Delta H_2 - \Delta H_1 + 1/3\ \Delta H_5 = 18.107\ kcal\ mol^{-1}$. The $\Delta_f H°(298.15\ K)$ of BeB_2O_4 was calculated from reaction 6 with auxiliary JANAF values.

Heat Capacity and Entropy

The vibrational frequencies estimated by comparison with B_2O_3 were adjusted to obtain an S°(1500 K) = 127 obtained in a manner analogous with $\Delta_f H°$ from the report of Blackburn and Buchler (1). The frequencies listed are not in point group order. All other molecular constants were estimated by comparison with related boron oxide molecules. The principal moments of inertia are $I_A = 8.5573 \times 10^{-39}\ g\ cm^2$, $I_B = 75.1462 \times 10^{-39}\ g\ cm^2$ and $I_C = 83.7035 \times 10^{-39}\ g\ cm^2$.

References

1. P. E. Blackburn and A. Buchler, Tech. Report No. 1 (1965).
2. JANAF Thermochemical Tables: $B_2O_3(\ell)$ and $B_2O_3(g)$, 6-30-71.

Beryllium Borate ($Be(BO_2)_2$) $B_2Be_1O_4(g)$

Enthalpy Reference Temperature = T_r = 298.15 K Standard State Pressure = p° = 0.1 MPa

T/K	$C_p°$ (J $K^{-1}mol^{-1}$)	S° (J $K^{-1}mol^{-1}$)	$-[G°-H°(T_r)]/T$ (J $K^{-1}mol^{-1}$)	$H°-H°(T_r)$ (kJ mol^{-1})	$\Delta_f H°$ (kJ mol^{-1})	$\Delta_f G°$ (kJ mol^{-1})	Log K_f
0	0.	0.	INFINITE	-18.639	-1348.345	-1348.345	INFINITE
100	52.679	250.301	396.063	-14.576	-1350.178	-1340.435	700.171
200	74.445	293.699	334.648	-8.190	-1350.857	-1330.401	347.465
250	83.774	311.333	328.252	-4.230	-1351.147	-1325.253	276.897
298.15	91.764	326.785	326.785	0.	-1351.432	-1320.240	231.301
300	92.053	327.354	326.787	0.170	-1351.443	-1320.047	229.841
350	99.380	342.106	327.935	4.960	-1351.739	-1314.791	196.222
400	105.839	355.807	330.573	10.094	-1352.029	-1309.492	171.002
450	111.510	368.608	334.096	15.530	-1352.309	-1304.158	151.383
500	116.475	380.619	338.154	21.233	-1352.572	-1298.794	135.684
600	124.603	402.607	347.099	33.305	-1353.048	-1287.993	112.130
700	130.814	422.303	356.460	46.090	-1353.475	-1277.116	95.300
800	135.589	440.096	365.821	59.420	-1353.880	-1266.180	82.673
900	139.296	456.289	374.987	73.172	-1354.296	-1255.193	72.849
1000	142.208	471.122	383.869	87.253	-1354.746	-1244.158	64.988
1100	144.522	484.788	392.430	101.594	-1355.254	-1233.074	58.554
1200	146.384	497.446	400.661	116.142	-1355.839	-1221.942	53.190
1300	147.899	509.225	408.564	130.859	-1356.517	-1210.757	48.649
1400	149.144	520.233	416.152	145.713	-1357.302	-1199.516	44.754
1500	150.179	530.559	423.438	160.681	-1358.206	-1188.215	41.377
1600	151.047	540.280	430.440	175.743	-1373.776	-1176.324	38.403
1700	151.781	549.459	437.174	190.886	-1374.547	-1163.960	35.764
1800	152.406	558.153	443.655	206.096	-1375.398	-1151.549	33.417
1900	152.943	566.408	449.901	221.364	-1376.333	-1139.087	31.316
2000	153.407	574.265	455.924	236.682	-1377.357	-1126.575	29.423
2100	153.811	581.760	461.739	252.043	-1378.476	-1114.008	27.709
2200	154.164	588.923	467.359	267.443	-1379.693	-1101.385	26.150
2300	154.475	595.783	472.794	282.875	-1381.011	-1088.707	24.725
2400	154.750	602.364	478.057	298.336	-1482.951	-1073.831	23.371
2500	154.994	608.686	483.156	313.824	-1484.633	-1056.750	22.080
2600	155.211	614.769	488.102	329.334	-1486.357	-1039.600	20.886
2700	155.406	620.631	492.903	344.865	-1488.124	-1022.383	19.779
2800	155.581	626.286	497.566	360.415	-1780.891	-998.878	18.634
2900	155.739	631.748	502.099	375.981	-1781.683	-970.936	17.488
3000	155.882	637.030	506.509	391.562	-1782.504	-942.965	16.418
3100	156.012	642.144	510.803	407.157	-1783.354	-914.965	15.417
3200	156.130	647.099	514.985	422.764	-1784.236	-886.940	14.478
3300	156.237	651.905	519.062	438.382	-1785.150	-858.886	13.595
3400	156.336	656.570	523.038	454.011	-1786.098	-830.802	12.764
3500	156.427	661.103	526.918	469.649	-1787.082	-802.692	11.980
3600	156.510	665.511	530.707	485.296	-1788.103	-774.552	11.238
3700	156.587	669.801	534.408	500.951	-1789.161	-746.384	10.537
3800	156.658	673.977	538.027	516.613	-1790.260	-718.184	9.872
3900	156.723	678.048	541.565	532.283	-1791.400	-689.957	9.241
4000	156.784	682.016	545.027	547.958	-1792.584	-661.701	8.641
4100	156.841	685.888	548.415	563.639	-1793.812	-633.413	8.070
4200	156.893	689.668	551.734	579.326	-2754.787	-591.047	7.351
4300	156.942	693.361	554.985	595.018	-2753.937	-539.539	6.554
4400	156.988	696.969	558.171	610.714	-2753.143	-488.052	5.794
4500	157.031	700.498	561.294	626.415	-2752.406	-436.582	5.068
4600	157.071	703.950	564.358	642.121	-2751.729	-385.127	4.373
4700	157.109	707.328	567.364	657.830	-2751.115	-333.685	3.708
4800	157.144	710.636	570.315	673.542	-2750.568	-282.256	3.072
4900	157.177	713.877	573.212	689.258	-2750.089	-230.837	2.461
5000	157.209	717.052	576.057	704.978	-2749.682	-179.428	1.874
5100	157.238	720.166	578.852	720.700	-2749.350	-128.026	1.311
5200	157.266	723.219	581.599	736.425	-2749.096	-76.632	0.770
5300	157.292	726.215	584.300	752.153	-2748.923	-25.238	0.249
5400	157.317	729.156	586.955	767.884	-2748.834	26.150	-0.253
5500	157.340	732.042	589.567	783.616	-2748.832	77.537	-0.736
5600	157.363	734.878	592.136	799.352	-2748.920	128.928	-1.203
5700	157.384	737.663	594.665	815.089	-2749.101	180.319	-1.652
5800	157.404	740.401	597.154	830.828	-2749.378	231.716	-2.087
5900	157.423	743.091	599.605	846.570	-2749.754	283.114	-2.507
6000	157.441	745.737	602.019	862.313	-2750.233	334.525	-2.912

PREVIOUS: June 1965 (1 atm) CURRENT: June 1965 (1 bar)

Beryllium Borate ($Be(BO_2)_2$) $B_2Be_1O_4(g)$

BERYLLIUM BORATE ($Be_3B_2O_6$) CRYSTAL M_r = 144.65294

$S°(298.15\ K)$ = [100 ± 13] J K^{-1} mol^{-1}

T_{fus} = 1768 ± 5 K

$\Delta_fH°(0\ K)$ = Unknown

$\Delta_fH°(298.15\ K)$ = -3104 ± 8 kJ mol^{-1}

$\Delta_{fus}H°$ = Unknown

Enthalpy of Formation

The adopted $\Delta_fH°(298.15\ K)$ = -741.96 ± 2 kcal mol^{-1} was calculated from $\Delta_rH°(298.15\ K)$ = -13.38 ± 0.5 kcal mol^{-1} for 3BeO(cr) + B_2O_3(g, ℓ) = $Be_3B_2O_6$(cr) measured calorimetrically by Gross (1). From equilibrium data relating to the same reaction at 1500 K, Blackburn and Buchler (2) report enthalpies of reaction of -21 ± 2 and -28.5 ± 5 kcal mol^{-1} which correspond to -21.7 and -29.2 kcal mol^{-1} at 298.15 K. The discrepancy of 8 to 16 kcal mol^{-1} is attributed primarily to the equilibrium values. These values result from differences between 2nd law enthalpies of reaction derived from the variation with temperature of ion intensities for $B_2O_3^+$ and $Be(BO_2)_2^+$, respectively, over the systems B_2O_3(ℓ) - $Be_3B_2O_6$(cr) and $Be_3B_2O_6$(cr) - BeO(cr). The reported reactions and their heats are:

	Reaction	$\Delta_rH°(1500\ K)$ kcal mol^{-1}
(a)	B_2O_3(ℓ) → B_2O_3(g)	91 ± 0.5
(b)	$Be_3B_2O_6$(cr) → 3 BeO(cr) + B_2O_3(g)	112 ± 1
(c)	1/3 $Be_3B_2O_6$(cr) + 2/3 B_2O_3(ℓ) → $Be(BO_2)$(g)	118 ± 2
(d)	$Be_3B_2O_6$(cr) → 2 BeO(cr) + $Be(BO_2)_2$(g)	137 ± 3

Heat Capacity and Entropy

The heat capacity was estimated from that for BeO(cr) and B_2O_3(cr). The adopted $S°(298.15\ K)$ = 24 ± 3 eu is an average value from the following three estimates. Blackburn and Buchler (2) report Kp = 2.7 x 10^{-5} atm and 9.3 x 10^{-6} atm at 1500 K for reactions a and b (see enthalpy of formation), based on vacuum balance data and ion intensities, respectively. Three Kp's yield $\Delta_rG°(1500\ K)$ = -3.17 kcal mol^{-1} for 3 BeO(cr) + B_2O_3(ℓ) → $Be_3B_2O_6$(cr). Combining this with the JANAF value for $\Delta_rH°(1500\ K)$ = -12.64 kcal mol^{-1}, we get $\Delta_rS°(1500\ K)$ = -6.3 and S°($Be_3B_2O_6$, cr, 298.15 K) = 22.9 cal K^{-1} mol^{-1}. This entropy agrees with 23 cal K^{-1} mol^{-1} estimated from the sum of entropies for the component oxides; however, the entropy for $Ca_3B_2O_6$(cr) observed by Kelley et al. (3) is larger than that of the component oxides by 2.5 cal K^{-1} mol^{-1}. Use of this increment leads to 25.5 cal K^{-1} mol^{-1} for $Be_3B_2O_6$(cr).

Fusion Data

T_{fus} was obtained from Blackburn and Buchler (2).

References

1. P. Gross, Fulmer Research Institute, Report No. 5 (1966).
2. P. E. Blackburn and A. Buchler, J. Phys. Chem. 69, 4250 (1965).
3. K. K. Kelley, S. S. Todd and C. H. Shomate, J. Am. Chem. Soc. 70, 1350 (1948).

Beryllium Borate ($Be_3B_2O_6$) $B_2Be_3O_6$(cr)

Enthalpy Reference Temperature = T_r = 298.15 K Standard State Pressure = p° = 0.1 MPa

T/K	$C_p°$ (J $K^{-1}mol^{-1}$)	S° (J $K^{-1}mol^{-1}$)	$-[G°-H°(T_r)]/T$ (J $K^{-1}mol^{-1}$)	$H°-H°(T_r)$ (kJ mol^{-1})	$\Delta_fH°$ (kJ mol^{-1})	$\Delta_fG°$ (kJ mol^{-1})	Log K_f
0							
100							
200							
298.15	139.746	100.416	100.416	0.	-3104.361	-2938.883	514.880
300	140.415	101.282	100.419	0.259	-3104.398	-2937.857	511.526
400	175.017	146.578	106.360	16.087	-3105.734	-2882.104	376.364
500	202.840	188.696	118.669	35.014	-3105.754	-2826.167	295.248
600	226.605	227.822	133.639	56.510	-3104.501	-2770.347	241.180
700	247.567	264.375	149.733	80.249	-3102.009	-2714.835	202.583
800	264.805	298.608	166.227	105.905	-3098.420	-2659.758	173.664
900	277.567	330.574	182.733	133.058	-3094.025	-2605.183	151.201
1000	286.269	360.294	199.021	161.272	-3089.170	-2551.124	133.257
1100	292.211	387.871	214.951	190.211	-3084.132	-2497.562	118.599
1200	296.395	413.483	230.442	219.650	-3079.092	-2444.461	106.405
1300	299.574	437.336	245.450	249.453	-3074.156	-2391.776	96.103
1400	302.210	459.635	259.961	279.544	-3069.382	-2339.466	87.287
1500	304.553	480.567	273.977	309.884	-3064.796	-2287.491	79.658
1600	306.729	500.292	287.512	340.448	-3104.023	-2234.237	72.940
1700	308.905	518.951	300.582	371.229	-3098.644	-2180.040	66.984
1800	311.206	536.674	313.210	402.235	-3093.254	-2126.161	61.700
1900	313.382	553.558	325.419	433.465	-3087.848	-2072.581	56.979
2000	315.474	569.686	337.232	464.907	-3082.437	-2019.285	52.738
2100	317.649	585.131	348.672	496.565	-3077.013	-1966.261	48.908
2200	319.658	599.955	359.759	528.431	-3071.583	-1913.494	45.432
2300	321.582	614.207	370.514	560.494	-3066.155	-1860.976	42.264
2400	323.423	627.932	380.956	592.745	-3161.256	-1806.557	39.319
2500	325.180	641.170	391.101	625.173	-3156.012	-1750.218	36.569
2600	327.105	653.962	400.966	657.789	-3150.710	-1694.091	34.035
2700	328.904	666.340	410.567	690.588	-3145.352	-1638.168	31.692
2800	330.829	678.337	419.917	723.575	-4012.804	-1563.777	29.173
2900	332.754	689.979	429.030	756.754	-4004.125	-1476.466	26.594
3000	334.720	701.293	437.918	790.127	-3995.326	-1389.454	24.193

PREVIOUS: CURRENT: September 1966

Beryllium Borate ($Be_3B_2O_6$) $B_2Be_3O_6$(cr)

DICHLOROBORANE $((BCl_2)_2)$ IDEAL GAS M_r = 163.432

$\Delta_fH°(0\ K)$ = Unknown

$S°(298.15\ K) = 358.96\ K\ J^{-1}\ mol^{-1}$ $\Delta_fH°(298.15\ K) = -489.1 \pm 5.0\ kJ\ mol^{-1}$

Vibrational Frequencies and Degeneracies

ν, cm^{-1}	ν, cm^{-1}	ν, cm^{-1}
1131(1)	Rotation	917(2)
401(1)	730(1)	617(2)
225(1)	291(1)	180(2)

Ground State Quantum Weight = 1 σ = 4

Point Group = D_{2d}

Bond Distances: B-Cl = 1.73 ± 0.02 Å B-B = 1.75 ± 0.01 Å

Bond Angles: Cl-B-Cl = 120 ± 2°

Dihedral angle between two BCl_2 planes = 90°

Product of the Moments of Inertia: $I_AI_BI_C = 5.336257 \times 10^{-112}\ g^3\ cm^6$

Enthalpy of Formation

Gunn et al. (1) determined the enthalpy change ($\Delta_rH°$) of the reaction $B_2Cl_4(\ell) + Cl_2(g) = 2\ BCl_3(g)$. From the value, $\Delta_rH°(298.15\ K) = -67.3 \pm 0.6\ kcal\ mol^{-1}$, the enthalpy of formation for $B_2Cl_4(\ell)$ was calculated as $-125.32 \pm 1.20\ kcal\ mol^{-1}$. Based on $\Delta_{vap}H°(338.7\ K) = 8.029\ kcal\ mol^{-1}$ reported by Urry et al. (2), the enthalpy of formation for $B_2Cl_4(g)$ was derived to be $-116.9 \pm 1.2\ kcal\ mol^{-1}$.

Heat Capacity and Entropy

The vibrational frequencies, molecular structure and constants were taken from Mann and Fano (3). The barrier restricting internal rotation about the B-B single bond is fixed at $1.53 \pm 0.60\ kcal\ mol^{-1}$ obtained from Mason and Kreevoy (4). The reduced moment of inertia for BCl_2 top is $I_r = 13.2149 \times 10^{-39}\ g\ cm^2$. The principal moments of inertia are: $I_A = 52.8594 \times 10^{-39}\ g\ cm^2$, $I_B = I_C = 100.4748 \times 10^{-39}\ g\ cm^2$.

References

1. S. R. Gunn, L. G. Green, and A. I. Von Egidy, J. Phys. Chem. 63, 1787 (1959).
2. G. Urry, L. Wartik, R. E. Moore and H. I. Schlesinger, J. Am. Chem. Soc. 76, 5293 (1954).
3. D. E. Mann and L. Fano, J. Chem. Phys. 26, 1665 (1957).
4. E. A. Mason and M. M. Kreevoy, J. Am. Chem. Soc. 77, 5808 (1955).

Dichloroborane $((BCl_2)_2)$ $B_2Cl_4(g)$

Enthalpy Reference Temperature = T_r = 298.15 K Standard State Pressure = p° = 0.1 MPa

T/K	$C_p°$ (J K⁻¹mol⁻¹)	S° (J K⁻¹mol⁻¹)	$-[G°-H°(T_r)]/T$ (J K⁻¹mol⁻¹)	$H°-H°(T_r)$ (kJ mol⁻¹)	$\Delta_fH°$ (kJ mol⁻¹)	$\Delta_fG°$ (kJ mol⁻¹)	Log K_f
0							
100	59.208	275.789	433.035	-15.725	-489.916	-479.632	250.534
200	80.655	323.936	367.210	-8.655	-489.588	-469.422	122.600
298.15	94.839	358.956	358.956	0.	-489.110	-459.631	80.526
300	95.056	359.543	358.958	0.176	-489.102	-459.448	79.997
400	104.801	388.301	362.812	10.195	-488.747	-449.622	58.715
500	111.307	412.425	370.388	21.019	-488.528	-439.868	45.953
600	115.683	433.124	379.160	32.378	-488.395	-430.149	37.448
700	118.541	451.179	388.187	44.095	-488.333	-420.448	31.374
800	120.792	467.162	397.079	56.066	-488.326	-410.751	26.819
900	122.242	481.475	405.675	68.220	-488.377	-401.052	23.276
1000	123.474	494.420	413.913	80.507	-488.487	-391.343	20.442
1100	124.199	506.223	421.776	92.892	-488.663	-381.621	18.122
1200	124.851	517.058	429.271	105.345	-488.911	-371.880	16.188
1300	125.430	527.075	436.414	117.859	-489.224	-362.115	14.550
1400	125.936	536.389	443.226	130.428	-489.598	-352.323	13.145
1500	126.369	545.093	449.730	143.044	-490.033	-342.503	11.927
1600	126.621	553.257	455.948	155.694	-490.535	-332.652	10.860
1700	126.854	560.940	461.900	168.368	-491.111	-322.767	9.917
1800	127.069	568.197	467.606	181.064	-491.759	-312.846	9.079
1900	127.266	575.073	473.083	193.781	-492.480	-302.887	8.327
2000	127.445	581.605	478.347	206.516	-493.274	-292.889	7.649
2100	127.548	587.826	483.413	219.266	-494.147	-282.848	7.035
2200	127.645	593.762	488.295	232.026	-495.102	-272.764	6.476
2300	127.738	599.438	493.005	244.795	-496.143	-262.635	5.965
2400	127.825	604.876	497.554	257.573	-597.789	-250.321	5.448
2500	127.907	610.096	501.952	270.360	-599.162	-235.815	4.927
2600	127.984	615.114	506.208	283.155	-600.561	-221.254	4.445
2700	128.056	619.946	510.332	295.957	-601.989	-206.637	3.998
2800	128.122	624.604	514.330	308.766	-603.448	-191.968	3.581
2900	128.184	629.101	518.211	321.581	-604.939	-177.246	3.193
3000	128.240	633.448	521.980	334.402	-606.464	-162.473	2.829
3100	128.273	637.653	525.644	347.228	-608.022	-147.647	2.488
3200	128.305	641.726	529.208	360.057	-609.615	-132.771	2.167
3300	128.336	645.675	532.678	372.889	-611.241	-117.844	1.865
3400	128.365	649.506	536.058	385.724	-612.897	-102.869	1.580
3500	128.393	653.228	539.353	398.562	-614.580	-87.843	1.311
3600	128.419	656.845	542.567	411.402	-616.287	-72.769	1.056
3700	128.444	660.364	545.703	424.245	-618.014	-57.648	0.814
3800	128.468	663.790	548.766	437.091	-619.756	-42.480	0.584
3900	128.491	667.127	551.758	449.939	-621.509	-27.266	0.365
4000	128.512	670.380	554.683	462.789	-623.268	-12.006	0.157
4100	128.527	673.554	557.544	475.641	-625.027	3.298	-0.042
4200	128.543	676.651	560.343	488.495	-1586.482	32.692	-0.407
4300	128.557	679.676	563.083	501.350	-1586.054	71.239	-0.865
4400	128.572	682.632	565.767	514.206	-1585.615	109.775	-1.303
4500	128.585	685.521	568.396	527.064	-1585.161	148.302	-1.721
4600	128.598	688.347	570.973	539.923	-1584.688	186.818	-2.121
4700	128.611	691.113	573.500	552.784	-1584.192	225.323	-2.504
4800	128.623	693.821	575.978	565.645	-1583.669	263.819	-2.871
4900	128.635	696.473	578.410	578.508	-1583.116	302.302	-3.223
5000	128.645	699.072	580.798	591.372	-1582.530	340.774	-3.560
5100	128.656	701.620	583.142	604.237	-1581.908	379.234	-3.884
5200	128.666	704.118	585.444	617.103	-1581.248	417.681	-4.196
5300	128.675	706.569	587.707	629.970	-1580.547	456.116	-4.495
5400	128.684	708.974	589.930	642.838	-1579.806	494.537	-4.784
5500	128.692	711.336	592.116	655.707	-1579.021	532.944	-5.061
5600	128.700	713.655	594.266	668.577	-1578.192	571.335	-5.329
5700	128.707	715.933	596.381	681.447	-1577.318	609.712	-5.587
5800	128.713	718.171	598.461	694.318	-1576.399	648.073	-5.837
5900	128.719	720.371	600.509	707.190	-1575.435	686.418	-6.077
6000	128.725	722.535	602.525	720.062	-1574.424	724.745	-6.309

PREVIOUS: December 1964 (1 atm) CURRENT: December

Dichloroborane $((BCl_2)_2)$ $B_2Cl_4(g)$

DIFLUOROBORANE $((BF_2)_2)$ IDEAL GAS M_r = 97.613612

$\Delta_fH°$(0 K) = Unknown

S°(298.15 K) = [318.47] J K^{-1} mol^{-1}

$\Delta_fH°$(298.15 K) = -1431.8 ± 4.2 kJ mol^{-1}

Vibrational Frequencies and Degeneracies

ν, cm^{-1}	ν, cm^{-1}	ν, cm^{-1}
1380 (1)	1155(1)	325(1)
[630](1)	542(1)	1378(1)
[304](1)	1386(1)	662(1)
Rotation	657(1)	325(1)

Ground State Quantum Weight: 1 σ = 4

Point Group: D_{2d}

Bond Distances: B-B = 1.67 Å B-F = 1.32 Å

Bond Angles: F-B-F = 120° F-B-B = 120°

Dihedral angle between two BF_2 planes = 90°

Product of the Moments of Inertia: $I_AI_BI_C$ = 2.502146 x 10^{-113} g^3 cm^6

Enthalpy of Formation

The enthalpy of formation was calculated based on the value, $\Delta_rH°$ = -82.0 ± 1.0 kcal mol^{-1} for the reaction $B_2F_4(g) + Cl_2(g) = 0.92\ BF_3(g) + 0.46\ BF_2Cl(g) + 0.32\ BFCl_2(g) + 0.30\ BCl_3(g)$ reported by Gunn and Green (1).

Heat Capacity and Entropy

The molecular and spectroscopic constants were obtained from Gayles and Self (2). The vibrational frequencies were corrected to the average isotropic species. The molecular configuration was reported to be staggered, analogous to that observed for $B_2Cl_4(g)$. The angle between two B-BF_2 planes is approximately 90°. The thermodynamic functions were evaluated based on an assumption that there is free rotation at the B-B bond. The reduced moment of inertia for the BF_2 top is I_r = 4.1231 x 10^{-39} g cm^2. The principal moments of inertia are: I_A = 16.4908 x 10^{-39} g cm^2, $I_B = I_C$ = 38.9529 x 10^{-39} g cm^2.

References

1. S. R. Gunn and L. G. Green, J. Phys. Chem. 65, 178 (1961).
2. J. N. Gayles and J. Self, J. Chem. Phys. 40, 3530 (1964).

Difluoroborane $((BF_2)_2)$ $B_2F_4(g)$

Enthalpy Reference Temperature = T_r = 298.15 K Standard State Pressure = p° = 0.1 MPa

T/K	$C_p°$ (J $K^{-1}mol^{-1}$)	S° (J $K^{-1}mol^{-1}$)	-[G°-H°(T_r)]/T (J $K^{-1}mol^{-1}$)	H°-H°(T_r) (kJ mol^{-1})	$\Delta_fH°$ (kJ mol^{-1})	$\Delta_fG°$ (kJ mol^{-1})	Log K_f
0							
100	43.322	255.051	376.204	-12.115	-1429.664	-1421.034	742.272
200	61.947	290.817	325.053	-6.847	-1430.920	-1411.855	368.739
250	70.075	305.535	319.702	-3.542	-1431.362	-1407.037	293.984
298.15	76.891	318.473	318.473	0.	-1431.765	-1402.315	245.680
300	77.135	318.950	318.475	0.142	-1431.780	-1402.133	244.133
350	83.308	331.314	319.437	4.157	-1432.198	-1397.158	208.514
400	88.716	342.799	321.648	8.461	-1432.626	-1392.124	181.793
450	93.438	353.528	324.601	13.017	-1433.059	-1387.035	161.003
500	97.541	363.590	328.002	17.794	-1433.485	-1381.898	144.366
600	104.163	381.988	335.496	27.895	-1434.294	-1371.504	119.400
700	109.105	398.435	343.334	38.571	-1435.032	-1360.979	101.558
800	112.815	413.257	351.163	49.675	-1435.712	-1350.354	88.169
900	115.634	426.715	358.822	61.104	-1436.355	-1339.645	77.751
1000	117.806	439.016	366.235	72.781	-1436.982	-1328.866	69.413
1100	119.505	450.327	373.372	84.650	-1437.613	-1318.024	62.588
1200	120.855	460.785	380.227	96.670	-1438.262	-1307.123	56.898
1300	121.940	470.503	386.802	108.812	-1438.942	-1296.168	52.081
1400	122.824	479.573	393.108	121.051	-1439.663	-1285.159	47.950
1500	123.553	488.073	399.158	133.371	-1440.433	-1274.096	44.368
1600	124.161	496.067	404.968	145.758	-1441.256	-1262.980	41.232
1700	124.671	503.610	410.551	158.200	-1442.137	-1251.811	38.463
1800	125.104	510.748	415.921	170.690	-1443.077	-1240.589	36.001
1900	125.475	517.522	421.091	183.219	-1444.074	-1229.313	33.796
2000	125.794	523.967	426.075	195.783	-1445.127	-1217.982	31.810
2100	126.070	530.111	430.884	208.376	-1446.233	-1206.598	30.012
2200	126.311	535.982	435.529	220.996	-1447.387	-1195.160	28.377
2300	126.523	541.601	440.019	233.638	-1448.583	-1183.668	26.882
2400	126.710	546.990	444.365	246.300	-1550.335	-1169.986	25.464
2500	126.875	552.166	448.574	258.979	-1551.754	-1154.109	24.114
2600	127.023	557.145	452.655	271.674	-1553.134	-1138.176	22.866
2700	127.154	561.941	456.614	284.383	-1554.468	-1122.191	21.710
2800	127.273	566.568	460.459	297.104	-1555.751	-1106.157	20.636
2900	127.379	571.036	464.195	309.837	-1556.980	-1090.078	19.634
3000	127.475	575.356	467.829	322.580	-1558.150	-1073.958	18.699
3100	127.563	579.537	471.366	335.332	-1559.257	-1057.800	17.824
3200	127.642	583.588	474.810	348.092	-1560.299	-1041.607	17.002
3300	127.715	587.517	478.166	360.860	-1561.272	-1025.383	16.230
3400	127.781	591.331	481.438	373.635	-1562.176	-1009.130	15.503
3500	127.842	595.036	484.631	386.416	-1563.008	-992.851	14.817
3600	127.898	598.638	487.748	399.203	-1563.768	-976.550	14.169
3700	127.949	602.143	490.793	411.996	-1564.455	-960.229	13.556
3800	127.997	605.556	493.768	424.793	-1565.069	-943.890	12.975
3900	128.040	608.881	496.678	437.595	-1565.610	-927.536	12.423
4000	128.081	612.124	499.523	450.401	-1566.079	-911.169	11.899
4100	128.119	615.287	502.308	463.211	-1566.475	-894.791	11.400
4200	128.154	618.374	505.035	476.025	-2526.500	-864.356	10.750
4300	128.187	621.390	507.706	488.842	-2524.581	-824.804	10.019
4400	128.218	624.338	510.324	501.662	-2522.597	-785.297	9.323
4500	128.246	627.219	512.889	514.485	-2520.551	-745.836	8.657
4600	128.273	630.038	515.406	527.311	-2518.443	-706.421	8.022
4700	128.298	632.797	517.874	540.140	-2516.276	-667.052	7.413
4800	128.322	635.499	520.296	552.971	-2514.052	-627.730	6.831
4900	128.344	638.145	522.675	565.804	-2511.773	-588.456	6.273
5000	128.365	640.738	525.010	578.639	-2509.442	-549.228	5.738
5100	128.384	643.280	527.304	591.477	-2507.060	-510.047	5.224
5200	128.403	645.773	529.559	604.316	-2504.631	-470.913	4.730
5300	128.420	648.219	531.774	617.157	-2502.156	-431.827	4.256
5400	128.437	650.620	533.953	630.000	-2499.638	-392.788	3.799
5500	128.452	652.977	536.096	642.845	-2497.079	-353.796	3.360
5600	128.467	655.291	538.204	655.691	-2494.481	-314.850	2.937
5700	128.481	657.565	540.278	668.538	-2491.847	-275.952	2.529
5800	128.495	659.800	542.319	681.387	-2489.178	-237.100	2.135
5900	128.507	661.997	544.329	694.237	-2486.478	-198.294	1.756
6000	128.519	664.157	546.308	707.088	-2483.748	-159.535	1.389

PREVIOUS: December 1964 (1 atm) CURRENT: December 1964 (1 bar)

Difluoroborane $((BF_2)_2)$ $B_2F_4(g)$

DIFLUOROBORANE OXIDE ($O(BF_2)_2$) IDEAL GAS M_r = 113.613012

$\Delta_f H°$(0 K) = Unknown

$S°$(298.15 K) = [337.52 ± 8.4] K J^{-1} mol^{-1} $\Delta_f H°$(298.15 K) = -1898.07 ± 8.4 kJ mol^{-1}

Vibrational Frequencies and Degeneracies

ν, cm^{-1}	ν, cm^{-1}	ν, cm^{-1}
[1400](1)	[145](1)	[1370](1)
[670](1)	[800](1)	[145](1)
[320](1)	[660](1)	[1800](1)
rotation	[660](1)	[1155](1)
[1370](1)	[1600](1)	[540](1)

Ground State Quantum Weight: 1 σ = 4

Point Group: D_{2h}

Bond Distances: B-O = [1.4] Å B-F = [1.32] Å

Bond Angles: B-O-B = [180]° F-B-F = [120]°

Dihedral angle between the two BF_2 planes = 90°

Product of the Moments of Inertia: $I_A I_B I_C$ = [7.701231 x 10^{-113}] g^3 cm^6

Enthalpy of Formation

Bidinosti and Coatsworth (1) have reported $\Delta_r H°$(1100 K) = 18.5 ± 3 kcal mol^{-1} for the reaction 4 BF_3(g) + B_2O_3(ℓ) = 3 B_2OF_4(g). This value was obtained from a mass spectrometric study of the variation of the ion intensities with temperature of the BF_3^+ and $B_2OF_4^+$ species. Employing JANAF auxiliary values (2) we obtain $\Delta_f H°$(B_2F_4O, g, 1100 K) = -455.75 ± 2 kcal mol^{-1} which reduces to $\Delta_f H°$(298.15 K) = -453.65 ± 2 kcal mol^{-1}.

Heat Capacity and Entropy

By analogy with B_2F_4(g) (3) free rotation of the BF_2 groups is assumed. Because of the large separation of the two BF_2 groups the molecule is considered to be planar. The B-O distance is assumed to be that in BF_2O(g) (2) and the B-F distance that in B_2F_4(g) (2). The vibrational frequencies are estimated by comparison with those for B_2F_4 (3); the values are given above in the order $3A_g$, A_u, $2B_{1g}$, $2B_{1u}$, B_{2g}, $3B_{2u}$, $3B_{3u}$. The thermodynamic functions were evaluated based on an assumption that there is free rotation at the two B-O bonds. The reduced moment of inertia for the BF_2 top is I_r = 4.1227 x 10^{-39} g cm^2. The principal moments of inertia are I_A = 16.4908 x 10^{-39} g cm^2, I_B = 60.5877 x 10^{-39} g cm^2 and I_C = 77.0785 x 10^{-39} g cm^2.

References

1. D. R. Bidinosti and L. L. Coatsworth, Can. J. Chem. 48, 2484 (1970).
2. JANAF Thermochemical Tables: BF_3(g), 6-30-69; B_2O_3(ℓ), 6-30-71; BF_2O(g), 12-31-66; B_2F_4(g), 12-31-64.
3. L. A. Nimon, K. S. Seshadri, D. White and R. C. Taylor, AD-693614, 1969.

Difluoroborane Oxide ($O(BF_2)_2$) $B_2F_4O_1$(g)

Enthalpy Reference Temperature = T_r = 298.15 K Standard State Pressure = p° = 0.1 MPa

T/K	$C_p°$ (J $K^{-1}mol^{-1}$)	S° (J $K^{-1}mol^{-1}$)	-[G°-H°(T_r)]/T (J $K^{-1}mol^{-1}$)	H°-H°(T_r) (kJ mol^{-1})	$\Delta_f H°$ (kJ mol^{-1})	$\Delta_f G°$ (kJ mol^{-1})	Log K_f
0							
100	51.315	267.752	399.768	-13.202	-1894.168	-1878.142	981.041
200	66.790	307.999	344.533	-7.307	-1896.253	-1861.276	486.115
250	74.652	323.753	338.831	-3.770	-1897.191	-1852.422	387.042
298.15	81.853	337.523	337.523	0.	-1898.072	-1843.720	323.012
300	82.121	338.031	337.525	0.152	-1898.105	-1843.382	320.961
350	89.101	351.220	338.551	4.434	-1898.993	-1834.191	273.738
400	95.547	363.545	340.913	9.053	-1899.853	-1824.875	238.304
450	101.427	375.144	344.078	13.980	-1900.674	-1815.453	210.732
500	106.733	386.110	347.738	19.186	-1901.443	-1805.942	188.665
600	115.686	406.393	355.852	30.325	-1902.793	-1786.711	155.547
700	122.698	424.775	364.406	42.258	-1903.900	-1767.273	131.876
800	128.153	441.531	373.015	54.812	-1904.799	-1747.692	114.113
900	132.412	456.881	381.493	67.849	-1905.537	-1728.007	100.291
1000	135.763	471.012	389.748	81.265	-1906.157	-1708.248	89.230
1100	138.428	484.082	397.737	94.979	-1906.696	-1688.430	80.177
1200	140.570	496.222	405.444	108.933	-1907.187	-1668.566	72.631
1300	142.311	507.544	412.868	123.080	-1907.653	-1648.662	66.244
1400	143.742	518.145	420.013	137.385	-1908.116	-1628.722	60.768
1500	144.929	528.104	426.890	151.820	-1908.591	-1608.749	56.022
1600	145.923	537.490	433.512	166.364	-1909.090	-1588.743	51.867
1700	146.763	546.362	439.892	180.999	-1909.624	-1568.706	48.200
1800	147.479	554.772	446.043	195.712	-1910.197	-1548.635	44.940
1900	148.093	562.762	451.977	210.492	-1910.815	-1528.532	42.022
2000	148.623	570.372	457.708	225.328	-1911.477	-1508.395	39.395
2100	149.084	577.635	463.247	240.214	-1912.183	-1488.224	37.018
2200	149.487	584.580	468.606	255.143	-1912.931	-1468.018	34.855
2300	149.841	591.233	473.794	270.110	-1913.718	-1447.777	32.880
2400	150.154	597.617	478.821	285.110	-2015.058	-1425.363	31.022
2500	150.432	603.752	483.696	300.139	-2016.065	-1400.772	29.268
2600	150.679	609.657	488.428	315.195	-2017.031	-1376.141	27.647
2700	150.901	615.348	493.024	330.274	-2017.954	-1351.474	26.146
2800	151.100	620.840	497.492	345.375	-2018.827	-1326.773	24.751
2900	151.280	626.145	501.837	360.494	-2019.648	-1302.042	23.452
3000	151.442	631.277	506.067	375.630	-2020.413	-1277.284	22.239
3100	151.590	636.245	510.186	390.782	-2021.118	-1252.501	21.105
3200	151.724	641.060	514.201	405.948	-2021.762	-1227.697	20.040
3300	151.847	645.730	518.116	421.126	-2022.340	-1202.873	19.040
3400	151.959	650.265	521.937	436.317	-2022.852	-1178.033	18.098
3500	152.062	654.672	525.667	451.518	-2023.296	-1153.179	17.210
3600	152.156	658.957	529.310	466.729	-2023.672	-1128.313	16.371
3700	152.244	663.127	532.870	481.949	-2023.978	-1103.437	15.578
3800	152.324	667.188	536.352	497.177	-2024.215	-1078.555	14.826
3900	152.399	671.146	539.758	512.413	-2024.383	-1053.666	14.112
4000	152.468	675.005	543.091	527.657	-2024.482	-1028.774	13.434
4100	152.532	678.770	546.354	542.907	-2024.513	-1003.882	12.790
4200	152.592	682.447	549.551	558.163	-2984.176	-964.940	12.001
4300	152.647	686.038	552.683	573.425	-2981.899	-916.890	11.138
4400	152.699	689.548	555.754	588.692	-2979.561	-868.893	10.315
4500	152.748	692.980	558.766	603.965	-2977.164	-820.950	9.529
4600	152.793	696.338	561.720	619.242	-2974.710	-773.062	8.778
4700	152.836	699.624	564.619	634.523	-2972.202	-725.227	8.060
4800	152.876	702.842	567.466	649.809	-2969.640	-677.445	7.372
4900	152.914	705.995	570.261	665.098	-2967.029	-629.718	6.713
5000	152.949	709.085	573.006	680.392	-2964.371	-582.045	6.081
5100	152.983	712.114	575.704	695.688	-2961.667	-534.425	5.474
5200	153.014	715.085	578.356	710.988	-2958.922	-486.859	4.891
5300	153.044	718.000	580.964	726.291	-2956.136	-439.345	4.330
5400	153.072	720.861	583.528	741.597	-2953.314	-391.886	3.791
5500	153.099	723.670	586.050	756.905	-2950.458	-344.479	3.272
5600	153.124	726.428	588.533	772.216	-2947.571	-297.122	2.771
5700	153.148	729.139	590.976	787.530	-2944.655	-249.820	2.289
5800	153.171	731.803	593.381	802.846	-2941.713	-202.567	1.824
5900	153.192	734.421	595.749	818.164	-2938.749	-155.366	1.376
6000	153.213	736.996	598.082	833.484	-2935.764	-108.215	0.942

PREVIOUS: June 1971 (1 atm) CURRENT: June 1971 (1 bar)

Difluoroborane Oxide ($O(BF_2)_2$) $B_2F_4O_1$(g)

DIHYDROXYBORANE ($(B(OH)_2)_2$) CRYSTAL M_r = 89.64936

Dihydroxyborane ($(B(OH)_2)_2$) $B_2H_4O_4$(cr)

$S°(298.15\ K)$ = [126 ± 4] K J^{-1} mol^{-1}

T_{fus} = Unknown

$\Delta_f H°(0\ K)$ = Unknown

$\Delta_f H°(298.15\ K)$ = -1410 ± 8 kJ mol^{-1}

$T_{fus}H°$ = Unknown

Enthalpy of Formation

The enthalpy change, $\Delta_r H°(298.15\ K)$ = -91.3 ± 1 kcal mol^{-1}, for the reaction $B_2(OH)_4(cr)$ + (m + 2)·$AgNO_3(aq)$ + (n + 2)·$H_2O(\ell)$ = 2 Ag(cr) + [2 H_3BO_3 + 2 HNO_3 + m$AgNO_3$]·n$H_2O(aq)$ was determined electrically by Finch et al. (1). The enthalpy of formation of $B_2(OH)_4(cr)$ was reported to be -335.1 kcal mol^{-1}. This value was corrected to be -337.1 ± 2 kcal mol^{-1} by use of more recent data on the $\Delta_f H°(298.15\ K)$ values used for calculation.

Heat Capacity and Entropy

The heat capacity for $B_2(OH)_4(cr)$ was estimated by comparison with those for $B(OH)_3(cr)$. $S°(298.15\ K)$ was calculated based on 3 cal $K^{-1}mol^{-1}$ per atom for $B_2(OH)_4(cr)$, which was obtained from the corresponding values for $HBO_2(cr)$ and $B(OH)_3(cr)$.

Sublimation Data

The value, $\Delta_{sub}H°(298.15\ K)$, was calculated as the difference between $\Delta_f H°(298.15\ K)$ for $B_2(OH)_4(g)$ and $B_2(OH)_4(cr)$.

Reference

1. A. Finch, P. J. Gardner, and I. J. Hyams, Trans. Faraday Soc. 61, 649 (1965).

Enthalpy Reference Temperature = T_r = 298.15 K   Standard State Pressure = p° = 0.1 MPa

	J $K^{-1}mol^{-1}$			kJ mol^{-1}			
T/K	$C_p°$	S°	$-[G°-H°(T_r)]/T$	$H°-H°(T_r)$	$\Delta_f H°$	$\Delta_f G°$	Log K_f
0							
100							
200							
298.15	116.190	125.520	125.520	0.	-1410.426	-1244.118	217.964
300	116.775	126.241	125.522	0.215	-1410.468	-1243.086	216.441
400	143.135	163.582	130.435	13.259	-1411.904	-1187.032	155.010
500	163.762	197.787	140.535	28.626	-1411.962	-1130.781	118.132
600	181.711	229.271	152.735	45.922	-1410.806	-1074.634	93.555
700	197.025	258.456	165.779	64.874	-1408.549	-1018.770	76.022
800	210.413	285.665	179.083	85.266	-1405.302	-963.298	62.897
900	221.459	311.098	192.354	106.869	-1401.222	-908.284	52.715
1000	231.333	334.953	205.434	129.519	-1396.428	-853.764	44.596
1100	239.994	357.417	218.239	153.095	-1391.008	-799.755	37.977
1200	247.442	378.626	230.729	177.476	-1385.059	-746.264	32.484
1300	253.801	398.691	242.885	202.548	-1378.678	-693.287	27.857
1400	258.990	417.697	254.699	228.197	-1371.963	-640.814	23.909
1500	263.006	435.709	266.171	254.307	-1365.021	-588.831	20.505

PREVIOUS: CURRENT: March 1966

Dihydroxyborane ($(B(OH)_2)_2$) $B_2H_4O_4$(cr)

DIHYDROXYBORANE ((B(OH)$_2$)$_2$) IDEAL GAS M_r = 89.64936

$S°$(298.15 K) = [348.93] J K^{-1} mol^{-1}

$\Delta_f H°$(0 K) = Unknown
$\Delta_f H°$(298.15 K) = [-1284 ±20] kJ mol^{-1}

Vibrational Frequencies and Degeneracies

ν, cm^{-1}	ν, cm^{-1}	ν, cm^{-1}
[1380](1)	[1386](1)	[824](2)
[630](1)	[657](1)	[209](2)
[304](1)	[325](1)	[3200](4)
Rotation	[1378](1)	[1200](2)
[1155](1)	[662](1)	[880](2)
[542](1)	[325](1)	

Ground State Quantum Weight: 1 σ = 2

Point Group: [C_2]

Bond Distances: B-B = [1.67] Å B-O = [1.36] Å O-H = [0.96] Å

Bond Angles: H-O-B = [114]° O-B-O = [120]° O-B-B = [120]°

Dihedral angle between two $B(OH)_2$ planes = 90°

Product of the Moments of Inertia: $I_A I_B I_C$ = [2.299983 x 10^{-113}] g^3 cm^6

Enthalpy of Formation

The enthalpy of formation, $\Delta_f H°(B_2(OH)_4$, g, 298.15 K), is calculated using bond energies, D(B-OH) = 132 and D(B-B) = 78 kcal mol^{-1}. The bond energy D(B-OH) was taken from that in $B(OH)_3$(g) and D(B-B) was estimated by comparison with that in B_2F_4(g).

Heat Capacity and Entropy

The molecular structure was assumed to be the same as that for B_2F_4(g), i.e. the two $B(OH)_2$ units have staggered configuration, and free rotation about B-B bond. The B-B bond distance was taken as that in B_2F_4 molecule. Those for B-O and B-H bond and bond angles were adopted from H_3BO_3 molecule. Eleven vibratonal frequencies were assumed to be the same as those for B_2F_4(g). In addition to these were added four OH torsion frequencies, 824(2) and 209(2), four OH stretching, 3200(4), and four B-O-H bending frequencies, 1200(2) and 880(2). The thermodynamic functions were evaluated based on an assumption that there is free rotation at the B-B bond. The reduced moment of inertia for the $B(OH)_2$ top is I_r = 4.1012 x 10^{-39} g cm^{-2}. The principal moments of inertia are: I_A = 16.4048 x 10^{-39}, I_B = 37.4430 x 10^{-39}, and I_C = 37.4441 x 10^{-39} g cm^2.

Dihydroxyborane ((B(OH)$_2$)$_2$) $B_2H_4O_4$(g)

Enthalpy Reference Temperature = T_r = 298.15 K Standard State Pressure = p° = 0.1 MPa

T/K	$C_p°$	S°	-[G°-H°(T_r)]/T	H°-H°(T_r)	$\Delta_f H°$	$\Delta_f G°$	Log K_f
	J K^{-1}mol^{-1}			kJ mol^{-1}			
0							
100	51.574	268.590	423.555	-15.496	-1275.111	-1247.102	651.419
200	78.745	312.793	357.611	-8.964	-1280.454	-1216.886	317.818
250	91.679	331.764	350.567	-4.701	-1282.641	-1200.736	250.880
298.15	103.433	348.929	348.929	0.	-1284.488	-1184.789	207.570
300	103.867	349.571	348.931	0.192	-1284.554	-1184.170	206.182
350	114.997	366.432	350.238	5.668	-1286.183	-1167.307	174.211
400	124.923	382.450	353.272	11.671	-1287.553	-1150.228	150.204
450	133.648	397.679	357.367	18.140	-1288.695	-1132.992	131.514
500	141.275	412.164	362.128	25.018	-1289.632	-1115.639	116.550
600	153.812	439.078	372.747	39.799	-1290.991	-1080.703	94.084
700	163.658	463.556	383.998	55.691	-1291.794	-1045.585	78.022
800	171.644	485.948	395.362	72.469	-1292.160	-1010.383	65.971
900	178.299	506.561	406.588	89.976	-1292.177	-975.156	56.597
1000	183.944	525.646	417.551	108.095	-1291.914	-939.943	49.098
1100	188.786	543.411	428.195	126.738	-1291.427	-904.768	42.964
1200	192.966	560.021	438.496	145.830	-1290.767	-869.645	37.855
1300	196.593	575.613	448.450	165.312	-1289.975	-834.583	33.534
1400	199.750	590.300	458.062	185.133	-1289.089	-799.585	29.833
1500	202.506	604.178	467.345	205.249	-1288.140	-764.654	26.628
1600	204.921	617.326	476.312	225.623	-1287.156	-729.787	23.825
1700	207.043	629.814	484.977	246.224	-1286.158	-694.982	21.354
1800	208.912	641.703	493.356	267.023	-1285.167	-660.236	19.160
1900	210.566	653.043	501.465	287.999	-1284.199	-625.544	17.197
2000	212.032	663.882	509.317	309.130	-1283.268	-590.902	15.433
2100	213.337	674.259	516.926	330.400	-1282.385	-556.305	13.837
2200	214.502	684.211	524.305	351.793	-1281.560	-521.749	12.388
2300	215.545	693.769	531.467	373.296	-1280.802	-487.230	11.065
2400	216.482	702.963	538.422	394.899	-1380.634	-450.605	9.807
2500	217.326	711.818	545.182	416.590	-1380.183	-411.863	8.605
2600	218.089	720.356	551.756	438.361	-1379.750	-373.139	7.496
2700	218.779	728.600	558.154	460.205	-1379.339	-334.430	6.470
2800	219.406	736.568	564.384	482.115	-1378.953	-295.736	5.517
2900	219.977	744.278	570.455	504.084	-1378.597	-257.058	4.630
3000	220.498	751.744	576.375	526.109	-1378.271	-218.389	3.802
3100	220.975	758.982	582.149	548.183	-1377.978	-179.730	3.028
3200	221.412	766.005	587.785	570.302	-1377.720	-141.083	2.303
3300	221.814	772.824	593.290	592.464	-1377.497	-102.442	1.622
3400	222.184	779.452	598.668	614.664	-1377.311	-63.806	0.980
3500	222.525	785.897	603.926	636.900	-1377.163	-25.176	0.376
3600	222.841	792.170	609.068	659.168	-1377.053	13.451	-0.195
3700	223.133	798.280	614.100	681.467	-1376.981	52.074	-0.735
3800	223.404	804.234	619.025	703.794	-1376.950	90.699	-1.247
3900	223.656	810.041	623.849	726.147	-1376.957	129.321	-1.732
4000	223.890	815.706	628.575	748.525	-1377.005	167.943	-2.193
4100	224.109	821.237	633.207	770.925	-1377.094	206.568	-2.632
4200	224.313	826.640	637.748	793.346	-2336.923	259.244	-3.224
4300	224.504	831.921	642.203	815.787	-2334.920	321.035	-3.900
4400	224.682	837.084	646.573	838.246	-2332.962	382.777	-4.544
4500	224.850	842.135	650.863	860.723	-2331.051	444.476	-5.159
4600	225.007	847.079	655.075	883.216	-2329.187	506.133	-5.747
4700	225.155	851.919	659.212	905.724	-2327.371	567.751	-6.310
4800	225.294	856.661	663.276	928.247	-2325.605	629.331	-6.849
4900	225.425	861.308	667.271	950.783	-2323.889	690.875	-7.365
5000	225.548	865.863	671.197	973.331	-2322.225	752.385	-7.860
5100	225.665	870.331	675.058	995.892	-2320.613	813.861	-8.336
5200	225.775	874.714	678.856	1018.464	-2319.054	875.304	-8.793
5300	225.880	879.016	682.592	1041.047	-2317.551	936.722	-9.232
5400	225.978	883.239	686.268	1063.640	-2316.104	998.108	-9.655
5500	226.072	887.386	689.888	1086.242	-2314.714	1059.468	-10.062
5600	226.161	891.460	693.451	1108.854	-2313.383	1120.807	-10.454
5700	226.245	895.464	696.960	1131.474	-2312.112	1182.120	-10.833
5800	226.325	899.400	700.416	1154.103	-2310.902	1243.413	-11.198
5900	226.401	903.269	703.822	1176.739	-2309.754	1304.680	-11.551
6000	226.474	907.075	707.178	1199.383	-2308.671	1365.935	-11.892

PREVIOUS: March 1966 (1 atm) CURRENT: March 1966 (1 bar)

Dihydroxyborane ((B(OH)$_2$)$_2$) $B_2H_4O_4$(g)

DIBORANE (B_2H_6) IDEAL GAS M_r = 27.66764

$S°(298.15\ K) = 233.17\ J\ K^{-1}\ mol^{-1}$

$\Delta_fH°(0\ K) = 56.7 \pm 16.7\ kJ\ mol^{-1}$
$\Delta_fH°(298.15\ K) = 41.0 \pm 16.7\ kJ\ mol^{-1}$

Vibrational Frequencies and Degeneracies

ν, cm^{-1}	ν, cm^{-1}	ν, cm^{-1}	ν, cm^{-1}
2532(1)	2600(1)	1026(1)	2615(1)
2109(1)	919(1)	584(1)	1601(1)
1184(1)	1755(1)	1905(1)	1174(1)
788(1)	942(1)	973(1)	
2612(1)	950(1)	368(1)	

Ground State Quantum Weight: [1] σ = 4

Point Group: D_{2h}

Bond Distances: B-B = 1.775 ± 0.004 Å B-H_t = 1.196 ± 0.016 Å B-H_b = 1.339 ± 0.013 Å

H_t = terminal hydrogen, H_b = bridge hydrogens.

Bond Angles: H_t - B - H_t = 120.2 ± 1.6°

Product of the Moments of Inertia: $I_AI_BI_C = 2.459830 \times 10^{-116}\ g^3\ cm^6$

Enthalpy of Formation

$\Delta_fH°(298.15\ K)$ is a weighted average of the following data. The older data of Roth et al. (8) deviates from the rest of the investigations and is not included.

Source	Reaction	$\Delta_rH°(298.15\ K)$ kcal mol^{-1}	$\Delta_fH°(298.15\ K)$ kcal mol^{-1}
(1)	$B_2H_6(g) \rightarrow 2\ B(am) + 3\ H_2(g)$	-5.0	7.4
(2)	$B_2H_6(g) \rightarrow 2\ B(am) + 3\ H_2(g)$	-6.73	9.13
(3)	$B_2H_6(g) + 6\ Cl_2 \rightarrow 2\ BCl_3(g) + 6\ HCl(g)$	-342.3	17.3
(4)	$B_2H_6(g) + 6\ H_2O(\ell) + [2000\ H_2O](\ell) \rightarrow 2[H_3BO_3 + 1000\ H_2O](soln) + 6\ H_2(g)$	-111.46 ± 0.54	9.32
(7)[a]	$B_2H_6(g) + 2006\ H_2O \rightarrow 2(H_3BO_3 + 1000\ H_2O) + 6\ H_2(g)$	-112.22 ± 0.10	10.08
(7)[b]	$BCl_3(\ell) + 130.3\ H_2O \rightarrow (H_3BO_3 + 1000\ H_2O) + 3(HCl + 100\ H_2O)$	-68.14	--
(a) - 2(b) = (c)	$B_2H_6(g) + 6(HCl{\cdot}100\ H_2O) \rightarrow 2\ BCl(\ell) + 6\ H_2(g)$	+24.06	10.06
(9)	$(CH_3)_3N(g) + 1/2\ B_2H_6(g) \rightarrow (CH_3)_3N{:}BH_3(cr)$	-32.33 ± 0.06	8.3
(13)	$(CH_3)_3N(g) + 1/2\ B_2H_6(g) \rightarrow (CH_3)_3N{:}BH_3(cr)$	-31.40	6.5

Sources of Auxiliary Data: From the JANAF tables, $\Delta_fH°(298.15\ K)$ = 1.2, -96.31, -22.063 kcal mol^{-1} of B(am), $BCl_3(g)$ and HCl(g), respectively. From Rossini (11), the $\Delta_{vap}H°(298.15\ K)$ = 5.6 kcal mol^{-1} of BCl_3. From Wagman (12), the $\Delta_fH°(298.15\ K)$ of (HCl + 1000 H_2O) = -39.657 kcal mol^{-1}. From Good et al. (5), the $\Delta_fH°(H_3BO_3$, cr, 298.15 K) = -261.47 ± 0.20 kcal mol^{-1} and the H_3BO_3 solution data of Fasolino (6). $\Delta_fH°((CH_3)_3N$, g, 298.15 K) = -5.9 ± 0.4 kcal mol^{-1} from Jaffe (10).

Heat Capacity and Entropy

The molecular constants reported by Carroll (14), were selected. They agree well with those reported by Hedberg and Schomaker (15). The principal moments of inertia are $I_A = 1.0562 \times 10^{-39}$, $I_B = 4.6381 \times 10^{-39}$, and $I_C = 5.0213 \times 10^{-39}\ g\ cm^2$.

The vibrational frequencies for $B_2{}^{11}H_6$ were those listed by Wu (16) except for ν_8, ν_{10} and ν_{14} which were taken from the calculated frequency column of Table VIII-1 page 55 of the Thesis.

References

1. S. R. Gunn and L. G. Green, J. Phys. Chem. 65, 779-83 (1961); J. Chem. Phys. 36, 1118 (1962). Explosive decomposition while mixed with stibine.
2. E. J. Prosen, W. H. Johnson and F. Y. Pergiel, J. Res. Nat. Bur. Stand. 61, 247-250 (1958). Decomposition by heating in a flow-type calorimeter.
3. J. R. Lacher, R. E. Scruby, and J. D. Park, J. Am. Chem. Soc. 74, 5292-4 (1952). A vapor phase chlorination in a flow-type calorimeter.
4. E. J. Prosen, W. H. Johnson, and F. Y. Pergiel, J. Res. Nat. Bur. Stand. 62, 43-47 (1959). Heat of reaction of diborane with water in a flow-type calorimeter.
5. W. D. Good, M. Mansson, and J. P. McCullough, "Thermochemistry of Boron and Some of its Compoudns. The heats of Formation of Trimethylamineborane and Orthaboric Acid", Symp. on Thermodynamics and Thermochemistry, Lund, Sweden, July 1963.
6. L. G. Fasolino, "Heats of Formation of B_2O_3(cr) and B_2O_3(amp)", Special Report NONR-360, National Research Corporation, Cambridge, Mass., June 2, 1964.
7. S. R. Gunn and L. G. Green, J. Phys. Chem. 64, 61-63 (1960). Heats of hydrolysis of diborane and boron trichloride in a sealed rocking bomb type calorimeter.
8. W. A. Roth, E. Borger, and A. Bertram, Ber. 70B, 49 (1937); Ber. 70B, 971 (1937) and W. A. Roth, Z. Naturforsch, 1, 574 (1946); heat of hydrolysis in a bomb type calorimeter.
9. S. R. Gunn, paper presented at 19th Annual Calorimetry Conference, Washington, DC and Bethesda, MD, Oct. 13-16, 1964.
10. I. Jaffe, Thesis, University of Maryland, 1958.
11. U. S. Nat. Bur. Stand. Circ. 500, 1952.
12. W. H. Evans, U. S. Nat. Bur. Stand. Report 8504, Appendix III, 1964.
13. R. E. McCoy and S. H. Bauer, J. Am. Chem. Soc. 78, 2061 (1956).
14. B. L. Carroll, Thesis, Iowa State University of Science and Technology, 1963.
15. K. Hedberg and V. Schomaker, J. Chem. Phys. 22, 985 (1954).
16. E. C. Wu, Thesis, University of Minnesota, 1962.

Diborane (B_2H_6) $B_2H_6(g)$

Enthalpy Reference Temperature = T_r = 298.15 K Standard State Pressure = p° = 0.1 MPa

T/K	$C_p°$ (J $K^{-1}mol^{-1}$)	S° (J $K^{-1}mol^{-1}$)	$-[G°-H°(T_r)]/T$ (J $K^{-1}mol^{-1}$)	$H°-H°(T_r)$ (kJ mol^{-1})	$\Delta_fH°$ (kJ mol^{-1})	$\Delta_fG°$ (kJ mol^{-1})	Log K_f
0	0.	0.	INFINITE	-12.149	56.684	56.684	INFINITE
100	34.592	186.917	274.911	-8.799	50.988	62.576	-32.686
200	43.499	213.191	237.985	-4.959	46.081	76.058	-19.864
250	50.526	223.629	234.083	-2.613	43.496	83.850	-17.519
298.15	58.100	233.170	233.170	0.	41.003	91.849	-16.092
300	58.397	233.530	233.171	0.108	40.909	92.165	-16.047
350	66.435	243.137	233.911	3.229	38.429	100.905	-15.059
400	74.255	252.522	235.653	6.748	36.106	109.991	-14.363
450	81.690	261.701	238.039	10.648	33.970	119.357	-13.855
500	88.683	270.674	240.856	14.909	32.037	128.950	-13.471
600	101.331	287.988	247.280	24.425	28.803	148.650	-12.941
700	112.275	304.451	254.282	35.119	26.373	168.828	-12.598
800	121.655	320.072	261.538	46.827	24.659	189.303	-12.360
900	129.635	334.875	268.871	59.403	23.547	209.956	-12.186
1000	136.394	348.893	276.179	72.714	22.922	230.706	-12.051
1100	142.112	362.168	283.398	86.647	22.678	251.499	-11.943
1200	146.952	374.747	290.491	101.107	22.725	272.300	-11.853
1300	151.059	386.676	297.435	116.013	22.986	293.089	-11.776
1400	154.558	398.002	304.217	131.298	23.400	313.851	-11.710
1500	157.550	408.770	310.832	146.908	23.917	334.580	-11.651
1600	160.122	419.022	317.276	162.794	24.497	355.272	-11.598
1700	162.342	428.798	323.551	178.920	25.110	375.927	-11.551
1800	164.269	438.133	329.659	195.253	25.732	396.546	-11.507
1900	165.949	447.060	335.605	211.766	26.343	417.130	-11.468
2000	167.421	455.611	341.393	228.436	26.928	437.682	-11.431
2100	168.716	463.811	347.028	245.244	27.475	458.206	-11.397
2200	169.860	471.687	352.517	262.174	27.975	478.705	-11.366
2300	170.876	479.260	357.864	279.212	28.418	499.183	-11.337
2400	171.780	486.552	363.075	296.345	-71.719	521.780	-11.356
2500	172.589	493.581	368.155	313.565	-71.560	546.506	-11.419
2600	173.315	500.365	373.111	330.861	-71.407	571.225	-11.476
2700	173.969	506.918	377.946	348.225	-71.265	595.939	-11.529
2800	174.559	513.256	382.666	365.652	-71.137	620.648	-11.578
2900	175.094	519.391	387.275	383.135	-71.025	645.353	-11.624
3000	175.581	525.335	391.779	400.669	-70.932	670.054	-11.667
3100	176.023	531.100	396.180	418.250	-70.862	694.752	-11.706
3200	176.428	536.695	400.485	435.873	-70.815	719.448	-11.744
3300	176.798	542.130	404.695	453.534	-70.793	744.144	-11.779
3400	177.139	547.413	408.815	471.232	-70.799	768.839	-11.812
3500	177.451	552.552	412.849	488.961	-70.834	793.535	-11.843
3600	177.740	557.555	416.799	506.721	-70.899	818.232	-11.872
3700	178.006	562.429	420.670	524.509	-70.996	842.931	-11.900
3800	178.253	567.179	424.463	542.322	-71.125	867.634	-11.926
3900	178.482	571.812	428.182	560.159	-71.288	892.340	-11.952
4000	178.694	576.334	431.829	578.017	-71.486	917.050	-11.975
4100	178.892	580.749	435.408	595.897	-71.719	941.767	-11.998
4200	179.076	585.062	438.920	613.795	-1031.687	980.538	-12.195
4300	179.248	589.278	442.368	631.712	-1029.819	1028.426	-12.493
4400	179.409	593.400	445.754	649.645	-1027.991	1076.270	-12.777
4500	179.559	597.434	449.080	667.593	-1026.205	1124.074	-13.048
4600	179.700	601.382	452.348	685.556	-1024.461	1171.838	-13.307
4700	179.833	605.248	455.560	703.533	-1022.759	1219.566	-13.554
4800	179.957	609.035	458.718	721.522	-1021.099	1267.258	-13.791
4900	180.075	612.747	461.824	739.524	-1019.481	1314.914	-14.017
5000	180.185	616.386	464.879	757.537	-1017.906	1362.540	-14.234
5100	180.289	619.955	467.885	775.561	-1016.373	1410.133	-14.443
5200	180.387	623.457	470.843	793.595	-1014.881	1457.697	-14.643
5300	180.480	626.894	473.755	811.638	-1013.430	1505.233	-14.835
5400	180.568	630.269	476.622	829.691	-1012.018	1552.742	-15.020
5500	180.651	633.583	479.446	847.752	-1010.645	1600.224	-15.198
5600	180.730	636.838	482.228	865.821	-1009.310	1647.683	-15.369
5700	180.805	640.038	484.968	883.898	-1008.010	1695.118	-15.534
5800	180.876	643.183	487.669	901.982	-1006.745	1742.530	-15.693
5900	180.944	646.276	490.331	920.073	-1005.513	1789.920	-15.847
6000	181.008	649.317	492.956	938.170	-1004.313	1837.290	-15.995

PREVIOUS: December 1964 (1 atm) CURRENT: December 1964 (1 bar)

Diborane (B_2H_6) $B_2H_6(g)$

MAGNESIUM BORIDE (MgB_2) CRYSTAL M_r = 45.925

$\Delta_f H°(0\ K) = -91.3 \pm 8.4\ kJ\ mol^{-1}$
$\Delta_f H°(298.15\ K) = -92.0 \pm 8.4\ kJ\ mol^{-1}$

$S°(298.15\ K) = 35.98 \pm 0.17\ J\ K^{-1}\ mol^{-1}$

T_{dcm} = [1320] K

Enthalpy of Formation

The equilibrium constants (993 - 1108 K) for the reaction 2 MgB_2(cr) = Mg(g) + MgB_4(cr) were determined by Wright and Walsh (1). The corresponding 2nd and 3rd law values of $\Delta_f H°$(298.15 K) for this reaction were derived as 53.1 ± 4.7 and 54.1 kcal mol^{-1}, respectively. Using the 3rd law value obtained, i.e., $\Delta_r H°(298.15\ K) = 54.1\ kcal\ mol^{-1}$, for the reaction, $\Delta_f H°(MgB_2$, cr, 298.15 K) was evaluated as -22.0 ± 2.0 kcal mol^{-1}.

Heat Capacity and Entropy

The low temperature (21.12 - 304.22 K) heat capacities were measured by Swift and White (2). Above 298.15 K, the values of C_p were estimated with those of other related borides. S°(298.15 K) was reported by Swift and White (2) using S°(extrap, 20 K) = 0.005 cal K^{-1} mol^{-1}.

Decomposition Data

T_{dcm} was taken from Markovsky et al. (3). On heating to high temperatures, MgB_2(cr) decomposes into boride phases, A, B, and C with the separation of Mg. The composition of phase A is close to MgB_6 and that of C is close to MgB_{12}.

References

1. M. Wright and P. N. Walsh, "The Vaporization of MgB_4(cr)", Ohio State University Research Foundation, Tech. Research Report OMCC-HEF-55 (1958).
2. R. M. Swift and D. White, J. Am. Chem. Soc. 79, 3641 (1957).
3. L. Y. Markovsky, Y. D. Kondrashev, and G. V. Kaputovskaya, Zhur. Obshchei Khim. 25, 409 (1955).

Magnesium Boride (MgB_2) B_2Mg_1(cr)

Enthalpy Reference Temperature = T_r = 298.15 K   Standard State Pressure = p° = 0.1 MPa

	J K⁻¹mol⁻¹			kJ mol⁻¹			
T/K	$C_p°$	S°	-[G°-H°(T_r)]/T	H°-H°(T_r)	$\Delta_f H°$	$\Delta_f G°$	Log K_f
0	0.	0.	INFINITE	-6.799	-91.337	-91.337	INFINITE
100	11.171	4.025	68.961	-6.494	-91.722	-91.112	47.592
200	34.392	19.205	40.125	-4.184	-92.085	-90.330	23.592
298.15	47.823	35.982	35.982	0.	-91.964	-89.473	15.675
300	47.865	36.278	35.983	0.089	-91.964	-89.457	15.576
400	54.392	51.002	37.946	5.222	-92.110	-88.609	11.571
500	58.409	63.590	41.848	10.871	-92.590	-87.684	9.160
600	61.379	74.507	46.402	16.863	-93.328	-86.638	7.542
700	64.015	84.168	51.120	23.134	-94.239	-85.452	6.377
800	66.526	92.879	55.804	29.660	-95.267	-84.128	5.493
900	69.162	100.867	60.373	36.445	-96.364	-82.670	4.798
1000	71.714	108.288	64.798	43.490	-106.100	-80.376	4.198
1100	74.078	115.233	69.070	50.779	-107.301	-77.744	3.692
1200	76.442	121.780	73.193	58.305	-108.384	-75.008	3.265
1300	78.795	127.992	77.171	66.067	-109.336	-72.188	2.901
1400	81.170	133.918	81.014	74.065	-237.558	-66.131	2.467
1500	83.643	139.603	84.732	82.307	-236.863	-53.910	1.877
1600	85.981	145.077	88.333	90.789	-236.012	-41.740	1.363
1700	88.162	150.355	91.827	99.497	-235.015	-29.628	0.910
1800	90.249	155.454	95.221	108.419	-233.881	-17.578	0.510
1900	92.242	160.387	98.522	117.544	-232.618	-5.595	0.154
2000	94.140	165.167	101.735	126.864	-231.232	6.318	-0.165

PREVIOUS: March 1963   CURRENT: March 1965

Magnesium Boride (MgB_2) B_2Mg_1(cr)

BORON OXIDE (B_2O) IDEAL GAS M_r = 37.6194

$S°(298.15\ K) = [227.7]\ J\ K^{-1}\ mol^{-1}$

$\Delta_f H°(0\ K) = 93 \pm 105\ kJ\ mol^{-1}$
$\Delta_f H°(298.15\ K) = 96 \pm 105\ kJ\ mol^{-1}$

Vibrational Frequencies and Degeneracies

ν, cm^{-1}
[1250](1)
[600](1)
[1800](1)

Ground State Quantum Weight: [1]

Point Group: $[C_{2v}]$ σ = [2]

Bond Distance: B–O = [1.26] Å

Bond Angle: B–O–B = [150]°

Product of the Moments of Inertia: $I_A I_B I_C = [4.732389 \times 10^{-117}]\ g^3\ cm^6$

Enthalpy of Formation

The enthalpy of formation was calculated from the estimated enthalpy of reaction $\Delta_r H°(0\ K) = 300 \pm 25\ kcal\ mol^{-1}$ for $B_2O(g) \rightarrow 2B(g) + O(g)$, using the JANAF auxiliary data (2) for B(g) and O(g). The value of $\Delta_r H°(0\ K)$ was assumed to be the sum of the enthalpies of dissociation, 120 kcal mol^{-1} for $B_2O(g) \rightarrow BO(g) + B(g)$ and 180 kcal mol^{-1} for $BO(g) \rightarrow B(g) + O(g)$. The former was estimated and the latter was derived from the JANAF values (2).

The possible existence of $B_2O(g)$ in the burning of boron under reducing conditions was discussed by Bahn (1). He estimated the enthalpy of formation to lie between -27 and +23 kcal mol^{-1} but favored the value of +23 kcal mol^{-1} in the theoretical calculations of boron burning in air.

Heat Capacity and Entropy

The bent molecular structure and the molecular constants were estimated by comparison with those in BO(g), AlO(g) and $Al_2O(g)$. The principal moments of inertia are $I_A = 0.1624 \times 10^{-39}$, $I_B = 5.3179 \times 10^{-39}$ and $I_C = 5.4803 \times 10^{-39}\ g\ cm^2$.

References

1. G. S. Bahn, Combustion Institute, University of Denver (1966).
2. JANAF Thermochemical Tables: B(g), 12-31-64; and O(g), 6-30-62.

Boron Oxide (B_2O) $B_2O_1(g)$

Enthalpy Reference Temperature = T_r = 298.15 K Standard State Pressure = $p°$ = 0.1 MPa

T/K	$C_p°$ (J K⁻¹mol⁻¹)	S° (J K⁻¹mol⁻¹)	$-[G°-H°(T_r)]/T$ (J K⁻¹mol⁻¹)	$H°-H°(T_r)$ (kJ mol⁻¹)	$\Delta_f H°$ (kJ mol⁻¹)	$\Delta_f G°$ (kJ mol⁻¹)	Log K_f
0	0.	0.	INFINITE	-10.375	92.626	92.626	INFINITE
100	33.368	189.391	259.874	-7.048	94.453	84.241	-44.003
200	35.469	213.037	231.161	-3.625	95.755	73.464	-19.187
250	36.956	221.110	228.368	-1.814	96.094	67.847	-14.176
298.15	38.410	227.744	227.744	0.	96.232	62.391	-10.931
300	38.466	227.982	227.745	0.071	96.234	62.181	-10.827
350	39.966	234.024	228.218	2.032	96.202	56.506	-8.433
400	41.433	239.457	229.289	4.067	96.019	50.846	-6.640
450	42.841	244.419	230.698	6.174	95.712	45.217	-5.249
500	44.168	249.002	232.303	8.350	95.311	39.627	-4.140
600	46.530	257.270	235.790	12.888	94.307	28.581	-2.488
700	48.483	264.595	239.391	17.642	93.123	17.718	-1.322
800	50.064	271.176	242.960	22.572	91.819	7.034	-0.459
900	51.336	277.149	246.432	27.645	90.425	-3.482	0.202
1000	52.359	282.612	249.781	32.831	88.957	-13.838	0.723
1100	53.186	287.643	252.997	38.110	87.421	-24.043	1.142
1200	53.861	292.300	256.081	43.464	85.822	-34.106	1.485
1300	54.414	296.634	259.036	48.878	84.162	-44.033	1.769
1400	54.873	300.684	261.867	54.343	82.441	-53.831	2.008
1500	55.256	304.483	264.583	59.850	80.659	-63.503	2.211
1600	55.579	308.060	267.190	65.392	78.815	-73.053	2.385
1700	55.853	311.438	269.694	70.964	76.909	-82.487	2.535
1800	56.088	314.637	272.103	76.562	74.939	-91.807	2.664
1900	56.289	317.675	274.422	82.181	72.905	-101.015	2.777
2000	56.464	320.567	276.658	87.819	70.807	-110.115	2.876
2100	56.616	323.326	278.815	93.473	68.643	-119.108	2.963
2200	56.750	325.963	280.898	99.141	66.413	-127.997	3.039
2300	56.867	328.488	282.913	104.822	64.117	-136.782	3.106
2400	56.971	330.911	284.863	110.514	-38.764	-143.329	3.119
2500	57.064	333.238	286.752	116.216	-41.349	-147.633	3.085
2600	57.147	335.478	288.583	121.927	-43.937	-151.833	3.050
2700	57.220	337.636	290.360	127.645	-46.527	-155.934	3.017
2800	57.287	339.718	292.086	133.371	-49.120	-159.938	2.984
2900	57.347	341.730	293.763	139.103	-51.718	-163.851	2.951
3000	57.401	343.675	295.395	144.840	-54.319	-167.673	2.919
3100	57.451	345.558	296.983	150.583	-56.924	-171.408	2.888
3200	57.496	347.382	298.529	156.330	-59.533	-175.060	2.858
3300	57.537	349.152	300.037	162.082	-62.147	-178.630	2.827
3400	57.575	350.870	301.506	167.837	-64.765	-182.120	2.798
3500	57.609	352.540	302.941	173.597	-67.388	-185.534	2.769
3600	57.641	354.163	304.341	179.359	-70.015	-188.872	2.740
3700	57.670	355.743	305.709	185.125	-72.647	-192.138	2.712
3800	57.697	357.281	307.046	190.893	-75.283	-195.331	2.685
3900	57.722	358.780	308.354	196.664	-77.923	-198.456	2.658
4000	57.746	360.242	309.633	202.437	-80.567	-201.513	2.631
4100	57.767	361.668	310.884	208.213	-83.216	-204.504	2.605
4200	57.787	363.060	312.110	213.991	-1045.567	-193.382	2.405
4300	57.806	364.420	313.311	219.770	-1046.050	-173.086	2.103
4400	57.824	365.750	314.488	225.552	-1046.541	-152.779	1.814
4500	57.840	367.049	315.641	231.335	-1047.040	-132.461	1.538
4600	57.855	368.321	316.773	237.120	-1047.548	-112.132	1.273
4700	57.870	369.565	317.883	242.906	-1048.066	-91.791	1.020
4800	57.883	370.784	318.972	248.694	-1048.595	-71.439	0.777
4900	57.896	371.977	320.042	254 483	-1049.135	-51.076	0.544
5000	57.908	373.147	321.092	260.273	-1049.687	-30.701	0.321
5100	57.919	374.294	322.124	266.064	-1050.252	-10.316	0.106
5200	57.930	375.419	323.138	271.857	-1050.832	10.080	-0.101
5300	57.940	376.522	324.135	277.650	-1051.426	30.488	-0.300
5400	57.949	377.605	325.115	283.445	-1052.037	50.907	-0.492
5500	57.958	378.669	326.080	289.240	-1052.665	71.338	-0.678
5600	57.967	379.713	327.028	295.036	-1053.312	91.780	-0.856
5700	57.975	380.739	327.961	300.833	-1053.978	112.234	-1.029
5800	57.982	381.747	328.880	306.631	-1054.665	132.700	-1.195
5900	57.990	382.739	329.784	312.430	-1055.375	153.177	-1.356
6000	57.997	383.713	330.675	318.229	-1056.108	173.668	-1.512

PREVIOUS: June 1966 (1 atm) CURRENT: June 1966 (1 bar)

Boron Oxide (B_2O) $B_2O_1(g)$

BOROON OXIDE ((BO)$_2$) IDEAL GAS M_r = 53.6188

$S°(298.15\ K) = 242.60\ J\ K^{-1}\ mol^{-1}$

$\Delta_f H°(0\ K) = -457.3 \pm 8.4\ kJ\ mol^{-1}$
$\Delta_f H°(298.15\ K) = -456.1 \pm 8.4\ kJ\ mol^{-1}$

Vibrational Frequencies and Degeneracies

ν, cm^{-1}	ν, cm^{-1}
2065(1)	565(2)
570(1)	285(2)
1910(1)	

Ground State Quantum Weight: 1 σ = 2
Point Group: $D_{\infty h}$
Bond Distances: O-B = 1.20 Å B-B = 1.70 Å
Bond Angle: O-B-B = 180°
Rotational Constant: B_0 = 0.112313 cm^{-1}

Enthalpy of Formation

$\Delta_f H°(B_2O_2$, g, 298.15 K) adopted is the weighted average of three $\Delta_f H°$(298.15 K) values derived from two different chemical reactions. The chemical reactions related to the production of B_2O_2(g) have been studied by several investigators. Using the equilibrium vapor pressure data, the respective values of $\Delta_r H°$(298.15 K) were evaluated by both the 2nd and 3rd law methods. The results obtained are based on the latter and presented as follows:

Source	Reaction	$\Delta_r H°$(298.15 K), kcal mol^{-1} 2nd law	3rd law	$\Delta_f H°$(298.15 K) kcal mol^{-1}
(1)	2/3 B(cr) + 2/3 B_2O_3(ℓ) = B_2O_2(g)	94.0	91.8 ± 2.2	-105.5 ± 0.7*
	2/3 B(cr) + 2/3 B_2O_3(g) = B_2O_2(g)	24.8	25.4 ± 0.2	-108.0 ± 0.5
(2)	2 MgO(cr) + 2 B(cr) = 2 Mg(g) + B_2O_2(g)	255.4	--	-102.6 ± 10*
(3)	2/3 B(cr) + 2/3 B_2O_3(ℓ) = B_2O_2(g)	89.3	77.9 ± 0.2	-110.2 ± 0.9
(4)	B_2O_3(ℓ) + 3 C(cr) = 3 CO(g) + 2 B(cr)	--	--	-108.5 ± 1.5
	B_2O_3(ℓ) + C(cr) = B_2O_2(g) + CO(g)	--	--	

*The value not used for the calculation of the weighted average of $\Delta_f H°(B_2O_2$, g, 298.15 K).

Heat Capacity and Entropy

The molecular structure and constants, and vibrational frequencies, corrected to the average isotopic species, were taken from Sommer et al. (5).

References

1. M. G. Inghram, R. F. Porter and W. A. Chupka, J. Chem. Phys. 25, 498 (1956).
2. A. W. Searcy and C. E. Myers, J. Phys. Chem. 61, 957 (1957).
3. M. D. Scheer, J. Phys. Chem. 62, 490 (1958).
4. P. Rentzepis, D. White and P. N. Walsh, J. Phys. Chem. 64 (1960). The value of the enthalpy of formation was obtained from derived log $K_f(B_2O_2$, g) by the 3rd law method
5. O. Sommer, D. White, M. J. Linevsky, and D. E. Mann, J. Phys. Chem. 38, 87 (1963).

Boron Oxide ((BO)$_2$) B_2O_2(g)

Enthalpy Reference Temperature = T_r = 298.15 K Standard State Pressure = p° = 0.1 MPa

T/K	$C_p°$ (J K^{-1}mol^{-1})	S°	$-[G°-H°(T_r)]/T$	$H°-H°(T_r)$ (kJ mol^{-1})	$\Delta_f H°$	$\Delta_f G°$	Log K_f
0	0.	0.	INFINITE	-12.398	-457.343	-457.343	INFINITE
100	34.367	193.295	286.969	-9.367	-457.265	-459.202	239.863
200	48.231	221.483	247.582	-5.220	-456.694	-461.326	120.486
250	53.477	232.836	243.520	-2.671	-456.345	-462.525	96.639
298.15	57.297	242.597	242.597	0.	-456.056	-463.743	81.246
300	57.424	242.952	242.598	0.106	-456.046	-463.791	80.753
350	60.445	252.040	243.309	3.056	-455.827	-465.100	69.412
400	62.871	260.275	244.923	6.141	-455.708	-466.434	60.910
450	64.921	267.802	247.053	9.337	-455.684	-467.777	54.298
500	66.722	274.737	249.479	12.629	-455.741	-469.118	49.008
600	69.820	287.184	254.750	19.461	-456.031	-471.770	41.071
700	72.409	298.147	260.182	26.576	-456.480	-474.359	35.397
800	74.575	307.962	265.551	33.928	-457.030	-476.877	31.137
900	76.378	316.853	270.765	41.479	-457.649	-479.321	27.819
1000	77.874	324.980	275.786	49.194	-458.320	-481.694	25.161
1100	79.116	332.462	280.603	57.045	-459.038	-483.996	22.983
1200	80.149	339.392	285.217	65.010	-459.800	-486.232	21.165
1300	81.013	345.842	289.635	73.069	-460.607	-488.402	19.624
1400	81.739	351.873	293.868	81.208	-461.461	-490.509	18.301
1500	82.353	357.534	297.925	89.413	-462.365	-492.552	17.152
1600	82.876	362.866	301.819	97.675	-463.323	-494.534	16.145
1700	83.323	367.904	305.560	105.986	-464.337	-496.454	15.254
1800	83.709	372.678	309.157	114.338	-465.410	-498.313	14.461
1900	84.042	377.213	312.621	122.726	-466.544	-500.110	13.749
2000	84.333	381.532	315.959	131.145	-467.743	-501.846	13.107
2100	84.588	385.653	319.180	139.591	-469.007	-503.520	12.524
2200	84.812	389.593	322.292	148.062	-470.339	-505.132	11.993
2300	85.010	393.367	325.301	156.553	-471.740	-506.683	11.507
2400	85.185	396.989	328.213	165.063	-573.730	-506.034	11.014
2500	85.342	400.470	331.034	173.589	-575.428	-503.178	10.513
2600	85.482	403.820	333.770	182.131	-577.133	-500.254	10.050
2700	85.608	407.048	336.424	190.685	-578.846	-497.265	9.620
2800	85.722	410.164	339.002	199.252	-580.567	-494.211	9.220
2900	85.824	413.174	341.508	207.829	-582.297	-491.098	8.846
3000	85.917	416.085	343.946	216.416	-584.037	-487.923	8.495
3100	86.002	418.903	346.319	225.012	-585.787	-484.690	8.167
3200	86.079	421.635	348.630	233.617	-587.546	-481.401	7.858
3300	86.150	424.285	350.883	242.228	-589.316	-478.057	7.567
3400	86.214	426.858	353.079	250.846	-591.095	-474.658	7.292
3500	86.274	429.358	355.223	259.471	-592.885	-471.208	7.032
3600	86.328	431.789	357.317	268.101	-594.684	-467.706	6.786
3700	86.379	434.155	359.361	276.736	-596.492	-464.154	6.553
3800	86.426	436.459	361.360	285.377	-598.311	-460.552	6.331
3900	86.469	438.705	363.315	294.021	-600.138	-456.903	6.120
4000	86.509	440.894	365.227	302.670	-601.975	-453.208	5.918
4100	86.547	443.031	367.099	311.323	-603.820	-449.466	5.726
4200	86.581	445.117	368.931	319.979	-1565.374	-431.630	5.368
4300	86.614	447.155	370.727	328.639	-1565.063	-404.640	4.915
4400	86.644	449.146	372.487	337.302	-1564.766	-377.657	4.483
4500	86.672	451.094	374.212	345.968	-1564.481	-350.681	4.071
4600	86.699	452.999	375.904	354.637	-1564.211	-323.711	3.676
4700	86.724	454.864	377.564	363.308	-1563.955	-296.746	3.298
4800	86.747	456.690	379.194	371.981	-1563.715	-269.787	2.936
4900	86.769	458.479	380.794	380.657	-1563.492	-242.831	2.589
5000	86.790	460.232	382.365	389.335	-1563.287	-215.881	2.255
5100	86.810	461.951	383.909	398.015	-1563.101	-188.935	1.935
5200	86.828	463.637	385.426	406.697	-1562.935	-161.993	1.627
5300	86.846	465.291	386.917	415.381	-1562.791	-135.052	1.331
5400	86.862	466.914	388.384	424.066	-1562.670	-108.116	1.046
5500	86.878	468.508	389.826	432.753	-1562.573	-81.181	0.771
5600	86.893	470.074	391.245	441.442	-1562.503	-54.246	0.506
5700	86.907	471.612	392.641	450.132	-1562.461	-27.314	0.250
5800	86.920	473.124	394.016	458.823	-1562.449	-0.381	0.003
5900	86.933	474.609	395.370	467.516	-1562.468	26.550	-0.235
6000	86.945	476.071	396.702	476.210	-1562.521	53.484	-0.466

PREVIOUS: December 1964 (1 atm) CURRENT: December 1964 (1 bar)

Boron Oxide ((BO)$_2$) B_2O_2(g)

BORON OXIDE (B_2O_3) CRYSTAL M_r = 69.6182

S°(298.15 K) = 53.95 ± 0.4 J K^{-1} mol^{-1}
T_{fus} = 723 ± 0.5 K

$\Delta_f H°$(0 K) = -1265.8 ± 2.1 kJ mol^{-1}
$\Delta_f H°$(298.15 K) = -1271.9 ± 2.1 kJ mol^{-1}
$\Delta_{fus} H°$ = 24.071 ± 0.4 kJ mol^{-1}

Enthalpy of Formation

The adopted $\Delta_f H°$(B_2O_3, cr, 298.15 K) = -304.0 ± 0.5 kcal mol^{-1} is calculated from JANAF $\Delta_f H°$(BF_3, g, 298.15 K) = -271.42 ± 0.4 kcal mol^{-1} (1) and $\Delta_c H°$(298.15 K) = -238.82 ± 0.39 kcal mol^{-1} for B_2O_3(cr) + 3 F_2(g) → 2 BF_3(g) + 1.5 O_2(g). The enthalpy of combustion of B_2O_3(cr) in fluorine was determined by Johnson and Hubbard (2). The direct combusion of boron in oxygen has led to $\Delta_f H°$(glass, 298.15 K) values which range from -280 to -368 kcal mol^{-1} (3, 4, 5, 6, 7, 8, 9) due to incomplete combustions or ill-defined states of combustion products. These values are not reliable.

A reliable enthalpy of formation of boric oxide can be derived from enthalpies of solution of H_3BO_3(cr) and B_2O_3(cr) which are listed below. Since the enthalpies of dilution of aqueous H_3BO_3 (10) are relatively small, no correction has been applied to the reported enthalpy of solution data. The derived $\Delta_f H°$ values are in very good agreement with the values adopted.

Source	Enthalpy of Solution, kcal mol^{-1}: $\Delta_{sol}H°(B_2O_3, cr)^{(a)}$	$\Delta_{sol}H°(H_3BO_3, cr)^{(b)}$	$\Delta_r H°(cr)^{(c)}$ kcal mol^{-1}	$\Delta_f H°$(B_2O_3, cr, 298.15 K) kcal mol^{-1}
(11)	-3.56*	+5.3	-14.16	-303.83
(12)	-3.41*	+5.10	-13.61	-304.38
(13)	-3.49*	+5.3	-14.09	-303.90
(14)	-3.48	+5.27	-14.02	-303.97
(15)	-3.49	+5.17	-13.83	-304.16
(16)	-3.45	+5.45	-14.35	-303.64

*The enthalpy of solution of B_2O_3(am) was combined with $\Delta_r H°$(25°C) = 4.44 kcal mol^{-1} (2) to obtain the $\Delta_{sol}H°$(B_2O_3, cr)

(a) B_2O_3(cr) + 3 H_2O(ℓ) → 2 H_3BO_3(aq) (b) H_3BO_3(cr) → H_3BO_3(aq) (c) B_2O_3(cr) + 3 H_2O(ℓ) → 2 H_3BO_3(cr)

Other solution calorimetric measurements (17, 18, 19) which involve the hydrolysis of B_2H_6(g) and BCl_3(ℓ) lead to enthalpies of formation of boric oxide (cr) with a relatively large uncertainty since the techniques were indirect, and many questionable auxiliary data were employed in the derivation.

Heat Capacity and Entropy

The adopted heat capacities are derived from C_p data of Kerr, Hersh and Johnston (20) and Shmidt (21) in the temperature regions 18-296.6 K and 303-703 K, respectively. These two sets of C_p data were measured adiabatically and are smoothly joined at 298.15 K by a polynomial curve fitting technique. The derived entropy, S°(298.15 K), is based on S°(18.08 K) = 0.025 cal K^{-1} mol^{-1}.

Kelley (22) also measured low temperature C_p from 53 to 295 K which are in very good agreement with the values adopted. Southard (23) determined high temperature enthalpy data from 350 to 721.5 K by drop calorimetry. The deviations of the observed enthalpy data from the adopted values are about 5% at 360 K, 1% at 570 K and 0.5% at 720 K.

Fusion Data

The melting point, 723 ± 0.5 K, was determined by Shmidt (21) in excellent agreement with the value 723 ± 2 K reported by Kracek et al. (24). The adopted enthalpy of fusion, $\Delta_{fus}H°$ = 5.75 ± 0.1 kcal mol^{-1}, is derived from $\Delta_r H°$(298.15 K) = 4.44 kcal mol^{-1} (2) for B_2O_3(crystal) → B_2O_3(am) and the adopted relative enthalpies of the two forms at 723 K. This value is reasonable agreement with $\Delta_{fus}H°$(723 K) = 5.87 ± 0.1 kcal mol^{-1} determined calorimetrically by Schmidt (21).

References

1. JANAF Thermochemical Tables: BF_3(g), 6-30-69 and H_3BO_3(cr), 12-31-64.
2. G. K. Johnson and W. N. Hubbard, J. Chem. Thermodynamics 1, 459 (1969).
3. W. A. Roth, Z. Natur. 1, 574 (1946).
4. W. Eggersgluess, A. G. Monroe and W. G. Parker, Trans. Faraday Soc. 45, 661 (1949).
5. J. A. Robertson, Thesis, Cornell University, 1944.
6. B. J. Todd and R. R. Miller, J. Amer. Chem. Soc. 68, 530 (1946).
7. C. C. Nathan, Ph.D. Thesis, University of Pittsburgh, 1948.
8. B. H. Eckstein and E. R. Van Artsdalen, J. Amer. Chem. Soc. 80, 1352 (1958).
9. G. L. Galchenko, A. N. Kornilov and S. M. Skruratov, Russ. J. Inorg. Chem. 5, 1039 (1960).
10. U. S. Nat. Bur. Stand. Tech. Note 270-3, 1968.
11. M. Von Stackelberg, F. Quatram and J. Dressel, Z. Elektrochem. 43, 14 (1937).
12. W. Roth, E. Borger and A. Bertram, Ber. 70, 97 (1937).
13. C. Katz, M. S Thesis, Cornell University, 1949.
14. J. Smisko and L. S. Masson, J. Amer. Chem. Soc. 72, 3679 (1950).
15. E. R. Van Artsdalen and K. P. Anderson, J. Amer. Chem. Soc. 73, 579 (1951).
16. L. G. Fasolino, J. Chem. Eng. Data 10, 373 (1965).
17. E. J. Prosen, W. H. Johnson and F. Y. Pergiel, J. Res. Nat. Bur. Stand. 62, 43 (1959).
18. W. H. Johnson, R. G. Miller and E. J. Prosen, J. Res. Nat. Bur. Stand. 62, 213 (1959).
19. S. R. Gunn and L. G. Green, J. Phys. Chem. 64, 61 (1960).
20. E. C. Kerr, H. N. Hersh and H. L. Johnston, J. Amer. Chem. Soc. 72, 4738 (1950).
21. N. E. Schmidt, Russ. J. Inorg. Chem. 11, 241 (1966).
22. K. K. Kelley, J. Amer. Chem. Soc. 63, 1137 (1941).
23. J. C. Southard, J. Amer. Chem. Soc. 63, 3147 (1941).
24. F. C. Kracek, G. W. Morey and H. E. Merwin, Am. J. Sci. 35, 143 (1938).

Boron Oxide (B_2O_3) B_2O_3(cr)

Enthalpy Reference Temperature = T_r = 298.15 K Standard State Pressure = p° = 0.1 MPa

T/K	$C_p°$ (J $K^{-1}mol^{-1}$)	S°	$-[G°-H°(T_r)]/T$	$H°-H°(T_r)$ (kJ mol^{-1})	$\Delta_f H°$	$\Delta_f G°$	Log K_f
0	0.	0.	INFINITE	-9.293	-1265.776	-1265.776	INFINITE
100	20.740	10.983	96.353	-8.537	-1269.425	-1244.466	650.042
200	43.844	32.869	59.093	-5.245	-1271.165	-1218.726	318.299
298.15	62.588	53.953	53.953	0.	-1271.936	-1192.796	208.973
300	62.927	54.341	53.954	0.116	-1271.944	-1192.305	207.599
400	77.948	74.600	56.614	7.194	-1272.047	-1165.729	152.229
500	89.287	93.269	62.105	15.582	-1271.710	-1139.180	119.009
600	98.115	110.360	68.744	24.969	-1271.024	-1112.733	96.872
700	105.228	126.039	75.825	35.149	-1270.036	-1086.426	81.070
723.000	106.692	129.464	77.477	37.587	------ CRYSTAL <--> LIQUID -----		
800	111.169	140.489	83.017	45.978	-1268.779	-1060.279	69.229
900	116.190	153.879	90.156	57.351	-1267.277	-1034.304	60.029
1000	120.625	166.356	97.159	69.197	-1265.548	-1008.509	52.679
1100	124.516	178.039	103.986	81.458	-1263.611	-982.897	46.674
1200	128.072	189.029	110.620	94.092	-1261.478	-957.470	41.678
1300	131.294	199.409	117.054	107.061	-1259.167	-932.228	37.457
1400	134.306	209.251	123.291	120.344	-1256.684	-907.171	33.847
1500	137.110	218.614	129.336	133.916	-1254.042	-882.298	30.724

PREVIOUS: December 1964 CURRENT: June 1971

Boron Oxide (B_2O_3) B_2O_3(cr)

BOROXIDE (B2O3) LIQUID M_r = 69.6182

Boron Oxide (B_2O_3) $B_2O_3(l)$

$S°(298.15\ K) = 78.44 \pm 0.4\ J\ K^{-1}\ mol^{-1}$

$T_{fus} = 723 \pm 0.5\ K$

$\Delta_f H°(298.15\ K) = -1253.4 \pm 2.5\ kJ\ mol^{-1}$

$\Delta_{fus}H° = 24.071 \pm 0.4\ kJ\ mol^{-1}$

Enthalpy of Formation

$\Delta_f H°(l,\ 298.15\ K)$, is calculated from $\Delta_r H°(25°C) = 4.44 \pm 0.06\ kcal\ mol^{-1}$ (1) for $B_2O_3(cr) \rightarrow B_2O_3(glass)$, using the JANAF value $\Delta_f H°(B_2O_3,\ cr,\ 298.15\ K) = -304.0 \pm 0.5\ kcal\ mol^{-1}$. The value of $\Delta_r H°$ is the difference between the enthalpies of crystalline and glassy B_2O_3 determined by Johnson and Hubbard (1).

Heat Capacity and Entropy

Heat capacities of B_2O_3(glass) have been measured adiabatically in the temperature regions 59.6-295 K by Turdakin and Tarasov (2), and 306.7-910.8 K by Schmidt (3). Thomas and Parks (4) also measured isothermally the heat capacities from 308 to 594 K in a radiation calorimeter. Both Schmidt (3) and Thomas and Parks (4) observed a rapid rise in heat capacity in the region 500-580 K. This rise indicates a glass transition in liquid B_2O_3.

Southard (5) and Krasovitskaya et al. (6) determined enthalpy changes by drop calorimetry in the temperature regions 381.7-1776.8 K and 1015-2154 K, respectively.

We have adopted the C_p data of Turdakin (60-295 K) and Shmidt (307-600 K) which are smoothly joined at 298.15 K by a polynomial curve fitting technique. The drop calorimetric data of Southard in this region 381 to 600 K are less reliable since the final state of the sample was dependent upon previous thermal history through the glass transition region. Above the melting point, we have chosen a constant C_p, 31.0 cal $K^{-1}\ mol^{-1}$, which is a compromise between the adiabatic C_p and drop calorimetric data. The uncertainty of this choice, ±0.6 cal $K^{-1}\ mol^{-1}$, is within the experimental errors of both techniques.

$S°(298.15\ K)$ is calculated from that of the crystal by adding $\Delta_{fus}H°/T_{fus}$ and the difference in entropy, $S°(723\ K)-S°(298.15\ K)$, between the crystal and liquid.

Fusion Data

See the crystal table for details.

Vaporization Data

The boiling point is calculated as the temperature at which the fugacity is 1 bar for $B_2O_3(l) \rightarrow B_2O_3(g)$. The enthalpy of vaporization is the difference in $\Delta_f H°$ at the boiling point between liquid and gas.

References

1. G. K. Johnson and W. N. Hubbard, J. Chem. Thermodynamics 1, 459 (1969).
2. V. A. Turdakin and V. V. Tarasov, Russ. J. Inorg. Chem. 11, 501 (1966).
3. N. E. Schmidt, Russ. J. Inorg. Chem. 11, 241 (1966).
4. S. B. Thomas and G. S. Parks, J. Phys. Chem. 35, 2091 (1931).
5. J. C. Southard, J. Amer. Chem. Soc. 63, 3147 (1941).
6. R. M. Krasovitskaya, P. B. Kantor, L. S. Kan, V. V. Kandyba, L. M. Kutsyna and E. N. Fomichev, Russ. J. Phys. Chem. 35, 737 (1961).

Enthalpy Reference Temperature = T_r = 298.15 K Standard State Pressure = p° = 0.1 MPa

T/K	$C_p°$	S°	$-[G°-H°(T_r)]/T$	$H°-H°(T_r)$	$\Delta_f H°$	$\Delta_f G°$	Log K_f
	J K⁻¹mol⁻¹			kJ mol⁻¹			
0	0.	16.617	INFINITE	-10.060	-1247.966	-1247.966	INFINITE
100	25.786	32.674	122.342	-8.967	-1251.278	-1228.488	641.696
200	45.915	56.849	83.667	-5.363	-1252.707	-1205.064	314.730
298.15	62.802	78.443	78.443	0.	-1253.359	-1181.521	206.998
300	63.095	78.832	78.444	0.116	-1253.366	-1181.075	205.644
400	77.404	98.998	81.097	7.160	-1253.504	-1156.945	151.082
500	89.203	117.606	86.563	15.522	-1253.193	-1132.832	118.346
600	132.842	138.348	93.368	26.988	-1250.428	-1108.930	96.541
700	129.788	158.562	101.280	40.097	-1246.511	-1085.668	81.014
723.000	129.704	162.757	103.170	43.081	------ CRYSTAL <--> LIQUID -----		
800	129.704	175.883	109.547	53.069	-1243.111	-1062.926	69.402
900	129.704	191.160	117.783	66.039	-1240.012	-1040.592	60.394
1000	129.704	204.826	125.816	79.009	-1237.159	-1018.589	53.206
1100	129.704	217.188	133.570	91.980	-1234.512	-996.861	47.337
1200	129.704	228.474	141.015	104.950	-1232.043	-975.367	42.457
1300	129.704	238.855	148.147	117.921	-1229.730	-954.072	38.335
1400	129.704	248.468	154.974	130.891	-1227.560	-932.950	34.809
1500	129.704	257.416	161.509	143.861	-1225.520	-911.979	31.758
1600	129.704	265.787	167.767	156.832	-1223.603	-891.139	29.093
1700	129.704	273.650	173.767	169.802	-1221.802	-870.416	26.745
1800	129.704	281.064	179.524	182.773	-1220.115	-849.796	24.660
1900	129.704	288.077	185.054	195.743	-1218.537	-829.266	22.798
2000	129.704	294.730	190.373	208.713	-1217.065	-808.817	21.124
2100	129.704	301.058	195.494	221.684	-1215.698	-788.438	19.611
2200	129.704	307.092	200.431	234.654	-1214.434	-768.122	18.238
2300	129.704	312.857	205.195	247.625	-1213.271	-747.862	16.984
2400	129.704	318.378	209.796	260.595	-1312.727	-725.513	15.790
2500	129.704	323.672	214.246	273.565	-1311.919	-701.062	14.648
2600	129.704	328.759	218.553	286.536	-1311.143	-676.643	13.594
2700	129.704	333.654	222.726	299.506	-1310.398	-652.253	12.619
2800	129.704	338.371	226.773	312.477	-1309.685	-627.890	11.713
2900	129.704	342.923	230.700	325.447	-1309.001	-603.554	10.871
3000	129.704	347.320	234.514	338.417	-1308.346	-579.238	10.085

PREVIOUS: December 1964 CURRENT: June 1971

Boron Oxide (B_2O_3) $B_2O_3(l)$

BORON OXIDE (B_2O_3) CRYSTAL–LIQUID M_r = 69.6182

0 to 723 K crystal
above 723 K liquid

Refer to the individual tables for details.

Boron Oxide (B_2O_3) B_2O_3(cr,l)

Enthalpy Reference Temperature = T_r = 298.15 K Standard State Pressure = $p°$ = 0.1 MPa

T/K	C_p° (J K⁻¹mol⁻¹)	S° (J K⁻¹mol⁻¹)	$-[G^\circ-H^\circ(T_r)]/T$ (J K⁻¹mol⁻¹)	$H^\circ-H^\circ(T_r)$ (kJ mol⁻¹)	Δ_fH° (kJ mol⁻¹)	Δ_fG° (kJ mol⁻¹)	Log K_f
0	0.	0.	INFINITE	-9.293	-1265.776	-1265.776	INFINITE
100	20.740	10.983	96.353	-8.537	-1269.425	-1244.466	650.042
200	43.844	32.869	59.093	-5.245	-1271.165	-1218.726	318.299
298.15	62.588	53.953	53.953	0.	-1271.936	-1192.796	208.973
300	62.927	54.341	53.954	0.116	-1271.944	-1192.305	207.599
400	77.948	74.600	56.614	7.194	-1272.047	-1165.729	152.229
500	89.287	93.269	62.105	15.582	-1271.710	-1139.180	119.009
600	98.115	110.360	68.744	24.969	-1271.024	-1112.733	96.872
700	105.228	126.039	75.825	35.149	-1270.036	-1086.426	81.070
723.000	106.692	129.464	77.477	37.587	CRYSTAL <--> LIQUID		
723.000	129.704	162.757	77.477	61.657	TRANSITION		
800	129.704	175.883	86.328	71.644	-1243.112	-1062.927	69.402
900	129.704	191.160	97.144	84.615	-1240.014	-1040.593	60.394
1000	129.704	204.826	107.241	97.585	-1237.160	-1018.590	53.206
1100	129.704	217.188	116.683	110.556	-1234.514	-996.863	47.337
1200	129.704	228.474	125.535	123.526	-1232.044	-975.368	42.457
1300	129.704	238.855	133.858	136.496	-1229.732	-954.074	38.335
1400	129.704	248.468	141.706	149.467	-1227.561	-932.951	34.809
1500	129.704	257.416	149.125	162.437	-1225.521	-911.980	31.758
1600	129.704	265.787	156.157	175.408	-1223.604	-891.141	29.093
1700	129.704	273.650	162.840	188.378	-1221.804	-870.417	26.745
1800	129.704	281.064	169.204	201.348	-1220.116	-849.797	24.660
1900	129.704	288.077	175.277	214.319	-1218.538	-829.267	22.798
2000	129.704	294.730	181.085	227.289	-1217.066	-808.818	21.124
2100	129.704	301.058	186.649	240.260	-1215.699	-788.439	19.611
2200	129.704	307.092	191.987	253.230	-1214.435	-768.123	18.238
2300	129.704	312.857	197.118	266.200	-1213.272	-747.863	16.985
2400	129.704	318.378	202.056	279.171	-1312.728	-725.514	15.790
2500	129.704	323.672	206.816	292.141	-1311.920	-701.063	14.648
2600	129.704	328.759	211.409	305.112	-1311.144	-676.644	13.594
2700	129.704	333.654	215.846	318.082	-1310.400	-652.254	12.619
2800	129.704	338.372	220.139	331.052	-1309.686	-627.891	11.713
2900	129.704	342.923	224.294	344.023	-1309.002	-603.555	10.871
3000	129.704	347.320	228.322	356.993	-1308.347	-579.240	10.085

PREVIOUS: CURRENT: June 1971

Boron Oxide (B_2O_3) B_2O_3(cr,l)

BOROXIDE (B$_2$O$_3$) IDEAL GAS M_r = 69.6182

$\Delta_f H°(0\ K) = -834.8 \pm 4.2\ kJ\ mol^{-1}$

$S°(298.15\ K) = [283.8 \pm 4.2]\ J\ K^{-1}\ mol^{-1}$ $\Delta_f H°(298.15\ K) = -836.0 \pm 4.2\ kJ\ mol^{-1}$

Vibrational Frequencies (all single degenerate)

ν, cm^{-1}	ν, cm^{-1}	ν, cm^{-1}	ν, cm^{-1}	ν, cm^{-1}	ν, cm^{-1}	ν, cm^{-1}	ν, cm^{-1}	ν, cm^{-1}
2073	730	521	[172]	[460]	2073	1240	457	480

Ground State Quantum Weight: 1 σ = 2

Point Group: C_{2v}

Bond Distances: O-B = 1.36 ± 0.02 Å B-O = 1.20 ± 0.03 Å

Bond Angles: O-B-O = 180° B-O-B = [120]°

Product of the Moments of Inertia: $I_A I_B I_C = 3.153764 \times 10^{-114}\ g^3\ cm^6$

Enthalpy of Formation

The vapor pressure of $B_2O_3(\ell)$ was measured by many investigators using various methods. 2nd and 3rd law analyses of these data (1-11) are given below. We have adopted the 3rd law enthalpy of vaporization, $\Delta_{vap}H°(298.15\ K) = 99.75\ kcal\ mol^{-1}$, to derive $\Delta_f H°(B_2O_3, g, 298.15\ K) = -199.8 \pm 1\ kcal\ mol^{-1}$, using $\Delta_f H°(B_2O_3, \ell, 298.15\ K) = -299.56\ kcal\ mol^{-1}$. The adopted $\Delta_{vap}H°(298.15\ K)$ is calculated from Hildenbrand's data (1) and this value also serves as an average of all derived enthalpies of vaporization except that of Cole (11). The uncertainty is assigned to be at least ±1 kcal mol^{-1} since the bending frequency was estimated to be either 172 or 260 cm^{-1} (12) which could lead to a difference ±1 kcal mol^{-1} in the $\Delta_{vap}H°$ calculation through the Gibbs energy functions of $B_2O_3(g)$.

Source	T/K	Method	Data Points	$\Delta_{vap}H°(298.15\ K)$, kcal mol^{-1} 2nd law	3rd law	Drift cal K^{-1} mol^{-1}	$\Delta_f H°(298.15\ K)$ kcal mol^{-1}
(1)	1436-1584	Torsion-effusion	24	100.97± 1.34	99.75±0.35	-0.8 ±0.7	-199.81
(2)	1946-2419	Transpiration and direct boiling point	16	92.55± 2.52	102.80±1.8	4.6 ±1.2	-196.76
(3)	1300	Mass spectrometric	--	100.5 ± 3	--	--	-199.06
(4)	1325-1547	Knudsen-effusion	9	93.08± 1.37	99.08±0.74	4.2 ±1.0	-200.48
(5)	1228-1641	Mass spectrometric	16	92.32± 0.94	99.17±1.40	4.6 ±0.6	-200.39
(6)	1229-1515	Knudsen-effusion	12	82.65± 2.85	98.52±1.70	11.1 ±2.0	-201.04
(7)	1414-1621	Torsion-effusion	19	95.55± 0.91	101.50±0.56	3.80±0.62	-198.06
(8)	1501-1566	Effusion	6	101.67±12.46	98.31±0.79	-2.2 ±8.1	-201.25
(9)	1567-1808	Carrier gas	6	66.17± 7.8	98.00±5.06	18.8 ±4.9	-201.56
(10)	1331-1642	Knudsen-effusion	22[a]	89.64± 1.62	99.06±1.44	6.4 ±1.1	-200.50
(11)	1473-1673	Carrier gas	6	73.27± 2.28	87.36±1.48	9.0 ±1.4	-212.20

[a] Four points were rejected due to failure of a statistical test.

Heat Capacity and Entropy

The V shaped molecular structure for B_2O_3 was proposed by many investigators on the basis of infrared spectrometry (12, 13) and electron diffraction (14). This structure has been reconfirmed by recent electric deflection (15) and electron diffraction (16) studies which definitely rule out any non-polar structure such as the bipyramidal model as suggested by Hanst et al. (17). Thus the assignment of vibrational frequencies, based on the V shaped structure, should be valid. The adopted vibrational frequencies (including two estimated values) were obtained from Sommer et al. (12) in their infrared spectrometric studies. Their values have been adjusted to the average isotopic species. Other spectrometric data (13, 18) are in reasonable agreement with those adopted in the tabulation.

Akishin and Spiridonov (14) determined the bond distances and bond angle (B-O-B = 95°) by electron diffraction. We have adopted their bond distances in the tabulation, but tentatiely select an apex angle B-O-B = 120° based on force-constant calculations of Weltner and Warn (13) and Sommer et al. (12). The principal moments of inertia are $I_A = 2.9763 \times 10^{-39}$, $I_B = 31.0977 \times 10^{-39}$ and $I_C = 34.0740 \times 10^{-39}\ g\ cm^2$.

References

1. D. L. Hildenbrand, W. F. Hall and N. D. Potter, J. Chem. Phys. 39, 296 (1963).
2. F. T. Greene and J. L. Margrave, J. Phys. Chem. 70, 2112 (1966).
3. A. Buchler and J. B. Berkowitz-Mattuck, J. Chem. Phys. 39, 286 (1963).
4. P. E. Blackburn, "Research on Thermodynamics of the Al-B-O and Be-B-O Systems", Quar. Rep. No. 4, 1963, A. D. Little, Inc., P. E. Blackburn, A. Buchler and J. L. Stauffer, J. Phys. Chem. 70, 2469 (1966).
5. A. Sommer, Ph.D. Thesis, The Ohio State University, 1962. His vapor pressure data superceded those values given by D. White et al., J. Phys. Chem. 65, 1404 (1961) who have revised their values because of errors in some of the temperature.
6. A. N. Nesmeyanov and L. P. Firsova, Russ. J. Phys. Chem. 34, 490 (1960).
7. M. D. Scheer, J. Phys. Chem. 61, 1184 (1957).
8. A. W. Searcy and C. E. Myers, J. Phys. Chem. 61, 957 (1957).
9. J. R. Soulen, P. Sthapitanonda and J. L. Margrave, J. Phys. Chem. 59, 132 (1955).
10. R. Speiser, S. Naiditch and H. L. Johnston, J. Am. Chem. Soc. 72, 2578 (1950).
11. S. S. Cole and N. W. Taylor, J. Amer. Chem. Soc. 18, 82 (1935).
12. A. Sommer, D. White, M. J. Linevsky and D. E. Mann, J. Chem. Phys. 38, 87 (1963).
13. W. Weltner and J. R. W. Warn, J. Chem. Phys. 37, 292 (1962).
14. P. A. Akishin and V. P. Spiridonov, Doklady, Akad. Nauk. SSSR 131, 557 (1960).
15. F. W. Kaiser, J. S. Muenter and W. Klemperer, J. Chem. Phys. 48, 3339 (1968).
16. Y. S. Ezhov, S. M. Tolmachev, V. P. Spiridonov and N. G. Rambidi, Teplofizika Vysokikh Temp. Akad. Nauk SSSR 6, 68 (1968).
17. P. L. Hanst, V. H. Early and W. Klemperer, J. Chem. Phys. 42, 1097 (1965).
18. V. S. Nikiten and A. A. Mal'tsev, Vest. Mosk. Khim. 11, 22 (1970).

Boron Oxide (B$_2$O$_3$) $B_2O_3(g)$

Enthalpy Reference Temperature = T_r = 298.15 K Standard State Pressure = p° = 0.1 MPa

T/K	$C_p°$ (J K^{-1}mol^{-1})	S°	$-[G°-H°(T_r)]/T$	$H°-H°(T_r)$ (kJ mol^{-1})	$\Delta_f H°$	$\Delta_f G°$	Log K_f
0	0.	0.	INFINITE	-14.337	-834.847	-834.847	INFINITE
100	40.042	226.986	334.961	-10.797	-835.713	-832.354	434.777
200	55.330	259.361	289.548	-6.037	-835.985	-828.844	216.472
250	61.832	272.432	284.842	-3.102	-835.979	-827.059	172.805
298.15	66.855	283.768	283.768	0.	-835.963	-825.343	144.597
300	67.028	284.182	283.769	0.124	-835.963	-825.277	143.694
350	71.228	294.841	284.601	3.584	-835.972	-823.496	122.900
400	74.727	304.587	286.499	7.235	-836.033	-821.710	107.304
450	77.728	313.566	289.014	11.048	-836.152	-819.913	95.173
500	80.362	321.894	291.891	15.002	-836.317	-818.100	85.466
600	84.817	336.954	298.173	23.269	-836.752	-814.418	70.901
700	88.439	350.310	304.685	31.938	-837.275	-810.655	60.492
800	91.403	362.320	311.151	40.935	-837.849	-806.814	52.680
900	93.830	373.230	317.452	50.200	-838.455	-802.898	46.599
1000	95.822	383.223	323.537	59.686	-839.087	-798.914	41.731
1100	97.462	392.435	329.387	69.353	-839.743	-794.864	37.745
1200	98.819	400.975	335.001	79.169	-840.428	-790.755	34.421
1300	99.950	408.931	340.386	89.109	-841.146	-786.586	31.605
1400	100.897	416.374	345.550	99.153	-841.902	-782.361	29.190
1500	101.696	423.363	350.507	109.284	-842.701	-778.081	27.095
1600	102.375	429.949	355.269	119.488	-843.550	-773.746	25.260
1700	102.955	436.173	359.846	129.755	-844.453	-769.355	23.639
1800	103.454	442.072	364.252	140.076	-845.415	-764.911	22.197
1900	103.886	447.677	368.496	150.444	-846.440	-760.411	20.905
2000	104.262	453.016	372.590	160.852	-847.531	-755.855	19.741
2100	104.591	458.111	376.542	171.295	-848.691	-751.243	18.686
2200	104.880	462.983	380.361	181.769	-849.924	-746.573	17.726
2300	105.136	467.651	384.056	192.270	-851.230	-741.847	16.848
2400	105.363	472.131	387.633	202.795	-953.131	-734.924	15.995
2500	105.565	476.436	391.099	213.342	-954.747	-725.799	15.165
2600	105.746	480.580	394.462	223.907	-956.376	-716.609	14.397
2700	105.908	484.574	397.726	234.490	-958.019	-707.356	13.685
2800	106.054	488.428	400.897	245.088	-959.677	-698.041	13.022
2900	106.186	492.152	403.980	255.700	-961.351	-688.669	12.404
3000	106.306	495.754	406.979	266.325	-963.042	-679.237	11.827
3100	106.415	499.242	409.899	276.961	-964.749	-669.748	11.285
3200	106.515	502.622	412.744	287.608	-966.474	-660.205	10.777
3300	106.606	505.901	415.518	298.264	-968.214	-650.608	10.298
3400	106.689	509.085	418.223	308.929	-969.971	-640.956	9.847
3500	106.765	512.178	420.864	319.602	-971.744	-631.254	9.421
3600	106.836	515.187	423.442	330.282	-973.532	-621.500	9.018
3700	106.901	518.115	425.961	340.969	-975.337	-611.697	8.636
3800	106.961	520.967	428.424	351.662	-977.156	-601.843	8.273
3900	107.017	523.746	430.833	362.361	-978.991	-591.943	7.928
4000	107.069	526.456	433.190	373.065	-980.840	-581.996	7.600
4100	107.117	529.100	435.497	383.774	-982.703	-572.001	7.287
4200	107.161	531.682	437.756	394.488	-1944.280	-547.913	6.814
4300	107.203	534.204	439.970	405.206	-1943.998	-514.669	6.252
4400	107.242	536.669	442.140	415.929	-1943.734	-481.433	5.715
4500	107.279	539.080	444.267	426.655	-1943.488	-448.202	5.203
4600	107.313	541.438	446.354	437.384	-1943.261	-414.977	4.712
4700	107.345	543.746	448.402	448.117	-1943.055	-381.755	4.243
4800	107.375	546.006	450.412	458.853	-1942.871	-348.538	3.793
4900	107.403	548.221	452.386	469.592	-1942.709	-315.323	3.361
5000	107.430	550.391	454.324	480.334	-1942.570	-282.113	2.947
5100	107.455	552.518	456.229	491.078	-1942.457	-248.904	2.549
5200	107.479	554.605	458.101	501.825	-1942.370	-215.700	2.167
5300	107.502	556.653	459.941	512.574	-1942.312	-182.494	1.799
5400	107.523	558.662	461.750	523.325	-1942.285	-149.291	1.444
5500	107.543	560.636	463.530	534.078	-1942.289	-116.088	1.103
5600	107.562	562.574	465.282	544.834	-1942.328	-82.883	0.773
5700	107.580	564.477	467.005	555.591	-1942.403	-49.678	0.455
5800	107.597	566.349	468.702	566.350	-1942.517	-16.470	0.148
5900	107.613	568.188	470.373	577.110	-1942.672	16.736	-0.148
6000	107.629	569.997	472.018	587.872	-1942.870	49.949	-0.435

PREVIOUS: June 1971 (1 atm) CURRENT: June 1971 (1 bar)

Boron Oxide (B$_2$O$_3$) $B_2O_3(g)$

LEAD BORATE (PbB_2O_4) CRYSTAL M_r = 292.8176

$S°(298.15\ K) = [130.5 \pm 12]\ J\ K^{-1}\ mol^{-1}$

$\Delta_fH°(0\ K)$ = Unknown
$\Delta_fH°(298.15\ K) = -1556 \pm 6\ kJ\ mol^{-1}$

Enthalpy of Formation

Evans (1) analyzed the enthalpies of formation of the lead borate system. Shartsis and Newman measured enthalpies of solution of the lead borate glasses in 2 N HNO_3. In their phase studies, Mazzetti and De Carli (3) established that the compounds formed were $PbO{\cdot}B_2O_3$, $PbO{\cdot}2B_2O_3$ and 2 $PbO{\cdot}5B_2O_3$. From the Shartsis and Newman (2) paper, Evans (1) obtained the following:

$$PbO(cr + B_2O_3(cr) \rightarrow PbO{\cdot}B_2O_3(gls) \qquad \Delta_rH° = -16.2\ kcal\ mol^{-1}$$

and estimated a $\Delta_fH°(298.15\ K)$ of 7.0 kcal mol^{-1} for the conversion to the $PbO{\cdot}B_2O_3(cr)$. With auxiliary JANAF values, the $\Delta_fH°(298.15\ K)$ was calculated to be -372 kcal mol^{-1}. The results of Lepinskikh and Esin (4) studying the lead borate system at 1000 K with cells of the type $Pb(\ell)/PbO,B_2O_3/Pt/O_2(g)$ were in fair agreement with the results of Shartsis and Newman (2).

Heat Capacity and Entropy

The heat capacity employed in this table was obtained by graphically smoothing the summation of the heat capacities for B_2O_3 and PbO.

The entropy at 298.15 was obtained from a consideration of the entropies of sodium and calcium borates and those of the constituent oxides, as given by Kelley and King (5). These indicate for the reaction MO + B_2O_3 → $Mo{\cdot}B_2O_3$ that the simple addition of entropies will give values at least 2.7 cal K^{-1} mol^{-1} too low for the reaction. The reason for this is probably structural, the $B_2O_4^{--}$ ion is a flexible chain. Therefore, $S°(PbB_2O_4$, cr, 298.15 K) was obtained from the summation of S°(298.15 K) for constituent oxides and the addition of 2.7 cal K^{-1} mol^{-1}.

References

1. W. H. Evans, U. S. Nat. Bur. Stand. Report 7192, 1961.
2. L. Shartsis and E. Newman, J. Am. Ceram. Soc. 31, 213 (1948).
3. C. Mazzetti and F. De Carli, Gazz. chim. ital. 56, 19 (1926).
4. B. M. Lepinskikh and O. A. Esin, Zhur. Neorg. Khim. 6, 1223 (1961).
5. K. K. Kelley and E. G. King, U. S. Bur. Mines Bull. 592, 1961.

Lead Borate (PbB_2O_4) $B_2O_4Pb_1(cr)$

Enthalpy Reference Temperature = T_r = 298.15 K Standard State Pressure = p° = 0.1 MPa

	J K⁻¹mol⁻¹			kJ mol⁻¹			
T/K	$C_p°$	S°	$-[G°-H°(T_r)]/T$	$H°-H°(T_r)$	$\Delta_fH°$	$\Delta_fG°$	Log K_f
0							
100							
200							
298.15	107.110	130.541	130.541	0.	-1556.448	-1450.245	254.077
300	107.529	131.205	130.543	0.199	-1556.450	-1449.586	252.395
400	129.704	165.276	135.038	12.095	-1555.950	-1414.009	184.651
500	147.695	196.213	144.227	25.993	-1554.445	-1378.679	144.030
600	162.339	224.474	155.281	41.516	-1552.093	-1343.734	116.982
700	174.473	250.436	167.046	58.374	-1553.877	-1308.443	97.637
800	184.514	274.413	178.987	76.340	-1550.166	-1273.627	83.159
900	192.464	296.613	190.839	95.196	-1545.870	-1239.312	71.928
1000	199.577	317.281	202.462	114.819	-1541.050	-1205.503	62.969
1100	204.179	336.522	213.785	135.011	-1535.862	-1172.197	55.663
1200	208.363	354.479	224.770	155.652	-1530.390	-1139.376	49.596
1300	210.874	371.262	235.400	176.620	-1524.741	-1107.020	44.481
1400	212.547	386.945	245.671	197.784	-1519.046	-1075.101	40.113
1500	215.058	401.700	255.586	219.172	-1513.274	-1043.592	36.341
1600	216.731	415.631	265.157	240.758	-1507.448	-1012.470	33.054
1700	218.823	428.833	274.400	262.535	-1501.572	-981.713	30.164
1800	220.915	441.402	283.332	284.526	-1495.624	-951.305	27.606
1900	222.589	453.391	291.969	306.701	-1489.630	-921.228	25.326
2000	224.262	464.851	300.329	329.044	-1483.608	-891.468	23.283

PREVIOUS: December 1962 CURRENT: March 1965

Lead Borate (PbB_2O_4) $B_2O_4Pb_1(cr)$

TITANIUM BORIDE (TiB_2) CRYSTAL M_r = 69.50

$S°(298.15\ K) = 28.48 \pm 0.4\ J\ K^{-1}\ mol^{-1}$

$T_{fus} = 3193\ K$

$\Delta_f H°(0\ K) = -277.8 \pm 16.7\ kJ\ mol^{-1}$

$\Delta_f H°(298.15\ K) = -279.5 \pm 16.7\ kJ\ mol^{-1}$

$\Delta_{fus} H° = [100.4]\ kJ\ mol^{-1}$

Enthalpy of Formation

Kibler et al. (1) studied the vapor pressure of the reaction $TiB_2(cr) + 1/2\ C(cr) = Ti(g) + 1/2\ B_4C(cr)$ by comparing the absorption intensity of the 3371.45 Å Ti resonance line of the above reaction at temperature T_1 to Ti(g) over pure titanium metal at temperature T_2. Using JANAF auxiliary data, $\Delta G°(2475\ K) = 81.612\ kcal\ mol^{-1}$ for the above reaction and $\Delta_f H°(TiB_2, cr, 298.15\ K) = -66.9\ kcal\ mol^{-1}$ were calculated.

Williams (2) found 2150 ± 25 K to be the temperature at which $TiB_2(cr)$ was formed from a mixture of TiN(cr) and BN(cr). If $\Delta G°(2150\ K) = 0$ is assumed for the reaction $TiN(cr) + 2\ BN(cr) = TiB_2(cr) + 3/2\ N_2(g)$, then $\Delta_f H°(TiB_2, cr, 298.15\ K) = -70.7\ kcal\ mol^{-1}$. If, however, $N_2(g)$ is 1/2 atm, then $\Delta G°(2150\ K) = 4.4\ kcal\ mol^{-1}$ and $\Delta_f H°(TiB_2, cr, 298.15\ K) = -66.3\ kcal\ mol^{-1}$. He has also shown by stability comparisons that ΔG and ΔH>0 for the reaction $TiB_2 + 3/2\ C = TiC + 1/2\ B_4C$. Thus $\Delta_f H°(TiB_2, 298.15\ K) < -51\ kcal\ mol^{-1}$.

Schissel and Trulson (3) used a mass spectrometer with Knudsen cells to study the vaporization of the titanium-boron system. Only the set of measurements without excess boron or titanium was used. The value $\Delta G°(2261\ K) = 194.9\ kcal\ mol^{-1}$ for the reaction $TiB_2(cr) = Ti(g) + 2\ B(g)$ was obtained. This corresponds to $\Delta_f H°(TiB_2, cr, 298.15\ K) = -62.9\ kcal\ mol^{-1}$ using JANAF auxiliary data.

Epel'baum and M. I. Starostina (4) report $\Delta_f H°(TiB_2, cr, 298.15\ K) = -66.85 \pm 2.7\ kcal\ mol^{-1}$ from combustion calorimetry.

Lowell and Williams (5) used high temperature calorimetry to obtain the enthalpy of formation of $TiB_2(cr)$. From their data and JANAF auxiliary data, $\Delta_f H°(TiB_2, cr, 298.15\ K) = -48.2 \pm 5\ kcal\ mol^{-1}$ was calculated. We question the authenticity of the experiment and believe their value is an upper limit.

$\Delta_f H°(TiB_2, cr, 298.15\ K) = -66.8 \pm 4\ kcal\ mol^{-1}$ was chosen as representative of the first four investigations.

Heat Capacity and Entropy

The heat capacity of $TiB_2(cr)$ has been determined by the following investigators.

Source	Method	T/K
(6)	Adiabatic Calorimetry	5- 350
(7)	Drop calorimeter	373- 977
(8)	Drop type ice calorimeter	586-2689
(9)	Copper-block drop calorimeter	420-1180
(10)	Arc-imaging furnace	1300-2150
(11)	Pulse-method	1733-2417

The low temperature data of Westrum (1) and the medium range data of Walker et al. (7) were used to obtain the C_p's of this tabulation. Their data were fitted to a Shomate plot and extrapolated to 4000 K.

$S°(298.15\ K) = 6.808\ cal\ K^{-1}\ mol^{-1}$ was determined by Westrum (1).

Fusion Data

The melting point has been reported as 3063 K, 3193 K, and 3203 K by Glaser (12), Post et al. (13), and Samsonov and Petrash (14), respectively. The enthalpy of fusion is derived from an estimated $\Delta_{fus} S = 2.5\ cal\ K^{-1}\ g\ atom^{-1}$ at 3193 K. It should be pointed out that an error of $0.5\ cal\ K^{-1}\ mol^{-1}\ g\ atom^{-1}$ in the estimation of $\Delta_{fus} S°$ is 5 kcal mol^{-1} in $\Delta_{fus} H°$.

References

1. G. M. Kibler, T. F. Lyon, M. J. Linevsky, and V. J. DeSantis, General Electric Co., WADD-TR-60-646, Part IV (1964).
2. W. S. Williams, J. Phys. Chem. 65, 2213 (1961).
3. P. O. Schissel and O. C. Trulson, J. Phys. Chem. 66, 1492 (1962).
4. V. A. Epel'baum and M. I. Starostina, Bor Trudy Konf. Khim. Bora i Ego Soedinenii 97 (1955, pub. 1958).
5. C. E. Lowell and W. S. Williams, Rev. Sci. Inst. 32, 1120 (1961).
6. L. Kaufman and E. V. Clougherty, Technical Documentary Report No. RTD-TDR-63-4096, Part 1, p. 239 (1963).
7. B. E. Walker, C. T. Ewing, and R. R. Miller, J. Phys. Chem. 61, 1682 (1957).
8. Southern Research Institute, Technical Documentary Report No. ASD-TDR-62-675, p. 253 (1963).
9. R. Mezaki, M.S. Thesis, University of Wisconsin (1961).
10. H. Prophet, Annual Report for 1962 of The Dow Chemical Co. to ARPA under Contract AF 04(611)7554.
11. R. J. Barriault et al., "Thermodynamics of Certain Refractory Compounds", AVCO Corporation, report ASD-TR-61-260, Part I, Vol. 1 (1962).
12. F. W. Glaser, Trans. AIME 194, 391 (1952).
13. B. Post, F. W. Glaser, and D. Moskowitz, Acta. Met. 2, 20 (1954).
14. G. V. Samsonov and G. V. Petrash, Metallovedi Obrabotka Metalov No. 4, 19 (1955).

Titanium Boride (TiB_2) $B_2Ti_1(cr)$

Enthalpy Reference Temperature = T_r = 298.15 K   Standard State Pressure = p° = 0.1 MPa

	J K⁻¹mol⁻¹			kJ mol⁻¹			
T/K	$C_p°$	S°	$-[G°-H°(T_r)]/T$	$H°-H°(T_r)$	$\Delta_f H°$	$\Delta_f G°$	Log K_f
0	0.	0.	INFINITE	-5.577	-277.811	-277.811	INFINITE
100	7.523	2.518	56.366	-5.385	-278.227	-277.591	144.999
200	28.125	14.046	32.058	-3.602	-279.027	-276.623	72.247
298.15	44.279	28.485	28.485	0.	-279.491	-275.334	48.237
300	44.522	28.759	28.486	0.082	-279.498	-275.308	47.935
400	54.890	43.136	30.375	5.104	-279.814	-273.864	35.763
500	61.668	56.156	34.253	10.952	-280.169	-272.336	28.451
600	66.166	67.815	38.893	17.353	-280.556	-270.734	23.569
700	69.496	78.274	43.785	24.142	-280.964	-269.064	20.078
800	72.069	87.723	48.696	31.222	-281.376	-267.336	17.455
900	74.534	96.355	53.519	38.553	-281.799	-265.556	15.412
1000	76.889	104.331	58.206	46.125	-282.273	-263.727	13.776
1100	79.136	111.765	62.741	53.927	-282.845	-261.846	12.434
1200	81.283	118.744	67.120	61.949	-287.496	-259.788	11.308
1300	83.320	125.331	71.347	70.180	-287.529	-257.477	10.346
1400	85.249	131.577	75.428	78.609	-287.537	-255.165	9.520
1500	87.069	137.521	79.371	87.226	-287.539	-252.853	8.805
1600	88.780	143.196	83.184	96.019	-287.553	-250.540	8.179
1700	90.383	148.627	86.875	104.978	-287.599	-248.225	7.627
1800	91.876	153.836	90.451	114.092	-287.698	-245.907	7.136
1900	93.261	158.841	93.920	123.350	-287.872	-243.581	6.696
2000	94.537	163.657	97.287	132.741	-302.828	-240.789	6.289
2100	95.705	168.298	100.558	142.254	-303.964	-237.659	5.911
2200	96.763	172.775	103.740	151.878	-305.057	-234.476	5.567
2300	97.713	177.098	106.836	161.603	-306.118	-231.244	5.252
2400	98.554	181.275	109.851	171.416	-407.675	-225.829	4.915
2500	99.391	185.315	112.789	181.314	-408.851	-218.227	4.560
2600	100.228	189.229	115.654	191.295	-409.944	-210.581	4.231
2700	101.065	193.028	118.450	201.359	-410.953	-202.893	3.925
2800	101.901	196.718	121.180	211.507	-411.879	-195.170	3.641
2900	102.738	200.309	123.847	221.739	-412.721	-187.415	3.376
3000	103.575	203.806	126.454	232.055	-413.479	-179.633	3.128
3100	104.412	207.215	129.004	242.454	-414.153	-171.827	2.895
3193.000	105.190	210.313	131.328	252.201	------ CRYSTAL <--> LIQUID -----		
3200	105.249	210.544	131.501	252.937	-414.744	-164.000	2.677
3300	106.085	213.795	133.945	263.504	-415.251	-156.157	2.472
3400	106.922	216.975	136.341	274.154	-415.674	-148.299	2.278
3500	107.759	220.086	138.689	284.889	-416.014	-140.429	2.096
3600	108.596	223.133	140.993	295.706	-416.270	-132.552	1.923
3700	109.433	226.120	143.253	306.608	-825.541	-116.880	1.650
3800	110.269	229.050	145.473	317.593	-824.395	-97.742	1.344
3900	111.106	231.925	147.653	328.662	-823.221	-78.635	1.053
4000	111.943	234.748	149.795	339.814	-822.017	-59.558	0.778

PREVIOUS: December 1962   CURRENT: June 1965

Titanium Boride (TiB_2) $B_2Ti_1(cr)$

TITANIUM BORIDE (TiB_2) LIQUID M_r = 69.50

$S°(298.15\ K) = [56.365]\ J\ K^{-1}\ mol^{-1}$ $\Delta_f H°(298.15\ K) = [-188.029]\ kJ\ mol^{-1}$

$T_{fus} = 3193\ K$ $\Delta_{fus}H° = [100.4]\ kJ\ mol^{-1}$

$T_{dcm} = [4250]\ K$

Enthalpy of Formation

$\Delta_f H°(TiB_2, \ell, 298.15\ K)$ is calculated from $\Delta_f H°(TiB_2, cr, 298.15\ K)$ by adding the enthalpy of fusion, $\Delta_{fus}H°$, and the difference in enthalpy, $H°(3193\ K)-H°(298.15\ K)$, between the crystal and liquid.

Heat Capacity and Entropy

A glass transition is assumed at 2100 K. Heat capacities at and below 2100 K are assumed to be equal to those of TiB_2(cr). Above 2100 K the heat capacities are estimated to be constant.

$S°(TiB_2, \ell, 298.15\ K)$ is calculated in a manner analogous to that used for the enthalpy of formation.

Fusion Data

See the crystal table for details.

Decomposition Data

T_{dcm} is estimated as the temperature for which $\Delta_r G° = 0$ for the reaction $TiB_2(\ell) \rightarrow Ti(g) + 2\ B(g)$.

Titanium Boride (TiB_2) $B_2Ti_1(l)$

Enthalpy Reference Temperature = T_r = 298.15 K Standard State Pressure = p° = 0.1 MPa

T/K	$C_p°$ (J K⁻¹mol⁻¹)	S° (J K⁻¹mol⁻¹)	$-[G°-H°(T_r)]/T$ (J K⁻¹mol⁻¹)	$H°-H°(T_r)$ (kJ mol⁻¹)	$\Delta_f H°$ (kJ mol⁻¹)	$\Delta_f G°$ (kJ mol⁻¹)	Log K_f
0							
100							
200							
298.15	44.279	56.365	56.365	0.	-188.029	-192.185	33.670
300	44.522	56.640	56.366	0.082	-188.036	-192.211	33.467
400	54.890	71.016	58.256	5.104	-188.352	-193.554	25.276
500	61.668	84.036	62.133	10.952	-188.707	-194.814	20.352
600	66.166	95.695	66.773	17.353	-189.094	-196.000	17.063
700	69.496	106.154	71.665	24.142	-189.502	-197.119	14.709
800	72.069	115.603	76.576	31.222	-189.914	-198.179	12.940
900	74.534	124.235	81.399	38.553	-190.337	-199.187	11.561
1000	76.889	132.211	86.086	46.125	-190.811	-200.145	10.455
1100	79.136	139.646	90.621	53.927	-191.383	-201.052	9.547
1200	81.283	146.624	95.000	61.949	-196.034	-201.782	8.783
1300	83.320	153.211	99.227	70.180	-196.067	-202.260	8.127
1400	85.249	159.457	103.308	78.609	-196.075	-202.736	7.564
1500	87.069	165.402	107.251	87.226	-196.077	-203.211	7.076
1600	88.780	171.076	111.064	96.019	-196.091	-203.687	6.650
1700	90.383	176.507	114.755	104.978	-196.137	-204.160	6.273
1800	91.876	181.716	118.331	114.092	-196.236	-204.630	5.938
1900	93.261	186.721	121.800	123.350	-196.410	-205.091	5.638
2000	94.537	191.538	125.167	132.741	-211.366	-205.088	5.356
2100	95.705	196.179	128.439	142.254	-212.502	-204.746	5.093
2100.000	95.705	196.179	128.439	142.254	GLASS <--> LIQUID		
2100.000	108.784	196.179	128.439	142.254	TRANSITION		
2200	108.784	201.240	131.634	153.132	-212.341	-204.381	4.853
2300	108.784	206.075	134.766	164.011	-212.248	-204.022	4.633
2400	108.784	210.705	137.835	174.889	-312.740	-201.527	4.386
2500	108.784	215.146	140.839	185.768	-312.936	-196.889	4.114
2600	108.784	219.412	143.779	196.646	-313.131	-192.244	3.862
2700	108.784	223.518	146.657	207.524	-313.326	-187.590	3.629
2800	108.784	227.474	149.473	218.403	-313.522	-182.930	3.413
2900	108.784	231.292	152.229	229.281	-313.717	-178.262	3.211
3000	108.784	234.979	154.926	240.160	-313.912	-173.588	3.022
3100	108.784	238.546	157.567	251.038	-314.108	-168.908	2.846
3193.000	108.784	241.762	159.972	261.155	------ CRYSTAL <--> LIQUID -----		
3200	108.784	242.000	160.151	261.916	-314.303	-164.221	2.681
3300	108.784	245.348	162.683	272.795	-314.498	-159.527	2.525
3400	108.784	248.595	165.162	283.673	-314.694	-154.828	2.379
3500	108.784	251.749	167.591	294.552	-314.889	-150.124	2.240
3600	108.784	254.813	169.971	305.430	-315.084	-145.413	2.110
3700	108.784	257.794	172.305	316.308	-724.378	-132.909	1.876
3800	108.784	260.695	174.593	327.187	-723.339	-116.938	1.607
3900	108.784	263.521	176.837	338.065	-722.355	-100.993	1.353
4000	108.784	266.275	179.039	348.944	-721.426	-85.072	1.111
4100	108.784	268.961	181.199	359.822	-720.549	-69.174	0.881
4200	108.784	271.582	183.320	370.700	-1679.422	-39.249	0.488
4300	108.784	274.142	185.403	381.579	-1676.471	-0.232	0.003
4400	108.784	276.643	187.448	392.457	-1673.571	38.717	-0.460
4500	108.784	279.088	189.457	403.336	-1670.720	77.600	-0.901
4600	108.784	281.479	191.432	414.214	-1667.918	116.421	-1.322
4700	108.784	283.818	193.373	425.092	-1665.163	155.181	-1.725
4800	108.784	286.108	195.281	435.971	-1662.454	193.883	-2.110
4900	108.784	288.351	197.158	446.849	-1659.788	232.529	-2.479
5000	108.784	290.549	199.004	457.728	-1657.165	271.121	-2.832
5100	108.784	292.703	200.820	468.606	-1654.583	309.661	-3.172
5200	108.784	294.816	202.607	479.484	-1652.041	348.150	-3.497
5300	108.784	296.888	204.367	490.363	-1649.538	386.592	-3.810
5400	108.784	298.921	206.099	501.241	-1647.072	424.986	-4.111
5500	108.784	300.917	207.805	512.120	-1644.642	463.335	-4.400
5600	108.784	302.877	209.485	522.998	-1642.247	501.640	-4.679
5700	108.784	304.803	211.140	533.876	-1639.886	539.903	-4.948
5800	108.784	306.695	212.772	544.755	-1637.558	578.125	-5.207
5900	108.784	308.554	214.379	555.633	-1635.262	616.306	-5.456
6000	108.784	310.383	215.964	566.512	-1632.997	654.449	-5.697

PREVIOUS: December 1962 CURRENT: June 1965

Titanium Boride (TiB_2) $B_2Ti_1(l)$

TITANIUM BORIDE (TiB_2) CRYSTAL-LIQUID M_r = 69.50

0 to 3193 K crystal
above 3193 K liquid

Refer to the individual tables for details.

Titanium Boride (TiB_2) B_2Ti_1(cr,l)

Enthalpy Reference Temperature = T_r = 298.15 K Standard State Pressure = p° = 0.1 MPa

T/K	C_p° (J K⁻¹mol⁻¹)	S° (J K⁻¹mol⁻¹)	$-[G^\circ - H^\circ(T_r)]/T$ (J K⁻¹mol⁻¹)	$H^\circ - H^\circ(T_r)$ (kJ mol⁻¹)	$\Delta_f H^\circ$ (kJ mol⁻¹)	$\Delta_f G^\circ$ (kJ mol⁻¹)	Log K_f
0	0.	0.	INFINITE	-5.577	-277.811	-277.811	INFINITE
100	7.523	2.518	56.366	-5.385	-278.227	-277.591	144.999
200	28.125	14.046	32.058	-3.602	-279.027	-276.623	72.247
298.15	44.279	28.485	28.485	0.	-279.491	-275.334	48.237
300	44.522	28.759	28.486	0.082	-279.498	-275.308	47.935
400	54.890	43.136	30.375	5.104	-279.814	-273.864	35.763
500	61.668	56.156	34.253	10.952	-280.169	-272.336	28.451
600	66.166	67.815	38.893	17.353	-280.556	-270.734	23.569
700	69.496	78.274	43.785	24.142	-280.964	-269.064	20.078
800	72.069	87.723	48.696	31.222	-281.376	-267.336	17.455
900	74.534	96.355	53.519	38.553	-281.799	-265.556	15.412
1000	76.889	104.331	58.206	46.125	-282.273	-263.727	13.776
1100	79.136	111.765	62.741	53.927	-282.845	-261.846	12.434
1200	81.283	118.744	67.120	61.949	-287.496	-259.788	11.308
1300	83.320	125.331	71.347	70.180	-287.529	-257.477	10.346
1400	85.249	131.577	75.428	78.609	-287.537	-255.165	9.520
1500	87.069	137.521	79.371	87.226	-287.539	-252.853	8.805
1600	88.780	143.196	83.184	96.019	-287.553	-250.540	8.179
1700	90.383	148.627	86.875	104.978	-287.599	-248.225	7.627
1800	91.876	153.836	90.451	114.092	-287.698	-245.907	7.136
1900	93.261	158.841	93.920	123.350	-287.872	-243.581	6.696
2000	94.537	163.657	97.287	132.741	-302.828	-240.789	6.289
2100	95.705	168.298	100.558	142.254	-303.964	-237.659	5.911
2200	96.763	172.775	103.740	151.878	-305.057	-234.476	5.567
2300	97.713	177.098	106.836	161.603	-306.118	-231.244	5.252
2400	98.554	181.275	109.851	171.416	-407.675	-225.829	4.915
2500	99.391	185.315	112.789	181.314	-408.851	-218.227	4.560
2600	100.228	189.229	115.654	191.295	-409.944	-210.581	4.231
2700	101.065	193.028	118.450	201.359	-410.953	-202.893	3.925
2800	101.901	196.718	121.180	211.507	-411.879	-195.170	3.641
2900	102.738	200.309	123.847	221.739	-412.721	-187.415	3.376
3000	103.575	203.806	126.454	232.055	-413.479	-179.633	3.128
3100	104.412	207.215	129.004	242.454	-414.153	-171.827	2.895
3193.000	105.190	210.313	131.328	252.201	CRYSTAL <--> LIQUID TRANSITION		
3193.000	108.784	241.762	131.328	352.617			
3200	108.784	242.000	131.569	353.378	-314.303	-164.220	2.681
3300	108.784	245.348	134.967	364.257	-314.498	-159.527	2.525
3400	108.784	248.595	138.261	375.135	-314.693	-154.828	2.379
3500	108.784	251.749	141.459	386.014	-314.889	-150.124	2.240
3600	108.784	254.813	144.565	396.892	-315.084	-145.413	2.110
3700	108.784	257.794	147.586	407.770	-724.378	-132.909	1.876
3800	108.784	260.695	150.524	418.649	-723.339	-116.937	1.607
3900	108.784	263.521	153.385	429.527	-722.355	-100.993	1.353
4000	108.784	266.275	156.173	440.406	-721.426	-85.072	1.111
4100	108.784	268.961	158.892	451.284	-720.549	-69.174	0.881
4200	108.784	271.582	161.544	462.162	-1679.422	-39.249	0.488
4300	108.784	274.142	164.133	473.041	-1676.470	-0.232	0.003
4400	108.784	276.643	166.661	483.919	-1673.570	38.717	-0.460
4500	108.784	279.088	169.133	494.798	-1670.720	77.600	-0.901
4600	108.784	281.479	171.549	505.676	-1667.918	116.421	-1.322
4700	108.784	283.818	173.913	516.554	-1665.163	155.181	-1.725
4800	108.784	286.108	176.227	527.433	-1662.453	193.883	-2.110
4900	108.784	288.351	178.492	538.311	-1659.788	232.529	-2.479
5000	108.784	290.549	180.711	549.190	-1657.165	271.121	-2.832
5100	108.784	292.703	182.886	560.068	-1654.583	309.661	-3.172
5200	108.784	294.816	185.018	570.946	-1652.041	348.150	-3.497
5300	108.784	296.888	187.110	581.825	-1649.538	386.592	-3.810
5400	108.784	298.921	189.161	592.703	-1647.072	424.986	-4.111
5500	108.784	300.917	191.175	603.582	-1644.642	463.335	-4.400
5600	108.784	302.877	193.152	614.460	-1642.247	501.640	-4.679
5700	108.784	304.803	195.094	625.338	-1639.886	539.903	-4.948
5800	108.784	306.695	197.002	636.217	-1637.558	578.125	-5.207
5900	108.784	308.554	198.877	647.095	-1635.262	616.306	-5.456
6000	108.784	310.383	200.721	657.974	-1632.997	654.449	-5.697

PREVIOUS: CURRENT: June 1965

Titanium Boride (TiB_2) B_2Ti_1(cr,l)

ZIRCONIUM BORIDE (ZrB_2) CRYSTAL M_r = 112.84

$S°(298.15\ K) = 35.94 \pm 0.04\ J\ K^{-1}\ mol^{-1}$

$T_{fus} = 3323\ K$

$\Delta_f H°(0\ K) = -321.3 \pm 6.7\ kJ\ mol^{-1}$

$\Delta_f H°(298.15\ K) = -322.6 \pm 6.7\ kJ\ mol^{-1}$

$\Delta_{fus} H° = [104.6]\ kJ\ mol^{-1}$

Enthalpy of Formation

Johnson (1) reported $\Delta_f H°(ZrB_{1.999\pm0.006}$, cr, 298.15 K) = $-78.7 \pm 1.5\ kcal\ mol^{-1}$. Their value was obtained by fluorine bomb calorimetry and was based on their recently redetermined value of $\Delta_f H°(BF_3$, g, 298.15 K) = $-271.80 \pm 0.20\ kcal\ mol^{-1}$.

Kibler et al. (2) studied the vaporization of $ZrB_{1.931}$ by the Knudsen effusion method. Over the range 2260 to 2500 K, log K_p = 22.038 − 10.164 x 10^4/T for the reaction $ZrB_{1.931}(cr) = Zr(g) + 1.931\ B(g)$. JANAF auxiliary data were used to calcuate $\Delta_r H°(298.15\ K) = 478.5 \pm 1.5\ kcal\ mol^{-1}$ by the 3rd law method and $\Delta_f H°(ZrB_{1.932}$, cr, 298.15 K) = $-76.7 \pm 3.2\ kcal\ mol^{-1}$.

Trulson and Goldstein (3) investigated the decomposition vapor pressure of the reaction $ZrB_2(cr) = Zr(g) + 2\ B(g)$ and reported $\Delta_f H°(ZRB_2$, cr, 298.15 K) = $-72.1 \pm 5.3\ kcal\ mol^{-1}$. They also reported $\Delta_r H°(298.15\ K) = 289.7\ kcal\ mol^{-1}$ for the reaction $ZrB_2(cr) + C(cr) = ZrC(cr) + 2\ B(g)$. Using the JANAF $\Delta_f H°(ZrC$, cr, 298.15 K) = $-47.0\pm3\ kcal\ mol^{-1}$, $\Delta_f H°(ZrB_2$, cr, 298.15 K) = $-76.1\pm3\ kcal\ mol^{-1}$ was calculated. JANAF values for $\Delta_{vap} H°(B$, g, 298.15 K) and Zr(g) were not used since their independent determinations serve to cancel instrumental constants.

From oxygen bomb calorimetry, Huber et al. (4) reported $\Delta_f H°(ZrB_2$, cr, 298.15 K) = $-77.2\pm1.2\ kcal\ mol^{-1}$.

$\Delta_f H°(ZrB_2$, cr, 298.15 K) = $-77.1\pm1.6\ kcal\ mol^{-1}$ was chosen as a value representative of the above.

Heat Capacity and Entropy

The heat capacity of $ZrB_2(cr)$ has been determined by the following investigators.

Source	Method	T/K
(5)	Adiabatic calorimetry	5-350
(6)	Copper-block drop calorimeter	410-1125
(7)	Copper-block drop calorimeter	430-1170
(8)	Arc-imaging furnace	1300-2150
(9)	Pulse-method	1740-2520
(10)	Drop type ice calorimeter	500-2500

Westrum and Feick's (5) values were adopted to 350 K. The disparities in heat capacities above room temperatures seem too great to be errors in measurement and are probably due to sample differences. Since the samples used by Westrum and Feick (5) and by Valentine et al. (6) were from the same batch, Valentine's data were used above 350 K. The two sets of data were fitted to a Shomate plot which was extrapolated to 4000 K. However, in consideration of the results of Mezsaki (7), Prophet (8) and Neel et al. (10), the C_p extrapolation ran somewhat higher than a direct extrapolation of Valentine's data would yield.

$S°(ZrB_2$, 298.15 K) = $8.59\ cal\ K^{-1}\ mol^{-1}$ was determined by Westrum (5).

Fusion Data

The melting point has been reported as 3313, 3323 and 3265 K by Glaser and Post (11), Post et al. (12) and Greenwood (13), respectively. The enthalpy of fusion is derived from an estimated $\Delta_{fus} S° = 2.5\ cal\ K^{-1}\ g\text{-atom}^{-1}$ at 3323 K. It should be pointed out that an error of $0.5\ cal\ K^{-1}\ g\text{-atom}^{-1}$ in the estimation of $\Delta_{fus} S°$ is $5\ kcal\ mol^{-1}$ in $\Delta_{fus} H°$.

References

1. G. K. Johnson, Argonne National Laboratory, personal communication, May 1965.
2. G. M. Kibler, T. F. Lyon, M. J. Linevsky, and V. J. De Santis, General Electric Co., WADD-TR-60-646, Part IV (1964).
3. O. C. Trulson and H. W. Goldstein, Union Carbide Res. Inst., report No. C-25 (1964).
4. E. J. Huber, Jr., E. L. Head, and C. E. Holley, Jr., J. Phys. Chem. 68, 3040 (1964).
5. E. F. Westrum, Jr. and G. Feick, J. Chem. Eng. Data 8, 193 (1963).
6. R. H. Valentine, T. F. Jambois and J. L. Margrave, J. Chem. Eng. Data 9, 182 (1964).
7. R. Mezaki, M.S. Thesis, University of Wisconsin, 1961.
8. H. Prophet, Annual Report for 1962 of the Dow Chemical Co. to ARPA under Contract AF 04(611)7554.
9. R. J. Barriault et al., Thermodynamics of Certain Refractory Compounds. AVCO Corporation, ASD-TR-61-260, Part I, Vol. 1 (1962).
10. D. Neel, C. Pears, and S. Oglesby, Southern Research Institute, ASD-TRD-62-675, 1962.
11. F. Glaser and B. Post, J. of Metals, 5, 1117 (1953).
12. B. Post, F. Glaser, and D. W. Moskowitz, Acta. Met. 2, 70 (1954).
13. H. Greenwood, Engineer 187, 349 (1949).

Zirconium Boride (ZrB_2) $B_2Zr_1(cr)$

Enthalpy Reference Temperature = T_r = 298.15 K   Standard State Pressure = p° = 0.1 MPa

T/K	$C_p°$ (J K⁻¹mol⁻¹)	S° (J K⁻¹mol⁻¹)	−[G°−H°(T_r)]/T (J K⁻¹mol⁻¹)	H°−H°(T_r) (kJ mol⁻¹)	$\Delta_f H°$ (kJ mol⁻¹)	$\Delta_f G°$ (kJ mol⁻¹)	Log K_f
0	0.	0.	INFINITE	-6.653	-321.314	-321.314	INFINITE
100	12.008	4.926	67.811	-6.289	-321.884	-320.909	167.626
200	32.803	19.732	39.899	-4.033	-322.488	-319.657	83.486
298.15	48.242	35.941	35.941	0.	-322.586	-318.234	55.753
300	48.488	36.240	35.941	0.089	-322.586	-318.207	55.405
400	57.505	51.582	37.972	5.444	-322.516	-316.760	41.365
500	62.634	65.010	42.068	11.471	-322.575	-315.317	32.941
600	65.810	76.725	46.889	17.902	-322.798	-313.847	27.323
700	68.028	87.043	51.903	24.598	-323.179	-312.327	23.306
800	69.710	96.240	56.881	31.487	-323.705	-310.743	20.289
900	71.090	104.534	61.723	38.530	-324.373	-309.084	17.939
1000	72.153	112.079	66.387	45.693	-325.191	-307.343	16.054
1100	73.174	119.004	70.860	52.959	-326.156	-305.512	14.508
1200	74.195	125.415	75.142	60.327	-330.967	-303.366	13.205
1300	75.216	131.394	79.242	67.798	-331.648	-301.039	12.096
1400	76.237	137.005	83.169	75.370	-332.366	-298.658	11.143
1500	77.258	142.300	86.937	83.045	-333.123	-296.224	10.315
1600	78.278	147.319	90.555	90.822	-333.926	-293.738	9.590
1700	79.299	152.095	94.036	98.701	-334.780	-291.200	8.947
1800	80.320	156.656	97.389	106.682	-335.694	-288.611	8.375
1900	81.341	161.026	100.624	114.765	-336.675	-285.969	7.862
2000	82.362	165.225	103.750	122.950	-337.731	-283.273	7.398
2100	83.383	169.268	106.774	131.237	-338.872	-280.522	6.978
2200	84.404	173.170	109.704	139.627	-361.430	-276.970	6.576
2300	85.425	176.945	112.546	148.118	-363.185	-273.091	6.202
2400	86.446	180.602	115.306	156.712	-465.422	-266.999	5.811
2500	87.467	184.152	117.989	165.407	-467.260	-258.694	5.405
2600	88.487	187.602	120.600	174.205	-468.997	-250.316	5.029
2700	89.508	190.961	123.144	183.105	-470.631	-241.874	4.679
2800	90.529	194.234	125.625	192.107	-472.163	-233.373	4.354
2900	91.550	197.429	128.046	201.211	-473.593	-224.820	4.049
3000	92.571	200.550	130.411	210.417	-474.921	-216.218	3.765
3100	93.592	203.602	132.723	219.725	-476.147	-207.575	3.498
3200	94.613	206.589	134.985	229.135	-477.271	-198.893	3.247
3300	95.634	209.516	137.199	238.647	-478.292	-190.177	3.010
3323.000	95.868	210.181	137.702	240.850	------ CRYSTAL <--> LIQUID -----		
3400	96.655	212.387	139.368	248.262	-479.212	-181.433	2.787
3500	97.675	215.203	141.495	257.978	-480.029	-172.662	2.577
3600	98.696	217.969	143.581	267.797	-480.745	-163.870	2.378
3700	99.717	220.687	145.628	277.717	-481.358	-155.059	2.189
3800	100.738	223.360	147.639	287.740	-481.869	-146.233	2.010
3900	101.759	225.990	149.614	297.865	-482.278	-137.395	1.840
4000	102.780	228.579	151.556	308.092	-482.585	-128.548	1.679

PREVIOUS: March 1963   CURRENT: June 1965

Zirconium Boride (ZrB_2) $B_2Zr_1(cr)$

ZIRCONIUM BORIDE (ZrB_2) LIQUID M_r = 112.84

$S°(298.15\ K) = [64.742]\ J\ K^{-1}\ mol^{-1}$

$T_{fus} = 3323\ K$

$T_{dcm} = [4466]\ K$

$\Delta_f H°(298.15\ K) = [-224.832]\ kJ\ mol^{-1}$

$\Delta_{fus} H° = [104.6]\ kJ\ mol^{-1}$

Enthalpy of Formation

$\Delta_f H°(ZrB_2, \ell, 298.15\ K)$ is calculated from $\Delta_f H°(ZrB_2, cr, 298.15\ K)$ by adding the enthalpy of fusion, $\Delta_{fus} H°$, and the difference in enthalpy, $H°(3323\ K)-H°(298.15\ K)$, between the crystal and liquid.

Heat Capacity and Entropy

A glass transition is assumed at 2200 K. The heat capacities at and below 2200 K are assumed to be equal to those of ZrB_2(cr). Above 2200 K the heat capacities were estimated to be constant.

$S°(ZrB_2, \ell, 298.15\ K)$ is calculated in a manner analogous to that used for the enthalpy of formation.

Fusion Data

See the crystal table for details.

Decomposition Data

T_{dcm} is estimated as the temperature for which $\Delta_r G° = 0$ for the reaction $ZrB_2(\ell) = Zr(g) + 2\ B(g)$.

Zirconium Boride (ZrB_2) $B_2Zr_1(l)$

Enthalpy Reference Temperature = T_r = 298.15 K Standard State Pressure = $p°$ = 0.1 MPa

	J K⁻¹mol⁻¹			kJ mol⁻¹			
T/K	$C°_p$	S°	$-[G°-H°(T_r)]/T$	$H°-H°(T_r)$	$\Delta_f H°$	$\Delta_f G°$	Log K_f
0							
100							
200							
298.15	48.242	64.742	64.742	0.	-224.832	-229.067	40.132
300	48.488	65.041	64.743	0.089	-224.831	-229.093	39.889
400	57.505	80.383	66.774	5.444	-224.761	-230.526	30.104
500	62.634	93.812	70.869	11.471	-224.820	-231.963	24.233
600	65.810	105.527	75.691	17.902	-225.044	-233.373	20.317
700	68.028	115.845	80.705	24.598	-225.424	-234.734	17.516
800	69.710	125.042	85.683	31.487	-225.950	-236.029	15.411
900	71.090	133.335	90.524	38.530	-226.619	-237.251	13.770
1000	72.153	140.881	95.188	45.693	-227.436	-238.390	12.452
1100	73.174	147.806	99.661	52.959	-228.401	-239.439	11.370
1200	74.195	154.216	103.944	60.327	-233.212	-240.174	10.454
1300	75.216	160.196	108.043	67.798	-233.893	-240.726	9.673
1400	76.237	165.807	111.971	75.370	-234.611	-241.225	9.000
1500	77.258	171.102	115.738	83.045	-235.368	-241.672	8.416
1600	78.278	176.120	119.357	90.822	-236.171	-242.066	7.903
1700	79.299	180.896	122.837	98.701	-237.025	-242.408	7.448
1800	80.320	185.458	126.190	106.682	-237.939	-242.699	7.043
1900	81.341	189.828	129.425	114.765	-238.920	-242.937	6.679
2000	82.362	194.026	132.551	122.950	-239.976	-243.121	6.350
2100	83.383	198.069	135.575	131.237	-241.118	-243.251	6.051
2200	84.404	201.972	138.505	139.627	-263.676	-242.578	5.760
2200.000	84.404	201.972	138.505	139.627	GLASS <--> LIQUID	TRANSITION	
2200.000	96.232	201.972	138.505	139.627			
2300	96.232	206.250	141.358	149.250	-264.298	-241.606	5.487
2400	96.232	210.345	144.148	158.873	-365.506	-238.467	5.190
2500	96.232	214.274	146.875	168.496	-366.417	-233.155	4.872
2600	96.232	218.048	149.540	178.119	-367.327	-227.807	4.577
2700	96.232	221.680	152.145	187.743	-368.238	-222.423	4.303
2800	96.232	225.179	154.692	197.366	-369.149	-217.006	4.048
2900	96.232	228.556	157.181	206.989	-370.060	-211.556	3.811
3000	96.232	231.819	159.615	216.612	-370.971	-206.075	3.588
3100	96.232	234.974	161.995	226.235	-371.881	-200.563	3.379
3200	96.232	238.029	164.324	235.859	-372.792	-195.022	3.183
3300	96.232	240.991	166.602	245.482	-373.703	-189.453	2.999
3323.000	96.232	241.659	167.119	247.695	------ CRYSTAL <--> LIQUID -----		
3400	96.232	243.863	168.833	255.105	-374.614	-183.856	2.825
3500	96.232	246.653	171.016	264.728	-375.525	-178.232	2.660
3600	96.232	249.364	173.155	274.351	-376.435	-172.582	2.504
3700	96.232	252.001	175.251	283.975	-377.346	-166.907	2.356
3800	96.232	254.567	177.304	293.598	-378.257	-161.208	2.216
3900	96.232	257.067	179.318	303.221	-379.168	-155.484	2.082
4000	96.232	259.503	181.292	312.844	-380.079	-149.737	1.955
4100	96.232	261.879	183.229	322.467	-380.989	-143.967	1.834
4200	96.232	264.198	185.129	332.091	-1341.599	-124.127	1.544
4300	96.232	266.463	186.994	341.714	-1340.335	-95.154	1.156
4400	96.232	268.675	188.826	351.337	-1339.075	-66.211	0.786
4500	96.232	270.838	190.624	360.960	-1337.819	-37.297	0.433
4600	96.232	272.953	192.391	370.583	-1336.568	-8.410	0.095
4700	96.232	275.022	194.127	380.207	-1335.322	20.450	-0.227
4800	96.232	277.048	195.834	389.830	-1895.018	60.900	-0.663
4900	96.232	279.032	197.511	399.453	-1893.436	101.632	-1.083
5000	96.232	280.977	199.161	409.076	-1891.877	142.332	-1.487

PREVIOUS: December 1962 CURRENT: June 1965

Zirconium Boride (ZrB_2) $B_2Zr_1(l)$

ZIRCONIUM BORIDE (ZrB_2) CRYSTAL-LIQUID M_r = 112.84

0 to 3323 K crystal
above 3323 K liquid

Refer to the individual tables for details.

Zirconium Boride (ZrB_2) B_2Zr_1(cr,l)

Enthalpy Reference Temperature = T_r = 298.15 K Standard State Pressure = p° = 0.1 MPa

T/K	C_p° (J K⁻¹mol⁻¹)	S° (J K⁻¹mol⁻¹)	−[G°−H°(T_r)]/T (J K⁻¹mol⁻¹)	H°−H°(T_r) (kJ mol⁻¹)	$\Delta_f H^\circ$ (kJ mol⁻¹)	$\Delta_f G^\circ$ (kJ mol⁻¹)	Log K_f
0	0.	0.	INFINITE	-6.653	-321.314	-321.314	INFINITE
100	12.008	4.926	67.811	-6.289	-321.884	-320.909	167.626
200	32.803	19.732	39.899	-4.033	-322.488	-319.657	83.486
298.15	48.242	35.941	35.941	0.	-322.586	-318.234	55.753
300	48.488	36.240	35.941	0.089	-322.586	-318.207	55.405
400	57.505	51.582	37.972	5.444	-322.516	-316.760	41.365
500	62.634	65.010	42.068	11.471	-322.575	-315.317	32.941
600	65.810	76.725	46.889	17.902	-322.798	-313.847	27.323
700	68.028	87.043	51.903	24.598	-323.179	-312.327	23.306
800	69.710	96.240	56.881	31.487	-323.705	-310.743	20.289
900	71.090	104.534	61.723	38.530	-324.373	-309.084	17.939
1000	72.153	112.079	66.387	45.693	-325.191	-307.343	16.054
1100	73.174	119.004	70.860	52.959	-326.156	-305.512	14.508
1200	74.195	125.415	75.142	60.327	-330.967	-303.366	13.205
1300	75.216	131.394	79.242	67.798	-331.648	-301.039	12.096
1400	76.237	137.005	83.169	75.370	-332.366	-298.658	11.143
1500	77.258	142.300	86.937	83.045	-333.123	-296.224	10.315
1600	78.278	147.319	90.555	90.822	-333.926	-293.738	9.590
1700	79.299	152.095	94.036	98.701	-334.780	-291.200	8.947
1800	80.320	156.656	97.389	106.682	-335.694	-288.611	8.375
1900	81.341	161.026	100.624	114.765	-336.675	-285.969	7.862
2000	82.362	165.225	103.750	122.950	-337.731	-283.273	7.398
2100	83.383	169.268	106.774	131.237	-338.872	-280.522	6.978
2200	84.404	173.170	109.704	139.627	-361.430	-276.970	6.576
2300	85.425	176.945	112.546	148.118	-363.185	-273.091	6.202
2400	86.446	180.602	115.306	156.712	-465.422	-266.999	5.811
2500	87.467	184.152	117.989	165.407	-467.260	-258.694	5.405
2600	88.487	187.602	120.600	174.205	-468.997	-250.316	5.029
2700	89.508	190.961	123.144	183.105	-470.631	-241.874	4.679
2800	90.529	194.234	125.625	192.107	-472.163	-233.373	4.354
2900	91.550	197.429	128.046	201.211	-473.593	-224.820	4.049
3000	92.571	200.550	130.411	210.417	-474.921	-216.218	3.765
3100	93.592	203.602	132.723	219.725	-476.147	-207.575	3.498
3200	94.613	206.589	134.985	229.135	-477.271	-198.893	3.247
3300	95.634	209.516	137.199	238.647	-478.292	-190.177	3.010
3323.000	95.868	210.181	137.702	240.850	CRYSTAL <--> LIQUID		
3323.000	96.232	241.659	137.702	345.450	TRANSITION		
3400	96.232	243.863	140.081	352.859	-374.614	-183.856	2.825
3500	96.232	246.653	143.087	362.483	-375.525	-178.233	2.660
3600	96.232	249.364	146.001	372.106	-376.436	-172.583	2.504
3700	96.232	252.001	148.831	381.729	-377.346	-166.908	2.356
3800	96.232	254.567	151.580	391.352	-378.257	-161.208	2.216
3900	96.232	257.067	154.252	400.975	-379.168	-155.484	2.082
4000	96.232	259.503	156.853	410.599	-380.079	-149.737	1.955
4100	96.232	261.879	159.386	420.222	-380.990	-143.967	1.834
4200	96.232	264.198	161.854	429.845	-1341.599	-124.127	1.544
4300	96.232	266.463	164.261	439.468	-1340.336	-95.154	1.156
4400	96.232	268.675	166.609	449.091	-1339.076	-66.211	0.786
4500	96.232	270.837	168.901	458.715	-1337.820	-37.297	0.433
4600	96.232	272.953	171.140	468.338	-1336.568	-8.410	0.095
4700	96.232	275.022	173.328	477.961	-1335.322	20.449	-0.227
4800	96.232	277.048	175.468	487.584	-1895.018	60.900	-0.663
4900	96.232	279.032	177.561	497.207	-1893.436	101.632	-1.083
5000	96.232	280.977	179.610	506.831	-1891.877	142.331	-1.487

PREVIOUS: CURRENT: June 1965

Zirconium Boride (ZrB_2) B_2Zr_1(cr,l)

TRICHLOROBOROXIN $((BOCl)_3)$ IDEAL GAS M_r = 186.7872

S°(298.15 K) = [382.38] J K^{-1} mol^{-1}

$\Delta_fH°$(0 K) = -1627 ± 8 kJ mol^{-1}

$\Delta_fH°$(298.15 K) = -1632 ± 8 kJ mol^{-1}

Vibrational Frequencies and Degeneracies

ν, cm^{-1}	ν, cm^{-1}	ν, cm^{-1}
[807] (1)	[980] (2)	[600] (1)
[690] (1)	[1300] (2)	[140] (1)
[333] (1)	[920] (2)	[400] (2)
[1037] (1)	[390] (2)	[120] (2)
[400] (1)	[150] (2)	

Ground State Quantum Weight: [1] σ = 6

Point Group: $[D_{3h}]$

Bond Distances: B-Cl = [1.75] Å B-O = [1.36] Å

Bond Angles: Cl-B-O = [120]° B-O-B = [120]°

Product of the Moments of Inertia: $I_AI_BI_C$ = [1.868800 x 10^{-111}] g^3 cm^6

Enthalpy of Formation

The equilibrium constants for the reaction B_2O_3(l) + BCl_3(g) = $(BOCl)_3$(g) were determined by Blauer and Farber in the temperature range 536-825 K by a transpiration method (1), and in the temperature range 1234-1309 K by an effusion method (2). Using the reported equilibrium constants, the enthalpy changes for this reaction were evaluated by both the 2nd and 3rd law methods.

Method Used	$\Delta_rH°$(298.15 K), kcal mol^{-1} 3rd Law	2nd Law	$\Delta_fH°$(293.15 K) kcal mol^{-1}
Transpiration	5.31	5.07 ± 0.4	-390 ± 2
Effusion	12.57	5.85 ± 3.3	-386 ± 3

By a mass spectroscopic method the enthalpy change $\Delta_rH°$(1000 K) = 16.6 ± 2.5 kcal mol^{-1}, for the reaction BCl_3(g) + B_2O_3(l) = $(BOCl)_3$(g) was determined by Porter and Gupta (3). The $\Delta_fH°$(298.15 K) value for $(BOCl)_3$(g) was derived as -377 ± 3 kcal mol^{-1}. However, it is possible that the intensity of $B_3O_3Cl_3^+$, employed to plot the ion ratios $I(B_3O_3Cl_3^+)/I(BCl_2^+)$ versus 1/T, does not fully reflect the concentration of $B_3O_3Cl_3$ especially at the highest temperatures, due to changes in the fragmentation pattern with temperature.

The value of $\Delta_fH°$(298.15 K) adopted for $(BOCl)_3$(g) is 390 ± 2 kcal mol^{-1} derived from transpiration data. The effusion data were measured in the temperature range where Porter and Gupta (3) found the presence of $(BOCl)_4$(g). However, this point was not taken into consideration in the report. Hence, the corresponding $\Delta_fH°$(298.15 K) value was not adopted.

Heat Capacity and Entropy

The vibration frequencies are estimated by comparison with related molecules such as $B_3O_3F_3$(g) and $(BH_2O)_3$(g). The B-Cl bond distance is assumed to be the same as that in BCl_3(g). The B-O bond distance and all bond angles are estimated. The principal moments of inertia are: $I_A = I_B$ = 97.7637 x 10^{-39} and I_C = 195.5274 x 10^{-39} g cm^2.

References

1. J. Blauer and M. Farber, J. Chem. Phys. 39, 158 (1963).
2. J. Blauer and M. Farber, Trans. Faraday Soc. 60, 301 (1964).
3. R. F. Porter and S. K. Gupta, J. Phys. Chem. 68, 280 (1964).

Trichloroboroxin $(B_3O_3Cl_3)$ $B_3Cl_3O_3$(g)

Enthalpy Reference Temperature = T_r = 298.15 K Standard State Pressure = p° = 0.1 MPa

T/K	$C_p°$ (J $K^{-1}mol^{-1}$)	S°	-[G°-H°(T_r)]/T	H°-H°(T_r) (kJ mol^{-1})	$\Delta_fH°$	$\Delta_fG°$	Log K_f
0	0.	0.	INFINITE	-25.342	-1626.664	-1626.664	INFINITE
100	70.079	275.129	480.665	-20.554	-1630.672	-1603.746	837.711
200	105.432	335.190	393.615	-11.685	-1631.726	-1576.278	411.682
250	119.582	360.271	384.479	-6.052	-1631.820	-1562.401	326.446
298.15	131.525	382.378	382.378	0.	-1631.760	-1549.034	271.384
300	131.953	383.193	382.381	0.244	-1631.755	-1548.521	269.622
350	142.628	404.357	384.028	7.115	-1631.558	-1534.663	229.036
400	151.707	424.012	387.812	14.480	-1631.268	-1520.839	198.601
450	159.351	442.335	392.864	22.262	-1630.909	-1507.057	174.935
500	165.756	459.466	398.677	30.394	-1630.492	-1493.317	156.006
600	175.618	490.611	411.458	47.492	-1629.526	-1465.970	127.624
700	182.610	518.238	424.777	65.423	-1628.450	-1438.795	107.364
800	187.665	542.970	438.032	83.950	-1627.326	-1411.778	92.180
900	191.400	565.300	450.953	102.912	-1626.198	-1384.903	80.378
1000	194.220	585.618	463.419	122.199	-1625.095	-1358.152	70.943
1100	196.392	604.236	475.386	141.735	-1624.038	-1331.509	63.228
1200	198.095	621.400	486.848	161.462	-1623.043	-1304.960	56.803
1300	199.452	637.311	497.818	181.342	-1622.121	-1278.491	51.370
1400	200.549	652.134	508.317	201.344	-1621.281	-1252.090	46.716
1500	201.448	666.002	518.372	221.445	-1620.531	-1225.746	42.684
1600	202.193	679.028	528.010	241.628	-1619.878	-1199.449	39.158
1700	202.817	691.305	537.258	261.880	-1619.325	-1173.189	36.048
1800	203.344	702.913	546.141	282.188	-1618.881	-1146.960	33.284
1900	203.793	713.919	554.685	302.546	-1618.548	-1120.751	30.812
2000	204.179	724.383	562.910	322.945	-1618.331	-1094.558	28.587
2100	204.513	734.353	570.839	343.380	-1618.235	-1068.372	26.574
2200	204.804	743.874	578.489	363.846	-1618.263	-1042.187	24.745
2300	205.059	752.983	585.879	384.340	-1618.419	-1015.999	23.074
2400	205.283	761.715	593.025	404.857	-1769.485	-986.595	21.473
2500	205.482	770.100	599.941	425.395	-1770.141	-953.961	19.932
2600	205.658	778.162	606.642	445.953	-1770.836	-921.301	18.509
2700	205.816	785.927	613.139	466.526	-1771.573	-888.612	17.191
2800	205.957	793.415	619.445	487.115	-1772.354	-855.895	15.967
2900	206.085	800.644	625.569	507.717	-1773.181	-823.151	14.827
3000	206.200	807.633	631.522	528.332	-1774.054	-790.376	13.762
3100	206.304	814.396	637.313	548.957	-1774.973	-757.571	12.765
3200	206.399	820.947	642.949	569.592	-1775.937	-724.738	11.830
3300	206.485	827.300	648.440	590.236	-1776.945	-691.872	10.951
3400	206.564	833.465	653.792	610.889	-1777.995	-658.975	10.124
3500	206.636	839.454	659.011	631.549	-1779.084	-626.048	9.343
3600	206.703	845.276	664.105	652.216	-1780.209	-593.087	8.605
3700	206.764	850.940	669.078	672.889	-1781.367	-560.097	7.907
3800	206.821	856.455	673.937	693.569	-1782.554	-527.072	7.245
3900	206.873	861.828	678.686	714.253	-1783.765	-494.017	6.617
4000	206.921	867.066	683.330	734.943	-1784.998	-460.932	6.019
4100	206.966	872.176	687.874	755.638	-1786.247	-427.814	5.450
4200	207.008	877.164	692.322	776.336	-3227.058	-373.594	4.646
4300	207.047	882.035	696.678	797.039	-3225.067	-305.678	3.713
4400	207.083	886.796	700.945	817.746	-3223.088	-237.809	2.823
4500	207.117	891.450	705.126	838.456	-3221.115	-169.984	1.973
4600	207.149	896.003	709.227	859.169	-3219.148	-102.203	1.161
4700	207.179	900.458	713.248	879.885	-3217.183	-34.465	0.383
4800	207.207	904.820	717.194	900.605	-3215.219	33.232	-0.362
4900	207.233	909.093	721.067	921.327	-3213.255	100.889	-1.075
5000	207.258	913.280	724.869	942.051	-3211.289	168.504	-1.760
5100	207.281	917.384	728.604	962.778	-3209.321	236.082	-2.418
5200	207.303	921.409	732.273	983.508	-3207.349	303.617	-3.050
5300	207.324	925.358	735.879	1004.239	-3205.375	371.118	-3.658
5400	207.343	929.234	739.424	1024.972	-3203.398	438.580	-4.242
5500	207.362	933.039	742.910	1045.708	-3201.419	506.006	-4.806
5600	207.380	936.775	746.338	1066.445	-3199.440	573.397	-5.348
5700	207.396	940.446	749.712	1087.183	-3197.462	640.750	-5.872
5800	207.412	944.053	753.031	1107.924	-3195.485	708.072	-6.377
5900	207.427	947.599	756.299	1128.666	-3193.512	775.355	-6.864
6000	207.441	951.085	759.517	1149.409	-3191.547	842.608	-7.336

PREVIOUS: March 1965 (1 atm) CURRENT: March 1965 (1 bar)

Trichloroboroxin $(B_3O_3Cl_3)$ $B_3Cl_3O_3$(g)

FLUOROBOROXIN ($B_3O_3H_2F$) IDEAL GAS M_r = 101.442483

Fluoroboroxin ($B_3H_2O_3F$) $B_3F_1H_2O_3(g)$

$S°(298.15\ K)$ = [314.84] J K^{-1} mol^{-1}

$\Delta_fH°(0\ K)$ = -1587 ± 25 kJ mol^{-1}
$\Delta_fH°(298.15\ K)$ = -1598 ± 25 kJ mol^{-1}

Vibrational Frequencies and Degeneracies

ν, cm^{-1}	ν, cm^{-1}	ν, cm^{-1}
[2100] (1)	[880] (1)	[350] (1)
[870] (1)	[1460] (1)	[225] (1)
[670] (1)	[520] (1)	[800] (1)
[2200] (1)	[2230] (1)	[225] (1)
[1390] (1)	[1390] (1)	[840] (1)
[1200] (1)	[1200] (1)	[800] (1)
[880] (1)	[350] (1)	[260] (1)

Ground State Quantum Weight: [1] σ = 2
Point Group: [C_{2v}]
Bond Distances: B–O = [1.38] Å B–H = [1.18] Å B–F = [1.41] Å
Bond Angles: B–O–B = [120]° O–B–H = [120]° O–B–F = [120]°
Product of the Moments of Inertia: $I_AI_BI_C$ = [2.164977 x 10^{-113}] g^3 cm^6

Enthalpy of Formation

The equilibrium constants (K), 1246-1326 K, for the two reactions $2B_3O_3F_2H(g) = B_3O_3FH_2(g) + B_3O_3F_3(g)$ and $2B_3O_3FH_2(g) = B_3O_3H_3(g) + B_3O_3F_2H(g)$ were reported by Porter and Sholette (1). The $\Delta_fH°(298.15\ K)$ for $B_3O_3H_2F(g)$ adopted was evaluated by use of K values slightly outside the ranges given by Porter and Sholette. See the table for $B_3O_3F_2H(g)$ for details.

Heat Capacity and Entropy

The structure is assumed to be a planar six membered ring structure of C_{2v} symmetry, which has 21 vibrations of the type $8A_1$, $7B_1$, $2A_2$, and $4B_2$. These frequencies are estimated from the values for boroxin and trifluoroboroxin, the symmetry being reduced from D_{3h} as follows: $3A_1' + 3A_1$, $2A_2' + 2B_1$, $5E' + 5A_1 + 5B_1$, $2A_2 + 2B_2$ and $2E'' + 2A_2 + 2B_2$.

The B-F distance is taken as approximately equal to that in $(CH_3)_2O \cdot BF_3$ reported by Bauer, Findlay and Laubengayer (2). The other bond lengths and angles are taken equal to those in boroxin. The principal moments of inertia are: I_A = 13.9711 x 10^{-39}, I_B = 32.9946 x 10^{-39}, and I_C= 46.9657 x 10^{-39} g cm^2.

References

1. R. F. Porter and W. P. Sholette, J. Chem. Phys. 37, 198 (1962).
2. S. H. Bauer, G. R. Findlay, and A. W. Laubengayer, J. Am. Chem. Soc. 67, 339 (1945).

Enthalpy Reference Temperature = T_r = 298.15 K; Standard State Pressure = p° = 0.1 MPa

T/K	$C_p°$ (J $K^{-1}mol^{-1}$)	S° (J $K^{-1}mol^{-1}$)	$-[G°-H°(T_r)]/T$ (J $K^{-1}mol^{-1}$)	$H°-H°(T_r)$ (kJ mol^{-1})	$\Delta_fH°$ (kJ mol^{-1})	$\Delta_fG°$ (kJ mol^{-1})	Log K_f
0	0.	0.	INFINITE	-17.877	-1586.619	-1586.619	INFINITE
100	46.702	241.315	383.279	-14.196	-1591.820	-1571.271	820.748
200	72.105	281.675	322.817	-8.228	-1595.375	-1549.266	404.627
250	84.135	299.063	316.347	-4.321	-1596.915	-1537.558	321.255
298.15	95.245	314.841	314.841	0.	-1598.288	-1526.000	267.349
300	95.659	315.432	314.843	0.177	-1598.338	-1525.551	265.622
350	106.362	330.993	316.047	5.231	-1599.624	-1513.316	225.850
400	116.099	345.842	318.851	10.797	-1600.778	-1500.906	195.998
450	124.856	360.032	322.644	16.825	-1601.803	-1488.359	172.764
500	132.685	373.600	327.066	23.267	-1602.695	-1475.705	154.166
600	145.879	399.004	336.970	37.220	-1604.090	-1450.168	126.248
700	156.339	422.308	347.520	52.351	-1605.020	-1424.435	106.293
800	164.653	443.748	358.227	68.417	-1605.565	-1398.596	91.319
900	171.295	463.539	368.843	85.226	-1605.811	-1372.708	79.670
1000	176.637	481.873	379.241	102.633	-1605.833	-1346.804	70.350
1100	180.969	498.919	389.355	120.520	-1605.698	-1320.907	62.725
1200	184.511	514.822	399.155	138.800	-1605.459	-1295.026	56.371
1300	187.433	529.710	408.631	157.402	-1605.162	-1269.169	50.996
1400	189.864	543.692	417.784	176.270	-1604.842	-1243.335	46.389
1500	191.902	556.862	426.621	195.362	-1604.529	-1217.525	42.398
1600	193.625	569.304	435.154	214.640	-1604.246	-1191.734	38.906
1700	195.093	581.088	443.395	234.078	-1604.010	-1165.959	35.826
1800	196.351	592.275	451.358	253.652	-1603.838	-1140.197	33.088
1900	197.436	602.921	459.057	273.343	-1603.740	-1114.442	30.638
2000	198.379	613.073	466.506	293.134	-1603.725	-1088.690	28.434
2100	199.202	622.772	473.718	313.014	-1603.800	-1062.937	26.439
2200	199.924	632.056	480.706	332.971	-1603.970	-1037.177	24.626
2300	200.562	640.958	487.481	352.996	-1604.238	-1011.409	22.970
2400	201.126	649.506	494.055	373.081	-1755.386	-982.420	21.382
2500	201.629	657.726	500.439	393.220	-1756.092	-950.199	19.853
2600	202.078	665.643	506.641	413.405	-1756.805	-917.949	18.442
2700	202.481	673.277	512.672	433.634	-1757.525	-885.671	17.134
2800	202.844	680.648	518.541	453.900	-1758.253	-853.367	15.920
2900	203.172	687.772	524.254	474.201	-1758.990	-821.037	14.788
3000	203.469	694.665	529.820	494.534	-1759.736	-788.681	13.732
3100	203.739	701.341	535.246	514.894	-1760.491	-756.300	12.744
3200	203.985	707.813	540.538	535.280	-1761.256	-723.895	11.816
3300	204.209	714.094	545.703	555.690	-1762.029	-691.466	10.945
3400	204.415	720.193	550.745	576.122	-1762.810	-659.012	10.124
3500	204.605	726.121	555.672	596.573	-1763.601	-626.537	9.351
3600	204.779	731.888	560.487	617.042	-1764.399	-594.038	8.619
3700	204.939	737.501	565.196	637.528	-1765.206	-561.517	7.927
3800	205.088	742.968	569.802	658.030	-1766.022	-528.973	7.271
3900	205.225	748.297	574.311	678.545	-1766.846	-496.408	6.649
4000	205.353	753.495	578.726	699.074	-1767.679	-463.823	6.057
4100	205.471	758.567	583.051	719.616	-1768.521	-431.216	5.494
4200	205.582	763.519	587.289	740.168	-3208.919	-377.516	4.695
4300	205.685	768.358	591.444	760.732	-3206.518	-310.129	3.767
4400	205.781	773.088	595.518	781.305	-3204.132	-242.800	2.882
4500	205.871	777.713	599.516	801.888	-3201.763	-175.524	2.037
4600	205.955	782.239	603.439	822.479	-3199.411	-108.301	1.230
4700	206.034	786.669	607.291	843.079	-3197.080	-41.127	0.457
4800	206.109	791.008	611.073	863.686	-3194.769	25.996	-0.283
4900	206.178	795.258	614.789	884.300	-3192.481	93.073	-0.992
5000	206.244	799.424	618.440	904.921	-3190.217	160.102	-1.673
5100	206.306	803.509	622.029	925.549	-3187.980	227.086	-2.326
5200	206.365	807.516	625.558	946.182	-3185.771	294.025	-2.954
5300	206.420	811.447	629.028	966.822	-3183.591	360.925	-3.557
5400	206.472	815.306	632.442	987.466	-3181.445	427.782	-4.138
5500	206.522	819.095	635.801	1008.116	-3179.332	494.599	-4.697
5600	206.569	822.817	639.108	1028.771	-3177.256	561.380	-5.236
5700	206.613	826.473	642.363	1049.430	-3175.219	628.123	-5.756
5800	206.655	830.067	645.568	1070.093	-3173.222	694.832	-6.258
5900	206.696	833.600	648.725	1090.761	-3171.268	761.503	-6.742
6000	206.734	837.074	651.836	1111.432	-3169.360	828.145	-7.210

PREVIOUS: December 1965 (1 atm) CURRENT: December 1965 (1 bar)

DIFLUOROBOROXIN ($B_3O_3HF_2$) IDEAL GAS M_r = 119.432946

S°(298.15 K) = [328.99] J K^{-1} mol^{-1}

$\Delta_f H°$(0 K) = -1977 ± 21 kJ mol^{-1}
$\Delta_f H°$(298.15 K) = -1987 ± 21 kJ mol^{-1}

Vibrational Frequencies and Degeneracies

ν, cm^{-1}	ν, cm^{-1}	ν, cm^{-1}
[1650] (1)	[660] (1)	[300] (1)
[830] (1)	[1370] (1)	[220] (1)
[610] (1)	[490] (1)	[710] (1)
[1800] (1)	[1840] (1)	[220] (1)
[1385] (1)	[1385] (1)	[780] (1)
[1080] (1)	[1080] (1)	[700] (1)
[660] (1)	[310] (1)	[230] (1)

Ground State Quantum Weight: [1] σ = 2
Point Group: [C_{2v}]
Bond Distances: B-O = [1.36] Å B-H = [1.18] Å B-F = [1.41] Å
Bond Angles: B-O-B = [120]° O-B-H = [120]° O-B-F = [120]°
Product of the Moments of Inertia: $I_A I_B I_C$ = [7.586853 x 10^{-113}] g^3 cm^6

Enthalpy of Formation

The stabilities of gaseous boroxin ($B_3O_3H_3$) and its fluoro derivatives have been studied by Porter and Sholette (1). From the mass spectra of the gaseous reaction products generated by reaction of an H_2-BF_3 mixture on B_2O_3 at 1250 K, the equilibrium constants (K) for the following proposed reactions were reported.

	Reaction	T/K	K	$\Delta_r H°$(298.15 K), kcal mol^{-1}
(1)	$H_2(g) + B_3O_3F_3(g) = HF(g) + B_3O_3F_2H(g)$	1246 - 1326	4.2 x 10^{-3} - 2.3 x 10^{-2}	20.45 - 17.17
(2)	$2B_3O_3F_2H(g) = B_3O_3FH_2(g) + B_3O_3F_3(g)$	1246 - 1326	0.62 - 0.46	1.28 - 2.15
(3)	$2B_3O_3FH_2(g) = B_3O_3H_3(g) + B_3O_3F_2H(g)$	1246 - 1326	1.0 - 0.63	-2.56 - -1.51

Using these reported K values the corresponding enthalpy changes are evaluated by the 3rd law method. The results obtained are listed in the above table. Based on $\Delta_f H°$(298.15 K) = -565.3 and -64.8 kcal mol^{-1} for $B_3O_3F_3$(g) and HF(g), respectively, the value of $\Delta_f H°$(298.15 K) for $B_3O_3F_2H$(g) is calculated from Reaction (1) as -482 kcal mol^{-1}. From Reaction (2), the $\Delta_f H°$(298.15 K) value for $B_3O_3FH_2$(g) is calculated as -397 kcal mol^{-1}. However, employing $\Delta_f H°$($B_3O_3H_3$, g, 298.15 K) = -291 kcal mol^{-1}, the values of $\Delta_f H°$(298.15 K) for $B_3O_3FH_2$(g) and $B_3O_3F_2H$(g) thus obtained do not fit properly for Reaction (3). Therefore, the $\Delta_f H°$(298.15 K) for $B_3O_3FH_2$(g) and $B_3O_3F_2H$(g) adopted are evaluated by use of K values slightly outside the ranges given by Porter and Sholette (1).

Heat Capacity and Entropy

The vibrational frequencies are obtained by comparison with $B_3O_3F_3$(g) and $B_3H_3O_3$(g). The types of vibration and symmetry are identical to those given for $B_3O_3H_2F$(g). The bond angles and lengths are taken equal to those for $B_3O_3H_2F$(g). The principal moments of inertia are: I_A = 22.0494 x 10^{-39}, I_B = 48.6611 x 10^{-39}, and I_C = 70.7105 x 10^{-39} g cm^2.

Reference

1. R. F. Porter and W. P. Sholette, J. Chem. Phys. 37, 198 (1962).

Difluoroboroxin ($B_3HO_3F_2$) $B_3F_2H_1O_3$(g)

Enthalpy Reference Temperature = T_r = 298.15 K; Standard State Pressure = p° = 0.1 MPa

T/K	$C_p°$ (J $K^{-1}mol^{-1}$)	S° (J $K^{-1}mol^{-1}$)	-[G°-H°(T_r)]/T (J $K^{-1}mol^{-1}$)	H°-H°(T_r) (kJ mol^{-1})	$\Delta_f H°$ (kJ mol^{-1})	$\Delta_f G°$ (kJ mol^{-1})	Log K_f
0	0.	0.	INFINITE	-19.127	-1976.801	-1976.801	INFINITE
100	49.248	249.646	403.289	-15.364	-1981.875	-1958.677	1023.108
200	78.340	292.875	337.680	-8.961	-1985.112	-1934.119	505.140
250	91.721	311.813	330.633	-4.705	-1986.357	-1921.223	401.418
298.15	103.528	328.994	328.994	0.	-1987.400	-1908.582	334.376
300	103.959	329.636	328.996	0.192	-1987.437	-1908.093	332.228
350	114.983	346.504	330.303	5.670	-1988.368	-1894.793	282.782
400	124.838	362.514	333.337	11.671	-1989.175	-1881.368	245.682
450	133.599	377.734	337.432	18.136	-1989.866	-1867.850	216.814
500	141.353	392.220	342.192	25.014	-1990.440	-1854.261	193.713
600	154.206	419.177	352.812	39.819	-1991.256	-1826.941	159.049
700	164.136	443.727	364.073	55.758	-1991.685	-1799.516	134.282
800	171.816	466.168	375.453	72.572	-1991.811	-1772.052	115.703
900	177.795	486.764	386.693	90.064	-1991.713	-1744.586	101.253
1000	182.495	505.750	397.662	108.088	-1991.461	-1717.140	89.694
1100	186.231	523.325	408.297	126.531	-1991.110	-1689.724	80.238
1200	189.236	539.663	418.571	145.310	-1990.706	-1662.343	72.360
1300	191.679	554.909	428.479	164.360	-1990.286	-1634.997	65.695
1400	193.686	569.190	438.025	183.631	-1989.876	-1607.682	59.983
1500	195.353	582.612	447.221	203.086	-1989.501	-1580.396	55.034
1600	196.750	595.265	456.082	222.693	-1989.178	-1553.133	50.705
1700	197.930	607.230	464.625	242.428	-1988.921	-1525.889	46.885
1800	198.936	618.572	472.865	262.273	-1988.740	-1498.658	43.490
1900	199.799	629.352	480.820	282.211	-1988.645	-1471.434	40.453
2000	200.544	639.620	488.505	302.229	-1988.639	-1444.212	37.719
2100	201.193	649.420	495.936	322.317	-1988.727	-1416.990	35.246
2200	201.760	658.793	503.127	342.465	-1988.911	-1389.759	32.997
2300	202.258	667.773	510.092	362.666	-1989.192	-1362.520	30.944
2400	202.699	676.391	516.843	382.915	-2140.347	-1332.059	28.992
2500	203.090	684.673	523.391	403.204	-2141.054	-1298.366	27.128
2600	203.439	692.645	529.749	423.531	-2141.758	-1264.645	25.407
2700	203.751	700.329	535.925	443.891	-2142.458	-1230.896	23.813
2800	204.032	707.744	541.930	464.280	-2143.152	-1197.121	22.333
2900	204.285	714.909	547.772	484.696	-2143.840	-1163.323	20.954
3000	204.513	721.838	553.459	505.136	-2144.520	-1129.500	19.666
3100	204.721	728.548	559.000	525.598	-2145.191	-1095.654	18.462
3200	204.910	735.050	564.400	546.080	-2145.852	-1061.789	17.332
3300	205.083	741.358	569.667	566.580	-2146.502	-1027.902	16.270
3400	205.241	747.483	574.808	587.096	-2147.138	-993.995	15.271
3500	205.386	753.435	579.827	607.628	-2147.762	-960.070	14.328
3600	205.519	759.222	584.730	628.173	-2148.371	-926.127	13.438
3700	205.642	764.855	589.522	648.731	-2148.966	-892.168	12.595
3800	205.756	770.341	594.209	669.301	-2149.547	-858.192	11.797
3900	205.861	775.687	598.794	689.882	-2150.112	-824.201	11.039
4000	205.958	780.900	603.282	710.473	-2150.663	-790.197	10.319
4100	206.049	785.987	607.676	731.073	-2151.199	-756.179	9.634
4200	206.133	790.953	611.981	751.683	-3591.269	-701.075	8.719
4300	206.211	795.804	616.200	772.300	-3588.516	-632.293	7.681
4400	206.285	800.546	620.336	792.925	-3585.755	-563.576	6.691
4500	206.353	805.182	624.392	813.557	-3582.988	-494.922	5.745
4600	206.417	809.719	628.372	834.195	-3580.217	-426.329	4.841
4700	206.477	814.158	632.278	854.840	-3577.444	-357.796	3.976
4800	206.534	818.506	636.112	875.490	-3574.670	-289.322	3.148
4900	206.587	822.765	639.878	896.147	-3571.898	-220.905	2.355
5000	206.637	826.939	643.578	916.808	-3569.129	-152.545	1.594
5100	206.684	831.032	647.213	937.474	-3566.367	-84.241	0.863
5200	206.729	835.046	650.787	958.145	-3563.614	-15.992	0.161
5300	206.771	838.984	654.301	978.819	-3560.872	52.207	-0.515
5400	206.810	842.849	657.757	999.499	-3558.145	120.352	-1.164
5500	206.848	846.644	661.157	1020.181	-3555.435	188.446	-1.790
5600	206.883	850.372	664.502	1040.868	-3552.744	256.494	-2.392
5700	206.917	854.034	667.796	1061.558	-3550.076	324.491	-2.974
5800	206.949	857.633	671.038	1082.251	-3547.434	392.444	-3.534
5900	206.980	861.171	674.230	1102.948	-3544.820	460.348	-4.076
6000	207.009	864.650	677.375	1123.647	-3542.238	528.212	-4.598

PREVIOUS: December 1965 (1 atm) CURRENT: December 1965 (1 bar)

Difluoroboroxin ($B_3HO_3F_2$) $B_3F_2H_1O_3$(g)

TRIFLUOROBOROXIN $((BOF)_3)$ CRYSTAL M_r = 137.423409

$S°(298.15\ K) = [217.6]\ J\ K^{-1}\ mol^{-1}$

$T_{fus} = 353\ K$

$\Delta_f H°(0\ K)$ = Unknown

$\Delta_f H°(298.15\ K) = -2454 \pm 12\ kJ\ mol^{-1}$

$\Delta_{fus} H°$ = Unknown

Enthalpy of Formation

The enthalpies of solution of the compounds: $B_2O_3(gl)$, $BF_3(g)$ and $(BOF)_3(cr)$ in water and methanol, respectively, were reported by Magee (1). The corresponding enthalpy changes for the reaction $B_2O_3(gl) + BF_3(g) = (BOF)_3(cr)$ were derived as -19.8 and -14.4 kcal mol^{-1}, yielding $\Delta_f H°(298.15\ K)$ values for $(BOF)_3(cr)$ as -589.18 and -583.78 kcal mol^{-1}, respectively. The adopted value of $\Delta_f H°(298.15\ K)$ for $(BOF)_3(cr)$ is the average of these two values obtained.

Heat Capacity and Entropy

The heat capacities for $(BOF)_3(cr)$ are estimated from the C_p values for $B_2O_3(cr)$ and $BF_3(g)$, according to the method suggested by Kubaschewski and Evans (2).

$S°(298.15\ K)$ for $(BOF)_3(cr)$ was estimated by Magee (1).

Fusion Data

The value of T_{fus} was reported by Magee (1).

Sublimation Data

The enthalpy of sublimation for $(BOF)_3(cr)$ is calculated as the difference in $\Delta_f H°(298.15\ K)$ between the crystal and gas.

References

1. E. M. Magee, J. Inorg. Nucl. Chem. 22, 156 (1961).
2. O. Kubaschewski and E. L. Evans, "Metallurgical Thermochemistry," Pergamon Press, New York, 1958.

Trifluoroboroxin $(B_3O_3F_3)$ $B_3F_3O_3(cr)$

Enthalpy Reference Temperature = T_r = 298.15 K Standard State Pressure = p° = 0.1 MPa

	J K⁻¹mol⁻¹			kJ mol⁻¹			
T/K	$C_p°$	S°	$-[G°-H°(T_r)]/T$	$H°-H°(T_r)$	$\Delta_f H°$	$\Delta_f G°$	Log K_f
0							
100							
200							
298.15	125.520	217.568	217.568	0.	-2453.916	-2331.126	408.404
300	125.874	218.346	217.570	0.233	-2453.915	-2330.364	405.753
400	144.766	257.156	222.735	13.769	-2453.752	-2289.199	298.939
500	163.176	291.484	233.111	29.187	-2453.164	-2248.116	234.859
600	178.761	322.641	245.476	46.299	-2451.944	-2207.210	192.155
700	192.464	351.250	258.570	64.876	-2450.042	-2166.562	161.671
800	204.284	377.743	271.831	84.729	-2447.457	-2126.234	138.829
900	214.221	402.396	284.985	105.670	-2444.251	-2086.267	121.084
1000	222.170	425.393	297.889	127.504	-2440.532	-2046.687	106.908
1100	228.446	446.875	310.468	150.048	-2436.420	-2007.498	95.328
1200	232.840	466.950	322.681	173.123	-2432.054	-1968.696	85.695
1300	235.978	485.720	334.508	196.575	-2427.556	-1930.265	77.559
1400	237.860	503.284	345.943	220.277	-2423.029	-1892.182	70.598
1500	238.488	519.723	356.987	244.105	-2418.582	-1854.421	64.577

PREVIOUS: September 1962 CURRENT: March 1965

Trifluoroboroxin $(B_3O_3F_3)$ $B_3F_3O_3(cr)$

TRIFLUOROBOROXIN ((BOF)$_3$) IDEAL GAS M_r = 137.423409

$S°(298.15\ K)$ = [342.45] J K^{-1} mol^{-1}

$\Delta_fH°(0\ K) = -2357.0 \pm 4.2$ kJ mol^{-1}
$\Delta_fH°(298.15\ K) = -2365.2 \pm 4.2$ kJ mol^{-1}

Vibrational Frequencies and Degeneracies

ν, cm^{-1}	ν, cm^{-1}	ν, cm^{-1}
[1233] (1)	1450 (2)	714 (1)
[790] (1)	1381 (2)	[170] (1)
[570] (1)	966 (2)	[630] (2)
[1280] (1)	[420] (2)	[185] (2)
[440] (1)	[220] (2)	

Ground State Quantum Weight: [1] $\sigma = 6$
Point Group: [D_{3h}]
Bond Distances: B-F = [1.3] Å B-O = [1.36] Å
Bond Angles: F-B-O = [120]° B-O-B = [120]° O-B-O = [120]°
Product of the Moments of Inertia: $I_AI_BI_C = [1.925790 \times 10^{-113}]$ g^3 cm^6

Enthalpy of Formation

The enthalpy changes of the reaction $B_2O_3(l) + BF_3(g) = (BOF)_3(g)$ are derived by both the 2nd and 3rd law methods, using the equilibrium constant data reported by several investigators.

Source	$\Delta_rH°(298.15\ K)$, kcal mol^{-1} 3rd law	2nd law	$\Delta_fH°(298.15\ K)$ kcal mol^{-1}
Hildenbrand, et al. (1)	4.06	4.01 ± 0.3	-565.32 ± 1.0
Fisher, et al. (2)	3.41	6.45 ± 0.6	-565.97 ± 1.5
Smith and Lawrence (3)	4.03	---	-565.35 ± 1.3
Farber (4)	5.25	---	-565.13 ± 2.0
	2.56	7.58 ± 0.2	-565.82 ± 2.0
Porter, et al. (5)	--	4.33 ± 0.30	-565.05 ± 1.0

The adopted value of $\Delta_fH°(298.15\ K)$ for $(BOF)_3(cr)$ is the weighted average of the six $\Delta_fH°(298.15\ K)$ values listed above.

Heat Capacity and Entropy

The infra-red spectrum has been reported by Smith and Lawrence (3) and by Fisher, Lehmann, and Shapiro (6) who also made a partial assignment. The unassigned bands at 1280 and 1233 from Fisher, et al. (6), can be assigned to the A_2' and A_1' species which may be observed if the selection rules are not rigorously obeyed in the solid. All the unobserved frequencies were estimated by comparison with $B_3O_3(OH)_3$, $B_3O_3(OD)_3$ from Parsons (7) and $B_3O_3H_3$ and $B_3O_3D_3$ from Gupta and Porter (8) and Lee, Bauer and Wiberley (9). The D_{3h} assumed symmetry has 14 vibrations of the type $3A_1'$, $2A_2'$, $5E'$, $2A_2''$ and $2E''$ which are given in order above. The B-F bond distance was assumed to be the same as that in BF_3. The B-O bond distance was estimated by comparison with other related compounds. The principal moments of inertia are: $I_A = I_B = 45.8345 \times 10^{-39}$ and $I_C = 91.6691 \times 10^{-39}$ g cm^2.

References

1. D. L. Hildenbrand, L. P. Theard and A. M. Saul, J. Chem. Phys. 39, 1973 (1963).
2. H. D. Fisher, J. Kiehl and A. Cane, Report HTC-61-90, June 1961, Hughes Tool Company, Culver City, California.
3. D. D. Smith and R. W. Lawrence, Aerojet Report No. 1952, February 1961, Aerojet-General Corporation.
4. M. Farber, J. Chem. Phys. 36, 661 (1962). The first $\Delta_rH°(298.15\ K)$ value was evaluated based on K (1000 K) = 11.8. The second $\Delta_rH°(298.15\ K)$ values refer to the reaction $B_2O_3(cr) + BF_3(g) = (BOF)_3(g)$.
5. R. F. Porter, D. R. Bidinosti and K. F. Watterson, J. Chem. Phys. 36, 2104 (1962). The $\Delta_rH°(298.15\ K)$ value was derived from $\Delta_rH°(700\ K) = 3.46 \pm 0.30$ kcal mol^{-1}.
6. H. D. Fisher, W. J. Lehmann, and I. Shapiro, J. Phys. Chem. 65, 1166 (1961).
7. J. L. Parsons, J. Chem. Phys. 33, 1860 (1960).
8. S. K. Gupta and R. F. Porter, J. Phys. Chem. 67, 1286 (1963).
9. G. H. Lee, W. H. Bauer, and S. E. Wiberley, J. Phys. Chem. 67, 1742 (1963).

Trifluoroboroxin ($B_3O_3F_3$) $B_3F_3O_3(g)$

Enthalpy Reference Temperature = T_r = 298.15 K Standard State Pressure = p° = 0.1 MPa

T/K	$C_p°$ (J $K^{-1}mol^{-1}$)	S° (J $K^{-1}mol^{-1}$)	$-[G°-H°(T_r)]/T$ (J $K^{-1}mol^{-1}$)	$H°-H°(T_r)$ (kJ mol^{-1})	$\Delta_fH°$ (kJ mol^{-1})	$\Delta_fG°$ (kJ mol^{-1})	Log K_f
0	0.	0.	INFINITE	-21.691	-2357.002	-2357.002	INFINITE
100	57.998	251.201	426.907	-17.571	-2361.671	-2335.147	1219.756
200	89.987	301.683	352.202	-10.104	-2363.963	-2307.608	602.685
250	103.398	323.231	344.283	-5.263	-2364.676	-2293.432	479.187
298.15	115.012	342.454	342.454	0.	-2365.215	-2279.660	399.387
300	115.435	343.167	342.456	0.213	-2365.234	-2279.129	396.832
350	126.216	361.786	343.902	6.260	-2365.667	-2264.743	337.994
400	135.793	379.278	347.242	12.815	-2366.005	-2250.301	293.859
450	144.217	395.770	351.726	19.820	-2366.260	-2235.822	259.527
500	151.566	411.354	356.917	27.218	-2366.431	-2221.318	232.060
600	163.450	440.091	368.427	42.998	-2366.545	-2192.281	190.855
700	172.330	465.987	380.546	59.809	-2366.409	-2163.244	161.423
800	178.998	489.454	392.716	77.391	-2366.095	-2134.241	139.352
900	184.064	510.843	404.671	95.555	-2365.665	-2105.284	122.187
1000	187.969	530.447	416.282	114.165	-2365.170	-2076.379	108.459
1100	191.023	548.511	427.492	133.120	-2364.647	-2047.524	97.229
1200	193.448	565.240	438.283	152.349	-2364.128	-2018.718	87.873
1300	195.399	580.804	448.655	171.794	-2363.636	-1989.954	79.957
1400	196.988	595.345	458.619	191.416	-2363.189	-1961.227	73.174
1500	198.299	608.982	468.193	211.183	-2362.803	-1932.529	67.297
1600	199.390	621.815	477.397	231.069	-2362.490	-1903.855	62.154
1700	200.307	633.932	486.252	251.055	-2362.258	-1875.197	57.618
1800	201.086	645.404	494.778	271.126	-2362.115	-1846.552	53.586
1900	201.751	656.294	502.995	291.268	-2362.064	-1817.911	49.978
2000	202.325	666.657	510.921	311.473	-2362.107	-1789.271	46.731
2100	202.822	676.541	518.574	331.731	-2362.246	-1760.626	43.793
2200	203.255	685.987	525.971	352.035	-2362.479	-1731.972	41.122
2300	203.636	695.030	533.126	372.380	-2362.805	-1703.306	38.683
2400	203.972	703.704	540.054	392.761	-2513.999	-1671.417	36.377
2500	204.269	712.037	546.768	413.173	-2514.735	-1636.294	34.189
2600	204.534	720.054	553.279	433.613	-2515.458	-1601.143	32.167
2700	204.771	727.777	559.600	454.079	-2516.163	-1565.963	30.295
2800	204.983	735.228	565.740	474.567	-2516.848	-1530.757	28.557
2900	205.175	742.425	571.709	495.075	-2517.510	-1495.529	26.937
3000	205.348	749.384	577.517	515.601	-2518.146	-1460.277	25.426
3100	205.505	756.120	583.170	536.144	-2518.753	-1425.004	24.011
3200	205.647	762.646	588.677	556.702	-2519.330	-1389.713	22.685
3300	205.778	768.976	594.045	577.273	-2519.874	-1354.405	21.438
3400	205.897	775.121	599.281	597.857	-2520.383	-1319.079	20.265
3500	206.006	781.091	604.391	618.452	-2520.857	-1283.740	19.159
3600	206.106	786.896	609.380	639.058	-2521.293	-1248.387	18.114
3700	206.199	792.545	614.255	659.673	-2521.690	-1213.024	17.125
3800	206.284	798.045	619.019	680.297	-2522.050	-1177.648	16.188
3900	206.363	803.404	623.678	700.930	-2522.370	-1142.264	15.299
4000	206.436	808.630	628.237	721.570	-2522.652	-1106.874	14.454
4100	206.504	813.728	632.699	742.217	-2522.895	-1071.477	13.651
4200	206.568	818.705	637.069	762.870	-3962.649	-1015.001	12.623
4300	206.627	823.566	641.350	783.530	-3959.555	-944.855	11.478
4400	206.682	828.317	645.545	804.195	-3956.430	-874.783	10.385
4500	206.733	832.962	649.659	824.866	-3953.276	-804.781	9.342
4600	206.781	837.507	653.693	845.542	-3950.095	-734.850	8.344
4700	206.826	841.954	657.652	866.222	-3946.890	-664.987	7.391
4800	206.869	846.309	661.537	886.907	-3943.662	-595.194	6.477
4900	206.908	850.575	665.351	907.596	-3940.414	-525.467	5.602
5000	206.946	854.756	669.098	928.289	-3937.150	-455.807	4.762
5100	206.981	858.854	672.778	948.985	-3933.872	-386.212	3.956
5200	207.014	862.873	676.396	969.685	-3930.583	-316.684	3.181
5300	207.046	866.817	679.951	990.388	-3927.286	-247.216	2.436
5400	207.076	870.687	683.448	1011.094	-3923.986	-177.812	1.720
5500	207.104	874.487	686.887	1031.803	-3920.684	-108.470	1.030
5600	207.130	878.219	690.270	1052.515	-3917.386	-39.185	0.366
5700	207.156	881.886	693.600	1073.229	-3914.094	30.038	-0.275
5800	207.180	885.489	696.877	1093.946	-3910.813	99.206	-0.893
5900	207.202	889.030	700.104	1114.665	-3907.545	168.314	-1.490
6000	207.224	892.513	703.282	1135.386	-3904.296	237.370	-2.066

PREVIOUS: March 1965 (1 atm) CURRENT: March 1965 (1 bar)

Trifluoroboroxin ($B_3O_3F_3$) $B_3F_3O_3(g)$

BOROXIN ($B_3O_3H_3$) CRYSTAL M_r = 83.45202

S°(298.15 K) = [167 ± 42] J K^{-1} mol^{-1}

T_{fus} = Unknown

$\Delta_f H°$(0 K) = Unknown

$\Delta_f H°$(298.15 K) = -1262.3 ± 42 kJ mol^{-1}

$\Delta_m H°$ = Unknown

Enthalpy of Formation

The value of $\Delta_f H°$(298.15 K) for $B_3O_3H_3$(cr) is calculated from the $\Delta_f H°$(298.15 K) for $B_3O_3H_3$(g) and $\Delta_{sub}H°$(298.15 K) for $B_3O_3H_3$(cr) reported by Porter and Gupta (1).

Heat Capacity and Entropy

The heat capacities were estimated from those from B_2O_3(cr) and 1/2 B_2H_6(g), according to the method suggested by Kubaschewski and Evans (2). The value of S°(298.15 K) for $B_3O_3H_3$(cr) is calculated based on $\Delta_r S°$(298.15 K) = 6.6 cal K^{-1} mol^{-1} for the reaction $B_3O_3H_3$(cr) = B_2O_3(am) + 1/2 B_2H_6(g), which was obtained from the calculated $\Delta_r H°$(298.15 K) and measured $\Delta_r G°$(298.15 K) evaluated from the vapor pressure reported by Sholette and Porter (3).

Sublimation Data

$\Delta_{sub}H°$(298.15 K) is taken from Porter and Gupta (3).

References

1. R. F. Porter and S. K. Gupta, J. Phys. Chem. 68, 280 (1964).
2. O. Kubaschewski and E. L. Evans, "Metallurgical Thermochemistry," Pergamon Press, New York, 1958.
3. W. P. Sholette and R. F. Porter, J. Phys. Chem. 67, 177 (1963).

Boroxin ($B_3H_3O_3$) $B_3H_3O_3$(cr)

Enthalpy Reference Temperature = T_r = 298.15 K Standard State Pressure = p° = 0.1 MPa

T/K	$C_p°$ (J $K^{-1}mol^{-1}$)	S° (J $K^{-1}mol^{-1}$)	$-[G°-H°(T_r)]/T$ (J $K^{-1}mol^{-1}$)	$H°-H°(T_r)$ (kJ mol^{-1})	$\Delta_f H°$ (kJ mol^{-1})	$\Delta_f G°$ (kJ mol^{-1})	Log K_f
0							
100							
200							
298.15	98.324	167.360	167.360	0.	-1262.313	-1156.803	202.667
300	98.742	167.969	167.362	0.182	-1262.355	-1156.148	201.303
400	120.081	199.288	171.488	11.120	-1264.321	-1120.431	146.313
500	141.838	228.436	179.991	24.223	-1265.383	-1084.312	113.277
600	162.758	256.203	190.396	39.484	-1265.199	-1048.093	91.245
700	179.912	282.612	201.700	56.638	-1263.798	-1012.004	75.517
800	194.556	307.628	213.392	75.389	-1261.331	-976.191	63.739
900	205.591	331.204	225.186	95.416	-1258.019	-940.740	54.599
1000	214.221	353.333	236.906	116.427	-1254.093	-905.692	47.309
1100	219.730	374.016	248.442	138.132	-1249.799	-871.057	41.363
1200	224.262	393.337	259.720	160.340	-1245.299	-836.824	36.426
1300	226.786	411.389	270.701	182.895	-1240.725	-802.970	32.264
1400	229.022	428.279	281.360	205.688	-1236.168	-769.467	28.709
1500	230.970	444.148	291.688	228.690	-1231.642	-736.290	25.640
1600	232.630	459.109	301.689	251.872	-1227.166	-703.413	22.964
1700	234.003	473.255	311.369	275.206	-1222.759	-670.814	20.612
1800	235.088	486.662	320.738	298.663	-1218.440	-638.472	18.528
1900	235.886	499.396	329.809	322.214	-1214.232	-606.367	16.670
2000	236.396	511.509	338.594	345.831	-1210.157	-574.479	15.004

PREVIOUS: CURRENT: March 1965

Boroxin ($B_3H_3O_3$) $B_3H_3O_3$(cr)

BOROXIN ($B_3O_3H_3$) IDEAL GAS M_r = 83.45202

$S°(298.15\ K) = [291.91]\ J\ K^{-1}\ mol^{-1}$

$\Delta_f H°(0\ K) = -1205 \pm 42\ kJ\ mol^{-1}$

$\Delta_f H°(298.15\ K) = -1218 \pm 42\ kJ\ mol^{-1}$

Vibrational Frequencies and Degeneracies

ν, cm^{-1}	ν, cm^{-1}	ν, cm^{-1}
2530 (1)	2620 (2)	903 (1)
[920] (1)	1404 (2)	[300] (1)
[735] (1)	1335 (2)	1115 (2)
[1560] (1)	940 (2)	[230] (2)
[550] (1)	[400] (2)	

Ground State Quantum Weight: [1] σ = [6]

Point Group: [D_{3h}]

Bond Distances: B–O = [1.36] Å B–H = [1.18] Å

Bond Angles: B–O–B = [120]° O–B–H = [120]°

Product of the Moments of Inertia: $I_A I_B I_C = [5.454021 \times 10^{-114}]\ g^3\ cm^6$

Enthalpy of Formation

The equilibrium constants (1290-1481 K) for the reaction (1) 3/2 H_2(g) + B(am) + B_2O_3(l) = $B_3O_3H_3$(g) were determined by Sholette and Porter (1). The equilibrium pressures of B_2H_6(g) and $B_3O_3H_3$(g) for the reaction (2) $B_3O_3H_3$(g) = 1/2 B_2H_6(g) + B_2O_3(cr) were measured at 300 K by Porter and Gupta (2). The equilibrium constants (1246-1326 K) for the reaction (3) $2B_3O_3FH_2$(g) = $B_3O_3H_3$(g) + $B_3O_3F_2H$(g) were reported by Porter and Sholette (3). Using these data the corresponding enthalpy changes were evaluated by both the 2nd and 3rd law methods.

Reaction	$\Delta_r H°$(298.15 K), kcal mol^{-1} 3rd law	2nd law	$\Delta_f H°$(298.15 K) kcal mol^{-1}
(1)	-5.90	+6.52 ± 4.5	-304.0 ± 5.0
(2)**	-11.11	---	-287.6 ± 2.0
	-12.21	---	-272.5 ± 2.0
(3)	-2.0	---	-314 ± 12.0

**The initial system for the first set of measurements was B_2H_6(g) + $B(OH)_3$(cr). That for the second set was B_2H_6(g) + B_2O_3(gl).

The value of $\Delta_f H°$(298.15 K) for $B_3O_3H_3$(g) adopted is a weighted average of these $\Delta_f H°$(298.15 K) values listed.

Heat Capacity and Entropy

The infrared spectrum os solid boroxin has been determined by Gupta and Porter (4) and a partial assignment was made. The infrared spectra of the gas has also been reported by Lee, Bauer, and Wiberley (5) and is in essential agreement with the solid. The assignment of the frequencies has been changed slightly due to the non-appearance of the band at 1115 cm^{-1} in the gas phase. The assumed D_{3h} symmetry has 14 vibrations of the type $3A_1'$, $2A_2'$, 5E', $2A_2''$, and 2E", and 7 estimated frequencies were obtained by comparison with $B_3O_6H_3$, by Parsons (6).

The point group, bond angles and bond lengths were assumed to be the same as for $(BOF)_3$(g). The B-H bond length was taken as the sum of the single covalent radii of the atoms. The principal moments of inertia are: $I_A = I_B = 13.9710 \times 10^{-39}$, and I_C = 27.9421 x 10^{-39} g cm^2.

References

1. W. P. Sholette and R. F. Porter, J. Phys. Chem. 67, 177 (1963).
2. R. F. Porter and S. K. Gupta, J. Phys. Chem. 68, 280 (1964).
3. R. F. Porter and W. P. Sholette, J. Phys. Chem. 37, 198 (1962).
4. S. K. Gupta and R. F. Porter, J. Phys. Chem. 67, 1286 (1963).
5. G. H. Lee, W. H. Bauer, and S. E. Wiberley, J. Phys. Chem. 67, 1742 (1963).
6. J. L. Parsons, J. Phys. Chem. 33, 1860 (1960).

Boroxin ($B_3H_3O_3$) $B_3H_3O_3$(g)

Enthalpy Reference Temperature = T_r = 298.15 K; Standard State Pressure = p° = 0.1 MPa

T/K	$C_p°$ (J K⁻¹mol⁻¹)	S°	-[G°-H°(T_r)]/T	H°-H°(T_r) (kJ mol⁻¹)	$\Delta_f H°$	$\Delta_f G°$	Log K_f
0	0.	0.	INFINITE	-16.846	-1205.022	-1205.022	INFINITE
100	44.531	223.177	355.472	-13.230	-1210.334	-1191.454	622.352
200	67.194	261.217	299.274	-7.612	-1214.122	-1171.045	305.846
250	77.761	277.347	293.296	-3.987	-1215.892	-1160.070	242.383
298.15	87.831	291.907	291.907	0.	-1217.544	-1149.168	201.329
300	88.213	292.451	291.909	0.163	-1217.606	-1148.743	200.014
350	98.298	306.813	293.020	4.828	-1219.211	-1137.138	169.708
400	107.750	320.564	295.609	9.982	-1220.690	-1125.310	146.950
450	116.426	333.765	299.120	15.590	-1222.032	-1113.306	129.229
500	124.290	346.445	303.223	21.611	-1223.226	-1101.160	115.037
600	137.736	370.338	312.444	34.736	-1225.177	-1076.554	93.722
700	148.586	392.415	322.312	49.072	-1226.596	-1051.664	78.476
800	157.375	412.850	332.369	64.385	-1227.565	-1026.603	67.030
900	164.538	431.814	342.377	80.493	-1228.173	-1001.443	58.122
1000	170.415	449.464	352.214	97.250	-1228.501	-976.231	50.993
1100	175.270	465.941	361.812	114.542	-1228.621	-950.996	45.159
1200	179.307	481.370	371.139	132.277	-1228.594	-925.758	40.297
1300	182.687	495.859	380.182	150.381	-1228.469	-900.526	36.184
1400	185.535	509.505	388.936	168.796	-1228.290	-875.306	32.658
1500	187.952	522.390	397.408	187.474	-1228.089	-850.100	29.603
1600	190.014	534.588	405.604	206.375	-1227.895	-824.908	26.930
1700	191.786	546.162	413.534	225.467	-1227.729	-799.726	24.573
1800	193.316	557.169	421.211	244.724	-1227.611	-774.553	22.477
1900	194.646	567.657	428.645	264.124	-1227.554	-749.385	20.602
2000	195.807	577.671	435.848	283.648	-1227.571	-724.218	18.915
2100	196.826	587.250	442.831	303.280	-1227.673	-699.048	17.388
2200	197.725	596.428	449.605	323.009	-1227.866	-673.871	16.000
2300	198.521	605.235	456.182	342.822	-1228.157	-648.684	14.732
2400	199.228	613.699	462.570	362.710	-1379.331	-620.275	13.500
2500	199.860	621.845	468.779	382.665	-1380.068	-588.633	12.299
2600	200.427	629.695	474.818	402.680	-1380.819	-556.961	11.189
2700	200.936	637.269	480.695	422.748	-1381.588	-525.259	10.162
2800	201.396	644.585	486.418	442.865	-1382.377	-493.528	9.207
2900	201.812	651.659	491.995	463.026	-1383.188	-461.770	8.317
3000	202.190	658.507	497.432	483.227	-1384.024	-429.983	7.487
3100	202.534	665.143	502.735	503.463	-1384.886	-398.167	6.709
3200	202.848	671.578	507.912	523.732	-1385.775	-366.324	5.980
3300	203.136	677.825	512.966	544.032	-1386.692	-334.452	5.294
3400	203.399	683.893	517.905	564.359	-1387.639	-302.551	4.648
3500	203.642	689.792	522.732	584.711	-1388.614	-270.623	4.039
3600	203.865	695.532	527.453	605.086	-1389.620	-238.666	3.463
3700	204.071	701.121	532.071	625.483	-1390.656	-206.682	2.918
3800	204.262	706.566	536.592	645.900	-1391.723	-174.667	2.401
3900	204.439	711.874	541.019	666.335	-1392.821	-142.624	1.910
4000	204.604	717.052	545.355	686.788	-1393.951	-110.555	1.444
4100	204.756	722.106	549.605	707.256	-1395.112	-78.455	1.000
4200	204.899	727.042	553.771	727.739	-2835.853	-25.255	0.314
4300	205.032	731.865	557.857	748.235	-2833.817	41.641	-0.506
4400	205.156	736.580	561.865	768.745	-2831.818	108.488	-1.288
4500	205.272	741.192	565.799	789.266	-2829.858	175.290	-2.035
4600	205.381	745.704	569.661	809.799	-2827.938	242.049	-2.749
4700	205.484	750.122	573.454	830.342	-2826.059	308.768	-3.432
4800	205.580	754.450	577.180	850.895	-2824.222	375.448	-4.086
4900	205.670	758.689	580.841	871.458	-2822.428	442.089	-4.713
5000	205.755	762.845	584.440	892.029	-2820.679	508.695	-5.314
5100	205.836	766.921	587.978	912.609	-2818.975	575.265	-5.892
5200	205.912	770.918	591.458	933.196	-2817.319	641.801	-6.447
5300	205.983	774.841	594.881	953.791	-2815.711	708.309	-6.981
5400	206.051	778.692	598.249	974.393	-2814.154	774.784	-7.495
5500	206.115	782.474	601.564	995.001	-2812.647	841.231	-7.989
5600	206.176	786.188	604.828	1015.616	-2811.193	907.654	-8.466
5700	206.234	789.838	608.042	1036.236	-2809.794	974.049	-8.926
5800	206.289	793.425	611.207	1056.862	-2808.450	1040.422	-9.370
5900	206.341	796.952	614.326	1077.494	-2807.163	1106.767	-9.799
6000	206.391	800.420	617.399	1098.131	-2805.936	1173.096	-10.213

PREVIOUS: March 1965 (1 atm) CURRENT: March 1965 (1 bar)

Boroxin ($B_3H_3O_3$) $B_3H_3O_3$(g)

BORIC ACID $((HBO_2)_3)$ IDEAL GAS M_r = 131.45022

$\Delta_f H°(0\ K)$ = [-2252 ± 13] kJ mol^{-1}

$S°(298.15\ K)$ = [347.57] J K^{-1} mol^{-1} $\Delta_f H°(298.15\ K)$ = [-2272 ± 13] kJ mol^{-1}

Vibrational Frequencies and Degeneracies

ν, cm^{-1}	ν, cm^{-1}	ν, cm^{-1}
[550] (1)	[250] (1)	[600] (2)
[1150] (1)	[500] (1)	[450] (2)
[950] (1)	[350] (1)	[3500] (2)
[3500] (1)	[750] (2)	[350] (2)
[1100] (1)	[900] (2)	[200] (2)
[600] (1)	[1300] (2)	[1000] (2)
[500] (1)	[1000] (2)	

Ground State Quantum Weight: 1 σ = 3

Point Group: C_{3h}

Bond Distances: B-O = [1.36] Å O-H = [1.1] Å

Bond Angles: H-O-B = [120]° O-B-O = [120]° B-O-B = [120]°

Product of the Moments of Inertia: $I_A I_B I_C$ = [1.776533 x 10^{-112}] g^3 cm^6

Enthalpy of Formation

The equilibrium constant, K_f (1451 K) = 0.9 for the reaction $H_3BO_3(l) + B_2O_3(l) = (HBO_2)_3(g)$ was reported by Meschi, Chupka and Berkowitz (1), which was determined by means of mass spectrometry. The equilibrium constants of the reactions between water vapor and liquid B_2O_3 were determined at 1147 K by Blauer and Farber (2) by use of a transpiration method. From the reported value, 1.5 K_2 + 0.5 K_3 = 0.0342 ± 0.0095, the equilibrium constant, K_2, for the reaction 1.5 $B_2O_3(l)$ + 1.5 $H_2O(g) = (HBO_2)_3$ derived as 0.0168, using K_3 = 0.0181 for the reaction 0.5 $B_2O_3(l)$ + 1.5 $H_2O(g) = H_3BO_3(g)$ evaluated by use of JANAF values of $\Delta_f G°$ for $B_2O_3(l)$, $H_2O(g)$ and $H_3BO_3(g)$. Employing the equilibrium constants obtained, the respective enthalpy changes for the two reactions were calculated by the 3rd law method. The corresponding values of $\Delta_f H°(298.15\ K)$ for $(HBO_2)_3(g)$ were also derived.

Source	Chemical Reaction	$\Delta_r H°(298.15\ K)$ kcal mol^{-1}	$\Delta_f H°(298.15\ K)$ kcal mol^{-1}
Meschi, et al. (1)	$H_3BO_3(l) + B_2O_3(l) = (HBO_2)_3(g)$	-7.31	-543.8
Blauer and Farber (2)	$1.5\ B_2O_3(l) + 1.5\ H_2O(g) = (HBO_2)_3(g)$	-6.67	-542.3

The adopted value of $\Delta_f H°(298.15\ K)$ for $(HBO_2)_3(g)$ is the average of these two listed in the above table. The equilibrium constants for the same reactions between $B_2O_3(l)$ and $H_2O(g)$ were also determined by a transpiration method in the temperature region 1000 - 1273 K by Randall and Margrave (3). The reported value, $\Delta_f H°(0\ K)$ = -537.5 ± 3 kcal mol^{-1} for $(HBO_2)_3(g)$, is in good agreement with the value adopted.

Heat Capacity and Entropy

All the molecular and spectroscopic constants were estimated by White, Mann, Walsh, and Sommer (4). The principal moments of inertia are: I_A = 43.6915 x 10^{-39}, I_B = 45.5585 x 10^{-39}, and I_C = 89.2499 x 10^{-39} g cm^2. The infrared absorption spectra for $B_3O_3(OH)_3(cr)$ in the 430 to 4000 cm^{-1} was reported by Parsons (5). A partial Raman spectrum was also given.

References

1. D. J. Meschi, W. A. Chupka, and J. Berkowitz, J. Chem. Phys. 33, 530 (1960).
2. J. A. Blauer and M. Farber, J. Phys. Chem. 68, 2357 (1964).
3. S. P. Randall and J. L. Margrave, J. Inorg. Nucl. Chem. 16, 29 (1960).
4. D. White, D. E. Mann, P. N. Walsh, and A. Sommer, J. Chem. Phys. 32, 488 (1960).
5. J. L. Parsons, J. Chem. Phys. 33, 1860 (1960).

Boric Acid $((HBO_2)_3)$ $B_3H_3O_6(g)$

Enthalpy Reference Temperature = T_r = 298.15 K Standard State Pressure = p° = 0.1 MPa

T/K	$C_p°$ (J K^{-1}mol^{-1})	S° (J K^{-1}mol^{-1})	$-[G°-H°(T_r)]/T$ (J K^{-1}mol^{-1})	$H°-H°(T_r)$ (kJ mol^{-1})	$\Delta_f H°$ (kJ mol^{-1})	$\Delta_f G°$ (kJ mol^{-1})	Log K_f
0	0.	0.	INFINITE	-22.763	-2252.282	-2252.282	INFINITE
100	51.541	251.641	441.169	-18.953	-2261.757	-2219.727	1159.467
200	96.849	301.091	358.876	-11.557	-2268.134	-2174.986	568.048
250	118.479	325.060	349.724	-6.166	-2270.324	-2151.434	449.518
298.15	137.265	347.567	347.567	0.	-2271.912	-2128.384	372.884
300	137.941	348.418	347.570	0.255	-2271.964	-2127.493	370.430
350	154.896	370.988	349.313	7.586	-2273.117	-2103.317	313.903
400	169.352	392.642	353.386	15.702	-2273.876	-2079.004	271.490
450	181.535	413.313	358.906	24.483	-2274.320	-2054.616	238.494
500	191.764	432.985	365.338	33.823	-2274.508	-2030.192	212.093
600	207.660	469.433	379.702	53.839	-2274.309	-1981.336	172.491
700	219.256	502.359	394.912	75.213	-2273.571	-1932.560	144.209
800	228.061	532.236	410.239	97.597	-2272.474	-1883.916	123.007
900	235.010	559.514	425.332	120.764	-2271.132	-1835.424	106.525
1000	240.666	584.577	440.020	144.557	-2269.617	-1787.093	93.348
1100	245.378	607.742	454.228	168.866	-2267.983	-1738.918	82.574
1200	249.367	629.269	467.929	193.608	-2266.271	-1690.896	73.603
1300	252.782	649.367	481.121	218.720	-2264.515	-1643.019	66.017
1400	255.730	668.211	493.818	244.149	-2262.741	-1595.278	59.521
1500	258.293	685.944	506.042	269.853	-2260.976	-1547.665	53.894
1600	260.531	702.686	517.813	295.797	-2259.240	-1500.167	48.975
1700	262.495	718.541	529.159	321.950	-2257.550	-1452.777	44.638
1800	264.225	733.595	540.102	348.288	-2255.925	-1405.486	40.786
1900	265.756	747.923	550.666	374.788	-2254.376	-1358.282	37.342
2000	267.115	761.590	560.873	401.433	-2252.917	-1311.156	34.244
2100	268.325	774.652	570.744	428.206	-2251.556	-1264.102	31.443
2200	269.406	787.160	580.299	455.094	-2250.302	-1217.109	28.898
2300	270.376	799.157	589.555	482.084	-2249.163	-1170.173	26.575
2400	271.248	810.683	598.531	509.166	-2398.922	-1120.077	24.378
2500	272.034	821.772	607.240	536.331	-2398.262	-1066.805	22.290
2600	272.745	832.456	615.698	563.570	-2397.633	-1013.560	20.363
2700	273.389	842.761	623.918	590.877	-2397.039	-960.337	18.579
2800	273.975	852.715	631.912	618.246	-2396.482	-907.135	16.923
2900	274.509	862.338	639.693	645.671	-2395.966	-853.955	15.381
3000	274.997	871.653	647.271	673.146	-2395.492	-800.790	13.943
3100	275.444	880.677	654.655	700.669	-2395.062	-747.639	12.598
3200	275.854	889.429	661.856	728.234	-2394.676	-694.505	11.337
3300	276.231	897.923	668.881	755.839	-2394.335	-641.380	10.152
3400	276.578	906.175	675.740	783.479	-2394.038	-588.264	9.038
3500	276.899	914.197	682.439	811.153	-2393.788	-535.157	7.987
3600	277.196	922.002	688.985	838.858	-2393.583	-482.056	6.994
3700	277.471	929.600	695.386	866.592	-2393.424	-428.962	6.056
3800	277.726	937.003	701.648	894.352	-2393.311	-375.867	5.167
3900	277.963	944.220	707.775	922.136	-2393.242	-322.776	4.323
4000	278.184	951.261	713.775	949.944	-2393.220	-269.690	3.522
4100	278.390	958.132	719.651	977.773	-2393.243	-216.601	2.760
4200	278.583	964.843	725.410	1005.621	-3832.860	-142.439	1.771
4300	278.763	971.401	731.054	1033.489	-3829.713	-54.608	0.663
4400	278.931	977.811	736.590	1061.374	-3826.619	33.148	-0.394
4500	279.089	984.081	742.020	1089.275	-3823.576	120.834	-1.403
4600	279.238	990.217	747.349	1117.191	-3820.587	208.453	-2.367
4700	279.378	996.224	752.581	1145.122	-3817.655	296.011	-3.290
4800	279.509	1002.107	757.718	1173.066	-3814.779	383.505	-4.173
4900	279.633	1007.872	762.765	1201.024	-3811.962	470.942	-5.020
5000	279.750	1013.522	767.724	1228.993	-3809.206	558.321	-5.833
5100	279.860	1019.063	772.598	1256.973	-3806.514	645.645	-6.613
5200	279.964	1024.498	777.390	1284.965	-3803.886	732.913	-7.362
5300	280.063	1029.832	782.103	1312.966	-3801.326	820.138	-8.083
5400	280.157	1035.068	786.739	1340.977	-3798.837	907.310	-8.776
5500	280.245	1040.209	791.301	1368.997	-3796.420	994.437	-9.444
5600	280.329	1045.260	795.791	1397.026	-3794.079	1081.525	-10.088
5700	280.409	1050.222	800.211	1425.063	-3791.816	1168.568	-10.709
5800	280.485	1055.100	804.564	1453.108	-3789.635	1255.576	-11.308
5900	280.557	1059.895	808.851	1481.160	-3787.539	1342.538	-11.886
6000	280.626	1064.611	813.075	1509.219	-3785.531	1429.475	-12.445

PREVIOUS: December 1964 (1 atm) CURRENT: December 1964 (1 bar)

Boric Acid $((HBO_2)_3)$ $B_3H_3O_6(g)$

BORAZINE ($B_3N_3H_6$) IDEAL GAS M_r = 80.49774

S°(298.15 K) = 288.72 J K^{-1} mol^{-1}

$\Delta_f H°$(0 K) = -484.5 ± 13 kJ mol^{-1}

$\Delta_f H°$(298.15 K) = -510.0 ± 13 kJ mol^{-1}

Vibrational Frequencies and Degeneracies

ν, cm^{-1}	ν, cm^{-1}
3450 (1)	3400 (2)
[554] (1)	[339] (1)
[266] (1)	528 (1)

Ground State Quantum Weight: 1 σ = 6

Point Group: D_{3h}

Bond Distances: B–N = 1.44 Å B–H = 1.20 Å H–N = 1.02 Å

Bond Angles: N–B–N = 120° B–N–B = 120° N–B–H = 120°

Product of the Moments of Inertia: $I_A I_B I_C$ = 8.326395 x 10^{-114} g^3 cm^6

Enthalpy of Formation

The enthalpy of combustion for the reaction $B_3N_3H_6$(l) + 15/4 O_2(g) + 3/2 H_2O(l) = 3 H_3BO_3(c) + 3/2 N_2(g) was reported to be -552.9 ± 3.0 kcal mol^{-1} by Kilday, Johnson and Prosen (1). $\Delta_f H°$(H_3BO_3, cr, 298.15 K) = -261.47 kcal mol^{-1} from JANAF tables were used to calculate $\Delta_f H°$($B_3N_3H_6$, l, 298.15 K) = -129.0 ± 3.0 kcal mol^{-1}. Vapor pressure data on $B_3N_3H_6$(l) reported by Stock and Pohland (3) were fitted to a linear least squares. ΔC_p(l-g) for borazine was approximated by benzene to calculate a 2nd law $\Delta_{vap}H°$(298.15 K) = 7.1 ± 0.1 kcal mol^{-1}.

Heat Capacity and Entropy

The fundamental frequency assignment was made by Crawford and Edsall (4). The three frequencies enclosed by brackets are calculated values. The principal moments of inertia are calculated from molecular constants given by Bauer (5). The principal moments of inertia are: $I_A = I_B$ = 16.0870 x 10^{-39} and I_C = 32.1740 x 10^{-39} g cm^2.

References

1. M. V. Kilday, W. H. Johnson, and E. J. Prosen, J. Res. Nat. Bur. Stand. 65A, 101 (1961).
2. U. S. Nat. Bur. Stand. Circ. 500, 1962.
3. A. Stock and E. Pohland, Ber. 59, 2215 (1926).
4. B. L. Crawford, Jr. and J. T. Edsall, J. Chem. Phys. 7, 223 (1939).
5. S. H. Bauer, J. Am. Chem. Soc. 60, 524 (1938).

Borazine ($B_3H_6N_3$) $B_3H_6N_3$(g)

Enthalpy Reference Temperature = T_r = 298.15 K Standard State Pressure = p° = 0.1 MPa

T/K	$C_p°$	S°	-[G°-H°(T_r)]/T	H°-H°(T_r)	$\Delta_f H°$	$\Delta_f G°$	Log K_f
	J $K^{-1}$$mol^{-1}$			kJ mol^{-1}			
0	0.	0.	INFINITE	-16.515	-484.495	-484.495	INFINITE
100	39.386	222.416	352.981	-13.056	-494.460	-462.420	241.543
200	65.053	256.820	296.554	-7.947	-502.796	-427.066	111.538
250	81.262	273.063	290.231	-4.292	-506.627	-407.685	85.181
298.15	96.927	288.723	288.723	0.	-510.030	-388.315	68.031
300	97.516	289.324	288.725	0.180	-510.154	-387.560	67.480
350	112.826	305.522	289.969	5.444	-513.304	-366.873	54.753
400	126.782	321.515	292.916	11.440	-516.075	-345.761	45.152
450	139.307	337.185	296.967	18.098	-518.480	-324.323	37.646
500	150.485	352.452	301.756	25.348	-520.537	-302.637	31.616
600	169.391	381.623	312.662	41.377	-523.714	-258.739	22.525
700	184.667	408.922	324.486	59.105	-525.828	-214.396	15.998
800	197.230	434.427	336.652	78.220	-527.084	-169.811	11.088
900	207.712	458.281	348.856	98.482	-527.656	-125.111	7.261
1000	216.552	480.636	360.928	119.708	-527.689	-80.378	4.198
1100	224.064	501.637	372.775	141.748	-527.300	-35.662	1.693
1200	230.486	521.416	384.346	164.484	-526.590	9.003	-0.392
1300	236.003	540.088	395.614	187.815	-525.636	53.598	-2.154
1400	240.763	557.756	406.571	211.659	-524.504	98.113	-3.661
1500	244.887	574.511	417.213	235.947	-523.246	142.543	-4.964
1600	248.475	590.433	427.546	260.619	-521.905	186.885	-6.101
1700	251.610	605.593	437.577	285.627	-520.514	231.142	-7.102
1800	254.359	620.054	447.316	310.928	-519.102	275.316	-7.989
1900	256.779	633.872	456.774	336.488	-517.691	319.412	-8.781
2000	258.918	647.099	465.962	362.274	-516.299	363.434	-9.492
2100	260.815	659.778	474.891	388.263	-514.942	407.387	-10.133
2200	262.503	671.951	483.574	414.430	-513.632	451.277	-10.715
2300	264.011	683.654	492.020	440.758	-512.379	495.108	-11.244
2400	265.363	694.919	500.241	467.228	-661.971	542.093	-11.798
2500	266.579	705.777	508.247	493.826	-661.093	592.244	-12.374
2600	267.675	716.254	516.047	520.539	-660.198	642.360	-12.905
2700	268.666	726.375	523.650	547.357	-659.291	692.441	-13.396
2800	269.565	736.162	531.066	574.269	-658.378	742.488	-13.851
2900	270.382	745.636	538.303	601.267	-657.465	792.503	-14.275
3000	271.128	754.815	545.367	628.343	-656.555	842.487	-14.669
3100	271.809	763.717	552.268	655.491	-655.652	892.439	-15.038
3200	272.433	772.356	559.012	682.703	-654.759	942.364	-15.383
3300	273.006	780.748	565.604	709.976	-653.880	992.261	-15.706
3400	273.534	788.906	572.053	737.303	-653.018	1042.130	-16.010
3500	274.020	796.843	578.362	764.681	-652.174	1091.975	-16.297
3600	274.469	804.568	584.539	792.106	-651.351	1141.796	-16.567
3700	274.885	812.094	590.588	819.574	-650.551	1191.595	-16.822
3800	275.270	819.430	596.514	847.082	-649.777	1241.373	-17.064
3900	275.628	826.585	602.322	874.627	-649.030	1291.130	-17.293
4000	275.961	833.568	608.016	902.207	-648.311	1340.868	-17.510
4100	276.272	840.386	613.601	929.818	-647.623	1390.589	-17.716
4200	276.561	847.047	619.080	957.460	-2086.514	1461.366	-18.175
4300	276.832	853.558	624.458	985.130	-2082.627	1545.794	-18.778
4400	277.086	859.925	629.737	1012.826	-2078.778	1630.131	-19.352
4500	277.323	866.154	634.922	1040.547	-2074.970	1714.381	-19.900
4600	277.546	872.252	640.015	1068.290	-2071.202	1798.547	-20.423
4700	277.756	878.223	645.020	1096.055	-2067.477	1882.632	-20.923
4800	277.953	884.073	649.940	1123.841	-2063.794	1966.639	-21.401
4900	278.138	889.806	654.776	1151.646	-2060.155	2050.567	-21.859
5000	278.313	895.427	659.533	1179.468	-2056.561	2134.424	-22.298
5100	278.478	900.940	664.213	1207.308	-2053.011	2218.207	-22.719
5200	278.635	906.349	668.818	1235.164	-2049.505	2301.923	-23.123
5300	278.782	911.658	673.350	1263.035	-2046.044	2385.571	-23.511
5400	278.922	916.870	677.811	1290.920	-2042.627	2469.155	-23.884
5500	279.055	921.990	682.204	1318.819	-2039.254	2552.674	-24.243
5600	279.180	927.019	686.531	1346.730	-2035.924	2636.134	-24.589
5700	279.299	931.961	690.794	1374.655	-2032.637	2719.535	-24.922
5800	279.413	936.820	694.994	1402.590	-2029.392	2802.878	-25.243
5900	279.521	941.597	699.133	1430.537	-2026.189	2886.164	-25.552
6000	279.623	946.296	703.214	1458.494	-2023.026	2969.398	-25.851

PREVIOUS: March 1965 (1 atm) CURRENT: March 1965 (1 bar)

Borazine ($B_3H_6N_3$) $B_3H_6N_3$(g)

POTASSIUM BORATE ($K_2B_4O_7$) CRYSTAL M_r = 233.4324

$\Delta_f H°(0\ K)$ = Unknown
$S°(298.15\ K) = [208.4 \pm 6.3]\ J\ K^{-1}\ mol^{-1}$ $\Delta_f H°(298.15\ K) = -3334.2\ kJ\ mol^{-1}$
T_{fus} = 1088 K $\Delta_{fus}H° = 104.5 \pm 4.2\ kJ\ mol^{-1}$

Enthalpy of Formation

$\Delta_f H°(298.15\ K)$ was recalculated from the experimental data of Shartsis and Capps (1). They studied the enthalpy of solution of various alkali borates in 2N nitric acid. Average molecular weights were used for physical mixtures of the oxides B_2O_3 and R_2O (where R is either Li, Na, or K) rather than for the borate compounds $R_2O{\cdot}nB_2O_3$ present. These incorrect molecular weights led to spurious correlations and faulty conclusions. In reinterpreting the data, the steps followed were: (1) The calculation of the correct molecular weights and molar enthalpies of reaction; (2) The establishment that the enthalpy of formation of the borates from the oxides is linear with the mole fraction of B_2O_3 in the total original moles of oxides for the reactions

$$R_2O(cr) + nB_2O_3(gl) \rightarrow R_2O{\cdot}nB_2O_3(gls) \quad (I)$$
and
$$R_2O(cr) + nB_2O_3(gl) \rightarrow R_2O{\cdot}nB_2O_3(cr) \quad (II)$$

With the enthalpy of formation being zero at 100 percent R_2C; (3) The evaluation of the least square fit of the line through the enthalpy of reaction data for reaction I and for reaction II: (4) The use of the proper weighting factors in applying least squares to adjust for both the sample size and the molecular weight of the mixture of borates used in each solution experiment.

Heat Capacity and Entropy

The enthalpy measurements of Smith (2), in the range 298-973 K were smoothed graphically and were used to obtain a smooth set of C_p data. The heat capacity was extrapolated smoothly abcve this region. The entropy at 298.15 K is estimated from that of $Na_2B_4O_7(cr)$ by using an average entropy change of $4.5 \pm 1.5\ cal\ K^{-1}\ mol^{-1}$ on substituting Na_2 by K_2.

Fusion Data

T_{fus} was given by Rollet (3). The enthalpy of melting was obtained from the enthalpies of formation of the crystal and glass at 298.15 K combined with the enthalpy differences between 298.15 K and 1088 K. This involves a slight extrapolation of the crystal data and an uncertainty of about 1.0 kcal mol^{-1}.

References

1. L. Shartsis and W. Capps, J. Am. Ceram. Soc. 37, 27 (1954).
2. G. S. Smith, Ph. D. Thesis, Penn. State Univ., Univ. Microfilms, Order No. Mic 59-2916 (1959).
3. A. P. Rollet, Comp. Rend. 200, 1763 (1935).

Potassium Borate ($K_2B_4O_7$) $B_4K_2O_7(cr)$

Enthalpy Reference Temperature = T_r = 298.15 K Standard State Pressure = p° = 0.1 MPa

T/K	$C_p°$ (J K⁻¹mol⁻¹)	S°	$-[G°-H°(T_r)]/T$	$H°-H°(T_r)$ (kJ mol⁻¹)	$\Delta_f H°$	$\Delta_f G°$	Log K_f
0							
100							
200							
298.15	170.498	208.363	208.363	0.	-3334.230	-3136.756	549.547
300	171.335	209.420	208.366	0.316	-3334.297	-3135.531	545.944
400	206.271	263.696	215.530	19.267	-3342.152	-3067.880	400.624
500	232.798	312.737	230.150	41.293	-3343.971	-2999.075	313.311
600	250.538	356.839	247.657	65.509	-3344.817	-2930.004	255.080
700	262.379	396.388	266.130	91.181	-3345.149	-2860.838	213.478
800	271.123	432.022	284.677	117.876	-3345.218	-2791.644	182.276
900	277.399	464.322	302.871	145.306	-3345.200	-2722.448	158.007
1000	283.257	493.860	320.514	173.346	-3345.151	-2653.256	138.592
1088.000	287.723	517.940	335.520	198.474	------ CRYSTAL <--> LIQUID -----		
1100	288.282	521.099	337.527	201.930	-3503.006	-2574.848	122.269
1200	292.478	546.369	353.890	230.975	-3500.895	-2490.563	108.411
1300	295.842	569.918	369.612	260.398	-3498.737	-2406.456	96.693
1400	298.374	591.940	384.715	290.115	-3496.583	-2322.515	86.654
1500	300.076	612.589	399.226	320.045	-3494.495	-2238.727	77.959

PREVIOUS: CURRENT: March 1963

Potassium Borate ($K_2B_4O_7$) $B_4K_2O_7(cr)$

POTASSIUM BORATE ($K_2B_4O_7$) LIQUID M_r = 233.4324

S°(298.15 K) = [208.363] J K^{-1} mol^{-1}

T_{fus} = 1088 K

Δ_fH°(298.15 K) = [-3334.230] kJ mol^{-1}

$\Delta_{fus}H$° = 104.5 ± 4.2 kJ mol^{-1}

Enthalpy of Formation

The enthalpy of formation is obtained form reaction (1), as given on the crystal table.

Heat Capacity and Entropy

The enthalpy measurements of Smith (1) in the range 298 to 1373 K on the glass and liquid were smoothed graphically and used to determine C_p. There is a discontinuity (glass transition) in the C_p curve at 676 K. Above 1373 K the C_p values were extrapolated smoothly.

The entropy at 298.15 K is calculated from that of the crystal by adding $\Delta_{fus}S$° and the entropy differences, S°(1088 K) - S°(298.15 K), between the crystal and liquid.

Fusion Data

Refer to the crystal table for details.

Reference

1. G. S. Smith, Ph. D. Thesis, Penn. State Univ., Univ. Microfilms, Order No. Mic 59-2916 (1959).

Potassium Borate ($K_2B_4O_7$) $B_4K_2O_7$(l)

Enthalpy Reference Temperature = T_r = 298.15 K Standard State Pressure = p° = 0.1 MPa

T/K	C_p° (J K⁻¹mol⁻¹)	S°	-[G°-H°(T_r)]/T	H°-H°(T_r) (kJ mol⁻¹)	Δ_fH°	Δ_fG°	Log K_f
0							
100							
200							
298.15	170.498	237.120	237.120	0.	-3289.042	-3100.143	543.132
300	171.335	238.177	237.124	0.316	-3289.110	-3098.971	539.579
400	206.271	292.452	244.287	19.266	-3296.965	-3034.195	396.225
500	232.840	341.499	258.908	41.296	-3298.781	-2968.266	310.093
600	250.538	385.610	276.416	65 516	-3299.623	-2902.072	252.648
676.000	259.821	416.068	290.428	84.933	GLASS <--> LIQUID TRANSITION		
676.000	369.237	416.068	290.428	84.933			
700	375.221	429.053	294.959	93.866	-3297.277	-2835.830	211.612
800	402.501	480.977	315.000	132.781	-3285.126	-2770.715	180.909
900	426.266	529.790	336.181	174.249	-3271.070	-2707.239	157.124
1000	446.558	575.818	357.866	217.952	-3255.357	-2645.420	138.183
1088.000	457.879	613.958	377.052	257.753	CRYSTAL <--> LIQUID		
1100	459.361	618.988	379.664	263.257	-3396.492	-2576.011	122.325
1200	471.143	659.470	401.312	309.790	-3376.892	-2502.281	108.921
1300	481.913	697.614	422.651	357.452	-3356.496	-2430.219	97.647
1400	491.670	733.691	443.591	406.139	-3335.372	-2359.755	88.044
1500	500.406	767.916	464.082	455.752	-3313.601	-2290.825	79.774
1600	507.268	800.434	484.096	506.140	-3291.314	-2223.364	72.585
1700	513.544	831.378	503.622	557.186	-3268.619	-2157.310	66.286
1800	519.234	860.896	522.657	608.830	-3245.569	-2092.603	60.726
1900	524.339	889.108	541.207	661.013	-3222.218	-2029.185	55.786
2000	528.858	916.121	559.282	713.678	-3198.621	-1967.003	51.373
2100	532.188	942.006	576.895	766.732	-3174.868	-1906.006	47.409
2200	535.234	966.835	594.059	820.106	-3151.032	-1846.140	43.833
2300	537.995	990.689	610.789	873.770	-3127.143	-1787.363	40.592
2400	540.472	1013.639	627.099	927.696	-3304.266	-1725.350	37.551
2500	542.665	1035.747	643.005	981.855	-3280.672	-1660.046	34.685
2600	544.573	1057.069	658.523	1036.219	-3256.986	-1595.688	32.058
2700	546.196	1077.653	673.668	1090.760	-3233.240	-1532.242	29.643
2800	547.535	1097.542	688.453	1145.449	-3209.468	-1469.676	27.417
2900	548.589	1116.775	702.893	1200.257	-3185.701	-1407.963	25.360
3000	549.359	1135.387	717.001	1255.157	-3161.975	-1347.065	23.454

PREVIOUS. CURRENT: March 1963

Potassium Borate ($K_2B_4O_7$) $B_4K_2O_7$(l)

POTASSIUM BORATE ($K_2B_4O_7$) CRYSTAL-LIQUID M_r = 233.4324

298.15 to 1088 K crystal
above 1088 K liquid

Refer to the individual tables for details.

Potassium Borate ($K_2B_4O_7$) $B_4K_2O_7$(cr,l)

Enthalpy Reference Temperature = T_r = 298.15 K; Standard State Pressure = p° = 0.1 MPa

T/K	C_p° (J K⁻¹mol⁻¹)	S° (J K⁻¹mol⁻¹)	$-[G^\circ-H^\circ(T_r)]/T$ (J K⁻¹mol⁻¹)	$H^\circ-H^\circ(T_r)$ (kJ mol⁻¹)	Δ_fH° (kJ mol⁻¹)	Δ_fG° (kJ mol⁻¹)	Log K_f
0							
100							
200							
298.15	170.498	208.363	208.363	0.	-3334.230	-3136.756	549.547
300	171.335	209.420	208.366	0.316	-3334.297	-3135.531	545.944
400	206.271	263.696	215.530	19.267	-3342.152	-3067.880	400.624
500	232.798	312.737	230.150	41.293	-3343.971	-2999.075	313.311
600	250.538	356.839	247.657	65.509	-3344.817	-2930.004	255.080
700	262.379	396.388	266.130	91.181	-3345.149	-2860.838	213.478
800	271.123	432.022	284.677	117.876	-3345.218	-2791.644	182.276
900	277.399	464.322	302.871	145.306	-3345.200	-2722.448	158.007
1000	283.257	493.860	320.514	173.346	-3345.151	-2653.256	138.592
1088.000	287.723	517.940	335.520	198.474	CRYSTAL <--> LIQUID		
1088.000	457.879	613.958	335.520	302.940	TRANSITION		
1100	459.361	618.988	338.585	308.444	-3396.492	-2576.011	122.325
1200	471.143	659.470	363.656	354.978	-3376.892	-2502.281	108.921
1300	481.913	697.614	387.892	402.639	-3356.496	-2430.219	97.647
1400	491.670	733.691	411.315	451.327	-3335.372	-2359.755	88.044
1500	500.406	767.916	433.957	500.939	-3313.601	-2290.825	79.774
1600	507.268	800.434	455.854	551.327	-3291.314	-2223.364	72.585
1700	513.544	831.378	477.041	602.373	-3268.619	-2157.310	66.286
1800	519.234	860.896	497.553	654.017	-3245.569	-2092.603	60.726
1900	524.339	889.108	517.424	706.200	-3222.218	-2029.185	55.786
2000	528.858	916.121	536.688	758.865	-3198.621	-1967.003	51.373
2100	532.188	942.006	555.377	811.920	-3174.868	-1906.006	47.409
2200	535.234	966.835	573.519	865.293	-3151.033	-1846.140	43.833
2300	537.995	990.689	591.142	918.957	-3127.143	-1787.363	40.592
2400	540.472	1013.639	608.271	972.883	-3304.266	-1725.350	37.551
2500	542.665	1035.747	624.931	1027.042	-3280.672	-1660.046	34.685
2600	544.573	1057.069	641.144	1081.406	-3256.986	-1595.688	32.058
2700	546.196	1077.653	656.932	1135.947	-3233.240	-1532.242	29.643
2800	547.535	1097.542	672.315	1190.636	-3209.468	-1469.676	27.417
2900	548.589	1116.775	687.311	1245.445	-3185.701	-1407.963	25.360
3000	549.359	1135.387	701.938	1300.344	-3161.975	-1347.065	23.454

PREVIOUS: CURRENT: March 1963

Potassium Borate ($K_2B_4O_7$) $B_4K_2O_7$(cr,l)

LITHIUM BORATE ($Li_2B_4O_7$) CRYSTAL M_r = 169.1178

$S°(298.15 K) = [155.6 \pm 4.2]\ J\ K^{-1}\ mol^{-1}$
$T_{fus} = 1190 \pm 2\ K$

$\Delta_f H°(0\ K)$ = Unknown
$\Delta_f H°(298.15\ K) = -3362.3 \pm 6.3\ kJ\ mol^{-1}$
$\Delta_{fus} H° = 120.5 \pm 6.3\ kJ\ mol^{-1}$

Enthalpy of Formation

The enthalpy of formation was calculated from $\Delta_r H°(298.15\ K) = -12.2$ kcal for the reaction $Li_2B_4O_7(cr) + 2\ HNO_3(aq) + 5\ H_2O(l) = 4\ H_3BO_3(aq) + 2\ LiNO_3(aq)$. The enthalpy of solution was interpolated from the enthalpy of solution data of Shartsis and Capps (1) Auxilliary data are from National Bureau of Standards Circular 500 (2).

Heat Capacity and Entropy

The enthalpy measurements of Smith (3) in the range 298 to 1190 K were smoothed graphically and used to determine C_p. A linear extrapolation of C_p was used from 1190 to 2000 K.

S°(298.15 K) was estimated by three routes: (1) $S°(Li_2B_4O_7$, cr, 298.15 K) = 37.3 cal $K^{-1}\ mol^{-1}$ based on JANAF $S°(Na_2B_4O_7$, cr, 298.15 K) = 45.3 and $\Delta S°(2Na - 2Li$, 298.15 K) = 8 cal $K^{-1}\ mol^{-1}$ from Latimer (4); (2) $S°(Li_2B_4O_7$, cr, 298.15 K) = 36.7 cal $K^{-1}\ mol^{-1}$ in a similar manner based on $\Delta_f S°(2Na^+ - 2Li^+$, 298.15 K) = 8.6 cal $K^{-1}\ mol^{-1}$ from Kelley (5); and (3) $S°(Li_2B_4O_7$, cr, 298.15 K) = 37.6 cal $K^{-1}\ mol^{-1}$ based on addition of the JANAF entropies for $2LiBO_2$ and B_2O_3. The value adopted is an average of these estimates.

Fusion Data

The value for T_{fus} is from Sastry and Hummel (6). Rollet and Bouaziz (7) earlier reported T_{fus} = 1188 K. $\Delta_{fus}H°$ = 28.8 kcal mol^{-1} is from Smith (3).

References

1. L. Shartsis and W. Capps, J. Am. Ceram. Soc. 37, 27 (1954).
2. U. S. Nat. Bur. Stand. Circ. 500, 1952.
3. G. S. Smith, Ph. D. Thesis, Penn. State Univ., Univ. Microfilms, Order No. Mic 59-2916 (1959).
4. W. M. Latimer, J. Am. Chem. Soc. 73, 1480 (1951).
5. K. K. Kelley, U. s. Bur. Mines, personal communication, June, 1960.
6. B. S. R. Sastry and F. A. Hummel, J. Am. Chem. Soc. 42, 216 (1959).
7. A. P. Rollet and R. Bouaziz, Comp. Rend. 240. 2417 (1955).

Lithium Borate ($Li_2B_4O_7$) $B_4Li_2O_7(cr)$

Enthalpy Reference Temperature = T_r = 298.15 K Standard State Pressure = p° = 0.1 MPa

T/K	$C_p°$	S°	$-[G°-H°(T_r)]/T$	$H°-H°(T_r)$	$\Delta_f H°$	$\Delta_f G°$	Log K_f
	J K⁻¹mol⁻¹			kJ mol⁻¹			
0							
100							
200							
298.15	183.050	155.645	155.645	0.	-3362.262	-3170.291	555.422
300	183.096	156.777	155.648	0.339	-3362.289	-3169.100	551.789
400	197.648	211.259	162.980	19.311	-3364.367	-3104.421	405.396
500	219.940	257.778	177.378	40.200	-3372.974	-3038.519	317.432
600	241.103	299.795	194.332	63.278	-3374.846	-2971.434	258.686
700	259.199	338.357	212.190	88.317	-3375.669	-2904.120	216.708
800	274.412	373.983	230.216	115.013	-3375.575	-2836.753	185.221
900	287.717	407.078	248.050	143.125	-3374.692	-2769.445	160.734
1000	300.424	438.042	265.517	172.525	-3373.030	-2702.277	141.152
1100	314.018	467.306	282.543	203.239	-3370.479	-2635.316	125.141
1190.000	327.010	492.501	297.475	232.080	------ CRYSTAL <--> LIQUID -----		
1200	327.180	495.378	299.118	235.511	-3366.736	-2568.641	111.810
1300	340.343	522.086	315.249	268.887	-3362.211	-2502.310	100.544
1400	353.506	547.790	330.947	303.580	-3356.657	-2436.365	90.902
1500	366.669	572.628	346.236	339.589	-3350.045	-2370.856	82.561
1600	379.832	596.712	361.141	376.914	-3342.359	-2305.821	75.277
1700	392.995	620.135	375.690	415.555	-3624.059	-2226.948	68.426
1800	406.158	642.970	389.908	455.513	-3612.686	-2145.089	62.249
1900	419.320	665.283	403.816	496.786	-3600.229	-2063.890	56.740
2000	432.483	687.126	417.437	539.377	-3586.685	-1983.375	51.800

PREVIOUS: December 1962 CURRENT: June 1965

Lithium Borate ($Li_2B_4O_7$) $B_4Li_2O_7(cr)$

LITHIUM BORATE ($Li_2B_4O_7$) LIQUID M_r = 169.1178

$S°(298.15\ K) = [173.375]\ J\ K^{-1}\ mol^{-1}$ $\Delta_f H°(298.15\ K) = [-3317.715]\ kJ\ mol^{-1}$

$T_{fus} = 1190 \pm 2\ K$ $\Delta_{fus}H° = 120.5 \pm 6.3\ kJ\ mol^{-1}$

Enthalpy of Formation

$\Delta_f H°(\ell,\ 298.15\ K)$ is calculated from $\Delta_f H°(cr,\ 298.15\ K)$ by adding the enthalpy of fusion and the difference in enthalpy, $H°(1190\ K) - H°(298.15\ K)$, between the crystal and liquid.

Heat Capacity and Entropy

The enthalpy measurements of Smith (1) in the range 296 to 1373 K were smoothed graphically and used to determine C_p. The C_p data have a discontinuity at about 756 K. A linear extrapolation of C_p was used from 1373 to 3000 K.

The entropy at 298.15 K is calculated in a manner analogous to that used for the enthalpy of formation.

Fusion Data

Refer to the crystal table for details.

Reference

1. G. S. Smith, Ph. D. Thesis, Penn. State Univ., Univ. Microfilms, Order No. Mic. 59-2916 (1959).

Lithium Borate ($Li_2B_4O_7$) $B_4Li_2O_7(l)$

Enthalpy Reference Temperature = T_r = 298.15 K Standard State Pressure = p° = 0.1 MPa

T/K	$C_p°$ (J K⁻¹mol⁻¹)	S° (J K⁻¹mol⁻¹)	$-[G°-H°(T_r)]/T$ (J K⁻¹mol⁻¹)	$H°-H°(T_r)$ (kJ mol⁻¹)	$\Delta_f H°$ (kJ mol⁻¹)	$\Delta_f G°$ (kJ mol⁻¹)	Log K_f
0							
100							
200							
298.15	181.146	173.375	173.375	0.	-3317.715	-3131.030	548.544
300	181.586	174.497	173.379	0.336	-3317.745	-3129.872	544.959
400	205.225	229.974	180.784	19.676	-3319.455	-3066.995	400.508
500	228.869	278.308	195.547	41.381	-3327.246	-3003.056	313.727
600	252.513	322.126	213.043	65.450	-3328.127	-2938.114	255.786
700	276.157	362.827	231.565	91.883	-3327.555	-2873.135	214.396
756.115	289.425	384.627	242.119	107.752	GLASS <--> LIQUID	TRANSITION	
756.115	455.516	384.627	242.119	107.752			
800	456.746	410.361	250.649	127.769	-3318.272	-2808.552	183.380
900	459.549	464.319	271.448	173.584	-3299.686	-2745.956	159.371
1000	462.352	512.883	293.204	219.679	-3281.329	-2685.416	140.272
1100	465.156	557.081	315.213	266.055	-3263.117	-2626.707	124.732
1190.000	467.680	593.760	334.910	308.032	------ CRYSTAL <--> LIQUID -----		
1200	467.959	597.675	337.083	312.710	-3244.989	-2569.652	111.854
1300	470.762	635.243	358.592	359.646	-3226.905	-2514.108	101.018
1400	473.444	670.231	379.616	406.861	-3208.829	-2459.955	91.782
1500	475.836	702.977	400.094	454.325	-3190.762	-2407.096	83.823
1600	478.294	733.765	419.995	502.031	-3172.695	-2355.440	76.897
1700	480.818	762.836	439.315	549.986	-3445.081	-2290.563	70.380
1800	483.408	790.392	458.061	598.196	-3425.455	-2223.218	64.516
1900	486.063	816.600	476.247	646.669	-3405.799	-2156.962	59.299
2000	488.785	841.600	493.895	695.411	-3386.104	-2091.743	54.631
2100	491.572	865.515	511.026	744.429	-3366.359	-2027.510	50.432
2200	494.425	888.449	527.663	793.728	-3346.557	-1964.217	46.636
2300	497.344	910.491	543.832	843.316	-3326.690	-1901.830	43.192
2400	500.328	931.721	559.554	893.199	-3507.788	-1836.032	39.960
2500	503.379	952.207	574.853	943.384	-3488.083	-1766.780	36.915
2600	506.495	972.010	589.750	993.877	-3468.163	-1698.320	34.120
2700	509.677	991.185	604.264	1044.685	-3448.023	-1630.630	31.546
2800	512.925	1009.779	618.417	1095.814	-3427.657	-1563.690	29.171
2900	516.239	1027.837	632.225	1147.275	-3407.059	-1497.485	26.973
3000	519.297	1045.390	645.706	1199.054	-3386.242	-1431.991	24.933

PREVIOUS: December 1962 CURRENT: June 1965

Lithium Borate ($Li_2B_4O_7$) $B_4Li_2O_7(l)$

LITHIUM BORATE ($Li_2B_4O_7$) CRYSTAL-LIQUID M_r = 169.1178

298.15 to 1190 K crystal
above 1190 K liquid

Refer to the individual tables for details.

Lithium Borate ($Li_2B_4O_7$) $B_4Li_2O_7$(cr,l)

Enthalpy Reference Temperature = T_r = 298.15 K Standard State Pressure = p° = 0.1 MPa

T/K	C_p° (J K⁻¹mol⁻¹)	S° (J K⁻¹mol⁻¹)	−[G°−H°(T_r)]/T (J K⁻¹mol⁻¹)	H°−H°(T_r) (kJ mol⁻¹)	$\Delta_f H^\circ$ (kJ mol⁻¹)	$\Delta_f G^\circ$ (kJ mol⁻¹)	Log K_f
0							
100							
200							
298.15	183.050	155.645	155.645	0.	-3362.262	-3170.291	555.422
300	183.096	156.777	155.648	0.339	-3362.289	-3169.100	551.789
400	197.648	211.259	162.980	19.311	-3364.367	-3104.421	405.396
500	219.940	257.778	177.378	40.200	-3372.974	-3038.519	317.432
600	241.103	299.795	194.332	63.278	-3374.846	-2971.434	258.686
700	259.199	338.357	212.190	88.317	-3375.669	-2904.120	216.708
800	274.412	373.983	230.216	115.013	-3375.575	-2836.753	185.221
900	287.717	407.078	248.050	143.125	-3374.692	-2769.445	160.734
1000	300.424	438.042	265.517	172.525	-3373.030	-2702.277	141.152
1100	314.018	467.306	282.543	203.239	-3370.479	-2635.316	125.141
1190.000	327.010	492.501	297.475	232.080	CRYSTAL <--> LIQUID		
1190.000	467.680	593.760	297.475	352.579	TRANSITION		
1200	467.960	597.675	299.961	357.257	-3244.989	-2569.652	111.854
1300	470.762	635.243	324.324	404.194	-3226.905	-2514.108	101.018
1400	473.444	670.231	347.797	451.408	-3208.828	-2459.955	91.782
1500	475.836	702.977	370.396	498.872	-3190.762	-2407.096	83.823
1600	478.294	733.765	392.154	546.578	-3172.695	-2355.440	76.897
1700	480.818	762.836	413.111	594.533	-3445.081	-2290.563	70.380
1800	483.408	790.392	433.312	642.744	-3425.455	-2223.218	64.516
1900	486.063	816.600	452.801	691.217	-3405.799	-2156.962	59.299
2000	488.785	841.600	471.621	739.958	-3386.104	-2091.743	54.631
2100	491.572	865.515	489.813	788.976	-3366.359	-2027.510	50.432
2200	494.425	888.449	507.415	838.275	-3346.557	-1964.217	46.636
2300	497.344	910.491	524.464	887.863	-3326.690	-1901.830	43.192
2400	500.328	931.721	540.993	937.746	-3507.788	-1836.032	39.960
2500	503.379	952.207	557.034	987.931	-3488.083	-1766.780	36.915
2600	506.495	972.010	572.616	1038.424	-3468.163	-1698.320	34.120
2700	509.677	991.185	587.766	1089.232	-3448.023	-1630.630	31.546
2800	512.925	1009.779	602.507	1140.361	-3427.657	-1563.690	29.171
2900	516.239	1027.837	616.864	1191.822	-3407.059	-1497.485	26.973
3000	519.297	1045.390	630.857	1243.601	-3386.242	-1431.991	24.933

PREVIOUS: CURRENT: June 1965

Lithium Borate ($Li_2B_4O_7$) $B_4Li_2O_7$(cr,l)

MAGNESIUM BORIDE (MgB_4) CRYSTAL M_r = 67.545

$S°(298.15\ K) = 51.92\ J\ K^{-1}\ mol^{-1}$

$T_{dcm} = [1100]\ K$

$\Delta_f H°(0\ K) = -104.7 \pm 8.4\ kJ\ mol^{-1}$

$\Delta_f H°(298.15\ K) = -105.0 \pm 8.4\ kJ\ mol^{-1}$

Enthalpy of Formation

The equilibrium pressures (1169-1177 K) of the reaction $MgB_4(cr) \rightarrow Mg(g) + 4B(cr)$ were measured by Wright and Walsh (1). The 3rd law value of $\Delta_r H°(298.15\ K)$ for this reaction is derived as $60.40 \pm 1.5\ kcal\ mol^{-1}$. This leads to a $\Delta_f H°(298.15\ K) = -25.1 \pm 2.0\ kcal\ mol^{-1}$ for $MgB_4(cr)$. The unit of equilibrium pressures reported should be in atmosphere not in mm Hg as printed. The diameter of the hole drilled through the center of the lid served as the effusion orifice should be 1/16 inch rather than the reported value 1/6 inch.

Heat Capacity and Entropy

The low temperature (17.34-299.53 K) heat capacities were measured by Swift and White (2). Above 298.15 K the C_p values are estimated by comparison with those of other related borides. S°(298.15 K) was reported by Swift and White (2) using S°(extrap., 20 K) = $0.022\ cal\ K^{-1}\ mol^{-1}$.

Decomposition Data

T_{dcm} was estimated from the decomposition reaction reported by Wright and Walsh (1).

References

1. M. Wright and P. N. Walsh, Ohio State Univ. Research Foundation, 'Technical Research Report OMCC-HEF-55, Jan. 9, 1958.
2. R. M. Swift and D. White, J. Am. Chem. Soc. 79, 3641 (1957).

Magnesium Boride (MgB_4) $B_4Mg_1(cr)$

Enthalpy Reference Temperature = T_r = 298.15 K Standard State Pressure = p° = 0.1 MPa

T/K	$C_p°$	S°	$-[G°-H°(T_r)]/T$	$H°-H°(T_r)$	$\Delta_f H°$	$\Delta_f G°$	Log K_f
	J K⁻¹mol⁻¹			kJ mol⁻¹			
0	0.	0.	INFINITE	-9.519	-104.683	-104.683	INFINITE
100	17.238	8.033	97.487	-8.945	-104.848	-104.578	54.626
200	46.275	28.702	57.635	-5.786	-105.028	-104.205	27.215
298.15	70.333	51.923	51.923	0.	-105.018	-103.801	18.185
300	70.710	52.360	51.925	0.130	-105.018	-103.793	18.072
400	79.496	73.935	54.807	7.651	-105.503	-103.339	13.495
500	87.027	92.491	60.529	15.981	-106.763	-102.664	10.725
600	94.140	109.006	67.256	25.050	-108.386	-101.695	8.853
700	99.956	123.955	74.305	34.755	-110.174	-100.440	7.495
800	105.730	137.684	81.380	45.043	-112.005	-98.924	6.459
900	111.043	150.450	88.354	55.887	-113.807	-97.180	5.640
1000	115.729	162.403	95.167	67.235	-124.164	-94.526	4.938
1100	118.512	173.561	101.793	78.945	-126.005	-91.473	4.344
1200	121.629	184.004	108.213	90.949	-127.787	-88.254	3.842
1300	125.525	193.897	114.427	103.311	-129.422	-84.892	3.411
1400	128.867	203.325	120.443	116.036	-258.297	-78.244	2.919
1500	131.587	212.311	126.270	129.062	-258.286	-65.382	2.277
1600	133.888	220.879	131.917	142.339	-258.193	-52.525	1.715
1700	135.562	229.046	137.393	155.811	-258.064	-39.674	1.219
1800	137.235	236.842	142.703	169.451	-257.921	-26.832	0.779
1900	138.909	244.307	147.855	183.258	-257.759	-13.998	0.385
2000	140.582	251.475	152.858	197.233	-257.573	-1.173	0.031

PREVIOUS: March 1963 CURRENT: March 1965

Magnesium Boride (MgB_4) $B_4Mg_1(cr)$

$S°(298.15\ K) = 189.49\ J\ K^{-1}\ mol^{-1}$

$T_{fus} = 1015.6\ K$

$\Delta_f H°(0\ K) = -3259.0 \pm 8\ k\ J\ mol^{-1}$

$\Delta_f H°(298.15\ K) = -3276.7 \pm 8\ kJ\ mol^{-1}$

$\Delta_{fus} H° = 76.9 \pm 6.3\ kJ\ mol^{-1}$

Enthalpy of Formation

The enthalpy of the reaction $Na_2O(cr) + 2B_2O_3(l) \rightarrow Na_2B_4O_7(cr)$ was reported by Shartsis and Capps (1) from their enthalpy of solution measurements. Their values for the enthalpy of reaction of the oxides have been converted to the enthalpy of formation using JANAF values (2) for $B_2O_3(l)$ and for $Na_2O(cr)$. The above $\Delta_r H° = -85.2\ kcal\ mol^{-1}$.

Heat Capacity and Entropy

In the low temperature region (5-344 K) the C_p values are those determined by Westrum and Grenier (3). The enthalpy measurements of Smith (4) in the range 296 to 1015.6 K were smoothed grpahically and used to determine C_p; the values did not join smoothly with those of Westrum and Grenier and the low temperature values were given greater weight. Above T_{fus} the heat capacity was extrapolated. S°(298.15 K) was derived from low temperature heat capacities.

Fusion Data

The value for T_{fus} is from Morey and Merwin (5). Other reported values are: 1008 ± 5 K by Coles, Scholes, and Amberg (6); 1011 K by Menzel (7); 1005 K by Ponomareff (8); 1015 K by Day and Allen (9); and 1064 K by Burgess and Holt (10). $\Delta_{fus} H°$ was obtained from the values of $\Delta_f H°$ for $Na_2B_4O_7(cr)$ and $Na_2B_4O_7(l)$ at 298.15 K as described above and the difference between H°(1015 K) - H°(298.15 K) for crystal and liquid.

References

1. L. Shartsis and W. Capps, J. Am. Ceram. Soc. 37, 27 (1954).
2. JANAF Thermochemical Tables: $B_2O_3(l)$, 12-31-64; $Na_2O(cr)$, 6-30-62.
3. E. F. Westrum, Jr. and G. Grenier, J. Am. Chem. Soc. 79, 1799 (1957).
4. G. S. Smith, Ph. D. Thesis, Penn. Stae Univ., Univ. Microfilms, Order No. Mic 59-2916 (1959).
5. G. W. Morey and H. E. Merwin, J. Am. Chem. Soc. 58, 2248 (1936).
6. S. S. Coles, S. R. Scholes, and C. R. Amberg, J. Am. Ceram. Soc. 18, 58 (1935).
7. H. Menzel, Z. anorg. Chem. 224, 1 (1935).
8. J. F. Ponomareff, Z. anorg. Chem. 89, 383 (1924).
9. A. L. Day and E. T. Allen, Carnegie Inst. of Washington, Publ. No. 31, 29 (1905).
10. C. H. Burgess and Q. Holt, Proc. Roy. Soc. (London) 74, 285 (1904).

Enthalpy Reference Temperature = T_r = 298.15 K   Standard State Pressure = p° = 0.1 MPa

	J K⁻¹mol⁻¹			kJ mol⁻¹			
T/K	$C_p°$	S°	$-[G°-H°(T_r)]/T$	$H°-H°(T_r)$	$\Delta_f H°$	$\Delta_f G°$	Log K_f
0	0.	0.	INFINITE	-30.382	-3258.983	-3258.983	INFINITE
100	76.525	50.814	321.976	-27.116	-3268.658	-3208.196	1675.790
200	139.746	124.566	205.130	-16.113	-3274.087	-3145.347	821.480
298.15	186.774	189.490	189.490	0.	-3276.741	-3081.522	539.870
300	187.447	190.648	189.494	0.346	-3276.774	-3080.310	536.330
400	213.802	248.460	197.196	20.505	-3283.707	-3014.160	393.609
500	228.534	297.784	212.510	42.637	-3285.408	-2946.574	307.826
600	241.500	340.624	230.368	66.154	-3286.907	-2878.661	250.610
700	252.697	378.715	248.889	90.878	-3288.098	-2810.522	209.724
800	262.128	413.077	267.299	116.622	-3288.983	-2742.233	179.049
900	271.215	444.482	285.266	143.294	-3289.531	-2673.853	155.186
1000	279.700	473.502	302.657	170.845	-3289.721	-2605.430	136.094
1015.600	280.970	477.841	305.314	175.218	------ CRYSTAL <--> LIQUID -----		
1100	287.583	500.534	319.431	199.214	-3289.570	-2537.005	120.472
1200	294.863	525.874	335.589	228.342	-3482.640	-2463.728	107.243
1300	301.541	549.743	351.153	258.167	-3480.079	-2378.921	95.586
1400	307.616	572.316	366.152	288.630	-3477.180	-2294.323	85.602
1500	313.089	593.718	380.616	319.653	-3473.995	-2209.943	76.957
1600	320.013	614.144	394.577	351.307	-3470.459	-2125.786	69.400
1700	327.017	633.755	408.074	383.658	-3466.432	-2041.866	62.739
1800	334.105	652.647	421.139	416.714	-3461.958	-1958.196	56.825
1900	341.276	670.903	433.807	450.482	-3457.003	-1874.786	51.541
2000	348.527	688.592	446.106	484.972	-3451.552	-1791.650	46.793

PREVIOUS: March 1963   CURRENT: March 1965

SODIUM BORATE ($Na_2B_4O_7$) LIQUID M_r = 201.21534

$S°(298.15\ K) = [194.636]\ J\ K^{-1}\ mol^{-1}$

$T_{fus} = 1015.6\ K$

$\Delta_fH°(298.15\ K) = [-3256.658]\ kJ\ mol^{-1}$

$\Delta_{fus}H° = 76.9 \pm 6.3\ kJ\ mol^{-1}$

Enthalpy of Formation

The enthalpy of reaction $Na_2O(cr) + 2B_2O_3(l) \rightarrow Na_2B_4O_7(l)$ was reported by Shartsis and Capps (1) from their enthalpy of solution measurements. Their values for the enthalpy of reaction of the oxides have been converted to the enthalpies of formation using JANAF values (2) for $B_2O_3(l)$ and $Na_2O(cr)$. The above $\Delta_rH° = -80.4\ kcal\ mol^{-1}$.

Heat Capacity and Entropy

The enthalpy measurements of Smith (3) in the range 298 to 1373 K, were smoothed graphically and used to determine C_p. These were joined smoothly with the low temperature data of Westrum and Grenier (4). At 743 K a glass transition was observed and above this temperature the C_p adopted as constant.

The entropy at 298.15 K is calculated from that of the crystal by adding $\Delta_{fus}S°$ and the difference in entropy, S°(1015 K) - S°(298.15 K), between the crystal and liquid.

Fusion Data

See the crystal table for details.

References

1. L. Shartsis and W. Capps, J. Am. Ceram. Soc. 37, 27 (1954).
2. JANAF Thermochemical Tables: $B_2O_3(\ell)$, 12-31-64; $Na_2O(cr)$, 6-30-62.
3. G. S. Smith, Ph. D. Thesis, Penn. State Univ., Univ. Microfilms, Order No. Mic 59-2916 (1959).
4. E. F. Westrum, Jr. and G. Grenier, J. Am. Chem. Soc. 79, 1799 (1957).

Sodium Borate ($Na_2B_4O_7$) $B_4Na_2O_7(l)$

Enthalpy Reference Temperature = T_r = 298.15 K — Standard State Pressure = p° = 0.1 MPa

T/K	$C_p°$ (J K⁻¹mol⁻¹)	S° (J K⁻¹mol⁻¹)	$-[G°-H°(T_r)]/T$ (J K⁻¹mol⁻¹)	$H°-H°(T_r)$ (kJ mol⁻¹)	$\Delta_fH°$ (kJ mol⁻¹)	$\Delta_fG°$ (kJ mol⁻¹)	Log K_f
0							
100							
200							
298.15	187.443	194.636	194.636	0.	-3256.658	-3062.973	536.620
300	188.141	195.798	194.640	0.347	-3256.690	-3061.771	533.102
400	221.752	254.719	202.445	20.909	-3263.220	-2996.176	391.260
500	247.274	307.021	218.230	44.396	-3263.566	-2929.350	306.027
600	268.613	353.997	237.008	70.194	-3262.784	-2862.562	249.208
700	289.533	396.980	256.831	98.104	-3260.789	-2795.998	208.640
743.010	298.402	414.507	265.454	110.748	GLASS <--> LIQUID TRANSITION		
743.010	444.885	414.507	265.454	110.748			
800	444.885	447.385	277.257	136.102	-3249.420	-2730.116	178.258
900	444.885	499.785	299.129	180.591	-3232.152	-2666.246	154.745
1000	444.885	546.658	321.579	225.079	-3215.404	-2604.270	136.033
1015.600	444.885	553.544	325.089	232.019	CRYSTAL <--> LIQUID		
1100	444.885	589.060	343.999	269.567	-3199.134	-2543.947	120.802
1200	444.885	627.770	366.057	314.056	-3376.842	-2480.207	107.961
1300	444.885	663.380	387.577	358.544	-3359.618	-2406.188	96.682
1400	444.885	696.350	408.469	403.033	-3342.693	-2333.483	87.063
1500	444.885	727.043	428.696	447.521	-3326.043	-2261.980	78.769
1600	444.885	755.756	448.249	492.010	-3309.653	-2191.579	71.548
1700	444.885	782.727	467.139	536.498	-3293.508	-2122.194	65.207
1800	444.885	808.156	485.385	580.987	-3277.602	-2053.755	59.598
1900	444.885	832.209	503.012	625.475	-3261.926	-1986.192	54.604
2000	444.885	855.029	520.047	669.964	-3246.476	-1919.448	50.131
2100	444.885	876.735	536.519	714.452	-3231.250	-1853.472	46.103
2200	444.885	897.431	552.458	758.941	-3216.243	-1788.214	42.458
2300	444.885	917.207	567.890	803.429	-3201.454	-1723.637	39.145
2400	444.885	936.141	582.842	847.918	-3387.918	-1655.423	36.029
2500	444.885	954.302	597.340	892.406	-3373.872	-1583.525	33.086
2600	444.885	971.751	611.407	936.895	-3359.910	-1512.187	30.380
2700	444.885	988.541	625.066	981.383	-3346.033	-1441.383	27.885
2800	444.885	1004.720	638.337	1025.872	-3332.239	-1371.092	25.578
2900	444.885	1020.332	651.242	1070.360	-3318.532	-1301.296	23.439
3000	444.885	1035.414	663.798	1114.848	-3304.911	-1231.967	21.450

PREVIOUS: March 1963 CURRENT: March 1965

Sodium Borate ($Na_2B_4O_7$) $B_4Na_2O_7(l)$

SODIUM BORATE ($Na_2B_4O_7$) CRYSTAL-LIQUID M_r = 201.21534

0 to 1015.6 K crystal
above 1015.6 K liquid

Refer to the individual tables for details.

Sodium Borate ($Na_2B_4O_7$) $B_4Na_2O_7$(cr,l)

Enthalpy Reference Temperature = T_r = 298.15 K Standard State Pressure = p° = 0.1 MPa

T/K	C_p° (J K⁻¹mol⁻¹)	S° (J K⁻¹mol⁻¹)	−[G°−H°(T_r)]/T (J K⁻¹mol⁻¹)	H°−H°(T_r) (kJ mol⁻¹)	$\Delta_f H^\circ$ (kJ mol⁻¹)	$\Delta_f G^\circ$ (kJ mol⁻¹)	Log K_f
0	0.	0.	INFINITE	-30.382	-3258.983	-3258.983	INFINITE
100	76.525	50.814	321.976	-27.116	-3268.658	-3208.196	1675.790
200	139.746	124.566	205.130	-16.113	-3274.087	-3145.347	821.480
298.15	186.774	189.490	189.490	0.	-3276.741	-3081.522	539.870
300	187.447	190.648	189.494	0.346	-3276.774	-3080.310	536.330
400	213.802	248.460	197.196	20.505	-3283.707	-3014.160	393.609
500	228.534	297.784	212.510	42.637	-3285.408	-2946.574	307.826
600	241.500	340.624	230.368	66.154	-3286.907	-2878.661	250.610
700	252.697	378.715	248.889	90.878	-3288.098	-2810.522	209.724
800	262.128	413.077	267.299	116.622	-3288.983	-2742.233	179.049
900	271.215	444.482	285.266	143.294	-3289.531	-2673.853	155.186
1000	279.700	473.502	302.657	170.845	-3289.721	-2605.430	136.094
1015.600	280.970	477.841	305.314	175.218	CRYSTAL <--> LIQUID	TRANSITION	
1015.600	444.885	553.544	305.314	252.102			
1100	444.885	589.060	325.741	289.651	-3199.134	-2543.946	120.802
1200	444.885	627.770	349.320	334.139	-3376.842	-2480.206	107.961
1300	444.885	663.379	372.127	378.628	-3359.618	-2406.187	96.682
1400	444.893	696.349	394.123	423.117	-3342.693	-2333.483	87.063
1500	444.885	727.043	415.306	467.605	-3326.043	-2261.980	78.769
1600	444.885	755.756	435.697	512.094	-3309.652	-2191.578	71.548
1700	444.885	782.727	455.325	556.582	-3293.507	-2122.194	65.207
1800	444.885	808.156	474.227	601.071	-3277.601	-2053.754	59.598
1900	444.885	832.209	492.441	645.559	-3261.925	-1986.191	54.604
2000	444.885	855.029	510.005	690.048	-3246.476	-1919.447	50.131
2100	444.885	876.735	526.956	734.536	-3231.249	-1853.471	46.103
2200	444.885	897.431	543.329	779.025	-3216.242	-1788.213	42.458
2300	444.885	917.207	559.158	823.513	-3201.453	-1723.636	39.145
2400	444.885	936.141	574.473	868.002	-3387.917	-1655.422	36.029
2500	444.885	954.302	589.306	912.490	-3373.871	-1583.524	33.086
2600	444.885	971.751	603.682	956.979	-3359.909	-1512.186	30.380
2700	444.885	988.541	617.627	1001.467	-3346.032	-1441.382	27.885
2800	444.885	1004.720	631.165	1045.956	-3332.239	-1371.091	25.578
2900	444.885	1020.332	644.317	1090.444	-3318.531	-1301.295	23.439
3000	444.885	1035.414	657.103	1134.933	-3304.910	-1231.967	21.450

PREVIOUS: CURRENT: March 1965

Sodium Borate ($Na_2B_4O_7$) $B_4Na_2O_7$(cr,l)

LEAD BORATE (PbB_4O_7) CRYSTAL M_r = 362.4358

$S°(298.15\ K) = [166.9 \pm 13]\ J\ K^{-1}\ mol^{-1}$

$\Delta_f H°(0\ K)$ = Unknown

$\Delta_f H°(298.15\ K) = [-2858 \pm 6]\ kJ\ mol^{-1}$

Enthalpy of Formation

See the PbB_2O_4(cr) table for the complete writeup. Evans (1) analysed the enthalpies of formation of the lead borate system based upon Shartsis and Newman's enthalpies of solution and Mazzetti and DeCarli's phase studies. From the Shartsis and Newman paper Evans obtained the following:

$$PbO(cr) + 2B_2O_3(cr) \rightarrow PbO \cdot 2B_2O_3(gl) \qquad \Delta_r H = -11.9\ kcal\ mol^{-1}$$

and estimated a $\Delta_f H°(298.15\ K)$ value of 11.4 kcal mol^{-1} for the conversion to the $PbO \cdot 2B_2O_3$(cr). With auxiliary JANAF values $\Delta_f H°(298.15\ K)$ is calculated to be -683.00 kcal mol^{-1}.

Heat Capacity and Entropy

The heat capacity employed in this table is obtained by graphically smoothing the summation of the heat capacities of the constituent oxides.

The entropy at 298.15 K is obtained from a consideration of the entropies of sodium and calcium borates and those of the constituent oxides, as given by Kelley and King (2). These indicate for the reaction:

$$MO \cdot B_2O_3 + B_2O_3 \rightarrow MO \cdot 2B_2O_3$$

that the simple addition of entropies will give values at least 2.7 cal $K^{-1}\ mol^{-1}$ in case of Na borates and 5.7 cal $K^{-1}\ mol^{-1}$ for Ca borates too high for the reaction. The reason for this is probably structural, the $B_4O_7^{--}$ ion is a rigid chain of rings. It was assumed that additional B_2O_3 would extend the ring structure into a rigid plane and thus the average entropy difference of 4.2 for the reaction was assumed to hold for each additional mole of B_2O_3. Therefore the entropy at 298.15 K is obtained from $S°(298.15\ K)$ for PbB_2O_4 and B_2O_3 and the subtraction of 4.2 cal $K^{-1}\ mol^{-1}$.

References

1. W. H. Evans, U. S. Nat. Bur. Stand. Report 1792, July 1, 1961.
2. K. K. Kelly and E. G. King, U. S. Bur. Mines, Bull. 592, 1961.

Lead Borate (PbB_4O_7) $B_4O_7Pb_1$(cr)

Enthalpy Reference Temperature = T_r = 298.15 K Standard State Pressure = p° = 0.1 MPa

T/K	$C_p°$ (J K⁻¹mol⁻¹)	S° (J K⁻¹mol⁻¹)	$-[G°-H°(T_r)]/T$ (J K⁻¹mol⁻¹)	$H°-H°(T_r)$ (kJ mol⁻¹)	$\Delta_f H°$ (kJ mol⁻¹)	$\Delta_f G°$ (kJ mol⁻¹)	Log K_f
0							
100							
200							
298.15	168.197	166.942	166.942	0.	-2857.672	-2667.096	467.264
300	169.034	167.985	166.945	0.312	-2857.684	-2665.914	464.177
400	206.522	221.877	174.039	19.135	-2857.439	-2601.980	339.784
500	238.572	271.498	188.634	41.432	-2855.586	-2538.298	265.174
600	265.558	317.462	206.323	66.683	-2852.207	-2475.133	215.480
700	287.148	360.073	225.282	94.354	-2852.370	-2411.845	179.974
800	304.512	399.590	244.632	123.967	-2846.584	-2349.295	153.393
900	318.277	436.271	263.912	155.123	-2839.860	-2287.529	132.765
1000	330.034	470.433	282.876	187.558	-2832.345	-2226.555	116.303
1100	339.490	502.348	301.393	221.050	-2824.179	-2166.366	102.872
1200	347.021	532.225	319.398	255.393	-2815.507	-2106.944	91.713
1300	352.502	560.226	336.858	290.378	-2806.499	-2048.261	82.300
1400	356.853	586.512	353.761	325.851	-2797.295	-1990.279	74.258
1500	360.577	611.264	370.111	361.730	-2787.962	-1932.961	67.312
1600	363.422	634.629	385.920	397.935	-2778.570	-1876.267	61.254
1700	365.682	656.732	401.206	434.395	-2769.182	-1820.160	55.927
1800	367.397	677.685	415.989	471.052	-2759.850	-1764.606	51.208
1900	368.694	697.585	430.290	507.859	-2750.617	-1709.567	46.999
2000	369.740	716.524	444.132	544.782	-2741.512	-1655.011	43.224

PREVIOUS: December 1962 CURRENT: March 1965

Lead Borate (PbB_4O_7) $B_4O_7Pb_1$(cr)

PENTABORANE (B_5H_9) LIQUID M_r = 63.12146

$S°(298.15 K) = 184.33 J K^{-1} mol^{-1}$ $\Delta_f H°(298.15 K) = 42.84 \pm 6.7 kJ mol^{-1}$

$T_{fus} = 226.34 K$ $\Delta_{fus}H° = 6.134 kJ mol^{-1}$

Enthalpy of Formation

The vapor pressures of $B_5H_9(l)$ were measured by Wirth and Palmer (1) and Johnston, Kerr, Clarke and Hallett (2). By use of both the 2nd and 3rd law methods, the respective enthalpies of vaporization are evaluated. The $\Delta_f H°(298.15 K)$ values for $B_5H_9(l)$ are calculated from $\Delta_f H°(298.15 K)$ for $B_5H_9(g)$ and $\Delta_{vap}H°(298.15 K)$ for $B_5H_9(l)$ obtained previously.

Source	$\Delta_r H°(298.15 K)$, kcal mol^{-1} 2nd law	3rd law	$\Delta_f H°(298.15 K)$ kcal mol^{-1}
Wirth and Palmer (1)	7.26 ± 0.02	7.26	10.24 ± 1.6
Johnston, et al. (2)	7.23 ± 0.04	7.25	10.23 ± 1.6

The adopted value is the average of the two $\Delta_f H°(298.15 K)$ values obtained.

Heat Capacity and Entropy

The heat capacities, 231.34 - 290.19 K, for $B_5H_9(l)$ were measured by Johnston, Kerr, Clarke and Hallett (2). The C_p values above 290.19 K are estimated by graphical extrapolation of the C_p curve plotted using the experimental data. The value of S°(298.15 K) was derived based on the low temperature C_p, 13.29 - 221.61 K, and $\Delta_{fus}H°$ for $B_5H_9(cr)$ reported by Johnston, Kerr, Clarke and Hallett (2) and C_p values, 226.34 - 298.15 K, for $B_5H_9(l)$, using $S°(13.29 K) = 0.236 cal K^{-1} mol^{-1}$ for $B_5H_9(cr)$.

Fusion Data

The values of T_{fus} and $\Delta_{fus}H°$ were taken from Johnston, Kerr, Clarke and Hallett (2).

Vaporization Data

The boiling point is calculated as the temperature at which the Gibbs energy change of the reaction $B_5H_9(l) \rightarrow B_5H_9(g)$ is zero. The enthalpy change of the reaction at T_{vap} is enthalpy of vaporization.

References

1. H. E. Wirth and E. D. Palmer, J. Phys. Chem. 60, 914 (1956).
2. H. L. Johnston, E. c. Kerr, J. T. Clarke, and N. Cl Hallet, The Ohio State University, TR-6, July 8, 1949.

Pentaborane (B_5H_9) $B_5H_9(l)$

Enthalpy Reference Temperature = T_r = 298.15 K Standard State Pressure = p° = 0.1 MPa

T/K	$C_p°$ (J $K^{-1}mol^{-1}$)	S°	$-[G°-H°(T_r)]/T$	$H°-H°(T_r)$ (kJ mol^{-1})	$\Delta_f H°$	$\Delta_f G°$	Log K_f
0							
100							
200							
298.15	151.126	184.330	184.330	0.	42.844	171.913	-30.118
300	152.298	185.269	184.333	0.281	42.780	172.714	-30.072
400	185.351	233.939	190.744	17.278	39.888	216.489	-28.271
500	190.372	275.566	203.682	35.942	36.745	260.981	-27.264
600	210.037	311.791	218.723	55.841	33.557	306.142	-26.652
700	244.346	347.030	234.537	78.745	32.465	351.697	-26.244
800	256.479	380.470	250.719	103.801	32.820	397.284	-25.940
900	267.096	411.315	266.871	130.000	33.724	442.793	-25.699
1000	275.307	439.903	282.763	157.140	35.036	488.177	-25.500
1100	280.272	466.381	298.267	184.925	36.497	533.421	-25.330
1200	284.512	490.955	313.313	213.170	37.945	578.532	-25.183
1300	288.027	513.872	327.869	241.803	39.327	623.524	-25.054
1400	290.816	535.323	341.929	270.751	40.588	668.414	-24.939
1500	292.880	555.462	355.500	299.942	41.671	713.219	-24.837

PREVIOUS: December 1962 CURRENT: March 1965

PENTABORANE (B_5H_9) IDEAL GAS M_r = 63.12146

S°(298.15 K) = 275.41 J K^{-1} mol^{-1}

$\Delta_fH°$(0 K) = 102.23 ± 6.7 kJ mol^{-1}
$\Delta_fH°$(298.15 K) = 73.22 ± 6.7 kJ mol^{-1}

Vibrational Frequencies and Degeneracies

ν, cm^{-1}	ν, cm^{-1}	ν, cm^{-1}	ν, cm^{-1}
2600 (1)	[1450](1)	[1500](1)	1449 (2)
2600 (1)	[500](1)	[1100](1)	1034 (2)
1844 (1)	1870 (1)	738 (1)	882 (2)
1413 (1)	1387 (1)	470 (1)	700 (2)
1126 (1)	[900](1)	2598 (2)	605 (2)
985 (1)	782 (1)	1802 (2)	568 (2)
799 (1)	2600 (1)	1621 (2)	

Ground State Quantum Weight: 1 σ = 4
Point Group: C_{4v}
Bond Distances: B–B = 1.800 ± 0.003 Å B-B(apex) = 1.687 ± 0.005 Å
B–H = B(apex)–H = 1.22 Å B-H(bridge) = 1.35 ± 0.02 Å
Bond Angles: B(apex)–B–H = 136° 10' ± 30' H–B(apex)–B = 131°
B–H(bridge)–B = 83° 37' B–B–B = 90°
Angle between B(apex)BB plane and BBH(bridge) plane = 196 ± 2°

Product of the Moments of Inertia: $I_AI_BI_C$ = 2.366493 x 10^{-114} g^3 cm^6

Enthalpy of Formation

The enthalpy change (2) of the reaction B_5H_9(g) + 5 B(am) + 9/2 H_2(g) was reported to be -12.99 ± 0.39 and -9.9 ± 1 kcal mol^{-1} by Prosen, Johnson and Pergiel (1) and Gunn and Green (2), respectively. Using $\Delta_rH°$(298.15 K) = 1.2 ± 0.4 kcal mol^{-1} for the reaction B(cr) → B(am), the corresponding $\Delta_fH°$(298.15 K) values for B_5H_9(g) were derived as 18.99 and 15.9 kcal mol^{-1}. The adopted value is the average of these two.

Heat Capacity and Entropy

The molecular structure, a tetragonal pyramid of boron atoms, and constants were obtained from Hrostowski and Myers (3). The vibrational frequencies were reported by Hrostowski and Pimentel (4). The earlier models of pentaborane were assumed to have a hydrocarbon-like structure of low symmetry by Bauer and Pauling (5), and Pitzer (6). However, the electron diffraction and X-ray data by Hedberg, Jones and Schomaker (7), and Dulmage and Lipscomb (8) revealed an unusual pyramidal structure of C_{4v} symmetry. The Raman spectrum of liquid B_5H_9 and the infrared spectrum of gaseous B_5H_9 from 3 to 25 microns have been obtained by Taylor, Beckett, Tung, Holden and Johnston (9). The near infrared spectrum of B_5H_9(l) have been examined at high resolution by Pondy and Beachell (10). Ten bands were observed. The principal moments of inertia are: I_A = 16.9101 x 10^{-39} and I_B = I_C = 11.8299 x 10^{-39} g cm^2.

References

1. E. J. Prosen, W. H. Johnson, and F. Y. Pergiel, J. Res. Nat. Bur. Stand. 61, 247 (1958).
2. S. R. Gunn and L. G. Green, J. Phys. Chem. 65, 2173 (1961).
3. H. J. Hrostowski and R. J. Myers, J. Chem. Phys. 22, 262 (1954).
4. H. J. Hrostowski and G. C. Pimental, J. Am. Chem. Soc. 76, 998 (1954).
5. S. H. Bauer and L. Pauling, J. Am. Chem. Soc. 58, 2403 (1936).
6. K. S. Pitzer, J. Am. Chem. Soc. 67, 1126 (1945).
7. K. Hedberg, M. E. Jones, and V. Schomaker, Proc. Natl. Acad. Sci., U.S. 38, 679 (1952).
8. W. J. Dulmage and W. N. Liscomb, Acta. Cryst. 5, 260 (1952).
9. W. C. Taylor, C. W. Beckett, J. Y. Tung, R. B. Holden, and H. L. Johnston, Phys. Rev. 79, 234 (1950).
10. P. R. Pondy and H. C. Beachill, J. Chem. Phys. 25, 238 (1956).

Pentaborane (B_5H_9) B_5H_9(g)

Enthalpy Reference Temperature = T_r = 298.15 K Standard State Pressure = p° = 0.1 MPa

T/K	$C_p°$ (J $K^{-1}mol^{-1}$)	S° (J $K^{-1}mol^{-1}$)	-[G°-H°(T_r)]/T (J $K^{-1}mol^{-1}$)	H°-H°(T_r) (kJ mol^{-1})	$\Delta_fH°$ (kJ mol^{-1})	$\Delta_fG°$ (kJ mol^{-1})	Log K_f
0	0.	0.	INFINITE	-15.163	102.230	102.230	INFINITE
100	34.678	216.114	334.297	-11.818	91.958	115.828	-60.502
200	57.782	245.727	282.800	-7.415	82.575	143.319	-37.431
250	75.771	260.522	276.853	-4.083	77.768	159.061	-33.234
298.15	93.818	275.410	275.410	0.	73.220	175.133	-30.683
300	94.509	275.993	275.412	0.174	73.049	175.766	-30.604
350	112.830	291.947	276.630	5.361	68.579	193.243	-28.840
400	130.205	308.160	279.557	11.441	64.427	211.339	-27.598
450	146.409	324.443	283.639	18.362	60.637	229.935	-26.690
500	161.367	340.654	288.531	26.061	57.240	248.932	-26.006
600	187.614	372.467	299.884	43.550	51.642	287.822	-25.057
700	209.395	403.077	312.455	63.435	47.531	327.530	-24.441
800	227.365	432.249	325.622	85.302	44.697	367.737	-24.011
900	242.175	459.912	339.020	108.803	42.903	408.234	-23.693
1000	254.403	486.081	352.430	133.651	41.923	448.886	-23.447
1100	264.538	510.819	365.715	159.614	41.561	489.604	-23.249
1200	272.980	534.209	378.791	186.502	41.653	530.334	-23.085
1300	280.052	556.347	391.605	214.164	42.064	571.043	-22.945
1400	286.013	577.325	404.128	242.476	42.688	611.711	-22.823
1500	291.069	597.235	416.344	271.337	43.441	652.330	-22.716
1600	295.384	616.162	428.246	300.665	44.258	692.895	-22.621
1700	299.088	634.183	439.834	330.393	45.084	733.410	-22.535
1800	302.287	651.371	451.113	360.466	45.882	773.877	-22.457
1900	305.064	667.791	462.088	390.836	46.617	814.300	-22.387
2000	307.488	683.502	472.769	421.467	47.264	854.688	-22.322
2100	309.615	698.557	483.165	452.324	47.805	895.045	-22.263
2200	311.489	713.005	493.286	483.381	48.222	935.379	-22.209
2300	313.149	726.888	503.143	514.615	48.503	975.698	-22.159
2400	314.625	740.247	512.745	546.005	-202.660	1021.352	-22.229
2500	315.942	753.118	522.104	577.535	-203.071	1072.361	-22.406
2600	317.122	765.533	531.230	609.189	-203.482	1123.386	-22.569
2700	318.183	777.522	540.131	640.955	-203.899	1174.427	-22.721
2800	319.141	789.111	548.817	672.822	-204.330	1225.484	-22.862
2900	320.007	800.325	557.298	704.780	-204.779	1276.557	-22.993
3000	320.793	811.187	565.581	736.821	-205.251	1327.645	-23.116
3100	321.509	821.718	573.674	768.936	-205.750	1378.749	-23.232
3200	322.163	831.936	581.586	801.121	-206.279	1429.871	-23.340
3300	322.761	841.859	589.323	833.367	-206.843	1481.010	-23.442
3400	323.309	851.502	596.893	865.671	-207.444	1532.165	-23.539
3500	323.814	860.882	604.302	898.028	-208.084	1583.339	-23.630
3600	324.278	870.010	611.557	930.432	-208.766	1634.532	-23.716
3700	324.707	878.901	618.663	962.882	-209.493	1685.745	-23.798
3800	325.104	887.566	625.626	995.373	-210.265	1736.978	-23.876
3900	325.472	896.016	632.451	1027.902	-211.086	1788.232	-23.951
4000	325.814	904.260	639.144	1060.466	-211.957	1839.507	-24.021
4100	326.131	912.309	645.708	1093.064	-212.878	1890.806	-24.089
4200	326.427	920.172	652.150	1125.692	-2613.099	1977.248	-24.591
4300	326.704	927.856	658.473	1158.349	-2608.690	2086.489	-25.346
4400	326.962	935.370	664.681	1191.032	-2604.344	2195.629	-26.065
4500	327.203	942.720	670.778	1223.740	-2600.062	2304.670	-26.752
4600	327.430	949.914	676.768	1256.472	-2595.847	2413.618	-27.407
4700	327.642	956.959	682.655	1289.226	-2591.698	2522.474	-28.034
4800	327.842	963.859	688.442	1322.000	-2587.617	2631.245	-28.634
4900	328.030	970.620	694.132	1354.794	-2583.605	2739.928	-29.208
5000	328.207	977.249	699.728	1387.606	-2579.661	2848.533	-29.758
5100	328.374	983.750	705.234	1420.435	-2575.787	2957.056	-30.286
5200	328.531	990.128	710.651	1453.280	-2571.983	3065.507	-30.793
5300	328.680	996.388	715.984	1486.141	-2568.247	3173.885	-31.281
5400	328.821	1002.533	721.233	1519.016	-2564.579	3282.193	-31.749
5500	328.955	1008.568	726.403	1551.905	-2560.978	3390.432	-32.200
5600	329.081	1014.496	731.495	1584.807	-2557.444	3498.608	-32.634
5700	329.201	1020.322	736.511	1617.721	-2553.975	3606.721	-33.052
5800	329.315	1026.048	741.454	1650.647	-2550.569	3714.775	-33.455
5900	329.423	1031.678	746.325	1683.584	-2547.226	3822.767	-33.844
6000	329.526	1037.216	751.127	1716.531	-2543.942	3930.705	-34.220

PREVIOUS: March 1965 (1 atm) CURRENT: March 1965 (1 bar)

Pentaborane (B_5H_9) B_5H_9(g)

POTASSIUM BORATE ($K_2B_6O_{10}$) CRYSTAL M_r = 303.0506

$S°(298.15\ K)$ = [251 ± 13] $J\ K^{-1}\ mol^{-1}$

T_{dcm} = 1098 K

$\Delta_f H°(0\ K)$ = Unknown

$\Delta_f H°(298.15\ K)$ = -4633.53 ± 10.4 $kJ\ mol^{-1}$

Enthalpy of Formation

The enthalpy of reaction $K_2O(cr) + 3B_2O_3(l) \rightarrow K_2B_6O_{10}(cr)$ was reported by Shartsis and Capps (1) from their enthalpy of solution measurements. Their values for the enthalpy of reaction of the oxides have been converted to the enthalpies of formation using JANAF values (2) for $B_2O_3(l)$ and for $K_2O(cr)$. The above $\Delta_r H°$ = -122.8 $kcal\ mol^{-1}$.

Heat Capacity and Entropy

The enthalpy measurements of Smith (3) in the range 298 to 973 K were graphically smoothed and used to determine C_p. Above this range C_p is extrapolated.

The entropy at 298.15 K is obtained from a consideration of the entropies of sodium and calcium borates and those of the constituent oxides, as given by Kelley and King (4). These indicate that for the reactions

$$MO + B_2O_3 \rightarrow MO.B_2O_3 \quad (1)$$

$$MO.B_2O_3 + B_2O_3 \rightarrow MO.2B_2O_3 \quad (2)$$

the simple addition of entropies will give values at least 2.7 $cal\ K^{-1}mol^{-1}$ too low for the reaction (1) and at least 2.7 $cal\ K^{-1}\ mol^{-1}$ too high for reaction (2). The reason for this is probably structural, the $B_2O_4^{--}$ ion being a flexible chain while the $B_4O_7^{--}$ ion probably is a rigid chain of rings. It was assumed that additional B_2O_3 would extend the ring structure into a rigid plane and thus the entropy difference for reaction (2) was assumed to hold for each additional mole of B_2O_3 added. Thus a value for $Na_2B_6O_{10}$ was estimated to be 55.5 ± 1.5 and then using an average value of 4.5 ± 1.5 $cal\ K^{-1}mol^{-1}$ for the difference between K_2 and Na_2 gave 60 ± 3 $cal\ K^{-1}\ mol^{-1}$ for the entropy fo $K_2B_6O_{10}$.

Decomposition Data

The incongruent melting point was given by Rollet (5); the decomposition products being $K_2B_4O_7(l)$ and $K_2B_8O_{13}(cr)$.

References

1. L. Shartsis and W. Capps, J. Am. Ceram. Soc. 37, 27 (1954).
2. JANAF Thermochemical Tables: $B_2O_3(l)$, 12-31-64; $K_2O(cr)$, 6-30-63.
3. G. S. Smith, Ph.D. Thesis, Penn. State Univ., Univ. Microfilms, Order No. Mic 59-2916 (1959).
4. K. K. Kelley and E. G. King, U.S. Bur. Mines, Bull. 592, 1961.
5. A.-P. Rollet, Comp. rend. 200, 1763 (1935).

Potassium Borate ($K_2B_6O_{10}$) $B_6K_2O_{10}(cr)$

Enthalpy Reference Temperature = T_r = 298.15 K Standard State Pressure = p° = 0.1 MPa

	$J\ K^{-1}mol^{-1}$			$kJ\ mol^{-1}$			
T/K	$C_p°$	S°	$-[G°-H°(T_r)]/T$	$H°-H°(T_r)$	$\Delta_f H°$	$\Delta_f G°$	Log K_f
0							
100							
200							
298.15	261.918	251.040	251.040	0.	-4633.529	-4353.554	762.725
300	262.337	252.661	251.045	0.485	-4633.551	-4351.817	757.719
400	285.767	331.362	261.617	27.898	-4640.126	-4256.761	555.876
500	308.277	397.560	282.348	57.606	-4642.314	-4160.664	434.662
600	330.118	455.703	306.487	89.530	-4644.154	-4064.153	353.816
700	351.456	508.194	331.610	123.608	-4645.271	-3967.385	296.050
800	372.794	556.524	356.742	159.826	-4645.388	-3870.524	252.719
900	393.463	601.639	381.473	198.150	-4644.347	-3773.716	219.021
1000	412.835	644.176	405.636	238.539	-4642.067	-3677.091	192.072
1100	423.295	684.043	429.155	280.376	-4796.992	-3571.514	169.597
1200	430.073	721.173	451.961	323.055	-4791.748	-3460.337	150.625
1300	435.680	755.823	474.018	366.347	-4786.378	-3349.604	134.589
1400	440.659	788.294	495.318	410.167	-4780.923	-3239.287	120.859
1500	445.387	818.859	515.879	454.471	-4775.390	-3129.364	108.974
1600	449.864	847.747	535.726	499.234	-4769.782	-3019.811	98.587
1700	454.299	875.154	554.893	544.443	-4764.094	-2910.611	89.432
1800	458.622	901.244	573.416	590.090	-4758.324	-2801.750	81.305
1900	462.834	926.153	591.330	636.164	-4752.475	-2693.210	74.042
2000	466.934	949.998	608.672	682.653	-4746.552	-2584.980	67.513

PREVIOUS: March 1963 CURRENT: March 1965

Potassium Borate ($K_2B_6O_{10}$) $B_6K_2O_{10}(cr)$

LITHIUM BORATE ($Li_2B_6O_{10}$) CRYSTAL M_r = 238.7360

$S°(298.15\ K)$ = [188.3 ± 4] J $K^{-1}mol^{-1}$

T_{dcm} = 1107 ± 4 K

$\Delta_f H°(0\ K)$ = Unknown

$\Delta_f H°(298.15\ K)$ = -4659.89 ± 5.6 kJ mol^{-1}

Enthalpy of Formation

The enthalpy of reaction $Li_2O(cr) + 3B_2O_3(l) \rightarrow Li_2B_6O_{10}(cr)$ was reported by Shartsis and Capps (1) from their enthalpy of solution measurements. Their values for the enthalpy of reaction of the oxides have been converted to the enthalpies of formation using JANAF values (2) for $B_2O_3(l)$ and $Li_2O(cr)$. The above $\Delta_r H°$ = -72.80 kcal mol^{-1}.

Heat Capacity and Entropy

The enthalpy measurement of Smith (3) in the range 298 to 823 K were used to determine C_p. Above this range C_p is extrapolated. Entropy at 298.15 K is estimated from that of the component oxides and the difference from the oxides observed for twice $LiBO_2(cr)$.

Decomposition Data

The incongruent melting point reported by Sastry and Hummel (4) is used here.

References

1. L. Shartsis and W. Capps, J. Am. Ceram. Soc. 37, 27 (1954).
2. JANAF Thermochemical Tables: $B_2O_3(l)$ 12-31-64; $Li_2O(cr)$, 3-31-64.
3. G. S. Smith, Ph.D. Thesis, Penn. State Univ., Univ. Microfilms, Order No. 59-2916 (1959).
4. B.S.R. Sastry and F. A. Hummel, J. Am. Ceram. Soc. 42, 216 (1959).

Lithium Borate ($Li_2B_6O_{10}$) $B_6Li_2O_{10}(cr)$

Enthalpy Reference Temperature = T_r = 298.15 K Standard State Pressure = p° = 0.1 MPa

T/K	$C_p°$ (J $K^{-1}mol^{-1}$)	S° (J $K^{-1}mol^{-1}$)	$-[G°-H°(T_r)]/T$ (J $K^{-1}mol^{-1}$)	$H°-H°(T_r)$ (kJ mol^{-1})	$\Delta_f H°$ (kJ mol^{-1})	$\Delta_f G°$ (kJ mol^{-1})	Log K_f
0							
100							
200							
298.15	293.215	188.280	188.280	0.	-4659.888	-4382.421	767.782
300	293.633	190.095	188.286	0.543	-4659.835	-4380.700	762.748
400	316.269	277.724	200.080	31.057	-4657.552	-4288.034	559.960
500	336.561	350.530	223.085	63.723	-4662.433	-4195.190	438.268
600	354.008	413.481	249.687	98.277	-4661.531	-4101.824	357.096
700	368.401	469.162	277.139	134.416	-4660.445	-4008.622	299.127
800	380.493	519.165	304.320	171.876	-4659.158	-3915.591	255.662
900	390.744	564.586	330.753	210.450	-4657.685	-3822.731	221.866
1000	399.572	606.219	356.247	249.972	-4656.018	-3730.046	194.838
1100	406.456	644.627	380.740	290.275	-4654.202	-3637.535	172.732
1200	413.157	680.281	404.234	331.257	-4652.249	-3545.196	154.318
1300	419.675	713.610	426.764	372.900	-4650.116	-3453.027	138.744
1400	426.012	744.944	448.383	415.186	-4647.768	-3361.029	125.402
1500	432.165	774.547	469.149	458.096	-4645.185	-3269.209	113.844
1600	438.137	802.630	489.122	501.613	-4642.361	-3177.568	103.737
1700	443.926	829.367	508.356	545.718	-4929.767	-3071.759	94.384
1800	449.532	854.900	526.905	590.392	-4924.960	-2962.603	85.973
1900	454.956	879.351	544.815	635.618	-4919.944	-2853.719	78.454
2000	460.198	902.822	562.133	681.377	-4914.730	-2745.104	71.695

PREVIOUS: December 1962 CURRENT: March 1965

Lithium Borate ($Li_2B_6O_{10}$) $B_6Li_2O_{10}(cr)$

SODIUM BORATE ($Na_2B_6O_{10}$) CRYSTAL M_r = 270.83354

$\Delta_fH°$(0 K) = Unknown

$S°$(298.15 K) = [232.2 ± 6.3] J K^{-1} mol^{-1} $\Delta_fH°$(298.15 K) = -4580.48 ± 9.2 kJ mol^{-1}

T_{dcm} = 1039 K

Enthalpy of Formation

The enthalpy of reaction Na_2O(cr) + 3 B_2O_3(l) → $Na_2B_6O_{10}$(cr) was reported by Shartsis and Capps (1) from their enthalpy of solution measurements. Their values for the enthalpy of reaction of the oxides have been converted to the enthalpies of formation using JANAF values (2) for B_2O_3(l) and Na_2O(cr). The above $\Delta_rH°$ = -97.52 kcal mol^{-1}.

Heat Capacity and Entropy

The enthalpy measurements of Smith (3) in the range 298 to 1039 K were smoothed graphically and used to determine C_p. Above this range C_p was extrapolated. Entropy at 298.15 K was estimated as described in detail for $K_2B_6O_{10}$(cr).

Decomposition Data

The incongruent melting point was reported by Morey and Merwin (4); the products are $Na_2B_4O_7$(l) and $Na_2B_8O_{13}$(cr).

References

1. L. Shartsis and W. Capps, J. Am. Ceram. Soc. 37, 27 (1954).
2. JANAF Thermochemical Tables: B_2O_3(l), 12-31-64; Na_2O(cr), 6-30-62.
3. G. S. Smith, Ph.D. Thesis, Penn. State Univ., Univ. Microfilms, Order No. 59-2916 (1959).
4. G. W. Morey and H. E. Merwin, J. Am. Chem. Soc. 58, 2248 (1936).

Sodium Borate ($Na_2B_6O_{10}$) $B_6Na_2O_{10}$(cr)

Enthalpy Reference Temperature = T_r = 298.15 K   Standard State Pressure = p° = 0.1 MPa

T/K	$C_p°$ (J $K^{-1}mol^{-1}$)	$S°$ (J $K^{-1}mol^{-1}$)	$-[G°-H°(T_r)]/T$ (J $K^{-1}mol^{-1}$)	$H°-H°(T_r)$ (kJ mol^{-1})	$\Delta_fH°$ (kJ mol^{-1})	$\Delta_fG°$ (kJ mol^{-1})	Log K_f
0							
100							
200							
298.15	243.509	232.212	232.212	0.	-4580.476	-4302.768	753.827
300	244.346	233.721	232.217	0.451	-4580.527	-4301.045	748.878
400	283.173	309.380	242.270	26.844	-4588.408	-4207.071	549.387
500	319.992	376.581	262.525	57.028	-4590.107	-4111.507	429.526
600	353.757	437.950	286.718	90.739	-4590.114	-4015.757	349.603
700	384.677	494.845	312.434	127.687	-4588.273	-3920.150	292.525
800	412.375	548.063	338.597	167.572	-4584.587	-3824.935	249.743
900	436.182	598.047	364.677	210.033	-4579.219	-3730.284	216.500
1000	456.056	645.071	390.389	254.682	-4572.428	-3636.311	189.941
1100	471.453	689.291	415.572	301.090	-4564.561	-3543.071	168.246
1200	482.917	730.824	440.131	348.831	-4749.519	-3445.703	149.988
1300	491.746	769.840	464.008	397.582	-4738.691	-3337.488	134.102
1400	498.482	806.540	487.177	447.108	-4727.528	-3230.120	120.517
1500	503.461	841.109	509.632	497.217	-4716.188	-3123.559	108.772
1600	507.017	873.720	531.378	547.747	-4704.808	-3017.754	98.520
1700	509.737	904.543	552.431	598.590	-4693.479	-2912.661	89.495
1800	511.829	933.739	572.811	649.672	-4682.263	-2808.233	81.493
1900	513.544	961.459	592.542	700.941	-4671.198	-2704.422	74.350
2000	515.134	987.841	611.653	752.376	-4660.301	-2601.189	67.936

PREVIOUS: March 1963   CURRENT: March 1965

Sodium Borate ($Na_2B_6O_{10}$) $B_6Na_2O_{10}$(cr)

LEAD BORATE (PbB_6O_{10}) CRYSTAL M_r = 432.0540

S°(298.15 K) = [203.3 ± 13] J K^{-1} mol^{-1}

Δ_fH°(0 K) = Unknown

Δ_fH°(298.15 K) = -4197 ± 8 kJ mol^{-1}

Enthalpy of Formation

See the PbB_2O_4(cr) table for the complete writeup. Evans (1) analysed the enthalpies of formation of the lead borate system based upon Shartsis and Newman's enthalpies of solution and Mazzetti and De Carli's phase studies. From the Shartsis and Newman paper Evans obtained the following:

PbO(cr) + $3B_2O_3$(cr) → PbO.$3B_2O_3$(gls) Δ_rH = -24.5 kcal mol^{-1}

and estimated a Δ_rH°(298.15 K) of 16.0 kcal mol^{-1} for the conversion to the PbO.$3B_2O_3$(cr). With auxiliary JANAF values the Δ_fH°(298.15 K) is calculated to be -1003.00 kcal mol^{-1}.

Heat Capacity and Entropy

The heat capactiy employed in this table was obtained by graphically smoothing the summation of heat capacities of the constituent oxides.

The entropy at 298.15 K was obtained in the manner decribed in the PbB_4O_7 table from S°(298.15 K) for PbB_4O_7 and B_2O_3 and the subtraction of 4.2 cal K^{-1} mol^{-1}.

References

1. W. H. Evans, U.S. Nat. Bur. Stand., Report 7192, July 1, 1961.

Lead Borate (PbB_6O_{10}) $B_6O_{10}Pb_1$(cr)

Enthalpy Reference Temperature = T_r = 298.15 K Standard State Pressure = p° = 0.1 MPa

T/K	C_p° (J $K^{-1}mol^{-1}$)	S° (J $K^{-1}mol^{-1}$)	-[G°-H°(T_r)]/T (J $K^{-1}mol^{-1}$)	H°-H°(T_r) (kJ mol^{-1})	Δ_fH° (kJ mol^{-1})	Δ_fG° (kJ mol^{-1})	Log K_f
0							
100							
200							
298.15	230.957	203.342	203.342	0.	-4196.552	-3921.603	687.049
300	231.794	204.774	203.347	0.428	-4196.572	-3919.897	682.515
400	286.186	279.058	213.108	26.380	-4196.379	-3827.635	499.838
500	334.218	348.197	233.290	57.454	-4193.800	-3735.697	390.266
600	375.849	412.869	257.889	92.988	-4188.839	-3644.504	317.282
700	413.714	473.741	284.419	132.526	-4186.327	-3553.532	265.168
800	444.466	531.083	311.705	175.503	-4176.749	-3463.763	226.161
900	467.060	584.797	339.098	221.129	-4165.426	-3375.304	195.898
1000	483.629	634.906	366.203	268.703	-4152.890	-3288.176	171.757
1100	495.469	681.590	392.778	317.694	-4139.548	-3202.344	152.067
1200	503.000	725.044	418.678	367.640	-4125.774	-3117.750	135.712
1300	507.812	765.511	443.819	418.199	-4111.850	-3034.313	121.920
1400	510.406	803.246	468.160	469.120	-4097.997	-2951.946	110.138
1500	511.787	838.510	491.688	520.234	-4084.360	-2870.563	99.962
1600	512.707	871.570	514.408	571.459	-4071.001	-2790.081	91.087
1700	513.502	902.678	536.342	622.771	-4057.931	-2710.424	83.281
1800	514.130	932.047	557.517	674.154	-4045.156	-2631.531	76.365
1900	514.590	959.857	577.968	725.591	-4032.686	-2553.337	70.196
2000	514.967	986.262	597.728	777.069	-4020.525	-2475.792	64.661

PREVIOUS: December 1962 CURRENT: March 1965

Lead Borate (PbB_6O_{10}) $B_6O_{10}Pb_1$(cr)

POTASSIUM BORATE ($K_2B_8O_{13}$) CRYSTAL M_r = 372.6688

$S°(298.15\ K) = [293.7 \pm 16.7]\ J\ K^{-1}\ mol^{-1}$

$T_{fus} = 1130\ K$

$\Delta_f H°(0\ K)$ = Unknown

$\Delta_f H°(298.15) = -5945.13 \pm 5.9\ kJ\ mol^{-1}$

$\Delta_{fus} H° = 149.8 \pm 4\ kJ\ mol^{-1}$

Enthalpy of Formation

The enthalpy of reaction $K_2O(cr) + 4\ B_2O_3(l) \rightarrow K_2B_8O_{13}(cr)$ was reported by Shartsis and Capps (1) from their enthalpy of solution measurements. Their values for the enthalpy of reaction of the oxides have been converted to the enthalpies of formation using JANAF values (2) for $B_2O_3(l)$ and $K_2O(cr)$. The above $\Delta_r H° = -137.00\ kcal\ mol^{-1}$.

Heat Capacity and Entropy

The enthalpy measurements of Smith (3) in the range 298 to 1130 K were smoothed graphically and used to determine C_p. Above this range C_p is extrapolated.

$S°(298.15\ K)$ is obtained by estimating the entropy of $Na_2B_8O_{13}(cr)$ as $65.7 \pm 2.5\ cal\ K^{-1}\ mol^{-1}$ as described on the table for $K_2B_6O_{10}(cr)$ and then using an average value of $4.5 \pm 1.5\ cal\ K^{-1}\ mol^{-1}$ for the replacement of Na_2 by K_2.

Fusion Data

T_{fus} is given by Rollet (4). $\Delta_{fus}H°$ was obtained from the values of $\Delta_f H°$ for $K_2B_8O_{13}(cr)$ and $K_2B_8O_{13}(l)$ at 298.15 K reported above and the enthalpy of the crystal and liquid at T_{fus}.

References

1. L. Shartsis and W. Capps, J. Am. Ceram. Soc. 37, 27 (1954).
2. JANAF Thermochemical Tables: $B_2O_3(l)$, 12-31-64; $K_2O(cr)$, 6-30-63.
3. G. S. Smith, Ph.D. Thesis, Penn State Univ., Univ. Microfilms, Order No. 59-2916 (1959).
4. A.-P. Rollet, Comp. rend. 200, 1763 (1935).

Potassium Borate ($K_2B_8O_{13}$) $B_8K_2O_{13}(cr)$

Enthalpy Reference Temperature = T_r = 298.15 K Standard State Pressure = p° = 0.1 MPa

T/K	$C_p°$ (J K⁻¹mol⁻¹)	S° (J K⁻¹mol⁻¹)	$-[G°-H°(T_r)]/T$ (J K⁻¹mol⁻¹)	$H°-H°(T_r)$ (kJ mol⁻¹)	$\Delta_f H°$ (kJ mol⁻¹)	$\Delta_f G°$ (kJ mol⁻¹)	Log K_f
0							
100							
200							
298.15	321.289	293.717	293.717	0.	-5945.129	-5582.653	978.058
300	322.042	295.707	293.723	0.595	-5945.165	-5580.404	971.635
400	360.493	393.657	306.819	34.735	-5952.194	-5457.589	712.688
500	397.313	478.036	332.801	72.617	-5954.258	-5333.683	557.206
600	435.136	553.841	363.420	114.252	-5955.089	-5209.463	453.524
700	471.453	623.739	395.677	159.643	-5954.085	-5085.245	379.466
800	500.406	688.630	428.291	208.272	-5951.363	-4961.293	323.939
900	525.092	749.032	460.612	259.578	-5947.212	-4837.770	280.777
1000	546.012	805.494	492.308	313.186	-5941.830	-4714.777	246.275
1100	560.656	858.271	523.206	368.571	-6093.530	-4583.146	217.635
1130.000	563.825	873.400	532.303	385.440	------ CRYSTAL <--> LIQUID -----		
1200	574.045	907.597	553.205	425.270	-6084.767	-4446.220	193.539
1300	586.178	954.032	582.269	483.292	-6075.326	-4310.053	173.180
1400	597.057	997.879	610.404	542.464	-6065.317	-4174.635	155.758
1500	606.680	1039.407	637.632	602.662	-6054.821	-4039.951	140.684
1600	615.048	1078.835	663.986	663.758	-6043.933	-3905.979	127.517
1700	622.161	1116.342	689.501	725.629	-6032.754	-3772.698	115.921
1800	628.018	1152.075	714.215	788.149	-6021.394	-3640.083	105.632
1900	632.621	1186.159	738.164	851.191	-6009.968	-3508.098	96.444
2000	635.968	1218.699	761.383	914.631	-5998.593	-3376.715	88.191

PREVIOUS: March 1963 CURRENT: March 1965

Potassium Borate ($K_2B_8O_{13}$) $B_8K_2O_{13}(cr)$

POTASSIUM BORATE ($K_2B_8O_{13}$) LIQUID M_r = 372.6688

$S°(298.15\ K)$ = [293.717] J K^{-1} mol^{-1}

T_{fus} = 1130 K

$\Delta_fH°(0\ K)$ = [-5945.129] kJ mol^{-1}

$\Delta_{fus}H°$ = 149.8 ± 4 kJ mol^{-1}

Enthalpy of Formation

The enthalpy of reaction $K_2O(cr) + 4B_2O_3(l) \rightarrow K_2B_8O_{13}(l)$ was reported by Shartsis and Capps (1) from their enthalpies of solution measurments. Their values for the enthalpy of reaction of the oxides have been converted to the enthalpies of formation using JANAF values(2) for $B_2O_3(l)$ and for $K_2O(cr)$. The above $\Delta_rH°$ = -119.65 kcal mol^{-1}.

Heat Capacity and Entropy

The enthalpy measurements of Smith (3) in the range 298 to 1373 K on the glass and liquid were smoothed graphically and used to determine C_p. There is a discontinuity in the C_p curve, a glass transition at about 673 K. Above 673 K the C_p values were estimated to be constant.

The entropy is calculated from that of the crystal by adding $\Delta_{fus}S°$ and differences in entropy, S°(1130 K)-S°(298.15 K), between the crystal and liquid.

Fusion Data

See the crystal table for details.

References

1. L. Shartsis and W. Capps, J. Am. Ceram. Soc. 37, 27 (1954).
2. JANAF Thermochemical Tables: $B_2O_3(l)$, 12-31-64; $K_2O(cr)$, 6-30-63.
3. G. S. Smith, Ph.D. Thesis, Penn State Univ., Univ. Microfilms, Order No. 59-2916 (1959).

Potassium Borate ($K_2B_8O_{13}$) $B_8K_2O_{13}(l)$

Enthalpy Reference Temperature = T_r = 298.15 K Standard State Pressure = p° = 0.1 MPa

T/K	$C_p°$	S°	$-[G°-H°(T_r)]/T$	$H°-H°(T_r)$	$\Delta_fH°$	$\Delta_fG°$	Log K_f
	J $K^{-1}mol^{-1}$			kJ mol^{-1}			
0							
100							
200							
298.15	323.088	324.085	324.085	0.	-5872.537	-5519.114	966.926
300	324.511	326.088	324.091	0.599	-5872.569	-5516.922	960.581
400	393.798	429.241	337.686	36.622	-5877.715	-5397.344	704.821
500	450.198	523.754	365.589	79.082	-5875.200	-5277.484	551.335
600	475.302	608.529	399.165	125.618	-5871.130	-5158.318	449.071
673.000	473.866	663.180	424.897	160.364	GLASS <--> LIQUID TRANSITION		
673.000	661.490	663.180	424.897	160.364			
700	661.490	689.200	434.593	178.225	-5862.911	-5039.893	376.081
800	661.490	777.530	472.062	244.374	-5842.669	-4923.718	321.486
900	661.490	855.442	510.417	310.523	-5823.675	-4810.002	279.165
1000	661.490	925.137	548.465	376.672	-5805.752	-4698.341	245.416
1100	661.490	988.184	585.619	442.821	-5946.688	-4579.208	217.448
1130.000	661.490	1005.983	596.544	462.666	------ CRYSTAL <--> LIQUID -----		
1200	661.490	1045.741	621.599	508.970	-5928.475	-4455.701	193.952
1300	661.490	1098.688	656.289	575.119	-5910.907	-4333.687	174.130
1400	661.490	1147.710	689.661	641.268	-5893.921	-4213.003	157.189
1500	661.490	1193.348	721.737	707.417	-5877.474	-4093.515	142.549
1600	661.490	1236.040	752.561	773.566	-5861.533	-3975.107	129.774
1700	661.490	1276.142	782.192	839.715	-5846.076	-3857.681	118.532
1800	661.490	1313.952	810.694	905.864	-5831.086	-3741.153	108.565
1900	661.490	1349.717	838.131	972.013	-5816.554	-3625.444	99.670
2000	661.490	1383.647	864.566	1038.162	-5802.470	-3510.488	91.685
2100	661.490	1415.921	890.059	1104.311	-5788.830	-3396 226	84.477
2200	661.490	1446.694	914.666	1170.460	-5775.631	-3282.600	77.939
2300	661.490	1476.098	938.442	1236.609	-5762.871	-3169.571	71.983
2400	661.490	1504.251	961.435	1302.758	-6152.624	-3048.536	66.350
2500	661.490	1531.254	983.691	1368.907	-6141.365	-2919.430	60.998
2600	661.490	1557.199	1005.254	1435.056	-6130.282	-2790.772	56.067
2700	661.490	1582.163	1026.161	1501.205	-6119.380	-2662.535	51.510
2800	661.490	1606.220	1046.451	1567.355	-6108.661	-2534.699	47.285
2900	661.490	1629.433	1066.156	1633.504	-6098.127	-2407.249	43.359
3000	661.490	1651.858	1085.307	1699.653	-6087.782	-2280.152	39.701

PREVIOUS: March 1963 CURRENT: March 1965

Potassium Borate ($K_2B_8O_{13}$) $B_8K_2O_{13}(l)$

POTASSIUM BORATE ($K_2B_8O_{13}$) CRYSTAL-LIQUID M_r = 372.6688

298.15 to 1130 K crystal
above 1130 K liquid

Refer to the individual tables for details.

Potassium Borate ($K_2B_8O_{13}$) $B_8K_2O_{13}$(cr,l)

Enthalpy Reference Temperature = T_r = 298.15 K Standard State Pressure = $p°$ = 0.1 MPa

T/K	$C_p°$ (J K⁻¹mol⁻¹)	$S°$ (J K⁻¹mol⁻¹)	$-[G°-H°(T_r)]/T$ (J K⁻¹mol⁻¹)	$H°-H°(T_r)$ (kJ mol⁻¹)	$\Delta_fH°$ (kJ mol⁻¹)	$\Delta_fG°$ (kJ mol⁻¹)	Log K_f
0							
100							
200							
298.15	321.289	293.717	293.717	0.	-5945.129	-5582.653	978.058
300	322.042	295.707	293.723	0.595	-5945.165	-5580.404	971.635
400	360.493	393.657	306.819	34.735	-5952.194	-5457.589	712.688
500	397.313	478.036	332.801	72.617	-5954.258	-5333.683	557.206
600	435.136	553.841	363.420	114.252	-5955.089	-5209.463	453.524
700	471.453	623.739	395.677	159.643	-5954.085	-5085.245	379.466
800	500.406	688.630	428.291	208.272	-5951.363	-4961.293	323.939
900	525.092	749.032	460.612	259.578	-5947.212	-4837.770	280.777
1000	546.012	805.494	492.308	313.186	-5941.830	-4714.777	246.275
1100	560.656	858.271	523.206	368.571	-6093.530	-4583.146	217.635
1130.000	563.825	873.400	532.303	385.440	CRYSTAL <--> LIQUID	TRANSITION	
1130.000	661.490	1005.983	532.303	535.258			
1200	661.490	1045.741	561.106	581.562	-5928.475	-4455.701	193.952
1300	661.490	1098.688	600.449	647.711	-5910.907	-4333.687	174.130
1400	661.490	1147.710	637.810	713.860	-5893.921	-4213.003	157.189
1500	661.490	1193.348	673.342	780.009	-5877.474	-4093.515	142.549
1600	661.490	1236.040	707.191	846.158	-5861.533	-3975.107	129.774
1700	661.490	1276.142	739.491	912.307	-5846.076	-3857.681	118.532
1800	661.490	1313.952	770.365	978.457	-5831.086	-3741.153	108.565
1900	661.490	1349.717	799.925	1044.606	-5816.554	-3625.444	99.670
2000	661.490	1383.647	828.270	1110.755	-5802.470	-3510.488	91.685
2100	661.490	1415.921	855.491	1176.904	-5788.830	-3396.226	84.477
2200	661.490	1446.694	881.670	1243.053	-5775.631	-3282.600	77.939
2300	661.490	1476.098	906.880	1309.202	-5762.871	-3169.571	71.983
2400	661.490	1504.251	931.188	1375.351	-6152.624	-3048.536	66.350
2500	661.490	1531.254	954.654	1441.500	-6141.365	-2919.430	60.998
2600	661.490	1557.199	977.334	1507.649	-6130.282	-2790.772	56.067
2700	661.490	1582.163	999.275	1573.798	-6119.380	-2662.535	51.510
2800	661.490	1606.220	1020.525	1639.947	-6108.661	-2534.699	47.285
2900	661.490	1629.433	1041.124	1706.096	-6098.127	-2407.249	43.359
3000	661.490	1651.858	1061.110	1772.245	-6087.782	-2280.152	39.701

PREVIOUS: CURRENT: March 1965

Potassium Borate ($K_2B_8O_{13}$) $B_8K_2O_{13}$(cr,l)

LITHIUM BORATE ($Li_2B_8O_{13}$)　　CRYSTAL　　M_r = 308.3542

$\Delta_f H°(0\ K)$ = Unknown

$S°(298.15\ K) = [265.3 \pm 4]\ J\ K^{-1}\ mol^{-1}$　　$\Delta_f H°(298.15) = -5914.38 \pm 6.8\ kJ\ mol^{-1}$

$T_{dcm} = 908 \pm 10\ K$

Enthalpy of Formation

The enthalpy of the reaction $Li_2O(cr) + 4\ B_2O_3(l) \rightarrow Li_2B_8O_{13}(cr)$ was reported by Shartsis and Capps (1) from their enthalpy of solution measurements. Their values for the enthalpy of reaction of the oxides have been converted to the enthalpies of formation using JANAF values (2) for $B_2O_3(l)$ and for $Li_2C(cr)$. The above $\Delta_r H° = -73.35\ kcal\ mol^{-1}$.

Heat Capacity and Entropy

The enthalpy measurements of Smith (3) in the range 298 to 823 K were used to determine C_p. Above this range C_p is extrapolated. The entropy at 298.15 K is estimated from that of the component oxides and the difference from the oxides observed for twice $LiBO_2(cr)$.

Decomposition Data

The incongruent melting point reported by Sastry and Hummel (4) is used here.

References

1. L. Shartsis and W. Capps, J. Am. Ceram. Soc. 37, 27 (1954).
2. JANAF Thermochemical Tables: $B_2O_3(l)$, 12-31-64; $Li_2O(cr)$, 3-31-64.
3. G. S. Smith, Ph.D. Thesis, Penn State Univ., Univ. Microfilms, Order No. 59-2916 (1959).
4. B. S. R. Sastry and T. A. Hummel, J. Am. Ceram. Soc. 42, 216 (1959).

Lithium Borate ($Li_2B_8O_{13}$)　　$B_8Li_2O_{13}(cr)$

Enthalpy Reference Temperature = T_r = 298.15 K　　Standard State Pressure = p° = 0.1 MPa

T/K	$C_p°$ (J K⁻¹mol⁻¹)	S° (J K⁻¹mol⁻¹)	−[G°−H°(T_r)]/T (J K⁻¹mol⁻¹)	H°−H°(T_r) (kJ mol⁻¹)	$\Delta_f H°$ (kJ mol⁻¹)	$\Delta_f G°$ (kJ mol⁻¹)	Log K_f
0							
100							
200							
298.15	325.515	265.266	265.266	0.	-5914.377	-5564.637	974.901
300	326.770	267.283	265.272	0.603	-5914.386	-5562.467	968.512
400	390.367	370.606	278.926	36.672	-5913.731	-5445.208	711.071
500	428.567	462.069	306.606	77.732	-5918.268	-5327.630	556.574
600	453.211	542.480	339.359	121.873	-5916.480	-5209.666	453.542
700	471.537	613.760	373.565	168.137	-5914.462	-5092.021	379.971
800	486.683	677.748	407.656	216.074	-5912.270	-4974.678	324.813
900	498.733	735.788	440.941	265.362	-5909.954	-4857.617	281.929
1000	508.649	788.860	473.117	315.743	-5907.545	-4740.819	247.635
1100	517.142	837.742	504.072	367.037	-5905.062	-4624.264	219.588
1200	524.623	883.062	533.789	419.127	-5902.502	-4507.941	196.226
1300	531.915	925.343	562.300	471.956	-5899.841	-4391.835	176.466
1400	539.016	965.023	589.664	525.504	-5897.031	-4275.938	159.537
1500	545.928	1002.448	615.947	579.752	-5894.040	-4160.251	144.873
1600	552.651	1037.897	641.221	634.683	-5890.855	-4044.767	132.048
1700	559.183	1071.599	665.554	690.276	-6177.943	-3915.137	120.298
1800	565.526	1103.741	689.012	746.513	-6172.856	-3782.179	109.756
1900	571.679	1134.483	711.654	803.375	-6167.596	-3649.505	100.332
2000	577.643	1163.959	733.538	860.843	-6162.172	-3517.113	91.858

PREVIOUS: December 1962　　CURRENT: March 1965

Lithium Borate ($Li_2B_8O_{13}$)　　$B_8Li_2O_{13}(cr)$

DECABORANE ($B_{10}H_{14}$) CRYSTAL M_r = 122.21116

$S°(298.15\ K) = 176.56\ J\ K^{-1}\ mol^{-1}$

$T_{fus} = 371.93\ K$

$\Delta_f H°(0\ K) = 14.3 \pm 19\ kJ\ mol^{-1}$

$\Delta_f H°(298.15) = -28.9 \pm 19\ kJ\ mol^{-1}$

$\Delta_{fus} H° = 21.966 \pm 0.04\ kJ\ mol^{-1}$

Enthalpy of Formation

The enthalpy change of the reaction $B_{10}H_{14}(cr) + 10\ B(am) + 7H_2(g)$ was reported to be 19.8 ± 1.4 and 18.0 ± 1.0 kcal mol^{-1} by Johnson, Kilday and Prosen (1) and Gal'chenko, Timofeev and Skuratov (2), respectively. Using $\Delta_r H°(298.15\ K) = 1.2 \pm 0.4$ kcal mol^{-1} for the reaction B(cr) → B(am), the corresponding $\Delta_f H°(298.15\ K)$ values for $B_{10}H_{14}(cr)$ are calculated to be -7.8 and -6.0 kcal mol^{-1}. The adopted value $\Delta_f H°$ (298.15 K) for $B_{10}H_{14}(cr)$ is the average of these two.

Heat Capacity and Entropy

C_p(60 to 371.93 K) was given by Furukawa and Park (3). The heat capacity above 371.93 K is estimated by comparison with that of the liquid decarborane. Heat capacity (14 to 205 K) was also reported by Kerr, Hallett and Johnston (4). The value of S°(298.15 K) was taken from Furukawa and Park (3).

Fusion Data

T_{fus} and $\Delta_{fus}H°$ were obtained from Furukawa and Park (3).

Sublimation Data

$\Delta_{sub}H°(298.15\ K)$ is calculated from Gibbs energy functions and vapor pressure reported by Furukawa and Park (3). See the ideal gas table for details.

References

1. W. H. Johnson, M. V. Kilday, and E. J. Prosen, J. Res. Nat. Bur. Stand. 64A, 521 (1960).
2. G. L. Gal'chenko, B. I. Timofeev, and S. M. Skuratov, Dokl. Akad. Nauk SSSR 142, 1077 (1962).
3. G. T. Furukawa and R. P. Park, J. Res. Nat. Bur. Stand. 55, 255 (1955).
4. E. C. Kerr, N. C. Hallett, and H. L. Johnston, J. Am. Chem, Soc. 73, 1117 (1951).

Decaborane ($B_{10}H_{14}$) $B_{10}H_{14}$(cr)

Enthalpy Reference Temperature = T_r = 298.15 K Standard State Pressure = p° = 0.1 MPa

T/K	$C_p°$ (J K⁻¹mol⁻¹)	S° (J K⁻¹mol⁻¹)	$-[G°-H°(T_r)]/T$ (J K⁻¹mol⁻¹)	$H°-H°(T_r)$ (kJ mol⁻¹)	$\Delta_f H°$ (kJ mol⁻¹)	$\Delta_f G°$ (kJ mol⁻¹)	Log K_f
0	0.	0.	INFINITE	-28.292	14.248	14.248	INFINITE
100	55.580	53.212	305.968	-25.276	-3.969	61.527	-32.138
200	123.386	109.725	193.426	-16.740	-17.619	132.450	-34.592
298.15	217.953	176.565	176.565	0.	-28.870	208.617	-36.549
300	219.706	177.918	176.569	0.405	-29.048	210.091	-36.580
371.930	293.483	232.548	182.044	18.784	------ CRYSTAL <--> LIQUID -----		
400	305.014	254.416	186.360	27.222	-36.197	291.017	-38.003
500	326.352	324.505	207.161	58.672	-42.516	373.532	-39.023
600	364.845	387.498	232.030	93.281	-48.223	457.307	-39.812
700	397.480	446.265	258.485	131.446	-52.174	541.898	-40.437
800	421.538	500.946	285.418	172.423	-54.694	626.951	-40.936
900	442.458	551.835	312.225	215.649	-56.109	712.253	-41.338
1000	460.240	599.400	338.590	260.810	-56.596	797.660	-41.665
1100	474.884	643.978	364.348	307.592	-56.385	883.080	-41.934
1200	486.390	685.814	389.412	355.682	-55.733	968.459	-42.156
1300	494.758	725.097	413.739	404.766	-54.909	1053.776	-42.341
1400	499.988	761.974	437.310	454.529	-54.192	1139.030	-42.498
1500	502.080	796.559	460.119	504.659	-53.862	1224.246	-42.632

PREVIOUS: September 1962 CURRENT: December 1964

Decaborane ($B_{10}H_{14}$) $B_{10}H_{14}$(cr)

DECABORANE ($B_{10}H_{14}$) LIQUID M_r = 122.21116

S°(298.15 K) = [234.935] J K^{-1} mol^{-1} $\Delta_f H°$(298.15 K) = [-7.133] kJ mol^{-1}

T_{fus} = 371.93 K $\Delta_{fus}H°$ = 21.966 ± 0.04 kJ mol^{-1}

Enthalpy of Formation

$\Delta_f H°$($B_{10}H_{14}$, cl, 298.15 K) is calculated from $\Delta_f H°$($B_{10}H_{14}$, cr, 298.15 K), by adding $\Delta_{fus}H°$ reported by Furukawa and Park (1) and the enthalpy difference, H°(371.93 K)-H°(298.15 K), between the crystal and liquid.

Heat Capacity and Entropy

C_p (371.93 to 380 K) was measured by Furukawa and Park (1). Heat capacities below 371.93 and above 380 K were estimated by comparison with that of the gaseous decarborane. S°(298.15 K) is calculated from that of crystal in a manner analogous to the used for the enthalpy of formation.

Fusion and Vaporization Data

T_{fus} and $\Delta_{fus}H°$ were obtained from Furukawa and Park (1). T_{vap} and $\Delta_{vap}H°$ are calculated from functions for $B_{10}H_{14}$(l) and $B_{10}H_{14}$(g).

References

1. G. T. Furukawa and R. P. Park, J. Res. Nat. Bur. Stand. 55, 255 (1955).

Decaborane ($B_{10}H_{14}$) $B_{10}H_{14}$(l)

Enthalpy Reference Temperature = T_r = 298.15 K Standard State Pressure = p° = 0.1 MPa

T/K	$C_p°$ (J K⁻¹mol⁻¹)	S° (J K⁻¹mol⁻¹)	-[G°-H°(T_r)]/T (J K⁻¹mol⁻¹)	H°-H°(T_r) (kJ mol⁻¹)	$\Delta_f H°$ (kJ mol⁻¹)	$\Delta_f G°$ (kJ mol⁻¹)	Log K_f
0							
100							
200							
298.15	221.208	234.935	234.935	0.	-7.133	212.950	-37.308
300	222.924	236.309	234.939	0.411	-7.306	214.316	-37.316
371.930	296.720	291.607	240.487	19.013	------ CRYSTAL <--> LIQUID -----		
400	323.005	314.790	244.892	27.959	-13.724	289.341	-37.784
500	347.272	389.393	266.520	61.436	-18.015	365.589	-38.193
600	389.949	456.569	292.659	98.346	-21.421	442.666	-38.537
700	426.768	519.520	320.619	139.231	-22.652	520.141	-38.813
800	456.056	578.465	349.207	183.407	-21.973	597.656	-39.023
900	481.160	633.672	377.780	230.302	-19.718	674.990	-39.175
1000	502.080	685.486	405.987	279.499	-16.169	752.000	-39.280
1100	518.816	734.158	433.631	330.579	-11.661	828.605	-39.347
1200	531.368	779.869	460.600	383.123	-6.555	904.771	-39.384
1300	539.736	822.760	486.826	436.713	-1.225	980.499	-39.397
1400	543.920	862.937	512.272	490.931	3.947	1055.820	-39.393
1500	543.920	900.488	536.916	545.358	8.574	1130.788	-39.378

PREVIOUS: September 1962 CURRENT: December 1964

Decaborane ($B_{10}H_{14}$) $B_{10}H_{14}$(l)

DECABORANE ($B_{10}H_{14}$) CRYSTAL–LIQUID M_r = 122.21116 Decaborane ($B_{10}H_{14}$) $B_{10}H_{14}$(cr,l)

0 to 371.93 K crystal
above 371.93 K liquid

Refer to the individual tables for details.

Enthalpy Reference Temperature = T_r = 298.15 K Standard State Pressure = p° = 0.1 MPa

T/K	C_p° (J K⁻¹mol⁻¹)	S° (J K⁻¹mol⁻¹)	−[G°−H°(T_r)]/T (J K⁻¹mol⁻¹)	H°−H°(T_r) (kJ mol⁻¹)	Δ_fH° (kJ mol⁻¹)	Δ_fG° (kJ mol⁻¹)	Log K_f
0	0.	0.	INFINITE	-28.292	14.248	14.248	INFINITE
100	55.580	53.212	305.968	-25.276	-3.969	61.527	-32.138
200	123.386	109.725	193.426	-16.740	-17.619	132.450	-34.592
298.15	217.953	176.565	176.565	0.	-28.870	208.617	-36.549
300	219.706	177.918	176.569	0.405	-29.048	210.091	-36.580
371.930	293.483	232.548	182.044	18.784	CRYSTAL <--> LIQUID		
371.930	296.720	291.607	182.044	40.750	TRANSITION		
400	323.005	314.790	190.551	49 696	-13.724	289.340	-37.784
500	347.272	389.393	223.047	83.173	-18.015	365.589	-38.193
600	389.949	456.569	256.431	120.083	-21.421	442.666	-38.537
700	426.768	519.520	289.567	160.967	-22.653	520.141	-38.813
800	456.056	578.465	322.036	205.143	-21.973	597.656	-39.023
900	481.160	633.672	353.628	252.039	-19.718	674.990	-39.175
1000	502.080	685.486	384.250	301.236	-16.170	752.000	-39.280
1100	518.816	734.158	413.871	352.316	-11.661	828.605	-39.347
1200	531.368	779.869	442.486	404.860	-6.555	904.771	-39.384
1300	539.736	822.760	470.106	458.450	-1.225	980.499	-39.397
1400	543.920	862.937	496.746	512.667	3.946	1055.819	-39.393
1500	543.920	900.488	522.425	567.094	8.574	1130.787	-39.378

PREVIOUS: CURRENT: December 1964

Decaborane ($B_{10}H_{14}$) $B_{10}H_{14}$(cr,l)

DECARBORANE ($B_{10}H_{14}$) IDEAL GAS M_r = 122.21116

$\Delta_fH°(0\ K)$ = 93.8 ± 18.8 kJ mol^{-1}
$S°(298.15\ K)$ = 352.16 J K^{-1} mol^{-1} $\Delta_fH°(298.15\ K)$ = 47.3 ± 18.8 kJ mol^{-1}

Vibrational Frequencies and Degeneracies

ν, cm^{-1}

230(1)	295(1)	344(1)	350(1)	387(1)	434(1)	446(1)	452(1)
464(1)	508(1)	588(1)	618(1)	635(1)	651(1)	690(1)	701(1)
709(1)	720(1)	747(1)	752(1)	767(1)	773(1)	814(1)	822(1)
832(1)	859(1)	862(1)	903(1)	915(1)	921(1)	924(1)	938(1)
942(1)	960(1)	967(1)	972(1)	1008(1)	1037(1)	1060(1)	1104(1)
1160(1)	1190(1)	1210(1)	1250(1)	1300(1)	1360(1)	1104(1)	1410(1)
1460(1)	1500(1)	1555(1)	1590(1)	1620(1)	1665(1)	1880(2)	1929(2)

Ground State Quantum Weight: [1] σ = 2

Point Group: C_{2V}

Bond Distances: B–B = 1.762 Å B–B(apex) = 1.773 Å
B–H(bridge) = 1.394 Å B–H = 1.256 Å

Bond Angles: B–B–B = 108° H–B(apex)–B = 59° 34' B–B–B(apex) = 60° 13'
B–H(bridge)–B = 46° 17' B(apex–B–H) = 87° 42' B(center)–B(center)–H = 120°

Product of the Moments of Inertia: $I_AI_BI_C$ = 8.657643 x 10^{-113} g^3 cm^6

Enthalpy of Formation

The vapor pressures of decaborane were measured by Furukawa and Park (1) and Miller (2). The enthalpies of vaporization were derived by both the 2nd and 3rd law methods. Using the average value of $\Delta_{vap}H°$, the respective of enthalpies of formation for $B_{10}H_{14}(g)$ are calculated.

Source	Reaction	$\Delta_vH°(298.15)$, kcal mol^{-1} 2nd law	3rd law	$\Delta_fH°(298.15)$ kcal mol^{-1}
(1)	$B_{10}H_{14}(cr) \rightarrow B_{10}H_{14}(g)$	16.63 ± 0.93	18.20	10.52 ± 4.5
	$B_{10}H_{14}(l) \rightarrow B_{10}H_{14}(g)$	13.28 ± 0.06	13.06	11.47 ± 4.5
(2)	$B_{10}H_{14}(cr) \rightarrow B_{10}H_{14}(g)$	17.81 ± 0.01	18.27	11.14 ± 4.5

The adopted value of $\Delta_fH°(298.15\ K)$ for $B_{10}H_{14}(g)$ is the weighted average of the three $\Delta_fH°(298.15\ K)$ values listed in the above table.

Heat Capacity and Entropy

The molecular structure of $B_{10}H_{14}(c)$ were reported by Kasper, Lucht and Harker (3), Moore, Dickerson, and Lipscomb (4), and Lipscomb (5). For the calculation of the moments of inertia for $B_{10}H_{14}(g)$ a simplified molecular model is adopted. The molecule is assumed to be composed of two regular pentagonal pyramids with a common base edge. The angle between the base planes of those pyramids is 73° 30'. Each of 10 H atoms is attached to a single B atom above the base plane. The remaining 4 H atoms, each bridge two B atoms. The vibrational frequencies were obtained from Evans (6), which were derived based on the infrared and Raman spectra reported by Stewart (7) and Keller and Johnston (8). The principal moments of inertia are: I_A = 35.1322 x 10^{-39}, I_B = 45.3594 x 10^{-39} and I_C = 54.3284 x 10^{-39} g cm^2.

References

1. G. T. Furukawa and R. P. Park, J. Res. Nat. Bur. Stand. 55, 255 (1955).
2. G. A. Miller, J. Phys. Chem. 67, 1363 (1963).
3. J. S. Kasper, C. M. Lucht, and D. Harker, Acta Cryst. 3, 436 (1950).
4. E. B. Moore, Jr., R. E. Dickerson, and W. N. Lipscomb, J. Chem. Phys. 27, 209 (1957).
5. W. N. Lipscomb, "Boron Hydrides," W. A. Benjamin, Inc., New York, 1963.
6. W. H. Evans, U.S. Nat. Bur. Stand., Report 7093, Jan. 1, 1961.
7. J. E. Stewart, U.S. Nat. Bur. Stand., personal communication, 1957.
8. W. E. Keller and H. L. Johnston, J. Chem. Phys. 20, 1749 (1952)

Decaborane ($B_{10}H_{14}$) $B_{10}H_{14}(g)$

Enthalpy Reference Temperature = T_r = 298.15 K — Standard State Pressure = p° = 0.1 MPa

T/K	$C°_p$ (J $K^{-1}mol^{-1}$)	S°	$-[G°-H°(T_r)]/T$	$H°-H°(T_r)$ (kJ mol^{-1})	$\Delta_fH°$	$\Delta_fG°$	Log K_f
0	0.	0.	INFINITE	-24.839	93.850	93.850	INFINITE
100	46.005	248.258	460.646	-21.239	76.216	122.208	-63.835
200	105.137	296.347	366.164	-13.963	61.306	174.051	-45.457
250	142.942	323.835	354.911	-7.769	54.016	203.084	-42.432
298.15	179.581	352.159	352.159	0.	47.279	232.412	-40.718
300	180.961	353.275	352.163	0.333	47.029	233.561	-40.667
350	216.949	383.907	354.500	10.292	40.573	265.169	-39.574
400	249.946	415.066	360.121	21.978	34.707	297.661	-38.871
450	279.729	446.256	367.960	35.233	29.462	330.852	-38.404
500	306.427	477.136	377.337	49.900	24.860	364.593	-38.089
600	351.663	537.156	399.010	82.888	17.533	433.268	-37.719
700	387.932	594.191	422.857	119.934	12.463	502.987	-37.533
800	417.186	647.970	447.669	160.241	9.273	573.299	-37.433
900	440.941	698.527	472.762	203.188	7.580	643.918	-37.372
1000	460.360	746.023	497.738	248.285	7.028	714.660	-37.330
1100	476.345	790.674	522.360	295.146	7.318	785.416	-37.296
1200	489.595	832.708	546.488	343.463	8.197	856.117	-37.266
1300	500.656	872.346	570.044	392.992	9.466	926.728	-37.236
1400	509.954	909.799	592.987	443.536	10.963	997.231	-37.207
1500	517.824	945.258	615.301	494.935	12.564	1067.623	-37.178
1600	524.530	978.897	636.983	547.062	14.169	1137.907	-37.149
1700	530.279	1010.873	658.044	599.809	15.700	1208.094	-37.120
1800	535.238	1041.327	678.498	653.091	17.099	1278.193	-37.092
1900	539.541	1070.384	698.365	706.835	18.318	1348.219	-37.065
2000	543.294	1098.156	717.666	760.981	19.318	1418.188	-37.039
2100	546.584	1124.745	736.422	815.479	20.073	1488.112	-37.015
2200	549.483	1150.240	754.656	870.285	20.557	1558.005	-36.992
2300	552.049	1174.723	772.391	925.364	20.752	1627.884	-36.970
2400	554.329	1198.268	789.649	980.685	-481.950	1708.449	-37.183
2500	556.363	1220.938	806.450	1036.222	-483.155	1799.741	-37.604
2600	558.185	1242.795	822.814	1091.951	-484.360	1891.079	-37.992
2700	559.824	1263.893	838.762	1147.853	-485.577	1982.465	-38.353
2800	561.301	1284.279	854.312	1203.910	-486.816	2073.897	-38.689
2900	562.638	1304.000	869.480	1260.108	-488.084	2165.374	-39.003
3000	563.852	1323.095	884.284	1316.434	-489.389	2256.895	-39.296
3100	564.956	1341.602	898.739	1372.875	-490.738	2348.459	-39.571
3200	565.964	1359.555	912.860	1429.422	-492.135	2440.067	-39.830
3300	566.886	1376.985	926.662	1486.065	-493.585	2531.722	-40.074
3400	567.732	1393.921	940.157	1542.796	-495.094	2623.420	-40.304
3500	568.510	1410.389	953.358	1599.609	-496.665	2715.164	-40.522
3600	569.226	1426.415	966.277	1656.496	-498.303	2806.953	-40.728
3700	569.887	1442.020	978.925	1713.452	-500.009	2898.790	-40.924
3800	570.499	1457.226	991.313	1770.472	-501.788	2990.673	-41.110
3900	571.066	1472.053	1003.450	1827.551	-503.643	3082.604	-41.287
4000	571.593	1486.518	1015.347	1884.684	-505.575	3174.582	-41.456
4100	572.082	1500.638	1027.011	1941.868	-507.587	3266.612	-41.617
4200	572.538	1514.429	1038.453	1999.099	-5308.175	3428.933	-42.645
4300	572.964	1527.906	1049.680	2056.375	-5299.480	3636.856	-44.179
4400	573.362	1541.083	1060.699	2113.691	-5290.888	3844.578	-45.641
4500	573.734	1553.972	1071.518	2171.046	-5282.400	4052.105	-47.036
4600	574.083	1566.586	1082.143	2228.437	-5274.020	4259.446	-48.368
4700	574.411	1578.936	1092.582	2285.862	-5265.751	4466.606	-49.641
4800	574.718	1591.033	1102.841	2343.319	-5257.594	4673.594	-50.859
4900	575.008	1602.886	1112.926	2400.805	-5249.551	4880.407	-52.026
5000	575.280	1614.505	1122.842	2458.320	-5241.623	5087.063	-53.144
5100	575.537	1625.900	1132.594	2515.861	-5233.812	5293.556	-54.217
5200	575.780	1637.078	1142.189	2573.427	-5226.117	5499.902	-55.247
5300	576.010	1648.048	1151.630	2631.016	-5218.540	5706.099	-56.237
5400	576.227	1658.817	1160.923	2688.628	-5211.080	5912.155	-57.189
5500	576.432	1669.392	1170.072	2746.261	-5203.735	6118.070	-58.105
5600	576.627	1679.780	1179.081	2803.914	-5196.507	6323.857	-58.987
5700	576.812	1689.988	1187.955	2861.586	-5189.393	6529.516	-59.836
5800	576.987	1700.021	1196.698	2919.276	-5182.391	6735.050	-60.656
5900	577.154	1709.886	1205.313	2976.983	-5175.501	6940.461	-61.446
6000	577.312	1719.588	1213.803	3034.707	-5168.720	7145.758	-62.209

PREVIOUS: December 1964 (1 atm) CURRENT: December 1964 (1 bar)

Decaborane ($B_{10}H_{14}$) $B_{10}H_{14}(g)$

LEAD BORATE ($Pb_2B_{10}O_{17}$) CRYSTAL M_r = 794.4898

$S°(298.15\ K) = [352.7 \pm 13]\ J\ K^{-1}\ mol^{-1}$

$\Delta_f H°(0\ K)$ = Unknown

$\Delta_f H°(298.15) = -7087.7 \pm 13\ kJ\ mol^{-1}$

Enthalpy of Formation

See the PbB_2O_4(cr) table for the complete writeup. Evans (1) analysed the enthalpies of formation of the lead borate system based upon Shartsis and Newman's enthalpies of solution and Mazzetti and DeCarli's phase studies. From the Shartsis and Newman paper Evans obtained the following:

$$2\ PbO(cr) + 5\ B_2O_3(cr) \rightarrow 2\ PbO.5B_2O_3(gls) \qquad \Delta_r H = -45.2\ kcal\ mol^{-1}$$

and estimated a ΔH_{298} of 26.8 kcal mol^{-1} for the conversion to the $2PbO.5B_2O_3$(cr). With auxiliary JANAF values the $\Delta_f H°$(298.15 K) is calculated to be -1694.00 kcal mol^{-1}.

Heat Capacity and Entropy

The heat capacity employed in this table was obtained by graphically smoothing the summation of the heat capacities of the constituent oxides.

The entropy at 298.15 K was obtained in the manner described in the PbB_4O_7 table from the entropies of $PbO\cdot 2B_2O_3$ and $Pb\cdot 3B_2O_3$ and the subtraction of 4.2 cal K^{-1} mol^{-1}.

References

1. W. H. Evans, U.S. Nat. Bur. Stand. Report 7192, July 1, 1961.

Lead Borate ($Pb_2B_{10}O_{17}$) $B_{10}O_{17}Pb_2$(cr)

Enthalpy Reference Temperature = T_r = 298.15 K Standard State Pressure = p° = 0.1 MPa

	J K⁻¹mol⁻¹			kJ mol⁻¹			
T/K	$C_p°$	S°	$-[G°-H°(T_r)]/T$	$H°-H°(T_r)$	$\Delta_f H°$	$\Delta_f G°$	Log K_f
0							
100							
200							
298.15	406.685	352.711	352.711	0.	-7087.696	-6616.932	1159.259
300	408.358	355.232	352.719	0.754	-7087.714	-6614.011	1151.602
400	502.917	486.150	369.928	46.489	-7086.317	-6456.200	843.093
500	576.974	606.499	405.379	100.560	-7081.183	-6299.195	658.072
600	641.826	717.604	448.275	161.597	-7072.593	-6143.547	534.843
700	694.962	820.629	494.201	228.499	-7070.549	-5987.999	446.830
800	740.568	916.489	541.065	300.339	-7055.936	-5834.313	380.942
900	778.642	1005.987	587.807	376.362	-7038.648	-5682.623	329.811
1000	809.186	1089.665	633.856	455.809	-7019.159	-5532.980	289.013
1100	833.034	1167.960	678.891	537.976	-6997.967	-5385.374	255.730
1200	850.189	1241.225	722.733	622.189	-6975.596	-5239.755	228.081
1300	861.067	1309.734	765.282	707.787	-6952.611	-5096.032	204.761
1400	867.762	1373.811	806.485	794.256	-6929.479	-4954.086	184.839
1500	871.109	1433.795	846.328	881.200	-6906.558	-4813.790	167.631
1600	874.456	1490.122	884.823	968.478	-6883.959	-4675.011	152.623
1700	877.803	1543.236	922.005	1056.091	-6861.659	-4537.635	139.425
1800	881.150	1593.504	957.927	1144.039	-6839.645	-4401.567	127.730
1900	884.498	1641.235	992.645	1232.321	-6817.903	-4266.712	117.300
2000	887.845	1686.689	1026.220	1320.938	-6796.422	-4132.994	107.943

PREVIOUS: December 1962 CURRENT: March 1965

Lead Borate ($Pb_2B_{10}O_{17}$) $B_{10}O_{17}Pb_2$(cr)

BARIUM (Ba) REFERENCE STATE A_r = 137.33

0 to 582.53	K	crystal, alpha
582.53 to 768.13	K	crystal, beta
768.13 to 1000	K	crystal, gamma
1000 to 2118.599	K	liquid
above 2118.599	K	ideal monatomic gas

Refer to the individual tables for details.

Barium (Ba) Ba_1(ref)

Enthalpy Reference Temperature = T_r = 298.15 K Standard State Pressure = p° = 0.1 MPa

T/K	C_p° (J K⁻¹mol⁻¹)	S° (J K⁻¹mol⁻¹)	$-[G^\circ-H^\circ(T_r)]/T$ (J K⁻¹mol⁻¹)	$H^\circ-H^\circ(T_r)$ (kJ mol⁻¹)	Δ_fH° (kJ mol⁻¹)	Δ_fG° (kJ mol⁻¹)	Log K_f
0	0.	0.	INFINITE	-6.912	0.	0.	0.
100	24.301	34.083	86.220	-5.214	0.	0.	0.
200	26.368	51.643	64.988	-2.669	0.	0.	0.
298.15	28.096	62.475	62.475	0.	0.	0.	0.
300	28.142	62.649	62.476	0.052	0.	0.	0.
400	33.229	71.295	63.622	3.069	0.	0.	0.
500	43.744	79.771	65.997	6.887	0.	0.	0.
582.530	54.400	87.232	68.471	10.929		ALPHA <--> BETA	
582.530	32.468	87.232	68.471	10.929		TRANSITION	
600	33.941	88.213	69.031	11.509	0.	0.	0.
700	42.363	94.077	72.186	15.324	0.	0.	0.
768.130	48.103	98.274	74.313	18.406		BETA <--> GAMMA	
768.130	39.066	98.274	74.313	18.406		TRANSITION	
800	39.066	99.863	75.299	19.651	0.	0.	0.
900	39.066	104.464	78.289	23.557	0.	0.	0.
1000	39.066	108.580	81.116	27.464	0.	0.	0.
1000.000	39.066	108.580	81.116	27.464		GAMMA <--> LIQUID	
1000.000	43.304	116.592	81.116	35.476		TRANSITION	
1100	42.007	120.657	84.530	39.741	0.	0.	0.
1200	40.878	124.262	87.693	43.883	0.	0.	0.
1300	39.999	127.498	90.633	47.925	0.	0.	0.
1400	39.330	130.437	93.373	51.890	0.	0.	0.
1500	38.869	133.133	95.935	55.798	0.	0.	0.
1600	38.618	135.633	98.339	59.670	0.	0.	0.
1700	38.576	137.971	100.602	63.528	0.	0.	0.
1800	38.740	140.180	102.740	67.392	0.	0.	0.
1900	39.116	142.284	104.766	71.283	0.	0.	0.
2000	39.706	144.304	106.693	75.223	0.	0.	0.
2100	40.518	146.260	108.531	79.232	0.	0.	0.
2118.599	40.692	146.618	108.863	79.987		LIQUID <--> IDEAL GAS	
2118.599	30.938	212.831	108.863	220.265		FUGACITY = 1 bar	
2200	32.665	214.029	112.732	222.853	0.	0.	0.
2300	34.873	215.530	117.169	226.230	0.	0.	0.
2400	37.123	217.062	121.299	229.830	0.	0.	0.
2500	39.355	218.622	125.161	233.654	0.	0.	0.
2600	41.512	220.208	128.786	237.698	0.	0.	0.
2700	43.543	221.814	132.202	241.952	0.	0.	0.
2800	45.406	223.431	135.431	246.401	0.	0.	0.
2900	47.068	225.054	138.493	251.026	0.	0.	0.
3000	48.509	226.675	141.406	255.807	0.	0.	0.
3100	49.718	228.286	144.182	260.720	0.	0.	0.
3200	50.695	229.880	146.836	265.743	0.	0.	0.
3300	51.410	231.450	149.376	270.844	0.	0.	0.
3400	51.931	232.992	151.813	276.012	0.	0.	0.
3500	52.262	234.503	154.154	281.223	0.	0.	0.
3600	52.420	235.978	156.406	286.458	0.	0.	0.
3700	52.431	237.415	158.576	291.702	0.	0.	0.
3800	52.319	238.812	160.669	296.940	0.	0.	0.
3900	52.105	240.168	162.691	302.162	0.	0.	0.
4000	51.814	241.483	164.644	307.358	0.	0.	0.
4200	51.068	243.993	168.363	317.645	0.	0.	0.
4400	50.234	246.350	171.855	327.776	0.	0.	0.
4600	49.384	248.563	175.143	337.732	0.	0.	0.
4800	48.545	250.640	178.245	347.499	0.	0.	0.
5000	47.832	252.606	181.180	357.129	0.	0.	0.
5200	47.194	254.465	183.963	366.608	0.	0.	0.
5400	46.697	256.236	186.607	375.995	0.	0.	0.
5600	46.176	257.908	189.122	385.200	0.	0.	0.
5800	45.812	259.517	191.522	394.376	0.	0.	0.
6000	45.369	261.036	193.810	403.351	0.	0.	0.

Previous: December 1970 (1 atm) Current: December 1970 (1 bar)

Barium (Ba) Ba_1(ref)

BARIUM ALPHA-BETA-GAMMA (Ba) CRYSTAL A_r = 137.33

$S°(298.15\ K) = 62.475 \pm 0.8\ J\ K^{-1}\ mol^{-1}$
T_{trs} = [582.53] K (α→β)
T_{trs} = [768.13] K (β→γ)
T_{fus} = 1000 ± 3 K

$\Delta_fH°\ (0\ K) = 0\ kJ\ mol^{-1}$
$\Delta_fH°(298.15\ K) = 0\ kJ\ mol^{-1}$
$\Delta_{trs}H°$ = [0.0] kJ mol^{-1}
$\Delta_{trs}H°$ = [0.0] kJ mol^{-1}
$\Delta_{fus}H° = 8.012 \pm 0.62\ kJ\ mol^{-1}$

Enthalpy of Formation

Zero by definition.

Heat Capacity and Entropy

The adopted values below 298.15 K are based on $C_p°$ (18-370 K) of Furukawa and Ishihara (1) and $C_p°$(1.5-20 K) of Roberts (2). S°(298.15 K) is calculated from $C_p°$ based on an extrapolation of about 0.001 cal K^{-1} mol^{-1} below 1.5 K. We estimate that S°(298.15 K) is uncertain by about 0.2 cal K^{-1} mol^{-1}, due mainly to possible effects of impurities. Furukawa and Ishihara (1) reported impurities (in mole %) of about 3% BaO, 1% Sr and 0.2% Ca. Further details on adjustment of the data for impurities are given in the original paper (1).

Relative enthalpies were measured by Jauch (3), Shpil'rain (4) and Ditmars and Douglas (5). Earlier reviewers dismissed the abnormally high $C_p°$ values of Jauch as due to a very impure sample. The new data confirm this decision, but at least one new study (5) may also involve bias due to impurities. Separate portions from a common sample consisting of three Ba rods were used in the heat capacity study (1), the enthalpy study (5) and the analytical studies (1, 5). The calorimetric data suggest variability in the impurity contents of the separate portions. The 2nd law study (4) used Ba which was not analyzed for O or N which are the most likely contaminants. Since the premelting effect in the enthalpy at 979 K appears to be 1/4 as large in (4) as in (5), the sample of (4) may have been more pure; however, such a conclusion would be quite speculative. Liquid enthalpies are in reasonable agreement, but the crystal enthalpies of (4) are 4 to 20% lower than those of (5). Another major difference is that (4) suggests a single "transition" near 650 K, while (5) suggests two "transitions" near 580 and 770 K. Further confusion arises because the enthalpies reported (4) for CsF(c, ℓ) are unreasonably large.

We tentatively adopt the smoothed $C_p°$ of Ditmars and Douglas (5), pending resolution of the transition discrepancy. The following alternative functions reproduce the enthalpies of Shpil'rain (4) and provide estimates of the possible bias in the adopted functions, Ba(α): $C_p°$/cal K^{-1} mol^{-1} = 5.023 ± 0.005657 T (298-650 K), $\Delta_{trs}H°$ = 0.345 kcal mol^{-1} at 650 K; Ba(γ): $C_p°$/cal K^{-1} mol^{-1} = 9.4 (650-999.6 K), $\Delta_{fus}H°$ = 2.048 kcal mol^{-1} at 999.6 K; Ba(ℓ): $C_p°$/cal K^{-1} mol^{-1} = 10.0 (999.6-1300 K). There is reasonable agreement between the alternative and adopted $C_p°$ for Ba(ℓ) and Ba(γ), suggesting that the γ-phases are identical in the two studies. $C_p°$(α) from the equation is a reasonable extension of the data below 298.15 K (1); $C_p°$ does not show the abnormally steep rise just above 300 K which appears in the enthalpy data (5) and to a lesser degree, in the $C_p°$ data (1). The alternative functions suggest that the adopted γ-phase entropy may be biased by about +0.6 cal K^{-1} mol^{-1} (about +6% in S°(T) - S°(298.15 K) due to impurities. Ditmars and Douglas stressed the possibility of such systematic errors (5) and estimated errors of up to 5% in $C_p°$ and H°. Our alternative functions suggest the maximum error in $C_p°$(α) may be much larger than 5%.

Transition Data

Recent data for Sr and Ca indicate that the pure metals exist in the fcc and bcc polymorphs but that impurities probably stabilize the hcp form. Although enthalpy data (4, 5) suggest a similar controversy for Ba, the pure metal is usually reported to have the bcc form at all temperatures at atmospheric pressure (6-9). Several high-pressure polymorphs are known (10-12). Earlier literature (13) contains much evidence for a "transition" near 648 K at one atmosphere; this is consistent with new enthalpy data (4) and could correspond to the "transition" found by Bridgman (13) near 17 kbar at room temperature. Bridgman's "transition" is in dispute; it seems to involve a very small volume contraction (12, 14) and no obvious change in X-ray pattern (15). We speculate that these two "transitions" are identical; however, evidence for their being a property of pure Ba is inconclusive. The main support comes from the relatively small premelting effect in enthalpy (4). The enthalpy data (4) are insufficient to distinguish whether the "transitions" is first or second order.

We tentatively adopt the two "transitions" with zero enthalpies selected by Ditmars and Douglas (5). These values are placed in brackets to emphasize that there is no confirmatory evidence whatsoever. The authors suggest that these "transitions" may have resulted wholly from impurities. We speculate further the T_{trs} = 582 K could be due to impurity lowering of T_{trs} = 650 K, while T_{trs} = 768 K could result from crossing of a solid-solubility limit below the eutectic temperature of the Ba-BaO system. This case would favor the alternative functions given above. It is also possible that the Sr and Ca impurities could nucleate other polymorphs.

The adopted functions include $\Delta_{trs}H°$ = 0.97 cal mol^{-1} at 57 K from a small λ-anomaly observed between 50 and 60 K (1).

Fusion and Sublimation Data

See the liquid and ideal gas tables for details.

References

1. G. T. Furukawa and S. Ishihara, U. S. Nat. Bur. Stand., Report 10326, chap. 2, July 1, 1970.
2. L. M. Roberts, Proc. Phys. Soc. (London) B70, 738 (1957).
3. R. Jauch, quoted by O. Kubaschewski, Z. Elektrochem. 54, 275 (1950); Z. Metallkunde 41, 445 (1950).
4. E. E. Shpil'rain and D. N. Kagan, Teplofiz. Vys. Temp. 7, 577 (1969); English transl., High Temp. 7, 525 (1969).
5. D. A. Ditmars and T. B. Douglas, U. S. Nat. Bur. Stand., Report 10326, chap. 3, July 1, 1970.
6. R. G. Hirst, A. J. King and F. A. Kanda, J. Phys. Chem. 60, 302 (1956).
7. D. V. Keller, F. A. Kanda and A. J. King, J. Phys. Chem. 62, 732 (1958).
8. D. T. Peterson and M. Indig, J. Amer. Chem. Soc. 82, 5645 (1960).
9. F. A. Kanda, R. M. Stevens and D. V. Keller, J. Phys. Chem. 69, 3867 (1965).
10. J. P. Bastide, C. Susse and R. Epain, C. R. Acad. Sci., Paris, Ser. C 267, 857 (1968).
11. H. G. Drickamer, in "Solid State Physics," Vol. 17, pp. 101-3, Academic Press, New York, 1965.
12. A. Jayaraman, W. Klement and G. C. Kennedy, Phys. Rev. Letters 10, 387 (1963).
13. Gmelin's Handbuch, Barium Erg. 30, 206-7 and 220-224 (1960).
14. F. F. Voronov and O. V. Stal'gorova, Soviet Physics JETP 22, 524 (1966).
15. J. D. Barnett, R. B. Bennion and H. T. Hall, Science 141 (3580) 534 (1963).

Barium (Ba) Ba_1(cr)

Enthalpy Reference Temperature = T_r = 298.15 K Standard State Pressure = p° = 0.1 MPa

T/K	$C_p°$	S°	$-[G°-H°(T_r)]/T$	$H°-H°(T_r)$	$\Delta_fH°$	$\Delta_fG°$	Log K_f
	J K^{-1}mol^{-1}			kJ mol^{-1}			
0	0.	0.	INFINITE	-6.912	0.	0.	0.
100	24.301	34.083	86.220	-5.214	0.	0.	0.
200	26.368	51.643	64.988	-2.669	0.	0.	0.
250	27.196	57.613	62.933	-1.330	0.	0.	0.
298.15	28.096	62.475	62.475	0.	0.	0.	0.
300	28.142	62.649	62.476	0.052	0.	0.	0.
350	29.898	67.105	62.824	1.498	0.	0.	0.
400	33.229	71.295	63.622	3.069	0.	0.	0.
450	38.031	75.475	64.706	4.846	0.	0.	0.
500	43.744	79.771	65.997	6.887	0.	0.	0.
582.530	54.400	87.232	68.471	10.929	ALPHA <--> BETA TRANSITION		
582.530	32.468	87.232	68.471	10.929			
600	33.941	88.213	69.031	11.509	0.	0.	0.
700	42.363	94.077	72.186	15.324	0.	0.	0.
768.130	48.103	98.274	74.313	18.406	BETA <--> GAMMA TRANSITION		
768.130	39.066	98.274	74.313	18.406			
800	39.066	99.863	75.299	19.651	0.	0.	0.
900	39.066	104.464	78.289	23.557	0.	0.	0.
1000	39.066	108.580	81.116	27.464	0.	0.	0.
1000.000	39.066	108.580	81.116	27.464	------- GAMMA <--> LIQUID ------		
1100	39.066	112.303	83.784	31.371	-8.370	0.820	-0.039
1200	39.066	115.702	86.305	35.277	-8.606	1.666	-0.073
1300	39.066	118.829	88.688	39.184	-8.741	2.528	-0.102
1400	39.066	121.724	90.946	43.090	-8.799	3.398	-0.127
1500	39.066	124.420	93.088	46.997	-8.801	4.269	-0.149

Previous: December 1970 Current: December 1970

Barium (Ba) Ba_1(cr)

$S°(298.15 K) = [66.328] J K^{-1} mol^{-1}$

T_{fus} = 1000 ± 3 K

$\Delta_f H°(298.15 K) = [4.985] kJ mol^{-1}$

$\Delta_{fus}H° = 8.012 \pm 0.62 kJ mol^{-1}$

Enthalpy of Formation

The enthalpy of formation is calculated from that of the crystal by adding $\Delta_{fus}H°$ and the difference in enthalpy, H°(1000 K) - H°(298.15 K), between the crystal and liqiud.

Heat Capacity and Entropy

$C_p^°$ is based on the parabolic equation derived by Douglas and Krause (1) from enthalpy data of Ditmars and Douglas (2). The enthalpy data (1003 to 1173 K) may be biased due to impurities in the sample (2); however, enthalpy data (1035 to 1253°K) of Shpil'rain (3) are in reasonable agreement. This suggests that the large discrepancies for Ba(cr) are partially compensated by different values of $\Delta_{fus}H°$. Detailed discussions of the discrepancies and possible alternative functions are given on the crystal table. The alternative functions, derived from the possibly purer sample of Shpil'rain (3), suggest that the adopted liquid entropies may be biased by about +0.4 cal K^{-1} mol^{-1}.

Douglas and Krause emphasized the uncertain nature (1) of the parabolic extrapolation of $C_p^°$ from 1200 to 2000 K. The extrapolation is reasonable provided that the temperature coefficient of $C_p^°$ is not grossly biased due to effects of impurities on the observed enthalpies. Such bias seems unlikely because of the agreement in enthalpy between samples of different purity (2, 3) and because the impurities (2) are present in amounts much below their solubility limits (2, 4, 5) in Ba(ℓ). We note also that S°(2000 K) is not significantly changed when the parabolic fit is replaced by a two-constant fit in which $C_p^°$ approaches a constant (8.48 cal K^{-1} mol^{-1}) at high temperature.

The parabolic $C_p^°$ is extrapolated to 580 K, the assumed glass transition temperature. $C_p^°$ below 580 K is taken to be the same as the crystal. S°(298.15 K) is calculated in a manner analogous to that of the enthalpy of formation.

Melting Data

Ditmars and Douglas (2) considered early melting point data and arbitrarily selected 991 K. Recent data yield higher values of 998 ± 1 (6, 7) and 1002 ± 1 K (8, 10). These values represent more carefully purified samples. We adopt 1000 ± 3 K and adjust $\Delta_{fus}H°$ slightly so that the result is consistent with the smoothed enthalpy (2) of the liquid. Alternative functions based on enthalpies of Shpil'rain (3) suggest that the adopted $\Delta_{fus}H°$ may be too small by about 0.15 kcal mol^{-1} (see the crystal table).

Vaporization Data

T_{vap} = 2118.599 K is calculated as the temperature at which $\Delta_r G° = 0$ for Ba(ℓ) → Ba(g). $\Delta_{vap}H° = 140.278 kJ mol^{-1}$ is the difference between $\Delta_f H°(g)$ and $\Delta_f H°(ℓ)$ at T_{vap}. Bohdansky and Schins (11) reported T_{vap} = 2063 K(1 atm) based on extrapolation of their vapor pressure data from 2024 K. Our calculated T_{vap} is some 60 K higher due to the entropy difference between the adopted functions and vapor pressure data (11) and to adoption of the mean 3rd law $\Delta_{sub}H°$ rather than the apparent value at 2024 K. See the ideal gas table for further details.

References

1. T. B. Douglas and R. F. Krause, U. S. Nat. Bur. Stand., Report 10326, chap. 4, July 1, 1970.
2. D. A. Ditmars and T. B. Douglas, U. S. Nat. Bur. Stand., Report 10326, chap. 3, July 1, 1970.
3. E. E. Shplil'rain and D. N. Kagan, Teplofiz. Vys. Temp. 7, 577 (1969); English transl., High Temp. 7, 525 (1969).
4. M. Schriel, Z. Anorg. Chem. 231, 313 (1937).
5. R. G. Hirst, A. J. King and F. A. Kanda, J. Phys. Chem. 60, 302 (1956).
6. A. S. Dworkin, H. R. Bronstein and M. A. Bredig, J. Phys. Chem. 72, 1892 (1968).
7. D. V. Keller, F. A. Kanda and A. J. King, J. Phys. Chem. 62, 732 (1958).
8. F. A. Kanda, R. M. Stevens and D. V. Keller, J. Phys. Chem. 69, 3867 (1965).
9. D. T. Peterson and M. Indig, J. Amer. Chem. Soc. 82, 5645 (1960).
10. D. T. Peterson and J. A. Hinkebein, J. Phys. Chem. 63, 1360 (1959).
11. J. Bohdansky and H. E. J. Schins, J. Phys. Chem. 71, 215 (1967).

Enthalpy Reference Temperature = T_r = 298.15 K   Standard State Pressure = p° = 0.1 MPa

T/K	$C_p^°$ (J K⁻¹mol⁻¹)	S° (J K⁻¹mol⁻¹)	−[G°−H°(T_r)]/T (J K⁻¹mol⁻¹)	H°−H°(T_r) (kJ mol⁻¹)	$\Delta_f H°$ (kJ mol⁻¹)	$\Delta_f G°$ (kJ mol⁻¹)	Log K_f
0							
100							
200							
298.15	28.096	66.328	66.328	0.	4.985	3.837	-0.672
300	28.142	66.502	66.329	0.052	4.985	3.829	-0.667
350	29.898	70.958	66.677	1.498	4.985	3.636	-0.543
400	33.229	75.147	67.475	3.069	4.985	3.444	-0.450
450	38.031	79.327	68.559	4.846	4.985	3.251	-0.377
500	43.744	83.623	69.849	6.887	4.985	3.059	-0.320
580.000	54.057	90.849	72.242	10.792	GLASS <--> LIQUID TRANSITION		
580.000	51.120	90.849	72.242	10.792			
600	50.664	92.574	72.891	11.809	5.286	2.669	-0.232
700	48.509	100.219	76.267	16.766	6.428	2.128	-0.159
800	46.564	106.567	79.670	21.518	6.853	1.489	-0.097
900	44.832	111.950	82.965	26.086	7.514	0.777	-0.045
1000	43.304	116.592	86.101	30.491	8.012	0.	0.
1000.000	43.304	116.592	86.101	30.491	GAMMA <--> LIQUID		
1100	42.007	120.657	89.062	34.755	0.	0.	0.
1200	40.878	124.262	91.848	38.898	0.	0.	0.
1300	39.999	127.498	94.468	42.940	0.	0.	0.
1400	39.330	130.437	96.933	46.904	0.	0.	0.
1500	38.869	133.133	99.258	50.813	0.	0.	0.
1600	38.618	135.633	101.454	54.685	0.	0.	0.
1700	38.576	137.971	103.534	58.543	0.	0.	0.
1800	38.740	140.180	105.509	62.407	0.	0.	0.
1900	39.116	142.284	107.390	66.298	0.	0.	0.
2000	39.706	144.304	109.185	70.238	0.	0.	0.
2100	40.518	146.260	110.905	74.247	0.	0.	0.
2118.599	40.692	146.618	111.217	75.002	FUGACITY = 1 bar		
2200	41.539	148.168	112.555	78.348	-139.520	5.375	-0.128
2300	42.769	150.041	114.144	82.562	-138.683	11.942	-0.271
2400	44.208	151.891	115.679	86.909	-137.935	18.475	-0.402
2500	45.857	153.728	117.164	91.410	-137.258	24.978	-0.522
2600	47.708	155.562	118.605	96.087	-136.626	31.455	-0.632
2700	49.771	157.400	120.008	100.959	-136.008	37.907	-0.733
2800	52.047	159.251	121.377	106.048	-135.368	44.337	-0.827
2900	54.536	161.120	122.715	111.376	-134.666	50.743	-0.914
3000	57.237	163.014	124.027	116.962	-133.860	57.123	-0.995
3100	60.151	164.938	125.315	122.830	-132.905	63.474	-1.070
3200	63.277	166.896	126.584	129.000	-131.758	69.791	-1.139
3300	66.616	168.894	127.836	135.493	-130.366	76.068	-1.204
3400	70.167	170.935	129.073	142.330	-128.696	82.299	-1.264
3500	73.931	173.023	130.299	149.533	-126.704	88.477	-1.320

Previous: December 1970   Current: December 1970

BARIUM (Ba) ALPHA-BETA-GAMMA-LIQUID A_r = 137.33

0	to	582.53	K	crystal, alpha
582.53	to	768.13	K	crystal, beta
768.13	to	1000	K	crystal, gamma
above		1000	K	liquid

Refer to the individual tables for details.

Barium (Ba) $Ba_1(cr,l)$

Enthalpy Reference Temperature = T_r = 298.15 K Standard State Pressure = $p°$ = 0.1 MPa

	J K^{-1}mol^{-1}			kJ mol^{-1}			
T/K	C_p°	S°	$-[G°-H°(T_r)]/T$	$H°-H°(T_r)$	$\Delta_fH°$	$\Delta_fG°$	Log K_f
0	0.	0.	INFINITE	-6.912	0.	0.	0.
100	24.301	34.083	86.220	-5.214	0.	0.	0.
200	26.368	51.643	64.988	-2.669	0.	0.	0.
250	27.196	57.613	62.933	-1.330	0.	0.	0.
298.15	28.096	62.475	62.475	0.	0.	0.	0.
300	28.142	62.649	62.476	0.052	0.	0.	0.
350	29.898	67.105	62.824	1.498	0.	0.	0.
400	33.229	71.295	63.622	3.069	0.	0.	0.
450	38.031	75.475	64.706	4.846	0.	0.	0.
500	43.744	79.771	65.997	6.887	0.	0.	0.
582.530	54.400	87.232	68.471	10.929	ALPHA <--> BETA TRANSITION		
582.530	32.468	87.232	68.471	10.929			
600	33.941	88.213	69.031	11.509	0.	0.	0.
700	42.363	94.077	72.186	15.324	0.	0.	0.
768.130	48.103	98.274	74.313	18.406	BETA <--> GAMMA TRANSITION		
768.130	39.066	98.274	74.313	18.406			
800	39.066	99.863	75.299	19.651	0.	0.	0.
900	39.066	104.464	78.289	23.557	0.	0.	0.
1000	39.066	108.580	81.116	27.464	0.	0.	0.
1000.000	39.066	108.580	81.116	27.464	GAMMA <--> LIQUID TRANSITION		
1000.000	43.304	116.592	81.116	35.476			
1100	42.007	120.657	84.530	39.741	0.	0.	0.
1200	40.878	124.262	87.693	43.883	0.	0.	0.
1300	39.999	127.498	90.633	47.925	0.	0.	0.
1400	39.330	130.437	93.373	51.890	0.	0.	0.
1500	38.869	133.133	95.935	55.798	0.	0.	0.
1600	38.618	135.633	98.339	59.670	0.	0.	0.
1700	38.576	137.971	100.602	63.528	0.	0.	0.
1800	38.740	140.180	102.740	67.392	0.	0.	0.
1900	39.116	142.284	104.766	71.283	0.	0.	0.
2000	39.706	144.304	106.693	75.223	0.	0.	0.
2100	40.518	146.260	108.531	79.232	0.	0.	0.
2118.599	40.692	146.618	108.863	79.987	------- FUGACITY = 1 bar -------		
2200	41.539	148.168	110.289	83.333	-139.520	5.375	-0.128
2300	42.769	150.041	111.977	87.547	-138.683	11.942	-0.271
2400	44.208	151.891	113.601	91.894	-137.935	18.475	-0.402
2500	45.857	153.728	115.170	96.396	-137.258	24.978	-0.522
2600	47.708	155.562	116.688	101.072	-136.626	31.455	-0.632
2700	49.771	157.400	118.162	105.944	-136.008	37.908	-0.733
2800	52.047	159.251	119.596	111.033	-135.368	44.337	-0.827
2900	54.536	161.120	120.996	116.361	-134.666	50.743	-0.914
3000	57.237	163.014	122.365	121.948	-133.860	57.123	-0.995
3100	60.151	164.938	123.707	127.815	-132.905	63.474	-1.070
3200	63.277	166.896	125.026	133.985	-131.758	69.791	-1.139
3300	66.616	168.894	126.325	140.478	-130.366	76.068	-1.204
3400	70.167	170.935	127.607	147.315	-128.696	82.299	-1.264
3500	73.931	173.023	128.874	154.518	-126.704	88.477	-1.320

Previous: December 1970 Current: December 1970

Barium (Ba) $Ba_1(cr,l)$

BARIUM (Ba) IDEAL GAS A_r = 137.33

IP (Ba,g) = 42035.14 ± 0.05 cm^{-1}

S°(298.15 K) = 170.245 ± 0.01 J K^{-1} mol^{-1}

$\Delta_fH°$(0 K) = 179.8 ± 5.0 kJ mol^{-1}

$\Delta_fH°$(298.15 K) = 179.1 ± 5.0 kJ mol^{-1}

Electronic Levels and Quantum Weights

ϵ_i, cm^{-1}	g_i	ϵ_i, cm^{-1}	g_i	ϵ_i, cm^{-1}	g_i	ϵ_i, cm^{-1}	g_i	ϵ_i, cm^{-1}	g_i	ϵ_i, cm^{-1}	g_i
0.0	1	22947	7	25704	3	33188	15	39765	[21]	[42000]	33
9034.0	3	23757	9	25957	5	33837	8	39905	[9]	[36440]	1
9215.5	5	23068	10	26160	3	35142	10	40736	3	[40000]	63
9596.6	7	23209	1	26816	7	34649	28	40893	15	[20000]	21
11395.4	5	23480	3	28230	1	35658	20	41097	9	[24000]	9
12266.0	1	23919	5	28554	3	35893	3	42012	3	[25000]	1
12636.6	3	24192	3	30237	5	36400	55	42436	9	[19000]	20
13514.7	5	24532	5	30771	15	36974	56	43264	3	[39000]	20
18060	3	24980	7	30903	9	37425	58	35500	9	[42000]	120
22065	5	25642	1	32547	3	38031	37	37700	9		

Enthalpy of Formation

$\Delta_fH°$ is $\Delta_{sub}H°$, 42.8 ± 1.2 kcal mol^{-1}, selected from 3rd law analyses of pressure data tubulated below. The adopted value is from the recent boiling-point study of Bohdansky and Schins (1). Except for the very scattered effusion data of Zavitsanos (4), the other studies differ by unusually large amounts: -2 (2), -6 (3) and +3 kcal mol^{-1} (5). Data of Ruff and Hartmann (3) are readily dismissed due to the large entropy discrepancy; their data for other metals have similar discrepancies. Data of Rudberg and Lempert (5) for Ca and Ba show almost identical biases, presumably due to poor effusion geometry or to impurities. A similar bias seems to exist in the effusion-resonance-fluorescence data of Hinnov and Ohlendorf (6): our analysis is not tabulated below since the graphical data include large discrepancies in temperature. Hartmann and Scheider's Ba data (2) deviate considereably more from the selected value than do their data for Li, Mg, Ca and Sr. The differences probably arise from impurity effects. Since Ba is the least volatile of the reactive metals studied (1, 2), purification by distillation presents special problems. Even Bohdansky and Schins (1) noted that their Ba data may be biased by impurities.

Analyses of the pressure data assume that the monatomic gas is the only significant species in the vapor and that gas implerfection is negligible. Evidence supporting the unimportance of the dimer was reviewed by Douglas and Krause (7). Estimated bond energies for the dimers of Mg, Ca, Sr nd Ba are less than 8.6 kcal mol^{-1} (8, 10).

Source	Method	T/K[a]	$\Delta_fH°$(298.15 K), kcal mol^{-1} 2nd law	3rd law	δS cal K^{-1} mol^{-1}
(1)	Boiling Point	1498-2027	46.7±0.3	42.78±0.04	2.2±0.1
(2)	Boiling Point	1334-1421	41.4±3.3	40.76±0.33	0.4±2.4
(3)	Boiling Point	1204-1404	92.6±3.0	37.0 ±6.1	43 ±2.3
(4)	Effusion	1120-1210	65 ±10	43.8 ±1.6	18 ±8.8
(4)	Torsion Effusion	1103-1216	36 ±10	43.5 ±1.9	-6.3±8.4
(5)	Effusion	798-1024	43.2±1.2	45.91±0.77	-3.0±1.3

[a]Temperature adjusted to IPTS-68 assuming published values to be IPTS-48.

Heat Capacity and Entropy

Observed energy levels and quantum weights are from Moore (11) as modified by Garton et al. (12, 13). Additional levels above the cutoffs have been reported (14, 15). We adopt an energy-level cutoff which is about kTmax (Tmax = 6000 K) below each series limit. For Ba I this corresponds to omitting levels above 9s, 8p, 7d and 5f. Energies of unobserved but predicted terms (11) are estimated by comparison of Ca I, Sr I, Ba I and their isoelectronic ions. The most important are terms of the $5d^2$ configuration (20000-25000 cm^{-1}) and those of the 5d6s' configuration (estimated at 19000 ± 10000 cm^{-1}). Levels above 30000 cm^{-1} are averaged. The adopted functions are essentially identical with earlier tables up to 2000 K; however, the entropies at 6000 K differ as follows: Hilsenrath et al. (16), 62.26; JANAF, 63.20; Gurvich et al. (17), 63.65 cal K^{-1} mol^{-1}. These differences arise from different methods of cutoff and different degrees of accounting for the unobserved terms.

References

1. J. Bohdansky and H. E. J. Schins, J. Phys. Chem. 71, 215 (1967).
2. H. Hartmann and R. Schneider, Z. Anorg. Chem. 180, 275 (1929).
3. O. Ruff and H. Hartmann, Z. Anorg. Chem. 133, 29 (1924).
4. P. D. Zavitsanos, G. E. Technical Report 68 SO 331, General Electric Co., Valley Forge, Pa., Nov. 1968; data quoted by Douglas and Krause (7).
5. E. Rudberg and J. Lempert, J. Chem. Phys. 3, 627 (1935).
6. E. Hinnov and W. Ohlendorf, J. Chem. Phys. 50, 3005 (1969).
7. T. B. Douglas and R. F. Krause, U. S. Nat. Bur. Stand., Report 10326, chap. 4, July 1, 1970.
8. K. A. Gingerich and H. C. Finkbeiner, J. Chem. Phys. 52, 2956 (1970).
9. A. M. Mellor, J. Chem. Phys. 51, 1678 (1969).
10. R. H. Ewing and A. M. Mellor, J. Chem. Phys. 53, 2983 (1970).
11. C. E. Moore, "Atomic Energy Levels," Vol. 3, U. S. Nat. Bur. Stand. Circ. 467, 1958.
12. W. R. S. Garton and K. Codling, Proc. Phys. Soc. 75, 87 (1960).
13. W. R. S. Garton, W. H. Parkinson and E. M. Reeves, Proc. Phys. Soc. 80, 860 (1962).
14. W. R. S. Garton, F. S. Tomkins, Astrophys. J. 158, 1219 (1969).
15. R. D. Hudson, V. L. Carter and P. A. Young, Phys. Rev. A2, 643 (1970).
16. J. Hilsenrath, C. G. Messina and W. H. Evans, AD606163, 1964.
17. L. V. Gurvich et al., "Thermodynamic Properties of Individual Substances," Vol. I and II, Moscow, 1962; English transl., AD65659, March, 1967.

Barium (Ba) $Ba_1(g)$

Enthalpy Reference Temperature = T_r = 298.15 K; Standard State Pressure = p° = 0.1 MPa

T/K	$C_p°$ (J $K^{-1}mol^{-1}$)	S°	$-[G°-H°(T_r)]/T$	$H°-H°(T_r)$ (kJ mol^{-1})	$\Delta_fH°$	$\Delta_fG°$	Log K_f
0	0.	0.	INFINITE	-6.197	179.790	179.790	INFINITE
100	20.786	147.537	188.725	-4.119	180.170	168.825	-88.185
200	20.786	161.945	172.146	-2.040	179.704	157.644	-41.172
250	20.786	166.584	170.587	-1.001	179.404	152.162	-31.792
298.15	20.786	170.245	170.245	0.	179.075	146.944	-25.744
300	20.786	170.373	170.245	0.038	179.062	146.744	-25.550
350	20.786	173.577	170.498	1.078	178.654	141.389	-21.101
400	20.786	176.353	171.060	2.117	178.123	136.100	-17.773
450	20.786	178.801	171.787	3.156	177.386	130.889	-15.193
500	20.786	180.991	172.600	4.196	176.384	125.774	-13.139
600	20.786	184.781	174.324	6.274	173.841	115.900	-10.090
700	20.786	187.985	176.053	8.353	172.104	106.368	-7.937
800	20.788	190.761	177.721	10.432	169.856	97.137	-6.342
900	20.796	193.210	179.309	12.511	168.028	88.157	-5.117
1000	20.821	195.402	180.811	14.591	166.202	79.380	-4.146
1100	20.883	197.389	182.229	16.676	156.011	71.606	-3.400
1200	21.014	199.211	183.569	18.770	153.963	64.024	-2.787
1300	21.251	200.901	184.838	20.882	152.033	56.608	-2.275
1400	21.635	202.489	186.043	23.025	150.211	49.337	-1.841
1500	22.204	204.000	187.190	25.216	148.493	42.192	-1.469
1600	22.993	205.457	188.286	27.473	146.878	35.159	-1.148
1700	24.024	206.881	189.338	29.822	145.369	28.223	-0.867
1800	25.307	208.289	190.352	32.287	143.969	21.373	-0.620
1900	26.839	209.697	191.333	34.892	142.684	14.598	-0.401
2000	28.599	211.118	192.287	37.662	141.514	7.887	-0.206
2100	30.556	212.560	193.218	40.618	140.461	1.232	-0.031
2118.599	30.938	212.831	193.389	41.190	------- FUGACITY = 1 bar -------		
2200	32.665	214.029	194.130	43.778	0.	0.	0.
2300	34.873	215.530	195.028	47.155	0.	0.	0.
2400	37.123	217.062	195.914	50.754	0.	0.	0.
2500	39.355	218.622	196.791	54.579	0.	0.	0.
2600	41.512	220.208	197.661	58.623	0.	0.	0.
2700	43.543	221.814	198.526	62.877	0.	0.	0.
2800	45.406	223.431	199.386	67.326	0.	0.	0.
2900	47.068	225.054	200.243	71.951	0.	0.	0.
3000	48.509	226.675	201.098	76.732	0.	0.	0.
3100	49.718	228.286	201.949	81.645	0.	0.	0.
3200	50.695	229.880	202.797	86.668	0.	0.	0.
3300	51.410	231.450	203.641	91.769	0.	0.	0.
3400	51.931	232.992	204.482	96.936	0.	0.	0.
3500	52.262	234.503	205.318	102.148	0.	0.	0.
3600	52.420	235.978	206.149	107.383	0.	0.	0.
3700	52.431	237.415	206.975	112.627	0.	0.	0.
3800	52.319	238.812	207.794	117.865	0.	0.	0.
3900	52.105	240.168	208.607	123.087	0.	0.	0.
4000	51.814	241.483	209.413	128.283	0.	0.	0.
4100	51.463	242.758	210.211	133.445	0.	0.	0.
4200	51.068	243.993	211.000	138.570	0.	0.	0.
4300	50.656	245.190	211.782	143.656	0.	0.	0.
4400	50.234	246.350	212.554	148.701	0.	0.	0.
4500	49.812	247.474	213.318	153.703	0.	0.	0.
4600	49.384	248.563	214.072	158.657	0.	0.	0.
4700	48.932	249.614	214.817	163.550	0.	0.	0.
4800	48.545	250.640	215.552	168.423	0.	0.	0.
4900	48.171	251.636	216.278	173.254	0.	0.	0.
5000	47.832	252.606	216.995	178.054	0.	0.	0.
5100	47.483	253.545	217.702	182.799	0.	0.	0.
5200	47.194	254.465	218.401	187.533	0.	0.	0.
5300	46.933	255.361	219.089	192.239	0.	0.	0.
5400	46.697	256.236	219.769	196.920	0.	0.	0.
5500	46.442	257.084	220.439	201.547	0.	0.	0.
5600	46.176	257.908	221.100	206.125	0.	0.	0.
5700	46.000	258.724	221.753	210.734	0.	0.	0.
5800	45.812	259.517	222.397	215.301	0.	0.	0.
5900	45.508	260.272	223.029	219.733	0.	0.	0.
6000	45.369	261.036	223.656	224.276	0.	0.	0.

Previous: December 1970 (1 atm) Current: December 1970 (1 bar)

Barium (Ba) $Ba_1(g)$

BARIUM, ION (Ba^+) IDEAL GAS M_r = 137.32945

$IP(Ba^+, g) = 80686.87 \pm 0.05\ cm^{-1}$

$S°(298.15\ K) = 176.008 \pm 0.05\ J\ K^{-1}\ mol^{-1}$

$\Delta_f H°(0\ K) = 682.6 \pm 5.0\ kJ\ mol^{-1}$

$\Delta_f H°(298.15\ K) = [688.126]\ kJ\ mol^{-1}$

Electronic Levels and Quantum Weights

State	$\varepsilon_i\ cm^{-1}$	g_i
$^2S_{1/2}$	0	2
$^2D_{3/2}$	4873.85	4
$^2D_{5/2}$	5674.824	6
$^2P_{1/2}$	20261.562	2
$^2P_{3/2}$	21952.422	4
.	.	.
.	.	.
.	.	.
.	.	.
$^2G_{9/2}$	77628	10
IP	80686.87	

Enthalpy of Formation

$\Delta_f H°(Ba^+, g, 0\ K)$ is calculated from $\Delta_f H°(Ba, g, 0\ K)$ (1) using the spectroscopic value of $IP(Ba) = 42035.14 \pm 0.05\ cm^{-1}$ ($502.8521 \pm 0.0006\ kJ\ mol^{-1}$) from Moore (2). The ionization limit is converted from cm^{-1} to $kJ\ mol^{-1}$ using the factor, $1\ cm^{-1} = 0.01196266\ kJ\ cm^{-1}$, which is derived from the 1973 CODATA fundamental constants (3). Rosenstock et al. (4) and Levin and Lias (5) have summarized additional ionization and appearance potential data.

$\Delta_f H°(Ba^+, g, 298.15\ K)$ is calculated from $\Delta_f H°(Ba, g, 0\ K)$ by using IP(Ba) with JANAF (1) enthalpies, H°(0 K) − H°(298.15 K), for Ba(g), $Ba^+(g)$, and $e^-(g)$. $\Delta_f H°(Ba \rightarrow Ba^+ + e^-, 298.15\ K)$ differs from a room temperature threshold energy due to inclusion of these enthalpies and to threshold effects discussed by Rosenstock et al. (4). Δ H°(298.15 K) should be changed by $-6.197\ kJ\ mol^{-1}$ of it is to be used in the ion convention that excludes the enthalpy of the electron.

Heat Capacity and Entropy

The information on electronic energy levels and quantum weights, given by Moore (2,6), is incomplete because many theoretically predicted levels have not been observed. Although we have listed only the ground, the first four excited states, the highest observed excited state, and the ionization potential for $Ba^+(g)$, all levels listed by Moore (2,6), as well as estimated levels, are used in the calculation. The observed levels are too numerous to list completely. The calculations indicate that for $Ba^+(g)$, the thermodynamic functions are independent of the estimated missing levels (for n = 6), the cut-off procedure, and the inclusion of n=12 levels up to 6000 K. The reported uncertainty in S°(298.15 K) is due to uncertainties in the relative ionic mass, and the fundamental constants. Extension of these calculations above 6000 K may require consideration of the higher excited states (n>12), and use of different fill and cut-off procedures (7).

References

1. JANAF Thermochemical Tables: Ba(g), 9-30-83; $e^-(g)$, 3-31-82.
2. C. E. Moore, U.S. Nat. Bur. Stand., NSRDS-NBS-34, 1970.
3. E. R. Cohen and B. N. Taylor, J. Phys. Chem. Ref. Data 2, 663(1973).
4. H.M. Rosenstock, K. Draxl et al., J. Phys. Chem. Ref. Data 6, Supp. 1 (1977).
5. R. D. Levin and S. G. Lias, U.S. Nat. Bur. Stand., NSRDS-NBS-71, 1982.
6. C. E. Moore, U.S. Nat. Bur. Stand., NSRDS-NBS-35, Volume III, 1970 [Reprint of NBS Circular 467, Volume III, 1958].
7. J. R. Downey, Jr., The Dow Chemical Company, AFOSR-TR-78-0960, Contract No. F44620-75-1-0048, 1978.

Barium, Ion (Ba^+) $Ba_1^+(g)$

Enthalpy Reference Temperature = T_r = 298.15 K Standard State Pressure = p° = 0.1 MPa

T/K	$C_p°$ (J K⁻¹mol⁻¹)	S° (J K⁻¹mol⁻¹)	−[G°−H°(T_r)]/T (J K⁻¹mol⁻¹)	H°−H°(T_r) (kJ mol⁻¹)	$\Delta_f H°$ (kJ mol⁻¹)	$\Delta_f G°$ (kJ mol⁻¹)	Log K_f
0	0.	0.	INFINITE	-6.197	682.643		
100	20.786	153.301	194.488	-4.119			
200	20.786	167.708	177.909	-2.040			
250	20.786	172.347	176.350	-1.001			
298.15	20.786	176.008	176.008	0.	688.126	648.021	-113.531
300	20.786	176.136	176.008	0.038	688.151	647.772	-112.787
350	20.786	179.341	176.261	1.078	688.783	640.991	-95.663
400	20.786	182.116	176.823	2.117	689.291	634.127	-82.809
450	20.787	184.564	177.550	3.156	689.593	627.211	-72.805
500	20.789	186.755	178.363	4.196	689.630	620.275	-64.800
600	20.811	190.546	180.087	6.275	689.166	606.457	-52.797
700	20.890	193.759	181.817	8.360	689.514	592.634	-44.223
800	21.081	196.560	183.488	10.457	689.363	578.810	-37.792
900	21.437	199.061	185.082	12.582	689.660	564.973	-32.790
1000	21.989	201.347	186.596	14.751	690.001	551.102	-28.787
1100	22.738	203.476	188.034	16.986	682.038	538.012	-25.548
1200	23.660	205.493	189.406	19.305	682.293	524.909	-22.849
1300	24.710	207.428	190.718	21.722	682.748	511.777	-20.563
1400	25.834	209.300	191.979	24.249	683.388	498.602	-18.603
1500	26.975	211.121	193.195	26.890	684.199	485.376	-16.902
1600	28.082	212.898	194.371	29.643	685.159	472.090	-15.412
1700	29.109	214.631	195.512	32.503	686.240	458.741	-14.095
1800	30.026	216.322	196.621	35.461	687.412	445.325	-12.923
1900	30.811	217.967	197.701	38.504	688.643	431.843	-11.872
2000	31.452	219.564	198.755	41.619	689.896	418.295	-10.925
2100	31.949	221.111	199.783	44.790	691.137	404.684	-10.066
2200	32.305	222.606	200.787	48.004	552.808	396.390	-9.412
2300	32.533	224.048	201.767	51.247	554.753	389.236	-8.840
2400	32.643	225.435	202.724	54.506	556.492	382.002	-8.314
2500	32.653	226.768	203.659	57.772	558.011	374.699	-7.829
2600	32.576	228.048	204.573	61.034	559.308	367.340	-7.380
2700	32.429	229.275	205.465	64.285	560.383	359.936	-6.963
2800	32.224	230.450	206.337	67.518	561.246	352.495	-6.576
2900	31.975	231.577	207.188	70.728	561.909	345.028	-6.215
3000	31.693	232.656	208.019	73.912	562.391	337.540	-5.877
3100	31.387	233.691	208.831	77.066	562.710	330.040	-5.561
3200	31.066	234.682	209.623	80.189	562.889	322.531	-5.265
3300	30.737	235.633	210.397	83.279	562.957	315.018	-4.986
3400	30.405	236.546	211.153	86.336	562.925	307.505	-4.724
3500	30.074	237.422	211.891	89.360	562.817	299.994	-4.477
3600	29.749	238.265	212.612	92.351	562.651	292.487	-4.244
3700	29.431	239.076	213.316	95.310	562.445	284.985	-4.023
3800	29.124	239.856	214.004	98.238	562.212	277.489	-3.814
3900	28.827	240.609	214.677	101.135	561.967	270.000	-3.616
4000	28.543	241.335	215.334	104.004	561.718	262.517	-3.428
4100	28.273	242.037	215.977	106.844	561.475	255.040	-3.249
4200	28.015	242.715	216.606	109.659	561.243	247.568	-3.079
4300	27.771	243.371	217.221	112.448	561.024	240.102	-2.917
4400	27.541	244.007	217.822	115.213	560.824	232.641	-2.762
4500	27.323	244.623	218.411	117.956	560.643	225.185	-2.614
4600	27.119	245.222	218.987	120.678	560.490	217.731	-2.472
4700	26.927	245.803	219.552	123.381	560.378	210.279	-2.337
4800	26.747	246.368	220.104	126.064	560.267	202.831	-2.207
4900	26.579	246.918	220.646	128.730	560.181	195.385	-2.083
5000	26.422	247.453	221.177	131.380	560.110	187.941	-1.963
5100	26.275	247.975	221.697	134.015	560.078	180.496	-1.849
5200	26.139	248.484	222.208	136.636	560.043	173.053	-1.738
5300	26.012	248.980	222.708	139.243	560.023	165.611	-1.632
5400	25.895	249.465	223.199	141.838	560.016	158.170	-1.530
5500	25.786	249.940	223.681	144.422	560.052	150.724	-1.431
5600	25.686	250.403	224.154	146.996	560.126	143.274	-1.336
5700	25.593	250.857	224.618	149.560	560.160	135.830	-1.245
5800	25.509	251.301	225.075	152.115	560.226	128.382	-1.156
5900	25.431	251.737	225.523	154.662	560.420	120.916	-1.071
6000	25.361	252.164	225.963	157.201	560.494	113.467	-0.988

Previous: Current: September 1983 (1 bar)

Barium, Ion (Ba^+) $Ba_1^+(g)$

BARIUM BROMIDE (BaBr) IDEAL GAS M_r = 217.234

$\Delta_f H°(0\ K) = -101.6 \pm 41.8\ kJ\ mol^{-1}$

$S°(298.15\ K) = 270.4 \pm 0.4\ J\ K^{-1}\ mol^{-1}$

$\Delta_f H°(298.15\ K) = -110.6 \pm 41.8\ kJ\ mol^{-1}$

Electronic Levels and Quantum Weights

State	ε_i, cm^{-1}	g_i	State	ε_i, cm^{-1}	g_i
$X^2\Sigma^+$	0	2	$D^2\Sigma$	25670.9	2
$[A^2\Pi]$	[13000]	[4]	$E^2\Sigma$	26865.9	2
$[B^2\Sigma]$	[13500]	[2]	$[F^2\Pi]$	[28000]	[4]
$C_1{}^2\Pi_{1/2}$	18650.9	2	$[G_1{}^2\Delta]$	[29000]	[4]
$C_2{}^2\Pi_{3/2}$	19192.5	2	$[H^2\Sigma]$	[30000]	[2]

$\omega_e = 193.2\ cm^{-1}$ $\omega_e x_e = 0.42\ cm^{-1}$ $\sigma = 1$

$B_e = [0.04052]\ cm^{-1}$ $\alpha_e = [0.000113]\ cm^{-1}$ $r_e = [2.87]$ Å

Enthalpy of Formation

The selected value, $\Delta_f H°(0\ K) = -24.3\ kcal\ mol^{-1}$, is obtained from an analysis of spectroscopic data. Herzberg (1) suggested $D_0^0 = 2.8$ eV for BaBr(g) which was derived from a linear Birge-Sponer extrapolation of the ground state vibrational levels. Our adopted vibrational constants give 2.74 eV by a similar extrapolation. Based on the ionicity correction developed by Hildenbrand (2), this value adjusts to 4.14 eV (95.5 kcal mol^{-1}) which is adopted. Gaydon (3) has claimed that the spectroscopic data for BaBr(g) are insufficient to obtain a reliable extrapolation. However, our adopted value for D_0^0 gives $D_0^0(BaBr)/\Delta_{at}H°(BaBr_2) = 0.48$ which is quite consistent with values of this ratio for other alkaline earth halide systems (4). $\Delta_f H°(298.15\ K)$ corresponds to $-26.4\ kcal\ mol^{-1}$.

Ionic model calculations (5, 6) have led to D_0^0 values of 4.9 eV (5) and 3.76 eV (6). The latter result is believed to represent a minimum value. Two other experimental values, which bracket the selected value, have been reported. Flame studies (7) gave $D_0^0 = 3.79$ eV, and chemiluminescence (8) from reaction of Ba atoms with Br_2 gave a lower limit to D_0^0 of 4.54 eV. We assign an uncertainty of ± 10 kcal mol^{-1} to $\Delta_f H°(0\ K)$ to include the possibility that these studies are correct.

Heat Capacity and Entropy

The value of r_e is obtained from that for gaseous $BaBr_2$(9) with $r_e(BaBr)/r_e(BaBr_2) = 0.96$. This value for the ratio is calculated from bond lengths (4) for several other alkaline earth halide systems. Two other estimates (5, 6) of r_e agree with the adopted value to within 0.05 Å. The rotational constant is calculated from the estimated value for r_e. The value of α_e is obtained from a Morse potential function.

The vibrational constants are taken from the compilation of Rosen (10) and are corrected for the natural isotopic abundances of the elements. The observed electronic levels for BaBr(g) are from Rosen (10). Also included are A, B, F, G, and H states. Their energies are estimated by analogy with those for SrBr(g) (9).

References

1. G. Herzberg, "Spectra of Diatomic Molecules," 2nd ed., D. Van Nostrand Company, New York, 1950.
2. D. L. Hildenbrand, "Advances in High Temperature Chemistry," Vol. 1, Ed. by Eyring, Academic Press, New York, 1967.
3. A. G. Gaydon, "Dissociation Energies and Spectra of Diatomic Molecules," Chapman and Hall, Ltd., London, 1968.
4. JANAF Thermochemical Tables.
5. J. L. Margrave, J. Phys. Chem. 58, 258 (1954).
6. K. S. Krasnov and N. V. Karaseva, Optics and Spectroscopy 19, 14 (1965).
7. A. N. Khitrov, V. G. Ryabova, and L. V. Gurvich, Teplofiz. Vys. Temp. 11, 1126 (1973).
8. M. Menzinger, Can. J. Chem. 52, 1988 (1974).
9. JANAF Thermochemical Tables: $BaBr_2$(g) and SrBr(g), 13-31-74.
10. B. Rosen, Ed., "Spectroscopic Data Relative to Diatomic Molecules," Pergamon Press, New York, 1970.

Barium Bromide (BaBr) $Ba_1Br_1(g)$

Enthalpy Reference Temperature = T_r = 298.15 K Standard State Pressure = p° = 0.1 MPa

T/K	$C_p°$ (J K^{-1}mol^{-1})	S° (J K^{-1}mol^{-1})	$-[G°-H°(T_r)]/T$ (J K^{-1}mol^{-1})	$H°-H°(T_r)$ (kJ mol^{-1})	$\Delta_f H°$ (kJ mol^{-1})	$\Delta_f G°$ (kJ mol^{-1})	Log K_f
0	0.	0.	INFINITE	-10.200	-101.621	-101.621	INFINITE
100	33.690	231.368	302.718	-7.135	-101.651	-118.687	61.996
200	36.304	255.748	273.762	-3.603	-103.114	-135.193	35.309
250	36.742	263.900	271.003	-1.776	-104.033	-143.109	29.901
298.15	36.996	270.395	270.395	0.	-110.587	-149.888	26.260
300	37.004	270.624	270.396	0.068	-110.641	-150.132	26.140
350	37.177	276.342	270.847	1.923	-126.558	-155.829	23.256
400	37.300	281.315	271.852	3.785	-127.181	-159.970	20.890
450	37.394	285.714	273.152	5.653	-128.011	-164.022	19.039
500	37.469	289.658	274.609	7.524	-129.106	-167.967	17.547
600	37.586	296.500	277.704	11.277	-131.834	-175.474	15.276
700	37.677	302.301	280.814	15.041	-133.755	-182.607	13.626
800	37.756	307.337	283.822	18.812	-136.187	-189.413	12.367
900	37.826	311.788	286.687	22.592	-138.196	-195.946	11.372
1000	37.891	315.777	289.400	26.378	-140.204	-202.255	10.565
1100	37.954	319.392	291.964	30.170	-150.580	-207.543	9.855
1200	38.015	322.697	294.390	33.968	-152.819	-212.622	9.255
1300	38.075	325.742	296.686	37.773	-154.957	-217.518	8.740
1400	38.136	328.566	298.863	41.583	-157.016	-222.253	8.292
1500	38.199	331.199	300.932	45.400	-159.017	-226.843	7.899
1600	38.268	333.666	302.902	49.223	-160.981	-231.300	7.551
1700	38.343	335.989	304.780	53.054	-162.929	-235.635	7.240
1800	38.428	338.183	306.576	56.892	-164.882	-239.856	6.960
1900	38.526	340.263	308.294	60.740	-166.859	-243.967	6.707
2000	38.639	342.242	309.943	64.598	-168.883	-247.973	6.476
2100	38.769	344.130	311.526	68.468	-170.975	-251.877	6.265
2200	38.919	345.937	313.049	72.352	-312.674	-250.304	5.943
2300	39.091	347.671	314.517	76.253	-314.122	-247.437	5.619
2400	39.286	349.338	315.934	80.171	-315.787	-244.503	5.321
2500	39.504	350.946	317.302	84.111	-317.667	-241.495	5.046
2600	39.746	352.500	318.626	88.073	-319.756	-238.407	4.790
2700	40.013	354.005	319.909	92.061	-322.040	-235.236	4.551
2800	40.304	355.466	321.153	96.076	-324.501	-231.976	4.328
2900	40.619	356.886	322.361	100.122	-327.119	-228.626	4.118
3000	40.956	358.268	323.535	104.201	-329.869	-225.183	3.921
3100	41.314	359.617	324.677	108.314	-332.725	-221.647	3.735
3200	41.692	360.935	325.789	112.464	-335.660	-218.017	3.559
3300	42.089	362.223	326.874	116.653	-338.641	-214.295	3.392
3400	42.502	363.486	327.932	120.883	-341.652	-210.482	3.234
3500	42.930	364.724	328.966	125.154	-344.668	-206.580	3.083
3600	43.372	365.940	329.976	129.469	-347.666	-202.592	2.940
3700	43.824	367.134	330.964	133.829	-350.628	-198.521	2.803
3800	44.286	368.309	331.932	138.234	-353.539	-194.371	2.672
3900	44.756	369.466	332.879	142.687	-356.384	-190.145	2.547
4000	45.231	370.605	333.808	147.186	-359.153	-185.848	2.427
4100	45.710	371.727	334.719	151.733	-361.836	-181.482	2.312
4200	46.191	372.835	335.614	156.328	-364.428	-177.051	2.202
4300	46.673	373.927	336.492	160.971	-366.928	-172.560	2.096
4400	47.154	375.006	337.355	165.662	-369.330	-168.012	1.995
4500	47.632	376.071	338.204	170.402	-371.634	-163.411	1.897
4600	48.106	377.123	339.038	175.189	-373.834	-158.760	1.803
4700	48.575	378.162	339.860	180.023	-375.916	-154.064	1.712
4800	49.037	379.190	340.668	184.903	-377.924	-149.322	1.625
4900	49.491	380.206	341.465	189.830	-379.832	-144.540	1.541
5000	49.937	381.210	342.250	194.801	-381.655	-139.720	1.460
5100	50.373	382.203	343.024	199.817	-383.368	-134.866	1.381
5200	50.798	383.186	343.786	204.876	-385.016	-129.977	1.306
5300	51.211	384.157	344.539	209.976	-386.584	-125.057	1.233
5400	51.612	385.118	345.282	215.117	-388.076	-120.109	1.162
5500	52.001	386.069	346.015	220.298	-389.462	-115.137	1.093
5600	52.376	387.009	346.738	225.517	-390.750	-110.145	1.027
5700	52.737	387.939	347.453	230.773	-392.022	-105.122	0.963
5800	53.084	388.860	348.159	236.064	-393.205	-100.082	0.901
5900	53.417	389.770	348.856	241.389	-394.209	-95.038	0.841
6000	53.735	390.670	349.546	246.747	-395.282	-89.958	0.783

PREVIOUS: December 1974 (1 atm) CURRENT: December 1974 (1 bar)

Barium Bromide (BaBr) $Ba_1Br_1(g)$

BARIUM BROMIDE ($BaBr_2$) CRYSTAL M_r = 297.138

$S°(298.15\ K) = [148.5 \pm 4.2]\ J\ K^{-1}\ mol^{-1}$

$T_{fus} = 1130 \pm 2\ K$

$\Delta_f H°(0\ K)$ = Unknown

$\Delta_f H°(298.15\ K) = -757.7 \pm 1.7\ kJ\ mol^{-1}$

$\Delta_{fus} H° = 31.96 \pm 0.63\ kJ\ mol^{-1}$

Enthalpy of Formation

The selected value is based on results obtained from solution calorimetry performed in aqueous acid media. Ehrlich et al. (1) reported measurements of the enthalpies of solution of Ba(cr) and $BaBr_2$(cr) in HBr · 555 H_2O. These results lead to $\Delta_f H°(BaBr_2$, cr, 298.15 K) = $-181.1 \pm 0.4\ kcal\ mol^{-1}$ ($-757.7 \pm 1.7\ kJ\ mol^{-1}$) when combined with the most recent thermal data (2, 3) for aqueous HBr. This value is adopted and is essentially that ($-181.0\ kcal\ mol^{-1}$) which has been selected by NBS (4).

Two independent calorimetric values in $kcal\ mol^{-1}$ of −180.7 (5) and −181.7 (6) bracket our adopted value. These values are based on results of early enthalpy of solution measurements (5, 6) which are combined with $\Delta_f H°(Ba^{+2}$, aq, 298.15 K) = −128.5 $kcal\ mol^{-1}$ (4) and $\Delta_f H°(Br^-$, aq, 298.15 K) = $-29.039 \pm 0.035\ kcal\ mol^{-1}$ (2). Use of the newly derived value (see BaO(cr) table) for $\Delta_f H°$ of Ba^{+2}(aq) makes these values less negative by 0.5 $kcal\ mol^{-1}$.

Heat Capacity and Entropy

$C_p°$ data needed to define S°(298.15 K) are unavailable. The adopted value S°(298.15 K) = $35.5 \pm 1.0\ cal\ K^{-1}\ mol^{-1}$ is a compromise between several estimates. JANAF (7) entropies for the other three barium dihalides suggest a value of S°(298.15 K) near 35.3 $cal\ K^{-1}\ mol^{-1}$ for the bromide. Application of the Berthelot principle (8) to the process $SrBr_2$(cr) + Ba(cr) → $BaBr_2$(cr) + Sr(cr) gives S°(298.15 K) = $36.7 \pm 1.7\ cal\ K^{-1}\ mol^{-1}$. Literature estimates which were considered are (in $cal\ K^{-1}\ mol^{-1}$) 35.0 (4), 35.5±5.0 (9), and 38 (10).

The adopted $C_p°$ data are obtained from JANAF curve fits of the relative enthalpies (825–1130 K) reported by Efremova and Matizen (11). For 27 enthalpy points the average deviation of the fit is ±0.4%; the maximum deviation is −1.0% at 847 K. Two points (1129 and 1130 K) are omitted from the fit, since these points presumably involve pre-melting effects. The derived value of $C_p°$(298.15 K) = 18.41 $cal\ K^{-1}\ mol^{-1}$ is supported by the value (18.48 $cal\ K^{-1} mol^{-1}$) obtained from the process $BaCl_2$(cr) + 2 KBr(cr) → $BaBr_2$(cr) + 2 KCl(cr) with $\Delta C_p° = 0$. Relative enthalpies (487–1126 K) reported in equation form by Janz et al. (12) suggest a somewhat lower value (17.8 $cal\ K^{-1}\ mol^{-1}$) for $C_p°$ at 298.15 K. Their smoothed enthalpies show deviations from our adopted ones that are generally less than ±0.3% but show maximum deviations of ~±1.0% at 500 K and 900 K. The enthalpy measurement of Dworkin and Bredig (13) at the melting point (T_{fus} = 1130 K) is roughly 1% lower than that which is adopted. The values of $C_p°$ above T_{fus} are obtained from the adopted enthalpy equation.

Fusion Data

See the liquid table for details.

Sublimation Data

$\Delta_{sub} H°$(298.15 K) is the difference in the values of $\Delta_f H°$ for the gas and crystal at 298.15 K.

References

1. P. Ehrlich, K. Peik, and E. Koch, Z. Anorg. Chem. 324, 113 (1963).
2. ICSU – CODATA Task Group, J. Chem. Thermodynamics 4, 331 (1972).
3. U. S. Nat. Bur. Stand. Tech. Note 270-3, 1968.
4. U. S. Nat. Bur. Stand. Tech. Note 270-6, 1971.
5. J. Thomsen, "Thermochemische Untersuchungen," Vol. III, Barth, Leipzig, 1883.
6. R. Varet, Ann. Chim. Phys. 8, 240 (1896).
7. JANAF Thermochemical Tables: BaO(cr), 6-30-74; BaF_2(cr) and $BaCl_2$(cr), 12-31-72; BaI_2(cr), 6-30-74.
8. N. N. Drozin, Zhur. Fiz. Khim. 35, 879 (1961).
9. O. Kubaschewski, E. Ll. Evans, and C. B. Alcock, "Metallurgical Thermochemistry," 4th ed., Pergamon Press, New York, 1967.
10. L. Brewer, G. R. Somayajulu, and E. Brackett, Chem. Revs. 63, 111 (1963).
11. R. I. Efremova and E. V. Matizen, Izv. Sib. Otd. Akad. Nauk SSSR, Ser. Khim. Nauk (1), 144 (1970).
12. G. J. Janz, F. J. Kelley, and J. L. Perano, Trans. Faraday Soc. 59, 2718 (1963).
13. A. S. Dworkin and M. A. Bredig, J. Phys. Chem. 67, 697 (1963).

Barium Bromide ($BaBr_2$) Ba_1Br_2(cr)

Enthalpy Reference Temperature = T_r = 298.15 K   Standard State Pressure = p° = 0.1 MPa

T/K	$C_p°$ (J K⁻¹mol⁻¹)	S° (J K⁻¹mol⁻¹)	−[G°−H°(T_r)]/T (J K⁻¹mol⁻¹)	H°−H°(T_r) (kJ mol⁻¹)	$\Delta_f H°$ (kJ mol⁻¹)	$\Delta_f G°$ (kJ mol⁻¹)	Log K_f
0							
100							
200							
298.15	77.015	148.532	148.532	0.	-757.722	-738.000	129.295
300	77.053	149.009	148.533	0.143	-757.772	-737.877	128.476
400	79.216	171.471	151.582	7.956	-787.455	-725.089	94.687
500	81.379	189.380	157.409	15.986	-786.935	-709.575	74.129
600	83.542	204.409	164.022	24.232	-787.031	-694.082	60.425
700	85.701	217.449	170.743	32.694	-786.122	-678.671	50.643
800	87.864	229.034	177.319	41.372	-785.523	-663.354	43.313
900	90.027	239.507	183.656	50.267	-784.299	-648.154	37.618
1000	92.190	249.105	189.727	59.377	-782.869	-633.101	33.070
1100	94.353	257.993	195.534	68.705	-789.601	-617.382	29.317
1130.000	95.002	260.540	197.226	71.545	------ CRYSTAL <--> LIQUID -----		
1200	96.512	266.295	201.089	78.248	-787.992	-601.795	26.195
1300	98.675	274.106	206.408	88.007	-786.076	-586.354	23.560
1400	100.839	281.498	211.510	97.983	-783.874	-571.071	21.307
1500	103.002	288.528	216.412	108.175	-781.409	-555.956	19.360

PREVIOUS: CURRENT: December 1974

Barium Bromide ($BaBr_2$) Ba_1Br_2(cr)

BARIUM BROMIDE ($BaBr_2$) LIQUID M_r = 297.138

$S°(298.15\ K)$ = [172.108] $J\ K^{-1}\ mol^{-1}$ $\Delta_f H°(298.15\ K)$ = [-730.192] $kJ\ mol^{-1}$

T_{fus} = 1130 ± 2 K $\Delta_{fus}H°$ = 31.96 ± 0.63 $kJ\ mol^{-1}$

Enthalpy of Formation

$\Delta_f H°$ of the liquid is calculated from that of the crystal by adding $\Delta_{fus}H°$ and the difference in enthalpy, H°(1130 K) - H°(298.15 K), between the crystal and liquid.

Heat Capacity and Entropy

$C_p°$ for the liquid is taken equal to 25.06 cal $K^{-1}mol^{-1}$ above the assumed glass transition temperature of 800 K. The constant value of $C_p°$ is derived from a curve fit of the relative enthalpies (1134 - 1963 K) reported by Efremova and Matizen (1). For 11 enthalpy points the average deviation of the fit is ±0.3%; the maximum deviation occurs at 1386 K and amounts to -0.7%. Our adopted value for $C_p°$ is supported by the enthalpy measurements of Dworkin and Bredig (2) who found $C_p°$ (ℓ) = 25.6 cal K^{-1} mol^{-1}. A considerably higher value (31 cal K^{-1} mol^{-1}) for $C_p°$(ℓ) is obtained from yet another enthalpy study (3); however, these latter measurements extend over a relatively short temperature interval (~100 K) above T_{fus} and are probably insufficient to accurately establish $C_p°$ of the liquid. $C_p°$ data below the glass transition are those of the crystal.

S°(298.15 K) is calculated in a manner analogous to that used for the enthalpy of formation.

Fusion Data

T_{fus} is the value determined in two independent drop-calorimetric investigations (1, 2). Nine other reported values for T_{fus} lie within 11 K of the adopted value. These other values are 1126 K (3-5), 1119 K (6, 7), 1128 K (8) and 1120 K (9-11).

$\Delta_{fus}H°$ is calculated as the difference between JANAF values for the enthalpies of the liquid and crystal at T_{fus}. Our value for $\Delta_{fus}H°$ is in excellent agreement with another drop-calorimetric value ($\Delta_{fus}H°$ = 7.63 kcal/mol, 2) and is further supported by a value ($\Delta_{fus}H°$ = 7.7 kcal mol^{-1}, 4) obtained by dynamic differential calorimetry. Other published values include (in kcal mol^{-1}) 7.75 ± 0.08 (1), 7.25 ± 0.19 (3), 7.15 ± 0.4 (4), and 7.0 (5).

Vaporization Data

T_{vap} is the temperature at which $\Delta_r G°$ = 0 for the process $BaBr_2$(ℓ) = $BaBr_2$(g) at p = 1 bar. $\Delta_{vap}H°$ is the corresponding difference in the $\Delta_f H°$ values for the gas and liquid at T_{vap}. Petersen and Hutchison (12) calculated T_{vap} = 2122 K (p = 1 atm) from their measured vapor pressure data which covered the range 1175 - 1321 K. This value is considered somewhat uncertain due to the rather long extrapolation.

References

1. R. I. Efremova and E. V. Matizen, Izv. Sib. Otd. Akad. Nauk SSSR, Ser. Khin. Nauk (1), 144 (1970).
2. A. S. Dworkin and M. A. Bredig, J. Phys. Chem. 67, 697 (1963).
3. G. J. Janz, F. J. Kelley, and J. L. Perano, Trans. Faraday Soc. 59, 2718 (1963).
4. H. H. Emons and B. Loffelholz, Wiss. Z. Tech. Hochsch. Chem., Leuna - Merseburg 6, 261 (1964).
5. M. Blanc, Compt. rend. 249, 2012 (1959).
6. E. D. Eastman, D. D. Cubicciotti, and C. D. Thurmond in "Chemistry and Metallurgy of Miscellaneous Materials: Thermodynamics," L. L. Quill, Ed., National Nuclear Energy Series IV-19B, McGraw - Hill Book Co., New York, 1950.
7. H. H. Emons, Z. Anorg. Chem. 323, 114 (1963).
8. R. Riccardi, C. Sinistri, G. V. Campari, and A. Magistris, Z. Naturforsch. 25a, 781 (1970).
9. G. Kellner, Z. Anorg. Chem 99, 137 (1917).
10. G. Kellner, J. Chem. Soc. 112, 469 (1917).
11. H. Schinke and F. Sauerwald, Z. Anorg. Chem. 304, 25 (1960).
12. D. T. Peterson and J. F. Hutchison, J. Chem. Eng. Data 15, 320 (1970).

Barium Bromide ($BaBr_2$) Ba_1Br_2(l)

Enthalpy Reference Temperature = T_r = 298.15 K Standard State Pressure = p° = 0.1 MPa

T/K	$C_p°$ (J K⁻¹mol⁻¹)	S° (J K⁻¹mol⁻¹)	-[G°-H°(T_r)]/T (J K⁻¹mol⁻¹)	H°-H°(T_r) (kJ mol⁻¹)	$\Delta_f H°$ (kJ mol⁻¹)	$\Delta_f G°$ (kJ mol⁻¹)	Log K_f
0							
100							
200							
298.15	77.015	172.108	172.108	0.000	-730.193	-717.500	125.703
300	77.053	172.585	172.109	0.143	-730.242	-717.421	124.914
400	79.216	195.047	175.158	7.956	-759.926	-706.990	92.323
500	81.379	212.956	180.985	15.986	-759.406	-693.834	72.484
600	83.542	227.985	187.598	24.232	-759.501	-680.698	59.260
700	85.701	241.025	194.319	32.694	-758.593	-667.645	49.820
800	87.866	252.610	200.895	41.372	-757.994	-654.685	42.747
800.000	87.866	252.610	200.895	41.372	GLASS <--> LIQUID TRANSITION		
800.000	104.851	252.610	200.895	41.372			
900	104.851	264.959	207.340	51.857	-755.179	-641.940	37.257
1000	104.851	276.007	213.664	62.342	-752.375	-629.509	32.882
1100	104.851	286.000	219.793	72.827	-757.949	-616.538	29.277
1130.000	104.851	288.821	221.589	75.973	------ CRYSTAL <--> LIQUID -----		
1200	104.851	295.123	225.696	83.313	-755.398	-603.794	26.282
1300	104.851	303.516	231.364	93.798	-752.756	-591.267	23.757
1400	104.851	311.286	236.798	104.283	-750.045	-578.946	21.601
1500	104.851	318.520	242.008	114.768	-747.287	-566.821	19.738
1600	104.851	325.287	247.004	125.253	-744.504	-554.881	18.115
1700	104.851	331.643	251.798	135.738	-741.717	-543.114	16.688
1800	104.851	337.637	256.402	146.223	-738.950	-531.512	15.424
1900	104.851	343.306	260.828	156.708	-736.225	-520.062	14.298
2000	104.851	348.684	265.087	167.193	-733.565	-508.754	13.287
2100	104.851	353.799	269.191	177.678	-730.993	-497.577	12.377
2200	104.851	358.677	273.148	188.164	-868.053	-481.145	11.424
2300	104.851	363.338	276.969	198.649	-864.890	-463.630	10.529
2400	104.851	367.800	280.661	209.134	-861.972	-446.248	9.712
2500	104.851	372.081	284.233	219.619	-859.302	-428.981	8.963
2600	104.851	376.193	287.691	230.104	-856.874	-411.818	8.274
2700	104.851	380.150	291.043	240.589	-854.678	-394.742	7.637
2800	104.851	383.963	294.294	251.074	-852.699	-377.745	7.047
2900	104.851	387.643	297.450	261.559	-850.916	-360.814	6.499
3000	104.851	391.197	300.516	272.044	-849.307	-343.942	5.989

PREVIOUS: CURRENT: December 1974

Barium Bromide ($BaBr_2$) Ba_1Br_2(l)

BARIUM BROMIDE ($BaBr_2$) CRYSTAL-LIQUID M_r = 297.138

0 to 1130 K crystal
above 1130 K liquid

Refer to the individual tables for details.

Enthalpy Reference Temperature = T_r = 298.15 K Standard State Pressure = $p°$ = 0.1 MPa

T/K	$C_p°$ (J K⁻¹mol⁻¹)	S° (J K⁻¹mol⁻¹)	$-[G°-H°(T_r)]/T$ (J K⁻¹mol⁻¹)	$H°-H°(T_r)$ (kJ mol⁻¹)	$\Delta_fH°$ (kJ mol⁻¹)	$\Delta_fG°$ (kJ mol⁻¹)	Log K_f
0							
100							
200							
298.15	77.015	148.532	148.532	0.	-757.722	-738.000	129.295
300	77.053	149.009	148.533	0.143	-757.772	-737.877	128.476
400	79.216	171.471	151.582	7.956	-787.455	-725.089	94.687
500	81.379	189.380	157.409	15.986	-786.935	-709.575	74.129
600	83.542	204.409	164.022	24.232	-787.031	-694.082	60.425
700	85.701	217.449	170.743	32.694	-786.122	-678.671	50.643
800	87.864	229.034	177.319	41.372	-785.523	-663.354	43.313
900	90.027	239.507	183.656	50.267	-784.299	-648.154	37.618
1000	92.190	249.105	189.727	59.377	-782.869	-633.101	33.070
1100	94.353	257.993	195.534	68.705	-789.602	-617.382	29.317
1130.000	95.002	260.540	197.226	71.545	CRYSTAL <--> LIQUID		
1130.000	104.851	288.821	197.226	103.502	TRANSITION		
1200	104.851	295.123	202.755	110.842	-755.398	-603.794	26.282
1300	104.851	303.516	210.187	121.327	-752.756	-591.267	23.757
1400	104.851	311.286	217.134	131.812	-750.045	-578.946	21.601
1500	104.851	318.520	223.655	142.297	-747.287	-566.821	19.738
1600	104.851	325.287	229.798	152.782	-744.504	-554.881	18.115
1700	104.851	331.644	235.604	163.268	-741.717	-543.114	16.688
1800	104.851	337.637	241.107	173.753	-738.950	-531.512	15.424
1900	104.851	343.306	246.338	184.238	-736.225	-520.062	14.298
2000	104.851	348.684	251.322	194.723	-733.565	-508.754	13.287
2100	104.851	353.799	256.081	205.208	-730.993	-497.577	12.377
2200	104.851	358.677	260.635	215.693	-868.053	-481.145	11.424
2300	104.851	363.338	265.000	226.178	-864.890	-463.630	10.529
2400	104.851	367.800	269.191	236.663	-861.972	-446.248	9.712
2500	104.851	372.081	273.221	247.148	-859.302	-428.982	8.963
2600	104.851	376.193	277.103	257.633	-856.874	-411.818	8.274
2700	104.851	380.150	280.847	268.119	-854.678	-394.742	7.637
2800	104.851	383.963	284.462	278.604	-852.699	-377.745	7.047
2900	104.851	387.643	287.957	289.089	-850.916	-360.814	6.499
3000	104.851	391.197	291.339	299.574	-849.307	-343.942	5.989

PREVIOUS: CURRENT: December 1974

BARIUM BROMIDE ($BaBr_2$) IDEAL GAS M_r = 297.138

$\Delta_f H°(0\ K) = -408.5 \pm 12.6\ kJ\ mol^{-1}$

$S°(298.15\ K) = [342.1 \pm 8.4]\ J\ K^{-1}\ mol^{-1}$ $\Delta_f H°(298.15\ K) = -424.7 \pm 12.6\ kJ\ mol^{-1}$

Vibrational Fequencies and Degeneracies

ν, cm^{-1}

[160] (1)
[28] (1)
[223] (1)

Ground State Quantum Weight: [1] σ = 2

Point Group: C_{2v}

Bond Distance: Ba-Br = 2.99 ± 0.03 Å

Bond Angle: Br-Ba-Br = [150±30]°

Product of Moments of Inertia: $I_A I_B I_C = [3.718284 \times 10^{-112}]\ g^3\ cm^6$

Enthalpy of Formation

The enthalpy of formation is calculated from the selected value for the enthalpy of vaporization, ($\Delta_{vap}H°(298.15\ K)$ = 73.04 ± 0.24 kcal mol^{-1}, and the adopted value for $\Delta_f H°(\ell)$. The selected value of $\Delta_{vap}H°$ is obtained from a 3rd law analysis of the vapor pressures for the liquid (1175-1321 K) reported by Peterson and Hutchison (1, 2). These measurements were made by the Knudsen effusion method on a sample of anhydrous $BaBr_2$ which had been prepared by direct union of high purity elements. The 2nd law $\Delta_{vap}H°$ = 73.7 kcal mol is in agreement with the 3rd law value; thus, the entropy deviation is small ($\Delta_{vap}S°$(2nd law) $-\Delta_{vap}S°$(3rd law) = 0.5 ± 0.8 cal $K^{-1}\ mol^{-1}$). The dominant vapor species is assumed to be the dihalide monomer. This assumption is supported by the fact that negligible concentrations of polymeric species have been observed for some of the other alkaline-earth dihalides by use of mass spectrometry (see $CaBr_2$(g) table (9)).

A 3rd law analysis of a single sublimation pressure reported by Stock and Heynemann (3) leads to $\Delta_f H°(298.15\ K)$ = 104.3 kcal mol^{-1}. This result suggests a possible uncertainty in the adopted value for $\Delta_f H°$ of as much as ± 3 kcal mol^{-1}. This also brings our value in agreement with the selection (-105 kcal mol^{-1}) of NBS (4).

Heat Capacity and Entropy

Experimental evidence which has been presented in favor of a linear or nonlinear structure for $BaBr_2$ is conflicting. Electron diffraction analysis by Akishin and Spiridonov (5) showed $BaBr_2$ to be linear (180 ± 30°). Later studies by Klemperer et al. (6, 7), using electric quadrupole deflection of molecular beams, have shown that this molecule possesses a permanent dipole moment and therefore must be bent. A simple explanation of the observed trends in the geometries of the alkaline-earth dihalides as established by the electric deflection experiments (6, 7) has been presented by Hayes (8), and it appears quite likely that $BaBr_2$ is probably bent. The bond angle is arbitrarily taken to be greater than that for $BaCl_2$ (100°, 9) but less than that for BaI_2 (170°, 9). The bond distance is that measured in the gas phase by an electron diffraction study of Akishin and Spiridonov (5). The principal moments of inertia are: $I_A = 228.6976 \times 10^{-39}$, $I_B = 221.3525 \times 10^{-39}$, and $I_C = 7.3451 \times 10^{-39}\ g\ cm^2$.

The vibrational frequencies are calculated from estimated force constants by the valence force method (10). The stretching force constant (K = 1.117×10^5 dynes/cm) is taken equal to that for BaBr (9). The ratio of the stretching to bending force constants is estimated as 125. This value for the ratio is to be compared with that for BaF_2 (93, 9) and $BaCl_2$ (119, 9). The uncertainty in the adopted value for ν_2 is believed to be no greater than ± 20 cm^{-1} which corresponds to roughly 1 cal $K^{-1}\ mol^{-1}$ in the value of S°(298.15 K). Other sets (11-13) of estimated vibrational frequencies compare favorably with ours; the maximum deviation is 35 cm^{-1} in ν_3 (13). The ground state quantum weight of one is assigned by analogy with that for $BaCl_2$ (9). Our Gibbs energy functions differ by roughly 2 cal $K^{-1}\ mol^{-1}$ in the range 298-2000 K from those given by Brewer et al. (11). However, their values are based on a linear structure for the bromide which now appears to be incorrect.

References

1. D. T. Peterson and J. F. Hutchison, J. Chem. Eng. Data 15, 320 (1970).
2. J. F. Hutchison, Ames Laboratory, Iowa State University, AEC Contract No. W-7405-eng-82, Report No. IS-T-50, November, 1965.
3. A. Stock and H. Heynemann, Chem. Ber. 42, 4088 (1909).
4. U. S. Nat. Bur. Stand. Tech. Note 270-6, 1971.
5. P. A. Akishin and V. P. Spiridonov, Kristallografiya 2, 475 (1957).
6. L. Wharton, R. A. Berg, and W. Klemperer, J. Chem. Phys. 39, 2023 (1963).
7. A. Buchler, J. L. Stauffer, and W. Klemperer, J. Amer. Chem. Soc. 86, 4544 (1964).
8. E. F. Hayes, J. Phys. Chem. 70, 3740 (1966).
9. JANAF Thermochemical Tables: $BaCl_2$(g) and BaF_2(g), 12-31-72; BaI_2(g), 6-30-74; BaBr(g), 12-31-74; $CaBr_2$(g), 6-30-74.
10. G. Herzberg, "Infrared and Raman Spectra of Polyatomic Molecules," D. Van Nostrand Company, Inc., New York, 1962.
11. L. Brewer, G. R. Somayajulu, and E. Brackett, Chem. Revs. 63, 111 (1963).
12. K. S. Krasnov and V. I. Svettsov, Izv. Vysshikh Uchebn. Zavedenii, Khim. i Khim. Tekhnol. 6, 167 (1963).
13. V. I. Baikov, Opt. Spektrosk. 27, 502 (1969).

Barium Bromide ($BaBr_2$) $Ba_1Br_2(g)$

Enthalpy Reference Temperature = T_r = 298.15 K Standard State Pressure = p° = 0.1 MPa

T/K	$C_p°$ (J K⁻¹mol⁻¹)	S° (J K⁻¹mol⁻¹)	$-[G°-H°(T_r)]/T$ (J K⁻¹mol⁻¹)	$H°-H°(T_r)$ (kJ mol⁻¹)	$\Delta_f H°$ (kJ mol⁻¹)	$\Delta_f G°$ (kJ mol⁻¹)	Log K_f
0	0.	0.	INFINITE	-15.255	-408.509	-408.509	INFINITE
100	50.661	282.411	391.697	-10.929	-408.675	-428.123	223.628
200	55.737	319.525	347.252	-5.545	-410.737	-446.830	116.700
250	56.566	332.061	343.004	-2.736	-412.082	-455.703	95.214
298.15	57.029	342.067	342.067	0.	-424.676	-462.656	81.055
300	57.043	342.420	342.068	0.106	-424.762	-462.891	80.597
350	57.340	351.237	342.764	2.966	-455.999	-467.523	69.774
400	57.536	358.908	344.313	5.838	-456.527	-469.135	61.263
450	57.673	365.693	346.319	8.718	-457.265	-470.669	54.634
500	57.772	371.775	348.566	11.604	-458.270	-472.107	49.321
600	57.901	382.320	353.339	17.389	-460.827	-474.626	41.320
700	57.980	391.252	358.133	23.183	-462.587	-476.798	35.579
800	58.031	398.998	362.768	28.984	-464.865	-478.667	31.254
900	58.067	405.835	367.181	34.789	-466.731	-480.280	27.875
1000	58.092	411.954	371.357	40.597	-468.603	-481.685	25.161
1100	58.111	417.492	375.304	46.407	-478.853	-482.083	22.892
1200	58.125	422.549	379.033	52.219	-480.975	-482.282	20.993
1300	58.136	427.202	382.562	58.032	-483.005	-482.308	19.379
1400	58.145	431.511	385.906	63.846	-484.965	-482.180	17.990
1500	58.152	435.522	389.082	69.661	-486.877	-481.915	16.782
1600	58.158	439.276	392.103	75.477	-488.763	-481.522	15.720
1700	58.163	442.802	394.983	81.293	-490.646	-481.012	14.780
1800	58.167	446.126	397.732	87.109	-492.547	-480.390	13.941
1900	58.171	449.271	400.363	92.926	-494.490	-479.662	13.187
2000	58.174	452.255	402.884	98.743	-496.498	-478.830	12.506
2100	58.176	455.094	405.303	104.561	-498.594	-477.896	11.887
2200	58.178	457.800	407.628	110.378	-640.321	-471.484	11.194
2300	58.180	460.386	409.866	116.196	-641.825	-463.777	10.533
2400	58.182	462.862	412.023	122.015	-643.574	-455.999	9.925
2500	58.183	465.237	414.104	127.833	-645.571	-448.143	9.363
2600	58.185	467.520	416.115	133.651	-647.810	-440.202	8.844
2700	58.186	469.715	418.060	139.470	-650.280	-432.171	8.361
2800	58.187	471.832	419.943	145.288	-652.968	-424.045	7.911
2900	58.188	473.873	421.768	151.107	-655.851	-415.818	7.490
3000	58.189	475.846	423.537	156.926	-658.908	-407.490	7.095
3100	58.190	477.754	425.256	162.745	-662.114	-399.057	6.724
3200	58.190	479.602	426.925	168.564	-665.444	-390.518	6.375
3300	58.191	481.392	428.549	174.383	-668.863	-381.874	6.045
3400	58.191	483.129	430.129	180.202	-672.357	-373.126	5.732
3500	58.192	484.816	431.667	186.021	-675.902	-364.273	5.436
3600	58.192	486.456	433.167	191.840	-679.474	-355.319	5.156
3700	58.193	488.050	434.628	197.660	-683.055	-346.265	4.888
3800	58.193	489.602	436.055	203.479	-686.629	-337.115	4.634
3900	58.194	491.113	437.447	209.298	-690.182	-327.871	4.391
4000	58.194	492.587	438.807	215.118	-693.703	-318.536	4.160
4100	58.194	494.024	440.137	220.937	-697.182	-309.114	3.938
4200	58.195	495.426	441.436	226.757	-700.612	-299.608	3.726
4300	58.195	496.795	442.708	232.576	-703.991	-290.020	3.523
4400	58.195	498.133	443.953	238.396	-707.315	-280.354	3.328
4500	58.195	499.441	445.171	244.215	-710.580	-270.614	3.141
4600	58.196	500.720	446.365	250.035	-713.780	-260.802	2.961
4700	58.196	501.972	447.535	255.854	-716.901	-250.923	2.789
4800	58.196	503.197	448.682	261.674	-719.984	-240.976	2.622
4900	58.196	504.397	449.806	267.493	-723.003	-230.966	2.462
5000	58.196	505.573	450.910	273.313	-725.972	-220.894	2.308
5100	58.197	506.725	451.993	279.133	-728.865	-210.765	2.159
5200	58.197	507.855	453.057	284.952	-731.725	-200.579	2.015
5300	58.197	508.964	454.101	290.772	-734.536	-190.337	1.876
5400	58.197	510.052	455.127	296.592	-737.300	-180.043	1.742
5500	58.197	511.119	456.136	302.412	-739.988	-169.703	1.612
5600	58.197	512.168	457.127	308.231	-742.605	-159.317	1.486
5700	58.198	513.198	458.101	314.051	-745.231	-148.878	1.364
5800	58.198	514.210	459.060	319.871	-747.794	-138.397	1.246
5900	58.198	515.205	460.003	325.691	-750.200	-127.888	1.132
6000	58.198	516.183	460.932	331.510	-752.697	-117.319	1.021

PREVIOUS: December 1974 (1 atm) CURRENT: December 1974 (1 bar)

Barium Bromide ($BaBr_2$) $Ba_1Br_2(g)$

BARIUM CHLORIDE (BaCl) IDEAL GAS M_r = 172.783

$\Delta_f H°(0\ K) = -140.6 \pm 12.6\ kJ\ mol^{-1}$

$S°(298.15\ K) = 258.64 \pm 0.4\ J\ K^{-1}\ mol^{-1}$ $\Delta_f H°(298.15\ K) = -142.3 \pm 12.6\ kJ\ mol^{-1}$

Electronic Levels and Quantum Weights

State	ε_i, cm^{-1}	g_i	State	ε_i, cm^{-1}	g_i
$X\ ^2\Sigma^+$	0.0	2	$D\ ^2\Sigma$	25471.5	2
$A\ ^2\Pi$	10351.1	2	$E\ ^2\Sigma$	27064.6	2
	10995.3	2	$F\ ^2\Sigma$	29493.6	2
$B\ ^2\Sigma^+$	11880.0		$G\ ^2\Sigma$	32511.4	2
$C\ ^2\Pi$	19062.9	2			
	19450.1	2			

$\omega_e = 277.9\ cm^{-1}$ $\omega_e x_e = 0.88\ cm^{-1}$ $\sigma = 1$

$B_e = [0.08518 \pm 0.003]\ cm^{-1}$ $\alpha_e = [0.00035]\ cm^{-1}$ $r_e = [2.65 \pm 0.05]$ Å

Enthalpy of Formation

$\Delta_f H°$ is based mainly on two sets of gas-phase equilibria measured mass spectrometrically by Hildenbrand (1) and, to a lesser extent, on the analogous data of Potter et al. (2). Analyses of the mass-spectrometric data (1-3) are summarized below. The results from Potter et al. (2) and Zmbov (3) are less negative by about 4 kcal mol^{-1}, suggesting the approximate uncertainty in $\Delta_f H°$. From the adopted $\Delta_f H°$ we derive D_0^0 = 106.4 ± 3 kcal mol^{-1}. Independent methods yield D_0^0 = 103 ± 5 (4), ≥110 (5), 106.8 ± 2 (6) and 105 ± 6 kcal mol^{-1} (6).

Electron-impact data (4) gave D_0^0 (ClBa-Cl) = 115.8 ± 3.3 kcal mol^{-1} and we use $\Delta_{at}H°(BaCl_2)$ = 219 ± 4 kcal mol^{-1} to derive D_0^0 = 103 ± 5 kcal mol^{-1}. Chemiluminescence from crossed beams of Ba atoms with Cl_2 gave an approximate lower limit to D_0^0 of 110 kcal mol^{-1} (5). Flame studies (6) gave D_0^0 = 106.8 ± 2 kcal mol^{-1} and D_0^0(ClBa-Cl) = 114 ± 4; we derive D_0^0 = 105 ± 6 kcal mol^{-1} from the latter value using $\Delta_{at}H°(BaCl_2)$ = 219 kcal mol^{-1}. Ryabova (6) considered the flames to involve $BaCl_2$, BaOHCl, and BaOH as well as BaCl. Schofield and Sugden (7) inferred that the flames involved mainly $BaCl_2$ rather than BaCl as assumed in earlier flame studies. Gaydon (8) discounted the values derived from electronic spectra, but Hildenbrand (1) showed that ionicity corrections bring the Birge-Sponer extrapolation into reasonable agreement with D_0^0.

Source	Reaction[a]	T/K	$\Delta_r H°$(298.15 K), kcal mol^{-1} 2nd law	3rd law	δS cal K^{-1} mol^{-1}	$\Delta_f H°$(298.15 K) kcal mol^{-1}	D_0^0
(1)	A	1234-1380	3.3±3.6	6.76±0.8	-2.6±2.7	-34.82±2.2	107.2
(2)	A	1507-1702	13.0±5.4	13.34±1.0	-0.2±3.3	-31.53±2.2	103.9
(3)[b]	A	974-1052	21.8±9.4	5.89±1.3	5.9±9.3	-30.26±2.2	102.7
(1)	B	1234-1380	10.3±5.1	13.21±1.1	-2.2±3.9	-34.29±2.6	106.7
(2)	B	1507-1702	9.7±7.3	17.72±1.5	-5.0±4.5	-29.78±2.8	102.2

a Reactions: (A) $Ba(g) + BaCl_2(g) \rightarrow 2BaCl(g)$; (B) $Ba(g) + AlCl(g) \rightarrow BaCl(g) + Al(g)$.

b Data given in Table 2 (3) are assumed to be 10^3 x Kp instead of 10^4 x Kp.

Heat Capacity and Entropy

Electronic levels and vibrational constants are taken from the review of Rosen (9). The level at 10351.1 cm^{-1} is T_0 rather than T_e. Vibrational constants are adjusted for natural isotopic abundances. B_e is calculated from r_e which is estimated by comparison of the bond lengths of MX_2(M=Mg, Ca, Sr, Ba and X=F, Cl) with those of MgCl, Cacl and MF(M=Mg, Ca, Sr, Ba). The value of α_e is estimated from the Morse potential function.

References

1. D. L. Hildenbrand, J. Chem. Phys. 52, 5751 (1970).
2. N. D. Potter, M. H. Boyer, F. Ju, D. L. Hildenbrand, E. Murad and W. F. Hall, AD-715567, 17-26 (1970); Aeronutronic Div., Philco Corp., Report U-4859, AFOSR 70-2311 TR, Aug. 31, 1970.
3. K. F. Zmbov, Chem. Phys. Letters 4, 191 (1969).
4. D. L. Hildenbrand, Int. J. Mass Spectrom. Ion Phys. 4, 75 (1970).
5. C. D. Jonah and R. N. Zare, Chem. Phys. Letters 9, 65 (1971).
6. V. G. Ryabova, Teplofiz. Vys. Temp. 10, 744 (1972).
7. K. Schofield and T. M. Sugden, Trans. Faraday Soc. 67, 1054 (1971).
8. A. G. Gaydon, "Dissociation Energies and Spectra of Diatomic Molecules," 3rd Edition, Chapman and Hall, Ltd., London, 1968.
9. B. Rosen, Ed., "Spectroscopic Data Relative to Diatomic Molecules," Pergamon Press, New York, 1970.

Barium Chloride (BaCl) $Ba_1Cl_1(g)$

Enthalpy Reference Temperature = T_r = 298.15 K Standard State Pressure = p° = 0.1 MPa

T/K	$C_p°$ (J K^{-1}mol^{-1})	S°	$-[G°-H°(T_r)]/T$	$H°-H°(T_r)$ (kJ mol^{-1})	$\Delta_f H°$	$\Delta_f G°$	Log K_f
0	0.	0.	INFINITE	-9.878	-140.631	-140.631	INFINITE
100	31.691	221.045	290.084	-6.904	-140.812	-150.058	78.382
200	35.235	244.313	261.945	-3.526	-141.498	-159.035	41.536
250	35.986	252.263	259.241	-1.744	-141.864	-163.379	34.136
298.15	36.436	258.643	258.643	0.	-142.256	-167.488	29.343
300	36.450	258.868	258.644	0.067	-142.272	-167.644	29.189
350	36.757	264.512	259.089	1.898	-142.747	-171.837	25.645
400	36.973	269.435	260.081	3.742	-143.350	-175.953	22.977
450	37.132	273.799	261.367	5.594	-144.162	-179.983	20.892
500	37.256	277.718	262.810	7.454	-145.241	-183.907	19.213
600	37.438	284.528	265.879	11.189	-147.944	-191.378	16.661
700	37.570	290.309	268.966	14.940	-149.844	-198.478	14.811
800	37.676	295.333	271.955	18.703	-152.258	-205.255	13.402
900	37.765	299.776	274.804	22.475	-154.253	-211.760	12.290
1000	37.846	303.759	277.504	26.255	-156.247	-218.043	11.389
1100	37.921	307.370	280.057	30.044	-166.610	-223.306	10.604
1200	37.995	310.673	282.473	33.840	-168.837	-228.361	9.940
1300	38.072	313.717	284.761	37.643	-170.962	-233.235	9.371
1400	38.156	316.541	286.931	41.454	-173.007	-237.948	8.878
1500	38.252	319.177	288.994	45.275	-174.990	-242.517	8.445
1600	38.364	321.649	290.959	49.105	-176.932	-246.955	8.062
1700	38.496	323.979	292.833	52.948	-178.851	-251.273	7.721
1800	38.652	326.184	294.625	56.805	-180.767	-255.478	7.414
1900	38.835	328.278	296.342	60.679	-182.698	-259.576	7.136
2000	39.046	330.275	297.989	64.573	-184.662	-263.571	6.884
2100	39.287	332.186	299.572	68.489	-186.679	-267.467	6.653
2200	39.557	334.020	301.097	72.431	-328.289	-265.891	6.313
2300	39.856	335.785	302.567	76.402	-329.632	-263.025	5.973
2400	40.183	337.488	303.986	80.404	-331.174	-260.097	5.661
2500	40.535	339.135	305.360	84.439	-332.915	-257.100	5.372
2600	40.909	340.732	306.689	88.511	-334.848	-254.030	5.104
2700	41.304	342.284	307.979	92.622	-336.962	-250.882	4.854
2800	41.716	343.793	309.231	96.773	-339.239	-247.653	4.620
2900	42.143	345.264	310.449	100.966	-341.661	-244.339	4.401
3000	42.581	346.700	311.633	105.202	-344.205	-240.940	4.195
3100	43.027	348.104	312.787	109.482	-346.846	-237.455	4.001
3200	43.478	349.477	313.912	113.807	-349.561	-233.883	3.818
3300	43.933	350.822	315.010	118.178	-352.319	-230.226	3.644
3400	44.387	352.140	316.083	122.594	-355.106	-226.485	3.480
3500	44.839	353.433	317.132	127.055	-357.898	-222.661	3.323
3600	45.287	354.703	318.158	131.561	-360.677	-218.758	3.174
3700	45.728	355.950	319.163	136.112	-363.425	-214.777	3.032
3800	46.161	357.175	320.147	140.707	-366.128	-210.724	2.897
3900	46.585	358.380	321.112	145.344	-368.775	-206.599	2.767
4000	46.998	359.564	322.058	150.023	-371.356	-202.408	2.643
4100	47.399	360.730	322.987	154.743	-373.864	-198.153	2 525
4200	47.787	361.877	323.900	159.503	-376.294	-193.838	2.411
4300	48.162	363.005	324.796	164.300	-378.646	-189.466	2.302
4400	48.523	364.117	325.677	169.135	-380.916	-185.040	2.197
4500	48.869	365.211	326.543	174.005	-383.103	-180.564	2.096
4600	49.200	366.289	327.396	178.908	-385.203	-176.040	1.999
4700	49.516	367.350	328.235	183.844	-387.203	-171.474	1.906
4800	49.817	368.396	329.060	188.811	-389.145	-166.863	1.816
4900	50.103	369.426	329.874	193.807	-391.005	-162.213	1.729
5000	50.374	370.441	330.675	198.831	-392.798	-157.526	1.646
5100	50.631	371.441	331.465	203.881	-394.499	-152.806	1.565
5200	50.873	372.427	332.243	208.957	-396.152	-148.050	1.487
5300	51.101	373.398	333.010	214.055	-397.741	-143.264	1.412
5400	51.315	374.355	333.767	219.176	-399.272	-138.448	1.339
5500	51.515	375.299	334.514	224.318	-400.714	-133.608	1.269
5600	51.703	376.229	335.250	229.479	-402.074	-128.746	1.201
5700	51.878	377.145	335.977	234.658	-403.433	-123.853	1.135
5800	52.040	378.049	336.695	239.854	-404.719	-118.940	1.071
5900	52.191	378.940	337.403	245.066	-405.840	-114.022	1.009
6000	52.331	379.818	338.103	250.292	-407.044	-109.066	0.950

PREVIOUS: December 1972 (1 atm) CURRENT: December 1972 (1 bar)

Barium Chloride (BaCl) $Ba_1Cl_1(g)$

BARIUM CHLORIDE ($BaCl_2$) CRYSTAL M_r = 208.236

$S°(298.15\ K) = 123.666 \pm 0.13\ J\ K^{-1}\ mol^{-1}$
$T_{trs} = 1198 \pm 1\ K$
$T_{fus} = 1235 \pm 1\ K$

$\Delta_f H°(0\ K) = -859.17 \pm 13\ (or \pm 2.5)\ kJ\ mol^{-1}$
$\Delta_f H°(298.15\ K) = -858.6 \pm 13\ (or \pm 2.5)\ kJ\ mol^{-1}$
$\Delta_{trs} H° = 16.90 \pm 0.21\ kJ\ mol^{-1}$
$\Delta_{fus} H° = 15.991 \pm 0.33\ kJ\ mol^{-1}$

Enthalpy of Formation

$\Delta_f H°$ is from NBS Technical Note 270-6 (1) and its selection was discussed in detail by Parker (2). The adopted value was derived by Ehrlich et al. (3) from their measurements of ΔH_{sol} of Ba(cr) and $BaCl_2$(cr) in HCl (550 H_2O). Earlier solution data for Ba (4) and $BaCl_2$ (5, 6) are in good agreement (2), but a recent study by Vorob'ev et al. (7) of Ba in HCl (240 and 400 H_2O) yields $\Delta_f H° = -201.8\ kcal\ mol^{-1}$. The deviation (+3.4 kcal mol^{-1}) from the adopted value is confirmed by direct measurements (7) of $\Delta_f H°$ and $\Delta_{sol} H°$ of BaH_2(cr), yielding -202.1 kcal mol^{-1} for $BaCl_2$(cr). All authors reported difficulties in obtaining or maintaining Ba and BaH_2 free of impurities; this provides conceivable reasons for bias in either of the opposing results. We assign an uncertainty of ±3 kcal mol^{-1} to include the possibility that Vorob'ev et al. (7) are correct but add an alternative of ±0.6 kcal mol^{-1} if their data are biased.

Parker (2) rejected $\Delta_f H° = -219.3\ kcal\ mol^{-1}$ obtained by Siemonsen (8) by direct chlorination and $\Delta_f H° = -212.0$ kcal mol^{-1} which would result from BaO(cr) based on Mah's direct combustion (9). Parker noted that these more negative values are incompatible with data for BaO_2, $Ba(OH)_2$ and $BaCO_3$. They are also incompatible with calorimetric data for BaH_2 noted above and equilibrium data as discussed on the tables for $BaCl_2$(ℓ and g).

Heat Capacity and Entropy

$C_p°$ is based on adiabatic calorimetry (6-346 K) of Goodman and Westrum (10). S° is calculated from $C_p°$ using S°(6 K) = 0.012 cal $K^{-1}\ mol^{-1}$. $C_p°$ above 300 K is from constrained fitting of enthalpy data (340-1197 K for α-phase and 1201-1231 K for β-phase) measured in an ice calorimeter by Gardner and Taylor (11). The sample studied at high temperatures was a portion of that used at low temperatures. Deviations of the H data (11) from the adopted values are <0.1% for most points but show a maximum of -0.3% for two points near 400 K.

Other enthalpy studies include those of Dworkin and Bredig (12), Janz et al. (13) and Gant (14). Deviations from the adopted values are about +2% for the smooth values (~1100-1230 K) of (12), +1.8 to +2.9% for the equations (892-1235 K) of (13), and -1.7 to +6.7% for the data (406-1230 K) of (14). Popov and Gal'chenko (15) used a continuous heating method to obtain $C_p°$ values which deviate by +4 to +9% (443-973 K).

Transition Data

T_{trs} = 1198 K is from the enthalpy study of Gardner and Taylor (11). Other enthalpy studies gave 1193 (12) and 1198 K (13), while thermal analysis gave 1199 (16) and 1197 K (17). $\Delta_{trs}H°$ is calculated from our adopted fits of the enthalpy data (11). Published values include 4.04 (11), 4.10 (12) and 4.15±0.16 (13).

Fusion Data

See the liquid table for details.

Sublimation Data

See the ideal gas table for details.

References

1. V. B. Parker, D. D. Wagman and W. H. Evans, U. S. Nat. Bur. Stand. Tech. Note 270-6, 1971.
2. V. B. Parker, U. S. Nat. Bur. Stand. Rept. 10074, 164 (1969).
3. P. Ehrlich, K. Peik and E. Koch, Z. Anorg. Allg. Chem. 324, 113 (1963).
4. A. Guntz and F. Benoit, Ann. Chim. (Paris) [9] 20, 5 (1923).
5. O. Ya. Samoilov, Izv. Akad. Nauk SSSR, Otdel. Khim. Nauk 1952, 627; 1950, 337.
6. F. E. Young, J. Amer. Chem. Soc. 66, 773 (1944).
7. A. F. Vorob'ev, A. S. Monaenkova and S. M. Skuratov, Vestn. Mosk. Univ., Ser. II 22 (6), 3 (1967); Dokl. Akad. Nauk SSSR 179, 1129 (1968); Izv. Vyssh. Ucheb. Zaved., Khim. i Khim. Tekhnol. 15, 191 (1972).
8. H. Siemonson, Z. Elektrochem. 55, 327 (1951).
9. A. D. Mah, U. S. Bur. Mines RI-6171, 1963.
10. R. M. Goodman and E. F. Westrum, J. Chem. Eng. Data 11, 294 (1966).
11. T. E. Gardner and A. R. Taylor, J. Chem. Eng. Data 14, 281 (1969).
12. A. S. Dworkin and M. A. Bredig, J. Phys. Chem. 67, 697 (1963).
13. G. J. Janz, F. J. Kelly and J. L. Perano, Trans. Faraday Soc. 59, 2718 (1963).
14. F. A. Gant, Ph.D. Thesis, Univ. Alabama, 1967; Diss. Abstr. B28, 133 (1967).
15. M. M. Popov and G. L. Gal'chenko, Zh. Obshch. Khim. 21, 2220 (1951).
16. K. Grjotheim, H. G. Nebell and J. Krogh-Moe, Acta Chem. Scand. 22, 1159 (1968).
17. C. Krohn, Acta Chem. Scand. 20, 255 (1966).

Barium Chloride ($BaCl_2$) Ba_1Cl_2(cr)

Enthalpy Reference Temperature = T_r = 298.15 K; Standard State Pressure = p° = 0.1 MPa

	J $K^{-1}mol^{-1}$			kJ mol^{-1}			
T/K	$C_p°$	S°	$-[G°-H°(T_r)]/T$	$H°-H°(T_r)$	$\Delta_f H°$	$\Delta_f G°$	Log K_f
0	0.	0.	INFINITE	-16.707	-859.171	-859.171	INFINITE
100	56.961	49.346	187.175	-13.783	-860.857	-843.483	440.590
200	71.038	94.429	130.426	-7.199	-859.856	-826.420	215.839
298.15	75.140	123.666	123.666	0.	-858.557	-810.290	141.959
300	75.195	124.131	123.668	0.139	-858.533	-809.990	141.032
400	77.329	146.082	126.646	7.774	-857.384	-793.994	103.685
500	78.857	163.505	132.335	15.585	-856.963	-778.216	81.300
600	80.425	178.018	138.772	23.547	-857.255	-762.428	66.375
700	82.233	190.548	145.293	31.678	-856.611	-746.684	55.718
800	84.299	201.661	151.657	40.003	-856.313	-730.992	47.729
900	86.624	211.721	157.780	48.546	-855.396	-715.380	41.520
1000	89.471	220.988	163.643	57.345	-854.241	-699.882	36.558
1100	93.531	229.692	169.256	66.480	-861.132	-683.694	32.466
1198.010	100.015	237.924	174.538	75.937	ALPHA <--> BETA	TRANSITION	
1198.010	123.846	252.033	174.538	92.840			
1200	123.846	252.239	174.666	93.087	-842.429	-667.664	29.063
1235.000	123.846	255.799	176.915	97.422	------- BETA	<--> LIQUID -------	
1300	123.846	262.152	181.020	105.472	-837.859	-653.284	26.249
1400	123.846	271.330	187.147	117.856	-833.221	-639.260	23.851
1500	123.846	279.874	193.047	130.241	-828.535	-625.569	21.784
1600	123.846	287.867	198.726	142.625	-823.823	-612.191	19.986
1700	123.846	295.375	204.193	155.010	-819.105	-599.109	18.408
1800	123.846	302.454	209.457	167.395	-814.402	-586.304	17.014
1900	123.846	309.150	214.529	179.779	-809.736	-573.759	15.774
2000	123.846	315.503	219.421	192.164	-805.128	-561.459	14.664

PREVIOUS: CURRENT: December 1972

Barium Chloride ($BaCl_2$) Ba_1Cl_2(cr)

BARIUM CHLORIDE ($BaCl_2$) LIQUID M_r = 208.236

$S°(298.15\ K) = [143.518]\ J\ K^{-1}\ mol^{-1}$ $\Delta_f H°(298.15\ K) = [-832.468]\ kJ\ mol^{-1}$

$T_{fus} = 1235 \pm 1\ K$ $\Delta_{fus}H° = 15.991 \pm 0.33\ kJ\ mol^{-1}$

Enthalpy of Formation

$\Delta_f H°$ is calculated from that of the crystal by addition of $\Delta_{fus}H°$ and the difference in enthalpy, H°(1235 K) −H°(298.15 K), between crystal and liquid. Independent values of $\Delta_f H°$ for liquid (and crystal) may be derived from equilibrium data (1, 2) for Ca(ℓ) + $BaCl_2$(ℓ) → Ba(ℓ) + $CaCl_2$(ℓ). Our 3rd-law analysis is summarized below; it shows that the data of Feschotte-Ostertag (1) are consistent with the adopted $\Delta_f H°(cr) = -205.2\ kcal\ mol^{-1}$ from Ehrlich et al. (cf. $BaCl_2$, cr), while the data of Peterson and Hinkebein (2) are more consistent with the alternative $\Delta_f H°(cr) = -201.8\ kcal\ mol^{-1}$ from Vorob'ev et al. Both are incompatible with $\Delta_f H°(cr) = -219.3$ from direct chlorination and $\Delta_f H°(cr) = -212.0$ derived from Mah's ΔHc value for $\Delta_f H°$ of BaO(cr).

Source	Remarks	T/K	δS cal K^{-1} mol^{-1}	$\Delta_r H°$(298.15 K) kcal mol^{-1}	$\Delta_f H°$(ℓ) kcal mol^{-1}	$\Delta_f H°$(cr) kcal mol^{-1}
(1)	Direct measurement	1273		13.24±1.6	-198.3±2	-204.5±2
(1)	Indirect via $SrCl_2$	1273		14.12±1.8	-199.1±2	-205.4±2
(2)	Direct measurement	1173-1223	-2.5±8.1	11.47±0.6	-196.5±1	-202.7±1

Heat Capacity and Entropy

The constant $C_p° = 26.0\ cal\ K^{-1}\ mol^{-1}$ is based on enthalpy data (1238-1347 K) measured in an ice calorimeter by Gardner and Taylor (3). Deviations of the data from the adopted fit are all less than 0.05%. Smoothed enthalpies of Dworkin and Bredig (4) deviate by +1.2% and those of Janz et al. (5) by +2.7 to 2.1%. Enthalpy data of Gant (6) deviate by +1.5%. $C_p°$ is taken equal to that of the crystal from 298 to 800 K, the assumed glass transition. S° is calculated in a manner analogous to that used for $\Delta_f H°$.

Fusion Data

T_{fus} is from the enthalpy study of Gardner and Taylor (3). Some other reported values include 1233 (4) and 1235 K (5, 6) from enthalpy studies and 1234 (7) and 1235 K (8) from thermal analysis. $\Delta_{fus}H°$ is obtained from the difference in the adopted enthalpy fits for liquid and β-phases. Published values include 3.82 (3), 3.90 (4) and 4.13 ± 0.12 (5) kcal mol^{-1}.

Vaporization Data

T_{vap} is calculated as the temperature at which $\Delta_r G° = 0$ for $BaCl_2$(ℓ) → $BaCl_2$(g) at p=1 bar. $\Delta_{vap}H°$ is calculated as the corresponding $\Delta_r H°$.

References

1. H. Feschotte-Ostertag, Ann. Chim. (Paris) 6, 791 (1961).
2. D. T. Peterson and J. A. Hinkebein, J. Phys. Chem. 63, 1360 (1959).
3. T. E. Gardner and A. R. Taylor, J. Chem. Eng. Data 14, 281 (1969).
4. A. S. Dworkin and M. A. Bredig, J. Phys. Chem. 67, 697 (1963).
5. G. J. Janz, F. J. Kelly and J. L. Perano, Trans. Faraday Soc. 59, 2718 (1963).
6. F. A. Gant, Ph.D. Thesis, Univ. Alabama, 1967; Diss. Abstr. B28, 133 (1967).
7. C. Krohn, Acta Chem. Scand. 20, 255 (1966).
8. K. Grjotheim, H. G. Nebell and J. Krogh-Moe, Acta Chem. Scand. 22, 1159 (1968).

Barium Chloride ($BaCl_2$) Ba_1Cl_2(l)

Enthalpy Reference Temperature = T_r = 298.15 K Standard State Pressure = p° = 0.1 MPa

T/K	$C_p°$ (J K⁻¹mol⁻¹)	S° (J K⁻¹mol⁻¹)	$-[G°-H°(T_r)]/T$ (J K⁻¹mol⁻¹)	$H°-H°(T_r)$ (kJ mol⁻¹)	$\Delta_f H°$ (kJ mol⁻¹)	$\Delta_f G°$ (kJ mol⁻¹)	Log K_f
0							
100							
200							
298.15	75.140	143.518	143.518	0.	-832.467	-790.119	138.426
300	75.195	143.983	143.520	0.139	-832.443	-789.857	137.526
400	77.329	165.934	146.498	7.774	-831.295	-775.846	101.315
500	78.857	183.357	152.187	15.585	-830.874	-762.052	79.611
600	80.425	197.869	158.624	23.547	-831.165	-748.249	65.141
700	82.233	210.399	165.145	31.678	-830.522	-734.491	54.808
800	84.299	221.512	171.509	40.003	-830.223	-720.784	47.062
800.010	84.299	221.513	171.509	40.004	GLASS <--> LIQUID TRANSITION		
800.010	108.784	221.513	171.509	40.004			
900	108.784	234.325	177.791	50.881	-826.972	-707.300	41.051
1000	108.784	245.786	184.027	61.759	-823.737	-694.176	36.260
1100	108.784	256.155	190.120	72.638	-828.885	-680.557	32.317
1200	108.784	265.620	196.023	83.516	-825.911	-667.203	29.043
1235.000	108.784	268.748	198.040	87.324	------- BETA <--> LIQUID -------		
1300	108.784	274.328	201.716	94.395	-822.847	-654.101	26.282
1400	108.784	282.389	207.194	105.273	-819.715	-641.237	23.925
1500	108.784	289.895	212.460	116.151	-816.536	-628.600	21.890
1600	108.784	296.915	217.522	127.030	-813.330	-616.175	20.116
1700	108.784	303.510	222.388	137.908	-810.118	-603.951	18.557
1800	108.784	309.728	227.069	148.787	-806.921	-591.916	17.177
1900	108.784	315.610	231.576	159.665	-803.761	-580.058	15.947
2000	108.784	321.190	235.918	170.543	-800.659	-568.365	14.844
2100	108.784	326.497	240.106	181.422	-797.639	-556.824	13.850
2200	108.784	331.558	244.149	192.300	-934.242	-540.051	12.822
2300	108.784	336.394	248.055	203.179	-930.614	-522.215	11.860
2400	108.784	341.023	251.833	214.057	-927.225	-504.532	10.981
2500	108.784	345.464	255.490	224.935	-924.075	-486.986	10.175
2600	108.784	349.731	259.033	235.814	-921.163	-469.560	9.434
2700	108.784	353.836	262.469	246.692	-918.478	-452.242	8.749
2800	108.784	357.793	265.803	257.571	-916.008	-435.020	8.115
2900	108.784	361.610	269.041	268.449	-913.733	-417.883	7.527
3000	108.784	365.298	272.189	279.327	-911.633	-400.821	6.979

PREVIOUS: CURRENT: December 1972

Barium Chloride ($BaCl_2$) Ba_1Cl_2(l)

BARIUM CHLORIDE ($BaCl_2$) ALPHA - BETA - LIQUID M_r = 208.236

0 to 1198 K		crystal, alpha
1198 K to 1235 K		crystal, beta
above 1235 K		liquid

Refer to the individual tables for details.

Barium Chloride ($BaCl_2$) Ba_1Cl_2(cr,l)

Enthalpy Reference Temperature = T_r = 298.15 K   Standard State Pressure = p° = 0.1 MPa

T/K	C_p° (J K⁻¹mol⁻¹)	S° (J K⁻¹mol⁻¹)	$-[G^\circ-H^\circ(T_r)]/T$ (J K⁻¹mol⁻¹)	$H^\circ-H^\circ(T_r)$ (kJ mol⁻¹)	Δ_fH° (kJ mol⁻¹)	Δ_fG° (kJ mol⁻¹)	Log K_f
0	0.	0.	INFINITE	-16.707	-859.171	-859.171	INFINITE
100	56.961	49.346	187.175	-13.783	-860.857	-843.483	440.590
200	71.038	94.429	130.426	-7.199	-859.856	-826.420	215.839
298.15	75.140	123.666	123.666	0.	-858.557	-810.290	141.959
300	75.195	124.131	123.668	0.139	-858.533	-809.990	141.032
400	77.329	146.082	126.646	7.774	-857.384	-793.994	103.685
500	78.857	163.505	132.335	15.585	-856.963	-778.216	81.300
600	80.425	178.018	138.772	23.547	-857.255	-762.428	66.375
700	82.233	190.548	145.293	31.678	-856.611	-746.684	55.718
800	84.299	201.661	151.657	40.003	-856.313	-730.992	47.729
900	86.624	211.721	157.780	48.546	-855.396	-715.380	41.520
1000	89.471	220.988	163.643	57.345	-854.241	-699.882	36.558
1100	93.531	229.692	169.256	66.480	-861.132	-683.694	32.466
1198.010	100.015	237.924	174.538	75.937		ALPHA <--> BETA	
1198.010	123.846	252.033	174.538	92.840		TRANSITION	
1200	123.846	252.239	174.666	93.087	-842.429	-667.664	29.063
1235.000	123.846	255.799	176.915	97.422		BETA <--> LIQUID	
1235.000	108.784	268.748	176.915	113.413		TRANSITION	
1300	108.784	274.328	181.648	120.484	-822.847	-654.101	26.282
1400	108.784	282.389	188.559	131.362	-819.715	-641.237	23.925
1500	108.784	289.895	195.068	142.241	-816.536	-628.600	21.890
1600	108.784	296.915	201.216	153.119	-813.330	-616.175	20.116
1700	108.784	303.510	207.041	163.997	-810.118	-603.951	18.557
1800	108.784	309.728	212.575	174.876	-806.921	-591.916	17.177
1900	108.784	315.610	217.845	185.754	-803.761	-580.058	15.947
2000	108.784	321.190	222.874	196.633	-800.659	-568.365	14.844
2100	108.784	326.497	227.683	207.511	-797.639	-556.825	13.850
2200	108.784	331.558	232.290	218.389	-934.242	-540.051	12.822
2300	108.784	336.394	236.712	229.268	-930.615	-522.215	11.860
2400	108.784	341.023	240.963	240.146	-927.225	-504.532	10.981
2500	108.784	345.464	245.054	251.025	-924.075	-486.986	10.175
2600	108.784	349.731	248.999	261.903	-921.163	-469.560	9.434
2700	108.784	353.836	252.806	272.781	-918.479	-452.243	8.749
2800	108.784	357.793	256.486	283.660	-916.008	-435.021	8.115
2900	108.784	361.610	260.045	294.538	-913.733	-417.883	7.527
3000	108.784	365.298	263.492	305.417	-911.634	-400.821	6.979

PREVIOUS: CURRENT: December 1972

BARIUM CHLORIDE ($BaCl_2$) IDEAL GAS M_r = 208.236

$\Delta_f H°(0\ K) = -497.3 \pm 17$ (or $\pm$ 6.7) kJ mol^{-1}

S°(298.15 K) = 325.74 ± 5.0 J K^{-1} mol^{-1} $\Delta_f H°(298.15\ K) = -498.7 \pm 17$(or $\pm$ 6.7) kJ mol^{-1}

Vibrational Fequencies and Degeneracies

ν, cm^{-1}	ν, cm^{-1}	ν, cm^{-1}
255 (1)	[36] (1)	265 (1)

Ground State Quantum Weight: [1] σ = 2

Point Group: C_{2v}

Bond Distance: Ba-Cl = 2.82 ± 0.03 Å

Bond Angle: Cl-Ba-Cl = [100 ± 20]°

Product of the Moments of Inertia: $I_A I_B I_C = 1.128013 \times 10^{-112}\ g^3 cm^6$

Enthalpy of Formation

$\Delta_f H°(298.15\ K)$ is calculated from $\Delta_f H°(cr)$ by addition of $\Delta_{sub}H° = 86.0 \pm 1.5$ kcal mol^{-1} derived from 3rd law analysis of the vapor-pressure equation of Hildenbrand et al. (1, 2). The equation (1) summarized five torsion-effusion runs (83 total points) and is reasonably consistent with five boiling pressures in the mmHg range measured by Novikov (3). Other vapor pressure data (4-6) reviewed below are considered less reliable. $\Delta_f H°(g)$ is assigned alternative uncertainties to reflect the discrepancies in $\Delta_f H°(cr)$. Equilibria derived from flame studies (7) yield $\Delta_f H°(g) = -119.1 \pm 3$ and $\Delta_f H°(cr) = -205.1 \pm 3$ kcal mol^{-1}, which confirm the adopted values by an independent path. Mass spectra (8) at 1230 K showed that the saturated vapor consists predominantly of $BaCl_2$ with considerably less than 1% of dimer.

Source	Method	Points	T/K	$\Delta_{sub}H°$(298.15 K), kcal mol^{-1} 2nd law	3rd law	δS cal K^{-1}mol^{-1}
(1, 2)	Torsion effusion	83[a]	1235-1440	85.1	86.01	-0.7
(3)	Boiling Point	5	1588-1710	75.6±4.2	86.30±0.7	-6.4±2.5
(4)	Effusion	3	1244-1277	78.5±1.6	85.3 ±0.2	-5.4±1.3
(5)	Boiling point	1	1819		>81	
(6)	Static	4	1343-1487	85.4±3.8	76.6±0.9	6.1±2.7
(7)	Flame Kp[b]	8	2134-2535	-9.5±8.7[b]	-13.5±2.4[b]	1.7±3.7

[a]Pressures obtained from an equation summarizing 5 runs (83 total points).

[b]$\Delta_r H°$ for $Ba(g) + 2\ HCl(g) \rightarrow BaCl_2(g) + 2\ H(g)$ yields $\Delta_f H°(298.15\ K) = 119.1 \pm 3$ kcal mol^{-1}.

Heat Capacity and Entropy

Vibrations ν_1 and ν_2 are the values quoted by Hildenbrand (9) and derived from IR spectra of molecules in matrix isolation (10). The bending mode was not observed; its value was calculated (10) from force-constant correlations as described for BaF_2 (11). Hastie et al. (12) assigned the bending mode to a band which they observed at 61 cm^{-1} in argon and 62 cm^{-1} in neon. This value would reduce S° by about 1.0 cal K^{-1} mol^{-1}. Our assigned uncertainty includes this possibility. Agreement is better for ν_1 = 255 (9, 10) or 262 cm^{-1} (12, in Ne) and for ν_3 = 265±5 (13, gas), 260 (9, 10) or 268 cm^{-1} (12, in Ne). We adopt the gas-phase value (13) for ν_3.

Bond length is from gas-phase electron diffraction (14) which also gave the bond angle 180° ± 40°. Later studies (15) of electric and magnetic deflection of molecular beams revealed a permanent electric dipole moment and suggested a bond angle of the order of roughly 120°. Hastie et al. (12) estimated an angle of 120° ± 10° from relative intensities of ν_1 and ν_3. The matrix spectra (9, 10) were first interpreted to give an angle of 120° but reanalysis by Hemple (10) gave the angle 95.4°, estimated from the three fundamentals. We adopt 100° ± 20° and note that S° changes by only 0.2 cal K^{-1} mol^{-1} in the range 95° to 120°. The singlet ground state is based on magnetic deflection data (15). The principal moments of inertia are $I_A = 25.5142 \times 10^{-39}$, $I_B = 54.9471 \times 10^{-39}$ and $I_C = 80.4613 \times 10^{-39}$ g cm^2.

References

1. D. L. Hildenbrand, Stanford Res. Inst., personal communication, 1973; D. L. Hildenbrand, E. Murad, N. D. Potter, L. P. Theard and W. F. Hall, Aeronutronic Division, Philco Corp., Report No. U-3183, Contract AF 49(838)-1397, June 30, 1965.
2. D. L. Hildenbrand, Proc. 2nd Meeting, ICRPG Working Group on Thermochemistry, CPIA Publ. No. 54U, pp. 79-80, Aug., 1964.
3. G. I. Novikov and F. G. Gavryuchenkov, Russ. J. Inorg. Chem. 9, 260 (1964).
4. J. A. Van Westenburg, Ph.D. Thesis, Iowa State Univ., 1964; Diss. Abstr. 25, 1129 (1964).
5. H. von Wartenburg and O. Bosse, Z. Elektrochem. 28, 384 (1922).
6. C. G. Maier, USBM Tech. Paper 360, 25 (1925).
7. K. Schofield and T. M. Sugden, Trans. Faraday Soc., 67, 1054 (1971).
8. D. L. Hildenbrand, Int. J. Mass Spectrom. Ion Phys. 4, 75 (1970)
9. D. L. Hildenbrand, J. Chem. Phys. 52, 5751 (1970).
10. David White, Univ. of Pennsylvania, personal communication, 1972; S. Hemple, U. S. Army Foreign Science and Technolog;y Center, Charlottesville, Va., personal comm., 1973.
11. V. Calder, D. E. Mann, K. S. Seshadri, M. Allavena and D. White, J. Chem. Phys. 51, 2093 (1969).
12. J. W. Hastie, R. H. Hauge and J. L. Margrave, High Temp. Sci. 3, 56 (1971).
13. V. I. Baikov, Optics and Spectroscopy 27, 502 (1969).
14. P. A. Akishin, V. P. Spiridonov, G. A. Solobev and V. A. Naumov, Zh. Fiz. Khim. 32, 58 (1958); Kristallografiya 2, 475 (1957).
15. L. Wharton, R. A. Berg and W. Klemperer, J. Chem. Phys. 39, 2023 (1963); A. Buchler, J. L. Stauffer and W. Klemperer, J. Amer. Chem. Soc. 86, 4544 (1964).

Barium Chloride ($BaCl_2$) $Ba_1Cl_2(g)$

Enthalpy Reference Temperature = T_r = 298.15 K Standard State Pressure = p° = 0.1 MPa

T/K	$C_p°$ (J K^{-1}mol^{-1})	S°	$-[G°-H°(T_r)]/T$	$H°-H°(T_r)$ (kJ mol^{-1})	$\Delta_f H°$	$\Delta_f G°$	Log K_f
0	0.	0.	INFINITE	-14.667	-497.307	-497.307	INFINITE
100	47.186	268.409	373.964	-10.556	-497.806	-502.338	262.394
200	54.049	303.691	330.820	-5.426	-498.258	-506.675	132.330
250	55.387	315.909	326.656	-2.687	-498.477	-508.755	106.299
298.15	56.160	325.736	325.736	0.	-498.733	-510.713	89.475
300	56.183	326.083	325.737	0.104	-498.744	-510.787	88.936
350	56.689	334.784	326.423	2.927	-499.086	-512.769	76.527
400	57.028	342.378	327.953	5.770	-499.565	-514.693	67.212
450	57.266	349.109	329.936	8.628	-500.260	-516.545	59.959
500	57.439	355.152	332.161	11.496	-501.228	-518.305	54.147
600	57.667	365.647	336.893	17.252	-503.726	-521.477	45.399
700	57.807	374.547	341.653	23.026	-505.439	-524.311	39.125
800	57.898	382.273	346.258	28.812	-507.680	-526.849	34.400
900	57.961	389.096	350.646	34.605	-509.514	-529.135	30 710
1000	58.006	395.205	354.802	40.403	-511.358	-531.216	27.748
1100	58.040	400.735	358.730	46.206	-521.583	-532.293	25.276
1200	58.066	405.787	362.444	52.011	-523.681	-533.173	23.208
1300	58.085	410.435	365.959	57.819	-525.688	-533.882	21.452
1400	58.101	414.740	369.292	63.628	-527.625	-534.439	19.940
1500	58.114	418.749	372.457	69.439	-529.513	-534.860	18.625
1600	58.125	422.500	375.469	75.251	-531.374	-535.155	17.471
1700	58.133	426.024	378.340	81.064	-533.228	-535.335	16.449
1800	58.141	429.348	381.082	86.877	-535.096	-535.405	15.537
1900	58.147	432.491	383.706	92.692	-536.999	-535.371	14.718
2000	58.152	435.474	386.221	98.507	-538.961	-535.235	13.979
2100	58.157	438.311	388.634	104.322	-541.004	-534.998	13.307
2200	58.160	441.017	390.954	110.138	-682.670	-529.288	12 567
2300	58.164	443.602	393.187	115.954	-684.104	-522.284	11.861
2400	58.167	446.078	395.340	121.771	-685.776	-515.214	11.213
2500	58.170	448.452	397.417	127.587	-687.688	-508.069	10.616
2600	58.172	450.734	399.424	133.405	-689.837	-500.842	10.062
2700	58.174	452.929	401.366	139.222	-692.214	-493.529	9.548
2800	58.176	455.045	403.245	145.039	-694.804	-486.123	9.069
2900	58.178	457.086	405.067	150.857	-697.590	-478.622	8.621
3000	58.179	459.059	406.834	156.675	-700.551	-471.021	8.201
3100	58.180	460.966	408.549	162.493	-703.664	-463.319	7.807
3200	58.182	462.814	410.216	168.311	-706.904	-455.515	7.436
3300	58.183	464.604	411.838	174.129	-710.241	-447.608	7.085
3400	58.184	466.341	413.415	179.948	-713.660	-439.598	6.754
3500	58.185	468.028	414.952	185.766	-717.139	-431.487	6.440
3600	58.186	469.667	416.449	191.585	-720.654	-423.276	6.142
3700	58.187	471.261	417.909	197.403	-724.190	-414.966	5.858
3800	58.187	472.813	419.333	203.222	-727.728	-406.562	5.589
3900	58.188	474.324	420.724	209.041	-731.256	-398.064	5.331
4000	58.189	475.797	422.082	214.859	-734.763	-389.476	5.086
4100	58.189	477.234	423.410	220.678	-738.238	-380.801	4.851
4200	58.190	478.636	424.708	226.497	-741.673	-372.041	4.627
4300	58.190	480.006	425.979	232.316	-745.066	-363.200	4.412
4400	58.191	481.343	427.222	238.135	-748.410	-354.281	4.206
4500	58.191	482.651	428.439	243.954	-751.704	-345.287	4.008
4600	58.192	483.930	429.631	249.774	-754.938	-336.220	3.818
4700	58.192	485.182	430.800	255.593	-758.097	-327.085	3.635
4800	58.192	486.407	431.946	261.412	-761.221	-317.882	3.459
4900	58.193	487.607	433.070	267.231	-764.285	-308.614	3.290
5000	58.193	488.782	434.172	273.051	-767.299	-299.284	3.127
5100	58.193	489.935	435.254	278.870	-770.236	-289.897	2.969
5200	58.194	491.065	436.317	284.689	-773.140	-280.450	2.817
5300	58.194	492.173	437.360	290.509	-775.992	-270.948	2.670
5400	58.194	493.261	438.385	296.328	-778.794	-261.393	2.528
5500	58.194	494.329	439.393	302.147	-781.515	-251.789	2.391
5600	58.195	495.377	440.383	307.967	-784.160	-242.141	2.259
5700	58.195	496.407	441.357	313.786	-786.808	-232.439	2.130
5800	58.195	497.419	442.315	319.606	-789.386	-222.693	2.006
5900	58.195	498.414	443.257	325.425	-791.800	-212.921	1.885
6000	58.195	499.392	444.185	331.245	-794.296	-203.088	1.768

PREVIOUS: December 1972 (1 atm) CURRENT: December 1972 (1 bar)

Barium Chloride ($BaCl_2$) $Ba_1Cl_2(g)$

$\Delta_fH°(0\ K) = -320.2 \pm 8.4\ kJ\ mol^{-1}$

$S°(298.15\ K) = 246.23 \pm 0.21\ J\ K^{-1}\ mol^{-1}$ $\Delta_fH°(298.15\ K) = -322.2 \pm 8.4\ kJ\ mol^{-1}$

Electronic Levels and Quantum Weights

State	ε_i, cm^{-1}	g_i	State	ε_i, cm^{-1}	g_i
X $^2\Sigma^+$	0	2	D $^2\Sigma^+$	24156.8	2
A $^2\Pi_{1/2}$	11646.9	2	D' $^2\Sigma^+$	26222.3	2
A $^2\Pi_{3/2}$	12278.2	2	E $^2\Sigma^+$	28139.7	2
B $^2\Sigma^+$	14062.5	2	F $^2\Sigma^+$	29411.3	2
C $^2\Pi$	19998.2	2	G $^2\Sigma^+$	31451.9	2
	20197.2	2	H $^2\Sigma^+$	31582.0	2

$\omega_e = 468.9\ cm^{-1}$ $\omega_ex_e = 1.79\ cm^{-1}$ $\sigma = 1$

$B_e = 0.2164\ cm^{-1}$ $\alpha_e = [0.00112]\ cm^{-1}$ $r_e = 2.152\ Å$

Enthalpy of Formation

$\Delta_fH°$ is obtained from a 3rd law analysis of Kp data for three gaseous dissociation equilibria. The Kp data were derived from observations made on the condensed phase systems BaF_2 - B (1) and BaF_2 - Al (2) by the mass spectrometric-Knudsen effusion method. Results of a 2nd and 3rd law analysis of these data are given below. Also included below are results derived from two mass spectrometric studies (2, 3) of the heterogeneous reaction (D), and a value for the heat of dissociation of BaF which was determined from spectrophotometric studies (4, 5) of reaction equilibria in H_2-O_2-N_2 flames.

Source	Reaction	T/K	Points	$\Delta_rH°(298.15\ K)$, kcal mol^{-1} 2nd law	3rd law	Drift cal K^{-1} mol^{-1}	$\Delta_fH°(298.15\ K)$ kcal mol^{-1}
(1)	A	1418-1512	8	- 1.0	-6.6±0.5	-3.8±2.2	-78.0±1.5
(1)	B	1418-1512	8	8.6	-3.5±1.1	-3.5±5.7	-77.3±3.6
(2)	C	1175-1255	8[a]	26.0	23.0±0.2	-2.5±0.9	-75.6±3.0
(2)	D	1175-1255	8	-118.8	-96.4±1.1	18.4±3.6	-74.9±1.7
(3)	D	1428		-103.5			-71.3±3.0
(4, 5)	E			145±6			-83.3±8.0

Reactions: (A) $Ba(g) + BaF_2(g) \rightarrow 2\ BaF(g)$ (D) $2\ BaF(g) \rightarrow Ba(g) + BaF_2(cr)$
(B) $2Ba(g) + BF_3(g) \rightarrow 2\ BaF(g) + BF(g)$ (E) $BaF(g) \rightarrow Ba(g) + F(g)$
(C) $Ba(g) + AlF(g) \rightarrow BaF(g) + Al(g)$

[a]One point rejected due to failure of a statistical test.

It is believed that the Kp data for the three homogeneous reactions (A, B, and C) are more reliable than those for the heterogeneous reaction (D) due to the need for a machine calibration constant which appears in the equilibrium expression for the latter. Ehlert et al. (2) determined this constant from vaporization experiments performed with silver contained in their Knudsen cell. Further support for this belief is provided by the large positive drift that arises in the 3rd law analysis of these Kp's. Also, it is felt that the results otained from the flame-spectrophotometric studies (4, 5) are somewhat more uncertain than those derived from the Kp data for the three homogeneous reactions. Therefore, we choose to adopt a median value $\Delta_fH°(BaF, g, 298.15\ K) = -77.0 \pm 2.0\ kcal\ mol^{-1}$ of the first three results. The adopted value corresponds to a dissociation energy of $D_0^0(Ba\text{-}F) = 137.9\ kcal\ mol^{-1}$ (5.98 eV).

Gaydon (7) has reported a spectroscopic value for D_0^0 of 3.8 eV which is considerably lower than the adopted thermochemical value. The spectroscopic value is based on a linear Birge-Sponer extrapolation of the thirteen known vibrational levels for the ground state ($^2\Sigma^+$). However, Gaydon (7) and later Hildenbrand (8) have shown that the linear extrapolation generally under-estimates D_0^0 when the bonding in the molecule is partially ionic. Correcting D_0^0(LBX) for the ionic character of BaF with the ionicity parameter suggested by Hildenbrand (8), we obtain D_0^0(Ba-F) = 6.9 eV which is somewhat high. Other values for D_0^0 that have been reported are 5.94 (1) and 5.68 eV (9). Both of these values were calculated from the Rittner ionic model (10) for polarizable ions and show remarkably good agreement with the adopted value.

Heat Capacity and Entropy

The spectroscopic data, except for α_e, are given by Rosen (11) for ^{138}BaF. α_e is calculated from the Morse potential function. Corrections to these data to account for the natural isotopic abundances of barium were found to be negligible. The electronic states and levels are also taken from the compilation of Rosen (11).

References

1. D. L. Hildenbrand, J. Chem. Phys. 48, 3657 (1968).
2. T. C. Ehlert, G. D. Blue, J. W. Green, and J. L. Margrave, J. Chem. Phys. 41, 2250 (1964).
3. K. F. Zmbov and J. L. Margrave, J. Chem. Phys. 45, 3167 (1966).
4. L. V. Gurvich and V. G. Ryabova, Teplofiz. Vysokikh Temperatur, Akad. Nauk SSSR 2, 401 (1964).
5. L. V. Gurvich and V. G. Ryabova, Teplofiz. Vysokikh Temperatur, Akad. Nauk SSSR 3, 652 (1965).
6. JANAF Thermochemical Tables: BaF_2(cr, g), 12-31-72; Ba(g), 12-31-70; BF_3(g) and AlF(g), 6-30-69; BF(g), 12-31-64; Al(g), 12-31-65.
7. A. G. Gaydon, "Dissociation Energies and Spectra of Diatomic Molecules," Third Ed., Chapman and Hall Ltd, London, 1968.
8. D. L. Hildenbrand in "Advances in High Temperature Chemistry," Vol. 1, Ed. by L. Eyring, Academic Press, New York, 1967.
9. K. S. Krasnov and N. V. Karaseva, Optics and Spectroscopy (Eng. Transl.) 19, 14 (1965).
10. E. S. Rittner, J. Chem. Phys. 19, 1030 (1951).
11. B. Rosen, "Spectroscopic Data Relative to Diatomic Molecules," Pergamon Press, New York 1970.

Enthalpy Reference Temperature = T_r = 298.15 K   Standard State Pressure = p° = 0.1 MPa

T/K	$C_p°$ (J $K^{-1}mol^{-1}$)	S°	$-[G°-H°(T_r)]/T$	$H°-H°(T_r)$ (kJ mol^{-1})	$\Delta_fH°$	$\Delta_fG°$	Log K_f
0	0.	0.	INFINITE	-9.345	-320.188	-320.188	INFINITE
100	29.574	211.330	275.613	-6.428	-320.424	-329.630	172.181
200	32.669	232.752	249.349	-3.319	-321.324	-338.481	88.402
250	33.896	240.181	246.796	-1.654	-321.748	-342.721	71.608
298.15	34.748	246.228	246.228	0.	-322.168	-346.723	60.744
300	34.776	246.443	246.229	0.064	-322.185	-346.875	60.396
350	35.405	251.854	246.655	1.820	-322.670	-350.953	52.377
400	35.863	256.613	247.608	3.602	-323.274	-354.954	46.352
450	36.206	260.858	248.849	5.404	-324.082	-358.868	41.656
500	36.468	264.687	250.244	7.221	-325.155	-362.679	37.889
600	36.839	271.371	253.224	10.888	-327.847	-369.921	32.205
700	37.088	277.069	256.234	14.585	-329.741	-376.796	28.117
800	37.268	282.034	259.155	18.303	-332.154	-383 347	25.030
900	37.406	286.432	261.946	22.037	-334.154	-389.626	22.613
1000	37.517	290.379	264.595	25.783	-336.159	-395.682	20.668
1100	37.611	293.959	267.105	29.540	-346.539	-400.718	19.029
1200	37.694	297.235	269.481	33.305	-348.789	-405.543	17.653
1300	37.770	300.255	271.734	37.078	-350.943	-410.185	16.481
1400	37.843	303.057	273.872	40.859	-353.022	-414.663	15.471
1500	37.917	305.671	275.906	44.647	-355.048	-418.995	14.591
1600	37.995	308.120	277.843	48.443	-357.039	-423.193	13.816
1700	38.081	310.426	279.693	52.246	-359.016	-427.267	13.128
1800	38.179	312.605	281.461	56.059	-360.997	-431.225	12.514
1900	38.292	314.673	283.155	59.883	-363.001	-435.072	11.961
2000	38.421	316.640	284.781	63.718	-365.045	-438.813	11.461
2100	38.570	318.518	286.343	67.568	-367.149	-442.450	11.005
2200	38.740	320.316	287.847	71.433	-508.849	-440.610	10.461
2300	38.933	322.042	289.296	75.316	-510.286	-437.476	9.935
2400	39.148	323.704	290.695	79.220	-511.924	-434.276	9.452
2500	39.387	325.307	292.048	83.147	-513.759	-431.004	9.005
2600	39.649	326.857	293.357	87.098	-515.782	-427.655	8.592
2700	39.934	328.358	294.626	91.077	-517.981	-424.224	8.207
2800	40.240	329.816	295.857	95.086	-520.335	-420.709	7.848
2900	40.566	331.234	297.052	99.126	-522.823	-417.107	7.513
3000	40.911	332.615	298.215	103.200	-525.421	-413.418	7.198
3100	41.273	333.962	299.346	107.309	-528.102	-409.640	6.902
3200	41.651	335.278	300.449	111.455	-530.842	-405.775	6.624
3300	42.042	336.566	301.524	115.639	-533.606	-401.825	6.360
3400	42.445	337.827	302.573	119.864	-536.382	-397.790	6.111
3500	42.859	339.063	303.598	124.129	-539.144	-393.673	5.875
3600	43.280	340.276	304.600	128.436	-541.871	-389.478	5.651
3700	43.708	341.468	305.580	132.785	-544.548	-385.208	5.438
3800	44.140	342.639	306.540	137.177	-547.159	-380.867	5.235
3900	44.575	343.792	307.481	141.613	-549.693	-376.457	5.042
4000	45.012	344.926	308.403	146.092	-552.141	-371.984	4.858
4100	45.448	346.043	309.307	150.615	-554.495	-367.451	4.681
4200	45.883	347.143	310.195	155.182	-556.750	-362.861	4.513
4300	46.315	348.228	311.067	159.792	-558.907	-358.219	4.352
4400	46.742	349.297	311.923	164.445	-560.963	-353.528	4.197
4500	47.164	350.352	312.766	169.140	-562.919	-348.791	4.049
4600	47.579	351.394	313.594	173.877	-564.768	-344.013	3.906
4700	47.986	352.421	314.409	178.656	-566.499	-339.197	3.770
4800	48.385	353.436	315.212	183.474	-568.157	-334.344	3.638
4900	48.775	354.437	316.002	188.332	-569.716	-329.457	3.512
5000	49.155	355.427	316.781	193.229	-571.193	-324.538	3.390
5100	49.524	356.404	317.548	198.163	-572.563	-319.594	3.273
5200	49.882	357.369	318.305	203.133	-573.872	-314.621	3.160
5300	50.228	358.322	319.051	208.139	-575.105	-309.623	3.052
5400	50.563	359.264	319.787	213.179	-576.266	-304.603	2.946
5500	50.884	360.195	320.513	218.251	-577.328	-299.566	2.845
5600	51.193	361.115	321.230	223.355	-578.297	-294.514	2.747
5700	51.490	362.023	321.938	228.489	-579.255	-289.438	2.652
5800	51.773	362.921	322.636	233.653	-580.131	-284.349	2.561
5900	52.043	363.809	323.327	238.844	-580.833	-279.261	2.472
6000	52.299	364.686	324.009	244.061	-581.611	-274.144	2.387

PREVIOUS: December 1972 (1 atm)   CURRENT: December 1972 (1 bar)

BARIUM FLUORIDE, ION (BaF^+) IDEAL GAS M_r = 156.327854

$S°(298.15\ K) = [239.0 \pm 8]\ J\ K^{-1}\ mol^{-1}$

$\Delta_f H°(0\ K) = 146.1 \pm 41.8\ kJ\ mol^{-1}$
$\Delta_f H°(298.15\ K) = 150.2 \pm 41.8\ kJ\ mol^{-1}$

Electronic Levels and Quantum Weights

State	ε_i, cm^{-1}	g_i
$X^1\Sigma^+$	0	1
$a^3\Pi$	[9000]	6
$A'^1\Pi$	[11000]	2
$B'^1\Delta$	[13000]	2
$b^3\Sigma^-$	[15000]	3
$A^1\Sigma^+$	[16000]	1
$B^1\Sigma^+$	[30000]	1

$\omega_e = [540]\ cm^{-1}$ $\omega_e x_e = [2.06]\ cm^{-1}$ $\sigma = 1$
$B_e = [0.2371]\ cm^{-1}$ $\alpha_e = [0.00122]\ cm^{-1}$ $r_e = [2.064]$ Å

Enthalpy of Formation

The appearance potential (AP) of BaF^+ from BaF(g) has been measured mass spectrometrically as 4.9 eV (1) and 4.8 ± 0.3 eV (2). This latter value has been confirmed by Zmbov and Margrave (3). Combining these results with $\Delta_f H°$(BaF, g, 0 K) = -76.5 ± 2.0 kcal mol^{-1} (4), we obtain $\Delta_f H°(BaF^+$, g, 0 K) equal to 36.5 and 34.2 ± 9.0 kcal mol^{-1}.

Hildenbrand (2) also reported an AP for BaF^+ from BaF_2(g) as 13.5 ± 1.0 eV. Green et al. (5) obtained an ionization efficiency curve for BaF^+ which extrapolates (high-energy portion) to 12.9 eV at zero ion current. Assuming the process to be $e^- + BaF_2(g) \rightarrow BaF^+(g) + F(g) + 2e^-$, we obtain $\Delta_f H°(BaF^+$, g, 0 K) values of 101.3 ± 25 and 87.5 kcal mol^{-1} by combining the above results with $\Delta_f H°(BaF_2$, g, 0 K) = -191.6 ± 1.5 kcal mol^{-1} and $\Delta_f H°$(F, g, 0 K) = 18.36 ± 0.4 kcal mol^{-1} (4).

The two lower values for Δ H°(0 K) are preferred here, since the dissociative ionization of BaF (g) most likely involved unknown kinetic energy factors. We choose to adopt $\Delta_f H°(BaF^+$, g, 0 K) = 35.0 ± 10.0 kcal mol^{-1} which gives $\Delta_f H°(BaF^+$,g, 298.15 K) = 35.9 ± 10.0 kcal mol^{-1}. The adopted heat of formation corresponds to an ionization potential for BaF(g) of 4.83 eV.

Heat Capacity and Entropy

The dissociation energy of $BaF^+(D_0^{\,0}$ = 145.4 kcal mol^{-1}) is slightly greater than that for BaF($D_0^{\,0}$ = 137.9 kcal mol^{-1}) (4) which implies somewhat stronger bonding in the ion. Thus, one would expect the internuclear distance (r_e) for the ground state of BaF^+ to be less than that for BaF. Barrow and Beale (6) postulated from a rotational analysis of the $F^2\Sigma^+ - X^2\Sigma^+$ band system of SrF a r_e for SrF^+ which was some 4% lower than that for SrF. We assume a similar decrease in r_e for BaF^+ and obtain r_e = 2.064 Å from r_e(BaF) = 2.152 Å (4). ω_e is calculated from Badger's rule (7) which can be written in the form $\omega_e^{\,2} = 3.159 \times 10^6/\mu(r_e - d_{ij})^3$. Molecular data for CsF (4), BaF (4), and BaO (8) were used to determine the constant d_{ij}. The product $x_e\mu^{1/2}$ has been shown to be constant within a group of similar molecules by Barrow and Caunt (9). Since $\mu^{1/2}(BaF^+) \cong \mu^{1/2}(BaF)$, we assume $x_e(BaF^+) = x_e(BaF)$ and obtain $\omega_e x_e$ equal to 2.06 cm^{-1} with our estimated value for ω_e. α_e is obtained from the other constants assuming a Morse potential function.

BaF has eight valence electrons; thus, the ground state configuration ($^1\Sigma^+$) and electronic levels for the A and B states are estimated from those for the isoelectronic molecule BaS (10). By analogy with the gaseous alkaline earth oxides (11), C_2 (4), and BeS (12), low-lying triplet states are to be expected. The two estimated triplet levels are based on those which have been observed directly in C_2 (4), BeS (12), and indirectly (through perturbative effects) in BaS (10). Also included are a $^1\Pi$ state at 11,000 cm^{-1} and $^1\Delta$ state at 13,000 cm^{-1}. These levels are also estimated by analogy with those for C_2(4) and BeS (12). It is believed that our adopted electronic entropies for BaF^+ are most likely minimal values. Probable upper limits for the electronic contributions can be established by decreasing the triplet levels to 1000 cm^{-1} ($^3\Pi$) and 5000 cm^{-1} ($^3\Sigma$) and neglecting the other excited states. This increases the entropy (in cal $K^{-1}mol^{-1}$) by 3.8 and 2.1 at 2000 K and 4000 K, respectively.

References

1. T. C. Ehlert, G. D. Blue, J. W. Green, and J. L. Margrave, J. Chem. Phys. 41, 2250 (1964).
2. D. L. Hildenbrand, J. Chem. Phys. 48, 3657 (1968).
3. K. F. Zmbov and J. L. Margrave, J. Chem. Phys. 45, 3167 (1966).
4. JANAF Thermochemical Tables: BaF(g) and BaF_2(g) dated 12-31-72; F(g) dated 9-30-65; CsF(g) dated 6-30-68; C_2(g) dated 12-31-69.
5. J. W. Green, G. D. Blue, T. C. Ehlert, and J. L. Margrave, J. Chem. Phys. 41, 2245 (1964).
6. R. F. Barrow and J. R. Beale, Chem. Comm., 1967, 606 (1967).
7. R. M. Badger, J. Chem. Phys. 2, 128 (1934).
8. M. Marcano and R. F. Barrow, Trans. Faraday Soc. 66, 1917 (1970).
9. R. F. Barow and A. D. Caunt, Proc. Roy. Soc (London) 219A, 120 (1953).
10. R. F. Barrow, W. G. Burton, and P. A. Jones, Trans. Faraday Soc. 67, 902 (1971).
11. L. Brewer and G. M. Rosenblatt in "Advances in High Temperature Chemistry,", Vol. 2, Ed. by L. Eyring, Academic Press, New York, 1969.
12. C. J. Cheetham, W. J. M. Gissane, and R. F. Barrow, Trans. Faraday Soc. 61, 1308 (1965).

Barium Fluoride, Ion (BaF^+) $Ba_1F_1^{\ +}$(g)

Enthalpy Reference Temperature = T_r = 298.15 K Standard State Pressure = p° = 0.1 MPa

	J $K^{-1}mol^{-1}$			kJ mol^{-1}			
T/K	$C_p°$	S°	$-[G°-H°(T_r)]/T$	$H°-H°(T_r)$	$\Delta_f H°$	$\Delta_f G°$	Log K_f
0	0.	0.	INFINITE	-9.205	146.127		
100	29.331	204.761	267.690	-6.293			
200	31.866	225.783	242.009	-3.245			
250	33.131	233.035	239.511	-1.619			
298.15	34.073	238.955	238.955	0.	150.206	121.564	-21.298
300	34.105	239.166	238.956	0.063	150.226	121.387	-21.135
350	34.831	244.480	239.374	1.787	150.749	116.538	-17.392
400	35.375	249.169	240.311	3.543	151.158	111.620	-14.576
450	35.787	253.360	241.532	5.323	151.367	106.663	-12.381
500	36.105	257.148	242.907	7.120	151.313	101.696	-10.624
600	36.557	263.773	245.848	10.755	150.668	91.843	-7.996
700	36.859	269.433	248.823	14.427	150.827	82.015	-6.120
800	37.074	274.369	251.714	18.124	150.471	72.215	-4.715
900	37.240	278.746	254.479	21.840	150.532	62.429	-3.623
1000	37.381	282.677	257.106	25.571	150.591	52.636	-2.749
1100	37.520	286.246	259.595	29.316	142.278	43.658	-2.073
1200	37.678	289.518	261.954	33.076	142.101	34.701	-1.510
1300	37.877	292.541	264.192	36.853	142.030	25.754	-1.035
1400	38.139	295.357	266.319	40.653	142.049	16.810	-0.627
1500	38.481	298.000	268.344	44.484	142.144	7.861	-0.274
1600	38.919	300.497	270.276	48.353	142.305	-1.096	0.036
1700	39.461	302.872	272.124	52.271	142.521	-10.065	0.309
1800	40.114	305.145	273.896	56.249	142.783	-19.048	0.553
1900	40.875	307.334	275.599	60.297	143.084	-28.046	0.771
2000	41.741	309.452	277.239	64.427	143.412	-37.061	0.968
2100	42.700	311.512	278.822	68.648	143.759	-46.094	1.147
2200	43.739	313.522	280.354	72.970	4.593	-49.767	1.182
2300	44.843	315.490	281.839	77.398	5.780	-52.265	1.187
2400	45.992	317.423	283.281	81.940	6.859	-54.813	1.193
2500	47.167	319.324	284.685	86.598	7.833	-57.403	1.199
2600	48.349	321.197	286.053	91.374	8.713	-60.030	1.206
2700	49.518	323.044	287.389	96.267	9.508	-62.689	1.213
2800	50.656	324.865	288.695	101.276	10.233	-65.376	1.220
2900	51.747	326.662	289.973	106.397	10.903	-68.089	1.226
3000	52.776	328.434	291.226	111.623	11.537	-70.823	1.233
3100	53.732	330.180	292.454	116.949	12.151	-73.579	1.240
3200	54.605	331.900	293.660	122.367	12.762	-76.354	1.246
3300	55.389	333.593	294.845	127.868	13.392	-79.149	1.253
3400	56.079	335.257	296.009	133.442	14.045	-81.963	1.259
3500	56.673	336.891	297.154	139.080	14.735	-84.797	1.266
3600	57.171	338.495	298.280	144.773	15.472	-87.651	1.272
3700	57.576	340.067	299.388	150.511	16.263	-90.526	1.278
3800	57.890	341.607	300.479	156.285	17.112	-93.424	1.284
3900	58.119	343.114	301.553	162.087	18.022	-96.344	1.290
4000	58.266	344.587	302.610	167.906	18.993	-99.289	1.297
4100	58.340	346.027	303.652	173.737	20.026	-102.259	1.303
4200	58.345	347.433	304.678	179.572	21.117	-105.255	1.309
4300	58.289	348.805	305.688	185.404	22.261	-108.277	1.315
4400	58.179	350.144	306.683	191.228	23.455	-111.326	1.322
4500	58.020	351.450	307.663	197.038	24.693	-114.403	1.328
4600	57.819	352.723	308.629	202.831	25.978	-117.509	1.334
4700	57.583	353.964	309.581	208.601	27.317	-120.645	1.341
4800	57.316	355.173	310.518	214.346	28.665	-123.807	1.347
4900	57.024	356.352	311.441	220.063	30.043	-126.999	1.354
5000	56.712	357.501	312.351	225.750	31.435	-130.218	1.360
5100	56.383	358.621	313.247	231.405	32.864	-133.467	1.367
5200	56.042	359.713	314.131	237.027	34.285	-136.742	1.374
5300	55.692	360.777	315.001	242.613	35.712	-140.045	1.380
5400	55.336	361.815	315.858	248.165	37.141	-143.374	1.387
5500	54.977	362.827	316.703	253.680	38.601	-146.735	1.394
5600	54.618	363.814	317.535	259.160	40.086	-150.124	1.400
5700	54.259	364.778	318.356	264.604	41.516	-153.534	1.407
5800	53.903	365.718	319.164	270.012	42.964	-156.972	1.414
5900	53.551	366.637	319.961	275.385	44.522	-160.451	1.421
6000	53.204	367.534	320.747	280.722	45.943	-163.937	1.427

PREVIOUS: December 1972 (1 atm) CURRENT: December 1972 (1 bar)

Barium Fluoride, Ion (BaF^+) $Ba_1F_1^{\ +}$(g)

BARIUM FLUORIDE (BaF_2) CRYSTAL M_r = 175.326806

$S°(298.15\ K) = 96.40 \pm 0.42\ J\ K^{-1}\ mol^{-1}$

T_{trs} = 1240 K ($\alpha \to \beta$)

T_{trs} = 1480 K ($\beta \to \gamma$)

T_{fus} = 1641 K

$\Delta_f H°(0\ K) = -1207.5 \pm 4.2\ kJ\ mol^{-1}$

$\Delta_f H°(298.15\ K) = -1208.8 \pm 4.2\ kJ\ mol^{-1}$

$\Delta_{trs} H° = 0.0\ kJ\ mol^{-1}$

$\Delta_{trs} H° = 2.674 \pm 0.837\ kJ\ mol^{-1}$

$\Delta_{fus} H° = 23.359 \pm 0.628\ kJ\ mol^{-1}$

Enthalpy of Formation

Guntz (1) measured the heat of neutralization of $Ba(OH)_2$(aq) with HF(aq) as ΔH = -34.8 kcal mol^{-1} at 284 K. We estimate ΔC_p = 106 cal K^{-1} mol^{-1} (2) for $Ba(OH)_2 \cdot 660\ H_2O + 2(HF \cdot 110\ H_2O) \to BaF_2(cr) + 2\ H_2O(\ell)$ which gives $\Delta_r H°(298.15\ K)$ = -33.4 kcal mol^{-1}. This leads to $\Delta_f H°(298.15\ K)$ = -288.3 kcal mol^{-1} when combined with auxiliary data (3, 4, 5). Petersen (6) investigated the same reaction but with HF · 200 H_2O and found that 16% of the BaF_2 remained dissolved. After correcting for incomplete precipitation his final result was ΔH°(292.15 K) = -33.92 kcal mol^{-1} which gives $\Delta_f H°(298.15\ K)$ = -288.2 kcal mol^{-1}.

In the same study, Petersen (6) measured the heat of precipitation of BaF_2 from $BaCl_2$(aq) with AgF(aq) as ΔH°(292 K) = -35.02 kcal mol^{-1}. We estimate ΔC_p = 108 cal K^{-1} mol^{-1} (2) for 2 (AgF · 200 H_2O) + $BaCl_2$ · 400 $H_2O \to BaF_2(cr)$ + 2AgCl(cr) which gives ΔH°(298.15 K) = -34.4 kcal mol^{-1}. Using auxiliary data (4, 7), we obtain $\Delta_f H°(BaF_2$, cr, 298.15 K) = -288.4 kcal mol^{-1}. Petersen (6) also reported the heat of solution of BaF_2 as $\Delta_{sol}H°$ = 1.88 kcal mol^{-1}. From this result, we obtain $\Delta_f H°(298.15\ K)$ = -290.3 kcal mol^{-1}.

Stephen and Stephen (9) have selected a "best" value for the solubility of BaF_2 of 1.614 g/ℓ at 298.15 K which corresponds to $K_{sp} = 3.1196 \times 10^{-6}$. Using $\Delta_f G°(Ba^{+2}$, ∞ aq, 298.15 K) = -134.02 (4) and $\Delta_f G°(F^-$, ∞ aq, 298.15 K) = -66.96 kcal mol^{-1} (5), we calculate $\Delta_f G°(298.15\ K)$ = -275.45 kcal mol^{-1}. Based on the adopted functions, $\Delta_f H°(298.15\ K)$ = -287.5 kcal mol^{-1}.

Two equilibrium studies (10, 11) have also been reported. 2nd and 3rd law analyses of these data are given below.

Source	Reaction	T/K	No. of Points	$\Delta_f H°(298.15\ K)$, kcal mol^{-1} 2nd law	3rd law	Drift cal K^{-1} mol^{-1}	$\Delta_f H°(298.15\ K)$ kcal mol^{-1}
Domange (10)	A	1173-1373	3	41.0	62.5±3.4	16.9±1.0	-267.3
Hood and Woyski (11)	B	477-1033	9	5.9	5.8±1.0	-0.1±0.1	-279.8

Reaction: (A) $BaF_2(cr) + H_2O(g) \to 2\ HF(g) + BaO(cr)$ (B) $HF(g) + 0.5\ BaCl_2(cr) \to HCl(g) + 0.5\ BaF_2(cr)$

The large discrepancy which arises in the $\Delta_f H°$values obtained from the equilibrium and calorimetric results is not well understood. Domange (10) has measured similar equilibria for CoF_2, MgF_2, and CaF_2 which are very consistent with JANAF heats of formation. Very recently, Finch et al. (12) attempted to determine the heat of precipitation of BaF_2 form $BaCl_2$(aq) by dissolving a slight excess of NaF. X-ray diffraction studies of the precipitate showed the formation of BaClF and Ba_2ClF_3 along with BaF_2. We believe the equilibrium results of Hood and Woyski (11) may have been influenced by the formation of mixed halides. Also, it seems quite likely that the calorimetric results of Petersen (6) on the precipitation of BaF_2 from $BaCl_2$(aq) with AgF(aq) would be similarly influenced. We adopt $\Delta_f H°(298.15\ K) = -288.9 \pm 1.0$ kcal mol^{-1} which is a weighted value of the calorimetric results of Guntz (1) and Petersen (6) on the heat of neutralization of $Ba(OH)_2$(aq), the heat of solution of BaF_2(cr) (6) and the K_{sp} value obtained from the solubility data of Stephen and Stephen (9). Barany et al. (8) determined the heat of solution of BaO(cr) in HF(aq), which contained some dissolved silica, as ΔH° = -73.04 kcal mol^{-1}. We assume this value refers to the reaction BaO(cr) + 2 HF · 4.41 $H_2O \to BaF_2(cr) + H_2O(\ell)$, and we calculate $\Delta_f H°(298.15\ K) = -290.2 \pm 2.0$ kcal mol^{-1} with $\Delta_f H°$(BaO, cr, 298.15 K) = - 132.3 kcal mol^{-1} (5). These results substantiate our adopted $\Delta_f H°$ value.

Heat Capacity and Entropy

C_p below 298 K is based on the low temperature (14-300 K) heat capacities of Pitzer et al. (13). These data show an unusual leveling off of C_p above 274 K, and values in the range 274-300 K are obtained by graphical extrapolation of their low temperature data. S°(298.15 K) is based on the adopted C_p's with S°(15 K) = 0.09 cal K^{-1} mol^{-1}.

Relative enthalpies have been reported by Efremova and Matizen (757-2053 K, 14) and Krestovnikov and Karetnikov (288-1273 K, 15). The enthalpy data of Efremova and Matizen (14) indicated the existence of a second order phase transition at 1240 K. We adopt their smoothed C_p data for 800-1480 K. C_p's below 800 K are estimated graphically by comparison with those for $BaCl_2$, CaF_2, and SrF_2 (5). C_p (BaF_2, γ) are based on a JANAF curve fit of their relative enthalpies (1482-1596 K). The rapid rise in the enthalpy near the melting point is assumed to be due to premelting, and these points were not used in the fit. The maximum deviation of our adopted functions from those used in the fit is 0.5%. Kelley (16) analyzed the enthalpy data of Krestovnikov and Karetnikov (15). His smoothed enthalpies are consistently 2-3% higher than those adopted. No second order phase transition was observed. These data are judged to be less reliable and are not used in our evaluation.

Transition Data

The transition temperatures are those reported by Efremova and Matizen (14). We assume the transition at 1240 K is of the second order type, since no discontinuity appears in their enthalpy data at this temperature. The heat of the β-γ transition is calculated as the difference between the adopted enthalpies for these two phases at 1480 K.

Fusion Data

See the liquid table for details.

Sublimation Data

See the ideal gas table for details.

References

1. M. Guntz, Ann. Chim. Phys. 3, 5 (1884).
2. $\Phi_c(Ba(OH)_2 \cdot 660\ H_2O)$ = [-60] cal K^{-1} mol^{-1}; $\Phi_c(HF \cdot 110\ H_2O)$ = 3.7 cal K^{-1} mol^{-1}; $\Phi_c(AgF \cdot 200\ H_2O)$ = -[3.0] cal K^{-1} mol^{-1}; $\Phi_c(BaCl_2 \cdot 400\ H_2O)$ = [-60] cal K^{-1} mol^{-1}.
3. U. S. Natl. Bur. Std. Tech. Note 270-3, 1968.
4. U. S. Natl. Bur. Std. Tech. Note 270-6, 1971.
5. JANAF Thermochemical Tables: Consistent with HF(g), 12-31-68; BaO(cr), $BaCl_2$(cr), SrF_2(cr), 12-31-72; CaF_2(cr), 12-31-68.
6. E. Petersen, Z. Physik Chem. 4, 384 (1889).
7. U. S. Natl. Bur. Std. Tech. Note 270-4, 1969.
8. R. Barany, E. G. King, and S. S. Todd, J. Amer. Chem. Soc. 79, 3639 (1957).
9. H. Stephen and T. Stephen, "Solubilities of Inorganic and Organic Compounds," Vol. I, Pergamon Press, Ltd., London, 1963.
10. L. Domange, Ann. Chim. 7, 225 (1937).
11. G. C. Hood and M. M. Woyski, J. Amer. Chem. Soc 73, 2738 (1951).
12. A. Finch, P. J. Gardner, and C. J. Steadman, Can. J. Chem. 46, 3447 (1965).
13. K. S. Pitzer, W. V. Smith, and W. M. Latimer, J. Amer. Chem. Soc. 60, 1826 (1938).
14. R. I Efremova and E. V. Matizen, Izv. Sib. Otd. Akad. Nauk SSSR, Ser. Khim. Nauk, (1), 3 (1970).
15. A. N. Krestovnikov and G. A. Karetnikov, Legkie Metal., 3, 29 (1934).
16. K. K. Kelley, U. S. Bur. Mines Bulletin 584, 1960.

Barium Fluoride (BaF_2) Ba_1F_2(cr)

Enthalpy Reference Temperature = T_r = 298.15 K Standard State Pressure = p° = 0.1 MPa

T/K	$C_p°$ (J $K^{-1}mol^{-1}$)	S°	-[G°-H°(T_r)]/T	H°-H°(T_r) (kJ mol^{-1})	$\Delta_f H°$	$\Delta_f G°$	Log K_f
0	0.	0.	INFINITE	-14.454	-1207.475	-1207.475	INFINITE
100	43.897	29.894	155.326	-12.543	-1210.169	-1192.713	623.010
200	65.796	68.764	102.826	-6.812	-1209.912	-1175.206	306.932
298.15	72.204	96.397	96.397	0.	-1208.758	-1158.410	202.949
300	72.304	96.844	96.398	0.134	-1208.734	-1158.097	201.643
400	75.940	118.170	99.281	7.556	-1207.548	-1141.405	149.052
500	78.450	135.397	104.837	15.280	-1207.007	-1124.951	117.523
600	80.333	149.864	111.167	23.218	-1207.165	-1108.515	96.505
700	82.383	162.395	117.610	31.350	-1206.400	-1092.143	81.497
800	84.935	173.549	123.917	39.705	-1205.980	-1075.841	70.245
900	88.701	183.750	130.005	48.370	-1204.877	-1059.636	61.500
1000	94.558	193.369	135.864	57.505	-1203.338	-1043.576	54.511
1100	104.182	202.789	141.520	67.396	-1209.444	-1026.882	48.763
1200	130.541	212.702	147.030	78.807	-1205.921	-1010.421	43.983
1240.000	194.914	217.601	149.222	84.790	ALPHA <--> BETA	TRANSITION	
1240.000	194.713	217.601	149.222	84.790			
1300	137.654	225.182	152.565	94.403	-1198.136	-994.440	39.957
1400	114.014	234.302	158.091	106.695	-1193.600	-978.956	36.525
1480.000	107.104	240.427	162.380	115.511	BETA <--> GAMMA	TRANSITION	
1480.000	107.654	242.234	162.380	118.184			
1500	107.654	243.679	163.454	120.337	-1187.676	-963.778	33.562
1600	107.654	250.627	168.687	131.103	-1184.612	-948.951	30.980
1641.000	107.654	253.351	170.769	135.517	GAMMA <--> LIQUID		
1700	107.654	257.153	173.701	141.868	-1181.550	-934.316	28.708
1800	107.654	263.307	178.510	152.634	-1178.508	-919.861	26.694
1900	107.654	269.127	183.128	163.399	-1175.506	-905.574	24.896
2000	107.654	274.649	187.567	174.164	-1172.561	-891.443	23.282

PREVIOUS: CURRENT: December 1972

BARIUM FLUORIDE (BaF_2) LIQUID M_r = 175.326806

S°(298.15 K) = [121.254] J K^{-1} mol^{-1} $\Delta_f H°$(298.15 K) = [-1171.294] kJ mol^{-1}

T_{fus} = 1641 K $\Delta_{fus}H°$ = 23.359 ± 0.628 kJ mol^{-1}

Enthalpy of Formation

The enthalpy of formation of the liquid is calculated from that of the crystal by addition of $\Delta_{fus}H°$ and the difference in enthalpy, H°(1641 K) - H°(298.15 K), between the crystal and liquid.

Heat Capacity and Entropy

A glass transition is assumed at 1100 K below which C_p° follows that of the crystal. C_p° is assumed constant in the temperature range 1100-3500 K and is based on a JANAF curve fit of the relative enthalpies of Efremova and Matizen (1) for the liquid (1641-1900 K). Some scatter was observed in these data at the higher temperatures, and these points were not included in the fit. The maximum deviation of the observed enthalpies which were used in the fit from those adopted is 0.29%.

S°(298.15 K) is calculated in a manner analogous to that used for the enthalpy of formation.

Fusion Data

T_{fus} is the value selected by Efremova and Matizen (1) from their relative enthalpy data. These data show premelting starting near 1620 K and complete melting occurring at 1646 K. Other values for T_{fus} that have been reported are 1627 (2), 1628 (3), 1627 (4), 1617 (5), and 1593 K (6). The selected T_{fus} was obtained from measurements made on samples of BaF_2 contained in steel and molybdenum capsules. There were no indications of any reaction between the sample and container. Three of the lower values given above were determined from measurements on BaF_2 contained in either graphite (3, 4) or tantalum (2) crucibles. It has now been established by mass spectroscopy (7, 8) that the alkaline earth difluorides can be appreciably reduced by these materials near 1400 K. Thus, we prefer the higher value.

$\Delta_{fus}H°$ is obtained as the difference in the adopted functions for the liquid and crystal at T_{fus}.

Vaporization Data

T_{vap} is the calculated temperature at which $\Delta_r G°$ = 0 for $BaF_2(\ell) \rightarrow BaF_2(g)$. $\Delta_{vap}H°$ is calculated from the difference in $\Delta_f H°(T_{vap})$ for the gas and liquid. Ruff and Le Boucher (9) derived T_{vap} = 2533 K and $\Delta_{vap}H°$ = 69.8 kcal mol^{-1} from an analysis of their vapor pressures (1960-2206 K).

References

1. R. I. Efremova and E. V. Matizen, Izv. Sib. Otd. Akal. Nauk SSSR, Ser. Khim. Nauk, (1), 3 (1970).
2. A. S. Dworkin and M. A. Bredig, J. Phys. Chem. 75, 2340 (1971).
3. H. Kojima, S. G. Whiteway, and C. R. Masson, Can. J. Chem. 46, 2968 (1968).
4. B. Porter and E. A. Brown, J. Amer. Ceram. Soc. 45, 49 (1962).
5. G. Petit and F. Delbove, Compt. rend. 254, 1388 (1962).
6. P. E. Hart and A. W. Searcy, J. Phys. Chem. 70, 2763 (1966).
7. D. L. Hildenbrand, J. Chem. Phys. 48, 3657 (1968).
8. J. W. Green, G. D. Blue, T. C. Ehlert, and J. L. Margrave, J. Chem. Phys. 41, 2245 (1964).
9. O. Ruff and L. Le Boucher, Z. Anorg. Chem. 219, 376 (1934).

Barium Fluoride (BaF_2) $Ba_1F_2(l)$

Enthalpy Reference Temperature = T_r = 298.15 K Standard State Pressure = p° = 0.1 MPa

T/K	C_p° (J $K^{-1}mol^{-1}$)	S° (J $K^{-1}mol^{-1}$)	$-[G°-H°(T_r)]/T$ (J $K^{-1}mol^{-1}$)	$H°-H°(T_r)$ (kJ mol^{-1})	$\Delta_f H°$ (kJ mol^{-1})	$\Delta_f G°$ (kJ mol^{-1})	Log K_f
0							
100							
200							
298.15	72.203	121.254	121.254	0.000	-1171.294	-1128.357	197.684
300	72.304	121.701	121.256	0.134	-1171.270	-1128.091	196.418
400	75.940	143.027	124.139	7.555	-1170.085	-1113.885	145.458
500	78.450	160.253	129.694	15.280	-1169.544	-1099.916	114.907
600	80.333	174.720	136.024	23.218	-1169.702	-1085.965	94.542
700	82.383	187.251	142.467	31.349	-1168.937	-1072.079	80.000
800	84.935	198.405	148.774	39.705	-1168.517	-1058.263	69.097
900	88.701	208.606	154.862	48.369	-1167.414	-1044.544	60.624
1000	94.558	218.222	160.721	57.501	-1165.879	-1030.969	53.852
1100	104.182	227.655	166.377	67.407	-1171.970	-1016.761	48.282
1100.000	104.182	227.655	166.377	67.407	———	GLASS <--> LIQUID	———
1100.000	99.826	227.655	166.377	67.407		TRANSITION	
1200	99.826	236.341	171.850	77.389	-1169.875	-1002.742	43.648
1300	99.826	244.332	177.123	87.372	-1167.704	-988.902	39.735
1400	99.826	251.730	182.191	97.354	-1165.476	-975.232	36.386
1500	99.826	258.617	187.059	107.337	-1163.213	-961.722	33.490
1600	99.826	265.060	191.735	117.320	-1160.931	-948.363	30.961
1641.000	99.826	267.585	193.598	121.413	-------	GAMMA <--> LIQUID	------
1700	99.826	271.111	196.228	127.302	-1158.652	-935.147	28.734
1800	99.826	276.817	200.548	137.285	-1156.393	-922.065	26.758
1900	99.826	282.215	204.705	147.267	-1154.174	-909.108	24.993
2000	99.826	287.335	208.710	157.250	-1152.011	-896.266	23.408
2100	99.826	292.206	212.571	167.233	-1149.925	-883.531	21.977
2200	99.826	296.850	216.297	177.215	-1287.453	-865.517	20.550
2300	99.826	301.287	219.897	187.198	-1284.735	-846.400	19.222
2400	99.826	305.536	223.377	197.181	-1282.236	-827.396	18.008
2500	99.826	309.611	226.745	207.163	-1279.952	-808.492	16.893
2600	99.826	313.526	230.008	217.146	-1277.876	-789.675	15.865
2700	99.826	317.293	233.172	227.128	-1275.994	-770.935	14.915
2800	99.826	320.924	236.241	237.111	-1274.288	-752.260	14.034
2900	99.826	324.427	239.222	247.094	-1272.736	-733.644	13.214
3000	99.826	327.811	242.119	257.076	-1271.316	-715.080	12.451
3100	99.826	331.084	244.936	267.059	-1270.001	-696.561	11.737
3200	99.826	334.254	247.678	277.041	-1268.767	-678.082	11.069
3300	99.826	337.326	250.349	287.024	-1267.581	-659.642	10.441
3400	99.826	340.306	252.951	297.007	-1266.430	-641.237	9.851
3500	99.826	343.199	255.488	306.989	-1265.291	-622.866	9.296

PREVIOUS: CURRENT: December 1972

Barium Fluoride (BaF_2) $Ba_1F_2(l)$

BARIUM FLUORIDE (BaF_2)

ALPHA -BETA - GAMMA - LIQUID

0 to 1240 K	crystal, alpha
1240 K to 1480 K	crystal, beta
1480 K to 1641 K	crystal, gamma
above 1641 K	liquid

Refer to the individual tables for details.

M_r = 175.326806

Barium Fluoride (BaF_2) Ba_1F_2(cr,l)

Enthalpy Reference Temperature = T_r = 298.15 K Standard State Pressure = p° = 0.1 MPa

T/K	C_p° (J K⁻¹mol⁻¹)	S° (J K⁻¹mol⁻¹)	$-[G^\circ-H^\circ(T_r)]/T$ (J K⁻¹mol⁻¹)	$H^\circ-H^\circ(T_r)$ (kJ mol⁻¹)	Δ_fH° (kJ mol⁻¹)	Δ_fG° (kJ mol⁻¹)	Log K_f
0	0.	0.	INFINITE	-14.454	-1207.475	-1207.475	INFINITE
100	43.897	29.894	155.326	-12.543	-1210.169	-1192.713	623.010
200	65.796	68.764	102.826	-6.812	-1209.912	-1175.206	306.932
298.15	72.204	96.397	96.397	0.	-1208.758	-1158.410	202.949
300	72.304	96.844	96.398	0.134	-1208.734	-1158.097	201.643
400	75.940	118.170	99.281	7.556	-1207.548	-1141.405	149.052
500	78.450	135.397	104.837	15.280	-1207.007	-1124.951	117.523
600	80.333	149.864	111.167	23.218	-1207.165	-1108.515	96.505
700	82.383	162.395	117.610	31.350	-1206.400	-1092.143	81.497
800	84.935	173.549	123.917	39.705	-1205.980	-1075.841	70.245
900	88.701	183.750	130.005	48.370	-1204.877	-1059.636	61.500
1000	94.558	193.369	135.864	57.505	-1203.338	-1043.576	54.511
1100	104.182	202.789	141.520	67.396	-1209.444	-1026.882	48.763
1200	130.541	212.702	147.030	78.807	-1205.921	-1010.421	43.983
1240.000	194.914	217.601	149.222	84.790	ALPHA <--> BETA		
1240.000	194.713	217.601	149.222	84.790	TRANSITION		
1300	137.654	225.182	152.565	94.403	-1198.136	-994.440	39.957
1400	114.014	234.302	158.091	106.695	-1193.600	-978.956	36.525
1480.000	107.104	240.427	162.380	115.511	BETA <--> GAMMA		
1480.000	107.654	242.234	162.380	118.184	TRANSITION		
1500	107.654	243.679	163.454	120.337	-1187.676	-963.778	33.562
1600	107.654	250.627	168.687	131.103	-1184.612	-948.951	30.980
1641.000	107.654	253.351	170.769	135.517	GAMMA <--> LIQUID		
1641.000	99.826	267.585	170.769	158.876	TRANSITION		
1700	99.826	271.111	174.191	164.766	-1158.652	-935.148	28.734
1800	99.826	276.817	179.735	174.748	-1156.393	-922.066	26.758
1900	99.826	282.215	184.988	184.731	-1154.174	-909.108	24.993
2000	99.826	287.335	189.978	194.713	-1152.012	-896.266	23.408
2100	99.826	292.206	194.731	204.696	-1149.925	-883.531	21.977
2200	99.826	296.850	199.268	214.679	-1287.454	-865.517	20.550
2300	99.826	301.287	203.608	224.661	-1284.736	-846.400	19.222
2400	99.826	305.536	207.767	234.644	-1282.236	-827.396	18.008
2500	99.826	309.611	211.760	244.626	-1279.952	-808.492	16.893
2600	99.826	313.526	215.599	254.609	-1277.876	-789.675	15.865
2700	99.826	317.293	219.296	264.592	-1275.994	-770.935	14.915
2800	99.826	320.924	222.862	274.574	-1274.288	-752.261	14.034
2900	99.826	324.427	226.304	284.557	-1272.736	-733.645	13.214
3000	99.826	327.811	229.631	294.539	-1271.316	-715.080	12.451
3100	99.826	331.084	232.851	304.522	-1270.001	-696.561	11.737
3200	99.826	334.254	235.971	314.505	-1268.767	-678.082	11.069
3300	99.826	337.326	238.996	324.487	-1267.581	-659.642	10.441
3400	99.826	340.306	241.932	334.470	-1266.430	-641.237	9.851
3500	99.826	343.199	244.784	344.452	-1265.291	-622.866	9.296

PREVIOUS: CURRENT: December 1972

Barium Fluoride (BaF_2) Ba_1F_2(cr,l)

BARIUM FLUORIDE (BaF_2) IDEAL GAS M_r = 175.326806

$\Delta_f H°(0\ K) = -801.7 \pm 6.3\ kJ\ mol^{-1}$

$S°(298.15\ K) = 301.30 \pm 2.09\ J\ K^{-1}\ mol^{-1}$ $\Delta_f H°(298.15\ K) = -803.7 \pm 6.3\ kJ\ mol^{-1}$

Vibrational Frequencies, ν/cm^{-1}, and Degeneracies

413(1) [64](1) 390(1)

Ground State Quantum Weight: 1 $\sigma = 2$

Point Group: C_{2v}

Bond Distance: Ba-F = 2.32 ± 0.03 Å

Bond Angle: F-Ba-F = [95]°

Product of Moments of Inertia: $I_A I_B I_C = [6.858791 \times 10^{-114}]\ g^3 cm^6$

Enthalpy of Formation

$\Delta_f H°(BaF_2, g, 298.15\ K) = -192.1 \pm 1.5\ kcal\ mol^{-1}$ ($-803.7 \pm 6.3\ kJ\ mol^{-1}$) is obtained from that of the crystal by addition of $\Delta_{sub}H° = 96.83 \pm 0.50\ kcal\ mol^{-1}$. The selected $\Delta_{sub}H°$ is based on an analysis of six sets of vapor pressure data (1 - 5). The results of our analysis are given below. A weighted value of these six results is adopted.

Source	Method	T/K	Data Points	$\Delta_{sub}H°$(298.15 K), kcal mol^{-1} 2nd law	3rd law	Drift cal K^{-1} mol^{-1}	$\Delta_f H°$(298.15 K) kcal mol^{-1}
Ruff and Le Boucher (1)	Dynamic B.P.	2061-2206	9	97.0[a]	86.79±0.88[a]	-4.9±0.8	-193.2±2.0
Green et al. (2)	Knudsen-Mass Spec.	1232-1503	20	96.5	98.17±0.45	1.2±0.6	-190.7±1.5
Bautista and Margrave (3)	Langmuir	1130-1250	10[b]	96.4	97.68±0.19	1.1±0.9	-191.2±1.2
Hart and Searcy (4)	Torsion-Effusion	1261-1548	49[c]	93.2	96.12±0.68	2.1±0.7	-192.8±1.7
Hart and Searcy (4)	Torsion-Langmuir	1315-1492	25[d]	93.7	96.47±0.38	2.0±0.8	-192.4±1.4
Hildenbrand et al. (5)	Torsion-Effusion	1265-1550	Equation	94.2	94.74±0.13	0.4	-194.2±1.1

[a] $\Delta_{vap}H°$, 2[b], 4[c], and 3[d] points rejected due to failure of a statistical test.

Heat Capacity and Entropy

The bond length is taken from the high-temperature electron diffraction studies of Akishin and Spiridonov (6). Although Akishin and Spiridonov (6) interpreted their results in terms of a linear configuration for BaF_2, recent electric-deflection experiments (7) and matrix isolation infrared studies (8, 9) clearly indicate that BaF_2 is bent. Calder et al. (8) estimated the bond angle as 100° from the changes in the bond angles with mass of the central atom for other alkaline earth difluorides. Baikov (10) estimated the angle as 115°. We prefer a slightly lower value (95°) than these, since there are indications that other heavy metal halides have bond angles below 100°. For example, bond angles for PbF_2 and $PbCl_2$ have been reported as 90° (11) and 96 ± 3° (12), respectively.

The symmetric (ν_1) and antisymmetric (ν_3) stretching frequencies are those observed by Calder et al. (8) in their study of infrared spectra of BaF_2 trapped in a krypton matrix. Other matrix frequencies in excellent agreement with the adopted ones have been reported by Snelson (9). Further confirmation of the adopted ν_3 is provided by observations of Baikov (13). He reported $\nu_3 = 415 \pm 7\ cm^{-1}$ which was observed in the infrared spectra of BaF_2 vapor at 1950 K. The bending frequency (ν_2) is that estimated by Calder et al. (8). This value is considerably less than that observed by Baikov (10) ($\nu_2 = 100\ cm^{-1}$). However, only the high frequency portion of this rather broad band was recorded. We have used our adopted frequencies in a valence bond calculation with the result $k/(k\ \delta/\ell^2) = 11.8$. This ratio of the stretching to bending force constants appears quite reasonable in comparison with those for other alkaline earth difluorides (14). Also, our adopted frequencies give thermodynamic functions which show good compatibility with the vapor pressure data (see Enthalpy of Formation section).

The principal moments of inertia are: $I_A = 30.6016 \times 10^{-39}$, $I_B = 18.4604 \times 10^{-39}$, and $I_C = 12.1412 \times 10^{-39}\ g\ cm^2$.

References

1. O. Ruff and L. Le Boucher, Z. Anorg. Chem. 219, 376 (1934).
2. J. W. Green, G. D. Blue, T. C. Ehlert, and J. L. Margrave, J. Chem. Phys. 41, 2245 (1964).
3. R. G. Bautista and J. L. Margrave, J. Phys. Chem. 69, 1770 (1965).
4. P. E. Hart and A. W. Searcy, J. Phys. Chem. 70, 2763 (1966).
5. D. L. Hildenbrand, E. Murad, N. D. Potter, L. P. Theard, and W. F. Hall, Aeronutronic Report No. U-3183, Ford Motor Company, June 1965.
6. P. A. Akishin and V. P. Spiridonov, Kristallografiya, 2, 475 (1957).
7. L. Wharton, R. A. Berg, and W. Klemperer, J. Chem. Phys. 39, 2023 (1963).
8. V. Calder, D. E. Mann, K. S. Seshadri, M. Allavena, and D. White, J. Chem. Phys. 51, 2093 (1969).
9. A. Snelson, J. Phys. Chem. 70, 3208 (1966).
10. V. I. Baikov, Opt. Spektrosk. 25, 356 (1967).
11. J. W. Hastie, R. H. Hauge, and J. L. Margrave, in "Annual Review of Physical Chemistry," Vol. 21, Ed. by H. Eyring, Annual Reviews, Inc., Palo Alto, California, 1970.
12. J. W. Hastie, R. H. Hauge and J. L. Margrave, High Temp. Sci. 3, 56 (1971).
13. V. I. Baikov, Opt. Spektrosk. 27, 502 (1969).
14. JANAF Thermochemical Tables: BeF_2(g), 6-30-70; MgF_2(g), 3-31-66; CaF_2(g), 12-31-68; SrF_2(g), 12-31-72.

Barium Fluoride (BaF_2) Ba_1F_2(g)

Enthalpy Reference Temperature = T_r = 298.15 K Standard State Pressure = p° = 0.1 MPa

T/K	$C_p°$ (J K^{-1}mol^{-1})	S°	$-[G°-H°(T_r)]/T$	$H°-H°(T_r)$ (kJ mol^{-1})	$\Delta_f H°$	$\Delta_f G°$	Log K_f
0	0.	0.	INFINITE	-13.649	-801.659	-801.659	INFINITE
100	42.748	248.473	346.341	-9.787	-802.402	-806.804	421.431
200	50.089	280.512	306.118	-5.121	-803.210	-810.853	211.773
250	52.332	291.947	302.175	-2.557	-803.487	-812.732	169.811
298.15	53.780	301.297	301.297	0.	-803.746	-814.489	142.695
300	53.826	301.630	301.298	0.100	-803.757	-814.556	141.827
350	54.840	310.008	301.957	2.818	-804.074	-816.332	121.831
400	55.551	317.380	303.434	5.579	-804.514	-818.055	106.827
450	56.064	323.954	305.355	8.370	-805.167	-819.711	95.150
500	56.444	329.882	307.516	11.183	-806.094	-821.280	85.799
600	56.957	340.222	312.130	16.855	-808.517	-824.081	71.743
700	57.276	349.027	316.788	22.568	-810.171	-826.557	61.678
800	57.487	356.690	321.307	28.307	-812.368	-828.742	54.111
900	57.634	363.470	325.622	34.063	-814.172	-830.680	48.211
1000	57.740	369.548	329.716	39.832	-816.000	-832.417	43.481
1100	57.819	375.055	333.592	45.610	-826.218	-833.150	39.563
1200	57.879	380.089	337.260	51.395	-828.321	-833.686	36.289
1300	57.926	384.724	340.735	57.185	-830.342	-834.051	33.513
1400	57.964	389.018	344.032	62.980	-832.303	-834.262	31.127
1500	57.994	393.018	347.166	68.778	-834.224	-834.335	29.054
1600	58.019	396.762	350.150	74.579	-836.125	-834.280	27.236
1700	58.040	400.280	352.997	80.382	-838.025	-834.107	25.629
1800	58.057	403.598	355.716	86.186	-839.944	-833.821	24.197
1900	58.072	406.737	358.320	91.993	-841.900	-833.428	22.913
2000	58.084	409.716	360.816	97.801	-843.913	-832.930	21.754
2100	58.095	412.550	363.212	103.610	-846.001	-832.330	20.703
2200	58.104	415.253	365.517	109.420	-987.701	-826.253	19.618
2300	58.113	417.836	367.736	115.231	-989.155	-818.883	18.597
2400	58.120	420.310	369.875	121.042	-990.827	-811.444	17.661
2500	58.126	422.682	371.941	126.854	-992.713	-803.932	16.797
2600	58.132	424.962	373.936	132.667	-994.807	-796.340	15.999
2700	58.137	427.156	375.867	138.481	-997.094	-788.664	15.258
2800	58.141	429.271	377.737	144.295	-999.556	-780.900	14.568
2900	58.145	431.311	379.549	150.109	-1002.173	-773.045	13.924
3000	58.149	433.282	381.308	155.924	-1004.920	-765.098	13.322
3100	58.152	435.189	383.015	161.739	-1007.773	-757.057	12.756
3200	58.155	437.035	384.675	167.554	-1010.706	-748.923	12.225
3300	58.158	438.825	386.288	173.370	-1013.687	-740.696	11.724
3400	58.160	440.561	387.859	179.186	-1016.703	-732.379	11.252
3500	58.163	442.247	389.389	185.002	-1019.730	-723.972	10.805
3600	58.165	443.885	390.880	190.818	-1022.748	-715.479	10.381
3700	58.167	445.479	392.335	196.635	-1025.740	-706.902	9.980
3800	58.168	447.030	393.754	202.452	-1028.692	-698.246	9.598
3900	58.170	448.541	395.139	208.268	-1031.593	-689.512	9.235
4000	58.172	450.014	396.493	214.086	-1034.434	-680.704	8.889
4100	58.173	451.451	397.816	219.903	-1037.208	-671.827	8.559
4200	58.174	452.852	399.110	225.720	-1039.910	-662.882	8.244
4300	58.176	454.221	400.375	231.538	-1042.540	-653.874	7.943
4400	58.177	455.559	401.614	237.355	-1045.096	-644.806	7.655
4500	58.178	456.866	402.828	243.173	-1047.577	-635.680	7.379
4600	58.179	458.145	404.016	248.991	-1049.978	-626.501	7.114
4700	58.180	459.396	405.181	254.809	-1052.287	-617.272	6.860
4800	58.181	460.621	406.324	260.627	-1054.547	-607.992	6.616
4900	58.181	461.821	407.444	266.445	-1056.734	-598.667	6.382
5000	58.182	462.996	408.543	272.263	-1058.862	-589.297	6.156
5100	58.183	464.148	409.622	278.081	-1060.907	-579.888	5.939
5200	58.184	465.278	410.682	283.900	-1062.914	-570.436	5.730
5300	58.184	466.386	411.723	289.718	-1064.866	-560.947	5.528
5400	58.185	467.474	412.745	295.536	-1066.768	-551.421	5.334
5500	58.185	468.542	413.750	301.355	-1068.591	-541.864	5.146
5600	58.186	469.590	414.738	307.174	-1070.341	-532.278	4.965
5700	58.186	470.620	415.709	312.992	-1072.098	-522.655	4.790
5800	58.187	471.632	416.664	318.811	-1073.791	-513.004	4.620
5900	58.187	472.626	417.605	324.630	-1075.327	-503.340	4.456
6000	58.188	473.604	418.530	330.448	-1076.953	-493.632	4.297

PREVIOUS: December 1972 (1 atm) CURRENT: December 1972 (1 bar)

Barium Fluoride (BaF_2) Ba_1F_2(g)

BARIUM HYDROXIDE (BaOH) IDEAL GAS M_r = 154.3734

$\Delta_f H°(0\ K) = -221.75 \pm 29.3\ kJ\ mol^{-1}$

$S°(298.15\ K) = [252.9 \pm 8]\ J\ K^{-1}\ mol^{-1}$ $\Delta_f H°(298.15\ K) = -226.44 \pm 29.3\ kJ\ mol^{-1}$

Electronic Levels and Quantum Weights				Vibrational Frequencies and Degeneracies
ε_i, cm^{-1}	g_i	ε_i, cm^{-1}	g_i	ν, cm^{-1}
0	[2]	[19500]	[2]	[469] (1)
[11800]	[4]	[20500]	[2]	[431] (2)
[13500]	[2]			[3650] (1)

Bond Distances: Ba–O = [2.17] Å σ = [1]
O–H = [0.96] Å
Bond Angle: Ba–O–H = [180]°
Rotational Constant: B_0 = [0.222044]cm^{-1}

Enthalpy of Formation

The adopted $\Delta_f H°$(BaOH, g, 0 K) = −53.0±7.0 kcal mol^{-1} is based on an assessment of dissociation energies derived from Knudsen-cell mass-spectrometric measurements of Stafford and Berkowitz (1) and Newbury (2) and flame spectral work (3-6).

Source	Reaction	No. Points	T/K	$\Delta_r H°$(298.15 K), kcal mol^{-1} 3rd law	Drift cal K^{-1} mol^{-1}	$\Delta_r H°$(298.15 K) kcal mol^{-1}	D_0^0(Ba-OH) kcal mol^{-1}
	1	2	1755–1785	115.23±3.55	-83.6	-51.58±6	102.8
(1)	2	1	1785	62.11	--	-55.82	108.1
(2)	3	1	1851	77.33	--	-53.67	104.8
(5)	4	2	1570–1800	2.40±2.39	7.3	-64.70	115.9

Reaction 1) BaO(cr) + BaO(g) + H_2O(g) → 2BaOH(g) + 0.5 O_2(g) 3) BaO(cr) + 0.5 H_2(g) = BaOH(g)
2) 2BaO(g) + H_2O(g) → 2BaOH(g) + O(g) 4) Ba(g) + H_2O(g) → BaOH(g) + H(g)

Ryabova and Gurvich (3) considered the reaction Ba(g) + H_2O(g) → BaOH(g) + H(g) to be dominant and derived D_0^0 (as defined in the table above) = 111±8 kcal mol^{-1}. Sugden and Schofield (4) considered the reaction Ba(g) + $2H_2O$(g) → Ba$(OH)_2$(g) + 2H(g) to be dominant. Cotton and Jenkins (5) found both the mono- and dihydroxide to be present in significant amounts in fuel-rich hydrogen-oxygen-nitrogen flames and derived D_0^0 = 114±5 kcal mol^{-1}. Cotton and Jenkins (5) recalculated the work of Ryabova and Gurvich (3) and of Sugden and Schofield (4) considering both BaOH and Ba$(OH)_2$ to be present and obtained recalculated values of D_0^0 = 110 and 114 kcal mol^{-1}, respectively. Ryabova et al. (6) made further measurements, considered BaOH as dominant, and reported D_0^0 = 109±3 kcal mol^{-1}. The D_0^0 calculation from the 3rd law analysis of Cotton and Jenkins (5) data above, using auxiliary data (7), is about 2 kcal mol^{-1} higher than theirs. Applying this correction to the values from Ryabova and Gurvich (3) and Sugden and Schofield (4) as recalculated by Cotton and Jenkins (5), gives 112 and 116 kcal mol^{-1}, respectively. The value D_0^0 = 109±3 kcal mol^{-1} given by Ryabova et al. (6) is an average of four calculations with two sets of molecular constants assumed for BaOH. This value, if recalculated with the present auxiliary data, would probably be higher.

The dissociation energies derived from flame spectra are higher, as a group, than those derived from the Knudsen cell mass spectrometric measurements. The ratio of D_0^0 (monohydroxide)/D_0^0(dihydroxide), where D_0^0 of the dihydroxide is defined by the reaction Ba$(OH)_2$(g) → Ba(g) + 2 OH(g) and is 209.6 kcal mol^{-1} (7), is approximately 0.53-0.55 for the flame spectral measurements and 0.50 for the average D_0^0 = 105.3 kcal mol^{-1} for the Knudsen-cell mass spectrometric measurements. The ratio for the barium mono- and difluoride (7) is 0.51, and the ratio for the other alkaline earth halides fall in the 0.40 to 0.51 range (7) with an average close to 0.46 (8). The similarity of the halides and hydroxides, particularly the fluorides, has been established (1, 9-11). D_0^0 = 105.3 kcal mol^{-1} is adopted from which is calculated $\Delta_f H°$(0 K) = −53.0±7.0 kcal mol^{-1}.

Heat Capacity and Entropy

The molecular configuration is assumed to be linear in accordance with the prediction of Walsh (12) and the evidence that gaseous alkali metal hydroxides are linear (13-15). The ground state is assumed to be $^2\Sigma^+$ by analogy with BaF and BaCl (7). The electronic levels are estimated from the observed band spectra (16-20) and the comparison with BaF and BaCl (7).

The Ba-O bond distance is estimated to be slightly larger, 0.02 Å, than the Ba-F distance (7) after noting the close similarity in bond distance of the alkali metal fluorides and hydroxides. The O-H bond distance is that in water (7). The Ba-O stretching frequency, 469 cm^{-1}, is estimated to be the same as in BaF (7, 11). The O-H stretching frequency, 3650 cm^{-1}, is estimated from the alkali hydroxide series. The bending frequency, 431 cm^{-1}, is estimated by assuming that the ratio of the bending force constant to the stretching force constant is 0.022, which is the average ratio found in the alkali metal monohydroxides by Acquista and Abramowitz (11, 15).

The entropy in the present table is lower by 1.25 cal K^{-1} mol^{-1} at 298 K and 1.77 cal K^{-1} mol^{-1} at 1000 K than that proposed by Jackson (11); the vibrational frequencies and electronic levels adopted here differ from those of Jackson (11). The moment of inertia in Jackson (11) appears to be in error.

References

1. F. E. Stafford and J. Berkowitz, J. Chem. Phys. 40, 2963 (1964).
2. R. S. Newbury, U. S. At. Energy Comm. UCRL-12225-T, 1965.
3. V. G. Ryabova and L. V. Gurvich, Teplofiz. Vysokikh Temperatur, Akad. Nauk SSSR 3 318 (1965).
4. T. M. Sugden and K. Schofield, Trans. Faraday Soc. 62, 566 (1966).
5. D. H. Cotton and D. R. Jenkins, Trans. Faraday Soc. 64, 2988 (1968).
6. V. G. Ryabova, A. N. Khitrov, and L. V. Gurvich, Teplofiz. Vys. Temp. 10, 744 (1972).
7. JANAF Thermochemical Tables: BaO(cr), BaO(ℓ), H(g), O(g) 6-30-74; H_2(g), H_2O(g), 3-31-61; O_2(g), 9-30-65; Ba(g), 12-30-70; BaCl(g), BaF(g), BaF_2(g) 12-31-72.
8. G. D. Blue, J. W. Green, T. C. Ehlert, and J. L. Margrave, Nature 199, 804 (1963).
9. R. C. Schoonmaker and R. F. Porter, J. Chem. Phys. 31, 830 (1959).
10. J. Berkowitz, D. J. Meschi, and W. A. Chupka, J. Chem. Phys. 33, 533 (1960).
11. D. D. Jackson, Lawrence Livermore Laboratory, Univ. of Calif., Report UCRL-51137, Dec. 8, 1971.
12. A. D. Walsh, J. Chem. Soc. 1953, 2288 (1953).
13. N. Acquista, S. Abramowitz, and R. Lide, J. Chem. Phys. 49, 780 (1968).
14. R. L. Kuczkowski and R. Lide, J. Chem. Phys. 44, 3131 (1966).
15. N. Acquista and S. Abramowitz, U. S. Nat. Bur. Stand., Report 9905, July, 1968.
16. C. G. James and T. M. Sugden, Nature 175, 333 (1955).
17. E. M. Bulewicz, Nature 177, 670 (1956).
18. M. Charton and A. G. Gaydon, Proc. Phys. Soc. (London) 69A, 520 (1956).
19. V. F. Zhitkevich, A. I. Lyutyi, N. A. Nesterko, V. S. Rossikhin, and I. L. Tsikora, Optika i Spektroskopiya 14, 35 (1963).
20. J. Van der Hurk, Tj. Hollander, and C. Th. J. Alkemade, J. Quant. Spectrosc. Radiat. Transfer 13, 273 (1973).

Barium Hydroxide (BaOH) $Ba_1H_1O_1$(g)

Enthalpy Reference Temperature = T_r = 298.15 K Standard State Pressure = p° = 0.1 MPa

T/K	$C_p°$ (J K^{-1}mol^{-1})	S°	−[G°−H°(T_r)]/T	H°−H°(T_r) (kJ mol^{-1})	$\Delta_f H°$	$\Delta_f G°$	Log K_f
0	0.	0.	INFINITE	-10.799	-221.752	-221.752	INFINITE
100	30.847	211.167	289.791	-7.862	-223.465	-227.472	118.819
200	40.469	235.500	256.974	-4.295	-225.245	-230.727	60.260
250	43.984	244.931	253.644	-2.178	-225.894	-232.020	48.478
298.15	46.356	252.893	252.893	0.	-226.440	-233.150	40.847
300	46.432	253.180	252.894	0.086	-226.460	-233.191	40.602
350	48.141	260.474	253.466	2.453	-227.003	-234.271	34.963
400	49.361	266.986	254.757	4.892	-227.610	-235.269	30.723
450	50.257	272.854	256.447	7.383	-228.384	-236.182	27.415
500	50.937	278.186	258.358	9.914	-229.396	-236.997	24.759
600	51.916	287.565	262.466	15.060	-231.917	-238.270	20.743
700	52.641	295.625	266.641	20.289	-233.599	-239.207	17.850
800	53.266	302.695	270.715	25.585	-235.775	-239.854	15.661
900	53.855	309.003	274.625	30.941	-237.515	-240.259	13.944
1000	54.427	314.707	278.352	36.355	-239.241	-240.471	12.561
1100	54.983	319.921	281.898	41.826	-249.320	-239.691	11.382
1200	55.519	324.728	285.269	47.351	-251.251	-238.730	10.392
1300	56.029	329.192	288.478	52.929	-253.067	-237.612	9.547
1400	56.512	333.363	291.537	58.556	-254.793	-236.358	8.819
1500	56.968	337.277	294.457	64.230	-256.452	-234.983	8.183
1600	57.398	340.968	297.250	69.949	-258.065	-233.499	7.623
1700	57.807	344.460	299.925	75.709	-259.656	-231.915	7.126
1800	58.199	347.775	302.492	81.509	-261.244	-230.237	6.681
1900	58.578	350.932	304.959	87.348	-262.852	-228.471	6.281
2000	58.951	353.946	307.334	93.225	-264.501	-226.619	5.919
2100	59.320	356.831	309.622	99.138	-266.213	-224.683	5.589
2200	59.689	359.599	311.832	105.089	-407.527	-217.288	5.159
2300	60.063	362.261	313.967	111.076	-408.587	-208.618	4.738
2400	60.442	364.825	316.033	117.102	-409.858	-199.898	4.351
2500	60.828	367.300	318.034	123.165	-411.342	-191.120	3.993
2600	61.223	369.694	319.975	129.268	-413.030	-182.278	3.662
2700	61.627	372.012	321.860	135.410	-414.913	-173.368	3.354
2800	62.040	374.260	323.691	141.593	-416.972	-164.384	3.067
2900	62.460	376.445	325.473	147.818	-419.188	-155.325	2.798
3000	62.888	378.570	327.208	154.086	-421.539	-146.186	2.545
3100	63.321	380.639	328.898	160.396	-423.999	-136.967	2.308
3200	63.759	382.656	330.547	166.750	-426.546	-127.668	2.084
3300	64.200	384.625	332.156	173.148	-429.146	-118.289	1.872
3400	64.642	386.548	333.727	179.590	-431.787	-108.829	1.672
3500	65.083	388.428	335.263	186.076	-434.447	-99.292	1.482
3600	65.522	390.268	336.766	192.607	-437.104	-89.678	1.301
3700	65.957	392.069	338.236	199.181	-439.743	-79.991	1.129
3800	66.386	393.833	339.676	205.798	-442.351	-70.233	0.965
3900	66.807	395.563	341.087	212.458	-444.915	-60.407	0.809
4000	67.220	397.260	342.470	219.159	-447.428	-50.516	0.660
4100	67.622	398.925	343.827	225.901	-449.884	-40.562	0.517
4200	68.013	400.559	345.158	232.683	-452.277	-30.550	0.380
4300	68.391	402.164	346.465	239.503	-454.608	-20.481	0.249
4400	68.755	403.740	347.749	246.361	-456.876	-10.359	0.123
4500	69.104	405.289	349.011	253.254	-459.081	-0.186	0.002
4600	69.438	406.812	350.251	260.181	-461.217	10.034	-0.114
4700	69.756	408.309	351.470	267.141	-463.275	20.299	-0.226
4800	70.057	409.780	352.670	274.132	-465.298	30.610	-0.333
4900	70.342	411.228	353.850	281.152	-467.262	40.961	-0.437
5000	70.610	412.652	355.012	288.200	-469.182	51.353	-0.536
5100	70.860	414.052	356.156	295.273	-471.036	61.779	-0.633
5200	71.093	415.431	357.282	302.371	-472.869	72.244	-0.726
5300	71.309	416.787	358.392	309.491	-474.665	82.745	-0.816
5400	71.509	418.122	359.486	316.632	-476.430	93.279	-0.902
5500	71.691	419.436	360.564	323.793	-478.136	103.840	-0.986
5600	71.857	420.729	361.627	330.970	-479.790	114.430	-1.067
5700	72.007	422.002	362.675	338.163	-481.473	125.056	-1.146
5800	72.142	423.256	363.709	345.371	-483.114	135.709	-1.222
5900	72.261	424.490	364.729	352.591	-484.622	146.371	-1.296
6000	72.366	425.705	365.735	359.823	-486.244	157.080	-1.368

PREVIOUS: December 1975 (1 atm) CURRENT: December 1975 (1 bar)

Barium Hydroxide (BaOH) $Ba_1H_1O_1$(g)

BARIUM HYDROXIDE, ION ($BaOH^+$) IDEAL GAS $M_r = 154.336791$

$S°(298.15\ K) = [247.26 \pm 8.4]\ J\ K^{-1}\ mol^{-1}$

$\Delta_f H°(0\ K) = 269.36 \pm 62.8\ kJ\ mol^{-1}$
$\Delta_f H°(298.15\ K) = 270.90 \pm 62.8\ kJ\ mol^{-1}$

Vibrational Fequencies and Degeneracies

ν, cm^{-1}
[460] (1)
[430] (2)
[3650] (1)

Ground State Quantum Weight: [1]
Point Group: [$C_{\infty v}$]
Bond Distance: Ba-O = [2.17] Å O-H = [0.96] Å
Bond Angle: Ba-O-H = [180]°
Rotational Constant: B_0 = [0.222053] cm^{-1}
σ = [2]

Enthalpy of Formation

The ionization potential of BaOH(g) was deduced by Kelly and Padley (1) to be 5.25 ± 0.1 eV. These authors quantitatively examined the total positive ion concentrations produced from Ba aqueous salt additives in fuel rich, premixed $H_2 + O_2 + N_2$ flames. Using current JANAF auxiliary data (2), we recalculate the ionization potential to be 5.36 eV.

Jensen (3) determined the enthalpy of reaction $\Delta_r H° = 6 \pm 10\ kcal\ mol^{-1}$ for $Ba(g) + OH(g) = BaOH^+(g) + e^-$ in atmospheric pressure $H_2 + O_2 + N_2$ flames using the microwave cavity resonance method. This value was calculated assuming a bent molecule for $BaOH^+$. The value is not significantly changed, within the uncertainty, by the change in the configuration. Using auxiliary data (2), we derive an ionization potential of 4.83 eV, which is not in good agreement with the value derived from the data of Kelly and Padley (1).

We adopt an ionization potential of 5.09 eV (117.38 kcal mol^{-1}) which is an average of the above two studies (1, 3). This leads to $\Delta_f H°(0\ K) = 64.38\ kcal\ mol^{-1}$ and $\Delta_f H°(298.15\ K) = 64.75\ kcal\ mol^{-1}$ for $BaOH^+(g)$. We assign an uncertainty of ±15 kcal mol^{-1}.

For comparison, the ionization potential of BaF(g) is 4.83 eV (2) while that of Ba(g) is 5.21 eV (4). The average of these two values is very close to the ionization potential adopted for $BaOH^+$.

Heat Capacity and Entropy

The molecular configuration is assumed to be linear, since experimental evidence indicates that the gaseous alkali metal hydroxides are linear (2, 5, 6, 7). In addition, Walsh (8) had predicted tha BAH molecules (H = hydrogen atom) with ten or less valence electrons ($BaOH^+$ has eight valence electrons) will be linear in their ground state. The molecule $BaOH^+$ is isoelectronic with CsOH.

The bond dissociation energy for $BaOH^+$ ($D_0^0 = 108.2\ kcal\ mol^{-1}$, 4, 2) for the process $BaOH^+(g) = Ba^+(g) + OH(g)$ is nearly the same as that for BaOH ($D_0^0 = 105.3\ kcal\ mol^{-1}$, 2). This suggests that the bonding in these two molecules is quite similar. Thus, the bond distances are assumed to be the same as those adopted for BaOH(g, 2). The vibrational frequencies are assumed to be similar to those adopted for BaOH(g, 2). The ground state quantum weight is assumed to be the same as that of CsOH(g, 2).

References

1. R. Kelly and P. J. Padley, Trans. Faraday Soc. 67, 1384 (1971).
2. JANAF Thermochemical Tables: e^- (ref st), 3-31-65; H_2(g) and H_2O(g), 3-31-61; OH(g) and Ba(g), 12-31-70; H(g), 6-30-74; BaOH(g) and $Ba(OH)_2$(g), 12-31-75; BaF(g), 12-31-72; CsOH(g), 6-30-71.
3. D. E. Jensen, Combustion and Flame 12, 261 (1968).
4. C. E. Moore, U. S. Nat. Bur. Stand., Circ. 467, Vol. III (1958).
5. N. Acquista, S. Abramowitz, and D. R. Lide, J. Chem. Phys. 49, 780 (1968).
6. R. L. Kuczkowski and D. R. Lide, J. Chem. Phys. 44, 3131 (1966).
7. N. Acquista and S. Abramowitz, U. S. Nat. Bur. Stand., Report 9905, July, 1968.
8. A. D. Walsh, J. Chem. Soc. 1953, 2288 (1953).

Barium Hydroxide, Ion ($BaOH^+$) $Ba_1H_1O_1^+(g)$

Enthalpy Reference Temperature = T_r = 298.15 K Standard State Pressure = p° = 0.1 MPa

T/K	$C_p°$ (J K⁻¹mol⁻¹)	S° (J K⁻¹mol⁻¹)	$-[G°-H°(T_r)]/T$ (J K⁻¹mol⁻¹)	$H°-H°(T_r)$ (kJ mol⁻¹)	$\Delta_f H°$ (kJ mol⁻¹)	$\Delta_f G°$ (kJ mol⁻¹)	Log K_f
0	0.	0.	INFINITE	-10.823	269.364		
100	30.902	205.415	284.269	-7.885			
200	40.603	229.824	251.356	-4.306			
250	44.104	239.283	248.018	-2.184			
298.15	46.458	247.265	247.265	0.	270.897	259.611	-45.483
300	46.533	247.552	247.266	0.086	270.916	259.541	-45.190
350	48.225	254.860	247.839	2.457	271.417	257.605	-38.445
400	49.430	261.383	249.132	4.900	271.853	255.601	-33.378
450	50.315	267.258	250.825	7.395	272.122	253.551	-29.431
500	50.986	272.596	252.739	9.928	272.151	251.483	-26.272
600	51.952	281.983	256.852	15.078	271.713	247.401	-21.538
700	52.668	290.047	261.032	20.311	272.113	243.306	-18.156
800	53.287	297.121	265.110	25.609	272.018	239.203	-15.618
900	53.872	303.431	269.023	30.967	272.358	235.081	-13.644
1000	54.440	309.137	272.754	36.383	272.713	230.920	-12.062
1100	54.994	314.351	276.302	41.855	264.713	227.543	-10.805
1200	55.526	319.159	279.675	47.381	264.862	224.158	-9.757
1300	56.031	323.624	282.886	52.959	265.125	220.757	-8.870
1400	56.505	327.794	285.947	58.586	265.477	217.331	-8.109
1500	56.945	331.707	288.868	64.259	265.895	213.877	-7.448
1600	57.351	335.396	291.662	69.974	266.357	210.395	-6.869
1700	57.723	338.884	294.338	75.728	266.839	206.882	-6.357
1800	58.064	342.193	296.906	81.517	267.319	203.342	-5.901
1900	58.375	345.341	299.373	87.339	267.772	199.775	-5.492
2000	58.658	348.343	301.747	93.191	268.177	196.185	-5.124
2100	58.917	351.211	304.034	99.070	268.510	192.577	-4.790
2200	59.152	353.957	306.242	104.974	129.227	194.331	-4.614
2300	59.366	356.591	308.374	110.900	130.185	197.267	-4.480
2400	59.562	359.122	310.436	116.846	130.913	200.167	-4.357
2500	59.740	361.557	312.433	122.812	131.410	203.042	-4.242
2600	59.903	363.903	314.367	128.794	131.680	205.901	-4.137
2700	60.053	366.167	316.244	134.792	131.732	208.754	-4.039
2800	60.190	368.354	318.066	140.804	131.580	211.609	-3.948
2900	60.316	370.468	319.837	146.830	131.243	214.473	-3.863
3000	60.432	372.515	321.559	152.867	130.741	217.351	-3.784
3100	60.539	374.498	323.235	158.916	130.097	220.248	-3.711
3200	60.637	376.422	324.867	164.975	129.334	223.168	-3.643
3300	60.729	378.289	326.458	171.043	128.483	226.113	-3.579
3400	60.813	380.103	328.009	177.120	127.555	229.085	-3.519
3500	60.891	381.867	329.523	183.205	126.573	232.085	-3.464
3600	60.964	383.584	331.001	189.298	125.557	235.114	-3.411
3700	61.032	385.255	332.445	195.398	124.523	238.172	-3.362
3800	61.094	386.883	333.856	201.504	123.483	241.258	-3.316
3900	61.153	388.471	335.236	207.617	122.450	244.371	-3.273
4000	61.208	390.020	336.586	213.735	121.431	247.509	-3.232
4100	61.259	391.532	337.908	219.858	120.436	250.674	-3.194
4200	61.307	393.009	339.203	225.987	119.468	253.862	-3.157
4300	61.352	394.452	340.471	232.120	118.528	257.073	-3.123
4400	61.395	395.863	341.714	238.257	117.619	260.305	-3.090
4500	61.434	397.243	342.932	244.398	116.741	263.558	-3.059
4600	61.472	398.594	344.128	250.544	115.901	266.829	-3.030
4700	61.507	399.916	345.301	256.693	115.111	270.117	-3.002
4800	61.540	401.211	346.452	262.845	114.329	273.423	-2.975
4900	61.571	402.481	347.583	269.001	113.578	276.745	-2.950
5000	61.601	403.725	348.693	275.159	112.848	280.083	-2.926
5100	61.629	404.945	349.784	281.321	112.160	283.432	-2.903
5200	61.655	406.142	350.856	287.485	111.473	286.796	-2.881
5300	61.680	407.317	351.911	293.652	110.801	290.175	-2.860
5400	61.704	408.470	352.947	299.821	110.143	293.565	-2.840
5500	61.727	409.602	353.967	305.992	109.527	296.964	-2.820
5600	61.748	410.715	354.971	312.166	108.948	300.371	-2.802
5700	61.768	411.808	355.958	318.342	108.326	303.794	-2.784
5800	61.788	412.882	356.930	324.520	107.734	307.226	-2.767
5900	61.806	413.939	357.888	330.700	107.264	310.650	-2.750
6000	61.824	414.977	358.831	336.881	106.670	314.102	-2.735

PREVIOUS: June 1976 (1 atm) CURRENT: June 1976 (1 bar)

Barium Hydroxide, Ion ($BaOH^+$) $Ba_1H_1O_1^+(g)$

BARIUM HYDROXIDE, ALPHA (α-$Ba(OH)_2$) CRYSTAL M_r = 171.4468

$S°(298.15\ K) = [107.11 \pm 8.4]\ J\ K^{-1}\ mol^{-1}$
$T_{fus} = 681.15 \pm 1\ K$
$T_{trs} = 521 \pm 2\ K\ (\beta-\alpha)$

$\Delta_f H°(0\ K)$ = Unknown
$\Delta_f H°(298.15\ K) = [-946.30]\ kJ\ mol^{-1}$
$\Delta_{fus} H° = 16.74 \pm 1.26\ kJ\ mol^{-1}$
$\Delta_{trs} H°$ = Unknown

Enthalpy of Formation

The adopted $\Delta_f H°(298.15\ K) = -226.17 \pm 1.5\ kcal\ mol^{-1}$ is obtained from the enthalpy of formation of the liquid (1) minus the enthalpy of fusion and the enthalpy difference of the liquid and the crystal between the melting point and 298.15 K. Using auxiliary data (1), our 2nd and 3rd law analysis of the dissociation pressure equation of Kondakov et al. (2) for the reaction $Ba(OH)_2(cr) = BaO(cr) + H_2O(g)$ is given below.

T/K	$\Delta_r H°(298.15\ K)$, $kcal^{-1}\ mol^{-1}$ 2nd law	3rd law	Drift $cal\ K^{-1}\ mol^{-1}$	$\Delta_f H°(298.15\ K)$ $kcal\ mol^{-1}$
559-682	46.4	38.04±1.63	-13.5 ± 0.2	-226.84 ± 2

While this is in good agreement with the adopted value, there is an inconsistency in the measurements of Kondakov et al. (2) for the crystal and the liquid and the adopted enthalpy of fusion. The liquid state dissociation pressures are in good agreement with other measurements (see the liquid table (1)) and are judged more reliable than the solid state measurements. The adopted value for the alpha crystal is also in good agreement with -225.8 kcal from another recent evaluation (4).

Heat Capacity and Entropy

The heat capacities at 298.15 K and above are derived from the drop-ice-calorimeter enthalpy measurements of Powers and Blalock (3, 425-681 K) combined with a graphical comparison of the heat capacities of $Mg(OH)_2(cr)$ and $Ca(OH)_2(cr)$ (1). From the information referenced in the Transition Data Section below, it is concluded tha the α-form is maintained on cooling to 273.15 K. This JANAF Table is a single-phase alpha-crystal table.

The adopted entropy, $S°(298.15\ K) = [25.6 \pm 2]\ cal\ K^{-1}\ mol^{-1}$, is calculated from Kelley's additive entropy constants for cations and anions (5).

Fusion Data

Seward (6) determined a melting point of 408 ± 1 °C and a cryoscopic enthalpy of fusion of $3400 \pm 100\ cal\ mol^{-1}$. Powers and Blalock (3) chose a melting point of 395°C where they found $\Delta_{fus}H°$ = 24 cal/g (4113 $cal\ mol^{-1}$) by drop calorimetry. However, their enthalpy measurements show a possible melting range of 395 to 421°C. (At the adopted temperature of melting, 408°C, our smoothing of Powers and Blalock's experimental enthalpies (3) leads to $\Delta_{fus}H° = 4051\ cal\ mol^{-1}$). Kondakov et al. observed barium hydroxide to melt at 682 K (ca. 409°C) while investigating the dissociation pressures of the crystal and liquid states. Michaud (7, 8) determined a melting point of 409 ± 1 °C and a cryoscopic enthalpy of fusion of $3720 \pm 200\ cal\ mol^{-1}$ (8). From a consideration of all of these data, a melting point of 681.15 ± 1 K (408 ± 1 °C) and a enthalpy of fusion of 4000 ± 300 $cal\ mol^{-1}$ are adopted.

Transition Data

Michaud has observed a β to α crystal transition, on heating anhydrous barium dihydroxide, at 250°C (7) and at 246°C (8) by both thermal and dilatometric techniques. In the absence of water vapor, the high temperature α-form is maintained on cooling to room temperature (7). This "metastable" state has been confirmed by others (9, 10). This present JANAF Table considers only the α-form. A transition temperature of 521 ± 2 K (248 ± 2 °C) is adopted.

References

1. JANAF Thermochemical Tables: $Ba(OH)_2(\ell)$, 12-31-75; BaO(cr), 6-30-74; $H_2O(g)$, 3-31-61; $Mg(OH)_2(cr)$, 12-31-75; $Ca(OH)_2(cr)$, 12-31-75.
2. B. V. Kondakov, P. V. Kovtunenko, and A. A. Bundel, Zh. Fiz. Khim. 38, 190 (1964).
3. W. D. Powers and G. C. Blalock, U. S. Atomic Energy Comm., ORNL-1653 (1954).
4. U. S. Nat. Bur. Stand. Tech. Note 270-6, 1971.
5. K. K. Kelley in Chap. 13, D. R. Stull and H. Prophet, "The Characterization of High Temperature Vapors," J. L. Margrave, ed., John Wiley and Sons, New York, 1967.
6. R. P. Seward, J. Am. Chem. Soc. 67, 1189 (1945).
7. M. Michaud, Compt. Rend. Ser. C 262, 1143 (1966).
8. M. Michaud, Rev. Chim. Miner. 5, 89 (1968).
9. P. Buck and H. Baernighausen, Acta Crystallogr., Sect. B 24, 1705 (1968).
10. H. D. Lutz, R. Heider, and R. A. Becker, Z. Naturforsch. B 24, 1657 (1969).

Barium Hydroxide, Alpha ($Ba(OH)_2$) $Ba_1H_2O_2(cr)$

Enthalpy Reference Temperature = T_r = 298.15 K Standard State Pressure = p° = 0.1 MPa

T/K	$C_p°$ (J K⁻¹mol⁻¹)	S°	$-[G°-H°(T_r)]/T$	$H°-H°(T_r)$ (kJ mol⁻¹)	$\Delta_f H°$	$\Delta_f G°$	Log K_f
0							
100							
200							
298.15	101.629	107.110	107.110	0.	-946.295	-859.476	150.577
300	101.797	107.740	107.112	0.188	-946.267	-858.938	149.554
400	112.633	138.679	111.257	10.969	-944.380	-830.099	108.400
500	118.658	164.520	119.399	22.560	-942.589	-801.748	83.758
600	122.717	186.510	128.797	34.628	-941.231	-773.693	67.356
681.150	126.022	202.280	136.628	44.719	------ CRYSTAL <--> LIQUID -----		
700	126.775	205.731	138.443	47.102	-938.765	-745.972	55.665
800	130.834	222.924	147.947	59.982	-936.501	-718.575	46.918
900	134.892	238.569	157.159	73.268	-933.501	-691.511	40.134
1000	138.951	252.991	166.030	86.961	-930.182	-664.799	34.726

PREVIOUS: CURRENT: December 1975

Barium Hydroxide, Alpha ($Ba(OH)_2$) $Ba_1H_2O_2(cr)$

BARIUM HYDROXIDE ($Ba(OH)_2$) LIQUID M_r = 171.4468

S°(298.15 K) = [123.357] J K^{-1} mol^{-1} $\Delta_fH°$(298.15 K) = [-934.123 kJ mol^{-1}

T_{fus} = 681.15 ± 1 K $\Delta_{fus}H°$ = 16.74 ± 1.26 kJ mol^{-1}

Enthalpy of Formation

The adopted $\Delta_fH°$(298.15 K) = -223.26 ± 1.0 kcal mol^{-1} is the average of values derived from the 3rd law analysis of dissociation pressure measurements of Tamaru and Shiomi (1) and of Kondakov et al. (2). Auxiliary data used in the analysis are from the JANAF Thermochemical Tables (3). Johnston (4) has also determined pressures for the dissociation reaction $Ba(OH)_2(\ell) = BaO(cr) + H_2O(g)$.

Source	No. of Points	T/K	$\Delta_rH°$(298.15 K), kcal mol^{-1} 2nd law	3rd law	Drift cal K^{-1} mol^{-1}	$\Delta_fH°$(298.15) kcal mol^{-1}
Tamaru and Shiomi (1)	7	788-1018	32.1	34.56±0.48	2.8 ± 0.5	-223.35±1
Kondakov et al. (2)	Equation	682- 993	34.0	34.37±0.21	0.4	-223.17±0.7
Johnston (4)	11	918-1263	34.4	33.86±0.38	-0.5±0.5	-222.66±0.9

The enthalpy of formation derived from Johnston's measurements is in good agreement with the adopted value but is not averaged in because, as indicated by Kondakov et al. (2), the measurements are believed less accurate.

Heat Capacity and Entropy

Cp° is based on the enthalpy measurements of Powers and Blalock (5) and is assumed constant at 33.7 cal K^{-1} mol^{-1} over the range of 450 to 1600 K. A glass transition is assumed at 450 K below which Cp° is that of the alpha crystal. The entropy at 298.15 K is derived from the value adopted for the alpha crystal.

Decomposition Data

T_{dcm} is calculated as the temperature at which $\Delta_rG° = 0$ for the reaction $Ba(OH)_2(\ell) = BaO(cr) + H_2O(g)$.

References

1. S. Tamaru and K. Shiomi, Z. physik. Chem. A171, 221 (1935).
2. B. V. Kondakov, P. V. Kovtunenko, and A. A. Bundel, Zh. Fiz. Khim. 38, 190 (1964).
3. JANAF Thermochemical Tables: BaO(cr), 6-30-74; $H_2O(g)$, 3-31-61.
4. J. Johnston, Z. physik. Chem. 62, 330 (1908).
5. W. D. Powers and G. C. Blalock, U. S. Atomic Energy Comm., ORNL-1653 (1954).

Barium Hydroxide ($Ba(OH)_2$) $Ba_1H_2O_2(l)$

Enthalpy Reference Temperature = T_r = 298.15 K Standard State Pressure = p° = 0.1 MPa

T/K	$C_p°$ (J $K^{-1}mol^{-1}$)	S°	$-[G°-H°(T_r)]/T$	$H°-H°(T_r)$ (kJ mol^{-1})	$\Delta_fH°$	$\Delta_fG°$	Log K_f
0							
100							
200							
298.15	101.629	123.357	123.357	0.000	-934.123	-852.148	149.293
300	101.797	123.987	123.359	0.188	-934.095	-851.640	148.284
400	112.633	154.926	127.504	10.969	-932.208	-824.426	107.659
450.010	116.065	168.404	131.312	16.692	GLASS <--> LIQUID TRANSITION		
450.010	141.001	168.404	131.312	16.692			
500	141.001	183.256	135.776	23.740	-929.236	-797.764	83.342
600	141.001	208.964	145.896	37.841	-925.847	-771.780	67.189
681.150	141.001	226.850	154.498	49.283	CRYSTAL <--> LIQUID		
700	141.001	230.699	156.498	51.941	-921.754	-746.439	55.700
800	141.001	249.527	166.976	66.041	-918.270	-721.627	47.117
900	141.001	266.135	177.090	80.141	-914.457	-697.277	40.469
1000	141.001	280.991	186.750	94.241	-910.729	-673.346	35.172
1100	141.001	294.430	195.938	108.341	-915.454	-648.965	30.817
1200	141.001	306.698	204.664	122.441	-912.123	-624.886	27.201
1300	141.001	317.984	212.953	136.541	-908.769	-601.085	24.152
1400	141.001	328.434	220.833	150.641	-905.411	-577.543	21.548
1500	141.001	338.162	228.334	164.741	-902.068	-554.241	19.300
1600	141.001	347.262	235.486	178.841	-898.759	-531.161	17.341
1700	141.001	355.810	242.315	192.941	-895.502	-508.286	15.618
1800	141.001	363.869	248.846	207.041	-892.316	-485.601	14.092
1900	141.001	371.493	255.102	221.142	-889.219	-463.090	12.731
2000	141.001	378.725	261.104	235.242	-886.231	-440.740	11.511

PREVIOUS: CURRENT: December 1975

Barium Hydroxide ($Ba(OH)_2$) $Ba_1H_2O_2(l)$

BARIUM HYDROXIDE ($Ba(OH)_2$) ALPHA - LIQUID M_r = 171.34468

0 to 681.15 K crystal, alpha*
above 681.15 K liquid

*A low temperature beta form does exist but is not included in this table, $T_{trs}(\beta \rightarrow \alpha)$ = 521 ± 2 K.

Refer to the individual tables for details.

Barium Hydroxide ($Ba(OH)_2$) $Ba_1H_2O_2(cr,l)$

Enthalpy Reference Temperature = T_r = 298.15 K Standard State Pressure = p° = 0.1 MPa

T/K	C_p° (J K⁻¹mol⁻¹)	S° (J K⁻¹mol⁻¹)	−[G°−H°(T_r)]/T (J K⁻¹mol⁻¹)	H°−H°(T_r) (kJ mol⁻¹)	Δ_fH° (kJ mol⁻¹)	Δ_fG° (kJ mol⁻¹)	Log K_f
0							
100							
200							
298.15	101.629	107.110	107.110	0.000	-946.295	-859.476	150.577
300	101.797	107.740	107.112	0.188	-946.267	-858.938	149.554
400	112.633	138.679	111.257	10.969	-944.380	-830.099	108.400
500	118.658	164.520	119.399	22.560	-942.589	-801.748	83.758
600	122.717	186.510	128.797	34.628	-941.231	-773.693	67.356
681.150	126.022	202.280	136.628	44.719	CRYSTAL <--> LIQUID	TRANSITION	
681.150	141.001	226.850	136.628	61.455			
700	141.001	230.699	139.110	64.113	-921.754	-746.439	55.700
800	141.001	249.527	151.761	78.213	-918.270	-721.627	47.117
900	141.001	266.135	163.565	92.313	-914.457	-697.277	40.469
1000	141.001	280.991	174.578	106.413	-910.729	-673.346	35.172
1100	141.001	294.430	184.872	120.513	-915.454	-648.965	30.817
1200	141.001	306.698	194.521	134.613	-912.123	-624.886	27.201
1300	141.001	317.984	203.590	148.713	-908.769	-601.085	24.152
1400	141.001	328.434	212.138	162.813	-905.411	-577.543	21.548
1500	141.001	338.162	220.219	176.913	-902.068	-554.241	19.300
1600	141.001	347.262	227.878	191.013	-898.759	-531.161	17.341
1700	141.001	355.810	235.155	205.113	-895.502	-508.286	15.618
1800	141.001	363.869	242.084	219.214	-892.316	-485.601	14.092
1900	141.001	371.493	248.696	233.314	-889.219	-463.090	12.731
2000	141.001	378.725	255.018	247.414	-886.231	-440.740	11.511

PREVIOUS: CURRENT December 1975

Barium Hydroxide ($Ba(OH)_2$) $Ba_1H_2O_2(cr,l)$

BARIUM HYDROXIDE ($Ba(OH)_2$) IDEAL GAS M_r = 171.4468

$S°(298.15\ K) = [315.0 \pm 12.6]\ J\ K^{-1}\ mol^{-1}$

$\Delta_fH°(0\ K) = -619.09 \pm 37.7\ kJ\ mol^{-1}$

$\Delta_fH°(298.15\ K) = -626.56 \pm 37.7\ kJ\ mol^{-1}$

Vibrational Fequencies and Degeneracies

ν, cm^{-1}	ν, cm^{-1}
[413] (1)	[3650] (2)
[64] (1)	[431] (4)
[390] (1)	

Ground State Quantum Weight: [1] σ = [2]

Point Group: [C_{2v}]

Bond Distances: Ba-O = [2.34] Å O-H = [0.96] Å

Bond Angles: O-Ba-O = [95]° Ba-O-H = [180]°

Product of the Moments of Inertia: $I_AI_BI_C = [6.366076 \times 10^{-114}]\ g^3\ cm^6$

Enthalpy of Formation

The adopted $\Delta_fH°(298.15\ K) = -149.75 \pm 9.0\ kcal\ mol^{-1}$ is derived from the Knudsen-cell mass-spectrometric equilibrium study of Newbury (1). Tabulated below are the results of a 2nd and 3rd law analysis of this study (1), the Knudsen-cell mass-spectrometric work of Stafford and Berkowitz (2), and the flame-spectral work of Cotton and Jenkins (3).

Source	Rxn No.	No. of Points	T/K	$\Delta_rH°(298.15\ K)$, kcal mol^{-1} 2nd law	3rd law	Drift cal K^{-1} mol^{-1}	$\Delta_fH°(298.15\ K)$ kcal/mol^{-1}	$D_0^{\,o}$ kcal/mol^{-1}
Stafford and Berkowitz (2)	A	24[a]	1485-1727	56.4	45.37±1.82	-6.7 ± 1.7	-143.43±2.5	203.3
Newbury (1)	A	44[a]	1428-1899	39.0	39.05±0.17	0.0 ± 0.1	-149.75±0.7	209.6
Cotton and Jenkin (3)	B	2	1570-1800	8.1	23.04±2.98	8.8 ± 0.0	-153.96±4.0	213.8

Reactions: A) $BaO(cr) + H_2O(g) = Ba(OH)_2(g)$ B) $Ba(g) + 2H_2O(g) = Ba(OH)_2(g) + 2H(g)$

a. Two points rejected by a statistical test.

b. $D_0^{\,o}$ is the dissociation energy for the reaction $Ba(OH)_2(g) = Ba(g) + 2OH(g)$.

In addition to the three investigations above, dissociation energies ($D_0^{\,o}$ as defined in the table above) have been derived from flame spectra by Ryabova and Gurvich (5) and by Sugden and Schofield (6). Ryabova and Gurvich (5) believed the dominant reaction to be $Ba(g) + H_2O(g) = BaOH(g) + H(g)$, but they also considered the possibility that the reaction $Ba(g) + 2H_2O(g) = Ba(OH) + 2H(g)$ was dominant and derived $D_0^{\,o} = 205 \pm 20\ kcal\ mol^{-1}$. Sugden and Schofield (6) considered the dihydroxide to be the domiant product and derived $D_0^{\,o} = 229 \pm 12\ kcal\ mol^{-1}$. Cotton and jenkins (3) found both BaOH and $Ba(OH)_2$ to be present and obtained recalculated $D_0^{\,o}$ values of 208 and 212 kcal mol^{-1}, respectively.

The 3rd law analysis of the data of Cotton and Jenkins (3) above combined with current JANAF Thermochemical Table data (4) leads to $D_0^{\,o} = 213.8\ kcal\ mol^{-1}$ which is 0.8 kcal mol^{-1} higher than the 213 kcal mol^{-1} derived by Cotton and Jenkins (3). Applying this difference to the values of Ryabova and Gurvich (5) and of Sugden and Schofield (6) as recalculated by Cotton and Jenkins (3) gives $D_0^{\,o}$ = 208.8 and 212.8 kcal mol^{-1}, respectively.

The average of the dissociation energies from the flame-spectral measurements (208.8, 212.8, and 213.8) and the dissociation energy from the Knudsen-cell mass-spectrometric work of Stafford and Berkowitz (203.3) is 209.7 kcal mol^{-1} which happens to be in agreement with the 209.6 kcal mol^{-1} calculated from Newbury's data.

$\Delta_fH°(298.15\ K) = -149.75 \pm 9.0\ kcal\ mol^{-1}$ is adopted because of the excellent agreement in the 2nd and 3rd law enthalpies of reaction from Newbury's data (1) and because of the larger uncertainty associated with dissociation energies derived from flame spectra.

The enthalpy of dissociation listed by Jackson (9) leads to $\Delta_fH°(298.15\ K) = -151.22\ kcal\ mol^{-1}$. Another recent compilation (10) lists $\Delta_fH°(298.15\ K) = -140\ kcal\ mol^{-1}$.

Heat Capacity and Entropy

The analogy between gaseous mono- and dihydroxides and gaseous mono- and dihalides, particularly the mono- and difluorides, has been recognized (2, 7-9). The O-Ba-O bond angle is assumed to be the same as the F-Ba-F bond angle (4); the Ba-O-H bond angle is considered to be linear as in BaOH (4). The Ba-O bond distance is estimated to slightly larger, 0.02 Å, than the Ba-F distance in BaF_2(4) after noting the close similarily in the bond distance of the alkale metal fluorides and hydroxides. The O-H bond distance is taken to be the same as in water (4).

The vibrational frequencies are assumed to be the same as in BaF_2 (4) (O-Ba-O symmetrical and asymmetrical stretch, and bend) and as in BaOH (4) (O-H stretch and Ba-O-H bend). The principal moments of inertial are: $I_A = 29.8004 \times 10^{-39}$, $I_B = 17.7971 \times 10^{-39}$, and $I_C = 12.0033 \times 10^{-39}\ g\ cm^2$.

Jackson (9) used a different molecular configuration and different vibrational frequencies to estimate $S°(298.15\ K) = 78.045\ cal\ K^{-1}\ mol^{-1}$. We assign uncertainty of ± 3 cal K^{-1} mol^{-1} to the adopted entropy.

References

1. R. A. Newbury, U. S. At. Energy Comm. UCRL-12225-T (1965).
2. F. E. Stafford and J. Berkowitz, J. Chem. Phys. 40, 2963 (1964).
3. D. H. Cotton and D. R. Jenkins, Trans. Faraday Soc. 64, 2988 (1968).
4. JANAF Thermochemical Tables: BaO(cr), H(g), 6-30-74; H_2O(g), 3-31-61; BaOH(g), 12-31-75; Ba(g), HO(g), 12-31-70; BaF_2(g), 12-31-72.
5. V. G. Ryabova and L. V. Gurvich, Teplofiz. Vysokikh Temperature, Akad. Nauk SSR 3, 318 (1965).
6. T. M. Sugden and K. Schofield, Trans. Faraday Soc. 62, 566 (1966).
7. R. C. Schoonmaker and R. F. Porter, J. Chem. Phys. 31, 830 (1959).
8. J. Berkowitz, D. J. Meschi, and W. A. Chupka, J. Chem. Phys. 33, 533 (1960).
9. D. D. Jackson, Lawrence Livermore Lab., Univ. of Calif., Report USRL-51137, Dec. 8, 1971.
10. U. S. Nat. Bur. Stand. Tech. Note 270-6, 1971.

Barium Hydroxide ($Ba(OH)_2$) $Ba_1H_2O_2(g)$

Enthalpy Reference Temperature = T_r = 298.15 K Standard State Pressure = p° = 0.1 MPa

T/K	C_p^o (J $K^{-1}mol^{-1}$)	S°	$-[G°-H°(T_r)]/T$	$H°-H°(T_r)$ (kJ mol^{-1})	$\Delta_fH°$	$\Delta_fG°$	Log K_f
0	0.	0.	INFINITE	-16.594	-619.094	-619.094	INFINITE
100	45.351	248.363	375.259	-12.690	-622.792	-616.816	322.192
200	65.873	286.310	321.778	-7.094	-625.345	-609.699	159.237
250	72.732	301.794	316.265	-3.618	-626.062	-605.700	126.554
298.15	77.256	315.016	315.016	0.	-626.562	-601.731	105.421
300	77.399	315.494	315.017	0.143	-626.579	-601.576	104.744
350	80.612	327.681	315.973	4.098	-626.996	-597.376	89.153
400	82.884	338.602	318.131	8.189	-627.428	-593.116	77.453
450	84.542	348.465	320.962	12.376	-627.994	-588.795	68.346
500	85.798	357.440	324.168	16.636	-628.780	-584.399	61.052
600	87.607	373.254	331.067	25.312	-630.814	-575.322	50.086
700	88.958	386.863	338.088	34.143	-631.991	-565.991	42.235
800	90.141	398.820	344.947	43.098	-633.652	-556.443	36.332
900	91.271	409.503	351.537	52.169	-634.868	-546.718	31.731
1000	92.380	419.177	357.825	61.352	-636.057	-536.860	28.043
1100	93.465	428.033	363.810	70.644	-645.589	-526.064	24.981
1200	94.513	436.210	369.507	80.044	-646.959	-515.137	22.423
1300	95.510	443.815	374.934	89.545	-648.204	-504.100	20.255
1400	96.447	450.928	380.111	99.144	-649.348	-492.972	18.393
1500	97.319	457.612	385.057	108.833	-650.416	-481.765	16.777
1600	98.124	463.919	389.791	118.605	-651.435	-470.488	15.360
1700	98.864	469.890	394.328	128.455	-652.428	-459.148	14.108
1800	99.540	475.561	398.685	138.376	-653.421	-447.750	12.993
1900	100.158	480.959	402.874	148.361	-654.439	-436.297	11.995
2000	100.721	486.111	406.908	158.406	-655.506	-424.788	11.094
2100	101.234	491.038	410.798	168.504	-656.648	-413.224	10.278
2200	101.702	495.759	414.554	178.651	-797.410	-396.228	9.408
2300	102.128	500.289	418.183	188.843	-797.936	-377.982	8.584
2400	102.517	504.644	421.696	199.075	-798.697	-359.709	7.829
2500	102.873	508.836	425.098	209.345	-799.697	-341.397	7.133
2600	103.198	512.877	428.397	219.649	-800.931	-323.042	6.490
2700	103.495	516.778	431.598	229.984	-802.391	-304.634	5.894
2800	103.768	520.547	434.708	240.347	-804.064	-286.169	5.339
2900	104.019	524.192	437.731	250.737	-805.932	-267.641	4.821
3000	104.250	527.723	440.673	261.150	-807.973	-249.044	4.336
3100	104.463	531.145	443.536	271.586	-810.166	-230.377	3.882
3200	104.659	534.464	446.326	282.042	-812.488	-211.638	3.455
3300	104.841	537.688	449.046	292.517	-814.908	-192.825	3.052
3400	105.009	540.820	451.699	303.010	-817.416	-173.936	2.672
3500	105.165	543.866	454.289	313.519	-819.987	-154.973	2.313
3600	105.309	546.831	456.819	324.043	-822.602	-135.936	1.972
3700	105.444	549.718	459.291	334.580	-825.247	-116.826	1.649
3800	105.569	552.532	461.708	345.131	-827.907	-97.643	1.342
3900	105.686	555.275	464.072	355.694	-830.571	-78.390	1.050
4000	105.796	557.953	466.386	366.268	-833.231	-59.071	0.771
4100	105.898	560.566	468.651	376.853	-835.879	-39.684	0.506
4200	105.993	563.119	470.870	387.447	-838.509	-20.233	0.252
4300	106.083	565.614	473.044	398.051	-841.123	-0.718	0.009
4400	106.167	568.054	475.176	408.664	-843.716	18.855	-0.224
4500	106.246	570.441	477.267	419.285	-846.289	38.488	-0.447
4600	106.321	572.777	479.318	429.913	-848.834	58.176	-0.661
4700	106.391	575.064	481.330	440.549	-851.342	77.919	-0.866
4800	106.457	577.305	483.307	451.191	-853.851	97.718	-1.063
4900	106.519	579.501	485.248	461.840	-856.341	117.568	-1.253
5000	106.578	581.653	487.154	472.495	-858.822	137.469	-1.436
5100	106.634	583.764	489.028	483.155	-861.271	157.417	-1.612
5200	106.687	585.835	490.870	493.821	-863.732	177.414	-1.782
5300	106.736	587.868	492.681	504.492	-866.188	197.461	-1.946
5400	106.784	589.864	494.462	515.169	-868.644	217.552	-2.104
5500	106.829	591.823	496.214	525.849	-871.068	237.684	-2.257
5600	106.872	593.749	497.939	536.534	-873.468	257.860	-2.405
5700	106.912	595.641	499.636	547.223	-875.922	278.083	-2.548
5800	106.951	597.500	501.308	557.917	-878.359	298.348	-2.687
5900	106.988	599.329	502.954	568.614	-880.687	318.635	-2.821
6000	107.023	601.127	504.575	579.314	-883.151	338.985	-2.951

PREVIOUS: December 1975 (1 atm) CURRENT: December 1975 (1 bar)

Barium Hydroxide ($Ba(OH)_2$) $Ba_1H_2O_2(g)$

BARIUM IODIDE (BaI) IDEAL GAS M_r = 264.345

$\Delta_f H°(0\ K) = -39.29 \pm 84\ kJ\ mol^{-1}$
$S°(298.15\ K) = 278.74 \pm 0.21\ J\ K^{-1}\ mol^{-1}$ $\Delta_f H°(298.15\ K) = -42.42 \pm 84\ kJ\ mol^{-1}$

Electronic Levels and Quantum Weights

State	ϵ_i, cm^{-1}	g_i
$X[^2\Sigma^+]$	0	2
$C[^2\Pi]$	17816.3	2
	18568.8	2
$D[^2\Sigma^+]$	25764.4	2
$E[^2\Sigma^+]$	26753.3	2

$\omega_e = 152.2\ cm^{-1}$ $\omega_e x_e = 0.29\ cm^{-1}$ $\sigma = 1$
$B_e = [0.0250]\ cm^{-1}$ $\alpha_e = [0.000059]\ cm^{-1}$ $r_e = [3.20]$ Å

Enthalpy of Formation

The adopted $\Delta_f H°(0\ K) = -9.39 \pm 20\ kcal\ mol^{-1}$ is calculated from $D_0^0 = 3.37$ eV (77.7 kcal mol^{-1}) calculated by a linear Birge-Sponer extrapolation of $\omega_e x_e$ data (1) with a correction for the ionic character of the molecule as described by Hildenbrand (2). Other values, in kcal mol^{-1}, are 55 (3) predicted from a correlation of binding energies of Group II and Group III diatomic hydrides and halides, 71.7 (4) as a lower bound from a consideration of ionic bonding forces and 66 (5) derived as a lower bound from a consideration of energy conservation and measured reaction threshold relative kinetic energy from crossed molecular beam experiments. D_0^0 = 77.7 kcal mol^{-1} is adopted because the ratio of $D_0^0(BaI)/\Delta_{at}H°(BaI_2) = 0.47$ which is nearly the same as the ratio of 0.46 found for a series of mono- and difluorides (6) and for other alkaline earth halides systems (8).

$\Delta_f H°(298.15\ K) = -10.14 \pm 20\ kcal\ mol^{-1}$ is calculated from the adopted $\Delta_f H°(0\ K) = -9.39 \pm 20\ kcal\ mol^{-1}$.

Heat Capacity and Entropy

The ground state vibrational constants are from the compilation of Rosen (1). The adopted value of r_e = 3.20 Å was estimated by Margrave (3) and is also the bond distance in $BaI_2(g)$ (7). B_e is calculated from the adopted r_e; α_e is calculated assuming a Morse potential function.

The electronic levels and their probable designation are those given by Rosen (1).

References

1. B. Rosen, Ed., "Spectroscopic Data Relative to Diatomic Molecules," Pergamon Press, New York, 1970.
2. D. L. Hildenbrand in "Advances in High Temperature Chemistry," Vol. 1, L. Eyring, Ed., Academic Press, New York, 1967, pp. 193-215.
3. J. L. Margrave, J. Phys. Chem. 58, 25 (1954).
4. K. S. Krasnov and N. V. Karaseva, Optics and Spectrosc. 19, 14 (1963).
5. C. A. Mims, S. Lin, and R. R. Herm, J. Chem. Phys. 57, 3099 (1972).
6. G. D. Blue, J. W. Green, T. C. Ehlert, and J. L. Margrave, Nature 199, 804 (1963).
7. P. A. Akishin, V. P. Spiridonov, G. A. Sobolev, and V. A. Naumov, Zhur. Fiz. Khim. 32, 58 (1958).
8. JANAF Thermochemical Tables; refer to all alkaline earth halide tables.

Barium Iodide (BaI) $Ba_1I_1(g)$

Enthalpy Reference Temperature = T_r = 298.15 K Standard State Pressure = p° = 0.1 MPa

T/K	$C_p°$ (J K^{-1}mol^{-1})	S°	$-[G°-H°(T_r)]/T$	$H°-H°(T_r)$ (kJ mol^{-1})	$\Delta_f H°$	$\Delta_f G°$	Log K_f
0	0.	0.	INFINITE	-10.381	-39.287	-39.287	INFINITE
100	34.820	239.050	311.435	-7.239	-39.368	-56.815	29.677
200	36.747	263.961	282.130	-3.634	-40.774	-73.742	19.259
250	37.046	272.196	279.349	-1.788	-41.582	-81.893	17.111
298.15	37.221	278.737	278.737	0.	-42.417	-89.582	15.694
300	37.227	278.967	278.738	0.069	-42.451	-89.875	15.649
350	37.349	284.716	279.192	1.933	-43.439	-97.703	14.581
400	37.439	289.709	280.201	3.803	-52.553	-105.106	13.725
450	37.510	294.123	281.507	5.677	-54.472	-111.562	12.950
500	37.569	298.078	282.970	7.554	-76.718	-115.892	12.107
600	37.666	304.937	286.077	11.316	-79.455	-123.460	10.748
700	37.747	310.749	289.197	15.087	-81.383	-130.654	9.749
800	37.820	315.794	292.213	18.865	-83.821	-137.520	8.979
900	37.887	320.253	295.086	22.650	-85.838	-144.111	8.364
1000	37.952	324.248	297.806	26.442	-87.853	-150.478	7.860
1100	38.015	327.868	300.376	30.241	-98.239	-155.823	7.399
1200	38.076	331.178	302.807	34.045	-100.493	-160.957	7.006
1300	38.137	334.229	305.109	37.856	-102.652	-165.908	6.666
1400	38.197	337.057	307.291	41.673	-104.744	-170.695	6.369
1500	38.256	339.694	309.364	45.495	-106.793	-175.334	6.106
1600	38.315	342.165	311.338	49.324	-108.826	-179.837	5.871
1700	38.374	344.490	313.220	53.158	-110.867	-184.213	5.660
1800	38.433	346.685	315.019	56.999	-112.941	-188.468	5.469
1900	38.493	348.764	316.741	60.845	-115.071	-192.606	5.295
2000	38.554	350.740	318.392	64.697	-117.278	-196.630	5.135
2100	38.616	352.623	319.977	68.556	-119.583	-200.541	4.988
2200	38.680	354.421	321.502	72.421	-261.525	-198.965	4.724
2300	38.746	356.142	322.971	76.292	-263.243	-196.084	4.453
2400	38.816	357.792	324.388	80.170	-265.199	-193.123	4.203
2500	38.889	359.378	325.756	84.055	-267.391	-190.076	3.971
2600	38.968	360.905	327.079	87.948	-269.805	-186.936	3.756
2700	39.051	362.377	328.359	91.849	-272.426	-183.700	3.554
2800	39.140	363.799	329.600	95.758	-275.234	-180.363	3.365
2900	39.236	365.174	330.803	99.677	-278.203	-176.922	3.187
3000	39.339	366.506	331.971	103.606	-281.307	-173.377	3.019
3100	39.450	367.798	333.106	107.545	-284.517	-169.727	2.860
3200	39.569	369.052	334.209	111.496	-287.807	-165.972	2.709
3300	39.696	370.272	335.284	115.459	-291.141	-162.114	2.566
3400	39.832	371.459	336.330	119.436	-294.505	-158.153	2.430
3500	39.977	372.615	337.351	123.426	-297.872	-154.094	2.300
3600	40.130	373.744	338.346	127.431	-301.221	-149.938	2.176
3700	40.293	374.845	339.318	131.452	-304.534	-145.690	2.057
3800	40.465	375.922	340.267	135.490	-307.795	-141.353	1.943
3900	40.646	376.976	341.195	139.546	-310.992	-136.932	1.834
4000	40.835	378.007	342.102	143.620	-314.116	-132.429	1.729
4100	41.033	379.018	342.990	147.713	-317.157	-127.849	1.629
4200	41.239	380.009	343.860	151.826	-320.111	-123.196	1.532
4300	41.453	380.982	344.712	155.961	-322.976	-118.473	1.439
4400	41.674	381.937	345.547	160.117	-325.750	-113.685	1.350
4500	41.902	382.876	346.366	164.296	-328.432	-108.835	1.263
4600	42.137	383.800	347.170	168.498	-331.016	-103.927	1.180
4700	42.377	384.709	347.959	172.724	-333.489	-98.966	1.100
4800	42.623	385.603	348.734	176 974	-335.894	-93.950	1.022
4900	42.874	386.485	349.495	181.248	-338.206	-88.886	0.948
5000	43.130	387.354	350.244	185.549	-340.439	-83.776	0.875
5100	43.389	388.210	350.980	189.874	-342.570	-78.623	0.805
5200	43.651	389.055	351.704	194.226	-344.641	-73.427	0.738
5300	43.916	389.889	352.417	198.605	-346.638	-68.193	0.672
5400	44.184	390.713	353.118	203.010	-348.563	-62.921	0.609
5500	44.452	391.526	353.809	207.442	-350.388	-57.618	0.547
5600	44.722	392.329	354.490	211.900	-352.119	-52.286	0.488
5700	44.992	393.123	355.161	216.386	-353.836	-46.916	0.430
5800	45.262	393.908	355.822	220.899	-355.467	-41.521	0.374
5900	45.532	394.684	356.474	225.438	-356.920	-36.114	0.320
6000	45.800	395.452	357.117	230.005	-358.442	-30.664	0.267

PREVIOUS: June 1974 (1 atm) CURRENT: June 1974 (1 bar)

Barium Iodide (BaI) $Ba_1I_1(g)$

BARIUM IODIDE (BaI_2) CRYSTAL M_r = 391.139

$S°(298.15\ K) = 165.14 \pm 0.4\ J\ K^{-1}\ mol^{-1}$

$T_{fus} = 984 \pm 2\ K$

$\Delta_fH°(0\ K) = -604.5 \pm 3.3\ kJ\ mol^{-1}$

$\Delta_fH°(298.15\ K) = -605.4 \pm 3.3\ kJ\ mol^{-1}$

$\Delta_{fus}H° = 26.53 \pm 0.54\ kJ\ mol^{-1}$

Enthalpy of Formation

Ehrlich, Peik, and Koch (1) derived $\Delta_fH°(BaI_2$, cr, 298.15 K) = -144.7 ± 0.4 kcal mol^{-1} from enthalpy of solution measurements of Ba(cr) and BaI_2(cr) in 0.1 N HI. An auxiliary HI enthalpy of solution value, -13.22 kcal mol^{-1}, calculated from data in reference (2) was used in the derivation. This auxiliary value will be changed by incorporation of the accepted CODATA key value of $\Delta_fH°(I^-$, aq, 298.15 K) = -13.60 kcal mol^{-1} (3) in the $\Delta_fH°$(HI, aq, 298.15 K) table, so that the above derived enthalpy of formation of BaI_2(cr) will be approximately -145.1 kcal mol^{-1}.

Combining $\Delta_fH°(Ba^{+2}$, aq, 298.15 K) = -128.5 kcal mol^{-1}, selected by Parker (4), with the CODATA value for I^-(aq) (3) gives $\Delta_fH°(BaI_2$, aq, 298.15 K) = -155.7 kcal mol^{-1}. Further combination with the enthalpy of solution, -10.98 kcal mol^{-1} (5, 11), gives $\Delta_fH°(BaI_2$, cr, 298.15 K) = -144.7 kcal mol^{-1}. This enthalpy of solution value was based on several sets of data, one of which was that of Ehrlich et al (1, 11).

We adopt -144.7 ± 0.8 kcal mol^{-1} (-605.425 ± 3.347 kJ mol^{-1}) because of the correlation of $\Delta_fH°(Ba^{+2}$, aq, 298.15 K) with the evaluation of data for several barium compounds (4). See also the updated discussion in the enthalpy of formation section for BaO(cr) (12).

Heat Capacity and Entropy

$C_p°$(13.08 - 300.17 K) has been measured by Paukov, Khriplovich, and Smirnova (6). Our T^3 extrapolation agrees with their S°(13) = 0.619 cal K^{-1} mol^{-1} and H°(13 K)-H°(0 K) = 6.01 cal K^{-1} mol^{-1}, and our smoothing and integration of their $C_p°$ data agrees with their S°(298.15 K) = 39.47 ± 0.1 cal K^{-1} mol^{-1} which is adopted.

The low-temperature heat capacities join smoothly with a linear extrapolation from $C_p°$(298.15 K) = 18.52 cal K^{-1} mol^{-1} to Dworkin and Bredig's reported $C_p°$ of 21.8 cal K^{-1} mol^{-1} at the adopted T_{fus} of 984 K (7). This gives H°(984 K) - H°(298.15 K) = 13.83 kcal mol^{-1}, in fair agreement with Dworkin and Bredig's drop calorimeter value of 13.9 kcal mol^{-1} (7).

Fusion Data

Dworkin and Bredit (7) report $\Delta_{fus}H°$ = 6.34 kcal mol^{-1} (±2%) at T_{fus} of 984 K from drop calorimetry; Emons and Loeffelholz (8) found 6.050 kcal mol^{-1} (±5%) at a reported T_{fus} of 981 K by high temperature cryoscopy. Hutchison (9) determined T_{fus} to be 983 K, while Kubaschewski et al. (10) listed 985 K. We adopt $\Delta_{fus}H°$ = 6.34 ± 0.13 kcal mol^{-1} and T_{fus} = 984 ± 2 K.

Sublimation Data

$\Delta_{sub}H°$(298.15 K) is calculated as the difference between the enthalpies of formation of the ideal gas and the crystal at 298.15 K.

References

1. P. Ehrlich, K. Peik, and E. Koch, Z. Anorg. Allgem. Chem. 324, 113 (1963).
2. U. S. Nat. Bur. Stand. Circ. 500, 1952.
3. ICSU-CODATA Task Group, J. Chem. Thermodynamics 4, 331 (1972).
4. V. B. Parker, U. S. Nat. Bur. Stand., Rept. 10074, 164 (1969).
5. U. S. Nat. Bur. Stand. Tech Note 270-6, 1971.
6. I. E. Paukov, L. M. Khriplovich, and O. M. Smirnova, Russ. J. Phys. Chem. 45, 1204 (1971).
7. A. S. Dworkin and M. A. Bredig, J. Phys. Chem. 67, 697 (1963).
8. H. H. Emons and B. Loeffelholz, Wiss. Z. Hochsch. Chem. Leuna-Merseburg 6, 261 (1964).
9. J. F. Hutchison, U. S. At. Energy Comm. IS-T-50, 1965.
10. O. Kubaschewski, E. Ll. Evans, and C. B. Alcock, "Metallurgical Thermochemistry," Fourth Ed:, Pergamon Press, New York, 1967.
11. V. B. Parker, U. S. Nat. Bur. Stand., personal communication, May 1974.
12. JANAF Thermochemical Tables: BaO(cr), 6-30-74.

Barium Iodide (BaI_2) Ba_1I_2(cr)

Enthalpy Reference Temperature = T_r = 298.15 K Standard State Pressure = p° = 0.1 MPa

T/K	$C_p°$ (J K⁻¹mol⁻¹)	S°	$-[G°-H°(T_r)]/T$	$H°-H°(T_r)$ (kJ mol⁻¹)	$\Delta_fH°$	$\Delta_fG°$	Log K_f
0	0.	0.	INFINITE	-19.230	-604.544	-604.544	INFINITE
100	68.952	84.559	232.338	-14.778	-604.841	-603.789	315.387
200	75.228	134.767	172.276	-7.502	-605.042	-602.675	157.403
298.15	77.488	165.142	165.142	0.	-605.425	-601.407	105.364
300	77.530	165.622	165.144	0.143	-605.434	-601.382	104.710
400	79.538	188.201	168.209	7.997	-622.236	-599.374	78.270
500	81.546	206.166	174.063	16.051	-666.196	-589.434	61.578
600	83.513	221.207	180.700	24.304	-666.319	-574.057	49.976
700	85.521	234.231	187.437	32.756	-665.450	-558.758	41.695
800	87.529	245.782	194.022	41.408	-664.904	-543.545	35.490
900	89.538	256.207	200.361	50.262	-663.747	-528.443	30.670
984.000	91.211	264.271	205.477	57.853	------ CRYSTAL <--> LIQUID -----		
1000	91.546	265.743	206.429	59.314	-662.403	-513.479	26.821
1100	93.617	274.566	212.227	68.573	-669.237	-497.840	23.640
1200	95.653	282.799	217.769	78.036	-667.748	-482.322	20.995
1300	97.653	290.535	223.072	87.702	-665.980	-466.940	18.762
1400	99.617	297.844	228.154	97.566	-663.968	-451.703	16.853
1500	101.546	304.783	233.033	107.624	-661.745	-436.618	15.204

PREVIOUS: CURRENT: June 1974

Barium Iodide (BaI_2) Ba_1I_2(cr)

BARIUM IODIDE (BaI_2) LIQUID M_r = 391.139

$S°(298.15\ K) = [183.669]\ J\ K^{-1}\ mol^{-1}$

$T_{fus} = 984 \pm 2\ K$

$\Delta_f H°\ (298.15\ K) = [-585.884]\ kJ\ mol^{-1}$

$\Delta_{fus} H° = 26.53 \pm 0.54\ kJ\ mol^{-1}$

Enthalpy of Formation

$\Delta_f H°(BaI_2$, ℓ, 298.15 K) is calculated from that of the crystal by adding $\Delta_{fus}H°$ and the difference in ethalpy, H°(984 K) - H°(298.15 K), between the crystal and liquid.

Heat Capacity and Entropy

The liquid heat capacity near the melting point, 27.0 cal $K^{-1}\ mol^{-1}$, was determined by Dworkin and Bredig (1) from drop calorimetric measurements. This value is adopted and assumed constant over a 700-2800 K range. A glass transition is assumed at 700 K below which the heat capacity is that of the crystal.

S°(298.15 K) is calculated in a manner similar to that used for the enthalpy of formation.

Vaporization Data

The temperature at which $\Delta_r G° = 0$ for the reaction $BaI_2(ℓ) = BaI_2(g)$ is the adopted T_{vap}. Peterson and Hutchison (2) have extrapolated Knudsen effusion cell measurements in the 1154-1301 K range to obtain a normal boiling point (1 atm) of 2300 K.

$\Delta_{vap}H°(T_{vap})$ is the calculated difference between the enthalpies of formation of the ideal gas and the liquid at T_{vap}. See the ideal gas table for details.

References

1. A. S. Dworking and M. A. Bredig, J. Phys. Chem. 67, 697 (1963).
2. D. T. Peterson and J. F. Hutchison, J. Chem. Eng. Data 15, 320 (1970).

Barium Iodide (BaI_2) $Ba_1I_2(l)$

Enthalpy Reference Temperature = T_r = 298.15 K Standard State Pressure = p° = 0.1 MPa

T/K	$C_p°$ (J K⁻¹mol⁻¹)	S° (J K⁻¹mol⁻¹)	$-[G°-H°(T_r)]/T$ (J K⁻¹mol⁻¹)	$H°-H°(T_r)$ (kJ mol⁻¹)	$\Delta_f H°$ (kJ mol⁻¹)	$\Delta_f G°$ (kJ mol⁻¹)	Log K_f
0							
100							
200							
298.15	77.488	183.669	183.669	0.	-585.884	-587.390	102.908
300	77.530	184.149	183.671	0.143	-585.893	-587.399	102.275
400	79.538	206.728	186.736	7.997	-602.695	-587.244	76.686
500	81.546	224.693	192.590	16.051	-646.655	-579.156	60.504
600	83.513	239.733	199.227	24.304	-646.778	-565.632	49.243
699.990	85.521	252.756	205.963	32.754	GLASS <--> LIQUID		
699.990	112.968	252.756	205.963	32.754	TRANSITION		
700	112.968	252.758	205.964	32.756	-645.910	-552.186	41.205
800	112.968	267.842	212.777	44.052	-642.719	-539.009	35.194
900	112.968	281.148	219.649	55.349	-639.119	-526.261	30.543
984.000	112.968	291.228	225.336	64.838	CRYSTAL <--> LIQUID		
1000	112.968	293.051	226.405	66.646	-635.530	-513.914	26.844
1100	112.968	303.818	232.960	77.943	-640.325	-501.106	23.796
1200	112.968	313.647	239.281	89.240	-637.003	-488.595	21.268
1300	112.968	322.689	245.354	100.536	-633.605	-476.365	19.141
1400	112.968	331.061	251.180	111.833	-630.159	-464.399	17.327
1500	112.968	338.855	256.768	123.130	-626.698	-452.680	15.764
1600	112.968	346.146	262.129	134.427	-623.251	-441.191	14.403
1700	112.968	352.995	267.275	145.724	-619.848	-429.917	13.210
1800	112.968	359.452	272.218	157.020	-616.516	-418.841	12.154
1900	112.968	365.559	276.972	168.317	-613.281	-407.948	11.215
2000	112.968	371.354	281.547	179.614	-610.164	-397.222	10.374
2100	112.968	376.866	285.956	190.911	-607.184	-386.649	9.617
2200	112.968	382.121	290.208	202.208	-743.879	-370.839	8.805
2300	112.968	387.143	294.315	213.504	-740.384	-353.962	8.039
2400	112.968	391.950	298.283	224.801	-737.157	-337.231	7.340
2500	112.968	396.562	302.123	236.098	-734.189	-320.629	6.699
2600	112.968	400.993	305.841	247.395	-731.462	-304.141	6.110
2700	112.968	405.256	309.445	258.692	-728.956	-287.754	5.567
2800	112.968	409.365	312.940	269.988	-726.644	-271.456	5.064

PREVIOUS: CURRENT: June 1974

Barium Iodide (BaI_2) $Ba_1I_2(l)$

BARIUM IODIDE (BaI_2) CRYSTAL-LIQUID M_r = 391.139

0 to 984 K crystal
above 984 K liquid

Refer to the individual tables for details.

Barium Iodide (BaI_2) Ba_1I_2(cr,l)

Enthalpy Reference Temperature = T_r = 298.15 K Standard State Pressure = p° = 0.1 MPa

T/K	C_p° (J K⁻¹mol⁻¹)	S° (J K⁻¹mol⁻¹)	$-[G^\circ-H^\circ(T_r)]/T$ (J K⁻¹mol⁻¹)	$H^\circ-H^\circ(T_r)$ (kJ mol⁻¹)	Δ_fH° (kJ mol⁻¹)	Δ_fG° (kJ mol⁻¹)	Log K_f
0	0.	0.	INFINITE	-19.230	-604.544	-604.544	INFINITE
100	68.952	84.559	232.338	-14.778	-604.841	-603.789	315.387
200	75.228	134.767	172.276	-7.502	-605.042	-602.675	157.403
298.15	77.488	165.142	165.142	0.	-605.425	-601.407	105.364
300	77.530	165.622	165.144	0.143	-605.434	-601.382	104.710
400	79.538	188.201	168.209	7.997	-622.236	-599.374	78.270
500	81.546	206.166	174.063	16.051	-666.196	-589.434	61.578
600	83.513	221.207	180.700	24.304	-666.319	-574.057	49.976
700	85.521	234.231	187.437	32.756	-665.450	-558.758	41.695
800	87.529	245.782	194.022	41.408	-664.904	-543.545	35.490
900	89.538	256.207	200.361	50.262	-663.747	-528.443	30.670
984.000	91.211	264.271	205.477	57.853	CRYSTAL <--> LIQUID		
984.000	112.968	291.228	205.477	84.380	TRANSITION		
1000	112.968	293.051	206.863	86.187	-635.530	-513.913	26.844
1100	112.968	303.818	215.196	97.484	-640.325	-501.105	23.796
1200	112.968	313.647	222.996	108.781	-637.003	-488.595	21.268
1300	112.968	322.689	230.322	120.078	-633.604	-476.365	19.141
1400	112.968	331.061	237.222	131.374	-630.159	-464.399	17.327
1500	112.968	338.855	243.741	142.671	-626.698	-452.679	15.764
1600	112.968	346.146	249.916	153.968	-623.251	-441.191	14.403
1700	112.968	352.995	255.780	165.265	-619.848	-429.916	13.210
1800	112.968	359.452	261.362	176.562	-616.516	-418.841	12.154
1900	112.968	365.559	266.687	187.858	-613.280	-407.947	11.215
2000	112.968	371.354	271.776	199.155	-610.163	-397.222	10.374
2100	112.968	376.866	276.650	210.452	-607.184	-386.648	9.617
2200	112.968	382.121	281.326	221.749	-743.879	-370.839	8.805
2300	112.968	387.143	285.818	233.046	-740.384	-353.961	8.039
2400	112.968	391.950	290.141	244.342	-737.157	-337.231	7.340
2500	112.968	396.562	294.306	255.639	-734.188	-320.629	6.699
2600	112.968	400.993	298.325	266.936	-731.462	-304.141	6.110
2700	112.968	405.256	302.207	278.233	-728.956	-287.754	5.567
2800	112.968	409.365	305.961	289.530	-726.644	-271.456	5.064

PREVIOUS: CURRENT: June 1974

Barium Iodide (BaI_2) Ba_1I_2(cr,l)

$\Delta_f H°(0\ K) = -298.5 \pm 16.7\ kJ\ mol^{-1}$

$S°(298.15\ K) = [348.14 \pm 8.4]\ J\ K^{-1}\ mol^{-1}$ $\Delta_f H°(298.15\ K) = -302.9 \pm 16.7\ kJ\ mol^{-1}$

Vibrational Fequencies and Degeneracies

ν, cm^{-1}
[111] (1)
[26] (1)
[185] (1)

Ground State Quantum Weight: [1] $\sigma = 2$

Point Group: C_{2v}

Bond Distance: Ba-I = 3.20 ± 0.03 Å

Bond Angle: I-Ba-I = [170]°

Product of Moments of Inertia: $I_A I_B I_C = 2.117174 \times 10^{-112}\ g^3\ cm^6$

Enthalpy of Formation

Winchell (1) has reported ΔH, ΔG, and ΔS for the vaporization of $BaI_2(\ell)$ at 1150 K derived from a mass spectrometric Knudsen effusion cell investigation (28 unpublished points in the range of 1019-1278 K). The spectra showed no evidence of BaI_2 polymers; also it was consluded that $BaI_2(\ell) = BaI_2(g)$ is the important vaporization. Peterson and Hutchison (2, 3) used a weight loss Knudsen effusion technique to observe vapor presured at 18 temperatures in the range of 1154-1301 K. Their enthalpy of vaporization, 53.6 ± 1.3 kcal mol^{-1}, is in good agreement with Winchell's value, 54 ± 4 kcal mol^{-1}.

Source	No. of Points	T/K	$\Delta_r H°(298.15\ K)$, kcal mol^{-1} 2nd law	3rd law	Drift cal $K^{-1} mol^{-1}$
Winchell (1)	3[a]	1019-1278	62.16	68.07	5.1
Peterson and Hutchison (2)	18[b]	1154-1301	64.35	67.61	2.6
			±1.5	±0.5	±1.2

[a] ΔG calculated at 1019 K and 1278 K from ΔH, ΔG, and ΔS values given at 1150 K to permit use of 2nd law.

[b] one point rejected by statistical tests.

We select $\Delta_{vap}H°(298.15\ K) = 67.6 \pm 3\ kcal\ mol^{-1}$ and combine this with $\Delta_f H°(\ell,\ 298.15\ K) = -140.029 \pm 0.9\ kcal\ mol^{-1}$ to give the adopted $\Delta_f H°(g,\ 298.15\ K) = -72.4 \pm 4\ kcal\ mol^{-1}$.

Heat Capacity and Entropy

The bond distance, as determined from an electron diffraction study, is taken from the work of Akishin et al. (4) who also judged the bond angle to be 180 ± 20°. The later electric deflection study of Buechler et al. (5) showed that all of the barium dihalide molecules are bent. Noting the trend in the magnitude of the bond angle in the alkaline earth-halide series, we have assumed a bond angle of 170°. The principal moments of inertia are: $I_A = 429.4543 \times 10^{-39}$, $I_B = 428.3033 \times 10^{-39}$, and $I_C = 1.1510 \times 10^{-39}\ g\ cm^2$.

In estimating the vibrational frequencies, the stretching force constants for gaseous BaF, BaF_2, BaCl, and $BaCl_2$ are calculated from their vibrational frequencies (6), using the valence force model. The trend in the ratio $k(monohalide)/k_1(dihalide)$ indicates that $k(BaI)/k_1(BaI_2) = 1$ is a reasonable approximation, an approximation used by Brewer et al. (7). The stretching force constant for BaI(g) is calculsted from the ground state vibrational frequency given by Rosen (8). The bending force constant is assumed to be 0.01 times the stretching force constant (7). The adopted vibrational frequencies are calculated from these estimated force constants. Other estimates for ν_1, ν_2, and ν_3, where a linear molecule was assumed, are 109, 26, 184 (7) and 106, 37, 178 (9).

We assign an uncertainty of ± 2 cal $K^{-1}\ mol^{-1}$ to the entropy to allow for error in the estimation of bond angle and of the vibrational frequencies.

References

1. P. Winchell, Nature 206, 1252 (1965).
2. J. F. Hutchison, U. S. At. Energy Comm. IS-T-50, 1965.
3. D. T. Peterson and J. F. Hutchison, J. Chem. Eng. Data 15, 320 (1970).
4. P. A. Akishin, V. P. Spiridonov, G. A. Sobolev, and V. A. Naumov, Zhur. Fiz. Khim. 32, 58 (1958).
5. A. Buechler, J. L. Stauffer, and W. Klemperer, J. Amer. Chem. Soc. 86, 4544 (1964).
6. JANAF Thermochemical Tables: BaF(g), BaF_2(g), and $BaCl_2$(g), 12-31-72.
7. L. Brewer, G. R. Somayajulu, and E. Brackett, Chem. Rev. 63, 111 (1963).
8. B. Rosen, Ed., "Spectroscopic Data Relative to Diatomic Molecules," Pergamon Press, New York, 1970.
9. K. S. Krasnov and V. I. Svettsov, Izv. Vysshikh Uchebn. Zavedenii, Khim. i Khim. Tekhnol. 6, 167 (1963).

Enthalpy Reference Temperature = T_r = 298.15 K Standard State Pressure = p° = 0.1 MPa

T/K	$C_p°$ (J $K^{-1}mol^{-1}$)	S°	$-[G°-H°(T_r)]/T$	$H°-H°(T_r)$ (kJ mol^{-1})	$\Delta_f H°$	$\Delta_f G°$	Log K_f
0	0.	0.	INFINITE	-15.651	-298.463	-298.463	INFINITE
100	52.981	287.151	398.518	-11.137	-298.697	-317.904	166.056
200	56.623	325.343	353.375	-5.606	-300.643	-336.392	87.856
250	57.165	338.042	349.083	-2.760	-301.764	-345.202	72.126
298.15	57.463	348.139	348.139	0.	-302.922	-353.464	61.925
300	57.472	348.494	348.140	0.106	-302.968	-353.778	61.598
350	57.661	357.369	348.840	2.985	-304.348	-362.141	54.047
400	57.785	365.077	350.399	5.871	-321.858	-369.746	48.284
450	57.871	371.888	352.415	8.763	-324.777	-375.560	43.594
500	57.933	377.989	354.673	11.658	-368.086	-377.235	39.410
600	58.014	388.559	359.466	17.456	-370.664	-378.814	32.979
700	58.063	397.506	364.278	23.260	-372.443	-380.043	28.359
800	58.096	405.262	368.927	29.068	-374.741	-380.966	24.875
900	58.118	412.106	373.352	34.879	-376.627	-381.631	22.149
1000	58.133	418.230	377.539	40.691	-378.523	-382.086	19.958
1100	58.145	423.771	381.494	46.505	-388.801	-381.530	18.117
1200	58.154	428.831	385.231	52.320	-390.961	-380.773	16.575
1300	58.161	433.486	388.766	58.136	-393.043	-379.839	15.262
1400	58.166	437.796	392.116	63.952	-395.078	-378.747	14.131
1500	58.171	441.810	395.297	69.769	-397.097	-377.510	13.146
1600	58.174	445.564	398.322	75.586	-399.129	-376.138	12.280
1700	58.177	449.091	401.206	81.404	-401.205	-374.638	11.511
1800	58.180	452.416	403.960	87.222	-403.352	-373.014	10.825
1900	58.182	455.562	406.594	93.040	-405.596	-371.267	10.207
2000	58.184	458.546	409.117	98.858	-407.957	-369.400	9.648
2100	58.186	461.385	411.539	104.677	-410.456	-367.411	9.139
2200	58.187	464.092	413.867	110.495	-552.629	-359.925	8.546
2300	58.188	466.679	416.107	116.314	-554.612	-351.122	7.974
2400	58.189	469.155	418.266	122.133	-556.863	-342.228	7.448
2500	58.190	471.530	420.350	127.952	-559.372	-333.234	6.963
2600	58.191	473.813	422.362	133.771	-562.124	-324.135	6.512
2700	58.192	476.009	424.309	139.590	-565.095	-314.925	6.093
2800	58.192	478.125	426.193	145.409	-568.261	-305.603	5.701
2900	58.193	480.167	428.019	151.229	-571.592	-296.164	5.334
3000	58.193	482.140	429.791	157.048	-575.056	-286.607	4.990
3100	58.194	484.048	431.510	162.867	-578.624	-276.934	4.666
3200	58.194	485.896	433.181	168.687	-582.263	-267.144	4.361
3300	58.195	487.687	434.806	174.506	-585.938	-257.240	4.072
3400	58.195	489.424	436.387	180.326	-589.631	-247.224	3.798
3500	58.195	491.111	437.926	186.145	-593.315	-237.100	3.539
3600	58.196	492.750	439.427	191.965	-596.969	-226.871	3.292
3700	58.196	494.345	440.890	197.784	-600.573	-216.540	3.057
3800	58.196	495.897	442.317	203.604	-604.113	-206.113	2.833
3900	58.196	497.408	443.710	209.423	-607.577	-195.594	2.620
4000	58.197	498.882	445.071	215.243	-610.957	-184.988	2.416
4100	58.197	500.319	446.401	221.063	-614.243	-174.298	2.221
4200	58.197	501.721	447.702	226.882	-617.433	-163.529	2.034
4300	58.197	503.091	448.974	232.702	-620.528	-152.685	1.855
4400	58.197	504.429	450.219	238.522	-623.524	-141.770	1.683
4500	58.198	505.736	451.438	244.342	-626.423	-130.788	1.518
4600	58.198	507.016	452.633	250.161	-629.221	-119.744	1.360
4700	58.198	508.267	453.803	255.981	-631.906	-108.641	1.207
4800	58.198	509.492	454.951	261.801	-634.522	-97.480	1.061
4900	58.198	510.692	456.076	267.621	-637.046	-86.265	0.920
5000	58.198	511.868	457.180	273.441	-639.493	-75.001	0.784
5100	58.198	513.021	458.264	279.260	-641.841	-63.689	0.652
5200	58.198	514.151	459.328	285.080	-644.134	-52.330	0.526
5300	58.198	515.259	460.373	290.900	-646.358	-40.929	0.403
5400	58.199	516.347	461.399	296.720	-648.518	-29.485	0.285
5500	58.199	517.415	462.408	302.540	-650.584	-18.006	0.171
5600	58.199	518.464	463.400	308.360	-652.565	-6.493	0.061
5700	58.199	519.494	464.375	314.180	-654.542	5.062	-0.046
5800	58.199	520.506	465.334	319.999	-656.443	16.647	-0.150
5900	58.199	521.501	466.277	325.819	-658.176	28.248	-0.250
6000	58.199	522.479	467.206	331.639	-659.990	39.898	-0.347

PREVIOUS: June 1974 (1 atm) CURRENT: June 1974 (1 bar)

Barium Iodide (BaI_2) $Ba_1I_2(g)$

BARIUM OXIDE (BaO) CRYSTAL M_r = 153.3294

$S°(298.15\ K) = 72.069 \pm 0.38\ J\ K^{-1}\ mol^{-1}$

$T_{fus} = 2286 \pm 15\ K$

$\Delta_f H°(0\ K) = -546.8 \pm 2.1\ kJ\ mol^{-1}$

$\Delta_f H°(298.15\ K) = -548.1 \pm 2.1\ kJ\ mol^{-1}$

$\Delta_{fus} H° = [59]\ kJ\ mol^{-1}$

Enthalpy of Formation

$\Delta_f H°$ is rounded from $-130.98 \pm 0.47\ kcal\ mol^{-1}$ derived by Fitzgibbon et al. (1) from calorimetric reactions of HCl(aq) with Ba(cr) and BaO(cr). Combination of their two reactions yields $\Delta_r H°(298.15\ K) = -62.66 \pm 0.47\ kcal\ mol^{-1}$ for Ba(cr) + $H_2O(\ell)$ + BaO(cr) + $H_2(g)$. Although the resulting $\Delta_f H°$ is 1.3 kcal mol^{-1} less negative than that of NBS Technical Note 270-6 (2), it confirms Parker's selection (2) of key values for BaO and $BaCl_2$ from calorimetry involving aqueous HCl (3-5) instead of discrepant values derived from other reactions (6-8). Earlier HCl calorimetry gave $\Delta_f H°(BaO, cr) = -132.0$ (3, 2) and -132.1 kcal mol^{-1} (4, 5). The new value (1) is preferred since both reactions were run in the same calorimeter on samples that were well characterized. The latter is particulaly important since impurities were a major problem in studies of BaO (3, 4, 9-14) and Ba (2, 5-7).

Direct measurements of $\Delta_c H$ of Ba(cr) by Mah (6) gave $-139.06 \pm 0.7\ kcal\ mol^{-1}$. Combustion studies of Fitzgibbon et al. (1) suggest that the negative bias of 8 kcal mol^{-1} was due to partial dissolution of the crucible lining (MgO) in molten BaO formed in the combustion. Dissolution was eliminated in combustions on sintered MgO disks, but large amounts of BaO_2 were formed both in this case and with the use of BaO crucibles (1). This led the authors (1) to dismiss combustion calorimetry in favor of solution calorimetry. Parker (2) noted earlier that the combustion result was incompatible with data for $BaCl_2$(aq), $Ba(OH)_2$(aq and cr), BaO_2(cr) and $BaCO_3$(cr); also a similar discrepancy existed for SrO(cr). We find that the combustion value is incompatible with equilibrium data (19) for $BaCl_2$(ℓ and g).

$\Delta_f H°(BaCl_2, cr)$ and $\Delta_f H°(Ba^{2+}, aq)$ deserve an updated discussion. We combine $\Delta_r H° = -127.95 \pm 0.41\ kcal\ mol^{-1}$ (1) for Ba(cr) + 2HCl(54.3 H_2O) = H_2(g) + $BaCl_2$ [in 149.9 HCl (55.0 H_2O)] with $\Delta_f H°(HCl, 54.3\ H_2O) = -39.52 \pm 0.02\ kcal\ mol^{-1}$ (16) and $\Delta_{sol} H° = -2.09 \pm 0.2\ kcal\ mol^{-1}$ (17, 18, 2) for $BaCl_2$(cr) in HCl(55 H_2O) to give $\Delta_f H°(BaCl_2, cr) = -204.9 \pm 0.5\ kcal\ mol^{-1}$. This confirms the JANAF (19) $\Delta_f H°(BaCl_2)$ of -205.2 kcal mol^{-1} based on the selection of Parker (2); the alternative uncertainty of $\pm$ 0.6 kcal mol^{-1} should now be used instead of $\pm$ 3 kcal mol^{-1} (19). The new $\Delta_f H°$ for $BaCl_2$ (1) yields $\Delta_f H°(Ba^{2+}, aq) = -128.0 \pm 0.6\ kcal\ mol^{-1}$, assuming $\Delta_{sol} H°(BaCl_2, cr) = -3.0 \pm 0.2\ kcal\ mol^{-1}$(many references) and $\Delta_f H°(Cl^-, aq) = -39.933 \pm 0.021\ kcal\ mol^{-1}$ (16). This is consistent with Parker's selection of $\Delta_f H°(Ba^{2+}, aq) = -128.5\ kcal\ mol^{-1}$ (2).

Heat Capacity and Entropy

Cp° is based on data (1.4-310 K) of Gmelin (12) who reported smoothed values only in the range 4 to 300 K. Based on Gmelin's graphs, we have corrected several typographical errors and resmoothed Cp° above 240 K where the scatter increases rapidly. The entropy is obtained by integration of our adopted Cp° using S° = 0.0015 cal K^{-1} mol^{-1} at 4 K. The value S° = 15.75 $\pm$ 0.02 cal K^{-1} mol^{-1} reported by Gmelin (12) corresponds to neither 273.15 K nor 298.15 K and must be erroneous. Cp° data (56-299 K) of Anderson (13) differ by $\pm$ 1.6% but average about 1.0% higher from 150 to 260 K.

Cp° above 300 K is from constrained fitting of enthalpy data (391-1299 K) of Lander (14). This calorimetry is subject to bias from a calibration based on Pt and from impurities in the two samples. Information is insufficient to assess the bias due to minor changes in the enthalpy of Pt (15). We apply crude corrections to the observed enthalpies for 27 mole % SrO in one sample and for about 2.8 mole % SiO_2 in the other sample. Deviations from the adopted curve are $\pm$ 0.9% (first sample) and $\pm$ 1.0% (second sample) except for -2.0% at 421 K.

Fusion Data

See the liquid table for details.

Sublimation Data

$\Delta_{sub} H°$ is calculated as the difference in $\Delta_f H°$ between gas and crystal.

References

1. G. C. Fitzgibbon, E. J. Huber and C. E. Holley, J. Chem. Thermodynamics 5, 577 (1973).
2. V. B. Parker, D. D. Wagman and W. H. Evans, U. S. Nat. Bur. Stand. Tech. Note 270-6, 1971; U. S. Nat. Bur. Stand. Report 10074, 164 (1969).
3. L. H. Adami and K. C. Conway, U. S. Bur. Mines RI 6822, 1966.
4. G. V. Flidlider, P. V. Kovtunenko and A. A. Bundel, Russ. J. Phys. Chem. 40, 1168 (1966).
5. A. Guntz and F. Benoit, Ann. Chim. (Paris) [9] 20, 5 (1923).
6. A. D. Mah, U. S. Bur. Mines RI 6171, 1963.
7. A. F. Vorob'ev, A. S. Monaenkova and S. M. Skuratov, Vestn. Mosk. Univ., Ser. II 22 (6), 3 (1967).
8. A. S. Monaenkova and a. F. Vorob'ev, Izv. Vyssh. Ucheb. Zaved., Khim. Khim. Tekhnol. 15, 191 (1972); Dokl. Akad. Nauk SSSR 179, 1129 (1968).
9. O. V. Kedrovskii, I. V. Kovtunenko, E. v. Kiseleva and A. A. Bundel, Russ. J. Phys. Chem. 41, 205 (1967).
10. M. D. Judd and M. I. Pope, J. Appl. Chem. 20, 384 (1970).
11. A. A. Bundel et al., Russ, J. Phys. Chem. 43, 835 (1969); 41, 726 (1967); Tr. Mosk. Khim.-Tekhnol. Inst. No. 62, 291 (1969); cf. chem. Abstr. 74, 130669.
12. E. Gmelin, Z. Naturforsch. 24A, 1794 (1969).
13. C. T. Anderson, J. Amer. Chem. Soc. 57, 429 (1935).
14. J. J. Lander, J. Amer. Chem. Soc. 73, 5794 (1951).
15. A. C. Macleod, J. Chem. Thermodynamics 4, 391 (1972).
16. ISCU-CODATA Task Group, J. Chem. Thermodynamics 4, 331 (1972); V. B. Parker, U. S. Nat. Bur. Stand. NSRDS-NBS2, 1965.
17. O. Ya. Samoilov, Bull. Acad. Sci. USSR, Div. Chem. Sci., 585 (1952); p. 627 (Russian edition).
18. F. E. Young, J. Amer. Chem. Soc. 66, 773 (1944).
19. JANAF Thermchemical Tables: $BaCl_2$(cr, ℓ, g), 12-31-72.

Barium Oxide (BaO) $Ba_1O_1(cr)$

Enthalpy Reference Temperature = T_r = 298.15 K — Standard State Pressure = p° = 0.1 MPa

T/K	$C_p°$	S°	$-[G°-H°(T_r)]/T$	$H°-H°(T_r)$	$\Delta_f H°$	$\Delta_f G°$	Log K_f
	J K⁻¹mol⁻¹			kJ mol⁻¹			
0	0.	0.	INFINITE	-9.983	-546.833	-546.833	INFINITE
100	31.941	27.301	111.119	-8.382	-548.383	-539.039	281.565
200	43.493	53.898	76.289	-4.478	-548.479	-529.582	138.313
298.15	47.279	72.069	72.069	0.	-548.104	-520.382	91.169
300	47.329	72.362	72.070	0.088	-548.096	-520.210	90.577
400	49.898	86.347	73.960	4.955	-547.731	-510.977	66.727
500	51.785	97.694	77.607	10.044	-547.990	-501.778	52.420
600	53.220	107.267	81.773	15.296	-548.939	-492.436	42.870
700	54.396	115.561	86.021	20.678	-548.999	-483.025	36.044
800	55.405	122.892	90.180	26.169	-549.503	-473.558	30.920
900	56.312	129.471	94.186	31.756	-549.526	-464.063	26.934
1000	57.153	135.448	98.018	37.429	-549.490	-454.569	23.744
1100	57.948	140.932	101.674	43.185	-557.766	-444.261	21.096
1200	58.706	146.007	105.159	49.018	-557.850	-433.937	18.889
1300	59.442	150.735	108.485	54.925	-557.776	-423.613	17.021
1400	60.162	155.167	111.663	60.905	-557.567	-413.300	15.420
1500	60.869	159.342	114.704	66.957	-557.244	-403.006	14.034
1600	61.563	163.292	117.618	73.079	-556.829	-392.737	12.822
1700	62.250	167.045	120.416	79.269	-556.342	-382.496	11.753
1800	62.927	170.623	123.107	85.528	-555.805	-372.285	10.803
1900	63.605	174.043	125.698	91.855	-555.239	-362.105	9.955
2000	64.275	177.323	128.198	98.249	-554.665	-351.955	9.192
2100	64.940	180.475	130.613	104.710	-554.107	-341.833	8.503
2200	65.605	183.511	132.949	111.237	-693.105	-326.362	7.749
2286.000	66.177	186.038	134.899	116.904	------ CRYSTAL <--> LIQUID -----		
2300	66.270	186.442	135.211	117.831	-691.803	-309.721	7.034
2400	66.927	189.276	137.405	124.491	-690.669	-293.134	6.380
2500	67.588	192.022	139.535	131.216	-689.705	-276.591	5.779
2600	68.245	194.686	141.605	138.008	-688.906	-260.082	5.225
2700	68.898	197.273	143.620	144.865	-688.261	-243.602	4.713
2800	69.555	199.791	145.581	151.788	-687.756	-227.142	4.237
2900	70.208	202.243	147.493	158.776	-687.372	-210.699	3.795
3000	70.860	204.634	149.358	165.830	-687.088	-194.267	3.382

PREVIOUS: December 1972 CURRENT: June 1974

Barium Oxide (BaO) $Ba_1O_1(cr)$

BARIUM OXIDE (BaO) LIQUID $M_r = 153.294$

$S°(298.15\ K) = [96.553]\ J\ K^{-1}\ mol^{-1}$ $\Delta_f H°(298.15\ K) = [-491.627]\ kJ\ mol^{-1}$

$T_{fus} = 2286 \pm 15\ K$ $\Delta_{fus}H° = [59]\ kJ\ mol^{-1}$

Enthalpy of Formation

$\Delta_f H°$ is calculated from that of the crystal by addition of $\Delta_{fus}H°$ and the difference in enthalpy, H°(2286 K) - H°(298.15 K), between crystal and liquid.

Heat Capacity and Entropy

Cp° of the liquid is estimated as 8 cal K^{-1} g-atom^{-1} by comparison with other oxides. Cp° is taken from the crystal in the range 298.15 K up to the glass transition assumed at 1600 K. S° is calculated in a manner analogous to that used for $\Delta_f H°$.

Fusion Data

Foex (1) obtained T_{fus} = 2010°C from a solar-furnace study using BaO as its own container. We adopt this result but increase it by 3° for conversion to IPTS-68. Foex showed that the much lower value (1918°C) of Schumacher (2, 3) probably resulted from contamination by WO_3 from tungsten supports of the sample. $\Delta_{fus}H°$ is estimated such that $\Delta_{fus}S°$ is 6 cal K^{-1} mol^{-1}.

Vaporization Data

Unlike the other alkaline-earth oxides, BaO(cr) vaporizes mainly to BaO(g) with traces of Ba_2O_2 and Ba_2O. Mass-spectrometric studies (4) suggest that decomposition to Ba(g) and O(g) is not a major contributor even though the observed intensities of Ba^+ are almost comparable with those of BaO^+. T_{vap} is the calculated temperature at which $\Delta_r G° = 0$ for BaO(ℓ) = BaO(g); $\Delta_{vap}H°$ is the corresponding difference in $\Delta_f H°$.

References - BaO(ℓ)

1. M. Foex, Solar Energy 9, 61 (1965).
2. E. E. Schumacher, J. Amer. Chem. Soc. 48, 396 (1926).
3. S. J. Schneider, U. S. Nat. Bur. Stand. Monograph 68, 1963.
4. JANAF Thermochemical Table: BaO(g) 6-30-74.

References

1. R. W. Field, J. Chem. Phys. 60, 2400 (1974).
2. B. Rosen, "Spectroscopic Data Relative to Diatomic Molecules," Pergamon Press, New York, 1970, p. 60.
3. J. Hoeft, F. J. Lovas, E. Tiemann and T. Torring, Z. Naturforsch. 25A, 1750 (1970).
4. L. Wharton, M. Kaufman and W. Klemperer, J. Chem. Phys. 37, 621 (1962); 38, 2705 (1963); 39, 240 (1963); 43, 3406 (1965).
5. JANAF Thermochemical Tables: BaO(cr, ℓ, g), SrO(g) 6-30-74; CaO(g), MgO(g), BeO(g), 12-31-74.
6. R. W. Field, A. D. English, T. Tanaka, D. O. Harris and D. A. Jennings, J. Chem. Phys. 59, 2191 (1973); Ber. Bunsen Ges. Phys. Chem. 78, 146 (1974).
7. P. J. Kalff and C. Th. J. Alkemade, J. Chem. Phys. 52, 1006 (1970); 59, 2572 (1973); 60, 1698 (1974).
8. R. Colin, P. Goldfinger and M. Jeunehomme, Trans. Faraday Soc. 60, 306 (1964).
9. G. A. Semenov, O. S. Popkov, A. I. Soloveichik and S. N. Persiyaninova, Russ. J. Phys. Chem. 46, 898 (1972).
10. R. S. Newbury, G. W. Barton and A. W. Searcy, J. Chem. Phys. 48, 793 (1968).
11. M. G. Inghram, W. A. Chupka and R. F. Porter, J. Chem. Phys. 23, 2159 (1955).
12. S. A. Shchukarev and G. A. Semenov, J. Inorg. Chem. (USSR) 2, (6), 1 (1957); p. 1217 (Russ. edition).
13. B. P. Nikinov and N. G. Otmakhova, Russ. J. Phys. Chem. 35, 734 (1961).
14. J. P. Blewett, H. A. Liebhafsky and E. f. hennelly, J. Chem. Phys. 7, 478 (1939).
15. G. Herrmann, Z. Phys. Chem. 35B, 298 (1937).
16. A. Claassen and C. F. Veenemans, Z. Physik 80, 342 (1933).
17. I. G. Panchenkov, A. V. Gusarov and L. N. Gorokhov, Russ. J. Phys. Chem. 47, 55 (1973).
18. C. D. Jonah, R. N. Zare and Ch. Ottinger, J. Chem. Phys. 56, 263 (1972); Chem. Phys. Lett. 5, 243 (1970).
19. K. Schofield, Chem. Rev. 67, 707 (1967).
20. L. Brewer and G. M. Rosenblatt, Adv. High Temp. Chem. 2, 1 (1969).

Barium Oxide (BaO) $Ba_1O_1(l)$

Enthalpy Reference Temperature = T_r = 298.15 K Standard State Pressure = p° = 0.1 MPa

T/K	$C_p°$ (J K⁻¹mol⁻¹)	S° (J K⁻¹mol⁻¹)	$-[G°-H°(T_r)]/T$ (J K⁻¹mol⁻¹)	$H°-H°(T_r)$ (kJ mol⁻¹)	$\Delta_f H°$ (kJ mol⁻¹)	$\Delta_f G°$ (kJ mol⁻¹)	Log K_f
0							
100							
200							
298.15	47.279	96.553	96.553	0.	-491.627	-471.205	82.553
300	47.329	96.845	96.554	0.088	-491.618	-471.078	82.022
400	49.898	110.831	98.443	4.955	-491.253	-464.293	60.631
500	51.785	122.178	102.090	10.044	-491.512	-457.543	47.799
600	53.220	131.750	106.257	15.296	-492.461	-450.649	39.232
700	54.396	140.045	110.504	20.678	-492.522	-443.686	33.108
800	55.405	147.375	114.664	26.169	-493.026	-436.668	28.511
900	56.312	153.954	118.670	31.756	-493.049	-429.621	24.935
1000	57.153	159.931	122.502	37.429	-493.013	-422.575	22.073
1100	57.948	165.416	126.157	43.185	-501.289	-414.716	19.693
1200	58.706	170.491	129.643	49.018	-501.372	-406.840	17.709
1300	59.442	175.219	132.969	54.925	-501.298	-398.965	16.031
1400	60.162	179.650	136.146	60.905	-501.090	-391.100	14.592
1500	60.869	183.825	139.187	66.957	-500.767	-383.254	13.346
1600	61.563	187.776	142.102	73.079	-500.351	-375.433	12.257
1600.010	61.563	187.776	142.102	73.079	GLASS <--> LIQUID TRANSITION		
1600.010	66.944	187.776	142.102	73.079			
1700	66.944	191.834	144.909	79.773	-499.361	-367.656	11.297
1800	66.944	195.661	147.623	86.468	-498.388	-359.937	10.445
1900	66.944	199.280	150.248	93.162	-497.455	-352.271	9.685
2000	66.944	202.714	152.786	99.856	-496.581	-344.653	9.001
2100	66.944	205.980	155.242	106.551	-495.789	-337.076	8.384
2200	66.944	209.094	157.619	113.245	-634.620	-324.159	7.697
2286.000	66.944	211.661	159.604	119.002	------ CRYSTAL <--> LIQUID -----		
2300	66.944	212.070	159.923	119.940	-633.217	-310.080	7.042
2400	66.944	214.919	162.155	126.634	-632.049	-296.057	6.444
2500	66.944	217.652	164.321	133.328	-631.116	-282.077	5.894
2600	66.944	220.278	166.423	140.023	-630.414	-268.130	5.387
2700	66.944	222.804	168.465	146.717	-629.932	-254.206	4.918
2800	66.944	225.239	170.449	153.412	-629.655	-240.296	4.483
2900	66.944	227.588	172.379	160.106	-629.565	-226.392	4.078
3000	66.944	229.857	174.257	166.800	-629.640	-212.489	3.700
3100	66.944	232.053	176.086	173.495	-629.857	-198.580	3.346
3200	66.944	234.178	177.869	180.189	-630.192	-184.663	3.014
3300	66.944	236.238	179.607	186.884	-630.614	-170.735	2.703
3400	66.944	238.236	181.302	193.578	-631.111	-156.792	2.409
3500	66.944	240.177	182.956	200.272	-631.660	-142.833	2.132
3600	66.944	242.063	184.572	206.967	-632.240	-128.859	1.870
3700	66.944	243.897	186.151	213.661	-632.837	-114.868	1.622
3800	66.944	245.682	187.694	220.356	-633.435	-100.860	1.386
3900	66.944	247.421	189.203	227.050	-634.023	-86.837	1.163
4000	66.944	249.116	190.680	233.744	-634.593	-72.800	0.951
4100	66.944	250.769	192.125	240.439	-635.135	-58.748	0.748
4200	66.944	252.382	193.541	247.133	-635.646	-44.684	0.556
4300	66.944	253.957	194.928	253.828	-636.125	-30.608	0.372
4400	66.944	255.496	196.287	260.522	-636.568	-16.521	0.196
4500	66.944	257.001	197.619	267.216	-636.975	-2.425	0.028
4600	66.944	258.472	198.926	273.911	-637.339	11.680	-0.133
4700	66.944	259.912	200.209	280.605	-637.649	25.790	-0.287
4800	66.944	261.321	201.467	287.300	-637.946	39.909	-0.434
4900	66.944	262.702	202.703	293.994	-638.206	54.034	-0.576
5000	66.944	264.054	203.916	300.688	-638.442	68.164	-0.712

PREVIOUS: December 1972 CURRENT: June 1974

Barium Oxide (BaO) $Ba_1O_1(l)$

BARIUM OXIDE (BaO) CRYSTAL–LIQUID M_r = 153.3294

0 to 2286 K crystal
above 2286 K liquid

Refer to the individual tables for details.

Barium Oxide (BaO) Ba_1O_1(cr,l)

Enthalpy Reference Temperature = T_r = 298.15 K   Standard State Pressure = p° = 0.1 MPa

	J K⁻¹mol⁻¹			kJ mol⁻¹			
T/K	C_p°	S°	$-[G^\circ - H^\circ(T_r)]/T$	$H^\circ - H^\circ(T_r)$	$\Delta_f H^\circ$	$\Delta_f G^\circ$	Log K_f
0	0.	0.	INFINITE	-9.983	-546.833	-546.833	INFINITE
100	31.941	27.301	111.119	-8.382	-548.383	-539.039	281.565
200	43.493	53.898	76.289	-4.478	-548.479	-529.582	138.313
298.15	47.279	72.069	72.069	0.	-548.104	-520.382	91.169
300	47.329	72.362	72.070	0.088	-548.096	-520.210	90.577
400	49.898	86.347	73.960	4.955	-547.731	-510.977	66.727
500	51.785	97.694	77.607	10.044	-547.990	-501.778	52.420
600	53.220	107.267	81.773	15.296	-548.939	-492.436	42.870
700	54.396	115.561	86.021	20.678	-548.999	-483.025	36.044
800	55.405	122.892	90.180	26.169	-549.503	-473.558	30.920
900	56.312	129.471	94.186	31.756	-549.526	-464.063	26.934
1000	57.153	135.448	98.018	37.429	-549.490	-454.569	23.744
1100	57.948	140.932	101.674	43.185	-557.766	-444.261	21.096
1200	58.706	146.007	105.159	49.018	-557.850	-433.937	18.889
1300	59.442	150.735	108.485	54.925	-557.776	-423.613	17.021
1400	60.162	155.167	111.663	60.905	-557.567	-413.300	15.420
1500	60.869	159.342	114.704	66.957	-557.244	-403.006	14.034
1600	61.563	163.292	117.618	73.079	-556.829	-392.737	12.822
1700	62.250	167.045	120.416	79.269	-556.342	-382.496	11.753
1800	62.927	170.623	123.107	85.528	-555.805	-372.285	10.803
1900	63.605	174.043	125.698	91.855	-555.239	-362.105	9.955
2000	64.275	177.323	128.198	98.249	-554.665	-351.955	9.192
2100	64.940	180.475	130.613	104.710	-554.107	-341.833	8.503
2200	65.605	183.511	132.949	111.237	-693.105	-326.362	7.749
2286.000	66.177	186.038	134.899	116.904	CRYSTAL <--> LIQUID		
2286.000	66.944	211.661	134.899	175.480	TRANSITION		
2300	66.944	212.070	135.367	176.417	-633.217	-310.080	7.042
2400	66.944	214.919	138.623	183.111	-632.049	-296.057	6.444
2500	66.944	217.652	141.730	189.806	-631.116	-282.077	5.894
2600	66.944	220.278	144.701	196.500	-630.414	-268.130	5.387
2700	66.944	222.804	147.547	203.194	-629.932	-254.206	4.918
2800	66.944	225.239	150.278	209.889	-629.655	-240.296	4.483
2900	66.944	227.588	152.904	216.583	-629.565	-226.392	4.078
3000	66.944	229.857	155.432	223.278	-629.640	-212.489	3.700
3100	66.944	232.053	157.868	229.972	-629.857	-198.580	3.346
3200	66.944	234.178	160.220	236.666	-630.192	-184.663	3.014
3300	66.944	236.238	162.492	243.361	-630.614	-170.735	2.703
3400	66.944	238.236	164.691	250.055	-631.111	-156.792	2.409
3500	66.944	240.177	166.820	256.750	-631.660	-142.833	2.132
3600	66.944	242.063	168.884	263.444	-632.240	-128.859	1.870
3700	66.944	243.897	170.887	270.138	-632.837	-114.868	1.622
3800	66.944	245.682	172.832	276.833	-633.435	-100.860	1.386
3900	66.944	247.421	174.722	283.527	-634.023	-86.837	1.163
4000	66.944	249.116	176.561	290.222	-634.593	-72.800	0.951
4100	66.944	250.769	178.351	296.916	-635.135	-58.748	0.748
4200	66.944	252.382	180.094	303.610	-635.646	-44.684	0.556
4300	66.944	253.957	181.794	310.305	-636.125	-30.608	0.372
4400	66.944	255.496	183.451	316.999	-636.568	-16.521	0.196
4500	66.944	257.001	185.069	323.694	-636.975	-2.425	0.028
4600	66.944	258.472	186.649	330.388	-637.339	11.680	-0.133
4700	66.944	259.912	188.192	337.082	-637.649	25.790	-0.287
4800	66.944	261.321	189.701	343.777	-637.946	39.909	-0.434
4900	66.944	262.702	191.177	350.471	-638.206	54.034	-0.576
5000	66.944	264.054	192.621	357.166	-638.442	68.163	-0.712

PREVIOUS: CURRENT: June 1974

Barium Oxide (BaO) Ba_1O_1(cr,l)

$S°(298.15\ K) = 235.457 \pm 0.33\ J\ K^{-1}\ mol^{-1}$

$\Delta_fH°(0\ K) = -121.6 \pm 8\ kJ\ mol^{-1}$

$\Delta_fH°(298.15\ K) = -123.8 \pm 8\ kJ\ mol^{-1}$

Electronic and Molecular Constants (σ = 1)

Source	State	ε_i, cm^{-1}	g_i	r_e, Å	B_e, cm^{-1}	α_e, cm^{-1}	ω_e, cm^{-1}	ω_ex_e, cm^{-1}
(1-4)	$X^1\Sigma$	0.0	1	1.9397	0.31261	0.00139	669.81	2.054
(1)	$a^3\Pi$	17372.	6	2.289	0.2244	0.0014	448.3	2.4
(1)	$A'^1\Pi$	17568.	2	2.289	0.2244	0.0014	448.3	2.4
(5)	$^3\Sigma^+$	[14000]	3	[2.133]	[0.2583]	[0.0011]	[499.7]	[1.64]
(1, 6, 2)	$A^1\Sigma^+$	16722.	1	2.133	0.25832	0.00107	499.7	1.64
(5)	$^3\Sigma^+$	[22000]	3	[2.26]	[0.230]	[0.001]	[480]	[2]
(5)	$^3\Delta$	[22000]	6	"	"	"	"	"
(5)	$^1\Delta$	[24000]	2	"	"	"	"	"
(5)	$^3\Sigma^-$	[24000]	3	"	"	"	"	"
(5)	$^1\Sigma^-$	[24000]	1	"	"	"	"	"
(5)	$^1\Sigma^+$	[28000]	1	"	"	"	"	"
(5)	$^3\Pi$	[32000]	6	[2.20]	[0.243]	"	[488]	[3.6]
(2)	$B^1\Pi$	32775.	2	"	"	"	488.	3.6

Enthalpy of Formation

We adopt D_0^0 = 131.0 ± 2 kcal mol^{-1} and $\Delta_fH°(298.15\ K)$ = -29.6 ± 2 kcal mol^{-1} based on equilibria and vaporization data analyzed below. The results are in excellent agreement, even though δS varies widely. The dissociation energy is confirmed also by values of 132 ± 5(electron impact, 17), ≥ 131.4 (crossed-beam chemiluminescence, 18) and ~131 kcal mol^{-1} (flame study, 1973, 7). We have adjusted the latter value to make it roughly consistent with our Gibbs energy functions. Less reliable data were reviewed by Schofield (19). Our analysis supports the conclusions of Brewer and Rosenblatt (20).

Source	Method	Reaction[a]	T/K	Points	δS cal K^{-1} mol^{-1}	$\Delta_rH°(298.15\ K)$, kcal mol^{-1} 2nd law	3rd law	$\Delta_fH°(298.15\ K)$, D_0^0 kcal mol^{-1}	
(7) Kalff (1970)	Flame photometric	A	~2680-2860	12	19. ±4	49. ±11	-4.7±2.5	-29.6±3	131.0
(8) Colin (1964)	Knudsen mass spec.	B	1934-2069	3	-11. ±10	-29. ±20	-6.8±2.0	-29.1±3	130.5
(9) Semenov (1972)	" " "	C	1423-1723	Eqn.	3.7	108.1	102.2±1.0	-28.8±2	130.2
(10) Newbury (1968)	Knudsen thermobal.	C	1563-1853	17	5.9±0.9	111.8±1.5	101.8±1.3	-29.2±2	130.6
	" " "	C	1503-1862	45	4.2±0.5	109.2±0.8	102.1±1.0	-28.9±2	130.3
	Knudsen mass spec.	C	1590-1892	11	4.8±0.2	110.1±0.3	101.7±1.1	-29.3±2	130.8
	" " "	C	1481-1917	38	5.6±0.3	111.4±0.5	101.8±1.4	-29.2±2	130.6
	" " "	C	1365-1910	29	6.1±0.4	112.1±0.6	102.1±1.6	-28.9±2	130.3
(11) Inghram (1955)	" " "	C	1530-1758	4	-2.8±6.1	99. ±10	103.2±1.7	-27.8±2	129.2
(12) Shchukarev (1957)	Langmuir mass spec.	C	~1173-1473	Eqn.	-6.1	93.3	101.4±0.6	-29.6±2	131.0
(13) Novikov (1961)	Knudsen effusion	C	~1140-1470	Eqn.	-2.0	97.2	99.8±1.5	-31.2±2	132.6
(14) Blewett (1939)	" "	C	1526-1800	12	-0.5±2.5	103.1±4.1	103.9±4.3	-27.1±3	128.5
(15) Herrmann (1937)	Langmuir	C	~1200-1560	Eqn.	16.	124.	101.8±1.6	-29.2±3	130.7
(16) Claassen (1933)	"	C	1223-1475	7	-4.5±1.5	95.9±2.1	101.9±1.0	-29.1±2	130.5

[a]Reactions: A) $Ba(g) + CO_2(g) = BaO(g) + CO(g)$; B) $Ba(g) + SO(g) = BaO(g) + S(g)$; C) $BaO(cr) = BaO(g)$.

Heat Capacity and Entropy

Electronic levels (T_o) and vibrational-rotational constants of the observed states are from Field (1) and Rosen (2). The long-sought $a^3\Pi$ and $A'^1\Pi$ states of CaO, SrO and BaO were characterized by Field (1) using a new method for assignment of perturbations. This study resolved the long-standing controversy over low-lying electronic levels and confirmed (1) that $X^1\Sigma^+$ is the ground state (4). We estimate the other potentially low-lying state ($^3\Sigma^+$) at 14000 cm^{-1} by assuming that it lies 3000 ± 2500 cm^{-1} (1, 5) below the isoconfigurational A state. Other predicted states and their vibrational-rotational constants are estimated in isoconfigurational groups by comaprison with BeO, MgO, CaO and SrO (5). Comparisons are facilitated by listing the states in the isoconfigurational order of MgO (5). Our thermodynamic functions correspond to an "effective" ground state with g = 1 instead of g = 6 (10, 19). Our functions are calculated using first-order anharmonic corrections to Q_r^i and Q_v^i in the partition function $Q = Q_t \Sigma Q_r^i Q_v^i g_i \exp(-c_2\varepsilon_i/T)$.

References

Reference are listed on the liquid table.

Enthalpy Reference Temperature = T_r = 298.15 K — Standard State Pressure = p° = 0.1 MPa

T/K	$C_p°$ (J K^{-1}mol^{-1})	S°	$-[G°-H°(T_r)]/T$	$H°-H°(T_r)$ (kJ mol^{-1})	$\Delta_fH°$	$\Delta_fG°$	Log K_f
0	0.	0.	INFINITE	-9.014	-121.607	-121.607	INFINITE
100	29.160	202.190	263.233	-6.104	-121.848	-129.993	67.901
200	30.735	222.767	238.400	-3.127	-122.870	-137.747	35.976
298.15	32.897	235.457	235.457	0.	-123.846	-144.839	25.375
300	32.932	235.661	235.458	0.061	-123.865	-144.969	25.241
400	34.443	245.359	236.770	3.435	-124.993	-151.844	19.829
500	35.385	253.154	239.293	6.930	-126.845	-158.363	16.544
600	35.988	259.662	242.160	10.501	-129.476	-164.410	14.313
700	36.393	265.242	245.068	14.122	-131.298	-170.100	12.693
800	36.679	270.121	247.901	17.776	-133.639	-175.478	11.458
900	36.889	274.454	250.616	21.455	-135.569	-180.592	10.481
1000	37.049	278.350	253.197	25.152	-137.510	-185.491	9.689
1100	37.177	281.887	255.647	28.864	-147.829	-189.375	8.993
1200	37.281	285.126	257.971	32.587	-150.023	-193.054	8.403
1300	37.370	288.114	260.176	36.319	-152.124	-196.554	7.898
1400	37.448	290.886	262.272	40.060	-154.154	-199.895	7.458
1500	37.520	293.472	264.267	43.809	-156.135	-203.093	7.072
1600	37.593	295.896	266.169	47.564	-158.085	-206.160	6.730
1700	37.671	298.178	267.985	51.327	-160.026	-209.105	6.425
1800	37.762	300.333	269.723	55.099	-161.977	-211.936	6.150
1900	37.873	302.378	271.388	58.880	-163.956	-214.658	5.901
2000	38.014	304.324	272.987	62.675	-165.982	-217.274	5.675
2100	38.195	306.183	274.523	66.485	-168.074	-219.787	5.467
2200	38.427	307.965	276.003	70.315	-309.769	-216.824	5.148
2300	38.720	309.679	277.430	74.172	-311.204	-212.568	4.828
2400	39.086	311.335	278.809	78.062	-312.841	-208.245	4.532
2500	39.535	312.939	280.142	81.992	-314.672	-203.850	4.259
2600	40.079	314.500	281.434	85.972	-316.684	-199.378	4.006
2700	40.725	316.024	282.687	90.011	-318.858	-194.825	3.769
2800	41.483	317.519	283.904	94.121	-321.166	-190.190	3.548
2900	42.357	318.989	285.089	98.312	-323.579	-185.470	3.341
3000	43.353	320.442	286.243	102.596	-326.064	-180.665	3.146
3100	44.471	321.881	287.369	106.986	-328.585	-175.777	2.962
3200	45.711	323.312	288.470	111.494	-331.106	-170.807	2.788
3300	47.069	324.739	289.548	116.132	-333.585	-165.760	2.624
3400	48.539	326.166	290.604	120.912	-335.997	-160.638	2.468
3500	50.112	327.595	291.640	125.844	-338.308	-155.447	2.320
3600	51.777	329.030	292.659	130.937	-340.490	-150.191	2.179
3700	53.519	330.472	293.661	136.201	-342.516	-144.877	2.045
3800	55.323	331.924	294.649	141.643	-344.367	-139.509	1.918
3900	57.172	333.384	295.624	147.268	-346.025	-134.096	1.796
4000	59.044	334.855	296.586	153.078	-347.479	-128.643	1.680
4100	60.922	336.337	297.537	159.077	-348.717	-123.157	1.569
4200	62.784	337.827	298.479	165.262	-349.737	-117.643	1.463
4300	64.609	339.326	299.411	171.632	-350.540	-112.107	1.362
4400	66.380	340.832	300.336	178.182	-351.127	-106.555	1.265
4500	68.076	342.342	301.252	184.906	-351.505	-100.992	1.172
4600	69.681	343.856	302.162	191.794	-351.675	-95.423	1.084
4700	71.180	345.371	303.065	198.838	-351.636	-89.855	0.999
4800	72.561	346.885	303.962	206.026	-351.439	-84.287	0.917
4900	73.812	348.394	304.854	213.346	-351.073	-78.725	0.839
5000	74.928	349.896	305.740	220.784	-350.565	-73.172	0.764
5100	75.901	351.390	306.620	228.327	-349.905	-67.632	0.693
5200	76.729	352.872	307.495	235.960	-349.151	-62.105	0.624
5300	77.413	354.340	308.365	243.668	-348.300	-56.593	0.558
5400	77.953	355.793	309.230	251.437	-347.371	-51.098	0.494
5500	78.354	357.227	310.090	259.254	-346.348	-45.624	0.433
5600	78.620	358.641	310.944	267.104	-345.252	-40.173	0.375
5700	78.759	360.034	311.793	274.974	-344.175	-34.735	0.318
5800	78.777	361.404	312.637	282.851	-343.058	-29.318	0.264
5900	78.684	362.750	313.475	290.725	-341.820	-23.939	0.212
6000	78.489	364.071	314.307	298.585	-340.718	-18.560	0.162

PREVIOUS: June 1974 (1 atm) CURRENT: June 1974 (1 bar)

BARIUM SULFIDE (BaS) CRYSTAL M_r = 169.39

$S°(298.15 K) = 78.4 \pm 1.3 J K^{-1} mol^{-1}$

$\Delta_fH°(0 K) = -462.7 \pm 2.1 kJ mol^{-1}$
$\Delta_fH°(298.15 K) = -463.6 \pm 2.1 kJ mol^{-1}$

Enthalpy of Formation

Sabatier (1) measured the enthalpy of solution of BaS in hydrochloric acid solution as $\Delta_rH°(284 K) = -27.0\pm1.0 kcal mol^{-1}$ for $BaS(cr) + 2HCl(100 H_2O, aq) \rightarrow BaCl_2(100 H_2O, aq) + H_2S(aq)$. We derive $\Delta_fH°(298.15 K) = -111.0\pm2.0 kcal mol^{-1}$ based on the following auxiliary data: $\Delta_fH°(HCl\cdot100 H_2O, aq, 298.15 K) = -39.657 kcal mol^{-1}$ (2), $\Delta_fH°(H_2S, aq, 298.15 K) = -9.5 kcal mol^{-1}$ (2), and $\Delta_fH°(BaCl_2\cdot100 H_2O, aq, 298.15 K) = -207.803 kcal mol^{-1}$ (3). Mourlot (4) also measured the enthalpy of solution of BaS in hydrochloric acid solution as $\Delta_rH° = -27.35 \pm 1.0 kcal mol^{-1}$ from which we calculate $\Delta_fH°(298.15 K) = -110.6 \pm 2 kcal mol^{-1}$ using the same auxiliary data. The uncertainties quoted reflect the correction to 298.15 K and the ambiguity of the exact solution composition.

Our analyses of equilibrium studies of Okuno (5), Culver and Hamdorf (6), Nikonov (7), Schenck and Hammerschmidt (8), and Colin et al. (9), are listed below. The calculated 3rd law $\Delta_rH°(298.15 K)$ may have an uncertainty of 0.5 kcal mol^{-1} since the JANAF Gibbs energy functions are partially based on the estimated $C_p°$ data (above 300 K). The enthalpy of formation of BaS(cr) derived from equilibrium studies is in good agreement with that derived from enthalpy of solution studies (1, 4). We discount the studies of Nikonov (7) and of Colin et al. (9). More information is given in the SrS(cr) table (10).

A weighted average, $\Delta_fH°(BaS, cr, 298.15 K) = -110.8\pm0.5 kcal mol^{-1}$, is adopted in the tabulation.

Source	Method	Reaction[a]	T/K	No. of Points	δS cal K^{-1}mol^{-1}	$\Delta_rH°(298.15 K)$, kcal mol^{-1} 2nd law	3rd law	$\Delta_fH°(298.15 K)$ kcal mol^{-1}
(5) Okuno (1935)	Equilibrium	A	973-1173	5	0.85 ± 0.75	-6.27±0.81	-7.18±0.25	-110.29±1.0
(6) Culver (1955)	Equilibrium	B	1073-1373	Equation	-2.89	6.14	9.68±0.75	-111.23±2.5
(7) Nikonov (1961)	Vaporization	C	1400-1600	Equation	2.2	-106.38	-109.74±0.6	-100.7 ±4
(8) Schenck (1933)	Equilibrium	C	1355-1399	2	-11.4	52.6	68.31±0.7	-111.05±4.0
(9) Colin et al. (1964)	Mass Spec	E	1846-2120	10	-17.6±7.5	-59.4±15.0	-24.2±4.2	-118.6 ±12

[a]Reactions:
A) $BaSO_4(cr) + 4CO(g) = BaS(cr) + 4CO_2(g)$
B) $BaSO_4(cr) + 4H_2(g) = BaS(cr) + 4H_2O(g)$
C) $BaS(g) = BaS(cr)$
D) $BaSO_4(cr) + S_2(g) = BaS(cr) + 2SO_2(g)$
E) $Ba(g) + 2S_2(g) = BaS(cr) + 3S(g)$

Heat Capacity and Entropy

King and Weller (11) measured the low temperature heat capacities from 54-296 K. Their smooth values are adopted in the tabulation. The entropy, $S°(298.15 K) = 18.7\pm0.2 cal K^{-1} mol^{-1}$, is based on $S°(51 K) = 2.60 cal K^{-1} mol^{-1}$ as extrapolated by King and Weller using Debye and Einstein functions (θ_D = 159, θ_E = 267). The $C_p°$ values above 300 K are estimated by graphical extrapolation combined with method B of Kubaschewski et al. (12).

Fusion Data

Literature melting data for BaS(cr) are not available. It has been reported (13, 14) that BaS melts above 2500 K but this value must be looked upon as a lower limit.

References

1. M. Sabatier, Ann. Chim. Phys. 22, 5-98 (1881).
2. U. S. Nat. Bur. Stand., Tech. Note 270-3, 1968.
3. U. S. Nat. Bur. Stand., Tech. Note 270-6, 1971.
4. A. Mourlot, Ann. Chim. Phys. 17, 510 (1899).
5. T. Okuno, J. Soc. Chem. Ind. Japan 38, 421 (1935).
6. R. V. Culver and C. J. Hamdorf, J. Appl. Chem. 5, 303 (1955).
7. B. P. Nikonov and N. G. Otmakhova, Zhur. Fiz. Khim. 35, 1494 (1961).
8. R. Schenck and F. Hammerschmidt, Z. anorg. allgem. Chem. 210, 305 (1933).
9. R. Colin, P. Goldfinger, and M. Jennehomme, Trans. Faraday Soc. 60, 306 (1964).
10. JANAF Thermochemical Tables: SrS(cr), 9-30-77.
11. E. G. King and W. W. Weller, U. S. Bur. Mines RI 5590, 1960.
12. O. Kubaschewski, E. L. Evans, and C. B. Alcock, "Metallurgical Thermochemistry," Fourth ed., Pergamon Press, Oxford, 1967.
13. K. C. Mills, "Thermodynamic Data for Inorganic Sulfides, Selenides, and Tellurides," Butterworths, London, 1974.
14. L. Brewer et al., U. S. Atomic Energy Commission Publ. AECD-2242(MB-LB-18-5) 1948.

Barium Sulfide (BaS) $Ba_1S_1(cr)$

Enthalpy Reference Temperature = T_r = 298.15 K Standard State Pressure = p° = 0.1 MPa

T/K	$C_p°$ (J K^{-1}mol^{-1})	S°	$-[G°-H°(T_r)]/T$	$H°-H°(T_r)$ (kJ mol^{-1})	$\Delta_fH°$	$\Delta_fG°$	Log K_f
0	0.	0.	INFINITE	-10.458	-462.722	-462.722	INFINITE
100	36.317	29.938	120.037	-9.010	-463.661	-461.995	241.321
200	46.610	59.176	82.843	-4.733	-463.571	-460.350	120.231
298.15	49.374	78.395	78.395	0.	-463.587	-458.776	80.376
300	49.407	78.701	78.396	0.091	-463.590	-458.746	79.875
400	51.128	93.153	80.355	5.119	-466.176	-457.002	59.678
500	52.718	104.737	84.110	10.314	-468.728	-454.445	47.476
600	54.057	114.473	88.380	15.655	-471.593	-451.302	39.289
700	55.061	122.882	92.722	21.112	-473.298	-447.793	33.415
800	55.982	130.296	96.965	26.665	-475.283	-444.006	28.991
900	56.777	136.935	101.043	32.302	-529.809	-438.930	25.475
1000	57.656	142.963	104.938	38.025	-529.849	-428.829	22.400
1100	58.430	148.494	108.650	43.829	-538.193	-417.908	19.845
1200	59.204	153.612	112.186	49.710	-538.341	-406.965	17.715
1300	59.983	158.381	115.558	55.670	-538.326	-396.017	15.912
1400	60.752	162.855	118.778	61.707	-538.172	-385.075	14.367
1500	61.500	167.071	121.859	67.819	-537.901	-374.148	13.029
1600	62.258	171.065	124.810	74.007	-537.534	-363.243	11.859
1700	63.037	174.862	127.644	80.272	-537.090	-352.363	10.827
1800	63.806	178.487	130.369	86.614	-536.588	-341.511	9.910
1900	64.554	181.957	132.993	93.032	-536.051	-330.688	9.091
2000	65.312	185.288	135.525	99.525	-535.498	-319.894	8.355
2100	66.086	188.493	137.971	106.095	-534.950	-309.127	7.689
2200	66.860	191.585	140.339	112.742	-673.947	-293.011	6.957
2300	67.640	194.574	142.632	119.467	-672.630	-275.725	6.262
2400	68.408	197.469	144.857	126.270	-671.468	-258.494	5.626
2500	69.156	200.277	147.018	133.148	-670.463	-241.308	5.042
2600	69.915	203.004	149.119	140.102	-669.609	-224.159	4.503
2700	70.694	205.657	151.164	147.132	-668.895	-207.041	4.005
2800	71.463	208.242	153.157	154.240	-668.304	-189.946	3.543
2900	72.221	210.763	155.100	161.424	-667.818	-172.871	3.114
3000	72.969	213.224	156.996	168.684	-667.418	-155.811	2.713

PREVIOUS: CURRENT: September 1977

Barium Sulfide (BaS) $Ba_1S_1(cr)$

$\Delta_f H°(0\ K) = 39.55 \pm 12.6\ kJ\ mol^{-1}$

$S°(298.15\ K) = 248.861 \pm 0.42\ J\ K^{-1}\ mol^{-1}$ $\Delta_f H°(298.15\ K) = 37.78 \pm 12.6\ kJ\ mol^{-1}$

Electronic and Molecular Constants (σ = 1)

Source	State	ε_i, cm^{-1}	g_i	r_e, Å	B_e, cm^{-1}	α_e, cm^{-1}	ω_e, cm^{-1}	$\omega_e x_e$, cm^{-1}
(1,2)	$X^1\Sigma^+$	0.0	1	2.5067	0.103160	0.000314	379.117	0.8828
(8)	$^3\Sigma^+$	[12000.]	3	[2.635]	[0.09339]	[0.000713]	[294.1]	[3.075]
(1,8)	$^3\Pi$	[14000.]	6	[2.88]	[0.0784]	[0.0004]	[234.8]	[0.9]
(8)	$^1\Pi$	[14200.]	2	[2.88]	[0.0784]	[0.0004]	[234.8]	[0.9]
(1)	$A^1\Sigma^+$	14450.	1	2.635	0.09339	0.000713	294.06	3.075
(1)	$B^1\Sigma^+$	26997.74	1	2.7468	0.08591	0.000439	253.90	0.4373

Enthalpy of Formation

The adopted value of the enthalpy of formation, $\Delta_f H°(BaS,\ g,\ 298.15\ K) = 9.03 \pm 3.0\ kcal\ mol^{-1}$, is based on the Knudsen mass-spectrometric study of Colin et al. (3). Our reanalysis of their ion intensity data is given below. Although the large drift would suggest a larger uncertainty in the derived quantities ($\Delta_f H°$, D_0^0), our experience with related work on other alkaline earth sulfides and oxides by the same authors indicates that results derived from a 3rd law analysis are preferred. Our value of the enthalpy of formation is intermediate between $\Delta_f H°(298.15\ K) = 7.3 \pm 5\ kcal\ mol^{-1}$ obtained by Mills (6) (using different Gibbs energy functions) in a recent critical compilation and $\Delta_f H°(298.15\ K) = 12.0\ kcal\ mol^{-1}$ preferred by NBS (7). Using auxiliary JANAF data (8), and the recommended value for $D_0^0(S_2,\ g) = 100.69 \pm 0.01\ kcal\ mol^{-1}$ (8), we calculate D_0^0(BaS, g) = 99.3 ± 3 kcal mol^{-1}. A linear Birge-Sponer extrapolation of the $X^1\Sigma^+$ ground state based on the data of Barrow et al. (1), after an ionic correction due to Hildenbrand (4), yields the dissociation energy of 99.8 kcal mol^{-1} for the products Ba(^{1}S) + S(^{1}D). The $X^1\Sigma^+$ ground state cannot dissociate to ground state atoms. If the dissociation products are instead Ba(^{3}D) + S(^{3}P), analogous to the work of Field et al. (5) on BaO, the dissociation energy becomes D_0^0 = 100.4 kcal mol^{-1}. Both of these values are uncertain due to the very long vibrational extrapolation.

Source	Method	Reaction[a]	T/K	No. of Points	δS cal K^{-1}mol^{-1}	$\Delta_r H°$(298.15 K), kcal mol^{-1} 2nd law	3rd law	$\Delta_f H°$(298.15 K), D_0^0 kcal mol^{-1}	
(3) Colin (1964)	Mass Spec.	A	1846-2120	10	-2.5±4.9	-3.37±9.8	1.68±2.5	9.03±3.0	99.2±3.0

[a]Reaction: A) $Ba(g) + S_2(g) = BaS(g) + S(g)$

Heat Capacity and Entropy

Electronic levels (T_{00}) and vibrational-rotational constants of the observed states are from the optical study of Barrow et al. (1) and the microwave work of Tiemann et al. (2). Other low-lying electronic states and their vibrational-rotational constants are estimated in isoconfigurational groups by analogy with BaO (8) and from trends observed in the known states of the other alkaline-earth oxides and sulfides. Thermodynamic functions are calculated using first-order anharmonic corrections to Q_r^i and Q_v^i in the partition function $Q = Q_t \Sigma Q_r^i Q_v^i g_i \exp(-c_2\varepsilon_i/T)$. Uncertainty in the energy and molecular constants for the estimated states may contribute as much as 2-3 cal K^{-1} mol^{-1} to S° at 3000 K. The molecular constants have been corrected to the natural isotopic abundances.

References

1. R. F. Barrow, W. G. Burton, and P. A. Jones, Trans. Faraday Soc 67, 902 (1971).
2. E. Tiemann, Ch. Ryzlewics, and T. Torring, Z. Naturforsch. 31A, 128 (1976).
3. R. Colin, P. Goldfinger, and M. Jeunehomme, Trans. Faraday Soc. 60, 306 (1964).
4. D. L. Hildenbrand, "Advance in High Temperature Chemistry," Vol. 1, L. Eyring (ed.) pp. 198-206, Academic Press, New York, 1967.
5. R. W. Field, C. R. Jones, and H. P. Broida, J. Chem. Phys. 60, 4377 (1974).
6. K. C. Mills, "Thermodynamic Data for Inorganic Sulphides, Selenides, and Tellurides," Butterworths, London, 1974.
7. U. S. Nat. Bur. Stand. Tech. Note 270-6, 1971.
8. JANAF Thermochemical Tables: Ba(g), 12-31-70; S_2(g), 12-31-65; S(g), 6-30-71; BaO(g), 6-30-74.

Enthalpy Reference Temperature = T_r = 298.15 K Standard State Pressure = p° = 0.1 MPa

T/K	$C_p°$ (J K^{-1}mol^{-1})	S°	$-[G°-H°(T_r)]/T$	$H°-H°(T_r)$ (kJ mol^{-1})	$\Delta_f H°$	$\Delta_f G°$	Log K_f
0	0.	0.	INFINITE	-9.556	39.549	39.549	INFINITE
100	30.195	212.875	279.140	-6.627	40.091	23.464	-12.256
200	33.796	234.997	252.065	-3.414	39.117	7.174	-1.874
298.15	35.531	248.861	248.861	0.	37.782	-8.232	1.442
300	35.552	249.081	248.862	0.066	37.753	-8.517	1.483
400	36.360	259.433	250.267	3.666	33.740	-23.599	3.082
500	36.789	267.597	252.946	7.326	29.653	-37.494	3.917
600	37.047	274.329	255.965	11.018	25.139	-50.484	4.395
700	37.219	280.053	259.007	14.732	21.691	-62.824	4.688
800	37.342	285.032	261.956	18.461	17.881	-74.630	4.873
900	37.438	289.436	264.769	22.200	-38.543	-84.914	4.928
1000	37.515	293.384	267.437	25.947	-40.558	-89.959	4.699
1100	37.581	296.963	269.961	29.702	-50.951	-93.981	4.463
1200	37.643	300.236	272.349	33.464	-53.219	-97.792	4.257
1300	37.705	303.251	274.612	37.231	-55.396	-101.418	4.075
1400	37.776	306.048	276.759	41.005	-57.505	-104.879	3.913
1500	37.867	308.657	278.799	44.787	-59.565	-108.190	3.768
1600	37.990	311.105	280.743	48.579	-61.593	-111.366	3.636
1700	38.161	313.413	282.597	52.386	-63.607	-114.415	3.516
1800	38.399	315.600	284.370	56.214	-65.620	-117.346	3.405
1900	38.721	317.685	286.069	60.069	-67.645	-120.164	3.304
2000	39.146	319.681	287.700	63.961	-69.693	-122.875	3.209
2100	39.691	321.604	289.269	67.902	-71.774	-125.484	3.121
2200	40.370	323.465	290.782	71.904	-213.416	-122.616	2.911
2300	41.193	325.277	292.242	75.981	-214.748	-118.460	2.690
2400	42.168	327.051	293.656	80.148	-216.221	-114.242	2.486
2500	43.296	328.794	295.026	84.420	-217.822	-109.961	2.298
2600	44.576	330.517	296.358	88.812	-219.530	-105.613	2.122
2700	45.998	332.225	297.655	93.340	-221.318	-101.198	1.958
2800	47.551	333.926	298.920	98.016	-223.159	-96.715	1.804
2900	49.216	335.623	300.157	102.853	-225.020	-92.167	1.660
3000	50.972	337.321	301.367	107.862	-226.871	-87.554	1.524
3100	52.795	339.022	302.554	113.050	-228.680	-82.880	1.397
3200	54.657	340.728	303.721	118.422	-230.418	-78.149	1.276
3300	56.529	342.438	304.868	123.982	-232.051	-73.365	1.161
3400	58.384	344.153	305.998	129.728	-233.569	-68.534	1.053
3500	60.193	345.872	307.113	135.657	-234.951	-63.659	0.950
3600	61.930	347.592	308.213	141.764	-236.184	-58.747	0.852
3700	63.570	349.312	309.301	148.040	-237.261	-53.803	0.760
3800	65.093	351.027	310.376	154.474	-238.178	-48.832	0.671
3900	66.482	352.736	311.441	161.054	-238.936	-43.839	0.587
4000	67.723	354.436	312.494	167.765	-239.543	-38.829	0.507
4100	68.809	356.122	313.538	174.593	-240.004	-33.805	0.431
4200	69.733	357.791	314.572	181.522	-240.331	-28.772	0.358
4300	70.494	359.441	315.596	188.534	-240.541	-23.732	0.288
4400	71.094	361.069	316.611	195.615	-240.647	-18.688	0.222
4500	71.539	362.672	317.617	202.748	-240.663	-13.644	0.158
4600	71.835	364.248	318.613	209.918	-240.600	-8.600	0.098
4700	71.991	365.795	319.601	217.110	-240.459	-3.560	0.040
4800	72.020	367.311	320.579	224.312	-240.296	1.479	-0.016
4900	71.931	368.795	321.548	231.510	-240.099	6.513	-0.069
5000	71.737	370.246	322.508	238.695	-239.892	11.544	-0.121
5100	71.451	371.664	323.458	245.855	-239.662	16.568	-0.170
5200	71.085	373.048	324.398	252.982	-239.460	21.590	-0.217
5300	70.650	374.398	325.329	260.069	-239.277	26.609	-0.262
5400	70.158	375.715	326.250	267.110	-239.123	31.624	-0.306
5500	69.619	376.997	327.161	274.100	-238.973	36.632	-0.348
5600	69.043	378.246	328.062	281.033	-238.837	41.635	-0.388
5700	68.439	379.463	328.953	287.907	-238.799	46.643	-0.427
5800	67.814	380.648	329.834	294.720	-238.788	51.648	-0.465
5900	67.175	381.802	330.705	301.470	-238.713	56.636	-0.501
6000	66.529	382.925	331.566	308.155	-238.821	61.643	-0.537

PREVIOUS: September 1977 (1 atm) CURRENT: September 1977 (1 bar)

BERYLLIUM (Be) REFERENCE STATE A_r = 9.01218

0 to 1527	K	crystal, alpha
1527 to 1560	K	crystal, beta
1560 to 2741.437	K	liquid
above 2741.437	K	ideal monatomic gas

Refer to the individual tables for details.

Beryllium (Be) Be_1(ref)

Enthalpy Reference Temperature = T_r = 298.15 K — Standard State Pressure = p° = 0.1 MPa

T/K	C_p° (J K⁻¹mol⁻¹)	S° (J K⁻¹mol⁻¹)	$-[G°-H°(T_r)]/T$ (J K⁻¹mol⁻¹)	$H°-H°(T_r)$ (kJ mol⁻¹)	$\Delta_fH°$ (kJ mol⁻¹)	$\Delta_fG°$ (kJ mol⁻¹)	Log K_f
0	0.	0.	INFINITE	-1.932	0.	0.	0.
100	1.819	0.503	19.434	-1.893	0.	0.	0.
200	9.984	4.174	10.751	-1.315	0.	0.	0.
298.15	16.380	9.440	9.440	0.	0.	0.	0.
300	16.472	9.542	9.440	0.030	0.	0.	0.
400	19.965	14.817	10.136	1.872	0.	0.	0.
500	21.943	19.501	11.551	3.975	0.	0.	0.
600	23.336	23.630	13.227	6.242	0.	0.	0.
700	24.463	27.314	14.981	8.634	0.	0.	0.
800	25.458	30.647	16.734	11.130	0.	0.	0.
900	26.384	33.699	18.452	13.723	0.	0.	0.
1000	27.274	36.525	20.119	16.406	0.	0.	0.
1100	28.147	39.166	21.732	19.177	0.	0.	0.
1200	29.016	41.652	23.290	22.035	0.	0.	0.
1300	29.886	44.009	24.793	24.980	0.	0.	0.
1400	30.763	46.256	26.247	28.013	0.	0.	0.
1500	31.650	48.409	27.653	31.133	0.	0.	0.
1527.000	31.892	48.975	28.025	31.991	ALPHA <--> BETA		
1527.000	30.000	53.461	28.025	38.840	TRANSITION		
1560.000	30.000	54.102	28.570	39.830	BETA <--> LIQUID		
1560.000	28.788	59.163	28.570	47.725	TRANSITION		
1600	28.874	59.893	29.344	48.879	0.	0.	0.
1700	29.089	61.650	31.193	51.777	0.	0.	0.
1800	29.304	63.319	32.932	54.696	0.	0.	0.
1900	29.519	64.909	34.573	57.638	0.	0.	0.
2000	29.734	66.429	36.128	60.600	0.	0.	0.
2100	29.949	67.884	37.606	63.584	0.	0.	0.
2200	30.164	69.283	39.014	66.590	0.	0.	0.
2300	30.379	70.628	40.360	69.617	0.	0.	0.
2400	30.594	71.926	41.648	72.666	0.	0.	0.
2500	30.809	73.179	42.885	75.736	0.	0.	0.
2600	31.024	74.391	44.073	78.828	0.	0.	0.
2700	31.239	75.566	45.218	81.941	0.	0.	0.
2741.437	31.328	76.043	45.680	83.237	LIQUID <--> IDEAL GAS		
2741.437	20.883	182.400	45.680	374.809	FUGACITY = 1 bar		
2800	20.905	182.842	48.545	376.032	0.	0.	0.
2900	20.949	183.576	53.188	378.125	0.	0.	0.
3000	21.006	184.287	57.546	380.223	0.	0.	0.
3100	21.075	184.977	61.646	382.327	0.	0.	0.
3200	21.159	185.648	65.511	384.438	0.	0.	0.
3300	21.259	186.300	69.161	386.559	0.	0.	0.
3400	21.376	186.937	72.616	388.691	0.	0.	0.
3500	21.512	187.558	75.891	390.835	0.	0.	0.
3600	21.668	188.166	79.001	392.994	0.	0.	0.
3700	21.844	188.762	81.960	395.169	0.	0.	0.
3800	22.041	189.347	84.778	397.363	0.	0.	0.
3900	22.259	189.923	87.467	399.578	0.	0.	0.
4000	22.500	190.489	90.035	401.816	0.	0.	0.
4100	22.761	191.048	92.492	404.079	0.	0.	0.
4200	23.045	191.600	94.845	406.369	0.	0.	0.
4300	23.349	192.146	97.102	408.688	0.	0.	0.
4400	23.674	192.686	99.268	411.039	0.	0.	0.
4500	24.019	193.222	101.350	413.424	0.	0.	0.
4600	24.383	193.754	103.353	415.844	0.	0.	0.
4700	24.764	194.282	105.282	418.301	0.	0.	0.
4800	25.162	194.808	107.142	420.797	0.	0.	0.
4900	25.576	195.331	108.936	423.334	0.	0.	0.
5000	26.003	195.852	110.669	425.913	0.	0.	0.
5200	26.894	196.889	113.965	431.202	0.	0.	0.
5400	27.822	197.921	117.056	436.673	0.	0.	0.
5600	28.776	198.950	119.962	442.332	0.	0.	0.
5800	29.743	199.977	122.704	448.184	0.	0.	0.
6000	30.711	201.001	125.297	454.229	0.	0.	0.

Previous: September 1961 (1 atm) Current: Septmeber 1983 (1 bar)

Beryllium (Be) Be_1(ref)

BERYLLIUM (Be) ALPHA-BETA $A_r = 9.01218$

$S°(298.15\ K) = 9.44 \pm 0.15\ J\ K^{-1}\ mol^{-1}$
$\Delta_{trs}H° = 1527 \pm 25\ K$
$T_{fus} = 1560 \pm 5\ K$

$\Delta_f H°(0\ K) = 0\ kJ\ mol^{-1}$
$\Delta_f H°(298.15\ K)\ 0\ kJ\ mol$
$\Delta_{trs}H° = 6.849 \pm 0.5\ kJ\ mol^{-1}$
$\Delta_{fus}H° = 7.895 \pm 0.5\ kJ\ mol^{-1}$

Enthalpy of Formation

Zero by definition.

Heat Capacity and Entropy

The adopted thermal functions for Be(α,β,cr) are derived from the studies of Ahlers (1.4-30 K, 1), Hill and Smith (4-300 K, 2), Ginnings et al. (367-1169 K, 3), and Kantor et al. (600-3200 K, 4). The mathematical and graphical treatment of these four studies yields a continuous and smooth heat capacity curve for α-Be(cr).

There are no heat capacity or enthalpy measurements for β-Be(cr). We assume a constant value of Cp = 30.0 $J\ K^{-1}\ mol^{-1}$ for the β-region.

The entropy at 298.15 K is 0.06 $J\ K^{-1}\ mol^{-1}$ less than that recommended by CODATA (5). This is undoubtedly due to the manner in whcih the data of Ahlers (1) and Hill and Smith (2) were merged. A detailed discussion of the heat capacities of all alkaline earth metals is in preparation.

Phase and Fusion Data

Cannon (6) has summarized the data on the polymorphic phase changes and the variation of the melting temperature of beryllium with pressure. His critical evaluation results in the recommendation that at ambient pressure there are two stable phases. His recommendations are entirely attributable to the resistance studies of Francois and Contre (7). The phases are denoted alpha(hcp) and beta(bcc).

The melting point of beryllium is fairly well defined. We adopt T_{fus} = 1560 ± 5 K based on the studies of Kantor et al. (4), Martin and Moore (8), Teitel and Cohen (9), and Loasby and Dearden (10).

The alpha-beta transition is ill-defined. There are many conflicting studies as to the value of the transition temperature and the value and sign of the derivative (dT/dP) for this transition. We adopt T_{trs} = 1527 ± 25 K based on the study of Ammonenko, et al. (11).

The enthalpies of transition and fusion are also poorly defined. Kantor, et al. (4) has measured the enthalpy of the crystal and liquid. No data points were obtained in the beta phase region. As a result, we can only derive a combined value for the enthalpy of transition and fusion. Following the suggestion of Abey (12), the enthalpy values of Loasby and Dearden (10) and Abey (12) have the same value for the ratio of the transition enthalpy to fusion enthalpy, even though the actual values for each enthalpy are quite different. Thus, using this ratio and the sum of the enthalpies, as determined by Kantor et al. (4), we calculate the individual enthalpies of transition and fusion. A much more detailed discussion is in progress.

Sublimation Data

See the ideal gas table for details.

References

1. G. Ahlers, Phys. Rev. 145, 419 (1966).
2. R. W. Hill and P. L. Smith, Phil. Mag. 44, 636 (1953).
3. D. C. Ginnings, T. B. Douglas, and A. F. Ball, J. Am. Chem. Soc. 73, 1236 (1951).
4. P. B. Kantor, R. M. Krasovitskaya, and A. N. Kisel, Fiz. Metal. i. Metalloved., Akad. Nauk SSSR 10, 835 (1960) [Phys. Metals Metallog. 10(6), 42 (1960)].
5. J. D. Cox, Chairman, ICSU-CODATA Task Group on Key Values for Thermodynamics, J. Chem. Thermodynamics 10, 903 (1978).
6. J. F. Cannon, J. Phys. Chem. Ref. Data 3, 781 (1974).
7. M. Francois and M. Contre, Conf. Int. Met. Beryllium, Commun., 3rd., pp. 201-11, 1965.
8. A. J. Martin and A. Moore, J. Less-Common Metals 1, 85 (1959).
9. R. J. Teital and M. Cohen, J. Metl. 185, 285 (1949).
10. R. G. Loasley and D. Dearden, J. Less-Common Metals 52, 137 (1977).
11. V. M. Amonenko, V. E. Ivanov, et al., Phys. Metals Metallog, 12, 77 (1962).
12. A. Abey, Lawrence Livermore National Laboratory, UCRL-53567, 1984.

Beryllium (Be) $Be_1(cr)$

Enthalpy Reference Temperature = T_r = 298.15 K — Standard State Pressure = p° = 0.1 MPa

T/K	$C_p°$ (J K⁻¹mol⁻¹)	S° (J K⁻¹mol⁻¹)	$-[G°-H°(T_r)]/T$ (J K⁻¹mol⁻¹)	$H°-H°(T_r)$ (kJ mol⁻¹)	$\Delta_f H°$ (kJ mol⁻¹)	$\Delta_f G°$ (kJ mol⁻¹)	Log K_f
0	0.	0.	INFINITE	-1.932	0.	0.	0.
100	1.819	0.503	19.434	-1.893	0.	0.	0.
200	9.984	4.174	10.751	-1.315	0.	0.	0.
250	13.574	6.801	9.695	-0.723	0.	0.	0.
298.15	16.380	9.440	9.440	0.	0.	0.	0.
300	16.472	9.542	9.440	0.030	0.	0.	0.
350	18.520	12.245	9.649	0.909	0.	0.	0.
400	19.965	14.817	10.136	1.872	0.	0.	0.
450	21.061	17.235	10.792	2.899	0.	0.	0.
500	21.943	19.501	11.551	3.975	0.	0.	0.
600	23.336	23.630	13.227	6.242	0.	0.	0.
700	24.463	27.314	14.981	8.634	0.	0.	0.
800	25.458	30.647	16.734	11.130	0.	0.	0.
900	26.384	33.699	18.452	13.723	0.	0.	0.
1000	27.274	36.525	20.119	16.406	0.	0.	0.
1100	28.147	39.166	21.732	19.177	0.	0.	0.
1200	29.016	41.652	23.290	22.035	0.	0.	0.
1300	29.886	44.009	24.793	24.980	0.	0.	0.
1400	30.763	46.256	26.247	28.013	0.	0.	0.
1500	31.650	48.409	27.653	31.133	0.	0.	0.
1527.000	31.892	48.975	28.025	31.991	ALPHA <--> BETA TRANSITION		
1527.000	30.000	53.461	28.025	38.840			
1560.000	30.000	54.102	28.570	39.830	BETA <--> LIQUID		
1600	30.000	54.862	29.218	41.030	-7.848	0.202	-0.007
1700	30.000	56.680	30.780	44.030	-7.747	0.702	-0.022
1800	30.000	58.395	32.267	47.030	-7.666	1.196	-0.035
1900	30.000	60.017	33.685	50.030	-7.607	1.687	-0.046
2000	30.000	61.556	35.041	53.030	-7.570	2.175	-0.057

Previous: September 1961 Current: September 1983

Beryllium (Be) $Be_1(cr)$

BERYLLIUM (Be) LIQUID $A_r = 9.01218$

$S°(298.15 K) = [19.539] J K^{-1} mol^{-1}$

$T_{fus} = 1560 \pm 5 K$

$\Delta_f H°(298.15 K) = [15.508] kJ mol^{-1}$

$\Delta_{fus} H° = 7.895 \pm 0.05 kJ mol^{-1}$

Enthalpy of Formation

The enthalpy of formation of liquid beryllium is calculated from that of the crystal by adding $\Delta_{fus}H°$ and the difference in enthalpy, H°(1560 K) - H°(298.15 K), between the crystal and liquid.

Heat Capacity and Entropy

The heat capacity of the liquid is derived from the measured enthalpy data of Kantor et al. (1), 17 data points in the liquid region, 1560 to 2200 K. The data was presented graphically. We adopt the reported equation: $Cp° = 6.079 + 5.138 \times 10^{-4} T$ cal $K^{-1} mol^{-1}$. This equation is used to calculate extrapolated values from 2200 K to 3500 K and from 1150 K to 1560 K.

A glass transition temperature is assumed at 1150 K, below which the heat capacity values of crystalline beryllium are adopted. The entropy at 298.15 K is calculated in a mannner analogous to that used for the enthalpy of formation.

Transition and Fusion Data

See the crystal table for details.

Vaporization Data

There is only one reported vaporization study for Be(ℓ): Baur and Brunner (926 -1283 K, 2). This study was not considered in the analysis as it had an incorrect temperature dependence, as was also noted by Hultgren et al. (3).

The boiling point (f = 1 bar), T_{vap}= 2741.437 K, is calculated as the temperature for which $\Delta_r G° = 0$ for Be(ℓ) = Be(g). The difference in the enthalpy of formation ofa Be(ℓ) and Be(g) at T_{vap} is the enthalpy of vaporization, $\Delta_{vap}H° = 291.572 kJ mol^{-1}$.

References

1. P. B. Kantor, R. M. Krasovitskaya, and A. N. Kisel, Fiz. Metal and Metalloved., Akad. Nauk SSSR. 10, 835 (1960).
2. E. Baur and R. Brunner, Helv. Chim. Acta 17, 958 (1934).
3. R. Hultgren, P. D. Desai, et al., "Selected Values of the Thermodynamic Properties of the Elements," American Society for Metals, Metals Park, Ohio, 1973.

Beryllium (Be) $Be_1(l)$

Enthalpy Reference Temperature = T_r = 298.15 K   Standard State Pressure = p° = 0.1 MPa

T/K	$C_p°$ (J K⁻¹mol⁻¹)	S°	$-[G°-H°(T_r)]/T$	$H°-H°(T_r)$ (kJ mol⁻¹)	$\Delta_f H°$	$\Delta_f G°$	Log K_f
0							
100							
200							
250							
298.15	16.380	19.539	19.539	0.	15.508	12.497	-2.189
300	16.472	19.641	19.540	0.030	15.508	12.478	-2.173
350	18.520	22.344	19.749	0.909	15.508	11.973	-1.787
400	19.965	24.917	20.235	1.872	15.508	11.468	-1.498
450	21.061	27.334	20.891	2.899	15.508	10.963	-1.273
500	21.943	29.600	21.650	3.975	15.508	10.458	-1.093
600	23.336	33.730	23.326	6.242	15.508	9.448	-0.823
700	24.463	37.414	25.080	8.634	15.508	8.438	-0.630
800	25.458	40.746	26.833	11.130	15.508	7.428	-0.485
900	26.384	43.799	28.551	13.723	15.508	6.418	-0.372
1000	27.274	46.625	30.219	16.406	15.508	5.408	-0.282
1100	28.147	49.265	31.832	19.177	15.508	4.398	-0.209
1150.000	28.582	50.526	32.617	20.595	GLASS <--> LIQUID TRANSITION		
1150.000	27.907	50.526	32.617	20.595			
1200	28.014	51.716	33.388	21.993	15.466	3.389	-0.148
1300	28.229	53.967	34.886	24.805	15.333	2.388	-0.096
1400	28.444	56.067	36.325	27.639	15.134	1.399	-0.052
1500	28.659	58.037	37.707	30.494	14.869	0.427	-0.015
1560.000	28.788	59.163	38.511	32.218	------- BETA <--> LIQUID -------		
1600	28.874	59.893	39.036	33.371	0.	0.	0.
1700	29.089	61.650	40.315	36.269	0.	0.	0.
1800	29.304	63.319	41.547	39.189	0.	0.	0.
1900	29.519	64.909	42.735	42.130	0.	0.	0.
2000	29.734	66.429	43.882	45.093	0.	0.	0.
2100	29.949	67.884	44.991	48.077	0.	0.	0.
2200	30.164	69.283	46.063	51.082	0.	0.	0.
2300	30.379	70.628	47.102	54.110	0.	0.	0.
2400	30.594	71.926	48.110	57.158	0.	0.	0.
2500	30.809	73.179	49.088	60.228	0.	0.	0.
2600	31.024	74.391	50.038	63.320	0.	0.	0.
2700	31.239	75.566	50.962	66.433	0.	0.	0.
2741.437	31.328	76.043	51.337	67.729	------- FUGACITY = 1 bar -------		
2800	31.454	76.706	51.861	69.568	-290.957	6.223	-0.116
2900	31.669	77.814	52.737	72.724	-289.894	16.817	-0.303
3000	31.884	78.891	53.591	75.901	-288.814	27.375	-0.477
3100	32.099	79.940	54.424	79.101	-287.718	37.897	-0.639
3200	32.314	80.963	55.237	82.321	-286.609	48.383	-0.790
3300	32.529	81.960	56.032	85.563	-285.488	58.834	-0.931
3400	32.744	82.934	56.809	88.827	-284.356	69.251	-1.064
3500	32.959	83.887	57.569	92.112	-283.215	79.635	-1.188

Previous: September 1961   Current: September 1983

Beryllium (Be) $Be_1(l)$

BERYLLIUM (Be) ALPHA-BETA-LIQUID A_r = 9.01218 Beryllium (Be) Be_1(cr,l)

0 to 1527 K crystal, alpha
1527 to 1560 K crystal, beta
above 1560 K liquid

Refer to the individual tables for details.

Enthalpy Reference Temperature = T_r = 298.15 K; Standard State Pressure = p° = 0.1 MPa

T/K	C_p° (J K⁻¹mol⁻¹)	S° (J K⁻¹mol⁻¹)	$-[G^\circ - H^\circ(T_r)]/T$ (J K⁻¹mol⁻¹)	$H^\circ - H^\circ(T_r)$ (kJ mol⁻¹)	$\Delta_f H^\circ$ (kJ mol⁻¹)	$\Delta_f G^\circ$ (kJ mol⁻¹)	Log K_f
0	0.	0.	INFINITE	-1.932	0.	0.	0.
100	1.819	0.503	19.434	-1.893	0.	0.	0.
200	9.984	4.174	10.751	-1.315	0.	0.	0.
250	13.574	6.801	9.695	-0.723	0.	0.	0.
298.15	16.380	9.440	9.440	0.	0.	0.	0.
300	16.472	9.542	9.440	0.030	0.	0.	0.
350	18.520	12.245	9.649	0.909	0.	0.	0.
400	19.965	14.817	10.136	1.872	0.	0.	0.
450	21.061	17.235	10.792	2.899	0.	0.	0.
500	21.943	19.501	11.551	3.975	0.	0.	0.
600	23.336	23.630	13.227	6.242	0.	0.	0.
700	24.463	27.314	14.981	8.634	0.	0.	0.
800	25.458	30.647	16.734	11.130	0.	0.	0.
900	26.384	33.699	18.452	13.723	0.	0.	0.
1000	27.274	36.525	20.119	16.406	0.	0.	0.
1100	28.147	39.166	21.732	19.177	0.	0.	0.
1200	29.016	41.652	23.290	22.035	0.	0.	0.
1300	29.886	44.009	24.793	24.980	0.	0.	0.
1400	30.763	46.256	26.247	28.013	0.	0.	0.
1500	31.650	48.409	27.653	31.133	0.	0.	0.
1527.000	31.892	48.975	28.025	31.991	ALPHA <--> BETA TRANSITION		
1527.000	30.000	53.461	28.025	38.840			
1560.000	30.000	54.102	28.570	39.830	BETA <--> LIQUID TRANSITION		
1560.000	28.788	59.163	28.570	47.725			
1600	28.874	59.893	29.344	48.879	0.	0.	0.
1700	29.089	61.650	31.193	51.777	0.	0.	0.
1800	29.304	63.319	32.932	54.696	0.	0.	0.
1900	29.519	64.909	34.573	57.638	0.	0.	0.
2000	29.734	66.429	36.128	60.600	0.	0.	0.
2100	29.949	67.884	37.606	63.584	0.	0.	0.
2200	30.164	69.283	39.014	66.590	0.	0.	0.
2300	30.379	70.628	40.360	69.617	0.	0.	0.
2400	30.594	71.926	41.648	72.666	0.	0.	0.
2500	30.809	73.179	42.885	75.736	0.	0.	0.
2600	31.024	74.391	44.073	78.828	0.	0.	0.
2700	31.239	75.566	45.218	81.941	0.	0.	0.
2741.437	31.328	76.043	45.680	83.237	------- FUGACITY = 1 bar -------		
2800	31.454	76.706	46.322	85.075	-290.957	6.223	-0.116
2900	31.669	77.814	47.389	88.232	-289.894	16.817	-0.303
3000	31.884	78.891	48.421	91.409	-288.814	27.375	-0.477
3100	32.099	79.940	49.421	94.608	-287.718	37.897	-0.639
3200	32.314	80.963	50.391	97.829	-286.609	48.383	-0.790
3300	32.529	81.960	51.333	101.071	-285.488	58.834	-0.931
3400	32.744	82.934	52.248	104.335	-284.356	69.251	-1.064
3500	32.959	83.887	53.138	107.620	-283.215	79.635	-1.188

Previous: Current: September 1983

Beryllium (Be) Be_1(cr,l)

BERYLLIUM (Be) IDEAL GAS A_r = 9.01218

IP(Be, g) = 75192.07 ± 0.1 cm^{-1} $\Delta_f H°$(0 K) = 320 ± 5 kJ mol^{-1}

S°(298.15 K) = 136.274 ± 0.020 J K^{-1} mol^{-1} $\Delta_f H°$(298.15 K) = 324 ± 5 kJ mol^{-1}

Electronic Levels and Quantum Weights

State	ε_i, cm^{-1}	g_i
$^1S_0$	0.0	1
$^3P_0$	21978.28	1
$^3P_1$	21978.92	3
$^3P_2$	21981.27	5
$^1P_1$	42565.35	3
.	.	.
.	.	.
.	.	.
$^1D_2$	74443.20	5
IP	75192.29	

Enthalpy of Formation

The enthalpy of formation is derived from the sublimation studies of Holden et al. (1172-1552 K, 1), Gulbranson and Andrew (1103-1229 K, 2), and Hildenbrand and Murad (1400-1500 K, 4). The adopted value, $\Delta_f H°$(Be, g, 298.15 K) = 324 ± 5 kJ mol^{-1}, is that recommended by CODATA (5). There are additional vapor pressure studies (~20), but they have been omitted in this analysis.

Heat Capacity and Entropy

The information on electronic energy levels and quantum weights, given by Moore (6, 7), is incomplete because many theoretically predicted levels have not been observed. Our calculations indicate that any reasonable method of filling in these missing levels and cutting off the summation in the partition function (8) has no effect on the thermodynamic functions to 6000 K. This is a result of the high energy of all levels other than the ground state and the three levels at ~22000 cm^{-1}; the next excited state is approximately 42565 cm^{-1} above the ground state. Since inclusion of these excited states has no effect on the thermodynamic functions (to 6000 K), we list only the ground state. The reported uncertainty in S°(298.15 K) is due to uncertainties in the relative ionic mass and the fundamental constants. Extension of these calculations above 6000 K may require consideration of the excited states and use of different fill and cutoff procedures (8).

The thermal functions at 298.15 K agree with the CODATA recommendations (5) except that the entropy differs by 0.1094 J K^{-1} mol^{-1}, since this table uses a standard state pressure of 1 bar (CODATA recommendations are based on 1 atm).

References

1. R. B. Holden, R. Speiser, and H. L. Johnston, J. Am. Chem. Soc. 70, 3897 (1948).
2. E. A. Gulbranson and R. F. Andrew, J. Electrochem. Soc. 97, 383 (1950).
3. G. P. Kovtun, A. A. Kruglykh, and V. S. Pavlov, Izvest. Akad. Nauk SSSR, Met. Gorn. Delo 2, 177 (1964).
4. D. L. Hildenbrand and E. Murad, J. Chem. Phys. 44, 1524 (1966).
5. J. D. Cox, chairman, ICSU-CODATA Task Group on Key Values for Thermodynamics, J. Chem. Thermodynamics 10, 903 (1978).
6. C. E. Moore, U. S. Nat. Bur. Stand., NSRDS-NBS-34, 1970.
7. C. E. Moore, U. S. Nat. Bur. Stand., NSRDS-NBS-35, Volume I, 1970 [Reprint of NBS Circular 467, Volume I, 1949].
8. J. R. Downey, Jr., The Dow Chemical Company, AFOSR-TR-78-0960, Contract No. F44620-75-1-0048, 1978.

Beryllium (Be) $Be_1(g)$

Enthalpy Reference Temperature = T_r = 298.15 K Standard State Pressure = p° = 0.1 MPa

T/K	$C_p°$ (J $K^{-1}mol^{-1}$)	S° (J $K^{-1}mol^{-1}$)	$-[G°-H°(T_r)]/T$ (J $K^{-1}mol^{-1}$)	$H°-H°(T_r)$ (kJ mol^{-1})	$\Delta_f H°$ (kJ mol^{-1})	$\Delta_f G°$ (kJ mol^{-1})	Log K_f
0	0.	0.	INFINITE	-6.197	319.735	319.735	INFINITE
100	20.786	113.567	154.755	-4.119	321.774	310.468	-162.172
200	20.786	127.975	138.176	-2.040	323.275	298.515	-77.964
250	20.786	132.613	136.617	-1.001	323.723	292.269	-61.066
298.15	20.786	136.274	136.274	0.	324.000	286.184	-50.138
300	20.786	136.403	136.275	0.038	324.008	285.950	-49.788
350	20.786	139.607	136.528	1.078	324.169	279.592	-41.727
400	20.786	142.383	137.090	2.117	324.245	273.218	-35.679
450	20.786	144.831	137.817	3.156	324.257	266.839	-30.974
500	20.786	147.021	138.630	4.196	324.221	260.460	-27.210
600	20.786	150.811	140.354	6.274	324.032	247.724	-21.566
700	20.786	154.015	142.082	8.353	323.719	235.029	-17.538
800	20.786	156.791	143.751	10.431	323.301	222.386	-14.520
900	20.786	159.239	145.339	12.510	322.787	209.802	-12.177
1000	20.786	161.429	146.840	14.589	322.183	197.279	-10.305
1100	20.786	163.410	148.258	16.667	321.490	184.822	-8.776
1200	20.786	165.219	149.597	18.746	320.711	172.431	-7.506
1300	20.786	166.882	150.863	20.824	319.844	160.109	-6.433
1400	20.786	168.423	152.063	22.903	318.890	147.857	-5.517
1500	20.786	169.857	153.202	24.982	317.848	135.676	-4.725
1600	20.786	171.198	154.286	27.060	302.182	124.093	-4.051
1700	20.786	172.458	155.318	29.139	301.362	112.988	-3.472
1800	20.787	173.647	156.303	31.218	300.521	101.931	-2.958
1900	20.787	174.770	157.246	33.296	299.659	90.922	-2.500
2000	20.789	175.837	158.149	35.375	298.775	79.958	-2.088
2100	20.791	176.851	159.016	37.454	297.870	69.040	-1.717
2200	20.795	177.818	159.849	39.533	296.943	58.165	-1.381
2300	20.801	178.743	160.650	41.613	295.996	47.332	-1.075
2400	20.811	179.628	161.423	43.694	295.028	36.541	-0.795
2500	20.824	180.478	162.168	45.775	294.039	25.791	-0.539
2600	20.844	181.295	162.888	47.859	293.031	15.081	-0.303
2700	20.870	182.082	163.584	49.944	292.003	4.410	-0.085
2741.437	20.883	182.400	163.867	50.809	------- FUGACITY = 1 bar ------		
2800	20.905	182.842	164.259	52.033	0.	0.	0.
2900	20.949	183.576	164.912	54.126	0.	0.	0.
3000	21.006	184.287	165.546	56.223	0.	0.	0.
3100	21.075	184.977	166.162	58.327	0.	0.	0.
3200	21.159	185.648	166.761	60.439	0.	0.	0.
3300	21.259	186.300	167.343	62.559	0.	0.	0.
3400	21.376	186.937	167.910	64.691	0.	0.	0.
3500	21.512	187.558	168.462	66.835	0.	0.	0.
3600	21.668	188.166	169.001	68.994	0.	0.	0.
3700	21.844	188.762	169.527	71.170	0.	0.	0.
3800	22.041	189.347	170.041	73.364	0.	0.	0.
3900	22.259	189.923	170.544	75.578	0.	0.	0.
4000	22.500	190.489	171.035	77.816	0.	0.	0.
4100	22.761	191.048	171.516	80.079	0.	0.	0.
4200	23.045	191.600	171.988	82.369	0.	0.	0.
4300	23.349	192.146	172.451	84.689	0.	0.	0.
4400	23.674	192.686	172.904	87.040	0.	0.	0.
4500	24.019	193.222	173.350	89.424	0.	0.	0.
4600	24.383	193.754	173.788	91.844	0.	0.	0.
4700	24.764	194.282	174.218	94.301	0.	0.	0.
4800	25.162	194.808	174.642	96.798	0.	0.	0.
4900	25.576	195.331	175.058	99.334	0.	0.	0.
5000	26.003	195.852	175.469	101.913	0.	0.	0.
5100	26.443	196.371	175.874	104.535	0.	0.	0.
5200	26.894	196.889	176.273	107.202	0.	0.	0.
5300	27.354	197.405	176.667	109.914	0.	0.	0.
5400	27.822	197.921	177.056	112.673	0.	0.	0.
5500	28.296	198.436	177.440	115.479	0.	0.	0.
5600	28.776	198.950	177.819	118.333	0.	0.	0.
5700	29.258	199.464	178.195	121.234	0.	0.	0.
5800	29.743	199.977	178.566	124.184	0.	0.	0.
5900	30.227	200.489	178.933	127.183	0.	0.	0.
6000	30.711	201.001	179.296	130.230	0.	0.	0.

Previous: September 1961 (1 atm) Current: Septmeber 1983 (1 bar)

Beryllium (Be) $Be_1(g)$

BERYLLIUM, ION (Be^+) IDEAL GAS M_r = 9.01163

Beryllium, Ion (Be^+) $Be_1^+(g)$

IP(Be^+, g) = 146882.86 ± 0.1 cm^{-1}

S°(298.15 K) = 142.037 ± 0.02 JK^{-1} mol^{-1}

$\Delta_f H°$(0 K) = 1208.23 ± 5.0 kJ mol^{-1}

$\Delta_f H°$(298.15 K) = [1219.232] kJ mol^{-1}

Electronic Levels and Quantum Weights

State	ε_i cm^{-1}	g_i
$^2S_{1/2}$	0	2
$^2P_{1/2}$	31928.8	2
$^2P_{3/2}$	31935.4	4

Enthalpy of Formation

$\Delta_f H°(Be^+$, g , 0 K) is calculated from $\Delta_f H°$(Be, g, 0 K) (1) using the spectroscopic value of IP(Be) = 75192.07 ± 0.1 cm^{-1} (899.497 ± 0.001 kJ mol^{-1}) from Moore (2). The ionization limit is converted from cm^{-1} to kJ mol^{-1} using the factor, 1 cm^{-1} = 0.01196266 kJ cm^{-1}, which is derived from the 1973 CODATA fundamental constants (3). Rossenstock et al. (4) and Levin and Lias (5) have summarized additional ionization and appearance potential data.

$\Delta_f H°(Be^+$, g, 298.15 K) is calculated from $\Delta_f H°$(Be, g, 0 K) by using IP(Be) with JANAF (1) enthalpies, H°(0 K) - H°(298.15 K), for Be(g), Be^+(g), and e^-(g). $\Delta_r H°(Be \rightarrow Be^+ e^-$, 298.15 K) differs from a room temperature threshold energy due to inclusion of these enthalpies and to threshold effects discussed by Rosenstock et al. (4). $\Delta_f H°$(298.15 K) should be changed by -6.197 kJ mol^{-1} if it is to be used in the ion convention that excludes the enthalpy of the electron.

Heat Capacity and Entropy

The information on electronic energy levels and quantum weights, given by Moore (2, 6), is incomplete because many theoretically predicted levels have not been observed. Our calculations indicate that any reasonable method of filling in these missing levels and cutting off the summation in the partition function (7) has no effect on the thermodynamic functions to 6000 K. This is a result of the high energy of all levels other than the ground state and the first two excited states; the next excited state is 88231.2 cm^{-1} above the ground state and the first two excited states. The reported uncertainty in S°(298.15 K) is due to uncertainties in the relative ionic mass and the fundamental constants. Extension of these calculations above 6000 K may require consideration of the excited states and use of different fill and cutoff procedures (7).

References

1. JANAF Thermochemical Tables: Be(g), 9-30-83; e^-(g), 3-31-82.
2. C. E. Moore, U. S. Nat. Bur. Stand., NSRDS-NBS 34, 1970.
3. E. R. Cohen and B. N. Taylor, J. Phys. Chem. Ref. Data 2, 663 (1973).
4. H. M. Rosenstock, K. Draxl et al., J. Phys. Chem. Ref. Data 6, Supp. 1 (1977).
5. R. D. Levin and S. G. Lias, U. S. Nat. Bur. Stand., NSRDS-NBS-71, 1982.
6. C. E. Moore, U. S. Nat. Bur. Stand., NSRDS-NBS-35, Volume I, 1970 [Reprint of NBS Circular 467, Volume I, 1949].
7. J. R. Downey, Jr., The Dow Chemical Company, AFOSR-TR-78-0960, Contract No. F44620-75-1-0048, 1978.

Enthalpy Reference Temperature = T_r = 298.15 K — Standard State Pressure = p° = 0.1 MPa

	J $K^{-1}mol^{-1}$			kJ mol^{-1}			
T/K	$C_p°$	S°	-[G°-H°(T_r)]/T	H°-H°(T_r)	$\Delta_f H°$	$\Delta_f G°$	Log K_f
0	0.	0.	INFINITE	-6.197	1208.770		
100	20.786	119.330	160.517	-4.119			
200	20.786	133.737	143.938	-2.040			
250	20.786	138.376	142.379	-1.001			
298.15	20.786	142.037	142.037	0.	1219.232	1173.444	-205.582
300	20.786	142.165	142.037	0.038	1219.279	1173.159	-204.265
350	20.786	145.370	142.290	1.078	1220.479	1165.376	-173.923
400	20.786	148.145	142.852	2.117	1221.594	1157.428	-151.145
450	20.786	150.593	143.579	3.156	1222.646	1149.343	-133.412
500	20.786	152.783	144.392	4.196	1223.648	1141.144	-119.215
600	20.786	156.573	146.116	6.274	1225.539	1124.464	-97.893
700	20.786	159.777	147.845	8.353	1227.304	1107.477	-82.641
800	20.786	162.553	149.514	10.431	1228.965	1090.244	-71.186
900	20.786	165.001	151.101	12.510	1230.529	1072.809	-62.264
1000	20.786	167.191	152.602	14.589	1232.003	1055.205	-55.118
1100	20.786	169.172	154.020	16.667	1233.390	1037.457	-49.265
1200	20.786	170.981	155.359	18.746	1234.689	1019.587	-44.381
1300	20.786	172.645	156.626	20.824	1235.901	1001.612	-40.245
1400	20.786	174.185	157.826	22.903	1237.026	983.547	-36.697
1500	20.786	175.619	158.965	24.982	1238.062	965.405	-33.618
1600	20.786	176.961	160.048	27.060	1224.474	947.722	-30.940
1700	20.786	178.221	161.080	29.139	1225.733	930.386	-28.587
1800	20.786	179.409	162.066	31.217	1226.971	912.977	-26.494
1900	20.786	180.533	163.008	33.296	1228.187	895.500	-24.619
2000	20.786	181.599	163.912	35.375	1229.381	877.959	-22.930
2100	20.786	182.613	164.778	37.453	1230.554	860.359	-21.400
2200	20.786	183.580	165.611	39.532	1231.706	842.703	-20.008
2300	20.786	184.504	166.412	41.610	1232.836	824.996	-18.736
2400	20.786	185.389	167.185	43.689	1233.944	807.240	-17.569
2500	20.786	186.237	167.930	45.768	1235.031	789.438	-16.494
2600	20.786	187.052	168.650	47.846	1236.097	771.594	-15.502
2700	20.786	187.837	169.346	49.925	1237.141	753.708	-14.581
2800	20.787	188.593	170.020	52.004	947.207	742.007	-13.842
2900	20.787	189.322	170.673	54.082	949.271	734.642	-13.232
3000	20.787	190.027	171.307	56.161	951.331	727.206	-12.662
3100	20.788	190.709	171.922	58.240	953.384	719.701	-12.127
3200	20.789	191.369	172.519	60.319	955.430	712.130	-11.624
3300	20.790	192.008	173.100	62.398	957.467	704.496	-11.151
3400	20.792	192.629	173.665	64.477	959.493	696.799	-10.705
3500	20.795	193.232	174.216	66.556	961.507	689.044	-10.283
3600	20.798	193.818	174.752	68.636	963.506	681.231	-9.884
3700	20.802	194.388	175.275	70.716	965.490	673.362	-9.506
3800	20.806	194.942	175.785	72.796	967.454	665.441	-9.147
3900	20.812	195.483	176.284	74.877	969.399	657.468	-8.806
4000	20.820	196.010	176.770	76.958	971.322	649.445	-8.481
4100	20.829	196.524	177.246	79.041	973.220	641.374	-8.171
4200	20.839	197.026	177.711	81.124	975.092	633.258	-7.876
4300	20.851	197.517	178.166	83.209	976.935	625.097	-7.593
4400	20.865	197.996	178.611	85.295	978.749	616.894	-7.323
4500	20.882	198.465	179.047	87.382	980.530	608.650	-7.065
4600	20.900	198.924	179.474	89.471	982.278	600.366	-6.817
4700	20.921	199.374	179.893	91.562	983.990	592.045	-6.580
4800	20.945	199.815	180.303	93.655	985.666	583.688	-6.352
4900	20.972	200.247	180.706	95.751	987.304	575.297	-6.133
5000	21.001	200.671	181.101	97.850	988.902	566.872	-5.922
5100	21.033	201.087	181.489	99.951	990.460	558.416	-5.719
5200	21.069	201.496	181.870	102.057	991.977	549.930	-5.524
5300	21.108	201.898	182.244	104.165	993.452	541.415	-5.336
5400	21.150	202.292	182.611	106.278	994.885	532.872	-5.155
5500	21.195	202.681	182.973	108.396	996.275	524.304	-4.979
5600	21.244	203.063	183.328	110.517	997.622	515.710	-4.810
5700	21.297	203.440	183.678	112.644	998.926	507.093	-4.647
5800	21.353	203.811	184.022	114.777	1000.187	498.453	-4.489
5900	21.412	204.176	184.360	116.915	1001.405	489.792	-4.336
6000	21.475	204.537	184.693	119.059	1002.581	481.111	-4.188

Previous: June 1965 (1 atm) Current: September 1983 (1 bar)

Beryllium, Ion (Be^+) $Be_1^+(g)$

BERYLLIUM BROMIDE (BeBr) IDEAL GAS M_r = 88.91618

$\Delta_f H°(0\ K)$ = 125.3 ± 41.8 kJ mol^{-1}

S°(298.15 K) = 228.5 ± 0.4 J K^{-1} mol^{-1} $\Delta_f H°(298.15\ K)$ = 120.1 ± 41.8 kJ mol^{-1}

Electronic Levels and Quantum Weights

State	ε_i, cm^{-1}	g_i
$X^2\Sigma^+$	0	2
$A_1{}^2\Pi_{1/2}$	26353.2	2
$A_2{}^2\Pi_{3/2}$	26550.2	2
$B[^2\Sigma]$	[27000]	[2]
$C[^2\Pi]$	[40000]	[4]

ω_e = 715.06 cm^{-1} $\omega_e x_e$ = 4.30 cm^{-1} σ = 1

B_e = [0.62155] cm^{-1} α_e = [0.005285] cm^{-1} r_e = [1.83] Å

Enthalpy of Formation

No thermochemical measurement of the enthalpy of formation has been made. The selected value, $\Delta_f H°$(BeBr, g, 0 K) = 30.0 ± 10.0 kcal mol^{-1}, is obtained from an analysis of spectroscopic data. The adopted values for the ground state vibrational constants give D_0^0 = 3.64_1 eV by a linear Birge-Sponer extrapolation (1). Based on the ionicity correction developed by Hildenbrand (2), this value adjusts to D_0^0 = 3.27_2 eV (75.46 kcal mol^{-1}) which is adopted. $\Delta_f H°$(298.15 K) corresponds to 28.7 kcal mol^{-1}.

Ionic model calculations (4, 6) and bond energy correlations (5) have led to D_0^0 values of (in eV) 1.2 (4), 4.1 (5), and 2.1 (6). The latter value which is based on a Rittner potential (7) is believed to represent a minimum value for D_0^0. Additional support for the adopted value of $\Delta_f H°$ is provided by a comparison of values for the ratio D_0^0 (MX, 298.15 K)/$\Delta_{at}H°$(MX_2, 298.15 K) for the alkaline-earth halides (3). We find D_0^0(BeBr, 298.15 K)/$\Delta_{at}H°$($BeBr_2$, 298.15 K) = 0.41 which is quite consistent with values of this ratio for other alkaline-earth halides.

Heat Capacity and Entropy

We estimate r_e to be 0.08 Å less than that for $BeBr_2$ (3). This estimation is made based on the observation that the ratio $r_e(MX)/r_e(MX_2)$ is near 0.96 for several alkaline-earth halides (3). The value of B_e is calculated from r_e. The value of α_e is obtained from the other constants by assuming a Morse potential function.

The ground state vibrational constants are taken from the results of a recent vibrational analysis of the $A^2\Pi$-$X^2\Sigma$ system by Reddy et al. (8). Other values (9) for these constants are slightly different than ours, but we consider them to be less accurate. It should also be noted that the values for $\omega_e x_e$ tabulated by Rosen (9) for the ground and first excited states have been inadvertently reversed. The electronic level for the $A_1{}^2\Pi_{1/2}$ state is also taken from Reddy et al. (8). The doublet splitting of this state is assumed to be 197 cm^{-1} as was observed in the emission spectrum of BeBr by Reddy and Rao (10, 11). We also include a $B^2\Sigma$ level at 27000 cm^{-1} and a $C^2\Pi$ level at 40000 cm^{-1}. These levels are estimated by analogy with those for CaBr, SrBr, and BaBr (3).

References

1. A. G. Gaydon, "Dissociation Energies and Spectra of Diatomic Molecules," Chapman and Hall, Ltd., London, 1968.
2. D. L. Hildenbrand in "Advances in High Temperature Chemistry," Vol. 1, Ed. by L. Eyring, Academic Press, New York, 1967.
3. JANAF Thermochemical Tables: $BeBr_2$(g), 6-30-75; CaBr(g), SrBr(g), and BaBr(g), 12-31-74.
4. J. L. Margrave, J. Phys. Chem. 58, 258 (1954).
5. S. P. Tandon and K. Tandon, Indian J. Phys. 38, 460 (1964).
6. K. S. Krasnov and N. V. Karaseva, Optics and Spectroscopy 19, 14 (1965).
7. E. S. Rittner, J. Chem. Phys. 19, 1030 (1951).
8. B. R. Reddy, Y. P. Reddy, and P. T. Rao, J. Phys. B, 3, L1 (1970).
9. B. Rosen, Ed., "Spectroscopic Data Relative to Diatomic Molecules," Pergamon Press, New York, 1970.
10. Y. P. Reddy and P. T. Rao, Proc. Phys. Soc., London, At. Mol. Phys. 1, 482 (1968).
11. Y. P. Reddy and P. T. Rao, Curr. Sci. 36, 399 (1967).

Beryllium Bromide (BeBr) Be_1Br_1(g)

Enthalpy Reference Temperature = T_r = 298.15 K Standard State Pressure = p° = 0.1 MPa

T/K	$C_p°$ (J $K^{-1}mol^{-1}$)	S° (J $K^{-1}mol^{-1}$)	$-[G°-H°(T_r)]/T$ (J $K^{-1}mol^{-1}$)	$H°-H°(T_r)$ (kJ mol^{-1})	$\Delta_f H°$ (kJ mol^{-1})	$\Delta_f G°$ (kJ mol^{-1})	Log K_f
0	0.	0.	INFINITE	-8.971	125.326	125.326	INFINITE
100	29.145	195.462	256.089	-6.063	126.799	109.995	-57.455
200	30.486	215.960	231.450	-3.098	126.735	93.120	-24.320
250	31.597	222.881	229.065	-1.546	126.287	84.763	-17.710
298.15	32.598	228.534	228.534	0.	120.110	77.477	-13.574
300	32.633	228.736	228.535	0.060	120.070	77.213	-13.444
350	33.508	233.834	228.935	1.715	104.521	70.926	-10.585
400	34.218	238.357	229.836	3.408	104.336	66.139	-8.637
450	34.787	242.421	231.012	5.134	104.114	61.377	-7.124
500	35.246	246.111	232.340	6.885	103.865	56.642	-5.917
600	35.922	252.601	235.191	10.446	103.298	47.249	-4.113
700	36.386	258.175	238.086	14.063	102.654	37.957	-2.832
800	36.721	263.057	240.908	17.719	101.938	28.763	-1.878
900	36.973	267.397	243.615	21.404	101.148	19.663	-1.141
1000	37.170	271.303	246.192	25.112	100.286	10.654	-0.557
1100	37.330	274.854	248.638	28.837	99.348	1.736	-0.082
1200	37.465	278.108	250.960	32.577	98.334	-7.094	0.309
1300	37.580	281.111	253.166	36.329	97.241	-15.836	0.636
1400	37.681	283.900	255.262	40.092	96.067	-24.490	0.914
1500	37.772	286.503	257.259	43.865	94.810	-33.058	1.151
1600	37.854	288.943	259.164	47.646	78.931	-41.014	1.339
1700	37.931	291.240	260.984	51.436	77.902	-48.479	1.490
1800	38.002	293.410	262.726	55.232	76.852	-55.883	1.622
1900	38.069	295.467	264.395	59.036	75.780	-63.228	1.738
2000	38.134	297.421	265.998	62.846	74.685	-70.516	1.842
2100	38.195	299.283	267.539	66.663	73.565	-77.748	1.934
2200	38.255	301.061	269.023	70.485	72.420	-84.927	2.016
2300	38.312	302.763	270.453	74.313	71.248	-92.053	2.091
2400	38.369	304.395	271.834	78.147	70.050	-99.128	2.157
2500	38.424	305.962	273.168	81.987	68.824	-106.152	2.218
2600	38.479	307.470	274.458	85.832	67.571	-113.126	2.273
2700	38.533	308.924	275.708	89.683	66.291	-120.052	2.323
2800	38.586	310.326	276.919	93.539	-225.973	-120.707	2.252
2900	38.640	311.681	278.095	97.400	-226.243	-116.943	2.106
3000	38.694	312.992	279.236	101.267	-226.522	-113.169	1.970
3100	38.749	314.262	280.346	105.139	-226.809	-109.386	1.843
3200	38.805	315.493	281.425	109.017	-227.106	-105.593	1.724
3300	38.862	316.688	282.475	112.900	-227.412	-101.791	1.611
3400	38.921	317.849	283.499	116.789	-227.727	-97.980	1.505
3500	38.982	318.978	284.497	120.684	-228.052	-94.159	1.405
3600	39.045	320.077	285.470	124.586	-228.388	-90.329	1.311
3700	39.111	321.148	286.420	128.494	-228.734	-86.489	1.221
3800	39.180	322.191	287.347	132.408	-229.091	-82.640	1.136
3900	39.252	323.210	288.254	136.330	-229.459	-78.781	1.055
4000	39.328	324.205	289.140	140.259	-229.840	-74.913	0.978
4100	39.408	325.177	290.007	144.195	-230.234	-71.035	0.905
4200	39.492	326.128	290.856	148.140	-230.642	-67.147	0.835
4300	39.581	327.058	291.687	152.094	-231.064	-63.249	0.768
4400	39.674	327.969	292.501	156.057	-231.502	-59.342	0.704
4500	39.772	328.862	293.300	160.029	-231.955	-55.424	0.643
4600	39.876	329.737	294.082	164.011	-232.426	-51.496	0.585
4700	39.984	330.596	294.850	168.004	-232.914	-47.557	0.529
4800	40.098	331.439	295.603	172.008	-233.420	-43.608	0.475
4900	40.218	332.267	296.343	176.024	-233.946	-39.648	0.423
5000	40.342	333.080	297.070	180.052	-234.491	-35.677	0.373
5100	40.473	333.880	297.784	184.093	-235.056	-31.696	0.325
5200	40.609	334.668	298.486	188.147	-235.642	-27.703	0.278
5300	40.751	335.443	299.176	192.215	-236.248	-23.698	0.234
5400	40.898	336.206	299.854	196.297	-236.876	-19.682	0.190
5500	41.051	336.957	300.522	200.395	-237.525	-15.654	0.149
5600	41.209	337.699	301.179	204.507	-238.195	-11.614	0.108
5700	41.372	338.429	301.826	208.636	-238.886	-7.561	0.069
5800	41.541	339.150	302.464	212.782	-239.598	-3.497	0.031
5900	41.714	339.862	303.092	216.945	-240.331	0.580	-0.005
6000	41.892	340.564	303.710	221.125	-241.084	4.670	-0.041

PREVIOUS: June 1975 (1 atm) CURRENT: June 1975 (1 bar)

Beryllium Bromide (BeBr) Be_1Br_1(g)

BERYLLIUM BROMIDE ($BeBr_2$) CRYSTAL M_r = 168.82018

$S°(298.15 K) = [100.4 \pm 4.2] J K^{-1} mol^{-1}$

$T_{fus} = 781 \pm 15 K$

$\Delta_f H°(0 K)$ = Unknown

$\Delta_f H°(298.15 K) = [-355.6 \pm 12.6] kJ mol^{-1}$

$\Delta_{fus} H° = [9.83 \pm 8.4] kJ mol^{-1}$

Enthalpy of Formation

A direct measurement of the enthalpy of formation has not been made. An estimate of $\Delta_f H°$ is obtained by a method suggested by Parker (1). Biltz and Messerknecht (2) have measured the enthalpies of solution of $BeCl_2$(cr) and $BeBr_2$(cr) in aqueous HCl (18.69%). Samples (3) of the dihalides were prepared from reactions of BeO-C mixtures with the halogens at elevated temperatures. We assume that the two dihalides had similar structures, i.e. α form. $\Delta_f H°(BeBr_2$, in 18.7% HCl, 298.15 K) is estimated as -140.1 kcal mol^{-1} by combining $\Delta_{sol} H°$ of $BeCl_2$(cr) with $\Delta_f H°(\alpha\text{-}BeCl_2$, cr, 298.15 K) = -117.3±0.8 kcal mol^{-1} (4) and twice the difference in $\Delta_f H°(HCl \cdot 8.8H_2O$, 298.15 K) and $\Delta_f H°(HBr \cdot 8.8H_2O$, 298.15 K) which is -21.2 kcal mol^{-1} (5). The heat associated with the interchange of the aqueous anions is assumed to be negligible. Combination of the estimated value for $\Delta_f H°(BeBr_2$, in 18.7% HCl, 298.15 K) with $\Delta_{sol} H°$ = -55.7 kcal mol^{-1} (2) gives $\Delta_f H°(BeBr_2$, cr, 298.15 K) = -84.4 kcal mol^{-1} which is essentially the value adopted by NBS (6). Subsequent measurements by Biltz et al. (7) in less concentrated HCl(1.48%) solutions lead to $\Delta_f H°$(298.15 K) equal to -85.6 kcal mol^{-1} by a similar route. We adopt an average value of -85.0 kcal mol^{-1} (-355.64 kJ mol^{-1}) but emphasize that the uncertainty in $\Delta_f H°$ is much greater than that indicated (~1.0 kcal mol^{-1}) by the agreement of these two results. A combined total uncertainty of ±3.0 kcal mol^{-1} (±12.552 kJ mol^{-1}) is believed to be more realistic.

Heat Capacity and Entropy

No low-temperature heat capacity or high-temperature enthalpy measurements have been reported. We estimate $C_p°$(298.15 K) = 15.78 cal K^{-1} mol^{-1} from the reaction BeF_2(cr) + 2LiBr(cr) = $BeBr_2$(cr) + 2LiF(cr) by assuming $\Delta C_p° = 0$. Comparison of this value with $C_p°$ data for α,β-$BeCl_2$ (4) suggests that our estimate is reasonable. $C_p°$ data above 298 K are estimated graphically by comparison with those for α,β-$BeCl_2$ and $MgCl_2$(4).

Several methods of estimation predict that the value of S°(298.15 K) should lie near 24.0 cal K^{-1} mol^{-1}. Application of the Berthelot principle (8) to the process $SrBr_2$(cr) + Be(cr) = Sr(cr) + $BeBr_2$(cr) and additive entropy constants (9) give values of 24.1 and 24.0 cal K^{-1} mol^{-1}, respectively. A graphical comparison of the standard entropies for other alkaline-earth dihalides (4) suggests values for S°(298.15 K) in the range 23.5-24.5 cal K^{-1} mol^{-1}. Literature estimates have included (in cal K^{-1} mol^{-1}) 29 (10) and 22 (11). We adopt S°(298.15 K) = 24.0±1.0 cal K^{-1} mol^{-1} (100.416±4.184 J K^{-1} mol^{-1}).

Fusion Data

The reported melting points for $BeBr_2$ show an unusual amount of scatter. Values of T_{fus} include 761 K (12), 763 K (13), and 779-781 K (14). Semenenko and Naumova (14) reported encountering difficulties in handling $BeBr_2$ due to its hygroscopic nature and susceptibility to oxidation by traces of oxygen. These factors could account for the discrepancies in the T_{fus} values. We tentatively adopt the highest value of 781 K (14) but believe that T_{fus} is more uncertain (~15 K) than the accuracy (±5 K) claimed in their temperature measurements.

A thermal analysis (14) of $BeBr_2$ up to temperatures near the melting point showed no polymorphic modifications. Also, the interpretation of electron-diffraction patterns for freshly sublimed $BeBr_2$ indicated that it is isostructural with the α form of $BeCl_2$. Thus, we estimate $\Delta_{fus} H°$ = 2.35±2.0 kcal mol^{-1} from $\Delta_{fus} S°$ = 3.0 cal K^{-1} mol^{-1} which is calculated for α-$BeCl_2$ (4).

Sublimation Data

$\Delta_{sub} H°$(298.15 K) is obtained from an analysis of the sublimation pressures reported by Rahlfs and Fischer (12). Further details of the analysis are given on the gas-phase table. The value, T_{sub} = 752 K, is the temperature at which ΔG approaches zero for the process $BeBr_2$(cr) = $BeBr_2$(g). T_{sub} has been measured as 746 K (12). The good agreement between the calculated and observed values of T_{sub} is believed to indicate the presence of only small amounts of dimer in the saturated vapor of $BeBr_2$ near the melting point. However, this conflicts with other evidence which is discussed on the gas-phase table. Our results show that $T_{sub} < T_{fus}$ which implies that the liquid phase is thermodynamically unstable under ordinary conditions.

References

1. V. B. Parker, personal communication, U. S. Nat. Bur. Stand., March 1975.
2. W. Biltz and C. Messerknecht, Z. Anorg. Chem. 148, 157 (1925).
3. C. Messerknecht and W. Biltz, Z. Anorg. Chem. 148, 152 (1925).
4. JANAF Thermochemical Tables: α,β-$BeCl_2$(cr), 6-30-65; $MgCl_2$(cr), 12-31-65.
5. U. S. Nat. Bur. Stand. Tech. Note 270-3, 1968.
6. U. S. Nat. Bur. Stand. Tech. Note 270-6, 1971.
7. W. Biltz, K. A. Klatte, and E. Rahlfs, Z. Anorg. Chem. 166, 339 (1927).
8. N. N. Drozin, Zhur. Fiz. Khim. 35, 879 (1961).
9. D. R. Stull and H. Prophet in "The Characterization of High Temperature Vapors," J. L. Margrave, Ed., John Wiley and Sons, Inc., New York, 1966.
10. L. Brewer, L. A. Bromley, P. W. Gilles, and N. L. Lofgren in "Chemistry and Metallurgy of Miscellaneous Materials: Thermodynamics," L. L. Quill, Ed., National Nuclear Energy Series IV-19B, McGraw-Hill Book Co., New York, 1950.
11. L. Brewer, G. R. Somayajulu, and E. Brackett, Chem. Revs. 63, 111 (1963).
12. O. Rahlfs and W. Fsicher, Z. Anorg. Chem. 211, 349 (1933).
13. P. Lebeau, Ann. Chim. Phys. 16, 472 (1899).
14. K. N. Semenenko and T. N. Naumova, Russ. J. Structural Chem. 4, 59 (1963).

Beryllium Bromide ($BeBr_2$) Be_1Br_2(cr)

Enthalpy Reference Temperature = T_r = 298.15 K Standard State Pressure = p° = 0.1 MPa

	J K⁻¹mol⁻¹			kJ mol⁻¹			
T/K	$C_p°$	S°	-[G°-H°(T_r)]/T	H°-H°(T_r)	$\Delta_f H°$	$\Delta_f G°$	Log K_f
0							
100							
200							
298.15	66.024	100.416	100.416	0.	-355.640	-337.384	59.108
300	66.149	100.825	100.417	0.122	-355.688	-337.271	58.724
400	70.626	120.475	103.066	6.963	-385.169	-324.995	42.440
500	74.810	136.710	108.216	14.247	-383.680	-310.120	32.398
600	77.613	150.608	114.151	21.874	-382.039	-295.560	25.731
700	79.705	162.735	120.244	29.744	-380.299	-281.283	20.990
800	81.337	173.487	126.240	37.798	-378.495	-267.260	17.450
900	82.634	183.144	132.035	45.998	-376.651	-253.467	14.711
1000	83.764	191.909	137.591	54.318	-374.788	-239.879	12.530
1100	84.810	199.942	142.899	62.747	-372.913	-226.479	10.755
1200	85.856	207.366	147.966	71.280	-371.030	-213.250	9.283
1300	86.902	214.280	152.804	79.919	-369.137	-200.178	8.043
1400	87.864	220.755	157.429	88.656	-367.241	-187.252	6.986
1500	88.910	226.852	161.856	97.494	-365.343	-174.462	6.075

PREVIOUS: June 1965 CURRENT: June 1975

Beryllium Bromide ($BeBr_2$) Be_1Br_2(cr)

BERYLLIUM BROMIDE ($BeBr_2$) IDEAL GAS M_r = 168.82018

$S°(298.15 K) = 273.9 \pm 2.1 J K^{-1} mol^{-1}$

$\Delta_f H°(0 K) = [-215.8 \pm 16.7] kJ mol^{-1}$
$\Delta_f H°(298.15 K) = [-229.3 \pm 16.7] kJ mol^{-1}$

Vibrational Frequencies and Degeneracies

ν, cm^{-1}
[227] (1)
220 (2)
1010 (1)

Ground State Quantum Weight: [1] $\sigma = 2$
Point Group: $D_{\infty h}$
Bond Distance: Be-Br = 1.91 ± 0.02 Å
Bond Angle: Br-Be-Br = 180°
Rotational Constant: B_0 = [0.028916] cm^{-1}

Enthalpy of Formation

Rahlfs and Fischer (1) have reported measurements of the sublimation pressures (624-695 K) and vapor densities (736-799 K) for $BeBr_2$. Both measurements were complicated by significant reaction of the dibromide with the quartz apparatus. Assuming the reaction to be $2BeBr_2(cr) + SiO_2(cr) = SiBr_4(g) + 2BeO(cr)$, Rahlfs and Fischer (1) corrected their measured total pressure for the partial pressure of the tetrabromide. Results of a 2nd and 3rd law analyses of their data are tabulated below.

Measurement	No. of Points	T/K	$\Delta_{sub}H°$(298.15 K), kcal mol^{-1} 2nd law	3rd law	Drift cal K^{-1} mol^{-1}	$\Delta_f H°$(298.15 K) kcal mol^{-1}
Sublimation Pressures[a]	8[b]	624-695	31.0	30.2±0.1	-1.0±0.8	-54.8
Vapor Densities-Total Pressures[c]	4	736-799	23.9	33.4±1.0	12.4±9.0	-51.6

[a]Pressures assigned to monomer.
[b]One point rejected due to failure of a statistical test.
[c]Pressures corrected for dimer.

The vapor density measurements (1) predict about 50% dimerization at temperatures near the melting point (T_{fus} = 781 K). This value appears to be unusually high when compared with dimerization data which are available for $BeCl_2$, $MgBr_2$, $MgCl_2$, and MgF_2 (2). Mass spectral studies of these alkaline-earth dihalides show the presence of only small amounts (1-5%) of dimer in the saturated vapors up to temperatures near their melting points. Electron-diffraction experiments on $BeBr_2$ vapor are also inconsistent with high concentrations of dimer. It is quite likely that the monomer and dimer pressures reported by Rahlfs and Fischer (1) from their vapor density measurements are unreliable. Thus, we choose to adopt $\Delta_f H°$(298.15 K) = -54.8 kcal mol^{-1} rather than the average value (-53.2 kcal mol^{-1}) of these two results. The sublimation pressures (1) are assigned solely to the monomer pending resolution of the dimerization problem. Also, $\Delta_f H°$(g) is placed in brackets to emphasize the uncertainties in $\Delta_f H°$(cr).

Heat Capacity and Entropy

Information available on the structure of $BeBr_2$ tend to indicate that the molecule is linear. Electron-diffraction patterns (3, 4) for $BeBr_2$ vapor have been successfully interpreted in terms of a linear configuration. Buchler et al. (5), using electric quadrupole deflection of molecular beams to detect permanent dipole moments in molecules, have shown that for the gaseous alkaline-earth dihalides the linear form is favored by a light metal-heavy halogen combination. The matrix-isolation spectrum (6) of $BeBr_2$ shows the two infrared-active vibrations which would be expected for a linear molecule. We adopt the linear configuration. The bond length is taken from the electron-diffraction studies of Akishin (3, 4).

The asymmetric stretching (ν_3) and bending (ν_2) frequencies ahve been observed in the matrix-isolation spectrum of $BeBr_2$ by Snelson (6). The adopted frequencies are gas-phase values which were estimated by Snelson (6) from the matrix data. The infrared-inactive symmetrical stretching frequency (ν_1) is calculated from an estimated force constant by a simple valence force field approximation (7). The stretching force constant (K = 2.430 x 10^5 dynes/cm) is assumed equal to that for BeBr (2). Literature estimates of ν_1 include (in cm^{-1}) 230 (6, 8) and 209 (9). The ground state quantum weight of one is assigned by analogy with $BaCl_2$ (2).

Brewer et al. (8) have tabulated Gibbs energy functions for $BeBr_2$ up to 1500 K. Their values are consistently lower than ours by about 2.5 cal K^{-1} mol^{-1}. These differences are due entirely to the higher bending frequency (ν_2 = 446 cm^{-1}) which was used by Brewer et al. (8). This estimated value is now known to be incorrect.

References

1. O. Rahlfs and W. Fischer, Z. Anorg, Chem. 211, 349 (1933).
2. JANAF Thermochemical Tables: Be_2Cl_4(g), 6-30-65; Mg_2Br_4(g), 6-30-74; Mg_2Cl_4(g), 12-31-69; Mg_2F_4(g), 3-31-66; BeBr(g), 6-30-75; $BaCl_2$(g), 12-31-72.
3. P. A. Akishin and V. P. Spiridonov, Kristallografiya 2, 472 (1957).
4. P. A. Akishin, V. P. Spiridonov, and G. A. Sobalev, Dokl. Akad. Nauk SSSR 118, 1134 (1958).
5. A. Buchler, J. L. Stauffer, and W. Klemperer, J. Amer. Chem. Soc. 86, 4544 (1964).
6. A. Snelson, J. Phys. Chem. 72, 250 (1968).
7. G. Herzberg, "Infrared and Raman Spectra of Polyatomic Molecules," D. Van Nostrand Company, Inc., New York, 1962.
8. L. Brewer, G. R. Somayajulu, and E. Brackett, Chem. Revs. 63. 111 (1963).
9. K. S. Krasnov and V. I. Svettsov, Izv. Vysshikh Uchekn. Zavedenii, Khim. i Khim. Tekhnol. 6, 167 (1963).

Beryllium Bromide ($BeBr_2$) Be_1Br_2(g)

Enthalpy Reference Temperature = T_r = 298.15 K Standard State Pressure = p° = 0.1 MPa

T/K	$C_p°$ (J $K^{-1}mol^{-1}$)	S°	-[G°-H°(T_r)]/T	H°-H°(T_r) (kJ mol^{-1})	$\Delta_f H°$	$\Delta_f G°$	Log K_f
0	0.	0.	INFINITE	-12.917	-215.758	-215.758	INFINITE
100	40.423	221.826	318.498	-9.667	-215.341	-232.089	121.231
200	49.650	253.339	278.680	-5.068	-216.221	-248.570	64.920
250	51.746	264.655	274.778	-2.531	-217.091	-256.564	53.606
298.15	53.315	273.909	273.909	0.	-229.283	-262.754	46.033
300	53.369	274.239	273.910	0.099	-229.355	-262.962	45.786
350	54.705	282.569	274.564	2.802	-260.180	-266.871	39.828
400	55.816	289.949	276.035	5.565	-260.210	-267.826	34.974
450	56.740	296.578	277.955	8.380	-260.264	-268.774	31.199
500	57.507	302.597	280.123	11.237	-260.333	-269.716	28.177
600	58.673	313.191	284.776	17.049	-260.507	-271.578	23.643
700	59.490	322.301	289.501	22.960	-260.726	-273.406	20.402
800	60.075	330.285	294.110	28.940	-260.996	-275.200	17.969
900	60.504	337.387	298.531	34.970	-261.322	-276.956	16.074
1000	60.826	343.779	302.742	41.037	-261.712	-278.673	14.556
1100	61.072	349.588	306.741	47.132	-262.171	-280.347	13.313
1200	61.265	354.911	310.536	53.250	-262.704	-281.977	12.274
1300	61.418	359.821	314.141	59.384	-263.315	-283.559	11.394
1400	61.542	364.377	317.568	65.532	-264.009	-285.091	10.637
1500	61.643	368.627	320.832	71.692	-264.789	-286.570	9.979
1600	61.726	372.608	323.945	77.860	-280.195	-287.468	9.385
1700	61.796	376.352	326.919	84.037	-280.757	-287.906	8.846
1800	61.855	379.886	329.764	90.219	-281.348	-288.309	8.367
1900	61.906	383.232	332.491	96.407	-281.970	-288.679	7.936
2000	61.949	386.408	335.108	102.600	-282.626	-289.015	7.548
2100	61.986	389.432	337.624	108.797	-283.317	-289.318	7.196
2200	62.018	392.316	340.045	114.997	-284.047	-289.587	6.876
2300	62.047	395.073	342.378	121.200	-284.816	-289.822	6.582
2400	62.072	397.715	344.629	127.406	-285.626	-290.022	6.312
2500	62.094	400.249	346.803	133.615	-286.478	-290.188	6.063
2600	62.114	402.685	348.906	139.825	-287.372	-290.319	5.833
2700	62.131	405.029	350.941	146.037	-288.309	-290.414	5.618
2800	62.147	407.289	352.914	152.251	-580.243	-284.252	5.303
2900	62.161	409.470	354.827	158.467	-580.198	-273.682	4.930
3000	62.174	411.578	356.683	164.683	-580.174	-263.113	4.581
3100	62.185	413.617	358.487	170.901	-580.171	-252.544	4.255
3200	62.196	415.591	360.241	177.120	-580.189	-241.975	3.950
3300	62.205	417.505	361.947	183.341	-580.227	-231.406	3.663
3400	62.214	419.362	363.609	189.562	-580.284	-220.835	3.393
3500	62.222	421.166	365.228	195.783	-580.359	-210.262	3.138
3600	62.230	422.919	366.806	202.006	-580.451	-199.686	2.897
3700	62.236	424.624	368.346	208.229	-580.560	-189.108	2.670
3800	62.243	426.284	369.849	214.453	-580.685	-178.526	2.454
3900	62.249	427.901	371.317	220.678	-580.826	-167.942	2.249
4000	62.254	429.477	372.751	226.903	-580.983	-157.353	2.055
4100	62.259	431.014	374.153	233.129	-581.156	-146.760	1.870
4200	62.264	432.514	375.525	239.355	-581.345	-136.163	1.693
4300	62.268	433.979	376.868	245.581	-581.550	-125.561	1.525
4400	62.272	435.411	378.182	251.808	-581.772	-114.954	1.365
4500	62.276	436.810	379.469	258.036	-582.012	-104.341	1.211
4600	62.279	438.179	380.731	264.263	-582.270	-93.724	1.064
4700	62.283	439.519	381.967	270.491	-582.547	-83.100	0.924
4800	62.286	440.830	383.180	276.720	-582.843	-72.471	0.789
4900	62.289	442.114	384.370	282.949	-583.160	-61.835	0.659
5000	62.291	443.373	385.537	289.178	-583.499	-51.192	0.535
5100	62.294	444.606	386.683	295.407	-583.859	-40.542	0.415
5200	62.296	445.816	387.809	301.636	-584.242	-29.885	0.300
5300	62.299	447.003	388.915	307.866	-584.649	-19.220	0.189
5400	62.301	448.167	390.001	314.096	-585.081	-8.548	0.083
5500	62.303	449.310	391.069	320.326	-585.537	2.133	-0.020
5600	62.305	450.433	392.119	326.557	-586.019	12.822	-0.120
5700	62.307	451.536	393.152	332.787	-586.528	23.520	-0.216
5800	62.308	452.619	394.168	339.018	-587.062	34.227	-0.308
5900	62.310	453.684	395.168	345.249	-587.624	44.944	-0.398
6000	62.312	454.732	396.152	351.480	-588.213	55.670	-0.485

PREVIOUS: June 1975 (1 atm) CURRENT: June 1975 (1 bar)

Beryllium Bromide ($BeBr_2$) Be_1Br_2(g)

BERYLLIUM CHLORIDE (BeCl) IDEAL GAS $M_r = 44.46518$

$S°(298.15\ K) = 217.6\ J\ K^{-1}\ mol^{-1}$

$\Delta_f H°(0\ K) = 58.3 \pm 12.6\ kJ\ mol^{-1}$
$\Delta_f H°(298.15\ K) = 60.7 \pm 12.6\ kJ\ mol^{-1}$

Electronic levels and Quantum Weight

State	ε_i, cm^{-1}	g_i
$X^2\Sigma^+$	0	2
$A^2\pi$	27970	4

$\omega_e = 845.29\ cm^{-1}$ $\omega_e x_e = 4.831\ cm^{-1}$ $\sigma = 1$
$B_e = [0.769]\ cm^{-1}$ $\alpha_e = [0.0063]$ $r_e = [1.75]$ Å

Enthalpy of Formation

Novikov (1) measured 39 bands in the ultraviolet spectrum of BeCl(g). A linear extrapolation leads to a ground state dissociation energy of 36,555 cm^{-1}. The authors assumed that the non-crossing rule applied and that the ground and first excited states had a common dissociation limit. However, in the analogous case of BeF, treated in the same way by the above authors, it was shown by Mann (2) that these assumptions were not correct. If it is assumed that the excited state dissociates $A^2\Pi \to Be(^3P) + F(^2P)$ and the ground state $X^2\Sigma \to Be(^1S_0) + F(^2P)$ then ν_{oo} + D' should equal D" + Excitation energy of Be atom. We find ν_{oo} + D' = 27,960 + 30,390 = 58,350 and D" + E_{Be} = 36,555 + 21,981 = 58,536 which is excellent agreement. Thus, the dissociation energy to ground state atoms is taken as 104.5 kcal mol^{-1} which leads to a enthalpy of formation of 3 kcal mol^{-1} using reaction a.

Greenbaum et al. (3) report equilibrium constants for reaction b in the temperature range 1573-1724 K using the molecular flow effusion method. Also Hildenbrand et al. (4) report equilibrium constants for reaction c and d which were calculated from mass spectrometric ion intensities. The $\Delta_r H°(298.15\ K)$ values by 2nd and 3rd law analysis for these data and the corresponding $\Delta_f H°(298.15\ K)$ values are listed below.

Author	Method	Reaction	T/K	$\Delta_r H°(298.15\ K)$, kcal mol^{-1} 2nd law	3rd law	Drift cal K^{-1} mol^{-1}	$\Delta_f H°(298.15\ K)$ kcal mol^{-1}
Novikov	Spectroscopic	(a) BeCl(g) → Be(g) + Cl(g)		104.5			2.677
Greenbaum	Kp	(b) $BeCl_2$(g) + Be(l) → 2 BeCl(g)	1573-1724	93.56±7	85.87	-4.521	5:2 1:3
Hildenbrand	Kp from Mass	(c) Be(g) + AlCl(g) → BeCl(g) + Al(g)	1382-1554	27.5 ±2			16.5 ± 3
	Spectroscopy				26.25	- .779	15.3 ± 2
		(d) Be(g) + BeCl(g) → 2 BeCl(g)	1382-1536	33.41±8			12.8 ± 5
					36.08	2.198	14.1 ± 2.5

The equilibrium data of Greenbaum et al. (3) was not weighted, since calculations indicated the possiblility of reaction of BeCl(g) with the Al_2O_3 tube at the temperatures and pressures used. The data of Hildenbrand et al. (4) which reflect the true gas phase equilibria and appear consistent with each other were adopted yielding $\Delta_f H°(298.15\ K) = 14.5$ kcal mol^{-1}.

Heat Capacity and Entropy

The vibrational constants used to calculate the functions were derived by Novikov (1) from a two constant equation $[\omega_o'' - (2\nu'' + 1)\omega_o''x_o'' = \Delta G_o''(\nu'')]$. The electronic levels were obtained from Herzberg (5). B_e was obtained from the estimated bond length, which was taken to be slightly less than that in $BeCl_2$(g) and α_e was estimated from the Morse potential function.

References

1. M. M. Novikov and L. N. Tunitskii, Optics and Spectroscopy, 8, 396 (1960).
2. D. E. Mann, Nat. Bur. Stand. Report 7587, July 1, 1962.
3. M. A. Greenbaum, M. L. Arin, M. Wong, and M. Farber, J. Phys. Chem. 68, 791-5 (1964).
4. D. L. Hildenbrand, E. Murad, L. P. Theard and F. Ju, Philco Report U-3197, contract Af 04(611) - 10743, July 30, 1965.
5. G. Herzberg, "Molecular Spectra and Molecular Structure of Diatomic Molecules", 2nd ed., D. Van Nostrand Company, Inc., New York, 1950.

Beryllium Chloride (BeCl) $Be_1Cl_1(g)$

Enthalpy Reference Temperature = T_r = 298.15 K Standard State Pressure = p° = 0.1 MPa

T/K	$C_p°$ (J $K^{-1}mol^{-1}$)	S°	$-[G°-H°(T_r)]/T$	$H°-H°(T_r)$ (kJ mol^{-1})	$\Delta_f H°$	$\Delta_f G°$	Log K_f
0	0.	0.	INFINITE	-8.860	58.331	58.331	INFINITE
100	29.118	185.044	244.568	-5.952	59.743	50.739	-26.504
200	29.867	205.369	220.449	-3.016	60.583	41.340	-10.797
250	30.735	212.124	218.131	-1.502	60.696	36.513	-7.629
298.15	31.642	217.615	217.615	0.	60.668	31.856	-5.581
300	31.677	217.811	217.615	0.059	60.665	31.677	-5.516
350	32.554	222.761	218.004	1.665	60.533	26.856	-4.008
400	33.316	227.159	218.879	3.312	60.341	22.057	-2.880
450	33.957	231.121	220.023	4.994	60.109	17.285	-2.006
500	34.490	234.727	221.315	6.706	59.847	12.541	-1.310
600	35.301	241.092	224.095	10.198	59.256	3.134	-0.273
700	35.872	246.579	226.924	13.758	58.589	-6.168	0.460
800	36.285	251.397	229.688	17.367	57.851	-15.370	1.004
900	36.596	255.690	232.343	21.012	57.043	-24.475	1.420
1000	36.836	259.558	234.875	24.684	56.164	-33.486	1.749
1100	37.029	263.079	237.281	28.378	55.211	-42.405	2.014
1200	37.187	266.308	239.567	32.089	54.183	-51.235	2.230
1300	37.319	269.289	241.740	35.814	53.077	-59.975	2.410
1400	37.433	272.059	243.808	39.552	51.892	-68.627	2.561
1500	37.533	274.646	245.779	43.300	50.624	-77.192	2.688
1600	37.622	277.071	247.659	47.058	34.737	-85.144	2.780
1700	37.702	279.354	249.457	50.824	33.700	-92.605	2.845
1800	37.776	281.511	251.179	54.598	32.646	-100.004	2.902
1900	37.844	283.555	252.829	58.379	31.572	-107.345	2.951
2000	37.908	285.498	254.415	62.167	30.478	-114.628	2.994
2100	37.968	287.349	255.939	65.961	29.364	-121.856	3.031
2200	38.025	289.117	257.407	69.760	28.228	-129.030	3.064
2300	38.080	290.808	258.823	73.566	27.069	-136.152	3.092
2400	38.133	292.430	260.190	77.376	25.886	-143.224	3.117
2500	38.184	293.988	261.511	81.192	24.680	-150.245	3.139
2600	38.233	295.486	262.789	85.013	23.448	-157.218	3.159
2700	38.282	296.930	264.027	88.839	22.190	-164.143	3.176
2800	38.329	298.323	265.227	92.669	-270.050	-164.798	3.074
2900	38.376	299.669	266.392	96.505	-270.297	-161.035	2.901
3000	38.423	300.971	267.523	100.345	-270.553	-157.263	2.738
3100	38.469	302.232	268.622	104.189	-270.821	-153.482	2.586
3200	38.515	303.454	269.692	108.038	-271.101	-149.693	2.443
3300	38.562	304.639	270.733	111.892	-271.395	-145.894	2.309
3400	38.609	305.791	271.747	115.751	-271.704	-142.087	2.183
3500	38.656	306.911	272.736	119.614	-272.027	-138.270	2.064
3600	38.705	308.001	273.700	123.482	-272.368	-134.443	1.951
3700	38.754	309.062	274.642	127.355	-272.725	-130.607	1.844
3800	38.805	310.096	275.561	131.233	-273.101	-126.761	1.742
3900	38.857	311.105	276.460	135.116	-273.495	-122.905	1.646
4000	38.911	312.089	277.338	139.004	-273.909	-119.038	1.554
4100	38.967	313.051	278.198	142.898	-274.343	-115.161	1.467
4200	39.025	313.991	279.039	146.798	-274.799	-111.273	1.384
4300	39.085	314.910	279.862	150.703	-275.276	-107.374	1.304
4400	39.148	315.809	280.669	154.615	-275.775	-103.464	1.228
4500	39.212	316.689	281.460	158.533	-276.296	-99.542	1.155
4600	39.280	317.552	282.235	162.458	-276.841	-95.608	1.086
4700	39.350	318.397	282.995	166.389	-277.410	-91.662	1.019
4800	39.423	319.227	283.742	170.328	-278.002	-87.704	0.954
4900	39.499	320.040	284.474	174.274	-278.619	-83.733	0.893
5000	39.578	320.839	285.193	178.228	-279.261	-79.749	0.833
5100	39.660	321.624	285.900	182.190	-279.927	-75.752	0.776
5200	39.745	322.394	286.595	186.160	-280.618	-71.742	0.721
5300	39.833	323.152	287.277	190.139	-281.334	-67.719	0.667
5400	39.923	323.898	287.948	194.126	-282.075	-63.681	0.616
5500	40.017	324.631	288.609	198.123	-282.841	-59.630	0.566
5600	40.114	325.353	289.258	202.130	-283.631	-55.564	0.518
5700	40.213	326.064	289.898	206.146	-284.446	-51.485	0.472
5800	40.316	326.764	290.528	210.173	-285.285	-47.390	0.427
5900	40.421	327.454	291.148	214.210	-286.147	-43.281	0.383
6000	40.529	328.135	291.758	218.257	-287.032	-39.158	0.341

PREVIOUS: September 1966 (1 atm) CURRENT: September 1966 (1 bar)

Beryllium Chloride (BeCl) $Be_1Cl_1(g)$

BERYLLIUM CHLORIDE, ION ($BeCl^+$) IDEAL GAS M_r = 44.464631

$S°$(298.15 K) = [213.0 ± 8] J K^{-1} mol^{-1}

$\Delta_f H°$(0 K) = [970 ± 84] kJ mol^{-1}
$\Delta_f H°$(298.15 K) = [979 ± 84] kJ mol^{-1}

Electronic Levels and Quantum Weights

State	ϵ_i, cm^{-1}	g_i
$^1\Sigma_g^+$	0	1
$^3\Pi_u$	[10000]	6
$^1\Pi_u$	[15000]	2
$^3\Sigma_g^-$	[30000]	3
$^1\Delta$	[35000]	2
$^1\Sigma^+$	[40000]	1

ω_e = [700] cm^{-1} $\omega_e x_e$ = [4.8] cm^{-1} σ = 1
B_e = [0.7241] cm^{-1} α_e = [0.0071] cm^{-1} r_e = [1.8] Å

Enthalpy of Formation

Krasnow (1) has reported calculations of the ionization potentials (IP) for the monohalides (MX, where X = F, Cl, Br, and I) of Group IIA elements (M = Mg, Ca, Sr, and Ba). Based on his data we obtain graphically the value IP(BeCl, g) = 8.9 ± 0.5 eV. The value of $\Delta_f H°$(298.15 K) is derived as 219 ± 15 kcal mol^{-1}, using $\Delta_f H°$(BeCl, g, O K) = 13.9 kcal mol^{-1}. Beckett (2) reported the ionization potential of BeCl(g) to be approximately 9 ± 1 eV, yielding $\Delta_f H°$(298.15 K) = 222 ± 23 kcal mol^{-1}. Hildenbrand (9) measured the appearance potential (AP) of $BeCl^+$(g) to be 9.5 ± 0.7 cal $K^{-1}mol^{-1}$ which is on the same order of the above IP values, and assumed that IP($BeCl^+$, g) = AP($BeCl^+$, g) = 9.5 ± 0.7 eV. Using this IP value, we obtain $\Delta_f H°$(298.15 K) = 234 ± 20 kcal mol^{-1} which is adopted.

Heat Capacity and Entropy

The six electronic states are obtained from Fougere (3). The electronic levels are estimated by comparison with those of the isoelectronic molecules, BeS (4), BeO (5, 6), and MgO (5, 7). The values of ω_e, $\omega_e x_e$ are estimated by comparison with those for BeCl(g) and LiCl(g). B_e and α_e are calculated using the method suggested by Herzberg (5). The bond distance is estimated (8).

References

1. K. S. Krasnov, Teplofiz. Vysokikh Temperatur, Akad. Nauk SSSR, 3, 927 (1965).
2. C. W. Beckett and E. C. Cassidy, U. S. Nat. Bur. Stand. Report 8628, January 1965.
3. P. F. Fougers and R. K. Nesbet, J. Chem. Phys. 44, 285 (1965).
4. G. Verhaegen and W. G. Richards, Proc. Phys. Soc. 90, 579 (1967).
5. G. Herzberg, "Spectra of Diatomic Molecules", D. Van Nostrand Co., Inc., New York, 1950.
6. G. Verhaegen and W. G. Richards, J. Chem. Phys. 45, 1828 (1966).
7. L. Brewer and S. Trajmar, J. Chem. Phys. 36, 1585 (1962).
8. K. S. Krasnov, Zh. Strukt. Khim. 1, 209 (1960).
9. D. L. Hildenbrand, E. Murad, L. P. Theard, and F. Ju, Aeronutronic, Philco Corporation, Report No. U-3197, July 30, 1965.

Beryllium Chloride, Ion ($BeCl^+$) $Be_1Cl_1^+(g)$

Enthalpy Reference Temperature = T_r = 298.15 K Standard State Pressure = p° = 0.1 MPa

T/K	$C_p°$ (J $K^{-1}mol^{-1}$)	S° (J $K^{-1}mol^{-1}$)	$-[G°-H°(T_r)]/T$ (J $K^{-1}mol^{-1}$)	$H°-H°(T_r)$ (kJ mol^{-1})	$\Delta_f H°$ (kJ mol^{-1})	$\Delta_f G°$ (kJ mol^{-1})	Log K_f
0	0.	0.	INFINITE	-8.990	970.392		
100	29.155	179.796	240.610	-6.081			
200	30.594	200.327	215.883	-3.111			
250	31.734	207.277	213.488	-1.553			
298.15	32.744	212.955	212.955	0.	979.056	945.379	-165.626
300	32.780	213.157	212.955	0.061	979.093	945.170	-164.569
350	33.652	218.278	213.358	1.722	980.057	939.439	-140.204
400	34.356	222.820	214.262	3.423	980.957	933.574	-121.912
450	34.918	226.900	215.443	5.155	981.814	927.599	-107.673
500	35.370	230.603	216.777	6.913	982.638	921.531	-96.272
600	36.036	237.115	219.639	10.486	984.206	909.161	-79.149
700	36.495	242.706	222.544	14.113	985.685	896.535	-66.900
800	36.828	247.602	225.377	17.780	987.084	883.703	-57.700
900	37.082	251.955	228.093	21.476	988.406	870.700	-50.534
1000	37.287	255.873	230.678	25.195	989.652	857.554	-44.794
1100	37.463	259.435	233.133	28.933	990.822	844.287	-40.092
1200	37.628	262.702	235.463	32.687	991.916	830.917	-36.169
1300	37.800	265.721	237.676	36.459	992.934	817.459	-32.846
1400	37.993	268.529	239.780	40.248	993.879	803.925	-29.995
1500	38.224	271.158	241.785	44.059	994.752	790.326	-27.522
1600	38.503	273.633	243.699	47.894	981.021	777.197	-25.373
1700	38.842	275.977	245.530	51.761	982.164	764.422	-23.488
1800	39.247	278.209	247.284	55.665	983.318	751.581	-21.810
1900	39.722	280.343	248.968	59.613	984.490	738.675	-20.308
2000	40.267	282.394	250.588	63.612	985.686	725.706	-18.954
2100	40.881	284.373	252.150	67.669	986.913	712.677	-17.727
2200	41.558	286.290	253.659	71.790	988.177	699.589	-16.610
2300	42.290	288.154	255.118	75.982	989.483	686.442	-15.590
2400	43.068	289.970	256.533	80.250	990.836	673.238	-14.653
2500	43.883	291.744	257.906	84.597	992.240	659.976	-13.789
2600	44.722	293.482	259.241	89.027	993.696	646.657	-12.991
2700	45.575	295.186	260.541	93.542	995.206	633.281	-12.252
2800	46.431	296.859	261.808	98.142	705.814	626.070	-11.679
2900	47.278	298.503	263.045	102.828	708.496	623.175	-11.225
3000	48.108	300.120	264.254	107.597	711.248	620.186	-10.798
3100	48.910	301.710	265.437	112.448	714.065	617.104	-10.398
3200	49.677	303.275	266.595	117.378	716.944	613.931	-10.021
3300	50.403	304.815	267.730	122.382	719.879	610.666	-9.666
3400	51.082	306.330	268.843	127.457	722.866	607.312	-9.330
3500	51.709	307.820	269.935	132.597	725.897	603.870	-9.012
3600	52.283	309.285	271.008	137.797	728.968	600.340	-8.711
3700	52.802	310.724	272.062	143.052	732.070	596.724	-8.424
3800	53.264	312.139	273.098	148.355	735.199	593.024	-8.152
3900	53.670	313.528	274.117	153.703	738.348	589.242	-7.892
4000	54.022	314.891	275.119	159.088	741.509	585.378	-7.644
4100	54.320	316.229	276.106	164.505	744.677	581.436	-7.408
4200	54.568	317.541	277.077	169.950	747.845	577.416	-7.181
4300	54.768	318.827	278.033	175.417	751.009	573.321	-6.964
4400	54.922	320.088	278.974	180.902	754.161	569.152	-6.757
4500	55.034	321.324	279.902	186.400	757.298	564.912	-6.557
4600	55.108	322.534	280.815	191.908	760.415	560.602	-6.366
4700	55.147	323.720	281.716	197.421	763.507	556.225	-6.182
4800	55.154	324.881	282.603	202.936	766.569	551.782	-6.005
4900	55.132	326.018	283.477	208.450	769.599	547.276	-5.834
5000	55.085	327.132	284.339	213.962	772.594	542.709	-5.670
5100	55.016	328.222	285.189	219.467	775.550	538.082	-5.511
5200	54.928	329.289	286.027	224.964	778.464	533.397	-5.358
5300	54.822	330.335	286.853	230.452	781.336	528.656	-5.210
5400	54.703	331.358	287.668	235.928	784.162	523.862	-5.067
5500	54.572	332.361	288.471	241.392	786.941	519.016	-4.929
5600	54.431	333.343	289.264	246.842	789.673	514.120	-4.796
5700	54.282	334.305	290.046	252.278	792.357	509.176	-4.666
5800	54.127	335.248	290.817	257.698	794.991	504.185	-4.541
5900	53.968	336.172	291.578	263.103	797.575	499.148	-4.419
6000	53.805	337.077	292.329	268.492	800.109	494.069	-4.301

PREVIOUS: June 1968 (1 atm) CURRENT: June 1968 (1 bar)

Beryllium Chloride, Ion ($BeCl^+$) $Be_1Cl_1^+(g)$

BERYLLIUM CHLORIDE FLUORIDE (BeClF) IDEAL GAS M_r = 63.63583

$S°(298.15\ K) = [246.49]\ J\ K^{-1}\ mol^{-1}$

$\Delta_f H°(0\ K) = [574 \pm 42]\ kJ\ mol^{-1}$
$\Delta_f H°(298.15\ K) = [573 \pm 42]\ kJ\ mol^{-1}$

Vibrational Frequencies and Degeneracies

ν, cm^{-1}
[500] (1)
[270] (2)
[1300] (1)

Ground State Quantum Weight: 1 $\sigma = 1$
Point Group: $C_{\infty v}$
Bond Distances: Be-Cl = [1.77] Å Be-F = [1.43] Å
Bond Angle: Cl-Be-F = [180]°
Rotational Constant: B_o = [0.129706] cm^{-1}

Enthalpy of Formation

The enthalpy of formation was estimated by assuming $\Delta_r H° = 0$ for the reaction $BeF_2 + BeCl_2 \rightarrow 2BeClF$.

Heat Capacity and Entropy

The structure, bond lengths and frequencies were all estimated from the properties of $BeCl_2$ and BeF_2.

Beryllium Chloride Fluoride (BeClF) $Be_1Cl_1F_1(g)$

Enthalpy Reference Temperature = T_r = 298.15 K Standard State Pressure = p° = 0.1 MPa

T/K	$C_p°$	S°	$-[G°-H°(T_r)]/T$	$H°-H°(T_r)$	$\Delta_f H°$	$\Delta_f G°$	Log K_f
	J K⁻¹mol⁻¹			kJ mol⁻¹			
0	0.	0.	INFINITE	-11.704	-573.977	-573.977	INFINITE
100	34.801	200.224	286.766	-8.654	-573.876	-575.879	300.808
200	44.537	227.702	250.869	-4.633	-573.416	-578.060	150.974
250	47.364	237.961	247.289	-2.332	-573.267	-579.240	121.026
298.15	49.413	246.486	246.486	0.	-573.208	-580.397	101.683
300	49.483	246.792	246.487	0.091	-573.207	-580.442	101.064
350	51.183	254.552	247.096	2.609	-573.221	-581.647	86.806
400	52.601	261.482	248.469	5.205	-573.280	-582.847	76.112
450	53.805	267.748	250.268	7.866	-573.368	-584.038	67.793
500	54.832	273.472	252.307	10.583	-573.474	-585.218	61.137
600	56.463	283.621	256.701	16.152	-573.725	-587.544	51.150
700	57.663	292.419	261.189	21.861	-574.019	-589.824	44.013
800	58.556	300.180	265.588	27.674	-574.357	-592.060	38.658
900	59.231	307.118	269.824	33.565	-574.746	-594.250	34.489
1000	59.749	313.387	273.872	39.515	-575.192	-596.394	31.152
1100	60.152	319.101	277.727	45.511	-575.702	-598.490	28.420
1200	60.472	324.349	281.397	51.543	-576.283	-600.536	26.141
1300	60.729	329.200	284.890	57.603	-576.938	-602.531	24.210
1400	60.938	333.708	288.218	63.687	-577.673	-604.473	22.553
1500	61.109	337.919	291.392	69.789	-578.491	-606.359	21.115
1600	61.252	341.867	294.425	75.908	-593.933	-607.662	19.838
1700	61.373	345.584	297.326	82.039	-594.527	-608.502	18.697
1800	61.474	349.095	300.106	88.182	-595.142	-609.307	17.682
1900	61.562	352.421	302.772	94.334	-595.781	-610.077	16.772
2000	61.636	355.581	305.334	100.494	-596.443	-610.812	15.953
2100	61.701	358.590	307.799	106.661	-597.128	-611.514	15.211
2200	61.758	361.462	310.174	112.834	-597.836	-612.182	14.535
2300	61.808	364.208	312.464	119.012	-598.566	-612.818	13.918
2400	61.851	366.840	314.675	125.195	-599.318	-613.422	13.351
2500	61.890	369.365	316.812	131.382	-600.090	-613.993	12.829
2600	61.925	371.793	318.881	137.573	-600.883	-614.534	12.346
2700	61.956	374.131	320.884	143.767	-601.696	-615.044	11.899
2800	61.983	376.385	322.826	149.964	-893.483	-609.300	11.367
2900	62.008	378.560	324.711	156.163	-893.268	-599.155	10.792
3000	62.031	380.663	326.541	162.365	-893.054	-589.017	10.256
3100	62.051	382.697	328.320	168.570	-892.839	-578.886	9.754
3200	62.070	384.667	330.050	174.776	-892.626	-568.761	9.284
3300	62.087	386.578	331.734	180.984	-892.413	-558.644	8.843
3400	62.102	388.431	333.375	187.193	-892.203	-548.533	8.427
3500	62.117	390.232	334.973	193.404	-891.995	-538.428	8.036
3600	62.130	391.982	336.533	199.616	-891.790	-528.329	7.666
3700	62.142	393.684	338.055	205.830	-891.589	-518.235	7.316
3800	62.153	395.342	339.540	212.045	-891.393	-508.148	6.985
3900	62.163	396.956	340.992	218.261	-891.203	-498.065	6.671
4000	62.173	398.530	342.411	224.477	-891.020	-487.986	6.372
4100	62.182	400.066	343.798	230.695	-890.845	-477.913	6.089
4200	62.190	401.564	345.156	236.914	-890.678	-467.843	5.818
4300	62.198	403.028	346.485	243.133	-890.522	-457.778	5.561
4400	62.205	404.457	347.786	249.353	-890.377	-447.716	5.315
4500	62.211	405.855	349.061	255.574	-890.244	-437.657	5.080
4600	62.218	407.223	350.311	261.795	-890.125	-427.600	4.856
4700	62.224	408.561	351.536	268.017	-890.020	-417.547	4.641
4800	62.229	409.871	352.738	274.240	-889.930	-407.495	4.434
4900	62.234	411.154	353.917	280.463	-889.858	-397.445	4.237
5000	62.239	412.412	355.074	286.687	-889.802	-387.397	4.047
5100	62.244	413.644	356.211	292.911	-889.765	-377.349	3.865
5200	62.248	414.853	357.327	299.136	-889.748	-367.302	3.690
5300	62.252	416.039	358.423	305.361	-889.750	-357.255	3.521
5400	62.256	417.202	359.501	311.586	-889.773	-347.207	3.359
5500	62.260	418.345	360.561	317.812	-889.817	-337.160	3.202
5600	62.263	419.467	361.603	324.038	-889.883	-327.111	3.051
5700	62.266	420.569	362.627	330.264	-889.971	-317.061	2.906
5800	62.270	421.652	363.636	336.491	-890.082	-307.008	2.765
5900	62.272	422.716	364.628	342.718	-890.215	-296.954	2.629
6000	62.275	423.763	365.605	348.946	-890.372	-286.898	2.498

PREVIOUS: June 1965 (1 atm) CURRENT: June 1965 (1 bar)

Beryllium Chloride Fluoride (BeClF) $Be_1Cl_1F_1(g)$

BERYLLIUM CHLORIDE, ALPLHA (α-$BeCl_2$) CRYSTAL M_r = 79.91818

$S°(298.15\ K) = 82.68 \pm 0.25\ J\ K^{-1}\ mol^{-1}$

$T_{fus} = 688\ K$

$\Delta_f H°(0\ K) = -491.8 \pm 3.3\ kJ\ mol^{-1}$

$\Delta_f H°(298.15\ K) = -490.9 \pm 3.3\ kJ\ mol^{-1}$

$\Delta_{fus} H° = 8.66 \pm 0.25\ kJ\ mol^{-1}$

Enthalpy of Formation

Johnson and Gilliland (1) have directly reacted beryllium and chlorine and obtained $\Delta_f H° = -118.03 \pm 0.56\ kcal\ mol^{-1}$ though the crystal form of the product was unknown. Gross, Heyman, Greene and Bingham (2) also by direct combustion of the elements obtain $\Delta_f H° = -117.1 \pm 0.4\ kcal\ mol^{-1}$. The crystal form of the sample was deduced to be α' by a comparison experiment, the assumption is made here that α and α' are identical. Thompson, Sinke and Stull (3) by solution calorimetry report a $\Delta_f H° = -118.25 \pm 0.5\ kcal\ mol^{-1}$. Initially this sample was reported to be of unknown crystal form. However, Sinke (4) reports that an x-ray of the sample used has been reinterpreted, in the light of more recent data, as being approximately 50% α' and 50% β. Using the enthalpies of transition adopted for these tables, enthalpies of formation of $-117.3 \pm 0.8\ kcal\ mol^{-1}$ for α and $-118.6 \pm 0.8\ kcal\ mol^{-1}$ for β have been adopted as being the most consistent with all measurements.

Heat Capacity and Entropy

McDonald and Oetting (5) have measured the heat capacity of the α' form from 13 and 304 K and the enthalpy (relative to the α' form) from 676 to 688 K. The assumption has been made that the α and α' forms are identical and the heat capacity curve between 304 and 676 K has been estimated graphically. The entropy was obtained by integration of the heat capacity curve assuming $S°(13\ K) = 0.16\ cal\ K^{-1}\ mol^{-1}$. The sample was identified as α' by x-ray diffraction both before and after the experiments.

Fusion Data

The temperature and enthalpy of melting were reported by McDonald and Oetting (5). Several investigations of the melting point are in disagreement mainly due to the uncertainty of the crystal form of the material and a solid state transition just below the pure α-liquid melting point.

Sublimation Data

The adopted enthalpy of sublimation was calculated from that of the β form and enthalpy of the α-β transition at 298.15 K.

References

1. W. H. Johnson and A. A. Gilliland, J. Res. Nat. Bur. Stand. 65A, 59 (1961).
2. P. Gross, C. Hayman, P. D. Greene and J. T. Bingham, Fulmer Research Institute, Report R. 163/SR.1/Sept. 1964 under Contract AF61 (052)-447.
3. C. J. Thompson, G. C. Sinke and D. R. Stull, J. Chem. Eng. Data 7, 380 (1962).
4. G. C. Sinke, personal communication, February 1965.
5. R. A. McDonald and F. L. Oetting, J. Phys. Chem., Novembr 1965.

Beryllium Chloride, Alpha ($BeCl_2$) $Be_1Cl_2(cr)$

Enthalpy Reference Temperature = T_r = 298.15 K; Standard State Pressure = p° = 0.1 MPa

T/K	$C_p°$	S°	$-[G°-H°(T_r)]/T$	$H°-H°(T_r)$	$\Delta_f H°$	$\Delta_f G°$	Log K_f
	J K⁻¹mol⁻¹			kJ mol⁻¹			
0	0.	0.	INFINITE	-11.979	-491.795	-491.795	INFINITE
100	33.660	28.974	131.942	-10.297	-493.064	-477.011	249.165
200	53.555	59.036	88.261	-5.845	-492.228	-461.208	120.455
298.15	64.852	82.676	82.676	0.	-490.930	-446.254	78.182
300	65.019	83.078	82.677	0.120	-490.903	-445.977	77.651
400	72.760	102.907	85.329	7.031	-489.303	-431.234	56.313
500	77.530	119.678	90.566	14.556	-487.453	-416.927	43.556
600	80.082	134.040	96.644	22.437	-485.471	-403.007	35.085
688.000	81.278	145.083	102.148	29.539	------- ALPHA <--> LIQUID ------		
700	81.358	146.489	102.896	30.515	-483.457	-389.422	29.059
800	82.257	157.413	109.042	38.697	-481.471	-376.125	24.558
900	83.023	167.146	114.967	46.961	-479.519	-363.074	21.072
1000	83.680	175.929	120.631	55.297	-477.603	-350.239	18.295
1100	84.230	183.931	126.027	63.694	-475.728	-337.593	16.031
1200	84.672	191.280	131.163	72.140	-473.902	-325.117	14.152
1300	85.007	198.071	136.052	80.625	-472.134	-312.790	12.568
1400	85.234	204.380	140.710	89.138	-470.435	-300.597	11.215
1500	85.354	210.265	145.153	97.668	-468.817	-288.523	10.047

PREVIOUS: CURRENT: June 1965

Beryllium Chloride, Alpha ($BeCl_2$) $Be_1Cl_2(cr)$

BERYLLIUM CHLORIDE, BETA (β-$BeCl_2$) CRYSTAL M_r = 79.91818

$S°(298.15\ K) = 75.81 \pm 0.21\ J\ K^{-1}\ mol^{-1}$

$T_{fus} = [682\ K]°$

$T_{trs} = 676\ K\ (\beta \rightarrow \alpha)$

$\Delta_f H°(0\ K) = -496.5 \pm 3.3\ kJ\ mol^{-1}$

$\Delta_f H°(298.15\ K) = -496.2 \pm 3.3\ kJ\ mol^{-1}$

$\Delta_{fus} H° = [15.27 \pm 0.63]\ kJ\ mol^{-1}$

$\Delta_{trs} H° = [6.82 \pm 0.63]\ kJ\ mol^{-1}$

Enthalpy of Formation

See the α-crystal table for details.

Heat Capacity and Entropy

McDonald and Oetting (1) have measured the heat capacity of the β-form from 13 to 304 K and the enthalpy of the β-form from 298 to 676 K. The entropy was obtained by integration of the heat capacity curve based on $S°(13\ K) = 0.06\ cal\ K^{-1}\ mol^{-1}$. The sample was identified as β-$BeCl_2$ both before and after the low temperature experiments.

Fusion Data

The temperature and enthalpy of melting were obtained from cross over point of the tables for β-$BeCl_2$ and liquid $BeCl_2$.

Transition Data

The temperature of the $\beta \rightarrow \alpha$ transition was obtained from McDonald and Oetting (1). The enthalpy of the transiton is a function of the heat capacity curve chosen for α-$BeCl_2$ between 298 and 676 K. A straight line interpolation was used by McDonald and Oetting, which gives a different result ($1.49\ kcal\ mol^{-1}$) than the curve employed here. It should also be noted that the transitional enthalpy is not included in this table, which is for β-$BeCl_2$ only.

Sublimation Data

The enthalpy of sublimation was determined by 2nd and 3rd law analysis of vapor pressure data as described fully on the table for $BeCl_2$(g).

References

1. R. A. McDonald and F. L. Oetting, J. Phys. Chem., November 1965.

Beryllium Chloride, Beta ($BeCl_2$) Be_1Cl_2(cr)

Enthalpy Reference Temperature = T_r = 298.15 K Standard State Pressure = p° = 0.1 MPa

T/K	$C_p°$ (J K⁻¹mol⁻¹)	S° (J K⁻¹mol⁻¹)	$-[G°-H°(T_r)]/T$ (J K⁻¹mol⁻¹)	$H°-H°(T_r)$ (kJ mol⁻¹)	$\Delta_f H°$ (kJ mol⁻¹)	$\Delta_f G°$ (kJ mol⁻¹)	Log K_f
0	0.	0.	INFINITE	-11.418	-496.527	-496.527	INFINITE
100	31.777	24.104	127.616	-10.351	-498.412	-481.871	251.704
200	51.798	52.928	81.211	-5.657	-497.332	-465.090	121.469
298.15	62.425	75.814	75.814	0.	-496.222	-449.501	78.751
300	62.593	76.201	75.815	0.116	-496.200	-449.211	78.215
400	68.701	95.099	78.350	6.700	-494.928	-433.736	56.640
500	72.802	110.900	83.324	13.788	-493.514	-418.599	43.731
600	75.814	124.449	89.075	21.224	-491.977	-403.758	35.150
700	78.157	136.327	94.995	28.933	-490.331	-389.184	29.041
800	79.956	146.885	100.833	36.841	-488.619	-374.851	24.475
900	81.379	156.387	106.487	44.910	-486.863	-360.735	20.936
1000	82.550	165.025	111.915	53.109	-485.084	-346.816	18.116
1100	83.387	172.934	117.108	61.408	-483.306	-333.075	15.816
1200	84.015	180.217	122.068	69.780	-481.555	-319.495	13.907
1300	84.475	186.961	126.803	78.205	-479.846	-306.060	12.298
1400	84.782	193.234	131.327	86.669	-478.196	-292.754	10.923
1500	84.935	199.089	135.652	95.157	-476.621	-279.564	9.735

PREVIOUS: CURRENT: June 1965

Beryllium Chloride, Beta ($BeCl_2$) Be_1Cl_2(cr)

BERYLLIUM CHLORIDE ($BeCl_2$) LIQUID M_r = 79.91818

S°(298.15 K) = [79.727] J K^{-1} mol^{-1} $\Delta_f H°$(298.15 K) = [-491.313] kj mol^{-1}

T_{fus} = 688 K $\Delta_{fus}H°$ = 8.66 ± 0.25 kJ mol^{-1}

Enthalpy of Formation

The enthalpy of formation is calculated from that of the α-$BeCl_2$ by adding $\Delta_{fus}H°$ and the difference in enthalpy, H°(688 K)-H°(298.15 K), between the crystal and liquid.

Heat Capacity and Entropy

McDonald and Oetting (1) have measured the enthalpy of the liquid, relative to α-$BeCl_2$, from 688 to 713 K. The heat capacity was assumed constant above this point and also below the melting point to 460 K where a glass transition was assumed. Below 416 K the heat capacity was assumed to be that of β-$BeCl_2$. The entropy is calculated in a manner analagous to that used for the enthalpy of formation.

Fusion

See α-crystal table for details.

Vaporization Data

The boiling point and enthalpy of vaporization are calculated from the cross over point of the $BeCl_2$ (ℓ) and (g) tables. The analysis of the liquid vapor pressure data is given on the table for $BeCl_2$(g).

References

1. R. A. McDonald and F. L. Oetting, J. Phys. Chem., November 1965.

Beryllium Chloride ($BeCl_2$) Be_1Cl_2(l)

Enthalpy Reference Temperature = T_r = 298.15 K Standard State Pressure = p° = 0.1 MPa

T/K	$C_p°$	S° (J $K^{-1}mol^{-1}$)	$-[G°-H°(T_r)]/T$ (J $K^{-1}mol^{-1}$)	$H°-H°(T_r)$ (kJ mol^{-1})	$\Delta_f H°$ (kJ mol^{-1})	$\Delta_f G°$ (kJ mol^{-1})	Log K_f
0							
100							
200							
298.15	62.425	79.727	79.727	0.	-491.313	-445.758	78.095
300	62.593	80.114	79.728	0.116	-491.291	-445.475	77.564
400	68.701	99.012	82.263	6.700	-490.019	-430.392	56.203
460.100	71.356	108.819	85.103	10.912	GLASS <--> LIQUID TRANSITION		
460.100	121.420	108.819	85.103	10.912			
500	121.420	118.917	87.404	15.757	-486.636	-415.729	43.431
600	121.420	141.054	94.557	27.898	-480.393	-402.138	35.009
688.000	121.420	157.671	101.591	38.583	------- ALPHA <--> LIQUID ------		
700	121.420	159.771	102.570	40.040	-474.314	-389.578	29.071
800	121.420	175.984	110.756	52.182	-468.369	-377.880	24.673
900	121.420	190.286	118.814	64.324	-462.540	-366.920	21.295
1000	121.420	203.078	126.612	76.466	-456.817	-356.603	18.627
1100	121.420	214.651	134.098	88.608	-451.197	-346.854	16.471
1200	121.420	225.216	141.257	100.750	-445.675	-337.613	14.696
1300	121.420	234.935	148.094	112.892	-440.250	-328.829	13.213
1400	121.420	243.933	154.623	125.034	-434.922	-320.458	11.956
1500	121.420	252.310	160.859	137.176	-429.692	-312.465	10.881
1600	121.420	260.146	166.822	149.318	-439.095	-304.293	9.934
1700	121.420	267.507	172.531	161.460	-433.660	-296.034	9.096
1800	121.420	274.447	178.002	173.602	-428.255	-288.095	8.360
1900	121.420	281.012	183.252	185.744	-422.881	-280.455	7.710
2000	121.420	287.240	188.297	197.886	-417.540	-273.097	7.133

PREVIOUS: September 1961 CURRENT: June 1965

BERYLLIUM CHLORIDE ($BeCl_2$)

BETA-ALPHA-LIQUID

M_r = 79.91818

Beryllium Chloride ($BeCl_2$)

$Be_1Cl_2(cr,l)$

0 to 676 K	crystal, beta
676 to 688	crystal, alpha
above 688	liquid

Refer to the individual tables for details.

Enthalpy Reference Temperature = T_r = 298.15 K   Standard State Pressure = $p°$ = 0.1 MPa

T/K	$C_p°$ (J K⁻¹mol⁻¹)	$S°$ (J K⁻¹mol⁻¹)	$-[G°-H°(T_r)]/T$ (J K⁻¹mol⁻¹)	$H°-H°(T_r)$ (kJ mol⁻¹)	$\Delta_fH°$ (kJ mol⁻¹)	$\Delta_fG°$ (kJ mol⁻¹)	Log K_f
0	0.	0.	INFINITE	-11.979	-491.795	-491.795	INFINITE
100	33.660	28.974	131.942	-10.297	-493.064	-477.011	249.165
200	53.555	59.036	88.261	-5.845	-492.228	-461.208	120.455
298.15	64.852	82.676	82.676	0.	-490.930	-446.254	78.182
300	65.019	83.078	82.677	0.120	-490.903	-445.977	77.651
400	72.760	102.907	85.329	7.031	-489.303	-431.234	56.313
500	77.530	119.678	90.566	14.556	-487.453	-416.927	43.556
600	80.082	134.040	96.644	22.437	-485.471	-403.007	35.085
688.000	81.278	145.083	102.148	29.539	ALPHA <--> LIQUID TRANSITION		
688.000	121.420	157.671	102.148	38.200			
700	121.420	159.771	103.118	39.657	-474.314	-389.578	29.071
800	121.420	175.984	111.235	51.799	-468.369	-377.880	24.673
900	121.420	190.286	119.240	63.941	-462.540	-366.920	21.295
1000	121.420	203.078	126.995	76.083	-456.817	-356.603	18.627
1100	121.420	214.651	134.446	88.225	-451.197	-346.854	16.471
1200	121.420	225.216	141.577	100.367	-445.675	-337.613	14.696
1300	121.420	234.935	148.389	112.509	-440.250	-328.829	13.212
1400	121.420	243.933	154.896	124.651	-434.922	-320.458	11.956
1500	121.420	252.310	161.115	136.793	-429.692	-312.465	10.881
1600	121.420	260.146	167.062	148.935	-439.095	-304.293	9.934
1700	121.420	267.507	172.756	161.077	-433.660	-296.034	9.096
1800	121.420	274.447	178.215	173.219	-428.255	-288.095	8.360
1900	121.420	281.012	183.454	185.361	-422.881	-280.455	7.710
2000	121.420	287.240	188.489	197.503	-417.540	-273.097	7.133

PREVIOUS: CURRENT: June 1965

Beryllium Chloride ($BeCl_2$) $Be_1Cl_2(cr,l)$

BERYLLIUM CHLORIDE ($BeCl_2$) IDEAL GAS M_r = 79.91818

$S°(298.15\ K) = 252 \pm 4\ J\ K^{-1}\ mol^{-1}$

$\Delta_fH°(0\ K) = -361.5 \pm 10.5\ kJ\ mol^{-1}$
$\Delta_fH°(298.15\ K) = -360.2 \pm 10.5\ kJ\ mol^{-1}$

Vibrational Frequencies and Degeneracies

ν, cm^{-1}
[375] (1)
[230] (2)
1113 (1)

Ground State Quantum Weight: 1 $\sigma = 2$
Point Group: $D_{\infty h}$
Bond Distance: Be-Cl = [1.77] Å
Bond Angle: Cl-Be-F = [180]°
Rotational Constant: $B_0 = 0.0759\ cm^{-1}$

Enthalpy of Formation

The vapor pressure data over the crystal was assumed to be for the β-form. The 2nd and 3rd law analyses of the data are summarized below after conversion to a common process. The data were also analyzed using gaseous Gibbs energy functions based on a bending frequency of 170 cm^{-1} but no significant improvement of the results was noted.

Source	T/K	$\Delta_{sub}H°(298.15\ K)$, kcal mol^{-1} 2nd law	3rd law	Drift cal K^{-1} mol^{-1}
(1)	680-740*	33.5 ± 0.6	31.38	-3.4 ± 0.7
(2)	680-740	34.3	31.22	-4.7
(1)	613-663	31.6 ± 1.6	30.91	-1.3 ± 2.6
(2)	638-668	35.6	31.07	-7.0
(3)	441-518	31.4 ± 0.8	32.45	2.5 ± 1.7
(3)	510-600*	31.8 ± 0.3	32.07	0.40± 0.5
(4)	460-504*	32.45± 0.4	32.48	0.14± 0.8

*Points rejected due to failure of a statistical test.

It is obvious that there is no real agreement between the data sets, and some of this is undoubtedly due to the difficulty of keeping a single crystal form throughout a series of measurements. It is also doubtful that water vapor was adequately excluded in some experiments. We have adopted the only piece of work which shows adequate 2nd and 3rd law agreement, and no 3rd law trend, that is $\Delta_{sub}H° = 32.5 \pm 1$ kcal mol^{-1} from (4). It is interesting to note that all the other data can be made much more precise and its drift removed if a constant pressure is added to all points. The addition which varied from 20-30% of the lowest pressure in each data set serves to indicate the magnitude of possible systematic errors.

Heat Capacity and Entropy

The linear structure of $BeCl_2$(g) has been confirmed by the electric deflection experiments of Buchler (5). The bond length was obtained from the electron diffraction data of Akishin, Spirodinov and Sobelev (6). Buchler and Klemperer (7) have reported the infra-red spectrum and assigned 1113 as the asymmetric stretch ν_3 and 482 as the bending frequency ν_2. However, Snelson (8) using matrix isolation experiments, has shown this second band to be due to beryllium chloride solid. The unobserved frequencies were estimated from a valence force field treatment by assuming that the ratio of stretching to bending force constant was 50:1, the same as for BeF_2(g). This gives 230 cm^{-1} for ν_2 and 375 cm^{-1} for the symmetric stretch.

References

1. O. Rahlfs and W. Fischer, Z. Anorg. Allgem. Chem. 211, 349 (1933).
2. W. Fischer, T. Petzel and S. Lauter, Z. Anorg. Allgem. Chem 333, 226 (1964).
3. M. A. Greenbaum et al., Rocket Power, Inc., 7th Quarterly Report March 31, 1963 under Contract AF 04(611)-7414.
4. D. L. Hildenbrand, L. P. Theard, E. Murad and F. Ju, Ford Motor Co., Aeronutronic Div., Report U-3068, April 1965 under Contract AF (04) 611-8523.
5. A. Buchler, A. D. Little, Inc., Progress Report No. 8, Dec. 1963 under contract No. DA-19-020-ORD-5584.
6. P. A. Akishin, V. P. Spirodinov and G. A. Sobelev, Dokl. Akad. Nauk SSSR 118, 1134 (1958).
7. A. Buchler and W. Klemperer, J. Chem. Phys. 29, 121 (1958).
8. A. Snelson, IIT Research Institute, Report No. IITRI-C6013-4, May 1964 under Contract No. DA-31-124-ARO (O)-111.

Beryllium Chloride ($BeCl_2$) Be_1Cl_2(g)

Enthalpy Reference Temperature = T_r = 298.15 K Standard State Pressure = p° = 0.1 MPa

T/K	$C_p°$ (J $K^{-1}mol^{-1}$)	S°	$-[G°-H°(T_r)]/T$	$H°-H°(T_r)$ (kJ mol^{-1})	$\Delta_fH°$	$\Delta_fG°$	Log K_f
0	0.	0.	INFINITE	-12.322	-361.451	-361.451	INFINITE
100	37.378	202.989	294.816	-9.183	-361.263	-362.611	189.409
200	47.267	232.467	256.831	-4.873	-360.568	-364.234	95.128
250	49.785	243.300	253.072	-2.443	-360.350	-365.179	76.300
298.15	51.622	252.232	252.232	0.	-360.242	-366.120	64.143
300	51.685	252.552	252.233	0.096	-360.240	-366.156	63.753
350	53.221	260.638	252.868	2.719	-360.213	-367.145	54.793
400	54.497	267.831	254.297	5.413	-360.234	-368.135	48.073
450	55.563	274.313	256.167	8.166	-360.285	-369.119	42.846
500	56.455	280.215	258.281	10.967	-360.355	-370.098	38.664
600	57.830	290.636	262.828	16.685	-360.536	-372.030	32.388
700	58.809	299.629	267.458	22.520	-360.764	-373.928	27.903
800	59.517	307.530	271.983	28.438	-361.043	-375.790	24.537
900	60.042	314.572	276.331	34.417	-361.376	-377.614	21.916
1000	60.438	320.920	280.478	40.442	-361.771	-379.398	19.818
1100	60.744	326.695	284.421	46.502	-362.233	-381.139	18.099
1200	60.983	331.991	288.167	52.588	-362.766	-382.835	16.664
1300	61.174	336.880	291.729	58.697	-363.375	-384.483	15.449
1400	61.329	341.419	295.118	64.822	-364.064	-386.081	14.405
1500	61.455	345.655	298.347	70.961	-364.836	-387.628	13.498
1600	61.560	349.625	301.430	77.112	-380.230	-388.594	12.686
1700	61.648	353.360	304.376	83.273	-380.776	-389.100	11.956
1800	61.722	356.885	307.196	89.441	-381.345	-389.573	11.305
1900	61.785	360.224	309.900	95.617	-381.938	-390.014	10.722
2000	61.840	363.395	312.496	101.798	-382.557	-390.424	10.197
2100	61.887	366.413	314.992	107.985	-383.203	-390.801	9.721
2200	61.928	369.293	317.395	114.175	-383.879	-391.147	9.287
2300	61.964	372.047	319.712	120.370	-384.585	-391.462	8.890
2400	61.995	374.685	321.948	126.568	-385.325	-391.745	8.526
2500	62.023	377.216	324.108	132.769	-386.099	-391.997	8.190
2600	62.048	379.649	326.198	138.972	-386.909	-392.217	7.880
2700	62.070	381.991	328.221	145.178	-387.756	-392.405	7.592
2800	62.090	384.249	330.182	151.386	-679.599	-386.339	7.207
2900	62.108	386.428	332.084	157.596	-679.460	-375.868	6.770
3000	62.124	388.534	333.931	163.808	-679.343	-365.401	6.362
3100	62.139	390.571	335.726	170.021	-679.251	-354.938	5.981
3200	62.152	392.544	337.470	176.236	-679.184	-344.477	5.623
3300	62.164	394.457	339.168	182.451	-679.143	-334.018	5.287
3400	62.176	396.313	340.822	188.668	-679.128	-323.560	4.971
3500	62.186	398.115	342.433	194.887	-679.140	-313.102	4.673
3600	62.195	399.867	344.005	201.106	-679.179	-302.644	4.391
3700	62.204	401.571	345.537	207.326	-679.244	-292.183	4.125
3800	62.212	403.230	347.034	213.546	-679.336	-281.721	3.873
3900	62.219	404.846	348.496	219.768	-679.455	-271.256	3.633
4000	62.226	406.422	349.924	225.990	-679.600	-260.788	3.406
4100	62.232	407.958	351.321	232.213	-679.771	-250.315	3.189
4200	62.238	409.458	352.687	238.436	-679.967	-239.838	2.983
4300	62.244	410.923	354.025	244.661	-680.188	-229.357	2.786
4400	62.249	412.354	355.334	250.885	-680.433	-218.869	2.598
4500	62.253	413.753	356.617	257.110	-680.703	-208.376	2.419
4600	62.258	415.121	357.874	263.336	-680.997	-197.877	2.247
4700	62.262	416.460	359.106	269.562	-681.314	-187.371	2.082
4800	62.266	417.771	360.315	275.788	-681.653	-176.858	1.925
4900	62.270	419.055	361.501	282.015	-682.016	-166.338	1.773
5000	62.273	420.313	362.664	288.242	-682.400	-155.810	1.628
5100	62.277	421.546	363.807	294.470	-682.807	-145.274	1.488
5200	62.280	422.755	364.929	300.698	-683.235	-134.730	1.353
5300	62.283	423.942	366.031	306.926	-683.684	-124.178	1.224
5400	62.285	425.106	367.114	313.154	-684.155	-113.617	1.099
5500	62.288	426.249	368.179	319.383	-684.646	-103.046	0.979
5600	62.290	427.371	369.226	325.612	-685.157	-92.467	0.863
5700	62.293	428.474	370.256	331.841	-685.688	-81.879	0.750
5800	62.295	429.557	371.269	338.070	-686.239	-71.281	0.642
5900	62.297	430.622	372.266	344.300	-686.810	-60.674	0.537
6000	62.299	431.669	373.247	350.530	-687.399	-50.056	0.436

PREVIOUS: June 1965 (1 atm) CURRENT: June 1965 (1 bar)

Beryllium Chloride ($BeCl_2$) Be_1Cl_2(g)

BERYLLIUM FLUORIDE (BeF) IDEAL GAS $M_r = 28.010583$

Ground State Configuration: $^2\Sigma^+$

$S°(298.15\ K) = [205.75 \pm 0.04]\ J\ K^{-1}\ mol^{-1}$

$\Delta_f H°(0\ K) = -172.2 \pm 8\ kJ\ mol^{-1}$

$\Delta_f H°(298.15\ K) = -169.9 \pm 8\ kJ\ mol^{-1}$

Electronic Levels and Quantum Weights

ε_i, cm^{-1}	g_i	ε_i, cm^{-1}	g_i
0.0	2	50419	2
33180	4	63916	[4]
49573	2	66373	[4]

$\omega_e = 1265.9\ cm^{-1}$ $\omega_e x_e = 9.35\ cm^{-1}$ $\sigma = 1$

$B_e = 1.489\ cm^{-1}$ $\alpha_e = 0.0176\ cm^{-1}$ $r_e = 1.361$ Å

Enthalpy of Formation

$\Delta_f H°$ is based on 3rd law analysis of Kp data for three gas-phase reactions observed mass spectrometrically by Hildenbrand et al. (1, 2). Our analysis is summarized below. Also included are data for heterogeneous reaction of Be(cr, ℓ) with $BeF_2(g)$, observed with a molecular-flow-effusion method by Greenbaum et al. (3). The mutual consistency of the gas-phase results prompts us to adopt a weighted average of -40.6 ± 2 kcal mol^{-1} rather than the mean value of -53.1 kcal mol^{-1} derived from the heterogeneous reaction.

Analogous studies of BeCl(g) gave a similar discrepancy which was reaffirmed by new transport studies of Gross and Lewin (4). These authors suggested that transport via $(BeCl)_2$ provides a possible, but inconclusive, explanation of the discrepancy. This explanation may also apply to BeF, but new data would be needed to establish this.

The adopted $\Delta_f H°$ yields $D_0^0 = 5.98 \pm 0.1$ eV which is consistent with spectroscopic results within their probable uncertainties (5). Spectroscopic values (1, 5) include 6.3 or 6.0 ± 0.5 eV (from $A^2\Pi$) and 5.3 cal K^{-1} mol^{-1} (from $X^2\Sigma$). Rationales have been given (5-8) for the lowness of the value derived from the ground state. Recent calculations (13) showed that the ground-state potential energy curve is consistent with $D_0^0 = 5.85$ eV. The flow-effusion data (3) yield 6.52 cal K^{-1} mol^{-1}.

Source	Method	Reaction[a]	T/K	$\Delta_r H°(298.15\ K)$, kcal mol^{-1} 2nd law	3rd law	δS cal K^{-1} mol^{-1}	$\Delta_f H°(298.15\ K)$ kcal mol^{-1}
Hildenbrand (1)	Mass Spec.	A	1458-1609	25.9±9	28.28±2	-1.6±5.7	-41.86±1.3
		A	1321-1518	31.2±2	29.09±2	1.5±1.5	-41.46±1.3
		A	1423-1443	-4.1	30.4 ±3	24.	-40.8 ±1.7
		B	1321-1503	-6.7±3	-7.86±2	-0.8±2.1	-37.46±3.0
		B	1423-1443	-38.	-8.7 ±3	20.	-38.3 ±3.8
		C	1423-1443	12.7	24.2 ±3	8.	-38.9 ±3.4
Greenbaum (3)	Flow Eff.	D	1420-1525	97. ±10	84.7 ±3	8.2±7.3	-52.8 ±1.6
		E	1570-1675	83. ±6	80.7 ±3	-1.2±3.7	-53.4 ±1.6

[a] A) $Be(g) + BeF_2(g) \rightarrow 2BeF(g)$; B) $Be(g) + CaF(g) \rightarrow BeF(g) + Ca(g)$; C) $Be(g) + AlF(g) \rightarrow BeF(g) + Al(g)$; D) $BeF_2(g) + Be(cr) \rightarrow 2BeF(g)$; E) $BeF_2(g) + Be(\ell) \rightarrow 2BeF(g)$.

Heat Capacity and Entropy

Rotational constants are from a new analysis of the A-X system by Walker and Barrow (9). Vibrational constants are those derived by Novikov and Gurvich (10) from data of Tatevskii et. al. (11). Excited states are from Novikov and Gurvich (10). We assume that the two highest states are $^2\Pi$ as postulated by Walker and Barrow (9) from comparison with quantum calculations (12).

References

1. D. L. Hildenbrand and E. Murad, J. Chem. Phys. 44, 1524 (1966).
2. D. L. Hildenbrand and L. P. Theard, J. Chem. Phys. 42, 3230 (1965).
3. M. A. Greenbaum, R. E. yates, M. L. Arin, M. Arshadi, J. Weiher and M. Farber, J. Phys. Chem. 67, 703 (1963).
4. P. Gross and R. H. Lewin, Fulmer Res. Inst. Report R.163/SR.8/July 1971.
5. A. G. Gaydon, "Dissociation Energies," 3rd ed., Chapman and hall Ltd., London, 1968.
6. D. L. Hildenbrand, pp. 193-217 in "Advance in High Temperature Chemistry," L. Eyring (editor), Vol. 1, Academic Press, New York, 1967.
7. K. S. Krasnov and N. V. Karaseva, Opt. Spectrosc. (Eng. Trans.) 19, 14 (1965); Opt. Spektrosk. 19, 30 (1965).
8. D. E. Mann, U. S. Nat. Bur. Stand. Report 7587, 109 (1962).
9. T. E. H. Walker and R. F. BArrow, J. Phys. B2, 102 (1969).
10. M. M. Novikov and L. V. Gurvich, Opt. Spectrosc. (Eng. Trans.) 23, 173 (1967); Opt. Spektrosk. 23, 323 (1967).
11. V. M. Tatevskii, L. N. Tunitskii and M. M. Novikov, Opt. Spektrosk. 5, 521 (1958).
12. T. E. Walker and W. G. Richards, Proc. Phys. Soc. 92, 285 (1967).

Beryllium Fluoride (BeF) $Be_1F_1(g)$

Enthalpy Reference Temperature = T_r = 298.15 K Standard State Pressure = $p°$ = 0.1 MPa

T/K	$C_p°$ (J $K^{-1}mol^{-1}$)	$S°$	$-[G°-H°(T_r)]/T$	$H°-H°(T_r)$ (kJ mol^{-1})	$\Delta_f H°$	$\Delta_f G°$	Log K_f
0	0.	0.	INFINITE	-8.711	-172.237((	-172.237	INFINITE
100	29.109	173.800	231.865	-5.807	-170.825	-179.636	93.832
200	29.204	193.990	208.457	-2.893	-169.954	-188.852	49.323
250	29.454	200.530	206.241	-1.428	-169.831	-193.594	40.449
298.15	29.872	205.751	205.751	0.	-169.870	-198.170	34.719
300	29.891	205.936	205.752	0.055	-169.874	-198.345	34.535
350	30.461	210.586	206.118	1.564	-170.039	-203.079	30.308
400	31.091	214.694	206.938	3.102	-170.279	-207.783	27.134
450	31.725	218.393	208.009	4.673	-170.569	-212.454	24.661
500	32.327	221.767	209.218	6.274	-170.893	-217.091	22.679
600	33.379	227.757	211.821	9.561	-171.609	-226.265	19.698
700	34.217	232.968	214.478	12.943	-172.395	-235.313	17.559
800	34.872	237.581	217.083	16.399	-173.241	-244.244	15.948
900	35.387	241.720	219.595	19.913	-174.146	-253.066	14.688
1000	35.795	245.470	221.998	23.472	-175.115	-261.783	13.674
1100	36.124	248.898	224.289	27.069	-176.149	-270.401	12.840
1200	36.393	252.053	226.473	30.695	-177.254	-278.921	12.141
1300	36.617	254.975	228.555	34.346	-178.433	-287.346	11.546
1400	36.806	257.696	230.540	38.018	-179.689	-295.677	11.032
1500	36.968	260.241	232.436	41.707	-181.026	-303.916	10.583
1600	37.109	262.631	234.250	45.411	-196.982	-311.537	10.171
1700	37.233	264.885	235.986	49.128	-198.085	-318.663	9.791
1800	37.343	267.016	237.651	52.857	-199.206	-325.723	9.452
1900	37.442	269.038	239.250	56.596	-200.344	-332.721	9.147
2000	37.532	270.961	240.788	60.345	-201.498	-339.659	8.871
2100	37.615	272.794	242.269	64.102	-202.668	-346.538	8.620
2200	37.691	274.545	243.697	67.868	-203.853	-353.361	8.390
2300	37.762	276.222	245.075	71.640	-205.052	-360.130	8.179
2400	37.829	277.831	246.406	75.420	-206.263	-366.847	7.984
2500	37.892	279.377	247.694	79.206	-207.484	-373.513	7.804
2600	37.951	280.864	248.942	82.998	-208.715	-380.130	7.637
2700	38.008	282.297	250.151	86.796	-209.953	-386.699	7.481
2800	38.063	283.681	251.323	90.600	-502.155	-387.000	7.220
2900	38.115	285.017	252.462	94.409	-502.341	-382.884	6.896
3000	38.166	286.310	253.569	98.223	-502.516	-378.762	6.595
3100	38.215	287.562	254.646	102.042	-502.678	-374.634	6.313
3200	38.263	288.776	255.693	105.866	-502.828	-370.501	6.048
3300	38.310	289.955	256.714	109.694	-502.968	-366.364	5.799
3400	38.356	291.099	257.708	113.528	-503.099	-362.222	5.565
3500	38.402	292.211	258.678	117.366	-503.221	-358.077	5.344
3600	38.446	293.294	259.625	121.208	-503.337	-353.928	5.135
3700	38.491	294.348	260.549	125.055	-503.448	-349.776	4.938
3800	38.535	295.375	261.452	128.906	-503.556	-345.622	4.751
3900	38.579	296.377	262.335	132.762	-503.663	-341.464	4.573
4000	38.623	297.354	263.198	136.622	-503.772	-337.304	4.405
4100	38.667	298.308	264.043	140.487	-503.884	-333.141	4.244
4200	38.712	299.240	264.870	144.356	-504.003	-328.975	4.091
4300	38.758	300.152	265.680	148.229	-504.130	-324.806	3.946
4400	38.804	301.043	266.474	152.107	-504.267	-320.634	3.806
4500	38.850	301.916	267.252	155.990	-504.417	-316.459	3.673
4600	38.898	302.770	268.014	159.877	-504.582	-312.280	3.546
4700	38.947	303.607	268.763	163.769	-504.764	-308.098	3.424
4800	38.997	304.428	269.497	167.667	-504.965	-303.911	3.307
4900	39.049	305.233	270.218	171.569	-505.187	-299.721	3.195
5000	39.102	306.022	270.927	175.477	-505.431	-295.525	3.087
5100	39.157	306.797	271.622	179.389	-505.700	-291.324	2.984
5200	39.213	307.558	272.306	183.308	-505.994	-287.118	2.884
5300	39.271	308.305	272.978	187.232	-506.314	-282.906	2.788
5400	39.331	309.040	273.639	191.162	-506.662	-278.687	2.696
5500	39.393	309.762	274.290	195.098	-507.039	-274.462	2.607
5600	39.458	310.472	274.929	199.041	-507.446	-270.230	2.521
5700	39.524	311.171	275.559	202.990	-507.882	-265.990	2.438
5800	39.593	311.859	276.179	206.946	-508.348	-261.742	2.357
5900	39.664	312.537	276.790	210.909	-508.845	-257.486	2.280
6000	39.738	313.204	277.391	214.879	-509.373	-253.222	2.204

PREVIOUS: December 1971 (1 atm) CURRENT: December 1971 (1 bar)

Beryllium Fluoride (BeF) $Be_1F_1(g)$

BERYLLIUM FLUORIDE (BeF_2) CRYSTAL M_r = 47.008986

$S°(298.15\ K) = 53.354 \pm 0.17\ J\ K^{-1}\ mol^{-1}$
T_{trs} = 500 K (low quartz → high quartz)
T_{fus} = 825 K

$\Delta_fH°(0\ K) = -1024.5 \pm 4.2\ kJ\ mol^{-1}$
$\Delta_fH°(298.15\ K) = -1026.8 \pm 4.2\ kJ\ mol^{-1}$
$\Delta_{trs}H° = 0.2201\ kJ\ mol^{-1}$
$\Delta_{fus}H° = 4.757\ kJ\ mol^{-1}$

Enthalpy of Formation

$\Delta_fH°$ is the value selected by Parker (1), based on a thorough review of data for BeF_2(cr, vitreous, aq) and BeO(cr). Parker's analysis remains essentially unchanged for BeO(cr) and BeF_2(cr), even though the results for BeF_2(HF, aq) become more negative by 0.93 kcal mol^{-1} due to JANAF selections for HF(aq). Three values derived by Parker for BeF_2(cr) from PbF_2(cr) are changed due to JANF auxiliary data; these values become -246.9, -246.0 and -247.2 for the three paths used by Parker.

The adopted $\Delta_fH°$ was derived by combination of calorimetric data for Be(cr) + F_2(g) → BeF_2(vitreous) and BeF_2(cr) → BeF_2 (vitreous). Churney and Armstrong (2) obtained $\Delta_fH°(298.15\ K) = -244.32 \pm 0.8$ kcal mol^{-1} for vitreous BeF_2 by combustion in fluorine of polytetrafluoroethylene and mixtures of Be(cr) with polytetrafluoroethylene. $\Delta_rH°$(298.15 K, cr → vitreous) = 1.125 ± 0.016 kcal mol^{-1} was measured via solution calorimetry by Taylor and Gardner (3).

Heat Capacity and Entropy

Cp° is based on JANAF curve fits of measurements by adiabatic calorimetry (8-300 K) and by ice calorimetry (354-776 K) as reported by Taylor and Gardner (3). The sample was estimated as approximately 99.5% BeF_2. Analyses indicated that the major impurity was about 0.3% oxygen, mainly in the form of absorbed water. Petrographic examination and X-ray diffraction showed the sample to be composed almost completely of quartz-type crystals. The entropy is derived from the smoothed Cp° based on the extrapolation $S°(8\ K) = 0.0065$ cal $K^{-1}\ mol^{-1}$. The adopted $C_p°$ above 300 K agrees closely with the relative enthalpy data and shows a remarkable similarity to that of SiO_2 (low quartz, high quartz).

Transition Data

T_{trs} is the value selected by Taylor and Gardner (3) from their relative enthalpy data. $\Delta_{trs}H°$ is the corresponding difference in relative enthalpy from the JANAF curve fits for the two crystalline forms. Levina (4) studied the polymorphism of BeF_2 by observing the peaks in Cp near the transition temperatures. For low quartz → high quartz, Levina's peak began near 438 K, was complete near 503 K, and gave $\Delta_{trs}H° = 0.17$ kcal mol^{-1}. Levina found a 2nd transition attributed to a form analogous to tridymite; this began above 670 K, was complete by 740 K, and gave $\Delta_{trs}H° = 0.24$ kcal mol^{-1}. There is no evidence for this transition in the relative enthalpy data, so we suspect that its occurrence may be due to impurities in Levina's sample. Levina also found that the cristobalite-like phase transformed in the region 380 to 420 K with $\Delta_{trs}H° = 1.27$ kcal mol^{-1}.

Fusion Data

See the liquid table for details.

Sublimation Data

See the ideal gas table for details.

References

1. V. B. Parker, Nat. Bur. Stand. Report 10004, Chapter 3, p. 58, issued Jan. 1, 1969.
2. K. L. Churney and G. T. Armstrong, J. Res. Nat. Bur. Stand. 73A, 281 (1969).
3. A. R. Taylor and T. Estelle Gardner, U. S. Bur. Mines RI-6664, 1965.
4. M. E. Levina, Izv. Vyssh. Ucheb. Zaved., Khim. Khim. Teknol. 8 (2), 177 (1965).

Beryllium Fluoride (BeF_2) Be_1F_2(cr)

Enthalpy Reference Temperature = T_r = 298.15 K Standard State Pressure = p° = 0.1 MPa

T/K	$C_p°$ (J K⁻¹mol⁻¹)	S° (J K⁻¹mol⁻¹)	$-[G°-H°(T_r)]/T$ (J K⁻¹mol⁻¹)	$H°-H°(T_r)$ (kJ mol⁻¹)	$\Delta_fH°$ (kJ mol⁻¹)	$\Delta_fG°$ (kJ mol⁻¹)	Log K_f
0	0.	0.	INFINITE	-8.468	-1024.465	-1024.465	INFINITE
100	21.004	14.753	90.253	-7.550	-1026.493	-1010.881	528.030
200	39.083	35.271	57.702	-4.486	-1026.935	-995.024	259.874
298.15	51.819	53.354	53.354	0.	-1026.754	-979.385	171.584
300	52.028	53.676	53.355	0.096	-1026.746	-979.091	170.475
400	62.522	70.077	55.522	5.822	-1026.081	-963.292	125.793
500	76.048	85.392	59.971	12.710	-1024.661	-947.738	99.009
500.000	76.048	85.832	59.971	12.930		II <--> I	
500.000	64.099	85.832	59.971	12.930		TRANSITION	
600	67.446	97.814	65.302	19.508	-1023.604	-932.474	81.179
700	70.793	108.463	70.720	26.420	-1022.636	-917.360	68.454
800	74.140	118.134	76.051	33.666	-1021.495	-902.396	58.920
825.000	74.977	120.428	77.362	35.530	---------	I <--> LIQUID	--------
900	77.488	127.060	81.229	41.248	-1020.160	-887.587	51.514
1000	80.835	135.397	86.234	49.164	-1018.617	-872.937	45.598
1100	84.182	143.259	91.064	57.415	-1016.858	-858.453	40.765
1200	87.529	150.727	95.727	66.000	-1014.876	-844.138	36.744
1300	90.876	157.865	100.234	74.920	-1012.670	-829.998	33.350

PREVIOUS: June 1964 CURRENT: June 1970

Beryllium Fluoride (BeF_2) Be_1F_2(cr)

BERYLLIUM FLUORIDE (BeF_2) LIQUID M_r = 47.008986

S°(298.15 K) = [59.920] $J\ K^{-1}\ mol^{-1}$ $\Delta_fH°$(298.15 K) = [-1022.048] $kJ\ mol^{-1}$

T_{fus} = 825 K $\Delta_{fus}H°$ = 4.757 $kJ\ mol^{-1}$

Enthalpy of Formation

$\Delta_fH°$ is calculated from that of the crystal by means of $\Delta_rH°$ = 1.125 ± 0.016 kcal mol^{-1} for BeF_2(cr) → BeF_2(vitreous) at 298.15 K. Further details are given in the section on Fusion Data. $\Delta_fH°$(ℓ) actually is based on the direct determination of $\Delta_fH°$(virteous) by Churney and Armstrong (1) as discussed on the crystal table.

Heat Capacity and Entropy

Cp° is based on a JANAF curve fit of relative enthalpies of Taylor and Gardner (2) for the liquid (828-1183 K) and vitreous (320-473 K) forms of BeF_2. The combined data yeild a normal Cp° curve and reveal no need for a glass transition. Scatter in the data is somewhat larger than for the crystal, presumably due to minor variations in the vitreous state at the conclusion of each drop. S°(298.15 K, ℓ) is calculated from that of the crystal by addition of $\Delta_{fus}S°$ and the difference for crystal and liquid of S°(825 K) - S°(298.15 K).

Fusion Data

T_{fus} is the value selected by Taylor and Gardner (2) from their relative enthalpy data. These showed premelting starting near 804 K and complete melting above 828 K. Cooling of the liquid produced the vitreous state. T_{fus} has also been reported as 815 (3), 821 (4), 818 (5), and 816 ± 5 K (6).

Solution calorimetry (2) gave $\Delta_rH°$ = 1.125 ± 0.016 kcal mol^{-1} for BeF_2(low quartz) → BeF_2(vitreous) at 298.15 K. Confirmation of ths result came from transposed temperature-drop calorimetry (7) and calorimetric conversion (8). These gave 1.1 ± 0.2 and 1.1 ± 0.5 kcal mol^{-1}, respectively. Combination of $\Delta_rH°$ = 1.125 with JANAF enthalpies yields $\Delta_{vap}H°$ = 1.137 kcal mol^{-1}.

Vaporization Data

T_{vap} is the calculated temperature at which $\Delta_rG°$ = 0 for BeF_2(ℓ) → BeF_2(g); $\Delta_{vap}H°$ is the corresponding value of $\Delta_rH°$ at T_{vap}. These calculations neglect the presence of dimer, which is assumed to be negligible based on extrapolation of the equations of Belousov (9). Cantor (10) obtained T_{vap} = 1442 K by extrapolation of his vapor pressures (1146-1372 K).

References

1. K. L. Churney and G. T. Armstrong, J. Res. Nat. Bur. Stand. 73A, 281 (1969).
2. A. R. Taylor and T. E. Gardner, U. S. Bur. Mines RI-6664, 1965.
3. M. A. Greenbaum, J. N. Foster, M. L. Arin and M. Farber, J. Phys. Chem. 67, 36 (1963).
4. R. E. Thoma, H. Insley, H. A. Friedman and C. F. Weaver, J. Phys. Chem. 64, 865 (1960).
5. K. A. Sense and R. W. Stone, J. Phys. Chem. 62, 453 (1958).
6. D. M. Roy, R. Roy and E. F. Osborn, J. Amer. Ceram. Soc. 36, 185 (1953).
7. M. E. Melnichak, unpublished result quoted by J. L. Holm and O. J. Kleppa, Inorg, Chem. 8, 207 (1969).
8. P. Gross, Fulmer Res. Inst. Report R.163/19/April 1965, Stokes Poges, England.
9. V. I. Belousov, L. N. Sidorov, S. A. Komarov and P. A. Akishin, Zh. Fiz. Khim. 41, 2969 (1967); Russ. J. Phys. Chem. 41, 1598 (1967).
10. S. Cantor, J. Chem. Eng. Data 10, 237 (1965).

Beryllium Fluoride (BeF_2) Be_1F_2(l)

Enthalpy Reference Temperature = T_r = 298.15 K Standard State Pressure = p° = 0.1 MPa

T/K	$C_p°$ (J K⁻¹mol⁻¹)	S° (J K⁻¹mol⁻¹)	-[G°-H°(T_r)]/T (J K⁻¹mol⁻¹)	H°-H°(T_r) (kJ mol⁻¹)	$\Delta_fH°$ (kJ mol⁻¹)	$\Delta_fG°$ (kJ mol⁻¹)	Log K_f
0							
100							
200							
298.15	49.903	59.920	59.920	0.	-1022.048	-976.637	171.103
300	50.112	60.229	59.921	0.093	-1022.044	-976.355	169.998
400	59.208	75.979	62.006	5.589	-1021.608	-961.180	125.517
500	65.559	89.906	66.222	11.842	-1020.823	-946.157	98.844
600	70.551	102.314	71.222	18.655	-1019.751	-931.320	81.079
700	74.810	113.516	76.477	25.928	-1018.421	-916.684	68.404
800	78.647	123.761	81.756	33.604	-1016.851	-902.254	58.911
825.000	79.538	126.195	83.066	35.582	---------	I <--> LIQUID	--------
900	82.228	133.232	86.956	41.648	-1015.054	-888.035	51.540
1000	85.642	142.073	92.030	50.043	-1013.032	-874.028	45.655
1100	88.943	150.391	96.961	58.773	-1010.793	-860.234	40.849
1200	92.165	158.269	101.745	67.829	-1008.342	-846.654	36.854
1300	95.332	165.771	106.383	77.204	-1005.681	-833.286	33.482
1400	98.458	172.950	110.884	86.894	-1002.814	-820.131	30.599
1500	101.550	179.849	115.253	96.894	-999.744	-807.188	28.109
1600	104.621	186.501	119.499	107.203	-1011.010	-793.931	25.919
1700	107.671	192.935	123.631	117.818	-1007.139	-780.481	23.981
1800	110.709	199.175	127.655	128.737	-1002.999	-767.266	22.265
1900	113.734	205.242	131.580	139.959	-998.590	-754.289	20.737
2000	116.750	211.153	135.411	151.483	-993.909	-741.551	19.367

PREVIOUS: June 1964 CURRENT: June 1970

Beryllium Fluoride (BeF_2) Be_1F_2(l)

BERYLLIUM FLUORIDE (BeF_2) CRYSTAL - CRYSTAL - LIQUID M_r = 47.008986

0 to 500 K	crystal, low quartz
500 K to 825 K	crystal, high quartz
above 825 K	liquid

Refer to the individual tables for details.

Beryllium Fluoride (BeF_2) Be_1F_2(cr,l)

Enthalpy Reference Temperature = T_r = 298.15 K Standard State Pressure = p° = 0.1 MPa

	J K⁻¹mol⁻¹			kJ mol⁻¹			
T/K	C_p°	S°	$-[G^\circ-H^\circ(T_r)]/T$	$H^\circ-H^\circ(T_r)$	Δ_fH°	Δ_fG°	Log K_f
0	0.	0.	INFINITE	-8.468	-1024.465	-1024.465	INFINITE
100	21.004	14.753	90.253	-7.550	-1026.493	-1010.881	528.030
200	39.083	35.271	57.702	-4.486	-1026.935	-995.024	259.874
298.15	51.819	53.354	53.354	0.	-1026.754	-979.385	171.584
300	52.028	53.676	53.355	0.096	-1026.746	-979.091	170.475
400	62.522	70.077	55.522	5.822	-1026.081	-963.292	125.793
500	76.048	85.392	59.971	12.710	-1024.661	-947.738	99.009
500.000	76.048	85.832	59.971	12.930		II <--> I	
500.000	64.099	85.832	59.971	12.930		TRANSITION	
600	67.446	97.814	65.302	19.508	-1023.604	-932.474	81.179
700	70.793	108.463	70.720	26.420	-1022.636	-917.360	68.454
800	74.140	118.134	76.051	33.666	-1021.495	-902.396	58.920
825.000	74.977	120.428	77.362	35.530		I <--> LIQUID	
825.000	79.538	126.195	77.362	40.287		TRANSITION	
900	82.228	133.232	81.727	46.354	-1015.054	-888.035	51.540
1000	85.642	142.073	87.324	54.749	-1013.032	-874.028	45.655
1100	88.943	150.391	92.683	63.479	-1010.793	-860.234	40.849
1200	92.165	158.268	97.823	72.535	-1008.342	-846.653	36.854
1300	95.332	165.771	102.763	81.910	-1005.681	-833.286	33.482
1400	98.458	172.950	107.522	91.600	-1002.814	-820.131	30.599
1500	101.550	179.849	112.115	101.600	-999.744	-807.188	28.109
1600	104.621	186.501	116.558	111.909	-1011.010	-793.931	25.919
1700	107.671	192.935	120.862	122.524	-1007.139	-780.481	23.981
1800	110.709	199.175	125.041	133.443	-1002.999	-767.266	22.265
1900	113.734	205.242	129.103	144.665	-998.590	-754.289	20.737
2000	116.750	211.153	133.058	156.189	-993.909	-741.551	19.367

PREVIOUS: CURRENT: June 1970

Beryllium Fluoride (BeF_2) Be_1F_2(cr,l)

BERYLLIUM FLUORIDE (BeF_2) IDEAL GAS M_r = 47.008986

$\Delta_f H°$(0 K) = -796.19 ± 4.2 kJ mol^{-1}

S°(298.15 K) = 227.56 ± 1.3 J K^{-1} mol^{-1} $\Delta_f H°$(298.15 K) = -796.01 ± 4.2 kJ mol^{-1}

Vibrational Frequencies and Degeneracies

ν, cm^{-1}	ν, cm^{-1}	ν, cm^{-1}
[670](1)	345 (2)	1530 (1)

Ground State Quantum Weight: 1 σ = 2

Point Group: $D_{\infty h}$

Bond Distance: Be-F = 1.40 ± 0.03 Å

Bond Angle: F-Be-F = 180°

Rotational Constant: B_0 = [0.226357] cm^{-1}

Enthalpy of Formation

$\Delta_f H°$ is calculated from that of the crystal by means of $\Delta_f H°$(298.15 K) = 55.15 ± 0.1 kcal mol^{-1}. The selected value of $\Delta_{sub}H°$ is an average of those obtained by 3rd law analysis of vapor pressures measured by manometry (3), torsion effusion (4) and entrainment (8-10). Mass-spectrometric data (1, 2) are in reasonable agreement and so is the average of two Knudsen-effusion studies (6, 7). Several techniques (2, 4, 6) were used to show that the amount of dimer in the vapor is negligible up to at least 1000 K. JANAF analyses of the vaporization data are summarized as follows:

Source	Method	T/K	δS cal K^{-1} mol^{-1}	$\Delta_{sub}H°$(298.15 K), kcal mol^{-1} 2nd law	3rd law
(1) Efimenko (1968)	Mass Spectrometry	705- 819	---	56.14±0.43	---
(2) Belousov (1967)	Mass Spectrometry	815- 977	-1.0	54.51	55.44
(3) Cantor (1965)	Manometry	1146-1372	0.65±0.08	55.96±0.10	55.14
(4) Hildenbrand (1965)	Torsion Effusion				
	Cell 13	891- 942	4.2 ±0.4	58.87±0.34	54.98
	Cell 27	872- 907	0.9 ±0.3	55.82±0.22	54.98
	Cell 12	821- 921	-0.8 ±0.3	54.42±0.27	55.12
	Cell 14	840- 921	3.4 ±0.3	58.13±0.29	55.13
(5) Blauer (1965)	Torsion Effusion				
	Cells A & B	713- 795	6.2 ±0.9	58.07±0.66	53.44
(6) Greenbaum (1963)	Effusion	823-1053	2.4 ±0.6	56.75±0.59	54.50
(7) Khandamirova (1959)	Effusion	846- 950	3.4 ±2.4	58.82±2.20	55.75
(6) Greenbaum (1963)	Entrainment	1123-1223	1.5 ±3.0	55.88±3.50	54.17
(8) Ozhigov (1961)	Entrainment	1019-1241	1.4 ±0.8	56.64±0.94	55.04
(9) Sense (1958)	Entrainment	1075-1298	-0.06±0.43	55.11±0.51	55.18
(10) Novoselova (1958)	Entrainment	1095-1376	-2.1 ±0.5	52.64±0.56	55.25

Heat Capacity and Entropy

The bond length and angle are those derived by Akishin et al. (11) from electron-diffraction data for the vapor. Linearity of BeF_2 was confirmed by electric-deflection studies (12) of the vapor and by infrared studies (13) of matrix-isolated BeF_2. Infrared absorptions were found near 330 and 1540 cm^{-1} in neon matrices, leading to gas-phase estimates of 345 and 1555 cm^{-1} (13). The latter absorption was observed (14) at 1520 cm^{-1} in the infrared spectra of the vapor at 1000°C. We adopt a compromise of 1530 cm^{-1} and use this value in the valence-force-field approximation to estimate 670 cm^{-1} for the symmetric stretching mode.

References

1. J. Efimenko, J. Res. Nat. Bur. Stand. 72A, 75 (1968).
2. V. I. Belousov, L. N. Sidorov, S. A. Komarov and P. A. Akishin, Zh. Fiz. Khim. 41, 2969 (1967); Russ. J. Phys. Chem. 41, 1598 (1967).
3. S. Cantor, J. Chem. Eng. Data 10, 237 (1965).
4. D. L. Hildenbrand and L. P. Theard, J. Chem. Phys. 42, 3230 (1965).
5. J. A. Blauer, M. A. Greenbaum, and M. Farber, J. Phys. Chem. 69, 1069 (1965).
6. M. A. Greenbaum, J. N. Foster, M. L. Arin and M. Farber, J. Phys. Chem. 67, 36 (1963).
7. N. E. Khandamirova, A. M. Evseev, G. V. Pozharskaya, E. A. Borisov, An. N. Nesmeyanov and Ya. I. Gerasimov, Zh. Neorg. Khim. 4, 2192 (1959).
8. E. P. Ozhigov and A. I. Zatsarin, Tr. Dal'nevost. Filiala, Akad. Nauk SSSR, Ser. Khim. 1961, No. 5, 24.
9. K. A. Sense and R. W. Stone, J. Phys. Chem. 62, 453 (1958).
10. A. V. Novoselova, F. Sh. Muratov, L. P. Reshetnikova and I. V. Gordeev, Vestn. Mosk. Univ., Ser. Mat., Mekh., Astron., Fiz. Khim. 13, No., 6, 181 (1958).
11. P. A. Akishin, V. P. Spiridonov and G. A. Sobolev, Dokl. Akad. Nauk SSSR 118, 1134 (1958); Proc. Acad. Sci. USSR (Eng. Trans.), Phys. Chem. Sec. 118, 111 (1958).
12. L. Wharton, R. A. Berg, and W. Klemperer, J. Chem. Phys. 39, 2023 (1963); A. Buchler, J. L. Stauffer and W. Klemperer, J. Amer. Chem. Soc. 86, 1554 (1964).
13. A. Snelson, J. Phys. Chem. 72, 250 (1968); 70, 3208 (1966).
14. A. Buchler and W. Klemperer, J. Chem. Phys. 29, 121 (1958).

Beryllium Fluoride (BeF_2) Be_1F_2(g)

Enthalpy Reference Temperature = T_r = 298.15 K Standard State Pressure = p° = 0.1 MPa

T/K	$C_p°$ (J $K^{-1}mol^{-1}$)	S° (J $K^{-1}mol^{-1}$)	$-[G°-H°(T_r)]/T$ (J $K^{-1}mol^{-1}$)	$H°-H°(T_r)$ (kJ mol^{-1})	$\Delta_f H°$ (kJ mol^{-1})	$\Delta_f G°$ (kJ mol^{-1})	Log K_f
0	0.	0.	INFINITE	-10.942	-796.191	-796.191	INFINITE
100	32.054	185.038	264.782	-7.974	-796.169	-797.586	416.616
200	40.895	210.117	231.640	-4.305	-796.006	-799.064	208.694
250	44.021	219.594	228.306	-2.178	-795.974	-799.834	167.116
298.15	46.364	227.556	227.556	0.	-796.006	-800.576	140.258
300	46.444	227.843	227.556	0.086	-796.008	-800.604	139.398
350	48.393	235.153	228.129	2.458	-796.103	-801.363	119.597
400	50.020	241.725	229.425	4.920	-796.236	-802.106	104.744
450	51.407	247.698	231.128	7.456	-796.393	-802.830	93.190
500	52.606	253.178	233.063	10.057	-796.566	-803.536	83.945
600	54.553	262.950	237.250	15.420	-796.945	-804.895	70.072
700	56.034	271.475	241.543	20.953	-797.355	-806.188	60.159
800	57.168	279.035	245.766	26.615	-797.799	-807.420	52.719
900	58.044	285.821	249.846	32.378	-798.283	-808.595	46.930
1000	58.728	291.973	253.756	38.217	-798.816	-809.712	42.295
1100	59.270	297.597	257.490	44.118	-799.406	-810.774	38.500
1200	59.704	302.774	261.050	50.068	-800.061	-811.779	35.336
1300	60.055	307.567	264.447	56.056	-800.786	-812.726	32.656
1400	60.344	312.028	267.688	62.077	-801.589	-813.615	30.356
1500	60.582	316.200	270.784	68.124	-802.473	-814.444	28.361
1600	60.782	320.116	273.746	74.192	-817.979	-814.685	26.597
1700	60.951	323.807	276.584	80.279	-818.636	-814.459	25.025
1800	61.094	327.295	279.305	86.381	-819.313	-814.194	23.627
1900	61.217	330.601	281.918	92.497	-820.010	-813.891	22.375
2000	61.323	333.744	284.432	98.624	-820.727	-813.550	21.248
2100	61.415	336.738	286.852	104.761	-821.461	-813.174	20.227
2200	61.496	339.597	289.185	110.907	-822.210	-812.762	19.297
2300	61.566	342.332	291.437	117.060	-822.973	-812.315	18.448
2400	61.629	344.954	293.612	123.220	-823.745	-811.835	17.669
2500	61.684	347.471	295.717	129.386	-824.524	-811.323	16.952
2600	61.734	349.891	297.754	135.556	-825.307	-810.779	16.289
2700	61.778	352.222	299.728	141.732	-826.091	-810.206	15.674
2800	61.818	354.469	301.644	147.912	-1117.830	-803.380	14.987
2900	61.853	356.639	303.503	154.095	-1117.545	-792.155	14.268
3000	61.886	358.737	305.309	160.282	-1117.237	-780.940	13.597
3100	61.915	360.766	307.066	166.473	-1116.905	-769.736	12.970
3200	61.942	362.732	308.775	172.665	-1116.550	-758.542	12.382
3300	61.966	364.639	310.439	178.861	-1116.171	-747.360	11.830
3400	61.989	366.489	312.060	185.059	-1115.769	-736.190	11.310
3500	62.009	368.286	313.641	191.259	-1115.345	-725.032	10.821
3600	62.028	370.033	315.183	197.460	-1114.901	-713.886	10.358
3700	62.045	371.733	316.689	203.664	-1114.437	-702.753	9.921
3800	62.061	373.388	318.159	209.869	-1113.956	-691.633	9.507
3900	62.076	375.000	319.596	216.076	-1113.461	-680.526	9.115
4000	62.090	376.572	321.001	222.285	-1112.952	-669.431	8.742
4100	62.103	378.105	322.375	228.494	-1112.434	-658.350	8.387
4200	62.115	379.602	323.720	234.705	-1111.908	-647.281	8.050
4300	62.126	381.064	325.037	240.917	-1111.377	-636.225	7.729
4400	62.136	382.492	326.326	247.130	-1110.844	-625.181	7.422
4500	62.146	383.889	327.590	253.344	-1110.311	-614.149	7.129
4600	62.155	385.255	328.829	259.559	-1109.781	-603.129	6.849
4700	62.163	386.591	330.044	265.775	-1109.256	-592.120	6.581
4800	62.171	387.900	331.235	271.992	-1108.740	-581.123	6.324
4900	62.179	389.182	332.405	278.209	-1108.234	-570.136	6.078
5000	62.186	390.439	333.553	284.428	-1107.740	-559.160	5.842
5100	62.192	391.670	334.681	290.647	-1107.262	-548.193	5.615
5200	62.199	392.878	335.788	296.866	-1106.800	-537.235	5.397
5300	62.205	394.063	336.877	303.086	-1106.357	-526.286	5.187
5400	62.210	395.225	337.946	309.307	-1105.935	-515.346	4.985
5500	62.215	396.367	338.998	315.528	-1105.534	-504.413	4.791
5600	62.220	397.488	340.033	321.750	-1105.156	-493.486	4.603
5700	62.225	398.589	341.050	327.972	-1104.802	-482.567	4.422
5800	62.230	399.672	342.052	334.195	-1104.474	-471.654	4.248
5900	62.234	400.735	343.037	340.418	-1104.172	-460.745	4.079
6000	62.238	401.781	344.008	346.642	-1103.897	-449.843	3.916

PREVIOUS: June 1970 (1 atm) CURRENT: June 1970 (1 bar)

Beryllium Fluoride (BeF_2) Be_1F_2(g)

LITHIUM TETRAFLUOROBERYLLATE ($LiBeF_3$) CRYSTAL M_r = 72.98389

$S°(298.15\ K) = 89.24 \pm 4.2\ J\ K^{-1}\ mol^{-1}$

$T_{fus} = [650\ K]$

$\Delta_f H°(0\ K)$ = Unknown

$\Delta_f H°(298.15\ K) = -1651.8 \pm 6.3\ kJ\ mol^{-1}$

$\Delta_{fus} H° = [27.196]\ kJ\ mol^{-1}$

Enthalpy of Formation

Interpolation of the calorimetric enthalpies of mixing of Holm and Kleppa (1) yields $\Delta_r H° = -0.56 \pm 0.1\ kcal\ mol^{-1}$ for $LiF(\ell) + BeF_2(\ell) \rightarrow LiBeF_3(\ell)$ at 1135 K. Reduction with the estimated JANAF functions yields $\Delta_r H° = -3.07 \pm 1.0\ kcal\ mol^{-1}$ at 298.15 K for $LiF(cr) + BeF_2(\ell) \rightarrow LiBeF_3(cr)$, where $BeF_2(\ell)$ is the supercooled liquid. $\Delta_r H°(298.15\ K)$ for the latter reaction may also be calculated as $-2.77 \pm 0.2\ kcal\ mol^{-1}$ from the difference of two calorimetric reactions measured by Gross (2). These results at 298.15 K were $\Delta_r H° = -2.28 \pm 0.12\ kcal\ mol^{-1}$ for $LiF(cr) + LiBeF_3(cr) \rightarrow Li_2BeF_4(cr)$ and $\Delta_r H° = -5.06 \pm 0.06\ kcal\ mol^{-1}$ for $2LiF(cr) + BeF_2(\ell) \rightarrow Li_2BeF_4(cr)$. We adopt the high-temperature results of Holm and Kleppa since the entropy must be calculated by combination of $\Delta_r H°$ with $\Delta_r G°$ obtained at high temperature.

Heat Capacity and Entropy

Cp° is estimated from Cp° of $Li_2BeF_4(cr)$ minus Cp° of LiF(cr). Existing information leads to three values for S° at 298.15 K; namely 21.41 cal $K^{-1}\ mol^{-1}$ from $\Delta_r G°$ and $\Delta_r H°$ for mixing of molten LiF and BeF_2, 21.33 cal $K^{-1}\ mol^{-1}$ from $\Delta_r G° = 0$ at 552 K for disproportionation of $LiBeF_3(cr)$ to $Li_2BeF_4(cr)$ and $BeF_2(cr)$, and 21.28 cal $K^{-1}\ mol^{-1}$ estimated from S° of LiF(cr) and $BeF_2(cr)$. We adopt the intermediate value but emphasize that S°(298.15 K) is much more uncertain than indicated by the consistency of the three values. There is less uncertainty in S°(900 K) since it does not involve large contributions from the estimated values of Cp° and $\Delta_{fus}H°$.

Hitch and Baes (3) reported emf data and activity coefficients which yield $\Delta_r G° = -5.91\ kcal\ mol^{-1}$ at 900 K for $LiF(\ell) + BeF_2(\ell) \rightarrow LiBeF_3(\ell)$. Combining $\Delta_r G°$ with $\Delta_r H° = -1.05\ kcal\ mol^{-1}$ derived with JANAF functions from data of Holm and Kleppa (1), we obtain $\Delta_r S° = 5.4$ and $S°(LiBeF_3, \ell) = 64.46\ cal\ K^{-1}\ mol^{-1}$, both at 900 K. The corresponding values from the adopted tables are $\Delta_r G° = -5.84\ kcal\ mol^{-1}$ and $S°(LiBeF_3, \ell) = 64.38\ cal\ K^{-1}\ mol^{-1}$.

Disproportionation Data

No transitions are known between room temperature and T_{fus}, but $LiBeF_3(cr)$ does disproportionate into $Li_2BeF_4(cr)$ and BeF_2 (high quartz). Disproportionation occurs at 553 K according to the latest phase diagrams (4, 5) or at 573 K according to Roy et al. (6). The adopted tables yield $\Delta_r G° = 0$ at 552 K, but this temperature is very sensitive to the value selected for S°, which in turn depends on the estimates of Cp° and $\Delta_{fus}H°$.

Fusion Data

See the liquid table for details.

References

1. J. L. Holm and O. J. Kleppa, Inorg. Chem. 8, 207 (1969).
2. P. Gross, Fulmer Res. Inst. Report, R.163/23/May 1966 and R.163/22/Jan. 1966, Contract AF 61(052)-863; R.163/18/Dec. 1964, Contract AF 61(052)-447.
3. B. F. Hitch and C. F. Baes, Inorg. Chem. 8, 201 (1969); U. S. Atomic Energy Comm., ORNL-4257, 46 pp., 1968.
4. K. A. Romberger, J. Braunstein and R. E. Thoma, J. Phys. Chem., to be published.
5. R. Thoma, H. Insley, H. A. Friedman, and G. M. Hebert, J. Nucl. Materials 27, 166 (1968).
6. D. M. Roy, R. Roy and E. F. Osborn, J. Amer. Ceram. Soc. 37, 300 (1954).

Lithium Trifluoroberyllate ($LiBeF_3$) $Be_1F_3Li_1(cr)$

Enthalpy Reference Temperature = T_r = 298.15 K; Standard State Pressure = p° = 0.1 MPa

T/K	$C_p°$	S°	$-[G°-H°(T_r)]/T$	$H°-H°(T_r)$	$\Delta_f H°$	$\Delta_f G°$	Log K_f
	J K⁻¹mol⁻¹			kJ mol⁻¹			
0							
100							
200							
298.15	91.816	89.245	89.245	0.	-1651.843	-1576.273	276.156
300	92.048	89.813	89.246	0.170	-1651.836	-1575.804	274.372
400	104.600	118.013	93.007	10.002	-1651.275	-1550.535	202.479
500	117.152	142.702	100.522	21.090	-1653.272	-1525.148	159.331
600	129.704	165.171	109.450	33.433	-1651.389	-1499.682	130.559
650.000	135.980	175.801	114.147	40.075	------ CRYSTAL <--> LIQUID -----		
700	142.256	186.108	118.921	47.031	-1648.435	-1474.616	110.037
800	154.808	205.923	128.568	61.884	-1644.389	-1450.048	94.678
900	167.360	224.881	138.223	77.992	-1639.246	-1426.053	82.766
1000	179.912	243.164	147.808	95.356	-1632.986	-1402.690	73.269
1100	192.464	260.900	157.286	113.975	-1625.600	-1380.009	65.531
1200	205.016	278.185	166.644	133.849	-1617.079	-1358.049	59.114

PREVIOUS: June 1966 CURRENT: December 1971

Lithium Trifluoroberyllate ($LiBeF_3$) $Be_1F_3Li_1(cr)$

LITHIUM TETRAFLUOROBERYLLATE ($LiBeF_3$) LIQUID M_r = 72.948389

$S°(298.15\ K) = [111.681]\ J\ K^{-1}\ mol^{-1}$ $\Delta_f H°(298.15\ K) = [-1634.323]\ kJ\ mol^{-1}$

$T_{fus} = [650]\ K$ $\Delta_{fus}H° = [27.2]\ kJ\ mol^{-1}$

Enthalpy of Formation

$\Delta_f H°$ is calculated from that of the crystal by addition of $\Delta_{fus}H°$ and the difference in enthalpy, H°(650 K) - H°(298.15 K), between the crystal and liquid. Note that the result is based on direct measurements of the enthalpy of mixing of molten LiF and BeF_2 at 1135 K (1).

Heat Capacity and Entropy

Cp° is estimated on the basis of 7.6 cal $K^{-1}\ mol^{-1}$ by comparison with the liquid phases of LiF, BeF_2 and Li_2BeF_4. A glass transition is assumed at 400 K and below this temperature Cp° is obtained from that of the crystal. The entropy is calculated in a manner analogous to that used for $\Delta_f H°$. Note that the resulting value of S°(900) differs by only 0.08 cal $K^{-1}\ mol^{-1}$ from that based on measurements of $\Delta_r H°(1135)$ (1) and $\Delta_r G°(900)$ (2) for mixing of molten LiF and BeF_2.

Fusion Data

The composition corresponding to $LiBeF_3$ melts incongruently (3), starting with the eutectic composition of 53 mole percent BeF_2 at 636.6 ± 0.5 K and finishing at a liquidus temperature near 653 K. Just below the melting region, the solid consists of an equimolar mixture of Li_2BeF_4(cr) and BeF_2(high quartz). We arbitrarily adopt 650 K as the hypothetical, congruent melting point of metastable $LiBeF_3$. We estimate $\Delta_{fus}H°$ on the basis of $\Delta_{fus}S° = 2.0$ cal K^{-1} g-atom^{-1} by comparison with Li_2BeF_4.

Vaporization Data

Mass-spectrometric data (4, 5) for the LiF-BeF_2 system suggest that the vapor consists mainly of BeF_2(g), some $LiBeF_3$(g), and traces of other molecules.

References

1. J. L. Holm and O. J. Kleppa, Inorg. Chem. 8, 207 (1969).
2. B. F. Hitch and C. F. Baes, Inorg. Chem. 8, 201 (1969); U. S. Atomic Energy Comm., ORNL-4257, 46 pp., 1968.
3. K. A. Romberger, J. Braunstien and R. E. Thoma, J. Phys. Chem., to be published.
4. A. Buchler and J. L. Stauffer, "Thermodynamics, Proc. Symp., Vienna, 1965," 1, 271, Intl. At. Energy Agency, Vienna, 1966.
5. J. Berkowitz, and W. A. Chupka, Ann. New York Acad. Sci. 79, Art. 11, 1073 (1960).

Lithium Trifluoroberyllate ($LiBeF_3$) $Be_1F_3Li_1(l)$

Enthalpy Reference Temperature = T_r = 298.15 K Standard State Pressure = p° = 0.1 MPa

T/K	$C_p°$ (J K⁻¹mol⁻¹)	S°	$-[G°-H°(T_r)]/T$	$H°-H°(T_r)$ (kJ mol⁻¹)	$\Delta_f H°$	$\Delta_f G°$	Log K_f
0							
100							
200							
298.15	91.816	111.681	111.681	0.	-1634.323	-1565.442	274.259
300	92.048	112.249	111.683	0.170	-1634.315	-1565.014	272.493
400	104.600	140.449	115.443	10.002	-1633.755	-1541.988	201.363
400.001	104.600	140.449	115.443	10.003	GLASS <--> LIQUID		
400.001	158.992	140.449	115.443	10.003	TRANSITION		
500	158.992	175.927	124.124	25.902	-1630.939	-1519.428	158.734
600	158.992	204.915	135.247	41.801	-1625.501	-1497.640	130.381
650.000	158.992	217.641	141.102	49.750	CRYSTAL <--> LIQUID		
700	158.992	229.423	146.995	57.700	-1620.246	-1476.748	110.196
800	158.992	250.654	158.655	73.599	-1615.154	-1456.597	95.106
900	158.992	269.380	169.938	89.498	-1610.219	-1437.076	83.406
1000	158.992	286.132	180.734	105.398	-1605.424	-1418.096	74.074
1100	158.992	301.285	191.015	121.297	-1600.757	-1399.590	66.461
1200	158.992	315.119	200.789	137.196	-1596.211	-1381.503	60.135
1300	158.992	327.846	210.080	153.095	-1591.783	-1363.792	54.798
1400	158.992	339.628	218.918	168.994	-1587.468	-1346.417	50.235
1500	158.992	350.597	227.335	184.894	-1583.263	-1329.346	46.292
1600	158.992	360.859	235.363	200.793	-1593.702	-1312.027	42.833
1700	158.992	370.497	243.032	216.692	-1734.546	-1287.380	39.556
1800	158.992	379.585	250.368	232.591	-1729.437	-1261.224	36.600
1900	158.992	388.181	257.397	248.490	-1724.368	-1235.351	33.962
2000	158.992	396.337	264.142	264.390	-1719.337	-1209.743	31.595

PREVIOUS: CURRENT: December 1971

Lithium Trifluoroberyllate ($LiBeF_3$) $Be_1F_3Li_1(l)$

LITHIUM TRIFLUOROBERYLLATE ($LiBeF_3$) CRYSTAL–LIQUID M_r = 72.948389

0 to 650 K crystal
above 650 K liquid

Refer to the individual tables for details.

Lithium Trifluoroberyllate ($LiBeF_3$) $Be_1F_3Li_1$(cr,l)

Enthalpy Reference Temperature = T_r = 298.15 K   Standard State Pressure = p° = 0.1 MPa

T/K	C_p° (J K⁻¹mol⁻¹)	S°	−[G°−H°(T_r)]/T	H°−H°(T_r) (kJ mol⁻¹)	Δ_fH°	Δ_fG°	Log K_f
0							
100							
200							
298.15	91.816	89.245	89.245	0.	-1651.843	-1576.273	276.156
300	92.048	89.813	89.246	0.170	-1651.836	-1575.804	274.372
400	104.600	118.013	93.007	10.002	-1651.275	-1550.535	202.479
500	117.152	142.702	100.522	21.090	-1653.272	-1525.148	159.331
600	129.704	165.171	109.450	33.433	-1651.389	-1499.682	130.559
650.000	135.980	175.801	114.147	40.075	CRYSTAL <--> LIQUID TRANSITION		
650.000	158.992	217.641	114.147	67.271			
700	158.992	229.423	121.965	75.221	-1620.246	-1476.748	110.196
800	158.992	250.654	136.754	91.120	-1615.154	-1456.597	95.106
900	158.992	269.380	150.470	107.019	-1610.219	-1437.076	83.406
1000	158.992	286.132	163.214	122.918	-1605.424	-1418.096	74.074
1100	158.992	301.285	175.088	138.817	-1600.757	-1399.590	66.461
1200	158.992	315.119	186.189	154.717	-1596.211	-1381.503	60.135
1300	158.992	327.846	196.603	170.616	-1591.783	-1363.792	54.798
1400	158.992	339.628	206.403	186.515	-1587.468	-1346.417	50.235
1500	158.992	350.597	215.655	202.414	-1583.263	-1329.346	46.292
1600	158.992	360.859	224.413	218.313	-1593.702	-1312.027	42.833
1700	158.992	370.497	232.725	234.213	-1734.546	-1287.380	39.556
1800	158.992	379.585	240.634	250.112	-1729.437	-1261.224	36.600
1900	158.992	388.181	248.176	266.011	-1724.368	-1235.351	33.962
2000	158.992	396.337	255.382	281.910	-1719.337	-1209.743	31.595

PREVIOUS: CURRENT: December 1971

Lithium Trifluoroberyllate ($LiBeF_3$) $Be_1F_3Li_1$(cr,l)

LITHIUM TRIFLUOROBERYLLATE ($LiBeF_3$) IDEAL GAS M_r = 72.948389

Lithium Trifluoroberyllate ($LiBeF_3$) $Be_1F_3Li_1(g)$

$S°(298.15\ K) = [267.42]\ J\ K^{-1}\ mol^{-1}$

$\Delta_f H°(0\ K) = -880\ kJ\ mol^{-1}$
$\Delta_f H°(298.15\ K) = -887\ kJ\ mol^{-1}$

Vibrational Frequencies and Degeneracies

ν, cm^{-1}	ν, cm^{-1}
[1530] (1)	[485] (2)
[820] (2)	[800] (3)
[700] (1)	

Point Group: $[C_{3v}]$ σ = [3]
Bond Distances: Be-F = [1.40] Å Li-Be = [1.209] Å
Li-F = [1.85] Å
Bond Angles: F-Be F = [120]° Li-Be-F = [90]°
Product of the Moments of Inertia: $I_AI_BI_C = 2.1635 \times 10^{-114}\ g^3\ cm^6$

Enthalpy of Formation

$\Delta_f H°(298.15\ K)$ is calculated from $\Delta_r H°(900\ K) = 53\ kcal\ mol^{-1}$ for the reaction: $LiF(g) + BeF_2(g) = LiBeF_3(g)$ reported by Hildenbrand, et al (1).

Heat Capacity and Entropy

The molecular structure was assumed to be pyramidal with Li atom at the top and BeF_3 located at the bottom. Beryllium atom was taken at the center of the equilateral triangle formed by the three F atoms. The Be-F distance was estimated from that of BeF(g). The bond distance of Li-F was taken from Porter and Zeller (2). Vibrational frequencies were estimated by comparison with related fluorides. The principal moments of inertia are: $I_A = I_B = 10.7996 \times 10^{-39}$ and $I_C = 18.5503 \times 10^{-39}\ g\ cm^2$.

References

1. D. L. Hildenbrand, et. al., Aeronutronic, Division of Ford Motor Co., Publication No. U-1734, June 1962.
2. R. F. Porter and E. E. Zeller, J. Chem. Phys., 33, 858 (1960).

Enthalpy Reference Temperature = T_r = 298.15 K   Standard State Pressure = p° = 0.1 MPa

T/K	$C_p°$ (J K⁻¹mol⁻¹)	S° (J K⁻¹mol⁻¹)	$-[G°-H°(T_r)]/T$ (J K⁻¹mol⁻¹)	$H°-H°(T_r)$ (kJ mol⁻¹)	$\Delta_f H°$ (kJ mol⁻¹)	$\Delta_f G°$ (kJ mol⁻¹)	Log K_f
0	0.	0.	INFINITE	-12.459	-879.675	-879.675	INFINITE
100	34.100	219.860	311.078	-9.122	-881.233	-876.878	458.034
200	45.467	246.303	272.513	-5.242	-884.166	-871.430	227.594
250	53.629	257.324	268.384	-2.765	-885.662	-868.072	181.374
298.15	61.090	267.419	267.419	0.	-887.008	-864.560	151.467
300	61.360	267.798	267.420	0.113	-887.058	-864.421	150.509
350	68.105	277.777	268.193	3.354	-888.341	-860.545	128.429
400	73.789	287.253	269.988	6.906	-889.536	-856.492	111.846
450	78.517	296.226	272.410	10.717	-890.676	-852.292	98.932
500	82.442	304.708	275.219	14.744	-894.782	-847.661	88.555
600	88.434	320.301	281.458	23.306	-896.681	-838.052	72.959
700	92.663	334.269	288.022	32.373	-898.258	-828.152	61.797
800	95.721	346.853	294.603	41.800	-899.638	-818.041	53.413
900	97.986	358.264	301.052	51.491	-900.912	-807.764	46.881
1000	99.700	368.681	307.302	61.379	-902.128	-797.349	41.649
1100	101.024	378.248	313.323	71.418	-903.322	-786.813	37.363
1200	102.064	387.085	319.106	81.574	-904.518	-776.169	33.786
1300	102.895	395.288	324.655	91.824	-905.740	-765.424	30.755
1400	103.568	402.939	329.976	102.148	-907.000	-754.583	28.154
1500	104.121	410.104	335.082	112.533	-908.309	-743.651	25.896
1600	104.579	416.839	339.983	122.969	-924.211	-732.104	23.901
1700	104.963	423.191	344.693	133.447	-1070.477	-712.890	21.904
1800	105.288	429.200	349.222	143.960	-1070.753	-691.847	20.077
1900	105.566	434.900	353.583	154.503	-1071.041	-670.789	18.441
2000	105.804	440.321	357.785	165.071	-1071.340	-649.716	16.969
2100	106.011	445.488	361.839	175.662	-1071.650	-628.627	15.636
2200	106.190	450.424	365.755	186.273	-1071.970	-607.523	14.424
2300	106.348	455.148	369.539	196.900	-1072.298	-586.404	13.318
2400	106.487	459.677	373.201	207.542	-1072.631	-565.271	12.303
2500	106.610	464.027	376.748	218.197	-1072.967	-544.125	11.369
2600	106.719	468.210	380.186	228.863	-1073.303	-522.964	10.506
2700	106.817	472.240	383.521	239.540	-1073.638	-501.791	9.708
2800	106.905	476.126	386.759	250.226	-1364.926	-474.383	8.850
2900	106.984	479.879	389.906	260.921	-1364.188	-442.591	7.972
3000	107.055	483.507	392.966	271.623	-1363.426	-410.825	7.153
3100	107.120	487.018	395.944	282.332	-1362.642	-379.084	6.388
3200	107.179	490.420	398.843	293.047	-1361.835	-347.369	5.670
3300	107.232	493.719	401.668	303.767	-1361.006	-315.680	4.997
3400	107.281	496.921	404.423	314.493	-1360.159	-284.016	4.363
3500	107.326	500.032	407.111	325.223	-1359.295	-252.377	3.767
3600	107.368	503.056	409.734	335.958	-1358.417	-220.763	3.203
3700	107.406	505.998	412.296	346.697	-1357.527	-189.174	2.671
3800	107.441	508.863	414.800	357.439	-1356.630	-157.609	2.166
3900	107.473	511.654	417.248	368.185	-1355.728	-126.068	1.688
4000	107.503	514.375	419.642	378.934	-1354.827	-94.549	1.235
4100	107.531	517.030	421.985	389.685	-1353.929	-63.054	0.803
4200	107.557	519.622	424.279	400.440	-1353.040	-31.579	0.393
4300	107.581	522.153	426.526	411.197	-1352.164	-0.127	0.002
4400	107.604	524.626	428.727	421.956	-1351.284	31.304	-0.372
4500	107.625	527.045	430.885	432.718	-1350.437	62.717	-0.728
4600	107.645	529.411	433.002	443.481	-1349.618	94.111	-1.069
4700	107.663	531.726	435.078	454.247	-1348.828	125.489	-1.395
4800	107.681	533.993	437.115	465.014	-1348.071	156.849	-1.707
4900	107.697	536.213	439.115	475.783	-1347.353	188.194	-2.006
5000	107.712	538.389	441.078	486.553	-1346.677	219.524	-2.293
5100	107.727	540.522	443.007	497.325	-1346.041	250.842	-2.569
5200	107.741	542.614	444.903	508.099	-1345.460	282.148	-2.834
5300	107.754	544.667	446.766	518.873	-1344.934	313.443	-3.089
5400	107.766	546.681	448.598	529.649	-1344.464	344.728	-3.335
5500	107.777	548.658	450.399	540.426	-1344.057	376.005	-3.571
5600	107.788	550.600	452.171	551.205	-1343.713	407.276	-3.799
5700	107.799	552.508	453.915	561.984	-1343.438	438.541	-4.019
5800	107.808	554.383	455.631	572.764	-1343.233	469.802	-4.231
5900	107.818	556.226	457.320	583.546	-1343.101	501.060	-4.436
6000	107.827	558.038	458.984	594.328	-1343.046	532.316	-4.634

PREVIOUS: December 1962 (1 atm)   CURRENT: December 1962 (1 bar)

Lithium Trifluoroberyllate ($LiBeF_3$) $Be_1F_3Li_1(g)$

LITHIUM TETRAFLUOROBERYLLATE (Li_2BeF_4) CRYSTAL M_r = 98.887792

$\Delta_f H°(0\ K)$ = Unknown

$S°(298.15\ K) = 130.58 \pm 2.1\ J\ K^{-1}\ mol^{-1}$ $\Delta_f H°(298.15\ K) = -2273.6 \pm 5.0\ kJ\ mol^{-1}$

$T_{fus} = 732.2 \pm 0.2\ K$ $\Delta_{fus}H° = 43.995 \pm 2.09\ kJ\ mol^{-1}$

Enthalpy of Formation

Interpolation of the calorimetric enthalpies of mixing of Holm and Kleppa (1) yields $\Delta_r H° = -3.2 \pm 0.1\ kcal\ mol^{-1}$ for $2LiF(\ell) + BeF_2(\ell) \rightarrow Li_2BeF_4(\ell)$ at 1135 K. Reduction with the estimated JANAF functions yields $\Delta_r H° = -4.24 \pm 0.2\ kcal\ mol^{-1}$ at 298.15 K for $2LiF(cr) + BeF_2(\ell) \rightarrow Li_2BeF_4(cr)$, where $BeF_2(\ell)$ is the supercooled liquid. Gross (2) studied the latter reaction calorimetrically and obtained $\Delta_r H° = -5.05 \pm 0.06\ kcal\ mol^{-1}$ at 298.15 K. The difference of 0.8 kcal mol^{-1} between the two results is greater than the combined uncertainties but probably is not unreasonable in this case. We adopt the high-temperature results of Holm and Kleppa since the entropy must be calculated by combination of $\Delta_r H°$ with $\Delta_r G°$ obtained at high temperature

Heat Capacity and Entropy

Cp° is taken from the enthalpy study (323-873 K) of Douglas and Payne (3). Their sample was only 98.6% pure and extensive premelting corrections were necessary above 650 K. Thus, Cp° may be more uncertain than usual in this region. The premelting corrections could also be affected by new phase data which show that pure Li_2BeF_4 melts congruently at a temperature 12° lower than assumed in the premelting analysis. The magnitude of the adopted Cp° is confirmed by data by Levina et al. (4) for samples prepared by fusion.

Hitch and Baes (5) reported emf data and activity coefficients which yield $\Delta_r G° = -9.33\ kcal\ mol^{-1}$ at 900 K for $2LiF(\ell) + BeF_2(\ell) \rightarrow Li_2BeF_4(\ell)$. Combining $\Delta_r G°$ with $\Delta_r H° = -4.19\ kcal\ mol^{-1}$ derived from data of Holm and Kleppa (1), we obtain $\Delta_r S° = 5.71\ cal\ K^{-1}\ mol^{-1}$ at 900 K and $S° = 31.21 \pm 0.5\ cal\ K^{-1}\ mol^{-1}$ for $Li_2BeF_4(cr)$ at 298.15 K. This experimental entropy may be compared with $29.8 \pm 2\ cal\ K^{-1}\ mol^{-1}$ estimated by assuming $\Delta_r S° = 0$ for formation form LiF(cr) and BeF_2(cr).

Transition Data

Li_2BeF_4 has a hexagonal unit cell (space group R$\bar{3}$) isotypic with phenacite (6). This structure shows no phase transitions between room temperature and T_{fus} (3, 4, 7), but exposure to traces of moisture at elevated temperatures causes decomposition to LiF and BeO (7, 8). Metastable forms have been reported (4) but not confirmed or identified.

Fusion Data

New electrochemical measurements (9) of the liquidus in the LiF-BeF_2 system show that Li_2BeF_4 melts congruently at 459.1 ± 0.2°C with the eutectic point at 458.9 ± 0.2°C and 32.8 mole percent BeF_2. The close proximity of the eutectic explains previous conflicting reports of incongruent (8) and congruent (4, 8) melting. We adopt $T_{fus} = 732.2 \pm 0.2\ K$ and derive $\Delta_{fus}H° = 10.515\ kcal\ mol^{-1}$ from the difference in smoothed enthalpies for liquid and crystal (3). The uncertainty estimated for $\Delta_{fus}H°$ emphasized the nature of the premelting corrections.

References

1. J. L. Holm and O. J. Kleppa, Inorg. Chem. 8, 207 (1969).
2. P. Gross, Fulmer Res. inst. Report R.163/23/May 1966 and R.163/22/Jan. 1966, Contract AF 61(052)-863; R.163/18/Dec. 1964, Contract AF 61(052)-447.
3. T. B. Douglas and W. H. Payne, J. Res. Nat. Bur. Stand. 73A, 479 (1969).
4. M. W. Levina, V. I. Kalitin and V. T. Kalinnikov, Vestnik Moskov. Univ., Khim. Ser. II 16, No. 4, 43 (1961).
5. B. F. Hitch and C. F. Baes, Inorg. Chem. 8, 201 (1969); U. S. Atomic Energy Comm., ORNL-4257, 46 pp., 1968.
6. J. H. Burns and E. K. Gordon, Acta Cryst. 20, 135 (1966).
7. T. Hahn, H. Bielen, W. Eysel and F. Weber, Chem. Erde 22, 175 (1962).
8. R. Thoma, H. Insley, H. A. Friedman, and G. M. Hebert, J. Nucl. Materials 27, 166 (1968).
9. K. A. Romberger, J. Braunstein and R. E. Thoma, J. Phys. Chem., to be published.

Lithium Tetrafluoroberyllate (Li_2BeF_4) $Be_1F_4Li_2(cr)$

Enthalpy Reference Temperature = T_r = 298.15 K Standard State Pressure = p° = 0.1 MPa

T/K	$C_p°$ (J K⁻¹mol⁻¹)	S° (J K⁻¹mol⁻¹)	−[G°−H°(T_r)]/T (J K⁻¹mol⁻¹)	H°−H°(T_r) (kJ mol⁻¹)	$\Delta_f H°$ (kJ mol⁻¹)	$\Delta_f G°$ (kJ mol⁻¹)	Log K_f
0							
100							
200							
298.15	135.269	130.583	130.583	0.	-2273.586	-2171.438	380.427
300	135.520	131.420	130.585	0.250	-2273.573	-2170.804	377.971
400	150.457	172.453	136.079	14.550	-2272.755	-2136.665	279.020
500	165.352	207.627	146.947	30.340	-2277.665	-2102.147	219 610
600	180.247	239.090	159.724	47.620	-2275.565	-2067.220	179.967
700	195.184	267.997	173.152	66.391	-2272.139	-2032.747	151.685
732.200	199.995	276.882	177.519	72.753	------ CRYSTAL <--> LIQUID -----		
800	210.100	295.034	186.715	86.656	-2267.383	-1998.858	130.512
900	225.004	320.641	200.184	108.411	-2261.310	-1965.643	114.083
1000	239.911	345.119	213.462	131.657	-2253.900	-1933.176	100.979
1100	254.819	368.684	226.509	156.393	-2245.137	-1901.516	90.295
1200	269.729	391.495	239.312	182.620	-2235.011	-1870.714	81.430
1300	284.640	413.674	251.875	210.339	-2223.521	-1840.813	73.965
1400	299.554	435.314	264.208	239.549	-2210.657	-1811.848	67.601
1500	314.469	456.490	276.323	270.250	-2196.413	-1783.850	62.119

PREVIOUS: June 1966 CURRENT: December 1971

Lithium Tetrafluoroberyllate (Li_2BeF_4) $Be_1F_4Li_2(cr)$

LITHIUM TETRAFLUOROBERYLLATE (Li_2BeF_4) LIQUID M_r = 98.887792

S°(298.15 K) = [171.395] J K^{-1} mol^{-1}

T_{fus} = 732.2 ± 0.2 K

$\Delta_f H°$(298.15 K) = [-2241.068] kJ mol^{-1}

$\Delta_{fus}H°$ = 43.995 ± 2.09 kJ mol^{-1}

Enthalpy of Formation

$\Delta_f H°$ is calculated from that of the crystal by addition of $\Delta_{fus}H°$ and the difference in enthalpy, H°(732.2 K) - H°(298.15 K), between the crystal and liquid. Note that the result is based on direct measurements of the enthalpy of mixing of molten LiF and BeF_2 at 1135 K (1).

Heat Capacity and Entropy

The constant Cp° is derived from enthalpy data (743-873 K) of Douglas and Payne (2). A glass transition is assumed at 500 K and below this temperature Cp° is obtained from that of the crystal. The entropy is calculated in a manner analogous to that used for $\Delta_f H°$. Note that the result is based on measurements of $\Delta_r H°$(1135 K) (1) and $\Delta_r G°$(900 K) (3) for mixing of molten LiF and BeF_2.

Fusion Data

See the crystal table for details.

Vaporization Data

Mass spectra of vapor effusing from twin-crucible Knudsen cells (4) suggest that the vapor consists of BeF_2, $LiBeF_3$ and traces of other molecules.

References

1. J. L. Holm and O. J. Kleppa, Inorg. Chem. 8, 207 (1969).
2. T. B. Douglas and W. H. Payne, J. Res. Nat. Bur. Stand. 73A, 479 (1969).
3. B. F. Hitch and C. F. Baes, Inorg. Chem. 8, 201 (1969); U. S. Atomic Energy Comm., ORNL-4257, 46 pp., 1968.
4. A. Buchler and J. L. Stauffer, "Thermodynamics, Proc. Symp., Vienna, 1965," 1, 271, Intl. At. Energy Agency, Vienna, 1966.

Lithium Tetrafluoroberyllate (Li_2BeF_4) $Be_1F_4Li_2(l)$

Enthalpy Reference Temperature = T_r = 298.15 K Standard State Pressure = p° = 0.1 MPa

T/K	$C_p°$ (J $K^{-1}mol^{-1}$)	S° (J $K^{-1}mol^{-1}$)	-[G°-H°(T_r)]/T (J $K^{-1}mol^{-1}$)	H°-H°(T_r) (kJ mol^{-1})	$\Delta_f H°$ (kJ mol^{-1})	$\Delta_f G°$ (kJ mol^{-1})	Log K_f
0							
100							
200							
298.15	135.269	171.395	171.395	0.	-2241.068	-2151.088	376.862
300	135.520	172.232	171.397	0.250	-2241.055	-2150.530	374.441
400	150.457	213.265	176.891	14.550	-2240.238	-2120.472	276.905
500	165.352	248.440	187.759	30.340	-2245.147	-2090.036	218.345
500.001	165.352	248.440	187.759	30.341	GLASS <--> LIQUID		
500.001	232.086	248.440	187.759	30.341	TRANSITION		
600	232.086	290.754	201.506	53.549	-2237.119	-2059.771	179.319
700	232.086	326.530	216.877	76.758	-2229.255	-2030.837	151.543
732.200	232.086	336.968	221.930	84.231	CRYSTAL <--> LIQUID		
800	232.086	357.521	232.563	99.966	-2221.555	-2003.019	130.784
900	232.086	384.857	247.996	123.175	-2214.029	-1976.156	114.693
1000	232.086	409.310	262.926	146.384	-2206.655	-1950.122	101.864
1100	232.086	431.430	277.255	169.592	-2199.420	-1924.820	91.402
1200	232.086	451.624	290.957	192.801	-2192.313	-1900.171	82.712
1300	232.086	470.201	304.040	216.010	-2185.333	-1876.110	75.383
1400	232.086	487.400	316.530	239.218	-2178.470	-1852.582	69.121
1500	232.086	503.413	328.461	262.427	-2171.718	-1829.540	63.710
1600	232.086	518.391	339.869	285.635	-2179.611	-1806.420	58.974
1700	232.086	532.461	350.788	308.844	-2463.146	-1768.956	54.353
1800	232.086	545.727	361.253	332.053	-2454.737	-1728.364	50.156
1900	232.086	558.275	371.296	355.261	-2446.377	-1688.238	46.413
2000	232.086	570.180	380.945	378.470	-2438.061	-1648.551	43.056
2100	232.086	581.503	390.228	401.679	-2429.783	-1609.280	40.029
2200	232.086	592.300	399.169	424.887	-2421.540	-1570.402	37.286
2300	232.086	602.617	407.792	448.096	-2413.325	-1531.897	34.790
2400	232.086	612.494	416.117	471.305	-2405.134	-1493.750	32.511
2500	232.086	621.968	424.163	494.513	-2396.963	-1455.944	30.420

PREVIOUS: June 1966 CURRENT: December 1971

Lithium Tetrafluoroberyllate (Li_2BeF_4) $Be_1F_4Li_2(l)$

LITHIUM TETRAFLUOROBERYLLATE (Li_2BeF_4) CRYSTAL-LIQUID M_r = 98.887792

0 to 732.2 K	crystal
above 732.2 K	liquid

Refer to the individual tables for details.

Lithium Tetrafluoroberyllate (Li_2BeF_4) $Be_1F_4Li_2$(cr,l)

Enthalpy Reference Temperature = T_r = 298.15 K   Standard State Pressure = p° = 0.1 MPa

	J K⁻¹mol⁻¹			kJ mol⁻¹			
T/K	C_p°	S°	$-[G^\circ-H^\circ(T_r)]/T$	$H^\circ-H^\circ(T_r)$	Δ_fH°	Δ_fG°	Log K_f
0							
100							
200							
298.15	135.269	130.583	130.583	0.	-2273.586	-2171.438	380.427
300	135.520	131.420	130.585	0.250	-2273.573	-2170.804	377.971
400	150.457	172.453	136.079	14.550	-2272.755	-2136.665	279.020
500	165.352	207.627	146.947	30.340	-2277.665	-2102.147	219.610
600	180.247	239.090	159.724	47.620	-2275.565	-2067.220	179.967
700	195.184	267.997	173.152	66.391	-2272.139	-2032.747	151.685
732.200	199.995	276.882	177.519	72.753	CRYSTAL <--> LIQUID		
732.200	232.086	336.968	177.519	116.748	TRANSITION		
800	232.086	357.521	191.917	132.484	-2221.555	-2003.019	130.784
900	232.086	384.857	211.866	155.692	-2214.029	-1976.156	114.693
1000	232.086	409.310	230.409	178.901	-2206.655	-1950.122	101.864
1100	232.086	431.430	247.694	202.110	-2199.420	-1924.820	91.402
1200	232.086	451.624	263.859	225.318	-2192.313	-1900.171	82.712
1300	232.086	470.201	279.026	248.527	-2185.333	-1876.110	75.383
1400	232.086	487.400	293.304	271.736	-2178.470	-1852.582	69.121
1500	232.086	503.413	306.783	294.944	-2171.718	-1829.540	63.710
1600	232.086	518.391	319.546	318.153	-2179.611	-1806.420	58.974
1700	232.086	532.461	331.660	341.361	-2463.146	-1768.956	54.353
1800	232.086	545.727	343.188	364.570	-2454.737	-1728.364	50.156
1900	232.086	558.275	354.181	387.779	-2446.377	-1688.238	46.413
2000	232.086	570.180	364.686	410.987	-2438.061	-1648.551	43.056
2100	232.086	581.503	374.743	434.196	-2429.783	-1609.280	40.029
2200	232.086	592.300	384.389	457.405	-2421.540	-1570.402	37.286
2300	232.086	602.617	393.654	480.613	-2413.325	-1531.897	34.790
2400	232.086	612.494	402.568	503.822	-2405.134	-1493.750	32.511
2500	232.086	621.968	411.156	527.031	-2396.963	-1455.944	30.420

PREVIOUS: CURRENT: December 1971

Lithium Tetrafluoroberyllate (Li_2BeF_4) $Be_1F_4Li_2$(cr,l)

BERYLLIUM HYDRIDE (BeH) IDEAL GAS M_r = 10.02012

$S°(298.15 K) = 176.822 J K^{-1} mol^{-1}$

$\Delta_f H°(0 K) = 319 \pm 29 kJ mol^{-1}$
$\Delta_f H°(298.15 K) = 321 \pm 29 kJ mol^{-1}$

Electronic Levels and Quantum Weight

State	ϵ_i, cm^{-1}	g_i
$^2\Sigma^+$	0	2

$\omega_e = 2058.6 cm^{-1}$ $\omega_e x_e = 35.5 cm^{-1}$ $\sigma = 1$
$B_e = 10.308 cm^{-1}$ $\alpha_e = 0.300 cm^{-1}$ $r_e = 1.3431 Å$

Enthalpy of Formation

Gaydon (1) selected a value of 2.3 ± 0.3 eV (53 ± 7 kcal mol^{-1}) for D_0^0. Herzberg (2) estimated D_0^0 as being 2.2 eV (50.7 kcal mol^{-1}). Gaydon's (1) value was used in this table.

Heat Capacity and Entropy

Molecular constants were found in Herzberg (2).

References

1. Gaydon, "Dissociation Energies and Spectra of Diatomic Molecules", Chapman and Hall Ltd., London, 1953.
2. Herzberg, "Molecular Spectra and Molecular Structure I. Spectra of Diatomic Molecules", D. Van Nostrand Co. Inc., New York, 1950.

Beryllium Hydride (BeH) $Be_1H_1(g)$

Enthalpy Reference Temperature = T_r = 298.15 K   Standard State Pressure = p° = 0.1 MPa

T/K	$C_p°$ (J $K^{-1}mol^{-1}$)	S°	$-[G°-H°(T_r)]/T$	$H°-H°(T_r)$ (kJ mol^{-1})	$\Delta_f H°$	$\Delta_f G°$	Log K_f
0	0.	0.	INFINITE	-8.648	318.716	318.716	INFINITE
100	29.128	144.978	202.751	-5.777	320.047	310.636	-162.260
200	29.149	165.173	179.492	-2.864	321.036	300.778	-78.555
250	29.171	171.680	177.303	-1.406	321.204	295.689	-61.781
298.15	29.224	176.822	176.822	0.	321.197	290.773	-50.942
300	29.227	177.003	176.822	0.054	321.194	290.585	-50.595
350	29.344	181.516	177.179	1.518	321.056	285.493	-42.607
400	29.538	185.446	177.972	2.990	320.835	280.426	-36.620
450	29.809	188.940	179.000	4.473	320.561	275.391	-31.967
500	30.144	192.098	180.154	5.972	320.253	270.389	-28.247
600	30.932	197.661	182.620	9.025	319.574	260.479	-22.677
700	31.769	202.492	185.122	12.160	318.849	250.686	-18.706
800	32.570	206.788	187.566	15.377	318.093	241.000	-15.736
900	33.296	210.667	189.921	18.671	317.307	231.410	-13.431
1000	33.937	214.209	192.175	22.033	316.485	221.910	-11.591
1100	34.496	217.470	194.328	25.456	315.617	212.494	-10.090
1200	34.982	220.493	196.384	28.930	314.694	203.159	-8.843
1300	35.404	223.310	198.348	32.450	313.708	193.904	-7.791
1400	35.774	225.947	200.227	36.009	312.653	184.728	-6.892
1500	36.098	228.427	202.025	39.603	311.522	175.629	-6.116
1600	36.385	230.766	203.749	43.228	295.776	167.134	-5.456
1700	36.641	232.980	205.404	46.879	294.882	159.121	-4.889
1800	36.871	235.081	206.994	50.555	293.972	151.162	-4.387
1900	37.079	237.080	208.526	54.253	293.042	143.253	-3.938
2000	37.269	238.987	210.001	57.970	292.092	135.394	-3.536
2100	37.443	240.809	211.425	61.706	291.120	127.583	-3.173
2200	37.603	242.555	212.801	65.458	290.128	119.819	-2.845
2300	37.752	244.230	214.131	69.226	289.113	112.100	-2.546
2400	37.892	245.839	215.419	73.008	288.076	104.426	-2.273
2500	38.023	247.389	216.667	76.804	287.017	96.795	-2.022
2600	38.146	248.883	217.878	80.613	285.935	89.208	-1.792
2700	38.264	250.324	219.053	84.433	284.830	81.662	-1.580
2800	38.376	251.718	220.195	88.265	-7.255	80.381	-1.500
2900	38.483	253.067	221.305	92.108	-7.341	83.512	-1.504
3000	38.585	254.373	222.386	95.962	-7.434	86.646	-1.509
3100	38.684	255.640	223.438	99.825	-7.534	89.784	-1.513
3200	38.780	256.869	224.464	103.698	-7.644	92.925	-1.517
3300	38.872	258.064	225.464	107.581	-7.764	96.070	-1.521
3400	38.962	259.226	226.440	111.473	-7.896	99.218	-1.524
3500	39.049	260.357	227.393	115.373	-8.042	102.371	-1.528
3600	39.135	261.458	228.324	119.283	-8.204	105.527	-1.531
3700	39.218	262.531	229.234	123.200	-8.384	108.689	-1.534
3800	39.300	263.578	230.124	127.126	-8.584	111.856	-1.538
3900	39.380	264.600	230.995	131.060	-8.806	115.028	-1.541
4000	39.459	265.598	231.848	135.002	-9.053	118.206	-1.544
4100	39.536	266.573	232.683	138.952	-9.327	121.391	-1.547
4200	39.613	267.527	233.501	142.909	-9.629	124.583	-1.549
4300	39.688	268.460	234.303	146.874	-9.962	127.783	-1.552
4400	39.762	269.373	235.090	150.847	-10.329	130.990	-1.555
4500	39.835	270.268	235.862	154.827	-10.730	134.206	-1.558
4600	39.908	271.144	236.619	158.814	-11.168	137.432	-1.561
4700	39.980	272.003	237.363	162.808	-11.644	140.667	-1.563
4800	40.051	272.846	238.093	166.810	-12.161	143.914	-1.566
4900	40.121	273.672	238.811	170.819	-12.719	147.171	-1.569
5000	40.191	274.483	239.517	174.834	-13.319	150.440	-1.572
5100	40.261	275.280	240.210	178.857	-13.964	153.721	-1.574
5200	40.329	276.062	240.892	182.886	-14.654	157.016	-1.577
5300	40.398	276.831	241.563	186.923	-15.389	160.324	-1.580
5400	40.466	277.587	242.223	190.966	-16.170	163.647	-1.583
5500	40.533	278.330	242.873	195.016	-16.998	166.984	-1.586
5600	40.600	279.061	243.512	199.072	-17.873	170.337	-1.589
5700	40.667	279.780	244.142	203.136	-18.794	173.706	-1.592
5800	40.733	280.488	244.763	207.206	-19.762	177.092	-1.595
5900	40.800	281.185	245.374	211.282	-20.777	180.494	-1.598
6000	40.865	281.871	245.977	215.366	-21.837	183.915	-1.601

PREVIOUS: March 1963 (1 atm)   CURRENT March 1963 (1 bar)

Beryllium Hydride (BeH) $Be_1H_1(g)$

BERYLLIUM HYDRIDE, ION (BeH^+) IDEAL GAS M_r = 10.019571

$S°(298.15\ K) = 170.65\ J\ K^{-1}\ mol^{-1}$

$\Delta_f H°(0\ K) = 1147.8 \pm 42\ kJ\ mol^{-1}$
$\Delta_f H°(298.15\ K) = 1156.5 \pm 42\ kJ\ mol^{-1}$

Electronic Levels and Quantum Weights

State	ε_i, cm^{-1}	g_i
$^1\Sigma$	0	1
$^3\Sigma$	[20000]	[3]

$\omega_e = 2221.7\ cm^{-1}$ $\omega_e x_e = 39.79\ cm^{-1}$ $\sigma = 1$
$B_e = 10.7996\ cm^{-1}$ $\alpha_e = 0.2935\ cm^{-1}$ $r_e = 1.31216$ Å

Enthalpy of Formation

The enthalpy of formation is calculated from the equation $BeH(g) \rightarrow BeH^+(g) + e^-$ with the JANAF auxiliary data for BeH(g) (1) using an ionization potential of 8.6 ± 0.4 eV (198.34 ± 9.23 kcal mol^{-1}) obtained from Beckett (2).

The dissociation energy, $D_0^0(Be^+\text{-}H)$ = 3.2 ± 0.2 eV (73.8 kcal mol^{-1}), has been reported by Gaydon (3). This yields $\Delta_f H°(BeH^+, g, 0\ K)$ = 270.0 kcal mol^{-1} which is in agreement with the value adopted.

Heat Capacity and Entropy

The molecular constants were obtained from the ultraviolet spectroscopic studies by Watson and Humphrey (4). The ground state configuration was given by Herzberg (5). The estimated electronic level and quantum weight were obtained by assuming a $^3\Sigma$ level lies below the first observed excited state ($^1\Sigma$) at 40,000 cm^{-1}.

References

1. JANAF Thermochemical Tables: BeH(g), 3-31-63.
2. C. W. Beckett, NBS Report 8628, Jan. 1965.
3. A. G. Gaydon, "Dissociation Energies", 2nd. Ed., Chapman and Hall Ltd., London, 1953.
4. W. W. Watson and R. F. Humphreys, Phys. Rev. 52, 318 (1937).
5. G. Herzberg, "Spectra of Diatomic Molecules", D. Van Nostrand Company, New York, 1950.

Beryllium Hydride, Ion (BeH^+) $Be_1H_1^+(g)$

Enthalpy Reference Temperature = T_r = 298.15 K Standard State Pressure = p° = 0.1 MPa

T/K	$C_p°$ (J K⁻¹mol⁻¹)	S° (J K⁻¹mol⁻¹)	$-[G°-H°(T_r)]/T$ (J K⁻¹mol⁻¹)	$H°-H°(T_r)$ (kJ mol⁻¹)	$\Delta_f H°$ (kJ mol⁻¹)	$\Delta_f G°$ (kJ mol⁻¹)	Log K_f
0	0.	0.	INFINITE	-8.644	1147.782		
100	29.126	138.816	196.573	-5.776			
200	29.144	159.009	173.321	-2.862			
250	29.159	165.514	171.133	-1.405			
298.15	29.193	170.653	170.653	0.	1156.458	1121.618	-196.503
300	29.195	170.833	170.653	0.054	1156.493	1121.402	-195.253
350	29.274	175.339	171.009	1.515	1157.391	1115.481	-166.476
400	29.413	179.256	171.800	2.982	1158.205	1109.438	-144.878
450	29.621	182.732	172.825	4.458	1158.963	1103.296	-128.067
500	29.892	185.866	173.975	5.946	1159.683	1097.072	-114.610
600	30.571	191.374	176.428	8.968	1161.052	1084.419	-94.407
700	31.339	196.144	178.911	12.063	1162.365	1071.543	-79.960
800	32.105	200.379	181.335	15.235	1163.643	1058.481	-69.112
900	32.823	204.203	183.667	18.482	1164.889	1045.260	-60.665
1000	33.471	207.695	185.898	21.797	1166.098	1031.903	-53.901
1100	34.046	210.913	188.027	25.174	1167.262	1018.427	-48.361
1200	34.551	213.897	190.060	28.604	1168.374	1004.847	-43.740
1300	34.994	216.681	192.002	32.082	1169.425	991.177	-39.826
1400	35.384	219.289	193.859	35.601	1170.408	977.428	-36.468
1500	35.728	221.742	195.637	39.157	1171.318	963.611	-33.556
1600	36.033	224.057	197.341	42.746	1157.614	950.262	-31.023
1700	36.304	226.250	198.978	46.363	1158.765	937.267	-28.799
1800	36.549	228.332	200.552	50.006	1159.900	924.205	-26.820
1900	36.770	230.314	202.066	53.672	1161.017	911.081	-25.047
2000	36.971	232.206	203.526	57.359	1162.115	897.898	-23.451
2100	37.156	234.014	204.935	61.065	1163.193	884.660	-22.005
2200	37.328	235.747	206.297	64.790	1164.251	871.372	-20.689
2300	37.489	237.409	207.614	68.531	1165.288	858.036	-19.487
2400	37.642	239.008	208.889	72.287	1166.304	844.655	-18.383
2500	37.788	240.548	210.124	76.059	1167.299	831.233	-17.368
2600	37.930	242.033	211.323	79.845	1168.273	817.771	-16.429
2700	38.069	243.467	212.487	83.645	1169.226	804.272	-15.560
2800	38.207	244.854	213.619	87.458	879.202	796.960	-14.867
2900	38.345	246.197	214.719	91.286	881.179	793.988	-14.301
3000	38.484	247.499	215.790	95.127	883.153	790.948	-13.772
3100	38.627	248.763	216.833	98.983	885.123	787.842	-13.275
3200	38.773	249.992	217.851	102.853	887.089	784.672	-12.808
3300	38.924	251.187	218.843	106.738	889.050	781.442	-12.369
3400	39.080	252.352	219.811	110.638	891.005	778.151	-11.955
3500	39.242	253.487	220.757	114.554	892.953	774.804	-11.563
3600	39.410	254.595	221.682	118.486	894.893	771.401	-11.193
3700	39.586	255.677	222.586	122.436	896.823	767.944	-10.841
3800	39.768	256.735	223.471	126.404	898.743	764.435	-10.508
3900	39.958	257.770	224.337	130.390	900.652	760.875	-10.191
4000	40.154	258.784	225.186	134.395	902.547	757.267	-9.889
4100	40.358	259.778	226.017	138.421	904.428	753.612	-9.601
4200	40.568	260.754	226.833	142.467	906.293	749.911	-9.327
4300	40.785	261.711	227.633	146.535	908.141	746.166	-9.064
4400	41.008	262.651	228.418	150.624	909.970	742.378	-8.813
4500	41.237	263.575	229.189	154.737	911.780	738.548	-8.573
4600	41.472	264.484	229.946	158.872	913.569	734.679	-8.343
4700	41.712	265.378	230.691	163.031	915.336	730.771	-8.122
4800	41.956	266.259	231.423	167.214	917.080	726.826	-7.909
4900	42.204	267.127	232.142	171.422	918.800	722.844	-7.706
5000	42.455	267.982	232.851	175.655	920.495	718.828	-7.510
5100	42.709	268.825	233.548	179.913	922.164	714.778	-7.321
5200	42.966	269.657	234.234	184.197	923.807	710.695	-7.139
5300	43.224	270.478	234.910	188.507	925.424	706.581	-6.964
5400	43.484	271.288	235.577	192.842	927.013	702.437	-6.795
5500	43.744	272.088	236.233	197.204	928.576	698.264	-6.632
5600	44.004	272.879	236.881	201.591	930.110	694.063	-6.474
5700	44.264	273.660	237.519	206.004	931.618	689.834	-6.322
5800	44.522	274.432	238.149	210.444	933.097	685.580	-6.174
5900	44.779	275.195	238.770	214.909	934.550	681.299	-6.032
6000	45.035	275.950	239.384	219.399	935.976	676.995	-5.894

PREVIOUS: September 1966 (1 atm) CURRENT: September 1966 (1 bar)

BERYLLIUM HYDROXIDE (BeOH)　　IDEAL GAS　　M_r = 26.01952

$S°(298.15\ K) = [209.59 \pm 12.6]\ J\ K^{-1}\ mol^{-1}$

$\Delta_f H°(0\ K) = -113.7 \pm 37.7\ kJ\ mol^{-1}$
$\Delta_f H°(298.15\ K) = -114.6 \pm 37.7\ kJ\ mol^{-1}$

Electronic Levels and Quantum Weights

ϵ_i, cm	g_i
0	[2]
[30000]	[4]

Point Group: $[C_{\infty v}]$　　σ = 1
Bond Distances: Be-O = [1.38] Å　　O-H = [0.96] Å
Bond Angle: Be-O-H = [180]°
Rotational Constant: $B_0 = [1.29312]\ cm^{-1}$

Vibrational Frequencies and Degeneracies

ν, cm^{-1}
[1266](1)
[606](2)
[3650](1)

Enthalpy of Formation

The adopted $\Delta_f H°(298.15\ K) = -27.4 \pm 10\ kcal\ mol^{-1}$, which corresponds to $D_0^0(Be\text{-}OH) = 113.9 \pm 10\ kcal\ mol^{-1}$, is derived from 3rd law analysis of Knudsen-cell mass-spectrometric studies and Knudsen-cell weight-loss studies of several BeOH(g) producing reactions (1-3). The earlier Knudsen-cell mass-spectrometric measurement by Hildenbrand et al. (4) had indicated a minimum absolute value of $\Delta_f H°(BeOH, g, 298.15\ K) = -25 \pm 10\ kcal\ mol^{-1}$.

Source	Reaction	Data Points	T/K	$\Delta_r H°(298.15\ K)$, 3rd law kcal mol^{-1}	Drift cal K^{-1} mol^{-1}	$\Delta_f H°(298.15\ K)$ kcal mol^{-1}
Inami and Ju (1)	A	3	2422-2485	9.73±0.93	-13.0 ± 2.0	-27.2
	B	3	2422-2485	0.82±0.59	6.2 ± 5.5	-37.3
	C	3	2422-2485	75.56±1.3	-5.3 ± 17	-16.1
	D	3	2422-2485	12.09±1.8	3.6 ± 25	-29.0
Inami and Ju (2)	E	3	2425-2518	22.11±2.1	-21.3 ± 8.6	-27.1
Ko, Greenbaum, and Farber (3)	F	3	2107-2368	106.01±1.40	5.1 ± 1.7	-39.4

Reactions:
A) $BeO(g) + OH(g) = BeOH(g) + O(g)$
B) $Be_2O(g) + OH(g) = BeOH(g) + BeO(g)$
C) $BeO(cr, \beta) + H(g) = BeOH(g)$
D) $Be_2O(g) + H(g) = BeOH(g) + Be(g)$
E) $BeCl(g) + OH(g) = BeOH(g) + Cl(g)$
F) $BeO(cr, \alpha) + 0.5H_2(g) = BeOH(g)$

The average $\Delta_f H°(BeOH, g, 298.15\ K)$ from the measurements of Inami and Ju (1, 2) is $-27.4\ kcal\ mol^{-1}$. The $\Delta_{at}H°(Be\text{-}OH, g, 298.15\ K) = 115\ kcal\ mol^{-1}$ from $\Delta_f H°(298.15\ K)$ of $-27.4\ kcal\ mol^{-1}$ appears more reasonable than a $\Delta_{at}H°(Be\text{-}OH, g, 298.15\ K) = 127\ kcal\ mol^{-1}$ from the $\Delta_f H°(298.15\ K)$ of $-39.4\ kcal\ mol^{-1}$, derived from Ko, Greenbaum, and Farber (3) above. This is based on a comparison of the trends in D_0^0 values of the alkaline earth monofluorides and monochlorides and in the adopted D_0^0 values for the other alkaline earth monohydroxides (5). The analogy between gaseous monohydroxides and monohalides, particularly the monofluorides, has already been recognized (6-9). Accordingly, $\Delta_f H°(BeOH, g, 298.15\ K) = -27.4 \pm 10\ kcal\ mol^{-1}$ is adopted.

Heat Capacity and Entropy

The molecular configuration is assumed to be linear in accordance with the prediction of Walsh (10) and the evidence that the gaseous alkali metal monohydroxides are linear (11-13). By analogy with BeF and BeCl (5) the ground state is assumed to be $^2\Sigma^+$ and the first excited state is estimated at 30000 cm^{-1}.

The Be-O bond distance is estimated to be slightly larger, 0.02 Å, than the Be-F bond distance (5) after noting the close similiarity in bond distance of the alkali metal fluorides and hydroxides. The O-H bond distance is that in water (5).

The Be-O stretching frequency, 1266 cm^{-1}, is estimated to be the same as the Be-F stretching frequency (5, 9). The O-H stretching frequency, 3650 cm^{-1}, is estimated from the alkali metal hydroxides series. The bending frequency, 606 cm^{-1}, is estimated by assuming that the ratio of the bending force constant to the stretching force constant is 0.022, which is the average ratio found in the alkali metal monohydroxides, by Acquista and Abramowitz (9, 13).

The entropy in the present table is higher by 1.83 cal K^{-1} mol^{-1} at 298.15 K and 1.76 cal K^{-1} mol^{-1} at 1000 K than that proposed by Jackson (9); the moment of inertia as calculated by Jackson (9) appears to be in error.

References

1. Y. H. Inami and F. Ju, Philco-Ford Corp., Aeronutronic Div., Report U-4307, Contract F 04611-68-C-0012, Feb 1, 1968.
2. Y. H. Inami and F. Ju, Philco-Ford Corp., Aeronutronic Div., Report U-4352, Contract F 06411-68-C-0012, May 1, 1968.
3. H. C. Ko, M. A. Greenbaum, and M. Farber, J. Phys. Chem. 71, 1815 (1967).
4. D. L. Hildenbrand, L. P. Theard, E. Murad, and F. Ju, Philco Corp., Aeronutronic Div., Report U-3068, Contract AF 04(611) -8523, April 1, 1965.
5. JANAF Thermochemical Tables: Be(g), 9-30-61; BeCl(g), 9-30-66; BeO(cr, α), BeO(cr, β), BeO(g), 12-31-74; $Be_2O(g)$, 9-30-63 Cl(g), 6-30-72; H(g), O(g), 6-30-74; OH(g), 12-31-70; BaOH(g), CaOH(g), MgOH(g), SrOH(g), 12-31-75; BeF(g), 12-31-71; BeCl(g), 9-30-66; $H_2O(g)$, 3-31-61.
6. R. C. Schoonmaker, and R. F. Porter, J. Chem. Phys. 31, 830 (1959).
7. J. Berkowitz, D. J. Meschi, and N. A. Chupka, J. Chem. Phys. 33, 533 (1960).
8. F. E. Stafford and J. Berkowitz, J. Chem. Phys. 40, 2963 (1964).
9. D. D. Jackson, Lawrence Livermore Laboratory, Univ. of Calif., Report UCRL-51137, Contract No. W-7405-Eng.-48, Dec. 8, 1971.
10. A. D. Walsh, J. Chem. Soc. 1953, 2288 (1953).
11. N. Acquista, S. Abramowitz, and R. Lide, J. Chem. Phys. 49, 780 (1968).
12. R. L. Kuczkowski and R. Lide, J. Chem. Phys. 44, 3131 (1966).
13. N. Acquista and S. Abramowitz, U. S. Nat. Bur. Stand. Report 9905, July, 1968.

Beryllium Hydroxide (BeOH)　　$Be_1H_1O_1(g)$

Enthalpy Reference Temperature = T_r = 298.15 K　　Standard State Pressure = p° = 0.1 MPa

T/K	$C_p°$ (J $K^{-1}mol^{-1}$)	S°	$-[G°-H°(T_r)]/T$	$H°-H°(T_r)$ (kJ mol^{-1})	$\Delta_f H°$	$\Delta_f G°$	Log K_f
0	0.	0.	INFINITE	-9.528	-113.662	-113.662	INFINITE
100	29.307	174.021	240.225	-6.620	-113.745	-117.396	61.321
200	33.323	195.328	212.944	-3.523	-114.028	-120.970	31.594
250	35.983	203.052	210.212	-1.790	-114.314	-122.673	25.631
298.15	38.321	209.594	209.594	0.	-114.642	-124.254	21.769
300	38.405	209.831	209.595	0.071	-114.655	-124.314	21.645
350	40.511	215.914	210.070	2.045	-115.022	-125.895	18.789
400	42.314	221.444	211.152	4.117	-115.390	-127.423	16.640
450	43.849	226.519	212.581	6.272	-115.750	-128.905	14.963
500	45.157	231.208	214.212	8.498	-116.102	-130.348	13.617
600	47.250	239.636	217.763	13.123	-116.788	-133.133	11.590
700	48.862	247.045	221.428	17.932	-117.467	-135.803	10.134
800	50.175	253.658	225.051	22.886	-118.155	-138.376	9.035
900	51.295	259.634	228.566	27.961	-118.862	-140.861	8.175
1000	52.281	265.090	231.950	33.140	-119.598	-143.267	7.483
1100	53.162	270.115	235.194	38.413	-120.371	-145.596	6.914
1200	53.956	274.776	238.301	43.770	-121.186	-147.854	6.436
1300	54.672	279.123	241.276	49.202	-122.051	-150.041	6.029
1400	55.319	283.199	244.126	54.702	-122.972	-152.160	5.677
1500	55.901	287.036	246.860	60.264	-123.955	-154.211	5.370
1600	56.426	290.661	249.485	65.881	-139.543	-155.669	5.082
1700	56.898	294.096	252.009	71.547	-140.267	-156.655	4.813
1800	57.324	297.360	254.439	77.259	-141.000	-157.598	4.573
1900	57.708	300.470	256.780	83.011	-141.746	-158.499	4.357
2000	58.053	303.439	259.040	88.799	-142.506	-159.361	4.162
2100	58.366	306.279	261.222	94.620	-143.285	-160.185	3.984
2200	58.648	309.001	263.332	100.471	-144.083	-160.971	3.822
2300	58.904	311.614	265.375	106.349	-144.903	-161.721	3.673
2400	59.136	314.126	267.355	112.251	-145.747	-162.434	3.535
2500	59.347	316.544	269.274	118.175	-146.615	-163.111	3.408
2600	59.540	318.876	271.137	124.120	-147.509	-163.754	3.290
2700	59.715	321.126	272.947	130.083	-148.430	-164.361	3.180
2800	59.876	323.301	274.707	136.063	-440.336	-158.711	2.961
2900	60.024	325.404	276.419	142.058	-440.249	-148.655	2.678
3000	60.159	327.442	278.086	148.067	-440.174	-138.601	2.413
3100	60.284	329.416	279.710	154.089	-440.114	-128.550	2.166
3200	60.400	331.332	281.294	160.123	-440.069	-118.500	1 934
3300	60.507	333.192	282.838	166.169	-440.042	-108.452	1.717
3400	60.606	335.000	284.346	172.224	-440.034	-98.404	1.512
3500	60.700	336.758	285.818	178.290	-440.047	-88.356	1.319
3600	60.787	338.470	287.257	184.364	-440.083	-78.307	1.136
3700	60.869	340.136	288.664	190.447	-440.145	-68.257	0.964
3800	60.947	341.760	290.040	196.538	-440.235	-58.204	0.800
3900	61.021	343.345	291.387	202.636	-440.354	-48.149	0.645
4000	61.092	344.890	292.705	208.742	-440.505	-38.091	0.497
4100	61.160	346.400	293.996	214.855	-440.690	-28.028	0.357
4200	61.226	347.874	295.262	220.974	-440.911	-17.961	0.223
4300	61.290	349.316	296.502	227.100	-441.170	-7.887	0.096
4400	61.353	350.726	297.718	233.232	-441.470	2.192	-0.026
4500	61.415	352.105	298.912	239.370	-441.811	12.279	-0.143
4600	61.477	353.456	300.083	245.515	-442.197	22.373	-0.254
4700	61.538	354.778	301.232	251.666	-442.628	32.478	-0.361
4800	61.598	356.075	302.362	257.823	-443.107	42.591	-0.463
4900	61.660	357.345	303.471	263.985	-443.635	52.716	-0.562
5000	61.721	358.592	304.561	270.154	-444.212	62.851	-0.657
5100	61.784	359.814	305.632	276.330	-444.842	72.999	-0.748
5200	61.847	361.015	306.686	282.511	-445.524	83.158	-0.835
5300	61.911	362.194	307.722	288.699	-446.259	93.333	-0.920
5400	61.977	363.351	308.741	294.893	-447.048	103.521	-1.001
5500	62.043	364.489	309.745	301.094	-447.892	113.724	-1.080
5600	62.111	365.608	310.732	307.302	-448.791	123.944	-1.156
5700	62.181	366.708	311.705	313.517	-449.746	134.179	-1.230
5800	62.252	367.790	312.662	319.738	-450.756	144.433	-1.301
5900	62.324	368.854	313.606	325.967	-451.822	154.703	-1.370
6000	62.399	369.903	314.535	332.203	-452.943	164.993	-1.436

PREVIOUS: December 1975 (1 atm)　　CURRENT: December 1975 (1 bar)

Beryllium Hydroxide (BeOH)　　$Be_1H_1O_1(g)$

BERYLLIUM HYDROXIDE, ION ($BeOH^+$) IDEAL GAS M_r = 26.018971

$S°(298.15 K) = [203.9 \pm 6.3]\ J\ K^{-1}\ mol^{-1}$

$\Delta_f H°(0\ K) = 754.70 \pm 50\ kJ\ mol^{-1}$

$\Delta_f H°(298.15\ K) = 759.93 \pm 50\ kJ\ mol^{-1}$

Vibrational Frequencies and Degeneracies

ν, cm^{-1}
[1260] (1)
[600] (2)
[3650] (1)

Ground State Quantum Weight: [1] $\sigma = 1$

Point Group: $[C_{\infty v}]$

Bond Distances: Be-O = [1.38] Å

Bond Angle: Be-O-H = [180]°

Rotational Constant: $B_0 = [1.2932]\ cm^{-1}$

Enthalpy of Formation

Using mass spectrometric techniques, Inami and Ju (1) determined the appearance potential of BeOH(g) to be 9.0 ± 0.5 eV (207.55 ± 11.53 kcal mol^{-1}). In the same study, Porter (2) was reported to have said that the appearance potential of a metal hydroxide is expected to be about the same as the corresponding fluoride. In this case, the appearance potential of BeF(g) is 9.1 eV (1), which is within 0.1 eV of the value for BeOH(g). It is also interesting to note that the ionization potential of Be(g) is 9.32 eV (3). We adopt the experimentally determined appearance potential of 9.0 ± 0.5 eV which refers to the process $BeOH(g) + e^- = BeOH^+(g) + 2e^-$. Using auxiliary data (3), we calculate $\Delta_f H°(0\ K) = 180.39 \pm 12.0\ kcal\ mol^{-1}$ for $BeOH^+(g)$. This leads to $\Delta_f H°(298.15\ K) = 181.63 \pm 12.0\ kcal\ mol^{-1}$.

Heat Capacity and Entropy

The molecular configuration is assumed to be linear since experimental evidence indicates that the gaseous alkali metal hydroxides are linear (3, 4, 5, 6). In addition, Walsh (7) had predicted that BAH molecules (H=hydrogen atom) with ten or less valence electrons ($BeOH^+$ has 8 valence electrons) will be linear in their ground state. The molecule $BeOH^+$ is isoelectronic with LiOH.

The bond dissociation energy for $BeOH^+$(121.2 kcal mol^{-1}, 3) for the process $BeOH^+(g) = Be^+(g) + OH(g)$ is fairly close to that for BeOH(113.9 kcal mol^{-1}, 3). This suggests a similar bonding in these two molecules. Thus, bond distances are assumed to be the same as those adopted for BeOH(g, 3). The vibrational frequencies are assumed to be similar to those adopted for BeOH(g, 3). The ground state quantum weight is assumed to be the same as that of LiOH(g, 3).

References

1. Y. H. Inami and F. Ju, Philco-Ford Corp., Aeronutronic Dev., Report U-4307, Contract F 04611-68-C-0012, Feb. 1, 1968.
2. R. F. Porter, personal communication, as referenced in (1).
3. JANAF Thermochemical Tables: e^-(ref), 3-31-65; Be^+(g), 6-30-65; LiOH(g), 6-30-71; BeOH(g), 12-31-75.
4. N. Acquista, S. Abramowitz, and D. R. Lide, J. Chem. Phys. 49, 780 (1968).
5. R. L. Kuczkowski and d. R. Lide, J. Chem. Phys. 44, 3131 (1966).
6. N. Acquista, and S. Abramowitz, U. S. Nat. Bur. Stand. Report 9905, July, 1968.
7. A. D. Walsh, J. Chem. Soc. 1953, 2288 (1953).

Beryllium Hydroxide, Ion ($BeOH^+$) $Be_1H_1O_1^+(g)$

Enthalpy Reference Temperature = T_r = 298.15 K   Standard State Pressure = p° = 0.1 MPa

T/K	$C_p°$ (J $K^{-1}mol^{-1}$)	S° (J $K^{-1}mol^{-1}$)	$-[G°-H°(T_r)]/T$ (J $K^{-1}mol^{-1}$)	$H°-H°(T_r)$ (kJ mol^{-1})	$\Delta_fH°$ (kJ mol^{-1})	$\Delta_fG°$ (kJ mol^{-1})	Log K_f
0	0.	0.	INFINITE	-9.546	754.698		
100	29.321	168.260	234.642	-6.638			
200	33.428	189.606	207.280	-3.535			
250	36.103	197.356	204.539	-1.796			
298.15	38.441	203.919	203.919	0.	759.934	745.758	-130.654
300	38.525	204.157	203.920	0.071	759.959	745.670	-129.833
350	40.625	210.257	204.397	2.051	760.638	743.235	-110.922
400	42.420	215.803	205.481	4.129	761.314	740.703	-96.726
450	43.946	220.889	206.914	6.289	761.998	738.085	-85.675
500	45.245	225.589	208.550	8.519	762.690	735.391	-76.826
600	47.323	234.031	212.109	13.153	764.091	729.800	-63.535
700	48.922	241.450	215.782	17.968	765.497	723.974	-54.024
800	50.224	248.071	219.411	22.927	766.894	717.946	-46.877
900	51.336	254.052	222.933	28.007	768.269	711.745	-41.309
1000	52.316	259.512	226.322	33.190	769.615	705.392	-36.846
1100	53.192	264.540	229.571	38.466	770.925	698.906	-33.188
1200	53.982	269.203	232.682	43.826	772.191	692.303	-30.135
1300	54.695	273.553	235.660	49.260	773.407	685.596	-27.548
1400	55.338	277.630	238.514	54.762	774.567	678.797	-25.326
1500	55.918	281.468	241.251	60.326	775.664	671.918	-23.398
1600	56.441	285.094	243.879	65.944	762.156	665.493	-21.726
1700	56.912	288.530	246.405	71.612	763.512	659.410	-20.261
1800	57.336	291.795	248.837	77.325	764.859	653.248	-18.957
1900	57.719	294.906	251.180	83.078	766.193	647.010	-17.788
2000	58.063	297.875	253.441	88.868	767.512	640.704	-16.733
2100	58.375	300.716	255.625	94.690	768.813	634.331	-15.778
2200	58.657	303.438	257.737	100.542	770.095	627.897	-14.908
2300	58.912	306.051	259.781	106.420	771.354	621.405	-14.113
2400	59.143	308.563	261.762	112.323	772.590	614.859	-13.382
2500	59.354	310.982	263.683	118.248	773.801	608.262	-12.709
2600	59.546	313.314	265.547	124.193	774.986	601.617	-12.087
2700	59.721	315.564	267.358	130.157	776.144	594.927	-11.510
2800	59.881	317.739	269.119	136.137	486.318	594.417	-11.089
2900	60.027	319.843	270.832	142.132	488.483	598.239	-10.775
3000	60.162	321.880	272.500	148.142	490.637	601.987	-10.482
3100	60.285	323.855	274.125	154.164	492.776	605.664	-10.205
3200	60.399	325.771	275.709	160.199	494.900	609.271	-9.945
3300	60.505	327.631	277.254	166.244	497.005	612.813	-9.700
3400	60.602	329.439	278.763	172.299	499.092	616.291	-9.468
3500	60.692	331.197	280.236	178.364	501.157	619.708	-9.249
3600	60.775	332.908	281.675	184.438	503.198	623.066	-9.040
3700	60.853	334.574	283.082	190.519	505.213	626.368	-8.843
3800	60.925	336.198	284.459	196.608	507.200	629.616	-8.655
3900	60.992	337.781	285.806	202.704	509.157	632.812	-8.476
4000	61.055	339.326	287.125	208.806	511.081	635.958	-8.305
4100	61.113	340.835	288.416	214.915	512.971	639.056	-8.142
4200	61.168	342.308	289.682	221.029	514.823	642.109	-7.986
4300	61.220	343.748	290.923	227.148	516.636	645.118	-7.837
4400	61.268	345.156	292.139	233.273	518.407	648.085	-7.694
4500	61.313	346.533	293.333	239.402	520.135	651.013	-7.557
4600	61.355	347.881	294.504	245.535	521.817	653.902	-7.425
4700	61.395	349.201	295.654	251.673	523.451	656.756	-7.299
4800	61.433	350.494	296.783	257.814	525.035	659.576	-7.178
4900	61.469	351.761	297.892	263.959	526.569	662.363	-7.061
5000	61.502	353.003	298.982	270.108	528.049	665.119	-6.948
5100	61.534	354.222	300.053	276.260	529.475	667.846	-6.840
5200	61.564	355.417	301.106	282.415	530.845	670.546	-6.736
5300	61.593	356.590	302.142	288.572	532.158	673.220	-6.635
5400	61.620	357.741	303.161	294.733	533.414	675.870	-6.538
5500	61.645	358.872	304.164	300.896	534.611	678.496	-6.444
5600	61.670	359.983	305.151	307.062	535.748	681.102	-6.353
5700	61.693	361.075	306.122	313.230	536.826	683.688	-6.265
5800	61.715	362.148	307.079	319.401	537.843	686.256	-6.180
5900	61.735	363.203	308.021	325.573	538.800	688.806	-6.098
6000	61.755	364.241	308.950	331.748	539.695	691.341	-6.019

PREVIOUS: December 1975 (1 atm)   CURRENT. December 1975 (1 bar)

Beryllium Hydroxide, Ion ($BeOH^+$) $Be_1H_1O_1^+(g)$

BERYLLIUM HYDRIDE (BeH_2) IDEAL GAS M_r = 11.02806

$S°(298.15\ K) = 173.1\ J\ K^{-1}\ mol^{-1}$

$\Delta_f H°(0\ K) = [127.21]\ kJ\ mol^{-1}$
$\Delta_f H°(298.15\ K) = [125.52]\ kJ\ mol^{-1}$

Vibrational Frequencies and Multiplicities

ν, cm^{-1}
[2200] (1)
[1500] (2)
[1300] (1)

Ground State Quantum Weight: 1 $\sigma = 2$
Point Group: $[D_{\infty h}]$
Bond Distance: Be-H: [1.34]Å
Bond Angle: H-Be-H: [180]°
Rotational constant: $B_0 = [4.657172]\ cm^{-1}$

Enthalpy of Formation

Estimated by O'Brien, Perrin and Perrine (1).

Heat Capacity and Entropy

Molecular and spectroscopic parameters estimated.

References

1. C. J. O'Brien and J. R. Perrin, Marquardt Corp., Van Nuys, Calif. and J. Perrine, Olin Mathieson Chemical Corp. New York 22, N. Y., "Estimation of the Heats of Formation of Gaseous Combustion Product Molecules", October, 1959.

Beryllium Hydride (BeH_2) $Be_1H_2(g)$

Enthalpy Reference Temperature = T_r = 298.15 K Standard State Pressure = p° = 0.1 MPa

T/K	$C_p°$ (J K⁻¹mol⁻¹)	S°	$-[G°-H°(T_r)]/T$	$H°-H°(T_r)$ (kJ mol⁻¹)	$\Delta_f H°$	$\Delta_f G°$	Log K_f
0	0.	0.	INFINITE	-8.714	127.206	127.206	INFINITE
100	29.101	141.109	199.331	-5.822	127.059	123.071	-64.286
200	29.204	161.292	175.842	-2.910	126.700	119.159	-31.121
250	29.588	167.843	173.610	-1.442	126.180	117.329	-24.515
298.15	30.370	173.115	173.115	0.	125.520	115.682	-20.267
300	30.408	173.303	173.115	0.056	125.492	115.622	-20.131
350	31.659	178.079	173.490	1.606	124.715	114.037	-17.019
400	33.233	182.406	174.338	3.227	123.916	112.566	-14.700
450	34.999	186.421	175.460	4.933	123.134	111.195	-12.907
500	36.842	190.204	176.747	6 729	122.392	109.908	-11.482
600	40.446	197.244	179 585	10.595	121.062	107.540	-9.362
700	43.678	203.727	182.577	14.805	119.943	105.378	-7.863
800	46.434	209.744	185.601	19.315	119.003	103.363	-6.749
900	48.733	215.351	188.599	24.077	118.198	101.458	-5.888
1000	50.634	220.587	191.539	29.048	117.482	99.637	-5.204
1100	52.204	225.489	194.405	34.192	116.816	97.885	-4.648
1200	53.505	230.088	197.189	39.480	116.167	96.192	-4.187
1300	54.589	234.415	199.887	44.886	115.508	94.554	-3.799
1400	55.496	238.495	202.501	50.392	114.817	92.968	-3.469
1500	56.262	242.351	205.030	55.981	114.078	91.433	-3.184
1600	56.912	246.003	207.478	61.640	98.740	90.475	-2.954
1700	57.467	249.470	209.847	67.360	98.268	89.973	-2.765
1800	57.944	252.769	212.140	73.131	97.786	89.499	-2.597
1900	58.357	255.913	214.362	78.947	97.287	89.052	-2.448
2000	58.716	258.916	216 515	84.801	96.769	88.632	-2.315
2100	59.030	261.788	218.603	90.688	96 227	88.238	-2.195
2200	59.306	264.541	220.629	96.605	95.659	87.871	-2.086
2300	59.550	267.183	222.596	102.548	95.065	87.530	-1.988
2400	59.766	269.722	224.507	108.514	94.441	87.215	-1.898
2500	59.958	272.165	226.365	114.501	93 787	86.928	-1.816
2600	60.131	274.520	228.172	120.505	93.102	86.667	-1.741
2700	60.285	276.793	229.931	126.526	92.386	86.433	-1.672
2800	60.425	278.988	231.644	132.562	-199.320	92.448	-1.725
2900	60.551	281.110	233.313	138.611	-199.037	102.864	-1.853
3000	60.665	283.165	234.941	144.672	-198.771	113.270	-1.972
3100	60.768	285.156	236.529	150.743	-198.524	123.667	-2.084
3200	60.863	287.087	238.079	156.825	-198.295	134.056	-2.188
3300	60.950	288.961	239.593	162.916	-198.089	144.439	-2.286
3400	61.029	290.782	241.071	169.015	-197.906	154.816	-2.378
3500	61.102	292.552	242.517	175.121	-197.748	165.188	-2.465
3600	61.169	294.274	243.931	181.235	-197.619	175.555	-2.547
3700	61.231	295.951	245.314	187.355	-197.519	185.920	-2.625
3800	61.288	297.585	246.669	193.481	-197.451	196.282	-2.698
3900	61.341	299.177	247.995	199.612	-197.418	206.643	-2.768
4000	61.390	300.731	249.294	205.749	-197.421	217.003	-2.834
4100	61.436	302.247	250.567	211.890	-197.463	227.365	-2.897
4200	61.478	303.728	251.815	218.036	-197.547	237.727	-2.957
4300	61.518	305.175	253.039	224.186	-197.674	248.092	-3.014
4400	61.555	306.590	254.240	230.339	-197.847	258.461	-3.068
4500	61.590	307.974	255.419	236.497	-198.067	268.834	-3 121
4600	61.622	309.328	256.576	242.657	-198.337	279.212	-3.171
4700	61.653	310.653	257.713	248.821	-198.658	289.597	-3.219
4800	61.681	311.952	258.829	254.988	-199.031	299 990	-3.265
4900	61.708	313.224	259.926	261.157	-199.458	310.390	-3.309
5000	61.734	314.471	261.005	267.329	-199.940	320.800	-3.351
5100	61.758	315.693	262.065	273.504	-200.478	331.220	-3.392
5200	61.780	316.893	263.108	279.681	-201.072	341.651	-3.432
5300	61.802	318.070	264.134	285.860	-201.724	352.094	-3.470
5400	61.822	319.225	265.144	292.041	-202.433	362.550	-3.507
5500	61.841	320.360	266.137	298.224	-203.200	373.019	-3.543
5600	61.859	321.474	267.115	304.409	-204.024	383.504	-3.577
5700	61.876	322.569	268.079	310.596	-204.905	394.003	-3.611
5800	61.893	323.646	269.028	316.785	-205.842	404.519	-3.643
5900	61.908	324.704	269.962	322.975	-206.836	415.050	-3.675
6000	61.923	325.744	270.883	329.166	-207.884	425.599	-3.705

PREVIOUS: December 1960 (1 atm) CURRENT. December 1960 (1 bar)

Beryllium Hydride (BeH_2) $Be_1H_2(g)$

BERYLLIUM HYDROXIDE, ALPHA (α-$Be(OH)_2$) CRYSTAL M_r = 43.02686

$\Delta_f H°$(0 K) = Unknown

$S°$(298.15 K) = [53.6 ± 8] J K^{-1} mol^{-1} $\Delta_f H°$(298.15 K) = -902.9 ± 2.0 kJ mol^{-1}

Enthalpy of Formation

Bear and Turnbull (1) measured the enthalpy of solution of α-$Be(OH)_2$ and Be in 22.6% HF(aq) at 21°C. Combining these results, adjusting to 25°C, and correcting for the small HF concentration differences in the two systems, we calculate $\Delta_r H°$(298.15 K) = -79.16 kcal mol^{-1} for the reaction Be(cr) + $2H_2O(\ell)$ = $Be(OH)_2$(α, cr) + H_2(g). Using auxiliary data (2, 3), we derive $\Delta_f H°$(298.15 K) = -215.8 kcal mol^{-1} for α-$Be(OH)_2$. Since this adopted value is calorimetrically determined and based on well characterized samples, we assign an uncertainty of ± 0.5 kcal mol^{-1}.

Parker (4) recently reviewed data pertaining to the enthalpy of formation of the beryllium hydroxides. Her analyses were intended to give additional support to the thermochemical values chosen for BeO(cr). Our interpretation would differ slightly due to the fact that our $\Delta_f H°$ value for BeO(cr) is different (by 0.3 kcal mol^{-1} at 298.15 K) from that adopted by Parker (4). Discussed in her review were the enthalpy of solution studies by Fricks and Wullhorst (5) [involving BeO and α-$Be(OH)_2$ in 11.59% HF] and Matignon and Marchal (6, 7) [amorphous $Be(OH)_2$ in 30% HF]. In addition there is also an earlier study by Mulert (8) involving amorphous $Be(OH)_2$ in 20% HF. All these studies are in fair agreement with our adopted value for $\Delta_f H°$(298.15 K). These latter studies are thought to be less reliable, however, due to incomplete characterization of the beryllium compounds.

Heat Capacity and Entropy

The heat capacity is assumed to be identical to that of the β-phase. The entropy difference between the α and β phases may be obtained from the solubility data of Fricke and Humme (9). Both crystalline forms of $Be(OH)_2$ were dissolved in varying concentrations of aqueous NaOH. By relating the quantities dissolved, we calculate $\Delta_r G°$(303 K) = -0.5 (± 0.2) kcal mol^{-1} for the process α-$Be(OH)_2$= β-$Be(OH)_2$. Using auxiliary data (2), we find $\Delta_r S°$(303 K) = -0.8 cal K^{-1} mol^{-1}, from which we calculate and adopt $S°$(298.15 K) = 12.8 cal K^{-1} mol^{-1} for α-$Be(OH)_2$. We assign an uncertainty of ± 2 cal K^{-1} mol^{-1} to reflect the possible error in this calculation of $\Delta_r S°$(303 K).

Phase Data

The structural information for α-$Be(OH)_2$ via x-ray techniques is not complete (1, 9, 10, 11). A tetragonal unit cell has been suggested by Guillemat and Lecocq (11) without supporting data. Bear and Turnbull (1), following this suggestion and using their x-ray data, calculated a crystal density which was consistent with the two experimentally determined values of Fricke and Severin (12, helium densitometry and a pycnometric method). This lends support to the premise that α-$Be(OH)_2$ has a tetragonal unit cell. The alpha phase is metastable.

Decomposition Data

T_{dcm} is calculated as the temperature at which $\Delta_r G°$ is zero for the reaction $Be(OH)_2$(α, cr) = BeO(α, cr) + H_2O(g).

References

1. I. J. Bear and A. G. Turnbull, J. Phys. Chem. 69, 2828 (1965).
2. JANAF Thermochemical Tables: Be(cr), 9-30-61; H_2(g) and H_2O(g), 3-31-61; BeO(aq, cr), 6-30-75; $Be(OH)_2$(β, cr), 12-31-75.
3. U. S. Nat. Bur. Stand. Tech. Note 270-3, 1968.
4. V. B. Parker, J. Res. Nat. Bur. Stand. 77A, 227 (1973); U. S. Nat. Bur. Stand. Report 10004, pp. 58-81, Jan., 1969.
5. R. Fricke and B. Wullhorst, Z. anorg. allgem. Chem. 205, 127 (1932).
6. C. Matignon and G. Marchal, Bull. Soc. Chim. France 39, 167 (1926).
7. C. Matignon and G. Marchal, Compt. Rend. 181, 859 (1925).
8. O. Mulert, Z. anorg. allgem. Chem. 75, 198 (1912).
9. R. Fricke and H. Humme, Z. anorg. allgem. Chem. 178, 400 (1929).
10. M. Bernier, Commission Energie Atomique (France) Rapport CEA2326 (1963).
11. A. Guillemat and A. Lecocq, Compt. Rend. 257, 1260 (1963).
12. R. Fricke and H. Severin, Z. anorg. allgem. Chem. 205, 287 (1932).

Beryllium Hydroxide, Alpha ($Be(OH)_2$) $Be_1H_2O_2$(cr)

Enthalpy Reference Temperature = T_r = 298.15 K Standard State Pressure = p° = 0.1 MPa

T/K	$C_p°$ (J $K^{-1}mol^{-1}$)	$S°$ (J $K^{-1}mol^{-1}$)	$-[G°-H°(T_r)]/T$ (J $K^{-1}mol^{-1}$)	$H°-H°(T_r)$ (kJ mol^{-1})	$\Delta_f H°$ (kJ mol^{-1})	$\Delta_f G°$ (kJ mol^{-1})	Log K_f
0							
100							
200							
298.15	65.701	53.555	53.555	0.	-902.907	-815.933	142.948
300	66.053	53.963	53.556	0.122	-902.923	-815.394	141.973
400	82.902	75.434	56.369	7.626	-903.139	-786.150	102.661
500	92.994	95.086	62.183	16.451	-902.397	-756.975	79.081
600	99.387	112.633	69.157	26.086	-901.119	-728.004	63.378
700	103.956	128.310	76.508	36.261	-899.527	-699.274	52.180
800	107.495	142.430	83.881	46.839	-897.735	-670.787	43.798
900	110.399	155.262	91.110	57.737	-895.810	-642.532	37.292
1000	112.905	167.027	98.121	68.906	-893.790	-614.498	32.098

PREVIOUS: December 1966 CURRENT: December 1975

Beryllium Hydroxide, Alpha ($Be(OH)_2$) $Be_1H_2O_2$(cr)

BERYLLIUM HYDROXIDE, BETA (β-$Be(OH)_2$) CRYSTAL M_r = 43.02686

$S°(298.15 K) = [50.2 \pm 4.2] J K^{-1} mol^{-1}$

T_{dcm} = [366] K

$\Delta_f H°(0 K)$ = Unknown

$\Delta_f H°(298.15 K) = -905.8 \pm 2.1 kJ mol^{-1}$

Enthalpy of Formation

Bear and Turnbull (1) measured the enthalpy of solution of β-$Be(OH)_2$ and Be in 22.6% HF(aq) at 21°C. Combining these results, adjusting to 25°C, and correcting for the small HF concentration differences in the two systems, we calculate $\Delta_r H°(298.15 K) = -79.89 kcal mol^{-1}$ for the reaction Be(cr) + $2H_2O(\ell)$ = $Be(OH)_2(\beta, cr)$ + $H_2(g)$. Using auxiliary data (2, 3), we derive $\Delta_f H°(298.15 K) = -216.5 kcal mol^{-1}$ for β-$Be(OH)_2$. Since this adopted value is calorimetrically determined and based on well characterized samples, we assign an uncertainty of ± 0.5 kcal mol^{-1}.

Parker (4) recently reviewed data pertaining to the enthalpy of formation of the beryllium hydroxides. Her analyses were intended to give additional support to the thermochemical values chosen for BeO(cr). Our interpretation would differ slightly due to the fact that our $\Delta_f H°$ value for BeO(cr) is different (by 0.3 kcal mol^{-1} at 298.15 K) from that adopted by Parker (4). Discussed in her review were the enthalpy of solution studies by Fricke and Wullhorst (5) [involving BeO and β-$Be(OH)_2$ in 11.59% HF] and Matignon and Marchal (6, 7) [amorphous $Be(OH)_2$ in 30% HF]. In addition there is also an earlier study by Mulert (8) involving amorphous $Be(OH)_2$ in 20% HF. These latter three studies are in fair agreement with our adopted value for $\Delta_f H°(298.15 K)$ but are thought to be less reliable, however, due to incomplete characterization of the beryllium compounds.

Fricke and Severin (9) and Baur and Lecocq (10) measured the deomposition pressure of water vapor over β-$Be(OH)_2$. Fricke and Severin (9) observed a decomposition temperature at 105°C at a pressure of 100 mm Hg while Baur and Lecocq (10) observed 223°C for the equilibrium with saturated water vapor at 24.2 atm. We reduce the latter data to a standard state value $\Delta_r G°(496 K) = -3028 cal mol^{-1}$ for the decomposition reaction $Be(OH)_2(\beta, cr)$ → BeO(α, cr) + $H_2O(g)$. The 3rd law analyses for these two decomposition studies are given below where $\Delta_f H°(298.15 K)$ refers to the enthalpy of formation of β-$Be(OH)_2$.

Source	T/K	Method	$\Delta_r H°(298.15 K)$ (kcal mol^{-1})	$\Delta_f H°(298.15 K)$ (kcal mol^{-1})
Fricke and Severin (9)	378	Kp	15.27	≥ -218.5
Baur and Lecocq (10)	496	Kp	14.88	≥ -218.1

These values are within 2 kcal mol^{-1} of our adopted $\Delta_f H°(298.15 K)$ value. The difference may be due to the formation of metastable BeO. Fricke and Severin (9) reported that BeO had a distorted lattice which would suggest a $\Delta_f H°$ value more positive than -145.4 kcal mol^{-1}. Thus, we conclude that these two studies (9, 10) are consistent with our adopted value but are not sufficiently definitive for further consideration.

Heat Capacity and Entropy

The heat capacity is estimated from that for $Mg(OH)_2$ by subtracting the values for MgO(cr) and adding those for BeO(α, cr) (2). The entropy is estimated to be $S°(298.15 K) = 12.0 cal K^{-1} mol^{-1}$. This value is a compromise of values suggested by two different paths. Additive entropy constants of Kelley (11) yield $S°(298.15 K) = 13.20 cal K^{-1} mol^{-1}$ whereas a comparison of entropy differences for the alkali metal and alkaline earth chlorides and hydroxides sugggest values in the range 10.2 to 12.4 cal $K^{-1} mol^{-1}$. Our adopted value is the same as that selected by NBS (12).

Phase Data

Bear and Turnbull (1) and Seitz et al. (13), using x-ray powder techniques, determined that β-$Be(OH)_2$ has an orthorhombic structure, the ϵ-$Zn(OH)_2$ structure. The β-structure is the stable crystalline form for $Be(OH)_2$ whereas the α-structure is metastable.

Decomposition Data

T_{dcm} is calculated as the temperature at which $\Delta_r G°$ is zero for the reaction $Be(OH)_2(\beta, cr)$ → BeO(α, cr) + $H_2O(g)$.

References

1. I. J. Bear and A. G. Turnbull, J. Phys. Chem. 69, 2828 (1965).
2. JANAF Thermochemical Tables: Be(cr), 9-30-61; $H_2(g)$ and $H_2O(g)$, 3-31-61; BeO(α, cr), 6-30-75; BeO(β, cr) and MgO(cr), 12-31-74; $Mg(OH)_2(cr)$, 12-31-75.
3. U. S. Nat. Bur. Stand. Tech. Note 270-3, 1968.
4. V. B. Parker, J. Res. Nat. Bur. Stand 77A, 227 (1973); U. S. Nat. Bur. Stand. Report 10004, pp. 58-81, Jan., 1969.
5. R. Fricke and B. Wullhorst, Z. anorg. allgem. Che. 205, 127 (1932).
6. C. Matignon and G. Marchal, Bull Soc. Chim. France 39, 167 (1926).
7. C. Matignon and G. Marchal, Compt. Rend. 181, 859 (1925).
8. O. Mulert, Z. anorg, allgem. Chem. 178, 198 (1912).
9. R. Fricke and H. Severin, Z. anorg. allgem. 205, 287 (1932).
10. A Baur and A. Lecocq, Comm. Energie At. (France), Rappt. CEA-R2611, 17 pp. (1964).
11. D. R. Stull and H. Prophet in "The Characterization of High Temperature Vapors", J. L. Margrave, ed., John Wiley and Sons Inc., New York 1966.
12. U. S. Nat. Bur Stand. Tech. Note 270-6, 1971.
13. A. Seitz, U. Rosler, and K. Schubert, Z. anorg. allgem. Chem. 261, 94 (1960).

Beryllium Hydroxide, Beta ($Be(OH)_2$) $Be_1H_2O_2(cr)$

Enthalpy Reference Temperature = T_r = 298.15 K; Standard State Pressure = p° = 0.1 MPa

T/K	$C_p°$ (J $K^{-1}mol^{-1}$)	S° (J $K^{-1}mol^{-1}$)	$-[G°-H°(T_r)]/T$ (J $K^{-1}mol^{-1}$)	$H°-H°(T_r)$ (kJ mol^{-1})	$\Delta_f H°$ (kJ mol^{-1})	$\Delta_f G°$ (kJ mol^{-1})	Log K_f
0							
100							
200							
298.15	65.701	50.208	50.208	0.	-905.836	-817.864	143.286
300	66.053	50.615	50.209	0.122	-905.852	-817.318	142.308
400	82.902	72.087	53.022	7.626	-906.067	-787.740	102.868
500	92.994	91.739	58.836	16.451	-905.326	-758.230	79.212
600	99.387	109.286	65.810	26.086	-904.047	-728.924	63.458
700	103.956	124.963	73.161	36.261	-902.456	-699.859	52.224
800	107.495	139.082	80.533	46.839	-900.664	-671.038	43.814
900	110.399	151.915	87.763	57.737	-898.739	-642.449	37.287
1000	112.905	163.680	94.774	68.906	-896.719	-614.080	32.076

PREVIOUS: December 1966 CURRENT: December 1975

Beryllium Hydroxide, Beta ($Be(OH)_2$) $Be_1H_2O_2(cr)$

BERYLLIUM HYDROXIDE ($Be(OH)_2$) IDEAL GAS M_r = 43.02686

$S°(298.15\ K)$ = [233.97 ± 12.6] J K^{-1} mol^{-1}

$\Delta_f H°(0\ K)$ = -670 ± 38 kJ mol^{-1}

$\Delta_f H°(298.15\ K)$ = -677 ± 38 kJ mol^{-1}

Vibrational Frequencies and Degeneracies

ν, cm^{-1}	ν, cm^{-1}
[670](1)	[3650](2)
[345](2)	[606](4)
[1530](1)	

Ground State Quantum Weight: [1] σ = 2

Point Group: [$C_{\infty v}$]

Bond Distances: Be-O = [1.42] Å O-H = [0.96] Å

Bond Angles: Be-O-H = [180]° O-Be-O = [180]°

Rotational Constant: B_0 = [0.22198] cm^{-1}

Enthalpy of Formation

The equilibrium reaction BeO(α, cr) + $H_2O(g)$ = $Be(OH)_2(g)$ has been studied by several investigators (1-6); all inferred that $Be(OH)_2$ was the major product, but there was no direct proof of this. A 2nd and 3rd law analysis is given below.

Investigator	Data Points	T/K	$\Delta_r H°(298.15\ K)$, kcal mol^{-1} 2nd law	3rd law	Drift cal K^{-1} mol^{-1}	$\Delta_f H°(298.15\ K)$ kcal mol^{-1}
Stuart and Price (1)	15**	1338-1653	40.66±0.70	41.76±0.26	0.7 ± 0.4	-161.74
Young (2)	7	1576-1839	42.25±1.46	41.58±0.43	-0.4 ± 0.9	-161.92
Grossweiner and Seifert (3)	24**	1472-1873	42.23±0.17	42.27±0.69	0.0 ± 0.7	-161.23
	11	1673		41.09±1.47		-161.81
Blauer et al. (4)	4	1567-1808	38.2 ±4.3	35.26±0.82	-1.8 ± 2.5	-168.24
Brewer and Elliott (5)	6	1527-1583	74.8 ±68.8	44.12±4.4	-20 ± 44	-159.38
Morize et al. (6)	eqn		49.1	37.7	-6.9	-165.80

**One (or two) point(s) rejected due to a statistical test.

Three of these studies (1-3) were conducted under similar conditions. Each of these investigations (1-3) varied the $H_2O(g)$ flow rate without a detectable change in Kp value. However, one study (3) used a considerable larger flow rate than the others (1, 2). The residence times in all cases were quite comparable. Blauer et al. (4) studied the same system using molecular flow effusion techniques. They reported a larger variation of Kp with surface area, operated at pressures near the upper limits of the molecular flow regime, and encountered unexpected variations of the blank experiments with sample size. Brewer and Elliot (5) also studied the same system but details of the sample are not available and the results show considerable drift. Baur and Lecocq (7) cited a study by Morize et al. (6). No details are given and the results are presented in equation form. It is presumed that the results were obtained in the vicinity of 1673 K.

We adopt $\Delta_f H°(298.15\ K)$ = -161.7 kcal mol^{-1} for $Be(OH)_2(g)$ which is an average value of three studies (1-3). We assign an uncertainty of ± 9.0 kcal mol^{-1}. The remaining studies are in reasonable agreement with our adopted value (and its uncertainty). As is suggested by Blauer et al. (4), the results of the transpiration studies are greatly affected by surface effects. The larger $\Delta_f H°$ value derived from Blauer et al. (4) is due to a consideration of sample size and surface effects. Thus, it is possible that the earlier studies (1, 2, 3) had not achieved equilibrium.

Hildenbrand et al. (8) examined the BeO-H_2O system at 2300°C mass-spectrometrically and detected BeOH and $Be(OH)_2$. They reported a maximum value for the equilibrium constant for the reaction BeO(g) + $H_2O(g)$ = $Be(OH)_2(g)$. From this maximum value, we calculate $\Delta_f H°(298.15\ K) \geq$ -161.3 kcal mol^{-1} for $Be(OH)_2(g)$. This gives additional support for our adopted value. Qualitative support is found by comparing the dissociation energies of BeOH(g) and $Be(OH)_2(g)$ into Be(g) and OH(g); the ratio for the mono- to dihydroxide dissociation is (113.9/256.1) = 0.445; a value which is consistent with similar ratios for the other alkaline earth hydroxides and halides.

Heat Capacity and Entropy

The analogy between gaseous mono- and dihydroxides and gaseous mono- and dihalides, particularly the mono- and difluorides, has been recognized (9-11). The O-Be-O angle is assumed to be the same as the F-Be-F angle (12), i.e. 180°; the Be-O-H angle is assumed to be linear as in MgOH(g) (12). The Be-O bond distance is estimated to be slightly larger, 0.02 Å, than the Be-F distance in BeF_2 (12) after noting the close similarity in the bond distance of the alkali metal fluorides and hydroxides. The O-H bond distance is taken to be the same as in water (12).

The vibrational frequencies are estimated to be the same as in BeF_2 (12) (O-Be-O symmetrical and antisymmetrical stretch and bend) and as in BeOH (12) (O-H stretch and Be-O-H bend).

Jackson (13) has used a different molecular configuration and different vibrational frequencies to estimate $S°(298.15\ K)$ = 59.076 cal K^{-1} mol^{-1}. We assign an uncertainty of ± 3 cal K^{-1} mol^{-1} to the adopted entropy.

One vibrational frequency, 649 cm^{-1}, was reported by Farber et al. (14) based on infrared spectral data. We do not adopt this value for the following reasons. It is within 21 cm^{-1} of an adopted value and will not drastically change the entropy. For consistency we estimate all the vibrational frequencies for the alkaline earth dihydroxides in the same manner since the result of Farber et al. (14) is the only reported value.

References

1. W. I. Stuart and G. H. Price, J. Nucl. Mater. 14, 417 (1964).
2. W. A. Young, J. Phys. Chem. 64, 1003 (1960).
3. L. I. Grossweiner and R. L. Seifert, J. Amer. Chem. Soc. 74, 2701 (1952).
4. J. Blauer, M. A. Greenbaum, and M. Farber, J. Phys. Chem. 70, 973 (1966).
5. L. Brewer and G. R. B. Elliott, UCRL 1831 (1952).
6. Morize, Roudier, and Besson, DM/948.
7. A. Baur and A. Lecocq, Comm. Energie At. (France), Rappt. CEA-R2611 (1964).
8. D. F. Hildenbrand, L. P. Theard, and F. Ju., Aeronautronic, 3rd Quarterly Technical Report No. U-2231, Contract AF 03(611) -8523, 31 July 1963.
9. R. C. Schoonmaker and R. F. Porter, J. Chem. Phys. 31, 830 (1959).
10. J. Berkowitz, D. J. Mescki, and W. A. Chupka, J. Chem. Phys. 33, 533 (1960).
11. F. E. Stafford and J. Berkowitz, J. Chem. Phys. 40, 2963 (1964).
12. JANAF Thermochemical Tables: $BeF_2(g)$, 12-31-71; Mg(OH)(g) and Be(OH)(g), 12-31-75; $H_2O(g)$, 3-31-61.
13. D. D. Jackson, Lawrence Livermore Laboratory, Univ. of Calif., UCRL-51137, Contract No., W-7405-Eng.-48, Dec. 8, 1971.
14. M. Farber, et al., Rocket Power, Inc., AFRPL-TR-66-220, September, 1966.

Beryllium Hydroxide ($Be(OH)_2$) $Be_1H_2O_2(g)$

Enthalpy Reference Temperature = T_r = 298.15 K Standard State Pressure = p° = 0.1 MPa

T/K	$C_p°$ (J $K^{-1}mol^{-1}$)	S° (J $K^{-1}mol^{-1}$)	$-[G°-H°(T_r)]/T$ (J $K^{-1}mol^{-1}$)	$H°-H°(T_r)$ (kJ mol^{-1})	$\Delta_f H°$ (kJ mol^{-1})	$\Delta_f G°$ (kJ mol^{-1})	Log K_f
0	0.	0.	INFINITE	-12.588	-670.058	-670.058	INFINITE
100	32.467	184.150	280.300	-9.615	-673.028	-663.989	346.832
200	49.187	211.481	239.353	-5.575	-675.170	-654.052	170.821
250	57.182	223.343	234.979	-2.909	-675.951	-648.679	135.534
298.15	63.419	233.968	233.968	0.	-676.553	-643 369	112.716
300	63.631	234.361	233.970	0.118	-676.573	-643.163	111.985
350	68.713	244.567	234.764	3.431	-677.063	-637.554	95.150
400	72.738	254.016	236.587	6.971	-677.439	-631.883	82.516
450	75.976	262.776	239.017	10.692	-677.722	-626.171	72.684
500	78.628	270.923	241.805	14.559	-677.935	-620.431	64.816
600	82.730	285.641	247.911	22.638	-678.212	-608.902	53.010
700	85.823	298.636	254.248	31.072	-678.362	-597.337	44.574
800	88.333	310.265	260.536	39.783	-678.437	-585.756	38.246
900	90.482	320.796	266.655	48.726	-678.466	-574.169	33.324
1000	92.385	330.429	272.558	57.872	-678.470	-562.580	29.386
1100	94.095	339.316	278.228	67.197	-678.464	-550.991	26.164
1200	95.642	347.571	283.667	76.685	-678.461	-539.403	23.480
1300	97.044	355.283	288.882	86.321	-678.474	-527.814	21.208
1400	98.311	362.522	293.886	96.089	-678.515	-516.224	19.261
1500	99.456	369.344	298.692	105.979	-678.596	-504.629	17.573
1600	100.490	375.796	303.311	115.977	-693.262	-492.502	16.079
1700	101.421	381.917	307.756	126.073	-693.048	-479.961	14.747
1800	102.262	387.738	312.039	136.258	-692.833	-467.433	13.565
1900	103.019	393.288	316.171	146.523	-692.622	-454.916	12.507
2000	103.703	398.590	320.160	156.860	-692.420	-442.410	11.555
2100	104.321	403.665	324.016	167.261	-692.234	-429.914	10.694
2200	104.881	408.531	327.748	177.722	-692.066	-417.427	9.911
2300	105.387	413.204	331.363	188.236	-691.921	-404.947	9.197
2400	105.847	417.699	334.867	198.798	-691.801	-392.473	8.542
2500	106.266	422.029	338.267	209.404	-691.711	-380.002	7.940
2600	106.647	426.204	341.570	220.050	-691.650	-367.535	7.384
2700	106.995	430.236	344.780	230.732	-691.622	-355.070	6.869
2800	107.313	434.133	347.902	241.448	-982.586	-336.382	6.275
2900	107.604	437.904	350.940	252.194	-981.564	-313.322	5.644
3000	107.872	441.556	353.900	262.968	-980.562	-290.296	5.055
3100	108.118	445.097	356.785	273.767	-979.581	-267.303	4.504
3200	108.345	448.534	359.599	284.591	-978.625	-244.342	3.988
3300	108.554	451.871	362.345	295.436	-977.695	-221.411	3.505
3400	108.747	455.114	365.026	306.301	-976.794	-198.506	3.050
3500	108.926	458.269	367.645	317.185	-975.923	-175.628	2.621
3600	109.093	461.340	370.205	328.086	-975.085	-152.775	2.217
3700	109.247	464.331	372.709	339.003	-974.282	-129.944	1.834
3800	109.390	467.247	375.159	349.935	-973.517	-107.134	1.473
3900	109.524	470.090	377.556	360.881	-972.791	-84.343	1.130
4000	109.649	472.865	379.905	371.839	-972.108	-61.572	0.804
4100	109.766	475.573	382.205	382.810	-971.470	-38.817	0.495
4200	109.875	478.220	384.460	393.792	-970.878	-16.076	0 200
4300	109.977	480.807	386.670	404.785	-970.336	6.651	-0.081
4400	110.073	483.336	388.839	415.788	-969.846	29.365	-0.349
4500	110.163	485.811	390.966	426.799	-969.410	52.069	-0.604
4600	110.248	488.233	393.055	437.820	-969.030	74.764	-0.849
4700	110.327	490.605	395.105	448.849	-968.708	97.452	-1.083
4800	110.402	492.928	397.119	459.885	-968.446	120.134	-1.307
4900	110.473	495.205	399.097	470.929	-968.247	142.811	-1.522
5000	110.540	497.438	401.042	481.980	-968.111	165.484	-1.729
5100	110.603	499.628	402.954	493.037	-968.041	188.155	-1.927
5200	110.663	501.776	404.833	504.100	-968.037	210.825	-2.118
5300	110.720	503.884	406.683	515.169	-968.102	233.497	-2.301
5400	110.774	505.954	408.502	526.244	-968.236	256.169	-2.478
5500	110.825	507.987	410.292	537.324	-968.440	278.844	-2.648
5600	110.873	509.985	412.055	548.409	-968.715	301.526	-2.813
5700	110.919	511.948	413.790	559.499	-969.062	324.211	-2.971
5800	110.963	513.877	415.499	570.593	-969.482	346.905	-3.124
5900	111.004	515.774	417.183	581.691	-969.974	369.603	-3.272
6000	111.044	517.640	418.841	592.793	-970.540	392.313	-3.415

PREVIOUS: December 1975 (1 atm) CURRENT December 1975 (1 bar)

Beryllium Hydroxide ($Be(OH)_2$) $Be_1H_2O_2(g)$

BERYLLIUM IODIDE (BeI) IDEAL GAS $M_r = 135.91668$

Beryllium Iodide (BeI) $Be_1I_1(g)$

$S°(298.15 K) = 237.30 \pm 0.21$ J K^{-1} mol^{-1}

$\Delta_fH°(0 K) = 169.5 \pm 41.8$ kJ mol^{-1}
$\Delta_fH°(298.15 K) = 170.0 \pm 41.8$ kJ mol^{-1}

Electronic Levels and Quantum Weight

State	ε_i, cm^{-1}	g_i
$X^2\Sigma^+$	0	2
$A_1{}^2\Pi_{1/2}$	23541	2
$A_2{}^2\Pi_{3/2}$	23898	2

$\omega_e = 611.7$ cm^{-1} $\omega_ex_e = [3.36]$ cm^{-1} $\sigma = 1$
$B_e = 0.4406$ cm^{-1} $\alpha_e = [0.0042]$ cm^{-1} $r_e = 2.132$ Å

Enthalpy of Formation

No thermochemical measurements of the enthalpy of formation has been made. The selected value, $\Delta_fH°(BeI, g, 0 K) = 40.5$ kcal mol^{-1}, is obtained from an analysis of spectroscopic data. The adopted values for the ground state vibrational constants give $D_0^0 = 3.41_4$ eV (78.73 kcal mol^{-1}) via a linear Berge-Sponer extrapolation (1). Based on the ionicity correction developed by Hildenbrand (2), this value adjusts to $D_0^0 = 2.70_7$ eV (62.43 kcal mol^{-1}). We adopt $D_0^0 = 62.4 \pm 10$ kcal mol^{-1} which corresponds to $\Delta_fH°(0 K) = 40.5 \pm 10.0$ kcal mol^{-1}.

Support for the adopted D_0^0 value is provided by an examination of the trends in the dissociation energies for all alkaline earth monohalides and a comparison of the values for the ratio $\Delta_{at}H°(MX, 298.15 K)/\Delta_{at}H°(MX_2, 298.15 K)$ for the alkaline earth halides (3). For the beryllium iodides this ratio is 0.44 which is consistent with the values of this ratio for other alkaline earth halides.

Heat Capacity and Entropy

The ground state vibrational and rotational constants are derived from spectroscopic studies by Murty and Rao (4, 5, 6). The reported value for ω_ex_e (1.6 cm^{-1}) appears to be inconsistent with the values obtained for other alkaline earth monohalides. By a comparison of (ω_e/ω_ex_e) and $x_e\mu^{1/2}$ values, we estimate $\omega_ex_e = 3.36$ cm^{-1} and adopt this value. The reported B_o value is converted to B_e by using a calculated value for α_e. We calculate α_e from the relation $\alpha_e/B_e = 13.2\ (B_e/\omega_e)$ as suggested by Calder and Ruedenberg (7). Initially we use the B_o value in this relationship and then iterate to obtain consistent values for B_e and α_e. We calculate $r_e = 2.132$ Å from B_e. This corresponds well with the 2.12 Å Be-I distance for $BeI_2(g)$ obtained via the electron diffraction study of Akishin et al. (8).

The electronic levels are also from the study by Murty and Rao (4, 5, 6). The possibility of additional levels ($^2\Pi$ and $^2\Sigma^+$) near 15000 cm^{-1} is suggested by a comparison with CaI, SrI, and BaI (3). Including these levels does not change the entropy at 298.15 K but does increase the entropy at 6000 K by 0.7 cal K^{-1} mol^{-1}.

References

1. A. G. Gaydon, "Dissociation Energies and Spectra of Diatomic Molecules", Chapman and Hall, Ltd., London, 1968.
2. D. L. Hildenbrand in "Advances in High Temperatures Chemistry", Vol. 1, Ed. by L. Eyring, Academic press, New York, 1967.
3. JANAF Thermochemical Tables: $BeI_2(g)$, 12-31-75; BaI(g), SrI(g), and CaI(g), 6-30-74.
4. P. S. Murty and P. T. Rao, Curr. Sci. 38, 187 (1969).
5. P. S. Murty and P. T. Rao, Curr. Sci. 38, 537 (1969).
6. P. S. Murty and P. T. Rao, Proc. Roy. Ir. Acad., Sect. A72, 71 (1972).
7. G. V. Calder and K. Ruedenberg, J. Chem. Phys. 49, 5399 (1968).
8. P. A. Akishin, V. P. Spiridonov, and G. A. Sobolev, Dokl. Akad. Nauk SSSR 118, 1134 (1958).

Enthalpy Reference Temperature = T_r = 298.15 K Standard State Pressure = $p°$ = 0.1 MPa

T/K	$C_p°$ (J $K^{-1}mol^{-1}$)	$S°$ (J $K^{-1}mol^{-1}$)	$-[G°-H°(T_r)]/T$ (J $K^{-1}mol^{-1}$)	$H°-H°(T_r)$ (kJ mol^{-1})	$\Delta_fH°$ (kJ mol^{-1})	$\Delta_fG°$ (kJ mol^{-1})	Log K_f
0	0.	0.	INFINITE	-9.098	169.452	169.452	INFINITE
100	29.217	203.629	265.506	-6.188	170.798	153.535	-80.198
200	31.228	224.384	240.293	-3.182	170.760	136.214	-35.575
250	32.473	231.489	237.843	-1.589	170.447	127.610	-26.663
298.15	33.475	237.297	237.297	0.	170.018	119.397	-20.918
300	33.509	237.505	237.298	0.062	170.000	119.083	-20.734
350	34.321	242.734	237.709	1.759	169.412	110.641	-16.512
400	34.949	247.359	238.632	3.491	160.767	102.563	-13.393
450	35.437	251.505	239.836	5.251	159.484	95.364	-11.070
500	35.821	255.259	241.193	7.033	138.109	90.209	-9.424
600	36.377	261.843	244.102	10.645	137.576	80.678	-7.024
700	36.755	267.481	247.048	14.302	136.959	71.242	-5.316
800	37.028	272.407	249.917	17.992	136.262	61.901	-4.042
900	37.236	276.781	252.663	21.706	135.488	52.652	-3.056
1000	37.401	280.713	255.275	25.438	134.636	43.492	-2.272
1100	37.538	284.284	257.752	29.185	133.705	34.422	-1.635
1200	37.655	287.556	260.102	32.945	132.690	25.441	-1.107
1300	37.758	290.574	262.331	36.716	131.588	16.547	-0.665
1400	37.850	293.376	264.450	40.496	130.393	7.742	-0.289
1500	37.934	295.990	266.466	44.286	129.098	-0.974	0.034
1600	38.012	298.441	268.389	48.083	113.161	-9.074	0.296
1700	38.085	300.747	270.225	51.888	112.050	-16.680	0.513
1800	38.155	302.926	271.982	55.700	110.892	-24.219	0.703
1900	38.222	304.991	273.665	59.519	109.684	-31.693	0.871
2000	38.286	306.953	275.281	63.344	108.427	-39.102	1.021
2100	38.349	308.823	276.834	67.176	107.120	-46.446	1.155
2200	38.410	310.608	278.329	71.014	105.767	-53.727	1.276
2300	38.470	312.317	279.770	74.858	104.372	-60.946	1.384
2400	38.530	313.955	281.160	78.708	102.938	-68.103	1.482
2500	38.588	315.529	282.504	82.564	101.472	-75.200	1.571
2600	38.647	317.044	283.803	86.426	99.979	-82.237	1.652
2700	38.706	318.504	285.062	90.293	98.465	-89.217	1.726
2800	38.765	319.912	286.281	94.167	-194.021	-89.917	1.677
2900	38.825	321.274	287.465	98.046	-194.497	-86.191	1.552
3000	38.886	322.591	288.614	101.932	-194.960	-82.448	1.436
3100	38.949	323.867	289.730	105.824	-195.409	-78.691	1.326
3200	39.014	325.105	290.817	109.722	-195.841	-74.919	1.223
3300	39.080	326.306	291.874	113.626	-196.253	-71.133	1.126
3400	39.149	327.474	292.904	117.538	-196.646	-67.336	1.034
3500	39.221	328.610	293.908	121.456	-197.018	-63.527	0.948
3600	39.295	329.716	294.887	125.382	-197.370	-59.708	0.866
3700	39.373	330.793	295.843	129.315	-197.702	-55.880	0.789
3800	39.454	331.844	296.777	133.257	-198.015	-52.042	0.715
3900	39.539	332.870	297.689	137.207	-198.312	-48.197	0.646
4000	39.628	333.872	298.581	141.165	-198.592	-44.344	0.579
4100	39.721	334.852	299.454	145.132	-198.860	-40.485	0.516
4200	39.818	335.810	300.308	149.109	-199.116	-36.619	0.455
4300	39.919	336.749	301.145	153.096	-199.362	-32.747	0.398
4400	40.024	337.667	301.964	157.093	-199.602	-28.869	0.343
4500	40.134	338.568	302.768	161.101	-199.837	-24.986	0.290
4600	40.247	339.451	303.556	165.120	-200.070	-21.098	0.240
4700	40.365	340.318	304.329	169.151	-200.302	-17.205	0.191
4800	40.486	341.169	305.087	173.193	-200.537	-13.307	0.145
4900	40.612	342.005	305.832	177.248	-200.776	-9.404	0.100
5000	40.742	342.827	306.564	181.316	-201.020	-5.496	0.057
5100	40.875	343.635	307.283	185.396	-201.273	-1.583	0.016
5200	41.012	344.430	307.990	189.491	-201.535	2.335	-0.023
5300	41.152	345.213	308.685	193.599	-201.808	6.258	-0.062
5400	41.295	345.983	309.368	197.721	-202.093	10.187	-0.099
5500	41.442	346.743	310.041	201.858	-202.392	14.121	-0.134
5600	41.591	347.491	310.703	206.010	-202.706	18.060	-0.168
5700	41.742	348.228	311.355	210.176	-203.035	22.005	-0.202
5800	41.896	348.955	311.997	214.358	-203.380	25.956	-0.234
5900	42.052	349.673	312.630	218.556	-203.741	29.913	-0.265
6000	42.210	350.381	313.253	222.769	-204.120	33.877	-0.295

PREVIOUS: December 1975 (1 atm) CURRENT: December 1975 (1 bar)

Beryllium Iodide (BeI) $Be_1I_1(g)$

BERYLLIUM IODIDE (BeI_2) CRYSTAL M_r = 262.82118

$S°(298.15 K) = [120.5 \pm 4.2] J K^{-1} mol^{-1}$
$T_{fus} = 753 \pm 15 K$

$\Delta_f H°(0 K)$ = Unknown
$\Delta_f H°(298.15 K) = [-188.7 \pm 21] kJ mol^{-1}$
$\Delta_{fus} H° = [20.9 \pm 12.6] kJ mol^{-1}$

Enthalpy of Formation

A direct measurement of the enthalpy of formation has not been made. An estimate of $\Delta_f H°$ is obtained by a method suggested by Parker (1) and used in the $BeBr_2$(cr) table (2). Biltz and Messerknecht (3) have measured the enthalpies of solution of $BeCl_2$(cr) and BeI_2(cr) is aqueous HCl (18.69%). Samples (4) of the dihalides were prepared from reactions of BeO-C mixtures with the halogens at elevated temperatures. We assume that the two dihalides had similar structures, i.e. the α-form (orthorhombic). $\Delta_f H°(BeI_2$, in 18.7% HCl, 298.15 K) is estimated at $-109.1 kcal mol^{-1}$ by combing $\Delta_f H°$ of $BeCl_2$(cr) with $\Delta_f H°(BeCl_2$, cr, α, 298.15 K) = $-117.3 \pm 0.8 kcal mol^{-1}$ (2) and twice the difference in $\Delta_f H°(HCl \cdot 8H_2O$, 298.15 K) and $\Delta_f H°$(298.15 K, $HI \cdot 8H_2O$) which is $-52.3 kcal mol^{-1}$ (5). The enthalpy associated with the aqueous anions is assumed to be negligible. Combination of the estimated value for $\Delta_f H°(BeI_2$, in 18.8% HCl, 298.15 K) with $\Delta_f H° = -62.5 kcal mol^{-1}$ (3) gives $\Delta_f H°(BeI_2$, cr, 298.15 K) = $-46.6 kcal mol^{-1}$ which is within 0.6 $kcal mol^{-1}$ of the value suggested by NBS (6). Subsequent measurements by Biltz et al. (7) in less concentrated HCl (1.48%) solution leads to $\Delta_f H°(298.15 K) = -43.6 kcal mol^{-1}$ by a similar route. We adopt an average value of $-45.1 kcal mol^{-1}$ but emphasize that the uncertainty in $\Delta_f H°$ is much greater than indicated (3 $kcal mol^{-1}$) by the closeness of these two values. An uncertainty of ± 0.5 $kcal mol^{-1}$ is believed to be more realistic.

Heat Capacity and Entropy

No low-temperature heat capacity or high-temperature enthalpy measurements have been reported. We estimate Cp°(298.15 K) = 16.482 $cal K^{-1} mol^{-1}$ from the reaction $BeCl_2$(cr, α) + 2LiI(cr) = BeI_2(cr) + 2LiCl(cr) by assuming ΔCp° = 0. Similar results are obtained using BeF_2(cr). Cp° values above 298.15 K are assumed to parallel those for α-$BeCl_2$ (2).

Application of the Berthelot principle (8) to the process BaI_2(cr) + Be(cr) → Ba(cr) + BeI_2(cr) suggests S°(298.15 K) = 26.8 $cal K^{-1} mol^{-1}$. A graphical comparison of the standard entropies for other alkaline-earth dihalides (2) indicates that this value is reasonable. Other estimated values are given by additive entropy constants (9, 27.4 $cal K^{-1} mol^{-1}$), Brewer (10, 31.0 $cal K^{-1} mol^{-1}$), and Brewer et al. (11, 25 $cal K^{-1} mol^{-1}$).

For additional information of the heat capacity and entropy, refer to the enthalpy of formation discussion for gaseous BeI_2 (2).

Fusion Data

See the liquid table for details.

Phase Data

Semenenko and Naumova (12) studied the BeI_2 crystal modifications by thermal and x-ray techniques. The study was complicated by the extreme hygroscopicity, high vapor pressures near the melting point, and susceptibility to oxidation on heating by traces of oxygen. As stated by Semenenko and Naumova (12), the sequence of polymorphic conversion and the character of the resulting modifications of BeI_2, which are largely dependent on the heating and cooling conditions, parallel those of $BeCl_2$ (13). They detected thermally transitions at 290°C, 370°C, 470°C, and a melting at 490°C. However, this data does not appear to be fully consistent in terms of crystallographic structures with that proposed by Johnson, Staritzky, and Douglas (14) and Messerknecht and Biltz (15). These latter works suggested two structures with a transition at 350°C.

We assume the similarity with $BeCl_2$ with a transition in the vicinity of 350 - 370°C. Further study is necessary to resolve all possible phases and their structures.

Sublimation Data

See the ideal gas table for details.

References

1. V. B. Parker, personal communication, Nat. Bur. Stand., March, 1975.
2. JANAF Thermochemical Tables: $BeBr_2$(cr), 6-30-75; BeI_2(ℓ) and BeI_2(g), 12-31-75; $BeCl_2$(cr, α), 6-30-75.
3. W. Biltz and C. Messerknecht. Z. Anorg. Chem. 148, 157 (1925).
4. C. Messerknecht and W. Biltz, Z. Anorg. Chem. 148, 152 (1925).
5. U. S. Nat. Bur. Stand. Tech. Note 270-3, 1968.
6. U. S. Nat. Bur. Stand. Tech. Note 270-6, 1971.
7. W. Biltz, K. A. Klatte, and E. Rahlfs, Z. Anorg. Chem. 166, 339 (1927).
8. N. N. Drozin, Zhur. Fiz. Khim. 35, 879 (1961).
9. D. R. Stull and H. Prophet in "The Characterization of High Temperature Vapors," J. L. Margrave, Ed., John Wiley and Sons, Inc., New York, 1966.
10. L. Brewer, L. A. Bromley, P. W. Gilles, and N. L. Lofgren in "Chemistry and Metallurgy of Miscellaneous Materials: Thermodynamics," L. L. Quill, Ed., National Nuclear Energy Series IV-19B, McGraw-Hill Book Co., New York, 1950.
11. L. Brewer, G. R. Somayajulu, and E. Brackett, Chem. Revs. 63, 111 (1963).
12. K. N. Semenenko and T. N. Naumova, Russ. J. Structural Chem. 4, 59 (1963).
13. O. N. Kuvyrkin, O. N. Breusov, A. V. Novoselova, and K. N. Semenenko, Zhur. Fiz. Khim. 34, 343 (1960).
14. R. E. Johnson, E. Staritzky, and R. M. Douglas, J. Amer. Chem. Soc. 79, 2037 (1957).
15. C. Messerknecht and W. Biltz, Z. anorg. allgem. Chem. 48, 152 (1925).

Beryllium Iodide (BeI_2) Be_1I_2(cr)

Enthalpy Reference Temperature = T_r = 298.15 K Standard State Pressure = p° = 0.1 MPa

T/K	$C_p°$ (J K⁻¹mol⁻¹)	S° (J K⁻¹mol⁻¹)	$-[G°-H°(T_r)]/T$ (J K⁻¹mol⁻¹)	$H°-H°(T_r)$ (kJ mol⁻¹)	$\Delta_f H°$ (kJ mol⁻¹)	$\Delta_f G°$ (kJ mol⁻¹)	Log K_f
0							
100							
200							
298.15	68.961	120.499	120.499	0.	-188.698	-187.183	32.794
300	69.128	120.926	120.501	0.128	-188.702	-187.174	32.590
400	76.868	141.944	123.314	7.452	-204.858	-186.084	24.300
500	81.638	159.651	128.858	15.396	-247.213	-177.328	18.525
600	84.190	174.780	135.283	23.698	-244.932	-163.563	14.239
700	85.467	187.860	141.881	32.185	-242.604	-150.186	11.207
753.000	85.944	194.116	145.340	36.728	------ CRYSTAL <--> LIQUID -----		
800	86.366	199.331	148.360	40.777	-240.288	-137.142	8.954
900	87.132	209.549	154.602	49.453	-237.995	-124.386	7.219
1000	87.789	218.764	160.565	58.200	-235.733	-111.885	5.844
1100	88.337	227.158	166.243	67.007	-233.512	-99.608	4.730
1200	88.780	234.864	171.645	75.864	-231.346	-87.531	3.810
1300	89.115	241.984	176.785	84.759	-229.252	-75.632	3.039
1400	89.341	248.597	181.681	93.683	-227.247	-63.891	2.384
1500	89.462	254.766	186.350	102.624	-225.354	-52.289	1.821

PREVIOUS: June 1965 CURRENT: December 1975

Beryllium Iodide (BeI_2) Be_1I_2(cr)

BERYLLIUM IODIDE (BeI_2) LIQUID M_r = 262.82118

$S°(298.15\ K) = [128.999]\ J\ K^{-1}\ mol^{-1}$ $\Delta_f H°(298.15\ K) = [-178.377]\ kJ\ mol^{-1}$

$T_{fus} = 753 \pm 15\ K$ $\Delta_{fus}H° = [20.9 \pm 12.6]\ kJ\ mol^{-1}$

Enthalpy of Formation

The enthalpy of formation is calculated from that of the crystal by adding $\Delta_{fus}H°$ and the differnce in enthalpy, H°(753 K) - H°(298.15 K), between the crystal and liquid.

Heat Capacity and Entropy

The heat capacity is estimated by comparison with the measured value for $BeCl_2$ (1). A glass transition is assumed at 400 K below which the heat capacity is that of the crystal. The entropy is calculated in a manner analogous to that used for the enthalpy of formation.

Fusion Data

Rahlfs and Fischer (2) reported a melting point of 753 K. Semenenko and Naumova (3), using thermal analysis techniques, implied a melting point of 763 K. Since this latter value was derived from a thermogram, it is probable that this value represents a maximum, with the true melting point somewhat lower. We adopt T_{fus} = 753 ± 15 K.

We estimate the enthalpy of melting, $\Delta_{fus}H° = 5.0 \pm 3.0\ kcal\ mol^{-1}$. This estimate includes a contribution for melting (based on the entropy of melting for other alkaline earth dihalides) and a contribution for a phase transition. There is no experimental data available pertaining to the enthalpy of melting.

Vaporization Data

T_{vap} is the temperature at which $\Delta_r G° = 0$ for the process $BeI_2(\ell) = BeI_2(g)$. $\Delta_{vap}H°$ is the corresponding difference in the $\Delta_f H°$ values for the liquid and gas at T_{vap}.

References

1. JANAF Thermochemical Table: $BeCl_2(\ell)$, 6-30-65.
2. O. Rahlfs and W. Fischer, Z. Anorg. Chem. 211, 349 (1933).
3. K. N. Semenenko and J. N. Naumova, Russ. J. Structural Chem. 4, 59 (1963).

Beryllium Iodide (BeI_2) $Be_1I_2(l)$

Enthalpy Reference Temperature = T_r = 298.15 K   Standard State Pressure = p° = 0.1 MPa

T/K	$C_p°$ (J K⁻¹mol⁻¹)	S° (J K⁻¹mol⁻¹)	$-[G°-H°(T_r)]/T$ (J K⁻¹mol⁻¹)	$H°-H°(T_r)$ (kJ mol⁻¹)	$\Delta_f H°$ (kJ mol⁻¹)	$\Delta_f G°$ (kJ mol⁻¹)	Log K_f
0							
100							
200							
298.15	68.961	128.999	128.999	0.	-178.377	-179.396	31.429
300	69.128	129.426	129.000	0.128	-178.380	-179.402	31.237
400	76.868	150.434	131.812	7.449	-194.539	-179.162	23.396
400.000	76.868	150.434	131.812	7.449	GLASS <--> LIQUID TRANSITION		
400.000	112.968	150.434	131.812	7.449			
500	112.968	175.643	138.151	18.746	-233.542	-171.653	17.932
600	112.968	196.239	146.168	30.043	-228.266	-159.773	13.909
700	112.968	213.653	154.597	41.339	-223.128	-148.765	11.101
753.000	112.968	221.898	159.047	47.327	CRYSTAL <--> LIQUID		
800	112.968	228.738	162.943	52.636	-218.107	-138.486	9.042
900	112.968	242.044	171.007	63.933	-213.193	-128.830	7.477
1000	112.968	253.946	178.716	75.230	-208.381	-119.715	6.253
1100	112.968	264.713	186.053	86.527	-203.671	-111.077	5.275
1200	112.968	274.542	193.023	97.823	-199.065	-102.863	4.478
1300	112.968	283.585	199.646	109.120	-194.569	-95.030	3.818
1400	112.968	291.957	205.944	120.417	-190.192	-87.538	3.266
1500	112.968	299.751	211.941	131.714	-185.943	-80.355	2.798

PREVIOUS: June 1965   CURRENT: December 1975

BERYLLIUM IODIDE (BeI_2) CRYSTAL-LIQUID M_r = 262.82118

0 to 753 K crystal
above 753 K liquid

Refer to the individual tables for details.

Beryllium Iodide (BeI_2) $Be_1I_2(cr,l)$

Enthalpy Reference Temperature = T_r = 298.15 K Standard State Pressure = p° = 0.1 MPa

T/K	C_p° (J K⁻¹mol⁻¹)	S° (J K⁻¹mol⁻¹)	$-[G^\circ - H^\circ(T_r)]/T$ (J K⁻¹mol⁻¹)	$H^\circ - H^\circ(T_r)$ (kJ mol⁻¹)	$\Delta_f H^\circ$ (kJ mol⁻¹)	$\Delta_f G^\circ$ (kJ mol⁻¹)	Log K_f
0							
100							
200							
298.15	68.961	120.499	120.499	0.	-188.698	-187.183	32.794
300	69.128	120.926	120.501	0.128	-188.702	-187.174	32.590
400	76.868	141.944	123.314	7.452	-204.858	-186.084	24.300
500	81.638	159.651	128.858	15.396	-247.213	-177.328	18.525
600	84.190	174.780	135.283	23.698	-244.932	-163.563	14.239
700	85.467	187.860	141.881	32.185	-242.604	-150.186	11.207
753.000	85.944	194.116	145.340	36.728	CRYSTAL <--> LIQUID		
753.000	112.968	221.898	145.340	57.648	TRANSITION		
800	112.968	228.738	150.041	62.958	-218.107	-138.486	9.042
900	112.968	242.044	159.539	74.254	-213.193	-128.830	7.477
1000	112.968	253.946	168.395	85.551	-208.381	-119.715	6.253
1100	112.968	264.713	176.669	96.848	-203.671	-111.077	5.275
1200	112.968	274.542	184.422	108.145	-199.065	-102.863	4.478
1300	112.968	283.585	191.707	119.442	-194.569	-95.030	3.818
1400	112.968	291.957	198.572	130.738	-190.192	-87.538	3.266
1500	112.968	299.751	205.060	142.035	-185.943	-80.354	2.798

PREVIOUS: CURRENT: December 1975

Beryllium Iodide (BeI_2) $Be_1I_2(cr,l)$

BERYLLIUM IODIDE (BeI_2) IDEAL GAS M_r = 262.82118

S°(298.15 K) = [291.52 ± 2.1] J K^{-1} mol^{-1}

Δ_fH°(0 K) = -62.5 ± 33.5 kJ mol^{-1}
Δ_fH°(298.15 K) = -64.0 ± 33.5 kJ mol^{-1}

Vibrational Frequencies and Degeneracies

ν, cm^{-1}
[160] (1)
[175] (2)
873 (1)

Ground State Quantum Weight: [1] σ = 2
Point Group: $D_{\infty h}$
Bond Distance: Be-I = 2.12 ± 0.05 Å
Bond Angle: I-Be-I = 180° ± 10°
Rotational Constant: B_o = 0.01478 cm^{-1}

Enthalpy of Formation

Rahlfs and Fischer (1) have reported measurements of the sublimaiton pressures (578-703 K, 9 pts) for BeI_2. The measurements were complicated by reaction of the diiodide with the quartz apparatus. Assuming the reaction to be BeI_2(cr) + SiO_2(cr) = SiI_4(g) + 2BeO(cr), Rahlfs and Fischer (1) corrected their measured total pressure for the partial pressure of the tetraiodide. A 2nd and 3rd law analysis of their corrected data yields Δ_{sub}H°(298.15 K) = 29.81 ± 0.29 kcal mol^{-1} (3rd law) and 28.34 ± 0.74 kcal mol^{-1} (2nd law) with a drift of 2.2 ± 1.1 cal K^{-1} mol^{-1}. We adopt Δ_{sub}H°(298.15 K) = 29.8 ± 2.0 kcal mol^{-1}. We have assumed negligible dimer formation. Δ_fH°(298.15 K) = -15.3 kcal mol^{-1} for BeI_2(g) when the adopted Δ_{sub}H°(298.15 K) value is added to the Δ_fH°(298.15 K) value for BeI_2(cr).

The drift could be reduced by further adjusting the Gibbs energy functions for the crystal. These functions may be changed by altering the Cp° values and/or the S°(298.15 K) value. Such changes would yield values which are unreasonable when compared to other alkaline earth dihalides. Note also that a similar sublimation study for $BeCl_2$ (2) gives a drift of -1.3 ± 2.6 cal K^{-1} mol^{-1}. We tentatively assume the drift is due to the data rather than our choice of functions.

Heat Capacity and Entropy

Electron diffraction patterns for BeI_2 vapor (3, 4) have been interpreted in terms of a linear configuration, with the latter study reporting the Be-I bond distance as 2.12 A. Other studies also suggest a linear molecule. For example, the electric deflection of mass spectrometricaly detected molecular beams (5) showed that all the beryllium dihalides are linear.

Snelson (6) observed the infrared spectra of BeI_2 in the spectral range 4000-200 cm^{-1} using a matrix isolation technique. Assuming a linear geometry, ν_3 was assigned as 873 cm^{-1}. Snelson (6) estimated a value for ν_2 based on force constant values for the other three beryllium dihalides. The values of ν_1 were calculated using a simple valence force field approximation. We adopt the two estimated and one measured vibrational frequency as reported by Snelson (6). The ground state quantum weight of one is assigned by analogy with $BaCl_2$ (2).

Brewer et al. (7) have tabulated Gibbs energy functions for BeI_2 up to 2000 K. Their values are consistently lower than ours due primarily to their use of a higher bending frequency (ν_2 = 395 cm^{-1}).

References

1. O. Rahlfs and W. Fischer, Z. Anorg. Chem. 211, 349 (1933).
2. JANAF Thermochemical Tables: $BeCl_2$(ℓ), 6-30-65.
3. P. S. Akishin and V. P. Spiridonov, Kristallografiya 2, 472 (1957).
4. P. A. Akishin, V. P. Spiridonov and G. A. Sobalev, Dokl. Akad. Nauk SSSR 118, 1134 (1958).
5. A. Buchler, J. L. Stauffer, and W. Klemperer, J. Amer. Chem. Soc. 86, 4544 (1964).
6. A. Snelson, J. Phys. Chem. 72, 250 (1968).
7. L. Brewer, G. R. Somoyajulu, and E. Brackett, Chem. Revs. 63, 111 (1963).

Beryllium Iodide (BeI_2) Be_1I_2(g)

Enthalpy Reference Temperature = T_r = 298.15 K Standard State Pressure = p° = 0.1 MPa

T/K	C_p° (J $K^{-1}mol^{-1}$)	S° (J $K^{-1}mol^{-1}$)	-[G°-H°(T_r)]/T (J $K^{-1}mol^{-1}$)	H°-H°(T_r) (kJ mol^{-1})	Δ_fH° (kJ mol^{-1})	Δ_fG° (kJ mol^{-1})	Log K_f
0	0.	0.	INFINITE	-13.633	-62.517	-62.517	INFINITE
100	44.605	236.565	337.992	-10.143	-62.117	-79.624	41.591
200	51.767	270.220	296.456	-5.247	-62.731	-96.949	25.320
250	53.542	281.970	292.421	-2.613	-63.316	-105.439	22.030
298.15	54.939	291.524	291.524	0.	-64.015	-113.491	19.883
300	54.988	291.864	291.525	0.102	-64.045	-113.798	19.814
350	56.181	300.433	292.199	2.882	-64.955	-122.021	18.211
400	57.156	308.001	293.710	5.716	-81.910	-129.559	16.919
450	57.948	314.780	295.681	8.594	-84.092	-135.385	15.715
500	58.591	320.920	297.903	11.508	-126.417	-137.167	14.330
600	59.544	331.692	302.662	17.418	-126.529	-139.307	12.128
700	60.192	340.923	307.484	23.407	-126.699	-141.425	10.553
800	60.648	348.992	312.179	29.450	-126.931	-143.513	9.370
900	60.977	356.155	316.675	35.532	-127.232	-145.569	8.449
1000	61.222	362.593	320.950	41.643	-127.606	-147.587	7.709
1100	61.408	368.437	325.005	47.775	-128.061	-149.563	7.102
1200	61.552	373.787	328.851	53.923	-128.604	-151.495	6.594
1300	61.667	378.718	332.499	60.084	-129.243	-153.377	6.163
1400	61.759	383.292	335.966	66.256	-129.991	-155.206	5.791
1500	61.834	387.555	339.265	72.436	-130.859	-156.978	5.466
1600	61.895	391.548	342.409	78.622	-146.395	-158.162	5.163
1700	61.947	395.302	345.411	84.814	-147.137	-158.875	4.882
1800	61.990	398.844	348.282	91.011	-147.961	-159.542	4.630
1900	62.027	402.197	351.032	97.212	-148.871	-160.161	4.403
2000	62.059	405.379	353.671	103.417	-149.870	-160.730	4.198
2100	62.086	408.408	356.206	109.624	-150.955	-161.246	4.011
2200	62.110	411.296	358.645	115.834	-152.121	-161.710	3.839
2300	62.131	414.058	360.994	122.046	-153.361	-162.118	3.682
2400	62.149	416.702	363.261	128.260	-154.666	-162.471	3.536
2500	62.165	419.240	365.450	134.476	-156.025	-162.768	3.401
2600	62.180	421.678	367.566	140.693	-157.425	-163.010	3.275
2700	62.193	424.025	369.614	146.911	-158.856	-163.198	3.157
2800	62.204	426.287	371.598	153.131	-451.264	-157.110	2.931
2900	62.214	428.470	373.521	159.352	-451.660	-146.597	2.641
3000	62.224	430.580	375.388	165.574	-452.039	-136.071	2.369
3100	62.232	432.620	377.202	171.797	-452.394	-125.533	2.115
3200	62.240	434.596	378.965	178.021	-452.718	-114.984	1.877
3300	62.247	436.511	380.680	184.245	-453.008	-104.425	1.653
3400	62.253	438.370	382.349	190.470	-453.259	-93.858	1.442
3500	62.259	440.174	383.976	196.696	-453.471	-83.284	1.243
3600	62.265	441.928	385.561	202.922	-453.641	-72.705	1.055
3700	62.270	443.634	387.108	209.149	-453.769	-62.122	0.877
3800	62.274	445.295	388.617	215.376	-453.858	-51.535	0.708
3900	62.278	446.913	390.091	221.603	-453.907	-40.947	0.548
4000	62.282	448.489	391.532	227.831	-453.920	-30.358	0.396
4100	62.286	450.027	392.940	234.060	-453.898	-19.770	0.252
4200	62.289	451.528	394.317	240.289	-453.844	-9.182	0.114
4300	62.292	452.994	395.664	246.518	-453.763	1.404	-0.017
4400	62.295	454.426	396.984	252.747	-453.656	11.989	-0.142
4500	62.298	455.826	398.276	258.977	-453.528	22.570	-0.262
4600	62.301	457.196	399.542	265.207	-453.381	33.148	-0.376
4700	62.303	458.535	400.783	271.437	-453.220	43.723	-0.486
4800	62.305	459.847	402.000	277.667	-453.048	54.295	-0.591
4900	62.308	461.132	403.193	283.898	-452.867	64.863	-0.691
5000	62.310	462.391	404.365	290.129	-452.683	75.427	-0.788
5100	62.311	463.625	405.515	296.360	-452.496	85.987	-0.881
5200	62.313	464.835	406.644	302.591	-452.311	96.544	-0.970
5300	62.315	466.022	407.753	308.822	-452.129	107.097	-1.056
5400	62.316	467.186	408.843	315.054	-451.954	117.647	-1.138
5500	62.318	468.330	409.914	321.286	-451.788	128.193	-1.217
5600	62.319	469.453	410.967	327.518	-451.633	138.737	-1.294
5700	62.321	470.556	412.003	333.750	-451.490	149.278	-1.368
5800	62.322	471.640	413.022	339.982	-451.362	159.817	-1.439
5900	62.323	472.705	414.025	346.214	-451.250	170.353	-1.508
6000	62.324	473.752	415.011	352.446	-451.154	180.888	-1.575

PREVIOUS: December 1975 (1 atm) CURRENT: December 1975 (1 bar)

Beryllium Iodide (BeI_2) Be_1I_2(g)

BERYLLIUM NITRIDE (BeN) IDEAL GAS M_r = 23.01888

$S°(298.15\ K) = [208.77]\ J\ K^{-1}\ mol^{-1}$

$\Delta_f H°(0\ K) = [424]\ kJ\ mol^{-1}$

$\Delta_f H°(298.15\ K) = [427]\ kJ\ mol^{-1}$

Electronic Levels and Multiplicities

state	ε_i, cm^{-1}	g_i
$^2\Pi$	0	[4]

$\omega_e = [1194]\ cm^{-1}$ $\omega_e x_e = [12.4]\ cm^{-1}$ $\sigma = 1$

$B_e = [1.555]\ cm^{-1}$ $\alpha_e = [0.019]$ $r_e = [1.406]$ Å

Enthalpy of Formation

The enthalpy of formation was taken from Gordon (1).

Heat Capacity and Entropy

The molecular constants were taken from Gordon (1).

Reference

1. J. S. Gordon, personal communication, April, 1963.

Beryllium Nitride (BeN) $Be_1N_1(g)$

Enthalpy Reference Temperature = T_r = 298.15 K Standard State Pressure = p° = 0.1 MPa

T/K	$C_p°$ (J K⁻¹mol⁻¹)	S° (J K⁻¹mol⁻¹)	$-[G°-H°(T_r)]/T$ (J K⁻¹mol⁻¹)	$H°-H°(T_r)$ (kJ mol⁻¹)	$\Delta_f H°$ (kJ mol⁻¹)	$\Delta_f G°$ (kJ mol⁻¹)	Log K_f
0	0.	0.	INFINITE	-8.725	424.231	424.231	INFINITE
100	29.111	176.758	234.965	-5.821	425.645	416.010	-217.301
200	29.254	196.957	211.487	-2.906	426.527	405.969	-106.028
250	29.583	203.516	209.260	-1.436	426.677	400.808	-83.744
298.15	30.089	208.768	208.768	0.	426.688	395.823	-69.347
300	30.111	208.954	208.768	0.056	426.687	395.632	-68.886
350	30.757	213.643	209.137	1.577	426.602	390.461	-58.273
400	31.439	217.794	209.964	3.132	426.462	385.308	-50.316
450	32.102	221.536	211.046	4.721	426.292	380.173	-44.129
500	32.716	224.951	212.268	6.341	426.099	375.059	-39.182
600	33.761	231.012	214.900	9.667	425.667	364.891	-31.767
700	34.574	236.280	217.586	13.086	425.172	354.800	-26.475
800	35.203	240.939	220.219	16.576	424.611	344.784	-22.512
900	35.693	245.115	222.758	20.122	423.976	334.843	-19.434
1000	36.081	248.897	225.185	23.711	423.262	324.977	-16.975
1100	36.395	252.351	227.500	27.336	422.467	315.186	-14.967
1200	36.653	255.529	229.705	30.988	421.587	305.471	-13.297
1300	36.869	258.471	231.806	34.665	420.622	295.833	-11.887
1400	37.052	261.211	233.810	38.361	419.569	286.273	-10.681
1500	37.211	263.772	235.723	42.074	418.427	276.791	-9.639
1600	37.351	266.179	237.552	45.803	402.661	267.914	-8.747
1700	37.475	268.447	239.303	49.544	401.741	259.521	-7.974
1800	37.587	270.592	240.982	53.297	400.800	251.182	-7.289
1900	37.689	272.627	242.595	57.061	399.838	242.896	-6.678
2000	37.783	274.563	244.145	60.835	398.855	234.661	-6.129
2100	37.870	276.408	245.638	64.617	397.851	226.476	-5.633
2200	37.951	278.172	247.077	68.408	396.826	218.339	-5.184
2300	38.027	279.860	248.466	72.207	395.781	210.249	-4.775
2400	38.100	281.480	249.808	76.014	394.716	202.206	-4.401
2500	38.169	283.037	251.106	79.827	393.632	194.206	-4.058
2600	38.236	284.535	252.363	83.648	392.527	186.251	-3.742
2700	38.300	285.980	253.582	87.474	391.403	178.339	-3.450
2800	38.361	287.374	254.764	91.307	99.302	176.691	-3.296
2900	38.421	288.721	255.912	95.147	99.202	179.457	-3.232
3000	38.479	290.024	257.027	98.992	99.100	182.226	-3.173
3100	38.536	291.287	258.112	102.842	98.994	184.998	-3.117
3200	38.592	292.511	259.168	106.699	98.882	187.774	-3.065
3300	38.646	293.700	260.197	110.561	98.764	190.554	-3.016
3400	38.700	294.854	261.199	114.428	98.638	193.337	-2.970
3500	38.752	295.977	262.177	118.301	98.501	196.125	-2.927
3600	38.804	297.069	263.131	122.178	98.353	198.916	-2.886
3700	38.855	298.133	264.063	126.061	98.190	201.712	-2.848
3800	38.906	299.170	264.973	129.949	98.012	204.512	-2.811
3900	38.956	300.181	265.863	133.843	97.816	207.317	-2.777
4000	39.005	301.168	266.733	137.741	97.600	210.127	-2.744
4100	39.054	302.132	267.585	141.644	97.362	212.944	-2.713
4200	39.102	303.074	268.419	145.551	97.099	215.766	-2.683
4300	39.151	303.994	269.235	149.464	96.810	218.594	-2.655
4400	39.198	304.895	270.035	153.381	96.492	221.430	-2.629
4500	39.246	305.776	270.820	157.304	96.143	224.274	-2.603
4600	39.293	306.639	271.589	161.231	95.763	227.125	-2.579
4700	39.340	307.485	272.344	165.162	95.348	229.985	-2.556
4800	39.387	308.314	273.085	169.099	94.896	232.855	-2.534
4900	39.433	309.126	273.812	173.040	94.408	235.734	-2.513
5000	39.479	309.923	274.526	176.985	93.880	238.623	-2.493
5100	39.525	310.706	275.228	180.935	93.311	241.524	-2.474
5200	39.571	311.474	275.918	184.890	92.701	244.436	-2.455
5300	39.617	312.228	276.596	188.850	92.048	247.360	-2.438
5400	39.662	312.969	277.262	192.814	91.352	250.297	-2.421
5500	39.708	313.697	277.918	196.782	90.612	253.247	-2.405
5600	39.753	314.413	278.564	200.755	89.826	256.211	-2.390
5700	39.798	315.117	279.199	204.733	88.995	259.189	-2.375
5800	39.843	315.809	279.824	208.715	88.119	262.183	-2.361
5900	39.888	316.491	280.440	212.701	87.196	265.192	-2.348
6000	39.933	317.162	281.046	216.692	86.227	268.217	-2.335

PREVIOUS: June 1963 (1 atm) CURRENT: June 1963 (1 bar)

Beryllium Nitride (BeN) $Be_1N_1(g)$

BERYLLIUM OXIDE, ALPHA (α-BeO) CRYSTAL M_r = 25.01158

$S°(298.15\ K) = 13.770 \pm 0.21\ J\ K^{-1}\ mol^{-1}$
$T_{trs} = 2373 \pm 15\ K$ (α-β)
$T_{fus} = [2821.2 \pm 100]\ K$ (β-ℓ)
$T_{fus} = 2780 \pm 100\ K$ (α-ℓ)

$\Delta_f H°(0\ K) = -604.9 \pm 0.3\ kJ\ mol^{-1}$
$\Delta_f H°(298.15\ K) = -608.4 \pm 3.3\ kJ\ mol^{-1}$
$\Delta_{trs} H° = [6.694 \pm 1.7]\ kJ\ mol^{-1}$
$\Delta_{fus} H° = [79.057 \pm 6.3]\ kJ\ mol^{-1}$
$\Delta_{fus} H° = 84.9 \pm 6.3\ kJ\ mol^{-1}$

Enthalpy of Formation

Parker (1) thoroughly reviewed the data as of 1969 and selected -145.4 ± 0.8 kcal mol^{-1}. She later revised (1) this value to -145.7 ± 0.6 kcal mol^{-1} due to new HF-solution calorimtry on BeF_2(amorphous) performed by Kilday et al. (2). Values of $\Delta_f H°$ from Parker's revised analysis (1) are summarized below. We adopt -145.4 ± 0.8, rather than the revised selection, as a compromise between the indirect results based on Be(cr) and those based on BeF_2(am). The latter depend on $\Delta_f H°$ for HF(n H_2O). Use of the JANAF $\Delta_f H°$ (4) in place of the NBS value (3) causes a change of ± 0.9 kcal mol^{-1} in the results based on BeF_2(am). Recent data for HF suggest that the change could be even larger.

Direct Determination of $\Delta_f H°$

kcal mol^{-1}	Source
-136.2	Mielenz & Von Wartenberg (1921)
-134.4	Moose & Parr (1924)
-145.3	Neumann et al. (1934)
-147.3	Roth et al. (1938)
-143.1	Cosgrove & Snyder (1953)

Indirect Determinations of $\Delta_f H°$

kcal mol^{-1}	Source
-144.9	Neumann et al. (1934, 1932), $\Delta_c H°$ & $\Delta_f H°$ of Be_3N_2
-141.1	Smirnov & Chukreev (1958), emf data
-145.7±1.5	Kilday et al. (1969), Thomson et al. (1962)[a]
-143.9±1.5	Kilday et al. (1969), Armstrong & Coyle (1965)[b]
-145.7±0.6	Kolesov et al. (1959), Bear & Turnbull (1965)[b]
-146.2, -145.65, -145.65, -145.3±0.6	Kilday et al. (1959), Bear & Turnbull (1969)[b]
-144.7(-145.6)[d], -144.5(-145.4)[d]±1.2	Kilday et al. (1971), Churney & Armstrong (1969)[c]

[a]HCl-solution calorimetry of BeO(cr) and Be(cr). [b]HF-solution calorimetry of BeO(cr) and Be(cr). [c]HF-solution calorimetry of BeO(cr) and BeF_2(am) and $\Delta_f H°$ of the latter. [d]Values in parentheses based on $\Delta_f H°$ of HF(n H_2O) from NBS (3) instead of JANAF (4).

Heat Capacity and Entropy

The adopted Cp° and S° below 298.15 K are taken from Furukawa and Reilly (5) who measured Cp° from 16 to 370 K. The authors give a detailed comparison with the earlier data of Gmelin (6, 5 to 75 K) and Kelley (7, 56 to 292 K). Gmelin's results deviate by roughly +20% from 15 to 75 and Kelley's results deviate by about +50% at 56 K, +8% at 100 K and +1% at 200 K. The deviation of Kelley are probably due to the sensitivity limit of his calorimeter and to non-standard state effects of finely powdered BeO.

The adopted Cp° above 298.15 K is from a constrained fit of the Cp° data (5) and enthalpy data of Victor and Douglas (8, 323 to 1173 K), Conway and Hein (9, 2161 to 2365 K) and Shpil'rain et al. (10, 2023 to 2708 K). The latter data show no evidence of transition even though they extend more than 300 K above T_{trs} = 2373 K. Deviations of the data (8) from the adopted enthalpies are <±0.15 % above 373 K and -0.4% at 323 K. Deviations above 2000 K are -0.3 to +0.8% (9) and -1.8 to +0.4% (10) excluding the point at 2266 K (10). Enthalpy measurements not used in the fit deviate by -2.1 to +2.9% (11, 1142 to 2697 K), -17±8% (12, 2273 to 2523 K), -2.1±0.8% (13, 400 to 1100 K) and +0.4 to -0.7% (14, 363 to 1128 K). Cp° data from the cooling-rate method (15) deviate by <2% (1300 to 1700 K) but by -5% at 2000 K.

Transition Data

See the β-crystal table for details.

Fusion Data

See the liquid table for details.

References

1. V. B. Parker, J. Res. Nat. Bur. Stand. 77A, 227 (1973); U. S. Nat. Bur. Stand. Report 10004, pp. 58-81, Jan., 1969.
2. M. V. Kilday, E. J. Prosen and D. D. Wagman, J. Res. Nat. Bur. Stand. 77A, 217 (1973).
3. U. S. Nat. Bur. Stand. Tech. Note 270-3, 1968.
4. JANAF Thermochemical Tables: HF(g), 12-31-68; BeF_2(cr, ℓ), 6-30-70.
5. G. T. Furukawa and M. L. Reilly, U. S. Nat. Bur. Stand. Report 9905, pp. 28-47, July, 1968.
6. E. Gmelin, Compt. Rend. 262, 1452 (1966).
7. K. K. Kelley, J. Amer. Chem. Soc. 61, 1217 (1939).
8. A. C. Victor and T. B. Douglas, J. Res. Nat. Bur. Stand 67A, 325 (1963).
9. J. B. Conway and R. A. Hein, Nucleonics 22 (6), 71 (1964).
10. E. E. Shpil'rain, D. N. Kagan and L. S. Barkhatov, High Temperature (USSR) 9, 842 (1971).
11. V. V. Kandyba, P. B. Kantor, R. M. Krasovitskaya and E. N. Fomichev, Dokl. Akad. Nauk SSSR 131, 566 (1960).
12. M. A. Greenbaum, J. Weiher and M. Farber, J. Phys. Chem. 69, 4035 (1965).
13. B. E. Walker, C. T. Ewing and R. R. Miller, J. Chem. Eng. Data 7, 595 (1962); B. E. Walker, personal communication, U. S. Naval Res. Lab., Washington, April 18, 1963.
14. E. N. Rodigina and K. Z. Gomel'skii, Russ. J. Phys. Chem. 35, 898 (1961).
15. R. A. McDonald, F. L. Oetting and H. Prophet, CPIA Working Group on Thermochemistry, Proc. 1st Meeting, Nov., 1963, CPIA Publ. No 44(U), pp. 213-245, Feb., 1964.

Beryllium Oxide, Alpha (BeO) Be_1O_1(cr)

Enthalpy Reference Temperature = T_r = 298.15 K Standard State Pressure = p° = 0.1 MPa

T/K	$C_p°$	S°	$-[G°-H°(T_r)]/T$	$H°-H°(T_r)$	$\Delta_f H°$	$\Delta_f G°$	Log K_f
	J $K^{-1}mol^{-1}$			kJ mol^{-1}			
0	0.	0.	INFINITE	-2.835	-604.914	-604.914	INFINITE
100	2.636	0.824	28.539	-2.771	-606.343	-597.709	312.211
200	14.159	5.887	15.759	-1.974	-607.579	-588.573	153.719
298.15	25.560	13.770	13.770	0.	-608.354	-579.062	101.449
300	25.744	13.928	13.770	0.047	-608.364	-578.880	100.792
400	33.757	22.513	14.886	3.051	-608.688	-568.992	74.303
500	38.920	30.641	17.235	6.703	-608.668	-559.065	58.405
600	42.376	38.061	20.098	10.778	-608.440	-549.163	47.809
700	44.823	44.786	23.152	15.144	-608.092	-539.310	40.244
800	46.656	50.897	26.244	19.722	-607.680	-529.511	34.574
900	48.091	56.478	29.298	24.462	-607.235	-519.767	30.166
1000	49.262	61.607	32.276	29.331	-606.780	-510.073	26.643
1100	50.254	66.350	35.161	34.308	-606.329	-500.424	23.763
1200	51.116	70.760	37.946	39.377	-605.892	-490.815	21.365
1300	51.882	74.882	40.630	44.527	-605.478	-481.243	19.337
1400	52.580	78.753	43.216	49.751	-605.094	-471.701	17.599
1500	53.225	82.403	45.708	55.042	-604.744	-462.185	16.095
1600	53.827	85.857	48.111	60.394	-618.971	-452.166	14.762
1700	54.396	89.138	50.428	65.806	-618.303	-441.761	13.574
1800	54.944	92.262	52.666	71.273	-617.614	-431.396	12.519
1900	55.467	95.247	54.829	76.794	-616.904	-421.069	11.576
2000	55.974	98.105	56.922	82.366	-616.176	-410.781	10.729
2100	56.467	100.848	58.949	87.988	-615.430	-400.530	9.963
2200	56.948	103.486	60.914	93.659	-614.669	-390.314	9.267
2300	57.421	106.028	62.820	99.378	-613.893	-380.133	8.633
2373.001	57.756	107.827	64.177	103.582	-------- ALPHA <--> BETA -------		
2400	57.881	108.482	64.672	105.143	-613.103	-369.987	8.053
2500	58.338	110.854	66.472	110.954	-612.300	-359.873	7.519
2600	58.789	113.151	68.224	116.810	-611.483	-349.792	7.027
2700	59.233	115.378	69.929	122.711	-610.654	-339.743	6.573
2800	59.676	117.540	71.591	128.657	-900.769	-323.502	6.035
2900	60.112	119.641	73.212	134.646	-898.851	-302.919	5.456
3000	60.547	121.687	74.794	140.679	-896.904	-282.403	4.917
3100	60.978	123.679	76.339	146.755	-894.930	-261.952	4.414
3200	61.404	125.622	77.849	152.874	-892.929	-241.565	3.943
3300	61.831	127.518	79.325	159.036	-890.903	-221.242	3.502
3400	62.254	129.370	80.770	165.240	-888.855	-200.979	3.088
3500	62.676	131.181	82.184	171.487	-886.784	-180.778	2.698

PREVIOUS: December 1974 CURRENT: June 1975

Beryllium Oxide, Alpha (BeO) Be_1O_1(cr)

BERYLLIUM OXIDE, BETA (β-BeO) CRYSTAL M_r = 25.01158

$S°(298.15\ K) = [16.591]\ J\ K^{-1}\ mol^{-1}$

$T_{trs} = 2373 \pm 15\ K\ (\alpha\to\beta)$

$T_{fus} = [2821.2 \pm 100]\ K\ (\beta\to\ell)$

$\Delta_f H°(298.15\ K) = [-601.659]\ kJ\ mol^{-1}$

$\Delta_{trs}H° = 6.694 \pm 1.7\ kJ\ mol^{-1}$

$\Delta_{fus}H° = [79.057 \pm 6.3]\ kJ\ mol^{-1}$

Enthalpy of Formation

$\Delta_f H°$ is calculated from that of BeO(α, cr) by addition of $\Delta_{trs}H°$, since the difference in enthalpy, H°(2373 K) − H°(298.15 K), between the α-and β-phases is zero according to the adopted functions.

Heat Capacity and Entropy

Cp° is taken to be the same as that of BeO(α, cr). Enthalpy data for β-BeO (1) extend from 2377 to 2501 K, a range too short for obtaining an accurate Cp° curve. The enthalpy data deviate from the adopted functions by −0.2 to ± 1.4%. Other enthalpy data (2-4) show no obvious transition to β-BeO; this suggests that experimental uncertainties masked the transition or that these samples failed to transform. S° is calculated in a manner analogous to that used for $\Delta_f H°$.

Transition Data

Studies of x-ray diffraction (5-7), optical properties (6, 8), thermal expansion (1, 9), thermal analysis (10-14), enthalpy (1), and decrepitation of single crystals (6, 9, 15) indicate the existence of a reversible transition near 2100°C. β-BeO is tetragonal with a structure related to rutile (5), while α-BeO is hexagonal close-packed with a wurtzite-type structure (16, 17).

T_{trs} is lower on cooling than on heating. Earlier studies gave temperature differences of about 40° (11, 12), 50° (7) and 80° (14), but a recent DTA study (10) gave T_{trs} values on cooling which were only ~15° below those on heating. T_{trs} = 2107 ± 7°C (IPTS-68) was proposed as a DTA standard temperature because it is reproducible and relatively unchanged by oxide impurities (10). Reported values of T_{trs} on heating include 2100 ± 10°C (1), 2107 ± 7°C (10), 2095°C (11), 2075 ± 15°C (12), 2050-2100°C (7), 2100-2250°C (7), 2144 ± 40°C (14), and 2050 ± 25°C (6). We adopt 2100 ± 15°C.

Reported values of $\Delta_{trs}H°$ include 1.35 ± 0.1 (1), 1.40 ± 0.25 (10), 1.25 ± 0.25 (14), and 0.95 ± 0.3 kcal mol^{-1} (13). The first value is from enthalpy data and the others are from thermal analysis. It is not clear why the transition fails to appear in two other enthalpy studies (2, 3). We adopt $\Delta_{trs}H°$ = 1.60 ± 0.4 kcal mol^{-1} derived from the difference between the enthalpy data for β-BeO (1) and the JANAF enthalpy for α-BeO.

Fusion Data

See the liquid table for details.

References

1. J. B. Conway and R. A. Hein, Nucleonics 22 (6), 71 (1964).
2. E. E. Shpil'rain, D. N. Kagan and L. S. Barkahatov, High Temperature (USSR) 9, 842 (1971).
3. V. V. Kandyba, P. B. Kantor, R. M. Krasovitskaya and E. N. Fomichev, Dokl. Adad. Nauk SSSR 131, 566 (1960).
4. M. A. Greenbaum, J. Weiher and M. Farber, J. Phys. Chem. 69, 4035 (1965).
5. D. K. Smith, C. F. Cline and S. B. Austerman, Acta Crystallogr. 18, 393 (1965).
6. D. K. Smith, C. F. Cline and V. D. Frechette, J. Nucl. Mater. 6, 265 (1962).
7. T. W. Baker and P. J. Baldock, J. Nucl. Mater. 19, 210 (1966); Nature 193, 1172 (1962).
8. S. B. Austerman, J. Nucl. Mater. 6, 332 (1962).
9. C. J. Engberg and E. H. Zehms, J. Amer. Ceram. Soc. 42, 300 (1959).
10. T. Sata and T. Takahashi, Colloq. Int. Cent. Nat. Rech. Sci., No. 205, 331 (1972); Yogyo Kyokai Shi 80, 142 (1972).
11. R. E. Latta, E. C. Duderstadt and R. E. Fryxell, J. Nucl. Mater. 35, 350 (1970).
12. Yu. A. Kocherzhinskii, E. A. Shishkin and L. M. Yupko, Refractories (USSR) 1969, 314; Ogneupory 34 (5), 50 (1969).
13. A. I. Kaznoff and L. N. Grossman, in "Thermodynamics of Nuclear Materials 1967," pp. 25-39, Intl. At. Energy Agency, Veinna, 1968.
14. S. B. Austerman, U. S. At. Energy Comm. NAA-SR-7654, 14 pp. (1963); Bull. Amer. Phys. Soc. II 7, 28 (1962).
15. S. B. Austerman, U. S. At. Energy Comm. NAA-SR-6428 (1961).
16. D. K. Smith, H. W. Newkirk and J. S. Kahn, J. Electrochem. Soc. 111, 78 (1964).
17. P. J. Baldock, W. E. Spindler and T. W. Baker, J. Nucl. Mater. 19, 169 (1966).

Beryllium Oxide, Beta (BeO) $Be_1O_1(cr)$

Enthalpy Reference Temperature = T_r = 298.15 K Standard State Pressure = p° = 0.1 MPa

T/K	$C_p°$	S°	$-[G°-H°(T_r)]/T$	$H°-H°(T_r)$	$\Delta_f H°$	$\Delta_f G°$	Log K_f
	J $K^{-1}mol^{-1}$			kJ mol^{-1}			
0							
100							
200							
298.15	25.560	16.591	16.591	0.	-601.659	-573.209	100.424
300	25.744	16.749	16.591	0.047	-601.669	-573.032	99.774
400	33.757	25.335	17.707	3.051	-601.993	-563.426	73.576
500	38.920	33.462	20.056	6.703	-601.974	-553.781	57.853
600	42.376	40.882	22.919	10.778	-601.746	-544.161	47.373
700	44.823	47.608	25.973	15.144	-601.398	-534.590	39.892
800	46.656	53.718	29.065	19.722	-600.986	-525.074	34.284
900	48.091	59.299	32.119	24.462	-600.541	-515.611	29.925
1000	49.262	64.428	35.097	29.331	-600.086	-506.199	26.441
1100	50.254	69.171	37.982	34.308	-599.635	-496.832	23.593
1200	51.116	73.581	40.767	39.377	-599.198	-487.506	21.221
1300	51.882	77.703	43.451	44.527	-598.784	-478.216	19.215
1400	52.580	81.574	46.037	49.751	-598.399	-468.956	17.497
1500	53.225	85.224	48.529	55.042	-598.050	-459.722	16.009
1600	53.827	88.678	50.932	60.394	-612.276	-449.985	14.691
1700	54.396	91.959	53.249	65.806	-611.609	-439.862	13.515
1800	54.944	95.083	55.487	71.273	-610.919	-429.779	12.472
1900	55.467	98.068	57.650	76.794	-610.210	-419.735	11.539
2000	55.974	100.926	59.743	82.366	-609.481	-409.729	10.701
2100	56.467	103.669	61.770	87.988	-608.736	-399.760	9.943
2200	56.948	106.307	63.735	93.659	-607.975	-389.826	9.256
2300	57.421	108.849	65.642	99.378	-607.199	-379.927	8.628
2373.001	57.758	110.649	66.999	103.582	-------- ALPHA <--> BETA -------		
2400	57.881	111.303	67.493	105.143	-606.409	-370.063	8.054
2500	58.338	113.675	69.293	110.954	-605.605	-360.232	7.527
2600	58.789	115.972	71.045	116.810	-604.789	-350.433	7.040
2700	59.233	118.199	72.750	122.711	-603.959	-340.666	6.591
2800	59.676	120.361	74.412	128.657	-894.074	-324.707	6.057
2821.220	59.770	120.812	74.759	129.924	------- BETA <--> LIQUID -------		
2900	60.112	122.463	76.033	134.646	-892.156	-304.406	5.483
3000	60.547	124.508	77.615	140.679	-890.210	-284.172	4.948
3100	60.978	126.500	79.160	146.755	-888.235	-264.003	4.448
3200	61.404	128.443	80.670	152.874	-886.235	-243.899	3.981
3300	61.831	130.339	82.146	159.036	-884.209	-223.857	3.543
3400	62.254	132.191	83.591	165.240	-882.160	-203.877	3.132
3500	62.676	134.002	85.005	171.487	-880.090	-183.957	2.745

PREVIOUS: June 1971 CURRENT: December 1974

Beryllium Oxide, Beta (BeO) $Be_1O_1(cr)$

BERYLLIUM OXIDE (BeO) LIQUID M_r = 25.01158

$S°(298.15\ K) = [35.931]\ J\ K^{-1}\ mol^{-1}$

$T_{fus} = [2821.2 \pm 100]\ K\ (\beta-\ell)$

$T_{fus} = 2780 \pm 100\ K\ (\alpha+\ell)$

$\Delta_f H°(298.15\ K) = [-542.705]\ kJ\ mol^{-1}$

$\Delta_{fus} H° = [79.057 \pm 6.3]\ kJ\ mol^{-1}$

$\Delta_{fus} H° = 84.9 \pm 6.3\ kJ\ mol^{-1}$

<u>Enthalpy of Formation</u>

$\Delta_f H°$ is calculated from that of α-BeO by adding $\Delta_{fus}H°$ and the difference in enthalpy, H°(2780 K) - H°(298.15 K), between the α and liquid phases.

<u>Heat Capacity and Entropy</u>

Cp° is taken to be 19 cal K^{-1} mol^{-1} based on the lower two of three enthalpy points (2867-3159 K) measured by Shpil'rain et al. (<u>1</u>). The upper two points yield 24.9 cal K^{-1} mol^{-1}, which is too large in our opinion. Earlier enthalpy data (<u>2</u>, <u>3</u>) are inadequate for deriving Cp°. Below the assumed glass transition at 1900 K, Cp° is taken to be the same as that of the crystal. S° is calculated in a manner analogous to that used for $\Delta_f H°$.

<u>Fusion Data</u>

Conflicting evidence suggests that the melting points of α- and β-Beo are quite uncertain. Schneider's review (<u>4</u>) listed six values (T_{fus}/°C, IPTS-48, with dates in parentheses): 2410(1916), 2452(1956), 2508(1937), 2557(1930), 2570(1948) and 2573°C 1913). Subsequent T_{fus} values include 2430 ± 10 (<u>5</u>), 2560 ± 10 (<u>6</u>), 2444 ± 30 (<u>7</u>), 2547 ± 9 (<u>2</u>), 2450 ± 30 and 2470 ± 30°C (<u>8</u>). The reported values fall roughly into two groups near 2450°C and 2560°C. Three enthalpy studies (<u>1</u>-<u>3</u>) avoided the region from 2435°C to 2547°C. Part of the conflict arises because, in most cases, the measurement of T_{fus} of BeO was only an incidental part of the study. Such values are more likely to be in error due to temperature measurement, detection of melting, impurities, volatilization and reaction with the surroundings. The actual discrepancies may be even larger than they appear, since one of the higher T_{fus} values seems to refer to α-BeO which should melt ~40 below β-BeO. Enthalpy data of Kandyba et al. (<u>2</u>) suggest that their sample may have remained as α-BeO, yet they reported T_{fus} = 2547 ± 9°C. In contrast T_{fus} = 2430 ± 10°C was found for β-BeO by Latta et al. (<u>5</u>); their thermal analysis showed both T_{trs} and T_{fus}.

The conflict is epitomized by the two most recent values of T_{fus}, 2430 ± 10°C (<u>5</u>) and 2560 ± 10°C (<u>6</u>). These studies (<u>5</u>, <u>6</u>) seem to be more satisfactory than their predecessors yet at least one of them has a large bias. They agree (<u>5</u>, <u>6</u>) on T_{fus} for $A\ell_2O_3$, Mo and Ta but differ in opposite directions for BeO and UO_2. Non-stoichiometry could explain the lower $T_{fus}(UO_2)$ reported by Riley (<u>6</u>), but analogous evidence for BeO is lacking. As a compromise, we adopt $T_{fus}(\alpha)$ = 2507°C = 2780 K and calculate $T_{fus}(\beta)$ = 2548°C = 2821.2 K from $\Delta_r G° = 0$ for BeO(β) → BeO(ℓ).

$\Delta_{fus}H°(\alpha) = 20.3\ kcal\ mol^{-1}$ is calculated by difference from enthalpies (2867-3159 K) of Shpil'rain et al. (<u>1</u>) and the JANAF enthalpy for α-BeO. We assume that BeO(ℓ) reverted to α-BeO and that β-BeO did not form during the drop calorimetry (<u>1</u>). $\Delta_{fus}H°(\beta) = 13.895\ kcal\ mol^{-1}$ is calculated from BeO(β) → BeO(ℓ) at the corresponding T_{fus} using the adopted tables. Enthalpy data of Greenbaum et al. (<u>3</u>) gave $\Delta_{fus}H°$ 19.3 kcal mol^{-1} (phase uncertain) even though both the crystal and liquid data have large negative bias. Ohta and Sata (9) derived $\Delta_{fus}H° = 7.6 \pm 2.1\ kcal\ mol^{-1}$ from liquidus data in the binary system BeO-ThO_2. This can be discounted along with other binary data (<u>10</u>, <u>11</u>) due to the uncertainty in T_{fus}.

<u>Vaporization Data</u>

The vapor over BeO is composed mainly of trimer, tetramer and individual atoms, along with minor amounts of several other molecules.

<u>References</u>

1. E. E. Sphil'rain, D. N. Kagan and L. S. Barkhatov, High Temperatures (USSR) <u>9</u>, 842 (1971).
2. V. V. Kandyba, P. B. Kantor, R. M. Krasovitskaya and E. N. Fomichev, Dokl. Akad. Nauk SSSR <u>131</u>, 566 (1960).
3. M. A. Greenbaum, J. Weiher and M. Farber, J. Phys. Chem. <u>69</u>, 4035 (1965).
4. S. J. Schneider, U. S. Nat. Bur. Stand. Monograph 68, 1963.
5. R. E. Latta, E. C. Duderstadt and R. E. Fryxell, J. Nucl. Mater. <u>35</u>, 350 (1970).
6. B. Riley, Rev. Int. Hautes Temp. Refract. <u>3</u>, 327 (1966).
7. J. Kordis, J. Nucl. Mater. <u>14</u>, 322 (1964).
8. C. J. Engberg and E. H. Zehms, J. Amer. Ceram. Soc. <u>42</u>, 300 (1959).
9. T. Ohta and T. Sata, Yogya Kyokai Shi <u>82</u>, 397 (1974).
10. E. M. Levin, C. R. Robbins and H. F. McMurdie, "Phase Diagrams for Ceramists," 3rd ed. and 1969 Supplement, The American Ceramic Soc., Columbus, Ohio, 1964 and 1969.
11. N. A. Toropov, V. P. Barzakovskii, V. V. Lapin and N. N. Kurtseva, "Handbook of Phase Diagrams of Silicate Systems, Vol. I, Binary Systems," 2nd revised ed. (1969), Eng. Trans. (avail. NTIS), Israel Program for Scientific Translations, Jerusalem, 1972.

Beryllium Oxide (BeO) $Be_1O_1(l)$

Enthalpy Reference Temperature = T_r = 298.15 K Standard State Pressure = p° = 0.1 MPa

T/K	$C_p°$	S°	$-[G°-H°(T_r)]/T$	$H°-H°(T_r)$	$\Delta_f H°$	$\Delta_f G°$	Log K_f
	J $K^{-1}mol^{-1}$			kJ mol^{-1}			
0							
100							
200							
298.15	25.560	35.931	35.931	0.	-542.705	-520.021	91.105
300	25.744	36.089	35.931	0.047	-542.715	-519.880	90.519
400	33.757	44.675	37.047	3.051	-543.039	-512.208	66.887
500	38.920	52.802	39.396	6.703	-543.019	-504.497	52.704
600	42.376	60.222	42.259	10.778	-542.791	-496.811	43.251
700	44.823	66.948	45.313	15.144	-542.444	-489.174	36.503
800	46.656	73.058	48.405	19.722	-542.031	-481.591	31.445
900	48.091	78.639	51.459	24.462	-541.586	-474.063	27.514
1000	49.262	83.768	54.437	29.331	-541.131	-466.585	24.372
1100	50.254	88.511	57.322	34.308	-540.680	-459.152	21.803
1200	51.116	92.921	60.107	39.377	-540.243	-451.760	19.665
1300	51.882	97.043	62.791	44.527	-539.830	-444.403	17.856
1400	52.580	100.914	65.377	49.751	-539.445	-437.077	16.308
1500	53.225	104.564	67.869	55.042	-539.096	-429.778	14.966
1600	53.827	108.018	70.272	60.394	-553.322	-421.975	13.776
1700	54.396	111.299	72.589	65.806	-552.655	-413.786	12.714
1800	54.944	114.423	74.827	71.273	-551.965	-405.637	11.771
1900	55.467	117.408	76.990	76.794	-551.255	-397.527	10.929
1900.001	55.467	117.408	76.990	76.794	GLASS <--> LIQUID	TRANSITION	
1900.001	79.496	117.408	76.990	76.794			
2000	79.496	121.486	79.114	84.743	-548.149	-389.516	10.173
2100	79.496	125.364	81.225	92.693	-545.077	-381.660	9.493
2200	79.496	129.063	83.316	100.643	-542.037	-373.950	8.879
2300	79.496	132.596	85.382	108.592	-539.030	-366.377	8.321
2400	79.496	135.980	87.421	116.542	-536.055	-358.934	7.812
2500	79.496	139.225	89.428	124.491	-533.113	-351.615	7.347
2600	79.496	142.343	91.404	132.441	-530.203	-344.412	6.919
2700	79.496	145.343	93.346	140.391	-527.326	-337.321	6.526
2800	79.496	148.234	95.255	148.340	-815.436	-324.113	6.046
2821.220	79.496	148.834	95.656	150.027	------- BETA <--> LIQUID -------		
2900	79.496	151.024	97.131	156.290	-811.558	-306.635	5.523
3000	79.496	153.719	98.972	164.239	-807.695	-289.290	5.037
3100	79.496	156.325	100.780	172.189	-803.847	-272.073	4.584
3200	79.496	158.849	102.556	180.139	-800.016	-254.980	4.162
3300	79.496	161.295	104.299	188.088	-796.202	-238.007	3.767
3400	79.496	163.669	106.010	196.038	-792.408	-221.149	3.398
3500	79.496	165.973	107.691	203.987	-788.635	-204.402	3.051
3600	79.496	168.213	109.341	211.937	-784.884	-187.763	2.724
3700	79.496	170.391	110.962	219.887	-781.157	-171.228	2.417
3800	79.496	172.511	112.554	227.836	-777.455	-154.792	2.128
3900	79.496	174.576	114.118	235.786	-773.782	-138.454	1.854
4000	79.496	176.588	115.654	243.735	-770.138	-122.211	1.596
4100	79.496	178.551	117.165	251.685	-766.525	-106.057	1.351
4200	79.496	180.467	118.649	259.635	-762.946	-89.992	1.119
4300	79.496	182.337	120.109	267.584	-759.403	-74.011	0.899
4400	79.496	184.165	121.544	275.534	-755.897	-58.112	0.690
4500	79.496	185.952	122.955	283.483	-752.431	-42.293	0.491
4600	79.496	187.699	124.344	291.433	-749.007	-26.550	0.301
4700	79.496	189.408	125.710	299.383	-745.626	-10.881	0.121
4800	79.496	191.082	127.055	307.332	-742.290	4.717	-0.051
4900	79.496	192.721	128.378	315.282	-739.001	20.246	-0.216
5000	79.496	194.327	129.681	323.231	-735.761	35.708	-0.373

PREVIOUS: December 1971 CURRENT: December 1974

Beryllium Oxide (BeO) $Be_1O_1(l)$

BERYLLIUM OXIDE (BeO) ALPHA-BETA-LIQUID M_r = 25.01158

0 to 2373 K		crystal, alpha
2327 to 2780 K		crystal, beta
above 2780 K		liquid

Refer to the individual tables for details.

Beryllium Oxide (BeO) $Be_1O_1(cr,l)$

Enthalpy Reference Temperature = T_r = 298.15 K   Standard State Pressure = $p°$ = 0.1 MPa

T/K	$C_p°$ (J K⁻¹mol⁻¹)	S° (J K⁻¹mol⁻¹)	$-[G°-H°(T_r)]/T$ (J K⁻¹mol⁻¹)	$H°-H°(T_r)$ (kJ mol⁻¹)	$\Delta_fH°$ (kJ mol⁻¹)	$\Delta_fG°$ (kJ mol⁻¹)	Log K_f
0	0.	0.	INFINITE	-2.835	-604.914	-604.914	INFINITE
100	2.636	0.824	28.539	-2.771	-606.343	-597.709	312.211
200	14.159	5.887	15.759	-1.974	-607.579	-588.573	153.719
298.15	25.560	13.770	13.770	0.	-608.354	-579.062	101.449
300	25.744	13.928	13.770	0.047	-608.364	-578.880	100.792
400	33.757	22.513	14.886	3.051	-608.688	-568.992	74.303
500	38.920	30.641	17.235	6.703	-608.668	-559.065	58.405
600	42.376	38.061	20.098	10.778	-608.440	-549.163	47.809
700	44.823	44.786	23.152	15.144	-608.092	-539.310	40.244
800	46.656	50.897	26.244	19.722	-607.680	-529.511	34.574
900	48.091	56.478	29.298	24.462	-607.235	-519.767	30.166
1000	49.262	61.607	32.276	29.331	-606.780	-510.073	26.643
1100	50.254	66.350	35.161	34.308	-606.329	-500.424	23.763
1200	51.116	70.760	37.946	39.377	-605.892	-490.815	21.365
1300	51.882	74.882	40.630	44.527	-605.478	-481.243	19.337
1400	52.580	78.753	43.216	49.751	-605.094	-471.701	17.599
1500	53.225	82.403	45.708	55.042	-604.744	-462.185	16.095
1600	53.827	85.857	48.111	60.394	-618.971	-452.166	14.762
1700	54.396	89.138	50.428	65.806	-618.303	-441.761	13.574
1800	54.944	92.262	52.666	71.273	-617.614	-431.396	12.519
1900	55.467	95.247	54.829	76.794	-616.904	-421.069	11.576
2000	55.974	98.105	56.922	82.366	-616.176	-410.781	10.729
2100	56.467	100.848	58.949	87.988	-615.430	-400.530	9.963
2200	56.948	103.486	60.914	93.659	-614.669	-390.314	9.267
2300	57.421	106.028	62.820	99.378	-613.893	-380.133	8.633
2373.001	57.756	107.827	64.177	103.582	ALPHA <--> BETA TRANSITION		
2373.001	57.758	110.649	64.177	110.276			
2400	57.881	111.303	64.704	111.837	-606.409	-370.063	8.054
2500	58.338	113.675	66.616	117.648	-605.605	-360.232	7.527
2600	58.789	115.972	68.470	123.504	-604.789	-350.433	7.040
2700	59.233	118.199	70.271	129.406	-603.959	-340.665	6.591
2800	59.676	120.361	72.021	135.351	-894.074	-324.707	6.057
2821.220	59.770	120.812	72.386	136.618	BETA <--> LIQUID TRANSITION		
2821.220	79.496	148.834	72.386	215.676			
2900	79.496	151.024	74.493	221.939	-811.558	-306.635	5.523
3000	79.496	153.719	77.089	229.888	-807.695	-289.290	5.037
3100	79.496	156.325	79.603	237.838	-803.847	-272.073	4.584
3200	79.496	158.849	82.041	245.787	-800.016	-254.980	4.162
3300	79.496	161.295	84.405	253.737	-796.202	-238.007	3.767
3400	79.496	163.669	86.702	261.687	-792.408	-221.149	3.398
3500	79.496	165.973	88.934	269.636	-788.635	-204.402	3.051
3600	79.496	168.213	91.105	277.586	-784.884	-187.763	2.724
3700	79.496	170.391	93.219	285.535	-781.157	-171.228	2.417
3800	79.496	172.511	95.278	293.485	-777.455	-154.792	2.128
3900	79.496	174.576	97.285	301.435	-773.782	-138.454	1.854
4000	79.496	176.588	99.242	309.384	-770.138	-122.211	1.596
4100	79.496	178.551	101.153	317.334	-766.525	-106.057	1.351
4200	79.496	180.467	103.018	325.283	-762.946	-89.992	1.119
4300	79.496	182.337	104.841	333.233	-759.403	-74.011	0.899
4400	79.496	184.165	106.624	341.183	-755.897	-58.112	0.690
4500	79.496	185.952	108.367	349.132	-752.431	-42.293	0.491
4600	79.496	187.699	110.072	357.082	-749.007	-26.550	0.301
4700	79.496	189.408	111.742	365.031	-745.626	-10.881	0.121
4800	79.496	191.082	113.378	372.981	-742.290	4.717	-0.051
4900	79.496	192.721	114.980	380.931	-739.001	20.246	-0.216
5000	79.496	194.327	116.551	388.880	-735.761	35.708	-0.373

PREVIOUS: CURRENT: December 1974

Beryllium Oxide (BeO) $Be_1O_1(cr,l)$

$S°(298.15\ K) = 197.625 \pm 0.4\ J\ K^{-1}\ mol^{-1}$

$\Delta_f H°(0\ K) = 134.0 \pm 13\ kJ\ mol^{-1}$
$\Delta_f H°(298.15\ K) = 136.4 \pm 13\ kJ\ mol^{-1}$

Electronic and Molecular Constants (σ = 1)

Source	State	ε_i, cm^{-1}	g_i	r_e, Å	B_e, cm^{-1}	α_e, cm^{-1}	ω_e, cm^{-1}	$\omega_e x_e$, cm^{-1}
(1)	$X^1\Sigma^+$	0.0	1	1.3310	1.6510	0.0190	1487.3	11.83
(1-7)	$^3\Pi$	[8000]	6	[1.463]	[1.366]	[0.0163]	1130.8	8.2
(1)	$A^1\Pi$	9234.8	2	1.4632	1.3661	0.0163	1144.2	8.42
(4-7)	$^3\Sigma^+$	[16000]	3	[1.362]	[1.576]	[0.015]	[1370]	[7.8]
(1)	$B^1\Sigma^+$	21197.	1	1.3623	1.5758	0.0154	1370.8	7.75
(4)	$^3\Sigma^+$	[37000]	3	[1.49]	[1.31]	[0.01]	[1082]	[9.]
(4)	$^3\Delta$	[38000]	6	[1.49]	[1.31]	[0.01]	[1082]	[9.]
(1)	$D^1\Delta$	38918.	2	1.49	1.31	0.01	1081.5	9.1
(1, 4)	$^3\Sigma^-$	[40000]	3	[1.49]	[1.31]	[0.01]	1012.2	8.4
(1, 4)	$^1\Sigma^-$	38956.	1	[1.49]	[1.31]	[0.01]	[1082]	[9.]
(4)	$^1\Sigma^+$	[47000]	1	[1.49]	[1.31]	[0.01]	[1082]	[9.]

Enthalpy of Formation

We adopt D_0^0 = 104.2 ± 3 and $\Delta_f H°(298.15\ K)$ = 32.6 ± 3 kcal mol^{-1} based on two mass-spectrometric studies (8, 9) analyzed below. The adopted values correspond to the average of the two results for reaction A and also to the average for reaction B. Improved agreement between reactions A and B is due to recalculation of Kp of reaction B using the current table (4) for BeO(α). Our analysis should be compared to the analysis of Brewer and Rosenblatt (10), using "functions based upon calculated levels."

The controversy over D_0^0 of alkaline earth oxides has been reviewed in detail (10-12). Uncertainty in the electronic partition function of BeO due to triplet states now is much reduced (2). Gaydon (12) derived spectroscopic values for D_0^0 of 91 and 111 kcal mol^{-1} from the X and A states, respectively. D_0^0 = 101 kcal mol^{-1} was derived (13) by fitting an electronegativity potential function to the X state.

Source	Method	Reaction[a]	T/K	Data Points	δS cal K^{-1} mol^{-1}	$\Delta_r H°(298.15\ K)$, kcal mol^{-1} 2nd law	3rd law	$\Delta_f H°(298.15\ K)$ kcal mol^{-1}	D_0^0
(8) Chupka (1959)	Knudsen mass spec.	A	2100-2474	8	1.6±2.3	-8.4±5	-11.9±3	30.6	106.2
(9) Theard (1964)	Knudsen mass spec.	A	2380	1	-	-	-15.9±3	34.6	102.2
(8) Chupka (1959)	Knudsen mass spec.	B[b]	1914-2304	6	0.7±0.9	107.3±2	105.9±3	31.9	104.9
(9) Theard (1964)	Knudsen mass spec.	B[b]	2380	1	-	-	104.5±3	33.3	103.5

[a]Reactions: A) BeO(g) + O(g) = Be(g) + O_2(g); B) BeO(g) = Be(g) + O(g).
[b]Recalculated as in (8) assuming $P_{Be} = P_O$ and JANAF values for BeO(α) = Be(g) + O(g).

Heat Capacity and Entropy

Electronic levels (T_0) and vibrational-rotational constants of the observed states are from Rosen (1). Field (2) concluded that calculations (3, 5-7) of the isoconfigurational $A^1\Pi - {}^3\Pi$ separation should be adequate for estimating the low-lying $^3\Pi$ state. The adopted separation of 1200 cm^{-1} is consistent with analysis (3) of perturbations. We estimate $^3\Sigma^+$ at 16000 cm^{-1} by assuming that it lies 5200 ± 4000 cm^{-1} (4-7) below the isoconfigurational B state. Other predicted states and their vibrational-rotational constants are facilitated by listing the states in the isconfigurational order of MgO (4). Our thermodynamic functions correspond to an "effective" ground state with 1<g<2 instead of g = 3 or 6 (10, 11). This comparison is only approximate since our functions are calculated using first-order anharmonic corrections to Q_r^i and Q_v^i in the partition function $Q = Q_t^i \Sigma\ Q_r^i\ Q_v^i\ g_i\ \exp(-c_2\varepsilon_i/T)$.

Sublimation and Vaporization Data

Mass spectra (8, 9) at 1900-2400 K showed the vapor to consist mainly of Be, O, $(BeO)_3$, and $(BeO)_4$ with small amounts of O_2, Be_2O, BeO and other polymer of BeO. Tetramer, trimer and pentamer become dominant at higher temperatures.

References

1. B. Rosen, "Spectroscopic Data Relative to Diatomic Molecules," Pergamon Press, New York, 1970, pp. 65-6.
2. R. W. Field, J. Chem. Phys. 60, 2400 (1974).
3. W. M. Huo, K. F. Freed and W. Klemperer, J. Chem. Phys. 46, 3556 (1967).
4. JANAF Thermochemical Tables: BeO(α,β,g), MgO(g), CaO(g) 12-31-74; SrO(g), BaO(g) 6-30-74.
5. P. K. Pearson, S. V. O'Neil and H. F. Schaeffer, J. Chem. Phys. 56, 3938 (1972); 55, 176 (1971); Chem. Phys. Lett. 10, 404 (1971).
6. M. L. Sink, Ph. D. Thesis, Case Western Reserve Univ., 1972; Diss. Abstr. B33, 671 (1972).
7. G. Verhaegen and W. G. Richards, J. Chem. Phys. 45, 1828 (1966).
8. W. A. Chupka, J. Berkowitz and C. F. Giese, J. Chem. Phys. 30, 827 (1959).
9. L. P. Theard and D. L. Hildenbrand, J. Chem. Phys. 41, 3416 (1964).
10. L. Brewer and G. Rosenblatt, Advan. High Temp. Chem. 2, 1 (1969).
11. K. Schofield, Chem. Rev. 67, 707 (1967).
12. A. G. Gaydon, "Dissociation Energies and Spectra of Diatomic Molecules," 3rd ed., Chapman and Hall Ltd., London, 1969, pp. 236-244.
13. B. Rai and S. N. Rai, Indian J. Pure Appl. Phys. 10, 401 (1972).

Enthalpy Reference Temperature = T_r = 298.15 K   Standard State Pressure = p° = 0.1 MPa

T/K	$C_p°$ (J $K^{-1}mol^{-1}$)	S°	$-[G°-H°(T_r)]/T$	$H°-H°(T_r)$ (kJ mol^{-1})	$\Delta_f H°$	$\Delta_f G°$	Log K_f
0	0.	0.	INFINITE	-8.688	133.984	133.984	INFINITE
100	29.108	165.763	223.605	-5.784	135.397	127.536	-66.618
200	29.139	185.943	200.307	-2.873	136.275	119.270	-31.150
298.15	29.481	197.625	197.625	0.	136.398	110.873	-19.425
300	29.493	197.807	197.626	0.055	136.395	110.715	-19.277
400	30.348	206.400	198.791	3.043	136.056	102.198	-13.346
500	31.421	213.286	201.023	6.131	135.513	93.793	-9.799
600	32.454	219.108	203.564	9.326	134.860	85.509	-7.444
700	33.345	224.179	206.155	12.617	134.133	77.340	-5.771
800	34.084	228.682	208.694	15.990	133.340	69.280	-4.524
900	34.704	232.733	211.144	19.430	132.485	61.324	-3.559
1000	35.252	236.418	213.490	22.928	131.569	53.465	-2.793
1100	35.784	239.803	215.730	26.480	130.595	45.702	-2.170
1200	36.357	242.941	217.869	30.086	129.569	38.029	-1.655
1300	37.023	245.876	219.911	33.755	128.501	30.444	-1.223
1400	37.825	248.648	221.866	37.496	127.403	22.942	-0.856
1500	38.788	251.290	223.740	41.325	126.291	15.519	-0.540
1600	39.927	253.829	225.542	45.259	110.646	8.696	-0.284
1700	41.236	256.288	227.278	49.316	109.959	2.346	-0.072
1800	42.698	258.685	228.957	53.511	109.377	-3.967	0.115
1900	44.285	261.036	230.584	57.860	108.914	-10.250	0.282
2000	45.956	263.350	232.164	62.371	108.582	-16.513	0.431
2100	47.670	265.633	233.704	67.052	108.386	-22.762	0.566
2200	49.379	267.890	235.206	71.905	108.329	-29.005	0.689
2300	51.042	270.122	236.676	76.927	108.408	-35.249	0.801
2400	52.616	272.328	238.116	82.110	108.617	-41.499	0.903
2500	54.070	274.506	239.528	87.446	108.944	-47.761	0.998
2600	55.375	276.653	240.915	92.919	109.378	-54.037	1.086
2700	56.514	278.765	242.277	98.515	109.902	-60.332	1.167
2800	57.475	280.838	243.618	104.216	-180.457	-60.425	1.127
2900	58.255	282.869	244.936	110.004	-178.741	-56.168	1.012
3000	58.856	284.854	246.234	115.861	-176.970	-51.972	0.905
3100	59.286	286.792	247.511	121.770	-175.163	-47.834	0.806
3200	59.556	288.679	248.768	127.713	-173.338	-43.757	0.714
3300	59.681	290.514	250.006	133.676	-171.511	-39.735	0.629
3400	59.676	292.295	251.223	139.645	-169.698	-35.769	0.550
3500	59.559	294.024	252.422	145.608	-167.911	-31.856	0.475
3600	59.345	295.699	253.601	151.554	-166.164	-27.994	0.406
3700	59.051	297.321	254.760	157.474	-164.466	-24.180	0.341
3800	58.691	298.891	255.901	163.362	-162.827	-20.410	0.281
3900	58.279	300.410	257.023	169.210	-161.254	-16.682	0.223
4000	57.827	301.880	258.126	175.016	-159.754	-12.995	0.170
4100	57.347	303.302	259.211	180.775	-158.332	-9.344	0.119
4200	56.848	304.678	260.277	186.485	-156.993	-5.726	0.071
4300	56.337	306.010	261.325	192.144	-155.740	-2.139	0.026
4400	55.822	307.299	262.356	197.752	-154.576	1.419	-0.017
4500	55.308	308.548	263.368	203.309	-153.503	4.952	-0.057
4600	54.801	309.758	264.364	208.814	-152.523	8.462	-0.096
4700	54.303	310.931	265.342	214.269	-151.636	11.952	-0.133
4800	53.818	312.069	266.304	219.675	-150.844	15.424	-0.168
4900	53.348	313.174	267.249	225.033	-150.146	18.881	-0.201
5000	52.895	314.247	268.178	230.345	-149.544	22.324	-0.233
5100	52.460	315.291	269.092	235.613	-149.035	25.756	-0.264
5200	52.044	316.305	269.990	240.838	-148.621	29.179	-0.293
5300	51.647	317.293	270.873	246.022	-148.301	32.595	-0.321
5400	51.271	318.255	271.742	251.168	-148.073	36.006	-0.348
5500	50.914	319.192	272.596	256.277	-147.937	39.413	-0.374
5600	50.577	320.106	273.436	261.351	-147.892	42.819	-0.399
5700	50.259	320.999	274.263	266.393	-147.936	46.225	-0.424
5800	49.961	321.870	275.076	271.404	-148.069	49.633	-0.447
5900	49.681	322.722	275.877	276.386	-148.289	53.042	-0.470
6000	49.418	323.555	276.665	281.340	-148.596	56.458	-0.492

PREVIOUS: December 1974 (1 atm)   CURRENT: December 1974 (1 bar)

BERYLLIUM SULFATE, ALPHA (α-$BeSO_4$) CRYSTAL M_r = 105.06978

$S°(298.15\ K) = 77.971\ J\ K^{-1}\ mol^{-1}$

T_{trs1} = 863 K ($\alpha \rightarrow \beta$)

T_{trs2} = 908 K ($\beta \rightarrow \gamma$)

$\Delta_f H°(0\ K) = -1190.1 \pm 3.3\ kJ\ mol^{-1}$

$\Delta_f H°(298.15\ K) = -1200.8 \pm 3.3\ kJ\ mol^{-1}$

$\Delta_{trs1}H°$ = [1.113] kJ mol^{-1}

$\Delta_{trs2}H°$ = [19.552] kJ mol^{-1}

Enthalpy of Formation

Bear et al. (1) measured the enthalpies of solution of Be(cr) and $BeSO_4$(cr,α) in concentrated HF and the enthalpy of reaction of H_2SO_4(aq) with $BeSO_4$(aq HF). Combination of these yields reaction (a) below. Taylor et al. (2) measured enthalpies of solution of Be(cr) and $BeSO_4$(cr,α) in H_2SO_4(aq) which yields reaction (b). Marchal (3) measured the total vapor pressure of $BeSO_4$(cr) in the temperature range 863-1103 K. Partial pressures of SO_3(g) were calculated from Marchal's (3) data using JANAF (4) values for the SO_3 - SO_2 equilibrium. 3rd law analysis of these data yields $\Delta_r H°$ for reaction (c) as given below. This was converted to $\Delta_f H°$ for α-$BeSO_4$ using JANAF (4) enthalpies of transitions.

Source	Method	Reaction	T/K	$\Delta_{fus}H$ kcal/mol	$\Delta_f H°$(298.15 K) kcal/mol
(1)	Calorimetric	(a) Be(cr) + H_2SO_4(109 H_2O) = H_2(g) + α-$BeSO_4$	294	-74.47	-286.65 ± 0.5
(2)	Calorimetric	(b) Be(cr) + H_2SO_4(47.849 H_2O) = H_2(g) + α-$BeSO_4$	298	-76.38	-287.55 ± 0.13
(3)	Kp (973-1083 K)	(c) $BeSO_4$(γ) → BeO(cr) + SO_3(g)	298	43.92	-286.55 ± 0.5

The adopted $\Delta_f H°(298.15\ K) = -287.0 \pm 0.8$ is an average value of reaction (a) and (b). This value is for relatively finely divided crystals of $BeSO_4$ as was indicated by Bear et al. (1).

Heat Capacity and Entropy

Low temperature heat capacities (12-301 K) and high temperature enthalpies (366-864 K) were measured by Taylor, Gardner and Smith (5). These data were joined smoothly at 298.15 K and extrapolated graphically above 864 K. The entropy was obtained from the heat capacities using $S°(12.56\ K) = 0.034\ cal\ K^{-1}\ mol^{-1}$.

Transition Data

Bosik, Novoselova and Simanov (6) observed two endothermic effects at 863-883 K and 908-913 K on the heating and cooling curves of $BeSO_4$(cr); thus we have adopted $T_{trs1}(\alpha \rightarrow \beta)$ = 863 K and $T_{trs2}(\beta \rightarrow \alpha)$ = 908 K. $\Delta_{trs1}H°(\alpha \rightarrow \beta) = 0.266\ kcal\ mol^{-1}$ was obtained from the single enthalpy point at 863.9 K observed by Taylor, Gardner and Smith (5). $\Delta_{trs2}H°(\beta \rightarrow \gamma) = 4.673\ kcal\ mol^{-1}$ was chosen in order to give good agreement with the equilibrium data of Marchal (3).

References

1. I. J. Bear and A. G. Turnbull, J. Phys. Chem. 70, 711-17 (1966).
2. A. R. Taylor, Jr., B. B. Letson and D. F. Smith, U. S. Bur. Mines, Report No. 6724, 1966.
3. G. Marchal, J. Chim. Phys. 22, 502 (1925).
4. JANAF Thermochemical Tables: SO_2(g), 6-30-61; SO_3(g), 9-30-65.
5. A. R. Taylor, Jr., T. E. Gardner and D. F. Smith, U. S. Bur. Mines, Report No. 6240, 1963.
6. I. I. Bosik, A. V. Novoselova and Yu. P. Simanov, Russ. J. Inorg. Chem. 6, 1295 (1961).

Beryllium Sulfate, Alpha ($BeSO_4$) $Be_1O_4S_1$(cr)

Enthalpy Reference Temperature = T_r = 298.15 K Standard State Pressure = p° = 0.1 MPa

T/K	$C_p°$ (J K⁻¹mol⁻¹)	S° (J K⁻¹mol⁻¹)	$-[G°-H°(T_r)]/T$ (J K⁻¹mol⁻¹)	$H°-H°(T_r)$ (kJ mol⁻¹)	$\Delta_f H°$ (kJ mol⁻¹)	$\Delta_f G°$ (kJ mol⁻¹)	Log K_f
0	0.	0.	INFINITE	-12.979	-1190.076	-1190.076	INFINITE
100	29.320	18.630	136.399	-11.777	-1195.412	-1161.311	606.607
200	60.675	48.883	85.043	-7.232	-1198.909	-1125.730	294.010
298.15	85.697	77.971	77.971	0.	-1200.808	-1089.354	190.850
300	86.111	78.502	77.973	0.159	-1200.830	-1088.662	189.553
400	103.880	105.869	81.583	9.714	-1203.656	-1051.063	137.255
500	116.357	130.457	88.944	20.756	-1204.763	-1012.781	105.805
600	126.817	152.591	97.737	32.913	-1204.777	-974.366	84.826
700	137.988	172.967	107.046	46.145	-1203.793	-936.028	69.847
800	149.787	192.159	116.496	60.531	-1201.788	-897.901	58.627
900	162.046	210.506	125.928	76.120	-1251.860	-859.007	49.855
1000	174.431	228.245	135.278	92.967	-1246.475	-815.635	42.604
1100	183.088	245.296	144.510	110.864	-1240.239	-772.847	36.699
1200	189.209	261.502	153.591	129.494	-1233.452	-730.653	31.805
1300	193.543	276.827	162.487	148.643	-1226.317	-689.041	27.686
1400	196.581	291.287	171.175	168.157	-1218.980	-647.986	24.177
1500	198.656	304.925	179.642	187.925	-1211.549	-607.460	21.154
1600	200.004	317.792	187.878	207.863	-1218.639	-566.905	18.508
1700	200.706	329.941	195.881	227.903	-1210.843	-526.410	16.175
1800	200.832	341.418	203.650	247.981	-1203.093	-486.374	14.114
1900	200.832	352.276	211.189	268.065	-1195.419	-446.766	12.282
2000	200.832	362.577	218.504	288.148	-1187.825	-407.560	10.644
2100	200.832	372.376	225.599	308.231	-1180.309	-368.732	9.172
2200	200.832	381.719	232.485	328.314	-1172.870	-330.259	7.841
2300	200.832	390.646	239.169	348.397	-1165.508	-292.124	6.634

PREVIOUS: CURRENT: September 1966

Beryllium Sulfate, Alpha ($BeSO_4$) $Be_1O_4S_1$(cr)

BERYLLIUM SULFATE, BETA (β-$BeSO_4$) CRYSTAL M_r = 63.63583

Beryllium Sulfate, Beta ($BeSO_4$) $Be_1O_4S_1$(cr)

$S°$(298.15 K) = [79.258] J K^{-1} mol^{-1}

T_{trs1} = 863 K ($\alpha \rightarrow \beta$)

T_{trs2} = 908 K ($\beta \rightarrow \gamma$)

$\Delta_f H°$(0 K) = Unknown

$\Delta_f H°$(298.15 K) = [-1199.695] kJ mol^{-1}

$\Delta_{trs1} H°$ = [1.113] kJ mol^{-1}

$\Delta_{trs2} H°$ = [19.552] kJ mol^{-1}

Enthalpy of Formation

The enthalpy of formation is calculated from that of the alpha phase by adding $\Delta_{trs1} H°$; since the heat capacities of both phases are assumed the same.

Heat Capacity and Entropy

The heat capacity was assumed to be the same as that of alpha phase. The entropy is calculated in a manner analogous to that used for the enthalpy of formation.

Transition Data

See α-crystal table for details.

Enthalpy Reference Temperature = T_r = 298.15 K   Standard State Pressure = p° = 0.1 MPa

T/K	$C_p°$	S°	$-[G°-H°(T_r)]/T$	$H°-H°(T_r)$	$\Delta_f H°$	$\Delta_f G°$	Log K_f
	J $K^{-1}mol^{-1}$			kJ mol^{-1}			
0							
100							
200							
298.15	85.697	79.258	79.258	0.	-1199.695	-1088.624	190.722
300	86.111	79.789	79.259	0.159	-1199.717	-1087.935	189.426
400	103.880	107.156	82.870	9.714	-1202.543	-1050.464	137.177
500	116.357	131.743	90.231	20.756	-1203.650	-1012.312	105.755
600	126.817	153.878	99.024	32.913	-1203.664	-974.025	84.796
700	137.988	174.253	108.332	46.145	-1202.680	-935.816	69.831
800	149.787	193.446	117.782	60.531	-1200.675	-897.818	58.621
900	162.046	211.792	127.215	76.120	-1250.747	-859.052	49.858
1000	174.431	229.532	136.564	92.967	-1245.362	-815.809	42.613
1100	183.088	246.583	145.797	110.864	-1239.126	-773.149	36.714
1200	189.209	262.789	154.877	129.494	-1232.339	-731.084	31.823
1300	193.543	278.114	163.773	148.643	-1225.204	-689.601	27.708
1400	196.581	292.574	172.462	168.157	-1217.867	-648.675	24.202
1500	198.656	306.211	180.928	187.925	-1210.436	-608.277	21.182
1600	200.004	319.079	189.164	207.863	-1217.526	-567.851	18.538
1700	200.706	331.228	197.167	227.903	-1209.730	-527.485	16.208
1800	200.832	342.704	204.937	247.981	-1201.980	-487.577	14.149
1900	200.832	353.563	212.476	268.065	-1194.306	-448.098	12.319
2000	200.832	363.864	219.790	288.148	-1186.712	-409.020	10.683
2100	200.832	373.663	226.886	308.231	-1179.196	-370.321	9.211
2200	200.832	383.005	233.772	328.314	-1171.757	-331.976	7.882
2300	200.832	391.933	240.456	348.397	-1164.395	-293.970	6.676

PREVIOUS: CURRENT: September 1966

Beryllium Sulfate, Beta ($BeSO_4$) $Be_1O_4S_1$(cr)

BERYLLIUM SULFATE, GAMMA (γ-$BeSO_4$) CRYSTAL M_r = 105.06978

$S°(298.15 K) = [100.784] J K^{-1} mol^{-1}$

T_{trs1} = 863 K ($\alpha \rightarrow \beta$)

T_{trs2} = 908 K ($\beta \rightarrow \gamma$)

$\Delta_f H°(0 K)$ = Unknown

$\Delta_f H°(298.15 K) = [-1180.143] kJ mol^{-1}$

$\Delta_{trs1} H° = [1.113] kJ mol^{-1}$

$\Delta_{trs2} H° = [19.552] kJ mol^{-1}$

Enthalpy of Formation

The enthalpy of formation is calculated from that of the beta phase by adding $\Delta_{trs2}H°$; since the heat capacities of both phases are assumed the same.

Heat Capacity and Entropy

The heat capacity was assumed to be the same as that of alpha phase. The entropy is calculated in a manner analogous to that used for the enthalpy of formation.

Transition Data

See α-crystal table for details.

Beryllium Sulfate, Gamma ($BeSO_4$) $Be_1O_4S_1(cr)$

Enthalpy Reference Temperature = T_r = 298.15 K   Standard State Pressure = $p°$ = 0.1 MPa

T/K	$C_p°$ (J K⁻¹mol⁻¹)	$S°$ (J K⁻¹mol⁻¹)	$-[G°-H°(T_r)]/T$ (J K⁻¹mol⁻¹)	$H°-H°(T_r)$ (kJ mol⁻¹)	$\Delta_f H°$ (kJ mol⁻¹)	$\Delta_f G°$ (kJ mol⁻¹)	Log K_f
0							
100							
200							
298.15	85.697	100.784	100.784	0.	-1180.143	-1075.491	188.422
300	86.111	101.316	100.786	0.159	-1180.166	-1074.841	187.147
400	103.880	128.682	104.396	9.714	-1182.991	-1039.523	135.748
500	116.357	153.270	111.757	20.756	-1184.098	-1003.523	104.837
600	126.817	175.405	120.550	32.913	-1184.113	-967.389	84.219
700	137.988	195.780	129.859	46.145	-1183.128	-931.333	69.497
800	149.787	214.973	139.309	60.531	-1181.123	-895.487	58.469
900	162.046	233.319	148.742	76.120	-1231.195	-858.874	49.848
1000	174.431	251.058	158.091	92.967	-1225.810	-817.784	42.717
1100	183.088	268.110	167.324	110.864	-1219.574	-777.277	36.910
1200	189.209	284.316	176.404	129.494	-1212.787	-737.364	32.097
1300	193.543	299.640	185.300	148.643	-1205.652	-698.033	28.047
1400	196.581	314.101	193.988	168.157	-1198.316	-659.260	24.597
1500	198.656	327.738	202.455	187.925	-1190.884	-621.016	21.626
1600	200.004	340.606	210.691	207.863	-1197.974	-582.742	19.025
1700	200.706	352.755	218.694	227.903	-1190.178	-544.528	16.731
1800	200.832	364.231	226.464	247.981	-1182.428	-506.773	14.706
1900	200.832	375.089	234.003	268.065	-1174.754	-469.446	12.906
2000	200.832	385.391	241.317	288.148	-1167.160	-432.522	11.296
2100	200.832	395.189	248.413	308.231	-1159.644	-395.975	9.849
2200	200.832	404.532	255.298	328.314	-1152.206	-359.783	8.542
2300	200.832	413.459	261.982	348.397	-1144.843	-323.930	7.357

PREVIOUS: CURRENT: September 1966

Beryllium Sulfate, Gamma ($BeSO_4$) $Be_1O_4S_1(cr)$

BERYLLIUM TUNGSTEN OXIDE ($BeWO_4$) CRYSTAL M_r = 256.85978

$S°(298.15\ K)$ = [88.37 ± 8.4] J K^{-1} mol^{-1}
T_{fus} = Unknown

$\Delta_f H°(0\ K)$ = Unknown
$\Delta_f H°(298.15\ K)$ = [-1513.4 ± 63] kJ mol^{-1}
$\Delta_{fus} H°$ = Unknown

Enthalpy of Formation

The enthalpy of formation, $\Delta_f H°(BeWO_4$, cr, 298.15 K) = -361.7 kcal mol^{-1}, is calculated from $\Delta_r H°(298.15\ K)$ = -17.1 kcal mol^{-1} for BeO(cr) + WO_3(cr) → $BeWO_4$(cr). The value of $\Delta_r H°(298.15\ K)$ is assumed to be the same as that for MgO(cr) + WO_3(cr) → $MgWO_4$(cr), using -143.7, -143.1, -201.5, and -362.3 kcal mol^{-1} for the enthalpies of formation of MgO(cr), BeO(cr), WO_3(cr) and $MgWO_4$(cr), respectively.

Heat Capacity and Entropy

Heat capacities are estimated by comparison with those of $CaWO_4$(cr) (1), BeO(cr) (2), and CaO(cr) (3).

The entropy, $S°(298.15\ K)$ = 21.12 cal K^{-1} mol^{-1}, is calculated from $\Delta S°(298.15\ K)$ = -0.4 cal K^{-1} mol^{-1} for BeO(cr) + WO_3(cr) → $BeWO_4$(cr). The value of $\Delta S°(298.15\ K)$ is assumed to be the same as that for MgO(cr) + WO_3(cr) → $MgWO_4$(cr), using 6.44, 3.38, 18.14 and 24.18 cal K^{-1} mol^{-1} for MgO(cr), BeO(cr), WO_3(cr) and $MgWO_4$(cr), respectively.

References

1. R. A. Yakovleva and T. N. Rezukhina, Russ. J. Phys. Chem. 34, 390 (1960).
2. JANAF Thermochemical Tables: BeO (cr), 9-30-65.
3. K. K. Kelley, U. S. Bur. Mines. Bull. 584 (1960).

Beryllium Tungsten Oxide ($BeWO_4$) $Be_1O_4W_1$(cr)

Enthalpy Reference Temperature = T_r = 298.15 K Standard State Pressure = p° = 0.1 MPa

T/K	$C_p°$	S° (J $K^{-1}mol^{-1}$)	$-[G°-H°(T_r)]/T$	$H°-H°(T_r)$ (kJ mol^{-1})	$\Delta_f H°$	$\Delta_f G°$	Log K_f
0							
100							
200							
298.15	97.278	88.366	88.366	0.	-1513.353	-1404.818	246.118
300	97.696	88.969	88.368	0.180	-1513.357	-1404.145	244.483
400	112.968	119.279	92.397	10.753	-1513.032	-1367.762	178.611
500	123.512	145.670	100.471	22.600	-1511.920	-1331.561	139.107
600	131.336	168.909	109.980	35.357	-1510.306	-1295.634	112.795
700	137.319	189.609	119.904	48.794	-1508.371	-1260.005	94.023
800	142.884	208.312	129.804	62.806	-1506.174	-1224.671	79.963
900	148.114	225.447	139.492	77.359	-1503.714	-1189.628	69.044
1000	153.009	241.307	148.890	92.417	-1500.997	-1154.873	60.324
1100	157.737	256.114	157.972	107.956	-1498.026	-1120.401	53.203
1200	162.256	270.033	166.736	123.956	-1494.806	-1086.212	47.282
1300	166.649	283.193	175.193	140.401	-1491.342	-1052.302	42.282
1400	171.126	295.708	183.358	157.291	-1487.628	-1018.667	38.007
1500	175.435	307.661	191.249	174.618	-1483.667	-985.307	34.311
1600	179.912	319.126	198.885	192.386	-1493.988	-951.693	31.070
1700	184.305	330.165	206.285	210.596	-1489.116	-917.947	28.205
1800	188.698	340.824	213.465	229.247	-1483.922	-884.498	25.667
1900	193.092	351.144	220.441	248.336	-1478.405	-851.345	23.405
2000	197.485	361.160	227.228	267.865	-1472.565	-818.492	21.377

PREVIOUS: June 1963 CURRENT: March 1967

Beryllium Tungsten Oxide ($BeWO_4$) $Be_1O_4W_1$(cr)

$S°(298.15 K) = [37.03 \pm 4.2]\ J\ K^{-1}\ mol^{-1}$

T_{trs} = 863 K

$\Delta_f H°(0\ K)$ = Unknown

$\Delta_f H°(298.15\ K) = -234.3 \pm 8.8\ kJ\ mol^{-1}$

Enthalpy of Formation

Von Wartenberg (1) has measured the enthalpy of solution of Be and BeS in H_2SO_4. The difference between these two enthalpies gives $\Delta_r H°(353\ K) = -51.35 \pm 2.1\ kcal\ mol^{-1}$ for the reaction $Be(cr) + H_2S(g) \rightarrow BeS(cr) + H_2(g)$. Using auxiliary JANAF (2) data we calculate $\Delta_f H°$(BeS, cr, 298.15 K) = $-56.0 \pm 2.1\ kcal\ mol^{-1}$. This is in agreement with the value for the enthalpy of formation preferred by NBS (3).

Heat Capacity and Entropy

No low-temperature heat capacity or high-temperature enthalpy measurements have been reported. We estimate $C_p°(298.15\ K) = 8.13\ cal\ K^{-1}\ mol^{-1}$ from the reaction BeO(cr, α) + MgS(cr) = BeS(cr) by assuming $\Delta C_p° = 0$ (2). Comparison of this value with $C_p°$ data for all the alkaline earth oxides and sulfides (2) suggests that our estimate is reasonable. $C_p°$ data above 298.15 K are estimated graphically by comparison with α- and β-BeO (2).

Several methods of estimation predict that the value of S°(298.15 K) should lie near 8.85 cal $K^{-1}\ mol^{-1}$. A graphical comparison of the standard entropies for other alkaline-earth oxides and sulfides suggest values for S°(298.15 K) in the range 9-10 cal $K^{-1}\ mol^{-1}$ while additive entropy constants (4) give 6.8 (Kelly's) and 9.3 (Latimer's) cal $K^{-1}\ mol^{-1}$. Literature estimates have included (in cal $K^{-1}\ mol^{-1}$) 7.4 (5), 8.0 (6), and 8.4 (7). We adopt S°(298.15 K) = 8.85 ± 1.0 cal $K^{-1}\ mol^{-1}$ based on cation increments for a series of beryllium and magnesium compounds.

Fusion Data

No literature melting data are available. BeO and BeS have a different crystal structure (cubic, zincblende type) than the remaining alkaline-earth oxides and sulfides (cubic, NaCl type).

References

1. H. Von Wartenberg, Z. Anorg. Chem. 252, 136 (1943).
2. JANAF Thermochemical Tables: $H_2S(g)$, 6-30-77; $H_2(g)$, 3-31-74; MgO(cr), 12-31-74; SrO(cr), 12-31-72; CaO(cr), 6-30-73; BaO(cr), 8-30-74; BeO(cr), 12-31-74; BaS(cr), SrS(cr), CaS(cr), and MgS(cr), 9-30-77.
3. U. S. Nat. Bur. Stand. Tech. Note 270-6, 1971.
4. D. R. Stull and H. Prophet in "The Characterization of High Temperature Vapors," J. L. Margrave, Ed., John Wiley Sons, Inc., New York, 1966.
5. V. A. Kireev, Zh. Obsh. Khim. 16, 1569 (1946).
6. P. J. Spencer, Atomic Energy Reviews No. 4 Beryllium, International Atomic Energy Agency, Vienna, 1973.
7. M. V. H. Karapet'yants and M. L. Karapet'yants, "Thermodynamic Constants of Inorganic and Organic Compounds," Eng. Trans. by J. Schmorak, Ann Arbor, 1970.

Enthalpy Reference Temperature = T_r = 298.15 K Standard State Pressure = p° = 0.1 MPa

	J K⁻¹mol⁻¹			kJ mol⁻¹			
T/K	$C_p°$	S°	$-[G°-H°(T_r)]/T$	$H°-H°(T_r)$	$\Delta_f H°$	$\Delta_f G°$	Log K_f
0							
100							
200							
298.15	34.016	37.028	37.028	0.	-234.304	-232.972	40.816
300	34.309	37.240	37.029	0.063	-234.313	-232.964	40.563
400	42.551	48.313	38.483	3.932	-236.884	-232.365	30.344
500	47.698	58.394	41.475	8.459	-238.387	-231.067	24.139
600	51.045	67.402	45.060	13.405	-239.293	-229.509	19.981
700	53.346	75.450	48.837	18.629	-239.807	-227.834	17.001
800	55.145	82.694	52.624	24.056	-240.088	-226.102	14.763
900	56.651	89.278	56.336	29.648	-293.346	-223.264	12.958
1000	57.907	95.314	59.936	35.377	-292.155	-215.540	11.259
1100	58.968	100.883	63.409	41.222	-290.953	-207.937	9.874
1200	59.915	106.056	66.750	47.167	-289.753	-200.443	8.725
1300	60.762	110.886	69.961	53.202	-288.566	-193.049	7.757
1400	61.463	115.415	73.048	59.315	-287.404	-185.745	6.930
1500	61.955	119.673	76.016	65.486	-286.287	-178.523	6.217
1600	62.425	123.686	78.871	71.705	-299.761	-170.848	5.578
1700	62.859	127.484	81.620	77.969	-298.358	-162.834	5.003
1800	63.304	131.089	84.269	84.277	-296.946	-154.903	4.495
1900	63.769	134.525	86.824	90.631	-295.523	-147.050	4.043
2000	64.224	137.807	89.292	97.031	-294.087	-139.273	3.637
2100	64.664	140.951	91.678	103.475	-292.639	-131.568	3.273
2200	65.103	143.970	93.986	109.963	-291.179	-123.932	2.943
2300	65.542	146.873	96.223	116.496	-289.706	-116.363	2.643
2400	65.982	149.672	98.392	123.072	-288.219	-108.858	2.369
2500	66.421	152.374	100.498	129.692	-286.718	-101.415	2.119
2600	66.860	154.988	102.543	136.356	-285.201	-94.033	1.889
2700	67.300	157.520	104.533	143.064	-283.668	-86.710	1.677
2800	67.739	159.975	106.469	149.816	-573.076	-73.220	1.366
2900	68.178	162.360	108.356	156.612	-570.446	-55.415	0.998
3000	68.618	164.679	110.195	163.452	-567.783	-37.700	0.656

PREVIOUS: CURRENT: September 1977

$\Delta_f H°(0\ K)$ = [261 ± 63] kJ mol^{-1}

S°(298.15 K) = 210.29 ± 0.42 J K^{-1} mol^{-1} $\Delta_f H°(298.15\ K)$ = [264 ± 63] kJ mol^{-1}

Electronic and Molecular Constants (σ = 1)

Source	State	ε_i, cm^{-1}	g_i	r_e, Å	B_e, cm^{-1}	α_e, cm^{-1}	ω_e, cm^{-1}	$\omega_e x_e$, cm^{-1}
(1)	$X^1\Sigma^+$	0.0	1	1.7415	0.79059	0.0064	997.94	6.137
(1, 2, 7)	$^3\Pi$	[6600.]	6	[1.9075]	[0.6590]	[0.00605]	[762.46]	[4.12]
(1)	$A^1\Pi$	7842.9	2	1.9075	0.6590	0.00605	762.46	4.12
(1)	$^1\Delta$	[13048.4]	2	[2.000]	[0.5997]	[0.0069]	[671.6]	[5.48]
(1, 2)	$^3\Sigma^+$	[21000.]	3	[1.8137]	[0.72894]	[0.00604]	[851.35]	[4.85]
(1)	$B^1\Sigma^+$	25868.6	1	1.8137	0.72894	0.00604	851.35	4.85

Enthalpy of Formation

No experimental data are available on BeS(g). We estimate D_0^0(BeS, g) by the method of Hauge and Margrave (1) based on data involving known bond energies and force constants of the gaseous diatomic oxides and sulfides. Based on dissociation energies of the alkaline earth monoxides, taken from Srivastava (2), dissociation energies of the alkaline earth sulfides and force constants calculated from JANAF data (3), we calculate D_0^0(BeS, g) = 80 ± 10 kcal mol^{-1}. A normal linear-Birge-Sponer extrapolation of the ground state yields D_0^0(BeS, g) = 115 kcal mol^{-1}. After correcting for ionic character (8) and for excited state products D_0^0(BeS, g) = 76 kcal mol^{-1}. These values are very uncertain. See the discussion of MgS(g) (3). We adopte D_0^0(BeS, g) = 80 ± 15 kcal mol^{-1}. Using auxiliary JANAF data (3), we calculate $\Delta_f H°$(BeS, g, 298.15 K) = 63 ± 15 kcal mol^{-1}.

Heat Capacity and Entropy

The spectroscopic constants of the observed states taken from Cheetham et al. (4) are corrected to account for the natural abundances of the elements. Other low-lying electronic states and their vibrational-rotational constants are estimated in isoconfigurational groups by analogy with BeO (5) and from trends observed in the known states of the other alkaline-earth oxides and sulfides (3). Uncertainty in the energy and molecular constants for the estimated states may contribute as much as 2-3 cal K^{-1} mol^{-1} to the entropy above 3000 K.

The splitting between the $A^1\Pi$ and a $^3\Pi$ states, ~1200 cm^{-1}, is based on trends in the triplet-singlet splitting observed by Field (6) for the alkaline-earth oxides. The Hartree-Fock calculations of Verhaegen and Richards (7) support this. Their theoretical signlet-triplet splitting should be reasonably accurate even though the absolute energies are biased. The thermodynamic functions are calculated using first-order anharmonic corrections to Q_r and Q_v in the partition function Q = $Q_t^i \Sigma\ Q_r^i\ Q_v^i\ g_i\ \exp(-c_2\varepsilon_i/T)$.

References

1. R. H. Hauge and J. L. Margrave, High Temp. Sci. 4, 170 (1972).
2. R. D. Srivastava, High Temp. Sci. 8, 225 (1976).
3. JANAF Thermochemical Tables: Be(g), 9-30-61; S(g) 6-30-71; S_2(g), 12-31-65; BaO(g) and SrO(g), 6-30-74; BeO(g), MgO(g), and CaO(g), 12-31-74; MgS(g), CaS(g), SrS(g), and BaS(g), 9-30-77.
4. C. J. Cheetham, W. J. M. Gissane, and R. F. Barrow, Trans. Faraday Soc. 61, 1308 (1965).
5. B. Rosen, "Spectroscopic Data Relative to Diatomic Molecules," Pergamon Press, New York, 1970.
6. R. W. Field, J. Chem. Phys. 60, 2400 (1974).
7. G. Verhaegen and W. G. Richards, Proc. Phys. Soc. (London) 90, 579 (1967).
8. D. L. Hildenbrand, "Advances in High Temperature Chemistry," Vol. 1, L. Eyring (ed.), pp. 198-206, Academic Press, New York, 1967.

Enthalpy Reference Temperature = T_r = 298.15 K Standard State Pressure = p° = 0.1 MPa

T/K	$C_p°$ (J $K^{-1}mol^{-1}$)	S° (J $K^{-1}mol^{-1}$)	$-[G°-H°(T_r)]/T$ (J $K^{-1}mol^{-1}$)	$H°-H°(T_r)$ (kJ mol^{-1})	$\Delta_f H°$ (kJ mol^{-1})	$\Delta_f G°$ (kJ mol^{-1})	Log K_f
0	0.	0.	INFINITE	-8.780	261.156	261.156	INFINITE
100	29.109	178.064	236.789	-5.872	263.335	246.831	-128.931
200	29.468	198.299	213.056	-2.951	264.036	229.939	-60.054
298.15	30.786	210.288	210.288	0.	263.592	213.267	-37.363
300	30.816	210.478	210.289	0.057	263.577	212.954	-37.079
400	32.371	219.561	211.516	3.218	260.298	196.318	-25.636
500	33.636	226.926	213.884	6.521	257.571	180.624	-18.870
600	34.575	233.146	216.590	9.934	255.132	165.469	-14.405
700	35.273	238.531	219.348	13.428	252.888	150.704	-11.246
800	35.834	243.279	222.049	16.984	250.736	136.254	-8.896
900	36.363	247.530	224.648	20.594	195.496	123.152	-7.148
1000	36.960	251.391	227.132	24.259	194.622	115.160	-6.015
1100	37.713	254.947	229.501	27.991	193.712	107.258	-5.093
1200	38.688	258.269	231.761	31.809	192.784	99.439	-4.328
1300	39.919	261.412	233.922	35.737	191.865	91.698	-3.684
1400	41.407	264.423	235.994	39.801	190.979	84.026	-3.135
1500	43.124	267.338	237.987	44.026	190.150	76.416	-2.661
1600	45.017	270.180	239.910	48.432	174.862	69.385	-2.265
1700	47.016	272.969	241.773	53.033	174.602	62.801	-1.930
1800	49.044	275.714	243.583	57.836	174.509	56.228	-1.632
1900	51.023	278.419	245.345	62.840	174.583	49.656	-1.365
2000	52.883	281.084	247.066	68.037	174.815	43.075	-1.125
2100	54.566	283.706	248.748	73.411	175.193	36.480	-0.907
2200	56.030	286.279	250.396	78.943	175.696	29.863	-0.709
2300	57.248	288.797	252.011	84.609	176.303	23.221	-0.527
2400	58.209	291.255	253.595	90.384	176.989	16.551	-0.360
2500	58.918	293.647	255.150	96.242	177.728	9.851	-0.206
2600	59.386	295.967	256.675	102.159	178.498	3.120	-0.063
2700	59.635	298.214	258.172	108.112	179.276	-3.640	0.070
2800	59.690	300.384	259.641	114.080	-110.916	-4.206	0.078
2900	59.580	302.477	261.083	120.045	-109.117	-0.427	0.008
3000	59.333	304.493	262.496	125.991	-107.347	3.291	-0.057
3100	58.975	306.433	263.882	131.907	-105.618	6.950	-0.117
3200	58.531	308.299	265.241	137.783	-103.942	10.555	-0.172
3300	58.023	310.092	266.574	143.611	-102.326	14.108	-0.223
3400	57.471	311.816	267.879	149.386	-100.779	17.613	-0.271
3500	56.890	313.474	269.158	155.105	-99.305	21.073	-0.314
3600	56.293	315.068	270.412	160.764	-97.909	24.492	-0.355
3700	55.693	316.602	271.639	166.363	-96.594	27.874	-0.394
3800	55.097	318.080	272.842	171.903	-95.361	31.222	-0.429
3900	54.512	319.503	274.020	177.383	-94.213	34.538	-0.463
4000	53.943	320.876	275.175	182.806	-93.150	37.825	-0.494
4100	53.393	322.201	276.306	188.172	-92.173	41.087	-0.523
4200	52.866	323.482	277.414	193.485	-91.281	44.326	-0.551
4300	52.363	324.720	278.500	198.746	-90.476	47.546	-0.578
4400	51.884	325.918	279.564	203.958	-89.756	50.747	-0.602
4500	51.430	327.079	280.607	209.124	-89.122	53.933	-0.626
4600	51.001	328.204	281.629	214.245	-88.574	57.106	-0.648
4700	50.597	329.297	282.632	219.325	-88.110	60.267	-0.670
4800	50.217	330.358	283.615	224.365	-87.731	63.420	-0.690
4900	49.859	331.390	284.580	229.369	-87.435	66.566	-0.710
5000	49.524	332.394	285.526	234.338	-87.222	69.707	-0.728
5100	49.210	333.371	286.455	239.274	-87.093	72.844	-0.746
5200	48.916	334.324	287.366	244.181	-87.044	75.979	-0.763
5300	48.641	335.253	288.261	249.058	-87.077	79.114	-0.780
5400	48.384	336.160	289.140	253.909	-87.190	82.251	-0.796
5500	48.143	337.045	290.003	258.736	-87.383	85.390	-0.811
5600	47.919	337.911	290.850	263.539	-87.653	88.534	-0.826
5700	47.709	338.757	291.683	268.320	-88.001	91.683	-0.840
5800	47.513	339.585	292.502	273.081	-88.425	94.839	-0.854
5900	47.331	340.396	293.307	277.823	-88.923	98.003	-0.868
6000	47.160	341.190	294.099	282.547	-89.496	101.176	-0.881

PREVIOUS: September 1977 (1 atm) CURRENT: September 1977 (1 bar)

BERYLLIUM (Be_2) IDEAL GAS M_r = 18.02436

D_0^0 = 8.535 ± 0.879 kJ mol^{-1} $\Delta_f H°(0\ K)$ = 631.1 ± 6.3 kJ mol^{-1}

S°(298.15 K) = 206.4 ± 4.0 J K^{-1} mol^{-1} $\Delta_f H°(298.15\ K)$ = 637.2 ± 7.0 kJ mol^{-1}

Spectroscopic Constants for Be_2 in cm^{-1} (ground state)

v/	0	1	2	3	4	5	6	7	8
G(v)-G(0)	0	224	328.6	405.1	459.7	508	547.9	578.2	600.7
B(v)*	0.548	0.457	0.366	0.317	0.296	0.275	0.254	0.233	0.212
10^6 x D(v)*	10.13	10.13	10.13	31.64	31.64	31.64	31.64	31.64	31.64

*Estimated as explained below.

Enthalpy of Formation

The adopted value of the ground state dissociation energy for $Be_2(g)$, D_0^0 = 8.535 ± 0.879 kJ mol^{-1}, was determined via a high level quantum mechanical calculation by Lengsfield, et al. (1). Using this value and the recommended value for $\Delta_f H°$(Be, g, 0 K) = 319.8 kJ mol^{-1} from CODATA (2), we calculate the adopted value of $\Delta_f H°(Be_2$, g, 0 K) = 631.1 ± 6.3 kJ mol^{-1}.

Heat Capacity and Entropy

The thermal functions were calculated using a direct summation technique analogous to the alkali dimers and diatomic halogens (12). As in the diatomic fluorine case, the values of G(v), B_v, and D_v were directly input into the calculation. Values of 9 bound vibrational energy levels for the $^1\Sigma_g^+$ ground state were determined from a high level *ab initio* quantum mechanical calculation (including full configuration interaction) (3). However, a recent spectroscopic determination by Bondybey (4) gave a value for $\Delta G_{1/2}$ = 224 cm^{-1} which is 1.21 times larger than the corresponding value from the theoretical calculation. We therefore adopt $\Delta G_{1/2}$ = 224 cm^{-1} and increase the rest of the theoretically determined vibrational spacings by the same factor. These scaled G(v) are given in the table above. Since the theoretical binding energy of Harrison and Handy (3) may also be too small, we adopt the value calculated by Lengsfield et al. (1). The adopted values of B_v are estimated using the correlation of Calder and Ruedenberg (5). Since a plot of $\Delta G_{v+1/2}$ vs. v exhibits two linear regions ($0 \le v \le 3$ and $4 \le v \le 8$), the value of B_e used to estimate α_e in the $4 \le v \le 8$ region is reduced from 0.593 to 0.39. This is done in order to assure a smoothly decreasing B_v as v increases. The D_v parameters are determined from the estimation method of Herzberg (6) where we use the first order D_e term only.

The rotational levels are extrapolated to high J values according to the method of Khachkuruzov (7) who proposed a simpler form of Wooley's method (8). The rotational levels are weighted in accordance with the nuclear spin-rotation interaction (9) as follows: even J weight = 0.625, odd J weight = 0.375. The so-called quasibound rotational levels were not included in the calculation. No suitable spectroscopic information exists for any of the excited states of Be_2. It should be noted that the inclusion of $^3\Sigma_u^+$ and $^3\Pi_g$ states in the range Te ≥ 5000 cm^{-1} as proposed by Jones (10) would cause our calculated thermal functions to be deficient at temperatures greater than about 2000 K (13). Due to the extremely shallow well of the ground state, most Be_2 molecules will probably exist in an excited state at temperatures T ≥ 1000 K. The calculated S°(298.15 K) is 5.41 J K^{-1} mol^{-1} smaller than the value adopted by Glushko, et al. (11).

References

1. B. H. Lengsfield III, A. D. McLean, M. Yoshimine and B. Liu, J. Chem. Phys. 79, 1891 (1983).
2. J. D. Cox, J. Chem. Thermodynamics 10, 903 (1977).
3. R. J. Harrison and N. C. Handy, Chem. Phys. Lett., 98, 97 (1983).
4. V. Bondybey, Bell Laboratories, submitted for publication, 1983.
5. G. V. Calder and K. Ruedenberg, J. Chem. Phys. 49, 5399 (1968).
6. G. Herzberg, "Spectra of Diatomic Molecules," 2nd Ed., D. Van Nostrand Co., Inc., New York, Eqn. III-123, 1950.
7. G. A. Khachkuruzov, Opt. Spectrosk. 30, 455 (1971).
8. H. W. Wooley, R. B. Scott, and F. G. Brickwedde, J. Res. Nat. Bur. Stand. 41, 379 (1948).
9. Ref. 6, Chapter 5.
10. R. O. Jones, J. Chem. Phys. 71, 1300 (1979).
11. V. P. Glushko, V. Medvedev, et al., "Thermal Constants of Substances", Volume IX, Moscow, 1979.
12. JANAF Thermochemical Tables, $Li_2(g)$, 12-31-83; $F_2(r)$, 6-30-82.
13. See for example JANAF Thermochemical Tables, $Mg_2(g)$, 9-30-83.

Beryllium (Be_2) $Be_2(g)$

Enthalpy Reference Temperature = T_r = 298.15 K Standard State Pressure = p° = 0.1 MPa

T/K	$C_p°$ (J $K^{-1}mol^{-1}$)	S° (J $K^{-1}mol^{-1}$)	$-[G°-H°(T_r)]/T$ (J $K^{-1}mol^{-1}$)	$H°-H°(T_r)$ (kJ mol^{-1})	$\Delta_f H°$ (kJ mol^{-1})	$\Delta_f G°$ (kJ mol^{-1})	Log K_f
0	0.	0.	INFINITE	-9.960	631.065	631.065	INFINITE
100	37.037	167.734	235.643	-6.791	634.156	617.483	-322.540
200	34.597	194.091	209.116	-3.005	636.787	599.638	-156.609
298.15	27.492	206.410	206.410	0.	637.161	581.249	-101.832
300	27.405	206.580	206.411	0.051	637.151	580.902	-101.144
400	24.295	213.975	207.440	2.614	636.030	562.294	-73.428
500	22.893	219.227	209.297	4.965	634.176	544.063	-56.838
600	22.173	223.331	211.306	7.215	631.891	526.249	-45.814
700	21.763	226.715	213.273	9.410	629.303	508.843	-37.970
800	21.509	229.603	215.138	11.572	626.472	491.825	-32.113
900	21.341	232.126	216.888	13.714	623.429	475.175	-27.578
1000	21.226	234.368	218.526	15.842	620.191	458.874	-23.969
1100	21.142	236.387	220.060	17.960	616.767	442.906	-21.032
1200	21.081	238.224	221.498	20.071	613.162	427.258	-18.598
1300	21.033	239.910	222.850	22.177	609.377	411.919	-16.551
1400	20.997	241.467	224.125	24.278	605.414	396.877	-14.808
1500	20.968	242.915	225.330	26.377	601.271	382.125	-13.307
1600	20.944	244.267	226.472	28.472	567.876	368.707	-12.037
1700	20.925	245.536	227.556	30.566	564.173	356.372	-10.950
1800	20.909	246.732	228.589	32.657	560.425	344.256	-9.990
1900	20.896	247.862	229.574	34.747	556.633	332.350	-9.137
2000	20.884	248.933	230.515	36.836	552.797	320.644	-8.374
2100	20.875	249.952	231.417	38.924	548.916	309.132	-7.689
2200	20.866	250.923	232.281	41.011	544.992	297.805	-7.071
2300	20.859	251.850	233.112	43.098	541.024	286.658	-6.510
2400	20.853	252.738	233.912	45.183	537.012	275.685	-6.000
2500	20.847	253.589	234.682	47.268	532.957	264.879	-5.534
2600	20.843	254.407	235.425	49.353	528.858	254.237	-5.108
2700	20.838	255.193	236.142	51.437	524.716	243.753	-4.716
2800	20.834	255.951	236.836	53.521	-61.384	245.869	-4.587
2900	20.831	256.682	237.508	55.604	-63.485	256.879	-4.627
3000	20.828	257.388	238.159	57.687	-65.598	267.962	-4.666
3100	20.825	258.071	238.791	59.769	-67.723	279.116	-4.703
3200	20.823	258.732	239.403	61.852	-69.864	290.338	-4.739
3300	20.820	259.373	239.999	63.934	-72.023	301.628	-4.774
3400	20.818	259.994	240.578	66.016	-74.204	312.983	-4.808
3500	20.816	260.598	241.141	68.098	-76.411	324.403	-4.841
3600	20.815	261.184	241.690	70.179	-78.647	335.887	-4.874
3700	20.813	261.755	242.225	72.261	-80.917	347.432	-4.905
3800	20.812	262.310	242.746	74.342	-83.223	359.040	-4.935
3900	20.810	262.850	243.255	76.423	-85.572	370.709	-4.965
4000	20.809	263.377	243.751	78.504	-87.967	382.439	-4.994
4100	20.808	263.891	244.236	80.585	-90.411	394.230	-5.023
4200	20.807	264.392	244.710	82.666	-92.911	406.080	-5.050
4300	20.806	264.882	245.173	84.746	-95.469	417.991	-5.078
4400	20.805	265.360	245.627	86.827	-98.091	429.962	-5.104
4500	20.804	265.828	246.071	88.907	-100.779	441.994	-5.131
4600	20.803	266.285	246.505	90.988	-103.539	454.085	-5.156
4700	20.803	266.732	246.931	93.068	-106.373	466.238	-5.182
4800	20.802	267.170	247.348	95.148	-109.285	478.452	-5.207
4900	20.801	267.599	247.757	97.228	-112.279	490.727	-5.231
5000	20.801	268.019	248.158	99.308	-115.356	503.064	-5.255
5100	20.800	268.431	248.551	101.388	-118.520	515.464	-5.279
5200	20.799	268.835	248.937	103.468	-121.774	527.927	-5.303
5300	20.799	269.231	249.317	105.548	-125.119	540.453	-5.326
5400	20.798	269.620	249.689	107.628	-128.556	553.043	-5.350
5500	20.798	270.002	250.055	109.708	-132.088	565.697	-5.373
5600	20.798	270.377	250.414	111.788	-135.715	578.417	-5.395
5700	20.797	270.745	250.768	113.868	-139.439	591.203	-5.418
5800	20.797	271.106	251.115	115.947	-143.259	604.054	-5.440
5900	20.796	271.462	251.457	118.027	-147.177	616.972	-5.462
6000	20.796	271.811	251.794	120.107	-151.191	629.958	-5.484

PREVIOUS: CURRENT: September 1983 (1 bar)

Beryllium (Be_2) $Be_2(g)$

BERYLLIUM CHLORIDE $((BeCl_2)_2)$ IDEAL GAS M_r = 159.83636

$\Delta_fH°(0\ K)$ = [-752] kJ mol^{-1}

$S°(298.15\ K)$ = [381.47] J K^{-1} mol^{-1} $\Delta_fH°(298.15\ K)$ = [-749] kJ mol^{-1}

Vibrational Frequencies and Degeneracies

ν, cm^{-1}	ν, cm^{-1}	ν, cm^{-1}
[220] (1)	[185] (1)	[250] (1)
[110] (1)	[100] (1)	[200] (1)
[180] (1)	860 (1)	[400] (1)
[105] (1)	[300] (1)	640 (1)

Ground State Quantum Weight: 1 σ = 4

Point Group: $[D_{2h}]$

Bond Distances: Be-Cl = [1.7]Å Be-Cl' (in ring) = [1.8]Å

Bond Angles: Cl'-Be-Cl' (in ring) = [88]° Cl'-Be-Cl = [136]°

Product of the Moments of Inertia: $I_AI_BI_C$ = [2.6276 X 10^{-112}] g^3 cm^6

Enthalpy of Formation

The enthalpy of formation was estimated by assuming that the concentration of dimer in equilibrium with the monomer was ~5% at 500 K as reported by Hildenbrand, Theard, Murad and Ju (1), giving an effective K_p = 5.5 x 10^{-4} for $2BeCl_2(g) \rightarrow Be_2Cl_4(g)$ at 500 K.

Heat Capacity and Entropy

The structure was assumed planar with a D_{2h} symmetry, the bond lengths were estimated from the values for $BeCl_2(g)$, a lengthening being assumed in the ring. The frequencies were estimated using K_2Cl_2 as a basis for the ring vibrations. Two polymeric frequencies had been observed by Snelson (2). These were observed in matrix isolated $BeCl_2$ gas and were ascribed to polymeric species on account of thier warm up behavior. They are here assigned to the B_{3u} asymmetric stretching mode and the B_{1u} symmetric out of plane wag of the non-ring chlorines. The order of the frequencies listed above is arbitrary and not related to their species type. The principal moments of inertia are: I_A = [18.4088 x 10^{-39}] g cm^2, I_B = [110.6214 x 10^{-39}] g cm^2 and I_C = [129.0303 x 10^{-39}] g cm^2.

References

1. D. L. Hildenbrand, L. P. Theard, E. Murad and F. Ju, Aeronutronic Div. of Philco Corp. Report U-3068, April 1965, under contract AF 04 (611)-8523.

2. A. Snelson, IIT Research Institute, Report No. IITRI-C 6013-4, May 1964, under contract No. DA-31-124-ARO (D)-111.

Beryllium Chloride $((BeCl_2)_2)$ $Be_2Cl_4(g)$

Enthalpy Reference Temperature = T_r = 298.15 K Standard State Pressure = p° = 0.1 MPa

T/K	$C_p°$	S° (J $K^{-1}mol^{-1}$)	$-[G°-H°(T_r)]/T$	$H°-H°(T_r)$ (kJ mol^{-1})	$\Delta_fH°$	$\Delta_fG°$	Log K_f
0	0.	0.	INFINITE	-25.305	-752.015	-752.015	INFINITE
100	77.971	273.932	475.560	-20.163	-752.775	-742.266	387.720
200	104.193	337.542	391.721	-10.836	-750.678	-732.532	191.318
250	110.809	361.547	383.350	-5.451	-749.715	-728.110	152.130
298.15	115.368	381.475	381.475	0.	-748.936	-724.021	126.846
300	115.517	382.189	381.477	0.214	-748.909	-723.867	126.036
350	118.970	400.268	382.897	6.080	-748.236	-719.748	107.417
400	121.557	416.331	386.091	12.096	-747.650	-715.719	93.463
450	123.529	430.768	390.266	18.225	-747.127	-711.760	82.619
500	125.056	443.865	394.982	24.442	-746.652	-707.856	73.949
600	127.217	466.871	405.099	37.063	-745.829	-700.176	60.956
700	128.625	486.595	415.366	49.860	-745.159	-692.623	51.684
800	129.587	503.837	425.370	62.774	-744.638	-685.155	44.736
900	130.270	519.142	434.955	75.769	-744.269	-677.743	39.335
1000	130.771	532.894	444.073	88.822	-744.055	-670.364	35.016
1100	131.149	545.377	452.723	101.919	-744.001	-662.999	31.483
1200	131.440	556.801	460.927	115.049	-744.111	-655.632	28.539
1300	131.669	567.331	468.712	128.205	-744.390	-648.248	26.047
1400	131.852	577.096	476.110	141.381	-744.842	-640.837	23.910
1500	132.001	586.198	483.149	154.574	-745.472	-633.387	22.057
1600	132.123	594.721	489.859	167.780	-775.356	-624.838	20.399
1700	132.225	602.734	496.265	180.998	-775.552	-615.425	18.910
1800	132.311	610.295	502.392	194.225	-775.800	-605.999	17.586
1900	132.384	617.450	508.261	207.459	-776.102	-596.558	16.401
2000	132.446	624.242	513.892	220.701	-776.460	-587.099	15.333
2100	132.500	630.706	519.302	233.948	-776.878	-577.621	14.368
2200	132.546	636.871	524.507	247.201	-777.359	-568.121	13.489
2300	132.587	642.763	529.521	260.457	-777.904	-558.598	12.686
2400	132.623	648.407	534.358	273.718	-778.519	-549.051	11.950
2500	132.655	653.822	539.029	286.982	-779.205	-539.476	11.272
2600	132.683	659.025	543.545	300.249	-779.965	-529.872	10.645
2700	132.708	664.033	547.915	313.518	-780.802	-520.237	10.065
2800	132.731	668.860	552.149	326.790	-1363.631	-498.125	9.293
2900	132.751	673.518	556.254	340.064	-1362.499	-467.234	8.416
3000	132.769	678.019	560.238	353.340	-1361.413	-436.381	7.598
3100	132.786	682.372	564.108	366.618	-1360.378	-405.564	6.834
3200	132.801	686.588	567.870	379.897	-1359.393	-374.780	6.118
3300	132.814	690.675	571.530	393.178	-1358.462	-344.025	5.445
3400	132.827	694.640	575.093	406.460	-1357.584	-313.298	4.813
3500	132.838	698.491	578.564	419.744	-1356.760	-282.596	4.218
3600	132.849	702.233	581.947	433.028	-1355.991	-251.916	3.655
3700	132.858	705.873	585.248	446.313	-1355.277	-221.257	3.124
3800	132.867	709.416	588.469	459.600	-1354.617	-190.617	2.620
3900	132.875	712.868	591.615	472.887	-1354.010	-159.994	2.143
4000	132.883	716.232	594.688	486.175	-1353.456	-129.385	1.690
4100	132.890	719.513	597.693	499.463	-1352.955	-98.789	1.259
4200	132.897	722.715	600.632	512.753	-1352.505	-68.206	0.848
4300	132.903	725.843	603.507	526.043	-1352.105	-37.632	0.457
4400	132.909	728.898	606.322	539.333	-1351.755	-7.067	0.084
4500	132.914	731.885	609.080	552.624	-1351.454	23.491	-0.273
4600	132.919	734.806	611.781	565.916	-1351.200	54.042	-0.614
4700	132.924	737.665	614.429	579.208	-1350.994	84.588	-0.940
4800	132.928	740.464	617.026	592.501	-1350.834	115.131	-1.253
4900	132.932	743.204	619.573	605.794	-1350.719	145.671	-1.553
5000	132.936	745.890	622.073	619.087	-1350.649	176.208	-1.841
5100	132.940	748.523	624.526	632.381	-1350.623	206.745	-2.118
5200	132.943	751.104	626.936	645.675	-1350.641	237.282	-2.384
5300	132.946	753.636	629.303	658.970	-1350.702	267.819	-2.640
5400	132.950	756.122	631.628	672.264	-1350.804	298.358	-2.886
5500	132.952	758.561	633.914	685.559	-1350.948	328.900	-3.124
5600	132.955	760.957	636.161	698.855	-1351.134	359.444	-3.353
5700	132.958	763.310	638.371	712.150	-1351.359	389.992	-3.574
5800	132.960	765.622	640.545	725.446	-1351.624	420.545	-3.787
5900	132.963	767.895	642.685	738.743	-1351.927	451.102	-3.994
6000	132.965	770.130	644.790	752.039	-1352.269	481.664	-4.193

PREVIOUS: June 1965 (1 atm) CURRENT: June 1965 (1 bar)

Beryllium Chloride $((BeCl_2)_2)$ $Be_2Cl_4(g)$

BERYLLIUM FLUORIDE OXIDE (O(BeF)2) IDEAL GAS M_r = 72.020566

$\Delta_f H°(0\ K) = -1204.1 \pm 21\ kJ\ mol^{-1}$

$S°(298.15\ K) = [298.9]\ J\ K^{-1}\ mol^{-1}$ $\Delta_f H°(298.15\ K) = -1204.6 \pm 21\ kJ\ mol^{-1}$

Vibrational Frequencies and Degeneracies

ν, cm^{-1}	ν, cm^{-1}
[1200] (1)	[1500] (1)
[600] (1)	[900] (1)
[400] (1)	[300] (1)
[100] (1)	[300] (1)
[300] (1)	

Ground State Quantum Weight: [1] σ = [2]

Point Group: [C_{2v}]

Bond Distances: Be-O = [1.33] Å Be-F = [1.43] Å

Bond Angles: O-Be-F = [180]° Be-O-Be = [150]°

Product of the Moments of Inertia: $I_A I_B I_C = [2.5549 \times 10^{-114}]\ g^3\ cm^6$

Enthalpy of Formation

Efimenko (1) has studied mass spectrometrically the reaction $BeO(cr) + BeF_2(g) \rightarrow Be_2OF_2(g)$ and also reported two sets of equilibrium constants. Using JANAF thermal functions, the enthalpy of reaction has been calculated from those reported equilibrium constants by the 2nd law and the 3rd law methods.

	Number of Measurments	T/K	$\Delta_r H°(298.15\ K)$, kcal/mol 2nd law	3rd law	Drift cal K^{-1} mol^{-1}
1st Set	13	1570-1891	41.448 ± 3.04	41.981	0.408 ± 1.726
2nd Set	21*	1872-2195	42.553 ± 0.70	42.714	0.015 ± 0.359

*There are 4 points rejected due to failure of statistical test.

A weighted average value, $\Delta_r H°(298.15\ K) = 42.70\ kcal\ mol^{-1}$, was chosen for the enthalpy of reaction. The enthalpy of formation $\Delta_f H°(Be_2OF_2,\ g,\ 298.15\ K) = 287.9 \pm 5\ kcal\ mol^{-1}$ was calculated from the selected enthalpy of reaction, using $\Delta_f H°(BeO,\ cr,\ 298.15\ K) = 143.1\ kcal\ mol^{-1}$ and $\Delta_f H°(BeF_2,\ g,\ 298.15\ K) = 187.5\ kcal\ mol^{-1}$.

Heat Capacity and Entropy

The adopted "V" shape molecular structure with nine fundamental vibrational frequencies was obtained from an estimation by Mann (3). The bond distance Be-O and Be-F were estimated to be the same as those in BeO(g) and BeF_2(g), respectively. The principal moments of inertia are: $I_A = 1.0103 \times 10^{-39}$, $I_B = 49.7842 \times 10^{-39}$ and $I_C = 50.7946 \times 10^{-39}\ g\ cm^2$.

References

1. J. Efimenko, NBS Report 8186, Jan., 1964.
2. D. E. Mann, NBS Report 8186, Jan., 1964.

Beryllium Fluoride Oxide ($O(BeF)_2$) $Be_2F_2O_1(g)$

Enthalpy Reference Temperature = T_r = 298.15 K Standard State Pressure = p° = 0.1 MPa

T/K	$C_p°$ (J K⁻¹mol⁻¹)	S°	$-[G°-H°(T_r)]/T$	$H°-H°(T_r)$ (kJ mol⁻¹)	$\Delta_f H°$	$\Delta_f G°$	Log K_f
0	0.	0.	INFINITE	-16.585	-1204.127	-1204.127	INFINITE
100	47.638	231.587	358.833	-12.725	-1204.705	-1202.060	627.892
200	65.525	270.625	305.569	-6.989	-1204.509	-1199.485	313.273
250	71.486	285.913	300.144	-3.558	-1204.493	-1198.233	250.357
298.15	76.151	298.915	298.915	0.	-1204.574	-1197.022	209.713
300	76.314	299.387	298.916	0.141	-1204.578	-1196.975	208.412
350	80.365	311.463	299.861	4.061	-1204.742	-1195.696	178.448
400	83.817	322.426	302.007	8.168	-1204.941	-1194.390	155.971
450	86.772	332.473	304.842	12.434	-1205.153	-1193.059	138.487
500	89.304	341.750	308.074	16.838	-1205.371	-1191.703	124.496
600	93.336	358.407	315.107	25.980	-1205.816	-1188.928	103.505
700	96.312	373.030	322.359	35.470	-1206.288	-1186.077	88.506
800	98.534	386.043	329.521	45.217	-1206.812	-1183.154	77.252
900	100.216	397.750	336.462	55.159	-1207.413	-1180.162	68.495
1000	101.510	408.378	343.131	65.248	-1208.111	-1177.098	61.485
1100	102.521	418.103	349.511	75.451	-1208.924	-1173.958	55.747
1200	103.324	427.059	355.605	85.745	-1209.867	-1170.738	50.961
1300	103.969	435.356	361.424	96.111	-1210.952	-1167.435	46.908
1400	104.494	443.080	366.984	106.535	-1212.190	-1164.042	43.431
1500	104.927	450.305	372.301	117.007	-1213.590	-1160.555	40.414
1600	105.288	457.089	377.390	127.518	-1244.232	-1155.918	37.737
1700	105.591	463.481	382.268	138.062	-1245.176	-1150.369	35.347
1800	105.849	469.524	386.949	148.635	-1246.160	-1144.764	33.220
1900	106.069	475.253	391.447	159.231	-1247.188	-1139.104	31.316
2000	106.258	480.699	395.775	169.847	-1248.259	-1133.387	29.601
2100	106.422	485.887	399.943	180.482	-1249.373	-1127.617	28.048
2200	106.566	490.841	403.963	191.131	-1250.528	-1121.791	26.635
2300	106.691	495.581	407.844	201.794	-1251.723	-1115.913	25.343
2400	106.802	500.124	411.595	212.469	-1252.956	-1109.982	24.158
2500	106.901	504.486	415.224	223.154	-1254.223	-1103.999	23.067
2600	106.988	508.681	418.739	233.849	-1255.522	-1097.965	22.058
2700	107.066	512.720	422.145	244.551	-1256.850	-1091.880	21.124
2800	107.137	516.615	425.450	255.262	-1840.120	-1073.300	20.023
2900	107.200	520.376	428.659	265.979	-1839.373	-1045.927	18.839
3000	107.257	524.011	431.777	276.701	-1838.615	-1018.579	17.735
3100	107.309	527.529	434.809	287.430	-1837.847	-991.257	16.703
3200	107.356	530.936	437.760	298.163	-1837.069	-963.961	15.735
3300	107.399	534.240	440.634	308.901	-1836.284	-936.688	14.827
3400	107.438	537.447	443.435	319.643	-1835.494	-909.439	13.972
3500	107.475	540.562	446.165	330.388	-1834.700	-882.214	13.166
3600	107.508	543.590	448.830	341.138	-1833.907	-855.011	12.406
3700	107.538	546.536	451.431	351.890	-1833.117	-827.831	11.687
3800	107.566	549.404	453.972	362.645	-1832.335	-800.671	11.006
3900	107.593	552.199	456.455	373.403	-1831.564	-773.532	10.360
4000	107.617	554.923	458.882	384.164	-1830.809	-746.413	9.747
4100	107.639	557.581	461.257	394.926	-1830.075	-719.312	9.164
4200	107.660	560.175	463.582	405.691	-1829.365	-692.229	8.609
4300	107.680	562.708	465.858	416.458	-1828.686	-665.162	8.080
4400	107.698	565.184	468.087	427.227	-1828.041	-638.111	7.575
4500	107.715	567.605	470.272	437.998	-1827.434	-611.074	7.093
4600	107.731	569.972	472.414	448.770	-1826.872	-584.051	6.632
4700	107.746	572.289	474.514	459.544	-1826.358	-557.038	6.191
4800	107.760	574.558	476.575	470.319	-1825.897	-530.036	5.768
4900	107.773	576.780	478.597	481.096	-1825.493	-503.043	5.363
5000	107.785	578.957	480.583	491.874	-1825.149	-476.058	4.973
5100	107.797	581.092	482.533	502.653	-1824.870	-449.079	4.600
5200	107.808	583.185	484.448	513.433	-1824.658	-422.105	4.240
5300	107.818	585.239	486.331	524.214	-1824.518	-395.134	3.894
5400	107.828	587.254	488.181	534.997	-1824.452	-368.166	3.561
5500	107.837	589.233	490.000	545.780	-1824.462	-341.198	3.240
5600	107.846	591.176	491.790	556.564	-1824.551	-314.228	2.931
5700	107.854	593.085	493.550	567.349	-1824.721	-287.257	2.632
5800	107.862	594.961	495.282	578.135	-1824.973	-260.282	2.344
5900	107.870	596.805	496.988	588.922	-1825.310	-233.302	2.065
6000	107.877	598.618	498.666	599.709	-1825.732	-206.315	1.796

PREVIOUS: June 1966 (1 atm) CURRENT: June 1966 (1 bar)

Beryllium Fluoride Oxide ($O(BeF)_2$) $Be_2F_2O_1(g)$

BERYLLIUM OXIDE (Be_2O) IDEAL GAS M_r = 34.02376

$\Delta_f H°(0\ K) = -64 \pm 42\ kJ\ mol^{-1}$

$S°(298.15\ K) = [220.8]\ J\ K^{-1}\ mol^{-1}$ $\Delta_f H°(298.15\ K) = -63 \pm 42\ kJ\ mol^{-1}$

Vibrational Frequencies and Degeneracies

ν, cm^{-1}	g_i
[1100]	(1)
[500]	(2)
[1600]	(1)

Ground State Quantum Weight: [3] $\sigma = 2$

Point Group: $D_{\infty h}$

Bond Distance: Be-O = [1.4] Å

Bond Angle: Be-O-Be = [180]°

Rotational Constant: B_0 = [0.477178] cm^{-1}

Enthalpy of Formation

Hildenbrand, Theard, and Ju (1) have reported the enthalpy of formation. Equilibrium constants were determined in a mass spectrometer for the reactions 2BeO(cr) + Be_2O(g) + O(g) - (1); 2BeO(g) + Be_2O(g) + O(g) - (2); and Be_2O_2(g) + Be_2O(g) + O(g) - (3). Using our own Gibbs energy functions we obtain $\Delta_f H°(Be_2O, g, 298.15\ K)$ = 13.1, -14.5, and -16.7 kcal mol^{-1}, all ± 10; the 2nd law treatment of reaction 1 gives -18.3 ± 10 kcal mol^{-1}. We adopt a median value -15 ± 10 kcal mol^{-1}.

Heat Capacity and Entropy

Buchler, Stauffer, Klemperer and Wharton (2) have shown Li_2O to be linear. Also Walsh (3) predicted that AB_2 molecules with less than 16 bonding electrons would be linear. Thus Be_2O was taken to be linear and symmetrical, with a Be-O bond length of 1.4 Å, which is a little longer than in BeO(g). The vibrational frequencies were estimated by taking ν_3 to be larger than the stretch in BeO(g) and then calculating ν_1 from the valence force field assumption. The bending force constant was estimated as 1/20th of the stretching force constant. The two outermost electrons were considered unpaired and so gave a $^3\Sigma$ ground state.

References

1. D. L. Hildenbrand, L. P. Theard, and F. Ju, Ford Motor Co., Aeronutronic Division, Third Quarterly Technical Report U-2231, April-June 1963.
2. A. Buchler, J. L. Stauffer, W. Klemperer and C. Wharton, A. D. Little, Inc., Interim Technical Report No. 5, 1963.
3. A. D. Walsh, J. Chem. Soc. London 2269 (1953).

Beryllium Oxide (Be_2O) $Be_2O_1(g)$

Enthalpy Reference Temperature = T_r = 298.15 K Standard State Pressure = p° = 0.1 MPa

T/K	$C_p°$ (J K⁻¹mol⁻¹)	S° (J K⁻¹mol⁻¹)	-[G°-H°(T_r)]/T (J K⁻¹mol⁻¹)	H°-H°(T_r) (kJ mol⁻¹)	$\Delta_f H°$ (kJ mol⁻¹)	$\Delta_f G°$ (kJ mol⁻¹)	Log K_f
0	0.	0.	INFINITE	-9.925	-64.479	-64.479	INFINITE
100	29.748	183.340	253.419	-7.008	-63.092	-72.660	37.954
200	35.536	205.584	224.408	-3.765	-62.460	-82.559	21.562
250	38.466	213.835	221.488	-1.913	-62.521	-87.581	18.299
298.15	40.952	220.827	220.827	0.	-62.760	-92.388	16.186
300	41.041	221.081	220.828	0.076	-62.772	-92.572	16.118
350	43.322	227.582	221.336	2.186	-63.156	-97.510	14.553
400	45.354	233.502	222.492	4.404	-63.614	-102.387	13.370
450	47.163	238.951	224.022	6.718	-64.112	-107.204	12.444
500	48.767	244.004	225.771	9.117	-64.636	-111.964	11.697
600	51.429	253.141	229.588	14.132	-65.734	-121.327	10.562
700	53.482	261.230	233.541	19.382	-66.894	-130.502	9.738
800	55.063	268.480	237.463	24.813	-68.126	-139.506	9.109
900	56.290	275.039	241.280	30.383	-69.443	-148.350	8.610
1000	57.250	281.021	244.959	36.062	-70.861	-157.043	8.203
1100	58.011	286.515	248.491	41.826	-72.394	-165.588	7.863
1200	58.620	291.590	251.874	47.659	-74.052	-173.988	7.573
1300	59.115	296.302	255.112	53.547	-75.846	-182.244	7.323
1400	59.520	300.698	258.213	59.479	-77.785	-190.357	7.102
1500	59.856	304.816	261.184	65.448	-79.877	-198.326	6.906
1600	60.138	308.689	264.033	71.449	-111.202	-205.098	6.696
1700	60.375	312.342	266.769	77.474	-112.818	-210.917	6.481
1800	60.577	315.799	269.397	83.522	-114.467	-216.641	6.287
1900	60.750	319.079	271.926	89.589	-116.153	-222.271	6.111
2000	60.899	322.198	274.363	95.671	-117.877	-227.812	5.950
2100	61.029	325.173	276.712	101.768	-119.641	-233.265	5.802
2200	61.142	328.015	278.980	107.877	-121.448	-238.634	5.666
2300	61.242	330.735	281.171	113.996	-123.298	-243.919	5.540
2400	61.330	333.343	283.291	120.125	-125.193	-249.123	5.422
2500	61.408	335.848	285.344	126.262	-127.134	-254.247	5.312
2600	61.478	338.258	287.333	132.406	-129.121	-259.293	5.209
2700	61.540	340.580	289.262	138.557	-131.155	-264.261	5.112
2800	61.596	342.819	291.135	144.714	-715.150	-256.707	4.789
2900	61.647	344.981	292.955	150.876	-715.152	-240.335	4.329
3000	61.692	347.072	294.724	157.043	-715.169	-223.961	3.900
3100	61.734	349.095	296.446	163.214	-715.203	-207.587	3.498
3200	61.771	351.056	298.122	169.390	-715.258	-191.212	3.121
3300	61.806	352.957	299.755	175.569	-715.336	-174.834	2.767
3400	61.837	354.803	301.347	181.751	-715.441	-158.454	2.434
3500	61.866	356.596	302.900	187.936	-715.576	-142.070	2.120
3600	61.893	358.339	304.416	194.124	-715.746	-125.682	1.824
3700	61.917	360.035	305.896	200.314	-715.953	-109.288	1.543
3800	61.940	361.687	307.343	206.507	-716.203	-92.888	1.277
3900	61.961	363.296	308.757	212.702	-716.498	-76.481	1.024
4000	61.980	364.865	310.140	218.899	-716.845	-60.067	0.784
4100	61.998	366.395	311.493	225.098	-717.246	-43.642	0.556
4200	62.015	367.890	312.818	231.299	-717.706	-27.208	0.338
4300	62.030	369.349	314.116	237.501	-718.230	-10.761	0.131
4400	62.045	370.775	315.388	243.705	-718.821	5.698	-0.068
4500	62.059	372.170	316.634	249.910	-719.484	22.172	-0.257
4600	62.071	373.534	317.856	256.117	-720.222	38.661	-0.439
4700	62.083	374.869	319.055	262.324	-721.040	55.167	-0.613
4800	62.095	376.176	320.232	268.533	-721.941	71.691	-0.780
4900	62.105	377.457	321.387	274.743	-722.929	88.236	-0.941
5000	62.115	378.711	322.521	280.954	-724.006	104.801	-1.095
5100	62.124	379.941	323.634	287.166	-725.175	121.389	-1.243
5200	62.133	381.148	324.729	293.379	-726.440	138.000	-1.386
5300	62.142	382.332	325.805	299.593	-727.802	154.637	-1.524
5400	62.149	383.493	326.862	305.807	-729.265	171.300	-1.657
5500	62.157	384.634	327.902	312.023	-730.828	187.991	-1.785
5600	62.164	385.754	328.925	318.239	-732.495	204.712	-1.909
5700	62.171	386.854	329.932	324.455	-734.266	221.463	-2.029
5800	62.177	387.935	330.923	330.673	-736.142	238.247	-2.146
5900	62.183	388.998	331.898	336.891	-738.125	255.063	-2.258
6000	62.189	390.043	332.858	343.109	-740.214	271.915	-2.367

PREVIOUS: September 1963 (1 atm) CURRENT: September 1963 (1 bar)

Beryllium Oxide (Be_2O) $Be_2O_1(g)$

BERYLLIUM OXIDE ((BeO)2) IDEAL GAS M_r = 50.02316

$S°(298.15\ K) = [247.61]\ J\ K^{-1}\ mol^{-1}$

$\Delta_f H°(0\ K) = -409 \pm 50\ kJ\ mol^{-1}$

$\Delta_f H°(298.15\ K) = -410 \pm 50\ kJ\ mol^{-1}$

Vibrational Frequencies and Degeneracies

ν, cm^{-1}	ν, cm^{-1}
[1120] (1)	[460] (1)
[630] (1)	[920] (1)
[760] (1)	[1480] (1)

Ground State Quantum Weight: [1] $\sigma = 4$

Point Group: $[V_h]$

Bond Distance: Be-O = [1.63] Å

Bond Angles: Be-O-Be = [90]° O-Be-O = [90]°

Product of the Moments of Inertia: $I_A I_B I_C = [3.0972 \times 10^{-115}]\ g^3\ cm^6$

Enthalpy of Formation

Chupka, Berkowitz, and Giese (1) studied the temperature dependence of the $(BeO)_2^+$ ion in a mass spectrometer and obtained a 2nd law enthalpy of vaporization of 172 ± 8 kcal mol^{-1} at 2150 K. No correction was made for fragmentation of higher polymers and this is reflected in the uncertainty. We adopted an enthalpy of sublimation of 178 kcal mol^{-1} in order to match more closely the equilibrium constant found by Chupka, et al. (1). However, it was not possible to alter either the enthalpy or entropy sufficiently to obtain exact agreement, the predicted pressure being too high by a factor of 20.

Heat Capacity and Entropy.

A planar ring configuration was assumed which has six vibrational frequencies of the type $2A_{1g}$, B_{1g}, B_{1u}, B_{2u}, and B_{3u}. These were estimated by comparison with $(LiO)_2$ for which values have been calculated by White, Seshadri, Dever, Mann, and Linevsky (2). It seems unlikely that any other structure than that of a ring is possible, as, in order to approach the observed equilibrium constant an even lower entropy is needed. A non-linear chain would have a higher entropy due to the change in symmetry number and the two odd electrons. A linear chain would have approximately the same entropy due to cancellation of opposite effects. The vibrational frequencies were also made as high as possible by normalizing them using B_{3u} equal to the BeO stretch in BeO(g). The bond length was taken as the sum of the covalent radii of O and Be, which is in general agreement with other polymeric structures which generally increase in bond length by 10-20% in going from monomer to polymer. The bond angle was chosen arbitrarily, however, changes in the angle would change the entropy less than 1 cal K^{-1} mol^{-1}.

The principal moments of inertia are: $I_A = 7.0589 \times 10^{-39}$, $I_B = 3.9761 \times 10^{-39}$, and $I_C = 11.0350 \times 10^{-39}\ g\ cm^2$.

References

1. W. A. Chupka, J. Berkowitz and C. F. Giese, J. Chem. Phys. 30, 827 (1959).
2. D. White, K. Seshadri, D. F. Dever, D. E. Mann, and M. J. Linevsky, prepublication communication, July 1963.

Beryllium Oxide $((BeO)_2)$ $Be_2O_2(g)$

Enthalpy Reference Temperature = T_r = 298.15 K Standard State Pressure = p° = 0.1 MPa

T/K	$C_p°$ (J K⁻¹mol⁻¹)	S° (J K⁻¹mol⁻¹)	$-[G°-H°(T_r)]/T$ (J K⁻¹mol⁻¹)	$H°-H°(T_r)$ (kJ mol⁻¹)	$\Delta_f H°$ (kJ mol⁻¹)	$\Delta_f G°$ (kJ mol⁻¹)	Log K_f
0	0.	0.	INFINITE	-11.409	-408.893	-408.893	INFINITE
100	33.845	204.646	285.396	-8.075	-408.542	-411.575	214.985
200	40.459	229.794	251.839	-4.409	-408.942	-414.535	108.265
250	45.019	239.309	248.400	-2.273	-409.448	-415.877	86.893
298.15	49.339	247.611	247.611	0.	-410.032	-417.064	73.068
300	49.499	247.917	247.612	0.091	-410.056	-417.107	72.625
350	53.618	255.862	248.230	2.671	-410.709	-418.231	62.418
400	57.274	263.266	249.652	5.445	-411.357	-419.261	54.750
450	60.456	270.200	251.554	8.391	-411.982	-420.211	48.777
500	63.197	276.715	253.748	11.484	-412.583	-421.093	43.991
600	67.555	288.643	258.589	18.033	-413.728	-422.687	36.798
700	70.753	299.310	263.658	24.956	-414.842	-424.092	31.646
800	73.124	308.920	268.725	32.156	-415.973	-425.337	27.772
900	74.910	317.641	273.683	39.561	-417.157	-426.437	24.750
1000	76.277	325.607	278.483	47.124	-418.423	-427.401	22.325
1100	77.342	332.929	283.105	54.807	-419.791	-428.233	20.335
1200	78.185	339.696	287.542	62.585	-421.278	-428.936	18.671
1300	78.861	345.982	291.799	70.438	-422.898	-429.510	17.258
1400	79.410	351.847	295.881	78.353	-424.662	-429.953	16.042
1500	79.863	357.342	299.797	86.317	-426.580	-430.265	14.983
1600	80.239	362.508	303.556	94.323	-457.732	-429.393	14.018
1700	80.555	367.382	307.169	102.363	-459.180	-427.577	13.138
1800	80.823	371.994	310.643	110.432	-460.666	-425.676	12.353
1900	81.051	376.371	313.988	118.526	-462.194	-423.690	11.648
2000	81.248	380.533	317.212	126.641	-463.767	-421.623	11.012
2100	81.419	384.501	320.323	134.775	-465.387	-419.476	10.434
2200	81.568	388.293	323.327	142.924	-467.057	-417.251	9.907
2300	81.698	391.921	326.231	151.088	-468.778	-414.949	9.424
2400	81.813	395.401	329.041	159.263	-470.553	-412.571	8.979
2500	81.915	398.743	331.763	167.450	-472.382	-410.118	8.569
2600	82.006	401.957	334.401	175.646	-474.265	-407.590	8.189
2700	82.087	405.054	336.961	183.851	-476.204	-404.989	7.835
2800	82.160	408.041	339.446	192.063	-1060.112	-389.869	7.273
2900	82.226	410.925	341.862	200.283	-1060.036	-365.934	6.591
3000	82.285	413.713	344.211	208.508	-1059.983	-342.000	5.955
3100	82.339	416.412	346.496	216.740	-1059.955	-318.068	5.359
3200	82.388	419.027	348.722	224.976	-1059.955	-294.137	4.801
3300	82.432	421.563	350.891	233.217	-1059.987	-270.205	4.277
3400	82.473	424.025	353.006	241.462	-1060.053	-246.270	3.783
3500	82.510	426.416	355.070	249.711	-1060.156	-222.335	3.318
3600	82.545	428.741	357.084	257.964	-1060.300	-198.395	2.879
3700	82.576	431.003	359.051	266.220	-1060.489	-174.451	2.463
3800	82.605	433.205	360.974	274.479	-1060.726	-150.500	2.069
3900	82.632	435.351	362.854	282.741	-1061.016	-126.543	1.695
4000	82.657	437.444	364.692	291.006	-1061.363	-102.578	1.340
4100	82.681	439.485	366.492	299.273	-1061.770	-78.604	1.001
4200	82.702	441.478	368.254	307.542	-1062.242	-54.618	0.679
4300	82.722	443.424	369.979	315.813	-1062.784	-30.621	0.372
4400	82.741	445.326	371.670	324.086	-1063.399	-6.610	0.078
4500	82.759	447.186	373.328	332.361	-1064.091	17.415	-0.202
4600	82.775	449.005	374.953	340.638	-1064.864	41.456	-0.471
4700	82.791	450.785	376.548	348.916	-1065.723	65.516	-0.728
4800	82.805	452.528	378.113	357.196	-1066.670	89.595	-0.975
4900	82.819	454.236	379.649	365.477	-1067.710	113.695	-1.212
5000	82.832	455.909	381.157	373.760	-1068.846	137.817	-1.440
5100	82.844	457.550	382.639	382.044	-1070.081	161.962	-1.659
5200	82.855	459.158	384.095	390.329	-1071.418	186.132	-1.870
5300	82.866	460.737	385.526	398.615	-1072.860	210.330	-2.073
5400	82.876	462.286	386.934	406.902	-1074.409	234.555	-2.269
5500	82.886	463.807	388.318	415.190	-1076.066	258.810	-2.458
5600	82.895	465.300	389.679	423.479	-1077.836	283.097	-2.641
5700	82.903	466.767	391.019	431.769	-1079.718	307.415	-2.817
5800	82.912	468.209	392.337	440.060	-1081.715	331.769	-2.988
5900	82.919	469.627	393.635	448.351	-1083.828	356.156	-3.153
6000	82.927	471.020	394.913	456.643	-1086.057	380.582	-3.313

PREVIOUS: September 1963 (1 atm) CURRENT: September 1963 (1 bar)

Beryllium Oxide $((BeO)_2)$ $Be_2O_2(g)$

BERYLLIUM SILICATE (Be_2SiO_4) CRYSTAL M_r = 110.10746

$\Delta_fH°$(0 K) = [-2105 ± 29] kJ mol^{-1}
$\Delta_fH°$(298.15 K)= [-2117 ± 29] kJ mol^{-1}

S°(298.15 K) = 64.190 ± 0.4 J K^{-1} mol^{-1}
T_{fus} = 1833 K

Enthalpy of Formation

The $\Delta_fH°$(298.15 K) was estimated from the $\Delta_rH°$(298.15 K) of Mg, Ca and Ba oxides with SiO_2 given by Kelley (1) and from the $\Delta_fH°$(298.15 K) of BeO(cr) and SiO_2(cr) in the JANAF tables (2). The $\Delta_rH°$(298.15 K) = 0 ± 7 kcal mol^{-1} was estimated for the reaction 2BeO(cr) + SiO_2(cr) → Be_2SiO_4(cr).

Heat Capacity and Entropy

At low temperatures C_p was used from Kelley (4). Above 298.15 K C_p values were estimated by summation of the component oxides using the JANAF tables (2), BeO and SiO_2. The entropy was calculated at 54.7 K using the Debye and Planck-Einstein functions D(422/T) + 2E(527/T) given by Kelley (4). The value of S°(54.7 K) = 13.59 cal K^{-1} mol^{-1}.

Fusion Data

The incongruent melting point was taken from Kubaschewski and Evans (5).

References

1. K. K. Kelley, U.S. Bur. Mines, RI-5901, 1962.
2. JANAF Thermochemical Table: BeO(cr), 9-30-63; SiO2(cr), 12-30-62.
3. K. K. Kelley, J. Am. Chem. Soc., 61, 1217 (1939).
4. Kubaschewski and Evans, "Metallurgical Thermochemistry" 3rd, edition, Pergamon Press., New York 1958.

Beryllium Silicate (Be_2SiO_4) $Be_2O_4Si_1$(cr)

Enthalpy Reference Temperature = T_r = 298.15 K   Standard State Pressure = p° = 0.1 MPa

T/K	$C_p°$	S°	$-[G°-H°(T_r)]/T$	$H°-H°(T_r)$	$\Delta_fH°$	$\Delta_fG°$	Log K_f
	J $K^{-1}mol^{-1}$			kJ mol^{-1}			
0	0.	0.	INFINITE	-12.227	-2105.003	-2105.003	INFINITE
100	18.794	7.470	124.264	-11.679	-2110.609	-2076.211	1084.501
200	60.433	33.245	71.847	-7.720	-2114.806	-2040.059	532.809
298.15	95.563	64.190	64.190	0.	-2117.225	-2002.794	350.881
300	96.098	64.782	64.191	0.177	-2117.255	-2002.084	348.594
400	120.750	96.007	68.281	11.090	-2118.090	-1963.528	256.411
500	136.942	124.764	76.748	24.008	-2117.773	-1924.906	201.094
600	149.160	150.850	86.960	38.334	-2116.676	-1886.425	164.228
700	158.866	174.605	97.810	53.757	-2114.993	-1848.175	137.913
800	166.021	196.312	108.786	70.021	-2112.905	-1810.198	118.194
900	170.749	216.154	119.630	86.872	-2110.613	-1772.496	102.873
1000	174.054	234.324	130.204	104.120	-2108.265	-1735.053	90.630
1100	176.356	251.025	140.439	121.644	-2105.957	-1697.844	80.624
1200	178.197	266.451	150.305	139.375	-2103.739	-1660.841	72.295
1300	179.636	280.773	159.797	157.268	-2101.643	-1624.019	65.254
1400	180.916	294.134	168.921	175.298	-2099.687	-1587.353	59.225
1500	181.887	306.649	177.690	193.439	-2097.894	-1550.822	54.004
1600	182.849	318.418	186.121	211.676	-2125.349	-1513.353	49.406
1700	183.803	329.532	194.233	230.008	-2173.281	-1474.725	45.313
1800	184.749	340.065	202.045	248.436	-2170.844	-1433.704	41.605
1900	185.686	350.079	209.575	266.958	-2168.403	-1392.819	38.291
2000	186.606	359.627	216.841	285.572	-2165.958	-1352.061	35.312
2100	187.493	368.753	223.859	304.277	-2163.512	-1311.427	32.620
2200	188.380	377.496	230.645	323.071	-2161.066	-1270.907	30.175
2300	189.267	385.889	237.213	341.954	-2158.618	-1230.501	27.946
2400	190.154	393.963	243.578	360.925	-2156.170	-1190.202	25.904
2500	191.041	401.743	249.750	379.984	-2153.720	-1150.003	24.028
2600	191.928	409.253	255.741	399.133	-2151.267	-1109.903	22.298
2700	192.815	416.513	261.562	418.370	-2148.810	-1069.896	20.698
2800	193.702	423.542	267.222	437.696	-2728.262	-1017.536	18.982
2900	194.589	430.354	272.730	457.111	-2723.668	-956.521	17.229
3000	195.476	436.966	278.095	476.614	-2719.034	-895.663	15.595

PREVIOUS: September 1961   CURRENT: June 1965

Beryllium Silicate (Be_2SiO_4) $Be_2O_4Si_1$(cr)

BERYLLIUM NITRIDE ALPHA (α-Be_3N_2) CRYSTAL M_r = 55.04994

$S°(298.15\ K) = 34.129\ J\ K^{-1}\ mol^{-1}$

$T_{fus} = 2473\ K$

$\Delta_f H°(0\ K) = -580.9 \pm 1.3\ kJ\ mol^{-1}$

$\Delta_f H°(298.15\ K) = -588.3 \pm 1.3\ kJ\ mol^{-1}$

$\Delta_{fus} H° = [129.3]\ kJ\ mol^{-1}$

Enthalpy of Formation

Gross, Hayman, Greene and Bingham (1) have measured the enthalpy of Chlorination of α-Be_3N_2 to α-$BeCl_2$, and the enthalpy of reaction of Be with ammonia. The chlorination reaction (A) α-Be_3N_2(cr) + $3Cl_2$(g) → 3 α-$BeCl_2$(cr) + N_2(g) had a $\Delta_r H°$(298.15 K) = -210.3 ± 0.4 kcal mol^{-1} which, combined with the value of -117.1 + 0.4 kcal mol^{-1} for α-$BeCl_2$ reported in the same paper, yields $\Delta_f H°$(α-Be_3N_2, cr, 298.15 K) = -141.0 ± 1.6 kcal mol^{-1}. Reaction (B) 3Be(cr) + 2 NH_3(g) → α-Be_3N_2(cr) + 3 H_2(g) had a $\Delta_r H°$(298.15 K) = -214.6 ± 0.3 kcal mol^{-1}, which, combined with the JANAF (2) value for NH_3(g), yields $\Delta_f H°$(α-Be_3N_2, cr, 298.15 K) = -140.3 ± 0.3 kcal mol^{-1}. A weighted mean value of the above two determinations is adopted.

The decomposition Be_3N_2(cr) = 3 Be(g) + N_2(g) has been studied by Yates, Greenbaum and Farber (3) using the torsion effusion technique. The same decomposition has also been investigated by Hoenig (4) using the Knudsen technique. An analysis of these data sets by 2nd and 3rd law methods is given below.

Source	T/K	Points	$\Delta_r H°$(298.15 K) 2nd law	kcal mol^{-1} 3rd law	Drift cal $K^{-1} mol^{-1}$	$\Delta_f H°$(298.15 K) kcal mol^{-1}
(3)-1	1478-1592	9*	364.5±5.9	374.00	6.0±3.8	-139.23
(3)-2	1541-1652	15*	369.6±8.3	376.51	4.0±5.3	-141.74
(3)-3	1438-1661	14	347.7±3.7	374.54	16.9±2.4	-139.77
(3)-4	1450-1646	17*	366.7±8.9	375.28	4.8±5.7	-140.51
(3)	1473-1661	52*	358.9±5.1	375.41	10.3±3.3	-140.64
(4)-1	1767-1946	14	355.8±3.3	377.77	11.9±1.8	-143.00
(4)-2	1660-1891	10	395.0±5.6	378.69	-9.1±3.1	-143.92
(4)-3	1648-1893	10	416.9±8.4	383.15	-19.5±4.6	-148.38

*Points rejected due to failure of statistical test.

The data of Yates et al. (3) are seen to be in good agreement with the adopted $\Delta_f H°$(298.15 K).

Heat Capacity and Entropy

The low temperature heat capacity of α-Be_3N_2 has been measured, 25 - 310 K, by Justice (5). These measurements are joined smoothly with the high temperature enthalpies reported by Douglas and Payne (6). Above 1200 K the values were smoothly extrapolated to a limiting value of 6.2 cal K^{-1} g-$atom^{-1}$. The entropy at 298.15 K was calculated by integration of the low temperature data based on S°(25 K) = 0.002 cal K^{-1} mol^{-1}.

Fusion Data

See the liquid table for details.

References

1. P. Gross, C. Hayman, P. D. Greene and J. T. Bingham, Trans. Faraday Soc. 62, 2719 (1966).
2. JANAF Thermochemical Tables: NH_3(g), 9-30-65.
3. R. E. Yates, M. A. Greenbaum and M. Farber, J. Phys. Chem. 68, 2682 (1964).
4. C. L. Hoenig, UCRL-7521, April 1964.
5. B. H. Justice, The Dow Chemical Co., 1st. Quarterly Report under contract AFO4611-67-C-0009, April 1967.
6. T. B. Douglas and W. H. Payne, U. S. Nat. Bur. Stand. Report 7587, July 1962.

Beryllium Nitride, Alpha (Be_3N_2) Be_3N_2(cr)

Enthalpy Reference Temperature = T_r = 298.15 K Standard State Pressure = p° = 0.1 MPa

T/K	$C_p°$ (J $K^{-1} mol^{-1}$)	S° (J $K^{-1} mol^{-1}$)	-[G°-H°(T_r)]/T (J $K^{-1} mol^{-1}$)	H°-H°(T_r) (kJ mol^{-1})	$\Delta_f H°$ (kJ mol^{-1})	$\Delta_f G°$ (kJ mol^{-1})	Log K_f
0	0.	0.	INFINITE	-7.075	-580.879	-580.879	INFINITE
100	6.109	1.845	71.174	-6.933	-583.756	-567.809	296.593
200	35.489	14.313	39.125	-4.962	-586.429	-550.790	143.852
298.15	64.358	34.129	34.129	0.	-588.270	-532.874	93.357
300	64.668	34.528	34.130	0.119	-588.296	-532.530	92.722
400	84.433	56.036	36.928	7.643	-589.216	-513.778	67.092
500	97.512	76.374	42.811	16.782	-589.325	-494.891	51.701
600	106.516	94.995	49.981	27.009	-588.882	-476.039	41.443
700	112.922	111.922	57.640	37.998	-588.110	-457.290	34.123
800	117.591	127.321	65.402	49.535	-587.172	-438.664	28.642
900	121.043	141.381	73.075	61.475	-586.186	-420.159	24.385
1000	123.612	154.273	80.559	73.714	-585.237	-401.763	20.986
1100	125.520	166.148	87.807	86.174	-584.387	-383.458	18.209
1200	126.922	177.132	94.799	98.800	-583.685	-365.224	15.898
1300	127.926	187.333	101.530	111.545	-583.169	-347.041	13.944
1400	128.616	196.840	108.002	124.374	-582.871	-328.890	12.271
1500	129.047	205.730	114.224	137.259	-582.816	-310.752	10.821
1600	129.269	214.066	120.206	150.176	-626.634	-291.033	9.501
1700	129.319	221.905	125.960	163.106	-625.924	-270.080	8.299
1800	129.366	229.298	131.498	176.040	-625.298	-249.166	7.231
1900	129.411	236.293	136.831	188.979	-624.753	-228.285	6.276
2000	129.452	242.932	141.971	201.922	-624.286	-207.431	5.418
2100	129.490	249.249	146.931	214.869	-623.896	-186.598	4.641
2200	129.526	255.274	151.720	227.820	-623.582	-165.782	3.936
2300	129.559	261.032	156.348	240.774	-623.343	-144.978	3.293
2400	129.588	266.547	160.826	253.731	-623.176	-124.184	2.703
2473.000	129.606	270.430	164.004	263.192	------ CRYSTAL <--> LIQUID -----		
2500	129.615	271.838	165.161	266.692	-623.083	-103.395	2.160
2600	129.639	276.922	169.362	279.654	-623.062	-82.608	1.660
2700	129.659	281.815	173.437	292.619	-623.112	-61.820	1.196
2800	129.677	286.530	177.393	305.586	-1496.105	-22.362	0.417
2900	129.692	291.081	181.235	318.555	-1493.106	30.219	-0.544
3000	129.704	295.478	184.970	331.524	-1490.129	82.696	-1.440

PREVIOUS: September 1964 CURRENT: March 1967

Beryllium Nitride, Alpha (Be_3N_2) Be_3N_2(cr)

BERYLLIUM NITRIDE (Be_3N_2) LIQUID M_r = 55.04994

$S°(298.15\ K)$ = [39.457] $J\ K^{-1}\ mol^{-1}$

T_{fus} = 2473 K

$\Delta_f H°(298.15\ K)$ = [-486.979] $kJ\ mol^{-1}$

$\Delta_{fus}H°$ = [129.3] $kJ\ mol^{-1}$

Enthalpy of Formation

The enthalpy of formation of $Be_3N_2(\ell)$ is calculated from that of the crystal by adding $\Delta_{fus}H°$ and the difference in enthalpy, H°(2473 K) - H°(298.15 K), between the crystal and liquid.

Heat Capacity and Entropy

The heat capacity is estimated to be constant and slightly greater than the maximum for Be_3N_2(cr). The entropy at 298.15 K is calculated in a manner analogous to that used for the enthalpy of formation.

Fusion Data

The melting point was reported by Fichter and Brunner (1). The enthalpy of melting is calculated on the assumption that each g-atom contributed 2.5 cal $K^{-1}\ mol^{-1}$ to $\Delta_{fus}S°$.

References

1. F. Fichter and E. Brunner, Z. Anorg. Allgem. Chem. 93, 84 (1915).

Beryllium Nitride (Be_3N_2) Be_3N_2(l)

Enthalpy Reference Temperature = T_r = 298.15 K   Standard State Pressure = p° = 0.1 MPa

T/K	$C_p°$	S°	$-[G°-H°(T_r)]/T$	$H°-H°(T_r)$	$\Delta_f H°$	$\Delta_f G°$	Log K_f
	J K⁻¹mol⁻¹			kJ mol⁻¹			
0							
100							
200							
298.15	133.888	39.457	39.457	0.	-486.979	-433.172	75.890
300	133.888	40.285	39.459	0.248	-486.877	-432.838	75.364
400	133.888	78.802	44.711	13.636	-481.931	-415.600	54.272
500	133.888	108.678	54.628	27.025	-477.790	-399.509	41.736
600	133.888	133.089	65.732	40.414	-474.185	-384.199	33.447
700	133.888	153.728	76.867	53.803	-471.014	-369.457	27.569
800	133.888	171.606	87.617	67.192	-468.225	-355.144	23.189
900	133.888	187.376	97.842	80.580	-465.790	-341.159	19.800
1000	133.888	201.482	107.513	93.969	-463.691	-327.426	17.103
1100	133.888	214.243	116.645	107.358	-461.913	-313.889	14.905
1200	133.888	225.893	125.271	120.747	-460.447	-300.499	13.080
1300	133.888	236.610	133.429	134.136	-459.287	-287.219	11.541
1400	133.888	246.532	141.157	147.524	-458.429	-274.017	10.224
1500	133.888	255.769	148.494	160.913	-457.870	-260.866	9.084
1600	133.888	264.410	155.471	174.302	-501.217	-246.166	8.036
1700	133.888	272.527	162.121	187.691	-500.048	-230.261	7.075
1800	133.888	280.180	168.469	201.080	-498.967	-214.423	6.222
1900	133.888	287.419	174.541	214.468	-497.972	-198.643	5.461
2000	133.888	294.287	180.358	227.857	-497.060	-182.913	4.777
2100	133.888	300.819	185.940	241.246	-496.228	-167.226	4.160
2200	133.888	307.047	191.304	254.635	-495.476	-151.577	3.599
2300	133.888	312.999	196.467	268.024	-494.802	-135.961	3.088
2400	133.888	318.697	201.442	281.412	-494.204	-120.372	2.620
2473.000	133.888	322.709	204.963	291.186	------ CRYSTAL <--> LIQUID -----		
2500	133.888	324.163	206.242	294.801	-493.682	-104.807	2.190
2600	133.888	329.414	210.879	308.190	-493.235	-89.261	1.793
2700	133.888	334.467	215.364	321.579	-492.861	-73.731	1.426
2800	133.888	339.336	219.705	334.968	-1365.432	-39.545	0.738
2900	133.888	344.034	223.911	348.356	-1362.013	7.747	-0.140
3000	133.888	348.573	227.992	361.745	-1358.617	54.923	-0.956
3100	133.888	352.964	231.953	375.134	-1355.246	101.985	-1.718
3200	133.888	357.214	235.801	388.523	-1351.905	148.939	-2.431
3300	133.888	361.334	239.543	401.912	-1348.597	195.789	-3.099
3400	133.888	365.331	243.184	415.300	-1345.327	242.539	-3.726
3500	133.888	369.212	246.730	428.689	-1342.101	289.194	-4.316
3600	133.888	372.984	250.185	442.078	-1338.923	335.758	-4.872
3700	133.888	376.652	253.553	455.467	-1335.801	382.234	-5.396
3800	133.888	380.223	256.840	468.856	-1332.738	428.626	-5.892
3900	133.888	383.701	260.048	482.244	-1329.743	474.938	-6.361
4000	133.888	387.091	263.182	495.633	-1326.820	521.175	-6.806

PREVIOUS: September 1964   CURRENT: March 1967

Beryllium Nitride (Be_3N_2) Be_3N_2(l)

BERYLLIUM NITRIDE (Be_3N_2) ALPHA-LIQUID M_r = 55.04994

0 to 2473 K crystal, alpha
above 2473 K liquid

Refer to the individual tables for details.

Beryllium Nitride (Be_3N_2) Be_3N_2(cr,l)

Enthalpy Reference Temperature = T_r = 298.15 K Standard State Pressure = $p°$ = 0.1 MPa

T/K	$C_p°$ (J K⁻¹mol⁻¹)	$S°$ (J K⁻¹mol⁻¹)	$-[G°-H°(T_r)]/T$ (J K⁻¹mol⁻¹)	$H°-H°(T_r)$ (kJ mol⁻¹)	$\Delta_f H°$ (kJ mol⁻¹)	$\Delta_f G°$ (kJ mol⁻¹)	Log K_f
0	0.	0.	INFINITE	-7.075	-580.879	-580.879	INFINITE
100	6.109	1.845	71.174	-6.933	-583.756	-567.809	296.593
200	35.489	14.313	39.125	-4.962	-586.429	-550.790	143.852
298.15	64.358	34.129	34.129	0.	-588.270	-532.874	93.357
300	64.668	34.528	34.130	0.119	-588.296	-532.530	92.722
400	84.433	56.036	36.928	7.643	-589.216	-513.778	67.092
500	97.512	76.374	42.811	16.782	-589.325	-494.891	51.701
600	106.516	94.995	49.981	27.009	-588.882	-476.039	41.443
700	112.922	111.922	57.640	37.998	-588.110	-457.290	34.123
800	117.591	127.321	65.402	49.535	-587.172	-438.664	28.642
900	121.043	141.381	73.075	61.475	-586.186	-420.159	24.385
1000	123.612	154.273	80.559	73.714	-585.237	-401.763	20.986
1100	125.520	166.148	87.807	86.174	-584.387	-383.458	18.209
1200	126.922	177.132	94.799	98.800	-583.685	-365.224	15.898
1300	127.926	187.333	101.530	111.545	-583.169	-347.041	13.944
1400	128.616	196.840	108.002	124.374	-582.871	-328.890	12.271
1500	129.047	205.730	114.224	137.259	-582.816	-310.752	10.821
1600	129.269	214.066	120.206	150.176	-626.634	-291.033	9.501
1700	129.319	221.905	125.960	163.106	-625.924	-270.080	8.299
1800	129.366	229.298	131.498	176.040	-625.298	-249.166	7.231
1900	129.411	236.293	136.831	188.979	-624.753	-228.285	6.276
2000	129.452	242.932	141.971	201.922	-624.286	-207.431	5.418
2100	129.490	249.249	146.931	214.869	-623.896	-186.598	4.641
2200	129.525	255.274	151.720	227.820	-623.582	-165.782	3.936
2300	129.558	261.032	156.348	240.774	-623.343	-144.978	3.293
2400	129.587	266.547	160.826	253.731	-623.177	-124.184	2.703
2473.000	129.606	270.430	164.004	263.192	CRYSTAL <--> LIQUID		
2473.000	133.888	322.709	164.004	392.477	TRANSITION		
2500	133.888	324.163	165.726	396.092	-493.682	-104.807	2.190
2600	133.888	329.414	171.921	409.481	-493.235	-89.261	1.793
2700	133.888	334.467	177.848	422.870	-492.861	-73.731	1.426
2800	133.888	339.336	183.529	436.259	-1365.432	-39.545	0.738
2900	133.888	344.034	188.984	449.647	-1362.013	7.747	-0.140
3000	133.888	348.573	194.228	463.036	-1358.617	54.923	-0.956
3100	133.888	352.964	199.278	476.425	-1355.246	101.985	-1.718
3200	133.888	357.214	204.148	489.814	-1351.905	148.939	-2.431
3300	133.888	361.334	208.849	503.203	-1348.597	195.789	-3.099
3400	133.888	365.331	213.393	516.591	-1345.327	242.539	-3.726
3500	133.888	369.212	217.789	529.980	-1342.101	289.194	-4.316
3600	133.888	372.984	222.048	543.369	-1338.923	335.758	-4.872
3700	133.888	376.653	226.177	556.758	-1335.801	382.233	-5.396
3800	133.888	380.223	230.184	570.147	-1332.738	428.626	-5.892
3900	133.888	383.701	234.076	583.535	-1329.743	474.938	-6.361
4000	133.888	387.091	237.860	596.924	-1326.820	521.175	-6.806

PREVIOUS: CURRENT: March 1967

Beryllium Nitride (Be_3N_2) Be_3N_2(cr,l)

BERYLLIUM OXIDE ((BeO)$_3$) IDEAL GAS M_r = 75.03474

S°(298.15 K) = [273.3] J K^{-1} mol^{-1}

$\Delta_fH°$(0 K) = -1049 ± 38 kJ mol^{-1}
$\Delta_fH°$(298.15 K) = -1054 ± 38 kJ mol^{-1}

Vibrational Frequencies and Degeneracies

ν, cm^{-1}	ν, cm^{-1}
[730] (1)	[1380] (2)
[990] (1)	[1450] (2)
[1480] (1)	[610] (2)
[1100] (1)	[320] (2)

Ground State Quantum Weight: [1] σ = [6]

Point Group: [D_{3h}]

Bond Distance: Be-O = [1.63] Å

Bond Angles: Be-O-Be = [120]° O-Be-O [120]°

Product of the Moments of Inertia: $I_AI_BI_C$ = [9.0703 x 10^{-114}] g^3 cm^6

Enthalpy of Formation

Chupka, Berkowitz, and Giese (1) measured the change in the intensity of the $(BeO)_3^+$ ion with temperature and obtained a 2nd law enthalpy of vaporization of 161 ± 6 kcal mol^{-1} at 2150 K. The uncertainty is large because of an approximate correction of 8 kcal mol^{-1} and the neglect of fragmentation effects. We adopted an enthalpy of sublimation at 2150 K of 163 kcal mol^{-1} in order to bring the equilibrium vapor pressure into agreement with that measured by Chupka, et al. (1).

Heat Capacity and Entropy

The assumed D_{3h} symmetry implies eight vibrational frequencies of the types $2A_1'$, A_2', A_2'', 3E' and E_2''. The frequencies were estimated by comparison with the values for the $B_3O_3(O^-)_3$ skeleton as given by Parsons (2) and the partial assignment of Boroxine by Gupta and Porter (3). The bond length was taken as that in Be_2O_2(g), while the bond angle was arbitrarily taken as 120°. The principal moments of inertia are: $I_A = I_B$ = 16.5525 x 10^{-39} and I_C = 33.1050 x 10^{-39} g cm^2.

References

1. W. A. Chupka, J. Berkowitz, and C. F. Giese, J. Chem. Phys. 30, 827 (1959).
2. J. L. Parsons, J. Chem. Phys. 33, 1860 (1960).
3. S. K. Gupta and R. F. Porter, J. Phys. Chem. 67, 1286 (1963).

Beryllium Oxide ((BeO)$_3$) Be_3O_3(g)

Enthalpy Reference Temperature = T_r = 298.15 K Standard State Pressure = p° = 0.1 MPa

T/K	$C_p°$ (J K^{-1}mol^{-1})	S° (J K^{-1}mol^{-1})	-[G°-H°(T_r)]/T (J K^{-1}mol^{-1})	H°-H°(T_r) (kJ mol^{-1})	$\Delta_fH°$ (kJ mol^{-1})	$\Delta_fG°$ (kJ mol^{-1})	Log K_f
0	0.	0.	INFINITE	-13.325	-1048.871	-1048.871	INFINITE
100	37.082	221.244	320.437	-9.919	-1049.940	-1045.917	546.331
200	50.124	250.834	278.710	-5.575	-1051.695	-1041.313	271.963
250	56.935	262.750	274.345	-2.899	-1052.982	-1038.572	216.998
298.15	63.458	273.337	273.337	0.	-1054.368	-1035.673	181.446
300	63.706	273.731	273.339	0.118	-1054.423	-1035.557	180.306
350	70.315	284.051	274.139	3.469	-1055.921	-1032.294	154.061
400	76.597	293.854	275.995	7.144	-1057.380	-1028.818	134.350
450	82.418	303.217	278.504	11.121	-1058.758	-1025.165	118.998
500	87.701	312.179	281.426	15.376	-1060.043	-1021.362	106.701
600	96.611	328.989	287.973	24.610	-1062.350	-1013.404	88.225
700	103.530	344.425	294.950	34.632	-1064.384	-1005.082	75.000
800	108.857	358.612	302.034	45.263	-1066.249	-996.482	65.064
900	112.975	371.682	309.056	56.363	-1068.035	-987.653	57.322
1000	116.187	383.758	315.930	67.828	-1069.812	-978.627	51.118
1100	118.723	394.955	322.612	79.578	-1071.639	-969.420	46.034
1200	120.750	405.376	329.079	91.556	-1073.559	-960.044	41.790
1300	122.389	415.108	335.327	103.715	-1075.609	-950.502	38.192
1400	123.730	424.229	341.355	116.024	-1077.818	-940.797	35.102
1500	124.840	432.804	347.168	128.454	-1080.212	-930.927	32.418
1600	125.766	440.892	352.776	140.985	-1126.417	-919.315	30.013
1700	126.547	448.540	358.186	153.602	-1128.033	-906.321	27.848
1800	127.210	455.793	363.409	166.291	-1129.677	-893.233	25.921
1900	127.778	462.686	368.454	179.041	-1131.359	-880.051	24.194
2000	128.268	469.253	373.331	191.844	-1133.088	-866.780	22.638
2100	128.694	475.522	378.049	204.692	-1134.870	-853.421	21.228
2200	129.065	481.517	382.617	217.581	-1136.711	-839.974	19.944
2300	129.391	487.262	387.043	230.504	-1138.615	-826.445	18.769
2400	129.679	492.775	391.334	243.458	-1140.587	-812.830	17.691
2500	129.935	498.074	395.499	256.439	-1142.629	-799.132	16.697
2600	130.162	503.175	399.542	269.444	-1144.743	-785.351	15.778
2700	130.366	508.091	403.472	282.470	-1146.932	-771.486	14.925
2800	130.549	512.835	407.294	295.516	-2022.067	-738.872	13.784
2900	130.713	517.419	411.013	308.579	-2021.218	-693.059	12.483
3000	130.862	521.853	414.634	321.658	-2020.398	-647.274	11.270
3100	130.997	526.147	418.162	334.751	-2019.610	-601.516	10.135
3200	131.120	530.307	421.602	347.857	-2018.860	-555.784	9.072
3300	131.232	534.344	424.958	360.975	-2018.151	-510.075	8.074
3400	131.335	538.263	428.233	374.104	-2017.489	-464.384	7.134
3500	131.429	542.072	431.431	387.242	-2016.879	-418.715	6.249
3600	131.516	545.775	434.556	400.389	-2016.327	-373.060	5.413
3700	131.595	549.380	437.611	413.545	-2015.839	-327.422	4.622
3800	131.669	552.890	440.599	426.708	-2015.420	-281.793	3.874
3900	131.737	556.311	443.522	439.878	-2015.078	-236.176	3.163
4000	131.800	559.647	446.384	453.055	-2014.817	-190.567	2.489
4100	131.859	562.903	449.186	466.238	-2014.646	-144.963	1.847
4200	131.914	566.081	451.932	479.427	-2014.569	-99.362	1.236
4300	131.964	569.185	454.622	492.621	-2014.594	-53.761	0.653
4400	132.012	572.220	457.261	505.820	-2014.728	-8.160	0.097
4500	132.056	575.187	459.848	519.023	-2014.975	37.446	-0.435
4600	132.098	578.090	462.388	532.231	-2015.342	83.058	-0.943
4700	132.137	580.931	464.880	545.443	-2015.836	128.682	-1.430
4800	132.173	583.714	467.326	558.658	-2016.461	174.316	-1.897
4900	132.208	586.439	469.730	571.877	-2017.224	219.966	-2.345
5000	132.240	589.111	472.091	585.100	-2018.130	265.632	-2.775
5100	132.271	591.730	474.411	598.325	-2019.182	311.318	-3.189
5200	132.299	594.298	476.692	611.554	-2020.387	357.023	-3.586
5300	132.326	596.819	478.935	624.785	-2021.747	402.757	-3.969
5400	132.352	599.292	481.141	638.019	-2023.266	448.516	-4.339
5500	132.376	601.721	483.311	651.255	-2024.949	494.304	-4.695
5600	132.399	604.106	485.447	664.494	-2026.798	540.126	-5.038
5700	132.421	606.450	487.549	677.735	-2028.815	585.981	-5.370
5800	132.442	608.753	489.619	690.978	-2031.003	631.875	-5.691
5900	132.462	611.017	491.658	704.224	-2033.364	677.804	-6.001
6000	132.480	613.244	493.665	717.471	-2035.900	723.779	-6.301

PREVIOUS: September 1963 (1 atm) CURRENT: September 1963 (1 bar)

Beryllium Oxide ((BeO)$_3$) Be_3O_3(g)

BERYLLIUM OXIDE $((BeO)_4)$ IDEAL GAS M_r = 100.04632

$S°(298.15\ K)$ = [302.6] $J\ K^{-1}\ mol^{-1}$

$\Delta_f H°(0\ K) = -1583 \pm 50\ kJ\ mol^{-1}$

$\Delta_f H°(298.15\ K) = -1590 \pm 50\ kJ\ mol^{-1}$

Vibrational Frequencies and Degeneracies

ν, cm^{-1}	ν, cm^{-1}
[700] (1)	[200] (2)
[850] (1)	[900] (1)
[1400] (1)	[400] (1)
[1000] (1)	[500] (1)
[800] (1)	[1150] (2)
[1000] (1)	[1200] (2)
[800] (1)	[600] (2)

Ground State Quantum Weight: [1] σ = [8]

Point Group: $[D_{4h}]$

Bond Distance: Be–O = [1.63] Å

Bond Angles: Be–O–Be = [115]° O–Be–O = [155]°

Product of the Moments of Inertia: $I_A I_B I_C = [1.1171 \times 10^{-112}]\ g^3\ cm^6$

Enthalpy of Formation

Chupka, Berkowitz, and Giese (1) measured the variation of intensity of the $(BeO)_4^+$ ion with temperature in a mass spectrometer. They reported a 2nd law enthalpy of vaporization of 177 ± 8 kcal mol^{-1} at 2150 K, which included an approximate correction of 9 kcal mol^{-1}, and this value was adopted here.

Heat Capacity and Entropy

The assumed D_{4h} structure has 14 vibrational frequencies of the type $2A_{1g}$, A_{2g}, $2B_{2g}$, E_g, A_{2u}, B_{2u}, B_{1u}, and $3E_u$. The frequencies were estimated by comparison with $(BeO)_3$ and cyclobutane from Rathjens, Freeman, Gwinn, and Pitzer (2). The estimated structure and frequencies gave an entropy which was in excellent agreement with the experimental enthalpy and equilibrium constant determined by Chupka, et al. (1).

The bond lengths were taken equal to those in $Be_2O_2(g)$ and the angles were chosen so that O–Be–O was almost linear. The principal moments of inertia are: $I_A = I_B = 38.2258 \times 10^{-39}$ and $I_C = 76.4515 \times 10^{-39}\ g\ cm^2$.

References

1. W. A. Chupka, J. Berkowitz, and C. F. Giese, J. Chem. Phys. 30, 827 (1959).
2. G. W. Rathjens, N. K. Freeman, W. D. Gwinn, and K. S. Pitzer, J. Am. Chem. Soc. 75, 5634 (1953).

Beryllium Oxide $((BeO)_4)$ $Be_4O_4(g)$

Enthalpy Reference Temperature = T_r = 298.15 K Standard State Pressure = p° = 0.1 MPa

T/K	$C_p°$ (J K⁻¹mol⁻¹)	S° (J K⁻¹mol⁻¹)	$-[G°-H°(T_r)]/T$ (J K⁻¹mol⁻¹)	$H°-H°(T_r)$ (kJ mol⁻¹)	$\Delta_f H°$ (kJ mol⁻¹)	$\Delta_f G°$ (kJ mol⁻¹)	Log K_f
0	0.	0.	INFINITE	-16.529	-1583.286	-1583.286	INFINITE
100	43.449	235.990	364.943	-12.895	-1585.579	-1574.365	822.364
200	64.574	272.167	309.947	-7.556	-1587.794	-1562.330	408.039
250	77.225	287.926	303.979	-4.013	-1588.943	-1555.829	325.073
298.15	89.389	302.575	302.575	0.	-1589.920	-1549.360	271.441
300	89.844	303.129	302.577	0.166	-1589.954	-1549.108	269.724
350	101.591	317.876	303.714	4.956	-1590.751	-1542.234	230.166
400	112.062	332.140	306.381	10.303	-1591.285	-1535.263	200.485
450	121.155	345.877	310.012	16.140	-1591.563	-1528.241	177.394
500	128.942	359.056	314.262	22.397	-1591.617	-1521.201	158.919
600	141.187	383.708	323.813	35.937	-1591.197	-1507.147	131.209
700	150.034	406.173	333.999	50.522	-1590.296	-1493.205	111.424
800	156.507	426.652	344.320	65.865	-1589.117	-1479.413	96.596
900	161.327	445.377	354.524	81.768	-1587.802	-1465.778	85.072
1000	164.987	462.573	364.481	98.092	-1586.452	-1452.292	75.860
1100	167.817	478.436	374.129	114.738	-1585.137	-1438.940	68.330
1200	170.043	493.137	383.441	131.635	-1583.912	-1425.704	62.059
1300	171.821	506.821	392.412	148.732	-1582.817	-1412.566	56.758
1400	173.262	519.608	401.045	165.988	-1581.884	-1399.505	52.216
1500	174.443	531.604	409.354	183.375	-1581.141	-1386.506	48.282
1600	175.423	542.894	417.350	200.870	-1624.217	-1371.972	44.790
1700	176.245	553.555	425.052	218.455	-1622.711	-1356.253	41.673
1800	176.940	563.649	432.474	236.115	-1621.241	-1340.622	38.904
1900	177.533	573.232	439.632	253.839	-1619.819	-1325.071	36.429
2000	178.043	582.351	446.542	271.619	-1618.453	-1309.593	34.203
2100	178.484	591.049	453.218	289.446	-1617.149	-1294.182	32.191
2200	178.869	599.361	459.673	307.314	-1615.915	-1278.831	30.363
2300	179.205	607.320	465.921	325.218	-1614.753	-1263.536	28.696
2400	179.502	614.953	471.972	343.154	-1613.669	-1248.290	27.168
2500	179.765	622.286	477.839	361.117	-1612.666	-1233.087	25.764
2600	179.998	629.341	483.531	379.106	-1611.745	-1217.922	24.468
2700	180.207	636.138	489.058	397.116	-1610.909	-1202.790	23.269
2800	180.394	642.696	494.429	415.146	-2483.029	-1169.021	21.808
2900	180.563	649.029	499.652	433.194	-2479.174	-1122.162	20.212
3000	180.715	655.153	504.733	451.258	-2475.357	-1075.433	18.725
3100	180.853	661.081	509.682	469.337	-2471.581	-1028.830	17.336
3200	180.979	666.825	514.503	487.429	-2467.852	-982.350	16.035
3300	181.093	672.395	519.204	505.532	-2464.172	-935.986	14.815
3400	181.198	677.803	523.789	523.647	-2460.548	-889.731	13.669
3500	181.294	683.057	528.265	541.772	-2456.984	-843.583	12.590
3600	181.382	688.166	532.636	559.905	-2453.485	-797.535	11.572
3700	181.463	693.136	536.907	578.048	-2450.057	-751.585	10.610
3800	181.538	697.977	541.082	596.198	-2446.706	-705.724	9.701
3900	181.607	702.693	545.166	614.355	-2443.437	-659.951	8.839
4000	181.671	707.292	549.162	632.519	-2440.258	-614.262	8.021
4100	181.731	711.779	553.074	650.689	-2437.174	-568.650	7.245
4200	181.786	716.158	556.905	668.865	-2434.190	-523.113	6.506
4300	181.838	720.437	560.658	687.047	-2431.315	-477.644	5.802
4400	181.886	724.618	564.337	705.233	-2428.554	-432.243	5.131
4500	181.931	728.706	567.945	723.424	-2425.912	-386.904	4.491
4600	181.974	732.705	571.483	741.619	-2423.397	-341.621	3.879
4700	182.013	736.619	574.955	759.818	-2421.015	-296.390	3.294
4800	182.050	740.451	578.363	778.021	-2418.770	-251.209	2.734
4900	182.085	744.205	581.710	796.228	-2416.669	-206.072	2.197
5000	182.118	747.884	584.996	814.438	-2414.717	-160.978	1.682
5100	182.149	751.491	588.226	832.652	-2412.919	-115.921	1.187
5200	182.178	755.028	591.400	850.868	-2411.280	-70.900	0.712
5300	182.205	758.498	594.520	869.087	-2409.804	-25.905	0.255
5400	182.231	761.905	597.588	887.309	-2408.495	19.060	-0.184
5500	182.256	765.249	600.606	905.533	-2407.356	64.003	-0.608
5600	182.279	768.533	603.576	923.760	-2406.393	108.930	-1.016
5700	182.301	771.759	606.498	941.989	-2405.607	153.838	-1.410
5800	182.322	774.930	609.375	960.220	-2405.001	198.737	-1.790
5900	182.342	778.047	612.207	978.454	-2404.577	243.622	-2.157
6000	182.361	781.112	614.997	996.689	-2404.339	288.506	-2.512

PREVIOUS: September 1963 (1 atm) CURRENT: September 1963 (1 bar)

BERYLLIUM OXIDE $((BeO)_5)$ IDEAL GAS $M_r = 125.0579$

$S°(298.15\ K) = [323.28]\ J\ K^{-1}\ mol^{-1}$

$\Delta_f H°(0\ K) = -2101 \pm 75\ kJ\ mol^{-1}$
$\Delta_f H°(298.15\ K) = -2113 \pm 75\ kJ\ mol^{-1}$

Vibrational Frequencies and Degeneracies

ν, cm^{-1}	ν, cm^{-1}
[700] (1)	[1200] (2)
[850] (1)	[600] (2)
[1400] (1)	[500] (2)
[1100] (2)	[900] (1)
[900] (2)	[200] (2)
[1000] (2)	[400] (2)
[1200] (2)	[500] (2)

Ground State Quantum Weight: [1] $\sigma = [10]$
Point Group: $[D_{5h}]$
Bond Distance: Be-O = [1.63] Å
Bond Angles: Be-O-Be = [110]° O-Be-O = [178]°
Product of the Moments of Inertia: $I_A I_B I_C = [6.9690 \times 10^{-112}]\ g^3\ cm^6$

Enthalpy of Formation

Chupka, Berkowitz, and Giese (1) measured the variation in intensity of the $(BeO)_5^+$ ion with temperature in a mass spectrometer. They calculated a 2nd law enthalpy of vaporization of 193 ± 12 kcal mol^{-1} at 2150 K; the uncertainty reflects the inclusion of an approximate correction of 10 kcal mol^{-1}. This value was adopted here.

Heat Capacity and Entropy

The assumed D_{5h} structure has 14 vibrations of the type $2A_1'$, A_2', $3E_1'$, $4E_2'$, A_2'', E_1'', and $2E_2''$. These were estimated by comparison with similar motions in other ring structures including $(BeO)_3$ and $(BeO)_4$. The estimated structure and frequencies gave an entropy in good agreement with the experimental enthalpy and equilibrium constant determined by Chupka, et al. (1).

The bond length was assumed to be the same as in $Be_2O_2(g)$ and the angles were chosen so as to make O-Be-O almost linear. The principal moments of inertia are: $I_A = I_B = 70.3689 \times 10^{-39}$ and $I_C = 140.7378 \times 10^{-39}\ g\ cm^2$.

References

1. W. A. Chupka, J. Berkowitz, and C. F. Giese, J. Chem. Phys. 30, 827 (1959).

Beryllium Oxide $((BeO)_5)$ $Be_5O_5(g)$

Enthalpy Reference Temperature = T_r = 298.15 K Standard State Pressure = p° = 0.1 MPa

T/K	$C_p°$	S°	$-[G°-H°(T_r)]/T$	$H°-H°(T_r)$	$\Delta_f H°$	$\Delta_f G°$	Log K_f
	J K⁻¹mol⁻¹			kJ mol⁻¹			
0	0.	0.	INFINITE	-19.090	-2100.640	-2100.640	INFINITE
100	45.283	244.859	399.133	-15.427	-2104.435	-2085.343	1089.271
200	77.909	285.714	332.441	-9.345	-2108.519	-2064.746	539.256
250	95.670	305.013	325.031	-5.005	-2110.783	-2053.541	429.064
298.15	112.018	323.278	323.278	0.	-2112.920	-2042.321	357.806
300	112.622	323.973	323.280	0.208	-2113.000	-2041.883	355.523
350	128.074	342.516	324.709	6.232	-2115.057	-2029.864	302.941
400	141.704	360.528	328.067	12.985	-2116.862	-2017.567	263.467
450	153.478	377.916	332.646	20.372	-2118.401	-2005.060	232.741
500	163.529	394.621	338.014	28.304	-2119.703	-1992.395	208.144
600	179.296	425.909	350.095	45.488	-2121.752	-1966.731	171.219
700	190.665	454.448	362.996	64.017	-2123.317	-1940.765	144.822
800	198.973	480.478	376.079	83.520	-2124.641	-1914.595	125.010
900	205.156	504.288	389.021	103.741	-2125.896	-1888.263	109.592
1000	209.847	526.157	401.656	124.501	-2127.206	-1861.791	97.250
1100	213.474	546.335	413.903	145.675	-2128.660	-1835.180	87.145
1200	216.326	565.037	425.728	167.170	-2130.328	-1808.428	78.719
1300	218.603	582.445	437.121	188.921	-2132.260	-1781.527	71.583
1400	220.447	598.715	448.089	210.877	-2134.500	-1754.465	65.460
1500	221.960	613.978	458.645	232.999	-2137.083	-1727.232	60.148
1600	223.215	628.344	468.807	255.260	-2212.718	-1697.186	55.407
1700	224.267	641.909	478.594	277.636	-2214.062	-1664.925	51.157
1800	225.157	654.754	488.027	300.108	-2215.478	-1632.583	47.376
1900	225.916	666.948	497.126	322.663	-2216.978	-1600.159	43.991
2000	226.568	678.553	505.909	345.288	-2218.572	-1567.654	40.943
2100	227.132	689.621	514.396	367.973	-2220.271	-1535.067	38.183
2200	227.624	700.199	522.603	390.712	-2222.081	-1502.394	35.671
2300	228.055	710.327	530.546	413.496	-2224.009	-1469.640	33.377
2400	228.435	720.041	538.241	436.321	-2226.060	-1436.798	31.271
2500	228.771	729.373	545.701	459.182	-2228.237	-1403.868	29.332
2600	229.070	738.352	552.939	482.074	-2230.544	-1370.848	27.541
2700	229.337	747.002	559.967	504.995	-2232.982	-1337.736	25.880
2800	229.577	755.347	566.797	527.941	-3690.339	-1273.419	23.756
2900	229.792	763.407	573.438	550.909	-3687.727	-1187.149	21.383
3000	229.987	771.201	579.901	573.898	-3685.169	-1100.965	19.169
3100	230.164	778.745	586.194	596.906	-3682.670	-1014.865	17.100
3200	230.324	786.055	592.326	619.931	-3680.238	-928.847	15.162
3300	230.471	793.145	598.305	642.971	-3677.879	-842.904	13.342
3400	230.605	800.027	604.137	666.024	-3675.602	-757.028	11.630
3500	230.727	806.713	609.830	689.091	-3673.417	-671.221	10.017
3600	230.840	813.215	615.390	712.170	-3671.330	-585.473	8.495
3700	230.944	819.541	620.822	735.259	-3669.353	-499.784	7.056
3800	231.040	825.701	626.133	758.358	-3667.496	-414.142	5.693
3900	231.128	831.704	631.328	781.466	-3665.767	-328.550	4.400
4000	231.210	837.556	636.410	804.583	-3664.178	-243.002	3.173
4100	231.287	843.266	641.386	827.708	-3662.738	-157.491	2.006
4200	231.357	848.841	646.260	850.841	-3661.460	-72.013	0.896
4300	231.424	854.285	651.034	873.980	-3660.353	13.438	-0.163
4400	231.485	859.606	655.714	897.125	-3659.427	98.862	-1.174
4500	231.543	864.809	660.303	920.277	-3658.693	184.268	-2.139
4600	231.597	869.899	664.805	943.434	-3658.162	269.660	-3.062
4700	231.647	874.880	669.222	966.596	-3657.842	355.045	-3.946
4800	231.695	879.758	673.557	989.763	-3657.743	440.424	-4.793
4900	231.739	884.536	677.814	1012.935	-3657.875	525.805	-5.605
5000	231.781	889.218	681.996	1036.111	-3658.245	611.190	-6.385
5100	231.821	893.808	686.104	1059.291	-3658.862	696.585	-7.134
5200	231.858	898.310	690.142	1082.475	-3659.732	781.991	-7.855
5300	231.893	902.727	694.111	1105.662	-3660.864	867.422	-8.549
5400	231.926	907.062	698.015	1128.853	-3662.262	952.872	-9.217
5500	231.958	911.317	701.854	1152.048	-3663.933	1038.351	-9.861
5600	231.988	915.497	705.632	1175.245	-3665.882	1123.867	-10.483
5700	232.016	919.604	709.350	1198.445	-3668.112	1209.416	-11.083
5800	232.043	923.639	713.010	1221.648	-3670.628	1295.011	-11.663
5900	232.068	927.606	716.614	1244.854	-3673.433	1380.645	-12.223
6000	232.093	931.506	720.163	1268.062	-3676.530	1466.335	-12.766

PREVIOUS: September 1963 (1 atm) CURRENT: September 1963 (1 bar)

Beryllium Oxide $((BeO)_5)$ $Be_5O_5(g)$

BERYLLIUM OXIDE $((BeO)_6)$ IDEAL GAS M_r = 150.06948

$\Delta_f H°(0\ K) = -2645 \pm 92\ kJ\ mol^{-1}$

$S°(298.15\ K) = [343.2]\ J\ K^{-1} mol^{-1}$ $\Delta_f H°(298.15\ K) = -2661 \pm 92\ kJ\ mol^{-1}$

Vibrational Frequencies and Degeneracies

ν, cm^{-1}	ν, cm^{-1}	ν, cm^{-1}	ν, cm^{-1}
[700] (1)	[600] (2)	[900] (1)	[1100] (2)
[850] (1)	[750] (2)	[1100] (1)	[1200] (2)
1400] (1)	[850] (2)	[1000] (1)	[500] (2)
[500] (1)	[850] (2)	[800] (1)	[200] (2)
[400] (1)	[950] (2)	[1400] (1)	[250] (2)

Ground State Quantum Weight: [1] σ = [12]

Point Group: $[D_{6h}]$

Bond Distance: Be-O = [1.63] Å

Bond Angles: Be-O-Be = [110]° O-Be-O = [190]°

Product of the Moments of Inertia: $I_A I_B I_C = [3.1282 \times 10^{-111}]\ g^3\ cm^6$

Enthalpy of Formation

Chupka, Berkowitz, and Giese (1) measured the temperature dependence of the $(BeO)_6^+$ ion in a mass spectrometer. They calculated a 2nd law enthalpy of vaporizaion of 203 ± 15 kcal mol^{-1} which was adopted here. The uncertainty reflects in part an approximate correction of 10 kcal mol^{-1} and the small observed intensities.

Heat Capacity and Entropy

The assumed symmetry D_{6h} has 20 vibrational frequencies of the type $2A_{1g}$, A_{2g}, B_{1g}, B_{2g}, E_{1g}, $4E_{2g}$, A_{2u}, $2B_{1u}$, $2B_{2u}$, $3E_{1u}$, and $2E_{2u}$. The frequencies were estimated by comparison with similar motions in other ring molecules including benzene, $(BeO)_3$ and $(BeO)_4$. The assumed structure and vibrational frequencies gave an entropy in good agreement with the experimental enthalpy and equilibrium constant determined by Chupka, et al (1). The bond length was taken equal to that in $Be_2O_2(g)$. The angles were chosen so that the O-Be-O angle was almost linear but with the Be-O-Be angle less than 120°. The principal moments of inertia are: $I_A = I_B = 116.0793 \times 10^{-39}$ and $I_C = 232.1586 \times 10^{-39}\ g\ cm^2$.

References

1. W. A. Chupka, J. Berkowitz, and C. F. Giese, J. Chem. Phys. 30, 827 (1959).

Beryllium Oxide $((BeO)_6)$ $Be_6O_6(g)$

Enthalpy Reference Temperature = T_r = 298.15 K Standard State Pressure = p° = 0.1 MPa

T/K	$C_p°$ (J K⁻¹mol⁻¹)	S° (J K⁻¹mol⁻¹)	$-[G°-H°(T_r)]/T$ (J K⁻¹mol⁻¹)	$H°-H°(T_r)$ (kJ mol⁻¹)	$\Delta_f H°$ (kJ mol⁻¹)	$\Delta_f G°$ (kJ mol⁻¹)	Log K_f
0	0.	0.	INFINITE	-21.479	-2644.860	-2644.860	INFINITE
100	50.386	253.784	430.456	-17.667	-2649.997	-2623.081	1370.157
200	88.526	299.761	353.847	-10.817	-2655.345	-2594.199	677.535
250	110.694	321.887	345.242	-5.839	-2658.293	-2578.570	538.763
298.15	131.624	343.191	343.191	0.	-2661.024	-2562.965	449.021
300	132.400	344.008	343.194	0.244	-2661.125	-2562.357	446.146
350	152.248	365.937	344.880	7.370	-2663.698	-2545.687	379.923
400	169.665	387.432	348.861	15.428	-2665.907	-2528.673	330.211
450	184.604	408.302	354.313	24.295	-2667.752	-2511.405	291.516
500	197.270	428.426	360.724	33.851	-2669.277	-2493.948	260.541
600	216.981	466.236	375.205	54.619	-2671.589	-2458.651	214.045
700	231.084	500.802	390.716	77.061	-2673.260	-2423.023	180.808
800	241.334	532.364	406.479	100.708	-2674.605	-2387.181	155.867
900	248.936	561.249	422.094	125.240	-2675.844	-2351.178	136.459
1000	254.690	587.789	437.355	150.434	-2677.134	-2315.036	120.925
1100	259.130	612.280	452.158	176.134	-2678.588	-2278.757	108.209
1200	262.617	634.983	466.459	202.228	-2680.289	-2242.336	97.606
1300	265.398	656.117	480.245	228.634	-2682.303	-2205.761	88.629
1400	267.649	675.871	493.520	255.290	-2684.682	-2169.017	80.927
1500	269.494	694.401	506.301	282.150	-2687.469	-2132.090	74.246
1600	271.024	711.844	518.608	309.179	-2777.915	-2091.808	68.291
1700	272.306	728.315	530.464	336.347	-2779.211	-2048.886	62.955
1800	273.389	743.911	541.892	363.633	-2780.590	-2005.888	58.209
1900	274.313	758.718	552.918	391.019	-2782.069	-1962.809	53.961
2000	275.107	772.809	563.563	418.491	-2783.660	-1919.649	50.136
2100	275.795	786.248	573.850	446.037	-2785.376	-1876.407	46.673
2200	276.393	799.092	583.798	473.647	-2787.224	-1833.078	43.523
2300	276.918	811.390	593.428	501.314	-2789.213	-1789.665	40.645
2400	277.379	823.186	602.757	529.029	-2791.348	-1746.162	38.004
2500	277.788	834.518	611.802	556.788	-2793.635	-1702.566	35.573
2600	278.152	845.420	620.579	584.585	-2796.076	-1658.875	33.327
2700	278.477	855.924	629.103	612.417	-2798.675	-1615.087	31.246
2800	278.768	866.056	637.385	640.279	-4547.176	-1533.865	28.615
2900	279.030	875.843	645.440	668.169	-4543.714	-1426.310	25.691
3000	279.267	885.307	653.279	696.084	-4540.316	-1318.870	22.964
3100	279.482	894.468	660.912	724.022	-4536.989	-1211.542	20.414
3200	279.677	903.344	668.350	751.980	-4533.742	-1104.324	18.026
3300	279.855	911.953	675.602	779.957	-4530.582	-997.205	15.784
3400	280.018	920.310	682.677	807.951	-4527.521	-890.177	13.676
3500	280.167	928.429	689.583	835.960	-4524.569	-783.241	11.689
3600	280.304	936.324	696.328	863.984	-4521.736	-676.385	9.814
3700	280.430	944.005	702.919	892.021	-4519.034	-569.609	8.041
3800	280.547	951.486	709.362	920.069	-4516.475	-462.900	6.363
3900	280.654	958.774	715.664	948.130	-4514.070	-356.257	4.772
4000	280.754	965.881	721.831	976.200	-4511.833	-249.677	3.260
4100	280.847	972.815	727.868	1004.280	-4509.776	-143.149	1.824
4200	280.933	979.584	733.781	1032.369	-4507.911	-36.669	0.456
4300	281.013	986.195	739.575	1060.467	-4506.252	69.770	-0.848
4400	281.088	992.656	745.254	1088.572	-4504.811	176.171	-2.091
4500	281.158	998.974	750.822	1116.684	-4503.599	282.541	-3.280
4600	281.224	1005.154	756.284	1144.803	-4502.631	388.888	-4.416
4700	281.285	1011.203	761.644	1172.929	-4501.916	495.218	-5.504
4800	281.343	1017.126	766.905	1201.060	-4501.467	601.535	-6.546
4900	281.397	1022.927	772.071	1229.197	-4501.294	707.847	-7.546
5000	281.448	1028.613	777.145	1257.339	-4501.407	814.157	-8.505
5100	281.496	1034.187	782.130	1285.487	-4501.817	920.473	-9.428
5200	281.541	1039.653	787.030	1313.638	-4502.530	1026.795	-10.314
5300	281.584	1045.016	791.848	1341.795	-4503.556	1133.141	-11.168
5400	281.624	1050.280	796.585	1369.955	-4504.904	1239.504	-11.990
5500	281.663	1055.448	801.245	1398.120	-4506.577	1345.895	-12.782
5600	281.699	1060.524	805.829	1426.288	-4508.584	1452.324	-13.547
5700	281.733	1065.510	810.342	1454.459	-4510.930	1558.787	-14.285
5800	281.766	1070.410	814.783	1482.634	-4513.617	1665.299	-14.998
5900	281.797	1075.227	819.157	1510.812	-4516.652	1771.853	-15.687
6000	281.826	1079.963	823.464	1538.994	-4520.037	1878.469	-16.354

PREVIOUS: September 1963 (1 atm) CURRENT: September 1963 (1 bar)

Beryllium Oxide $((BeO)_6)$ $Be_6O_6(g)$

BROMINE (Br) IDEAL GAS M_r = 79.904

IP(Br, g) = 95284.8 ± 0.5 cm^{-1}
S°(298.15 K) = 175.017 ± 0.003 J K^{-1} mol^{-1}

$\Delta_f H°$(0 K) = 117.92 ± 0.06 KJ mol^{-1}
$\Delta_f H°$(298.15 K) = 111.86 ± 0.06 kJ mol^{-1}

Electronic Levels and Quantum Weights

State	ε_i, cm^{-1}	g_i
$^2P°_{3/2}$	0.0	4
$^2P°_{1/2}$	3685.24	2

Enthalpy of Formation

The adopted value for the enthalpy of formation of Br(g) is derived from the spectroscopically determined dissociation energy of Br_2(g) as reported by Barrow et al. (1, 2). This value is also recommended by Brewer and Winn (3) and Huber and Herzberg (4). The adopted value, D_0^0 = 15895.6±0.02 cm^{-1} (190.154±0.001 kJ mol^{-1}), has been corrected for the natural isotopic abundance from the dissociation energies of $^{79}Br_2$ and $^{81}Br_2$. The convergence limit in the absorption spectrum of the $^3\Pi_{0^+u}$ - $^1\Sigma_g^+$ system corresponds to dissociation to a ground state bromine atom ($^3P_{3/2}$) and a bromine atom in the $^3P_{1/2}$ excited state.

Using earlier data, the CODATA recommended value (5) was calculated using D_0^0 (^{79}Br ^{81}Br) = 15893.1±2 cm^{-1} based on Horsley and Barrow (6) and Sitterly (7). LeRoy (8) recalculated the dissociation energy from these data; his result is correct only if Horsley and Barrow (6) missed many upper rotational levels of higher vibrational states in their interpretation of the absorption spectra. Subsequent to these initial CODATA evaluation, LeRoy and Bernstein (9), using the band origins obtained by Horsley and Barrow (6), derived D_0^0 (^{79}Br ^{81}Br) = 15895.5±0.5 cm^{-1}.

Heat Capacity and Entropy

Information on the electronic energy levels and quantum weights is taken from Moore (10, 11). Only two levels are included in the calculation. All other levels (observed and predicted) lie above 63430 cm^{-1}. Our calculations indicate that any reasonable method of filling in the missing levels and cutting off the summation in the partition function (12) has no effect on the thermodynamic properties to 6000 K. This is undoubtedly a result of the high energy of these levels. Therefore, we list only the ground state and the first excited state. Extension to higher temperatures may require consideration of the higher excited states and utilization of different fill and cutoff procedures.

The thermodynamic functions at 298.15 K are in agreement with recent CODATA recommendations (13) except for two minor differences. First, the entropy differs by 0.1094 J K^{-1} mol^{-1} because this table uses a standard-state pressure of 1 bar, whereas the CODATA recommendations are based on 1 atm. Second, entropy differences of the order of 0.001-0.004 J K^{-1} mol^{-1} for the monatomic halogens arise due to the use of slightly different values for the relative atomic mass and for R; this table uses R = 8.31441 J K^{-1} mol^{-1}. Considering these minor changes, this table agrees within the estimated uncertainty with those by Hultgren et al. (14), Gurvich et al. (15) and Wagman et al. (16). The estimated uncertainty is due to uncertainties in the relative atomic mass and fundamental constants which are based on the 1981 scale (17) and the 1973 values (18), respectively.

References

1. R. F. Barrow, D. F. Broyd, L. B. Pederson and K. K. Yee, Chem. Phys. Letters 18, 357 (1973).
2. R. F. Barrow, Clark, Coxon and K. K. Yee, J. Mol. Spectrosc. 51, 428 (1974).
3. L. Brewer and J. S. Winn, Faraday Symp. Chem. Soc. (GB) No. 14, 126 (1980).
4. K. P. Huber and G. Herzberg, "Constants of Diatomic Molecules", Van Nostrand Reinhold Company, New York, N.Y. (1979).
5. J. D. Cox, chairman, ICSU-CODATA Task Group on Key Values for Thermodynamics, J. Chem. Thermodynamics 4, 331 (1972).
6. T. A. Horsley and R. F. Barrow, Trans. Faraday Soc. 63, 32 (1967).
7. C. M. Sitterly, U. S. Nat. Bur. Stand., personal communication to ISCU-CODATA Task Group on Key Value for Thermodynamics.
8. R. J. LeRoy, J. Chem. Phys. 52, 2678 (1970).
9. R. J. LeRoy and R. B. Bernstein, J. Chem. Phys. 52, 2678, 3869 (1970).
10. C. E. Moore, U. S. Nat. Bur. Stand., NSRDS-NBS 34, 1970.
11. C. E. Moore, U. S. Nat. Bur. Stand., NSRDS-NBS-35, Volume II, 1970 [Reprint of NBS Circular 467, Volume II, 1952].
12. J. R. Downey, Jr., The Dow Chemical Co., AFOSR-TR-78-0960, Contract No. F44620-75-1-0048, 1978.
13. J. D. Cox, chairman, ICSU-CODATA Task Group on Key Values for Thermodynamics, J. Chem. Thermodynamics, 10, 903 (1978).
14. R. Hultgren, P. D. Desai et al., "Selected Values of the Thermodynamic Properties of the Elements", American Society for Metals, Metals Park, Ohio, 1973.
15. L. V. Gurvich, I. V. Veits et al., "Thermodynamic Properties of Individual Substances," 3rd ed., Vol. 1, Nauka, Moscow, 1978.
16. D. D. Wagman, W. H. Evans et al., J. Phys. Chem. Ref. Data 11, Supp. 2 (1982).
17. N. E. Holden and R. L. Martin, Pure Appl. Chem. 55, 1101 (1983).
18. E. R. Cohen and B. N. Taylor, J. Phys. Chem. Ref. Data 2, 663 (1973).

Bromine (Br) Br_1(g)

Enthalpy Reference Temperature = T_r = 298.15 K   Standard State Pressure = p° = 0.1 MPa

T/K	$C_p°$ (J $K^{-1}mol^{-1}$)	S° (J $K^{-1}mol^{-1}$)	-[G°-H°(T_r)]/T (J $K^{-1}mol^{-1}$)	H°-H°(T_r) (kJ mol^{-1})	$\Delta_f H°$ (kJ mol^{-1})	$\Delta_f G°$ (kJ mol^{-1})	Log K_f
0	0.	0.	INFINITE	-6.197	117.917	117.917	INFINITE
100	20.786	152.310	193.497	-4.119	118.599	106.061	-55.400
200	20.786	166.717	176.918	-2.040	118.227	93.626	-24.453
250	20.786	171.356	175.359	-1.001	117.859	87.515	-18.285
298.15	20.786	175.017	175.017	0.	111.860	82.369	-14.431
300	20.786	175.145	175.017	0.038	111.828	82.186	-14.310
350	20.786	178.350	175.270	1.078	96.543	78.082	-11.653
400	20.787	181.125	175.832	2.117	96.667	75.436	-9.851
450	20.790	183.574	176.559	3.157	96.786	72.775	-8.447
500	20.798	185.765	177.372	4.196	96.900	70.100	-7.323
600	20.833	189.559	179.097	6.277	97.122	64.720	-5.634
700	20.908	192.776	180.827	8.364	97.339	59.302	-4.425
800	21.027	195.575	182.499	10.461	97.560	53.853	-3.516
900	21.184	198.060	184.093	12.571	97.788	48.376	-2.808
1000	21.365	200.301	185.603	14.698	98.028	42.873	-2.239
1100	21.559	202.347	187.034	16.844	98.283	37.345	-1.773
1200	21.752	204.231	188.389	19.010	98.552	31.794	-1.384
1300	21.937	205.979	189.676	21.195	98.837	26.219	-1.054
1400	22.107	207.611	190.899	23.397	99.134	20.622	-0.769
1500	22.258	209.142	192.065	25.615	99.443	15.004	-0.522
1600	22.388	210.583	193.178	27.848	99.761	9.364	-0.306
1700	22.497	211.943	194.242	30.092	100.085	3.704	-0.114
1800	22.586	213.232	195.262	32.346	100.412	-1.975	0.057
1900	22.657	214.455	196.240	34.609	100.740	-7.672	0.211
2000	22.710	215.619	197.180	36.877	101.066	-13.386	0.350
2100	22.748	216.728	198.085	39.150	101.387	-19.117	0.476
2200	22.772	217.786	198.956	41.426	101.701	-24.862	0.590
2300	22.785	218.799	199.797	43.704	102.006	-30.622	0.695
2400	22.788	219.769	200.609	45.983	102.301	-36.395	0.792
2500	22.782	220.699	201.394	48.262	102.585	-42.180	0.881
2600	22.769	221.592	202.154	50.539	102.856	-47.976	0.964
2700	22.750	222.451	202.890	52.815	103.114	-53.782	1.040
2800	22.727	223.278	203.603	55.089	103.360	-59.598	1.112
2900	22.699	224.075	204.296	57.360	103.592	-65.421	1.178
3000	22.669	224.844	204.968	59.629	103.813	-71.253	1.241
3100	22.635	225.587	205.621	61.894	104.023	-77.092	1.299
3200	22.600	226.305	206.256	64.156	104.222	-82.938	1.354
3300	22.564	227.000	206.875	66.414	104.411	-88.790	1.405
3400	22.526	227.673	207.476	68.669	104.593	-94.647	1.454
3500	22.489	228.325	208.063	70.919	104.767	-100.510	1.500
3600	22.450	228.958	208.634	73.166	104.936	-106.377	1.543
3700	22.412	229.573	209.192	75.409	105.101	-112.249	1.585
3800	22.374	230.170	209.736	77.649	105.263	-118.126	1.624
3900	22.336	230.751	210.268	79.884	105.423	-124.006	1.661
4000	22.298	231.316	210.787	82.116	105.582	-129.891	1.696
4100	22.261	231.866	211.294	84.344	105.743	-135.780	1.730
4200	22.225	232.402	211.791	86.568	105.904	-141.673	1.762
4300	22.189	232.925	212.276	88.789	106.069	-147.569	1.793
4400	22.154	233.434	212.751	91.006	106.237	-153.470	1.822
4500	22.120	233.932	213.216	93.220	106.409	-159.374	1.850
4600	22.087	234.418	213.672	95.430	106.587	-165.283	1.877
4700	22.055	234.892	214.118	97.637	106.770	-171.195	1.903
4800	22.023	235.356	214.556	99.841	106.960	-177.111	1.927
4900	21.992	235.810	214.985	102.042	107.156	-183.031	1.951
5000	21.962	236.254	215.406	104.239	107.359	-188.955	1.974
5100	21.933	236.689	215.819	106.434	107.570	-194.883	1.996
5200	21.905	237.114	216.225	108.626	107.789	-200.816	2.017
5300	21.878	237.531	216.623	110.815	108.016	-206.753	2.038
5400	21.851	237.940	217.014	113.002	108.251	-212.694	2.057
5500	21.825	238.341	217.398	115.186	108.495	-218.640	2.076
5600	21.800	238.734	217.775	117.367	108.746	-224.590	2.095
5700	21.776	239.119	218.146	119.546	109.007	-230.545	2.113
5800	21.752	239.498	218.511	121.722	109.275	-236.504	2.130
5900	21.730	239.870	218.870	123.896	109.552	-242.468	2.147
6000	21.708	240.235	219.223	126.068	109.838	-248.437	2.163

PREVIOUS: June 1974 (1 atm)   CURRENT: June 1982 (1 bar)

Bromine (Br) Br_1(g)

BROMINE, ION (Br^+) IDEAL GAS M_r = 79.90345

IP(Br^+, g) = 175870 ± 10 cm^{-1}
S°(298.15 K) = 176.872 ± 0.005 J K^{-1} mol^{-1}

$\Delta_f H°$(0 K) = 1256.95 ± 0.13 kJ mol^{-1}
$\Delta_f H°$(298.15 K) = [1257.091] kJ mol^{-1}

Electronic Levels and Quantum Weights

State	ϵ_i, cm^{-1}	g_i
$^3P_2$	0.0	5
$^3P_1$	3136.4	3
$^3P_0$	3837.5	1
$^1D_2$	12089.1	5
$^1S_0$	27867.1	1

Enthalpy of Formation

$\Delta_f H°$(Br^+, g, 0 K) is calculated from $\Delta_f H°$(Br, g, 0 K) (1) using the spectroscopic value of IP(Br) = 95284.8±0.5 cm^{-1} (1139.86±0.01 kJ mol^{-1}) from Moore (2). The ionization limit is converted from cm^{-1} to kJ mol^{-1} using the factor, 1 cm^{-1} = 0.01196266 kJ mol^{-1}, which is derived from the 1973 CODATA fundamental constants (3). Rosenstock et al. (4) and Levin and Lias (5) have summarized additional ionization potential and appearance potential data.

$\Delta_f H°$(Br^+, g, 298.15 K) is calculated from $\Delta_f H°$(Br, g, 0 K) by using IP(Br) with JANAF (1) enthalpies, H°(0 K)-H°(298.15 K), for Br(g), Br^+(g), and e^-(g). $\Delta_f H°$(Br → Br^+ + e^-, 298.15 K) differs from a room temperature threshold energy due to inclusion of these enthalpies and to threshold effects discussed by Rosenstock et al. (4). $\Delta_f H°$(298.15 K) should be changed by -6.197 kJ mol^{-1} if it is to be used in the ion convention that excludes the enthalpy of the electron.

Heat Capacity and Entropy

The information on electronic energy levels and quantum weights, given by Moore (2, 6), is incomplete because many theoretically predicted levels have not been observed. Our calculations indicate that any reasonable method of filling in these missing levels and cutting off the summation in the partition function (7) has no effect on the thermodynamic functions to 6000 K. This is a result of the high energy of all levels other than the ground state and the lowest four excited states; the next lowest level is approximately 93927 cm^{-1} above the ground state. Since inclusion of these upper levels has no effect on the thermodynamic functions (to 6000 K), we list only the ground state and four excited states, with the energy of these states taken from recent study by Moore (2). The reported uncertainty in S°(298.15 K) is due to uncertainties in the relative ionic mass, fundamental constants, and the position of the four lowest excited states. Extension of these calculations above 6000 K may require consideration of the higher excited states and use of different fill and cutoff procedures (7).

References

1. JANAF Thermochemical Tables: Br(g), 3-31-82; e^-(r), 3-31-82.
2. C. E. Moore, U. S. Nat. Bur. Stand., NSRDS-NBS-34, 1970.
3. E. R. Cohen and B. N. Taylor, J. Phys. Chem. Ref. Data 2, 663 (1973).
4. H. M. Rosenstock, K. Draxl et al., J. Phys. Chem. Ref. Data 6, Supp. 1 (1977).
5. R. D. Levin and S. G. Lias, U. S. Nat. Bur. Stand., NSRDS-NBS-71, 1982.
6. C. E. Moore, U. S. Nat. Bur. Stand., NSRDS-NBS-35, Volume II, 1970 [reprint of NBS Circular 467, Volume II, 1952].
7. J. R. Downey, Jr., The Dow Chemical Company, AFOSR-TR-78-0960, Contract No. F44620-75-1-0048, 1978.

Bromine, Ion (Br^+) Br_1^+(g)

Enthalpy Reference Temperature = T_r = 298.15 K Standard State Pressure = p° = 0.1 MPa

T/K	$C_p°$ (J $K^{-1}mol^{-1}$)	S° (J $K^{-1}mol^{-1}$)	-[G°-H°(T_r)]/T (J $K^{-1}mol^{-1}$)	H°-H°(T_r) (kJ mol^{-1})	$\Delta_f H°$ (kJ mol^{-1})	$\Delta_f G°$ (kJ mol^{-1})	Log K_f
0	0.	0.	INFINITE	-6.197	1256.951		
100	20.786	154.165	195.352	-4.119			
200	20.786	168.573	178.773	-2.040			
250	20.786	173.211	177.214	-1.001			
298.15	20.786	176.872	176.872	0.	1257.091	1220.792	-213.878
300	20.786	177.001	176.872	0.038	1257.098	1220.567	-212.520
350	20.788	180.205	177.125	1.078	1242.852	1215.232	-181.364
400	20.794	182.981	177.688	2.117	1244.016	1211.207	-158.167
450	20.809	185.431	178.415	3.157	1245.175	1207.037	-140.109
500	20.838	187.625	179.228	4.199	1246.330	1202.737	-125.649
600	20.953	191.433	180.954	6.287	1248.638	1193.802	-103.930
700	21.153	194.677	182.688	8.392	1250.952	1184.480	-88.387
800	21.426	197.519	184.368	10.520	1253.282	1174.825	-76.708
900	21.746	200.060	185.973	12.679	1255.637	1164.877	-67.608
1000	22.084	202.369	187.499	14.870	1258.020	1154.665	-60.313
1100	22.415	204.490	188.948	17.095	1260.432	1144.212	-54.334
1200	22.720	206.453	190.326	19.352	1262.872	1133.540	-49.342
1300	22.990	208.283	191.638	21.638	1265.336	1122.662	-45.109
1400	23.220	209.995	192.889	23.949	1267.821	1111.594	-41.474
1500	23.409	211.604	194.083	26.281	1270.322	1100.348	-38.318
1600	23.560	213.120	195.226	28.629	1272.835	1088.934	-35.550
1700	23.677	214.552	196.321	30.992	1275.355	1077.363	-33.103
1800	23.765	215.908	197.372	33.364	1277.879	1065.644	-30.924
1900	23.829	217.194	198.382	35.744	1280.403	1053.784	-28.971
2000	23.874	218.418	199.353	38.129	1282.924	1041.792	-27.209
2100	23.904	219.583	200.289	40.518	1285.440	1029.673	-25.612
2200	23.923	220.696	201.192	42.909	1287.948	1017.435	-24.157
2300	23.935	221.760	202.063	45.302	1290.447	1005.083	-22.826
2400	23.942	222.778	202.905	47.696	1292.935	992.623	-21.604
2500	23.946	223.756	203.720	50.091	1295.413	980.059	-20.477
2600	23.948	224.695	204.508	52.485	1297.880	967.396	-19.435
2700	23.950	225.599	205.273	54.880	1300.336	954.638	-18.469
2800	23.952	226.470	206.014	57.275	1302.781	941.790	-17.569
2900	23.955	227.310	206.734	59.671	1305.216	928.854	-16.730
3000	23.959	228.123	207.434	62.066	1307.643	915.835	-15.946
3100	23.964	228.908	208.114	64.462	1310.062	902.735	-15.211
3200	23.971	229.669	208.776	66.859	1312.474	889.557	-14.521
3300	23.979	230.407	209.420	69.257	1314.882	876.303	-13.871
3400	23.988	231.123	210.048	71 655	1317.286	862.977	-13.258
3500	23.997	231.818	210.660	74.054	1319.687	849.580	-12.679
3600	24.008	232.495	211.257	76.455	1322.088	836.114	-12.132
3700	24.018	233.153	211.840	78.856	1324.490	822.582	-11.613
3800	24.029	233.793	212.409	81.258	1326.893	808.985	-11.120
3900	24.040	234.418	212.966	83.662	1329.300	795.324	-10.652
4000	24.051	235.026	213.510	86.066	1331.711	781.602	-10.207
4100	24.061	235.620	214.042	88.472	1334.127	767.819	-9.782
4200	24.071	236.200	214.562	90.878	1336.550	753.977	-9.377
4300	24.080	236.767	215.072	93.286	1338.980	740.078	-8.990
4400	24.088	237.320	215.572	95.694	1341.418	726.122	-8.620
4500	24.095	237.862	216.061	98.104	1343.864	712.110	-8.266
4600	24.101	238.391	216.541	100.513	1346.320	698.044	-7.927
4700	24.107	238.910	217.011	102.924	1348.785	683.925	-7.601
4800	24.111	239.417	217.473	105.335	1351.260	669.753	-7.288
4900	24.114	239.915	217.926	107.746	1353.746	655.529	-6.988
5000	24.116	240.402	218.370	110.157	1356.242	641.254	-6.699
5100	24.116	240.879	218.807	112.569	1358.748	626.930	-6.421
5200	24.116	241.348	219.236	114.981	1361.265	612.556	-6.153
5300	24.114	241.807	219.658	117.392	1363.793	598.133	-5.895
5400	24.111	242.258	220.072	119.803	1366.332	583.663	-5.646
5500	24.107	242.700	220.479	122.214	1368.881	569.146	-5.405
5600	24.102	243.134	220.880	124.625	1371.440	554.582	-5.173
5700	24.096	243.561	221.274	127.035	1374.010	539.973	-4.948
5800	24.089	243.980	221.662	129.444	1376.590	525.318	-4.731
5900	24.081	244.392	222.044	131.853	1379.181	510.618	-4.521
6000	24.072	244.796	222.420	134.260	1381.780	495.875	-4.317

PREVIOUS: CURRENT: June 1982 (1 bar)

Bromine, Ion (Br^+) Br_1^+(g)

BROMINE, ION (Br^-) IDEAL GAS M_r = 79.90455

EA(Br, g) = 3.365 ± 0.003 eV
S°(298.15 K) = 163.491 ± 0.005 J K^{-1} mol^{-1}

$\Delta_f H°$(0 K) = -206.8 ± 0.5 kJ mol^{-1}
$\Delta_f H°$(298.15 K) = [-219.008] kJ mol^{-1}

Electronic Level and Quantum Weight

State	ε_i, cm^{-1}	g_i
$^1S_0$	0.0	1

Enthalpy of Formation

$\Delta_f H°(Br^-, g, 0 K)$ is calculated from $\Delta_f H°$(Br, g, 0 K) (1) using the adopted electron affinity of EA(Br) = 3.363±0.003 eV (324.671±0.289 kJ mol^{-1}). This value, recommended by Hotop and Lineberger (10), is based on plasma absorption and plasma emission studies (5, 6). Additional information on Br^-(g) may be obtained in the critical discussions of Hotop and Lineberger (2, 10), Rosenstock et al. (3) and Massey (4).

$\Delta_f H°(Br^-, g, 298.15 K)$ is obtained from $\Delta_f H°$(Br, g, 0 K) by using EA(Br) with JANAF (1) enthalpies, H°(0 K) - H°(298.15 K), for Br^-(g), Br(g), and e^-(g). $\Delta_r H°(Br^- \rightarrow Br + e^-$, 298.15 K) differs from a room-temperature threshold energy due to inclusion of these enthalpies and to threshold effects discussed by Rosenstock et al. (3). $\Delta_f H°$(298.15 K) should be changed by +6.197 kJ mol^{-1} if it is to be used in the ion convention that excludes the enthalpy of the electron.

Heat Capacity and Entropy

The ground state configuration for Br^-(g) is given by Hotop and Lineberger (2, 10), Rosenstock et al. (3) and Massey (4). Lacking any experimental evidence as to the stability of any excited states, we assume that no stable excited states exist.

The entropy at all temperatures (0-6000 K) agrees within 0.001 J K^{-1} mol^{-1} with the values of Gurvich et al. (8) except for one minor difference. The entropy differs by 0.1094 J K^{-1} mol^{-1} because this table uses a standard-state pressure of 1 bar, whereas the tabulation of Gurvich et al. (8) is based on 1 atm. The estimated uncertainty is due to uncertainties in the relative ionic mass and the fundamental constants which are based on the 1981 scale (9) and the 1973 values (7), respectively.

References

1. JANAF Thermochemical Tables: Br(g), 6-30-82; e^-(g), 3-31-82.
2. H. Hotop and W. C. Lineberger, J. Phys. Chem. Ref. Data 4, 539 (1975).
3. H. M. Rosenstock, K. Draxl et al., J. Phys. Chem. Ref. Data 6, Supp. 1 (1977).
4. H. S. W. Massey, "Negative Ions," 3rd ed., Cambridge University Press, Cambridge, 1976.
5. R. S. Berry and C. W. Reimann, J. Chem. Phys. 38, 1540 (1963).
6. H. Frank, M. Neiger and H.-P. Popp, Z. Naturforsch. 25a, 1617 (1970).
7. E. R. Cohen and B. N. Taylor, J. Phys. Chem. Ref. Data 2, 663 (1973).
8. L. V. Gurvich, I. V. Veits et al., "Thermodynamic Properties of Individual Substances", 3rd ed., Vol. 1, Nauka, Moscow, 1978.
9. N. E. Holden and R. L. Martin, Pure Appl. Chem. 55, 1101 (1983).
10. H. Hotop and W. C. Lineberger, J. Phys. Chem. Ref. Data, to be published, 1985.

Bromine, Ion (Br^-) Br_1^-(g)

Enthalpy Reference Temperature = T_r = 298.15 K Standard State Pressure = p° = 0.1 MPa

T/K	$C_p°$ (J $K^{-1}mol^{-1}$)	S° (J $K^{-1}mol^{-1}$)	-[G°-H°(T_r)]/T (J $K^{-1}mol^{-1}$)	H°-H°(T_r) (kJ mol^{-1})	$\Delta_f H°$ (kJ mol^{-1})	$\Delta_f G°$ (kJ mol^{-1})	Log K_f
0	0.	0.	INFINITE	-6.197	-206.753		
100	20.786	140.783	181.971	-4.119			
200	20.786	155.191	165.392	-2.040			
250	20.786	159.829	163.833	-1.001			
298.15	20.786	163.491	163.491	0.	-219.008	-238.808	41.838
300	20.786	163.619	163.491	0.038	-219.078	-238.930	41.601
350	20.786	166.823	163.744	1.078	-235.403	-241.321	36.015
400	20.786	169.599	164.306	2.117	-236.318	-242.104	31.616
450	20.786	172.047	165.033	3.156	-237.239	-242.772	28.180
500	20.786	174.237	165.846	4.196	-238.164	-243.337	25.421
600	20.786	178.027	167.570	6.274	-240.023	-244.198	21.259
700	20.786	181.231	169.299	8.353	-241.893	-244.746	18.263
800	20.786	184.007	170.967	10.431	-243.769	-245.025	15.999
900	20.786	186.455	172.555	12.510	-245.651	-245.069	14.223
1000	20.786	188.645	174.056	14.589	-247.538	-244.904	12.792
1100	20.786	190.626	175.474	16.667	-249.429	-244.549	11.613
1200	20.786	192.435	176.813	18.746	-251.325	-244.021	10.622
1300	20.786	194.099	178.080	20.824	-253.226	-243.335	9.777
1400	20.786	195.639	179.280	22.903	-255.130	-242.503	9.048
1500	20.786	197.073	180.419	24.982	-257.040	-241.535	8.411
1600	20.786	198.415	181.502	27.060	-258.954	-240.439	7.850
1700	20.786	199.675	182.534	29.139	-260.875	-239.223	7.350
1800	20.786	200.863	183.520	31.217	-262.802	-237.893	6.903
1900	20.786	201.987	184.462	33.296	-264.736	-236.457	6.501
2000	20.786	203.053	185.365	35.375	-266.679	-234.918	6.135
2100	20.786	204.067	186.232	37.453	-268.631	-233.282	5.803
2200	20.786	205.034	187.065	39.532	-270.593	-231.553	5.498
2300	20.786	205.958	187.866	41.610	-272.566	-229.734	5.217
2400	20.786	206.843	188.639	43.689	-274.550	-227.830	4.959
2500	20.786	207.691	189.384	45.768	-276.545	-225.842	4.719
2600	20.786	208.506	190.104	47.846	-278.551	-223.774	4.496
2700	20.786	209.291	190.800	49.925	-280.569	-221.629	4.288
2800	20.786	210.047	191.474	52.004	-282.597	-219.409	4.093
2900	20.786	210.776	192.127	54.082	-284.636	-217.117	3.911
3000	20.786	211.481	192.761	56.161	-286.684	-214.754	3.739
3100	20.786	212.162	193.375	58.239	-288.739	-212.322	3.578
3200	20.786	212.822	193.973	60.318	-290.802	-209.824	3.425
3300	20.786	213.462	194.554	62.397	-292.871	-207.261	3.281
3400	20.786	214.082	195.119	64.475	-294.944	-204.636	3.144
3500	20.786	214.685	195.670	66.554	-297.020	-201.950	3.014
3600	20.786	215.271	196.206	68.632	-299.098	-199.204	2.890
3700	20.786	215.840	196.729	70.711	-301.176	-196.401	2.773
3800	20.786	216.394	197.239	72.790	-303.254	-193.541	2.660
3900	20.786	216.934	197.737	74.868	-305.329	-190.627	2.553
4000	20.786	217.461	198.224	76.947	-307.401	-187.659	2.451
4100	20.786	217.974	198.699	79.025	-309.469	-184.640	2.352
4200	20.786	218.475	199.164	81.104	-311.532	-181.571	2.258
4300	20.786	218.964	199.619	83.183	-313.588	-178.452	2.168
4400	20.786	219.442	200.064	85.261	-315.637	-175.286	2.081
4500	20.786	219.909	200.500	87.340	-317.678	-172.073	1.997
4600	20.786	220.366	200.927	89.418	-319.711	-168.815	1.917
4700	20.786	220.813	201.345	91.497	-321.735	-165.513	1.839
4800	20.786	221.250	201.755	93.576	-323.749	-162.167	1.765
4900	20.786	221.679	202.158	95.654	-325.754	-158.780	1.693
5000	20.786	222.099	202.552	97.733	-327.748	-155.352	1.623
5100	20.786	222.510	202.940	99.811	-329.732	-151.885	1.556
5200	20.786	222.914	203.320	101.890	-331.705	-148.378	1.490
5300	20.786	223.310	203.693	103.969	-333.667	-144.834	1.427
5400	20.786	223.699	204.060	106.047	-335.618	-141.253	1.366
5500	20.786	224.080	204.421	108.126	-337.559	-137.635	1.307
5600	20.786	224.454	204.775	110.204	-339.488	-133.983	1.250
5700	20.786	224.822	205.124	112.283	-341.407	-130.296	1.194
5800	20.786	225.184	205.466	114.362	-343.314	-126.576	1.140
5900	20.786	225.539	205.804	116.440	-345.212	-122.823	1.087
6000	20.786	225.889	206.135	118.519	-347.098	-119.038	1.036

PREVIOUS: CURRENT: June 1982 (1 bar)

Bromine, Ion (Br^-) Br_1^-(g)

$S°(298.15\ K) = 252.89 \pm 0.21\ J\ K^{-1}\ mol^{-1}$

$\Delta_f H°(0\ K) = -41.3 \pm 41.8\ kJ\ mol^{-1}$
$\Delta_f H°(298.15\ K) = -49.4 \pm 41.8\ kJ\ mol^{-1}$

Electronic Levels and Quantum Weights

State	ε_i, cm^{-1}	g_i	State	ε_i, cm^{-1}	g_i
$X\ ^2\Sigma^+$	0	2	$D\ ^2\Sigma$	30190.6	2
$A_1\ ^2\Pi_{1/2}$	15922.5	2	$E\ ^2\Sigma$	33942.2	2
$A_2\ ^2\Pi_{3/2}$	15985.8	2	$[F\ ^2\Pi]$	[35000]	[4]
$B\ ^2\Sigma$	16380.0	2	$[G\ ^2\Delta]$	[36000]	[4]
$C_1\ ^2\Pi_{1/2}$	25314.0	2	$[H\ ^2\Sigma]$	36798.7	[2]
$C_2\ ^2\Pi_{3/2}$	25537.5	2			

$\omega_e = 284.56\ cm^{-1}$ $\omega_e x_e = 0.86\ cm^{-1}$ $\sigma = 1$

$B_e = [0.09637]\ cm^{-1}$ $\alpha_e = [0.000389]\ cm^{-1}$ $r_e = [2.56]$ Å

Enthalpy of Formation

The selected value, $\Delta_f H°(0\ K) = -41.258\ kJ\ mol^{-1}$, is obtained from an analysis of spectroscopic data. The adopted values of the ground state vibrational constants, ω_e and $\omega_e x_e$, give $D_0^0 = 2.90$ eV for CaBr(g) by a linear Birge-Sponer extrapolation. Based on the ionicity correction suggested by Hildenbrand (1), this value adjusts to 3.50 eV (337.696 kJ mol^{-1}) which is adopted. The adopted value for D_0^0 gives $D_0^0(CaBr)/\Delta_{at}H°(CaBr_2) = 0.43$ which is quite consistent with values of this ratio for other alkaline earth halide systems (2). Also, Hildenbrand (3, 4) found that the ionicity parameter brings thermochemical and spectroscopic dissociation energies for CaF(g) and CaCl(g) into reasonable agreement. $\Delta_f H°(298.15\ K)$ corresponds to $-49.4\ kJ\ mol^{-1}$.

Ionic model calculations have led to D_0^0 values of 5.29 eV (5) and 3.4 eV (6). The latter result is believed to represent a minimum value for D_0^0. Two other experimental values, which bracket the selected value, have also been reported. Flame studies (7) gave $D_0^0 = 3.29$ eV and chemiluminescence (8) from reaction of Ca atoms with Br gave a lower limit to D_0^0 of 4.22 eV. We assign an uncertainty of $\pm 41.8\ kJ\ mol^{-1}$ to $\Delta_f H°(0\ K)$ to include the possibility that these studies are correct.

Heat Capacity and Entropy

The value of r_e is obtained from that for gaseous $CaBr_2$ (9) with $r_e(CaBr)/r_e(CaBr_2) = 0.96$. The value of this ratio is calculated from bond lengths (2) for several other alkaline earth halide systems. Our adopted value for r_e is supported by an estimate (r_e = 2.6 Å) of Krasnov and Karaseva (6) while another etimated value (5) is only 0.14 Å larger than ours. The rotational constant is calculated from the estimated value of r_e. The value of α_e is obtained from a Morse potential function.

The vibrational constants are those recently determined from a complete vibrational analysis of the $D\ ^2\Sigma - X\ ^2\Sigma$ system by Shah (10). These two values are corrected for the natural isotopic abundances of Br. The electronic levels with the exception of those for the D, F, G, and H states are from the compilation of Rosen (11). The D state energy is from Shah (10). The two doublet states (F and G) are estimated by analogy with those for SrBr (9). Recently, Reddy et al. (12) reported observing a new system of bands in the visible emission spectrum of CaBr which was associated with $^2\Sigma - A\ ^2\Pi$ transition. The upper state of this system is assigned the $H\ ^2\Sigma$ state by analogy with SrBr.

References

1. D. L. Hildenbrand in "Advances in High Temperature Chemistry," Vol. 1, Ed. by L. Eyring, Academic Press, New York, 1967.
2. JANAF Thermochemical Tables: refer to all alkaline earth halide gases, MX and MX2.
3. D. L. Hildenbrand, J. Chem. Phys. 48, 3657 (1968).
4. D. L. Hildenbrand, J. Chem. Phys. 52, 5751 (1970).
5. J. L. Margrave, J. Phys. Chem. 58, 258 (1954).
6. K. S. Krasnov and N. V. Karaseva, Optics and Spectroscopy 19, 14 (1965).
7. A. N. Khitrov, V. G. Ryabova, and L. V. Gurvich, Teplofiz. Vys. Temp. 11, 1126 (1973).
8. M. Menzinger, Can. J. Chem. 52, 1688 (1974).
9. JANAF Thermochemical Tables: $CaBr_2$(g), 6-30-74; SrBr(g), 12-31-74.
10. S. G. Shah, Indian J. Pure Appl. Phys. 8, 118 (1970).
11. B. Rosen, Ed., "Spectroscopic Data Relative to Diatomic Molecules," Pergamon Press, New York, 1970.
12. B. R. K. Reddy, Y. P. Reddy, and P. T. Rao, Curr. Sci. 39, 485 (1970).

Enthalpy Reference Temperature = T_r = 298.15 K   Standard State Pressure = p° = 0.1 MPa

	J K⁻¹mol⁻¹			kJ mol⁻¹			
T/K	$C_p°$	S°	$-[G°-H°(T_r)]/T$	$H°-H°(T_r)$	$\Delta_f H°$	$\Delta_f G°$	Log K_f
0	0.	0.	INFINITE	-9.854	-41.259	-41.259	INFINITE
100	31.560	215.410	284.258	-6.885	-40.667	-57.919	30.254
200	35.141	238.594	256.190	-3.519	-42.021	-74.700	19.510
250	35.915	246.526	253.491	-1.741	-42.901	-82.771	17.294
298.15	36.381	252.894	252.894	0.	-49.396	-89.707	15.716
300	36.395	253.119	252.895	0.067	-49.447	-89.957	15.663
350	36.713	258.755	253.339	1.896	-65.250	-95.832	14.302
400	36.935	263.673	254.330	3.737	-65.652	-100.174	13.081
450	37.100	268.033	255.615	5.588	-66.084	-104.463	12.126
500	37.227	271.949	257.056	7.446	-66.554	-108.703	11.356
600	37.413	278.754	260.122	11.179	-67.624	-117.035	10.189
700	37.548	284.532	263.207	14.927	-68.883	-125.173	9.341
800	37.654	289.553	266.193	18.688	-71.193	-133.014	8.685
900	37.743	293.993	269.040	22.458	-72.831	-140.646	8.163
1000	37.822	297.974	271.738	26.236	-74.760	-148.080	7.735
1100	37.894	301.582	274.289	30.022	-76.988	-155.307	7.375
1200	37.962	304.882	276.703	33.815	-87.252	-161.695	7.038
1300	38.026	307.923	278.989	37.614	-88.852	-167.834	6.744
1400	38.088	310.744	281.158	41.420	-90.451	-173.849	6.486
1500	38.149	313.373	283.219	45.232	-92.049	-179.750	6.259
1600	38.209	315.837	285.181	49.050	-93.646	-185.545	6.057
1700	38.269	318.156	287.054	52.873	-95.242	-191.240	5.876
1800	38.331	320.345	288.843	56.703	-245.513	-194.630	5.648
1900	38.395	322.419	290.556	60.540	-245.698	-191.798	5.273
2000	38.463	324.390	292.199	64.383	-245.889	-188.956	4.935
2100	38.536	326.268	293.777	68.232	-246.091	-186.104	4.629
2200	38.615	328.063	295.295	72.090	-246.306	-183.243	4.351
2300	38.703	329.781	296.757	75.956	-246.539	-180.371	4.096
2400	38.800	331.430	298.168	79.831	-246.792	-177.489	3.863
2500	38.908	333.016	299.530	83.716	-247.071	-174.596	3.648
2600	39.028	334.545	300.848	87.613	-247.381	-171.691	3.449
2700	39.161	336.020	302.123	91.522	-247.726	-168.773	3.265
2800	39.307	337.447	303.359	95.445	-248.113	-165.842	3.094
2900	39.468	338.829	304.559	99.384	-248.545	-162.896	2.934
3000	39.644	340.170	305.723	103.340	-249.030	-159.935	2.785
3100	39.835	341.473	306.856	107.313	-249.573	-156.956	2.645
3200	40.041	342.741	307.957	111.307	-250.178	-153.959	2.513
3300	40.262	343.976	309.030	115.322	-250.852	-150.942	2.389
3400	40.498	345.182	310.076	119.360	-251.600	-147.903	2.272
3500	40.748	346.359	311.096	123.422	-252.427	-144.842	2.162
3600	41.013	347.511	312.091	127.510	-253.338	-141.755	2.057
3700	41.291	348.638	313.064	131.625	-254.336	-138.642	1.957
3800	41.582	349.743	314.015	135.769	-255.426	-135.501	1.863
3900	41.885	350.827	314.945	139.942	-256.610	-132.330	1.772
4000	42.199	351.892	315.855	144.146	-257.892	-129.127	1.686
4100	42.524	352.938	316.747	148.382	-259.272	-125.891	1.604
4200	42.858	353.966	317.621	152.651	-260.751	-122.620	1.525
4300	43.201	354.979	318.478	156.954	-262.332	-119.313	1.449
4400	43.552	355.976	319.319	161.291	-264.009	-115.968	1.377
4500	43.910	356.959	320.144	165.664	-265.770	-112.584	1.307
4600	44.274	357.928	320.955	170.074	-267.634	-109.160	1.240
4700	44.643	358.884	321.752	174.519	-269.594	-105.694	1.175
4800	45.016	359.828	322.536	179.002	-271.643	-102.185	1.112
4900	45.394	360.760	323.306	183.523	-273.778	-98.633	1.051
5000	45.773	361.681	324.064	188.081	-275.993	-95.036	0.993
5100	46.155	362.591	324.811	192.677	-278.285	-91.395	0.936
5200	46.538	363.491	325.546	197.312	-280.642	-87.708	0.881
5300	46.921	364.381	326.271	201.985	-283.068	-83.974	0.828
5400	47.303	365.261	326.984	206.696	-285.553	-80.194	0.776
5500	47.685	366.133	327.688	211.446	-288.090	-76.368	0.725
5600	48.065	366.996	328.383	216.233	-290.664	-72.496	0.676
5700	48.442	367.850	329.067	221.059	-293.286	-68.577	0.628
5800	48.817	368.695	329.743	225.921	-295.942	-64.612	0.582
5900	49.188	369.533	330.411	230.822	-298.626	-60.601	0.537
6000	49.555	370.363	331.070	235.759	-301.330	-56.544	0.492

PREVIOUS: December 1974 (1 atm)   CURRENT: December 1974 (1 bar)

BROMINE CHLORIDE (BrCl) IDEAL GAS M_r = 115.357

$S°(298.15\ K)$ = 240.001 $J\ K^{-1}\ mol^{-1}$

$\Delta_f H°(0\ K)$ = 22.087 ± 1.25 $kJ\ mol^{-1}$
$\Delta_f H°(298.15\ K)$ = 14.644 ± 1.25 $kJ\ mol^{-1}$

Electronic Level and Quantum Weight

State	ε_i, cm^{-1}	g_i
$^1\Sigma^+$	0	1

ω_e = 443.1 cm^{-1} $\omega_e x_e$ = 1.8 cm^{-1} σ = 1

B_e = 0.150797 cm^{-1} α_e = 0.0007597 cm^{-1} r_e = 2.136 Å

Enthalpy of Formation

The average $\Delta_r H°(298.15\ K)$ = 0.406 kcal mol^{-1} for the reaction $Br_2(g) + Cl_2(g) = 2BrCl(g)$ and the $\Delta_f H°(298.15\ K)$ = 3.694 kcal mol^{-1} for 1/2 $Br_2(g)$ from JANAF tables were used to calculate the $\Delta_f H°(298.15\ K)$ = 3.5 ± 0.3 kcal mol^{-1} (14.644±1.25 kJ mol^{-1}). Other equilibrium measurements not included below have been summarized by Beeson and Yost (3).

Source	T, K	Method	$\Delta_r H°(298.15\ K)$, kcal mol^{-1}
(1)	301	Light absorption	.51
(2)	298	Light absorption	0.34
(3)	372-492	Total pressure Run III	0.39
(3)	372-492	Total pressure Run II	0.47
(4)	313±5	Mass spectrometric	0.32
		Average	0.406

Heat Capacity and Entropy

All spectroscopic and molecular constants were obtained from Evans et al. (5) except r_e which was calculated from B_e. The value of r_e was reported to be 2.13 Å and 2.11 Å by Beeson and Yost (3) and Cole and Elverum (6), respectively. Ground state configuration was taken from Herzberg (7).

References

1. H. G. Vesper and G. K. Rollefson, J. Am. Chem. Soc. 56, 620 (1934).
2. G. Braune and E. Victor, Z. Elektrochem. 41, 508 (1935).
3. C. M. Beeson and D. M. Yost, Am. Chem. Soc. 61, 1432 (1939).
4. H. C. Mattraw, C. F. Pachucki and N. J. Hawkins, General Electric Co., Knolls Atomic Power Laboratory, Schenectady, New York Report No. KAPL-1016 (1953).
5. W. H. Evans, T. R. Munson and D. D. Wagman, J. Res. Natl. Bur. of Standards, 55, 147 (1955).
6. L. G. Cole and G. W. Elverum Jr., J. Chem. Phys., 20, 1543 (1952).
7. G. Herzberg, "Spectra of Diatomic Molecules", D. Van Nostrand Co., New York, 1950.

Bromine Chloride (BrCl) $Br_1Cl_1(g)$

Enthalpy Reference Temperature = T_r = 298.15 K Standard State Pressure = p° = 0.1 MPa

T/K	$C_p°$ (J $K^{-1}mol^{-1}$)	S° (J $K^{-1}mol^{-1}$)	$-[G°-H°(T_r)]/T$ (J $K^{-1}mol^{-1}$)	$H°-H°(T_r)$ (kJ mol^{-1})	$\Delta_f H°$ (kJ mol^{-1})	$\Delta_f G°$ (kJ mol^{-1})	Log K_f
0	0.	0.	INFINITE	-9.402	22.087	22.087	INFINITE
100	29.709	204.809	269.641	-6.483	22.153	13.815	-7.216
200	32.987	226.411	243.147	-3.347	21.320	5.776	-1.509
250	34.180	233.907	240.573	-1.666	20.783	1.950	-0.407
298.15	34.990	240.001	240.001	0.	14.644	-0.967	0.169
300	35.015	240.218	240.002	0.065	14.607	-1.063	0.185
350	35.604	245.662	240.430	1.831	-0.811	-2.829	0.422
400	36.029	250.445	241.389	3.622	-0.810	-3.117	0.407
450	36.344	254.708	242.637	5.432	-0.809	-3.405	0.395
500	36.586	258.550	244.039	7.256	-0.808	-3.694	0.386
600	36.927	265.253	247.032	10.932	-0.807	-4.271	0.372
700	37.157	270.964	250.053	14.637	-0.807	-4.848	0.362
800	37.323	275.937	252.984	18.362	-0.809	-5.426	0.354
900	37.452	280.340	255.784	22.101	-0.812	-6.003	0.348
1000	37.557	284.292	258.441	25.851	-0.817	-6.579	0.344
1100	37.646	287.876	260.956	29.612	-0.823	-7.155	0.340
1200	37.724	291.155	263.338	33.380	-0.832	-7.730	0.336
1300	37.793	294.177	265.596	37.156	-0.842	-8.305	0.334
1400	37.858	296.980	267.738	40.939	-0.855	-8.878	0.331
1500	37.918	299.594	269.776	44.727	-0.871	-9.451	0.329
1600	37.974	302.043	271.717	48.522	-0.891	-10.022	0.327
1700	38.028	304.347	273.569	52.322	-0.916	-10.592	0.325
1800	38.081	306.522	275.340	56.128	-0.946	-11.160	0.324
1900	38.131	308.583	277.036	59.938	-0.984	-11.727	0.322
2000	38.180	310.540	278.663	63.754	-1.029	-12.291	0.321
2100	38.229	312.404	280.225	67.574	-1.085	-12.853	0.320
2200	38.276	314.183	281.729	71.400	-1.152	-13.412	0.318
2300	38.323	315.886	283.177	75.230	-1.232	-13.967	0.317
2400	38.369	317.518	284.574	79.064	-1.326	-14.519	0.316
2500	38.414	319.085	285.923	82.903	-1.434	-15.066	0.315
2600	38.460	320.592	287.228	86.747	-1.558	-15.609	0.314
2700	38.504	322.045	288.491	90.595	-1.698	-16.147	0.312
2800	38.549	323.446	289.714	94.448	-1.853	-16.680	0.311
2900	38.593	324.799	290.901	98.305	-2.023	-17.206	0.310
3000	38.637	326.108	292.053	102.166	-2.208	-17.726	0.309
3100	38.681	327.376	293.172	106.032	-2.407	-18.240	0.307
3200	38.724	328.605	294.260	109.903	-2.617	-18.748	0.306
3300	38.768	329.797	295.319	113.777	-2.838	-19.248	0.305
3400	38.811	330.955	296.350	117.656	-3.068	-19.742	0.303
3500	38.854	332.081	297.355	121.539	-3.303	-20.230	0.302
3600	38.897	333.176	298.335	125.427	-3.543	-20.710	0.300
3700	38.940	334.242	299.291	129.319	-3.784	-21.183	0.299
3800	38.983	335.281	300.225	133.215	-4.025	-21.650	0.298
3900	39.026	336.294	301.137	137.115	-4.263	-22.111	0.296
4000	39.068	337.283	302.028	141.020	-4.495	-22.566	0.295
4100	39.111	338.248	302.900	144.929	-4.719	-23.015	0.293
4200	39.154	339.191	303.752	148.842	-4.933	-23.458	0.292
4300	39.196	340.113	304.587	152.760	-5.135	-23.897	0.290
4400	39.239	341.015	305.405	156.682	-5.322	-24.331	0.289
4500	39.281	341.897	306.206	160.608	-5.493	-24.761	0.287
4600	39.323	342.761	306.992	164.538	-5.645	-25.187	0.286
4700	39.366	343.607	307.762	168.472	-5.777	-25.611	0.285
4800	39.408	344.436	308.517	172.411	-5.888	-26.032	0.283
4900	39.450	345.249	309.258	176.354	-5.975	-26.450	0.282
5000	39.492	346.046	309.986	180.301	-6.039	-26.868	0.281
5100	39.535	346.829	310.701	184.252	-6.077	-27.284	0.279
5200	39.577	347.597	311.403	188.208	-6.089	-27.700	0.278
5300	39.619	348.351	312.093	192.168	-6.074	-28.115	0.277
5400	39.661	349.092	312.772	196.132	-6.032	-28.531	0.276
5500	39.703	349.820	313.439	200.100	-5.961	-28.948	0.275
5600	39.745	350.536	314.095	204.072	-5.861	-29.367	0.274
5700	39.788	351.240	314.740	208.049	-5.732	-29.788	0.273
5800	39.830	351.932	315.375	212.030	-5.574	-30.212	0.272
5900	39.872	352.614	316.001	216.015	-5.387	-30.638	0.271
6000	39.914	353.284	316.617	220.004	-5.170	-31.068	0.270

PREVIOUS: September 1965 (1 atm) CURRENT: September 1965 (1 bar)

Bromine Chloride (BrCl) $Br_1Cl_1(g)$

BROMINE FLUORIDE (BrF) IDEAL GAS M_r = 98.902403

$S°(298.15\ K) = 228.967\ J\ K^{-1}\ mol^{-1}$

$\Delta_f H°(0\ K) = -50.815 \pm 1.7\ kJ\ mol^{-1}$
$\Delta_f H°(298.15\ K) = -58.463 \pm 1.7\ kJ\ mol^{-1}$

Electronic Level and Quantum Weight

State	ε_i, cm^{-1}	g_i
$^1\Sigma^+$	0	1

$\omega_e = 673\ cm^{-1}$ $\omega_e x_e = 4\ cm^{-1}$ $\sigma = 1$

$B_e = 0.356319\ cm^{-1}$ $\alpha_e = 0.005206\ cm^{-1}$ $r_e = 1.7555$ Å

Enthalpy of Formation

The equilibrium constant for the reaction $Br_2(g) + BrF_3(g) = 3BrF(g)$ has been measured over the temperature range 328-380 K by Steunenberg et al. (1). 3rd law analysis of the data yields a $\Delta_r H°(298.15\ K) = 11.785\pm0.2\ kcal\ mol^{-1}$, with a drift of $-0.6\pm1.2\ cal\ K^{-1}\ mol^{-1}$. 2nd law analysis yields $\Delta_r H°(298.15\ K) = 11.98\pm0.45\ kcal\ mol^{-1}$, thus it is obvious that the data is thermodynamically self consistent. There is an uncertainty in the entropy of $BrF_3(g)$ of ±1 cal $K^{-1}\ mol^{-1}$ but this corresponds only to ±0.35 kcal mol^{-1} in $\Delta_r H°(298.15\ K)$, thus a best value of $\Delta_r H°(298.15\ K) = 11.78\pm0.5\ kcal\ mol^{-1}$ is adopted. This yields $\Delta_f H°(BrF, g, 298.15\ K) = -13.973\pm0.4\ kcal\ mol^{-1}$ ($-58.463\pm1.7\ kJ\ mol^{-1}$), the uncertainty in the enthalpy of formation being a maximum value comprising the sum of the uncertainties in Br_2, BrF_3 and $\Delta_r H°$.

Attempts have been made to analyze the spectra of BrF to obtain a dissociation energy. Brodersen and Sicre (2), obtained two dissociation limits for BrF of 65.62 kcal mol^{-1} and 54.98 kcal mol^{-1}. The difference corresponds to the $Br(^2P_{1/2}) - Br(^2P_{3/2})$ excitation energy, indicating that the upper state goes to $F(^2P_{3/2})$ and $Br(^2P_{1/2})$. This yields an enthalpy of formation of BrF of -10.23 kcal mol^{-1}. Since this value is well outside the possible uncertainty limits of the equilibrium data we must conclude that the extrapolation is in error. Evans et al. (3) have also analyzed the spectra and conclude that an enthalpy of formation of -14.4 kcal mol^{-1} is possible. However, Evans (4) now believes that the dissociating state was too strongly perturbed for meaningful extrapolation.

Heat Capacity and Entropy

All molecular constants were obtained from Evans et al. (3), except r_e and the ground state configuration which were taken from Herzberg (5).

References

1. R. K. Steunenberg, R. C. Vogel and J. Fischer, J. Amer. Chem. Soc. 79 1320 (1957).
2. P. H. Brodersen and J. E. Sicre, Z. Physik 141, 515 (1955).
3. W. H. Evans, T. R. Munson and D. D. Wagman, J. Res. Nat. Bur. Stand. 55, 147 (1955).
4. W. H. Evans, U. S. Nat. Bur. Stand., personal communication, Nov. 1965.
5. G. Herzberg, "Spectra of Diatomic Molecules", D. Van Nostrand Co., New York, 1950.

Bromine Fluoride (BrF) $Br_1F_1(g)$

Enthalpy Reference Temperature = T_r = 298.15 K Standard State Pressure = p° = 0.1 MPa

	J K⁻¹mol⁻¹			kJ mol⁻¹			
T/K	$C_p°$	S°	$-[G°-H°(T_r)]/T$	$H°-H°(T_r)$	$\Delta_f H°$	$\Delta_f G°$	Log K_f
0	0.	0.	INFINITE	-9.019	-50.815	-50.815	INFINITE
100	29.162	195.677	256.772	-6.109	-50.755	-59.112	30.877
200	30.760	216.261	231.914	-3.131	-51.692	-67.137	17.534
250	31.940	223.252	229.504	-1.563	-52.283	-70.931	14.820
298.15	32.956	228.967	228.967	0.	-58.463	-73.809	12.931
300	32.991	229.171	228.968	0.061	-58.501	-73.904	12.868
350	33.852	234.324	229.373	1.733	-73.949	-75.622	11.286
400	34.537	238.891	230.283	3.443	-73.968	-75.859	9.906
450	35.080	242.991	231.471	5.184	-73.982	-76.095	8.833
500	35.513	246.710	232.812	6.949	-73.991	-76.329	7.974
600	36.148	253.245	235.688	10.535	-74.002	-76.796	6.686
700	36.584	258.852	238.606	14.172	-74.009	-77.261	5.765
800	36.900	263.759	241.450	17.847	-74.015	-77.725	5.075
900	37.140	268.119	244.175	21.550	-74.022	-78.188	4.538
1000	37.330	272.043	246.769	25.274	-74.030	-78.651	4.108
1100	37.487	275.608	249.231	29.015	-74.041	-79.112	3.757
1200	37.620	278.876	251.567	32.770	-74.054	-79.573	3.464
1300	37.736	281.892	253.785	36.538	-74.071	-80.032	3.216
1400	37.840	284.692	255.894	40.317	-74.092	-80.490	3.003
1500	37.934	287.306	257.902	44.106	-74.118	-80.946	2.819
1600	38.021	289.757	259.817	47.903	-74.149	-81.400	2.657
1700	38.102	292.064	261.647	51.710	-74.186	-81.852	2.515
1800	38.179	294.244	263.398	55.524	-74.229	-82.302	2.388
1900	38.252	296.311	265.076	59.345	-74.278	-82.749	2.275
2000	38.323	298.275	266.688	63.174	-74.332	-83.193	2.173
2100	38.391	300.146	268.237	67.010	-74.392	-83.635	2.080
2200	38.457	301.933	269.728	70.852	-74.457	-84.073	1.996
2300	38.522	303.644	271.166	74.701	-74.524	-84.509	1.919
2400	38.585	305.285	272.553	78.556	-74.595	-84.942	1.849
2500	38.647	306.862	273.894	82.418	-74.665	-85.371	1.784
2600	38.708	308.379	275.192	86.286	-74.735	-85.798	1.724
2700	38.769	309.841	276.448	90.160	-74.802	-86.222	1.668
2800	38.828	311.252	277.666	94.040	-74.865	-86.644	1.616
2900	38.887	312.615	278.848	97.925	-74.920	-87.064	1.568
3000	38.946	313.934	279.995	101.817	-74.967	-87.482	1.523
3100	39.003	315.212	281.111	105.714	-75.003	-87.898	1.481
3200	39.061	316.452	282.196	109.618	-75.025	-88.314	1.442
3300	39.118	317.654	283.252	113.527	-75.033	-88.729	1.404
3400	39.175	318.823	284.282	117.441	-75.023	-89.144	1.370
3500	39.232	319.960	285.285	121.362	-74.995	-89.560	1.337
3600	39.288	321.065	286.263	125.288	-74.946	-89.977	1.306
3700	39.344	322.143	287.219	129.219	-74.875	-90.395	1.276
3800	39.400	323.193	288.152	133.156	-74.781	-90.816	1.248
3900	39.455	324.217	289.063	137.099	-74.662	-91.239	1.222
4000	39.511	325.216	289.955	141.047	-74.517	-91.666	1.197
4100	39.566	326.193	290.827	145.001	-74.345	-92.097	1.173
4200	39.622	327.147	291.680	148.961	-74.145	-92.532	1.151
4300	39.677	328.080	292.516	152.926	-73.917	-92.973	1.129
4400	39.732	328.993	293.334	156.896	-73.660	-93.419	1.109
4500	39.786	329.886	294.137	160.872	-73.374	-93.871	1.090
4600	39.841	330.761	294.923	164.853	-73.058	-94.330	1.071
4700	39.896	331.619	295.695	168.840	-72.712	-94.796	1.054
4800	39.951	332.459	296.452	172.832	-72.336	-95.270	1.037
4900	40.005	333.283	297.196	176.830	-71.930	-95.752	1.021
5000	40.060	334.092	297.925	180.833	-71.494	-96.242	1.005
5100	40.114	334.886	298.642	184.842	-71.029	-96.742	0.991
5200	40.168	335.665	299.347	188.856	-70.533	-97.251	0.977
5300	40.223	336.431	300.039	192.876	-70.008	-97.769	0.964
5400	40.277	337.183	300.720	196.901	-69.454	-98.298	0.951
5500	40.331	337.923	301.390	200.931	-68.871	-98.838	0.939
5600	40.385	338.650	302.049	204.967	-68.260	-99.388	0.927
5700	40.439	339.365	302.697	209.008	-67.621	-99.950	0.916
5800	40.494	340.069	303.336	213.055	-66.955	-100.523	0.905
5900	40.548	340.762	303.964	217.107	-66.261	-101.107	0.895
6000	40.602	341.444	304.583	221.164	-65.541	-101.704	0.885

PREVIOUS: September 1965 (1 atm) CURRENT: September 1965 (1 bar)

Bromine Fluoride (BrF) $Br_1F_1(g)$

BROMINE FLUORIDE (BrF_3) IDEAL GAS M_r = 136.899209

$S°(298.15\ K) = 292.397\ J\ K^{-1}\ mol^{-1}$

$\Delta_fH°(0\ K) = -244.37 \pm 2.9\ kJ\ mol^{-1}$

$\Delta_fH°(298.15\ K) = -255.59 \pm 2.9\ kJ\ mol^{-1}$

Vibrational Frequencies and Degeneracies

ν, cm^{-1}	ν, cm^{-1}
674 (1)	613 (1)
[528](1)	[384](1)
[300](1)	[289](1)

Ground State Quantum Weight: 1 $\sigma = 2$

Point Group: C_{2v}

Bond Distances: $Br-F_1 = Br-F_3 = 1.8061$ Å $Br-F_2 = 1.721$ Å

Bond Angles: $F_1-Br-F_2 = 86°\ 12.6'$ $F_2-Br-F_3 = 86°\ 12.6'$ $F_3-Br-F_1 - 187°\ 34.8'$

Product of the Moment of Inertia: $I_AI_BI_C = 4.48703 \times 10^{-114}\ g^3\ cm^6$

Enthalpy for Formation

$\Delta_fH°(298.15\ K) = -255.588 \pm 2.9\ kJ\ mol^{-1}$ is from Stein (1) and was derived from enthaply of reaction of F_2 with Br_2 which were measured in an adiabatic calorimeter. Stein's value for the $\Delta_fH°(298.15\ K) = -271.040\ kJ\ mol^{-1}$ was based upon gaseous Br_2, which has been adjusted to the liquid Br_2 reference state at 298.15 K.

Heat Capacity and Entropy

Magnuson (2) determined the moments of inertia and quadrupole coupling coefficients of both $^{79}BrF_3$ and $^{81}BrF_3$ from a number of microwave transitions. From the moment of inertia the bromine trifluoride molecule was found to have a distorted "T" structure with one short, 1.721 Å, and two long, 1.810 Å Br-F bonds. The FBrF angle was found to be 86° 12.6'.

Claasen et al. (3), observed two of the six vibrational levels in the vapor, and calculated the other four levels from a normal coordinate treatment of the molecule. They also obtained a measured entropy of BrF_3(g), from low temperature data combined with vapor pressure data and an approximate correction for dimerization. This value is 300.9 $J\ K^{-1}\ mol^{-1}$ at 316.27 K compared with a value of 296.42 $J\ K^{-1}\ mol^{-1}$ from this table, the difference is probably within the uncertainty of the data. The principal moments of inertia are: $I_A = 7.7525 \times 10^{-39}$, $I_B = 20.4920 \times 10^{-39}$ and $I_C = 28.2445 \times 10^{-39}\ g\ cm^2$.

References

1. L. Stein, J. Phys. Chem. 66, 288 (1962).
2. D. W. Magnuson, Carbide and Carbon Chemicals Co., K-25 Plant, Oak Ridge, Tennessee, Report No. K-1179, December, 1954.
3. H. H. Classen, B. Weinstock, and J. G. Malm, J. Chem. Phys. 28, 285 (1958).

Bromine Fluoride (BrF_3) $Br_1F_3(g)$

Enthalpy Reference Temperature = T_r = 298.15 K Standard State Pressure = p° = 0.1 MPa

T/K	$C_p°$ (J K⁻¹mol⁻¹)	S° (J K⁻¹mol⁻¹)	−[G°−H°(T_r)]/T (J K⁻¹mol⁻¹)	H°−H°(T_r) (kJ mol⁻¹)	$\Delta_fH°$ (kJ mol⁻¹)	$\Delta_fG°$ (kJ mol⁻¹)	Log K_f
0	0.	0.	INFINITE	-14.275	-244.370	-244.370	INFINITE
100	39.106	235.530	343.752	-10.822	-246.675	-241.980	126.398
200	55.795	267.920	298.179	-6.052	-248.749	-236.395	61.740
250	62.017	281.071	293.468	-3.099	-249.458	-233.222	48.729
298.15	66.497	292.397	292.397	0.	-255.588	-229.383	40.187
300	66.644	292.808	292.398	0.123	-255.622	-229.221	39.911
350	70.048	303.351	293.223	3.545	-270.908	-223.983	33.428
400	72.571	312.877	295.094	7.113	-270.700	-217.293	28.376
450	74.468	321.539	297.558	10.791	-270.444	-210.632	24.450
500	75.918	329.464	300.358	14.553	-270.155	-204.002	21.312
600	77.935	343.498	306.409	22.253	-269.525	-190.829	16.613
700	79.228	355.616	312.593	30.116	-268.859	-177.765	13.265
800	80.100	366.255	318.649	38.085	-268.180	-164.798	10.760
900	80.713	375.727	324.475	46.127	-267.501	-151.916	8.817
1000	81.160	384.256	330.034	54.222	-266.828	-139.110	7.266
1100	81.495	392.007	335.321	62.355	-266.166	-126.370	6.001
1200	81.752	399.110	340.345	70.518	-265.518	-113.690	4.949
1300	81.954	405.662	345.120	78.704	-264.887	-101.063	4.061
1400	82.115	411.741	349.664	86.908	-264.274	-88.485	3.301
1500	82.246	417.411	353.994	95.126	-263.680	-75.949	2.645
1600	82.353	422.723	358.125	103.356	-263.108	-63.453	2.072
1700	82.443	427.718	362.073	111.596	-262.557	-50.991	1.567
1800	82.518	432.433	365.853	119.844	-262.025	-38.561	1.119
1900	82.581	436.896	369.475	128.099	-261.512	-26.161	0.719
2000	82.636	441.133	372.953	136.360	-261.016	-13.787	0.360
2100	82.683	445.166	376.297	144.626	-260.533	-1.438	0.036
2200	82.723	449.014	379.515	152.897	-260.058	10.889	-0.259
2300	82.759	452.692	382.617	161.171	-259.589	23.195	-0.527
2400	82.790	456.214	385.611	169.448	-259.121	35.480	-0.772
2500	82.818	459.595	388.503	177.729	-258.647	47.745	-0.998
2600	82.842	462.843	391.300	186.012	-258.164	59.991	-1.205
2700	82.864	465.970	394.008	194.297	-257.666	72.218	-1.397
2800	82.884	468.984	396.633	202.584	-257.149	84.427	-1.575
2900	82.901	471.893	399.178	210.874	-256.606	96.617	-1.740
3000	82.917	474.704	401.649	219.165	-256.035	108.787	-1.894
3100	82.931	477.423	404.050	227.457	-255.430	120.937	-2.038
3200	82.944	480.056	406.384	235.751	-254.788	133.068	-2.172
3300	82.956	482.609	408.655	244.046	-254.105	145.179	-2.298
3400	82.967	485.085	410.867	252.342	-253.379	157.267	-2.416
3500	82.977	487.490	413.022	260.639	-252.605	169.334	-2.527
3600	82.986	489.828	415.123	268.937	-251.783	181.378	-2.632
3700	82.995	492.102	417.173	277.236	-250.909	193.398	-2.730
3800	83.002	494.315	419.174	285.536	-249.983	205.394	-2.823
3900	83.010	496.471	421.129	293.837	-249.002	217.365	-2.911
4000	83.016	498.573	423.039	302.138	-247.966	229.310	-2.994
4100	83.022	500.623	424.906	310.440	-246.875	241.229	-3.073
4200	83.028	502.624	426.733	318.743	-245.727	253.120	-3.148
4300	83.033	504.578	428.520	327.046	-244.522	264.983	-3.219
4400	83.038	506.487	430.271	335.349	-243.261	276.817	-3.286
4500	83.043	508.353	431.985	343.653	-241.943	288.623	-3.350
4600	83.047	510.178	433.665	351.958	-240.569	300.398	-3.411
4700	83.051	511.964	435.312	360.263	-239.139	312.143	-3.469
4800	83.055	513.713	436.928	368.568	-237.654	323.857	-3.524
4900	83.059	515.425	438.512	376.874	-236.115	335.539	-3.577
5000	83.062	517.103	440.067	385.180	-234.522	347.189	-3.627
5100	83.065	518.748	441.594	393.486	-232.877	358.807	-3.675
5200	83.068	520.361	443.093	401.793	-231.180	370.393	-3.721
5300	83.071	521.943	444.566	410.100	-229.433	381.945	-3.764
5400	83.074	523.496	446.013	418.407	-227.636	393.464	-3.806
5500	83.076	525.021	447.436	426.715	-225.790	404.948	-3.846
5600	83.079	526.517	448.835	435.023	-223.898	416.399	-3.884
5700	83.081	527.988	450.211	443.331	-221.959	427.816	-3.920
5800	83.083	529.433	451.564	451.639	-219.975	439.198	-3.955
5900	83.085	530.853	452.896	459.947	-217.948	450.546	-3.989
6000	83.087	532.250	454.207	468.256	-215.878	461.859	-4.021

PREVIOUS: September 1965 (1 atm) CURRENT: September 1965 (1 bar)

Bromine Fluoride (BrF_3) $Br_1F_3(g)$

BROMINE FLUORIDE (BrF_5) IDEAL GAS M_r = 174.896015

$\Delta_f H°(0\ K) = -413.58 \pm 2.1\ kJ\ mol^{-1}$

$S°(298.15\ K) = 323.682\ J\ K^{-1}\ mol^{-1}$ $\Delta_f H°(298.15\ K) = -428.72 \pm 2.1\ kJ\ mol^{-1}$

Vibrational Frequencies and Degeneracies

ν, cm⁻¹	ν, cm⁻¹	ν, cm⁻¹
683 (1)	535 (1)	644 (2)
587 (1)	[281](1)	415 (2)
369 (1)	312 (1)	245 (2)

Ground State Quantum Weight: 1 σ = 4

Point Group: C_{4v}

Bond Distances: Br-F = 1.79 Å Br-F* = 1.68 Å

Bond Angles: F-Br-F = [90]° and F*-Br-F = [90]°

(* axial)

Product of the Moments of Inertia: $I_A I_B I_C = 3.20483 \times 10^{-113}\ g^3\ cm^6$

Enthalpy of Formation

$\Delta_f H°(298.15\ K) = -428.718 \pm 2.1\ kJ\ mol^{-1}$ is from Stein (1) and was derived from enthalpies of reaction of F_2 with Br_2 which were measured in an adiabatic calorimeter. Stein's value for the $\Delta_f H°(298.15\ K) = -444.17\ kJ\ mol^{-1}$ was based upon gaseous Br_2 which has been adjusted to the liquid Br_2 reference state at 298.15 K.

Heat Capacity and Entropy

Begun et al. (2) report the infrared (gas phase) and Raman (liquid phase) spectra and correlate the observations by means of normal coordinate calculations for the similar molecules BrF_5, IF_5, ClF_5 and $XeOF_4$. The fundamental frequencies are taken from Begun et al. (2) except for ν_9 which was beyond the range of their infrared measurements. The value $\nu_9 = 245\ cm^{-1}$ for the gas has been observed by McDowell and Asprey (3) and is confirmed by 237 cm^{-1} found in the Raman spectra of the liquid. Raman values from the liquid were used for the infrared inactive fundamentals ν_4 and ν_6. One Raman active frequency, persumably ν_5, is not observed in any of the four molecules. The value $\nu_5 = 281\ cm^{-1}$ was obtained from the normal coordinate calculation. Previous Raman studies were reported by Stephenson and Jones (4).

A tetragonal pyramidal structure (C_{4v} symmetry) with the above parameters was used by Begun et al. (2) in the analysis of the spectra. This symmetry is consistent with the NMR spectra of Gutowsky and Hoffman (5), with the dipole moment determined by Rogers et al. (6), and with the X-ray crystal structure obtained by Robinson and Bensey (7). Bond distances were based on those in the crystal and the bond angle was assumed to be 90°, slightly larger than those in the crystal (80.5-86.5°). Begun et al. (2) indicate that the angle in the gaseous molecule is probably not exactly 90° but that the normal coordinate calculations are not sensitive to this angle. The three principal moments of inertia are: $I_A = 40.4333 \times 10^{-39}$ and $I_B = I_C = 28.1536 \times 10^{-39}\ g\ cm^2$.

References

1. L. Stein, J. Phys. Chem. 66, 288 (1962).
2. G. M. Begun, W. H. Fletcher and D. F. Smith, J. Chem. Phys. 42, 2236 (1965).
3. R. S. McDowell and L. B. Asprey, J. Chem. Phys. 37, 165 (1962).
4. C. V. Stephenson and E. A. Jones, J. Chem. Phys. 20, 1830 (1952).
5. H. S. Gutowsky and C. J. Hoffman, J. Chem. Phys. 19, 1259 (1951).
6. M. T. Rogers, R. D. Pruett, H. B. Thompson and J. L. Speirs, J. Amer. Chem. Soc. 78, 44 (1956).
7. R. D. Robinson and F. N. Bensey, Jr., J. Chem. Phys. 27, 982 (1957).

Enthalpy Reference Temperature = T_r = 298.15 K Standard State Pressure = p° = 0.1 MPa

	J K⁻¹mol⁻¹			kJ mol⁻¹			
T/K	$C_p°$	S°	−[G°−H°(T_r)]/T	H°−H°(T_r)	$\Delta_f H°$	$\Delta_f G°$	Log K_f
0	0.	0.	INFINITE	-19.180	-413.581	-413.581	INFINITE
100	47.334	243.616	398.823	-15.521	-418.586	-397.662	207.717
200	80.797	287.176	332.382	-9.041	-421.879	-375.246	98.004
250	92.816	306.562	325.307	-4.686	-422.688	-363.487	75.946
298.15	101.404	323.682	323.682	0.	-428.718	-351.379	61.560
300	101.685	324.310	323.684	0.188	-428.745	-350.899	61.097
350	108.182	340.499	324.948	5.443	-443.787	-337.105	50.310
400	112.985	355.274	327.829	10.978	-443.243	-321.901	42.036
450	116.590	368.799	331.640	16.721	-442.588	-306.771	35.609
500	119.343	381.232	335.986	22.623	-441.858	-291.719	30.476
600	123.169	403.356	345.417	34.763	-440.261	-261.839	22.795
700	125.618	422.540	355.095	47.211	-438.561	-232.235	17.330
800	127.270	439.429	364.603	59.861	-436.812	-202.879	13.247
900	128.431	454.490	373.768	72.649	-435.041	-173.744	10.084
1000	129.277	468.067	382.531	85.537	-433.265	-144.806	7.564
1100	129.911	480.420	390.877	98.497	-431.496	-116.046	5.511
1200	130.398	491.745	398.817	111.514	-429.740	-87.447	3.806
1300	130.780	502.198	406.373	124.573	-428.004	-58.993	2.370
1400	131.085	511.902	413.568	137.667	-426.291	-30.672	1.144
1500	131.332	520.954	420.429	150.789	-424.606	-2.472	0.086
1600	131.535	529.437	426.979	163.932	-422.948	25.616	-0.836
1700	131.704	537.416	433.243	177.094	-421.320	53.601	-1.647
1800	131.846	544.949	439.242	190.272	-419.719	81.491	-2.365
1900	131.966	552.080	444.995	203.463	-418.142	109.293	-3.005
2000	132.069	558.852	450.520	216.665	-416.586	137.013	-3.578
2100	132.158	565.298	455.833	229.876	-415.044	164.654	-4.096
2200	132.235	571.448	460.950	243.096	-413.510	192.223	-4.564
2300	132.302	577.327	465.883	256.323	-411.976	219.723	-4.990
2400	132.361	582.959	470.644	269.556	-410.435	247.155	-5.379
2500	132.413	588.364	475.246	282.795	-408.878	274.522	-5.736
2600	132.460	593.558	479.697	296.038	-407.297	301.827	-6.064
2700	132.501	598.558	484.007	309.287	-405.682	329.071	-6.366
2800	132.538	603.377	488.185	322.539	-404.028	356.253	-6.646
2900	132.571	608.029	492.238	335.794	-402.325	383.377	-6.905
3000	132.601	612.524	496.173	349.053	-400.567	410.440	-7.146
3100	132.628	616.872	499.997	362.314	-398.748	437.443	-7.371
3200	132.653	621.083	503.715	375.578	-396.861	464.387	-7.580
3300	132.676	625.166	507.334	388.845	-394.903	491.271	-7.776
3400	132.696	629.127	510.858	402.113	-392.868	518.095	-7.960
3500	132.715	632.973	514.292	415.384	-390.753	544.857	-8.132
3600	132.732	636.712	517.641	428.656	-388.555	571.559	-8.293
3700	132.748	640.349	520.909	441.930	-386.271	598.197	-8.445
3800	132.763	643.890	524.099	455.206	-383.900	624.772	-8.588
3900	132.776	647.338	527.215	468.483	-381.439	651.285	-8.723
4000	132.789	650.700	530.260	481.761	-378.889	677.733	-8.850
4100	132.800	653.979	533.238	495.040	-376.248	704.115	-8.971
4200	132.811	657.180	536.151	508.321	-373.516	730.433	-9.084
4300	132.821	660.305	539.002	521.603	-370.695	756.684	-9.192
4400	132.831	663.358	541.794	534.885	-367.783	782.869	-9.294
4500	132.839	666.344	544.528	548.169	-364.782	808.987	-9.390
4600	132.848	669.263	547.208	561.453	-361.693	835.036	-9.482
4700	132.855	672.120	549.836	574.738	-358.518	861.018	-9.569
4800	132.862	674.918	552.413	588.024	-355.256	886.932	-9.652
4900	132.869	677.657	554.941	601.311	-351.911	912.776	-9.730
5000	132.876	680.342	557.422	614.598	-348.483	938.551	-9.805
5100	132.882	682.973	559.858	627.886	-344.975	964.257	-9.876
5200	132.887	685.553	562.250	641.174	-341.387	989.894	-9.944
5300	132.893	688.085	564.601	654.463	-337.723	1015.461	-10.008
5400	132.898	690.569	566.911	667.753	-333.983	1040.957	-10.069
5500	132.902	693.007	569.181	681.043	-330.169	1066.383	-10.128
5600	132.907	695.402	571.414	694.333	-326.285	1091.741	-10.183
5700	132.911	697.754	573.610	707.624	-322.330	1117.027	-10.236
5800	132.915	700.066	575.770	720.916	-318.308	1142.243	-10.287
5900	132.919	702.338	577.896	734.207	-314.220	1167.392	-10.335
6000	132.923	704.572	579.989	747.499	-310.068	1192.468	-10.381

PREVIOUS: September 1965 (1 atm) CURRENT: September 1965 (1 bar)

SULFUR BROMIDE FLUORIDE ($BrSF_5$) IDEAL GAS M_r = 206.956015

$S°(298.15 K) = [333.648 \pm 2.1] J K^{-1} mol^{-1}$

$\Delta_f H°(0 K) = [-953.5 \pm 59] kJ mol^{-1}$

$\Delta_f H°(298.15 K) = [-972.8 \pm 59] kJ mol^{-1}$

Vibrational Frequencies and Degeneracies

ν, cm^{-1}	ν, cm^{-1}	ν, cm^{-1}
848 (1)	621 (1)	580 (2)
694 (1)	[325](1)	423 (2)
597 (1)	502 (1)	225 (2)
279 (1)	892 (2)	

Ground State Quantum Weight: 1 σ = 4

Point Group: C_{4v}

Bond Distances: S-F = 1.597 Å S-Br = 2.1902 Å

Bond Angles: Br-S-F* = [92°] F*-S-F* = 90°

(* equatorial)

Product of Moments of Inertia: $I_A I_B I_C = [1.66563 \times 10^{-112}] g^3 cm^6$

Enthalpy of Formation

$\Delta_f H°(BrSF_5, g, 0 K) = -953.529 \pm 59 kJ mol^{-1}$ is calculated from a S-Br bond energy of $173.0 \pm 42 kJ mol^{-1}$ by combining $D_0^{\,0}$ with JANAF enthalpies of formation (1) for SF_5 and Br. We estimate the strength of the S-Br bond in $BrSF_5$ from bond energy correlations. In the case of related sulfur chlorine molecules, we calculate that the ratio of the S-Cl bond strength in $ClSF_5$ (1) is 1.1 times the mean S-Cl bond energy in $SOCl_2$ (2). Assuming that this relationship holds for $BrSF_5$ and $SOBr_2$, we obtain the adopted value for $D_0^{\,0}(SF_5\text{-}Br)$ from $D_0^{\,0}(SOBr\text{-}Br)$, 157.3 kJ mol^{-1}.

The enthalpy of atomization and mean S-F bond energy are calculated to be $1730.5 \pm 67 kJ mol^{-1}$ and 311.3 kJ mol^{-1}, respectively. The latter value is identical with that found for $ClSF_5$, indicating that the sulfur fluorine bonding is very similar in these two molecules. This conclusion is supported by spectroscopic data (1, 3, 4).

Heat Capacity and Entropy

Microwave (3) and vibrational spectroscopic (4) measurements indicate that $BrSF_5$ is a symmetrical top molecule possessing C_{4v} symmetry. We adopt the structural parameters proposed by Neuvar and Jache (3) from an analysis of the observed rotational spectrum. The adopted bond lengths were calculated assuming that the Br-S-F* bond angle was 92° which was estimated from the value (91.6°) for the Cl-S-F* angle in $ClSF_5$ reported by Kewley et al. (5). A more plausible value for the Cl-S-F* angle is 90.7±0.2 (1), indicating that the bond lengths reported by Neuvar and Jache (3) may be slightly biased. We place the values of S°(298.15 K) and $I_A I_B I_C$ in brackets to emphasize this uncertainty.

The vibrational frequencies are based on the infrared and laser-excited Raman spectra of $BrSF_5$ recorded under matrix (argon) isolation conditions by Smardzewski et al. (4). We adjust their reported frequencies for a matrix effect which is estimated as 5 cm^{-1} from gas phase (1) and argon-matrix (4) spectral data for $ClSF_5$. Evidence available for ClS_5 (1) indicates that ν_6 = 220 cm^{-1} by Smardzewski et al. (4) should be reassigned. We believe that this Raman band is the low frequency mode ν_{11} (S-Br wag) and that ν6 has not been resolved. We estimate ν_6 = 325 cm^{-1} from the force-field value for $ClSF_5$(1).

The principal moments of inertia are: $I_A = 32.1450 \times 10^{-39} g cm^2$ and $I_B = I_C = 71.9835 \times 10^{-39} g cm^2$.

References

1. JANAF Thermochemical Tables: SF_5(g) and $ClSF_5$(g), 12-31-77; Br(g), 6-30-74.
2. U. S. Nat. Bur. Stand. Tech. Note 270-3, 1968.
3. E. W. Neuvar and A. W. Jache, J. Chem. Phys. 39, 596 (1963).
4. R. R. Smardzewski, R. E. Noftle, and W. B. Fox, J. Mol. Spectrosc. 62, 449 (1976).
5. R. Kewley, K. S. R. Murty, and T. M. Sugden, Trans. Faraday Soc. 56, 1732 (1960).

Sulfur Bromide Fluoride ($SBrF_5$) $Br_1F_5S_1$(g)

Enthalpy Reference Temperature = T_r = 298.15 K Standard State Pressure = p° = 0.1 MPa

	J $K^{-1}mol^{-1}$			kJ mol^{-1}			
T/K	$C_p°$	S°	-[G°-H°(T_r)]/T	H°-H°(T_r)	$\Delta_f H°$	$\Delta_f G°$	Log K_f
0	0.	0.	INFINITE	-19.433	-953.526	-953.526	INFINITE
100	47.160	252.860	410.417	-15.756	-959.203	-937.951	489.936
200	81.206	296.049	342.686	-9.327	-964.189	-914.603	238.870
250	95.727	315.781	335.351	-4.893	-965.938	-901.996	188.462
298.15	107.073	333.648	333.648	0.	-972.822	-888.897	155.731
300	107.459	334.312	333.650	0.198	-972.880	-888.376	154.680
350	116.682	351.599	334.995	5.811	-988.731	-873.408	130.349
400	123.876	367.669	338.087	11.833	-991.131	-856.830	111.891
450	129.504	382.598	342.213	18.173	-991.805	-840.009	97.506
500	133.940	396.481	346.954	24.763	-992.388	-823.108	85.990
600	140.315	421.507	357.341	38.499	-992.781	-789.203	68.706
700	144.531	443.474	368.109	52.755	-992.620	-755.280	56.360
800	147.434	462.974	378.772	67.362	-992.125	-721.405	47.103
900	149.507	480.466	389.116	82.215	-1044.546	-686.525	39.845
1000	151.033	496.301	399.055	97.245	-1042.483	-646.855	33.788
1100	152.186	510.752	408.562	112.409	-1040.383	-607.393	28.843
1200	153.077	524.034	417.639	127.674	-1038.265	-568.124	24.730
1300	153.779	536.315	426.302	143.018	-1036.147	-529.031	21.257
1400	154.341	547.733	434.572	158.425	-1034.040	-490.102	18.286
1500	154.798	558.397	442.476	173.882	-1031.951	-451.322	15.716
1600	155.175	568.400	450.037	189.382	-1029.886	-412.681	13.473
1700	155.488	577.817	457.279	204.915	-1027.849	-374.168	11.497
1800	155.753	586.713	464.225	220.478	-1025.840	-335.774	9.744
1900	155.977	595.140	470.895	236.065	-1023.857	-297.492	8.179
2000	156.169	603.145	477.309	251.672	-1021.896	-259.312	6.773
2100	156.335	610.769	483.485	267.297	-1019.953	-221.232	5.503
2200	156.479	618.045	489.437	282.938	-1018.020	-183.243	4.351
2300	156.605	625.004	495.181	298.593	-1016.091	-145.341	3.301
2400	156.715	631.671	500.730	314.259	-1014.158	-107.524	2.340
2500	156.813	638.071	506.096	329.935	-1012.212	-69.787	1.458
2600	156.900	644.223	511.291	345.621	-1010.244	-32.129	0.645
2700	156.977	650.146	516.325	361.315	-1008.246	5.453	-0.105
2800	157.047	655.856	521.207	377.016	-1006.210	42.960	-0.801
2900	157.109	661.368	525.946	392.724	-1004.128	80.394	-1.448
3000	157.166	666.695	530.549	408.438	-1001.994	117.755	-2.050
3100	157.217	671.849	535.024	424.157	-999.800	155.044	-2.612
3200	157.263	676.841	539.379	439.881	-997.541	192.261	-3.138
3300	157.305	681.681	543.618	455.609	-995.213	229.407	-3.631
3400	157.344	686.378	547.748	471.342	-992.810	266.480	-4.094
3500	157.379	690.940	551.774	487.078	-990.330	303.482	-4.529
3600	157.412	695.373	555.702	502.818	-987.769	340.413	-4.939
3700	157.441	699.687	559.535	518.560	-985.125	377.271	-5.326
3800	157.469	703.886	563.279	534.306	-982.397	414.055	-5.692
3900	157.494	707.977	566.937	550.054	-979.582	450.768	-6.037
4000	157.518	711.964	570.513	565.805	-976.681	487.407	-6.365
4100	157.540	715.854	574.011	581.558	-973.693	523.972	-6.675
4200	157.560	719.651	577.433	597.313	-970.618	560.464	-6.970
4300	157.579	723.358	580.784	613.070	-967.457	596.880	-7.251
4400	157.597	726.981	584.066	628.828	-964.211	633.223	-7.517
4500	157.613	730.523	587.281	644.589	-960.880	669.491	-7.771
4600	157.629	733.987	590.433	660.351	-957.467	705.684	-8.013
4700	157.643	737.378	593.523	676.115	-953.971	741.802	-8.244
4800	157.657	740.697	596.555	691.880	-950.396	777.845	-8.465
4900	157.670	743.948	599.530	707.646	-946.742	813.812	-8.675
5000	157.682	747.133	602.450	723.414	-943.011	849.704	-8.877
5100	157.693	750.256	605.318	739.182	-939.207	885.521	-9.070
5200	157.704	753.318	608.135	754.952	-935.329	921.262	-9.254
5300	157.714	756.322	610.902	770.723	-931.381	956.928	-9.431
5400	157.723	759.270	613.623	786.495	-927.364	992.519	-9.601
5500	157.732	762.164	616.297	802.268	-923.280	1028.034	-9.763
5600	157.741	765.006	618.928	818.041	-919.132	1063.476	-9.920
5700	157.749	767.798	621.515	833.816	-914.922	1098.842	-10.070
5800	157.757	770.542	624.061	849.591	-910.650	1134.133	-10.214
5900	157.764	773.239	626.566	865.367	-906.320	1169.352	-10.353
6000	157.771	775.890	629.033	881.144	-901.933	1204.496	-10.486

PREVIOUS: December 1977 (1 atm) CURRENT: December 1977 (1 bar)

Sulfur Bromide Fluoride ($SBrF_5$) $Br_1F_5S_1$(g)

HYDROGEN BROMIDE (HBr) IDEAL GAS M_r = 80.91194

$\Delta_f H°(0\ K) = -28.60 \pm 0.25\ kJ\ mol^{-1}$

$S°(298.15\ K) = 198.699 \pm 0.033\ J\ K^{-1}\ mol^{-1}$ $\Delta_f H°(298.15\ K) = -36.44 \pm 0.17\ kJ\ mol^{-1}$

Electronic Level and Quantum Weight

State	ε_i, cm^{-1}	g_i
$^1\Sigma^+$	0	1

$\omega_e = 2649.1816 \pm 0.007\ cm^{-1}$ $\omega_e x_e = 45.2245 \pm 0.003\ cm^{-1}$ $\sigma = 1$

$B_e = 8.466203 \pm 0.000025\ cm^{-1}$ $\alpha_e = 0.233332 \pm 0.00004\ cm^{-1}$ $r_e = 1.414435$ Å

Enthaply of Formation

For HBr(g), CODATA (1) recommended the key value $\Delta_f H°(298.15\ K) = -36.38 \pm 0.17\ kJ\ mol^{-1}$ which is adopted. References 2-9 were cited by CODATA.

Heat Capacity and Entropy

The values of ω_e, $\omega_e x_e$, B_e, and α_e are taken from Rank et al. (10) after correction to the averge isotopic species. See also James and Thibault (11), Jones and Gordy (12), Plyler (13), and Mould et al. (14). The ground state configuration is from Huber and Herzberg (15) where molecular constants for HBr(g) are also given.

Our evaluations lead to $S°(298.15\ K) = 198.589\ J\ K^{-1}\ mol^{-1}$ in agreement with the CODATA recommended value (1) of 198.585 ± 0.033 $J\ K^{-1}\ mol^{-1}$, both values based on a reference pressure of one atm. The value adopted, $S°(298.15\ K) = 198.694 \pm 0.033\ J\ K^{-1}\ mol^{-1}$, is the CODATA value corrected, by 0.109 $J\ K^{-1} mol^{-1}$, to a reference pressure of one bar.

References

1. ISCU - CODATA Task Group, "CODATA Recommended Key Values for Thermodynamics, 1977," CODATA Bulletin No. 28, 1978; J. Chem. Thermodynamics 10, 903 (1978).
2. W. H. Johnson and S. Sunner, Acta Chem. Scand. 17, 1917 (1963).
3. W. H. Johnson and J. R. Ambrose, J. Res. Nat. Bur. Stand 67A, 427 (1963).
4. S. Sunner and S. Thoren, Acta Chem. Scand. 18, 1328 (1964).
5. J. R. Lacher, L. Casali, and J. D. Park, J. Phys. Chem. 60, 607 (1956).
6. P. B. Howard and H. A. Skinner, J. Chem. Soc. 1966, 1536 (1966).
7. C. E. Vanderzee and J. D. Nutter, J. Phys. Chem. 67, 2521 (1963).
8. W. A. Roth and A. Bertram, Z. Elektrochem. 43, 376 (1937).
9. J. Thomsen, "Thermochemische Untersuchungen," J. Barth, Leipzig, 1884.
10. D. H. Rank, U. Fink, and T. A. Wiggins, J. Mol. Spectrosc. 18, 170 (1965).
11. T. C. James and R. J. Thibault, J. Chem. Phys. 46, 1450 (1965).
12. G. Jones and W. Gordy, Phys. Rev. 136, A 1229 (1964).
13. E. K. Plyler, J. Res. Nat. Bur. Stand. 64, 377 (1960).
14. H. M. Mould, W. C. Price, and G. R. Wilkinson, Spectrochim. Acta 16, 479 (1960).
15. K. P. Huber and G. Herzberg, "Constants of Diatomic Molecules," Van Nostrand Reinhold, New York, 1979.

Hydrogen Bromide (HBr) $Br_1H_1(g)$

Enthalpy Reference Temperature = T_r = 298.15 K Standard State Pressure = p° = 0.1 MPa

T/K	$C_p°$ (J K⁻¹mol⁻¹)	S° (J K⁻¹mol⁻¹)	$-[G°-H°(T_r)]/T$ (J K⁻¹mol⁻¹)	$H°-H°(T_r)$ (kJ mol⁻¹)	$\Delta_f H°$ (kJ mol⁻¹)	$\Delta_f G°$ (kJ mol⁻¹)	Log K_f
0	0.	0.	INFINITE	-8.648	-28.603	-28.603	INFINITE
100	29.115	166.884	224.595	-5.771	-28.622	-37.581	19.630
200	29.124	187.067	201.363	-2.859	-29.507	-46.237	12.076
250	29.130	193.567	199.178	-1.403	-30.157	-50.348	10.520
298.15	29.141	198.699	198.699	0.	-36.443	-53.513	9.375
300	29.141	198.879	198.699	0.054	-36.485	-53.619	9.336
350	29.167	203.373	199.054	1.512	-52.077	-55.615	8.300
400	29.220	207.270	199.843	2.971	-52.261	-56.108	7.327
450	29.313	210.717	200.863	4.434	-52.449	-56.577	6.567
500	29.453	213.812	202.006	5.903	-52.636	-57.026	5.957
600	29.870	219.216	204.437	8.868	-52.996	-57.869	5.038
700	30.427	223.861	206.887	11.882	-53.320	-58.655	4.377
800	31.055	227.965	209.270	14.956	-53.599	-59.398	3.878
900	31.698	231.660	211.556	18.093	-53.830	-60.109	3.489
1000	32.319	235.032	213.737	21.294	-54.018	-60.796	3.176
1100	32.897	238.140	215.816	24.556	-54.168	-61.466	2.919
1200	33.425	241.025	217.798	27.872	-54.287	-62.124	2.704
1300	33.902	243.720	219.690	31.239	-54.381	-62.773	2.522
1400	34.329	246.248	221.497	34.651	-54.455	-63.416	2.366
1500	34.711	248.630	223.228	38.103	-54.517	-64.054	2.231
1600	35.053	250.881	224.886	41.592	-54.568	-64.688	2.112
1700	35.360	253.015	226.479	45.113	-54.614	-65.319	2.007
1800	35.635	255.044	228.010	48.663	-54.658	-65.947	1.914
1900	35.884	256.978	229.484	52.239	-54.703	-66.573	1.830
2000	36.109	258.824	230.905	55.838	-54.751	-67.196	1.755
2100	36.313	260.591	232.277	59.460	-54.805	-67.817	1.687
2200	36.500	262.285	233.603	63.101	-54.865	-68.436	1.625
2300	36.672	263.911	234.885	66.759	-54.935	-69.051	1.568
2400	36.830	265.475	236.128	70.434	-55.014	-69.663	1.516
2500	36.976	266.982	237.332	74.125	-55.104	-70.272	1.468
2600	37.111	268.435	238.500	77.829	-55.205	-70.877	1.424
2700	37.238	269.838	239.635	81.547	-55.317	-71.477	1.383
2800	37.357	271.194	240.738	85.277	-55.440	-72.074	1.345
2900	37.468	272.507	241.811	89.018	-55.574	-72.665	1.309
3000	37.573	273.779	242.856	92.770	-55.719	-73.252	1.275
3100	37.673	275.013	243.873	96.532	-55.872	-73.834	1.244
3200	37.767	276.210	244.865	100.304	-56.034	-74.411	1.215
3300	37.857	277.374	245.833	104.086	-56.203	-74.983	1.187
3400	37.943	278.505	246.777	107.876	-56.378	-75.549	1.161
3500	38.025	279.606	247.699	111.674	-56.558	-76.110	1.136
3600	38.104	280.678	248.601	115.481	-56.742	-76.666	1.112
3700	38.181	281.724	249.482	119.295	-56.928	-77.217	1.090
3800	38.254	282.743	250.344	123.117	-57.116	-77.763	1.069
3900	38.325	283.737	251.187	126.946	-57.304	-78.304	1.049
4000	38.394	284.709	252.013	130.782	-57.492	-78.840	1.030
4100	38.461	285.657	252.822	134.624	-57.677	-79.372	1.011
4200	38.526	286.585	253.615	138.474	-57.860	-79.898	0.994
4300	38.589	287.492	254.392	142.329	-58.039	-80.421	0.977
4400	38.651	288.380	255.155	146.191	-58.214	-80.940	0.961
4500	38.711	289.249	255.903	150.059	-58.384	-81.454	0.945
4600	38.770	290.101	256.637	153.934	-58.548	-81.965	0.931
4700	38.828	290.935	257.358	157.814	-58.705	-82.472	0.917
4800	38.885	291.753	258.066	161.699	-58.856	-82.976	0.903
4900	38.941	292.556	258.762	165.591	-58.999	-83.478	0.890
5000	38.996	293.343	259.446	169.487	-59.134	-83.975	0.877
5100	39.050	294.116	260.118	173.390	-59.260	-84.471	0.865
5200	39.103	294.875	260.779	177.297	-59.378	-84.964	0.853
5300	39.155	295.620	261.429	181.210	-59.487	-85.455	0.842
5400	39.207	296.352	262.069	185.128	-59.585	-85.944	0.831
5500	39.258	297.072	262.699	189.052	-59.675	-86.432	0.821
5600	39.309	297.780	263.319	192.980	-59.753	-86.917	0.811
5700	39.359	298.476	263.930	196.913	-59.822	-87.402	0.801
5800	39.408	299.161	264.531	200.852	-59.879	-87.885	0.791
5900	39.457	299.835	265.124	204.795	-59.925	-88.368	0.782
6000	39.506	300.499	265.708	208.743	-59.960	-88.850	0.774

PREVIOUS: September 1965 (1 atm) CURRENT: September 1965 (1 bar)

Hydrogen Bromide (HBr) $Br_1H_1(g)$

$S°(298.15\ K) = 262.47 \pm 0.21\ J\ K^{-1}\ mol^{-1}$

$\Delta_fH°(0\ K) = [-61.84 \pm 17]\ kJ\ mol^{-1}$
$\Delta_fH°(298.15\ K) = [-78.24 \pm 17]\ kJ\ mol^{-1}$

Vibrational Frequencies and Degeneracies

ν, cm^{-1}	ν, cm^{-1}
2200 (1)	2209.2(2)
930.3(1)	946.3(2)
430.6(1)	632.6(2)

Ground State Quantum Weight: [1]   σ = 3
Point Group: C_{3v}
Bond Distances: Si-H = 1.486 ± 0.008 Å Si-Br = 2.210 Å
Bond Angles: H-Si-H = 110.42° ± 0.5° H-Si-Br = 108.5° ± 0.5°
Product of the Moments of Inertia: $I_AI_BI_C = 3.79963 \times 10^{-115}\ g^3\ cm^6$

Enthalpy for Formation

$\Delta_fH°(298.15\ K)$ is estimated by linear interpolation between the values (1) of $SiBr_4(g)$ and $SiH_4(g)$. The only experimental value for $SiHBr_3(g)$ appears to be too uncertain to justify a nonlinear interpolation such as that adopted for SiH_3Cl (1). Normally we would seek a comparison with $\Delta_fH°$ values of CH_nBr_{4-n}. These were recently reviewed by Kudchadker and Kudchadker (2). They adopted a cubic variation of $\Delta_fH°$ with n, based on $\Delta_fH°$ values of $CHBr_3$ and CBr_4 selected by Wagman et al. (3). We presume that Wagman's values are calculated rather than experimental. Uncertainty in these values (4-6) precludes their use as a comparison for SiH_nBr_{4-n}. We conclude, as did Hunt and Sirtl (15), that the available data justify only linear interpolation of $\Delta_fH°$.

Heat Capacity and Entropy

The molecular structure is based on microwave data of Kewley et al. (7) for isotopic forms of the four monohalosilanes. The authors concluded that the silyl group is effectively the same in all four halides. We adopt effective (r_o) structural parameters rather than substitutional (r_s) parameters. These are in reasonable agreement with previous structural determinations (8). The principal moments of inertia are $I_A = 0.9971 \times 10^{-39}$ and $I_B = I_C = 19.5210 \times 10^{-39}\ g\ cm^2$.

Vibrational frequencies are from gas-phase infrared spectra (9) measured with a grating spectrometer by Buerger et al. Assignments were based on analysis of band contours and rotational fine structure. These results supersede Shimanouchi's selected frequencies (10) which derive from lower resolution spectra. The new data agree with unpublished data cited by Ball et al. (11).

We neglect excited states and assume the electronic ground state to be $^1A_1$ by analogy with CH_3Br (12). SiH_3Br has several high-lying electronic absorptions and a diffuse band with a maximum at ~52000 cm^{-1} (13). Comparison (12, 13) with CH_3Br (12,14) suggests that the lowest excited state in SiH_3Br is near 40000 cm^{-1}. Contributions due to excited states should be unimportant.

References

1. JANAF Thermochemical Tables: $Br_4Si(g)$, $Br_3HSi(g)$, $ClH_3Si(g)$, 12-31-76; $H_4Si(g)$ 6-30-76.
2. S. A. Kudchadker and A. P. Kudchadker, J. Phys. Chem. Ref. Data 4, 457 (1975).
3. D. D. Wagman et al., U. S. Natl. Bur. Std. Tech. Note 270-3, 1968.
4. H. J. Bernstein, J. Phys. Chem. 69, 1550 (1965).
5. S. Furuyama, D. M. Golden and S. W. Benson, J. Amer. Chem. Soc. 91, 7564 (1969).
6. K. D. King, D. M. Golden and S. W. Benson, J. Phys. Chem. 75, 987 (1971).
7. R. Kewley, P. M. McKinney and A. G. Robiette, J. Mol. Spectrosc. 34, 390 (1970).
8. L. E. Sutton, "Interatomic Distances," Spec. Publ. No. 11, 1958; "Interatomic Distances Supplement," Spec. Publ. No. 18, The Chemical Society, London, 1965.
9. H. Buerger, J. Cichon and A. Ruoff, Spectrochim. Acta A30, 223 (1974); Z. Naturforsch. B26, 1068 (1971).
10. T. Shimanouchi, NSRDS-NBS 39, 30 (1972).
11. D. F. Ball, M. J. Buttler and D. C. McKean, Spectrochim. Acta 21, 451 (1965).
12. G. Herzberg, "Electronic Spectra of Polyatomic Molecules," D. Van Nostrand Co., Inc., New York, 1966.
13. G. C. Causley, J. B. Clark and B. R. Russell, Chem. Phys. Lett. 38, 602 (1976).
14. G. C. Causley and B. R. Russell, J. Chem. Phys. 62, 848 (1975).
15. L. P. Hunt and E. Sirtl, J. Electrochem. Soc. 120, 806 (1973).

Enthalpy Reference Temperature = T_r = 298.15 K   Standard State Pressure = p° = 0.1 MPa

T/K	$C_p°$ (J $K^{-1}mol^{-1}$)	S°	$-[G°-H°(T_r)]/T$	$H°-H°(T_r)$ (kJ mol^{-1})	$\Delta_fH°$	$\Delta_fG°$	Log K_f
0	0.	0.	INFINITE	-11.769	-61.836	-61.836	INFINITE
100	34.071	217.875	302.187	-8.431	-64.660	-68.263	35.657
200	42.240	243.595	266.964	-4.674	-68.572	-70.393	18.385
250	47.756	253.612	263.308	-2.424	-70.675	-70.606	14.752
298.15	52.865	262.466	262.466	0.	-78.241	-69.751	12.220
300	53.052	262.794	262.467	0.098	-78.330	-69.698	12.135
350	57.817	271.337	263.131	2.872	-95.089	-67.304	10.045
400	62.019	279.337	264.662	5.870	-96.278	-63.252	8.260
450	65.736	286.861	266.715	9.066	-97.318	-59.060	6.856
500	69.060	293.962	269.088	12.437	-98.219	-54.760	5.721
600	74.795	307.077	274.345	19.639	-99.646	-45.926	3.998
700	79.577	318.976	279.884	27.365	-100.645	-36.888	2.753
800	83.586	329.872	285.461	35.529	-101.295	-27.732	1.811
900	86.942	339.917	290.961	44.060	-101.669	-18.512	1.074
1000	89.747	349.227	296.328	52.899	-101.834	-9.261	0.484
1100	92.092	357.894	301.535	61.994	-101.845	-0.002	0.
1200	94.058	365.993	306.573	71.304	-101.747	9.252	-0.403
1300	95.712	373.589	311.439	80.795	-101.577	18.496	-0.743
1400	97.111	380.735	316.136	90.438	-101.368	27.724	-1.034
1500	98.299	387.476	320.669	100.210	-101.141	36.937	-1.286
1600	99.315	393.854	325.046	110.092	-100.917	46.135	-1.506
1700	100.188	399.902	329.273	120.069	-150.886	55.766	-1.713
1800	100.942	405.650	333.358	130.126	-150.476	67 911	-1.971
1900	101.597	411.126	337.308	140.254	-150.062	80.032	-2.200
2000	102.169	416.352	341.130	150.443	-149.650	92.132	-2.406
2100	102.671	421.349	344.832	160.685	-149.247	104.211	-2.592
2200	103.114	426.136	348.420	170.975	-148.858	116.271	-2.761
2300	103.505	430.728	351.899	181.306	-148.485	128.314	-2.914
2400	103.854	435.141	355.276	191.675	-148.132	140.341	-3.054
2500	104.165	439.387	358.556	202.076	-147.801	152.354	-3.183
2600	104.443	443.478	361.744	212.506	-147.493	164.353	-3.302
2700	104.694	447.424	364.845	222.964	-147.209	176.342	-3.412
2800	104.920	451.236	367.863	233.444	-146.951	188.321	-3.513
2900	105.124	454.921	370.802	243.947	-146.717	200.291	-3.608
3000	105.310	458.488	373.665	254.469	-146.508	212.253	-3.696
3100	105.478	461.944	376.458	265.008	-146.324	224.208	-3.778
3200	105.633	465.295	379.182	275.564	-146.164	236.158	-3.855
3300	105.774	468.548	381.841	286.134	-146.027	248.104	-3.927
3400	105.903	471.708	384.438	296.718	-145.912	260.045	-3.995
3500	106.022	474.779	386.975	307.314	-145.818	271.984	-4.059
3600	106.132	477.767	389.456	317.922	-529.897	294.381	-4.271
3700	106.233	480.677	391.882	328.540	-529.430	317.271	-4.479
3800	106.326	483.511	394.256	339.168	-528.983	340.148	-4.676
3900	106.413	486.274	396.580	349.806	-528.555	363.015	-4.862
4000	106.494	488.969	398.857	360.451	-528.145	385.870	-5.039
4100	106.569	491.600	401.087	371.104	-527.752	408.716	-5.207
4200	106.638	494.169	403.272	381.764	-527.374	431.552	-5.367
4300	106.704	496.679	405.416	392.432	-527.010	454.379	-5.520
4400	106.764	499.133	407.518	403.105	-526.660	477.198	-5.665
4500	106.821	501.532	409.580	413.784	-526.323	500.009	-5.804
4600	106.875	503.881	411.605	424.469	-525.997	522.813	-5.937
4700	106.925	506.180	413.593	435.159	-525.683	545.610	-6.064
4800	106.972	508.432	415.545	445.854	-525.378	568.400	-6.185
4900	107.016	510.638	417.463	456.553	-525.082	591.184	-6.302
5000	107.058	512.800	419.349	467.257	-524.795	613.962	-6.414
5100	107.097	514.921	421.202	477.965	-524.515	636.734	-6.521
5200	107.134	517.000	423.024	488.677	-524.243	659.501	-6.625
5300	107.169	519.042	424.817	499.392	-523.976	682.263	-6.724
5400	107.203	521.045	426.580	510.110	-523.715	705.020	-6.820
5500	107.234	523.012	428.316	520.832	-523.457	727.771	-6.912
5600	107.264	524.945	430.024	531.557	-523.204	750.519	-7.001
5700	107.292	526.844	431.706	542.285	-522.953	773.262	-7.086
5800	107.319	528.710	433.362	553.016	-522.705	796.001	-7.169
5900	107.345	530.545	434.994	563.749	-522.457	818.734	-7.249
6000	107.369	532.349	436.602	574.484	-522.209	841.464	-7.326

PREVIOUS. December 1976 (1 atm)   CURRENT: December 1976 (1 bar)

AMMONIUM BROMIDE (NH_4Br) CRYSTAL M_r = 97.94246

$S°(298.15\ K) = 112.84\ J\ K^{-1}\ mol^{-1}$
$T_{trs} = 411\ K$
$T_{fus} = 815.2\ K$

$\Delta_f H°(0\ K) = -255.0 \pm 1.3\ kJ\ mol^{-1}$
$\Delta_f H°(298.15\ K) = -271.5 \pm 1.3\ kJ\ mol^{-1}$
$\Delta_{trs} H° = 3.222\ kJ\ mol^{-1}$
$\Delta_{fus} H°$ = Unknown

Enthalpy of Formation

The equilibrium pressures for the reaction $NH_4Br(cr) \rightarrow NH_3(g) + HBr(g)$ have been measured by several investigators. Using the densi-tensimeter, Smits and Purcell (2), determined both the equilibrium pressures and vapor densities of the decomposition products, simultaneously, at several temperatures. In order to check the density values obtained, the authors applied the "extrapolation method", from which it was calculated that the corresponding vapor density agreed with complete dissociation. Based on this conclusion, the reported equilibrium pressures were employed to evaluate the enthalpy changes ($\Delta_r H°$) of the decomposition reaction by both the 2nd and 3rd law methods. The results obtained are presented as follows.

Source	T/K	$\Delta_r H°(298.15\ K)$, kcal mol^{-1} 2nd Law	3rd Law	Equipment Used
Smith and Calvert (1)	576.40-676.02	45.12±0.13	44.93	Isoteniscope
Smits and Purcell (2)	604.95-668.15	45.50±0.17	45.08	Densi-tensimeter
Johnson (3)	573.15-667.75	45.41±0.44	44.85	Spiral Manometer

The value of $\Delta_r H°(298.15\ K)$ adopted is 45.2±0.2 kcal mol^{-1}. Using $\Delta_f H°(298.15\ K) = -10.97$ and -8.71 kcal mol^{-1} for $NH_3(g)$ and HBr(g), respectively, the value of $\Delta_f H°(298.15\ K)$ for $NH_4Br(cr)$ is evaluated as -64.9±0.3 kcal mol^{-1} (-271.542±1.3 kJ mol^{-1}). The corresponding $\Delta_r H°(298.15\ K)$ value evaluated from solution data, selected by Parker (4) is 45.5±0.2 kcal mol^{-1}.

The vapor densities (573-661 K) of dissociation products of $NH_4Br(cr)$ were also measured by Smith and Lombard (5). Using the equilibrium pressures reported by Smith and Calvert (1), they derived the corresponding degrees of dissociation of $NH_4Br(cr)$ at different temperatures. The calculated results indicated that only 48% of $NH_4Br(cr)$ decomposed at 573 K and as reaction temperatures increased to 661 K the degree of dissocation decreased to 10%. This conclusion was stated to be erroneous by Smits and Purcell (2) probably because of incorrect density determinations.

Heat Capacity and Entropy

The low temperature (13-305 K) heat capacities and S°(298.15 K) were obtained from Sorai et al. (6). The $C_p°$ values above 305 K were estimated by comparison with those for NaCl(cr), NaBr(cr) and $NH_4Cl(cr)$. The low temperature heat capacities were also measured by Ewald (7), 139-301 K, and by Simon et al. (8), 201.2-277.1 K.

Transition Data

The temperature (T_{trs}) and enthalpy of transition ($\Delta_{trs}H°$) were given by Bridgman (9). T_{trs} = 419.5±0.6 K was reported Markowitz and Boryta (10), which was determined by differential thermal analysis. According to Sorai et al. (6), there are three solid phase transitions, namely Phase IV (CsCl) → Phase III (tetragonal) at 108 K, Phase III (tetragonal) → Phase II (CsCl) at 235 K and Phase II (CsCl) → Phase I (NaCl) at 411 K.

Fusion Data

The value of T_{fus} was taken from Rassow (11), determined under pressure.

Decomposition Data

The temperature of decomposition was calculated as 670 K, i.e. the temperature at which the total pressure of the decomposition products for the reaction $NH_4Br(cr) \rightarrow NH_3(g) + HBr(g)$ equals 1 atm. Values of 674 and 668.4 K were reported by Luft (12) and Markowitz and Boryta (10), respectively.

References

1. A. Smith and R. P. Calvert, J. Amer. Chem. Soc. 36, 1363 (1914).
2. A. Smits and R. Purcell, J. Chem. Soc. 2944 (1928).
3. F. M. G. Johnson, Z. Physik. Chem. 65, 38 (1909).
4. V. B. Parker, U.S. Nat. Bur. Stand., NSRDS-NBS 2, 1965.
5. A. Smith and R. H. Lombard, J. Amer. Chem. Soc. 37, 38 (1915).
6. M. Sorai, H. Suga and S. Seki, Bull. Chem. Soc. Japan, 38, 1125 (1965).
7. R. Ewald, Ann. Physik, 44, 1213 (1914).
8. F. Simon, C. V. Simson and M. Ruhemann, Z. Phys. Chem. 129, 339 (1927).
9. P. W. Bridgman, Proc. Am. Acad. Arts Sci. 52, 91 (1916-17).
10. M. M. Markowitz and D. A. Boryta, J. Phys. Chem. 66, 1477 (1962).
11. H. Rassow, Z. Anorg. Chem. 114, 117 (1920).
12. N. W. Luft, Ind. Chemist, 31, 502 (1955).

Ammonium Bromide (NH_4Br) $Br_1H_4N_1(cr)$

Enthalpy Reference Temperature = T_r = 298.15 K; Standard State Pressure = p° = 0.1 MPa

T/K	$C_p°$ (J $K^{-1}mol^{-1}$)	S°	$-[G°-H°(T_r)]/T$	$H°-H°(T_r)$ (kJ mol^{-1})	$\Delta_f H°$	$\Delta_f G°$	Log K_f
0	0.	0.	INFINITE	-16.954	-254.971	-254.971	INFINITE
100	44.283	35.690	183.887	-14.820	-261.683	-234.424	122.450
200	78.680	77.864	120.813	-8.590	-264.747	-205.814	53.753
298.15	88.676	112.842	112.842	0.	-271.542	-176.007	30.836
300	88.893	113.392	112.844	0.164	-271.581	-175.414	30.542
400	92.717	139.666	116.401	9.306	-286.950	-140.189	18.307
411.000	92.844	142.183	117.057	10.327		II <--> I	
411.000	72.867	150.022	117.057	13.548		TRANSITION	
500	87.195	165.676	124.323	20.677	-284.740	-104.074	10.873
600	102.424	182.959	132.660	30.179	-284.447	-67.944	5.915
700	115.060	199.708	141.050	41.061	-282.832	-31.971	2.386
800	126.775	215.856	149.397	53.167	-280.062	3.690	-0.241
900	136.817	231.376	157.650	66.353	-276.295	38.943	-2.260
1000	146.022	246.276	165.772	80.503	-271.660	73.728	-3.851
1100	154.102	260.577	173.745	95.515	-266.266	108.012	-5.129
1200	161.502	274.308	181.557	111.301	-260.207	141.775	-6.171
1300	168.127	287.500	189.203	127.787	-253.560	175.007	-7.032
1400	174.264	300.187	196.680	144.910	-246.386	207.708	-7.750
1500	179.912	312.406	203.990	162.623	-238.733	239.879	-8.353

PREVIOUS: March 1962 CURRENT: September 1965

Ammonium Bromide (NH_4Br) $Br_1H_4N_1(cr)$

MERCURY BROMIDE (HgBr) IDEAL GAS $M_r = 280.494$

$S°(298.15\ K) = 271.538\ J\ K^{-1}\ mol^{-1}$

$\Delta_f H°(0\ K) = 115.5 \pm 38\ kJ\ mol^{-1}$
$\Delta_f H°(298.15\ K) = 104.2 \pm 38\ kJ\ mol^{-1}$

Electronic Level and Quantum Weight

State	ε_i, cm^{-1}	g_i
$^2\Sigma^+$	0	2

$\omega_e = 187.29\ cm^{-1}$ $\omega_e x_e = 0.9665\ cm^{-1}$ $\sigma = 1$

$B_e = [0.0545]\ cm^{-1}$ $\alpha_e = [0.0003]\ cm^{-1}$ $r_e = [2.33]$ Å

Enthalpy of Formation

Gaydon (1) gives 0.7±0.4 eV for the dissociation energy into gaseous atoms.

Heat Capacity and Entropy

Vibrational constants from Wieland (2) were weighted according to their natural isotopic abundances. The bond length was estimated to be 0.965 of the mercuric bond length by analogy with the mercury chlorides. Rotational constants were estimated (3).

References

1. A. G. Gaydon, "Dissociation Energies and Spectra of Diatomic Molecules," 2nd Ed., Chapman and Hall, London, 1953.
2. K. Wieland, Z. Elektrochem. 64, 761 (1960).
3. G. Herzberg, "Spectra of Diatomic Molecules," van Nostrand, New York, 1950; Eqns. III, 115, p. 106 and III, 123, p. 108.

Mercury Bromide (HgBr) $Br_1Hg_1(g)$

Enthalpy Reference Temperature = T_r = 298.15 K Standard State Pressure = p° = 0.1 MPa

T/K	$C_p°$ (J K⁻¹mol⁻¹)	S° (J K⁻¹mol⁻¹)	$-[G°-H°(T_r)]/T$ (J K⁻¹mol⁻¹)	$H°-H°(T_r)$ (kJ mol⁻¹)	$\Delta_f H°$ (kJ mol⁻¹)	$\Delta_f G°$ (kJ mol⁻¹)	Log K_f
0	0.	0.	INFINITE	-10.252	115.527	115.527	INFINITE
100	33.918	232.289	304.036	-7.175	115.502	98.688	-51.549
200	36.493	256.811	274.923	-3.622	114.019	82.403	-21.522
250	36.940	265.007	272.149	-1.786	110.751	74.762	-15.621
298.15	37.210	271.538	271.538	0.	104.182	68.581	-12.015
300	37.218	271.768	271.538	0.069	104.129	68.360	-11.902
350	37.413	277.521	271.992	1.935	88.280	63.279	-9.444
400	37.561	282.526	273.003	3.809	87.864	59.736	-7.801
450	37.681	286.958	274.312	5.690	87.457	56.245	-6.529
500	37.783	290.933	275.779	7.577	87.058	52.798	-5.516
600	37.956	297.837	278.897	11.364	86.271	46.020	-4.006
700	38.105	303.700	282.032	15.168	26.732	45.945	-3.428
800	38.242	308.797	285.066	18.985	26.594	48.699	-3.180
900	38.372	313.309	287.958	22.816	26.464	51.470	-2.987
1000	38.497	317.358	290.699	26.659	26.342	54.255	-2.834
1100	38.619	321.033	293.292	30.515	26.228	57.051	-2.709
1200	38.738	324.399	295.746	34.383	26.121	59.858	-2.606
1300	38.857	327.504	298.071	38.263	26.022	62.674	-2.518
1400	38.974	330.388	300.278	42.154	25.930	65.497	-2.444
1500	39.090	333.081	302.376	46.057	25.845	68.326	-2.379
1600	39.206	335.607	304.375	49.972	25.767	71.161	-2.323
1700	39.322	337.988	306.283	53.899	25.694	74.000	-2.274
1800	39.437	340.238	308.107	57.837	25.627	76.843	-2.230
1900	39.552	342.374	309.855	61.786	25.563	79.690	-2.191
2000	39.666	344.405	311.532	65.747	25.503	82.541	-2.156
2100	39.780	346.343	313.144	69.719	25.444	85.394	-2.124
2200	39.895	348.197	314.695	73.703	25.388	88.250	-2.095
2300	40.009	349.973	316.191	77.698	25.331	91.109	-2.069
2400	40.123	351.678	317.634	81.705	25.276	93.970	-2.045
2500	40.237	353.318	319.029	85.723	25.220	96.834	-2.023
2600	40.350	354.898	320.378	89.752	25.164	99.699	-2.003
2700	40.464	356.423	321.685	93.793	25.108	102.567	-1.984
2800	40.578	357.897	322.952	97.845	25.054	105.437	-1.967
2900	40.691	359.323	324.182	101.908	25.000	108.309	-1.951
3000	40.805	360.704	325.377	105.983	24.948	111.182	-1.936
3100	40.919	362.044	326.538	110.069	24.900	114.057	-1.922
3200	41.032	363.345	327.668	114.167	24.856	116.934	-1.909
3300	41.146	364.609	328.768	118.276	24.818	119.812	-1.896
3400	41.259	365.839	329.841	122.396	24.787	122.691	-1.885
3500	41.372	367.037	330.886	126.527	24.763	125.571	-1.874
3600	41.486	368.204	331.907	130.670	24.750	128.452	-1.864
3700	41.599	369.342	332.903	134.825	24.747	131.332	-1.854
3800	41.713	370.453	333.877	138.990	24.756	134.213	-1.845
3900	41.826	371.538	334.829	143.167	24.778	137.093	-1.836
4000	41.939	372.599	335.760	147.355	24.815	139.972	-1.828
4100	42.053	373.635	336.671	151.555	24.868	142.851	-1.820
4200	42.166	374.650	337.563	155.766	24.936	145.728	-1.812
4300	42.279	375.644	338.437	159.988	25.022	148.603	-1.805
4400	42.393	376.617	339.294	164.222	25.126	151.475	-1.798
4500	42.506	377.571	340.134	168.467	25.249	154.346	-1.792
4600	42.619	378.506	340.958	172.723	25.391	157.213	-1.785
4700	42.733	379.424	341.767	176.991	25.552	160.077	-1.779
4800	42.846	380.325	342.561	181.270	25.733	162.937	-1.773
4900	42.959	381.210	343.340	185.560	25.934	165.794	-1.767
5000	43.072	382.079	344.106	189.861	26.156	168.646	-1.762
5100	43.186	382.933	344.859	194.174	26.398	171.493	-1.756
5200	43.299	383.772	345.600	198.499	26.660	174.336	-1.751
5300	43.412	384.598	346.328	202.834	26.942	177.173	-1.746
5400	43.525	385.411	347.044	207.181	27.244	180.005	-1.741
5500	43.639	386.211	347.749	211.539	27.565	182.831	-1.736
5600	43.752	386.998	348.443	215.909	27.906	185.651	-1.732
5700	43.865	387.773	349.126	220.290	28.265	188.464	-1.727
5800	43.978	388.537	349.799	224.682	28.643	191.271	-1.723
5900	44.092	389.290	350.462	229.085	29.039	194.072	-1.718
6000	44.205	390.032	351.115	233.500	29.452	196.866	-1.714

PREVIOUS: December 1961 (1 atm) CURRENT: December 1961 (1 bar)

Mercury Bromide (HgBr) $Br_1Hg_1(g)$

$S°(298.15\ K) = 258.95\ J\ K^{-1}\ mol^{-1}$

$\Delta_f H°(0\ K) = 49.815 \pm 0.08\ kJ\ mol^{-1}$

$\Delta_f H°(298.15\ K) = 40.878 \pm 0.08\ kJ\ mol^{-1}$

Electronic Levels and Quantum Weights

State	ε_i, cm^{-1}	g_i	State	ε_i, cm^{-1}	g_i
X $^1\Sigma^+$	0	1	B' 0^+	16880	1
A $^3\Pi_1$	12230	6	C ($\Pi_{1,2}$)	[51677]	2
B $^3\Pi_{0^+}$	[16155]	5	D ($\Pi_{1,0}$)	[56349]	2

$\omega_e = 267.38\ cm^{-1}$ $\omega_e x_e = 0.774\ cm^{-1}$ $\sigma = 1$

$B_e = 0.05568\ cm^{-1}$ $\alpha_e = 0.00034\ cm^{-1}$ $r_e = 2.485$ Å

Enthalpy for Formation

The equilibrium constants for the reaction (A) $I_2(g) + Br_2(g) = 2\ IBr(g)$ were determined to be 0.01124 and 0.01303 at 547.7 and 578 K, respectively, by Muller (1) who studied the rate of reaction between $H_2(g)$ and $Br_2(g)$ in the presence of $I_2(g)$. Bodenstein and Schmidt (2) derived two values of the equilibrium constants at 1495 K for the same reaction by vapor density studies on the system I_2, I, Br_2, Br and BrI. The equilibrium between $CuBr_2(cr)$, $I_2(g)$, CuBr(cr) and IBr(g) was investigated by McMorris and Yost (3) and the equilibrium constants at 115.0, 151.2 and 176.0°C for reaction (A) were evaluated. Based on these reported equilibrium constants, the corresponding enthalpy changes for reaction (A) are calculated by both the 2nd and 3rd law methods.

Badger and Yost (4) observed the infrared bands of IBr and classified them as the $A^3\Pi_2$-X'Σ transition. They have shown that the dissociation products of the upper state are normal atoms. Brown (5) assigned a faint set of bands in the red as the $B^3\Pi_0 \leftarrow$ X'Σ system. By means of a vibrational analysis he found that the absorption spectrum of IBr is analogous to that of ICl. The dissociation energy for IBr was evaluated to be 14,660±5 cm^{-1} or 1.817 eV. Hence the enthalpy change for the reaction (B) IBr(g) = I(g) + Br(g) is calculated as 41.92 kcal mol^{-1}.

The enthaplies of formation for IBr(g) derived from the enthalpy changes for reaction (A) and (B) are presented in the following table. The value of $\Delta_f H°$(IBr, g, 298.15 K) adopted is 9.77±0.02 kcal mol^{-1} (40.878±0.08 kJ mol^{-1}).

Source	Reaction	T/K	Data Points	$\Delta_r H°$(298.15 K), kcal mol^{-1} 2nd Law	3rd Law	Drift cal K^{-1} mol^{-1}	$\Delta_f H°$(298.15 K) kcal mol^{-1}
Muller (1)	A	547.7, 578	2	-3.07	-3.38	-0.55	9.47
Bodenstein and Schmidt (2)	A	1495	1	-	-2.92	-	9.70
McMorris and Yost (3)	A	388.2-449.2	14	-2.46	-2.85	-0.9±0.4	9.73
Brown (5)	B	298	-	-	41.92**	-	9.77**

**The value is derived from D_0^0(I-Br).

Heat Capacity and Entropy

The ground state configuration, electronic levels, quantum weights, ω_e and $\omega_e x_e$ are obtained from Herzberg (6). The values of B_e, α_e(corrected to the average isotopic species) and r_e were reported by Jaseja (7) who analyzed the microwave spectrum of IBr, corresponding to the transitions J = 4 → 5 and J = 5 → 6, for the two isotopic species, $^{127}I^{79}Br$ and $^{127}I\ ^{81}Br$.

The infrared absorption spectrum of IBr at 8000 - 6800 Å, associated with a $^3\Pi_1 \leftarrow X^1\Sigma^+$ transition, was observed by Selin (8). The derived rotational constants are different from those reported by Jaseja (7).

References

1. W. Muller, Z. Physik, Chem. 123, 1 (1926).
2. M. Bodenstein and A. Schmidt, Z. Physik. Chem. 123, 28 (1926).
3. J. McMorris and D. M. Yost, J. Amer. Chem. Soc. 53, 2625 (1931).
4. R. M. Badger and D. M. Yost, Phys. Rev. 37, 1548 (1931).
5. W. G. Brown, Phys. Rev. 42, 355 (1932).
6. G. Herzberg, "Spectra of Diatomic Molecules," D. Van Nostrand, Inc., New York, 1950.
7. T. S. Jaseja, J. Mol. Spectrosc. 5, 445 (1960).
8. L. E. Selin, Arkiv Fysik 21, 479 (1962).

Enthalpy Reference Temperature = T_r = 298.15 K   Standard State Pressure = p° = 0.1 MPa

T/K	$C_p°$ (J $K^{-1}mol^{-1}$)	S°	-[G°-H°(T_r)]/T	H°-H°(T_r) (kJ mol^{-1})	$\Delta_f H°$	$\Delta_f G°$	Log K_f
0	0.	0.	INFINITE	-9.917	49.815	49.815	INFINITE
100	31.904	221.168	290.519	-6.935	49.875	33.500	-17.499
200	35.387	244.574	262.265	-3.538	48.355	17.678	-4.617
250	36.103	252.554	259.552	-1.749	47.422	10.114	-2.113
298.15	36.531	258.952	258.952	0.	40.878	3.675	-0.644
300	36.544	259.178	258.953	0.068	40.825	3.444	-0.600
350	36.836	264.835	259.399	1.903	24.929	-1.903	0.284
400	37.042	269.768	260.394	3.750	16.448	-5.428	0.709
450	37.196	274.140	261.683	5.606	15.367	-8.098	0.940
500	37.316	278.066	263.128	7.469	-5.777	-8.748	0.914
600	37.494	284.886	266.203	11.210	-5.773	-9.342	0.813
700	37.627	290.676	269.296	14.966	-5.769	-9.937	0.742
800	37.734	295.708	272.290	18.734	-5.767	-10.533	0.688
900	37.827	300.158	275.144	22.513	-5.766	-11.129	0.646
1000	37.911	304.147	277.848	26.300	-5.767	-11.724	0.612
1100	37.989	307.764	280.406	30.095	-5.771	-12.320	0.585
1200	38.066	311.073	282.825	33.897	-5.781	-12.915	0.562
1300	38.143	314.123	285.117	37.708	-5.798	-13.509	0.543
1400	38.227	316.953	287.291	41.526	-5.828	-14.101	0.526
1500	38.323	319.593	289.358	45.354	-5.874	-14.690	0.512
1600	38.437	322.070	291.326	49.191	-5.939	-15.276	0.499
1700	38.576	324.405	293.203	53.042	-6.027	-15.857	0.487
1800	38.748	326.614	294.999	56.908	-6.139	-16.432	0.477
1900	38.962	328.715	296.718	60.793	-6.273	-17.000	0.467
2000	39.223	330.720	298.369	64.702	-6.427	-17.561	0.459
2100	39.538	332.641	299.955	68.639	-6.596	-18.114	0.451
2200	39.911	334.488	301.483	72.611	-6.771	-18.658	0.443
2300	40.347	336.272	302.957	76.624	-6.944	-19.194	0.436
2400	40.845	337.999	304.382	80.683	-7.104	-19.724	0.429
2500	41.408	339.678	305.760	84.795	-7.239	-20.247	0.423
2600	42.033	341.314	307.096	88.966	-7.337	-20.765	0.417
2700	42.718	342.913	308.393	93.203	-7.386	-21.280	0.412
2800	43.457	344.480	309.654	97.512	-7.374	-21.795	0.407
2900	44.245	346.018	310.882	101.896	-7.290	-22.311	0.402
3000	45.076	347.532	312.078	106.362	-7.124	-22.832	0.398
3100	45.942	349.024	313.246	110.913	-6.866	-23.359	0.394
3200	46.834	350.497	314.387	115.551	-6.508	-23.897	0.390
3300	47.745	351.952	315.503	120.280	-6.044	-24.447	0.387
3400	48.664	353.391	316.597	125.101	-5.469	-25.013	0.384
3500	49.585	354.815	317.668	130.013	-4.779	-25.597	0.382
3600	50.497	356.224	318.720	135.017	-3.972	-26.203	0.380
3700	51.393	357.620	319.752	140.112	-3.046	-26.833	0.379
3800	52.264	359.002	320.767	145.295	-2.001	-27.490	0.378
3900	53.105	360.371	321.765	150.564	-0.839	-28.176	0.377
4000	53.909	361.726	322.747	155.915	0.439	-28.893	0.377
4100	54.671	363.066	323.714	161.344	1.829	-29.643	0.378
4200	55.386	364.392	324.667	166.847	3.327	-30.429	0.378
4300	56.050	365.704	325.606	172.420	4.929	-31.251	0.380
4400	56.661	366.999	326.532	178.056	6.630	-32.112	0.381
4500	57.217	368.279	327.446	183.750	8.425	-33.012	0.383
4600	57.717	369.542	328.347	189.497	10.307	-33.954	0.386
4700	58.161	370.788	329.237	195.292	12.272	-34.937	0.388
4800	58.548	372.017	330.115	201.127	14.312	-35.963	0.391
4900	58.881	373.228	330.983	206.999	16.423	-37.032	0.395
5000	59.159	374.420	331.840	212.902	18.598	-38.145	0.398
5100	59.387	375.594	332.686	218.829	20.830	-39.302	0.403
5200	59.565	376.749	333.522	224.777	23.116	-40.503	0.407
5300	59.696	377.885	334.349	230.741	25.448	-41.749	0.411
5400	59.782	379.001	335.165	236.715	27.822	-43.039	0.416
5500	59.828	380.099	335.972	242.696	30.233	-44.373	0.421
5600	59.836	381.177	336.770	248.679	32.675	-45.751	0.427
5700	59.809	382.236	337.558	254.662	35.145	-47.174	0.432
5800	59.749	383.276	338.338	260.640	37.639	-48.640	0.438
5900	59.661	384.296	339.108	266.611	40.152	-50.149	0.444
6000	59.546	385.298	339.869	272.571	42.681	-51.701	0.450

PREVIOUS: December 1966 (1 atm)   CURRENT: December 1966 (1 bar)

POTASSIUM BROMIDE (KBr) CRYSTAL M_r = 119.0023

$S°(298.15\ K) = 95.94 \pm 0.04\ J\ K^{-1}\ mol^{-1}$

$T_{fus} = 1007\ K$

$\Delta_f H°(0\ K) = -386.674 \pm 0.4\ kJ\ mol^{-1}$

$\Delta_f H°(298.15\ K) = -393.798 \pm 0.4\ kJ\ mol^{-1}$

$\Delta_{fus} H° = 25.5\ kJ\ mol^{-1}$

Enthalpy of Formation

The enthalpy of solution, $\Delta_{sol}H$, of KBr(cr) in $H_2O(\ell)$ has been measured by many investigators. The results were reviewed in detail by Parker (1). Eight pertinent $\Delta_{sol}H°(aq, \infty)$ values are reproduced in the table below. Adopting the best value, $\Delta_r H°(298.15\ K) = 4.75 \pm 0.02\ kcal\ mol^{-1}$, for the reaction $KBr(cr) = K^+(aq, \infty) + Br^-(aq, \infty)$, we derive $\Delta_f H°(298.15\ K) = -94.12\ kcal\ mol^{-1}$ for KBr(cr) ($-393.798\ kJ\ mol^{-1}$). The auxiliary values $\Delta_f H°(298.15\ K) = -60.32$ and $-29.05\ kcal\ mol^{-1}$ for $K^+(aq, \infty)$ and $Br^-(\infty H_2O)$, respectively, are obtained from reference (2).

$\Delta_{sol}H°(aq, \infty)$* kcal mol^{-1}	T/°C	m**	Source
4.727	16.6	0.28	Walden (3)
4.777	25	0.37	Wust and Lange (4)
7.753	25	0.09	Chipman et al. (5)
4.684	20	0.19	Popov et al. (6)
4.889	23.5	0.14	Fedorov and Sil'chenko (7)
4.783	25	0.13	Lange and Martin (8)
4.542	20.5	0.12	Popov et al. (9)
4.739	25	0.28	Hietala (10)

*Values are adjusted to 298.15 K and to infinite dilution.
**Lowest experimental molality.

Heat Capacity and Entropy

The low temperature heat capacities, 2.9 - 272.2 K, were measured by Berg and Morrison (11) using an adiabatic calorimeter. The high temperature heat capacities, 325.25 - 711.75 K, are taken from Mustajoki (12). These two sets of $C°_p$ data are joined smoothly at 298.15 K. The $C°_p$ values at temperatures, 720 - 1007 K, are extrapolated graphically, so that the derived enthalpy change $H°(1007\ K)-H°(298.15\ K) = 9.9\ kcal\ mol^{-1}$ agrees with that determined by Dworkin (13). The derived enthalpies at 700, 800 and 900 K are also in good agreement with those measured by Skuratov and Lapushkin (14). The $C°_p$ values above 1007 K are obtained by graphical extrapolation.

The S°(298.15 K) is calculated from the adopted low temperature $C°_p$, based on $S°(2.9\ K) = 0.006276\ J\ K^{-1}\ mol^{-1}$.

The heat capacities of KBr(cr) were also measured by Nernst and Lindemann (15), 78.7 - 89.2 K; Koref (16), 137 - 234 K; and Clusius et al. (17). The enthalpies of KBr(cr) were determined by Magnus (18) and by Cooper (19). The results are in reasonable agreement with the adopted functions except for those reported by Cooper (14).

Fusion Data

See the liquid table for details.

Sublimation Data

The heats of sublimation to monomer and dimer are calculated as the enthalpy changes for the following two reactions: KBr(cr) = KBr(g) and 2 KBr(cr) = $K_2Br_2(g)$.

References

1. V. B. Parker, U. S. Nat. Bur. Stand., NSRDS-NBS 2, 1965.
2. U. S. Nat. Bur. Stand. Tech. Note 270-1, 1965.
3. P. T. Walden, Z. Physik. Chem. 58, 479 (1907).
4. J. Wust and E. Lange, Z. Physik. Chem. 116, 161 (1925).
5. H. R. Chipman, F. M. G. Johnson, and O. Maass, Proc. Trans. Nova Scotian Inst. Sci. 17, 149 (1929).
6. M. M. Popov, A. Bundel, and V. Choller, Z. Physik. Chem. A147, 302 (1930).
7. A. S. Fedorov and G. F. Sil'chenko, Ukrain. Khim. Zhur. 12, 53 (1937).
8. E. Lange and W. Martin, Z. Physik. Chem. A180, 233 (1937).
9. M. M. Popov, S. M. Skuratov, and M. M. Strel'tsova, Zhur. Obshchei Khim. 10, 2023 (1940).
10. J. Hietala, Ann. Acad. Sci. Fennicae Ser. A VI No. 63, 1 (1960).
11. W. T. Berg and J. A. Morrison, Proc. Roy. Soc. (London) A242, 467 (1957).
12. A. Mustajoki, Ann. Acad. Sci. Fennicae, Ser. A I, 98, 7 (1951).
13. A. S. Dworkin, Oak Ridge National Laboratory, Oak Ridge, Tennessee, personal communication, Dec. 1, 1964.
14. S. M. Skuratov and S. A. Lapushkin, J. Gen. Chem. USSR (Eng. Transl.) 21, 2485 (1951).
15. W. Nernst and F. A. Lindemann, Z. Elektrochem. 17, 817 (1911).
16. F. Koref, Ann. Physik 4, 49 (1911).
17. K. Clusius, J. Goldmann, and A. Perlick, Z. Naturforsch. 4A, 424 (1949).
18. A. Magnus, Phys. Z. 14, 5 (1913).
19. C. B. Cooper, J. Chem. Phys. 21, 777 (1953).

Potassium Bromide (KBr) $Br_1K_1(cr)$

Enthalpy Reference Temperature = T_r = 298.15 K; Standard State Pressure = p° = 0.1 MPa

T/K	$C°_p$ (J K⁻¹mol⁻¹)	S° (J K⁻¹mol⁻¹)	$-[G°-H°(T_r)]/T$ (J K⁻¹mol⁻¹)	$H°-H°(T_r)$ (kJ mol⁻¹)	$\Delta_f H°$ (kJ mol⁻¹)	$\Delta_f G°$ (kJ mol⁻¹)	Log K_f
0	0.	0.	INFINITE	-12.213	-386.674	-386.674	INFINITE
100	43.145	42.949	140.561	-9.761	-387.351	-385.394	201.309
200	49.873	75.509	100.634	-5.025	-387.660	-383.321	100.113
298.15	52.379	95.939	95.939	0.	-393.798	-380.431	66.650
300	52.300	96.263	95.940	0.097	-393.826	-380.348	66.224
400	53.806	111.525	98.011	5.406	-411.235	-372.297	48.617
500	55.250	123.669	101.968	10.850	-410.744	-362.615	37.882
600	56.358	133.837	106.456	16.429	-410.065	-353.050	30.736
700	58.084	142.647	111.010	22.146	-409.212	-343.612	25.641
800	60.417	150.546	115.466	28.064	-408.148	-334.311	21.828
900	63.687	157.840	119.773	34.260	-406.816	-325.158	18.872
1000	68.011	164.764	123.928	40.836	-405.141	-316.171	16.515
1007.000	68.367	165.240	124.214	41.313	------ CRYSTAL <--> LIQUID -----		
1100	73.053	171.491	127.948	47.897	-481.989	-302.762	14.377
1200	76.860	178.018	131.850	55.402	-478.460	-286.622	12.476
1300	79.622	184.283	135.644	63.231	-474.610	-270.790	10.880
1400	81.797	190.265	139.333	71.304	-470.520	-255.263	9.524
1500	83.680	195.974	142.921	79.580	-466.233	-240.036	8.359

PREVIOUS: December 1961 CURRENT: March 1967

Potassium Bromide (KBr) $Br_1K_1(cr)$

$S°(298.15\ K) = [105.540]\ J\ K^{-1}\ mol^{-1}$

$T_{fus} = 1007\ K$

$\Delta_f H°(298.15\ K) = [-376.492]\ kJ\ mol^{-1}$

$\Delta_{fus} H° = 25.522\ kJ\ mol^{-1}$

Enthalpy of Formation

$\Delta_f H°(298.15\ K)$ is calculated from that of the crystal by adding $\Delta_{fus}H°$ and the difference in enthalpy, H°(1007 K)-H°(298.15 K), between the crystal and liquid.

Toguri et al. (1) studied the chemical equilibrium for the reaction KBr(ℓ) + HCl(g) = KCl(ℓ) + HBr(g). They obtained $\Delta_r G°(1073\ K) = 3.50 \pm 0.08$ and 3.54 kcal mol^{-1} from partial pressure data and emf data, respectively. The corresponding value calculated from these tables is 3.78 kcal mol^{-1}. Combination of the experimental results with JANAF auxiliary data leads to -89.70 and -89.74 kcal mol^{-1} for $\Delta_f H°(298.15\ K)$.

Heat Capacity and Entropy

The heat capacity of 16.7 cal K^{-1} mol^{-1} (69.873 J K^{-1} mol^{-1}) was derived by Dworkin (2) from enthalpy data in the range 1010 -1100 K. This $C_p°$ value is adopted here for the temperature range 298 - 2500 K.

The entropy is calculated in a manner analogous to that used for the enthalpy of formation.

Fusion Data

The adopted T_{fus} = 1007 K and $\Delta_{fus}H°$ = 6.1 kcal mol^{-1} (25.522 kJ mol^{-1}) were determined by Johnson and Bredig (3) and Dworkin and Bredig (4), respectively.

Enthalpies of melting of 6.2 kcal mol^{-1} from phase diagram studies and 6.7 kcal mol^{-1} from calorimetric measurements were obtained by Aukrust et al. (5).

Vaporization Data

The boiling point, T_{vap} = 1671 K, is calculated as the temperature at which the sum of the partial pressures of KBr(g) and K_2Br_2(g) over KBr(ℓ) equals one atmosphere. The value of $\Delta_{vap}H°$ is calculated as the enthalpy required to produce one mole of vapor mixture, which contains 19.5% dimer (K_2Br_2) and 80.5% monomer (KBr).

T_{vap} was reported to be 1648 and 1668 K by von Wartenberg and Albrecht (6) and Ruff and Mugdan (7), respectively.

References

1. J. Toguri, H. Flood and T. Forland, Acta Chem. Scand. 17, 1502 (1963).
2. A. S. Dworkin, Oak Ridge National Laboratory, Oak Ridge, Tennessee, personal communication, Dec. 1, 1964.
3. J. W. Johnson and M. A. Bredig, J. Phys. Chem. 62, 604 (1958).
4. A. S. Dworkin and M. A. Bredig, J. Phys. Chem. 64, 269 (1960).
5. E. Aukrust, B. Bjorge, H. Flood, and T. Forland, Annals of the New York Academy of Sciences 79, Art. 11, 830 (1960).
6. H. von Wartenberg and P. Albrecht, Z. Elektrochem. 27, 162 (1921).
7. O. Ruff and S. Mugdan, Z. Anorg. Chem. 117, 147 (1921).

Enthalpy Reference Temperature = T_r = 298.15 K — Standard State Pressure = p° = 0.1 MPa

T/K	$C_p°$ (J K⁻¹mol⁻¹)	S° (J K⁻¹mol⁻¹)	$-[G°-H°(T_r)]/T$ (J K⁻¹mol⁻¹)	$H°-H°(T_r)$ (kJ mol⁻¹)	$\Delta_f H°$ (kJ mol⁻¹)	$\Delta_f G°$ (kJ mol⁻¹)	Log K_f
0							
100							
200							
298.15	69.873	105.540	105.540	0.	-376.492	-365.987	64.119
300	69.873	105.972	105.541	0.129	-376.487	-365.922	63.713
400	69.873	126.073	108.282	7.117	-392.218	-359.099	46.894
500	69.873	141.665	113.457	14.104	-390.185	-351.053	36.674
600	69.873	154.404	119.252	21.091	-388.097	-343.422	29.898
700	69.873	165.175	125.063	28.078	-385.974	-336.144	25.083
800	69.873	174.505	130.673	35.066	-383.840	-329.171	21.493
900	69.873	182.735	136.010	42.053	-381.718	-322.465	18.715
1000	69.873	190.097	141.057	49.040	-379.630	-315.994	16.506
1007.000	69.873	190.585	141.399	49.529	------ CRYSTAL <--> LIQUID ------		
1100	69.873	196.757	145.823	56.027	-456.553	-305.119	14.489
1200	69.873	202.836	150.324	63.015	-453.541	-291.485	12.688
1300	69.873	208.429	154.581	70.002	-450.532	-278.103	11.174
1400	69.873	213.607	158.615	76.989	-447.529	-264.951	9.885
1500	69.873	218.428	162.444	83.977	-444.531	-252.015	8.776
1600	69.873	222.938	166.085	90.964	-441.539	-239.278	7.812
1700	69.873	227.174	169.555	97.951	-438.554	-226.728	6.967
1800	69.873	231.167	172.868	104.938	-435.579	-214.354	6.220
1900	69.873	234.945	176.037	111.926	-432.615	-202.144	5.557
2000	69.873	238.529	179.073	118.913	-429.665	-190.091	4.965
2100	69.873	241.938	181.986	125.900	-426.731	-178.184	4.432
2200	69.873	245.189	184.785	132.888	-423.816	-166.417	3.951
2300	69.873	248.295	187.480	139.875	-420.923	-154.782	3.515
2400	69.873	251.269	190.076	146.862	-418.055	-143.272	3.118
2500	69.873	254.121	192.581	153.849	-415.215	-131.881	2.756

PREVIOUS: December 1961 CURRENT: March 1967

POTASSIUM BROMIDE (KBr)　　CRYSTAL - LIQUID　　M_r = 119.0023

0 to 1007 K　crystal

above　1007 K　liquid

Refer to the individual tables for details.

Potassium Bromide (KBr)　　$Br_1K_1(cr,l)$

Enthalpy Reference Temperature = T_r = 298.15 K　　Standard State Pressure = p° = 0.1 MPa

T/K	C_p° (J K⁻¹mol⁻¹)	S° (J K⁻¹mol⁻¹)	$-[G^\circ - H^\circ(T_r)]/T$ (J K⁻¹mol⁻¹)	$H^\circ - H^\circ(T_r)$ (kJ mol⁻¹)	$\Delta_f H^\circ$ (kJ mol⁻¹)	$\Delta_f G^\circ$ (kJ mol⁻¹)	Log K_f
0	0.	0.	INFINITE	-12.213	-386.674	-386.674	INFINITE
100	43.145	42.949	140.561	-9.761	-387.351	-385.394	201.309
200	49.873	75.509	100.634	-5.025	-387.660	-383.321	100.113
298.15	52.379	95.939	95.939	0.	-393.798	-380.431	66.650
300	52.300	96.263	95.940	0.097	-393.826	-380.348	66.224
400	53.806	111.525	98.011	5.406	-411.235	-372.297	48.617
500	55.250	123.669	101.968	10.850	-410.744	-362.615	37.882
600	56.358	133.837	106.456	16.429	-410.065	-353.050	30.736
700	58.084	142.647	111.010	22.146	-409.212	-343.612	25.641
800	60.417	150.546	115.466	28.064	-408.148	-334.311	21.828
900	63.687	157.840	119.773	34.260	-406.816	-325.158	18.872
1000	68.011	164.764	123.928	40.836	-405.141	-316.171	16.515
1007.000	68.367	165.240	124.214	41.313	CRYSTAL <--> LIQUID		
1007.000	69.873	190.585	124.214	66.835	TRANSITION		
1100	69.873	196.757	130.090	73.333	-456.553	-305.119	14.489
1200	69.873	202.836	135.902	80.321	-453.541	-291.485	12.688
1300	69.873	208.429	141.269	87.308	-450.532	-278.103	11.174
1400	69.873	213.607	146.254	94.295	-447.529	-264.951	9.885
1500	69.873	218.428	150.906	101.283	-444.531	-252.015	8.776
1600	69.873	222.938	155.269	108.270	-441.539	-239.278	7.812
1700	69.873	227.174	159.375	115.257	-438.554	-226.728	6.967
1800	69.873	231.167	163.254	122.244	-435.579	-214.354	6.220
1900	69.873	234.945	166.929	129.232	-432.615	-202.144	5.557
2000	69.873	238.529	170.420	136.219	-429.665	-190.091	4.965
2100	69.873	241.938	173.745	143.206	-426.731	-178.184	4.432
2200	69.873	245.189	176.919	150.194	-423.816	-166.417	3.951
2300	69.873	248.295	179.955	157.181	-420.923	-154.782	3.515
2400	69.873	251.269	182.865	164.168	-418.055	-143.272	3.118
2500	69.873	254.121	185.659	171.155	-415.215	-131.881	2.756

PREVIOUS:　　CURRENT: March 1967

Potassium Bromide (KBr)　　$Br_1K_1(cr,l)$

POTASSIUM BROMIDE (KBr) IDEAL GAS M_r = 119.0023

$S°(298.15) = 250.532\ J\ K^{-1}\ mol^{-1}$

$\Delta_f H°(0\ K) = -170.85 \pm 2.1\ kJ\ mol^{-1}$

$\Delta_f H°(298.15\ K) = -180.08 \pm 2.1\ kJ\ mol^{-1}$

Electronic Level and Quantum Weight

State	ε_i, cm^{-1}	g_i
$^1\Sigma^+$	0	1

$\omega_e = 218.47\ cm^{-1}$ $\omega_e x_e = 0.7533\ cm^{-1}$ $\sigma = 1$

$B_e = 0.0807\ cm^{-1}$ $\alpha_e = 0.000401\ cm^{-1}$ $r_e = 2.821$ Å

Enthalpy of Formation

The enthalpy of formation is derived from the sublimation and vaporization data analyzed below. Numerous investigators have obtained total pressures by static methods or "apparent" pressures by effusion or transpiration. These pressures are converted to monomer pressures by use of functions [see $K_2Br_2(g)$ table] which are consistent with the dimer-monomer equilibrium data of Hagemark et al. (6). 2nd and 3rd law analyses of the monomer pressures show remarkable consistency over temperatures from 637 to 1668 K. The selected value of $\Delta_f H°(298\ K) = -43.04 \pm 0.5\ kcal\ mol^{-1}$ ($-180.079 \pm 2.1\ kJ\ mol^{-1}$) may be compared with $-43.8\ kcal\ mol^{-1}$ derived from $D_0^0 = 3.95$ eV which is reported by Gaydon (1)] based on atomic fluorescence data.

Source	Reaction	T/K	Data Points	$\Delta_r H°(298.15\ K)$, kcal mol^{-1} 2nd Law	3rd Law	Drift cal K^{-1}mol^{-1}	$\Delta_f H°(298.15\ K)$ kcal mol^{-1}
Wartenberg-Albrecht (1)	A	1368.15-1654.15	14	48.28±0.31	46.70	-1.1±0.2	-43.28
Ruff-Mugdan (2)	"	1361.15-1668.15	10	49.06±0.56	47.16	-1.4±0.4	-42.82
Fiock-Rodebush (3)	"	1179.15-1335.75	10	47.80±0.05	46.95	-0.7±0.1	-43.03
Hintz-Jellinek (4)	"	1523.15	1	-	47.34	-	-42.64
Murgulescu-Marta (5)	"	1373.15-1473.15	5	46.20±1.15	47.69	1.1±0.8	-42.29
Hagemark et al. (6)	"	1138.95-1416.02	26	47.15±0.49	46.94	-0.2±0.4	-43.04
Niwa (7)	B	823.15-923.15	6	48.04±0.51	51.67	4.1±0.6	-42.45
Mayer-Wintner (8)	"	884.90-929.40	6	56.15±3.17	51.49	-5.2±3.5	-42.63
Zimm-Mayer (9)	"	636.94-900.09	15	50.71±0.21	50.58	-0.1±0.2	-43.54

Reactions: (A) KBr(ℓ) = KBr(g) (B) KBr(cr) = KBr(g)

Heat Capacity and Entropy

Rusk and Gordy (10) have investigated the pure rotational spectra of KBr in the 1.5 to 5.0 mm range of the microwave region by millimeter wave molecular beam spectroscopy. The reported values of ω_e, $\omega_e x_e$, B_e and r_e are adopted here and corrected to the average isotopic species. The value of α_e is obtained from Fabricand et al. (11). Molecular constants for KBr(g) have also been reported by Hertzberg (12) and Rice and Klemperer (13). They are in good agreement with the values adopted. Herzberg (12) also lists an A state in 31770 cm^{-1} based on observed electronic spectra.

References

1. H. von Wartenberg and P. Albrecht, Z. Elektrochem. 27, 162 (1921).
2. O. Ruff and S. Mugdan, Z. Anorg. Chem. 117, 147 (1921).
3. E. F. Fiock and W. H. Rodebush, J. Am. Chem. Soc. 48, 2522 (1926).
4. H. Hintz and K. Jellinek, Z. Elektrochem. 42, 187 (1936).
5. I. G. Murgulescu and L. Marta, Rev. Roumaine Chim. 11, 457 (1966). The authors adopted total vapor pressures reported by Wartenberg and Albrecht, loc. cit., for the evaluation of the partial pressures of KBr(g) and $K_2Br_2(g)$. We use the KBr partial pressures derived from JANAF table thermodynamic functions for evaluation.
6. K. I. Hagemark, M. Blander and E. B. Luchsinger, J. Phys. Chem. 70, 276 (1966), and K. I. Hagemark, 3M Company, St. Paul, Minnesota, personal communication, Feb. 6, 1967.
7. K. Niwa, "Vapor Pressures of Alkali-Halides," J. Fac. Sci., Hokkaido Imperial Univ., Ser. III, 2, 201 (1938).
8. J. E. Mayer and I. H. Wintner, J. Chem. Phys. 6, 301 (1938).
9. B. H. Zimm and J. E. Mayer, J. Chem. Phys. 12, 362 (1944).
10. J. R. Rusk and W. Gordy, Phys. Rev. 127, 817 (1962).
11. P. B. Fabricand, R. O. Carlson, C. A. Lee, and I. I. Rabi, Phys. Rev. 91, 1403 (1953).
12. G. Herzberg, "Spectra of Diatomic Molecules," D. Van Nostrand Co., New York, 1950.
13. S. A. Rice and W. Klemperer, J. Chem. Phys. 27, 573 (1957).
14. A. G. Gaydon, "Dissociation Energies," Chapman and Hall, Ltd., London, 1953.

Potassium Bromide (KBr) $Br_1K_1(g)$

Enthalpy Reference Temperature = T_r = 298.15 K Standard State Pressure = p° = 0.1 MPa

T/K	$C_p°$ (J K^{-1}mol^{-1})	S°	$-[G°-H°(T_r)]/T$	$H°-H°(T_r)$ (kJ mol^{-1})	$\Delta_f H°$	$\Delta_f G°$	Log K_f
0	0.	0.	INFINITE	-10.109	-170.852	-170.852	INFINITE
100	33.058	211.876	282.656	-7.078	-170.949	-185.884	97.096
200	36.056	235.946	253.888	-3.588	-172.505	-200.253	52.301
250	36.600	244.056	251.138	-1.771	-173.469	-207.082	43.267
298.15	36.922	250.532	250.532	0.	-180.079	-212.804	37.282
300	36.932	250.761	250.533	0.068	-180.136	-213.007	37.088
350	37.155	256.471	250.983	1.921	-198.500	-217.498	32.460
400	37.316	261.444	251.987	3.783	-199.139	-220.168	28.751
450	37.440	265.846	253.287	5.652	-199.755	-222.760	25.857
500	37.541	269.796	254.744	7.526	-200.349	-225.284	23.535
600	37.701	276.656	257.841	11.289	-201.486	-230.163	20.037
700	37.829	282.477	260.955	15.066	-202.575	-234.856	17.525
800	37.940	287.536	263.968	18.854	-203.639	-239.394	15.631
900	38.041	292.011	266.840	22.653	-204.705	-243.800	14.150
1000	38.136	296.024	269.562	26.462	-205.796	-248.086	12.959
1100	38.226	299.663	272.135	30.280	-285.888	-247.650	11.760
1200	38.314	302.992	274.570	34.107	-286.036	-244.167	10.628
1300	38.400	306.063	276.876	37.943	-286.179	-240.672	9.670
1400	38.484	308.911	279.063	41.787	-286.318	-237.167	8.849
1500	38.568	311.569	281.143	45.640	-286.455	-233.651	8.136
1600	38.650	314.061	283.123	49.501	-286.589	-230.126	7.513
1700	38.732	316.407	285.013	53.370	-286.723	-226.593	6.962
1800	38.814	318.623	286.819	57.247	-286.858	-223.052	6.473
1900	38.895	320.724	288.549	61.132	-286.996	-219.504	6.035
2000	38.975	322.721	290.208	65.026	-287.139	-215.948	5.640
2100	39.056	324.624	291.802	68.927	-287.291	-212.385	5.283
2200	39.136	326.443	293.335	72.837	-287.454	-208.814	4.958
2300	39.216	328.184	294.813	76.755	-287.631	-205.236	4.661
2400	39.296	329.855	296.238	80.680	-287.824	-201.649	4.389
2500	39.376	331.461	297.615	84.614	-288.037	-198.054	4.138
2600	39.456	333.007	298.947	88.556	-288.273	-194.450	3.907
2700	39.535	334.497	300.236	92.505	-288.533	-190.837	3.692
2800	39.615	335.937	301.486	96.463	-288.822	-187.213	3.493
2900	39.694	337.328	302.698	100.428	-289.140	-183.579	3.307
3000	39.774	338.675	303.875	104.401	-289.491	-179.933	3.133
3100	39.853	339.981	305.018	108.383	-289.877	-176.275	2.970
3200	39.932	341.247	306.131	112.372	-290.301	-172.603	2.817
3300	40.012	342.477	307.214	116.369	-290.757	-168.919	2.674
3400	40.091	343.673	308.269	120.374	-291.260	-165.219	2.538
3500	40.170	344.836	309.297	124.387	-291.806	-161.505	2.410
3600	40.249	345.969	310.300	128.408	-292.401	-157.773	2.289
3700	40.328	347.073	311.279	132.437	-293.047	-154.025	2.174
3800	40.408	348.149	312.235	136.474	-293.747	-150.259	2.065
3900	40.487	349.200	313.169	140.519	-294.504	-146.473	1.962
4000	40.566	350.226	314.083	144.571	-295.315	-142.667	1.863
4100	40.645	351.229	314.977	148.632	-296.192	-138.841	1.769
4200	40.724	352.209	315.852	152.700	-297.140	-134.992	1.679
4300	40.803	353.168	316.708	156.777	-298.160	-131.119	1.593
4400	40.882	354.107	317.548	160.861	-298.978	-127.249	1.511
4500	40.961	355.027	318.370	164.953	-300.084	-123.334	1.432
4600	41.040	355.928	319.177	169.053	-301.262	-119.393	1.356
4700	41.119	356.811	319.969	173.161	-302.492	-115.428	1.283
4800	41.198	357.678	320.745	177.277	-303.687	-111.448	1.213
4900	41.277	358.528	321.508	181.401	-305.060	-107.429	1.145
5000	41.356	359.363	322.256	185.532	-306.507	-103.381	1.080
5100	41.435	360.183	322.992	189.672	-308.031	-99.304	1.017
5200	41.514	360.988	323.715	193.819	-309.630	-95.196	0.956
5300	41.593	361.779	324.426	197.975	-311.305	-91.056	0.897
5400	41.672	362.558	325.125	202.138	-313.057	-86.884	0.840
5500	41.751	363.323	325.812	206.309	-314.885	-82.679	0.785
5600	41.830	364.076	326.489	210.488	-316.787	-78.440	0.732
5700	41.909	364.817	327.155	214.675	-318.418	-74.214	0.680
5800	41.988	365.547	327.810	218.870	-320.421	-69.913	0.630
5900	42.067	366.265	328.456	223.073	-322.491	-65.576	0.581
6000	42.146	366.973	329.092	227.283	-324.625	-61.204	0.533

PREVIOUS: March 1967 (1 atm) CURRENT: March 1967 (1 bar)

Potassium Bromide (KBr) $Br_1K_1(g)$

LITHIUM BROMIDE (LiBr) CRYSTAL M_r = 86.845

$S°(298.15 K) = [74.06] J K^{-1} mol^{-1}$

T_{fus} = 823 K

$\Delta_f H°(0 K)$ = unknown

$\Delta_f H°(298.15 K) = -350.9 \pm 0.4 kJ mol^{-1}$

$\Delta_{fus} H° = 17.656 kJ mol^{-1}$

Enthalpy of Formation

$\Delta_f H°$(LiBr, cr, 298.15 K) = -83.87 kcal mol^{-1}(-350.912 kJ mol^{-1}) was calculated from the enthalpy of solution of lithium bromide (cr) at infinite dilution and the ionic enthalpies of formation of Li^+(aq,∞) and Br^-(aq,∞).

Parker (1) reviewed the enthalpy of solution data in the literature and gave the "best" value, $\Delta_{sol} H^{\infty}$ = -11,670 ± 50 cal mol^{-1} for LiBr(cr) + LiBr(aq,∞) at 298.15 K. The ionic enthalpy of formation, $\Delta_f H°(Li^+$, aq,∞ 298.15 K) = -66.49 kcal mol^{-1}, was calculated from $\Delta_f H°$(LiOH, aq,∞ 298.15 K) = -121.46 kcal mol^{-1} (2) with $\Delta_f H°(OH^-$,aq,∞ 298.15 K) = -54.97 kcal mol^{-1} (3). The $\nu\Delta_f H°$(Br, aq,∞, 298.15 K) = -29.05 kcal mol^{-1} was also obtained from (3). Combination of the ionic enthalpies of formation of Li^+(aq,∞) and Br^-(aq,∞) gives the enthalpy of formation of LiBr(aq,∞), $\Delta_f H°$(298.15 K) = -95.54 kcal mol^{-1}.

Heat Capacity and Entropy

Dworkin (4) has measured the enthalpy changes by the drop method (723-883 K), yielding H°(823;298) = 7.0 kcal mol^{-1} for the crystal at the melting point. Heat capacities derived from his data were $C_p°$(773 K) = 15.1 and $C_p°$(853 K) = 15.6 cal K^{-1} mol^{-1} for the crystal and liquid, respectively. The tabulated heat capacities were estimated based on these values and on the heat capacities of LiCl(cr), NaCl(cr) and NaBr(cr).

S°(298.15 K) = 17.7 cal K^{-1} mol^{-1}(74.06 J K^{-1} mol^{-1}) was estimated by adding the entropy difference between NaBr(cr) and NaCl(cr) to the entropy of lithium chloride (cr) at 298.15 K. Comparisons with other alkali halides give results within ±0.5 cal K^{-1} mol^{-1} of this value. Kelley (5) has estimated $C_p° = 11.50 + 3.02 \times 10^{-3} T$ cal K^{-1} mol^{-1} and S°(298.15 K) = 16.0 ± 0.5 cal K^{-1} mol^{-1}.

Fusion Data

The selected enthalpy of fusion, $\Delta_{fus} H°$(823 K) = 4.22 kcal mol^{-1}(17.656 kJ mol^{-1}), was obtained from enthalpy measurements in a drop calorimeter by Dworkin and Bredig (6).

Kelley (7) reviewed some phase diagram studies of the lithium bromide system in the literature, and gave the enthalpy of fusion $\Delta_{fus} H°$(825 K) = 2.90 kcal mol^{-1}. Blanc (8) reported the enthalpy of fusion $\Delta_{fus} H°$(819 K) = 3.095 kcal mol^{-1} by a cryoscopic method. Both values are too low.

References

1. V. B. Parker, U.S. Nat. Bur. Stand., NSRDS-NBS 2, 1965.
2. JANAF Thermochemical Tables: LiOH(cr), 3-31-66.
3. U. S. Nat. Bur. Stand. Tech. Note 270-1, 1965.
4. A. S. Dworkin, Oak Ridge National Laboratory, Oak Ridge, Tennessee, personal communication, Dec. 1964.
5. K. K. Kelley, U. S. Bur. Mines Bull. 584, 1960 and 592, 1961.
6. A. S. Dworkin and M. A. Bredig, J. Phys. Chem. 64, 269 (1960).
7. K. K. Kelley, U. S. Bur. Mines Bull. 393, 1936.
8. M. Blanc, Compt. rend. 246, 570 (1958).

Lithium Bromide (LiBr) $Br_1Li_1(cr)$

Enthalpy Reference Temperature = T_r = 298.15 K — Standard State Pressure = p° = 0.1 MPa

T/K	$C_p°$	S°	$-[G°-H°(T_r)]/T$	$H°-H°(T_r)$	$\Delta_f H°$	$\Delta_f G°$	Log K_f
	J $K^{-1}mol^{-1}$			kJ mol^{-1}			
0							
100							
200							
298.15	48.919	74.057	74.057	0.	-350.912	-341.630	59.852
300	48.995	74.360	74.058	0.091	-350.937	-341.573	59.473
400	51.233	88.767	76.007	5.104	-365.764	-335.373	43.795
500	53.430	100.430	79.760	10.335	-368.312	-327.524	34.216
600	56.087	110.394	84.054	15.804	-367.686	-319.419	27.808
700	59.768	119.300	88.463	21.586	-366.698	-311.447	23.240
800	64.455	127.583	92.840	27.794	-365.263	-303.647	19.826
823.000	65.333	129.422	93.837	29.287	------ CRYSTAL <--> LIQUID -----		
900	68.471	135.414	97.139	34.447	-363.383	-296.053	17.182
1000	71.379	142.787	101.339	41.448	-361.155	-288.689	15.080
1100	73.555	149.695	105.425	48.698	-358.678	-281.560	13.370
1200	75.228	156.172	109.387	56.142	-356.005	-274.666	11.956
1300	76.442	162.244	113.222	63.730	-353.191	-268.001	10.768
1400	77.278	167.942	116.929	71.418	-350.273	-261.557	9.759
1500	77.822	173.293	120.510	79.175	-347.283	-255.325	8.891
1600	78.199	178.329	123.968	86.977	-344.245	-249.293	8.139
1700	78.408	183.076	127.307	94.808	-486.412	-236.276	7.260
1800	78.576	187.563	130.531	102.658	-482.570	-221.673	6.433
1900	78.701	191.815	133.645	110.522	-478.721	-207.284	5.699
2000	78.743	195.853	136.655	118.395	-474.875	-193.097	5.043

PREVIOUS: September 1961 CURRENT: June 1966

Lithium Bromide (LiBr) $Br_1Li_1(cr)$

LITHIUM BROMIDE (LiBr) LIQUID M_r = 86.845

$S°(298.15 K) = [84.603] J K^{-1} mol^{-1}$ $\Delta_f H°(298.15 K) = [-338.226] kJ mol^{-1}$

T_{fus} = 823 K $\Delta_{fus}H° = 17.656 kJ mol^{-1}$

Enthalpy of Formation

$\Delta_f H°$(298.15 K) is calculated from that of the crystal by adding the $\Delta_{fus}H°$ and the difference in enthalpy, H°(823 K) - H°(298.15 K), between the crystal and liquid.

Heat Capacity and Entropy

Dworkin (1) has derived the heat capacity Cp = 15.6 cal K^{-1} mol^{-1} from enthalpy measurements (823 - 883 K) by the drop method. The liquid heat capacity was assumed to be constant as 15.6 cal K^{-1} mol^{-1}. S°(298.15 K) is obtained in a manner analogous to that used for the enthalpy of formation.

Fusion Data

See the crystal table for details.

Vaporization Data

T_{vap}(ℓ→monomer) is taken as the temperature at which the calculated Gibbs energy change is zero for LiBr(ℓ) → LiBr(g), while $\Delta_{vap}H°$(ℓ→monomer) is the corresponding enthalpy of vaporization.

T_{vap}(ℓ→equilibrium mixture) is taken as the temperature at which the sum of the calculated partial vapor pressures of LiBr(g) and Li_2Br_2(g) reaches one bar (trimer and higher polymers have been neglected in calculation). This value (1562 K) is in good agreement with the boiling point (at 1 atm) of 1583 K obtained from total vapor pressure measurements by von Wartenberg and Schulz (2) and also the boiling point of 1538 K by Ruff and Mugdan (3). $\Delta_{vap}H°$(ℓ→equilibrium mixture) at the boiling point is calculated as the heat of vaporization of one mole of liquid to vapor containing 29.23 mole percent of dimer. For detailed information see LiBr(g) and Li_2Br_2(g) tables.

References

1. A. S. Dworkin, Oak Ridge National Laboratory, Oak Ridge, Tennessee, personal communication, Dec. 1964.
2. H. von Wartenberg and H. Schulz, Z. Elektrochem. 27, 568 (1921).
3. O. Ruff and S. Mugdan, Z. Anorg. Chem. 117, 147 (1921).

Lithium Bromide (LiBr) $Br_1Li_1(l)$

Enthalpy Reference Temperature = T_r = 298.15 K   Standard State Pressure = p° = 0.1 MPa

T/K	$C_p°$	S°	$-[G°-H°(T_r)]/T$	$H°-H°(T_r)$	$\Delta_f H°$	$\Delta_f G°$	Log K_f
	J K⁻¹mol⁻¹			kJ mol⁻¹			
0							
100							
200							
298.15	65.270	84.603	84.603	0.	-338.226	-332.088	58.181
300	65.270	85.007	84.605	0.121	-338.221	-332.050	57.815
400	65.270	103.784	87.165	6.648	-351.534	-327.150	42.721
500	65.270	118.349	91.999	13.175	-352.786	-320.957	33.530
600	65.270	130.249	97.413	19.702	-351.102	-314.748	27.401
700	65.270	140.311	102.841	26.229	-349.369	-308.825	23.045
800	65.270	149.026	108.081	32.756	-347.614	-303.153	19.794
823.000	65.270	150.876	109.251	34.257	------ CRYSTAL <--> LIQUID -----		
900	65.270	156.714	113.066	39.283	-345.861	-297.701	17.278
1000	65.270	163.591	117.781	45.810	-344.107	-292.444	15.276
1100	65.270	169.812	122.233	52.337	-342.352	-287.363	13.646
1200	65.270	175.491	126.438	58.864	-340.597	-282.441	12.294
1300	65.270	180.715	130.415	65.391	-338.843	-277.666	11.157
1400	65.270	185.552	134.182	71.918	-337.086	-273.026	10.187
1500	65.270	190.056	137.759	78.445	-335.326	-268.511	9.350
1600	65.270	194.268	141.160	84.972	-333.563	-264.115	8.622
1700	65.270	198.225	144.402	91.499	-477.034	-252.652	7.763
1800	65.270	201.956	147.497	98.026	-474.514	-239.526	6.951
1900	65.270	205.485	150.457	104.553	-472.004	-226.539	6.228
2000	65.270	208.833	153.293	111.080	-469.503	-213.685	5.581

PREVIOUS: September 1961   CURRENT: June 1966

Lithium Bromide (LiBr) $Br_1Li_1(l)$

LITHIUM BROMIDE (LiBr) CRYSTAL - LIQUID M_r = 86.845 Lithium Bromide (LiBr) $Br_1Li_1(cr,l)$

298.15 to 823 K crystal

above 823 K liquid

Refer to the individual tables for details.

Enthalpy Reference Temperature = T_r = 298.15 K Standard State Pressure = $p°$ = 0.1 MPa

T/K	$C_p°$ (J K⁻¹mol⁻¹)	$S°$ (J K⁻¹mol⁻¹)	$-[G°-H°(T_r)]/T$ (J K⁻¹mol⁻¹)	$H°-H°(T_r)$ (kJ mol⁻¹)	$\Delta_fH°$ (kJ mol⁻¹)	$\Delta_fG°$ (kJ mol⁻¹)	Log K_f
0							
100							
200							
298.15	48.919	74.057	74.057	0.	-350.912	-341.630	59.852
300	48.995	74.360	74.058	0.091	-350.937	-341.573	59.473
400	51.233	88.767	76.007	5.104	-365.764	-335.373	43.795
500	53.430	100.430	79.760	10.335	-368.312	-327.524	34.216
600	56.087	110.394	84.054	15.804	-367.686	-319.419	27.808
700	59.768	119.300	88.463	21.586	-366.698	-311.447	23.240
800	64.455	127.583	92.840	27.794	-365.263	-303.647	19.826
823.000	65.333	129.422	93.837	29.287	CRYSTAL <--> LIQUID		
823.000	65.270	150.876	93.837	46.944	TRANSITION		
900	65.270	156.714	98.970	51.969	-345.861	-297.701	17.278
1000	65.270	163.591	105.094	58.496	-344.107	-292.444	15.276
1100	65.270	169.812	110.700	65.023	-342.352	-287.363	13.646
1200	65.270	175.491	115.866	71.550	-340.597	-282.441	12.294
1300	65.270	180.715	120.656	78.078	-338.843	-277.666	11.157
1400	65.270	185.552	125.121	84.605	-337.086	-273.026	10.187
1500	65.270	190.056	129.301	91.132	-335.326	-268.511	9.350
1600	65.270	194.268	133.231	97.659	-333.563	-264.115	8.622
1700	65.270	198.225	136.939	104.186	-477.034	-252.652	7.763
1800	65.270	201.956	140.449	110.713	-474.514	-239.526	6.951
1900	65.270	205.485	143.780	117.240	-472.004	-226.539	6.228
2000	65.270	208.833	146.949	123.767	-469.503	-213.685	5.581

PREVIOUS: CURRENT: June 1966

Lithium Bromide (LiBr) $Br_1Li_1(cr,l)$

LITHIUM BROMIDE (LiBr) IDEAL GAS M_r = 86.845

$S°(298.15\ K) = 224.345\ J\ K^{-1}\ mol^{-1}$

$\Delta_f H°(0\ K) = -146.3 \pm 13\ kJ\ mol^{-1}$
$\Delta_f H°(298.15\ K) = -154.0 \pm 13\ kJ\ mol^{-1}$

Electronic Level and Quantum Weight

State	ε_i, cm^{-1}	g_i
$^1\Sigma$	0	1

$\omega_e = 563.16\ cm^{-1}$ $\omega_e x_e = 3.53\ cm^{-1}$ $\sigma = 1$

$B_e = 0.56112\ cm^{-1}$ $\alpha_e = 0.00573\ cm^{-1}$ $r_e = 2.1704$ Å

Enthalpy of Formation

$\Delta_f H°$(LiBr, g, 298.15 K) = $-36.8\pm3\ kcal\ mol^{-1}$ ($-153.971\pm13\ kJ\ mol^{-1}$) is calculated from the selected enthalpy of vaporization and the enthalpy of formation for lithium bromide (ℓ). Lithium bromide vaporizes to a mixture of monomeric and dimeric gases. (Higher polymers have been neglected in the calculation.) The enthalpies of vaporization to monomer and to dimer were chosen to satisfy (1) the total vapor pressure data measured by von Wartenberg and Schulz (1) and by Ruff and Mugdan (2); the partial vapor pressures of monomer and dimer derived from Miller and Kusch (3) in an analysis of the velocity distribution of molecules in alkali halide vapor. The selected enthalpies of vaporization are $\Delta_{vap}H°$(ℓ→monomer, 298.15 K) = $44.0\pm3\ kcal\ mol^{-1}$ and $\Delta_{vap}H°$(ℓ→dimer, 298.15 K) = $42.0\pm5\ kcal\ mol^{-1}$ which combine with the enthalpy of formation of lithium bromide (ℓ), $\Delta_f H°$(298.15 K) = $-80.84\ kcal\ mol^{-1}$, to give the standard enthaply of formation of LiBr(g) and Li_2Br_2(g) -36.8 and -119.7 $kcal\ mol^{-1}$ (-153.971 and -500.825 $kJ\ mol^{-1}$), respectively. The derived enthaply of dissociation is $\Delta_d H°$(298.15 K) = 46.0 $kcal\ mol^{-1}$ for $Li_2Br_2(g) \rightarrow 2LiBr(g)$.

Berkowitz et al. (4) have measured mass-spectrometrically the enthaply of dissociation $\Delta_r H°$(850 K) = 45.9 $kcal\ mol^{-1}$ ($\Delta_r H°$(298.15 K) = 46.9 $kcal\ mol^{-1}$) for $Li_2Br_2(g) \rightarrow 2LiBr(g)$ in a double oven apparatus by the 2nd Law method. Hildenbrand et al. (5) have determined mass-spectrometrically the enthaply of sublimation of lithium bromide by the 2nd Law method as $\Delta_{sub}H°$(cr→monomer, 740 K) = 46.9 and $\Delta_{sub}H°$(cr→dimer, 740 K) = 45.0 $kcal\ mol^{-1}$ (equivalent to $\Delta_{vap}H°$(ℓ→monomer, 298.15 K) = 45.82 $kcal\ mol^{-1}$ and $\Delta_{vap}H°$(ℓ→dimer, 298.15 K) = 41.93 $kcal\ mol^{-1}$). These values are in reasonable agreement with those selected in the tabulation.

Heat Capacity and Entropy

The bond distance (r_e) was obtained from the microwave studies by Honig et al. (6). The vibrational constants (ω_e and $\omega_e x_e$) were determined from the infrared spectrum by Klemperer et al. (7). The rotational constants which have been corrected to the average isotopic species were obtained by Herbert et al. (8) using microwave spectra. Their data are in good agreement with those reported by Honig et al. (6) and Rusk and Gordy (9).

The tabulated thermodynamic functions are in reasonable agreement with those calculated by Wilkins (10) who used slightly different molecular constants.

References

1. H. von Wartenberg and H. Schulz, Z. Elektrochem. 27, 568 (1921).
2. O. Ruff and S. Mugdan, Z. Anorg. Chem. 117, 147 (1921).
3. R. C. Miller and P. Kusch, J. Chem. Phys. 25, 860 (1956), 28, 981 (1958).
4. J. Berkowitz, H. A. Tasman and W. A. Chupka, J. Chem. Phys. 36, 2170 (1962).
5. D. L. Hildenbrand, L. P. Theard, W. F. Hall and N. D. Potter, Philco Report U-2289, Contract NOw61-0905-C, Sept. 15, 1963.
6. A. Honig, M. Mandel, M. L. Stitch and C. H. Townes, Phys. Rev. 96, 629 (1954).
7. W. Klemperer, W. G. Norris, A. Buchler, and A. G. Emslie, J. Chem. Phys. 33, 1534 (1960).
8. A. J. Herbert, F. W. Breivogel, Jr., and K. Street, Jr., J. Chem. Phys. 41, 2368 (1964).
9. J. R. Rusk and W. Gordy, Phys. Rev. 127, 817 (1962).
10. R. L. Wilkins, J. Chem. Eng. Data 5, 337 (1960).

Lithium Bromide (LiBr) $Br_1Li_1(g)$

Enthalpy Reference Temperature = T_r = 298.15 K Standard State Pressure = p° = 0.1 MPa

	J K⁻¹mol⁻¹			kJ mol⁻¹			
T/K	$C_p°$	S°	$-[G°-H°(T_r)]/T$	$H°-H°(T_r)$	$\Delta_f H°$	$\Delta_f G°$	Log K_f
0	0.	0.	INFINITE	-9.177	-146.271	-146.271	INFINITE
100	29.297	190.290	252.948	-6.266	-145.252	-160.854	84.022
200	31.695	211.235	227.385	-3.230	-146.508	-176.048	45.979
250	32.974	218.450	224.899	-1.612	-147.426	-183.330	38.305
298.15	33.952	224.345	224.345	0.	-153.971	-189.498	33.199
300	33.985	224.555	224.345	0.063	-154.024	-189.718	33.033
350	34.753	229.854	224.762	1.782	-169.893	-194.785	29.070
400	35.336	234.534	225.697	3.535	-170.392	-198.308	25.896
450	35.785	238.724	226.916	5.314	-170.958	-201.764	23.420
500	36.136	242.513	228.289	7.112	-174.594	-204.847	21.400
600	36.646	249.149	231.228	10.753	-175.796	-210.783	18.350
700	36.995	254.826	234.204	14.436	-176.908	-216.525	16.157
800	37.252	259.784	237.098	18.149	-177.967	-222.112	14.502
900	37.451	264.183	239.867	21.884	-179.005	-227.568	13.208
1000	37.614	268.138	242.500	25.638	-180.024	-232.909	12.166
1100	37.752	271.729	244.996	29.406	-181.029	-238.148	11.309
1200	37.873	275.020	247.363	33.188	-182.019	-243.297	10.590
1300	37.981	278.055	249.609	36.980	-182.999	-248.364	9.979
1400	38.081	280.874	251.743	40.784	-183.966	-253.356	9.453
1500	38.174	283.504	253.773	44.596	-184.921	-258.279	8.994
1600	38.261	285.971	255.709	48.418	-185.863	-263.139	8.591
1700	38.345	288.293	257.559	52.248	-332.030	-260.764	8.012
1800	38.425	290.487	259.327	56.087	-332.199	-256.567	7.445
1900	38.503	292.567	261.023	59.933	-332.369	-252.360	6.938
2000	38.579	294.543	262.650	63.788	-332.541	-248.145	6.481
2100	38.653	296.428	264.214	67.649	-332.718	-243.920	6.067
2200	38.726	298.227	265.719	71.518	-332.901	-239.688	5.691
2300	38.798	299.950	267.170	75.394	-333.093	-235.447	5.347
2400	38.869	301.603	268.571	79.278	-333.295	-231.197	5.032
2500	38.939	303.191	269.924	83.168	-333.508	-226.938	4.742
2600	39.008	304.720	271.233	87.065	-333.736	-222.671	4.474
2700	39.077	306.193	272.501	90.970	-333.978	-218.395	4.225
2800	39.145	307.616	273.730	94.881	-334.236	-214.109	3.994
2900	39.213	308.990	274.922	98.799	-334.512	-209.814	3.779
3000	39.280	310.321	276.080	102.723	-334.806	-205.509	3.578
3100	39.347	311.610	277.205	106.655	-335.119	-201.194	3.390
3200	39.414	312.860	278.300	110.593	-335.451	-196.869	3.214
3300	39.480	314.074	279.366	114.537	-335.803	-192.533	3.048
3400	39.546	315.254	280.404	118.489	-336.175	-188.186	2.891
3500	39.612	316.401	281.416	122.447	-336.568	-183.828	2.743
3600	39.678	317.518	282.404	126.411	-336.981	-179.458	2.604
3700	39.744	318.606	283.367	130.382	-337.415	-175.076	2.472
3800	39.810	319.667	284.309	134.360	-337.870	-170.683	2.346
3900	39.875	320.702	285.229	138.344	-338.346	-166.277	2.227
4000	39.940	321.712	286.128	142.335	-338.843	-161.858	2.114
4100	40.005	322.699	287.008	146.332	-339.363	-157.427	2.006
4200	40.071	323.664	287.870	150.336	-339.905	-152.983	1.903
4300	40.136	324.607	288.713	154.346	-340.470	-148.526	1.804
4400	40.201	325.531	289.539	158.363	-341.038	-144.057	1.710
4500	40.265	326.435	290.349	162.386	-341.641	-139.574	1.620
4600	40.330	327.321	291.143	166.416	-342.270	-135.077	1.534
4700	40.395	328.189	291.922	170.452	-342.925	-130.566	1.451
4800	40.460	329.040	292.687	174.495	-343.605	-126.040	1.372
4900	40.524	329.875	293.437	178.544	-344.311	-121.500	1.295
5000	40.589	330.694	294.174	182.600	-345.046	-116.946	1.222
5100	40.653	331.499	294.898	186.662	-345.805	-112.377	1.151
5200	40.718	332.289	295.610	190.731	-346.598	-107.792	1.083
5300	40.782	333.065	296.309	194.806	-347.424	-103.192	1.017
5400	40.847	333.828	296.997	198.887	-348.283	-98.575	0.954
5500	40.911	334.578	297.673	202.975	-349.177	-93.943	0.892
5600	40.976	335.316	298.339	207.069	-350.108	-89.294	0.833
5700	41.040	336.041	298.994	211.170	-351.077	-84.628	0.776
5800	41.104	336.756	299.639	215.277	-352.086	-79.945	0.720
5900	41.169	337.459	300.274	219.391	-353.137	-75.244	0.666
6000	41.233	338.151	300.900	223.511	-354.231	-70.525	0.614

PREVIOUS: June 1966 (1 atm) CURRENT: June 1966 (1 bar)

Lithium Bromide (LiBr) $Br_1Li_1(g)$

MAGNESIUM BROMIDE (MgBr) IDEAL GAS M_r = 104.209

$\Delta_fH°(0\ K) = -27.7 \pm 41.8\ kJ\ mol^{-1}$

$S°(298.15\ K) = 244.95 \pm 0.21\ J\ K^{-1}\ mol^{-1}$ $\Delta_fH°(298.15\ K) = -35.3 \pm 41.8\ kJ\ mol^{-1}$

Electronic Levels and Quantum Weights

State	ε_i, cm^{-1}	g_i
$X\ ^2\Sigma^+$	0	2
$A_1\ ^2\Pi_{1/2}$	25765.2	2
$A_2\ ^2\Pi_{3/2}$	25876.3	2
$B\ [^2\Sigma]$	[26500]	2
$C\ [^2\Sigma]$	39285.9	2

$\omega_e = 373.2\ cm^{-1}$ $\omega_ex_e = 1.34\ cm^{-1}$ $\sigma = 1$

$B_e = 0.16241\ cm^{-1}$ $\alpha_e = [0.00079]\ cm^{-1}$ $r_e = 2.36 \pm 0.10\ Å$

Enthaply of Formation

No thermochemical measurement of the enthaply of formation has been made. The selected value, $\Delta_fH°$(MgBr, g, 0 K) = -6.611±10.0 kcal mol^{-1} (-27.660 kJ mol^{-1}), is based on an analysis of spectroscopic data. Herzberg (1) obtained the value D_0^0< 3.35 eV from predissociation which sets in above v = 3 of $A^2\Pi$ state. Gaydon (2) recommended D_0^0 = 3.2±1.0 eV which was obtained from a linear Birge-Sponer extrapolation of the ground state vibrational levels (v, 0-6). Our adopted vibrational constants give this same value by a similar extrapolation. The linear Birge-Sponer D_0^0 value adjusts to 3.03 eV (69.79 kcal mol^{-1}) when corrected for the ionic character of the Mg-Br bond by the method suggested by Hildenbrand (3). This adjusted D_0^0 value is adopted and corresponds to $\Delta_fH°$(298.15 K) of -35.3 kJ mol^{-1}.

Two lower values of D_0^0 have been reported from results of ionic model calculations (4, 5). Margrave (4) calculated an ionic binding energy of 135 kcal mol^{-1} which gives D_0^0 = 1.75 eV. Krasnov and Karaseva (5), using a Rittner potential function (6), found D_0^0 = 2.39 eV which probably represents a minimum value. In addition, we find D_0^0(MgBr, g, 298.15 K)/$\Delta_{at}H°$($MgBr_2$, g, 298.15 K) = 0.44 which is quite consistent with values of this ratio for other alkaline-earth halides (7). This consistency provides further support for our adopted results. An estimated uncertainty of ±10.0 kcal mol^{-1} (41.8 kJ mol^{-1}) is believed to be realistic.

Heat Capacity and Entropy

Values for the ground state vibrational constants and bond length are taken from the tabulation of Rosen (8). The adopted value of r_e which was obtained from a rotational analysis of the (0,0) bands of the $A^2\Pi$-$X^2\Sigma$ system by Patel and Patel (9, 10) gives $r_e(MgBr)/r_e(MgBr_2)$ = 1.01. Comparison of values for this ratio for several alkaline-earth halides (7) shows that $r_e(MX)/r_e(MX_2)$ is generally slightly less than one (~0.96). This suggests that the uncertainty in r_e(MgBr) may be as high as 0.1 Å, assuming r_e for $MgBr_2$ is correct (7). The value of B_e is calculated from r_e. α_e is obtained from the other constants assuming a Morse potential function.

The electronic levels except for the two uppe most states are from Rosen (8). We estimate a $^2\Sigma$ state to lie at 26500 cm^{-1} by analogy with those observed for CaBr and SrBr (7). The assignment o the level at 39285.9 cm^{-1} is rather uncertain. Rosen (8) has assigned this level as a $C^2\Pi$ state. Very recently, Reddy and Rao (11) observed that the bands were single-headed, and they attributed the system to a $C^2\Sigma$ - $X^2\Sigma$ transition by analogy with that for MgF. Comparison of the observed spectra for MgCl, CaBr, SrBr, and BaBr (7) suggests yet another assignment. It appears likely that the observed level near 40000 cm^{-1} arises from a $D^2\Sigma$-X $^2\Sigma$ transition, and the $C^2\Pi$ state, estimated to lie near 30000 cm^{-1}, has gone unobserved. We tentatively adopt the assignment of Reddy and Rao (11). However, thermodynamic functions based on the alternate assignments are not significantly different from those adopted, below 4500 K.

References

1. G. Hezberg, "Spectra of Diatomic Molecules," 2nd Ed., D. Van Nostrand Company, New York, 1950.
2. A. G. Gaydon, "Dissociation Energies and Spectra of Diatomic Molecules," Chapman and Hall, Ltd., London, 1968.
3. D. L. Hildenbrand in "Advances in High Temperature Chemistry," Vol. 1, Ed. by L. Eyring, Academic Press, New York, 1967.
4. J. L. Margrave, J. Phys. Chem. 58, 258 (1954).
5. K. S. Krasnov and N. V. Karaseva, Optics and Spectroscopy 19, 14 (1965).
6. E. S. Rittner, J. Chem. Phys. 19, 1030 (1951).
7. JANAF Thermochemical Tables: CaBr(g), SrBr(g), and BaBr(g), 12-31-74; MgCl(g), 3-31-66; $MgBr_2$(g), 6-30-74.
8. B. Rosen, Ed., "Spectroscopic Data Relative to Diatomic Molecules," Pergamon Press, New York, 1970.
9. M. M. Patel and P. D. Patel, Indian J. Phys. 42, 419 (1968).
10. M. M. Patel and P. D. Patel, J. Phys. B (Atom. Molec. Phys.) 2, 515 (1969).
11. B. R. K. Reddy and P. T. Rao, Curr. Sci. 39, 509 (1970).

Magnesium Bromide (MgBr) $Br_1Mg_1(g)$

Enthalpy Reference Temperature = T_r = 298.15 K Standard State Pressure = p° = 0.1 MPa

T/K	$C_p°$ (J $K^{-1}mol^{-1}$)	S° (J $K^{-1}mol^{-1}$)	-[G°-H°(T_r)]/T (J $K^{-1}mol^{-1}$)	H°-H°(T_r) (kJ mol^{-1})	$\Delta_fH°$ (kJ mol^{-1})	$\Delta_fG°$ (kJ mol^{-1})	Log K_f
0	0.	0.	INFINITE	-9.580	-27.670	-27.670	INFINITE
100	30.272	208.838	275.326	-6.649	-26.777	-44.018	22.993
200	33.916	231.036	248.160	-3.425	-28.011	-60.847	15.892
250	34.967	238.726	245.529	-1.701	-28.865	-68.960	14.408
298.15	35.639	244.946	244.946	0.	-35.342	-75.942	13.305
300	35.660	245.166	244.947	0.066	-35.392	-76.194	13.266
350	36.132	250.701	245.383	1.861	-51.184	-82.118	12.255
400	36.467	255.549	246.357	3.677	-51.577	-86.511	11.297
450	36.715	259.859	247.622	5.507	-51.988	-90.853	10.546
500	36.904	263.738	249.043	7.347	-52.418	-95.148	9.940
600	37.173	270.491	252.071	11.052	-53.341	-103.609	9.020
700	37.359	276.236	255.123	14.779	-54.355	-111.908	8.351
800	37.498	281.235	258.081	18.522	-55.476	-120.055	7.839
900	37.609	285.658	260.904	22.278	-56.721	-128.054	7.432
1000	37.703	289.625	263.582	26.044	-66.699	-135.195	7.062
1100	37.785	293.223	266.115	29.818	-68.247	-141.969	6.742
1200	37.860	296.514	268.513	33.601	-69.791	-148.603	6.469
1300	37.928	299.547	270.785	37.390	-71.333	-155.108	6.232
1400	37.993	302.360	272.941	41.186	-200.282	-158.328	5.907
1500	38.054	304.983	274.991	44.989	-200.467	-155.324	5.409
1600	38.113	307.441	276.943	48.797	-200.652	-152.309	4.972
1700	38.171	309.754	278.806	52.611	-200.837	-149.282	4.587
1800	38.227	311.937	280.586	56.431	-201.023	-146.244	4.244
1900	38.281	314.005	282.291	60.256	-201.210	-143.195	3.937
2000	38.335	315.970	283.927	64.087	-201.401	-140.137	3.660
2100	38.389	317.842	285.497	67.923	-201.596	-137.069	3.409
2200	38.441	319.629	287.009	71.765	-201.796	-133.992	3.181
2300	38.493	321.339	288.464	75.612	-202.001	-130.905	2.973
2400	38.545	322.978	289.868	79.464	-202.214	-127.809	2.782
2500	38.597	324.553	291.225	83.321	-202.434	-124.705	2.606
2600	38.649	326.068	292.536	87.183	-202.662	-121.591	2.443
2700	38.702	327.527	293.805	91.051	-202.898	-118.468	2.292
2800	38.755	328.936	295.034	94.924	-203.142	-115.337	2.152
2900	38.808	330.297	296.227	98.802	-203.396	-112.197	2.021
3000	38.863	331.613	297.385	102.685	-203.658	-109.047	1.899
3100	38.919	332.888	298.510	106.574	-203.930	-105.889	1.784
3200	38.976	334.125	299.603	110.469	-204.211	-102.722	1.677
3300	39.035	335.325	300.668	114.370	-204.501	-99.546	1.576
3400	39.097	336.491	301.704	118.276	-204.801	-96.361	1.480
3500	39.161	337.626	302.715	122.189	-205.111	-93.168	1.390
3600	39.228	338.730	303.700	126.108	-205.431	-89.965	1.305
3700	39.298	339.806	304.661	130.035	-205.761	-86.753	1.225
3800	39.372	340.855	305.600	133.968	-206.103	-83.532	1.148
3900	39.450	341.878	306.517	137.909	-206.457	-80.302	1.076
4000	39.531	342.878	307.413	141.858	-206.823	-77.062	1.006
4100	39.618	343.855	308.290	145.816	-207.203	-73.813	0.940
4200	39.708	344.811	309.149	149.782	-207.597	-70.555	0.877
4300	39.804	345.746	309.989	153.757	-208.006	-67.288	0.817
4400	39.905	346.663	310.812	157.743	-208.432	-64.010	0.760
4500	40.011	347.561	311.619	161.739	-208.874	-60.723	0.705
4600	40.123	348.441	312.410	165.745	-209.334	-57.426	0.652
4700	40.239	349.305	313.186	169.763	-209.814	-54.118	0.601
4800	40.362	350.154	313.947	173.793	-210.314	-50.800	0.553
4900	40.490	350.987	314.694	177.836	-210.835	-47.472	0.506
5000	40.623	351.807	315.428	181.891	-211.379	-44.132	0.461
5100	40.762	352.612	316.150	185.961	-211.946	-40.782	0.418
5200	40.907	353.405	316.858	190.044	-212.538	-37.420	0.376
5300	41.056	354.186	317.555	194.142	-213.155	-34.046	0.336
5400	41.212	354.955	318.241	198.256	-213.799	-30.661	0.297
5500	41.372	355.713	318.915	202.385	-214.470	-27.263	0.259
5600	41.537	356.459	319.579	206.530	-215.170	-23.853	0.222
5700	41.707	357.196	320.233	210.692	-215.899	-20.430	0.187
5800	41.882	357.923	320.876	214.872	-216.659	-16.995	0.153
5900	42.062	358.641	321.510	219.069	-217.450	-13.546	0.120
6000	42.245	359.349	322.135	223.284	-218.273	-10.083	0.088

PREVIOUS: June 1975 (1 atm) CURRENT: June 1975 (1 bar)

Magnesium Bromide (MgBr) $Br_1Mg_1(g)$

MOLYBDENUM BROMIDE (MoBr) IDEAL GAS M_r = 175.844

$S°(298.15) = [269.52 \pm 8]\ J\ K^{-1}\ mol^{-1}$

$\Delta_f H°(0\ K) = [464.4 \pm 42]\ kJ\ mol^{-1}$
$\Delta_f H°(298.15) = [457.3 \pm 42]\ kJ\ mol^{-1}$

Electronic Levels and Quantum Weights

ε_i, cm^{-1}	g_i	ε_i, cm^{-1}	g_i	ε_i, cm^{-1}	g_i
0	[6]	[15199]	[6]	[15890]	[2]
[11783]	[2]	[15331]	[8]	[16796]	[2]
[12034]	[4]	[15428]	[10]	[16947]	[8]
[12417]	[6]	[15447]	[12]	[17174]	[4]
[12900]	[8]	[15691]	[6]	[17344]	[6]
[13461]	[10]	[15699]	[4]		

$\omega_e = [300]\ cm^{-1}$ $\omega_e x_e = [0.7]\ cm^{-1}$ $\sigma = 1$

$B_e = [0.066]\ cm^{-1}$ $\alpha_e = [0.0002]\ cm^{-1}$ $r_e = [2.42]$ Å

Enthaply of Formation

The adopted value of $\Delta_f H°(298.15\ K)$ is that estimated by Brewer (1) using bonding models. The value of $\Delta_f H°(0\ K)$ combined with JANAF data (2) for Mo(g) and Br(g) gives $D_0^{\circ}(Mo\text{-}Br) = 310.87 \pm 42\ kJ\ mol^{-1}$. This value of the bond dissociation energy agrees favorably with the mean Mo-Br bond energy $\Delta_{at}H°(MoBr_4,\ g,\ 0\ K)/4 = 317.98\ kJ\ mol^{-1}$ (2).

Enthaply Capacity and Entropy

The Mo-Br bond length of 2.42 Å and the ground state vibrational frequency of 300 cm^{-1} are estimates from Brewer (1). The electronic contributions are assumed to be the same as those of the free gaseous ion, Mo^+, as suggested by Brewer (1). All states up to 17,344 cm^{-1} as listed by Moore (3) are included in the calculation. Our thermodynamic functions differ slightly from those reported by Brewer (1) due to the use of non-zero values for $\omega_e x_e$ and α_e. Estimates for α_e and $\omega_e x_e$ are based on the empirical correlations of Calder and Ruedenberg (4).

References

1. L. Brewer, Materials and Molecular Research Division, Lawrence Berkeley Laboratory, University of California, Berkeley; personal communication of Sept. 29, 1978; preliminary draft of review to be submitted for publication in Atomic Energy Review, International Atomic Energy Agency, Vienna, Austria.
2. JANAF Thermochemical Tables: Mo(g), 3-31-78; Br(g), 9-30-61; $MoBr_4$(g), 9-30-78.
3. C. E. Moore, U. S. Nat. Bur. Stand., NSRDS-NBS 34, 1970; NSRDS-NBS 35, Vol. III, 1971 [reprint of NBS Circular 467, Volume III, 1958).
4. G. V. Calder and K. Ruedenberg, J. Chem. Phys. 49, 5399 (1968).

Molybdenum Bromide (MoBr) $Br_1Mo_1(g)$

Enthalpy Reference Temperature = T_r = 298.15 K Standard State Pressure = p° = 0.1 MPa

	J $K^{-1}mol^{-1}$			kJ mol^{-1}			
T/K	$C_p°$	S°	$-[G°-H°(T_r)]/T$	$H°-H°(T_r)$	$\Delta_f H°$	$\Delta_f G°$	Log K_f
0	0.	0.	INFINITE	-9.796	464.355	464.355	INFINITE
100	31.264	232.325	300.685	-6.836	465.427	445.589	-232.752
200	34.897	255.303	272.802	-3.500	464.470	426.047	-111.272
250	35.718	263.186	270.117	-1.733	463.711	416.526	-87.028
298.15	36.214	269.523	269.523	0.	457.311	408.172	-71.510
300	36.229	269.747	269.524	0.067	457.264	407.867	-71.016
350	36.566	275.359	269.966	1.887	441.545	400.504	-59.772
400	36.800	280.258	270.953	3.722	441.223	394.663	-51.538
450	36.971	284.603	272.233	5.566	440.882	388.863	-45.138
500	37.101	288.505	273.668	7.418	440.525	383.102	-40.022
600	37.285	295.287	276.723	11.138	439.769	371.687	-32.358
700	37.413	301.044	279.796	14.874	438.962	360.404	-26.894
800	37.509	306.047	282.772	18.620	438.111	349.239	-22.803
900	37.587	310.469	285.608	22.375	437.217	338.183	-19.628
1000	37.653	314.433	288.296	26.137	436.280	327.229	-17.093
1100	37.712	318.024	290.838	29.905	435.294	316.371	-15.023
1200	37.768	321.308	293.242	33.679	434.254	305.605	-13.303
1300	37.822	324.333	295.519	37.459	433.152	294.929	-11.850
1400	37.880	327.138	297.679	41.244	431.982	284.340	-10.609
1500	37.945	329.754	299.731	45.035	430.739	273.837	-9.536
1600	38.024	332.205	301.685	48.833	429.415	263.419	-8.600
1700	38.123	334.513	303.549	52.640	428.007	253.087	-7.776
1800	38.251	336.696	305.330	56.459	426.507	242.841	-7.047
1900	38.416	338.768	307.036	60.292	424.911	232.680	-6.397
2000	38.627	340.744	308.672	64.144	423.213	222.606	-5.814
2100	38.891	342.635	310.245	68.019	421.410	212.620	-5.289
2200	39.216	344.451	311.759	71.924	419.497	202.722	-4.813
2300	39.610	346.203	313.218	75.864	417.471	192.913	-4.381
2400	40.075	347.898	314.628	79.848	415.328	183.195	-3.987
2500	40.617	349.545	315.992	83.882	413.063	173.569	-3.627
2600	41.236	351.150	317.313	87.974	410.663	164.036	-3.296
2700	41.931	352.719	318.596	92.132	408.096	154.599	-2.991
2800	42.701	354.257	319.842	96.363	405.307	145.260	-2.710
2900	43.541	355.770	321.055	100.674	366.340	136.076	-2.451
3000	44.445	357.261	322.237	105.073	364.925	128.160	-2.231
3100	45.406	358.734	323.391	109.565	363.596	120.290	-2.027
3200	46.414	360.192	324.518	114.156	362.358	112.461	-1.836
3300	47.460	361.636	325.621	118.849	361.217	104.670	-1.657
3400	48.534	363.069	326.701	123.649	360.178	96.912	-1.489
3500	49.623	364.491	327.761	128.557	359.244	89.182	-1.331
3600	50.717	365.904	328.801	133.574	358.418	81.478	-1.182
3700	51.805	367.309	329.822	138.700	357.700	73.796	-1.042
3800	52.875	368.705	330.827	143.934	357.091	66.131	-0.909
3900	53.916	370.092	331.816	149.274	356.590	58.481	-0.783
4000	54.920	371.469	332.790	154.716	356.194	50.842	-0.664
4100	55.877	372.837	333.751	160.256	355.901	43.212	-0.551
4200	56.781	374.195	334.697	165.890	355.707	35.588	-0.443
4300	57.624	375.541	335.632	171.611	355.606	27.968	-0.340
4400	58.401	376.875	336.554	177.412	355.593	20.349	-0.242
4500	59.109	378.195	337.465	183.289	355.662	12.729	-0.148
4600	59.744	379.502	338.364	189.232	355.807	5.107	-0.058
4700	60.306	380.793	339.253	195.235	356.020	-2.520	0.028
4800	60.793	382.067	340.132	201.290	356.296	-10.151	0.110
4900	61.205	383.325	341.001	207.391	356.626	-17.788	0.190
5000	61.545	384.565	341.860	213.529	-234.380	-19.702	0.206
5100	61.815	385.787	342.709	219.698	-235.641	-15.396	0.158
5200	62.016	386.989	343.549	225.890	-237.013	-11.064	0.111
5300	62.153	388.172	344.380	232.099	-238.501	-6.705	0.066
5400	62.229	389.335	345.202	238.318	-240.111	-2.317	0.022
5500	62.248	390.477	346.014	244.543	-241.844	2.102	-0.020
5600	62.214	391.598	346.818	250.766	-243.706	6.554	-0.061
5700	62.131	392.699	347.614	256.984	-245.697	11.041	-0.101
5800	62.004	393.778	348.400	263.191	-247.820	15.563	-0.140
5900	61.837	394.837	349.178	269.383	-250.074	20.123	-0.178
6000	61.634	395.874	349.948	275.557	-252.464	24.722	-0.215

PREVIOUS: September 1978 (1 atm) CURRENT: September 1978 (1 bar)

Molybdenum Bromide (MoBr) $Br_1Mo_1(g)$

BROMOIMIDOGEN (NBr) IDEAL GAS M_r = 93.9107

$S°(298.15\ K) = 235.48 \pm 0.4\ J\ K^{-1}\ mol^{-1}$

$\Delta_f H°(0\ K) = 308.4 \pm 21\ kJ\ mol^{-1}$
$\Delta_f H°(298.15\ K) = 300.8 \pm 21\ kJ\ mol^{-1}$

Electronic Level and Quantum Weight

State	ε_i, cm^{-1}	g_i
$^3\Sigma^-$	0	3

$\omega_e = 699.1\ cm^{-1}$ $\omega_e x_e = 4.7\ cm^{-1}$ $\sigma = 1$

$B_e = 0.443\ cm^{-1}$ $\alpha_e = 0.00399\ cm^{-1}$ $r_e = 1.79\ Å$

Enthaply of Formation

Miller and Dunford (1) have extrapolated the ground state vibrational levels and obtain $D_0^0 = 67 \pm 5\ kcal\ mol^{-1}$ or $\Delta_f H°(0\ K) = 73.7 \pm 5\ kcal\ mol^{-1}$ ($308.361 \pm 21\ kJ\ mol^{-1}$).

Heat Capacity and Entropy

Miller and Dunford (1) give all the vibrational and rotational constants, which have been adjusted to the normally occurring isotopic mixture for bromine.

Reference

1. R. V. Miller and H. B. Dunford, J. Chem. Phys. 35, 1202 (1961).

Enthalpy Reference Temperature = T_r = 298.15 K Standard State Pressure = p° = 0.1 MPa

T/K	$C_p°$ (J K⁻¹mol⁻¹)	S° (J K⁻¹mol⁻¹)	$-[G°-H°(T_r)]/T$ (J K⁻¹mol⁻¹)	$H°-H°(T_r)$ (kJ mol⁻¹)	$\Delta_f H°$ (kJ mol⁻¹)	$\Delta_f G°$ (kJ mol⁻¹)	Log K_f
0	0.	0.	INFINITE	-8.989	308.359	308.359	INFINITE
100	29.149	202.331	263.129	-6.080	308.421	298.871	-156.114
200	30.586	222.859	238.409	-3.110	307.484	289.653	-75.650
250	31.724	229.806	236.015	-1.552	306.907	285.261	-59.602
298.15	32.730	235.482	235.482	0.	300.758	281.804	-49.371
300	32.766	235.685	235.483	0.061	300.722	281.686	-49.046
350	33.634	240.803	235.885	1.721	285.329	279.359	-41.692
400	34.334	245.342	236.789	3.421	285.384	278.502	-36.369
450	34.892	249.419	237.969	5.152	285.462	277.638	-32.227
500	35.339	253.119	239.302	6.909	285.556	276.763	-28.913
600	35.997	259.625	242.162	10.478	285.774	274.985	-23.940
700	36.449	265.209	245.065	14.101	286.007	273.168	-20.384
800	36.774	270.099	247.895	17.763	286.237	271.318	-17.715
900	37.019	274.445	250.608	21.453	286.457	269.440	-15.638
1000	37.211	278.356	253.190	25.165	286.662	267.538	-13.975
1100	37.368	281.910	255.642	28.894	286.851	265.617	-12.613
1200	37.499	285.167	257.969	32.638	287.024	263.678	-11.478
1300	37.613	288.173	260.178	36.394	287.183	261.726	-10.516
1400	37.713	290.964	262.279	40.160	287.328	259.763	-9.692
1500	37.803	293.569	264.279	43.936	287.460	257.789	-8.977
1600	37.885	296.012	266.187	47.720	287.580	255.807	-8.351
1700	37.961	298.311	268.009	51.513	287.690	253.818	-7.799
1800	38.032	300.483	269.754	55.312	287.788	251.822	-7.308
1900	38.100	302.541	271.425	59.119	287.875	249.822	-6.868
2000	38.165	304.497	273.031	62.932	287.951	247.817	-6.472
2100	38.226	306.360	274.574	66.752	288.016	245.809	-6.114
2200	38.286	308.140	276.059	70.577	288.070	243.798	-5.788
2300	38.345	309.843	277.491	74.409	288.112	241.784	-5.491
2400	38.401	311.476	278.874	78.246	288.143	239.769	-5.218
2500	38.457	313.045	280.209	82.089	288.163	237.753	-4.968
2600	38.511	314.554	281.501	85.938	288.171	235.737	-4.736
2700	38.565	316.009	282.753	89.791	288.170	233.720	-4.522
2800	38.617	317.412	283.966	93.651	288.158	231.703	-4.322
2900	38.670	318.768	285.142	97.515	288.138	229.688	-4.137
3000	38.721	320.080	286.285	101.384	288.110	227.672	-3.964
3100	38.772	321.351	287.396	105.259	288.076	225.658	-3.802
3200	38.823	322.582	288.476	109.139	288.036	223.646	-3.651
3300	38.873	323.778	289.528	113.024	287.993	221.634	-3.508
3400	38.923	324.939	290.553	116.913	287.948	219.624	-3.374
3500	38.972	326.068	291.551	120.808	287.901	217.615	-3.248
3600	39.021	327.167	292.525	124.708	287.856	215.607	-3.128
3700	39.070	328.236	293.476	128.612	287.812	213.601	-3.016
3800	39.119	329.279	294.405	132.522	287.772	211.596	-2.909
3900	39.168	330.296	295.312	136.436	287.737	209.592	-2.807
4000	39.216	331.288	296.199	140.355	287.707	207.589	-2.711
4100	39.264	332.257	297.067	144.279	287.685	205.586	-2.619
4200	39.312	333.204	297.916	148.208	287.671	203.584	-2.532
4300	39.360	334.129	298.747	152.142	287.666	201.582	-2.449
4400	39.408	335.035	299.562	156.080	287.671	199.580	-2.369
4500	39.456	335.921	300.360	160.023	287.687	197.577	-2.293
4600	39.503	336.789	301.143	163.971	287.714	195.575	-2.221
4700	39.551	337.639	301.910	167.924	287.753	193.571	-2.151
4800	39.598	338.472	302.663	171.882	287.805	191.567	-2.085
4900	39.646	339.289	303.402	175.844	287.870	189.561	-2.021
5000	39.693	340.090	304.128	179.811	287.948	187.554	-1.959
5100	39.740	340.877	304.841	183.782	288.039	185.546	-1.900
5200	39.787	341.649	305.541	187.759	288.144	183.535	-1.844
5300	39.834	342.407	306.230	191.740	288.263	181.522	-1.789
5400	39.881	343.152	306.907	195.726	288.396	179.507	-1.736
5500	39.928	343.884	307.572	199.716	288.543	177.489	-1.686
5600	39.975	344.604	308.227	203.711	288.704	175.468	-1.637
5700	40.022	345.312	308.872	207.711	288.879	173.445	-1.589
5800	40.069	346.009	309.506	211.716	289.067	171.418	-1.544
5900	40.116	346.694	310.130	215.725	289.268	169.388	-1.500
6000	40.163	347.369	310.746	219.739	289.483	167.354	-1.457

PREVIOUS: December 1962 (1 atm) CURRENT: December 1962 (1 bar)

NITROSYL BROMIDE (NOBr) IDEAL GAS M_r = 109.9101

S°(298.15 K) = 273.52 J K^{-1} mol^{-1}

$\Delta_f H°$(0 K) = 91.41 ± 0.8 kJ mol^{-1}
$\Delta_f H°$(298.15 K) = 82.13 ± 0.8 kJ mol^{-1}

Vibrational Frequencies and Degeneracies

ν, cm^{-1}
1801 (1)
542 (1)
[265] (1)

Ground State Quantum Weight: 1 σ = 1
Point Group: C_s
Bond Distances: O-N = 1.15 ± 0.04 Å N-Br = 2.14 ± 0.02 Å
Bond Angle: O-N-Br = 117 ± 3°
Product of the Moments of Inertia: $I_A I_B I_C = 5.16414 \times 10^{-115}\ g^3\ cm^6$

Enthaply of Formation

The chemical equilibrium of the reaction 2NO(g) + Br_2(g) = 2NOBr(g) has been studied by Trautz and Dalal (1), Blair et al. (2), and Krauss (3). Using the reported equilibrium constants, the enthalpy changes $\Delta_r H°$(298.15 K), of this reaction are evaluated by both the 2nd and 3rd law methods. Based on the 3rd law $\Delta_r H°$(298.15 K) values, the enthaplies of formation for NOBr(g) are derived. The results obtained are presented in the table below. The value of $\Delta_f H°$(NOBr, g, 298.15 K) adopted is 19.63±0.2 kcal mol^{-1} (82.132±0.8 kJ mol^{-1}).

Source	Data Points	T/K	$\Delta_r H°$(298.15 K), kcal mol^{-1} 2nd Law	3rd Law	Drift cal $K^{-1}mol^{-1}$	$\Delta_f H°$(298.15 K) kcal mol^{-1}
Trautz and Dalal (1)	*	258.2 - 603.2	-11.11	-11.30	-0.4	19.62
Blair et al. (2)	30	296.9 - 502.9	-11.74 + 0.07	-11.26	1.3 ± 0.2	19.64
Krauss (3)	23	264.0 - 290.1	-13.46 + 0.30	- 9.62	-13.7 ± 1.1	20.46

*The data points employed for evaluation are calculated from a given equation.

Heat Capacity and Entropy

The molecular structure of NOBr(g) has been determined by electron diffraction by Ketelaar and Palmer (4). The results were confirmed later by Weatherly and Williams (5), who studied the microwave spectrum of NOBr(g) in the region 20000 - 40000 Mc/sec and analyzed the J = 2 → 3 transition. The values of bond length and angle adopted are obtained from Ketelaar and Palmer (4). The infrared absorption spectrum of NOBr(g) has been examined from 400 to 5303 cm^{-1} by Burns and Bernstein (6). The authors observed the first two fundamental vibrational frequencies and obtained the third from combination and overtones. These assignments are adopted. The principal moments of inertia are: $I_A = 0.9405 \times 10^{-39}$, $I_B = 22.9668 \times 10^{-39}$ and $I_C = 23.9093 \times 10^{-39}$ g cm^2.

References

1. M. Trautz and V. P. Dalal, Z. Anorg. Chem. 110, 1 (1920).
2. C. M. Blair, Jr., P. D. Brass and D. M. Yost, J. Am. Chem. Soc. 56, 1916 (1934).
3. W. Krauss, Z. Physik. Chem. A175, 295 (1936).
4. J. A. A. Ketelaar and K. J. Palmer, J. Amer. Chem. Soc. 59, 2629 (1937).
5. T. L. Weatherly and Q. Williams, J. Chem. Phys. 25, 717 (1956).
6. W. G. Burns and H. J. Bernstein, J. Chem. Phys. 18, 1669 (1950).

Nitrosyl Bromide (ONBr) $Br_1N_1O_1$(g)

Enthalpy Reference Temperature = T_r = 298.15 K Standard State Pressure = p° = 0.1 MPa

T/K	$C_p°$ (J $K^{-1}mol^{-1}$)	S°	$-[G°-H°(T_r)]/T$	$H°-H°(T_r)$ (kJ mol^{-1})	$\Delta_f H°$	$\Delta_f G°$	Log K_f
0	0.	0.	INFINITE	-11.654	91.410	91.410	INFINITE
100	36.257	228.950	311.488	-8.254	90.510	86.963	-45.425
200	42.124	256.017	277.590	-4.315	89.087	83.973	-21.931
250	44.073	265.637	274.266	-2.157	88.380	82.775	-17.295
298.15	45.473	273.525	273.525	0.	82.132	82.417	-14.439
300	45.520	273.806	273.526	0.084	82.092	82.419	-14.350
350	46.660	280.912	274.084	2.390	66.605	83.326	-12.436
400	47.623	287.207	275.339	4.747	66.571	85.717	-11.194
450	48.480	292.866	276.977	7.150	66.562	88.111	-10.228
500	49.264	298.015	278.827	9.594	66.573	90.505	-9.455
600	50.656	307.124	282.804	14.592	66.639	95.286	-8.295
700	51.831	315.023	286.855	19.718	66.748	100.053	-7.466
800	52.807	322.010	290.821	24.951	66.881	104.802	-6.843
900	53.609	328.278	294.640	30.274	67.031	109.533	-6.357
1000	54.266	333.961	298.293	35.668	67.187	114.247	-5.968
1100	54.804	339.159	301.775	41.123	67.347	118.945	-5.648
1200	55.248	343.947	305.092	46.626	67.506	123.629	-5.381
1300	55.615	348.385	308.254	52.170	67.660	128.300	-5.155
1400	55.922	352.518	311.270	57.747	67.809	132.959	-4.961
1500	56.179	356.385	314.150	63.352	67.951	137.607	-4.792
1600	56.397	360.018	316.904	68.982	68.082	142.247	-4.644
1700	56.583	363.443	319.542	74.631	68.202	146.878	-4.513
1800	56.742	366.681	322.072	80.297	68.309	151.503	-4.396
1900	56.880	369.753	324.501	85.979	68.402	156.122	-4.292
2000	57.000	372.674	326.837	91.673	68.477	160.737	-4.198
2100	57.104	375.457	329.087	97.378	68.535	165.349	-4.113
2200	57.196	378.116	331.255	103.093	68.575	169.958	-4.035
2300	57.276	380.660	333.349	108.817	68.594	174.566	-3.965
2400	57.348	383.099	335.371	114.548	68.592	179.173	-3.900
2500	57.412	385.442	337.327	120.286	68.569	183.781	-3.840
2600	57.469	387.695	339.222	126.030	68.525	188.390	-3.785
2700	57.520	389.864	341.057	131.780	68.461	193.002	-3.734
2800	57.566	391.957	342.838	137.534	68.375	197.616	-3.687
2900	57.607	393.978	344.567	143.293	68.271	202.233	-3.643
3000	57.645	395.932	346.247	149.055	68.147	206.855	-3.602
3100	57.679	397.822	347.880	154.821	68.007	211.481	-3.563
3200	57.710	399.654	349.469	160.591	67.850	216.111	-3.528
3300	57.739	401.430	351.017	166.363	67.679	220.747	-3.494
3400	57.765	403.154	352.525	172.139	67.495	225.388	-3.463
3500	57.789	404.829	353.996	177.916	67.300	230.034	-3.433
3600	57.811	406.457	355.431	183.696	67.095	234.687	-3.405
3700	57.831	408.042	356.831	189.478	66.882	239.345	-3.379
3800	57.850	409.584	358.199	195.262	66.662	244.010	-3.354
3900	57.867	411.087	359.536	201.048	66.437	248.680	-3.331
4000	57.884	412.552	360.843	206.836	66.208	253.355	-3.308
4100	57.899	413.982	362.122	212.625	65.977	258.037	-3.287
4200	57.913	415.377	363.374	218.416	65.744	262.724	-3.267
4300	57.926	416.740	364.599	224.207	65.510	267.417	-3.248
4400	57.938	418.072	365.799	230.001	65.277	272.115	-3.230
4500	57.949	419.374	366.975	235.795	65.045	276.818	-3.213
4600	57.960	420.648	368.128	241.590	64.815	281.527	-3.197
4700	57.970	421.895	369.259	247.387	64.587	286.240	-3.181
4800	57.979	423.115	370.368	253.184	64.361	290.959	-3.166
4900	57.988	424.311	371.457	258.983	64.138	295.682	-3.152
5000	57.996	425.482	372.526	264.782	63.918	300.410	-3.138
5100	58.004	426.631	373.576	270.582	63.701	305.142	-3.125
5200	58.012	427.757	374.607	276.383	63.486	309.878	-3.113
5300	58.019	428.862	375.620	282.184	63.274	314.619	-3.101
5400	58.025	429.947	376.616	287.986	63.064	319.363	-3.089
5500	58.032	431.012	377.595	293.789	62.856	324.111	-3.078
5600	58.037	432.057	378.559	299.593	62.650	328.863	-3.068
5700	58.043	433.085	379.506	305.397	62.444	333.618	-3.057
5800	58.048	434.094	380.439	311.201	62.238	338.378	-3.047
5900	58.054	435.086	381.357	317.006	62.032	343.140	-3.038
6000	58.058	436.062	382.260	322.812	61.825	347.907	-3.029

PREVIOUS: December 1966 (1 atm) CURRENT: December 1966 (1 bar)

Nitrosyl Bromide (ONBr) $Br_1N_1O_1$(g)

SODIUM BROMIDE (NaBr) CRYSTAL M_r = 102.89377

$S°(298.15\ K) = 86.82 \pm 0.25\ J\ K^{-1}\ mol^{-1}$
$T_{fus} = 1020\ K$

$\Delta_f H°(0\ K) = -354.30 \pm 0.42\ kJ\ mol^{-1}$
$\Delta_f H°(298.15\ K) = -361.41 \pm 0.42\ kJ\ mol^{-1}$
$\Delta_{fus} H° = 26.108\ kJ\ mol^{-1}$

Enthalpy of Formation

The enthalpy of solution of NaBr(cr) in water has been measured by Askew et al. (1) and Wallace (2) to be +0.01 and -0.144 kcal mol^{-1}, respectively. Using $\Delta_f H°(Na^+, aq, 298.15\ K) = -57.39$ and $\Delta_f H°(Br^-, aq, 298.15\ K) = -29.05$ kcal mol^{-1}, from Wagman (3)., the respective value of $\Delta_f H°(298.15\ K)$ for NaBr(cr) was found to be -86.3 and -86.45 kcal mol^{-1}. The adopted value is the average of these two.

Heat Capacity and Entropy

The low temperature heat capacities (7.21 - 301.73 K) were measured by Gardner and Taylor (4). The high temperature enthalpy changes (290 - 645, 290 - 816 K) were determined by Magnus (5). Based upon the latter data, the high temperature heat capacities were derived. The two sets of $C_p°$ data were joined smoothly at 298.15 K by graphical method. S°(298.15 K) was obtained from Gardner and Taylor (4) using $S°(10\ K) = 0.034\ cal\ K^{-1}\ mol^{-1}$.

Fusion Data

T_{fus} and $\Delta_{fus}H°$ were reported by Dworkin and Bredig (6). The values, T_{fus} = 1014 K and $\Delta_{fus}H°$ = 5.52 kcal mol^{-1}, reported by Blanc (7) were not used.

Sublimation Data

The value of $\Delta_{sub}H°(298.15\ K)$ was derived from six sets of vapor pressure data by both the 2nd and 3rd law methods. See the ideal gas table for details.

References

1. F. A. Askew, E. Bullock, H. T. Smith, R. K. Tinkler, O. Gatty and J. H. Wolfenden, J. Chem. Soc. 1368 (1934).
2. W. E. Wallace, J. Amer. Chem. Soc. 71, 2485 (1949).
3. D. D. Wagman, U. S. Nat. Bur. Stand., personal communication, July 2, 1964.
4. T. E. Gardner and A. R. Taylor,Jr., U. S. Bur. Mines RI 6435 (1964).
5. A. Magnus, Phys. Z. 14, 5 (1913).
6. A. S. Dworkin and M. A. Bredig, J. Phys. Chem. 64, 269 (1960).
7. M. Blanc, Compt. Rend., 246, 570 (1958).

Sodium Bromide (NaBr) $Br_1Na_1(cr)$

Enthalpy Reference Temperature = T_r = 298.15 K; Standard State Pressure = p° = 0.1 MPa

T/K	$C_p°$ (J K⁻¹mol⁻¹)	S° (J K⁻¹mol⁻¹)	$-[G°-H°(T_r)]/T$ (J K⁻¹mol⁻¹)	$H°-H°(T_r)$ (kJ mol⁻¹)	$\Delta_f H°$ (kJ mol⁻¹)	$\Delta_f G°$ (kJ mol⁻¹)	Log K_f
0	0.	0.	INFINITE	-11.590	-354.302	-354.302	INFINITE
100	40.166	35.606	130.332	-9.473	-354.921	-353.408	184.601
200	48.744	66.818	91.420	-4.920	-355.275	-351.758	91.870
298.15	51.400	86.818	86.818	0.	-361.414	-349.267	61.190
300	51.442	87.136	86.819	0.095	-361.441	-349.192	60.800
400	53.451	102.226	88.863	5.345	-379.052	-341.804	44.635
500	54.852	114.311	92.784	10.764	-378.581	-332.542	34.740
600	56.066	124.419	97.236	16.310	-377.911	-323.394	28.154
700	57.321	133.156	101.757	21.979	-377.063	-314.372	23.459
800	58.576	140.892	106.175	27.774	-376.054	-305.484	19.946
900	59.831	147.864	110.426	33.694	-374.903	-296.730	17.222
1000	61.086	154.232	114.492	39.740	-373.631	-288.111	15.049
1020.000	61.337	155.445	115.284	40.964	------ CRYSTAL <--> LIQUID -----		
1100	62.342	160.113	118.376	45.911	-372.260	-279.625	13.278
1200	63.597	165.592	122.085	52.208	-467.569	-268.827	11.702
1300	64.852	170.732	125.631	58.631	-465.125	-252.364	10.140
1400	66.107	175.584	129.027	65.179	-462.561	-236.093	8.809
1500	67.345	180.187	132.286	71.851	-459.877	-220.010	7.661
1600	68.618	184.573	135.418	78.649	-457.072	-204.109	6.663
1700	69.925	188.772	138.434	85.576	-454.144	-188.388	5.788
1800	71.267	192.807	141.343	92.635	-451.091	-172.843	5.016
1900	72.645	196.697	144.155	99.830	-447.910	-157.471	4.329
2000	74.057	200.459	146.876	107.165	-444.598	-142.270	3.716

PREVIOUS: CURRENT: September 1964

Sodium Bromide (NaBr) $Br_1Na_1(cr)$

SODIUM BROMIDE (NaBr) LIQUID M_r = 102.89377

$S°(298.15\ K) = [104.363]\ J\ K^{-1}\ mol^{-1}$ $\Delta_f H°(298.15\ K) = [-339.343]\ kJ\ mol^{-1}$

$T_{fus} = 1020\ K$ $\Delta_{fus}H° = 26.108\ kJ\ mol^{-1}$

Enthalpy of Formation

$\Delta_f H°(\ell, 298.15\ K)$ is calculated from $\Delta_f H°(cr, 298.15\ K)$ by adding $\Delta_{fus}H°$ and the difference in enthalpy, H°(1020 K)-H°(298.15 K), between the crystal and liquid. The equilibrium constant of the reaction NaBr(ℓ) + HCl(g) = NaCl(ℓ) + HBr(g) has been determined by Toguri et al. (1). From the reported data, ln K(800°C) = -1.19 ± 0.01, the enthalpy of reaction ($\Delta_r H°$(298.15 K)) was derived to be 2.11 kcal mol^{-1}. Using $\Delta_f H°$(HCl, g, 298.15 K) = -22.062, $\Delta_f H°$(NaCl, ℓ, 298.15 K) = -92.24 and $\Delta_f H°$(HBr, g, 298.15 K) = 8.70 kcal mol^{-1}, the value of $\Delta_f H°$(298.15 K) for NaBr(ℓ) was evaluated to be -81.0 ± 0.5 kcal mol^{-1} which is in reasonable agreement with the value adpted. The values of $\Delta_f H°$(298.15 K) for HCl(g) and HBr(g) were obtained from Wagman (2).

Heat Capacity and Entropy

Heat capacity for NaBr(ℓ) is estimated by comparison with those for NaCl(cr), NaCl(ℓ), AgCl(cr), and AgCl(ℓ). The entropy is calculated in a manner analogous to that used for the enthalpy of formation.

Fusion Data

T_{fus} and $\Delta_{fus}H°$ were reported by Dworkin and Bredig (3). The values, T_{fus} = 1014 K and $\Delta_{fus}H°$ = 5.52 kcal mol^{-1}, reported by Blanc (4), were not used.

Vaporization Data

The boiling point, T_{vap}(ℓ → monomer), is calculated as the temperature at which the value of $\Delta_r G° = 0$ for the reaction NaBr(ℓ) → NaBr(g). The difference between $\Delta_f H°$ for NaBr(ℓ) and NaBr(g) at T_{vap}(ℓ → monomer) is the enthalpy of vaporization, $\Delta_{vap}H°$(ℓ → monomer).

References

1. J. Toguri, H. Flood and T. Forland, Acta Chem. Scand. 17, 1502 (1963).
2. D. D. Wagman, U. S. Nat. Bur. Stand., personal communication, July 2, 1964.
3. A. S. Dworkin and M. A. Bredig, J. Phys. Chem. 64, 269 (1960).
4. M. Blanc, Compt. Rend., 246, 570 (1958).

Sodium Bromide (NaBr) $Br_1Na_1(l)$

Enthalpy Reference Temperature = T_r = 298.15 K Standard State Pressure = p° = 0.1 MPa

T/K	$C_p°$	S°	$-[G°-H°(T_r)]/T$	$H°-H°(T_r)$	$\Delta_f H°$	$\Delta_f G°$	Log K_f
	J K⁻¹mol⁻¹			kJ mol⁻¹			
0							
100							
200							
298.15	62 342	104.363	104.363	0.	-339.343	-332.427	58.240
300	62.342	104.749	104.364	0.115	-339.350	-332.384	57.873
400	62.342	122.683	106.810	6.349	-355.977	-326.912	42.690
500	62.342	136.594	111.427	12.584	-354.690	-319.792	33.408
600	62.342	147.961	116.598	18.818	-353.332	-312.939	27.244
700	62.342	157.571	121.782	25.052	-351.919	-306.318	22.858
800	62.342	165.895	126.787	31.286	-350.470	-299.903	19.582
900	62.342	173.238	131.549	37.520	-349.005	-293.670	17.044
1000	62.342	179.806	136.052	43.754	-347.545	-287.600	15.023
1020.000	62.342	181.041	136.922	45.001	------ CRYSTAL <--> LIQUID -----		
1100	62.342	185.748	140.304	49.989	-346.111	-281.675	13.376
1200	62.342	191.172	144.320	56.223	-441.483	-273.439	11.902
1300	62.342	196.162	148.119	62.457	-439.228	-259 526	10.428
1400	62.342	200.782	151.717	68.691	-436.977	-245.788	9.170
1500	62.342	205.084	155.133	74.925	-434.731	-232.210	8.086
1600	62.342	209.107	158.382	81.159	-432.490	-218.781	7.142
1700	62.342	212.886	161.478	87.394	-430.256	-205.493	6.314
1800	62.342	216.450	164.434	93.628	-428.027	-192.336	5.581
1900	62.342	219.820	167.262	99.862	-425.807	-179.303	4.929
2000	62.342	223.018	169.970	106.096	-423.596	-166.386	4.346
2100	62.342	226.060	172.569	112.330	-421.395	-153.580	3.820
2200	62.342	228.960	175.067	118.564	-419.205	-140.878	3.345
2300	62.342	231.731	177.471	124.799	-417.028	-128.276	2.913
2400	62.342	234.384	179.787	131.033	-414.864	-115.768	2.520
2500	62.342	236.929	182.023	137.267	-412.716	-103.349	2.159

PREVIOUS: CURRENT: September 1964

Sodium Bromide (NaBr) $Br_1Na_1(l)$

SODIUM BROMIDE (NaBr) CRYSTAL - LIQUID M_r = 102.89377

0 to 1020 K crystal

above 1020 K liquid

Refer to the individual tables for details.

Sodium Bromide (NaBr) $Br_1Na_1(cr,l)$

Enthalpy Reference Temperature = T_r = 298.15 K Standard State Pressure = $p°$ = 0.1 MPa

T/K	$C_p°$ (J K⁻¹mol⁻¹)	$S°$ (J K⁻¹mol⁻¹)	$-[G°-H°(T_r)]/T$ (J K⁻¹mol⁻¹)	$H°-H°(T_r)$ (kJ mol⁻¹)	$\Delta_fH°$ (kJ mol⁻¹)	$\Delta_fG°$ (kJ mol⁻¹)	Log K_f
0	0.	0.	INFINITE	-11.590	-354.302	-354.302	INFINITE
100	40.166	35.606	130.332	-9.473	-354.921	-353.408	184.601
200	48.744	66.818	91.420	-4.920	-355.275	-351.758	91.870
298.15	51.400	86.818	86.818	0.	-361.414	-349.267	61.190
300	51.442	87.136	86.819	0.095	-361.441	-349.192	60.800
400	53.451	102.226	88.863	5.345	-379.052	-341.804	44.635
500	54.852	114.311	92.784	10.764	-378.581	-332.542	34.740
600	56.066	124.419	97.236	16.310	-377.911	-323.394	28.154
700	57.321	133.156	101.757	21.979	-377.063	-314.372	23.459
800	58.576	140.892	106.175	27.774	-376.054	-305.484	19.946
900	59.831	147.864	110.426	33.694	-374.903	-296.730	17.222
1000	61.086	154.232	114.492	39.740	-373.631	-288.111	15.049
1020.000	61.337	155.445	115.284	40.964	CRYSTAL <--> LIQUID		
1020.000	62.342	181.041	115.284	67.072	TRANSITION		
1100	62.342	185.748	120.239	72.060	-346.111	-281.675	13.376
1200	62.342	191.172	125.928	78.294	-441.483	-273.439	11.902
1300	62.342	196.162	131.141	84.528	-439.228	-259.526	10.428
1400	62.342	200.782	135.952	90.762	-436.977	-245.788	9.170
1500	62.342	205.084	140.419	96.996	-434.731	-232.210	8.086
1600	62.342	209.107	144.588	103.230	-432.490	-218.781	7.142
1700	62.342	212.886	148.496	109.465	-430.256	-205.493	6.314
1800	62.342	216.450	152.173	115.699	-428.027	-192.336	5.581
1900	62.342	219.820	155.645	121.933	-425.807	-179.303	4.929
2000	62.342	223.018	158.935	128.167	-423.596	-166.386	4.346
2100	62.342	226.060	162.059	134.401	-421.395	-153.580	3.820
2200	62.342	228.960	165.035	140.635	-419.205	-140.878	3.345
2300	62.342	231.731	167.875	146.870	-417.028	-128.276	2.913
2400	62.342	234.384	170.591	153.104	-414.864	-115.768	2.520
2500	62.342	236.929	173.194	159.338	-412.716	-103.349	2.159

PREVIOUS: CURRENT: September 1964

Sodium Bromide (NaBr) $Br_1Na_1(cr,l)$

SODIUM BROMIDE (NaBr) IDEAL GAS M_r = 102.89377

$S°(298.15\ K) = 241.218\ J\ K^{-1}\ mol^{-1}$

$\Delta_f H°(0\ K) = -135.04 \pm 2.09\ kJ\ mol^{-1}$
$\Delta_f H°(298.15\ K) = -143.93 \pm 2.09\ kJ\ mol^{-1}$

Electronic Level and Quantum Weight

State	ε_i, cm^{-1}	g_i
$^1\Sigma^+$	0	1

$\omega_e = 302.1 \pm 4\ cm^{-1}$ $\omega_e x_e = 1.50\ cm^{-1}$ $\sigma = 1$
$B_e = 0.1495\ cm^{-1}$ $\alpha_e = 0.000939\ cm^{-1}$ $r_e = 2.5020 \pm 0.0001$ Å

Enthalpy of Formation

$\Delta_f H°(298.15\ K)$ was calculated from $\Delta_f H°(298.15\ K)$ and $\Delta_{sub}H°(298.15\ K)$ for NaBr(cr). The latter was derived by both the 2nd and 3rd law methods from six sets of vapor pressure data, corrected for the presence of dimer species in the vapor. The results are listed as follows.

Source	Reaction	$\Delta_r H°(298.15\ K)$, kcal mol^{-1} 2nd Law	3rd Law	$\Delta_f H°(298.15\ K)^a$ kcal mol^{-1}
Niwa (1)	NaBr(cr) → NaBr(g)	50.31	51.97	-35.24
Cogin and Kimball (2)	NaBr(cr) → NaBr(g)	52.14	52.26	-34.18
Mayer and Wintner (3)	NaBr(cr) → NaBr(g)	65.30	52.59	-33.79b
Ruff and Mugdan (4)	NaBr(ℓ) → NaBr(g)	47.48	46.43	-34.15
Wartenberg and Albrecht (5)	NaBr(ℓ) → NaBr(g)	47.54	46.38	-34.15
Bloom et al. (6)	NaBr(ℓ) → NaBr(g)	51.53	46.23	-34.88b

a Based on the average of the 2nd and 3rd law values.
b Only the 3rd law value being used.

The value of $\Delta_f H°$(NaBr, g, 298.15 K) = -34.40±0.50 kcal mol^{-1} (-143.930 kJ mol^{-1}) adopted is the average value of the six $\Delta_f H°(298.15\ K)$ values listed in the above table. The dissociation energy was calculated to be 86.28 kcal mol^{-1} or 3.74 eV which is in good agreement with the values, 3.84 ± 1 and 3.85 eV, reported by Gaydon (7) and Herzberg (8), respectively. According to Brewer (9), Gaydon's original reported value, D_0^0 = 3.8, is an average of the value of 3.84 eV from atomic fluorescence and the value of 3.75 eV calculated for Gaydon by Brewer from pressure data available at that time.

Heat Capacity and Entropy

The values of ω_e, $\omega_e x_e$, B_e and α_e were taken from Rice and Klemperer (10). The adopted bond distance (r_e) reported by Honing et al. (11) was derived from microwave spectrum. By electron-diffraction method, the Na-Br bond distance was determined as 2.64 ± 0.01 Å by Maxwell et al. (12). The discrepancy may be due to the presence of a large proportion of dimer at the higher pressures used in the electron-diffraction determination. The ground state configuration was reported by G. Herzberg (8).

References

1. K. Niwa, J. Fac. Sci., Hokkaido Univ. Ser. III, 2, 201 (1938).
2. G. E. Cogin and G. E. Kimball, J. Chem. Phys. 16, 1035 (1948).
3. J. E. Mayer and I. H. Wintner, J. Chem. Phys. 6, 301 (1938).
4. O. Ruff and S. Mugdan, Z. Anorg. Chem., 117, 147 (1921).
5. H. von Wartenberg and P. Albrecht, Z. Electrochem., 27, 162 (1921).
6. H. Bloom, J. O'M. Bockris, N. E. Richards and R. G. Taylor, J. Amer. Chem. Soc. 80, 2044 (1958).
7. A. G. Gaydon, "Dissociation Energies," Chapman and Hall Ltd., London, 1953.
8. G. Herzberg, "Spectra of Diatomic Molecules," D. Van Nostrand Company, New York, 1950.
9. L. Brewer, Univ. of Calif., Berkeley, personal communication, Oct. 23, 1964.
10. S. A. Rice and W. Klemperer, J. Chem. Phys. 27, 573 (1957).
11. A. Honig, M. Mandel, M. L. Stitch and C. H. Townes, Phys. Rev. 96, 629 (1954).
12. L. R. Maxwell, S. B. Hendricks and V. M. Mosley, Phys. Rev. 52, 968 (1937).

Sodium Bromide (NaBr) $Br_1Na_1(g)$

Enthalpy Reference Temperature = T_r = 298.15 K Standard State Pressure = p° = 0.1 MPa

T/K	$C_p°$ (J $K^{-1}mol^{-1}$)	S°	$-[G°-H°(T_r)]/T$	$H°-H°(T_r)$ (kJ mol^{-1})	$\Delta_f H°$	$\Delta_f G°$	Log K_f
0	0.	0.	INFINITE	-9.807	-135.035	-135.035	INFINITE
100	31.273	203.967	272.441	-6.847	-134.812	-150.135	78.422
200	34.956	226.965	244.505	-3.508	-136.379	-164.890	43.065
250	35.803	234.864	241.813	-1.737	-137.340	-171.910	35.919
298.15	36.323	241.218	241.218	0.	-143.930	-177.817	31.153
300	36.339	241.443	241.219	0.067	-143.984	-178.027	30.997
350	36.699	247.073	241.663	1.894	-159.936	-182.814	27.283
400	36.957	251.991	242.653	3.735	-163.178	-185.836	24.268
450	37.150	256.356	243.937	5.588	-163.808	-188.630	21.896
500	37.301	260.278	245.379	7.450	-164.411	-191.355	19.991
600	37.529	267.100	248.447	11.192	-165.545	-196.636	17.119
700	37.699	272.899	251.536	14.954	-166.604	-201.733	15.054
800	37.837	277.942	254.529	18.731	-167.612	-206.682	13.495
900	37.957	282.405	257.383	22.520	-168.592	-211.507	12.276
1000	38.065	286.410	260.089	26.322	-169.565	-216.223	11.294
1100	38.165	290.043	262.649	30.133	-170.553	-220.841	10.487
1200	38.260	293.368	265.073	33.954	-268.339	-222.928	9.704
1300	38.351	296.434	267.369	37.785	-268.487	-219.138	8.805
1400	38.440	299.279	269.547	41.625	-268.631	-215.337	8.034
1500	38.526	301.934	271.619	45.473	-268.770	-211.525	7.366
1600	38.611	304.423	273.592	49.330	-268.907	-207.704	6.781
1700	38.695	306.767	275.476	53.195	-269.041	-203.875	6.264
1800	38.778	308.981	277.276	57.069	-269.173	-200.038	5.805
1900	38.860	311.080	279.000	60.950	-269.305	-196.193	5.394
2000	38.941	313.075	280.655	64.840	-269.438	-192.342	5.023
2100	39.022	314.977	282.244	68.739	-269.573	-188.484	4.688
2200	39.103	316.794	283.774	72.645	-269.711	-184.619	4.383
2300	39.183	318.534	285.247	76.559	-269.854	-180.748	4.105
2400	39.263	320.203	286.669	80.481	-270.002	-176.871	3.849
2500	39.342	321.808	288.043	84.412	-270.158	-172.987	3.614
2600	39.422	323.352	289.371	88.350	-270.321	-169.097	3.397
2700	39.501	324.841	290.658	92.296	-270.492	-165.201	3.196
2800	39.580	326.279	291.904	96.250	-270.674	-161.298	3.009
2900	39.660	327.670	293.114	100.212	-270.865	-157.388	2.835
3000	39.739	329.016	294.288	104.182	-271.066	-153.472	2.672
3100	39.817	330.320	295.430	108.160	-271.279	-149.548	2.520
3200	39.896	331.585	296.540	112.146	-271.502	-145.618	2.377
3300	39.975	332.814	297.621	116.139	-271.737	-141.681	2.243
3400	40.054	334.009	298.673	120.141	-271.985	-137.736	2.116
3500	40.132	335.171	299.700	124.150	-272.246	-133.784	1.997
3600	40.211	336.303	300.701	128.167	-272.519	-129.824	1.884
3700	40.289	337.405	301.678	132.192	-272.807	-125.856	1.777
3800	40.368	338.481	302.632	136.225	-273.110	-121.880	1.675
3900	40.446	339.530	303.565	140.265	-273.429	-117.897	1.579
4000	40.524	340.555	304.477	144.314	-273.766	-113.904	1.487
4100	40.603	341.557	305.369	148.370	-274.121	-109.903	1.400
4200	40.681	342.536	306.243	152.434	-274.494	-105.894	1.317
4300	40.759	343.495	307.098	156.506	-274.892	-101.875	1.238
4400	40.837	344.433	307.936	160.586	-275.271	-97.850	1.162
4500	40.916	345.351	308.757	164.674	-275.695	-93.813	1.089
4600	40.994	346.251	309.562	168.769	-276.153	-89.767	1.019
4700	41.072	347.134	310.352	172.873	-276.638	-85.710	0.953
4800	41.150	347.999	311.128	176.984	-277.153	-81.642	0.888
4900	41.228	348.849	311.889	181.103	-277.700	-77.563	0.827
5000	41.307	349.682	312.636	185.230	-278.273	-73.474	0.768
5100	41.385	350.501	313.371	189.364	-278.888	-69.371	0.711
5200	41.463	351.305	314.093	193.506	-279.543	-65.257	0.656
5300	41.541	352.096	314.802	197.657	-280.239	-61.129	0.602
5400	41.619	352.873	315.500	201.815	-280.979	-56.988	0.551
5500	41.697	353.638	316.187	205.981	-281.766	-52.833	0.502
5600	41.775	354.390	316.862	210.154	-282.601	-48.664	0.454
5700	41.853	355.130	317.527	214.336	-283.488	-44.478	0.408
5800	41.931	355.858	318.182	218.525	-284.428	-40.277	0.363
5900	42.010	356.576	318.826	222.722	-285.424	-36.059	0.319
6000	42.088	357.282	319.461	226.927	-286.319	-31.842	0.277

PREVIOUS: September 1964 (1 atm) CURRENT: September 1964 (1 bar)

Sodium Bromide (NaBr) $Br_1Na_1(g)$

PHOSPHORUS BROMIDE (PBr) IDEAL GAS M_r = 110.87776

$S°(298.15\ K) = [249.2]\ J\ K^{-1}\ mol^{-1}$

$\Delta_f H°(0\ K) = [170.7]\ kJ\ mol^{-1}$
$\Delta_f H°(298.15\ K) = [162.6]\ kJ\ mol^{-1}$

Electronic Level and Quantum Weight

State	ε_i, cm^{-1}	g_i
$^3\Sigma$	0	[3]

$\omega_e = [400]\ cm^{-1}$ $\quad \omega_e x_e = [1.5]\ cm^{-1}$ $\quad \sigma = 1$
$B_e = [0.1519]\ cm^{-1}$ $\quad \alpha_e = [0.0008]\ cm^{-1}$ $\quad r_e = [2.23]$ Å

Enthalpy of Formation

$\Delta_f H°(0\ K)$ was estimated as 40.8 kcal mol^{-1} by Gordon (1) with white (α) phosphorus as the reference state.

Heat Capacity and Entropy

Molecular constants were estimated by Gordon (1).

Reference

(1) J. S. Gordon, Astrosystems International, Livingston, New Jersey, personal communication, August 25, 1961.

Phosphorus Bromide (PBr) $Br_1P_1(g)$

Enthalpy Reference Temperature = T_r = 298.15 K Standard State Pressure = p° = 0.1 MPa

T/K	$C_p°$ (J K^{-1}mol^{-1})	S° (J K^{-1}mol^{-1})	$-[G°-H°(T_r)]/T$ (J K^{-1}mol^{-1})	$H°-H°(T_r)$ (kJ mol^{-1})	$\Delta_f H°$ (kJ mol^{-1})	$\Delta_f G°$ (kJ mol^{-1})	Log K_f
0	0.	0.	INFINITE	-9.508	170.707	170.707	INFINITE
100	30.017	213.468	279.294	-6.583	171.349	154.463	-80.683
200	33.551	235.417	252.392	-3.395	169.823	137.906	-36.017
250	34.666	243.032	249.783	-1.688	169.030	130.016	-27.165
298.15	35.394	249.204	249.204	0.	162.601	123.238	-21.591
300	35.417	249.423	249.205	0.065	162.552	122.994	-21.415
350	35.935	254.924	249.638	1.850	146.076	117.345	-17.513
400	36.304	259.748	250.606	3.657	145.651	113.270	-14.791
450	36.578	264.040	251.865	5.479	145.237	109.247	-12.681
500	36.786	267.905	253.279	7.313	144.830	105.270	-10.997
600	37.083	274.640	256.294	11.008	144.032	97.433	-8.482
700	37.285	280.373	259.335	14.727	143.249	89.729	-6.696
800	37.434	285.362	262.283	18.463	142.476	82.136	-5.363
900	37.552	289.778	265.097	22.212	141.711	74.639	-4.332
1000	37.649	293.739	267.767	25.973	140.952	67.228	-3.512
1100	37.734	297.332	270.294	29.742	140.197	59.892	-2.844
1200	37.810	300.618	272.686	33.519	75.873	53.703	-2.338
1300	37.879	303.648	274.953	37.304	75.900	51.854	-2.084
1400	37.943	306.457	277.104	41.095	75.924	50.004	-1.866
1500	38.004	309.077	279.149	44.892	75.945	48.152	-1.677
1600	38.062	311.532	281.097	48.695	75.963	46.298	-1.511
1700	38.118	313.841	282.956	52.504	75.978	44.444	-1.366
1800	38.173	316.021	284.733	56.319	75.989	42.588	-1.236
1900	38.226	318.086	286.434	60.139	75.996	40.732	-1.120
2000	38.278	320.048	288.066	63.964	75.997	38.876	-1.015
2100	38.329	321.917	289.634	67.794	75.992	37.021	-0.921
2200	38.380	323.701	291.142	71.630	75.980	35.165	-0.835
2300	38.430	325.409	292.595	75.470	75.961	33.310	-0.756
2400	38.479	327.045	293.997	79.316	75.933	31.456	-0.685
2500	38.529	328.617	295.351	83.166	75.898	29.604	-0.619
2600	38.577	330.129	296.659	87.021	75.854	27.753	-0.558
2700	38.626	331.586	297.926	90.882	75.802	25.904	-0.501
2800	38.674	332.992	299.153	94.747	75.743	24.057	-0.449
2900	38.722	334.350	300.344	98.616	75.677	22.212	-0.400
3000	38.770	335.663	301.499	102.491	75.605	20.370	-0.355
3100	38.817	336.935	302.622	106.370	75.528	18.530	-0.312
3200	38.865	338.168	303.714	110.254	75.448	16.692	-0.272
3300	38.912	339.365	304.776	114.143	75.366	14.857	-0.235
3400	38.959	340.527	305.810	118.037	75.282	13.025	-0.200
3500	39.007	341.657	306.819	121.935	75.198	11.195	-0.167
3600	39.054	342.757	307.802	125.838	75.116	9.368	-0.136
3700	39.100	343.827	308.761	129.746	75.038	7.543	-0.106
3800	39.147	344.871	309.698	133.658	74.963	5.719	-0.079
3900	39.194	345.888	310.613	137.575	74.894	3.898	-0.052
4000	39.241	346.881	311.507	141.497	74.831	2.078	-0.027
4100	39.288	347.851	312.382	145.424	74.776	0.260	-0.003
4200	39.334	348.798	313.237	149.355	74.731	-1.557	0.019
4300	39.381	349.724	314.075	153.290	74.694	-3.373	0.041
4400	39.427	350.630	314.896	157.231	74.668	-5.188	0.062
4500	39.474	351.517	315.700	161.176	74.654	-7.002	0.081
4600	39.520	352.385	316.488	165.126	74.651	-8.817	0.100
4700	39.567	353.235	317.261	169.080	74.661	-10.632	0.118
4800	39.613	354.069	318.019	173.039	74.684	-12.447	0.135
4900	39.660	354.886	318.763	177.003	74.720	-14.262	0.152
5000	39.706	355.688	319.493	180.971	74.770	-16.078	0.168
5100	39.753	356.474	320.211	184.944	74.834	-17.896	0.183
5200	39.799	357.247	320.916	188.922	74.912	-19.715	0.198
5300	39.845	358.005	321.608	192.904	75.004	-21.536	0.212
5400	39.892	358.750	322.289	196.891	75.110	-23.358	0.226
5500	39.938	359.483	322.959	200.882	75.231	-25.183	0.239
5600	39.984	360.203	323.618	204.878	75.367	-27.010	0.252
5700	40.030	360.911	324.266	208.879	75.516	-28.839	0.264
5800	40.077	361.608	324.903	212.884	75.680	-30.671	0.276
5900	40.123	362.293	325.531	216.894	75.858	-32.507	0.288
6000	40.169	362.968	326.150	220.909	76.051	-34.345	0.299

PREVIOUS: December 1963 (1 atm) CURRENT: December 1963 (1 bar)

Phosphorus Bromide (PBr) $Br_1P_1(g)$

LEAD BROMIDE (PbBr) IDEAL GAS M_r = 287.104

$\Delta_f H°(0\ K)$ = 79.91 ± 42 kJ mol^{-1}

S°(298.15 K) = 272.47 ± 0.21 J K^{-1} mol^{-1} $\Delta_f H°(298.15\ K)$ = 70.92 ± 42 kJ mol^{-1}

Electronic Levels and Quantum Weights

State	ε_i, cm^{-1}	g_i
X $^2\Pi_{1/2}$	0	2
$^2\Pi_{3/2}$	[8300]	2
A $[^2\Sigma]$	20884.3	[2]
B $[^2\Sigma]$	34523.7	[2]

ω_e = 206.56 cm^{-1} $\omega_e x_e$ = 0.495 cm^{-1} σ = 1

B_e = 0.0451 cm^{-1} α_e = [0.00014] cm^{-1} r_e = 2.546 Å

Enthalpy of Formation

$\Delta_f H°(0\ K)$ = 19.10±10 kcal mol^{-1} (79.914±42 kJ mol^{-1}) is calculated and rounded from the D_0^0 = 2.43 eV (56.04 kcal mol^{-1}) determined by Singh from emission spectra (1), $\Delta_f H°$(Pb, g, 0 K) = 46.91±0.13 kcal mol^{-1} (2), and $\Delta_f H°$(Br, g, 0K) = 28.188 kcal mol^{-1} (2). Rosen (3) and Gaydon (4) recommended 57±9 kcal mol^{-1} and 58±9 kcal mol^{-1}, respectively, for D_0^0. A linear Birge-Sponer extrapolation of the ground state vibrational constants, as derived from absorption spectra (ν, 0-11) observed by Morgan (5), gives a dissociation energy of 2.66 eV. Correction for the ionic character of PbBr as proposed by Hildenbrand (6) gives D_0^0 = 2.44 eV which is in good agreement with the value determined by Singh (1). $\Delta_f H°(298.15\ K)$ = 16.95±10 kcal mol^{-1} is calculated from the adopted $\Delta_f H°(0\ K)$ = 19.10±10 kcal mol^{-1}.

Heat Capacity and Entropy

The vibrational constants are from the compilation of Rosen (3), corrected for the natural isotopic abundance of ^{79}Br and ^{81}Br. Analogous to PbF (2) and PbCl (2), a $^2\Pi_{3/2}$ state at 8300 cm^{-1} is added to the electronic states and levels listed by Rosen (3). The assignment of the first excited state remains uncertain; the choice appears to be between $^2\Sigma$ state, tentatively assigned to PbF (2) and PbCl (2), and $^2\Delta$, tentatively assigned to PbI (2). The $^2\Sigma$ state is adopted. The entropy is increased by 0.009 cal K^{-1} mol^{-1} at 4000 K and 0.012 cal K^{-1} mol^{-1} at 6000 K if the $^2\Delta$ state is used. The adopted value of B_e was determined by Lal and Khanna (7) and is adjusted for the natural abundance of the bromine isotopes. The value of r_e is calculated from the adopted B_e. The value of α_e is calculated assuming a Morse potential function.

References

1. S. P. Singh, Indian J. Pure Appl. Phys. 1968, 384 (1968).
2. JANAF Thermochemical Tables: Pb(g), 3-31-62; Br(g), 9-30-61; PbF(g), 12-31-73; PbCl(g), 6-30-73; PbI(g), 12-31-73.
3. B. Rosen, Ed., "Spectroscopic Data Relative to Diatomic Molecules," Pergamon Press, New York, 1970.
4. A. G. Gaydon, "Dissociation Energies and Spectra of Diatomic Molecules," 3rd ed., Chapman and Hall, London, 1968.
5. F. Morgan, Phys. Rev. 49, 47 (1936).
6. D. L. Hildenbrand in "Advances in High Temperature Chemistry," Vol. 1, L. Eyring, Ed., Academic Press, New York, 1967, pp. 193-215.
7. K. M. Lal and B. M. Khanna, Can. J. Phys. 46, 1991 (1968).

Lead Bromide (PbBr) $Br_1Pb_1(g)$

Enthalpy Reference Temperature = T_r = 298.15 K Standard State Pressure = p° = 0.1 MPa

T/K	$C_p°$ (J $K^{-1}mol^{-1}$)	S° (J $K^{-1}mol^{-1}$)	$-[G°-H°(T_r)]/T$ (J $K^{-1}mol^{-1}$)	$H°-H°(T_r)$ (kJ mol^{-1})	$\Delta_f H°$ (kJ mol^{-1})	$\Delta_f G°$ (kJ mol^{-1})	Log K_f
0	0.	0.	INFINITE	-10.146	79.910	79.910	INFINITE
100	33.337	233.663	304.665	-7.100	79.793	62.802	-32.804
200	36.150	257.867	275.829	-3.592	78.326	46.349	-12.105
250	36.638	265.991	273.077	-1.771	77.433	38.455	-8.035
298.15	36.922	272.471	272.471	0.	70.923	31.692	-5.552
300	36.930	272.699	272.471	0.068	70.872	31.448	-5.476
350	37.123	278.407	272.922	1.920	55.044	25.756	-3.844
400	37.260	283.374	273.925	3.780	54.614	21.602	-2.821
450	37.363	287.769	275.223	5.645	54.162	17.502	-2.032
500	37.446	291.710	276.678	7.516	53.691	13.454	-1.405
600	37.572	298.549	279.770	11.267	52.693	5.499	-0.479
700	37.671	304.348	282.878	15.029	46.764	-1.485	0.111
800	37.754	309.384	285.883	18.801	45.642	-8.301	0.542
900	37.830	313.835	288.746	22.580	44.554	-14.978	0.869
1000	37.904	317.825	291.458	26.367	43.499	-21.536	1.125
1100	37.980	321.441	294.022	30.161	42.477	-27.990	1.329
1200	38.062	324.749	296.447	33.963	41.489	-34.352	1.495
1300	38.155	327.799	298.743	37.773	40.529	-40.633	1.633
1400	38.262	330.631	300.920	41.594	39.582	-46.841	1.748
1500	38.385	333.274	302.990	45.426	38.641	-52.981	1.845
1600	38.524	335.756	304.961	49.272	37.699	-59.058	1.928
1700	38.680	338.096	306.842	53.132	36.753	-65.077	2.000
1800	38.852	340.312	308.641	57.008	35.801	-71.039	2.062
1900	39.038	342.417	310.363	60.903	34.841	-76.949	2.115
2000	39.237	344.425	312.017	64.816	33.872	-82.807	2.163
2100	39.446	346.344	313.606	68.750	-144.328	-81.502	2.027
2200	39.663	348.184	315.136	72.706	-144.927	-78.497	1.864
2300	39.886	349.952	316.612	76.683	-145.603	-75.462	1.714
2400	40.112	351.655	318.037	80.683	-146.359	-72.397	1.576
2500	40.339	353.297	319.414	84.706	-147.195	-69.298	1.448
2600	40.566	354.883	320.748	88.751	-148.109	-66.165	1.329
2700	40.791	356.418	322.041	92.819	-149.101	-62.994	1.219
2800	41.012	357.906	323.295	96.909	-150.165	-59.786	1.115
2900	41.230	359.349	324.514	101.021	-151.297	-56.538	1.018
3000	41.441	360.750	325.699	105.155	-152.492	-53.251	0.927
3100	41.648	362.112	326.851	109.309	-153.743	-49.922	0.841
3200	41.848	363.438	327.974	113.484	-155.042	-46.553	0.760
3300	42.041	364.728	329.068	117.678	-156.382	-43.142	0.683
3400	42.228	365.986	330.136	121.892	-157.756	-39.689	0.610
3500	42.407	367.213	331.178	126.124	-159.155	-36.197	0.540
3600	42.581	368.410	332.195	130.373	-160.574	-32.663	0.474
3700	42.747	369.579	333.190	134.640	-162.003	-29.091	0.411
3800	42.907	370.721	334.163	138.922	-163.438	-25.479	0.350
3900	43.061	371.838	335.114	143.221	-164.870	-21.830	0.292
4000	43.210	372.930	336.046	147.534	-166.295	-18.144	0.237
4100	43.352	373.999	336.959	151.863	-167.707	-14.423	0.184
4200	43.489	375.045	337.853	156.205	-169.101	-10.668	0.133
4300	43.621	376.070	338.730	160.560	-170.473	-6.879	0.084
4400	43.749	377.074	339.590	164.929	-171.820	-3.059	0.036
4500	43.872	378.059	340.434	169.310	-173.137	0.791	-0.009
4600	43.991	379.024	341.263	173.703	-174.422	4.671	-0.053
4700	44.106	379.971	342.076	178.108	-175.674	8.578	-0.095
4800	44.217	380.901	342.875	182.524	-176.889	12.511	-0.136
4900	44.325	381.814	343.661	186.951	-178.067	16.469	-0.176
5000	44.430	382.711	344.433	191.389	-179.207	20.451	-0.214
5100	44.531	383.591	345.192	195.837	-180.308	24.455	-0.250
5200	44.631	384.457	345.939	200.295	-181.369	28.480	-0.286
5300	44.727	385.308	346.674	204.763	-182.390	32.526	-0.321
5400	44.821	386.145	347.397	209.240	-183.372	36.590	-0.354
5500	44.914	386.968	348.109	213.727	-184.315	40.672	-0.386
5600	45.004	387.778	348.810	218.223	-185.219	44.771	-0.418
5700	45.092	388.576	349.501	222.728	-186.085	48.886	-0.448
5800	45.178	389.361	350.181	227.241	-186.914	53.015	-0.477
5900	45.263	390.134	350.852	231.763	-187.706	57.159	-0.506
6000	45.346	390.895	351.513	236.294	-188.463	61.315	-0.534

PREVIOUS: December 1973 (1 atm) CURRENT: December 1973 (1 bar)

Lead Bromide (PbBr) $Br_1Pb_1(g)$

S°(298.15 K) = 247.42 ± 0.21 J K^{-1} mol^{-1}

$\Delta_f H°$(0 K) = 240.8 ± 46.0 kJ mol^{-1}
$\Delta_f H°$(298.15 K) = 235.3 ± 46.0 kJ mol^{-1}

Electronic States and Molecular Constants

State	ε_i, cm^{-1}	g_i	ω_e, cm^{-1}	$\omega_e x_e$, cm^{-1}	B_e, cm	α_e, cm^{-1}	r_e, Å	Source
X $^2\Pi_{1/2}$	0	2	423.2	1.5	0.1669	0.0019	2.210	(5)
$^2\Pi_{3/2}$	419.2	2	423.2	1.5	0.1669	0.0019	2.210	(5)
A $^2\Sigma$	20937.6	2	249.6	0.5	[0.1669]	[0.0019]	[2.210]	(6)
B' $^2\Delta$	23920	4	394	4	[0.1669]	[0.0019]	[2.210]	(1)
B $^2\Sigma$	33572.7	2	573.6	4.0	0.1793	[0.0019]	2.132	(5)
C $^2\Pi$	41057	4	529.2	2.0	[0.1669]	[0.0019]	[2.210]	(10)

Enthalpy of Formation

Kuznetsova and Kuzyakov (1) suggested a value of D_0^0 = 85.8±11.4 kcal mol^{-1} based on linear Birge-Sponer extrapolations of the ground state and B' $^2\Delta$ state vibrational data. We correct this value for the ionic character of the bond according to Hildenbrand (2) and obtain D_0^0 = 77.3±10.3 kcal mol^{-1}. With auxiliary JANAF data (3) this yields the adopted $\Delta_f H°$(0 K) = 57.543 kcal mol^{-1} (240.760 kJ mol^{-1}). Tandon and Tandon (4) obtained D_0^0 = 85.3 kcal mol^{-1} from a theoretical treatment based on Sutherland's potential function in good agreement with the adopted value. The value of D_0^0 may be compared to the average (per bond) enthalpy of atomization of 80.58 and 77.89 kcal mol^{-1} for $SiBr_2$(g) and $SiBr_4$(g), respectively (3).

Heat Capacity and Entropy

Molecular constant data for the ground state are from Mishra and Khanna (5). The vibrational data is in good agreement with data from Rao and Haranath (6), Kuznetsova et al. (7), and Jevons and Bashford (8). Data for the A $^2\Sigma$ and B' $^2\Delta$ states are from Rao and Haranath (6) and Kuznetsova and Kuzyakov (1), respectively. The B $^2\Sigma$ state data are from Mishra and Khanna (5) and the vibrational data agree well with Jevons and Bashford (8). Slightly different values for the rotational constants for both this state and the ground state have been obtained by Kuznetsova and Kuzyakov (9). Data for the C $^2\Pi$ state is from Oldershaw and Robinson (10) who also reported data for several higher states; these are not included due to uncertainties in their assignments and degeneracies. Splittings of 18 and 12 cm^{-1} for the B' $^2\Delta$ and C $^2\Pi$ states (1, 10), respectively, have not been included in our calculations. The $^4\Sigma$ state observed for SiF and SiI has not been observed for SiBr. It is expected to lie near 24000 cm^{-1} by analogy with SiF and SiI (3) which is very near the B' $^2\Delta$ state. This raises the possibility that this state may have been missed, or the $^2\Delta$ state misassigned, due to the large number of lines in this region (1). A high resolution re-examination of this region of the spectrum is desirable to clarify the situation. We have assigned an uncertainity of ±0.05 cal K^{-1} mol^{-1} to S°(298.15 K) in view of these uncertainties. All molecular constant data has been corrected to the natural abundances of Si and Br assuming the observed data was for ^{28}Si ^{79}Br. The ground state is treated as two distinct levels due to the splitting of this state as expressed by the spin coupling constant (A = 419.2 cm^{-1}). This approximation gives slightly biased results at low temperature: the stated uncertainty in S°(298.15 K) should account for this.

References

1. L. A. Kuznetsova and Y. Y. Kuzyakov, Zhur. Prikl. Spectrosc., 10, 413 (1969).
2. D. L. Hildenbrand, in "Advances in High Temperature Chemistry," Vol. 1, L. Eyring, Ed. Academic Press, New York, 1967.
3. JANAF Thermochemical Tables: Si(g), 3-31-67; SiF(g), SiI(g), 12-31-76; Br(g), 6-30-74; $SiBr_2$(g), $SiBr_4$(g), 12-31-76.
4. S. P. Tandon and K. Tandon, Indian J. Phys., 38, 460 (1964).
5. R. K. Mishra and B. N. Khanna, Indian J. Phys., 46, 1 (1972).
6. K. B. Rao and P. B. V. Haranath, Proc. Phys. Soc. London., (At. Mol. Phys.), 2, 1381 (1969).
7. L. A. Kuznetsova, N. E. Kuzmenko, and Y. Y. Kuzyakov, Vestn. Mosk. Univ. Khim., 23, 30 (1968).
8. W. Jevons and L. A. Bashford, Proc. Phys. Soc. (London), 49, 554 (1937).
9. L. A. Kuznetsova and Y. Y. Kuzyakov, Vestn. Mosk. Univ. Khim., 24, 103 (1969).
10. G. A. Oldershaw and K. Robinson, Trans. Faraday Soc., 67, 1870 (1971).

Enthalpy Reference Temperature = T_r = 298.15 K Standard State Pressure = p° = 0.1 MPa

T/K	$C_p°$ (J $K^{-1}mol^{-1}$)	S° (J $K^{-1}mol^{-1}$)	-[G°-H°(T_r)]/T (J $K^{-1}mol^{-1}$)	H°-H°(T_r) (kJ mol^{-1})	$\Delta_f H°$ (kJ mol^{-1})	$\Delta_f G°$ (kJ mol^{-1})	Log K_f
0	0.	0.	INFINITE	-10.040	240.758	240.758	INFINITE
100	30.564	209.105	280.168	-7.106	242.029	224.194	-117.107
200	36.635	232.296	250.917	-3.724	241.782	206.398	-53.906
298.15	38.731	247.418	247.418	0.	235.325	189.859	-33.262
300	38.746	247.658	247.419	0.072	235.290	189.577	-33.008
400	39.025	258.862	248.945	3.967	219.824	177.510	-23.180
500	38.903	267.559	251.831	7.864	219.597	166.955	-17.442
600	38.742	274.638	255.060	11.746	219.244	156.458	-13.621
700	38.615	280.600	258.294	15.614	218.794	146.028	-10.897
800	38.529	285.750	261.412	19.471	218.266	135.668	-8.858
900	38.477	290.285	264.373	23.321	217.672	125.379	-7.277
1000	38.449	294.337	267.170	27.167	217.020	115.158	-6.015
1100	38.440	298.001	269.809	31.011	216.317	105.006	-4.986
1200	38.446	301.346	272.300	34.855	215.566	94.919	-4.132
1300	38.462	304.424	274.654	38.701	214.771	84.897	-3.411
1400	38.486	307.275	276.884	42.548	213.931	74.938	-2.796
1500	38.517	309.931	278.999	46.398	213.048	65.041	-2.265
1600	38.552	312.418	281.011	50.252	212.120	55.204	-1.802
1700	38.591	314.757	282.928	54.109	160.972	45.873	-1.409
1800	38.634	316.964	284.758	57.970	160.187	39.125	-1.135
1900	38.679	319.054	286.509	61.836	159.398	32.421	-0.891
2000	38.726	321.039	288.186	65.706	158.606	25.758	-0.673
2100	38.775	322.930	289.796	69.581	157.810	19.135	-0.476
2200	38.826	324.735	291.343	73.461	157.008	12.550	-0.298
2300	38.879	326.462	292.833	77.346	156.201	6.002	-0.136
2400	38.933	328.117	294.269	81.237	155.388	-0.511	0.011
2500	38.989	329.708	295.655	85.133	154.570	-6.990	0.146
2600	39.047	331.238	296.994	89.035	153.745	-13.436	0.270
2700	39.107	332.713	298.290	92.942	152.916	-19.850	0.384
2800	39.170	334.136	299.545	96.856	152.082	-26.234	0.489
2900	39.235	335.512	300.761	100.777	151.244	-32.587	0.587
3000	39.304	336.843	301.942	104.704	150.403	-38.912	0.678
3100	39.376	338.133	303.089	108.638	149.562	-45.209	0.762
3200	39.451	339.385	304.204	112.579	148.721	-51.478	0.840
3300	39.530	340.600	305.288	116.528	147.882	-57.721	0.914
3400	39.614	341.781	306.344	120.485	147.046	-63.939	0.982
3500	39.702	342.931	307.373	124.451	146.216	-70.133	1.047
3600	39.796	344.050	308.377	128.426	-238.758	-65.841	0.955
3700	39.894	345.142	309.356	132.410	-239.158	-61.033	0.862
3800	39.998	346.207	310.311	136.405	-239.548	-56.213	0.773
3900	40.107	347.248	311.245	140.410	-239.927	-51.384	0.688
4000	40.223	348.265	312.158	144.426	-240.292	-46.545	0.608
4100	40.344	349.259	313.051	148.455	-240.642	-41.697	0.531
4200	40.471	350.233	313.925	152.495	-240.975	-36.840	0.458
4300	40.604	351.187	314.780	156.549	-241.289	-31.976	0.388
4400	40.743	352.122	315.618	160.616	-241.582	-27.105	0.322
4500	40.888	353.039	316.440	164.698	-241.853	-22.227	0.258
4600	41.040	353.939	317.245	168.794	-242.100	-17.344	0.197
4700	41.197	354.824	318.035	172.906	-242.323	-12.456	0.138
4800	41.360	355.693	318.811	177.034	-242.520	-7.563	0.082
4900	41.529	356.547	319.572	181.178	-242.690	-2.666	0.028
5000	41.703	357.388	320.320	185.340	-242.832	2.234	-0.023
5100	41.883	358.216	321.055	189.519	-242.945	7.137	-0.073
5200	42.069	359.031	321.778	193.717	-243.029	12.041	-0.121
5300	42.259	359.834	322.488	197.933	-243.084	16.947	-0.167
5400	42.454	360.626	323.187	202.169	-243.108	21.853	-0.211
5500	42.654	361.406	323.875	206.424	-243.101	26.760	-0.254
5600	42.859	362.177	324.552	210.700	-243.064	31.666	-0.295
5700	43.067	362.937	325.219	214.996	-242.995	36.572	-0.335
5800	43.280	363.688	325.875	219.313	-242.896	41.475	-0.374
5900	43.496	364.430	326.523	223.652	-242.766	46.377	-0.411
6000	43.716	365.163	327.161	228.013	-242.603	51.276	-0.446

PREVIOUS: December 1976 (1 atm) CURRENT: December 1976 (1 bar)

STRONTIUM BROMIDE (SrBr) IDEAL GAS M_r = 167.524

$\Delta_f H°$(0 K) = -80.4 ± 42 kJ mol^{-1}

S°(298.15 K) = 263.8 ± 0.4 J K^{-1} mol^{-1} $\Delta_f H°$(298.15 K) = -89.1 ± 42 kJ mol^{-1}

Electronic Levels and Quantum Weights

State	ε_i, cm^{-1}	g_i	State	ε_i, cm^{-1}	g_i
X $^2\Sigma^+$	0	2	E $^2\Sigma$	32052.5	2
A_1 $^2\Pi_{1/21}$	14699.4	2	F_1 $^2\Pi_{1/2}$	33131.7	2
A_2 $^2\Pi_{3/2}$	15000.7	2	F_2 $^2\Pi_{3/2}$	33215.0	2
B $^2\Sigma$	15352.0	2	G_1 $^2\Delta_{3/2}$	34257.0	2
C_1 $^2\Pi_{1/2}$	24343.7	2	G_2 $^2\Delta_{3/2}$	34282.8	2
C_2 $^2\Pi_{3/2}$	24665.8	2	H $^2\Sigma$	34357.7	2
D $^2\Sigma$	28958.2	2			

ω_e = 216.5 cm^{-1} $\omega_e x_e$ = 0.51 cm^{-1} σ = 1

B_e = [0.054924] cm^{-1} α_e = [0.000171] cm^{-1} r_e = [2.71] Å

Enthalpy of Formation

The selected value, $\Delta_f H°$(298.15 K) = -21.3 kcal mol^{-1} (-89.119 kJ mol^{-1}), is obtained from an analysis of spectroscopic data. Herzberg (1) suggested D_0^0 = 2.8 eV for SrBr(g) which was derived from a linear Birge-Sponer extrapolation of the ground state vibrational levels. Later, Gaydon (2) claimed that this value is unreliable and suggested that the true value may be much higher. The adopted ground state vibrational constants give D_0^0 = 2.84 eV by a similar extrapolation. We note that JANAF analyses (3) of the spectroscopic and thermochemical data for SrF(g) and SrCl(g) show that the ionicity corrections of Hildenbrand (4) bring the Birge-Sponer extrapolations into reasonable agreement with adopted D_0^0 values. Based on this correction for SrBr(g), we obtain D_0^0 = 3.76 eV (86.7 kcal mol^{-1}) which is adopted. The adopted value of D_0^0 corresponds to $\Delta_f H°$(298.15 K) = 21.3 kcal mol^{-1}. We also find D_0^0 (SrBr)/$\Delta_{at}H°(SrBr_2)$ = 0.46 which is quite consistent with values of this ratio for other alkaline earth halide systems (5).

Ionic model calculations (6, 7) have led to D_0^0 values of 5.07 eV (6) and 3.53 eV (7). The latter result is believed to represent a minimum value for D_0^0. Two other experimental values for D_0^0 which bracket the selected value, have been reported. Flame studies (8) gave D_0^0 = 3.4 eV, and chemiluminescence (9) from reaction of Sr atoms with Br_2 gave a lower limit to D_0^0 of 4.1 eV. We assign an uncertainty of ± 10 kcal mol^{-1} to $\Delta_f H°$(298.15 K) to include the possibility that these studies are correct.

Heat Capacity and Entropy

The value of r_e is obtained from that for gaseous $SrBr_2$ (3) with $r_e(SrBr)/r_e(SrBr_2)$ = 0.96. This value for the ratio is calculated from bond lengths (5) for several other alkaline earth halide systems. Our adopted value for r_e agrees with that (2.74 Å) estimated by Krasnov and Karaseva (7), while two other estimated values (6) 10] lie somewhat higher (~0.2 Å). The rotational constant is calculated from the estimated value for r_e. The value of α_e is obtained from a Morse potential function.

The vibrational constants and first seven electronic states and levels are taken from the compilation of Rosen (11). The E state energy has been measured by Reddy and Rao (12), while the F, G, and H state energies are due to Reddy et al. (13). The five uppermost states were associated with transitions between the excited states of SrBr, and their assignments (13) were made by analogy with the observed spectrum for SrCl(g).

References

1. G. Herzberg, "Spectra of Diatomic Molecules," 2nd ed., D. Van Nostrand Company, New York, 1950.
2. A. G. Gaydon, "Dissociation Energies and Spectra of Diatomic Molecules," Chapman and Hall, Ltd., London, 1968.
3. JANAF Thermochemical Tables: SrF(g) and SrCl(g), 12-31-72; $SrBr_2$(g), 6-30-74.
4. D. L. Hildenbrand in "Advances in High Temperature Chemistry," Vol. 1, L. Eyring, Ed., Academic Press, New York, 1967.
5. JANAF Thermochemical Tables.
6. J. L. Margrave, J. Phys. Chem. 58, 258 (1954).
7. K. S. Krasnov and N. V. Karaseva, Optics and Spectroscopy 19, 14 (1973).
8. A. N. Khitrov, V. G. Ryabova, and L. V. Gurvich, Teplofiz. Vys. Temp. 11, 1126 (1973).
9. M. Menzinger, Can. J. Chem. 52, 1688 (1974).
10. K. M. Guggenheimer, Proc. Phys. Soc. London 58, 456 (1946).
11. B. Rosen, Ed., "Spectroscopic Data Relative to Diatomic Molecules," Pergamon Press, New York, 1970.
12. Y. P. Reddy and P. T. Rao, Indian J. Pure and Appl. Phys. 4, 251 (1966).
13. B. R. K. Reddy, T. P. Reddy, and P. T. Rao, Curr. Sci. 40, 317 (1971).

Strontium Bromide (SrBr) Br_1Sr_1(g)

Enthalpy Reference Temperature = T_r = 298.15 K Standard State Pressure = p° = 0.1 MPa

	J K^{-1}mol^{-1}			kJ mol^{-1}			
T/K	$C_p°$	S°	-[G°-H°(T_r)]/T	H°-H°(T_r)	$\Delta_f H°$	$\Delta_f G°$	Log K_f
0	0.	0.	INFINITE	-10.105	-80.401	-80.401	INFINITE
100	33.079	225.114	295.842	-7.073	-80.265	-97.279	50.813
200	36.028	249.184	267.101	-3.584	-81.715	-113.767	29.713
250	36.551	257.285	264.356	-1.768	-82.609	-121.685	25.425
298.15	36.856	263.751	263.751	0.	-89.119	-128.461	22.506
300	36.866	263.979	263.751	0.068	-89.171	-128.705	22.410
350	37.073	269.678	264.201	1.917	-105.000	-134.416	20.060
400	37.219	274.639	265.202	3.774	-105.433	-138.589	18.098
450	37.329	279.029	266.500	5.638	-105.892	-142.706	16.565
500	37.416	282.967	267.953	7.507	-106.375	-146.771	15.333
600	37.548	289.801	271.042	11.255	-107.416	-154.754	13.473
700	37.649	295.597	274.146	15.015	-108.559	-162.554	12.130
800	37.733	300.630	277.149	18.785	-109.805	-170.184	11.112
900	37.808	305.078	280.010	22.562	-111.784	-177.579	10.306
1000	37.877	309.065	282.719	26.346	-112.921	-184.829	9.655
1100	37.941	312.678	285.281	30.137	-121.958	-191.600	9.098
1200	38.003	315.982	287.704	33.934	-124.003	-197.841	8.612
1300	38.063	319.027	289.998	37.738	-126.046	-203.911	8.193
1400	38.122	321.849	292.173	41.547	-128.088	-209.824	7.829
1500	38.181	324.482	294.240	45.362	-130.128	-215.591	7.508
1600	38.241	326.948	296.208	49.183	-132.168	-221.222	7.222
1700	38.303	329.268	298.085	53.010	-271.122	-225.546	6.930
1800	38.368	331.459	299.879	56.844	-271.301	-222.860	6.467
1900	38.438	333.535	301.596	60.684	-271.487	-220.163	6.053
2000	38.515	335.509	303.243	64.531	-271.682	-217.457	5.679
2100	38.601	337.390	304.825	68.387	-271.890	-214.741	5.341
2200	38.696	339.188	306.346	72.252	-272.115	-212.014	5.034
2300	38.803	340.910	307.812	76.127	-272.363	-209.277	4.753
2400	38.924	342.564	309.226	80.013	-272.640	-206.528	4.495
2500	39.059	344.156	310.591	83.912	-272.951	-203.767	4.257
2600	39.209	345.691	311.912	87.825	-273.305	-200.993	4.038
2700	39.376	347.173	313.190	91.754	-273.706	-198.205	3.834
2800	39.559	348.609	314.430	95.701	-274.164	-195.400	3.645
2900	39.760	350.000	315.633	99.667	-274.686	-192.578	3.469
3000	39.977	351.352	316.801	103.653	-275.279	-189.737	3.304
3100	40.212	352.667	317.937	107.663	-275.951	-186.875	3.149
3200	40.463	353.947	319.042	111.696	-276.709	-183.989	3.003
3300	40.730	355.196	320.119	115.756	-277.559	-181.079	2.866
3400	41.012	356.416	321.168	119.843	-278.508	-178.141	2.737
3500	41.309	357.609	322.193	123.959	-279.562	-175.174	2.614
3600	41.619	358.777	323.193	128.105	-280.725	-172.175	2.498
3700	41.942	359.922	324.170	132.283	-282.000	-169.143	2.388
3800	42.277	361.045	325.126	136.494	-283.392	-166.074	2.283
3900	42.622	362.148	326.061	140.739	-284.901	-162.968	2.183
4000	42.977	363.231	326.977	145.018	-286.528	-159.821	2.087
4100	43.341	364.297	327.874	149.334	-288.274	-156.632	1.996
4200	43.712	365.346	328.754	153.687	-290.132	-153.399	1.908
4300	44.089	366.379	329.617	158.077	-292.108	-150.120	1.824
4400	44.472	367.397	330.464	162.505	-294.194	-146.794	1.743
4500	44.859	368.400	331.296	166.971	-296.386	-143.420	1.665
4600	45.250	369.391	332.113	171.477	-298.614	-140.001	1.590
4700	45.643	370.368	332.917	176.021	-300.984	-136.527	1.517
4800	46.037	371.333	333.707	180.605	-303.437	-133.002	1.447
4900	46.433	372.286	334.485	185.229	-305.969	-129.426	1.380
5000	46.828	373.228	335.250	189.892	-308.564	-125.797	1.314
5100	47.222	374.160	336.004	194.594	-311.201	-122.117	1.251
5200	47.614	375.080	336.747	199.336	-313.909	-118.384	1.189
5300	48.004	375.991	337.478	204.117	-316.661	-114.597	1.129
5400	48.391	376.892	338.200	208.937	-319.447	-110.758	1.071
5500	48.774	377.783	338.912	213.795	-322.260	-106.868	1.015
5600	49.153	378.666	339.614	218.692	-325.092	-102.926	0.960
5700	49.527	379.539	340.306	223.626	-327.934	-98.934	0.907
5800	49.895	380.404	340.990	228.597	-330.780	-94.892	0.855
5900	50.257	381.260	341.666	233.605	-333.411	-90.824	0.804
6000	50.613	382.107	342.333	238.648	-336.211	-86.689	0.755

PREVIOUS: December 1974 (1 atm) CURRENT: December 1974 (1 bar)

Strontium Bromide (SrBr) Br_1Sr_1(g)

TITANIUM BROMIDE (TiBr) IDEAL GAS M_r = 127.784

$\Delta_f H°(0\ K)$ = [219.7 ± 42] kJ mol^{-1}

$S°(298.15\ K)$ = [260.25 ± 8.4] J K^{-1} mol^{-1} $\Delta_f H°(298.15\ K)$ = [212.5 ± 42] kJ mol^{-1}

Electronic Levels and Quantum Weights

State	ϵ_i, cm^{-1}	g_i
$[^4\Sigma]$	0	[4]
	[1000]	[4]
	[2000]	[4]
	[4000]	[4]
	[6000]	[4]
	[8000]	[4]
	[10000]	[4]

ω_e = [295] cm^{-1} $\omega_e x_e$ = [2.21] cm^{-1} σ = 1

B_e = [0.0904] cm^{-1} α_e = [0.00067] cm^{-1} r_e = [2.5] Å

Enthalpy of Formation

$\Delta_f H°$(TiBr, g, 298.15 K) is calculated from the dissociation energy which is estimated as 104 kcal mol^{-1}. This estimate is obtained from the relation $\Delta_{at}H°(TiBr_4) < D_0^0(TiBr) < \Delta_{at}H°(TiBr_3)$, which is valid for the titanium fluorides. The dissociation energy of TiF(g) from which the relation is derived was estimated relative to that of TiF_2(g) by Zmbov and Margrave (1).

Heat Capacity and Entropy

The vibrational frequency, ω_e, and the anharmonic vibrational term, $\omega_e x_e$, are estimated from those of TiCl(g) and comparisons of the mercury and alkali monohalides. The interatomic distance is estimated from Guggenhemier's relation (2). B_e is calculated from r_e. The ground state term and electronic levels are estimated from the ground state multiplet of Ti^+(3). α_e is estimated from the Morse potential function.

References

1. K. F. Zmbov and J. L. Margrave, J. Phys. Chem. 71, 2893 (1967).
2. K. M. Guggenheimer, Proc. Phys. Soc. 58, 456 (1946).
3. C. E. Moore, U. S. Nat. Bur. Stand. Circ. 467, 1949.

Titanium Bromide (TiBr) $Br_1Ti_1(g)$

Enthalpy Reference Temperature = T_r = 298.15 K Standard State Pressure = p° = 0.1 MPa

	J $K^{-1}mol^{-1}$			kJ mol^{-1}			
T/K	$C_p°$	S°	$-[G°-H°(T_r)]/T$	$H°-H°(T_r)$	$\Delta_f H°$	$\Delta_f G°$	Log K_f
0	0.	0.	INFINITE	-9.938	219.694	219.694	INFINITE
100	31.427	222.442	292.172	-6.973	220.702	201.976	-105.501
200	35.424	245.604	263.646	-3.608	219.699	183.565	-47.942
250	36.809	253.660	260.868	-1.802	218.934	174.618	-36.484
298.15	38.020	260.249	260.249	0.	212.547	166.815	-29.225
300	38.065	260.484	260.249	0.070	212.501	166.531	-28.996
350	39.192	266.438	260.717	2.002	196.821	159.737	-23.839
400	40.173	271.737	261.769	3.987	196.564	154.457	-20.170
450	41.000	276.518	263.147	6.017	196.316	149.209	-17.320
500	41.678	280.874	264.705	8.085	196.076	143.988	-15.042
600	42.651	288.566	268.058	12.305	195.611	133.614	-11.632
700	43.232	295.189	271.472	16.602	195.150	123.317	-9.202
800	43.559	300.985	274.806	20.943	194.690	113.087	-7.384
900	43.734	306.127	278.006	25.309	194.183	102.916	-5.973
1000	43.829	310.740	281.053	29.687	193.553	92.807	-4.848
1100	43.887	314.920	283.945	34.073	192.733	82.771	-3.930
1200	43.933	318.741	286.687	38.464	187.733	72.938	-3.175
1300	43.980	322.259	289.290	42.860	187.248	63.391	-2.547
1400	44.034	325.520	291.763	47.260	186.685	53.884	-2.010
1500	44.097	328.561	294.116	51.667	186.032	44.420	-1.547
1600	44.169	331.409	296.359	56.080	185.276	35.003	-1.143
1700	44.249	334.089	298.500	60.501	184.404	25.637	-0.788
1800	44.336	336.621	300.548	64.930	183.402	16.326	-0.474
1900	44.426	339.020	302.510	69.368	182.258	7.075	-0.195
2000	44.519	341.301	304.393	73.816	166.270	-1.659	0.043
2100	44.612	343.475	306.203	78.272	164.051	-10.001	0.249
2200	44.704	345.553	307.945	82.738	161.831	-18.237	0.433
2300	44.794	347.542	309.624	87.213	159.610	-26.372	0.599
2400	44.881	349.450	311.244	91.697	157.386	-34.410	0.749
2500	44.965	351.284	312.809	96.189	155.159	-42.356	0.885
2600	45.044	353.049	314.323	100.689	152.929	-50.213	1.009
2700	45.119	354.751	315.789	105.198	150.696	-57.984	1.122
2800	45.190	356.393	317.210	109.713	148.460	-65.672	1.225
2900	45.256	357.980	318.588	114.235	146.220	-73.280	1.320
3000	45.318	359.515	319.927	118.764	143.977	-80.810	1.407
3100	45.376	361.002	321.228	123.299	141.732	-88.266	1.487
3200	45.430	362.444	322.494	127.839	139.486	-95.650	1.561
3300	45.481	363.842	323.726	132.385	137.239	-102.963	1.630
3400	45.528	365.201	324.926	136.935	134.993	-110.208	1.693
3500	45.572	366.521	326.095	141.490	132.748	-117.387	1.752
3600	45.614	367.806	327.236	146.050	130.506	-124.502	1.806
3700	45.653	369.056	328.350	150.613	-280.831	-123.767	1.747
3800	45.690	370.274	329.437	155.180	-281.831	-119.508	1.643
3900	45.726	371.461	330.499	159.751	-282.880	-115.223	1.543
4000	45.760	372.619	331.538	164.325	-283.977	-110.910	1.448
4100	45.793	373.750	332.554	168.903	-285.119	-106.570	1.358
4200	45.825	374.854	333.548	173.484	-286.303	-102.201	1.271
4300	45.856	375.932	334.521	178.068	-287.527	-97.803	1.188
4400	45.887	376.987	335.474	182.655	-288.788	-93.376	1.109
4500	45.918	378.018	336.408	187.245	-290.084	-88.920	1.032
4600	45.949	379.028	337.324	191.839	-291.412	-84.435	0.959
4700	45.980	380.016	338.222	196.435	-292.769	-79.921	0.888
4800	46.011	380.985	339.103	201.035	-294.154	-75.378	0.820
4900	46.042	381.934	339.967	205.637	-295.564	-70.806	0.755
5000	46.074	382.864	340.816	210.243	-296.996	-66.204	0.692
5100	46.106	383.777	341.649	214.852	-298.447	-61.574	0.631
5200	46.140	384.673	342.468	219.464	-299.917	-56.915	0.572
5300	46.173	385.552	343.273	224.080	-301.402	-52.228	0.515
5400	46.208	386.415	344.064	228.699	-302.901	-47.513	0.460
5500	46.244	387.263	344.841	233.322	-304.411	-42.769	0.406
5600	46.280	388.097	345.606	237.948	-305.930	-37.998	0.354
5700	46.318	388.916	346.359	242.578	-307.458	-33.200	0.304
5800	46.356	389.722	347.100	247.212	-308.991	-28.376	0.256
5900	46.395	390.515	347.829	251.849	-310.528	-23.524	0.208
6000	46.436	391.295	348.547	256.491	-312.068	-18.647	0.162

PREVIOUS: June 1968 (1 atm) CURRENT: June 1968 (1 bar)

Titanium Bromide (TiBr) $Br_1Ti_1(g)$

$S°(298.15 K) = [272.6] J K^{-1} mol^{-1}$

$\Delta_f H°(0 K) = [593.7 \pm 84] kJ mol^{-1}$
$\Delta_f H°(298.15 K) = [586.2 \pm 84] kJ mol^{-1}$

Electronic Levels and Quantum Weights

State	ε_i, cm^{-1}	g_i
[$^2\Delta$]	0	[4]
	[5000]	[2]
	[15000]	[20]

$\omega_e = [321.4] cm^{-1}$ $\omega_e x_e = [0.80] cm^{-1}$ $\sigma = 1$

$B_e = [0.05254] cm^{-1}$ $\alpha_e = [0.00015] cm^{-1}$ $r_e = [2.40] Å$

Enthalpy of Formation

$\Delta_f H°$(WBr, g, 298.15 K) = 140.1 kcal mol^{-1} (586.178 kJ mol^{-1}) is calculated from the bond dissociation energy, $\Delta_{at}H°$(W-Br, 298.15 K) = 90±20 kcal mol^{-1}. This value is estimated to be slightly higher than the average bond dissociation energy of WBr_6(g), by analogy with the WCl_x system.

Heat Capacity and Entropy

The bond distance is estimated to be the same as that in WBr_6(g). The distance is then used with Guggenheimer's relation (1) for polar molecules to calculate the fundamental vibrational freqency, ω_e. The anharmonicity correction x_e is estimated roughly by assuming $x_e = \omega_e/(D° + 0.5\ \omega_e) = 0.0025$. The rotational constant B_e is calculated from the estimated bond distance. The value of α_e is calculated from the Morse potential function.

The ground state configuration, low lying electronic levels and their quantum weights are assumed to be the same as those estimated for WF(g) (2).

References

1. K. M. Guggenheimer, Proc. Phys. Soc. (London) 58, 456 (1946).
2. JANAF Thermochemical Table: WF(g), 3-31-67.

Enthalpy Reference Temperature = T_r = 298.15 K — Standard State Pressure = p° = 0.1 MPa

T/K	$C_p°$ (J K^{-1}mol^{-1})	S° (J K^{-1}mol^{-1})	$-[G°-H°(T_r)]/T$ (J K^{-1}mol^{-1})	$H°-H°(T_r)$ (kJ mol^{-1})	$\Delta_f H°$ (kJ mol^{-1})	$\Delta_f G°$ (kJ mol^{-1})	Log K_f
0	0.	0.	INFINITE	-9.726	593.680	593.680	INFINITE
100	30.915	235.768	303.535	-6.777	594.581	574.657	-300.170
200	34.595	258.493	275.876	-3.477	593.422	555.121	-144.983
250	35.487	266.317	273.208	-1.723	592.612	545.636	-114.004
298.15	36.032	272.617	272.617	0.	586.178	537.325	-94.137
300	36.049	272.840	272.618	0.067	586.130	537.022	-93.504
350	36.421	278.427	273.058	1.879	570.393	529.709	-79.055
400	36.681	283.308	274.041	3.707	570.067	523.919	-68.417
450	36.870	287.640	275.316	5.546	569.733	518.171	-60.148
500	37.013	291.533	276.746	7.393	569.393	512.460	-53.536
600	37.218	298.300	279.791	11.105	568.688	501.139	-43.628
700	37.366	304.049	282.856	14.835	567.947	489.939	-36.560
800	37.494	309.047	285.824	18.578	567.169	478.847	-31.266
900	37.622	313.471	288.655	22.334	566.354	467.856	-27.154
1000	37.761	317.441	291.339	26.103	565.502	456.957	-23.869
1100	37.914	321.048	293.878	29.886	564.615	446.145	-21.186
1200	38.082	324.354	296.282	33.686	563.695	435.415	-18.953
1300	38.260	327.409	298.560	37.503	562.741	424.764	-17.067
1400	38.445	330.251	300.723	41.338	561.756	414.187	-15.453
1500	38.635	332.910	302.782	45.192	560.738	403.681	-14.057
1600	38.824	335.409	304.743	49.065	559.686	393.245	-12.838
1700	39.013	337.769	306.617	52.957	558.600	382.876	-11.764
1800	39.200	340.004	308.411	56.868	557.480	372.571	-10.812
1900	39.386	342.128	310.130	60.797	556.322	362.330	-9.961
2000	39.572	344.153	311.781	64.745	555.126	352.150	-9.197
2100	39.760	346.088	313.369	68.712	553.891	342.032	-8.508
2200	39.952	347.943	314.898	72.697	552.616	331.973	-7.882
2300	40.151	349.723	316.374	76.702	551.299	321.973	-7.312
2400	40.358	351.436	317.799	80.728	549.942	312.031	-6.791
2500	40.576	353.088	319.178	84.774	548.546	302.147	-6.313
2600	40.806	354.684	320.513	88.843	547.113	292.319	-5.873
2700	41.050	356.228	321.808	92.936	545.618	282.548	-5.466
2800	41.307	357.726	323.064	97.054	544.029	272.833	-5.090
2900	41.578	359.180	324.284	101.198	542.308	263.177	-4.740
3000	41.863	360.594	325.471	105.370	540.431	253.584	-4.415
3100	42.162	361.972	326.626	109.571	538.356	244.056	-4.112
3200	42.473	363.315	327.752	113.803	536.055	234.598	-3.829
3300	42.795	364.627	328.850	118.066	533.504	225.216	-3.565
3400	43.126	365.910	329.921	122.362	530.671	215.916	-3.317
3500	43.465	367.165	330.967	126.691	527.494	206.704	-3.085
3600	43.810	368.394	331.990	131.055	523.879	197.589	-2.867
3700	44.159	369.599	332.990	135.453	484.961	188.771	-2.665
3800	44.510	370.781	333.969	139.887	483.760	180.782	-2.485
3900	44.861	371.942	334.928	144.355	482.597	172.824	-2.315
4000	45.210	373.082	335.867	148.859	481.472	164.896	-2.153
4100	45.556	374.203	336.789	153.397	480.386	156.995	-2.000
4200	45.896	375.305	337.693	157.970	479.340	149.120	-1.855
4300	46.230	376.389	338.580	162.576	478.334	141.270	-1.716
4400	46.555	377.455	339.452	167.216	477.368	133.442	-1.584
4500	46.870	378.505	340.308	171.887	476.441	125.636	-1.458
4600	47.174	379.538	341.149	176.589	475.554	117.851	-1.338
4700	47.466	380.556	341.977	181.321	474.706	110.084	-1.223
4800	47.746	381.558	342.791	186.082	473.896	102.334	-1.114
4900	48.012	382.546	343.592	190.870	473.123	94.602	-1.008
5000	48.264	383.518	344.381	195.684	472.386	86.884	-0.908
5100	48.501	384.476	345.158	200.522	471.685	79.181	-0.811
5200	48.723	385.420	345.923	205.384	471.017	71.491	-0.718
5300	48.931	386.350	346.677	210.267	470.381	63.815	-0.629
5400	49.123	387.267	347.421	215.169	469.776	56.149	-0.543
5500	49.301	388.170	348.153	220.091	469.200	48.495	-0.461
5600	49.463	389.060	348.876	225.029	468.653	40.851	-0.381
5700	49.611	389.936	349.588	229.983	468.132	33.216	-0.304
5800	49.744	390.800	350.292	234.951	467.635	25.590	-0.230
5900	49.863	391.652	350.985	239.931	467.162	17.973	-0.159
6000	49.969	392.491	351.670	244.923	466.711	10.363	-0.090

PREVIOUS: June 1967 (1 atm) CURRENT: June 1967 (1 bar)

ZIRCONIUM BROMIDE (ZrBr) IDEAL GAS $M_r = 171.124$

$\Delta_f H°(0\ K) = [309.3 \pm 42]\ kJ\ mol^{-1}$

$S°(298.15\ K) = [265.6 \pm 8]\ J\ K^{-1}\ mol^{-1}$ $\Delta_f H°(298.15\ K) = [301.2 \pm 2]\ kJ\ mol^{-1}$

Electronic Levels and Quantum Weights

State	ε_i, cm^{-1}	g_i
$[^4\Sigma]$	0	[4]
	[3000]	[4]
	[6000]	[4]
	[12000]	[4]
	[18000]	[4]
	[24000]	[4]
	[30000]	[4]

$\omega_e = [317]\ cm^{-1}$ $\omega_e x_e = [0.69]\ cm^{-1}$ $\sigma = 1$

$B_e = [0.065]\ cm^{-1}$ $\alpha_e = [0.00018]\ cm^{-1}$ $r_e = [2.47]\ Å$

Enthalpy of Formation

$\Delta_f H°(ZrBr, g, 298.15\ K) = 72\ kcal\ mol^{-1}$ ($301.248\ kJ\ mol^{-1}$) is derived from the estimated bond dissociation energy, $\Delta_{at}H°(Zr\text{-}Br, 298.15\ K) = 103.5 \pm 10\ kcal\ mol^{-1}$. This value of is calculated from the relation $\Delta_{at}H°(Zr\text{-}Br, 298.15\ K)/\Delta_{at}H°(Ti\text{-}Br, 298.15\ K) = \Delta atH°(ZrBr4, 298.15\ K)/\Delta at\ H°(TiBr4, 298.15\ K)$ where the average bond dissociation energies, $\Delta_{at}H°(ZrBr_4, 298.15\ K) = 102.2\ kcal\ mol^{-1}$, $\Delta_{at}H°(TiBr_4, 298.15\ K) = 87.9\ kcal\ mol^{-1}$ and $\Delta_{at}H°(TiBr, 298.15\ K) = 89.0\ kcal\ mol^{-1}$, all calculated from JANAF $\Delta_f H°(298.15\ K)$ for $ZrBr_4(g)$, $TiBr_4(g)$, TiBr(g), Zr(g), Ti(g) and Br(g).

Heat Capacity and Entropy

The bond distance is assumed to be the same as that in $ZrBr_4(g)$, which was estimated as 2.47 Å by Godnev et al. (1). The estimated bond distance is then used with Guggenheimer's (2) relation for polar molecules to calculate the fundamental frequency ω_e. The anharmonicity correction x_e is estimated roughly by assuming $x_e = \omega_e/4\ (\Delta_{at}H° + 0.5\ \omega_e) = 0.0021$. The rotational constant B_e is calculated from the estimated bond distance. The value of α_e is calculated from the Morse potential function.

The ground state configuration is taken from the ground state multiplet of Zr^+ reported by Moore (3). The electronic levels and quantum weights are estimated to be the same as those of ZrF(g).

References

1. I. N. Godnev, A. M. Aleksandrovskaia and I. V. Rigina, Optics and Spectroscopy, 7, 172 (1959).
2. K. M. Guggenheimer, Proc. Phys. Soc. (London) 58, 416 (1946).
3. C. E. Moore, U. S. Nat. Bur. Stand. Circ. 467, Volume II, 1952.

Zirconium Bromide (ZrBr) $Br_1Zr_1(g)$

Enthalpy Reference Temperature = T_r = 298.15 K Standard State Pressure = p° = 0.1 MPa

T/K	$C_p°$ (J K⁻¹mol⁻¹)	S° (J K⁻¹mol⁻¹)	$-[G°-H°(T_r)]/T$ (J K⁻¹mol⁻¹)	$H°-H°(T_r)$ (kJ mol⁻¹)	$\Delta_f H°$ (kJ mol⁻¹)	$\Delta_f G°$ (kJ mol⁻¹)	Log K_f
0	0.	0.	INFINITE	-9.739	309.261	309.261	INFINITE
100	30.981	228.641	296.521	-6.788	309.929	291.163	-152.088
200	34.653	251.415	268.819	-3.481	308.592	272.862	-71.264
250	35.530	259.250	266.148	-1.725	307.725	264.027	-55.165
298.15	36.065	265.557	265.557	0.	301.248	256.351	-44.912
300	36.082	265.780	265.557	0.067	301.198	256.073	-44.586
350	36.451	271.371	265.998	1.881	285.416	249.427	-37.225
400	36.719	276.257	266.982	3.710	285.042	244.311	-31.904
450	36.936	280.595	268.258	5.552	284.658	239.243	-27.771
500	37.133	284.496	269.690	7.403	284.265	234.218	-24.469
600	37.542	291.302	272.741	11.137	283.447	224.284	-19.526
700	38.018	297.124	275.818	14.914	282.589	214.491	-16.006
800	38.558	302.235	278.807	18.743	281.691	204.824	-13.374
900	39.134	306.810	281.669	22.627	280.746	195.272	-11.333
1000	39.714	310.963	284.394	26.569	279.746	185.828	-9.707
1100	40.274	314.775	286.985	30.569	278.682	176.487	-8.381
1200	40.798	318.302	289.449	34.623	273.839	167.468	-7.290
1300	41.277	321.587	291.796	38.727	273.174	158.630	-6.374
1400	41.705	324.662	294.035	42.877	272.508	149.844	-5.591
1500	42.082	327.552	296.174	47.067	271.825	141.106	-4.914
1600	42.410	330.279	298.222	51.292	271.108	132.415	-4.323
1700	42.691	332.858	300.184	55.547	270.342	123.769	-3.803
1800	42.929	335.306	302.068	59.828	269.511	115.171	-3.342
1900	43.129	337.632	303.879	64.131	268.599	106.621	-2.931
2000	43.295	339.849	305.622	68.453	267.591	98.122	-2.563
2100	43.432	341.964	307.303	72.789	266.472	89.675	-2.231
2200	43.543	343.988	308.925	77.138	243.906	82.031	-1.948
2300	43.633	345.925	310.492	81.497	242.108	74.713	-1.697
2400	43.705	347.784	312.007	85.864	240.308	67.473	-1.469
2500	43.763	349.569	313.474	90.238	238.502	60.309	-1.260
2600	43.808	351.286	314.895	94.616	236.690	53.217	-1.069
2700	43.844	352.940	316.274	98.999	234.871	46.195	-0.894
2800	43.872	354.535	317.612	103.385	233.044	39.240	-0.732
2900	43.894	356.075	318.912	107.773	231.210	32.351	-0.583
3000	43.911	357.564	320.176	112.164	229.369	25.525	-0.444
3100	43.926	359.004	321.405	116.555	227.521	18.761	-0.316
3200	43.938	360.399	322.602	120.949	225.667	12.056	-0.197
3300	43.948	361.751	323.768	125.343	223.809	5.410	-0.086
3400	43.958	363.063	324.905	129.738	221.947	-1.181	0.018
3500	43.968	364.337	326.013	134.134	220.083	-7.716	0.115
3600	43.977	365.576	327.095	138.532	218.219	-14.198	0.206
3700	43.988	366.781	328.151	142.930	216.355	-20.628	0.291
3800	43.999	367.954	329.183	147.329	214.492	-27.009	0.371
3900	44.011	369.097	330.192	151.730	212.633	-33.340	0.447
4000	44.024	370.212	331.179	156.131	210.779	-39.623	0.517
4100	44.038	371.299	332.144	160.535	208.930	-45.860	0.584
4200	44.053	372.360	333.089	164.939	207.088	-52.052	0.647
4300	44.070	373.397	334.015	169.345	205.254	-58.201	0.707
4400	44.088	374.411	334.921	173.753	203.429	-64.306	0.763
4500	44.107	375.402	335.810	178.163	201.613	-70.371	0.817
4600	44.127	376.371	336.681	182.574	199.808	-76.395	0.867
4700	44.148	377.320	337.536	186.988	198.014	-82.380	0.916
4800	44.171	378.250	338.374	191.404	-364.705	-76.710	0.835
4900	44.194	379.161	339.197	195.822	-366.127	-70.695	0.754
5000	44.219	380.054	340.006	200.243	-367.553	-64.651	0.675
5100	44.244	380.930	340.799	204.666	-368.982	-58.579	0.600
5200	44.270	381.789	341.580	209.092	-370.412	-52.479	0.527
5300	44.296	382.633	342.346	213.520	-371.842	-46.351	0.457
5400	44.323	383.461	343.100	217.951	-373.269	-40.196	0.389
5500	44.351	384.275	343.841	222.385	-374.694	-34.015	0.323
5600	44.379	385.074	344.570	226.821	-376.113	-27.808	0.259
5700	44.408	385.860	345.288	231.261	-377.527	-21.576	0.198
5800	44.437	386.633	345.994	235.703	-378.933	-15.319	0.138
5900	44.466	387.392	346.689	240.148	-380.331	-9.038	0.080
6000	44.495	388.140	347.374	244.596	-381.720	-2.733	0.024

PREVIOUS: June 1970 (1 atm) CURRENT: June 1970 (1 bar)

Zirconium Bromide (ZrBr) $Br_1Zr_1(g)$

BROMINE (Br_2) REFERENCE STATE M_r = 159.808

0 to 265.90 K	crystal	
265.90 to 332.503 K	liquid	
above 332.503 K	diatomic ideal gas	

Refer to the individual tables for details.

Bromine (Br_2) Br_2(ref)

Enthalpy Reference Temperature = T_r = 298.15 K Standard State Pressure = $p°$ = 0.1 MPa

	J K⁻¹mol⁻¹			kJ mol⁻¹			
T/K	$C_p°$	S°	$-[G°-H°(T_r)]/T$	$H°-H°(T_r)$	$\Delta_f H°$	$\Delta_f G°$	Log K_f
0	0.	0.	INFINITE	-24.509	0.	0.	0.
100	43.587	53.847	271.009	-21.716	0.	0.	0.
200	53.761	87.421	171.496	-16.815	0.	0.	0.
265.900	61.630	103.687	152.725	-13.039	CRYSTAL <--> LIQUID TRANSITION		
265.900	77.724	143.445	152.725	-2.468			
298.15	75.674	152.206	152.206	0.	0.	0.	0.
300	75.624	152.674	152.208	0.140	0.	0.	0.
332.503	75.302	160.429	152.639	2.590	LIQUID <--> IDEAL GAS FUGACITY = 1 bar		
332.503	36.332	249.341	152.639	32.153			
400	36.723	256.093	169.543	34.620	0.	0.	0.
500	37.077	264.329	187.706	38.312	0.	0.	0.
600	37.301	271.110	201.058	42.031	0.	0.	0.
700	37.460	276.873	211.488	45.770	0.	0.	0.
800	37.586	281.883	219.981	49.522	0.	0.	0.
900	37.692	286.316	227.110	53.286	0.	0.	0.
1000	37.787	290.293	233.233	57.060	0.	0.	0.
1100	37.876	293.898	238.586	60.843	0.	0.	0.
1200	37.962	297.198	243.335	64.635	0.	0.	0.
1300	38.049	300.240	247.597	68.436	0.	0.	0.
1400	38.140	303.063	251.459	72.245	0.	0.	0.
1500	38.239	305.697	254.988	76.064	0.	0.	0.
1600	38.348	308.169	258.236	79.893	0.	0.	0.
1700	38.471	310.497	261.242	83.734	0.	0.	0.
1800	38.611	312.700	264.040	87.588	0.	0.	0.
1900	38.769	314.792	266.657	91.457	0.	0.	0.
2000	38.945	316.785	269.114	95.342	0.	0.	0.
2100	39.137	318.690	271.429	99.246	0.	0.	0.
2200	39.345	320.515	273.619	103.170	0.	0.	0.
2300	39.563	322.269	275.697	107.116	0.	0.	0.
2400	39.789	323.957	277.673	111.083	0.	0.	0.
2500	40.018	325.586	279.557	115.074	0.	0.	0.
2600	40.244	327.160	281.358	119.087	0.	0.	0.
2700	40.463	328.683	283.082	123.122	0.	0.	0.
2800	40.672	330.158	284.737	127.179	0.	0.	0.
2900	40.864	331.589	286.328	131.256	0.	0.	0.
3000	41.038	332.977	287.860	135.351	0.	0.	0.
3100	41.190	334.326	289.338	139.463	0.	0.	0.
3200	41.318	335.635	290.764	143.589	0.	0.	0.
3300	41.420	336.908	292.143	147.726	0.	0.	0.
3400	41.496	338.146	293.478	151.872	0.	0.	0.
3500	41.545	339.350	294.772	156.024	0.	0.	0.
3600	41.567	340.521	296.026	160.180	0.	0.	0.
3700	41.563	341.659	297.244	164.337	0.	0.	0.
3800	41.533	342.767	298.428	168.492	0.	0.	0.
3900	41.480	343.846	299.578	172.642	0.	0.	0.
4000	41.404	344.895	300.698	176.787	0.	0.	0.
4100	41.306	345.916	301.789	180.922	0.	0.	0.
4200	41.190	346.910	302.851	185.047	0.	0.	0.
4300	41.056	347.878	303.887	189.160	0.	0.	0.
4400	40.907	348.820	304.898	193.258	0.	0.	0.
4500	40.744	349.738	305.884	197.341	0.	0.	0.
4600	40.569	350.631	306.847	201.407	0.	0.	0.
4700	40.383	351.502	307.788	205.454	0.	0.	0.
4800	40.189	352.350	308.708	209.483	0.	0.	0.
4900	39.987	353.176	309.607	213.492	0.	0.	0.
5000	39.780	353.982	310.486	217.480	0.	0.	0.
5100	39.568	354.768	311.347	221.448	0.	0.	0.
5200	39.353	355.534	312.189	225.394	0.	0.	0.
5300	39.136	356.282	313.014	229.318	0.	0.	0.
5400	38.917	357.011	313.822	233.221	0.	0.	0.
5500	38.698	357.723	314.614	237.102	0.	0.	0.
5600	38.480	358.419	315.390	240.961	0.	0.	0.
5700	38.263	359.098	316.151	244.798	0.	0.	0.
5800	38.047	359.761	316.897	248.613	0.	0.	0.
5900	37.834	360.410	317.629	252.407	0.	0.	0.
6000	37.624	361.044	318.347	256.180	0.	0.	0.

PREVIOUS: September 1961 (1 atm) CURRENT: June 1982 (1 bar)

Bromine (Br_2) Br_2(ref)

$\Delta_f H°(0\ K) = 0\ kJ\ mol^{-1}$

$S°(298.15\ K) = 152.21 \pm 0.04\ J\ K^{-1}\ mol^{-1}$

$\Delta_f H°(298.15\ K) = 0\ kJ\ mol^{-1}$

$T_{fus} = 265.90 \pm 0.05\ K$

$\Delta_{fus} H° = 10.571 \pm 0.008\ kJ\ mol^{-1}$

Enthalpy of Formation

Zero by definition.

Heat Capacity and Entropy

The heat capacity for crystalline and liquid bromine was measured by Hildenbrand et al. (1), 179 data points in the range 15-310 K. The reported values are corrected for the current relative molecular weight and converted to joules. The liquid phase heat capacity is extrapolated to a constant value of 18.0 cal $K^{-1}\ mol^{-1}$ for temperatures above 320 K. The value recommended by CODATA (2) for S°(Br_2, ℓ, 298.15 K) is the same as adopted here and is based on the same study. References to earlier work are given by Hildenbrand et al. (1). They are not used in this analysis.

Fusion Data

The adopted values for the fusion point and the enthalpy of fusion are from the study of Hildenbrand et al. (1).

Vaporization Data

The enthalpy of vaporization of bromine at 25°C was measured by Hildenbrand et al. (1). Six vaporization determinations were made using 14 to 18 gram samples of bromine. Correcting their reported values for the current relative atomic mass and converting to Joules yields $\Delta_{vap}H°(Br_2, ℓ, 298.15\ K) = 30.91 \pm 0.11\ kJ\ mol^{-1}$. CODATA (2) has recommended the same value based on the same study.

Vapor pressure studies include: Fischer and Bingle (298-389 K, 3), Frey and Gregory (177-195 K, 4), Henglein, et al. (177-241 K, 5), Ramsay and Young (256-329 K, 6) and Scheffer and Voogd (253-362 K, 7). For the liquid region, the data of Fischer and Bingle (3) and Scheffer and Voogd (7) show a decided decrease of $\Delta_{vap}H°$ with increasing temperature. The data of Ramsey and Young (6) show only a small temperature trend and are in good agreement with the adopted calorimetric study. For the crystal region, the data are too scattered to be conclusive.

References

1. D. L. Hildenbrand, W. R. Kramer, R. A. McDonald, and D. R. Stull, J. Amer. Chem. Soc. 80, 4129 (1958).
2. J. D. Cox, Chairman, ICSU-CODATA Task Group on Key Values for Thermodynamics, J. Chem. Thermodynamics 10, 903 (1978).
3. J. Fischer and J. Bingle, J. Am. Chem. Soc. 77, 6511 (1955).
4. M. B. Frey and N. W. Gregory, J. Am. Chem. Cos. 82, 1068 (1960).
5. F. A. Henglein, G. von Rosenberg, and A. Muchlinski, Z. Physik 11, 1 (1922).
6. J. Ramsay and J. Young, J. Chem. Soc. 49, 453 (1886).
7. F. E. C. Scheffer and M. Voogd, Rec. Trav. Chem. 45, 214 (1926).

Enthalpy Reference Temperature = T_r = 298.15 K — Standard State Pressure = p° = 0.1 MPa

T/K	$C_p°$ (J K⁻¹mol⁻¹)	S° (J K⁻¹mol⁻¹)	−[G°−H°(T_r)]/T (J K⁻¹mol⁻¹)	H°−H°(T_r) (kJ mol⁻¹)	$\Delta_f H°$ (kJ mol⁻¹)	$\Delta_f G°$ (kJ mol⁻¹)	Log K_f
0	0.	0.	INFINITE	-24.509	0.	0.	0.
100	43.587	53.847	271.009	-21.716	0.	0.	0.
200	53.761	87.421	171.496	-16.815	0.	0.	0.
250	59.225	99.965	155.962	-13.999	0.	0.	0.
265.900	61.630	103.687	152.725	-13.039	CRYSTAL <--> LIQUID	TRANSITION	
265.900	77.724	143.445	152.725	-2.468			
298.15	75.674	152.206	152.206	0.	0.	0.	0.
300	75.624	152.674	152.208	0.140	0.	0.	0.
332.503	75.302	160.429	152.639	2.590	FUGACITY = 1 bar		
350	75.302	164.291	153.126	3.908	-28.882	1.538	-0.230
400	75.302	174.347	155.164	7.673	-26.947	5.752	-0.751
450	75.302	183.216	157.798	11.438	-25.023	9.723	-1.129
500	75.302	191.150	160.743	15.203	-23.108	13.481	-1.408

PREVIOUS: September 1961 CURRENT: June 1982

BROMINE (Br_2) IDEAL GAS M_r = 159.808

D_0^0 = 191.154 ± 0.001 kJ mol^{-1} (natural abundance)
S°(298.15 K) = 245.394 ± 0.050 J K^{-1} mol^{-1}

$\Delta_f H°$(0 K) = 45.70 ± 0.121 kJ mol^{-1}
$\Delta_f H°$(298.15 K) = 30.911 ± 0.11 kJ mol^{-1}

Spectroscopic Data for $^{79}Br_2$ in cm^{-1}

State	T_e	D°	ω_e	$\omega_e x_e$	B_e	$10^3 \alpha_e$	$10^7 D_e$	r_e, Å	Reference
$^1\Sigma_g^+$	0	15894.6	325.3213	1.07742[a]	0.082107	0.31873	0.2092	2.28107	(1)
$^3\Pi_{2u}$	13400	15894.6	160	2.4	0.057	0.5	0.29	2.74	(2)
$^3\Pi_{1u}$	13904	15894.6	153.8	2.78	0.0588	0.72	0.34	2.695	(3)
$^3\Pi_{o_u}^-$	15850	15894.6	70	6.2	0.06	2.8	1.76	2.67	(2)
$^3\Pi_{o_u}^+$	15902.47	19579.76	166.546	1.61517[b]	0.059589	0.489095[c]	0.3013	2.6776	(4)[d]

(a) $-2.29798 \times 10^{-3}(v + 1/2)^3$
(b) $-7.53588 \times 10^{-3}(v + 1/2)^3 - 2.44989 \times 10^{-4}(v + 1/2)^4 + 0.991688 \times 10^{-5}(v + 1/2)^5 - 6.97853 \times 10^{-8}(v + 1/2)^6$
(c) $-6.6369 \times 10^{-6}(v + 1/2)^2$
(d) $^3\Pi_{o_u}^+$ constants for natural abundance Br_2, actual coefficients for $^{79}Br_2$ will differ slightly from these according to the scaling method of Herzberg (5).
(e) The dissociation energy of each state is relative to the lowest level of the ground state.

Enthalpy of Formation

We adopt the enthalpy of formation values of CODATA (6) as follows: $\Delta_f H°$(298.15 K) = 30.91 ± 0.11 kJ mol^{-1} and $\Delta_f H°$ (0 K) = 45.71 ± 0.12 kJ mol^{-1}. These values were based on the heat capacity and enthalpy of vaporization measurements of Hildenbrand et al (7). Note that the latter value of $\Delta_f H°$ is deduced from the former using the CODATA values of H°(298.15 K) - H°(0 K) for Br_2(g) and Br_2(ℓ). Our calculated value of H°(298.15 K) - H°(0 K) for Br_2(g) is only 1.3 J mol^{-1} smaller than the recommended value of CODATA. Our adopted dissociation energy for the natural abundance mixture is derived from the isotopic values reported by Barrow et al. (1) as given by Huber and Herzberg (see reference (3)). The CODATA value for D_0^0(Br_2, g), as deduced from the $\Delta_f H°$(0 K) data for Br_2(g) and Br(g), is 190.14 kJ mol^{-1}. The 0.01 kJ mol^{-1} discrepancy is due to CODATA'S adoption of D_0^0 for $^{79}Br^{81}Br$ based on earlier work of Horsley and Barrow (8). Refer to the monatomic bromine gas table for additional discussion of the dissociation energy.

Heat Capacity and Entropy

The thermal functions are calculated using a direct summation technique. Included in the calculation are the ground state plus the four lowest lying excited states. Spectroscopic constants for the ground state are taken from Barrow, et al. (1) for $^{79}Br_2$ while the excited state constants are from Gurvich et al. (2-4). Note that Barrow, et al. (1) presented constants for the $^3\Pi_{o_u}^+$ state of $^{79}Br_2$ for v≤10 only. Therefore for this state we adopted the constants of Gurvich, et al. (2) which are fit to the experimental data up to the dissociation limit. For the ground state and the $^3\Pi_{1u}$ and $^3\Pi_{0^+u}$ excited states, the calculation is performed including the quasi-bound rotational levels above the dissociation limit (see Cl_2(g) table) using J_{lim} and v_{max} given by Gurvich et al. (2).

Splitting of the rotational levels due to the rotational-electronic interaction in the $^3\Pi_{2u}$ and $^3\Pi_{1u}$ states is taken into account only insofar as the statistical weight of the rotational levels is doubled. The rotational levels are extrapolated to high J values according to the method of Khachkuruzov (9) who proposed a simpler form of Wooley's method (10).

The spectroscopic constants for the hypothetical natural abundance species are determined from the $^{79}Br_2$constants using a standard reduced mass scaling scheme. The calculated value of S°(298.15 K) is 0.07 J mol^{-1} K^{-1} smaller than that adopted by CODATA (6) and also by Gurvich et al. (2). For this comparison, the entropy value reported by CODATA and Gurvich is increased by 0.1094 J K^{-1} mol^{-1} to account for a change in the standard state pressure from 1 atm to 1 bar.

References

1. R. F. Barrow, T. C. Clark, J. A. Coxon, and K. K. Yee, J. Mol. Spectrosc. 51, 428 (1974).
2. L. V. Gurvich, I. V. Veits et al., "Thermodynamic Properties of Individual Substances," Nauka, Moscow, USSR (1978), Vol I, Part 1, pp. 183-187. See also Fig. 10.1, p. 201.
3. Taken from reference (2). Data are in agreement with the values adopted by K. P. Huber and G. Herzberg, "Constants of Diatomic Molecules," Van Nostrand-Reinhold Co., New York, 1979.
4. Taken from reference (2), Table 9.4, p. 186.
5. G. Herzberg, "Spectra of Diatomic Molecules, 2nd. ed., D. Van Nostrand Co., Inc., New York (1950), pp. 141-145.
6. J. D. Cox, Chairman, ICSU-CODATA Task Group on Key Values for Thermodynamics, J. Chem. Thermodynamics 10, 903 (1978).
7. D. L. Hildenbrand, W. R. Kramer, R. A. McDonald, and D. R. Stull, J. Am. Chem. Soc. 80, 4129 (1958).
8. T. A. Horsley and R. F. Barrow, Trans. Faraday Soc. 63, 32 (1967).
9. G. A. Khachkuruzov, Opt. Spectrosc. 30, 455 (1971).
10. H. W. Woolley, R. B. Scott, and F. G. Brickwedde, J. Res. Nat. Bur. Stand. 41, 379 (1948).

Bromine (Br_2) Br_2(g)

Enthalpy Reference Temperature = T_r = 298.15 K Standard State Pressure = p° = 0.1 MPa

T/K	$C_p°$ (J $K^{-1}mol^{-1}$)	S°	-[G°-H°(T_r)]/T	H°-H°(T_r) (kJ mol^{-1})	$\Delta_f H°$	$\Delta_f G°$	Log K_f
0	0.	0.	INFINITE	-9.722	45.697	45.697	INFINITE
100	30.884	208.563	276.306	-6.774	45.852	30.380	-15.869
200	34.582	231.269	248.653	-3.477	44.248	15.479	-4.043
250	35.489	239.092	245.984	-1.723	43.186	8.404	-1.756
298.15	36.048	245.394	245.394	0.	30.910	3.126	-0.548
300	36.066	245.617	245.394	0.067	30.837	2.954	-0.514
332.503	36.332	249.341	245.601	1.243	------- FUGACITY = 1 bar -------		
350	36.451	251.207	245.835	1.880	0.	0.	0.
400	36.723	256.093	246.818	3.710	0.	0.	0.
450	36.923	260.431	248.094	5.551	0.	0.	0.
500	37.077	264.329	249.526	7.402	0.	0.	0.
600	37.301	271.110	252.575	11.121	0.	0.	0.
700	37.460	276.873	255.645	14.860	0.	0.	0.
800	37.586	281.883	258.618	18.612	0.	0.	0.
900	37.692	286.316	261.454	22.376	0.	0.	0.
1000	37.787	290.293	264.143	26.150	0.	0.	0.
1100	37.876	293.898	266.686	29.933	0.	0.	0.
1200	37.962	297.198	269.093	33.725	0.	0.	0.
1300	38.049	300.240	271.374	37.526	0.	0.	0.
1400	38.140	303.063	273.538	41.335	0.	0.	0.
1500	38.239	305.697	275.595	45.154	0.	0.	0.
1600	38.348	308.169	277.554	48.983	0.	0.	0.
1700	38.471	310.497	279.424	52.824	0.	0.	0.
1800	38.611	312.700	281.212	56.678	0.	0.	0.
1900	38.769	314.792	282.925	60.547	0.	0.	0.
2000	38.945	316.785	284.569	64.432	0.	0.	0.
2100	39.137	318.690	286.148	68.336	0.	0.	0.
2200	39.345	320.515	287.669	72.260	0.	0.	0.
2300	39.563	322.269	289.136	76.206	0.	0.	0.
2400	39.789	323.957	290.552	80.173	0.	0.	0.
2500	40.018	325.586	291.921	84.164	0.	0.	0.
2600	40.244	327.160	293.246	88.177	0.	0.	0.
2700	40.463	328.683	294.530	92.212	0.	0.	0.
2800	40.672	330.158	295.777	96.269	0.	0.	0.
2900	40.864	331.589	296.987	100.346	0.	0.	0.
3000	41.038	332.977	298.164	104.441	0.	0.	0.
3100	41.190	334.326	299.308	108.553	0.	0.	0.
3200	41.318	335.635	300.423	112.679	0.	0.	0.
3300	41.420	336.908	301.510	116.816	0.	0.	0.
3400	41.496	338.146	302.569	120.962	0.	0.	0.
3500	41.545	339.350	303.603	125.114	0.	0.	0.
3600	41.567	340.521	304.612	129.270	0.	0.	0.
3700	41.563	341.659	305.598	133.427	0.	0.	0.
3800	41.533	342.767	306.562	137.582	0.	0.	0.
3900	41.480	343.846	307.504	141.732	0.	0.	0.
4000	41.404	344.895	308.426	145.877	0.	0.	0.
4100	41.306	345.916	309.328	150.012	0.	0.	0.
4200	41.190	346.910	310.211	154.137	0.	0.	0.
4300	41.056	347.878	311.076	158.250	0.	0.	0.
4400	40.907	348.820	311.923	162.348	0.	0.	0.
4500	40.744	349.738	312.753	166.431	0.	0.	0.
4600	40.569	350.631	313.567	170.497	0.	0.	0.
4700	40.383	351.502	314.365	174.544	0.	0.	0.
4800	40.189	352.350	315.147	178.573	0.	0.	0.
4900	39.987	353.176	315.915	182.582	0.	0.	0.
5000	39.780	353.982	316.668	186.570	0.	0.	0.
5100	39.568	354.768	317.408	190.538	0.	0.	0.
5200	39.353	355.534	318.133	194.484	0.	0.	0.
5300	39.136	356.282	318.846	198.408	0.	0.	0.
5400	38.917	357.011	319.546	202.311	0.	0.	0.
5500	38.698	357.723	320.234	206.192	0.	0.	0.
5600	38.480	358.419	320.910	210.051	0.	0.	0.
5700	38.263	359.098	321.574	213.888	0.	0.	0.
5800	38.047	359.761	322.226	217.703	0.	0.	0.
5900	37.834	360.410	322.868	221.497	0.	0.	0.
6000	37.624	361.044	323.499	225.270	0.	0.	0.

PREVIOUS: December 1961 (1 atm) CURRENT: June 1982 (1 bar)

Bromine (Br_2) Br_2(g)

CALCIUM BROMIDE ($CaBr_2$) CRYSTAL M_r = 199.888

$\Delta_fH°(0\ K)$ = Unknown

S°(298.15 K) = [129.7 ± 4.2] J K^{-1} mol^{-1} $\Delta_fH°(298.15\ K)$ = -683.2 ± 4.2 kJ mol^{-1}

T_{fus} = 1015 K $\Delta_{fus}H°$ = 29.079 ± 0.63 kJ mol^{-1}

Enthalpy of Formation

The adopted enthalpy of formation is the mean of three values derived from two independent paths. One path involves the enthalpy of solution of Ca(cr) and $CaBr_2$(cr) in HBr·555 H_2O which were measured by Ehrlich et al. (1). Recalculation of their results using recent thermal data (2) 3) for aqueous HBr yields $\Delta_fH°(CaBr_2$, cr, 298.15 K) = -164.0 kcal mol^{-1}. The uncertainty in this value could approach ±1.0 kcal mol^{-1} due primarily to the rather impure metal (99.3% free Ca) used in these measurements.

An alternate path used to obtain values for $\Delta_fH°$ involves the results of measurements on the enthalpy of solution of $CaBr_2$ (cr) in aqueous solution. The work (4-6) reported in this area has been reviewed by Bichowsky and Rossini (7). Recalculation of these results using $\Delta_fH°(Ca^{+2}$, ∞ aq, 298.15 K) = -129.74 kcal mol^{-1} (8) and $\Delta_fH°(Br^-$, ∞ aq, 298.15 K) = -29.039±0.036 kcal mol^{-1} (2) yields values for $\Delta_fH°$ of $CaBr_2$(cr) in kcal mol^{-1} of -162.3 (4), -160.2 (5), and -163.6 (6). The disagreement in these values is disturbing but can be mostly accounted for when one considers the difficulties involved in the preparation (6) of "pure" anhydrous $CaBr_2$.

A mean value (-163.3 kcal mol^{-1}, 683.247 kJ mol^{-1}) for $\Delta_fH°(CaBr_2$, cr, 298.15 K) of three results (1, 4, and 6) is preferred because of the difficulties reported in obtaining and keeping samples free of impurities. NBS (8) has selected essentially the same value (-163.2 kcal mol^{-1}) from their analysis of the available data.

Heat Capacity and Entropy

No low temperature $C_p°$ data have been reported for $CaBr_2$; thus, the value of S°(298.15 K) must be estimated. Application of the Berthelot principle (9) to the process $SrBr_2$(cr) + Ca(cr) = $CaBr_2$(cr) + Sr(cr) gives S°(298.15 K) = 31.7±1.5 cal K^{-1} mol^{-1} for $CaBr_2$. Auxiliary entropies are taken from the JANAF Tables (10). Also, JANAF entropies (10) for the other three calcium dihalides suggest a value near 30.5 cal K^{-1} mol^{-1} for $CaBr_2$. We adopt S°(298.15 K) = 31±1 cal K^{-1} mol^{-1} which is the value selected by NBS (8). Other reported estimates for S°(298.15 K) in cal K^{-1} mol^{-1} are 32.0±2.8 (11), 32.2±0.7 (12), and 33 (13).

$C_p°$ at 298.15 K has not been measured. However, Janz et al. (14) measured relative enthalpies (434-1013 K) for $CaBr_2$ and reported the results in equation form. Their equation gives $C_p°$ = 16.3 cal K^{-1} mol^{-1} at 298.15 K which appears to be too low by about 2 units when compared with JANAF $C_p°$ data for several other alkaline earth dihalides. The only other enthalpy study that has been reported is that of Dworkin and Bredig (15) who measured H°(1015 K)-H°(298.15 K) = 14.00 kcal mol^{-1} in a copper block drop-calorimeter. We adopt this result. JANAF data (10) are used to obtain $C_p°$ at 298.15 K from the process $CaCl_2$(cr) + 2 KBr(cr) = $CaBr_2$(cr) + 2 KCl(cr) by assuming $\Delta C_p°$ = 0. Data in the temperature range 300-1013 K are estimated by comparison with those for CaF_2 and $CaCl_2$ (10). More weight is given to the results for $CaCl_2$, since it has the same crystal structure (15) as $CaBr_2$ and nearly the same melting temperature (1045 K) (10) and enthalpy H°(1045 K)-H°(298.15 K) = 14.2 kcal mol^{-1} (10) at T_{fus} as does $CaBr_2$. The $C_p°$ estimates are also made so as to reproduce as closely as possible the adopted enthalpy at 1015 K. Our estimated $C_p°$ data give a value of H°(1015 K)-H°(298.15 K) = 13.994 kcal mol^{-1} which agrees with the measured value (15) to within 6 cal mol^{-1}. $C_p°$ data above 1015 K are obtained by graphical extrapolation. The enthalpies obtained from the equation of Janz et al. (14) are systematically lower than our adopted values. The deviation is -0.4% at 1015 K; it increases at lower temperatures and reaches a maximum value of -2.7% at 600 K.

Fusion Data

See the liquid table for details.

Sublimation Data

$\Delta_{sub}H°$(298.15 K) is calculated as the difference in the adopted enthalpies of formation for the ideal gas and crystal at 298.15 K.

References

1. P. Ehrlich, K. Peik, and E. Koch, Z. Anorg. Chem. 324, 113 (1963).
2. ICSU-CODATA Task Group, J. Chem. Thermodynamics 4, 331 (1972).
3. U. S. Nat. Bur. Stand. Tech. Note 270-3, 1968.
4. J. Thomsen, "Thermochemische Untersuchungen," Volume III, Barth, Leipzig, 1883.
5. R. Varet, Ann. Chim. Phys. 8, 240 (1896).
6. S. U. Pickering, J. Chem. Soc. 53, 865 (1888).
7. F. R. Bichowsky and F. D. Rossini, "The Thermochemistry of The Chemical Substances," Reinhold Publishing Corporation, New York, 1936.
8. U. S. Nat. Bur. Stand. Tech. Note 270-6, 1971.
9. N. N. Drozin, Zhur. Fiz. Khim. 35, 879 (1961).
10. JANAF Thermochemical Tables: $SrBr_2$(cr), Ca(cr), and CaI_2(cr), 6-30-74; CaF_2(cr), 12-31-68; Sr(cr), 12-31-70; $CaCl_2$(cr), 6-30-70; KCl(cr), 3-31-66; KBr(cr), 3-31-67.
11. O. Kubaschewski, E. Ll. Evans, and C. B. Alcock, "Metallurigcal Thermochemistry," 4th ed., Pergamon Press, New York, 1967.
12. K. K. Kelley and E. G. King, U. S. Bur. Mines Bull. 592, 1961.
13. L. Brewer, G. R. Somayajulu, and E. Brackett, Chem. Revs. 63, 111 (1963).
14. G. J. Janz, F. J. Kelley, and J. L. Perano, Trans. Faraday Soc. 59, 2718 (1963).
15. A. S. Dworkin and M. A. Bredig, J. Phys. Chem. 67, 697 (1963).

Calcium Bromide ($CaBr_2$) $Br_2Ca_1(cr)$

Enthalpy Reference Temperature = T_r = 298.15 K; Standard State Pressure = p° = 0.1 MPa

T/K	$C_p°$ (J $K^{-1}mol^{-1}$)	S° (J $K^{-1}mol^{-1}$)	$-[G°-H°(T_r)]/T$ (J $K^{-1}mol^{-1}$)	$H°-H°(T_r)$ (kJ mol^{-1})	$\Delta_fH°$ (kJ mol^{-1})	$\Delta_fG°$ (kJ mol^{-1})	Log K_f
0							
100							
200							
298.15	75.040	129.704	129.704	0.	-683.247	-664.139	116.354
300	75.124	130.168	129.705	0.139	-683.296	-664.020	115.616
400	77.990	152.206	132.691	7.806	-712.744	-651.460	85.072
500	79.454	169.774	138.410	15.682	-711.325	-636.304	66.474
600	80.500	184.352	144.887	23.679	-709.990	-621.427	54.100
700	81.630	196.836	151.437	31.780	-708.767	-606.765	45.277
800	83.471	207.851	157.813	40.030	-708.463	-592.168	38.665
900	85.856	217.816	163.935	48.493	-707.289	-577.703	33.529
1000	88.617	227.004	169.788	57.216	-706.160	-563.365	29.427
1015.000	89.036	228.327	170.643	58.549	------ CRYSTAL <--> LIQUID -----		
1100	91.922	235.601	175.384	66.238	-705.045	-549.139	26.076
1200	95.563	243.753	180.745	75.611	-711.624	-534.394	23.262
1300	99.412	251.554	185.893	85.358	-709.177	-519.722	20.883
1400	103.387	259.067	190.853	95.499	-706.346	-505.252	18.851
1500	107.320	266.334	195.644	106.034	-703.129	-490.998	17.098

PREVIOUS: CURRENT: June 1974

Calcium Bromide ($CaBr_2$) $Br_2Ca_1(cr)$

CALCIUM BROMIDE ($CaBr_2$) LIQUID M_r = 199.888

$S°(298.15\ K)$ = [147.861] $J\ K^{-1}\ mol^{-1}$

T_{fus} = 1015 K

$\Delta_f H°(298.15\ K)$ = [-662.989] $kJ\ mol^{-1}$

$\Delta_{fus}H°$ = 29.079 ± 0.63 $kJ\ mol^{-1}$

Enthalpy of Formation

$\Delta_f H°(298.15\ K)$ is calculated from that of the crystal by adding of $\Delta_{fus}H°$ and the difference in enthalpy, H°(1015 K) - H°(298.15 K), between the crystal and liquid. Toguri et al. (1) have reported a K_p value for the anion exchange equilibrium 0.5 $CaBr_2(\ell)$ + HCl(g) = 0.5 $CaCl_2(\ell)$ + HBr(g) at 1073 K. A 3rd law analysis of this single value gives $\Delta_r H°(298.15\ K)$ = -0.22 kcal mol^{-1} with JANAF functions. The 3rd law enthalpy leads to $\Delta_f H°(\ell,\ 298.15\ K)$ = -157.9 kcal mol^{-1} which supports our adopted value by an independent path.

Heat Capacity and Entropy

$C_p°$ of the liquid above 700 K is the value (27.0 cal $K^{-1}\ mol^{-1}$) determined by Dworkin and Bredig (2) from enthalpy measurements in the vicinity of the melting point. A similar enthalpy study (1013-1132 K) by Janz et al. (3) suggests only a slightly higher value (27.38 cal $K^{-1}\ mol^{-1}$) for $C_p°$ of $CaBr_2(\ell)$. A glass transition is assumed at 700 K below which $C_p°$ is that of the crystal.

The entropy at 298.15 K is calculated in a manner analogous to that used for the enthalpy of formation.

Fusion Data

The adopted value of T_{fus} is that measured calorimetrically by Dworkin and Bredig (2). Similar measurements by Janz et al. (3) gave essentially the same value (1013 K). Other reported values for T_{fus} are 1018 K (4), 1003 K (5-7), and 1016 K (8).

$\Delta_{fus}H°$ is that measured by Dworkin and Bredig (2) in a copper block drop-calorimeter. Another calorimetrically determined value (3)($\Delta_{fus}H°$ = 6.86 ± 0.09 kcal mol^{-1})] and one based on cryoscopic measurements (4)($\Delta_{fus}H°$ = 7.03 ± 0.35 kcal mol^{-1})] bracket the selected value.

Vaporization Data

T_{vap} (f = 1 bar) is the temperature at which $\Delta_r G°$ for the process $CaBr_2(\ell) = CaBr_2(g)$ approaches zero. $\Delta_{vap}H°$ is the difference in the enthalpies of formation of the gas and liquid at T_{vap}. Peterson and Hutchison (9) obtained T_{vap} = 2088 K (at 1 atm) from vapor pressure measurements on the liquid (1149-1321 K). This value is nearly identical to our result.

References

1. J. Toguri, H. Flood, and T. Forland, Acta Chem. Scand. 17, 1502 (1963).
2. A. S. Dworkin and M. A. Bredig, J. Phys. Chem. 67, 697 (1963).
3. G. J. Janz, F. J. Kelly, and J. L. Pérano, Trans. Faraday Soc. 59, 2718 (1963).
4. H. H. Emons and B. Loffelholz, Wiss. Z. Tech. Hochsch. Chem., Leuna-Merseburg 6, 261 (1964).
5. G. Kellner, J. Chem. Soc. 112, 469 (1917).
6. G. Kellner, Z. Anorg. Chem. 99, 137 (1917).
7. H. H. Emons, Z. Anorg. Chem. 323, 114 (1963).
8. H. Schinke and F. Sauerwald, Z. Anorg. Chem. 304, 25 (1960).
9. D. T. Peterson and J. F. Hutchison, J. Chem. Eng. Data 15, 320 (1970).

Calcium Bromide ($CaBr_2$) $Br_2Ca_1(l)$

Enthalpy Reference Temperature = T_r = 298.15 K Standard State Pressure = p° = 0.1 MPa

T/K	$C_p°$ (J K⁻¹mol⁻¹)	S° (J K⁻¹mol⁻¹)	$-[G°-H°(T_r)]/T$ (J K⁻¹mol⁻¹)	$H°-H°(T_r)$ (kJ mol⁻¹)	$\Delta_f H°$ (kJ mol⁻¹)	$\Delta_f G°$ (kJ mol⁻¹)	Log K_f
0							
100							
200							
298.15	75.040	147.861	147.861	0.	-662.989	-649.294	113.754
300	75.124	148.325	147.862	0.139	-663.038	-649.209	113.037
400	77.990	170.363	150.847	7.806	-692.486	-638.465	83.375
500	79.454	187.931	156.567	15.682	-691.067	-625.125	65.306
600	80.500	202.508	163.043	23.679	-689.732	-612.063	53.285
700	81.631	215.001	169.594	31.785	-688.503	-599.217	44.714
700.000	81.631	215.001	169.594	31.785	GLASS <--> LIQUID TRANSITION		
700.000	112.968	215.001	169.594	31.785			
800	112.968	230.086	176.234	43.082	-685.153	-586.647	38.304
900	112.968	243.392	182.971	54.378	-681.146	-574.578	33.348
1000	112.968	255.294	189.619	65.675	-677.444	-562.938	29.405
1015.000	112.968	256.976	190.602	67.370	CRYSTAL <--> LIQUID		
1100	112.968	266.061	196.087	76.972	-674.053	-551.654	26.196
1200	112.968	275.891	202.333	88.269	-678.708	-540.043	23.507
1300	112.968	284.933	208.344	99.566	-674.712	-528.650	21.241
1400	112.968	293.305	214.117	110.862	-670.724	-517.564	19.311
1500	112.968	301.099	219.659	122.159	-666.746	-506.763	17.647
1600	112.968	308.389	224.979	133.456	-662.779	-496.226	16.200
1700	112.968	315.238	230.089	144.753	-658.823	-485.938	14.931
1800	112.968	321.695	235.001	156.050	-803.554	-473.671	13.746
1900	112.968	327.803	239.726	167.346	-798.212	-455.490	12.522
2000	112.968	333.598	244.276	178.643	-792.893	-437.590	11.429
2100	112.968	339.109	248.662	189.940	-787.600	-419.955	10.446
2200	112.968	344.365	252.893	201.237	-782.338	-402.572	9.558
2300	112.968	349.386	256.980	212.534	-777.112	-385.427	8.753
2400	112.968	354.194	260.931	223.830	-771.927	-368.508	8.020
2500	112.968	358.806	264.755	235.127	-766.790	-351.805	7.351

PREVIOUS: CURRENT: June 1974

Calcium Bromide ($CaBr_2$) $Br_2Ca_1(l)$

CALCIUM BROMIDE ($CaBr_2$) CRYSTAL - LIQUID M_r = 199.888

209.15 to 1015 K crystal

above 1015 K liquid

Refer to the individual tables for details.

Calcium Bromide ($CaBr_2$) $Br_2Ca_1(cr,l)$

Enthalpy Reference Temperature = T_r = 298.15 K Standard State Pressure = $p°$ = 0.1 MPa

T/K	$C_p°$ (J K⁻¹mol⁻¹)	$S°$ (J K⁻¹mol⁻¹)	$-[G°-H°(T_r)]/T$ (J K⁻¹mol⁻¹)	$H°-H°(T_r)$ (kJ mol⁻¹)	$\Delta_fH°$ (kJ mol⁻¹)	$\Delta_fG°$ (kJ mol⁻¹)	Log K_f
0							
100							
200							
298.15	75.040	129.704	129.704	0.	-683.247	-664.139	116.354
300	75.124	130.168	129.705	0.139	-683.296	-664.020	115.616
400	77.990	152.206	132.691	7.806	-712.744	-651.460	85.072
500	79.454	169.774	138.410	15.682	-711.325	-636.304	66.474
600	80.500	184.352	144.887	23.679	-709.990	-621.427	54.100
700	81.630	196.836	151.437	31.780	-708.767	-606.765	45.277
800	83.471	207.851	157.813	40.030	-708.463	-592.168	38.665
900	85.856	217.816	163.935	48.493	-707.289	-577.703	33.529
1000	88.617	227.004	169.788	57.216	-706.160	-563.365	29.427
1015.000	89.036	228.327	170.643	58.549	CRYSTAL <--> LIQUID		
1015.000	112.968	256.976	170.643	87.628	TRANSITION		
1100	112.968	266.061	177.670	97.230	-674.053	-551.654	26.196
1200	112.968	275.891	185.452	108.527	-678.708	-540.043	23.507
1300	112.968	284.933	192.761	119.824	-674.712	-528.650	21.241
1400	112.968	293.305	199.647	131.120	-670.724	-517.564	19.311
1500	112.968	301.099	206.154	142.417	-666.746	-506.763	17.647
1600	112.968	308.389	212.318	153.714	-662.779	-496.226	16.200
1700	112.968	315.238	218.173	165.011	-658.823	-485.938	14.931
1800	112.968	321.695	223.747	176.308	-803.554	-473.671	13.746
1900	112.968	327.803	229.064	187.604	-798.212	-455.490	12.522
2000	112.968	333.598	234.147	198.901	-792.893	-437.590	11.429
2100	112.968	339.109	239.015	210.198	-787.600	-419.955	10.446
2200	112.968	344.365	243.685	221.495	-782.338	-402.572	9.558
2300	112.968	349.386	248.172	232.792	-777.112	-385.427	8.753
2400	112.968	354.194	252.491	244.088	-771.927	-368.508	8.020
2500	112.968	358.806	256.652	255.385	-766.790	-351.805	7.351

PREVIOUS: CURRENT: June 1974

Calcium Bromide ($CaBr_2$) $Br_2Ca_1(cr,l)$

CALCIUM BROMIDE ($CaBr_2$) IDEAL GAS M_r = 199.888

$S°(298.15 K) = [314.8 \pm 8.4]\ J\ K^{-1}\ mol^{-1}$

$\Delta_f H°(0 K) = -370.5 \pm 8.4\ kJ\ mol^{-1}$
$\Delta_f H°(298.15 K) = -384.9 \pm 8.4\ kJ\ mol^{-1}$

Vibrational Frequencies and Degeneracies

ν, cm^{-1}
[148](1)
[47](2)
330 (1)

Ground State Quantum Weight: [1] σ = 2

Point Group: $D_{\infty h}$

Bond Distance: Ca-Br = 2.67 ± 0.03 Å Bond Angle: Br-Ca-Br = 180°

Rotational Constant: $B_0 = 0.014797\ cm^{-1}$

Enthalpy of Formation

The selection of $\Delta_f H°(298.15 K) = -92.0 \pm 2.0\ kcal\ mol^{-1}$ ($-384.928 \pm 8.4\ kJ\ mol^{-1}$) for $CaBr_2(g)$ is based on results derived from two independent means. Peterson and Hutchison (1, 2) have reported results of an extensive study of the vapor pressures (1149-1321 K) for liquid $CaBr_2$. Measurements were made by the Knudsen effusion method on a sample which had been prepared by direct union of high purity elements. X-ray diffraction patterns on the final product showed no metal or oxide lines. Schofield and Sugden (3) investigated the equilibrium $Ca(g) + 2\ HBr(g) = CaBr_2(g) + 2\ H(g)$ in H_2-O_2-N_2 flames and reported K_p data for temperatures in the range 2137-2532 K. Results of our 2nd and 3rd law analysis of these data are tabulated below. We assume that the dibromide monomer is the only vapor species produced in the volatilization of $CaBr_2(\ell)$. Support for this assumption is provided by mass spectra for some of the other alkaline earth dihalides (4). These studies indicate that the saturated vapors consist predominately of the metal dihalide monomer and agree with predictions which were made by Brewer et al. (5) for these dihalide molecules. Included in the table for comparison are results of our analysis of older vapor pressure data (6),7) which have been reviewed by Brewer et al. (5).

Source	Method	Reaction	Data Points	T/K	$\Delta_r H°$, kcal mol^{-1} 2nd law	3rd Law	Drift cal $K^{-1}mol^{-1}$	$\Delta_f H°(298.15 K)$ kcal mol^{-1}
Peterson and Hutchison (1, 2)	Knudsen Effusion	A	18(a)	1149-1321	65.7	67.70 ± 0.60	1.6 ±1.3	-90.8 ± 1.6
Brewer (5, 6)		B	1	990		68.9		-94.4
Brewer (5, 7)	Boiling Point	A	2	1180;1480	62.5	64.76 ±0.7	1.7	-93.7 ± 1.7
Schofield and Sugden (3)	Flame K_p	C	8(a)	2137-2532	-9.4	-14.38 ±1.7	-2.1 ±3.2	-93.2 ± 2.0

Reactions: (A) $CaBr_2(\ell) = CaBr_2(g)$ (B) $CaBr_2(cr) = CaBr_2(g)$ (C) $Ca(g) + 2\ HBr(g) = CaBr_2(g) + 2\ H(g)$

(a) One point rejected due to failure of a statistical test.

In the selection of $\Delta_f H°$, no weight is given to those measurements made prior to 1965. Also, we note that Schofield and Sugden (3) reported in the same paper similar results for $SrCl_2$ and $BaCl_2$ which lead to $\Delta_f H°$ values that are quite consistent with JANAF enthalpies of formation (4). We, therefore, choose to select an average value of the results of Peterson and Hutchison (1, 2) and Schofield and Sugden (3). An earlier analysis (14) of the vapor pressure data (1, 2) with older functions (5) gave $\Delta_f H°$ of $-95.2\ kcal\ mol^{-1}$.

Heat Capacity and Entropy

The bond length is from the high-temperature electron diffraction work of Akishin and Spiridonov (8). Electron diffraction patterns (8) were satisfactorily explained on the basis of a linear model (180° ± 10°). Later studies by Wharton et al. (9), using electric deflection of molecular beams to detect dipole moments, showed no polarity in the $CaBr_2$ molecule. The absence of dipolar character is most reasonably explained by a linear and centrosymmetric configuration.

The antisymmetric stretching frequency (ν_3) was observed in the high-temperature infrared spectra of $CaBr_2$ vapor by Baikov (10). The symmetric stretching (ν_1) and bending (ν_2) frequencies are calculated from force constants by the valence force method (11). The observed value of ν_3 gives $k = 1.028 \times 10^5$ dynes/cm which leads to $\nu_1 = 148\ cm^{-1}$. Other estimates that have been reported for ν_1 (in cm^{-1}) are 148 (10), 164 (5), and 172 (13). A comparison of values for the ratio of the stretch to bend force constants for the linear molecules $CaCl_2$ (4)(ratio = 81)] and $SrBr_2$ (4)(ratio = [100])] suggests a value near 100 for $CaBr_2$. This value for the ratio gives $k\delta/\ell^2 = 1.028 \times 10^3$ dynes/cm, or $\nu_2 = 47\ cm^{-1}$. However, similar data for $BeBr_2$ (12) indicate that the value of the ratio could be as low as 41. This leads to a $\nu_2 = 73\ cm^{-1}$ which agrees with the value (85 cm^{-1}) recommended by Krasnov and Svettsov (13). Brewer et al. (5) choose $\nu_2 = 52\ cm^{-1}$. We prefer the lowest value of ν_2 (47 cm^{-1}) at this time but include in the uncertainty (± 2.0 cal $K^{-1}mol^{-1}$) assigned to the value of S°(298.15 K) the possibility that the highest value (85 cm^{-1}) is correct. The singlet ground state is assigned by analogy with that for $BaCl_2$ (4).

References

1. D. T. Peterson and J. F. Hutchison, J. Chem. Eng. Data 15, 320 (1970).
2. J. F. Hutchison, Ames Laboratory, Iowa State University, M. S. Thesis, AEC Contract No. W-7405-eng-82, November, 1965.
3. K. Schofield and T. M. Sugden, Trans. Faraday Soc. 67, 1054 (1971).
4. JANAF Thermochemical Tables: $BaI_2(g)$, $SrBr_2(g)$, 6-30-74; $BaCl_2(g)$, $SrCl_2(g)$, 12-31-72; $MgCl_2(g)$, 12-31-69; $BeF_2(g)$, $CaCl_2(g)$, 6-30-70.
5. L. Brewer, G. R. Somayajulu, and E. Brackett, Chem. Revs. 63, 111 (1963).
6. A. Stock and H. Heynemann, Chem. Ber. 42, 4088 (1909).
7. H. von Wartenberg and O. Bosse, Z. Elektrochem. 28, 384 (1922).
8. P. A. Akishin and V. P. Spiridonov, Kristallografiya 2, 472 (1952).
9. L. Wharton, R. A. Berg, and W. Klemperer, J. Chem. Phys. 39, 2023 (1963).
10. V. I. Baikov, Opt. Spektrosk. 27, 502 (1969).
11. G. Herzberg, "Infrared and Raman Spectra of Polyatomic Molecules," D. Van Nostrand Company, Inc., New York, 1962.
12. A. Snelson, J. Phys. Chem. 72, 250 (1968).
13. K. S. Krasnov and V. I. Svettsov, Izv. Vysshikh Uchebn. Zavendenii, Khim. i Khim. Tekhnol. 6, 167 (1963).
14. U. S. Nat. Bur. Stand. Tech. Note 270-6, 1971.

Calcium Bromide ($CaBr_2$) $Br_2Ca_1(g)$

Enthalpy Reference Temperature = T_r = 298.15 K Standard State Pressure = p° = 0.1 MPa

T/K	$C_p°$ (J $K^{-1}mol^{-1}$)	S°	$-[G°-H°(T_r)]/T$	$H°-H°(T_r)$ (kJ mol^{-1})	$\Delta_f H°$	$\Delta_f G°$	Log K_f
0	0.	0.	INFINITE	-15.799	-370.482	-370.482	INFINITE
100	52.538	252.373	366.898	-11.452	-369.908	-388.164	202.756
200	58.448	290.991	320.226	-5.847	-371.473	-405.889	106.007
250	59.666	304.176	315.742	-2.892	-372.583	-414.370	86.578
298.15	60.387	314.752	314.752	0.	-384.928	-420.991	73.756
300	60.409	315.125	314.753	0.112	-385.004	-421.215	73.340
350	60.889	324.476	315.490	3.145	-415.927	-425.550	63.510
400	61.214	332.629	317.134	6.198	-416.033	-426.918	55.750
450	61.443	339.853	319.264	9.265	-416.170	-428.271	49.712
500	61.611	346.336	321.653	12.341	-416.346	-429.606	44.881
600	61.833	357.590	326.732	18.515	-416.835	-432.216	37.628
700	61.970	367.133	331.840	24.705	-417.522	-434.728	32.440
800	62.059	375.414	336.780	30.907	-419.266	-437.023	28.535
900	62.121	382.727	341.487	37.116	-420.347	-439.181	25.489
1000	62.166	389.275	345.944	43.331	-421.727	-441.202	23.046
1100	62.199	395.201	350.157	49.549	-423.415	-443.070	21.040
1200	62.224	400.615	354.139	55.770	-433.145	-444.149	19.333
1300	62.244	405.596	357.909	61.994	-434.222	-445.022	17.881
1400	62.260	410.209	361.482	68.219	-435.307	-445.813	16.633
1500	62.272	414.505	364.875	74.445	-436.399	-446.525	15.549
1600	62.283	418.525	368.104	80.673	-437.500	-447.164	14.598
1700	62.291	422.301	371.182	86.902	-438.613	-447.734	13.757
1800	62.298	425.861	374.122	93.131	-588.411	-446.027	12.943
1900	62.304	429.230	376.934	99.362	-588.136	-438.125	12.045
2000	62.310	432.426	379.630	105.592	-587.883	-430.236	11.237
2100	62.314	435.466	382.217	111.823	-587.655	-422.359	10.506
2200	62.318	438.365	384.704	118.055	-587.458	-414.493	9.841
2300	62.321	441.135	387.097	124.287	-587.297	-406.635	9.235
2400	62.324	443.788	389.405	130.519	-587.177	-398.782	8.679
2500	62.327	446.332	391.631	136.752	-587.104	-390.934	8.168
2600	62.329	448.776	393.782	142.985	-587.084	-383.088	7.696
2700	62.331	451.129	395.863	149.218	-587.124	-375.241	7.259
2800	62.333	453.396	397.877	155.451	-587.228	-367.392	6.854
2900	62.335	455.583	399.830	161.684	-587.405	-359.538	6.476
3000	62.337	457.696	401.724	167.918	-587.659	-351.677	6.123
3100	62.338	459.740	403.562	174.152	-587.997	-343.805	5.793
3200	62.339	461.720	405.349	180.386	-588.425	-335.921	5.483
3300	62.340	463.638	407.086	186.620	-588.949	-328.023	5.192
3400	62.341	465.499	408.777	192.854	-589.574	-320.107	4.918
3500	62.342	467.306	410.424	199.088	-590.305	-312.171	4.659
3600	62.343	469.062	412.028	205.322	-591.147	-304.212	4.414
3700	62.344	470.770	413.593	211.556	-592.105	-296.229	4.182
3800	62.345	472.433	415.120	217.791	-593.181	-288.219	3.962
3900	62.345	474.052	416.610	224.025	-594.380	-280.178	3.753
4000	62.346	475.631	418.066	230.260	-595.703	-272.105	3.553
4100	62.347	477.170	419.489	236.495	-597.152	-263.997	3.363
4200	62.347	478.673	420.880	242.729	-598.728	-255.853	3.182
4300	62.348	480.140	422.241	248.964	-600.433	-247.669	3.009
4400	62.348	481.573	423.573	255.199	-602.262	-239.445	2.843
4500	62.348	482.974	424.878	261.434	-604.203	-231.178	2.683
4600	62.349	484.345	426.156	267.669	-606.274	-222.866	2.531
4700	62.349	485.686	427.408	273.903	-608.469	-214.508	2.384
4800	62.350	486.998	428.636	280.138	-610.780	-206.102	2.243
4900	62.350	488.284	429.840	286.373	-613.205	-197.646	2.107
5000	62.350	489.544	431.022	292.608	-615.738	-189.140	1.976
5100	62.351	490.778	432.181	298.843	-618.374	-180.581	1.850
5200	62.351	491.989	433.320	305.078	-621.105	-171.971	1.727
5300	62.351	493.177	434.438	311.314	-623.930	-163.307	1.609
5400	62.351	494.342	435.537	317.549	-626.842	-154.589	1.495
5500	62.352	495.486	436.616	323.784	-629.834	-145.816	1.385
5600	62.352	496.610	437.678	330.019	-632.890	-136.989	1.278
5700	62.352	497.713	438.721	336.254	-636.021	-128.106	1.174
5800	62.352	498.798	439.748	342.489	-639.212	-119.168	1.073
5900	62.352	499.864	440.758	348.725	-642.458	-110.174	0.975
6000	62.353	500.912	441.752	354.960	-645.751	-101.124	0.880

PREVIOUS: June 1974 (1 atm) CURRENT: June 1974 (1 bar)

Calcium Bromide ($CaBr_2$) $Br_2Ca_1(g)$

IRON BROMIDE ($FeBr_2$) CRYSTAL M_r = 215.655

$S°(298.15\ K) = 140.67 \pm 1.3\ J\ K^{-1}\ mol^{-1}$
T_{trs} = 650 K
T_{fus} = 964 K

$\Delta_f H°(0\ K)$ = Unknown
$\Delta_f H°(298.15\ K) = -248.9 \pm 2.1\ kJ\ mol^{-1}$
$\Delta_{trs} H° = 0.4184\ kJ\ mol^{-1}$
$\Delta_{fus} H° = [50.208 \pm 13]\ kJ\ mol^{-1}$

Enthalpy of Formation

The enthalpies of solution of Fe(cr), $Br_2(\ell)$ and $FeBr_2$(cr) in Br_2-KBr aqueous solution were measured by use of an ice calorimeter by Hieber and Woerner (1). From the results obtained, the enthalpy of formation for $FeBr_2$(cr) was reported to be $-59.87\ kcal\ mol^{-1}$. The enthalpies of solution of $FeBr_2$(cr) in water were determined by Li and Gregory (2) and by Paoletti (3) to be -20.1±0.4 and -20.06±0.06 kcal mol^{-1}, respectively. Using $\Delta_f H°(298.15\ K)$ = -29.05 and -21.3 kcal mol^{-1} for Br^-(aq, ∞) and Fe^{++}(aq, ∞), the corresponding $\Delta_f H°(298.15\ K)$ for $FeBr_2$(cr) were calculated as -59.3 and -59.34 kcal mol^{-1}. The value of $\Delta_f H°(Br^-$, aq, ∞, 298.15 K) was taken from (4). The value of $\Delta_f H°(Fe^{++}$, aq, ∞, 298.15 K) was derived from enthalpies of solution and formation for $FeCl_2$(cr) using $\Delta_f H°(Cl^-$, aq, ∞, 298.15 K) = -39.952 kcal mol^{-1} from (4).

The value of $\Delta_f H°(FeBr_2$, cr, 298.15 K) adopted is -59.5±0.5 kcal mol^{-1} (-248.948±2.1 kJ mol^{-1}).

Heat Capacity and Entropy

The heat capacities, 323-633 K, were measured by Gregory and O'Neal (5). Those for temperatures below 232 K and above 633 K were estimated by linear extrapolation. Low temperature $C_p°$, 12-110 K, were reported by Miljutin and Nachimowitsch (6). However, these data are inadequate for obtaining an accurate entropy at 298.15 K. The adopted S°(298.15 K) value was reported by Gregory and MacLaren (7) and was obtained from Westrum (8). A magnetic transition at 11 K was reported by Bizette et al. (9).

Transition Data

T_{trs} and $\Delta_{trs}H°$ were taken from Gregory and O'Neal (5).

Fusion Data

T_{fus} adopted was reported by MacLaren and Gregory (10). The value was calculated from the vapor pressure equations and is slightly above 684°C (957 K) reported by Ferrari et al. (11). However, it agrees well with 962±2 K obtained by the differential thermocouple cooling curve analysis (10). The $\Delta_{fus}H°$ value was estimated from the vapor pressure data in order to obtain good agreement between the 2nd and 3rd law heats of vaporization. See the $FeBr_2$(g) table for details.

Sublimation Data

$\Delta_{sub}H°$(monomer → 298.15 K) is calculated as the difference between $\Delta_f H°(298.15\ K)$ for $FeBr_2$(g) and $FeBr_2$(cr).
$\Delta_{sub}H°$(dimer → 298.15 K) is calculated as the difference between those for Fe_2Br_4(g) and 2 $FeBr_2$(cr).

References

1. W. Hieber and A. Woerner, Z. Elektrochem. 40, 287 (1934).
2. J. C. M. Li and N. W. Gregory, J. Am. Chem. Soc. 74, 4670 (1952).
3. P. Paoletti, Trans. Faraday Soc. 61, 219 (1965).
4. U. S. Nat. Bur. Stand. Tech. Note 270-1, 1965.
5. N. W. Gregory and H. E. O'Neal, J. Am. Chem. Soc. 81, 2649 (1959).
6. G. Miljutin and N. Nachimowitsch, Phys. Trans. Ukrain. Acad. Sci. 9, 71 (1940).
7. N. W. Gregory and R. O. MacLaren, J. Phys. Chem. 59, 110 (1955).
8. E. F. Westrum, Jr., University of Michigan, Ann Arbor, Michigan, personal communication to Gregory and MacLaren (7).
9. H. Bizette, C. Terrier, and B. Tasi, Compt. Rend. 245, 507 (1957).
10. R. O. MacLaren and N. W. Gregory, J. Phys. Chem. 59, 184 (1955).
11. A. Ferrari, A. Celeri and F. Giorgi, Atti Accad. Nazl. Lincei 9, 782,1134 (1929).

Iron Bromide ($FeBr_2$) Br_2Fe_1(cr)

Enthalpy Reference Temperature = T_r = 298.15 K Standard State Pressure = p° = 0.1 MPa

T/K	$C_p°$ (J K⁻¹mol⁻¹)	S° (J K⁻¹mol⁻¹)	$-[G°-H°(T_r)]/T$ (J K⁻¹mol⁻¹)	$H°-H°(T_r)$ (kJ mol⁻¹)	$\Delta_f H°$ (kJ mol⁻¹)	$\Delta_f G°$ (kJ mol⁻¹)	Log K_f
0							
100							
200							
298.15	80.232	140.666	140.666	0.	-248.948	-237.362	41.585
300	80.274	141.163	140.668	0.148	-248.986	-237.290	41.316
400	82.500	164.560	143.843	8.287	-277.955	-227.334	29.687
500	84.726	183.209	149.912	16.648	-276.138	-214.890	22.449
600	86.952	198.853	156.799	25.232	-274.360	-202.807	17.656
650.000	88.061	205.857	160.306	29.608		I <--> II	
650.000	88.069	206.500	160.306	30.026		TRANSITION	
700	89.178	213.067	163.843	34.457	-272.199	-191.053	14.257
800	91.404	225.121	170.763	43.486	-270.537	-179.575	11.725
900	93.630	236.015	177.417	52.738	-269.078	-168.295	9.768
964.000	95.054	242.495	181.525	58.776	--------	II <--> LIQUID	--------
1000	95.855	245.995	183.783	62.212	-268.125	-157.156	8.209
1100	98.081	255.236	189.864	71.909	-267.854	-146.056	6.936
1200	100.307	263.865	195.675	81.828	-266.817	-135.047	5.878
1300	102.533	271.982	201.235	91.970	-263.919	-124.183	4.990
1400	104.759	279.662	206.565	102.335	-260.891	-113.547	4.236
1500	106.985	286.965	211.684	112.922	-257.733	-103.132	3.591

PREVIOUS: CURRENT: September 1966

Iron Bromide ($FeBr_2$) Br_2Fe_1(cr)

IRON BROMIDE ($FeBr_2$) LIQUID M_r = 215.655

$S°(298.15\ K) = [169.376]\ J\ K^{-1}\ mol^{-1}$

$\Delta_f H°(298.15\ K) = [-211.005]\ kJ\ mol^{-1}$

$T_{fus} = 964\ K$

$\Delta_{fus}H° = [50.208 \pm 13]\ kJ\ mol^{-1}$

Enthalpy of Formation

$\Delta_f H°(\ell,\ 298.15\ K)$ is calculated from $\Delta_f H°(cr,\ 298.15\ K)$ by adding $\Delta_{fus}H°$ and the difference in enthalpy, H°(964 K) - H°(298.15 K), between $FeBr_2(cr)$ and $FeBr_2(\ell)$.

Heat Capacity and Entropy

The heat capacity is assumed to be constant in the temperature range from 298.15 to 2000 K and is estimated on the basis of 8.5 cal $K^{-1}\ mol^{-1}$ per atom. S°(298.15 K) is calculated in a manner similar to that used for the enthalpy of formation.

Fusion Data

See the crystal table for details.

Vaporization Data

T_{vap} is calculated as the temperature at which the total pressure of $FeBr_2(g)$ and $Fe_2Br_4(g)$ over $FeBr_2(\ell)$ equals one bar. Based on the vapor composition and the values of $\Delta_{vap}H°$ for both the monomer and dimer at T_{vap} for both the monomer and dimer at T_{vap}, the enthalpy of vaporization is derived.

Iron Bromide ($FeBr_2$) $Br_2Fe_1(l)$

Enthalpy Reference Temperature = T_r = 298.15 K Standard State Pressure = p° = 0.1 MPa

T/K	$C_p°$	S°	$-[G°-H°(T_r)]/T$	$H°-H°(T_r)$	$\Delta_f H°$	$\Delta_f G°$	Log K_f
	J K⁻¹mol⁻¹			kJ mol⁻¹			
0							
100							
200							
298.15	106.692	169.376	169.376	0.	-211.005	-207.978	36.437
300	106.692	170.036	169.378	0.197	-210.994	-207.960	36.209
400	106.692	200.729	173.563	10.867	-237.433	-201.279	26.284
500	106.692	224.537	181.465	21.536	-233.308	-192.723	20.134
600	106.692	243.989	190.314	32.205	-229.444	-184.974	16.103
700	106.692	260.436	199.187	42.874	-225.839	-177.850	13.271
800	106.692	274.683	207.753	53.543	-222.537	-171.224	11.180
900	106.692	287.249	215.902	64.213	-219.661	-164.988	9.576
964.000	106.692	294.578	220.885	71.041	-------- II <--> LIQUID --------		
1000	106.692	298.490	223.608	74.882	-217.513	-159.039	8.307
1100	106.692	308.659	230.885	85.551	-216.269	-153.237	7.277
1200	106.692	317.942	237.759	96.220	-214.482	-147.605	6.425
1300	106.692	326.482	244.260	106.889	-211.057	-142.172	5.713
1400	106.692	334.389	250.419	117.559	-207.724	-136.998	5.111
1500	106.692	341.750	256.265	128.228	-204.485	-132.060	4.599
1600	106.692	348.636	261.825	138.897	-201.339	-127.335	4.157
1700	106.692	355.104	267.124	149.566	-199.239	-122.785	3.773
1800	106.692	361.202	272.183	160.235	-196.620	-118.364	3.435
1900	106.692	366.971	277.021	170.905	-208.198	-113.383	3.117
2000	106.692	372.443	281.657	181.574	-206.017	-108.449	2.832

PREVIOUS: CURRENT: September 1966

Iron Bromide ($FeBr_2$) $Br_2Fe_1(l)$

IRON BROMIDE ($FeBr_2$) CRYSTAL - CRYSTAL - LIQUID M_r = 215.655

298.15 to 650 K		crystal, I
650	to 964 K	crystal, II
above	964 K	liquid

Refer to the individual tables for details.

Iron Bromide ($FeBr_2$) Br_2Fe_1(cr,l)

Enthalpy Reference Temperature = T_r = 298.15 K Standard State Pressure = p° = 0.1 MPa

T/K	C_p° (J K⁻¹mol⁻¹)	S° (J K⁻¹mol⁻¹)	$-[G^\circ - H^\circ(T_r)]/T$ (J K⁻¹mol⁻¹)	$H^\circ - H^\circ(T_r)$ (kJ mol⁻¹)	$\Delta_f H^\circ$ (kJ mol⁻¹)	$\Delta_f G^\circ$ (kJ mol⁻¹)	Log K_f
0							
100							
200							
298.15	80.232	140.666	140.666	0.	-248.948	-237.362	41.585
300	80.274	141.163	140.668	0.148	-248.986	-237.290	41.316
400	82.500	164.560	143.843	8.287	-277.955	-227.334	29.687
500	84.726	183.209	149.912	16.648	-276.138	-214.890	22.449
600	86.952	198.853	156.799	25.232	-274.360	-202.807	17.656
650.000	88.061	205.857	160.306	29.608		I <--> II	
650.000	88.069	206.500	160.306	30.026		TRANSITION	
700	89.178	213.067	163.843	34.457	-272.199	-191.053	14.257
800	91.404	225.121	170.763	43.486	-270.537	-179.575	11.725
900	93.630	236.015	177.417	52.738	-269.078	-168.295	9.768
964.000	95.054	242.495	181.525	58.776		II <--> LIQUID	
964.000	106.692	294.578	181.525	108.984		TRANSITION	
1000	106.692	298.490	185.666	112.825	-217.513	-159.039	8.307
1100	106.692	308.659	196.392	123.494	-216.269	-153.237	7.277
1200	106.692	317.942	206.140	134.163	-214.482	-147.605	6.425
1300	106.692	326.482	215.073	144.832	-211.057	-142.172	5.713
1400	106.692	334.389	223.317	155.501	-207.724	-136.998	5.111
1500	106.692	341.750	230.970	166.171	-204.485	-132.060	4.599
1600	106.692	348.636	238.111	176.840	-201.339	-127.335	4.157
1700	106.692	355.104	244.805	187.509	-199.239	-122.785	3.773
1800	106.692	361.202	251.103	198.178	-196.620	-118.364	3.435
1900	106.692	366.971	257.051	208.847	-208.198	-113.383	3.117
2000	106.692	372.443	262.685	219.517	-206.017	-108.449	2.832

PREVIOUS: CURRENT: September 1966

Iron Bromide ($FeBr_2$) Br_2Fe_1(cr,l)

IRON BROMIDE ($FeBr_2$) IDEAL GAS M_r = 215.655

$S°(298.15 K) = [337.4] J K^{-1} mol^{-1}$

$\Delta_f H°(0 K) = [-28.43 \pm 2.1] kJ mol^{-1}$
$\Delta_f H°(298.15 K) = [-41.42 \pm 2.1] kJ mol^{-1}$

Electronic Levels and Quantum Weights

ϵ_i, cm^{-1}	g_i
0	[10]
[4450]	[10]
[6900]	[5]

Vibrational Frequencies and Degeneracies

ν, cm^{-1}
[150](1)
[35](2)
[300](1)

Point Group $[D_{\infty h}]$ σ = [2]
Bond Distance: Fe-Br = [2.24] Å
Bond Angle: Br-Fe-Br = [180]°
Rotational Constant: B_o = [0.021023] cm^{-1}

Enthalpy of Formation

The vapor pressures of $FeBr_2(cr)$ and $FeBr_2(\ell)$ have been measured in the temperature range from 623.15 to 962.15 K (1) 2). Using the reported data the corresponding enthalpy changes for the reactions (A) $FeBr_2(cr) = FeBr_2(g)$ and (B) $FeBr_2(\ell) = FeBr_2(g)$ were evaluated by both the 2nd and 3rd law methods. The results obtained are presented in the following table. The values of $\Delta_f H°(298.15 K)$ for $FeBr_2(g)$ are calculated based on the 3rd law values for $\Delta_r H°(298.15 K)$. The value of $\Delta_f H°(298.15 K)$ for $FeBr_2(g)$ adopted is -9.9 ± 0.5 kcal mol^{-1} (-41.422 kJ mol^{-1}).

Source	Method	Reaction	T/K	$\Delta_r H°(298.15 K)$, kcal mol^{-1} 2nd law	3rd law	Drift cal K^{-1} mol^{-1}	$\Delta_f H°(298.15 K)$ kcal mol^{-1}
(1)	Transpiration	(A)	673.15- 962.15	50.04	49.56	- 0.75	-9.94
(1)	Effusion	(A)	623.15- 718.15	49.21	50.14	+ 1.37	-9.36
(1)	Diaphragm	(A)	873.15- 962.15	59.20	49.95	-10.09	-9.55
(2)	Torsion-Effusion	(A)	670.0 - 740.0	49.11	49.59	+ 0.71	-9.91
(1)	Diaphragm	(B)	962.15-1182.15	39.77	40.84	+ 0.93	-9.59

Heat Capacity and Entropy

The molecular structure was assumed to be linear. The electronic levels and quantum weights were estimated by comparison with those for $FeCl_2(g)$ reported by DeKock and Gruen (3). The bond distance was estimated by Brewer et al. (4). The vibrtional frequencies were estimated in order to obtain good agreement between the 2nd and 3rd law enthalpies of sublimation and vaporization.

References

1. R. O. MacLaren and N. W. Gregory, J. Phys. Chem. 59, 184 (1955).
2. R. J. Sime and N. W. Gregory, J. Phys. Chem. 64, 86 (1960).
3. C. W. DeKock and D. M. Gruen, J. Chem. Phys. 44, 4387 (1966).
4. L. Brewer, G. R. Somayajulu and E. Brackett, Chem. Rev. 63, 111 (1963).

Iron Bromide ($FeBr_2$) $Br_2Fe_1(g)$

Enthalpy Reference Temperature = T_r = 298.15 K Standard State Pressure = p° = 0.1 MPa

T/K	$C_p°$ (J $K^{-1}mol^{-1}$)	S°	$-[G°-H°(T_r)]/T$	$H°-H°(T_r)$ (kJ mol^{-1})	$\Delta_f H°$	$\Delta_f G°$	Log K_f
0	0.	0.	INFINITE	-16.019	-28.425	-28.425	INFINITE
100	53.226	274.411	389.883	-11.547	-27.169	-48.619	25.396
200	58.914	313.454	342.863	-5.882	-28.174	-69.790	18.227
250	60.014	326.729	338.354	-2.906	-29.151	-80.088	16.733
298.15	60.653	337.359	337.359	0.	-41.422	-88.479	15.501
300	60.672	337.734	337.360	0.112	-41.496	-88.771	15.456
350	61.093	347.120	338.100	3.157	-72.387	-94.953	14.171
400	61.376	355.298	339.750	6.219	-72.497	-98.170	12.820
450	61.575	362.539	341.887	9.293	-72.661	-101.370	11.767
500	61.722	369.035	344.283	12.376	-72.884	-104.548	10.922
600	61.932	380.307	349.375	18.559	-73.506	-110.827	9.648
700	62.102	389.867	354.494	24.761	-74.369	-116.982	8.729
800	62.284	398.171	359.446	30.980	-75.517	-122.995	8.031
900	62.506	405.520	364.165	37.219	-77.071	-128.842	7.478
1000	62.777	412.119	368.636	43.483	-79.328	-134.483	7.025
1100	63.094	418.117	372.866	49.776	-82.461	-139.832	6.640
1200	63.445	423.622	376.869	56.103	-85.016	-144.954	6.310
1300	63.819	428.715	380.664	62.466	-85.897	-149.913	6.024
1400	64.202	433.458	384.267	68.867	-86.832	-154.803	5.776
1500	64.582	437.901	387.696	75.306	-87.823	-159.624	5.559
1600	64.949	442.080	390.966	81.783	-88.870	-164.377	5.366
1700	65.295	446.028	394.090	88.295	-90.926	-169.043	5.194
1800	65.614	449.770	397.080	94.841	-92.431	-173.596	5.038
1900	65.902	453.325	399.948	101.417	-108.102	-177.360	4.876
2000	66.158	456.712	402.702	108.020	-109.987	-180.956	4.726
2100	66.380	459.945	405.351	114.648	-111.866	-184.459	4.588
2200	66.570	463.038	407.904	121.295	-113.744	-187.871	4.461
2300	66.728	466.001	410.366	127.960	-115.627	-191.199	4.342
2400	66.855	468.843	412.743	134.640	-117.518	-194.444	4.232
2500	66.955	471.575	415.042	141.331	-119.420	-197.610	4.129
2600	67.029	474.202	417.268	148.030	-121.336	-200.700	4.032
2700	67.080	476.733	419.423	154.736	-123.268	-203.716	3.941
2800	67.110	479.173	421.514	161.445	-125.218	-206.660	3.855
2900	67.121	481.528	423.543	168.157	-127.185	-209.534	3.774
3000	67.116	483.804	425.514	174.869	-129.171	-212.340	3.697
3100	67.096	486.004	427.430	181.580	-131.174	-215.079	3.624
3200	67.065	488.134	429.294	188.288	-481.494	-210.330	3.433
3300	67.023	490.197	431.108	194.992	-481.628	-201.854	3.195
3400	66.971	492.197	432.876	201.692	-481.809	-193.373	2.971
3500	66.913	494.138	434.599	208.386	-482.035	-184.887	2.759
3600	66.848	496.022	436.279	215.074	-482.305	-176.393	2.559
3700	66.778	497.852	437.918	221.756	-482.617	-167.891	2.370
3800	66.704	499.632	439.519	228.430	-482.970	-159.380	2.191
3900	66.626	501.364	441.083	235.096	-483.364	-150.860	2.021
4000	66.546	503.050	442.611	241.755	-483.796	-142.328	1.859
4100	66.465	504.692	444.105	248.405	-484.265	-133.786	1.704
4200	66.382	506.293	445.567	255.048	-484.772	-125.232	1.557
4300	66.299	507.854	446.997	261.682	-485.315	-116.665	1.417
4400	66.215	509.377	448.398	268.308	-485.894	-108.085	1.283
4500	66.132	510.864	449.769	274.925	-486.507	-99.492	1.155
4600	66.049	512.317	451.113	281.534	-487.156	-90.884	1.032
4700	65.966	513.736	452.431	288.135	-487.840	-82.262	0.914
4800	65.885	515.124	453.723	294.727	-488.558	-73.625	0.801
4900	65.805	516.482	454.990	301.312	-489.312	-64.973	0.693
5000	65.726	517.810	456.233	307.888	-490.102	-56.305	0.588
5100	65.648	519.111	457.453	314.457	-490.928	-47.621	0.488
5200	65.572	520.385	458.651	321.018	-491.791	-38.921	0.391
5300	65.497	521.633	459.828	327.571	-492.692	-30.203	0.298
5400	65.424	522.857	460.983	334.117	-493.634	-21.468	0.208
5500	65.352	524.057	462.119	340.656	-494.616	-12.715	0.121
5600	65.283	525.234	463.236	347.188	-495.641	-3.944	0.037
5700	65.215	526.389	464.334	353.713	-496.711	4.846	-0.044
5800	65.148	527.522	465.414	360.231	-497.828	13.654	-0.123
5900	65.083	528.635	466.476	366.742	-498.994	22.483	-0.199
6000	65.020	529.729	467.521	373.248	-500.213	31.332	-0.273

PREVIOUS: September 1966 (1 atm) CURRENT: September 1966 (1 bar)

Iron Bromide ($FeBr_2$) $Br_2Fe_1(g)$

DIBROMOSILANE (SiH_2Br_2) IDEAL GAS M_r = 189.90938

$\Delta_fH°(0\ K)$ = [-168.4 ± 17] kJ mol^{-1}

S°(298.15 K) = [310.05 ± 1.3] J K^{-1} mol^{-1} $\Delta_fH°(298.15\ K)$ = [-190.4 ± 17] kJ mol^{-1}

Vibrational Frequencies and Degeneracies

ν, cm^{-1}	ν, cm^{-1}
2206(1)	2232(1)
942(1)	556(1)
407(1)	843(1)
122(1)	471(1)
688(1)	

Ground State Quantum Weight: [1] σ = [2]

Point Group: [C_{2v}]

Bond Distances: Si-H = [1.49] Å Si-Br = [2.19] Å

Bond Angles: H-Si-H = [111]° Br-Si-Br = [110]°

Product of the Moments of Inertia: $I_AI_BI_C$ = [6.36266 x 10^{-113}] g^3 cm^6

Enthalpy of Formation

$\Delta_fH°$(298.15 K) is estimated by linear interpolation between the values (1) of $SiBr_4$(g) and SiH_4(g). The only experimental $\Delta_fH°$ for $SiHBr_3$(g) appears to be too uncertain to justify a nonlinear interpolation such as that adopted for SiH_3Cl (1). Normally we would seek a comparison with $\Delta_fH°$ values of CH_nBr_{4-n}. These were recently reviewed by Kudchadker and Kudchadker (2). They adopted a cubic variation of $\Delta_fH°$ with n, based on $\Delta_fH°$ values of $CHBr_3$ and CBr_4 selected by Wagman et al. (3). We presume that Wagman's values are calculated rather than experimental. Uncertainty in these values (4 - 6) precludes their use as a comparison for SiH_nBr_{4-n}. We conclude, as did Hunt and Sirtl (7), that the available data justify only linear interpolation of $\Delta_fH°$.

Heat Capacity and Entropy

The molecular structure is estimated by comparison with SiH_3Br, $SiHBr_3$ and the chlorosilanes (1). Bond distances are assumed to be intermediate between those of SiH_3Br and $SiHBr_3$ (1). The principal moments of inertia are I_A = 7.9748 x 10^{-39}, I_B = 85.9075 x 10^{-39} and I_C = 92.8728 x 10^{-39} g cm^2.

Vibrational frequencies are those selected by Shimanouchi (8) based on gas-phase infrared spectra (9) and liquid-phase Raman spectra (10). Gas-phase frequencies are adopted except for ν_4 = 122 cm^{-1} and infrared inactive ν_5 = 688 cm^{-1}.

We neglect excited states and assume the electronic ground state to be a singlet. We expect that contributions from excited states should be unimportant, as discussed on the tables for SiH_3Br and SiH_2Cl_2 (1).

References

1. JANAF Thermochemical Tables: Br_4Si(g), Br_3HSi(g), BrH_3Si(g), ClH_3Si(g), Cl_2H_2Si(g), Cl_3HSi(g) 12-31-76; H_4Si(g) 6-30-76.
2. S. A. Kudchadker and A. P. Kudchadker, J. Phys. Chem. Ref. Data 4, 457 (1975).
3. U. S. Nat. Bur. Stand. Tech. Note 270-3, 1968.
4. H. J. Bernstein, J. Phys. Chem. 69, 1550 (1965).
5. S. Furuyama, D. M. Golden and S. W. Benson, J. Am. Chem. Soc. 91, 7564 (1969).
6. K. D. King, D. M. Golden and S. W. Benson, J. Phys. Chem. 75, 987 (1971).
7. L. P. Hunt and E. Sirtl, J. Electrochem. Soc. 120, 806 (1973).
8. T. Shimanouchi, J. Phys. Chem. Ref. Data 2, 135 (1973).
9. D. W. Mayo, H. E. Opitz and J. S. Peake, J. Chem. Phys. 23, 1344 (1955).
10. F. Francois and M. B. Buisset, Compt. Rend. 230, 1946 (1950).

Dibromosilane (SiH_2Br_2) $Br_2H_2Si_1$(g)

Enthalpy Reference Temperature = T_r = 298.15 K Standard State Pressure = p° = 0.1 MPa

T/K	$C_p°$ (J K^{-1}mol^{-1})	S° (J K^{-1}mol^{-1})	$-[G°-H°(T_r)]/T$ (J K^{-1}mol^{-1})	$H°-H°(T_r)$ (kJ mol^{-1})	$\Delta_fH°$ (kJ mol^{-1})	$\Delta_fG°$ (kJ mol^{-1})	Log K_f
0	0.	0.	INFINITE	-14.272	-168.449	-168.449	INFINITE
100	41.215	253.984	360.165	-10.618	-170.854	-180.412	94.237
200	53.911	286.243	315.696	-5.891	-174.900	-188.449	49.218
250	60.273	298.971	311.102	-3.033	-177.104	-191.584	40.029
298.15	65.545	310.051	310.051	0.	-190.372	-192.860	33.788
300	65.730	310.457	310.053	0.121	-190.481	-192.875	33.583
350	70.258	320.941	310.870	3.525	-222.212	-191.509	28.581
400	74.012	330.575	312.739	7.134	-222.976	-187.069	24.429
450	77.170	339.480	315.222	10.916	-223.620	-182.541	21.189
500	79.882	347.754	318.066	14.844	-224.158	-177.947	18.590
600	84.357	362.730	324.289	23.065	-224.962	-168.624	14.680
700	87.937	376.013	330.747	31.686	-225.465	-159.190	11.879
800	90.864	387.952	337.164	40.631	-225.734	-149.702	9.775
900	93.276	398.798	343.419	49.841	-225.824	-140.191	8.136
1000	95.273	408.733	349.461	59.272	-225.782	-130.677	6.826
1100	96.931	417.893	355.271	68.885	-225.647	-121.172	5.754
1200	98.315	426.389	360.848	78.649	-225.452	-111.683	4.861
1300	99.476	434.305	366.197	88.540	-225.222	-102.211	4.107
1400	100.455	441.714	371.329	98.538	-224.981	-92.758	3.461
1500	101.285	448.674	376.256	108.626	-224.743	-83.322	2.902
1600	101.994	455.234	380.989	118.791	-224.525	-73.901	2.413
1700	102.603	461.436	385.541	129.022	-274.514	-64.046	1.968
1800	103.128	467.316	389.922	139.309	-274.134	-51.677	1.500
1900	103.584	472.904	394.143	149.645	-273.759	-39.328	1.081
2000	103.982	478.228	398.216	160.024	-273.396	-26.999	0.705
2100	104.331	483.309	402.148	170.440	-273.049	-14.688	0.365
2200	104.638	488.170	405.948	180.889	-272.722	-2.393	0.057
2300	104.910	492.828	409.625	191.366	-272.420	9.888	-0.225
2400	105.152	497.298	413.185	201.870	-272.145	22.156	-0.482
2500	105.368	501.595	416.636	212.396	-271.899	34.414	-0.719
2600	105.561	505.731	419.984	222.943	-271.683	46.662	-0.937
2700	105.735	509.718	423.234	233.508	-271.497	58.902	-1.140
2800	105.892	513.567	426.392	244.089	-271.342	71.136	-1.327
2900	106.034	517.285	429.462	254.686	-271.216	83.366	-1.502
3000	106.162	520.882	432.450	265.295	-271.118	95.591	-1.664
3100	106.280	524.365	435.359	275.918	-271.047	107.813	-1.817
3200	106.387	527.741	438.194	286.551	-271.001	120.033	-1.959
3300	106.484	531.016	440.957	297.195	-270.977	132.253	-2.093
3400	106.574	534.196	443.653	307.848	-270.974	144.472	-2.220
3500	106.656	537.287	446.284	318.509	-270.989	156.691	-2.338
3600	106.732	540.293	448.854	329.179	-655.172	179.372	-2.603
3700	106.803	543.218	451.365	339.855	-654.802	202.548	-2.859
3800	106.867	546.067	453.820	350.539	-654.445	225.715	-3.103
3900	106.928	548.844	456.221	361.229	-654.098	248.873	-3.333
4000	106.983	551.552	458.571	371.924	-653.759	272.021	-3.552
4100	107.035	554.194	460.871	382.625	-653.425	295.162	-3.760
4200	107.084	556.774	463.124	393.331	-653.095	318.294	-3.959
4300	107.129	559.294	465.331	404.042	-652.765	341.418	-4.147
4400	107.171	561.757	467.494	414.757	-652.435	364.535	-4.328
4500	107.210	564.166	469.616	425.476	-652.103	387.644	-4.500
4600	107.247	566.523	471.697	436.199	-651.767	410.746	-4.664
4700	107.282	568.830	473.739	446.926	-651.426	433.840	-4.822
4800	107.315	571.089	475.744	457.655	-651.078	456.928	-4.972
4900	107.345	573.302	477.713	468.388	-650.724	480.007	-5.117
5000	107.374	575.471	479.646	479.124	-650.361	503.080	-5.256
5100	107.402	577.598	481.546	489.863	-649.989	526.145	-5.389
5200	107.427	579.683	483.413	500.605	-649.607	549.203	-5.517
5300	107.452	581.730	485.249	511.349	-649.215	572.254	-5.640
5400	107.475	583.739	487.054	522.095	-648.811	595.296	-5.758
5500	107.496	585.711	488.830	532.844	-648.395	618.331	-5.872
5600	107.517	587.648	490.578	543.594	-647.968	641.359	-5.982
5700	107.537	589.551	492.297	554.347	-647.528	664.379	-6.088
5800	107.555	591.422	493.990	565.102	-647.075	687.391	-6.191
5900	107.573	593.260	495.657	575.858	-646.609	710.394	-6.289
6000	107.590	595.068	497.299	586.616	-646.128	733.390	-6.385

PREVIOUS: December 1976 (1 atm) CURRENT: December 1976 (1 bar)

Dibromosilane (SiH_2Br_2) $Br_2H_2Si_1$(g)

MERCURY BROMIDE ($HgBr_2$) CRYSTAL M_r = 360.398

$S°(298.15\ K) = [170.314 \pm 6.3]\ J\ K^{-1}\ mol^{-1}$

$T_{fus} = 514\ K$

$\Delta_f H°(0\ K)$ = Unknown

$\Delta_f H°(298.15\ K) = -169.45 \pm 2.1\ kJ\ mol^{-1}$

$\Delta_{fus} H° = 17.908 \pm 0.33\ kJ\ mol^{-1}$

Enthalpy of Formation

$\Delta_f H°(298.15\ K)$ is taken from (1).

Heat Capacity and Entropy

$C°_p$ is assumed to be a linear function of temperature and is fitted to the data of Guinchant (2) and Janz and Goodkin (3). The entropy is estimated by adjusting its value until the melting, sublimation, and vaporization data are in agreement.

Fusion Data

T_{fus} is from (1); $\Delta_{fus}H°$ is from Janz and Goodkin (3).

References

1. U. S. Nat. Bur. Stand. Circ. 500, 1952.
2. M. Guinchant, Comp. Rend. 145, 320 (1907).
3. G. J. Janz and J. Goodkin, J. Phys. Chem. 63, 1975 (1959).

Mercury Bromide ($HgBr_2$) $Br_2Hg_1(cr)$

Enthalpy Reference Temperature = T_r = 298.15 K Standard State Pressure = p° = 0.1 MPa

T/K	$C°_p$ (J K⁻¹mol⁻¹)	S° (J K⁻¹mol⁻¹)	−[G°−H°(T_r)]/T (J K⁻¹mol⁻¹)	H°−H°(T_r) (kJ mol⁻¹)	$\Delta_f H°$ (kJ mol⁻¹)	$\Delta_f G°$ (kJ mol⁻¹)	Log K_f
0							
100							
200							
298.15	75.312	170.314	170.314	0.	-169.452	-152.183	26.662
300	75.366	170.780	170.315	0.139	-169.504	-152.076	26.479
400	78.321	192.861	173.306	7.822	-199.068	-140.110	18.297
500	81.206	210.688	179.035	15.816	-197.492	-125.547	13.116
514.000	81.437	212.933	179.948	16.955	------ CRYSTAL <--> LIQUID	-----	
600	84.157	225.738	185.612	24.076	-195.666	-111.325	9.692
700	87.086	238.931	192.305	32.638	-252.316	-90.860	6.780
800	90.015	250.751	198.885	41.493	-249.292	-67.998	4.440
900	92.943	261.522	205.255	50.641	-245.987	-45.531	2.643
1000	95.872	271.466	211.385	60.082	-242.399	-23.448	1.225
1100	98.801	280.741	217.273	69.815	-238.527	-1.738	0.083
1200	101.730	289.464	222.929	79.842	-234.371	19.607	-0.853
1300	104.659	297.722	228.367	90.161	-229.931	40.593	-1.631
1400	107.587	305.585	233.604	100.774	-225.206	61.227	-2.284
1500	110.516	313.108	238.655	111.679	-220.199	81.514	-2.839

PREVIOUS: CURRENT: March 1962

Mercury Bromide ($HgBr_2$) $Br_2Hg_1(cr)$

MERCURY BROMIDE ($HgBr_2$) LIQUID M_r = 360.398

$S°(298.15\ K) = [192.172]\ J\ K^{-1}\ mol^{-1}$

$\Delta_f H°(298.15\ K) = [-156.626]\ kJ\ mol^{-1}$

$T_{fus} = 514\ K$

$\Delta_{fus} H° = 17.908 \pm 0.33\ kJ\ mol^{-1}$

Enthalpy of Formation

$\Delta_f H°(298.15\ K)$ is calculated from that of the crystal by adding $\Delta_{fus} H°$ and the difference in enthalpy, H°(514 K)-H°(298.15 K), between the crystal and liquid.

Heat Capacity and Entropy

$C°_p$ is obtained from the data of Janz and Goodkin (1) in the range 507-544 K; it is assumed constant above and below this range. S°(298.15 K) is calculated in a manner similar to that used for the enthalpy of formation.

Fusion and Vaporization Data

T_{fus} and T_{vap} (1 atm) are from (2). $\Delta_{fus} H°$ is the value given by Janz and Goodkin (1), and $\Delta_{vap} H$ (1 atm) is obtained from the data of Prideaux (3) and Johnson (4).

References

1. G. J. Janz and J. Goodkin, J. Phys. Chem. 63, 1975 (1959).
2. U. S. Nat. Bur. Stand. Circ. 500, 1952.
3. E. B. R. Prideaux, J. Chem. Soc. (London) 97, 2032 (1910).
4. F. M. G. Johnson, J. Am. Chem. Soc. 33, 777 (1911).

Mercury Bromide ($HgBr_2$) $Br_2Hg_1(l)$

Enthalpy Reference Temperature = T_r = 298.15 K; Standard State Pressure = p° = 0.1 MPa

T/K	$C°_p$ (J K⁻¹mol⁻¹)	S° (J K⁻¹mol⁻¹)	$-[G°-H°(T_r)]/T$ (J K⁻¹mol⁻¹)	$H°-H°(T_r)$ (kJ mol⁻¹)	$\Delta_f H°$ (kJ mol⁻¹)	$\Delta_f G°$ (kJ mol⁻¹)	Log K_f
0							
100							
200							
298.15	102.090	192.172	192.172	0.	-156.626	-145.874	25.557
300	102.090	192.804	192.174	0.189	-156.629	-145.807	25.387
400	102.090	222.173	196.179	10.398	-183.665	-136.433	17.816
500	102.090	244.954	203.740	20.607	-179.875	-125.064	13.065
514.000	102.090	247.773	204.901	22.036	------ CRYSTAL <--> LIQUID -----		
600	102.090	263.567	212.207	30.816	-176.100	-114.456	9.964
700	102.090	279.304	220.697	41.025	-231.103	-97.908	7.306
800	102.090	292.936	228.894	51.234	-226.726	-79.179	5.170
900	102.090	304.961	236.691	61.443	-222.359	-60.998	3.540
1000	102.090	315.717	244.065	71.652	-218.003	-43.303	2.262
1100	102.090	325.447	251.028	81.861	-213.656	-26.043	1.237
1200	102.090	334.330	257.605	92.070	-209.317	-9.179	0.400
1300	102.090	342.502	263.826	102.278	-204.987	7.323	-0.294
1400	102.090	350.067	269.719	112.487	-200.666	23.493	-0.877
1500	102.090	357.111	275.313	122.696	-196.355	39.354	-1.370

PREVIOUS: CURRENT: March 1962

Mercury Bromide ($HgBr_2$) $Br_2Hg_1(l)$

MERCURY BROMIDE ($HgBr_2$) CRYSTAL - LIQUID M_r = 360.398 Mercury Bromide ($HgBr_2$) Br_2Hg_1(cr,l)

298.15 to 514 K crystal

above 514 K liquid

Refer to the individual tables for details.

Enthalpy Reference Temperature = T_r = 298.15 K Standard State Pressure = p° = 0.1 MPa

T/K	C_p° (J K⁻¹mol⁻¹)	S° (J K⁻¹mol⁻¹)	−[G°−H°(T_r)]/T (J K⁻¹mol⁻¹)	H°−H°(T_r) (kJ mol⁻¹)	Δ_fH° (kJ mol⁻¹)	Δ_fG° (kJ mol⁻¹)	Log K_f
0							
100							
200							
298.15	75.312	170.314	170.314	0.	-169.452	-152.183	26.662
300	75.366	170.780	170.315	0.139	-169.504	-152.076	26.479
400	78.321	192.861	173.306	7.822	-199.068	-140.110	18.297
500	81.206	210.688	179.055	15.816	-197.492	-125.547	13.116
514.000	81.437	212.933	179.948	16.955	CRYSTAL <--> LIQUID		
514.000	102.090	247.773	179.948	34.862	TRANSITION		
600	102.090	263.567	190.830	43.642	-176.100	-114.456	9.964
700	102.090	279.304	202.374	53.851	-231.103	-97.908	7.306
800	102.090	292.936	212.861	64.060	-226.726	-79.179	5.170
900	102.090	304.961	222.440	74.269	-222.359	-60.998	3.540
1000	102.090	315.717	231.239	84.478	-218.003	-43.303	2.262
1100	102.090	325.447	239.368	94.687	-213.656	-26.043	1.237
1200	102.090	334.330	246.917	104.896	-209.317	-9.179	0.400
1300	102.090	342.502	253.960	115.105	-204.987	7.323	-0.294
1400	102.090	350.067	260.557	125.314	-200.666	23.493	-0.877
1500	102.090	357.111	266.762	135.523	-196.355	39.354	-1.370

PREVIOUS: CURRENT: March 1962

Mercury Bromide ($HgBr_2$) Br_2Hg_1(cr,l)

MERCURY BROMIDE ($HgBr_2$) IDEAL GAS M_r = 360.398

$S°(298.15\ K) = 320.223\ J\ K^{-1}\ mol^{-1}$

$\Delta_f H°(0\ K) = -67.26 \pm 84\ kJ\ mol^{-1}$
$\Delta_f H°(298.15\ K) = -85.45 \pm 8.4\ kJ\ mol^{-1}$

Vibrational Frequencies and Degeneracies

ν, cm^{-1}
225(1)
41(2)
293(1)

Ground State Quantum Weight: 1 $\sigma = 2$
Point Group: $D_{\infty h}$
Bond Distance: Hg-Br = 2.4 A
Bond Angle: Br-Hg-Br = 180°
Rotational Constant: $B_0 = 10.018162\ cm^{-1}$

Enthalpy of Formation

$\Delta_f H°$(cr, 298.15 K) (1) is combined with the 3rd law $\Delta_{sub} H°$(298.15 K) derived from the data of Niwa and Shibata (2) and Johnson (3).

Heat Capacity and Entropy

The vibrational constants are those given by Klemperer and Lindeman (4). The bond length is an average of the values given by Braune and Knoke (5), Gregg et al. (6), and Akishin et al. (7).

References

1. U. S. Nat. Bur. Stand. Circ. 500, 1952.
2. K. Niwa and Z. Shibata, J. Fac. Sci. Hokkaido, Imp. Univ. Ser. III 2, 183 (1938).
3. F. M. G. Johnson, J. Am. Chem. Soc. 33, 777 (1911).
4. W. Klemperer and L. Lindeman, J. Chem. Phys. 25, 397 (1956).
5. H. Braune and S. Knoke, Z. Phys. Chem. B23, 163 (1933).
6. A. H. Gregg, G. C. Hampson, G. I. Jenkins, P. L. F. Jones, and L. E. Sutton, Trans. Faraday Soc. 33, 852 (1937).
7. P. A. Akishin, V. P. Spiridinov, and A. N. Khodchenkov, Zh. Fiz. Khim. 33, 20 (1959).

Mercury Bromide ($HgBr_2$) $Br_2Hg_1(g)$

Enthalpy Reference Temperature = T_r = 298.15 K Standard State Pressure = p° = 0.1 MPa

T/K	$C_p°$ (J K⁻¹mol⁻¹)	S° (J K⁻¹mol⁻¹)	$-[G°-H°(T_r)]/T$ (J K⁻¹mol⁻¹)	$H°-H°(T_r)$ (kJ mol⁻¹)	$\Delta_f H°$ (kJ mol⁻¹)	$\Delta_f G°$ (kJ mol⁻¹)	Log K_f
0	0.	0.	INFINITE	-15.658	-67.260	-67.260	INFINITE
100	51.210	258.441	372.093	-11.365	-67.466	-84.203	43.983
200	58.139	296.540	325.684	-5.829	-69.415	-100.235	26.179
250	59.492	309.672	321.212	-2.885	-72.984	-107.644	22.491
298.15	60.276	320.223	320.223	0.	-85.454	-112.881	19.776
300	60.300	320.596	320.225	0.112	-85.534	-113.051	19.684
350	60.814	329.933	320.961	3.140	-116.545	-115.929	17.302
400	61.160	338.078	322.602	6.190	-116.701	-115.831	15.126
450	61.403	345.296	324.730	9.255	-116.845	-115.713	13.432
500	61.579	351.775	327.116	12.329	-116.981	-115.580	12.075
600	61.812	363.025	332.191	18.500	-117.244	-115.275	10.036
700	61.955	372.565	337.295	24.689	-176.267	-108.354	8.086
800	62.048	380.844	342.232	30.889	-175.898	-98.678	6.443
900	62.113	388.156	346.937	37.097	-175.532	-89.047	5.168
1000	62.159	394.703	351.392	43.311	-175.171	-79.457	4.150
1100	62.193	400.629	355.603	49.529	-174.815	-69.903	3.319
1200	62.219	406.042	359.584	55.750	-174.465	-60.381	2.628
1300	62.240	411.023	363.351	61.973	-174.121	-50.888	2.045
1400	62.256	415.636	366.923	68.197	-173.785	-41.421	1.545
1500	62.269	419.932	370.316	74.424	-173.456	-31.978	1.114
1600	62.280	423.951	373.544	80.651	-173.136	-22.557	0.736
1700	62.289	427.727	376.621	86.880	-172.827	-13.155	0.404
1800	62.296	431.287	379.560	93.109	-172.531	-3.772	0.109
1900	62.303	434.656	382.372	99.339	-172.248	5.596	-0.154
2000	62.308	437.851	385.067	105.569	-171.982	14.950	-0.390
2100	62.313	440.891	387.653	111.800	-171.733	24.290	-0.604
2200	62.317	443.790	390.139	118.032	-171.504	33.619	-0.798
2300	62.320	446.561	392.533	124.264	-171.296	42.938	-0.975
2400	62.323	449.213	394.840	130.496	-171.110	52.248	-1.137
2500	62.326	451.757	397.066	136.728	-170.947	61.552	-1.286
2600	62.328	454.202	399.217	142.961	-170.806	70.849	-1.423
2700	62.331	456.554	401.297	149.194	-170.687	80.141	-1.550
2800	62.333	458.821	403.311	155.427	-170.589	89.429	-1.668
2900	62.334	461.008	405.263	161.661	-170.511	98.714	-1.778
3000	62.336	463.121	407.157	167.894	-170.452	107.996	-1.880
3100	62.337	465.165	408.995	174.128	-170.408	117.277	-1.976
3200	62.338	467.145	410.782	180.362	-170.379	126.557	-2.066
3300	62.340	469.063	412.519	186.595	-170.361	135.836	-2.150
3400	62.341	470.924	414.209	192.829	-170.351	145.114	-2.229
3500	62.342	472.731	415.856	199.064	-170.348	154.392	-2.304
3600	62.343	474.487	417.460	205.298	-170.349	163.671	-2.375
3700	62.343	476.195	419.025	211.532	-170.350	172.949	-2.442
3800	62.344	477.858	420.551	217.766	-170.349	182.227	-2.505
3900	62.345	479.477	422.041	224.001	-170.345	191.506	-2.565
4000	62.346	481.056	423.497	230.235	-170.334	200.784	-2.622
4100	62.346	482.595	424.920	236.470	-170.314	210.062	-2.676
4200	62.347	484.098	426.311	242.705	-170.284	219.339	-2.728
4300	62.347	485.565	427.672	248.939	-170.242	228.615	-2.777
4400	62.348	486.998	429.004	255.174	-170.186	237.890	-2.824
4500	62.348	488.399	430.308	261.409	-170.115	247.164	-2.869
4600	62.349	489.770	431.586	267.644	-170.028	256.436	-2.912
4700	62.349	491.111	432.838	273.879	-169.923	265.706	-2.953
4800	62.349	492.423	434.066	280.114	-169.800	274.973	-2.992
4900	62.350	493.709	435.270	286.349	-169.658	284.238	-3.030
5000	62.350	494.968	436.452	292.584	-169.498	293.499	-3.066
5100	62.350	496.203	437.611	298.819	-169.318	302.758	-3.101
5200	62.351	497.414	438.750	305.054	-169.118	312.012	-3.134
5300	62.351	498.602	439.868	311.289	-168.898	321.263	-3.166
5400	62.351	499.767	440.966	317.524	-168.660	330.509	-3.197
5500	62.351	500.911	442.046	323.759	-168.402	339.750	-3.227
5600	62.352	502.035	443.107	329.994	-168.125	348.987	-3.255
5700	62.352	503.138	444.151	336.229	-167.829	358.218	-3.283
5800	62.352	504.223	445.177	342.464	-167.516	367.444	-3.309
5900	62.352	505.288	446.187	348.700	-167.186	376.665	-3.335
6000	62.352	506.336	447.181	354.935	-166.839	385.880	-3.359

PREVIOUS: March 1962 (1 atm) CURRENT: March 1962 (1 bar)

Mercury Bromide ($HgBr_2$) $Br_2Hg_1(g)$

MERCURY BROMIDE (Hg_2Br_2) CRYSTAL M_r = 560.988

$S°(298.15\ K) = 218.75 \pm 3.2\ J\ K^{-1}\ mol^{-1}$

$\Delta_f H°(0\ K)$ = Unknown

$\Delta_f H°(298.15\ K) = -204.18 \pm 0.8\ kJ\ mol^{-1}$

Enthalpy of Formation

The average of the values obtained from the cell measurements of Dakin and Ewing (1) and Ishikawa and Ueda (2) is adopted.

Heat Capacity and Entropy

The heat capacity is estimated by analogy with Hg_2Cl_2(cr). The entropy was obtained from the Gibbs energy and enthalpy of formation (1, 2).

Sublimation Data

According to Jung and Ziegler (3), the vapor pressure over Hg_2Br_2(cr) reaches one atmosphere at 666 K where the vapor is dissociated into Hg(g) and $HgBr_2$(g).

References

1. T. W. Dakin and D. T. Ewing, J. Am. Chem. Soc. 62, 2280 (1940).
2. F. Ishikawa and Y. Ueda, J. Chem. Soc. Japan 51, 59 (1930).
3. G. Jung and W. Ziegler, Z. Phys. Chem. A150, 139 (1930).

Mercury Bromide (Hg_2Br_2) Br_2Hg_2(cr)

Enthalpy Reference Temperature = T_r = 298.15 K Standard State Pressure = p° = 0.1 MPa

T/K	$C_p°$ (J K⁻¹mol⁻¹)	S° (J K⁻¹mol⁻¹)	$-[G°-H°(T_r)]/T$ (J K⁻¹mol⁻¹)	$H°-H°(T_r)$ (kJ mol⁻¹)	$\Delta_f H°$ (kJ mol⁻¹)	$\Delta_f G°$ (kJ mol⁻¹)	Log K_f
0							
100							
200							
298.15	104.600	218.752	218.752	0.	-204.179	-178.684	31.305
300	104.703	219.399	218.754	0.194	-204.229	-178.526	31.084
400	109.621	250.226	222.924	10.921	-233.513	-163.837	21.395
500	113.052	275.074	230.948	22.063	-231.518	-146.642	15.320
600	115.604	295.921	240.086	33.501	-229.227	-129.878	11.307
700	117.529	313.890	249.376	45.160	-344.254	-100.372	7.490
800	119.118	329.691	258.447	56.995	-340.329	-65.797	4.296
900	120.416	343.797	267.161	68.973	-336.272	-31.723	1.841
1000	121.545	356.544	275.472	81.072	-332.104	1.893	-0.099
1100	122.539	368.176	283.379	93.277	-327.840	35.087	-1.666
1200	123.470	378.878	290.897	105.578	-323.488	67.889	-2.955
1300	124.209	388.789	298.050	117.960	-319.064	100.325	-4.031
1400	125.143	398.027	304.865	130.426	-314.564	132.417	-4.941
1500	126.273	406.698	311.368	142.995	-309.971	164.185	-5.717

PREVIOUS: CURRENT: March 1962

Mercury Bromide (Hg_2Br_2) Br_2Hg_2(cr)

POTASSIUM BROMIDE ((KBr)$_2$) IDEAL GAS M_r = 238.0046

S°(298.15 K) = [376.2] J K^{-1} mol^{-1}

Δ_fH°(0 K) = -522.9 ± 17 kJ mol^{-1}
Δ_fH°(298.15 K) = -540.6 ± 17 kJ mol^{-1}

Vibrational Frequencies and Degeneracies

ν, cm^{-1}	ν, cm^{-1}
[135](1)	[80](1)
[80](1)	[125](1)
[130](1)	[150](1)

Ground State Quantum Weight: 1 σ = [4]
Point Group: [D_{2h}]
Bond Distances: K-Br = [3.07] Å K-K = [1.94] Å Br-Br = [2.39] Å
Bond Angles: Br-K-Br = [102]° K-Br-K = [78]°
Product of the Moments of Inertia: $I_AI_BI_C$ = [1.46082 x 10^{-111}] g^3 cm^6

Enthalpy of Formation

Vapor densities and vapor pressures of KBr were measured, using a liquid gold isoteniscope, by Hagemark et al. (1), and Hagemark (2). Based on the equilibrium constants for the reaction 2KBr(g) = K_2Br_2(g) in the temperature range 1267-1434 K, the enthalpy change Δ_rH°(298.15 K) of this reaction is evaluated by the 2nd and 3rd law methods to be -42.80±4.26 and -43.12 kcal mol^{-1}, respectively. The enthalpy of formation for K_2Br_2(g) is calculated to be -129.2±4 kcal mol^{-1}, using the 3rd law Δ_rH°(298.15 K) value and Δ_fH°(KBr, g, 298.15 K) = -43.04 kcal mol^{-1}. The drift in the 3rd law Δ_rH°(298.15 K) values is 0.01±3.2 cal K^{-1} mol^{-1}.

Heat Capacity and Entropy

The molecular structure and bond distances were estimated by Berkowitz (3), and are tentatively adopted. Berkowitz (4) has calculated vibrational frequencies on the basis of the potential function for an ionic model. These values are adjusted so that the 3rd law Δ_rH°(298.15 K) value for the reaction 2KBr(g) = K_2Br_2(g) agrees with the 2nd law value. The prinicpal moments of inertia are: I_A = 48.4690 x 10^{-39}, I_B = 151.0555 x 10^{-39} and I_C = 199.5245 x 10^{-39} g cm^2.

References

1. K. I. Hagemark, M. Blander, and E. B. Luchsinger, J. Phys. Chem. 70, 276 (1966).
2. K. I. Hagemark, 3M Company, St. Paul, Minnesota, personal communication, Feb. 6, 1967.
3. J. Berkowitz, J. Chem. Phys. 29, 1386 (1958).
4. J. Berkowitz, J. Chem. Phys. 32, 1519 (1960).

Potassium Bromide ((KBr)$_2$) Br_2K_2(g)

Enthalpy Reference Temperature = T_r = 298.15 K Standard State Pressure = p° = 0.1 MPa

	J K^{-1}mol^{-1}			kJ mol^{-1}			
T/K	C_p^o	S°	-[G°-H°(T_r)]/T	H°-H°(T_r)	Δ_fH°	Δ_fG°	Log K_f
0	0.	0.	INFINITE	-21.014	-522.913	-522.913	INFINITE
100	72.670	290.395	447.575	-15.718	-523.875	-540.409	282.281
200	80.188	343.855	383.672	-7.964	-526.211	-556.100	145.238
250	81.223	361.872	377.574	-3.926	-527.737	-563.402	117.717
298.15	81.782	376.230	376.230	0.	-540.573	-568.802	99.652
300	81.799	376.736	376.231	0.151	-540.671	-568.977	99.068
350	82.151	389.374	377.229	4.251	-577.004	-571.752	85.329
400	82.381	400.359	379.448	8.364	-577.894	-570.940	74.557
450	82.540	410.072	382.322	12.488	-578.740	-570.020	66.166
500	82.654	418.775	385.540	16.618	-579.548	-569.007	59.444
600	82.803	433.859	392.374	24.891	-581.073	-566.754	49.340
700	82.893	446.630	399.236	33.176	-582.518	-564.253	42.105
800	82.952	457.703	405.867	41.469	-583.932	-561.547	36.665
900	82.992	467.476	412.180	49.766	-585.364	-558.663	32.424
1000	83.021	476.222	418.155	58.067	-586.863	-555.617	29.022
1100	83.042	484.135	423.799	66.370	-746.380	-543.196	25.794
1200	83.058	491.362	429.132	74.675	-746.024	-524.740	22.841
1300	83.071	498.010	434.178	82.982	-745.676	-506.314	20.344
1400	83.081	504.167	438.960	91.289	-745.336	-487.914	18.204
1500	83.089	509.899	443.501	99.598	-745.006	-469.539	16.351
1600	83.096	515.262	447.820	107.907	-744.687	-451.185	14.730
1700	83.101	520.300	451.937	116.217	-744.383	-432.850	13.300
1800	83.106	525.050	455.868	124.527	-744.097	-414.533	12.029
1900	83.110	529.543	459.629	132.838	-743.833	-396.231	10.893
2000	83.113	533.806	463.232	141.149	-743.596	-377.942	9.871
2100	83.116	537.862	466.690	149.461	-743.391	-359.665	8.946
2200	83.119	541.728	470.013	157.772	-743.224	-341.396	8.106
2300	83.121	545.423	473.212	166.084	-743.101	-323.135	7.339
2400	83.123	548.961	476.295	174.397	-743.027	-304.877	6.635
2500	83.124	552.354	479.270	182.709	-743.008	-286.621	5.989
2600	83.126	555.614	482.144	191.021	-743.050	-268.365	5.392
2700	83.127	558.751	484.924	199.334	-743.157	-250.106	4.839
2800	83.128	561.775	487.615	207.647	-743.336	-231.842	4.325
2900	83.129	564.692	490.223	215.960	-743.590	-213.570	3.847
3000	83.130	567.510	492.752	224.273	-743.926	-195.288	3.400
3100	83.131	570.236	495.208	232.586	-744.348	-176.994	2.982
3200	83.132	572.875	497.594	240.899	-744.862	-158.684	2.590
3300	83.133	575.433	499.914	249.212	-745.453	-140.358	2.222
3400	83.133	577.915	502.172	257.526	-746.157	-122.011	1.874
3500	83.134	580.325	504.371	265.839	-746.961	-103.643	1.547
3600	83.135	582.667	506.513	274.152	-747.880	-85.250	1.237
3700	83.135	584.945	508.602	282.466	-748.916	-66.829	0.943
3800	83.136	587.162	510.641	290.779	-750.076	-48.380	0.665
3900	83.136	589.321	512.631	299.093	-751.366	-29.898	0.400
4000	83.136	591.426	514.574	307.407	-752.781	-11.381	0.149
4100	83.137	593.479	516.474	315.720	-754.341	7.172	-0.091
4200	83.137	595.482	518.331	324.034	-756.061	25.766	-0.320
4300	83.137	597.438	520.148	332.348	-757.939	44.403	-0.539
4400	83.138	599.350	521.927	340.661	-759.431	63.031	-0.748
4500	83.138	601.218	523.668	348.975	-761.514	81.746	-0.949
4600	83.138	603.045	525.374	357.289	-763.755	100.510	-1.141
4700	83.139	604.833	527.046	365.603	-766.117	119.319	-1.326
4800	83.139	606.584	528.684	373.917	-768.425	138.157	-1.503
4900	83.139	608.298	530.292	382.231	-771.105	157.072	-1.674
5000	83.139	609.978	531.869	390.544	-773.949	176.043	-1.839
5100	83.139	611.624	533.416	398.858	-776.961	195.073	-1.998
5200	83.140	613.238	534.936	407.172	-780.140	214.162	-2.151
5300	83.140	614.822	536.428	415.486	-783.488	233.316	-2.299
5400	83.140	616.376	537.895	423.800	-787.005	252.533	-2.443
5500	83.140	617.902	539.335	432.114	-790.688	271.818	-2.582
5600	83.140	619.400	540.752	440.428	-794.536	291.170	-2.716
5700	83.140	620.871	542.145	448.742	-797.858	310.498	-2.845
5800	83.140	622.317	543.514	457.056	-801.941	329.978	-2.972
5900	83.141	623.738	544.862	465.370	-806.172	349.530	-3.095
6000	83.141	625.136	546.188	473.684	-810.546	369.155	-3.214

PREVIOUS: March 1967 (1 atm) CURRENT: March 1967 (1 bar)

Potassium Bromide ((KBr)$_2$) Br_2K_2(g)

LITHIUM BROMIDE $((LiBr)_2)$ IDEAL GAS M_r = 173.690

$S°(298.15\ K) = [314.53]\ J\ K^{-1}\ mol^{-1}$

$\Delta_f H°(0\ K) = -483.8 \pm 21\ kJ\ mol^{-1}$
$\Delta_f H°(298.15\ K) = -500.8 \pm 21\ kJ\ mol^{-1}$

Vibrational Frequencies and Degeneracies

ν, cm^{-1}	ν, cm^{-1}
[202](1)	[327](1)
[203](1)	413 (1)
[353](1)	295 (1)

Ground State Quantum Weight: 1 $\sigma = 4$

Point Group: D^2h

Bond Distance: Li-Br = 2.35 Å Li-Li = 2.70 Å

Bond Angle: Br-Li-Br = 110°

Product of the Moments of Inertia: $I_A I_B I_C = 4.22259 \times 10^{-113}\ g^3\ cm^6$

Enthalpy of Formation

This was obtained from the enthalpy of formation of the liquid and the selected enthalpy of vaporization to the dimer, which derivation has been given in the LiBr(g) table.

Heat Capacity and Entropy

Berkowitz (1) has calculated the molecular structure and vibrational frequencies based on an ionic model. The planar rhombic structure for dimeric lithium bromide, proposed by Berkowitz, has been confirmed by the lack of polarity in electric deflection by Buchler et al. (2). The selected bond distances and angle were obtained from the electron diffraction studies of monomer-dimer vapor by Akischin and Rambidi (3). The bond distances (r_{Br-Br} = 4.20 Å and r_{Li-Li} = 2.60 Å) calculated by Berkowitz are in reasonable agreement with those selected. The principal moments of inertia are: $I_A = 4.1882 \times 10^{-39}$, $I_B = 98.3375 \times 10^{-39}$ and $I_C = 102.5257 \times 10^{-39}\ g\ cm^2$.

Bauer et al. (4) have estimated six vibrational frequencies (576, 576, 202, 230, 329 and 365 cm^{-1}) for $Li_2Br_2(g)$ in the electron diffraction studies of $Li_2Cl_2(g)$. Klemperer and Norris (5) have observed two fundamental vibrational frequencies (413 and 295 cm^{-1}) in the infrared spectrum and tentatively assigned them as B_{2u} and B_{3u} modes; and these have been adopted in the tabulation. The remaining four vibrational frequencies were obtained from Berkowitz (1) because his model and derivation are self-consistent.

References

1. J. Berkowitz, J. Chem. Phys. 29, 1386 (1958), 32, 1519 (1960).
2. A. Buchler, J. L. Stauffer and W. Klemperer, J. Am. Chem. Soc. 86, 4544 (1964).
3. P. A. Akischin and N. G. Rambidi, Z. Phys. Chem. 213, 111 (1960).
4. S. H. Bauer, T. Ino and R. F. Porter, J. Chem. Phys. 33, 685 (1960).
5. W. Klemperer and W. G. Norris, J. Chem. Phys. 34, 1071 (1961).

Lithium Bromide $((LiBr)_2)$ $Br_2Li_2(g)$

Enthalpy Reference Temperature = T_r = 298.15 K; Standard State Pressure = p° = 0.1 MPa

T/K	$C_p°$	S°	$-[G°-H°(T_r)]/T$	$H°-H°(T_r)$	$\Delta_f H°$	$\Delta_f G°$	Log K_f
	J K⁻¹mol⁻¹			kJ mol⁻¹			
0	0.	0.	INFINITE	-16.696	-483.767	-483.767	INFINITE
100	47.860	245.752	375.407	-12.966	-483.820	-501.542	261.979
200	67.533	285.942	321.206	-7.053	-486.491	-518.265	135.357
250	72.201	301.555	315.755	-3.550	-488.060	-526.031	109.908
298.15	75.049	314.534	314.534	0.	-500.825	-531.879	93.183
300	75.136	314.998	314.535	0.139	-500.917	-532.072	92.642
350	77.065	326.735	315.458	3.947	-532.285	-535.530	79.923
400	78.388	337.117	317.529	7.835	-532.902	-535.953	69.988
450	79.329	346.407	320.231	11.780	-533.645	-536.291	62.251
500	80.020	354.803	323.275	15.764	-540.530	-535.925	55.988
600	80.943	369.481	329.787	23.816	-542.164	-534.847	46.563
700	81.513	382.005	336.374	31.941	-543.628	-533.510	39.811
800	81.888	392.915	342.775	40.112	-545.002	-531.970	34.734
900	82.148	402.576	348.893	48.315	-546.345	-530.260	30.775
1000	82.335	411.242	354.702	56.540	-547.667	-528.402	27.601
1100	82.474	419.096	360.205	64.780	-548.972	-526.412	24.997
1200	82.580	426.277	365.416	73.033	-550.263	-524.304	22.822
1300	82.663	432.890	370.355	81.296	-551.545	-522.089	20.978
1400	82.729	439.018	375.043	89.565	-552.817	-519.775	19.393
1500	82.782	444.728	379.501	97.841	-554.076	-517.371	18.016
1600	82.826	450.072	383.746	106.121	-555.323	-514.883	16.809
1700	82.862	455.094	387.797	114.406	-847.035	-497.965	15.301
1800	82.892	459.832	391.669	122.693	-846.762	-477.440	13.855
1900	82.918	464.314	395.375	130.984	-846.503	-456.929	12.562
2000	82.940	468.568	398.929	139.277	-846.263	-436.431	11.398
2100	82.959	472.615	402.343	147.572	-846.045	-415.945	10.346
2200	82.975	476.475	405.625	155.869	-845.853	-395.469	9.390
2300	82.990	480.163	408.786	164.167	-845.690	-375.001	8.517
2400	83.002	483.696	411.834	172.466	-845.561	-354.539	7.716
2500	83.013	487.084	414.777	180.767	-845.468	-334.082	6.980
2600	83.023	490.340	417.621	189.069	-845.416	-313.628	6.301
2700	83.032	493.474	420.373	197.372	-845.406	-293.174	5.672
2800	83.040	496.493	423.038	205.675	-845.441	-272.721	5.088
2900	83.047	499.407	425.621	213.980	-845.524	-252.265	4.544
3000	83.053	502.223	428.128	222.285	-845.656	-231.806	4.036
3100	83.059	504.946	430.562	230.590	-845.839	-211.342	3.561
3200	83.064	507.584	432.928	238.897	-846.074	-190.870	3.116
3300	83.069	510.140	435.230	247.203	-846.360	-170.391	2.697
3400	83.073	512.620	437.469	255.510	-846.700	-149.902	2.303
3500	83.077	515.028	439.651	263.818	-847.094	-129.403	1.931
3600	83.081	517.368	441.778	272.126	-847.541	-108.891	1.580
3700	83.084	519.644	443.851	280.434	-848.043	-88.366	1.248
3800	83.087	521.860	445.875	288.743	-848.599	-67.827	0.932
3900	83.090	524.018	447.851	297.052	-849.211	-47.272	0.633
4000	83.093	526.122	449.782	305.361	-849.878	-26.701	0.349
4100	83.095	528.174	451.669	313.670	-850.603	-6.113	0.078
4200	83.098	530.176	453.515	321.980	-851.384	14.494	-0.180
4300	83.100	532.132	455.320	330.290	-852.225	35.119	-0.427
4400	83.102	534.042	457.088	338.600	-853.084	55.762	-0.662
4500	83.104	535.910	458.819	346.910	-854.027	76.428	-0.887
4600	83.105	537.736	460.514	355.220	-855.035	97.116	-1.103
4700	83.107	539.524	462.177	363.531	-856.105	117.826	-1.309
4800	83.109	541.273	463.806	371.842	-857.240	138.560	-1.508
4900	83.110	542.987	465.405	380.153	-858.441	159.318	-1.698
5000	83.111	544.666	466.973	388.464	-859.711	180.102	-1.882
5100	83.113	546.312	468.513	396.775	-861.041	200.910	-2.058
5200	83.114	547.926	470.025	405.086	-862.454	221.746	-2.227
5300	83.115	549.509	471.509	413.398	-863.944	242.611	-2.391
5400	83.116	551.063	472.968	421.709	-865.513	263.504	-2.549
5500	83.117	552.588	474.402	430.021	-867.166	284.426	-2.701
5600	83.118	554.085	475.812	438.333	-868.904	305.380	-2.848
5700	83.119	555.557	477.198	446.645	-870.732	326.366	-2.991
5800	83.120	557.002	478.561	454.957	-872.653	347.383	-3.129
5900	83.121	558.423	479.903	463.269	-874.670	368.436	-3.262
6000	83.121	559.820	481.223	471.581	-876.786	389.523	-3.391

PREVIOUS: June 1966 (1 atm) CURRENT: June 1966 (1 bar)

Lithium Bromide $((LiBr)_2)$ $Br_2Li_2(g)$

MAGNESIUM BROMIDE ($MgBr_2$) CRYSTAL M_r = 184.113

$\Delta_f H°(0\ K)$ = Unknown

$S°(298.15\ K) = [117.2 \pm 4.2]\ J\ K^{-1}\ mol^{-1}$ $\Delta_f H°(298.15\ K) = -524.26 \pm 2.1\ kJ\ mol^{-1}$

$T_{fus} = 984 \pm 15\ K$ $\Delta_{fus}H° = [39.330 \pm 4.2]\ kJ\ mol^{-1}$

Enthalpy of Formation

The selected value for $\Delta_f H°(298.15\ K)$ is that given by NBS (1). This value is derived from the solution calorimetric measurements of Finch et al. (2). Their enthalpies for the reaction of crystalline MgO with aqueous HBr and solution of anhydrous $MgBr_2$ in the same solvent are combined to give $\Delta_r H°(298.15\ K) = 7.77 \pm 0.40\ kcal\ mol^{-1}$ for the process $MgO(cr) + 2\ HBr(aq) = MgBr_2(cr) + H_2O(\ell)$. When this result is combined with auxiliary data (1, 3, and 4), the selected value is obtained. This value is confirmed by an independent means. Bichowsky and Rossini (5) give the enthalpy of solution of $MgBr_2$ in 800 H_2O as $-43.3\ kcal\ mol^{-1}$ based on measurements by Beketoff (6). Correction of this value to 298.15 K and infinite dilution gives a value for $\Delta_{sol}H°$ which leads to $\Delta_f H°(298.15\ K) = -125.4\ kcal\ mol^{-1}$ when combined with $\Delta_f H°(Mg^{+2}, aq, \infty, 298.15\ K) = -111.58\ kcal\ mol^{-1}$ (1) and $\Delta_f H°(Br^-, aq, \infty, 298.15\ K) = -29.039 \pm 0.035\ kcal\ mol^{-1}$ (4).

Two emf studies (7, 8) on the potential of the magnesium electrode in ether solution give results in disagreement with the calorimetric values. From these cell data we derive $\Delta_f H°(298.15\ K)$ values of $-118.8\ kcal\ mol^{-1}$ (7) and $-115.0\ kcal\ mol^{-1}$ (8). These rather large discrepancies are taken as an indication that the electrodes are probably irreversible under these conditions. In fact one study (8) reported that from 10 to 20 hours were required before stable emf values were obtained.

Heat Capacity and Entropy

$C°_p$ at 298.15 K is obtained from the process $MgCl_2(cr) + 2\ NaBr(cr) = MgBr_2(cr) + 2\ NaCl(cr)$ by assuming $\Delta C°_p$ is zero. A comparison of the $C°_p$ value (17.49 cal $K^{-1}\ mol^{-1}$) with those for other alkaline earth dihalides shows that it is reasonable. $C°_p$ data above 300 K are estimated by comparison with similar data for $MgCl_2$ and CaI_2 (3).

Several estimates (9-12) lying in the range 27-30 cal $K^{-1}\ mol^{-1}$ for S°(298.15 K) have been reported. Available experimental data for several other alkaline earth dihalides suggest a value for S°(298.15 K) of $MgBr_2(cr)$ near 28 cal $K^{-1}\ mol^{-1}$ (117.152 J $K^{-1}\ mol^{-1}$). We tentatively adopt this value. It is also the value recommended by Kelley and King (10) and NBS (1).

Fusion Data

The value of T_{fus} is that determined from cooling curves by Kellner (13, 14). The uncertainty in these measurements is probably near ± 15 K.

$\Delta_{fus}H°$ is calculated from a $\Delta_{fus}S° = 9.5\ cal\ K^{-1}\ mol^{-1}$ by multiplication of T_{fus}. $\Delta_{fus}S°$ is assumed the same as that for CaI_2 (3) which also has the hexagonal CdI_2 structure (15). Kelley (16) obtained a value of $\Delta_{fus}H°$ equal to 8.3 kcal mol^{-1} from an analysis of phase diagrams for the systems $MgBr_2$–MBr(M = Na, K) (14) and $MgBr_2$–LiBr (17). However, he implied that the value was somewhat uncertain. Furthermore, it is noted that several $\Delta_{fus}H°$ values listed by Kelley for other alkaline earth dihalides deviate considerably (1-3 kcal mol^{-1}) from those measured calorimetrically (18). Thus, we prefer the estimated value at this time. (See also $MgBr_2(\ell)$ table Enthalpy of Formation Section).

Sublimation Data

See the ideal gas table for details.

References

1. U. S. Nat. Bur. Stand. Tech. Note 270-6, 1971.
2. A. Finch, P. J. Gardner, and C. J. Steadman, Can. J. Chem. 46, 3447 (1968).
3. JANAF Thermochemical Tables: MgO(cr) and $MgCl_2$(cr), 12-31-65; CaI_2(cr), 6-30-74.
4. ICSU-CODATA Task Group on Key Values for Thermodynamics, J. Chem. Thermodynamics 4, 331 (1972).
5. F. R. Bichowsky and F. D. Rossini, "The Thermochemistry of the Chemical Substances," Reinhold Publishing Corporation, New York, 1936.
6. N. Beketoff, Bull. Acad. Sci. USSR 34, 291 (1892).
7. G. A. Scherer and R. F. Newton, J. Amer. Chem. Soc. 56, 18 (1934).
8. J. A. Ricketts and P. Brown, Proc. Indiana Acad. Sci. 75, 105 (1965).
9. L. Brewer, G. R. Somayajulu, and E. Brackett, Chem. Revs. 63, 111 (1963).
10. K. K. Kelley and E. G. King, U. S. Bur. Mines Bull. 592, 1961.
11. L. Brewer, L. A. Bromley, P. W. Gilles, and N. L. Lofgren in "Chemistry and Metallurgy of Miscellaneous Materials: Thermodynamics," L. L. Quill, Ed., National Nuclear Energy Series IV-19B, McGraw-Hill Book Co., New York, 1950.
12. W. M. Latimer, "The Oxidation States of the Elements and Their Potentials in Aqueous Solutions," 2nd. ed., Prentice-Hall, Englewood Cliffs, N. J., 1952.
13. G. Kellner, J. Chem. Soc. 112, 469 (1917).
14. G. Kellner, Z. Anorg. Chem. 99, 137 (1917).
15. O. Kubaschewski, E. LL. Evans and C. B. Alcock, "Metallurgical Thermochemistry," 4th ed., Pergamon Press, New York, 1967.
16. K. K. Kelley, U. S. Bur. Mines Bull. 393, 1936.
17. A. Ferrari and C. Colla, Atti accad. Leucei 13, 78 (1931).
18. A. S. Dworkin and M. A. Bredig, J. Phys. Chem. 67, 697 (1963).

Magnesium Bromide ($MgBr_2$) $Br_2Mg_1(cr)$

Enthalpy Reference Temperature = T_r = 298.15 K Standard State Pressure = p° = 0.1 MPa

T/K	$C°_p$ (J K⁻¹mol⁻¹)	S° (J K⁻¹mol⁻¹)	$-[G°-H°(T_r)]/T$ (J K⁻¹mol⁻¹)	$H°-H°(T_r)$ (kJ mol⁻¹)	$\Delta_f H°$ (kJ mol⁻¹)	$\Delta_f G°$ (kJ mol⁻¹)	Log K_f
0							
100							
200							
298.15	73.161	117.152	117.152	0.	-524.255	-504.063	88.310
300	73.387	117.605	117.153	0.136	-524.306	-503.937	87.743
400	77.278	139.284	120.083	7.681	-553.796	-491.005	64.119
500	79.743	156.808	125.731	15.538	-552.296	-475.479	49.673
600	81.429	171.498	132.168	23.598	-550.723	-460.263	40.069
700	82.969	184.165	138.712	31.818	-549.114	-445.313	33.230
800	84.554	195.346	145.106	40.193	-547.480	-430.595	28.115
900	86.270	205.403	151.256	48.733	-545.822	-416.084	24.149
984.000	87.811	213.168	156.213	56.044	------ CRYSTAL <--> LIQUID -----		
1000	88.136	214.586	157.136	57.450	-552.735	-401.046	20.948
1100	90.144	223.079	162.749	66.363	-551.036	-385.957	18.328
1200	92.257	231.013	168.111	75.483	-549.140	-371.032	16.151
1300	94.433	238.484	173.240	84.817	-547.036	-356.274	14.315
1400	96.567	245.560	178.155	94.367	-672.136	-338.518	12.630
1500	98.701	252.295	182.875	104.131	-668.270	-314.822	10.963

PREVIOUS: June 1966 CURRENT: June 1974

MAGNESIUM BROMIDE ($MgBr_2$) LIQUID M_r = 184.113

$S°(298.15\ K)$ = [150.500] $J\ K^{-1}\ mol^{-1}$ $\Delta_f H°(298.15\ K)$ = [-490.408] $kJ\ mol^{-1}$

T_{fus} = 984 ± 15 K $\Delta_{fus}H°$ = [39.330 ± 4.2] $kJ\ mol^{-1}$

Enthalpy of Formation

$\Delta_f H°(298.15\ K)$ is calculated from that of the crystal by adding $\Delta_{fus}H°$ and the difference in enthalpy, H°(984 K)-H°(298.15 K), between the crystal and liquid. The derived value is supported by results obtained from an equilibrium study (1) of the anion exchange reaction 0.5 $MgBr_2(\ell)$ + HCl(g) = 0.5 $MgCl_2(\ell)$ + HBr(g). A third law analysis of the K_p value reported for the equilibrium at 1073 K gives $\Delta_r H°(298.15\ K)$ = 0.05 kcal mol^{-1} which leads to $\Delta_f H°(\ell,\ 298.15\ K)$ = -117.2 kcal mol^{-1}. In the same paper, Toguri et al. (1) reported similar data for the chloride-bromide salts of Ca, K, and Na. These results lead to $\Delta_f H°(298.15\ K)$ values for the molten bromides which show deviations from adopted JANAF values (2) of no more than 0.6 kcal mol^{-1}. Thus, although the exact agreement in the $\Delta_f H°(298.15\ K)$ values for $MgBr_2(\ell)$ is fortuitous, it does tend to substantiate our thermal data.

Heat Capacity and Entropy

$C°_p$ for the liquid is estimated as 25.0 cal $K^{-1}\ mol^{-1}$ by analogy with measured $C°_p$ data for molten $CaBr_2$ and $SrBr_2$ (2). The value is taken to be constant above the assumed glass transition at 700 K. Below 700 K, $C°_p$ is that of the crystal.

S°(298.15 K) is calculated in a manner analogous to that used for the enthalpy of formation.

Fusion Data

See the crystal table for details.

Vaporization Data

T_{vap} is the temperature at which $\Delta_r G°$ =0 at 1 bar for the process $MgBr_2(\ell)$ = $MgBr_2(g)$. $\Delta_{vap}H°$ is the difference in the enthalpies of formation of the gas and liquid at T_{vap}.

References

1. J. Toguri, H. Flood, and T. Forland, Acta Chem. Scand. 17, 502 (1963).
2. JANAF Thermochemical Tables: $CaBr_2(\ell)$ and $SrBr_2(\ell)$, 6-30-74; Kbr(ℓ), 3-31-67; NaBr(ℓ), 6-30-64.

Magnesium Bromide ($MgBr_2$) $Br_2Mg_1(l)$

Enthalpy Reference Temperature = T_r = 298.15 K Standard State Pressure = p° = 0.1 MPa

T/K	$C°_p$	S°	$-[G°-H°(T_r)]/T$	$H°-H°(T_r)$	$\Delta_f H°$	$\Delta_f G°$	Log K_f
	J $K^{-1}mol^{-1}$			kJ mol^{-1}			
0							
100							
200							
298.15	73.161	150.500	150.500	0.	-490.408	-480.158	84.122
300	73.387	150.954	150.502	0.136	-490.458	-480.094	83.592
400	77.278	172.633	153.431	7.681	-519.948	-470.497	61.441
500	79.743	190.156	159.080	15.538	-518.449	-458.306	47.879
600	81.429	204.846	165.516	23.598	-516.876	-446.424	38.865
700	82.970	217.516	172.060	31.819	-515.265	-434.809	32.446
700.000	82.970	217.516	172.060	31.819	GLASS <--> LIQUID		
700.000	104.600	217.516	172.060	31.819	TRANSITION		
800	104.600	231.484	178.635	42.279	-511.546	-423.571	27.656
900	104.600	243.804	185.205	52.739	-507.968	-412.790	23.958
984.000	104.600	253.137	190.611	61.526	------ CRYSTAL <--> LIQUID -----		
1000	104.600	254.824	191.625	63.199	-513.139	-401.687	20.982
1100	104.600	264.794	197.831	73.659	-509.893	-390.699	18.553
1200	104.600	273.895	203.796	84.119	-506.655	-380.006	16.541
1300	104.600	282.268	209.514	94.579	-503.427	-369.583	14.850
1400	104.600	290.019	214.991	105.039	-627.616	-356.242	13.292
1500	104.600	297.236	220.237	115.499	-623.054	-337.017	11.736
1600	104.600	303.987	225.262	125.959	-618.502	-318.096	10.385
1700	104.600	310.328	230.082	136.419	-613.961	-299.460	9.201
1800	104.600	316.307	234.707	146.879	-609.434	-281.091	8.157
1900	104.600	321.962	239.152	157.339	-604.921	-262.972	7.230
2000	104.600	327.328	243.428	167.799	-600.426	-245.091	6.401

PREVIOUS: June 1966 CURRENT June 1974

Magnesium Bromide ($MgBr_2$) $Br_2Mg_1(l)$

MAGNESIUM BROMIDE ($MgBr_2$) CRYSTAL - LIQUID

298.15 to 984 K crystal

above 984 K liquid

Refer to the individual tables for details.

M_r = 184.113 Magnesium Bromide ($MgBr_2$) Br_2Mg_1(cr,l)

Enthalpy Reference Temperature = T_r = 298.15 K Standard State Pressure = p° = 0.1 MPa

T/K	C_p° (J K⁻¹mol⁻¹)	S° (J K⁻¹mol⁻¹)	$-[G°-H°(T_r)]/T$ (J K⁻¹mol⁻¹)	$H°-H°(T_r)$ (kJ mol⁻¹)	$\Delta_f H°$ (kJ mol⁻¹)	$\Delta_f G°$ (kJ mol⁻¹)	Log K_f
0							
100							
200							
298.15	73.161	117.152	117.152	0.	-524.255	-504.063	88.310
300	73.387	117.605	117.153	0.136	-524.306	-503.937	87.743
400	77.278	139.284	120.083	7.681	-553.796	-491.005	64.119
500	79.743	156.808	125.731	15.538	-552.296	-475.479	49.673
600	81.429	171.498	132.168	23.598	-550.723	-460.263	40.069
700	82.969	184.165	138.712	31.818	-549.114	-445.313	33.230
800	84.554	195.346	145.106	40.193	-547.480	-430.595	28.115
900	86.270	205.403	151.256	48.733	-545.822	-416.084	24.149
984.000	87.811	213.168	156.213	56.044	CRYSTAL <--> LIQUID		
984.000	104.600	253.137	156.213	95.373	TRANSITION		
1000	104.600	254.824	157.778	97.047	-513.139	-401.687	20.982
1100	104.600	264.794	167.060	107.507	-509.893	-390.699	18.553
1200	104.600	273.895	175.590	117.967	-506.655	-380.006	16.541
1300	104.600	282.268	183.478	128.427	-503.427	-369.583	14.850
1400	104.600	290.019	190.815	138.887	-627.616	-356.242	13.292
1500	104.600	297.236	197.672	149.347	-623.054	-337.017	11.736
1600	104.600	303.987	204.108	159.807	-618.502	-318.096	10.385
1700	104.600	310.328	210.171	170.267	-613.961	-299.460	9.201
1800	104.600	316.307	215.903	180.727	-609.434	-281.091	8.157
1900	104.600	321.962	221.338	191.187	-604.921	-262.973	7.230
2000	104.600	327.328	226.504	201.647	-600.426	-245.091	6.401

PREVIOUS: CURRENT: June 1974

Magnesium Bromide ($MgBr_2$) Br_2Mg_1(cr,l)

MAGNESIUM BROMIDE ($MgBr_2$) IDEAL GAS M_r = 184.113

$S°(298.15\ K) = [301.0 \pm 8.4]\ J\ K^{-1}\ mol^{-1}$

$\Delta_f H°(0\ K) = -288.4 \pm 10.5\ kJ\ mol^{-1}$
$\Delta_f H°(298.15\ K) = -302.9 \pm 10.5\ kJ\ mol^{-1}$

Vibrational Frequencies and Degeneracies

ν, cm^{-1}
[179](1)
[70](2)
490 (1)

Ground State Quantum Weight: 1 $\sigma = 2$

Point group: $[D_{\infty h}]$

Bond Distance: Mg-Br = 2.34 ± 0.03 Å Bond Angle: Br-Mg-Br = [180]°

Rotational Constant: B_0 = [0.019265] cm^{-1}

Enthalpy of Formation

$\Delta_f H°(298.15\ K)$ of the gas is obtained from that of the crystal by addition of $\Delta_{sub}H°(298.15\ K) = 52.9 \pm 2.0\ kcal\ mol^{-1}$ (221.334±8.4 kJ mol^{-1}). The selected value for the enthalpy of sublimation is derived from results of a mass spectrometric-Knudsen effusion study by Berkowitz and Marquart (1). The presence of ~2% dimer in the vapor phase over the solid at near 800 K was revealed by the mass spectral results. Berkowitz and Marquart (1). using $MgBr_2^+$-ion intensity as a measure of the monomer concentration, obtained a 2nd law $\Delta_{sub}H$ = 50.3 kcal mol^{-1} at 751 K. This value when corrected to 298.15 K gives $\Delta_{sub}H°(298.15\ K)$ = 52.4 kcal mol^{-1}. Absolute partial pressures were also determined at 798 and 842 K by integrating the ion current during complete volatilization of the sample. A 3rd law analysis of these two pressures gives a enthalpy of sublimation equal to 53.4 kcal mol^{-1} which is in good agreement with the value obtained from the ion intensities by the 2nd law. The drift in the 3rd law analysis is 1.0 cal K^{-1} mol^{-1}. The selected value of $\Delta_{sub}H°(298.15\ K)$ is an average of these two results.

Heat Capacity and Entropy

The bond length is from the electron-diffraction measurements of Akishin and Spiridonov (3). Their diffraction patterns for $MgBr_2$ were satisfactorily explained on the basis of a linear model. Klemperer et al. (4, 5) have performed electronic quadrupole deflection experiments on several symmetrical triatomic dihalides in an attempt to determine which molecules possess permanent dipole moments. Their results on the geometrics of the alkaline earth dihalides have been summarized by Hayes (6) who pointed out that the linear form is favored by the light metal-heavy halogen combination. Thus, one would expect $MgBr_2$ to be linear, and we adopt this configuration.

The antisymmetric stretching frequency (ν_3 = 490 cm^{-1}) has been observed in the infrared absorption spectrum of gaseous $MgBr_2$ by Randall et al. (7). The symmetric stretching (ν_1) and bending (ν_2) frequencies are calculated from force constants by the valence force method (8). The observed ν_3 frequency leads to k = 1.5 x 10^5 dynes/cm, or ν_1 = 179 cm^{-1}. A comparison of values for the ratio of the stretch to bend force constants for the linear molecules $MgCl_2$ (9)(ratio = 88)], $CaBr_2$ (9)(ratio = [100])], and $SrBr_2$ (9)(ratio = [100])] suggests a value near 100 for $MgBr_2$. This value for the ratio gives $k\delta/\ell^2$ = 1.5 x 10^3 dynes/cm, or ν_2 = 70 cm^{-1}. Based on similar data for linear $BeBr_2$ (10), it is possible that the value of the ratio could be as low as 41. This ratio corresponds to a bending frequency of 109 cm^{-1}. We prefer the lower value of ν_2, since it is consistent with the observed value for $MgCl_2$(9)(ν_2 = 88 cm^{-1})]. However, we do include in the uncertainty of the value for S°(298.15 K) the possibility that the higher value is correct. Brewer et al. (2) used a value for the ratio of 10 and obtained ν_2 = 220 cm^{-1}. By comparison with the observed frequencies for $MgCl_2$ and $BeBr_2$(10)(ν_2 = 220 cm^{-1})] it is unlikely that the value for $MgBr_2$ is this high. Krasnov and Svettsov (11) have reported ν_1 = 188 cm^{-1} and ν_2 = 148 cm^{-1} based on force constant correlations. The ground state is assumed to be singlet by analogy with that for $BaCl_2$ (9).

References

1. J. Berkowitz and J. R. Marquart, J. Chem. Phys. 37, 1853 (1962).
2. L. Brewer, G. R. Somayajulu, and E. Brackett, Chem. Revs. 63, 111 (1963).
3. P. A. Akishin and V. P. Spiridonov, Kristallografiya 2, 472 (1957).
4. L. Wharton, R. A. Berg, and W. Klemperer, J. Chem. Phys. 39, 2023 (1963).
5. A. Buchler, J. L. Stauffer, and W. Klemperer, J. Am. Chem. Soc. 86, 4544 (1964).
6. E. F. Hayes, J. Phys. Chem. 70, 3740 (1966).
7. S. P. Randall, F. T. Greene, and J. L. Margrave, J. Phys. Chem. 63, 758 (1959).
8. G. Herzberg, "Infrared and Raman Spectra of Polyatomic Molecules," D. Van Nostrand Co., Inc., New York, 1962.
9. JANAF Thermochemical Tables: $MgCl_2$(g), 12-31-69; $CaBr_2$(g) and $SrBr_2$(g), 6-30-74; $BaCl_2$(g), 12-31-72.
10. A. Snelson, J. Phys. Chem. 72, 250 (1968).
11. K. S. Krasnov and V. I. Svettsov, Izv. Vysshikh Uchebn. Zavedenii, Khim. i Khim. Tekhnol. 6, 167 (1963).

Magnesium Bromide ($MgBr_2$) $Br_2Mg_1(g)$

Enthalpy Reference Temperature = T_r = 298.15 K Standard State Pressure = p° = 0.1 MPa

T/K	$C_p°$	S°	$-[G°-H°(T_r)]/T$	$H°-H°(T_r)$	$\Delta_f H°$	$\Delta_f G°$	Log K_f
	J $K^{-1}mol^{-1}$			kJ mol^{-1}			
0	0.	0.	INFINITE	-15.012	-288.427	-288.427	INFINITE
100	49.669	241.523	351.095	-10.957	-287.807	-305.624	159.641
200	55.867	278.131	306.332	-5.640	-289.398	-322.911	84.336
250	57.579	290.793	301.998	-2.801	-290.546	-331.162	69.192
298.15	58.711	301.038	301.038	0.	-302.922	-337.555	59.138
300	58.747	301.401	301.039	0.109	-302.999	-337.769	58.811
350	59.557	310.521	301.757	3.067	-333.953	-341.862	51.020
400	60.134	318.514	303.363	6.061	-334.082	-342.984	44.789
450	60.555	325.622	305.448	9.078	-334.226	-344.088	39.941
500	60.870	332.019	307.791	12.114	-334.387	-345.175	36.060
600	61.299	343.158	312.784	18.224	-334.763	-347.299	30.235
700	61.568	352.629	317.817	24.369	-335.230	-349.353	26.069
800	61.747	360.863	322.694	30.535	-335.804	-351.332	22.940
900	61.872	368.143	327.347	36.716	-336.505	-353.233	20.501
1000	61.962	374.667	331.759	42.908	-345.943	-354.335	18.509
1100	62.030	380.576	335.932	49.108	-346.958	-355.125	16.863
1200	62.081	385.976	339.881	55.314	-347.975	-355.822	15.489
1300	62.122	390.946	343.620	61.524	-348.996	-356.435	14.322
1400	62.154	395.551	347.167	67.738	-477.432	-353.802	13.201
1500	62.180	399.841	350.538	73.954	-477.113	-344.982	12.013
1600	62.201	403.854	353.746	80.174	-476.801	-336.184	10.975
1700	62.219	407.626	356.805	86.395	-476.500	-327.404	10.060
1800	62.234	411.182	359.728	92.617	-476.210	-318.643	9.247
1900	62.247	414.548	362.526	98.841	-475.933	-309.896	8.520
2000	62.258	417.741	365.207	105.067	-475.672	-301.164	7.866
2100	62.267	420.779	367.782	111.293	-475.429	-292.445	7.274
2200	62.275	423.675	370.257	117.520	-475.205	-283.737	6.737
2300	62.282	426.444	372.640	123.748	-475.003	-275.038	6.246
2400	62.288	429.095	374.938	129.976	-474.822	-266.348	5.797
2500	62.294	431.637	377.155	136.205	-474.665	-257.665	5.384
2600	62.298	434.081	379.298	142.435	-474.532	-248.988	5.002
2700	62.303	436.432	381.371	148.665	-474.424	-240.315	4.649
2800	62.307	438.698	383.378	154.896	-474.339	-231.646	4.321
2900	62.310	440.884	385.324	161.126	-474.278	-222.980	4.016
3000	62.313	442.997	387.211	167.358	-474.241	-214.315	3.732
3100	62.316	445.040	389.044	173.589	-474.226	-205.651	3.465
3200	62.319	447.019	390.825	179.821	-474.233	-196.987	3.215
3300	62.321	448.936	392.557	186.053	-474.260	-188.323	2.981
3400	62.323	450.797	394.242	192.285	-474.307	-179.657	2.760
3500	62.325	452.603	395.884	198.517	-474.373	-170.990	2.552
3600	62.327	454.359	397.484	204.750	-474.458	-162.321	2.355
3700	62.329	456.067	399.045	210.983	-474.561	-153.649	2.169
3800	62.330	457.729	400.567	217.216	-474.681	-144.975	1.993
3900	62.332	459.348	402.054	223.449	-474.818	-136.296	1.825
4000	62.333	460.926	403.506	229.682	-474.972	-127.614	1.666
4100	62.334	462.465	404.925	235.915	-475.144	-118.928	1.515
4200	62.335	463.968	406.313	242.149	-475.333	-110.238	1.371
4300	62.336	465.434	407.671	248.382	-475.541	-101.543	1.233
4400	62.337	466.867	409.000	254.616	-475.767	-92.842	1.102
4500	62.338	468.268	410.302	260.850	-476.012	-84.137	0.977
4600	62.339	469.639	411.577	267.084	-476.278	-75.426	0.856
4700	62.340	470.979	412.827	273.318	-476.566	-66.708	0.741
4800	62.341	472.292	414.052	279.552	-476.876	-57.985	0.631
4900	62.341	473.577	415.253	285.786	-477.210	-49.254	0.525
5000	62.342	474.837	416.433	292.020	-477.570	-40.517	0.423
5100	62.343	476.071	417.590	298.254	-477.956	-31.772	0.325
5200	62.343	477.282	418.726	304.488	-478.370	-23.020	0.231
5300	62.344	478.469	419.842	310.723	-478.813	-14.258	0.141
5400	62.344	479.635	420.939	316.957	-479.287	-5.489	0.053
5500	62.345	480.779	422.016	323.192	-479.794	3.290	-0.031
5600	62.345	481.902	423.076	329.426	-480.334	12.078	-0.113
5700	62.346	483.005	424.118	335.661	-480.909	20.876	-0.191
5800	62.346	484.090	425.142	341.895	-481.522	29.684	-0.267
5900	62.346	485.155	426.150	348.130	-482.172	38.504	-0.341
6000	62.347	486.203	427.143	354.364	-482.862	47.334	-0.412

PREVIOUS: June 1974 (1 atm) CURRENT: June 1974 (1 bar)

MAGNESIUM BROMIDE, ION ($MgBr_2^+$) IDEAL GAS M_r = 184.112451

$\Delta_fH°(0\ K)$ = 739.2 ± 25.1 kJ mol^{-1}

S°(298.15 K) = [321.7 ± 12.6] J K^{-1} mol^{-1} $\Delta_fH°(298.15\ K)$ = [731.363] kJ mol^{-1}

Electronic Levels and Quantum Weights

State	ε_i, cm^{-1}	g_i
$^2\Pi$	0	4
$^2\Pi$	[25000]	4
$^2\Sigma$	[30000]	2

Vibrational Frequencies and Degeneracies

ν, cm^{-1}
[150](1)
[50](2)
[450](1)

Point Group: [$D_{\infty h}$] σ = 2
Bond Distance: Mg-Br = [2.6] Å
Bond Angle: Br-Mg-Br = [180]°
Rotational Constant: B_0 = [0.015605] cm^{-1}

Enthalpy of Formation

Berkowitz and Marquart (1) have reported the appearance potential of $MgBr_2^+$ as 10.6±0.1$_5$ eV(245.6 kcal mol^{-1}). Assuming this value refers to the direct ionization process $MgBr_2(g) + e^- = MgBr_2^+(g) + 2e^-$, we obtain $\Delta_fH°(MgBr_2^+$, g, 0 K) = 176.7±6.0 kcal mol^{-1} by combining the appearance potential with $\Delta_fH°(MgBr_2$, g, 0 K) = -68.9±2.5 kcal mol^{-1} (2). The former value corresponds to $\Delta_fH°(298.15\ K)$ = 174.8 kcal mol^{-1} (731.363 kJ mol^{-1}) which is adopted.

Heat Capacity and Entropy

A comparison of the dissociation energy, $\Delta_{at}H°$= 91.0 kcal mol^{-1}, for $MgBr_2^+$ with that for $MgBr_2$ (2), $\Delta_{at}H°$= 160.3 kcal mol^{-1}, suggests the existence of somewhat weaker bonding in the ion relative to the molecule. Thus, one would expect the bond length in the ion to be greater than that for $MgBr_2$. We assume a 10% increase in r_e for the ion. The correlation diagrams of Walsh (3) predict a linear configuration for $MgBr_2^+$(fifteen valence electrons). This prediction is supported by the fact that several other fifteen valence electron molecules (BO_2, N_3, NCO, and NNO^+) are now known to be linear (2). We adopt a linear configuration for the ion.

The vibrational frequencies are estimated to be slightly less than those for $MgBr_2$ (2). The ground state electronic configuration is assumed to be $^2\Pi$ by analogy with those for the isoelectronic molecules (2) BO_2, N_3, NCO, and N_2O^+. Two excited states are also included based on those observed for BO_2 (2). The enthalpy at 0 K is -15.441 kJ mol^{-1}.

References

1. J. Berkowitz and J. R. Marquart, J. Chem. Phys. 37, 1853 (1962).
2. JANAF Thermochemical Tables: $MgBr_2$(g), 6-30-74; BO_2(g), 6-30-68; CNO(g), N_2O^+(g), and N_3(g), 12-31-70.
3. A. D. Walsh, J. Chem. Soc. 1953, 2266 (1953).

Magnesium Bromide, Ion ($MgBr_2^+$) $Br_2Mg_1^+$(g)

Enthalpy Reference Temperature = T_r = 298.15 K Standard State Pressure = p° = 0.1 MPa

	J $K^{-1}mol^{-1}$			kJ mol^{-1}			
T/K	$C_p°$	S°	$-[G°-H°(T_r)]/T$	$H°-H°(T_r)$	$\Delta_fH°$	$\Delta_fG°$	Log K_f
0	0.	0.	INFINITE	-15.442	739.231		
100	51.289	261.045	372.577	-11.153			
200	56.811	298.524	327.095	-5.714			
250	58.322	311.375	322.708	-2.833			
298.15	59.303	321.737	321.737	0.	731.363	684.304	-119.887
300	59.334	322.104	321.738	0.110	731.325	684.012	-119.097
350	60.027	331.305	322.463	3.095	701.437	677.744	-101.148
400	60.514	339.354	324.082	6.109	702.368	674.296	-88.054
450	60.868	346.503	326.183	9.144	703.280	670.732	-77.857
500	61.131	352.931	328.542	12.194	704.174	667.067	-69.688
600	61.487	364.110	333.566	18.327	705.898	659.482	-57.413
700	61.710	373.607	338.625	24.487	707.527	651.616	-48.624
800	61.857	381.857	343.524	30.666	709.043	643.524	-42.018
900	61.960	389.149	348.196	36.857	710.431	635.249	-36.869
1000	62.034	395.681	352.624	43.057	703.079	627.540	-32.779
1100	62.090	401.596	356.811	49.264	704.150	619.935	-29.438
1200	62.132	407.001	360.772	55.475	705.217	612.232	-26.650
1300	62.165	411.975	364.522	61.690	706.279	604.440	-24.287
1400	62.191	416.583	368.078	67.908	579.926	599.734	-22.376
1500	62.213	420.875	371.456	74.128	582.327	601.064	-20.931
1600	62.230	424.891	374.672	80.350	584.720	602.236	-19.661
1700	62.245	428.664	377.738	86.574	587.103	603.257	-18.536
1800	62.257	432.222	380.667	92.799	589.474	604.139	-17.532
1900	62.267	435.588	383.470	99.025	591.831	604.889	-16.630
2000	62.276	438.782	386.156	105.252	594.173	605.516	-15.814
2100	62.284	441.821	388.735	111.480	596.496	606.026	-15.074
2200	62.290	444.719	391.214	117.709	598.800	606.426	-14.398
2300	62.296	447.488	393.601	123.938	601.083	606.721	-13.779
2400	62.302	450.139	395.902	130.168	603.343	606.918	-13.209
2500	62.306	452.682	398.123	136.398	605.580	607.021	-12.683
2600	62.311	455.126	400.269	142.629	607.793	607.035	-12.195
2700	62.315	457.478	402.344	148.861	609.982	606.964	-11.742
2800	62.320	459.744	404.354	155.092	612.146	606.812	-11.320
2900	62.324	461.931	406.302	161.325	614.287	606.584	-10.926
3000	62.329	464.044	408.192	167.557	616.404	606.282	-10.556
3100	62.335	466.088	410.027	173.791	618.500	605.910	-10.210
3200	62.341	468.067	411.810	180.024	620.574	605.471	-9.883
3300	62.348	469.986	413.544	186.259	622.627	604.967	-9.576
3400	62.355	471.847	415.231	192.494	624.662	604.401	-9.285
3500	62.364	473.655	416.875	198.730	626.677	603.776	-9.011
3600	62.374	475.412	418.476	204.967	628.676	603.094	-8.751
3700	62.386	477.121	420.038	211.205	630.657	602.356	-8.504
3800	62.399	478.785	421.563	217.444	632.622	601.564	-8.269
3900	62.414	480.406	423.051	223.685	634.571	600.721	-8.046
4000	62.431	481.986	424.504	229.927	636.504	599.829	-7.833
4100	62.450	483.528	425.925	236.171	638.422	598.888	-7.630
4200	62.471	485.033	427.315	242.417	640.324	597.901	-7.436
4300	62.494	486.503	428.674	248.665	642.209	596.868	-7.251
4400	62.519	487.940	430.005	254.916	644.079	595.793	-7.073
4500	62.547	489.346	431.308	261.169	645.932	594.674	-6.903
4600	62.577	490.721	432.585	267.425	647.766	593.515	-6.740
4700	62.610	492.067	433.836	273.685	649.583	592.316	-6.583
4800	62.645	493.385	435.063	279.947	651.380	591.078	-6.432
4900	62.683	494.677	436.266	286.214	653.157	589.804	-6.287
5000	62.723	495.944	437.447	292.484	654.912	588.493	-6.148
5100	62.765	497.187	438.607	298.758	656.644	587.148	-6.014
5200	62.810	498.406	439.745	305.037	658.353	585.768	-5.884
5300	62.857	499.603	440.863	311.320	660.038	584.356	-5.759
5400	62.907	500.778	441.962	317.609	661.696	582.913	-5.639
5500	62.959	501.933	443.042	323.902	663.327	581.439	-5.522
5600	63.013	503.068	444.103	330.201	664.930	579.935	-5.409
5700	63.069	504.184	445.148	336.505	666.503	578.403	-5.300
5800	63.127	505.281	446.175	342.814	668.044	576.844	-5.195
5900	63.188	506.361	447.186	349.130	669.553	575.259	-5.093
6000	63.250	507.423	448.181	355.452	671.029	573.648	-4.994

PREVIOUS: June 1974 (1 atm) CURRENT: June 1974 (1 bar)

Magnesium Bromide, Ion ($MgBr_2^+$) $Br_2Mg_1^+$(g)

MOLYBDENUM BROMIDE ($MoBr_2$) CRYSTAL M_r = 255.748

$\Delta_f H°$(0 K) = unknown

S°(298.15 K) = [124.7 ± 17] J K^{-1} mol^{-1} $\Delta_f H°$(298.15 K) = -202.9 ± 25 kJ mol^{-1}

Enthalpy of Formation

Shukurov et al. (1)(determined the enthalpies of combustion of $MoBr_2$(cr) and $MoBr_3$(cr). Dellien et al. (2) used the combustion values of Shukurov to obtain $\Delta_f H°$($MoBr_2$, cr, 298.15 K) = -54 kcal mol^{-1} and $\Delta_f H°$($MoBr_3$, cr, 298.15 K) = -64 kcal mol^{-1}. Oppermann (3) carried out equilibrium experiments on the mixed phases $MoBr_2$(cr) + $MoBr_3$(cr). Brewer (4) re-evaluated the data of Oppermann (see the discussion in the $MoBr_4$(g) table (5)) and supplemented the combustion data with other data on $MoBr_4$(cr). For this reason we believe Brewer's value of $\Delta_f H°$($MoBr_2$, cr, 298.15 K) = -48.5±6 kcal mol^{-1}(-202.924±25 kJ mol^{-1}) to be the most reliable among all the data and is our adopted value. Note that the enthalpy of formation listed in NBS 270-4 (6), $\Delta_f H°$(298.15 K) = -62.4 kcal mol^{-1}, is much more negative.

Heat Capacity and Entropy

Brewer (4) estimated S°(298.15 K) = 29.8±4 cal K^{-1} mol^{-1} (124.683±17 J K^{-1} mol^{-1}) and derived $C_p°$ = 18.30 + 2.50 x 10^{-3}T cal K^{-1} mol^{-1} (298-1353 K) from the heat capacities estimated by Opperman (3). Other estimates are $C_p°$(298.15 K) = 17.1 cal K^{-1} mol^{-1} from the ionic contributions of Kubaschewski and Unal (7) and S°(298.15 K) = 30.4 cal K^{-1} mol^{-1} from the additive entropy constants of Stull and Prophet (8). The heat capacity at 298.15 K listed in NBS 270-4 (6) is 18.3 cal K^{-1} mol^{-1}.

Decomposition Data

T_{dcm} is the calculated temperature at which the pressure of $MoBr_4$(g) is one atmosphere for the reaction 2 $MoBr_2$(cr) = Mo(cr) + $MoBr_4$(g).

References

1. A. Shukurov, T. Nikol'skaya, I. V. Vasil'kova and S. A. Shchukarev, Izv. Akad. Nauk Tadzhsk. SSR, Otd. Geol. - Khim. Tekhn. Nauk, No. 4, 23 (1961).
2. I. Dellien, F. M. Hall, and L. G. Hepler, Chem. Rev. 76, 283 (1976).
3. H. Oppermann, Z. Anorg. Allg. Chem. 395, 249 (1973).
4. L. Brewer, Materials and Molecular Research Division, Lawrence Berkeley Laboratory, University of California, Berkeley; personal communication, Sept. 29, 1978; preliminary draft of review to be submitted for publication in Atomic Energy Review, International Atomic Energy Agency, Vienna, Austria.
5. JANAF Thermochemical Tables: $MoBr_4$(g), 9-30-78.
6. U. S. Nat. Bur. Stand. Tech. Note 270-4, 1969.
7. O. Kubaschewski and H. Unal, High Temp.-High Pressures 9, 361 (1977).
8. D. R. Stull and H. Prophet, Chap. 13 in "The Characterization of High Temperature Vapors" Ed. by J. L. Margrave, John Wiley, 1967.

Molybdenum Bromide ($MoBr_2$) Br_2Mo_1(cr)

Enthalpy Reference Temperature = T_r = 298.15 K Standard State Pressure = p° = 0.1 MPa

T/K	$C_p°$ (J $K^{-1}mol^{-1}$)	S° (J $K^{-1}mol^{-1}$)	-[G°-H°(T_r)]/T (J $K^{-1}mol^{-1}$)	H°-H°(T_r) (kJ mol^{-1})	$\Delta_f H°$ (kJ mol^{-1})	$\Delta_f G°$ (kJ mol^{-1})	Log K_f
0							
100							
200							
298.15	79.684	124.683	124.683	0.	-202.924	-186.189	32.620
300	79.705	125.176	124.685	0.147	-202.961	-186.085	32.400
400	80.751	148.249	127.824	8.170	-231.874	-174.412	22.776
500	81.797	166.381	133.785	16.298	-229.986	-160.265	16.743
600	82.843	181.387	140.504	24.530	-228.091	-146.499	12.754
700	83.889	194.235	147.284	32.866	-226.165	-133.052	9.928
800	84.935	205.506	153.871	41.308	-224.197	-119.884	7.828
900	85.981	215.570	160.177	49.853	-222.182	-106.965	6.208
1000	87.027	224.683	166.179	58.504	-220.118	-94.273	4.924
1100	88.073	233.027	171.882	67.259	-218.009	-81.790	3.884
1200	89.119	240.735	177.303	76.118	-215.860	-69.501	3.025
1300	90.165	247.909	182.461	85.083	-213.677	-57.393	2.306
1400	91.211	254.630	187.379	94.152	-211.468	-45.454	1.696
1500	92.257	260.958	192.075	103.325	-209.238	-33.674	1.173
1600	93.303	266.946	196.569	112.603	-206.997	-22.043	0.720
1700	94.349	272.634	200.878	121.986	-204.750	-10.552	0.324
1800	95.395	278.056	205.016	131.473	-202.508	0.807	-0.023

PREVIOUS: CURRENT: September 1978

Molybdenum Bromide ($MoBr_2$) Br_2Mo_1(cr)

MOLYBDENUM BROMIDE ($MoBr_2$) IDEAL GAS M_r = 255.748

$S°(298.15\ K) = [321.0 \pm 8]\ J\ K^{-1}\ mol^{-1}$

$\Delta_f H°(0\ K) = [179.2 \pm 84]\ kJ\ mol^{-1}$
$\Delta_f H°(298.15\ K) = [167.4 \pm 84]\ kJ\ mol^{-1}$

Electronic Levels and Quantum Weights

ε_i, cm^{-1}	g_i
0	[1]
[243.10]	[3]
[669.60]	[5]
[1225.20]	[7]
[1873.80]	[9]
[11271]	[1]
[12510]	[3]
[12630]	[9]
[13201]	[11]
[13742]	[13]

Vibrational Frequencies and Degeneracies

ν_i, cm^{-1}
[300](1)
[60](2)
[300](1)

Ground State Quantum Weight: 1 σ = [2]
Point Group: [$D_{\infty h}$]
Bond Distance: Mo-Br = [2.42] Å
Bond Angle: Br-Mo-Br = [180]°
Rotational Constant: B_o = [0.018012] cm^{-1}

Enthalpy of Formation

The adopted value of $\Delta_f H°(298.15\ K)$ = 40.0±20 kcal mol^{-1} (167.360±84 kJ mol^{-1}) is that estimated by Brewer (1) using bonding models. The value is consistent with sublimation data for $MoBr_2$ at 1023 K and 1273 K reported by Schafer, et al. (2). Brewer gave no weight to the partial pressures of $MoBr_2$ obtained by Oppermann (3) by difference (see the discussion on the $MoBr_4$(g) table (4)). The value of $\Delta_f H°(0\ K)$ combined with JANAF data (4) for Mo(g) and Br(g) gives $\Delta_{at}H°(MoBr_2, g)$ = 171±20 kcal mol^{-1} (715.5±84 kJ mol^{-1}). This yields a mean bond energy $\Delta_{at}H°(MoBr_2, g)/2$ = 85.5 kcal mol^{-1}, somewhat higher than that found in $MoBr_4$(g), 76.0 kcal mol^{-1} (4)

Heat Capacity and Entropy

The linear structure of $MoBr_2$(g), bond length, and the ground state vibrational frequencies are estimates from Brewer (1). The electronic contributions are assumed to be the same as the free ion, Mo^{++}, as suggested by Brewer (1). All states up to 13,742 cm^{-1} as listed by Rico (5) are included in the calculation.

References

1. L. Brewer, Materials and Molecular Research Division, Lawrence Berkeley Laboratory, University of California, Berkeley; personal communication of Sept. 29, 1978; preliminary draft of review to be submitted for publication in Atomic Energy Review, International Atomic Energy Agency, Vienna, Austria.
2. H. Schafer, T. Grofe, and M. Trenkel, J. Solid State Chem. 8, 14 (1973).
3. H. Oppermann, Z. Anorg. Allg. Chem. 395, 249 (1973).
4. JANAF Thermochemical Tables: Mo(g), 3-31-78; Br(g), 9-30-61, $MoBr_4$(g), 9-30-78.
5. F. R. Rico, Anales. Real Soc. Espan. Fys. Quim., Ser. A, 61, 103 (1965).

Molybdenum Bromide ($MoBr_2$) $Br_2Mo_1(g)$

Enthalpy Reference Temperature = T_r = 298.15 K — Standard State Pressure = p° = 0.1 MPa

T/K	$C_p°$ (J $K^{-1}mol^{-1}$)	S°	$-[G°-H°(T_r)]/T$	$H°-H°(T_r)$ (kJ mol^{-1})	$\Delta_f H°$	$\Delta_f G°$	Log K_f
0	0.	0.	INFINITE	-17.293	179.161	179.161	INFINITE
100	56.960	250.449	380.393	-12.994	180.175	161.218	-84.211
200	66.592	293.789	327.229	-6.688	179.738	142.360	-37.181
250	68.260	308.843	322.096	-3.313	179.179	133.075	-27.805
298.15	69.291	320.960	320.960	0.	167.360	125.575	-22.000
300	69.322	321.389	320.962	0.128	167.304	125.315	-21.819
350	70.005	332.131	321.809	3.613	136.924	119.974	-17.905
400	70.409	341.508	323.697	7.124	137.364	117.522	-15.347
450	70.590	349.813	326.147	10.650	137.784	115.017	-13.351
500	70.590	357.252	328.892	14.180	138.180	112.466	-11.749
600	70.206	370.094	334.722	21.223	138.887	107.254	-9.337
700	69.536	380.868	340.566	28.212	139.465	101.935	-7.606
800	68.771	390.104	346.195	35.127	139.906	96.541	-6.304
900	68.020	398.160	351.531	41.966	140.215	91.101	-5.287
1000	67.333	405.291	356.557	48.733	140.395	85.633	-4.473
1100	66.728	411.679	361.283	55.436	140.452	80.153	-3.806
1200	66.204	417.462	365.727	62.082	140.387	74.673	-3.250
1300	65.756	422.743	369.913	68.679	140.203	69.204	-2.781
1400	65.375	427.602	373.862	75.235	139.900	63.753	-2.379
1500	65.053	432.101	377.597	81.756	139.477	58.328	-2.031
1600	64.783	436.290	381.136	88.247	138.932	52.935	-1.728
1700	64.559	440.211	384.497	94.714	138.262	47.580	-1.462
1800	64.377	443.896	387.695	101.161	137.464	42.268	-1.227
1900	64.232	447.372	390.745	107.591	136.530	37.004	-1.017
2000	64.122	450.664	393.660	114.008	135.456	31.793	-0.830
2100	64.043	453.790	396.449	120.416	134.233	26.640	-0.663
2200	63.993	456.769	399.124	126.818	132.855	21.548	-0.512
2300	63.970	459.613	401.693	133.216	131.313	16.522	-0.375
2400	63.972	462.335	404.163	139.613	129.599	11.567	-0.252
2500	63.996	464.947	406.543	146.011	127.704	6.688	-0.140
2600	64.040	467.458	408.837	152.413	125.607	1.888	-0.038
2700	64.103	469.876	411.054	158.820	123.271	-2.827	0.055
2800	64.182	472.208	413.196	165.234	120.637	-7.451	0.139
2900	64.275	474.462	415.270	171.657	81.743	-11.924	0.215
3000	64.380	476.643	417.280	178.089	80.314	-15.129	0.263
3100	64.496	478.756	419.229	184.533	78.881	-18.287	0.308
3200	64.620	480.805	421.121	190.989	77.445	-21.399	0.349
3300	64.750	482.796	422.960	197.457	76.011	-24.465	0.387
3400	64.886	484.731	424.749	203.939	74.581	-27.489	0.422
3500	65.025	486.614	426.490	210.434	73.159	-30.470	0.455
3600	65.167	488.448	428.185	216.944	71.747	-33.411	0.485
3700	65.309	490.235	429.838	223.468	70.349	-36.312	0.513
3800	65.452	491.979	431.451	230.006	68.966	-39.177	0.539
3900	65.593	493.681	433.025	236.558	67.602	-42.005	0.563
4000	65.732	495.343	434.562	243.124	66.258	-44.798	0.585
4100	65.868	496.968	436.064	249.704	64.937	-47.558	0.606
4200	66.000	498.557	437.533	256.298	63.640	-50.286	0.625
4300	66.128	500.111	438.971	262.904	62.368	-52.984	0.644
4400	66.252	501.633	440.377	269.523	61.123	-55.652	0.661
4500	66.370	503.123	441.755	276.154	59.906	-58.292	0.677
4600	66.483	504.583	443.105	282.797	58.717	-60.905	0.692
4700	66.590	506.014	444.429	289.451	57.558	-63.493	0.706
4800	66.691	507.417	445.726	296.115	56.428	-66.057	0.719
4900	66.786	508.793	446.999	302.789	55.327	-68.597	0.731
5000	66.875	510.143	448.249	309.472	-537.129	-65.384	0.683
5100	66.958	511.468	449.475	316.163	-539.850	-55.922	0.573
5200	67.034	512.769	450.680	322.863	-542.688	-46.406	0.466
5300	67.105	514.047	451.864	329.570	-545.640	-36.834	0.363
5400	67.169	515.302	453.027	336.284	-548.707	-27.205	0.263
5500	67.227	516.535	454.170	343.004	-551.885	-17.519	0.166
5600	67.280	517.747	455.295	349.729	-555.175	-7.774	0.073
5700	67.327	518.938	456.401	356.460	-558.571	2.031	-0.019
5800	67.369	520.109	457.489	363.194	-562.074	11.896	-0.107
5900	67.405	521.261	458.561	369.933	-565.679	21.823	-0.193
6000	67.436	522.394	459.615	376.675	-569.388	31.812	-0.277

PREVIOUS: September 1978 (1 atm) CURRENT: September 1978 (1 bar)

Molybdenum Bromide ($MoBr_2$) $Br_2Mo_1(g)$

SODIUM BROMIDE $((NaBr)_2)$ IDEAL GAS M_r = 205.78754

$\Delta_f H°(0\ K)$ = -468.33 ± 3.77 kJ mol^{-1}

$S°(298.15\ K)$ = [348.949] J K^{-1} mol^{-1} $\Delta_f H°(298.15\ K)$ = -486.35 ± 3.77 kJ mol^{-1}

Vibrational Frequencies and Degeneracies

ν, cm^{-1}	ν, cm^{-1}
[182](1)	[134](1)
[120](1)	[188](1)
[207](1)	[228](1)

Ground State Quantum Weight: [1] σ = 4

Point Group: $[D_{2h}]$

Bond Distance: Na-Br = [2.75] Å

Bond Angles: Na-Br-Na = [71]° Br-Na-Br = [109]°

Product of the Moments of Inertia: $I_A I_B I_C$ = [3.94920 x 10^{-112}] g^3 cm^6

Enthalpy of Formation

The temperature dependence of the equilibrium constants (1293-1434 K) for the reaction $(NaBr)_2(g) = 2NaBr(g)$ has been studied by Datz et al. (1). The dissociation energy ($\Delta_r E°$(1300 K) was evaluated to be 42.9±1.3 kcal mol^{-1}. The vapor mixture contains 37 to 17% of dimers in the temperature range of 1293 to 1434 K. Hence the vapor pressure data on NaBr(g) reported by the previous investigators were reexamined. In other words, the reported vapor pressures of NaBr(g) is the sum of the partial pressures of both the monomer (NaBr) and the dimer (Na_2Br_2). The partial pressures of $Na_2Br_2(g)$ thus obtained were used to evaluate the enthaply of vaporization by both the 2nd and 3rd law methods. The values of $\Delta_f H°(Na_2Br_2$, g, 298.15 K) were then calculated. The results are as follows.

Source	Reaction	$\Delta_r H°$(298.15 K), kcal mol^{-1} 2nd Law	3rd Law	$\Delta_f H°$(298.15 K)* kcal mol^{-1}
Niwa (2)	2NaBr(cr) + Na_2Br_2(g)	54.84	56.51	-117.08
Cogin and Kimball (3)	2NaBr(cr) + Na_2Br_2(g)	56.69	56.80	-116.01
Mayer and Wintner (4)	2NaBr(cr) + Na_2Br_2(g)	68.78	57.13	-115.63**
Ruff and Mugdan (5)	2NaBr(ℓ) + Na_2Br_2(g)	46.71	45.70	-116.00
Wartenberg and Albrecht (6)	2NaBr(ℓ) + Na_2Br_2(g)	46.77	45.64	-116.00
Bloom et al. (7)	2NaBr(ℓ) + Na_2Br_2(g)	51.17	45.49	-116.72**

*Based on the average of the 2nd and 3rd law values.

**Only the third law value being used.

The value of $\Delta_f H°(Na_2Br_2$, g, 298.15 K) = -116.24 kcal mol^{-1} (-486.348 kJ mol^{-1}) adopted is the average of the six $\Delta_f H°$(298.15 K) values listed in the above table.

Heat Capacity and Entropy

Bond distance and angles were calculated based on the related data reported by Berkowitz (8). The vibrational frequencies were calculated by Berkowitz (9), from an assumed model. The ground state quantum weight was estimated. The principal moments of inertia are I_A = 19.4713 x 10^{-39}, I_B = 133.0124 x 10^{-39}, and I_C = 152.4836 x 10^{-39} g cm^2.

References

1. S. Datz, W. T. Smith, Jr. and E. H. Taylor, J. Chem. Phys. 34, 558 (1961).
2. K. Niwa, J. Fac. Sci., Hokkaido Univ., Ser. III, 2, 201 (1938).
3. G. E. Cogin nd G. E. Kimball, J. Chem. Phys. 16, 1035 (1948).
4. J. E. Mayer and I. H. Wintner, J. Chem. Phys. 6, 301 (1938).
5. O. Ruff and S. Mugdan, Z. Anorg. Allgem. Chem. 117, 147 (1921).
6. H. von Wartenberg and P. Albrecht, Z. Electrochem. 27, 162 (1921).
7. H. Bloom, J. O'M. Bockris, N. E. Richards and R. G. Taylor, J. Am. Chem. Soc. 80, 2044 (1958).
8. J. Berkowitz, J. Chem. Phys. 29, 1386 (1958).
9. J. Berkowitz, J. Chem. Phys. 32, 1519 (1960).

Sodium Bromide $((NaBr)_2)$ $Br_2Na_2(g)$

Enthalpy Reference Temperature = T_r = 298.15 K Standard State Pressure = p° = 0.1 MPa

T/K	$C_p°$ (J $K^{-1}mol^{-1}$)	S°	$-[G°-H°(T_r)]/T$	$H°-H°(T_r)$ (kJ mol^{-1})	$\Delta_f H°$	$\Delta_f G°$	Log K_f
0	0.	0.	INFINITE	-19.383	-468.328	-468.328	INFINITE
100	63.244	268.362	417.381	-14.902	-469.320	-486.008	253.865
200	76.762	317.552	356.196	-7.729	-471.959	-501.707	131.032
250	78.921	334.937	350.262	-3.831	-473.526	-508.967	106.343
298.15	80.121	348.949	348.949	0.	-486.348	-514.324	90.108
300	80.157	349.445	348.950	0.148	-486.444	-514.498	89.582
350	80.925	361.863	349.929	4.177	-517.971	-517.428	77.222
400	81.432	372.704	352.113	8.236	-524.079	-516.883	67.498
450	81.785	382.317	354.945	12.317	-524.965	-515.930	59.888
500	82.039	390.948	358.121	16.413	-525.797	-514.881	53.789
600	82.373	405.937	364.879	24.635	-527.326	-512.551	44.622
700	82.576	418.652	371.675	32.884	-528.721	-509.977	38.055
800	82.708	429.687	378.252	41.148	-530.027	-507.210	33.117
900	82.799	439.435	384.519	49.424	-531.290	-504.281	29.268
1000	82.865	448.162	390.455	57.707	-532.555	-501.213	26.181
1100	82.913	456.062	396.066	65.996	-533.866	-498.016	23.649
1200	82.950	463.278	401.370	74.289	-728.786	-489.816	21.321
1300	82.978	469.919	406.391	82.586	-728.447	-469.916	18.881
1400	83.001	476.069	411.151	90.885	-728.114	-450.041	16.791
1500	83.020	481.796	415.672	99.186	-727.790	-430.190	14.981
1600	83.035	487.155	419.974	107.489	-727.473	-410.361	13.397
1700	83.047	492.189	424.076	115.793	-727.168	-390.550	12.000
1800	83.058	496.936	427.993	124.098	-726.875	-370.758	10.759
1900	83.066	501.427	431.741	132.404	-726.596	-350.981	9.649
2000	83.074	505.688	435.332	140.711	-726.335	-331.219	8.651
2100	83.081	509.741	438.780	149.019	-726.093	-311.469	7.747
2200	83.086	513.607	442.094	157.327	-725.873	-291.730	6.927
2300	83.091	517.300	445.284	165.636	-725.679	-272.001	6.177
2400	83.095	520.836	448.359	173.946	-725.511	-252.280	5.491
2500	83.099	524.229	451.326	182.255	-725.372	-232.565	4.859
2600	83.103	527.488	454.193	190.565	-725.265	-212.855	4.276
2700	83.106	530.624	456.967	198.876	-725.190	-193.148	3.737
2800	83.108	533.647	459.651	207.187	-725.150	-173.444	3.236
2900	83.111	536.563	462.254	215.498	-725.145	-153.740	2.769
3000	83.113	539.381	464.778	223.809	-725.176	-134.036	2.334
3100	83.115	542.106	467.229	232.120	-725.246	-114.330	1.926
3200	83.117	544.745	469.610	240.432	-725.353	-94.622	1.545
3300	83.118	547.302	471.926	248.743	-725.499	-74.909	1.186
3400	83.120	549.784	474.179	257.055	-725.685	-55.192	0.848
3500	83.121	552.193	476.374	265.367	-725.912	-35.468	0.529
3600	83.122	554.535	478.513	273.680	-726.182	-15.737	0.228
3700	83.124	556.812	480.598	281.992	-726.495	4.001	-0.056
3800	83.125	559.029	482.633	290.304	-726.855	23.749	-0.326
3900	83.126	561.188	484.620	298.617	-727.262	43.507	-0.583
4000	83.127	563.293	486.561	306.929	-727.719	63.276	-0.826
4100	83.127	565.346	488.457	315.242	-728.229	83.057	-1.058
4200	83.128	567.349	490.312	323.555	-728.791	102.851	-1.279
4300	83.129	569.305	492.126	331.868	-729.417	122.659	-1.490
4400	83.130	571.216	493.902	340.181	-730.022	142.476	-1.691
4500	83.130	573.084	495.641	348.494	-730.734	162.312	-1.884
4600	83.131	574.911	497.344	356.807	-731.527	182.166	-2.069
4700	83.131	576.699	499.014	365.120	-732.391	202.038	-2.245
4800	83.132	578.449	500.651	373.433	-733.329	221.929	-2.415
4900	83.132	580.163	502.256	381.746	-734.347	241.841	-2.578
5000	83.133	581.843	503.831	390.059	-735.434	261.773	-2.735
5100	83.133	583.489	505.377	398.373	-736.621	281.729	-2.885
5200	83.134	585.103	506.895	406.686	-737.902	301.708	-3.031
5300	83.134	586.687	508.385	415.000	-739.282	321.714	-3.171
5400	83.134	588.241	509.850	423.313	-740.764	341.747	-3.306
5500	83.135	589.766	511.289	431.626	-742.356	361.808	-3.436
5600	83.135	591.264	512.704	439.940	-744.060	381.899	-3.562
5700	83.135	592.736	514.095	448.253	-745.882	402.021	-3.684
5800	83.136	594.182	515.463	456.567	-747.827	422.176	-3.802
5900	83.136	595.603	516.809	464.881	-749.899	442.366	-3.916
6000	83.136	597.000	518.134	473.194	-751.786	462.556	-4.027

PREVIOUS: September 1964 (1 atm) CURRENT: September 1964 (1 bar)

Sodium Bromide $((NaBr)_2)$ $Br_2Na_2(g)$

LEAD BROMIDE ($PbBr_2$) CRYSTAL M_r = 367.008

$S°(298.15\ K) = 161.13 \pm 2.1\ J\ K^{-1}\ mol^{-1}$

$T_{fus} = 644 \pm 5\ K$

$\Delta_f H°(0\ K) = -265.2 \pm 2.5\ kJ\ mol^{-1}$

$\Delta_f H°(298.15\ K) = -277.4 \pm 2.5\ kJ\ mol^{-1}$

$\Delta_{fus} H° = 16.443 \pm 0.8\ kJ\ mol^{-1}$

Enthaply of Formation

$\Delta_f H°(PbBr_2$, cr, 298.15 K) = -66.3±0.6 kcal mol^{-1} (-277.399 kJ mol^{-1}) is the rounded average of -66.350 kcal mol^{-1} calorimetrically determined by Braune and Koref (1) at about 293.15 K and five other values obtained from a 2nd and 3rd law analysis of electromotive force data (2, 3, 4). The auxiliary value $\Delta_f H°$(AgBr, cr, 298.15 K) = -23.99 kcal mol^{-1} (5) is used in the analysis.

Source	Reaction*	Data Points	T/K	$\Delta_r H°(298.15\ K)$, kcal mol^{-1} 2nd Law	3rd Law	Drift cal K^{-1} mol^{-1}	$\Delta_f H°(298.15\ K)$, kcal mol^{-1} 2nd Law	3rd Law
Krahmer (2)	A	5	273-292	17.88±0.06	18.52±0.04	-2.2±0.2	-65.86	-66.50
Cann and Sumner (3)	A	1	298		18.57			-66.57
Jahn-Held and Jellinek (4)	B	3	288-308	66.68±0.06	66.09±0.04	2.0±0.2	-66.68	-66.09

*Reactions: (A) Pb(cr) + 2 AgBr(cr) = $PbBr_2$(cr) + 2 Ag(cr) (B) Pb(cr) + Br_2(ℓ) = $PbBr_2$(cr)

Heat Capacity and Entropy

$C_p°$ (18.4 - 297.0 K) has been measured by Latimer and Hoenshel (6). The enthalpy data of Ehrhardt (273-786 K) (7) and of Goodwin and Kalmus (298-860 K) (8) are discarded because the observed high T_{fus} shows their samples were not pure. (See melting data discussion.) Linsey (16) measured the enthalpy (39 points) of $PbBr_2$ in the range 319-424 K using an ice calorimeter. The heat capacity data of Latimer and Hoenshel (6) is smoothed graphically. The enthalpy data was analyzed by Linsey (16) to yield heat capacity values from 273-644 K. Both sets of heat capacity values are adjusted graphically in the region 200-400 K so as to yield a smooth curve in the vicinity of 298 K. The adjustments are quite small such that the resulting enthalpy at 600 K differs by about 7 cal mol^{-1} from the smooth enthalpies reported by Linsey (16). The data of Linsey (16) did not indicate any transitions other than the solid-liquid transition at 644 K.

Latimer and Hoenshel (6) obtained S°(15.85 K) = 2.4 cal K^{-1} mol^{-1} after allowing for an approach to the T^3 rule. Kelley and King (10) later derived S°(17.80 K) = 1.64 cal K^{-1} mol^{-1} from the same data; this value is adopted. Combining this with S°(298.15;17.80) = 36.871 cal K^{-1} mol^{-1} derived from the adopted $C_p°$ value gives S°(298.15 K) = 38.511±0.5 cal K^{-1} mol^{-1} (161.128 J K^{-1} mol^{-1}). A graphical extrapolation and integration of $C_p°$ gives H°(17.80;0) = 0.0252 kcal mol^{-1}. This value may not be entirely consistent with the adopted S°(17.80 K) value, but the error involved is undoubtedly small.

Fusion Data

The older reported values of T_{fus} have been summarized by Mellor (11). Seven determinations fall in the 636-653 K range and four more range from 753 to 772 K. Included in the high melting range are T_{fus} = 763 K by Ehrhardt (7) and T_{fus} = 761 K by Goodwin and Kalmus (8). Knowles (12), suggested that contamination of the $PbBr_2$ by PbO probably caused the high melting temperatures. The enthalpy measurements of Ehrhardt (7) and of Goodwin and Kalmus (8) are therefore judged not to represent $PbBr_2$. More recently, Blanc and Petit (13) have reported T_{fus} = 640 K and $\Delta_{fus}H$ = 4.96 kcal mol^{-1} from a cryoscopic investigation. A visual polytherm technique was employed by Il'yasov (14) who found T_{fus} = 649 K. Cola et al. (15) determined T_{fus} = 644 K by differential thermal analysis and $\Delta_{fus}H°$ = 4.41±0.07 kcal mol^{-1} by differential scanning calorimetry. The enthalpy study by Linsey (16) gave T_{fus} = 644 K and $\Delta_{fus}H°$ = 3.93 kcal mol^{-1} (16.443 kJ mol^{-1}). We adopt these latter two values.

Sublimation Data

$\Delta_{sub}H°$(298.15 K) = 41.35 kcal mol^{-1} (173.008 kJ mol^{-1}) is the sum of the heat of fusion, the enthalpy difference of crystal and liquid between the melting point and 298.15 K, and $\Delta_{vap}H°$(298.15 K). See the ideal gas table for details.

References

1. H. Braune and F. Koref, Z. Anorg. Chem. 87, 175 (1914).
2. C. Krahmer, Z. Elektrochem. 26, 97 (1920).
3. J. Y. Cann and R. A. Sumner, J. Phys. Chem. 36, 2615 (1932).
4. W. Jahn-Held and K. Jellinek, Z. Elektrochem. 42, 401 (1936).
5. U. S. Nat. Bur. Stand. Tech. Note 270-4, 1969.
6. W. M. Latimer and H. D. Hoenshel, J. Am. Chem. Soc. 48, 19 (1926).
7. O. Ehrhardt, Wied. Ann. 24, 215 (1885).
8. H. M. Goodwin and H. T. Kalmus, Phys. Rev. 28, 1 (1909).
9. JANAF Thermochemical Tables: $PbCl_2$(cr), 6-30-73.
10. K. K. Kelley and E. G. King, U. S. Bur. Mines Bull. 592, 1961.
11. J. W. Mellor, "A Comprehensive Treatise on Inorganic and Theoretical Chemistry," Vol. VII, p. 747, Longmans, Green, and Co., New York, 1927.
12. L. M. Knowles, J. Am. Chem. Soc. 72, 4817 (1954) and J. Chem. Phys. 19, 1128 (1951).
13. M. Blanc and G. Petit, Comp. rend. 248, 1305 (1959).
14. I. I. Il'yasov, Russ. J. Inorg. Chem. 13, 866 (1968).
15. M. Cola, V. Massarotti, R. Riccardi and C. Sinistri, Z. Naturforsch. 26a, 1328 (1971).
16. C. W. Linsey, Ph.D. Dissertation, North Texas State University, 1970.

Lead Bromide ($PbBr_2$) Br_2Pb_1(cr)

Enthalpy Reference Temperature = T_r = 298.15 K — Standard State Pressure = p° = 0.1 MPa

T/K	$C_p°$ (J $K^{-1}mol^{-1}$)	S°	$-[G°-H°(T_r)]/T$	$H°-H°(T_r)$ (kJ mol^{-1})	$\Delta_f H°$	$\Delta_f G°$	Log K_f
0	0.	0.	INFINITE	-19.235	-265.247	-265.247	INFINITE
100	68.199	79.592	229.504	-14.991	-265.562	-264.454	138.136
200	76.567	129.902	168.330	-7.686	-265.682	-263.324	68.773
298.15	79.578	161.128	161.128	0.	-277.399	-260.743	45.681
300	79.613	161.620	161.129	0.147	-277.442	-260.640	45.381
400	81.253	184.747	164.272	8.190	-306.608	-248.951	32.510
500	84.684	203.229	170.271	16.479	-304.824	-234.739	24.523
600	88.784	219.026	177.110	25.149	-302.762	-220.910	19.232
644.000	90.667	225.375	180.193	29.097	------ CRYSTAL <--> LIQUID -----		
700	93.061	233.033	184.116	34.242	-305.230	-206.654	15.421
800	97.341	245.739	191.037	43.762	-302.480	-192.754	12.586
900	101.621	257.451	197.774	53.710	-299.281	-179.226	10.402
1000	105.901	268.380	204.294	64.086	-295.634	-166.078	8.675

PREVIOUS: March 1962 CURRENT: December 1973

Lead Bromide ($PbBr_2$) Br_2Pb_1(cr)

LEAD BROMIDE ($PbBr_2$) LIQUID M_r = 367.008

$S°(298.15\ K) = [173.878]\ J\ K^{-1}\ mol^{-1}$

$T_{fus} = 644 \pm 5\ K$

$\Delta_f H°(298.15\ K) = [-267.412]\ kJ\ mol^{-1}$

$\Delta_{fus}H° = 16.443 \pm 0.8\ kJ\ mol^{-1}$

Enthalpy of Formation

$\Delta_f H°(PbBr_2, \ell, 298.15\ K)$ is calculated from that of the crystal by adding $\Delta_{fus}H°$ and the difference in enthalpy, H°(644 K)-H°(298.15 K), between the crystal and liquid. $\Delta_f H°(\ell)$ can also be obtained from a 2nd and 3rd law analysis of emf measurements for the cell reaction $Pb(\ell) + Br_2(g) = PbBr_2(\ell)$. Results from three investigations are tabulated below. The auxiliary values, $\Delta_f H°(Pb, \ell, 298.15\ K) = 1.025\ kcal\ mol^{-1}$ (1) and $\Delta_f H°(Br_2, g, 298.15\ K) = 7.387\ kcal\ mol^{-1}$ (1), are used. The average result or $\Delta_f H°(\ell, 298.15\ K)$ is $-63.80\ kcal\ mol^{-1}$, in good agreement with the adopted value.

Source	Data Points	T/K	$-\Delta_r H°(298.15\ K)$, kcal mol 2nd Law	3rd Law	Drift cal K^{-1} mol^{-1}	$\Delta_f H°(298.15\ K)$, kcal mol^{-1} 2nd Law	3rd Law
Salstrom and Hildebrand (2)	16	711-849	73.60±0.10	72.10±0.13	1.9±0.1	-65.19	-63.69
Lantratov and Shevlyakova (3)	1	862		72.36			-63.95
Bloom and Welch (4)	4	723-1000	74.22±0.33	68.79±1.63	6.3±0.4	-65.80	-60.38

Heat Capacity and Entropy

Ehrhardt (5) and Goodwin and Kalmus (6) have reported enthalpy measurements in the liquid temperature range. Their data are rejected because the high melting points observed indicated the samples were not entirely $PbBr_2$. See the $PbBr_2$(cr) table discussion. Bizouard and Paulty (7) have measured 26.3±0.6 cal K^{-1} mol^{-1} for the heat capacity of the molten salt. Linsey (9) reported a heat capacity value of 26.8 cal K^{-1} mol^{-1} based on enthalpy measurements from T_{fus} to 924 K. This value is adopted and is assumed to represent the heat capacity up to 2000 K. A glass transition is assumed at 400 K below which the heat capacity is that of the crystal.

$S°(PbBr_2, \ell, 298.15\ K)$ is calculated in a manner similar to that used for the enthalpy of formation.

Vaporization Data

T_{vap} = 1185 K is calculated as the temperature at which $\Delta_r G° = 0$ for the reaction $PbBr_2(\ell) = PbBr_2(g)$ at one atm. $\Delta_{vap}H°$ = 28.23 kcal mol^{-1} (118.114 kJ mol^{-1}) is calculated as the difference between $\Delta_f H°$ at T_{vap} for the gas and the liquid. Bloom and Anthony (8) found an average value of $\Delta_{vap}H°$ = 29.8±2.0 kcal mol^{-1} from mass spectrometric studies in the 660-735 K range. The present table gives $\Delta_{vap}H°(700)$ = 34.50 kcal mol^{-1}.

References

1. JANAF Thermochemical Tables: Pb(ℓ), 3-31-62; Br_2(g), 12-31-61.
2. E. J. Salstrom and J. H. Hildebrand, J. Am. Chem. Soc. 52, 4641 (1930).
3. M. F. Lantratov and T. N. Shevlyakova, Russ. J. Inorg. Chem. 6, 95 (1961).
4. H. Bloom and B. J. Welch, Trans. Faraday Soc. 59, 410 (1963).
5. O. Ehrhardt, Wied. Ann. 24, 215 (1885).
6. H. M. Goodwin and H. T. Kalmus, Phys. Rev. 28, 1 (1909).
7. M. Bizouard and F. Pauty, Comp. Rend. 252, 51 (1961).
8. H. Bloom and R. G. Anthony, Aust. J. Chem. 24, 2001 (1971).
9. C. W. Linsey, Ph.D. Dissertation, North Texas State University, 1970.

Lead Bromide ($PbBr_2$) Br_2Pb_1(l)

Enthalpy Reference Temperature = T_r = 298.15 K Standard State Pressure = p° = 0.1 MPa

T/K	$C_p°$ (J $K^{-1}mol^{-1}$)	S°	$-[G°-H°(T_r)]/T$	$H°-H°(T_r)$ (kJ mol^{-1})	$\Delta_f H°$	$\Delta_f G°$	Log K_f
0							
100							
200							
298.15	79.580	173.878	173.878	0.	-267.412	-254.558	44.598
300	79.613	174.370	173.879	0.147	-267.455	-254.478	44.309
400	81.254	197.507	177.023	8.194	-296.617	-244.065	31.872
400.000	81.254	197.507	177.023	8.194	GLASS <--> LIQUID TRANSITION		
400.000	112.131	197.507	177.023	8.194			
500	112.131	222.528	183.715	19.407	-291.910	-231.474	24.182
600	112.131	242.972	191.939	30.620	-287.305	-219.820	19.137
644.000	112.131	250.908	195.700	35.554	CRYSTAL <--> LIQUID		
700	112.131	260.257	200.496	41.833	-287.652	-208.133	15.531
800	112.131	275.230	208.923	53.046	-283.209	-197.076	12.868
900	112.131	288.437	217.038	64.259	-278.745	-186.577	10.829
1000	112.131	300.252	224.779	75.472	-274.261	-176.576	9.223
1100	112.131	310.939	232.134	86.685	-269.755	-167.026	7.931
1200	112.131	320.696	239.113	97.899	-265.228	-157.886	6.873
1300	112.131	329.671	245.739	109.112	-260.686	-149.125	5.992
1400	112.131	337.981	252.034	120.325	-256.145	-140.714	5.250
1500	112.131	345.717	258.025	131.538	-251.615	-132.627	4.618
1600	112.131	352.954	263.734	142.751	-247.104	-124.842	4.076
1700	112.131	359.752	269.184	153.964	-242.617	-117.338	3.605
1800	112.131	366.161	274.396	165.177	-238.159	-110.098	3.195
1900	112.131	372.223	279.386	176.390	-233.735	-103.104	2.835
2000	112.131	377.975	284.173	187.604	-229.347	-96.342	2.516

PREVIOUS: March 1962 CURRENT: December 1973

Lead Bromide ($PbBr_2$) Br_2Pb_1(l)

LEAD BROMIDE ($PbBr_2$) CRYSTAL - LIQUID M_r = 367.008 Lead Bromide ($PbBr_2$) $Br_2Pb_1(cr,l)$

0 to 644 K crystal

above 644 K liquid

Refer to the individual tables for details.

Enthalpy Reference Temperature = T_r = 298.15 K Standard State Pressure = $p°$ = 0.1 MPa

T/K	$C_p°$ (J K⁻¹mol⁻¹)	$S°$ (J K⁻¹mol⁻¹)	$-[G°-H°(T_r)]/T$ (J K⁻¹mol⁻¹)	$H°-H°(T_r)$ (kJ mol⁻¹)	$\Delta_f H°$ (kJ mol⁻¹)	$\Delta_f G°$ (kJ mol⁻¹)	Log K_f
0	0.	0.	INFINITE	-19.235	-265.247	-265.247	INFINITE
100	68.199	79.592	229.504	-14.991	-265.562	-264.454	138.136
200	76.567	129.902	168.330	-7.686	-265.682	-263.324	68.773
298.15	79.578	161.128	161.128	0.	-277.399	-260.743	45.681
300	79.613	161.620	161.129	0.147	-277.442	-260.640	45.381
400	81.253	184.747	164.272	8.190	-306.608	-248.951	32.510
500	84.684	203.229	170.271	16.479	-304.824	-234.739	24.523
600	88.784	219.026	177.110	25.149	-302.762	-220.910	19.232
644.000	90.667	225.375	180.193	29.097	CRYSTAL <--> LIQUID		
644.000	112.131	250.908	180.193	45.540	TRANSITION		
700	112.131	260.257	186.229	51.820	-287.652	-208.133	15.531
800	112.131	275.230	196.439	63.033	-283.209	-197.076	12.868
900	112.131	288.437	205.942	74.246	-278.745	-186.577	10.829
1000	112.131	300.252	214.792	85.459	-274.261	-176.576	9.223
1100	112.131	310.939	223.055	96.672	-269.755	-167.026	7.931
1200	112.131	320.696	230.791	107.885	-265.228	-157.886	6.873
1300	112.131	329.671	238.057	119.098	-260.686	-149.125	5.992
1400	112.131	337.981	244.901	130.312	-256.145	-140.714	5.250
1500	112.131	345.717	251.367	141.525	-251.615	-132.627	4.618
1600	112.131	352.954	257.493	152.738	-247.104	-124.842	4.076
1700	112.131	359.752	263.310	163.951	-242.617	-117.338	3.605
1800	112.131	366.161	268.847	175.164	-238.159	-110.098	3.195
1900	112.131	372.223	274.130	186.377	-233.735	-103.104	2.835
2000	112.131	377.975	279.180	197.590	-229.347	-96.342	2.516

PREVIOUS: CURRENT: December 1973

Lead Bromide ($PbBr_2$) $Br_2Pb_1(cr,l)$

LEAD BROMIDE ($PbBr_2$) IDEAL GAS M_r = 367.008

$\Delta_f H°(0\ K) = -87.99 \pm 6.3\ kJ\ mol^{-1}$

$S°(298.15\ K) = 339.39 \pm 2.9\ J\ K^{-1}\ mol^{-1}$ $\Delta_f H°(298.15\ K) = -104.39 \pm 6.3\ kJ\ mol^{-1}$

Vibrational Freqencies and Degeneracies

ν, cm^{-1}
208 (1)
64 (1)
189 (1)

Ground State Quantum Weight : [1] σ = 2

Point Group: C_{2v}

Bond Distance: Pb-Br = 2.6 ± 0.3 Å

Bond Angle: Br-Pb-Br = [95]°

Product of Moments of Inertia: $I_A I_B I_C = [6.47916 \times 10^{-112}]\ g^3\ cm^6$

Enthalpy of Formation

$\Delta_f H°(298.15\ K) = -24.95 \pm 1.5\ kcal\ mol^{-1}$ ($-104.391\ kJ\ mol^{-1}$) is the sum of $\Delta_f H°(\ell, 298.15\ K) = -63.913 \pm 0.8\ kcal\ mol^{-1}$ and $\Delta_{vap}H°(\ell, 298.15\ K) = 39.96\ kcal\ mol^{-1}$. The adopted enthaply of vaporization is from the 2nd and 3rd law analyses, tabulated below, of seven sets of liquid vapor pressure measurements, $PbBr_2(\ell) = PbBr_2(g)$. The overall average for $\Delta_{vap}H°(298.15\ K)$ is 39.23 kcal mol^{-1}. The average value of the 3rd law $\Delta_{vap}H°(298.15\ K)$ is 38.96 kcal mol^{-1}. This is in good agreement with both the 2nd and 3rd law values from the data of Wartenberg and Bosse (1). The adopted enthalpy of formation for $PbBr_2(g)$ yields an atomization energy of $\Delta_{at}H°$ = 5.42 eV for the process $PbBr_2(g) = Pb(g) + 2\ Br(g)$.

Source	Data Points	T/K	$\Delta_{vap}H°(298.15\ K)$, kcal mol^{-1} 2nd law	3rd law	Drift cal K^{-1} mol^{-1}
Wartenberg and Bosse (1)	7[a]	1008-1191	38.05±0.19	39.02±0.12	0.9±0.2
Volmer (2)	11	684-841	39.36±0.11	38.93±0.08	-0.6±0.1
Greiner and Jellinek (3)	1	1043		38.90	
Jahn-Held and Jellinek (4)	3	1045-1153	41.85±6.54	38.68±0.73	-2.9±5.9
Bloom, et al. (5)	Equation	796-1133	39.56±0.48	38.81±0.30	-0.8±0.5
Murgulescu and Marta (6)	1	973		39.13	
Bloom and Hastie (7)	1	973		39.27	

[a]One point rejected by statistical test.

Heat Capacity and Entropy

Molecular dimensions are those given by Sutton (8). Beattie and Perry (9) observed the frequencies $\nu_1 = 200\ cm^{-1}$ and $\nu_2 = 64\ cm^{-1}$ in a gas phase Raman study of $PbBr_2$ in the presence of excess bromine; the $PbBr_2$ spectrum was superimposed on the resonance fluorescence spectrum of Br_2. The matrix isolation laser Raman spectroscopic study by Ozin and Vander Voet (10) gave $\nu_1 = 208\ cm^{-1}$ and $\nu_3 = 189\ cm^{-1}$. Experimental conditions prevented observations below 90 cm^{-1} so ν_2 expected about 65 cm^{-1} was not confirmed. The principal moments of inertia are: $I_A = 46.2255 \times 10^{-39}$, $I_B = 97.5131 \times 10^{-39}$, and $I_C = 143.7386 \times 10^{-39}\ g\ cm^2$.

References

1. H. von Wartenberg and O. Bosse, Z. Elektrochem. 28, 384 (1922).
2. F. Volmer, Phys. Z. 30, 590 (1929).
3. B. Greiner and K. Jellinek, Z. Physik. Chem. A165, 97 (1933).
4. W. Jahn-Held and K. Jellinek, Z. Elektrochem. 43, 491 (1937).
5. H. Bloom, J. O'M. Bockris, N. E. Richards, and R. G. Taylor, J. Amer. Chem. Soc. 80, 2044 (1958).
6. I. G. Murgulescu and L. Marta, Rev. Chim., Acad. Rep. Populaire Roumaine 6, 317 (1961).
7. H. Bloom and J. W. Hastie, Aust. J. Chem. 21, 583 (1968).
8. L. E. Sutton, Ed., "Tables of Interatomic Distances and Configuration in Molecules and Ions," The Chemical Society, London, 1958.
9. I. R. Beattie and R. O. Perry, J. Chem. Soc. A1970, 2428 (1970).
10. G. A. Ozin and A. Vander Voet, J. Chem. Phys. 56, 4768 (1972).

Lead Bromide ($PbBr_2$) $Br_2Pb_1(g)$

Enthalpy Reference Temperature = T_r = 298.15 K Standard State Pressure = p° = 0.1 MPa

T/K	$C_p°$ (J $K^{-1}mol^{-1}$)	S°	$-[G°-H°(T_r)]/T$	$H°-H°(T_r)$ (kJ mol^{-1})	$\Delta_f H°$	$\Delta_f G°$	Log K_f
0	0.	0.	INFINITE	-14.989	-87.992	-87.992	INFINITE
100	49.793	280.145	388.803	-10.866	-88.428	-107.375	56.087
200	55.493	316.912	344.558	-5.529	-90.518	-125.561	32.793
250	56.408	329.404	340.321	-2.729	-91.839	-134.174	28.034
298.15	56.918	339.386	339.386	0.	-104.391	-140.883	24.682
300	56.933	339.739	339.388	0.105	-104.475	-141.109	24.569
350	57.259	348.541	340.082	2.961	-135.624	-145.497	21.714
400	57.474	356.202	341.629	5.829	-135.960	-146.885	19.181
450	57.624	362.981	343.632	8.707	-136.320	-148.229	17.206
500	57.732	369.058	345.876	11.591	-136.704	-149.532	15.622
600	57.873	379.597	350.644	17.372	-137.531	-152.022	13.235
700	57.960	388.525	355.434	23.164	-143.300	-153.568	11.459
800	58.016	396.269	360.065	28.963	-144.271	-154.968	10.118
900	58.054	403.104	364.475	34.766	-145.216	-156.248	9.068
1000	58.082	409.222	368.649	40.573	-146.138	-157.425	8.223
1100	58.103	414.759	372.593	46.383	-147.036	-158.509	7.527
1200	58.118	419.816	376.321	52.194	-147.911	-159.514	6.943
1300	58.131	424.468	379.848	58.006	-148.770	-160.446	6.447
1400	58.140	428.776	383.191	63.820	-149.628	-161.312	6.019
1500	58.148	432.788	386.365	69.634	-150.497	-162.116	5.645
1600	58.154	436.541	389.385	75.449	-151.384	-162.862	5.317
1700	58.160	440.067	392.264	81.265	-152.294	-163.551	5.025
1800	58.164	443.391	395.013	87.081	-153.234	-164.187	4.765
1900	58.168	446.536	397.642	92.898	-154.206	-164.769	4.530
2000	58.171	449.520	400.162	98.715	-155.215	-165.299	4.317
2100	58.174	452.358	402.581	104.532	-333.483	-158.662	3.947
2200	58.176	455.064	404.905	110.350	-334.182	-150.321	3.569
2300	58.178	457.650	407.143	116.167	-334.991	-141.946	3.224
2400	58.180	460.126	409.299	121.985	-335.912	-133.534	2.906
2500	58.182	462.501	411.380	127.803	-336.947	-125.080	2.613
2600	58.183	464.783	413.390	133.622	-338.096	-116.584	2.342
2700	58.185	466.979	415.335	139.440	-339.354	-108.040	2.090
2800	58.186	469.095	417.217	145.258	-340.719	-99.449	1.855
2900	58.187	471.137	419.042	151.077	-342.183	-90.806	1.636
3000	58.188	473.110	420.811	156.896	-343.740	-82.112	1.430
3100	58.188	475.018	422.529	162.715	-345.382	-73.364	1.236
3200	58.189	476.865	424.198	168.534	-347.100	-64.562	1.054
3300	58.190	478.656	425.822	174.352	-348.884	-55.705	0.882
3400	58.191	480.393	427.401	180.172	-350.726	-46.793	0.719
3500	58.191	482.080	428.940	185.991	-352.614	-37.827	0.565
3600	58.192	483.719	430.439	191.810	-354.541	-28.806	0.418
3700	58.192	485.313	431.900	197.629	-356.496	-19.730	0.279
3800	58.193	486.865	433.326	203.448	-358.471	-10.603	0.146
3900	58.193	488.377	434.719	209.267	-360.459	-1.422	0.019
4000	58.193	489.850	436.079	215.087	-362.450	7.809	-0.102
4100	58.194	491.287	437.408	220.906	-364.438	17.090	-0.218
4200	58.194	492.689	438.707	226.726	-366.418	26.420	-0.329
4300	58.194	494.059	439.979	232.545	-368.382	35.796	-0.435
4400	58.195	495.397	441.223	238.364	-370.327	45.218	-0.537
4500	58.195	496.705	442.441	244.184	-372.247	54.684	-0.635
4600	58.195	497.984	443.635	250.003	-374.139	64.193	-0.729
4700	58.195	499.235	444.805	255.823	-376.000	73.742	-0.820
4800	58.196	500.460	445.951	261.643	-377.826	83.330	-0.907
4900	58.196	501.660	447.076	267.462	-379.616	92.956	-0.991
5000	58.196	502.836	448.180	273.282	-381.368	102.618	-1.072
5100	58.196	503.988	449.263	279.101	-383.081	112.315	-1.150
5200	58.196	505.119	450.326	284.921	-384.754	122.045	-1.226
5300	58.197	506.227	451.370	290.741	-386.386	131.807	-1.299
5400	58.197	507.315	452.396	296.560	-387.977	141.599	-1.370
5500	58.197	508.383	453.405	302.380	-389.527	151.421	-1.438
5600	58.197	509.431	454.396	308.200	-391.036	161.269	-1.504
5700	58.197	510.461	455.370	314.019	-392.506	171.145	-1.568
5800	58.197	511.474	456.329	319.839	-393.937	181.046	-1.630
5900	58.197	512.468	457.272	325.659	-395.328	190.972	-1.691
6000	58.198	513.447	458.200	331.479	-396.682	200.920	-1.749

PREVIOUS: December 1973 (1 atm) CURRENT: December 1973 (1 bar)

Lead Bromide ($PbBr_2$) $Br_2Pb_1(g)$

$S°(298.15\ K)$ = [305.22 ± 2.13] J K^{-1} mol^{-1}

$\Delta_f H°(0\ K)$ = -37.9 ± 16.7 kJ mol^{-1}
$\Delta_f H°(298.15\ K)$ = -52.3 ± 16.7 kJ mol^{-1}

Electronic Levels and Quantum Weights

State	ε_i, cm^{-1}	g_i
[$^1A_1$]	0	[1]
[$^3B_1$]	[20000]	[3]

Vibrational Frequencies and Degeneracies

ν, cm^{-1}
402 (1)
[120](1)
398.9(1)

Point Group: C_{2v} σ = 2
Bond Distance: Si-Br - [2.19] Å
Bond Angle: Br-Si-Br = 109° ± 3°
Product of the Moments of Inertia: $I_AI_BI_C$ = [4.91229 x 10^{-113}] g^3 cm^6

Enthalpy of Formation

The equilibrium reaction Si(cr) + $SiBr_4$(g) = 2 $SiBr_2$(g) was studied by Schafer et al. (1) and Wolf and Herbst (2). Both investigators used a flow technique in an argon atmosphere. A 2nd and 3rd law analysis of their results (reported only in equation form) is summarized below. As in the case of $SiCl_2$ (3), there is acceptable agreement between these two studies. Using the mean $\Delta_r H°(298.15\ K)$ from the 3rd law results and auxiliary data (3), we calculate and adopt $\Delta_f H°(298.15\ K)$ = -12.5 kcal mol^{-1} (52.300 kJ mol^{-1}) for $SiBr_2$(g).

Source	T/K	$\Delta_r H°(298.15\ K)$, kcal mol^{-1} 2nd law	3rd law	Drift cal K^{-1} mol^{-1}
Schafer et al. (1)	1320-1475 K	71.61	73.26	1.18
Wolf and Herbst (2)	1200-1550 K	77.33	75.21	-1.54

The adopted enthalpy of formation leads to a $\Delta_{at}H°$ value which implies that the average bond energy is roughly 10% larger in $SiBr_2$(g) than in SiBr(g) or $SiBr_4$(g) (3).

Heat Capacity and Entropy

Maass, Hauge, and Margrave (4) observed the infrared spectra of $SiBr_2$ in nitrogen and argon matrices in the Si-Br stretching region. Attempts to observe the bending fundamental in the region between 90 cm^{-1} and 150 cm^{-1} were unsuccessful. Maass et al. (4) measured and assigned ν_1 and ν_3. The bond angle was calculated to be 109±3° from the isotopic splittings due to three silicon isotopes. The authors also assumed ν_2 = 120 cm^{-1}. This bending frequency is consistent with the analogous $SiBr_2$ "bending mode" in SiH_2Br_2 (3). We adopt the frequencies and bond angles as suggested by Maass et al. (4).

We assume the Si-Br bond distance to be the same as that in SiH_2Br_2 (3). By observing the trends in SiF_2 and $SiCl_2$ (3), we assume a $^1A_1$ ground electronic state and a $^3B_1$ excited state at ~20000 cm^{-1}. Additional support for this assignment comes from the luminescense spectrum of a glow discharge in $SiBr_4$ vapors observed by Kuznetsova and Kuzyokov (5). They states that the complex spectral structure of the 16800-23500 cm^{-1} band could be attributed to the oscillation and deformation of the nonlinear $SiBr_2$ molecule. The inclusion of this triplet electronic state increases the entropy by 0.0004 cal K^{-1} mol^{-1} at 3000K. The principal moments of inertia are: I_A = 6.4154 x 10^{-39}, I_B = 84.3558 x 10^{-39}, and I_C = 90.7711 x 10^{-39}g cm^2.

References
1. H. Schafer, H. Bruderreck and B. Morcher, Z. Anorg. Allg. Chem. 352, 122 (1967).
2. E. Wolf and C. Herbst, Z. Anorg. Allg. Chem. 347, 113 (1966).
3. JANAF Thermochemical Tables: $SiBr_4$(g), SiBr(g), and SiH_2Br_2(g), 12-31-76; Si(g), 3-31-67; Br(g), 6-30-74; $SiCl_2$(g) and SiF_2(g), 12-31-77.
4. G. Maass, R. H. Hauge and J. L. Margrave, Z. Anorg. Allg. Chem. 392, 295 (1972).
5. L. A. Kusnetsova and Yu. Ya. Kuzyokov, Zh. Prikl Spektrosk. 10, 413 (1969).

Enthalpy Reference Temperature = T_r = 298.15 K Standard State Pressure = p° = 0.1 MPa

T/K	$C_p°$ (J $K^{-1}mol^{-1}$)	S°	$-[G°-H°(T_r)]/T$	$H°-H°(T_r)$ (kJ mol^{-1})	$\Delta_f H°$	$\Delta_f G°$	Log K_f
0	0.	0.	INFINITE	-13.368	-37.941	-37.941	INFINITE
100	41.529	252.970	349.983	-9.701	-37.333	-56.862	29.702
200	49.762	284.529	310.029	-5.100	-38.812	-75.900	19.823
250	52.126	295.906	306.100	-2.549	-39.926	-85.050	17.770
298.15	53.637	305.225	305.225	0.	-52.300	-92.311	16.173
300	53.684	305.557	305.226	0.099	-52.378	-92.559	16.116
350	54.738	313.916	305.884	2.811	-83.351	-97.553	14.559
400	55.473	321.276	307.357	5.568	-83.511	-99.571	13.003
450	56.003	327.843	309.275	8.355	-83.689	-101.568	11.790
500	56.395	333.764	311.433	11.166	-83.882	-103.545	10.817
600	56.923	344.097	316.040	16.834	-84.310	-107.438	9.353
700	57.251	352.898	320.693	22.544	-84.786	-111.255	8.302
800	57.468	360.558	325.208	28.280	-85.311	-115.001	7.509
900	57.619	367.336	329.519	34.035	-85.882	-118.679	6.888
1000	57.728	373.413	333.610	39.803	-86.499	-122.290	6.388
1100	57.809	378.919	337.483	45.580	-87.162	-125.838	5.976
1200	57.871	383.952	341.148	51.364	-87.868	-129.323	5.629
1300	57.919	388.586	344.621	57.154	-88.619	-132.747	5.334
1400	57.958	392.879	347.917	62.947	-89.417	-136.112	5.078
1500	57.989	396.879	351.049	68.745	-90.263	-139.418	4.855
1600	58.014	400.623	354.032	74.545	-91.158	-142.666	4.658
1700	58.036	404.140	356.877	80.348	-142.281	-145.410	4.468
1800	58.054	407.458	359.596	86.152	-143.051	-145.572	4.224
1900	58.070	410.597	362.198	91.958	-143.833	-145.691	4.005
2000	58.084	413.576	364.693	97.766	-144.630	-145.768	3.807
2100	58.098	416.411	367.089	103.575	-145.445	-145.805	3.627
2200	58.111	419.114	369.393	109.385	-146.278	-145.803	3.462
2300	58.125	421.697	371.611	115.197	-147.131	-145.762	3.310
2400	58.140	424.171	373.750	121.010	-148.005	-145.684	3.171
2500	58.157	426.545	375.815	126.825	-148.900	-145.569	3.041
2600	58.178	428.826	377.810	132.642	-149.816	-145.418	2.921
2700	58.202	431.022	379.740	138.461	-150.752	-145.231	2.810
2800	58.230	433.139	381.610	144.283	-151.707	-145.009	2.705
2900	58.264	435.183	383.422	150.107	-152.679	-144.752	2.607
3000	58.304	437.159	385.181	155.936	-153.665	-144.462	2.515
3100	58.351	439.072	386.888	161.768	-154.664	-144.139	2.429
3200	58.405	440.925	388.548	167.606	-155.672	-143.784	2.347
3300	58.466	442.723	390.163	173.449	-156.685	-143.396	2.270
3400	58.536	444.470	391.735	179.299	-157.700	-142.979	2.197
3500	58.614	446.168	393.266	185.157	-158.715	-142.531	2.127
3600	58.701	447.820	394.758	191.023	-543.876	-131.593	1.909
3700	58.797	449.430	396.214	196.897	-544.464	-120.133	1.696
3800	58.901	450.999	397.635	202.782	-545.041	-108.657	1.494
3900	59.014	452.530	399.023	208.678	-545.605	-97.166	1.301
4000	59.136	454.026	400.380	214.585	-546.152	-85.660	1.119
4100	59.266	455.488	401.706	220.505	-546.678	-74.141	0.945
4200	59.405	456.918	403.004	226.439	-547.181	-62.610	0.779
4300	59.551	458.317	404.274	232.387	-547.657	-51.067	0.620
4400	59.704	459.688	405.518	238.349	-548.104	-39.513	0.469
4500	59.864	461.031	406.736	244.328	-548.519	-27.949	0.324
4600	60.031	462.349	407.931	250.322	-548.901	-16.377	0.186
4700	60.203	463.642	409.103	256.334	-549.248	-4.797	0.053
4800	60.380	464.911	410.252	262.363	-549.557	6.791	-0.074
4900	60.562	466.158	411.381	268.410	-549.829	18.384	-0.196
5000	60.749	467.383	412.488	274.476	-550.061	29.983	-0.313
5100	60.938	468.588	413.577	280.560	-550.253	41.586	-0.426
5200	61.131	469.773	414.646	286.663	-550.405	53.192	-0.534
5300	61.325	470.940	415.697	292.786	-550.515	64.801	-0.639
5400	61.522	472.088	416.731	298.929	-550.584	76.412	-0.739
5500	61.719	473.219	417.748	305.091	-550.611	88.023	-0.836
5600	61.917	474.332	418.748	311.272	-550.597	99.634	-0.929
5700	62.114	475.430	419.733	317.474	-550.542	111.245	-1.019
5800	62.311	476.512	420.703	323.695	-550.446	122.854	-1.106
5900	62.507	477.579	421.658	329.936	-550.311	134.462	-1.190
6000	62.702	478.631	422.598	336.197	-550.135	146.067	-1.272

PREVIOUS: December 1977 (1 atm) CURRENT: December 1977 (1 bar)

STRONTIUM BROMIDE ($SrBr_2$) CRYSTAL M_r = 247.428

$S°(298.15\ K) = 143.4 \pm 4.2\ J\ K^{-1}\ mol^{-1}$

T_{trs} = 918 K

T_{fus} = 930 K

$\Delta_fH°(0\ K) = -704.96 \pm 1.7\ kJ\ mol^{-1}$

$\Delta_fH°(298.15\ K) = -717.97 \pm 1.7\ kJ\ mol^{-1}$

$\Delta_{trs}H° = 12.217 \pm 0.21\ kJ\ mol^{-1}$

$\Delta_{fus}H° = 10.125 \pm 0.21\ kJ\ mol^{-1}$

Enthalpy of Formation

The selected value for $\Delta_fH°$ is derived from results of solution calorimetric measurements in aqueous acids performed by Ehrlich et al. (1). Their results on the enthalpy of solution of Sr(cr) and $SrBr_2$(cr) in HBr·555 H_2O are combined with data for HBr(aq) (2, 3) in a thermochemical cycle to give $\Delta_fH°(SrBr_2$, cr, 298.15 K) = -171.6±0.4 kcal mol^{-1} (-717.974 ± 1.7 kJ mol^{-1}). In the same paper, Ehrlich et al. (1) reported similar results for three (Ca, Sr, and Ba) of the alkaline earth dichlorides; these results have been the basis for adopted JANAF enthalpies of formation (4) for these dichlorides.

Independent values for $\Delta_fH°$ can be obtained from measurements of the enthalpy of solution of $SrBr_2$(cr) in aqueous solution (5-7) as summarized by Bichowsky and Rossini (8). Combining these results with $\Delta_fH°(Sr^{+2}$, ∞ aq, 298.15 K) = -130.45 kcal mol^{-1} (9) and $\Delta_fH°(Br^-$, ∞ aq, 298.15 K) = -29.039±0.036 kcal mol^{-1} (2), we derive $\Delta_fH°$ values for $SrBr_2$(cr) in kcal mol^{-1} of -171.5 (5), -172.2 (6), and -171.6 (7). Deviations from our adopted value are at worst only 0.6 kcal mol^{-1}, and the results of two studies (5, 7) provide additional support for the selected value of $\Delta_fH°$. Also, our value for $\Delta_fH°$ is essentially that (-171.5 kcal mol^{-1}) selected by NBS (9).

Heat Capacity and Entropy

$C_p°$ data below 300 K are based primarily on the adiabatic calorimetry (60-302 K) of Taylor and Smith (10]. These $C_p°$ data show an unusual leveling off at near 18.0 cal K^{-1} mol^{-1} above 245 K. We adopt their $C_p°$ data in the temperature range 60-245 K; values above 245 K are estimated by comparison with similar data for $SrCl_2$ and $BaCl_2$ (4]. Our value for $C_p°$ at 298.15 K (18.37 cal K^{-1} mol^{-1}) is roughly 0.4 cal K^{-1} mol^{-1} higher than the measured value of Taylor and Smith (10]. Huttig and Slonim (11] have reported mean heat capacity values over three temperature intervals (85-198, 196-271, and 276-368 K). Their measurements suggest a slightly higher value (18.8 cal K^{-1} mol^{-1}) for $C_p°$ at 298.15 K than is adopted.

Taylor and Smith (10] reported S°(298.15 K) = 32.29 cal K^{-1} mol^{-1} based on S°(60 K) = 6.68 cal K^{-1} mol^{-1} obtained from Debye-Einstein functions which represented their $C_p°$ data to only ±1.8 percent from 60 to 100 K. A comparison of their extrapolated $C_p°$ data with those which have been measured for $SrCl_2$, $BaCl_2$, and CaI_2 (4] indicates that the values for $SrBr_2$ decrease much more rapidly with temperature below 50 K than would be expected. We have made our own extrapolation to 0 K for $SrBr_2$ by comparison with the measured data for $SrCl_2$ (7.4 to 60 K), $BaCl_2$(6 to 60 K), and CaI (13.1 to 60 K). Our extrapolation gives S°(60 K) = 8.7 cal K^{-1} mol^{-1}, or S°(298.15 K) = 34.28 cal K^{-1} mol^{-1} (143.438 J $K^{-1}mol^{-1}$) based on the adopted $C_p°$ data (60-300 K). Our value for S°(298.15 K) is consistent with the estimates of 35 cal K^{-1} mol^{-1} (12] and 33.8 cal K^{-1} mol^{-1} (13], suggesting a possible uncertainty of ±1 cal K^{-1} mol^{-1} in the adopted value.

Taylor and Smith (10] measured relative enthalpy data (293-902 K) on a portion of the same sample used for their $C_p°$ measurements. This sample was reported to contain less than 0.1% oxide and was investigated in a Bunsen ice calorimeter. Our analysis of their relative enthalpies by curve fitting techniques reveals the existence of considerable scatter in the data, and no weight is given to their results. The average deviation of 36 points in a Shomate type fit is 2.5%; the maximum deviation is -9.3% at 397 K. Also, these workers (10] were apparently unaware of the transition in $SrBr_2$ at 918 K, since their data set contains only two enthalpy points between 902-927 K. Relative enthalpies have also been measured by Dworkin and Bredig (14], and the results were reported in graphical form. The short temperature interval (846-902 K) over which these measurements were made for the α phase preclude derivation of accurate heat capacities. Dworkin and Bredig (14) also reported in the same paper enthalpy data for three (Ca, Sr, and Ba) of the alkaline earth dichlorides. A comparison of these enthalpies with those adopted by JANAF (4) indicates that their results are probably reliable to better than ±2%. We adopt their measured value for H°(918;298.15) = 12.37 kcal mol^{-1} of the α phase at the transition temperature. $C_p°$ values (300-918 K) are then estimated by comparison with those for $SrCl_2$ and $BaCl_2$ (4). These estimates are made so as to reproduce as closely as possible the adopted enthalpy at 918 K. Our estimated $C_p°$ data give a value of H°(918;298.15) = 12.35 kcal mol^{-1} which agrees with the measured value of Dworkin and Bredig to within 20 calories. The enthalpies of Taylor and Smith (10) are roughly 2.5% higher than the adopted values at temperatures near the transition. $C_p°$ data for the β phase (918-930 K) are taken from Dworkin and Bredig (14) and extrapolated above the melting point.

Transition Data

Existence of two forms of $SrBr_2$ has been shown by thermal analysis (15-18) and drop calorimetry (14). Values of T_{trs} reported are 915 K (15-16), 920 K (17), 919 K (18), and 918 K (14). We adopt the last value which is based on the drop calorimetry of Dworkin and Bredig (14). Several other investigators (10, 19-21) have mistakenly interpreted the transition as due to melting.

$\Delta_{trs}H°$ of 2.92 kcal mol^{-1} (12.217 kJ mol^{-1}) is derived from the enthalpy data of Dworkin and Bredig (14).

Fusion Data

See the liquid table for details.

Sublimation Data

$\Delta_{sub}H°$(298.15 K) is calculated as the difference in the adopted gas and crystal enthalpies of formation at 298.15 K.

References

The references are listed on the liquid table.

Strontium Bromide ($SrBr_2$) Br_2Sr_1(cr)

Enthalpy Reference Temperature = T_r = 298.15 K Standard State Pressure = p° = 0.1 MPa

	J K⁻¹mol⁻¹			kJ mol⁻¹			
T/K	$C_p°$	S°	$-[G°-H°(T_r)]/T$	$H°-H°(T_r)$	$\Delta_fH°$	$\Delta_fG°$	Log K_f
0	0.	0.	INFINITE	-18.064	-704.960	-704.960	INFINITE
100	64.337	64.952	209.316	-14.436	-705.626	-703.931	367.696
200	73.645	113.388	150.374	-7.397	-705.976	-702.127	183.377
298.15	76.853	143.438	143.438	0.	-717.974	-698.755	122.419
300	76.902	143.913	143.439	0.142	-718.022	-698.636	121.643
400	79.036	166.337	146.482	7.942	-747.431	-686.047	89.589
500	80.835	184.166	152.295	15.935	-745.957	-670.871	70.085
600	82.676	199.060	158.881	24.107	-744.434	-655.995	57.109
700	84.893	211.966	165.563	32.482	-742.832	-641.381	47.860
800	87.571	223.468	172.094	41.099	-741.107	-627.004	40.939
900	91.002	233.971	178.394	50.020	-739.825	-612.781	35.565
918.000	91.717	235.781	179.501	51.664	ALPHA <--> BETA TRANSITION		
918.000	115.060	249.098	179.501	63.890			
930.000	115.060	250.593	180.409	65.271	------- BETA <--> LIQUID -------		
1000	115.060	258.943	185.617	73.325	-723.327	-599.967	31.339
1100	115.060	269.909	192.790	84.831	-726.540	-587.492	27.898
1200	115.060	279.920	199.640	96.337	-722.773	-575.018	25.030
1300	115.060	289.130	206.174	107.843	-719.013	-562.857	22.616
1400	115.060	297.657	212.408	119.349	-715.263	-550.986	20.558
1500	115.060	305.595	218.359	130.855	-711.522	-539.383	18.783

PREVIOUS: CURRENT: June 1974

Strontium Bromide ($SrBr_2$) Br_2Sr_1(cr)

STRONTIUM BROMIDE ($SrBr_2$) LIQUID M_r = 247.428

$S°(298.15\ K)$ = [154.843] J K^{-1} mol^{-1} $\Delta_f H°(298.15\ K)$ = [-705.098] kJ mol^{-1}

T_{fus} = 930 K $\Delta_{fus}H°$ = 10.125 ± 0.21 kJ mol^{-1}

Enthalpy of Formation

$\Delta_f H°(SrBr_2, \ell)$ is calculated from that of the crystal by adding $\Delta_{fus}H°$ and the difference in enthalpy, H°(930 K)-H°(298.15 K), betweeen the crystal and liquid.

Heat Capacity and Entropy

$C_p°$ for the liquid in the temperature range 600-3000 K is assumed constant at 27.818 cal K^{-1} mol^{-1}. This value is based on JANAF curve fits (deviations ± 0.1%) of the relative enthalpies (931-1002 K) reported by Dworkin and Bredig (1). Taylor and Smith (2) have also reported enthalpies (927-1118 K) for the liquid. These results suggest a somewhat higher value (28.36 cal K^{-1} mol^{-1}) for $C_p°$; however, the data are considered less reliable. (See $SrBr_2$(cr) table for further discussion). The average deviation of the data of Taylor and Smith (2) from our adopted enthalpies is ±0.4%; the maximum deviation is 1.2% at 1067 K. A glass transition is assumed at 600 K below which $C_p°$ is that of the crystal.

S°(298.15 K) is calculated in a manner analogous to that used for the enthalpy of formation.

Fusion Data

T_{fus} is that observed by Dworkin and Bredig (1) from drop calorimetry. Other reported values for T_{fus} are 926 K (3), 931 K (4), 924 K (5), and 928 K (6). Four other investigators (2, 7-10) have reported values of T_{fus} near 916 K, but they have mistakenly interpreted the transition in $SrBr_2$ as that due to melting.

$\Delta_{fus}H°$ is calculated as the difference in the enthalpies of the liquid and β phase at T_{fus}. The enthalpy of the liquid at T_{fus} is based on JANAF curve fits of the enthalpy data reported by Dworkin and Bredig (1). We adopted their measured value of 15.6 kcal mol^{-1} for H°(930;298.15) of the β phase at T_{fus}. A cryoscopic (4) determination of $\Delta_{fus}H°$ gave 2.6±1.3 kcal mol^{-1} which agrees well with the adopted calorimetric value. Taylor and Smith (2) reported a calorimetric value of $\Delta_{fus}H°$ = 5.25 kcal mol^{-1}; however, this value also includes the heat of transition. Addition of our values for $\Delta_{trs}H°$ and $\Delta_{vap}H°$ gives 5.34 kcal mol^{-1}.

Vaporization Data

T_{vap} is the temperature at which $\Delta_r G°$ for the process $SrBr_2(\ell) = SrBr_2(g)$ approaches zero. $\Delta_{vap}H°$ is the difference in the heats of formation of the gas and liquid at T_{vap}. Peterson and Hutchinson (10) obtained T_{vap} = 2318 K from vapor pressure measurements on the liquid (1114-1304 K). However, the value is based on a rather long extrapolation and is probably uncertain to at least ± 100 K.

References - $SrBr_2(l)$

1. A. S. Dworkin and M. A. Bredig, J. Phys. Chem. 67, 697 (1963).
2. A. R. Taylor, Jr., and D. F. Smith, U.S. Bur. Mines RI 5967, 1962.
3. E. D. Eastman, D. D. Cubicciotti and C. D. Thurmond in "Chemistry and Metallurgy of Miscellaneous Materials: Thermodynamics," L. L. Quill, Ed., National Nuclear Energy Series IV-19B, McGraw-Hill Book Co., New York, 1950.
4. H. H. Emons and B. Loffelholz, Wiss. Z. Tech. Hochsch. Chem., Leuna-Merseburg 6, 261 (1964).
5. H. Schinke and F. Sauerwald, Z. Anorg. Chem. 304, 25 (1960).
6. R. Riccardi, C. Sinistri, G. V. Campari, and A. Magistris, Z. Naturforsch. 25a, 781 (1970).
7. H. H. Emons, Z. Anorg. Chem. 323, 114 (1963).
8. G. Kellner, J. Chem. Soc. 112, 469 (1917).
9. G. Kellner, Z. Anorg. Chem. 99, 137 (1917).
10. D. T. Peterson and J. F. Hutchison, J. Chem. Eng. Data 15, 320 (1970).

References - $SrBr_2$(cr)

1. P. Ehrlich, K. Peik, and E. Koch, Z. Anorg. Chem. 324, 113 (1963).
2. ICSU-CODATA Task Group, J. Chem. Thermodynamics 4, 331 (1972).
3. U.S. Nat. Bur. Stand. Tech. Note 270-3, 1968.
4. JANAF Thermochemical Tables: $SrCl_2$(cr) and $BaCl_2$(cr), 12-31-72; $CaCl_2$(cr), 6-30-70; CaI_2(cr), 6-30-74.
5. J. Thomsen, "Thermochemische Untersuchungen," Vol. III, Barth, Zeipzig, 1883.
6. E. Tassilly, Ann. Chim. Phys. 17, 38 (1899).
7. G. F. Huttig and Ch. Slonim, Z. Anorg. Chem. 181, 65 (1929).
8. F. R. Bichowsky and F. D. Rossini, "The Thermochemistry of the Chemical Substances," Reinhold Publishing Corporation, New York, 1936.
9. U.S. Nat. Bur. Stand. Tech. Note 270-6, 1971.
10. A. R. Taylor, Jr., and D. F. Smith, U.S. Bur. Mines RI 5967, 1962.
11. G. F. Huttig and Ch. Slonim, Z. Anorg. Chem. 181, 74 (1929).
12. K. K. Kelley and E. G. King, U.S. Bur. Mines Bull. 592, 1961.
13. W. M. Latimer, "The Oxidation States of the Elements and Their Potentials in Aqueous Solutions," 2nd ed., Prentice-Hall, Englewood Cliffs, N.J., 1952.
14. A. S. Dworkin and M. A. Bredig, J. Phys. Chem. 67, 697 (1963).
15. E. D. Eastman, D. D. Cubicciotti and C. D. Thurmond in "Chemistry and Metallurgy of Miscellaneous Materials: Thermodyanics," L. L. Quill, Ed., National Nuclear Energy Series IV-19B, McGraw-Hill Book Co., New York, 1950.
16. H. H. Emons and B. Loffelholz, Wiss. Z. Tech. Hochsch. Chem., Leuna-Merseburg 6, 261 (1964).
17. H. Schinke and F. Sauerwald, Z. Anorg. Chem. 304, 25 (1960).
18. R. Riccardi, C. Sinistri, G. V. Campari, and A. Magistris Z. Naturforsch. 25a, 781 (1970).
19. G. Kellner, Z. Anorg. Chem. 99, 137 (1917).
20. H. H. Emons, Z. Anorg. Chem. 323, 114 (1963).
21. G. Kellner, J. Chem. Soc. 112, 469 (1917).

Strontium Bromide ($SrBr_2$) $Br_2Sr_1(l)$

Enthalpy Reference Temperature = T_r = 298.15 K Standard State Pressure = p° = 0.1 MPa

	J K^{-1} mol^{-1}			kJ mol^{-1}			
T/K	$C_p°$	S°	$-[G°-H°(T_r)]/T$	$H°-H°(T_r)$	$\Delta_f H°$	$\Delta_f G°$	Log K_f
0							
100							
200							
298.15	76.852	154.843	154.843	0.	-705.097	-689.278	120.759
300	76.902	155.319	154.845	0.142	-705.144	-689.180	119.997
400	79.036	177.742	157.888	7.942	-734.554	-677.732	88.503
500	80.835	195.571	163.701	15.935	-733.080	-663.696	69.336
600	82.678	210.471	170.287	24.110	-731.554	-649.961	56.584
600.000	82.678	210.471	170.287	24.110	GLASS <--> LIQUID	TRANSITION	
600.000	116.391	210.471	170.287	24.110			
700	116.391	228.413	177.342	35.749	-726.687	-636.749	47.515
800	116.391	243.955	184.719	47.388	-721.940	-624.226	40.758
900	116.391	257.664	192.077	59.027	-717.940	-612.219	35.532
930.000	116.391	261.480	194.255	62.519	------- BETA <--> LIQUID -------		
1000	116.391	269.926	199.260	70.667	-713.108	-600.732	31.379
1100	116.391	281.020	206.196	82.306	-716.188	-589.362	27.986
1200	116.391	291.147	212.860	93.945	-712.288	-578.004	25.160
1300	116.391	300.463	219.245	105.584	-708.395	-566.972	22.781
1400	116.391	309.089	225.358	117.223	-704.512	-556.239	20.754
1500	116.391	317.119	231.211	128.862	-700.638	-545.784	19.006
1600	116.391	324.630	236.817	140.501	-696.774	-535.586	17.485
1700	116.391	331.687	242.193	152.140	-829.837	-524.450	16.114
1800	116.391	338.339	247.351	163.779	-824.138	-506.651	14.703
1900	116.391	344.632	252.307	175.418	-818.459	-489.167	13.448
2000	116.391	350.602	257.074	187.057	-812.805	-471.982	12.327
2100	116.391	356.281	261.664	198.696	-807.182	-455.080	11.319
2200	116.391	361.695	266.089	210.335	-801.595	-438.444	10.410
2300	116.391	366.869	270.359	221.974	-796.051	-422.062	9.585
2400	116.391	371.823	274.484	233.613	-790.559	-405.919	8.835
2500	116.391	376.574	278.473	245.252	-785.126	-390.004	8.149
2600	116.391	381.139	282.335	256.891	-779.760	-374.306	7.520
2700	116.391	385.532	286.076	268.530	-774.469	-358.812	6.942
2800	116.391	389.764	289.704	280.169	-769.263	-343.513	6.408
2900	116.391	393.849	293.225	291.809	-764.150	-328.398	5.915
3000	116.391	397.795	296.645	303.448	-759.138	-313.458	5.458

PREVIOUS: CURRENT: June 1974

Strontium Bromide ($SrBr_2$) $Br_2Sr_1(l)$

STRONTIUM BROMIDE ($SrBr_2$) I-II-LIQUID M_r = 247.428

0 to 918 K		crystal, I
918 to 930 K		crystal, II
above	930 K	liquid

Refer to the individual tables for details.

Strontium Bromide ($SrBr_2$) $Br_2Sr_1(cr,l)$

Enthalpy Reference Temperature = T_r = 298.15 K Standard State Pressure = $p°$ = 0.1 MPa

	J K⁻¹mol⁻¹			kJ mol⁻¹			
T/K	$C_p°$	S°	$-[G°-H°(T_r)]/T$	$H°-H°(T_r)$	$\Delta_fH°$	$\Delta_fG°$	Log K_f
0	0.	0.	INFINITE	-18.064	-704.960	-704.960	INFINITE
100	64.337	64.952	209.316	-14.436	-705.626	-703.931	367.696
200	73.645	113.388	150.374	-7.397	-705.976	-702.127	183.377
298.15	76.853	143.438	143.438	0.	-717.974	-698.755	122.419
300	76.902	143.913	143.439	0.142	-718.022	-698.636	121.643
400	79.036	166.337	146.482	7.942	-747.431	-686.047	89.589
500	80.835	184.166	152.295	15.935	-745.957	-670.871	70.085
600	82.676	199.060	158.881	24.107	-744.434	-655.995	57.109
700	84.893	211.966	165.563	32.482	-742.832	-641.381	47.860
800	87.571	223.468	172.094	41.099	-741.107	-627.004	40.939
900	91.002	233.971	178.394	50.020	-739.825	-612.781	35.565
918.000	91.717	235.781	179.501	51.664	ALPHA <--> BETA		
918.000	115.060	249.098	179.501	63.890	TRANSITION		
930.000	115.060	250.593	180.409	65.271	BETA <--> LIQUID		
930.000	116.391	261.480	180.409	75.396	TRANSITION		
1000	116.391	269.926	186.383	83.543	-713.108	-600.732	31.379
1100	116.391	281.020	194.490	95.182	-716.189	-589.362	27.986
1200	116.391	291.147	202.129	106.822	-712.288	-578.005	25.160
1300	116.391	300.463	209.340	118.461	-708.396	-566.973	22.781
1400	116.391	309.089	216.160	130.100	-704.512	-556.240	20.754
1500	116.391	317.119	222.626	141.739	-700.639	-545.784	19.006
1600	116.391	324.630	228.769	153.378	-696.775	-535.586	17.485
1700	116.391	331.687	234.618	165.017	-829.837	-524.450	16.114
1800	116.391	338.339	240.197	176.656	-824.138	-506.651	14.703
1900	116.391	344.632	245.530	188.295	-818.459	-489.168	13.448
2000	116.391	350.602	250.635	199.934	-812.805	-471.983	12.327
2100	116.391	356.281	255.532	211.573	-807.182	-455.080	11.320
2200	116.391	361.695	260.235	223.212	-801.595	-438.445	10.410
2300	116.391	366.869	264.760	234.851	-796.052	-422.062	9.585
2400	116.391	371.823	269.118	246.490	-790.560	-405.920	8.835
2500	116.391	376.574	273.322	258.129	-785.126	-390.005	8.149
2600	116.391	381.139	277.382	269.768	-779.760	-374.306	7.520
2700	116.391	385.532	281.307	281.407	-774.470	-358.812	6.942
2800	116.391	389.764	285.105	293.046	-769.264	-343.513	6.408
2900	116.391	393.849	288.785	304.685	-764.151	-328.398	5.915
3000	116.391	397.795	292.353	316.324	-759.139	-313.458	5.458

PREVIOUS: CURRENT: June 1974

Strontium Bromide ($SrBr_2$) $Br_2Sr_1(cr,l)$

STRONTIUM BROMIDE ($SrBr_2$) IDEAL GAS M_r = 247.428

$S°(298.15 K) = [323.4 ± 8.4] J K^{-1} mol^{-1}$

$\Delta_f H°(0 K) = -392.1 ± 12.6 kJ mol^{-1}$
$\Delta_f H°(298.15 K) = -407.1 ± 12.6 kJ mol^{-1}$

Vibrational Frequencies and Degeneracies

ν, cm^{-1}
[157](1)
[37](2)
[263](1)

Ground State Quantum Weight: [1] σ = 2
Point Group: $D_{\infty h}$
Bond Distance: Sr-Br = 2.82 ± 0.03 Å
Bond Angle: Br-Sr-Br = 180°
Rotational Constant: B_0 = 0.013265 cm^{-1}

Enthalpy of Formation

Peterson and Hutchison (1, 2) have measured vapor pressures (1114-1304 K) for liquid $SrBr_2$ by the Knudsen effusion method. Results of 2nd and 3rd law analysis of these pressures are tabulated below. We assume that the dibromide monomer was the only vapor species present. Mass spectra for some of the other alkaline earth dihalides (3) indicate that the saturated vapor consist predominantly of the metal dihalide monomer which agrees with predictions which were made by Brewer et al. (4) for the dihalide molecules. Included in the table is a value for $\Delta_f H°$ derived from a 3rd law analysis of a single vapor pressure point determined by Stock and Heynemann (5) and tabulated by Brewer et al. (4).

Source	Reaction[a]	Data Points	T/K	Δ H°(298.15 K), kcal mol^{-1} 2nd law	3rd law	Drift cal K^{-1} mol^{-1}	$\Delta_f H°$(298.15 K) kcal mol^{-1}
(1, 2)	A	13[a]	1147-1304	69.9	71.24±0.86	1.1±2.5	97.3±2.5
(4, 5)	A	1	1040		66.9		101.6

[a]Reaction: (A) $SrBr_2(\ell) = SrBr_2(g)$; [b]Two points rejected due to failure of a statistical test.

The value $\Delta_f H°(SrBr_2, g, 298.15 K) = -97.3±3.0$ kcal mol^{-1} (-407.103±12.6 kJ mol^{-1}) is preferred here rather than an average (-99.5 kcal mol^{-1}) of the two results, since it is impossible to assess the reliability of the measurement by Stock and Heynemann (5). However, our value does compare favorably with that (-98.0 kcal mol^{-1}) selected by NBS (6).

Heat Capacity and Entropy

The value of the bond length is that measured by Akishin and Spiridonov (7) in a high-temperature electron diffraction study. Electron diffraction patterns (7) for $SrBr_2$ were satisfactorily explained on the basis of a linear model (180° ± 10°). Later studies by Wharton et al. (8), using electric deflection of molecular beams to detect dipole moments, showed no polarity in the $SrBr_2$ molecule; this is most reasonably explained by a linear and centrosymmetric configuration.

Few infrared spectral studies (9-11) have been reported for the gaseous alkaline earth dibromides. The antisymmetric stretch (ν_3) and bending (ν_2) frequencies for $BeBr_2$ were observed by Snelson (9) in a neon matrix. Baikov (10) observed a band at 330±5 cm^{-1} in the high-temperature infrared spectra of $CaBr_2$ vapor and assigned this band to ν_3. In a similar study on $MgBr_2$(g) by Randall et al. (11) ν_3 was observed at 490 cm^{-1}. The only frequencies available for $SrBr_2$ are estimates (4, 10, and 12) which have been obtained by force constant calculations. The adopted vibrational frequencies are derived from force constants estimated by the valence force method (13). A comparison of values for the ratio of the stretching force constants for diatomic and triatomic strontium fluorides (ratio = 1.33, 3) and chlorides (ratio = 1.15, 3) suggests a value of near unity for the ratio of the bromides. Therefore, the stretching force constant for $SrBr_2$ is taken equal to that for SrBr(k = 1.154 x 10^5 dynes/cm) (4). Our values for ν_1 and ν_3 compare favorably with those estimated by a similar procedure (4) (ν_1 = 156 cm^{-1} and ν_3 = 262 cm^{-1}) and with those given by Krasnov and Svettsov (12, ν_1 = 158 cm^{-1} and ν_3 = 265 cm^{-1}). The estimated frequencies of Baikov (10) are roughly 12% lower than ours. A comparison of values for the ratio of the stretch to bend force constants for the bent molecules SrF_2(ratio = 50, 3) and $SrCl_2$(ratio = 78, 3) suggests a value near 100 for $SrBr_2$. Brewer (4) used this same value in his force constant calculations. This value for the ratio gives $k\delta/\ell^2 = 1.154 \times 10^3$ dynes/cm, or ν_2 = 37 cm^{-1}. A comparison of similar data for the linear molecule $BeBr_2$ (9) indicates that the value of the ratio could be as low as 41. This leads to a ν_2 = 58 cm^{-1} which is in excellent agreement with the value (56 cm^{-1}) recommended by Krasnov and Svettsov (12). We prefer the lower value of ν_2 at this time but include in the uncertainty ±2.0 cal K^{-1} mol^{-1} assigned to the value of S°(298.15 K) the possibility that the higher value is the correct one. The singlet ground state is assigned by analogy with that for $BaCl_2$ (3). Our adopted functions agree with those tabulated by Brewer et al. (4) to within 0.1 cal K^{-1} mol^{-1} in the range 298-2000 K.

References

1. D. T. Peterson and J. F. Hutchison, J. Chem. Eng. Data 15, 320 (1970).
2. J. T. Hutchison, Ames Laboratory, Iowa State University, M.S. Thesis, AEC Contract No. W-7405-eng-82, November, 1965.
3. JANAF Thermochemical Tables: $BaCl_2$(g), $SrCl_2$(g), SrCl(g), SrF_2(g), and SrF(g), 12-31-72; $MgCl_2$(g), 12-31-69; BeF_2(g), 6-30-70; $SrBr_2(\ell)$, and BaI_2(g), 6-30-74.
4. L. Brewer, G. R. Somayajulu, and E. Brackett, Chem. Revs. 63, 111 (1963).
5. A. Stock and H. Heynemann, Chem. Ber. 42, 4088 (1909).
6. U.S. Nat. Bur. Stand. Tech. Note 270-6, 1971.
7. P. A. Akishin and V. P. Spiridonov, Kristallografiya 2, 472 (1952).
8. L. Wharton, R. A. Berg, and W. Klemperer, J. Chem. Phys. 39, 2023 (1963).
9. A. Snelson, J. Phys. Chem. 72, 250 (1968).
10. V. I. Baikov, Opt. Spektrosk. 27, 502 (1969).
11. S. P. Randall, F. T. Greene, and J. L. Margrave, J. Phys. Chem. 63, 758 (1959).
12. K. S. Krasnov and V. I. Svettsov, Izv. Vysshikh Uchebn. Zavedenii, Khim. i Khim. Tekhnol. 6, 167 (1963).
13. G. Herzberg, "Infrared and Raman Spectra of Polyatomic Molecules," D. Van Nostrand Company, Inc., New York, 1962.

Strontium Bromide ($SrBr_2$) $Br_2Sr_1(g)$

Enthalpy Reference Temperature = T_r = 298.15 K; Standard State Pressure = p° = 0.1 MPa

T/K	$C_p°$ (J $K^{-1}mol^{-1}$)	S°	-[G°-H°(T_r)]/T	H°-H°(T_r) (kJ mol^{-1})	$\Delta_f H°$	$\Delta_f G°$	Log K_f
0	0.	0.	INFINITE	-16.092	-392.118	-392.118	INFINITE
100	53.702	259.975	376.278	-11.630	-391.949	-409.756	214.035
200	59.338	299.377	328.945	-5.914	-393.621	-426.971	111.513
250	60.331	312.735	324.413	-2.920	-394.746	-435.188	90.928
298.15	60.896	323.413	323.413	0.	-407.103	-441.544	77.357
300	60.913	323.790	323.415	0.113	-407.180	-441.757	76.917
350	61.279	333.209	324.158	3.168	-438.128	-445.818	66.535
400	61.523	341.409	325.813	6.239	-438.263	-446.908	58.360
450	61.693	348.666	327.957	9.319	-438.425	-447.979	52.000
500	61.817	355.173	330.358	12.407	-438.614	-449.031	46.910
600	61.979	366.459	335.463	18.598	-439.073	-451.073	39.269
700	62.079	376.021	340.591	24.801	-439.642	-453.029	33.805
800	62.144	384.315	345.550	31.012	-440.323	-454.897	29.702
900	62.188	391.637	350.272	37.229	-441.744	-456.600	26.500
1000	62.220	398.191	354.742	43.449	-442.331	-458.220	23.935
1100	62.244	404.123	358.965	49.673	-450.827	-459.414	21.816
1200	62.262	409.539	362.958	55.898	-452.340	-460.128	20.029
1300	62.276	414.523	366.735	62.125	-453.860	-460.715	18.512
1400	62.288	419.139	370.315	68.353	-455.388	-461.185	17.207
1500	62.297	423.437	373.715	74.583	-456.923	-461.546	16.072
1600	62.304	427.458	376.950	80.813	-458.469	-461.804	15.076
1700	62.310	431.235	380.033	87.043	-596.940	-460.785	14.158
1800	62.315	434.797	382.977	93.275	-596.648	-452.784	13.139
1900	62.320	438.166	385.794	99.506	-596.377	-444.799	12.228
2000	62.323	441.363	388.493	105.739	-596.130	-436.828	11.409
2100	62.327	444.404	391.084	111.971	-595.913	-428.868	10.668
2200	62.329	447.303	393.574	118.204	-595.732	-420.918	9.994
2300	62.332	450.074	395.971	124.437	-595.595	-412.976	9.379
2400	62.334	452.727	398.281	130.670	-595.508	-405.038	8.815
2500	62.336	455.271	400.510	136.904	-595.480	-397.102	8.297
2600	62.338	457.716	402.663	143.137	-595.520	-389.167	7.818
2700	62.339	460.069	404.746	149.371	-595.635	-381.228	7.375
2800	62.340	462.336	406.763	155.605	-595.834	-373.284	6.964
2900	62.342	464.524	408.717	161.839	-596.125	-365.330	6.580
3000	62.343	466.637	410.613	168.074	-596.519	-357.366	6.222
3100	62.344	468.681	412.453	174.308	-597.021	-349.386	5.887
3200	62.345	470.661	414.241	180.542	-597.641	-341.388	5.573
3300	62.345	472.579	415.980	186.777	-598.385	-333.369	5.277
3400	62.346	474.440	417.672	193.011	-599.259	-325.326	4.998
3500	62.347	476.248	419.320	199.246	-600.270	-317.254	4.735
3600	62.347	478.004	420.926	205.481	-601.423	-309.152	4.486
3700	62.348	479.712	422.492	211.715	-602.720	-301.016	4.250
3800	62.348	481.375	424.020	217.950	-604.165	-292.843	4.025
3900	62.349	482.994	425.511	224.185	-605.759	-284.630	3.812
4000	62.349	484.573	426.968	230.420	-607.504	-276.374	3.609
4100	62.350	486.113	428.392	236.655	-609.398	-268.072	3.415
4200	62.350	487.615	429.784	242.890	-611.437	-259.723	3.230
4300	62.351	489.082	431.146	249.125	-613.623	-251.324	3.053
4400	62.351	490.516	432.479	255.360	-615.951	-242.871	2.883
4500	62.351	491.917	433.785	261.595	-618.416	-234.365	2.720
4600	62.352	493.287	435.063	267.830	-620.948	-225.807	2.564
4700	62.352	494.628	436.316	274.066	-623.651	-217.188	2.414
4800	62.352	495.941	437.545	280.301	-626.467	-208.510	2.269
4900	62.352	497.227	438.750	286.536	-629.391	-199.773	2.130
5000	62.353	498.486	439.932	292.771	-632.409	-190.975	1.995
5100	62.353	499.721	441.092	299.006	-635.497	-182.118	1.865
5200	62.353	500.932	442.231	305.242	-638.685	-173.197	1.740
5300	62.353	502.120	443.350	311.477	-641.944	-164.214	1.618
5400	62.353	503.285	444.449	317.712	-645.266	-155.169	1.501
5500	62.353	504.429	445.530	323.948	-648.643	-146.063	1.387
5600	62.354	505.553	446.591	330.183	-652.065	-136.894	1.277
5700	62.354	506.656	447.636	336.418	-655.524	-127.664	1.170
5800	62.354	507.741	448.662	342.654	-659.013	-118.373	1.066
5900	62.354	508.807	449.673	348.889	-662.314	-109.045	0.965
6000	62.354	509.855	450.667	355.125	-665.808	-99.639	0.867

PREVIOUS: June 1974 (1 atm) CURRENT: June 1974 (1 bar)

Strontium Bromide ($SrBr_2$) $Br_2Sr_1(g)$

TITANIUM BROMIDE ($TiBr_2$) CRYSTAL M_r = 207.688

$\Delta_f H°$(0 K) = unknown

$S°$(298.15 K) = [108.4 ± 8.4] J K^{-1} mol^{-1} $\Delta_f H°$(298.15 K) = -405.43 ± 20.9 kJ mol^{-1}

Enthalpy of Formation

The enthalpy of formation of $TiBr_2$ crystal is derived from the data of Hall and Blocher (1) for the reaction 2 $TiBr_3$(cr) = $TiBr_4$(g) + $TiBr_2$(cr). 2nd and 3rd law analyses of these data are not useful because of the formation of solid solutions in the above process. Hall and Blocher (1) obtained a value of $\Delta_r H°$(800 K) of 31.3 kcal mol^{-1} for the reaction by integrating the incremental Gibbs energy changes over varying compositions of the solid solutions. The corresponding $\Delta_r H°$(298.15 K) and $\Delta_f H°$(298.15 K) are calculated as 34.6 kcal mol^{-1} and -96.9 kcal mol^{-1} (-405.430 kJ mol^{-1}), respectively, using auxiliary JANAF enthalpy of formation and enthalpy data.

Heat Capacity and Entropy

The heat capacity of $TiBr_2$(cr) was estimated by Kelley (2). The value of $S°$(298.15 K) is estimated from that of $TiCl_2$(cr) and the difference between ionic entropy contribution of Cl^- and Br^-.

Sublimation Data

The enthalpy sublimation of $TiBr_2$(cr) is taken as the difference in the enthalpies of formation of $TiBr_2$(cr) and $TiBr_2$(g) at the sublimation temperature. The sublimation temperature is taken as the point at which $\Delta_r G°$ = 0 for the reaction $TiBr_2$(cr) $TiBr_2$(g). The enthalpy of formation of $TiBr_2$(g), upon which these quantities depend, is an estimated quantity.

References

1. E. H. Hall and J. M. Blocher, Jr., J. Phys. Chem. 63, 1525 (1959).
2. K. K. Kelley, U.S. Bur. Mines Bull. 584, 1960.

Titanium Bromide ($TiBr_2$) Br_2Ti_1(cr)

Enthalpy Reference Temperature = T_r = 298.15 K Standard State Pressure = p° = 0.1 MPa

T/K	$C_p°$	$S°$	$-[G°-H°(T_r)]/T$	$H°-H°(T_r)$	$\Delta_f H°$	$\Delta_f G°$	Log K_f
	J $K^{-1}mol^{-1}$			kJ mol^{-1}			
0							
100							
200							
298.15	78.688	108.366	108.366	0.	-405.430	-383.188	67.133
300	78.709	108.852	108.367	0.146	-405.471	-383.050	66.695
400	79.856	131.653	111.468	8.074	-434.636	-369.491	48.251
500	81.002	149.595	117.362	16.117	-433.025	-353.391	36.919
600	82.149	164.465	124.008	24.274	-431.412	-337.615	29.392
700	83.295	177.214	130.719	32.546	-429.767	-322.112	24.036
800	84.441	188.412	137.245	40.933	-428.057	-306.849	20.035
900	85.588	198.424	143.496	49.435	-426.311	-291.802	16.936
1000	86.734	207.501	149.450	58.051	-424.590	-276.950	14.466
1100	87.881	215.821	155.110	66.782	-422.957	-262.266	12.454
1200	89.027	223.517	160.494	75.627	-425.398	-247.606	10.778
1300	90.174	230.688	165.621	84.587	-423.219	-232.878	9.357
1400	91.320	237.413	170.511	93.662	-421.013	-218.319	8.146
1500	92.466	243.752	175.185	102.851	-418.793	-203.919	7.101

PREVIOUS: June 1964 CURRENT: June 1968

Titanium Bromide ($TiBr_2$) Br_2Ti_1(cr)

$\Delta_f H°(0\ K)$ = [-164.9 ± 20.9] kJ mol^{-1}

S°(298.15 K) = [308.7 ± 12.6] J K^{-1} mol^{-1}

$\Delta_f H°(298.15\ K)$ = [-179.1 ± 20.9] kJ mol^{-1}

Electronic Levels and Quantum Weights

ε_i, cm^{-1}	g_i
0	[3]
[7000]	[6]
[17000]	[6]
[22000]	[15]

Vibrational Frequences and Degeneracies

ν, cm^{-1}
[130](1)
[105](2)
[333](1)

Ground State Quantum Weight: [3] σ = 2

Point Group: [$D_{\infty h}$]

Bond Distance: Ti-Br = [2.4] Å

Bond Angle: Br-Ti-Br = [180°]

Rotational Constant: B_0 = [0.018314] cm^{-1}

Enthalpy of Formation

The enthalpy of formation of $TiBr_2(g)$ is calculated from the estimated Ti-Br bond energy. The bond energy is estimated as being the same as the average Ti-Cl bond energy of $TiCl_2(g)$. This estimate is used because the measured average bond energies of $TiBr_3(g)$ and $TiCl_3(g)$ and those of $TiBr_4(g)$ and $TiCl_4(g)$ are very nearly equal.

Heat Capacity and Entropy

The interatomic distances are estimated from those of $TiCl_2(g)$, $TiCl_4(g)$ and $TiBr_4(g)$. The vibrational frequencies are estimated from a valence force field model. The force constant k is estimated as 1.2 millidynes Å^{-1}, and the constant k_δ/ℓ^2 is assumed to be 0.06 millidynes Å^{-1}. These values are derived by a correlation with other dihalides.

The electronic levels are assumed to be the same as $TiCl_2(g)$. The levels of $TiCl_2(g)$ are estimated by assuming that they correspond to the inverted states of $NiCl_2(g)$ (1). The linear configuration is adopted since experimental evidence indicates that other transition metal dihalides are linear (2, 3, 4, 5).

References

1. C. W. DeKock and D. M. Gruen, J. Chem. Phys. 44, 4387 (1966).
2. R. A. Berg and O. Sinanoglu, J. Chem. Phys. 32, 1082 (1960).
3. J. T. Hougen, G. E. Leroi and T. C. James, J. Chem. Phys. 34, 1670 (1961).
4. G. E. Leroi, T. C. James, J. T. Hougen and W. Klemperer, J. Chem. Phys. 36, 2879 (1962).
5. A. Buchler, J. L. Stauffer and W. Klemperer, J. Chem. Phys. 40, 3471 (1964).

Enthalpy Reference Temperature = T_r = 298.15 K   Standard State Pressure = p° = 0.1 MPa

T/K	$C_p°$ (J K^{-1}mol^{-1})	S° (J K^{-1}mol^{-1})	$-[G°-H°(T_r)]/T$ (J K^{-1}mol^{-1})	$H°-H°(T_r)$ (kJ mol^{-1})	$\Delta_f H°$ (kJ mol^{-1})	$\Delta_f G°$ (kJ mol^{-1})	Log K_f
0	0.	0.	INFINITE	-15.127	-164.863	-164.863	INFINITE
100	49.952	247.576	360.164	-11.259	-164.348	-182.895	95.535
200	57.678	285.140	314.116	-5.795	-165.703	-201.001	52.496
250	59.160	298.185	309.669	-2.871	-166.758	-209.710	43.816
298.15	60.026	308.685	308.685	0.	-179.075	-216.558	37.940
300	60.052	309.056	308.686	0.111	-179.151	-216.791	37.747
350	60.624	318.359	309.419	3.129	-210.070	-221.365	33.037
400	61.009	326.481	311.055	6.170	-210.185	-222.971	29.117
450	61.281	333.684	313.177	9.228	-210.326	-224.561	26.066
500	61.479	340.151	315.556	12.297	-210.490	-226.134	23.624
600	61.741	351.385	320.619	18.459	-210.873	-229.228	19.956
700	61.904	360.916	325.712	24.642	-211.317	-232.253	17.331
800	62.016	369.189	330.641	30.839	-211.798	-235.211	15.358
900	62.109	376.499	335.338	37.045	-212.346	-238.106	13.819
1000	62.204	383.048	339.788	43.260	-213.026	-240.933	12.585
1100	62.319	388.982	343.994	49.486	-213.898	-243.684	11.572
1200	62.466	394.410	347.973	55.725	-218.946	-246.226	10.718
1300	62.655	399.417	351.740	61.981	-219.471	-248.478	9.984
1400	62.887	404.069	355.313	68.258	-220.063	-250.688	9.353
1500	63.163	408.417	358.710	74.560	-220.730	-252.853	8.805
1600	63.478	412.503	361.946	80.892	-221.482	-254.970	8.324
1700	63.825	416.362	365.034	87.256	-222.330	-257.038	7.898
1800	64.198	420.020	367.988	93.657	-223.287	-259.052	7.518
1900	64.587	423.502	370.819	100.097	-224.365	-261.010	7.176
2000	64.986	426.825	373.537	106.575	-240.264	-262.455	6.855
2100	65.387	430.005	376.151	113.094	-242.373	-263.512	6.555
2200	65.784	433.056	378.669	119.653	-244.462	-264.470	6.279
2300	66.171	435.989	381.097	126.250	-246.533	-265.333	6.026
2400	66.546	438.813	383.444	132.886	-248.588	-266.106	5.792
2500	66.904	441.537	385.713	139.559	-250.630	-266.794	5.574
2600	67.246	444.168	387.911	146.267	-252.659	-267.401	5.372
2700	67.568	446.712	390.042	153.007	-254.677	-267.929	5.183
2800	67.872	449.175	392.110	159.780	-256.686	-268.384	5.007
2900	68.157	451.561	394.119	166.581	-258.685	-268.766	4.841
3000	68.425	453.876	396.073	173.410	-260.675	-269.080	4.685
3100	68.675	456.124	397.974	180.266	-262.655	-269.327	4.538
3200	68.911	458.308	399.826	187.145	-264.625	-269.511	4.399
3300	69.132	460.432	401.630	194.047	-266.583	-269.633	4.268
3400	69.341	462.499	403.390	200.971	-268.530	-269.696	4.143
3500	69.538	464.512	405.108	207.915	-270.462	-269.703	4.025
3600	69.726	466.474	406.785	214.878	-272.378	-269.654	3.913
3700	69.904	468.387	408.424	221.860	-683.375	-261.763	3.695
3800	70.074	470.253	410.027	228.859	-684.021	-250.360	3.441
3900	70.238	472.075	411.595	235.874	-684.701	-238.939	3.200
4000	70.395	473.856	413.129	242.906	-685.412	-227.500	2.971
4100	70.547	475.596	414.632	249.953	-686.152	-216.044	2.752
4200	70.694	477.298	416.103	257.015	-686.917	-204.568	2.544
4300	70.836	478.963	417.546	264.092	-687.705	-193.075	2.345
4400	70.974	480.593	418.960	271.183	-688.512	-181.562	2.155
4500	71.108	482.189	420.348	278.287	-689.335	-170.031	1.974
4600	71.237	483.754	421.709	285.404	-690.172	-158.482	1.800
4700	71.362	485.287	423.046	292.534	-691.020	-146.915	1.633
4800	71.484	486.791	424.358	299.676	-691.876	-135.329	1.473
4900	71.601	488.266	425.647	306.831	-692.739	-123.725	1.319
5000	71.714	489.713	426.914	313.996	-693.605	-112.104	1.171
5100	71.822	491.135	428.160	321.173	-694.473	-100.465	1.029
5200	71.926	492.530	429.384	328.361	-695.340	-88.810	0.892
5300	72.026	493.901	430.588	335.558	-696.205	-77.137	0.760
5400	72.120	495.249	431.773	342.766	-697.067	-65.448	0.633
5500	72.210	496.573	432.940	349.982	-697.924	-53.744	0.510
5600	72.294	497.875	434.088	357.207	-698.774	-42.024	0.392
5700	72.373	499.155	435.218	364.441	-699.616	-30.289	0.278
5800	72.447	500.414	436.331	371.682	-700.449	-18.539	0.167
5900	72.516	501.653	437.428	378.930	-701.273	-6.775	0.060
6000	72.579	502.873	438.508	386.185	-702.086	5.003	-0.044

PREVIOUS: June 1968 (1 atm)   CURRENT: June 1968 (1 bar)

ZIRCONIUM BROMIDE ($ZrBr_2$) CRYSTAL M_r = 251.028

$S°(298.15 K)$ = [115.9 ± 13] $J K^{-1} mol^{-1}$

T_{fus} = [900] K

$\Delta_f H°(0 K)$ = unknown

$\Delta_f H°(298.15 K)$ = [-404.6 ± 42] $kJ mol^{-1}$

$\Delta_{fus}H°$ = [62.760 ± 42] $kJ mol^{-1}$

Enthalpy of Formation

The value of $\Delta_f H°(ZrBr_2$, cr, 298.15 K) = -96.7 $kcal mol^{-1}$ (-404.593 $kJ mol^{-1}$) is calculated from the estimated $\Delta_{sub}H°(298.15 K)$ = 55±5 $kcal mol^{-1}$ for $ZrBr_2(cr) \rightarrow ZrBr_2(g)$, using $\Delta_f H°(ZrBr_2$, g, 298.15 K) = -41.7 $kcal mol^{-1}$ (1). The estimated $\Delta_{sub}H°(298.15 K)$ is obtained as an average of the enthalpies of sublimation at 298°K for $CrBr_2$, $MnBr_2$, $CoBr_2$ and $NiBr_2$ given by Brewer, Somayajulu and Brackett (2) and $TiBr_2$ and $FeBr_2$ listed in JANAF tables (1).

Heat Capacity and Entropy

The heat capacities of $ZrBr_2(cr)$ are estimated from $TiBr_2(cr)$ (1), based on the assumption $C_p°(ZrBr_2) = C_p°(TiBr_2)$ + 3/2 $R \ln[A_r$ of Zr)/(A_r of Ti)]. The entropy, S°(298.15 K) = 27.7 $cal K^{-1} mol^{-1}$ (115.897 $J K^{-1} mol^{-1}$) is estimated in a similar manner based on $TiCl_2(cr)$ (1) and $TiBr_2(cr)$ (1).

Fusion Data

The enthalpy of fusion is roughly estimated, so that the entropy of vaporization is about 20 $cal K^{-1} mol^{-1}$. The melting point was estimated by Brewer (3).

References

1. JANAF Thermochemical Tables: $ZrBr_2(g)$, 6-30-70; $TiBr_2(cr)$, $FeBr_2(cr)$, 9-30-66; $TiCl_2(cr)$, 12-31-68.
2. L. Brewer, G. R. Somayajulu and E. Brackett, Chem. Rev. 63, 111 (1963).
3. L. Brewer, "The Chemistry and Metallurgy of Miscellaneous Materials: Thermodynamics," L. L. Quill Ed., McGraw-Hill Book Company, New York, 1949.

Zirconium Bromide ($ZrBr_2$) $Br_2Zr_1(cr)$

Enthalpy Reference Temperature = T_r = 298.15 K Standard State Pressure = p° = 0.1 MPa

T/K	$C_p°$ (J K⁻¹mol⁻¹)	S° (J K⁻¹mol⁻¹)	-[G°-H°(T_r)]/T (J K⁻¹mol⁻¹)	H°-H°(T_r) (kJ mol⁻¹)	$\Delta_f H°$ (kJ mol⁻¹)	$\Delta_f G°$ (kJ mol⁻¹)	Log K_f
0							
100							
200							
298.15	86.722	115.897	115.897	0.	-404.593	-382.178	66.956
300	86.743	116.433	115.898	0.160	-404.619	-382.039	66.519
400	87.889	141.545	119.315	8.892	-432.927	-368.554	48.128
500	89.036	161.280	125.803	17.738	-430.397	-352.753	36.852
600	90.182	177.614	133.116	26.699	-427.847	-337.464	29.379
700	91.328	191.602	140.495	35.775	-425.276	-322.603	24.073
800	92.475	203.872	147.666	44.965	-422.688	-308.111	20.118
900	93.621	214.830	154.530	54.270	-420.095	-293.945	17.060
900.000	93.621	214.830	154.530	54.270	------ CRYSTAL <--> LIQUID -----		
1000	94.768	224.753	161.064	63.689	-417.506	-280.067	14.629
1100	95.914	233.839	167.273	73.223	-414.926	-266.448	12.653
1200	97.060	242.234	173.174	82.872	-416.071	-252.842	11.006
1300	98.207	250.048	178.791	92.635	-412.977	-239.365	9.618
1400	99.353	257.368	184.145	102.513	-409.819	-226.128	8.437
1500	100.500	264.262	189.258	112.506	-406.609	-213.119	7.421

PREVIOUS: March 1962 CURRENT: June 1970

ZIRCONIUM BROMIDE ($ZrBr_2$) LIQUID M_r = 251.028

$S°(298.15\ K)$ = [184.024] $J\ K^{-1}\ mol^{-1}$

T_{fus} = [900] K

$\Delta_f H°(298.15\ K)$ = [-342.333] $kJ\ mol^{-1}$

$\Delta_{fus}H°$ = [62.760 ± 42] $kJ\ mol^{-1}$

Enthalpy of Formation

$\Delta_f H°(ZrBr_2$, ℓ, 298.15 K) is calculated from that of the crystal by adding $\Delta_{fus}H°$ and the difference in enthalpy, H°(900 K) - H°(298.15 K), between the crystal and liquid.

Heat Capacity and Entropy

The heat capacity is assumed to be constant at 7.25 cal K^{-1} g-atom^{-1}. The entropy is calculated in a manner analogous to that used for the enthalpy of formation.

Fusion Data

See the crystal table for details.

Vaporization Data

T_{vap} is the temperature at which the Gibbs energy change approaches zero for the process $ZrBr_2$(ℓ) → $ZrBr_2$(g) at one bar. The difference between $\Delta_f H°$ for ZrBr(ℓ) and $ZrBr_2$(g) at T_{vap} is $\Delta_{vap}H°$.

Zirconium Bromide ($ZrBr_2$) Br_2Zr_1(l)

Enthalpy Reference Temperature = T_r = 298.15 K Standard State Pressure = p° = 0.1 MPa

T/K	$C_p°$	S°	$-[G°-H°(T_r)]/T$	$H°-H°(T_r)$	$\Delta_f H°$	$\Delta_f G°$	Log K_f
	J K⁻¹mol⁻¹			kJ mol⁻¹			
0							
100							
200							
298.15	91.002	184.024	184.024	0.	-342.333	-340.230	59.607
300	91.002	184.587	184.026	0.168	-342.351	-340.217	59.237
400	91.002	210.767	187.596	9.269	-370.290	-333.606	43.565
500	91.002	231.074	194.336	18.369	-367.506	-324.760	33.927
600	91.002	247.665	201.884	27.469	-364.817	-316.464	27.551
700	91.002	261.693	209.452	36.569	-362.222	-308.612	23.029
800	91.002	273.845	216.758	45.669	-359.724	-301.125	19.661
900	91.002	284.563	223.708	54.770	-357.335	-293.945	17.060
900.000	91.002	284.563	223.708	54.770	CRYSTAL <--> LIQUID		
900.000	91.002	284.563	223.708	54.770	TRANSITION		
1000	91.002	294.151	230.282	63.870	-355.065	-287.025	14.993
1100	91.002	302.825	236.488	72.970	-352.919	-280.325	13.312
1200	91.002	310.743	242.351	82.070	-354.613	-273.594	11.909
1300	91.002	318.027	247.896	91.170	-352.182	-266.942	10.726
1400	91.002	324.771	253.149	100.271	-349.802	-260.475	9.718
1500	91.002	331.049	258.136	109.371	-347.484	-254.175	8.851
1600	91.002	336.923	262.878	118.471	-345.240	-248.028	8.097
1700	91.002	342.440	267.398	127.571	-343.081	-242.019	7.436
1800	91.002	347.641	271.713	136.671	-341.020	-236.134	6.852
1900	91.002	352.561	275.839	145.772	-339.070	-230.361	6.333
2000	91.002	357.229	279.793	154.872	-337.242	-224.687	5.868

PREVIOUS: March 1962 CURRENT: June 1970

Zirconium Bromide ($ZrBr_2$) Br_2Zr_1(l)

ZIRCONIUM BROMIDE ($ZrBr_2$) CRYSTAL - LIQUID M_r = 251.028

298.15 to 900 K crystal

above 900 K liquid

Refer to the individual tables for details.

Zirconium Bromide ($ZrBr_2$) Br_2Zr_1(cr,l)

Enthalpy Reference Temperature = T_r = 298.15 K Standard State Pressure = p° = 0.1 MPa

T/K	C_p° (J K⁻¹mol⁻¹)	S° (J K⁻¹mol⁻¹)	−[G°−H°(T_r)]/T (J K⁻¹mol⁻¹)	H°−H°(T_r) (kJ mol⁻¹)	$\Delta_f H^\circ$ (kJ mol⁻¹)	$\Delta_f G^\circ$ (kJ mol⁻¹)	Log K_f
0							
100							
200							
298.15	86.722	115.897	115.897	0.	-404.593	-382.178	66.956
300	86.743	116.433	115.898	0.160	-404.619	-382.039	66.519
400	87.889	141.545	119.315	8.892	-432.927	-368.554	48.128
500	89.036	161.280	125.803	17.738	-430.397	-352.753	36.852
600	90.182	177.614	133.116	26.699	-427.847	-337.464	29.379
700	91.328	191.602	140.495	35.775	-425.276	-322.603	24.073
800	92.475	203.872	147.666	44.965	-422.688	-308.111	20.118
900	93.621	214.830	154.530	54.270	-420.095	-293.945	17.060
900.000	93.621	284.563	154.530	117.030	CRYSTAL <--> LIQUID	TRANSITION	
900.000	91.002	284.563	154.530	117.030			
1000	91.002	294.151	168.022	126.130	-355.065	-287.025	14.993
1100	91.002	302.825	179.888	135.230	-352.919	-280.325	13.312
1200	91.002	310.743	190.468	144.330	-354.613	-273.594	11.909
1300	91.002	318.027	200.004	153.430	-352.182	-266.942	10.726
1400	91.002	324.771	208.678	162.531	-349.802	-260.475	9.718
1500	91.002	331.049	216.629	171.631	-347.484	-254.175	8.851
1600	91.002	336.923	223.966	180.731	-345.240	-248.028	8.097
1700	91.002	342.440	230.774	189.831	-343.081	-242.019	7.436
1800	91.002	347.641	237.124	198.931	-341.020	-236.134	6.852
1900	91.002	352.561	243.071	208.032	-339.070	-230.361	6.333
2000	91.002	357.229	248.663	217.132	-337.242	-224.687	5.868

PREVIOUS: CURRENT: June 1970

Zirconium Bromide ($ZrBr_2$) Br_2Zr_1(cr,l)

ZIRCONIUM BROMIDE ($ZrBr_2$) IDEAL GAS M_r = 251.028

$\Delta_f H°(0\ K)$ = [-160.0 ± 42] kJ mol^{-1}

S°(298.15 K) = [316.8 ± 13] J K^{-1} mol^{-1} $\Delta_f H°(298.15\ K)$ = [-174.5 ± 42] kJ mol^{-1}

Electronic Levels and Quantum Weights

ε_i, cm^{-1}	g_i
0	[3]
[7000]	[6]
[17000]	[6]
[22000]	[15]

Vibrational Frequencies and Degeneracies

ν cm^{-1}
[160](1)
[84](2)
[265](1)

Ground State Quantum Weight: [3] σ = [2]

Point Group: [$D_{\infty h}$]

Bond Distance: Zr-Br = [2.47] Å

Bond Angle: Br-Zr-Br = [180]°

Rotational Constant: B_0 = [0.017290] cm^{-1}

Enthalpy of Formation

$\Delta_f H°(ZrBr_2$, g, 298.15 K) = -41.7 kcal mol^{-1} (-174.473 kJ mol^{-1}) is derived from the estimated $\Delta_r H°$(298.15 K) = 243.5±10 kcal mol^{-1} for $ZrBr_2$(g) → Zr(g) + 2 Br(g). The value of $\Delta_r H°$(298.15 K) is twice the average bond dissociation energy, $\Delta_{at}H°$(298.15 K)/2 = 121.75 kcal mol^{-1}, which is calculated from the proportion of the average bond dissociation energies at 298.15 K for $ZrBr_2/TiBr_2 = ZrBr_4/TiBr_4$. The average bond dissociation energies at 298.15 K, $\Delta_{at}H°(ZrBr_4)/_4$ = 102.2 kcal mol^{-1}, $\Delta_{at}H°(TiBr_4)/_4$ = 87.9 kcal mol^{-1}, and $\Delta_{at}H°(TiBr_2)/_2$ = 104.6 kcal mol^{-1}, are all calculated from JANAF $\Delta_f H°$(298.15 K) for $ZrBr_4$(g), $TiBr_4$(g), $TiBr_2$(g), Zr(g), Ti(g), and Br(g).

Heat Capacity and Entropy

The molecular configuration is assumed to be linear, since experimental evidence indicates that the transition metal dihalides are generally linear (1, 2, 3, 4, 5) even though a few fluorides are bent (6). The bond distance is assumed to be the same as that in $ZrBr_4$(g) which was estimated as 2.47 Å by Godnev et al. (7).

The vibrational frequencies are calculated from a valence force field model. The stretching force constant is estimated to be 1.2 millidyne Å^{-1} and the bending force constant 0.06 millidyne Å^{-1}. These values are derived from force constants of transition metal dihalides listed by Brewer, Somayajulu and Brackett (8). The electronic levels and quantum weights are estimated to be the same as those of gaseous $TiCl_2$ (9).

References

1. R. A. Berg and O. Sinanoglu, J. Chem. Phys. 32, 1082 (1960).
2. J. T. Hougen, G. E. Leroi and T. C. James, J. Chem. Phys. 34, 1670 (1961).
3. A. Buchler, J. L. Stauffer and W. Klemperer, J. Chem. Phys. 40, 3471 (1964).
4. M. E. Jacox and D. E. Milligan, J. Chem. Phys. 51, 4143 (1969).
5. K. R. Thompson and K. D. Carlson, J. Chem. Phys. 49, 4379 (1968).
6. J. W. Hastie, R. H. Hauge and J. L. Margrave, J. Chem. Phys. 51, 2648 (1969).
7. I. N. Godnev, A. M. Aleksandrovskaia and I. V. Rigina, Optics and Spectroscopy 7, 172 (1969).
8. L. Brewer, G. R. Somayajulu and E. Brackett, Chem. Rev. 63, 111 (1963).
9. JANAF Thermochemical Tables: $TiCl_2$(g), 12-31-68.

Zirconium Bromide ($ZrBr_2$) Br_2Zr_1(g)

Enthalpy Reference Temperature = T_r = 298.15 K Standard State Pressure = p° = 0.1 MPa

T/K	$C_p°$	S°	$-[G°-H°(T_r)]/T$	$H°-H°(T_r)$	$\Delta_f H°$	$\Delta_f G°$	Log K_f
	J K^{-1}mol^{-1}			kJ mol^{-1}			
0	0.	0.	INFINITE	-15.564	-160.030	-160.030	INFINITE
100	52.080	254.108	369.245	-11.514	-159.659	-178.280	93.124
200	58.882	292.867	322.284	-5.883	-161.123	-196.402	51.295
250	60.034	306.143	317.773	-2.908	-162.179	-205.105	42.854
298.15	60.685	316.777	316.777	0.	-174.473	-211.951	37.133
300	60.704	317.152	316.778	0.112	-174.547	-212.183	36.944
350	61.125	326.544	317.519	3.159	-205.422	-216.760	32.350
400	61.404	334.725	319.169	6.222	-205.477	-218.376	28.517
450	61.599	341.970	321.308	9.298	-205.547	-219.984	25.535
500	61.741	348.468	323.705	12.381	-205.634	-221.584	23.149
600	61.927	359.742	328.799	18.566	-205.860	-224.754	19.567
700	62.042	369.298	333.920	24.765	-206.166	-227.880	17.005
800	62.123	377.588	338.871	30.973	-206.560	-230.956	15.080
900	62.193	384.909	343.588	37.189	-207.056	-233.977	13.580
1000	62.273	391.465	348.053	43.412	-207.663	-236.936	12.376
1100	62.375	397.405	352.274	49.644	-208.385	-239.830	11.389
1200	62.514	402.838	356.265	55.888	-212.935	-242.431	10.553
1300	62.695	407.849	360.042	62.148	-213.344	-244.872	9.839
1400	62.922	412.503	363.625	68.429	-213.783	-247.281	9.226
1500	63.194	416.853	367.030	74.734	-214.260	-249.658	8.694
1600	63.505	420.941	370.273	81.069	-214.782	-252.001	8.227
1700	63.849	424.802	373.368	87.436	-215.356	-254.310	7.814
1800	64.219	428.461	376.328	93.840	-215.992	-256.583	7.446
1900	64.607	431.944	379.165	100.281	-216.701	-258.819	7.115
2000	65.003	435.268	381.887	106.761	-217.493	-261.016	6.817
2100	65.403	438.449	384.505	113.282	-218.380	-263.170	6.546
2200	65.798	441.501	387.027	119.842	-240.697	-264.535	6.281
2300	66.184	444.434	389.460	126.441	-242.227	-265.584	6.032
2400	66.558	447.259	391.810	133.078	-243.741	-266.567	5.802
2500	66.916	449.983	394.082	139.752	-245.242	-267.487	5.589
2600	67.256	452.614	396.283	146.461	-246.730	-268.348	5.391
2700	67.578	455.159	398.417	153.202	-248.208	-269.151	5.207
2800	67.881	457.622	400.488	159.975	-249.676	-269.900	5.035
2900	68.165	460.009	402.499	166.778	-251.134	-270.596	4.874
3000	68.432	462.324	404.455	173.608	-252.583	-271.243	4.723
3100	68.683	464.572	406.358	180.464	-254.023	-271.841	4.580
3200	68.918	466.757	408.212	187.344	-255.453	-272.393	4.446
3300	69.139	468.881	410.018	194.247	-256.871	-272.900	4.320
3400	69.347	470.948	411.780	201.171	-258.276	-273.365	4.200
3500	69.544	472.961	413.499	208.116	-259.668	-273.788	4.086
3600	69.731	474.923	415.178	215.080	-261.044	-274.172	3.978
3700	69.909	476.836	416.819	222.062	-262.403	-274.518	3.875
3800	70.079	478.702	418.423	229.061	-263.742	-274.828	3.778
3900	70.243	480.525	419.992	236.077	-265.061	-275.102	3.685
4000	70.400	482.305	421.528	243.110	-266.357	-275.343	3.596
4100	70.551	484.046	423.032	250.157	-267.629	-275.552	3.511
4200	70.698	485.747	424.505	257.220	-268.876	-275.730	3.429
4300	70.840	487.413	425.948	264.297	-270.095	-275.879	3.351
4400	70.978	489.043	427.364	271.388	-271.287	-275.999	3.277
4500	71.111	490.639	428.752	278.492	-272.449	-276.093	3.205
4600	71.240	492.204	430.115	285.610	-273.581	-276.162	3.136
4700	71.366	493.737	431.452	292.740	-274.682	-276.206	3.070
4800	71.487	495.241	432.765	299.883	-836.689	-264.610	2.880
4900	71.604	496.716	434.056	307.037	-837.379	-252.684	2.694
5000	71.717	498.164	435.323	314.203	-838.054	-240.745	2.515
5100	71.825	499.585	436.569	321.380	-838.713	-228.792	2.343
5200	71.929	500.981	437.795	328.568	-839.354	-216.827	2.178
5300	72.028	502.352	439.000	335.766	-839.976	-204.849	2.019
5400	72.123	503.699	440.186	342.974	-840.578	-192.860	1.866
5500	72.212	505.023	441.352	350.190	-841.160	-180.859	1.718
5600	72.296	506.325	442.501	357.416	-841.720	-168.849	1.575
5700	72.376	507.606	443.632	364.649	-842.258	-156.828	1.437
5800	72.449	508.865	444.746	371.891	-842.773	-144.799	1.304
5900	72.518	510.104	445.843	379.139	-843.265	-132.761	1.175
6000	72.581	511.323	446.924	386.394	-843.733	-120.714	1.051

PREVIOUS: June 1970 (1 atm) CURRENT: June 1970 (1 bar)

Zirconium Bromide ($ZrBr_2$) Br_2Zr_1(g)

$S°(298.15 K) = 348.05 \pm 0.4 J K^{-1} mol^{-1}$

$\Delta_f H°(0 K) = [-276.5 \pm 17] kJ mol^{-1}$
$\Delta_f H°(298.15 K) = [-302.9 \pm 17] kJ mol^{-1}$

Vibrational Frequencies and Degeneracies

ν, cm^{-1}	ν, cm^{-1}
2239.2(1)	774. (2)
362. (1)	483.5(2)
168.5(1)	116.8(2)

Ground State Quantum Weight: [1] $\sigma = 3$

Point Group: C_{3v}

Bond Distances: Si-H = 1.494 Å Si-Br = 2.170 Å

Bond Angles: H-Si-Br = 107.51° Br-Si-Br = 111.36°

Product of the Moments of Inertia: $I_A I_B I_C = 1.31228 \times 10^{-11}$

Enthalpy of Formation

$\Delta_f H°$(298.15 K) is estimated by linear interpolation between the values (1) of $SiBr_4$(g) and SiH_4(g). In adopting linear estimate for SiH_nBr_{4-n} (n=1,2,3), we dismiss the solution calorimetry of Wolf et al. (2) because of possible negative bias. These authors reported $\Delta_f H$(g, 298.15 K) = -69.3±2.0 kcal mol^{-1} derived from enthalpies of solution in dilute aqueous NaOH. Wagman et al. (3) changed this value to -75.9 kcal mol^{-1} due to revised auxiliary data, especially $\Delta_f H°$ of Na_2SiO_3(cr) and Na_2SiO_5(sln). The calorimetric result is 3.5 kcal mol^{-1} more negative than our linear estimate. This could imply a cubic variation of $\Delta_f H°(SiH_nBr_{4-n})$ with n, such as we adopt for the chlorosilanes (1). We tentatively reject this hypothesis because the solution results (2) for $SiHCl_3$(g), $SiCl_4$(g), $SiBr_4$(ℓ) and SiI_4(cr) are also more negative than our adopted values (1). It is conceivable that the hydrolysis reactions did not proceed to equal degrees in the different calorimetric studies. We conclude, as did Hunt and Sirtl (4), that the available data justify only linear interpolation of $\Delta_f H°$.

Heat Capacity and Entropy

The molecular structure is based on microwave data of Mitzlaff et al. (5) for eight isotopic forms of $SiHBr_3$ and $SiDBr_3$. Structural parameters are presumably substitutional (r_s) values. The principal moments of inertia are: $I_A = I_B = 87.7371 \times 10^{-39}$ and $I_C = 170.4749 \times 10^{-39}$ g cm^2. Vibrational frequencies are from gas-phase infrared spectra of Buerger and Cichon (6). Assignments are based on band contour analysis and are consistent with Shimanouchi's values (7) derived from liquid-phase Raman spectra (8).

We neglect excited states and assume the electronic ground state to be a singlet. We expect that contributions from excited states should be unimportant, as discussed on the tables for SiH_3Br and SiH_2Cl_2 (1).

References

1. JANAF Thermochemical Tables: Br_4Si(ℓ, g), Br_2H_2Si(g), BrH_3Si(g), ClH_3Si(g), Cl_2H_2Si(g), Cl_3HSi(g) 12-31-76; H_4Si(g), I_4Si(cr) 6-30-76.
2. E. Wolf, W. Stahn and M. Schonherr, Z. Anorg. Allgem. Chem. 319, 168 (1962); 313, 228 (1961); Z. Chem. 7, 283 (1967).
3. D. D. Wagman et al., U.S. Nat. Bur. Stand. Tech. Note 270-3, 1968.
4. L. P. Hunt and E. Sirtl, J. Electrochem. Soc. 120, 806 (1973).
5. M. Mitzlaff, R. Holm and H. Hartmann, Z. Naturforsch. A23, 65 (1968).
6. H. Buerger and J. Cichon, Spectrochim. Acta A27, 2191 (1971).
7. T. Shimanouchi, J. Phys. Chem. Ref. Data 1, 205 (1972).
8. M. L. Delwaulle and M. F. Francois, Compt. Rend. 230, 743 (1950).

Enthalpy Reference Temperature = T_r = 298.15 K — Standard State Pressure = p° = 0.1 MPa

T/K	$C_p°$	S°	$-[G°-H°(T_r)]/T$	$H°-H°(T_r)$	$\Delta_f H°$	$\Delta_f G°$	Log K_f
	J $K^{-1}mol^{-1}$			kJ mol^{-1}			
0	0.	0.	INFINITE	-17.823	-276.529	-276.529	INFINITE
100	53.723	275.599	411.866	-13.627	-278.288	-292.351	152.709
200	69.766	318.044	355.097	-7.411	-281.949	-305.058	79.673
250	75.812	334.287	349.349	-3.765	-284.076	-310.594	64.895
298.15	80.410	348.049	348.049	0.	-302.922	-313.530	54.929
300	80.565	348.547	348.051	0.149	-303.046	-313.595	54.602
350	84.261	361.256	349.046	4.273	-349.657	-312.784	46.681
400	87.163	372.704	351.299	8.562	-349.928	-307.498	40.155
450	89.488	383.109	354.265	12.980	-350.126	-302.181	35.076
500	91.398	392.640	357.632	17.504	-350.263	-296.846	31.011
600	94.380	409.581	364.914	26.800	-350.386	-286.148	24.911
700	96.632	424.307	372.370	36.356	-350.355	-275.443	20.554
800	98.399	437.330	379.691	46.111	-350.214	-264.749	17.286
900	99.818	449.004	386.755	56.024	-349.996	-254.079	14.746
1000	100.970	459.583	393.518	66.065	-349.728	-243.435	12.716
1100	101.915	469.252	399.969	76.211	-349.433	-232.820	11.056
1200	102.697	478.155	406.119	86.443	-349.127	-222.232	9.674
1300	103.348	486.401	411.981	96.746	-348.825	-211.670	8.505
1400	103.894	494.081	417.574	107.109	-348.541	-201.131	7.504
1500	104.356	501.265	422.917	117.522	-348.284	-190.611	6.638
1600	104.748	508.013	428.026	127.978	-348.064	-180.106	5.880
1700	105.085	514.373	432.920	138.470	-398.065	-169.168	5.198
1800	105.374	520.388	437.614	148.993	-397.709	-155.714	4.519
1900	105.625	526.092	442.122	159.544	-397.368	-142.279	3.912
2000	105.844	531.516	446.457	170.117	-397.047	-128.861	3.366
2100	106.036	536.685	450.632	180.712	-396.751	-115.459	2.872
2200	106.205	541.622	454.656	191.324	-396.484	-102.071	2.423
2300	106.354	546.346	458.541	201.952	-396.249	-88.694	2.014
2400	106.486	550.875	462.294	212.594	-396.049	-75.326	1.639
2500	106.604	555.225	465.925	223.249	-395.884	-61.966	1.295
2600	106.710	559.408	469.441	233.914	-395.757	-48.612	0.977
2700	106.805	563.437	472.848	244.590	-395.666	-35.263	0.682
2800	106.891	567.323	476.153	255.275	-395.610	-21.916	0.409
2900	106.968	571.075	479.362	265.968	-395.589	-8.570	0.154
3000	107.039	574.703	482.480	276.669	-395.600	4.776	-0.083
3100	107.103	578.213	485.512	287.376	-395.640	18.122	-0.305
3200	107.161	581.615	488.462	298.089	-395.706	31.470	-0.514
3300	107.214	584.913	491.335	308.808	-395.794	44.821	-0.709
3400	107.263	588.115	494.135	319.532	-395.901	58.175	-0.894
3500	107.308	591.224	496.864	330.260	-396.023	71.531	-1.068
3600	107.350	594.248	499.528	340.993	-780.307	95.353	-1.384
3700	107.388	597.190	502.128	351.730	-780.033	119.673	-1.689
3800	107.423	600.054	504.667	362.471	-779.765	143.986	-1.979
3900	107.456	602.845	507.149	373.215	-779.497	168.291	-2.254
4000	107.486	605.566	509.575	383.962	-779.228	192.590	-2.515
4100	107.515	608.220	511.949	394.712	-778.952	216.882	-2.763
4200	107.541	610.812	514.272	405.465	-778.667	241 167	-2.999
4300	107.566	613.342	516.547	416.220	-778.371	265.446	-3.225
4400	107.589	615.815	518.775	426.978	-778.059	289.717	-3.439
4500	107.610	618.234	520.958	437.738	-777.731	313.981	-3.645
4600	107.630	620.599	523.099	448.500	-777.384	338.237	-3.841
4700	107.649	622.914	525.198	459.264	-777.015	362.486	-4.029
4800	107.667	625.180	527.258	470.030	-776.625	386.727	-4.208
4900	107.684	627.401	529.279	480.797	-776.210	410.959	-4.381
5000	107.699	629.576	531.263	491.566	-775.771	435.183	-4.546
5100	107.714	631.709	533.212	502.337	-775.305	459.397	-4.705
5200	107.728	633.801	535.126	513.109	-774.813	483.602	-4.858
5300	107.742	635.853	537.007	523.883	-774.294	507.798	-5.005
5400	107.754	637.867	538.856	534.657	-773.748	531.983	-5.146
5500	107.766	639.844	540.675	545.433	-773.173	556.158	-5.282
5600	107.777	641.786	542.463	556.211	-772.571	580.322	-5.413
5700	107.788	643.694	544.222	566.989	-771.941	604.476	-5.539
5800	107.798	645.569	545.953	577.768	-771.283	628.617	-5.661
5900	107.808	647.412	547.658	588.548	-770.598	652.748	-5.779
6000	107.817	649.224	549.335	599.330	-769.884	676.866	-5.893

PREVIOUS December 1976 (1 atm) CURRENT. December 1976 (1 bar)

MOLYBDENUM BROMIDE ($MoBr_3$) CRYSTAL M_r = 335.652

$\Delta_f H°$(0 K) = unknown

S°(298.15 K) = [174.5 ± 17] J K^{-1} mol^{-1} $\Delta_f H°$(298.15 K) = -283.7 ± 25 kJ mol^{-1}

Enthalpy of Formation

Shukurov et al. (1) determined the enthalpy of combustion of $MoBr_3$(cr). Dellien et al. (2) obtained $\Delta_f H°$(298.15 K) = -64±5 kcal mol^{-1} from the calorimetric result. Brewer (3) re-examined the equilibrium measurements of Oppermann (4) on the mixed solids $MoBr_2$ + $MoBr_3$ (see the discussion on the $MoBr_4$(g) table (5)) and obtained $\Delta_f H°$(298.15 K) = -67.8±6 kcal mol^1 (-283.675±25 kJ mol^{-1}) which is the value adopted here.

Heat Capacity and Entropy

Brewer (3) estimated S°(298.15 K) = 41.7±4 cal K^{-1} mol^{-1} (174.473±17 J K^{-1} mol^{-1}) and derived $C_p°$ = 24.14 + 3.50 x 10^{-3}T cal K^{-1} mol^{-1} (298-963 K) from the heat capacities estimated by Opperman (4). Other estimates are S°(298.15 K) = 40.9 cal K^{-1} mol^{-1} from the additive entropy constants of Stull and Prophet (6) and $C_p°$(298.15 K) = 23.3 cal K^{-1} mol $^{-1}$ from the ionic contributions of Kubaschewski and Unal (7). We adopt the estimates of Brewer (3).

Decomposition Data

T_{dcm} is the calculated temperature at which $\Delta_r G°$ = 0 for the reaction 2 $MoBr_3$(cr) = $MoBr_2$(cr) + $MoBr_4$(g) at one bar.

References

1. A. Shukurov, T. Nikol'skaya, I. V. Vasil'kova, and S. A. Shchukarev, Izv. Akad. Nauk Tadzhsk. SSR, Otd. Geol. - Khim. Tekhn. Nauk, No. 4, 23 (1961).
2. I. Dellien, M. F. Hall, and L. G. Hepler, Chem. Rev. 76, 283 (1976).
3. L. Brewer, Materials and Molecular Research Division, Lawrence Berkeley Laboratory, University of California, Berkeley; personal communication, Sept. 29, 1978; preliminary draft of review to be submitted for publication in Atomic Energy Review, International Atomic Energy Agency, Vienna, Austria.
4. H. Oppermann, Z. Anorg. Allg. Chem. 395, 249 (1973).
5. JANAF Thermochemical Tables: $MoBr_4$(g), 9-30-78.
6. D. R. Stull and H. Prophet, in "The Characterization of High-Temperature Vapors," Ed. by J. L. Margrave, John Wiley, 1967.
7. O. Kubaschewski and H. Unal, High Temp.-High Pressures 9, 361 (1977).

Molybdenum Bromide ($MoBr_3$) Br_3Mo_1(cr)

Enthalpy Reference Temperature = T_r = 298.15 K Standard State Pressure = p° = 0.1 MPa

T/K	$C_p°$ (J $K^{-1}mol^{-1}$)	S° (J $K^{-1}mol^{-1}$)	$-[G°-H°(T_r)]/T$ (J $K^{-1}mol^{-1}$)	$H°-H°(T_r)$ (kJ mol^{-1})	$\Delta_f H°$ (kJ mol^{-1})	$\Delta_f G°$ (kJ mol^{-1})	Log K_f
0							
100							
200							
298.15	105.370	174.473	174.473	0.	-283.675	-259.095	45.392
300	105.395	175.125	174.475	0.195	-283.734	-258.942	45.086
400	106.859	205.645	178.626	10.808	-327.297	-241.576	31.547
500	108.324	229.648	186.514	21.567	-324.624	-220.454	23.031
600	109.788	249.527	195.406	32.472	-321.915	-199.874	17.401
700	111.253	266.561	204.383	43.524	-319.143	-179.752	13.413
800	112.717	281.512	213.108	54.723	-316.294	-160.032	10.449
900	114.181	294.873	221.464	66.068	-313.362	-140.675	8.165
1000	115.646	306.979	229.420	77.559	-310.344	-121.648	6.354
1100	117.110	318.070	236.982	89.197	-307.244	-102.928	4.888
1200	118.575	328.322	244.171	100.981	-304.066	-84.494	3.678
1300	120.039	337.871	251.016	112.912	-300.817	-66.327	2.665
1400	121.503	346.821	257.543	124.989	-297.504	-48.414	1.806
1500	122.968	355.254	263.778	137.213	-294.134	-30.739	1.070

PREVIOUS: CURRENT: September 1978

Molybdenum Bromide ($MoBr_3$) Br_3Mo_1(cr)

MOLYBDENUM BROMIDE ($MoBr_3$) IDEAL GAS $M_r = 335.652$

$S°(298.15\ K) = [375.0 \pm 8]\ J\ K^{-1}\ mol^{-1}$

$\Delta_f H°(0\ K) = [13.1 \pm 33]\ kJ\ mol^{-1}$
$\Delta_f H°(298.15\ K) = [-8.4 \pm 33]\ kJ\ mol^{-1}$

Electronic Levels and Quantum Weights

ε_i, cm^{-1}	g_i
0	[2]
[3505.43]	[2]
[10860.95]	[4]
[11062]	[4]

Vibrational Frequencies and Degeneracies

ν, cm^{-1}	ν, cm^{-1}
[170](1)	[260](2)
[95](1)	[90](2)

Point Group: $[D_{3h}]$ $\sigma = [6]$
Bond Distance: Mo–Br = [2.42] Å
Bond Angle: Br–Mo–Br = [120]°
Product of the Moments of Inertia: $I_A I_B I_C = [3.16715 \times 10^{-111}]\ g^3\ cm^6$

Enthalpy of Formation

The adopted value of $\Delta_f H°(298.15\ K) = -2.0 \pm 8\ kcal\ mol^{-1}$ ($-8.368 \pm 33\ kJ\ mol^{-1}$) is that estimated by Brewer (1) who used bonding models to estimate the enthalpy of formation of MoBr(g), $MoBr_2$(g), and $MoBr_3$(g). The value of $\Delta_f H°(0\ K)$ combined with JANAF data (2) for Mo(g) and Br(g) leads to $\Delta_{at}H°(MoBr_3,\ g) = 238.6 \pm 8\ kcal\ mol^{-1}$. The mean bond energy $\Delta_{at}H°(MoBr_3,\ g)/3 = 79.5 \pm 3\ kcal\ mol^{-1}$ compared to $\Delta_{at}H°(MoBr_4,\ g)/4 = 76.0\ kcal\ mol^{-1}$ (4). The enthalpy of formation of $MoBr_4$(g) is based on the experimental measurements of Oppermann (3), providing some confidence in the estimate of Brewer.

Heat Capacity and Entropy

Brewer (1) used a slight modification of the vibrational frequencies and internuclear distance estimated by Lofgren (4). The Mo-Br distance was taken as 2.42 Å. Drake and Rosenblatt (5) predicted that $MoBr_3$ would be planar. The electronic contributions are taken to be the same as those for MoF_3(g) (2) which are estimated from TaO(g) (2). The principal moments of inertia are $I_A = I_B = 116.5591 \times 10^{-39}\ g\ cm^2$ and $I_C = 233.1183 \times 10^{-39}\ g\ cm^2$

References

1. L. Brewer, Materials and Molecular Research Division, Lawrence Berkeley Laboratory, University of California, Berkeley; personal communication of Sept. 29, 1978, preliminary draft of review to be submitted for publication in Atomic Energy Review, International Atomic Energy Agency, Vienna, Austria.
2. JANAF Thermochemical Tables: $MoBr_4$(g) and MoF_3(g), 9-30-78; TaO(g), 12-31-73, Br(g), 9-30-61; Mo(g), 3-31-78.
3. H. Oppermann, Z. Anorg. Allg. Chem. 395, 249 (1973).
4. N. L. Lofgren, private communication to L. Brewer, 1967.
5. M. C. Drake and G. M. Rosenblatt, paper 388, The Electrochemical Society, Atlanta, Georgia, October, 1977.

Molybdenum Bromide ($MoBr_3$) Br_3Mo_1(g)

Enthalpy Reference Temperature = T_r = 298.15 K Standard State Pressure = p° = 0.1 MPa

T/K	$C_p°$	S°	$-[G°-H°(T_r)]/T$	$H°-H°(T_r)$	$\Delta_f H°$	$\Delta_f G°$	Log K_f
	J K^{-1}mol^{-1}			kJ mol^{-1}			
0	0.	0.	INFINITE	-19.910	13.071	13.071	INFINITE
100	65.839	293.440	443.828	-15.039	13.261	-7.303	3.815
200	77.191	343.488	382.257	-7.754	11.352	-27.224	7.110
250	79.157	360.945	376.306	-3.840	9.924	-36.710	7.670
298.15	80.270	374.990	374.990	0.	-8.368	-43.572	7.634
300	80.303	375.487	374.992	0.149	-8.474	-43.790	7.625
350	81.025	387.924	375.972	4.183	-54.628	-47.145	7.036
400	81.509	398.777	378.159	8.247	-54.551	-46.082	6.018
450	81.853	408.398	380.995	12.332	-54.493	-45.027	5.227
500	82.117	417.036	384.174	16.431	-54.453	-43.977	4.594
600	82.533	432.046	390.939	24.664	-54.416	-41.886	3.647
700	82.916	444.797	397.745	32.937	-54.423	-39.798	2.970
800	83.325	455.895	404.335	41.248	-54.462	-37.706	2.462
900	83.766	465.735	410.621	49.603	-54.520	-35.609	2.067
1000	84.222	474.584	416.582	58.002	-54.594	-33.504	1.750
1100	84.672	482.633	422.226	66.447	-54.686	-31.390	1.491
1200	85.099	490.019	427.572	74.936	-54.804	-29.267	1.274
1300	85.491	496.846	432.642	83.466	-54.956	-27.133	1.090
1400	85.845	503.195	437.457	92.033	-55.153	-24.986	0.932
1500	86.161	509.128	442.039	100.633	-55.406	-22.823	0.795
1600	86.445	514.698	446.408	109.264	-55.726	-20.641	0.674
1700	86.703	519.947	450.581	117.921	-56.125	-18.436	0.566
1800	86.943	524.909	454.574	126.604	-56.615	-16.206	0.470
1900	87.172	529.616	458.401	135.310	-57.207	-13.945	0.383
2000	87.395	534.093	462.074	144.038	-57.914	-11.650	0.304
2100	87.617	538.363	465.606	152.789	-58.746	-9.317	0.232
2200	87.841	542.444	469.007	161.562	-59.714	-6.941	0.165
2300	88.067	546.354	472.285	170.357	-60.832	-4.518	0.103
2400	88.297	550.106	475.450	179.175	-62.108	-2.043	0.044
2500	88.531	553.716	478.509	188.016	-63.556	0.489	-0.010
2600	88.766	557.192	481.469	196.881	-65.196	3.083	-0.062
2700	89.001	560.547	484.336	205.770	-67.068	5.743	-0.111
2800	89.235	563.788	487.116	214.681	-69.233	8.479	-0.158
2900	89.464	566.923	489.814	223.616	-107.653	11.347	-0.204
3000	89.688	569.960	492.435	232.574	-108.604	15.467	-0.269
3100	89.904	572.905	494.984	241.554	-109.558	19.618	-0.331
3200	90.110	575.762	497.464	250.555	-110.511	23.800	-0.388
3300	90.305	578.538	499.879	259.576	-111.461	28.012	-0.443
3400	90.487	581.237	502.232	268.615	-112.406	32.253	-0.496
3500	90.656	583.862	504.527	277.673	-113.343	36.521	-0.545
3600	90.810	586.418	506.766	286.746	-114.269	40.816	-0.592
3700	90.949	588.908	508.953	295.834	-115.182	45.137	-0.637
3800	91.072	591.335	511.089	304.935	-116.079	49.481	-0.680
3900	91.180	593.702	513.177	314.048	-116.958	53.850	-0.721
4000	91.272	596.012	515.219	323.171	-117.817	58.241	-0.761
4100	91.350	598.267	517.218	332.302	-118.655	62.652	-0.798
4200	91.412	600.469	519.174	341.440	-119.470	67.084	-0.834
4300	91.459	602.620	521.089	350.584	-120.260	71.535	-0.869
4400	91.493	604.723	522.966	359.731	-121.026	76.005	-0.902
4500	91.513	606.780	524.806	368.882	-121.765	80.491	-0.934
4600	91.521	608.791	526.610	378.034	-122.477	84.994	-0.965
4700	91.517	610.759	528.380	387.186	-123.162	89.511	-0.995
4800	91.502	612.686	530.116	396.337	-123.820	94.043	-1.023
4900	91.476	614.573	531.820	405.486	-124.450	98.589	-1.051
5000	91.441	616.420	533.494	414.632	-716.437	108.878	-1.137
5100	91.397	618.231	535.138	423.774	-718.691	125.407	-1.284
5200	91.345	620.005	536.753	432.911	-721.065	141.980	-1.426
5300	91.285	621.744	538.340	442.042	-723.555	158.601	-1.563
5400	91.219	623.450	539.900	451.168	-726.162	175.270	-1.695
5500	91.147	625.123	541.435	460.286	-728.882	191.988	-1.823
5600	91.069	626.765	542.944	469.397	-731.715	208.756	-1.947
5700	90.987	628.376	544.429	478.500	-734.658	225.577	-2.067
5800	90.900	629.958	545.890	487.594	-737.709	242.448	-2.183
5900	90.810	631.511	547.328	496.680	-740.864	259.375	-2.296
6000	90.716	633.036	548.744	505.756	-744.125	276.355	-2.406

PREVIOUS: September 1978 (1 atm) CURRENT: September 1978 (1 bar)

Molybdenum Bromide ($MoBr_3$) Br_3Mo_1(g)

PHOSPHORYL BROMIDE ($POBr_3$) IDEAL GAS M_r = 286.68516

$S°(298.15\ K) = 359.83\ J\ K^{-1}\ mol^{-1}$

$\Delta_f H°(0\ K) = [-379.4]\ kJ\ mol^{-1}$
$\Delta_f H°(298.15\ K) = [-406.0]\ kJ\ mol^{-1}$

Vibrational Frequencies and Degeneracies

$\nu,\ cm^{-1}$	$\nu,\ cm^{-1}$
118 (2)	340 (1)
173 (1)	488 (2)
267 (2)	1261 (1)

Ground State Quantum Weight: [1] $\sigma = 3$
Point Group: C_{3v}
Bond Distances: P–O = 1.41 ± 0.07 Å P–Br = 2.06 ± 0.03 Å
Bond Angle: Br–P–Br = 108° ± 3°
Product of the Moments of Inertia: $I_A I_B I_C = 1.11121 \times 10^{-111}\ g^3\ cm^6$

Enthalpy of Formation

The $\Delta_f H°(POBr_3$, g, 298.15 K) is calculated from the enthalpy of formation of the crystal and the enthalpy of sublimation. The enthalpy of sublimation is obtained by combining the $\Delta_{vap}H°$ and an estimated $\Delta_{fus}H°$. The enthalpy of formation of the crystal is derived from the enthalpies observed for the reaction: $POBr_3(cr) + (n+3)H_2O(\ell) \rightarrow [H_3PO_4 + 3\ HBr]$ in $nH_2O(\ell)$. The data used in these calculations may be summarized as follows:

Data	Source
$\Delta_f H°(H_3PO_4$, aq, 298.15 K)	Obtained graphically from Holmes (1)
$\Delta_f H°$(HBr, aq, 298.15 K)	Obtained graphically from (2)
$\Delta_f H°(H_2O$, ℓ, 298.15 K) = −68.3174 kcal mol^{-1}	Obtained from (2)
$\Delta_f H°(POBr_3$, cr, 298.15 K) = −109.1 kcal mol^{-1}	Recalculated from data of Charnley and Skinner (3)
$\Delta_{fus}H°$(298.15 K) = [3.7] kcal mol^{-1}	Estimated from comparison with $POCl_3$
$\Delta_{vap}H°$(298.15 K) = [12.1] kcal mol^{-1}	Estimated from the $\Delta_{vap}H°$ at T_{vap}(10.6 kcal mol^{-1} at 464.9 K). $\Delta_{vap}H$ was calculated from a least squares treatment of the vapor pressure data of Driel (4)

Heat Capacity and Entropy

The molecular constants used were calculated from electron diffraction measurements by Secrist and Brockway (5). The principal moments of inertia are: $I_A = I_B = 86.8222 \times 10^{-39}\ g\ cm^2$ and $I_C = 147.4125 \times 10^{-39}\ g\ cm^2$.

From the Raman spectra of $POBr_3$ the vibrational frequencies were measured and assigned by Gerding and Driel (6). The Raman spectra was studied and the vibrational frequencies were listed by Delwaulle and Francois (7).

References

1. W. S. Holmes, Trans. Faraday Soc. 58, 1916 (1962).
2. U. S. Nat. Bur. Stand. Circ. 500, 1952.
3. T. Charnley and H. A. Skinner, J. Chem. Soc. 450 (1953).
4. M. van Driel, Rec. Trav. Chim. 61, 748 (1942).
5. J. H. Secrist and L. O. Brockway, J. Am. Chem. Soc. 66, 1941 (1944).
6. H. Gerding and M. van Driel, Rec. Trav. Chim. 61, 419 (1942).
7. M. L. Delwaulle and F. Francois, Compt. Rend. 220, 817 (1945); ibid. 222, 550 (1946); and J. Chim. Phys. 46, 87 (1949).

Phosphoryl Bromide ($POBr_3$) $Br_3O_1P_1(g)$

Enthalpy Reference Temperature = T_r = 298.15 K Standard State Pressure = p° = 0.1 MPa

T/K	$C°_p$ (J $K^{-1}mol^{-1}$)	S° (J $K^{-1}mol^{-1}$)	−[G°−H°(T_r)]/T (J $K^{-1}mol^{-1}$)	H°−H°(T_r) (kJ mol^{-1})	$\Delta_f H°$ (kJ mol^{-1})	$\Delta_f G°$ (kJ mol^{-1})	Log K_f
0	0.	0.	INFINITE	-19.850	-379.406	-379.406	INFINITE
100	59.241	277.278	432.506	-15.523	-381.608	-390.825	204.146
200	80.331	325.763	367.792	-8.406	-385.562	-398.714	104.134
250	85.998	344.337	361.291	-4.239	-387.439	-401.789	83.949
298.15	89.862	359.832	359.832	0.	-406.022	-402.406	70.500
300	89.989	360.388	359.834	0.166	-406.136	-402.383	70.061
350	92.960	374.494	360.941	4.743	-453.209	-399.138	59.568
400	95.267	387.063	363.435	9.451	-453.309	-391.405	51.112
450	97.107	398.394	366.700	14.262	-453.335	-383.665	44.535
500	98.602	408.706	370.393	19.156	-453.303	-375.925	39.273
600	100.851	426.894	378.334	29.136	-453.116	-360.465	31.381
700	102.423	442.566	386.417	39.304	-452.815	-345.046	25.748
800	103.555	456.320	394.313	49.606	-452.443	-329.675	21.526
900	104.391	468.568	401.895	60.005	-452.025	-314.354	18.245
1000	105.022	479.601	409.123	70.477	-451.577	-299.081	15.622
1100	105.508	489.634	415.993	81.005	-451.112	-283.854	13.479
1200	105.890	498.832	422.519	91.576	-514.208	-267.591	11.648
1300	106.194	507.320	428.719	102.180	-512.953	-247.090	9.928
1400	106.441	515.199	434.618	112.813	-511.705	-226.686	8.458
1500	106.643	522.550	440.238	123.467	-510.466	-206.371	7.186
1600	106.810	529.438	445.600	134.140	-509.241	-186.138	6.077
1700	106.950	535.917	450.724	144.828	-508.033	-165.981	5.100
1800	107.069	542.034	455.629	155.529	-506.847	-145.895	4.234
1900	107.170	547.826	460.330	166.241	-505.687	-125.874	3.461
2000	107.257	553.325	464.844	176.963	-504.556	-105.913	2.766
2100	107.332	558.560	469.183	187.692	-503.459	-86.008	2.139
2200	107.398	563.555	473.360	198.429	-502.398	-66.154	1.571
2300	107.455	568.330	477.386	209.172	-501.376	-46.348	1.053
2400	107.506	572.904	481.271	219.920	-500.395	-26.585	0.579
2500	107.551	577.294	485.025	230.673	-499.456	-6.863	0.143
2600	107.590	581.513	488.655	241.430	-498.559	12.823	-0.258
2700	107.626	585.574	492.170	252.191	-497.704	32.475	-0.628
2800	107.658	589.489	495.576	262.955	-496.890	52.097	-0.972
2900	107.687	593.267	498.880	273.722	-496.114	71.689	-1.291
3000	107.712	596.918	502.088	284.492	-495.374	91.256	-1.589
3100	107.736	600.450	505.204	295.264	-494.667	110.799	-1.867
3200	107.757	603.871	508.234	306.039	-493.990	130.319	-2.127
3300	107.777	607.187	511.183	316.816	-493.337	149.818	-2.371
3400	107.795	610.405	514.054	327.594	-492.706	169.299	-2.601
3500	107.811	613.530	516.852	338.375	-492.091	188.760	-2.817
3600	107.826	616.567	519.580	349.157	-491.490	208.205	-3.021
3700	107.840	619.522	522.241	359.940	-490.897	227.632	-3.214
3800	107.853	622.398	524.839	370.725	-490.308	247.044	-3.396
3900	107.864	625.200	527.377	381.510	-489.721	266.441	-3.569
4000	107.875	627.931	529.856	392.297	-489.130	285.822	-3.732
4100	107.885	630.595	532.281	403.085	-488.533	305.188	-3.888
4200	107.895	633.195	534.653	413.874	-487.927	324.540	-4.036
4300	107.904	635.733	536.974	424.664	-487.308	343.877	-4.177
4400	107.912	638.214	539.247	435.455	-486.675	363.200	-4.312
4500	107.920	640.639	541.473	446.247	-486.025	382.507	-4.440
4600	107.927	643.011	543.655	457.039	-485.355	401.801	-4.563
4700	107.933	645.333	545.794	467.832	-484.666	421.079	-4.680
4800	107.940	647.605	547.891	478.626	-483.955	440.343	-4.792
4900	107.946	649.831	549.949	489.420	-483.221	459.592	-4.899
5000	107.951	652.012	551.969	500.215	-482.463	478.826	-5.002
5100	107.957	654.149	553.951	511.010	-481.681	498.044	-5.101
5200	107.961	656.246	555.898	521.806	-480.876	517.246	-5.196
5300	107.966	658.302	557.811	532.603	-480.045	536.433	-5.287
5400	107.971	660.320	559.691	543.399	-479.190	555.604	-5.374
5500	107.975	662.302	561.539	554.197	-478.312	574.758	-5.459
5600	107.979	664.247	563.355	564.994	-477.409	593.897	-5.540
5700	107.983	666.158	565.142	575.792	-476.484	613.020	-5.618
5800	107.986	668.036	566.900	586.591	-475.536	632.125	-5.693
5900	107.989	669.882	568.630	597.390	-474.567	651.214	-5.765
6000	107.993	671.697	570.333	608.189	-473.577	670.287	-5.835

PREVIOUS: December 1963 (1 atm) CURRENT: December 1963 (1 bar)

Phosphoryl Bromide ($POBr_3$) $Br_3O_1P_1(g)$

PHOSPHORUS BROMIDE (PBr_3) IDEAL GAS M_r = 270.68576

$\Delta_f H°(0\ K) = -121.7 \pm 6.3\ kJ\ mol^{-1}$

$S°(298.15\ K) = 348.234\ J\ K^{-1}\ mol^{-1}$ $\Delta_f H°(298.15\ K) = -146.0 \pm 6.3\ kJ\ mol^{-1}$

Vibrational Frequencies and Degeneracies

ν, cm^{-1}	ν, cm^{-1}
392 (1)	392.2 (2)
161.3 (1)	115.7 (2)

Ground State Quantum Weight: [1] $\sigma = 3$

Point Group: C_{3v}

Bond Distance: P-Br = 2.20 ± 0.03 Å

Bond Angle: Br-P-Br = 106° ± 3°

Product of the Moments of Inertia: $I_A I_B I_C = 1.18987 \times 10^{-111}\ g^3\ cm^6$

Enthalpy of Formation

The enthalpy of formation of PBr gas from white (α) phosphorus, $-34.9\ kcal\ mol^{-1}$, was obtained by Charnley and Skinner (1). To obtain their value ($-34.9\ kcal\ mol^{-1}$), Charnley and Skinner (1) measured the enthalpy of formation of the liquid, and combined it and their estimate of the liquid enthalpy with the enthalpy of vaporization ($9.5\ kcal\ mol^{-1}$) calculated by Driel and Gerding (2). For the liquid enthalpy of formation Charnley and Skinner (1) observed the $\Delta_r H° = -67.2 \pm 0.6\ kcal\ mol^{-1}$ for $PBr_3(l) + (n+3)H_2O(l) \rightarrow [H_3PO_3 + 3HBr]$ in $nH_2O(l)$. Driel and Gerding (2) measured the vapor pressure over $PBr_3(l)$ and calculated the enthalpy of vaporization with the aid of a modified Claperyron equation. The uncertainties on the enthalpy of formation are estimates.

Heat Capacity and Entropy

Williams and Gordy (3) have reported moments of inertia from the microwave spectrum of PBr_3. From this data Whiffen (4) has calculated the selected molecular constants. The principal moments of inertia are: $I_A = I_B = 85.2188 \times 10^{-39}\ g\ cm^2$ and $I_C = 163.8427 \times 10^{-39}\ g\ cm^2$.

Lister and Sutton (5) have reported P-Br = 2.23 ± 0.04 Å and the angle Br-P-Br = 100±2. Swingle (6) has measured P-Br = 2.18±0.03 Å and the angle Br-P-Br = 101.5±1.5. Swingle's data as well as that of Lister and Sutton were obtained by electron diffraction. The 106° angle was selected since 101.5° and 100° give imaginary force constants with the ranges allowed for the type A, 392 cm^{-1} frequency and the form of the potential function assumed by Davis and Oetjen (7).

The vibrational frequencies were measured and assigned in the infrared by Davis and Oetjen (7). Raman measurements and assignments by Cabannes and Rausset (8) were also listed by Delwaulle (9). The Raman assignments were as follows: type A, 380 cm^{-1} and 162 cm^{-1} and type E, 400 cm^{-1} and 116 cm^{-1}.

References

1. T. Charnley and H. A. Skinner, J. Chem. Soc. 450 (1953).
2. M. van Driel and H. Gerding, Rec. Trav. Chem. 60, 943 (1941).
3. Q. Williams and W. Gordy, Phys. Rev. 79, 225 (1950).
4. D. H. Whiffen in "Tables of Interatomic Distances and Configuration in Molecules and Ions", Scientific Editor, L. E. Sutton, Special Publication No. 11, The Chemical Society (London) 1958.
5. M. Lister and L. E. Sutton, Trans. Faraday Soc. 37, 393 (1941).
6. S. M. Swingle, personal communication from L. Pauling and V. Schomaker, California Institute of Technology to P. W. Allen and L. E. Sutton, Acta. Cryst. 3, 46 (1950).
7. P. W. Davis and R. A. Oetjen, J. Molecular Spec. 2, 253 (1958).
8. J. Cabannes and A. Rausset, Ann. Phys. (Paris) 19, 229 (1933).
9. M. L. Delwaulle, Compt. Rend. 222, 1391 (1946), Compt. Rend. 224, 389 (1947).

Phosphorus Bromide (PBr_3) $Br_3P_1(g)$

Enthalpy Reference Temperature = T_r = 298.15 K   Standard State Pressure = p° = 0.1 MPa

T/K	$C_p°$ (J K⁻¹mol⁻¹)	S° (J K⁻¹mol⁻¹)	$-[G°-H°(T_r)]/T$ (J K⁻¹mol⁻¹)	$H°-H°(T_r)$ (kJ mol⁻¹)	$\Delta_f H°$ (kJ mol⁻¹)	$\Delta_f G°$ (kJ mol⁻¹)	Log K_f
0	0.	0.	INFINITE	-17.753	-121.651	-121.651	INFINITE
100	54.768	275.812	411.024	-13.521	-122.496	-140.231	73.249
200	69.764	319.032	355.025	-7.199	-125.788	-156.943	40.989
250	73.619	335.046	349.473	-3.607	-127.512	-164.539	34.378
298.15	76.033	348.234	348.234	0.	-146.022	-169.530	29.701
300	76.108	348.704	348.235	0.141	-146.135	-169.675	29.543
350	77.770	360.569	349.168	3.990	-193.196	-170.981	25.518
400	78.922	371.034	351.260	7.909	-193.338	-167.797	21.912
450	79.748	380.380	353.986	11.877	-193.449	-164.598	19.106
500	80.357	388.815	357.054	15.881	-193.537	-161.387	16.860
600	81.176	403.545	363.610	23.961	-193.669	-154.944	13.489
700	81.683	416.099	370.234	32.105	-193.764	-148.482	11.080
800	82.018	427.029	376.665	40.292	-193.840	-142.007	9.272
900	82.250	436.704	382.809	48.506	-193.904	-135.525	7.866
1000	82.417	445.379	388.640	56.739	-193.964	-129.035	6.740
1100	82.542	453.240	394.161	64.988	-194.023	-122.539	5.819
1200	82.637	460.427	399.388	73.247	-257.656	-114.959	5.004
1300	82.711	467.044	404.341	81.514	-256.947	-103.097	4.142
1400	82.771	473.176	409.041	89.789	-256.250	-91.288	3.406
1500	82.818	478.888	413.509	98.068	-255.566	-79.529	2.769
1600	82.858	484.234	417.764	106.352	-254.896	-67.816	2.214
1700	82.890	489.259	421.824	114.639	-254.243	-56.143	1.725
1800	82.918	493.997	425.703	122.930	-253.610	-44.508	1.292
1900	82.941	498.481	429.416	131.223	-252.999	-32.908	0.905
2000	82.960	502.736	432.977	139.518	-252.414	-21.340	0.557
2100	82.977	506.784	436.396	147.815	-251.856	-9.800	0.244
2200	82.992	510.644	439.684	156.113	-251.329	1.714	-0.041
2300	83.005	514.334	442.850	164.413	-250.835	13.205	-0.300
2400	83.016	517.867	445.903	172.714	-250.374	24.675	-0.537
2500	83.026	521.256	448.849	181.016	-249.948	36.126	-0.755
2600	83.035	524.512	451.697	189.319	-249.558	47.561	-0.956
2700	83.043	527.646	454.453	197.623	-249.201	58.982	-1.141
2800	83.050	530.667	457.121	205.928	-248.877	70.390	-1.313
2900	83.057	533.581	459.707	214.233	-248.585	81.788	-1.473
3000	83.062	536.397	462.217	222.539	-248.320	93.175	-1.622
3100	83.067	539.121	464.654	230.846	-248.081	104.554	-1.762
3200	83.072	541.758	467.023	239.153	-247.864	115.926	-1.892
3300	83.076	544.314	469.326	247.460	-247.666	127.291	-2.015
3400	83.080	546.794	471.568	255.768	-247.481	138.651	-2.130
3500	83.084	549.203	473.752	264.076	-247.307	150.005	-2.239
3600	83.087	551.543	475.881	272.385	-247.139	161.355	-2.341
3700	83.090	553.820	477.957	280.694	-246.974	172.700	-2.438
3800	83.093	556.036	479.982	289.003	-246.806	184.039	-2.530
3900	83.096	558.194	481.960	297.312	-246.634	195.375	-2.617
4000	83.098	560.298	483.893	305.622	-246.453	206.707	-2.699
4100	83.100	562.350	485.781	313.932	-246.260	218.033	-2.778
4200	83.102	564.353	487.628	322.242	-246.052	229.355	-2.852
4300	83.104	566.308	489.435	330.552	-245.826	240.671	-2.924
4400	83.106	568.219	491.204	338.863	-245.580	251.983	-2.991
4500	83.108	570.086	492.936	347.174	-245.312	263.288	-3.056
4600	83.109	571.913	494.634	355.485	-245.019	274.587	-3.118
4700	83.111	573.700	496.297	363.796	-244.700	285.879	-3.177
4800	83.112	575.450	497.928	372.107	-244.354	297.165	-3.234
4900	83.113	577.164	499.527	380.418	-243.979	308.442	-3.288
5000	83.115	578.843	501.097	388.729	-243.574	319.712	-3.340
5100	83.116	580.489	502.638	397.041	-243.139	330.974	-3.390
5200	83.117	582.103	504.150	405.353	-242.673	342.226	-3.438
5300	83.118	583.686	505.636	413.664	-242.176	353.470	-3.484
5400	83.119	585.240	507.096	421.976	-241.647	364.704	-3.528
5500	83.120	586.765	508.531	430.288	-241.087	375.927	-3.570
5600	83.121	588.262	509.941	438.600	-240.494	387.140	-3.611
5700	83.121	589.734	511.328	446.912	-239.871	398.343	-3.650
5800	83.122	591.179	512.692	455.224	-239.215	409.533	-3.688
5900	83.123	592.600	514.035	463.537	-238.529	420.713	-3.725
6000	83.124	593.997	515.356	471.849	-237.812	431.880	-3.760

PREVIOUS: December 1963 (1 atm)   CURRENT: December 1963 (1 bar)

Phosphorus Bromide (PBr_3) $Br_3P_1(g)$

THIOPHOSPHORYL BROMIDE ($PSBr_3$) IDEAL GAS M_r = 302.74576

$S°(298.15 K) = 372.828 J K^{-1} mol^{-1}$

$\Delta_f H°(0 K) = [-255.7] kJ mol^{-1}$
$\Delta_f H°(298.15 K) = [-281.1] kJ mol^{-1}$

Vibrational Frequencies and Degeneracies

ν, cm^{-1}
299 (1)
438 (2)
718 (1)
115 (2)
165 (1)
179 (2)

Ground State Quantum Weight: [1] $\sigma = 3$

Point Group: C_{3v}

Bond Distance: P–Br = 2.13 ± 0.03 Å P–S = 1.89 ± 0.06 Å

Bond Angle: Br–P–Br = 106° ± 3°

Product of the Moments of Inertia: $I_A I_B I_C = 1.94531 \times 10^{-111} g^3 cm^6$

Enthalpy of Formation

The $\Delta_f H°(PSBr_3$, g, 298.15 K) was calculated from the $\Delta_f H°(POBr_3$, g, 298.15 K) and estimated strengths for P–S and P–O bonds.

The difference in the P–S and P–O bond strengths was assumed to be the same for $POF_3 - PSF_3$ and $POBr_3 - PSBr_3$. For the P–S bond strength in PSF_3, Henderson and Scheffee (1) estimated 91 kcal from a consideration of unpublished data. The P–O bond was estimated to be 115 kcal by Neale and Williams (2) and Neale et al. (3).

Auxiliary data for $POBr_3$(g), O(g), and S(g) are from the JANAF tables (4).

Heat Capacity and Entropy

The molecular constants were determined from electron diffraction data by Secrist and Brockway (5). The principal moments of inertia are: $I_A = I_B = 112.5445 \times 10^{-39} g cm^2$ and $I_C = 153.5823 \times 10^{-39} g cm^2$.

The vibrational frequencies obtained from the Raman spectra of $PSBr_3$ were reported by Delwaulle and Francois (6).

References

1. C. B. Henderson and R. S. Scheffee, "Survey of Thermochemical Data," Atlantic Research Corp., Alexandria, Virginia, January, 1960.
2. E. Neale and L. T. D. Williams, J. Chem. Soc. 2485 (1955).
3. E. Neale, L. T. D. Williams, and V. T. Moores, J. Chem. Soc., 422 (1956).
4. JANAF Thermochemical Tables: $POBr_3$(g), 3-31-63; O(g), 6-30-62; S(g), 3-31-61.
5. J. H. Secrist and L. O. Brockway, J. Am. Chem. Soc. 66, 1941 (1944).
6. M. L. Delwaulle and F. Francois, Compt. Rend. 224, 1422 (1947).

Thiophosphoryl Bromide ($PSBr_3$) $Br_3P_1S_1(g)$

Enthalpy Reference Temperature = T_r = 298.15 K Standard State Pressure = p° = 0.1 MPa

T/K	$C_p°$ (J K⁻¹mol⁻¹)	S° (J K⁻¹mol⁻¹)	$-[G°-H°(T_r)]/T$ (J K⁻¹mol⁻¹)	$H°-H°(T_r)$ (kJ mol⁻¹)	$\Delta_f H°$ (kJ mol⁻¹)	$\Delta_f G°$ (kJ mol⁻¹)	Log K_f
0	0.	0.	INFINITE	-21.191	-255.707	-255.707	INFINITE
100	65.087	284.297	450.202	-16.591	-256.873	-274.205	143.230
200	85.612	336.689	381.258	-8.914	-260.454	-290.412	75.848
250	91.175	356.429	374.371	-4.486	-262.360	-297.688	62.198
298.15	94.934	372.828	372.828	0.	-281.052	-302.335	52.968
300	95.055	373.415	372.829	0.176	-281.172	-302.467	52.664
350	97.818	388.287	373.998	5.001	-328.424	-303.383	45.277
400	99.831	401.487	376.624	9.945	-330.972	-299.695	39.136
450	101.330	413.336	380.056	14.976	-331.945	-295.735	34.328
500	102.468	424.074	383.930	20.072	-332.943	-291.656	30.469
600	104.045	442.907	392.234	30.404	-334.408	-283.253	24.659
700	105.051	459.026	400.652	40.862	-335.536	-274.635	20.494
800	105.728	473.101	408.847	51.403	-336.468	-265.869	17.359
900	106.204	485.583	416.693	62.001	-390.405	-255.914	14.853
1000	106.551	496.792	424.152	72.640	-389.916	-240.997	12.588
1100	106.810	506.960	431.225	83.309	-389.427	-226.128	10.738
1200	107.010	516.262	437.929	94.000	-452.515	-210.226	9.151
1300	107.166	524.834	444.289	104.709	-451.267	-190.086	7.638
1400	107.291	532.781	450.329	115.432	-450.038	-170.041	6.344
1500	107.392	540.187	456.076	126.167	-448.833	-150.084	5.226
1600	107.475	547.120	461.552	136.910	-447.652	-130.206	4.251
1700	107.544	553.638	466.779	147.661	-446.498	-110.401	3.392
1800	107.602	559.787	471.777	158.419	-445.374	-90.663	2.631
1900	107.652	565.606	476.563	169.181	-444.283	-70.987	1.952
2000	107.694	571.129	481.155	179.949	-443.227	-51.367	1.342
2100	107.730	576.384	485.565	190.720	-442.207	-31.799	0.791
2200	107.762	581.397	489.808	201.495	-441.227	-12.279	0.292
2300	107.789	586.187	493.895	212.272	-440.286	7.197	-0.163
2400	107.813	590.775	497.837	223.052	-439.387	26.633	-0.580
2500	107.835	595.177	501.643	233.835	-438.530	46.033	-0.962
2600	107.854	599.407	505.322	244.619	-437.714	65.399	-1.314
2700	107.871	603.477	508.883	255.405	-436.937	84.735	-1.639
2800	107.886	607.401	512.332	266.193	-436.198	104.042	-1.941
2900	107.899	611.187	515.676	276.982	-435.495	123.324	-2.221
3000	107.912	614.845	518.921	287.773	-434.825	142.582	-2.483
3100	107.923	618.384	522.072	298.565	-434.184	161.819	-2.727
3200	107.933	621.810	525.136	309.358	-433.569	181.035	-2.955
3300	107.942	625.132	528.116	320.151	-432 975	200.232	-3.169
3400	107.950	628.354	531.017	330.946	-432.400	219.411	-3.371
3500	107.958	631.483	533.843	341.741	-431.839	238.574	-3.561
3600	107.965	634.525	536.598	352.538	-431.288	257.721	-3.739
3700	107.972	637.483	539.285	363.334	-430.743	276.853	-3.908
3800	107.978	640.363	541.907	374.132	-430.201	295.969	-4.068
3900	107.983	643.167	544.467	384.930	-429.657	315.072	-4.220
4000	107.988	645.901	546.969	395.728	-429.109	334.161	-4.364
4100	107.993	648.568	549.415	406.528	-428.553	353.235	-4.500
4200	107.998	651.170	551.807	417.327	-427.987	372.297	-4.630
4300	108.002	653.712	554.147	428.127	-427.407	391.344	-4.754
4400	108.005	656.195	556.438	438.927	-426.813	410.378	-4.872
4500	108.009	658.622	558.682	449.728	-426.202	429.398	-4.984
4600	108.012	660.996	560.881	460.529	-425.571	448.405	-5.092
4700	108.016	663.319	563.036	471.331	-424.921	467.397	-5.195
4800	108.019	665.593	565.149	482.132	-424.249	486.376	-5.293
4900	108.021	667.820	567.221	492.934	-423.555	505.340	-5.387
5000	108.024	670.002	569.255	503.737	-422.837	524.290	-5.477
5100	108.026	672.142	571.252	514.539	-422.095	543.225	-5.564
5200	108.029	674.239	573.212	525.342	-421.330	562.145	-5.647
5300	108.031	676.297	575.138	536.145	-420.540	581.051	-5.727
5400	108.033	678.316	577.030	546.948	-419.725	599.942	-5.803
5500	108.035	680.299	578.889	557.751	-418.885	618.817	-5.877
5600	108.037	682.245	580.718	568.555	-418.021	637.676	-5.948
5700	108.039	684.158	582.516	579.359	-417.133	656.520	-6.016
5800	108.040	686.037	584.284	590.163	-416.221	675.347	-6.082
5900	108.042	687.884	586.025	600.967	-415.285	694.160	-6.146
6000	108.043	689.699	587.738	611.771	-414.326	712.956	-6.207

PREVIOUS: December 1963 (1 atm) CURRENT: December 1963 (1 bar)

Thiophosphoryl Bromide ($PSBr_3$) $Br_3P_1S_1(g)$

TRIBROMOSILYL ($SiBr_3$) IDEAL GAS M_r = 267.7975

Tribromosilyl ($SiBr_3$) $Br_3Si_1(g)$

$S°(298.15\ K) = [351.77 \pm 8.4]\ J\ K^{-1}\ mol^{-1}$

$\Delta_f H°(0\ K) = [-179.1 \pm 63]\ kJ\ mol^{-1}$

$\Delta_f H°(298.15\ K) = [-201.7 \pm 63]\ kJ\ mol^{-1}$

Vibrational Frequencies and Degeneracies

ν, cm^{-1}
[362] (1)
[168] (1)
[483.5](2)
[116.8](2)

Ground State Quantum Weight: [2] σ = [3]

Point Group: [C_{3v}]

Bond distance: Si-Br = [2.17]Å

Bond Angle: Br-Si-Br = [111.36]°

Product of the Moments of Inertia: $I_AI_BI_C = [1.29082 \times 10^{111}]\ g^3\ cm^6$

Enthalpy of Formation

The enthalpy of formation of $SiBr_3(g)$ is based on an assumed average bond energy of 78±5 kcal mol^{-1} (1). This average bond energy is that of $SiBr_4(g)$, i.e. $\Delta_a H°(SiBr_4, g,)/4$. The rationale for this assumption is based on the same relationship existing for the silicon chloride and fluoride species (1).

Heat Capacity and Entropy

The molecular structure is assumed to be identical to the $SiBr_3$ group in $SiHBr_3$ (1). From this structure we estimate the following principal moments of inertia: $I_A = I_B = 87.0168 \times 10^{-39}$, and $I_C = 170.479 \times 10^{-39}\ g\ cm^2$. The vibrational frequencies are assumed to be those of the $SiBr_3$ group in $SiHBr_3(g)$.

Reference

1. JANAF Thermochemical Tables: $SiCl_3(g)$ and $SiF_3(g)$, 12-31-77; $SiBr_4(g)$ and $SiHBr_3(g)$, 12-31-76.

Enthalpy Reference Temperature = T_r = 298.15 K Standard State Pressure = p° = 0.1 MPa

T/K	$C_p°$ (J K⁻¹mol⁻¹)	S° (J K⁻¹mol⁻¹)	$-[G°-H°(T_r)]/T$ (J K⁻¹mol⁻¹)	$H°-H°(T_r)$ (kJ mol⁻¹)	$\Delta_f H°$ (kJ mol⁻¹)	$\Delta_f G°$ (kJ mol⁻¹)	Log K_f
0	0.	0.	INFINITE	-17.370	-179.082	-179.082	INFINITE
100	53.692	281.244	412.980	-13.174	-179.341	-199.005	103.950
200	67.782	323.273	358.414	-7.028	-181.727	-217.822	56.889
250	71.881	338.868	352.988	-3.530	-183.302	-226.671	47.360
298.15	74.578	351.774	351.774	0.	-201.694	-232.894	40.802
300	74.663	352.236	351.776	0.138	-201.803	-233.087	40.584
350	76.584	363.898	352.692	3.922	-248.029	-235.764	35.186
400	77.944	374.218	354.750	7.787	-247.995	-234.014	30.559
450	78.934	383.459	357.436	11.710	-247.958	-232.268	26.961
500	79.673	391.816	360.463	15.676	-247.921	-230.527	24.083
600	80.677	406.438	366.941	23.698	-247.855	-227.055	19.767
700	81.305	418.925	373.497	31.799	-247.809	-223.592	16.685
800	81.723	429.811	379.871	39.952	-247.794	-220.134	14.373
900	82.014	439.454	385.966	48.140	-247.815	-216.675	12.576
1000	82.224	448.107	391.754	56.352	-247.874	-213.213	11.137
1100	82.381	455.951	397.240	64.583	-247.974	-209.742	9.960
1200	82.501	463.125	402.435	72.827	-248.116	-206.260	8.978
1300	82.595	469.732	407.361	81.082	-248.302	-202.765	8.147
1400	82.670	475.856	412.038	89.346	-248.536	-199.254	7.434
1500	82.731	481.562	416.485	97.616	-248.818	-195.724	6.816
1600	82.780	486.903	420.721	105.892	-249.152	-192.174	6.274
1700	82.822	491.923	424.763	114.172	-299.718	-188.154	5.781
1800	82.856	496.658	428.627	122.456	-299.935	-181.586	5.269
1900	82.886	501.138	432.326	130.743	-300.171	-175.004	4.811
2000	82.911	505.390	435.874	139.033	-300.429	-168.410	4.398
2100	82.932	509.436	439.281	147.325	-300.712	-161.802	4.025
2200	82.951	513.294	442.559	155.619	-301.023	-155.180	3.684
2300	82.967	516.982	445.715	163.915	-301.365	-148.543	3.374
2400	82.982	520.514	448.758	172.212	-301.739	-141.891	3.088
2500	82.994	523.901	451.697	180.511	-302.145	-135.222	2.825
2600	83.006	527.157	454.537	188.811	-302.584	-128.537	2.582
2700	83.016	530.289	457.285	197.112	-303.056	-121.834	2.357
2800	83.025	533.309	459.946	205.414	-303.559	-115.113	2.147
2900	83.033	536.222	462.527	213.717	-304.091	-108.373	1.952
3000	83.040	539.037	465.030	222.021	-304.650	-101.615	1.769
3100	83.047	541.760	467.462	230.325	-305.232	-94.838	1.598
3200	83.053	544.397	469.825	238.630	-305.835	-88.041	1.437
3300	83.058	546.953	472.124	246.936	-306.455	-81.225	1.286
3400	83.063	549.432	474.361	255.242	-307.088	-74.391	1.143
3500	83.068	551.840	476.541	263.548	-307.729	-67.538	1.008
3600	83.072	554.180	478.665	271.855	-692.528	-50.205	0.728
3700	83.076	556.457	480.737	280.163	-692.761	-32.359	0.457
3800	83.079	558.672	482.759	288.470	-692.993	-14.508	0.199
3900	83.082	560.830	484.733	296.778	-693.220	3.350	-0.045
4000	83.086	562.934	486.662	305.087	-693.438	21.214	-0.277
4100	83.088	564.985	488.547	313.396	-693.643	39.082	-0.498
4200	83.091	566.988	490.391	321.705	-693.833	56.956	-0.708
4300	83.093	568.943	492.195	330.014	-694.003	74.834	-0.909
4400	83.096	570.853	493.961	338.323	-694.153	92.716	-1.101
4500	83.098	572.720	495.691	346.633	-694.278	110.601	-1.284
4600	83.100	574.547	497.385	354.943	-694.378	128.488	-1.459
4700	83.102	576.334	499.046	363.253	-694.450	146.377	-1.627
4800	83.103	578.084	500.675	371.563	-694.493	164.268	-1.788
4900	83.105	579.797	502.272	379.874	-694.505	182.159	-1.942
5000	83.107	581.476	503.839	388.184	-694.487	200.050	-2.090
5100	83.108	583.122	505.378	396.495	-694.436	217.940	-2.232
5200	83.109	584.736	506.889	404.806	-694.353	235.829	-2.369
5300	83.111	586.319	508.372	413.117	-694.238	253.716	-2.501
5400	83.112	587.872	509.830	421.428	-694.089	271.601	-2.627
5500	83.113	589.397	511.263	429.739	-693.907	289.482	-2.749
5600	83.114	590.895	512.672	438.050	-693.693	307.360	-2.867
5700	83.115	592.366	514.057	446.362	-693.446	325.234	-2.980
5800	83.116	593.812	515.420	454.674	-693.168	343.103	-3.090
5900	83.117	595.232	516.760	462.985	-692.859	360.967	-3.196
6000	83.118	596.629	518.080	471.297	-692.518	378.825	-3.298

PREVIOUS: December 1977 (1 atm) CURRENT: December 1977 (1 bar)

Tribromosilyl ($SiBr_3$) $Br_3Si_1(g)$

TITANIUM BROMIDE ($TiBr_3$) CRYSTAL M_r = 287.592

$S°(298.15\ K) = 176.4 \pm 3.3\ J\ K^{-1}\ mol^{-1}$

$T_{trs} = 179.9\ K$

$\Delta_f H°(0\ K) = -531.6 \pm 8.4\ kJ\ mol^{-1}$

$\Delta_f H°(298.15\ K) = -550.2 \pm 8.4\ kJ\ mol^{-1}$

$\Delta_{trs} H° = 0\ kJ\ mol^{-1}$

Enthalpy of Formation

The enthalpy of formation of $TiBr_3(cr)$ is based on the vapor pressure data reported by Hall and Blocher (1) for the process $1/2Hg_2Br_2(cr) + TiBr_3(cr) = TiBr_4(g) + Hg(\ell)$ over the temperature range 430 to 546 K. 2nd and 3rd law analyses of these data give values for the enthalpy of reaction of 24.2±0.3 and 24.4 kcal mol^{-1}, respectively, for seventy three points, with ten points rejected due to failure of a statistical test. The 3rd law drift in the data is calculated as 0.6±0.8 cal K^{-1} mol^{-1}. The 3rd law value of 24.4 kcal mol^{-1} is combined with the JANAF enthalpies of formation of $TiBr_4(g)$ and $HgBr_2(cr)$ to obtain the adopted enthalpy of formation, $\Delta_f H°(cr,\ 298.15\ K) = -131.5 \pm 2.0\ kcal\ mol^{-1}$ ($-550.196 \pm 8.4\ kJ\ mol^{-1}$).

Heat Capacity and Entropy

The heat capacity and entropy of $TiBr_3(cr)$ have been measured over the temperature range 51 to 800 K by King et al. (2). Heat capacities above 800 K are estimated from graphical extrapolation. The value of S°(298.15 K) is derived from these data, based on $S°(51\ K) = 8.60\ cal\ K^{-1}\ mol^{-1}$. The value of S°(51 K) is estimated from a Debye-Einstein extrapolation of the measured heat capacities, the equation being $C_p° = D(70.0/T) + E(120/T) + 2E(306/T)$. It is assumed that all electronic entropy is contained within the measured and extrapolated heat capacities.

Transition Data

A second order transition at 179.9 K was observed by King et al. (2). The heat capacity at this temperature is in excess of 56.1 cal K^{-1} mol^{-1}. King et al. (2) measured the value of H°(188 K) −H°(160 K) as 0.773 kcal mol^{-1}.

Sublimation Data

The enthalpy of sublimation is calculated from the enthalpies of formation of $TiBr_3(cr)$ and $TiBr_3(g)$ at the sublimation temperature. The sublimation temperature is taken as the point at which $\Delta_r G = 0$ for the process $TiBr_3(cr) = TiBr_3(g)$ at one bar.

References

1. E. H. Hall and J. M. Blocher, Jr., J. Electrochem. Soc. 105, 40 (1958).
2. E. G. King, W. W. Weller, A. U. Christensen and K. K. Kelley, U. S. Bur. Mines RI5799, 1961.

Titanium Bromide ($TiBr_3$) $Br_3Ti_1(cr)$

Enthalpy Reference Temperature = T_r = 298.15 K — Standard State Pressure = p° = 0.1 MPa

T/K	$C_p°$ (J K⁻¹mol⁻¹)	S° (J K⁻¹mol⁻¹)	−[G°−H°(T_r)]/T (J K⁻¹mol⁻¹)	H°−H°(T_r) (kJ mol⁻¹)	$\Delta_f H°$ (kJ mol⁻¹)	$\Delta_f G°$ (kJ mol⁻¹)	Log K_f
0	0.	0.	INFINITE	-22.970	-531.572	-531.572	INFINITE
100	70.655	74.136	264.466	-19.033	-532.385	-530.896	277.312
200	97.818	137.244	185.506	-9.652	-532.274	-529.251	138.226
298.15	101.705	176.443	176.443	0.	-550.196	-525.561	92.076
300	101.776	177.073	176.445	0.188	-550.264	-525.408	91.482
400	105.818	206.789	180.471	10.527	-594.259	-507.949	66.331
500	114.709	231.322	188.246	21.538	-591.526	-486.673	50.842
600	125.507	253.176	197.274	33.541	-587.928	-466.024	40.571
700	137.210	273.423	206.722	46.691	-583.274	-446.060	33.285
800	147.277	292.463	216.262	60.961	-577.557	-426.837	27.870
900	151.879	310.075	225.722	75.918	-571.237	-408.372	23.701
1000	156.673	326.329	234.980	91.349	-564.588	-390.630	20.404
1100	161.084	341.471	243.980	107.240	-557.687	-373.567	17.739
1200	165.111	355.663	252.702	123.553	-554.556	-357.021	15.541
1300	168.755	369.025	261.141	140.250	-546.541	-340.883	13.697
1400	172.015	381.653	269.302	157.291	-538.272	-325.372	12.140
1500	174.891	393.621	277.195	174.640	-529.802	-310.459	10.811

PREVIOUS: June 1964 CURRENT: June 1968

Titanium Bromide ($TiBr_3$) $Br_3Ti_1(cr)$

TITANIUM BROMIDE ($TiBr_3$) IDEAL GAS M_r = 287.592

$S°(298.15 K) = [359.1 \pm 5.0] J K^{-1} mol^{-1}$

$\Delta_f H°(0 K) = -351.4 \pm 10.5 kJ mol^{-1}$
$\Delta_f H°(298.15 K) = -374.9 \pm 10.5 kJ mol^{-1}$

Electronic Levels and Quantum Weights

ϵ_i, cm^{-1}	g_i	ϵ_i, cm^{-1}	g_i
0	[2]	[4000]	[2]
[600]	[2]	[10000]	[2]
[1500]	[2]		

Vibrational Frequencies and Degeneracies

ν, cm^{-1}	ν, cm^{-1}
[380](1)	[427](2)
[162](1)	[103](2)

Point Group: C_{3v} σ = 3
Bond Distance: Ti-Br = [2.4] Å
Bond Angle: Br-Ti-Br = [100]°
Product of the Moments of Inertia: $I_A I_B I_C = [1.72305 \times 10^{-111}] g^3 cm^6$

Enthalpy of Formation

The enthalpy of formation of $TiBr_3(g)$ is calculated from the enthalpy of sublimation of $TiBr_3(cr)$. A tentative equation for the vapor pressure of $TiBr_3(cr)$ was reported by Hall and Blocher (1). 2nd and 3rd law analyses of the equation over the temperature range 700 to 900 K give values of $\Delta_{sub}H°(298.15 K)$ of 45.2 and 41.9 kcal mol^{-1}, respectively, with a 3nd law drift of -4.2 cal $K^{-1} mol^{-1}$. The chosen value of $\Delta_f H°(298.15 K)$, -89.6±2.5 kcal mol^{-1} (-374.886±10.5 kJ mol^{-1}), is based on the 3rd law $\Delta_{sub}H°(298.15 K)$.

Heat Capacity and Entropy

The interatomic distance is estimated from those of $TiCl_4$, $TiCl_3$ and $TiBr_4$. The pyramidal bond angle is estimated assuming that $TiBr_3(g)$ is similar to the group V trihalides. The principal moments of inertia are: $I_A = I_B = 98.0035 \times 10^{-39}$ g cm^2 and $I_C = 179.3970 \times 10^{-39}$ g cm^2. The vibrational frequencies are estimated from valence force field predictions and comparisons with group V trihalides. The electronic levels are estimated from the levels of Ti^{+3} (2).

References

1. E. H. Hall and J. M Blocher, J. Phys. Chem. 63, 1525 (1959).
2. C. E. Moore, U. S. Nat. Bur. Stand., Circ. 467, 1949.

Titanium Bromide ($TiBr_3$) $Br_3Ti_1(g)$

Enthalpy Reference Temperature = T_r = 298.15 K Standard State Pressure = p° = 0.1 MPa

T/K	$C_p°$ (J $K^{-1}mol^{-1}$)	S° (J $K^{-1}mol^{-1}$)	$-[G°-H°(T_r)]/T$ (J $K^{-1}mol^{-1}$)	$H°-H°(T_r)$ (kJ mol^{-1})	$\Delta_f H°$ (kJ mol^{-1})	$\Delta_f G°$ (kJ mol^{-1})	Log K_f
0	0.	0.	INFINITE	-18.122	-351.415	-351.415	INFINITE
100	55.060	285.335	423.540	-13.820	-351.863	-371.494	194.048
200	71.226	328.902	366.106	-7.441	-354.752	-390.061	101.873
250	76.150	345.361	360.354	-3.748	-356.447	-398.697	83.303
298.15	79.334	359.064	359.064	0.	-374.886	-404.700	70.902
300	79.433	359.555	359.066	0.147	-374.996	-404.885	70.497
350	81.623	371.976	360.042	4.177	-421.229	-407.328	60.790
400	83.104	382.978	362.234	8.297	-421.179	-405.345	52.933
450	84.118	392.829	365.096	12.480	-421.117	-403.370	46.822
500	84.816	401.730	368.322	16.704	-421.050	-401.401	41.934
600	85.616	417.274	375.222	25.231	-420.928	-397.483	34.604
700	85.968	430.502	382.198	33.813	-420.842	-393.583	29.370
800	86.103	441.992	388.970	42.417	-420.791	-389.693	25.444
900	86.137	452.136	395.436	51.030	-420.815	-385.806	22.392
1000	86.128	461.211	401.568	59.643	-420.984	-381.908	19.949
1100	86.102	469.419	407.369	68.255	-421.362	-377.984	17.949
1200	86.068	476.909	412.856	76.863	-425.936	-373.896	16.275
1300	86.033	483.797	418.052	85.469	-426.013	-369.557	14.849
1400	85.996	490.171	422.978	94.070	-426.184	-365.209	13.626
1500	85.958	496.103	427.658	102.668	-426.465	-360.844	12.566
1600	85.921	501.649	432.111	111.262	-426.870	-356.457	11.637
1700	85.884	506.857	436.356	119.852	-427.413	-352.040	10.817
1800	85.847	511.765	440.411	128.438	-428.111	-347.587	10.087
1900	85.812	516.406	444.289	137.021	-428.980	-343.091	9.432
2000	85.779	520.806	448.006	145.601	-444.720	-338.090	8.830
2100	85.748	524.991	451.573	154.177	-446.724	-332.710	8.276
2200	85.718	528.979	455.002	162.750	-448.760	-327.233	7.769
2300	85.691	532.789	458.302	171.321	-450.832	-321.663	7.305
2400	85.665	536.435	461.482	179.889	-452.939	-316.001	6.878
2500	85.641	539.932	464.550	188.454	-455.083	-310.252	6.482
2600	85.619	543.290	467.515	197.017	-457.263	-304.416	6.116
2700	85.598	546.521	470.381	205.578	-459.479	-298.495	5.775
2800	85.578	549.634	473.156	214.137	-461.730	-292.492	5.457
2900	85.560	552.637	475.846	222.693	-464.012	-286.407	5.159
3000	85.542	555.537	478.454	231.249	-466.323	-280.244	4.879
3100	85.525	558.342	480.986	239.802	-468.661	-274.002	4.617
3200	85.508	561.057	483.446	248.354	-471.022	-267.686	4.370
3300	85.492	563.688	485.838	256.904	-473.401	-261.294	4.136
3400	85.475	566.239	488.165	265.452	-475.796	-254.831	3.915
3500	85.459	568.717	490.432	273.999	-478.201	-248.297	3.706
3600	85.443	571.124	492.640	282.544	-480.614	-241.694	3.507
3700	85.426	573.465	494.793	291.087	-892.127	-227.236	3.208
3800	85.410	575.743	496.893	299.629	-893.307	-209.251	2.876
3900	85.393	577.961	498.944	308.169	-894.538	-191.233	2.561
4000	85.376	580.123	500.946	316.707	-895.815	-173.183	2.262
4100	85.358	582.231	502.903	325.244	-897.134	-155.101	1.976
4200	85.340	584.288	504.816	333.779	-898.489	-136.987	1.704
4300	85.322	586.296	506.688	342.312	-899.876	-118.839	1.444
4400	85.304	588.257	508.520	350.844	-901.291	-100.659	1.195
4500	85.285	590.174	510.313	359.373	-902.730	-82.447	0.957
4600	85.266	592.048	512.070	367.901	-904.190	-64.203	0.729
4700	85.247	593.882	513.791	376.426	-905.666	-45.926	0.510
4800	85.228	595.676	515.478	384.950	-907.155	-27.618	0.301
4900	85.208	597.433	517.133	393.472	-908.655	-9.279	0.099
5000	85.188	599.154	518.756	401.992	-910.161	9.091	-0.095
5100	85.168	600.841	520.349	410.509	-911.671	27.492	-0.282
5200	85.147	602.495	521.913	419.025	-913.184	45.921	-0.461
5300	85.127	604.116	523.449	427.539	-914.695	64.380	-0.635
5400	85.106	605.707	524.957	436.051	-916.204	82.867	-0.802
5500	85.086	607.269	526.440	444.560	-917.708	101.382	-0.963
5600	85.065	608.802	527.897	453.068	-919.205	119.925	-1.119
5700	85.044	610.307	529.330	461.573	-920.694	138.494	-1.269
5800	85.023	611.786	530.738	470.076	-922.173	157.088	-1.415
5900	85.002	613.239	532.125	478.578	-923.640	175.709	-1.556
6000	84.981	614.668	533.488	487.077	-925.095	194.354	-1.692

PREVIOUS: June 1968 (1 atm) CURRENT: June 1968 (1 bar)

Titanium Bromide ($TiBr_3$) $Br_3Ti_1(g)$

ZIRCONIUM BROMIDE ($ZrBr_3$) CRYSTAL M_r = 330.932

$\Delta_f H°(0\ K)$ = [-616.1 ± 67] kJ mol^{-1}

S°(298.15 K) = [172.05] J K^{-1} mol^{-1} $\Delta_f H°(298.15\ K)$ = [-636.0 ± 67] kJ mol^{-1}

Enthalpy of Formation

Larsen and Leddy (1) studied the reaction Zr(cr) + 3 $ZrBr_4$(g) = 4 $ZrBr_3$(cr) in the temperature range 473 to 973 K and pressure range 5 to 15 atm. They present a plot of the fractional reaction, for a series of temperatures, as a function of the reaction time; however, they express doubt concerning the attainment of thermodynamic equilibrium.

The Gibbs energy function change for this reaction was calculated at 100 K invervals from 500 to 1100 K. At each temperature, the equilibrium constant was assumed to be 1, and $\Delta_r H$(298.15 K) accordingly calculated. These values and the enthalpy of formation of $ZrBr_4$(g) [see $ZrBr_4$ table] were used to compute a series of values for the standard enthalpy of formation of $ZrBr_3$(cr). The following table gives representative values at four temperatures.

T/K	K_p	$\Delta_r H°$(298.15 K)	$\Delta_f H°$($ZrBr_3$, cr, 298.15 K)
500	1	-70.26 kcal mol^{-1}	-132.8 kcal mol^{-1}
700	1	-96.48 kcal mol^{-1}	-139.3 kcal mol^{-1}
900	1	-121.75 kcal mol^{-1}	-145.6 kcal mol^{-1}
1100	1	-146.28 kcal mol^{-1}	-151.8 kcal mol^{-1}

If one assumes an initial pressure of from 0 to 10 atm for $ZrBr_4$ and stoichiometric amounts of reactants, then, on the basis of zirconium, the reaction must proceed to the extent of 80 to 90% for the pressure of $ZrBr_4$ to attain a value of 1 atm and hence an equilibrium constant of 1. The data of Larsen and Leddy (1) indicate about 82% reaction at 973 K. It was assumed that around 1100 K the equilibrium constant attains a value of 1 giving -152 kcal mol^{-1} (-635.968 kJ mol^{-1}) for the enthaply of formation of $ZrBr_3$(cr) at 298.15 K. The limits of error were taken to be ±16 kcal mol^{-1} which corresponds to a ±500 K temperature spread in the above table.

A 2nd law calculation of their data was carried out but the results are of doubtful value in view of the uncertainty in the attainment of equilibrium at the lower temperatures.

Heat Capacity and Entropy

The heat capacity was estimated in the same manner as for $ZrBr_4$(cr) [see $ZrBr_4$(cr) table]. The values for θ_D and θ_E were taken to be the same as those estimated for $ZrBr_4$(cr). The internal contribution was obtained from the estimated $ZrBr_3$ vibrational frequencies and the anharmonicity factor "a" was taken to be 2.5 x 10^{-3}. The specific heat above 300 K was obtained by graphical extrapolation.

It was assumed, in the above estimation, that the crystalline lattice is made up of $ZrBr_3$ molecules. However, Holze (2) concluded that crystalline ZrI_3 was composed of a chain lattice of $(Zr_2I_6)_n$ units. A chain lattice of $[Zr_2Br_6)_n$ units is probably a better representation of the solid state structure of $ZrBr_3$. Until more quantitative data becomes available, however, it is felt that the above analysis gives a fair approximation to the heat capacity of $ZrBr_3$.

Disproportionation Data

As detailed above, this was assumed to be 1100 ± 500 K.

Liquid Phase

It is assumed that the liquid phase is thermodynamically unstable under ordinary conditions.

Sublimation Data

The enthalpy of sublimation at 298.15 K was obtained from the difference in the enthalpy of formation of the gas and solid at 298.15 K. The sublimation point was obtained from the Gibbs energy crossover between gas and solid at one bar.

References

1. E. M. Larsen and J. J. Leddy, J. Am. Chem. Soc. 78, 5983 (1956).
2. E. Holze in R. F. Rolsten, "Iodide Metals and Metal Iodides," John Wiley & Sons, Inc., New York, 1961. page 46.

Zirconium Bromide ($ZrBr_3$) Br_3Zr_1(cr)

Enthalpy Reference Temperature = T_r = 298.15 K Standard State Pressure = p° = 0.1 MPa

T/K	$C_p°$ (J K^{-1}mol^{-1})	S°	-[G°-H°(T_r)]/T	H°-H°(T_r) (kJ mol^{-1})	$\Delta_f H°$	$\Delta_f G°$	Log K_f
0	0.	0.	INFINITE	-22.372	-616.078	-616.078	INFINITE
100	74.421	74.693	255.609	-18.092	-616.874	-614.861	321.170
200	93.232	133.499	180.987	-9.498	-617.825	-612.489	159.966
298.15	99.483	172.054	172.054	0.	-635.968	-607.607	106.450
300	99.562	172.670	172.056	0.184	-636.040	-607.430	105.763
400	102.508	201.756	176.002	10.302	-680.202	-588.696	76.876
500	103.763	224.773	183.535	20.619	-678.047	-566.068	59.137
600	104.600	243.771	192.037	31.040	-675.896	-543.874	47.348
700	105.018	259.927	200.611	41.521	-673.790	-522.038	38.955
800	105.437	273.978	208.922	52.044	-671.745	-500.500	32.679
900	105.855	286.422	216.855	62.610	-669.773	-479.213	27.813
1000	106.064	297.587	224.380	73.207	-667.892	-458.142	23.931
1100	106.148	307.700	231.502	83.818	-666.128	-437.253	20.763
1200	106.232	316.940	238.242	94.437	-668.199	-416.298	18.121
1300	106.315	325.447	244.628	105.065	-666.140	-395.390	15.887
1400	106.357	333.327	250.685	115.698	-664.132	-374.639	13.978
1500	106.399	340.666	256.442	126.336	-662.185	-354.029	12.328
1600	106.441	347.534	261.923	136.978	-660.314	-333.547	10.889
1700	106.483	353.989	267.151	147.624	-658.530	-313.179	9.623
1800	106.525	360.076	272.146	158.275	-656.846	-292.913	8.500
1900	106.566	365.837	276.927	168.929	-655.276	-272.738	7.498
2000	106.608	371.304	281.510	179.588	-653.832	-252.642	6.598

PREVIOUS: March 1962 CURRENT: June 1964

Zirconium Bromide ($ZrBr_3$) Br_3Zr_1(cr)

ZIRCONIUM BROMIDE ($ZrBr_3$) IDEAL GAS $M_r = 330.932$

$S°(298.15\ K) = [371.58]\ J\ K^{-1}\ mol^{-1}$

$\Delta_fH°(0\ K) = [-408]\ kJ\ mol^{-1}$
$\Delta_fH°(298.15\ K) = [-431]\ kJ\ mol^{-1}$

Vibrational Frequencies and Degeneracies

ν, cm^{-1}

[255](1)
[120](1)
[273](2)
[95](2)

Ground State Quantum Weight: [2] $\sigma = [3]$
Point Group: [C_{3v}]
Bond Distance: Zr-Br = [2.39] Å
Bond Angle: Br-Zr-Br = [98]°
Product of the Moments of Inertia: $I_AI_BI_C = [1.77605 \times 10^{-111}]\ gm^3\ cm^6$

Enthalpy of Formation

The enthalpies of formation, from the gaseous atoms, of the gaseous zirconium tetrahalides are computed from data issued in these JANAF tables. The zirconium-halide bond energy, taken as 1/4 of this enthaply of formation, is found to be linear with internuclear separation. From gaseous titanium tri- and tetrachloride, it is found that the bond energy of the trichloride is around 7 kcal mol^{-1} greater than that for titanium tetrachloride. This amount is added to the bond energy of zirconium tetrachloride to get that of zirconium trichloride which, when used with its estimated internuclear distance, is found to lie almost exactly on the bond energy versus internuclear distance curve for the tetrahalides. The bond energy for each of the gaseous trihalides of zirconium is determined from this curve and their estimated bond distances.

For zirconium tribromide, the enthaply of formation from the gaseous atoms is -329 kcal mol^{-1} and from the elements in their standard state, -103 kcal mol^{-1} (-430.952 kJ mol^{-1}).

Heat Capacity and Entropy

The measured internuclear distances for the tribromides of P, As, and Sb are plotted as a function of the atomic weight of these atoms and a smooth curve is drawn through the points. The internuclear distance for $ZrBr_3$ is taken from this plot. The bond angle is assumed to be intermediate between the bond angles of $AsBr_3$ and $SbBr_3$; it is taken to be 98°. The principal moments of inertia are: $I_A = I_B = 101.4161 \times 10^{-39}\ gm\ cm^2$, $I_C = 172.6792 \times 10^{-39}\ g\ cm^2$.

The measured vibrational frequencies for the tribromides of phosphorus, arsenic, and antimony are plotted as a function of internuclear distance. Smooth curves through these points are obtained for ν_1, ν_2, and ν_4; these curves and the estimated $ZrBr_3$ bond distance are used to obtain correlation values of ν_1, ν_2, and ν_4 for $ZrBr_3$. Force constants, assuming a valence force field and a Br-Zr-Br angle of 98°, are derived from the above values for ν_1 and ν_2. These force constants are then used to derive values for ν_3 and ν_4. By a process of successive approximation, a set of force constants are obtained which give frequencies in good agreement with the three correlation values. The three correlation frequencies and the derived frequency are used in this table.

Zirconium Bromide ($ZrBr_3$) $Br_3Zr_1(g)$

Enthalpy Reference Temperature = T_r = 298.15 K Standard State Pressure = p° = 0.1 MPa

T/K	$C_p°$ (J K⁻¹mol⁻¹)	S° (J K⁻¹mol⁻¹)	$-[G°-H°(T_r)]/T$ (J K⁻¹mol⁻¹)	$H°-H°(T_r)$ (kJ mol⁻¹)	$\Delta_fH°$ (kJ mol⁻¹)	$\Delta_fG°$ (kJ mol⁻¹)	Log K_f
0	0.	0.	INFINITE	-19.302	-407.993	-407.993	INFINITE
100	62.307	292.306	439.075	-14.677	-408.444	-428.192	223.664
200	75.575	340.557	378.756	-7.640	-410.951	-447.026	116.751
250	78.032	357.710	372.886	-3.794	-412.545	-455.867	95.248
298.15	79.443	371.585	371.585	0.	-430.952	-462.081	80.955
300	79.486	372.076	371.586	0.147	-431.062	-462.274	80.489
350	80.406	384.403	372.558	4.146	-477.309	-464.937	69.388
400	81.022	395.183	374.726	8.183	-477.306	-463.169	60.484
450	81.453	404.752	377.541	12.245	-477.310	-461.402	53.558
500	81.766	413.351	380.699	16.326	-477.324	-459.634	48.018
600	82.180	428.299	387.424	24.525	-477.396	-456.090	39.706
700	82.432	440.987	394.192	32.757	-477.538	-452.529	33.768
800	82.597	452.006	400.745	41.009	-477.765	-448.942	29.313
900	82.711	461.742	406.992	49.274	-478.093	-445.321	25.846
1000	82.793	470.460	412.911	57.550	-478.534	-441.657	23.070
1100	82.853	478.354	418.507	65.832	-479.098	-437.943	20.796
1200	82.900	485.566	423.799	74.120	-483.500	-433.950	18.889
1300	82.936	492.203	428.809	82.412	-483.777	-429.810	17.270
1400	82.964	498.350	433.559	90.707	-484.107	-425.647	15.881
1500	82.987	504.075	438.072	99.004	-484.501	-421.458	14.676
1600	83.006	509.431	442.366	107.304	-484.972	-417.240	13.621
1700	83.022	514.464	446.461	115.606	-485.533	-412.990	12.690
1800	83.035	519.210	450.372	123.909	-486.197	-408.704	11.860
1900	83.046	523.700	454.114	132.213	-486.977	-404.378	11.117
2000	83.056	527.960	457.701	140.518	-487.886	-400.008	10.447
2100	83.064	532.012	461.144	148.824	-488.940	-395.589	9.840
2200	83.071	535.876	464.453	157.130	-511.472	-390.370	9.269
2300	83.077	539.569	467.640	165.438	-513.267	-384.825	8.740
2400	83.083	543.105	470.711	173.746	-515.094	-379.202	8.253
2500	83.088	546.497	473.675	182.054	-516.955	-373.501	7.804
2600	83.092	549.756	476.539	190.363	-518.850	-367.726	7.388
2700	83.096	552.892	479.309	198.673	-520.778	-361.877	7.001
2800	83.099	555.914	481.991	206.983	-522.737	-355.956	6.640
2900	83.102	558.830	484.591	215.293	-524.727	-349.965	6.304
3000	83.105	561.647	487.113	223.603	-526.743	-343.904	5.988
3100	83.107	564.372	489.561	231.914	-528.784	-337.776	5.691
3200	83.110	567.011	491.941	240.224	-530.846	-331.582	5.413
3300	83.112	569.568	494.254	248.535	-532.924	-325.322	5.149
3400	83.114	572.049	496.506	256.847	-535.016	-319.000	4.901
3500	83.115	574.459	498.699	265.158	-537.117	-312.616	4.666
3600	83.117	576.800	500.836	273.470	-539.223	-306.172	4.442
3700	83.118	579.077	502.920	281.782	-541.330	-299.669	4.231
3800	83.120	581.294	504.954	290.093	-543.435	-293.110	4.029
3900	83.121	583.453	506.939	298.405	-545.533	-286.495	3.837
4000	83.122	585.557	508.878	306.718	-547.622	-279.826	3.654
4100	83.123	587.610	510.773	315.030	-549.697	-273.106	3.479
4200	83.124	589.613	512.627	323.342	-551.756	-266.335	3.312
4300	83.125	591.569	514.440	331.655	-553.796	-259.515	3.152
4400	83.126	593.480	516.215	339.967	-555.815	-252.648	2.999
4500	83.127	595.348	517.953	348.280	-557.811	-245.735	2.852
4600	83.127	597.175	519.655	356.593	-559.781	-238.778	2.711
4700	83.128	598.963	521.324	364.905	-561.723	-231.779	2.576
4800	83.129	600.713	522.959	373.218	-1124.574	-213.121	2.319
4900	83.129	602.427	524.564	381.531	-1126.110	-194.117	2.069
5000	83.130	604.107	526.138	389.844	-1127.632	-175.081	1.829
5100	83.131	605.753	527.683	398.157	-1129.139	-156.015	1.598
5200	83.131	607.367	529.200	406.470	-1130.628	-136.920	1.375
5300	83.132	608.951	530.690	414.783	-1132.097	-117.796	1.161
5400	83.132	610.504	532.153	423.096	-1133.545	-98.644	0.954
5500	83.132	612.030	533.592	431.410	-1134.971	-79.466	0.755
5600	83.133	613.528	535.006	439.723	-1136.372	-60.263	0.562
5700	83.133	614.999	536.396	448.036	-1137.749	-41.034	0.376
5800	83.134	616.445	537.764	456.350	-1139.100	-21.782	0.196
5900	83.134	617.866	539.110	464.663	-1140.424	-2.506	0.022
6000	83.134	619.263	540.434	472.976	-1141.720	16.791	-0.146

PREVIOUS: June 1964 (1 atm) CURRENT: June 1964 (1 bar)

Zirconium Bromide ($ZrBr_3$) $Br_3Zr_1(g)$

IRON BROMIDE $((FeBr_2)_2)$ IDEAL GAS $M_r = 431.310$

$\Delta_f H°(0\ K) = -228.3 \pm 8\ kJ\ mol^{-1}$

$S°(298.15\ K) = [516.0]\ J\ K^{-1}\ mol^{-1}$ $\Delta_f H°(298.15\ K) = -253.1 \pm 8\ kJ\ mol^{-1}$

Electronic Levels and Quantum Weights

ε_i, cm^{-1}	g_i
0	[5]
[4450]	[5]
[6900]	[5]

Vibrational Frequencies and Degeneracies

ν, cm^{-1}	ν, cm^{-1}	ν, cm^{-1}
[110](1)	[75](1)	[95](2)
[70](1)	[100](1)	[24](2)
[110](1)	[115](1)	[190](2)

Point Group: $[D_{2h}]$ $\sigma = [4]$

Bond Distance: Fe-Br = [2.24] Å

Bond Angles: $Fe\text{-}Br_{bridge}\text{-}Fe = [90]°$ $Br_{bridge}\text{-}Fe\text{-}Br_{bridge} = [90]°$ $Br\text{-}Fe\text{-}Br_{bridge} = [135]°$

Product of the Moments of Inertia: $I_A I_B I_C = [1.44991 \times 10^{-110}]\ g^3\ cm^6$

Enthaply of Formation

The chemical equilibria, 622-665 K, for the reaction $Fe_2Br_4(g) = 2\ FeBr_2(g)$ were studied by Porter and Schoonmaker (1), using a mass spectrometer. Based on the reported partial pressures for the two species, the enthalpy change $\Delta_r H°(298.15\ K)$ for the reaction is evaluated by both the 2nd and 3rd law method to be 40.36±2.0 and 40.75±0.2 kcal mol^{-1}, respectively. Using the 3rd law $\Delta_r H°$ value and $\Delta_f H°(FeBr_2, g, 298.15\ K) = -9.9 \pm 0.5$ kcal mol^{-1}, the enthaply of formation $\Delta_f H°(Fe_2Br_4, g, 298.15\ K)$ is calculated as -60.5±2.0 kcal mol^{-1} (-253.132 kJ mol^{-1}) and is adopted.

Heat Capacity and Entropy

The molecular structure is assumed to be planar. The two Fe atoms are at the two opposite corners of a square. The other two corners of the square are occupied by two Br atoms. The two remaining Br atoms are situated outside the square on a straight line joining the two Fe atoms. The Fe-Br bond distance was estimated to be the same as that in the $FeBr_2(g)$ molecule. The first six vibrational frequencies were estimated by comparison with those for $K_2Br_2(g)$ calculated by Berkowitz (2). The last three degenerate frequencies were assigned arbitrarily in order to give good second and third law agreement for the heats of dissociation of dimer to monomer. The order of the frequencies listed above is arbitrary and not related to their species types.

The electronic levels and quantum weights were estimated by comparison with those for $FeBr_2(g)$. The principal moments of inertia are: $I_A = 66.5764 \times 10^{-39}$, $I_B = 434.5679 \times 10^{-39}$ and $I_C = 501.1443 \times 10^{-39}\ g\ cm^2$.

References

1. R. F. Porter and R. C. Schoonmaker, J. Phys. Chem. 63, 626 (1959).
2. J. Berkowitz, J. Chem. Phys. 32, 1519 (1960).

Iron Bromide $((FeBr_2)_2)$ $Br_4Fe_2(g)$

Enthalpy Reference Temperature = T_r = 298.15 K Standard State Pressure = $p°$ = 0.1 MPa

T/K	$C_p°$ (J K⁻¹mol⁻¹)	$S°$ (J K⁻¹mol⁻¹)	$-[G°-H°(T_r)]/T$ (J K⁻¹mol⁻¹)	$H°-H°(T_r)$ (kJ mol⁻¹)	$\Delta_f H°$ (kJ mol⁻¹)	$\Delta_f G°$ (kJ mol⁻¹)	Log K_f
0	0.	0.	INFINITE	-33.210	-228.310	-228.310	INFINITE
100	115.643	379.131	629.897	-25.077	-226.608	-252.539	131.913
200	127.972	464.315	527.906	-12.718	-227.590	-278.305	72.686
250	129.727	493.080	518.165	-6.271	-229.050	-290.829	60.765
298.15	130.683	516.018	516.018	0.	-253.132	-299.931	52.547
300	130.711	516.827	516.021	0.242	-253.263	-300.221	52.273
350	131.315	537.025	517.614	6.794	-314.584	-304.690	45.473
400	131.711	554.587	521.162	13.370	-314.351	-303.294	39.606
450	131.986	570.117	525.755	19.963	-314.236	-301.920	35.046
500	132.185	584.034	530.900	26.567	-314.242	-300.553	31.399
600	132.461	608.160	541.826	39.801	-314.620	-297.787	25.925
700	132.671	628.595	552.799	53.057	-315.491	-294.920	22.007
800	132.880	646.325	563.406	66.335	-316.948	-291.890	19.058
900	133.125	661.989	573.506	79.635	-319.234	-288.631	16.752
1000	133.420	676.030	583.069	92.961	-322.950	-285.051	14.890
1100	133.765	688.762	592.108	106.320	-328.442	-280.967	13.342
1200	134.155	700.418	600.655	119.716	-332.810	-276.496	12.036
1300	134.577	711.172	608.748	133.152	-333.862	-271.761	10.920
1400	135.018	721.162	616.425	146.632	-335.056	-266.941	9.960
1500	135.466	730.492	623.722	160.156	-336.391	-262.030	9.125
1600	135.908	739.249	630.671	173.725	-337.869	-257.025	8.391
1700	136.335	747.501	637.303	187.337	-341.395	-251.885	7.739
1800	136.737	755.306	643.644	200.991	-343.842	-246.550	7.155
1900	137.110	762.709	649.717	214.684	-374.643	-239.671	6.589
2000	137.448	769.750	655.544	228.412	-377.891	-232.483	6.072
2100	137.749	776.464	661.144	242.172	-381.144	-225.132	5.600
2200	138.012	782.878	666.533	255.960	-384.408	-217.627	5.167
2300	138.238	789.018	671.726	269.773	-387.691	-209.973	4.769
2400	138.426	794.906	676.736	283.607	-390.997	-202.175	4.400
2500	138.579	800.560	681.577	297.457	-394.332	-194.239	4.058
2600	138.699	805.997	686.258	311.321	-397.699	-186.169	3.740
2700	138.789	811.234	690.791	325.196	-401.100	-177.968	3.443
2800	138.850	816.282	695.183	339.078	-404.536	-169.642	3.165
2900	138.886	821.155	699.443	352.965	-408.008	-161.191	2.903
3000	138.899	825.864	703.579	366.855	-411.514	-152.621	2.657
3100	138.892	830.418	707.598	380.744	-415.052	-143.933	2.425
3200	138.867	834.828	711.505	394.632	-1115.220	-120.283	1.963
3300	138.828	839.100	715.307	408.517	-1115.013	-89.194	1.412
3400	138.774	843.244	719.009	422.397	-1114.894	-58.111	0.893
3500	138.710	847.266	722.617	436.272	-1114.860	-27.030	0.403
3600	138.636	851.172	726.134	450.139	-1114.908	4.052	-0.059
3700	138.555	854.970	729.565	463.999	-1115.036	35.136	-0.496
3800	138.467	858.664	732.914	477.850	-1115.239	66.224	-0.910
3900	138.373	862.259	736.184	491.692	-1115.517	97.318	-1.303
4000	138.276	865.761	739.380	505.524	-1115.865	128.421	-1.677
4100	138.175	869.174	742.504	519.347	-1116.284	159.533	-2.032
4200	138.072	872.503	745.560	533.159	-1116.769	190.656	-2.371
4300	137.968	875.751	748.550	546.961	-1117.321	221.792	-2.694
4400	137.863	878.921	751.477	560.753	-1117.938	252.941	-3.003
4500	137.757	882.018	754.344	574.534	-1118.620	284.105	-3.298
4600	137.652	885.045	757.153	588.304	-1119.365	315.285	-3.580
4700	137.547	888.004	759.905	602.064	-1120.173	346.481	-3.851
4800	137.443	890.899	762.604	615.814	-1121.046	377.696	-4.110
4900	137.341	893.732	765.252	629.553	-1121.984	408.930	-4.359
5000	137.240	896.505	767.849	643.282	-1122.987	440.183	-4.599
5100	137.140	899.222	770.398	657.001	-1124.057	471.457	-4.829
5200	137.042	901.884	772.901	670.710	-1125.196	502.752	-5.050
5300	136.946	904.494	775.360	684.410	-1126.407	534.071	-5.264
5400	136.852	907.053	777.775	698.099	-1127.691	565.412	-5.469
5500	136.760	909.563	780.148	711.780	-1129.052	596.779	-5.668
5600	136.671	912.026	782.481	725.452	-1130.495	628.170	-5.859
5700	136.583	914.444	784.775	739.114	-1132.021	659.589	-6.044
5800	136.497	916.819	787.032	752.768	-1133.637	691.033	-6.223
5900	136.414	919.152	789.251	766.414	-1135.348	722.508	-6.397
6000	136.333	921.444	791.435	780.051	-1137.158	754.012	-6.564

PREVIOUS: September 1966 (1 atm) CURRENT: September 1966 (1 bar)

Iron Bromide $((FeBr_2)_2)$ $Br_4Fe_2(g)$

MAGNESIUM BROMIDE ($(MgBr_2)_2$) IDEAL GAS M_r = 368.226

$\Delta_fH°(0\ K) = -739.7 \pm 2.09\ kJ\ mol^{-1}$

$S°(298.15\ K) = [461.3 \pm 20.9]\ J\ K^{-1}\ mol^{-1}$ $\Delta_fH°(298.15\ K) = -767.8 \pm 20.9\ kJ\ mol^{-1}$

Vibrational Frequencies and Degeneracies

ν, cm^{-1}	ν, cm^{-1}	ν, cm^{-1}
[182](1)	[188](1)	[100](1)
[120](1)	[228](1)	[100](1)
[207](1)	[125](1)	[75](1)
[134](1)	[125](1)	[75](1)

Ground State Quantum Weight: [1] σ = 4

Point Group: [D_{2h}]

Bond Distances: Mg-Br = [2.34] Å Mg-Br'(bridge) = [2.60] Å

Bond Angles: Mg-Br'(bridge)-Mg = [90]° Br'(bridge)-Mg-Br'(bridge) = [90]°

Br-Mg-Br'(bridge) = [135]°

Product of Moments of Inertia: $I_AI_BI_C = [2.55371 \times 10^{-110}]\ g^3\ cm^6$

Enthalpy of Formation

$\Delta_fH°$ of the dimer is based on an analysis of mass spectrometric data reported by Berkowitz and Marquart (1) These studies have revealed the presence of approximately 2% dimer in the equilibrium vapor over $MgBr_2$(cr) at around 800 K. The existence of the dimer was inferred from the observed ion intensities for $Mg_2Br_3^+$. Berkowitz and Marquart (1) reported a 2nd law Δ_rH(803 K) of 12.9 kcal mol^{-1} for the process $MgBr_2(cr) + MgBr_2(g) = Mg_2Br_4(g)$. When corrected to 298.15 K this value gives $\Delta_rH°$ = 14.0 kcal mol^{-1} which leads to $\Delta_fH°(Mg_2Br_4$, g, 298.15 K) = -183.7 kcal mol^{-1}. The absolute pressure of Mg_2Br_4(g) was also measured at 798 K by integrating the ion current during complete volatilization of a previously weighed sample. A 3rd law analysis of the reported pressure gives $\Delta_{sub}H°$(298.15 K) = 67.4 kcal mol^{-1} for 2 $MgBr_2(cr) = Mg_2Br_4(g)$. This value for the enthalpy of sublimation leads to $\Delta_fH°(Mg_2Br_4$, g, 298.15 K) = -183.2 kcal mol^{-1} which is in excellent agreement with the 2nd law result. We adopt an average value (-183.5 kcal mol^{-1}, -767.764 kJ mol^{-1}) of these two results. The uncertainty in $\Delta_fH°$ is estimated as ±5.0 kcal mol^{-1}, and the adopted $\Delta_fH°$ value value corresponds to a dimerization energy for $MgBr_2$(g) of 38.7 kcal mol^{-1}.

Heat Capacity and Entropy

The dimer molecule is assumed to have a bridge-bond structure of D_{2h} symmetry similar to that suggested by Thompson and Carlson (2) for the dimers of several transition metal dichlorides. The two outer Mg-Br bond lengths are assumed to be the same as that for $MgBr_2$(3)(2.34 Å). The four ring Mg-Br'(bridge) bond lengths are taken to be somewhat longer (2.6 Å). The four atoms which lie in the ring form a square. The Br'(bridge)-Mg-Br bond angle is estimated as 135°. The principal moments of inertia are: $I_A = 89.6957 \times 10^{-39}$, $I_B = 490.6138 \times 10^{-39}$, and $I_C = 580.3094 \times 10^{-39}\ g\ cm^2$.

The vibratonal frequencies of the ring (first six listed) are taken equal to those for Na_2Br_2 (3). These frequencies were obtained by Berkowitz (4) from ionic model calculations. The remaining six frequencies are estimated by analogy with those for $MgBr_2$ and Mg_2Cl_4 (3). Following the observations made by Thompson and Carlson (2), all of the dimer frequencies are assumed to lie above the bending frequency (ν_2 = 70 cm^{-1}) of the monomer (3). Our adopted functions for the monomer and dimer reproduce the pressures measured by Berkowitz and Marquart (1) and show that the dimer becomes an increasingly important vapor species with rising temperatures.

References

1. J. Berkowitz and J. R. Marquart, J. Chem. Phys. 37, 1853 (1962).
2. K. R. Thompson and K. D. Carlson, J. Chem. Phys. 49, 4379 (1968).
3. JANAF Thermochemical Tables: $MgBr_2$(g), 6-30-74; Na_2Br_2(g), 9-30-64; Mg_2Cl_4(g), 12-31-69.
4. J. Berkowitz, J. Chem. Phys. 32, 1519 (1960).

Magnesium Bromide ($(MgBr_2)_2$) Br_4Mg_2(g)

Enthalpy Reference Temperature = T_r = 298.15 K Standard State Pressure = p° = 0.1 MPa

T/K	$C_p°$ (J K⁻¹mol⁻¹)	S°	-[G°-H°(T_r)]/T	H°-H°(T_r) (kJ mol⁻¹)	$\Delta_fH°$	$\Delta_fG°$	Log K_f
0	0.	0.	INFINITE	-30.967	-739.716	-739.716	INFINITE
100	105.171	329.984	572.231	-24.225	-739.844	-760.172	397.073
200	124.475	410.634	473.038	-12.481	-741.917	-779.818	203.668
250	127.401	438.757	463.463	-6.176	-743.586	-789.111	164.876
298.15	129.013	461.347	461.347	0.	-767.764	-795.072	139.293
300	129.061	462.145	461.349	0.239	-767.897	-795.241	138.464
350	130.087	482.123	462.925	6.719	-829.242	-796.439	118.862
400	130.763	499.540	466.436	13.242	-828.965	-791.773	103.395
450	131.231	514.971	470.988	19.792	-828.738	-787.138	91.369
500	131.569	528.816	476.090	26.363	-828.560	-782.526	81.750
600	132.011	552.846	486.940	39.544	-828.353	-773.342	67.325
700	132.280	573.218	497.847	52.759	-828.359	-764.175	57.023
800	132.455	590.893	508.398	65.996	-828.602	-754.993	49.296
900	132.575	606.502	518.448	79.248	-829.116	-745.764	43.283
1000	132.661	620.474	527.964	92.510	-847.114	-735.037	38.394
1100	132.725	633.122	536.958	105.780	-848.273	-723.773	34.369
1200	132.774	644.672	545.460	119.055	-849.443	-712.403	31.010
1300	132.812	655.302	553.506	132.334	-850.626	-700.935	28.164
1400	132.842	665.145	561.133	145.617	-1106.643	-683.042	25.485
1500	132.866	674.311	568.376	158.903	-1105.152	-652.837	22.734
1600	132.886	682.887	575.268	172.190	-1103.681	-622.731	20.330
1700	132.903	690.943	581.838	185.480	-1102.230	-592.716	18.212
1800	132.916	698.540	588.112	198.771	-1100.804	-562.786	16.332
1900	132.928	705.727	594.115	212.063	-1099.407	-532.934	14.651
2000	132.938	712.546	599.868	225.356	-1098.042	-503.155	13.141
2100	132.947	719.032	605.389	238.650	-1096.714	-473.443	11.776
2200	132.954	725.217	610.696	251.945	-1095.426	-443.794	10.537
2300	132.961	731.127	615.805	265.241	-1094.180	-414.203	9.407
2400	132.966	736.786	620.729	278.538	-1092.980	-384.664	8.372
2500	132.971	742.214	625.480	291.834	-1091.828	-355.175	7.421
2600	132.976	747.429	630.071	305.132	-1090.724	-325.731	6.544
2700	132.980	752.448	634.511	318.430	-1089.669	-296.328	5.733
2800	132.983	757.284	638.810	331.728	-1088.663	-266.964	4.980
2900	132.987	761.951	642.976	345.026	-1087.704	-237.634	4.280
3000	132.989	766.459	647.018	358.325	-1086.793	-208.337	3.627
3100	132.992	770.820	650.941	371.624	-1085.927	-179.070	3.017
3200	132.994	775.042	654.754	384.923	-1085.104	-149.830	2.446
3300	132.997	779.135	658.461	398.223	-1084.323	-120.614	1.909
3400	132.999	783.105	662.069	411.523	-1083.582	-91.422	1.405
3500	133.000	786.961	665.583	424.823	-1082.880	-62.252	0.929
3600	133.002	790.707	669.006	438.123	-1082.214	-33.100	0.480
3700	133.004	794.351	672.345	451.423	-1081.585	-3.966	0.056
3800	133.005	797.898	675.603	464.724	-1080.990	25.150	-0.346
3900	133.006	801.353	678.783	478.024	-1080.430	54.251	-0.727
4000	133.007	804.721	681.890	491.325	-1079.905	83.339	-1.088
4100	133.009	808.005	684.926	504.626	-1079.414	112.414	-1.432
4200	133.010	811.210	687.894	517.926	-1078.959	141.477	-1.760
4300	133.011	814.340	690.799	531.227	-1078.540	170.530	-2.072
4400	133.011	817.398	693.641	544.529	-1078.158	199.574	-2.369
4500	133.012	820.387	696.425	557.830	-1077.815	228.610	-2.654
4600	133.013	823.311	699.152	571.131	-1077.514	257.638	-2.926
4700	133.014	826.171	701.824	584.432	-1077.255	286.660	-3.186
4800	133.014	828.972	704.444	597.734	-1077.043	315.677	-3.435
4900	133.015	831.714	707.013	611.035	-1076.878	344.690	-3.674
5000	133.016	834.401	709.534	624.337	-1076.764	373.701	-3.904
5100	133.016	837.036	712.008	637.638	-1076.703	402.710	-4.125
5200	133.017	839.618	714.438	650.940	-1076.698	431.717	-4.337
5300	133.017	842.152	716.824	664.242	-1076.751	460.726	-4.541
5400	133.018	844.639	719.168	677.544	-1076.866	489.736	-4.737
5500	133.018	847.079	721.471	690.845	-1077.046	518.749	-4.927
5600	133.019	849.476	723.736	704.147	-1077.293	547.765	-5.109
5700	133.019	851.831	725.962	717.449	-1077.611	576.787	-5.286
5800	133.020	854.144	728.152	730.751	-1078.003	605.814	-5.456
5900	133.020	856.418	730.307	744.053	-1078.471	634.849	-5.621
6000	133.020	858.654	732.428	757.355	-1079.019	663.893	-5.780

PREVIOUS: June 1974 (1 atm) CURRENT: June 1974 (1 bar)

Magnesium Bromide ($(MgBr_2)_2$) Br_4Mg_2(g)

MOLYBDENUM BROMIDE ($MoBr_4$) CRYSTAL M_r = 415.556

$\Delta_f H°(0\ K)$ = Unknown

$S°(298.15\ K) = [209.2 \pm 17]\ J\ K^{-1}\ mol^{-1}$ $\Delta_f H°(298.15\ K) = -304.2 \pm 8\ kJ\ mol^{-1}$

Enthalpy of Formation

Shchukarev et al. (1) have measured the enthalpy of solution of $MoBr_4(cr)$ from which Dellien et al. (2) have obtained $\Delta_f H°(MoBr_4,\ cr,\ 298.15\ K) = -70.35\ kcal\ mol^{-1}$. Brewer (3) has reviewed the equilibrium data of Oppermann (4) (see the discussion on the $MoBr_4(g)$ table (5)) as well as the above measurements and obtained $\Delta_f H°(298.15\ K) = -72.7 \pm 2\ kcal\ mol^{-1}$ ($-304.177\ kJ\ mol^{-1}$) which is the value adopted here. This differs from the value $\Delta_f H°(298.15\ K) = -76.8\ kcal\ mol^{-1}$ adopted by NBS (6).

Heat Capacity and Entropy

Brewer (3) estimated $C_p°(298.15\ K) = 31.8 \pm 2\ cal\ K^{-1}\ mol^{-1}$ and $S°(298.15\ K) = 50.0\ cal\ K^{-1}\ mol^{-1}$ ($209.200\ kJ\ mol^{-1}$). Other estimates are $C_p°(298.15\ K) = 29.5\ cal\ K^{-1}\ mol^{-1}$ from the ionic contributions of Kubaschewski and Unal (7) and $S°(298.15\ K) = 51.4\ cal\ K^{-1}\ mol^{-1}$ from the additive entropy constants of Stull and Prophet (8). We adopt Brewer's values (3) of $C_p°$ and S° at 298.15 K and estimate $C_p°$ at higher temperatures by comparison with $MoBr_2(cr)$ and $MoBr_3(cr)$ (5).

Decomposition Data

T_{dcm} is the calculated temperature at which $\Delta_r G° = 0$ for the reaction $MoBr_4(cr) = MoBr_3(cr) + 0.5\ Br_2(g)$. Carnell et al. (9) have given conditions for the formation of $MoBr_4(cr)$ and its dissociation to $MoBr_3(cr)$.

References

1. S. A. Shchukarev, I. V. Vasil'kova, and N. D. Zaitseva, Vestn. Leningrad Univ. Ser. Fiz. Khim. 4, 127 (1961); Chem. Abstr. 56, 9510g (1962).
2. I. Dellien, M. F. Hall, and L. G. Hepler, Chem. Rev. 76, 283 (1976).
3. L. Brewer, Materials and Molecular Research Division, Lawrence Berkeley Laboratory, University of California, Berkeley; personal communication, Sept. 29, 1978; preliminary draft of review to be submitted for publication in Atomic Energy Review, International Atomic Energy Agency, Vienna, Austria.
4. H. Oppermann, Z. Anorg. Allg. Chem. 395, 249 (1973).
5. JANAF Thermochemical Tables: $MoBr_4(g)$, $MoBr_3(cr)$ and $MoBr_2(cr)$, 9-30-78.
6. U. S. Nat. Bur. Stand. Tech. Note 270-4, 1969.
7. O. Kubaschewski and H. Unal, Hign Temp.-High Pressure 9, 361 (1977).
8. D. R. Stull and H. Prophet, Chap. 13 in "The Characterization of High-Temperature Vapors," Ed. by J. L. Margrave, John Wiley, 1967.
9. P. J. H. Carnell, R. E. McCarley, and R. D. Hogue, Inorg. Syn. 10, 49 (1967).

Molybdenum Bromide ($MoBr_4$) $Br_4Mo_1(cr)$

Enthalpy Reference Temperature = T_r = 298.15 K Standard State Pressure = p° = 0.1 MPa

	J K⁻¹mol⁻¹			kJ mol⁻¹			
T/K	$C_p°$	S°	$-[G°-H°(T_r)]/T$	$H°-H°(T_r)$	$\Delta_f H°$	$\Delta_f G°$	Log K_f
0							
100							
200							
298.15	133.051	209.200	209.200	0.	-304.177	-267.261	46.823
300	133.093	210.023	209.203	0.246	-304.255	-267.031	46.494
400	135.143	248.601	214.449	13.661	-362.256	-242.498	31.667
500	136.817	278.930	224.418	27.256	-358.593	-212.982	22.250
600	138.909	304.061	235.656	41.043	-354.861	-184.208	16.037
700	140.844	325.621	247.004	55.032	-351.022	-156.068	11.646
800	142.674	344.549	258.038	69.208	-347.071	-128.486	8.389
900	144.400	361.454	268.606	83.563	-343.011	-101.405	5.885
1000	146.022	376.753	278.668	98.085	-338.850	-74.782	3.906

PREVIOUS: CURRENT: September 1978

MOLYBDENUM BROMIDE ($MoBr_4$) IDEAL GAS M_r = 415.556

$\Delta_f H°(0\ K) = -142.4 \pm 8\ kJ\ mol^{-1}$

$S°(298.15\ K) = 418.89 \pm 8\ J\ K^{-1}\ mol^{-1}$ $\Delta_f H°(298.15\ K) = -171.1 \pm 8\ kJ\ mol^{-1}$

Electronic Levels and Quantum Weights

ϵ_i, cm^{-1}	g_i
0	[3]
[15000]	[11]
[20000]	[9]

Vibrational Frequencies and Degeneracies

ν, cm^{-1}	ν, cm^{-1}
[230](1)	[80](3)
[70](2)	[390](3)

Point Group: T_d
Bond Distance: Mo-Br = 2.39 ± 0.02
Bond Angle: Br-Mo-Br = 109.47°
Product of the Moments of Inertia: $I_A I_B I_C = 8.25596 \times 10^{-111}\ g^3\ cm^6$

Enthalpy of Formation

Oppermann (1) made careful pressure measurements on the systems $MoBr_3(cr) + Br_2(g)$, $MoBr_{2+x}(cr) + MoBr_{3-y}(cr)$ and $MoBr_2(cr) + Mo(cr)$. Brewer (2), by comparison with the tungsten-bromide system, indicated the possibility of intermediate phases between $MoBr_2$ and $MoBr_3$ and thus modified the treatment of the data. Brewer used a 3rd law treatment combined with independent data on the enthalpies of formation of $MoBr_2(cr)$, $MoBr_3(cr)$ and $MoBr_4(cr)$ and removed the restriction that x=0 and y=0. For this reason we adopt the value of $\Delta_f H°(298.15\ K)$ obtained by Brewer, -40.9±2 kcal mol^{-1} (-171.126 kJ mol^{-1}), rather than the value of -36.7 kcal mol^{-1} reported by Dellien et al. (7).

Heat Capacity and Entropy

Brewer (2) used a slight modification of the vibrational frequencies and internuclear distance estimated by Lofgren (3). He assumed an Mo-Br distance of 2.42 Å in all the molybdenum bromides. The electron diffraction study of Spiridonov and Romanov (4) established the structure of $MoBr_4(g)$ to be a regular tetrahedron with an Mo-Br distance of 2.39±0.02 Å (5). This is the value adopted here. Our value for the product of the moments of inertia $I_A I_B I_C = 8.256 \times 10^{-111}\ g^3\ cm^6$ is 8% lower than Brewer's value ($8.898 \times 10^{-111}\ g^3\ cm^6$). This reduces S° by 0.1 cal $K^{-1}\ mol^{-1}$ at all temperatures. Electronic levels are assumed to be the same as the MoF_4 levels (6), which Brewer [2] estimated from crystal field theory.

References

1. H. Oppermann, Z. Anorg. Allg. Chem. 305, 249 (1973).
2. L. Brewer, Materials and Molecular Research Division, Lawrence Berkeley Laboratory, University of California, Berkeley; personal communication, Sept. 29, 1978; a preliminary draft of review to be submitted for publication in Atomic Energy Review, International Atomic Energy Agency, Vienna, Austria.
3. N. L. Lofgren, personal communication to L. Brewer, 1967.
4. V. P. Spiridonov and G. V. Romanov, Vestn. Mosk. Univ., Ser II, 22, 118 (1967); Chem. Abstr. 69, 21045 (1968).
5. B. Beagley, in "Molecular Structure by Diffraction Methods," Vol. 1, Chemical Society Specialist Periodical Report, Burlington House, London, 1973, p. 151.
6. JANAF Thermochemical Tables: $MoCl_6(g)$, $MoCl_5(g)$ and $MoCl_4(g)$, 12-31-68; $MoF_4(g)$, 12-30-78.
7. I. Dellien, F. M. Hall, and L. G. Hepler, Chem. Rev. 76, 283 (1976).

Molybdenum Bromide ($MoBr_4$) $Br_4Mo_1(g)$

Enthalpy Reference Temperature = T_r = 298.15 K Standard State Pressure = p° = 0.1 MPa

T/K	$C_p°$ (J K⁻¹mol⁻¹)	S° (J K⁻¹mol⁻¹)	$-[G°-H°(T_r)]/T$ (J K⁻¹mol⁻¹)	$H°-H°(T_r)$ (kJ mol⁻¹)	$\Delta_f H°$ (kJ mol⁻¹)	$\Delta_f G°$ (kJ mol⁻¹)	Log K_f
0	0.	0.	INFINITE	-24.839	-142.361	-142.361	INFINITE
100	80.869	316.427	506.136	-18.971	-142.571	-162.741	85.007
200	97.616	378.758	428.168	-9.882	-145.126	-182.014	47.537
250	100.942	400.931	420.574	-4.911	-146.904	-191.039	39.915
298.15	102.884	418.889	418.889	0.	-171.126	-196.728	34.466
300	102.942	419.526	418.891	0.190	-171.259	-196.887	34.281
350	104.222	435.498	420.150	5.372	-232.592	-197.799	29.520
400	105.084	449.474	422.960	10.606	-232.260	-192.851	25.184
450	105.690	461.888	426.609	15.876	-231.937	-187.944	21.816
500	106.131	473.048	430.704	21.172	-231.625	-183.073	19.126
600	106.716	492.454	439.426	31.817	-231.036	-173.419	15.097
700	107.074	508.934	448.208	42.508	-230.495	-163.859	12.227
800	107.308	523.248	456.713	53.228	-230.001	-154.374	10.080
900	107.470	535.897	464.822	63.967	-229.556	-144.948	8.413
1000	107.587	547.226	472.506	74.720	-229.164	-135.569	7.081
1100	107.673	557.485	479.772	85.483	-228.829	-126.226	5.994
1200	107.739	566.856	486.645	96.254	-228.561	-116.911	5.089
1300	107.791	575.482	493.151	107.031	-228.366	-107.615	4.324
1400	107.832	583.472	499.321	117.812	-228.254	-98.331	3.669
1500	107.867	590.913	505.181	128.597	-228.232	-89.053	3.101
1600	107.899	597.875	510.760	139.385	-228.309	-79.772	2.604
1700	107.929	604.418	516.078	150.177	-228.495	-70.483	2.166
1800	107.960	610.587	521.159	160.971	-228.800	-61.180	1.775
1900	107.995	616.426	526.021	171.769	-229.234	-51.857	1.426
2000	108.037	621.966	530.681	182.570	-229.810	-42.507	1.110
2100	108.089	627.238	535.154	193.377	-230.538	-33.125	0.824
2200	108.153	632.268	539.455	204.189	-231.430	-23.704	0.563
2300	108.232	637.077	543.596	215.008	-232.496	-14.238	0.323
2400	108.328	641.686	547.587	225.836	-233.747	-4.723	0.103
2500	108.444	646.110	551.440	236.674	-235.192	4.849	-0.101
2600	108.581	650.366	555.164	247.525	-236.853	14.482	-0.291
2700	108.741	654.467	558.766	258.391	-238.765	24.185	-0.468
2800	108.924	658.425	562.255	269.274	-240.987	33.963	-0.634
2900	109.132	662.250	565.638	280.177	-279.479	43.877	-0.790
3000	109.365	665.954	568.920	291.101	-280.510	55.045	-0.958
3100	109.623	669.544	572.109	302.050	-281.550	66.247	-1.116
3200	109.904	673.029	575.208	313.027	-282.591	77.483	-1.265
3300	110.209	676.416	578.224	324.032	-283.625	88.751	-1.405
3400	110.536	679.710	581.161	335.069	-284.646	100.051	-1.537
3500	110.884	682.920	584.022	346.140	-285.645	111.380	-1.662
3600	111.251	686.048	586.813	357.247	-286.616	122.737	-1.781
3700	111.635	689.102	589.537	368.391	-287.550	134.121	-1.893
3800	112.035	692.084	592.196	379.574	-288.443	145.529	-2.000
3900	112.447	695.000	594.795	390.798	-289.286	156.961	-2.102
4000	112.870	697.852	597.336	402.064	-290.075	168.413	-2.199
4100	113.301	700.644	599.822	413.372	-290.803	179.885	-2.292
4200	113.738	703.380	602.255	424.724	-291.467	191.373	-2.380
4300	114.178	706.061	604.638	436.120	-292.062	202.876	-2.464
4400	114.620	708.691	606.973	447.560	-292.584	214.393	-2.545
4500	115.061	711.272	609.262	459.044	-293.031	225.920	-2.622
4600	115.498	713.806	611.507	470.572	-293.400	237.457	-2.696
4700	115.931	716.294	613.710	482.144	-293.689	249.000	-2.767
4800	116.356	718.739	615.873	493.758	-293.898	260.549	-2.835
4900	116.772	721.143	617.997	505.414	-294.025	272.102	-2.901
5000	117.177	723.506	620.084	517.112	-885.455	289.387	-3.023
5100	117.570	725.830	622.134	528.849	-887.097	312.901	-3.205
5200	117.949	728.117	624.151	540.625	-888.805	336.446	-3.380
5300	118.313	730.367	626.134	552.439	-890.575	360.026	-3.548
5400	118.662	732.582	628.084	564.288	-892.410	383.639	-3.711
5500	118.993	734.762	630.004	576.170	-894.306	407.287	-3.868
5600	119.307	736.909	631.894	588.086	-896.265	430.970	-4.020
5700	119.603	739.024	633.755	600.031	-898.283	454.689	-4.167
5800	119.880	741.106	635.588	612.006	-900.362	478.442	-4.309
5900	120.139	743.158	637.394	624.007	-902.498	502.232	-4.446
6000	120.378	745.179	639.174	636.033	-904.696	526.060	-4.580

PREVIOUS: September 1978 (1 atm) CURRENT: September 1978 (1 bar)

LEAD BROMIDE ($PbBr_4$) IDEAL GAS M_r = 526.816

$\Delta_fH°(0\ K)$ = [-426.4 ± 84] kJ mol^{-1}
$S°(298.15\ K)$ = [426.20 ± 12.6] J K^{-1} mol^{-1} $\Delta_fH°(298.15\ K)$ = [-456.4 ± 84] kJ mol^{-1}

Vibrational Frequencies and Degeneracies

ν, cm^{-1}
[207](1)
[59](2)
[231](3)
[73](3)

Ground State Quantum Weight: [1] σ = 12
Point Group: [T_d]
Bond Distance: Pb-Br = [2.58] Å
Bond Angle: Br-Pb-Br = [109° 28']
Product of Moments of Inertia: $I_AI_BI_C$ = [1.30647 x 10^{-110}] g^3 cm^6

Enthalpy of Formation

The adopted $\Delta_fH°(0\ K)$ = -101.9±20 kcal mol^{-1} (-426.350±84 kJ mol^{-1}) is derived from the dissociation energy for the process $PbBr_4(g) = Pb(g) + 4\ Br(g)$. D_0^0 = 11.34 eV is estimated from an intercomparison of the lead mono-, di-, and tetra-halides (1). This adopted D_0^0 value for $PbBr_4(g)$ is greater than the D_0^0 value for $PbBr_2(g)$ by a factor of 2.09. The auxiliary values $\Delta_fH°$(Pb, g, 0 K)) = 46.91 kcal mol^{-1} (1) and $\Delta_fH°$(Br, g, 0 K) = 28.188 kcal mol^{-1} (1) are used. $\Delta_fH°$($PbBr_4$, g, 298.15 K) = -109.1±20 kcal mol^{-1} (-456.361±84 kJ mol^{-1}) is calculated from $\Delta_fH°(0\ K)$.

Heat Capacity and Entropy

The molecular constants and vibrational frequencies are those calculated by Aleksandrovskaya et al. (2). Later works (3, 4, 5) mistakenly cite this work as observed data and use a bond distance of 2.54 Å rather than the 2.58 Å originally published (2).

The principal moments of inertia are $I_A = I_B = I_C$ = 23.5227 x 10^{-39} g cm^2.

References

1. JANAF Thermochemical Tables: PbF(g), PbF_2(g), PbF_4(g), PbBr(g), $PbBr_2$(g), PbI(g), and PbI_2(g), 12-31-73; PbCl(g) and $PbCl_2$(g), 6-30-73; Pb(g), 3-31-62; Br(g), 9-30-61.
2. A. M. Aleksandrovskaya, I. V. Rigina, and I. N. Godnev, Optics and Spectrosc. 7, 495 (1959).
3. G. Nagarajan, Bull. Soc. Chim. Belges 71, 119 (1962).
4. M. Radhakrishnan, Indian J. Pure Appl Phys. 1, 402 (1963).
5. S. P. Singh, Labdev, Part A 7, 185 (1969).

Lead Bromide ($PbBr_4$) $Br_4Pb_1(g)$

Enthalpy Reference Temperature = T_r = 298.15 K; Standard State Pressure = p° = 0.1 MPa

T/K	$C_p°$ (J $K^{-1}mol^{-1}$)	S°	$-[G°-H°(T_r)]/T$	$H°-H°(T_r)$ (kJ mol^{-1})	$\Delta_fH°$	$\Delta_fG°$	Log K_f
0	0.	0.	INFINITE	-25.895	-426.359	-426.359	INFINITE
100	86.541	319.504	516.117	-19.661	-427.478	-444.976	232.432
200	100.868	385.092	435.680	-10.118	-430.262	-461.457	120.520
250	103.277	407.885	427.917	-5.008	-432.089	-469.054	98.003
298.15	104.631	426.201	426.201	0.	-456.361	-473.357	82.930
300	104.671	426.849	426.203	0.194	-456.497	-473.462	82.437
350	105.543	443.054	427.481	5.451	-517.895	-472.925	70.580
400	106.122	457.188	430.330	10.743	-517.637	-466.519	60.921
450	106.525	469.711	434.023	16.060	-517.400	-460.144	53.412
500	106.816	480.951	438.163	21.394	-517.183	-453.794	47.408
600	107.199	500.463	446.969	32.096	-516.809	-441.152	38.406
700	107.433	517.006	455.823	42.829	-521.375	-427.769	31.921
800	107.585	531.363	464.388	53.580	-521.146	-414.412	27.058
900	107.690	544.041	472.547	64.344	-520.895	-401.085	23.278
1000	107.765	555.391	480.274	75.117	-520.625	-387.788	20.256
1100	107.821	565.665	487.577	85.896	-520.337	-374.518	17.784
1200	107.863	575.048	494.481	96.681	-520.030	-361.275	15.726
1300	107.896	583.683	501.015	107.469	-519.714	-348.058	13.985
1400	107.922	591.680	507.209	118.260	-519.404	-334.865	12.494
1500	107.944	599.127	513.092	129.053	-519.113	-321.694	11.202
1600	107.961	606.094	518.689	139.848	-518.849	-308.542	10.073
1700	107.975	612.640	524.025	150.645	-518.619	-295.405	9.077
1800	107.987	618.812	529.121	161.443	-518.430	-282.280	8.192
1900	107.998	624.651	533.997	172.243	-518.289	-269.165	7.400
2000	108.006	630.190	538.669	183.043	-518.200	-256.055	6.687
2100	108.014	635.460	543.154	193.844	-695.388	-235.834	5.866
2200	108.020	640.485	547.464	204.645	-695.027	-213.960	5.080
2300	108.026	645.287	551.614	215.448	-694.797	-192.099	4.363
2400	108.031	649.885	555.613	226.251	-694.701	-170.245	3.705
2500	108.036	654.295	559.473	237.054	-694.741	-148.392	3.100
2600	108.039	658.532	563.202	247.858	-694.917	-126.535	2.542
2700	108.043	662.610	566.809	258.662	-695.225	-104.669	2.025
2800	108.046	666.539	570.301	269.466	-695.660	-82.789	1.544
2900	108.049	670.330	573.685	280.271	-696.216	-60.891	1.097
3000	108.051	673.993	576.968	291.076	-696.882	-38.972	0.679
3100	108.054	677.537	580.155	301.881	-697.649	-17.030	0.287
3200	108.056	680.967	583.252	312.687	-698.506	4.939	-0.081
3300	108.058	684.292	586.264	323.493	-699.441	26.936	-0.426
3400	108.059	687.518	589.195	334.298	-700.441	48.963	-0.752
3500	108.061	690.650	592.049	345.104	-701.495	71.019	-1.060
3600	108.062	693.695	594.831	355.911	-702.590	93.107	-1.351
3700	108.064	696.655	597.543	366.717	-703.715	115.225	-1.627
3800	108.065	699.537	600.189	377.523	-704.858	137.373	-1.888
3900	108.066	702.344	602.773	388.330	-706.009	159.552	-2.137
4000	108.067	705.080	605.296	399.136	-707.158	181.761	-2.374
4100	108.068	707.749	607.763	409.943	-708.294	203.998	-2.599
4200	108.069	710.353	610.174	420.750	-709.411	226.262	-2.814
4300	108.070	712.896	612.534	431.557	-710.501	248.553	-3.019
4400	108.071	715.380	614.843	442.364	-711.556	270.869	-3.216
4500	108.071	717.809	617.104	453.171	-712.571	293.208	-3.403
4600	108.072	720.184	619.320	463.978	-713.541	315.570	-3.583
4700	108.073	722.509	621.490	474.786	-714.462	337.952	-3.756
4800	108.073	724.784	623.619	485.593	-715.329	360.353	-3.921
4900	108.074	727.012	625.706	496.400	-716.141	382.771	-4.080
5000	108.074	729.196	627.754	507.208	-716.893	405.206	-4.233
5100	108.075	731.336	629.764	518.015	-717.586	427.656	-4.380
5200	108.075	733.435	631.738	528.823	-718.217	450.117	-4.521
5300	108.076	735.493	633.676	539.630	-718.785	472.591	-4.658
5400	108.076	737.513	635.580	550.438	-719.291	495.075	-4.789
5500	108.077	739.496	637.452	561.245	-719.734	517.567	-4.915
5600	108.077	741.444	639.292	572.053	-720.114	540.067	-5.038
5700	108.077	743.357	641.100	582.861	-720.433	562.573	-5.155
5800	108.078	745.236	642.880	593.669	-720.691	585.083	-5.269
5900	108.078	747.084	644.630	604.476	-720.888	607.599	-5.379
6000	108.078	748.900	646.353	615.284	-721.028	630.116	-5.486

PREVIOUS: December 1973 (1 atm) CURRENT: December 1973 (1 bar)

TETRABROMOSILANE ($SiBr_4$) LIQUID M_r = 347.7015

Tetrabromosilane ($SiBr_4$) Br_4Si_1(l)

$S°$(298.15 K) = [278.24 ± 1.3] J K^{-1} mol^{-1}

T_{fus} = 278.4 ± 1.0 K

$\Delta_f H°$(298.15 K) = -457.31 ± 8.4 kJ mol^{-1}

$\Delta_{fus}H°$ = unknown

Enthalpy of Formation

Schafer and Heine (1) measured the enthalpy of solution of Si(cr) in an HF solution containing AgF. For the overall reaction $SiBr_4$(ℓ) + 4 Ag(cr) = Si(cr) + 4 AgBr(cr), they reported $\Delta_r H°$(298.15 K) = +13.3 kcal mol^{-1}. Using auxiliary data for AgBr (2), we calculate $\Delta_f H°$(298.15 K) = -109.3 kcal mol^{-1} (-457.311 kJ mol^{-1}) for $SiBr_4$(ℓ). We adopt this value and assign an uncertainty of ±2.0 kcal mol^{-1}.

Wolf et al. (3) studied the enthalpies of solution of $SiBr_4$(ℓ) and Na_2SiO_3(cr) in caustic solution. The net reaction of interest, $SiBr_4$(ℓ) + 6 NaOH(cr) = Na_2SiO_3(cr) + 4 NaBr(cr) + 3 H_2O(ℓ), yielded $\Delta_r H°$(298.15 K) = -199.32 kcal mol^{-1} based on the appropriate combination of results from five solution studies. Using current auxiliary data (4, 5), we update this value to $\Delta_r H°$(298.15 K) = -198.06 kcal mol^{-1} and calculate $\Delta_f H°$(298.15 K) = -114.8 kcal mol^{-1} for $SiBr_4$(ℓ). This data is suspect due to uncertainties in the conversion of data from Na_2SiO_3 (calorimetric solution) to Na_2SiO_3(cr). The final state of the calorimetric solution is not well defined or well known and thus large uncertainties result.

Heat Capacity and Entropy

The heat capacity from 25°-100°C was determined calorimetrically within 2% by Sladkov (6). The constant value of 35.0 cal K^{-1} mol^{-1} is extrapolated to 800 K. The entropy at 298.15 K is chosen so as to bring the 2nd and 3rd law analysis of the vapor pressure data into agreement.

Fusion Data

Pohland (7) reported the melting point of $SiBr_4$ to be 5.2 °C (±1 °C). No data are available on the enthalpy of fusion

Vaporization Data

T_{vap} is calculated as that temperature for which $\Delta_r G°$ = 0 for the process $SiBr_4$(ℓ) = $SiBr_4$(g) [f = 1 bar]. $\Delta_{vap}H°$ is calculated as the difference between the $\Delta_f H°$ values for the ideal gas and the liquid at T_{vap}. Normal boiling points (p = 1 atm) reported in the literature are 426.0 K (7) and 426.6 K (8), and 427.8 K (766 mm Hg, 9). A normal boiling point (p = 1 bar) should be slightly lower than our calculated T_{vap} which corresponds to f = 1 bar. Refer to the $SiBr_4$(g) table for a possible explanation of this discrepancy.

References

1. H. Schafer and H. Heine, Z. Anorg. Allgem. Chem. 332, 25 (1964).
2. U. S. Nat. Bur. Stand. Tech. Note 270-4, 1969.
3. E. Wolf, W. Stahn, and M. Schonherr, Z. Anorg. Allgem. Chem. 319, 168 (1962).
4. ICSU-CODATA Task Group, J. Chem. Thermodynamics 4, 331 (1972).
5. JANAF Thermochemical Tables: NaBr(cr), 9-30-64; NaOH(cr), 12-31-70; Na_2SiO_3(cr), 9-30-67.
6. D. B. Sladkov, Russ. J. Phys. Chem. 48, 1117 (1974).
7. E. Pohland, Z. Anorg. Allgem. Chem. 201, 265 (1931).
8. W. C. Schumb and R. C. Young, J. Am. Chem. Soc. 52, 1464 (1930).
9. W. Biltz, A. Sapper, and E. Wunnenberg, Z. Anorg. Allgem. Chem. 203, 277 (1932).

Enthalpy Reference Temperature = T_r = 298.15 K   Standard State Pressure = p° = 0.1 MPa

T/K	$C_p°$ (J $K^{-1}mol^{-1}$)	S° (J $K^{-1}mol^{-1}$)	$-[G°-H°(T_r)]/T$ (J $K^{-1}mol^{-1}$)	$H°-H°(T_r)$ (kJ mol^{-1})	$\Delta_f H°$ (kJ mol^{-1})	$\Delta_f G°$ (kJ mol^{-1})	Log K_f
0							
100							
200							
298.15	146.440	278.236	278.236	0.	-457.311	-443.896	77.769
300	146.440	279.142	278.239	0.271	-457.357	-443.812	77.275
400	146.440	321.270	283.983	14.915	-513.795	-427.415	55.815
500	146.440	353.947	294.829	29.559	-508.812	-406.401	42.456
600	146.440	380.646	306.975	44.203	-503.983	-386.374	33.637
700	146.440	403.220	319.153	58.847	-499.264	-367.147	27.397
800	146.440	422.774	330.911	73.491	-494.634	-348.591	22.761

PREVIOUS: CURRENT: December 1976

Tetrabromosilane ($SiBr_4$) Br_4Si_1(l)

TETRABROMOSILANE ($SiBr_4$) IDEAL GAS M_r = 347.7015

S°(298.15 K) = 379.36 ± 0.8 J K^{-1} mol^{-1}

$\Delta_f H°$(0 K) = -385.55 ± 16.7 kJ mol^{-1}
$\Delta_f H°$(298.15 K) = -415.47 ± 16.7 kJ mol^{-1}

Vibrational Frequencies and Degeneracies

ν, cm^{-1}	ν, cm^{-1}
246.7(1)	494 (3)
84.8(2)	133.6(3)

Ground State Quantum Weight: [1] σ = 12
Point Group: T_d
Bond Distance: Si-Br = 2.15 ± 0.02 Å
Bond Angle: Br-Si-Br = 109.47122°
Product of the Moments of Inertia: $I_A I_B I_C$ = 4.37533 x 10^{-111} g^3 cm^6

Enthalpy of Formation Pohland (1) has studied the vaporization of $SiBr_4$ from 260-426 K (0.8-755.5 mm Hg). This vapor pressure data (20 points) is corrected for vapor non-ideality by means of the equation ΔG°/T = -R ln p - Bp/T. The Berthelot equation of state and critical constants T_c = 656 K and P_c = 41.3 atm as suggested in the review article by Lapidus et al. (2) are used to calculate B. The recent study on orthobaric densities by Nisel'son et al. (3) reported T_c = 663 K. The use of this latter value does not signficantly affect the results.

A 2nd and 3rd law analysis of this corrected data (1) yields $\Delta_{vap}H°$(298.15 K) = 10.0 kcal mol^{-1}. The entropy of the liquid at 298.15 K is adjusted so as to bring the 2nd and 3rd law results into agreement. This treatment suggests that the five lowest pressure data points (260-345 K, 0.8-60.3 mm Hg) are biased. These five points are not included in the above analysis. Any vapor pressure equation used to represent this data will be significantly altered in neglecting these five data points.

We adopt $\Delta_{vap}H°$(298.15 K) = 10.0 kcal mol^{-1} which leads to $\Delta_f H°$(298.15 K) = -99.3 kcal mol^{-1} (-415.471 kJ mol^{-1}) (4). Using auxiliary data (4), we calculate $\Delta_{at}H°$($SiBr_4$, g) = 311.54 kcal mol^{-1} (1303.48 kJ mol^{-1}). This value is 4.03 times the dissociation energy of SiBr(g) (4).

Heat Capacity and Entropy

The adopted vibrational frequencies are from the work of Clark and Rippon (5), who recorded the Raman spectra in the vapor phase (210°C). The spectral data were interpreted in terms of a tetrahedral structure. This structure is consistent with the electron diffraction data of Lister and Sutton (6), Yamasaki et al. (7), and Spitzer et al. (8), which suggested a tetrahedral structure with a Si-Br bond distance of 2.15±0.02 Å. We adopt this bond distance. The principal moments of inertia are: $I_A = I_B = I_C$ = 163.5574x 10^{-39} g cm^2. Shimanouchi, in a recent compilation of molecular vibrational frequencies (9), suggested somewhat different values (249, 90, 487, 137 cm^{-1}) based on earlier infrared and Raman studies by Trumpy (10), Delwaulle (11), and Radhakrishnan (12), and Long et al. (13). These frequencies would lead to a S°(298.15 K) value of 90.3 cal K^{-1} mol^{-1} (377.9 J K^{-1} mol^{-1}). Much literature has been published on the inter-relationships between force constants and vibrational frequencies. Since the majority of this work was published prior to the Clark and Rippon study (5), it will not be further discussed or referenced. The same situation exists for temperature dependent thermochemical tabulations of $SiBr_4$(g).

References

1. E. Pohland, Z. Anorg. Allg. Chem. 201, 265 (1931).
2. I. I. Lapidus, L. A. Nisel'son, and A. L. Seifer in "Thermophysical Properties of Gases and Liquids;" Vol. 1, V. A. Rabinovish, Ed., Israel Program for Scientific Translations, Jerusalem, 1970, pp 102-135.
3. L. A. Nisel'son, T. D. Sokolova, and I. I. Lapidus, Zh. Neorg. Khim. 12, 1423 (1967).
4. JANAF Thermochemical Tables: SiBr(g) and $SiBr_4$(cr), 12-31-76; Si(g), 3-31-67; Br(g), 6-30-74.
5. R. J. H. Clark and D. M. Rippon, J. Mol. Spectrosc. 44, 479 (1972).
6. M. W. Lister and L. E. Sutton, Trans. Faraday Soc. 37, 393 (1941).
7. K. Yamasaki, A. Kotera, M. Yokoi, and M. Iwasaki, J. Chem. Phys. 17, 1355 (1949).
8. R. Spitzer, W. J. Howell, Jr., and V. Schomaker, J. Amer. Chem. Soc. 64, 62 (1942).
9. T. Shimanouchi, NSRDS-NBS 39, 1972.
10. B. Trumpy. Z. Physik 68, 675 (1931).
11. M. L. Delwaulle, J. Phys. Chem. 56, 355 (1952).
12. M. Radhakrishnan, Z. Phys. Chem. 35, 247 (1962).
13. D. A. Long, T. V. Spencer, D. N. Waters, and L. A. Woodward, Proc. Roy. Soc. (London) A240, 499 (1957).

Tetrabromosilane ($SiBr_4$) Br_4Si_1(g)

Enthalpy Reference Temperature = T_r = 298.15 K Standard State Pressure = p° = 0.1 MPa

T/K	$C_p°$ (J $K^{-1}mol^{-1}$)	S° (J $K^{-1}mol^{-1}$)	-[G°-H°(T_r)]/T (J $K^{-1}mol^{-1}$)	H°-H°(T_r) (kJ mol^{-1})	$\Delta_f H°$ (kJ mol^{-1})	$\Delta_f G°$ (kJ mol^{-1})	Log K_f
0	0.	0.	INFINITE	-22.313	-385.548	-385.548	INFINITE
100	70.691	287.079	459.223	-17.214	-386.301	-403.857	210.953
200	88.479	342.240	388.009	-9.154	-389.222	-420.368	109.789
250	93.599	362.569	380.944	-4.594	-391.143	-427.942	89.414
298.15	97.009	379.365	379.365	0.	-415.471	-432.207	75.721
300	97.117	379.965	379.367	0.180	-415.609	-432.311	75.272
350	99.568	395.131	380.559	5.100	-477.023	-431.728	64.432
400	101.316	408.547	383.235	10.125	-476.745	-425.276	55.535
450	102.594	420.558	386.726	15.224	-476.453	-418.860	48.620
500	103.552	431.419	390.661	20.379	-476.152	-412.477	43.091
600	104.857	450.424	399.083	30.805	-475.541	-399.799	34.806
700	105.676	466.654	407.605	41.334	-474.936	-387.224	28.895
800	106.222	480.803	415.889	51.931	-474.354	-374.733	24.468
900	106.603	493.337	423.811	62.573	-473.801	-362.314	21.028
1000	106.879	504.584	431.336	73.248	-473.285	-349.955	18.280
1100	107.085	514.781	438.466	83.947	-472.809	-337.645	16.033
1200	107.243	524.106	445.219	94.664	-472.375	-325.377	14.163
1300	107.366	532.695	451.622	105.394	-471.986	-313.143	12.582
1400	107.464	540.655	457.701	116.136	-471.645	-300.938	11.228
1500	107.544	548.072	463.481	126.886	-471.357	-288.755	10.055
1600	107.609	555.015	468.987	137.644	-471.123	-276.590	9.030
1700	107.663	561.540	474.242	148.408	-521.126	-263.990	8.111
1800	107.709	567.696	479.264	159.177	-520.785	-248.874	7.222
1900	107.747	573.520	484.073	169.949	-520.470	-233.777	6.427
2000	107.780	579.048	488.685	180.726	-520.184	-218.695	5.712
2100	107.809	584.307	493.114	191.505	-519.932	-203.627	5.065
2200	107.833	589.323	497.374	202.287	-519.717	-188.570	4.477
2300	107.855	594.117	501.477	213.072	-519.543	-173.522	3.941
2400	107.874	598.707	505.433	223.858	-519.412	-158.480	3.449
2500	107.890	603.111	509.253	234.647	-519.324	-143.443	2.997
2600	107.905	607.343	512.945	245.436	-519.280	-128.410	2.580
2700	107.918	611.416	516.517	256.228	-519.279	-113.376	2.193
2800	107.930	615.341	519.977	267.020	-519.320	-98.342	1.835
2900	107.941	619.128	523.331	277.814	-519.400	-83.306	1.500
3000	107.950	622.788	526.585	288.608	-519.515	-68.267	1.189
3100	107.959	626.328	529.746	299.404	-519.663	-53.222	0.897
3200	107.967	629.756	532.818	310.200	-519.837	-38.173	0.623
3300	107.974	633.078	535.806	320.997	-520.034	-23.118	0.366
3400	107.981	636.301	538.715	331.795	-520.248	-8.057	0.124
3500	107.987	639.432	541.548	342.593	-520.474	7.010	-0.105
3600	107.992	642.474	544.309	353.392	-904.858	32.546	-0.472
3700	107.997	645.433	547.003	364.192	-904.677	58.583	-0.827
3800	108.002	648.313	549.631	374.992	-904.495	84.614	-1.163
3900	108.006	651.118	552.197	385.792	-904.305	110.640	-1.482
4000	108.010	653.853	554.705	396.593	-904.103	136.662	-1.785
4100	108.014	656.520	557.156	407.394	-903.883	162.679	-2.073
4200	108.017	659.123	559.552	418.196	-903.643	188.689	-2.347
4300	108.021	661.665	561.898	428.998	-903.377	214.694	-2.608
4400	108.024	664.148	564.193	439.800	-903.083	240.693	-2.857
4500	108.026	666.576	566.442	450.602	-902.756	266.684	-3.096
4600	108.029	668.950	568.644	461.405	-902.396	292.667	-3.323
4700	108.031	671.273	570.803	472.208	-901.999	318.643	-3.541
4800	108.034	673.548	572.920	483.011	-901.563	344.610	-3.750
4900	108.036	675.775	574.997	493.815	-901.087	370.567	-3.950
5000	108.038	677.958	577.034	504.618	-900.570	396.514	-4.142
5100	108.040	680.097	579.034	515.422	-900.010	422.450	-4.327
5200	108.042	682.195	580.998	526.226	-899.407	448.374	-4.504
5300	108.043	684.253	582.927	537.031	-898.760	474.288	-4.674
5400	108.045	686.273	584.822	547.835	-898.070	500.188	-4.838
5500	108.047	688.255	586.685	558.640	-897.335	526.075	-4.996
5600	108.048	690.202	588.516	569.444	-896.557	551.948	-5.148
5700	108.049	692.115	590.317	580.249	-895.735	577.807	-5.295
5800	108.051	693.994	592.088	591.054	-894.871	603.650	-5.436
5900	108.052	695.841	593.831	601.859	-893.966	629.479	-5.573
6000	108.053	697.657	595.546	612.665	-893.018	655.292	-5.705

PREVIOUS: December 1976 (1 atm) CURRENT: December 1976 (1 bar)

Tetrabromosilane ($SiBr_4$) Br_4Si_1(g)

TITANIUM BROMIDE ($TiBr_4$) CRYSTAL M_r = 367.496

$\Delta_f H°(0\ K)$ = -592.68 ± 4.6 kJ mol^{-1}
$\Delta_f H°(298.15\ K)$ = -617.98 ± 4.2 kJ mol^{-1}
$\Delta_{fus} H°$ = 12.887 kJ mol^{-1}

S°(298.15 K) = 243.6 ± 6.7 J K^{-1} mol^{-1}
T_{fus} = 311.4 K

Enthalpy of Formation

$\Delta_f H°(TiBr_4$, cr, 298.15 K) has been determined calorimetrically by several investigators as follows.

Source	Reaction	$\Delta_f H°(298.15\ K)$ kcal mol^{-1}
Nelson et al. (1)	$Ti(cr) + 2\ Br_2(g) = TiBr_4(cr)$	-147.3 ± 1.1
Johnson et al. (2)	$TiCl_4(\ell) + 2\ Br_2(\ell) = TiBr_4(cr) + 2\ Cl_2(g)$	-147.6 ± 1.3
Schlafer and Schmidtke (3)	$Ti(cr) + 2\ Br_2(\ell) = TiBr_4(cr)$	-147.7 ± 0.3
Gross, et al. (4)	$Ti(cr) + 2\ Br_2(\ell) = TiBr_4(cr)$	-148.1 ± 0.3

The chosen value of $\Delta_f H°(298.15\ K)$ = -147.7 ± 1.0 kcal mol^{-1} (617.977 ±4.2 kJ mol^{-1}) is the average of these determinations. The value from the work of Nelson et al. (1) is obtained from their enthalpy of reaction and the JANAF value for the enthalpy of vaporization of bromine. The value obtained from Johnson et al. (2) is a combination of their enthalpy of reaction and the JANAF value for the enthalpy of formation of $TiCl_4(\ell)$.

Heat Capacity and Entropy

The heat capacity of $TiBr_4$(cr) has been measured over the temperature range 51 to 298.15 K by King et al. (5). They reported a value of 42.74 cal K^{-1} mol^{-1} for S°(298.15 K)-S°(51 K) based on their measurements. The value of S°(51 K) is estimated as 15.46 cal K^{-1} mol^{-1}. King et al. (5) reported an estimate of 14.75±1.60 for the same quantity. The former estimate is used so that the values of ΔsubH°(298.15 K) obtained by both 2nd and 3rd law methods are in agreement. See $TiBr_4$(g) table for details. The value of H°(51 K)-H°(0 K) is estimated as 0.372 kcal mol^{-1}. This estimate is based on a Debye-Einstein extrapolation of the measured heat capacity with vibrational contributions removed.

Fusion Data

The melting temperature and heat of melting were reported by King et al. (5).

References

1. R. A. Nelson, W. H. Johnson and E. J. Prosen, J. Res. Nat. Bur. Stand. 62, 67 (1959).
2. W. H. Johnson, A. A. Gilliland and E. J. Prosen, J. Res. Nat. Bur. Stand. 63, 161 (1959).
3. H. L. Schlafer and H. H. Schmidtke, Z. Physik. Chem. 11, 297 (1959).
4. P. Gross, C. Hayman and D. L. Levi, Trans. Faraday Soc. 53, 1601 (1957).
5. E. G. King, W. W. Weller, A. U. Christensen and K. K. Kelley, U. S. Bur. Mines RI 5799, 1961.

Titanium Bromide ($TiBr_4$) $Br_4Ti_1(cr)$

Enthalpy Reference Temperature = T_r = 298.15 K — Standard State Pressure = p° = 0.1 MPa

T/K	$C_p°$ (J $K^{-1}mol^{-1}$)	S° (J $K^{-1}mol^{-1}$)	$-[G°-H°(T_r)]/T$ (J $K^{-1}mol^{-1}$)	$H°-H°(T_r)$ (kJ mol^{-1})	$\Delta_f H°$ (kJ mol^{-1})	$\Delta_f G°$ (kJ mol^{-1})	Log K_f
0	0.	0.	INFINITE	-28.556	-592.684	-592.684	INFINITE
100	94.098	120.922	349.954	-22.903	-593.178	-593.675	310.104
200	117.273	194.146	255.191	-12.209	-594.203	-593.819	155.090
298.15	131.503	243.622	243.622	0.	-617.977	-590.681	103.485
300	131.815	244.436	243.624	0.244	-618.060	-590.512	102.817
311.400	133.779	249.388	243.745	1.757	------ CRYSTAL <--> LIQUID -----		
400	133.779	282.886	248.860	13.611	-676.266	-569.177	74.327
500	133.779	312.738	258.761	26.989	-673.012	-542.785	56.704
600	133.779	337.129	269.852	40.366	-669.899	-517.034	45.012
700	133.779	357.751	280.974	53.744	-666.885	-491.796	36.698
800	133.779	375.615	291.712	67.122	-663.937	-466.985	30.491
900	133.779	391.372	301.927	80.500	-661.078	-442.539	25.684
1000	133.779	405.467	311.589	93.878	-658.370	-418.404	21.855
1100	133.779	418.218	320.712	107.256	-655.873	-394.530	18.735
1200	133.779	429.858	329.329	120.634	-657.574	-370.753	16.138
1300	133.779	440.566	337.480	134.012	-654.777	-346.966	13.941
1400	133.779	450.480	345.202	147.390	-652.077	-323.390	12.066
1500	133.779	459.710	352.531	160.768	-649.487	-300.004	10.447

PREVIOUS: June 1964 CURRENT: June 1968

Titanium Bromide ($TiBr_4$) $Br_4Ti_1(cr)$

TITANIUM BROMIDE ($TiBr_4$) LIQUID M_r = 367.496

$S°(298.15\ K) = [284.168]\ J\ K^{-1}\ mol^{-1}$ $\Delta_f H°(298.15\ K) = [-605.345]\ kJ\ mol^{-1}$

$T_{fus} = 311.4\ K$ $\Delta_{fus}H° = 12.887\ kJ\ mol^{-1}$

Enthalpy of Formation

$\Delta_f H°(298.15\ K)$ is calculated from that of the crystal by adding $\Delta_{fus}H°$ and the difference in enthalpy, H°(311.4 K) - H°(298.15 K), between the crystal and liquid.

Heat Capacity and Entropy

King et al. (1) reported the heat capacity of $TiBr_4(\ell)$ between the melting point and normal boiling point. $S°(TiBr_4, \ell, 298.15\ K)$ is calculated in a manner similar to that used for the enthalpy of formation.

Fusion Data

The melting temperature and heat of melting were reported by King et al. (1).

Vaporization Data

The boiling temperature, T_{vap}, is taken as the point at which $K_p = 1$ for the reaction $TiBr_4(\ell) = TiBr_4(g)$ at one bar. The enthalpy of vaporization is calculated as the difference between $\Delta_f H°$ of the liquid and gas at the boiling temperature. The vapor pressure data are discussed in the table for $TiBr_4(g)$.

Reference

1. E. G. King, W. W. Weller, A. U. Christensen and K. K. Kelley, U. S. Bur. Mines RI 5799, 1961.

Titanium Bromide ($TiBr_4$) $Br_4Ti_1(l)$

Enthalpy Reference Temperature = T_r = 298.15 K Standard State Pressure = p° = 0.1 MPa

	J K⁻¹mol⁻¹			kJ mol⁻¹			
T/K	$C_p°$	S°	$-[G°-H°(T_r)]/T$	$H°-H°(T_r)$	$\Delta_f H°$	$\Delta_f G°$	Log K_f
0							
100							
200							
298.15	151.879	284.168	284.168	0.	-605.345	-590.138	103.390
300	151.879	285.107	284.171	0.281	-605.391	-590.044	102.736
311.400	151.879	290.772	284.309	2.012	------ CRYSTAL <--> LIQUID -----		
400	151.879	328.800	290.128	15.469	-661.776	-573.053	74.833
500	151.879	362.691	301.377	30.657	-656.712	-551.461	57.611
600	151.879	390.382	313.974	45.845	-651.789	-530.876	46.217
700	151.879	413.794	326.605	61.033	-646.966	-511.106	38.139
800	151.879	434.075	338.799	76.221	-642.207	-492.023	32.126
900	151.879	451.964	350.399	91.408	-637.539	-473.531	27.483
1000	151.879	467.966	361.369	106.596	-633.020	-455.552	23.796

PREVIOUS: June 1964 CURRENT: June 1968

Titanium Bromide ($TiBr_4$) $Br_4Ti_1(l)$

TITANIUM BROMIDE ($TiBr_4$) CRYSTAL - LIQUID M_r = 367.496

0 to 311.4 K crystal

above 311.4 K liquid

Refer to the individual tables for details.

Titanium Bromide ($TiBr_4$) $Br_4Ti_1(cr,l)$

Enthalpy Reference Temperature = T_r = 298.15 K Standard State Pressure = $p°$ = 0.1 MPa

T/K	$C_p°$	S°	$-[G°-H°(T_r)]/T$	$H°-H°(T_r)$	$\Delta_fH°$	$\Delta_fG°$	Log K_f
	J K⁻¹mol⁻¹			kJ mol⁻¹			
0	0.	0.	INFINITE	-28.556	-592.684	-592.684	INFINITE
100	94.098	120.922	349.954	-22.903	-593.178	-593.675	310.104
200	117.273	194.146	255.191	-12.209	-594.203	-593.819	155.090
298.15	131.503	243.622	243.622	0.	-617.977	-590.681	103.485
300	131.815	244.436	243.624	0.244	-618.060	-590.512	102.817
311.400	133.779	249.388	243.745	1.757	CRYSTAL <--> LIQUID		
311.400	151.879	290.772	243.745	14.644	TRANSITION		
400	151.879	328.800	258.549	28.101	-661.776	-573.053	74.833
500	151.879	362.691	276.114	43.289	-656.712	-551.461	57.611
600	151.879	390.382	292.921	58.476	-651.789	-530.876	46.217
700	151.879	413.794	308.559	73.664	-646.966	-511.106	38.139
800	151.879	434.075	323.009	88.852	-642.207	-492.023	32.126
900	151.879	451.964	336.363	104.040	-637.539	-473.531	27.483
1000	151.879	467.966	348.738	119.228	-633.020	-455.552	23.796

PREVIOUS: CURRENT: June 1968

Titanium Bromide ($TiBr_4$) $Br_4Ti_1(cr,l)$

$\Delta_f H°(0\ K) = -520.3 \pm 5.0\ kJ\ mol^{-1}$

$S°(298.15\ K) = 398.60 \pm 4.2\ J\ K^{-1}\ mol^{-1}$ $\Delta_f H°(298.15\ K) = -550.2 \pm 5.0\ kJ\ mol^{-1}$

Vibrational Frequencies and Degeneracies

ν, cm^{-1}	ν, cm^{-1}
229.5(1)	382 (3)
74 (2)	90.5(3)

Ground State Quantum Weight: 1 σ = 12

Point Group: T_d

Bond Distance: Ti-Br = 2.31 Å

Bond Angle: Br-Ti-Br = 109° 28'

Product of the Moments of Inertia: $I_A I_B I_C = 6.73056 \times 10^{-111}\ g^3\ cm^6$

Enthalpy of Formation

The enthalpy of formation of $TiBr_4$ (g) is calculated from those of $TiBr_4$ (cr) and $TiBr_4$ (ℓ) and the enthalpies of reaction for the processes (A) $TiBr_4(cr) = TiBr_4(g)$ and (B) $TiBr_4(ℓ) = TiBr_4(g)$. 2nd and 3rd law analyses of the vapor pressure data for these processes yield the following results. The first four investigations employed the spoon gauge method, the last investigation being a manometric determination.

Source	Reaction	Data Points	T/K	$\Delta_r H$(298.15 K), kcal mol^{-1} 2nd Law	3rd Law	Drift cal $K^{-1} mol^{-1}$	$\Delta_f H°$(298.15 K) kcal mol^{-1}
Boni, 1966 (1)	A	13	275-311	16.4±0.2	16.19	-0.6±0.5	-131.5
Funaki et al., 1961 (2)	B	14	385-493	12.2±0.3	12.92	1.6±0.3	-131.8
Seki, 1941 (3)	A	12	286-306	15.4±0.5	15.90	1.6±1.6	-131.8
Hall et al., 1958 (4)	B	89*	341-499	13.2±0.1	13.14	-0.1±0.1	-131.5
Keavney & Smith, 1960 (5)	A	14	287-310	16.1±0.3	16.14	0.2±0.8	-131.6

* Four points rejected due to failure of a statistical test.

The adopted value of -131.5 kcal mol^{-1} (-550.196 kJ mol^{-1}) is that obtained from the data of Hall et al. (4), whose 3rd law drift was adjusted to zero by changing the entropy of the crystal within its uncertainty.

Heat Capacity and Entropy

The interatomic distance was reported by Lister and Sutton (6). The tetrahedral configuration is confirmed by spectroscopic studies (7, 8). The principal moments of inertia are: $I_A = I_B = I_C = 188.8066 \times 10^{-39}\ g\ cm^2$.

The vibrational frequencies have been reported by Delwaulle and Francois (7) and by Miller and Carlson (8). These values are based primarily on the Raman and infrared spectra of $TiBr_4$(ℓ), the value of ν_3 being the only frequency measured for $TiBr_4$(g).

The electronic spectra of $TiBr_4$ in solution have been measured by Dijkgraaf (9). Di Sipio et al. (10) have reported the near ultraviolet spectra of $TiBr_4$(g). Both of these studies indicate that $TiBr_4$ has no low lying electronic levels which would contribute significantly to the entropy.

References

1. A. A. Boni, J. Electrochem. Soc. 113, 1089 (1966).
2. K. Funaki, K. Uchimura and Y. Kuniya, Kogyo Kagaku Zasshi 64, 1914 (1961).
3. S. Seki, J. Chem. Soc. Japan 62, 789 (1941).
4. E. H. Hall, J. M. Blocher and I. E. Campbell, J. Electrochem. Soc. 105, 271 (1958).
5. J. J. Keavney and N. O. Smith, J. Phys. Chem. 64, 737 (1960).
6. M. W. Lister and L. E. Sutton, Trans. Faraday Soc. 37, 393 (1941).
7. M. L. Delwaulle and F. Francois, Compt. Rend. 220, 173 (1945).
8. F. A. Miller and G. L. Carlson, Spectrochim. Acta 16, 6 (1960).
9. C. Dijkgraaf, Spectrochim. Acta 21, 769 (1965).
10. L. Di Sipio, G. DeMichelis, E. Tondello and L. Oleari, Gazz. Chim. Ital. 96, 1785 (1966).

Enthalpy Reference Temperature = T_r = 298.15 K Standard State Pressure = p° = 0.1 MPa

T/K	$C_p°$ (J $K^{-1}mol^{-1}$)	S°	-[G°-H°(T_r)]/T	H°-H°(T_r) (kJ mol^{-1})	$\Delta_f H°$	$\Delta_f G°$	Log K_f
0	0.	0.	INFINITE	-23.904	-520.252	-520.252	INFINITE
100	76.813	300.230	482.949	-18.272	-520.766	-539.194	281.646
200	94.011	359.619	407.640	-9.604	-523.818	-556.528	145.350
250	98.156	381.078	400.246	-4.792	-525.801	-564.484	117.943
298.15	100.711	398.601	398.601	0.	-550.196	-569.107	99.705
300	100.790	399.224	398.603	0.186	-550.336	-569.224	99.111
350	102.533	414.900	399.837	5.272	-611.838	-569.000	84.919
400	103.734	428.675	402.598	10.431	-611.665	-562.892	73.506
450	104.591	440.945	406.189	15.640	-611.496	-556805	64.632
500	105.222	451.999	410.227	20.886	-611.333	-550.737	57.535
600	106.066	471.264	418.841	31.454	-611.030	-538.647	46.893
700	106.588	487.657	427.530	42.089	-610.760	-526.605	39.296
800	106.932	501.914	435.956	52.766	-610.513	-514.599	33.600
900	107.171	514.523	443.999	63.472	-610.326	-502.623	29.171
1000	107.343	525.824	451.626	74.198	-610.269	-490.660	25.629
1100	107.471	536.061	458.844	84.939	-610.410	-478.694	22.731
1200	107.568	545.417	465.674	95.691	-614.736	-466.586	20.310
1300	107.644	554.030	472.144	106.452	-614.557	-454.248	18.252
1400	107.705	562.009	478.281	117.219	-614.467	-441.921	16.488
1500	107.754	569.442	484.114	127.992	-614.482	-429.597	14.960
1600	107.794	576.398	489.667	138.770	-614.618	-417.267	13.622
1700	107.827	582.934	494.963	149.551	-614.891	-404.925	12.442
1800	107.855	589.098	500.023	160.335	-615.318	-392.563	11.392
1900	107.879	594.930	504.866	171.122	-615.917	-380.172	10.452
2000	107.899	600.464	509.508	181.911	-631.391	-367.291	9.593
2100	107.917	605.729	513.966	192.702	-633.132	-354.043	8.806
2200	107.932	610.749	518.252	203.494	-634.911	-340.712	8.090
2300	107.945	615.547	522.379	214.288	-636.732	-327.299	7.433
2400	107.957	620.142	526.357	225.083	-638.596	-313.805	6.830
2500	107.967	624.549	530.197	235.879	-640.504	-300.233	6.273
2600	107.976	628.784	533.908	246.676	-642.457	-286.584	5.758
2700	107.984	632.859	537.498	257.474	-644.453	-272.859	5.279
2800	107.991	636.786	540.974	268.273	-646.492	-259.059	4.833
2900	107.998	640.576	544.344	279.073	-648.570	-245.185	4.416
3000	108.004	644.237	547.613	289.873	-650.684	-231.240	4.026
3100	108.009	647.779	550.787	300.673	-652.831	-217.223	3.660
3200	108.014	651.208	553.872	311.475	-655.005	-203.136	3.316
3300	108.018	654.532	556.872	322.276	-657.201	-188.981	2.991
3400	108.022	657.756	559.792	333.078	-659.415	-174.760	2.685
3500	108.026	660.888	562.636	343.881	-661.641	-160.473	2.395
3600	108.029	663.931	565.408	354.683	-663.873	-146.122	2.120
3700	108.032	666.891	568.111	365.486	-1075.206	-123.921	1.749
3800	108.035	669.772	570.748	376.290	-1076.202	-98.198	1.350
3900	108.038	672.578	573.324	387.093	-1077.245	-72.447	0.970
4000	108.040	675.314	575.839	397.897	-1078.329	-46.669	0.609
4100	108.042	677.982	578.298	408.701	-1079.447	-20.864	0.266
4200	108.045	680.585	580.703	419.506	-1080.595	4.969	-0.062
4300	108.047	683.127	583.055	430.310	-1081.767	30.829	-0.375
4400	108.048	685.611	585.358	441.115	-1082.958	56.718	-0.673
4500	108.050	688.040	587.613	451.920	-1084.163	82.633	-0.959
4600	108.052	690.414	589.822	462.725	-1085.378	108.575	-1.233
4700	108.053	692.738	591.987	473.530	-1086.599	134.544	-1.495
4800	108.055	695.013	594.110	484.336	-1087.821	160.538	-1.747
4900	108.056	697.241	596.192	495.141	-1089.041	186.558	-1.989
5000	108.057	699.424	598.235	505.947	-1090.255	212.604	-2.221
5100	108.058	701.564	600.240	516.753	-1091.462	238.673	-2.445
5200	108.059	703.662	602.209	527.559	-1092.657	264.765	-2.660
5300	108.060	705.721	604.142	538.365	-1093.838	290.881	-2.867
5400	108.061	707.741	606.042	549.171	-1095.004	317.019	-3.067
5500	108.062	709.723	607.909	559.977	-1096.151	343.178	-3.259
5600	108.063	711.671	609.745	570.783	-1097.279	369.358	-3.445
5700	108.064	713.583	611.550	581.590	-1098.386	395.558	-3.625
5800	108.065	715.463	613.325	592.396	-1099.469	421.776	-3.799
5900	108.066	717.310	615.072	603.202	-1100.528	448.014	-3.966
6000	108.066	719.126	616.791	614.009	-1101.562	474.269	-4.129

PREVIOUS: June 1968 (1 atm) CURRENT: June 1968 (1 bar)

ZIRCONIUM BROMIDE ($ZrBr_4$) CRYSTAL M_r = 410.836

$S°(298.15\ K) = [224.7 \pm 4.2]\ J\ K^{-1}\ mol^{-1}$
$\Delta_f H°(0\ K) = [-734.3 \pm 8.4]\ kJ\ mol^{-1}$
$\Delta_f H°(298.15\ K) = -760.7 \pm 8.4\ kJ\ mol^{-1}$
$T_{fus} = 723 \pm 1\ K$
$\Delta_{fus}H°$ = Unknown

Enthalpy of Formation

Turnbull (1) measured the enthalpies of reaction for the dissolution of the zirconium tetrahalides in caustic and in water. The reported enthalpies of reaction and the corresponding reactions may be combined to yield the following:

$$ZrBr_4(cr) + 4NaCl(aq) = ZrCl_4(cr) + 4NaBr(aq),\ \Delta_r H°(298.15\ K) = -8.10 \pm 0.9\ kcal\ mol^{-1}$$
$$ZrBr_4(cr) + 4Cl^-(aq) = ZrCl_4(cr) + 4Br^-(aq),\ \Delta_r H°(298.15\ K) = -9.83 \pm 0.6\ kcal\ mol^{-1}$$

Using auxiliary data (2, 4), we calculate $\Delta_f H°(298.15\ K) = -182.67$ and $-181.02\ kcal\ mol^{-1}$ for $ZrBr_4(cr)$ from these two reactions. We adopt a mean of these two values, $\Delta_f H°(298.15\ K) = -181.8\ kcal\ mol^{-1}$ ($-760.651\ kJ\ mol^{-1}$), and assign an uncertainty of $\pm 1.5\ kcal\ mol^{-1}$. This same value was suggested by NBS (3).

Heat Capacity and Entropy

There are no heat capacity and enthalpy data reported in the literature for $ZrBr_4(cr)$. The adopted heat capacity values are estimated so as to give reasonable trends in comparison with $ZrCl_4$ and ZrI_4 and to be consistent with the existing sublimation data.

The crystal data compilation of Donnay and Ondik (5) tabulated both $ZrCl_4$ and $ZrBr_4$ as cubic structures. Thus, the adopted heat capacity values are estimated so as to parallel those for $ZrCl_4$. The heat capacity values below 300 K are calculated by summing contributions due to hindered translations, librations, and internal vibrations of the crystal. The parameters used in the calculations are determined by a correlation with corresponding parameters for $ZrCl_4$ (6) and a consideration of the sublimation data for $ZrBr_4$ (6). The high temperature heat capacities are obtained graphically.

Fusion Data

The melting point was observed by Rahlfs and Fischer (7) to be 723±1 K and by Nisel'son (8) to be 723±0.5 K.

Sublimation Data

The sublimation data is treated in the $ZrBr_4(g)$ table (6). The enthalpy of sublimation is adopted as $\Delta H°(298.15\ K) = 27.7 \pm 0.3\ kcal\ mol^{-1}$ ($115.897 \pm 1.3\ kJ\ mol^{-1}$). The sublimation temperture, T_{sub}, is calculated as that temperature for which $\Delta_r G° = 0$ for the process $ZrBr_4(cr) = ZrBr_4(g)$ at one bar. Since T_{sub} is less than T_{fus}, the liquid phase is thermodynamically unstable at a pressure of one atmosphere.

References

1. A. G. Turnbull, J. Phys. Chem. 65, 1652 (1961).
2. JANAF Thermochemical Tables: $ZrCl_4(cr)$, 6-30-75.
3. U. S. Nat. Bur. Stand. Tech Note 270-5, 1971.
4. ICSU-CODATA Task Group, J. Chem. Thermodynamics 4, 331 (1972).
5. J. D. H. Donnay and H. M. Ondik, "Crystal Data, Determinative Tables," 3rd edition, Vol. 2, Nat. Bur. Stand. and Joint Committee on Powder Diffraction Standards, 1973.
6. JANAF Thermochemical Tables: $ZrCl_4(cr)$ and $ZrBr_4(g)$, 6-30-75.
7. O. Rahlfs and W. Fischer, Z. Anorg. Allgem. Chem. 211, 349 (1933).
8. L. A. Nisel'son, Russ. J. Inorg. Chem. 7, 354 (1962).

Zirconium Bromide ($ZrBr_4$) $Br_4Zr_1(cr)$

Enthalpy Reference Temperature = T_r = 298.15 K Standard State Pressure = p° = 0.1 MPa

T/K	$C_p°$ (J K⁻¹mol⁻¹)	S° (J K⁻¹mol⁻¹)	$-[G°-H°(T_r)]/T$ (J K⁻¹mol⁻¹)	$H°-H°(T_r)$ (kJ mol⁻¹)	$\Delta_f H°$ (kJ mol⁻¹)	$\Delta_f G°$ (kJ mol⁻¹)	Log K_f
0	0.	0.	INFINITE	-28.162	-734.298	-734.298	INFINITE
100	93.625	103.324	328.883	-22.556	-735.164	-733.321	383.048
200	116.077	176.515	235.865	-11.870	-736.473	-730.998	190.917
298.15	124.805	224.689	224.689	0.	-760.651	-725.293	127.068
300	124.917	225.462	224.692	0.231	-760.747	-725.073	126.246
400	129.286	262.045	229.650	12.958	-819.539	-700.929	91.532
500	131.629	291.157	239.139	26.009	-816.496	-671.627	70.164
600	133.344	315.313	249.879	39.260	-813.375	-642.945	55.973
700	134.725	335.974	260.738	52.666	-810.213	-614.789	45.876
800	135.896	354.041	271.295	66.197	-807.037	-587.089	38.333
900	137.068	370.115	281.398	79.845	-803.864	-559.786	32.489
1000	138.239	384.617	291.007	93.610	-800.703	-532.835	27.832

PREVIOUS: March 1964 CURRENT: June 1975

Zirconium Bromide ($ZrBr_4$) $Br_4Zr_1(cr)$

ZIRCONIUM BROMIDE ($ZrBr_4$) IDEAL GAS M_r = 410.836

$S°(298.15\ K) = 414.49 \pm 0.4\ J\ K^{-1}\ mol^{-1}$

$\Delta_fH°(0\ K) = -615.2 \pm 8.4\ kJ\ mol^{-1}$
$\Delta_fH°(298.15\ K) = -644.8 \pm 8.4\ kJ\ mol^{-1}$

Vibrational Frequencies and Degeneracies

ν, cm^{-1}
223 (1)
60 (2)
315 (3)
72 (3)

Ground State Quantum Weight: [1] σ = 12
Point Group: T_d
Bond Distance: Zr–Br = 2.44 ± 0.02 Å
Bond Angle: Br–Zr–Br = 109° 28'
Product of the Moments of Inertia: $I_AI_BI_C = 9.34800 \times 10^{-111}\ g^3\ cm^6$

Enthalpy of Formation

The enthaply of formation for $ZrBr_4$(g) is calculated from the enthaplies of formation and sublimation of $ZrBr_4$(cr) at 298.15 K. The adopted value for the enthaply of sublimation, $\Delta_{sub}H°(298.15\ K) = 27.7\pm0.3\ kcal\ mol^{-1}$, is based on the mean of the 3rd law results from the following sublimation data.

In analyzing the vapor pressure data for the four sublimation studies (1, 2, 3, 4), corrections were made for non-ideality by means of the equation $\Delta G°/T = -R\ln p - Bp/T$. The Berthelot equation of state and the critical constants T_c = 805.15 K and P_c = 42.9 atm, as reported by Nisel'son and Sokolova (5), are used to calculate B.

Source	Method	Data Points	T/K	$\Delta_{sub}H°$(298.15 K),kcal mol^{-1} 2nd Law	3rd Law	Drift cal K^{-1} mol^{-1}	$\Delta_fH°$(298.15 K) kcal mol^{-1}
Rahlfs and Fischer (1)	static	15	538-633	27.85±0.24	27.78±0.09	-0.1±0.4	-154.02
Schlafer and Skoludek (2)	static	eqn	494-620	28.59	27.66	-1.7	-154.14
Berdonosova et al. (3)		17	489-606	28.87±0.12	28.03±0.12	-1.5±0.2	-153.77
Normanton and Shelton (4)	effusion	eqn	400-500	26.32	27.30	2.2	-154.50

For the enthalpy of sublimation, we adopt the mean of the 3rd law results and assign an uncertainty of $\pm0.3\ kcal\ mol^{-1}$. Combining the adopted $\Delta_{sub}H°$(298.15 K) value with the enthalpy of formation of $ZrBr_4$(cr), $\Delta_fH°(298.15\ K) = -181.8\pm1.5\ kcal\ mol^{-1}$ (6), we calculate $\Delta_fH°(298.15\ K) = -154.1\ kcal\ mol^{-1}$ ($-644.754\ kJ\ mol^{-1}$) for $ZrBr_4$(g) and assign an uncertainty of $\pm2.0\ kcal\ mol^{-1}$ ($\pm8.4\ kJ\ mol^{-1}$).

Heat Capacity and Entropy

The adopted vibrational frequencies are from the work of Clark et al. (7, 8, 9), who recorded the Raman spectra of $ZrBr_4$ in the vapor phase (380-420°C). These studies by Clark et al. (7, 8, 9) indicated that $ZrBr_4$ is a tetrahedral monomer in the vapor phase. Shimanouchi, in his compilation of molecular vibrational frequencies (10), also adopted the values of Clark et al. (7, 8, 9) for $ZrBr_4$(g). Rahlfs and Fischer (1), through vapor density measurements, had earlier concluded that $ZrBr_4$ was monomeric in the vapor phase.

Berdonosov et al. (11) referenced an electron diffraction study by Cherkasov (12) which showed that, in the vapor phase, the Zr-Br internuclear distance was 2.44±0.02 Å. We adopt this value. The principal moments of inertia are: $I_A = I_B = I_C = 210.6556 \times 10^{-39}\ g\ cm^2$.

Much literature has been published on the inter-relationships between force constants and vibrational frequencies. Since the majority of these articles are based on estimated frequencies, they will not be further discussed or referenced. The same situation is true for thermodynamic tabulations of $ZrBr_4$(g). One exception is that Clark et al. (7) calculated thermodynamic properties based on their experimental vibrational frequencies. Their tabulation is very similar to ours in the range 100-1000 K; the difference in entropy being less than 0.02 cal K^{-1} mol^{-1} in this range.

References

1. O. Rahlfs and W. Fischer, Z. Anorg. Allgem. Chem. 211, 349 (1933).
2. H. L. Schlafer and H. Skoludek, Z. Elektrochem. 66, 367 (1962).
3. S. S. Berdonosova, V. I. Tsirel'nikov, and A. V. Lapitskii, Vestn. Mosk. Univ. Ser. II, Khim. 20, 26 (1965).
4. A. S. Normanton and R. A. J. Shelton, Trans. Faraday Soc. 66, 33 (1970).
5. L. A. Nisel'son and T. D. Sokolova, Russ. J. Inorg. Chem. 7, 1382 (1962).
6. JANAF Thermochemical Tables: $ZrBr_4$(cr), 6-30-75.
7. R. J. H. Clark, B. K. Hunter, and D. M. Rippon, Chem. Ind. (London) 1971, 787 (1971); Inorg. Chem. 11, 56 (1971).
8. R. J. H. Clark and D. M. Rippon, J. Mol. Spectrosc. 44, 479 (1972).
9. R. J. H. Clark and D. M. Rippon, Advan. Raman Spectrosc. 1, 493 (1972).
10. T. Shimanouchi, J. Phys. Chem. Ref. Data 3, 269 (1974).
11. S. S. Berdonosov, D. H. Berdonosova, A. V. Lapitskii, and L. G. Vlasov, Russ. J. Inorg. Chem. 8, 277 (1963).
12. I. A. Cherkasov, Diploma Thesis, Moscow State University, 1961 [as referenced in (11)].

Zirconium Bromide ($ZrBr_4$) Br_4Zr_1(g)

Enthalpy Reference Temperature = T_r = 298.15 K Standard State Pressure = p° = 0.1 MPa

T/K	$C_p°$ (J $K^{-1}mol^{-1}$)	S°	$-[G°-H°(T_r)]/T$	$H°-H°(T_r)$ (kJ mol^{-1})	$\Delta_fH°$	$\Delta_fG°$	Log K_f
0	0.	0.	INFINITE	-24.954	-615.192	-615.192	INFINITE
100	81.107	312.254	501.497	-18.924	-615.635	-634.686	331.526
200	97.300	374.463	423.743	-9.856	-618.562	-652.677	170.462
250	100.676	396.571	416.167	-4.899	-620.453	-660.995	138.107
298.15	102.669	414.487	414.487	0.	-644.754	-665.984	116.678
300	102.730	415.122	414.489	0.190	-644.891	-666.115	115.981
350	104.052	431.064	415.745	5.362	-706.291	-666.288	99.438
400	104.947	445.021	418.550	10.588	-706.012	-660.593	86.265
450	105.577	457.420	422.193	15.852	-705.736	-654.932	76.022
500	106.038	468.569	426.283	21.143	-705.465	-649.301	67.832
600	106.649	487.960	434.994	31.780	-704.959	-638.117	55.553
700	107.023	504.431	443.766	42.465	-704.517	-627.013	46.788
800	107.269	518.739	452.263	53.180	-704.156	-615.966	40.218
900	107.439	531.384	460.366	63.916	-703.896	-604.959	35.111
1000	107.561	542.710	468.044	74.667	-703.750	-593.976	31.026
1100	107.652	552.966	475.305	85.428	-703.727	-583.000	27.684
1200	107.721	562.336	482.173	96.196	-707.543	-571.800	24.890
1300	107.775	570.961	488.675	106.971	-707.238	-560.500	22.521
1400	107.818	578.949	494.842	117.751	-706.988	-549.223	20.492
1500	107.852	586.389	500.700	128.534	-706.806	-537.961	18.733
1600	107.881	593.351	506.275	139.321	-706.704	-526.708	17.195
1700	107.904	599.892	511.592	150.110	-706.698	-515.459	15.838
1800	107.924	606.060	516.670	160.902	-706.800	-504.208	14.632
1900	107.941	611.896	521.530	171.695	-707.025	-492.947	13.552
2000	107.955	617.433	526.188	182.490	-707.388	-481.671	12.580
2100	107.967	622.700	530.659	193.286	-707.903	-470.373	11.700
2200	107.978	627.723	534.958	204.083	-729.907	-458.301	10.881
2300	107.987	632.523	539.096	214.882	-731.183	-445.927	10.127
2400	107.995	637.119	543.085	225.681	-732.503	-433.496	9.435
2500	108.003	641.528	546.936	236.481	-733.868	-421.009	8.797
2600	108.009	645.764	550.656	247.281	-735.278	-408.468	8.206
2700	108.015	649.840	554.254	258.082	-736.732	-395.870	7.659
2800	108.020	653.769	557.739	268.884	-738.228	-383.219	7.149
2900	108.024	657.559	561.116	279.686	-739.763	-370.513	6.674
3000	108.028	661.221	564.392	290.489	-741.335	-357.754	6.229
3100	108.032	664.764	567.573	301.292	-742.939	-344.941	5.812
3200	108.036	668.194	570.664	312.095	-744.571	-332.077	5.421
3300	108.039	671.518	573.670	322.899	-746.226	-319.160	5.052
3400	108.041	674.743	576.596	333.703	-747.898	-306.194	4.704
3500	108.044	677.875	579.445	344.507	-749.582	-293.179	4.375
3600	108.046	680.919	582.221	355.312	-751.273	-280.114	4.064
3700	108.049	683.880	584.929	366.117	-752.966	-267.003	3.769
3800	108.051	686.761	587.571	376.922	-754.655	-253.847	3.489
3900	108.052	689.568	590.151	387.727	-756.336	-240.646	3.223
4000	108.054	692.303	592.670	398.532	-758.003	-227.401	2.970
4100	108.056	694.972	595.133	409.338	-759.653	-214.116	2.728
4200	108.057	697.575	597.541	420.143	-761.281	-200.791	2.497
4300	108.059	700.118	599.897	430.949	-762.884	-187.426	2.277
4400	108.060	702.602	602.203	441.755	-764.459	-174.025	2.066
4500	108.061	705.031	604.462	452.561	-766.002	-160.589	1.864
4600	108.062	707.406	606.674	463.367	-767.512	-147.119	1.671
4700	108.063	709.730	608.842	474.173	-768.985	-133.616	1.485
4800	108.064	712.005	610.967	484.980	-1331.356	-108.465	1.180
4900	108.065	714.233	613.052	495.786	-1332.403	-82.977	0.885
5000	108.066	716.416	615.098	506.593	-1333.426	-57.468	0.600
5100	108.067	718.556	617.106	517.399	-1334.423	-31.938	0.327
5200	108.068	720.655	619.077	528.206	-1335.391	-6.391	0.064
5300	108.068	722.713	621.013	539.013	-1336.329	19.177	-0.189
5400	108.069	724.733	622.915	549.820	-1337.234	44.761	-0.433
5500	108.070	726.716	624.784	560.627	-1338.107	70.361	-0.668
5600	108.070	728.664	626.622	571.434	-1338.944	95.977	-0.895
5700	108.071	730.576	628.429	582.241	-1339.746	121.608	-1.114
5800	108.072	732.456	630.206	593.048	-1340.510	147.252	-1.326
5900	108.072	734.303	631.955	603.855	-1341.237	172.909	-1.531
6000	108.073	736.120	633.676	614.662	-1341.926	198.579	-1.729

PREVIOUS: June 1975 (1 atm) CURRENT: June 1975 (1 bar)

Zirconium Bromide ($ZrBr_4$) Br_4Zr_1(g)

NIOBIUM BROMIDE ($NbBr_5$) CRYSTAL M_r = 492.4264

$\Delta_f H°$(0 K) = Unknown
$S°$(298.15 K) = [258.78 ± 6.3] J K^{-1} mol^{-1} $\Delta_f H°$(298.15 K) = -556.05 ± 12.6 kJ mol^{-1}
T_{fus} = 527 ± 3 K $\Delta_{fus} H°$ = 24.016 ± 6.3 kJ mol^{-1}

Enthalpy of Formation

The adopted value for the enthalpy of formation of $NbBr_5$(cr), $\Delta_f H°$(298.15 K) = -132.9±3.0 kcal mol^{-1} (-556.054±12.6 kJ mol^{-1}) is based on the study of the direct bromination of Nb(cr) by Gross et al. (1).

Shchukarev et al. (2) studied the hydrolysis of $NbBr_5$(cr). Using the experimental results reported by Shchukarev et al. (2) and auxiliary data (4, 5), we calculate, $\Delta_f H°$(298.15 K) = -135.5±1.2 kcal mol^{-1}. Schafer and Heine (3) also determined the enthaply of formation of $NbBr_5$(cr) via calorimetric measurements of the heats of solution of Nb(cr) and $NbBr_5$(cr) in hydrofluoric acid. Using their data and auxiliary results (4, 5), we calculate $\Delta_f H°$(298.15 K) = -131.86±1.0 kcal mol^{-1}. These two solution results (2, 3) are in reasonable agreement with the adopted value. The direct bromination (1) is thought to be the more reliable.

Heat Capacity and Entropy

The heat capacity and entropy are estimated in comparison with $NbCl_5$(cr) (4). As there is no low temperature heat capacity data reported in the literature, the entropy at 298.15 K is calculated from the equation $\Delta_{sub} S°$(298.15 K) = $S°$(g, 298.15 K) - $S°$(cr, 298.15 K); $S°$(g, 298.15 K) = 107.35 cal K^{-1} mol^{-1} as given in the $NbBr_5$(g) table (4) and $\Delta_{sub} S°$(298.15 K) = 45.5 cal K^{-1} mol^{-1} as suggested by comparison with $NbCl_5$ and $TaCl_5$ (4).

Fusion Data

See the liquid table for details.

Sublimation Data

The enthalpy of sublimation, $\Delta_{sub} H°$(298.15 K), is the difference between the $\Delta_f H°$(298.15 K) values for $NbBr_5$(g) and $NbBr_5$(cr). Two sublimation studies are summarized in the $NbBr_5$(g) table (4).

References

1. P. Gross, C. Hayman, D. L. Levi, and G. L. Wilson, Trans. Faraday Soc. 56, 318 (1960).
2. S. A. Shchukarev, E. K. Smirnova, I. V. Vasil'kova, and N. I. Borovkova, Russ. J. Inorg. Chem. 7, 625 (1962).
3. H. Schafer and H. Heine, Z. Anorg. Allg. Chem. 352, 258 (1967).
4. JANAF Thermochemical Tables: $NbCl_5$(cr), $TaCl_5$(ℓ), $NaBr_5$(ℓ) and $NbBr_5$(g), 12-31-74; Nb_2O_5, 12-31-72.
5. U. S. Nat. Bur. Stand. Tech. Note 270-3, 1968.

Niobium Bromide ($NbBr_5$) Br_5Nb_1(cr)

Enthalpy Reference Temperature = T_r = 298.15 K   Standard State Pressure = p° = 0.1 MPa

T/K	$C_p°$ (J $K^{-1}mol^{-1}$)	$S°$ (J $K^{-1}mol^{-1}$)	$-[G°-H°(T_r)]/T$ (J $K^{-1}mol^{-1}$)	$H°-H°(T_r)$ (kJ mol^{-1})	$\Delta_f H°$ (kJ mol^{-1})	$\Delta_f G°$ (kJ mol^{-1})	Log K_f
0							
100							
200							
298.15	147.904	258.780	258.780	0.	-556.054	-508.887	89.155
300	147.904	259.695	258.783	0.274	-556.176	-508.593	88.554
400	147.904	302.245	264.585	15.064	-630.093	-477.367	62.338
500	147.904	335.249	275.540	29.854	-627.096	-439.535	45.918
527.000	147.904	343.027	278.800	33.848	------ CRYSTAL <--> LIQUID -----		
600	147.904	362.215	287.807	44.645	-624.217	-402.295	35.023
700	147.904	385.014	300.107	59.435	-621.429	-365.530	27.276
800	147.904	404.764	311.982	74.226	-618.717	-329.159	21.492

PREVIOUS: CURRENT: December 1974

NIOBIUM BROMIDE ($NbBr_5$) LIQUID M_r = 492.4264

$S°(298.15\ K) = [274.005]\ J\ K^{-1}\ mol^{-1}$

$T_{fus}H° = 527 \pm 3\ K$

$\Delta_f H°(298.15\ K) = [-545.074]\ kJ\ mol^{-1}$

$\Delta_{fus}H° = 24.016 \pm 6.3\ kJ\ mol^{-1}$

Enthalpy of Formation

$\Delta_f H°(298.15)$ is calculated from that of $NbBr_5$(cr) by adding $\Delta_{fus}H°$ and the difference in enthalpy, H°(527 K)-H(298.15 K), between the crystal and liquid.

Heat Capacity and Entropy

The heat capacity is assumed to be identical with that of $NbCl_5(\ell)$ (3) including the assumed glass transition at 350 K. The entropy at 298.15 K is calcualted in a manner analogous to that used for $\Delta_f H°(298.15\ K)$.

Fusion Data

The adopted melting point, T_{fus} = 527±3 K (254°C), is based on the studies by Nisel'son et al. (1) and Berdonosov et al. (2). The melting point was determined by Nisel'son et al. (1) from cooling curves, T_{fus} = 255 C. Berdonosov et al. (2) determined the melting point by three methods: T_{fus} = 252.0±1.5 based on an analysis of their vapor pressure data, T_{fus} = 255±2 based on visual observation, and T_{fus} = 254±1 based on cooling curves.

The heat of melting is chosen to be $\Delta_{fus}H°$ = 5.74±1.5 kcal mol^{-1} (24.016±6.3 kJ mol^{-1}). This value is consistent with the vaporization data and the thermodynamic functions we have adopted. The entropy of melting, $\Delta_{fus}S°$ = 10.89 cal $K^{-1}\ mol^{-1}$ is somewhat lower than anticipated, based on the expected similarity with $NbCl_5$ and $TaCl_5$ as far as condensed phase dimerization is concerned (3).

Vaporization Data

T_{vap}, the normal boiling point, is calculated as that temperature for which the Gibbs energy approaches zero for the process $NbBr_5(\ell) \rightleftharpoons NbBr_5(g)$. $\Delta_{vap}H°$ is calculated as the difference between the $\Delta_f H°$ values for $NbBr_5$(g) and $NbBr_5(\ell)$ at T_{vap}. Two vaporization studies are summarized in the $NbBr_5$(g) table.

References

1. L. A. Nisel'son and J. D. Sokolova, Russ. J. Inorg. Chem. 9, 1117 (1964).
2. S. S. Berdonosov, A. V. Lapitskii, and E. K. Bakov, Russ. J. Inorg. Chem. 10, 173 (1965).
3. JANAF Thermochemical Tables: $NbCl_5(\ell)$, 12-31-74.

Niobium Bromide ($NbBr_5$) $Br_5Nb_1(l)$

Enthalpy Reference Temperature = T_r = 298.15 K Standard State Pressure = p° = 0.1 MPa

T/K	$C_p°$ (J K⁻¹mol⁻¹)	S° (J K⁻¹mol⁻¹)	$-[G°-H°(T_r)]/T$ (J K⁻¹mol⁻¹)	$H°-H°(T_r)$ (kJ mol⁻¹)	$\Delta_f H°$ (kJ mol⁻¹)	$\Delta_f G°$ (kJ mol⁻¹)	Log K_f
0							
100							
200							
298.15	147.904	274.005	274.005	0.	-545.074	-502.446	88.026
300	147.904	274.920	274.008	0.274	-545.196	-502.181	87.438
350	147.904	297.720	275.809	7.669	-620.672	-490.908	73.264
350.000	147.904	297.720	275.809	7.669	GLASS <--> LIQUID		
350.000	228.280	297.720	275.809	7.669	TRANSITION		
400	224.848	327.984	280.486	18.999	-615.178	-472.748	61.735
500	216.564	377.299	295.131	41.084	-604.887	-438.351	45.794
527.000	215.228	388.599	299.634	46.884	------ CRYSTAL <--> LIQUID -----		
600	206.397	415.912	312.167	62.247	-595.635	-405.932	35.339
700	194.430	446.854	329.276	82.304	-587.580	-374.969	27.981
800	180.611	471.921	345.594	101.061	-580.902	-345.069	22.531
900	166.218	492.368	360.805	118.407	-575.705	-315.918	18.335

PREVIOUS: CURRENT: December 1974

Niobium Bromide ($NbBr_5$) $Br_5Nb_1(l)$

NIOBIUM BROMIDE ($NbBr_5$) CRYSTAL - LIQUID M_r = 492.4264 Niobium Bromide ($NbBr_5$) Br_5Nb_1(cr,l)

298.15 to 527 K crystal

above 527 K liquid

Refer to the individual tables for details.

Enthalpy Reference Temperature = T_r = 298.15 K Standard State Pressure = p° = 0.1 MPa

T/K	C_p° (J K⁻¹mol⁻¹)	S° (J K⁻¹mol⁻¹)	$-[G^\circ-H^\circ(T_r)]/T$ (J K⁻¹mol⁻¹)	$H^\circ-H^\circ(T_r)$ (kJ mol⁻¹)	Δ_fH° (kJ mol⁻¹)	Δ_fG° (kJ mol⁻¹)	Log K_f
0							
100							
200							
298.15	147.904	258.780	258.780	0.	-556.054	-508.887	89.155
300	147.904	259.695	258.783	0.274	-556.176	-508.593	88.554
400	147.904	302.245	264.585	15.064	-630.093	-477.367	62.338
500	147.904	335.249	275.540	29.854	-627.096	-439.535	45.918
527.000	147.904	343.027	278.800	33.848	CRYSTAL <--> LIQUID	TRANSITION	
527.000	215.228	388.599	278.800	57.864			
600	206.397	415.912	293.867	73.227	-595.635	-405.932	35.339
700	194.430	446.854	313.591	93.284	-587.580	-374.969	27.981
800	180.611	471.921	331.870	112.041	-580.902	-345.069	22.531
900	166.218	492.368	348.605	129.387	-575.705	-315.918	18.335

PREVIOUS: CURRENT: December 1974

Niobium Bromide ($NbBr_5$) Br_5Nb_1(cr,l)

NIOBIUM BROMIDE ($NbBr_5$) IDEAL GAS $M_r = 492.4264$

$S°(298.15\ K) = 449.26 \pm 3.1\ J\ K^{-1}\ mol^{-1}$

$\Delta_fH°(0\ K) = -406.5 \pm 12.6\ kJ\ mol^{-1}$

$\Delta_fH°(298.15\ K) = -443.6 \pm 12.6\ kJ\ mol^{-1}$

Vibrational Frequencies and Degeneracies

ν, cm^{-1}	ν, cm^{-1}
234.0 (1)	[375.6](2)
178.0 (1)	119.0 (2)
[288.9](1)	67.0 (2)
[106.2](1)	101.0 (2)

Ground State Quantum Weight: [1]

Point group: D_{3h} $\sigma = 6$

Bond Distance: Nb-Br = 2.45 ± 0.02 Å

Bond Angles: Br*-Nb-Br* = 120° Br*-Nb-Br** = 90° Br**-Nb-Br** = 180°

(* - equatorial ** - axial)

Product of the Moments of Inertia: $I_AI_BI_C = 1.85664 \times 10^{-110}\ g^3\ cm^6$

Enthalpy of Formation

The vapor pressures over $NbBr_5$(cr, ℓ) have been measured by Alexander and Fairbrother (1) and Berdonosov et al. (2). A 2nd and 3rd law analysis of their data is given below. The enthalpy of formation for $NbBr_5$(g) is derived from the sublimation data of Berdonosov et al. (2). Our 3rd law analysis of their data gives $\Delta_{sub}H°(298.15\ K) = 26.88\ kcal\ mol^{-1}$ which leads to the adopted value, $\Delta_fH°(298.15\ K) = -106.02\ kcal\ mol^{-1}$ ($-443.588\ kJ\ mol^{-1}$) for $NbBr_5$(g). The sublimation data of Alexander and Fairbrother (1) is not acceptable as it leads to a large entropy drift, $-45\pm6\ cal\ K^{-1}\ mol^{-1}$.

The enthalpy of melting is chosen as $5.74\ kcal\ mol^{-1}$ so as to give reasonable entropy drifts for the vaporization data.

Source	Reaction[a]	Method	Data Points	T/K	$\Delta_rH°(298.15\ K)$, kcal mol^{-1} 2nd Law	3rd Law	Drift cal K^{-1} mol^{-1}
Alexander and Fairbrother (1)	A	static	4	480-517	50.82±2.94	27.77±1.57	-45±6
Berdonosov et al. (2)	A	static	12[b]	478-524	27.28±0.16	26.88±0.04	-0.8±0.3
Alexander and Fairbrother (1)	B	static	26	528-635	24.07±0.13	24.26±0.06	+0.3±0.2
Berdonosov et al. (2)	B	static	14	529-606	24.80±0.27	24.23±0.09	-1.0±0.5

[a]Reaction (A) $NbBr_5(\ell) = NbBr_5(g)$, (B) $NbBr_5(cr) = NbBr_5(g)$

[b]One point rejected due to a statistical test.

Heat Capacity and Entropy

Monomeric $NbBr_5$(g) was shown by Spiridonov and Romanov (4, 5), using electron diffraction techniques, to have a trigonal bipyramidal structure of D_{3h} symmetry: all the Nb-Br bond lengths being equal within expeirmental uncertainty, Nb-Br = 2.45±0.02 Å. Skinner and Sutton (3) earlier used electron diffraction techniques and had suggested the same structure although a square pyramidal structure was consistent with their experimental results. We adopt the results of Spiridonov and Romanov (4, 5). The principal moments of inertia are: $I_A = 238.9340 \times 10^{-39}$, $I_B = I_C = 278.7563 \times 10^{-39}\ g\ cm^2$. A normal coordinate treatment of $NbBr_5$(g) in the Urey-Bradley force fields was performed by So (7) using the reported vibrational frequencies of Beattie and Ozin (6). This work by So (7) was intended to check the correctness of the reported fundamental frequencies and predict those unobserved frequencies (ν_3, ν_4, ν_5). Beattie and Ozin (6) had recorded the gas phase Raman spectra of niobium and tantalum chloride and bromide. We adopt the results of So (7) which support the work of Beattie and Ozin (6).

References

1. K. A. Alexander and F. Fairbrother, J. Chem. Soc. (London) 1949, S223 (1949).
2. S. S. Berdonosov, A. V. Lapitskii, and E. K. Bakov, Russ. J. Inorg. Chem. 10, 173 (1965).
3. H. Skinner and L. Sutton, Trans. Faraday Soc. 36, 668 (1940).
4. V. P. Spiridonov and G. V. Romanov, Vestn. Mosk. Univ., Khim. 21, 109 (1966).
5. V. P. Spiridonov and G. V. Romanov, Vestn. Mosk. Univ., Khim. 24, 65 (1969).
6. I. R. Beattie and G. A. Ozin, J. Chem. Soc. (London) A1969, 1691 (1964).
7. S. P. So, J. Mol. Struct. 16, 311 (1973).

Niobium Bromide ($NbBr_5$) $Br_5Nb_1(g)$

Enthalpy Reference Temperature = T_r = 298.15 K Standard State Pressure = p° = 0.1 MPa

T/K	$C_p°$ (J $K^{-1}mol^{-1}$)	S°	$-[G°-H°(T_r)]/T$	$H°-H°(T_r)$ (kJ mol^{-1})	$\Delta_fH°$	$\Delta_fG°$	Log K_f
0	0.	0.	INFINITE	-29.426	-406.500	-406.500	INFINITE
100	95.964	325.261	555.381	-23.012	-407.850	-425.658	222.341
200	118.641	400.321	460.586	-12.053	-411.242	-442.220	115.496
250	123.151	427.323	451.317	-5.998	-413.411	-449.724	93.965
298.15	125.806	449.259	449.259	0.	-443.588	-453.212	79.401
300	125.886	450.037	449.261	0.233	-443.750	-453.271	78.922
350	127.648	469.585	450.801	6.574	-520.280	-450.669	67.259
400	128.840	486.712	454.242	12.988	-519.703	-440.764	57.558
450	129.681	501.939	458.711	19.452	-519.118	-430.932	50.021
500	130.295	515.635	463.730	25.953	-518.532	-421.165	43.999
600	131.109	539.469	474.425	39.026	-517.370	-401.801	34.980
700	131.610	559.720	485.200	52.164	-516.234	-382.629	28.552
800	131.938	577.317	495.639	65.343	-515.135	-363.618	23.742
900	132.165	592.871	505.595	78.548	-514.078	-344.743	20.008
1000	132.328	606.805	515.032	91.773	-513.067	-325.983	17.028
1100	132.449	619.423	523.957	105.012	-512.107	-307.321	14.593
1200	132.541	630.952	532.400	118.262	-511.197	-288.745	12.569
1300	132.613	641.564	540.394	131.520	-510.339	-270.242	10.858
1400	132.670	651.394	547.976	144.784	-509.536	-251.804	9.395
1500	132.717	660.548	555.179	158.054	-508.796	-233.421	8.128
1600	132.754	669.115	562.035	171.327	-508.128	-215.084	7.022
1700	132.786	677.164	568.573	184.604	-507.544	-196.787	6.047
1800	132.812	684.755	574.819	197.884	-507.055	-178.522	5.181
1900	132.835	691.936	580.796	211.167	-506.671	-160.281	4.406
2000	132.854	698.750	586.525	224.451	-506.404	-142.058	3.710
2100	132.870	705.233	592.024	237.737	-506.262	-123.844	3.080
2200	132.884	711.414	597.312	251.025	-506.257	-105.635	2.508
2300	132.897	717.321	602.402	264.314	-506.399	-87.422	1.985
2400	132.908	722.978	607.309	277.604	-506.702	-69.200	1.506
2500	132.917	728.403	612.045	290.896	-507.181	-50.961	1.065
2600	132.926	733.617	616.621	304.188	-507.862	-32.700	0.657
2700	132.933	738.633	621.048	317.481	-508.781	-14.408	0.279
2800	132.940	743.468	625.334	330.774	-536.348	4.407	-0.082
2900	132.946	748.133	629.489	344.069	-536.593	23.725	-0.427
3000	132.952	752.640	633.519	357.364	-536.884	43.051	-0.750
3100	132.957	757.000	637.432	370.659	-537.215	62.387	-1.051
3200	132.961	761.221	641.235	383.955	-537.580	81.735	-1.334
3300	132.965	765.313	644.933	397.251	-537.974	101.095	-1.600
3400	132.969	769.282	648.533	410.548	-538.389	120.467	-1.851
3500	132.973	773.137	652.038	423.845	-538.820	139.851	-2.087
3600	132.976	776.883	655.454	437.143	-539.260	159.248	-2.311
3700	132.979	780.526	658.785	450.440	-539.701	178.657	-2.522
3800	132.981	784.072	662.036	463.738	-540.138	198.077	-2.723
3900	132.984	787.527	665.210	477.037	-540.564	217.510	-2.913
4000	132.986	790.894	668.310	490.335	-540.973	236.953	-3.094
4100	132.988	794.177	671.340	503.634	-541.361	256.406	-3.267
4200	132.990	797.382	674.303	516.933	-541.722	275.868	-3.431
4300	132.992	800.511	677.202	530.232	-542.051	295.338	-3.588
4400	132.994	803.569	680.039	543.531	-542.344	314.816	-3.737
4500	132.996	806.558	682.818	556.831	-542.599	334.300	-3.880
4600	132.997	809.481	685.539	570.130	-542.811	353.789	-4.017
4700	132.998	812.341	688.207	583.430	-542.977	373.282	-4.149
4800	133.000	815.141	690.822	596.730	-543.096	392.779	-4.274
4900	133.001	817.884	693.388	610.030	-543.166	412.277	-4.395
5000	133.002	820.571	695.904	623.330	-543.184	431.776	-4.511
5100	133.003	823.204	698.375	636.631	-543.149	451.275	-4.622
5200	133.004	825.787	700.800	649.931	-1233.733	480.082	-4.822
5300	133.005	828.321	703.183	663.231	-1233.923	513.042	-5.056
5400	133.006	830.807	705.523	676.532	-1234.031	546.001	-5.282
5500	133.007	833.247	707.823	689.833	-1234.158	578.966	-5.499
5600	133.008	835.644	710.084	703.133	-1234.259	611.932	-5.708
5700	133.009	837.998	712.308	716.434	-1234.333	644.901	-5.910
5800	133.009	840.311	714.495	729.735	-1234.381	677.870	-6.105
5900	133.010	842.585	716.647	743.036	-1234.405	710.840	-6.293
6000	133.011	844.821	718.764	756.337	-1234.405	743.810	-6.475

PREVIOUS: December 1974 (1 atm) CURRENT: December 1974 (1 bar)

TUNGSTEN BROMIDE (WBr_5) CRYSTAL M_r = 583.370

$S°(298.15\ K) = [272 \pm 21]\ J\ K^{-1}\ mol^{-1}$

$T_{fus} = 559 \pm 10\ K$

$\Delta_fH°(0\ K)$ = Unknown

$\Delta_fH°(298.15\ K) = -311.7 \pm 13\ kJ\ mol^{-1}$

$\Delta_{fus}H° = 17.154\ kJ\ mol^{-1}$

Enthalpy of Formation

Shchukarev and Kokovin (1) have measured calorimetrically the enthalpy of reaction $\Delta_rH°(298.15\ K) = -187.475 \pm 0.9\ kcal\ mol^{-1}$ for $WBr_5(cr) + 1/2\ Br_2(\ell) + 8NaOH(77.4H_2O) \rightarrow Na_2WO_4(cr) + 6NaBr(cr) + 4H_2O(\ell)$. This leads to $\Delta_fH°(WBr_5$, cr, 298.15 K) = $-74.5 \pm 3\ kcal\ mol^{-1}$ ($-311.708 \pm 13\ kJ\ mol^{-1}$), using $\Delta_fH°(NaOH\cdot 77.4H_2O$, 298.15 K) = $-112.348\ kcal\ mol^{-1}$ (2), $\Delta_fH°(NaWO_4$, cr, 298.15 K) = $-369.2\ kcal\ mol^{-1}$ (3), $\Delta_fH°$(NaBr, cr, 298.15 K) = $-86.38\ kcal\ mol^{-1}$ (3) and $\Delta_fH°(H_2O$, ℓ, 298.15 K) = $-68.315\ kcal\ mol^{-1}$ (4).

Heat Capacity and Entropy

$C_p°(300\ K) = 6.2\ cal\ K^{-1}\ g\text{-}atom^{-1}$ and $C_p°(559\ K) = 7.25\ cal\ K^{-1}\ g\text{-}atom^{-1}$ are estimated using the method described by Kubaschewski and Evans (5). Between 300 K and 559 K, which is the melting point, the heat capacity is obtained by linear interpolation.

The entropy, $S°(298.15\ K) = 65\ cal\ K^{-1}\ mol^{-1}$ ($271.960\ J\ K^{-1}\ mol^{-1}$), is estimated from that of WCl_5(cr) (3) and the entropy difference between bromides and chlorides. The latter is estimated to be $13\ cal\ K^{-1}\ mol^{-1}$ based on an extrapolation to pentavalent compounds of the entropy values of bromides and chlorides given by Latimer (7). Both WBr_5 and WCl_5 are paramagnetic with an effective moment of about 1.1 Bohr magnetons as indicated by Figgis and Lewis (8).

Fusion Data

Shchukarev et al. (6) have found the melting point, 559 K, by the thermographic method and 568 K by the tensimetric method. The value of 559 K is adopted in the tabulation.

The heat of melting is calculated from the difference between the enthalpies of sublimation and vaporization at the melting point. The enthalpies are both calculated from a 2nd law analysis of the vapor pressure data reported by Shchukarev et al. (6), who gave the uncorrected enthalpy of fusion, $5\ kcal\ mol^{-1}$. This value, when corrected for $\Delta C_p°$ of vaporization and sublimation, is in good agreement with the value adopted in the tabulation.

References

1. S. A. Shchukarev and G. A. Kokovin, Zh. Neorg. Khim. 9, 1309 (1964).
2. The value $\Delta_fH°(NaOH\cdot 77.4H_2O$, 298.15 K) = $-112.348\ kcal\ mol^{-1}$ is calculated from $\Delta_fH°(NaOH\cdot\infty H_2O$, 298.15 K) = $-112.448\ kcal\ mol^{-1}$ and $\phi_L = 100\ cal\ mol^{-1}$ for $NaOH(\infty H_2O) \rightarrow NaOH(77\cdot 4H_2O)$. The former is obtained from JANAF NaOH(cr) table (Mar. 31, 1966) and the latter is obtained from V. B. Parker, U. S. Nat. Bur. Stand. NSRDS-NBS 2, 1965.
3. JANAF Thermochemcial Tables: Na_2WO4(cr), 6-30-67; NaBr(cr), 9-30-64; $WC\ell_5$(cr), 12-31-66.
4. U. S. Nat. Bur. Stand. Tech. Note 270-1, 1965.
5. O. Kubaschewski and E. Ll. Evans, "Metallurgical Thermochemistry," Pergamon Press, New York, 1958.
6. S. A. Shchukarev, G. I. Novikov and G. A. Kokovin, Zh. Neorg. Khim. 4, 2185 (1959).
7. W. M. Latimer, J. Amer. Chem. Soc. 73, 1480 (1951).
8. B. N. Figgis and J. L. Lewis, "Progress in Inorg. Chem.," Vol. 6, page 121, Interscience Publishers, New York, 1964.

Tungsten Bromide (WBr_5) Br_5W_1(cr)

Enthalpy Reference Temperature = T_r = 298.15 K — Standard State Pressure = p° = 0.1 MPa

T/K	$C_p°$ (J K⁻¹mol⁻¹)	S°	$-[G°-H°(T_r)]/T$	$H°-H°(T_r)$ (kJ mol⁻¹)	$\Delta_fH°$	$\Delta_fG°$	Log K_f
0							
100							
200							
298.15	155.461	271.960	271.960	0.	-311.708	-269.605	47.234
300	155.645	272.922	271.963	0.288	-311.815	-269.343	46.897
400	165.686	319.074	278.188	16.354	-384.412	-239.991	31.340
500	175.728	357.110	290.273	33.419	-379.091	-204.484	21.362
559.000	182.004	377.052	298.393	43.971	------ CRYSTAL <--> LIQUID -----		
600	185.770	390.077	304.218	51.516	-372.851	-170.133	14.811
700	195.811	419.468	318.618	70.595	-365.719	-136.898	10.215
800	205.853	446.270	332.923	90.678	-357.662	-104.748	6.839
900	215.894	471.096	346.912	111.765	-348.673	-73.664	4.275
1000	225.936	494.362	360.506	133.857	-338.750	-43.631	2.279

PREVIOUS: December 1962 CURRENT: June 1967

Tungsten Bromide (WBr_5) Br_5W_1(cr)

TUNGSTEN BROMIDE (WBr_5) LIQUID M_r = 583.370

$S°(298.15\ K)$ = [293.341] J K^{-1} mol^{-1} $\Delta_fH°(298.15\ K)$ = [-298.059] kJ mol^{-1}

T_{fus} = 559 ± 10 K $\Delta_{fus}H°$ = 17.154 kJ mol^{-1}

Enthalpy of Formation

$\Delta_fH°(WBr_5, \ell, 298.15\ K)$ is calculated from that of WBr_5(cr) by adding the $\Delta_{fus}H°$ and the difference in enthalpy, H°(559 K) - H°(298.15 K), between the crystal and liquid.

Heat Capacity and Entropy

The heat capacity is assumed to be a constant, 7.25 cal K^{-1} g-atm^{-1}.

S°(298.15 K) is calcualted in a manner analogous to that used for the enthalpy of formation.

Fusion Data

See the crystal table for details.

Vaporization Data

T_{vap} is calculated as the temperature at which the Gibbs energy of reaction $WBr_5(\ell) \rightarrow WBr_5(g)$ approaches zero. The difference between the enthalpies of formation for liquid and gas at the normal boiling point is $\Delta_{vap}H°$.

Shchukarev et al. (1) derived the boiling point 665 K and the enthalpy of vaporization 14.5±0.5 kcal mol^{-1} from vapor pressure data. Our 2nd law values, corrected for $\Delta C_p°$, are T_{vap} = 666 K and $\Delta_{vap}H°$ = 13.8 kcal mol^{-1}. The adopted values are significantly different because the 2nd law value of $\Delta_{vap}S$ is not adopted in the table.

Reference

1. S. A. Shchukarev, G. I. Novikov and G. A. Kokovin, Zh. Neorg. Khim. 4, 2185 (1959).

Tungsten Bromide (WBr_5) Br_5W_1(l)

Enthalpy Reference Temperature = T_r = 298.15 K Standard State Pressure = p° = 0.1 MPa

T/K	$C_p°$ (J $K^{-1}mol^{-1}$)	S° (J $K^{-1}mol^{-1}$)	$-[G°-H°(T_r)]/T$ (J $K^{-1}mol^{-1}$)	$H°-H°(T_r)$ (kJ mol^{-1})	$\Delta_fH°$ (kJ mol^{-1})	$\Delta_fG°$ (kJ mol^{-1})	Log K_f
0							
100							
200							
298.15	182.004	293.341	293.341	0.	-298.059	-262.330	45.959
300	182.004	294.467	293.344	0.337	-298.117	-262.108	45.637
400	182.004	346.826	300.483	18.537	-368.580	-235.260	30.722
500	182.004	387.439	313.964	36.738	-362.123	-202.680	21.174
559.000	182.004	407.740	322.810	47.476	------ CRYSTAL <--> LIQUID -----		
600	182.004	420.622	329.059	54.938	-355.779	-171.389	14.921
700	182.004	448.678	344.195	73.138	-349.526	-141.153	10.533
800	182.004	472.982	358.808	91.339	-343.351	-111.808	7.300
900	182.004	494.419	372.708	109.539	-337.250	-83.232	4.831
1000	182.004	513.595	385.855	127.740	-331.218	-55.332	2.890

PREVIOUS: December 1962 CURRENT: June 1967

TUNGSTEN BROMIDE (WBr_5) CRYSTAL - LIQUID M_r = 583.370

298.15 to 559 K crystal

above 559 K liquid

Refer to the individual tables for details.

Tungsten Bromide (WBr_5) Br_5W_1(cr,l)

Enthalpy Reference Temperature = T_r = 298.15 K Standard State Pressure = p° = 0.1 MPa

T/K	C_p° (J K⁻¹mol⁻¹)	S° (J K⁻¹mol⁻¹)	$-[G°-H°(T_r)]/T$ (J K⁻¹mol⁻¹)	$H°-H°(T_r)$ (kJ mol⁻¹)	$\Delta_fH°$ (kJ mol⁻¹)	$\Delta_fG°$ (kJ mol⁻¹)	Log K_f
0							
100							
200							
298.15	155.461	271.960	271.960	0.	-311.708	-269.605	47.234
300	155.645	272.922	271.963	0.288	-311.815	-269.343	46.897
400	165.686	319.074	278.188	16.354	-384.412	-239.991	31.340
500	175.728	357.110	290.273	33.419	-379.091	-204.484	21.362
559.000	182.004	377.052	298.393	43.971	CRYSTAL <--> LIQUID	TRANSITION	
559.000	182.004	407.740	298.393	61.125			
600	182.004	420.622	306.310	68.587	-355.779	-171.389	14.921
700	182.004	448.678	324.696	86.788	-349.526	-141.153	10.533
800	182.004	472.982	341.747	104.988	-343.351	-111.808	7.300
900	182.004	494.419	357.543	123.188	-337.250	-83.232	4.831
1000	182.004	513.595	372.206	141.389	-331.218	-55.332	2.890

PREVIOUS: CURRENT: June 1967

Tungsten Bromide (WBr_5) Br_5W_1(cr,l)

TUNGSTEN BROMIDE (WBr_5) IDEAL GAS M_r = 583.370

S°(298.15 K) = [461.4] J K^{-1} mol^{-1}

$\Delta_fH°(0\ K) = -162.9 \pm 21\ kJ\ mol^{-1}$
$\Delta_fH°(298.15\ K) = -199.2 \pm 21\ kJ\ mol^{-1}$

Eletronic Levels and Quantum Weights

ε_i, cm^{-1}	g_i
0	[2]
[7000]	[2]
[14000]	[2]

Vibrational Frequencies and Degeneracies

ν, cm^{-1}	ν, cm^{-1}
[300](1)	[240](2)
[250](1)	[100](2)
[235](1)	[70](2)
[90](1)	[110)(2)

Ground State Quantum Weight: [2] σ = [6]
Point Group: [D_{3h}]
Bond Distance: W-Br = [2.40] Å
Bond Angles: Br*-W-Br* = [120°] Br*-W-Br** = [90°]
Br**-W-Br** = [180°]
*Equatorial ** Axial
Product of the Moments of Inertia: $I_AI_BI_C$ = [1.6458] x 10^{-110} g^3 cm^6

Enthalpy of Formation

Shchukarev et al. (5) measured the vapor pressure of WBr_5(cr, ℓ) using a null manometer. Their results are given as below. The adopted enthalpy of formation is -47.6 kcal mol^{-1} (-199.158 kJ mol^{-1}).

Reactions	Data Points	T/K	$\Delta_rH°$(298.15 K), kcal mol^{-1} 2nd law	3rd law	Drift cal $K^{-1}$$mol^{-1}$	$\Delta_fH°$(298.15 K) kcal mol^{-1}
WBr_5(ℓ) → WBr_5(g)	13	573.3-657.5	18.37	24.11	9.4±0.1	-47.13
WBr_5(cr) → WBr_5(g)	6	443.5-496.5	21.61	26.46	9.4±0.4	-48.04

Heat Capacity and Entropy

The molecular configuration is assumed to be a trigonal bipyramid similar to that of $MoCl_5$ determined by electron diffraction (1). The bond distance is estimated to be the same as that in WBr_6(g). The principal moments of inertia are: I_A = 229.2811 x 10^{-39} and $I_B = I_C$ = 267.4946 x 10^{-39} g cm^2.

All vibrational frequencies are estimated from those of WCl_5(g) (2), using the average value of $\nu(WBr_5)/\nu(WCl_5)$ = 0.62 for modes which are independent of the central atom and 0.68 for modes involving the central atom. These average values of 0.62 and 0.68 are obtained from the ratios of corresponding vibrational frequencies of $ReX_6^=$, $SnX_6^=$ and $PtX_6^=$ summarized by Siebert (3) and Nakamoto (4).

The electronic levels and quantum weights are estimated to be the same as those in WCl_5(g) (2).

References

1. R. V. G. Ewens and M. W. Lister, Trans. Faraday Soc. 34, 1358 (1938).
2. JANAF Thermochemical Table: WCl_5(g), 12-31-66.
3. H. Siebert, "Anwendungen der Schwingunsspektroskopie in der Anorganischen Chemie," Springer-Verlag, Berlin, 1966.
4. K. Nakamoto, "Infrared Spectra of Inorganic and Coordination Compounds," John Wiley and Sons, Inc., New York, 1963.
5. S. A. Shchukarev, G. I. Novikov and G. A. Kokovin, Zh. Neorg. Khim. 4, 2185 (1959).

Tungsten Bromide (WBr_5) Br_5W_1(g)

Enthalpy Reference Temperature = T_r = 298.15 K Standard State Pressure = p° = 0.1 MPa

T/K	$C_p°$ (J $K^{-1}$$mol^{-1}$)	S°	-[G°-H°(T_r)]/T	H°-H°(T_r) (kJ mol^{-1})	$\Delta_fH°$	$\Delta_fG°$	Log K_f
0	0.	0.	INFINITE	-30.021	-162.933	-162.933	INFINITE
100	98.777	334.646	569.460	-23.481	-164.028	-183.070	95.626
200	121.036	411.719	472.895	-12.235	-167.043	-201.022	52.501
250	124.972	439.192	463.494	-6.076	-169.080	-209.292	43.729
298.15	127.212	461.411	461.411	0.	-199.158	-213.540	37.411
300	127.279	462.198	461.413	0.235	-199.318	-213.629	37.196
350	128.733	481.935	462.969	6.638	-275.765	-211.832	31.614
400	129.704	499.193	466.441	13.101	-275.116	-202.743	26.476
450	130.382	514.511	470.947	19.604	-274.468	-193.735	22.488
500	130.873	528.275	476.003	26.136	-273.824	-184.799	19.306
600	131.522	552.199	486.768	39.258	-272.558	-167.114	14.549
700	131.918	572.505	497.602	52.432	-271.332	-149.637	11.166
800	132.180	590.138	508.091	65.638	-270.152	-132.334	8.640
900	132.368	605.718	518.090	78.866	-269.023	-115.175	6.685
1000	132.517	619.673	527.562	92.110	-267.947	-98.139	5.126
1100	132.650	632.309	536.520	105.368	-266.926	-81.208	3.856
1200	132.781	643.857	544.990	118.640	-265.959	-64.367	2.802
1300	132.921	654.490	553.010	131.925	-265.045	-47.605	1.913
1400	133.073	664.346	560.614	145.225	-264.185	-30.912	1.153
1500	133.242	673.533	567.840	158.540	-263.379	-14.278	0.497
1600	133.426	682.138	574.717	171.873	-262.629	2.304	-0.075
1700	133.624	690.233	581.276	185.226	-261.936	18.841	-0.579
1800	133.835	697.877	587.544	198.599	-261.302	35.338	-1.025
1900	134.055	705.118	593.543	211.993	-260.733	51.803	-1.424
2000	134.281	712.000	599.295	225.410	-260.231	68.239	-1.782
2100	134.511	718.557	604.820	238.849	-259.801	84.652	-2.106
2200	134.743	724.820	610.133	252.312	-259.447	101.045	-2.399
2300	134.974	730.815	615.251	265.798	-259.173	117.425	-2.667
2400	135.202	736.564	620.186	279.307	-258.982	133.795	-2.912
2500	135.425	742.088	624.953	292.838	-258.875	150.158	-3.137
2600	135.643	747.404	629.561	306.392	-258.849	166.518	-3.345
2700	135.854	752.527	634.021	319.966	-258.933	182.880	-3.538
2800	136.057	757.471	638.342	333.562	-259.158	199.246	-3.717
2900	136.252	762.249	642.533	347.178	-259.561	215.625	-3.884
3000	136.438	766.871	646.601	360.812	-260.166	232.020	-4.040
3100	136.614	771.348	650.553	374.465	-261.013	248.439	-4.186
3200	136.781	775.688	654.396	388.135	-262.127	264.890	-4.324
3300	136.939	779.899	658.136	401.821	-263.529	281.381	-4.454
3400	137.087	783.990	661.777	415.522	-265.249	297.918	-4.577
3500	137.225	787.966	665.326	429.238	-267.344	314.511	-4.694
3600	137.353	791.833	668.787	442.967	-269.906	331.171	-4.805
3700	137.473	795.598	672.163	456.708	-307.794	348.100	-4.914
3800	137.583	799.266	675.460	470.461	-307.985	365.829	-5.029
3900	137.683	802.841	678.681	484.224	-308.156	383.563	-5.137
4000	137.775	806.328	681.828	497.997	-308.300	401.301	-5.240
4100	137.859	809.731	684.907	511.779	-308.414	419.043	-5.339
4200	137.934	813.054	687.918	525.569	-308.493	436.786	-5.432
4300	138.002	816.300	690.866	539.366	-308.533	454.532	-5.521
4400	138.061	819.474	693.753	553.169	-308.532	472.278	-5.607
4500	138.114	822.577	696.582	566.978	-308.487	490.023	-5.688
4600	138.159	825.613	699.354	580.791	-308.394	507.767	-5.766
4700	138.198	828.585	702.072	594.609	-308.251	525.508	-5.840
4800	138.230	831.495	704.738	608.431	-308.058	543.246	-5.912
4900	138.257	834.345	707.354	622.255	-307.812	560.979	-5.980
5000	138.277	837.138	709.922	636.082	-307.513	578.706	-6.046
5100	138.293	839.877	712.443	649.910	-307.159	596.428	-6.109
5200	138.303	842.562	714.920	663.740	-306.751	614.140	-6.169
5300	138.308	845.197	717.353	677.571	-306.288	631.846	-6.227
5400	138.309	847.782	719.745	691.402	-305.770	649.542	-6.283
5500	138.306	850.320	722.096	705.233	-305.198	667.228	-6.337
5600	138.299	852.812	724.408	719.063	-304.571	684.902	-6.389
5700	138.288	855.260	726.682	732.892	-303.891	702.566	-6.438
5800	138.274	857.665	728.920	746.720	-303.158	720.216	-6.486
5900	138.256	860.028	731.122	760.547	-302.373	737.854	-6.532
6000	138.236	862.352	733.290	774.371	-301.537	755.478	-6.577

PREVIOUS: June 1967 (1 atm) CURRENT: June 1967 (1 bar)

Tungsten Bromide (WBr_5) Br_5W_1(g)

TUNGSTEN BROMIDE (WBr_6) CRYSTAL M_r = 663.274

$S°(298.15\ K)$ = [313.8 ± 21] $J\ K^{-1}\ mol^{-1}$

T_{fus} = 582 K

$\Delta_f H°(0\ K)$ = Unknown

$\Delta_f H°(298.15\ K)$ = -343.088 $kJ\ mol^{-1}$

$\Delta_{fus}H°$ = Unknown

Enthalpy of Formation

Shchukarev and Kokovin (1) have measured calorimetrically the enthalpy of reaction $\Delta_r H°(298.15\ K)$ = -180.0±0.8 $kcal\ mol^{-1}$ for $WBr_6(cr) + 8NaOH(77.4H_2O) \rightarrow Na_2WO_4(cr) + 6NaBr(cr) + 4H_2O(\ell)$. This leads to $\Delta_f H°(WBr_6$, cr, 298.15 K) = -82.0±3 $kcal\ mol^{-1}$ (-343.088±13 $kJ\ mol^{-1}$), using $\Delta_f H°(NaOH{\cdot}77.4H_2O$, 298.15 K) = -112.348 $kcal\ mol^{-1}$ (2), $\Delta_f H°(Na_2WO_4$, cr, 298.15 K) = -369.2 $kcal\ mol^{-1}$ (3), $\Delta_f H°$(NaBr, cr, 298.15 K) = -86.38 $kcal\ mol^{-1}$ (3) and $\Delta_f H°(H_2O$, ℓ, 298.15 K) = -68.315 $kcal\ mol^{-1}$ (4).

Heat Capacity and Entropy

$C_p°(300\ K)$ = 6.2 cal K^{-1} g-atom^{-1} and $C_p°(582\ K)$ = 7.25 cal K^{-1} g-atom^{-1} are estimated using the method described by Kubaschewski and Evans (5). Between 300 and 582 K, which is the melting point, the heat capacity is obtained by linear interpolation.

The entropy, $S°(WBr_6$, cr, 298.15 K) = 75 cal $K^{-1}\ mol^{-1}$ (313.800 $J\ K^{-1}\ mol^{-1}$), is estimated by assuming $S°(WBr_6$, 298.15 K) = $S°(WCl_6$, 298.15 K) + 6[$S°(Br^-$, 298.15 K) - $S°(Cl^-$, 298.15 K)]. The value, $S°(WCl_6$, cr, 298.15 K) = 57 cal $K^{-1}\ mol^{-1}$, is obtained from WCl_6(cr, α) (3); and the value, $S°(Br^-$, 298.15 K) - $S°(Cl^-$, 298.15 K) = 3 cal $K^{-1}\ mol^{-1}$, is obtained from the average entropy values for Br^- and Cl^- ions given by Latimer (6) and Kelley (7).

Fusion Data

Shchukarev et al. (8) have found the melting point, 582 K, under 50 atm pressure of $Br_2(g)$ by the thermographic method. The heat of fusion is unknown.

References

1. S. A. Shchukarev and G. A. Kokovin, Zh. Neorg. Khim. 9, 1309 (1964).
2. This value, $\Delta_f H°(NaOH{\cdot}77.4H_2O$, 298.15 K) = -112.348 $kcal\ mol^{-1}$ is calculated from $\Delta_f H°(NaOH{\cdot}\infty H_2O$, 298.15 K) = 112.448 kcal mol^{-1} and ϕ_L = 100 $kcal\ mol^{-1}$ for $NaOH(\infty H_2O) \rightarrow NaOH(77.4H_2O)$. The former is obtained from JANAF NaOH(cr) table (Mar. 31, 1966) and the latter is obtained from V. B. Parker, U. S. Nat. Bur. Stand. NSRDS-NBS 2, 1965.
3. JANAF Thermochemical Tables: Na_2WO_4(cr), 6-30-67; NaBr(cr), 6-30-64; WCl_6(cr, α), 12-31-66.
4. U. S. Nat. Bur. Stand. Tech. Note 270-1, 1965.
5. O. Kubaschewski and E. Ll. Evans, "Metallurgical Thermochemistry," Pergamon Press, New York, 1958.
6. W. M. Latimer, J. Amer. Chem. Soc. 73, 1480 (1951).
7. K. K. Kelley, personal communication, June, 1960.
8. S. A. Shchukarev, G. I. Novikov and G. A. Kokovin, Zh. Neorg. Khim. 4, 2185 (1959).

Tungsten Bromide (WBr_6) Br_6W_1(cr)

Enthalpy Reference Temperature = T_r = 298.15 K — Standard State Pressure = p° = 0.1 MPa

T/K	$C_p°$	S°	$-[G°-H°(T_r)]/T$	$H°-H°(T_r)$	$\Delta_f H°$	$\Delta_f G°$	Log K_f
	J K⁻¹mol⁻¹			kJ mol⁻¹			
0							
100							
200							
298.15	181.385	313.800	313.800	0.	-343.088	-290.769	50.942
300	181.586	314.923	313.803	0.336	-343.217	-290.444	50.571
400	192.464	368.651	321.056	19.038	-430.418	-254.610	33.249
500	203.342	412.767	335.110	38.829	-424.217	-211.356	22.080
600	214.221	450.802	351.291	59.707	-417.055	-169.440	14.751
700	225.099	484.642	367.966	81.673	-408.905	-128.801	9.611
800	235.978	515.410	384.501	104.727	-399.754	-89.399	5.837
900	246.856	543.832	400.645	128.868	-389.593	-51.204	2.972
1000	257.734	570.404	416.306	154.098	-378.419	-14.195	0.741

PREVIOUS: December 1962 CURRENT: June 1967

Tungsten Bromide (WBr_6) Br_6W_1(cr)

TUNGSTEN BROMIDE (WBr_6) IDEAL GAS M_r = 663.274

$S°(298.15\ K) = [482.7 \pm 42]\ J\ K^{-1}\ mol^{-1}$

$\Delta_f H°(0\ K) = [-199.8 \pm 42]\ kJ\ mol^{-1}$
$\Delta_f H°(298.15\ K) = [-243.1 \pm 42]\ kJ\ mol^{-1}$

Vibrational Frequencies and Degeneracies

ν, cm^{-1}	ν, cm^{-1}
[250](1)	[110](3)
[200](2)	[130](3)
[250](3)	[70](3)

Ground State Quantum Weight: [1] σ = [24]
Point Group: [O_h]
Bond Distance: W-Br = [2.40] Å
Bond Angle: Br-W-Br = [90°]
Product of the Moments of Inertia: $I_AI_BI_C = [2.85707 \times 10^{-110}]\ g^3\ cm^6$

Enthalpy of Formation

$\Delta_f H°(WBr_6, g, 298.15\ K) = -58.1\ kcal\ mol^{-1}$ ($-243.090\ kJ\ mol^{-1}$) is calculated from that of the crystal plus the estimated enthalpy of sublimation $\Delta_{sub}H°(298.15\ K) = 23.9\ kcal\ mol^{-1}$ for $WBr_6(cr) \rightarrow WBr_6(g)$. The value of $\Delta_{sub}H°(298.15\ K)$ is assumed to be the same as that for $WCl_6(cr) \rightarrow WCl_6(g)$. See WCl_6 table (Dec. 31, 1966) for details.

Heat Capacity and Entropy

The molecular configuration is assumed to be an octahedron similar to those of $WF_6(g)$, and $WCl_6(g)$ determined by electron diffraction. The bond distance is estimated to be 2.40 Å by assuming $r_{W-Br} = r_{W-Cl} + (r_{Na-Br} - r_{Na-Cl})$. The bond distances, W-Cl, Na-Br and Na-Cl, are given in JANAF WCl(g) an NaBr(g) and NaCl(g) tables, as 2.26 Å, 2.50 Å and 2.36 Å, respectively. The principal moments of inertia are: $I_A = I_B = I_C = 305.7081 \times 10^{-39}\ g\ cm^2$.

All vibrational frequencies are estimated from those of $WCl_6(g)$ (1), using the average value of $\nu(WBr_6)/\nu(WCl_6) = 0.62$ for modes which are independent of the central atom and 0.68 for modes involving the central atom. These average values of 0.62 and 0.68 are obtained from the ratios of corresponding vibrational frequencies of $ReX_6^=$, $SnX_6^=$ and $PtX_6^=$ summarized by Siebert (2), and Nakamoto (3).

References

1. J. C. Evans and G. Y-S Lo, The Dow Chemical Company, personal communication, June 6, 1967, have obtained the six fundamental vibrational frequencies of 408, 312, 367, 165, 206 and 97 cm^{-1} for $WCl_6(g)$ by infrared spectrometry.
2. H. Siebert, "Anwendungen der Schwingungsspektroskopie in der Anorganischen Chemie," Springer-Verlag, Berlin, 1966.
3. K. Nakamoto, "Infrared Spectra of Inorganic and Coordination Compounds," John Wiley and Sons, Inc., New York, 1963.

Tungsten Bromide (WBr_6) $Br_6W_1(g)$

Enthalpy Reference Temperature = T_r = 298.15 K — Standard State Pressure = p° = 0.1 MPa

T/K	$C_p°$ (J K⁻¹mol⁻¹)	S° (J K⁻¹mol⁻¹)	−[G°−H°(T_r)]/T (J K⁻¹mol⁻¹)	H°−H°(T_r) (kJ mol⁻¹)	$\Delta_f H°$ (kJ mol⁻¹)	$\Delta_f G°$ (kJ mol⁻¹)	Log K_f
0	0.	0.	INFINITE	-35.204	-199.793	-199.793	INFINITE
100	116.477	331.931	611.326	-27.940	-201.561	-217.638	113.683
200	144.174	423.537	496.369	-14.566	-204.899	-232.499	60.722
250	148.782	456.254	485.179	-7.231	-207.167	-239.149	49.968
298.15	151.369	482.699	482.699	0.	-243.090	-241.129	42.245
300	151.446	483.636	482.702	0.280	-243.275	-241.116	41.982
350	153.112	507.116	484.553	7.897	-334.833	-235.752	35.184
400	154.218	527.638	488.683	15.582	-333.877	-221.663	28.946
450	154.988	545.849	494.042	23.313	-332.922	-207.694	24.108
500	155.544	562.209	500.055	31.077	-331.971	-193.830	20.249
600	156.277	590.639	512.853	46.672	-330.092	-166.379	14.485
700	156.723	614.765	525.732	62.323	-328.257	-139.240	10.390
800	157.014	635.713	538.199	78.011	-326.472	-112.360	7.336
900	157.214	654.219	550.082	93.723	-324.741	-85.700	4.974
1000	157.358	670.791	561.339	109.452	-323.068	-59.230	3.094
1100	157.464	685.794	571.981	125.193	-321.455	-32.925	1.563
1200	157.545	699.498	582.045	140.944	-319.904	-6.764	0.294
1300	157.608	712.111	591.571	156.702	-318.418	19.270	-0.774
1400	157.659	723.793	600.604	172.465	-316.999	45.192	-1.686
1500	157.699	734.672	609.183	188.233	-315.651	71.016	-2.473
1600	157.732	744.851	617.348	204.005	-314.376	96.752	-3.159
1700	157.760	754.414	625.132	219.779	-313.181	122.410	-3.761
1800	157.783	763.432	632.567	235.557	-312.071	148.001	-4.295
1900	157.802	771.963	639.681	251.336	-311.050	173.532	-4.771
2000	157.819	780.058	646.500	267.117	-310.127	199.012	-5.198
2100	157.834	787.758	653.044	282.900	-309.306	224.448	-5.583
2200	157.846	795.101	659.336	298.684	-308.593	249.848	-5.932
2300	157.857	802.118	665.392	314.469	-307.992	275.218	-6.250
2400	157.866	808.836	671.230	330.255	-307.507	300.564	-6.542
2500	157.875	815.281	676.864	346.042	-307.140	325.893	-6.809
2600	157.882	821.473	682.308	361.830	-306.887	351.208	-7.056
2700	157.889	827.432	687.573	377.618	-306.774	376.517	-7.284
2800	157.895	833.174	692.671	393.408	-306.833	401.824	-7.496
2900	157.900	838.715	697.612	409.197	-307.101	427.138	-7.694
3000	157.905	844.068	702.405	424.988	-307.598	452.464	-7.878
3100	157.909	849.246	707.059	440.778	-308.363	477.811	-8.051
3200	157.913	854.259	711.581	456.570	-309.418	503.188	-8.214
3300	157.917	859.119	715.979	472.361	-310.784	528.602	-8.367
3400	157.920	863.833	720.258	488.153	-312.486	554.063	-8.512
3500	157.923	868.411	724.426	503.945	-314.581	579.578	-8.650
3600	157.926	872.860	728.488	519.738	-317.157	605.162	-8.781
3700	157.929	877.187	732.449	535.530	-355.072	631.014	-8.908
3800	157.931	881.398	736.313	551.323	-355.301	657.668	-9.040
3900	157.933	885.501	740.086	567.116	-355.517	684.327	-9.166
4000	157.935	889.499	743.772	582.910	-355.713	710.993	-9.285
4100	157.937	893.399	747.374	598.703	-355.883	737.663	-9.398
4200	157.939	897.205	750.896	614.497	-356.020	764.336	-9.506
4300	157.940	900.921	754.342	630.291	-356.120	791.013	-9.609
4400	157.942	904.552	757.715	646.085	-356.177	817.691	-9.707
4500	157.943	908.102	761.017	661.880	-356.187	844.370	-9.801
4600	157.945	911.573	764.253	677.674	-356.146	871.049	-9.891
4700	157.946	914.970	767.423	693.468	-356.051	897.726	-9.977
4800	157.947	918.295	770.532	709.263	-355.899	924.401	-10.060
4900	157.948	921.552	773.581	725.058	-355.687	951.071	-10.139
5000	157.949	924.743	776.573	740.853	-355.414	977.738	-10.214
5100	157.950	927.871	779.509	756.648	-355.078	1004.398	-10.287
5200	157.951	930.938	782.391	772.443	-354.678	1031.050	-10.357
5300	157.952	933.947	785.223	788.238	-354.212	1057.695	-10.424
5400	157.953	936.899	788.004	804.033	-353.682	1084.330	-10.489
5500	157.953	939.797	790.738	819.828	-353.085	1110.954	-10.551
5600	157.954	942.643	793.425	835.624	-352.423	1137.567	-10.611
5700	157.955	945.439	796.067	851.419	-351.695	1164.168	-10.668
5800	157.955	948.186	798.667	867.215	-350.903	1190.754	-10.724
5900	157.956	950.886	801.224	883.010	-350.046	1217.327	-10.777
6000	157.957	953.541	803.740	898.806	-349.125	1243.885	-10.829

PREVIOUS: June 1967 (1 atm) CURRENT: June 1967 (1 bar)

Tungsten Bromide (WBr_6) $Br_6W_1(g)$

NIOBIUM CARBIDE ($NbC_{0.98}$) CRYSTAL M_r = 104.6772

$S°(298.15\ K) = 34.966 \pm 0.4\ J\ K^{-1}\ mol^{-1}$

$\Delta_f H°(0\ K) = -138.061 \pm 2.5\ kJ\ mol^{-1}$
$\Delta_f H°(298.15\ K) = -138.909 \pm 2.5\ kJ\ mol^{-1}$

Enthalpy of Formation

There are five oxygen bomb calorimetry studies on NbC reported in the literature (1-5) and one CO(g) equilibrium study (6). The adopted $\Delta_f H°(298.15\ K)$ value is derived from the calorimetric study of Huber et al. (3). These authors studied the combustion of eight NbC_x samples where x covers the range 0.686 to 0.984. For each composition six to eight combustions were performed. Huber et al. (3) reported that the combustions varied from 98.78 to 100% of completion. Corrections of the order 0.51% or less were made to the enthalpy of combustion value to account for impurities in seven of the eight samples. The correction was 3.33% for the x = 0.984 sample. The calculated heats of formation were fitted to a quadratic equation by the least squares method by Huber et al. (3). Each value was weighted inversely proportional to the square of its uncertainty. Using the quadratic equation for $\Delta_f H°(298.15\ K)$ and correcting for the JANAF value of $\Delta_f H°(Nb_2O_5$, cr, 298.15 K) (7), we interpolate to x = 0.98, calculate and adopt $\Delta_f H°(298.15\ K) = -33.20 \pm 0.6\ kcal\ mol^{-1}$. Due to the large error in the data points, a linear equation can be chosen to fit the data without making a statistically significant difference (3).

Huber et al. (3) and Storms (8) have reanalyzed much of the calorimetric and equilibrium data (1-6) for NbC_x (x = 0.686 to 0.984). A graphical comparison of the results (i.e., $\Delta_f H°(298.15\ K)$ values vs. x) indicated that the quadratic equation suggested by Huber et al. (3) is a satisfactory representation of the various $\Delta_f H°(298.15\ K)$ results for NbC_x. The shortcomings of the various studies have been discussed by Huber et al. (3) and Storms (8). For comparison, the oxygen combustion study by Mah (1) leads to $\Delta_f H°(298.15\ K) = -31.14 \pm 0.8\ kcal\ mol^{-1}$ for $NbC_{0.9445}$.

Heat Capacity and Entropy

There have been three low temperature heat capacity studies for NbC_x: Pankratz et al. (13) using a $NbC_{0.996}$ sample (52-296 K), Toth et al. (19) using five NbC_x samples where x = 0.77, 0.83, 0.86, 0.91, and 0.96 (1.5-18 K), and Sandernaw and Storms using three NbC_x samples where x = 0.702, 0.825, and 0.980 (7.5-320 K). In addition there are numerous high temperature heat capacity and enthalpy studies on various NbC_x samples (9-13), (15-18). The adopted heat capacity values are those suggested by Turchanin et al. (20). Their $C_p°$ values were represented by a five term equation for the range 0-3000 K and is a compromise of the reported data mentioned, all adjusted to the composition $NbC_{0.98}$.

Phase Data

NbC has a face centered cubic structure of the NaCl type (8). The NbC phase exists, following the phase diagram of Storms (8), from roughly $NbC_{0.70}$ to near $NbC_{0.99}$. At the NbC-C eutectic temperature of 3573±50 K, the phase limit is $NbC_{0.97}$.

References

1. A. D. Mah and B. J. Boyle, J. Amer. Chem. Soc. 77, 6512 (1955).
2. F. G. Kusenko and P. V. Gel'd, Izv. Siber. Otdel. Adad. Nauk SSSR, No. 2, 46 (1960).
3. E. J. Huber, Jr., E. L. Head, C. E. Holley, Jr., E. K. Storms, and N. H. Krikorian, J. Phys. Chem. 65, 1846 (1961).
4. A. N. Kornilov, V. Yo. Leonidov, and S. N. Skuratov, Vest. Mosk. Univ. Ser II: Khim. 17, 48 (1962).
5. A. N. Kornilov, I. D. Zaikin, S. M. Shuratov, and G. P. Shoveikin, Zhur. Fiz. Khim. 40, 1070 (1966).
6. W. L. Worrell and J. Chipman, J. Phys. Chem. 68, 860 (1964).
7. JANAF Thermochemical Tables: Nb_2O_5(cr), 12-31-72.
8. E. K. Storms, "Refractory Carbides," Academic Press, New York, 1967.
9. D. S. Neel, C. D. Peers, and S. Oglesby, Jr., WADD-TR-60-924, 1962 [AD 275 536].
10. P. V. Gel'd and F. G. Kusenko, Izv. Akad. Nauk SSSR, Otd. Tikhn. Nauk, Met. i Toplivo, No. 2, 79 (1960).
11. L. S. Levinson, J. Chem. Phys. 39, 1550 (1963).
12. R. A. McDonald, F. L. Oetting, and H. Prophet, CPIA-44, 1, 213 (1964).
13. L. B. Pankratz, W. W. Weller, and K. K. Kelley, U. S. Bur. Mines RI 6446, 1964.
14. T. A. Sandenaw and E. K. Storms, LA-3331 (1965); J. Phys. Chem. Solids 27, 217 (1966).
15. A. G. Turchanin, Tr. Nauch. Konf. Aspir. Inst. Probl. Materialoved. Akad. Nauk Ukr. SSR, 1st, 88 (1967).
16. V. V. Fresenko, A. G. Turchanin, and E. A. Guseva, Russ. J. Phys. Chem. 42, 1234 (1968).
17. P. B. Kantor and E. N. Formichiv, Teplofiz. Vys. Temp. 1, 48 (1967).
18. A. G. Turchanin, S. S. Ordan'yan, and V. V. Fesenko, Porosh. Met. 9, 23 (1967).
19. L. E. Toth, M. Ishikawa, and Y. A. Chang, Acta Met. 16, 1183 (1968).
20. A. G. Turchanin, E. A. Guseva, and V. V. Fesenko, Porosh. Met. 11, 45 (1971).

Niobium Carbide ($NbC_{0.98}$) $C_{0.98}Nb_1(cr)$

Enthalpy Reference Temperature = T_r = 298.15 K Standard State Pressure = p° = 0.1 MPa

	J K⁻¹mol⁻¹			kJ mol⁻¹			
T/K	$C_p°$	S°	$-[G°-H°(T_r)]/T$	$H°-H°(T_r)$	$\Delta_f H°$	$\Delta_f G°$	Log K_f
0	0.	0.	INFINITE	-5.422	-138.061	-138.061	INFINITE
100	14.426	7.983	56.725	-4.874	-138.352	-137.801	71.980
200	27.476	22.238	35.932	-2.739	-138.635	-137.103	35.808
298.15	36.233	34.966	34.966	0.	-138.909	-136.785	23.964
300	36.380	35.190	34.966	0.067	-138.903	-136.772	23.814
400	41.882	46.463	36.466	3.999	-138.481	-136.120	17.776
500	45.145	56.179	39.461	8.359	-137.986	-135.587	14.165
600	47.363	64.616	42.966	12.990	-137.514	-135.153	11.766
700	48.911	72.038	46.600	17.806	-137.091	-134.793	10.058
800	50.124	78.651	50.200	22.760	-136.718	-134.491	8.781
900	51.087	84.611	53.698	27.822	-136.389	-134.233	7.791
1000	51.923	90.038	57.065	32.973	-136.097	-134.009	7.000
1100	52.635	95.021	60.292	38.202	-135.839	-133.813	6.354
1200	53.304	99.630	63.380	43.499	-135.609	-133.640	5.817
1300	53.932	103.921	66.336	48.861	-135.398	-133.484	5.363
1400	54.504	107.939	69.165	54.283	-135.207	-133.344	4.975
1500	55.061	111.718	71.878	59.761	-135.038	-133.217	4.639
1600	55.605	115.289	74.480	65.295	-134.894	-133.101	4.345
1700	56.134	118.676	76.981	70.882	-134.781	-132.992	4.086
1800	56.651	121.900	79.388	76.521	-134.705	-132.889	3.856
1900	57.153	124.976	81.707	82.211	-134.671	-132.790	3.651
2000	57.656	127.920	83.944	87.952	-134.683	-132.691	3.466
2100	58.158	130.746	86.106	93.743	-134.745	-132.590	3.298
2200	58.660	133.463	88.197	99.583	-134.863	-132.484	3.146
2300	59.162	136.081	90.223	105.475	-135.042	-132.373	3.006
2400	59.664	138.610	92.187	111.416	-135.293	-132.251	2.878
2500	60.166	141.056	94.093	117.407	-135.630	-132.118	2.760
2600	60.668	143.425	95.945	123.449	-136.076	-131.969	2.651
2700	61.170	145.724	97.746	129.541	-136.669	-131.801	2.550
2800	61.672	147.958	99.500	135.683	-163.822	-131.122	2.446
2900	62.174	150.131	101.208	141.875	-163.565	-129.959	2.341
3000	62.676	152.247	102.874	148.118	-163.272	-128.805	2.243

PREVIOUS: CURRENT: December 1973

Niobium Carbide ($NbC_{0.98}$) $C_{0.98}Nb_1(cr)$

S°(298.15 K) = 5.74 ± 0.21 J K^{-1} mol^{-1}

$\Delta_f H°$(0 K) = 0 kJ mol^{-1}

$\Delta_f H°$(298.15 K)= 0 kJ mol^{-1}

Enthalpy of Formation

$\Delta_f H°$ is zero by definition for the adopted standard reference state of Achesson spectroscopic graphite, i.e. artificial graphite of spectroscopic grade which is graphitized near 2600°C in an Acheson-type electric furnace. This standard state was adopted by CODATA (1) in order to be more consistent with the samples used in determing $\Delta_f H°$(298.15) of CO_2(g). CODATA (1) cites references which show the effects of various graphite samples on $\Delta_f H°(CO_2)$. Their assigned uncertainty (1) of ±0.031 kcal/mol for $\Delta_f H°(CO_2)$ is a rough measure of the deviations of various high purity graphites from our standard state. Larger deviations occur in carbon samples which are impure or less completely graphitized or which have much disorder or high surface area. Gross deviations (up to several kcal/mol) have been found for other carbons; e.g., many "glassy" carbons appear to be more stable than graphite at elevated temperatures due to large $\Delta_f S$ values (2) which compensate for positive values of $\Delta_f H$.

Heat Capacity and Entropy

S°(298.15) and H°(298.15 K) - H°(0 K) are the values selected by CODATA (1) based on Cp° (13-301 K) of CS-grade Acheson graphite measured by DeSorbo and Tyler (3). DeSorbo's later study (3) gave minor corrections to Cp° and more information on his sample. Due to differences in Cp° of various graphites, we increase the uncertainty of S°(298.15) from ±0.029 (1) to ±0.05 cal K^{-1} mol^{-1}. The entropy extrapolation below 13 K was given (3) as 0.0041 cal K^{-1} mol^{-1}. Properties tabulated below 298.15 K are from our smoothing of the data (3). Cp° above 298.15 K is based on enthalpy data of McDonald (4) and West and Ishihara (5) and on Cp° data of Cezairliyan et al. (6, 7). Deviations of the enthalpy data from our table are -0.3 to +0.6% (4, omitting 341 and 1037 K) and 0.0 to -0.6% (5). Deviations of the Cp° equations (6, 7) from our table are much less than the estimated uncertainties of 3% (6) and 3 to 5% (7). The selected studies (4-7) show reasonable agreement in their regions of overlap even though they involve different grades of graphite. Together, they yield mutually consistent data extending up to 3800 K.

Discrepancies in other Cp data are of concern since no single type of graphite has been accurately measured over the whole temperature range. At all temperatures the many studies show unusually large differences, some of which are due to deviations of the graphite samples from our chosen standard state. For example, the Cp of natural (Ceylon) graphite (3) was found to be approximately 9% lower from 200 to 300 K, yielding values of S°(298.15) and H°(298.15 K) - H°(0 K) which are lower by 0.08 cal K^{-1} mol^{-1} and 21 cal mol^{-1}, respectively. Lutkov et al. (8) reported similar differences in Cp among two graphites and one pyrolytic carbon. Discrepancies are even larger below 30 K (9, 10) and above 3000 K (6, 10-15).

If there were a large bias in our Cp° at 300-1500 K or above 3000 K, it could affect the interpretation of data on vaporization and melting. New studies of various graphites by laser flash (16) and differential scanning calorimetry (17) tend to confirm our Cp° at 350-950 K, where the largest deviations (-3%) are roughly equal to the experimental error. More controversial are Cp (12) and enthalpy (13) studies which gave Cp curves that rise very rapidly near 3500 K. Recent pulse-heating data (7), however, gave a slowly rising Cp° which is linear up to 3800 K. Four Cp° studies (6, 7, 12, 14) and an enthalpy study (15) agree satisfactorily below 3500 K, so the discrepancy must be due to temperature-dependent differences in experimental methods or graphite samples. This is a region where some graphites degrade rapidly with emission of particles, leading to possible bias in Cp data. Moreover, Whittaker has proposed (18) that graphite is metastable in this region and slowly transforms to carbynes. If this is the case, then very rapid measurements (7) should yield Cp° data which are different and more appropriate to graphite. The rapidly rising enthalpy data (13) could be explained by Whittaker's proposal or by differences in the various graphite samples (13).

Sublimation Data

See the tables for C(g), C_2(g), C_3(g), C_4(g), etc. The most abundant species is C_3. Recent analyses gave sublimation temperature values (corresponding to 1 atm total pressure of all carbon species) of 3895-4020 K (19) from carbon arc data, ≤4070 K (20) from mass-loss data at 0.1-11 atm, and approximately 3915 K (21) from extrapolation of mass-spectrometric data at low pressure. Earlier values for this controversial point (see below) were summarized by Palmer(22).

Fusion Data and Phase Data

Controversy surrounds the solid-liquid-vapor triple point which recent studies place either at approximately 4000 K and approximately 1 atm (18, carbyne form) or at 4130±30 K and 120±10 atm (23, pyrolytic graphite). We doubt the latter triple point because the temperature is displaced by 400 to 800 kelvins from the sublimation curves (19-22 and ref. cited therein). The nature of the observed phenomenon (22-24) has been questioned (18) as well as the measurement of temperature (18, 19, 24). Also controversial is Whittaker's proposal (18) that several carbyne forms are more stable than graphite above 2600 K. Evidence includes X-ray data which were interpreted (18) to indicate slow forward transitions, graphite → carbynes, but rapid reverse transitions. These transitions might explain discrepancies observed in various Cp curves but alternative explanations are possible. We retain graphite as the reference state at all temperatures; this arbitrary choice avoids the need to switch phases at the proposed transition point, sublimation point or melting point, all of which are subject to change.

References

References are listed on the C(g) table.

Enthalpy Reference Temperature = T_r = 298.15 K Standard State Pressure = p° = 0.1 MPa

T/K	$C_p°$ (J $K^{-1}mol^{-1}$)	S° (J $K^{-1}mol^{-1}$)	$-[G°-H°(T_r)]/T$ (J $K^{-1}mol^{-1}$)	$H°-H°(T_r)$ (kJ mol^{-1})	$\Delta_f H°$ (kJ mol^{-1})	$\Delta_f G°$ (kJ mol^{-1})	Log K_f
0	0.	0.	INFINITE	-1.051	0.	0.	0.
100	1.674	0.952	10.867	-0.991	0.	0.	0.
200	5.006	3.082	6.407	-0.665	0.	0.	0.
250	6.816	4.394	5.871	-0.369	0.	0.	0.
298.15	8.517	5.740	5.740	0.	0.	0.	0.
300	8.581	5.793	5.741	0.016	0.	0.	0.
350	10.241	7.242	5.851	0.487	0.	0.	0.
400	11.817	8.713	6.117	1.039	0.	0.	0.
450	13.289	10.191	6.487	1.667	0.	0.	0.
500	14.623	11.662	6.932	2.365	0.	0.	0.
600	16.844	14.533	7.961	3.943	0.	0.	0.
700	18.537	17.263	9.097	5.716	0.	0.	0.
800	19.827	19.826	10.279	7.637	0.	0.	0.
900	20.824	22.221	11.475	9.672	0.	0.	0.
1000	21.610	24.457	12.662	11.795	0.	0.	0.
1100	22.244	26.548	13.831	13.989	0.	0.	0.
1200	22.766	28.506	14.973	16.240	0.	0.	0.
1300	23.204	30.346	16.085	18.539	0.	0.	0.
1400	23.578	32.080	17.167	20.879	0.	0.	0.
1500	23.904	33.718	18.216	23.253	0.	0.	0.
1600	24.191	35.270	19.234	25.658	0.	0.	0.
1700	24.448	36.744	20.221	28.090	0.	0.	0.
1800	24.681	38.149	21.178	30.547	0.	0.	0.
1900	24.895	39.489	22.107	33.026	0.	0.	0.
2000	25.094	40.771	23.008	35.525	0.	0.	0.
2100	25.278	42.000	23.883	38.044	0.	0.	0.
2200	25.453	43.180	24.734	40.581	0.	0.	0.
2300	25.618	44.315	25.561	43.134	0.	0.	0.
2400	25.775	45.408	26.365	45.704	0.	0.	0.
2500	25.926	46.464	27.148	48.289	0.	0.	0.
2600	26.071	47.483	27.911	50.889	0.	0.	0.
2700	26.212	48.470	28.654	53.503	0.	0.	0.
2800	26.348	49.426	29.379	56.131	0.	0.	0.
2900	26.481	50.353	30.086	58.773	0.	0.	0.
3000	26.611	51.253	30.777	61.427	0.	0.	0.
3100	26.738	52.127	31.451	64.095	0.	0.	0.
3200	26.863	52.978	32.111	66.775	0.	0.	0.
3300	26.986	53.807	32.756	69.467	0.	0.	0.
3400	27.106	54.614	33.387	72.172	0.	0.	0.
3500	27.225	55.401	34.005	74.889	0.	0.	0.
3600	27.342	56.170	34.610	77.617	0.	0.	0.
3700	27.459	56.921	35.203	80.357	0.	0.	0.
3800	27.574	57.655	35.784	83.109	0.	0.	0.
3900	27.688	58.372	36.354	85.872	0.	0.	0.
4000	27.801	59.075	36.913	88.646	0.	0.	0.
4100	27.913	59.763	37.462	91.432	0.	0.	0.
4200	28.024	60.437	38.001	94.229	0.	0.	0.
4300	28.134	61.097	38.531	97.037	0.	0.	0.
4400	28.245	61.745	39.051	99.856	0.	0.	0.
4500	28.354	62.381	39.562	102.685	0.	0.	0.
4600	28.462	63.006	40.065	105.526	0.	0.	0.
4700	28.570	63.619	40.560	108.378	0.	0.	0.
4800	28.678	64.222	41.047	111.240	0.	0.	0.
4900	28.785	64.814	41.526	114.114	0.	0.	0.
5000	28.893	65.397	41.997	116.997	0.	0.	0.
5100	28.999	65.970	42.462	119.892	0.	0.	0.
5200	29.106	66.534	42.919	122.797	0.	0.	0.
5300	29.211	67.089	43.370	125.713	0.	0.	0.
5400	29.317	67.636	43.814	128.640	0.	0.	0.
5500	29.422	68.175	44.252	131.577	0.	0.	0.
5600	29.528	68.706	44.684	134.524	0.	0.	0.
5700	29.632	69.230	45.110	137.482	0.	0.	0.
5800	29.737	69.746	45.531	140.451	0.	0.	0.
5900	29.842	70.255	45.945	143.429	0.	0.	0.
6000	29.946	70.758	46.355	146.419	0.	0.	0.

PREVIOUS: March 1978 (1 atm) CURRENT: March 1978 (1 bar)

CARBON (C) IDEAL GAS A_r = 12.011

IP(C, g) = 90820.42 ± 0.1 cm^{-1}
S°(298.15 K) = 158.100 ± 0.86 J K^{-1} mol^{-1}

$\Delta_f H°$(0 K) = 711.19 ± 0.46 kJ mol^{-1}
$\Delta_f H°$(298.15 K) = 716.67 ± 0.46 kJ mol^{-1}

Electronic Levels and Quantum Weights

ϵ_i, cm^{-1}	g_i
0.00	1
16.40	3
43.40	5
10192.63	5
21648.01	1
33735.20	5

Enthalpy of Formation

$\Delta_f H°$(298.15 K) = 171.29 ± 0.11 kcal mol^{-1} is the value selected by CODATA (1) based on D_0^0(CO) = 89595 ± 30 cm^{-1} (256.165 ± 0.086 kcal mol^{-1}) from Douglas and Moller (2). The conversion factor used by CODATA (1) corresponds to the negligibly different value of 256.163 kcal mol^{-1}; this yields the adopted $\Delta_f H°$ when combined with $\Delta_f H°$(CO, 0 K) = -27.201 ± 0.024 kcal mol^{-1} (1, 3). According to Krupenie (4) it is most likely that the atomic dissociation products are in their ground states as assumed in the adopted D_0^0(CO).

The enthalpy of sublimation, $\Delta_{sub}H° = \Delta_f H°$, is confirmed by Knudsen-effusion mass-spectrometric data of Wachi and Gilmartin (5) for graphite. They obtained absolute pressures from calibrations using gold and cobalt; pressures from the latter were higher by 8.74%. Our 3rd law analyses yield $\Delta_{sub}H°$(298.15 K) = 171.22 ± 0.61 and 170.78 ± 0.61 kcal mol^{-1} from gold and cobalt, respectively. 2nd law values are $\Delta_{sub}H°$(298.15 K) = 172.5 ± 3.3 kcal mol^{-1} and $\Delta_{sub}S°$(2nd law)- $\Delta_{sub}S°$(3rd law) = 0.5 ± 1.2 (gold) and 0.7 ± 1.2 (cobalt) cal K^{-1} mol^{-1}). The data consisted of 14 points in the range 2509-2753 K. None of the other studies of C(g) over pure graphite included calibrations for conversion of ion intensities to absolute pressures.

Heat Capacity and Entropy

The atomic energy levels are from Moore (6). S°(298.15 K) and H°(0 K) - H°(298.15 K) are the same as the values selected by CODATA (1). Observed and predicted levels above 60333 cm^{-1} are omitted since they do not change the thermodynamic functions at 6000 K. At 14000 K the BETHE-FILL cut-off method would yield a Gibbs energy function ~0.1 cal K^{-1} mol^{-1} larger than either the present cut-off method (TEMPER, 7).

Reference For C(g)

1. ICSU-CODATA Task Group on Key Values for Thermodynamics, J. Chem. Thermodyn. 3, 1 (1971).
2. A. E. Douglas and C. K. Moller, Can. J. Phys. 33, 125 (1955).
3. JANAF Thermochemical Tables: C(ref), 3-31-78; O(g), 3-31-77.
4. P. H. Krupenie, U. S. Natl. Bur. Stand., NSRDS-NBS 5, 1966.
5. F. M. Wachi and D. E. Gilmartin, High Temp. Sci. 4, 423 (1972).
6. C. E. Moore, U. S. Nat. Bur. Stand., NSRDS-NBS 3, Section 3, 1970.
7. J. R. Downey, The Dow Chemical Co., Rept. AFOSR-TR-78-0960, Contract No F44620-75-C-0048, Mar. 9, 1978.

References For C(ref. st., graphite, spectroscopic-grade Acheson)

1. ICSU-CODATA Task Group on Key Values for Thermodynamics, J. Chem. Thermody. 3, 1 (1971). S° and H° - H°(0 K) at 298.15 K.
2. S. K. Das and E. E. Hucke, Carbon 13, 33 (1975). Stability of "glassy" carbon.
3. W. Desorbo and W. W. Tyler, J. Chem. Phys. 21, 1660 (1953); W. DeSorbo, J. Amer. Chem. Soc. 77, 4713 (1955). Cp (13-301 K) of CS-grade Acheson graphite and of natural graphite from Ceylon.
4. R. A. McDonald, J. Chem. Eng. Data 10, 243 (1965); H (341-1723 K) of SPK-grade graphite.
5. E. D. West and S. Ishihara, "Third Symp. Thermophys. Properties," ASME 1965, p. 146-51. H(1191-2576 K) of CCH-grade graphite.
6. A. Cezairliyan and F. Righini, Rev. Int. Hautes Temp. Refract. 12, 124 (1975). Cp(1500-3000 K) of POCO graphite.
7. A. Cezairliyan and J. L. McClure, U. S. Natl. Bur. Stand., personal communication, Nov., 1977; "Seventh Symp. Thermophys. Properties," ASME, 1977, p. 112. Cp (2500-3800 K) of POCO graphite.
8. A. I. Lutkov, V. I. Volga and B. K. Dymov, Carbon 8, 753 (1970). Cp of various graphites.
9. Y. Takahashi and E. F. Westrum, Jr., J. Chem. Thermodyn. 2, 847 (1970). comparison of Cp.
10. Y. S. Touloukian and E. H. Buyco, "Thermophysical Properties of Matter," Vol. 5, IFI/Plenum, New York, 1970, pp. 9-14. Review of Cp.
11. A. T. D. Butland and R. J. Maddison, J. Nucl. Mater. 49, 45 (1973). Review and evaluation of Cp.
12. N. S. Rasor and J. D. McClelland, J. Phys. Chem. Solids 15, 17 (1960); Rev. Sci. Instrum. 31, 595 (1960). Cp(~1400-3900 K) of four types of graphite.
13. A. E. Sheindlin, I. S. Belevich and I. G. Kozhevnikov, High Temperature 10, 897 (1972). H (various ranges up to 3628 K) of 8 brands of graphite.
14. Ya. A. Kraftmakher and V. O. Shestopal, Zh. Prikl. Mekhan. i Tekhn. Fiz., 170 (1965). Cp (1750-2850 K).
15. L. M. Buchnev et al., High Temperature 11, 1072 (1973). H (500-3257 K).
16. J. Yokyama et al., Tanso, No. 65, 44 (1971). Cp (298-970 K) of artificial and natural graphite.
17. R. E. Taylor and H. Groot, Purde Univ., Rept. AFOSR-TR-78-1375, Contract No. AFOSR-77-3280, July, 1978. Cp (350-950 K) of POCO graphite.
18. A. G. Whittaker, Science 200, 111 (1978); Aerospace Corp., Rept. SAMSO-TR-78-121, Contract No. F04701-77-C-0078, Aug., 1978; Ext. Abstr. Program - Bienn. Conf. Carbon 13, 413 (1977). Transition to carbyne and triple point.
19. J. Abrahmson, Carbon 12, 111 (1974); 16, 341 (1978). Sublimation temperature from analysis of arc data.
20. J. H. Lundell and R. R. Dickey, Prog. Astronaut. Aeronaut. 56, 405 (1977). Sublimation temperature from mass-loss data.
21. J. R. Leider, O. H. Krikorian and D. A. Young, Carbon 11, 555 (1973). Sublimation and fusion temperature.
22. H. B. Palmer and M. Shelef, Chem. Phys. Carbon 4, 85 (1968); Carbon 8, 243 (1970). Vaporization review.
23. N. A. Gokcen et al., High Temp. Sci. 8, 81 (1976); Triple point of pyrolytic carbon.
24. D. M. Haaland, Carbon 14, 357 (1976). Triple point and plume effects.

Carbon (C) C_1(g)

Enthalpy Reference Temperature = T_r = 298.15 K Standard State Pressure = p° = 0.1 MPa

T/K	$C_p°$	S°	$-[G°-H°(T_r)]/T$	$H°-H°(T_r)$	$\Delta_f H°$	$\Delta_f G°$	Log K_f
	J $K^{-1}mol^{-1}$			kJ mol^{-1}			
0	0.	0.	INFINITE	-6.536	711.185	711.185	INFINITE
100	21.271	135.180	176.684	-4.150	713.511	700.088	-365.689
200	20.904	149.768	160.007	-2.048	715.287	685.950	-179.152
250	20.861	154.427	158.443	-1.004	716.035	678.527	-141.770
298.15	20.838	158.100	158.100	0.	716.670	671.244	-117.599
300	20.838	158.228	158.100	0.039	716.693	670.962	-116.825
350	20.824	161.439	158.354	1.080	717.263	663.294	-98.991
400	20.815	164.219	158.917	2.121	717.752	655.550	-85.606
450	20.809	166.671	159.645	3.162	718.165	647.749	-75.189
500	20.804	168.863	160.459	4.202	718.507	639.906	-66.851
600	20.799	172.655	162.185	6.282	719.009	624.135	-54.336
700	20.795	175.861	163.916	8.362	719.315	608.296	-45.392
800	20.793	178.638	165.587	10.441	719.474	592.424	-38.681
900	20.792	181.087	167.175	12.520	719.519	576.539	-33.461
1000	20.791	183.278	168.678	14.600	719.475	560.654	-29.286
1100	20.791	185.259	170.097	16.679	719.360	544 777	-25.869
1200	20.793	187.068	171.437	18.758	719.188	528.913	-23.023
1300	20.796	188.733	172.704	20.837	718.968	513.066	-20.615
1400	20.803	190.274	173.905	22.917	718.709	497.237	-18.552
1500	20.814	191.710	175.044	24.998	718.415	481.427	-16.765
1600	20.829	193.053	176.128	27.080	718.092	465.639	-15.202
1700	20.850	194.317	177.162	29.164	717.744	449.871	-13.823
1800	20.878	195.509	178.148	31.250	717.373	434.124	-12.598
1900	20.912	196.639	179.092	33.340	716.984	418.399	-11.503
2000	20.952	197.713	179.996	35.433	716.577	402.694	-10.517
2100	20.999	198.736	180.864	37.530	716.156	387.010	-9.626
2200	21.052	199.714	181.699	39.633	715.722	371.347	-8.817
2300	21.110	200.651	182.503	41.741	715.277	355.703	-8.078
2400	21.174	201.551	183.278	43.855	714.821	340.079	-7.402
2500	21.241	202.417	184.026	45.976	714.357	324.474	-6.780
2600	21.313	203.251	184.750	48.103	713.884	308.888	-6.206
2700	21.387	204.057	185.450	50.238	713.405	293.321	-5.675
2800	21.464	204.836	186.129	52.381	712.920	277.771	-5.182
2900	21.542	205.591	186.787	54.531	712.429	262.239	-4.723
3000	21.621	206.322	187.426	56.689	711.932	246.723	-4.296
3100	21.701	207.032	188.047	58.856	711.431	231.224	-3.896
3200	21.780	207.723	188.651	61.030	710.925	215.742	-3.522
3300	21.859	208.394	189.239	63.212	710.414	200.275	-3.170
3400	21.936	209.048	189.812	65.401	709.899	184.824	-2.839
3500	22.012	209.685	190.371	67.599	709.380	169.389	-2.528
3600	22.087	210.306	190.916	69.804	708.857	153.968	-2.234
3700	22.159	210.912	191.448	72.016	708.329	138.561	-1.956
3800	22.230	211.504	191.968	74.235	707.797	123.169	-1.693
3900	22.298	212.082	192.477	76.462	707.260	107.791	-1.444
4000	22.363	212.648	192.974	78.695	706.719	92.427	-1.207
4100	22.426	213.201	193.461	80.934	706.173	77.077	-0.982
4200	22.487	213.742	193.937	83.180	705.621	61.740	-0.768
4300	22.544	214.272	194.404	85.432	705.065	46.416	-0.564
4400	22.600	214.791	194.861	87.689	704.503	31.105	-0.369
4500	22.652	215.299	195.310	89.951	703.936	15.807	-0.183
4600	22.702	215.797	195.750	92.219	703.363	0.521	-0.006
4700	22.750	216.286	196.182	94.492	702.784	-14.752	0.164
4800	22.795	216.766	196.605	96.769	702.199	-30.012	0.327
4900	22.838	217.236	197.022	99.051	701.607	-45.261	0.482
5000	22.878	217.698	197.431	101.337	701.009	-60.497	0.632
5100	22.917	218.151	197.832	103.626	700.404	-75.721	0.776
5200	22.953	218.597	198.227	105.920	699.793	-90.933	0.913
5300	22.987	219.034	198.616	108.217	699.174	-106.133	1.046
5400	23.020	219.464	198.998	110.517	698.548	-121.322	1.174
5500	23.051	219.887	199.374	112.821	697.914	-136.499	1.296
5600	23.080	220.302	199.744	115.127	697.273	-151 664	1.415
5700	23.107	220.711	200.108	117.437	696.625	-166.818	1.529
5800	23.133	221.113	200.467	119.749	695.968	-181.961	1.639
5900	23.157	221.509	200.820	122.063	695.304	-197.092	1.745
6000	23.181	221.898	201.168	124.380	694.631	-212.211	1.847

PREVIOUS: March 1978 (1 atm) CURRENT: March 1983 (1 bar)

Carbon (C) C_1(g)

CARBON, ION (C^+) IDEAL GAS Mr = 12.01045

$IP(C^+, g) = 196664.7 \pm 0.5\ cm^{-1}$ $\Delta_f H°(0\ K) = 1797.639 \pm 0.46\ kJ\ mol^{-1}$

$S°(298.15\ K) = 154.662 \pm 0.02\ JK^{-1}\ mol^{-1}$ $\Delta_f H°(298.15\ K) = [1809.435]\ kJ\ mol^{-1}$

Electronic Levels and Quantum Weights

State	ε_i, cm^{-1}	g_i
$^2P_{1/2}$	0	2
$^2P_{3/2}$	63.42	4
$^4P_{1/2}$	43003.3	2
$^4P_{3/2}$	43025.3	4
$^4P_{5/2}$	43053.6	6

Enthalpy of Formation

$\Delta_f H°(C^+, g, 0\ K)$ is calculated from $\Delta_f H°(C, g, 0\ K)$ (1) using the spectroscopic value of IP(C) = 90820.42 ± 0.1 cm^{-1} (1086.454 ± 0.001 kJ mol^{-1}) from Moore (2). The ionization limit is converted from cm^{-1} to kJ mol^{-1} using the factor, 1 cm^{-1} = 0.01196266 kJ cm^{-1}, which is derived from the 1973 CODATA fundamental constants (3). Rosenstock et al. (4) and Levin and Lias (5) have summarized additional ionization and appearance potential data.

$\Delta_f H°(C^+, g, 298.15\ K)$ is calculated from $\Delta_f H°(C, g, 0\ K)$ by using IP(C) with JANAF (1) enthalpies, H°(0 K) − H°(298.15 K), for C(g), $C^+(g)$, and $e^-(g)$. $\Delta_f H°(C \rightarrow C^+ + e^-, 298.15\ K)$ differs from a room temperature threshold energy due to inclusion of these enthalpies and to threshold effects discussed by Rosenstock et al. (4). $\Delta_f H°(298.15\ K)$ should be changed by −6.197 kJ mol^{-1} if it is to be used in the ion convention that excludes the enthalpy of the electron.

Heat Capacity and Entropy

The information on electronic energy levels and quantum weights, given by Moore (2, 6) is incomplete because many theoritically predicted levels that any reasonable method of filling in these missing levels and cutting off the summation in the partition function (6) has no effect on the thermodynamic functions to 6000 K. This is a result of the high energy of all levels other than the ground state and the first four excited states; the next excited state is 74930.1 cm^{-1} above the ground state. Since inclusion of these excited states has no effect on the thermodynamic functions (to 6000 K), we list only the ground state and the first four excited states. The reported uncertainty in S°(298.15 K) is due to uncertainties in the relative ionic mass and the fundamental constants. Extension of these calculations above 6000 K may require consideration of the excited states and use of different fill and cutoff procedures (6).

References

1. JANAF Thermochemical Tables: C (g), 3-31-83; e^-(g), 3-31-82.
2. C. E. Moore, U. S. Nat. Bur. Stand. NSRDS-NBS-3, Section 3, 1970.
3. E. R. Cohen and B. N. Taylor, J. Phys. Chem. Ref. Data 2, 663 (1973).
4. H. M. Rosenstock, K. Draxl et al., J. Phys. Chem. Ref. Data 6, Supp. 1 (1977).
5. R. D. Levin and S. G. Lias, U. S. Nat. Bur. Stand., NSRDS-NBS-71, 1982.
6. J. R. Downey, Jr., The Dow Chemical Company, AFOSR-TR-0960, Contract No. F44620-75-1-0048, 1978.

Carbon, Ion (C^+) $C_1^+(g)$

Enthalpy Reference Temperature = T_r = 298.15 K Standard State Pressure = p° = 0.1 MPa

T/K	$C_p°$ (J K⁻¹mol⁻¹)	S° (J K⁻¹mol⁻¹)	−[G°−H°(T_r)]/T (J K⁻¹mol⁻¹)	H°−H°(T_r) (kJ mol⁻¹)	$\Delta_f H°$ (kJ mol⁻¹)	$\Delta_f G°$ (kJ mol⁻¹)	Log K_f
0	0.	0.	INFINITE	-6.649	1797.639		
100	22.496	131.192	173.519	-4.233			
200	21.213	146.246	156.586	-2.068			
250	21.056	150.961	155.008	-1.012			
298.15	20.974	154.662	154.662	0.	1809.435	1758.779	-308.131
300	20.971	154.792	154.662	0.039	1809.496	1758.464	-306.176
350	20.921	158.020	154.918	1.086	1811.112	1749.830	-261.148
400	20.889	160.812	155.484	2.131	1812.644	1740.970	-227.347
450	20.867	163.271	156.215	3.175	1814.099	1731.923	-201.036
500	20.851	165.468	157.032	4.218	1815.483	1722.717	-179.971
600	20.831	169.268	158.765	6.302	1818.067	1703.917	-148.339
700	20.819	172.478	160.500	8.384	1820.455	1684.701	-125.714
800	20.811	175.257	162.175	10.466	1822.695	1665.154	-108.723
900	20.806	177.708	163.768	12.547	1824.819	1645.332	-95.493
1000	20.802	179.900	165.273	14.627	1826.855	1625.279	-84.896
1100	20.799	181.883	166.694	16.707	1828.820	1605.026	-76.216
1200	20.797	183.692	168.037	18.787	1830 727	1584.597	-68.976
1300	20.795	185.357	169.306	20.866	1832.586	1564.010	-62.843
1400	20.794	186.898	170.508	22.946	1834.405	1543.282	-57.581
1500	20.793	188.333	171.649	25.025	1836.188	1522.425	-53.016
1600	20.792	189.675	172.734	27.105	1837.941	1501.450	-49.017
1700	20.791	190.935	173.768	29.184	1839.667	1480.366	-45.486
1800	20.791	192.123	174.755	31.263	1841.368	1459.182	-42.344
1900	20.790	193.247	175.699	33.342	1843.047	1437.903	-39.531
2000	20.790	194.314	176.603	35.421	1844.705	1416.537	-36.996
2100	20.790	195.328	177.471	37.500	1846.344	1395.088	-34.701
2200	20.789	196.295	178.305	39.579	1847.964	1373.562	-32.613
2300	20.789	197.219	179.107	41.658	1849 568	1351.962	-30.704
2400	20.789	198.104	179.881	43.737	1851.156	1330.293	-28.953
2500	20.789	198.953	180.627	45.815	1852.729	1308.558	-27.341
2600	20.788	199.768	181.347	47.894	1854.286	1286.761	-25.851
2700	20.788	200.553	182.044	49.973	1855.829	1264.903	-24.471
2800	20.788	201.309	182.719	52.052	1857.359	1242.989	-23.188
2900	20.788	202.038	183.372	54.131	1858.875	1221.021	-21.993
3000	20.788	202.743	184.006	56.209	1860.377	1199.000	-20.876
3100	20.788	203.425	184.622	58.288	1861.867	1176.929	-19.831
3200	20.788	204.085	185.220	60.367	1863.345	1154.811	-18.850
3300	20.787	204.724	185.801	62.446	1864.809	1132.647	-17.928
3400	20.787	205.345	186.367	64.525	1866.262	1110.438	-17.060
3500	20.787	205.947	186.918	66.603	1867.703	1088.187	-16.240
3600	20.787	206.533	187.455	68.682	1869.132	1065.895	-15.466
3700	20.787	207.103	187.978	70.761	1870.549	1043.563	-14.732
3800	20.787	207.657	188.489	72.839	1871.955	1021.193	-14.037
3900	20.788	208.197	188.987	74.918	1873.349	998.787	-13.377
4000	20.788	208.723	189.474	76.997	1874.732	976.344	-12.750
4100	20.788	209.236	189.950	79.076	1876.104	953.868	-12.152
4200	20.788	209.737	190.415	81 155	1877.464	931.358	-11.583
4300	20.789	210.227	190.870	83.233	1878.814	908.815	-11.040
4400	20.789	210.705	191.315	85.312	1880.153	886.242	-10.521
4500	20.790	211.172	191.751	87.391	1881.480	863.638	-10.025
4600	20.791	211.629	192.179	89.470	1882.797	841.005	-9.550
4700	20.792	212.076	192.597	91.550	1884.103	818.343	-9.095
4800	20.794	212.514	193.008	93 629	1885.399	795 653	-8.658
4900	20.795	212.942	193.410	95.708	1886.683	772.937	-8.240
5000	20.797	213.362	193.805	97.788	1887.958	750.195	-7.837
5100	20.800	213.774	194.192	99.868	1889.222	727.427	-7.450
5200	20.803	214.178	194.573	101.948	1890.475	704.634	-7.078
5300	20.806	214.575	194.947	104.028	1891.718	681.818	-6.720
5400	20.810	214.964	195.314	106.109	1892 951	658.978	-6.374
5500	20.814	215.345	195.674	108.190	1894.174	636.115	-6.041
5600	20.819	215.720	196.029	110.272	1895.387	613.230	-5.720
5700	20.824	216.089	196.378	112.354	1896.590	590.324	-5.410
5800	20.831	216.451	196.721	114.437	1897.782	567.397	-5.110
5900	20.838	216.807	197.058	116.520	1898.965	544.449	-4.820
6000	20.845	217.158	197.390	118.604	1900.139	521.481	-4.540

PREVIOUS: March 1978 (1 atm) CURRENT: March 1983 (1 bar)

Carbon, Ion (C^+) $C_1^+(g)$

CARBON, ION (C^-) IDEAL GAS M_r = 12.01155

EA(C, g) = 1.2629 ± 0.0003 eV

S°(298.15 K) = 151.384 ± 0.003 J K^{-1} mol^{-1}

$\Delta_f H°$(0 K) = 589.33 ± 2 kJ mol^{-1}

$\Delta_f H°$(298.15 K) = [588.284] kJ mol^{-1}

Electonic Levels and Quantum Weights

State	ε_i, cm^{-1}	g_i
$^4S_{3/2}$	0.0	4
^2D	9920	10

Enthalpy of Formation

$\Delta_f H°(C^-$, g, 0 K) is calculated from $\Delta_f H°$(C, g 0 K) (1) using the adopted electron affinity of EA(C) = 1.2629 + 0.0003 eV (121.850 ± 0.029 kJ mol^{-1}). This value, recommended by Hotop and Lineberger (2), is based on a laser photodetachment electron spectrometry study (3). Additional information on C^-(g) may be obtained in the critical discussions of Hotop and Lineberger (2, 4), Rosenstock et al. (5) and Massey (6).

$\Delta_f H°(C^-$, g, 298.15 K) is obtained from $\Delta_f H°$(C, g, 0 K) by using EA(C) with JANAF (1) enthalpies, H°(0 K) - H°(298.15 K), for C^-(g), C(g), and e^-(g). $\Delta_r H°(C^- \rightarrow C + e^-$, 298.15 K) differs from a room-temperature threshold energy due to inclusion of these enthalpies and to threshold effects discussed by Rosenstock et al. (5). $\Delta_f H°$(298.15 K) should be changed by +6.197 kJ mol^{-1} if it is to be used in the ion convention that excludes the enthalpy of the electron.

Heat Capacity and Entropy

Electronic energy levels are from the laser photodetachment study of Feldmann (3). He found that the $C^-(^2D)$ state has a binding energy of 0.033 ± 0.001 eV (266 ± 8 cm^{-1}) relative to ground state $C(^3P_0)$. The electronic energy of 9920 cm^{-1} is the difference in binding energy between ground and excited states. Spin-orbit splitting of the ^2D state was not resolved (3) but has been estimated (2, 4) to be 3 cm^{-1}. We neglect this splitting because its effect is negligible. $C^-(^2D)$ binding energies obtained by other mehtods (4, 7) are in satisfactory agreement.

Note that we include the excited state even though its binding energy is only 266 cm^{-1}. If this state were omitted, the Gibbs energy function (J K^{-1} mol^{-1}) would change by -0.004 (1700 K), -0.176 (3000 K) and -1.732 (6000 K).

References

1. JANAF Thermochemical Tables: C(g), 3-31-83; e^-(g), 3-31-82.
2. H. Hotop and W. kC. Lineberger, J. Phys. Chem. Ref. Data, to be published, 1985.
3. D. Feldmann, Chem. Phys. Lett. 47, 338 (1977).
4. H. Hotop and W. C. Lineberger, J. Phys. Chem. Ref. Data 4, 539 (1975).
5. H. M. Rosenstock, K. Draxl et al. J. Phys. Chem. Ref. Data 6, Supp. 1 (1977).
6. H. S. W. Massey, "Negative Ions", 3rd ed., Cambridge University Press, Cambridge, 1976.
7. V. A. Aparin et al., Soviet Phys. - JETP 39, 989 (1974).

Carbon, Ion (C^-) C_1^-(g)

Enthalpy Reference Temperature = T_r = 298.15 K Standard State Pressure = p° = 0.1 MPa

	J $K^{-1}mol^{-1}$			kJ mol^{-1}			
T/K	$C_p°$	S°	-[G°-H°(T_r)]/T	H°-H°(T_r)	$\Delta_f H°$	$\Delta_f G°$	Log K_f
0	0.	0.	INFINITE	-6.197	589.335		
100	20.786	128.676	169.864	-4.119			
200	20.786	143.084	153.285	-2.040			
250	20.786	147.722	151.726	-1.001			
298.15	20.786	151.384	151.384	0.	588.284	551.115	-96.553
300	20.786	151.512	151.384	0.038	588.268	550.885	-95.918
350	20.786	154.716	151.637	1.078	587.797	544.690	-81.291
400	20.786	157.492	152.199	2.117	587.245	538.569	-70.330
450	20.786	159.940	152.926	3.156	586.617	532.521	-61.813
500	20.786	162.130	153.739	4.196	585.919	526.547	-55.008
600	20.786	165.920	155.463	6.274	584.340	514.817	-44.819
700	20.786	169.124	157.191	8.353	582.568	503.368	-37.562
800	20.786	171.900	158.860	10.431	580.647	492.183	-32.136
900	20.787	174.348	160.448	12.510	578.612	481.247	-27.931
1000	20.789	176.538	161.949	14.589	576.489	470.541	-24.579
1100	20.794	178.520	163.367	16.668	574.296	460.052	-21.846
1200	20.806	180.330	164.706	18.748	572.046	449.765	-19.578
1300	20.829	181.996	165.973	20.830	569.750	439 668	-17.666
1400	20.867	183.541	167.173	22.914	567.416	429.749	-16.034
1500	20.925	184.982	168.313	25.004	565.053	419.998	-14.626
1600	21.007	186.335	169.398	27.100	562.665	410.405	-13.398
1700	21.117	187.612	170.432	29.206	560.261	400.963	-12.320
1800	21.256	188.822	171.420	31.324	557.844	391.662	-11.366
1900	21.425	189.976	172.367	33.458	555.420	382.496	-10.516
2000	21.625	191.080	173.275	35.610	552.994	373.457	-9.754
2100	21.853	192.141	174.148	37.784	550.570	364.540	-9.067
2200	22.108	193.163	174.989	39.982	548.153	355.738	-8.446
2300	22.385	194.152	175.801	42.206	545.745	347.046	-7.882
2400	22.683	195.111	176.586	44.459	543.350	338.458	-7.366
2500	22.996	196.043	177.346	46.743	540.970	329.970	-6.894
2600	23.320	196.951	178.082	49.059	538.608	321.577	-6.461
2700	23.653	197.837	178.798	51.408	536.263	313.274	-6.061
2800	23.989	198.704	179.493	53.790	533.939	305.058	-5.691
2900	24.325	199.551	180.170	56.205	531.634	296.924	-5.348
3000	24.658	200.382	180.830	58.655	529.350	288.870	-5.030
3100	24.984	201.195	181.474	61.137	527.086	280.891	-4.733
3200	25.301	201.994	182.103	63.651	524.842	272.985	-4 456
3300	25.607	202.777	182.717	66.197	522.616	265.149	-4.197
3400	25.899	203.546	183.319	68.772	520.409	257.380	-3.954
3500	26.176	204.301	183.908	71.376	518.217	249.676	-3.726
3600	26.437	205.042	184.484	74.007	516.041	242.034	-3.512
3700	26.680	205.769	185.050	76.663	513.879	234.453	-3.310
3800	26.905	206.484	185.604	79.342	511.728	226.929	-3.119
3900	27.112	207.186	186.149	82.043	509.587	219.463	-2.939
4000	27.300	207.874	186.683	84.764	507.455	212.051	-2.769
4100	27.469	208.551	187.209	87.502	505.329	204.692	-2.608
4200	27.621	209.214	187.725	90.257	503.208	197.385	-2.455
4300	27.754	209.866	188.232	93.026	501.091	190.128	-2.310
4400	27.870	210.505	188.731	95.807	498.974	182.921	-2.172
4500	27.969	211.133	189.222	98.599	496.858	175.762	-2.040
4600	28.052	211.748	189.705	101.401	494.740	168.650	-1.915
4700	28.120	212.352	190.180	104.209	492.618	161.584	-1.796
4800	28.173	212.945	190.648	107.024	490.492	154.563	-1.682
4900	28.212	213.526	191.109	109.843	488.359	147.587	-1.573
5000	28.239	214.097	191.563	112.666	486.220	140.654	-1.469
5100	28.253	214.656	192.011	115.491	484.071	133.764	-1.370
5200	28.256	215.205	192.452	118.316	481.913	126.916	-1.275
5300	28.249	215.743	192.886	121.142	479.744	120.110	-1.184
5400	28.232	216.271	193.314	123.966	477.563	113.345	-1.096
5500	28.207	216.789	193.736	126.788	475.369	106.620	-1.013
5600	28.173	217.297	194.152	129.607	473.162	99.935	-0.932
5700	28.132	217.795	194.563	132.422	470.941	93.290	-0.855
5800	28.084	218.284	194.968	135.233	468.705	86.684	-0.781
5900	28.030	218.763	195.367	138.039	466.453	80.117	-0.709
6000	27.971	219.234	195.761	140.839	464.185	73.588	-0.641

PREVIOUS: March 1978 (1 atm) CURRENT: March 1983 (1 bar)

Carbon, Ion (C^-) C_1^-(g)

$\Delta_f H°(0\ K) = [686.2 \pm 10]\ kJ\ mol^{-1}$

$S°(298.15\ K) = [223.4 \pm 4]\ J\ K^{-1}\ mol^{-1}$ $\Delta_f H°(298.15\ K) = [689.5 \pm 10]\ kJ\ mol^{-1}$

Electronic Level and Quantum Weight

State	ε_i, cm^{-1}	g_i
$^2\Pi$	0	4

$\omega_e = [760]\ cm^{-1}$ $\omega_e x_e = [5.0]\ cm^{-1}$ $\sigma = 1$

$B_e = [0.650]\ cm^{-1}$ $\alpha_e = [0.006]\ cm^{-1}$ $r_e = [1.77]$ Å

Enthalpy of Formation

$\Delta_f H°(0\ K)$ was calculated from D^0_0 = 83.0 kcal mol^{-1}, estimated by Gordon (1). The spectrum originally attributed to AlC(g) (2) was later shown to be due to Al_2(g) (3, 4).

Heat Capacity and Entropy

Molecular constants were estimated by Gordon (1). The recent review by Huber and Herzberg (5) did not indicate any experimental data available for the vibrational-rotational constants for AlC(g).

References

1. J. S. Gordon, Astrosystems, International, Livingston, N.J., personal communication, 1963.
2. P. Zeeman, Can. J. Phys. 32, 9 (1954).
3. D. S. Ginter, M. L. Ginter, and K. K. Innes, Astrophys. J. 139, 365 (1964).
4. H. Wooley, U. S. Nat. Bur. Stand. Report 6928, 1960.
5. K. P. Huber and G. Herzberg, "Constants of Diatomic Molecules", Van Nostrand Reinhold Company, New York, 1979.

Enthalpy Reference Temperature = T_r = 298.15 K — Standard State Pressure = p° = 0.1 MPa

	J K⁻¹mol⁻¹			kJ mol⁻¹			
T/K	$C°_p$	S°	$-[G°-H°(T_r)]/T$	$H°-H°(T_r)$	$\Delta_f H°$	$\Delta_f G°$	Log K_f
0	0.	0.	INFINITE	-8.928	686.185	686.185	INFINITE
100	29.131	190.573	250.776	-6.020	688.550	670.287	-350.122
200	30.241	210.997	226.335	-3.067	689.375	651.620	-170.186
250	31.276	217.855	223.975	-1.530	689.503	642.164	-134.173
298.15	32.256	223.449	223.449	0.	689.523	633.044	-110.907
300	32.292	223.649	223.450	0.060	689.522	632.693	-110.162
350	33.178	228.695	223.846	1.697	689.454	623.226	-93.011
400	33.913	233.175	224.738	3.375	689.307	613.774	-80.151
450	34.513	237.205	225.903	5.086	689.086	604.345	-70.151
500	35.001	240.868	227.219	6.824	688.797	594.944	-62.153
600	35.729	247.318	230.046	10.363	688.022	576.242	-50.166
700	36.233	252.866	232.919	13.963	687.001	557.689	-41.615
800	36.598	257.729	235.722	17.605	685.742	539.299	-35.213
900	36.872	262.056	238.412	21.279	684.240	521.082	-30.243
1000	37.087	265.952	240.975	24.978	671.900	503.805	-26.316
1100	37.260	269.496	243.409	28.695	670.249	487.075	-23.129
1200	37.405	272.744	245.720	32.429	668.556	470.497	-20.480
1300	37.528	275.743	247.916	36.176	666.829	454.062	-18.244
1400	37.636	278.528	250.004	39.934	665.072	437.761	-16.333
1500	37.733	281.128	251.993	43.703	663.291	421.586	-14.681
1600	37.820	283.566	253.891	47.480	661.489	405.531	-13.239
1700	37.900	285.862	255.705	51.266	659.668	389.589	-11.971
1800	37.975	288.030	257.441	55.060	657.830	373.755	-10.846
1900	38.045	290.085	259.106	58.861	655.977	358.024	-9.843
2000	38.112	292.038	260.704	62.669	654.110	342.391	-8.942
2100	38.176	293.899	262.241	66.484	652.231	326.851	-8.130
2200	38.237	295.677	263.720	70.304	650.340	311.401	-7.394
2300	38.296	297.378	265.147	74.131	648.438	296.037	-6.723
2400	38.354	299.009	266.524	77.963	646.526	280.756	-6.110
2500	38.410	300.576	267.855	81.802	644.604	265.555	-5.548
2600	38.465	302.083	269.143	85.645	642.672	250.431	-5.031
2700	38.519	303.536	270.390	89.495	640.732	235.382	-4.554
2800	38.571	304.938	271.599	93.349	344.883	221.372	-4.130
2900	38.623	306.292	272.772	97.209	344.021	216.976	-3.908
3000	38.675	307.602	273.911	101.074	343.152	212.610	-3.702
3100	38.726	308.871	275.018	104.944	342.275	208.273	-3.509
3200	38.776	310.102	276.096	108.819	341.390	203.965	-3.329
3300	38.826	311.296	277.144	112.699	340.497	199.684	-3.161
3400	38.875	312.455	278.166	116.584	339.596	195.431	-3.002
3500	38.924	313.583	279.162	120.474	338.688	191.204	-2.854
3600	38.973	314.680	280.133	124.369	337.771	187.003	-2.713
3700	39.021	315.749	281.081	128.269	336.847	182.827	-2.581
3800	39.069	316.790	282.008	132.173	335.915	178.677	-2.456
3900	39.117	317.805	282.912	136.082	334.974	174.552	-2.338
4000	39.165	318.796	283.797	139.997	334.024	170.450	-2.226
4100	39.213	319.764	284.663	143.915	333.066	166.373	-2.120
4200	39.260	320.710	285.510	147.839	332.098	162.319	-2.019
4300	39.307	321.634	286.339	151.767	331.120	158.288	-1.923
4400	39.354	322.538	287.152	155.700	330.132	154.280	-1.832
4500	39.401	323.423	287.948	159.638	329.135	150.295	-1.745
4600	39.448	324.290	288.729	163.581	328.125	146.331	-1.662
4700	39.495	325.138	289.494	167.528	327.104	142.390	-1.582
4800	39.541	325.970	290.245	171.480	326.069	138.471	-1.507
4900	39.588	326.786	290.983	175.436	325.022	134.574	-1.435
5000	39.634	327.586	291.707	179.397	323.959	130.698	-1.365
5100	39.680	328.372	292.418	183.363	322.882	126.843	-1.299
5200	39.727	329.143	293.117	187.333	321.788	123.010	-1.236
5300	39.773	329.900	293.804	191.308	320.677	119.198	-1.175
5400	39.819	330.644	294.479	195.288	319.549	115.407	-1.116
5500	39.865	331.375	295.144	199.272	318.401	111.637	-1.060
5600	39.911	332.094	295.797	203.261	317.233	107.888	-1.006
5700	39.957	332.800	296.440	207.254	316.043	104.160	-0.955
5800	40.003	333.496	297.073	211.252	314.834	100.453	-0.905
5900	40.049	334.180	297.696	215.255	313.599	96.768	-0.857
6000	40.095	334.853	298.310	219.262	312.362	93.101	-0.811

PREVIOUS: June 1963 (1 atm) CURRENT: June 1963 (1 bar)

BORON CARBIDE (BC) IDEAL GAS M_r = 22.821

$S°(298.15\ K) = [208.56]\ J\ K^{-1}\ mol^{-1}$

$\Delta_f H°(0\ K) = 822.0 \pm 41\ kJ\ mol^{-1}$

$\Delta_f H°(298.15\ K) = 828.4 \pm 41\ kJ\ mol^{-1}$

Electronic Level and Quantum Weight

State	ε_i, cm^{-1}	g_i
$^4\Sigma^-$	0	4

$\omega_e = [1350]\ cm^{-1}$ $\omega_e x_e = [12.0]\ cm^{-1}$ $\sigma = 1$

$B_e = [1.555]\ cm^{-1}$ $\alpha_e = [0.018]\ cm^{-1}$

Enthalpy of Formation

The enthalpy of formation at 298.15 K was calculated from $\Delta_f H°(0\ K) = 105 \pm 10\ kcal\ mol^{-1}$ for the reaction BC(g) = B(g) + C(g) reported by Verhaegen et al. (1, 2).

Heat Capacity and Entropy

All spectroscopic constants were obtained from Gordon (3). Theoretical calculations by Kruba (4) predict a $^4\Sigma^-$ ground state. The recent review by Huber and Herzberg (5) did not indicate any experimental data available on the vibrational-rotational constants for BC(g).

References

1. D. G. Verhaegen, F. E. Stafford, M. Ackerman, and J. Drowart, Nature 193, 1280 (1962).
2. D. G. Verhaegen, F. E. Stafford and J. Drowart, J. Chem. Phys. 40, 1622 (1964).
3. J. S. Gordon, Astrosystems, International, West Caldwell, New Jersey, personal communication, November 2, 1962.
4. J. E. Kouba and V. Ohrn, J. Chem. Phys. 53, 3923 (1970).
5. K. P. Huber and G. Herzberg, "Constants of Diatomic Molecules", Van Nostrand Reinhold Company, New York, 1979.

Boron Carbide (BC) $C_1B_1(g)$

Enthalpy Reference Temperature = T_r = 298.15 K Standard State Pressure = p° = 0.1 MPa

T/K	$C_p°$ (J K⁻¹mol⁻¹)	S° (J K⁻¹mol⁻¹)	$-[G°-H°(T_r)]/T$ (J K⁻¹mol⁻¹)	$H°-H°(T_r)$ (kJ mol⁻¹)	$\Delta_f H°$ (kJ mol⁻¹)	$\Delta_f G°$ (kJ mol⁻¹)	Log K_f
0	0.	0.	INFINITE	-8.700	821.997	821.997	INFINITE
100	29.109	176.645	234.607	-5.796	824.817	807.279	-421.679
200	29.172	196.831	211.250	-2.884	827.070	788.804	-206.015
250	29.358	203.357	209.042	-1.421	827.866	779.141	-162.793
298.15	29.700	208.555	208.555	0.	828.432	769.702	-134.849
300	29.716	208.739	208.556	0.055	828.450	769.338	-133.954
350	30.212	213.355	208.919	1.553	828.850	759.452	-113.342
400	30.788	217.427	209.733	3.077	829.087	749.520	-97.877
450	31.388	221.088	210.795	4.632	829.186	739.567	-85.847
500	31.975	224.425	211.993	6.216	829.168	729.609	-76.222
600	33.032	230.351	214.571	9.468	828.861	709.720	-61.787
700	33.898	235.510	217.202	12.816	828.281	689.906	-51.481
800	34.589	240.084	219.782	16.242	827.503	670.190	-43.759
900	35.138	244.191	222.270	19.729	826.574	650.580	-37.759
1000	35.578	247.917	224.651	23.266	825.525	631.080	-32.964
1100	35.934	251.325	226.923	26.842	824.378	611.691	-29.047
1200	36.227	254.464	229.089	30.450	823.146	592.409	-25.787
1300	36.471	257.374	231.154	34.086	821.841	573.234	-23.033
1400	36.677	260.085	233.125	37.743	820.469	554.161	-20.676
1500	36.854	262.621	235.008	41.420	819.037	535.189	-18.637
1600	37.007	265.005	236.809	45.113	817.549	516.314	-16.856
1700	37.142	267.252	238.534	48.821	816.008	497.533	-15.287
1800	37.262	269.379	240.189	52.541	814.417	478.845	-13.896
1900	37.369	271.396	241.779	56.273	812.778	460.247	-12.653
2000	37.466	273.316	243.308	60.015	811.093	441.736	-11.537
2100	37.555	275.146	244.781	63.766	809.363	423.310	-10.529
2200	37.637	276.895	246.201	67.526	807.589	404.968	-9.615
2300	37.713	278.570	247.573	71.293	805.772	386.708	-8.782
2400	37.784	280.176	248.898	75.068	753.654	369.596	-8.044
2500	37.851	281.720	250.180	78.850	751.676	353.634	-7.389
2600	37.914	283.206	251.422	82.638	749.689	337.752	-6.786
2700	37.974	284.638	252.626	86.432	747.694	321.946	-6.228
2800	38.032	286.020	253.794	90.233	745.692	306.214	-5.712
2900	38.087	287.355	254.928	94.039	743.681	290.554	-5.233
3000	38.140	288.647	256.031	97.850	741.663	274.963	-4.788
3100	38.191	289.899	257.103	101.667	739.637	259.440	-4.372
3200	38.241	291.112	258.147	105.488	737.603	243.982	-3.983
3300	38.289	292.290	259.164	109.315	735.562	228.588	-3.618
3400	38.336	293.433	260.155	113.146	733.514	213.256	-3.276
3500	38.382	294.545	261.122	116.982	731.458	197.984	-2.955
3600	38.427	295.627	262.066	120.822	729.395	182.771	-2.652
3700	38.471	296.681	262.987	124.667	727.325	167.616	-2.366
3800	38.514	297.707	263.887	128.516	725.248	152.516	-2.096
3900	38.557	298.708	264.767	132.370	723.163	137.471	-1.841
4000	38.598	299.685	265.628	136.228	721.072	122.480	-1.599
4100	38.640	300.639	266.470	140.090	718.973	107.542	-1.370
4200	38.681	301.570	267.295	143.956	237.018	99.678	-1.240
4300	38.721	302.481	268.103	147.826	235.992	96.420	-1.171
4400	38.761	303.371	268.894	151.700	234.958	93.1.6	-1.106
4500	38.800	304.243	269.670	155.578	233.914	89.976	-1.044
4600	38.839	305.096	270.431	159.460	232.861	86.789	-0.986
4700	38.878	305.932	271.177	163.346	231.799	83.625	-0.929
4800	38.916	306.751	271.910	167.235	230.727	80.484	-0.876
4900	38.954	307.554	272.629	171.129	229.645	77.365	-0.825
5000	38.992	308.341	273.336	175.026	228.552	74.268	-0.776
5100	39.030	309.113	274.030	178.927	227.449	71.193	-0.729
5200	39.067	309.872	274.712	182.832	226.335	68.140	-0.684
5300	39.104	310.616	275.382	186.741	225.209	65.109	-0.642
5400	39.141	311.347	276.041	190.653	224.072	62.099	-0.601
5500	39.178	312.066	276.690	194.569	222.923	59.110	-0.561
5600	39.215	312.772	277.328	198.489	221.761	56.142	-0.524
5700	39.251	313.467	277.956	202.412	220.587	53.195	-0.487
5800	39.287	314.150	278.574	206.339	219.400	50.268	-0.453
5900	39.324	314.822	279.183	210.269	218.199	47.363	-0.419
6000	39.360	315.483	279.782	214.204	216.985	44.477	-0.387

PREVIOUS: June 1963 (1 atm) CURRENT: June 1963 (1 bar)

Boron Carbide (BC) $C_1B_1(g)$

BORON CARBIDE (B_4C) CRYSTAL M_r = 55.251

$S°(298.15\ K) = 27.175 \pm 0.13\ J\ K^{-1}\ mol^{-1}$

$T_{fus} = 2743 \pm 20\ K$

$\Delta_f H°(0\ K) = -62.40 \pm 6.28\ kJ\ mol^{-1}$

$\Delta_f H°(298.15\ K) = -62.68 \pm 6.28\ kJ\ mol^{-1}$

$\Delta_{fus} H° = [104.6 \pm 10]\ kJ\ mol^{-1}$

Enthalpy of Formation

Using a high temperature solution calorimetric technique, Hong and Kleppa (1) measured $\Delta_f H°(B_4C$, cr, 1320 K) = -15.98±0.84 kcal mol^{-1}. We adopt this value and correct to 298.15 K using tabulated enthalpy data (2) to obtain $\Delta_f H°(B_4C$, cr, 298.15 K) = -14.98±1.5 kcal mol^{-1} which is a correction to that reported by Hong and Kleppa (1). The uncertainty limit is that recommended by Hong and Kleppa.

Two additional experimental determinations of $\Delta_f H°(B_4C$, cr, 298.15 K) have been made. Smith, Dworkin, and Van Artsdalen (3) measured a value of $\Delta_r H°$(298.15 K) = -683.2±2.2 kcal mol^{-1} for the reaction $B_4C(cr) + 4\ O_2(g) \rightarrow 2\ B_2O_3(am) + CO_2(g)$. Due to the amorphous nature of the B_2O_3 product, a value of $\Delta_f H°$(298.15 K) derived from this reaction may be unreliable. Domalski and Armstrong (4) determined $\Delta_f H°$(298.15 K) = -17.1±2.7 kcal mol^{-1} based on careful fluorine bomb calorimetry of the non-stoichiometric compound $B_{4.222}C(cr)$. The data were corrected to the stoichiometric $B_4C(cr)$ assuming the presence of excess free elemental boron. Note that the Hong and Kleppa result lies within the stated uncertainty limits of the Domalski and Armstrong value. In a review of similar data, Gurvich et al. (5) adopted a value for $\Delta_f H°$(298.15 K) of -14.82 ± 1.43 kcal mol^{-1}.

Heat Capacity and Entropy

The low temperature heat capacities, 54-294 K, were taken from Kelley (6). Above 298 K at least four enthalpy determinations have been made. The enthalpy data of King (7) in the range 298 to 1726 K were used to determine the equation: $C_p°/cal\ K^{-1}\ mol^{-1} = 22.99 + 5.40 \times 10^{-3}T - 10.72 \times 10^5 T^{-2}$. Similar measurements of Ancey-Moret and Deniel (8) lead to the equation $C_p°/cal\ K^{-1}\ mol^{-1} = 23.077 + 5.24 \times 10^{-3}T - 10.75 \times 10^5 T^{-2}$. Enthalpy measurements of Mathieu, Durand, and Bonnier (9) in the range 661-1014 K also support the latter two studies. We adopt the high temperature equation of King which joins smoothly at 298 K with the low temperature data of Kelley. The S°(298.15 K) is calculated based on the low temperature heat capacities of Kelley using S°(53.1 K) = 0.047 cal $K^{-1}\ mol^{-1}$. The heat capacities adopted by Gurvich et al. (5) are in essential agreement with those adopted here up to 1500 K but are larger in the high temperature range (T>1500 K). This is due to the adoption (by Gurvich et al.) of the heat capacity results of Sheindlin et al. (10) which can be represented by the equation $C_p°/cal\ K^{-1}\ mol^{-1} = 22.65 + 5.27 \times 10^{-3}T - 12.82 \times 10^5 T^{-2} + 6.45 \times 10^{13}\ T^{-2}\exp(-33 \times 10^3/T)$. Note the additional exponential term which causes the high temperature disparity. Due to the relatively high impurity content of the material used (5.6 mole%) and also the thermodynamic propensity and endothermicity of the reaction by $B_4C(cr) = 4\ B(cr) + C(cr)$ at temperatures greater than 1500 K, we are in doubt as to the exponential behavior of $C_p°$ as reported in the Sheindlin study.

Fusion Data

Of the several reported values of the fusion temperature, T_{fus}/K, [2623(11, 13), 2723(12), 2713±20(10), 2633(14), 2743±20(15)], we adopt the value of Dolloff (15). The value for $\Delta_{fus}H°$ is estimated. Note that $B_4C(cr)$ decomposes at moderate temperatures to form B(cr, g) + C(cr) (2).

References

1. K. C. Hong and O. J. Kleppa, J. Chem. Thermodynamics 10, 797 (1978).
2. JANAF Thermochemical Tables: B(ref), 3-31-82; C(ref), 3-31-82; B(g), 3-31-82.
3. D. Smith, A. S. Dworkin and E. R. Van Artsdalen, J. Am. Chem. Soc. 77, 2654 (1955).
4. E. S. Domalski and G. T. Armstrong, J. Res. Nat. Bur. Stand. 72A, 133 (1968).
5. L. V. Gurvich, I. V. Veits et al., "Thermodynamic Properties of Individual Substances," 3rd ed., Vol. III, Nauka, Moscow, 1981.
6. K. K. Kelley, J. Am. Chem. Soc. 63, 1137 (1941).
7. E. G. King, Ind. Eng. Chem. 41, 1298 (1949).
8. M. F. Ancey-Moret and M. Y. Deniel, IRSID AMC-188 (1971).
9. J. C. Mathieu, F. Durand, and E. Bonnier, Thermodynamics IAEA Proc. Symp., Vienna, 75 (1966).
10. A. E. Sheindlin, J. S. Belevich, and I. G. Kozhevnikov, Teplofiz Vys. Temp. 10, 421 (1972).
11. G. A. Meerson and G. V. Samsonov, Repts. Sector for Phys. Chem. Anal. 22, 92 (1953); see G. N. Makarenko in "Boron and Refractory Borides," V I. Matkovich, ed., pp. 310-330, Springer-Verlag, Berlin, 1977.
12. R. Kieffer, E. Gugel, G. Laimer, and P. Ettmayer, Ber. Deut. Keram. Ges. 48, 385 (1971).
13. J. Flinta, Vacuum 2, 257 (1952).
14. G. Vuillard, Compt. Rend. 257, 3927 (1963).
15. R. T. Dollof, U. S. Dept. Com., Office Tech. Serv., P. B. Rept. 171, 365 (1960).

Boron Carbide (B_4C) $C_1B_4(cr)$

Enthalpy Reference Temperature = T_r = 298.15 K — Standard State Pressure = p° = 0.1 MPa

T/K	$C_p°$ (J K⁻¹mol⁻¹)	S° (J K⁻¹mol⁻¹)	$-[G°-H°(T_r)]/T$ (J K⁻¹mol⁻¹)	$H°-H°(T_r)$ (kJ mol⁻¹)	$\Delta_f H°$ (kJ mol⁻¹)	$\Delta_f G°$ (kJ mol⁻¹)	Log K_f
0	0.	0.	INFINITE	-5.635	-62.404	-62.404	INFINITE
100	5.071	1.634	56.731	-5.510	-62.435	-62.380	32.584
200	27.426	11.311	31.236	-3.985	-62.568	-62.279	16.266
298.15	53.764	27.175	27.175	0.	-62.676	-62.109	10.881
300	54.266	27.509	27.176	0.100	-62.676	-62.106	10.814
400	76.358	46.353	29.595	6.703	-62.546	-61.931	8.087
500	89.789	64.925	34.822	15.051	-62.448	-61.791	6.455
600	98.366	82.103	41.293	24.486	-62.517	-61.657	5.368
700	103.680	97.682	48.254	34.599	-62.796	-61.494	4.589
800	107.654	111.793	55.329	45.171	-63.277	-61.277	4.001
900	110.989	124.666	62.328	56.104	-63.906	-60.991	3.540
1000	114.299	136.533	69.163	67.370	-64.611	-60.630	3.167
1100	117.336	147.571	75.795	78.954	-65.341	-60.196	2.858
1200	120.190	157.904	82.211	90.831	-66.071	-59.696	2.599
1300	122.909	167.632	88.411	102.987	-66.781	-59.136	2.376
1400	125.533	176.837	94.402	115.410	-67.456	-58.522	2.183
1500	128.089	185.585	100.191	128.091	-68.086	-57.862	2.015
1600	130.587	193.932	105.791	141.025	-68.663	-57.161	1.866
1700	133.047	201.923	111.213	154.207	-69.178	-56.426	1.734
1800	135.474	209.596	116.467	167.633	-69.626	-55.663	1.615
1900	137.875	216.986	121.564	181.301	-70.004	-54.877	1.509
2000	140.256	224.118	126.514	195.208	-70.307	-54.072	1.412
2100	142.620	231.019	131.327	209.352	-70.532	-53.255	1.325
2200	144.971	237.708	136.012	223.731	-70.677	-52.428	1.245
2300	147.306	244.203	140.575	238.345	-70.739	-51.597	1.172
2400	149.637	250.522	145.025	253.193	-271.756	-46.490	1.012
2500	151.955	256.677	149.369	268.272	-271.961	-37.099	0.775
2600	154.268	262.682	153.612	283.583	-271.950	-27.704	0.557
2700	156.578	268.548	157.760	299.126	-271.722	-18.313	0.354
2743.000	157.569	271.030	159.517	305.880	------ CRYSTAL <--> LIQUID -----		
2800	158.887	274.284	161.820	314.899	-271.276	-8.935	0.167
2900	161.197	279.900	165.795	330.903	-270.614	0.423	-0.008
3000	163.507	285.403	169.691	347.138	-269.733	9.755	-0.170
3100	165.816	290.802	173.511	363.604	-268.635	19.054	-0.321
3200	168.126	296.103	177.259	380.301	-267.318	28.313	-0.462

PREVIOUS: December 1965 CURRENT: June 1983

Boron Carbide (B_4C) $C_1B_4(cr)$

BORON CARBIDE (B_4C) LIQUID M_r = 55.251

$S°(298.15\ K) = [69.426]\ J\ K^{-1}\ mol^{-1}$ $\Delta_f H°(298.15\ K) = [51.885]\ kJ\ mol^{-1}$

$T_{fus} = 2743 \pm 20\ K$ $\Delta_{fus}H° = [104.6 \pm 10]\ kJ\ mol^{-1}$

Enthalpy of Formation

$\Delta_f H°(B_4C, \ell, 298.15\ K)$ is calculated from $\Delta_f H°(B_4C, cr, 298.15\ K)$ by adding the enthalpy of fusion, $\Delta_{fus}H°$, and the difference in enthalpy, $H°(2743\ K) - H°(298.15\ K)$, between the crystal and liquid.

Heat Capacity and Entropy

A glass transition is assumed at 1750 K. The heat capacity below 1750 K is assumed to be that of the crystal. Above 1750 K the heat capacity is assumed constant and estimated as 135.98 $J\ K^{-1}\ mol^{-1}$ or 27.196 $J\ K^{-1}$ g-atom^{-1} (6.5 cal K^{-1} g-atom^{-1}).

The entropy is calculated in a manner analogous to that used for the enthalpy of formation.

Fusion Data

See the crystal table for details.

Boron Carbide (B_4C) $C_1B_4(l)$

Enthalpy Reference Temperature = T_r = 298.15 K Standard State Pressure = p° = 0.1 MPa

T/K	$C_p°$ (J K⁻¹mol⁻¹)	S° (J K⁻¹mol⁻¹)	$-[G°-H°(T_r)]/T$ (J K⁻¹mol⁻¹)	$H°-H°(T_r)$ (kJ mol⁻¹)	$\Delta_f H°$ (kJ mol⁻¹)	$\Delta_f G°$ (kJ mol⁻¹)	Log K_f
0							
100							
200							
298.15	53.764	69.426	69.426	0.000	51.885	39.855	-6.982
300	54.266	69.760	69.427	0.100	51.885	39.780	-6.926
400	76.358	88.604	71.846	6.703	52.016	35.730	-4.666
500	89.789	107.176	77.074	15.051	52.113	31.645	-3.306
600	98.366	124.354	83.544	24.486	52.045	27.554	-2.399
700	103.680	139.933	90.506	34.599	51.765	23.491	-1.753
800	107.654	154.045	97.580	45.171	51.285	19.483	-1.272
900	110.989	166.917	104.580	56.104	50.656	15.544	-0.902
1000	114.299	178.785	111.414	67.370	49.950	11.681	-0.610
1100	117.336	189.822	118.046	78.954	49.220	7.889	-0.375
1200	120.190	200.155	124.463	90.831	48.491	4.164	-0.181
1300	122.909	209.884	130.663	102.987	47.781	0.499	-0.020
1400	125.533	219.089	136.653	115.410	47.105	-3.113	0.116
1500	128.089	227.837	142.443	128.091	46.475	-6.677	0.233
1600	130.587	236.184	148.043	141.025	45.899	-10.202	0 333
1700	133.047	244.174	153.464	154.207	45.384	-13.692	0.421
1750.000	134.265	248.049	156.112	160.890	GLASS <--> LIQUID TRANSITION		
1750.000	135.980	248.049	156.112	160.890			
1800	135.980	251.879	158.719	167.689	44.991	-17.155	0.498
1900	135.980	259.231	163.817	181.287	44.544	-20.596	0.566
2000	135.980	266.206	168.764	194.885	43.932	-24.009	0.627
2100	135.980	272.841	173.563	208.483	43.161	-27.388	0.681
2200	135.980	279.167	178.221	222.081	42.235	-30.726	0.730
2300	135.980	285.211	182.742	235.679	41.156	-34.019	0.773
2400	135.980	290.998	187.133	249.277	-161.109	-32.987	0.718
2500	135.980	296.549	191.399	262.875	-162.796	-27.614	0.577
2600	135.980	301.883	195.547	276.473	-164.498	-22.173	0.445
2700	135.980	307.014	199.581	290.071	-166.215	-16.666	0.322
2743.000	135.980	309.163	201.282	295.918	CRYSTAL <--> LIQUID		
2800	135.980	311.960	203.507	303.669	-167.945	-11.096	0.207
2900	135.980	316.731	207.329	317.267	-169.688	-5.463	0.098
3000	135.980	321.341	211.053	330.865	-171.445	0.230	-0.004
3100	135.980	325.800	214.683	344.463	-173.214	5.981	-0.101
3200	135.980	330.117	218.223	358.061	-174.996	11.790	-0.192
3300	135.980	334.302	221 678	371.659	-176.791	17.655	-0.279
3400	135.980	338.361	225.050	385.257	-178.597	23.574	-0.362
3500	135.980	342.303	228.344	398.855	-180.416	29.547	-0.441
3600	135.980	346.133	231.563	412.453	-182.246	35.572	-0.516
3700	135.980	349.859	234.710	426.051	-184.088	41.648	-0.588
3800	135.980	353.486	237.788	439.649	-185.942	47.773	-0.657
3900	135.980	357.018	240.801	453.247	-187.807	53.948	-0.723
4000	135.980	360.460	243.749	466.845	-189.684	60.171	-0.786
4100	135.980	363.818	246.637	480.443	-191.571	66.440	-0.846
4200	135.980	367.095	249.466	494.041	-2112.868	100.853	-1.254
4300	135.980	370.295	252.239	507.639	-2110.429	153.532	-1.865
4400	135.980	373.421	254.958	521.237	-2108.008	206.154	-2.447
4500	135.980	376.477	257.624	534.835	-2105.607	258.721	-3.003

PREVIOUS: December 1965 CURRENT: June 1983

Boron Carbide (B_4C) $C_1B_4(l)$

BORON CARBIDE (B_4C) CRYSTAL–LIQUID M_r = 55.251 Boron Carbide (B_4C) C_1B_4(cr,l)

0 - 2743 K Crystal
2743 - 4500 K Liquid

Refer to the individual tables for the details.

Enthalpy Reference Temperature = T_r = 298.15 K Standard State Pressure = p° = 0.1 MPa

T/K	C_p° (J K⁻¹mol⁻¹)	S° (J K⁻¹mol⁻¹)	−[G°−H°(T_r)]/T (J K⁻¹mol⁻¹)	H°−H°(T_r) (kJ mol⁻¹)	$\Delta_f H^\circ$ (kJ mol⁻¹)	$\Delta_f G^\circ$ (kJ mol⁻¹)	Log K_f
0	0.	0.	INFINITE	-5.635	-62.404	-62.404	INFINITE
100	5.071	1.634	56.731	-5.510	-62.435	-62.380	32.584
200	27.426	11.311	31.236	-3.985	-62.568	-62.279	16.266
298.15	53.764	27.175	27.175	0.	-62.676	-62.109	10.881
300	54.266	27.509	27.176	0.100	-62.676	-62.106	10.814
400	76.358	46.353	29.595	6.703	-62.546	-61.931	8.087
500	89.789	64.925	34.822	15.051	-62.448	-61.791	6.455
600	98.366	82.103	41.293	24.486	-62.517	-61.657	5.368
700	103.680	97.682	48.254	34.599	-62.796	-61.494	4.589
800	107.654	111.793	55.329	45.171	-63.277	-61.277	4.001
900	110.989	124.666	62.328	56.104	-63.906	-60.991	3.540
1000	114.299	136.533	69.163	67.370	-64.611	-60.630	3.167
1100	117.336	147.571	75.795	78.954	-65.341	-60.196	2.858
1200	120.190	157.904	82.211	90.831	-66.071	-59.696	2.599
1300	122.909	167.632	88.411	102.987	-66.781	-59.136	2.376
1400	125.533	176.837	94.402	115.410	-67.456	-58.522	2.183
1500	128.089	185.585	100.191	128.091	-68.086	-57.862	2.015
1600	130.587	193.932	105.791	141.025	-68.663	-57.161	1.866
1700	133.047	201.923	111.213	154.207	-69.178	-56.426	1.734
1800	135.474	209.596	116.467	167.633	-69.626	-55.663	1.615
1900	137.875	216.986	121.564	181.301	-70.004	-54.877	1.509
2000	140.256	224.118	126.514	195.208	-70.307	-54.072	1.412
2100	142.620	231.019	131.327	209.352	-70.532	-53.255	1.325
2200	144.971	237.708	136.012	223.731	-70.677	-52.428	1.245
2300	147.306	244.203	140.575	238.345	-70.739	-51.597	1.172
2400	149.637	250.522	145.025	253.193	-271.756	-46.490	1.012
2500	151.955	256.677	149.369	268.272	-271.961	-37.099	0.775
2600	154.268	262.682	153.612	283.583	-271.950	-27.704	0.557
2700	156.578	268.548	157.760	299.126	-271.722	-18.313	0.354
2743.000	157.569	271.030	159.517	305.880	CRYSTAL <--> LIQUID		
2743.000	135.980	309.163	159.517	410.480	TRANSITION		
2800	135.980	311.960	162.592	418.231	-167.945	-11.096	0.207
2900	135.980	316.731	167.825	431.829	-169.688	-5.463	0.098
3000	135.980	321.341	172.866	445.427	-171.445	0.230	-0.004
3100	135.980	325.800	177.728	459.025	-173.214	5.981	-0.101
3200	135.980	330.117	182.423	472.623	-174.996	11.790	-0.192
3300	135.980	334.302	186.962	486.221	-176.791	17.655	-0.279
3400	135.980	338.361	191.356	499.819	-178.597	23.574	-0.362
3500	135.980	342.303	195.612	513.417	-180.416	29.547	-0.441
3600	135.980	346.133	199.741	527.015	-182.246	35.572	-0.516
3700	135.980	349.859	203.748	540.613	-184.088	41.648	-0.588
3800	135.980	353.486	207.641	554.211	-185.942	47.773	-0.657
3900	135.980	357.018	211.426	567.809	-187.807	53.948	-0.723
4000	135.980	360.460	215.109	581.407	-189.684	60.171	-0.786
4100	135.980	363.818	218.695	595.005	-191.571	66.440	-0.846
4200	135.980	367.095	222.189	608.603	-2112.868	100.853	-1.254
4300	135.980	370.295	225.597	622.201	-2110.429	153.532	-1.865
4400	135.980	373.421	228.921	635.799	-2108.008	206.154	-2.447
4500	135.980	376.477	232.166	649.397	-2105.607	258.721	-3.003

PREVIOUS: CURRENT: June 1983

Boron Carbide (B_4C) C_1B_4(cr,l)

BERYLLIUM CARBIDE (Be_2C) CRYSTAL M_r = 30.03536

$\Delta_f H°(0\ K)$ = Unknown

$S°(298.15\ K) = [16.3 \pm 4.2]\ J\ K^{-1}\ mol^{-1}$ $\Delta_f H°(298.15\ K) = -90.79 \pm 10.5\ kJ\ mol^{-1}$

T_{fus} = [2400] K $\Delta_{fus}H° = [75.312]\ kJ\ mol^{-1}$

Enthalpy of Formation

The selected $\Delta_f H°(298.15\ K)$ was calculated by the 3rd law method using vapor pressures reported by Pollock (1). Pressures for the reaction $Be_2C(cr) \rightarrow 2Be(g) + C(graphite)$ were measured by Pollock in the temperature range 1430-1669 K by Knudsen technique. Vapor pressure data for this decomposition reaction have also been reported by Baboin (2), Quirk (3) and Muratov and Novoselova (4). The calculated $\Delta_f H°(298.15\ K)$ values are summarized as follows:

Calc. Method	Number of Pressures	T/K	$\Delta_r H°(298.15\ K)$ kcal mol^{-1}	$\Delta_f H°(298.15\ K)$ kcal mol^{-1}	Source
3rd law	14	1430-1669	89.117	-21.7	Pollock (1)
2nd law	14	1430-1669	90.8	-25.1	Least squares fit of Pollock's v.p. to the equation: log P (atm) = 6.946-19,593/T
3rd law	6	1673-1953	71.411	+13.7	Muratov and Novoselova (4)
3rd law	8	1900-2600	81.278	- 6.1	Quirk (3)
3rd law	8	1600-2300	96.266	-36.0	Baboin (2)

Schick et al. (5) have listed the measured vapor pressures. The vapor pressures reported by Baboin (2) used in the 3rd law calculation were those listed by Schick, et al. (5).

Heat Capacity and Entropy

The heat capacity values, from 298.15 to 1200 K, were derived from; the mean $C°_p$ data measured by Neely et al. (6). Neely, et al., estimated an overall error of 15% in their mean heat capacity data. The powdered Be_2C material which was used analyzed 80 to 74% Be_2C. The impurities were mainly oxide and nitride of Be. No corrections were made for the effect of these impurities.

From 1200-2400 K the $C°_p$ was extrapolated so as to approach the value, 22 cal $K^{-1}\ mol^{-1}$ calculated from Krikorian (7). The $S°(298.15\ K) = 3.9\ cal\ K^{-1}\ mol^{-1}$, was estimated by Krikorian (7).

Fusion Data

Schick et al. (5), have summarized the available melting data i.e.: Oishi and Hamona (8) have reported decomposition upon heating as low as 1873 K. Brewer et al. (9) and Gaev (10) reported melting or dissociation between 2373 and 2423 K.

The melting point is assumed to be 2400 K as estimated by Douglas and Victor (11) and Schick et al. (5). From the estimated melting point, 2400 K, and an estimated 2.5 cal g-atom^{-1} for the entropy of melting, the heat of melting was calculated.

References

1. B. D. Pollock, J. Phys. Chem. 63, 587 (1959).
2. J. Baboin, French Patents 1.193.790 (1959).
3. J. Quirk, Reactor Handbook 3, 98 (1955).
4. F. Muratov and A. Novoselova, Dokl. Akad. Nauk SSSR 129, 334 (1959).
5. H. L. Schick, D. F. Anthrop, P. L. Hanst, R. E. Dreikorn, and M. B. Panish, "Thermodyanmics of Certain Refractory Compounds, Part II", March 1, 1963 to May 31, 1963.
6. J. J. Neely, E. Teeter, Jr., and J. B. Trice, J. Amer. Ceram. Soc. 33, 363 (1950).
7. O. H. Krikorian, "Estimation of High-Temperature Heat Capacities of Carbides," University of California, UCRL 6785, February 6, 1962.
8. Y. Oishi and Y. Hamona, Osaka Kogyo Gijutsu Shikenjo Kiho 8, 89 (1957).
9. K. L. Brewer, L. Bromley, P. Gilles, and N. Lofgren, "The Chemistry and Metallurgy of Miscellaneous Materials," in "Thermodynamics", McGraw-Hill, New York (1950).
10. I. S. Gaev, AEC TR 3036 Transl. from Zh. Neorg. Khim. 1, 196 (1956).
11. T. B. Douglas and A. C. Victor, U. S. Nat. Bur. Stand. Preliminary Report 6645, January 1960.

Beryllium Carbide (Be_2C) $C_1Be_2(cr)$

Enthalpy Reference Temperature = T_r = 298.15 K **Standard State Pressure = p° = 0.1 MPa**

T/K	$C°_p$ (J $K^{-1}mol^{-1}$)	S° (J $K^{-1}mol^{-1}$)	$-[G°-H°(T_r)]/T$ (J $K^{-1}mol^{-1}$)	$H°-H°(T_r)$ (kJ mol^{-1})	$\Delta_f H°$ (kJ mol^{-1})	$\Delta_f G°$ (kJ mol^{-1})	Log K_f
0							
100							
200							
298.15	43.250	16.318	16.318	0.	-90.793	-88.317	15.473
300	43.329	16.585	16.318	0.080	-90.789	-88.302	15.375
400	47.614	29.689	18.076	4.645	-90.931	-87.468	11.422
500	49.752	40.543	21.516	9.513	-91.595	-86.535	9.040
600	51.890	49.706	25.469	14.542	-92.678	-85.426	7.437
700	60.434	58.401	29.552	20.194	-93.582	-84.139	6.279
800	64.706	66.750	33.686	26.451	-94.240	-82.744	5.403
900	68.977	74.617	37.801	33.135	-94.775	-81.273	4.717
1000	73.249	82.106	41.860	40.246	-95.153	-79.751	4.166
1100	77.525	89.288	45.847	47.785	-95.351	-78.200	3.713
1200	81.797	96.224	49.758	55.760	-95.343	-76.639	3.336
1300	85.011	102.905	53.591	64.108	-95.184	-75.087	3.017
1400	87.278	109.292	57.343	72.728	-94.968	-73.549	2.744
1500	88.881	115.371	61.011	81.540	-94.772	-72.026	2.508
1600	90.023	121.145	64.590	90.488	-123.720	-69.463	2.268
1700	90.843	126.629	68.080	99.534	-122.903	-66.096	2.031
1800	91.349	131.837	71.478	108.645	-122.087	-62.778	1.822
1900	91.621	136.784	74.786	117.795	-121.299	-59.505	1.636
2000	91.722	141.486	78.005	126.963	-120.556	-56.272	1.470
2100	91.797	145.963	81.135	136.139	-119.867	-53.075	1.320
2200	91.885	150.235	84.180	145.323	-119.231	-49.910	1.185
2300	91.964	154.322	87.141	154.515	-118.646	-46.772	1.062
2400	92.048	158.237	90.023	163.716	-118.113	-43.659	0.950
2400.000	92.048	158.237	90.023	163.716	------ CRYSTAL <--> LIQUID -----		
2500	92.153	161.997	92.827	172.926	-117.628	-40.567	0.848
2600	92.253	165.613	95.557	182.146	-117.191	-37.493	0.753
2700	92.350	169.097	98.217	191.376	-116.802	-34.435	0.666
2800	92.443	172.457	100.809	200.616	-698.373	-18.946	0.353
2900	92.534	175.703	103.335	209.865	-695.951	5.276	-0.095
3000	92.621	178.841	105.800	219.122	-693.543	29.416	-0.512
3100	92.705	181.879	108.206	228.389	-691.152	53.475	-0.901
3200	92.786	184.824	110.554	237.663	-688.781	77.457	-1.264
3300	92.864	187.680	112.848	246.946	-686.432	101.366	-1.604
3400	92.939	190.454	115.090	256.236	-684.110	125.203	-1.924
3500	93.010	193.149	117.282	265.534	-681.817	148.973	-2.223

PREVIOUS: March 1962 CURRENT: June 1963

Beryllium Carbide (Be_2C) $C_1Be_2(cr)$

BERYLLIUM CARBIDE (Be_2C) LIQUID M_r = 30.03536

S°(298.15 K) = [47.467] J K^{-1} mol^{-1}

T_{fus} = [2400] K

$\Delta_f H°$(298.15 K) = [15.893] kJ mol^{-1}

$\Delta_{fus}H°$ = [75.312] kJ mol^{-1}

Enthalpy of Formation

$\Delta_f H°$(298.15 K) is calculated from that of the crystal by adding $\Delta_{fus}H°$ and the difference in enthalpy, H°(2400 K) - H°(298.15 K), between the crystal and liquid.

Heat Capacity and Entropy

The heat capacity is assumed constant and equal to the heat capacity of solid Be_2C at the estimated melting point. S°(298.15 K) is calculated in a manner similar to that used for the enthalpy of formation.

Fusion Data

The U. S. Nat. Bur. Stand. (1), states that Be_2C seems to decompose easily only at temperatures above its reported melting point. For this reason they extended their Be_2C table into the liquid region. For detailed fusion data see the crystal table.

Reference

1. U. S. Nat. Bur. Stand. Report 6645, 1960.

Beryllium Carbide (Be_2C) $C_1Be_2(l)$

Enthalpy Reference Temperature = T_r = 298.15 K Standard State Pressure = p° = 0.1 MPa

T/K	$C_p°$ (J K⁻¹mol⁻¹)	S° (J K⁻¹mol⁻¹)	-[G°-H°(T_r)]/T (J K⁻¹mol⁻¹)	H°-H°(T_r) (kJ mol⁻¹)	$\Delta_f H°$ (kJ mol⁻¹)	$\Delta_f G°$ (kJ mol⁻¹)	Log K_f
0							
100							
200							
298.15	43.250	47.467	47.467	0.	-15.893	-22.704	3.978
300	43.329	47.735	47.468	0.080	-15.889	-22.747	3.961
400	47.614	60.838	49.225	4.645	-16.031	-25.027	3.268
500	49.752	71.692	52.665	9.513	-16.694	-27.209	2.843
600	51.890	80.855	56.618	14.542	-17.778	-29.215	2.543
700	60.434	89.550	60.702	20.194	-18.682	-31.043	2.316
800	64.706	97.899	64.835	26.451	-19.340	-32.763	2.139
900	68.977	105.767	68.950	33.135	-19.875	-34.407	1.997
1000	73.249	113.255	73.009	40.246	-20.253	-36.000	1.880
1100	77.525	120.437	76.996	47.785	-20.450	-37.564	1.784
1200	81.797	127.374	80.907	55.760	-20.443	-39.118	1.703
1300	85.011	134.054	84.740	64.108	-20.284	-40.680	1.635
1400	87.278	140.441	88.492	72.728	-20.068	-42.257	1.577
1500	88.881	146.520	92.160	81.540	-19.872	-43.849	1.527
1600	90.023	152.295	95.739	90.489	-48.819	-44.401	1.450
1600.000	90.023	152.295	95.739	90.489	GLASS <--> LIQUID TRANSITION		
1600.000	92.048	152.295	95.739	90.489			
1700	92.048	157.876	99.232	99.694	-47.843	-44.155	1.357
1800	92.048	163.137	102.638	108.899	-46.934	-43.965	1.276
1900	92.048	168.114	105.954	118.104	-46.090	-43.823	1.205
2000	92.048	172.835	109.181	127.308	-45.310	-43.724	1.142
2100	92.048	177.326	112.320	136.513	-44.592	-43.663	1.086
2200	92.048	181.608	115.373	145.718	-43.936	-43.634	1.036
2300	92.048	185.700	118.342	154.923	-43.339	-43.634	0.991
2400	92.048	189.617	121.231	164.128	-42.801	-43.659	0.950
2400.000	92.048	189.617	121.231	164.128	------ CRYSTAL <--> LIQUID -----		
2500	92.048	193.375	124.042	173.332	-42.321	-43.705	0.913
2600	92.048	196.985	126.779	182.537	-41.900	-43.769	0.879
2700	92.048	200.459	129.444	191.742	-41.535	-43.848	0.848
2800	92.048	203.807	132.040	200.947	-623.142	-31.494	0.588
2900	92.048	207.037	134.571	210.152	-620.764	-10.406	0.187
3000	92.048	210.157	137.039	219.356	-618.409	10.601	-0.185
3100	92.048	213.176	139.446	228.561	-616.080	31.529	-0.531
3200	92.048	216.098	141.796	237.766	-613.778	52.383	-0.855
3300	92.048	218.930	144.091	246.971	-611.507	73.166	-1.158
3400	92.048	221.678	146.333	256.176	-609.270	93.880	-1.442
3500	92.048	224.347	148.524	265.380	-607.070	114.528	-1.709

PREVIOUS: March 1962 CURRENT: June 1963

Beryllium Carbide (Be_2C) $C_1Be_2(l)$

BERYLLIUM CARBIDE (Be_2C) CRYSTAL–LIQUID M_r = 30.03536 Beryllium Carbide (Be_2C) C_1Be_2(cr,l)

0 to 2400 K crystal
above 2400 K liquid

Refer to the individual tables for details.

Enthalpy Reference Temperature = T_r = 298.15 K Standard State Pressure = p° = 0.1 MPa

T/K	C_p° (J K⁻¹mol⁻¹)	S° (J K⁻¹mol⁻¹)	−[G°−H°(T_r)]/T (J K⁻¹mol⁻¹)	H°−H°(T_r) (kJ mol⁻¹)	$\Delta_f H^\circ$ (kJ mol⁻¹)	$\Delta_f G^\circ$ (kJ mol⁻¹)	Log K_f
0							
100							
200							
298.15	43.250	16.318	16.318	0.	-90.793	-88.317	15.473
300	43.329	16.585	16.318	0.080	-90.789	-88.302	15.375
400	47.614	29.689	18.076	4.645	-90.931	-87.468	11.422
500	49.752	40.543	21.516	9.513	-91.595	-86.535	9.040
600	51.890	49.706	25.469	14.542	-92.678	-85.426	7.437
700	60.434	58.401	29.552	20.194	-93.582	-84.139	6.279
800	64.706	66.750	33.686	26.451	-94.240	-82.744	5.403
900	68.977	74.617	37.801	33.135	-94.775	-81.273	4.717
1000	73.249	82.106	41.860	40.246	-95.153	-79.751	4.166
1100	77.525	89.288	45.847	47.785	-95.351	-78.200	3.713
1200	81.797	96.224	49.758	55.760	-95.343	-76.639	3.336
1300	85.011	102.905	53.591	64.108	-95.184	-75.087	3.017
1400	87.278	109.292	57.343	72.728	-94.968	-73.549	2.744
1500	88.881	115.371	61.011	81.540	-94.772	-72.026	2.508
1600	90.023	121.145	64.590	90.488	-123.720	-69.463	2.268
1700	90.843	126.629	68.080	99.534	-122.903	-66.096	2.031
1800	91.349	131.837	71.478	108.645	-122.087	-62.778	1.822
1900	91.621	136.784	74.786	117.795	-121.299	-59.505	1.636
2000	91.722	141.486	78.005	126.963	-120.556	-56.272	1.470
2100	91.797	145.963	81.135	136.139	-119.867	-53.075	1.320
2200	91.885	150.235	84.180	145.323	-119.231	-49.910	1.185
2300	91.964	154.322	87.141	154.515	-118.646	-46.772	1.062
2400	92.048	158.237	90.023	163.716	-118.113	-43.659	0.950
2400.000	92.048	189.617	90.023	239.028	CRYSTAL <--> LIQUID		
2400.000	92.048	189.617	90.023	239.028	TRANSITION		
2500	92.048	193.375	94.082	248.233	-42.321	-43.705	0.913
2600	92.048	196.985	97.971	257.437	-41.900	-43.769	0.879
2700	92.048	200.459	101.703	266.642	-41.535	-43.848	0.848
2800	92.048	203.807	105.290	275.847	-623.142	-31.494	0.588
2900	92.048	207.037	108.743	285.052	-620.764	-10.406	0.187
3000	92.048	210.157	112.072	294.257	-618.409	10.601	-0.185
3100	92.048	213.176	115.285	303.461	-616.080	31.529	-0.531
3200	92.048	216.098	118.390	312.666	-613.778	52.383	-0.855
3300	92.048	218.930	121.394	321.871	-611.507	73.166	-1.158
3400	92.048	221.678	124.303	331.076	-609.270	93.880	-1.442
3500	92.048	224.347	127.124	340.281	-607.070	114.528	-1.709

PREVIOUS: CURRENT: June 1963

Beryllium Carbide (Be_2C) C_1Be_2(cr,l)

BROMOMETHYLIDYNE (CBr) IDEAL GAS M_r = 91.915

$S°(298.15\ K)$ = [233.45] $J\ K^{-1}\ mol^{-1}$

$\Delta_f H°(0\ K) = 514.3 \pm 63\ kJ\ mol^{-1}$
$\Delta_f H°(298.15\ K) = 510.4 \pm 63\ kJ\ mol^{-1}$

Electronic Levels and Quantum Weights

State	ε_i, cm^{-1}	g_i
$X_2\ ^2\Pi_{1/2}$	0	2
$X_1\ ^2\Pi_{3/3}$	465	2
$A_1\ ^2\Delta_{3/2}$	33163	2
$A_2\ ^2\Delta_{5/2}$	33218	2
	[40000]	[2]

ω_e = [736] cm^{-1} $\omega_e x_e$ = [4.0] cm^{-1} σ = 1

B_e = 0.4890 cm^{-1} α_e = [0.0036] cm^{-1} r_e = 1.8170 Å

Enthalpy of Formation

Dixon and Kroto (1) have observed predissociation of CBr(g) which indicates that the dissociation energy D_0^0 is less than 94.7 kcal mol^{-1}. They evaluated the kinetic data of Simons and Yarwood (2), obtaining a value of D_0^0 = 97±5 kcal mol^{-1}, based on the enthalpy of formation of bromoform and the dissociation energy of HBr, and concluded that the true value of D_0^0 is very close to the upper limit of 94.7 kcal mol^{-1}. Comparisons of C-X bond energies of CX_4 and CX compounds, however, indicate that this value is too high to be consistent with other carbon halides. Therefore the value of D_0^0 adopted is 75±15 kcal mol^{-1}. The corresponding $\Delta_f H°$(CBr, g, 298.15 K) is 122±15 kcal mol^{-1} (510.448±63 kJ mol^{-1}).

Heat Capacity and Entropy

The rotational structure of the near UV spectra of CBr(g) has been investigated by Dixon and Kroto (1) in the 3000 Å region. They assigned the above doublet electronic levels to the ground ($^2\Pi$) and first excited ($^2\Delta$) states and reported a value of B_0 = 0.4872±0.0003 cm^{-1} for the ground state. An additional electronic level ($^2\Sigma$) is expected to be close to the first excited state by analogy with CF(g) and CCl(g).

The fundamental vibrational frequency ω_e is estimated from Guggenheimer's relation for single bonded molecules (3). The anharmonic vibrational term $\omega_e x_e$ is estimated by assuming $\omega_e x_e = \omega_e^2/(4D_0 + 2\omega_e)$. The value of α_e is calculated from the Morse potential function. The rotational constant B_e is calculated from $B_e = B_0 + \alpha_e/2$. The bond distance is calculated from B_e.

References

1. R. N. Dixon and H. W. Kroto, Trans. Faraday Soc. 59, 1484 (1963).
2. J. P. Simons and A. J. Yarwood, Trans. Faraday Soc. 57, 2167 (1961).
3. K. M. Guggenheimer, Proc. Phys. Soc. (London) 58, 456 (1946).

Bromomethylidyne (CBr) $C_1Br_1(g)$

Enthalpy Reference Temperature = T_r = 298.15 K Standard State Pressure = p° = 0.1 MPa

T/K	$C_p°$ (J $K^{-1}mol^{-1}$)	S° (J $K^{-1}mol^{-1}$)	$-[G°-H°(T_r)]/T$ (J $K^{-1}mol^{-1}$)	$H°-H°(T_r)$ (kJ mol^{-1})	$\Delta_f H°$ (kJ mol^{-1})	$\Delta_f G°$ (kJ mol^{-1})	Log K_f
0	0.	0.	INFINITE	-9.482	514.272	514.272	INFINITE
100	29.596	197.941	263.600	-6.566	515.732	498.725	-260.507
200	33.413	219.553	236.680	-3.425	516.095	481.543	-125.766
250	35.011	227.191	234.040	-1.712	516.105	472.901	-98.807
298.15	36.041	233.452	233.452	0.	510.448	465.246	-81.509
300	36.072	233.675	233.453	0.067	510.429	464.965	-80.958
350	36.733	239.290	233.895	1.888	495.454	458.199	-68.382
400	37.139	244.223	234.884	3.736	495.835	452.850	-59.136
450	37.388	248.613	236.170	5.599	496.150	447.457	-51.939
500	37.544	252.561	237.615	7.473	496.400	442.033	-46.179
600	37.708	259.423	240.695	11.237	496.726	431.125	-37.533
700	37.785	265.242	243.796	15.012	496.859	420.179	-31.354
800	37.829	270.291	246.800	18.793	496.842	409.224	-26.720
900	37.861	274.748	249.662	22.577	496.711	398.279	-23.115
1000	37.889	278.739	252.374	26.365	496.488	387.353	-20.233
1100	37.916	282.351	254.937	30.155	496.193	376.453	-17.876
1200	37.944	285.651	257.361	33.948	495.839	365.583	-15.913
1300	37.973	288.690	259.656	37.744	495.435	354.745	-14.254
1400	38.002	291.505	261.831	41.543	494.989	343.939	-12.833
1500	38.033	294.128	263.898	45.344	494.507	333.166	-11.602
1600	38.065	296.583	265.865	49.149	493.993	322.426	-10.526
1700	38.097	298.892	267.741	52.957	493.448	311.720	-9.578
1800	38.130	301.070	269.532	56.769	492.876	301.047	-8.736
1900	38.163	303.133	271.247	60.583	492.277	290.406	-7.984
2000	38.197	305.091	272.891	64.401	491.653	279.797	-7.308
2100	38.232	306.956	274.469	68.223	491.004	269.220	-6.696
2200	38.267	308.735	275.986	72.048	490.330	258.674	-6.142
2300	38.302	310.437	277.447	75.876	489.632	248.160	-5.636
2400	38.337	312.068	278.856	79.708	488.910	237.677	-5.173
2500	38.373	313.633	280.216	83.544	488.166	227.224	-4.748
2600	38.408	315.139	281.530	87.383	487.398	216.801	-4.356
2700	38.444	316.589	282.802	91.225	486.609	206.409	-3.993
2800	38.480	317.988	284.034	95.071	485.799	196.046	-3.657
2900	38.517	319.339	285.228	98.921	484.969	185.712	-3.345
3000	38.553	320.646	286.387	102.775	484.120	175.407	-3.054
3100	38.590	321.910	287.513	106.632	483.254	165.131	-2.782
3200	38.626	323.136	288.607	110.493	482.372	154.883	-2.528
3300	38.663	324.325	289.671	114.357	481.475	144.663	-2.290
3400	38.700	325.480	290.708	118.225	480.566	134.470	-2.066
3500	38.737	326.602	291.717	122.097	479.645	124.304	-1.855
3600	38.775	327.694	292.702	125.973	478.714	114.165	-1.656
3700	38.813	328.757	293.662	129.852	477.775	104.051	-1.469
3800	38.851	329.793	294.599	133.735	476.829	93.963	-1.292
3900	38.889	330.802	295.514	137.622	475.878	83.900	-1.124
4000	38.928	331.787	296.409	141.513	474.922	73.862	-0.965
4100	38.968	332.749	297.284	145.408	473.963	63.847	-0.813
4200	39.008	333.689	298.139	149.307	473.003	53.856	-0.670
4300	39.048	334.607	298.977	153.210	472.041	43.887	-0.533
4400	39.089	335.505	299.797	157.117	471.080	33.941	-0.403
4500	39.131	336.384	300.600	161.028	470.120	24.017	-0.279
4600	39.174	337.244	301.387	164.943	469.161	14.115	-0.160
4700	39.217	338.087	302.159	168.862	468.205	4.233	-0.047
4800	39.261	338.914	302.916	172.786	467.252	-5.629	0.061
4900	39.306	339.724	303.659	176.715	466.303	-15.471	0.165
5000	39.353	340.518	304.389	180.647	465.358	-25.293	0.264
5100	39.400	341.298	305.105	184.585	464.417	-35.097	0.359
5200	39.448	342.063	305.808	188.528	463.481	-44.882	0.451
5300	39.498	342.815	306.499	192.475	462.551	-54.650	0.539
5400	39.549	343.554	307.179	196.427	461.625	-64.399	0.623
5500	39.601	344.280	307.847	200.385	460.705	-74.132	0.704
5600	39.654	344.994	308.504	204.347	459.791	-83.848	0.782
5700	39.709	345.697	309.150	208.315	458.883	-93.548	0.857
5800	39.765	346.388	309.786	212.289	457.980	-103.232	0.930
5900	39.822	347.068	310.412	216.268	457.083	-112.900	1.000
6000	39.881	347.738	311.029	220.254	456.193	-122.554	1.067

PREVIOUS: June 1967 (1 atm) CURRENT: June 1967 (1 bar)

Bromomethylidyne (CBr) $C_1Br_1(g)$

BROMOTRIFLUOROMETHANE ($CBrF_3$) IDEAL GAS M_r = 148.910209

$S°(298.15 K) = 297.83 \pm 0.50\ J\ K^{-1}\ mol^{-1}$

$\Delta_f H°(0 K) = -636.9 \pm 2.9\ kJ\ mol^{-1}$
$\Delta_f H°(298.15 K) = -648.9 \pm 2.9\ kJ\ mol^{-1}$

Vibrational Frequencies and Degeneracies

ν, cm^{-1}	ν, cm^{-1}
1086 (1)	1206 (2)
761 (1)	549 (2)
348 (1)	300 (2)

Ground State Quantum Weight: 1 σ = 4
Point Group: C_{3v}
Bond Distance: C-F = 1.328 Å C-Br = 1.909 Å
Bond Angle: F-C-F = 108.6 ± 1° F-C-Br = 110.33°
Product of the Moments of Inertia: $I_A I_B I_C = 2.340709 \times 10^{-113}\ g^3\ cm^6$

Enthalpy of Formation

The selected value is obtained from least squares, simultaneous adjustment of the enthalpies of formation of CF_3, CF_3X (X = H, Cl, Br, I, CF_3) and C_2F_4. Details of the input data and the adjustment are given in (1). Coomber and Whittle (2, 3, 4) have determined equilibrium data for $CClF_3(g) + BrCl(g) = CBrF_3(g) + Cl_2(g)$ and for $CF_3X(g) + Br_2(g) = CF_3Br(g) + BrX(g)$ where X = H, Cl, and CF . Equilibrium data for $CF_3I(g) + BrI(g) = CBrF_3(g) + I_2(g)$ were obtained by Lord et al. (5). These data link $CBrF_3$ with C_2F_6, CHF_3, $CClF_3$ and CF_3I.

Heat Capacity and Entropy

The molecular structure has been derived from electron-diffraction studies by Sharbaugh et al. (6), Bowen (7), and Anderson (8) and from microwave data by Sheridan and Gordy (9). The adoped parameters are an average of those reported by (8, 9). Infrared spectra of the gas were observed by several investigators (10, 11, 12) and Raman spectra of the liquid were obtained by Edgell and May (12). The vibrational assignments of (11, 12) are in close agreement. These assignments are adopted but preference is given to frequencies observed for the gas. The value for ν_6 is taken as 300 cm^{-1} which is between the values of 305 and 297 ± 3 cm^{-1} derived from Raman spectra and IR (combination and overtone) spectra, respectively.

The thermodynamic functions of this table are essentially the same as those calculated by (11). Uncertainty in the entropy is estimated to vary from 0.12 to 0.5 cal/(K mol) over the range from 300 to 1000 K. Uncertainty in ν_6 and neglect of anharmonicity are the major sources of error. Discussions of the magnitude of anharmonic effects appear on the tables for $CClF_3(g)$ and $CF_4(g)$. The principal moments of inertia are: $I_A = 14.6768 \times 10^{-39}$ and $I_B = I_C = 39.9354 \times 10^{-39}\ g\ cm^2$.

References

1. A. N. Syverud, Quarterly Tech. Report T-0009-2Q-69 under AFRPL contract NR. F04611-67-C-0009, Dow Chemical Co., July, 1969.
2. J. W. Coomber and E. Whittle, Trans. Faraday Soc. 63, 608 (1967).
3. J. W. Coomber and E. Whittle, Trans. Faraday Soc. 63, 1394 (1967).
4. J. W. commber and E. Whittle, Trans. Faraday Soc. 63, 2656 (1967).
5. A. Lord, C. A. Goy and H. O. Pritchard, J. Phys. Chem. 71, 2705 (1967).
6. A. H. Sharbaugh, B. S. Pritchard and T. C. Madison, Phys. Rev. 77, 302 (1950).
7. H. J. M. Bowen, Trans. Faraday Soc. 50, 444 (1954).
8. R. E. Anderson, Dissertation Abstracts 19, 50 (1958).
9. J. Sheridan and M. Gordy, J. Chem. Phys. 20, 591 (1952).
10. E. K. Plyler and N. Acquista, J. Res. Nat. Bur. Stand. 48, 92 (1952).
11. P. R. McGee, F. F. Cleveland, A. G. Meister and C. E. Decker, J. Chem. Phys. 21, 242 (1953).
12. W. F. edgell and C. E. May, J. Chem. Phys. 22, 1808 (1954).

Bromotrifluoromethane ($CBrF_3$) $C_1Br_1F_3(g)$

Enthalpy Reference Temperature = T_r = 298.15 K Standard State Pressure = p° = 0.1 MPa

T/K	$C_p°$ (J K⁻¹mol⁻¹)	S° (J K⁻¹mol⁻¹)	$-[G°-H°(T_r)]/T$ (J K⁻¹mol⁻¹)	$H°-H°(T_r)$ (kJ mol⁻¹)	$\Delta_f H°$ (kJ mol⁻¹)	$\Delta_f G°$ (kJ mol⁻¹)	Log K_f
0	0.	0.	INFINITE	-14.468	-636.863	-636.863	INFINITE
100	39.323	240.114	350.236	-11.012	-639.224	-634.892	331.634
200	56.420	272.770	303.778	-6.202	-641.584	-629.584	164.431
250	63.460	286.136	298.936	-3.200	-642.540	-626.472	130.894
298.15	69.330	297.827	297.827	0.	-648.938	-622.641	109.084
300	69.540	298.256	297.828	0.128	-648.983	-622.478	108.383
350	74.784	309.380	298.695	3.740	-664.551	-617.201	92.112
400	79.273	319.667	300.681	7.594	-664.608	-610.432	79.714
450	83.084	329.230	303.329	11.656	-664.596	-603.660	70.071
500	86.305	338.156	306.370	15.893	-664.531	-596.892	62.357
600	91.313	354.359	313.046	24.788	-664.284	-583.385	50.788
700	94.902	368.720	319.994	34.108	-663.933	-569.929	42.529
800	97.514	381.572	326.902	43.736	-663.517	-556.527	36.338
900	99.454	393.175	333.631	53.589	-663.062	-543.181	31.525
1000	100.923	403.733	340.122	63.611	-662.585	-529.886	27.678
1100	102.058	413.407	346.350	73.762	-662.099	-516.640	24.533
1200	102.949	422.327	352.315	84.014	-661.613	-503.437	21.914
1300	103.660	430.596	358.022	94.346	-661.134	-490.276	19.700
1400	104.236	438.300	363.485	104.742	-660.669	-477.150	17.803
1500	104.708	445.508	368.715	115.190	-660.220	-464.058	16.160
1600	105.100	452.279	373.728	125.681	-659.792	-450.994	14.723
1700	105.428	458.661	378.539	136.208	-659.386	-437.957	13.457
1800	105.705	464.695	383.159	146.765	-659.002	-424.943	12.332
1900	105.942	470.417	387.602	157.348	-658.640	-411.950	11.325
2000	106.145	475.856	391.880	167.952	-658.300	-398.974	10.420
2100	106.321	481.039	396.003	178.576	-657.978	-386.017	9.602
2200	106.475	485.989	399.982	189.216	-657.670	-373.073	8.858
2300	106.609	490.725	403.825	199.870	-657.375	-360.143	8.179
2400	106.727	495.265	407.541	210.537	-657.086	-347.227	7.557
2500	106.832	499.624	411.138	221.215	-656.800	-334.322	6.985
2600	106.925	503.816	414.622	231.903	-656.512	-321.428	6.458
2700	107.008	507.853	418.001	242.600	-656.217	-308.546	5.969
2800	107.083	511.746	421.280	253.304	-655.910	-295.675	5.516
2900	107.150	515.504	424.464	264.016	-655.587	-282.815	5.094
3000	107.211	519.138	427.560	274.734	-655.243	-269.966	4.701
3100	107.266	522.654	430.571	285.458	-654.874	-257.130	4.333
3200	107.316	526.061	433.502	296.187	-654.477	-244.306	3.988
3300	107.361	529.364	436.357	306.921	-654.048	-231.494	3.664
3400	107.403	532.569	439.140	317.659	-653.584	-218.697	3.360
3500	107.441	535.683	441.854	328.401	-653.082	-205.913	3.073
3600	107.476	538.711	444.503	339.147	-652.540	-193.144	2.802
3700	107.508	541.656	447.089	349.896	-651.956	-180.391	2.547
3800	107.538	544.523	449.616	360.649	-651.329	-167.655	2.305
3900	107.566	547.317	452.085	371.404	-650.657	-154.935	2.075
4000	107.591	550.041	454.500	382.162	-649.939	-142.233	1.857
4100	107.615	552.698	456.863	392.922	-649.175	-129.550	1.650
4200	107.637	555.291	459.176	403.685	-648.363	-116.886	1.454
4300	107.658	557.824	461.440	414.450	-647.505	-104.242	1.266
4400	107.677	560.299	463.659	425.216	-646.599	-91.618	1.088
4500	107.695	562.719	465.834	435.985	-645.647	-79.015	0.917
4600	107.712	565.087	467.966	446.755	-644.648	-66.435	0.754
4700	107.727	567.403	470.057	457.527	-643.603	-53.876	0.599
4800	107.742	569.671	472.109	468.301	-642.512	-41.340	0.450
4900	107.756	571.893	474.122	479.076	-641.377	-28.827	0.307
5000	107.769	574.070	476.100	489.852	-640.198	-16.338	0.171
5100	107.781	576.204	478.042	500.630	-638.976	-3.873	0.040
5200	107.793	578.297	479.950	511.408	-637.713	8.568	-0.086
5300	107.804	580.351	481.825	522.188	-636.408	20.984	-0.207
5400	107.814	582.366	483.668	532.969	-635.064	33.375	-0.323
5500	107.824	584.344	485.481	543.751	-633.681	45.741	-0.434
5600	107.833	586.287	487.263	554.534	-632.261	58.082	-0.542
5700	107.842	588.196	489.017	565.318	-630.804	70.396	-0.645
5800	107.851	590.072	490.744	576.102	-629.313	82.684	-0.745
5900	107.859	591.915	492.443	586.888	-627.787	94.948	-0.841
6000	107.866	593.728	494.116	597.674	-626.229	107.184	-0.933

PREVIOUS: June 1969 (1 atm) CURRENT: June 1969 (1 bar)

Bromotrifluoromethane ($CBrF_3$) $C_1Br_1F_3(g)$

CYANOGEN BROMIDE (BrCN) IDEAL GAS Mr = 105.9217

$\Delta_f H°(0\ K) = 192.75 \pm 6.3\ kJ\ mol^{-1}$

$S°(298.15\ K) = 248.363 \pm 0.040\ J\ K^{-1}\ mol^{-1}$ $\Delta_f H°(298.15\ K) = 186.19 \pm 6.3\ kJ\ mol^{-1}$

Vibrational Frequencies and Degeneracies

ν, cm^{-1}
575 (1)
342.5 (2)
2200 (1)

Ground State Quantum Weight: 1 $\sigma = 1$

Point Group: $C_{\infty v}$

Bond Distances: C-Br = 1.789 Å C-N = 1.158 Å

Bond Angle: Br-C-N = 180°

Rotational Constants: $B_0 = 0.137172\ cm^{-1}$ $D_0 = 2.929 \times 10^{-8}\ cm^{-1}$

Spectroscopic constants used in calculating corrections to rigid-rotator-harmonic oscillator approximation (cm^{-1}).

$\alpha_1 = 5.17 \times 10^{-4}$	$x_{11} = [-3.0]$	$x_{12} = [-4.0]$
$\alpha_2 = -3.84 \times 10^{-4}$	$x_{22} = -0.86$	$x_{23} = -6.19$
$\alpha_3 = 6.77 \times 10^{-4}$	$x_{33} = [-5.0]$	$x_{13} = [-3.0]$
	$g_{22} = 1.14$	

Enthalpy of Formation

The enthalpy of formation is calculated from $\Delta_f H°$(BrCN, cr, 298.15 K) = 33.58 kcal mol^{-1} (1) and $\Delta_{sub}H°$ (298.15 K) = 10.92 ± 0.1 kcal mol^{-1}. This latter is obtained from an anlysis of the vapor pressure data of Lord and Woof (2) and Baxter et al. (3).

The enthalpy of formation of the crystal is calculated from $\Delta_r H°$(298.15 K) = -56.0 kcal mol^{-1} obtained by Lord and Woolf (2) for the reaction

$$BrCN(cr) + 2NaOH(56\ H_2O) = NaCNO(2800\ H_2O) + NaBr(2800\ H_2O) + H_2O(\ell).$$

The auxiliary data used were $\Delta_f H°$(NaOH.56 H_2O, 298.15 K) = -112.26 kcal mol^{-1} (1, 4), $\Delta_f H$(NaCNO·2800 H_2O, 298.15 K) = -92.25 kcal mol^{-1} (1, 4) and $\Delta_f H°$(NaBr·2800 H_2O, 298.15 K) = -70.54 kcal mol^{-1} (1, 4). The uncertainty in the final value reflects the uncertainty in $\Delta_f H$(NaCNO·2800 H_2O) which was determined by Lord and Woolf (2) to be -93.6 kcal mol^{-1} in excess NaOH·56 H_2O.

Heat Capacity and Entropy

Gordon (5) used the constants given above to calculate the adopted table, using the method of Pennington and Kobe (6) for anharmonicity, vibration rotation interaction, and centrifugal stretching corrections.

The bond lengths and angle were reported by Tyler and Sheridan (7).

References

1. U. S. Nat. Bur. Stand. Tech. Note 270-3, 1968.
2. G. Lord and A. A. Woolf, J. Chem. Soc. (London), 2546 (1954).
3. G. P. Baxter, T. K. Bezzenburger and C. H. Wilson, J. Amer. Chem. Soc. 42, 1386 (1920).
4. V. B. Parker, U. S. Nat. Bur. Stand., NSRDS-NBS 2, 1965.
5. J. S. Gordon, J. Chem. Eng. Data 11, 553 (1966).
6. R. E. Pennington and K. A. Kobe, J. Chem. Phys. 22, 1442 (1954).
7. J. K. Tyler and J. Sheridan, Trans. Faraday Soc. 59, 2661 (1963).

Cyanogen Bromide (BrCN) $C_1Br_1N_1(g)$

Enthalpy Reference Temperature = T_r = 298.15 K Standard State Pressure = p° = 0.1 MPa

T/K	$C_p°$ (J K⁻¹mol⁻¹)	S° (J K⁻¹mol⁻¹)	$-[G°-H°(T_r)]/T$ (J K⁻¹mol⁻¹)	$H°-H°(T_r)$ (kJ mol⁻¹)	$\Delta_f H°$ (kJ mol⁻¹)	$\Delta_f G°$ (kJ mol⁻¹)	Log K_f
0	0.	0.	INFINITE	-11.075	192.754	192.754	INFINITE
100	32.213	205.109	286.153	-8.104	192.817	183.084	-95.633
200	41.714	230.527	252.430	-4.381	192.308	173.560	-45.329
298.15	47.124	248.363	248.363	0.	186.188	165.104	-28.926
300	47.196	248.655	248.364	0.087	186.163	164.974	-28.724
400	50.137	262.658	250.253	4.962	171.316	160.993	-21.023
500	52.095	274.064	253.910	10.077	171.789	158.355	-16.543
600	53.664	283.704	258.092	15.367	172.149	155.632	-13.549
700	55.020	292.080	262.362	20.802	172.421	152.857	-11.406
800	56.204	299.506	266.550	26.365	172.631	150.047	-9.797
900	57.237	306.187	270.589	32.038	172.800	147.214	-8.544
1000	58.137	312.265	274.457	37.808	172.939	144.363	-7.541
1100	58.915	317.843	278.151	43.661	173.059	141.500	-6.719
1200	59.597	322.999	281.676	49.587	173.163	138.626	-6.034
1300	60.191	327.793	285.042	55.577	173.257	135.744	-5.454
1400	60.714	332.273	288.257	61.623	173.342	132.856	-4.957
1500	61.174	336.478	291.333	67.718	173.418	129.961	-4.526
1600	61.588	340.440	294.280	73.856	173.488	127.062	-4.148
1700	61.961	344.185	297.106	80.034	173.550	124.158	-3.815
1800	62.300	347.736	299.821	86.247	173.605	121.251	-3.519
1900	62.614	351.113	302.433	92.493	173.653	118.341	-3.253
2000	62.898	354.332	304.948	98.769	173.692	115.429	-3.015
2100	63.162	357.407	307.373	105.072	173.722	112.515	-2.799
2200	63.409	360.351	309.715	111.401	173.742	109.600	-2.602
2300	63.643	363.175	311.978	117.753	173.752	106.684	-2.423
2400	63.860	365.888	314.168	124.129	173.751	103.768	-2.258
2500	64.070	368.500	316.290	130.525	173.739	100.852	-2.107
2600	64.270	371.016	318.346	136.942	173.716	97.937	-1.968
2700	64.459	373.446	320.342	143.379	173.683	95.023	-1.838
2800	64.643	375.793	322.281	149.834	173.640	92.110	-1.718
2900	64.819	378.065	324.166	156.307	173.587	89.200	-1.607
3000	64.990	380.265	325.999	162.797	173.525	86.291	-1.502
3100	65.157	382.399	327.784	169.305	173.456	83.384	-1.405
3200	65.321	384.470	329.524	175.829	173.381	80.480	-1.314
3300	65.480	386.482	331.219	182.369	173.300	77.578	-1.228
3400	65.634	388.440	332.873	188.925	173.216	74.678	-1.147
3500	65.785	390.344	334.488	195.496	173.130	71.781	-1.071
3600	65.931	392.200	336.066	202.081	173.042	68.887	-1.000
3700	66.078	394.008	337.608	208.682	172.954	65.995	-0.932
3800	66.224	395.772	339.115	215.297	172.868	63.105	-0.867
3900	66.362	397.494	340.590	221.926	172.785	60.218	-0.807
4000	66.505	399.176	342.034	228.570	172.705	57.333	-0.749
4100	66.643	400.820	343.448	235.227	172.630	54.449	-0.694
4200	66.781	402.428	344.833	241.898	172.562	51.568	-0.641
4300	66.915	404.001	346.191	248.583	172.500	48.688	-0.591
4400	67.049	405.540	347.522	255.281	172.446	45.809	-0.544
4500	67.182	407.049	348.828	261.993	172.400	42.931	-0.498
4600	67.312	408.527	350.110	268.718	172.363	40.055	-0.455
4700	67.442	409.976	351.368	275.455	172.336	37.179	-0.413
4800	67.572	411.397	352.604	282.206	172.319	34.303	-0.373
4900	67.701	412.792	353.818	288.970	172.312	31.428	-0.335
5000	67.831	414.161	355.011	295.746	172.315	28.553	-0.298
5100	67.957	415.505	356.184	302.536	172.330	25.678	-0.263
5200	68.086	416.826	357.338	309.338	172.356	22.802	-0.229
5300	68.212	418.124	358.473	316.153	172.393	19.925	-0.196
5400	68.337	419.400	359.589	322.980	172.441	17.048	-0.165
5500	68.463	420.655	360.688	329.820	172.500	14.170	-0.135
5600	68.584	421.890	361.770	336.673	172.571	11.291	-0.105
5700	68.710	423.105	362.835	343.537	172.652	8.410	-0.077
5800	68.835	424.301	363.885	350.414	172.745	5.528	-0.050
5900	68.956	425.479	364.919	357.304	172.848	2.644	-0.023
6000	69.078	426.639	365.938	364.206	172.961	-0.242	0.002

PREVIOUS: June 1969 (1 atm) CURRENT: June 1969 (1 bar)

Cyanogen Bromide (BrCN) $C_1Br_1N_1(g)$

TETRABROMOMETHANE (CBr_4) IDEAL GAS M_r = 331.627

$S°(298.15\ K) = 358.095\ J\ K^{-1}\ mol^{-1}$

$\Delta_f H°(0\ K) = [79.890]\ kJ\ mol^{-1}$

$\Delta_f H°(298.15\ K) = [50.208]\ kJ\ mol^{-1}$

Vibrational Frequencies and Degeneracies

ν, cm^{-1}	ν, cm^{-1}
267.2(1)	670.7(3)
122.5(2)	182.5(3)

Ground State Quantum Weight: [1] σ = 12

Point Group: T_d

Bond Distance: C-Br = 1.942 Å

Bond Angle: Br-C-Br = 109° 28'

Product of the Moments of Inertia: $I_A I_B I_C = 2.376154 \times 10^{-111}\ g^3\ cm^6$

Enthalpy of Formation

$\Delta_f H°(298.15\ K)$ was calculated and reported by Rossini et al. (1).

Heat Capacity and Entropy

The molecular constants, bond distance, and bond angle were determined by electron diffraction by Finbak et al. (2, 3). The constants reported by Finbak et al. (3) in 1943 were selected. These constants are listed in "Tables of Interatomic Distances and Configuration in Molecules and Ions," Special Publication No. 11, The Chemical Society (London), 1958. The principal moments of inertia calculated from these constants are: $I_A = I_B = I_C = 133.4417 \times 10^{-39}\ g\ cm^2$.

The vibrational frequencies used were assigned by Meister et al. (4). They measured the infrared and Raman spectrum, considered the previous literature, calculated force constants and made most probable value assignments for the frequencies. A more recent paper by Moszynska and Mathieu (5) made assignments from Raman and infrared spectra which agreed with all of Meister et al's. (4) assignments except for ν_2 which they found was 126 cm^{-1} rather than 122 cm^{-1}.

References

1. U. S. Nat. Bur. Stand. Circ. 500, 1952.
2. C. Finbak, O. Hassel and B. Ottar, Arch. Math. Naturv. B44, No. 13 (1941).
3. C. Finbak, O. Hassel and O. Olaussen, Tidsskr. Kjemi Bergv. 3, 13 (1943).
4. A. O. Meister, S. E. Rosser, and F. F. Cleveland, J. Chem. Phys. 18, 346 (1950).
5. B. Moszynska and J. P. Mathieu, Compt. Rend 256, 2820 (1963).

Tetrabromomethane (CBr_4) $C_1Br_4(g)$

Enthalpy Reference Temperature = T_r = 298.15 K Standard State Pressure = p° = 0.1 MPa

T/K	$C_p°$	S°	$-[G°-H°(T_r)]/T$	$H°-H°(T_r)$	$\Delta_f H°$	$\Delta_f G°$	Log K_f
	J $K^{-1}mol^{-1}$			kJ mol^{-1}			
0	0.	0.	INFINITE	-20.388	79.890	79.890	INFINITE
100	63.538	273.277	432.002	-15.873	78.759	62.296	-32.540
200	81.403	323.603	366.156	-8.511	75.992	46.857	-12.238
250	87.009	342.397	359.574	-4.294	74.282	39.763	-8.308
298.15	91.180	358.095	358.095	0.	50.208	35.914	-6.292
300	91.319	358.659	358.096	0.169	50.081	35.826	-6.238
350	94.594	372.993	359.221	4.820	-11.039	36.793	-5.491
400	97.081	385.795	361.757	9.615	-10.455	43.587	-5.692
450	98.987	397.344	365.080	14.519	-9.863	50.306	-5.839
500	100.464	407.853	368.840	19.507	-9.274	56.960	-5.951
600	102.548	426.368	376.926	29.665	-8.132	70.099	-6.103
700	103.902	442.285	385.153	39.992	-7.055	83.051	-6.197
800	104.823	456.223	393.184	50.431	-6.042	95.854	-6.259
900	105.475	468.609	400.889	60.948	-5.088	108.533	-6.299
1000	105.951	479.748	408.227	71.520	-4.187	121.108	-6.326
1100	106.310	489.864	415.196	82.134	-3.333	133.596	-6.344
1200	106.586	499.126	421.810	92.780	-2.523	146.008	-6.356
1300	106.802	507.666	428.090	103.449	-1.753	158.354	-6.363
1400	106.976	515.588	434.060	114.139	-1.022	170.642	-6.367
1500	107.116	522.973	439.744	124.844	-0.329	182.880	-6.368
1600	107.232	529.890	445.165	135.561	0.325	195.072	-6.368
1700	107.328	536.394	450.342	146.289	0.939	207.225	-6.367
1800	107.409	542.531	455.295	157.026	1.511	219.343	-6.365
1900	107.478	548.341	460.040	167.771	2.039	231.430	-6.362
2000	107.537	553.855	464.594	178.522	2.519	243.491	-6.359
2100	107.588	559.103	468.971	189.278	2.949	255.528	-6.356
2200	107.632	564.109	473.182	200.039	3.325	267.547	-6.352
2300	107.670	568.894	477.240	210.804	3.646	279.549	-6.349
2400	107.704	573.477	481.155	221.573	3.910	291.540	-6.345
2500	107.734	577.875	484.937	232.345	4.116	303.520	-6.342
2600	107.760	582.101	488.593	243.119	4.265	315.492	-6.338
2700	107.784	586.168	492.132	253.897	4.357	327.461	-6.335
2800	107.805	590.088	495.561	264.676	4.395	339.427	-6.332
2900	107.824	593.872	498.886	275.458	4.381	351.392	-6.329
3000	107.841	597.527	502.114	286.241	4.319	363.359	-6.327
3100	107.857	601.064	505.249	297.026	4.213	375.329	-6.324
3200	107.871	604.488	508.297	307.812	4.068	387.302	-6.322
3300	107.884	607.808	511.262	318.600	3.889	399.281	-6.320
3400	107.896	611.029	514.150	329.389	3.681	411.266	-6.318
3500	107.906	614.156	516.962	340.179	3.450	423.257	-6.317
3600	107.916	617.196	519.705	350.970	3.202	435.255	-6.315
3700	107.925	620.153	522.380	361.762	2.940	447.260	-6.314
3800	107.934	623.032	524.991	372.555	2.671	459.272	-6.313
3900	107.941	625.835	527.541	383.349	2.400	471.291	-6.312
4000	107.949	628.568	530.032	394.143	2.132	483.318	-6.311
4100	107.955	631.234	532.468	404.939	1.870	495.351	-6.311
4200	107.962	633.835	534.851	415.735	1.619	507.390	-6.310
4300	107.967	636.376	537.183	426.531	1.383	519.435	-6.310
4400	107.973	638.858	539.465	437.328	1.164	531.485	-6.310
4500	107.978	641.285	541.701	448.126	0.966	543.540	-6.309
4600	107.982	643.658	543.892	458.924	0.792	555.599	-6.309
4700	107.987	645.980	546.039	469.722	0.644	567.662	-6.309
4800	107.991	648.254	548.145	480.521	0.523	579.728	-6.309
4900	107.995	650.480	550.211	491.320	0.431	591.795	-6.309
5000	107.998	652.662	552.238	502.120	0.370	603.865	-6.309
5100	108.002	654.801	554.228	512.920	0.340	615.935	-6.308
5200	108.005	656.898	556.183	523.720	0.343	628.005	-6.308
5300	108.008	658.956	558.103	534.521	0.379	640.076	-6.308
5400	108.011	660.974	559.989	545.322	0.448	652.145	-6.308
5500	108.014	662.956	561.843	556.123	0.551	664.213	-6.308
5600	108.016	664.903	563.666	566.925	0.687	676.278	-6.308
5700	108.019	666.815	565.459	577.726	0.857	688.341	-6.308
5800	108.021	668.693	567.223	588.528	1.059	700.399	-6.308
5900	108.023	670.540	568.958	599.331	1.295	712.455	-6.308
6000	108.026	672.355	570.667	610.133	1.562	724.506	-6.307

PREVIOUS: December 1963 (1 atm) CURRENT: December 1963 (1 bar)

Tetrabromomethane (CBr_4) $C_1Br_4(g)$

CHLOROMETHYLIDYNE (CCl) IDEAL GAS $M_r = 47.464$

$S°(298.15\ K) = 224.52 \pm 0.4\ J\ K^{-1}\ mol^{-1}$

$\Delta_f H°(0\ K) = 498.3 \pm 20\ kJ\ mol^{-1}$
$\Delta_f H°(298.15\ K) = 502.1 \pm 20\ kJ\ mol^{-1}$

Electronic Levels and Quantum Weights

State	ε_i, cm^{-1}	g_i
$X^2\Pi_{1/2}$	0	2
$^2\Pi_{3/2}$	134.92	2
$[^4\Sigma]$	[20000]	[4]
$A^2\Delta$	36004	4

$\omega_e = 876.4\ cm^{-1}$ $\omega_e x_e = 6.15\ cm^{-1}$ $\sigma = 1$
$B_e = 0.69666\ cm^{-1}$ $\alpha_e = [0.0066]\ cm^{-1}$ $r_e = 1.642\ Å$

Enthalpy of Formation

The data of Miller and Palmer (1) provide a value of $D_0^0(CCl) = 80 \pm 5$ kcal mol^{-1}, which appears to be reasonably substantiated. They studied the chemiluminescence from flames of $CHCl_3$ and CCl_4 burning in potassium vapor. These flames exhibit C_2 Swan band radiation with major departures from vibrational and rotational equilibrium. The spectra are interpreted as due to the reaction $C + CCl \rightarrow C_2 + Cl$ where the carbon atom is either ^3D ground state or ^1D first excited state. The separation between the ground and first excited state of the carbon atom is matched closely by the difference in preferential population of the excited C_2 levels. The assumption is made that all the exothermicity of the reaction is retained as vibrational excitation of the C_2, thus the populated levels are measures of the heat of reaction. In this manner we obtain a heat of reaction of 67±2.5 kcal for the reaction $C + CCl \rightarrow C_2 + Cl$ which, using JANAF auxiliary data, gives $\Delta_f H°(CCl, g) = 124 \pm 2.5$ kcal mol^{-1}; however, this represents the initial state of the CCl which in the flame is presumed to have about 4 kcal of vibrational energy at roughly 2000 K. Thus, we obtain $\Delta_f H°(CCl, g) = 120 \pm 5$ kcal mol^{-1} (502.080±20 kJ mol^{-1}) which is adopted.

The enthalpy of formation can also be estimated from the dissociation energy calculated from the linear Birge-Sponer extrapolation of the ground state vibrational constants. This yields $D^0_0 = 89$ kcal or $\Delta_f H°(298.15\ K) = 111$ kcal mol^{-1}. There are several indications that this dissociation energy is probably a maximum value, for example, Hildenbrand (5) has applied an ionic-covalent correction to the linear extrapolation which gives $D^0_0 = 72$ kcal. We would expect this enthalpy of formation to be a minimum value. A real minimum for the dissociation energy can be obtained from the knowledge that the first vibrational level of the $^2\Delta$ state is occupied and so the dissociation energy of this state must be greater than 850 cm^{-1}. This yields a minimum dissociation energy of the ground state of 76 kcal, assuming the dissociation products are $C(^1\Delta)$ and $Cl(^2P)$, and $\Delta_f H°(CCl, g, 298.15\ K) = 124$ kcal mol^{-1} which is the maximum value of the heat of formation.

Heat Capacity and Entropy

The rotational and electronic constants are those of Verma and Mulliken (2). The vibrational constants are those derived by Ovcharenko et al. (3) from the data of Gordon and King (4). The anharmonicities listed by Verma and Mulliken are not used, since these are based only on the 0,0 and 0,1 transitions. The $^4\Sigma$ level is estimated by analogy with CF(g). The value for α_e was obtained from the other constants assuming a Morse potential function.

References

1. W. J. Miller and H. B. Palmer, J. Chem. Phys. 40, 3701 (1964).
2. R. D. Verma and R. S. Mulliken, J. Mol. Spectrosc. 6, 419 (1961).
3. I. E. Ovcharenko, Yu. Ya. Kuzyakov, and V. M. Tatievskii, Opt. Spectrosc. (Eng. Transl.) 19, 294 (1965).
4. R. D. Gordon and G. W. King, Can. J. Phys. 39, 252 (1961).
5. D. L. Hildenbrand, Douglas Advanced Research Laboratories, Huntington Beach, Calif., private communication, Dec. 30, 1969.

Chloromethylidyne (CCl) $C_1Cl_1(g)$

Enthalpy Reference Temperature = T_r = 298.15 K Standard State Pressure = $p°$ = 0.1 MPa

	J $K^{-1}mol^{-1}$			kJ mol^{-1}			
T/K	$C_p°$	S°	$-[G°-H°(T_r)]/T$	$H°-H°(T_r)$	$\Delta_f H°$	$\Delta_f G°$	Log K_f
0	0.	0.	INFINITE	-9.395	498.327	498.327	INFINITE
100	32.553	189.824	252.667	-6.284	499.922	490.485	-256.203
200	31.329	211.872	227.428	-3.111	501.249	480.488	-125.491
250	31.663	218.890	225.043	-1.538	501.717	475.241	-99.296
298.15	32.254	224.516	224.516	0.	502.080	470.108	-82.361
300	32.279	224.715	224.516	0.060	502.092	469.909	-81.818
350	32.956	229.742	224.912	1.691	502.393	464.521	-69.326
400	33.594	234.185	225.799	3.355	502.630	459.093	-59.951
450	34.156	238.175	226.956	5.049	502.807	453.640	-52.657
500	34.637	241.799	228.262	6.769	502.932	448.170	-46.820
600	35.391	248.185	231.064	10.272	503.041	437.205	-38.062
700	35.934	253.684	233.912	13.840	503.000	426.233	-31.806
800	36.335	258.509	236.691	17.455	502.843	415.276	-27.115
900	36.639	262.807	239.359	21.104	502.598	404.344	-23.468
1000	36.877	266.680	241.900	24.780	502.283	393.443	-20.551
1100	37.068	270.204	244.316	28.478	501.911	382.577	-18.167
1200	37.226	273.437	246.610	32.193	501.494	371.747	-16.182
1300	37.360	276.422	248.789	35.922	501.038	360.953	-14.503
1400	37.476	279.195	250.863	39.664	500.550	350.195	-13.066
1500	37.577	281.784	252.839	43.417	500.033	339.473	-11.822
1600	37.668	284.212	254.725	47.179	499.490	328.787	-10.734
1700	37.751	286.498	256.528	50.950	498.925	318.135	-9.775
1800	37.827	288.658	258.253	54.729	498.338	307.517	-8.924
1900	37.898	290.705	259.908	58.515	497.732	296.932	-8.163
2000	37.965	292.651	261.497	62.309	497.107	286.380	-7.479
2100	38.029	294.505	263.025	66.108	496.464	275.859	-6.862
2200	38.090	296.275	264.496	69.914	495.803	265.370	-6.301
2300	38.149	297.970	265.915	73.726	495.124	254.911	-5.789
2400	38.207	299.595	267.285	77.544	494.428	244.481	-5.321
2500	38.264	301.155	268.608	81.367	493.714	234.082	-4.891
2600	38.321	302.657	269.889	85.197	492.982	223.711	-4.494
2700	38.378	304.105	271.130	89.032	492.233	213.368	-4.128
2800	38.436	305.501	272.333	92.872	491.466	203.054	-3.788
2900	38.494	306.851	273.500	96.719	490.682	192.767	-3.472
3000	38.555	308.157	274.633	100.571	489.881	182.508	-3.178
3100	38.616	309.422	275.735	104.430	489.063	172.276	-2.903
3200	38.680	310.649	276.807	108.295	488.230	162.070	-2.646
3300	38.746	311.841	277.851	112.166	487.382	151.891	-2.404
3400	38.814	312.998	278.868	116.044	486.520	141.737	-2.178
3500	38.885	314.124	279.859	119.929	485.646	131.609	-1.964
3600	38.959	315.221	280.826	123.821	484.760	121.506	-1.763
3700	39.035	316.289	281.770	127.721	483.865	111.428	-1.573
3800	39.114	317.331	282.692	131.628	482.961	101.374	-1.393
3900	39.196	318.348	283.594	135.543	482.051	91.344	-1.223
4000	39.281	319.342	284.475	139.467	481.135	81.338	-1.062
4100	39.369	320.313	285.337	143.400	480.217	71.355	-0.909
4200	39.459	321.263	286.181	147.341	479.296	61.393	-0.764
4300	39.553	322.192	287.008	151.292	478.376	51.454	-0.625
4400	39.649	323.103	287.818	155.252	477.458	41.536	-0.493
4500	39.748	323.995	288.612	159.222	476.543	31.639	-0.367
4600	39.849	324.870	289.391	163.201	475.632	21.763	-0.247
4700	39.953	325.728	290.155	167.192	474.728	11.906	-0.132
4800	40.059	326.570	290.905	171.192	473.831	2.068	-0.023
4900	40.167	327.397	291.641	175.203	472.943	-7.751	0.083
5000	40.277	328.210	292.364	179.226	472.065	-17.553	0.183
5100	40.390	329.008	293.075	183.259	471.197	-27.336	0.280
5200	40.503	329.794	293.774	187.304	470.342	-37.103	0.373
5300	40.619	330.566	294.461	191.360	469.500	-46.854	0.462
5400	40.736	331.327	295.136	195.427	468.671	-56.588	0.547
5500	40.854	332.075	295.801	199.507	467.857	-66.308	0.630
5600	40.974	332.812	296.456	203.598	467.057	-76.012	0.709
5700	41.095	333.539	297.100	207.702	466.273	-85.703	0.785
5800	41.216	334.254	297.734	211.817	465.505	-95.380	0.859
5900	41.338	334.960	298.359	215.945	464.753	-105.044	0.930
6000	41.462	335.656	298.975	220.085	464.018	-114.696	0.999

PREVIOUS: December 1969 (1 atm) CURRENT: December 1969 (1 bar)

CARBONIC CHLORIDE FLUORIDE (COClF) IDEAL GAS $M_r = 82.461803$

$\Delta_f H°(0\ K) = [-424.3 \pm 33]\ kJ\ mol^{-1}$

$S°(298.15\ K) = [277.012]\ J\ K^{-1}\ mol^{-1}$ $\Delta_f H°(298.15\ K) = [-426.8 \pm 33]\ kJ\ mol^{-1}$

Vibrational Frequencies and Degeneracies

ν, cm^{-1}	ν, cm^{-1}
776(1)	1095(1)
1868(1)	415(1)
501(1)	667(1)

Ground State Quantum Weight: 1 $\sigma = 1$

Point Group: C_s

Product of the Moments of Inertia: $I_A I_B I_C = 2.7911 \times 10^{-114}\ g^3\ cm^6$

Enthalpy of Formation

The enthalpies of formation of $COCl_2$ and COF_2 have been averaged to obtain that of COClF.

Heat Capacity and Entropy

The molecular constants are from Lovell et al (1). The principal moments of inertia are: $I_A = 7.480 \times 10^{-39}$, $I_B = 16.008 \times 10^{-39}$ and $I_C = 23.31 \times 10^{-39}\ g\ cm^2$.

Reference

1. R. J. Lovell, C. V. Stephenson, and E. A. Jones, J. Chem. Phys. 22, 1953 (1954).

Carbonic Chloride Fluoride (COClF) $C_1Cl_1F_1O_1(g)$

Enthalpy Reference Temperature = T_r = 298.15 K Standard State Pressure = $p°$ = 0.1 MPa

T/K	$C_p°$ (J K⁻¹mol⁻¹)	S° (J K⁻¹mol⁻¹)	$-[G°-H°(T_r)]/T$ (J K⁻¹mol⁻¹)	$H°-H°(T_r)$ (kJ mol⁻¹)	$\Delta_f H°$ (kJ mol⁻¹)	$\Delta_f G°$ (kJ mol⁻¹)	Log K_f
0	0.	0.	INFINITE	-11.904	-424.276	-424.276	INFINITE
100	34.405	231.656	317.256	-8.560	-425.354	-421.790	220.320
200	43.300	257.957	281.526	-4.714	-426.273	-417.838	109.128
250	48.207	268.154	277.852	-2.424	-426.569	-415.693	86.854
298.15	52.396	277.012	277.012	0.	-426.768	-413.579	72.457
300	52.545	277.337	277.013	0.097	-426.774	-413.497	71.996
350	56.256	285.723	277.667	2.820	-426.915	-411.272	61.379
400	59.407	293.446	279.163	5.713	-427.011	-409.031	53.414
450	62.091	300.602	281.152	8.753	-427.080	-406.779	47.218
500	64.390	307.266	283.434	11.916	-427.133	-404.520	42.260
600	68.086	319.349	288.434	18.549	-427.211	-399.990	34.822
700	70.879	330.064	293.631	25.503	-427.268	-395.448	29.509
800	73.020	339.674	298.796	32.703	-427.313	-390.899	25.523
900	74.684	348.375	303.829	40.091	-427.349	-386.345	22.423
1000	75.993	356.314	308.686	47.628	-427.380	-381.788	19.943
1100	77.036	363.608	313.352	55.281	-427.410	-377.227	17.913
1200	77.876	370.348	317.825	63.028	-427.442	-372.664	16.222
1300	78.561	376.610	322.109	70.851	-427.481	-368.097	14.790
1400	79.124	382.453	326.213	78.736	-427.528	-363.528	13.563
1500	79.593	387.928	330.146	86.673	-427.587	-358.954	12.500
1600	79.987	393.078	333.920	94.653	-427.660	-354.377	11.569
1700	80.320	397.938	337.544	102.668	-427.750	-349.794	10.748
1800	80.605	402.537	341.028	110.715	-427.856	-345.205	10.018
1900	80.849	406.902	344.382	118.788	-427.982	-340.610	9.364
2000	81.060	411.054	347.612	126.884	-428.126	-336.008	8.776
2100	81.244	415.014	350.728	134.999	-428.290	-331.398	8.243
2200	81.405	418.797	353.737	143.132	-428.473	-326.780	7.759
2300	81.546	422.419	356.645	151.279	-428.675	-322.153	7.316
2400	81.672	425.892	359.458	159.440	-428.897	-317.517	6.911
2500	81.783	429.228	362.183	167.613	-429.136	-312.871	6.537
2600	81.882	432.438	364.824	175.797	-429.393	-308.216	6.192
2700	81.971	435.530	367.385	183.989	-429.666	-303.550	5.873
2800	82.051	438.512	369.873	192.191	-429.955	-298.874	5.576
2900	82.123	441.393	372.290	200.399	-430.259	-294.187	5.299
3000	82.188	444.178	374.640	208.615	-430.576	-289.489	5.040
3100	82.247	446.874	376.927	216.837	-430.905	-284.781	4.799
3200	82.301	449.486	379.153	225.064	-431.245	-280.062	4.572
3300	82.350	452.019	381.323	233.297	-431.596	-275.332	4.358
3400	82.395	454.478	383.439	241.534	-431.954	-270.592	4.157
3500	82.437	456.867	385.503	249.776	-432.320	-265.841	3.967
3600	82.475	459.190	387.518	258.021	-432.691	-261.079	3.788
3700	82.510	461.451	389.485	266.271	-433.066	-256.307	3.618
3800	82.542	463.651	391.408	274.523	-433.444	-251.524	3.457
3900	82.572	465.796	393.288	282.779	-433.823	-246.731	3.305
4000	82.600	467.887	395.127	291.038	-434.203	-241.930	3.159
4100	82.626	469.927	396.927	299.299	-434.581	-237.118	3.021
4200	82.650	471.918	398.689	307.563	-434.956	-232.297	2.889
4300	82.672	473.863	400.415	315.829	-435.329	-227.468	2.763
4400	82.693	475.764	402.105	324.097	-435.696	-222.630	2.643
4500	82.713	477.623	403.763	332.368	-436.058	-217.783	2.528
4600	82.731	479.441	405.389	340.640	-436.414	-212.929	2.418
4700	82.748	481.220	406.983	348.914	-436.763	-208.067	2.312
4800	82.765	482.962	408.548	357.189	-437.104	-203.197	2.211
4900	82.780	484.669	410.084	365.467	-437.438	-198.320	2.114
5000	82.794	486.342	411.593	373.745	-437.763	-193.437	2.021
5100	82.808	487.981	413.074	382.025	-438.080	-188.547	1.931
5200	82.820	489.589	414.530	390.307	-438.388	-183.652	1.845
5300	82.832	491.167	415.962	398.589	-438.688	-178.750	1.762
5400	82.844	492.715	417.369	406.873	-438.979	-173.843	1.682
5500	82.854	494.236	418.752	415.158	-439.262	-168.931	1.604
5600	82.865	495.729	420.114	423.444	-439.538	-164.012	1.530
5700	82.874	497.195	421.453	431.731	-439.807	-159.090	1.458
5800	82.883	498.637	422.772	440.019	-440.068	-154.163	1.388
5900	82.892	500.054	424.069	448.308	-440.324	-149.232	1.321
6000	82.900	501.447	425.347	456.597	-440.575	-144.296	1.256

PREVIOUS: June 1961 (1 atm) CURRENT: June 1961 (1 bar)

Carbonic Chloride Fluoride (COClF) $C_1Cl_1F_1O_1(g)$

CHLOROTRIFLUOROMETHANE ($CClF_3$) IDEAL GAS M_r = 104.459209

$S°(298.15\ K) = 285.35 \pm 0.4\ J\ K^{-1}\ mol^{-1}$

$\Delta_f H°(0\ K) = -702.84 \pm 3.3\ kJ\ mol^{-1}$
$\Delta_f H°(298.15\ K) = -707.93 \pm 3.3\ kJ\ mol^{-1}$

Vibrational Frequencies and Degeneracies

ν, cm^{-1}	ν, cm^{-1}
1106(1)	1217(2)
782(1)	560(2)
474(1)	350(2)

Ground State Quantum Weight: 1 $\sigma = 3$
Point Group: C_{3v}
Bond Distances: C–F = 1.328 Å C–Cl = 1.751 Å
Bond Angles: F–C–F = 108.6 ± 0.5° F–C–Cl = 110.33°
Product of the Moments of Inertia: $I_A I_B I_C = 9.444771 \times 10^{-114}\ g^3\ cm^6$

Enthalpy of Formation

The selected value is obtained from least squares, simultaneous adjustment of the heats of formation of CF_3, CF_3X (X = H, Cl, Br, I, CF_3) and C_2F_4. Details of the input data and the adjustment are given in (1). Coomber and Whittle (2) have determined equilibrium data for $CClF_3(g) + BrX(g) = CBrF_3(g) + ClX(g)$ where X = Cl and Br. Lord et al. (3) studied the analogous equilibria $CF_3I(g) + ClI(g) = CClF_3(g) + I_2(g)$. These data link $CClF_3$ with $CBrF_3$ and CF_3I.

Other data which were omitted from the adjustment include those of Baibuz (4, 5), Mears (6), Kolesov (7), Kirkbride (8), and von Wartenberg (9). Baibuz (4, 5) derived $\Delta_f H° = -166.2 \pm 2.2\ kcal\ mol^{-1}$ by applying the method of explosion in a spherical bomb to mixtures of H_2, O_2 and $CClF_3$. By comparison with similar data for CF_4 and CCl_4, we estimate the bias in Baibuz' value to be about +2 kcal mol^{-1}. Mears (6) gave a preliminary report of values of $\Delta_r H$ at 300°C for reactions involving HF, HCl and CCl_nF_{4-n} (n = 1, 2, 3, 4). We derive $\Delta_f H° = -172$ and $-173\ kcal\ mol^{-1}$ from two different combinations of reactions; however, no detailed account of the data has been published. The other authors (7, 8, 9) measured $\Delta_r H$ for reaction of $CClF_3$ with an alkali metal (Na or K) to form amorphous carbon and mixtures of the fluoride and chloride. Discrepancies of many kcal mol^{-1} exist in these studies, presumably due to inadequate characterization of the thermochemical reactions.

Heat Capacity and Entropy

The molecular structure has been derived from electron-diffraction studies by Bowen (10) and Bartell (11). Results from Bartell are adopted. These parameters are consistent with the microwave data of Coles (12). The principal moments of inertia are: $I_A = 14.6768 \times 10^{-39}$ and $I_B = I_C = 25.3677 \times 10^{-39}\ g\ cm^2$. Vibrational assignments are based on the work of Claassen (13) and the gas-phase Raman spectra observed by Holzer and Moser (14).

The largest uncertainty in $C°_p$ and S° probably results from the neglect of anharmonicity. Barho (15) and Albright (16) have given estimates for the anharmonic contributions. These vary from 0.1 to 0.3 cal $K^{-1}\ mol^{-1}$ in both $C°_p$ and S° over the range from 300 to 1000 K, respectively. Somewhat larger contributions are possible based on the anharmonic functions calculated by McBride (17) for $CF_4(g)$. We estimate the overall uncertainty in the entropy of this JANAF table to be 0.1 cal $K^{-1}\ mol^{-1}$ at 300 K and 0.4 cal $K^{-1}\ mol^{-1}$ at 1000 K. $C°_p$ is consistent with $C°_v$ measured at 300 K by the wire-ribbon method (18).

References

1. A. N. Syverud, Quarterly Tech. Report T-0009-20-69, AFRPL Contract No. F04611-67-C-0009, Dow Chemical Co., July, 1969.
2. J. W. Coomber and E. Whittle, Trans. Faraday Soc. 63, 2656 (1967).
3. A. Lord, A. A. Goy and H. O. Pritchard, J. Phys. Chem. 71, 2705 (1967).
4. V. F. Baibuz, Dokl. Akad. Nauk SSSR 140 (6), 1358 (1961); English translation, 786 (1961).
5. V. F. Baibuz and V. A. Medvedev, Tr., Gos. Inst. Prikl. Khim. No. 49, 84 (1962).
6. W. H. Mears and R. F. Stahl, I.U.P.A.C. Thermochemical Bulletin 2, 5 (1956).
7. V. P. Kolesov, I. D. Zenkov and S. M. Skuratov, Russ. J. Phys. Chem. 36 (9), 1120 (1962); 37 (3), 378 (1963).
8. F. W. Kirkbride and F. G. Davidson, Nature 174, 79 (1954).
9. H. von Wartenberg and J. Schiefer, Z. Anorg. Allg. Chem. 278, 326 (1955).
10. H. J. M. Bowen, Trans. Faraday Soc. 50, 444 (1954).
11. L. S. Bartell and L. O. Brockway, J. Chem. Phys. 23, 1860 (1955).
12. D. K. Coles and R. H. Hughes, Phys. Rev. 76, 858 (1949); recalculated by J. Sheridan and W. Gordy, J. Chem. Phys. 20, 591 (1952).
13. H. H. Claassen, J. Chem. Phys. 22, 50 (1954).
14. W. Holzer and H. Moser, J. Mol. Spectrosc. 20, 188 (1966).
14. W. Barho, Kaltetechnik 17, 219 (1965).
15. L. Albright, W. C. Calegar and K. K. Innes, J. Amer. Chem. Soc. 76, 6017 (1954).
16. B. J. McBride, S. Heimel, J. G. Ehlers and S. Gordon, NASA SP-3001, 156 (1963).
17. W. N. Vanderkooi and T. DeVries, J. Phys. Chem. 60, 636 (1956).

Chlorotrifluoromethane ($CClF_3$) $C_1Cl_1F_3(g)$

Enthalpy Reference Temperature = T_r = 298.15 K Standard State Pressure = p° = 0.1 MPa

T/K	$C°_p$ (J K⁻¹mol⁻¹)	S° (J K⁻¹mol⁻¹)	$-[G°-H°(T_r)]/T$ (J K⁻¹mol⁻¹)	$H°-H°(T_r)$ (kJ mol⁻¹)	$\Delta_f H°$ (kJ mol⁻¹)	$\Delta_f G°$ (kJ mol⁻¹)	Log K_f
0	0.	0.	INFINITE	-13.782	-702.836	-702.836	INFINITE
100	36.815	231.109	335.022	-10.391	-705.321	-693.331	362.159
200	53.051	261.466	291.051	-5.917	-707.086	-680.571	177.747
250	60.573	274.131	286.420	-3.072	-707.600	-673.878	140.799
298.15	66.887	285.353	285.353	0.	-707.933	-667.351	116.917
300	67.112	285.767	285.354	0.124	-707.943	-667.099	116.152
350	72.738	296.546	286.192	3.624	-708.157	-660.274	98.540
400	77.534	306.581	288.121	7.384	-708.270	-653.424	85.328
450	81.595	315.954	290.699	11.365	-708.305	-646.566	75.051
500	85.019	324.733	293.668	15.533	-708.282	-639.707	66.830
600	90.333	340.731	300.206	24.315	-708.103	-626.006	54.499
700	94.135	354.958	307.030	33.549	-707.806	-612.345	45.694
800	96.901	367.718	313.833	43.108	-707.432	-598.733	39.093
900	98.953	379.255	320.471	52.906	-707.010	-585.170	33.962
1000	100.508	389.765	326.883	62.882	-706.560	-571.657	29.860
1100	101.709	399.403	333.044	72.996	-706.096	-558.189	26.506
1200	102.652	408.295	338.949	83.215	-705.627	-544.763	23.713
1300	103.404	416.542	344.604	93.520	-705.162	-531.377	21.351
1400	104.014	424.229	350.020	103.891	-704.706	-518.026	19.328
1500	104.513	431.422	355.210	114.319	-704.265	-504.707	17.575
1600	104.927	438.181	360.187	124.791	-703.840	-491.417	16.043
1700	105.274	444.553	364.964	135.302	-703.434	-478.153	14.692
1800	105.568	450.579	369.554	145.844	-703.047	-464.913	13.491
1900	105.818	456.294	373.970	156.414	-702.678	-451.694	12.418
2000	106.033	461.727	378.224	167.007	-702.325	-438.493	11.452
2100	106.219	466.905	382.324	177.620	-701.985	-425.310	10.579
2200	106.381	471.850	386.282	188.250	-701.656	-412.143	9.786
2300	106.523	476.582	390.106	198.895	-701.334	-398.990	9.061
2400	106.648	481.118	393.804	209.554	-701.014	-385.851	8.398
2500	106.759	485.474	397.384	220.224	-700.693	-372.726	7.788
2600	106.858	489.663	400.854	230.905	-700.367	-359.614	7.225
2700	106.946	493.698	404.218	241.596	-700.030	-346.514	6.704
2800	107.025	497.589	407.484	252.294	-699.680	-333.428	6.220
2900	107.096	501.346	410.656	263.000	-699.313	-320.353	5.770
3000	107.160	504.977	413.740	273.713	-698.926	-307.293	5.350
3100	107.218	508.492	416.740	284.432	-698.515	-294.245	4.958
3200	107.271	511.897	419.661	295.156	-698.078	-281.211	4.590
3300	107.319	515.199	422.506	305.886	-697.611	-268.191	4.245
3400	107.363	518.403	425.280	316.620	-697.113	-255.186	3.920
3500	107.404	521.516	427.985	327.359	-696.582	-242.195	3.615
3600	107.441	524.542	430.625	338.101	-696.015	-229.220	3.326
3700	107.475	527.486	433.203	348.847	-695.411	-216.262	3.053
3800	107.507	530.353	435.722	359.596	-694.769	-203.321	2.795
3900	107.536	533.146	438.185	370.348	-694.088	-190.397	2.550
4000	107.563	535.869	440.593	381.103	-693.365	-177.491	2.318
4100	107.588	538.525	442.949	391.860	-692.601	-164.603	2.097
4200	107.611	541.118	445.256	402.620	-691.795	-151.735	1.887
4300	107.633	543.650	447.515	413.382	-690.946	-138.887	1.687
4400	107.653	546.125	449.728	424.147	-690.053	-126.058	1.496
4500	107.672	548.544	451.897	434.913	-689.117	-113.250	1.315
4600	107.690	550.911	454.024	445.681	-688.136	-100.464	1.141
4700	107.707	553.227	456.110	456.451	-687.112	-87.700	0.975
4800	107.722	555.495	458.157	467.222	-686.045	-74.957	0.816
4900	107.737	557.716	460.166	477.995	-684.933	-62.238	0.663
5000	107.751	559.893	462.139	488.770	-683.778	-49.541	0.518
5100	107.764	562.027	464.077	499.546	-682.580	-36.868	0.378
5200	107.776	564.120	465.981	510.323	-681.340	-24.219	0.243
5300	107.788	566.173	467.852	521.101	-680.058	-11.594	0.114
5400	107.799	568.188	469.691	531.880	-678.734	1.006	-0.010
5500	107.809	570.166	471.500	542.660	-677.369	13.581	-0.129
5600	107.819	572.108	473.280	553.442	-675.964	26.131	-0.244
5700	107.828	574.017	475.030	564.224	-674.520	38.655	-0.354
5800	107.837	575.892	476.753	575.007	-673.037	51.154	-0.461
5900	107.845	577.736	478.449	585.792	-671.516	63.628	-0.563
6000	107.853	579.548	480.119	596.576	-669.959	76.074	-0.662

PREVIOUS: June 1969 (1 atm) CURRENT: June 1969 (1 bar)

Chlorotrifluoromethane ($CClF_3$) $C_1Cl_1F_3(g)$

CYANOGEN CHLORIDE (ClCN) IDEAL GAS $M_r = 61.4707$

$\Delta_f H°(0\ K) = 137.24\ kJ\ mol^{-1}$

$S°(298.15\ K) = 236.334\ J\ K^{-1}\ mol^{-1}$ $\Delta_f H°(298.15\ K) = 137.95\ kJ\ mol^{-1}$

Vibrational Frequencies and Degeneracies

ν, cm^{-1}
714.52(1)
378.3 (2)
2215.5 (1)

Ground State Quantum Weight: 1 $\sigma = 1$

Point Group: $C_{\infty v}$

Bond Distances: Cl–C = 1.631 Å C–N = 1.162 Å

Bond Angle: Cl-C-N = 180°

Rotational Constants: $B_o = 0.19817\ cm^{-1}$ $D_o = 5.503 \times 10^{-8}\ cm^{-1}$

Spectroscopic constants used in calculating corrections to rigid rotator-harmonic oscillator approximation (cm^{-1}):

$\alpha_1 = 8.25 \times 10^{-4}$ $x_{11} = [-4]$ $x_{12} = [-6.8]$ $g_{22} = 0.95$

$\alpha_2 = -5.46 \times 10^{-4}$ $x_{22} = -0.65$ $x_{23} = -7.236$

$\alpha_3 = 1.06 \times 10^{-3}$ $x_{33} = [-7]$ $x_{13} = -2.8$

Enthalpy of Formation

The enthalpy of formation was taken from Wagman et al. (1).

Heat Capacity and Entropy

Gordon (2) has used the constants listed above to calculate C_p from 298.15 to 6000 K using the method of Pennington and Kobe (3) which takes anharmonicity, vibration-rotation interaction, and centrifugal stretching into account. The constants B_o and D_o were taken from Burrus and Gordy (4), ω_1, x_1, x_3, x_{22}, x_{23}, x_{13}, and g_{22} were obtained from Lafferty et al. (5), α_2 has been taken from Townes et al. (6) and x_{11}, x_{33} and x_{12} were estimated by Gordon (2). The bond distances and bond angle were taken from Lafferty et al. (5).

References

1. U. S. Nat. Bur. Stand. Tech. Note 270-1, 1965.
2. J. S. Gordon, Astrosystems International, Inc., Fairfield, NJ, personal communication, Feb., 1966.
3. R. E. Pennington and K. A. Kobe, J. Chem. Phys. 22, 1442 (1954).
4. C. A. Burrus and W. Gordy, Phys. Rev. 101, 599 (1956).
5. W. J. Lafferty, D. R. Lide and R. A. Toth, J. Chem. Phys. 43, 2063 (1965).
6. C. H. Townes, A. N. Holden and F. R. Merritt, Phys. Rev. 74, 1113 (1948).

Cyanogen Chloride (ClCN) $C_1Cl_1N_1(g)$

Enthalpy Reference Temperature = T_r = 298.15 K Standard State Pressure = p° = 0.1 MPa

	J $K^{-1}mol^{-1}$			kJ mol^{-1}			
T/K	$C_p°$	S°	$-[G°-H°(T_r)]/T$	$H°-H°(T_r)$	$\Delta_fH°$	$\Delta_fG°$	Log K_f
0	0.	0.	INFINITE	-10.686	137.237	137.237	INFINITE
100	31.246	195.017	272.421	-7.740	137.216	135.250	-70.648
200	39.643	219.305	240.225	-4.184	137.472	133.222	-34.794
298.15	45.271	236.334	236.334	0.	137.946	131.015	-22.953
300	45.346	236.614	236.335	0.084	137.956	130.972	-22.804
400	48.718	250.147	238.157	4.796	138.452	128.567	-16.789
500	51.024	261.275	241.701	9.787	138.861	126.046	-13.168
600	52.831	270.742	245.771	14.982	139.170	123.453	-10.748
700	54.354	279.003	249.941	20.343	139.401	120.814	-9.015
800	55.656	286.348	254.041	25.845	139.578	118.146	-7.714
900	56.781	292.970	258.005	31.469	139.718	115.459	-6.701
1000	57.748	299.003	261.807	37.196	139.834	112.757	-5.890
1100	58.504	304.543	265.444	43.009	139.929	110.044	-5.226
1200	59.204	309.664	268.918	48.895	140.009	107.324	-4.672
1300	59.847	314.429	272.238	54.848	140.079	104.597	-4.203
1400	60.432	318.886	275.412	60.862	140.147	101.865	-3.801
1500	60.961	323.073	278.452	66.932	140.213	99.129	-3.452
1600	61.365	327.021	281.365	73.049	140.275	96.388	-3.147
1700	61.745	330.752	284.161	79.205	140.331	93.643	-2.877
1800	62.103	334.292	286.849	85.397	140.384	90.895	-2.638
1900	62.437	337.659	289.435	91.624	140.433	88.144	-2.423
2000	62.747	340.869	291.927	97.884	140.480	85.391	-2.230
2100	63.010	343.937	294.331	104.172	140.523	82.636	-2.055
2200	63.262	346.874	296.653	110.485	140.560	79.878	-1.897
2300	63.503	349.692	298.899	116.824	140.591	77.119	-1.751
2400	63.733	352.399	301.072	123.186	140.616	74.359	-1.618
2500	63.952	355.005	303.177	129.570	140.635	71.598	-1.496
2600	64.150	357.518	305.219	135.975	140.646	68.836	-1.383
2700	64.342	359.942	307.201	142.400	140.648	66.074	-1.278
2800	64.529	362.286	309.127	148.844	140.642	63.312	-1.181
2900	64.710	364.553	310.999	155.306	140.628	60.551	-1.091
3000	64.885	366.750	312.821	161.785	140.604	57.790	-1.006
3100	65.050	368.880	314.596	168.282	140.572	55.030	-0.927
3200	65.212	370.948	316.324	174.795	140.530	52.271	-0.853
3300	65.371	372.957	318.010	181.325	140.481	49.514	-0.784
3400	65.527	374.911	319.655	187.870	140.424	46.758	-0.718
3500	65.680	376.813	321.261	194.430	140.360	44.004	-0.657
3600	65.828	378.665	322.830	201.005	140.290	41.252	-0.599
3700	65.973	380.471	324.364	207.595	140.215	38.502	-0.544
3800	66.117	382.232	325.864	214.200	140.137	35.754	-0.491
3900	66.259	383.951	327.331	220.819	140.055	33.008	-0.442
4000	66.400	385.631	328.768	227.452	139.973	30.264	-0.395
4100	66.536	387.272	330.175	234.099	139.890	27.523	-0.351
4200	66.670	388.877	331.553	240.759	139.808	24.783	-0.308
4300	66.804	390.447	332.905	247.433	139.729	22.045	-0.268
4400	66.937	391.984	334.230	254.120	139.653	19.309	-0.229
4500	67.070	393.490	335.530	260.820	139.582	16.575	-0.192
4600	67.201	394.966	336.806	267.534	139.518	13.842	-0.157
4700	67.331	396.412	338.059	274.260	139.460	11.111	-0.123
4800	67.461	397.831	339.290	281.000	139.411	8.380	-0.091
4900	67.590	399.224	340.499	287.752	139.372	5.651	-0.060
5000	67.718	400.590	341.687	294.518	139.342	2.922	-0.031
5100	67.845	401.933	342.855	301.296	139.323	0.194	-0.002
5200	67.972	403.251	344.004	308.087	139.316	-2.534	0.025
5300	68.097	404.547	345.134	314.890	139.321	-5.262	0.052
5400	68.222	405.821	346.246	321.706	139.339	-7.990	0.077
5500	68.346	407.074	347.341	328.535	139.370	-10.718	0.102
5600	68.469	408.307	348.418	335.375	139.415	-13.448	0.125
5700	68.591	409.520	349.480	342.228	139.474	-16.178	0.148
5800	68.712	410.714	350.525	349.093	139.547	-18.909	0.170
5900	68.833	411.889	351.555	355.971	139.634	-21.642	0.192
6000	68.952	413.047	352.571	362.860	139.735	-24.376	0.212

PREVIOUS: June 1966 (1 atm) CURRENT: June 1966 (1 bar)

Cyanogen Chloride (ClCN) $C_1Cl_1N_1(g)$

CARBONYL CHLORIDE (COCl) IDEAL GAS Mr = 63.4634

$S°(298.15\ K) = [265.965]\ J\ K^{-1}\ mol^{-1}$

$\Delta_f H°(0\ K) = [-64.3 \pm 42]\ kJ\ mol^{-1}$
$\Delta_f H°(298.15\ K) = [-62.8 \pm 42]\ kJ\ mol^{-1}$

Vibrational Frequencies and Degeneracies

ν, cm^{-1}
570 (1)
281 (1)
1880 (1)

Ground State Quantum Weight: 2 $\sigma = 1$
Point Group: C_s
Bond Distance: C–Cl = [1.75] Å C–O = [1.17] Å
Bond Angle: Cl–C–O = [120]°
Product of the Moments of Inertia: $I_A I_B I_C = [1.276664 \times 10^{-115}]\ g^3\ cm^6$

Enthaply of Formation

The enthalpy change ($\Delta_r H°$) for the reaction $COCl_2(g) = CO(g) + 2Cl(g)$ is calculated to be 82 kcal mol^{-1}, which is the sum of two bond energies, i.e. D(Cl–COCl) + D(Cl–CO). By comparison with values, $D(H_3C\text{–}COCH_3)/D(H_3C\text{–}CO) = 72/18 = 4.0$ for $CH_3COCH_3(g)$ and $D(H_3C\text{–}COCl)/D(Cl\text{–}CO) = 4.3$ for $CH_3COCl(g)$, the ratio of D(Cl–COCl)/D(Cl–CO) is estimated to be 4.0 for $COCl_2(g)$. Hence the enthalpy change for the reaction $COCl_2(g) = COCl(g) + Cl(g)$ is calculated as 82 x (4/5) = 65.6 kcal mol^{-1}. Using the $\Delta_r H°$ value obtained, the enthalpy of formation for COCl(g) is derived to be -15±10 kcal mol^{-1}.

Heat Capacity and Entropy

The vibrational frequencies were reported by Jacox and Milligan (1). The bond distances and angle were estimated by Jacox and Milligan (1). Due to the presence of one unpaired electron on C atom, the ground state quantum weight is calculated to be 2. The principal moments of inertia are: $I_A = 0.7159 \times 10^{-39}$, $I_B = 13.0005 \times 10^{-39}$, and $I_C = 13.7165 \times 10^{-39}\ g\ cm^2$.

References

1. M. E. Jacoy and D. E. Milligan, J. Chem. Phys. 43, 866 (1965).

Carbonyl Chloride (COCl) $C_1Cl_1O_1(g)$

Enthalpy Reference Temperature = T_r = 298.15 K Standard State Pressure = p° = 0.1 MPa

T/K	$C_p°$ (J $K^{-1}mol^{-1}$)	S° (J $K^{-1}mol^{-1}$)	$-[G°-H°(T_r)]/T$ (J $K^{-1}mol^{-1}$)	$H°-H°(T_r)$ (kJ mol^{-1})	$\Delta_f H°$ (kJ mol^{-1})	$\Delta_f G°$ (kJ mol^{-1})	Log K_f
0	0.	0.	INFINITE	-11.551	-64.327	-64.327	INFINITE
100	35.882	221.891	303.521	-8.163	-63.908	-67.886	35.460
200	41.634	248.636	269.991	-4.271	-63.317	-72.083	18.826
250	43.627	258.152	266.699	-2.137	-63.016	-74.309	15.526
298.15	45.072	265.965	265.965	0.	-62.760	-76.508	13.404
300	45.120	266.244	265.966	0.083	-62.751	-76.593	13.336
350	46.289	273.290	266.520	2.370	-62.533	-78.918	11.778
400	47.265	279.537	267.764	4.709	-62.369	-81.271	10.613
450	48.125	285.154	269.389	7.094	-62.258	-83.641	9.709
500	48.908	290.266	271.225	9.520	-62.199	-86.020	8.987
600	50.301	299.309	275.171	14.483	-62.211	-90.786	7.904
700	51.487	307.155	279.192	19.574	-62.356	-95.539	7.129
800	52.484	314.097	283.130	24.774	-62.595	-100.264	6.547
900	53.312	320.328	286.923	30.065	-62.901	-104.955	6.091
1000	53.996	325.981	290.550	35.431	-63.257	-109.609	5.725
1100	54.560	331.155	294.010	40.860	-63.652	-114.225	5.424
1200	55.028	335.923	297.306	46.340	-64.079	-118.804	5.171
1300	55.417	340.344	300.449	51.863	-64.533	-123.346	4.956
1400	55.743	344.463	303.447	57.421	-65.011	-127.852	4.770
1500	56.018	348.318	306.312	63.010	-65.513	-132.324	4.608
1600	56.251	351.941	309.051	68.624	-66.038	-136.761	4.465
1700	56.450	355.357	311.676	74.259	-66.585	-141.164	4.337
1800	56.621	358.589	314.193	79.913	-67.155	-145.535	4.223
1900	56.769	361.654	316.611	85.582	-67.748	-149.874	4.120
2000	56.898	364.570	318.937	91.266	-68.364	-154.180	4.027
2100	57.010	367.348	321.176	96.961	-69.004	-158.455	3.941
2200	57.109	370.003	323.336	102.667	-69.669	-162.699	3.863
2300	57.196	372.543	325.420	108.383	-70.359	-166.913	3.791
2400	57.274	374.979	327.435	114.106	-71.076	-171.096	3.724
2500	57.343	377.319	329.384	119.837	-71.820	-175.248	3.662
2600	57.405	379.569	331.271	125.575	-72.592	-179.370	3.604
2700	57.460	381.737	333.100	131.318	-73.391	-183.461	3.549
2800	57.510	383.827	334.875	137.066	-74.219	-187.523	3.498
2900	57.555	385.846	336.598	142.820	-75.075	-191.555	3.450
3000	57.596	387.798	338.272	148.577	-75.960	-195.556	3.405
3100	57.633	389.687	339.900	154.339	-76.872	-199.527	3.362
3200	57.667	391.517	341.485	160.104	-77.812	-203.469	3.321
3300	57.698	393.292	343.028	165.872	-78.779	-207.381	3.283
3400	57.726	395.015	344.532	171.643	-79.772	-211.263	3.246
3500	57.752	396.689	345.998	177.417	-80.789	-215.116	3.210
3600	57.776	398.316	347.429	183.193	-81.830	-218.939	3.177
3700	57.798	399.900	348.826	188.972	-82.893	-222.734	3.144
3800	57.818	401.441	350.190	194.753	-83.978	-226.498	3.113
3900	57.837	402.943	351.524	200.536	-85.082	-230.234	3.084
4000	57.855	404.408	352.828	206.320	-86.204	-233.942	3.055
4100	57.871	405.837	354.103	212.107	-87.343	-237.621	3.027
4200	57.887	407.231	355.352	217.895	-88.497	-241.273	3.001
4300	57.901	408.594	356.574	223.684	-89.666	-244.896	2.975
4400	57.914	409.925	357.772	229.475	-90.846	-248.493	2.950
4500	57.926	411.227	358.945	235.267	-92.038	-252.063	2.926
4600	57.938	412.500	360.095	241.060	-93.241	-255.606	2.902
4700	57.949	413.746	361.224	246.854	-94.452	-259.122	2.880
4800	57.959	414.966	362.331	252.650	-95.672	-262.613	2.858
4900	57.969	416.161	363.417	258.446	-96.899	-266.078	2.836
5000	57.978	417.333	364.484	264.244	-98.132	-269.518	2.816
5100	57.986	418.481	365.531	270.042	-99.372	-272.933	2.795
5200	57.994	419.607	366.560	275.841	-100.616	-276.325	2.776
5300	58.002	420.712	367.572	281.641	-101.867	-279.691	2.757
5400	58.009	421.796	368.566	287.441	-103.122	-283.035	2.738
5500	58.016	422.860	369.543	293.242	-104.381	-286.355	2.720
5600	58.023	423.906	370.505	299.044	-105.646	-289.652	2.702
5700	58.029	424.933	371.451	304.847	-106.915	-292.927	2.684
5800	58.035	425.942	372.382	310.650	-108.189	-296.179	2.667
5900	58.040	426.934	373.298	316.454	-109.469	-299.410	2.651
6000	58.045	427.910	374.200	322.258	-110.754	-302.618	2.635

PREVIOUS: December 1965 (1 atm) CURRENT: December 1965 (1 bar)

Carbonyl Chloride (COCl) $C_1Cl_1O_1(g)$

DICHLOROMETHYLENE (CCl_2) IDEAL GAS M_r = 82.917

$S°(298.15\ K) = [265.346]\ J\ K^{-1}\ mol^{-1}$

$\Delta_fH°(0\ K) = 237.3 \pm 21\ kJ\ mol^{-1}$
$\Delta_fH°(298.15\ K) = 238.5 \pm 21\ kJ\ mol^{-1}$

Electronic Levels and Quantum Weights

ϵ_i, cm^{-1}	g_i
0	1
[4000]	[3]
[17500]	[1]

Vibrational Frequencies and Degeneracies

ν, cm^{-1}
721 (1)
[350](1)
748 (1)

Point Group: C_{2v} σ = 2
Bond Distance: C-Cl = [1.76] Å
Bond Angle: Cl-C-Cl = [100 ± 10]°
Product of the Moments of Inertia: $I_AI_BI_C = [1.101940 \times 10^{-114}]\ g^3\ cm^6$

Enthalpy of Formation

Shapiro and Lossing (1) have measured mass-spectrometrically the appearance potentials 12.2 and 14.7 eV for the processes $CHCl_3(g) + e^- \rightarrow CCl_2^+(g) + HCl(g) + 2e^-$ and $C_2Cl_4(g) + e^- \rightarrow CCl_2^+(g) + CCl_2(g) + 2e^-$, respectively. The difference in these two processes gives $\Delta_rH°$ = 2.5 eV (57.65 kcal mol^{-1}), for the reaction $C_2Cl_4(g) + HCl(g) \rightarrow CCl_2(g) + CHCl_3(g)$ which leads to $\Delta_fH°(CCl_2$, g, 298.15 K) = 57.32 kcal mol^{-1}, using $\Delta_fH°$(298.15 K) = -2.97, -22.063 and -24.7 kcal mol^{-1} for $C_2Cl_4(g)$, HCl(g) and $CHCl_3(g)$, respectively. Since the authors claimed that the experimental error was probably 5 kcal mol^{-1}, the value 57±5 kcal mol^{-1} (238.488±21 kJ mol^{-1}), is adopted for the enthalpy of formation.

Heat Capacity and Entropy

Milligan and Jacox (2) have found two stretching fundamentals (721 and 748 cm^{-1}) for the CCl_2 radical in the infrared and ultraviolet spectra by matrix-isolation, and proposed a bent molecular structure (the valence angle = 100 ± 10°) with a ground singlet state which was based on the analogy with the electronic absorption spectrum of CF_2. L. Andrews (3) has confirmed all these observations in his infrared spectrometric studies of dichlorocarbene in solid argon, and made vibrational assignment to the fundamentals and also eliminated the possibility of a linear molecular structure for CCl_2. The bending frequency (350 cm^{-1}) is estimated by the valence force method (4), using the same ratio of bending over stretching force constants from $CF_2(g)$ (5). The bond distance is estimated to be the same as that of $CCl_4(g)$ measured by Bartell et al. (6). The principal moments of inertia are $I_A = 2.1829 \times 10^{-39}$, $I_B = 21.4029 \times 10^{-39}$, and $I_C = 23.5858 \times 10^{-39}\ g\ cm^2$.

The triplet low lying electronic level of 4000 cm^{-1} is estimated by analogy with that of $CF_2(g)$. The other level of 17,5000 cm^{-1} is also estimated based on a weak band appearing between 4400 and 5600 Å in the spectra of CCl_2 observed by Milligan and Jacox (2).

References

1. J. S. Shapiro and F. P. Lossing, J. Phys. Chem. 72, 1552 (1968).
2. D. E. Milligan and M. E. Jacox, J. Chem. Phys. 47, 703 (1967).
3. L. Andrews, J. Chem. Phys. 48, 979 (1968).
4. G. Herzberg, "Infrared and Raman Spectra," D. Van Nostrand Co., Inc., New York, 1945.
5. JANAF Thermochemical Tables: $CF_2(g)$, 9-30-66.
6. L. S. Bartell, L. O. Brockway and R. H. Schwendeman, J. Chem. Phys. 23, 1854 (1955).

Dichloromethylene (CCl_2) $C_1Cl_2(g)$

Enthalpy Reference Temperature = T_r = 298.15 K; Standard State Pressure = p° = 0.1 MPa

T/K	$C_p°$ (J K⁻¹mol⁻¹)	S° (J K⁻¹mol⁻¹)	-[G°-H°(T_r)]/T (J K⁻¹mol⁻¹)	H°-H°(T_r) (kJ mol⁻¹)	$\Delta_fH°$ (kJ mol⁻¹)	$\Delta_fG°$ (kJ mol⁻¹)	Log K_f
0	0.	0.	INFINITE	-11.388	237.332	237.332	INFINITE
100	34.695	222.225	302.564	-8.034	237.714	234.488	-122.484
200	40.669	248.013	269.408	-4.279	238.105	231.112	-60.360
250	43.720	257.422	266.094	-2.168	238.302	229.341	-47.918
298.15	46.249	265.346	265.346	0.	238.488	227.598	-39.874
300	46.337	265.632	265.347	0.086	238.495	227.530	-39.617
350	48.459	272.941	265.919	2.458	238.677	225.688	-33.682
400	50.142	279.526	267.215	4.924	238.841	223.821	-29.228
450	51.476	285.511	268.921	7.466	238.978	221.935	-25.762
500	52.546	290.992	270.858	10.067	239.086	220.035	-22.987
600	54.180	300.725	275.045	15.408	239.216	216.211	-18.823
700	55.475	309.176	279.330	20.892	239.256	212.373	-15.847
800	56.676	316.663	283.538	26.500	239.243	208.533	-13.616
900	57.888	323.408	287.599	32.228	239.216	204.696	-11.880
1000	59.129	329.571	291.492	38.079	239.207	200.861	-10.492
1100	60.369	335.265	295.216	44.054	239.238	197.025	-9.356
1200	61.560	340.569	298.777	50.151	239.322	193.184	-8.409
1300	62.656	345.541	302.185	56.362	239.462	189.334	-7.608
1400	63.622	350.220	305.451	62.677	239.656	185.471	-6.920
1500	64.437	354.638	308.584	69.082	239.895	181.593	-6.324
1600	65.093	358.819	311.594	75.559	240.168	177.697	-5.801
1700	65.593	362.781	314.490	82.095	240.463	173.784	-5.340
1800	65.948	366.541	317.278	88.673	240.767	169.853	-4.929
1900	66.173	370.113	319.965	95.280	241.068	165.905	-4.561
2000	66.286	373.510	322.558	101.904	241.354	161.941	-4.229
2100	66.306	376.745	325.062	108.534	241.618	157.964	-3.929
2200	66.250	379.829	327.482	115.163	241.849	153.975	-3.656
2300	66.134	382.771	329.823	121.782	242.040	149.976	-3.406
2400	65.971	385.583	332.088	128.388	242.187	145.970	-3.177
2500	65.776	388.272	334.282	134.976	242.285	141.959	-2.966
2600	65.556	390.848	336.408	141.542	242.330	137.945	-2.771
2700	65.321	393.317	338.470	148.086	242.320	133.930	-2.591
2800	65.078	395.688	340.472	154.606	242.253	129.917	-2.424
2900	64.831	397.968	342.415	161.102	242.129	125.907	-2.268
3000	64.585	400.161	344.304	167.572	241.947	121.902	-2.123
3100	64.343	402.275	346.140	174.019	241.709	117.904	-1.987
3200	64.107	404.314	347.926	180.441	241.415	113.915	-1.859
3300	63.879	406.283	349.665	186.840	241.068	109.936	-1.740
3400	63.660	408.187	351.359	193.217	240.670	105.968	-1.628
3500	63.451	410.030	353.009	199.573	240.223	102.013	-1.522
3600	63.253	411.814	354.618	205.908	239.731	98.071	-1.423
3700	63.064	413.545	356.187	212.224	239.197	94.143	-1.329
3800	62.886	415.224	357.719	218.521	238.624	90.230	-1.240
3900	62.718	416.855	359.214	224.801	238.015	86.333	-1.156
4000	62.560	418.441	360.675	231.065	237.375	82.452	-1.077
4100	62.411	419.984	362.103	237.314	236.707	78.587	-1.001
4200	62.272	421.486	363.499	243.548	236.015	74.739	-0.930
4300	62.141	422.950	364.865	249.768	235.302	70.907	-0.861
4400	62.018	424.377	366.201	255.976	234.572	67.092	-0.796
4500	61.903	425.770	367.509	262.172	233.827	63.294	-0.735
4600	61.795	427.129	368.791	268.357	233.072	59.513	-0.676
4700	61.694	428.457	370.046	274.531	232.309	55.748	-0.620
4800	61.599	429.755	371.277	280.696	231.542	52.000	-0.566
4900	61.510	431.024	372.483	286.851	230.771	48.267	-0.515
5000	61.427	432.266	373.666	292.998	230.001	44.550	-0.465
5100	61.348	433.482	374.827	299.137	229.234	40.849	-0.418
5200	61.275	434.672	375.967	305.268	228.470	37.162	-0.373
5300	61.206	435.839	377.085	311.392	227.713	33.491	-0.330
5400	61.141	436.982	378.184	317.509	226.964	29.833	-0.289
5500	61.080	438.103	379.263	323.620	226.224	26.190	-0.249
5600	61.022	439.203	380.324	329.725	225.495	22.559	-0.210
5700	60.968	440.283	381.366	335.825	224.778	18.942	-0.174
5800	60.916	441.343	382.391	341.919	224.073	15.337	-0.138
5900	60.868	442.384	383.399	348.008	223.382	11.744	-0.104
6000	60.822	443.406	384.391	354.093	222.705	8.162	-0.071

PREVIOUS: December 1968 (1 atm) CURRENT: December 1968 (1 bar)

Dichloromethylene (CCl_2) $C_1Cl_2(g)$

DICHLORODIFLUOROMETHANE (CCl_2F_2) IDEAL GAS M_r = 120.913806

$\Delta_f H°(0\ K) = -487.4 \pm 8\ kJ\ mol^{-1}$

S°(298.15 K) = 300.897 ± 0.21 J K^{-1} mol^{-1} $\Delta_f H°(298.15\ K) = -491.6 \pm 8\ kJ\ mol^{-1}$

Vibrational Frequencies and Degeneracies

ν, cm^{-1}	ν, cm^{-1}	ν, cm^{-1}
1099 (1)	457.5(1)	436(1)
261.5(1)	322 (1)	1162(1)
667.2(1)	922 (1)	446(1)

Ground State Quantum Weight: 1 σ = 2

Point Group: C_{2v}

Bond Distances: C-Cl = 1.77 ± 0.02 Å C-F = 1.33 ± 0.02 Å

Bond Angles: F-C-F = 109.5 ± 3° Cl-C-Cl = 108.5 ± 2

F-C-F plane perpendicular to Cl-C-Cl plane

Product of the Moments of Inertia: $I_AI_BI_C = 2.537004 \times 10^{113}\ g^3\ cm^6$

Enthalpy of Formation

Mears and Stahl (1) investigated several heats of reaction involving CF_4, CF_3Cl, CF_2Cl_2, $CFCl_3$, CCl_4, HF and HCl. Using JANAF values for all compounds except CF_2Cl_2 and $CFCl_3$ we obtain the following observations:

1. $\Delta_f H°(CCl_3F,\ g,\ 298.15\ K) = -70.12 \pm 1.1\ kcal\ mol^{-1}$
2. $\Delta_f H°(CCl_2F_2,\ g,\ 298.15\ K) = -116.53 \pm 1.5\ kcal\ mol^{-1}$
3. $\Delta_f H°(CCl_2F_2,\ g,\ 298.15\ K) - \Delta_f H°(CCl_3F,\ g,\ 298.15\ K) = -49.39 \pm 1.0\ kcal\ mol^{-1}$
4. $\Delta_f H°(CCl_2F_2,\ g,\ 298.15\ K) - 2\ \Delta_f H°(CCl_3F,\ g,\ 298.15\ K) = 20.34 \pm 1.0\ kcal\ mol^{-1}$
5. $\Delta_f H°(CCl_3F,\ g,\ 298.15\ K) - 2\ \Delta_f H°(CCl_2F_2,\ g,\ 298.15\ K) = 165.47 \pm 1.0\ kcal\ mol^{-1}$

The sum of the squares of the deviations is smallest when $\Delta_f H°(CCl_2F_2,\ g,\ 298.15\ K) = -117.5\ kcal\ mol^{-1}$ and $\Delta_f H°(CCl_3F,\ g,\ 298.15\ K) = -69\ kcal\ mol^{-1}$, and all the values are within their uncertainty ranges.

Peterson and Pitzer (8) obtained an approximate equilibrium constant for the reaction 2 $CClF_3(g)$ + $CCl_4(g)$ = 3 $CCl_2F_2(g)$ of log K_{400} = -3.0. This on 3rd law reaction gives $\Delta_r H°(298.15\ K) = 7.64\ kcal\ mol^{-1}$ and with $\Delta_f H°(CClF_3,\ g,\ 298.15\ K) = -169.2\ kcal\ mol^{-1}$ and $\Delta_f H°(CCl_4,\ g,\ 298.15\ K) = -22.94\ kcal\ mol^{-1}$ this yields $\Delta_f H°(CCl_2F_2,\ g,\ 298.15\ K) = -117.9\ kcal\ mol^{-1}$.

Kirkbride and Davidson (2) investigated the reaction of potassium with the chlorofluoromethanes and obtained $\Delta_f H°(CCl_2F_2,\ g,\ 298.15\ K) = -115.1 \pm 2\ kcal\ mol^{-1}$ and $\Delta_f H°(CCl_3F,\ g,\ 298.15\ K) = -72.9 \pm 4\ kcal\ mol^{-1}$ after correction with JANAF auxiliary data. Wartenberg and Schiefer (3) investigated the same reaction and we obtain, after correction with JANAF auxiliary data, $\Delta_f H°(CCl_2F_2,\ g,\ 298.15\ K) = -112.7\ kcal\ mol^{-1}$ and $\Delta_f H°(CCl_3F,\ g,\ 298.15\ K) = -63.1\ kcal\ mol^{-1}$. Kolesov et al (4) have reported analogous data for the reaction with sodium which, after correction with JANAF auxiliary data, yields $\Delta_f H°(CCl_2F_2,\ g,\ 298.15\ K) = -114.4 \pm 3.5\ kcal\ mol^{-1}$.

We adopt the values calculated above from Mears and Stahl (1), since they have a high degree of internal consistency and also lead to a realistic progression in the heats of atomization in going from CF_4 to CCl_4. With the adopted heats of formation, the successive replacement of F by Cl changes the atomization heats by 43.8, 41.6, 38.5, and 36.0 kcal.

Heat Capacity and Entropy

Claassen (5) chose the frequencies given above from his own gas-phase Raman measurements and the infrared values of earlier investigators. The values are in good agreement with the assignment of Plyler and Benedict (6). Masi (9) has measured the gas phase heat capacities at 243, 273, 318 and 363 K, within this range our adopted heat capacities are from .3 to .7% high. The entropies agree to within 0.05% and further adjustment to obtain better fits was not considered necessary. The molecular structure was determined from the microwave measurements of Long et al. (7). The principal moments of inertia are: $I_A = 21.0747 \times 10^{-39}$, $I_B = 31.7398 \times 10^{-39}$, and $I_C = 37.9277 \times 10^{-39}\ g\ cm^2$.

References

1. W. H. Mears and R. F. Stahl, IUPAC Thermochemical Bulletin 2, 5 (1956).
2. F. W. Kirkbride and F. G. Davidson, Nature 174, 79 (1954).
3. H. Wartenberg and J. Schiefer, Z. Anorg. Chem. 278, 326 (1955).
4. V. P. Kolesov, I. D. Zenkov, and S. M. Skuratov, Russ. J. Phys. Chem. 37, 378 (1963).
5. H. H. Claassen, J. Chem. Phys. 22, 50 (1954).
6. E. K. Plyler and W. S. Benedict, J. Res. Nat. Bur. Stand. 47, 202 (1951).
7. M. W. Long, Q. Williams, and T. L. Weatherly, J. Chem. Phys. 33, 508 (1960).
8. D. E. Petersen and K. S. Pitzer, J. Phys. Chem. 61, 1252 (1957).
9. J. F. Masi, J. Amer. Chem. Soc. 74, 4738 (1952) and 75, 2276 (1953).

Dichlorodifluoromethane (CCl_2F_2) $C_1Cl_2F_2(g)$

Enthalpy Reference Temperature = T_r = 298.15 K Standard State Pressure = p° = 0.1 MPa

	J $K^{-1}mol^{-1}$			kJ mol^{-1}			
T/K	$C_p°$	S°	-[G°-H°(T_r)]/T	H°-H°(T_r)	$\Delta_f H°$	$\Delta_f G°$	Log K_f
0	0.	0.	INFINITE	-14.866	-487.429	-487.429	INFINITE
100	39.629	241.343	355.353	-11.401	-489.843	-477.944	249.652
200	58.584	274.768	307.109	-6.468	-491.203	-465.417	121.554
250	66.235	288.688	302.056	-3.342	-491.494	-458.933	95.889
298.15	72.404	300.897	300.897	0.	-491.620	-452.648	79.302
300	72.620	301.346	300.899	0.134	-491.622	-452.406	78.771
350	77.938	312.952	301.803	3.902	-491.633	-445.868	66.542
400	82.349	323.657	303.875	7.913	-491.556	-439.335	57.371
450	85.998	333.574	306.630	12.124	-491.416	-432.815	50.240
500	89.016	342.796	309.791	16.502	-491.230	-426.314	44.537
600	93.597	359.456	316.711	25.647	-490.769	-413.372	35.987
700	96.801	374.138	323.887	35.176	-490.236	-400.514	29.887
800	99.095	387.222	331.001	44.977	-489.666	-387.735	25.316
900	100.779	398.996	337.913	54.975	-489.077	-375.029	21.766
1000	102.044	409.683	344.564	65.119	-488.483	-362.389	18.929
1100	103.015	419.456	350.935	75.374	-487.891	-349.808	16.611
1200	103.773	428.453	357.025	85.715	-487.310	-337.281	14.681
1300	104.377	436.784	362.843	96.123	-486.741	-324.801	13.051
1400	104.864	444.538	368.405	106.586	-486.191	-312.365	11.654
1500	105.262	451.787	373.725	117.093	-485.659	-299.968	10.446
1600	105.592	458.591	378.819	127.636	-485.150	-287.605	9.389
1700	105.867	465.001	383.701	138.210	-484.663	-275.274	8.458
1800	106.100	471.059	388.388	148.808	-484.198	-262.970	7.631
1900	106.298	476.801	392.891	159.428	-483.756	-250.692	6.892
2000	106.469	482.258	397.224	170.067	-483.335	-238.436	6.227
2100	106.616	487.456	401.398	180.721	-482.935	-226.202	5.626
2200	106.744	492.419	405.424	191.390	-482.554	-213.985	5.081
2300	106.856	497.166	409.310	202.070	-482.190	-201.785	4.583
2400	106.955	501.716	413.066	212.760	-481.841	-189.602	4.127
2500	107.042	506.084	416.700	223.460	-481.505	-177.432	3.707
2600	107.120	510.284	420.219	234.169	-481.181	-165.275	3.320
2700	107.190	514.328	423.630	244.884	-480.866	-153.131	2.962
2800	107.252	518.228	426.940	255.606	-480.559	-140.998	2.630
2900	107.308	521.992	430.153	266.334	-480.256	-128.877	2.321
3000	107.358	525.631	433.275	277.068	-479.957	-116.765	2.033
3100	107.404	529.152	436.311	287.806	-479.658	-104.664	1.764
3200	107.446	532.563	439.266	298.548	-479.357	-92.572	1.511
3300	107.484	535.869	442.144	309.295	-479.053	-80.490	1.274
3400	107.518	539.079	444.948	320.045	-478.742	-68.417	1.051
3500	107.550	542.196	447.682	330.798	-478.422	-56.353	0.841
3600	107.579	545.226	450.350	341.555	-478.092	-44.298	0.643
3700	107.606	548.174	452.954	352.314	-477.747	-32.253	0.455
3800	107.631	551.044	455.498	363.076	-477.386	-20.218	0.278
3900	107.654	553.840	457.984	373.840	-477.007	-8.192	0.110
4000	107.675	556.566	460.414	384.607	-476.607	3.824	-0.050
4100	107.695	559.225	462.792	395.375	-476.183	15.830	-0.202
4200	107.713	561.820	465.119	406.146	-475.733	27.825	-0.346
4300	107.731	564.355	467.397	416.918	-475.256	39.808	-0.484
4400	107.746	566.832	469.629	427.692	-474.749	51.781	-0.615
4500	107.761	569.253	471.816	438.467	-474.211	63.742	-0.740
4600	107.775	571.622	473.960	449.244	-473.639	75.690	-0.859
4700	107.788	573.940	476.063	460.022	-473.032	87.625	-0.974
4800	107.801	576.210	478.126	470.802	-472.389	99.547	-1.083
4900	107.812	578.432	480.150	481.582	-471.709	111.456	-1.188
5000	107.823	580.611	482.138	492.364	-470.990	123.350	-1.289
5100	107.833	582.746	484.090	503.147	-470.232	135.229	-1.385
5200	107.843	584.840	486.007	513.931	-469.434	147.093	-1.478
5300	107.852	586.894	487.891	524.716	-468.595	158.942	-1.566
5400	107.861	588.910	489.743	535.501	-467.715	170.774	-1.652
5500	107.869	590.890	491.564	546.288	-466.794	182.589	-1.734
5600	107.877	592.833	493.356	557.075	-465.831	194.388	-1.813
5700	107.884	594.743	495.118	567.863	-464.827	206.168	-1.889
5800	107.891	596.619	496.851	578.652	-463.782	217.931	-1.963
5900	107.897	598.463	498.558	589.441	-462.695	229.676	-2.033
6000	107.904	600.277	500.238	600.231	-461.568	241.400	-2.102

PREVIOUS: June 1970 (1 atm) CURRENT: June 1970 (1 bar)

CARBONIC DICHLORIDE ($COCl_2$) IDEAL GAS Mr = 98.9164

$S°(298.15\ K) = 283.796\ J\ K^{-1}\ mol^{-1}$

$\Delta_f H°(0\ K) = -218.371 \pm 3.3\ kJ\ mol^{-1}$
$\Delta_f H°(298.15\ K) = -220.08 \pm 3.3\ kJ\ mol^{-1}$

Vibrational Frequencies and Degeneracies

ν, cm^{-1}	
[570]	(1)
1827	(2)
285	(1)
849	(1)
440	(1)
[585]	(1)

Point Group: C_{2v} $\sigma = 2$
Bond Distances: C-O = 1.166±0.002 Cl-O = 2.589±0.001 Cl-Cl = 2.884±0.001 C-Cl = 1.746±0.004
Bond Angle: Cl-C-Cl = 111.3±0.1°
Product of the Moments of Inertia: $I_A I_B I_C = 9.138289 \times 10^{-114}\ g^3\ cm^6$

Enthalpy of Formation

Thomsen (1) reports the enthalpy of alkaline hydrolysis of $COCl_2$, from which $\Delta_f H°(298.15\ K)$ is calculated to be -54.4±0.3 kcal mol^{-1}. From the enthalpy of combustion of $COCl_2$ in O_2 and H_2, reported by Thomsen (2), $\Delta_f H°(298.15\ K)$ is calculated to be -51.73 ± 0.43 kcal mol^{-1}. Bodenstein and Plaut (3), Christiansen (4), and Bodenstein and Dunant (5) report 10 values of K_p from 647 to 876 K for the reaction $CO + Cl_2 = COCl_2$. These yield -25.91±0.04 kcal for $\Delta_f H°(298.15\ K)$ for $COCl_2$ is -52.33±0.62 kcal mol^{-1}. The average of the 3 foregoing values of $\Delta_f H°(298.15\ K)$ weighted for the uncertainties, is adopted here. Values calculated from the degree of dissociation of $COCl_2$ from 374 to 778 K, reported by Atkinson, et al. (6), range from -47.06 to -53.11 kcal mol^{-1} and are ignored.

Heat Capacity and Entropy

The vibrational frequencies were assigned by Catalano and Pitzer (7), on the basis of their own infra-red spectra and other data. The molecular dimensions listed above were deduced from microwave spectra by Robinson (8), and are confirmed by electron-diffraction values compiled by Allen and Sutton (9). The moments of inertia are weighted for the isotopes and were calculated from Robinson's dimensions by Giauque and Ott (10). Giauque and Ott (10) determined S°(298.15 K) experimentally to be 67.81 cal K^{-1} mol^{-1}, with which the value calculated here agrees, i.e. 67.816 cal K^{-1} mol^{-1}. The principal moments of inertia are: $I_A = 10.6393 \times 10^{-39}$, $I_B = 24.4666 \times 10^{-39}$, and $I_C = 35.1059 \times 10^{39}\ g\ cm^2$.

References

1. J. Thomsen, Thermochemische Untersochungen, Barth, Leipzig Vol. 2, 361 (1882).
2. J. Thomsen, Thermochemische Untersochungen, Barth, Leipzig Vol. 2, 363 (1882).
3. M. Bodenstein and H. Plaut, Z. Phys. Chem. 110, 399 (1924).
4. J. A. Christiansen, Z. Phys. Chem. 103, 99 (1923).
5. M. Bodenstein and G. Dunant, Z. Phys. Chem. 61, 437 (1908).
6. R. H. Atkinson, C. T. Heyeock, and W. J. Pope, J. Chem. Soc. 117, 1410 (1920).
7. E. Catalano and K. S. Pitzer, J. Amer. Chem. Soc. 80, 1054 (1958).
8. G. W. Robinson, J. Chem. Phys. 21, 1741 (1953).
9. P. W. Allen and L. E. Sutton, Acta Cryst. 3, 46 (1950).
10. W. F. Giauque and J. B. Ott, J. Amer. Chem. Soc. 82, 2689 (1960).

Carbonic Dichloride ($COCl_2$) $C_1Cl_2O_1(g)$

Enthalpy Reference Temperature = T_r = 298.15 K Standard State Pressure = p° = 0.1 MPa

	J K⁻¹mol⁻¹			kJ mol⁻¹			
T/K	$C_p°$	S°	$-[G°-H°(T_r)]/T$	$H°-H°(T_r)$	$\Delta_f H°$	$\Delta_f G°$	Log K_f
0	0.	0.	INFINITE	-12.867	-218.372	-218.372	INFINITE
100	36.537	233.747	328.453	-9.471	-219.399	-215.113	112.363
200	48.362	262.618	288.800	-5.236	-219.985	-210.551	54.990
250	53.603	273.992	284.724	-2.683	-220.075	-208.180	43.497
298.15	57.695	283.796	283.796	0.	-220.078	-205.887	36.071
300	57.835	284.153	283.797	0.107	-220.077	-205.799	35.833
350	61.211	293.332	284.514	3.086	-220.026	-203.423	30.359
400	63.938	301.689	286.147	6.217	-219.945	-201.056	26.255
450	66.184	309.354	288.305	9.472	-219.853	-198.701	23.065
500	68.068	316.427	290.768	12.830	-219.760	-196.356	20.513
600	71.048	329.115	296.127	19.793	-219.587	-191.691	16.688
700	73.282	340.242	301.650	27.014	-219.438	-187.054	13.958
800	74.993	350.144	307.105	34.432	-219.309	-182.437	11.912
900	76.324	359.057	312.390	42.000	-219.198	-177.835	10.321
1000	77.373	367.155	317.468	49.687	-219.102	-173.244	9.049
1100	78.210	374.570	322.327	57.468	-219.020	-168.662	8.009
1200	78.886	381.406	326.969	65.324	-218.952	-164.087	7.143
1300	79.437	387.742	331.403	73.241	-218.898	-159.517	6.409
1400	79.891	393.646	335.641	81.208	-218.859	-154.951	5.781
1500	80.270	399.172	339.694	89.217	-218.836	-150.387	5.237
1600	80.588	404.363	343.575	97.260	-218.831	-145.824	4.761
1700	80.857	409.256	347.296	105.332	-218.845	-141.261	4.340
1800	81.086	413.885	350.868	113.430	-218.880	-136.697	3.967
1900	81.284	418.274	354.301	121.549	-218.937	-132.130	3.632
2000	81.455	422.448	357.605	129.686	-219.018	-127.559	3.331
2100	81.604	426.426	360.788	137.839	-219.125	-122.983	3.059
2200	81.734	430.225	363.859	146.006	-219.259	-118.402	2.811
2300	81.849	433.861	366.824	154.185	-219.423	-113.814	2.585
2400	81.950	437.347	369.690	162.375	-219.618	-109.218	2.377
2500	82.040	440.694	372.464	170.575	-219.846	-104.614	2.186
2600	82.120	443.913	375.150	178.783	-220.108	-99.999	2.009
2700	82.192	447.014	377.755	186.999	-220.405	-95.374	1.845
2800	82.257	450.004	380.282	195.221	-220.738	-90.737	1.693
2900	82.316	452.892	382.737	203.450	-221.108	-86.088	1.551
3000	82.368	455.683	385.122	211.684	-221.515	-81.425	1.418
3100	82.416	458.385	387.442	219.923	-221.958	-76.748	1.293
3200	82.460	461.002	389.700	228.167	-222.437	-72.057	1.176
3300	82.500	463.540	391.899	236.415	-222.951	-67.350	1.066
3400	82.537	466.004	394.043	244.667	-223.498	-62.626	0.962
3500	82.570	468.397	396.133	252.922	-224.076	-57.887	0.864
3600	82.601	470.723	398.173	261.181	-224.685	-53.130	0.771
3700	82.629	472.987	400.164	269.443	-225.320	-48.356	0.683
3800	82.656	475.191	402.110	277.707	-225.981	-43.564	0.599
3900	82.680	477.338	404.011	285.974	-226.664	-38.754	0.519
4000	82.703	479.432	405.871	294.243	-227.366	-33.927	0.443
4100	82.724	481.474	407.690	302.514	-228.085	-29.083	0.371
4200	82.743	483.468	409.471	310.787	-228.819	-24.220	0.301
4300	82.761	485.415	411.214	319.063	-229.564	-19.340	0.235
4400	82.778	487.318	412.922	327.340	-230.318	-14.442	0.171
4500	82.794	489.178	414.596	335.618	-231.079	-9.528	0.111
4600	82.809	490.998	416.237	343.898	-231.844	-4.596	0.052
4700	82.823	492.779	417.847	352.180	-232.611	0.352	-0.004
4800	82.836	494.523	419.426	360.463	-233.378	5.317	-0.058
4900	82.848	496.231	420.976	368.747	-234.143	10.299	-0.110
5000	82.860	497.905	422.498	377.033	-234.905	15.295	-0.160
5100	82.871	499.546	423.993	385.319	-235.662	20.306	-0.208
5200	82.881	501.155	425.462	393.607	-236.413	25.332	-0.254
5300	82.891	502.734	426.905	401.895	-237.157	30.373	-0.299
5400	82.900	504.284	428.323	410.185	-237.893	35.428	-0.343
5500	82.909	505.805	429.718	418.476	-238.620	40.496	-0.385
5600	82.917	507.299	431.090	426.767	-239.339	45.577	-0.425
5700	82.925	508.766	432.440	435.059	-240.048	50.671	-0.464
5800	82.932	510.209	433.769	443.352	-240.748	55.778	-0.502
5900	82.939	511.626	435.076	451.645	-241.438	60.896	-0.539
6000	82.946	513.020	436.364	459.940	-242.119	66.026	-0.575

PREVIOUS: June 1961 (1 atm) CURRENT: June 1961 (1 bar)

Carbonic Dichloride ($COCl_2$) $C_1Cl_2O_1(g)$

TRICHLOROMETHYL (CCl_3) IDEAL GAS Mr = 118.370

$\Delta_f H°(0\ K) = 80.1 \pm 8\ kJ\ mol^{-1}$

$S°(298.15\ K) = 296.83 \pm 6.3\ J\ K^{-1}\ mol^{-1}$ $\Delta_f H°(298.15\ K) = 79.5 \pm 8\ kJ\ mol^{-1}$

Vibrational Frequencies and Degeneracies

ν, cm^{-1}	
[450]	(1)
[460]	(1)
898	(2)
[240]	(2)

Ground State Quantum Weight: 2 $\sigma = 6$

Point Group: D_{3h}

Bond Distance: C-Cl = [1.74]Å

Bond Angle: Cl-C-Cl = [120]°

Product of the Moments of Inertia: $I_A I_B I_C = [3.822311 \times 10^{-113}]\ g^3\ cm^6$

Enthalpy of Formation

Benson (1) has reanalysed the kinetic data of Sullivan and Davidson (2) for the reaction $Br_2(g) + CHCl_3(g) = HBr(g) + CBrCl_3(g)$ and has obtained from it the enthalpy of the initial step, $Br(g) + CHCl_3(g) + CCl_3(g) + HBr(g)$, as 8.2 ± 1 kcal mol^{-1}. Substituting this value and $\Delta_f H°$(Br, g, 298. 15 K) = 26.74 kcal mol^{-1}, $\Delta_f H°(CHCl_3$, g, 298.15 K) = -24.66 kcal mol^{-1} and $\Delta_f H°$(HBr, g, 298.15 K) = -8.71 kcal mol^{-1}, we obtain $\Delta_f H°(CCl_3$, g, 298.15 K) = 19 ± 2 kcal mol^{-1} (79.496 ± 8 k J mol^{-1}), which is adopted here.

Heat Capacity and Entropy

Walsh (3) has predicted that CCl_3 is non-planar on the basis of his orbital correlation diagram. Andrews (4) indicates that his matrix infrared observations of the symmetric and antisymmetric stretches are consistent with a non-planar C_{3v} structure and a near tetrahedral angle. The antisymmetric stretch has been confirmed by Current and Burdett (5). They rationalize the absence of the symmetric stretch and claim support of the C_{3v} structure. Rogers et al. (6) have made a thorough search for the symmetric stretch and did not find it in the region where is was reported by Andrews (4). Thus, they conclude that since only the antisymmetric stretch has been confirmed, the possibility of a planar D_{3h} molecule cannot be excluded. Maget et al. (7) have observed the NMR spectrum of trapped CCl_3 and find it consistent with a planar molecule. The absence of the symmetric stretch strongly implies a planar configuration, and we adopt such a configuration until more definitive proof of the non-planarity can be obtained. The remaining vibrational frequencies are estimated by comparison with those of BCl_3. The bond length was estimated as 1.74 Å by Andrews (4) from a correlation with related molecules. The principal moments of inertia are: $I_A = I_B = 26.7361 \times 10^{-39}$ and $I_C = 53.4723 \times 10^{-39}\ g\ cm^2$.

References

1. S. W. Benson, J. Chem. Phys. 43, 2044 (1965).
2. J. H. Sullivan and N. Davidson, J. Chem. Phys. 19, 143 (1951).
3. A. D. Walsh, J. Chem. Soc. (London) 1953, 2301 (1953).
4. L. Andrews, J. Chem. Phys. 48, 972 (1968).
5. J. H. Current and J. K. Burdett, J. Phys. Chem. 73, 3504 (1969).
6. E. E. Rogers, S. Abramowitz, M. E. Jacox and D. E. Milligan, J. Chem. Phys. 52, 2198 (1970).
7. M. Magat, N. Leray and J. Roncin, Vysokomolekulyarnye Soedineniya, Vsesoyuz. Khim. Obshch. D. I. Mendeleeva, 11, 223 (1966).

Trichloromethyl (CCl_3) $C_1Cl_3(g)$

Enthalpy Reference Temperature = T_r = 298.15 K Standard State Pressure = p° = 0.1 MPa

T/K	$C_p°$ (J $K^{-1}mol^{-1}$)	S° (J $K^{-1}mol^{-1}$)	$-[G°-H°(T_r)]/T$ (J $K^{-1}mol^{-1}$)	$H°-H°(T_r)$ (kJ mol^{-1})	$\Delta_f H°$ (kJ mol^{-1})	$\Delta_f G°$ (kJ mol^{-1})	Log K_f
0	0.	0.	INFINITE	-14.194	80.125	80.125	INFINITE
100	40.983	240.296	346.941	-10.665	79.226	83.643	-43.691
200	54.687	273.207	302.392	-5.837	79.171	88.135	-23.018
250	59.708	285.968	297.860	-2.973	79.311	90.361	-18.880
298.15	63.649	296.833	296.833	0.	79.496	92.473	-16.201
300	63.784	297.228	296.835	0.118	79.504	92.554	-16.115
350	67.068	307.316	297.624	3.392	79.729	94.711	-14.135
400	69.696	316.450	299.415	6.814	79.972	96.835	-12.645
450	71.799	324.785	301.778	10.353	80.219	98.928	-11.483
500	73.488	332.440	304.466	13.987	80.462	100.994	-10.551
600	75.965	346.073	310.293	21.468	80.916	105.057	-9.146
700	77.636	357.916	316.269	29.153	81.321	109.048	-8.137
800	78.803	368.364	322.141	36.978	81.676	112.984	-7.377
900	79.643	377.697	327.805	44.903	81.985	116.878	-6.783
1000	80.266	386.122	333.222	52.900	82.254	120.741	-6.307
1100	80.739	393.795	338.385	60.951	82.486	124.578	-5.916
1200	81.106	400.837	343.300	69.044	82.685	128.395	-5.589
1300	81.396	407.341	347.979	77.170	82.853	132.197	-5.312
1400	81.629	413.381	352.438	85.321	82.993	135.988	-5.074
1500	81.818	419.020	356.690	93.494	83.105	139.769	-4.867
1600	81.975	424.306	360.753	101.684	83.190	143.544	-4.686
1700	82.105	429.279	364.639	109.888	83.249	147.314	-4.526
1800	82.215	433.975	368.362	118.104	83.282	151.081	-4.384
1900	82.309	438.423	371.933	126.331	83.288	154.847	-4.257
2000	82.389	442.647	375.364	134.566	83.268	158.614	-4.143
2100	82.458	446.669	378.665	142.808	83.219	162.383	-4.039
2200	82.518	450.506	381.844	151.057	83.140	166.154	-3.945
2300	82.571	454.175	384.909	159.311	83.030	169.930	-3.859
2400	82.617	457.690	387.869	167.571	82.886	173.711	-3.781
2500	82.658	461.064	390.730	175.835	82.708	177.499	-3.709
2600	82.694	464.306	393.498	184.102	82.492	181.295	-3.642
2700	82.726	467.428	396.179	192.373	82.239	185.100	-3.581
2800	82.756	470.437	398.777	200.647	81.947	188.914	-3.524
2900	82.782	473.342	401.299	208.924	81.615	192.740	-3.472
3000	82.805	476.148	403.747	217.204	81.243	196.578	-3.423
3100	82.827	478.864	406.127	225.485	80.831	200.430	-3.377
3200	82.846	481.494	408.441	233.769	80.381	204.294	-3.335
3300	82.864	484.043	410.694	242.054	79.893	208.175	-3.295
3400	82.880	486.517	412.887	250.342	79.370	212.069	-3.258
3500	82.895	488.920	415.026	258.630	78.814	215.980	-3.223
3600	82.908	491.255	417.111	266.921	78.228	219.908	-3.191
3700	82.921	493.527	419.146	275.212	77.614	223.852	-3.160
3800	82.932	495.739	421.132	283.505	76.976	227.812	-3.131
3900	82.943	497.893	423.073	291.798	76.319	231.790	-3.104
4000	82.953	499.993	424.970	300.093	75.645	235.785	-3.079
4100	82.962	502.042	426.825	308.389	74.959	239.797	-3.055
4200	82.971	504.041	428.640	316.686	74.265	243.826	-3.032
4300	82.979	505.993	430.416	324.983	73.566	247.872	-3.011
4400	82.986	507.901	432.155	333.281	72.866	251.933	-2.991
4500	82.993	509.766	433.859	341.580	72.170	256.011	-2.972
4600	82.999	511.590	435.529	349.880	71.480	260.104	-2.954
4700	83.005	513.375	437.167	358.180	70.800	264.212	-2.936
4800	83.011	515.123	438.773	366.481	70.133	268.334	-2.920
4900	83.016	516.835	440.348	374.782	69.483	272.470	-2.905
5000	83.022	518.512	441.895	383.084	68.852	276.619	-2.890
5100	83.026	520.156	443.413	391.387	68.242	280.780	-2.876
5200	83.031	521.768	444.905	399.689	67.656	284.953	-2.862
5300	83.035	523.350	446.370	407.993	67.095	289.138	-2.850
5400	83.039	524.902	447.810	416.296	66.562	293.332	-2.837
5500	83.043	526.426	449.226	424.601	66.059	297.537	-2.826
5600	83.046	527.922	450.617	432.905	65.586	301.749	-2.815
5700	83.050	529.392	451.987	441.210	65.144	305.970	-2.804
5800	83.053	530.836	453.334	449.515	64.736	310.199	-2.794
5900	83.056	532.256	454.659	457.820	64.360	314.435	-2.784
6000	83.059	533.652	455.964	466.126	64.019	318.676	-2.774

PREVIOUS: June 1970 (1 atm) CURRENT: June 1970 (1 bar)

Trichloromethyl (CCl_3) $C_1Cl_3(g)$

TRICHLOROFLUOROMETHANE (CCl_3F) IDEAL GAS M_r = 137.368403

$S°(298.15\ K) = 309.735 \pm 0.21\ J\ K^{-1}\ mol^{-1}$

$\Delta_fH°(0\ K) = -285.52 \pm 6.3\ kJ\ mol^{-1}$
$\Delta_fH°(298.15\ K) = -288.70 \pm 6.3\ kJ\ mol^{-1}$

Vibrational Frequencies and Degeneracies

ν, cm^{-1}	ν, cm^{-1}
349.5(1)	241(2)
535.3(1)	398(2)
1085 (1)	847(2)

Ground State Quantum Weight: 1 $\sigma = 3$
Point Group: C_{3v}
Bond Distances: C-Cl = 1.76 Å C-F = 1.33 Å
Bond Angles: Cl-C-Cl = 109°40' F-C-Cl = 109°28'
Product of the Moments of Inertia: $I_AI_BI_C = 5.736023 \times 10^{-113}\ g^3\ cm^6$

Enthalpy of Formation

Mears and Stahl (1) investigated several enthalpiess of reaction involving CF_4, CF_3Cl, CF_2Cl_2, $CFCl_3$, CCl_4, HF and HCl. Using JANAF values for all compounds except CF_2Cl_2 and $CFCl_3$, we obtain the following observations:

1. $\Delta_fH°(CCl_3F,\ g,\ 298.15\ K) = -70.12 \pm 1.1\ kcal\ mol^{-1}$
2. $\Delta_fH°(CCl_2F_2,\ g,\ 298.15\ K) = -116.53 \pm 1.5\ kcal\ mol^{-1}$
3. $\Delta_fH°(CCl_2F_2,\ g,\ 298.15\ K) - \Delta_fH°(CCl_3F,\ g,\ 298.15\ K) = -49.39 \pm 1.0\ kcal\ mol^{-1}$
4. $\Delta_fH°(CCl_2F_2,\ g,\ 298.15\ K) - 2\Delta_fH°(CCl_3F,\ g,\ 298.15\ K) = 20.34 \pm 1.0\ kcal\ mol^{-1}$
5. $\Delta_fH°(CCl_3F,\ g,\ 298.15\ K) - 2\Delta_fH°(CCl_2F_2,\ g,\ 298.15\ K) = 165.47 \pm 1.0\ kcal\ mol^{-1}$

The sum of the squares of the deviations is smallest when $\Delta_fH°(CCl_2F_2,\ g,\ 298.15\ K) = -117.5\ kcal\ mol^{-1}$ and $\Delta_fH°(CCl_3F,\ g,\ 298.15\ K) = -69\ kcal\ mol^{-1}$, and all the values are within their uncertainty ranges.

Petersen and Pitzer (9) obtained an approximate equilibrium constant for the reaction $CClF_3(g) + 2CCl_4(g) = 3CCl_3F(g)$ of $\log K_{400} = -4.5$. This on third law reduction gives $\Delta_rH°(298.15\ K) = 10.56$ kcal and with $\Delta_fH°(CClF_3,\ g,\ 298.15\ K) = -169.2\ kcal\ mol^{-1}$ and $\Delta_fH°(CCl_4,\ g,\ 298.15\ K) = -22.94\ kcal\ mol^{-1}$ this yields $\Delta_fH°(CCl3F,\ g,\ 298.15\ K) = -68.2\ kcal\ mol^{-1}$.

Kirkbride and Davidson (2) investigated the reaction of potassium with the chlorofluoromethanes and obtained $\Delta_fH°(CCl_2F_2,\ g,\ 298.15\ K) = -115.1 \pm 2\ kcal\ mol^{-1}$ and $\Delta_fH°(CCl_3F,\ g,\ 298.15\ K) = -72.9 \pm 4\ kcal\ mol^{-1}$ after correction with JANAF auxiliary data. Wartenberg and Schiefer (3) investigated the same reaction and we obtain, after correction with JANAF auxiliary data, $\Delta_fH°(CCl_2F_2,\ g,\ 298.15\ K) = -112.7\ kcal\ mol^{-1}$ and $\Delta_fH°(CCl_3F,\ g,\ 298.15\ K) = -63.1\ kcal\ mol^{-1}$. Baibuz (4) has reported $\Delta_fH°(CCl_3F,\ g,\ 298.15\ K) = -66.4 \pm 2.1$ from heat of explosion measurements. A correlation of his other results would indicate that this value is not negative enough, possibly by as much as 3 kcal.

We adopt the values calculated above from Mears and Stahl (1), since they have a high degree of internal consistency and also lead to a realistic progression in the heats of atomization in going from CF_4 to CCl_4. With the adopted enthalpies of formation, the successive replacement of F by Cl changes the atomization enthalpies by 43.8, 41.6, 38.5 and 36.0 kcal.

Heat Capacity and Entropy

Claassen (5) chose the frequencies given above from his own gas phase Raman measurements and the infrared values of earlier investigators. The values are in good agreement with the assignment of Plyler and Benedict (6), and the matrix studies of King (7). The molecular structure was determined from the microwave measurements of Long et al. (8). The principal moments of inertia are: $I_A = I_B = 34.3029 \times 10^{-39}$ and $I_C = 48.7471 \times 10^{-39}\ g\ cm^2$.

References

1. W. H. Mears and R. F. Stahl, IUPAC Thermochemical Bulletin 2, 5 (1956).
2. F. W. Kirkbride and F. G. Davidson, Nature 174, 79 (1954).
3. H. Wartenberg and J. Schiefer, Z. Anorg. Chem. 278, 326 (1955).
4. V. F. Baibuz, Dokl. Akad. Nauk SSSR, 140, 1358 (1961).
5. H. H. Claassen, J. Chem. Phys. 22, 50 (1954).
6. E. K. Plyler and W. S. Benedict, J. Res. Nat. Bur. Stand. 47, 202 (1951).
7. S. T. King, J. Chem. Phys. 49, 1321 (1968).
8. M. W. Long, Q. Williams, and T. L. Weatherly, J. Chem. Phys. 33, 508 (1960).
9. D. E. Petersen and K. S. Pitzer, J. Phys. Chem. 61, 1252 (1957).

Triclorofluormethane (CCl_3F) $C_1Cl_3F_1(g)$

Enthalpy Reference Temperature = T_r = 298.15 K Standard State Pressure = p° = 0.1 MPa

T/K	$C_p°$ (J K⁻¹mol⁻¹)	S°	$-[G°-H°(T_r)]/T$	$H°-H°(T_r)$ (kJ mol⁻¹)	$\Delta_fH°$	$\Delta_fG°$	Log K_f
0	0.	0.	INFINITE	-16.060	-285.522	-285.522	INFINITE
100	43.323	244.360	369.234	-12.487	-287.829	-275.301	143.802
200	64.411	281.288	316.474	-7.037	-288.727	-262.314	68.509
250	72.066	296.515	310.988	-3.618	-288.783	-255.700	53.426
298.15	78.018	309.735	309.735	0.	-288.696	-249.335	43.682
300	78.223	310.219	309.737	0.145	-288.691	-249.090	43.371
350	83.173	322.662	310.708	4.184	-288.494	-242.505	36.192
400	87.149	334.038	312.924	8.446	-288.227	-235.952	30.812
450	90.347	344.494	315.859	12.886	-287.912	-229.437	26.632
500	92.930	354.151	319.211	17.470	-287.569	-222.957	23.292
600	96.749	371.455	326.511	26.967	-286.835	-210.103	18.291
700	99.350	386.577	334.034	36.780	-286.079	-197.374	14.728
800	101.178	399.969	341.455	46.811	-285.322	-184.754	12.063
900	102.502	411.967	348.635	56.999	-284.577	-172.228	9.996
1000	103.488	422.820	355.520	67.300	-283.848	-159.784	8.346
1100	104.238	432.720	362.094	77.688	-283.139	-147.412	7.000
1200	104.822	441.816	368.364	88.143	-282.452	-135.104	5.881
1300	105.284	450.225	374.342	98.649	-281.788	-122.852	4.936
1400	105.655	458.042	380.044	109.196	-281.148	-110.650	4.128
1500	105.959	465.342	385.490	119.778	-280.533	-98.493	3.430
1600	106.209	472.189	390.697	130.386	-279.943	-86.377	2.820
1700	106.418	478.634	395.682	141.018	-279.379	-74.296	2.283
1800	106.594	484.722	400.461	151.669	-278.842	-62.248	1.806
1900	106.744	490.489	405.049	162.336	-278.330	-50.229	1.381
2000	106.872	495.968	409.459	173.017	-277.845	-38.237	0.999
2100	106.983	501.185	413.704	183.710	-277.387	-26.268	0.653
2200	107.080	506.164	417.794	194.413	-276.957	-14.320	0.340
2300	107.164	510.926	421.741	205.125	-276.553	-2.391	0.054
2400	107.239	515.488	425.552	215.846	-276.178	9.521	-0.207
2500	107.304	519.867	429.238	226.573	-275.830	21.418	-0.448
2600	107.363	524.077	432.805	237.306	-275.510	33.302	-0.669
2700	107.415	528.130	436.261	248.045	-275.219	45.174	-0.874
2800	107.462	532.037	439.612	258.789	-274.955	57.035	-1.064
2900	107.504	535.809	442.865	269.537	-274.719	68.888	-1.241
3000	107.542	539.454	446.024	280.290	-274.508	80.732	-1.406
3100	107.576	542.981	449.095	291.045	-274.323	92.571	-1.560
3200	107.607	546.397	452.083	301.805	-274.161	104.403	-1.704
3300	107.636	549.708	454.991	312.567	-274.020	116.231	-1.840
3400	107.662	552.922	457.824	323.332	-273.897	128.055	-1.967
3500	107.685	556.043	460.586	334.099	-273.791	139.875	-2.088
3600	107.707	559.077	463.280	344.869	-273.697	151.693	-2.201
3700	107.728	562.029	465.909	355.640	-273.613	163.509	-2.308
3800	107.746	564.902	468.477	366.414	-273.534	175.321	-2.410
3900	107.763	567.701	470.985	377.190	-273.458	187.132	-2.506
4000	107.779	570.429	473.437	387.967	-273.381	198.942	-2.598
4100	107.794	573.091	475.836	398.745	-273.298	210.749	-2.685
4200	107.808	575.688	478.182	409.526	-273.206	222.554	-2.768
4300	107.821	578.225	480.480	420.307	-273.102	234.356	-2.847
4400	107.832	580.704	482.729	431.090	-272.982	246.156	-2.922
4500	107.844	583.128	484.934	441.873	-272.842	257.953	-2.994
4600	107.854	585.498	487.094	452.658	-272.679	269.747	-3.063
4700	107.864	587.818	489.213	463.444	-272.490	281.537	-3.129
4800	107.873	590.089	491.291	474.231	-272.273	293.322	-3.192
4900	107.882	592.313	493.330	485.019	-272.024	305.103	-3.252
5000	107.890	594.493	495.331	495.807	-271.742	316.878	-3.310
5100	107.897	596.629	497.297	506.597	-271.424	328.648	-3.366
5200	107.905	598.724	499.227	517.387	-271.068	340.410	-3.419
5300	107.911	600.780	501.124	528.178	-270.674	352.166	-3.471
5400	107.918	602.797	502.988	538.969	-270.238	363.913	-3.520
5500	107.924	604.777	504.821	549.761	-269.761	375.653	-3.568
5600	107.930	606.722	506.623	560.554	-269.241	387.383	-3.613
5700	107.935	608.632	508.396	571.347	-268.678	399.103	-3.657
5800	107.940	610.510	510.140	582.141	-268.070	410.814	-3.700
5900	107.945	612.355	511.857	592.935	-267.418	422.513	-3.741
6000	107.950	614.169	513.547	603.730	-266.721	434.200	-3.780

PREVIOUS: June 1970 (1 atm) CURRENT: June 1970 (1 bar)

Triclorofluormethane (CCl_3F) $C_1Cl_3F_1(g)$

TETRACHLOROMETHANE (CCl_4) IDEAL GAS M_r = 153.823

$S°(298.15 K) = 309.809 \pm 0.04 J K^{-1} mol^{-1}$

$\Delta_f H°(0 K) = -93.81 \pm 2.1 kJ mol^{-1}$
$\Delta_f H°(298.15 K) = -95.98 \pm 2.1 kJ mol^{-1}$

Vibrational Frequencies and Degeneracies

ν, cm^{-1}	
458	(1)
218	(2)
776	(3)
314	(3)

Ground State Quantum Weight: 1 σ = 2
Point Group: T_d
Bond Distance: C-Cl = 1.760 Å
Bond Angle: Cl-C-Cl = 109.47°
Product of the Moments of Inertia: $I_A I_B I_C = 1.150034 \times 10^{-112} g^3 cm^6$

Enthalpy of Formation

Hu and Sinke (1) have measured the enthalpy of combustion of CCl_4(ℓ) in the rotating platinum bomb calorimeter, using As_2O_3 solution as the reducing agent. They reported $\Delta_c H°(298.15 K) = -86.02$ kcal mol^{-1} for $CCl_4(\ell) + 2H_2O(\ell) \rightarrow CO_2(g) + 4HCl (600H_2O, aq)$ which leads to $\Delta_f H°(CCl_4$, ℓ, 298.15 K) = -30.69 kcal mol^{-1}, using $\Delta_f H°$(298.15 K) = -94.051, -39.823 and -68.315 kcal mol^{-1} for CO_2(g), HCl(600H_2O, aq) and H_2O(ℓ), respectively (2). The standard deviation of five combustion runs was ±0.074 kcal mol^{-1}. Employing the enthalpy of vaporization of CCl_4(ℓ) at 298.15 K as 7.75 kcal mol^{-1} (3), we obtain $\Delta_f H°(CCl_4$, g, 298.15K) = -22.94 kcal mol^{-1} which is adopted in the tabulation.

Smith et al. (5), based on the combustion data of Eftring (4), re-evaluatd the enthalpy of combustion of carbon tetrachloride at 18.7°C as $\Delta_c E°/M = -567.5$ cal/g for $CCl_4(\ell) + 2 H_2O(\ell) \rightarrow CO_2(g) + 4HCl(600H_2O)$. This value needs to be further corrected for a change in the heat of oxidation of As_2O_3 (6) and calculated to 25°C to give $\Delta_c E°$(298.15 K) = -88.105 kcal mol^{-1}. $\Delta_c H°$(298.15 K) = -87.513 kcal mol^{-1}, $\Delta_f H°(CCl_4$, ℓ, 298.15 K) = -29.20 kcal mol^{-1} and $\Delta_f H°(CCl_4$, g, 298.15 K) = -21.45 kcal mol^{-1}.

Neugebauer (7) has measured the enthalpy of reaction of CCl_4 with Mg in presence of hydrogen in the stationary bomb calorimeter and then derived the enthalpy of formation, $\Delta_f H°(CCl_4$, ℓ, 298.15 K) = -32.3 ± 2 kcal mol^{-1} which yields $\Delta_f H°(CCl_4$, g, 298.15 K) = -24.55 ± 2 kcal mol^{-1}. Baibuz (8), using an explosion technique, obtained $\Delta_f H°(CCl_4$, g, 298.15 K) = -24.6±1.9 kcal mol^{-1}. Bodenstein and Gunther (9) have measured calorimetrically $\Delta_r H°$(293 K) = -62.57±0.35 kcal mol^{-1} for $CCl_4(g) + 2H_2(g) \rightarrow C(cr) + 4HCl(g)$. They calibrated their calorimeter by means of the reaction $H_2(g) + Cl_2(g) \rightarrow 2HCl(g)$ for which they took $\Delta_f H°$(293 K) = -44 kcal mol^{-1}. Their data have been re-evaluated to give $\Delta_f H°(CCl_4$, g, 298.15 K) = -25.68 kcal mol^{-1}, but no correction is applied for the enthalpy of formation of carbon which was not the reference state "graphite" formed in the reaction. Domalski and Armstrong (10) reviewed several cases of active carbon formed in similar reactions and quoted a enthalpy of formation of carbon from +1.9 to +3.95 kcal mol^{-1}. If this correction is applied to the data of Bodenstein, their enthalpy of formation of CCl_4(g) should be 2 to 4 kcal mol^{-1} less negative than -25.68 kcal mol^{-1}.

Heat Capacity and Entropy

Bartell, Brockway and Schwendeman (11) have measured the bond distance as 1.760 ± 0.004 Å by the electron diffrction measurement. The assigned fundamental vibrational frequencies are obtained from Shimanouchi's selection (12) in which they are essentially the same as those frequencies determined from Raman and infrared spectra by Classen (13). The principal moments of inertia are $I_A = I_B = I_C = 48.6299 \times 10^{-39}$ g cm^2.

References

1. A. T. Hu, G. C. Sinke, presented at the 23rd Calorimetry Conference, Midland, Mich., 1968.
2. U. S. Nat. Bur. Stand. Tech. Note 270-3, 1968.
3. D. L. Hildenbrand and R. A. McDonald, J. Phys. Chem. 63, 1521 (1959).
4. E. Eftring, Diss., University of Lund, 1938.
5. L. Smith, L. Bjellerup, S. Krook and H. Westermark, Acta Chem. Scand. 7, 65 (1953).
6. S. Sunner and S. Thoren, Acta Chem. Scand. 18, 1528 (1964).
7. C. A. Neugebauer, Diss. Abstr. 17, 1478 (1957).
8. V. F. Baibuz, Dokl. Akad. Nauk. SSSR 140, 1358 (1961).
9. M. Bodenstein and P. Gunther, 49th General Meeting of the Am. Electrochem. Soc., Chicago, April 22, 1926; Angew. Chem. 39, 875 (1926).
10. E. S. Domalski and G. T. Armstrong, J. Res. Nat. Bur. Stand. 71A, 105 (1967).
11. L. S. Bartell, L. O. Brockway and R. H. Schwendeman, J. Chem. Phys. 23, 1854 (1955).
12. T. Shimanouchi, U. S. Nat. Bur. Stand. NSRDS-NBS6, 1967.
13. H. W. Classen, J. Chem. Phys. 22, 50 (1954).

Tetrachloromethane (CCl_4) $C_1Cl_4(g)$

Enthalpy Reference Temperature = T_r = 298.15 K Standard State Pressure = p° = 0.1 MPa

T/K	$C_p°$ (J $K^{-1}mol^{-1}$)	S° (J $K^{-1}mol^{-1}$)	$-[G°-H°(T_r)]/T$ (J $K^{-1}mol^{-1}$)	$H°-H°(T_r)$ (kJ mol^{-1})	$\Delta_f H°$ (kJ mol^{-1})	$\Delta_f G°$ (kJ mol^{-1})	Log K_f
0	0.	0.	INFINITE	-17.238	-93.806	-93.806	INFINITE
100	47.244	238.738	374.168	-13.543	-95.995	-81.972	42.818
200	69.995	279.142	317.054	-7.582	-96.436	-67.662	17.672
250	77.656	295.619	311.152	-3.883	-96.270	-60.485	12.638
298.15	83.401	309.809	309.809	0.	-95.981	-53.617	9.393
300	83.594	310.326	309.811	0.154	-95.968	-53.354	9.290
350	88.174	323.571	310.846	4.454	-95.577	-46.282	6.907
400	91.707	335.586	313.200	8.954	-95.131	-39.270	5.128
450	94.451	346.553	316.305	13.611	-94.654	-32.315	3.751
500	96.602	356.620	319.840	18.390	-94.165	-25.415	2.655
600	99.676	374.525	327.500	28.215	-93.182	-11.757	1.024
700	101.698	390.052	335.352	38.290	-92.223	1.737	-0.130
800	103.086	403.728	343.061	48.534	-91.300	15.096	-0.986
900	104.074	415.930	350.492	58.894	-90.415	28.342	-1.645
1000	104.800	426.935	357.595	69.340	-89.566	41.492	-2.167
1100	105.348	436.950	364.361	79.848	-88.751	54.558	-2.591
1200	105.771	446.135	370.798	90.405	-87.969	67.551	-2.940
1300	106.103	454.615	376.923	100.999	-87.219	80.481	-3.234
1400	106.370	462.488	382.757	111.624	-86.498	93.353	-3.483
1500	106.586	469.835	388.320	122.272	-85.806	106.175	-3.697
1600	106.765	476.720	393.632	132.940	-85.142	118.952	-3.883
1700	106.913	483.197	398.712	143.624	-84.508	131.689	-4.046
1800	107.038	489.311	403.577	154.321	-83.902	144.389	-4.190
1900	107.144	495.101	408.243	165.031	-83.326	157.055	-4.318
2000	107.235	500.600	412.725	175.750	-82.781	169.693	-4.432
2100	107.313	505.834	417.035	186.477	-82.270	182.304	-4.535
2200	107.381	510.827	421.186	197.212	-81.793	194.891	-4.627
2300	107.441	515.602	425.188	207.953	-81.354	207.458	-4.712
2400	107.493	520.176	429.051	218.700	-80.954	220.006	-4.788
2500	107.539	524.565	432.784	229.452	-80.597	232.539	-4.859
2600	107.580	528.783	436.396	240.208	-80.285	245.058	-4.923
2700	107.617	532.844	439.893	250.968	-80.019	257.566	-4.983
2800	107.650	536.759	443.283	261.731	-79.802	270.065	-5.038
2900	107.679	540.537	446.572	272.497	-79.633	282.557	-5.089
3000	107.706	544.188	449.765	283.267	-79.514	295.044	-5.137
3100	107.730	547.720	452.869	294.038	-79.444	307.528	-5.182
3200	107.752	551.140	455.886	304.813	-79.422	320.010	-5.224
3300	107.772	554.456	458.823	315.589	-79.446	332.493	-5.263
3400	107.790	557.674	461.684	326.367	-79.513	344.977	-5.300
3500	107.807	560.799	464.471	337.147	-79.621	357.464	-5.335
3600	107.822	563.836	467.189	347.928	-79.766	369.954	-5.368
3700	107.836	566.790	469.841	358.711	-79.943	382.448	-5.399
3800	107.849	569.666	472.431	369.495	-80.148	394.948	-5.429
3900	107.861	572.468	474.960	380.281	-80.377	407.453	-5.457
4000	107.872	575.199	477.432	391.067	-80.623	419.965	-5.484
4100	107.883	577.863	479.849	401.855	-80.883	432.483	-5.510
4200	107.892	580.462	482.214	412.644	-81.150	445.007	-5.534
4300	107.901	583.001	484.528	423.434	-81.419	457.538	-5.558
4400	107.909	585.482	486.795	434.224	-81.686	470.075	-5.581
4500	107.917	587.907	489.015	445.015	-81.946	482.618	-5.602
4600	107.925	590.279	491.191	455.808	-82.193	495.167	-5.623
4700	107.931	592.600	493.324	466.600	-82.423	507.720	-5.643
4800	107.938	594.873	495.416	477.394	-82.632	520.279	-5.662
4900	107.944	597.098	497.468	488.188	-82.815	532.842	-5.680
5000	107.949	599.279	499.483	498.983	-82.971	545.407	-5.698
5100	107.955	601.417	501.460	509.778	-83.094	557.977	-5.715
5200	107.960	603.513	503.403	520.574	-83.181	570.547	-5.731
5300	107.965	605.570	505.311	531.370	-83.231	583.119	-5.747
5400	107.969	607.588	507.187	542.166	-83.241	595.692	-5.762
5500	107.973	609.569	509.030	552.964	-83.209	608.265	-5.777
5600	107.977	611.515	510.843	563.761	-83.132	620.837	-5.791
5700	107.981	613.426	512.626	574.559	-83.010	633.406	-5.805
5800	107.985	615.304	514.380	585.357	-82.841	645.974	-5.818
5900	107.988	617.150	516.106	596.156	-82.624	658.538	-5.830
6000	107.992	618.965	517.806	606.955	-82.358	671.097	-5.842

PREVIOUS: December 1968 (1 atm) CURRENT: December 1968 (1 bar)

Tetrachloromethane (CCl_4) $C_1Cl_4(g)$

COPPER CYANIDE (CuCN) CRYSTAL M_r = 89.5637

$S°(298.15 K) = 90.002 \pm 0.25\ J\ K^{-1}\ mol^{-1}$

T_{fus} = 746 K

$\Delta_f H°(0\ K) = 93.32 \pm 2.1\ kJ\ mol^{-1}$

$\Delta_f H°(298.15\ K) = 94.98 \pm 2.1\ kJ\ mol^{-1}$

$\Delta_{fus} H° = 12 \pm 4\ kJ\ mol^{-1}$

Enthalpy of Formation

Taylor, Brown and Taylor (1) determined by solution calorimetry the enthalpy of reaction, $\Delta_r H°(298.15\ K) = -13.437 \pm 100$ cal $K^{-1} mol^{-1}$ for NaCN(cr) + 0.5 Cu_2O(cr) + 0.5 H_2O(ℓ) → CuCN(cr) + NaOH(128.5 H_2O). This leads to $\Delta_f H°$(CuCN, cr, 298.15 K) = 22.70±0.5 kcal mol^{-1} using the following auxiliary data: $\Delta_f H°$(NaCN, cr, 298.15 K) = -21.68±0.3 kcal mol^{-1} (2), $\Delta_f H°$(Cu_2O, cr, 298.15 K) = -40.7±0.5 kcal mol^{-1} (3), $\Delta_f H°$(H_2O, ℓ, 298.15 K) = -68.315 kcal mol^{-1} (4) and $\Delta_f H°$(NaOH · 128.5 H_2O, aq, 298.15 K) = -112.321 kcal mol^{-1} (2, 4).

Heat Capacity and Entropy

The heat capacity of CuCN(cr) has been measured adiabatically over the temperature range from 8 to 402 K by Taylor, Brown and Taylor (1). The smooth $C°_p$ values are adopted in the tabulation. The value of S°(298.15 K) is calculated from the $C°_p$ data and is based on an extrapolation of S°(8 K) = 0.104 cal K^{-1} mol^{-1}. They also found an anomaly in the heat capacity at 300 K which is considered as a second order phase transition. Heat capacities above 402 K are estimated by extrapolation of the measured $C°_p$ data.

Fusion Data

The melting point 746 K was determined by Truthe (5). The heat of melting is calculated from the phase diagram studies of KCN - CuCN and NaCN - CuCN reported by Truthe (5).

References

1. A. R. Taylor, Jr., M. H. Brown and E. G. Taylor, U. S. Bur. Mines RI 7499, 1971.
2. JANAF Thermochemical Tables: NaCN(cr), 3-31-66; Cu_2O(cr), 6-30-66; NaOH(cr), 12-30-70.
3. U. S. Nat. Bur. Stand. Tech. Note 270-3, 1968.
4. JANAF NaOH(cr) table, dated Dec. 30, 1970. Heat of dilution data is obtained from V. B. Parker, "Thermal Properties of Aqueous Uni-univalent Electrolytes," NSRDS-NBS 2, 1965.
5. W. Truthe, Z. Anorg. Chem. 76, 144 (1912).

Copper Cyanide (CuCN) $C_1Cu_1N_1$(cr)

Enthalpy Reference Temperature = T_r = 298.15 K Standard State Pressure = p° = 0.1 MPa

T/K	$C°_p$ (J K⁻¹mol⁻¹)	S° (J K⁻¹mol⁻¹)	-[G°-H°(T_r)]/T (J K⁻¹mol⁻¹)	H°-H°(T_r) (kJ mol⁻¹)	$\Delta_f H°$ (kJ mol⁻¹)	$\Delta_f G°$ (kJ mol⁻¹)	Log K_f
0	0.	0.	INFINITE	-12.052	93.317	93.317	INFINITE
100	35.238	36.869	137.842	-10.097	93.093	98.495	-51.449
200	52.484	67.308	95.337	-5.606	93.790	103.689	-27.081
298.15	61.024	90.002	90.002	0.	94.977	108.306	-18.975
300	62.195	90.383	90.003	0.114	95.003	108.389	-18.872
400	66.735	108.930	92.500	6.572	96.486	112.628	-14.708
500	70.249	124.212	97.358	13.427	97.978	116.491	-12.170
600	73.136	137.282	102.949	20.600	99.462	120.054	-10.452
700	75.689	148.753	108.689	28.045	100.938	123.370	-9.206
800	78.032	159.016	114.350	35.733	102.427	126.473	-8.258
900	80.249	168.337	119.838	43.649	103.943	129.388	-7.509
1000	82.341	176.900	125.122	51.778	105.495	132.132	-6.902

PREVIOUS: CURRENT: December 1971

Copper Cyanide (CuCN) $C_1Cu_1N_1$(cr)

FLUOROMETHYLIDYNE (CF)   IDEAL GAS M_r = 31.009403

$S°(298.15\ K) = 213.03 \pm 0.04\ J\ K^{-1}\ mol^{-1}$

$\Delta_f H°(0\ K) = 251.6 \pm 8\ kJ\ mol^{-1}$
$\Delta_f H°(298.15\ K) = 255.2 \pm 8\ kJ\ mol^{-1}$

Electronic Levels and Quantum Weights

State	ε_i, cm^{-1}	g_i
$^2\Pi_r$	0	2
	77.11	2
[$^4\Sigma$]	[25000]	[4]
	42705	2
	49452	2

$\omega_e = 1308.1\ cm^{-1}$ $\omega_e x_e = 11.10\ cm^{-1}$ $\sigma = 1$

$B_e = 1.4172\ cm^{-1}$ $\alpha_e = 0.0184\ cm^{-1}$ $r_e = 1.2667$ Å

Enthalpy of Formation

Faber et al. (1) have studied the weight change of graphite in CF_4 at high temperatures. Our reanalysis of their data use the equilibrium constants for CF_4, CF_3 and CF_2 from the JANAF Tables (2); the only other C-F species considered is assumed to be CF. This treatment yields negative values for the CF pressure except at the highest temperature, where 3rd law analysis of the equilibrium constant yields $\Delta_f H°(CF, g, 298.15\ K) = 59.4\ kcal\ mol^{-1}$. Farber et al. (1) also investigated mass-spectrometrically the vapor species present at equilibrium and followed the intensities of CF^+ and F^+. From a 2nd law treatment of the intensities they report $\Delta_f H°(CF, g, 2300\ K) = 62.0\ kcal\ mol^{-1}$, or $\Delta_f H°(CF, g, 298.15\ K) = 64\ kcal\ mol^{-1}$.

Porter et al. (3) have investigated the A-X band system under high dispersion. From an analytical extrapolation of the 10 observed levels they reported a dissociation energy of 126.9±2.9 kcal mol^{-1}, which yields $\Delta_f H°(CF, g, 298.15\ K) = 61.9 \pm 2.9\ kcal\ mol^{-1}$.

The predissociation limit of CF(g) has been reported by Thrush and Zwolenik (4) as <132 kcal mol^{-1}; Kuzyakov and Tatievsky (5) determined the same quantity as <133 kcal mol^{-1}. These results yield $\Delta_f H°(CF, g, 298.15\ K) > 56\ kcal\ mol^{-1}$. Thrush and Zwolenik (4) also report a predissociation limit for the reaction $CF_2 \rightarrow CF + F$ which is <120 kcal mol^{-1}. This value has now been shown to be erroneous (see the CF_2(g) table for details).

Modica (6) has studied the equilibrium $CF_2 \rightarrow CF + F$ in a shock tube in the range 2680 to 3400 K. Since functions for CF_2, CF and F are well established, the data were analyzed using 2nd and 3rd law treatments. The 2nd law analysis gives $\Delta_r H°(298.15\ K) = 107.2 \pm 7.6\ kcal\ mol^{-1}$, in agreement with with value reported by Modica. The 3rd law analysis yields $\Delta_r H°(298.15\ K) = 134.9 \pm 7.2\ kcal\ mol^{-1}$ and a drift of 9.7±2.5 cal K^{-1} mol^{-1}. The excessive drift indicates that the reported equilibrium constants do not represent the stated reaction.

Andrews and Barrow (7) report a dissociation energy of 114 kcal from an extrapolation of four B state levels. This extrapolation is considered unreliable by Porter et al. (3) since this state is formed from an avoided crossing and is likely to be perturbed.

The adopted value of $\Delta_f H°(CF, g, 298.15\ K)$ is 61±2 kcal mol^{-1} (255.224±8 kJ mol^{-1}), which is within the uncertainties and ranges of the acceptable investigations.

Heat Capacity and Entropy

The vibrational and rotational constants are from Porter et al. (3); the higher electronic levels are those reported by Andrews and Barrow (7), except that the B state is now considered to be $^2\Sigma$. The estimated level at 25000 cm^{-1} is proposed in order to explain the low bond dissociation energy of C_2F_2, by analogy with the reasoning used by Simons [8] for C_2F_4. This level would be a $^4\Sigma$ level and could correlate with the ground state atoms.

References

1. M. Farber, M. A. Frisch, and H. C. Ko, Trans. Faraday Soc. 65, 3202 (1969).
2. JANAF Thermochemical Tables: CF_4(g), 6-30-69; CF_3(g), 6-30-69; CF_2(g), 7-30-70.
3. T. L. Porter, D. E. Mann and N. Acquista, J. Mol. Spectrosc. 16, 228 (1965).
4. B. A. Thrush and J. J. Zwolenik, Trans. Faraday Soc. 59, 582 (1963).
5. Yu. Ya. Kuzyakov and V. M. Tatievsky, Opt. Spektrosk. 5, 699 (1958).
6. A. P. Modica, J. Chem. Phys. 44, 1585 (1966).
7. E. B. Andrews and R. F. Barrow, Proc. Phys. Soc. (London), 64A, 481 (1951).
8. J. P. Simons, Nature 205, 1308 (1965).

Fluoromethylidyne (CF) C_1F_1(g)

Enthalpy Reference Temperature = T_r = 298.15 K   Standard State Pressure = p° = 0.1 MPa

T/K	$C_p°$	S°	$-[G°-H°(T_r)]/T$	$H°-H°(T_r)$	$\Delta_f H°$	$\Delta_f G°$	Log K_f
	J $K^{-1}mol^{-1}$			kJ mol^{-1}			
0	0.	0.	INFINITE	-9.082	251.605	251.605	INFINITE
100	31.017	180.140	239.625	-5.948	253.226	243.826	-127.361
200	29.778	201.126	215.766	-2.928	254.455	233.912	-61.092
250	29.792	207.766	213.526	-1.440	254.896	228.723	-47.789
298.15	30.060	213.033	213.033	0.	255.224	223.650	-39.183
300	30.074	213.219	213.034	0.056	255.235	223.455	-38.907
350	30.537	217.888	213.402	1.570	255.484	218.137	-32.555
400	31.094	222.002	214.224	3.111	255.658	212.789	-27.787
450	31.681	225.698	215.297	4.680	255.765	207.423	-24.077
500	32.253	229.066	216.508	6.279	255.816	202.049	-21.108
600	33.280	235.040	219.111	9.557	255.779	191.296	-16.654
700	34.116	240.235	221.766	12.928	255.602	180.561	-13.474
800	34.780	244.836	224.368	16.374	255.322	169.859	-11.091
900	35.305	248.964	226.875	19.880	254.966	159.197	-9.240
1000	35.725	252.706	229.274	23.432	254.550	148.577	-7.761
1100	36.064	256.127	231.562	27.022	254.086	138.002	-6.553
1200	36.343	259.278	233.742	30.643	253.583	127.471	-5.549
1300	36.576	262.196	235.820	34.289	253.046	116.983	-4.700
1400	36.773	264.914	237.802	37.957	252.479	106.538	-3.975
1500	36.942	267.457	239.695	41.643	251.885	96.134	-3.348
1600	37.088	269.846	241.506	45.344	251.267	85.771	-2.800
1700	37.218	272.098	243.240	49.060	250.628	75.447	-2.318
1800	37.333	274.229	244.903	52.787	249.969	65.161	-1.891
1900	37.436	276.250	246.500	56.526	249.292	54.913	-1.510
2000	37.530	278.173	248.036	60.274	248.601	44.700	-1.167
2100	37.617	280.006	249.515	64.032	247.896	34.522	-0.859
2200	37.697	281.758	250.941	67.797	247.180	24.378	-0.579
2300	37.771	283.435	252.318	71.571	246.456	14.267	-0.324
2400	37.841	285.044	253.648	75.352	245.725	4.187	-0.091
2500	37.907	286.591	254.935	79.139	244.990	-5.861	0.122
2600	37.970	288.079	256.181	82.933	244.253	-15.881	0.319
2700	38.030	289.513	257.389	86.733	243.516	-25.872	0.501
2800	38.089	290.897	258.561	90.539	242.780	-35.836	0.669
2900	38.145	292.234	259.700	94.351	242.047	-45.774	0.824
3000	38.200	293.528	260.806	98.168	241.319	-55.686	0.970
3100	38.254	294.782	261.882	101.991	240.597	-65.574	1.105
3200	38.307	295.997	262.929	105.819	239.882	-75.439	1.231
3300	38.360	297.177	263.949	109.652	239.175	-85.282	1.350
3400	38.413	298.323	264.943	113.491	238.477	-95.104	1.461
3500	38.466	299.437	265.913	117.335	237.789	-104.905	1.566
3600	38.519	300.521	266.859	121.184	237.110	-114.686	1.664
3700	38.572	301.578	267.783	125.038	236.442	-124.449	1.757
3800	38.627	302.607	268.686	128.898	235.785	-134.195	1.845
3900	38.682	303.611	269.569	132.764	235.139	-143.922	1.928
4000	38.738	304.591	270.432	136.635	234.505	-153.634	2.006
4100	38.796	305.548	271.277	140.511	233.882	-163.330	2.081
4200	38.855	306.484	272.104	144.394	233.270	-173.010	2.152
4300	38.915	307.399	272.915	148.282	232.670	-182.676	2.219
4400	38.977	308.294	273.709	152.177	232.081	-192.329	2.283
4500	39.040	309.171	274.487	156.078	231.504	-201.968	2.344
4600	39.105	310.030	275.250	159.985	230.938	-211.594	2.403
4700	39.172	310.871	275.999	163.899	230.383	-221.209	2.458
4800	39.241	311.697	276.734	167.820	229.839	-230.811	2.512
4900	39.311	312.507	277.456	171.747	229.306	-240.403	2.563
5000	39.383	313.301	278.165	175.682	228.784	-249.983	2.612
5100	39.457	314.082	278.862	179.624	228.272	-259.553	2.658
5200	39.532	314.849	279.546	183.573	227.771	-269.114	2.703
5300	39.610	315.603	280.220	187.530	227.279	-278.664	2.746
5400	39.689	316.344	280.882	191.495	226.798	-288.206	2.788
5500	39.769	317.073	281.533	195.468	226.327	-297.739	2.828
5600	39.851	317.790	282.174	199.449	225.865	-307.263	2.866
5700	39.935	318.496	282.805	203.438	225.413	-316.779	2.903
5800	40.021	319.192	283.427	207.436	224.970	-326.288	2.939
5900	40.107	319.876	284.039	211.443	224.536	-335.788	2.973
6000	40.196	320.551	284.642	215.458	224.111	-345.282	3.006

PREVIOUS: June 1970 (1 atm)   CURRENT: June 1970 (1 bar)

Fluoromethylidyne (CF) C_1F_1(g)

FLUOROMETHYLIDYNE, ION (CF^+)   IDEAL GAS M_r = 31.008854

$\Delta_f H°(0\ K)$ = 1140.0 ± 5.0 kJ mol^{-1}
$S°(298.15\ K)$ = 201.51 ± 2.1 J K^{-1} mol^{-1} $\Delta_f H°(298.15\ K)$ = 1149.4 ± 5.0 kJ mol^{-1}

Electronic Levels and Quantum Weights

State	ε_i, cm^{-1}	g_i
$X^1\Sigma$	0	1
$[a^3\Pi]$	[35000]	[6]

ω_e = [1380] cm^{-1} $\omega_e x_e$ = [11.6] cm^{-1} σ = 1
B_e = [1.4361] cm^{-1} α_e = [0.0193] cm^{-1} r_e = [1.263] Å

Enthalpy of Formation

There have been no direct determinations of the ionization potential but Johns and Barrow (1) have calculated a value by assuming the $B^2\Sigma^+$ state of CF to be the first of a Rydberg series. They report IP = 8.91 eV (205.464 $kcal^{-1}$) which yields $\Delta_f H°(CF^+, g, 0\ K)$ = 265.6 ± 10 kcal mol^{-1}. An identical value has been calculated by Carroll and Grennan (7) who assumed the $A^2\Sigma^+$ state of CF to be the start of a Rydberg series and estimated the termination by comparison with the same series in N_2, CO and NO.

Diebler and Liston (2) have measured the photoionization onset of the process FCN + hv → CF^+ + N + e^1 ($\Delta H°(0\ K)$ = 377.03 kcal mol^{-1}). Using JANAF auxiliary data (3), we obtain $\Delta_f H°(CF^+, g, 0\ K)$ = 273.0 ± 4 kcal mol^{-1}.

Walter et al. (4) have observed the photoionization onset of the process C_2F_4 + hv → CF^+ + CF_3 + e^- ($\Delta H°(0\ K)$ = 317.306 ± 0.23 kcal mol^{-1}. The JANAF values (3) for C_2F_4 and CF_3 are consistent since they are both derived from the same simultaneous solution, they yield $\Delta_f H°(CF^+, g, 0\ K)$ = 272.464 ± 1.2 kcal mol^{-1}(1139.989 ± 5.0 kJ mol^{-1}), which is adopted. This corresponds to an ionization potential of 212.328 kcal (9.21 eV).

Heat Capacity and Entropy

Spectra reported in the literature, Verma (5), were later shown to belong to BF by Verma (6). We estimated the molecular constants by comparison with CF(g) and BF(g) (3). The $^3\Pi$ electronic level was estimated by comparison with AlCl(g) (3). The enthalpy between 0 and 298.15 K is 2.072 kcal mol^{-1}.

References

1. J. W. C. Johns and R. F. Barrow, Proc. Phys. Soc. (London) 71A, 476 (1958).
2. V. H. Dibeler and S. K. Liston, J. Chem. Phys. 47, 4548 (1967).
3. JANAF Thermochemical Tables: FCN(g), 6-30-69; N(g), 3-31-61; C_2F_4(g), 6-30-69; CF_3(g), 6-30-69; CF(g), 6-30-70; BF(g), 12-31-64; AlCl(g), 6-30-70.
4. T. A. Walter, C. Livshitz, W. A. Chupka and J. Berkowitz, J. Chem. Phys. 51, 3531 (1969).
5. R. D. Verma, Can. J. Phys. 39, 1377 (1961).
6. R. D. Verma, Can. J. Phys. 40, 1852 (1962).
7. P. K. Carroll and T. P. Grennan, J. Phys. At. Mol. Phys. 3B, 865 (1970).

Fluoromethylidyne, Ion (CF^+) $C_1F_1^+(g)$

Enthalpy Reference Temperature = T_r = 298.15 K   Standard State Pressure = p° = 0.1 MPa

T/K	$C_p°$ (J $K^{-1}mol^{-1}$)	S° (J $K^{-1}mol^{-1}$)	$-[G°-H°(T_r)]/T$ (J $K^{-1}mol^{-1}$)	$H°-H°(T_r)$ (kJ mol^{-1})	$\Delta_f H°$ (kJ mol^{-1})	$\Delta_f G°$ (kJ mol^{-1})	Log K_f
0	0.	0.	INFINITE	-8.697	1139.985		
100	29.108	169.611	227.538	-5.793			
200	29.161	189.794	204.197	-2.881			
250	29.327	196.316	201.993	-1.419			
298.15	29.641	201.507	201.507	0.	1149.416	1115.024	-195.348
300	29.656	201.690	201.507	0.055	1149.464	1114.811	-194.106
350	30.125	206.295	201.870	1.549	1150.732	1108.934	-165.499
400	30.680	210.353	202.681	3.069	1151.925	1102.880	-144.021
450	31.266	214.001	203.740	4.617	1153.051	1096.682	-127.299
500	31.845	217.325	204.935	6.195	1154.120	1090.360	-113.909
600	32.900	223.227	207.504	9.434	1156.123	1077.418	-93.798
700	33.775	228.367	210.125	12.769	1157.988	1064.151	-79.408
800	34.477	232.924	212.695	16.183	1159.754	1050.624	-68.599
900	35.038	237.019	215.174	19.660	1161.448	1036.881	-60.179
1000	35.488	240.734	217.547	23.187	1163.086	1022.952	-53.434
1100	35.854	244.134	219.812	26.755	1164.678	1008.861	-47.907
1200	36.155	247.267	221.971	30.356	1166.234	994.627	-43.295
1300	36.405	250.172	224.030	33.984	1167.757	980.264	-39.388
1400	36.617	252.877	225.995	37.635	1169.252	965.785	-36.034
1500	36.798	255.410	227.873	41.306	1170.722	951.201	-33.124
1600	36.955	257.790	229.669	44.994	1172.169	936.519	-30.574
1700	37.093	260.035	231.390	48.697	1173.595	921.747	-28.322
1800	37.215	262.158	233.041	52.412	1175.003	906.891	-26.317
1900	37.325	264.174	234.627	56.139	1176.394	891.958	-24.522
2000	37.424	266.091	236.152	59.877	1177.770	876.952	-22.904
2100	37.514	267.919	237.622	63.624	1179.133	861.878	-21.438
2200	37.598	269.666	239.039	67.380	1180.486	846.739	-20.104
2300	37.675	271.339	240.407	71.143	1181.831	831.538	-18.885
2400	37.747	272.944	241.730	74.914	1183.169	816.279	-17.766
2500	37.814	274.486	243.009	78.693	1184.503	800.965	-16.735
2600	37.878	275.971	244.249	82.477	1185.836	785.597	-15.783
2700	37.939	277.401	245.450	86.268	1187.168	770.178	-14.900
2800	37.996	278.782	246.616	90.065	1188.501	754.710	-14.079
2900	38.052	280.116	247.748	93.867	1189.838	739.193	-13.314
3000	38.105	281.407	248.849	97.675	1191.179	723.631	-12.600
3100	38.157	282.658	249.919	101.488	1192.526	708.024	-11.930
3200	38.207	283.870	250.962	105.306	1193.880	692.373	-11.302
3300	38.256	285.046	251.977	109.130	1195.241	676.680	-10.711
3400	38.304	286.189	252.966	112.958	1196.611	660.946	-10.154
3500	38.352	287.300	253.931	116.790	1197.990	645.171	-9.629
3600	38.399	288.381	254.873	120.628	1199.379	629.356	-9.132
3700	38.446	289.434	255.793	124.470	1200.777	613.503	-8.661
3800	38.493	290.460	256.692	128.317	1202.186	597.612	-8.215
3900	38.540	291.460	257.571	132.169	1203.605	581.684	-7.791
4000	38.588	292.437	258.430	136.025	1205.034	565.719	-7.388
4100	38.637	293.390	259.271	139.886	1206.474	549.718	-7.003
4200	38.687	294.322	260.095	143.753	1207.925	533.682	-6.637
4300	38.738	295.233	260.901	147.624	1209.386	517.611	-6.288
4400	38.791	296.124	261.692	151.500	1210.858	501.507	-5.954
4500	38.846	296.996	262.467	155.382	1212.340	485.369	-5.634
4600	38.904	297.851	263.227	159.270	1213.833	469.197	-5.328
4700	38.964	298.688	263.972	163.163	1215.336	452.993	-5.034
4800	39.027	299.509	264.704	167.063	1216.849	436.757	-4.753
4900	39.094	300.314	265.423	170.969	1218.374	420.490	-4.482
5000	39.163	301.105	266.128	174.881	1219.908	404.191	-4.223
5100	39.237	301.881	266.822	178.801	1221.453	387.861	-3.973
5200	39.315	302.644	267.503	182.729	1223.008	371.501	-3.732
5300	39.396	303.393	268.174	186.665	1224.574	355.111	-3.500
5400	39.483	304.131	268.833	190.608	1226.150	338.691	-3.276
5500	39.574	304.856	269.481	194.561	1227.738	322.242	-3.060
5600	39.669	305.570	270.119	198.523	1229.336	305.764	-2.852
5700	39.770	306.273	270.747	202.495	1230.945	289.257	-2.651
5800	39.876	306.965	271.366	206.478	1232.565	272.722	-2.456
5900	39.988	307.648	271.975	210.471	1234.196	256.160	-2.268
6000	40.105	308.321	272.575	214.475	1235.839	239.569	-2.086

PREVIOUS: December 1970 (1 atm)   CURRENT: December 1970 (1 bar)

Fluoromethylidyne, Ion (CF^+) $C_1F_1^+(g)$

CYANOGEN FLUORIDE (FCN) IDEAL GAS $M_r = 45.016103$

$\Delta_f H°(0\ K) = 35.51 \pm 16.7\ kJ\ mol^{-1}$

$S°(298.15\ K) = 225.40 \pm 0.8\ J\ K^{-1}\ mol^{-1}$ $\Delta_f H°(298.15\ K) = 35.98 \pm 16.7\ kJ\ mol^{-1}$

Vibrational Frequencies and Degeneracies

ν, cm^{-1}
1077 (1)
[420] (2)
2290 (1)

Ground State Quantum Weight: 1 $\sigma = 1$

Point Group: $C_{\infty v}$

Bond Distances: C-F = 1.262 Å C-N = 1.159 Å

Bond Angle: F-C-N = 180°

Rotational Constant: $B_0 = 0.353106\ cm^{-1}$

Enthalpy of Formation

Davis and Okabe (1) have determined the dissociation energy of FCN to F(g) and CN(g) by measuring the threshold energies of incident photon radiation to produce $B^2\Sigma^+$ CN which is monitored by its fluorescence. The dissociation energy determined by this technique is $\leq$ 111 kcal mol^{-1} with an uncertainty of about 1 kcal in the determination of the limit.

Dibeler and Liston (2) have reported a dissociation energy of $\leq$ 115.6 kcal mol^{-1} from the threshold of the CN^+ ion as determined by photoionization.

We adopt a median value of 113.1±2.5 kcal mol^{-1} which leads to $\Delta_f H°$(FCN, g, 298.15 K) = 8.6±4 kcal mol^{-1} (35.982±16.7 kJ mol^{-1}) with auxiliary JANAF values for F(g) and CN(g).

Heat Capacity and Entropy

The molecular structure and bond lengths and angle were determined using microwave spectroscopy by Tyler and Sheridan (3). The vibrational frequencies were reported by Aynsley et. al. (4) from the infrared spectrum, except for the bending frequency which is estimated from the values for ClCN, BrCN and ICN, by comparison of bending force constants. The reasonable limits for this value as calculated from generous limits on the bending force constant are 405-450 cm^{-1}.

References

1. D. D. Davis and H. Okabe, J. Chem. Phys. 49, 5526 (1968).
2. V. H. Dibeler and S. K. Liston, J. Chem. Phys. 47, 4548 (1967).
3. J. K. Tyler and J. Sheridan, Trans. Faraday Soc. 59, 2661 (1963).
4. E. E. Aynsley, R. E. Dodd, and R. Little, Proc. Chem. Soc. (London) 1959, 265 (1959).

Cyanogen Fluoride (FCN) $C_1F_1N_1(g)$

Enthalpy Reference Temperature = T_r = 298.15 K Standard State Pressure = p° = 0.1 MPa

T/K	$C_p°$ (J $K^{-1}mol^{-1}$)	S°	$-[G°-H°(T_r)]/T$	$H°-H°(T_r)$ (kJ mol^{-1})	$\Delta_f H°$	$\Delta_f G°$	Log K_f
0	0.	0.	INFINITE	-10.272	35.509	35.509	INFINITE
100	30.550	186.140	259.533	-7.339	35.477	33.468	-17.482
200	37.491	209.468	229.132	-3.933	35.638	31.424	-8.207
250	40.199	218.136	226.088	-1.988	35.808	30.352	-6.342
298.15	42.312	225.403	225.403	0.	35.982	29.285	-5.131
300	42.385	225.665	225.404	0.078	35.989	29.243	-5.092
350	44.203	232.339	225.927	2.244	36.161	28 105	-4.194
400	45.750	238.345	227.110	4.494	36.314	26.944	-3.518
450	47.097	243.814	228.667	6.816	36.441	25.764	-2.991
500	48.291	248.839	230.436	9.201	36.542	24.573	-2.567
600	50.339	257.830	234.270	14.136	36.670	22.165	-1.930
700	52.041	265.722	238.211	19.258	36.721	19.743	-1.473
800	53.470	272.767	242.098	24.535	36.719	17.317	-1.131
900	54.669	279.136	245.865	29.944	36.677	14.894	-0.864
1000	55.674	284.950	249.487	35.463	36.608	12.477	-0.652
1100	56.518	290.297	252.957	41.073	36.516	10.068	-0.478
1200	57.227	295.246	256.278	46.762	36.406	7.668	-0.334
1300	57.826	299.851	259.454	52.515	36.279	5.279	-0.212
1400	58.333	304.155	262.495	58.324	36.136	2.899	-0.108
1500	58.765	308.195	265.408	64.179	35.977	0.531	-0.018
1600	59.135	311.999	268.203	70.075	35.804	-1.827	0.060
1700	59.453	315.594	270.886	76.005	35.616	-4.173	0.128
1800	59.729	319.000	273.465	81.964	35.415	-6.508	0.189
1900	59.969	322.236	275.947	87.949	35.200	-8.831	0.243
2000	60.178	325.318	278.339	93.957	34.973	-11.143	0.291
2100	60.362	328.258	280.647	99.984	34.736	-13.443	0.334
2200	60.524	331.070	282.876	106.028	34.489	-15.731	0.374
2300	60.668	333.764	285.030	112.088	34.234	-18.008	0.409
2400	60.796	336.349	287.115	118.162	33.974	-20.274	0.441
2500	60.910	338.833	289.134	124.247	33.708	-22.529	0.471
2600	61.013	341.224	291.092	130.343	33.440	-24.773	0.498
2700	61.105	343.528	292.992	136.449	33.171	-27.007	0.522
2800	61.188	345.752	294.836	142.564	32.902	-29.231	0.545
2900	61.264	347.901	296.629	148.687	32.634	-31.445	0.566
3000	61.332	349.979	298.373	154.816	32.369	-33.650	0.586
3100	61.394	351.991	300.071	160.953	32.107	-35.847	0.604
3200	61.451	353.941	301.724	167.095	31.850	-38.035	0.621
3300	61.503	355.833	303.335	173.243	31.598	-40.215	0.637
3400	61.551	357.669	304.906	179.395	31.352	-42.387	0.651
3500	61.595	359.454	306.439	185.553	31.112	-44.552	0.665
3600	61.635	361.190	307.936	191.714	30.878	-46.711	0.678
3700	61.672	362.879	309.398	197.880	30.652	-48.863	0.690
3800	61.707	364.524	310.827	204.049	30.432	-51.009	0.701
3900	61.739	366.128	312.225	210.221	30.218	-53.149	0.712
4000	61.769	367.691	313.592	216.396	30.012	-55.284	0.722
4100	61.796	369.217	314.930	222.575	29.812	-57.414	0.731
4200	61.822	370.706	316.241	228.756	29.618	-59.539	0.740
4300	61.846	372.161	317.524	234.939	29.430	-61.660	0.749
4400	61.869	373.583	318.782	241.125	29.248	-63.776	0.757
4500	61.890	374.974	320.015	247.313	29.072	-65.888	0.765
4600	61.909	376.334	321.225	253.503	28.901	-67.997	0.772
4700	61.928	377.666	322.412	259.695	28.735	-70.101	0.779
4800	61.945	378.970	323.577	265.888	28.572	-72.203	0.786
4900	61.962	380.247	324.720	272.084	28.414	-74.300	0.792
5000	61.977	381.499	325.843	278.281	28.259	-76.395	0.798
5100	61.992	382.727	326.947	284.479	28.108	-78.487	0.804
5200	62.005	383.931	328.031	290.679	27.959	-80.575	0.809
5300	62.018	385.112	329.097	296.880	27.812	-82.661	0.815
5400	62.031	386.271	330.145	303.083	27.667	-84.744	0.820
5500	62.042	387.410	331.176	309.286	27.523	-86.824	0.825
5600	62.053	388.528	332.190	315.491	27.380	-88.902	0.829
5700	62.064	389.626	333.188	321.697	27.238	-90.978	0.834
5800	62.074	390.706	334.170	327.904	27.095	-93.050	0.838
5900	62.083	391.767	335.138	334.112	26.952	-95.120	0.842
6000	62.092	392.810	336.090	340.320	26.808	-97.188	0.846

PREVIOUS: June 1969 (1 atm) CURRENT: June 1969 (1 bar)

Cyanogen Fluoride (FCN) $C_1F_1N_1(g)$

CARBONYL FLUORIDE (COF) IDEAL GAS M_r = 47.008803

$\Delta_f H°(0\ K)$ = [-172.1 ± 63] kJ mol^{-1}

S°(298.15 K) = [248.48] J K^{-1} mol^{-1} $\Delta_f H°(298.15\ K)$ = [-171.5 ± 63] kJ mol^{-1}

Vibrational Frequencies and Degeneracies

ν, cm^{-1}
1018(1)
626(1)
1855(1)

Ground State Quantum Weight: 2 σ = 1

Point Group: C_s

Bond Distances: C–F = [1.34] Å C–O = [1.18] Å

Bond Angle: F–C–O = [135]°

Product of the Moments of Inertia: $I_A I_B I_C$ = [2.199751 x 10^{-116}] g^3 cm^6

Enthalpy of Formation

The enthalpy change $\Delta_r H°$ for the reaction $COF_2(g) = CO(g) + 2F(g)$ was calculated as 160.5 kcal mol^{-1}. By analogy with $\Delta_{at}H°$(Cl–COCl, 0 K), $\Delta_{at}H°$(Cl–CO, 0 K), the value of $\Delta_{at}H°$(F–COF, 0 K), $\Delta_{at}H°$(F–CO, 0 K) was taken to be 4. Based on $\Delta_r H°$ = 160.5 x (4/5) = 128.4 kcal mol^{-1} for the reaction $COF_2(g) = COF(g) + F(g)$, the enthalpy of formation for COF(g) was derived as -41±15 kcal mol^{-1}.

Heat Capacity and Entropy

The vibrational frequencies were reported by Milligan et al. (1). The bond distances and angle were estimated by comparison with those for HFCO(g). Due to the presence of one unpaired electron on C atom, the ground state quantum weight is calculated to be 2. The principal moments of inertia are: I_A = 0.3399 x 10^{-39}, I_B = 7.8768 x 10^{-39}, and I_C = 8.2167 x 10^{-39} g cm^2.

Reference

1. D. E. Milligan, M. E. Jacox, A. M. Başs, J. J. Comeford and D. E. Mann, J. Chem. Phys. 42, 3187 (1965).

Carbonyl Fluoride (COF) $C_1F_1O_1(g)$

Enthalpy Reference Temperature = T_r = 298.15 K Standard State Pressure = p° = 0.1 MPa

T/K	$C_p°$ (J $K^{-1}mol^{-1}$)	S° (J $K^{-1}mol^{-1}$)	$-[G°-H°(T_r)]/T$ (J $K^{-1}mol^{-1}$)	$H°-H°(T_r)$ (kJ mol^{-1})	$\Delta_f H°$ (kJ mol^{-1})	$\Delta_f G°$ (kJ mol^{-1})	Log K_f
0	0.	0.	INFINITE	-10.393	-172.132	-172.132	INFINITE
100	33.341	210.080	280.740	-7.066	-171.770	-175.499	91.671
200	35.464	233.679	251.926	-3.649	-171.600	-179.306	46.830
250	37.207	241.777	249.110	-1.833	-171.560	-181.237	37.867
298.15	38.930	248.479	248.479	0.	-171.544	-183.103	32.079
300	38.995	248.720	248.479	0.072	-171.544	-183.175	31.894
350	40.702	254.860	248.960	2.065	-171.555	-185.113	27.627
400	42.284	260.400	250.050	4.140	-171.594	-187.048	24.426
450	43.733	265.465	251.485	6.291	-171.663	-188.975	21.936
500	45.051	270.142	253.120	8.511	-171.762	-190.894	19.943
600	47.318	278.564	256.674	13.134	-172.034	-194.696	16.950
700	49.151	286.001	260.343	17.960	-172.383	-198.447	14.808
800	50.621	292.663	263.974	22.952	-172.786	-202.143	13.199
900	51.799	298.696	267.502	28.075	-173.227	-205.787	11.944
1000	52.747	304.205	270.901	33.304	-173.697	-209.380	10.937
1100	53.514	309.269	274.162	38.618	-174.191	-212.924	10.111
1200	54.140	313.953	277.285	44.002	-174.706	-216.423	9.421
1300	54.655	318.308	280.275	49.443	-175.241	-219.877	8.835
1400	55.082	322.374	283.138	54.930	-175.795	-223.290	8.331
1500	55.440	326.187	285.883	60.457	-176.368	-226.663	7.893
1600	55.741	329.775	288.515	66.016	-176.962	-229.997	7.509
1700	55.997	333.162	291.042	71.603	-177.575	-233.293	7.168
1800	56.216	336.369	293.472	77.214	-178.209	-236.552	6.865
1900	56.405	339.414	295.811	82.846	-178.862	-239.776	6.592
2000	56.569	342.311	298.064	88.495	-179.535	-242.964	6.346
2100	56.711	345.075	300.237	94.159	-180.226	-246.119	6.122
2200	56.836	347.716	302.336	99.836	-180.934	-249.240	5.918
2300	56.947	350.245	304.364	105.526	-181.657	-252.329	5.731
2400	57.044	352.671	306.327	111.225	-182.396	-255.386	5.558
2500	57.131	355.001	308.228	116.934	-183.147	-258.412	5.399
2600	57.209	357.243	310.070	122.651	-183.909	-261.407	5.252
2700	57.278	359.404	311.857	128.375	-184.680	-264.373	5.115
2800	57.341	361.488	313.593	134.106	-185.460	-267.310	4.987
2900	57.397	363.501	315.279	139.843	-186.246	-270.220	4.867
3000	57.448	365.448	316.919	145.586	-187.038	-273.102	4.755
3100	57.495	367.332	318.515	151.333	-187.833	-275.958	4.650
3200	57.537	369.158	320.070	157.084	-188.631	-278.788	4.551
3300	57.576	370.930	321.584	162.840	-189.432	-281.593	4.457
3400	57.611	372.649	323.061	168.599	-190.233	-284.374	4.369
3500	57.643	374.319	324.502	174.362	-191.035	-287.131	4.285
3600	57.673	375.944	325.908	180.128	-191.836	-289.865	4.206
3700	57.701	377.524	327.282	185.897	-192.637	-292.577	4.130
3800	57.726	379.063	328.624	191.668	-193.436	-295.267	4.059
3900	57.750	380.563	329.937	197.442	-194.235	-297.936	3.990
4000	57.772	382.026	331.221	203.218	-195.032	-300.586	3.925
4100	57.792	383.452	332.478	208.996	-195.828	-303.215	3.863
4200	57.811	384.845	333.708	214.776	-196.623	-305.824	3.803
4300	57.829	386.206	334.913	220.558	-197.416	-308.415	3.746
4400	57.845	387.535	336.094	226.342	-198.209	-310.987	3.692
4500	57.861	388.835	337.252	232.127	-199.001	-313.541	3.639
4600	57.875	390.107	338.387	237.914	-199.792	-316.078	3.589
4700	57.889	391.352	339.501	243.702	-200.584	-318.597	3.541
4800	57.901	392.571	340.594	249.492	-201.377	-321.100	3.494
4900	57.913	393.765	341.667	255.283	-202.170	-323.586	3.449
5000	57.925	394.935	342.720	261.075	-202.966	-326.056	3.406
5100	57.935	396.082	343.755	266.868	-203.764	-328.509	3.365
5200	57.945	397.207	344.773	272.662	-204.565	-330.948	3.324
5300	57.955	398.311	345.772	278.457	-205.370	-333.370	3.286
5400	57.964	399.395	346.755	284.253	-206.179	-335.778	3.248
5500	57.972	400.458	347.722	290.049	-206.993	-338.171	3.212
5600	57.980	401.503	348.673	295.847	-207.814	-340.548	3.176
5700	57.988	402.529	349.609	301.645	-208.642	-342.911	3.142
5800	57.995	403.538	350.530	307.445	-209.477	-345.259	3.109
5900	58.002	404.529	351.437	313.244	-210.321	-347.594	3.077
6000	58.008	405.504	352.330	319.045	-211.175	-349.913	3.046

PREVIOUS: December 1965 (1 atm) CURRENT: December 1965 (1 bar)

Carbonyl Fluoride (COF) $C_1F_1O_1(g)$

$\Delta_fH°(0\ K) = -182.48 \pm 6.3\ kJ\ mol^{-1}$

$S°(298.15\ K) = 240.833 \pm 0.04\ J\ K^{-1}\ mol^{-1}$ $\Delta_fH°(298.15\ K) = -182.00 \pm 6.3\ kJ\ mol^{-1}$

Electronic Levels and Quantum Weights

State	ε_i, cm^{-1}	g_i
X^1A_1	0	1
a^3B_1	[16000]	3
A^1B_1	37226	1

Vibrational Frequencies and Degeneracies

ν_1, cm^{-1}	ν_2, cm^{-1}	ν_3, cm^{-1}
1222 (1)	665 (1)	1110 (1)

Point Group: C_{2v} $\sigma = 2$

Bond Distance: C-F = 1.300 Å

Bond Angle: F-C-F = 104.9°

Product of the Moments of Inertia: $I_AI_BI_C = 4.88015 \times 10^{-116}\ g^3\ cm^6$

Enthalpy of Formation

Thrush and Zwolenik (1) reported a predissociation limit for $CF_2 \rightarrow CF + F$; however, Marsigny et al. (2) have since observed in emission the assumed predissociated level, and others above it, which invalidates the reported limit.

Walter et al. (3) have obtained the photoionization threshold at 0°K for onset of the reaction $C_2F_4 + e^- \rightarrow CF_2^+ + CF_2 + 2e^-$, as 14.63 ± 0.04 eV (337.37 kcal). The ionization potential of CF_2 has been reported as 11.7 eV (4), 11.8 ± 0.3 eV (5) and 11.86 ± 0.1 eV (6). Since these values were obtained by electron impact methods, there may be excess energy involved. Thus, adopting the lowest value as 11.7 ± 0.2 eV, we obtain $\Delta_fH°(CF_2, g, 298.15\ K) = -44.4 \pm 2.5\ kcal\ mol^{-1}$ (7). All previous electron impact methods are considered superseded by this photoionization result.

Modica and LaGraff (8) determined equilibrium constants for the reaction (A) $C_2F_4 \rightarrow 2CF_2$ by use of a shock tube to achieve high temperatures and ultra violet absorption to detect the CF_2 concentration. Carlson (9) has studied the same reaction using a similar technique, and also Zmbov et al. (5) have investigated this equilibrium in a mass spectrometer. A 2nd and 3rd law analysis of all three sets of data is given below along with our recalculated equilibrium data of Farber et al. (10) for the reaction (B) $C + 2F \rightarrow CF_2$ from weight loss experiments in the system $C + CF_4$.

Kinetic methods were used by Edwards and Small (11) to study the equilibrium (C) $CHF_2Cl(g) \rightleftarrows CF_2(g) + HCl(g)$. They analysed their data by assuming that the only other equilibrium occurring was $2CF_2(g) \rightleftarrows C_2F_4(g)$, and fit their kinetic data by trial and error. In this manner they obtained activation energies of the forward and back reaction (C) of 55.79 ± 0.49 and 6.21 ± 3.0 kcal, thus ΔE_a = 49.6 ± 3.0 and ΔH(900 K) = 51.4 ± 3 kcal, which yields $\Delta_fH(CF_2, 298.15\ K) = -42 \pm 4\ kcal\ mol^{-1}$ (7). They also obtained the activation energy of reaction (A) as 70.36 ± 1.3 kcal. Dalby (12) has reported the activation energy of the reverse of (A) as 1.2 ± 0.2 kcal, thus, ΔE_a = 69.2 ± 1.3 and ΔH(900 K) = 71.0 ± 1.3 kcal, which yields $\Delta_fH(CF_2, 298.15\ K) = -42.6 \pm 1.0\ kcal\ mol^{-1}$ (7). Gozzo and Patrick (13) also investigated the kinetics of reaction C and obtained activation energies of 51.4 ± 2.5 and 14.4 ± 5 kcal, the latter value being corrected for the activation energy of CF_2 recombination (12). Thus, we obtain ΔE_a = 37 ± 7 and ΔH(1000 K) = 39 ± 7 kcal, which yields $\Delta_fH°(CF_2, 298.15\ K) = -54 \pm 9\ kcal\ mol^{-1}$ (7).

Modica and LaGraff (14) investigated the kinetics of CHF_3 decomposition by a mass spectrometer coupled to a shock tube. They investigated the pressure dependence of the rate and concluded that at the hgh pressure limit E_o = 58.4 ± 2.2 kcal. Taking this value as ΔH°(0 K) for the reaction (D) $CHF_3(g) \rightarrow CF_2(g) + HF(g)$, we obtain $\Delta_fH°(CF_2, 298.15\ K) = -41.3 \pm 2.5\ kcal\ mol^{-1}$ (7). However, any activation energy of the reverse reaction (D) will make this value more negative. Since values of the reverse activation energy of analogous reaction (C) range from 6.2 to 14.4 kcal, it appears likely that reverse reaction (D) will have a significant activation enrgy; thus, $\Delta_fH(CF_2, 298.15\ K) < -41.3 \pm 2.5$ kcal. Tschiukow-Roux (15) has investigated the same reaction in a shock tube, but was unable to obtain sufficient variation of rate with pressure to calculate the high pressure limit.

Reference and Reaction		Points	T/R	Δ_rH(298.15 K) kcal mol^{-1} 2nd law	3rd law	Drift eu	**$\Delta_fH(CF_2, g, 298.15\ K)$ kcal mol^{-1}
(8)	A	20*	1165-1620	70.89±2.7	66.68±2.28	-2.4±2.1	-45.4±2.0
(9)	A	35*	1240-1742	67.92±0.7	68.40±0.72	0.3±0.5	-44.5±0.7
(5)	A	11	1127-1244	77.3 ±2.1	75.13±0.5	-1.8±1.8	-41.15±1.5
(10)	B	8	2023-2474	-73.8 ±1.4	-80.04±1.0	-2.8±0.6	-42.3±1.0

*Points ommitted due to failure of a statistical test **Calculated from the 3rd law heat of reaction.

A median value of $\Delta_fH(CF_2, g, 298.15\ K)$ covering the area of maximum probability is adopted as $-43.5 \pm 1.5\ kcal\ mol^{-1}$.

Heat Capacity and Entropy

The structure of CF_2 has been reported by Powell and Lide (16) from microwave measurements and has been confirmed by Mathews (17) from the fine structure of the 2500Å absorption band. Mathews also reported ω_2, and Herr and Pimentel (18) reported ω_3, both from gas phase studies. Milligan et al. (19) have observed all three vibrations in matrix-isolated CF_2 as 1222, 668 and 1102 cm^{-1}. The electronic levels are from Mathews (17) with the triplet level estimated by Simons (20) to explain the low C=C bond energy of 70 kcal in C_2F_4.

References

1. B. A. Thrush and J. H. Zwolenik, Trans. Faraday Soc. 59, 582 (1963).
2. L. Marsigny, J. Ferran, J. Lebreton, and R. Lagrange, Comp. Rend. 266c, 507 (1968).
3. T. S. Walter, C. Livshitz, W. A. Chupka, and J. Berkowitz, J. Chem. Phys. 51, 3531 (1969).
4. I. P. Fisher, J. B. Homer, and F. P. Lossing, J. Amer. Chem. Soc. 87, 957 (1965).
5. K. F. Zmbov, O. M. Uy and J. L. Margrave, J. Amer. Chem. Soc. 90, 5090 (1968).
6. R. F. Pottie, J. Chem. Phys. 42, 2607 (1965).
7. JANAF Thermochemical Tables: C_2F_4 dated 6-30-69; $CHClF_2$ dated 6-30-70; CHF_3 dated 6-30-69.
8. A. P. Modica and J. E. LaGraff, J. Chem. Phys. 43, 3383 (1965).
9. G. A. Carlson, Sandia Laboratories, Albuquerque, N. M. private communication, April 7, 1970.
10. M. Farber, M. A. Frish, and H. C. Ko, Trans. Faraday Soc. 65, 3202 (1969).
11. J. W. Edwards and P. A. Small, Ind. Eng. Chem. Fundamentals 4, 396 (1965).
12. F. W. Dalby, J. Chem. Phys. 41, 2297 (1964).
13. F. Gozzo and C. R. Patrick, Nature 202, 80 (1964).
14. A. P. Modica and J. E. LaGraff, J. Chem. Phys. 44, 3375 (1966).
15. E. Tschuikow-Roux, J. Chem. Phys. 42, 3639 and 2049 (1965).
16. F. X. Powell and D. R. Lide, J. Chem. Phys. 45, 1067 (1966)
17. C. W. Mathews, J. Chem. Phys. 45, 1068 (1966) and Can. J. Phys. 45, 2355 (1967).
18. K. C. Herr and G. C. Pimentel, Appl. Opt. 4, 25 (1965).
19. D. E. Milligan, D. E. Mann, M. E. Jacox and R. A. Mitsch, J. Chem. Phys. 41, 1199 (1964) and D. E. Milligan and M. E. Jacox, J. Chem. Phys. 48, 2265 (1968).
20. J. P. Simons, Nature 205, 1308 (1965) and J. Chem. Soc. (London), 5406 (1965).

Enthalpy Reference Temperature = T_r = 298.15 K Standard State Pressure = p° = 0.1 MPa

T/K	$C_p°$ (J $K^{-1}mol^{-1}$)	S°	$-[G°-H°(T_r)]/T$	$H°-H°(T_r)$ (kJ mol^{-1})	$\Delta_fH°$	$\Delta_fG°$	Log K_f
0	0.	0.	INFINITE	-10.354	-182.482	-182.482	INFINITE
100	33.311	202.633	272.908	-7.028	-182.122	-185.253	96.766
200	35.154	226.123	244.265	-3.628	-181.979	-188.456	49.220
250	36.964	234.154	241.463	-1.827	-181.975	-190.077	39.714
298.15	38.948	240.833	240.833	0.	-182.004	-191.635	33.574
300	39.026	241.074	240.834	0.072	-182.006	-191.695	33.377
350	41.119	247.248	241.316	2.076	-182.061	-193.306	28.849
400	43.097	252.869	242.414	4.182	-182.138	-194.907	25.452
450	44.887	258.051	243.867	6.383	-182.232	-196.498	22.809
500	46.465	262.864	245.529	8.667	-182.345	-198.077	20.693
600	49.020	271.573	249.160	13.448	-182.616	-201.199	17.516
700	50.913	279.279	252.923	18.449	-182.939	-204.271	15.243
800	52.319	286.173	256.656	23.614	-183.305	-207.294	13.535
900	53.376	292.400	260.287	28.901	-183.706	-210.269	12.204
1000	54.184	298.067	263.786	34.281	-184.139	-213.197	11.136
1100	54.812	303.262	267.142	39.732	-184.602	-216.081	10.261
1200	55.308	308.053	270.354	45.239	-185.092	-218.921	9.529
1300	55.705	312.497	273.427	50.791	-185.609	-221.719	8.909
1400	56.028	316.637	276.367	56.378	-186.152	-224.477	8.375
1500	56.293	320.512	279.182	61.994	-186.721	-227.195	7.912
1600	56.515	324.152	281.880	67.635	-187.314	-229.874	7.505
1700	56.702	327.584	284.469	73.296	-187.930	-232.515	7.144
1800	56.864	330.830	286.955	78.975	-188.568	-235.119	6.823
1900	57.006	333.908	289.346	84.668	-189.225	-237.688	6.534
2000	57.134	336.836	291.648	90.375	-189.899	-240.221	6.274
2100	57.252	339.626	293.867	96.095	-190.585	-242.721	6.037
2200	57.365	342.292	296.008	101.825	-191.280	-245.187	5.821
2300	57.475	344.844	298.076	107.567	-191.980	-247.621	5.624
2400	57.586	347.293	300.076	113.321	-192.680	-250.026	5.442
2500	57.700	349.646	302.012	119.085	-193.376	-252.401	5.274
2600	57.820	351.911	303.888	124.861	-194.062	-254.748	5.118
2700	57.947	354.096	305.707	130.649	-194.734	-257.069	4.973
2800	58.083	356.206	307.473	136.451	-195.388	-259.366	4.839
2900	58.229	358.246	309.189	142.266	-196.020	-261.640	4.713
3000	58.384	360.223	310.858	148.097	-196.625	-263.892	4.595
3100	58.551	362.140	312.481	153.943	-197.201	-266.125	4.484
3200	58.728	364.002	314.062	159.807	-197.743	-268.339	4.380
3300	58.916	365.812	315.603	165.689	-198.249	-270.538	4.282
3400	59.115	367.574	317.106	171.591	-198.716	-272.721	4.190
3500	59.323	369.290	318.572	177.513	-199.143	-274.891	4.103
3600	59.541	370.964	320.004	183.456	-199.527	-277.050	4.020
3700	59.768	372.599	321.404	189.421	-199.866	-279.199	3.942
3800	60.002	374.196	322.772	195.410	-200.160	-281.339	3.867
3900	60.242	375.758	324.111	201.422	-200.407	-283.472	3.797
4000	60.488	377.286	325.421	207.458	-200.607	-285.599	3.730
4100	60.739	378.783	326.705	213.520	-200.760	-287.722	3.666
4200	60.993	380.249	327.962	219.606	-200.865	-289.841	3.605
4300	61.250	381.687	329.195	225.718	-200.922	-291.959	3.547
4400	61.508	383.098	330.404	231.856	-200.932	-294.076	3.491
4500	61.766	384.484	331.590	238.020	-200.895	-296.193	3.438
4600	62.024	385.844	332.755	244.209	-200.811	-298.312	3.387
4700	62.280	387.181	333.899	250.425	-200.682	-300.433	3.339
4800	62.533	388.495	335.023	256.665	-200.508	-302.557	3.292
4900	62.783	389.786	336.127	262.931	-200.290	-304.685	3.248
5000	63.029	391.057	337.213	269.222	-200.029	-306.818	3.205
5100	63.270	392.308	338.281	275.537	-199.727	-308.957	3.164
5200	63.506	393.539	339.332	281.875	-199.385	-311.102	3.125
5300	63.735	394.751	340.366	288.238	-199.003	-313.254	3.087
5400	63.958	395.944	341.384	294.622	-198.584	-315.413	3.051
5500	64.174	397.120	342.387	301.029	-198.129	-317.581	3.016
5600	64.383	398.278	343.375	307.457	-197.639	-319.757	2.983
5700	64.584	399.419	344.348	313.905	-197.116	-321.943	2.950
5800	64.776	400.544	345.307	320.373	-196.561	-324.138	2.919
5900	64.961	401.653	346.253	326.860	-195.976	-326.342	2.889
6000	65.137	402.746	347.185	333.365	-195.361	-328.557	2.860

PREVIOUS: June 1970 (1 atm) CURRENT: June 1970 (1 bar)

DIFLUOROMETHYLENE, ION (CF_2^+) IDEAL GAS Mr = 50.007257

$S°(298.15\ K) = [246.77 \pm 2.1]\ J\ K^{-1}\ mol^{-1}$

$\Delta_f H°(0\ K) = 935.12 \pm 12.6\ kJ\ mol^{-1}$
$\Delta_f H°(298.15\ K) = 941.79 \pm 12.6\ kJ\ mol^{-1}$

Electronic Levels and Quantum Weights

State	ν_i, cm^{-1}	g_i
$^2A_1$	0	2
$^2B_1$	[15000]	2

Vibrational Frequencies and Degeneracies

ν, cm^{-1}
[1100] (1)
[650] (1)
[1400] (1)

Point Group: C_{2v}
Bond Distance: C-F = [1.32] Å
Bond Angle: F-C-F = [116 ± 10°]
Product of the Moments of inertia: $I_A I_B I_C = [5.07002 \times 10^{-116}]\ g^3\ cm^6$

Enthalpy of Formation

The ionization potential has been reported as 11.7 eV (1), 11.8 eV (2,3), and 11.86 eV (4) from electron impact measurements. Since there may be excess energy involved we prefer the lowest value, 11.7 ± 0.2 eV. This leads to $\Delta_f H°(CF^+$, g, 0 K) = 226.2 ± 6 kcal mol^{-1} using the JANAF $\Delta_f H°(CF_2$, g, 0 K) = -43.61 ± 1.5 kcal mol^{-1} (5).

A more precise value is available from the photoionization measurements of Walter et al. (6), who measured the photoionization threshold the process. for

$$C_2F_4 + h\nu \rightarrow CF_2^+ + CF_2 + e^- \qquad \Delta H°(0\ K) = 337.4 \pm 0.9\ kcal\ mol^{-1}$$

Employing JANAF auxiliary enthalpies (5) we obtain $\Delta_f H°(CF_2^+$, g, 0 K) = 223.5 ± 3.0 kcal mol^{-1} which is adopted.

Heat Capacity and Entropy

The configuration of CF_2^+ is predicted to be bent by Walsh (7), since it has 17 electrons, thus, its ground state is $^2A_1$, a low lying electronic state is predicted by analogy with NO_2 (8). We estimate the angle to be 116 ± 10°, which is larger than CF_2 at 105°, this is expected from the Walsh correlations (7), but smaller than isoelectronic CO_2^- at 127° and NO_2 at 134°. The vibrational frequencies were estimated by comparison with those for CF_2, CO_2^- and COF (7) as 1100 ± 150 cm^{-1}, 650 ± 50 cm^{-1} and 1400 ± 200 cm^{-1}.

The principal moments of interia are: $I_A = 0.7415 \times 10^{-39}$, $I_B = 7.9066 \times 10^{-39}$ and $I_C = 8.6481 \times 10^{-39}\ g\ cm^2$.

References

1. I. P. Fisher, J. B. Homer, and F. P. Lossing, J. Amer. Chem. Soc. 87, 957 (1965).
2. K. F. Zmbov, O. M. Uy, and J. L. Margrave, J. Amer. Chem. Soc. 90, 5090 (1968).
3. W. J. Martin and H. A. McGee, J. Phys. Chem. 72, 738 (1968).
4. R. F. Pottie, J. Chem. Phys. 42, 2607 (1965).
5. JANAF Thermochemical Tables: CF_2(g), 6-30-70; C_2F_4(g), 6-30-69; CO_2^-(g), 12-31-66; COF(g), 12-31-65.
6. T. A. Walter, C. Livshitz, W. A. Chupka and J. Berkowitz, J. Chem. Phys. 51, 3531 (1969).
7. A. D. Walsh, J. Chem. Soc. (London), 2266 (1953).
8. G. Herzberg, "Electronic Spectra of Polyatomic Molecules", D. Van Nostrand Company, Inc., New York, 1966.

Difluoromethylene, Ion (CF_2^+) $C_1F_2^+(g)$

Enthalpy Reference Temperature = T_r = 298.15 K; Standard State Pressure = p° = 0.1 MPa

T/K	$C_p°$ (J K⁻¹mol⁻¹)	S° (J K⁻¹mol⁻¹)	$-[G°-H°(T_r)]/T$ (J K⁻¹mol⁻¹)	$H°-H°(T_r)$ (kJ mol⁻¹)	$\Delta_f H°$ (kJ mol⁻¹)	$\Delta_f G°$ (kJ mol⁻¹)	Log K_f
0	0.	0.	INFINITE	-10.354	935.118		
100	33.321	208.556	278.829	-7.027			
200	35.209	232.075	250.193	-3.624			
250	36.922	240.111	247.397	-1.822			
298.15	38.754	246.769	246.769	0.	941.793	924.137	-161.905
300	38.826	247.009	246.770	0.072	941.830	924.028	-160.888
350	40.758	253.140	247.250	2.062	942.799	920.984	-137.449
400	42.608	258.704	248.339	4.146	943.741	917.803	-119.853
450	44.312	263.823	249.779	6.320	944.658	914.505	-106.153
500	45.841	268.572	251.423	8.574	945.555	911.106	-95.183
600	48.379	277.165	255.013	13.291	947.299	904.052	-78.705
700	50.311	284.774	258.731	18.230	948.992	896.710	-66.913
800	51.776	291.592	262.420	23.338	950.648	889.127	-58.054
900	52.895	297.758	266.010	28.574	952.274	881.339	-51.152
1000	53.761	303.378	269.470	33.908	953.874	873.372	-45.620
1100	54.440	308.535	272.790	39.320	955.450	865.245	-41.087
1200	54.980	313.296	275.970	44.792	957.003	856.976	-37.303
1300	55.415	317.715	279.013	50.312	958.534	848.578	-34.096
1400	55.770	321.835	281.926	55.872	960.043	840.063	-31.343
1500	56.063	325.693	284.717	61.464	961.528	831.441	-28.953
1600	56.308	329.319	287.392	67.083	962.992	822.721	-26.859
1700	56.515	332.739	289.960	72.724	964.435	813.910	-25.008
1800	56.692	335.974	292.427	78.385	965.857	805.014	-23.361
1900	56.846	339.044	294.801	84.062	967.262	796.040	-21.885
2000	56.982	341.963	297.086	89.754	968.652	786.993	-20.554
2100	57.103	344.746	299.290	95.458	970.029	777.876	-19.349
2200	57.214	347.405	301.417	101.174	971.397	768.693	-18.251
2300	57.317	349.951	303.472	106.901	972.761	759.449	-17.248
2400	57.414	352.392	305.460	112.637	974.123	750.145	-16.326
2500	57.507	354.738	307.385	118.383	975.488	740.785	-15.478
2600	57.598	356.995	309.250	124.138	976.859	731.370	-14.693
2700	57.688	359.171	311.059	129.903	978.242	721.902	-13.966
2800	57.778	361.270	312.815	135.676	979.638	712.382	-13.290
2900	57.869	363.299	314.521	141.458	981.052	702.813	-12.659
3000	57.961	365.263	316.180	147.250	982.486	693.194	-12.070
3100	58.054	367.165	317.794	153.051	983.943	683.526	-11.517
3200	58.150	369.010	319.366	158.861	985.426	673.812	-10.999
3300	58.247	370.800	320.897	164.681	986.936	664.051	-10.511
3400	58.346	372.541	322.391	170.510	988.476	654.243	-10.051
3500	58.446	374.233	323.848	176.350	990.045	644.390	-9.617
3600	58.549	375.881	325.270	182.200	991.647	634.492	-9.206
3700	58.653	377.487	326.660	188.060	993.281	624.549	-8.817
3800	58.758	379.053	328.018	193.930	994.948	614.560	-8.448
3900	58.864	380.580	329.346	199.811	996.648	604.528	-8.097
4000	58.970	382.072	330.646	205.703	998.381	594.452	-7.763
4100	59.077	383.529	331.918	211.605	1000.148	584.332	-7.444
4200	59.184	384.954	333.164	217.518	1001.949	574.168	-7.141
4300	59.291	386.348	334.385	223.442	1003.782	563.961	-6.851
4400	59.397	387.712	335.581	229.377	1005.647	553.711	-6.573
4500	59.503	389.048	336.755	235.322	1007.544	543.419	-6.308
4600	59.607	390.357	337.906	241.277	1009.472	533.084	-6.053
4700	59.710	391.640	339.035	247.243	1011.431	522.707	-5.809
4800	59.812	392.898	340.144	253.219	1013.419	512.287	-5.575
4900	59.912	394.133	341.234	259.205	1015.436	501.826	-5.350
5000	60.010	395.344	342.304	265.201	1017.481	491.324	-5.133
5100	60.105	396.533	343.356	271.207	1019.552	480.780	-4.924
5200	60.199	397.702	344.389	277.223	1021.650	470.196	-4.723
5300	60.290	398.849	345.406	283.247	1023.772	459.570	-4.529
5400	60.378	399.977	346.406	289.280	1025.918	448.905	-4.342
5500	60.464	401.086	347.391	295.323	1028.088	438.199	-4.162
5600	60.547	402.176	348.359	301.373	1030.279	427.455	-3.987
5700	60.628	403.248	349.313	307.432	1032.491	416.670	-3.818
5800	60.705	404.303	350.252	313.499	1034.723	405.846	-3.655
5900	60.780	405.342	351.177	319.573	1036.975	394.985	-3.497
6000	60.851	406.364	352.088	325.655	1039.244	384.084	-3.344

PREVIOUS: December 1970 (1 atm) CURRENT: December 1970 (1 bar)

Difluoromethylene, Ion (CF_2^+) $C_1F_2^+(g)$

CARBONIC DIFLUORIDE (COF_2) IDEAL GAS Mr = 66.007206

$\Delta_f H°(0\ K) = -635.8 \pm 1.71\ kJ\ mol^{-1}$

$S°(298.15\ K) = 258.887 \pm 0.08\ J\ K^{-1}\ mol^{-1}$ $\Delta_f H°(298.15\ K) = -638.9 \pm 1.7\ kJ\ mol^{-1}$

Vibrational Frequencies and Degeneracies

ν, cm^{-1}	ν, cm^{-1}
965 (1)	1249 (1)
1928 (1)	584 (1)
626 (1)	774 (1)

Ground State Quantum Weight: 1 $\sigma = 2$

Point Group: C_{2v}

Bond Distances: C-F = 1.312 Å C-O = 1.174 Å

Bond Angles: F-C-F = 108° F-C-O = 126° 1'

Product of the Moments of Inertia: $I_A I_B I_C = 7.225667 \times 10^{-115}\ g^3\ cm^6$

Enthalpy of Formation

Ruff and Li (1) measured equilibrium constants for the reaction

$$2COF_2(g) \rightleftarrows CO_2(g) + CF_4(g)$$

in the temperature range 573-1273 K. These equilibrium data were recently analyzed by Stull, Westrum, and Sinke (2). Their 3rd law analysis of the data clearly indicates that the three low temperature points are in error and should not be used in calculating $\Delta_r H(298.15\ K)$. They reported $\Delta_r H(298.15\ K) = -12 \pm 3\ kcal\ mol^{-1}$ based upon an average of the five high temperature points. Combining this result with the enthalpy of formation of carbon dioxide and carbon tetrafluoride (3), we derive,

$$\Delta_f H°(COF_2,\ g,\ 298.15\ K) = -152.5 \pm 3.3\ kcal\ mol^{-1}.$$

Wartenberg and Riteris (4) measured the enthalpy of hydrolysis of CF_2O. They reported $\Delta_r H = -26.7 \pm 0.2\ kcal\ mol^{-1}$ for the reaction

$$COF_2(g) + H_2O(\ell) \rightarrow CO_2(g) + 2HF\cdot 300\ H_2O.$$

Combining this result with enthaply of formation data for carbon dioxide and HF($\Delta_f H°(HF\cdot 300\ H_2O,\ 298.15\ K) = -76.84\ kcal\ mol^{-1}$ (5)), we derive

$$\Delta_f H°(COF_2,\ g,\ 298.15\ K) = -152.7 \pm 0.4\ kcal\ mol^{-1}.$$

Armstrong, Coyle, and Krieger (6) measured the enthalpies of combustion of CH_4 in O_2-F_2 mixtures and found $\Delta_f H°(COF_2,\ g,\ 298.15\ K) = -154.8\ kcal\ mol^{-1}$. The value calculated from Wartenberg's data is adopted here.

Heat Capacity nd Entropy

The vibrational frequencies are from the infrared studies of Hopper et al. (7) and are in good agreement with the results of Lovell et al. (8) and Nielsen et al. (9). The molecular constants are from the microwave studies of Laurie et al. (10) and are in reasonable agreement with the electron-diffraction work of Brown and Livingston (11). The principal moments of inertia are: $I_A = 7.1086 \times 10^{-39}$, $I_B = 7.1359 \times 10^{-39}$, and $I_C = 14.2445 \times 10^{-39}\ g\ cm^2$.

References

1. O. Ruff and S. Li, Z. Anorg. Chem. 242, 272 (1939).
2. D. R. Stull, E. F. Westrum, Jr., and G. C. Sinke, "The Chemical Thermodynamics of Organic Compounds", John Wiley & Sons, Inc., New York, 1969, p. 80.
3. JANAF Thermochemical Tables: CF_4(g) and CO_2(g), 6-30-69.
4. H. von Wartenberg and G. Riteris, Z. Anorg. Chem. 258, 356 (1949).
5. Ref. 3, Dec. 31, 1968.
6. G. T. Armstrong, C. F. Coyle, and L. A. Krieger, WADC TR 58-541, October, 1958.
7. M. J. Hopper, J. W. Russell, and J. Overend, J. Chem. Phys. 48, 3765 (1968).
8. R. J. Lovell, C. V. Stephenson, E. A. Jones, J. Chem. Phys. 22, 1953 (1954).
9. A. H. Nielsen, T. G. Burke, P. J. H. Woltz, and E. A. Jones, J. Chem. Phys. 20, 596 (1952).
10. V. W. Laurie, D. T. Pence, and R. H. Jackson, J. Chem. Phys. 37, 2995 (1962).
11. T. T. Brown and R. L. Livingston, J. Amer. Chem. Soc. 74, 6084 (1952).

Carbonic Difluoride (COF_2) $C_1F_2O_1(g)$

Enthalpy Reference Temperature = T_r = 298.15 K Standard State Pressure = p° = 0.1 MPa

T/K	$C_p°$ ($J\ K^{-1} mol^{-1}$)	S° ($J\ K^{-1} mol^{-1}$)	$-[G°-H°(T_r)]/T$ ($J\ K^{-1} mol^{-1}$)	$H°-H°(T_r)$ ($kJ\ mol^{-1}$)	$\Delta_f H°$ ($kJ\ mol^{-1}$)	$\Delta_f G°$ ($kJ\ mol^{-1}$)	Log K_f
0	0.	0.	INFINITE	-11.119	-635.798	-635.798	INFINITE
100	33.489	217.322	295.224	-7.790	-636.888	-632.823	330.553
200	38.897	241.797	262.939	-4.228	-638.038	-628.302	164.096
250	43.168	250.931	259.642	-2.178	-638.514	-625.811	130.756
298.15	47.255	258.887	258.887	0.	-638.897	-623.329	109.205
300	47.407	259.180	258.888	0.088	-638.910	-623.232	108.514
350	51.305	266.786	259.480	2.557	-639.239	-620.592	92.618
400	54.778	273.868	260.841	5.211	-639.514	-617.909	80.691
450	57.831	280.500	262.661	8.028	-639.751	-615.194	71.410
500	60.502	286.735	264.759	10.988	-639.959	-612.454	63.983
600	64.875	298.170	269.393	17.266	-640.312	-606.918	52.837
700	68.225	308.434	274.250	23.929	-640.602	-601.329	44.872
800	70.811	317.720	279.112	30.886	-640.843	-595.701	38.895
900	72.827	326.181	283.879	38.072	-641.049	-590.046	34.245
1000	74.417	333.940	288.503	45.437	-641.228	-584.369	30.524
1100	75.686	341.094	292.963	52.945	-641.388	-578.676	27.479
1200	76.709	347.725	297.253	60.566	-641.538	-572.968	24.941
1300	77.544	353.899	301.376	68.280	-641.684	-567.247	22.792
1400	78.231	359.672	305.336	76.070	-641.831	-561.516	20.950
1500	78.804	365.089	309.141	83.923	-641.984	-555.774	19.354
1600	79.284	370.191	312.799	91.828	-642.147	-550.021	17.956
1700	79.691	375.010	316.318	99.777	-642.321	-544.258	16.723
1800	80.038	379.575	319.706	107.764	-642.508	-538.485	15.626
1900	80.337	383.911	322.973	115.783	-642.710	-532.700	14.645
2000	80.595	388.038	326.123	123.830	-642.925	-526.905	13.761
2100	80.820	391.976	329.166	131.901	-643.152	-521.098	12.962
2200	81.016	395.741	332.107	139.993	-643.390	-515.280	12.234
2300	81.189	399.346	334.953	148.103	-643.637	-509.452	11.570
2400	81.342	402.804	337.709	156.230	-643.890	-503.612	10.961
2500	81.478	406.128	340.379	164.371	-644.146	-497.762	10.400
2600	81.600	409.326	342.970	172.525	-644.402	-491.902	9.882
2700	81.708	412.408	345.485	180.691	-644.656	-486.031	9.403
2800	81.806	415.381	347.929	188.867	-644.904	-480.152	8.957
2900	81.894	418.253	350.304	197.052	-645.145	-474.264	8.542
3000	81.974	421.031	352.616	205.245	-645.376	-468.367	8.155
3100	82.046	423.720	354.866	213.446	-645.595	-462.463	7.792
3200	82.112	426.326	357.059	221.654	-645.800	-456.553	7.452
3300	82.173	428.854	359.196	229.869	-645.989	-450.636	7.133
3400	82.228	431.308	361.281	238.089	-646.162	-444.713	6.832
3500	82.278	433.692	363.316	246.314	-646.317	-438.786	6.549
3600	82.325	436.010	365.304	254.544	-646.453	-432.854	6.281
3700	82.368	438.267	367.245	262.779	-646.570	-426.920	6.027
3800	82.408	440.464	369.143	271.018	-646.668	-420.982	5.787
3900	82.444	442.605	371.000	279.260	-646.746	-415.041	5.559
4000	82.478	444.693	372.816	287.506	-646.804	-409.099	5.342
4100	82.510	446.730	374.594	295.756	-646.843	-403.156	5.136
4200	82.539	448.718	376.335	304.008	-646.863	-397.212	4.940
4300	82.567	450.661	378.041	312.264	-646.864	-391.268	4.753
4400	82.592	452.559	379.713	320.522	-646.846	-385.325	4.574
4500	82.616	454.415	381.353	328.782	-646.811	-379.381	4.404
4600	82.639	456.232	382.961	337.045	-646.760	-373.439	4.241
4700	82.660	458.009	384.539	345.310	-646.692	-367.498	4.084
4800	82.680	459.750	386.088	353.577	-646.609	-361.558	3.935
4900	82.698	461.455	387.608	361.846	-646.512	-355.621	3.791
5000	82.716	463.125	389.102	370.116	-646.402	-349.685	3.653
5100	82.732	464.764	390.570	378.389	-646.279	-343.752	3.521
5200	82.748	466.370	392.012	386.663	-646.146	-337.821	3.393
5300	82.762	467.947	393.430	394.938	-646.003	-331.893	3.271
5400	82.776	469.494	394.824	403.215	-645.851	-325.968	3.153
5500	82.789	471.013	396.196	411.493	-645.691	-320.047	3.040
5600	82.802	472.505	397.545	419.773	-645.525	-314.126	2.930
5700	82.814	473.970	398.873	428.054	-645.354	-308.211	2.824
5800	82.825	475.411	400.180	436.336	-645.179	-302.297	2.722
5900	82.836	476.827	401.467	444.619	-645.001	-296.387	2.624
6000	82.846	478.219	402.735	452.903	-644.822	-290.480	2.529

PREVIOUS: December 1969 (1 atm) CURRENT: December 1969 (1 bar)

Carbonic Difluoride (COF_2) $C_1F_2O_1(g)$

TRIFLUOROMETHYL (CF_3) IDEAL GAS Mr = 69.006209

$S°(298.15\ K) = 265.082 \pm 4.2\ J\ K^{-1}\ mol^{-1}$

$\Delta_f H°(0\ K) = -467.52 \pm 4.2\ kJ\ mol^{-1}$
$\Delta_f H°(298.15\ K) = -470.28 \pm 4.2\ kJ\ mol^{-1}$

Vibrational Frequencies and Degeneracies

ν, cm^{-1}	ν, cm^{-1}
1090 (1)	1259 (2)
701 (1)	[500](2)

Ground State Quantum Weight: 2 $\sigma = 3$
Point Group: C_{3v}
Bond Distance: C-F = [1.33] Å
Bond Angle: F-C-F = [112]°
Product of the Moments of Inertia: $I_A I_B I_C = [9.610547 \times 10^{-115}]\ g^3\ cm^6$

Enthalpy of Formation

The selected value is obtained from least squares, simultaneous adjustment of the enthalpies of formation of CF_3, CF_3X (X = H, Cl, Br, I, CF_3) and C_2F_4. Details of the input data and the adjustment are given in (1). Kinetic data provide the most consistent paths to $\Delta_f H°$ of $CF_3(g)$. Forward and reverse activation energies for the reactions $CHF_3(g) + X(g) \rightarrow CF_3(g) + HX(g)$ have been reported for X = Cl by Coomber (2), for X + Br by Amphlett (3), and for X = I by Goy (4). Increases of about 0.3 kcal mol^{-1} in the reverse activation energies for X = Cl and Br were proposed by Arthur (5) on the basis of additional data. Combining these values and reducing to 298.15 K, we get $\Delta_r H° = 2.93 \pm 0.7$ (X = Cl), 18.89 ± 0.7 (X = Br), and 35.5 ± 3 kcal mol^{-1} (X = I). Coomber (2) reviewed data for the analogous reactions with X = CH_3 and CD_3; these yield $\Delta_r H° = 0.6 \pm 2$ kcal mol^{-1} at 298.15 K for X = CH_3. Activation energies have also been reported (6) for $CF_3I(g) + I(g) \rightarrow CF_3I(g) + I_2(g)$ from which we derive the average value $\Delta_r H° = 17.1 \pm 2$ kcal mol^{-1}. The data of Tschuikow-Roux (7) for $C_2F_6(g) \rightarrow 2CF_3(g)$, recalculated as suggested by Coomber (8) but with JANAF functions, yield $\Delta_r H° = 97.6 \pm 6$ kcal mol^{-1}. Differences between the calculated and observed values of $\Delta_r H°$ are all less than half of the uncertainties assigned to the observed values.

The thorough discussion of photoionization data by Noutary (9) emphasized that CF_3^+ is formed with large amounts of excess energy both during ionization of CF_3 and during fragmentation of other fluorides. The amount of excess energy varies with the parent molecule. This may also be involved in electron-impact studies and could account for some of the discrepancies in bond-dissociation energies derived from appearance potentials and ionization potentials. Such data have been summarized by Coomber (10). Ionization data were omitted from the simultaneous adjustment due to uncertainty in the amounts of excess energy.

Also omitted was the mass-spectrometric study of $3C_2F_4(g) + 4CF_3(g) + 2C(graphite)$ by Ehlert (11) which gave $\Delta_f H° = -102 \pm 3$ kcal mol^{-1}. The discrepancy of 10 kcal mol^{-1} suggests that the data may not represent equilibrium for the above reaction. This reaction was one of many which resulted from heating MnF_3 in an effusion cell made of graphite.

Heat Capacity and Entropy

The bond length is assumed to be 1.33 Å by comparison with the molecules CF_3X, and the bond angle is estimated from molecular-orbital calculations of Morokuma (12) and Pople and coworkers (13). The nonplanar structure is consistent with electron-spin-resonance spectra (14) of CF_3 in krypton and xenon matrices, with infrared spectra of CF_3 in various matrices (15) and in the gaseous phase (16), and with photoionization spectra (17). The principal moments of Inertia are: $I_A = I_B = 7.9146 \times 10^{-39}$ and $I_C = 15.3421 \times 10^{-39}\ g\ cm^2$. The observed vibrational fundamentals of Carlson (16) are confirmed by the matrix spectra of Milligan (15). The latter estimated ν_4 to be in the range of 500 to 600 cm^{-1}. We choose the lower value by comparison with CF_3H, SiF_3H and SiF_3 (18).

The uncertainty in ν_4 corresponds to an uncertainty in the entropy of 0.4 cal K^{-1} mol^{-1} at 300 K and 0.7 cal K^{-1} mol^{-1} at 1000 K. Coomber (2) and Amphlett (19) used kinetic data to derive entropies of 63.2 cal K^{-1} mol^{-1} at 350 K and 71.5 cal K^{-1} mol^{-1} at 600 K. The calculated entropies from this table are higher by 2.0 and 1.7 cal K^{-1} mol^{-1}, respectively. These differences are only slightly larger than our estimate of the uncertainty in the kinetic results. We estimate the overall uncertainty in the calculated entropies as 1.0 cal K^{-1} mol^{-1} at 298.15 K and 1.6 cal K^{-1} mol^{-1} at 1000 K.

References

1. A. N. Syverud, Quarterly Tech. Report T-0009-2Q-69, AFRPL Contract Nr. F04611-67-C-0009, Dow Chemical Co., July, 1969.
2. J. W. Coomber and E. Whittle, Trans. Faraday Soc. 62, 2183 (1966).
3. J. C. Amphlett and E. Whittle, Trans. Faraday Soc. 64, 2130 (1968).
4. C. A. Goy, A. Lord and H. O. Pritchard, J. Phys. Chem. 71, 1086 (1967).
5. N. L. Arthur and P. Gray, Trans. Faraday Soc. 65, 434 (1969).
6. Reference (4), Table II; G. S. Laurence, Trans. Faraday Soc. 63, 1155 (1967).
7. E. Tschuikow-Roux, J. Chem. Phys. 43, 2251 (1965).
8. J. W. Coomber and E. Whittle, Trans. Faraday Soc. 63, 1394 (1967).
9. C. J. Noutary, J. Res. Nat. Bur. Stand. 72A, 479 (1968).
10. J. W. Coomber and E. Whittle, Trans. Faraday Soc. 63, 2656 (1967).
11. T. C. Ehlert, J. Phys. Chem. 73, 949 (1969).
12. K. Morokuma, L. Pedersen and M. Karplus, J. Chem. Phys. 48, 4801 (1968).
13. D. L. Beveridge, P. A. Dobosh and J. A. Pople, J. Chem. Phys. 48, 4803 (1968); M. S. Gordon and J. A. Pople, J. Chem. Phys. 49, 4643 (1968).
14. R. W. Fessenden and R. H. Schuler, J. Chem. Phys. 43, 2704 (1965).
15. D. E. Milligan, M. E. Jacox and J. J. Comeford, J. Chem. Phys. 44, 4058 (1966).
16. G. A. Carlson and G. C. Pimentel, J. Chem. Phys. 44, 4053 (1966).
17. C. Lifshitz and W. A. Chupka, J. Chem. Phys. 47, 3439 (1967).
18. D. E. Milligan and M. E. Jacox, J. Chem. Phys. 49, 5330 (1968).
19. J. C. Amphlett and E. Whittle, Trans. Faraday Soc. 63, 2695 (1967).

Trifluoromethyl (CF_3) $C_1F_3(g)$

Enthalpy Reference Temperature = T_r = 298.15 K Standard State Pressure = p° = 0.1 MPa

T/K	$C_p°$ (J K⁻¹mol⁻¹)	S° (J K⁻¹mol⁻¹)	$-[G°-H°(T_r)]/T$ (J K⁻¹mol⁻¹)	$H°-H°(T_r)$ (kJ mol⁻¹)	$\Delta_f H°$ (kJ mol⁻¹)	$\Delta_f G°$ (kJ mol⁻¹)	Log K_f
0	0.	0.	INFINITE	-11.525	-467.518	-467.518	INFINITE
100	33.941	221.531	303.427	-8.190	-468.603	-465.105	242.946
200	41.234	246.999	269.364	-4.473	-469.606	-461.194	120.452
250	45.682	256.679	265.879	-2.300	-469.982	-459.046	95.912
298.15	49.804	265.082	265.082	0.	-470.282	-456.912	80.049
300	49.957	265.390	265.082	0.092	-470.292	-456.829	79.541
350	53.911	273.393	265.705	2.691	-470.548	-454.564	67.840
400	57.467	280.828	267.136	5.477	-470.759	-452.266	59.060
450	60.598	287.782	269.048	8.430	-470.935	-449.943	52.228
500	63.315	294.311	271.251	11.530	-471.081	-447.602	46.761
600	67.663	306.259	276.110	18.090	-471.310	-442.884	38.556
700	70.862	316.942	281.193	25.024	-471.476	-438.132	32.694
800	73.232	326.567	286.274	32.235	-471.601	-433.360	28.296
900	75.013	335.300	291.243	39.651	-471.700	-428.574	24.874
1000	76.373	343.277	296.054	47.223	-471.786	-423.778	22.136
1100	77.431	350.608	300.684	54.915	-471.867	-418.973	19.895
1200	78.265	357.382	305.131	62.702	-471.951	-414.160	18.028
1300	78.934	363.674	309.395	70.563	-472.043	-409.341	16.448
1400	79.477	369.544	313.484	78.484	-472.147	-404.514	15.093
1500	79.923	375.043	317.407	86.455	-472.266	-399.679	13.918
1600	80.294	380.214	321.172	94.467	-472.403	-394.836	12.890
1700	80.605	385.091	324.790	102.512	-472.558	-389.983	11.983
1800	80.869	389.706	328.269	110.586	-472.730	-385.121	11.176
1900	81.094	394.085	331.619	118.685	-472.918	-380.249	10.454
2000	81.287	398.249	334.847	126.804	-473.120	-375.366	9.804
2100	81.455	402.219	337.962	134.941	-473.332	-370.473	9.215
2200	81.601	406.012	340.969	143.094	-473.550	-365.570	8.680
2300	81.729	409.642	343.877	151.261	-473.769	-360.657	8.191
2400	81.842	413.123	346.690	159.439	-473.985	-355.734	7.742
2500	81.942	416.466	349.415	167.629	-474.193	-350.803	7.330
2600	82.031	419.682	352.056	175.827	-474.388	-345.863	6.948
2700	82.110	422.779	354.618	184.034	-474.564	-340.917	6.595
2800	82.182	425.766	357.106	192.249	-474.719	-335.964	6.267
2900	82.246	428.651	359.524	200.470	-474.848	-331.006	5.962
3000	82.304	431.441	361.875	208.698	-474.947	-326.044	5.677
3100	82.356	434.140	364.163	216.931	-475.013	-321.080	5.410
3200	82.404	436.756	366.390	225.169	-475.044	-316.113	5.160
3300	82.448	439.292	368.561	233.412	-475.037	-311.147	4.925
3400	82.488	441.754	370.678	241.659	-474.992	-306.181	4.704
3500	82.524	444.146	372.743	249.909	-474.905	-301.217	4.495
3600	82.558	446.471	374.759	258.163	-474.777	-296.256	4.299
3700	82.589	448.733	376.728	266.421	-474.607	-291.299	4.112
3800	82.618	450.936	378.652	274.681	-474.394	-286.348	3.936
3900	82.644	453.083	380.533	282.944	-474.139	-281.403	3.769
4000	82.668	455.175	382.373	291.210	-473.841	-276.464	3.610
4100	82.691	457.217	384.174	299.478	-473.501	-271.534	3.459
4200	82.712	459.210	385.937	307.748	-473.120	-266.612	3.316
4300	82.732	461.156	387.663	316.020	-472.698	-261.701	3.179
4400	82.750	463.059	389.355	324.294	-472.236	-256.799	3.049
4500	82.768	464.918	391.014	332.570	-471.734	-251.908	2.924
4600	82.784	466.738	392.640	340.848	-471.195	-247.029	2.805
4700	82.799	468.518	394.236	349.127	-470.619	-242.162	2.691
4800	82.813	470.262	395.802	357.408	-470.007	-237.307	2.582
4900	82.826	471.969	397.339	365.690	-469.360	-232.466	2.478
5000	82.839	473.643	398.848	373.973	-468.680	-227.639	2.378
5100	82.851	475.283	400.331	382.257	-467.968	-222.825	2.282
5200	82.862	476.892	401.788	390.543	-467.224	-218.025	2.190
5300	82.872	478.471	403.220	398.830	-466.451	-213.240	2.102
5400	82.882	480.020	404.628	407.117	-465.648	-208.470	2.017
5500	82.892	481.541	406.012	415.406	-464.818	-203.716	1.935
5600	82.900	483.034	407.374	423.696	-463.962	-198.975	1.856
5700	82.909	484.502	408.715	431.986	-463.080	-194.252	1.780
5800	82.917	485.944	410.034	440.277	-462.174	-189.544	1.707
5900	82.924	487.361	411.332	448.569	-461.245	-184.850	1.637
6000	82.932	488.755	412.611	456.862	-460.294	-180.174	1.569

PREVIOUS: June 1969 (1 atm) CURRENT: June 1969 (1 bar)

Trifluoromethyl (CF_3) $C_1F_3(g)$

TRIFLUOROMETHYL, ION (CF_3^+) IDEAL GAS Mr = 69.005660

$S°(298.15\ K) = 254.53 \pm 1.7\ J\ K^{-1}\ mol^{-1}$

$\Delta_f H°(0\ K) = 417.55 \pm 11.7\ kJ\ mol^{-1}$

$\Delta_f H°(298.15\ K) = 421.01 \pm 11.7\ kJ\ mol^{-1}$

Vibrational Frequencies and Degeneracies

ν, cm^{-1}
[888] (1)
830 (1)
[1404] (2)
[480] (2)

Ground State Quantum Weight: 1 $\sigma = 1$

Point Group: D_{3h}

Bond Distance: C-F = [1.33] Å

Bond Angle: F-C-F = [120]°

Product of the Moments of Inertia: $I_A I_B I_C = [1.173095 \times 10^{-114}]\ g^3\ cm^6$

Enthalpy of Formation

The ionization potential of CF_3 has been carefully measured by Lifshitz and Chupka (1). They used a photoionization technique on CF_3 radicals prepared by heating CF_3CH_2ONO. Because of the low ionization cross-section near the threshold it is very difficult to determine the value exactly. The authors conclude that it is improbable that the ionization potential differs from 9.25 eV by as much as 0.2 eV. They choose to accept a value of 9.25 ± 0.04 eV (213.3 ± 0.9 kcal) based on their interpretation of the regular structure near the threshold. In a later paper, Walter, Lifshitz, Chupka and Berkowitz (9) use photodissociative ionization of CF_3I, CF_3BR and CF_3Cl to obtain an indirect ionization potential for CF_3 of 9.11 eV. We adopt a median value of 9.17 ± 0.08 eV (211.5 ± 1.8 kcal), which with $\Delta_f H°(CF_3, g, 0\ K) = -111.7 \pm 1.0\ kcmol^{-1}$ (2) gives $\Delta_f H°(CF_3^+, g, 0\ K) = 99.8 \pm 2.8\ kcal\ mol^{-1}$.

The above value supercedes all earlier values for the ionization potential of CF_3 which generally were obtained by electron impact methods, which do not have the energy resolution available in the photoionization technique. For further details of the older values see references 3, 4, 5, 6 and 7.

Heat Capacity and Entropy

The correlation rules proposed by Walsh (8) for AB_3 molecules indicate that molecules with less than 25 valence electrons are planar in their ground states. Since BF_3 is isoelectronic with CF_3^+, we estimate the vibrational frequencies by transferring the valence force field constants from BF_3 to CF_3^+. This gives 888, 674, 1404 and 480 cm^{-1}; however, Lifshitz and Chupka (1) conclude that the 674 cm^{-1} frequency is in fact at 830 cm^{-1}. This value is obtained from the spacings in the fine structure of the photoionization curve near threshold. They believe that the out-of-plane motion is most likely to be excited by the transition from pyramidal CF_3 to planar CF_3^+, thus, they ascribe the 830 cm^{-1} spacing to this motion. We adopt this one frequency in place of the calculated value. The bond lengths are assumed equal to those in CF_3. The principal moments of inertia are: $I_A = I_B = 8.3708 \times 10^{-39}$, and $I_C = 16.7416 \times 10^{-39}\ g\ cm^2$.

References

1. C. Lifshitz and W. A. Chupka, J. Chem. Phys. 47, 3439 (1967).
2. JANAF Thermochemical Tables: CF_3(g), 6-30-69.
3. J. B. Farmer, I. H. S. Henderson and F. P. Lossing, J. Chem. Phys. 23, 403 (1955).
4. R. I. Reed and W. Sneddon, Trans. Faraday Soc. 54, 301 (1958).
5. V. H. Dibeler, R. M. Reese and F. L. Mohler, J. Res. Nat. Bur. Stand. 57, 113 (1956).
6. F. P. Lossing, P. Kebarle and J. B. de Sousa, "Advances in Mass Spectrometry", J. Waldren, Editor, Pergamon Press, Inc., New York, 1959.
7. C. Lifshitz and F. A. Long, J. Phys. Chem 69, 3731 (1965).
8. A. D. Walsh, J. Chem. Soc. (London), 1953, 2301 (1953).
9. T. A. Walter, C. Lifshitz, W. A. Chupka and J. Berkowitz, J. Chem. Phys. 51, 353 (1969).

Trifluoromethyl, Ion (CF_3^+) $C_1F_3^+$(g)

Enthalpy Reference Temperature = T_r = 298.15 K Standard State Pressure = p° = 0.1 MPa

	J K⁻¹mol⁻¹			kJ mol⁻¹			
T/K	$C_p°$	S°	$-[G°-H°(T_r)]/T$	$H°-H°(T_r)$	$\Delta_f H°$	$\Delta_f G°$	Log K_f
0	0.	0.	INFINITE	-11.543	417.554		
100	34.065	210.860	292.920	-8.206			
200	41.352	236.450	258.810	-4.472			
250	45.673	246.144	255.327	-2.296			
298.15	49.645	254.532	254.532	0.	421.007	431.267	-75.556
300	49.792	254.839	254.533	0.092	421.034	431.331	-75.101
350	53.582	262.804	255.153	2.678	421.805	432.987	-64.620
400	56.993	270.186	256.576	5.444	422.613	434.529	-56.744
450	60.013	277.077	258.475	8.371	423.450	435.969	-50.606
500	62.659	283.540	260.662	11.439	424.312	437.314	-45.686
600	66.955	295.363	265.479	17.930	426.093	439.748	-38.283
700	70.176	305.937	270.517	24.794	427.935	441.879	-32.973
800	72.600	315.474	275.551	31.939	429.823	443.742	-28.973
900	74.445	324.136	280.475	39.295	431.742	445.367	-25.848
1000	75.869	332.057	285.243	46.814	433.682	446.777	-23.337
1100	76.983	339.342	289.835	54.458	435.631	447.992	-21.273
1200	77.869	346.080	294.244	62.203	437.584	449.030	-19.546
1300	78.582	352.342	298.475	70.026	439.534	449.904	-18.077
1400	79.163	358.188	302.534	77.915	441.475	450.629	-16.813
1500	79.642	363.666	306.429	85.856	443.404	451.216	-15.713
1600	80.041	368.819	310.169	93.840	445.320	451.674	-14.746
1700	80.377	373.682	313.763	101.862	447.219	452.013	-13.889
1800	80.662	378.285	317.221	109.914	449.104	452.240	-13.124
1900	80.906	382.652	320.551	117.993	450.974	452.363	-12.436
2000	81.116	386.808	323.761	126.094	452.833	452.389	-11.815
2100	81.298	390.770	326.858	134.215	454.683	452.321	-11.251
2200	81.457	394.556	329.850	142.353	456.529	452.165	-10.736
2300	81.596	398.180	332.742	150.506	458.374	451.925	-10.264
2400	81.719	401.655	335.542	158.672	460.224	451.605	-9.829
2500	81.828	404.993	338.254	166.849	462.083	451.208	-9.427
2600	81.925	408.205	340.883	175.037	463.956	450.736	-9.055
2700	82.012	411.298	343.434	183.234	465.848	450.192	-8.709
2800	82.090	414.282	345.911	191.439	467.762	449.577	-8.387
2900	82.160	417.164	348.319	199.651	469.704	448.893	-8.085
3000	82.223	419.950	350.660	207.871	471.675	448.142	-7.803
3100	82.281	422.647	352.939	216.096	473.679	447.325	-7.537
3200	82.333	425.261	355.158	224.327	475.720	446.442	-7.287
3300	82.381	427.795	357.321	232.562	477.798	445.495	-7.052
3400	82.424	430.255	359.430	240.803	479.916	444.485	-6.829
3500	82.464	432.645	361.488	249.047	482.074	443.411	-6.618
3600	82.501	434.968	363.497	257.295	484.275	442.276	-6.417
3700	82.535	437.229	365.460	265.547	486.519	441.078	-6.227
3800	82.566	439.431	367.377	273.802	488.805	439.819	-6.046
3900	82.595	441.576	369.253	282.060	491.134	438.500	-5.873
4000	82.622	443.667	371.087	290.321	493.505	437.121	-5.708
4100	82.647	445.708	372.882	298.585	495.919	435.681	-5.551
4200	82.670	447.700	374.640	306.851	498.375	434.182	-5.400
4300	82.692	449.645	376.362	315.119	500.872	432.624	-5.255
4400	82.712	451.546	378.049	323.389	503.408	431.008	-5.117
4500	82.731	453.405	379.703	331.661	505.984	429.334	-4.984
4600	82.748	455.224	381.325	339.935	508.598	427.601	-4.856
4700	82.765	457.004	382.916	348.211	511.250	425.812	-4.732
4800	82.780	458.746	384.478	356.488	513.937	423.966	-4.614
4900	82.795	460.453	386.011	364.767	516.659	422.064	-4.499
5000	82.809	462.126	387.517	373.047	519.415	420.105	-4.389
5100	82.822	463.766	388.996	381.328	522.203	418.091	-4.282
5200	82.834	465.374	390.449	389.611	525.022	416.023	-4.179
5300	82.846	466.952	391.878	397.895	527.872	413.899	-4.079
5400	82.856	468.501	393.282	406.180	530.750	411.722	-3.983
5500	82.867	470.021	394.664	414.466	533.656	409.491	-3.889
5600	82.876	471.515	396.023	422.754	536.589	407.207	-3.798
5700	82.886	472.982	397.360	431.042	539.547	404.870	-3.710
5800	82.895	474.423	398.677	439.331	542.529	402.481	-3.625
5900	82.903	475.840	399.972	447.621	545.535	400.041	-3.542
6000	82.911	477.234	401.249	455.911	548.562	397.549	-3.461

PREVIOUS: December 1971 (1 atm) CURRENT: December 1971 (1 bar)

Trifluoromethyl, Ion (CF_3^+) $C_1F_3^+$(g)

TRIFLUOROIODOMETHANE (CF_3I) IDEAL GAS Mr = 195.910709

$S°(298.15\ K) = 307.629 \pm 0.63\ J\ K^{-1}\ mol^{-1}$

$\Delta_f H°(0\ K) = -583.23 \pm 3.3\ kJ\ mol^{-1}$
$\Delta_f H°(298.15\ K) = -589.11 \pm 3.3\ kJ\ mol^{-1}$

Vibrational Frequencies and Degeneracies

ν, cm^{-1}	ν, cm^{-1}
1074 (1)	1185 (2)
742 (1)	539 (2)
284 (1)	260 (2)

Ground State Quantum Weight: 1 $\sigma = 3$
Point Group: C_{3v}
Bond Distances: C-F = 1.332 Å C-I = 2.130 Å
Bond Angles: F-C-F = 108.3° ± 1° F-C-I = 110.62°
Product of the Moments of Inertia: $I_A I_B I_C = 4.410587 \times 10^{-113}\ g^3\ cm^6$

Enthalpy of Formation

The selected value is obtained from least squares, simultaneous adjustment of the enthalpies of formation of CF_3, CF_3X (X = H, Cl, Br, I, CF_3) and C_2F_4. Details of the input data and the adjustment are given in (1). Lord, Goy and Pritchard (2, 3) report equilibrium data for the reactions $CF_3X(g) + I_2(g) = CF_3I(g) + XI(g)$ where X = H, Cl, and Br. In addition to these prime links, there is a weak link to CF_3 through kinetic studies of $CF_3I(g) + I(g) \rightarrow CF_3(g) + I_2(g)$. Data for this reaction (3, 12) yield the average value $\Delta_r H° = 17.1 \pm 2\ kcal\ mol^{-1}$ at 298.15 K.

Heat Capacity and Entropy

The molecular structure has been derived from electron-diffraction studies by Bowen (4), Anderson (5), and Wong and Schomaker (6). Differences among the data are small. The results of Anderson are adopted since they are intermediate between the others. Microwave data (7) are consistent with the adopted parameters. Infrared spectra of the gas were observed by several investigators (8, 9, 10) and Raman spectra of the liquid were obtained by Edgell and May (10). The vibrational assignments of (9, 10) are in close agreement. These assignments are adopted but preference is given to frequencies observed for the gas. In particular, ν_6 is taken as 260 cm^{-1} (from overtone and combination bands of the gas) rather than 265 cm^{-1} (from the Raman spectra of the liquid). The approximate values of ν_2, ν_3 abd ν_6 have been confirmed by analysis (11) of absorption spectra in the UV.

The thermodynamic functions of this table are essentially the same as those calculated by (9). Uncertainty in the entropy is estimated to vary from 0.15 to 0.6 cal K^{-1} mol^{-1} over the range from 300 to 1000 K. Uncertainty in ν_6 and neglect of anharmonicity are the major sources of error. Discussions of the magnitude of anharmonic effects appear on the tables for $CClF_3(g)$ and $CF_4(g)$. The principal moments of inertia are: $I_A = 14.7097 \times 10^{-39}$ and $I_B = I_C = 54.7578 \times 10^{-39}\ g\ cm^2$.

References

1. A. N. Lsyverud, Quarterly Tech. Report T-0009-2Q-69 under AFRPL contract Nr. F04611-67-C-0009, Dow Chemical Co., July, 1969.
2. A. Lord, C. A. Goy and H. O. Pritchard, J. Phys. Chem. 71, 2705 (1967).
3. C. A. Goy, A. Lord and H. O. Pritchard, J. Phys. Chem. 71, 1086 (1967).
4. H. J. M. Bowen, Trans. Faraday Soc. 50, 444 (1954).
5. R. E. Anderson, Dissertation Abstracts 19, 50 (1968).
6. C. Wong and V. Schomaker, J. Chem. Phys. 28, 1010 (1958).
7. J. Sheridan and M. Gordy, J. Chem. Phys. 20, 591 (1952).
8. E. K. Plyler and N. Acquista, J. Res. Nat. Bur. Stand. 48, 92 (1952).
9. P. R. McGee, F. F. Cleveland, A. G. Meister and C. E. Decker, J. Chem. Phys. 21, 242 (1953).
10. W. F. Edgell and C. E. May. J. Chem. Phys. 22, 1808 (1954).
11. L. H. Sutcliffe and A. D. Walsh, Trans. Faraday Soc. 57, 873 (1961).
12. G. S. Laurence, Trans. Faraday Soc. 63, 1155 (1967).

Trifluoriodomethane (CIF_3) $C_1F_3I_1(g)$

Enthalpy Reference Temperature = T_r = 298.15 K Standard State Pressure = p° = 0.1 MPa

	J $K^{-1}mol^{-1}$			kJ mol^{-1}			
T/K	$C_p°$	S°	$-[G°-H°(T_r)]/T$	$H°-H°(T_r)$	$\Delta_f H°$	$\Delta_f G°$	Log K_f
0	0.	0.	INFINITE	-15.008	-583.227	-583.227	INFINITE
100	41.917	247.280	361.949	-11.467	-585.632	-581.659	303.828
200	58.726	281.794	313.747	-6.391	-587.742	-576.793	150.643
250	65.362	295.629	308.766	-3.284	-588.499	-573.965	119.923
298.15	70.940	307.629	307.629	0.	-589.107	-571.109	100.056
300	71.139	308.068	307.630	0.131	-589.129	-570.997	99.419
350	76.154	319.420	308.515	3.817	-589.704	-567.930	84.759
400	80.462	329.878	310.541	7.735	-598.196	-564.510	73.717
450	84.124	339.573	313.234	11.852	-599.224	-560.237	65.031
500	87.220	348.601	316.325	16.138	-620.266	-553.952	57.871
600	92.033	364.953	323.096	25.114	-619.956	-540.717	47.074
700	95.478	379.413	330.129	34.499	-619.555	-527.541	39.366
800	97.982	392.335	337.111	44.179	-619.100	-514.427	33.589
900	99.840	403.988	343.905	54.074	-618.615	-501.372	29.099
1000	101.246	414.583	350.451	64.132	-618.117	-488.371	25.510
1100	102.331	424.286	356.729	74.313	-617.617	-475.421	22.576
1200	103.184	433.228	362.736	84.590	-617.126	-462.516	20.133
1300	103.863	441.515	368.481	94.944	-616.653	-449.651	18.067
1400	104.413	449.233	373.977	105.359	-616.208	-436.822	16.298
1500	104.864	456.452	379.237	115.823	-615.797	-424.023	14.766
1600	105.238	463.233	384.277	126.329	-615.428	-411.250	13.426
1700	105.551	469.622	389.111	136.869	-615.106	-398.499	12.244
1800	105.816	475.663	393.753	147.438	-614.834	-385.766	11.195
1900	106.042	481.390	398.216	158.031	-614.613	-373.046	10.256
2000	106.236	486.835	402.512	168.645	-614.440	-360.336	9.411
2100	106.403	492.022	406.652	179.277	-614.311	-347.635	8.647
2200	106.550	496.975	410.646	189.925	-614.220	-334.938	7.952
2300	106.678	501.715	414.503	200.586	-614.157	-322.245	7.318
2400	106.791	506.257	418.232	211.260	-614.113	-309.554	6.737
2500	106.890	510.619	421.841	221.944	-614.078	-296.865	6.203
2600	106.979	514.813	425.337	232.638	-614.041	-284.177	5.709
2700	107.059	518.852	428.726	243.339	-613.991	-271.491	5.252
2800	107.130	522.747	432.015	254.049	-613.919	-258.807	4.828
2900	107.194	526.507	435.209	264.765	-613.815	-246.126	4.433
3000	107.252	530.142	438.313	275.488	-613.671	-233.450	4.065
3100	107.304	533.660	441.332	286.215	-613.479	-220.779	3.720
3200	107.352	537.067	444.271	296.948	-613.233	-208.115	3.397
3300	107.395	540.371	447.133	307.686	-612.929	-195.459	3.094
3400	107.435	543.578	449.923	318.427	-612.561	-182.814	2.809
3500	107.471	546.693	452.643	329.172	-612.126	-170.181	2.540
3600	107.505	549.721	455.298	339.921	-611.622	-157.560	2.286
3700	107.536	552.667	457.890	350.673	-611.047	-144.955	2.046
3800	107.564	555.535	460.422	361.428	-610.400	-132.367	1.820
3900	107.590	558.329	462.897	372.186	-609.681	-119.796	1.604
4000	107.615	561.054	465.317	382.946	-608.890	-107.245	1.400
4100	107.637	563.711	467.685	393.709	-608.028	-94.715	1.207
4200	107.658	566.305	470.002	404.474	-607.094	-82.206	1.022
4300	107.678	568.839	472.271	415.241	-606.091	-69.721	0.847
4400	107.696	571.314	474.494	426.009	-605.021	-57.259	0.680
4500	107.713	573.735	476.673	436.780	-603.883	-44.822	0.520
4600	107.729	576.102	478.808	447.552	-602.681	-32.412	0.368
4700	107.744	578.419	480.903	458.326	-601.417	-20.028	0.223
4800	107.758	580.688	482.959	469.101	-600.091	-7.672	0.083
4900	107.772	582.910	484.976	479.877	-598.707	4.656	-0.050
5000	107.784	585.087	486.956	490.655	-597.265	16.955	-0.177
5100	107.796	587.222	488.901	501.434	-595.770	29.224	-0.299
5200	107.807	589.315	490.812	512.214	-594.221	41.465	-0.417
5300	107.817	591.369	492.690	522.995	-592.622	53.674	-0.529
5400	107.827	593.384	494.537	533.778	-590.974	65.853	-0.637
5500	107.837	595.363	496.352	544.561	-589.279	78.000	-0.741
5600	107.845	597.306	498.137	555.345	-587.540	90.118	-0.841
5700	107.854	599.215	499.894	566.130	-585.757	102.202	-0.937
5800	107.862	601.091	501.623	576.916	-583.934	114.255	-1.029
5900	107.869	602.935	503.324	587.702	-582.071	126.278	-1.118
6000	107.876	604.748	504.999	598.489	-580.170	138.268	-1.204

PREVIOUS: June 1969 (1 atm) CURRENT: June 1969 (1 bar)

Trifluoriodomethane (CIF_3) $C_1F_3I_1(g)$

$S°(298.15\ K) = 261.419 \pm 0.25\ J\ K^{-1}\ mol^{-1}$

$\Delta_f H°(0\ K) = -927.23 \pm 1.3\ kJ\ mol^{-1}$
$\Delta_f H°(298.15\ K) = -933.20 \pm 1.3\ kJ\ mol^{-1}$

Vibrational Frequencies and Degeneracies

ν, cm^{-1}	ν, cm^{-1}
908.5 (1)	1283.0 (3)
435.0 (2)	631.5 (3)

Ground State Quantum Weight: 1 $\sigma = 12$
Point Group: T_d
Bond Distance: C–F = 1.320 Å
Bond Angle: F–C–F = 109.4712°
Product of the Moments of Inertia: $I_A I_B I_C = 3.149690 \times 10^{-114}\ g^3\ cm^6$

Enthalpy of Formation

The selected value is obtained from least squares, simultaneous adjustment of the enthaplies of formation of HF(g), HF(50 H_2O) and five closely related fluorides. Adjustment of 23 selected observations relating the 7 variables gives the following enthalpies of formation in kcal mol^{-1} at 298.15 K: HF(g), −65.14 ± 0.2; HF(50 H_2O), −76.78 ± 0.1; NaF(cr), −137.52 ± 0.2; BF_3(g), −271.42 ± 0.4; CF_4(g), −223.04 ± 0.3; NF_3(g), −31.43 ± 0.3; and C_2F_4(polymer), −198.2 ± 0.7. The ± values are estimates of the overall uncertainty including systematic error. Further details of the simultaneous adjustment are given in (1).

The selected value is essentially the same as that obtained by Greenberg and Hubbard (2) in their thorough calorimetric study of the combustion in fluorine of natural and synthetic graphite. Their results is confirmed by the data of Domalski and Armstrong (3) for combustion of graphite-Teflon mixtures. Reliable data (1) also link CF_4(g) with HF(n H_2O), NF_3(g) and C_2F_4(polymer). Armstrong (4, 3) critically reviewed the more uncertain data which were omitted from the adjustment.

Heat Capacity and Entropy

The fundamental vibrations are those observed in the gas-phase Raman spectra by Monostori and Weber (5) and the gas-phase infrared spectra by Maki et al. (6) and Chalmers and McKean (7). The frequencies are essentially the same as those selected by Shimanouchi (8) who also lists earlier spectral studies. Electron-diffraction studies of the gas were reported by Thornton (9) and Hoffman and Livingston (10). The adopted bond length is the average of these two results which differ by only 0.006 Å. The principal moments of inertia are: $I_A = I_B = I_C = 14.6585 \times 10^{-39}\ g\ cm^2$.

The most significant uncertainty in the calculated thermodynamic functtions is due to neglect of anharmonic contributions. Estimates for these contributions were given by Barho (11) and Albright et al. (12), and functions calculated with anharmonic terms were given by McBride et al. (13). Differences between McBride and this JANAF table vary from 0.1 to 0.75 cal $K^{-1}\ mol^{-1}$ in Cp° and 0.05 to 0.49 cal $K^{-1}\ mol^{-1}$ in S° over the range from 300 to 1000 K. Measurements (14) of C_v(303–473 K) suggest that anharmonic contributions are significant, but the data are not sufficiently accurate to test the calculations of McBride (13). Heat capacity data (15) for the condensed phases were reviewed by Kelley and King (16) who derived S°(298.15 K) = 62.8 ± 0.5 cal $K^{-1}\ mol^{-1}$ for CF_4(g).

References

1. A. N. Syverud, Report AFRPL-TR-69-70, Jan., 1969.
2. E. Greenberg and W. N. Hubbard, J. Phys. Chem. 72, 222 (1968).
3. E. S. Domalski and G. T. Armstrong, J. Res. Nat. Bur. Stand. 71A, 105 (1967).
4. G. T. Armstrong, U. S. Nat. Bur. Stand., Report 9803, 1-18 (1968).
5. B. Monostori and A. Weber, J. Chem. Phys. 33, 1867 (1960).
6. A. Maki, E. K. Plyler and R. Thibault, J. Chem. Phys. 37, 1899 (1962).
7. A. A. Chalmers and D. C. McKean, Spectrochim. Acta 22, 251 (1966).
8. T. Shimanouchi, U. S. Nat. Bur. Stand. NSRDS-NBS 6, 11 (1967).
9. C. G. Thornton, Dissertation Abstracts 14, 604 (1954); R. L. Livingston, Ann. Rev. Phys. Chem. 5, 397 (1954).
10. C. W. W. Hoffman and R. L. Livingston, J. Chem. Phys. 21, 565 (1953).
11. W. Barho, Kaltetechnik 17, 219 (1965).
12. L. F. Albright, W. C. Calegar and K. K. Innes, J. Am. Chem. Soc. 76, 6017 (1954).
13. B. J. McBride, S. Heimel, J. G. Ehlers and S. Gordon, NASA SP-3001, 156 (1963).
14. Y. Hwang and J. J. Martin, A. I. Ch. E. J. 10, 89 (1964).
15. A. Eucken and E. Schroder, Z. Phys. Chem. 41B, 307 (1938).
16. K. K. Kelley and E. G. King, U. S. Bur. Mines Bull. 592, 1961.

Enthalpy Reference Temperature = T_r = 298.15 K Standard State Pressure = p° = 0.1 MPa

T/K	$C_p°$	S°	$-[G°-H°(T_r)]/T$	$H°-H°(T_r)$	$\Delta_f H°$	$\Delta_f G°$	Log K_f
	J $K^{-1}mol^{-1}$			kJ mol^{-1}			
0	0.	0.	INFINITE	-12.731	-927.229	-927.229	INFINITE
100	34.745	212.363	306.188	-9.383	-929.755	-916.822	478.899
200	47.370	239.882	266.577	-5.339	-931.896	-902.995	235.838
250	54.623	251.239	262.389	-2.788	-932.645	-895.679	187.142
298.15	61.054	261.419	261.419	0.	-933.199	-888.507	155.663
300	61.288	261.798	261.420	0.113	-933.218	-888.229	154.654
350	67.217	271.700	262.189	3.329	-933.651	-880.695	131.437
400	72.400	281.022	263.966	6.822	-933.970	-873.107	114.016
450	76.875	289.815	266.354	10.557	-934.198	-865.485	100.463
500	80.712	298.118	269.120	14.499	-934.351	-857.841	89.618
600	86.781	313.399	275.249	22.890	-934.485	-842.523	73.348
700	91.209	327.127	281.697	31.801	-934.451	-827.197	61.726
800	94.476	339.530	288.164	41.093	-934.298	-811.884	53.011
900	96.927	350.806	294.507	50.669	-934.065	-796.596	46.233
1000	98.797	361.119	300.660	60.459	-933.778	-781.337	40.813
1100	100.249	370.607	306.593	70.415	-933.456	-766.108	36.379
1200	101.394	379.380	312.298	80.499	-933.115	-750.909	32.686
1300	102.312	387.534	317.776	90.686	-932.766	-735.739	29.562
1400	103.057	395.144	323.033	100.956	-932.417	-720.597	26.886
1500	103.669	402.276	328.081	111.293	-932.075	-705.479	24.567
1600	104.178	408.983	332.930	121.686	-931.744	-690.383	22.539
1700	104.605	415.312	337.591	132.126	-931.428	-675.308	20.750
1800	104.966	421.302	342.077	142.605	-931.125	-660.251	19.160
1900	105.275	426.986	346.398	153.117	-930.835	-645.210	17.738
2000	105.540	432.392	350.563	163.658	-930.556	-630.184	16.459
2100	105.770	437.547	354.584	174.224	-930.283	-615.173	15.302
2200	105.970	442.473	358.467	184.811	-930.011	-600.174	14.250
2300	106.146	447.187	362.223	195.417	-929.735	-585.187	13.290
2400	106.301	451.708	365.858	206.040	-929.449	-570.213	12.410
2500	106.438	456.050	369.379	216.677	-929.146	-555.251	11.601
2600	106.560	460.227	372.794	227.327	-928.821	-540.302	10.855
2700	106.669	464.251	376.107	237.988	-928.466	-525.365	10.164
2800	106.767	468.132	379.325	248.660	-928.077	-510.442	9.522
2900	106.855	471.880	382.452	259.342	-927.649	-495.534	8.926
3000	106.935	475.504	385.494	270.031	-927.177	-480.642	8.369
3100	107.007	479.012	388.454	280.728	-926.656	-465.766	7.848
3200	107.073	482.410	391.337	291.432	-926.084	-450.907	7.360
3300	107.132	485.706	394.147	302.143	-925.458	-436.068	6.902
3400	107.187	488.905	396.888	312.859	-924.775	-421.248	6.472
3500	107.237	492.013	399.561	323.580	-924.034	-406.449	6.066
3600	107.283	495.034	402.171	334.306	-923.233	-391.671	5.683
3700	107.326	497.974	404.721	345.036	-922.372	-376.918	5.321
3800	107.365	500.837	407.213	355.771	-921.450	-362.188	4.979
3900	107.401	503.626	409.650	366.509	-920.468	-347.483	4.654
4000	107.435	506.346	412.033	377.251	-919.425	-332.804	4.346
4100	107.466	508.999	414.366	387.996	-918.322	-318.153	4.053
4200	107.495	511.589	416.650	398.744	-917.160	-303.528	3.775
4300	107.522	514.119	418.887	409.495	-915.940	-288.932	3.510
4400	107.547	516.591	421.080	420.249	-914.663	-274.365	3.257
4500	107.571	519.008	423.229	431.005	-913.330	-259.827	3.016
4600	107.593	521.373	425.337	441.763	-911.943	-245.321	2.786
4700	107.614	523.687	427.405	452.523	-910.503	-230.844	2.566
4800	107.633	525.953	429.435	463.285	-909.011	-216.399	2.355
4900	107.651	528.172	431.427	474.050	-907.470	-201.986	2.153
5000	107.669	530.347	433.384	484.816	-905.880	-187.604	1.960
5100	107.685	532.480	435.306	495.583	-904.243	-173.255	1.774
5200	107.700	534.571	437.195	506.353	-902.562	-158.937	1.597
5300	107.714	536.622	439.052	517.123	-900.837	-144.653	1.426
5400	107.728	538.636	440.877	527.895	-899.069	-130.403	1.261
5500	107.741	540.613	442.673	538.669	-897.262	-116.185	1.103
5600	107.753	542.554	444.439	549.444	-895.416	-102.000	0.951
5700	107.765	544.461	446.177	560.219	-893.532	-87.849	0.805
5800	107.776	546.336	447.888	570.997	-891.613	-73.731	0.664
5900	107.786	548.178	449.572	581.775	-889.659	-59.645	0.528
6000	107.796	549.990	451.231	592.554	-887.672	-45.595	0.397

PREVIOUS: June 1969 (1 atm) CURRENT: June 1969 (1 bar)

TRIFLUOROMETHYL HYPOFLUORITE (CF_3OF) IDEAL GAS Mr = 104.004012

$S°(298.15 K) = 322.53 \pm 2.9 J K^{-1} mol^{-1}$

$\Delta_fH°(0 K) = -756.4 \pm 13 kJ mol^{-1}$
$\Delta_fH°(298.15 K) = -764.8 \pm 13 kJ mol^{-1}$

Vibrational Frequencies and Degeneracies

ν, cm^{-1}	ν, cm^{-1}	ν, cm^{-1}
1294.4 (1)	679.0 (1)	1223.2 (1)
1262.5 (1)	584.1 (1)	607.5 (1)
947.3 (1)	433.7 (1)	250.0 (1)
882.1 (1)	227.5 (1)	internal rotation

Ground State Quantum Weight: 1 σ = 1
Point Group: C_s
Bond Distances: C-F = [1.33] Å F-O = [1.42] Å C-O = [1.43] Å
Bond Angles: C-O-F = [103.2]° F-C-F = [109.5]°
Product of the Moments of Inertia: $I_AI_BI_C = 1.151629 \times 10^{-113} g^3 cm^6$

Enthalpy of Formation

Porter and Cady (1) measured equilibrium constants for the reaction $CF_3OF(g) = COF_2(g) + F_2(g)$ in the temperature range 630-750 K. Equilibrium constants were also determined with excess fluorine added to the system. 2nd and 3rd law analyses of these data yield the following enthalpy of reaction:

	$\Delta_rH°$(298.15 K), kcal mol^{-1}		Drift
	2nd law	3rd law	cal K^{-1} mol^{-1}
Set 1	27.65 ± 0.2	30.10	3.6 ± 0.3
Set added fluorine	25.11 ± 0.3	30.09	7.2 ± 0.5

The 3rd law analyses indicate significant trends within each set of equilibrium data. If the entire error is assigned to the Gibbs energy functions, then the drifts in the 3rd law analysis imply that the Gibbs energy functions for CF_3OF should be increased by 3.6 or 7.2 cal K^{-1} mol^{-1}. However, the real error in these Gibbs energy functions is probably no larger than 1 cal K^{-1} mol^{-1} in the temperature range of the experimental data. Apparently, the equilibrium reaction investigated by Porter and Cady (1) was affected by side reactions possibly involving perfluorodimethyl peroxide. However, since the only other approach to the enthalpy of formation of CF_3OF involves bond energies, we adopt the 3rd law value, $\Delta_rH° = 30.1$ kcal mol^{-1}, and derive $\Delta_fH°(CF_3OF$, g, 298.15 K) = -182.8 ± 3 kcal mol^{-1} with $\Delta_fH°(COF_2$, g, 298.15 K) = -152.7 ± 0.4 kcal mol^{-1} (2).

Czarnowski et al. (3) recently determined the O-F bond energy in CF_3OF from kinetic studies on the thermal decomposition of CF_3OF in the presence of SO_3. They reported the value D_0^0(O-F) = 43.5 ± 0.5 kcal mol^{-1}. Combining this result with the following data,

$\Delta_fH°$(C, g, 298.15 K) = 170.89 ± kcal mol^{-1} (4)
$\Delta_fH°$(F, g, 298.15 K) = 18.86 ± 0.40 kcal mol^{-1} (5)
$\Delta_fH°$(O, g, 298.15 K) = 59.559 ± 0.03 kcal mol^{-1} (6)
D_0^0(C-F) = 117.4 kcal mol^{-1} (from $\Delta_fH°(CF_4$, g, 298.15 K) = -223.04 ± 0.3 kcal mol^{-1} (7)
D_0^0(C-O) = 84 kcal mol^{-1} (8)

we derive $\Delta_fH°(CF_3OF$, g, 298.15 K) = -173.8 kcal mol^{-1}.

The value determined from Porter and Cady's equilibrium data is adopted here.

Heat Capacity and Entropy

The vibrational frequencies are from the infrared and Raman studies of Wilt and Jones (9) and are in reasonable agreement with nine of the twelve expected fundamentals observed by Lagemann et al. (10). Bond lengths and angles are estimated quantities chosen by Wilt and Jones (9). The principal moments of inertia are: $I_A = 14.8190 \times 10^{-39}$, $I_B = 27.8485 \times 10^{-39}$, and $I_C = 27.9057 \times 10^{-39}$ g cm^2. The contributions of hindered internal rotation to the thermodynamic properties of CF_3OF are from the tables of Pitzer and Gwinn (11). The reduced moment of inertia, I_r, and the potential barrier, V_o, are from the work of Wilt and Jones (9) and are:

$I_r = 2.939 \times 10^{-39}$ g cm^{-1} $V_o = 1130$ kcal mol^{-1} ($\nu_{12} = 56$ cm^{-1})

References

1. R. S. Porter and G. H. Cady, J. Amer. Chem. Soc., 79, 5628 (1957).
2. JANAF Thermochemical Tables: COF_2(g), 12-31-69.
3. J. Czarnowski, E. Castellano, and H. J. Schumacher, Chem. Commun., 20 1255 (1968).
4. Ref. 2, March 31, 1961.
5. Ref. 2, Sept. 30, 1965.
6. Ref. 2, June 30, 1962.
7. Ref. 2, June 30, 1969.
8. E. S. Gould, "Inorganic Reactions and Structure", Revised Ed., Holt, Rinehart, and Winston, New York, 1962, P. 138.
9. P. M. Wilt and E. A. Jones, J. Inorg. Nucl. Chem., 30, 2933 (1968).
10. R. T. Lagemann, E. A. Jones, and P. J. H. Woltz. J. Chem. Phys., 20, 1768 (1952).
11. K. S. Pitzer and W. D. Gwinn, J. Chem. Phys., 10, 428 (1942).

Trifluoromethyl Hypofluorite (CF_3OF) $C_1F_4O_1(g)$

Enthalpy Reference Temperature = T_r = 298.15 K Standard State Pressure = p° = 0.1 MPa

T/K	$C_p°$	S°	$-[G°-H°(T_r)]/T$	$H°-H°(T_r)$	$\Delta_fH°$	$\Delta_fG°$	Log K_f
	J $K^{-1}$$mol^{-1}$			kJ mol^{-1}			
0	0.	0.	INFINITE	-14.644	-756.437	-756.437	INFINITE
100	50.170	253.856	383.560	-12.970	-762.089	-744.640	388.960
200	65.831	293.567	329.248	-7.136	-763.895	-726.383	189.712
298.15	79.354	322.533	322.533	0.	-764.835	-707.781	124.000
300	79.617	323.024	322.534	0.147	-764.847	-707.427	123.174
400	91.023	347.556	325.802	8.702	-765.239	-688.216	89.872
500	99.876	368.861	332.329	18.266	-765.263	-668.951	69.885
600	106.462	387.683	340.017	28.599	-765.034	-649.707	56.562
700	111.056	404.453	348.047	39.484	-764.652	-630.514	47.050
800	114.533	419.520	356.055	50.772	-764.174	-611.383	39.919
900	117.089	433.163	363.877	62.358	-763.633	-592.316	34.377
1000	119.039	445.604	371.437	74.167	-763.058	-573.312	29.947
1100	120.625	457.028	378.706	86.154	-762.459	-554.366	26.325
1200	121.779	467.574	385.678	98.275	-761.855	-535.475	23.309
1300	122.779	477.362	392.358	110.505	-761.254	-516.634	20.759
1400	123.591	486.492	398.760	122.825	-760.662	-497.840	18.575
1500	124.256	495.042	404.897	135.218	-760.085	-479.087	16.683
1600	124.805	503.080	410.784	147.672	-759.527	-460.372	15.030
1700	125.206	510.658	416.439	160.173	-758.995	-441.691	13.572
1800	125.599	517.826	421.874	172.714	-758.488	-423.041	12.276
1900	125.934	524.626	427.105	185.291	-758.004	-404.419	11.118
2000	126.219	531.093	432.144	197.899	-757.539	-385.821	10.077
2100	126.470	537.257	437.003	210.533	-757.090	-367.246	9.135
2200	126.687	543.146	441.695	223.192	-756.651	-348.692	8.279
2300	126.876	548.782	446.229	235.870	-756.218	-330.158	7.498
2400	127.043	554.185	450.616	248.566	-755.785	-311.644	6.783
2500	127.189	559.374	454.863	261.278	-755.345	-293.147	6.125
2600	127.261	564.364	458.979	274.001	-754.895	-274.668	5.518
2700	127.340	569.168	462.972	286.730	-754.431	-256.206	4.957
2800	127.445	573.801	466.848	299.470	-753.943	-237.762	4.436
2900	127.541	578.275	470.614	312.219	-753.425	-219.336	3.951
3000	127.629	582.601	474.275	324.978	-752.873	-200.929	3.498
3100	127.705	586.787	477.837	337.744	-752.280	-182.540	3.076
3200	127.775	590.842	481.305	350.518	-751.645	-164.172	2.680
3300	127.841	594.775	484.685	363.299	-750.964	-145.823	2.308
3400	127.901	598.593	487.979	376.086	-750.234	-127.497	1.959
3500	127.955	602.301	491.193	388.879	-749.453	-109.192	1.630
3600	128.005	605.906	494.329	401.677	-748.620	-90.910	1.319
3700	128.051	609.414	497.392	414.480	-747.734	-72.653	1.026
3800	128.093	612.830	500.386	427.287	-746.794	-54.420	0.748
3900	128.132	616.157	503.312	440.099	-745.799	-36.212	0.485
4000	128.168	619.402	506.173	452.914	-744.751	-18.031	0.235
4100	128.201	622.567	508.974	465.732	-743.649	0.123	-0.002
4200	128.231	625.657	511.715	478.554	-742.494	18.251	-0.227
4300	128.261	628.674	514.400	491.378	-741.287	36.349	-0.442
4400	128.289	631.623	517.031	504.206	-740.029	54.419	-0.646
4500	128.315	634.507	519.610	517.036	-738.721	72.461	-0.841
4600	128.340	637.327	522.138	529.869	-737.364	90.471	-1.027
4700	128.363	640.088	524.619	542.704	-735.961	108.453	-1.205
4800	128.384	642.790	527.053	555.541	-734.511	126.404	-1.376
4900	128.403	645.438	529.442	568.381	-733.019	144.325	-1.539
5000	128.420	648.032	531.788	581.222	-731.484	162.214	-1.695
5100	128.437	650.575	534.092	594.065	-729.909	180.073	-1.844
5200	128.454	653.069	536.356	606.909	-728.297	197.900	-1.988
5300	128.470	655.516	538.581	619.755	-726.647	215.696	-2.126
5400	128.485	657.918	540.769	632.603	-724.964	233.460	-2.258
5500	128.499	660.276	542.921	645.452	-723.248	251.192	-2.386
5600	128.512	662.591	545.037	658.303	-721.501	268.894	-2.508
5700	128.522	664.866	547.119	671.155	-719.727	286.563	-2.626
5800	128.533	667.101	549.169	684.007	-717.925	304.202	-2.740
5900	128.545	669.298	551.186	696.861	-716.099	321.810	-2.849
6000	128.558	671.459	553.173	709.716	-714.250	339.385	-2.955

PREVIOUS: December 1969 (1 atm) CURRENT: December 1969 (1 bar)

Trifluoromethyl Hypofluorite (CF_3OF) $C_1F_4O_1(g)$

PENTAFLUORO (TRIFLUOROMETHYL) SULFUR (CF_3SF_5) IDEAL GAS M_r = 196.058224

$S°(298.15\ K)$ = [379.97 ± 2.1] J K^{-1} mol^{-1}

$\Delta_fH°(0\ K)$ = [-1700.4 ± 63] kJ mol^{-1}
$\Delta_fH°(298.15\ K)$ = [-1717.1 ± 63] kJ mol^{-1}

Vibrational Frequencies and Degeneracies

ν, cm^{-1}	ν, cm^{-1}	ν, cm^{-1}
1168(1)	Free Rotation	591(2)
883(1)	627(1)	558(2)
755(1)	319(1)	425(2)
692(1)	501(1)	290(2)
612(1)	1256(2)	220(2)
325(1)	903(2)	

Ground State Quantum Weight: 1 σ (overall) = 12
Point Group: C_{4v}
Bond Distances: C-F = [1.32] Å S-F = [1.56] Å C-S = 1.92 Å
Bond Angles: S-C-F = [109.5°] F*-S-F* = 90°
F*-S-C = [90°] F**-S-C = 180°
(* - equatorial ** - axial)
Product of Moments of Inertia: $I_AI_BI_C$ = [2.659738 x 10^{-112}] g^3 cm^6
Reduced Moment of Inertia: I_{red} = [9.9216 x 10^{-39}] g cm^2

Enthalpy of Formation

We adopt $D°(CF_3-SF_5)$ = 80±10 kcal mol^{-1} from which $\Delta_fH°$ is calculated to be -406.4±15 kcal mol^{-1} at 0 K by combining D° with $\Delta_fH°(CF_3, g)$ = -111.7±1.0 kcal mol^{-1} and $\Delta_fH°(SF_5, g, 0\ K)$ = -214.7±3.6 kcal mol^{-1} (1). We estimate that the D°(C-S) in CF_3SF_5 of the C-C, S-S, and C-S bond lengths and the corresponding stretching frequencies for C_2F_6, S_2F_{10}, and CF_3SF_5 indicates that D°(C-S) should be intermediate between those for C-C and S-S. We calculate the heat of atomization ($\Delta_aH°$) and $\Delta_fH°$ value at 298.15 K as 788.6±19 kcal mol^{-1} and -410.385±15 kcal mol^{-1}, respectively.

Heat Capacity and Entropy

The microwave spectrum of CSF_8 observed by Kisliuk and Silvey (2) shows that the molecule is a symmetric rotor comprised of CF_3 and SF_5 groups joined by a C-S bond. The infrared (3, 4) and Raman (4) spectra obey the selection rules predicted for the point group C_{4v}. We estimate all structural data except for the C-S bond length by analogy with the CF_3 structure in C_2F_6 and the SF_5 structure in S_2F_{10} (1). We select a value for the C-S bond length which when combined with the estimated data reproduces the experimentally determined amount of inertia ($I_A = I_B$ = 7.656 x 10^{-38} g cm^2, 2) within the experimental error. The principal moment of inertia about the symmetry axis is I_C = 45.3692 x 10^{-39} g cm^2.

The vibrational frequencies are rounded values taken from the gas phase infrared and liquid phase Raman study of Griffiths (4). The infrared data are supported by results obtained in an earlier study (3). Griffiths (4) assigned the Raman band at 262 cm^{-1} to the b_1 SF_2 out-of-plane deformation mode (ν_9) by analogy with his assignment for $ClSF_5$ (ν_6 = 271 cm^{-1}). More recent information (1) on $ClSF_5$ indicates that the 271 cm^{-1} Raman band is the low-frequency SF_5 rocking fundamental π_{11}. By analogy with this new assignment for $ClSF_5$, we assign the infrared band observed at 290 cm^{-1} by Griffiths (4) to the SF_5 rocking fundamental (π_{16}). The Raman band at 319 cm^{-1} is believed to be π_9 since it is close to the values (332 cm^{-1} and 317 cm^{-1}) predicted for the analogous deformation mode (π_6) in $ClSF_5$ (1) by force field calculations. We assume that the internal rotation about the C-S bond is unrestricted based on spectroscopic evidence (3, 4) which indicates that the barrier is very small. Kisliuk and Silvey (2) have derived a potential barrier of 219 cm^{-1} (0.63 kcal mol^{-1}) from satellite lines observed in their microwave spectrum; however, these lines are probably due to the CF_3 rocking mode (π_{17}) rather than the torsional vibration (π_7), as indicated by Eggers et al. (3). The overall symmetry number is calculated as the product of the external (λ) and internal (×) symmetry numbers with × assumed equal to 12.

References

1. JANAF Thermochemical Tables: CF_3(g) and C_2F_6(g), 6-30-69; SF_5(g), S_2F_{10}(g), and $ClSF_5$(g), 12-31-77.
2. P. Kisliuk and G. A. Silvey, J. Chem. Phys. 20, 517 (1952).
3. D. F. Eggers, Jr., H. E. Wright and D. W. Robinson, J. Chem. Phys. 35, 1045 (1961).
4. J. E. Griffiths, Spectrochim Acta, 23A, 2145 (1967).

Pentafluoro(Trifluoromethyl)Sulfur (CF_3SF_5) $C_1F_8S_1$(g)

Enthalpy Reference Temperature = T_r = 298.15 K Standard State Pressure = p° = 0.1 MPa

T/K	$C_p°$ (J $K^{-1}mol^{-1}$)	S° (J $K^{-1}mol^{-1}$)	$-[G°-H°(T_r)]/T$ (J $K^{-1}mol^{-1}$)	$H°-H°(T_r)$ (kJ mol^{-1})	$\Delta_fH°$ (kJ mol^{-1})	$\Delta_fG°$ (kJ mol^{-1})	Log K_f
0	4.157	0.	INFINITE	-24.084	-1700.372	-1700.372	INFINITE
100	55.764	278.734	477.625	-19.889	-1708.555	-1666.933	870.716
200	102.475	331.780	391.607	-11.965	-1714.316	-1622.806	423.833
250	122.790	356.890	382.170	-6.320	-1715.995	-1599.720	334.243
298.15	139.222	379.967	379.967	0.	-1717.051	-1577.222	276.323
300	139.794	380.830	379.969	0.258	-1717.082	-1576.354	274.468
350	153.755	403.463	381.725	7.609	-1717.724	-1552.844	231.749
400	165.138	424.764	385.787	15.591	-1720.246	-1529.177	199.690
450	174.415	444.768	391.241	24.087	-1720.972	-1505.256	174.726
500	181.997	463.550	397.543	33.004	-1721.551	-1481.253	154.745
600	193.374	497.804	411.456	51.808	-1721.803	-1433.155	124.767
700	201.250	528.240	426.007	71.563	-1721.375	-1385.073	103.355
800	206.857	555.499	440.520	91.983	-1720.525	-1337.084	87.303
900	210.958	580.112	454.685	112.884	-1772.531	-1288.138	74.762
1000	214.031	602.505	468.364	134.141	-1770.014	-1234.451	64.481
1100	216.384	623.020	481.505	155.667	-1767.433	-1181.020	56.082
1200	218.223	641.930	494.096	177.401	-1764.821	-1127.824	49.093
1300	219.683	659.457	506.150	199.299	-1762.201	-1074.847	43.188
1400	220.861	675.781	517.690	221.328	-1759.592	-1022.072	38.134
1500	221.824	691.053	528.744	243.464	-1757.006	-969.483	33.760
1600	222.621	705.396	539.341	265.687	-1754.451	-917.064	29.939
1700	223.287	718.913	549.510	287.984	-1751.931	-864.805	26.572
1800	223.850	731.692	559.280	310.341	-1749.446	-812.693	23.584
1900	224.329	743.808	568.676	332.751	-1746.992	-760.718	20.914
2000	224.740	755.325	577.723	355.205	-1744.564	-708.872	18.514
2100	225.095	766.299	586.443	377.697	-1742.151	-657.147	16.346
2200	225.405	776.778	594.859	400.222	-1739.742	-605.537	14.377
2300	225.676	786.803	602.988	422.777	-1737.327	-554.035	12.583
2400	225.914	796.413	610.848	445.356	-1734.891	-502.640	10.940
2500	226.125	805.640	618.457	467.958	-1732.421	-451.347	9.430
2600	226.313	814.512	625.828	490.581	-1729.903	-400.154	8.039
2700	226.480	823.057	632.975	513.220	-1727.325	-349.058	6.753
2800	226.630	831.296	639.912	535.876	-1724.676	-298.059	5.560
2900	226.765	839.251	646.649	558.546	-1721.944	-247.156	4.452
3000	226.888	846.941	653.198	581.229	-1719.119	-196.350	3.419
3100	226.998	854.382	659.569	603.923	-1716.194	-145.640	2.454
3200	227.099	861.591	665.770	626.628	-1713.161	-95.024	1.551
3300	227.190	868.581	671.810	649.342	-1710.014	-44.506	0.704
3400	227.274	875.364	677.698	672.066	-1706.748	5.915	-0.091
3500	227.351	881.953	683.440	694.797	-1703.361	56.239	-0.839
3600	227.421	888.359	689.044	717.536	-1699.849	106.465	-1.545
3700	227.486	894.591	694.515	740.281	-1696.212	156.590	-2.211
3800	227.546	900.658	699.860	763.033	-1692.447	206.615	-2.840
3900	227.601	906.570	705.085	785.790	-1688.555	256.540	-3.436
4000	227.652	912.333	710.195	808.553	-1684.538	306.364	-4.001
4100	227.700	917.955	715.194	831.320	-1680.395	356.085	-4.537
4200	227.744	923.442	720.087	854.093	-1676.130	405.704	-5.046
4300	227.786	928.802	724.879	876.869	-1671.743	455.219	-5.530
4400	227.824	934.039	729.573	899.650	-1667.238	504.632	-5.991
4500	227.860	939.159	734.174	922.434	-1662.617	553.942	-6.430
4600	227.894	944.168	738.685	945.222	-1657.883	603.146	-6.849
4700	227.925	949.069	743.109	968.013	-1653.040	652.247	-7.249
4800	227.955	953.868	747.450	990.807	-1648.090	701.244	-7.631
4900	227.983	958.569	751.711	1013.603	-1643.036	750.136	-7.997
5000	228.009	963.175	755.894	1036.403	-1637.883	798.924	-8.346
5100	228.034	967.690	760.003	1059.205	-1632.632	847.608	-8.681
5200	228.057	972.118	764.040	1082.010	-1627.289	896.189	-9.002
5300	228.079	976.463	768.007	1104.816	-1621.856	944.666	-9.310
5400	228.099	980.726	771.907	1127.625	-1616.336	993.039	-9.606
5500	228.119	984.912	775.741	1150.436	-1610.733	1041.308	-9.890
5600	228.138	989.022	779.513	1173.249	-1605.049	1089.477	-10.162
5700	228.155	993.060	783.225	1196.064	-1599.288	1137.541	-10.424
5800	228.172	997.028	786.877	1218.880	-1593.453	1185.504	-10.677
5900	228.188	1000.929	790.472	1241.698	-1587.547	1233.369	-10.919
6000	228.203	1004.764	794.011	1264.518	-1581.573	1281.129	-11.153

PREVIOUS: December 1977 (1 atm) CURRENT: December 1977 (1 bar)

Pentafluoro(Trifluoromethyl)Sulfur (CF_3SF_5) $C_1F_8S_1$(g)

METHYLIDYNE (CH) IDEAL GAS M_r = 13.01894

$S°(298.15\ K) = 183.04 \pm 0.04\ J\ K^{-1}\ mol^{-1}$

$\Delta_f H°(0\ K) = 590.683 \pm 17.5\ kJ\ mol^{-1}$

$\Delta_f H°(298.15\ K) = 594.128 \pm 17.5\ kJ\ mol^{-1}$

Electronic Levels and Quantum Weights

State	ε_i, cm^{-1}	g_i
$^2\Pi_r$	0	2
	17.9	2
$^4\Sigma^-$	[4500]	[4]
	23150	4
	25949	2
	31821	2

$\omega_e = 2861.39\ cm^{-1}$ $\omega_e x_e = 64.29\ cm^{-1}$ $\sigma = 1$

$B_e = 14.455\ cm^{-1}$ $\alpha_e = 0.5339\ cm^{-1}$ $r_e = 1.1198\ Å$

Enthalpy of Formation

Herzberg (1) reanalyzed the rotational predissociation reported by Shidei (2) and concluded that D^0_0 = 80.0 kcal mol^{-1} (3.47 eV). This yields $\Delta_f H°$(CH, g, 298.15 K) = 142.0 kcal mol^{-1}(594.128 ± 17.5 kJ mol^{-1}), and is the adopted value.

Brewer and Kester (3) have confirmed the value as 142 ± 5 kcal mol^{-1}, from an equilibrium study of the temperature variation of the intensity of the $^2\Delta$+$^2\Pi$ emission band.

Linevsky (4) determined the concentration of CH radicals in equilibrium with 1 atm. of hydrogen gas, by application of a high resolution spectro-photographic technique to three electronic transitions in absorption. A third law analysis, using the present functions, of 27 determinations in the temperature range 3065 - 3155 K yielded $\Delta_f H°$(CH, g, 298.15 K) = 142.01 ± 1.28 kcal mol^{-1} with a negligible drift. This value is in excellent agreement with the spectroscopic predissociation value.

Heat Capacity and Entropy

The vibrational and rotational constants and electronic levels are from Herzberg (1) with the exception of the estimated $^4\Sigma$ level. All the constants have been adjusted to the normal isotopic abundances for both carbon and hydrogen. The $^4\Sigma$ level is estimated from the ab initio calculations of Hurley (5) which include electron correlation effects. Earlier LCAO calculations had predicted the $^4\Sigma^-$ level to be the ground state, but Higuchi (6) used LCAO-SCF calculations and obtained the $^4\Sigma$ level at 800 cm^{-1}. Krauss and Wehner (7), using essentially the same technique as Hurley (5), obtained $^4\Sigma$ at 3000 cm^{-1}.

References

1. G. Herzberg, "Spectra of Diatomic Molecules," D. Van Nostrand Co., Inc., New York (1950).
2. T. Shidei, Japan J. Phys. 11, 23 (1936).
3. R. G. Brewer and F. L. Kester, J. Chem. Phys. 40, 812 (1964).
4. M. Linevsky, J. Chem. Phys. 47, 3485 (1967).
5. A. C. Hurley, Proc. Roy. Soc. (London) 249A, 402 (1959).
6. J. Higuchi, J. Chem. Phys. 22, 1339 (1954).
7. M. Krauss and J. F. Wehner, J. Chem. Phys. 29, 1287 (1958).

Methylidyne (CH) $C_1H_1(g)$

Enthalpy Reference Temperature = T_r = 298.15 K Standard State Pressure = p° = 0.1 MPa

	J $K^{-1}mol^{-1}$			kJ mol^{-1}			
T/K	$C°_p$	S°	$-[G°-H°(T_r)]/T$	$H°-H°(T_r)$	$\Delta_f H°$	$\Delta_f G°$	Log K_f
0	0.	0.	INFINITE	-8.730	590.683	590.683	INFINITE
100	29.261	151.151	208.978	-5.783	592.071	582.087	-304.051
200	29.171	171.393	185.708	-2.863	593.317	571.596	-149.286
250	29.167	177.902	183.520	-1.404	593.782	566.110	-118.282
298.15	29.171	183.040	183.040	0.	594.128	560.747	-98.241
300	29.171	183.220	183.040	0.054	594.139	560.540	-97.599
350	29.187	187.718	183.395	1.513	594.403	554.918	-82.817
400	29.223	191.617	184.185	2.973	594.583	549.264	-71.727
450	29.291	195.063	185.206	4.436	594.687	543.592	-63.099
500	29.400	198.154	186.349	5.903	594.725	537.913	-56.195
600	29.757	203.543	188.778	8.859	594.638	526.555	-45.841
700	30.292	208.168	191.225	11.860	594.397	515.226	-38.447
800	30.967	212.256	193.603	14.922	594.062	503.937	-32.904
900	31.739	215.947	195.884	18.057	593.675	492.695	-28.595
1000	32.565	219.334	198.062	21.272	593.265	481.496	-25.151
1100	33.411	222.477	200.140	24.570	592.850	470.340	-22.335
1200	34.250	225.420	202.126	27.954	592.443	459.220	-19.989
1300	35.062	228.194	204.025	31.420	592.050	448.135	-18.006
1400	35.831	230.821	205.846	34.965	591.673	437.078	-16.308
1500	36.548	233.318	207.595	38.584	591.314	426.049	-14.836
1600	37.208	235.698	209.278	42.272	590.972	415.042	-13.550
1700	37.808	237.972	210.899	46.024	590.644	404.057	-12.415
1800	38.348	240.149	212.464	49.832	590.329	393.090	-11.407
1900	38.830	242.235	213.977	53.691	590.023	382.141	-10.506
2000	39.257	244.238	215.440	57.596	589.723	371.207	-9.695
2100	39.632	246.163	216.857	61.541	589.426	360.289	-8.962
2200	39.961	248.014	218.232	65.521	589.130	349.385	-8.295
2300	40.247	249.797	219.566	69.532	588.832	338.494	-7.687
2400	40.495	251.515	220.861	73.569	588.529	327.616	-7.130
2500	40.709	253.173	222.121	77.630	588.219	316.751	-6.618
2600	40.893	254.773	223.346	81.710	587.901	305.898	-6.146
2700	41.051	256.319	224.539	85.807	587.572	295.058	-5.708
2800	41.187	257.815	225.701	89.919	587.231	284.231	-5.302
2900	41.303	259.262	226.833	94.044	586.878	273.415	-4.925
3000	41.402	260.664	227.938	98.179	586.510	262.613	-4.572
3100	41.488	262.023	229.015	102.324	586.127	251.822	-4.243
3200	41.562	263.341	230.067	106.477	585.729	241.044	-3.935
3300	41.627	264.621	231.095	110.636	585.314	230.280	-3.645
3400	41.683	265.865	232.100	114.802	584.883	219.527	-3.373
3500	41.734	267.074	233.082	118.973	584.435	208.788	-3.116
3600	41.780	268.250	234.042	123.148	583.970	198.062	-2.874
3700	41.822	269.396	234.982	127.328	583.487	187.349	-2.645
3800	41.861	270.511	235.903	131.513	582.988	176.649	-2.428
3900	41.899	271.599	236.804	135.701	582.471	165.963	-2.223
4000	41.936	272.661	237.687	139.892	581.937	155.290	-2.028
4100	41.973	273.696	238.553	144.088	581.386	144.630	-1.843
4200	42.010	274.708	239.402	148.287	580.819	133.985	-1.666
4300	42.048	275.697	240.235	152.490	580.235	123.352	-1.498
4400	42.087	276.664	241.052	156.696	579.635	112.734	-1.338
4500	42.127	277.611	241.854	160.907	579.020	102.130	-1.185
4600	42.170	278.537	242.641	165.122	578.388	91.539	-1.039
4700	42.214	279.444	243.414	169.341	577.742	80.962	-0.900
4800	42.260	280.334	244.174	173.565	577.082	70.400	-0.766
4900	42.309	281.206	244.921	177.793	576.407	59.850	-0.638
5000	42.359	282.061	245.655	182.027	575.719	49.316	-0.515
5100	42.412	282.900	246.378	186.265	575.018	38.794	-0.397
5200	42.468	283.724	247.088	190.509	574.304	28.287	-0.284
5300	42.525	284.534	247.787	194.759	573.579	17.794	-0.175
5400	42.585	285.329	248.475	199.014	572.842	7.314	-0.071
5500	42.647	286.111	249.152	203.276	572.095	-3.152	0.030
5600	42.711	286.880	249.819	207.544	571.337	-13.604	0.127
5700	42.777	287.637	250.476	211.818	570.571	-24.043	0.220
5800	42.846	288.381	251.123	216.099	569.795	-34.468	0.310
5900	42.916	289.114	251.760	220.387	569.012	-44.879	0.397
6000	42.987	289.836	252.389	224.683	568.221	-55.278	0.481

PREVIOUS: December 1967 (1 atm) CURRENT: December 1967 (1 bar)

Methylidyne (CH) $C_1H_1(g)$

METHYLIDYNE, ION (CH^+) IDEAL GAS M_r = 13.01839

$S°(298.15\ K) = 171.67 \pm 0.21\ J\ K^{-1}\ mol^{-1}$

$\Delta_f H°(0\ K) = 1623.40 \pm 2.1\ kJ\ mol^{-1}$
$\Delta_f H°(298.15\ K) = 1626.74 \pm 2.1\ kJ\ mol^{-1}$

Electronic Levels and Quantum Weights

State	ε_i, cm^{-1}	g_i
$^1\Sigma^+$	0	1
$^1\Pi$	23596.94	2

$\omega_e = 2885.7\ cm^{-1}$ $\omega_e x_e = [73]\ cm^{-1}$ $\sigma = 1$

$B_e = 14.1776\ cm^{-1}$ $\alpha_e = 0.4917\ cm^{-1}$ $r_e = 1.3909$ Å

Enthalpy of Formation

Douglas and Herzberg (1) first observed the spectrum of CH^+ and concluded that $D_0^0(CH^+) = 3.61 \pm 0.22$ eV from a short extrapolation of the A state vibrational separations, this leads to IP = 11.13±0.22 eV. Later Douglas and Morton (2) showed that the basis for this extrapolation was incorrect and the value was thus invalidated. Herzberg and Johns (3) then reported a value of IP(CH) = 10.64 eV (245.36 kcal mol^{-1}) which was obtained from the limit of the Rydberg states of CH (see also Herzberg (4)). This value in conjuntion with the JANAF value for $\Delta_f H°(CH,\ g,\ 0\ K) = 141.178$ kcal mol^{-1} leads to $\Delta_f H°(CH^+,\ g,\ 0\ K) = 386.5 \pm 0.5$ kcal mol^{-1} (1617.116±2.1 kJ mol^{-1}) and is adopted.

Heat Capacity and Entropy

The molecular constants are those reported by Douglas and Morton (2). The value for ω_e is derived from the observed $\Delta G_{1/2}$ by adding $2\omega_e x_e$. The $\omega_e x_e$ value is that calculated by Moore et al. (5) from ab initio variational calculations of the $^1\Sigma$ and $^1\Pi$ states.

References

1. A. E. Douglas and G. Herzberg, Can. J. Res. 20A, 71 (1942).
2. A. E. Douglas and J. R. Morton, Astrophys. J. 131, 1 (1960).
3. G. Herzberg and J. W. C. Johns, Symposium on Molecular Structure and Spectra, Ohio State University, 1963.
4. G. Herzberg, "The Spectra and Structure of Simple Free Radicals", Cornell University Press, Ithaca (1971).
5. P. L. Moore, J. C. Brown and F. A. Matsen, J. Chem. Phys. 43, 903 (1965).

Methylidyne, Ion (CH^+) $C_1H_1^+(g)$

Enthalpy Reference Temperature = T_r = 298.15 K Standard State Pressure = p° = 0.1 MPa

	J $K^{-1}mol^{-1}$			kJ mol^{-1}			
T/K	$C_p°$	S°	$-[G°-H°(T_r)]/T$	$H°-H°(T_r)$	$\Delta_f H°$	$\Delta_f G°$	Log K_f
0	0.	0.	INFINITE	-8.628	1623.396		
100	29.124	139.844	197.578	-5.773			
200	29.135	160.035	174.337	-2.860			
250	29.143	166.537	172.151	-1.403			
298.15	29.153	171.671	171.671	0.	1626.739	1596.748	-279.744
300	29.154	171 851	171.671	0.054	1626.751	1596.562	-277.986
350	29.173	176.346	172.026	1.512	1627.013	1591.509	-237.520
400	29.211	180.244	172.815	2.972	1627.193	1586.423	-207.166
450	29.279	183.688	173.836	4.434	1627.296	1581.320	-183.555
500	29.385	186.778	174.978	5.900	1627.333	1576.209	-164.665
600	29.723	192.163	177.406	8.854	1627.244	1565.990	-136.332
700	30.205	196.780	179.852	11.850	1626.998	1555.798	-116.095
800	30.778	200.850	182.227	14.898	1626.649	1545.650	-100.921
900	31.390	204.510	184.503	18.006	1626.236	1535.549	-89.121
1000	32.000	207.849	186.673	21.176	1625.780	1525.497	-79.684
1100	32.585	210.927	188.740	24.406	1625.297	1515.491	-71.965
1200	33.132	213.786	190.709	27.692	1624.792	1505.531	-65.534
1300	33.634	216.458	192.588	31.031	1624.272	1495.614	-60.095
1400	34.092	218.967	194.384	34.417	1623.737	1485.737	-55.434
1500	34.507	221.334	196.102	37.848	1623.189	1475.899	-51.395
1600	34.884	223.573	197.750	41.317	1622.628	1466.098	-47.863
1700	35.225	225.698	199.332	44.823	1622.055	1456.332	-44.748
1800	35.535	227.721	200.853	48.361	1621.469	1446.601	-41.979
1900	35.817	229.650	202.319	51.929	1620.872	1436.902	-39.503
2000	36.074	231.494	203.732	55.524	1620.262	1427.235	-37.276
2100	36.311	233.259	205.096	59.143	1619.640	1417.599	-35.261
2200	36.528	234.954	206.415	62.785	1619.006	1407.993	-33.430
2300	36.729	236.582	207.691	66.448	1618.360	1398.416	-31.759
2400	36.916	238.149	208.928	70.131	1617.702	1388.867	-30.228
2500	37.090	239.660	210.127	73.831	1617.032	1379.346	-28.820
2600	37.254	241.117	211.291	77.548	1616.351	1369.852	-27.521
2700	37.408	242.526	212.422	81.282	1615.658	1360.384	-26.318
2800	37.555	243.890	213.522	85.030	1614.953	1350.943	-25.202
2900	37.695	245.210	214.592	88.792	1614.237	1341.526	-24.164
3000	37.830	246.490	215.634	92.569	1613.510	1332.135	-23.195
3100	37.960	247.733	216.649	96.358	1612.772	1322.768	-22.288
3200	38.087	248.940	217.640	100.161	1612.024	1313.425	-21.439
3300	38.211	250.114	218.606	103.976	1611.264	1304.105	-20.642
3400	38.334	251.256	219.550	107.803	1610.495	1294.809	-19.892
3500	38.455	252.369	220.471	111.642	1609.716	1285.535	-19.186
3600	38.576	253.454	221.373	115.494	1608.926	1276.284	-18.518
3700	38.698	254.513	222.254	119 358	1608.128	1267.055	-17.888
3800	38.820	255.546	223.117	123.233	1607.320	1257.848	-17.290
3900	38.943	256.556	223.961	127.122	1606.503	1248.662	-16.724
4000	39.067	257.544	224.788	131.022	1605.678	1239.497	-16.186
4100	39.193	258.510	225.599	134.935	1604.845	1230.353	-15.675
4200	39.322	259.456	226.394	138.861	1604.004	1221.229	-15.188
4300	39.453	260.383	227.174	142.800	1603.156	1212.125	-14.724
4400	39.586	261.291	227.939	146.751	1602.302	1203.041	-14.282
4500	39.723	262.183	228.690	150.717	1601.441	1193.977	-13.859
4600	39.862	263.057	229.428	154.696	1600.574	1184.932	-13.455
4700	40.003	263.916	230.152	158.689	1599.702	1175.905	-13.069
4800	40.148	264.760	230.865	162.697	1598.825	1166.898	-12.698
4900	40.296	265.589	231.565	166.719	1597.944	1157.908	-12.343
5000	40.446	266.405	232.253	170.756	1597.060	1148.937	-12.003
5100	40.599	267.207	232.931	174.808	1596.172	1139.983	-11.676
5200	40.755	267.997	233.598	178.876	1595.282	1131.047	-11.362
5300	40.913	268.775	234.254	182.959	1594.390	1122.128	-11.059
5400	41.074	269.541	234.900	187.059	1593.498	1113.226	-10.768
5500	41.237	270.296	235.537	191.174	1592.604	1104.340	-10.488
5600	41.402	271.041	236.165	195.306	1591.711	1095.471	-10.218
5700	41.569	271.775	236.783	199.455	1590.819	1086.617	-9.958
5800	41.738	272.499	237.392	203.620	1589.927	1077.779	-9.706
5900	41.908	273.214	237.994	207.803	1589.038	1068.957	-9.464
6000	42.080	273.920	238.586	212.002	1588.152	1060.149	-9.229

PREVIOUS: December 1971 (1 atm) CURRENT: December 1971 (1 bar)

Methylidyne, Ion (CH^+) $C_1H_1^+(g)$

CHLOROMETHYLENE (CHCl) IDEAL GAS M_r = 48.47194

$S°(298.15 K) = 234.91 \pm 0.21 J K^{-1} mol^{-1}$

$\Delta_f H°(0 K) = [334 \pm 42] kJ mol^{-1}$

$\Delta_f H°(298.15 K) = [335 \pm 42] kJ mol^{-1}$

Electronic Levels and Quantum Weights

ε_i, cm^{-1}	g_i
0	1
[3000]	3
12288	1

Vibrational Frequencies and Degeneracies

ν, cm^{-1}
[2800] (1)
1201 (1)
815 (1)

Ground State Quantum Weight: 1 σ = 1

Point Group: C_s

Bond Distances: C-H = 1.12 Å C-Cl = 1.689 Å

Bond Angle: H-C-Cl = 103.4°

Product of the Moments of Inertia: $I_A I_B I_C = 3.9883 \times 10^{-117} g^3 cm^6$

Enthalpy of Formation

The enthalpy of formation can be estimated as the average of those of the CH_2 and CCl_2 radicals, which gives $\Delta_f H°$(CHCl, g 298.15 K) = 74.5 ± 15 kcal mol^{-1}. A better value for the enthalpy of formation can be obtained by considering the reactions CHCl → CH + Cl and CHCl → CCl + H. When $\Delta_f H°$(CCl, g, 298.15 K) = 74.5 kcal mol^{-1} is substituted into the above equations it gives D_0^0(C-Cl) = 96 kcal and D_0^0(C-H) = 98 kcal. Although these values are not unreasonable, the D_0^0(C-H) value appears to be too high compared to that in CFH. Thus, we adopt $\Delta_f H°$(CHCl, g, 298.15 K) = 80 kcal mol^{-1} which gives D_0^0(C-Cl) = 91 kcal and D_0^0(C-H) = 92 kcal.

Heat Capacity and Entropy

The structure and electronic levels of the molecule have been established by Merer and Travis (1) from an analysis of the rotational structure of the visible absoorption bands. The estimated triplet state is based on the electronic levels of CCl_2 and CH_2 as given in these tables. The vibrational levels are from Jacox and Milligan (2) who investigated the infrared spectrum of matrix-isolated CHCl, and also estimated the missing frequency.

The principal moments of inertia are: $I_A = 0.1780 \times 10^{-39}$, $I_B = 4.6457 \times 10^{-39}$, and $I_C = 4.8237 \times 10^{-39}$ g cm^2.

References

1. A. J. Merer and D. N. Travis, Can. J. Phys. 44, 525 (1966).
2. M. E. Jacox and D. E. Milligan, J. Chem. Phys. 47, 1626 (1967).

Chloromethylene (CHCl) $C_1H_1Cl_1(g)$

Enthalpy Reference Temperature = T_r = 298.15 K Standard State Pressure = p° = 0.1 MPa

T/K	$C_p°$ (J $K^{-1}mol^{-1}$)	S° (J $K^{-1}mol^{-1}$)	$-[G°-H°(T_r)]/T$ (J $K^{-1}mol^{-1}$)	$H°-H°(T_r)$ (kJ mol^{-1})	$\Delta_f H°$ (kJ mol^{-1})	$\Delta_f G°$ (kJ mol^{-1})	Log K_f
0	0.	0.	INFINITE	-10.154	334.441	334.441	INFINITE
100	33.267	197.591	265.877	-6.829	334.751	329.574	-172.152
200	34.185	220.819	238.182	-3.473	334.915	324.305	-84.700
250	35.365	228.567	235.509	-1.735	334.849	321.659	-67.207
298.15	36.737	234.912	234.912	0.	334.720	319.129	-55.910
300	36.792	235.139	234.912	0.068	334.714	319.032	-55.548
350	38.276	240.922	235.366	1.945	334.536	316.432	-47.225
400	39.724	246.128	236.391	3.895	334.330	313.860	-40.986
450	41.113	250.888	237.741	5.916	334.105	311.315	-36.136
500	42.461	255.290	239.278	8.006	333.867	308.795	-32.260
600	45.130	263.267	242.625	12.385	333.388	303.826	-26.450
700	47.839	270.427	246.094	17.033	332.958	298.934	-22.307
800	50.542	276.992	249.551	21.953	332.631	294.097	-19.203
900	53.113	283.095	252.943	27.137	332.433	289.294	-16.790
1000	55.427	288.814	256.247	32.567	332.369	284.505	-14.861
1100	57.404	294.192	259.454	38.211	332.426	279.717	-13.283
1200	59.013	299.258	262.562	44.035	332.578	274.919	-11.967
1300	60.262	304.034	265.571	50.002	332.799	270.105	-10.853
1400	61.187	308.535	268.480	56.077	333.062	265.273	-9.897
1500	61.837	312.780	271.294	62.230	333.341	260.421	-9.069
1600	62.264	316.786	274.013	68.437	333.617	255.551	-8.343
1700	62.515	320.569	276.641	74.677	333.874	250.663	-7.702
1800	62.634	324.146	279.182	80.935	334.100	245.762	-7.132
1900	62.656	327.533	281.638	87.201	334.287	240.849	-6.621
2000	62.610	330.746	284.014	93.464	334.427	235.927	-6.162
2100	62.517	333.799	286.313	99.721	334.518	231.000	-5.746
2200	62.393	336.705	288.538	105.967	334.557	226.069	-5.368
2300	62.253	339.475	290.693	112.199	334.544	221.138	-5.022
2400	62.103	342.121	292.781	118.417	334.477	216.209	-4.706
2500	61.952	344.653	294.806	124.620	334.357	211.283	-4.415
2600	61.803	347.080	296.770	130.807	334.185	206.364	-4.146
2700	61.659	349.410	298.677	136.980	333.962	201.451	-3.897
2800	61.522	351.650	300.529	143.139	333.688	196.549	-3.667
2900	61.394	353.807	302.329	149.285	333.367	191.656	-3.452
3000	61.274	355.886	304.080	155.418	332.998	186.776	-3.252
3100	61.162	357.893	305.783	161.540	332.584	181.908	-3.065
3200	61.060	359.833	307.442	167.651	332.126	177.055	-2.890
3300	60.965	361.711	309.059	173.752	331.626	172.217	-2.726
3400	60.878	363.529	310.634	179.845	331.086	167.394	-2.572
3500	60.798	365.293	312.171	185.928	330.508	162.588	-2.426
3600	60.725	367.005	313.670	192.004	329.893	157.799	-2.290
3700	60.657	368.668	315.134	198.073	329.245	153.028	-2.160
3800	60.595	370.284	316.564	204.136	328.564	148.274	-2.038
3900	60.538	371.858	317.962	210.193	327.854	143.539	-1.922
4000	60.486	373.390	319.329	216.244	327.115	138.822	-1.813
4100	60.438	374.883	320.666	222.290	326.349	134.124	-1.709
4200	60.393	376.338	321.974	228.331	325.560	129.445	-1.610
4300	60.351	377.759	323.255	234.369	324.747	124.785	-1.516
4400	60.312	379.146	324.509	240.402	323.914	120.145	-1.426
4500	60.276	380.501	325.739	246.431	323.062	115.523	-1.341
4600	60.242	381.825	326.943	252.457	322.192	110.921	-1.260
4700	60.211	383.121	328.125	258.480	321.307	106.337	-1.182
4800	60.181	384.388	329.284	264.499	320.407	101.773	-1.108
4900	60.152	385.629	330.421	270.516	319.495	97.227	-1.036
5000	60.125	386.844	331.538	276.530	318.570	92.701	-0.968
5100	60.100	388.034	332.634	282.541	317.636	88.193	-0.903
5200	60.075	389.201	333.710	288.550	316.692	83.703	-0.841
5300	60.052	390.345	334.768	294.556	315.741	79.231	-0.781
5400	60.029	391.467	335.808	300.560	314.783	74.778	-0.723
5500	60.007	392.568	336.830	306.562	313.819	70.342	-0.668
5600	59.986	393.649	337.835	312.562	312.850	65.924	-0.615
5700	59.966	394.711	338.823	318.559	311.877	61.524	-0.564
5800	59.946	395.754	339.796	324.555	310.901	57.140	-0.515
5900	59.927	396.778	340.753	330.548	309.923	52.773	-0.467
6000	59.908	397.785	341.695	336.540	308.943	48.423	-0.422

PREVIOUS: December 1969 (1 atm) CURRENT: December 1969 (1 bar)

Chloromethylene (CHCl) $C_1H_1Cl_1(g)$

CHLORODIFLUOROMETHANE ($CHClF_2$) IDEAL GAS M_r = 86.468746

$\Delta_f H°(0\ K) = -475.3 \pm 13\ kJ\ mol^{-1}$

$S°(298.15\ K) = 280.97 \pm 0.8\ J\ K^{-1}\ mol^{-1}$ $\Delta_f H°(298.15\ K) = -481.6 \pm 13\ kJ\ mol^{-1}$

Vibrational Frequencies and Degeneracies

ν, cm^{-1}	ν, cm^{-1}	ν, cm^{-1}
3024(1)	812 (1)	1350(1)
1312(1)	598 (1)	1108(1)
836(1)	417 (1)	400(1)

Ground State Quantum Weight: 1 σ = 1

Point Group: C_s

Bond Distances: C-H = 1.09 Å C-Cl = 1.74 Å C-F = 1.35 Å

Bond Angles: H-C-Cl = 107° Cl-C-F = 110.5° F-C-F = 107°

Product of the Moments of Inertia: $I_A I_B I_C = 3.435053 \times 10^{-114}\ g^3\ cm^6$

Enthalpy of Formation

Edwards and Small (1) determined a single equilibrium constant at 600°C for the reaction $2CHClF_2(g) \rightarrow C_2F_4(g) + 2HCl(g)$. The value was not obtained directly but was obtained by extrapolation to eliminate the effects of side reactions. The equilibrium constant was used with JANAF free energy functions (2) to obtain $\Delta_r H°(298.15\ K)$ = 28.75 kcal, which yields $\Delta_f H°(CHClF_2, g, 298.15\ K) = -115.1\ kcal\ mol^{-1}$. An uncertainty of ±3 $kcal\ mol^{-1}$ corresponds to an uncertainty of about ±500% in the equilibrium constant.

Heat Capacity and Entropy

The gas phase vibrational frequencies have been observed in the Raman by Holzer and Moser (3); they do not see anything in the 1350 cm^{-1} region as reported in the liquid by Glockler and Edgell (4). Plyler and Benedict (5) in their assignment replaced an 831 cm^{-1} frequency by 1178 cm^{-1}. We essentially adopt Glockler and Edgell's assignment with updated gas phase values for the observed frequencies. The microwave spectrum has been analyzed by McLay and Mann (6) and Beeson et al. (7), who derive bond lengths and angles in excellent agreement with each other. The values given by Beeson et al. are adopted here.

Neilson and White (8) have measured the low temperature heat capacity and heat of vaporization and have reported an entropy in the gas phase at 232.5 K of 63.919±0.28 cal $K^{-1}\ mol^{-1}$. This compares with 63.959 cal $K^{-1}\ mol^{-1}$ calculated from our adopted functions.

The principal moments of inertia are: $I_A = 8.2033 \times 10^{-39}$, $I_B = 17.4031 \times 10^{-39}$ and $I_C = 24.0612 \times 10^{-39}\ g\ cm^2$.

References

1. J. W. Edwards and P. A. Small, Ind. Eng. Chem. Fundamentals 4, 396 (1965).
2. JANAF Thermochemical Tables: C_2F_4(g), 6-30-69; HCl(g), 9-30-64.
3. W. Holzer and H. Moser, J. Mol. Spectrosc. 20, 185 (1966) and 25, 123 (1968).
4. G. Glockler and W. F. Edgell, J. Chem. Phys. 9, 224 (1941).
5. E. K. Plyler and W. S. Benedict, J. Res. Nat. Bur. Stand. 47, 202 (1951).
6. D. B. McLay and C. R. Mann, Can. J. Phys. 40, 61 (1962).
7. E. L. Beeson, T. L. Weatherly and Q. Williams, J. Chem. Phys. 37, 2926 (1962).
8. E. F. Neilson and D. White, J. Amer. Chem. Soc. 79, 5618 (1957).

Chlorodifluoromethane ($CHClF_2$) $C_1H_1Cl_1F_2(g)$

Enthalpy Reference Temperature = T_r = 298.15 K Standard State Pressure = p° = 0.1 MPa

T/K	$C_p°$ (J K⁻¹mol⁻¹)	S° (J K⁻¹mol⁻¹)	-[G°-H°(T_r)]/T (J K⁻¹mol⁻¹)	H°-H°(T_r) (kJ mol⁻¹)	$\Delta_f H°$ (kJ mol⁻¹)	$\Delta_f G°$ (kJ mol⁻¹)	Log K_f
0	0.	0.	INFINITE	-12.413	-475.292	-475.292	INFINITE
100	35.011	233.244	323.826	-9.058	-477.859	-469.564	245.275
200	45.732	260.548	285.833	-5.057	-479.979	-460.404	120.245
250	51.680	271.395	281.879	-2.621	-480.848	-455.407	95.152
298.15	57.126	280.968	280.968	0.	-481.578	-450.439	78.915
300	57.328	281.322	280.969	0.106	-481.604	-450.246	78.395
350	62.496	290.554	281.687	3.104	-482.250	-444.967	66.408
400	67.107	299.207	283.342	6.346	-482.794	-439.603	57.406
450	71.156	307.350	285.561	9.805	-483.249	-434.176	50.398
500	74.683	315.034	288.126	13.453	-483.626	-428.702	44.786
600	80.420	329.181	293.813	21.221	-484.191	-417.660	36.361
700	84.813	341.923	299.791	29.492	-484.549	-406.541	30.336
800	88.253	353.481	305.791	38.152	-484.745	-395.382	25.816
900	91.004	364.041	311.685	47.120	-484.814	-384.206	22.299
1000	93.244	373.749	317.412	56.336	-484.781	-373.029	19.485
1100	95.093	382.725	322.947	65.756	-484.670	-361.858	17.183
1200	96.636	391.067	328.280	75.345	-484.498	-350.701	15.266
1300	97.935	398.855	333.413	85.075	-484.282	-339.560	13.644
1400	99.038	406.154	338.351	94.925	-484.036	-328.436	12.254
1500	99.979	413.020	343.102	104.877	-483.768	-317.331	11.050
1600	100.788	419.499	347.676	114.917	-483.488	-306.245	9.998
1700	101.488	425.631	352.083	125.031	-483.202	-295.176	9.070
1800	102.096	431.450	356.332	135.211	-482.914	-284.123	8.245
1900	102.627	436.984	360.433	145.448	-482.628	-273.088	7.508
2000	103.093	442.260	364.393	155.734	-482.346	-262.066	6.844
2100	103.505	447.300	368.222	166.065	-482.068	-251.059	6.245
2200	103.869	452.124	371.927	176.434	-481.795	-240.065	5.700
2300	104.192	456.748	375.515	186.837	-481.526	-229.083	5.203
2400	104.481	461.189	378.993	197.271	-481.260	-218.113	4.747
2500	104.740	465.460	382.367	207.732	-480.996	-207.154	4.328
2600	104.972	469.572	385.642	218.218	-480.732	-196.206	3.942
2700	105.182	473.538	388.824	228.726	-480.467	-185.268	3.584
2800	105.372	477.367	391.919	239.254	-480.199	-174.339	3.252
2900	105.544	481.067	394.929	249.800	-479.926	-163.420	2.944
3000	105.700	484.648	397.861	260.362	-479.647	-152.511	2.655
3100	105.843	488.116	400.716	270.939	-479.361	-141.612	2.386
3200	105.974	491.479	403.500	281.530	-479.065	-130.721	2.134
3300	106.093	494.742	406.216	292.134	-478.758	-119.840	1.897
3400	106.203	497.910	408.867	302.749	-478.440	-108.968	1.674
3500	106.305	500.990	411.455	313.374	-478.108	-98.106	1.464
3600	106.398	503.986	413.984	324.009	-477.761	-87.254	1.266
3700	106.484	506.903	416.456	334.654	-477.399	-76.411	1.079
3800	106.564	509.744	418.874	345.306	-477.021	-65.579	0.901
3900	106.639	512.513	421.239	355.966	-476.624	-54.757	0.733
4000	106.708	515.213	423.555	366.634	-476.209	-43.945	0.574
4100	106.772	517.849	425.823	377.308	-475.775	-33.144	0.422
4200	106.832	520.423	428.045	387.988	-475.321	-22.353	0.278
4300	106.888	522.937	430.222	398.674	-474.846	-11.574	0.141
4400	106.940	525.395	432.358	409.365	-474.349	-0.806	0.010
4500	106.989	527.799	434.452	420.062	-473.831	9.951	-0.116
4600	107.035	530.151	436.507	430.763	-473.291	20.696	-0.235
4700	107.078	532.453	438.524	441.468	-472.727	31.428	-0.349
4800	107.119	534.708	440.504	452.178	-472.141	42.149	-0.459
4900	107.157	536.917	442.449	462.892	-471.531	52.857	-0.563
5000	107.193	539.082	444.361	473.610	-470.898	63.553	-0.664
5100	107.227	541.206	446.239	484.331	-470.241	74.235	-0.760
5200	107.259	543.288	448.085	495.055	-469.560	84.905	-0.853
5300	107.289	545.331	449.901	505.782	-468.855	95.561	-0.942
5400	107.318	547.337	451.687	516.513	-468.126	106.203	-1.027
5500	107.345	549.307	453.444	527.246	-467.373	116.832	-1.110
5600	107.371	551.241	455.173	537.982	-466.596	127.447	-1.189
5700	107.395	553.142	456.875	548.720	-465.796	138.048	-1.265
5800	107.418	555.010	458.551	559.461	-464.971	148.634	-1.339
5900	107.441	556.846	460.201	570.204	-464.123	159.207	-1.410
6000	107.462	558.652	461.827	580.949	-463.251	169.764	-1.478

PREVIOUS: June 1970 (1 atm) CURRENT: June 1970 (1 bar)

Chlorodifluoromethane ($CHClF_2$) $C_1H_1Cl_1F_2(g)$

DICHLOROFLUOROMETHANE ($CHCl_2F$) IDEAL GAS M_r = 102.923343

$S°(298.15\ K) = 293.260 \pm 0.8\ J\ K^{-1}\ mol^{-1}$

$\Delta_f H°(0\ K) = [-277.7 \pm 13]\ kJ\ mol^{-1}$
$\Delta_f H°(298.15\ K) = [-283.3 \pm 13]\ kJ\ mol^{-1}$

Vibrational Frequencies and Degeneracies

ν, cm^{-1}	ν, cm^{-1}	ν, cm^{-1}
3023(1)	741 (1)	1242(1)
1313(1)	455 (1)	806(1)
1079(1)	270 (1)	368(1)

Ground State Quantum Weight: 1 $\sigma = 1$

Point Group: C_s

Bond Distances: C-H = 1.09 Å C-Cl = 1.75 Å C-F = 1.367 Å

Bond Angles: Cl-C-Cl = 112°12' F-C-Cl = 109° H-C-Cl = 109°45'

Product of the Moments of Inertia: $I_A I_B I_C = 1.110094 \times 10^{-113}\ g^3\ cm^6$

Enthalpy of Formation

The change in enthalpy of atomization by successive replacement of F by Cl in the CF_4 to CCl_4 series follows a regular pattern (see CCl_2F_2 table). A similar pattern is to be expected in the CHF_3 to $CHCl_3$ series. The first replacement of the series is, in fact, 4 kcal higher than the mean versus 3.8 kcal higher in the CF_4-CCl_4 series. Thus, we have selected the atomization energy of the second and third replacements to continue this pattern. In this manner we have selected the atomization energy of $CHCl_2F$ as 367.4 kcal which corresponds to $\Delta_f H°(CHCl_2F, g, 298.15\ K) = -67.7\ kcal\ mol^{-1}$. With this adopted value, the successive replacement of F by Cl changes the enthalpy of atomization by 41.4, 37.3 and 33.0 kcal.

Heat Capacity and Entropy

Holzer (1) has reported the gas phase Raman spectrum which is in excellent agreement with the assignments of Plyler and Benedict (2). The only discrepancy is in ν_7, which Plyler and Benedict assign at 1242 cm^{-1}, and Holzer (1) does not report a band in this region. Holzer does, however, report a band at 728 cm^{-1} which has not been reported in the earlier liquid studies. There is strong evidence from correlations with $CHCl_3$ and CHF_3 that two frequencies should occur between 1220 and 1376 cm^{-1}. On this basis we adopt the assignment of Plyler and Benedict for ν_7, while accepting the gas phase values of Holzer for the other frequencies. The extra band at 728 cm^{-1} is not used. It is probable that the bands at 741 and 728 cm^{-1} are due to Fermi resonance of ν_4 with $2\nu_9$; the unperturbed level would lie at 735 cm^{-1}. The symmetry classes of ν_4 and $2\nu_9$ are identical and thus Fermi resonance is allowed.

McLay (3) has reported the molecular structure from his microwave investigations. The principal moments of inertia are: $I_A = 12.0028 \times 10^{-39}$, $I_B = 25.7010 \times 10^{-39}$ and $I_C = 35.9855 \times 10^{-39}\ g\ cm^2$.

References

1. W. Holzer, J. Mol. Spectrosc. 25, 123 (1968).
2. E. K. Plyler and W. S. Benedict, J. Res. Nat. Bur. Stand. 47, 202 (1951).
3. D. B. McLay, Can. J. Phys. 42, 720 (1964).

Dichlorofluoromethane ($CHCl_2F$) $C_1H_1Cl_2F_1(g)$

Enthalpy Reference Temperature = T_r = 298.15 K Standard State Pressure = p° = 0.1 MPa

T/K	$C_p°$ (J K⁻¹mol⁻¹)	S° (J K⁻¹mol⁻¹)	$-[G°-H°(T_r)]/T$ (J K⁻¹mol⁻¹)	$H°-H°(T_r)$ (kJ mol⁻¹)	$\Delta_f H°$ (kJ mol⁻¹)	$\Delta_f G°$ (kJ mol⁻¹)	Log K_f
0	0.	0.	INFINITE	-13.293	-277.672	-277.672	INFINITE
100	37.673	241.161	339.849	-9.869	-280.172	-271.738	141.941
200	50.180	271.154	298.501	-5.469	-281.948	-262.564	68.575
250	55.897	282.973	294.236	-2.816	-282.659	-257.633	53.830
298.15	60.995	293.260	293.260	0.	-283.257	-252.758	44.282
300	61.182	293.638	293.261	0.113	-283.278	-252.569	43.976
350	65.973	303.436	294.024	3.294	-283.806	-247.407	36.924
400	70.228	312.529	295.776	6.701	-284.245	-242.177	31.625
450	73.950	321.021	298.114	10.308	-284.606	-236.896	27.498
500	77.184	328.983	300.807	14.088	-284.900	-231.578	24.193
600	82.432	343.542	306.740	22.081	-285.319	-220.871	19.229
700	86.446	356.563	312.944	30.534	-285.556	-210.108	15.678
800	89.594	368.320	319.143	39.342	-285.650	-199.322	13.014
900	92.119	379.024	325.211	48.432	-285.629	-188.532	10.942
1000	94.182	388.840	331.090	57.750	-285.517	-177.748	9.285
1100	95.891	397.899	336.757	67.256	-285.334	-166.980	7.929
1200	97.322	406.306	342.206	76.919	-285.097	-156.230	6.801
1300	98.530	414.145	347.442	86.713	-284.819	-145.502	5.846
1400	99.558	421.485	352.471	96.619	-284.512	-134.797	5.029
1500	100.438	428.385	357.305	106.620	-284.186	-124.115	4.322
1600	101.195	434.891	361.952	116.702	-283.848	-113.454	3.704
1700	101.851	441.047	366.426	126.856	-283.505	-102.815	3.159
1800	102.422	446.885	370.735	137.070	-283.162	-92.196	2.675
1900	102.922	452.436	374.890	147.338	-282.822	-81.596	2.243
2000	103.361	457.727	378.901	157.652	-282.490	-71.014	1.855
2100	103.748	462.779	382.775	168.008	-282.168	-60.448	1.504
2200	104.091	467.614	386.523	178.400	-281.857	-49.898	1.185
2300	104.397	472.248	390.150	188.825	-281.560	-39.361	0.894
2400	104.669	476.697	393.664	199.279	-281.277	-28.836	0.628
2500	104.914	480.974	397.071	209.758	-281.010	-18.324	0.383
2600	105.134	485.094	400.378	220.261	-280.759	-7.821	0.157
2700	105.332	489.065	403.590	230.784	-280.525	2.672	-0.052
2800	105.511	492.899	406.711	241.326	-280.308	13.157	-0.245
2900	105.674	496.605	409.747	251.886	-280.108	23.634	-0.426
3000	105.822	500.190	412.703	262.461	-279.925	34.105	-0.594
3100	105.958	503.662	415.581	273.050	-279.758	44.570	-0.751
3200	106.081	507.028	418.387	283.652	-279.606	55.029	-0.898
3300	106.195	510.294	421.122	294.266	-279.468	65.485	-1.037
3400	106.299	513.466	423.792	304.890	-279.343	75.936	-1.167
3500	106.395	516.548	426.398	315.525	-279.228	86.383	-1.289
3600	106.483	519.547	428.944	326.169	-279.123	96.828	-1.405
3700	106.565	522.465	431.433	336.822	-279.026	107.270	-1.514
3800	106.641	525.308	433.866	347.482	-278.933	117.709	-1.618
3900	106.711	528.079	436.246	358.150	-278.844	128.146	-1.716
4000	106.777	530.782	438.576	368.824	-278.755	138.580	-1.810
4100	106.838	533.419	440.857	379.505	-278.666	149.013	-1.898
4200	106.895	535.994	443.092	390.191	-278.572	159.443	-1.983
4300	106.948	538.510	445.282	400.884	-278.473	169.870	-2.064
4400	106.997	540.970	447.429	411.581	-278.366	180.296	-2.140
4500	107.044	543.375	449.534	422.283	-278.249	190.719	-2.214
4600	107.087	545.728	451.600	432.990	-278.120	201.139	-2.284
4700	107.128	548.031	453.627	443.700	-277.978	211.556	-2.351
4800	107.167	550.287	455.617	454.415	-277.819	221.970	-2.416
4900	107.203	552.497	457.572	465.134	-277.643	232.380	-2.477
5000	107.237	554.663	459.492	475.856	-277.449	242.787	-2.536
5100	107.269	556.787	461.379	486.581	-277.234	253.190	-2.593
5200	107.300	558.871	463.234	497.310	-276.998	263.588	-2.648
5300	107.329	560.915	465.058	508.041	-276.739	273.981	-2.700
5400	107.356	562.921	466.852	518.775	-276.457	284.370	-2.751
5500	107.382	564.891	468.616	529.512	-276.150	294.752	-2.799
5600	107.406	566.826	470.353	540.252	-275.818	305.130	-2.846
5700	107.430	568.728	472.062	550.993	-275.460	315.500	-2.891
5800	107.452	570.596	473.745	561.737	-275.075	325.865	-2.935
5900	107.473	572.433	475.402	572.484	-274.662	336.222	-2.977
6000	107.493	574.240	477.034	583.232	-274.222	346.572	-3.017

PREVIOUS: June 1970 (1 atm) CURRENT: June 1970 (1 bar)

Dichlorofluoromethane ($CHCl_2F$) $C_1H_1Cl_2F_1(g)$

TRICHLOROMETHANE ($CHCl_3$) IDEAL GAS M_r = 119.37794

$S°(298.15\ K) = 295.620 \pm 0.04\ J\ K^{-1}\ mol^{-1}$

$\Delta_f H°(0\ K) = -98.27 \pm 1.3\ kJ\ mol^{-1}$
$\Delta_f H°(298.15\ K) = -103.18 \pm 1.3\ kJ\ mol^{-1}$

Vibrational Frequencies and Degeneracies

ν, cm^{-1}	ν, cm^{-1}
3034(1)	1220(2)
680(1)	774(2)
363(1)	261(2)

Ground State Quantum Weight: 1 $\sigma = 3$

Point Group: C_{3v}

Bond Distances: C-H = 1.100 Å C-Cl = 1.758 Å

Bond Angles: Cl-C-Cl = 111°18' H-C-Cl = 107°34'

Product of the Moments of Inertia: $I_A I_B I_C = 3.282103 \times 10^{-113}\ g^3\ cm^6$

Enthalpy of Formation

Hu and Sinke (1) have measured the enthalpy of combustion of $CHCl_3$(ℓ) in the rotating bomb calorimeter, using As_2O_3 solution as the reducing agent. The reported $\Delta_c H°(298.15\ K) = -113.10\ kcal\ mol^{-1}$ for $CHCl_3$(ℓ) + 0.5O_2 + H_2O(ℓ) + CO_2(g) + 3HCl(aq, 600H_2O) which leads to $\Delta_f H°(CHCl_3$, ℓ, 298.15 K) = $-32.10\ kcal\ mol^{-1}$, using $\Delta_f H°$(298.15 K) = -94.051, -39.823, -68.315 $kcal\ mol^{-1}$ for CO_2(g), HCl(aq, 600H_2O) and H_2O(ℓ), respectively (2). The standard deviation of six combustion runs was 0.09 $kcal\ mol^{-1}$. Employing the heat of vaporization of chloroform at 298.15 K as 7.44 $kcal\ mol^{-1}$ (3), we obtain $\Delta_f H°(CHCl_3$, g, 298.15 K) = -24.66 $kcal\ mol^{-1}$ (-103.177 ± 1.3 $kJ\ mol^{-1}$) which is adopted in the tabulation.

Smith et al. (4), based on the combustion data of Eftring (5), re-evaluated the enthalpy of combustion of chloroform at 18.7°C as $\Delta_c E/M° = -947.7\ cal\ K^{-1}\ mol^{-1}$ for $CHCl_3$(ℓ) + 0.5O_2(g) + H_2O(ℓ) + CO_2(g) + 3HCl(aq, 600H_2O). This value needs to be further corrected for a change in the enthalpy of oxidation of As_2O_3 (6) and calculated to 25°C to give $\Delta_c H°(298.15\ K) = -113.44\ kcal\ mol^{-1}$, $\Delta_f H°(CHCl_3$, ℓ, 298.15 K) = $-31.76\ kcal\ mol^{-1}$ and $\Delta_f H°(CHCl_3$, g, 298.15 K) = $-24.32\ kcal\ mol^{-1}$.

Heat Capacity and Entropy

The assigned fundamental vibrational frequencies are obtained from Shimanouchi's selection (7), based on infrared and Raman spectra measured by Nielson and Ward (8), Madigan and Cleveland (9), Gibian and McKinney (10), and Stanevich and Yaroslaviskii (11).

The bond distances and angles were obtained from the microwave spectra by Jen and Lide (12). The principal moments of inertia are: $I_A = I_B = 25.7217 \times 10^{-39}$ and $I_C = 49.6081 \times 10^{-39}\ g\ cm^2$.

References

1. A. T. Hu and G. C. Sinke, presented at the 23rd Calorimetry Conference, Midland, Mich., 1968.
2. U. S. Nat. Bur. Stand. Tech. Note 270-3, 1968.
3. J. H. Mathews, J. Am. Chem. Soc. 48, 562 (1926).
4. L. Smith. L. Bjellerup, S. Krook and H. Westermark, Acta Chem. Scand. 7, 65 (1953).
5. E. Eftring, Ph. D. Thesis, University of Lund, Lund, Sweden, 1938.
6. S. Sunner and S. Thoren, Acta Chem. Scand. 18, 1528 (1964).
7. T. Shimanouchi, U. S. Nat. Bur. Stand. NSRDS-NBS 6, 1967.
8. J. R. Nielsen and N. E. Ward, J. Chem. Phys. 10, 81 (1941).
9. J. R. Madigan and F. F. Cleveland, J. Chem. Phys. 19, 119 (1951).
10. T. G. Gibian and D. S. McKinney, J. Am. Chem. Soc. 73, 1431 (1951).
11. A. E. Stanevich and N. G. Yaroslaviskii, Opt. Spectrosc. 9, 31 (1961).
12. M. Jen and D. R. Lide, Jr., J. Chem. Phys. 36, 2525 (1962).

Trichloromethane ($CHCl_3$) $C_1H_1Cl_3$(g)

Enthalpy Reference Temperature = T_r = 298.15 K Standard State Pressure = p° = 0.1 MPa

T/K	$C_p°$ (J K⁻¹mol⁻¹)	S°	-[G°-H°(T_r)]/T	H°-H°(T_r) (kJ mol⁻¹)	$\Delta_f H°$	$\Delta_f G°$	Log K_f
0	0.	0.	INFINITE	-14.153	-98.275	-98.275	INFINITE
100	40.321	239.353	345.889	-10.654	-100.702	-91.155	47.614
200	54.302	271.795	301.259	-5.893	-102.171	-80.983	21.151
250	60.243	284.561	296.668	-3.027	-102.727	-75.620	15.800
298.15	65.383	295.620	295.620	0.	-103.177	-70.357	12.326
300	65.568	296.025	295.621	0.121	-103.193	-70.154	12.215
350	70.235	306.492	296.437	3.519	-103.569	-64.616	9.643
400	74.255	316.139	298.305	7.134	-103.861	-59.031	7.709
450	77.685	325.089	300.790	10.935	-104.083	-53.413	6.200
500	80.609	333.429	303.641	14.894	-104.246	-47.774	4.991
600	85.262	348.558	309.893	23.199	-104.432	-36.459	3.174
700	88.772	361.977	316.394	31.908	-104.472	-25.125	1.875
800	91.513	374.017	322.857	40.928	-104.399	-13.794	0.901
900	93.716	384.927	329.157	50.193	-104.236	-2.477	0.144
1000	95.524	394.898	335.240	59.658	-104.001	8.817	-0.461
1100	97.031	404.075	341.086	69.288	-103.710	20.085	-0.954
1200	98.299	412.574	346.693	79.056	-103.375	31.325	-1.364
1300	99.376	420.485	352.069	88.941	-103.007	42.535	-1.709
1400	100.296	427.884	357.223	98.926	-102.617	53.717	-2.004
1500	101.086	434.832	362.168	108.996	-102.212	64.869	-2.259
1600	101.770	441.378	366.916	119.140	-101.798	75.994	-2.481
1700	102.363	447.566	371.479	129.347	-101.383	87.094	-2.676
1800	102.881	453.432	375.871	139.610	-100.970	98.169	-2.849
1900	103.336	459.007	380.101	149.921	-100.565	109.221	-3.003
2000	103.735	464.318	384.180	160.275	-100.172	120.252	-3.141
2100	104.089	469.388	388.118	170.667	-99.794	131.264	-3.265
2200	104.402	474.237	391.923	181.092	-99.436	142.258	-3.378
2300	104.682	478.885	395.604	191.546	-99.102	153.236	-3.480
2400	104.932	483.345	399.167	202.027	-98.795	164.201	-3.574
2500	105.156	487.633	402.621	212.532	-98.518	175.153	-3.660
2600	105.358	491.762	405.970	223.058	-98.273	186.095	-3.739
2700	105.540	495.741	409.222	233.603	-98.065	197.029	-3.812
2800	105.705	499.583	412.381	244.165	-97.893	207.955	-3.879
2900	105.855	503.295	415.452	254.743	-97.761	218.875	-3.942
3000	105.991	506.886	418.440	265.336	-97.669	229.792	-4.001
3100	106.116	510.363	421.350	275.941	-97.616	240.706	-4.056
3200	106.230	513.734	424.184	286.559	-97.604	251.620	-4.107
3300	106.335	517.004	426.948	297.187	-97.630	262.534	-4.156
3400	106.431	520.180	429.643	307.825	-97.695	273.448	-4.201
3500	106.519	523.267	432.275	318.473	-97.794	284.366	-4.244
3600	106.601	526.269	434.844	329.129	-97.927	295.286	-4.284
3700	106.677	529.190	437.355	339.793	-98.091	306.211	-4.323
3800	106.747	532.036	439.809	350.464	-98.282	317.140	-4.359
3900	106.812	534.810	442.209	361.142	-98.497	328.075	-4.394
4000	106.872	537.515	444.558	371.826	-98.732	339.016	-4.427
4100	106.929	540.155	446.858	382.516	-98.984	349.963	-4.459
4200	106.981	542.732	449.110	393.212	-99.249	360.916	-4.489
4300	107.030	545.250	451.317	403.913	-99.524	371.876	-4.517
4400	107.076	547.711	453.480	414.618	-99.804	382.842	-4.545
4500	107.119	550.118	455.601	425.328	-100.086	393.814	-4.571
4600	107.160	552.473	457.681	436.042	-100.367	404.793	-4.597
4700	107.198	554.778	459.722	446.760	-100.643	415.777	-4.621
4800	107.233	557.035	461.726	457.481	-100.910	426.768	-4.644
4900	107.267	559.246	463.694	468.206	-101.167	437.764	-4.667
5000	107.299	561.414	465.627	478.935	-101.410	448.765	-4.688
5100	107.328	563.539	467.526	489.666	-101.636	459.771	-4.709
5200	107.357	565.623	469.392	500.400	-101.843	470.781	-4.729
5300	107.383	567.668	471.227	511.137	-102.029	481.795	-4.748
5400	107.409	569.676	473.032	521.877	-102.191	492.812	-4.767
5500	107.433	571.647	474.807	532.619	-102.329	503.832	-4.785
5600	107.455	573.583	476.554	543.363	-102.440	514.854	-4.802
5700	107.477	575.485	478.273	554.110	-102.523	525.878	-4.819
5800	107.497	577.354	479.965	564.859	-102.576	536.903	-4.835
5900	107.517	579.192	481.631	575.609	-102.598	547.929	-4.851
6000	107.535	580.999	483.272	586.362	-102.590	558.954	-4.866

PREVIOUS: December 1968 (1 atm) CURRENT: December 1968 (1 bar)

Trichloromethane ($CHCl_3$) $C_1H_1Cl_3$(g)

FLUOROMETHYLENE (CHF) IDEAL GAS M_r = 32.017343

$S°(298.15\ K) = 223.35 \pm 0.21\ J\ K^{-1}\ mol^{-1}$

$\Delta_fH°(0\ K) = [125.2 \pm 29]\ kJ\ mol^{-1}$
$\Delta_fH°(298.15\ K) = [125.5 \pm 29]\ kJ\ mol^{-1}$

Electronic Levels and Quantum Weights

ϵ_i, cm^{-1}	g_i
0	1
[10000]	3
17287	1

Vibrational Frequencies and Degeneracies

ν, cm^{-1}
[3000](1)
1403 (1)
1182 (1)

Point Group: C_s $\sigma = 1$
Bond Distances: C-H = 1.121 Å C-F = 1.314 Å
Bond Angle: H-C-F = 101.6°
Product of the Moments of Inertia: $I_AI_BI_C = 1.0217 \times 10^{-117}\ g^3\ cm^6$

Enthalpy of Formation

The enthalpy of formation can be estimated as the average of those of the CH_2 and CF_2 radicals, which gives $\Delta_fH°$(CHF, g, 298.15 K) = 25±20 kcal mol^{-1}. A better value for the enthalpy of formation can be obtained by considering the bond energies CHF → CH + F and CHF → CF + H. When $\Delta_fH°$(CHF, g, 298.15 K) = 25 kcal mol^{-1} is substituted into the above equations it gives D_0^0(CF) = 136 kcal mol^{-1} and D_0^0(CH) = 86 kcal mol^{-1}. These are not unreasonable values but are probably close to the upper and lower limits respectively. A value of $\Delta_fH°$(CHF, g, 298.15 K) = 30 kcal mol^{-1} gives D_0^0(CF) = 131 kcal mol^{-1} and D_0^0(CH) = 81 kcal which are more probable since both fall within the bounds of normal bond strengths. The strongest established C-F bond is 135 kcal in CF_3 and the weakest established C-H bond is 80 kcal in CH. We adopt $\Delta_fH°$(CHF, g, 298.15 K) = 30±7 kcal mol^{-1} (125.520 kJ mol^{-1}).

Heat Capacity and Entropy

The structure and electronic levels of the molecule have been established by Merer and Travis (1) from an analysis of the rotational structure of the visible absorption bands. The estimated triplet state is based on the electronic levels of CF_2 and CH_2 as given in these tables. The vibrational levels are from Merer and Trvis (1) and Jacox and Milligan (2), the latter investigated the infrared spectrum of matrix-isolated HCF and also estimated the missing frequency.

The principal moments of inertia are: $I_A = 0.1795 \times 10^{-39}$, $I_B = 2.2980 \times 10^{-39}$, and $I_C = 2.4774 \times 10^{-39}\ g\ cm^2$.

References

1. A. J. Merer and D. N. Travis, Can. J. Phys. 44, 1541 (1966).
2. M. E. Jacox and D. E. Milligan, J. Chem. Phys. 50, 3252 (1969).

Fluoromethylene (CHF) $C_1H_1F_1(g)$

Enthalpy Reference Temperature = T_r = 298.15 K Standard State Pressure = p° = 0.1 MPa

T/K	$C_p°$ (J K⁻¹mol⁻¹)	S° (J K⁻¹mol⁻¹)	-[G°-H°(T_r)]/T (J K⁻¹mol⁻¹)	H°-H°(T_r) (kJ mol⁻¹)	$\Delta_fH°$ (kJ mol⁻¹)	$\Delta_fG°$ (kJ mol⁻¹)	Log K_f
0	0.	0.	INFINITE	-9.982	125.235	125.235	INFINITE
100	33.258	186.756	253.322	-6.657	125.548	120.522	-62.954
200	33.415	209.828	226.465	-3.327	125.739	115.396	-30.138
250	33.855	217.325	223.913	-1.647	125.675	112.816	-23.572
298.15	34.604	223.348	223.348	0.	125.520	110.352	-19.333
300	34.639	223.562	223.349	0.064	125.513	110.258	-19.198
350	35.681	228.978	223.774	1.821	125.280	107.733	-16.078
400	36.860	233.818	224.732	3.634	124.998	105.245	-13.744
450	38.079	238.230	225.990	5.508	124.679	102.795	-11.932
500	39.277	242.304	227.421	7.442	124.334	100.382	-10.487
600	41.503	249.666	230.528	11.483	123.596	95.660	-8.328
700	43.461	256.214	233.738	15.733	122.829	91.064	-6.795
800	45.165	262.132	236.923	20.167	122.060	86.579	-5.653
900	46.649	267.539	240.029	24.759	121.303	82.190	-4.770
1000	47.940	272.522	243.032	29.490	120.564	77.883	-4.068
1100	49.066	277.146	245.926	34.342	119.843	73.650	-3.497
1200	50.052	281.458	248.709	39.299	119.136	69.482	-3.024
1300	50.922	285.499	251.385	44.348	118.442	65.373	-2.627
1400	51.701	289.302	253.959	49.480	117.757	61.316	-2.288
1500	52.408	292.894	256.436	54.686	117.079	57.308	-1.996
1600	53.063	296.297	258.822	59.960	116.408	53.345	-1.742
1700	53.680	299.533	261.122	65.297	115.744	49.425	-1.519
1800	54.273	302.618	263.343	70.695	115.088	45.542	-1.322
1900	54.851	305.568	265.488	76.151	114.443	41.696	-1.146
2000	55.422	308.396	267.563	81.665	113.811	37.884	-0.989
2100	55.990	311.113	269.573	87.235	113.197	34.103	-0.848
2200	56.557	313.731	271.521	92.863	112.604	30.350	-0.721
2300	57.126	316.258	273.411	98.547	112.035	26.624	-0.605
2400	57.696	318.701	275.248	104.288	111.494	22.923	-0.499
2500	58.265	321.068	277.033	110.086	110.984	19.243	-0.402
2600	58.832	323.364	278.771	115.941	110.509	15.583	-0.313
2700	59.394	325.595	280.465	121.852	110.071	11.940	-0.231
2800	59.949	327.765	282.115	127.820	109.672	8.313	-0.155
2900	60.494	329.878	283.726	133.842	109.313	4.700	-0.085
3000	61.027	331.938	285.299	139.918	108.995	1.098	-0.019
3100	61.544	333.948	286.836	146.047	108.719	-2.494	0.042
3200	62.043	335.910	288.339	152.226	108.485	-6.078	0.099
3300	62.523	337.826	289.810	158.455	108.291	-9.655	0.153
3400	62.982	339.700	291.250	164.730	108.138	-13.227	0.203
3500	63.417	341.532	292.660	171.050	108.023	-16.794	0.251
3600	63.828	343.324	294.043	177.413	107.945	-20.359	0.295
3700	64.214	345.078	295.398	183.815	107.903	-23.923	0.338
3800	64.575	346.795	296.728	190.255	107.894	-27.486	0.378
3900	64.909	348.477	298.034	196.729	107.915	-31.048	0.416
4000	65.218	350.124	299.316	203.236	107.965	-34.612	0.452
4100	65.500	351.738	300.574	209.772	108.041	-38.177	0.486
4200	65.757	353.320	301.812	216.335	108.140	-41.745	0.519
4300	65.989	354.870	303.028	222.923	108.260	-45.315	0.550
4400	66.196	356.389	304.223	229.532	108.399	-48.888	0.580
4500	66.380	357.879	305.399	236.161	108.553	-52.464	0.609
4600	66.541	359.340	306.556	242.807	108.721	-56.044	0.636
4700	66.681	360.773	307.694	249.469	108.900	-59.628	0.663
4800	66.799	362.178	308.815	256.143	109.087	-63.216	0.688
4900	66.898	363.556	309.918	262.828	109.282	-66.807	0.712
5000	66.979	364.908	311.004	269.522	109.481	-70.403	0.735
5100	67.042	366.235	312.074	276.223	109.684	-74.003	0.758
5200	67.089	367.538	313.128	282.930	109.887	-77.606	0.780
5300	67.120	368.816	314.167	289.640	110.090	-81.214	0.800
5400	67.138	370.071	315.191	296.353	110.291	-84.825	0.821
5500	67.142	371.303	316.200	303.067	110.489	-88.441	0.840
5600	67.134	372.513	317.194	309.781	110.683	-92.059	0.859
5700	67.115	373.701	318.175	316.494	110.871	-95.681	0.877
5800	67.085	374.868	319.143	323.204	111.052	-99.307	0.894
5900	67.047	376.014	320.097	329.910	111.226	-102.935	0.911
6000	67.000	377.141	321.038	336.613	111.392	-106.566	0.928

PREVIOUS: December 1969 (1 atm) CURRENT: December 1969 (1 bar)

Fluoromethylene (CHF) $C_1H_1F_1(g)$

FORMYL FLUORIDE (HCOF) IDEAL GAS $M_r = 48.016743$

$S°(298.15 K) = 246.520 \pm 0.8 J K^{-1} mol^{-1}$

$\Delta_f H°(0 K) = [-372.964] kJ mol^{-1}$
$\Delta_f H°(298.15 K) = [-376.560] kJ mol^{-1}$

Vibrational Frequencies and Degeneracies

ν, cm^{-1}	ν, cm^{-1}
2981 (1)	1064.8(1)
1836.9(1)	662.5(1)
1342.5(1)	[1030] (1)

Ground State Quantum Weight: 1 $\sigma = 1$

Point Group: C_1

Bond Distances: C–F = 1.31 ± 0.02 Å C–O = 1.183 ± 0.003 Å C–H = 1.10 ± 0.02 Å

Bond Angles: F–C–O = 122.7 ± 0.05° H–C–F = 108.3 ± 3° H–C–O = 129 ± 3°

Product of the Moments of Inertia: $I_A I_B I_C = 4.96336 \times 10^{-116} g^3 cm^6$

Enthalpy of Formation

$\Delta_r H°$ for III was estimated to be the average of the $\Delta_r H$'s of I and II.

(I)	$CH_2O(g) = CO(g) + 2 H(g)$	106 kcal
(II)	$CF_2O(g) = CO(g) + 2 F(g)$	162 kcal
(III)	$CHFO(g) = CO(g) + F(g) + H(g)$	[134] kcal

The auxiliary $\Delta_f H°(298.15 K)$ of $CH_2O = -28$, H = 52, F = 19, CO = -26, $CF_2O = -150 kcal mol^{-1}$ was used. See the respective sheets for details. The estimated enthalpy of formation is consistent with the instability of CHFO(g) at room temperature.

Heat Capacities and Entropies

The molecular structural parameters were obtained by LeBlanc et al (1) from combining the results of microwave measurements with those of electron diffraction determined by Jones et al. (2).. Agreement within the assigned uncertainties was obtained by Stratton and Nielsen (3) on the basis of infrared measurements.

Five of the fundamentals were observed by Stratton and Nielsen (3) in agreement with those obtained by Morgan et al. (4). The estimated frequency was based upon the observed value of 857.4 cm^{-1} for the fundamental in the infrared spectrum of the deuterated compound and the application of the Teller-Redlich isotope rule. The principal moments of inertia are: $I_A = 0.9081 \times 10^{-39}$, $I_B = 6.9527 \times 10^{-39}$, and $I_C = 7.8608 \times 10^{-39} g cm^2$.

References

1. O. H. LeBlanc, V. W. Laurie and W. D. Gwinn, J. Chem. Phys. 33, 598 (1960).
2. M. E. Jones, K. Hedberg, and V. Schomaker, J. Amer. Chem. Soc. 77, 5278 (1955).
3. R. F. Stratton and A. H. Nielsen, J. Mol. Spectrosc. 4, 373 (1960).
4. H. W. Morgan, P. A. Staats and J. H. Goldstein, J. Chem. Phys. 25, 337 (1956).

Formyl Fluoride (HCOF) $C_1H_1F_1O_1(g)$

Enthalpy Reference Temperature = T_r = 298.15 K Standard State Pressure = p° = 0.1 MPa

T/K	$C_p°$ (J K⁻¹mol⁻¹)	S°	$-[G°-H°(T_r)]/T$	$H°-H°(T_r)$ (kJ mol⁻¹)	$\Delta_f H°$	$\Delta_f G°$	Log K_f
0	0.	0.	INFINITE	-10.443	-372.964	-372.964	INFINITE
100	33.314	207.960	279.124	-7.116	-374.103	-372.583	194.617
200	35.452	231.500	250.043	-3.709	-375.288	-370.617	96.795
250	37.759	239.647	247.170	-1.881	-375.934	-369.375	77.177
298.15	40.414	246.520	246.520	0.	-376.560	-368.054	64.482
300	40.521	246.771	246.521	0.075	-376.584	-368.002	64.075
350	43.451	253.236	247.025	2.174	-377.213	-366.521	54.700
400	46.351	259.229	248.180	4.419	-377.810	-364.952	47.658
450	49.110	264.849	249.723	6.807	-378.373	-363.311	42.172
500	51.678	270.158	251.504	9.327	-378.903	-361.609	37.777
600	56.197	279.992	255.446	14.728	-379.861	-358.058	31.172
700	59.958	288.946	259.601	20.541	-380.693	-354.356	26.442
800	63.083	297.163	263.790	26.698	-381.406	-350.544	22.888
900	65.689	304.748	267.925	33.141	-382.016	-346.649	20.119
1000	67.874	311.785	271.963	39.822	-382.535	-342.690	17.900
1100	69.714	318.343	275.885	46.704	-382.981	-338.684	16.083
1200	71.270	324.478	279.682	53.755	-383.367	-334.639	14.566
1300	72.594	330.236	283.351	60.950	-383.708	-330.564	13.282
1400	73.724	335.658	286.896	68.268	-384.014	-326.465	12.181
1500	74.694	340.779	290.319	75.690	-384.297	-322.344	11.225
1600	75.531	345.627	293.625	83.202	-384.563	-318.205	10.388
1700	76.256	350.228	296.821	90.792	-384.820	-314.050	9.650
1800	76.888	354.605	299.910	98.450	-385.074	-309.880	8.992
1900	77.441	358.777	302.900	106.167	-385.328	-305.696	8.404
2000	77.926	362.762	305.794	113.936	-385.585	-301.498	7.874
2100	78.355	366.575	308.598	121.751	-385.848	-297.287	7.395
2200	78.734	370.229	311.317	129.605	-386.118	-293.063	6.958
2300	79.072	373.736	313.955	137.496	-386.396	-288.827	6.559
2400	79.374	377.108	316.517	145.419	-386.682	-284.579	6.194
2500	79.644	380.354	319.006	153.370	-386.976	-280.319	5.o57
2600	79.887	383.482	321.426	161.347	-387.277	-276.047	5.546
2700	80.106	386.501	323.781	169.346	-387.585	-271.763	5.258
2800	80.304	389.418	326.073	177.367	-387.900	-267.467	4.990
2900	80.484	392.239	328.306	185.407	-388.220	-263.161	4.740
3000	80.648	394.971	330.483	193.463	-388.546	-258.842	4.507
3100	80.797	397.618	332.606	201.536	-388.876	-254.513	4.289
3200	80.934	400.185	334.678	209.622	-389.211	-250.174	4.084
3300	81.059	402.678	336.701	217.722	-389.549	-245.824	3.891
3400	81.174	405.099	338.677	225.834	-389.890	-241.463	3.710
3500	81.280	407.454	340.609	233.957	-390.234	-237.093	3.538
3600	81.377	409.745	342.498	242.089	-390.580	-232.713	3.377
3700	81.468	411.976	344.345	250.232	-390.930	-228.323	3.223
3800	81.551	414.149	346.154	258.383	-391.282	-223.923	3.078
3900	81.629	416.269	347.925	266.542	-391.637	-219.514	2.940
4000	81.701	418.336	349.659	274.708	-391.995	-215.096	2.809
4100	81.768	420.355	351.359	282.882	-392.356	-210.669	2.684
4200	81.831	422.326	353.025	291.062	-392.721	-206.233	2.565
4300	81.889	424.252	354.659	299.248	-393.089	-201.789	2.451
4400	81.944	426.135	356.263	307.440	-393.461	-197.336	2.343
4500	81.995	427.977	357.836	315.637	-393.838	-192.875	2.239
4600	82.043	429.780	359.380	323.838	-394.219	-188.405	2.139
4700	82.088	431.545	360.897	332.045	-394.606	-183.926	2.044
4800	82.131	433.274	362.387	340.256	-394.999	-179.439	1.953
4900	82.171	434.968	363.851	348.471	-395.399	-174.944	1.865
5000	82.208	436.628	365.290	356.690	-395.805	-170.441	1.781
5100	82.244	438.256	366.705	364.913	-396.218	-165.929	1.699
5200	82.278	439.854	368.096	373.139	-396.639	-161.410	1.621
5300	82.309	441.421	369.465	381.368	-397.069	-156.882	1.546
5400	82.339	442.960	370.812	389.601	-397.508	-152.346	1.474
5500	82.368	444.471	372.137	397.836	-397.956	-147.803	1.404
5600	82.395	445.955	373.442	406.074	-398.413	-143.250	1.336
5700	82.420	447.414	374.727	414.315	-398.882	-138.690	1.271
5800	82.445	448.848	375.993	422.558	-399.361	-134.120	1.208
5900	82.468	450.257	377.240	430.804	-399.852	-129.544	1.147
6000	82.490	451.643	378.468	439.052	-400.355	-124.958	1.088

PREVIOUS: June 1961 (1 atm) CURRENT: June 1961 (1 bar)

Formyl Fluoride (HCOF) $C_1H_1F_1O_1(g)$

TRIFLUOROMETHANE (CHF_3) IDEAL GAS M_r = 70.014149

$\Delta_f H°(0\ K) = -690.10 \pm 3.3\ kJ\ mol^{-1}$

$S°(298.15\ K) = 259.657 \pm 0.34\ J\ K^{-1}\ mol^{-1}$ $\Delta_f H°(298.15\ K) = -697.05 \pm 3.3\ kJ\ mol^{-1}$

Vibrational Frequencies and Degeneracies

ν, cm^{-1}	ν, cm^{-1}
3035(1)	1376(2)
1137(1)	1152(2)
700(1)	508(2)

Ground State Quantum Weight: 1 $\sigma = 3$

Point Group: C_{3v}

Bond Distances: C-F = 1.333 Å C-H = 1.098 Å

Bond Angles: F-C-F = 108.6 ± 0.5° F-C-H = 110.33°

Product of the Moments of Inertia: $I_A I_B I_C = 9.722739 \times 10^{-115}\ g^3\ cm^6$

Enthalpy of Formation

The selected value is obtained from least squares, simultaneous adjustment of the enthalpies of formation of CF_3, CF_3X (X = H, Cl, Br, I, CF_3) and C_2F_4. Details of the input data and the adjustment are given in (1). Trifluoromethane is linked to HF(aq, 21.5 H_2O) by calorimetric data, to $CBrF_3$(g) and CF_3I(g) by equilibrium data, and to CF_3(g) by kinetic data. Neugebauer (2) reported calorimetric data for $CHF_3(g) + 0.5\ O_2(g) + H_2O(\ell) = CO_2(g) + 3\ HF(aq, 21.5\ H_2O)$, where CF_4 was also formed in a competing reaction. Recalculation of the original data with $\Delta_f H°(CF_4, g, 298.15\ K) = -223.0\ kcal\ mol^{-1}$ yields the mean value of $\Delta_r H°(298.15\ K) = -90.35 \pm 0.7\ kcal\ mol^{-1}$, but the individual values of $\Delta_r H°$ show a disturbing trend with the amount of competing reaction. Possible causes for the trend include corrosion and error in the analysis of the reaction products. Calibration was done in the presence of HF in an attempt to eliminate the heat due to corrosion. Because of this unexplained trend, we increase the uncertainty to ±1.0 kcal mol^{-1}. Goy (3) and Coomber (4) determined equilibrium data for $CHF_3(g) + X_2(g) = CF_3X(g) + HX(g)$ where X = I and Br. Kinetic data pertinent to CHF_3 are noted on the table for CF_3(g).

Heat Capacity and Entropy

The molecular structure has been derived from electron-diffraction data by Thornton (5) and from microwave data by Ghosh (6). The results are in close agreement and average values are adopted. Vibrational frequencies are from the assignment of Long (7) who reviewed the spectral data and gave force-constant calculations. Preferences has been given to the gas-phase spectra, particularly the Raman data of Claassen (8). Shimanouchi (9) assigned $\nu_2 = 1117$ rather than 1137 cm^{-1}, but he apparently did not consider the gas-phase data of (8, 10). The principal moments of inertia are: $I_A = I_B = 8.1086 \times 10^{-39}$ and $I_C = 14.7875 \times 10^{-39}\ g\ cm^2$.

Valentine (1)] used low-temperature calorimetric data to derive the ideal gas entropy at the normal boiling point of 190.97 K. The experimental value of 57.18 cal K^{-1} mol^{-1} is in good agreement with 57.216 cal K^{-1} mol^{-1} obtained from this table. Schwing (12) measured $C_v°$ for CHF_3 vapor (25-100°C) and confirmed that the vibrational assignment is adequate. Uncertainty in $C_p°$ and S° increases at higher temperatures due to neglect of anharmonicity in the calculations. Discussions of anharmonic contributions are given on the tables for CF_4(g) and $CClF_3$(g). The error probably does not exceed 0.3 cal K^{-1} mol^{-1} in S° at 1000 K. $C_p°$ is consistent with $C_v°$ measured at 300 K by the wire-ribbon method (13).

References

1. A. N. Syverud, Quarterly Tech. Report T-0009-2Q-69 under AFRPL Contract Nr. F04611-67-C-0009, Dow Chemical Co., July, 1969.
2. C. A. Neugebauer and J. L. Margrave, J. Phys. Chem. 62, 1043 (1958).
3. C. A. Goy, A. Lord and H. O. Pritchard, J. Phys. Chem. 71, 1086 (1967).
4. J. W. Coomber and E. Whittle, Trans. Faraday Soc. 63, 608 (1967).
5. C. G. Thornton, Dissertation Abstracts 14, 604 (1954); R. L. Livingston, Ann. Rev. Phys. Chem. 5, 397 (1954).
6. S. N. Ghosh, R. Trambarulo and W. Gordy, J. Chem. Phys. 20, 605 (1952); C. A. Burrus and W. Gordy, ibid. 26, 391 (1957); M. Jen and D. R. Lide, ibid. 36, 2525 (1962).
7. D. A. Long, R. B. Gravenor and D. T. L. Jones, Trans. Faraday Soc. 60, 1509 (1964).
8. H. H. Claassen and J. R. Nielsen, J. Opt. Soc. Amer. 43, 352 (1953).
9. T. Shimanouchi, U. S. Nat. Bur. Stand. NSRDS-NBS 17, 17 (1967).
10. W. F. Edgell and C. May, J. Chem. Phys. 21, 1901 (1953); 22, 1808 (1954).
11. R. H. Valentine, G. E. Brodale and W. F. Giauque, J. Phys. Chem. 66, 392 (1962).
12. R. C. Schwing, Dissertation Abstracts 25, 2404 (1964).
13. W. N. Vanderkooi and T. DeVries, J. Phys. Chem. 60, 636 (1956).

Trifluoromethane (CHF_3) $C_1H_1F_3(g)$

Enthalpy Reference Temperature = T_r = 298.15 K Standard State Pressure = p° = 0.1 MPa

T/K	$C_p°$ (J $K^{-1}mol^{-1}$)	S° (J $K^{-1}mol^{-1}$)	$-[G°-H°(T_r)]/T$ (J $K^{-1}mol^{-1}$)	$H°-H°(T_r)$ (kJ mol^{-1})	$\Delta_f H°$ (kJ mol^{-1})	$\Delta_f G°$ (kJ mol^{-1})	Log K_f
0	0.	0.	INFINITE	-11.564	-690.096	-690.096	INFINITE
100	33.889	215.987	298.282	-8.230	-692.682	-683.593	357.072
200	41.227	241.391	264.002	-4.522	-695.041	-673.566	175.917
250	46.139	251.112	260.470	-2.339	-696.106	-668.072	139.586
298.15	51.040	259.657	259.657	0.	-697.054	-662.586	116.082
300	51.228	259.973	259.658	0.095	-697.089	-662.372	115.329
350	56.250	268.250	260.300	2.782	-697.980	-656.514	97.979
400	61.000	276.075	261.787	5.715	-698.773	-650.535	84.951
450	65.358	283.515	263.791	8.876	-699.472	-644.462	74.807
500	69.278	290.608	266.121	12.244	-700.081	-638.317	66.685
600	75.849	303.844	271.322	19.513	-701.065	-625.867	54.487
700	80.998	315.939	276.845	27.366	-701.781	-613.274	45.763
800	85.065	327.031	282.435	35.677	-702.282	-600.593	39.215
900	88.323	337.245	287.965	44.352	-702.610	-587.861	34.119
1000	90.969	346.692	293.371	53.321	-702.801	-575.100	30.040
1100	93.146	355.468	298.622	62.530	-702.885	-562.325	26.703
1200	94.954	363.652	303.704	71.938	-702.886	-549.546	23.921
1300	96.470	371.314	308.613	81.511	-702.826	-536.770	21.568
1400	97.751	378.512	313.352	91.224	-702.721	-524.000	19.551
1500	98.842	385.294	317.924	101.055	-702.584	-511.239	17.803
1600	99.777	391.704	322.337	110.987	-702.426	-498.488	16.274
1700	100.583	397.778	326.597	121.006	-702.253	-485.747	14.925
1800	101.282	403.547	330.713	131.100	-702.073	-473.016	13.727
1900	101.891	409.040	334.693	141.260	-701.887	-460.296	12.654
2000	102.425	414.280	338.542	151.476	-701.696	-447.585	11.690
2100	102.895	419.289	342.269	161.742	-701.502	-434.885	10.817
2200	103.310	424.085	345.879	172.053	-701.302	-422.193	10.024
2300	103.679	428.686	349.380	182.403	-701.093	-409.511	9.300
2400	104.008	433.106	352.777	192.788	-700.874	-396.838	8.637
2500	104.302	437.357	356.076	203.203	-700.640	-384.175	8.027
2600	104.567	441.453	359.282	213.647	-700.389	-371.521	7.464
2700	104.805	445.404	362.399	224.116	-700.116	-358.877	6.943
2800	105.020	449.220	365.432	234.607	-699.818	-346.244	6.459
2900	105.215	452.909	368.385	245.119	-699.493	-333.622	6.009
3000	105.393	456.479	371.262	255.650	-699.138	-321.012	5.589
3100	105.555	459.937	374.067	266.197	-698.750	-308.415	5.197
3200	105.703	463.291	376.803	276.760	-698.327	-295.830	4.829
3300	105.838	466.545	379.474	287.337	-697.868	-283.258	4.484
3400	105.962	469.707	382.081	297.928	-697.371	-270.702	4.159
3500	106.077	472.780	384.629	308.530	-696.835	-258.161	3.853
3600	106.183	475.770	387.119	319.143	-696.261	-245.635	3.564
3700	106.280	478.681	389.555	329.766	-695.647	-233.126	3.291
3800	106.371	481.516	391.938	340.398	-694.994	-220.634	3.033
3900	106.454	484.280	394.270	351.040	-694.302	-208.160	2.788
4000	106.532	486.976	396.554	361.689	-693.572	-195.704	2.556
4100	106.605	489.608	398.792	372.346	-692.803	-183.267	2.335
4200	106.673	492.178	400.985	383.010	-691.998	-170.848	2.125
4300	106.736	494.688	403.135	393.680	-691.156	-158.450	1.925
4400	106.795	497.143	405.244	404.357	-690.279	-146.072	1.734
4500	106.850	499.544	407.313	415.039	-689.368	-133.713	1.552
4600	106.902	501.893	409.343	425.727	-688.424	-121.376	1.378
4700	106.951	504.192	411.337	436.419	-687.448	-109.059	1.212
4800	106.996	506.444	413.295	447.117	-686.441	-96.763	1.053
4900	107.039	508.651	415.219	457.819	-685.405	-84.489	0.901
5000	107.080	510.814	417.109	468.525	-684.340	-72.236	0.755
5100	107.118	512.935	418.967	479.235	-683.247	-60.005	0.615
5200	107.154	515.015	420.794	489.948	-682.127	-47.795	0.480
5300	107.189	517.057	422.591	500.665	-680.983	-35.608	0.351
5400	107.221	519.060	424.359	511.386	-679.813	-23.442	0.227
5500	107.252	521.028	426.099	522.110	-678.620	-11.298	0.107
5600	107.281	522.961	427.812	532.836	-677.405	0.825	-0.008
5700	107.308	524.860	429.498	543.566	-676.167	12.924	-0.118
5800	107.334	526.726	431.158	554.298	-674.908	25.003	-0.225
5900	107.359	528.561	432.793	565.033	-673.629	37.060	-0.328
6000	107.383	530.366	434.404	575.770	-672.330	49.094	-0.427

PREVIOUS: June 1969 (1 atm) CURRENT: June 1969 (1 bar)

Trifluoromethane (CHF_3) $C_1H_1F_3(g)$

HYDROGEN CYANIDE (HCN) IDEAL GAS $M_r = 27.02564$

$S°(298.15\ K) = 201.828 \pm 0.04\ J\ K^{-1}\ mol^{-1}$

$\Delta_f H°(0\ K) = 135.53 \pm 8.4\ kJ\ mol^{-1}$
$\Delta_f H°(298.15\ K) = 135.14 \pm 8.4\ kJ\ mol^{-1}$

Vibrational Frequencies and Degeneracies

$\nu,\ cm^{-1}$
2096.3(1)
713.5(2)
3311.5(1)

Ground State Quantum Weights: 1 $\sigma = 1$
Point Group: $C_{\infty v}$
Bond Distances: C-H = 1.066 Å C-N = 1.153 Å
Bond Angle: H-C-N = 180°
Rotational Constant: $B_0 = 1.484360\ cm^{-1}$

Spectroscopic constants used in calculating corrections to the rigid-rotator harmonic oscillator approximation (cm^{-1}):

$\alpha_1 = 10.10 \times 10^{-3}$	$-x_{11} = 7.07$	$-x_{12} = 2.53$	$g_{22} = 5.16$
$\alpha_2 = -3.61 \times 10^{-3}$	$-x_{22} = 2.65$	$-x_{23} = 19.01$	
$\alpha_3 = 10.42 \times 10^{-3}$	$-x_{33} = 52.49$	$-x_{13} = 10.44$	

Enthalpy of Formation

The value adopted is from NBS (1). The data of Thomsen (2) who measured both C_2N_2 and HCN and Berthelot (3) were used in conjunction with the well established value for C_2N_2 to derive the adopted value (4).

Badger (5) studied the equilibrium $NH_3(g) + C(cr) = HCN(g) + H_2(g)$ from both sides; we obtain a 3rd law $\Delta_r H°(298.15\ K) = 46\pm0.9$ kcal and a 2nd law $\Delta_r H°(298.15\ K) = 34.5\pm19$ kcal, the 3rd law drift is 14±24 cal $K^{-1}\ mol^{-1}$. The 3rd law value leads to $\Delta_f H°(HCN,\ g,\ 298.15\ K) = 35\pm1\ kcal\ mol^{-1}$.

Horiuchi et al. (6) studied the equilibrium $2\ CO(g) + NH_3(g) = HCN(g) + CO_2(g) + H_2(g)$. We obtain $\Delta_r H°(298.15\ K) = 1.1\pm0.5$ kcal from a 3rd law analysis with a drift of -2±3 cal $K^{-1}\ mol^{-1}$, and a 2nd law value of 2.9±2.6 kcal. The 3rd law value leads to $\Delta_f H°(HCN,\ g,\ 298.15\ K) = 31.35\pm0.6\ kcal\ mol^{-1}$. The two sets of equilibrium data are self-consistent but disagree with each other and the direct combustions. We adopt an uncertainty of 2.0 kcal mol^{-1} to draw attention to this unsatisfactory situation.

Heat Capacity and Entropy

McBride et al. (7) used the constants listed above to calculate the adopted table which includes first and second corrections for anharmonicity, vibration-rotation interaction and centrifugal stretching. The bond lengths and molecular constants are those listed by Rank et al. (8). A more extensive set of anharmonic constants has recently been determined by Nakagawa and Morino (9), they are in generally good agreement with the values used.

References

1. U. S. Nat. Bur. Stand. Tech. Note 270-3, 1968.
2. J. Thomsen, "Thermochemische Untersuchungen," Barth, Leipzig, 1886.
3. M. Berthelot, Ann. Chim. Phys. 23, 252 (1881).
4. W. H. Evans, U. S. Nat. Bur. Stand., personal communication, May 1969.
5. R. M. Badger, J. Amer. Chem. Soc. 46, 2166 (1924).
6. J. Horiuchi, T. Yano, and K. Kanai, Shokubai 7, 8 (1951); CA. 49 6553e.
7. B. J. McBride, S. Heimel, J. G. Ehlers, and S. Gordon, NASA SP 3001 (1963).
8. D. H. Rank, G. Skorinko, D. P. Eastman, and T. A. Wiggins, J. Opt. Soc. 50, 421 (1960).
9. T. Nakagawa and Y. Morino, Bull. Chem. Soc. Japan 42, 2312 (1969).

Hydrogen Cyanide (HCN) $C_1H_1N_1(g)$

Enthalpy Reference Temperature = T_r = 298.15 K Standard State Pressure = p° = 0.1 MPa

	J K⁻¹mol⁻¹			kJ mol⁻¹			
T/K	$C_p°$	S°	$-[G°-H°(T_r)]/T$	$H°-H°(T_r)$	$\Delta_f H°$	$\Delta_f G°$	Log K_f
0	0.	0.	INFINITE	-9.234	135.529	135.529	INFINITE
100	29.167	167.608	230.911	-6.330	135.422	131.784	-68.837
200	31.719	188.377	204.966	-3.318	135.306	128.187	-33.479
298.15	35.857	201.828	201.828	0.	135.143	124.725	-21.851
300	35.928	202.050	201.829	0.066	135.140	124.660	-21.705
400	39.229	212.863	203.281	3.833	134.972	121.192	-15.826
500	41.731	221.894	206.125	7.885	134.766	117.769	-12.303
600	43.806	229.690	209.417	12.164	134.511	114.393	-9.959
700	45.643	236.583	212.815	16.638	134.222	111.063	-8.288
800	47.317	242.788	216.180	21.287	133.919	107.775	-7.037
900	48.844	248.451	219.456	26.096	133.618	104.525	-6.066
1000	50.229	253.670	222.619	31.051	133.328	101.308	-5.292
1100	51.484	258.517	225.665	36.137	133.052	98.120	-4.659
1200	52.610	263.046	228.594	41.343	132.793	94.955	-4.133
1300	53.618	267.298	231.409	46.655	132.549	91.812	-3.689
1400	54.522	271.305	234.117	52.063	132.318	88.687	-3.309
1500	55.329	275.095	236.724	57.556	132.099	85.579	-2.980
1600	56.049	278.689	239.235	63.126	131.889	82.484	-2.693
1700	56.697	282.107	241.657	68.764	131.685	79.403	-2.440
1800	57.275	285.364	243.996	74.463	131.486	76.333	-2.215
1900	57.794	288.475	246.255	80.217	131.289	73.274	-2.014
2000	58.262	291.451	248.441	86.020	131.094	70.226	-1.834
2100	58.689	294.304	250.558	91.868	130.898	67.187	-1.671
2200	59.070	297.043	252.609	97.756	130.700	64.158	-1.523
2300	59.421	299.677	254.598	103.681	130.499	61.138	-1.388
2400	59.739	302.213	256.530	109.639	130.294	58.127	-1.265
2500	60.028	304.657	258.406	115.628	130.085	55.124	-1.152
2600	60.296	307.017	260.231	121.644	129.869	52.130	-1.047
2700	60.542	309.297	262.006	127.686	129.647	49.144	-0.951
2800	60.764	311.503	263.735	133.752	129.417	46.167	-0.861
2900	60.973	313.639	265.419	139.838	129.180	43.198	-0.778
3000	61.170	315.710	267.061	145.946	128.934	40.237	-0.701
3100	61.347	317.718	268.663	152.072	128.679	37.284	-0.628
3200	61.513	319.669	270.226	158.215	128.415	34.340	-0.561
3300	61.669	321.564	271.754	164.374	128.141	31.405	-0.497
3400	61.814	323.407	273.246	170.548	127.856	28.478	-0.438
3500	61.948	325.201	274.705	176.736	127.560	25.559	-0.381
3600	62.074	326.948	276.132	182.938	127.253	22.649	-0.329
3700	62.190	328.650	277.528	189.151	126.934	19.748	-0.279
3800	62.300	330.310	278.896	195.376	126.603	16.855	-0.232
3900	62.401	331.930	280.235	201.611	126.259	13.971	-0.187
4000	62.496	333.511	281.547	207.855	125.902	11.097	-0.145
4100	62.587	335.055	282.833	214.110	125.532	8.231	-0.105
4200	62.672	336.564	284.095	220.373	125.148	5.375	-0.067
4300	62.750	338.040	285.332	226.644	124.750	2.528	-0.031
4400	62.823	339.483	286.546	232.923	124.338	-0.310	0.004
4500	62.890	340.896	287.739	239.208	123.911	-3.138	0.036
4600	62.952	342.279	288.909	245.500	123.469	-5.957	0.068
4700	63.011	343.633	290.059	251.799	123.013	-8.765	0.097
4800	63.065	344.961	291.189	258.102	122.541	-11.564	0.126
4900	63.115	346.261	292.300	264.412	122.054	-14.353	0.153
5000	63.162	347.537	293.392	270.725	121.552	-17.131	0.179
5100	63.206	348.788	294.466	277.044	121.034	-19.900	0.204
5200	63.245	350.016	295.522	283.366	120.501	-22.658	0.228
5300	63.278	351.221	296.562	289.693	119.952	-25.406	0.250
5400	63.308	352.404	297.585	296.022	119.388	-28.143	0.272
5500	63.335	353.566	298.593	302.354	118.808	-30.870	0.293
5600	63.358	354.707	299.584	308.689	118.212	-33.586	0.313
5700	63.378	355.829	300.561	315.026	117.601	-36.292	0.333
5800	63.395	356.931	301.524	321.364	116.975	-38.986	0.351
5900	63.407	358.015	302.472	327.705	116.333	-41.670	0.369
6000	63.417	359.081	303.407	334.046	115.675	-44.342	0.386

PREVIOUS: December 1969 (1 atm) CURRENT: December 1969 (1 bar)

Hydrogen Cyanide (HCN) $C_1H_1N_1(g)$

HYDROGEN ISOCYANATE (HNCO) IDEAL GAS $M_r = 43.02504$

$S°(298.15\ K) = 238.228 \pm 0.4\ J\ K^{-1}\ mol^{-1}$

$\Delta_f H°(0\ K) = -98.7 \pm 8\ kJ\ mol^{-1}$
$\Delta_f H°(298.15\ K) = -101.7 \pm 8\ kJ\ mol^{-1}$

Vibrational Frequencies and Degeneracies

ν, cm^{-1}	ν, cm^{-1}
3531(1)	659.8(1)
2274(1)	577.5(1)
1327(1)	777.1(1)

Ground State Quantum Weight: 1 $\sigma = 1$
Point Group: C_s
Bond Distances: H–N = 0.987 Å N–C = 1.207 Å C–O = 1.171 Å
Bong Angles: H–N–C = 128.1° N–C–O = 180°
Product of the Moments of Inertia: $I_A I_B I_C = 5.3166 \times 10^{-117}\ g^3\ cm^6$

Enthalpy of Formation

Okabe (1) has measured the onset of photodissociation for the process HNCO(g) → NH(c' Π) + CO(g) as 8.79±0.03 eV (202.7±0.7 kcal). Okabe and Lenzi (2) have also reported the onset of photodissociation for the process $NH_3(g) \rightarrow NH(c'\Pi) + H_2(g)$ as 9.35±0.06 eV (215.6±1.4 kcal). Elimination of the NH(c'Π) gives

$$NH_3(g) + CO(g) \rightarrow NHCO(g) + H_2(g) \qquad \Delta_f H°(0\ K) = 12.9 \pm 2\ kcal$$

which with JANAF auxiliary data (3) yields $\Delta_f H°$(HNCO, g, 0 K) = $-23.6\pm2\ kcal\ mol^{-1}$, or $\Delta_f H°$(HNCO, g, 298.15 K) = $-24.3\pm2\ kcal\ mol^{-1}$.

The heat of formation of ionized cyanic acid in aqueous solution is $-34.9\ kcal\ mol^{-1}$ (4). Since the ions from cyanic acid and isocyanic acid are identical in solution we can use this value for aqueous isocyanic acid. Assuming the heat of solution is $-3\pm3\ kcal\ mol^{-1}$, a value covering the range of heats of solution of most organic liquids (4), we obtain $\Delta_f H°$(HNCO, ℓ, 298.15 K) = $-32\pm3\ kcal\ mol^{-1}$. Using $\Delta_{vap}H° = 7.2\ kcal\ mol^{-1}$ (5) we obtain $\Delta_f H°$(HNCO, g, 298.15 K) = $-25\pm3\ kcal\ mol^{-1}$. If we assume that $\Delta_f H°$(HNCO, g, 298.15 K) − $\Delta_f H°$(HNCO, ionized aq, 298.15 K) = $\Delta_f H°$(HNCS, g, 298.15 K) − $\Delta_f H°$(HCNS, ionized aq, 298.15 K) then we obtain $\Delta_f H°$(HNCO, g, 298.15 K) = $-23\ kcal\ mol^{-1}$ employing values from (4). We adopt the value $\Delta_f H°$(HNCO, g, 298.15 K) = $-24.3\pm2\ kcal\ mol^{-1}$ ($-101.671\pm8\ kJ\ mol^{-1}$).

Heat Capacity and Entropy

The vibrational frequency determinations have been summarized by Herzberg and Reid (6). Ashby and Werner (7) have remeasured the three lower fundamentals under much higher resolution. They reverse the assignments of ν_5 and ν_6 of Herzberg and Reid. Dixon and Kirby (8) have discussed the corrections due to Coriolis interaction made by Ashby and Werner and conclude they are in the wrong direction. On this basis they reverse the ν_4 and ν_6 assignments of Herzberg and Reid.

The bond distances and angles are from the microwave measurements of Jones et al. (9) which give rotational constants in good agreement with the later microwave work of Kewley (10). The principal moments are: $I_A = 0.0912 \times 10^{-39}$, $I_B = 7.5910 \times 10^{-39}$, and $I_C = 7.6822 \times 10^{-39}\ g\ cm^2$.

References

1. H. Okabe, J. Chem. Phys. 53, 3507 (1970).
2. H. Okabe and M. Lenzi, J. Chem. Phys. 47, 5241 (1967).
3. JANAF Thermochemical Tables: $NH_3(g)$, 9-30-65; CO(g), 9-30-65; $H_2(g)$, 3-31-61.
4. U. S. Nat. Bur. Stand. Tech. Note 270-3, 1968.
5. U. S. Nat. Bur. Stand. Circ. 500, 1952.
6. G. Herzberg and C. Reid, Disc. Faraday Soc. 9, 92 (1950).
7. R. A. Ashby and R. L. Werner, J. Mol. Spectrosc. 18, 184 (1965).
8. R. N. Dixon and G. H. Kirby, Trans. Faraday Soc. 64, 2002 (1968).
9. L. H. Jones, J. N. Shoolery, R. G. Shulman, and D. M. Yost, J. Chem. Phys. 18, 990 (1950).
10. R. Kewley, K. V. L. N. Sastry and M. Winnewisser, J. Mol. Spectrosc. 10, 418 (1963).

Hydrogen Isocyanate (HNCO) $C_1H_1N_1O_1(g)$

Enthalpy Reference Temperature = T_r = 298.15 K Standard State Pressure = p° = 0.1 MPa

T/K	$C_p°$ (J K⁻¹mol⁻¹)	S° (J K⁻¹mol⁻¹)	$-[G°-H°(T_r)]/T$ (J K⁻¹mol⁻¹)	$H°-H°(T_r)$ (kJ mol⁻¹)	$\Delta_f H°$ (kJ mol⁻¹)	$\Delta_f G°$ (kJ mol⁻¹)	Log K_f
0	0.	0.	INFINITE	-10.961	-98.671	-98.671	INFINITE
100	33.470	197.325	273.654	-7.633	-99.805	-97.750	51.060
200	38.269	221.663	242.134	-4.094	-100.851	-95.279	24.884
250	41.813	230.583	238.952	-2.092	-101.299	-93.833	19.605
298.15	45.043	238.228	238.228	0.	-101.671	-92.360	16.181
300	45.160	238.507	238.229	0.083	-101.684	-92.302	16.071
350	48.133	245.697	238.790	2.417	-102.013	-90.712	13.538
400	50.729	252.297	240.071	4.890	-102.297	-89.077	11.632
450	53.001	258.407	241.773	7.485	-102.553	-87.409	10.146
500	55.007	264.097	243.724	10.186	-102.789	-85.714	8.954
600	58.410	274.437	247.999	15.863	-103.226	-82.257	7.161
700	61.228	283.659	252.446	21.849	-103.631	-78.730	5.875
800	63.627	291.995	256.877	28.094	-104.006	-75.147	4.907
900	65.704	299.612	261.208	34.563	-104.349	-71.519	4.151
1000	67.516	306.631	265.404	41.226	-104.662	-67.854	3.544
1100	69.104	313.142	269.452	48.059	-104.946	-64.159	3.047
1200	70.498	319.216	273.348	55.041	-105.204	-60.439	2.631
1300	71.724	324.908	277.098	62.153	-105.439	-56.699	2.278
1400	72.802	330.264	280.706	69.381	-105.657	-52.941	1.975
1500	73.752	335.320	284.180	76.709	-105.862	-49.169	1.712
1600	74.591	340.107	287.527	84.127	-106.057	-45.383	1.482
1700	75.333	344.652	290.755	91.624	-106.248	-41.585	1.278
1800	75.990	348.976	293.870	99.191	-106.437	-37.776	1.096
1900	76.575	353.101	296.880	106.820	-106.628	-33.956	0.934
2000	77.095	357.042	299.790	114.504	-106.824	-30.126	0.787
2100	77.560	360.815	302.607	122.237	-107.028	-26.287	0.654
2200	77.977	364.433	305.336	130.014	-107.241	-22.437	0.533
2300	78.351	367.908	307.981	137.831	-107.465	-18.577	0.422
2400	78.688	371.250	310.548	145.683	-107.702	-14.707	0.320
2500	78.992	374.468	313.041	153.568	-107.954	-10.828	0.226
2600	79.268	377.572	315.464	161.481	-108.220	-6.937	0.139
2700	79.518	380.568	317.820	169.420	-108.504	-3.036	0.059
2800	79.745	383.464	320.113	177.384	-108.804	0.875	-0.016
2900	79.953	386.266	322.346	185.369	-109.122	4.798	-0.086
3000	80.143	388.980	324.522	193.374	-109.459	8.732	-0.152
3100	80.316	391.611	326.644	201.397	-109.814	12.678	-0.214
3200	80.476	394.163	328.714	209.437	-110.189	16.635	-0.272
3300	80.623	396.642	330.735	217.492	-110.583	20.604	-0.326
3400	80.758	399.051	332.709	225.561	-110.997	24.585	-0.378
3500	80.883	401.393	334.638	233.643	-111.430	28.579	-0.427
3600	80.999	403.674	336.524	241.737	-111.884	32.586	-0.473
3700	81.106	405.894	338.369	249.842	-112.358	36.605	-0.517
3800	81.205	408.059	340.175	257.958	-112.852	40.638	-0.559
3900	81.298	410.169	341.943	266.083	-113.367	44.684	-0.598
4000	81.384	412.229	343.674	274.217	-113.903	48.744	-0.637
4100	81.464	414.239	345.371	282.360	-114.459	52.817	-0.673
4200	81.539	416.203	347.034	290.510	-115.037	56.904	-0.708
4300	81.610	418.123	348.665	298.667	-115.635	61.005	-0.741
4400	81.675	420.000	350.265	306.832	-116.255	65.119	-0.773
4500	81.737	421.836	351.835	315.002	-116.895	69.248	-0.804
4600	81.795	423.633	353.377	323.179	-117.558	73.392	-0.833
4700	81.850	425.393	354.890	331.361	-118.241	77.551	-0.862
4800	81.901	427.116	356.377	339.549	-118.947	81.724	-0.889
4900	81.949	428.806	357.838	347.741	-119.674	85.912	-0.916
5000	81.995	430.462	359.274	355.939	-120.424	90.116	-0.941
5100	82.038	432.086	360.686	364.140	-121.196	94.334	-0.966
5200	82.079	433.679	362.074	372.346	-121.990	98.567	-0.990
5300	82.117	435.243	363.440	380.556	-122.806	102.817	-1.013
5400	82.154	436.778	364.784	388.770	-123.646	107.082	-1.036
5500	82.188	438.286	366.107	396.987	-124.508	111.362	-1.058
5600	82.221	439.767	367.409	405.207	-125.393	115.659	-1.079
5700	82.252	441.223	368.691	413.431	-126.302	119.971	-1.099
5800	82.282	442.654	369.954	421.658	-127.234	124.300	-1.119
5900	82.310	444.060	371.198	429.887	-128.190	128.644	-1.139
6000	82.337	445.444	372.424	438.120	-129.170	133.006	-1.158

PREVIOUS: December 1970 (1 atm) CURRENT: December 1970 (1 bar)

Hydrogen Isocyanate (HNCO) $C_1H_1N_1O_1(g)$

FORMYL (HCO) IDEAL GAS M_r = 29.01834

$\Delta_f H°(0\ K) = 43.2 \pm 8\ kJ\ mol^{-1}$

$S°(298.15\ K) = 224.649 \pm 0.04\ J\ K^{-1}\ mol^{-1}$ $\Delta_f H°(298.15\ K) = 43.5 \pm 8\ kJ\ mol^{-1}$

Electronic Levels and Molecular Constants

State	ε_i, cm^{-1}	g_i	ν_1, cm^{-1}	ν_2, cm^{-1}	ν_3, cm^{-1}	B_0, cm^{-1}	$I_A I_B I_C$, $g^3\ cm^6$
X^2A'	0	2	2488 (1)	1083 (1)	1820 (1)	--	0.4655×10^{-117}
$\tilde{A}^2\Sigma^-$	9294	2	3316.2 (1)	802.3 (2)	1813.4 (1)	1.333284	--
$\tilde{B}^2A'$	38691	2	[2488] (1)	[1083] (1)	[1820] (1)	--	1.0087×10^{-117}
$\tilde{C}^2A''$	41270	2	[2488] (1)	[1083] (1)	[1820] (1)	--	$[0.4655] \times 10^{-117}$

Point Group: C_s $\sigma = 1$

Bond Distance (X^2A'): C-H 1.16 Å C-O = 1.17 Å

Bond Angle (X^2A'): H-C-O 123.8°

Enthalpy of Formation

The adopted $\Delta_f H°$(HCO, g, 298.15 K) = 10.4±2 kcal mol^{-1} is calculated from $\Delta_{at} H°$(OHC-CHO, g) = 71.5±2 kcal mol^{-1} and $\Delta_f H°$(glyoxal, g, 298.15 K) = -50.66±0.2 kcal mol^{-1}. The C-C bond dissociation energy in glyoxal was determined from chemiluminescent recombination of formyl radicals by Hartley (1). The heat of formation of glyoxal was determined from flame combustion calorimetry by Fletcher and Pilcher (2).

The other route to derive the enthalpy of formation of formyl radical was from $\Delta_{at} H°$(H-CHO) which has been studied by many investigators listed below. There are several conflicting values between 77 to 91 kcal for $\Delta_{at} H°$(H-CHO). Based on the adopted $\Delta_f H°$(HCO, g, 298.15 K) = 10.4±2 kcal mol^{-1}, we have derived $\Delta_{at} H°$(H-CHO) = 89.3±2 kcal mol^{-1} which is in essential agreement with the high values of $\Delta_{at} H°$(H-CHO). The low values were probably not reliable as pointed out by Calvert (3), Walsh and Benson (5), Haney and Franklin (9).

Source	Method	$\Delta_{at} H°$(H-CHO) kcal mol^{-1}
Calvert (3)	Kinetic	91
Klein and Schoen (4)	Kinetic	77
Walsh and Benson (5)	Kinetic	87 ± 1
Brand and Reed (6)	Spectroscopic	<82.3
Reed and Brand (7)	Electron Impact	75 ± 2.3
Shannon and Harrison (8)	Electron Impact	79 ± 3
Haney and Franklin (9)	Electron Impact	87 ± 3

Heat Capacity and Entropy

The thermodynamic functions were calculated by summing over the individual partition functions for the separate states. The observed molecular constants and electronic levels were taken from Herzberg and Ramsay (10, 11, 12) and Dixon (13). The vibrational frequency (ν_1 = 2488 cm^{-1}) was observed using the matrix isolation infrared technique by Milligan and Jacox (14). Using the same technique, Ewing, Thompson and Pimentel (15) found ν_2 = 1091 cm^{-1} and ν_3 = 1860 cm^{-1} which are in good agreement with the values adopted (10). These small differences are probably due to matrix shifts. The bond lengths and angle were calculated from the normal coordinate analyses by Shirk and Pimentel (16). These values are a better fit than those calculated by Johns, Priddle and Ramsay (12) who assumed a normal CH bond length of 1.08 Å.

References

1. D. B. Hartley, Chem. Comm. 1967, 1281 (1967).
2. R. A. Fletcher and G. Pilcher, Trans. Faraday Soc. 66, 794 (1970).
3. J. G. Calvert, J. Chem. Phys. 29, 954 (1958); J. Phys. Chem. 61, 1206 (1957).
4. R. Klein and L. J. Shoen, J. Chem. Phys. 29, 953 (1958); 24, 1094 (1956).
5. R. Walsh and S. W. Benson, J. Amer. Chem. Soc. 88, 4570 (1966).
6. J. C. D. Brand and R. I. Reed, J. Chem. Soc. 1957, 2386 (1957).
7. R. I. Reed and J. C. D. Brand, Trans. Faraday Soc. 54, 478 (1958).
8. T. W. Shannon and A. S. Harrison, Can. J. Chem. 39, 1392 (1961).
9. M. A. Haney and J. L. Franklin, Trans. Faraday Soc. 65, 1794 (1969).
10. G. Herzberg, "Electronic Spectra of Polyatomic Molecules," D. Van Nostrand Co., Inc., New York, 1966.
11. G. Herzberg and D. A. Ramsey, Proc. Roy. Soc. (London) A233, 34 (1955).
12. J. W. C. Johns, S. H. Priddle and D. A. Ramsay, Discussion Faraday Soc. 35, 90 (1963).
13. R. N. Dixon, Trans. Faraday Soc. 65, 3141 (1969).
14. D. E. Milligan and M. E. Jacox, J. Chem. Phys. 51, 277 (1969).
15. G. E. Ewing, W. E. Thompson and G. C. Pimentel, J. Chem. Phys. 32, 927 (1960).
16. J. S. Shirk and G. C. Pimentel, J. Amer. Chem. Soc. 90, 3349 (1968).

Formyl (HCO) $C_1H_1O_1(g)$

Enthalpy Reference Temperature = T_r = 298.15 K Standard State Pressure = p° = 0.1 MPa

T/K	$C_p°$ (J $K^{-1}mol^{-1}$)	S°	$-[G°-H°(T_r)]/T$	$H°-H°(T_r)$ (kJ mol^{-1})	$\Delta_f H°$	$\Delta_f G°$	Log K_f
0	0.	0.	INFINITE	-9.989	43.151	43.151	INFINITE
100	33.258	188.025	254.660	-6.664	43.465	38.459	-20.089
200	33.469	211.108	227.769	-3.332	43.667	33.352	-8.711
250	33.921	218.619	225.214	-1.649	43.628	30.776	-6.430
298.15	34.597	224.649	224.649	0.	43.514	28.309	-4.960
300	34.627	224.863	224.650	0.064	43.508	28.215	-4.913
350	35.518	230.266	225.074	1.817	43.327	25.680	-3.833
400	36.530	235.074	226.029	3.618	43.100	23.174	-3.026
450	37.614	239.438	227.280	5.471	42.837	20.698	-2.403
500	38.733	243.459	228.699	7.380	42.545	18.254	-1.907
600	40.965	250.719	231.777	11.365	41.908	13.455	-1.171
700	43.068	257.194	234.954	15.568	41.242	8.765	-0.654
800	44.962	263.071	238.106	19.972	40.580	4.171	-0.272
900	46.623	268.465	241.184	24.553	39.936	-0.341	0.020
1000	48.056	273.454	244.165	29.289	39.316	-4.783	0.250
1100	49.282	278.093	247.041	34.157	38.717	-9.164	0.435
1200	50.329	282.427	249.811	39.139	38.134	-13.491	0.587
1300	51.224	286.492	252.478	44.218	37.562	-17.770	0.714
1400	51.991	290.317	255.045	49.380	36.995	-22.004	0.821
1500	52.651	293.927	257.518	54.613	36.429	-26.199	0.912
1600	53.223	297.343	259.901	59.907	35.859	-30.356	0.991
1700	53.722	300.585	262.200	65.255	35.282	-34.477	1.059
1800	54.161	303.669	264.419	70.650	34.695	-38.563	1.119
1900	54.549	306.608	266.563	76.085	34.096	-42.617	1.172
2000	54.896	309.415	268.636	81.558	33.483	-46.638	1.218
2100	55.206	312.101	270.642	87.063	32.854	-50.629	1.259
2200	55.488	314.675	272.585	92.598	32.209	-54.589	1.296
2300	55.744	317.148	274.469	98.160	31.546	-58.520	1.329
2400	55.978	319.525	276.298	103.746	30.866	-62.422	1.359
2500	56.195	321.815	278.073	109.355	30.167	-66.294	1.385
2600	56.395	324.023	279.798	114.985	29.450	-70.139	1.409
2700	56.581	326.155	281.475	120.634	28.713	-73.955	1.431
2800	56.755	328.215	283.108	126.300	27.959	-77.744	1.450
2900	56.918	330.210	284.698	131.984	27.185	-81.505	1.468
3000	57.071	332.142	286.248	137.684	26.393	-85.240	1.484
3100	57.216	334.016	287.758	143.398	25.582	-88.947	1.499
3200	57.353	335.835	289.233	149.127	24.753	-92.629	1.512
3300	57.483	337.601	290.672	154.868	23.905	-96.284	1.524
3400	57.606	339.319	292.077	160.623	23.039	-99.913	1.535
3500	57.723	340.991	293.451	166.389	22.154	-103.517	1.545
3600	57.835	342.619	294.794	172.167	21.252	-107.094	1.554
3700	57.942	344.205	296.108	177.956	20.331	-110.647	1.562
3800	58.044	345.751	297.395	183.756	19.393	-114.174	1.569
3900	58.141	347.260	298.654	189.565	18.436	-117.676	1.576
4000	58.235	348.733	299.888	195.384	17.462	-121.154	1.582
4100	58.324	350.172	301.096	201.212	16.469	-124.607	1.588
4200	58.410	351.579	302.282	207.048	15.459	-128.036	1.592
4300	58.493	352.954	303.444	212.894	14.430	-131.440	1.597
4400	58.573	354.300	304.585	218.747	13.384	-134.821	1.601
4500	58.650	355.617	305.704	224.608	12.320	-138.178	1.604
4600	58.724	356.907	306.803	230.477	11.238	-141.510	1.607
4700	58.796	358.171	307.883	236.353	10.137	-144.819	1.609
4800	58.866	359.409	308.944	242.236	9.018	-148.104	1.612
4900	58.933	360.624	309.986	248.126	7.881	-151.365	1.614
5000	58.999	361.815	311.011	254.022	6.726	-154.603	1.615
5100	59.063	362.984	312.018	259.926	5.552	-157.818	1.616
5200	59.126	364.132	313.009	265.835	4.360	-161.010	1.617
5300	59.187	365.258	313.985	271.751	3.149	-164.179	1.618
5400	59.248	366.365	314.945	277.673	1.919	-167.325	1.619
5500	59.307	367.453	315.889	283.600	0.671	-170.448	1.619
5600	59.365	368.522	316.820	289.534	-0.596	-173.547	1.619
5700	59.423	369.573	317.736	295.473	-1.882	-176.624	1.619
5800	59.480	370.607	318.639	301.418	-3.188	-179.678	1.618
5900	59.536	371.625	319.528	307.369	-4.512	-182.711	1.618
6000	59.593	372.626	320.405	313.326	-5.855	-185.719	1.617

PREVIOUS: December 1970 (1 atm) CURRENT: December 1970 (1 bar)

Formyl (HCO) $C_1H_1O_1(g)$

FORMYL, ION (HCO^+) IDEAL GAS M_r = 29.01779

$S°(298.15 K) = [203.417 \pm 4]\ J\ K^{-1}\ mol^{-1}$

$\Delta_f H°(0 K) = 827.2 \pm 8\ kJ\ mol^{-1}$
$\Delta_f H°(298.15 K) = 833.0 \pm 8\ kJ\ mol^{-1}$

Vibrational Frequencies and Degeneracies

ν, cm^{-1}
[3223](1)
[707](2)
[2088](1)

Ground State Quantum Weight: 1 σ = 1
Point Group: $[C_{\infty v}]$
Bond Distances: C-H = [1.07] Å C-O = [1.17] Å
Bond Angles: H-C-O = [180]°
Rotational Constant: B_0 = [1.367073] cm^{-1}

Enthalpy of Formation

Using the photoionization technique, Matthews and Warneck (1) measured the appearance potentials of $HCO^+(g)$ from formaldehyde, formic acid and acetaldehyde as 11.95, 12.79 and 11.79 eV, respectively, whose average yields $\Delta_f H°(CHCO^+, g, 0 K)$ = 197.7±1.5 kcal mol^{-1}, based on the following $\Delta_f H°(0 K)$ data (in units of kcal mol^{-1}): 51.63 for H(g), 9.35 for OH(g), 35.62 for $CH_3(g)$, -26.78 for $H_2CO(g)$ (2), -88.74 for HCOOH(g) (3) and -37.14 for CH_3CHO (4). The appearance potentials obtained from photoionization are several tenths to 1 eV lower than the recent electron impact data (5, 6, 7, 8, 9). The appearance potentials determined by the electron impact method tend to be high because the fundamental nature of the process does not lead to a sharp onset in contrast to the step-function behavior of photoionization onset.

Heat Capacity and Entropy

The molecular configuration is assumed to be linear based on two reasons: (1) the molecule $HCO^+(g)$ is isoelectronic with HCN(g) which is linear; (2) A. D. Walsh (10) predicted that "HAB" molecules with 10 or less valence electrons will be linear in their ground state. $HCO^+(g)$ has 10 valence electrons.

The bond distances HC and CO were estimated to be the same as those in HCN and HCO molecules, respectively. The vibrational frequencies were calculated by the valence force method from the estimated stretching and bending force constants which were obtained from HCN(g).

References

1. C. S. Mathews and P. Warneck, J. Chem. Phys. 51, 854 (1969).
2. JANAF Thermochemical Tables: H(g), 9-30-65; OH(g), 12-31-70; $CH_3(g)$, 6-30-69; $H_2CO(g)$, 3-31-61.
3. J. H. S. Green, J. Chem. Soc. 1961, 2241 (1961).
4. U. S. Nat. Bur. Stand. Tech. Note 270-3, 1968.
5. M. A. Haney and J. L. Franklin, Trans. Faraday Soc. 65, 1794 (1969).
6. J. L. Franklin, J. G. Dillard, H. M. Rosenstock, J. T. Herron, K. Draxl and F. H. Field, "Ionization Potentials, Appearance Potentials, and Heats of Formation of Gaseous Positive Ions," NSRDS-NBS-26, 1969.
7. H. Pritchard and A. G. Harrison, J. Chem. Phys. 48, 2827 (1968); 50, 1043 (1969).
8. F. H. Dorman, J. Chem. Phys. 50, 1042 (1969).
9. J. R. Majer, C. R. Patrick and J. C. Robb, Trans. Faraday Soc. 57, 14 (1961).
10. A. D. Walsh, J. Chem. Soc. 1953, 2288 (1953).

Formyl, Ion (HCO^+) $C_1H_1O_1^+(g)$

Enthalpy Reference Temperature = T_r = 298.15 K Standard State Pressure = p° = 0.1 MPa

	J $K^{-1}mol^{-1}$			kJ mol^{-1}			
T/K	$C_p°$	S°	$-[G°-H°(T_r)]/T$	$H°-H°(T_r)$	$\Delta_f H°$	$\Delta_f G°$	Log K_f
0	0.	0.	INFINITE	-9.249	827.215		
100	29.166	169.137	232.573	-6.344			
200	31.793	189.930	206.575	-3.329			
250	33.980	197.259	203.998	-1.685			
298.15	35.963	203.417	203.417	0.	833.034	817.906	-143.294
300	36.035	203.639	203.417	0.067	833.070	817.812	-142.394
350	37.811	209.331	203.863	1.914	834.023	815.193	-121.661
400	39.328	214.482	204.873	3.843	834.964	812.439	-106.094
450	40.645	219.192	206.206	5.843	835.886	809.568	-93.972
500	41.819	223.536	207.725	7.905	836.787	806.595	-84.264
600	43.892	231.348	211.025	12.194	838.531	800.392	-69.680
700	45.727	238.254	214.431	16.676	840.223	793.901	-59.242
800	47.389	244.471	217.804	21.333	841.893	787.169	-51.397
900	48.895	250.141	221.087	26.149	843.563	780.229	-45.283
1000	50.251	255.364	224.256	31.107	845.244	773.102	-40.383
1100	51.463	260.211	227.307	36.194	846.942	765.805	-36.365
1200	52.541	264.736	230.240	41.395	848.657	758.354	-33.010
1300	53.496	268.980	233.058	46.698	850.387	750.759	-30.166
1400	54.339	272.976	235.768	52.091	852.130	743.029	-27.723
1500	55.084	276.751	238.376	57.563	853.881	735.175	-25.601
1600	55.741	280.327	240.887	63.105	855.638	727.204	-23.741
1700	56.322	283.725	243.308	68.708	857.395	719.124	-22.096
1800	56.837	286.959	245.644	74.367	859.151	710.939	-20.631
1900	57.293	290.044	247.900	80.074	860.901	702.657	-19.317
2000	57.700	292.993	250.082	85.824	862.644	694.284	-18.133
2100	58.062	295.818	252.193	91.612	864.377	685.823	-17.059
2200	58.386	298.526	254.238	97.435	866.098	677.280	-16.081
2300	58.677	301.128	256.220	103.288	867.806	668.659	-15.186
2400	58.939	303.631	258.144	109.169	869.499	659.964	-14.364
2500	59.175	306.042	260.012	115.075	871.175	651.199	-13.606
2600	59.388	308.367	261.827	121.004	872.835	642.367	-12.905
2700	59.581	310.612	263.593	126.952	874.478	633.472	-12.255
2800	59.757	312.782	265.311	132.919	876.102	624.516	-11.650
2900	59.917	314.882	266.984	138.903	877.707	615.502	-11.086
3000	60.063	316.916	268.615	144.902	879.293	606.434	-10.559
3100	60.197	318.887	270.205	150.915	880.859	597.313	-10.065
3200	60.320	320.800	271.756	156.941	882.406	588.141	-9.600
3300	60.433	322.658	273.271	162.979	883.933	578.921	-9.164
3400	60.537	324.464	274.750	169.028	885.439	569.656	-8.752
3500	60.633	326.220	276.196	175.086	886.925	560.347	-8.363
3600	60.722	327.930	277.609	181.154	888.391	550.995	-7.995
3700	60.804	329.594	278.992	187.230	889.837	541.603	-7.646
3800	60.880	331.217	280.345	193.314	891.262	532.172	-7.315
3900	60.951	332.799	281.670	199.406	892.666	522.705	-7.001
4000	61.017	334.343	282.967	205.504	894.050	513.200	-6.702
4100	61.078	335.851	284.239	211.609	895.413	503.662	-6.417
4200	61.136	337.323	285.485	217.720	896.755	494.091	-6.145
4300	61.190	338.762	286.707	223.836	898.077	484.488	-5.885
4400	61.240	340.170	287.907	229.958	899.377	474.855	-5.637
4500	61.287	341.546	289.083	236.084	900.657	465.192	-5.400
4600	61.331	342.894	290.239	242.215	901.915	455.501	-5.172
4700	61.373	344.213	291.373	248.350	903.152	445.783	-4.954
4800	61.412	345.506	292.487	254.490	904.368	436.039	-4.745
4900	61.449	346.773	293.582	260.633	905.563	426.270	-4.544
5000	61.484	348.014	294.659	266.779	906.736	416.476	-4.351
5100	61.517	349.232	295.717	272.929	907.888	406.660	-4.165
5200	61.548	350.427	296.757	279.083	909.018	396.820	-3.986
5300	61.577	351.600	297.781	285.239	910.127	386.960	-3.814
5400	61.605	352.751	298.788	291.398	911.213	377.079	-3.648
5500	61.632	353.882	299.780	297.560	912.277	367.177	-3.487
5600	61.657	354.992	300.756	303.724	913.319	357.257	-3.332
5700	61.680	356.084	301.717	309.891	914.339	347.318	-3.183
5800	61.703	357.157	302.664	316.060	915.337	337.362	-3.038
5900	61.724	358.212	303.596	322.232	916.312	327.388	-2.898
6000	61.745	359.249	304.515	328.405	917.264	317.398	-2.763

PREVIOUS: December 1970 (1 atm) CURRENT: December 1970 (1 bar)

Formyl, Ion (HCO^+) $C_1H_1O_1^+(g)$

METHINOPHOSPHIDE (CHP) IDEAL GAS M_r = 43.99270

$S°(298.15\ K) = 215.06 \pm 0.04\ J\ K^{-1}\ mol^{-1}$

$\Delta_f H°(0\ K) = [151.2 \pm 63]\ kJ\ mol^{-1}$
$\Delta_f H°(298.15\ K) = [149.9 \pm 63]\ kJ\ mol^{-1}$

Electronic Levels and Quantum Weights

State	ε_i, cm^{-1}	g_i
$X^1\Sigma^+$	0	1
$a^3\Sigma^+$	24440	3
$b^3\Pi$	30430	6
$c^3\Sigma^-$	31024	3
A^1A''	34746	1
$B^1\Pi$	35927	2
$d^3\Pi$	35976	6
C^1A'	40248	1

Vibrational Frequencies and Degeneracies

ν, cm^{-1}
3216.9(1)
674.7(2)
1278.4(1)

Spectroscopic constants used in calculating corrections to the rigid-rotator harmonic oscillator model (cm^{-1}):

$\alpha_1 = 3.4 \times 10^{-3}$ $-x_{11} = [3.7]$ $-x_{12} = [1.9]$
$\alpha_2 = -3.4 \times 10^{-4}$ $-x_{22} = [2.4]$ $-x_{23} = [18.0]$
$\alpha_3 = 3.1 \times 10^{-3}$ $-x_{23} = [51.0]$ $-x_{13} = [8.0]$

Point Group: $C_{\infty v}$ σ = 1
Bond Distances: C-H = 1.0667 Å C-P - 1.5421 Å
Bond Angle: H-C-P = 180°
Rotational Constant: $B_0 = 0.667545\ cm^{-1}$

Enthalpy of Formation

The bonding in HCP and HCN is very similar, as evidenced by the bond lengths and vibrational frequencies. We assume that the C-H bond strength is thus the same in these two molecules. For the reactions HCN + H + CN we derive $\Delta_r H°$ = 123.8 kcal from $\Delta_f H°$(HCN, 298.15 K) = 32.3, $\Delta_f H°$(H, 298.15 K) = 52.1 and $\Delta_f H°$(CN, 298.15 K) = 104 kcal mol^{-1}. Applying this to the reaction HCP + H + CP, we obtain $\Delta_f H°$(HCP, g, 298.15 K) = 40±15 kcal mol^{-1}, from $\Delta_f H°$(CP, 298.15 K) = 111.7 kcal mol^{-1}.

Heat Capacity and Entropy

The molecular configuration was first reported by Gier (1), and later the bond lengths were determined by Tyler (2) using microwave spectroscopy. The vibrational frequencies and electronic levels were obtained from an analysis of the ultraviolet spectrum by Johns et al. (3). Since the first excited state is above 24000 cm^{-1} and the first nonlinear state at 34000 cm^{-1}, there is negligible error introduced by assuming ground state molecular constants for all excited states. The anharmonic constants were estimated from those of HCN by Gordon (4) and the adopted table includes these corrections and the electronic contributions.

References

1. T. E. Gier, J. Amer. Chem. Soc. 83, 1769 (1961).
2. J. K. Tyler, J. Chem. Phys. 40, 1170 (1964).
3. J. W. C. Johns, H. F. Shurvell and J. K. Tyler, Can. J. Phys. 47, 893 (1969).
4. J. S. Gordon, Atlantic Research Corp., Alexandria, Va., personal communication, Jan. 12, 1970.

Methinophosphide (CHP) $C_1H_1P_1(g)$

Enthalpy Reference Temperature = T_r = 298.15 K Standard State Pressure = p° = 0.1 MPa

T/K	$C_p°$ (J K⁻¹mol⁻¹)	S° (J K⁻¹mol⁻¹)	$-[G°-H°(T_r)]/T$ (J K⁻¹mol⁻¹)	$H°-H°(T_r)$ (kJ mol⁻¹)	$\Delta_f H°$ (kJ mol⁻¹)	$\Delta_f G°$ (kJ mol⁻¹)	Log K_f
0	0.	0.	INFINITE	-9.353	151.192	151.192	INFINITE
100	29.196	180.303	244.749	-6.445	151.654	140.524	-73.402
200	32.275	201.236	218.271	-3.407	150.755	129.490	-33.819
298.15	37.166	215.062	215.062	0.	149.900	119.219	-20.887
300	37.254	215.293	215.063	0.069	149.882	119.029	-20.725
400	41.430	226.606	216.577	4.011	148.097	109.082	-14.245
500	44.593	236.205	219.567	8.319	146.984	99.458	-10.390
600	47.011	244.556	223.051	12.903	145.893	90.055	-7.840
700	48.936	251.951	226.661	17.703	144.819	80.834	-6.032
800	50.551	258.594	230.245	22.679	143.765	71.766	-4.686
900	51.944	264.630	233.735	27.806	142.737	62.828	-3.646
1000	53.166	270.167	237.105	33.062	141.736	54.003	-2.821
1100	54.254	275.287	240.346	38.434	140.762	45.277	-2.150
1200	55.225	280.050	243.459	43.909	76.242	37.717	-1.642
1300	56.095	284.505	246.447	49.476	76.091	34.513	-1.387
1400	56.869	288.691	249.316	55.125	75.956	31.319	-1.169
1500	57.568	292.639	252.074	60.847	75.833	28.136	-0.980
1600	58.195	296.374	254.727	66.636	75.721	24.959	-0.815
1700	58.764	299.920	257.282	72.484	75.617	21.790	-0.670
1800	59.279	303.293	259.745	78.387	75.519	18.627	-0.541
1900	59.743	306.511	262.123	84.338	75.426	15.468	-0.425
2000	60.170	309.586	264.419	90.334	75.337	12.315	-0.322
2100	60.563	312.532	266.641	96.371	75.249	9.166	-0.228
2200	60.923	315.358	268.791	102.445	75.162	6.021	-0.143
2300	61.254	318.073	270.876	108.554	75.074	2.881	-0.065
2400	61.568	320.687	272.897	114.696	74.986	-0.256	0.006
2500	61.852	323.206	274.859	120.867	74.896	-3.389	0.071
2600	62.128	325.637	276.766	127.066	74.804	-6.519	0.131
2700	62.383	327.987	278.619	133.292	74.709	-9.645	0.187
2800	62.626	330.260	280.423	139.542	74.611	-12.768	0.238
2900	62.856	332.462	282.180	145.816	74.510	-15.887	0.286
3000	63.078	334.596	283.892	152.113	74.405	-19.002	0.331
3100	63.300	336.668	285.561	158.432	74.296	-22.114	0.373
3200	63.509	338.681	287.190	164.773	74.184	-25.222	0.412
3300	63.718	340.639	288.780	171.134	74.068	-28.326	0.448
3400	63.923	342.544	290.333	177.516	73.949	-31.428	0.483
3500	64.132	344.400	291.852	183.919	73.827	-34.525	0.515
3600	64.337	346.209	293.337	190.342	73.703	-37.619	0.546
3700	64.551	347.975	294.790	196.787	73.577	-40.710	0.575
3800	64.772	349.700	296.212	203.253	73.450	-43.797	0.602
3900	64.990	351.385	297.605	209.741	73.322	-46.881	0.628
4000	65.220	353.033	298.970	216.251	73.195	-49.961	0.652
4100	65.471	354.647	300.309	222.786	73.070	-53.038	0.676
4200	65.714	356.227	301.621	229.345	72.948	-56.113	0.698
4300	65.977	357.777	302.909	235.929	72.830	-59.184	0.719
4400	66.258	359.297	304.174	242.541	72.718	-62.253	0.739
4500	66.547	360.789	305.415	249.181	72.613	-65.319	0.758
4600	66.852	362.255	306.635	255.851	72.518	-68.384	0.777
4700	67.178	363.696	307.834	262.552	72.433	-71.446	0.794
4800	67.513	365.114	309.012	269.287	72.361	-74.506	0.811
4900	67.873	366.509	310.172	276.056	72.304	-77.565	0.827
5000	68.249	367.884	311.312	282.862	72.265	-80.623	0.842
5100	68.647	369.240	312.435	289.706	72.244	-83.681	0.857
5200	69.065	370.577	313.540	296.592	72.245	-86.738	0.871
5300	69.496	371.896	314.629	303.520	72.270	-89.796	0.885
5400	69.948	373.200	315.701	310.492	72.321	-92.854	0.898
5500	70.425	374.487	316.758	317.510	72.401	-95.914	0.911
5600	70.919	375.761	317.801	324.577	72.511	-98.975	0.923
5700	71.433	377.021	318.829	331.695	72.655	-102.038	0.935
5800	71.965	378.268	319.843	338.865	72.834	-105.104	0.947
5900	72.521	379.502	320.843	346.089	73.052	-108.174	0.958
6000	73.090	380.726	321.831	353.369	73.311	-111.248	0.969

PREVIOUS: December 1969 (1 atm) CURRENT: December 1969 (1 bar)

Methinophosphide (CHP) $C_1H_1P_1(g)$

METHYLENE (CH_2) IDEAL GAS M_r = 14.02688

$\Delta_f H°(0\ K)$ = 385.92 ± 4.2 kJ mol^{-1}

$S°(298.15\ K)$ = 193.93 ± 0.5 J K^{-1} mol^{-1} $\Delta_f H°(298.15\ K)$ = 386.39 ± 4.2 kJ mol^{-1}

Electronic Levels, Quantum Weights and Molecular Constants

	ε_i, cm^{-1}	g_i	ν, cm^{-1}	C-H Distance	H-C-H Angle	$I_A I_B I_C$ $g^3 cm^6$
X $^3B_1$	0	3	[2954](1) [1056(1) [3123](1)	1.029 Å	136°	4.506 x 10^{-120}
a $^1A_1$	[2600]	1	[3026](1) [1398](1) [3074](1)	1.11 Å	102.4°	13.53 x 10^{-120}
b $^1B_1$	7100+[2600]	1	[2931](1) [761](1) [3110](1)	1.05 Å	140°	4.373 x 10^{-120}
c $^1A_1$	27700+[2600]	1	[2954](1) [1056](1) [3123](1)	[1.029] Å	[136°]	4.506 x 10^{-120}

Point Group = C_{2v}

Enthalpy of Formation

The electron impact data of Langer et al. (1) was reanalyzed by Prophet (2) to give $\Delta_f H°(CH_2, g, 298.15\ K)$ = 95±5 kcal mol^{-1}. Dibeler and Liston (3) reported the photoionization threshold for the process $CH_3CN \rightarrow CH_2^+ + HCN$ as 14.94 eV (344.55 kcal mol^{-1}). Using auxiliary data from (4) and (5), we obtain $\Delta_f H°(CH_2, g, 0\ K)$ = 95 kcal mol^{-1}, which is a maximum value as the process is likely to occur with excess energy. $\Delta_f H°(CH_2, g, 0\ K)$ values of 95.5 kcal mol^{-1} (a maximum) and 86.8±6.0 kcal mol^{-1} have been obtained by Chupka et al. from the appearance threshold of CH_2^+ from $C_2H_4(g)$ by photoionization (6), and from chemical equilibria measured in a mass spectrometer (7), respectively. Chupka (8) and Dibeler et al. (9) have determined the photoionization threshold for the production of CH_2^+ from CH_4. If the latter value is corrected for rotational energy effects as suggested by Chupka (8), then both studies yield $\Delta_f H°(CH_2, g, 0\ K)$ = 94.6 kcal mol^{-1}. Chupka (8) indicates that this is a maximum value since the process of H_2 elimation is likely to occur with excess energy in the products at the threshold.

Chupka and Lifshitz (10) have obtained the photoionization threshold for CH_2^+ from CH_3 as 15.09±0.03 eV. Combining this with the spectroscopic IP (4), they report $\Delta_{at}H°(CH_2\text{-}H)$ = 4.694±0.033 eV (108.25±0.77 kcal mol^{-1}). Using $\Delta_f H°(CH_3, g, 0\ K)$ = 35.64 kcal mol^{-1} and $\Delta_f H°(H, g, 0\ K)$ = 51.63 kcal mol^{-1} from these tables yields $\Delta_f H°(CH_2, g, 0\ K)$ = 92.24±1.0 kcal mol^{-1}.

Heat Capacity and Entropy

Bernheim et al. (11), in examining the EPR spectra of CH_2 which was produced by the photolysis of diazirine and trapped in xenon at 4 K, concluded that the ground state of CH_2 is a triplet state and it is bent. In similar followup work, they deduced a bond angle of 137.7±0.6 for CD_2 from a measurement of the isotropic component of the ^{13}C hyperfine interaction. Wasserman et al. (12), observing the EPR spectra on photolysis of diazomethane and diazirine at 4K, concluded that CH_2 has a triplet ground state with a probable bond angle of 136±8. Waserman et al. (13) also reported that their investigation of the EPR spectra of CD_2 and CHD supported their earlier conclusions (12). In these latter two works (12, 13), the bond angle was determined from the zero-field parameter. In more recent work by Wasserman et al. (14), the ^{13}C hyperfine structure was observed in the EPR spectrum of ground state $^{13}CD_2$ at 4 K. The authors conclude that ground state CH_2, by analogy, has an angle of 137-143°.

Herzberg and Johns (15) state that the electronic spectra of CH_2 and its isotopic derivatives is not only compatible with a linear ground state but it is also consistent with a bond angle of 136° if an ad hoc assumption of a heterogeneous predissociation of the excited state is made. Recent theoretical calculations indicate a bent ground state (16, 17). We adopt the triplet ground state, X $^3B_1$, as bent with an angle of 136±5° on the basis of the previously mentioned experimental work and weighted by the remarks of Herzberg and Johns (15). The bond length is 1.029 Å as reported by Herzberg (4). The bond lengths and angles for the a $^1A_1$ and b $^1B_1$ states are also from Herzberg and Johns (18).

The singlet electronic levels are firmly established (19). The unobserved singlet-triplet separation is likely to be less then 8000 cm^{-1}, as suggested by Herzberg (19). Many theoretical arguments yield energy separations within 10% of this value (16, 17, 20). Goldberg and Riter (24) estimate a value of 2600±2000 cm^{-1} by calculations which allow for correlation energy effects in different states, all assumed to be linear, however. Foner and Hudson (21) suggest a "small" separation due to the absence of resolved structure in the CH_2 ionization curve. Carr, Eder, and Topor (22) via ketone photolysis deduced an energy gap of roughly 350-700 cm^{-1}. Halberstadt and McNesby (23) suggest 880 cm^{-1} based on ketone photolysis in the presence of methane and propane. The value chosen for the singlet-triplet energy separation is 2600±2000 cm^{-1}. This value is intermediate between the experimental and theoretical suggestions.

The vibrational frequencies for the three lower lying states were calculated via the valence force field method using the force constants calculated by O'Neil, Schaeffer, and Bender (17). The geometry and vibrational levels for the c $^1A_1$ state were estimated. The uncertainty in the entropy mainly reflects the uncertain position of the low lying electronic levels.

References

1. A. Langer, J. A. Hipple, and D. P. Stevenson, J. Chem. Phys. 22, 1836 (1954).
2. H. Prophet, J. Chem. Phys. 38, 2345 (1963).
3. V. H. Dibeler and S. K. Liston, J. Chem. Phys. 48, 4765 (1968).
4. G. Herzberg, Proc. Roy. Soc. (London) A262, 291 (1961).
5. U. S. Nat. Bur. Stand. Tech. Note 270-3, 1968.
6. W. A. Chupka, J. Berkowitz, and K. M. A. Rafaey, J. Chem. Phys. 50, 1938 (1969).
7. W. A. Chupka, J. Berkowitz, and D. J. Meschi, "Advances in Mass Spectrometry," R. M. Elliott, Ed., Pergammon Press, Ltd., London 1963.
8. W. A. Chupka, J. Chem. Phys. 48, 2337 (1968).
9. V. H. Dibeler, M. Krauss, R. M. Reese, and F. N. Harllee, J. Chem. Phys. 42, 3791 (1965).
10. W. A. Chupka and C. Lifshitz, J. Chem. Phys. 48, 2337 (1968).
11. R. A. Bernheim, H. W. Bernard, P. S. Wang, L. S. Wood, and P. S. Skell, J. Chem. Phys. 53, 1280 (1970), 54, 3223 (1971).
12. E. Wasserman, W. A. Yager, and V. J. Kuck, Chem. Phys. Letters 7, 409 (1970).
13. E. Wasserman, V. J. Kuck, R. S. Hutton, and W. A. Yager, J. Am. Chem. Soc. 92, 7491 (1970).
14. E. Wasserman, V. J. Kuck, R. S. Hutton, E. D. Anderson, and W. A. Yager, J. Chem. Phys. 54, 4120 (1971).
15. G. Herzberg and J. W. C. Johns, J. Chem. Phys. 54, 2276 (1971).
16. J. F. Harrison, J. Am. Chem. Soc. 93, 4112 (1971).
17. S. V. O'Neil, H. F. Schaeffer, and C. F. Bender, J. Chem. Phys. 55, 162 (1971).
18. G. Herzberg and J. W. C. Johns, Proc. Roy. Soc. (London) A295, 107 (1966).
19. G. Herzberg, "Electronic Spectra of Polyatomic Molecules," D. Van Nostrand Co., Inc., New York, 1966.
20. J. M. Foster and S. F. Boys, Rev. Mod. Phys. 32, 305 (1960).
21. S. N. Foner and R. L. Hudson, J. Chem. Phys. 45, 49 (1966).
22. R. W. Carr, F. W. Eder, and M. G. Topor, J. Chem. Phys. 53, 4716 (1970).
23. M. L. Halberstadt and J. R. McNesby, J. Am. Chem. Soc. 89, 3417 (1967).
24. M. C. Goldberg and J. R. Riter, J. Chem. Phys. 50, 547 (1969).

Methylene (CH_2) $C_1H_2(g)$

Enthalpy Reference Temperature = T_r = 298.15 K Standard State Pressure = p° = 0.1 MPa

	J $K^{-1}mol^{-1}$			kJ mol^{-1}			
T/K	$C_p°$	S°	$-[G°-H°(T_r)]/T$	$H°-H°(T_r)$	$\Delta_f H°$	$\Delta_f G°$	Log K_f
0	0.	0.	INFINITE	-9.994	385.917	385.917	INFINITE
100	33.258	157.288	223.967	-6.668	386.184	380.623	-198.817
200	33.499	180.376	197.055	-3.336	386.496	374.920	-97.919
298.15	34.600	193.931	193.931	0.	386.392	369.245	-64.690
300	34.627	194 146	193.932	0.064	386.387	369.139	-64.273
400	36.142	204.313	195.308	3.602	385.996	363.443	-47.461
500	37.663	212.541	197.957	7.292	385.437	357.866	-37.386
600	39.206	219.544	200.985	11.135	384.773	352.413	-30.680
700	40.808	225.707	204.085	15.135	384.063	347.076	-25.899
800	42.442	231.263	207.140	19.298	383.351	341.841	-22.320
900	44.048	236.355	210.107	23.623	382.667	336.693	-19.541
1000	45.571	241.075	212.971	28.105	382.022	331.620	-17.322
1100	46.976	245.486	215.728	32.733	381.418	326.609	-15.509
1200	48.247	249.629	218.382	37.495	380.850	321.652	-14.001
1300	49.379	253.536	220.938	42.378	380.313	316.741	-12.727
1400	50.381	257.233	223.399	47.367	379.799	311.870	-11.636
1500	51.264	260.740	225.773	52.450	379.299	307.036	-10.692
1600	52.040	264.073	228.063	57.616	378.809	302.234	-9.867
1700	52.724	267.249	230.276	62.855	378.323	297.463	-9.140
1800	53.327	270.280	232.415	68.158	377.835	292.721	-8.495
1900	53.861	273.178	234.484	73.518	377.343	288.005	-7.918
2000	54.336	275.953	236.489	78.928	376.844	283.316	-7.399
2100	54.760	278.615	238.432	84.384	376.335	278.652	-6.931
2200	55.142	281.171	240.317	89.879	375.815	274.013	-6.506
2300	55.485	283.630	242.147	95.411	375.282	269.398	-6.118
2400	55.797	285.998	243.925	100.975	374.736	264.806	-5.763
2500	56.080	288.282	245.654	106.569	374.174	260.237	-5.437
2600	56.338	290.486	247.336	112.190	373.598	255.690	-5.137
2700	56.575	292.617	248.974	117.836	373.005	251.167	-4.859
2800	56.792	294.678	250.570	123.505	372.396	246.665	-4.602
2900	56.993	296.675	252.125	129.194	371.770	242.186	-4.362
3000	57.177	298.610	253.643	134.903	371.127	237.728	-4.139
3100	57.348	300.488	255.124	140.629	370.466	233.292	-3.931
3200	57.506	302.311	256.570	146.372	369.787	228.878	-3.736
3300	57.652	304.083	257.983	152.130	369.089	224.486	-3.553
3400	57.787	305.806	259.364	157.902	368.372	220.115	-3.382
3500	57.913	307.483	260.715	163.687	367.636	215.765	-3.220
3600	58.030	309.116	262.037	169.484	366.880	211.436	-3.068
3700	58.139	310.708	263.331	175.293	366.103	207.129	-2.924
3800	58.240	312.259	264.598	181.112	365.307	202.843	-2.788
3900	58.334	313.773	265.840	186.940	364.489	198 578	-2.660
4000	58.421	315.251	267.057	192.778	363.650	194.335	-2.538
4100	58.502	316.695	268.250	198.624	362.790	190.113	-2.422
4200	58.577	318.106	269.420	204.478	361.908	185.912	-2.312
4300	58.647	319.485	270.569	210.339	361.003	181.732	-2.208
4400	58.713	320.834	271.696	216.207	360.077	177.573	-2.108
4500	58.773	322.154	272.802	222.082	359.129	173.436	-2.013
4600	58.830	323.446	273.889	227.962	358.158	169.320	-1.923
4700	58.882	324.712	274.957	233.848	357.164	165.226	-1.836
4800	58.931	325.952	276.007	239.738	356.149	161.153	-1.754
4900	58.976	327.168	277.039	245.634	355.111	157.101	-1.675
5000	59.018	328.360	278.053	251.533	354.052	153.071	-1.599
5100	59.058	329.529	279.051	257.437	352.971	149.062	-1.527
5200	59.094	330.676	280.033	263.345	351.868	145.074	-1.457
5300	59.128	331.802	280.999	269.256	350.745	141.108	-1.391
5400	59.160	332.908	281.950	275.170	349.602	137.164	-1.327
5500	59.189	333.993	282.886	281.088	348.438	133.240	-1.265
5600	59.217	335.060	283.809	287.008	347.256	129.338	-1.206
5700	59.242	336.108	284.717	292.931	346.054	125.457	-1.150
5800	59.266	337.139	285.612	298.857	344.835	121.598	-1.095
5900	59.288	338.152	286.494	304.784	343.599	117.759	-1.043
6000	59.309	339.149	287.363	310.714	342.347	113.942	-0.992

PREVIOUS: December 1972 (1 atm) CURRENT: December 1972 (1 bar)

Methylene (CH_2) $C_1H_2(g)$

CHLOROFLUOROMETHANE (CH_2ClF) IDEAL GAS M_r = 68.478283

$\Delta_f H°(0\ K)$ = [-254.649 ± 13] kJ mol^{-1}

S°(298.15 K) = 264.421 ± 0.4 J K^{-1} mol^{-1} $\Delta_f H°(298.15\ K)$ = [-261.918 ± 13] kJ mol^{-1}

Vibrational Frequencies and Degeneracies

ν, cm^{-1}	ν, cm^{-1}	ν, cm^{-1}
2993(1)	1068(1)	3048(1)
1470(1)	760(1)	1236(1)
1351(1)	385(1)	1004(1)

Ground State Quantum Weight: 1 σ = 1

Point Group: C_s

Bond Distances: C-H = 1.078 Å C-Cl = 1.759 Å C-F = 1.378 Å

Bond Angles: H-C-H = 111°54' F-C-Cl = 110° H-C-Cl = 109°06'

Product of the Moments of Inertia: $I_A I_B I_C$ = 4.818560 x 10^{-115} g^3 cm^6

Enthalpy of Formation

The change in enthalpy of atomization by successive replacement of F by Cl, in the CF_4 to CCl_4 series and CHF_3 to $CHCl_3$ series, follows a regular pattern (see CCl_2F_2 and $CHCl_2F$ tables). We make the assumption that this same regular variation occurs also in the CH_2F_2 to CH_2Cl_2 series. Thus, the difference between the atomization energies of CH_2F_2 and CH_2Cl_2 is divided into two unequal parts which fall into the pattern established in the CF_4 and CHF_3 sequences. We adopt an enthalpy of atomization of 385.5 kcal which corresponds to a $\Delta_f H°(CH_2ClF$, g, 298.15 K) = -62.6 kcal mol^{-1}. The changes in kcal, in the enthalpy of atomization by successive replacement of F by Cl are tabulated, below for all three series.

CF_4		
43.8		
CF_3Cl	CHF_3	
41.6	41.4	
CF_2Cl_2	CHF_2Cl	CH_2F_2
38.5	37.3	35.0
$CFCl_3$	$CHFCl_2$	CH_2FCl
36.0	33.0	29.7
CCl_4	$CHCl_3$	CH_2Cl_2

Heat Capacity and Entropy

Plyler and Benedict (1) assigned the vibrational frequencies from the infrared and Raman data for the liquid. The molecular structure was obtained from the microwave measurements of Muller (2). The principal moments of inertia are: I_A = 2.0097 x 10^{-39}, I_B = 14.7640 x 10^{-39}, and I_C = 16.2396 x 10^{-39} g cm^2.

References

1. E. K. Plyler and W. S. Benedict, J. Res. Nat. Bur. Stand. 47, 202 (1951).
2. N. Muller, J. Amer. Chem. Soc. 75, 800 (1953).

Chlorofluoromethane (CH_2ClF) $C_1H_2Cl_1F_1(g)$

Enthalpy Reference Temperature = T_r = 298.15 K Standard State Pressure = p° = 0.1 MPa

T/K	$C_p°$ (J $K^{-1}mol^{-1}$)	S°	-[G°-H°(T_r)]/T	H°-H°(T_r) (kJ mol^{-1})	$\Delta_f H°$	$\Delta_f G°$	Log K_f
0	0.	0.	INFINITE	-11.251	-254.649	-254.649	INFINITE
100	34.287	222.046	301.119	-7.907	-257.273	-251.341	131.287
200	39.571	247.291	268.467	-4.235	-259.604	-244.502	63.857
250	43.143	256.493	265.173	-2.170	-260.791	-240.589	50.268
298.15	47.037	264.421	264.421	0.	-261.918	-236.595	41.451
300	47.192	264.712	264.422	0.087	-261.961	-236.438	41.168
350	51.421	272.304	265.012	2.552	-263.069	-232.096	34.638
400	55.568	279.443	266.374	5.228	-264.093	-227.600	29.722
450	59.472	286.216	268.205	8.105	-265.026	-222.981	25.883
500	63.064	292.671	270.331	11 170	-265.869	-218.264	22.802
600	69.292	304.738	275.075	17.798	-267.301	-208.603	18.160
700	74.417	315.817	280.115	24.991	-268.430	-198.726	14.829
800	78.676	326.041	285.225	32.652	-269.297	-188.707	12.321
900	82.258	335.520	290.293	40.704	-269.942	-178.592	10.365
1000	85.296	344.348	295.262	49.086	-270.401	-168.416	8.797
1100	87.887	352.603	300.104	57.748	-270.706	-158.201	7.512
1200	90.108	360.347	304.805	66.651	-270.887	-147.965	6.441
1300	92.019	367.637	309.361	75.760	-270.968	-137.717	5.534
1400	93.668	374.518	313.771	85.046	-270.972	-127.467	4.756
1500	95.097	381.031	318.040	94.486	-270.915	-117.218	4.082
1600	96.340	387.209	322.172	104.059	-270.812	-106.975	3.492
1700	97.425	393.083	326.172	113.749	-270.675	-96.739	2.972
1800	98.376	398.679	330.046	123.540	-270.514	-86.512	2.511
1900	99.212	404.021	333.799	133.420	-270.335	-76.295	2.097
2000	99.950	409.129	337.439	143.379	-270.145	-66.087	1.726
2100	100.604	414.021	340.970	153.407	-269.948	-55.889	1.390
2200	101.186	418.715	344.398	163.497	-269.749	-45.700	1.085
2300	101.705	423.225	347.728	173.642	-269.549	-35.521	0.807
2400	102.170	427.563	350.965	183.837	-269.352	-25.350	0.552
2500	102.587	431.743	354.113	194.075	-269.159	-15.187	0.317
2600	102.964	435.774	357.176	204.353	-268.971	-5.032	0.101
2700	103.304	439.666	360.160	214.666	-268.789	5.116	-0.099
2800	103.612	443.429	363.067	225.012	-268.614	15.258	-0.285
2900	103.892	447.069	365.901	235.388	-268.445	25.393	-0.457
3000	104.147	450.596	368.666	245.790	-268.285	35.523	-0.619
3100	104.381	454.015	371.364	256.216	-268.131	45.647	-0.769
3200	104.594	457.332	373.999	266.665	-267.985	55.766	-0.910
3300	104.790	460.554	376.573	277.135	-267.847	65.882	-1.043
3400	104.970	463.685	379.090	287.623	-267.715	75.992	-1.167
3500	105.136	466.730	381.550	298.128	-267.590	86.100	-1.285
3600	105.290	469.694	383.958	308.650	-267.470	96.203	-1.396
3700	105.432	472.581	386.314	319.186	-267.356	106.304	-1.501
3800	105.563	475.394	388.622	329.736	-267.247	116.401	-1.600
3900	105.686	478.138	390.882	340.298	-267.142	126.496	-1.694
4000	105.799	480.815	393.097	350.873	-267.040	136.588	-1.784
4100	105.905	483.429	395.268	361.458	-266.941	146.677	-1.869
4200	106.004	485.982	397.398	372.053	-266.843	156.765	-1.950
4300	106.096	488.477	399.487	382.658	-266.747	166.849	-2.027
4400	106.183	490.918	401.537	393.272	-266.652	176.932	-2.100
4500	106.263	493.305	403.550	403.895	-266.556	187.012	-2.171
4600	106.339	495.641	405.527	414.525	-266.459	197.090	-2.238
4700	106.411	497.929	407.469	425.162	-266.360	207.166	-2.302
4800	106.478	500.170	409.377	435.807	-266.259	217.241	-2.364
4900	106.541	502.366	411.252	446.458	-266.154	227.312	-2.423
5000	106.601	504.519	413.096	457.115	-266.046	237.382	-2.480
5100	106.657	506.631	414.909	467.778	-265.933	247.449	-2.534
5200	106.710	508.702	416.693	478.446	-265.814	257.514	-2.587
5300	106.760	510.735	418.448	489.120	-265.690	267.577	-2.637
5400	106.808	512.731	420.176	499.798	-265.560	277.638	-2.686
5500	106.853	514.692	421.877	510.481	-265.422	287.695	-2.732
5600	106.896	516.617	423.551	521.169	-265.276	297.751	-2.777
5700	106.936	518.510	425.201	531.860	-265.121	307.804	-2.821
5800	106.975	520.370	426.826	542.556	-264.958	317.854	-2.863
5900	107.011	522.199	428.427	553.255	-264.784	327.900	-2.903
6000	107.046	523.998	430.005	563.958	-264.600	337.944	-2.942

PREVIOUS: June 1970 (1 atm) CURRENT: June 1970 (1 bar)

Chlorofluoromethane (CH_2ClF) $C_1H_2Cl_1F_1(g)$

DICHLOROMETHANE (CH_2Cl_2) IDEAL GAS M_r = 84.93288

$\Delta_f H°(0\ K) = -88.663 \pm 1.3\ kJ\ mol^{-1}$

$S°(298.15\ K) = 270.293 \pm 0.04\ J\ K^{-1}\ mol^{-1}$ $\Delta_f H°(298.15\ K) = -95.521 \pm 1.3\ kJ\ mol^{-1}$

Vibrational Frequencies and Degeneracies

ν, cm⁻¹	ν, cm⁻¹	ν, cm⁻¹
2999 (1)	282 (1)	898 (1)
1467 (1)	1153 (1)	1268 (1)
717 (1)	3040 (1)	758 (1)

Ground State Quantum Weight: 1 σ = 2

Point Group: C_{2v}

Bond Distances: C-H = 1.068 Å C-Cl = 1.7724 Å

Bond Angles: H-C-H = 112° Cl-C-Cl = 111.8°

Product of the Moments of Inertia: $I_A I_B I_C = 1.869359 \times 10^{114}\ g^3\ cm^6$

Enthalpy of Formation

Hu and Sinke (1) have measured the enthalpy of combustion of $CH_2Cl_2(\ell)$ in the rotating bomb calorimeter, using As_2O_3 solution as the reducing agent. They reported $\Delta_c H°(298.15\ K) = -144.00\ kcal\ mol^{-1}$ for $CH_2Cl_2(\ell) + O_2(g) + CO_2(g) + 2\ HCl(aq, 600\ H_2O)$ which leads to $\Delta_f H°(CH_2Cl_2,\ \ell,\ 298.15\ K) = -29.70\ kcal\ mol^{-1}$, using $\Delta_f H°(298.15\ K) = -94.051$ and $-39.823\ kcal\ mol^{-1}$ for $CO_2(g)$ and HCl(aq, 600 H_2O), respectively (2). The standard deviation of five combustion runs was 0.12 kcal mol^{-1}. Employing the heat of vaporization of $CH_2Cl_2(\ell)$ at 298.15 K as 6.87 kcal mol^{-1} (3), we obtain $\Delta_f H°(CH_2Cl_2,\ g,\ 298.15\ K) = -22.83\ kcal\ mol^{-1}$ ($-95.521 \pm 1.3\ kJ\ mol^{-1}$) which is adopted in the tabulation.

L. Smith et al. (4), based on the combustion data of E. Eftring (5), re-evaluated the enthalpy of combustion of dichloromethane at 18.7°C as $\Delta_c E°/M = -1701.4$ cal/g for $CH_2Cl_2(\ell) + O_2(g) + CO_2(g) + 2\ HCl(aq,\ 600\ H_2O)$. This value needs to be further corrected for a change in the enthalpy of oxidation of As_2O_3 (6) and calculated to 25°C to give $\Delta_c E°(298.15\ K) = \Delta_c H°(298.15\ K) = -144.89\ kcal\ mol^{-1}$, $\Delta_f H°(CH_2Cl_2,\ \ell,\ 298.15\ K) = -28.80\ kcal\ mol^{-1}$ and $\Delta_f H°(CH_2Cl_2,\ g,\ 298.15\ K) = -21.93\ kcal\ mol^{-1}$.

J. Lacher et al. (7) have measured the enthalpy of hydrogenation of dichloromethane to methane and hydrogen chloride at 250°C as $-40.07\ kcal\ mol^{-1}$. Correction to 298.15 K gives $\Delta_f H°(CH_2Cl_2,\ g\ 298.15\ K) = -23.05\ kcal\ mol^{-1}$ which is in good agreement with the value adopted.

Heat Capacity and Entropy

The bond distances and angles have been determined from the microwave spectra by Meyers and Gwinn (8). The principal moments of inertia are: $I_A = 2.6307 \times 10^{-39}$, $I_B = 25.6246 \times 10^{-39}$ and $I_C = 27.7305 \times 10^{-39}\ g\ cm^2$.

The assigned fundamental vibrational frequencies are obtained from Shimanouchi's selection (9), based on infrared and Raman spectra by Shimanouchi and Syzuki (10), Welsh et al. (11), and Palma et al. (12).

References

1. A. T. Hu and G. C. Sinke, presented at the 23rd Calorimetry Conference, Midland, MI, 1968.
2. U. S. Nat. Bur. Stand. Tech. Note 270-3, 1968.
3. J. H. Mathews, J. Amer. Chem. Soc. 48, 562 (1926).
4. L. Smith, L. Bjellerup, S. Krook and H. Westermark, Acta Chem. Scand. 7, 65 (1953).
5. E. Eftring, Ph.D. Dissertation, University of Lund, Lund, Sweden, 1938.
6. S. Sunner and S. Thoren, Acta Chem. Scand. 18, 1528 (1964).
7. J. R. Lacher, A. Amador and J. D. Park, Trans. Faraday Soc. 63, 1608 (1967).
8. R. J. Myers and W. D. Gwinn, J. Chem. Phys. 20, 1420 (1952).
9. T. Shimanouchi, U. S. Nat. Bur. Stand. NSRDS-NBS 6, 1967.
10. T. Shimanouchi and I. Suzuki, J. Mol. Spectrosc. 8, 222 (1962).
11. F. E. Palma, E. A. Piotrowski, S. Sundram and F. F. Cleveland, J. Mol. Spectrosc. 13, 119 (1964).
12. H. L. Welsh, M. F. Crawford, T. R. Thomas and C. R. Love, Can. J. Phys. 30, 577 (1952).

Dichloromethane (CH_2Cl_2) $C_1H_2Cl_2(g)$

Enthalpy Reference Temperature = T_r = 298.15 K Standard State Pressure = p° = 0.1 MPa

T/K	$C_p°$ (J K⁻¹mol⁻¹)	S° (J K⁻¹mol⁻¹)	−[G°−H°(T_r)]/T (J K⁻¹mol⁻¹)	H°−H°(T_r) (kJ mol⁻¹)	$\Delta_f H°$ (kJ mol⁻¹)	$\Delta_f G°$ (kJ mol⁻¹)	Log K_f
0	0.	0.	INFINITE	-11.842	-88.663	-88.663	INFINITE
100	35.760	225.132	309.694	-8.456	-91.248	-84.693	44.239
200	42.366	251.822	274.664	-4.568	-93.419	-77.291	20.186
250	46.584	261.721	271.107	-2.347	-94 507	-73.133	15.280
298.15	50.896	270.293	270.293	0.	-95.521	-68.924	12.075
300	51.062	270.609	270.294	0.094	-95.558	-68.759	11.972
350	55.459	278.813	270.932	2.758	-96.533	-64.214	9.583
400	59.570	286.490	272.402	5.635	-97.416	-59.536	7.775
450	63.309	293.726	274.373	8.709	-98.207	-54.752	6.355
500	66.667	300.574	276.654	11.960	-98.912	-49.885	5.211
600	72.367	313.251	281.715	18.922	-100.090	-39.964	3.479
700	76.997	324.766	287.055	26.398	-100.996	-29.868	2.229
800	80.838	335.306	292.437	34.295	-101.672	-19.658	1.284
900	84.079	345.020	297.747	42.545	-102.151	-9.376	0.544
1000	86.842	354.025	302.930	51.095	-102.465	0.951	-0.050
1100	89.211	362.416	307.961	59.901	-102.643	11.302	-0.537
1200	91.252	370.268	312.829	68.926	-102.708	21.664	-0.943
1300	93.014	377.643	317.534	78.142	-102.685	32.028	-1.287
1400	94.542	384.593	322.078	87.521	-102.591	42.387	-1.581
1500	95.869	391.163	326.467	97.043	-102.442	52.738	-1.836
1600	97.026	397.388	330.707	106.690	-102.252	63.077	-2.059
1700	98.039	403.301	334.804	116.444	-102.032	73.404	-2.255
1800	98.927	408.930	338.767	126.293	-101.791	83.717	-2.429
1900	99.710	414.300	342.603	136.226	-101.537	94.016	-2.585
2000	100.403	419.433	346.317	146.232	-101.278	104.301	-2.724
2100	101.017	424.347	349.916	156.304	-101.019	114.574	-2.850
2200	101.564	429.059	353.407	166.433	-100.765	124.834	-2.964
2300	102.052	433.584	356.796	176.615	-100.523	135.083	-3.068
2400	102.490	437.937	360.086	186.842	-100.295	145.322	-3.163
2500	102.883	442.129	363.285	197.111	-100.086	155.552	-3.250
2600	103.238	446.171	366.395	207.417	-99.899	165.773	-3.330
2700	103.559	450.074	369.423	217.757	-99.738	175.988	-3.405
2800	103.849	453.845	372.371	228.128	-99.603	186.198	-3.474
2900	104.114	457.494	375.244	238.526	-99.499	196.403	-3.538
3000	104.355	461.028	378.044	248.950	-99.425	206.605	-3.597
3100	104.575	464.453	380.777	259.397	-99.383	216.806	-3.653
3200	104.777	467.777	383.444	269.864	-99.373	227.005	-3.705
3300	104.962	471.004	386.049	280.352	-99.395	237.205	-3.755
3400	105.133	474.140	388.594	290.856	-99.450	247.405	-3.801
3500	105.290	477.189	391.082	301.378	-99.536	257.608	-3.845
3600	105.435	480.158	393.515	311.914	-99.651	267.814	-3.886
3700	105.570	483.048	395.896	322.464	-99.796	278.023	-3.925
3800	105.694	485.865	398.226	333.028	-99.967	288.237	-3.962
3900	105.810	488.612	400.509	343.603	-100.164	298.455	-3.997
4000	105.917	491.293	402.745	354.189	-100.383	308.679	-4.031
4100	106.018	493.909	404.937	364.786	-100.624	318.908	-4.063
4200	106.111	496.465	407.086	375.393	-100.883	329.144	-4.094
4300	106.199	498.963	409.194	386.008	-101.159	339.386	-4.123
4400	106.280	501.405	411.262	396.632	-101.448	349.634	-4.151
4500	106.357	503.795	413.292	407.264	-101.750	359.889	-4.177
4600	106.429	506.133	415.284	417.903	-102.061	370.152	-4.203
4700	106 497	508.423	417.242	428.550	-102.379	380.420	-4.228
4800	106.560	510.665	419.165	439.203	-102.702	390.697	-4.252
4900	106.620	512.863	421.055	449.862	-103.028	400.979	-4.274
5000	106.677	515.018	422.913	460.527	-103.355	411.268	-4.296
5100	106.730	517.131	424.739	471.197	-103.682	421.564	-4.318
5200	106.780	519.204	426.536	481.872	-104.005	431.866	-4.338
5300	106.828	521.238	428.304	492.553	-104.325	442.174	-4.358
5400	106.873	523.236	430.043	503.238	-104.637	452.488	-4.377
5500	106.916	525.197	431.756	513.927	-104.943	462.808	-4.395
5600	106.956	527.124	433.441	524.621	-105.239	473.134	-4.413
5700	106.995	529.017	435.102	535.319	-105.524	483.464	-4.430
5800	107.031	530.878	436.737	546.020	-105.798	493.800	-4.447
5900	107.066	532.708	438.348	556.725	-106.058	504.140	-4.463
6000	107.099	534.508	439.936	567.433	-106.304	514.484	-4.479

PREVIOUS. December 1968 (1 atm) CURRENT: December 1968 (1 bar)

Dichloromethane (CH_2Cl_2) $C_1H_2Cl_2(g)$

DIFLUOROMETHANE (CH_2F_2) IDEAL GAS M_r = 52.023686

$S°(298.15\ K) = 246.698 \pm 0.04\ J\ K^{-1}\ mol^{-1}$

$\Delta_fH°(0\ K) = -443.005 \pm 1.7\ kJ\ mol^{-1}$

$\Delta_fH°(298.15\ K) = -450.659 \pm 1.7\ kJ\ mol^{-1}$

Vibrational Frequencies and Degeneracies

ν, cm⁻¹	ν, cm⁻¹	ν, cm⁻¹
2949(1)	529(1)	1176(1)
1508(1)	1262(1)	1435(1)
1116(1)	3013(1)	1090(1)

Ground State Quantum Weight: [1] σ = 2

Point Group: C_{2v}

Bond Distances: C-F = 1.358 ± 0.001 Å C-H = 1.092 ± 0.003 Å

Bond Angles: F-C-F = 108° 17' ± 6' H-C-H = 111° 52' ± 25'

Product of the Moments of Inertia: $I_AI_BI_C = 1.22816 \times 10^{-115}\ g^3\ cm^6$

Enthalpy of Formation

Neugebauer and Margrave (1) measured the enthalpy of combustion of CH_2F_2. Their result for the combustion reaction

$$CH_2F_2(g) + O_2(g) + CO_2(g) + 2HF \cdot 21.5\ H_2O$$

was $\Delta_fH(298.15\ K) = -139.84 \pm 0.22\ kcal\ mol^{-1}$. Using $\Delta_fH°(HF \cdot 21.5\ H_2O,\ 298.15\ K) = -76.75 \pm 0.1\ kcal\ mol^{-1}$ (2), obtained with dilution data compiled by Parker (3), we derive, $\Delta_fH°(CH_2F_2,\ g,\ 298.15\ K) = -107.7 \pm 0.4\ kcal\ mol^{-1}$ ($-450.659 \pm 1.7\ kJ\ mol^{-1}$) from Margrave's data.

Heat Capacity and Entropy

The vibrational frequencies are average values from the infrared studies of Stewart and Nielsen (4) and Plyler and Benedict (5). Principal moments of inertia are calculated from the rotational constants reported by Lide (6) with the result $I_A = 1.7088 \times 10^{-39}$, $I_B = 7.9171 \times 10^{-39}$, and $I_C = 9.0781 \times 10^{-39}\ g\ cm^2$. The molecular structure data are from the microwave studies of Lide (6) and are in excellent agreement with electron-diffraction studies of Brockway, Thornton, and Bartell whose results are reported in the Annual Review of Physical Chemistry (7).

References

1. C. A. Neugebauer and J. L. Margrave, J. Phys. Chem., 62, 1043 (1958).
2. JANAF Thermochemical Tables: HF(g), 12-31-68.
3. V. B. Parker, NSRDS-NBS 2, 1965.
4. H. B. Stewart and H. H. Nielsen, Phys. Rev., 75, 640 (1949).
5. E. K. Plyler and W. S. Benedict, J. Research Nat. Bur. Stand. 47, 202 (1951).
6. D. R. Lide, Jr., J. Amer. Chem. Soc., 74, 3548 (1952).
7. R. L. Livingston, Annual Review of Phys. Chem., 5, 397 (1954).

Difluoromethane (CH_2F_2) $C_1H_2F_2(g)$

Enthalpy Reference Temperature = T_r = 298.15 K Standard State Pressure = p° = 0.1 MPa

T/K	$C_p°$ (J K⁻¹mol⁻¹)	S° (J K⁻¹mol⁻¹)	−[G°−H°(T_r)]/T (J K⁻¹mol⁻¹)	H°−H°(T_r) (kJ mol⁻¹)	$\Delta_fH°$ (kJ mol⁻¹)	$\Delta_fG°$ (kJ mol⁻¹)	Log K_f
0	0.	0.	INFINITE	-10.690	-443.005	-443.005	INFINITE
100	33.494	206.992	280.603	-7.361	-445.642	-439.137	229.381
200	36.678	230.995	250.396	-3.880	-448.111	-431.680	112.743
250	39.472	239.465	247.384	-1.980	-449.405	-427.424	89.305
298.15	42.859	246.698	246.698	0.	-450.659	-423.076	74.121
300	43.000	246.964	246.699	0.079	-450.706	-422.905	73.634
350	46.984	253.888	247.237	2.328	-451.967	-418.171	62.409
400	51.113	260.431	248.481	4.780	-453.153	-413.261	53.966
450	55.154	266.686	250.158	7.438	-454.252	-408.208	47.384
500	58.973	272.697	252.114	10.292	-455.257	-403.037	42.105
600	65.767	284.068	256.503	16.539	-456.991	-392.424	34.164
700	71.458	294.646	261.206	23.408	-458.383	-381.549	28.472
800	76.212	304.508	266.009	30.799	-459.476	-370.494	24.191
900	80.205	313.722	270.804	38.625	-460.313	-359.319	20.854
1000	83.577	322.352	275.532	46.819	-460.936	-348.062	18.181
1100	86.438	330.455	280.161	55.324	-461.384	-336.752	15.991
1200	88.876	338.083	284.673	64.093	-461.691	-325.407	14.165
1300	90.962	345.282	289.061	73.087	-461.885	-314.042	12.618
1400	92.754	352.090	293.322	82.275	-461.991	-302.665	11.293
1500	94.301	358.543	297.457	91.630	-462.030	-291.282	10.143
1600	95.641	364.673	301.468	101.129	-462.016	-279.900	9.138
1700	96.807	370.507	305.359	110.752	-461.963	-268.519	8.251
1800	97.825	376.070	309.134	120.485	-461.881	-257.142	7.462
1900	98.719	381.384	312.798	130.313	-461.776	-245.770	6.757
2000	99.507	386.468	316.355	140.225	-461.655	-234.404	6.122
2100	100.203	391.340	319.811	150.212	-461.520	-223.045	5.548
2200	100.822	396.016	323.169	160.263	-461.373	-211.693	5.026
2300	101.373	400.510	326.435	170.374	-461.215	-200.347	4.550
2400	101.866	404.835	329.612	180.536	-461.047	-189.009	4.114
2500	102.308	409.003	332.704	190.745	-460.868	-177.677	3.712
2600	102.706	413.023	335.717	200.996	-460.677	-166.353	3.342
2700	103.065	416.906	338.652	211.285	-460.473	-155.037	2.999
2800	103.391	420.660	341.514	221.608	-460.255	-143.729	2.681
2900	103.686	424.294	344.307	231.962	-460.022	-132.428	2.385
3000	103.955	427.813	347.032	242.344	-459.773	-121.136	2.109
3100	104.201	431.226	349.693	252.752	-459.507	-109.853	1.851
3200	104.426	434.538	352.293	263.184	-459.223	-98.578	1.609
3300	104.632	437.754	354.834	273.637	-458.922	-87.312	1.382
3400	104.822	440.881	357.319	284.110	-458.602	-76.056	1.168
3500	104.997	443.922	359.750	294.601	-458.264	-64.810	0.967
3600	105.158	446.882	362.130	305.109	-457.908	-53.573	0.777
3700	105.307	449.765	364.459	315.632	-457.534	-42.347	0.598
3800	105.445	452.576	366.741	326.170	-457.143	-31.131	0.428
3900	105.573	455.316	368.978	336.721	-456.735	-19.925	0.267
4000	105.693	457.991	371.170	347.284	-456.310	-8.730	0.114
4100	105.804	460.602	373.319	357.859	-455.870	2.454	-0.031
4200	105.907	463 153	375.428	368.445	-455.415	13.627	-0.169
4300	106.004	465.646	377.497	379.040	-454.947	24.789	-0.301
4400	106.095	468.084	379.528	389.645	-454.465	35.940	-0.427
4500	106.180	470.469	381.523	400.259	-453.970	47.080	-0.546
4600	106.259	472.804	383.482	410.881	-453.465	58.209	-0.661
4700	106.334	475.090	385.407	421.511	-452.948	69.327	-0.770
4800	106.404	477.329	387.299	432.148	-452.421	80.434	-0.875
4900	106.471	479.524	389.158	442.791	-451.885	91.529	-0.976
5000	106.533	481.676	390.987	453.442	-451.340	102.614	-1.072
5100	106.592	483.786	392.786	464.098	-450.787	113.687	-1.164
5200	106.647	485.856	394.556	474.760	-450.226	124.750	-1.253
5300	106.700	487.888	396.298	485.427	-449.658	135.802	-1.338
5400	106.750	489.883	398.013	496.100	-449.083	146.843	-1.420
5500	106.797	491.842	399.701	506.777	-448.501	157.873	-1.499
5600	106.842	493.767	401.364	517.459	-447.912	168.893	-1.575
5700	106.884	495.658	403.001	528.145	-447.317	179.902	-1.649
5800	106.925	497.518	404.615	538.836	-446.716	190.901	-1.719
5900	106.963	499.346	406.205	549.530	-446.108	201.889	-1.787
6000	106.999	501.144	407.772	560.228	-445.494	212.866	-1.853

PREVIOUS: December 1969 (1 atm) CURRENT December 1969 (1 bar)

Difluoromethane (CH_2F_2) $C_1H_2F_2(g)$

FORMALDEHYDE (H_2CO) IDEAL GAS M_r = 30.02628

$S°(298.15 K) = 218.950 \pm 0.4\ J\ K^{-1}\ mol^{-1}$

$\Delta H_f°(0 K) = -112.057 \pm 6.3\ kJ\ mol^{-1}$
$\Delta_f H°(298.15 K) = -115.897 \pm 6.3\ kJ\ mol^{-1}$

Vibrational Levels and Degeneracies

ν, cm^{-1}	ν, cm^{-1}
2766.4 (1)	2843.4(1)
1746.07(1)	1247.4(1)
1500.6 (1)	1163.5(1)

Ground State Quantum Weight: 1 σ = 2
Point Group: C_{2v}
Bond Distances: C-H = 1.12 ± 0.01 Å C-O = 1.21 ± 0.01 Å
Bond Angles: H-C-H = 118 ± 2°
Product of the Moments of Inertia: $I_A I_B I_C = 1.6597 \times 10^{-117}\ g^3\ cm^6$

Enthalpy of Formation

Delepine and Badoche (1) measured the enthalpy of combustion of a formaldehyde polymer. Their value is revised to $\Delta_f H°(298.15 K) = -25.0\ kcal\ mol^{-1}$ based on the following thermochemical cycle:

I	$C(cr) + 1/2O_2(g) + H_2(g) = \frac{1}{n}(CH_2O)_n(cr)$	-42.3 kcal
II	$[1/n]\ (CH_2O)_n(cr) + H_2O = CH_2O(aq)$	2.4
III	$CH_2O(g) = CH_2O(aq)$	14.9

Equation I and II are based upon data listed by Delepine and Badoche (1). For reaction III, Delepine and Badoche obtained 15 $kcal\ mol^{-1}$. Whereas, Walker (2) obtained 14.8 $kcal\ mol^{-1}$. A value of -27.7 $kcal\ mol^{-1}$ is obtained from the combustion experiments of Wartenberg and Lerner-Steinberg (3). Corrections of -0.6 and 0.1 $kcal\ mol^{-1}$ were applied to obtain a enthalpy of reaction at constant pressure and at 298.15 K.

Equilibrium constant determinations of reaction (IV) by Newton and Dodge (4) lead to a $\Delta_f H°(298.15 K) = -29.0\ kcal\ mol^{-1}$.

IV $H_2(g) + CO(g) = H_2CO(g)$

The value obtained from the measurements of Wartenberg and Lerner-Steinberg (3) is believed to be the most reliable.

Heat Capacity and Entropy

Frequencies were measured by Blau and Nielson (5) in the I. R. Microwave determinations of Lawrence and Strandberg (6), and those of Erlandson (7) are in essential agreement. The principal moments of inertia are: $I_A = 0.3085 \times 10^{-39}$, $I_B = 2.1701 \times 10^{-39}$, and $I_C = 2.4786 \times 10^{-39}\ g\ cm^2$.

References

1. M. Delepine and M. Badoche, Compt. Rend. 214, 777 (1942).
2. F. Walker, J. Amer. Chem. Soc. 55, 2821 (1933).
3. H. von Wartenberg and B. Lerner-Steinberg, Z. Angew. Chem. 38, 591 (1925).
4. R. H. Newton and B. F. Dodge, J. Amer. Chem. Soc. 55, 4747 (1933).
5. H. H. Blau and H. H. Nielsen, J. Molec. Spectrosc. 1, 124 (1957).
6. R. B. Lawrence and M. P. W. Strandberg, Phys. Rev. 83, 363 (1951).
7. G. Erlandson, J. Chem. Phys. 25, 579 (1956).

Formaldehyde (H_2CO) $C_1H_2O_1(g)$

Enthalpy Reference Temperature = T_r = 298.15 K Standard State Pressure = p° = 0.1 MPa

T/K	$C_p°$ (J K⁻¹mol⁻¹)	S°	$-[G°-H°(T_r)]/T$	$H°-H°(T_r)$ (kJ mol⁻¹)	$\Delta_f H°$	$\Delta_f G°$	Log K_f
0	0.	0.	INFINITE	-10.021	-112.057	-112.057	INFINITE
100	33.258	182.208	249.156	-6.695	-113.243	-112.630	58.832
200	33.502	205.292	222.109	-3.363	-114.387	-111.598	29.146
250	34.194	212.831	219.525	-1.673	-115.118	-110.819	23.154
298.15	35.402	218.950	218.950	0.	-115.897	-109.921	19.258
300	35.458	219.169	218.951	0.066	-115.928	-109.884	19.132
350	37.203	224.761	219.388	1.880	-116.771	-108.810	16.239
400	39.264	229.861	220.383	3.791	-117.616	-107.615	14.053
450	41.489	234.613	221.703	5.810	-118.445	-106.314	12.341
500	43.766	239.102	223.220	7.941	-119.245	-104.923	10.961
600	48.216	247.478	226.575	12.542	-120.731	-101.917	8.873
700	52.319	255.224	230.121	17.572	-122.039	-98.675	7.363
800	55.983	262.455	233.716	22.991	-123.162	-95.259	6.220
900	59.199	269.239	237.290	28.754	-124.111	-91.712	5.323
1000	61.995	275.624	240.807	34.817	-124.906	-88.068	4.600
1100	64.411	281.649	244.249	41.140	-125.570	-84.351	4.005
1200	66.494	287.345	247.605	47.688	-126.127	-80.578	3.507
1300	68.290	292.740	250.871	54.429	-126.596	-76.763	3.084
1400	69.840	297.859	254.046	61.338	-126.998	-72.914	2.720
1500	71.182	302.724	257.131	68.390	-127.349	-69.039	2.404
1600	72.346	307.356	260.126	75.568	-127.661	-65.141	2.127
1700	73.360	311.773	263.035	82.855	-127.946	-61.225	1.881
1800	74.246	315.992	265.861	90.236	-128.213	-57.292	1.663
1900	75.023	320.028	268.606	97.700	-128.470	-53.345	1.467
2000	75.709	323.894	271.275	105.237	-128.724	-49.385	1.290
2100	76.314	327.602	273.869	112.839	-128.979	-45.411	1.130
2200	76.852	331.165	276.393	120.498	-129.240	-41.426	0.984
2300	77.331	334.592	278.850	128.208	-129.510	-37.428	0.850
2400	77.759	337.893	281.241	135.963	-129.792	-33.419	0.727
2500	78.143	341.075	283.572	143.758	-130.090	-29.397	0.614
2600	78.489	344.146	285.843	151.590	-130.404	-25.363	0.510
2700	78.801	347.115	288.057	159.455	-130.736	-21.317	0.412
2800	79.083	349.986	290.218	167.349	-131.088	-17.258	0.322
2900	79.339	352.765	292.327	175.270	-131.461	-13.186	0.238
3000	79.573	355.459	294.387	183.216	-131.855	-9.101	0.158
3100	79.786	358.072	296.399	191.184	-132.272	-5.002	0.084
3200	79.981	360.608	298.366	199.173	-132.713	-0.890	0.015
3300	80.159	363.072	300.290	207.180	-133.177	3.236	-0.051
3400	80.323	365.467	302.172	215.204	-133.666	7.377	-0.113
3500	80.475	367.798	304.014	223.244	-134.179	11.533	-0.172
3600	80.614	370.067	305.817	231.299	-134.717	15.704	-0.228
3700	80.743	372.277	307.584	239.367	-135.281	19.890	-0.281
3800	80.863	374.432	309.315	247.447	-135.871	24.092	-0.331
3900	80.974	376.534	311.011	255.539	-136.486	28.310	-0.379
4000	81.077	378.586	312.675	263.641	-137.128	32.543	-0.425
4100	81.173	380.589	314.307	271.754	-137.796	36.793	-0.469
4200	81.263	382.546	315.909	279.876	-138.491	41.060	-0.511
4300	81.346	384.459	317.481	288.006	-139.213	45.343	-0.551
4400	81.425	386.330	319.024	296.145	-139.962	49.643	-0.589
4500	81.498	388.161	320.541	304.291	-140.738	53.961	-0.626
4600	81.567	389.953	322.030	312.444	-141.540	58.296	-0.662
4700	81.631	391.708	323.494	320.604	-142.371	62.650	-0.696
4800	81.692	393.427	324.933	328.771	-143.228	67.021	-0.729
4900	81.749	395.112	326.348	336.943	-144.113	71.410	-0.761
5000	81.803	396.764	327.740	345.120	-145.025	75.818	-0.792
5100	81.854	398.384	329.109	353.303	-145.964	80.244	-0.822
5200	81.902	399.974	330.457	361.491	-146.930	84.689	-0.851
5300	81.948	401.535	331.783	369.683	-147.924	89.153	-0.879
5400	81.991	403.067	333.089	377.880	-148.944	93.636	-0.906
5500	82.031	404.572	334.375	386.082	-149.991	98.137	-0.932
5600	82.070	406.050	335.642	394.287	-151.064	102.659	-0.958
5700	82.107	407.503	336.890	402.495	-152.164	107.199	-0.982
5800	82.141	408.932	338.120	410.708	-153.290	111.759	-1.007
5900	82.175	410.336	339.332	418.924	-154.442	116.338	-1.030
6000	82.206	411.717	340.527	427.143	-155.619	120.938	-1.053

PREVIOUS: March 1961 (1 atm) CURRENT March 1961 (1 bar)

Formaldehyde (H_2CO) $C_1H_2O_1(g)$

METHYL (CH_3) IDEAL GAS M_r = 15.03482

$S°(298.15 K) = 194.171 \pm 1.3 J K^{-1} mol^{-1}$

$\Delta_f H°(0 K) = 149.031 \pm 0.8 kJ mol^{-1}$
$\Delta_f H°(298.15 K) = 145.687 \pm 0.8 kJ mol^{-1}$

Electronic Levels and Quantum Weights

ε_i, cm^{-1}	g_i
0	2
46205	2

Vibrational Frequencies and Degeneracies

ν, cm^{-1}
[3002] (1)
580 (1)
[3184] (2)
1383 (2)

Point Group: D_{3h} $\sigma = 6$
Bond Distance: C-H = 1.079 Å
Bond Angle: H-C-H = 120°
Product of the Moments of Inertia: $I_A I_B I_C = 4.99 \times 10^{-119} g^3 cm^6$

Enthalpy of Formation

W. A. Chupka (1) measured the photoionization spectrum of methane and from the onset of CH_3^+ production has reported $\Delta_{at}H°(CH_3\text{-}H) = 103.244 \pm 0.12$ kcal. This value was corrected for the thermal rotation energy of the fragment ions and is the value adopted here.

V. H. Dibeler et al. (2) had earlier reported $\Delta_{at}H°(CH_3\text{-}H) = 101.7 \pm 0.5$ kcal but they did not apply the rotational energy correction, which would bring them closer to Chupka.

The photoionization values are in good agreement with earlier electron impact determinations as summarized by Stevenson (3) who gives $\Delta_{at}H°(CH_3\text{-}H) = 101.9 \pm 1$ kcal. Kinetic determinations have been recently reviewed by Kerr (4) and yield $\Delta_{at}H°(CH_3\text{-}H) = 102.5 \pm 1$ kcal again in excellent agreement.

Heat Capacity and Entropy

The structure, bond length, angles and electronic levels are those reported by Herzberg (5). A set of vibrational frequencies have been reported by Andrews and Pimentel (6) in an argon matrix, of which only two were observed. Milligan and Jacox (7) observed $\nu_2 = 611 cm^{-1}$ in an argon matrix and disagreed with Andrews and Pimentel who reported 730 cm^{-1}. Later Tan and Pimentel (8) agreed that the earlier values were for a methyl radical interacting with an alkali halide molecule. This would affect the ν_2 mode strongest, thus, we adopt the remaining in plane frequencies which appear to be of the correct magnitudes. The ν_2 mode is taken from the analysis of Herzberg (5) since this represents the gas phase molecule and is in reasonable agreement with the matrix value of Milligan and Jacox. The uncertainty in the entropy reflects the uncertainties in the vibrational frequencies.

The principal moments of inertia are: $I_A = I_B = 0.2923 \times 10^{-39}$, $I_C = 0.5846 \times 10^{-39} g cm^2$.

References

1. W. A. Chupka, J. Chem. Phys. 48, 2337 (1968).
2. V. H. Dibeler, M. Krauss, R. M. Reese and F. N. Harllee, J. Chem. Phys. 42, 3791 (1965).
3. D. P. Stevenson, Discuss. Faraday Soc. 10, 35 (1951).
4. J. A. Kerr, Chem. Revs. 66, 465 (1966).
5. G. Herzberg, "Electronic Spectra of Polyatomic Molecules," D. Van Nostrand Co., Inc., New York, 1966.
6. L. Andrews and G. C. Pimentel, J. Chem. Phys. 47, 3637 (1967).
7. D. E. Milligan and M. E. Jacox, J. Chem. Phys. 47, 5146 (1967).
8. L. Y. Tan and G. C. Pimentel, J. Chem. Phys. 48, 5202 (1968).

Methyl (CH_3) $C_1H_3(g)$

Enthalpy Reference Temperature = T_r = 298.15 K Standard State Pressure = p° = 0.1 MPa

T/K	$C_p°$ (J K⁻¹mol⁻¹)	S° (J K⁻¹mol⁻¹)	$-[G°-H°(T_r)]/T$ (J K⁻¹mol⁻¹)	$H°-H°(T_r)$ (kJ mol⁻¹)	$\Delta_f H°$ (kJ mol⁻¹)	$\Delta_f G°$ (kJ mol⁻¹)	Log K_f
0	0.	0.	INFINITE	-10.407	149.031	149.031	INFINITE
100	33.395	155.665	226.465	-7.080	147.800	147.438	-77.014
200	35.638	179.380	197.609	-3.646	146.868	147.432	-38.505
250	37.162	187.495	194.799	-1.826	146.297	147.637	-30.847
298.15	38.693	194.170	194.170	0.	145.687	147.950	-25.920
300	38.753	194.410	194.171	0.072	145.663	147.964	-25.763
350	40.392	200.506	194.649	2.050	144.997	148.400	-22.147
400	42.041	206.008	195.730	4.111	144.320	148.932	-19.449
450	43.668	211.054	197.156	6.254	143.644	149.549	-17.359
500	45.252	215.737	198.783	8.477	142.976	150.241	-15.696
600	48.288	224.258	202 333	13.155	141.682	151.817	-13.217
700	51.174	231.921	206.021	18.130	140.477	153.602	-11.462
800	53.926	238.935	209.703	23.386	139.383	155.554	-10.157
900	56.527	245.438	213.316	28.910	138.411	157.635	-9.149
1000	58.954	251.521	216.836	34.686	137.557	159.817	-8.348
1100	61.187	257.247	220.252	40.694	136.814	162.080	-7.697
1200	63.217	262.659	223.562	46.916	136.167	164.406	-7.156
1300	65.048	267.793	226.769	53.331	135.602	166.783	-6.701
1400	66.690	272.675	229.875	59.919	135.105	169.201	-6.313
1500	68.156	277.327	232.885	66.663	134.662	171.652	-5.977
1600	69.463	281.768	235.802	73.545	134.262	174.131	-5.685
1700	70.628	286.015	238.632	80.551	133.896	176.634	-5.427
1800	71.667	290.082	241.378	87.667	133.554	179.158	-5.199
1900	72.594	293.982	244.045	94.881	133.230	181.701	-4.995
2000	73.422	297.727	246.636	102.182	132.917	184.260	-4.812
2100	74.165	301.327	249.155	109.562	132.610	186.835	-4.647
2200	74.831	304.793	251.606	117.013	132.305	189.424	-4.497
2300	75.430	308.133	253.991	124.526	131.999	192.027	-4.361
2400	75.971	311.355	256.315	132.097	131.688	194.644	-4.236
2500	76.459	314.466	258.579	139.719	131.369	197.273	-4.122
2600	76.902	317.474	260.787	147.387	131.041	199.916	-4 016
2700	77.304	320.384	262.940	155.098	130.701	202.571	-3.919
2800	77.670	323.202	265.042	162.847	130.348	205.239	-3.829
2900	78.003	325.933	267.095	170.630	129.980	207.921	-3.745
3000	78.309	328.583	269.101	178.446	129.595	210.615	-3.667
3100	78.588	331.155	271.061	186.291	129.193	213.322	-3.594
3200	78.845	333.655	272.979	194.163	128.772	216.042	-3.527
3300	79.082	336.084	274.854	202.060	128.331	218.777	-3.463
3400	79.300	338.449	276.690	209.979	127.869	221.524	-3.403
3500	79.501	340.750	278.488	217.919	127.385	224.286	-3.347
3600	79.687	342.992	280.248	225.879	126.879	227.061	-3.295
3700	79.860	345.178	281.974	233.856	126.349	229.852	-3.245
3800	80.020	347.310	283.665	241.850	125.796	232.656	-3.198
3900	80.169	349.391	285.324	249.860	125.217	235.476	-3.154
4000	80.308	351.422	286.951	257.884	124.613	238.311	-3.112
4100	80.438	353.407	288.548	265.921	123.984	241.161	-3.072
4200	80.559	355.346	290.115	273.971	123.328	244.027	-3.035
4300	80.672	357.243	291.654	282.033	122.645	246.909	-2.999
4400	80.778	359.099	293.166	290.105	121.936	249.807	-2.966
4500	80.878	360.916	294.652	298.188	121.199	252.721	-2 934
4600	80.972	362.694	296.112	306.281	120.436	255.652	-2.903
4700	81.060	364.437	297.547	314.382	119.645	258.600	-2.874
4800	81.143	366.144	298.958	322.492	118.827	261.565	-2.846
4900	81.221	367.818	300.347	330.611	117.982	264.547	-2.820
5000	81.295	369.460	301.712	338.737	117.111	267.548	-2.795
5100	81.365	371.070	303.057	346.870	116.214	270.565	-2.771
5200	81.432	372.651	304.380	355.009	115.292	273.601	-2.748
5300	81.495	374.203	305.683	363.156	114.345	276.654	-2.727
5400	81.554	375.726	306.966	371.308	113.373	279.726	-2.706
5500	81.611	377.223	308.230	379.467	112.379	282.815	-2.686
5600	81.666	378.694	309.475	387.630	111.362	285.923	-2.667
5700	81.718	380.140	310.702	395.800	110.324	289.050	-2.649
5800	81.767	381.562	311.911	403.974	109.266	292.195	-2.631
5900	81.815	382.960	313.104	412.153	108.188	295.357	-2.615
6000	81.860	384.336	314.280	420.337	107.094	298.539	-2.599

PREVIOUS: June 1969 (1 atm) CURRENT: June 1969 (1 bar)

Methyl (CH_3) $C_1H_3(g)$

CHLOROMETHANE (CH_3Cl) IDEAL GAS M_r = 50.48782

$\Delta_f H°(0\ K) = -75.753 \pm 2.1\ kJ\ mol^{-1}$

$S°(298.15\ K) = 234.367 \pm 0.42\ J\ K^{-1}\ mol^{-1}$ $\Delta_f H°(298.15\ K) = -83.680 \pm 2.1\ kJ\ mol^{-1}$

Vibrational Frequencies and Degeneracies

ν, cm^{-1}	ν, cm^{-1}
2968(1)	3054(2)
1355(1)	1455(2)
732(1)	1017(2)

Ground State Quantum Weight: 1 σ = 3

Point Group: C_{3v}

Bond Distances: C-H = 1.0959 Å C-Cl = 1.7812 Å

Bond Angles: H-C-H = 108° H-C-Cl = 110°55'

Product of the Moments of Inertia: $I_A I_B I_C = 2.14799 \times 10^{-116}\ g^3\ cm^6$

Enthalpy of Formation

The enthalpy of formation has been determined by hydrogenation of methyl chloride to give methane and hydrogen chloride by Lacher, et al. (1), and by Lacher, et al. (2). They reported an average enthalpy of hydrogenation at 248°C to be -19,665±45 cal K^{-1} mol^{-1}. Using this value, together with JANAF data (5) for H_2(g), CH_4(g) and HCl(g), the enthalpy of formation of methyl chloride is calculated to be $\Delta_f H°(298.15\ K) = -20.69\ kcal\ mol^{-1}$. Fowell, et al. [3], using a similar hydrogenation technique almost ten years later, report a value $\Delta_f H°(CH_3Cl, g, 298.15\ K) = -20.741\ kcal\ mol^{-1}$.

Fletcher and Pilcher (4) derived the enthalpy of formation by measuring the enthalpy of combustion of methyl chloride by flame calorimetry. They reported a value $\Delta_f H°(CH_3Cl, g, 298.15\ K) = -19.59 \pm 0.16\ kcal\ mol^{-1}$ where the error, according to the authors, is twice the standard deviation of the mean including the uncertainty in calibration, measurement, determination of ignition energy, and values of auxiliary data used.

In support of their results, Fletcher and Pilcher (4) have compared the enthalpies of formation of C_2H_5Cl(g) and i-C_3H_7Cl(g) as determined by equilibrium studies (2nd and 3rd law) with their work via flame calorimetry. The agreement is excellent (<0.2 kcal mol^{-1} difference) but in both cases results for hydrogenation studies are at least 1 kcal mol^{-1} less negative. In the case of CH_3Cl(g) no equilibrium results are currently available and the results of $\Delta_f H°$(298.15 K) from hydrogenation studies are approximately 1 kcal mol^{-1} more negative than the results from flame calorimetry.

The value chosen as the enthalpy of formation is $\Delta_f H°(CH_3Cl, g, 298.15\ K) = -20.0 \pm 0.5\ kcal\ mol^{-1}$. This value is close to the median value and is arbitrarily chosen as it is uncertain whether the flame calorimetry or hydrogenation studies are the more accurate. Slightly more weight was given to the flame calorimetry work due to favorable comparisons with equilibrium studies in related compounds. An error of ± 0.5 kcal mol^{-1} is chosen so as to encompass the current $\Delta_f H°$(298.15 K) experimental values even though the actual experimental error is considerably less.

Heat Capacity and Entropy

The vibrational frequencies are essentially typical values which were taken from the experimental work of Plyler and Benedict (6), Reichman and Overend (7), Morino and Nakamura (8), Jones, et al. (9), Flether (10), and King, et al. (11). The value of ω_1 as given by Plyler and Benedict (6) appears too low and was not included. Costain (12) has calculated the bond distances and angles based on a isotopic substitution method. From this latter data, the moments of inertia are calculated to be: $I_A = 0.5263 \times 10^{-39}$ and $I_B = I_C = 6.3887 \times 10^{-39}\ g\ cm^2$.

References

1. J. R. Lacher, A. Kianpour, F. Oetting, and J. D. Park, Trans. Faraday Soc. 52, 1500 (1956).
2. J. R. Lacher, E. Emery, E. Bohmfalk, and J. D. Park, J. Phys. Chem. 60, 492 (1956).
3. P. Fowell, J. R. Lacher, and J. D. Park, Trans. Faraday Soc. 61, 1324 (1965).
4. R. A. Fletcher and G. Pilcher, Trans. Faraday Soc. 67, 3191 (1971).
5. JANAF Thermochemical Tables: H_2(g), 3-31-61; CH_4(g), 3-31-61; HCl(g), 9-30-64.
6. E. K. Plyler and W. S. Benedict, J. Res. Nat. Bur. Stand. 47, 202 (1951).
7. S. Richman, and J. Overend, J. Chem. Phys. 48, 3095 (1968).
8. Y. Morino and J. Nakamura, Bull. Chem. Soc. Japan 38, 443 (1965).
9. E. W. Jones, R. J. L. Popplewell, and H. W. Thompson, Spectrochem. Acta 22, 669 (1966).
10. W. H. Fletcher, "Essays in Structural Chemistry," A. J. Downs, D. A. Long, and L. A. K. Staveley, eds., Ch. 3, pp. 73-74, Plenum Press, New York.
11. W. T. King, I. M. Mills, and B. Crawford, Jr., J. Chem. Phys. 27, 455 (1957).
12. C. C. Costain, J. Chem. Phys. 29, 864 (1958).

Chloromethane (CH_3Cl) $C_1H_3Cl_1$(g)

Enthalpy Reference Temperature = T_r = 298.15 K Standard State Pressure = p° = 0.1 MPa

T/K	$C_p°$ (J K⁻¹mol⁻¹)	S° (J K⁻¹mol⁻¹)	-[G°-H°(T_r)]/T (J K⁻¹mol⁻¹)	H°-H°(T_r) (kJ mol⁻¹)	$\Delta_f H°$ (kJ mol⁻¹)	$\Delta_f G°$ (kJ mol⁻¹)	Log K_f
0	0.	0.	INFINITE	-10.415	-75.753	-75.753	INFINITE
100	33.284	195.966	266.858	-7.089	-78.441	-73.383	38.331
200	35.151	219.389	237.889	-3.700	-80.938	-67.379	17.598
250	37.631	227.482	235.019	-1.884	-82.322	-63.831	13.337
298.15	40.731	234.367	234.367	0.	-83.680	-60.146	10.537
300	40.859	234.619	234.368	0.075	-83.732	-60.000	10.447
350	44.466	241.186	234.878	2.208	-85.103	-55.936	8.348
400	48.155	247.365	236.056	4.523	-86.401	-51.679	6.749
450	51.745	253.245	237.642	7.022	-87.609	-47.266	5.486
500	55.145	258.875	239.486	9.695	-88.725	-42.722	4.463
600	61.284	269.485	243.613	15.523	-90.685	-33.332	2.902
700	66.605	279.341	248.021	21.924	-92.300	-23.640	1.764
800	71.239	288.545	252.518	28.822	-93.602	-13.740	0.897
900	75.294	297.175	257.005	36.153	-94.627	-3.692	0.214
1000	78.845	305.296	261.432	43.864	-95.414	6.456	-0.337
1100	81 952	312.960	265.771	51.907	-95.998	16.672	-0.792
1200	84.669	320.209	270.009	60.241	-96.413	26.935	-1.172
1300	87.043	327.082	274.137	68.829	-96.691	37.226	-1.496
1400	89.119	333.611	278.154	77.640	-96.857	47.534	-1.774
1500	90.936	339.823	282.060	86.644	-96.935	57.851	-2.015
1600	92.529	345.743	285.856	95.819	-96.941	68.170	-2.226
1700	93.929	351.396	289.547	105.144	-96.894	78.488	-2 412
1800	95.162	356.800	293.134	114.599	-96.804	88.803	-2.577
1900	96.253	361.975	296.622	124.171	-96.684	99.111	-2.725
2000	97.219	366.937	300.015	133.846	-96.543	109.412	-2.858
2100	98.079	371.702	303.316	143.612	-96.388	119.706	-2.978
2200	98.845	376.283	306.529	153.459	-96.227	129.993	-3.086
2300	99.531	380.692	309.658	163.378	-96.064	140.272	-3.186
2400	100.147	384.941	312.707	173.362	-95.906	150.544	-3.277
2500	100.701	389.041	315.679	183.405	-95.755	160.810	-3.360
2600	101.201	393.000	318.577	193.501	-95.617	171.069	-3.437
2700	101 654	396.828	321.405	203.644	-95.495	181.324	-3.508
2800	102.065	400.533	324.165	213.830	-95.390	191.575	-3.574
2900	102.439	404.121	326.860	224.056	-95.306	201.822	-3.635
3000	102.780	407.600	329.494	234.317	-95.244	212.067	-3.692
3100	103.092	410.975	332.068	244.611	-95.206	222.309	-3.746
3200	103.378	414.253	334.586	254.934	-95.194	232.551	-3.796
3300	103.640	417.438	337.048	265.286	-95.207	242.794	-3.843
3400	103.882	420.535	339.458	275.662	-95.247	253.037	-3.887
3500	104.105	423.550	341.818	286.061	-95.314	263.281	-3.929
3600	104.311	426.486	344.129	296.482	-95.408	273.528	-3.969
3700	104.502	429.346	346.394	306.923	-95.530	283.778	-4.006
3800	104.678	432.135	348.614	317.382	-95.678	294.032	-4.042
3900	104.843	434.857	350.790	327.858	-95.852	304.289	-4.075
4000	104.996	437.513	352.925	338.350	-96.053	314.552	-4.108
4100	105.138	440.107	355.020	348.857	-96.278	324.820	-4.138
4200	105.271	442.643	357.076	359.378	-96.529	335.094	-4.167
4300	105.395	445.121	359.095	369.911	-96.802	345.373	-4.195
4400	105.512	447.545	361.078	380.456	-97.098	355.660	-4.222
4500	105.621	449.918	363.026	391.013	-97.416	365.953	-4.248
4600	105.723	452.240	364.940	401.580	-97.755	376.254	-4.272
4700	105.819	454.515	366.822	412.158	-98.113	386.562	-4.296
4800	105.910	456.744	368.672	422.744	-98.489	396.879	-4.319
4900	105.995	458.929	370.492	433.339	-98.883	407.203	-4.341
5000	106.076	461.071	372.282	443.943	-99.293	417.536	-4.362
5100	106.151	463.172	374.044	454.554	-99.717	427.876	-4.382
5200	106.223	465.234	375.778	465.173	-100.156	438.225	-4.402
5300	106.291	467.258	377.485	475.799	-100.606	448.583	-4.421
5400	106.355	469.246	379.166	486.431	-101.068	458.950	-4.439
5500	106.416	471.198	380.821	497.070	-101.539	469.324	-4.457
5600	106.474	473.116	382.452	507.714	-102.018	479.708	-4.475
5700	106.529	475.001	384.060	518.364	-102.505	490.101	-4.491
5800	106.581	476.854	385.643	529.020	-102.997	500.502	-4.508
5900	106.630	478.676	387.205	539.680	-103.493	510.910	-4.523
6000	106.677	480.469	388.744	550.346	-103.992	521.328	-4.539

PREVIOUS June 1972 (1 atm) CURRENT June 1972 (1 bar)

Chloromethane (CH_3Cl) $C_1H_3Cl_1$(g)

TRICHLOROMETHYLSILANE ($Si(CH_3)Cl_3$) IDEAL GAS M_r = 149.47932

$S°(298.15\ K) = 351.147\ J\ K^{-1}\ mol^{-1}$

$\Delta_f H°(0\ K)$ = Unknown

$\Delta_f H°(298.15\ K) = [-528.858]\ kJ\ mol^{-1}$

Enthalpy of Formation

Source unknown.

Heat Capacity and Entropy

$C_p°$ values are taken from Janz et al. (1).

Reference

1. G. J. Janz, Y. Mikawa, and F. Behnke, 'Silanes Vibrational Spectra, Molecular Parameters, and Thermodynamic Properties,' Rensselaer Polytechnic Institute, Troy, N. Y., 1960.

Trichloromethylsilane ($SiCH_3Cl_3$) $C_1H_3Cl_3Si_1(g)$

Enthalpy Reference Temperature = T_r = 298.15 K; Standard State Pressure = p° = 0.1 MPa

	J K⁻¹mol⁻¹			kJ mol⁻¹			
T/K	$C_p°$	S°	$-[G°-H°(T_r)]/T$	$H°-H°(T_r)$	$\Delta_f H°$	$\Delta_f G°$	Log K_f
0							
100							
200							
298.15	102.382	351.147	351.147	0.	-528.858	-409.577	71.756
300	102.634	351.781	351.149	0.190	-528.975	-408.836	71.185
400	114.725	383.033	355.328	11.082	-535.150	-367.848	48.036
500	123.888	409.653	363.596	23.029	-540.932	-325.346	33.989
600	131.127	432.901	373.250	35.791	-546.360	-281.713	24.525
700	137.026	453.570	383.275	49.206	-551.487	-237.196	17.700
800	141.963	472.197	393.245	63.161	-556.369	-191.962	12.534
900	146.231	489.170	402.974	77.576	-561.055	-146.128	8.481
1000	149.871	504.769	412.384	92.385	-565.597	-99.780	5.212
1100	153.051	519.206	421.447	107.535	-570.038	-52.983	2.516
1200	155.770	532.643	430.160	122.979	-574.422	-5.784	0.252
1300	158.155	545.207	438.532	138.678	-578.783	41.779	-1.679
1400	160.205	557.004	446.577	154.598	-583.150	89.678	-3.346
1500	162.004	568.120	454.313	170.711	-587.547	137.891	-4.802

PREVIOUS: December 1960 (1 atm) CURRENT: December 1960 (1 bar)

Trichloromethylsilane ($SiCH_3Cl_3$) $C_1H_3Cl_3Si_1(g)$

FLUOROMETHANE (CH_3F) IDEAL GAS M_r = 34.033223

$S°(298.15\ K) = 222.843\ J\ K^{-1}\ mol^{-1}$

$\Delta_fH°(0\ K) = -226.3 \pm 33\ kJ\ mol^{-1}$
$\Delta_fH°(298.15\ K) = -234.3 \pm 29\ kJ\ mol^{-1}$

Vibrational Frequencies and Degeneracies

ν, cm^{-1}	ν, cm^{-1}
2965 (1)	3006 (2)
1464 (1)	1467 (2)
1049 (1)	1182 (2)

Ground State Quantum Weight: 1 $\sigma = 3$

Point Group: C_{3v}

Bond Distances: C-F = 1.391 ± 0.005 Å C-H = 1.095 ± 0.010 Å

Bond Angle: H-C-H = 109.5° ± 2°

Product of the Moments of Inertia: $I_AI_BI_C = 5.8630 \times 10^{-117}\ g^3\ cm^6$

Enthalpy of Formation

Appearance potentials for CH_3^+ from CH_4 and CH_3F reported by Lossing, Ingold, and Henderson (1), the enthalpies of dissociation of hydrogen (104.20 kcal) and fluorine (37.72 kcal), and the enthalpy of formation of methane (-17.90 kcal) were used to calculate the enthalpy of formation of gaseous CH_3F. These authors measured an appearance potential for CH_3^+ from CH_3F of 14.6 ± 0.6 eV, which would give an uncertainty of about ±15 kcal mol^{-1} in the enthalpy of formation of methyl fluoride. However, a correlation of the enthalpy of formation from the atoms of the methyl fluorides and methyl chlorides indicates that the uncertainty can be reduced to ± 7 kcal mol^{-1}.

Heat Capacity and Entropy

Fundamental vibrational frequencies were taken from the summary reported by Smith and Mills (2). Moments of inertia from the microwave results of Gilliam, Edwards, and Gordy (3) are: $I_A = 0.5354 \times 10^{-39}$, $I_B = I_C = 3.3093 \times 10^{-39}$ g, cm^2. Bond parameters are the vlaues of Brockway, Thornton, and Bartell (4) from electron diffraction results. These values are in close agreement with the microwave parameters reported by Gilliam, Edwards, and Gordy (3).

References

1. F. P. Lossing, K. V. Ingold, and I. H. S. Henderson, J. Chem. Phys. 22, 1489 (1954).
2. W. L. Smith and I. M. Mills, J. Mol. Spectrosc. 11, 11 (1963).
3. O. R. Gilliam, H. D. Edwards, and W. Gordy, Phys. Rev. 75, 1014 (1949).
4. L. O. Brockway, C. G. Thornton, and L. S. Bartell, Ann. Rev. Phys. Chem. 5, 397 (1954).

Fluoromethane (CH_3F) $C_1H_3F_1(g)$

Enthalpy Reference Temperature = T_r = 298.15 K Standard State Pressure = p° = 0.1 MPa

T/K	$C_p°$	S°	$-[G°-H°(T_r)]/T$	$H°-H°(T_r)$	$\Delta_fH°$	$\Delta_fG°$	Log K_f
	J $K^{-1}mol^{-1}$			kJ mol^{-1}			
0	0.	0.	INFINITE	-10.135	-226.275	-226.275	INFINITE
100	33.258	185.647	253.738	-6.809	-228.961	-223.802	116.902
200	33.825	208.776	226.124	-3.470	-231.453	-217.703	56.858
250	35.227	216.456	223.446	-1.748	-232.872	-214.103	44.734
298.15	37.500	222.843	222.843	0.	-234.304	-210.359	36.854
300	37.603	223.075	222.843	0.069	-234.359	-210.210	36.601
350	40.695	229.097	223.312	2.025	-235.843	-206.068	30.754
400	44.167	234.755	224.392	4.145	-237.275	-201.716	26.341
450	47.751	240.164	225.846	6.443	-238.629	-197.188	22.889
500	51.280	245.378	227.540	8.919	-239.894	-192.515	20.112
600	57.871	255.322	231.350	14.383	-242.139	-188.822	15.916
700	63.705	264.690	235.450	20.468	-244.010	-172.783	12.893
800	68.812	273.538	239.663	27.100	-245.533	-162.500	10.610
900	73.269	281.906	243.896	34.209	-246.747	-152.045	8.824
1000	77.150	289.831	248.097	41.734	-247.695	-141.470	7.390
1100	80.526	297.347	252.236	49.622	-248.420	-130.810	6.212
1200	83.459	304.482	256.295	57.825	-248.959	-120.093	5.228
1300	86.009	311.265	260.265	66.301	-249.347	-109.338	4.393
1400	88.227	317.723	264.140	75.016	-249.614	-98.557	3.677
1500	90.161	323.877	267.919	83 937	-249.784	-87.761	3.056
1600	91.851	329.751	271.601	93.040	-249.878	-76.956	2.512
1700	93.332	335.365	275.188	102.300	-249.912	-66.147	2.032
1800	94.633	340.737	278.682	111.700	-249.900	-55.338	1.606
1900	95.780	345.885	282.084	121.222	-249.852	-44.530	1.224
2000	96.795	350.824	285.399	130.852	-249.777	-33.725	0.881
2100	97.696	355.569	288.628	140.577	-249.682	-22.925	0.570
2200	98.499	360.133	291.775	150.388	-249.571	-12.130	0.288
2300	99.216	364.527	294.843	160.274	-249.449	-1.339	0.030
2400	99.858	368.764	297.835	170.228	-249.318	9.445	-0.206
2500	100.436	372.852	300.755	180.243	-249.180	20.224	-0.423
2600	100.957	376.802	303.604	190.314	-249.038	30.998	-0.623
2700	101.429	380.621	306.386	200.433	-248.892	41.765	-0.808
2800	101.856	384.317	309.104	210.598	-248.743	52.528	-0.980
2900	102.245	387.899	311.760	220.803	-248.593	63.285	-1.140
3000	102.599	391.371	314.356	231.046	-248.441	74.037	-1.289
3100	102.923	394.741	316.895	241.322	-248.290	84.783	-1.429
3200	103.220	398.013	319.379	251.630	-248.138	95.525	-1.559
3300	103.492	401.193	321.810	261.965	-247.988	106.262	-1.682
3400	103.743	404.287	324.191	272.327	-247.839	116.995	-1.797
3500	103.974	407.297	326.522	282.713	-247.693	127.723	-1.906
3600	104.187	410.229	328.807	293.122	-247.550	138.447	-2.009
3700	104.385	413.087	331.046	303.550	-247.410	149.168	-2.106
3800	104.568	415.873	333.242	313.998	-247.276	159.884	-2.198
3900	104.738	418.591	335.396	324.463	-247.147	170.597	-2.285
4000	104.896	421.245	337.509	334.945	-247.024	181.307	-2.368
4100	105.044	423.837	339.583	345.442	-246.908	192.014	-2.446
4200	105.181	426.370	341.619	355.954	-246.800	202.718	-2.521
4300	105.310	428.847	343.619	366.478	-246.700	213.420	-2.593
4400	105.430	431.269	345.584	377.015	-246.609	224.119	-2.661
4500	105.543	433.640	347.514	387.564	-246.528	234.816	-2.726
4600	105.649	435.961	349.412	398.124	-246.457	245.512	-2.788
4700	105.748	438.234	351.278	408.694	-246.397	256.206	-2.847
4800	105.842	440.461	353.113	419.273	-246.348	266.900	-2.904
4900	105.930	442.644	354.918	429.862	-246.309	277.592	-2.959
5000	106.013	444.785	356.694	440.459	-246.282	288.284	-3.012
5100	106.091	446.885	358.442	451.064	-246.266	298.974	-3.062
5200	106.165	448.946	360.162	461.677	-246.261	309.666	-3.111
5300	106.235	450.969	361.857	472.297	-246.267	320.357	-3.157
5400	106.302	452.956	363.525	482.924	-246.283	331.048	-3.202
5500	106.364	454.907	365.169	493.557	-246.310	341.739	-3.246
5600	106.424	456.824	366.789	504.197	-246.347	352.431	-3.287
5700	106.481	458.708	368.385	514.842	-246.392	363.124	-3.328
5800	106.534	460.560	369.958	525.493	-246.446	373.818	-3.367
5900	106.585	462.382	371.509	536.149	-246.508	384.512	-3.404
6000	106.634	464.174	373.039	546.810	-246.576	395.208	-3.441

PREVIOUS: December 1963 (1 atm) CURRENT: December 1963 (1 bar)

Fluoromethane (CH_3F) $C_1H_3F_1(g)$

TRIFLUOROMETHYLSILANE ($Si(CH_3)F_3$) IDEAL GAS M_r = 100.115529

$S°$(298.15 K) = 314.2 J K^{-1} mol^{-1}

$\Delta_f H°$(0 K) = Unknown

$\Delta_f H°$(298.15 K) = [-1232.71] kJ mol^{-1}

Enthalpy of Formation

Source unknown.

Heat Capacity and Entropy

Data from Janz et al. (1).

Reference

1. G. J. Janz, Y. Mikawa, and F. Behnke, "Silanes. Vibrational Spectra, Molecular Parameters, and Thermodynamic Properties," Rensselaer Polytechnic Institute, Troy, N. Y., 1960.

Trifluoromethylsilane ($SiCH_3F_3$) $C_1H_3F_3Si_1$(g)

Enthalpy Reference Temperature = T_r = 298.15 K Standard State Pressure = p° = 0.1 MPa

	J $K^{-1}mol^{-1}$			kJ mol^{-1}			
T/K	$C_p°$	S°	$-[G°-H°(T_r)]/T$	$H°-H°(T_r)$	$\Delta_f H°$	$\Delta_f G°$	Log K_f
0	0.	0.	INFINITE	0.	-1189.803	-1189.803	INFINITE
100	52.770	238.288	382.810	-14.452	-1217.939	-1185.515	619.250
200	73.670	281.472	321.981	-8.102	-1225.568	-1150.071	300.368
298.15	90.876	314.202	314.202	0.	-1232.711	-1111.489	194.728
300	91.169	314.765	314.204	0.168	-1232.842	-1110.736	193.396
400	105.269	342.993	317.957	10.014	-1239.687	-1068.988	139.596
500	116.483	367.732	325.489	21.122	-1246.001	-1025.572	107.141
600	125.311	389.777	334.400	33.226	-1251.848	-980.930	85.398
700	132.382	409.640	343.753	46.121	-1257.315	-935.340	69.796
800	138.239	427.710	353.135	59.660	-1262.478	-888.988	58.045
900	143.135	444.282	362.355	73.735	-1267.405	-842.003	48.869
1000	147.277	459.582	371.322	88.260	-1272.160	-794.480	41.499
1100	150.875	473.793	379.999	103.173	-1276.794	-746.487	35.448
1200	153.888	487.052	388.374	118.414	-1281.357	-698.076	30.386
1300	156.523	499.476	396.448	133.937	-1285.889	-649.285	26 089
1400	158.825	511.161	404.228	149.706	-1290.420	-600.146	22.392
1500	160.917	522.192	411.728	165.695	-1294.970	-550.682	19.176

PREVIOUS: December 1960 (1 atm) CURRENT: December 1960 (1 bar)

Trifluoromethylsilane ($SiCH_3F_3$) $C_1H_3F_3Si_1$(g)

METHANE (CH_4) IDEAL GAS M_r = 16.04276

$\Delta_f H°$(0 K) = -66.911 ± 0.34 kJ mol^{-1}

S°(298.15 K) = 186.251 ± 0.04 J K mol^{-1} $\Delta_f H°$(298.15 K) = -74.873 ± 0.34 kJ mol^{-1}

Vibrational Frequencies and Degeneracies

ν, cm^{-1}
2916.5(1)
1534.0(2)
3018.7(3)
1306 (3)

Ground State Quantum Weight: 1 σ = 12

Point Group: T_d

Bond Distance: C-H = 1.091 ± 0.002 Å

Bond Angle: H-C-H = 109.4712°

Product of the Moments of Inertia: $I_A I_B I_C$ = 1.499 x 10^{-118} g^3 cm^6.

Enthalpy of Formation

Rossini (1) measured the enthalpy of combustion of methane gas. His value at 298.15 K was corrected to the presently accepted molecular weight of water. The enthalpy of formation was calculated using -68.3174 and -94.0540 kcal mol^{-1} for the enthalpy of formation of H_2O(ℓ) and CO_2(g), respectively.

Heat Capacity and Entropy

Stevenson and Ibers (2), calculated the bond distance from an analysis of available spectroscopic data. Vibrational frequencies listed by Jones and McDowell (3), are consistent with their results of a force constant calculation. The principal moments of inertia are: $I_A = I_B = I_C$ = 0.5313 x 10^{-39} g cm^2.

References

1. F. D. Rossini, J. Res. Nat. Bur. Stand. 6, 37 (1931).
2. D. P. Stevenson and J. A. Ibers, J. Chem. Phys. 33, 762 (1960).
3. L. H. Jones and R. S. McDowell, J. Mol. Spectrosc. 3, 632 (1959).

Methane (CH_4) C_1H_4(g)

Enthalpy Reference Temperature = T_r = 298.15 K Standard State Pressure = p° = 0.1 MPa

T/K	$C_p°$ (J $K^{-1}mol^{-1}$)	S° (J $K^{-1}mol^{-1}$)	$-[G°-H°(T_r)]/T$ (J $K^{-1}mol^{-1}$)	$H°-H°(T_r)$ (kJ mol^{-1})	$\Delta_f H°$ (kJ mol^{-1})	$\Delta_f G°$ (kJ mol^{-1})	Log K_f
0	0.	0.	INFINITE	-10.024	-66.911	-66.911	INFINITE
100	33.258	149.500	216.485	-6.698	-69.644	-64.353	33.615
200	33.473	172.577	189.418	-3.368	-72.027	-58.161	15.190
250	34.216	180.113	186.829	-1.679	-73.426	-54.536	11.395
298.15	35.639	186.251	186.251	0.	-74.873	-50.768	8.894
300	35.708	186.472	186.252	0.066	-74.929	-50.618	8.813
350	37.874	192.131	186.694	1.903	-76.461	-46.445	6.932
400	40.500	197.356	187.704	3.861	-77.969	-42.054	5.492
450	43.374	202.291	189.053	5.957	-79.422	-37.476	4.350
500	46.342	207.014	190.614	8.200	-80.802	-32.741	3.420
600	52.227	215.987	194.103	13.130	-83.308	-22.887	1.993
700	57.794	224.461	197.840	18.635	-85.452	-12.643	0.943
800	62.932	232.518	201.675	24.675	-87.238	-2.115	0.138
900	67.601	240.205	205.532	31.205	-88.692	8.616	-0.500
1000	71.795	247.549	209.370	38.179	-89.849	19.492	-1.018
1100	75.529	254.570	213.162	45 549	-90.750	30.472	-1.447
1200	78.833	261.287	216.895	53 270	-91.437	41.524	-1.807
1300	81.744	267.714	220.558	61.302	-91.945	52.626	-2.115
1400	84.305	273.868	224.148	69.608	-92.308	63.761	-2.379
1500	86.556	279.763	227.660	78.153	-92.553	74.918	-2.609
1600	88.537	285.413	231.095	86.910	-92.703	86.088	-2.810
1700	90.283	290.834	234.450	95.853	-92.780	97.265	-2.989
1800	91.824	296.039	237.728	104 960	-92.797	108.445	-3.147
1900	93.188	301.041	240.930	114.212	-92.770	119.624	-3.289
2000	94.399	305.853	244.057	123.592	-92.709	130.802	-3.416
2100	95.477	310.485	247.110	133.087	-92.624	141.975	-3.531
2200	96.439	314.949	250.093	142.684	-92.521	153.144	-3.636
2300	97.301	319.255	253.007	152.371	-92.409	164.308	-3.732
2400	98.075	323.413	255.854	162.141	-92.291	175.467	-3.819
2500	98.772	327.431	258.638	171.984	-92.174	186.622	-3.899
2600	99.401	331.317	261.359	181.893	-92.060	197.771	-3.973
2700	99.971	335.080	264.020	191.862	-91.954	208.916	-4.042
2800	100.489	338.725	266.623	201.885	-91.857	220.058	-4.105
2900	100.960	342.260	269.171	211.958	-91.773	231.196	-4.164
3000	101.389	345.690	271.664	222.076	-91.705	242.332	-4.219
3100	101.782	349.021	274.106	232.235	-91.653	253.465	-4.271
3200	102.143	352.258	276.498	242.431	-91.621	264.598	-4.319
3300	102.474	355.406	278.842	252.662	-91.609	275.730	-4.364
3400	102.778	358.470	281.139	262.925	-91.619	286.861	-4.407
3500	103.060	361.453	283.391	273.217	-91.654	297.993	-4.447
3600	103.319	364.360	285.600	283.536	-91.713	309.127	-4.485
3700	103.560	367.194	287.767	293.881	-91.798	320.262	-4.521
3800	103.783	369.959	289.894	304.248	-91.911	331.401	-4.555
3900	103.990	372.658	291.982	314.637	-92.051	342.542	-4.588
4000	104.183	375.293	294.032	325.045	-92.222	353.687	-4.619
4100	104.363	377.868	296.045	335.473	-92.422	364.838	-4.648
4200	104.531	380.385	298.023	345 918	-92.652	375.993	-4.676
4300	104.688	382.846	299.967	356.379	-92.914	387.155	-4.703
4400	104.834	385.255	301.879	366.855	-93.208	398.322	-4.729
4500	104 972	387.612	303.758	377.345	-93.533	409.497	-4.753
4600	105.101	389.921	305.606	387.849	-93.891	420.679	-4.777
4700	105.223	392.182	307.424	398.365	-94.281	431.869	-4.800
4800	105.337	394.399	309.213	408.893	-94.702	443.069	-4.822
4900	105.445	396.572	310.973	419.432	-95.156	454.277	-4.843
5000	105.546	398.703	312.707	429.982	-95.641	465.495	-4.863
5100	105.642	400 794	314.414	440.541	-96.157	476.722	-4.883
5200	105.733	402.847	316.095	451.110	-96.703	487.961	-4.902
5300	105.818	404.861	317.750	461.688	-97.278	499.210	-4.920
5400	105.899	406.840	319.382	472.274	-97.882	510.470	-4.938
5500	105.976	408.784	320.990	482.867	-98.513	521.741	-4.955
5600	106.049	410.694	322.575	493.469	-99.170	533.025	-4.972
5700	106.118	412.572	324.137	504.077	-99.852	544.320	-4.988
5800	106.184	414.418	325.678	514.692	-100.557	555.628	-5.004
5900	106.247	416.234	327.197	525.314	-101.284	566.946	-5.019
6000	106.306	418.020	328.696	535.942	-102.032	578.279	-5.034

PREVIOUS. March 1961 (1 atm) CURRENT: March 1961 (1 bar)

Methane (CH_4) C_1H_4(g)

CYANOGEN IODIDE (ICN) IDEAL GAS M_r = 152.9222

$\Delta_f H°(0\ K)$ = 226.5 ± 6.3 kJ mol^{-1}

S°(298.15 K) = 257.34 ± 0.04 J K^{-1} mol^{-1} $\Delta_f H°(298.15\ K)$ = 225.9 ± 6.3 kJ mol^{-1}

Vibrational Frequencies and Degeneracies

ν, cm^{-1}
485.8 (1)
304.5 (2)
2188 (1)

Ground State Quantum Weight: 1 σ = 1

Point Group: $C_{\infty v}$

Bond Distances: C–I = 1.994 Å C–N = 1.159 Å

Bond Angle: I–C–N = 180°

Rotational Constant: B_0 = 0.107604 cm^{-1}

Enthalpy of Formation

The enthalpy of formation is calculated from $\Delta_f H°$(ICN, cr, 298.15 K) = 39.71 kcal mol^{-1} (1) and $\Delta_{sub}H°$(298.15 K) = 14.29 ± 0.15 kcal mol^{-1}. This latter value is obtained from a 2nd and 3rd law analysis of the vapor pressure data of Yost and Stone (2) and Ketelaar and Kruyer (3), which are in good agreement after dropping the lowest point in the first set and the lowest 3 points in the second set.

The enthalpy of formation of the crystal is calculated from $\Delta_r H°$(298.15 K) = -46.3 ± 0.08 kcal obtained by Lord and Woolf (4) for the reaction ICN(cr) + 2 NaOH(56 H_2O) = NaCNO(2800 H_2O) + NaI(2800 H_2O) + H_2O(ℓ). The auxiliary data used were $\Delta_f H°$(NaOH·56 H_2O, 298.15 K) = -112.26 kcal mol^{-1} (1, 5); $\Delta_f H°$(NaCNO·2800 H_2O, 298.15 K) = -92.25 kcal mol^{-1} (1, 5) and $\Delta_f H°$(NaI·2800 H_2O, 298.15 K) = -70.54 kcal mol^{-1} (1, 5). The uncertainty in the final value reflects the uncertainty in $\Delta_f H°$(NaCNO·2800 H_2O, 298.15 K) which was determined by Lord and Woolf to be -93.6 kcal mol^{-1} in excess NaOH·56 H_2O.

Cyanogen Iodide (CNI) $C_1I_1N_1(g)$

Enthalpy Reference Temperature = T_r = 298.15 K Standard State Pressure = p° = 0.1 MPa

T/K	$C_p°$	S°	−[G°−H°(T_r)]/T	H°−H°(T_r)	$\Delta_f H°$	$\Delta_f G°$	Log K_f
	J $K^{-1}mol^{-1}$			kJ mol^{-1}			
0	0.	0.	INFINITE	-11.470	226.451	226.451	INFINITE
100	33.571	212.187	296.813	-8.463	226.423	216.340	-113.004
200	43.694	238.890	261.644	-4.551	226.087	206.419	-53.911
250	46.561	248.970	258.128	-2.289	226.011	201.512	-42.104
298.15	48.420	257.340	257.340	0.	225.936	196.800	-34.478
300	48.479	257.640	257.341	0.090	225.932	196.619	-34.234
350	49.827	265.220	257.937	2.549	225.787	191.743	-28.616
400	50.845	271.942	259.275	5.067	217.609	187.166	-24.441
450	51.676	277.980	261.024	7.631	216.795	183.409	-21.290
500	52.398	283.463	262.998	10.233	195.880	181.645	-18.976
600	53.662	293.131	267.235	15.537	196.238	178.762	-15.563
700	54.769	301.488	271.545	20.960	196.483	175.828	-13.120
800	55.742	308.866	275.758	26.487	196.644	172.866	-11.287
900	56.587	315.482	279.811	32.104	196.744	169.887	-9.860
1000	57.312	321.482	283.682	37.800	196.796	166.899	-8.718
1100	57.930	326.975	287.372	43.563	196.808	163.909	-7.783
1200	58.456	332.038	290.886	49.383	196.786	160.919	-7.005
1300	58.903	336.735	294.234	55.252	196.731	157.932	-6.346
1400	59.284	341.115	297.428	61.161	196.641	154.950	-5.781
1500	59.610	345.217	300.479	67.107	196.514	151.977	-5.292
1600	59.890	349.073	303.397	73.082	196.346	149.013	-4.865
1700	60.132	352.711	306.192	79.083	196.135	146.061	-4.488
1800	60.342	356.154	308.872	85.107	195.877	143.122	-4.153
1900	60.525	359.422	311.448	91.151	195.572	140.200	-3.854
2000	60.684	362.530	313.925	97.211	195.218	137.294	-3.586
2100	60.825	365.495	316.310	103.287	194.819	134.408	-3.343
2200	60.949	368.327	318.611	109.376	194.376	131.541	-3.123
2300	61.059	371.039	320.832	115.476	193.894	128.696	-2.923
2400	61.157	373.640	322.978	121.587	193.377	125.872	-2.740
2500	61.245	376.138	325.055	127.708	192.832	123.070	-2.571
2600	61.323	378.542	327.066	133.836	192.264	120.291	-2.417
2700	61.394	380.858	329.016	139.972	191.680	117.534	-2.274
2800	61.458	383.091	330.908	146.115	191.084	114.799	-2.142
2900	61.516	385.249	332.744	152.263	190.483	112.085	-2.019
3000	61.568	387.335	334.530	158.418	189.881	109.392	-1.905
3100	61.616	389.355	336.266	164.577	189.283	106.719	-1.798
3200	61.660	391.312	337.956	170.741	188.692	104.065	-1.699
3300	61.700	393.210	339.601	176.909	188.112	101.429	-1.605
3400	61.736	395.053	341.205	183.081	187.545	98.811	-1.518
3500	61.770	396.843	342.769	189.256	186.992	96.209	-1.436
3600	61.801	398.583	344.296	195.435	186.456	93.623	-1.358
3700	61.830	400.277	345.786	201.616	185.938	91.051	-1.285
3800	61.857	401.926	347.242	207.801	185.438	88.494	-1.216
3900	61.881	403.533	348.665	213.987	184.956	85.949	-1.151
4000	61.904	405.100	350.056	220.177	184.493	83.416	-1.089
4100	61.925	406.629	351.417	226.368	184.049	80.895	-1.031
4200	61.945	408.122	352.750	232.562	183.622	78.384	-0.975
4300	61.964	409.579	354.054	238.757	183.213	75.883	-0.922
4400	61.981	411.004	355.333	244.954	182.822	73.391	-0.871
4500	61.997	412.397	356.585	251.153	182.446	70.909	-0.823
4600	62.012	413.760	357.813	257.354	182.087	68.434	-0.777
4700	62.027	415.094	359.018	263.556	181.741	65.967	-0.733
4800	62.040	416.400	360.200	269.759	181.410	63.508	-0.691
4900	62.053	417.679	361.360	275.964	181.092	61.055	-0.651
5000	62.065	418.933	362.499	282.170	180.785	58.608	-0.612
5100	62.076	420.162	363.618	288.377	180.490	56.167	-0.575
5200	62.086	421.368	364.717	294.585	180.205	53.732	-0.540
5300	62.096	422.550	365.797	300.794	179.930	51.303	-0.506
5400	62.106	423.711	366.858	307.004	179.663	48.878	-0.473
5500	62.115	424.851	367.902	313.215	179.404	46.459	-0.441
5600	62.123	425.970	368.929	319.427	179.152	44.044	-0.411
5700	62.131	427.070	369.940	325.640	178.906	41.633	-0.382
5800	62.139	428.150	370.934	331.853	178.666	39.227	-0.353
5900	62.146	429.213	371.913	338.068	178.430	36.825	-0.326
6000	62.153	430.257	372.877	344.283	178.197	34.427	-0.300

PREVIOUS: June 1969 (1 atm) CURRENT: June 1969 (1 bar)

Cyanogen Iodide (CNI) $C_1I_1N_1(g)$

POTASSIUM CYANIDE (KCN) CRYSTAL M_r = 65.1160

$S°(298.15\ K) = 127.780 \pm 1.05\ J\ K^{-1}\ mol^{-1}$
$T_{trs} = 168.3 \pm 0.1\ K$
$T_{fus} = 895 \pm 2\ K$

$\Delta_f H°(0\ K) = -118.34 \pm 0.8\ kJ\ mol^{-1}$
$\Delta_f H°(298.15\ K) = -113.47 \pm 0.8\ kJ\ mol^{-1}$
$\Delta_{trs} H° = 1.167\ kJ\ mol^{-1}$
$\Delta_{fus} H° = 14.644\ kJ\ mol^{-1}$

Enthalpy of Formation

The enthalpy change, $\Delta_r H°(298.15\ K) = 2.80 \pm 0.10\ kcal\ mol^{-1}$, for the reaction KCN(cr) → KCN(aq, ∞) was reported by Parker (1). Incorporating these data with the enthalpy of formation for K^+(aq, ∞) and CN^-(aq, ∞), reported by Wagman et al. (2), the value of $\Delta_f H°$(KCN, cr, 298.15 K) was evaluated to be $-27.12 \pm 0.2\ kcal\ mol^{-1}$.

Heat Capacity and Entropy

The heat capacity for KCN(cr) has been measured with a heat conduction method by Messer and Ziegler (3), 101.6-345.8 K, and adiabatically by Suga et al. (4), 14.23-308.5 K. The data of Suga (4) were adopted after correction for curvature. The anomaleous region between 50-105 K has been treated as second order by inclusion of the enthalpy in the heat capacity. The $C_p°$ values above 310 K were estimated by graphical extrapolation. The value of S°(298.15 K) was evaluated based on S°(15 K) = 0.193 cal $K^{-1}\ mol^{-1}$. This extrapolated value appears more reasonable than the value, S°(15 K) = 0.307 cal $K^{-1}\ mol^{-1}$ reported by Suga et al (4).

Transition Data

T_{trs} was taken from Suga et al (4). $\Delta_{trs}H°$ was calculated as the difference between the total enthalpy change, H°(173.80 K) − H°(159.02 K) = 0.533 kcal mol^{-1} obtained from $C_p°$ values measured by Messer and Ziegler (3), and the enthalpy change, 0.254 kcal mol^{-1}, calculated from the arbitrarily smoothed $C_p°$ curve. Suga et al (4) reported $\Delta_{trs}H°$ = 0.339 kcal mol^{-1}; however, their $C_p°$ data do not adequately define the total enthalpy change through the transition region. The values of T_{trs} and $\Delta_{trs}H°$ reported by Messer and Ziegler (3) were 168.3±0.4 K and 0.30 kcal mol^{-1}, respectively. The lower transition, T_{trs} = 82.9±0.2 K and $\Delta_f H°$ = 0.095±0.001 kcal mol^{-1} reported by Suga et al. (4) was treatment as second order by inclusion of the enthalpy in the heat capacity.

Fusion Data

See the liquid table for details.

Sublimation Data

The difference between $\Delta_f H°$(298.15 K) for $(KCN)_2$(g) and 2 KCN(cr) is $\Delta_{sub}H°$(cr → dimer, 298.15 K) and for KCN(g) and KCN(cr) is $\Delta_{sub}H°$(cr → monomer, 298.15 K).

References

1. V. B. Parker, NSRDS-NBS 2, 1965.
2. U. S. Nat. Bur. Stand. Tech. Note 270-1, 1965.
3. C. E. Messer and W. T. Ziegler, J. Amer. Chem. Soc. 63, 2703 (1941).
4. H. Suga, T. Matsuo and S. Seki, Bull. Chem. Soc. Japan 38, 1115 (1965).

Potassium Cyanide (KCN) $C_1K_1N_1$(cr)

Enthalpy Reference Temperature = T_r = 298.15 K — Standard State Pressure = p° = 0.1 MPa

T/K	$C_p°$ (J K⁻¹mol⁻¹)	S° (J K⁻¹mol⁻¹)	−[G°−H°(T_r)]/T (J K⁻¹mol⁻¹)	H°−H°(T_r) (kJ mol⁻¹)	$\Delta_f H°$ (kJ mol⁻¹)	$\Delta_f G°$ (kJ mol⁻¹)	Log K_f
0	0.	0.	INFINITE	-17.343	-118.345	-118.345	INFINITE
100	55.483	49.213	191.980	-14.277	-118.522	-111.797	58.397
168.300	76.283	82.687	141.073	-9.826	I <--> II		
168.300	67.280	89.623	141.073	-8.659	TRANSITION		
200	66.828	101.196	133.862	-6.533	-115.154	-106.080	27.705
298.15	66.391	127.780	127.780	0.	-113.470	-102.011	17.872
300	66.383	128.191	127.781	0.123	-113.445	-101.940	17.749
400	66.316	147.269	130.383	6.755	-114.772	-97.829	12.775
500	66.354	162.072	135.295	13.388	-114.043	-93.681	9.787
600	66.392	174.173	140.797	20.025	-113.515	-89.662	7.806
700	66.431	184.410	146.315	26.667	-113.164	-85.716	6.396
800	66.471	193.283	151.644	33.312	-112.972	-81.810	5.342
895.000	66.509	200.744	156.467	39.628	I <--> LIQUID		
900	66.511	201.115	156.714	39.961	-112.928	-77.919	4.522
1000	66.551	208.124	161.510	46.614	-113.031	-74.025	3.867
1100	66.590	214.469	166.041	53.271	-192.235	-65.504	3.111
1200	66.629	220.265	170.322	59.932	-191.578	-54.012	2.351
1300	66.669	225.600	174.371	66.597	-190.988	-42.573	1.711
1400	66.708	230.542	178.209	73 266	-190.455	-31.177	1.163
1500	66.747	235.146	181.853	79.938	-189.970	-19.817	0.690
1600	66.787	239.455	185.320	86.615	-189.529	-8.488	0.277
1700	66.826	243.505	188.625	93.296	-189.125	2.815	-0.086
1800	66.865	247.325	191.781	99.980	-188.757	14.094	-0.409
1900	66.905	250.942	194.800	106.669	-188.422	25.355	-0.697
2000	66.944	254.375	197.694	113.361	-188.117	36.598	-0.956

PREVIOUS: June 1962 CURRENT: March 1966

Potassium Cyanide (KCN) $C_1K_1N_1$(cr)

POTASSIUM CYANIDE (KCN) LIQUID M_r = 65.1160

$S°(298.15\ K) = [134.321]\ J\ K^{-1}\ mol^{-1}$

$T_{fus} = 895 \pm 2\ K$

$\Delta_f H°(298.15\ K) = [-104.148]\ kJ\ mol^{-1}$

$\Delta_{fus} H° = 14.644\ kJ\ mol^{-1}$

Enthalpy of Formation

$\Delta_f H°(\ell, 298.15\ K)$ is calculated from $\Delta_f H°(cr, 298.15\ K)$ by adding $\Delta_{fus} H°$ and the difference in enthalpy, $H°(895\ K) - H°(298.15\ K)$, between the crystal and liquid.

Heat Capacity and Entropy

Heat capacity is estimated and assumed constant in the temperature range 298.15-3000 K. The entropy is calculated in a manner analogous to that used for the enthalpy of formation.

Fusion Data

The melting temperature was reported by Truthe (1). The value of $\Delta_{fus} H°$ was taken from Kelly (2) on phase diagrams of the systems KCN-AgCN, KCN-$Cu_2(CN)_2$, KCN-KCl, KCN-NaCN, and KCN-$Zn(CN)_2$ reported by Truthe (1).

Vaporization Data

T_{vap} is the temperature at which the Gibbs energy change for the reaction KCN(ℓ) = KCN(g) becomes zero. The difference between $\Delta_f H°$ for KCN(g) and KCN(ℓ) at T_{vap} is $\Delta_{vap} H°$.

References

1. W. Truthe, Z. Anorg. Chem. 76, 129 (1912).
2. K. K. Kelley, U. S. Bur. Mines Bull. 393, 1936.

Potassium Cyanide (KCN) $C_1K_1N_1(l)$

Enthalpy Reference Temperature = T_r = 298.15 K   Standard State Pressure = p° = 0.1 MPa

	J K⁻¹mol⁻¹			kJ mol⁻¹			
T/K	$C_p°$	S°	$-[G°-H°(T_r)]/T$	$H°-H°(T_r)$	$\Delta_f H°$	$\Delta_f G°$	Log K_f
0							
100							
200							
298.15	75.312	134.321	134.321	0.	-104.148	-94.638	16.580
300	75.312	134.787	134.323	0.139	-104.106	-94.580	16.468
400	75.312	156.453	137.277	7.671	-104.534	-91.264	11.918
500	75.312	173.258	142.855	15.202	-102.907	-88.139	9.208
600	75.312	186.989	149.101	22.733	-101.485	-85.322	7.428
700	75.312	198.599	155.364	30.264	-100.244	-82.729	6.173
800	75.312	208.655	161.411	37.795	-99.166	-80.302	5.243
895.000	75.312	217.106	166.883	44.950	--------- I	<--> LIQUID	--------
900	75.312	217.526	167.163	45.327	-98.240	-78.001	4.527
1000	75.312	225.461	172.603	52.858	-97.465	-75.795	3.959
1100	75.312	232.639	177.740	60.389	-175.795	-69.050	3.279
1200	75.312	239.192	182.592	67.920	-174.268	-59.414	2.586
1300	75.312	245.220	187.180	75.451	-172.811	-49.902	2.005
1400	75.312	250.801	191.528	82.983	-171.416	-40.500	1.511
1500	75.312	255.997	195.655	90.514	-170.073	-31.197	1.086
1600	75.312	260.858	199.580	98.045	-168.777	-21.981	0.718
1700	75.312	265.423	203.320	105.576	-167.523	-12.844	0.395
1800	75.312	269.728	206.891	113.107	-166.308	-3.781	0.110
1900	75.312	273.800	210.306	120.639	-165.130	5.216	-0.143
2000	75.312	277.663	213.578	128.170	-163.986	14.152	-0 370
2100	75.312	281.337	216.718	135.701	-162.878	23.031	-0.573
2200	75.312	284.841	219.735	143.232	-161.804	31.859	-0.756
2300	75.312	288.189	222.639	150.763	-160.764	40.638	-0.923
2400	75.312	291.394	225.438	158.295	-159.761	49.373	-1.075
2500	75.312	294.468	228.138	165.826	-158.794	58.067	-1.213
2600	75.312	297.422	230.746	173.357	-157.867	66.723	-1.340
2700	75.312	300.264	233.269	180.888	-156.980	75.344	-1.458
2800	75.312	303.003	235.711	188.419	-156.136	83.933	-1.566
2900	75.312	305.646	238.077	195.951	-155.338	92.493	-1.666
3000	75.312	308.199	240.372	203.482	-154.588	101.026	-1.759

PREVIOUS: June 1962   CURRENT: March 1966

Potassium Cyanide (KCN) $C_1K_1N_1(l)$

POTASSIUM CYANIDE (KCN) CRYSTAL–CRYSTAL–LIQUID M_r = 65.1160

0 to 168.3 K crystal, I
168.3 to 895 K crystal, II
above 895 K liquid

Refer to the individual tables for details.

Potassium Cyanide (KCN) $C_1K_1N_1$(cr,l)

Enthalpy Reference Temperature = T_r = 298.15 K Standard State Pressure = p° = 0.1 MPa

T/K	C_p° (J K⁻¹mol⁻¹)	S° (J K⁻¹mol⁻¹)	$-[G°-H°(T_r)]/T$ (J K⁻¹mol⁻¹)	$H°-H°(T_r)$ (kJ mol⁻¹)	$\Delta_fH°$ (kJ mol⁻¹)	$\Delta_fG°$ (kJ mol⁻¹)	Log K_f
0	0.	0.	INFINITE	-17.343	-118.345	-118.345	INFINITE
100	55.483	49.213	191.980	-14.277	-118.522	-111.797	58.397
168.300	76.283	82.687	141.073	-9.826		I <--> II	
168.300	67.280	89.623	141.073	-8.659		TRANSITION	
200	66.828	101.196	133.862	-6.533	-115.154	-106.080	27.705
298.15	66.391	127.780	127.780	0.	-113.470	-102.011	17.872
300	66.383	128.191	127.781	0.123	-113.445	-101.940	17.749
400	66.316	147.269	130.383	6.755	-114.772	-97.829	12.775
500	66.354	162.072	135.295	13.388	-114.043	-93.681	9.787
600	66.392	174.173	140.797	20.025	-113.515	-89.662	7.806
700	66.431	184.410	146.315	26.667	-113.164	-85.716	6.396
800	66.471	193.283	151.644	33.312	-112.972	-81.810	5.342
895.000	66.509	200.744	156.467	39.628		I <--> LIQUID	
895.000	75.312	217.106	156.467	54.272		TRANSITION	
900	75.312	217.526	156.805	54.649	-98.240	-78.001	4.527
1000	75.312	225.461	163.281	62.180	-97.465	-75.795	3.959
1100	75.312	232.639	169.265	69.711	-175.795	-69.050	3.279
1200	75.312	239.192	174.823	77.242	-174.268	-59.414	2.586
1300	75.312	245.220	180.009	84.774	-172.811	-49.902	2.005
1400	75.312	250.801	184.869	92.305	-171.416	-40.500	1.511
1500	75.312	255.997	189.440	99.836	-170.073	-31.197	1.086
1600	75.312	260.858	193.753	107.367	-168.777	-21.981	0.718
1700	75.312	265.423	197.836	114.898	-167.523	-12.844	0.395
1800	75.312	269.728	201.712	122.430	-166.308	-3.781	0.110
1900	75.312	273.800	205.400	129.961	-165.130	5.216	-0.143
2000	75.312	277.663	208.917	137.492	-163.986	14.152	-0.370
2100	75.312	281.337	212.279	145.023	-162.878	23.031	-0.573
2200	75.312	284.841	215.498	152.554	-161.804	31.859	-0.756
2300	75.312	288.189	218.586	160.086	-160.764	40.638	-0.923
2400	75.312	291.394	221.554	167.617	-159.761	49.373	-1.075
2500	75.312	294.468	224.409	175.148	-158.794	58.067	-1.213
2600	75.312	297.422	227.161	182.679	-157.867	66.723	-1.340
2700	75.312	300.264	229.816	190.210	-156.980	75.344	-1.458
2800	75.312	303.003	232.381	197.742	-156.136	83.933	-1.566
2900	75.312	305.646	234.862	205.273	-155.338	92.493	-1.666
3000	75.312	308.199	237.265	212.804	-154.588	101.026	-1.759

PREVIOUS: CURRENT: March 1966

Potassium Cyanide (KCN) $C_1K_1N_1$(cr,l)

POTASSIUM CYANIDE (KCN) IDEAL GAS M_r = 65.1160

$\Delta_f H°(0\ K)$ = [80 ± 21] kJ mol^{-1}

$S°(298.15\ K)$ = [253.15] J K^{-1} mol^{-1} $\Delta_f H°(298.15\ K)$ = [79 ± 21] kJ mol^{-1}

Vibrational Frequencies and Degeneracies

ν, cm^{-1}
[370] (1)
207 (2)
2158 (1)

Ground State Quantum Weight: 1 σ = 1

Point Group: $C_{\infty v}$

Bond Distances: C–N = 1.16 ± 0.01 Å K–C = [2.294] Å

Bond Angle: K–C–C = 180°

Rotational Constant: B_o = [0.118914] cm^{-1}

Enthalpy of Formation

The vapor pressure of KCN(g) at 870°C was estimated to be 1-10 mm Hg by Villars (1). Assuming the vapor pressure at this temperature to be 1 mm Hg, the enthalpy of vaporization, $\Delta_{vap}H°(298.15\ K)$, for KCN(ℓ) was evaluated as 44 kcal mol^{-1}. Based on $\Delta_f H°(298.15\ K)$ = -24.89 kcal mol^{-1} for KCN(ℓ), the value of $\Delta_f H°$(KCN, g, 298.15 K) is calculated to be 19 ± 5 kcal mol^{-1}.

Heat Capacity and Entropy

The molecular structure of KCN(g) has not been determined. It was assumed to be linear by comparison with that for HCN(g). The C-N bond distance was determined by Elliott and Hastings (2) by use of neutron diffraction measurements. The K-C bond distance was taken from Leroi (3), which was estimated. The vibrational frequencies, ν_2 and ν_3, were obtained from Leroi and Klemperer (4). The value of ν_1 was calculated by valence force method described by Herzberg (5) using appropriate force constants.

References

1. D. S. Villars, J. Am. Chem. Soc. 53, 405 (1931).
2. N. Elliott and J. Hastings, Acta Cryst. 14, 1018 (1961).
3. G. E. Leroi, Ph. D. Dissertation, Harvard University, 1961.
4. G. E. Leroi and W. Klemperer, J. Chem. Phys. 35, 774 (1961).
5. G. Herzberg, "Infrared and Raman Spectra", D. Van Nostrand, Inc., New York, 1945.

Potassium Cyanide (KCN) $C_1K_1N_1(g)$

Enthalpy Reference Temperature = T_r = 298.15 K Standard State Pressure = p° = 0.1 MPa

	J $K^{-1}mol^{-1}$			kJ mol^{-1}			
T/K	$C_p°$	S°	$-[G°-H°(T_r)]/T$	$H°-H°(T_r)$	$\Delta_f H°$	$\Delta_f G°$	Log K_f
0	0.	0.	INFINITE	-12.456	79.509	79.509	INFINITE
100	38.592	203.295	295.884	-9.259	79.462	70.779	-36.971
200	47.724	233.411	257.718	-4.861	79.484	62.115	-16.223
250	49.674	244.289	253.978	-2.422	79.525	57.767	-12.070
298.15	50.857	253.146	253.146	0.	79.496	53.578	-9.387
300	50.894	253.461	253.147	0.094	79.493	53.417	-9.301
350	51.734	261.373	253.770	2.661	76.969	49.177	-7.339
400	52.384	268.325	255.164	5.264	76.703	45.225	-5.906
450	52.944	274.528	256.977	7.898	76.412	41.307	-4.795
500	53.465	280.133	259.017	10.558	76.093	37.424	-3.910
600	54.456	289.969	263.378	15.955	75.380	29.755	-2.590
700	55.391	298.435	267.795	21.447	74.583	22.213	-1.658
800	56.246	305.888	272.100	27.030	73.713	14.790	-0.966
900	57.006	312.558	276.231	32.694	72.771	7.481	-0.434
1000	57.666	318.599	280.171	38.428	71.749	0.280	-0.015
1100	58.234	324.122	283.919	44.224	-8.316	-2.203	0.105
1200	58.719	329.211	287.484	50.072	-8.472	-1.641	0.071
1300	59.133	333.927	290.877	55.965	-8.654	-1.065	0.043
1400	59.487	338.323	294.111	61.896	-8.858	-0.473	0.018
1500	59.790	342.438	297.197	67.861	-9.082	0.133	-0.005
1600	60.051	346.305	300.147	73.853	-9.325	0.756	-0.025
1700	60.277	349.952	302.970	79.870	-9.585	1.393	-0.043
1800	60.472	353.403	305.677	85.907	-9.864	2.047	-0.059
1900	60.643	356.678	308.276	91.963	-10.161	2.717	-0.075
2000	60.792	359.792	310.774	98.035	-10.477	3.403	-0.089
2100	60.923	362.761	313.180	104.121	-10.814	4.105	-0.102
2200	61.039	365.598	315.499	110.219	-11.173	4.824	-0.115
2300	61.142	368.314	317.736	116.329	-11.555	5.559	-0.126
2400	61.234	370.918	319.898	122.448	-11.964	6.312	-0.137
2500	61.316	373.420	321.989	128.575	-12.401	7.083	-0.148
2600	61.389	375.826	324.014	134.710	-12.870	7.871	-0.158
2700	61.456	378.144	325.976	140.853	-13.372	8.678	-0.168
2800	61.515	380.380	327.880	147.001	-13.911	9.505	-0.177
2900	61.569	382.540	329.727	153.156	-14.489	10.351	-0.186
3000	61.619	384.628	331.523	159.315	-15.111	11.218	-0.195
3100	61.663	386.649	333.269	165.479	-15.779	12.106	-0.204
3200	61.704	388.607	334.967	171.648	-16.498	13.017	-0.212
3300	61.742	390.507	336.622	177.820	-17.261	13.951	-0.221
3400	61.776	392.350	338.234	183.996	-18.088	14.909	-0.229
3500	61.808	394.142	339.806	190.175	-18.972	15.892	-0.237
3600	61.837	395.883	341.339	196.357	-19.924	16.901	-0.245
3700	61.864	397.578	342.837	202.542	-20.946	17.938	-0.253
3800	61.888	399.228	344.299	208.730	-22.041	19.003	-0.261
3900	61.912	400.836	345.728	214.920	-23.215	20.099	-0.269
4000	61.933	402.404	347.126	221.112	-24.465	21.224	-0.277
4100	61.953	403.933	348.492	227.307	-25.804	22.383	-0.285
4200	61.971	405.426	349.830	233.503	-27.239	23.575	-0.293
4300	61.989	406.885	351.140	239.701	-28.772	24.803	-0.301
4400	62.005	408.310	352.423	245.900	-30.129	26.040	-0.309
4500	62.020	409.704	353.681	252.102	-31.800	27.336	-0.317
4600	62.034	411.067	354.914	258.304	-33.571	28.669	-0.326
4700	62.048	412.401	356.123	264.509	-35.422	30.040	-0.334
4800	62.060	413.708	357.309	270.714	-37.267	31.440	-0.342
4900	62.072	414.987	358.473	276.921	-39.319	32.892	-0.351
5000	62.083	416.241	359.616	283.128	-41.475	34.387	-0.359
5100	62.094	417.471	360.738	289.337	-43.736	35.927	-0.368
5200	62.104	418.677	361.841	295.547	-46.103	37.512	-0.377
5300	62.113	419.860	362.924	301.758	-48.577	39.143	-0.386
5400	62.122	421.021	363.990	307.970	-51.157	40.822	-0.395
5500	62.130	422.161	365.037	314.182	-53.843	42.550	-0.404
5600	62.138	423.280	366.067	320.396	-56.633	44.328	-0.413
5700	62.146	424.380	367.080	326.610	-59.183	46.109	-0.423
5800	62.153	425.461	368.078	332.825	-62.136	47.981	-0.432
5900	62.160	426.524	369.059	339.040	-65.186	49.906	-0.442
6000	62.166	427.569	370.026	345.257	-68.329	51.883	-0.452

PREVIOUS: March 1966 (1 atm) CURRENT: March 1966 (1 bar)

Potassium Cyanide (KCN) $C_1K_1N_1(g)$

POTASSIUM CARBONATE (K_2CO_3) CRYSTAL M_r = 138.2058

$S°(298.15\ K) = 155.519 \pm 0.4\ J\ K^{-1} mol^{-1}$

$T_{fus} = 1174 \pm 1\ K$

$\Delta_f H°(0\ K) = -1144.61 \pm 2.1\ kJ\ mol^{-1}$

$\Delta_f H°(298.15\ K) = -1150.18 \pm 2.1\ kJ\ mol^{-1}$

$\Delta_{fus} H° = 27.614 \pm 0.4\ kJ\ mol^{-1}$

Enthalpy of Formation

The enthalpy of formation was obtained from the enthalpy of solution $\Delta_\infty H° = -7.6 \pm 0.5$ kcal mol^{-1} and the enthalpies of formation for $K^+(aq)$ and $CO_3^{--}(aq)$, -60.32 and -161.84 kcal mol^{-1}, respectively. The standard enthalpy solution, $\Delta_\infty H°$ = -7.6 ± 0.5 kcal mol^{-1}, was calculated by extrapolation to infinite dilution of the calorimetric data of Benjamin (1), using the methods suggested by Rupert et al. (2). The equilibrium constant for the hydrolysis $CO_3^{--}(aq, \infty) + H_2O(\ell) + HCO_3^-(aq, \infty) + OH^-(\infty H_2O)$ was taken as K -2.1 x 10^{-4} based on the selected values of Wagman et al. (3). The resulting extrapolations were not so consistent as those of Rupert et al. (2) for Na_2CO_3. The ionic enthalpies of formation were taken from Wagman et al. (3). This procedure was adopted even though it may involve an inconsistency between the extrapolations for $\Delta_\infty H°$ and $CO_3^{--}(\infty H_2O)$. The alternative evaluation of $\Delta_f H°(298.15\ K)$ for $K_2CO_3(\sim 1900\ H_2O)$ from the enthalpies of mixing given by Muller (4) and Berthelot (5) may be subject to significant error if dissolved CO_2 was not excluded from the solutions. Presumably, this path was used by Wagman (6) in the selection of $\Delta_f H°(298.15\ K) = -274.3$ kcal mol^{-1}. This value is 0.6 kcal mol^{-1} less negative than the value adopted here.

Heat Capacity and Entropy

The low temperature heat capacity, 14.0-315 K was measured by Oetting (7). Janz et al. (8) measured the heat content by drop calorimetry in the temperature range 630-1250 K, and gave enthalpy and heat capacity equations based on their measurements. The above information was used in a Shomate analysis in order to smooth the enthalpy and calculate heat capacity above 298.15 K. The values from the low and high temperature sources join smoothly at 298.15 K. The heat capacity was graphically extrapolated above the melting point. The entropy at 14.0 K was calculated from the extrapolated low temperature $C_p°$ data, using the cubic law. The value of $S°(14\ K) = 0.109$ cal K^{-1} mol^{-1}.

Fusion Data

T_{fus} was obtained from Reisman (9) and $\Delta_{fus}H°$ was taken from Janz (8).

References

1. L. Benjamin, J. Chem. Eng. Data 7, 239-40 (1962).
2. J. P. Rupert, H. P. Hopkins, Jr., and C. A. Wulff, J. Phys. Chem. 69, 3059-62 (1965).
3. U. S. Nat. Bur. Stand. Tech. Note 270-1, 1965.
4. J. A. Muller, Ann. Chim. Phys. (6), 15, 517 (1888).
5. M. Berthelot, Ann. Chim. Phys. (4) 29, 433 (1873).
6. D. D. Wagman, U. S. Nat. Bur. Stand. Report 8628, 1965.
7. F. L. Oetting, unpublished data, Thermal Research Laboratory, Dow Chemical Company, Midland, Michigan, 1963.
8. G. I. Janz, E. Neuenschwander and F. J. Kelly, Trans. Faraday Soc. 59, 841 (1963).
9. A. Reisman, J. Amer. Chem. Soc. 80, 3558 (1958).

Potassium Carbonate (K_2CO_3) $C_1K_2O_3(cr)$

Enthalpy Reference Temperature = T_r = 298.15 K Standard State Pressure = p° = 0.1 MPa

	J $K^{-1}mol^{-1}$			kJ mol^{-1}			
T/K	$C_p°$	S°	$-[G°-H°(T_r)]/T$	$H°-H°(T_r)$	$\Delta_f H°$	$\Delta_f G°$	Log K_f
0	0.	0.	INFINITE	-22.673	-1144.614	-1144.614	INFINITE
100	72.383	53.095	246.061	-19.297	-1149.119	-1121.217	585.664
200	99.140	113.018	165.506	-10.498	-1150.202	-1092.747	285.396
298.15	114.437	155.519	155.519	0.	-1150.182	-1064.528	186.501
300	114.699	156.228	155.521	0.212	-1150.176	-1063.997	185.258
400	128.135	191.103	160.186	12.367	-1154.457	-1034.430	135.083
500	140.049	221.005	169.426	25.790	-1153 165	-1004.553	104.945
600	150.666	247.493	180.270	40.334	-1151.017	-975.020	84.883
700	160.483	271.462	191.611	55.896	-1148.102	-945.907	70.584
800	170.043	293.516	202.989	72.422	-1144.456	-917.262	59.891
900	179.515	314.092	214.202	89.901	-1140.085	-889.118	51.603
1000	188.949	333.495	225.170	108.325	-1135.003	-861.499	45.000
1100	198.321	351.944	235.863	127.688	-1287.134	-825.202	39.186
1174.000	205.259	365.079	243.596	142.621	------ CRYSTAL <--> LIQUID -----		
1200	207.695	369.602	246.277	147.989	-1278.565	-783.580	34.108
1300	217.066	386.596	256.421	169.227	-1269.158	-742.707	29.842
1400	226.438	403.026	266.309	191.403	-1258.901	-702.590	26.214
1500	235.810	418.968	275.958	214.515	-1247.784	-663.237	23.096
1600	245.182	434.486	285.383	238.565	-1235.802	-624.653	20.393
1700	254.555	449.631	294.601	263.552	-1222.949	-586.845	18.032
1800	263.927	464.446	303.626	289.476	-1209.226	-549.819	15.955
1900	273.299	478.967	312.474	316.337	-1194.631	-513.579	14.119
2000	282.671	493.224	321.156	344.135	-1179.164	-478.131	12.487

PREVIOUS: December 1961 CURRENT: March 1966

Potassium Carbonate (K_2CO_3) $C_1K_2O_3(cr)$

POTASSIUM CARBONATE (K_2CO_3) LIQUID M_r = 138.2058

$S°(298.15\ K) = [170.373]\ J\ K^{-1}\ mol^{-1}$ $\Delta_fH°(298.15\ K) = [-1130.609]\ kJ\ mol^{-1}$

$T_{fus} = 1174 \pm 1\ K$ $\Delta_{fus}H° = 27.614 \pm 0.4\ kJ\ mol^{-1}$

Enthalpy of Formation

$\Delta_fH°(\ell,\ 298.15\ K)$ is calculated from $\Delta_fH°(cr,\ 298.15\ K)$ by adding $\Delta_{fus}H°$ and the difference in enthalpy, H°(1174 K) - H°(298.15 K), between the crystal and liquid.

Heat Capacity and Entropy

A glass transition is assumed at 800 K. The heat capacity below 800 K is obtained from the heat capacity of the crystal. Above 800 K the heat capacity is adopted as 50.0 cal $K^{-1}\ mol^{-1}$, based on the enthalpy measurements in the range 1169-1250 K reported by Janz et al. (1). The entropy is calculated in a manner analogous to that used for the enthalpy of formation.

Fusion Data

See the crystal table for details.

Reference

1. G. J. Janz, E. Neuenschwander and F. J. Kelly, Trans. Faraday Soc. 59, 841 (1963).

Potassium Carbonate (K_2CO_3) $C_1K_2O_3(l)$

Enthalpy Reference Temperature = T_r = 298.15 K Standard State Pressure = p° = 0.1 MPa

T/K	$C_p°$ (J K⁻¹mol⁻¹)	S° (J K⁻¹mol⁻¹)	-[G°-H°(T_r)]/T (J K⁻¹mol⁻¹)	H°-H°(T_r) (kJ mol⁻¹)	$\Delta_fH°$ (kJ mol⁻¹)	$\Delta_fG°$ (kJ mol⁻¹)	Log K_f
0							
100							
200							
298.15	114.437	170.373	170.373	0.	-1130.609	-1049.385	183.848
300	114.699	171.082	170.375	0.212	-1130.604	-1048.881	182.626
400	128.135	205.957	175.039	12.367	-1134.885	-1020.799	133.303
500	140.049	235.859	184.280	25.790	-1133.593	-992.408	103.676
600	150.666	262.347	195.124	40.334	-1131.445	-964.360	83.955
700	160.483	286.316	206.465	55.896	-1128.530	-936.732	69.900
800	183.110	308.360	217.832	72.422	-1124.884	-909.565	59.388
800.000	183.110	308.360	217.832	72.422	GLASS <--> LIQUID		
800.000	209.200	308.360	217.832	72.422	TRANSITION		
900	209.200	333.000	229.286	93.342	-1117.071	-883.121	51.255
1000	209.200	355.041	240.779	114.262	-1109.493	-857.535	44.793
1100	209.200	374.980	252.087	135.182	-1260.068	-823.476	39.104
1174.000	209.200	388.601	260.267	150.663	CRYSTAL <--> LIQUID		
1200	209.200	393.183	263.098	156.102	-1250.880	-784.193	34.135
1300	209.200	409.928	273.757	177.022	-1241.791	-745.671	29.961
1400	209.200	425.431	284.044	197.942	-1232.789	-707.846	26.410
1500	209.200	439.865	293.957	218.862	-1223.865	-670.663	23.355
1600	209.200	453.366	303.502	239.782	-1215.012	-634.072	20.700
1700	209.200	466.049	312.695	260.702	-1206.226	-598.032	18.375
1800	209.200	478.006	321.550	281.622	-1197.507	-562.508	16.324
1900	209.200	489.317	330.084	302.542	-1188.853	-527.467	14.501
2000	209.200	500.048	338.317	323.462	-1180.265	-492.879	12.873
2100	209.200	510.255	346.263	344.382	-1171.745	-458.720	11.410
2200	209.200	519.987	353.940	365.302	-1163.295	-424.964	10.090
2300	209.200	529.286	361.363	386.222	-1154.918	-391.592	8.893
2400	209.200	538.189	368.547	407.142	-1146.618	-358.582	7.804
2500	209.200	546.729	375.504	428.062	-1138.398	-325.917	6.810

PREVIOUS: December 1961 CURRENT: March 1966

POTASSIUM CARBONATE (K_2CO_3)   CRYSTAL–LIQUID M_r = 138.2058

0 to 1174 K crystal
above 1174 K liquid

Refer to the individual tables for details.

Potassium Carbonate (K_2CO_3) $C_1K_2O_3$(cr,l)

Enthalpy Reference Temperature = T_r = 298.15 K   Standard State Pressure = p° = 0.1 MPa

T/K	C_p°	S°	$-[G^\circ-H^\circ(T_r)]/T$	$H^\circ-H^\circ(T_r)$	Δ_fH°	Δ_fG°	Log K_f
	J K⁻¹mol⁻¹			kJ mol⁻¹			
0	0.	0.	INFINITE	-22.673	-1144.614	-1144.614	INFINITE
100	72.383	53.095	246.061	-19.297	-1149.119	-1121.217	585.664
200	99.140	113.018	165.506	-10.498	-1150.202	-1092.747	285.396
298.15	114.437	155.519	155.519	0.	-1150 182	-1064.528	186.501
300	114.699	156.228	155.521	0.212	-1150.176	-1063.997	185.258
400	128.135	191.103	160.186	12.367	-1154.457	-1034.430	135.083
500	140.049	221.005	169.426	25.790	-1153.165	-1004.553	104.945
600	150.666	247.493	180.270	40.334	-1151.017	-975.020	84.883
700	160.483	271.462	191.611	55.896	-1148.102	-945.907	70.584
800	170.043	293.516	202.989	72.422	-1144.456	-917.262	59.891
900	179.515	314.092	214.202	89.901	-1140.085	-889.118	51.603
1000	188.949	333.495	225.170	108.325	-1135.003	-861.499	45.000
1100	198.321	351.944	235.863	127.688	-1287.134	-825.202	39.186
1174.000	205.259	365.079	243.596	142.621	CRYSTAL <--> LIQUID		
1174.000	209.200	388.601	243.596	170.235	TRANSITION		
1200	209.200	393.183	246.788	175.675	-1250.880	-784.193	34.135
1300	209.200	409.928	258.701	196.595	-1241.791	-745.671	29.961
1400	209.200	425.431	270.064	217.515	-1232.789	-707.846	26.410
1500	209.200	439.865	280.908	238.435	-1223.865	-670.663	23.355
1600	209.200	453.366	291.270	259.355	-1215.012	-634.072	20.700
1700	209.200	466.049	301.181	280.275	-1206.226	-598.032	18.375
1800	209.200	478.006	310.676	301.195	-1197.507	-562.508	16.324
1900	209.200	489.317	319.783	322.115	-1188.853	-527.467	14.501
2000	209.200	500.048	328.530	343.035	-1180.265	-492.879	12.873
2100	209.200	510.255	336.943	363.955	-1171.745	-458.720	11.410
2200	209.200	519.987	345.044	384.875	-1163.295	-424.964	10.090
2300	209.200	529.286	352.854	405.795	-1154.918	-391.592	8.893
2400	209.200	538.189	360.392	426.715	-1146.618	-358.582	7.804
2500	209.200	546.729	367.676	447.635	-1138.398	-325.917	6.810

PREVIOUS: CURRENT: March 1966

Potassium Carbonate (K_2CO_3) $C_1K_2O_3$(cr,l)

LITHIUM CARBONATE (Li_2CO_3) CRYSTAL M_r = 73.8912

S°(298.15 K) = 90.171 ± 0.21 J K^{-1} mol^{-1}

T_{trs1} = 623 K

T_{trs2} = 683 ± 3 K

T_{fus} = 993 ± 1 K

$\Delta_f H°$(0 K) = -1207.891 ± 0.17 kJ mol^{-1}

$\Delta_f H°$(298.15 K) = -1216.038 ± 0.17 kJ mol^{-1}

$\Delta_{trs1}H°$ = [0.561] kJ mol^{-1}

$\Delta_{trs2}H°$ = [2.238] kJ mol^{-1}

$\Delta_{fus}H°$ = 44.769 ± 0.4 kJ mol^{-1}

Enthalpy of Formation

The enthalpy of formation was derived from the enthalpy change of the following reactions:

	$\Delta_r H°$(298.15 K) kcal mol^{-1}	References
2 Li^+(aq, ∞) + CO_3^{--}(aq, ∞) = Li_2CO_3(cr)	4.20	[1]
2 Li(cr) + 2 H_2O(ℓ) = 2 Li^+(aq, ∞) + 2 OH^-(aq, ∞) + H_2(g)	-106.310	[2][3][4][5]

Summation of these two reactions gives

2 Li(cr) + 2 H_2O(ℓ) + CO_3^{--}(aq, ∞) = Li_2CO_3(cr) + 2 OH^-(aq, ∞) + H_2(g) -102.110

Using $\Delta_f H°$ (6) of H_2O(ℓ), CO_3^{--}(aq, ∞) and OH^-(aq, ∞) yields $\Delta_f H°$ (Li_2CO_3, cr, 298.15 K) = -290.64 kcal mol^{-1}.

Heat Capacity and Entropy

The low temperature heat capacity, 16.68-300.20 K, was measured by Brown and Latimer (1). Janz et al. (7) measured the heat content by drop calorimetry in the temperature range 560-1150 K and gave enthalpy and heat capacity equations based on their measurement. The above information was used in a Shomate analysis in order to smooth the enthalpy and calculate heat capacity. The values from the low and high temperature sources join smoothly at 298 K. The heat capacity was graphically extrapolated above the melting point. The entropy at 16.68 K was calculated from the extrapolated low temperature $C_p°$ data, using the cubic law. The value of S°(16.68 K) = 0.053 cal K^{-1} mol^{-1}.

Transition Data

T_{trs} were taken from Reisman (8). The enthalpies of transition were estimated from the relative areas of the transitions and the heat of fusion of the differential analysis curve reported by Reisman (8).

Fusion Data

T_{fus} was obtained from Reisman (8) and $\Delta_{fus}H°$ was taken from Janz et al. (7).

References

1. O. L. I. Brown and W. M. Latimer, J. Amer. Chem. Soc. 58, 2228 (1936).
2. JANAF Thermochemical tables: LiOH(cr), 3-31-66.
3. S. R. Gunn and L. G. Green, J. Amer. Chem. Soc. 80, 4782 (1958); S. R. Gunn, "Note Concerning an Electrical Heater Error in the Rocking-Bomb Solution Calorimeters," UCRL-7992, Oct. 1964.
4. C. B. Messer, L. G. Fasolino and C. E. Thalmayer, J. Amer. Chem. Soc. 77, 4524 (1955).
5. V. B. Parker, NSRDS-NBS 2, 1965.
6. U. S. Nat. Bur. Stand. Tech. Note 270-1, 1965.
7. G. I. Janz, E. Neuenschwander and F. J. Kelly, Trans. Faraday Soc. 59, 841 (1963).
8. A. Reisman, J. Amer. Chem. Soc. 80, 3558 (1958).

Lithium Carbonate (Li_2CO_3) $C_1Li_2O_3$(cr)

Enthalpy Reference Temperature = T_r = 298.15 K   Standard State Pressure = p° = 0.1 MPa

T/K	$C_p°$ (J $K^{-1}mol^{-1}$)	S°	-[G°-H°(T_r)]/T	H°-H°(T_r) (kJ mol^{-1})	$\Delta_f H°$	$\Delta_f G°$	Log K_f
0	0.	0.	INFINITE	-15.174	-1207.891	-1207.891	INFINITE
100	34.811	20.048	158.357	-13.831	-1211.955	-1186.400	619.712
200	72.425	56.513	98.271	-8.352	-1214.852	-1159.562	302.847
298.15	96.232	90.171	90.171	0.	-1216.038	-1132.120	198.343
300	96.567	90.767	90.172	0.178	-1216.048	-1131.600	197.029
400	112.173	120.752	94.156	10.638	-1216.269	-1103.403	144.090
500	128.951	147.525	102.190	22.668	-1222.020	-1074.589	112.262
600	149.369	172.780	111.863	36.550	-1220.421	-1045.224	90.995
623.150	154.643	178.533	114.233	40.069		I <--> II	
623.150	154.638	179.433	114.233	40.630		TRANSITION	
683.150	170.212	194.286	120.612	50.331		II <--> III	
683.150	138.782	197.563	120.612	52.569		TRANSITION	
700	140.875	200.969	122.505	54.925	-1214.552	-1016.396	75.844
800	158.992	220.962	133.565	69.917	-1212.279	-988.225	64.524
900	177.025	240.727	144.376	86.716	-1208.404	-960.434	55.742
993.150	194.096	258.989	154.270	104.001	-------- III	<--> LIQUID	-------
1000	194.723	260.350	154.993	105.357	-1202.852	-933.163	48.743
1100	205.936	279.471	165.446	125.428	-1196.001	-906.517	43.047
1200	212.673	297.699	175.714	146.382	-1188.373	-880.533	38.329
1300	216.564	314.883	185.765	167.853	-1180.320	-855.204	34.363
1400	219.325	331.040	195.571	189.656	-1172.009	-830.506	30.987
1500	221.083	346.233	205.114	211.678	-1163.538	-806.408	28.082
1600	222.589	360.553	214.386	233.868	-1154.954	-782.878	25.558
1700	223.426	374.072	223.385	256.168	-1436.778	-745.536	22.908
1800	224.262	386.867	232.115	278.553	-1426.584	-705.169	20.463
1900	225.099	399.014	240.582	301.021	-1416.368	-665.369	18.292
2000	225.936	410.582	248.795	323.572	-1406.127	-626.107	16.352

PREVIOUS: CURRENT: March 1966

Lithium Carbonate (Li_2CO_3) $C_1Li_2O_3$(cr)

LITHIUM CARBONATE (Li_2CO_3) LIQUID M_r = 73.8912

S°(298.15 K) = [130.567] J K^{-1} mol^{-1}

T_{fus} = 993.15 ± 1 K

$\Delta_f H°$(298.15 K) =[-1175.083] kJ mol^{-1}

$\Delta_{fus}H°$ = 44.769 ± 0.4 kJ mol^{-1}

Enthalpy of Formation

$\Delta_f H°$(ℓ, 298.15 K) is calculated from $\Delta_f H°$(cr, 298.15 K) by adding $\Delta_{fus}H°$ and the difference in enthalpy, H°(993.15 K) - H°(298.15 K), between the crystal and liquid.

Heat Capacity and Entropy

A glass transition is assumed at 683.15 K. The heat capacity below 683.15 K is obtained from the heat capacity of the crystal. Above 683.15 K the heat capacity is adopted as 44.32 cal deg^{-1} mol^{-1}, based on the enthalpy measurements in the range 996-1150 K reported by Janz et al. (1). The entropy is calculated in a manner analogous to that used for the enthalpy of formation.

Fusion Data

See the crystal table for details.

Reference

1. G. J. Janz, E. Neuenschwander and F. J. Kelly, Trans. Faraday Soc. 59, 841 (1963).

Lithium Carbonate (Li_2CO_3) $C_1Li_2O_3$(l)

Enthalpy Reference Temperature = T_r = 298.15 K Standard State Pressure = p° = 0.1 MPa

T/K	$C_p°$	S°	$-[G°-H°(T_r)]/T$	$H°-H°(T_r)$	$\Delta_f H°$	$\Delta_f G°$	Log K_f
	J $K^{-1}mol^{-1}$			kJ mol^{-1}			
0							
100							
200							
298.15	96.232	130.567	130.567	0	-1175.083	-1103.210	193.278
300	96.567	131.163	130.569	0.178	-1175.094	-1102.764	192.008
400	112.173	161.148	134.553	10.638	-1175.314	-1078.607	140.852
500	128.951	187.921	142.586	22.668	-1181.066	-1053.833	110.093
600	149.369	213.176	152.259	36.550	-1179.467	-1028.507	89.540
623.150	154.643	218.930	154.629	40.069		I <--> II	
623.150	154.638	219.829	154.629	40.630		TRANSITION	
683.150	170.212	234.683	161.008	50.331		II <--> III	
683.150	185.435	234.683	161.008	50.331		TRANSITION	
700	185.435	239.201	162.836	53.455	-1175.067	-1003.674	74.895
800	185.435	263.962	173.964	71.999	-1169.242	-979.589	63 961
900	185.435	285.803	185.201	90.542	-1163.624	-956.222	55.498
993.150	185.435	304.066	195.507	107.816	--------	III <--> LIQUID	-------
1000	185.435	305.341	196.255	109.086	-1158.169	-933.471	48.760
1100	185.435	323.015	206.988	127.629	-1152.845	-911.260	43.272
1200	185.435	339.150	217.339	146.173	-1147.628	-889.528	38.720
1300	185.435	353.992	227.288	164.716	-1142.503	-868.229	34.886
1400	185.435	367.735	236.835	183.260	-1137.451	-847.320	31.614
1500	185.435	380.528	245.993	201.803	-1132.459	-826.772	28.791
1600	185.435	392.496	254.779	220.347	-1127.520	-806.554	26.331
1700	185.435	403.738	263.214	238.890	-1413.101	-772.291	23.730
1800	185.435	414.337	271.318	257.434	-1406.749	-734.781	21.323
1900	185.435	424.363	279.112	275.977	-1400.457	-697.621	19.179
2000	185.435	433.875	286.614	294.521	-1394.224	-660.791	17.258

PREVIOUS: CURRENT: March 1966

Lithium Carbonate (Li_2CO_3) $C_1Li_2O_3$(l)

LITHIUM CARBONATE (Li_2CO_3) CRYSTAL–CRYSTAL–CRYSTAL–LIQUID

M_r = 73.8912

0 to 623 K crystal, I
623 to 683 K crystal, II
683 to 993 K crystal, III
above 993 K liquid

Refer to the individual tables for details.

Lithium Carbonate (Li_2CO_3) $C_1Li_2O_3$(cr,l)

Enthalpy Reference Temperature = T_r = 298.15 K Standard State Pressure = p° = 0.1 MPa

T/K	C_p° (J K⁻¹mol⁻¹)	S° (J K⁻¹mol⁻¹)	$-[G^\circ-H^\circ(T_r)]/T$ (J K⁻¹mol⁻¹)	$H^\circ-H^\circ(T_r)$ (kJ mol⁻¹)	Δ_fH° (kJ mol⁻¹)	Δ_fG° (kJ mol⁻¹)	Log K_f
0	0.	0.	INFINITE	-15.174	-1207.891	-1207.891	INFINITE
100	34.811	20.048	158.357	-13.831	-1211.955	-1186.400	619.712
200	72.425	56.513	98.271	-8.352	-1214.852	-1159.562	302.847
298.15	96.232	90.171	90.171	0.	-1216.038	-1132.120	198.343
300	96.567	90.767	90.172	0.178	-1216.048	-1131 600	197.029
400	112.173	120.752	94.156	10.638	-1216.269	-1103.403	144.090
500	128.951	147.525	102.190	22.668	-1222.020	-1074.589	112.262
600	149.369	172.780	111.863	36.550	-1220.421	-1045.224	90.995
623.150	154.643	178.533	114.233	40.069		I <--> II	
623.150	154.638	179.433	114.233	40.630		TRANSITION	
683.150	170.212	194.286	120.612	50.331		II <--> III	
683.150	138.782	197.563	120.612	52.569		TRANSITION	
700	140.875	200.969	122.505	54.925	-1214.552	-1016.396	75.844
800	158.992	220.962	133.565	69 917	-1212.279	-988.225	64.524
900	177.025	240.727	144.376	86.716	-1208.404	-960.434	55.742
993.150	194.096	258.989	154.270	104.001		III <--> LIQUID	
993.150	185.435	304.066	154.270	148.770		TRANSITION	
1000	185.435	305.341	155.301	150.040	-1158.169	-933.471	48.760
1100	185.435	323.015	169.757	168.584	-1152.845	-911.260	43.272
1200	185.435	339.150	183.210	187.127	-1147.628	-889.528	38.720
1300	185.435	353.992	195.784	205.671	-1142.503	-868.229	34.886
1400	185.435	367.735	207.582	224.214	-1137.451	-847.321	31.614
1500	185.435	380.528	218.690	242.758	-1132.459	-826.772	28.791
1600	185.435	392.496	229.183	261.301	-1127.520	-806.554	26.331
1700	185.435	403.738	239.124	279.845	-1413.101	-772.291	23.730
1800	185.435	414.337	248.566	298.388	-1406.749	-734.781	21.323
1900	185.435	424.363	257.557	316.932	-1400.457	-697.621	19.179
2000	185.435	433.875	266.137	335.475	-1394.224	-660.791	17.258

PREVIOUS: CURRENT: March 1966

Lithium Carbonate (Li_2CO_3) $C_1Li_2O_3$(cr,l)

MAGNESIUM CARBONATE ($MgCO_3$) CRYSTAL M_r = 84.3142

$S°(298.15\ K) = 65.854\ J\ K^{-1}\ mol^{-1}$

$\Delta_f H°(0\ K) = -1104.3 \pm 8\ kJ\ mol^{-1}$
$\Delta_f H°(298.15\ K) = -1111.7 \pm 8\ kJ\ mol^{-1}$

Enthalpy of Formation

The adopted $\Delta_f H°(298.15\ K) = -265.7\pm2\ kcal\ mol^{-1}$ ($-1111.689\pm8\ kJ\ mol^{-1}$) is the average value derived from the dissociation pressures of $MgCO_3$(cr) in the temperature range 671-782 K measured by Marc (1) and from the $\Delta_r H°(923\ K) = 25.9\pm0.6\ kcal\ mol^{-1}$ determined by Thomasson and Cunningham (2) using a high temperature differential calorimeter. The decomposition of $MgCO_3$ has been studied in a bomb calorimeter by Roth (3); this yields $\Delta_r H°(293\ K) = 23.8\ kcal\ mol^{-1}$ when corrected to constant pressure. The value is derived from the small difference between two large heats of combustion, so the uncertainty is probably large. Roth also reported $\Delta_r H°(363\ K) = 26.15\ kcal\ mol^{-1}$ from calorimetric data on the heat of solution in HCl. Berthelot (4) measured the heat of solution of $MgSO_4$(aq) with K_2CO_3(aq). Analyzing his data we find $\Delta_f H°(298.15\ K) = -268.6\ kcal\ mol^{-1}$.

Discrepancies are apparent in the wide range of the calorimetric results and in the 3rd law drift of the equilibrium data. Possible sources for the drift include lack of true equilibrium (e.g., excess surface energy for MgO) and uncertainty in the high temperature entropy of $MgCO_3$(see entropy write-up). Non-standard state behavior of the crystal phases may also be a problem in the calorimetric studies. The data of Thomasson and Cunningham (2), based on a high temperature differential thermal analysis technique, are probably the most reliable of the calorimetric values; however, there are uncertainties in the absolute energy calibration. The first result of Roth (3) may be dismissed as too discrepant and the remaining two calorimetric values are within ±3 kcal of the selected enthalpy of formation.

Source	Method	Reaction	T/K	$\Delta_r H°(298.15\ K)$ kcal mol		Drift cal K^{-1} mol^{-1}	$\Delta_f H°(298.15\ K)$ kcal mol^{-1}
Marc (1)	Kp	$MgCO_3(cr) = MgO(cr) + CO_2(g)$	671-782	-	28.3	4.8 ± 0.9	-266.04
Thomasson	Calorimetric	$MgCO_3(cr) = MgO(cr) + CO_2(g)$	923	25.9	27.7		-265.45
Roth (3)	Calorimetric	$MgCO_3(cr) = MgO(cr) + CO_2(g)$	293	23.8	23.8		-260.35
Roth (3)	Calorimetric	$MgCO_3(cr) = MgO(cr) + CO_2(g)$	363	26.15	26.10		-263.95
Berthelot (4)	Calorimetric	$K_2CO_3(\sim.5m) + MgSO_4(\sim.5m) = MgCO_3(cr) + K_2SO_4(\sim.5m)$	288	1.05	1.35		-268.6

Heat Capacity and Entropy

Low temperature heat capacities (56.3 - 291.6 K) were measured by Anderson (5). The high temperature heat capacities are derived by a Shomate analysis of smoothed enthalpies reported by Kelley (6). Kelley obtained the smoothed values from unpublished high temperature (400-743 K) enthalpy measurements by Shomate (7). The data were joined smoothly at 298.15 K and extrapolated graphically above 743 K. The entropy was obtained from the heat capacities using $S°(56.2) = 0.57\ cal\ K^{-1}\ mol^{-1}$.

Because of the drift in the decomposition pressure data, it is pertinent to assess the uncertainty in the functions. The greatest potential source for error is in the values above room temperature where the increment of entropy is about 20 cal K^{-1} mol^{-1} between 298.15 and 700 K. Unfortunately, no evaluation of the unpublished enthalpies is possible, but an increase of 10% in the entropy increment would reduce the drift to a tolerable level.

Decomposition Data

T_{dcm} is calculated as the temperature at which $\Delta_r G°$ equals zero for $MgCO_3(cr) = MgO(cr) + CO_2(g)$.

References

1. R. Marc and A. Simek, Z. Anorg. Chem. 82, 17 (1913).
2. C. V. Thomasson and D. A. Cunningham, J. Sci. Instr. 41, 308-10 (1964).
3. W. A. Roth, J. Prakt. Chem. 158, 117 (1941).
4. Berthelot, Ann. Chim. Phys. 4, 160-179 (1875).
5. C. T. Anderson, J. Amer. Chem. Soc. 56, 849 (1934).
6. K. K. Kelley, U. S. Bur. of Mines, 584, p. 114, 1960.
7. C. H. Shomate, U. S. Bur. of Mines, Berkeley Thermodynamics Laboratory, Region II, Berkeley, California, as referenced by Kelley (6).

Magnesium Carbonate ($MgCO_3$) $C_1Mg_1O_3(cr)$

Enthalpy Reference Temperature = T_r = 298.15 K Standard State Pressure = p° = 0.1 MPa

T/K	$C_p°$	S°	$-[G°-H°(T_r)]/T$	$H°-H°(T_r)$	$\Delta_f H°$	$\Delta_f G°$	Log K_f
	J $K^{-1}mol^{-1}$			kJ mol^{-1}			
0	0.	0.	INFINITE	-11.689	-1104.304	-1104.304	INFINITE
100	24.683	10.780	119.865	-10.909	-1108.582	-1082.618	565.502
200	57.572	39.145	72.274	-6.626	-1110.999	-1055.538	275.678
298.15	76.210	65.854	65.854	0.	-1111.689	-1028.124	180.123
300	76.501	66.327	65.856	0.141	-1111.691	-1027.606	178.922
400	89.855	90.253	69.028	8.490	-1111.377	-999.603	130 535
500	100.061	111.423	75.430	17.996	-1110.452	-971.756	101.519
600	109.035	130.486	83.043	28.466	-1109.068	-944.140	82.195
700	116.232	147.849	91.080	39.739	-1107.322	-916.786	68.411
800	122.256	163.771	99.185	51.669	-1105.306	-889.702	58.092
900	127.612	178.491	107.189	64.172	-1103.064	-862.883	50.080
1000	131.796	192.163	115.011	77.152	-1109.256	-835.612	43.648

PREVIOUS: December 1960 CURRENT: December 1966

Magnesium Carbonate ($MgCO_3$) $C_1Mg_1O_3(cr)$

CYANOGEN (CN) IDEAL GAS M_r = 26.0177

$S°(298.15\ K) = 202.64 \pm 0.04\ J\ K^{-1}\ mol^{-1}$

$\Delta_f H°(0\ K) = 436.8 \pm 10\ kJ\ mol^{-1}$
$\Delta_f H°(298.15\ K) = 435.1 \pm 10\ kJ\ mol^{-1}$

Electronic Levels and Quantum Weights

ϵ_i, cm^{-1}	g_i	ϵ_i, cm^{-1}	g_i
0	2	58960	2
9118	4	59683	4
25798	2	60939	4
53955	4	64782	4

$\omega_e = 2068.61\ cm^{-1}$ $\omega_e x_e = 13.114\ cm^{-1}$ $\sigma = 1$
$B_e = 1.8989\ cm^{-1}$ $\alpha_e = 0.0172\ cm^{-1}$ $r_e = 1.175$ Å

Enthalpy of Formation

The enthalpy of formation of CN has been the subject of considerable uncertainty for many years. Berkowitz (1) has reviewed the earlier work and he concludes that $\Delta_f H°$(CN, g, 0 K) = 109 kcal mol^{-1}. Since this review a considerable body of work has emerged which supports a value near 100 kcal mol^{-1}.

Tsang et al. (2) obtained $\Delta_f H°$(CN, g, 298.15 K) = 100 ± 4 kcal mol^{-1} from shock tube studies of the decomposition of cyanogen.

Dibeler and Liston (3, 4) have studied a series of cyanogen compounds and obtained their enthaplies of dissociation from photoionization threshold measurements. The values below are obtained from the halogen ion threshold measurements, except for HCN which is from the CN^+ threshold measurements. The value below is calculated using the IP from CNCl measurements (3).

Davis and Okabe (5) employed a direct photodissociation technique to obtain dissociation energies for many of the same molecules. The method determines the disappearance threshold for fluorescence due to decay of excited CN $B^2\Sigma^+ - X^2\Sigma^+$. This corresponds to the energy needed to just break the bond and form excited cyanogen in the $B^2\Sigma^+$ state, whose energy above ground is well known.

Setser and Steadman (6) used metastable excited argon atoms with a maximum electronic energy of 270 kcal mol^{-1} to break the bonds in HCN and $(CN)_2$. From the emission spectra of the excited $B^2\Sigma^+$ CN they were able to calculate the highest populated vibrational level and thus a maximum limit to the dissociation energy.

Berkowitz et al. (7) have recently examined the photoionization spectrum of HCN using the H^+ ion threshold. They also determined the threshold for the CN^+ ion which agrees exactly with that of Diebeler and Liston (4) when the latter is corrected for rotational energy contributions (0.7 kcal). Berkowitz et al. (7) suggest that the HCN limit of Setser and Steadman (6) should be ≤105 due to neglect of the significant vibrational population of the first harmonic of the bending mode. However, if the same allowance is made for the C_2N_2 data the limit is only ≤104 and this presumes significant occupation of the third harmonic of the bending mode of C_2N_2.

It is apparent from the data that considerable scatter still exists and several sources of error are probable. HCN and the cyanogen halides all have significant uncertainties in their enthalpy of formation, and the effects of rotational and vibrational contributions at 300 K are uncertain, but could be eliminated by using refrigerated gases. We choose to adopt a median value within the range of almost all determinations of $\Delta_f H°$(CN, g, 0 K) = 103.2 ± 2.5 kcal mol^{-1}.

X	$\Delta_r H°$(CNX → CN + X, 0 K), kcal mol^{-1}					$\Delta_f H°$(CN, g, 0 K), kcal mol^{-1}				
Reference	2	3, 4	5	6	7	2	3, 4	5	6	7
Cl		99.23±0.5	99.86±1.2				103.5±2.0	104.1±2.5		
Br		84.72±0.5	86.94±1.2				102.6±2.0	104.8±2.5		
I		72.95±0.5	71.72±1.0				101.4±2.0	100.2±2.5		
H		119.99±0.5	119.92±1.2	≤124	124.57±0.3		100.7±2.5	100.7±3.0	≤103	105.3±2.5
CN	125±8		129.61±1.2	≤133		100±4		101.5±0.8	≤103	

Heat Capacity and Entropy

The therodynamic functions are taken from Johnston et al. (8) calculated based on the direct summation of the energy levels. All the electronic levels and molecular constants listed above are from Carroll (9). Johnston et al. used only the first two excited electronic states but the contribution of the higher levels is negligible.

References
1. J. Berkowitz, J. Chem. Phys. 36, 2533 (1962).
2. W. Tsang, S. H. Bauer and M. Cowperthwaite, J. Chem. Phys. 36, 1768 (1962).
3. V. H. Dibeler and S. K. Liston, J. Chem. Phys. 47 4548 (1967).
4. V. H. Dibeler and S. K. Liston, J. Chem. Phys. 48, 4765 (1968).
5. D. D. Davis and H. Okabe, J. Chem. Phys. 49, 5526 (1968).
6. D. W. Setser and D. H. Steadman, J. Chem. Phys. 49, 467 (1968).
7. J. Berkowitz, W. A. Chupka, and T. A. Walter, J. Chem. Phys. 50, 1497 (1969).
8. H. L. Johnston, J. Belzer and L. Savedoff, "Thermodynamic Properties of CN between 1° and 6000° computed from spectroscopic data," TR 316-7 under contract N6onr-225, 1953. [AD 13734].
9. P. K. Carroll, Can. J. Phys. 34, 83 (1956).

Cyanogen (CN) $C_1N_1(g)$

Enthalpy Reference Temperature = T_r = 298.15 K   Standard State Pressure = p° = 0.1 MPa

T/K	$C_p°$	S°	$-[G°-H°(T_r)]/T$	$H°-H°(T_r)$	$\Delta_f H°$	$\Delta_f G°$	Log K_f
	J $K^{-1}mol^{-1}$			kJ mol^{-1}			
0	0.	0.	INFINITE	-8.672	436.848	436.848	INFINITE
100	29.104	170.835	228.528	-5.769	437.598	429.551	-224.374
200	29.108	191.009	205.304	-2.859	436.720	421.761	-110.153
298.15	29.154	202.638	202.638	0.	435.136	414.736	-72.660
300	29.157	202.819	202.639	0.054	435.101	414.609	-72.190
400	29.408	211.235	203.785	2.980	432.990	408.085	-53.290
500	29.936	217.849	205.959	5.945	430.493	402.140	-42.011
600	30.650	223.368	208.413	8.973	427.684	396.729	-34.538
700	31.423	228.151	210.898	12.077	424.621	391.809	-29.237
800	32.171	232.396	213.325	15.257	421.337	387.343	-25.291
900	32.852	236.226	215.660	18.509	417.848	383.301	-22.246
1000	33.456	239.719	217.894	21.825	405.565	380.371	-19.869
1100	33.987	242.933	220.026	25.198	401.664	378.041	-17.952
1200	34.457	245.911	222.060	28.620	397.730	376 066	-16.370
1300	34.882	248.686	224.003	32.088	393.770	374.422	-15.044
1400	35.277	251.285	225.860	35.596	262.382	376.249	-14.038
1500	35.656	253.732	227.637	39.143	259.742	384.475	-13.389
1600	36.031	256.046	229.341	42.727	257.093	392.877	-12.826
1700	36.408	258.241	230.977	46.349	254.441	401.444	-12.335
1800	36.795	260.333	232.550	50.009	251.791	410.168	-11.903
1900	37.194	262.333	234.066	53.708	249.148	419.039	-11.520
2000	37.608	264.251	235.527	57.448	246.516	428.049	-11.179
2100	38.033	266.097	236.939	61.230	243.897	437 190	-10.875
2200	38.474	267.876	238.305	65.055	241.297	446.455	-10.600
2300	38.921	269.596	239.629	68.925	238.716	455.839	-10.352
2400	39.374	271.262	240.912	72.840	236.158	465.334	-10.128
2500	39.829	272.879	242.159	76.800	233.623	474.935	-9.923
2600	40.282	274.450	243.371	80.806	231.112	484.637	-9.736
2700	40.730	275.978	244.550	84.856	228.625	494.435	-9.565
2800	41.169	277.467	245.699	88.951	226.161	504.325	-9.408
2900	41.596	278.920	246.820	93.090	223.718	514.303	-9.264
3000	42.008	280.337	247.913	97.270	221.296	524.364	-9.130
3100	42.403	281.721	248.982	101.491	218.891	534.506	-9.006
3200	42.779	283.073	250.026	105.750	216.501	544.726	-8.892
3300	43.135	284.395	251.048	110.046	214.123	555.019	-8.785
3400	43.470	285.688	252.048	114.376	211.753	565.385	-8.686
3500	43.783	286.952	253.027	118.739	209.388	575.821	-8.594
3600	44.073	288.190	253.986	123.132	207.024	586.323	-8.507
3700	44.341	289.401	254.927	127.553	204.656	596.892	-8.427
3800	44.586	290.587	255.850	131.999	202.281	607.525	-8.351
3900	44.810	291.748	256.756	136.469	199.894	618.220	-8.280
4000	45.013	292.885	257.645	140.961	197.491	628.977	-8.214
4100	45.196	293.999	258.518	145.471	195.068	639.794	-8.151
4200	45.359	295.090	259.376	149.999	192.621	650.671	-8.092
4300	45.504	296.159	260.219	154.542	190.145	661.606	-8.037
4400	45.631	297.206	261.048	159.099	187.638	672.599	-7.985
4500	45.742	298.233	261.863	163.668	185.094	683.649	-7.936
4600	45.838	299.240	262.664	168.247	182.510	694.757	-7.889
4700	45.920	300.226	263.453	172.835	179.883	705.921	-7.845
4800	45.988	301.194	264.229	177.431	177.209	717.141	-7.804
4900	46.044	302.143	264.993	182.032	174.486	728.418	-7.765
5000	46.090	303.073	265.746	186.639	171.708	739.751	-7.728
5100	46.126	303.986	266.486	191.250	168.875	751.140	-7.693
5200	46.153	304.882	267.216	195.864	165.984	762.585	-7.660
5300	46.171	305.762	267.935	200.480	163.031	774.086	-7.629
5400	46.183	306.625	268.644	205.098	160.015	785.643	-7.600
5500	46.187	307.472	269.342	209.717	156.934	797.257	-7.572
5600	46.186	308.305	270.030	214 335	153.784	808.928	-7.545
5700	46.175	309.122	270.709	218.953	150.565	820.655	-7.520
5800	46.166	309.925	271.378	223.571	147.273	832.440	-7.497
5900	46.145	310.714	272.038	228.186	143.908	844.282	-7.475
6000	46.129	311.489	272.689	232.800	140.468	856.181	-7.454

PREVIOUS: June 1969 (1 atm)   CURRENT: June 1969 (1 bar)

Cyanogen (CN) $C_1N_1(g)$

CYANOGEN, ION (CN^+) IDEAL GAS M_r = 26.01715

$S°(298.15\ K) = 213.44 \pm 6.3\ J\ K^{-1}\ mol^{-1}$

$\Delta_fH°(0\ K) = 1793.26 \pm 12.6\ kJ\ mol^{-1}$
$\Delta_fH°(298.15\ K) = 1802.76 \pm 12.6\ kJ\ mol^{-1}$

Electronic Levels and Quantum Weights

ε_i, cm^{-1}	g_i
0	[6]
[1500]	[1]
[6000]	[2]
[8000]	[3]
[15000]	[2]

ω_e = [1580] cm^{-1} ω_ex_e = [12.5] cm^{-1} σ = 1
B_e = [1.566] cm^{-1} α_e = [0.017] cm^{-1} r_e = [1.29] Å

Enthalpy of Formation

Dibeler and Liston (1) have measured the photoionization thresholds for the processes:

$ClCN + h\nu \rightarrow Cl^+ + CN + e^-$ $\Delta_rH°(0\ K) = 399.44 \pm 0.5$ kcal
$ClCN + h\nu \rightarrow Cl + CN^+ + e^-$ $\Delta_rH°(0\ K) = 426.65 \pm 0.5$ kcal

By subtraction we obtain

$Cl + CN^+ \rightarrow Cl^+ + CN$ $\Delta_rH°(0\ K) = -27.21 \pm 0.7$ kcal

Using JANAF vlaues for Cl, Cl^+ and CN we obtain $\Delta_fH°(CN^+,\ 0\ K) = 430.6 \pm 3.0\ kcal\ mol^{-1}$ and IP(CN) = $327.42 \pm 0.7\ kcal\ mol^{-1}$.
Berkowitz et al. (2) have measured the photoionization thresholds for the processes:

$HCN + h\nu \rightarrow H^+ + CN + e^-$ $\Delta_rH°(0\ K) = 438.18 \pm 0.3$ kcal
$HCN + h\nu \rightarrow H + CN^+$ $\Delta_rH°(0\ K) = 448.09 \pm 0.3$ kcal

By subtraction we obtain

$H + CN^+ \rightarrow H^+ + CN$ $\Delta_rH°(0\ K) = -9.91 \pm 0.4$ kcal

Using JANAF values for H, H^+ and CN we obtain $\Delta_fH°(CN^+,\ 0\ K) = 426.7 \pm 3.0\ kcal\ mol^{-1}$, and IP(CN) = $323.51 \pm 0.4\ kcal\ mol^{-1}$.
We adopt a median value for $\Delta_fH°(CN^+,\ 0\ K) = 428.6 \pm 3.0\ kcal\ mol^{-1}$ and $325.4 \pm 2\ kcal\ mol^{-1}$ for IP(CN).

Heat Capacity and Entropy

The molecule has eight valence electrons and thus we expect that its molecular constants should fall between those for C_2 and BN. The ground electronic state is either $^1\Sigma$ or $^3\Pi$ and the two levels are likely to be very close. As a consequence the thermodynamic functions are dominated by the $^3\Pi$ contribution and so we choose this to be the ground state. The other levels are estimated based on the values of Fougere and Nesbet (3) for C_2 and of Verhaegen et al. (4) for BN. The vibrational constants and bond length are estimated between those for the $^3\Pi$ states of BN and C_2. The rotational constants B_e is obtained from the estimated bond length and α_e is calculated from the other constants assuming a Morse potential function.

References

1. V. H. Dibeler and S. K. Liston, J. Chem. Phys. 47, 4548 (1967).
2. J. Berkowitz, W. A. Chupka and T. A. Walter, J. Chem. Phys. 50, 1497 (1969).
3. P. F. Fougere and R. K. Nesbet, J. Chem. Phys. 44, 285 (1966).
4. G. Verheagen, W. G. Richards and C. M. Moser, J. Chem. Phys. 46, 160 (1967).

Cyanogen, Ion (CN^+) $C_1N_1^+(g)$

Enthalpy Reference Temperature = T_r = 298.15 K Standard State Pressure = p° = 0.1 MPa

T/K	$C_p°$ (J K^{-1}mol^{-1})	S°	$-[G°-H°(T_r)]/T$	$H°-H°(T_r)$ (kJ mol^{-1})	$\Delta_fH°$	$\Delta_fG°$	Log K_f
0	0.	0.	INFINITE	-8.685	1793.264		
100	29.106	181.588	239.398	-5.781			
200	29.130	201.766	216.115	-2.870			
250	29.221	208.274	213.920	-1.411			
298.15	29.431	213.437	213.437	0.	1802.760	1763.145	-308.896
300	29.442	213.619	213.437	0.054	1802.810	1762.899	-306.948
350	29.809	218.183	213.797	1.535	1804.131	1756.142	-262.090
400	30.294	222.194	214.601	3.037	1805.390	1749.199	-228.422
450	30.852	225.794	215.648	4.566	1806.597	1742.102	-202.218
500	31.440	229.075	216.829	6.123	1807.758	1734.874	-181.241
600	32.585	234.910	219.368	9.325	1809.969	1720.087	-149.747
700	33.585	240.010	221.960	12.635	1812.063	1704.940	-127.224
800	34.412	244.551	224.505	16.036	1814.068	1689.498	-110.313
900	35.083	248.644	226.964	19.512	1815.999	1673.811	-97.145
1000	35.630	252.370	229.321	23.049	1817.871	1657.911	-86.600
1100	36.083	255.788	231.574	26.635	1819.694	1641.826	-77.964
1200	36.469	258.944	233.725	30.263	1821.475	1625.578	-70.760
1300	36.807	261.877	235.779	33.927	1823.221	1609.182	-64.658
1400	37.112	264.616	237.742	37.624	1824.940	1592.653	-59.423
1500	37.394	267.186	239.620	41.349	1826.635	1576.001	-54.881
1600	37.659	269.608	241.419	45.102	1828.312	1559.238	-50.904
1700	37.912	271.899	243.146	48.881	1829.975	1542.370	-47.391
1800	38.155	274.073	244.804	52.684	1831.626	1525.404	-44.266
1900	38.389	276.142	246.399	56.511	1833.267	1508.347	-41.467
2000	38.615	278.117	247.936	60.362	1834.903	1491.203	-38.946
2100	38.832	280.006	249.419	64.234	1836.532	1473.978	-36.663
2200	39.042	281.818	250.850	68.128	1838.158	1456.676	-34.586
2300	39.242	283.557	252.235	72.042	1839.781	1439.299	-32.688
2400	39.435	285.232	253.575	75.976	1841.401	1421.852	-30.946
2500	39.618	286.845	254.874	79.929	1843.019	1404.338	-29.342
2600	39.792	288.402	256.134	83.899	1844.635	1386.759	-27.860
2700	39.957	289.907	257.357	87.887	1846.249	1369.118	-26.487
2800	40.113	291.363	258.545	91.890	1847.861	1351.416	-25.211
2900	40.261	292.774	259.702	95.909	1849.471	1333.658	-24.022
3000	40.400	294.141	260.827	99.942	1851.078	1315.844	-22.911
3100	40.530	295.468	261.923	103.989	1852.683	1297.976	-21.871
3200	40.652	296.756	262.991	108.048	1854.284	1280.057	-20.895
3300	40.766	298.009	264.034	112.119	1855.882	1262.088	-19.977
3400	40.873	299.228	265.051	116.201	1857.476	1244.070	-19.113
3500	40.972	300.414	266.045	120.293	1859.065	1226.005	-18.297
3600	41.065	301.570	267.015	124.395	1860.650	1207.896	-17.526
3700	41.151	302.696	267.965	128.506	1862.230	1189.742	-16.796
3800	41.231	303.794	268.893	132.625	1863.804	1171.545	-16.104
3900	41.305	304.866	269.802	136.752	1865.372	1153.307	-15.447
4000	41.373	305.913	270.691	140.886	1866.933	1135.030	-14.822
4100	41.436	306.935	271.563	145.026	1868.488	1116.713	-14.227
4200	41.495	307.935	272.417	149.173	1870.036	1098.358	-13.660
4300	41.548	308.912	273.255	153.325	1871.576	1079.966	-13.119
4400	41.598	309.867	274.076	157.483	1873.109	1061.539	-12.602
4500	41.643	310.803	274.882	161.645	1874.634	1043.077	-12.108
4600	41.685	311.718	275.673	165.811	1876.150	1024.581	-11.634
4700	41.724	312.615	276.449	169.982	1877.658	1006.053	-11.181
4800	41.759	313.494	277.212	174.156	1879.157	987.492	-10.746
4900	41.791	314.355	277.961	178.333	1880.647	968.900	-10.329
5000	41.820	315.200	278.697	182.514	1882.128	950.278	-9.927
5100	41.846	316.028	279.421	186.697	1883.599	931.627	-9.542
5200	41.871	316.841	280.133	190.883	1885.060	912.946	-9.171
5300	41.893	317.639	280.833	195.071	1886.511	894.238	-8.813
5400	41.913	318.422	281.522	199.261	1887.952	875.502	-8.469
5500	41.931	319.192	282.200	203.454	1889.382	856.740	-8.137
5600	41.947	319.947	282.867	207.647	1890.803	837.952	-7.816
5700	41.962	320.690	283.524	211.843	1892.212	819.138	-7.507
5800	41.975	321.420	284.171	216.040	1893.610	800.300	-7.207
5900	41.987	322.137	284.809	220.238	1894.997	781.438	-6.918
6000	41.997	322.843	285.437	224.437	1896.373	762.553	-6.639

PREVIOUS: December 1970 (1 atm) CURRENT: December 1970 (1 bar)

Cyanogen, Ion (CN^+) $C_1N_1^+(g)$

CYANIDE (CN^-) IDEAL GAS M_r = 26.01825

$S°(298.15\ K) = 195.98 \pm 2.1\ J\ K^{-1}\ mol^{-1}$

$\Delta_f H°(0\ K) = 63.6 \pm 13\ kJ\ mol^{-1}$
$\Delta_f H°(298.15\ K) = 60.7 \pm 13\ kJ\ mol^{-1}$

Electronic Levels and Quantum Weights

State	ε_i, cm^{-1}	g_i
$[{}^1\Sigma^+]$	0	[1]
	[67000]	[2]

$\omega_e = [2250]\ cm^{-1}$ $\omega_e x_e = [13.8]\ cm^{-1}$ $\sigma = 1$
$B_e = [2.116]\ cm^{-1}$ $\alpha_e = [0.0177]\ cm^{-1}$ $r_e = [1.11]$ Å

Enthalpy of Formation

Berkowitz et al. (1) reported 88 ± 0.5 kcal mol^{-1} for the electron affinity of CN from the photoionization thresholds of the reactions $HCN \rightarrow N^+ + CN^-$ and $HCN \rightarrow H^+ = CN + e^-$.

Other values have been reported by Page (2), who reinterpreted his earlier data (3), to yield 73 ± 1 kcal mol^{-1}; Bakulina and Ionov (4) in conjunction with the data of Branscomb and Smith (5) reported 78 ± 5 kcal mol^{-1} and Inoue (6) who obtained 83 ± 7 kcal mol^{-1}.

We adopt the value given by Berkowitz et al. (1) as being the most well defined in terms of both the precision of measurements and the reactions being studied. This leads to $\Delta_f H°(CN^-, g, 0\ K) = 15.2 \pm 3$ kcal mol^{-1} or $\Delta_f H°(CN^-, g, 298.15\ K) = 14.5 \pm 3$ kcal mol^{-1}.

Heat Capacity and Entropy

The electronic structure and molecular constants are estimated by analogy with the isoelectronic species CO and N_2. The uncertainties in the constants are generally small since N_2 and CO are quite similar, thus the entropy has an uncertainty of only 0.5 cal K^{-1} mol^{-1}.

References

1. J. Berkowitz, W. A. Chupka, and T. A. Walter, J. Chem. Phys. 50, 1497 (1969).
2. F. M. Page, J. Chem. Phys. 49, 2466 (1968).
3. R. Napper and F. M. Page, Trans. Faraday Soc. 59, 1086 (1963).
4. I. N. Bakulina and N. I. Ionov, Russ. J. Phys. Chem. 33, 236 (1959).
5. L. M. Branscomb and S. J. Smith, J. Chem. Phys. 25, 598 (1956).
6. M. Inoue, J. Chim. Phys. 63, 1061 (1966).

Cyanide (CN^-) $C_1N_1^-(g)$

Enthalpy Reference Temperature = T_r = 298.15 K Standard State Pressure = p° = 0.1 MPa

	J $K^{-1}mol^{-1}$			kJ mol^{-1}			
T/K	$C_p°$	S°	$-[G°-H°(T_r)]/T$	$H°-H°(T_r)$	$\Delta_f H°$	$\Delta_f G°$	Log K_f
0	0.	0.	INFINITE	-8.670	63.581		
100	29.105	164.176	221.860	-5.768			
200	29.109	184.351	198.640	-2.858			
250	29.114	190.847	196.456	-1.402			
298.15	29.133	195.976	195.976	0.	60.668	38.768	-6.792
300	29.135	196.156	195.977	0.054	60.641	38.632	-6.726
350	29.190	200.651	196.332	1.512	59.860	35.025	-5.227
400	29.297	204.555	197.121	2.974	59.001	31.535	-4.118
450	29.466	208.015	198.143	4.443	58.069	28.157	-3.268
500	29.695	211.131	199.288	5.921	57.073	24.886	-2.600
600	30.291	216.595	201.730	8.919	54.923	18.647	-1.623
700	30.986	221.316	204.198	11.983	52.613	12.782	-0.954
800	31.693	225.500	206.604	15.117	50.193	7.256	-0.474
900	32.362	229.272	208.917	18.320	47.695	2.038	-0.118
1000	32.969	232.714	211.127	21.587	45.140	-2.898	0.151
1100	33.506	235.882	213.235	24.912	42.544	-7.576	0.360
1200	33.976	238.818	215.246	28.286	39.914	-12.017	0.523
1300	34.385	241.554	217.166	31.705	37.258	-16.237	0.652
1400	34.741	244.116	219.000	35.161	34.580	-20.252	0.756
1500	35.050	246.523	220.756	38.651	31.882	-24.074	0.838
1600	35.321	248.794	222.438	42.170	29.168	-27.716	0.905
1700	35.557	250.943	224.052	45.714	26.439	-31.188	0.958
1800	35.766	252.981	225.603	49.281	23.695	-34.499	1.001
1900	35.950	254.920	227.095	52.867	20.939	-37.657	1.035
2000	36.114	256.768	228.533	56.470	18.170	-40.669	1.062
2100	36.260	258.534	229.920	60.089	15.389	-43.542	1.083
2200	36.391	260.224	231.259	63.722	12.596	-46.283	1.099
2300	36.510	261.844	232.554	67.367	9.793	-48.897	1.110
2400	36.617	263.400	233.807	71.023	6.978	-51.388	1.118
2500	36.714	264.897	235.021	74.690	4.153	-53.762	1.123
2600	36.804	266.339	236.198	78.366	1.317	-56.023	1.126
2700	36.885	267.729	237.340	82.050	-1.529	-58.174	1.125
2800	36.961	269.072	238.450	85.743	-4.386	-60.220	1.123
2900	37.030	270.370	239.528	89.442	-7.252	-62.164	1.120
3000	37.095	271.627	240.577	93.149	-10.129	-64.008	1.114
3100	37.155	272.844	241.599	96.861	-13.016	-65.757	1.108
3200	37.212	274.025	242.593	100.579	-15.912	-67.412	1.100
3300	37.265	275.170	243.563	104.303	-18.819	-68.976	1.092
3400	37.314	276.284	244.509	108.032	-21.735	-70.452	1.082
3500	37.361	277.366	245.433	111.766	-24.661	-71.843	1.072
3600	37.406	278.419	246.335	115.504	-27.597	-73.149	1.061
3700	37.448	279.445	247.216	119.247	-30.543	-74.374	1.050
3800	37.488	280.444	248.077	122.994	-33.499	-75.519	1.038
3900	37.526	281.418	248.919	126.745	-36.464	-76.586	1.026
4000	37.563	282.369	249.744	130.499	-39.439	-77.577	1.013
4100	37.598	283.297	250.551	134.257	-42.424	-78.493	1.000
4200	37.632	284.203	251.341	138.019	-45.418	-79.337	0.987
4300	37.664	285.089	252.116	141.784	-48.422	-80.109	0.973
4400	37.696	285.955	252.875	145.552	-51.436	-80.811	0.959
4500	37.726	286.803	253.620	149.323	-54.460	-81.445	0.945
4600	37.755	287.632	254.350	153.097	-57.493	-82.011	0.931
4700	37.783	288.444	255.067	156.874	-60.536	-82.511	0.917
4800	37.811	289.240	255.771	160.653	-63.588	-82.947	0.903
4900	37.838	290.020	256.462	164.436	-66.650	-83.318	0.888
5000	37.864	290.785	257.141	168.221	-69.723	-83.627	0.874
5100	37.889	291.535	257.808	172.009	-72.805	-83.875	0.859
5200	37.914	292.271	258.463	175.799	-75.897	-84.062	0.844
5300	37.939	292.993	259.108	179.591	-78.998	-84.189	0.830
5400	37.962	293.703	259.742	183.386	-82.110	-84.258	0.815
5500	37.986	294.399	260.366	187.184	-85.231	-84.269	0.800
5600	38.009	295.084	260.980	190.984	-88.362	-84.223	0.786
5700	38.031	295.757	261.584	194.786	-91.504	-84.121	0.771
5800	38.053	296.419	262.179	198.590	-94.655	-83.964	0.756
5900	38.075	297.069	262.765	202.396	-97.817	-83.753	0.741
6000	38.096	297.709	263.342	206.205	-100.989	-83.488	0.727

PREVIOUS: December 1970 (1 atm) CURRENT: December 1970 (1 bar)

Cyanide (CN^-) $C_1N_1^-(g)$

SODIUM CYANIDE (NaCN) CRYSTAL M_r = 49.00747

$S°(298.15 K) = [118.47 \pm 2.1] J K^{-1} mol^{-1}$

T_{fus} = 835 K

$\Delta_f H°(0 K) = -98.3 \pm 1.3 kJ mol^{-1}$

$\Delta_f H°(298.15 K) = -90.7 \pm 1.3 kJ mol^{-1}$

$\Delta_{fus} H° = [8.786] kJ mol^{-1}$

Enthalpy of Formation

The enthalpy change for the reaction NaCN(cr) + NaCN(aq, ∞) was reported to be 0.29 ± 0.20 kcal mol^{-1} by Parker (1). Incorporating this data with the heats of formation for Na^+(aq, ∞), -57.39, and CN^-(aq, ∞), 36.0 kcal mol^{-1}, reported by Wagman et al. (2), the value of $\Delta_f H°$(298.15 K) for NaCN(cr) was derived and adopted.

Lord and Woolf (3) determined the value, $\Delta_f H°$(NaCN, cr, 298.15 K) = -21.4 kcal mol^{-1}, by solution calorimetry. This value is in agreement with the adopted one.

Heat Capacity and Entropy

The low temperature heat capacities, 101.6-345.8 K, were reported by Messer and Ziegler (4). The $C°_p$ values above 345.8 K were estimated by graphical extrapolation. S°(298.15 K) was derived from low temperature heat capacities, based on S°(100 K) = 8.4 cal deg^{-1} mol^{-1} which was estimated by comparison with the S°(298.15 K) values for NaCl, KCl, and KCN crystals.

Transition Data

The transition temperatures, T_{trs_2} and T_{trs_1}, were determined by Messer and Ziegler (4). The corresponding enthalpies of transition were reported to be 0.15 and 0.70 kcal mol^{-1}, respectively. However, these quantities were incorporated with $C°_p$ values used to evaluate H°(0 K) - H°(298.15 K). Therefore $\Delta_{trs} H°$ values are not listed.

Fusion Data

See the liquid table for details.

Sublimation Data

The difference between $\Delta_f H°$(298.15 K) for $(NaCN)_2$(g) and 2 NaCN(cr) is $\Delta_{sub} H°$(cr → dimer, 298.15 K); for NaCN(g) and NaCN(cr) is $\Delta_{sub} H°$(cr→ monomer, 298.15 K).

References

1. V. B. Parker, NSRDS-NBS 2, 1965.
2. U. S. Nat. Bur. Stand. Tech. Note 270-1, 1965.
3. G. Lord and A. A. Woolf, J. Chem. Soc. 2546 (1954).
4. C. E. Messer and W. T. Ziegler, J. Amer. Chem. Soc. 63, 2703 (1941).

Sodium Cyanide (NaCN) $C_1N_1Na_1(cr)$

Enthalpy Reference Temperature = T_r = 298.15 K Standard State Pressure = p° = 0.1 MPa

	J K⁻¹mol⁻¹			kJ mol⁻¹			
T/K	$C°_p$	S°	$-[G°-H°(T_r)]/T$	$H°-H°(T_r)$	$\Delta_f H°$	$\Delta_f G°$	Log K_f
0	0.	0.	INFINITE	-19.422	-98.299	-98.299	INFINITE
100	47.363	35.146	200.082	-16.494	-98.220	-91.268	47.673
172.100	93.722	69.586	138.127	-11.796		I <--> II	
172.100	93.724	69.588	138.127	-11.796		TRANSITION	
200	67.877	80.075	129.338	-9.853	-95.817	-85.077	22.220
288.50	1548.088	111.062	118.520	-2.152		II <--> III	
288.50	1548.129	111.063	118.520	-2.151		TRANSITION	
298.15	68.681	118.467	118.467	0.	-90.709	-80.413	14.088
300	68.575	118.892	118.469	0.127	-90.677	-80.349	13.990
400	68.659	138.659	121.165	6.998	-91.909	-76.931	10.046
500	68.752	153.990	126.254	13.868	-90.937	-73.303	7.658
600	68.848	166.534	131.953	20.748	-90.143	-69.854	6.081
700	68.945	177.154	137.671	27.638	-89.499	-66.526	4.964
800	69.046	186.367	143.195	34.537	-88.985	-63.280	4.132
835.000	69.082	189.324	145.067	36.955	--------	III <--> LIQUID	-------
900	69.145	194.505	148.453	41.447	-88.585	-60.092	3.488
1000	69.245	201.795	153.429	48.367	-88.296	-56.943	2.974
1100	69.354	208.400	158.131	55.297	-88.117	-53.817	2.556
1200	69.462	214.439	162.575	62.237	-184.812	-48.263	2.101
1300	69.570	220.004	166.781	69.189	-183.935	-36.920	1.483
1400	69.676	225.163	170.769	76.151	-183.108	-25.642	0.957
1500	69.781	229.974	174.558	83.124	-182.322	-14.422	0.502
1600	69.885	234.481	178.164	90.107	-181.572	-3.254	0.106
1700	69.988	238.721	181.602	97.101	-180.852	7.869	-0.242
1800	70.090	242.724	184.888	104.105	-180.158	18.950	-0.550
1900	70.191	246.516	188.032	111.119	-179.488	29.993	-0.825
2000	70.291	250.119	191.047	118.143	-178.837	41.002	-1.071

PREVIOUS: December 1960 CURRENT: March 1966

Sodium Cyanide (NaCN) $C_1N_1Na_1(cr)$

SODIUM CYANIDE (NaCN) LIQUID M_r = 49.00747

$S°(298.15\ K)$ = [125.507] $J\ K^{-1}\ mol^{-1}$ $\Delta_f H°(298.15\ K)$ = [-84.398] $kJ\ mol^{-1}$

T_{fus} = 835 K $\Delta_{fus}H°$ = [8.7] $kJ\ mol^{-1}$

Enthalpy of Formation

$\Delta_f H°(\ell,\ 298.15\ K)$ was obtained from $\Delta_f H°(cr,\ 298.15\ K)$ by adding $\Delta_{fus}H°$ and the difference in enthalpy, H°(835 K) - H°(298.15 K), between the crystal and liquid.

Heat Capacity and Entropy

The heat capacity is estimated by comparison with those for NaCN(cr), KCN(cr) and KCN(ℓ). A glass transition temperature is assumed at 600 K. The entropy is calculated in a manner analogous to that used for the enthalpy of formation.

Fusion Data

The melting temperature was reported by Truthe (1). $\Delta_{fus}H°$ is estimated such that the derived third law value for $\Delta_{vap}H°$(NaCN, ℓ) agrees with the 2nd law value. See the ideal gas table for details.

Vaporization Data

T_{vap} is the temperature at which the Gibbs energy change for the reaction NaCN(ℓ) = NaCN(g) becomes zero. The difference between $\Delta_f H°$ for NaCN(g) and NACN(ℓ) at T_{vap} is $\Delta_{vap}H°$.

Reference

1. W. Truthe, Z. Anorg. Chem. 76, 129 (1912).

Sodium Cyanide (NaCN) $C_1N_1Na_1(l)$

Enthalpy Reference Temperature = T_r = 298.15 K Standard State Pressure = p° = 0.1 MPa

	J K⁻¹mol⁻¹			kJ mol⁻¹			
T/K	$C_p°$	S°	$-[G°-H°(T_r)]/T$	$H°-H°(T_r)$	$\Delta_f H°$	$\Delta_f G°$	Log K_f
0							
100							
200							
298.15	68.680	125.507	125.507	0.	-84.398	-76.201	13.350
300	68.575	125.931	125.508	0.127	-84.366	-76.150	13.259
400	68.659	145.699	128.205	6.998	-85.598	-73.436	9.590
500	68.749	161.029	133.293	13.868	-84.626	-70.512	7.366
600	68.848	173.572	138.993	20.748	-83.832	-67.767	5.900
600.000	68.848	173.573	138.993	20.748	GLASS <--> LIQUID	TRANSITION	
600.000	79.496	173.573	138.993	20.748			
700	79.496	185.827	144.830	28.698	-82.128	-65.226	4.867
800	79.496	196.443	150.633	36.647	-80.563	-62.920	4.108
835.000	79.496	199.847	152.625	39.430	-------- III <-->	LIQUID -------	
900	79.496	205.806	156.254	44.597	-79.124	-60.802	3.529
1000	79.496	214.182	161.635	52.547	-77.805	-58.838	3.073
1100	79.496	221.758	166.762	60.496	-76.606	-57.000	2.707
1200	79.496	228.675	171.637	68.446	-172.292	-52.827	2.299
1300	79.496	235.039	176.273	76.395	-170.417	-42.948	1.726
1400	79.496	240.930	180.683	84.345	-168.603	-33.211	1.239
1500	79.496	246.414	184.885	92.295	-166.841	-23.601	0.822
1600	79.496	251.545	188.892	100.244	-165.124	-14.108	0.461
1700	79.496	256.364	192.721	108.194	-163.448	-4.722	0.145
1800	79.496	260.908	196.384	116.143	-161.809	4.568	-0.133
1900	79.496	265.206	199.894	124.093	-160.203	13.767	-0.378
2000	79.496	269.284	203.263	132.043	-158.627	22.883	-0.598
2100	79.496	273.163	206.500	139.992	-157.079	31.920	-0.794
2200	79.496	276.861	209.615	147.942	-155.559	40.884	-0.971
2300	79.496	280.395	212.616	155.891	-154.064	49.780	-1.131
2400	79.496	283.778	215.511	163.841	-152.593	58.611	-1.276
2500	79.496	287.023	218.307	171.791	-151.148	67.381	-1.408
2600	79.496	290.141	221.010	179.740	-149.726	76.094	-1.529
2700	79.496	293.141	223.626	187.690	-148.329	84.753	-1.640
2800	79.496	296.032	226.161	195.639	-146.956	93.360	-1.742
2900	79.496	298.822	228.619	203.589	-145.608	101.919	-1.836
3000	79.496	301.517	231.004	211.539	-144.287	110.432	-1.923

PREVIOUS: December 1960 CURRENT: March 1966

Sodium Cyanide (NaCN) $C_1N_1Na_1(l)$

SODIUM CYANIDE (NaCN) CRYSTAL–LIQUID M_r = 49.00747

0 to 835 K crystal
above 835 K liquid

Refer to the individual tables for details.

Sodium Cyanide (NaCN) $C_1N_1Na_1(cr,l)$

Enthalpy Reference Temperature = T_r = 298.15 K Standard State Pressure = p° = 0.1 MPa

T/K	C_p° (J K⁻¹mol⁻¹)	S° (J K⁻¹mol⁻¹)	$-[G°-H°(T_r)]/T$ (J K⁻¹mol⁻¹)	$H°-H°(T_r)$ (kJ mol⁻¹)	$\Delta_fH°$ (kJ mol⁻¹)	$\Delta_fG°$ (kJ mol⁻¹)	Log K_f
0	0.	0.	INFINITE	-19.422	-98.299	-98.299	INFINITE
100	47.363	35.146	200.082	-16.494	-98.220	-91.268	47.673
172.100	93.722	69.586	138.127	-11.796		I <--> II	
172.100	93.724	69.588	138.127	-11.796		TRANSITION	
200	67.877	80.075	129.338	-9.853	-95.817	-85.077	22.220
288.50	1548.088	111.062	118.520	-2.152		II <--> III	
288.50	1548.129	111.063	118.520	-2.151		TRANSITION	
298.15	68.681	118.467	118.467	0.	-90.709	-80.413	14.088
300	68.575	118.892	118.469	0.127	-90.677	-80.349	13.990
400	68.659	138.659	121.165	6.998	-91.909	-76.931	10.046
500	68.752	153.990	126.254	13.868	-90.937	-73.303	7.658
600	68.848	166.534	131.953	20.748	-90.143	-69.854	6.081
700	68.945	177.154	137.671	27.638	-89.499	-66.526	4.964
800	69.046	186.367	143.195	34.537	-88.985	-63.280	4.132
835.000	69.082	189.324	145.067	36.955		III <--> LIQUID	
835.000	79.496	199.847	145.067	45.741		TRANSITION	
900	79.496	205.806	149.241	50.908	-79.124	-60.802	3.529
1000	79.496	214.182	155.324	58.858	-77.805	-58.838	3.073
1100	79.496	221.758	161.024	66.807	-76.606	-57.000	2.707
1200	79.496	228.675	166.378	74.757	-172.292	-52.826	2.299
1300	79.496	235.039	171.418	82.707	-170.417	-42.947	1.726
1400	79.496	240.930	176.175	90.656	-168.603	-33.210	1.239
1500	79.496	246.414	180.677	98.606	-166.840	-23.601	0.822
1600	79.496	251.545	184.948	106.555	-165.124	-14.108	0.461
1700	79.496	256.364	189.009	114.505	-163.448	-4.721	0.145
1800	79.496	260.908	192.878	122.455	-161.809	4.568	-0.133
1900	79.496	265.206	196.573	130.404	-160.202	13.767	-0.378
2000	79.496	269.284	200.107	138.354	-158.627	22.883	-0.598
2100	79.496	273.163	203.494	146.303	-157.079	31.920	-0.794
2200	79.496	276.861	206.746	154.253	-155.559	40.884	-0.971
2300	79.496	280.395	209.872	162.203	-154.064	49.780	-1.131
2400	79.496	283.778	212.881	170.152	-152.593	58.611	-1.276
2500	79.496	287.023	215.782	178.102	-151.147	67.381	-1.408
2600	79.496	290.141	218.583	186.051	-149.726	76.094	-1.529
2700	79.496	293.141	221.289	194.001	-148.328	84.753	-1.640
2800	79.496	296.032	223.907	201.951	-146.956	93.361	-1.742
2900	79.496	298.822	226.442	209.900	-145.608	101.919	-1.836
3000	79.496	301.517	228.900	217.850	-144.287	110.432	-1.923

PREVIOUS: CURRENT: March 1966

Sodium Cyanide (NaCN) $C_1N_1Na_1(cr,l)$

SODIUM CYANIDE (NaCN) IDEAL GAS M_r = 49.00747

$S°(298.15\ K)$ = [243.37] J K^{-1} mol^{-1}

$\Delta_f H°(0\ K)$ = 93.97 ± 2.1 kJ mol^{-1}

$\Delta_f H°(298.15\ K)$ = 94.27 ± 2.1 kJ mol^{-1}

Vibrational Frequencies and Degeneracies

ν, cm^{-1}
[400] (1)
239 (2)
2176 (1)

Ground State Quantum Weight: 1 σ = 1

Point Group: $C_{\infty v}$

Bond Distances: Na-C = [1.992] Å C-N = [1.16] Å

Bond Angle: Na-C-N = [180]°

Rotational Constant: B_o = [0.182724] cm^{-1}

Enthalpy of Formation

The vapor pressures of NaCN(ℓ) at temperatures 805-1353 K were measured by Ingold (1). According to the investigation reported by Porter (2), the vapor consists of NaCN and $(NaCN)_2$. Based on his data the dimer/monomer ratios at temperatures 805-1353°C(1078.2-1626.2 K) were calculated to be 0.29-0.09. Hence the respective partial pressures for NaCN(g) and $(NaCN)_2$(g) were derived. By the 2nd and 3rd law methods, the values of $\Delta_{vap}H°$(ℓ + monomer, 298.15 K) were evaluated to be 43.43 and 42.69 kcal mol^{-1}, respectively. The enthalpy of formation for NaCN(g) is calculated from $\Delta_f H°$(NaCN, cr, 298.15 K) and the 3rd law value for $\Delta_{vap}H°$.

Heat Capacity and Entropy

The molecular structure of NaCN(g) has not been determined. It was assumed to be linear by comparison with that for HCN(g). The C-N bond distance was assumed to be the same as that in KCN(g) and the Na-C bond distance was taken from Leroi (3), which was calculated. The vibrational frequencies, ν_2 and ν_3, were measured by Leroi and Klemperer (4) from infrared absorption spectra. The value of ν_1 was calculated by valence force method described by Herzberg (5), using appropriate force constants.

References

1. C. K. Ingold, J. Chem. Soc. 123, 885 (1923).
2. R. F. Porter, J. Chem. Phys. 35, 318 (1961).
3. G. E. Leroi, Ph. D. Dissertation, Harvard University, 1960.
4. G. E. Leroi and W. Klemperer, J. Chem. Phys. 35, 774 (1961).
5. G. Herzberg, "Infrared and Raman Spectra", D. Van Nostrand, Inc., New York, 1945.

Sodium Cyanide (NaCN) $C_1N_1Na_1(g)$

Enthalpy Reference Temperature = T_r = 298.15 K Standard State Pressure = p° = 0.1 MPa

T/K	$C_p°$ (J K^{-1}mol^{-1})	S° (J K^{-1}mol^{-1})	$-[G°-H°(T_r)]/T$ (J K^{-1}mol^{-1})	$H°-H°(T_r)$ (kJ mol^{-1})	$\Delta_f H°$ (kJ mol^{-1})	$\Delta_f G°$ (kJ mol^{-1})	Log K_f
0	0.	0.	INFINITE	-12.125	93.973	93.973	INFINITE
100	36.717	195.016	285.125	-9.011	94.237	85.203	-44.505
200	46.524	224.002	247.863	-4.772	94.238	76.192	-19.899
250	48.786	234.647	244.186	-2.385	94.278	71.676	-14.976
298.15	50.178	243.367	243.367	0.	94.266	67.323	-11.795
300	50.222	243.678	243.368	0.093	94.263	67.155	-11.693
350	51.210	251.498	243.983	2.630	94.148	62.645	-9.349
400	51.962	258.387	245.362	5.210	91.278	58.365	-7.622
450	52.595	264.544	247.157	7.824	90.968	54.269	-6.299
500	53.168	270.116	249.179	10.469	90.638	50.209	-5.245
600	54.227	279.904	253.506	15.839	89.923	42.189	-3.673
700	55.203	288.338	257.893	21.311	89.149	34.294	-2.559
800	56.088	295.768	262.172	26.877	88.329	26.513	-1.731
900	56.870	302.420	266.281	32.525	87.468	18.837	-1.093
1000	57.548	308.448	270.201	38.247	86.559	11.259	-0.588
1100	58.130	313.961	273.932	44.032	85.593	3.776	-0.179
1200	58.627	319.041	277.483	49.870	-12.205	-1.177	0.051
1300	59.051	323.751	280.863	55.755	-12.395	-0.251	0.010
1400	59.414	328.141	284.085	61.678	-12.606	0.691	-0.026
1500	59.725	332.251	287.160	67.636	-12.836	1.649	-0.057
1600	59.992	336.114	290.100	73.622	-13.083	2.622	-0.086
1700	60.223	339.758	292.915	79.633	-13.346	3.612	-0.111
1800	60.424	343.206	295.614	85.666	-13.623	4.617	-0.134
1900	60.598	346.478	298.206	91.717	-13.915	5.639	-0.155
2000	60.752	349.590	300.698	97.784	-14.221	6.676	-0.174
2100	60.886	352.558	303.097	103.867	-14.542	7.728	-0.192
2200	61.005	355.393	305.410	109.961	-14.876	8.797	-0.209
2300	61.111	358.107	307.643	116.067	-15.224	9.880	-0.224
2400	61.205	360.710	309.800	122.183	-15.588	10.980	-0.239
2500	61.289	363.210	311.887	128.308	-15.967	12.094	-0.253
2600	61.364	365.615	313.907	134.440	-16.362	13.225	-0.266
2700	61.432	367.933	315.866	140.580	-16.774	14.370	-0.278
2800	61.493	370.168	317.765	146.727	-17.205	15.532	-0.290
2900	61.549	372.327	319.610	152.879	-17.655	16.709	-0.301
3000	61.599	374.414	321.402	159.036	-18.126	17.902	-0.312
3100	61.645	376.435	323.145	165.199	-18.619	19.110	-0.322
3200	61.687	378.393	324.841	171.365	-19.135	20.336	-0.332
3300	61.725	380.291	326.493	177.536	-19.676	21.577	-0.342
3400	61.761	382.135	328.102	183.710	-20.245	22.836	-0.351
3500	61.793	383.925	329.672	189.888	-20.842	24.112	-0.360
3600	61.823	385.667	331.203	196.069	-21.470	25.405	-0.369
3700	61.851	387.361	332.698	202.252	-22.131	26.716	-0.377
3800	61.876	389.011	334.158	208.439	-22.827	28.045	-0.386
3900	61.900	390.618	335.585	214.628	-23.559	29.394	-0.394
4000	61.922	392.186	336.981	220.819	-24.332	30.761	-0.402
4100	61.942	393.715	338.346	227.012	-25.147	32.149	-0.410
4200	61.961	395.208	339.682	233.207	-26.004	33.556	-0.417
4300	61.979	396.666	340.991	239.404	-26.910	34.985	-0.425
4400	61.996	398.091	342.272	245.603	-27.824	36.432	-0.432
4500	62.011	399.484	343.528	251.803	-28.811	37.902	-0.440
4600	62.026	400.848	344.759	258.005	-29.858	39.396	-0.447
4700	62.040	402.182	345.967	264.208	-30.961	40.914	-0.455
4800	62.052	403.488	347.152	270.413	-32.121	42.455	-0.462
4900	62.065	404.767	348.315	276.619	-33.343	44.021	-0.469
5000	62.076	406.021	349.456	282.826	-34.620	45.612	-0.477
5100	62.087	407.251	350.578	289.034	-35.970	47.230	-0.484
5200	62.097	408.457	351.679	295.243	-37.388	48.875	-0.491
5300	62.106	409.639	352.762	301.453	-38.878	50.548	-0.498
5400	62.116	410.800	353.826	307.664	-40.441	52.250	-0.505
5500	62.124	411.940	354.872	313.876	-42.081	53.982	-0.513
5600	62.132	413.060	355.901	320.089	-43.800	55.744	-0.520
5700	62.140	414.160	356.913	326.303	-45.601	57.537	-0.527
5800	62.147	415.240	357.910	332.517	-47.485	59.363	-0.535
5900	62.154	416.303	358.891	338.732	-49.455	61.222	-0.542
6000	62.161	417.347	359.856	344.948	-51.355	63.097	-0.549

PREVIOUS: March 1966 (1 atm) CURRENT: March 1966 (1 bar)

Sodium Cyanide (NaCN) $C_1N_1Na_1(g)$

NCO RADICAL (NCO) IDEAL GAS M_r = 42.0171

$S°(298.15\ K) = 232.2 \pm 0.8\ J\ K^{-1}\ mol^{-1}$

$\Delta_f H°(0\ K) = 159.0 \pm 10.5\ kJ\ mol^{-1}$

$\Delta_f H°(298.15\ K) = 159.4 \pm 10.5\ kJ\ mol^{-1}$

Electronic Levels and Quantum Weights

State	ε_i, cm^{-1}	g_i
$X^2\Pi_{3/2}$	0	2
$X^2\Pi_{1/2}$	95.6	2
$A^2\Sigma^+$	22802	2
$B^2\Pi$	31811	4

Vibrational Frequencies and Degeneracies

ν, cm^{-1}
1922 (1)
536 (2)
1275 (1)

Point Group: $C_{\infty v}$ σ = 1

Bond Distances: N–C = [1.23] Å C–O = [1.18] Å

Bond Angle: N–C–O = 108°

Rotational Constant: B_0 = 0.388394 cm^{-1}

Enthalpy of Formation

Okabe (1) has measured the onset of photodissociation for the process HNCO(g) + H(g) + NCO($A^2\Sigma$) as 7.73 ± 0.01 eV (178.25 ± 0.25 kcal). This result in combination with JANAF auxiliary data (2) results in $\Delta_f H°$(NCO, g, 0 K) = 38 ± 2.5 kcal mol^{-1}.

Dixon (3) has discussed the diffusencess in the $^2\Pi$-$^2\Pi$ bands of NCO above 33700 cm^{-1} and concludes this is due to either predissociation or unresolved groups of lines due to a perturbing state. Assuming the dissociation is to N(^{2}D) + CO($^1\Sigma$) yields D_0^0(N–CO) ≤ 41 kcal, or $\Delta_f H°$(NCO, 0 K) ≥ 44 kcal mol^{-1}; assuming the products CN($^2\Sigma$) + O(^{3}P) then D_0^0(NC–O) ≤ 97 kcal, or $\Delta_f H°$(NCO, 0 K) ≥ 65 kcal mol^{-1}. Milligan and Jacox (4) confirm the diffuseness and observe increased amounts of CO. Thus, they prefer the first dissociation mechanism which favors $\Delta_f H°$(NCO, 0 K) ≥ 44 kcal mol^{-1}. An upper limit to the enthalpy formation can be obtained from HNCO(g) (2) by assuming a maximum D_0^0(H–NCO) ≤ 122 kcal, equal to D_0^0(H–CN), which yields $\Delta_f H°$(NCO, 0 K) ≤ 45 kcal mol^{-1}. Another estimate may be obtained by making the reasonable assumption that the sum of the atomization energies of NCN + NNO equal those of NNN + NCO, where the central atoms are simply interchanged. Using respective atomization values (2) of 282, 264 and 242 kcal yields 304 kcal for NCO, or $\Delta_f H°$(NCO, 0 K) = 38 kcal mol^{-1}. It appears that the diffuseness in the spectrum is not due to predissociation, since all the estimates and measurements fall below the vlaues calculated on that basis. We adopt $\Delta_f H°$(NCO, g, 0 K) = 38 ± 2.5 kcal mol^{-1}.

Heat Capacity and Entropy

The electronic levels, structure, and ν_2 are from Dixon (3). The ν_1 and ν_3 vibrational frequencies are from the matrix isolation data of Milligan and Jacox (4). The individual bond lengths are those selected by Milligan and Jacox (4) to fit the overall length determined by Dixon (3).

References

1. H. Okabe, J. Chem. Phys. 53, 3507 (1970).
2. JANAF Thermochemical Tables: HNCO(g), 12-31-70; HCN(g), 12-31-69; NCN(g), 12-31-70; NNO(g), 12-31-64; NNN(g), 12-31-70; CN(g), 6-30-69; H(g), 9-30-65.
3. R. N. Dixon, Phil. Trans. Roy. Soc. (London) 252A, 165 (1960) and Can. J. Phys. 38, 10 (1960).
4. D. E. Milligan and M. E. Jacox, J. Chem. Phys. 47, 5157 (1967).

NCO Radical (NCO) $C_1N_1O_1(g)$

Enthalpy Reference Temperature = T_r = 298.15 K Standard State Pressure = p° = 0.1 MPa

T/K	$C_p°$ (J K^{-1}mol^{-1})	S°	−[G°−H°(T_r)]/T	H°−H°(T_r) (kJ mol^{-1})	$\Delta_f H°$	$\Delta_f G°$	Log K_f
0	0.	0.	INFINITE	-10.196	158.943	158 943	INFINITE
100	32.077	194.218	264.724	-7.051	159.125	156.454	-81.723
200	35.509	217.193	235.771	-3.716	159.222	153.747	-40.155
250	37.940	225.381	232.896	-1.879	159.307	152.369	-31.836
298.15	40.053	232.248	232.248	0.	159.410	151.024	-26.459
300	40.130	232.496	232.249	0.074	159.415	150.972	-26.286
350	42.089	238.832	232.744	2.131	159.533	149.555	-22.320
400	43.865	244.570	233.869	4.280	159.654	148.121	-19.343
450	45.485	249.831	235.355	6.514	159.768	146.673	-17.025
500	46.967	254.701	237.049	8.826	159.874	145.212	-15.170
600	49.543	263.500	240.740	13.656	160.054	142.262	-12.385
700	51.652	271.301	244.559	18.719	160.196	139.285	-10.394
800	53.361	278.314	248.347	23.973	160.305	136.290	-8.899
900	54.740	284.681	252.036	29.381	160.387	133.283	-7.736
1000	55.855	290.508	255.596	34.912	160.445	130.268	-6.805
1100	56.761	295.876	259.017	40.545	160.480	127.249	-6.043
1200	57.503	300.848	262.298	46.259	160.495	124.227	-5.407
1300	58.114	305 475	265.444	52.041	160.489	121.205	-4.870
1400	58.622	309.801	268.459	57.878	160.463	118.184	-4.409
1500	59.048	313.861	271.352	63.763	160.418	115.165	-4.010
1600	59.408	317.683	274.130	69.686	160.353	112.150	-3.661
1700	59.713	321.294	276.799	75.642	160.269	109.140	-3.353
1800	59.975	324.715	279.367	81.627	160.165	106.135	-3.080
1900	60.201	327.964	281.840	87.636	160.040	103.137	-2.835
2000	60.397	331.057	284.224	93.666	159.895	100.146	-2.616
2100	60.568	334.008	286.525	99.715	159.730	97.162	-2.417
2200	60.718	336.829	288.748	105.779	159.544	94.187	-2.236
2300	60.851	339.531	290.897	111.858	159.337	91.221	-2.072
2400	60.968	342.123	292.978	117.949	159 109	88.264	-1.921
2500	61.073	344.614	294.994	124.051	158.860	85.317	-1.783
2600	61.167	347.012	296.949	130.163	158.591	82.381	-1.655
2700	61.251	349.322	298.846	136.284	158.301	79.455	-1.537
2800	61.328	351.551	300.689	142.413	157.992	76.541	-1.428
2900	61.398	353.704	302.480	148.550	157.661	73.637	-1 326
3000	61.462	355.787	304.222	154.693	157.312	70.746	-1.232
3100	61.521	357.803	305.918	160.842	156.942	67.866	-1.144
3200	61.576	359.757	307.570	166.997	156.554	64.999	-1.061
3300	61.627	361.652	309.181	173.157	156.147	62.144	-0.984
3400	61.676	363.493	310.751	179.322	155.721	59.302	-0.911
3500	61.722	365.281	312.284	185.492	155.278	56.472	-0.843
3600	61.766	367.021	313.780	191.666	154.817	53.656	-0.779
3700	61.808	368.714	315.242	197.845	154.338	50.852	-0.718
3800	61.850	370.363	316.671	204.028	153.843	48.062	-0.661
3900	61.891	371.970	318.068	210.215	153.332	45.285	-0.607
4000	61.931	373.537	319.436	216.406	152.804	42.521	-0.555
4100	61.971	375.067	320.774	222.601	152.261	39.771	-0.507
4200	62.011	376.561	322.084	228.800	151.702	37 034	-0.461
4300	62.052	378.020	323.368	235.003	151.128	34.311	-0.417
4400	62.092	379.447	324.627	241.211	150.539	31.601	-0.375
4500	62.134	380.843	325.861	247.422	149.936	28.904	-0.336
4600	62.176	382.209	327.071	253.637	149 317	26.221	-0.298
4700	62.219	383.547	328.258	259.857	148.685	23.552	-0.262
4800	62.263	384.857	329.424	266.081	148.038	20.896	-0.227
4900	62.308	386.142	330.568	272.310	147.376	18.255	-0.195
5000	62.355	387.401	331.692	278.543	146.700	15.626	-0.163
5100	62.402	388.636	332.797	284.781	146.010	13.012	-0.133
5200	62.451	389.848	333.882	291.023	145.305	10.411	-0.105
5300	62.501	391.038	334.950	297.271	144.585	7.824	-0.077
5400	62.552	392.207	335.999	303.524	143.851	5.250	-0.051
5500	62.605	393.355	337.031	309.782	143.101	2.690	-0.026
5600	62.659	394.484	338.047	316.045	142.337	0.144	-0.001
5700	62.714	395.593	339.047	322.313	141.556	-2.388	0.022
5800	62.771	396.685	340.032	328.588	140.759	-4.906	0.044
5900	62.829	397.758	341.001	334.868	139.946	-7.412	0.066
6000	62.888	398.815	341.956	341.154	139.116	-9.902	0.086

PREVIOUS: December 1970 (1 atm) CURRENT: December 1970 (1 bar)

NCO Radical (NCO) $C_1N_1O_1(g)$

CNN RADICAL (CNN) IDEAL GAS M_r = 40.0244

$S°(298.15\ K) = 231.716\ J\ K^{-1}\ mol^{-1}$

$\Delta_f H°(0\ K) = 584 \pm 126\ kJ\ mol^{-1}$
$\Delta_f H°(298.15\ K) = 585 \pm 126\ kJ\ mol^{-1}$

Vibrational Frequencies and Degeneracies

ν, cm^{-1}
1241 (1)
393 (2)
2847 (1)

Ground State Quantum Weight: [3]
Point Group: $C_{\infty v}$ σ = 1
Bond Distances: C-N = [1.15] Å N-N = [1.25] Å
Bond Angle: C-N-N = 180°
Rotational Constant: B_0 = [0.450392] cm^{-1}

Enthalpy of Formation

The enthalpy of formation is calculated from the estimated enthalpy of reaction $\Delta_r H°(0\ K) = 255 \pm 30\ kcal\ mol^{-1}$ for CNN(g) → C(g) + 2 N(g). Since CNN radical can dissociate to give C + N_2 or CN + N, the molecule must have more energy than either CN or N_2. It is assumed that the extra bond will be quite weak, for this we estimate 30±30 kcal mol^{-1} which gives $\Delta_r H°(0\ K) = 255 \pm 30\ kcal\ mol^{-1}$ (583.856±126 kJ mol^{-1}) for the first path and $\Delta_r H° = 204 \pm 30\ kcal\ mol^{-1}$ for the second. We adopt the former value.

Heat Capacity and Entropy

Milligan and Jacox (1) have measured three fundamental vibrational frequencies for CNN radical (g) in the infrared and ultraviolet spectra by matrix-isolation, and proposed a linear molecular configuration with a $^3\Sigma$ ground state which has been confirmed by Wasserman et al. (2) in the electron paramagnetic resonance observations. Also Milligan and Jacox (1) suggested a "C≡N=N" bond structure, based on the agreement between the C-N and N-N stretching force constants (calculated from observed vibrational frequencies in the CNN radical) and the typical C≡N and N=N stretching force constants. In addition, they estimated the bond distances r(C-N) = 1.15 Å and r(N-N) = 1.25 Å from the proposed structure.

References

1. D. E. Milligan and M. E. Jacox, J. Chem. Phys. 44, 2850 (1966).
2. E. Wasserman, L. Barash and W. A. Yager, J. Amer. Chem. Soc. 87, 2075 (1965).

CNN Radical (CNN) $C_1N_2(g)$

Enthalpy Reference Temperature = T_r = 298.15 K Standard State Pressure = p° = 0.1 MPa

	J $K^{-1}mol^{-1}$			kJ mol^{-1}			
T/K	$C_p°$	S°	$-[G°-H°(T_r)]/T$	$H°-H°(T_r)$	$\Delta_f H°$	$\Delta_f G°$	Log K_f
0	0.	0.	INFINITE	-10.372	583.854	583.854	INFINITE
100	30.975	191.895	266.206	-7.431	583.833	580.720	-303.337
200	38.075	215.660	235.463	-3.961	584.066	577.548	-150.840
250	40.478	224.427	232.403	-1.994	584.282	575.894	-120.326
298.15	42.281	231.716	231.716	0.	584.505	574.258	-100.608
300	42.343	231.978	231.717	0.078	584.513	574.195	-99.976
350	43.883	238.624	232.238	2.235	584.742	572.457	-85.435
400	45.200	244.572	233.415	4.463	584.958	570.687	-74.524
450	46.353	249.964	234.959	6.752	585.154	568 891	-66.035
500	47.377	254.902	236.709	9.096	585.325	567.075	-59.242
600	49.142	263.700	240.493	13.925	585.592	563.398	-49.048
700	50.642	271.391	244.369	18.916	585.768	559.684	-41.764
800	51.950	278.241	248.182	24.047	585.868	555.950	-36.300
900	53.100	284.427	251.871	29.300	585.911	552.207	-32.049
1000	54.111	290.075	255.413	34.662	585.909	548.462	-28.649
1100	54.997	295.275	258.804	40.118	585.874	544.718	-25.867
1200	55.772	300.095	262.046	45.658	585.814	540.979	-23.548
1300	56.449	304.586	265.148	51.270	585.733	537.246	-21.587
1400	57.040	308.791	268.117	56.945	585.634	533.520	-19.906
1500	57.556	312.745	270.961	62.675	585.522	529.802	-18.449
1600	58.007	316.474	273.691	68.454	585.397	526 091	-17.175
1700	58.404	320.003	276.312	74.275	585.260	522.388	-16.051
1800	58.752	323.351	278.833	80.133	585.113	518.694	-15.052
1900	59.060	326.536	281.261	86.024	584.954	515.009	-14.159
2000	59.332	329.573	283.601	91.944	584.786	511.332	-13.355
2100	59.573	332.473	285.860	97.889	584.608	507.663	-12.627
2200	59.788	335.250	288.042	103.857	584.420	504.004	-11.967
2300	59.981	337.912	290.153	109.846	584.222	500.353	-11.363
2400	60.153	340.468	292.196	115.853	584.014	496.711	-10.811
2500	60.308	342.927	294.177	121.876	583.795	493.078	-10.302
2600	60.448	345.295	296.097	127.914	583.567	489.453	-9.833
2700	60.574	347.579	297.962	133 965	583.328	485.838	-9.399
2800	60.689	349.784	299.774	140.028	583.079	482.232	-8.996
2900	60.793	351.915	301.535	146.102	582.819	478.635	-8.621
3000	60.888	353.978	303.249	152.187	582.550	475.047	-8.271
3100	60.975	355.976	304.918	158.280	582.269	471.468	-7.944
3200	61.054	357.913	306.544	164.381	581.978	467.898	-7.638
3300	61.127	359.793	308.129	170.490	581.676	464 338	-7.350
3400	61.195	361.619	309.676	176.607	581.363	460.787	-7.079
3500	61.257	363.394	311.185	182.729	581.039	457.245	-6.824
3600	61.314	365.120	312.660	188.858	580.704	453.713	-6.583
3700	61.367	366.801	314.100	194.992	580.359	450.190	-6.356
3800	61.416	368.438	315.509	201.131	580.002	446.677	-6 140
3900	61.461	370.034	316.886	207.275	579.634	443.173	-5.936
4000	61.504	371.590	318.235	213.423	579.255	439.679	-5.742
4100	61.543	373.110	319.555	219.576	578.865	436.194	-5.557
4200	61.580	374.593	320.847	225.732	578.463	432.719	-5.382
4300	61.614	376.043	322.114	231.891	578.050	429.254	-5.214
4400	61.647	377.459	323.356	238.055	577.626	425.799	-5.055
4500	61.677	378.845	324.574	244.221	577.190	422.353	-4.903
4600	61.705	380.201	325.768	250.390	576.743	418.917	-4.757
4700	61.732	381.528	326.941	256.562	576.284	415.491	-4.618
4800	61.757	382.828	328.092	262.736	575.813	412.075	-4.484
4900	61.780	384.102	329.222	268.913	575.331	408.668	-4.356
5000	61.803	385.350	330.332	275.092	574.837	405.272	-4.234
5100	61.824	386.574	331.423	281.274	574.331	401.886	-4.116
5200	61.844	387.775	332.495	287.457	573.812	398.510	-4.003
5300	61.862	388.953	333.549	293.642	573.282	395.143	-3.894
5400	61.880	390.110	334.586	299.829	572.740	391.787	-3.790
5500	61.897	391.245	335.606	306.018	572.186	388.442	-3.689
5600	61.913	392.361	336.609	312.209	571.619	385.106	-3.592
5700	61.928	393.457	337.597	318.401	571.039	381.780	-3.499
5800	61.942	394.534	338.569	324.594	570.447	378.465	-3.408
5900	61.956	395.593	339.527	330.789	569.842	375.160	-3.321
6000	61.969	396.634	340.470	336.985	569.223	371.866	-3.237

PREVIOUS: June 1966 (1 atm) CURRENT: June 1966 (1 bar)

CNN Radical (CNN) $C_1N_2(g)$

NCN RADICAL (NCN) IDEAL GAS M_r = 40.0244

$S°(298.15\ K) = 226.21 \pm 0.42\ J\ K^{-1}\ mol^{-1}$

$\Delta_fH°(0\ K) = 472 \pm 21\ kJ\ mol^{-1}$
$\Delta_fH°(298.15\ K) = 473 \pm 21\ kJ\ mol^{-1}$

Electronic Levels and Quantum Weights

State	ε_i, cm^{-1}	g_i
$X^3\Sigma_g^-$	0	3
$A^3\Pi_u$	30384	6
$a^1\Delta$	[38000]	2

Vibrational Frequencies and Degeneracies

ν, cm^{-1}
1197 (1)
423 (2)
1475 (1)

Point Group: $D_{\infty h}$ σ = 2
Bond Distance: C-N = 1.232 Å
Bond Angle: N-C-N = 180°
Rotational Constant: $B_0 = 0.396469\ cm^{-1}$

Enthalpy of Formation

Okabe and Mele (1) have measured the photodissociation onset of the process

$CN{\cdot}N_3(g) \rightarrow NCN(g) + N_2(A^3\Sigma)$ $\Delta_rH°(0\ K) = 149.3 \pm 2.5$ kcal

Using $\Delta_fH°(N_2,\ A^3\Sigma,\ 298.15\ K) = 143.5\ kcal\ mol^{-1}$ (2) and $\Delta_fH°(CN{\cdot}N_3,\ g,\ 298.15\ K) = 107 \pm 3\ kcal\ mol^{-1}$ (3), we obtain $\Delta_fH°(NCN,\ g,\ 298.15\ K) = 113 \pm 5\ kcal\ mol^{-1}$.

Heat Capacity and Entropy

The X and A electronic levels, ground state structure, and rotational constants are from Herzberg and Travis (4), the $a^1\Delta$ level is estimated from the work of Schoen (5). The vibrational frequencies have been observed in the matrix-isolated species by Milligan and Jacox (6, 7). Wasserman et al. (8) have confirmed the ground state is linear and triplet by electron paramagnetic resonance.

References

1. H. Okabe and A. Mele, J. Chem. Phys. 51, 2100 (1969).
2. G. Herzberg, "Spectra of Diatomic Molecules", D. Van Nostrand Co., Inc., New York, 1950.
3. JANAF Thermochemical Tables: $N_3(g)$, 12-31-70.
4. G. Herzberg and D. N. Travis, Can. J. Phys. 42, 1658 (1964).
5. L. J. Schoen, J. Chem. Phys. 45, 2773 (1966).
6. D. E. Milligan, M. E. Jacox and A. M. Bass, J. Chem. Phys. 43, 3149 (1965).
7. D. E. Milligan and M. E. Jacox, J. Chem. Phys. 45, 1387 (1966).
8. E. Wasserman, L. Barash and W. A. Yager, J. Amer. Chem. Soc. 87, 2075 (1965).

NCN Radical (NCN) $C_1N_2(g)$

Enthalpy Reference Temperature = T_r = 298.15 K Standard State Pressure = p° = 0.1 MPa

	J K⁻¹mol⁻¹			kJ mol⁻¹			
T/K	$C_p°$	S°	$-[G°-H°(T_r)]/T$	$H°-H°(T_r)$	$\Delta_fH°$	$\Delta_fG°$	Log K_f
0	0.	0.	INFINITE	-10.244	472.269	472.269	INFINITE
100	30.508	187.072	260.196	-7.312	472.239	469.608	-245.298
200	37.333	210.338	229.916	-3.916	472.399	466.945	-121.953
250	40.004	218.965	226.885	-1.980	472.583	465.561	-97.274
298.15	42.192	226.203	226.203	0.	472.792	464.189	-81.324
300	42.271	226.464	226.203	0.078	472.800	464.136	-80.813
350	44.309	233.135	226.726	2.243	473.038	462.673	-69.050
400	46.171	239.175	227.910	4.506	473.288	461.176	-60.223
450	47.866	244.713	229.474	6.858	473.546	459.646	-53.354
500	49.391	249.837	231.257	9.290	473.806	458.088	-47.856
600	51.948	259.078	235.141	14.362	474.317	454.896	-39.602
700	53.928	267.241	239.155	19.660	474.799	451.621	-33.700
800	55.451	274.546	243.130	25.133	475.241	448.279	-29.270
900	56.627	281.148	246.993	30.739	475.636	444.884	-25.820
1000	57.545	287.163	250.714	36.449	475.984	441.448	-23.059
1100	58.270	292.683	254.282	42.242	476.285	437.980	-20.798
1200	58.849	297.779	257.697	48.099	476.542	434.486	-18.913
1300	59.317	302.509	260.964	54.008	476.758	430.972	-17.317
1400	59.700	306.919	264.091	59.959	476.936	427.443	-15.948
1500	60.017	311.049	267.086	65.945	477.080	423.903	-14.762
1600	60.282	314.931	269.956	71.961	477.191	420.354	-13.723
1700	60.505	318.593	272.710	78.000	477.273	416.799	-12.807
1800	60.695	322.057	275.356	84.061	477.328	413.240	-11.992
1900	60.857	325.343	277.901	90.138	477.356	409.678	-11.263
2000	60.997	328.468	280.352	96.231	477.361	406.116	-10.607
2100	61.119	331.447	282.715	102.337	477.344	402.554	-10.013
2200	61.225	334.293	284.995	108.455	477.304	398.994	-9.473
2300	61.318	337.016	287.198	114.582	477.245	395.435	-8.981
2400	61.400	339.628	289.329	120.718	477.166	391.880	-8.529
2500	61.473	342.136	291.391	126.862	477.068	388.329	-8.114
2600	61.538	344.548	293.390	133.012	476.952	384.781	-7.730
2700	61.597	346.872	295.328	139.169	476.819	381.239	-7.376
2800	61.649	349.113	297.209	145.331	476.669	377.701	-7.046
2900	61.697	351.277	299.036	151.499	476.503	374.170	-6.740
3000	61.740	353.369	300.812	157.671	476.321	370.644	-6.453
3100	61.779	355.394	302.541	163.847	476.123	367.125	-6.186
3200	61.816	357.356	304.223	170.026	475.910	363.612	-5.935
3300	61.849	359.259	305.862	176.210	475.682	360.106	-5.700
3400	61.881	361.106	307.460	182.396	475.439	356.607	-5.479
3500	61.910	362.900	309.019	188.586	475.183	353.116	-5.270
3600	61.938	364.645	310.540	194.778	474.912	349.632	-5.073
3700	61.966	366.342	312.025	200.973	474.628	346.156	-4.887
3800	61.992	367.995	313.476	207.171	474.329	342.688	-4.711
3900	62.018	369.606	314.895	213.372	474.018	339.228	-4.543
4000	62.043	371.176	316.282	219.575	473.694	335.776	-4.385
4100	62.069	372.708	317.640	225.780	473.357	332.332	-4.234
4200	62.095	374.204	318.969	231.989	473.007	328.896	-4.090
4300	62.122	375.666	320.271	238.199	472.645	325.469	-3.954
4400	62.150	377.094	321.546	244.413	472.272	322.051	-3.823
4500	62.179	378.491	322.796	250.629	471.886	318.641	-3.699
4600	62.209	379.858	324.022	256.849	471.489	315.240	-3.580
4700	62.241	381.196	325.224	263.071	471.081	311.848	-3.466
4800	62.274	382.507	326.404	269.297	470.662	308.464	-3.357
4900	62.309	383.792	327.562	275.526	470.231	305.090	-3.252
5000	62.347	385.051	328.699	281.759	469.791	301.724	-3.152
5100	62.386	386.286	329.816	287.996	469.340	298.367	-3.056
5200	62.428	387.498	330.914	294.236	468.879	295.019	-2.964
5300	62.472	388.687	331.993	300.481	468.408	291.680	-2.875
5400	62.519	389.855	333.053	306.731	467.929	288.350	-2.789
5500	62.568	391.003	334.097	312.985	467.440	285.029	-2.707
5600	62.620	392.131	335.123	319.244	466.942	281.717	-2.628
5700	62.674	393.240	336.133	325.509	466.435	278.414	-2.551
5800	62.732	394.330	337.127	331.779	465.919	275.119	-2.478
5900	62.792	395.403	338.105	338.056	465.395	271.834	-2.407
6000	62.854	396.459	339.069	344.338	464.863	268.558	-2.338

PREVIOUS: December 1970 (1 atm) CURRENT: December 1970 (1 bar)

NCN Radical (NCN) $C_1N_2(g)$

SODIUM CARBONATE (Na_2CO_3) CRYSTAL M_r = 105.98874

$S°(298.15\ K) = 138.797 \pm 0.8\ J\ K^{-1}\ mol^{-1}$

$T_{trs} = 723.15\ K$

$T_{fus} = 1123.15\ K$

$\Delta_f H°(0\ K) = -1124.610 \pm 0.21\ kJ\ mol^{-1}$

$\Delta_f H°(298.15\ K) = -1130.768 \pm 0.21\ kJ\ mol^{-1}$

$\Delta_{trs} H° = 0.690\ kJ\ mol^{-1}$

$\Delta_{fus} H° = 29.665\ kJ\ mol^{-1}$

Enthalpy of Formation

The enthalpy of formation was obtained from the enthalpy of solution $\Delta_{sol}H = -6.36\ kcal\ mol^{-1}$ and the enthalpies of formation for $Na^+(aq, \infty)$ and $CO_2^{--}(aq, \infty)$, -57.39 and -161.84 kcal mol^{-1}, respectively. The enthalpy of solution was taken from Rupert et al. (1). The ionic enthalpies of formation were taken from Wagman et al. (2).

A value of $\Delta_f H°(298.15\ K) = -270.9\ kcal\ mol^{-1}$ has been selected by Wagman (3).

Heat Capacity and Entropy

The low temperature heat capacities, 54.6-292.1 K, were measured by Anderson (4). In the temperature range 473.15-873.15 K, Popov and Galchenko (5) measured the heat capacities (by heat conduction calorimetry) and observed two transformations, one at about 623 K and the other at about 750 K. Popov and Ginzburg (6) measured enthalpies in the temperature range 480-1350 K and tabulated their data only in the range 476-872 K. Ginzburg (7) has given heat capacity equations based on the data from (5, 6) and has tabulated thermodynamic functions in the range 500-1500 K. There appear to be inconsistencies among the data, equations and functions. Rolin and Recapet (8) also measured the enthalpy in the temperature range 645-1322 K. Janz et al. (9) have given an enthalpy equation based on their heat content measured data in the range 707-1127 K. May (10) has tabulated smoothed measured enthalpy values in the range 400-1500 K. All the above information was used in a Shomate analysis in order to smooth the enthalpies and calculate heat capacities. The values from the low and high temperature sources join smoothly at 298 K. The heat capacity was graphically extrapolated above the melting point.

The entropy was calculated at 54.60 using the Debye and Einstein function D(138/T) + 2E(266/T) + 2D(737/T) given by Anderson (4). The value of $S°(54.6\ K) = 3.943\ cal\ deg^{-1}\ mole^{-1}$.

Transition Data

T_{trs} was taken from Ginzburg [7] and $\Delta_{trs}H°$ was obtained from the above reported enthalpy measurements by means of Shomate function analysis. The heat of transition at 593.15 K has been incorporated in the heat capacity.

Fusion Data

T_{fus} was taken from Ginzburg [7] and $\Delta_{fus}H°$ was obtained from the above reported enthalpy measurements by means of Shomate function analysis.

References

1. J. P. Rupert, H. P. Hopkins, Jr., and C. A. Wulff, J. Phys. Chem. 69, 3059-62 (1965).
2. U. S. Nat. Bur. Stand. Tech. Note 270-1, 1965.
3. D. D. Wagman, U. S. Nat. Bur. Stand. Report 8628, 87, 1965.
4. C. T. Anderson, J. Amer. Chem. Soc. 55, 3621 (1933).
5. M. M. Popov and G. L. Galchenko, Zh. Obshch. Khim. 21, 2220 (1951).
6. M. M. Popov and D. M. Ginzburg, Zh. Obshch. Khim. 26, 971-80 (1956).
7. D. M. Ginzburg, Zh. Obshch. Khim. 26, 968-70 (1956).
8. M. Rolin and J. M. Recapet, Bull. Soc. Chim., 2504 (1964).
9. G. J. Janz, E. Neuenschwander and F. J. Kelly, Trans. Faraday Soc. 59, 841 (1963).
10. M. N. May, Tappi, 35, 511 (1952).

Sodium Carbonate (Na_2CO_3) $C_1Na_2O_3(cr)$

Enthalpy Reference Temperature = T_r = 298.15 K; Standard State Pressure = p° = 0.1 MPa

T/K	$C_p°$ (J K⁻¹mol⁻¹)	S° (J K⁻¹mol⁻¹)	$-[G°-H°(T_r)]/T$ (J K⁻¹mol⁻¹)	$H°-H°(T_r)$ (kJ mol⁻¹)	$\Delta_f H°$ (kJ mol⁻¹)	$\Delta_f G°$ (kJ mol⁻¹)	Log K_f
0	0.	0.	INFINITE	-20.812	-1124.610	-1124.610	INFINITE
100	61.241	43.237	225.158	-18.192	-1129.086	-1102.556	575.916
200	94.140	97.766	148.470	-10.141	-1130.640	-1075.252	280.827
298.15	111.003	138.797	138.797	0.	-1130.768	-1048.009	183.607
300	111.253	139.484	138.799	0.206	-1130.764	-1047.496	182.385
400	125.102	173.307	143.317	11.996	-1135.696	-1019.370	133.116
500	142.214	203.007	152.332	25.337	-1134.473	-990.400	103.466
600	163.302	230.759	163.106	40.592	-1131.568	-961.826	83.734
700	187.569	257.722	174.702	58.115	-1126.604	-933.897	69.688
723.150	193.385	263.919	177.459	62.524		I <--> II	
723.150	143.762	264.873	177.459	63.214		TRANSITION	
800	153.344	279.866	186.580	74.628	-1122.835	-906.701	59.202
900	166.230	298.668	197.996	90.605	-1119.776	-879.857	51.066
1000	179.201	316.851	208.977	107.874	-1115.597	-853.412	44.578
1100	192.046	334.537	219.591	126.441	-1110.305	-827.441	39.292
1123.150	195.101	338.569	222.002	130.922	--------	II <--> LIQUID	--------
1200	200.748	351.719	229.891	146.194	-1297.547	-797.118	34.698
1300	206.188	368.015	239.895	166.556	-1289.016	-755.759	30.367
1400	209.744	383.430	249.602	187.359	-1280.130	-715.071	26.680
1500	212.547	398.000	259.014	208.478	-1271.004	-675.027	23.507
1600	214.765	411.789	268.136	229.845	-1261.701	-635.598	20.750
1700	216.899	424.875	276.974	251.431	-1252.242	-596.755	18.336
1800	218.656	437.323	285.539	273.210	-1242.651	-558.473	16.206
1900	220.204	449.186	293.843	295.153	-1232.955	-520.727	14.316
2000	221.752	460.521	301.895	317.251	-1223.161	-483.495	12.628

PREVIOUS: December 1960 CURRENT: March 1966

Sodium Carbonate (Na_2CO_3) $C_1Na_2O_3(cr)$

SODIUM CARBONATE (Na_2CO_3) LIQUID M_r = 105.98874

$S°(298.15\ K) = [156.423]\ J\ K^{-1}\ mol^{-1}$ $\Delta_f H°(298.15\ K) = [-1108.515]\ kJ\ mol^{-1}$

$T_{fus} = 1123.15\ K$ $\Delta_{fus}H° = 29.665\ kJ\ mol^{-1}$

Enthalpy of Formation

The $\Delta_f H°(\ell,\ 298.15\ K)$ is calculated from $\Delta_f H°(cr,\ 298.15\ K)$ by adding $\Delta_{fus}H°$ and the difference in enthalpy, $H°(1123.15\ K) - H°(298.15\ K)$, between the crystal and liquid.

Heat Capacity and Entropy

A glass transition is assumed at 723.15 K. The heat capacity below 723.15 K is obtained from the heat capacity of the crystal. Above 723.15 K the heat capacity is adopted as 45.30 cal $K^{-1}\ mol^{-1}$, based on the enthalpy measurements in the range 1127-1210 K reported by Janz, et al. [1]. The entropy is calculated in a manner analogous to that used for the enthalpy of formation.

Fusion Data

See the crystal table for details.

Reference

1. G. J. Janz, E. Neuenschwander and F. J. Kelly, Trans. Faraday Soc. 59, 841 (1963).

Sodium Carbonate (Na_2CO_3) $C_1Na_2O_3(l)$

Enthalpy Reference Temperature = T_r = 298.15 K Standard State Pressure = p° = 0.1 MPa

T/K	$C_p°$ (J K⁻¹mol⁻¹)	S° (J K⁻¹mol⁻¹)	$-[G°-H°(T_r)]/T$ (J K⁻¹mol⁻¹)	$H°-H°(T_r)$ (kJ mol⁻¹)	$\Delta_f H°$ (kJ mol⁻¹)	$\Delta_f G°$ (kJ mol⁻¹)	Log K_f
0							
100							
200							
298.15	111.002	156.423	156.423	0.	-1108.515	-1031.011	180.629
300	111.253	157.110	156.425	0.206	-1108.511	-1030.531	179.431
400	125.102	190.934	160.944	11.996	-1113.443	-1004.168	131.131
500	142.214	220.614	169.957	25.329	-1112.228	-976.960	102.062
600	163.302	248.359	180.729	40.578	-1109.329	-950.147	82.718
700	187.569	275.336	192.322	58.110	-1104.356	-923.979	68.948
723.150	193.387	281.533	195.079	62.519		I <--> II	
723.150	189.535	281.533	195.079	62.519		TRANSITION	
800	189.535	300.675	204.319	77.085	-1098.125	-898.639	58.675
900	189.535	322.999	216.289	96.039	-1092.089	-874.068	50.730
1000	189.535	342.969	227.976	114.992	-1086.226	-850.159	44.408
1100	189.535	361.033	239.264	133.946	-1080.547	-826.828	39.263
1123.150	189.535	364.981	241.815	138.333	--------	II <--> LIQUID	--------
1200	189.535	377.525	250.109	152.899	-1268.589	-799.127	34.785
1300	189.535	392.696	260.501	171.853	-1261.466	-760.294	30.549
1400	189.535	406.742	270.452	190.806	-1254.430	-722.007	26.938
1500	189.535	419.818	279.979	209.760	-1247.470	-684.221	23.827
1600	189.535	432.051	289.105	228.713	-1240.579	-646.896	21.119
1700	189.535	443.541	297.855	247.667	-1233.753	-610.000	18.743
1800	189.535	454.375	306.252	266.620	-1226.988	-573.504	16.643
1900	189.535	464.622	314.320	285.574	-1220.282	-537.383	14.774
2000	189.535	474.344	322.081	304.527	-1213.632	-501.613	13.101
2100	189.535	483.592	329.553	323.481	-1207.037	-466.174	11.595
2200	189.535	492.409	336.757	342.434	-1200.497	-431.048	10.234
2300	189.535	500.834	343.709	361.388	-1194.012	-396.219	8.998
2400	189.535	508.901	350.425	380.341	-1187.582	-361.671	7.872
2500	189.535	516.638	356.920	399.295	-1181.206	-327.389	6.840

PREVIOUS: December 1960 CURRENT: March 1966

Sodium Carbonate (Na_2CO_3) $C_1Na_2O_3(l)$

SODIUM CARBONATE (Na_2CO_3) CRYSTAL–CRYSTAL–LIQUID M_r 105.98874

0 to 723.15 K crystal, I
723.15 to 1123.15 K crystal, II
above 1123.15 K liquid

Refer to the individual tables for details.

Sodium Carbonate ($NaCO_3$) $C_1Na_2O_3$(cr,l)

Enthalpy Reference Temperature = T_r = 298.15 K   Standard State Pressure = $p°$ = 0.1 MPa

T/K	$C_p°$ (J K⁻¹mol⁻¹)	$S°$ (J K⁻¹mol⁻¹)	$-[G°-H°(T_r)]/T$ (J K⁻¹mol⁻¹)	$H°-H°(T_r)$ (kJ mol⁻¹)	$\Delta_f H°$ (kJ mol⁻¹)	$\Delta_f G°$ (kJ mol⁻¹)	Log K_f
0	0.	0.	INFINITE	-20.812	-1124.610	-1124.610	INFINITE
100	61.241	43.237	225.158	-18.192	-1129.086	-1102.556	575.916
200	94.140	97.766	148.470	-10.141	-1130.640	-1075.252	280.827
298.15	111.003	138.797	138.797	0.	-1130.768	-1048.009	183.607
300	111.253	139.484	138.799	0.206	-1130.764	-1047.496	182.385
400	125.102	173.307	143.317	11.996	-1135.696	-1019.370	133.116
500	142.214	203.007	152.332	25.337	-1134.473	-990.400	103.466
600	163.302	230.759	163.106	40.592	-1131.568	-961.826	83.734
700	187.569	257.722	174.702	58.115	-1126.604	-933.897	69.688
723.150	193.387	263.919	177.459	62.524		I <--> II	
723.150	143.761	264.873	177.459	63.214		TRANSITION	
800	153.344	279.866	186.580	74.628	-1122.835	-906.701	59.202
900	166.230	298.668	197.996	90.605	-1119.776	-879.857	51.066
1000	179.201	316.851	208.977	107.874	-1115.597	-853.412	44.578
1100	192.046	334.537	219.591	126.441	-1110.305	-827.441	39.292
1123.150	195.101	338.569	222.002	130.922		II <--> LIQUID	
1123.150	189.535	364.981	222.002	160.586		TRANSITION	
1200	189.535	377.525	231.565	175.152	-1268.589	-799.127	34.785
1300	189.535	392.696	243.384	194.106	-1261.466	-760.294	30.549
1400	189.535	406.742	254.557	213.059	-1254.430	-722.007	26.938
1500	189.535	419.818	265.143	232.013	-1247.470	-684.221	23.827
1600	189.535	432.051	275.197	250.966	-1240.579	-646.896	21.119
1700	189.535	443.541	284.765	269.920	-1233.753	-610.000	18.743
1800	189.535	454.375	293.890	288.873	-1226.988	-573.504	16.643
1900	189.535	464.622	302.608	307.827	-1220.282	-537.383	14.774
2000	189.535	474.344	310.954	326.780	-1213.632	-501.613	13.101
2100	189.535	483.592	318.957	345.734	-1207.037	-466.174	11.595
2200	189.535	492.409	326.642	364.687	-1200.497	-431.048	10.234
2300	189.535	500.834	334.034	383.641	-1194.012	-396.219	8.998
2400	189.535	508.901	341.153	402.594	-1187.582	-361.671	7.872
2500	189.535	516.638	348.019	421.548	-1181.206	-327.389	6.840

PREVIOUS: CURRENT: March 1966

Sodium Carbonate ($NaCO_3$) $C_1Na_2O_3$(cr,l)

$S°(298.15\ K) = 197.65 \pm 0.04\ J\ K^{-1}\ mol^{-1}$

$\Delta_f H°(0\ K) = -113.81 \pm 0.17\ kJ\ mol^{-1}$

$\Delta_f H°(298.15\ K) = -110.53 \pm 0.17\ kJ\ mol^{-1}$

Electronic Level and Quantum Weight

State	ε_i, cm^{-1}	g_i
$^1\Sigma$	0	1

$\omega_e = 2169.52\ cm^{-1}$ $\omega_e x_e = 13.453\ cm^{-1}$ $\sigma = 1$

$B_e = 1.9302\ cm^{-1}$ $\alpha_e = 0.01746\ cm^{-1}$ $r_e = 1.1281$ Å

Enthalpy of Formation

The enthalpy change $\Delta_r H°(298.15\ K)$ for the reaction: $CO(g) + 1/2\ O_2(g) = CO_2(g)$ was reported to be -67.636 ± 0.029 kcal mol^{-1}, based on molecular weight of CO_2 = 44.010, by Rossini (1). It was recalculated to be -67.638 ± 0.03 kcal mol^{-1}, using molecular weight of CO_2 = 44.011, for internal consistency. From the value of $\Delta_r H°(298.15\ K)$, the enthalpy of formation $\Delta_f H°(298.15\ K)$ for CO(g) was derived to be -26.417 ± 0.04 kcal mol^{-1}, which yields $D_0^0(CO)$ = 11.09 eV.

The $D_0^0(CO)$ value has been proposed to be 6.92 to 11.11 eV in the past few decades in order to explain data collected from spectroscopic, flame, shock-wave, detonation, and electron-impact studies. Recent evaluations by Brewer and Searcy (2), Fineman and Petrocelli (3), and Giese and Maier (4) favored the value 11.11 eV, reported by Gaydon (5).

The enthaply of combustion of CO(g) was also determined by Awbery and Griffiths (6), Fenning and Cotton (7), and Roth and Banse (8).

Heat Capacity and Entropy

The functions adopted here were obtained from Belzer et al. (9), assuming the thermodynamic functions for the naturally occurring isotopic mixture to be the same as those for $C^{12}O^{16}(g)$. The spectroscopic constants employed for calculation were reported by Herzberg (10). The tabulated functions include the second order corrections to the rigid-rotator and harmonic-oscillator molecular model for vibrational anharmonicity, rotational stretching and rotational-vibrational interaction.

The spectroscopic constants listed above are for the naturally occurring isotopic composition given by Strominger et al. (11).

Thermodynamic properties for CO from 70 to 300 K, with pressures to 300 atmospheres were reported by Hust et al. (12). Calculations of the vapor pressure and enthalpies of vaporization and sublimation of CO and CO_2 below one atmosphere were reported by Mullins et al. (13).

References

1. F. D. Rossini, J. Research Natl. Bur. Stand. 22, 407 (1939).
2. L. Brewer and A. Search, Ann. Rev. Phys. Chem. 7, 259 (1956).
3. M. A. Fineman and A. W. Petrocelli, J. Chem. Phys. 36, 25 (1962).
4. C. F. Giese and W. B. Maier II, J. Chem. Phys. 39, 197 (1963).
5. A. G. Gaydon, "Dissociation Energies", Chapman and Hall Ltd., 1953.
6. J. H. Awbery and E. Griffiths, Proc. Roy. Soc. (London) [A] 141, 1 (1933).
7. R. W. Fenning and F. T. Cotton, Proc. Roy. Soc. (London) [A] 141, 17 (1933).
8. W. A. Roth and H. Banse, Arch. Eisenhutten 6, 43 (1932-1933).
9. J. Belzer, L. G. Savedoff and H. L. Johnston, Ohio State University, TR 316-6, May 1953.
10. G. Herzberg, "Spectra of Diatomic Molecules", D. Van Nostrand Company, Inc., 1950.
11. D. Strominger, J. M. Hollander and G. T. Seaborg, Rev. Mod. Phys. 30, 585 (1958).
12. J. G. Hust and R. B. Stewart, U. S. Natl. Bur. of Stand., NBS-TN-202 1963.
13. J. C. Mullins, B. S. Kirk, and W. T. Ziegler, U.S. Atomic Energy Commission, NP-13862 1963.

Enthalpy Reference Temperature = T_r = 298.15 K   Standard State Pressure = p° = 0.1 MPa

	J K^{-1}mol^{-1}			kJ mol^{-1}			
T/K	$C_p°$	S°	$-[G°-H°(T_r)]/T$	$H°-H°(T_r)$	$\Delta_f H°$	$\Delta_f G°$	Log K_f
0	0.	0.	INFINITE	-8.671	-113.805	-113.805	INFINITE
100	29.104	165.850	223.539	-5.769	-112.415	-120.239	62.807
200	29.108	186.025	200.317	-2.858	-111.286	-128.526	33.568
298.15	29.142	197.653	197.653	0.	-110.527	-137.163	24.030
300	29.142	197.833	197.653	0.054	-110.516	-137.328	23.911
400	29.342	206.238	198.798	2.976	-110.102	-146.338	19.110
500	29.794	212.831	200.968	5.931	-110.003	-155.414	16.236
600	30.443	218.319	203.415	8.942	-110.150	-164.486	14.320
700	31.171	223.066	205.890	12.023	-110.469	-173.518	12.948
800	31.899	227.277	208.305	15.177	-110.905	-182.497	11.916
900	32.577	231.074	210.628	18.401	-111.418	-191.416	11.109
1000	33.183	234.538	212.848	21.690	-111.983	-200.275	10.461
1100	33.710	237.726	214.967	25.035	-112.586	-209.075	9.928
1200	34.175	240.679	216.988	28.430	-113.217	-217.819	9.481
1300	34.572	243.431	218.917	31.868	-113.870	-226.509	9.101
1400	34.920	246.006	220.761	35.343	-114.541	-235.149	8.774
1500	35.217	248.426	222.526	38.850	-115.229	-243.740	8.488
1600	35.480	250.707	224.216	42.385	-115.933	-252.284	8.236
1700	35.710	252.865	225.839	45.945	-116.651	-260.784	8.013
1800	35.911	254.912	227.398	49.526	-117.384	-269.242	7.813
1900	36.091	256.859	228.897	53.126	-118.133	-277.658	7.633
2000	36.250	258.714	230.342	56.744	-118.896	-286.034	7.470
2100	36.392	260.486	231.736	60.376	-119.675	-294.372	7.322
2200	36.518	262.182	233.081	64.021	-120.470	-302.672	7.186
2300	36.635	263.809	234.382	67.683	-121.278	-310.936	7.062
2400	36.321	265.359	235.641	71.324	-122.133	-319.164	6.946
2500	36.836	266.854	236.860	74.985	-122.994	-327.356	6.840
2600	36.924	268.300	238.041	78.673	-123.854	-335.514	6.741
2700	37.003	269.695	239.188	82.369	-124.731	-343.638	6.648
2800	37.083	271.042	240.302	86.074	-125.623	-351.729	6.562
2900	37.150	272.345	241.384	89.786	-126.532	-359.789	6.480
3000	37.217	273.605	242.437	93.504	-127.457	-367.816	6.404
3100	37.279	274.827	243.463	97.229	-128.397	-375.812	6.332
3200	37.338	276.011	244.461	100.960	-129.353	-383.778	6.265
3300	37.392	277.161	245.435	104.696	-130.325	-391.714	6.200
3400	37.443	278.278	246.385	108.438	-131.312	-399.620	6.139
3500	37.493	279.364	247.311	112.185	-132.313	-407.497	6.082
3600	37.543	280.421	248.216	115.937	-133.329	-415.345	6.027
3700	37.589	281.450	249.101	119.693	-134.360	-423.165	5.974
3800	37.631	282.453	249.965	123.454	-135.405	-430.956	5.924
3900	37.673	283.431	250.811	127.219	-136.464	-438.720	5.876
4000	37.715	284.386	251.638	130.989	-137.537	-446.457	5.830
4100	37.756	285.317	252.449	134.762	-138.623	-454.166	5.786
4200	37.794	286.228	253.242	138.540	-139.723	-461.849	5.744
4300	37.832	287.117	254.020	142.321	-140.836	-469.506	5.703
4400	37.869	287.988	254.782	146.106	-141.963	-477.136	5.664
4500	37.903	288.839	255.529	149.895	-143.103	-484.741	5.627
4600	37.941	289.673	256.262	153.687	-144.257	-492.321	5.590
4700	37.974	290.489	256.982	157.483	-145.424	-499.875	5.555
4800	38.007	291.289	257.688	161.282	-146.605	-507.404	5.522
4900	38.041	292.073	258.382	165.084	-147.800	-514.908	5.489
5000	38.074	292.842	259.064	168.890	-149.009	-522.387	5.457
5100	38.104	293.596	259.733	172.699	-150.231	-529.843	5.427
5200	38.137	294.336	260.392	176.511	-151.469	-537.275	5.397
5300	38.171	295.063	261.039	180.326	-152.721	-544.681	5.368
5400	38.200	295.777	261.676	184.146	-153.987	-552.065	5.340
5500	38.074	296.476	262.302	187.957	-155.279	-559.426	5.313
5600	38.263	297.164	262.919	191.775	-156.585	-566.762	5.287
5700	38.296	297.842	263.525	195.603	-157.899	-574.075	5.261
5800	38.325	298.508	264.123	199.434	-159.230	-581.364	5.236
5900	38.355	299.163	264.711	203.268	-160.579	-588.631	5.211
6000	38.388	299.808	265.291	207.106	-161.945	-595.875	5.188

PREVIOUS: September 1965 (1 atm)   CURRENT: September 1965 (1 bar)

CARBON OXIDE SULFIDE (COS) IDEAL GAS M_r = 60.0704

$S°(298.15\ K) = 231.581\ J\ K^{-1}\ mol^{-1}$

$\Delta_f H°(0\ K) = -138.531 \pm 1.05\ kJ\ mol^{-1}$
$\Delta_f H°(298.15\ K) = -138.407 \pm 1.05\ kJ\ mol^{-1}$

Vibrational Levels and Degeneracies

ν, cm^{-1}
859(1)
524(2)
2064(1)

Ground State Quantum Weight: 1 $\sigma = 1$
Point Group: $C_{\infty v}$
Rotational Constant: $B_{000} = 0.20287\ cm^{-1}$

Spectroscopic constants used in calculating corrections to rigid rotator-harmonic oscillator approximation (cm^{-1}):

$\alpha_1 = 0.0006044$	$X_{11} = -4.0$	$X_{12} = -6.8$	$A_{000} = 0$	$g_{22} = 3.2$
$\alpha_2 = 0.0003539$	$X_{22} = -0.4$	$X_{23} = -11.5$	$C_{000} = 0$	
$\alpha_3 = 0.001838$	$X_{33} = -7.0$	$X_{13} = -4.5$	$D_{000} = 4.37 \times 10^{-8}$	

Enthalpy of Formation

Terres and Wesemann (1), report equilibrium constants from 623 to 873 K for the reaction $CO_2(g) + H_2S(g) \rightleftarrows COS(g) + H_2O(g)$. These yield $\Delta_r H°(700\ K) = 8.19 \pm 0.23$ kcal, whence $\Delta_f H°(298.15\ K)$ for COS is $-33.08 \pm 0.25\ kcal\ mol^{-1}$. The same authors report equilibrium constants for the reaction $COS(g) + H_2S(g) \rightleftarrows CS_2(g) + H_2O(g)$, but regard these as less reliable than the others. The resulting value of $-26.39\ kcal\ mol^{-1}$ for $\Delta_f H°(298.15\ K)$ of COS has been disregarded.

Heat Capacity

Gordon (2) has used the constants listed above to calculate $C°_p$ from 298.15 to 6000 K by the method of Pennington and Kobe (3), which takes anharmonicity, vibration-rotation interaction, and centrifugal stretching into account. The constants have been taken from (4, 5, 6, 7, 8). The thermodynamic functions below 298.15 K have been calculated for the rigidly rotating harmonic oscillator.

References

1. E. Terres and H. Wesemann, Angew. Chem. 45, 795 (1932).
2. J. S. Gordon, personal communication, February, 1961.
3. R. E. Pennington and K. A. Kobe, J. Chem. Phys. 22, 1442 (1954).
4. C. A. Burrus and W. Gordy, Phys. Rev. 93, 897 (1954).
5. H. J. Callomon, D. C. McKean, and H. W. Thompson, Proc. Roy Soc. (London), A208, 341 (1951); A222, 431 (1954).
6. C. H. Townes, A. N. Holden, and F. R. Merritt, Phys. Rev. 74, 1113 (1948).
7. T. Wentink, J. Chem. Phys. 30, 105 (1959).
8. M. W. P. Strandberg, T. Wentink, and R. L. Kyhl, Phys. Rev. 75, 270 (1949).

Carbon Oxide Sulfide (COS) $C_1O_1S_1(g)$

Enthalpy Reference Temperature = T_r = 298.15 K   Standard State Pressure = $p°$ = 0.1 MPa

T/K	$C°_p$ (J K⁻¹mol⁻¹)	S°	$-[G°-H°(T_r)]/T$	$H°-H°(T_r)$ (kJ mol⁻¹)	$\Delta_f H°$	$\Delta_f G°$	Log K_f
0	0.	0.	INFINITE	-9.929	-138.531	-138.531	INFINITE
100	29.610	194.193	264.316	-7.012	-137.816	-147.223	76.901
200	35.472	216.276	235.229	-3.791	-138.018	-156.581	40.895
298.15	41.497	231.581	231.581	0.	-138.407	-165.601	29.013
300	41.589	231.838	231.582	0.077	-138.415	-165.770	28.863
400	45.852	244.415	233.268	4.459	-141.138	-174.727	22.817
500	48.903	254.988	236.582	9.203	-143.179	-182.902	19.108
600	51.250	264.118	240.428	14.214	-144.910	-190.679	16.600
700	53.137	272.164	244.398	19.436	-146.435	-198.184	14.789
800	54.677	279 363	248.327	24.829	-147.842	-205.480	13.416
900	55.940	285.878	252.143	30.362	-202.304	-211.523	12.276
1000	56.990	291.828	255.818	36.010	-202.366	-212.544	11.102
1100	57.861	297.302	259.344	41.754	-202.442	-213.559	10.141
1200	58.597	302.369	262.721	47.577	-202.531	-214.565	9.340
1300	59.216	307.084	265.954	53.469	-202.633	-215.564	8.661
1400	59.748	311.492	269.051	59.418	-202.748	-216.555	8.080
1500	60.208	315.631	272.020	65.416	-202.878	-217.536	7.575
1600	60.605	319.529	274.869	71.457	-203.024	-218.509	7.134
1700	60.957	323.214	277.605	77.535	-203.187	-219.472	6.744
1800	61.266	326.707	280.237	83.647	-203.366	-220.425	6.397
1900	61.547	330.027	282.771	89.788	-203.563	-221.367	6.086
2000	61.798	333.191	285.213	95.955	-203.778	-222.299	5.806
2100	62.024	336.211	287.570	102.146	-204.011	-223.219	5.552
2200	62.233	339.102	289.847	108.359	-204.261	-224.128	5.321
2300	62.423	341.872	292.050	114.592	-204.529	-225.025	5.110
2400	62.601	344.533	294.181	120.843	-204.815	-225.910	4.917
2500	62.764	347.092	296.247	127.112	-205.118	-226.783	4.738
2600	62.919	349.556	298.250	133.396	-205.437	-227.643	4.573
2700	63.065	351.934	300.195	139.695	-205.773	-228.491	4.420
2800	63.203	354.230	302.084	146.009	-206.124	-229.326	4.278
2900	63.333	356.450	303.920	152.336	-206.491	-230.148	4.145
3000	63.459	358.599	305.707	158.675	-206.873	-230.958	4.021
3100	63.576	360.682	307.447	165.027	-207.270	-231.754	3.905
3200	63.691	362.702	309.143	171.390	-207.681	-232.537	3.796
3300	63.802	364.664	310.795	177.765	-208.107	-233.308	3.693
3400	63.910	366.570	312.408	184.151	-208.546	-234.064	3.596
3500	64.015	368.424	313.982	190.547	-208.998	-234.809	3.504
3600	64.114	370.229	315.520	196.954	-209.464	-235.539	3.418
3700	64.211	371.987	317.022	203.370	-209.944	-236.257	3.335
3800	64.306	373.701	318.491	209.796	-210.436	-236.962	3.257
3900	64.400	375.372	319.928	216.231	-210.942	-237.653	3.183
4000	64.492	377.004	321.335	222.676	-211.462	-238.331	3.112
4100	64.582	378.597	322.712	229.129	-211.994	-238.997	3.045
4200	64.670	380.155	324.061	235.592	-212.540	-239.649	2.980
4300	64.757	381.677	325.384	242.063	-213.100	-240.287	2.919
4400	64.843	383.167	326.680	248.543	-213.673	-240.913	2.860
4500	64.927	384.625	327.952	255.032	-214.261	-241.526	2.804
4600	65.009	386.053	329.199	261.529	-214.863	-242.125	2.749
4700	65.090	387.452	330.424	268.034	-215.479	-242.711	2.697
4800	65.170	388.823	331.626	274.547	-216.111	-243.284	2.647
4900	65.250	390.168	332.807	281.068	-216.759	-243.843	2.599
5000	65.329	391.487	333.968	287.597	-217.422	-244.389	2.553
5100	65.407	392.781	335.108	294.133	-218.101	-244.922	2.509
5200	65.484	394.052	336.230	300.678	-218.797	-245.441	2.465
5300	65.561	395.300	337.332	307.230	-219.511	-245.946	2.424
5400	65.638	396.527	338.417	313.790	-220.242	-246.439	2.384
5500	65.714	397.732	339.485	320.358	-220.991	-246.917	2.345
5600	65.789	398.916	340.536	326.933	-221.759	-247.381	2.307
5700	65.864	400.082	341.570	333.516	-222.546	-247.832	2.271
5800	65.939	401.228	342.589	340.106	-223.353	-248.268	2.236
5900	66.013	402.355	343.592	346.704	-224.180	-248.692	2.202
6000	66.086	403.466	344.581	353.308	-225.028	-249.099	2.169

PREVIOUS: March 1961 (1 atm)   CURRENT: March 1961 (1 bar)

Carbon Oxide Sulfide (COS) $C_1O_1S_1(g)$

CARBON DIOXIDE (CO_2) IDEAL GAS M_r = 44.0098

$\Delta_f H°$(0 K) = -393.151 ± 0.05 kJ mol^{-1}

$S°$(298.15 K) = 213.795 ± 0.12 J K^{-1} mol^{-1} $\Delta_f H°$(298.15 K) = -393.522 ± 0.05 kJ mol^{-1}

Vibrational Frequencies and Degeneracies

ν, cm^{-1}
1384.86 (1)
667.30 (2)
2349.30 (1)

Point Group: $D_{\infty h}$ σ = 2

Bond Distances: C–O = 1.16 Å

Bond Angle: O–C–O = 108°

Rotational Constant: B_0 = 0.391514 cm^{-1}

Enthalpy of Formation

The enthalpy change of the reaction C(c, graphite) + O_2(g) = CO_2(g) has been measured by Dewey and Harper (1). Jessup (2), and Prosen and Rossini (3). Based on these data, the enthalpy of formation for CO_2(g) was reported to be -94.0518 ± 0.0108 cal mol^{-1}, using molecular weight of CO_2 = 44.010, by Prosen, Jessup anad Rossini (4). This value was recalculated to be -94.054 ± 0.011 kca mol^{-1}, based on molecular weight of CO_2 = 44.011, for internal consistency.

Heat Capacity and Entropy

The functions adopted here were obtained from Woolley (5), who calculated the thermodynamic functions by means of a direct summation for the naturally occurring isotopic composition. The spectroscopic constants used are essentially those selected by Wentnik (6). Slightly different sets of spectroscopic constants were obtained by Courtoy (7) and Stull, Wyatt and Plass (8). The high-resolution infrared spectrum of O^{18} - enriched CO_2 was examined in the region 5400-1620 cm^{-1}, using an Ebert grating Spectrometer with spectral slit widths ranging from 0.4 to 0.2 cm^{-1} by Berney (9).

The molecular structure was reported by Herzberg (10). The rotational constant, B_o, was obtained from Woolley (5). The value of bond distance, r_e, was calculated from B_e which was derived from B_o, using B_e-B_o = 0.0011 cm^{-1} given in Herzberg (10).

Heat capacities of CO_2(g) at high pressures were reported by Vukalovich, Altunin and Gureev (11), Krueger (12) and Vukalovich and Gureev (13).

References

1. P. H. Dewey and D. R. Harper, J. Res. Nat. Bur. Stand. 21, 457 (1938).
2. R. S. Jessup, J. Res. Nat. Bur. Stand. 21, 491 (1938).
3. E. J. Prosen and F. D. Rossini, J. Res. Nat. Bur. Stand. 33, 439 (1944).
4. E. J. Prosen, R. S. Jessup, and F. D. Rossini, J. Res. Nat. Bur. Stand. 33, 447 (1944).
5. H. W. Woolley, J. Res. Nat. Bur. Stand. 52, 289 (1954).
6. T. Wentnik, Jr., J. Chem. Phys. 30, 105 (1959).
7. C. P. Courtoy, Mem. Soc. Roy. Liege 18, 496 (1957).
8. V. R. Stull, P. J. Wyatt, and G. N. Plass, J. Chem. Phys. 37, 1442 (1962).
9. C. V. Berney, Ph. D. Dissertation, University of Washington, 1962.
10. G. Herzberg, "Infrared and Raman Spectra", D. Van Nostrand, Inc., New York, 1945.
11. M. P. Vukalovich, V. V. Altunin, and A. N. Gureev, Teploenergetika 12, 58 (1965).
12. K. Krueger, Ver. Deut. Ingr. Z. 106, 1620 (1964).
13. M. P. Vukalovich and A. N. Gureev, Teploenergetika, 11 80 (1964).

Carbon Dioxide (CO_2) C_1O_2(g)

Enthalpy Reference Temperature = T_r = 298.15 K   Standard State Pressure = p° = 0.1 MPa

T/K	$C_p°$	$S°$	$-[G°-H°(T_r)]/T$	$H°-H°(T_r)$	$\Delta_f H°$	$\Delta_f G°$	Log K_f
	J $K^{-1}mol^{-1}$			kJ mol^{-1}			
0	0.	0.	INFINITE	-9.364	-393.151	-393.151	INFINITE
100	29.208	179.009	243.568	-6.456	-393.208	-393.683	205.639
200	32.359	199.975	217.046	-3.414	-393.404	-394.085	102.924
298.15	37.129	213.795	213.795	0.	-393.522	-394.389	69.095
300	37.221	214.025	213.795	0.069	-393.523	-394.394	68.670
400	41.325	225.314	215.307	4.003	-393.583	-394.675	51.539
500	44.627	234.901	218.290	8.305	-393.666	-394.939	41.259
600	47.321	243.283	221.772	12.907	-393.803	-395.182	34.404
700	49.564	250.750	225.388	17.754	-393.983	-395.398	29.505
800	51.434	257.494	228.986	22.806	-394.188	-395.586	25.829
900	52.999	263.645	232.500	28.030	-394.405	-395.748	22.969
1000	54.308	269.299	235.901	33.397	-394.623	-395.886	20.679
1100	55.409	274.528	239.178	38.884	-394.838	-396.001	18.805
1200	56.342	279.390	242.329	44.473	-395.050	-396.098	17.242
1300	57.137	283.932	245.356	50.148	-395.257	-396.177	15.919
1400	57.802	288.191	248.265	55.896	-395.462	-396.240	14.784
1500	58.379	292.199	251.062	61.705	-395.668	-396.288	13.800
1600	58.886	295.983	253.753	67.569	-395.876	-396.323	12.939
1700	59.317	299.566	256.343	73.480	-396.090	-396.344	12.178
1800	59.701	302.968	258.840	79.431	-396.311	-396.353	11.502
1900	60.049	306.205	261.248	85.419	-396.542	-396.349	10.896
2000	60.350	309.293	263.574	91.439	-396.784	-396.333	10.351
2100	60.622	312.244	265.822	97.488	-397.039	-396.304	9.858
2200	60.865	315.070	267.996	103.562	-397.309	-396.262	9.408
2300	61.086	317.781	270.102	109.660	-397.596	-396.209	8.998
2400	61.287	320.385	272.144	115.779	-397.900	-396.142	8.622
2500	61.471	322.890	274.124	121.917	-398.222	-396.062	8.275
2600	61.647	325.305	276.046	128.073	-398.562	-395.969	7.955
2700	61.802	327.634	277.914	134.246	-398.921	-395.862	7.658
2800	61.952	329.885	279.730	140.433	-399.299	-395.742	7.383
2900	62.095	332.061	281.497	146.636	-399.695	-395.609	7.126
3000	62.229	334.169	283.218	152.852	-400.111	-395.461	6.886
3100	62.347	336.211	284.895	159.081	-400.545	-395.298	6.661
3200	62.462	338.192	286.529	165.321	-400.998	-395.122	6.450
3300	62.573	340.116	288.124	171.573	-401.470	-394.932	6.251
3400	62.681	341.986	289.681	177.836	-401.960	-394.726	6.064
3500	62.785	343.804	291.202	184.109	-402.467	-394.506	5.888
3600	62.884	345.574	292.687	190.393	-402.991	-394.271	5.721
3700	62.980	347.299	294.140	196.686	-403.532	-394.022	5.563
3800	63.074	348.979	295.561	202.989	-404.089	-393.756	5.413
3900	63.166	350.619	296.952	209.301	-404.662	-393.477	5.270
4000	63.254	352.219	298.314	215.622	-405.251	-393.183	5.134
4100	63.341	353.782	299.648	221.951	-405.856	-392.874	5.005
4200	63.426	355.310	300.955	228.290	-406.475	-392.550	4.882
4300	63.509	356.803	302.236	234.637	-407.110	-392.210	4.764
4400	63.588	358.264	303.493	240.991	-407.760	-391.857	4.652
4500	63.667	359.694	304.726	247.354	-408.426	-391.488	4.544
4600	63.745	361.094	305.937	253.725	-409.106	-391.105	4.441
4700	63.823	362.466	307.125	260.103	-409.802	-390.706	4.342
4800	63.893	363.810	308.292	266.489	-410.514	-390.292	4.247
4900	63.968	365.128	309.438	272.882	-411.242	-389.862	4.156
5000	64.046	366.422	310.565	279.283	-411.986	-389.419	4.068
5100	64.128	367.691	311.673	285.691	-412.746	-388.959	3.984
5200	64.220	368.937	312.762	292.109	-413.522	-388.486	3.902
5300	64.312	370.161	313.833	298.535	-414.314	-387.996	3.824
5400	64.404	371.364	314.888	304.971	-415.123	-387.493	3.748
5500	64.496	372.547	315.925	311.416	-415.949	-386.974	3.675
5600	64.588	373.709	316.947	317.870	-416.794	-386.439	3.605
5700	64.680	374.853	317.953	324.334	-417.658	-385.890	3.536
5800	64.772	375.979	318.944	330.806	-418.541	-385.324	3.470
5900	64.865	377.087	319.920	337.288	-419.445	-384.745	3.406
6000	64.957	378.178	320.882	343.779	-420.372	-384.148	3.344

PREVIOUS: September 1965 (1 atm)   CURRENT: September 1965 (1 bar)

Carbon Dioxide (CO_2) C_1O_2(g)

CARBON DIOXIDE, ION (CO_2^-) IDEAL GAS M_r = 44.010348

$S°(298.15\ K)$ = [240.65] J K^{-1} mol^{-1}

$\Delta_fH°(0\ K)$ = [-435.6 ± 113] kJ mol^{-1}
$\Delta_fH°(298.15\ K)$ = [-441.4 ± 113] kJ mol^{-1}

Vibrational Frequencies and Degeneracies

ν, cm^{-1}
[1400] (1)
[800] (1)
1671 (1)

Ground State Quantum Weight: [2] σ = 2
Point Group: [C_{2v}]
Bond Distances: C–O = [1.25] Å
Bond Angle: O–C–O = 127° ± 8°
Product of the Moments of Inertia: $I_AI_BI_C$ = [2.13011 x 10^{-116}]

Enthalpy of Formation

$\Delta_fH°(CO_2^-$, g, 298.15 K) = -105.5 ± 27 kcal mol^{-1} is estimated from $\Delta_rH°$(298.15 K) = 88.6 ± 10 kcal mol^{-1} for $HCO_2^-(g)$ + H(g) + $CO_2^-(g)$, using the $\Delta_fH°(HCO_2^-$, g, 298.15 K) = 142 ± 17 kcal mol^{-1} and $\Delta_fH°$(H, g, 298.15 K) = 52.1 kcal mol^{-1}. The value of $\Delta_rH°$(298.15 K) is assumed to be the same as that for HCOOH(g) + H(G) + COOH(g). The values of $\Delta_fH°$(HCOOH, g, 298.15 K) and $\Delta_fH°$(H, g, 298.15 K), - 90.5 and 52.1 kcal mol^{-1}, respectively, are from NBS (1), and the value of $\Delta_fH°$(COOH, g, 298.15 K) = -54 ± 3 kcal mol^{-1} is from Benson (2). The value of $\Delta_fH°(HCO_2^-$, g, 298.15 K) is estimated from an approximate lattice energy calculation for sodium formate. The derived electron affinity of CO_2(g) is 11.5 kcal mol^{-1} (0.5 eV) with an estimated uncertainty range of 0-40 kcal mol^{-1}.

Heat Capacity and Entropy

The correlations of Walsh (3) predict a bent molecular structure of CO_2^- based on its 17 valence electrons. Hartman and Hisatsune (4) measured the asymmetric C–O stretching frequency and derived the bond angle of 127° ± 8° from the infrared spectrum of matrix isolated CO_2^-. They also estimated the bond distance C–O and the symmetric stretching and the bending vibrational frequencies by the valence force method. These molecular constants are adopted in the tabulation. The principal moments of inertia are: I_A = 0.4511 x 10^{-39}, I_B = 6.6497 x 10^{-39} and I_C = 7.1008 x 10^{-39} g cm^2.

Ovenall and Whiffen (5) have derived a bond angle of 134° which is in good agreement with the value selected. This result was obtained from electron spin resonance for the CO_2^- radical trapped in sodium formate.

References

1. U. S. Nat. Bur. Stand. Tech. Note 270-1, 1965.
2. S. W. Benson, J. Chem. Ed. 42, 502 (1965).
3. A. D. Walsh, J. Chem. Soc. 2266 (1953).
4. K. O. Hartman and I. C. Hisatsune, J. Chem. Phys. 44, 1913 (1966).
5. D. W. Ovenall and D. H. Whiffen, Mol. Phys. 4, 135 (1961).

Carbon Dioxide, Ion (CO_2^-) $C_1O_2^-(g)$

Enthalpy Reference Temperature = T_r = 298.15 K Standard State Pressure = p° = 0.1 MPa

T/K	$C_p°$ (J $K^{-1}mol^{-1}$)	S°	-[G°-H°(T_r)]/T	H°-H°(T_r) (kJ mol^{-1})	$\Delta_fH°$	$\Delta_fG°$	Log K_f
0	0.	0.	INFINITE	-10.147	-435.628		
100	33.269	203.352	271.567	-6.822			
200	34.178	226.586	243.909	-3.465			
250	35.280	234.325	241.243	-1.730			
298.15	36.593	240.648	240.648	0.	-441.412	-444.030	77.792
300	36.647	240.875	240.649	0.068	-441.453	-444.047	77.315
350	38.163	246.636	241.100	1.938	-442.570	-444.391	66.322
400	39.737	251.835	242.122	3.885	-443.708	-444.573	58.055
450	41.295	256.606	243.470	5.911	-444.867	-444.612	51.609
500	42.785	261.034	245.008	8.013	-446.044	-444.521	46.439
600	45.454	269.078	248.363	12.429	-448.445	-443.992	38.653
700	47.653	276.256	251.844	17.088	-450.891	-443.057	33.061
800	49.420	282.739	255.307	21.945	-453.371	-441.769	28.845
900	50.828	288.644	258.688	26.960	-455.874	-440.168	25.547
1000	51.952	294.060	261.958	32.101	-458.397	-438.288	22.894
1100	52.854	299.055	265.106	37.343	-460.937	-436.154	20.711
1200	53.586	303.686	268.131	42.667	-463.492	-433.789	18.882
1300	54.183	308.000	271.034	48.056	-466.063	-431.209	17.326
1400	54.677	312.034	273.820	53.500	-468.651	-428.431	15.985
1500	55.088	315.821	276.495	58.989	-471.257	-425.468	14.816
1600	55.433	319.387	279.065	64.515	-473.881	-422.330	13.788
1700	55.725	322.757	281.537	70.073	-476.525	-419.027	12.875
1800	55.974	325.949	283.917	75.659	-479.191	-415.568	12.059
1900	56.188	328.981	286.209	81.267	-481.880	-411.960	11.326
2000	56.374	331.868	288.421	86.895	-484.592	-408.210	10.661
2100	56.535	334.623	290.556	92.541	-487.329	-404.324	10.057
2200	56.676	337.256	292.619	98.202	-490.092	-400.306	9.504
2300	56.800	339.778	294.615	103.876	-492.881	-396.163	8.997
2400	56.910	342.198	296.548	109.561	-495.697	-391.898	8.529
2500	57.008	344.523	298.420	115.257	-498.539	-387.514	8.097
2600	57.095	346.761	300.237	120.962	-501.409	-383.017	7.695
2700	57.173	348.917	302.000	126.676	-504.305	-378.408	7.321
2800	57.243	350.998	303.713	132.397	-507.229	-373.691	6.971
2900	57.306	353.008	305.378	138.124	-510.179	-368.871	6.644
3000	57.363	354.951	306.999	143.858	-513.156	-363.947	6.337
3100	57.415	356.833	308.576	149.597	-516.159	-358.924	6.048
3200	57.462	358.657	310.113	155.341	-519.187	-353.804	5.775
3300	57.505	360.426	311.611	161.089	-522.241	-348.589	5.518
3400	57.545	362.143	313.072	166.842	-525.319	-343.280	5.274
3500	57.581	363.811	314.498	172.598	-528.422	-337.881	5.043
3600	57.615	365.434	315.890	178.358	-531.548	-332.393	4.823
3700	57.645	367.013	317.251	184.121	-534.698	-326.818	4.614
3800	57.674	368.551	318.581	189.887	-537.871	-321.156	4.415
3900	57.700	370.049	319.881	195.655	-541.066	-315.411	4.224
4000	57.724	371.510	321.154	201.427	-544.283	-309.585	4.043
4100	57.747	372.936	322.399	207.200	-547.523	-303.677	3.869
4200	57.768	374.328	323.619	212.976	-550.783	-297.691	3.702
4300	57.788	375.687	324.814	218.754	-554.066	-291.625	3.543
4400	57.806	377.016	325.986	224.533	-557.370	-285.485	3.389
4500	57.823	378.315	327.134	230.315	-560.695	-279.268	3.242
4600	57.839	379.586	328.261	236.098	-564.041	-272.978	3.100
4700	57.854	380.830	329.366	241.883	-567.409	-266.614	2.963
4800	57.869	382.049	330.451	247.669	-570.799	-260.178	2.831
4900	57.882	383.242	331.516	253.456	-574.211	-253.671	2.704
5000	57.894	384.411	332.562	259.245	-577.646	-247.094	2.581
5100	57.906	385.558	333.590	265.035	-581.103	-240.449	2.463
5200	57.917	386.683	334.601	270.826	-584.584	-233.736	2.348
5300	57.928	387.786	335.594	276.619	-588.089	-226.955	2.237
5400	57.938	388.869	336.570	282.412	-591.620	-220.109	2.129
5500	57.947	389.932	337.531	288.206	-595.175	-213.197	2.025
5600	57.956	390.976	338.476	294.001	-598.757	-206.218	1.924
5700	57.965	392.002	339.406	299.797	-602.367	-199.177	1.825
5800	57.973	393.010	340.322	305.594	-606.005	-192.071	1.730
5900	57.980	394.001	341.223	311.392	-609.672	-184.905	1.637
6000	57.987	394.976	342.111	317.190	-613.369	-177.673	1.547

PREVIOUS: December 1966 (1 atm) CURRENT: December 1966 (1 bar)

Carbon Dioxide, Ion (CO_2^-) $C_1O_2^-(g)$

$S°(298.15\ K) = 216.26 \pm 0.2\ J\ K^{-1}\ mol^{-1}$

$\Delta_f H°(0\ K) = 447.6 \pm 9\ kJ\ mol^{-1}$
$\Delta_f H°(298.15\ K) = 449.9 \pm 9\ kJ\ mol^{-1}$

Electronic States and Molecular Constants.

State	T_e, cm^{-1}	ω_e, cm^{-1}	$\omega_e x_e$, cm^{-1}	B_e, cm^{-1}	α_e, cm^{-1}	D_e, cm^{-1}	r_e, Å
X $^2\Sigma^+$	0	1239.67	6.86	0.7986	0.00597	[1.33]	1.5622
A $^2\Pi_i$	7053.2 6894.9	1061.99	6.035	0.7135	0.0058	[1.3]	1.653
B $^2\Sigma^+$	29100.4	836.32	5.917	0.6829	0.00628	[1.82]	1.6894

Enthalpy of Formation

We adopt the value which Huber and Herzberg (1) recommended, D^0_0(CP, g) = 5.28 eV (509.4 kJ mol^{-1}), based on the mass-spectrometric studies of Gingerich (2), Smoes et al. (3), and Kordis and Gingerich (4). Smoes et al. (3) studied the equilibrium between graphite and mono- and di-atomic phosphorus vapor when gadolinium phosphide was heated in a graphite Knudsen cell in the absence of excess graphite powder (2120-2500 K). Kordis and Gingerich (4) studied the effusate from a graphite-lined tantalum Knudsen cell containing a mixture of graphite(excess)-TiP-Rh.

Heat Capacity and Entropy

The molecular constants for the three electronic states are taken from Huber and Herzberg (1) based on their review in 1976 of the data of Barwald et al. (5), Thakur and Singh (6), Chaudhry and Upadhya (7), and Wentink and Spindler (8).

References

1. K. P. Huber and G. Herzberg, "Molecular Spectra and Molecular Structure; IV. Constants of Diatomic Molecules," Van Nostrand Reinhold Company, New York, 1979.
2. K. A. Gingerich, Thermochim. Acta 2, 233 (1971).
3. S. Smoes, C. E. Myers and J. Drowart, Chem. Phys. Lett. 8, 10 (1971).
4. J. Kordis and K. A. Gingerich, J. Chem. Phys. 58, 5058 (1973).
5. H. Barwald, G. Herzberg and L. Herzberg, Ann. Physik (Leipzig) 20, 569 (1934).
6. S. N. Thakur and N. L. Singh, J. Sci. Res. Banaras Hindu Univ. 18, 253 (1967).
7. A K. Chaudhry and K. N. Upadhya, Indian J. Phys. 43, 83 (1969).
8. T. Wentink and R. J. Spindler, J. Quant. Spectrosc. Radiat. Transfer 10, 621 (1970).

Enthalpy Reference Temperature = T_r = 298.15 K   Standard State Pressure = p° = 0.1 MPa

T/K	$C_p°$ (J $K^{-1}mol^{-1}$)	S° (J $K^{-1}mol^{-1}$)	$-[G°-H°(T_r)]/T$ (J $K^{-1}mol^{-1}$)	$H°-H°(T_r)$ (kJ mol^{-1})	$\Delta_f H°$ (kJ mol^{-1})	$\Delta_f G°$ (kJ mol^{-1})	Log K_f
0	0.000	0.000	INFINITE	-8.715	447.589	447.589	INFINITE
100	29.105	184.299	242.380	-5.808	449.549	432.983	-226.167
200	29.207	204.487	218.963	-2.895	449.872	416.016	-108.652
250	29.473	211.029	216.745	-1.429	449.951	407.540	-85.151
298.15	29.910	216.255	216.255	0.000	449.893	399.375	-69.969
300	29.930	216.440	216.256	0.055	449.888	399.062	-69.483
350	30.515	221.097	216.622	1.566	448.992	390.674	-58.305
400	31.154	225.213	217.444	3.108	448.666	382.365	-49.932
450	31.789	228.919	218.516	4.681	448.295	374.099	-43.424
500	32.388	232.300	219.728	6.286	447.885	365.877	-38.223
600	33 423	238.300	222.336	9.578	446.967	349.559	-30.432
700	34.239	243.516	224.997	12.963	445.946	333.404	-24.879
800	34.872	248.131	227.606	16.420	444.849	317.401	-20.724
900	35.364	252.268	230.120	19.933	443.695	301.538	-17.501
1000	35.752	256.015	232.526	23.489	442.496	285.807	-14.929
1100	36.061	259.438	234.819	27.081	441.261	270.198	-12.831
1200	36.312	262.586	237.003	30.700	376.424	255.781	-11.134
1300	36.519	265.501	239.085	34.342	375.909	245.748	-9.874
1400	36.692	268.214	241.070	38.002	375.367	235.756	-8.796
1500	36.838	270.751	242.965	41.679	374.803	225.803	-7 863
1600	36.964	273.132	244.777	45.369	374.218	215.888	-7.048
1700	37.074	275.377	246.511	49.071	373.614	206.011	-6.330
1800	37.170	277.499	248.174	52.784	372.994	196.170	-5.693
1900	37.256	279.511	249.771	56.505	372.357	186.364	-5.123
2000	37.333	281.424	251.306	60.235	371.706	176.592	-4.612
2100	37.403	283.247	252.784	63.972	371.041	166.852	-4.150
2200	37.466	284.988	254.209	67.715	370.362	157.145	-3.731
2300	37.525	286.655	255.583	71.465	369.671	147.468	-3.349
2400	37.579	288.253	256.912	75.220	368.967	137.822	-3.000
2500	37.630	289.788	258.196	78.980	368.252	128.206	-2.679
2600	37.678	291.265	259.440	82.746	367.525	118.619	-2.383
2700	37.723	292.688	260.645	86.516	366.786	109.059	-2.110
2800	37.765	294.061	261.814	90.290	366.037	99.528	-1.857
2900	37.806	295.387	262.949	94.069	365.277	90.023	-1.621
3000	37.845	296.669	264.052	97.851	364.506	80.544	-1.402
3100	37.882	297.911	265.124	101.638	363.725	71.092	-1.198
3200	37.918	299.114	266.168	105.428	362.933	61.665	-1.007
3300	37.953	300.281	267.184	109.221	362.131	52.262	-0.827
3400	37.987	301.415	268.174	113.018	361.319	42 885	-0.659
3500	38.019	302.516	269.140	116.819	360.497	33.531	-0.500
3600	38.051	303.588	270.082	120.622	359.666	24.201	-0.351
3700	38.082	304.631	271.001	124.429	358.824	14.894	-0.210
3800	38.113	305.647	271.900	128.239	357.973	5.610	-0.077
3900	38.143	306.637	272.778	132.052	357.112	-3.652	0.049
4000	38.172	307.603	273.636	135.867	356.241	-12.891	0.168
4100	38.201	308.546	274.476	139.686	355.360	-22.109	0.282
4200	38.229	309.467	275.299	143.507	354.471	-31.304	0.389
4300	38.257	310.367	276.104	147.332	353.571	-40.479	0.492
4400	38.285	311.247	276.892	151.159	352.662	-49.632	0.589
4500	38.312	312.107	277.665	154.989	351.744	-58.765	0.682
4600	38.339	312.950	278.423	158.821	350.816	-67.877	0.771
4700	38.366	313.775	279.167	162.657	349.879	-76.969	0.855
4800	38.392	314.583	279.896	166.495	348.933	-86.041	0.936
4900	38.418	315.375	280.612	170.335	347.977	-95.093	1.014
5000	38.444	316.151	281.315	174.178	347.012	-104.126	1.088
5100	38.470	316.912	282.006	178.024	346.038	-113.139	1.159
5200	38.496	317.660	282.684	181.872	345.054	-122.133	1.227
5300	38.521	318.393	283.351	185.723	344.061	-131.108	1.292
5400	38.546	319.114	284.007	189.576	343.059	-140.064	1.355
5500	38.571	319.821	284.652	193.432	342.048	-149.001	1.415
5600	38.596	320.516	285.286	197.291	341.028	-157.920	1.473
5700	38.621	321.200	285.910	201.151	339.998	-166.821	1.529
5800	38.645	321.872	286.524	205.015	338.959	-175.703	1.582
5900	38.670	322.532	287.129	208.881	337.911	-184.568	1.634
6000	38.694	323.182	287.724	212.749	336.854	-193.414	1.684

PREVIOUS: June 1962 (1 atm)   CURRENT: June 1962 (1 bar)

CARBON SULFIDE (CS) IDEAL GAS M_r = 44.071

$\Delta_f H°(0\ K) = 277.1 \pm 25\ kJ\ mol^{-1}$

$S°(298.15\ K) = 210.55 \pm 0.04\ J\ K^{-1}\ mol^{-1}$ $\Delta_f H°(298.15\ K) = 280.3 \pm 25\ kJ\ mol^{-1}$

Electronic Levels and Quantum Weights

State	ε_i, cm^{-1}	g_i
$X^1\Sigma^+$	0	1
$a^3\pi$	27509	2
	27590	2
	27607	2
$a(^3\pi)$	38613.9	6
$A^1\pi$	38797.6	2
$a'^3\Sigma^+$	38851.8	3
$c^3\Sigma^-$	39160.8	3
$(^1\Delta)$	39170	2
$(^1\Delta)$	39950	2

$\omega_e = 1285.08\ cm^{-1}$ $\omega_e x_e = 6.46\ cm^{-1}$ $\sigma = 1$

$B_e = 0.820046\ cm^{-1}$ $\alpha_e = 0.005922\ cm^{-1}$ $r_e = 1.5349\ Å$

Enthalpy of Formation

Several values have been proposed for the enthalpy of formation or the dissociation energy of CS(g) (1 - 17). Many of these have been summarized by Hubin-Franksin et al. (16) although distinction was not made between $\Delta_f H°(0\ K)$ and $\Delta_f H°(298.15\ K)$. The values proposed or derivable for $\Delta_f H°(298.15\ K)$ vary from <32 kcal mol^{-1} (electron impact of COS)] to 77±4 kcal mol^{-1} (photodissociation of COS, 15) . We give more weight to the equilibrium measurements -- $\Delta_f H(298.15\ K)$ = 69 kcal mol^{-1} derived from (3), 58±5 and 60±6 kcal mol^{-1} (5), 55±1 kcal mol^{-1} (7), and 70±2 kcal mol^{-1} (11) (lack of data prohibits a complete re-analysis of these values) -- and adopt $\Delta_f H°$(CS, g, 298.15 K) = 67±6 kcal mol^{-1} (280.328±25 kJ mol^{-1}). This leads to $\Delta_f H°(0\ K)$ = 66.2±6 kcal mol^{-1} and to D^0_0 = 169.1±6 kcal mol^{-1} for the process CS(g) = C(g) + S(g), using JANAF values for the heats of formation of C(g) and S(g) (18).

Heat Capacity and Entropy

The molecular constants and electronic levels are those compiled by Rosen (19) except that the value of $\omega_e x_e$ = 8.46 cm^{-1} given by Rosen (19) and by Suchard (20) appears to be a transcription error for the original $\omega_e x_e$ = 6.46 cm^{-1} by Lagerqvist, et al. (4).

The microwave spectra of carbon monosulfide have been critically reviewed by Lovas and Krupenie (21). Their derived rotational constants are in agreement with those adopted.

References

1. H. D. Smyth and J. P. Blewett, Phys. Rev. 46, 276 (1934).
2. L. P. Blanchard and P. Le Goff, Can. J. Chem. 35, 89 (1957).
3. C. J. Fincham and R. A. Bergman, J. Metals. 9, 690 (1957).
4. A. Lagerqvist, H. Westerlund, C. V. Wright, and R. F. Barrow, Ark. Fys. 14, 387 (1958).
5. H. Schafer and H. Wiedemeier, Z. Anorg. Allg. Chem. 296, 242 (1958).
6. E. Gallegos and R. W. Kiser, J. Phys. Chem. 65, 1177 (1961).
7. H. Wiedemeier and H. Schafer, Z. Anorg. Allg. Chem. 326, 235 (1964).
8. K. P. R. Nair, R. B. Singh, and D. K. Rai, J. Chem. Phys. 43, 3570 (1965).
9. V. H. Dibeler and J. A. Walker, J. Opt. Soc. Amer. 57, 1007 (1967).
10. G. Hancock, C. Morley, and I. W. M. Smith, Chem. Phys. Lett. 12, 193 (1971).
11. D. L. Hildenbrand, Chem. Phys. Lett. 15, 379 (1972).
12. S. Bell, T. L. Ng, and C. Suggitt, J. Mol. Spectrosc. 44, 267 (1972).
13. H. Okaba, J. Chem. Phys. 56, 4381 (1972).
14. G. W. Taylor, J. Phys. Chem. 77, 124 (1973).
15. L. C. Lee and D. L. Judge, J. Chem. Phys. 63, 2782 (1975).
16. M. J. Hubin-Franksin, J. Katihabwa, and J. E. Collin, Int. J. Mass Spectrom. Ion Phys. 20, 285 (1976).
17. A. G. Gaydon, "Dissociation Energies and Spectra of Diatomic Molecules," 3rd ed., Chapman and Hall Ltd., London, 1968.
18. JANAF Thermochemical Tables: C(g), 3-31-61; S(g), 6-30-71.
19. B. Rosen, Ed., "Spectroscopic Data Relative to Diatomic Molecules," Pergamon Press, New York, 1970.
20. A. N. Suchard, The Aerospace Corporation, Air Force Report No. SAMSO-TR-74-82, Air Force Space and Missile Systems Organization Contract F04701-73-C-0074, March 29, 1974.
21. F. J. Lovas and P. H. Krupenie, J. Phys. Chem. Ref. Data 3, 245 (1974).

Carbon Sulfide (CS) $C_1S_1(g)$

Enthalpy Reference Temperature = T_r = 298.15 K Standard State Pressure = p° = 0.1 MPa

	J $K^{-1}mol^{-1}$			kJ mol^{-1}			
T/K	$C°_p$	S°	$-[G°-H°(T_r)]/T$	$H°-H°(T_r)$	$\Delta_f H°$	$\Delta_f G°$	Log K_f
0	0.	0.	INFINITE	-8.708	277.083	277.083	INFINITE
100	29.105	178.626	236.638	-5.801	279.240	262.725	-137.234
200	29.184	198.810	213.254	-2.889	280.184	245.766	-64.188
250	29.409	205.343	211.042	-1.425	280.333	237.141	-49.548
298.15	29.798	210.554	210.554	0.	280.328	228.820	-40.088
300	29.816	210.738	210.555	0.055	280.325	228.501	-39.786
350	30.358	215.374	210.920	1.559	280.192	219.872	-32.814
400	30.965	219.467	211.737	3.092	277.742	211.358	-27.601
450	31.583	223.150	212.804	4.656	276.753	203.110	-23.576
500	32.174	226.509	214.009	6.250	275.646	194.988	-20.370
600	33.214	232.470	216.601	9.521	273.754	179.039	-15.587
700	34.046	237.655	219.247	12.886	271.999	163.393	-12.193
800	34.698	242.245	221.840	16.324	270.305	147.994	-9.663
900	35.208	246.363	224.340	19.820	215.510	133.886	-7.771
1000	35.612	250.094	226.732	23.362	215.073	124.839	-6.521
1100	35.934	253.504	229.013	26.940	214.585	115.839	-5.501
1200	36.197	256.642	231.186	30.547	214.054	106.886	-4.653
1300	36.412	259.548	233.257	34.178	213.483	97.978	-3.937
1400	36.593	262.253	235.233	37.828	212.876	89.115	-3.325
1500	36.745	264.783	237.120	41.496	212.235	80.297	-2.796
1600	36.876	267.159	238.924	45.177	211.563	71.523	-2.335
1700	36.989	269.398	240.651	48.870	210.862	62.792	-1.929
1800	37.088	271.515	242.307	52.574	210.133	54.103	-1.570
1900	37.176	273.523	243.898	56.287	209.378	45.455	-1.250
2000	37.255	275.432	245.427	60.009	208.598	36.848	-0.962
2100	37.326	277.251	246.900	63.738	207.796	28.280	-0.703
2200	37.391	278.989	248.319	67.474	206.973	19.751	-0.469
2300	37.450	280.653	249.689	71.216	206.129	11.259	-0.256
2400	37.505	282.248	251.013	74.964	205.267	2.805	-0.061
2500	37.557	283.780	252.293	78.717	204.386	-5.613	0.117
2600	37.605	285.254	253.532	82.475	203.489	-13.995	0.281
2700	37.651	286.674	254.734	86.238	202.575	-22.342	0.432
2800	37.696	288.044	255.899	90.005	201.647	-30.656	0.572
2900	37.740	289.367	257.031	93.777	200.704	-38.936	0.701
3000	37.783	290.648	258.130	97.553	199.746	-47.183	0.822
3100	37.827	291.887	259.199	101.334	198.776	-55.398	0.933
3200	37.871	293.089	260.239	105.119	197.793	-63.581	1.038
3300	37.917	294.255	261.253	108.908	196.798	-71.734	1.135
3400	37.966	295.388	262.240	112.702	195.791	-79.856	1.227
3500	38.017	296.489	263.203	116.501	194.774	-87.949	1.313
3600	38.072	297.561	264.142	120.306	193.745	-96.012	1.393
3700	38.132	298.605	265.060	124.116	192.707	-104.046	1.469
3800	38.198	299.622	265.956	127.932	191.659	-112.053	1.540
3900	38.269	300.615	266.832	131.756	190.602	-120.031	1.608
4000	38.348	301.585	267.689	135.586	189.537	-127.983	1.671
4100	38.435	302.533	268.527	139.426	188.464	-135.908	1.731
4200	38.530	303.461	269.348	143.274	187.384	-143.806	1.788
4300	38.636	304.368	270.152	147.132	186.298	-151.679	1.843
4400	38.752	305.258	270.940	151.001	185.206	-159.526	1.894
4500	38.879	306.130	271.712	154.883	184.111	-167.348	1.943
4600	39.018	306.986	272.469	158.777	183.012	-175.147	1.989
4700	39.171	307.827	273.213	162.687	181.911	-182.921	2.033
4800	39.337	308.653	273.943	166.612	180.809	-190.671	2.075
4900	39.517	309.466	274.659	170.555	179.707	-198.399	2.115
5000	39.712	310.267	275.363	174.516	178.607	-206.105	2.153
5100	39.922	311.055	276.056	178.497	177.509	-213.788	2.190
5200	40.148	311.832	276.736	182.501	176.416	-221.450	2.224
5300	40.391	312.599	277.406	186.528	175.329	-229.091	2.258
5400	40.651	313.357	278.064	190.580	174.249	-236.711	2.290
5500	40.927	314.105	278.713	194.658	173.178	-244.312	2.320
5600	41.221	314.845	279.351	198.766	172.118	-251.893	2.350
5700	41.532	315.578	279.981	202.903	171.070	-259.455	2.378
5800	41.861	316.303	280.601	207.073	170.036	-266.999	2.405
5900	42.207	317.021	281.212	211.276	169.018	-274.525	2.430
6000	42.571	317.734	281.815	215.515	168.018	-282.035	2.455

PREVIOUS: December 1976 (1 atm) CURRENT: December 1976 (1 bar)

Carbon Sulfide (CS) $C_1S_1(g)$

CARBON DISULFIDE (CS_2) IDEAL GAS M_r = 76.131

$S°(298.15\ K) = 237.977 \pm 0.08\ J\ K^{-1}\ mol^{-1}$

$\Delta_f H°(0\ K) = 116.131 \pm 0.84\ kJ\ mol^{-1}$

$\Delta_f H°(298.15\ K) = 116.943 \pm 0.84\ kJ\ mol^{-1}$

Vibrational Frequencies, Degeneracies, and Anharmonicity and Rotational Constants

ν, cm^{-1}	x_{ii}, cm^{-1}	x_{ij}, cm^{-1}	α_i, cm^{-1}
657.98 (1)	x_{11} = -0.83	x_{12} = -1.75	α_1 = 0.0002144
395.93 (2)	x_{22} = 0.82	x_{23} = -6.49	α_2 = 0.0007130
1535.35 (1)	x_{33} = -6.54	x_{13} = -7.59	α_3 = -0.0002148

Point Group: $D_{\infty h}$ $\sigma = 2$

Bond Distance: C-S = 1.553 Å

Bond Angle: S-C-S = 180°

$B_e = 0.109277\ cm^{-1}$ $g_{22} = -0.74\ cm^{-1}$

$D_o = 1.05 \times 10^{-8}\ cm^{-1}$

Enthalpy of Formation

Good et al. (1) have determined $\Delta_f H°(CS_2, \ell, 298.15\ K) = 21.37 \pm 0.17\ kcal\ mol^{-1}$ by rotating bomb combustion calorimetry. They combined this with an unpublished measurement of $\Delta_{vap}H°(298.15\ K) = 6.61 \pm 0.02\ kcal\ mol^{-1}$ to obtain $\Delta_f H°(CS_2, g, 298.15\ K) = 27.98 \pm 0.19\ kcal\ mol^{-1}$. We adopt $\Delta_{vap}H°(298.15\ K) = 6.578 \pm 0.003\ kcal\ mol^{-1}$ determined by Weddington et al. (2) by vapor flow calorimetry and obtain the adopted $\Delta_f H°(CS_2, g, 298.15\ K) = 27.95 \pm 0.20\ kcal\ mol^{-1}$.

The resulting $\Delta_f H°(CS_2, g, 0\ K) = 27.75 \pm 0.20\ kcal\ mol^{-1}$ leads to $\Delta_a H°(0\ K) = 273.33 \pm 0.66\ kcal\ mol^{-1}$ for the process $CS_2(g) = C(g) + 2\ S(g)$. Auxiliary data are from the JANAF Tables (12).

Heat Capacity and Entropy

The fundamental ground state vibrational frequencies are those from a re-analysis of infra-red and Raman spectra by Guenther and Stoicheff, referenced by Kleman (3) as a private communication. Several sets of anharmonicity constants and/or rotational constants have been published (2, 4-9). Some are corrected for Fermi resonance to some degree in some manner (6, 7, 9), others are not (2, 4). We adopt the x_{ii}, x_{ij}, α_i, and g_{22} terms determined from the least-square estimates of the force constants in the general quartic force field by Giguere et al. (9). The rotational constants B_e and D_o are from McBride and Gordon (7). The bond distance, r_e, is a rounded average of 1.5529±0.0005 Å (4) and 1.5523±0.0005 Å (5) and is consistent with the adopted B_e. The bond angle of 180° is from Cross and Brockway (10).

Waddington et al. (2) have measured $C_p°$ of carbon disulfide gas in the 325.65-502.25 K range. At 325.65 K, the value calculated by Waddington et al. (2), where Fermi resonance was not considered explicitly, is approximately 0.03% lower than the observed value; at 300 K, the present table is higher than the table calculated by Waddington et al. (2) by about 0.3%. At 502.25 K, the present table is approximately 0.6% higher (0.074 cal K^{-1} mol^{-1}) than the measured value of 12.484 cal K^{-1} mol^{-1}. The value calculated by Waddington et al. (2) is 0.2% higher (0.025 cal K^{-1} mol^{-1}) than the observed value. At 6000 K, the value of $C_p°$ in the present table is approximately 1.1% higher (0.17 cal K^{-1} mol^{-1}) than the value calculated with the anharmonicity constants used by Waddington et al. (2).

The tabulated functions have been generated by the NASA-Lewis computer program (11). We adopted these functions directly for 0 to 298.15 K and reproduce the functions at higher temperatures by integration of $C_p°$.

The allowed error of ±0.02 cal K^{-1} mol^{-1} for S°(298.15 K) includes the values calculated using other anharmonicity correction constants (2, 6, 7).

References

1. W. D. Good, J. L. Lacina and J. P. McCullough, J. Phys. Chem. 65, 2229 (1961).
2. G. Waddington, J. C. Smith, K. D. Williamson, and D. W. Scott, J. Phys. Chem. 66, 1074 (1962).
3. B. Kleman, Can. J. Phys. 41, 2034 (1963). References unpublished work of A. H. Guenther and B. P. Stoicheff, 1962.
4. B. P. Stoicheff, Can. J. Phys. 36, 218 (1958).
5. A. H. Guenther, T. A. Wiggins, and D. H. Rank, J. Chem. Phys. 28, 682 (1958).
6. J. S. Gordon, J. Chem. Eng. Data 6, 390 (1961).
7. B. J. McBride and S. Gordon, J. Chem. Phys. 35, 2198 (1961).
8. W. J. Walker and A. Weber, J. Mol. Spectrosc. 39, 57 (1971).
9. J. Giguere, V. K. Wang, J. Overend, and A. Cabana, Spectrochim. Acta, Part A 29, 1197 (1973).
10. P. C. Cross and L. O. Brockway, J. Chem. Phys. 3, 821 (1935).
11. B. J. McBride and S. Gordon, NASA-TN-D-4097, 135pp., Aug. 1967; Scientific and Technical Aerospace Reports 5, p 3680, N67-35192 (1967); CA 68, 63364.
12. JANAF Thermochemical Tables: C(g), 3-31-61; S(g), 6-30-71.

Carbon Disulfide (CS_2) $C_1S_2(g)$

Enthalpy Reference Temperature = T_r = 298.15 K Standard State Pressure = p° = 0.1 MPa

	J $K^{-1}mol^{-1}$			kJ mol^{-1}			
T/K	$C_p°$	S°	$-[G°-H°(T_r)]/T$	$H°-H°(T_r)$	$\Delta_f H°$	$\Delta_f G°$	Log K_f
0	0.	0.	INFINITE	-10.687	116.131	116.131	INFINITE
100	30.995	196.787	274.228	-7.744	117.634	100.555	-52.525
200	39.648	220.929	241.981	-4.210	117.558	83.443	-21.793
298.15	45.664	237.977	237.977	0.	116.943	66.816	-11.706
300	45.751	238.260	237.978	0.085	116.927	66.505	-11.580
400	49.677	251.995	239.824	4.868	111.494	50.016	-6.531
500	52.494	263.398	243.431	9.984	107.427	35.091	-3.666
600	54.616	273.165	247.592	15.344	104.039	20.954	-1.824
700	56.242	281.711	251.868	20.890	101.119	7.342	-0.548
800	57.503	289.307	256.082	26.580	98.466	-5.871	0.383
900	58.493	296.140	260.160	32.382	-10.281	-16.601	0.963
1000	59.283	302.345	264.073	38.272	-10.226	-17.306	0.904
1100	59.923	308.026	267.814	44.233	-10.201	-18.016	0.855
1200	60.450	313.263	271.386	50.253	-10.207	-18.726	0.815
1300	60.891	318.120	274.796	56.320	-10.243	-19.435	0.781
1400	61.266	322.646	278.054	62.429	-10.311	-20.139	0.751
1500	61.588	326.884	281.170	68.572	-10.409	-20.838	0.726
1600	61.869	330.868	284.153	74.745	-10.537	-21.530	0.703
1700	62.118	334.627	287.012	80.944	-10.695	-22.212	0.682
1800	62.340	338.184	289.757	87.167	-10.882	-22.884	0.664
1900	62.539	341.560	292.396	93.412	-11.095	-23.545	0.647
2000	62.722	344.772	294.935	99.675	-11.335	-24.194	0.632
2100	62.889	347.837	297.382	105.955	-11.598	-24.831	0.618
2200	63.045	350.766	299.742	112.252	-11.882	-25.455	0.604
2300	63.189	353.571	302.022	118.564	-12.188	-26.065	0.592
2400	63.325	356.264	304.226	124.890	-12.514	-26.661	0.580
2500	63.453	358.851	306.360	131.229	-12.857	-27.244	0.569
2600	63.575	361.342	308.427	137.580	-13.217	-27.812	0.559
2700	63.691	363.744	310.432	143.943	-13.592	-28.366	0.549
2800	63.802	366.062	312.377	150.318	-13.981	-28.906	0.539
2900	63.909	368.303	314.267	156.704	-14.384	-29.432	0.530
3000	64.012	370.471	316.105	163.100	-14.800	-29.944	0.521
3100	64.111	372.572	317.893	169.506	-15.228	-30.442	0.513
3200	64.208	374.609	319.633	175.922	-15.668	-30.926	0.505
3300	64.303	376.586	321.329	182.348	-16.118	-31.396	0.497
3400	64.394	378.507	322.983	188.782	-16.580	-31.852	0.489
3500	64.484	380.375	324.596	195.226	-17.054	-32.294	0.482
3600	64.572	382.193	326.171	201.679	-17.538	-32.723	0.475
3700	64.658	383.963	327.709	208.141	-18.034	-33.138	0.468
3800	64.742	385.689	329.212	214.611	-18.541	-33.539	0.461
3900	64.825	387.371	330.682	221.089	-19.060	-33.927	0.454
4000	64.907	389.014	332.120	227.576	-19.591	-34.302	0.448
4100	64.988	390.617	333.527	234.070	-20.135	-34.663	0.442
4200	65.067	392.184	334.905	240.573	-20.692	-35.010	0.435
4300	65.145	393.716	336.255	247.084	-21.261	-35.344	0.429
4400	65.223	395.215	337.578	253.602	-21.845	-35.665	0.423
4500	65.299	396.682	338.875	260.128	-22.443	-35.973	0.418
4600	65.375	398.118	340.148	266.662	-23.056	-36.267	0.412
4700	65.450	399.524	341.396	273.203	-23.684	-36.547	0.406
4800	65.524	400.903	342.621	279.752	-24.327	-36.814	0.401
4900	65.598	402.255	343.825	286.308	-24.987	-37.067	0.395
5000	65.671	403.581	345.007	292.871	-25.663	-37.307	0.390
5100	65.743	404.882	346.168	299.442	-26.355	-37.533	0.384
5200	65.815	406.159	347.309	306.020	-27.065	-37.745	0.379
5300	65.886	407.414	348.432	312.605	-27.793	-37.944	0.374
5400	65.957	408.646	349.535	319.197	-28.538	-38.128	0.369
5500	66.027	409.857	350.621	325.796	-29.301	-38.299	0.364
5600	66.097	411.047	351.690	332.403	-30.082	-38.455	0.359
5700	66.166	412.218	352.741	339.016	-30.882	-38.598	0.354
5800	66.235	413.369	353.777	345.636	-31.700	-38.726	0.349
5900	66.303	414.502	354.796	352.263	-32.536	-38.840	0.344
6000	66.372	415.617	355.801	358.896	-33.392	-38.940	0.339

PREVIOUS: December 1976 (1 atm) CURRENT: December 1976 (1 bar)

Carbon Disulfide (CS_2) $C_1S_2(g)$

SILICON CARBIDE, ALPHA (SiC) CRYSTAL M_r = 40.0965

$S°(298.15\ K) = 16.485 \pm 0.13\ J\ K^{-1}\ mol^{-1}$

$\Delta_fH°(0\ K) = -70.530 \pm 6.3\ kJ\ mol^{-1}$

$\Delta_fH°(298.15\ K) = -71.546 \pm 6.3\ kJ\ mol^{-1}$

Enthalpy of Formation

The enthalpy of formation is calculated from $\Delta_rH°(298.15\ K) = -591.9\ kcal\ mol^{-1}$ for $SiC(\alpha) + 4F_2(g) \rightarrow SiF_4(g) + CF_4(g)$ based on the tentative value of $\Delta_fH°(298.15\ K) = -223\pm1\ kcal\ mol^{-1}$ for $CF_4(g)$. Greenberg (1) determined the enthalpy of reaction by fluorine bomb calorimetry. Data of Humphrey (2) obtained by oxygen bomb calorimetry yield a $\Delta_fH°$ which is over 2 kcal more negative; however, this difference is probably within the uncertainties due to use of large amounts of Ti as a kindling agent and to corrections for incomplete combustion. Decomposition pressure data obtained mass spectrometrically by Drowart (3) are in close agreement with the selected $\Delta_fH°$, but Knudsen weight loss data of Grieveson (4) and solubility-activity data of Kirkwood (5) yield values less negative by 2 and 3 kcal, respectively. The results are summarized below.

The data of Drowart (3) show a significant drift while those of Grieveson (4) do not, but the latter should be corrected for $C_2Si(g)$ and $CSi_2(g)$ which are present in amounts of about 4 and 3%, respectively, at 2000 K. Silicon carbide may have a small vaporization coefficient, since rates of free evaporation reported by Voronin (6) and Ghoshtagore (7) yield apparent pressures which are smaller by factors of 1/30 to 1/10000.

Source	Method	Reaction	T/K	Data Points	$\Delta_rH°(298.15\ K)$, kcal mol^{-1} 2nd Law	3rd Law	Drift cal K^{-1} mol^{-1}	$\Delta_fH°(298.15\ K)$ kcal mol^{-1}
Greenberg (1)	F_2 Calorimetry	A	298	-	-591.9		-	-17.1
Humphrey (2)	O_2Calorimetry	B	303	-	-292.0		-	-19.6
Drowart (3)	Mass Spec.	C	2149-2316	7	136.9±0.7	124.61	-5.6±0.3	-16.9(-17.6)*
Grieveson (4)	Knudsen Wt. Loss	C	1785-2004	10	122.9±0.7	122.66	-0.2±0.4	-15.0
Kirkwood (5)	Soly. in Pb	D	1693	1	-	25.66	-	-14.1

A) $SiC(\alpha) + 4F_2(g) \rightarrow SiF_4(g) + CF_4(g)$ C) $SiC(\alpha) \rightarrow C(graph) + Si(g)$

B) $SiC(\alpha) + 2O_2(g) \rightarrow SiO_2(low\text{-}quartz) + CO_2(g)$ D) $SiC(\alpha) \rightarrow C(graph) + Si(\ell)$

*Value in parentheses is based on authors' data for $Si(\ell) \rightarrow Si(g)$ rather than JANAF value.

Heat Capacity and Entropy

Low temperature heat capacities are from data (54-296 K) of Humphrey (2) and are in good agreement with earlier data (54-295 K) of Kelley (8). The entropy is obtained from $C_p°$ using $S°(50\ K) = 0.038\ cal\ K^{-1}\ mol^{-1}$. $C_p°$ above room temperature is based on enthalpy data of Humphrey (2) [389-1789 K], Walker (9) [373-1073 K] and Magnus (10) [372-1172 K]. Maximum deviations of the data from the selected values are 1.4, 2 and -1.7%, respectively; these occur in the range 370-621 K. Kirillin (11) obtained enthalpy data (1114-2843 K) for SiC mixed with 12% free carbon. When corrected for free carbon but not for 0.73% iron impurity, these data deviate by -1±1.5% from the selected values. Enthalpies (793-1790 K) of Fieldhouse (12) are high by 1-8% while those of Maksimenko (13) were not available for analysis.

Transition Data

SiC(α), also called hexagonal II or 6H, is one of the more common of many hexagonal forms which arise from various possible stacking sequences of the hexagonal SiC layers (14). The properties of these phases are so similar that they have not been adequately differentiated thermodynamically. It has frequently been assumed that cubic SiC(β) transforms to alpha at about 2300 K, but this seems unlikely since both phases have been prepared over temperature ranges of 1700-3000 K (14). Heat of formation and equilibrium data indicate that alpha is less stable up to 2000 K. The adopted functions suggest that this is the case at all temperatures; however, the stability difference is small.

Decomposition Data

The decomposition temperature, T_{dcm} = ~ 3245 K, is calculated as the value at which the total pressure reaches one atm for vapor consisting of C_2Si(44.3 mol %), Si(32.0%), CSi_2(21.2%), Si_2(1.9%), CSi(0.27%), and Si_3(0.25%). The tables predict decomposition to graphite and pure liquid silicon at about 3280 K; however, Scace (15) observed peritectic decomposition to graphite and solution containing 19 atom percent of carbon at 3103±40 K. In contrast, Dolloff (16) reported the peritectic at 2813±40 K and 27 percent carbon. Badami (17) has shown that decomposition in vacuum at lower temperatures (2400 K) leads to graphite.

References

1. E. Greenberg et al., Argonne National Laboratory, ANL-7175, 138 (1966); ANL-7020, 169 (1965).
2. G. L. Humphrey, S. S. Todd, J. P. Coughlin and E. G. King, U. S. Bur. Mines RI 4888, 1952.
3. J. Drowart, G. DeMaria and M. G. Inghram, J. Chem. Phys. 29, 1015 (1958); J. Drowart and G. DeMaria, pp 16-23 in (14).
4. P. Grieveson and C. B. Alcock, "Special Ceramics," Proc. Symposium Brit. Ceram. Research Assoc., Stoke-on-Trent 1959, 183-208, Haywood & Co., London, 1960.
5. D. H. Kirkwood and J. Chipman, J. Phys. Chem. 65, 1082 (1961).
6. N. I. Voronin, N. L. Makarova and B. F. Yudin, Issled. v Obl. Khim. Silikatov i Okislov, Akad. Nauk SSSR, Sb. Statei, 203-8 (1965).
7. R. N. Ghoshtagore, Solid-State Electronics 9, 178 (1966).
8. K. K. Kelley, J. Amer. Chem. Soc. 63, 1137 (1941).
9. B. E. Walker, C. T. Ewing and R. R. Miller, J. Chem. Eng. Data 7, 595 (1962).
10. A. Magnus, Ann. Physik 70, 303 (1923).
11. V. A. Kirillin, A. E. Sheindlin and V. Ya. Chekhovskoi, Teplofiz. Vys. Temp. 2, 9 (1964).
12. I. B. Fieldhouse and J. C. Hedge, AD-150954 (1958).
13. M. S. Maksimenko and A. S. Polubelova, Trudy Leningrad. Teknol. Inst. im. Lensoveta 33, 30 (1955).
14. J. R. O'Connor and J. Smiltens, editors, "Silicon Carbide," Pergamon Press, London, 1960.
15. R. Scace and G. A. Slack, J. Chem. Phys. 30, 1551 (1959); pp 24-30 in ref. (14).
16. R. T. Dolloff, U. S. Dept. Comm., Clearinghouse, PB Rept. 171, 365, 28 pp (1960).
17. D. V. Badami, Nature 193, 569 (1962).

Silicon Carbide, Alpha (SiC) $C_1Si_1(cr)$

Enthalpy Reference Temperature = T_r = 298.15 K Standard State Pressure = p° = 0.1 MPa

T/K	$C_p°$ (J K^{-1}mol^{-1})	S°	$-[G°-H°(T_r)]/T$	$H°-H°(T_r)$ (kJ mol^{-1})	$\Delta_fH°$	$\Delta_fG°$	Log K_f
0	0.	0.	INFINITE	-3.252	-70.530	-70.530	INFINITE
100	4.197	1.414	32.865	-3.145	-70.748	-70.411	36.779
200	16.246	7.954	18.613	-2.132	-71.240	-69.881	18.251
298.15	26.761	16.485	16.485	0.	-71.546	-69.139	12.113
300	26.932	16.651	16.485	0.050	-71.550	-69.124	12.036
400	34.100	25.466	17.639	3.131	-71.613	-68.301	8.919
500	38.610	33.593	20.030	6.781	-71.567	-67.477	7.049
600	41.723	40.922	22.912	10.806	-71.496	-66.666	5.804
700	44.007	47.534	25.965	15.098	-71.425	-65.866	4.915
800	45.769	53.530	29.042	19.590	-71.363	-65.076	4.249
900	47.170	59.004	32.071	24.239	-71.310	-64.294	3.732
1000	48.321	64.035	35.019	29.016	-71.268	-63.517	3.318
1100	49.283	68.687	37.871	33.897	-71.236	-62.743	2.979
1200	50.108	73.011	40.622	38.868	-71.216	-61.972	2.698
1300	50.823	77.051	43.270	43.915	-71.208	-61.202	2.459
1400	51.451	80.841	45.820	49.029	-71.216	-60.432	2.255
1500	52.011	84.410	48.275	54.203	-71.241	-59.661	2.078
1600	52.518	87.783	50.640	59.430	-71.285	-58.888	1.922
1700	52.974	90.981	52.919	64.704	-121.527	-57.665	1.772
1800	53.396	94.021	55.119	70.023	-121.385	-53.912	1.564
1900	53.781	96.918	57.243	75.382	-121.224	-50.168	1.379
2000	54.141	99.686	59.297	80.779	-121.047	-46.432	1.213
2100	54.476	102.336	61.284	86.210	-120.854	-42.706	1.062
2200	54.785	104.877	63.208	91.673	-120.647	-38.990	0.926
2300	55.082	107.319	65.073	97.166	-120.427	-35.283	0.801
2400	55.363	109.669	66.882	102.689	-120.194	-31.586	0.687
2500	55.630	111.935	68.640	108.238	-119.949	-27.899	0.583
2600	55.881	114.122	70.347	113.814	-119.693	-24.222	0.487
2700	56.124	116.235	72.008	119.414	-119.426	-20.555	0.398
2800	56.358	118.281	73.624	125.039	-119.149	-16.898	0.315
2900	56.580	120.262	75.198	130.686	-118.864	-13.251	0.239
3000	56.798	122.184	76.732	136.355	-118.569	-9.615	0.167
3100	57.007	124.050	78.229	142.045	-118.266	-5.988	0.101
3200	57.208	125.863	79.689	147.756	-117.955	-2.371	0.039
3300	57.404	127.626	81.115	153.486	-117.636	1.236	-0.020
3400	57.593	129.343	82.509	159.236	-117.310	4.833	-0.074
3500	57.781	131.015	83.871	165.005	-116.978	8.421	-0.126
3600	57.961	132.645	85.203	170.792	-500.790	22.460	-0.326
3700	58.137	134.236	86.507	176.597	-500.031	36.984	-0.522
3800	58.312	135.789	87.783	182.420	-499.268	51.488	-0.708
3900	58.480	137.305	89.034	188.259	-498.500	65.971	-0.884
4000	58.647	138.788	90.259	194.116	-497.728	80.435	-1.050

PREVIOUS: December 1962 CURRENT: March 1967

Silicon Carbide, Alpha (SiC) $C_1Si_1(cr)$

SILICON CARBIDE, BETA (SiC) CRYSTAL M_r = 40.0965

$S°(298.15\ K) = 16.610 \pm 0.13\ J\ K^{-1}\ mol^{-1}$

$\Delta_f H°(0\ K) = -72.221 \pm 6.3\ kJ\ mol^{-1}$

$\Delta_f H°(298.15\ K) = -73.220 \pm 6.3\ kJ\ mol^{-1}$

Enthalpy of Formation

The enthalpy of formation is calculated from $\Delta_r H°(298.15\ K) = -591.5\ kcal\ mol^{-1}$ for $SiC(\beta)+4F_2(g)\rightarrow SiF_4(g)+CF_4(g)$, based on the tentative value of $\Delta_f H°(CF_4, g, 298.15\ K) = -223\pm1\ kcal\ mol^{-1}$. Greenberg (1) determined the enthalpy of reaction by fluorine bomb calorimetry. Data of Humphrey (2) obtained by oxygen bomb calorimetry yield a $\Delta_f H°$ which is over 3 kcal more negative; however, this difference is probably within the uncertainties due to use of large amounts of Ti as a kindling agent and to corrections for incomplete combustion. Decomposition pressure data obtained with the Knudsen weight loss method by Davis (3) are in good agreement with the selected $\Delta_f H°$, but similar data of Grieveson (4) and solubility-activity data of Rein (5) and d'Entremont (6) yield values less negative by 1.6 to 2.5 kcal. The results ar summarized below. Other equilibrium data (5) relating SiC and SiO_2 are not included, pending the revision of tables for SiO_2.

The data of Davis show a serious drift, while those of Grieveson do not; but the latter should be corrrected for $C_2Si(g)$ and $CSi_2(g)$ which are present in amounts of about 4 and 3%, respectively, at 2000 K. Silicon carbide may have a small vaporization coefficient, since rates of free evaporation reported by Voronin (7) and Ghoshtagore (8) yield apparent pressures which are smaller by factors of 1/30 to 1/10000.

Source	Method	Reaction	T/K	Data Points	$\Delta_r H°$(298.15 K), kcal mol^{-1} 2nd law	3rd law	Drift cal K^{-1} mol^{-1}	$\Delta_f H°$(298.15 K) kcal mol^{-1}
Greenberg (1)	F_2 Calorimetry	A	298	-	-591.5		-	-17.5
Humphrey (2)	O_2 Calorimetry	B	303	-	-290.9		-	-20.6
Davis (3)	Knudsen Wt. Loss	C	2117-2171	8	167±28	126.10	-19±13	-18.4(-17.6)*
Grieveson (4)	" " "	C	1808-1973	6	123.3±0.3	122.65	-04±0.2	-15.0
Rein (5)	Soly. in C-Fe	D	1823-1873	2	32.0	27.43	- 2.5	-15.9
d'Entremont (6)	Soly. in Ag	D	1693	1	-	27.31	-	-15.7

A) $SiC(\beta) + 4F_2(g) \rightarrow SiF_4(g) + CF_4(g)$

B) $SiC(\beta) + 2O_2(g) \rightarrow SiO_2(\text{low quartz}) + CO_2(g)$

C) $SiC(\beta) \rightarrow C(graph) + Si(g)$

D) $SiC(\beta) \rightarrow C(graph) + Si(\ell)$

*Value in parentheses is based on author's data for $Si(\ell) \rightarrow Si(g)$ rather than JANAF value.

Heat Capacity and Entropy

Low temperature heat capacities are from data (54-296 K) of Humphrey (2). The entropy is obtained from $C_p°$ using S°(50 K) = 0.041 cal K^{-1} mol^{-1}. $C_p°$ above room temperature is based on Humphrey's enthalpy data (389-1693 K), which show a maximum deviation of ±1.2% from the selected functions. Humphrey's sample contained about 1% of α-SiC and about 0.6% of other impurities. Kirillin (9) obtained enthalpy data (1114-2843 K) for SiC mixed with 12% free carbon. When corrected for free carbon but not for 0.73% iron impurity, these data deviate by -1±2% from the selected values.

Transition Data

SiC(β), also called cubic or 3C, has a structure of the diamond or zinc blend type (10). The cubic phase differs from the many hexagonal forms only in the order of the stacking sequence of the hexagonal SiC layers. It has frequently been assumed that β transforms into α at about 2300 K, but this seems unlikely, since both phases have been prepared over temperature ranges of 1700-3000 K (10). Heat of formation and equilibrium data indicate that β is more stable up to 2000 K. The adopted functions suggest that this is the case at all temperatures; however, the stability difference is small.

Decomposition Data

The decomposition temperature, T_{dcm} = ~ 3259 K, is calculated as the value at which the total pressure reaches one atm for vapor consisting of C_2Si(45.4 mol %), Si(32.1%), CSi_2(20.1%), Si_2(1.8%), CSi(0.28%) and Si_3(0.23%). The tables predict decomposition to graphite and pure liquid silicon at about 3337 K; however, Scace (11) observed peritectic decomposition to graphite and solution containing 19 atom percent of carbon at 3103±40 K. In contrast, Dolloff (12) reported the peritectic at 2813±40 K and 27 percent carbon.

References

1. E. Greenberg et al., Argonne National Laboratory, ANL-7175, 138 (1966); ANL-7020, 169 (1965).
2. G. L. Humphrey, S. S. Todd, J. P. Coughlin and E. G. King, U. S. Bur. Mines RI 4888, 1952.
3. S. G. Davis, D. F. Anthrop and A. W. Searcy, J. Chem. Phys. 34, 659 (1961).
4. P. Grieveson and C. B. Alcock, "Special Ceramics," Proc. Symposium Brit. Ceram. Research Assoc., Stoke-on-Trent 1959, 183-208, Haywood & Co., London, 1960.
5. R. H. Rein and J. Chipman, J. Phys. Chem. 67, 839 (1963).
6. J. C. d'Entremont and J. Chipman, J. Phys. Chem. 67, 499 (1963).
7. N. I. Voronin, N. L. Makarova and B. F. Yudin, Issled. v Obl. Khim. Silikatov i Okislov, Akad. Nauk SSSR, Sb. Statei, 203-8, 1965.
8. R. N. Ghoshtagore, Solid-State Electronics 9, 178 (1966).
9. V. A Kirillin, A. E. Sheindlin and V. Ya. Chekhovskoi, Teplofiz. Vys. Temp. 2, 9 (1964).
10. J. R. O'Connor and J. Smiltens, editors, "Silicon Carbide," Pergamon Press, London, 1960.
11. R. I. Scace and G. A. Slack, J. Chem. Phys. 30, 1551 (1959); pp 24-30 in ref. 10.
12. R. T. Dolloff, U. S. Dept. Comm., Clearinghouse, PB Rept. 171, 365, 28 pp (1960).

Silicon Carbide, Beta (SiC) $C_1Si_1(cr)$

Enthalpy Reference Temperature = T_r = 298.15 K Standard State Pressure = p° = 0.1 MPa

T/K	$C_p°$ (J K^{-1}mol^{-1})	S°	$-[G°-H°(T_r)]/T$	$H°-H°(T_r)$ (kJ mol^{-1})	$\Delta_f H°$	$\Delta_f G°$	Log K_f
0	0.	0.	INFINITE	-3.270	-72.221	-72.221	INFINITE
100	4.263	1.460	33.062	-3.160	-72.437	-72.104	37.664
200	16.301	8.033	18.740	-2.141	-72.923	-71.580	18.695
298.15	26.840	16.610	16.610	0.	-73.220	-70.850	12.413
300	27.008	16.777	16.611	0.050	-73.223	-70.835	12.333
400	34.100	25.598	17.765	3.133	-73.284	-70.025	9.144
500	38.639	33.727	20.158	6.784	-73.237	-69.215	7.231
600	41.790	41.066	23.042	10.814	-73.161	-68.417	5.956
700	44.108	47.690	26.098	15.115	-73.082	-67.633	5.047
800	45.878	53.700	29.178	19.618	-73.009	-66.859	4.365
900	47.283	59.188	32.212	24.278	-72.945	-66.094	3.836
1000	48.421	64.231	35.165	29.065	-72.892	-65.336	3.413
1100	49.363	68.891	38.022	33.956	-72.851	-64.583	3.067
1200	50.158	73.221	40.777	38.932	-72.824	-63.832	2.779
1300	50.840	77.263	43.430	43.983	-72.813	-63.084	2.535
1400	51.430	81.053	45.984	49.097	-72.821	-62.335	2.326
1500	51.944	84.619	48.442	54.266	-72.851	-61.585	2.145
1600	52.400	87.986	50.809	59.484	-72.904	-60.832	1.986
1700	52.806	91.176	53.090	64.745	-123.160	-59.629	1.832
1800	53.170	94.204	55.291	70.044	-123.038	-55.895	1.622
1900	53.497	97.088	57.416	75.377	-122.903	-52.169	1.434
2000	53.798	99.840	59.469	80.742	-122.757	-48.450	1.265
2100	54.070	102.471	61.454	86.136	-122.601	-44.738	1.113
2200	54.321	104.992	63.376	91.556	-122.438	-41.034	0.974
2300	54.555	107.412	65.239	97.000	-122.267	-37.338	0.848
2400	54.769	109.739	67.045	102.466	-122.090	-33.649	0.732
2500	54.965	111.979	68.797	107.953	-121.908	-29.967	0.626
2600	55.153	114.138	70.500	113.459	-121.722	-26.294	0.528
2700	55.329	116.223	72.155	118.983	-121.531	-22.627	0.438
2800	55.492	118.238	73.765	124.524	-121.338	-18.967	0.354
2900	55.643	120.188	75.332	130.081	-121.142	-15.315	0.276
3000	55.789	122.077	76.859	135.653	-120.944	-11.669	0.203
3100	55.928	123.908	78.348	141.239	-120.746	-8.030	0.135
3200	56.057	125.686	79.799	146.838	-120.546	-4.397	0.072
3300	56.183	127.413	81.216	152.450	-120.346	-0.770	0.012
3400	56.300	129.092	82.600	158.074	-120.146	2.850	-0.044
3500	56.413	130.726	83.951	163.710	-119.946	6.465	-0.096
3600	56.522	132.316	85.273	169.357	-503.899	20.535	-0.298
3700	56.626	133.866	86.565	175.014	-503.288	35.094	-0.495
3800	56.722	135.378	87.830	180.681	-502.679	49.637	-0.682
3900	56.819	136.853	89.068	186.359	-502.074	64.163	-0.859
4000	56.911	138.292	90.281	192.045	-501.472	78.675	-1.027

PREVIOUS: CURRENT: March 1967

Silicon Carbide, Beta (SiC) $C_1Si_1(cr)$

SILICON CARBIDE (SiC) IDEAL GAS M_r = 40.0965

$S°(298.15\ K) = [213.04]\ J\ K^{-1}\ mol^{-1}$

$\Delta_f H°(0\ K) = 714.7 \pm 3.3\ kJ\ mol^{-1}$
$\Delta_f H°(298.15\ K) = 719.6 \pm 33\ kJ\ mol^{-1}$

Electronic Levels and Quantum Weights

State	ε_i, cm^{-1}	g_i
$[^1\Sigma]$	0	[1]
	[1000]	[6]
	[5000]	[3]
	[8000]	[2]
	[14000]	[2]
	[18000]	[1]

$\omega_e = [1226]\ cm^{-1}$ $\omega_e x_e = [5]\ cm^{-1}$ $\sigma = 1$
$B_e = [0.6933]\ cm^{-1}$ $\alpha_e = [0.006]\ cm^{-1}$ $r_e = [1.7]\ Å$

Enthalpy of Formation

Drowart and co-workers have used the Knudsen effusion-mass spectrometric technique to determine the vapor equilibrium over the systems SiC-graphite (1) and boron-silicon-graphite (2). 3rd law analysis of the partial pressures of Si and SiC yields the results summarized below. The adopted value, $\Delta_f H°(298.15\ K) = 172 \pm 8$ is the mean of the two results and the uncertainty reflects the possible effect of an error of up to 4 cal $K^{-1}\ mol^{-1}$ in the functions. The corresponding dissociation energy is $D^0_0 = 106.5 \pm 8\ kcal\ mol^{-1}$.

Source	Method	Reaction*	T/K	Data Points	$\Delta_r H°(298.15\ K)$, kcal mol^{-1} 2nd law	3rd law	Drift cal $K^{-1}\ mol^{-1}$	$\Delta_f H°(298.15\ K)$ kcal mol^{-1}
Drowart (1)	Mass Spec.	A	2181-2316	3	27.1±4	65.94	17±2	173.6
Verhaegen (2)	Mass Spec.	A	2249-2344	3	48.7±4	62.67	6±2	170.4

*Reaction: (A) C(graphite) + Si(g) = SiC(g)

Heat Capacity and Entropy

The ground state is assumed to be $^1\Sigma$, as suggested by Weltner and McLeod (3). By analogy with C_2, there should be a low-lying $^3\Pi$ excited state; this would dominate the thermodynamic functions at temperatures above 2000 K if the level lies below 5000 cm^{-1}. The $^3\Pi$ level is assumed to lie at 1000 cm^{-1}, intermediate between the value (4) observed for C_2 and that estimated for Si_2 (see JANAF table). Additional excited states, $^3\Sigma$, $^1\Pi$, $^1\Delta$ and $^1\Sigma$, are estimated by comparison with those observed or predicted (5) for C_2. The estimates for the electronic levels are relatively uncertain and probably yield an upper limit for the entropy at temperatures where SiC(g) is significant. A probable lower limit for the electronic contribution may be obtained by increasing the $^3\Pi$ level to 7000 cm^{-1} and omitting the other excited states. This would reduce the entropy by 3.5 and 2.0 cal $K^{-1}\ mol^{-1}$ at 2000 and 4000 K.

The vibrational frequency is that estimated by Weltner (3) from a valence bond calculation using $k = 7.44 \times 10^5$ dyn/cm obtained from C_2Si. The constants $\omega_e x_e$, α_e, and r_e are estimated by interpolation between those (4, 6) for the $^3\Pi_u$ states of Si_2 and C_2. Use of constants based on the $^3\Pi$ state is consistent with the assumption that this state dominates the electronic partition function at high temperatures. B_e is calculated from r_e.

References

1. J. Drowart, G. DeMaria and M. G. Inghram, J. Chem. Phys. 29, 1015 (1958).
2. G. Verhaegen, F. E. Stafford and J. Drowart, J. Chem. Phys. 40, 1622 (1964).
3. W. Weltner, Jr., and D. McLeod, Jr., J. Chem. Phys. 41, 235 (1964).
4. E. A. Ballik and D. A. Ramsay, Astrophys. J. 137, 84 (1963).
5. P. F. Fougere and R. K. Nesbet, J. Chem. Phys. 44, 285 (1966).
6. A. E. Douglas, Can. J. Phys. 33, 801 (1955).

Silicon Carbide (SiC) $C_1Si_1(g)$

Enthalpy Reference Temperature = T_r = 298.15 K Standard State Pressure = p° = 0.1 MPa

T/K	$C°_p$ (J $K^{-1}mol^{-1}$)	S°	$-[G°-H°(T_r)]/T$	$H°-H°(T_r)$ (kJ mol^{-1})	$\Delta_f H°$	$\Delta_f G°$	Log K_f
0	0.	0.	INFINITE	-9.266	714.651	714.651	INFINITE
100	29.111	178.849	242.437	-6.359	717.232	699.826	-365.552
200	31.133	199.342	216.303	-3.392	718.694	681.775	-178.061
250	34.526	206.629	213.654	-1.756	719.184	672.487	-140.508
298.15	38.417	213.044	213.044	0.	719.648	663.452	-116.234
300	38.562	213.282	213.044	0.071	719.666	663.103	-115.456
350	42.006	219.498	213.527	2.090	720.179	653.636	-97.550
400	44.241	225.267	214.639	4.251	720.702	644.093	-84.110
450	45.276	230.548	216.118	6.494	721.192	634.487	-73.649
500	45.426	235.333	217.804	8.764	721.610	624.830	-65.276
600	44.395	243.540	221.433	13.264	722.156	605.416	-52.706
700	42.946	250.274	225.087	17.631	722.302	585.943	-43.724
800	41.723	255.925	228.598	21.861	722.103	566.473	-36.987
900	40.839	260.785	231.911	25.987	721.631	547.045	-31.750
1000	40.250	265.055	235.016	30.039	720.950	527.681	-27.563
1100	39.884	268.872	237.923	34.044	720.105	508.394	-24.142
1200	39.677	272.333	240.649	38.021	719.132	489.190	-21.294
1300	39.581	275.504	243.210	41.983	718.055	470.071	-18.888
1400	39.561	278.437	245.622	45.940	716.889	451.038	-16.828
1500	39.591	281.167	247.902	49.897	715.648	432.092	-15.047
1600	39.654	283.724	250.062	53.859	714.339	413.231	-13.491
1700	39.737	286.130	252.113	57.828	662.791	394.900	-12.134
1800	39.831	288.404	254.067	61.807	661.593	379.176	-11.003
1900	39.931	290.560	255.931	65.795	660.383	363.519	-9.994
2000	40.031	292.611	257.715	69.793	659.162	347.926	-9.087
2100	40.129	294.567	259.423	73.801	657.931	332.395	-8.268
2200	40.224	296.436	261.063	77.819	656.693	316.922	-7.525
2300	40.314	298.226	262.641	81.846	655.447	301.506	-6.847
2400	40.398	299.943	264.159	85.881	654.193	286.144	-6.228
2500	40.477	301.594	265.624	89.925	652.932	270.834	-5.659
2600	40.550	303.183	267.038	93.976	651.664	255.575	-5.135
2700	40.618	304.714	268.405	98.035	650.389	240.366	-4.650
2800	40.680	306.193	269.729	102.100	649.106	225.203	-4.201
2900	40.738	307.621	271.011	106.171	647.816	210.086	-3.784
3000	40.791	309.003	272.254	110.247	646.518	195.014	-3.396
3100	40.839	310.342	273.461	114.329	645.212	179.986	-3.033
3200	40.884	311.639	274.634	118.415	643.899	164.999	-2.693
3300	40.925	312.898	275.775	122.505	642.577	150.054	-2.375
3400	40.963	314.120	276.885	126.600	641.248	135.149	-2.076
3500	40.998	315.308	277.966	130.698	639.910	120.283	-1.795
3600	41.030	316.463	279.019	134.799	254.411	115.917	-1.682
3700	41.060	317.588	280.046	138.904	253.470	112.082	-1.582
3800	41.087	318.683	281.049	143.011	252.518	108.274	-1.488
3900	41.112	319.751	282.027	147.121	251.557	104.490	-1.399
4000	41.135	320.792	282.984	151.234	250.585	100.732	-1.315
4100	41.157	321.808	283.918	155.348	249.603	96.998	-1.236
4200	41.176	322.800	284.832	159.465	248.612	93.288	-1.160
4300	41.194	323.769	285.727	163.583	247.611	89.601	-1.088
4400	41.211	324.716	286.602	167.704	246.601	85.938	-1.020
4500	41.227	325.643	287.459	171.826	245.582	82.298	-0.955
4600	41.241	326.549	288.299	175.949	244.554	78.681	-0.893
4700	41.254	327.436	289.123	180.074	243.516	75.086	-0.834
4800	41.266	328.305	289.930	184.200	242.470	71.514	-0.778
4900	41.277	329.156	290.722	188.327	241.414	67.963	-0.724
5000	41.287	329.990	291.499	192.455	240.349	64.434	-0.673
5100	41.296	330.807	292.261	196.584	239.274	60.926	-0.624
5200	41.305	331.609	293.010	200.714	238.190	57.440	-0.577
5300	41.312	332.396	293.746	204.845	237.097	53.974	-0.532
5400	41.319	333.168	294.469	208.977	235.993	50.529	-0.489
5500	41.326	333.927	295.180	213.109	234.881	47.105	-0.447
5600	41.331	334.671	295.878	217.242	233.757	43.701	-0.408
5700	41.337	335.403	296.565	221.375	232.623	40.317	-0.369
5800	41.341	336.122	297.241	225.509	231.478	36.954	-0.333
5900	41.346	336.829	297.906	229.643	230.323	33.610	-0.298
6000	41.350	337.524	298.561	233.778	229.156	30.285	-0.264

PREVIOUS: March 1967 (1 atm) CURRENT: March 1967 (1 bar)

Silicon Carbide (SiC) $C_1Si_1(g)$

SILICON CARBIDE (Si_2C) IDEAL GAS M_r = 68.1820

$\Delta_f H°(0\ K)$ = 531.7 ± 25 kJ mol^{-1}

$S°(298.15\ K)$ = [242.286] J K^{-1} mol^{-1} $\Delta_f H°(298.15\ K)$ = 535.6 ± 25 kJ mol^{-1}

Electronic Levels and Quantum Weights

ε_i, cm^{-1}	g_i
0	[1]
[12000]	[6]
[19000]	[2]

Vibrational Frequencies and Degeneracies

ν, cm^{-1}
[670](1)
[275](2)
[1600](1)

Ground State Quantum Weight: 1 σ = [2]
Point Group: [$D_{\infty h}$]
Bond Distance: C-Si = [1.75] Å
Bond Angle: C-Si-Si = [180]°
Rotational Constant: B_0 = [0.097996] cm^{-1}

Enthalpy of Formation

The selected value is an average based on the equilibrium data summarized below. Drowart and co-workers have used the Knudsen effusion-mass spectrometric technique to determine the vapor species over the SiC-graphite (1), SiC-silicon (2) and boron-carbon-silicon (3) systems. 3rd law analysis of the aprtial pressures of CSi_2 and Si yields the concordant values 127.4, 127.9 and 128.6 kcal mol^{-1}. Two of the drifts are rather large, but they are opposite in sign so that no consistent entropy difference is apparent between the data and the tabulated functions. The drifts may be related to the low condensation coefficient for CSi_2 suggested by Drowart (2). Rates of free evaporation reported by Voronin (4) yield apparent pressures which are lower by a factor of 1/30, thus confirming the existence of a low evaporation coefficient. The adopted value of $\Delta_f H°(298.15\ K)$ = 128±6 kcal mol^{-1} (535.552 ± 25 k J mol^{-1} includes allowance for an error in the entropy of up to 3 cal K^{-1} mol^{-1}.

Source	Method	Reaction*	T/K	Data Points	$\Delta_r H°(298.15\ K)$, kcal mol^{-1} 2nd Law	3rd Law	Drift cal K^{-1} mol^{-1}	$\Delta_f H°(298.15\ K)$ kcal mol^{-1}
Drowart (1958)	Mass Spec.	A	2149-2316	7	-104±7	-88.0	7.4±3.0	127.4
Drowart (1960)	Mass Spec.	B	1825-2160	7	141±3	134.4	-3.4±1.7	127.9
Verhaegen (1964)	Mass Spec.	A	2083-2344	6	- 68±25	-86.8	-9±12	128.6

*Reactions: A) C(graph) + 2 Si(g) = CSi_2(g) B) SiC(α) + Si(ℓ) = CSi_2(g)

Heat Capacity and Entropy

Weltner and McLeod (5) observed a band system in absorption at 5300 Å in matrix isolation and assigned this tentatively to the transition $^1\Pi_u \leftarrow X^1\Sigma_g^+$ of CSi_2. This assignment is adopted, although the evidence is not at all conclusive. A $^3\Pi_u$ excited state is assumed to lie at 12000 cm^{-1}, presumably arising from the same molecular orbital configuration as the $^1\Pi_u$ state. A linear, symmetric structure is assumed, as suggested by Weltner. The bond distance is estimated by comparison with Si_2, C_2 and C_3. Vibrational frequencies are estimated from a valence bond calculation using $k_1 = 7.46 \times 10$ (5) and $k\delta/l^2 = 0.11 \times 10^5$ dyn/cm. The stretching force constant is taken from C_2Si (5), while the bending force constant is from C_2Si and the $^1\Pi_u$ excited state of C_3 (5).

References

1. J. Drowart, G. DeMaria and M. G. Inghram, J. Chem. Phys. 29, 1015 (1958).
2. J. Drowart and G. DeMaria, pp. 16-23 in "Silicon Carbide," Edited by J. R. O'Connor and J. Smiltens, Pergamon Press, New York, 1960.
3. G. Verhaegen, F. E. Stafford and J. Drowart, J. Chem. Phys. 40, 1622 (1964).
4. N. I. Voronin, N. L. Makarova and B. F. Yudin, Issled. v Obl. Khim. Silikatov i Okislov, Akad. Nauk SSSR, Sb. Statei 203-8 (1965).
5. W. Weltner, Jr., and D. McLeod, Jr., J. Chem. Phys. 41, 235 (1964); 40, 1305 (1964); 45, 3096 (1966).

Silicon Carbide (Si_2C) $C_1Si_2(g)$

Enthalpy Reference Temperature = T_r = 298.15 K Standard State Pressure = $p°$ = 0.1 MPa

T/K	$C_p°$ (J K^{-1}mol^{-1})	$S°$	$-[G°-H°(T_r)]/T$	$H°-H°(T_r)$ (kJ mol^{-1})	$\Delta_f H°$	$\Delta_f G°$	Log K_f
0	0.	0.	INFINITE	-11.389	531.650	531.650	INFINITE
100	34.326	197.543	281.045	-8.350	534.097	515.204	-269.115
200	42.817	224.254	246.492	-4.448	535.316	495.747	-129.476
250	45.453	234.105	243.056	-2.238	535.530	485.825	-101.508
298.15	47.424	242.286	242.286	0.	535.552	476.248	-83.437
300	47.491	242.580	242.287	0.088	535.550	475.880	-82.858
350	49.147	250.029	242.871	2.505	535.427	465.943	-69.538
400	50.550	256.685	244.189	4.998	535.195	456.032	-59.552
450	51.770	262.711	245.918	7.557	534.877	446.155	-51.788
500	52.843	268.222	247.877	10.173	534.487	436.317	-45.582
600	54.631	278.022	252.104	15.550	533.534	416.769	-36.283
700	56.029	286.553	256.429	21.086	532.401	397.395	-29.654
800	57.122	294.108	260.676	26.746	531.122	378.195	-24.694
900	57.979	300.888	264.774	32.503	529.720	359.162	-20.845
1000	58.657	307.033	268.697	38.336	528.208	340.291	-17.775
1100	59.199	312.650	272.441	44.230	526.597	321.576	-15.270
1200	59.640	317.821	276.010	50.172	524.891	303.013	-13.190
1300	60.004	322.609	279.413	56.155	523.093	284.595	-11.435
1400	60.315	327.068	282.659	62.171	521.205	266.319	-9.936
1500	60.592	331.239	285.761	68.217	519.228	248.182	-8.642
1600	60.849	335.157	288.727	74.289	517.163	230.179	-7.515
1700	61.102	338.854	291.568	80.387	414.659	213.201	-6.551
1800	61.363	342.354	294.293	86.510	412.886	201.402	-5.845
1900	61.643	345.679	296.910	92.660	411.118	189.701	-5.215
2000	61.949	348.848	299.429	98.839	409.358	178.093	-4.651
2100	62.289	351.879	301.855	105.051	407.612	166.573	-4.143
2200	62.667	354.785	304.195	111.298	405.884	155.136	-3.683
2300	63.086	357.580	306.456	117.586	404.178	143.776	-3.265
2400	63.546	360.274	308.642	123.917	402.501	132.490	-2.884
2500	64.047	362.879	310.760	130.296	400.856	121.274	-2.534
2600	64.587	365.401	312.813	136.728	399.248	110.123	-2.212
2700	65.162	367.849	314.807	143.215	397.682	99.032	-1.916
2800	65.767	370.230	316.744	149.761	396.161	87.999	-1.642
2900	66.399	372.549	318.628	156.369	394.688	77.020	-1.387
3000	67.050	374.811	320.463	163.042	393.267	66.090	-1.151
3100	67.716	377.020	322.252	169.780	391.898	55.207	-0.930
3200	68.389	379.180	323.998	176.585	390.584	44.367	-0.724
3300	69.065	381.295	325.702	183.458	389.325	33.567	-0.531
3400	69.737	383.367	327.368	190.398	388.122	22.805	-0.350
3500	70.398	385.398	328.997	197.405	386.973	12.077	-0.180
3600	71.045	387.390	330.591	204.477	-382.426	22.303	-0.324
3700	71.672	389.346	332.153	211.613	-382.641	33.548	-0.474
3800	72.275	391.265	333.683	218.811	-382.810	44.799	-0.616
3900	72.849	393.150	335.184	226.067	-382.934	56.054	-0.751
4000	73.393	395.001	336.656	233.379	-383.016	67.311	-0.879
4100	73.902	396.820	338.102	240.744	-383.058	78.570	-1.001
4200	74.376	398.606	339.521	248.159	-383.062	89.829	-1.117
4300	74.812	400.362	340.915	255.618	-383.033	101.088	-1.228
4400	75.209	402.086	342.286	263.120	-382.973	112.346	-1.334
4500	75.568	403.780	343.634	270.659	-382.886	123.602	-1.435
4600	75.888	405.445	344.960	278.232	-382.776	134.856	-1.531
4700	76.169	407.080	346.264	285.835	-382.645	146.108	-1.624
4800	76.411	408.686	347.548	293.464	-382.499	157.356	-1.712
4900	76.617	410.264	348.812	301.116	-382.340	168.602	-1.797
5000	76.787	411.814	350.056	308.787	-382.172	179.844	-1.879
5100	76.923	413.336	351.282	316.472	-381.999	191.082	-1.957
5200	77.025	414.830	352.490	324.170	-381.824	202.318	-2.032
5300	77.097	416.298	353.680	331.876	-381.651	213.550	-2.105
5400	77.139	417.740	354.853	339.588	-381.483	224.778	-2.174
5500	77.153	419.155	356.009	347.303	-381.321	236.004	-2.241
5600	77.142	420.545	357.149	355.018	-381.171	247.226	-2.306
5700	77.108	421.911	358.274	362.731	-381.035	258.446	-2.368
5800	77.051	423.251	359.382	370.439	-380.916	269.664	-2.429
5900	76.974	424.568	360.476	378.140	-380.816	280.880	-2.487
6000	76.880	425.861	361.555	385.833	-380.736	292.095	-2.543

PREVIOUS: March 1967 (1 atm) CURRENT: March 1967 (1 bar)

Silicon Carbide (Si_2C) $C_1Si_2(g)$

TANTALUM CARBIDE (TaC) CRYSTAL M_r = 192.9589

$S°(298.15\ K) = 42.373 \pm 0.84\ J\ K^{-1}\ mol^{-1}$

$T_{fus} = 4273 \pm 200\ K$

$\Delta_f H°(0\ K) = -143.9 \pm 3.8\ kJ\ mol^{-1}$

$\Delta_f H°(298.15\ K) = -144.1 \pm 3.8\ kJ\ mol^{-1}$

$\Delta_{fus} H° = [104.6 \pm 20.9]\ kJ\ mol^{-1}$

Enthalpy of Formation

There are eight oxygen bomb calorimetric studies on TaC reported in the literature (1-8) and one CO(g) equilibrium study (9). The adopted $\Delta_f H°(298.15\ K)$ value is derived from the calorimetric study of Huber et al. (6). These authors studied the combustion of ten TaC_x samples where x covers the range 0.724 to 0.998. For each composition seven or eight combustions were performed. Huber et al. (6) reported that the combustions varied from 94.83% to 100% completion but were usually greater than 99%. Corrections of the order 0.4% or less were made to the heat of combustion value to account for impurities. The calculated heats of formation for the ten compositions were fitted to both a quadratic and linear equation by the least squares method by Huber et al. (6). Each value was weighted inversely proportional to the square of its uncertainty. Using the linear equation for $\Delta_f H°(298.15\ K)$ and correcting for the JANAF value of $\Delta_f H°(Ta_2O_5$, cr, 298.15 K) (10), we extrapolate to x = 1.0, calculate, and adopt $\Delta_f H°(298.15\ K) = -34.44 \pm 0.9\ kcal\ mol^{-1}$ for TaC. The quadratic equation leads to a value of $-33.94 \pm 1.2\ kcal\ mol^{-1}$ which is consistent with our adopted value.

Huber et al. (6) and Storms (11) have reanalyzed the calorimetric and equilibrium data (1-9) for TaC_x (x=0.724 to 0.998). A graphical comparison of the results (i.e., $\Delta_f H°(298.15\ K)$ values vs. x) indicated that the linear or quadratic equation suggested by Huber et al. (6) is a satisfactory representation of all the $\Delta_f H°(298.15\ K)$ results for TaC_x. The shortcomings of the various studies have been discussed by Huber et al. (6) and Storms (11). For comparison, the oxygen combustion study by Mah (8) led to a reported value of $\Delta_f H°(298.15\ K) = -35.4 \pm 0.5\ kcal\ mol^{-1}$ for $TaC_{0.996}$ based on six combustions.

Heat Capacity and Entropy

Kelley (12) measured the heat capacity of TaC in the region 54-294 K (32 points). The reported data refer to stoichiometric TaC with corrections for 0.03% free carbon having been made by Kelley (12, 13). The sample contained less than 0.02% impurities other than free carbon. Using the Debye function suggested by Kelley (12), we calculate and adopt $S°(50\ K) = 0.831\ cal\ K^{-1}\ mol^{-1}$ and $H°(50;0) = 30.10\ cal\ mol^{-1}$.

Of the numerous enthalpy and heat capacity studies on TaC (14-24), the adopted $C_p°$ values for T > 298 K are derived primarily from the studies of Pankratz et al. (21), Chang (23), Levinson (17), and Bolgar et al. (22). The heat capacity data for $200 \leq T \leq 298$ K are used to derive enthalpy values which in turn are used to constrain the high temperature enthalpy data. A polynomial curve fitting technique is used for the Pankratz et al. (21) and Levinson (17) data, with the above mentioned constraints, to obtain a smooth joining of the results near 298 K and to provide an equation suitable for extrapolation to T_{fus}. Pankratz et al. (21) determined the enthalpy of $TaC_{1.008}$ in the region 405-1700 K (14 data points) while Levinson (17) studied $TaC_{1.007}$ in the range 1289-2778 K (15 data points). Both sets of data are corrected to the stoichiometric composition (TaC) assuming excess free carbon. These corrections are roughly 20 cal mol^{-1} at 1000 K and 90 cal mol^{-1} at 2500 K (10). Chang (23) determined the enthalpy of a reported stoichiometric TaC sample for 324-985 K (11 data points). Bolgar et al. (22) reported an equation for $C_p°$ values as a function of T and x for the 1200-2400 K range. This equation was based on a study of four TaC_x compositions. We use this equation to extrapolate to the TaC composition.

The adopted tabulation yields enthalpy values for T < 1200 K which are very similar to those suggested by Pankratz et al. (21), the difference being no more than ±20 cal mol^{-1}. Above 1200 K, the data of Pankratz et al. (21) lies low approaching roughly 1% in the 1700-1800 K region. The Chang data (23) lies below the tabulated enthalpy values for T < 600 K. The average deviation is 0.94% with the maximum of -2.5% occurring at 324.4 K. The Levinson data (17) has an average deviation of 0.5% with considerable scatter. The $C_p°$ values of Bolgar et al. (24) has considerable scatter and deviates from our tabulated values by an average deviation of 6%.

Phase Data

See the liquid table for details.

Fusion Data

See the liquid table for details.

References

1. P. M. McKenna, Ind. Eng. Chem. 28, 767 (1936).
2. M. C. Sanz, North American Aviation, Inc., Report A1-203, May, 1947.
3. G. L. Humphrey, J. Amer. Chem. Soc. 76, 978 (1954).
4. V. I. Smirnova and B. F. Ormont, Dokl. Akad. Nauk SSSR 100, 127 (1955); Zhur. Fiz. Khim. 30, 1327 (1956).
5. A. N. Kornilov, V. Y. Leonidov, and S. M. Shuratov, Vestn. Most, Univ. Ser. II: Khim. 17, 48 (1962).
6. E. J. Huber, Jr., E. L. Head, C. E. Holley, Jr., and A. L. Bowman, J. Phys. Chem. 67, 793 (1963).
7. A. N. Kornilov, I. D. Zaikin, S. M. Shuratov, L. B. Dubrovskaya, and G. P. Shoeikin, Zhur. Fiz. Khim. 38, 702 (1964).
8. A. D. Mah, U. S. Bur. Mines RI 6663, 1965.
9. W. L. Worrell and J. Chipman, J. Phys. Chem. 68, 860 (1964).
10. JANAF Thermochemical Tables: Ta_2O_5(cr), 12-31-72; C(ref. st.), 3-31-61.
11. E. K. Storms, "The Refractory Carbides," Academic Press, New York, 1967.
12. K. K. Kelley, J. Amer. Chem. Soc. 62, 818 (1940).
13. R. Hultgren, R. L. Orr, and K. K. Kelley, "Supplement to Selected Values of Thermodynamic Properties of Metals and Alloys, C-Ta table, December, 1965.
14. D. S. Neel, C. D. Pears, and S. Oglesby, Jr., WADD TR 60-924, 1962 [AD 275 536].
15. M. A. Levinstein, ASD-TDR 62-201, 1962 [AD 283 967].
16. R. J. Barriault et al., ASD-TR 61-260, 1962 [AD 278 633].
17. L. S. Levinson, J. Chem. Phys. 42, 3342 (1965); ibid. 39, 1550 (1963).
18. R. Mezaki, E. W. Tilleux, T. F. Jambois, and J. L. Margrave, in "Advances in Thermophysical Properties at Extreme Temperatures and Pressures," ASME, 1965, pp. 138-45.
19. R. Mezaki, M. S. Thesis, Univ. of Wisconsin, 1961.
20. R. A. McDonald, F. L. Oetting, and H. Prophet, CPIA-44, 1, 213 (1964).
21. L. B. Pankratz, W. W. Weller, and E. G. King, U.S. Bur. Mines RI 6861, 1966.
22. A. S. Bolgar, E. A. Guseva, V. A. Gorbatyuk and V. V. Fesenko, Porosh. Met. 8, 60 (1968).
23. Y. A. Chang, Trans. Met. Soc. AIME 239, 1685 (1967).
24. A. E. Sheindlin, I. S. Belevick, and I. G. Kozhevnikov, Teplofiz. Vys. Temp. 10, 650 (1972).

Tantalum Carbide (TaC) $C_1Ta_1(cr)$

Enthalpy Reference Temperature = T_r = 298.15 K; Standard State Pressure = p° = 0.1 MPa

	J K⁻¹mol⁻¹			kJ mol⁻¹			
T/K	$C_p°$	S°	$-[G°-H°(T_r)]/T$	$H°-H°(T_r)$	$\Delta_f H°$	$\Delta_f G°$	Log K_f
0	0.	0.	INFINITE	-6.514	-143.879	-143.879	INFINITE
100	18.619	12.880	69.673	-5.679	-144.099	-143.677	75.049
200	29.192	29.187	45.519	-3.266	-144.264	-143.168	37.392
298.15	36.790	42.373	42.373	0.	-144.097	-142.654	24.992
300	36.903	42.601	42.374	0.068	-144.091	-142.645	24.837
400	41.664	53.915	43.884	4.012	-143.729	-142.215	18.571
500	44.530	63.537	46.877	8.330	-143.347	-141.882	14.822
600	46.447	71.833	50.361	12.883	-143.033	-141.620	12.329
700	47.890	79.104	53.959	17.602	-142.791	-141.404	10.552
800	49.074	85.578	57.514	22.451	-142.596	-141.220	9.221
900	50.116	91.419	60.962	27.411	-142.426	-141.058	8.187
1000	51.066	96.749	64.278	32.471	-142.270	-140.914	7.361
1100	51.928	101.656	67.456	37.621	-142.124	-140.786	6.685
1200	52.779	106.211	70.498	42.856	-141.987	-140.670	6.123
1300	53.616	110.469	73.411	48.176	-141.850	-140.566	5.648
1400	54.441	114.473	76.202	53.579	-141.697	-140.473	5.241
1500	55.254	118.257	78.881	59.064	-141.513	-140.392	4.889
1600	56.035	121.848	81.455	64.628	-141.291	-140.324	4.581
1700	56.813	125.268	83.932	70.271	-141.035	-140.271	4.310
1800	57.588	128.537	86.320	75.991	-140.761	-140.234	4.069
1900	58.360	131.672	88.625	81.788	-140.482	-140.213	3.855
2000	59.128	134.685	90.853	87.663	-140.200	-140.206	3.662
2100	59.890	137.588	93.010	93.614	-139.913	-140.213	3.488
2200	60.650	140.392	95.100	99.641	-139.621	-140.234	3.330
2300	61.409	143.104	97.129	105.744	-139.325	-140.269	3.186
2400	62.167	145.734	99.100	111.922	-139.030	-140.316	3.054
2500	62.923	148.287	101.016	118.177	-138.741	-140.376	2.933
2600	63.676	150.770	102.882	124.507	-138.467	-140.446	2.822
2700	64.429	153.187	104.701	130.912	-138.218	-140.527	2.719
2800	65.181	155.544	106.475	137.393	-138.006	-140.617	2.623
2900	65.933	157.844	108.207	143.948	-137.845	-140.713	2.535
3000	66.685	160.092	109.899	150.579	-137.753	-140.814	2.452
3100	67.436	162.291	111.554	157.285	-137.746	-140.916	2.374
3200	68.187	164.444	113.173	164.066	-137.846	-141.017	2.302
3300	68.938	166.553	114.759	170.923	-174.531	-140.642	2.226
3400	69.688	168.622	116.312	177.854	-174.488	-139.616	2.145
3500	70.438	170.653	117.836	184.860	-174.382	-138.592	2.068
3600	71.187	172.648	119.331	191.941	-174.213	-137.571	1.996
3700	71.935	174.609	120.799	199.098	-173.981	-136.557	1.928
3800	72.683	176.537	122.240	206.328	-173.686	-135.549	1.863
3900	73.431	178.435	123.657	213.634	-173.328	-134.550	1.802
4000	74.178	180.303	125.050	221.015	-172.905	-133.561	1.744
4100	74.925	182.144	126.420	228.470	-172.420	-132.583	1.689
4200	75.671	183.959	127.768	236.000	-171.871	-131.618	1.637
4273.000	76.216	185.267	128.739	241.544	------ CRYSTAL <--> LIQUID -----		
4300	76.418	185.748	129.096	243.604	-171.259	-130.667	1.587
4400	77.167	187.513	130.403	251.283	-170.582	-129.730	1.540
4500	77.914	189.256	131.692	259.037	-169.842	-128.810	1.495
4600	78.662	190.976	132.962	266.866	-169.038	-127.907	1.452
4700	79.408	192.676	134.215	274.770	-168.170	-127.022	1.412

PREVIOUS: CURRENT: December 1973

Tantalum Carbide (TaC) $C_1Ta_1(cr)$

TANTALUM CARBIDE (TaC) LIQUID M_r = 192.9589

$S°(298.15\ K) = [68.279]\ J\ K^{-1}\ mol^{-1}$ $\Delta_f H°(298.15\ K) = [-33.955]\ kJ\ mol^{-1}$

$T_{fus} = 4273 \pm 200\ K$ $\Delta_{fus}H° = [104.6 \pm 20.9]\ kJ\ mol^{-1}$

Enthalpy of Formation

The enthalpy of formation of TaC(ℓ) at 298.15 K is calculated from that of the crystal by adding $\Delta_{fus}H°$ and the difference in enthalpy, H°(4273 K) - H°(298.15 K), between TaC(cr) and TaC(ℓ).

Heat Capacity and Entropy

The heat capacity for TaC(ℓ) is assumed to be 16.0 cal $K^{-1}\ mol^{-1}$ for the entire liquid phase. A glass transition is assumed at 2800 K, below which the heat capacity values of the crystal are used. The entropy at 298.15 K is calculated in a manner analogous to that used for the enthalpy of formation.

Phase Data

TaC has a cubic structure of the NaCl type (11) and has a wide homogeneity range. From the phase diagram for the Ta-C system given by Storms (11), the temperature dependent homogeneity range extends from x=0.58 at 3603±100 K to x=0.998 or possibly x=1.00. The phase boundary at a eutectic temperature of 3673±50 K is $TaC_{0.98}$. A discussion of the phase diagram for TaC_x (x=0.00 to 1.6) is given by Storms (11).

Fusion Data

The phase diagram given by Storms (1) indicated a eutectic melting at 3673±50 K. A congruent melting point was estimated by Storms (1) at roughly 4273 K for $TaC_{0.88}$. As an approximation due to the uncertainties in this area of the phase diagram, we adopt T_{fus} = 4273±200 K and assign this melting point to TaC. We assume a heat of melting $\Delta_{fus}H°$ = 25.0±5.0 kcal mol^{-1} ($\Delta_{fus}S°$ = 5.85 cal $K^{-1}\ mol^{-1}$).

Decomposition Data

Using JANAF auxiliary data (2), we calculate T_{dcm} = ~ 5580 K, as that temperature at which $\Delta_r G°$ is zero for the reaction TaC(ℓ) = Ta(ℓ) + C(g).

References

1. E. K. Storms, "The Refractory Carbides," Academic Press, New York, 1967.
2. JANAF Thermochemical Tables: Ta(ℓ), 12-31-72; C(g), 3-31-61.

Tantalum Carbide (TaC) $C_1Ta_1(l)$

Enthalpy Reference Temperature = T_r = 298.15 K Standard State Pressure = p° = 0.1 MPa

T/K	$C_p°$ (J K⁻¹mol⁻¹)	S° (J K⁻¹mol⁻¹)	$-[G°-H°(T_r)]/T$ (J K⁻¹mol⁻¹)	$H°-H°(T_r)$ (kJ mol⁻¹)	$\Delta_f H°$ (kJ mol⁻¹)	$\Delta_f G°$ (kJ mol⁻¹)	Log K_f
0							
100							
200							
298.15	36.790	68.279	68.279	0.	-33.955	-40.236	7.049
300	36.903	68.507	68.279	0.068	-33.950	-40.275	7.013
400	41.664	79.820	69.790	4.012	-33.587	-42.436	5.542
500	44.530	89.443	72.783	8.330	-33.205	-44.693	4.669
600	46.447	97.738	76.267	12.883	-32.891	-47.021	4.094
700	47.890	105.010	79.865	17.602	-32.649	-49.396	3.686
800	49.074	111.484	83.420	22.451	-32.454	-51.802	3.382
900	50.116	117.325	86.868	27.411	-32.285	-54.231	3.148
1000	51.066	122.654	90.184	32.471	-32.128	-56.678	2.961
1100	51.928	127.562	93.362	37.621	-31.982	-59.140	2.808
1200	52.779	132.117	96.404	42.856	-31.846	-61.615	2.682
1300	53.616	136.375	99.316	48.176	-31.708	-64.102	2.576
1400	54.441	140.378	102.108	53.579	-31.555	-66.599	2.485
1500	55.254	144.162	104.786	59.064	-31.371	-69.109	2.407
1600	56.035	147.753	107.361	64.628	-31.149	-71.631	2.339
1700	56.813	151.174	109.838	70.271	-30.894	-74.169	2.279
1800	57.588	154.443	112.226	75.991	-30.620	-76.723	2.226
1900	58.360	157.577	114.531	81.788	-30.340	-79.292	2.180
2000	59.128	160.590	116.759	87.663	-30.058	-81.875	2.138
2100	59.890	163.494	118.916	93.614	-29.771	-84.473	2.101
2200	60.651	166.297	121.006	99.641	-29.479	-87.085	2.068
2300	61.410	169.010	123.035	105.744	-29.183	-89.710	2.037
2400	62.167	171.640	125.005	111.923	-28.888	-92.348	2.010
2500	62.923	174.193	126.922	118.177	-28.599	-94.998	1.985
2600	63.678	176.675	128.788	124.507	-28.325	-97.659	1.962
2700	64.431	179.093	130.607	130.913	-28.075	-100.331	1.941
2800	65.183	181.449	132.380	137.393	-27.863	-103.011	1.922
2800.000	65.183	181.449	132.380	137.393	GLASS <--> LIQUID TRANSITION		
2800.000	66.944	181.449	132.380	137.393			
2900	66.944	183.799	134.113	144.088	-27.564	-105.701	1.904
3000	66.944	186.068	135.807	150.782	-27.408	-108.398	1.887
3100	66.944	188.263	137.464	157.476	-27.413	-111.098	1.872
3200	66.944	190.389	139.085	164.171	-27.600	-113.795	1.858
3300	66.944	192.449	140.671	170.865	-64.446	-116.013	1.836
3400	66.944	194.447	142.224	177.560	-64.641	-117.572	1.806
3500	66.944	196.388	143.744	184.254	-64.847	-119.126	1.778
3600	66.944	198.273	145.232	190.948	-65.065	-120.674	1.751
3700	66.944	200.108	146.691	197.643	-65.294	-122.215	1.725
3800	66.944	201.893	148.120	204.337	-65.536	-123.751	1.701
3900	66.944	203.632	149.521	211.032	-65.788	-125.279	1.678
4000	66.944	205.327	150.895	217.726	-66.052	-126.801	1.656
4100	66.944	206.980	152.243	224.420	-66.328	-128.317	1.635
4200	66.944	208.593	153.566	231.115	-66.614	-129.825	1.615
4273.000	66.944	209.746	154.516	236.002	------ CRYSTAL <--> LIQUID -----		
4300	66.944	210.168	154.864	237.809	-66.912	-131.327	1.595
4400	66.944	211.707	156.138	244.504	-67.220	-132.821	1.577
4500	66.944	213.212	157.390	251.198	-67.540	-134.309	1.559
4600	66.944	214.683	158.619	257.892	-67.870	-135.789	1.542
4700	66.944	216.123	159.828	264.587	-68.211	-137.262	1.525
4800	66.944	217.532	161.015	271.281	-68.563	-138.727	1.510
4900	66.944	218.912	162.183	277.976	-68.926	-140.185	1.494
5000	66.944	220.265	163.331	284.670	-69.300	-141.636	1.480
5100	66.944	221.591	164.460	291.364	-69.684	-143.078	1.465
5200	66.944	222.890	165.571	298.059	-70.079	-144.514	1.452
5300	66.944	224.166	166.665	304.753	-70.484	-145.941	1.438
5400	66.944	225.417	167.741	311.448	-70.900	-147.361	1.425
5500	66.944	226.645	168.801	318.142	-71.327	-148.773	1.413
5600	66.944	227.852	169.845	324.836	-71.764	-150.177	1.401
5700	66.944	229.036	170.873	331.531	-72.211	-151.574	1.389
5800	66.944	230.201	171.886	338.225	-811.424	-150.161	1.352
5900	66.944	231.345	172.884	344.920	-812.106	-138.754	1.228
6000	66.944	232.470	173.868	351.614	-812.842	-127.335	1.109

PREVIOUS: CURRENT: December 1973

Tantalum Carbide (TaC) $C_1Ta_1(l)$

TANTALUM CARBIDE (TaC) CRYSTAL–LIQUID

0 to 4273 K crystal
above 4273 K liquid

Refer to the individual tables for details.

M_r = 192.9589

Tantalum Carbide (TaC) C_1Ta_1(cr,l)

Enthalpy Reference Temperature = T_r = 298.15 K Standard State Pressure = p° = 0.1 MPa

T/K	C_p° (J K⁻¹mol⁻¹)	S° (J K⁻¹mol⁻¹)	−[G°−H°(T_r)]/T (J K⁻¹mol⁻¹)	H°−H°(T_r) (kJ mol⁻¹)	$\Delta_f H^\circ$ (kJ mol⁻¹)	$\Delta_f G^\circ$ (kJ mol⁻¹)	Log K_f
0	0.	0.	INFINITE	-6.514	-143.879	-143.879	INFINITE
100	18.619	12.880	69.673	-5.679	-144.099	-143.677	75.049
200	29.192	29.187	45.519	-3.266	-144.264	-143.168	37.392
298.15	36.790	42.373	42.373	0.	-144.097	-142.654	24.992
300	36.903	42.601	42.374	0.068	-144.091	-142.645	24.837
400	41.664	53.915	43.884	4.012	-143.729	-142.215	18.571
500	44.530	63.537	46.877	8.330	-143.347	-141.882	14.822
600	46.447	71.833	50.361	12.883	-143.033	-141.620	12.329
700	47.890	79.104	53.959	17.602	-142.791	-141.404	10.552
800	49.074	85.578	57.514	22.451	-142.596	-141.220	9.221
900	50.116	91.419	60.962	27.411	-142.426	-141.058	8.187
1000	51.066	96.749	64.278	32.471	-142.270	-140.914	7.361
1100	51.928	101.656	67.456	37.621	-142.124	-140.786	6.685
1200	52.779	106.211	70.498	42.856	-141.987	-140.670	6.123
1300	53.616	110.469	73.411	48.176	-141.850	-140.566	5.648
1400	54.441	114.473	76.202	53.579	-141.697	-140.473	5.241
1500	55.254	118.257	78.881	59.064	-141.513	-140.392	4.889
1600	56.035	121.848	81.455	64.628	-141.291	-140.324	4.581
1700	56.813	125.268	83.932	70.271	-141.035	-140.271	4.310
1800	57.588	128.537	86.320	75.991	-140.761	-140.234	4.069
1900	58.360	131.672	88.625	81.788	-140.482	-140.213	3.855
2000	59.128	134.685	90.853	87.663	-140.200	-140.206	3.662
2100	59.890	137.588	93.010	93.614	-139.913	-140.213	3.488
2200	60.650	140.392	95.100	99.641	-139.621	-140.234	3.330
2300	61.409	143.104	97.129	105.744	-139.325	-140.269	3.186
2400	62.167	145.734	99.100	111.922	-139.030	-140.316	3.054
2500	62.923	148.287	101.016	118.177	-138.741	-140.376	2.933
2600	63.676	150.770	102.882	124.507	-138.467	-140.446	2.822
2700	64.429	153.187	104.701	130.912	-138.218	-140.527	2.719
2800	65.181	155.544	106.475	137.393	-138.006	-140.617	2.623
2900	65.933	157.844	108.207	143.948	-137.845	-140.713	2.535
3000	66.685	160.092	109.899	150.579	-137.753	-140.814	2.452
3100	67.436	162.291	111.554	157.285	-137.746	-140.916	2.374
3200	68.187	164.444	113.173	164.066	-137.846	-141.017	2.302
3300	68.938	166.553	114.759	170.923	-174.531	-140.642	2.226
3400	69.688	168.622	116.312	177.854	-174.488	-139.616	2.145
3500	70.438	170.653	117.836	184.860	-174.382	-138.592	2.068
3600	71.187	172.648	119.331	191.941	-174.213	-137.571	1.996
3700	71.935	174.609	120.799	199.098	-173.981	-136.557	1.928
3800	72.683	176.537	122.240	206.328	-173.686	-135.549	1.863
3900	73.431	178.435	123.657	213.634	-173.328	-134.550	1.802
4000	74.178	180.303	125.050	221.015	-172.905	-133.561	1.744
4100	74.925	182.144	126.420	228.470	-172.420	-132.583	1.689
4200	75.671	183.959	127.768	236.000	-171.871	-131.618	1.637
4273.000	76.216	185.267	128.739	241.544	CRYSTAL <--> LIQUID		
4273.000	66.944	209.746	128.739	346.144	TRANSITION		
4300	66.944	210.168	129.249	347.951	-66.912	-131.327	1.595
4400	66.944	211.707	131.106	354.645	-67.220	-132.821	1.577
4500	66.944	213.212	132.914	361.340	-67.540	-134.309	1.559
4600	66.944	214.683	134.676	368.034	-67.870	-135.789	1.542
4700	66.944	216.123	136.393	374.729	-68.211	-137.262	1.525
4800	66.944	217.532	138.069	381.423	-68.563	-138.727	1.510
4900	66.944	218.912	139.705	388.117	-68.926	-140.185	1.494
5000	66.944	220.265	141.302	394.812	-69.300	-141.636	1.480
5100	66.944	221.591	142.864	401.506	-69.684	-143.078	1.465
5200	66.944	222.890	144.390	408.201	-70.079	-144.514	1.452
5300	66.944	224.166	145.884	414.895	-70.484	-145.941	1.438
5400	66.944	225.417	147.345	421.589	-70.900	-147.361	1.425
5500	66.944	226.645	148.775	428.284	-71.327	-148.773	1.413
5600	66.944	227.852	150.177	434.978	-71.764	-150.177	1.401
5700	66.944	229.036	151.550	441.673	-72.211	-151.574	1.389
5800	66.944	230.201	152.896	448.367	-811.424	-150.161	1.352
5900	66.944	231.345	154.216	455.061	-812.106	-138.754	1.228
6000	66.944	232.470	155.511	461.756	-812.842	-127.335	1.109

PREVIOUS: CURRENT: December 1973

Tantalum Carbide (TaC) C_1Ta_1(cr,l)

TITANIUM CARBIDE (TiC) CRYSTAL M_r = 59.891

$S°(298.15\ K) = 24.229 \pm 0.21\ J\ K^{-1}\ mol^{-1}$

$T_{fus} = 3290 \pm 15\ K$

$\Delta_f H°(0\ K) = -182.8 \pm 4\ kJ\ mol^{-1}$

$\Delta_f H°(298.15\ K) = -184.1 \pm 4\ kJ\ mol^{-1}$

$\Delta_{fus} H° = [71.128]\ kJ\ mol^{-1}$

Enthalpy of Formation

The adopted $\Delta_f H°(298.15\ K) = -44.0\ kcal\ mol^{-1}$ ($-184.096\ kJ\ mol^{-1}$) is derived from $\Delta_r H°(298.15\ K) = -275.72\ kcal\ mol^{-1}$ for reaction (a) determined by Humphrey (1) who burned TiC(cr) in an oxygen bomb calorimeter. The author reported that the combustion product TiO_2 was >95 per cent rutile and <5 per cent anatase, the heat of formation is calculated assuming the above distribution. The uncertainty adopted is due mainly to the uncertainty in TiO_2(rutile).

Using atomic absorbtion spectroscopy Vidale (2) determined that the pressure ($\log P_{Ti} = -7.269$) of Ti(g) at 1666 K was equal to the pressure of TiC(cr) at 2220 K. Using this data for reaction (b) we obtain a $\Delta_f H°(298.15\ K) = -43.02\ kcal\ mol^{-1}$. Lowell and Williams (3) measured the $\Delta_f H°(2028\ K) = -43.3 \pm 4\ kcal\ mol^{-1}$ of TiC(cr) directly using a high temperature calorimeter, which reduces to $\Delta_f H°(298.15\ K) = -37.7 \pm 4\ kcal\ mol^{-1}$.

Fujishiro and Gokcen (4) determined the equilibrium pressure of Ti(g) in reaction (d) in the temperature range 2383-2593 K. 3rd law analysis of the data yields a $\Delta_f H°(298.15\ K) = -31.6\ kcal\ mol^{-1}$. Brantley and Beckman (5) measured equilibirum pressures of CO for reaction (e) in the temperature range 1278 - 1428 K. 3rd law analysis of the data yields a $\Delta_f H°(298.15\ K) = -61.9\ kcal\ mol^{-1}$.

Heat Capacity and Entropy

Low temperature heat capacities (55 - 295 K) have been taken from Kelley (6). Naylor (7) measured heat contents from 360 to 1738 K and Levinson (8) from 1274 to 2799 K. High temperature heat capacities are derived from the enthalpies by a fitting technique which constrains the curve to join smoothly with the low temperature values. Above 2799 K the heat capacity is graphically extrapolated. The entropy is based on $S°(50;0) = 0.065\ cal\ K^{-1}\ mol^{-1}$.

Fusion Data

See the liquid table for details.

Source	Method	Reaction	T/K	$\Delta_r H°(298.15\ K)$, kcal mol^{-1} 2nd Law	3rd Law	Drift cal K^{-1} mol^{-1}	$\Delta_f H°(298.15\ K)$ kcal mol^{-1}
(1)	Calorimetry	(a) $TiC(cr)+2O_2(g) \rightarrow CO_2(g)+TiO_2$(rutile)	298	-275.72			-44.00
(2)	Equilibrium	(b) TiC(cr)→Ti(g)+C(cr)			156.02		-43.02
(3)	Calorimetry	(c) Ti(cr)+C(cr)→TiC(cr)		-37.7±4			-37.7
(4)	Equilibrium	(d) TiC(cr)+C(graph)→Ti(g)	2383-2593	169.5±10.3	144.6	-10.0±4.1	-31.60
(5)	Equilibrium	(e) TiO_2(rutile)+3C(graph)→TiC(cr)+2CO(g)	1278-1428	48.07±1.83	111.07	46.7±1.36	-61.90

References

1. G. L. Humphrey, J. Amer. Chem. Soc. 73, 2261 (1951).
2. G. L. Vidale, General Electric Co., R61SD147, Contract AF 33(616)-6841, August, 1961.
3. C. E. Lowell and W. S. Williams, Rev. Sci. Instr. 32, 1120-1123 (1961).
4. S. Fujishiro and N. A. Gokcen, J. Phys. Chem. 65, 161 (1961).
5. L. R. Brandley and A. O. Bickman, J. Amer. Chem. Soc. 52, 3956 (1930).
6. K. K. Kelley, Ind. Eng. Chem. 36, 865 (1944).
7. B. F. Naylor, J. Amer. Chem. Soc. 68, 370 (1946).
8. L. S. Levinson, J. Chem. Phys. 42, 2891 (1965).

Titanium Carbide (TiC) $C_1Ti_1(cr)$

Enthalpy Reference Temperature = T_r = 298.15 K Standard State Pressure = p° = 0.1 MPa

T/K	$C_p°$ (J K^{-1}mol^{-1})	S°	$-[G°-H°(T_r)]/T$	$H°-H°(T_r)$ (kJ mol^{-1})	$\Delta_f H°$	$\Delta_f G°$	Log K_f
0	0.	0.	INFINITE	-4.607	-182.822	-182.822	INFINITE
100	7.335	2.602	46.702	-4.410	-183.245	-182.584	95.372
200	14.924	12.780	27.026	-2.849	-183.928	-181.622	47.435
298.15	33.807	24.229	24.229	0.	-184.096	-180.438	31.612
300	33.936	24.439	24.230	0.063	-184.096	-180.415	31.413
400	40.685	35.180	25.648	3.813	-183.982	-179.200	23.401
500	45.166	44.777	28.533	8.122	-183.740	-178.031	18.599
600	47.647	53.248	31.962	12.772	-183.494	-176.913	15.402
700	49.032	60.703	35.546	17.610	-183.316	-175.831	13.121
800	49.903	67.310	39.112	22.558	-183.214	-174.770	11.411
900	50.572	73.227	42.579	27.583	-183.215	-173.715	10.082
1000	51.179	78.586	45.916	32.670	-183.372	-172.653	9.018
1100	51.794	83.493	49.112	37.818	-183.732	-171.566	8.147
1200	52.446	88.027	52.169	43.030	-188.267	-170.317	7.414
1300	53.145	92.252	55.091	48.309	-188.267	-168.821	6.783
1400	53.894	96.218	57.889	53.661	-188.314	-167.324	6.243
1500	54.689	99.963	60.570	59.090	-188.410	-165.821	5.774
1600	55.530	103.519	63.144	64.600	-188.559	-164.311	5.364
1700	56.404	106.912	65.620	70.197	-188.767	-162.789	5.002
1800	57.317	110.161	68.005	75.882	-189.042	-161.254	4.679
1900	58.254	113.285	70.306	81.661	-189.390	-159.701	4.390
2000	59.220	116.298	72.531	87.534	-204.508	-157.674	4.118
2100	60.208	119.211	74.685	93.506	-205.780	-155.301	3.863
2200	61.216	122.035	76.773	99.577	-206.969	-152.869	3.630
2300	62.241	124.779	78.801	105.749	-208.073	-150.384	3.415
2400	63.283	127.450	80.773	112.025	-209.091	-147.854	3.218
2500	64.337	130.054	82.692	118.406	-210.019	-145.283	3.036
2600	65.400	132.598	84.563	124.893	-210.856	-142.677	2.866
2700	66.480	135.087	86.388	131.487	-211.600	-140.040	2.709
2800	67.563	137.524	88.171	138.189	-212.249	-137.378	2.563
2900	68.659	139.914	89.914	145.000	-212.803	-134.694	2.426
3000	69.760	142.260	91.620	151.921	-213.261	-131.992	2.298
3100	70.869	144.566	93.291	158.952	-213.621	-129.277	2.178
3200	71.982	146.833	94.929	166.095	-213.882	-126.552	2.066
3290.000	72.987	148.844	96.376	172.618	------ CRYSTAL <--> LIQUID -----		
3300	73.101	149.066	96.536	173.349	-214.044	-123.820	1.960
3400	74.224	151.264	98.113	180.715	-214.106	-121.085	1.860
3500	75.353	153.432	99.663	188.194	-214.068	-118.349	1.766
3600	76.485	155.571	101.186	195.786	-213.928	-115.616	1.678
3700	77.621	157.682	102.684	203.491	-622.785	-105.100	1.484
3800	78.760	159.767	104.159	211.310	-621.207	-91.129	1.253
3900	79.902	161.828	105.612	219.243	-619.582	-77.201	1.034
4000	81.048	163.865	107.042	227.291	-617.908	-63.315	0.827

PREVIOUS: December 1964 CURRENT: June 1968

Titanium Carbide (TiC) $C_1Ti_1(cr)$

TITANIUM CARBIDE (TiC) LIQUID M_r = 59.891

$S°(298.15\ K) = [47.401]\ J\ K^{-1}\ mol^{-1}$

$T_{fus} = 3290 \pm 15\ K$

$\Delta_f H°(298.15\ K) = [-108.334]\ kJ\ mol^{-1}$

$\Delta_{fus} H° = [71.128]\ kJ\ mol^{-1}$

Enthalpy of Formation

The enthalpy of formation is calculated from $\Delta_f H°$(cr, 298.15 K) by adding $\Delta_{fus}H°$ and the difference in enthalpy, H°(3290 K) - H°(198.15 K), between the crystal and liquid.

Heat Capacity and Entropy

A glass transition is assumed at 2200 K below which the heat capacity is obtained from that of the crystal; above 2200 K it is assumed constant and estimated as 15 cal K^{-1} mol^{-1}. The entropy is calculated in a manner analogous to that used for the enthalpy of formation.

Fusion Data

The adopted T_{fus} is taken from the phase diagram of the system Ti-C, at the composition 50% Ti and 50% C, reported by Rudy and Harmon (1). Earlier reports by Friederich and Sittig (2) and Moers (3) gave a T_{fus} of 3430±100 K and 3410±90 K, respectively. $\Delta_{fus}H°$ is derived from an estimated $\Delta_{fus}S°$ = 2.5 cal K^{-1} g-atom^{-1}, as suggested by Kubaschewski et al (3).

References

1. E. Rudy and D. P. Harmon, Aerojet General Corporation, BSPN No. 64-6899-735001, Contract AF-33 (6115), April 1, 1965.
2. E. Friederich and L. Sittig, Z. Anorg. Chem. 144, 171 (1925).
3. C. Agte and K. Moers, Z. Anorg. Chem. 198, 233 (1931).
4. O. Kubaschewski, E. Ll. Evans and C. B. Alcock, "Metallurgical Thermochemistry," Pergamon Press, New York, 1967.

Titanium Carbide (TiC) $C_1Ti_1(l)$

Enthalpy Reference Temperature = T_r = 298.15 K — Standard State Pressure = p° = 0.1 MPa

T/K	$C_p°$ (J K⁻¹mol⁻¹)	S° (J K⁻¹mol⁻¹)	−[G°−H°(T_r)]/T (J K⁻¹mol⁻¹)	H°−H°(T_r) (kJ mol⁻¹)	$\Delta_f H°$ (kJ mol⁻¹)	$\Delta_f G°$ (kJ mol⁻¹)	Log K_f
0							
100							
200							
298.15	33.807	47.401	47.401	0.	-108.334	-111.585	19.549
300	33.936	47.610	47.401	0.063	-108.334	-111.605	19.432
400	40.685	58.352	48.820	3.813	-108.221	-112.707	14.718
500	45.166	67.949	51.705	8.122	-107.978	-113.855	11.894
600	47.647	76.420	55.133	12.772	-107.732	-115.054	10.016
700	49.032	83.875	58.718	17.610	-107.555	-116.290	8.678
800	49.903	90.481	62.283	22.558	-107.452	-117.545	7.675
900	50.572	96.398	65.751	27.583	-107.454	-118.808	6.895
1000	51.179	101.758	69.088	32.670	-107.610	-120.063	6.271
1100	51.794	106.664	72.284	37.818	-107.971	-121.293	5.760
1200	52.446	111.199	75.340	43.030	-112.505	-122.361	5.326
1300	53.145	115.424	78.263	48.309	-112.506	-123.183	4.950
1400	53.894	119.389	81.060	53.661	-112.552	-124.002	4.627
1500	54.689	123.135	83.742	59.090	-112.648	-124.817	4.347
1600	55.530	126.691	86.316	64.600	-112.797	-125.624	4.101
1700	56.404	130.083	88.791	70.197	-113.006	-126.420	3.884
1800	57.317	133.333	91.176	75.882	-113.280	-127.201	3.691
1900	58.254	136.457	93.478	81.661	-113.629	-127.965	3.518
2000	59.220	139.470	95.702	87.534	-128.747	-128.256	3.350
2100	60.208	142.383	97.856	93.506	-130.018	-128.199	3.189
2200	61.216	145.207	99.945	99.577	-131.207	-128.085	3.041
2200.000	61.216	145.207	99.945	99.577	GLASS <--> LIQUID	TRANSITION	
2200.000	62.760	145.207	99.945	99.577			
2300	62.760	147.997	101.974	105.853	-132.209	-127.920	2.905
2400	62.760	150.668	103.947	112.129	-133.226	-127.712	2.780
2500	62.760	153.230	105.868	118.405	-134.259	-127.461	2.663
2600	62.760	155.691	107.737	124.681	-135.307	-127.169	2.555
2700	62.760	158.060	109.557	130.957	-136.369	-126.835	2.454
2800	62.760	160.342	111.330	137.233	-137.444	-126.463	2.359
2900	62.760	162.544	113.059	143.509	-138.534	-126.051	2.270
3000	62.760	164.672	114.744	149.785	-139.636	-125.602	2.187
3100	62.760	166.730	116.388	156.061	-140.751	-125.116	2.108
3200	62.760	168.723	117.992	162.337	-141.879	-124.594	2.034
3290.000	62.760	170.463	119.404	167.985	------ CRYSTAL <--> LIQUID -----		
3300	62.760	170.654	119.559	168.613	-143.019	-124.036	1.963
3400	62.760	172.527	121.089	174.889	-144.171	-123.443	1.896
3500	62.760	174.347	122.585	181.165	-145.336	-122.817	1.833
3600	62.760	176.115	124.048	187.441	-146.512	-122.157	1.772
3700	62.760	177.834	125.478	193.717	-556.798	-113.676	1.605
3800	62.760	179.508	126.878	199.993	-556.763	-101.700	1.398
3900	62.760	181.138	128.249	206.269	-556.795	-89.724	1.202
4000	62.760	182.727	129.591	212.545	-556.892	-77.747	1.015
4100	62.760	184.277	130.906	218.821	-557.053	-65.767	0.838
4200	62.760	185.789	132.195	225.097	-557.277	-53.781	0.669
4300	62.760	187.266	133.458	231.373	-557.560	-41.790	0.508
4400	62.760	188.709	134.698	237.649	-557.902	-29.792	0.354
4500	62.760	190.119	135.914	243.925	-558.301	-17.785	0.206

PREVIOUS: December 1964 CURRENT: June 1968

Titanium Carbide (TiC) $C_1Ti_1(l)$

TITANIUM CARBIDE (TiC) CRYSTAL-LIQUID M_r = 59.891

0 to 3290 K crystal
above 3290 K liquid

Refer to the individual tables for details.

Titanium Carbide (TiC) $C_1Ti_1(cr,l)$

Enthalpy Reference Temperature = T_r = 298.15 K Standard State Pressure = p° = 0.1 MPa

T/K	C_p°	S°	$-[G^\circ-H^\circ(T_r)]/T$	$H^\circ-H^\circ(T_r)$	Δ_fH°	Δ_fG°	Log K_f
	J K⁻¹mol⁻¹			kJ mol⁻¹			
0	0.	0.	INFINITE	-4.607	-182.822	-182.822	INFINITE
100	7.335	2.602	46.702	-4.410	-183.245	-182.584	95.372
200	14.924	12.780	27.026	-2.849	-183.928	-181.622	47.435
298.15	33.807	24.229	24.229	0.	-184.096	-180.438	31.612
300	33.936	24.439	24.230	0.063	-184.096	-180.415	31.413
400	40.685	35.180	25.648	3.813	-183.982	-179.200	23.401
500	45.166	44.777	28.533	8.122	-183.740	-178.031	18.599
600	47.647	53.248	31.962	12.772	-183.494	-176.913	15.402
700	49.032	60.703	35.546	17.610	-183.316	-175.831	13.121
800	49.903	67.310	39.112	22.558	-183.214	-174.770	11.411
900	50.572	73.227	42.579	27.583	-183.215	-173.715	10.082
1000	51.179	78.586	45.916	32.670	-183.372	-172.653	9.018
1100	51.794	83.493	49.112	37.818	-183.732	-171.566	8.147
1200	52.446	88.027	52.169	43.030	-188.267	-170.317	7.414
1300	53.145	92.252	55.091	48.309	-188.267	-168.821	6.783
1400	53.894	96.218	57.889	53.661	-188.314	-167.324	6.243
1500	54.689	99.963	60.570	59.090	-188.410	-165.821	5.774
1600	55.530	103.519	63.144	64.600	-188.559	-164.311	5.364
1700	56.404	106.912	65.620	70.197	-188.767	-162.789	5.002
1800	57.317	110.161	68.005	75.882	-189.042	-161.254	4.679
1900	58.254	113.285	70.306	81.661	-189.390	-159.701	4.390
2000	59.220	116.298	72.531	87.534	-204.508	-157.674	4.118
2100	60.208	119.211	74.685	93.506	-205.780	-155.301	3.863
2200	61.216	122.035	76.773	99.577	-206.969	-152.869	3.630
2300	62.241	124.779	78.801	105.749	-208.073	-150.384	3.415
2400	63.283	127.450	80.773	112.025	-209.091	-147.854	3.218
2500	64.337	130.054	82.692	118.406	-210.019	-145.283	3.036
2600	65.400	132.598	84.563	124.893	-210.856	-142.677	2.866
2700	66.480	135.087	86.388	131.487	-211.600	-140.040	2.709
2800	67.563	137.524	88.171	138.189	-212.249	-137.378	2.563
2900	68.659	139.914	89.914	145.000	-212.803	-134.694	2.426
3000	69.760	142.260	91.620	151.921	-213.261	-131.992	2.298
3100	70.869	144.566	93.291	158.952	-213.621	-129.277	2.178
3200	71.982	146.833	94.929	166.095	-213.882	-126.552	2.066
3290.000	72.987	148.844	96.376	172.618	CRYSTAL <--> LIQUID		
3290.000	62.760	170.463	96.376	243.746	TRANSITION		
3300	62.760	170.654	96.601	244.374	-143.019	-124.036	1.963
3400	62.760	172.527	98.807	250.650	-144.171	-123.443	1.896
3500	62.760	174.347	100.939	256.926	-145.336	-122.817	1.833
3600	62.760	176.115	103.003	263.202	-146.512	-122.157	1.772
3700	62.760	177.834	105.002	269.478	-556.798	-113.676	1.605
3800	62.760	179.508	106.941	275.754	-556.763	-101.700	1.398
3900	62.760	181.138	108.823	282.030	-556.795	-89.724	1.202
4000	62.760	182.727	110.650	288.306	-556.892	-77.747	1.015
4100	62.760	184.277	112.427	294.582	-557.053	-65.767	0.838
4200	62.760	185.789	114.156	300.858	-557.277	-53.781	0.669
4300	62.760	187.266	115.839	307.134	-557.560	-41.790	0.508
4400	62.760	188.709	117.479	313.410	-557.902	-29.792	0.354
4500	62.760	190.119	119.078	319.686	-558.301	-17.785	0.206

PREVIOUS: CURRENT: June 1968

Titanium Carbide (TiC) $C_1Ti_1(cr,l)$

ZIRCONIUM CARBIDE (ZrC) CRYSTAL M_r = 103.231

$S°(298.15\ K) = 33.321\ J\ K^{-1}\ mol^{-1}$

$T_{fus} = 3805 \pm 125\ K$

$\Delta_f H°(0\ K) = -196.0\ kJ\ mol^{-1}$

$\Delta_f H°(298.15\ K) = -196.6 \pm 13\ kJ\ mol^{-1}$

$\Delta_{fus} H° = [79.496]\ kJ\ mol^{-1}$

Enthalpy of Formation

$\Delta_f H°(298.15\ K) = -47.16\pm1.5\ kcal\ mol^{-1}$ was calculated from the standard Gibbs energy of formation of $-38.9\pm1.5\ kcal\ mol^{-1}$ at 2675 K reported by Pollock (1). Using spectrophotometric data reported by Vidale (2), we calculated $\Delta_f H°(298.15\ K) = -46.9\pm1\ kcal\ mol^{-1}$. These values were averaged with $\Delta_f H°(298.15\ K) = 47.0\pm0.6\ kcal\ mol^{-1}$ reported by Mah (3). The equilibria data reported by Prescott (4) and Kutsev (5) were considered but not used.

Heat Capacity and Entropy

Enthalpy data over the range 470.8 to 1,174.4 K were taken from a preliminary report by Mezaki et al. (6). These data were smoothed by means of the Shomate plot into low temperature data reported by Westrum and Feick (7). High temperature data reported by Neel et al. (8) and McDonald et al. (9) were then used to obtain the heat capacity above 1174 K.

Fusion Data

Melting points were measured by Agte and Alterthum (10) and Fredrich and Sittig (12). The value of 3805±125 K by Agte and Alterthum was chosen. Frederich and Sittig (11) obtained a measurement of 3300 K but estimated 3500 K because of the clouding formation of ZrO_2. Worrell (12) quoted a value of 3480°C taken from Storms (13). The enthalpy of melting is derived from an estimated $\Delta_{fus}S° = 2.5\ cal\ K^{-1}\ g\text{-}atom^{-1}$.

References

1. B. D. Pollock, J. Phys. Chem. 65, 731 (1961).
2. G. L. Vidale, Missle and Space Vehicle Department, General Electric R61SD147, August, 1961.
3. A. D. Mah, U. S. Bur. Mines RI 6518, 1964.
4. C. H. Prescott Jr., J. Amer. Chem. Soc. 48, 2534 (1926).
5. V. S. Kutsev, B. F. Ormont, and V. A. Epel'baum, Doklady Akad, Nauk SSSR 104, 567 (1955).
6. R. Mezaki, T. F. Jambois, A. K. Gangopadhyay, and J. L. Margrave, "Thermodynamic Properties of Inorganic Substances VIII.
7. E. F. Westrum Jr., and G. Feick, J. Chem. Eng. Data 8, 176 (1963).
The High Temperature Heat Contents of Zirconium Carbide and Tantalum Carbide", Univ. of Wisconsin, Madison, Wisconsin.
8. D. S. Neel, C. D. Pears, and S. Oglesby Jr., Southern Research Institute, Birmingham, Alabama, Technical Documentary Report No. WADD 60-924, p. 134, February 1962.
9. R. A. McDonald, F. L. Oetting, and H. Prophet, CPIA Publication No. 44(u), p. 213, February, 1964.
10. C. Agte and H. Alterthum, Z. Tech. Physik 11, 182 (1930).
11. E. Friederich and L. Sittig, Z. Anorg. Chem. 144, 169 (1925).
12. W. L. Worrell, J. Phys. Chem. 68, 954 (1964).
13. E. K. Storms, Los Alamos Scientific Lab Report LAMS-2674, (1962).

Zirconium Carbide (ZrC) $C_1Zr_1(cr)$

Enthalpy Reference Temperature = T_r = 298.15 K Standard State Pressure = p° = 0.1 MPa

T/K	$C_p°$	S°	$-[G°-H°(T_r)]/T$	$H°-H°(T_r)$	$\Delta_f H°$	$\Delta_f G°$	Log K_f
	J K⁻¹mol⁻¹			kJ mol⁻¹			
0	0.	0.	INFINITE	-5.862	-195.961	-195.961	INFINITE
100	12.782	5.832	60.224	-5.439	-196.485	-195.567	102.154
200	28.706	19.970	36.518	-3.310	-196.875	-194.442	50.783
298.15	37.899	33.321	33.321	0.	-196.648	-193.282	33.862
300	37.656	33.555	33.322	0.070	-196.641	-193.262	33.650
400	43.593	45.309	34.885	4.170	-196.123	-192.208	25.100
500	47.133	55.452	38.009	8.721	-195.523	-191.299	19.985
600	49.380	64.257	41.666	13.554	-194.959	-190.508	16.585
700	50.965	72.009	45.459	18.585	-194.468	-189.806	14.164
800	52.070	78.890	49.216	23.740	-194.084	-189.168	12.351
900	52.836	85.070	52.862	28.987	-193.819	-188.570	10.944
1000	53.388	90.662	56.367	34.296	-193.689	-187.995	9.820
1100	54.229	95.794	59.721	39.680	-193.670	-187.428	8.900
1200	54.580	100.528	62.927	45.121	-197.482	-186.635	8.124
1300	54.932	104.910	65.990	50.596	-197.174	-185.744	7.463
1400	55.283	108.994	68.917	56.107	-196.914	-184.875	6.898
1500	55.635	112.820	71.718	61.653	-196.706	-184.023	6.408
1600	55.986	116.422	74.401	67.234	-196.556	-183.182	5.980
1700	56.338	119.827	76.974	72.850	-196.473	-182.349	5.603
1800	56.689	123.057	79.445	78.502	-196.464	-181.519	5.268
1900	57.040	126.131	81.822	84.188	-196.538	-180.687	4.967
2000	57.392	129.066	84.111	89.910	-196.702	-179.849	4.697
2100	57.743	131.875	86.319	95.667	-196.968	-179.001	4.452
2200	58.095	134.569	88.451	101.458	-218.665	-177.391	4.212
2300	58.446	137.159	90.513	107.285	-219.576	-175.494	3.986
2400	58.798	139.654	92.509	113.148	-220.468	-173.559	3.777
2500	59.149	142.061	94.443	119.045	-221.339	-171.586	3.585
2600	59.501	144.388	96.320	124.978	-222.191	-169.579	3.407
2700	59.852	146.640	98.142	130.945	-223.021	-167.540	3.241
2800	60.204	148.823	99.913	136.948	-223.830	-165.470	3.087
2900	60.555	150.942	101.637	142.986	-224.618	-163.372	2.943
3000	60.906	153.001	103.315	149.059	-225.384	-161.247	2.808
3100	61.258	155.004	104.950	155.167	-226.127	-159.097	2.681
3200	61.609	156.954	106.545	161.311	-226.848	-156.923	2.562
3300	61.961	158.855	108.101	167.489	-227.546	-154.727	2.449
3400	62.312	160.710	109.621	173.703	-228.220	-152.510	2.343
3500	62.664	162.522	111.107	179.952	-228.872	-150.274	2.243
3600	63.015	164.292	112.560	186.235	-229.501	-148.019	2.148
3700	63.367	166.023	113.981	192.555	-230.106	-145.747	2.058
3800	63.718	167.718	115.373	198.909	-230.687	-143.459	1.972
3805.000	63.736	167.802	115.442	199.227	------ CRYSTAL <--> LIQUID -----		
3900	64.070	169.377	116.737	205.298	-231.245	-141.157	1.891
4000	64.421	171.004	118.073	211.723	-231.779	-138.840	1.813
4100	64.772	172.599	119.384	218.182	-232.289	-136.510	1.739
4200	65.124	174.164	120.669	224.677	-232.775	-134.168	1.669
4300	65.475	175.701	121.931	231.207	-233.237	-131.815	1.601
4400	65.827	177.210	123.171	237.772	-233.674	-129.451	1.537
4500	66.178	178.693	124.388	244.373	-234.088	-127.078	1.475
4600	66.530	180.151	125.584	251.008	-234.477	-124.695	1.416
4700	66.881	181.586	126.761	257.679	-234.843	-122.305	1.359
4800	67.233	182.998	127.918	264.384	-796.120	-108.289	1.178
4900	67.584	184.388	129.056	271.125	-796.088	-93.960	1.002
5000	67.936	185.757	130.176	277.901	-796.049	-79.631	0.832

PREVIOUS: June 1964 CURRENT: December 1964

Zirconium Carbide (ZrC) $C_1Zr_1(cr)$

ZIRCONIUM CARBIDE (ZrC) LIQUID M_r = 103.231

$S°(298.15 K) = [53.577] J K^{-1} mol^{-1}$

T_{fus} = 3805 ± 125 K

$\Delta_f H°(298.15 K) = [-118.877] kJ mol^{-1}$

$\Delta_{fus} H° = [79.496] kJ mol^{-1}$

Enthalpy of Formation

$\Delta_f H°$(ZrC, ℓ, 298.15 K) is calculated from $\Delta_f H°$(ZrC, cr, 298.15 K) by adding $\Delta_{fus} H°$ and the difference in enthalpy, H°(3805 K) - H°(298.15 K), between the crystal and liquid.

Heat Capacity and Entropy

$C_p°$(ℓ) at and below 2500 K is assumed to be equal to that of ZrC(cr). Above 2500 K the heat capacity is estimated to be constant. S°(ℓ, 298.15 K) is calculated in a manner similar to that used for the enthalpy of formation.

Fusion Data

See the crystal table for details.

Zirconium Carbide (ZrC) $C_1Zr_1(l)$

Enthalpy Reference Temperature = T_r = 298.15 K   Standard State Pressure = p° = 0.1 MPa

T/K	$C_p°$ (J K⁻¹mol⁻¹)	S° (J K⁻¹mol⁻¹)	-[G°-H°(T_r)]/T (J K⁻¹mol⁻¹)	H°-H°(T_r) (kJ mol⁻¹)	$\Delta_f H°$ (kJ mol⁻¹)	$\Delta_f G°$ (kJ mol⁻¹)	Log K_f
0							
100							
200							
298.15	37.899	53.577	53.577	0.	-118.877	-121.550	21.295
300	38.030	53.811	53.577	0.070	-118.869	-121.567	21.167
400	43.593	65.581	55.144	4.175	-118.347	-122.541	16.002
500	47.133	75.723	58.270	8.727	-117.746	-123.658	12.918
600	49.380	84.529	61.930	13.560	-117.183	-124.895	10.873
700	50.965	92.281	65.723	18.590	-116.692	-126.220	9.419
800	52.070	99.162	69.481	23.745	-116.308	-127.609	8.332
900	52.836	105.341	73.128	28.992	-116.043	-129.039	7.489
1000	53.388	110.934	76.633	34.301	-115.913	-130.491	6.816
1100	54.229	116.066	79.988	39.686	-115.893	-131.950	6.266
1200	54.580	120.799	83.194	45.126	-119.706	-133.185	5.797
1300	54.932	125.182	86.258	50.602	-119.398	-134.321	5.397
1400	55.283	129.266	89.186	56.112	-119.138	-135.479	5.055
1500	55.635	133.092	91.986	61.658	-118.929	-136.654	4.759
1600	55.986	136.694	94.669	67.239	-118.780	-137.841	4.500
1700	56.338	140.098	97.242	72.855	-118.697	-139.035	4.272
1800	56.689	143.329	99.714	78.507	-118.688	-140.232	4.069
1900	57.040	146.403	102.091	84.193	-118.761	-141.427	3.888
2000	57.392	149.338	104.380	89.915	-118.926	-142.617	3.725
2100	57.743	152.146	106.588	95.672	-119.191	-143.795	3.577
2200	58.095	154.841	108.721	101.464	-140.889	-144.213	3.424
2300	58.446	157.431	110.783	107.291	-141.800	-144.343	3.278
2400	58.798	159.926	112.779	113.153	-142.691	-144.435	3.144
2500	59.149	162.333	114.713	119.050	-143.563	-144.489	3.019
2500.000	59.149	162.333	114.713	119.051	GLASS <--> LIQUID TRANSITION		
2500.000	62.760	162.333	114.713	119.051			
2600	62.760	164.795	116.592	125.327	-144.071	-144.516	2.903
2700	62.760	167.163	118.422	131.603	-144.593	-144.524	2.796
2800	62.760	169.446	120.203	137.879	-145.129	-144.511	2.696
2900	62.760	171.648	121.940	144.155	-145.678	-144.480	2.602
3000	62.760	173.776	123.632	150.431	-146.241	-144.429	2.515
3100	62.760	175.834	125.283	156.707	-146.816	-144.359	2.432
3200	62.760	177.826	126.894	162.983	-147.405	-144.270	2.355
3300	62.760	179.757	128.467	169.259	-148.005	-144.163	2.282
3400	62.760	181.631	130.003	175.535	-148.618	-144.037	2.213
3500	62.760	183.450	131.504	181.811	-149.242	-143.894	2.147
3600	62.760	185.218	132.972	188.087	-149.878	-143.732	2.085
3700	62.760	186.938	134.407	194.363	-150.527	-143.552	2.027
3800	62.760	188.612	135.812	200.639	-151.186	-143.355	1.971
3805.000	62.760	188.694	135.881	200.952	------ CRYSTAL <--> LIQUID -----		
3900	62.760	190.242	137.187	206.915	-151.857	-143.140	1.917
4000	62.760	191.831	138.533	213.191	-152.540	-142.908	1.866
4100	62.760	193.380	139.852	219.467	-153.233	-142.659	1.817
4200	62.760	194.893	141.145	225.743	-153.938	-142.392	1.771
4300	62.760	196.370	142.412	232.019	-154.654	-142.109	1.726
4400	62.760	197.812	143.654	238.295	-155.381	-141.809	1.683
4500	62.760	199.223	144.874	244.571	-156.119	-141.492	1.642
4600	62.760	200.602	146.070	250.847	-156.868	-141.159	1.603
4700	62.760	201.952	147.245	257.123	-157.627	-140.809	1.565
4800	62.760	203.273	148.398	263.399	-719.334	-128.826	1.402
4900	62.760	204.567	149.532	269.675	-719.767	-116.519	1.242
5000	62.760	205.835	150.645	275.951	-720.228	-104.203	1.089
5100	62.760	207.078	151.739	282.227	-720.715	-91.878	0.941
5200	62.760	208.297	152.815	288.503	-721.227	-79.543	0.799
5300	62.760	209.492	153.874	294.779	-721.762	-67.198	0.662
5400	62.760	210.665	154.914	301.055	-722.320	-54.842	0.530
5500	62.760	211.817	155.939	307.331	-722.899	-42.476	0.403
5600	62.760	212.948	156.946	313.607	-723.497	-30.100	0.281
5700	62.760	214.058	157.939	319.883	-724.113	-17.712	0.162
5800	62.760	215.150	158.916	326.159	-724.746	-5.314	0.048
5900	62.760	216.223	159.878	332.435	-725.396	7.096	-0.063
6000	62.760	217.278	160.826	338.711	-726.059	19.517	-0.170

PREVIOUS: June 1964   CURRENT: December 1964

Zirconium Carbide (ZrC) $C_1Zr_1(l)$

ZIRCONIUM CARBIDE (ZrC) CRYSTAL-LIQUID M_r = 103.231 Zirconium Carbide (ZrC) $C_1Zr_1(cr,l)$

0 to 3805 K crystal
above 3805 K liquid

Refer to the individual tables for details.

Enthalpy Reference Temperature = T_r = 298.15 K; Standard State Pressure = $p°$ = 0.1 MPa

T/K	$C_p°$ (J K⁻¹mol⁻¹)	$S°$ (J K⁻¹mol⁻¹)	$-[G°-H°(T_r)]/T$ (J K⁻¹mol⁻¹)	$H°-H°(T_r)$ (kJ mol⁻¹)	$\Delta_fH°$ (kJ mol⁻¹)	$\Delta_fG°$ (kJ mol⁻¹)	Log K_f
0	0.	0.	INFINITE	-5.862	-195.961	-195.961	INFINITE
100	12.782	5.832	60.224	-5.439	-196.485	-195.567	102.154
200	28.706	19.970	36.518	-3.310	-196.875	-194.442	50.783
298.15	37.899	33.321	33.321	0.	-196.648	-193.282	33.862
300	37.656	33.555	33.322	0.070	-196.641	-193.262	33.650
400	43.593	45.309	34.885	4.170	-196.123	-192.208	25.100
500	47.133	55.452	38.009	8.721	-195.523	-191.299	19.985
600	49.380	64.257	41.666	13.554	-194.959	-190.508	16.585
700	50.965	72.009	45.459	18.585	-194.468	-189.806	14.164
800	52.070	78.890	49.216	23.740	-194.084	-189.168	12.351
900	52.836	85.070	52.862	28.987	-193.819	-188.570	10.944
1000	53.388	90.662	56.367	34.296	-193.689	-187.995	9.820
1100	54.229	95.794	59.721	39.680	-193.670	-187.428	8.900
1200	54.580	100.528	62.927	45.121	-197.482	-186.635	8.124
1300	54.932	104.910	65.990	50.596	-197.174	-185.744	7.463
1400	55.283	108.994	68.917	56.107	-196.914	-184.875	6.898
1500	55.635	112.820	71.718	61.653	-196.706	-184.023	6.408
1600	55.986	116.422	74.401	67.234	-196.556	-183.182	5.980
1700	56.338	119.827	76.974	72.850	-196.473	-182.349	5.603
1800	56.689	123.057	79.445	78.502	-196.464	-181.519	5.268
1900	57.040	126.131	81.822	84.188	-196.538	-180.687	4.967
2000	57.392	129.066	84.111	89.910	-196.702	-179.849	4.697
2100	57.743	131.875	86.319	95.667	-196.968	-179.001	4.452
2200	58.095	134.569	88.451	101.458	-218.665	-177.391	4.212
2300	58.446	137.159	90.513	107.285	-219.576	-175.494	3.986
2400	58.798	139.654	92.509	113.148	-220.468	-173.559	3.777
2500	59.149	142.061	94.443	119.045	-221.339	-171.586	3.585
2600	59.501	144.388	96.320	124.978	-222.191	-169.579	3.407
2700	59.852	146.640	98.142	130.945	-223.021	-167.540	3.241
2800	60.204	148.823	99.913	136.948	-223.830	-165.470	3.087
2900	60.555	150.942	101.637	142.986	-224.618	-163.372	2.943
3000	60.906	153.001	103.315	149.059	-225.384	-161.247	2.808
3100	61.258	155.004	104.950	155.167	-226.127	-159.097	2.681
3200	61.609	156.954	106.545	161.311	-226.848	-156.923	2.562
3300	61.961	158.855	108.101	167.489	-227.546	-154.727	2.449
3400	62.312	160.710	109.621	173.703	-228.220	-152.510	2.343
3500	62.664	162.522	111.107	179.952	-228.872	-150.274	2.243
3600	63.015	164.292	112.560	186.235	-229.501	-148.019	2.148
3700	63.367	166.023	113.981	192.555	-230.106	-145.747	2.058
3800	63.718	167.718	115.373	198.909	-230.687	-143.459	1.972
3805.000	63.736	167.802	115.442	199.227	CRYSTAL <--> LIQUID		
3805.000	62.760	188.694	115.442	278.723	TRANSITION		
3900	62.760	190.242	117.245	284.686	-151.857	-143.140	1.917
4000	62.760	191.831	119.090	290.962	-152.540	-142.908	1.866
4100	62.760	193.380	120.883	297.238	-153.233	-142.659	1.817
4200	62.760	194.893	122.628	303.514	-153.938	-142.392	1.771
4300	62.760	196.370	124.325	309.790	-154.654	-142.109	1.726
4400	62.760	197.812	125.979	316.066	-155.381	-141.809	1.683
4500	62.760	199.223	127.591	322.342	-156.119	-141.492	1.642
4600	62.760	200.602	129.164	328.618	-156.868	-141.159	1.603
4700	62.760	201.952	130.698	334.894	-157.627	-140.809	1.565
4800	62.760	203.273	132.196	341.170	-719.334	-128.826	1.402
4900	62.760	204.567	133.660	347.446	-719.767	-116.519	1.242
5000	62.760	205.835	135.091	353.722	-720.228	-104.203	1.089
5100	62.760	207.078	136.490	359.998	-720.715	-91.878	0.941
5200	62.760	208.297	137.859	366.274	-721.227	-79.543	0.799
5300	62.760	209.492	139.200	372.550	-721.762	-67.198	0.662
5400	62.760	210.665	140.512	378.826	-722.320	-54.842	0.530
5500	62.760	211.817	141.798	385.102	-722.899	-42.476	0.403
5600	62.760	212.948	143.059	391.378	-723.497	-30.100	0.281
5700	62.760	214.058	144.295	397.654	-724.113	-17.712	0.162
5800	62.760	215.150	145.507	403.930	-724.746	-5.314	0.048
5900	62.760	216.223	146.696	410.206	-725.396	7.096	-0.063
6000	62.760	217.278	147.864	416.482	-726.059	19.517	-0.170

PREVIOUS: CURRENT: December 1964

Zirconium Carbide (ZrC) $C_1Zr_1(cr,l)$

CARBON (C_2) IDEAL GAS M_r = 24.022

$S°(298.15\ K) = 199.38 \pm 0.04\ J\ K^{-1}\ mol^{-1}$

$\Delta_f H°(0\ K) = 829.3 \pm 3.8\ kJ\ mol^{-1}$
$\Delta_f H°(298.15\ K) = 837.7 \pm 3.8\ kJ\ mol^{-1}$

Electronic Levels and Molecular Constants (σ = 2)

State	ε_i, cm^{-1}	g_i	r_e, Å	B_e, cm^{-1}	α_e, cm^{-1}	ω_e, cm^{-1}	$\omega_e x_e$, cm^{-1}
$X^1\Sigma_g^+$	0	1	1.2425	1.81984	0.01765	1854.71	13.34
$X'^3\Pi_u$	610	6	1.3119	1.63246	0.01661	1641.35	11.67
$A'^3\Sigma_g^-$	6243.5	3	1.3693	1.49852	0.01634	1470.45	11.19
$b^1\Pi_u$	8268.33	2	1.3184	1.61634	0.01686	1608.35	12.078
$A''^3\Sigma_u^+$	13365	3	1.23	1.87	[0.018]	1961.6	13.65
$^1\Delta_g$	[15000]	2	[1.39]	[1.454]	[0.015]	[1510]	[11.1]
$^1\Sigma_g^+$	[18570]	1	[1.38]	[1.475]	[0.013]	[1510]	[10.0]
$A^3\Pi_g$	19916	6	1.266	1.7527	0.01608	1788.2	16.44
$^5\Pi_g$	[22260]	10	[1.46]	[1.324]	[0.012]	[1400]	[9.9]
$^5\Sigma_g^+$	[30320]	5	[1.35]	[1.557]	[0.015]	[1550]	[11.7]
$c^1\Pi_g$	34262	2	1.2552	1.7334	0.018	1809.1	15.81
$^3\Delta_u$	[34530]	6	[1.51]	[1.23]	[0.013]	[1160]	[9.2]
$^1\Sigma_u^-$	[34760]	1	[1.90]	[0.778]	[0.005]	[1025]	[5.8]
$^3\Phi_g$	[38020]	6	[1.53]	[1.20]	[0.011]	[1290]	[9.0]
$B^3\Pi_g$	40690	6	1.535	1.1922	0.0242	1106.56	39.26
$^3\Delta_u$	[40780]	6	[1.51]	[1.23]	[0.011]	[1380]	[9.2]
$^3\Pi_g$	[42370]	6	[1.49]	[1.265]	[0.012	[1340]	[9.5]
$^1\Phi_g$	[42860]	2	[1.51]	[1.232]	[0.011]	[1360]	[9.2]
$d^1\Sigma_u^+$	43240	1	1.2378	1.8334	0.0204	1829.57	13.97
$^3\Sigma_u^+$	[43360]	3	[1.44]	[1.355]	[0.011]	[1660]	[10.21]

Enthalpy of Formation

Messerle and Krauss (1) have reported the dissociation energy as 49300 ± 300 cm^{-1} from the variation of the rotational structure cut-off with vibrational level in the $c^1\Pi_g$ state. This state dissociates into $C(^3P) + C^3(P)$, due to an avoided crossing (2), these dissociation products are those of the ground state. The uncertainty in the determination is generous and thus we adopt D_0^0 = 140.95 ± 0.9 kcal or $\Delta_f H°(C_2, g, 0\ K) = 198.2 \pm 0.9\ kcal\ mol^{-1}$, using $\Delta_f H°(C, g, 0\ K) = 169.576\ kcal\ mol^{-1}$ (3).

Brewer et al. (4) have reviewed several other earlier determinations which confirm $\Delta_f H°(0\ K) = 195 \pm 5\ kcal\ mol^{-1}$. The extensive mass spectrometric data of Drowart et al. (5) yield a 3rd law $\Delta_f H°(0\ K) = 199.2 \pm 1.4\ kcal\ mol^{-1}$ in excellent agreement with the adopted value.

Heat Capacity and Entropy

The observed molecular constants and electronic levels were taken from Ballik and Ramsay (2). The estimated parameters are from the correlation calculations of Fougere and Nesbet (6) and Clementi (7), the electronic levels are T_o values. Not all predicted states were included since the contribution of levels above 40000 cm^{-1} is neligible. The functions were calculated from the partition function $Q = Q_t \Sigma Q_r^i Q_v^i g_i \exp(-c_2\varepsilon_i/T)$, the values for Q_r^i and Q_v^i were calculated with first order anharmonic corrections. The functions are similar to those reported by Altman (8) and Clementi (7) but include several additional states.

References

1. G. Messerle and L. Krauss, Z. Naturforsch. 22A, 2023 (1967), 22A 1744 (1967) and 22A, 2015 (1967).
2. E. A. Ballik and D. A. Ramsay, Astrophys. Journ. 137, 84 (1963).
3. JANAF Thermochemical Tables: C(g), 3-31-61.
4. L. Brewer, W. T. Hicks, and O. H Krikorian, J. Chem. Phys. 36, 182 (1962).
5. J. Drowart, R. P. Burns, G. DeMaria and M. G. Inghram, J. Chem. Phys. 31, 1131 (1959).
6. P. F. Fougere and R. K. Nesbet, J. Chem. Phys. 44, 285 (1966).
7. E. Clementi, Astrophys. Journ. 133, 303 (1961).
8. R. Altman, J. Chem. Phys. 32, 615 (1960).

Carbon (C_2) C_2(g)

Enthalpy Reference Temperature = T_r = 298.15 K Standard State Pressure = p° = 0.1 MPa

T/K	$C_p°$ (J $K^{-1}mol^{-1}$)	S° (J $K^{-1}mol^{-1}$)	$-[G°-H°(T_r)]/T$ (J $K^{-1}mol^{-1}$)	$H°-H°(T_r)$ (kJ mol^{-1})	$\Delta_f H°$ (kJ mol^{-1})	$\Delta_f G°$ (kJ mol^{-1})	Log K_f
0	0.	0.	INFINITE	-10.578	829.262	829.262	INFINITE
100	29.765	158.761	235.431	-7.667	832.053	816.367	-426.427
200	40.456	182.316	203.335	-4.204	834.863	799.633	-208.843
250	43.379	191.730	200.097	-2.092	836.384	790.649	-165.197
298.15	43.145	199.382	199.382	0.	837.737	781.714	-136.953
300	43.100	199.649	199.383	0.080	837.785	781.367	-136.048
350	41.443	206.176	199.902	2.196	838.960	771.867	-115.195
400	39.650	211.590	201.035	4.222	839.882	762.217	-99.535
450	38.191	216.172	202.469	6.166	840.570	752.465	-87.344
500	37.141	220.138	204.042	8.048	841.055	742.648	-77.584
600	36.000	226.793	207.299	11.696	841.547	722.911	-62.935
700	35.622	232.307	210.488	15.273	841.578	703.131	-52.468
800	35.603	237.060	213.520	18.833	841.295	683.369	-44.619
900	35.746	241.261	216.373	22.399	840.793	663.656	-38.518
1000	35.963	245.038	219.054	25.984	840.132	644.008	-33.640
1100	36.212	248.477	221.575	29.593	839.353	624.433	-29.652
1200	36.479	251.639	223.950	33.227	838.484	604.932	-26.332
1300	36.756	254.570	226.194	36.889	837.548	585.507	-23.526
1400	37.038	257.304	228.320	40.579	836.559	566.156	-21.124
1500	37.324	259.869	230.338	44.297	835.528	546.878	-19.044
1600	37.610	262.287	232.260	48.043	834.465	527.669	-17.227
1700	37.896	264.576	234.094	51.819	833.375	508.527	-15.625
1800	38.179	266.750	235.849	55.622	832.266	489.450	-14.203
1900	38.457	268.822	237.530	59.454	831.140	470.436	-12.933
2000	38.729	270.801	239.145	63.314	830.000	451.481	-11.791
2100	38.994	272.697	240.697	67.200	828.849	432.583	-10.760
2200	39.251	274.517	242.194	71.112	827.688	413.741	-9.823
2300	39.500	276.268	243.637	75.050	826.518	394.951	-8.970
2400	39.740	277.954	245.032	79.012	825.341	376.212	-8.188
2500	39.972	279.581	246.382	82.997	824.156	357.523	-7.470
2600	40.197	281.153	247.689	87.006	822.965	338.881	-6.808
2700	40.414	282.674	248.957	91.037	821.767	320.285	-6.196
2800	40.624	284.148	250.188	95.088	820.563	301.734	-5.629
2900	40.827	285.577	251.383	99.161	819.353	283.226	-5.101
3000	41.026	286.964	252.546	103.254	818.136	264.759	-4.610
3100	41.219	288.313	253.678	107.366	816.914	246.333	-4.151
3200	41.409	289.624	254.781	111.498	815.685	227.947	-3.721
3300	41.595	290.901	255.857	115.648	814.450	209.599	-3.318
3400	41.778	292.146	256.906	119.816	813.210	191.289	-2.939
3500	41.960	293.360	257.930	124.003	811.963	173.016	-2.582
3600	42.140	294.544	258.931	128.208	810.712	154.778	-2.246
3700	42.318	295.701	259.909	132.431	809.454	136.574	-1.928
3800	42.496	296.832	260.866	136.672	808.192	118.405	-1.628
3900	42.674	297.938	261.802	140.931	806.924	100.270	-1.343
4000	42.852	299.021	262.719	145.207	805.652	82.166	-1.073
4100	43.030	300.081	263.618	149.501	804.375	64.095	-0.817
4200	43.209	301.120	264.498	153.813	803.093	46.055	-0.573
4300	43.388	302.139	265.362	158.143	801.807	28.046	-0.341
4400	43.568	303.139	266.209	162.490	800.517	10.066	-0.119
4500	43.749	304.120	267.041	166.856	799.223	-7.884	0.092
4600	43.932	305.083	267.857	171.240	797.925	-25.806	0.293
4700	44.115	306.030	268.659	175.643	796.624	-43.699	0.486
4800	44.299	306.961	269.448	180.063	795.320	-61.564	0.670
4900	44.485	307.876	270.222	184.503	794.013	-79.402	0.846
5000	44.671	308.777	270.985	188.960	792.703	-97.214	1.016
5100	44.859	309.663	271.734	193.437	791.390	-114.999	1.178
5200	45.047	310.536	272.472	197.932	790.075	-132.759	1.334
5300	45.237	311.396	273.198	202.446	788.757	-150.493	1.483
5400	45.427	312.243	273.914	206.980	787.438	-168.202	1.627
5500	45.619	313.079	274.618	211.532	786.116	-185.887	1.765
5600	45.811	313.902	275.312	216.103	784.792	-203.548	1.899
5700	46.004	314.715	275.997	220.694	783.467	-221.185	2.027
5800	46.197	315.517	276.671	225.304	782.140	-238.799	2.151
5900	46.391	316.308	277.336	229.933	780.812	-256.390	2.270
6000	46.586	317.089	277.992	234.582	779.482	-273.959	2.385

PREVIOUS: December 1969 (1 atm) CURRENT: December 1969 (1 bar)

Carbon (C_2) C_2(g)

CARBON, ION (C_2^-) IDEAL GAS M_r = 24.022549

$S°(298.15\ K) = 196.60 \pm 0.21\ J\ K^{-1}\ mol^{-1}$

$\Delta_f H°(0\ K) = 443 \pm 84\ kJ\ mol^{-1}$
$\Delta_f H°(298.15\ K) = 444 \pm 84\ kJ\ mol^{-1}$

Electronic Levels and Quantum Weights

ε_i, cm^{-1}	g_i
0	2
[9000]	[4]
18484	[2]

$\omega_e = 1781.04\ cm^{-1}$ $\omega_e x_e = 11.585\ cm^{-1}$ $\sigma = 2$
$B_e = 1.74685\ cm^{-1}$ $\alpha_e = 0.0167\ cm^{-1}$ $r_e = 1.2682$ Å

Enthalpy of Formation

The electron affinity of C_2 has been obtained by Honig (1) as 4 ± 0.8 eV from the difference in enthalpies of sublimation of neutral C_2 and C_2^- and the work function of graphite. Thus, we have the reaction

$$C_2(g) + e^- \rightarrow C_2^-(g) \qquad \Delta H(0\ K) = -92.2 \pm 20\ kcal.$$

Using $\Delta_f H°(C_2,\ g,\ 0\ K) = 198.2\ kcal\ mol^{-1}$ (2), we obtain $\Delta_f H°(C_2^-,\ g,\ 0\ K) = 106 \pm 20\ kcal\ mol^{-1}$, which is adopted.

A value for the enthalpy of formation may be obtained from a linear Birge-Sponer extrapolation of the ground state vibrational levels, which gives 195.8 kcal for the reaction.

$$C_2^-(g) \rightarrow C(g) + C^-(g).$$

With auxiliary data (2) this yields $\Delta_f H°(C_2^-,\ g,\ 0\ K) = 114.5\ kcal\ mol^{-1}$, in good agreement with the adopted value.

Heat Capacity and Entropy

The molecular constants and electronic levels are from Herzberg and Lagerqvist (3). The values are not definitely ascribed to C_2^- by Herzberg and Lagerqvist, but there is a good deal of evidence to support this assignment. Recently Milligan and Jacox (4) have shown that matrix isolated bands previously attributed to $C_2(g)$ are enhanced in the presence of a photoelectron source. The matrix spectra correspond to the gas system reported by Harzberg and Lagerqvist (3) and leave little doubt that the species is C_2^-.

The estimated electronic level is assumed to be a $^2\Pi$ state and is obtained from the isoelectronic CN(g) (2).

References

1. R. E. Honig, J. Chem. Phys. 22, 126 (1954).
2. JANAF Thermochemical Tables: C_2(g), 12-31-69.
3. G. Herzberg and A. Lagerqvist, Can. J. Phys. 42, 2363 (1968).
4. D. E. Milligan and M. E. Jacox, J. Chem. Phys. 51, 1952 (1969).

Carbon, Ion (C_2^-) $C_2^-(g)$

Enthalpy Reference Temperature = T_r = 298.15 K Standard State Pressure = p° = 0.1 MPa

	J $K^{-1}mol^{-1}$			kJ mol^{-1}			
T/K	$C_p°$	S°	$-[G°-H°(T_r)]/T$	$H°-H°(T_r)$	$\Delta_f H°$	$\Delta_f G°$	Log K_f
0	0.	0.	INFINITE	-8.676	443.127		
100	29.106	164.780	222.507	-5.773			
200	29.115	184.957	199.266	-2.862			
250	29.148	191.457	197.078	-1.405			
298.15	29.241	196.597	196.597	0.	443.504	394.566	-69.126
300	29.246	196.778	196.598	0.054	443.488	394.263	-68.647
350	29.438	201.300	196.955	1.521	442.974	386.097	-57.622
400	29.726	205.248	197.750	2.999	442.309	378.015	-49.364
450	30.092	208.770	198.782	4.495	441.509	370.025	-42.951
500	30.511	211.962	199.943	6.010	440.588	362.131	-37.832
600	31.407	217.604	202.428	9.105	438.448	346.635	-30.177
700	32.272	222.511	204.954	12.290	436.008	331.522	-24.738
800	33.042	226.872	207.426	15.556	433.354	316.774	-20.683
900	33.703	230.803	209.809	18.895	430.545	302.369	-17.549
1000	34.263	234.383	212.090	22.294	427.619	288.283	-15.058
1100	34.741	237.672	214.268	25.744	424.604	274.495	-13.035
1200	35.154	240.713	216.347	29.240	421.518	260.985	-11.360
1300	35.522	243.541	218.331	32.774	418.375	247.734	-9.954
1400	35.861	246.186	220.227	36.343	415.187	234.727	-8.758
1500	36.185	248.672	222.041	39.945	411.962	221.950	-7.729
1600	36.503	251.017	223.780	43.580	408.707	209.388	-6.836
1700	36.824	253.240	225.448	47.246	405.431	197.031	-6.054
1800	37.152	255.354	227.051	50.945	402.137	184.867	-5.365
1900	37.490	257.372	228.594	54.677	398.833	172.886	-4.753
2000	37.840	259.303	230.082	58.443	395.522	161.080	-4.207
2100	38.199	261.158	231.518	62.245	392.208	149.440	-3.717
2200	38.568	262.944	232.906	66.083	388.894	137.957	-3.276
2300	38.943	264.666	234.249	69.959	385.584	126.625	-2.876
2400	39.323	266.332	235.552	73.872	382.279	115.437	-2.512
2500	39.704	267.945	236.815	77.823	378.982	104.386	-2.181
2600	40.083	269.509	238.043	81.813	375.692	93.467	-1.878
2700	40.459	271.029	239.237	85.840	372.413	82.675	-1.599
2800	40.828	272.507	240.399	89.904	369.142	72.003	-1.343
2900	41.189	273.946	241.531	94.005	365.882	61.449	-1.107
3000	41.539	275.349	242.635	98.142	362.630	51.007	-0.888
3100	41.876	276.716	243.712	102.313	359.388	40.673	-0.685
3200	42.200	278.051	244.764	106.517	356.153	30.443	-0.497
3300	42.510	279.354	245.793	110.752	352.925	20.315	-0.322
3400	42.803	280.628	246.799	115.018	349.703	10.284	-0.158
3500	43.081	281.872	247.783	119.312	346.486	0.348	-0.005
3600	43.342	283.090	248.747	123.634	343.272	-9.496	0.138
3700	43.587	284.281	249.691	127.980	340.059	-19.251	0.272
3800	43.815	285.446	250.617	132.351	336.848	-28.919	0.398
3900	44.027	286.587	251.525	136.743	333.635	-38.503	0.516
4000	44.223	287.704	252.415	141.155	330.420	-48.004	0.627
4100	44.403	288.798	253.289	145.587	327.202	-57.425	0.732
4200	44.567	289.870	254.148	150.036	323.978	-66.767	0.830
4300	44.717	290.921	254.991	154.500	320.748	-76.032	0.924
4400	44.853	291.951	255.819	158.979	317.510	-85.222	1.012
4500	44.976	292.960	256.633	163.470	314.263	-94.339	1.095
4600	45.085	293.950	257.434	167.973	311.006	-103.383	1.174
4700	45.182	294.920	258.221	172.487	307.738	-112.356	1.249
4800	45.268	295.873	258.996	177.009	304.457	-121.260	1.320
4900	45.343	296.807	259.758	181.540	301.163	-130.095	1.387
5000	45.407	297.723	260.508	186.077	297.854	-138.862	1.451
5100	45.462	298.623	261.246	190.621	294.529	-147.564	1.511
5200	45.507	299.506	261.974	195.169	291.189	-156.200	1.569
5300	45.545	300.374	262.690	199.722	287.831	-164.771	1.624
5400	45.574	301.225	263.396	204.278	284.456	-173.279	1.676
5500	45.596	302.062	264.091	208.837	281.062	-181.725	1.726
5600	45.611	302.883	264.777	213.397	277.649	-190.108	1.773
5700	45.620	303.691	265.452	217.959	274.216	-198.431	1.818
5800	45.624	304.484	266.118	222.521	270.762	-206.693	1.861
5900	45.622	305.264	266.775	227.083	267.288	-214.895	1.903
6000	45.615	306.031	267.423	231.645	263.793	-223.038	1.942

PREVIOUS: December 1969 (1 atm) CURRENT: December 1969 (1 bar)

Carbon, Ion (C_2^-) $C_2^-(g)$

BERYLLIUM CARBIDE (BeC_2) IDEAL GAS M_r = 33.03418

Beryllium Carbide (BeC_2) $C_2Be_1(g)$

$S°(298.15\ K) = [218.643]\ J\ K^{-1}\ mol^{-1}$

$\Delta_f H°(0\ K) = [558.678]\ kJ\ mol^{-1}$

$\Delta_f H°(298.15\ K) = [564.840]\ kJ\ mol^{-1}$

Vibrational Frequencies and Degeneracies

ν, cm^{-1}
[1580](1)
[423](2)
[1947](1)

Ground State Quantum Weight: 1 $\sigma = 1$

Point Group: $[C_{\infty v}]$

Bond Distances: Be–C = [1.407] Å C≡C = [1.2] Å

Bond Angle: Be–C–C = [180]°

Rotational Constant: B_0 = [0.481016] cm^{-1}

Enthalpy of Formation

$\Delta_f H°(298.15\ K)$ was estimated as a lower positive limit by U. S. Nat. Bur. Stand (1).

Heat Capacity and Entropy

The vibrational frequencies, bond angle, and configuration were estimated by Chupka, et al (2). Bond distances were calculated assuming the C–C bond distance was the same as in acetylene.

References

1. G. T. Armstrong and L. A. Krieger, U. S. Nat. Bur. Stand. Report 6645, Chapter A2, 1960.
2. Chupka, Berkowitz, Giese, and Inghram, J. Phys. Chem. 62, 611 (1958).

Enthalpy Reference Temperature = T_r = 298.15 K — Standard State Pressure = p° = 0.1 MPa

T/K	$C_p°$	S°	$-[G°-H°(T_r)]/T$	$H°-H°(T_r)$	$\Delta_f H°$	$\Delta_f G°$	Log K_f
	J $K^{-1}mol^{-1}$			kJ mol^{-1}			
0	0.	0.	INFINITE	-10.196	558.678	558.678	INFINITE
100	30.508	179.700	252.347	-7.265	561.451	543.722	-284.011
200	37.212	202.950	222.304	-3.871	563.615	525.092	-137.140
250	39.569	211.521	219.313	-1.948	564.354	515.371	-107.681
298.15	41.282	218.643	218.643	0.	564.840	505.889	-88.630
300	41.341	218.898	218.643	0.076	564.854	505.523	-88.019
350	42.814	225.384	219.152	2.181	565.139	495.610	-73.966
400	44.152	231.189	220.300	4.356	565.246	485.668	-63.422
450	45.426	236.464	221.808	6.595	565.202	475.722	-55.220
500	46.651	241.314	223.519	8.897	565.032	465.787	-48.660
600	48.933	250.025	227.227	13.679	564.390	445.993	-38.827
700	50.936	257.722	231.044	18.675	563.449	426.331	-31.813
800	52.634	264.638	234.818	23.856	562.291	406.819	-26.563
900	54.048	270.921	238.486	29.192	560.965	387.464	-22.488
1000	55.215	276.678	242.021	34.657	559.501	368.263	-19.236
1100	56.179	281.988	245.416	40.228	557.914	349.215	-16.583
1200	56.976	286.911	248.672	45.887	556.212	330.317	-14.378
1300	57.639	291.498	251.792	51.619	554.401	311.565	-12.519
1400	58.194	295.791	254.783	57.411	552.481	292.956	-10.930
1500	58.661	299.822	257.652	63.255	550.455	274.489	-9.559
1600	59.057	303.621	260.408	69.141	533.786	256.686	-8.380
1700	59.395	307.212	263.056	75.064	531.947	239.423	-7.357
1800	59.685	310.615	265.605	81.019	530.068	222.270	-6.450
1900	59.936	313.849	268.060	87.000	528.151	205.222	-5.642
2000	60.155	316.929	270.427	93.005	526.194	188.276	-4.917
2100	60.345	319.869	272.712	99.030	524.197	171.429	-4.264
2200	60.513	322.680	274.919	105.073	522.161	154.678	-3.673
2300	60.660	325.373	277.055	111.132	520.086	138.021	-3.135
2400	60.792	327.958	279.122	117.205	517.971	121.455	-2.643
2500	60.908	330.442	281.126	123.290	515.815	104.977	-2.193
2600	61.013	332.833	283.069	129.386	513.620	88.587	-1.780
2700	61.106	335.137	284.955	135.492	511.385	72.282	-1.398
2800	61.191	337.361	286.787	141.607	218.152	62.283	-1.162
2900	61.267	339.509	288.568	147.730	216.899	56.738	-1.022
3000	61.336	341.588	290.301	153.860	215.622	51.237	-0.892
3100	61.399	343.600	291.988	159.997	214.320	45.779	-0.771
3200	61.456	345.550	293.632	166.139	212.991	40.363	-0.659
3300	61.508	347.442	295.234	172.288	211.634	34.989	-0.554
3400	61.556	349.279	296.796	178.441	210.246	29.657	-0.456
3500	61.600	351.064	298.322	184.599	208.827	24.366	-0.364
3600	61.641	352.800	299.811	190.761	207.373	19.117	-0.277
3700	61.678	354.489	301.266	196.927	205.884	13.908	-0.196
3800	61.712	356.135	302.688	203.096	204.356	8.740	-0.120
3900	61.744	357.738	304.079	209.269	202.788	3.612	-0.048
4000	61.774	359.302	305.440	215.445	201.177	-1.475	0.019
4100	61.802	360.827	306.773	221.624	199.522	-6.520	0.083
4200	61.827	362.317	308.078	227.805	197.819	-11.525	0.143
4300	61.851	363.772	309.356	233.989	196.068	-16.489	0.200
4400	61.874	365.194	310.609	240.176	194.265	-21.412	0.254
4500	61.895	366.585	311.837	246.364	192.409	-26.292	0.305
4600	61.914	367.946	313.042	252.554	190.498	-31.132	0.354
4700	61.933	369.277	314.225	258.747	188.530	-35.928	0.399
4800	61.950	370.581	315.385	264.941	186.503	-40.683	0.443
4900	61.966	371.859	316.525	271.137	184.416	-45.394	0.484
5000	61.982	373.111	317.644	277.334	182.267	-50.063	0.523
5100	61.996	374.339	318.744	283.533	180.054	-54.687	0.560
5200	62.010	375.543	319.825	289.733	177.777	-59.268	0.595
5300	62.023	376.724	320.887	295.935	175.435	-63.804	0.629
5400	62.035	377.883	321.932	302.138	173.026	-68.296	0.661
5500	62.046	379.022	322.959	308.342	170.550	-72.742	0.691
5600	62.057	380.140	323.971	314.547	168.007	-77.143	0.720
5700	62.068	381.238	324.966	320.753	165.395	-81.498	0.747
5800	62.077	382.318	325.945	326.961	162.716	-85.806	0.773
5900	62.087	383.379	326.910	333.169	159.968	-90.067	0.797
6000	62.096	384.423	327.860	339.378	157.151	-94.281	0.821

PREVIOUS: December 1963 (1 atm) CURRENT: December 1963 (1 bar)

Beryllium Carbide (BeC_2) $C_2Be_1(g)$

DICHLOROETHYNE (C_2Cl_2) IDEAL GAS M_r = 94.928

$\Delta_fH°(0\ K)$ = [206.3 ± 42] kJ mol^{-1}

S°(298.15 K) = [272.027 ± 13] J K^{-1} mol^{-1} $\Delta_fH°(298.15\ K)$ = [209.6 ± 42] kJ mol^{-1}

Vibrational Frequencies and Degeneracies

ν, cm^{-1}
[410](1)
[2200](1)
[925](1)
[380](2)
[165](2)

Ground State Quantum Weight: 1 σ = 2

Point Group: $D_{\infty h}$

Bond Distances: C–C = 1.195 Å C–Cl = 1.64 Å

Bond Angle: Cl–C–C = 180°

Rotational Constant: B_0 = 0.046368 cm^{-1}

Enthalpy of Formation

The enthalpy of formation of dichloroacetylene is calculated from the estimated $\Delta_rH°(298.15\ K)$ = -53±10 kcal mol^{-1} for ClC≡CCl(g) + Cl_2(g) → $Cl_2C{=}CCl_2$(g), using $\Delta_fH°(C_2Cl_4$, g, 298.15 K) = -2.9 kcal mol^{-1} (1). The value of $\Delta_rH°(298.15\ K)$ is assumed to be the same as that for the reaction HC≡CH(g) + Cl_2(g) → CHCl=CHCl(g), using $\Delta_fH°$(CHCl=CHCl, g, 298.15 K) = 1.2 kcal mol^{-1} (2).

Heat Capacity and Entropy

The linear molecular structure and bond distances were determined from electron diffraction studies by Hassel and Viervoll (3).

All five fundamental vibrational frequencies are calculated by the modified valence force method (4) from estimated force constants which are interpolated from those of C_2I_2 (5), CH≡CF (6), CH≡CCl (6), CH≡CBr (6), and C_2D_2 (4).

References

1. U. S. Nat. Bur. Stand. Tech. Note 270-3, 1968.
2. This is an average value of 0.9 and 1.47 kcal mol^{-1} for cis- and trans- CHCl = CHCl(g), respectively, which are obtained from reference (1).
3. O. Hassel and H. Viervoll, Acta Chem. Scand. 1, 149 (1947).
4. G. Herzberg, "Infrared and Raman Spectra," pp. 188-189, D. Van Nostrand Co., Inc., New York, 1950.
5. A. G. Meister and F. F. Cleveland, J. Chem. Phys. 17, 212 (1949).
6. G. R. Hunt and M. K. Wilson, J. Chem. Phys. 34, 1301 (1961).

Dichloroethyne (C_2Cl_2) C_2Cl_2(g)

Enthalpy Reference Temperature = T_r = 298.15 K Standard State Pressure = p° = 0.1 MPa

T/K	$C_p°$ (J K^{-1}mol^{-1})	S°	-[G°-H°(T_r)]/T	H°-H°(T_r) (kJ mol^{-1})	$\Delta_fH°$	$\Delta_fG°$	Log K_f
0	0.	0.	INFINITE	-14.565	206.336	206.336	INFINITE
100	42.627	212.575	324.555	-11.198	206.672	204.506	-106.823
200	57.837	247.336	277.809	-6.095	208.085	201.843	-52.716
250	62.360	260.755	273.089	-3.084	208.886	200.190	-41.827
298.15	65.573	272.027	272.027	0.	209.618	198.448	-34.767
300	65.680	272.433	272.028	0.121	209.645	198.378	-34.541
350	68.213	282.756	272.837	3.471	210.335	196.445	-29.318
400	70.218	292.000	274.665	6.934	210.942	194.418	-25.388
450	71.861	300.368	277.063	10.487	211.463	192.320	-22.324
500	73.250	308.013	279.782	14.116	211.900	190.169	-19.867
600	75.515	321.577	285.646	21.559	212.554	185.757	-16.172
700	77.312	333.358	291.639	29.203	212.981	181.255	-13.525
800	78.775	343.780	297.517	37.010	213.246	176.703	-11.538
900	79.977	353.130	303.186	44.950	213.396	172.125	-9.990
1000	80.970	361.610	308.611	52.999	213.462	167.535	-8.751
1100	81.794	369.367	313.787	61.138	213.464	162.942	-7.737
1200	82.480	376.514	318.720	69.353	213.414	158.351	-6.893
1300	83.055	383.139	323.424	77.630	213.322	153.765	-6.178
1400	83.539	389.312	327.912	85.961	213.191	149.189	-5.566
1500	83.950	395.090	332.200	94.336	213.026	144.623	-5.036
1600	84.300	400.520	336.302	102.749	212.829	140.069	-4.573
1700	84.600	405.640	340.232	111.194	212.602	135.528	-4.164
1800	84.859	410.483	344.001	119.667	212.344	131.002	-3.802
1900	85.084	415.077	347.622	128.165	212.056	126.490	-3.477
2000	85.280	419.447	351.105	136.683	211.738	121.995	-3.186
2100	85.452	423.612	354.459	145.220	211.389	117.516	-2.923
2200	85.604	427.590	357.694	153.773	211.008	113.055	-2.684
2300	85.738	431.399	360.816	162.340	210.594	108.611	-2.467
2400	85.857	435.050	363.834	170.920	210.146	104.187	-2.268
2500	85.964	438.557	366.753	179.511	209.662	99.782	-2.085
2600	86.059	441.931	369.580	188.112	209.142	95.397	-1.917
2700	86.144	445.180	372.320	196.723	208.583	91.033	-1.761
2800	86.222	448.315	374.978	205.341	207.987	86.689	-1.617
2900	86.291	451.341	377.560	213.967	207.351	82.368	-1.484
3000	86.355	454.268	380.068	222.599	206.677	78.070	-1.359
3100	86.412	457.100	382.508	231.238	205.963	73.795	-1.243
3200	86.465	459.845	384.882	239.881	205.211	69.543	-1.135
3300	86.513	462.506	387.194	248.530	204.421	65.316	-1 034
3400	86.557	465.090	389.447	257.184	203.595	61.113	-0.939
3500	86.598	467.599	391.644	265.842	202.734	56.935	-0.850
3600	86.635	470.039	393.788	274.503	201.840	52.782	-0.766
3700	86.670	472.413	395.881	283.169	200.915	48.654	-0.687
3800	86.702	474.725	397.926	291.837	199.961	44 551	-0.612
3900	86.731	476.978	399.924	300.509	198.981	40.474	-0.542
4000	86.759	479.174	401.878	309.183	197.978	36.423	-0.476
4100	86.784	481.317	403.790	317.861	196.953	32.396	-0.413
4200	86.808	483.408	405.660	326.540	195.909	28.395	-0.353
4300	86.830	485.451	407.492	335.222	194.850	24.420	-0.297
4400	86.851	487.447	409.287	343.906	193.776	20.468	-0.243
4500	86.870	489.399	411.046	352.592	192.692	16.542	-0.192
4600	86.888	491.309	412.770	361.280	191.599	12.639	-0.144
4700	86.906	493.178	414.461	369.970	190.500	8.761	-0.097
4800	86.922	495.008	416.120	378.661	189.397	4.906	-0.053
4900	86.937	496.800	417.748	387.354	188.291	1.074	-0.011
5000	86.951	498.557	419.347	396.048	187.185	-2.736	0.029
5100	86.964	500.279	420.917	404.744	186.079	-6.523	0.067
5200	86.977	501.967	422.459	413.441	184.977	-10.289	0.103
5300	86.989	503.624	423.975	422.140	183.878	-14.034	0.138
5400	87.000	505.250	425.465	430.839	182.785	-17.758	0.172
5500	87.011	506.847	426.930	439.540	181.698	-21.461	0.204
5600	87.021	508.415	428.372	448.241	180.617	-25.146	0.235
5700	87.031	509.955	429.789	456.944	179.545	-28.811	0.264
5800	87.040	511.469	431.185	465.647	178.481	-32.456	0.292
5900	87.048	512.957	432.558	474.352	177.427	-36.084	0.319
6000	87.057	514.420	433.910	483.057	176.381	-39.695	0.346

PREVIOUS: December 1968 (1 atm) CURRENT: December 1968 (1 bar)

Dichloroethyne (C_2Cl_2) C_2Cl_2(g)

TETRACHLOROETHENE (C_2Cl_4) IDEAL GAS M_r = 165.834

$\Delta_fH°(0\ K) = -11.873 \pm 2.9\ kJ\ mol^{-1}$

$S°(298.15\ K) = 343.421\ J\ K^{-1}\ mol^{-1}$ $\Delta_fH°(298.15\ K) = -12.426 \pm 2.9\ kJ\ mol^{-1}$

Electronic Levels and Quantum Weights

ε_i, cm^{-1}	g_i	ε_i, cm^{-1}	g_i
0	1	63557	[1]
[37700]	[1]	70097	[1]
[50350]	[1]	72410	[1]
[59000]	[1]		

Vibrational Frequencies and Degeneracies

ν, cm^{-1}	ν, cm^{-1}	ν, cm^{-1}
1571 (1)	1000 (1)	918 (1)
447 (1)	347 (1)	176 (1)
235 (1)	324 (1)	777 (1)
110 (1)	512 (1)	288 (1)

Point Group: D_{2h} σ = 4
Bond Distances: Cl-C = 1.724 Å C-C = 1.327 Å
Bond Angle: Cl-C-Cl = 113.3°
Product of the Moments of Inertia: $I_AI_BI_C = 3.431926 \times 10^{-112}\ g^3\ cm^6$

Enthalpy of Formation

The enthalpy of combustion $\Delta E(Cl_2Cl_4$, ℓ, 293 K) has been determined by the "Quartz Wool" method by Smith et al. (1). The free chlorine produced from combustion was reduced to hydrogen chloride by a quartz spiral moistened with a solution of arsenious acid in the combustion bomb. Based on the reported results, the enthalpy change at 298.15 K for the reaction $C_2Cl_4(\ell) + O_2(g) + 2H_2O(\ell) = 2CO_2(g) + 4HCl(aq, 600H_2O)$ is evaluated as -198.40 kcal mol^{-1}, using recent auxiliary data from (2) and Parker (3) and $C_p°$ for $C_2Cl_4(\ell)$ from Kurbatov [4]. Employing $\Delta_fH°(298.15\ K)$ = -94.054, -39.823 and -68.315 kcal mol^{-1} for $CO_2(g)$, HCl(aq, $600H_2O$) and $H_2O(\ell)$, respectively, we calculate the enthalpy of formation for $C_2Cl_4(\ell)$ as -12.367 kcal mol^{-1}. Incorporating this value with the value $\Delta_{vap}H°(298.15\ K) = 9.4$ kcal mol^{-1} obtained from Eftring (5), we derive $\Delta_fH°(298.15\ K) = -2.97\pm0.7$ kcal mol^{-1} for $C_2Cl_4(g)$.

Heat Capacity and Entropy

The molecular structure of $C_2Cl_4(g)$ was investigated by the sector-visual method of electron diffraction by Ramachandra and Livingston (6). The reported bond distances and angle are in agreement with those measured by Karle and Karle (7), using the sector-microphotometer method, and Lipscomb (8). The molecular structure has also been determined by Brockway et al. (9) Laszlo (10), and Dornte (11) by use of electron diffraction. The values reported by Ramachandra and Livingston (6), are adopted.

The vibrational spectra of tetrachloroethylene have been studied by numerous investigators. The vibrational frequencies assigned by Mann et al. (12), are adopted here. These authors also reviewed the results of earlier investigations in detail.

The electronic levels and quantum weights are obtained from Herzberg (13).

The principal moments of inertia are: $I_A = 48.8379 \times 10^{-39}$, $I_B = 62.8935 \times 10^{-39}$ and $I_C = 111.7314 \times 10^{-39}\ g\ cm^2$.

References

1. L. Smith, L. Bjellerup, S. Krook and H. Westermark, Acta Chem. Scand. 7, 65 (1953).
2. U. S. Nat. Bur. Stand. Tech. Note 270-1, 1965.
3. V. B. Parker, NSRDS-NBS 2, 1965.
4. V. Y. Kurbatov, J. Gen. Chem. USSR 18, 372 (1948).
5. E. Eftring, Ph.D. Thesis, University of Lund, 1938.
6. C. H. Ramachandra and R. L. Livingston, Current Sci. 27, 330 (1958).
7. I. L. Karle and J. Karle, J. Chem. Phys. 20, 63 (1952).
8. W. N. Lipscomb, Ph.D. Thesis, California Institute of Technology, 1946.
9. L. O. Brockway, J. Y. Beach and L. Pauling, J. Amer. Chem. Soc. 57, 2693 (1935).
10. H. de Laszlo, Nature 135, 474 (1935).
11. R. W. Dornte, J. Chem. Phys. 1, 566 (1933).
12. D. E. Mann, N. Acquista and E. K. Plyler, J. Res. Nat. Bur. Stand. 52, 67 (1954).
13. G. Herzberg, "Electronic Spectra of Polyatomic Molecules", D. Van Nostrand Co., Inc., New York, 1966.

Enthalpy Reference Temperature = T_r = 298.15 K Standard State Pressure = p° = 0.1 MPa

T/K	$C_p°$ (J K^{-1}mol^{-1})	S° (J K^{-1}mol^{-1})	$-[G°-H°(T_r)]/T$ (J K^{-1}mol^{-1})	$H°-H°(T_r)$ (kJ mol^{-1})	$\Delta_fH°$ (kJ mol^{-1})	$\Delta_fG°$ (kJ mol^{-1})	Log K_f
0	0.	0.	INFINITE	-19.911	-11.873	-11.873	INFINITE
100	56.855	260.362	417.923	-15.756	-13.662	-1.706	0.891
200	81.269	308.108	351.744	-8.727	-13.361	10.236	-2.673
250	89.301	327.140	344.961	-4.455	-12.918	16.086	-3.361
298.15	95.546	343.421	343.421	0.	-12.426	21.628	-3.789
300	95.762	344.013	343.423	0.177	-12.407	21.839	-3.802
350	101.048	359.185	344.609	5.101	-11.861	27.504	-4.105
400	105.411	372.972	347.306	10.266	-11.303	33.089	-4.321
450	109.036	385.603	350.869	15.630	-10.747	38.605	-4.481
500	112.065	397.253	354.933	21.160	-10.205	44.059	-4.603
600	116.755	418.123	363.766	32.614	-9.171	54.814	-4.772
700	120.131	436.388	372.863	44.468	-8.208	65.401	-4.880
800	122.611	452.600	381.836	56.611	-7.306	75.854	-4.953
900	124.471	467.154	390.521	68.969	-6.457	86.198	-5.003
1000	125.893	480.345	398.854	81.490	-5.655	96.449	-5.038
1100	126.999	492.398	406.818	94.137	-4.897	106.623	-5.063
1200	127.874	503.487	414.418	106.883	-4.177	116.729	-5.081
1300	128.576	513.751	421.669	119.706	-3.496	126.776	-5.094
1400	129.147	523.301	428.592	132.593	-2.852	136.773	-5.103
1500	129.616	532.228	435.206	145.532	-2.244	146.725	-5.109
1600	130.007	540.606	441.535	158.514	-1.672	156.637	-5.114
1700	130.336	548.498	447.597	171.532	-1.135	166.515	-5.116
1800	130.614	555.956	453.412	184.579	-0.636	176.362	-5.118
1900	130.852	563.024	458.996	197.653	-0.175	186.182	-5.119
2000	131.057	569.741	464.367	210.749	0.247	195.979	-5.118
2100	131.234	576.140	469.538	223.863	0.627	205.757	-5.118
2200	131.389	582.249	474.524	236.995	0.964	215.517	-5.117
2300	131.525	588.092	479.335	250.141	1.254	225.262	-5.116
2400	131.645	593.692	483.984	263.299	1.496	234.996	-5.115
2500	131.751	599.069	488.481	276.469	1.686	244.721	-5.113
2600	131.845	604.238	492.834	289.649	1.822	254.440	-5.112
2700	131.930	609.215	497.053	302.838	1.903	264.155	-5.110
2800	132.005	614.015	501.145	316.035	1.925	273.867	-5.109
2900	132.074	618.648	505.118	329.239	1.890	283.580	-5.108
3000	132.135	623.127	508.977	342.449	1.796	293.295	-5.107
3100	132.191	627.460	512.730	355.666	1.643	303.013	-5.106
3200	132.242	631.658	516.381	368.887	1.432	312.738	-5.105
3300	132.289	635.728	519.936	382.114	1.166	322.471	-5.104
3400	132.331	639.678	523.400	395.345	0.847	332.212	-5.104
3500	132.370	643.515	526.777	408.580	0.478	341.963	-5.104
3600	132.406	647.244	530.072	421.819	0.063	351.726	-5.103
3700	132.440	650.872	533.288	435.061	-0.395	361.500	-5.103
3800	132.470	654.405	536.429	448.307	-0.891	371.287	-5.104
3900	132.499	657.846	539.498	461.555	-1.419	381.088	-5.104
4000	132.525	661.201	542.499	474.807	-1.976	390.904	-5.105
4100	132.550	664.474	545.434	488.060	-2.555	400.733	-5.105
4200	132.574	667.668	548.307	501.317	-3.151	410.576	-5.106
4300	132.596	670.788	551.119	514.575	-3.760	420.434	-5.107
4400	132.617	673.836	553.874	527.836	-4.376	430.306	-5.108
4500	132.636	676.817	556.573	541.098	-4.994	440.192	-5.110
4600	132.655	679.732	559.219	554.363	-5.609	450.092	-5.111
4700	132.674	682.585	561.813	567.629	-6.217	460.005	-5.112
4800	132.691	685.379	564.358	580.898	-6.814	469.931	-5.114
4900	132.708	688.115	566.856	594.168	-7.395	479.870	-5.115
5000	132.725	690.796	569.308	607.439	-7.957	489.819	-5.117
5100	132.741	693.425	571.716	620.713	-8.496	499.780	-5.119
5200	132.758	696.002	574.082	633.987	-9.010	509.751	-5.121
5300	132.774	698.531	576.406	647.264	-9.496	519.732	-5.122
5400	132.790	701.013	578.691	660.542	-9.951	529.722	-5.124
5500	132.807	703.450	580.937	673.822	-10.372	539.720	-5.126
5600	132.823	705.843	583.146	687.104	-10.759	549.725	-5.128
5700	132.840	708.194	585.319	700.387	-11.110	559.737	-5.129
5800	132.857	710.505	587.458	713.672	-11.422	569.755	-5.131
5900	132.875	712.776	589.563	726.958	-11.696	579.777	-5.133
6000	132.893	715.009	591.635	740.247	-11.930	589.804	-5.135

PREVIOUS: September 1967 (1 atm) CURRENT: September 1967 (1 bar)

HEXACHLOROETHANE (C_2Cl_6) IDEAL GAS M_4 = 236.740

$\Delta_fH°(0\ K) = -135.83 \pm 8.4\ kJ\ mol^{-1}$

$S°(298.15\ K) = 397.880\ J\ K^{-1}\ mol^{-1}$ $\Delta_fH°(298.15\ K) = -134.22 \pm 8.4\ kJ\ mol^{-1}$

Vibrational Frequencies and Degeneracies

ν, cm^{-1}	ν, cm^{-1}	ν, cm^{-1}
976(1)	678(1)	223(1)
431(1)	377(1)	778(2)
164(1)	858(1)	267(2)
Internal Rotation 340(2)		139(2)

Point Group: D_{3d}
Bond Distances: Cl-C = 1.74 Å C-C = 1.55 Å σ = 6
Bond Angles: Cl-C-Cl = 110° Cl-C-C = 109°
Product of the Moments of Inertia: $I_AI_BI_C = 9.144710 \times 10^{-112}\ g^3\ cm^6$

Enthalpy of Formation

Smith et al. (1) determined the heat of combustion of C_2Cl_6(cr) to be ΔE(293 K) = -718.7±2 cal/g (uncorrected) by use of the 'Quartz Wool' method. The free chlorine produced was reduced to HCl by employing the quartz spiral moistened iwth a solution of arsenious acid in the combustion bomb. Adopting the reported results and recent auxiliary data (2, 3, 4), we evaluate the values $\Delta_cH°$(298.15 K) = -173.525 and $\Delta_fH°$(298.15 K) = -48.58 kcal mol^{-1} for C_2Cl_6(cr). Kirkbride (5) measured the enthalpy changes as -32.0 and 4.7 kcal mol^{-1} for the following two reactions by solution calorimetry: $C_2Cl_4(\ell)+Cl_2(g) = C_2Cl_6(aq)$ and $C_2Cl_6(cr) = C_2Cl_6(aq)$. Hence we obtain $\Delta_fH°$(298.15 K) = -49.07 kcal mol^{-1} for C_2Cl_6(cr), using $\Delta_fH°(C_2Cl_4, \ell$, 298.15 K) = -12.37 kcal mol^{-1}. This value is in reasonable agreement with the adopted value -48.58 kcal mol^{-1}.

Hexachloroethane (cr) has three modifications, namely triclinic, monoclinic and cubic, with transition temperatures of 318 and 345 K (4). The sublimiation pressures over the different phases have been measured by Ivin (8), Lee (7), and Nitta (26). Based oin these data, values for $\Delta_{sub}H°$(triclinic, 298.15 K) are derived by a modified second and third law mthod to be 16.5, 17.1 and 14.8 kcal mol^{-1}, respectively. The analysis depends on the $C_p°$ and $\Delta_{trs}H°$ data determined by (4). The order of magnitude of the two $\Delta_{trs}H°$ values is confirmed by R. A. McDonald, private communication, Thermal Research Laboratory, The Dow Chemical Company, Midland, Michigan. Pressures reported by Nelson (6) are systematically lower than the above data by about 10 per cent; therefore, his data are not used. Third law analysis is not applicable to the pressure data due to lack of S°(triclinic, 298.15 K). Adopting $\Delta_{sub}H°$(298.15 K) = 16.5 and $\Delta_fH°$(298.15 K) = -48.58 kcal mol^{-1} for C_2Cl_6 (triclinic), we obtain $\Delta_fH°$(298.15 K) = -32.08 kcal mol^{-1} for C_2Cl_6(g).

The equilibrium constant at 776 K for the reaction $C_2Cl_6(g) = C_2Cl_4(g) + Cl_2(g)$ was reported as 0.8±0.1 (atm) by Puyo et al. (9). The corresponding enthalpy change at 298.15 K is evaluated to be 30.8 kcal mol^{-1} by the third law method, yielding $\Delta_fH°(C_2Cl_6$, g, 298.15 K) = -33.77 kcal mol^{-1} which is in fair agreement with the adopted one.

Heat Capacity and Entropy

The molecular structure of C_2Cl_6 has been studied previously by means of infrared and Raman spectroscopy (10), X-ray (11), and electron diffraction (12-18). The molecule was found to be in the staggered position in the vapor and solid states. The bond distances and angles adopted here are those reported by Swick et al. (16).

Incomplete assignments of vibrational frequencies for C_2Cl_6(g) were reported by many investigators (21-24). The adopted frequencies, except the torsional mode ν_4, are obtained from Carney et al. (19). The torsional mode is treated as a hindered internal rotation, and thermodynamic properties are calculated (25) from the barrier height of V_0 = 12.8 kcal mol^{-1} which was evaluated by Karle (20) from electron diffraction data. The value of S°(298.15 K) is calculated based on S°(400 K) = 105.355 cal K^{-1} mol^{-1} which is derived from the free rotor value S°(400 K) = 108.121 cal K^{-1} mol^{-1} by subtraction of the entropy difference between a free rotor and hindered rotor at 400 K. Internal rotation contributions to $C_p°$ at temperatures below 350 K could not be calculated since they are beyond the range of the values above 350 K and the corresponding harmonic oscillator values (ν_4 = 67 cm^{-1}) below 150 K. This calculation presumes that the internal rotation approaches harmonic oscillation as V/RT >>1.

The principal moments of inertia are: $I_A = 79.9217 \times 10^{-39}$, $I_B = 92.8302 \times 10^{-39}$ and $I_C = 123.2582 \times 10^{-39}\ g\ cm^2$.

References

1. L. Smith, L. Bjellerup, S. Krook and H. Westermark, Acta Chem. Scand. 7, 65 (1953).
2. U. S. Nat. Bur. Stand. Tech. Note 270-1, 1965.
3. V. B. Parker, NSRDS-NBS 2, 1965.
4. S. Seki and M. Momotani, BUll. Chem. Soc. Japan 23, 30 (1950).
5. F. W. Kirkbride, J. Appl. Chem. 6, 11 (1956).
6. O. A. Nelson, Ind. Eng. Chem. 22, 971 (1930).
7. P. J. v. d. Lee, Z. Anorg. Allgem. Chem. 223, 213 (1935); 346 - 458 K, 32 data points.
8. K. J. Ivin and F. S. Dainton, Trns. Faraday Soc. 43, 32 (1947); 304 - 448, 20 data points.
9. J. Puyo, D. Balesdent, M. Niclause and M. Dzierzynski, Compt. Rend. 256, 3471 (1963).
10. Landolt-Bornstein, Zahlenwerte and Funktionen aus Physik, Chemie, Astronomie, Geophysik and Technik, Springer-Verlag, Berlin, 1951.
11. Y. Sasada and M. Atoji, J. Chem. Phys. 21 , 145 (1953).
12. S. Mizushima, Y. Morino and S. Yamaguchi, Sci. Pap. Inst. Phys. Chem. Res. (Tokyo) 42, 5 (1944).
13. O. Hassel and H. Viervold, Acta Chem. Scand. 1, 149 (1947).
14. M. Kimura and Y. Morino, J. Chem. Soc. Japan 68, 78 (1947).
15. Y. Morino and M. Iwasaki, J. Chem. Phys. 17, 216 (1949).
16. D. A. Swick, I. L. Karle and J. Karle, J. Chem. Phys. 22, 1242 (1954).
17. Y. Morino and E. Hirota, J. Chem. Phys. 28, 185 (1958).
18. A. Almenningen, B. Andersena nd M. Traetteberg, Acta Chem. Scand. 18, 603 (1964).
19. R. A. Carney, E. A. Piotrowski, A. G. Meister, J. H. Braun and F. F. Cleveland, J. Mol. Spectry. 7, 209 (1961).
20. J. Karle, J. Chem. Phys. 45, 4149 (1966).
21. D. T. Hamilton and F. F. Cleveland, J. Chem. Phys. 12, 249 (1944).
22. H. Gerding and H. G. Haring, Rec. Trav. Chim. 65, 743 (1946).
23. S. Mizushima, Y. MOrino, T. Simanonti and K. Kuratani, J. Chem. Phys. 17, 838 (1949).
24. E. Hirota, Bull. Chem. Soc. Japan 31, 296 (1958).
25. G. N. Lewis and M. Randall, "Thermodynamics," revised by K. S. Pitzer and L. Brewer, McGraw-Hill Book Company, Inc., New York, 1961.
26. I. Nitta and S. Seki, J. Chem. Soc. Japan 62, 581 (1941) 290-343 K, 24 data points.

Hexachloroethane (C_2Cl_6) C_2Cl_6(g)

Enthalpy Reference Temperature = T_r = 298.15 K Standard State Pressure = p° = 0.1 MPa

T/K	$C_p°$	S°	$-[G°-H°(T_r)]/T$	$H°-H°(T_r)$	$\Delta_fH°$	$\Delta_fG°$	Log K_f
	J $K^{-1}mol^{-1}$			kJ mol^{-1}			
0							
100	76.371	280.567	504.581	-22.401	-135.834	-106.999	55.890
200	115.964	346.938	409.925	-12.597	-135.796	-77.973	20.364
298.15	138.934	397.880	397.880	0.	-134.223	-49.894	8.741
300	139.267	398.741	397.883	0.257	-134.186	-49.371	8.596
400	153.457	440.913	403.536	14.951	-131.948	-21.427	2.798
500	162.260	476.174	414.636	30.769	-129.496	5.921	-0.619
600	167.954	506.295	427.464	47.298	-127.020	32.772	-2.853
700	171.799	532.488	440.638	64.295	-124.584	59.211	-4.418
800	174.531	555.615	453.593	81.618	-122.203	85.304	-5.570
900	176.536	576.293	466.098	99.175	-119.875	111.102	-6.448
1000	178.033	594.973	478.066	116.907	-117.600	136.644	-7.138
1100	179.128	611.994	489.479	134.766	-115.378	161.960	-7.691
1200	180.012	627.620	500.349	152.725	-113.208	187.076	-8.143
1300	180.651	642.055	510.701	170.760	-111.088	212.014	-8.519
1400	181.150	655.462	520.568	188.851	-109.022	236.790	-8.835
1500	181.478	667.971	529.983	206.983	-107.011	261.420	-9.103
1600	181.732	679.692	538.977	225.144	-105.059	285.919	-9.334
1700	181.832	690.713	547.582	243.322	-103.171	310.296	-9.534
1800	181.910	701.108	555.825	261.510	-101.350	334.565	-9.709
1900	181.968	710.945	563.733	279.704	-99.596	358.734	-9.862
2000	182.004	720.280	571.329	297.902	-97.908	382.813	-9.998
2100	181.922	729.158	578.635	316.099	-96.295	406.809	-10.119
2200	181.841	737.619	585.671	334.287	-94.762	430.730	-10.227
2300	181.761	745.701	592.454	352.467	-93.311	454.584	-10.324
2400	181.681	753.435	599.002	370.639	-91.945	478.375	-10.412
2500	181.602	760.850	605.328	388.803	-90.666	502.112	-10.491
2600	181.523	767.971	611.448	406.960	-89.475	525.800	-10.563
2700	181.445	774.820	617.372	425.108	-88.375	549.443	-10.630
2800	181.367	781.417	623.114	443.249	-87.367	573.047	-10.690
2900	181.290	787.780	628.683	461.381	-86.452	596.617	-10.746
3000	181.213	793.925	634.089	479.506	-85.630	620.157	-10.798
3100	181.128	799.865	639.342	497.624	-84.899	643.670	-10.846
3200	181.045	805.615	644.448	515.732	-84.258	667.162	-10.890
3300	180.965	811.185	649.417	533.833	-83.704	690.636	-10.932
3400	180.888	816.586	654.255	551.925	-83.232	714.093	-10.971
3500	180.813	821.828	658.968	570.010	-82.837	737.538	-11.007
3600	180.740	826.921	663.563	588.088	-82.513	760.972	-11.041
3700	180.671	831.872	668.045	606.158	-82.253	784.399	-11.074
3800	180.603	836.689	672.420	624.222	-82.049	807.818	-11.104
3900	180.539	841.380	676.693	642.279	-81.894	831.234	-11.133
4000	180.477	845.950	680.867	660.330	-81.781	854.646	-11.161
4100	180.423	850.405	684.948	678.375	-81.699	878.056	-11.187
4200	180.370	854.753	688.940	696.414	-81.642	901.464	-11.211
4300	180.319	858.996	692.845	714.449	-81.600	924.871	-11.235
4400	180.269	863.141	696.669	732.478	-81.566	948.277	-11.257
4500	180.221	867.192	700.413	750.503	-81.533	971.682	-11.279
4600	180.174	871.152	704.082	768.522	-81.492	995.087	-11.300
4700	180.128	875.026	707.678	786.538	-81.437	1018.489	-11.319
4800	180.084	878.818	711.204	804.548	-81.362	1041.891	-11.338
4900	180.041	882.531	714.663	822.554	-81.259	1065.292	-11.356
5000	180.000	886.168	718.057	840.556	-81.123	1088.688	-11.373
5100	179.960	889.732	721.388	858.554	-80.950	1112.084	-11.390
5200	179.922	893.226	724.659	876.548	-80.734	1135.474	-11.406
5300	179.885	896.653	727.872	894.539	-80.471	1158.860	-11.421
5400	179.849	900.015	731.029	912.525	-80.157	1182.241	-11.436
5500	179.815	903.315	734.131	930.509	-79.789	1205.616	-11.450
5600	179.782	906.555	737.182	948.488	-79.365	1228.983	-11.463
5700	179.751	909.736	740.181	966.465	-78.881	1252.342	-11.476
5800	179.721	912.862	743.131	984.439	-78.335	1275.693	-11.489
5900	179.692	915.934	746.034	1002.409	-77.726	1299.033	-11.501
6000	179.665	918.954	748.891	1020.377	-77.053	1322.361	-11.512

PREVIOUS: September 1967 (1 atm) CURRENT: September 1967 (1 bar)

Hexachloroethane (C_2Cl_6) C_2Cl_6(g)

CHROMIUM CARBIDE (Cr_3C_2) CRYSTAL $M_r = 180.010$

$S°(298.15\ K) = 85.434 \pm 0.42\ J\ K^{-1}\ mol^{-1}$

$\Delta_f H°(0\ K) = -86.21 \pm 12.6\ kJ\ mol^{-1}$
$\Delta_f H°(298.15\ K) = -85.35 \pm 12.6\ kJ\ mol^{-1}$

Enthalpy of Formation

Mah (1) determined the enthalpy of combustion of $Cr_3C_2(cr)$ using oxygen bomb calorimetry. She reported $\Delta_c H°(298.15\ K) = -574.5 \pm 0.4\ kcal\ mol^{-1}$ based on the reaction $Cr_3C_2(cr) + 17/4\ O_2(g) = 3/2\ Cr_2O_3(cr) + 2\ CO_2(g)$. Using auxiliary JANAF data (2), we calculate $\Delta_f H°(298.15\ K) = -20.4\ kcal\ mol^{-1}$ for $Cr_3C_2(cr)$. Mah [1] reported that of the six combustion runs four were complete combustions whereas the other two were 99.17% and 98.57% complete. Corrections were made by Mah (1) for incomplete combustion, formation of small amounts of CO(g), and impurities. These corrections amount to less than 1.95% of the $\Delta_c E°_y$. There are many equilibrium studies involving $Cr_3C_2(cr)$ (3-13). Our analysis of a portion of these studies is tabulated. Reference

Source	Reaction	T/K	Method	$\Delta_r H°(298.15\ K)$, kcal mol^{-1} 2nd Law	3rd Law	Drift cal K^{-1} mol^{-1}	$\Delta_f H°(298.15\ K)$ kcal mol^{-1}
(4)	A(14 pts.)**	1244-1382	CO equilibrium pressure	528.39	533.26	3.7±3.4	-21.29
(8)	A(18 pts.)	1316-1366	CO equilibrium pressure	495.46	526.56	23.0±6.4	-24.64
(10)	B(eqn)	880-1100	emf (CaF_2)	-9.34	-12.60	-3.3	-12.60
(11)	B(eqn)	1073-1303	emf ($ThO_2 \cdot Y_2O_3$)	-12.70	-15.73	-2.5	-15.73
(12)	B(eqn)	885-1095	emf (CaF_2)	-15.42	-14.92	0.5	-14.92

Reactions: (A) $3\ Cr_2O_3(cr) + 13\ C(cr) = 2\ Cr_3C_2(cr) + 9\ CO(g)$. (B) $3\ Cr(cr) + 2\ C(cr) = Cr_3C_2(cr)$.

**1374 K point rejected due to statistical test.

The CO(g) equilibrium pressures measured by Heusler (3) in the range 1159-1369 K are higher than those of Kelley et al. (4) by a factor varying from 2.0 to 1.4 with increasing temperature. Three studies measured the chromium pressure in equilibrium with C(cr) and $Cr_3C_2(cr)$ (5, 7, 9). Reported heats of formation for Cr_3C_2 from these studies are $-23.1\ kcal\ mol^{-1}$ (5), $-14.2\ kcal\ mol^{-1}$ (7), and $-22\ kcal\ mol^{-1}$ (9). Alekseev and Shvartsman (6) studied the decarburization of the chromium carbides. Mah (1) analyzed their data and reported $\Delta_f H°(298.15\ K) = -16.6\ kcal\ mol^{-1}$ for $Cr_3C_2(cr)$.

The $\Delta_f H°$ results are quite scattered with the emf results being roughly 5 $kcal\ mol^{-1}$ more positive than the combustion and pressure equilibrium studies. Possible explanations for these discrepancies are discussed by Kulkarni and Worrell (16). We adopt $\Delta_f H°(298.15\ K) = -20.4 \pm 3.0\ kcal\ mol^{-1}$ ($-85.354 \pm 12.6\ kJ\ mol^{-1}$) for $Cr_3C_2(cr)$ based on the combustion study by Mah (1).

Heat Capacity and Entropy

The heat capacity of $Cr_3C_2(cr)$ has been measured by Kelley et al. (4) [53-296 K, 29 pts.) and DeSorbo (14) (12-301 K, 65 pts.). As reported by DeSorbo (14), the results of these two studies differ by no more than 0.1%. Using the Debye function 3D (300/T) as suggested by DeSorbo (14), we calculate and adopt $S°(12\ K) = 0.0297\ cal\ K^{-1}\ mol^{-1}$ and $H°(12;0) = 0.268\ cal\ mol^{-1}$. The enthalpy has been measured by Kelley et al. (4) [481-1576 K, 17 pts.] and Oriani and Murphy (15) [297-1188 K, 11 pts.]. As stated by Oriani and Murphy (15), their results are somewhat lower than those of Kelley et al. (4); the maximum difference is 1.2% at 970 K. For T > 298.15 K, we base our adopted values on the enthalpy data of Kelley et al. (4). The enthalpy data is subjected to a curve fit procedure with the constraint that it join smoothly in the 298 K region with the enthalpy derived from the heat capacity data of Kelley et al. (4). The deviations between the tabulated and observed enthalpy average 0.26% with a maximum of 0.78% at the lowest observed temperature, 480.7 K.

Phase Data

Storms (17) has summarized the phase information for the Cr-Cr_3C_2 system and reported that Cr_3C_2 has an orthorhombic structure. The range of homogeneity is unknown (17).

Decomposition Data

The tentative phase diagram for the Cr-Cr_3C_2 system proposed by Storms (17) indicated that Cr_3C_2 has an incongruent melting point at 2168 K.

References

1. A. D. Mah, U. S. Bur. Mines RI 7217, 1969.
2. JANAF Thermochemical Tables: $Cr_2O_3(cr)$, 12-31-73; $CO_2(g)$, 9-30-65.
3. O. Heusler, Z. Anorg. Chem. 154, 353 (1926).
4. K. K. Kelley, F. S. Boericke, G. E. Moore, E. H. Huffman, and W. M. Bangert, U. S. Bur. Mines Tech. Paper 662, 1949.
5. S. Fiyishiro and N. A. Gokcen, Trans. Met. Soc. AIME 221, 275 (1961).
6. V. I. Alekseev and L. A. Shvartsman, Fiz. Metallov i Metalloved 2, 545 (1961); Fiz. Khim. Osnovy Met. Protsessov. Komis, po Fiz. Khim. Osmovam Proiz. Stali. sb statei, pp. 414-21, 1964.
7. F. Z. Vintaikin, Fiz. Metallov i Metalloved. 16, 144 (1963).
8. M. Gleiser, J. Phys. Chem. 69, 1771 (1965).
9. A. S. Bolgar, V. V. Fesenko, and G. P. Gordinko, Poroshk. Met. 38, 100 (1966).
10. H. Kleykamp, Ber. Bunsenges., Physik. Chem. 73, 354 (1969).
11. H. Mabuchi, N. Sano, and Y. Matsushita, Met. Trans. 2, 1503 (1971).
12. H. Tanaka, Y. Kishida, A. Kawaguchi, and J. Moriyama, see ref. (11).
13. Y. A. Chang and D. Naujock, Met. Trans. 3, 1963 (1972).
14. W. DeSorbo, J. Am. Chem. Soc. 75, 1825 (1953).
15. R. A. Oriani and W. K. Murphy, J. Amer. Chem. Soc. 76, 343 (1954).
16. A. D. Kulkarni and W. L. Worrell, Met. Trans. 3, 2363 (1972).
17. E. K. Storms, "The Refractory Carbides", Academic Press, New York, 1967.

Chromium Carbide (Cr_3C_2) $C_2Cr_3(cr)$

Enthalpy Reference Temperature = T_r = 298.15 K Standard State Pressure = p° = 0.1 MPa

T/K	$C_p°$ (J K^{-1}mol^{-1})	S°	$-[G°-H°(T_r)]/T$	$H°-H°(T_r)$ (kJ mol^{-1})	$\Delta_f H°$	$\Delta_f G°$	Log K_f
0	0.	0.	INFINITE	-15.132	-86.211	-86.211	INFINITE
100	31.322	14.297	155.599	-14.130	-86.263	-86.216	45.035
200	74.718	50.620	93.807	-8.637	-86.175	-86.144	22.499
298.15	99.299	85.434	85.434	0.	-85.354	-86.277	15.115
300	99.663	86.049	85.436	0.184	-85.331	-86.283	15.023
400	115.478	117.037	89.553	10.994	-84.193	-86.719	11.324
500	125.549	143.964	97.805	23.079	-82.549	-87.540	9.145
600	131.486	167.406	107.496	35.946	-80.997	-88.687	7.721
700	135.658	187.997	117.555	49.309	-79.624	-90.079	6.722
800	139.101	206.340	127.528	63.050	-78.422	-91.657	5.985
900	142.260	222.907	137.220	77.118	-77.406	-93.374	5.419
1000	145.327	238.054	146.556	91.498	-76.621	-95.193	4.972
1100	148.377	252.048	155.518	106.183	-76.120	-97.077	4.610
1200	151.448	265.090	164.112	121.174	-75.926	-98.993	4.309
1300	154.553	277.335	172.355	136.474	-76.068	-100.912	4.055
1400	157.695	288.903	180.270	152.086	-76.569	-102.807	3.836
1500	160.871	299.891	187.882	168.014	-77.437	-104.653	3.644
1600	164.084	310.376	195.213	184.261	-78.678	-106.429	3.475
1700	167.322	320.420	202.284	200.831	-80.296	-108.115	3.322
1800	170.586	330.077	209.117	217.727	-82.294	-109.696	3.183
1900	173.870	339.388	215.730	234.949	-84.677	-111.155	3.056
2000	177.176	348.390	222.139	252.501	-87.445	-112.479	2.938
2100	180.489	357.115	228.360	270.385	-90.603	-113.655	2.827
2200	183.761	365.586	234.406	288.597	-152.162	-112.705	2.676
2300	187.075	373.828	240.289	307.139	-150.527	-110.947	2.520
2400	190.389	381.860	246.021	326.012	-148.592	-109.267	2.378
2500	193.702	389.699	251.612	345.217	-146.356	-107.673	2.250

PREVIOUS: CURRENT: December 1973

DIFLUOROETHYNE (C_2F_2) IDEAL GAS M_r = 62.018806

$\Delta_fH°(0\ K)$ = [19.4 ± 21] kJ mol^{-1}
S°(298.15 K) = [244.06 ± 63] J K^{-1} mol^{-1} $\Delta_fH°(298.15\ K)$ = [20.9 ± 21] kJ mol^{-1}

Vibrational Frequencies and Degeneracies

ν, cm^{-1}	ν, cm^{-1}
[1050](1)	[450](2)
[1900](1)	[300](2)
1149 (1)	

Ground State Quantum Weight: 1 σ = 2
Point Group: $[D_{\infty h}]$
Bond Distances: C-F = [1.28] Å C-C = [1.20] Å
Bond Angle: F-C-C = [180]°
Rotational Constant: B_0 = [0.117932] cm^{-1}

Enthalpy of Formation

Farber et al. (1), have obtained equilibrium-weight change data over the C-CF_4 system. Our reanalysis of these data assumes that the equilibrium constants for CF_4, CF_3 and CF_2 are those in the JANAF Tables, and that C_2F_2 was the only other C-F species present at equilibrium. This gives maximum possible concentrations of C_2F_2 and places a lower limit on the heat of formation. The $\Delta_fH°(C_2F_2$, g, 298.15 K) values so obtained range from +10 kcal mol^{-1} from data at 2023 K to -10 kcal mol^{-1} from data at 2373 K; we adopt a weighted mean of +5±5 kcal mol^{-1} (20.920 ± 21 kJ mol^{-1}). This value still predicts complete dimerization of CF at 1 atmosphere pressure below about 2200 K, but only CF would exist under experimental conditions such as the study by Farber.

Farber et al. (1) also investigated the mass spectrum of the vapor over the C-CF_4 system and did not detect any ion ascribable to C_2F_2. This is additional support for the value adopted.

The C-C bond dissociation energy in C_2F_2 is 113 kcal mol^{-1}, using JANAF CF(g) table (Dec. 31, 1967). This value is 115 kcal weaker than the bond dissociation energy of acetylene. In order to explain such a large decrease, a crossing of the repulsive approach of two $^2\Pi$ CF molecules with an attractive approach of two $^4\Sigma$ CF molecules may be postulated. This follows the reasoning used by Simons (2) to explain the bond dissociation energy of C_2F_4.

Heat Capacity and Entropy

The molecular structure is assumed to be the same as that for C_2H_2(g). The C-F and C-C bond distances are estimated from those in HC ≡ CF reported by Tyler and Sheridan (3). The assymmetric stretching frequency, ν_3 = 1149 cm^{-1}, was observed by Heicklen and Knight (4). The other vibrational frequencies were estimated by comparison with those for C_2H_2(g) and C_2HF(g) reported by Herzberg (5), Hunt and Wilson (6), and Brown and Tyler (7).

References

1. M. Farber, M. A. Greenbaum, M. A. Frisch, H. C. Ko, G. Grenier and E. Graper, Rocket Power Inc., 3rd Quarterly Report, AFRPL-TR-67-147, Contract F04611-67-C-0010, May 1967.
2. J. P. Simons, Nature 205, 1308 (1965)
3. J. K. Tyler and J. Sheridan, Trans. Faraday Soc. 59, 2661, 1936.
4. J. Heicklen and V. Knight, J. Phys. Chem. 69, 2484 (1965).
5. G. Herzberg, "Infrared and Raman Spectra", D. Van Nostrand Company, Inc., New York, 1945.
6. G. R. Hunt and M. K. Wilson, J. Chem. Phys. 34, 1301 (1961).
7. J. K. Brown and J. K. Tyler, Proc. Chem. Soc. 1961, 13 (1961).

Difluoroethyne (C_2F_2) C_2F_2(g)

Enthalpy Reference Temperature = T_r = 298.15 K Standard State Pressure = p° = 0.1 MPa

T/K	$C_p°$ (J $K^{-1}mol^{-1}$)	S°	$-[G°-H°(T_r)]/T$	$H°-H°(T_r)$ (kJ mol^{-1})	$\Delta_fH°$	$\Delta_fG°$	Log K_f
0	0.	0.	INFINITE	-12.403	19.444	19.444	INFINITE
100	34.428	194.599	288.398	-9.380	19.441	17.208	-8.989
200	48.351	223.037	249.016	-5.196	20.043	14.799	-3.865
250	53.162	234.364	244.977	-2.653	20.492	13.436	-2.807
298.15	56.935	244.060	244.060	0.	20.920	12.038	-2.109
300	57.068	244.412	244.061	0.105	20.936	11.983	-2.086
350	60.380	253.465	244.768	3.044	21.344	10.458	-1.561
400	63.249	261.719	246.379	6.136	21.702	8.878	-1.159
450	65.757	269.317	248.511	9.363	22.005	7.256	-0.842
500	67.957	276.361	250.948	12.707	22.253	5.604	-0.585
600	71.592	289.087	256.267	19.692	22.609	2.238	-0.195
700	74.412	300.344	261.775	26.998	22.818	-1.176	0.088
800	76.610	310.429	267.237	34.554	22.922	-4.612	0.301
900	78.336	319.556	272.552	42.304	22.949	-8.056	0.468
1000	79.704	327.883	277.674	50.209	22.917	-11.499	0.601
1100	80.799	335.533	282.591	58.236	22.837	-14.938	0.709
1200	81.685	342.603	287.302	66.362	22.714	-18.366	0.799
1300	82.410	349.171	291.811	74.568	22.553	-21.784	0.875
1400	83.007	355.301	296.130	82.839	22.355	-25.187	0.940
1500	83.506	361.045	300.268	91.166	22.122	-28.575	0.995
1600	83.924	366.448	304.237	99.538	21.855	-31.946	1.043
1700	84.279	371.547	308.048	107.949	21.556	-35.300	1.085
1800	84.582	376.373	311.711	116.392	21.227	-38.635	1.121
1900	84.843	380.953	315.236	124.864	20.868	-41.951	1.153
2000	85.069	385.311	318.631	133.360	20.484	-45.247	1.182
2100	85.265	389.466	321.906	141.876	20.077	-48.524	1.207
2200	85.437	393.437	325.068	150.412	19.649	-51.781	1.229
2300	85.588	397.238	328.124	158.963	19.205	-55.018	1.249
2400	85.722	400.884	331.080	167.529	18.748	-58.235	1.267
2500	85.841	404.386	333.943	176.107	18.282	-61.433	1.284
2600	85.948	407.754	336.717	184.697	17.809	-64.612	1.298
2700	86.043	411.000	339.409	193.296	17.334	-67.773	1.311
2800	86.128	414.131	342.022	201.905	16.859	-70.917	1.323
2900	86.205	417.154	344.561	210.522	16.387	-74.043	1.334
3000	86.275	420.078	347.030	219.146	15.920	-77.154	1.343
3100	86.339	422.908	349.432	227.776	15.462	-80.249	1.352
3200	86.397	425.650	351.771	236.413	15.012	-83.329	1.360
3300	86.449	428.310	354.050	245.056	14.574	-86.395	1.368
3400	86.498	430.891	356.273	253.703	14.148	-89.448	1.374
3500	86.542	433.399	358.440	262.355	13.735	-92.489	1.380
3600	86.583	435.838	360.557	271.011	13.336	-95.518	1.386
3700	86.620	438.210	362.624	279.671	12.951	-98.537	1.391
3800	86.655	440.521	364.643	288.335	12.581	-101.545	1.396
3900	86.687	442.772	366.618	297.002	12.226	-104.544	1.400
4000	86.717	444.967	368.549	305.673	11.885	-107.533	1.404
4100	86.745	447.109	370.439	314.346	11.558	-110.515	1.408
4200	86.771	449.200	372.290	323.022	11.246	-113.488	1.411
4300	86.795	451.242	374.102	331.700	10.947	-116.455	1.415
4400	86.817	453.237	375.878	340.381	10.661	-119.414	1.418
4500	86.838	455.189	377.619	349.063	10.387	-122.367	1.420
4600	86.858	457.097	379.326	357.748	10.125	-125.315	1.423
4700	86.876	458.966	381.001	366.435	9.875	-128.256	1.425
4800	86.894	460.795	382.644	375.123	9.634	-131.192	1.428
4900	86.910	462.587	384.257	383.814	9.403	-134.124	1.430
5000	86.925	464.343	385.842	392.505	9.181	-137.051	1.432
5100	86.940	466.064	387.398	401.199	8.967	-139.973	1.434
5200	86.954	467.752	388.927	409.893	8.760	-142.892	1.435
5300	86.966	469.409	390.430	418.589	8.559	-145.806	1.437
5400	86.979	471.035	391.907	427.287	8.364	-148.717	1.439
5500	86.990	472.631	393.361	435.985	8.174	-151.624	1.440
5600	87.001	474.198	394.790	444.685	7.989	-154.528	1.441
5700	87.011	475.738	396.197	453.385	7.806	-157.429	1.443
5800	87.021	477.252	397.581	462.087	7.626	-160.326	1.444
5900	87.031	478.739	398.944	470.789	7.448	-163.220	1.445
6000	87.039	480.202	400.287	479.493	7.271	-166.111	1.446

PREVIOUS: December 1967 (1 atm) CURRENT: December 1967 (1 bar)

Difluoroethyne (C_2F_2) C_2F_2(g)

TRIFLUOROACETONITRILE (CF_3CN) IDEAL GAS M_r = 95.023909

$\Delta_f H°(0\ K) = -491.61 \pm 2.9\ kJ\ mol^{-1}$

$S°(298.15\ K) = 298.533 \pm 0.4\ J\ K^{-1}\ mol^{-1}$ $\Delta_f H°(298.15\ K) = -495.39 \pm 2.9\ kJ\ mol^{-1}$

Vibrational Frequencies and Degeneracies

ν, cm^{-1}	ν, cm^{-1}
2271(1)	1215(2)
1228(1)	625(2)
801(1)	464(2)
521(1)	192(2)

Ground State Quantum Weight: [1] $\sigma = 3$

Point Group: C_{3v}

Bond Distances: C–F = 1.335 Å C–C = 1.461 Å CN = 1.153 Å

Bond Angles: F–C–F = 107.5° F–C–C = 111.38°

Product of the Moments of Inertia: $I_A I_B I_C = 1.186732 \times 10^{-113}\ g^3\ cm^6$

Enthalpy of Formation

The selected value is obtained from least squares, simultaneous adjustment of the enthalpies of formation of CF_3, C_2F_4, and CF_3X(X = H, Cl, Br, I, CN). Details of the input data and adjustment are given in (1). Input included a calorimetric link to NF_3(g) and CF_4(g) plus an equilibrium link to C_2F_6(g) and C_2N_2(g). The calorimetric study (2) gave $\Delta_r H° = -824.4\pm0.6\ kcal\ mol^{-1}$ at 298.15 K for the reaction $3CF_3CN(g) + 5NF_3(g) + 6CF_4(g) + 4N_2(g)$. Equilibrium data (3) in the range 865–925 K give $\Delta_r H°$ (3rd law) = 10.59 and $\Delta_r H°$ (2nd law) = 10.59±0.20 kcal mol^{-1} at 298.15 K for the reaction $C_2F_6(g) + C_2N_2(g) + 2CF_3CN(g)$. The entropy difference, $\Delta_r S°$ (2nd law) − $\Delta_r S°$ (3rd law), is 0.0±1.5 cal $K^{-1}\ mol^{-1}$. Deviations between the selected $\Delta_f H°$ and the two observations are 1.37 and 0.15 kcal mol^{-1}, respectively, both considerably less than the overall uncertainties including those in auxiliary values of $\Delta_f H°$. Subsequent to the adjustment, Curnutt and Sinke (4) reported $\Delta_r H° = -197.6\pm0.9\ kcal\ mol^{-1}$ from a calorimetric study of $CF_3CN(g) + 1.25\ O_2(g) + 1.5\ H_2O(\ell) + 2\ CO_2(g) + 0.5\ N_2(g) + 3HF(aq,\ 120\ H_2O)$. Their result of $\Delta_f H° = -118.4\pm1.2\ kcal\ mol^{-1}$ is confirmation of the selected value.

Heat Capacity and Entropy

The molecular structure has been derived from electron-diffraction data by Danford and Livingston [5] and from microwave data by Sheridan and Gordy (6) and Thomas et al. (7). These studies are in substantial agreement except for a difference of about 0.04 Å in the C–C bond distance. We adopt the results of Thomas et al. (7) since they are based on data for several isotopic species. Principal moments of inertia are: $I_A = 14.6267 \times 10^{-39}$ and $I_B = I_C = 28.4842 \times 10^{-39}\ g\ cm^2$. Vibrational frequencies are taken from the assignments of Edgell and Potter (8). All values except ν_8 are from infrared spectra of the vapor phase. Raman spectra of the liquid provide ν_8 and confirm the infrared data. Similar assignments have been given by Janz and Wait (9) based on spectra of Wait and Janz (10).

Pace and Bobka (11) used low temperature calorimetric data to derive the entropy of the ideal gas at the normal boiling point of 205.47 K. The experimental value of 65.01±0.2 cal $K^{-1}\ mol^{-1}$ is in good agreement with 65.03 cal $K^{-1}\ mol^{-1}$ obtained from this table. The largest uncertainty in $C_p°$ and S° probably arises from neglect of anharmonicity and rotational distortion. Uncertainty in the entropy is estimated to vary from 0.1 to 0.5 cal $K^{-1}mol^{-1}$ over the range 300 to 1000 K. Discussions of the neglect of anharmonicity are given on the tables for $CClF_3$(g) and CF_4(g).

References

1. A. N. Syverud, Quarterly Technical Report AFRPL-TR-69-187, Aug., 1969.
2. L. C. Walker, The Dow Chemical Co., personal communication, June, 1969.
3. D. J. Perettie, The Dow Chemical Co., personal communication, June, 1969.
4. J. L. Curnutt and G. C. Sinke, presented at the 24th Annual Calorimetry Conference, Portsmouth, New Hampshire, October, 1969.
5. M. D. Danford and R. L. Livingston, J. Amer. Chem. Soc. 77, 2944 (1955).
6. J. Sheridan and W. Gordy, J. Chem. Phys. 20, 591 (1952).
7. L. F. Thomas, J. S. Heeks and J. Sheridan, Z. Elektrochem. 61, 935 (1957).
8. W. F. Edgell and R. M. Potter, J. Chem. Phys. 24, 80 (1956).
9. G. J. Janz and S. C. Wait, J. Chem. Phys. 26, 1766 (1957).
10. S. C. Wait and G. J. Janz, J. Chem. Phys. 26, 1554 (1957).
11. E. L. Pace and R. J. Bobka, J. Chem. Phys. 35, 454 (1961).

Trifluoroacetonitrile (CF_3CN) $C_2F_3N_1(g)$

Enthalpy Reference Temperature = T_r = 298.15 K   Standard State Pressure = p° = 0.1 MPa

T/K	$C_p°$ (J K⁻¹mol⁻¹)	S°	−[G°−H°(T_r)]/T	H°−H°(T_r) (kJ mol⁻¹)	$\Delta_f H°$	$\Delta_f G°$	Log K_f
0	0.	0.	INFINITE	-15.897	-491.608	-491.608	INFINITE
100	43.761	234.503	356.929	-12.243	-493.884	-483.598	252.606
200	62.441	270.513	305.210	-6.939	-495.083	-472.759	123.472
250	71.031	285.389	299.782	-3.598	-495.314	-467.147	97.605
298.15	78.255	298.533	298.533	0.	-495.386	-461.713	80.890
300	78.513	299.017	298.534	0.145	-495.386	-461.505	80.355
350	84.951	311.616	299.514	4.236	-495.349	-455.859	68.033
400	90.470	323.330	301.768	8.625	-495.239	-450.225	58.793
450	95.200	334.266	304.778	13.270	-495.084	-444.607	51.609
500	99.263	344.512	308.244	18.134	-494.902	-439.008	45.863
600	105.795	363.216	315.878	28.402	-494.492	-427.867	37.249
700	110.730	379.912	323.855	39.240	-494.049	-416.797	31.102
800	114.522	394.956	331.818	50.510	-493.589	-405.793	26.496
900	117.482	408.623	339.604	62.116	-493.122	-394.846	22.916
1000	119.823	421.126	347.140	73.986	-492.653	-383.952	20.056
1100	121.700	432.638	354.396	86.066	-492.190	-373.104	17.717
1200	123.221	443.295	361.366	98.314	-491.737	-362.298	15.770
1300	124.468	453.208	368.054	110.701	-491.299	-351.529	14.125
1400	125.500	462.471	374.471	123.201	-490.881	-340.794	12.715
1500	126.362	471.160	380.630	135.795	-490.486	-330.087	11.495
1600	127.089	479.339	386.547	148.469	-490.115	-319.406	10.428
1700	127.706	487.063	392.234	161.209	-489.769	-308.748	9.487
1800	128.234	494.378	397.707	174.007	-489.449	-298.109	8.651
1900	128.690	501.324	402.980	186.854	-489.153	-287.487	7.904
2000	129.084	507.935	408.064	199.743	-488.879	-276.880	7.231
2100	129.428	514.241	412.971	212.669	-488.623	-266.287	6.624
2200	129.730	520.270	417.712	225.627	-488.382	-255.705	6.071
2300	129.996	526.042	422.297	238.614	-488.152	-245.134	5.567
2400	130.232	531.580	426.736	251.625	-487.927	-234.572	5.105
2500	130.441	536.901	431.037	264.659	-487.704	-224.021	4.681
2600	130.629	542.020	435.208	277.713	-487.477	-213.478	4.289
2700	130.797	546.954	439.256	290.784	-487.241	-202.944	3.926
2800	130.948	551.713	443.188	303.872	-486.993	-192.419	3.590
2900	131.084	556.311	447.010	316.973	-486.729	-181.903	3.276
3000	131.208	560.757	450.727	330.088	-486.445	-171.396	2.984
3100	131.320	565.061	454.347	343.214	-486.139	-160.900	2.711
3200	131.422	569.232	457.872	356.352	-485.807	-150.413	2.455
3300	131.516	573.277	461.308	369.499	-485.448	-139.938	2.215
3400	131.601	577.205	464.659	382.654	-485.060	-129.474	1.989
3500	131.680	581.021	467.930	395.819	-484.642	-119.021	1.776
3600	131.752	584.731	471.123	408.990	-484.192	-108.581	1.575
3700	131.819	588.342	474.242	422.169	-483.710	-98.154	1.386
3800	131.881	591.858	477.291	435.354	-483.197	-87.741	1.206
3900	131.938	595.285	480.273	448.545	-482.651	-77.340	1.036
4000	131.991	598.626	483.190	461.741	-482.073	-66.955	0.874
4100	132.040	601.885	486.046	474.943	-481.464	-56.585	0.721
4200	132.086	605.068	488.842	488.149	-480.824	-46.229	0.575
4300	132.128	608.176	491.581	501.360	-480.153	-35.890	0.436
4400	132.168	611.214	494.266	514.575	-479.454	-25.566	0.304
4500	132.206	614.185	496.898	527.793	-478.726	-15.259	0.177
4600	132.241	617.091	499.479	541.016	-477.970	-4.968	0.056
4700	132.274	619.936	502.012	554.241	-477.189	5.307	-0.059
4800	132.304	622.721	504.498	567.470	-476.382	15.564	-0.169
4900	132.333	625.449	506.938	580.702	-475.552	25.804	-0.275
5000	132.361	628.123	509.335	593.937	-474.699	36.027	-0.376
5100	132.386	630.744	511.690	607.174	-473.825	46.233	-0.474
5200	132.411	633.315	514.005	620.414	-472.930	56.422	-0.567
5300	132.434	635.837	516.280	633.656	-472.017	66.593	-0.656
5400	132.455	638.313	518.517	646.901	-471.086	76.747	-0.742
5500	132.476	640.744	520.717	660.147	-470.138	86.883	-0.825
5600	132.495	643.131	522.882	673.396	-469.175	97.002	-0.905
5700	132.514	645.476	525.012	686.646	-468.198	107.103	-0.981
5800	132.531	647.781	527.109	699.899	-467.208	117.187	-1.055
5900	132.548	650.047	529.173	713.153	-466.207	127.256	-1.127
6000	132.564	652.275	531.207	726.408	-465.195	137.305	-1.195

PREVIOUS: December 1969 (1 atm)   CURRENT: December 1969 (1 bar)

Trifluoroacetonitrile (CF_3CN) $C_2F_3N_1(g)$

TETRAFLUOROETHENE (C_2F_4) IDEAL GAS M_r = 100.015612

$S°(298.15\ K) = 300.015 \pm 0.4\ J\ K^{-1}\ mol^{-1}$

$\Delta_fH°(0\ K) = -655.14 \pm 2.9\ kJ\ mol^{-1}$
$\Delta_fH°(298.15\ K) = -658.56 \pm 2.9\ kJ\ mol^{-1}$

Vibrational Frequencies and Degeneracies

ν, cm^{-1}	ν, cm^{-1}	ν, cm^{-1}
1872(1)	1340(1)	1337(1)
778(1)	551(1)	218(1)
394(1)	406(1)	1186(1)
190(1)	508(1)	558(1)

Ground State Quantum Weight: 1 $\sigma = 4$
Point Group: D_{2h}
Bond Distances: C-F = 1.313 Å C-C = 1.313 Å
Bond Angle: F-C-F = 114°
Product of the Moments of Inertia: $I_AI_BI_C = 1.588028 \times 10^{-113}\ g^3\ cm^6$

Enthalpy of Formation

The selected value is obtained from least squares, simultaneous adjustment of the heats of formation of CF_3, CF_3X (X = H, Cl, Br, I, CF_3) and C_2F_4. Details of the input data and the adjustment are given in (1). In this case, the selected $\Delta_fH°$ corresponds to the weighted average of three calorimetric studies (2, 3) which may be reduced to the following:

Source	Reaction	$\Delta_rH°$(298.15 K) kcal mo^{-1}	$\Delta_fH°$(298.15 K) kcal mol^{-1}
(2)	$C_2F_4(g) + CF_4(g) + C(graphite)$	-65.4±0.5	-157.6±0.6
(2)	$C_2F_4(g) + 2H_2(g) \rightarrow 4HF(50\ H_2O) + 2C(graphite)$	-150.9±1.2	-156.2±1.3
(3)	$C_2F_4(g) + 4Na(cr) \rightarrow 4NaF(cr) + 2C(graphite)$	-392.9±1.3	-157.2±1.4

Amorphous carbon was formed in all three calorimetric reactions and separate combustions were made to refer the observed product to the standard state of graphite. Other calorimetric studies (4, 5, 6) are omitted because of apparent uncertainties in the reductions to the standard-state reactions.

Heat Capacity and Entropy

The molecular structure has been derived from electron-diffraction data by Karle (7), Broun (8) and Young (9). The resulting structures differ in minor respects, particularly in the exact location of the two carbons. Values from Karle (7) are adopted since they are intermediate between the others. Values from Young (9) and Broun (8) correspond to changes of ±2.2% in the product of the moments of inertia. Principal moments of inertia for the adopted structure are: $I_A = 15.3019 \times 10^{-39}$, $I_B = 25.4600 \times 10^{-39}$ and $I_C = 40.7619 \times 10^{-39}\ g\ cm^2$. Vibrational fundamentals are those selected by Shimanouchi (10) from the data of Mann (11), Monfils (12) and Nielsen (13).

Furukawa (14) reported low-temperature calorimetric data from which the ideal gas entropy may be derived as 64.54±0.09 cal $K^{-1}\ mol^{-1}$ at 197.53 K. The calculated entropy from this table is also 64.54 cal $K^{-1}\ mol^{-1}$. The range in values of $I_AI_BI_C$ corresponds to an entropy change of ±0.022 cal $K^{-1}\ mol^{-1}$.

References

1. A. N. Syverud, Quarterly Tech. Report T-0009-2Q-69 under AFRPL Contract Nr. F04611-67-C-0009, The Dow Chemical Co., July, 1969.
2. C. Neugebauer and J. L. Margrave, J. Phys. Chem. 60, 1318 (1956).
3. V. P. Kolesov, I. D. Zenkov and S. M. Skuratov, Russ. J. Phys. Chem. 36, 45 (1962).
4. H. C. Duus, Ind. Eng. Chem. 47, 1445 (1955).
5. F. W. Kirkbride and F. G. Davidson, Nature 174, 79 (1954).
6. H. von Wartenberg and J. Schiefer, Z. Anorg. Allg. Chem. 278, 326 (1955).
7. I. L. Karle and J. Karle, J. Chem. Phys. 18, 963 (1950).
8. T. T. Broun and R. L. Livingston, J. Amer. Chem. Soc. 74, 6084 (1952).
9. J. A. Young, Dissertation Abstracts 16, 460 (1956).
10. T. Shimanouchi, NSRDS-NBS 6, 25 (1967).
11. D. E. Mann, N. Acquista and E. K. Plyler, J. Res. Nat. Bur. Stand. 52, 67 (1954).
12. A. Monfils and J. Duchesne, J. Chem. Phys. 18, 1415 (1950).
13. J. R. Nielsen, H. H. Claassen and D. C. Smith, J. Chem. Phys. 18, 812 (1950).
14. G. T. Furukawa, R. McCoskey and M. L. Reilly, J. Res. Nat. Bur. Stand. 51, 69 (1953).

Tetrafluoroethene (C_2F_4) C_2F_4(g)

Enthalpy Reference Temperature = T_r = 298.15 K Standard State Pressure = p° = 0.1 MPa

T/K	$C_p°$ (J K^{-1}mol^{-1})	S° (J K^{-1}mol^{-1})	$-[G°-H°(T_r)]/T$ (J K^{-1}mol^{-1})	$H°-H°(T_r)$ (kJ mol^{-1})	$\Delta_fH°$ (kJ mol^{-1})	$\Delta_fG°$ (kJ mol^{-1})	Log K_f
0	0.	0.	INFINITE	-16.330	-655.140	-655.140	INFINITE
100	44.191	233.706	360.595	-12.689	-657.432	-646.538	337.717
200	65.202	270.959	306.921	-7.192	-658.446	-635.144	165.883
250	73.646	286.445	301.303	-3.715	-658.565	-629.302	131.485
298.15	80.458	300.015	300.015	0.	-658.562	-623.665	109.263
300	80.698	300.514	300.017	0.149	-658.560	-623.448	108.552
350	86.707	313.416	301.022	4.338	-658.491	-617.601	92.172
400	91.903	325.342	303.326	8.806	-658.387	-611.767	79.889
450	96.426	336.433	306.396	13.517	-658.267	-605.946	70.336
500	100.373	346.802	309.923	18.439	-658.138	-600.140	62.696
600	106.824	365.699	317.676	28.814	-657.867	-588.565	51.239
700	111.747	382.553	325.762	39.753	-657.577	-577.037	43.059
800	115.524	397.732	333.826	51.125	-657.266	-565.553	36.927
900	118.450	411.515	341.704	62.830	-656.939	-554.108	32.160
1000	120.743	424.118	349.324	74.794	-656.601	-542.700	28.348
1100	122.561	435.715	356.658	86.963	-656.259	-531.327	25.231
1200	124.020	446.444	363.699	99.294	-655.922	-519.984	22.634
1300	125.206	456.419	370.452	111.758	-655.595	-508.669	20.439
1400	126.179	465.735	376.929	124.328	-655.285	-497.380	18.557
1500	126.986	474.468	383.143	136.988	-654.995	-486.111	16.928
1600	127.662	482.686	389.110	149.721	-654.729	-474.860	15.503
1700	128.233	490.443	394.845	162.517	-654.489	-463.626	14.245
1800	128.720	497.787	400.362	175.365	-654.274	-452.405	13.128
1900	129.137	504.758	405.675	188.258	-654.082	-441.196	12.129
2000	129.498	511.391	410.796	201.191	-653.911	-429.995	11.230
2100	129.811	517.717	415.738	214.156	-653.757	-418.804	10.417
2200	130.085	523.762	420.512	227.152	-653.614	-407.619	9.678
2300	130.326	529.550	425.128	240.172	-653.477	-396.440	9.003
2400	130.539	535.102	429.595	253.216	-653.339	-385.268	8.385
2500	130.728	540.434	433.923	266.279	-653.195	-374.101	7.816
2600	130.896	545.565	438.119	279.361	-653.038	-362.940	7.292
2700	131.047	550.508	442.190	292.458	-652.862	-351.786	6.806
2800	131.182	555.276	446.144	305.570	-652.662	-340.639	6.355
2900	131.305	559.882	449.987	318.694	-652.432	-329.499	5.935
3000	131.415	564.335	453.725	331.830	-652.167	-318.368	5.543
3100	131.515	568.646	457.363	344.977	-651.865	-307.246	5.177
3200	131.607	572.823	460.906	358.133	-651.521	-296.135	4.834
3300	131.690	576.874	464.359	371.298	-651.132	-285.035	4.512
3400	131.766	580.806	467.727	384.471	-650.697	-273.947	4.209
3500	131.836	584.627	471.012	397.651	-650.214	-262.873	3.923
3600	131.901	588.342	474.220	410.838	-649.681	-251.813	3.654
3700	131.960	591.957	477.354	424.031	-649.097	-240.770	3.399
3800	132.015	595.476	480.416	437.230	-648.463	-229.743	3.158
3900	132.066	598.906	483.410	450.434	-647.778	-218.732	2.930
4000	132.113	602.250	486.340	463.643	-647.042	-207.740	2.713
4100	132.156	605.513	489.207	476.856	-646.257	-196.768	2.507
4200	132.197	608.698	492.014	490.074	-645.422	-185.814	2.311
4300	132.235	611.809	494.764	503.295	-644.539	-174.882	2.124
4400	132.270	614.850	497.459	516.521	-643.609	-163.970	1.947
4500	132.303	617.823	500.101	529.749	-642.633	-153.080	1.777
4600	132.334	620.731	502.692	542.981	-641.613	-142.212	1.615
4700	132.363	623.577	505.233	556.216	-640.550	-131.367	1.460
4800	132.391	626.364	507.728	569.454	-639.446	-120.545	1.312
4900	132.416	629.094	510.177	582.694	-638.301	-109.746	1.170
5000	132.440	631.770	512.582	595.937	-637.118	-98.971	1.034
5100	132.463	634.393	514.945	609.182	-635.899	-88.220	0.904
5200	132.484	636.965	517.267	622.430	-634.644	-77.493	0.778
5300	132.505	639.489	519.549	635.679	-633.356	-66.791	0.658
5400	132.524	641.966	521.793	648.930	-632.036	-56.114	0.543
5500	132.542	644.398	524.001	662.184	-630.686	-45.461	0.432
5600	132.559	646.786	526.172	675.439	-629.307	-34.833	0.325
5700	132.575	649.132	528.309	688.696	-627.900	-24.231	0.222
5800	132.591	651.438	530.412	701.954	-626.468	-13.652	0.123
5900	132.606	653.705	532.482	715.214	-625.011	-3.098	0.027
6000	132.620	655.934	534.521	728.475	-623.532	7.430	-0.065

PREVIOUS: June 1969 (1 atm) CURRENT: June 1969 (1 bar)

Tetrafluoroethene (C_2F_4) C_2F_4(g)

HEXAFLUOROETHANE (C_2F_6) IDEAL GAS M_r = 138.012418

$S°(298.15\ K) = 332.185 \pm 0.8\ J\ K^{-1}\ mol^{-1}$

$\Delta_f H°(0\ K) = -1335.6 \pm 5.0\ kJ\ mol^{-1}$
$\Delta_f H°(298.15\ K) = -1343.9 \pm 5.0\ kJ\ mol^{-1}$

Vibrational Frequencies and Degeneracies

ν, cm^{-1}	ν, cm^{-1}	ν, cm^{-1}
1228(1)	1117(1)	376(2)
807(1)	714(1)	1251(2)
348(1)	1243(2)	523(2)
Torsion	619(2)	220(2)

Ground State Quantum Weight: 1 $\sigma = 6$
Point Group: D_{3d}
Bond Distance: C–F = 1.32 Å C–C = 1.56 Å
Bond Angle: C–C–F = 109.5 ± 1.5°
Product of the Moments of Inertia: $I_A I_B I_C = 6.005449 \times 10^{-113}\ g^3\ cm^6$
Reduced Moment of Inertia: $I_{red} = 7.32 \times 10^{-39}\ g\ cm^2$
Barrier to Internal Rotation: $V = 3.96\ kcal\ mol^{-1}$

Enthalpy of Formation

The selected value is obtained from least squares, simultaneous adjustment of the heats of formation of CF_3, CF_3X (X = H, Cl, Br, I, CF_3) and C_2F_4. Details of the input data and the adjustment are given in (1). Data pertinent to C_2F_6 are summarized below. Calorimetric data (7) for the reaction of C_2F_6 with potassium were omitted from the adjustment due to insufficient characterization of the products.

Source	Method	Reaction	$\Delta_r H°(298.15\ K)$ $kcal^{-1} mol^{-1}$	$\Delta_f H°(298.15\ K)$ $kcal^{-1} mol^{-1}$
(2)	Calorimetric	$3\ C_2F_6(g) + 2\ NF_3(g) \rightarrow 6\ CF_4(g) + N_2(g)$	-311.6 ± 3.0	-321.3 ± 1.2
(3)	Calorimetric	$3\ CF_3CN(g) + 5\ NF_3(g) \rightarrow 6\ CF_4(g) + 4\ N_2(g)$	-824.4 ± 0.6	---
(4)	Equilibrium	$2\ CF_3CN(g) \rightarrow C_2F_6(g) + C_2N_2(g)$	-10.59± 0.5	-322.2 ± 1.8
(5)	Equilibrium	$C_2F_6(g) + Br_2(g) \rightarrow 2\ CBrF_3(g)$	3.30± 0.7	-320.9 ± 1.6
(6, 5)	Kinetic	$C_2F_6(g) \rightarrow 2\ CF_3(g)$	97.6 ± 6	-322.4 ± 6

Heat Capacity and Entropy

The adopted molecular structure was derived from electron-diffraction data by Swick and Karle (8). The vibrational frequencies are from the assignment of Carney et al. (9), who reviewed the extensive spectral data. Shimanouchi (10) has selected the same frequencies. Low-temperature calorimetric data were used by Pace and Aston (11) to derive a value for the entropy of the ideal gas of 69.88 ± 0.20 cal K^{-1} mol^{-1} at 194.87 K. This value is used to fix the barrier to internal rotation at 3960 cal/mol. Thermodynamic functions are calculated from these parameters, using hindered internal rotation in place of a torsional vibration. Principal moments of inertia are: $I_A = 29.3165 \times 10^{-39}$, and $I_B = I_C = 45.2602 \times 10^{-39}\ g\ cm^2$.

An uncertainty of ±0.2 cal K^{-1} mol^{-1} in the 3rd law entropy corresponds to a variation in the barrier to internal rotation of about ±700 cal/mol. This range easily includes the barrier of 4300 cal/mol derived by Karle (12) from analysis of the electron-diffraction data (8). Uncertainty in the entropy of this table should not exceed 0.5 cal K^{-1} mol^{-1} at 1000 K.

References

1. A. N. Syverud, Quarterly Tech. Report T-0009-2Q-69 under AFRPL Contract Nr. F04611-67-C-0009, Dow Chemical Co., July, 1969.
2. G. C. Sinke, J. Phys. Chem. 70, 1326 (1966).
3. L. C. Walker, Dow Chemical Co., Midland, Mich., personal communication, June, 1969.
4. D. J. Perettie, Dow Chemical Co., Midland, Mich., personal communication, June, 1969.
5. J. W. Coomber and E. Whittle, Trans. Faraday Soc. 63, 1394 (1967).
6. E. Tschuikow-Roux, J. Chem. Phys. 43, 2251 (1965).
7. F. W. Kirkbride and F. G. Davidson, Nature 174, 79 (1954).
8. D. A. Swick and I. L. Karle, J. Chem. Phys. 23, 1499 (1955).
9. R. A. Carney, E. A. Piotrowski, A. G. Meister, J. H. Braun and F. F. Cleveland, J. Mol. Spectrosc. 7, 209 (1961).
10. T. Shimanouchi, NSRDS-NBS 6, 29, 1967.
11. E. L. Pace and J. G. Aston, J. Amer. Chem. Soc. 70, 566 (1948).
12. J. Karle, J. Chem. Phys. 45, 4149 (1966).

Hexafluoroethane (C_2F_6) $C_2F_6(g)$

Enthalpy Reference Temperature = T_r = 298.15 K Standard State Pressure = p° = 0.1 MPa

T/K	$C_p°$ (J $K^{-1}mol^{-1}$)	S° (J $K^{-1}mol^{-1}$)	$-[G°-H°(T_r)]/T$ (J $K^{-1}mol^{-1}$)	$H°-H°(T_r)$ (kJ mol^{-1})	$\Delta_f H°$ (kJ mol^{-1})	$\Delta_f G°$ (kJ mol^{-1})	Log K_f
0	0.	0.	INFINITE	-20.255	-1335.579	-1335.579	INFINITE
100	53.137	248.355	409.522	-16.117	-1340.281	-1313.815	686.267
200	82.437	294.717	341.306	-9.318	-1342.922	-1286.241	335.932
298.15	106.407	332.185	332.185	0.	-1343.901	-1258.134	220.420
300	106.809	332.845	332.187	0.197	-1343.909	-1257.602	218.968
400	125.545	366.266	336.617	11.859	-1343.950	-1228.806	160.466
500	139.139	395.825	345.562	25.132	-1343.428	-1200.072	125.371
600	148.708	422.089	356.170	39.551	-1342.585	-1171.476	101.986
700	155.465	445.549	367.293	54.779	-1341.558	-1143.037	85.294
800	160.310	466.640	378.416	70.579	-1340.428	-1114.753	72.786
900	163.804	485.732	389.296	86.792	-1339.247	-1086.614	63.065
1000	166.486	503.134	399.823	103.311	-1338.044	-1058.608	55.296
1100	168.458	519.097	409.951	120.061	-1336.841	-1030.723	48.945
1200	170.088	533.827	419.668	136.991	-1335.652	-1002.946	43.657
1300	171.296	547.490	428.981	154.062	-1334.487	-975.268	39.187
1400	172.318	560.223	437.906	171.244	-1333.356	-947.679	35.358
1500	173.114	572.140	446.462	188.517	-1332.263	-920.169	32.043
1600	173.808	583.335	454.670	205.864	-1331.213	-892.731	29.145
1700	174.312	593.887	462.552	223.270	-1330.207	-865.356	26.589
1800	174.770	603.864	470.128	240.725	-1329.245	-838.040	24.319
1900	175.182	613.325	477.418	258.222	-1328.321	-810.777	22.290
2000	175.548	622.320	484.440	275.759	-1327.426	-783.560	20.465
2100	175.816	630.891	491.211	293.328	-1326.556	-756.389	18.814
2200	176.065	639.076	497.748	310.922	-1325.704	-729.258	17.315
2300	176.293	646.908	504.064	328.540	-1324.858	-702.166	15.947
2400	176.502	654.415	510.173	346.180	-1324.007	-675.110	14.693
2500	176.690	661.624	516.088	363.840	-1323.141	-648.091	13.541
2600	176.839	668.557	521.820	381.516	-1322.251	-621.106	12.478
2700	176.978	675.234	527.379	399.207	-1321.328	-594.156	11.495
2800	177.107	681.672	532.775	416.912	-1320.362	-567.242	10.582
2900	177.226	687.889	538.017	434.628	-1319.346	-540.362	9.733
3000	177.335	693.899	543.114	452.356	-1318.271	-513.519	8.941
3100	177.426	699.716	548.072	470.095	-1317.131	-486.713	8.201
3200	177.511	705.350	552.900	487.841	-1315.923	-459.944	7.508
3300	177.590	710.814	557.602	505.597	-1314.640	-433.214	6.857
3400	177.662	716.116	562.187	523.359	-1313.279	-406.525	6.245
3500	177.728	721.267	566.659	541.129	-1311.838	-379.876	5.669
3600	177.783	726.275	571.024	558.904	-1310.315	-353.270	5.126
3700	177.834	731.146	575.286	576.685	-1308.708	-326.707	4.612
3800	177.881	735.890	579.450	594.471	-1307.017	-300.190	4.126
3900	177.924	740.511	583.521	612.261	-1305.242	-273.717	3.666
4000	177.962	745.016	587.502	630.056	-1303.384	-247.291	3.229
4100	177.994	749.411	591.398	647.853	-1301.442	-220.913	2.814
4200	178.022	753.700	595.211	665.654	-1299.419	-194.582	2.420
4300	178.048	757.889	598.946	683.458	-1297.316	-168.302	2.044
4400	178.071	761.983	602.605	701.264	-1295.134	-142.071	1.687
4500	178.092	765.985	606.191	719.072	-1292.875	-115.889	1.345
4600	178.108	769.899	609.708	736.882	-1290.542	-89.760	1.019
4700	178.122	773.730	613.157	754.693	-1288.136	-63.682	0.708
4800	178.135	777.480	616.541	772.506	-1285.661	-37.655	0.410
4900	178.146	781.153	619.863	790.320	-1283.117	-11.682	0.125
5000	178.155	784.752	623.125	808.135	-1280.509	14.239	-0.149
5100	178.159	788.280	626.329	825.951	-1277.837	40.108	-0.411
5200	178.163	791.740	629.477	843.767	-1275.105	65.924	-0.662
5300	178.167	795.134	632.571	861.584	-1272.314	91.687	-0.904
5400	178.171	798.464	635.612	879.401	-1269.468	117.396	-1.136
5500	178.176	801.733	638.603	897.218	-1266.568	143.051	-1.359
5600	178.180	804.944	641.545	915.036	-1263.617	168.655	-1.573
5700	178.184	808.098	644.439	932.854	-1260.616	194.203	-1.780
5800	178.188	811.197	647.288	950.673	-1257.568	219.699	-1.979
5900	178.192	814.243	650.092	968.492	-1254.475	245.145	-2.170
6000	178.197	817.238	652.852	986.311	-1251.339	270.535	-2.355

PREVIOUS: June 1969 (1 atm) CURRENT: June 1969 (1 bar)

Hexafluoroethane (C_2F_6) $C_2F_6(g)$

ETHYNYL (C_2H) IDEAL GAS M_r = 25.02994

$\Delta_fH°$(0 K) = [473.9 ± 29] kJ mol^{-1}

S°(298.15 K) = [207.444 ± 5.4] J K^{-1} mol^{-1} $\Delta_fH°$(298.15 K) = [477.0 ± 29] kJ mol^{-1}

Electronic Levels and Quantum Weights

ε_i, cm^{-1}	g_i
0	[4]
[7000]	[2]

Vibrational Frequencies and Degeneracies

ν, cm^{-1}
[1920](1)
[640](2)
[3220](1)

Point Group: [$C_{\infty v}$] σ = 1

Bond Distances: C–C = [1.207] Å C–H = [1.061] Å

Bond Angle: C–C–H = [180]°

Rotational Constant: B_0 = [1.474960] cm^{-1}

Enthalpy of Formation

The enthalpy of formation of C_2H(g) is not well established; however, we can place reliable upper and lower limits on the value.

Cottrell (1), gives an upper limit for the C–H bond dissociation energy in acetylene, $\Delta_{at}H°$(HCC–H) <121 kcal. This was obtained from the fact that ultra violet light of 121 kcal mol^{-1} energy causes acetylene to from diacetylene and hydrogen; assuming that this process occurs through the CCH radical, the light energy is equated with $\Delta_{at}H°$(HCC–H). This value of $\Delta_{at}H°$ (HCC–H) gives an upper limit for $\Delta_fH°$(CCH, g, 298.15 K) <123 kcal mol^{-1}.

A lower limit for $\Delta_{at}H°$(HCC–H) is obtained from a consideration of the C–H bond dissociation energies in ethylene and ethane, $\Delta_{at}H°(C_2H_3–H)$ = 103 kcal and $\Delta_{at}H°(C_2H_5–H)$ = 98 kcal. The heats of formation of the C_2H_3 and C_2H_5 radicals and their parent compounds are taken from Benson (2). The vibrational frequency of the C–H stretch increases steadily form ethane to ethylene and acetylene. Similarly we expect the bond strength to increase, and so $\Delta_{at}H°$(HCC–H) >103 kcal; this yields $\Delta_fH°$(CCH, g, 298.15 K) >105 kcal mol^{-1}.

Coats and Anderson (3), obtained several values for $\Delta_fH°$(CCH, g, 298.15 K) from mass-spectroscopic measurements; their values range from 110 kcal mol^{-1} to 153 kcal mol^{-1}, but they chose to adopt 112± 3 kcal mol^{-1}.

Our adopted value of $\Delta_fH°$(CCH, g, 298.15 K), 114±7 kcal mol^{-1}, (476.976±29 kJ mol^{-1}), is 2 kcal smaller than that estimated by Tsang et al (4).

Heat Capacity and Entropy

The molecule is assumed to be linear; the bond lengths are assumed to be the same as in acetylene and were taken from Sutton (5).

The vibrational frequencies are those estimated by Plooster and Reed (6). It should be noted that the values of the free energy function given by Plooster and Reed (6) are incorrect due to the neglect of the electronic contributions.

The electronic levels, $X^2\Sigma$ and $A^2\Pi$, were estimated by analogy with those of CN(g) which is isoelectronic; this assumption holds very well in the parent molecules HCCH and HCN (7). The molecule $C_2H_2^+$, which is also isoelectronic with CCH, is predicted to have a $^2\Pi$ ground state by Herzberg (7).

References

1. T. L. Cottrell, "The Strengths of Chemical Bonds," 2nd Ed., Academic press, New York, 1958.
2. S. W. Benson, J. Chem. Educ. 42, 502 (1965).
3. F. H. Coats and R. C. Anderson, J. Amer. Chem. Soc. 79, 1340 (1957).
4. W. Tsang, S. H. Bauer and M. Cowperthwaite, J. Chem. Phys. 36, 1768 (1962).
5. L. E. Sutton, "Interatomic Distances Supplement," Special Publication No. 18, The Chemical Society, London, 1965.
6. M. N. Plooster and T. B. Reed, J. Chem. Phys. 31, 66 (1959).
7. G. Herzberg, "Electronic Spectra and Electronic Structure of Polyatomic Molecules," D. Van Nostrand Co., New York, 1966.

Ethynyl (C_2H) C_2H_1(g)

Enthalpy Reference Temperature = T_r = 298.15 K Standard State Pressure = p° = 0.1 MPa

T/K	$C_p°$ (J K^{-1}mol^{-1})	S° (J K^{-1}mol^{-1})	−[G°−H°(T_r)]/T (J K^{-1}mol^{-1})	H°−H°(T_r) (kJ mol^{-1})	$\Delta_fH°$ (kJ mol^{-1})	$\Delta_fG°$ (kJ mol^{-1})	Log K_f
0	0.	0.	INFINITE	-9.404	473.908	473.908	INFINITE
100	29.242	172.435	237.415	-6.498	475.195	463.178	-241.940
200	32.702	193.530	210.703	-3.435	476.258	450.726	-117.718
250	35.084	201.087	208.044	-1.739	476.664	444.295	-92.830
298.15	37.104	207.444	207.444	0.	476.976	438.031	-76.741
300	37.175	207.673	207.444	0.069	476.986	437.789	-76.226
350	38.922	213.539	207.904	1.972	477.224	431.236	-64.358
400	40.395	218.835	208.944	3.956	477.376	424.655	-55.454
450	41.675	223.669	210.316	6.009	477.441	418.060	-48.527
500	42.818	228.120	211.877	8.122	477.426	411.463	-42.985
600	44.837	236.109	215.265	12.507	477.190	398.288	-34.674
700	46.615	243.157	218.755	17.081	476.750	385.170	-28.742
800	48.212	249.487	222.208	21.824	476.174	372.125	-24.297
900	49.651	255.251	225.564	26.718	475.512	359.158	-20.845
1000	50.943	260.550	228.801	31.749	474.795	346.268	-18.087
1100	52.102	265.461	231.913	36.902	474.041	333.451	-15.834
1200	53.142	270.040	234.902	42.165	473.263	320.705	-13.960
1300	54.081	274.331	237.771	47.527	472.466	308.024	-12.377
1400	54.935	278.370	240.528	52.979	471.656	295.405	-11.022
1500	55.720	282.188	243.180	58.512	470.837	282.844	-9.849
1600	56.451	285.807	245.732	64.121	470.010	270.338	-8.826
1700	57.140	289.251	248.191	69.801	469.179	257 884	-7.924
1800	57.795	292.535	250.564	75.548	468.346	245.478	-7 124
1900	58.424	295.677	252.857	81.359	467.512	233.120	-6.409
2000	59.032	298.689	255.073	87.232	466.681	220.805	-5.767
2100	59.622	301.584	257.220	93.165	465.854	208.531	-5.187
2200	60.194	304.371	259.300	99.156	465.032	196.297	-4.661
2300	60.750	307.059	261.318	105.203	464.217	184.100	-4.181
2400	61.289	309.656	263.279	111.305	463.409	171.938	-3.742
2500	61.810	312.168	265.184	117.460	462.609	159.810	-3.339
2600	62.313	314.602	267.038	123.666	461.817	147.714	-2.968
2700	62.796	316.963	268.844	129.922	461.032	135.648	-2.624
2800	63.258	319.255	270.604	136.225	460.254	123.611	-2.306
2900	63.698	321.483	272.320	142.573	459.482	111.602	-2.010
3000	64.115	323.649	273.995	148.964	458.715	99.619	-1.735
3100	64.509	325.758	275.631	155.395	457.951	87.662	-1.477
3200	64.878	327.812	277.230	161.865	457.190	75.729	-1.236
3300	65.223	329.814	278.793	168.370	456.428	63.820	-1.010
3400	65.544	331.766	280.322	174.909	455.666	51.934	-0.798
3500	65.840	333.670	281.819	181.478	454.900	40.071	-0.598
3600	66.113	335.529	283.286	188.076	454.128	28.230	-0.410
3700	66.362	337.344	284.722	194.700	453.350	16.410	-0.232
3800	66.588	339.117	286.130	201.347	452.562	4.611	-0.063
3900	66.792	340.849	287.511	208.017	451.763	-7.166	0.096
4000	66.975	342.542	288.866	214.705	450.952	-18.924	0.247
4100	67.138	344.198	290.195	221.411	450.126	-30.660	0.391
4200	67.281	345.818	291.501	228.132	449.284	-42.376	0.527
4300	67.407	347.402	292.782	234 867	448.424	-54.073	0.657
4400	67.515	348.953	294.041	241.613	447.544	-65.748	0.781
4500	67.607	350.472	295.278	248.369	446.644	-77.404	0.898
4600	67.684	351.958	296.495	255.134	445.722	-89.040	1.011
4700	67.746	353.415	297.690	261.905	444.777	-100.655	1.119
4800	67.796	354.842	298.866	268.683	443.807	-112.249	1.222
4900	67.834	356.240	300.023	275.464	442.812	-123.824	1.320
5000	67.860	357.611	301.161	282.249	441.792	-135.377	1.414
5100	67.876	358.955	302.281	289.036	440.744	-146.910	1.505
5200	67.882	360.273	303.383	295.824	439.669	-158.422	1.591
5300	67.880	361.566	304.469	302.612	438.567	-169.914	1.675
5400	67.871	362.834	305.538	309.400	437.436	-181.384	1.755
5500	67.854	364.080	306.591	316.186	436.276	-192.833	1.831
5600	67.830	365.302	307.629	322.970	435.087	-204.261	1.905
5700	67.801	366.502	308.651	329.752	433.870	-215.667	1.976
5800	67.766	367.681	309.659	336.530	432.623	-227.051	2.045
5900	67.727	368.839	310.652	343.305	431.348	-238.414	2.111
6000	67.684	369.977	311.631	350.075	430.043	-249.755	2.174

PREVIOUS: March 1967 (1 atm) CURRENT: March 1967 (1 bar)

Ethynyl (C_2H) C_2H_1(g)

CHLOROETHYNE (C_2HCl) IDEAL GAS M_r = 60.48294

$S°(298.15\ K)$ = 241.996 J K^{-1} mol^{-1}

$\Delta_f H°(0\ K)$ = [212.9 ± 42] kJ mol^{-1}
$\Delta_f H°(298.15\ K)$ = [213.8 ± 42] kJ mol^{-1}

Vibrational Frequencies and Degeneracies

ν, cm^{-1}	ν, cm^{-1}
3340(1)	604(2)
2110(1)	326(2)
756(1)	

Ground State Quantum Weight: [1] σ = 1
Point Group: $C_{\infty v}$
Bond Distances: C-C = 1.204 Å C-H = 1.055 Å C-Cl = 1.637 Å
Bond Angles: Cl-C-C = 180° C-C-H = 180°
Rotational Constant: B_0 = 0.188645 cm^{-1}

Enthalpy of Formation

The value of $\Delta_f H°(298.15\ K)$ = 51.1 kcal mol^{-1} is calculated from the estimated $\Delta_r H°(298.15\ K)$ = -53±10 kcal mol^{-1} for HC≡CCl(g) + Cl_2(g) → CHCl=CCl_2(g), using $\Delta_f H°$(CHCl=CCl_2, g, 298.15 K) = -1.86 kcal mol^{-1} (1). The value of $\Delta_r H°(298.15\ K)$ is assumed to be the same as that for the reaction HC≡CH(g) + Cl_2(g) → CHCl=CHCl(g), using $\Delta_f H°$(C_2H_2, g, 298.15 K) = 54.19 kcal mol^{-1} (1) and $\Delta_f H°$(CHCl=CHCl, g, 298.15 K) = 1.2 kcal mol^{-1} (2).

Heat Capacity and Entropy

The linear molecular structure and bond distances were determined from the microwave spectrometrical studies by Tyler and Sheridan (3). Their results are in good agreement with those of Westenberg et al.(4).

The vibrational frequencies were observed in the infrared spectra by Hunt and Wilson (5), and Richardson and Goldstein (6).

References

1. U. S. Nat. Bur. Stand. Tech. Note 270-3, 1968.
2. This is an average value of 0.9 and 1.47 kcal mol^{-1} for cis- and trans- CHCl = CHCl(g), respectively which are obtained from reference (1).
3. J. K. Tyler and J. Sheridan, Trans. Faraday Soc. 59, 2661 (1963).
4. A. A. Westenberg, J. H. Goldstein and E. B. Wilson, Jr., J. Chem. Phys. 17, 1319 (1949).
5. G. R. Hunt and M. K. Wilson, J. Chem. Phys. 34, 1301 (1961).
6. W. S. Richardson and J. H. Goldstein, J. Chem. Phys. 18, 1314 (1950).

Chloroethyne (C_2HCl) $C_2H_1Cl_1(g)$

Enthalpy Reference Temperature = T_r = 298.15 K Standard State Pressure = p° = 0.1 MPa

	J $K^{-1}mol^{-1}$			kJ mol^{-1}			
T/K	$C_p°$	S°	$-[G°-H°(T_r)]/T$	$H°-H°(T_r)$	$\Delta_f H°$	$\Delta_f G°$	Log K_f
0	0.	0.	INFINITE	-11.787	212.941	212.941	INFINITE
100	32.752	195.662	283.695	-8.803	212.851	207.961	-108.628
200	45.079	222.136	246.697	-4.912	213.223	202.966	-53.009
250	50.310	232.778	242.869	-2.523	213.513	200.369	-41.865
298.15	54.319	241.996	241.996	0.	213.802	197.811	-34.656
300	54.455	242.332	241.997	0.101	213.813	197.712	-34.425
350	57.682	250.979	242.672	2.907	214.094	195.006	-29.103
400	60.227	258.854	244.210	5.857	214.337	192.261	-25.107
450	62.290	266.071	246.244	8.922	214.527	189.490	-21.995
500	64.017	272.725	248.564	12.081	214.660	186.701	-19.504
600	66.824	284.655	253.608	18.629	214.771	181.095	-15.766
700	69.104	295.133	258.807	25.428	214.720	175.485	-13.095
800	71.050	304.491	263.943	32.438	214.562	169.889	-11.093
900	72.751	312.959	268.926	39.630	214.337	164.318	-9.537
1000	74.248	320.704	273.722	46.982	214.072	158.775	-8.294
1100	75.570	327.844	278.322	54.474	213.782	153.259	-7.278
1200	76.735	334.470	282 728	62.091	213.476	147.770	-6.432
1300	77.762	340.654	286.948	69.817	213.157	142.308	-5.718
1400	78.667	346.450	290.994	77.639	212.828	136.870	-5.107
1500	79.466	351.905	294.874	85.546	212.487	131.456	-4.578
1600	80.170	357.057	298.601	93.529	212.134	126.065	-4.116
1700	80.793	361.936	302.185	101.578	211.767	120.697	-3.709
1800	81.345	366.570	305.634	109.685	211.386	115.351	-3.347
1900	81.835	370.982	308.958	117.845	210.987	110.027	-3.025
2000	82.271	375.191	312.165	126.050	210.570	104.724	-2.735
2100	82.660	379.214	315.263	134.297	210.133	99.442	-2.473
2200	83.009	383.068	318.258	142.581	209.673	94.182	-2.236
2300	83.322	386.765	321.157	150.898	209.191	88.943	-2.020
2400	83.603	390.317	323.965	159.244	208.683	83.725	-1.822
2500	83.857	393.735	326.688	167.618	208.148	78.530	-1.641
2600	84.087	397.029	329.330	176.015	207.586	73.356	-1.474
2700	84.295	400.206	331.897	184.434	206.995	68.205	-1.319
2800	84.485	403.275	334.392	192.874	206.374	63.075	-1.177
2900	84.658	406.243	336.818	201.331	205.722	57.969	-1.044
3000	84.816	409.116	339.181	209.805	205.039	52.886	-0.921
3100	84.960	411.899	341.482	218.293	204.324	47.826	-0.806
3200	85.093	414.598	343.725	226.796	203.578	42.789	-0.698
3300	85.215	417.219	345.912	235.312	202.800	37.777	-0.598
3400	85.327	419.764	348.047	243.839	201.990	32.788	-0.504
3500	85.431	422.239	350.132	252.377	201.150	27.823	-0.415
3600	85.527	424.647	352.168	260.925	200.279	22.884	-0.332
3700	85.616	426.992	354.159	269.482	199.379	17.968	-0.254
3800	85.698	429.276	356.106	278.048	198.450	13.078	-0.180
3900	85.775	431.503	358.011	286.621	197.493	8.212	-0.110
4000	85.846	433.676	359.875	295 202	196.510	3.371	-0.044
4100	85.913	435.796	361.701	303.790	195.500	-1.445	0.018
4200	85.975	437.868	363.490	312.385	194.467	-6.236	0.078
4300	86.033	439.891	365.244	320.985	193.410	-11.002	0.134
4400	86.088	441.870	366.963	329.591	192.331	-15.744	0.187
4500	86.139	443.805	368.649	338.203	191.230	-20.460	0.237
4600	86.187	445.699	370.303	346.819	190.110	-25.152	0.286
4700	86.232	447.553	371.927	355.440	188.972	-29.820	0.331
4800	86.275	449.369	373.522	364.065	187.815	-34.463	0.375
4900	86.315	451.148	375.088	372.695	186.642	-39.081	0.417
5000	86.352	452.892	376.627	381.328	185.454	-43.676	0.456
5100	86.388	454.603	378.139	389.965	184.251	-48.247	0.494
5200	86.422	456.280	379.625	398.606	183.034	-52.793	0.530
5300	86.454	457.927	381.087	407.249	181.804	-57.317	0.565
5400	86.484	459.543	382.525	415.896	180.562	-61.817	0.598
5500	86.512	461.130	383.940	424.546	179.309	-66.294	0.630
5600	86.540	462.689	385.332	433.199	178.046	-70.748	0.660
5700	86.565	464.221	386.703	441.854	176.772	-75.180	0.689
5800	86.590	465.727	388.053	450.512	175.490	-79.588	0.717
5900	86.613	467.207	389.382	459.172	174.199	-83.975	0.743
6000	86.635	468.663	390.691	467.834	172.900	-88.341	0.769

PREVIOUS: December 1968 (1 atm) CURRENT December 1968 (1 bar)

Chloroethyne (C_2HCl) $C_2H_1Cl_1(g)$

FLUOROETHYNE (C_2HF) IDEAL GAS M_r = 44.028343

$S°(298.15 K) = 231.611 ± 0.04 J K^{-1} mol^{-1}$

$\Delta_f H°(0 K) = [124.8 ± 63] kJ mol^{-1}$

$\Delta_f H°(298.15 K) = [125.5 ± 63] kJ mol^{-1}$

Vibrational Frequencies and Degeneracies

ν, cm^{-1}
3360(1)
2250(1)
1060(1)
580(2)
367(2)

Ground State Quantum Weight: 1 σ = 1

Point Group: $C_{\infty v}$

Bond Distances: C–F = 1.279 Å C–C = 1.198 Å C–H = 1.053 Å

Bond Angles: F–C–C = 180° C–C–H = 180°

Rotational Constant: $B_0 = 0.323577 cm^{-1}$

Enthalpy of Formation

The enthalpy of formation is estimated by assuming that the C–C bond strength is the average of those in C_2F_2(113 kcal) and C_2H_2(229 kcal). Combining the average value (171 kcal) with the enthalpies of formation of CF and CH from the JANAF Tables, yields $\Delta_f H°(C_2HF, g, 298.15 K) = 30±15 kcal mol^{-1}$.

Heat Capacity and Entropy

Hunt and Wilson (1) and Brown and Tyler (2), have reported the infrared spectrum and have assigned the fundamental frequencies. The frequencies used in the present table are median values of the two sets and are generally within 5 cm^{-1} of each set.

The molecule has been shown to be linear by Tyler and Sheridan (3), who reported the rotational constant B_0.

The electronic ground is assumed to be $^1\Sigma$ by analogy with C_2H_2.

References

1. G. R. Hunt and M. K. Wilson, J. Chem. Phys. 34, 1301 (1961).
2. J. K. Brown and J. K. Tyler, Proc. Chem. Soc. (London) 1961, 13 (1961).
3. J. K. Tyler an J. Sheridan, Trans. Faraday Soc. 59, 2661 (1963).

Fluoroethyne (C_2HF) $C_2H_1F_1(g)$

Enthalpy Reference Temperature = T_r = 298.15 K Standard State Pressure = p° = 0.1 MPa

T/K	$C_p°$ (J K⁻¹mol⁻¹)	S° (J K⁻¹mol⁻¹)	−[G°−H°(T_r)]/T (J K⁻¹mol⁻¹)	H°−H°(T_r) (kJ mol⁻¹)	$\Delta_f H°$ (kJ mol⁻¹)	$\Delta_f G°$ (kJ mol⁻¹)	Log K_f
0	0.	0.	INFINITE	-11.452	124.816	124.816	INFINITE
100	31.761	186.882	271.833	-8.495	124.701	119.758	-62.555
200	43.530	212.465	236.141	-4.735	124.996	114.742	-29.968
250	48.488	222.731	232.452	-2.430	125.261	112.149	-23.432
298.15	52.321	231.611	231.611	0.	125.520	109.600	-19.201
300	52.452	231.935	231.612	0.097	125.530	109.501	-19.066
350	55.619	240.268	232.263	2.802	125.774	106.810	-15.941
400	58.190	247.868	233.746	5.649	125.974	104.087	-13.592
450	60.324	254.849	235.708	8.613	126.118	101.342	-11.763
500	62.143	261.301	237.949	11.676	126.203	98.584	-10.299
600	65.147	272.907	242.830	18.046	126.216	93.055	-8.101
700	67.611	283.140	247.872	24.688	126.067	87.538	-6.532
800	69.723	292.309	252.863	31.557	125.813	82.051	-5.357
900	71.570	300.630	257.715	38.624	125.496	76.599	-4.446
1000	73.196	308.257	262.393	45.864	125.143	71.185	-3.718
1100	74.630	315.302	266.887	53.256	124.769	65.807	-3.125
1200	75.894	321.851	271.198	60.784	124.382	60.464	-2.632
1300	77.007	327.971	275.332	68.430	123.985	55.153	-2.216
1400	77.987	333.714	279.299	76.181	123.579	49.874	-1.861
1500	78.850	339.125	283.109	84.024	123.164	44.623	-1.554
1600	79.612	344.238	286.771	91.948	122.738	39.401	-1.286
1700	80.285	349.085	290.295	99.943	122.300	34.206	-1.051
1800	80.882	353.692	293.690	108.002	121.848	29.037	-0.843
1900	81.411	358.079	296.965	116.117	121.383	23.893	-0.657
2000	81.882	362.267	300.126	124.282	120.903	18.775	-0.490
2100	82.302	366.273	303.181	132.492	120.409	13.680	-0.340
2200	82.678	370.110	306.137	140.741	119.901	8.610	-0.204
2300	83.015	373.793	308.999	149.026	119.379	3.563	-0.081
2400	83.319	377.332	311.773	157.343	118.845	-1.461	0.032
2500	83.593	380.739	314.464	165.689	118.298	-6.463	0.135
2600	83.840	384.023	317.076	174.061	117.740	-11.442	0.230
2700	84.065	387.191	319.615	182.456	117.172	-16 400	0.317
2800	84.269	390.252	322.083	190.873	116.594	-21.337	0.398
2900	84.455	393.213	324.485	199.309	116.008	-26.252	0.473
3000	84.625	396.079	326.824	207.764	115.413	-31.148	0.542
3100	84.781	398.856	329.103	216.234	114.812	-36.024	0.607
3200	84.924	401.550	331.325	224.719	114.203	-40.879	0.667
3300	85.055	404.165	333.493	233.218	113.587	-45.716	0.724
3400	85.176	406.706	335.609	241.730	112.965	-50.534	0.776
3500	85.288	409.177	337.676	250.253	112.337	-55.334	0.826
3600	85.391	411.581	339.696	258.787	111.703	-60 115	0.872
3700	85.487	413.922	341.670	267.331	111.062	-64.879	0.916
3800	85.576	416.203	343.602	275.885	110.415	-69.626	0.957
3900	85.659	418.427	345.492	284.446	109.760	-74.355	0.996
4000	85.735	420.597	347.343	293.016	109.099	-79.068	1.033
4100	85.807	422.715	349.155	301.593	108.430	-83.764	1.067
4200	85.874	424.783	350.931	310.177	107.754	-88.443	1 100
4300	85.937	426.805	352.673	318.768	107.069	-93.107	1.131
4400	85.995	428.781	354.380	327.365	106.376	-97.754	1.160
4500	86.051	430.714	356.055	335.967	105.673	-102 385	1.188
4600	86.102	432.606	357.698	344.575	104.962	-107.001	1.215
4700	86.151	434.458	359.312	353.187	104.240	-111.601	1.240
4800	86.196	436.273	360.897	361.805	103.509	-116.186	1.264
4900	86.240	438.050	362.453	370.426	102.767	-120.755	1.287
5000	86.280	439.793	363.982	379.052	102.015	-125.309	1.309
5100	86.319	441.502	365.486	387.682	101.251	-129.848	1.330
5200	86.355	443.178	366.964	396.316	100.476	-134 372	1.350
5300	86.389	444.824	368.417	404.953	99.690	-138.881	1.369
5400	86.422	446.439	369.847	413.594	98.892	-143.374	1.387
5500	86.452	448.025	371.254	422.238	98.083	-147.854	1.404
5600	86.482	449.583	372.639	430.884	97.262	-152.318	1.421
5700	86.509	451.114	374.003	439.534	96.429	-156.767	1.437
5800	86.536	452.619	375.345	448.186	95.584	-161.202	1.452
5900	86.561	454.098	376.667	456.841	94.727	-165.622	1.466
6000	86.585	455.553	377.970	465.498	93.858	-170.027	1.480

PREVIOUS: December 1967 (1 atm) CURRENT: December 1967 (1 bar)

Fluoroethyne (C_2HF) $C_2H_1F_1(g)$

ETHYNE (C_2H_2) IDEAL GAS M_r = 26.03788

$\Delta_fH°$(0 K) = 235.755 ± 0.79 kJ mol^{-1}

$S°$(298.15 K) = 200.958 J K^{-1} mol^{-1} $\Delta_fH°$(298.15 K) = 226.731 ± 0.79 kJ mol^{-1}

Vibrational Frequencies and Degeneracies

ν, cm^{-1}
3373.7(1)
1973.8(1)
3281.9(1)
611.6(2)
729.3(2)

Ground State Quantum Weight: 1 σ = 2

Point Group: $D_{\infty h}$

Rotational Constant B_0 = 1.17684 ± 0.00016 cm^{-1}

Spectroscopic constants used in calculating corrections to rigid rotator-harmonic oscillator approximation (cm^{-1}):

X_{11} = -24.08	X_{22} = -7.92	X_{34} = -9.06	D_0 = 2.19 x 10^{-6}
X_{12} = -16.94	X_{23} = -1.38	X_{35} = -5.73	g_{44} = 1.1
X_{13} = -99.01	X_{24} = -6.15	X_{44} = 5.38	g_{55} = 2.49
X_{14} = -16.46	X_{25} = -0.85	X_{45} = -12.65	
X_{15} = -11.75	X_{33} = -25.69	X_{55} = -2.27	

Values of α_1 not available.

Enthalpy of Formation

Taken from Wagman et al. (1).

Heat Capacity and Entropy

Gordon (2) has used the constants listed above to calculate $C_p°$ from 298.15 K to 6000 K by the method of Pennington and Kobe (3). The constants are from Bell and Nielsen (4) and Allen et al. (5). Heat capacities below 298.15 K have been calculated for a rigidly rotating harmonic oscillator.

References

1. D. D. Wagman, J. E. Kilpatrick, K. S. Pitzer, and F. D. Rossini, J. Res. Nat. Bur. Stand. 35, 467 (1945).
2. J. S. Gordon, personal communication, February, 1961.
3. R. E. Pennington and K. A. Kobe, J. Chem. Phys. 22, 1442 (1954).
4. E. E. Bell and H. H. Nielsen, J. Chem. Phys. 18, 1382 (1950).
5. H. C. Allen, E. D. Tidewell and E. K. Plyler, J. Res. Nat. Bur. Stand. 57, 213 (1956).

Ethyne (C_2H_2) C_2H_2(g)

Enthalpy Reference Temperature = T_r = 298.15 K — Standard State Pressure = p° = 0.1 MPa

T/K	$C_p°$ (J $K^{-1}mol^{-1}$)	S° (J $K^{-1}mol^{-1}$)	-[G°-H°(T_r)]/T (J $K^{-1}mol^{-1}$)	H°-H°(T_r) (kJ mol^{-1})	$\Delta_fH°$ (kJ mol^{-1})	$\Delta_fG°$ (kJ mol^{-1})	Log K_f
0	0.	0.	INFINITE	-10.012	235.755	235.755	INFINITE
100	29.347	163.294	234.338	-7.104	232.546	236.552	-123.562
200	35.585	185.097	204.720	-3.925	229.685	241.663	-63.116
298.15	44.095	200.958	200.958	0.	226.731	248.163	-43.477
300	44.229	201.231	200.959	0.082	226.674	248.296	-43.232
400	50.480	214.856	202.774	4.833	223.568	255.969	-33.426
500	54.869	226.610	206.393	10.108	220.345	264.439	-27.626
600	58.287	236.924	210.640	15.771	216.993	273.571	-23.816
700	61.149	246.127	215.064	21.745	213.545	283.272	-21.138
800	63.760	254.466	219.476	27.992	210.046	293.471	-19.162
900	66.111	262.113	223.794	34.487	206.522	304.111	-17.650
1000	68.275	269.192	227.984	41.208	202.989	315.144	-16.461
1100	70.245	275.793	232.034	48.136	199.451	326.530	-15.506
1200	72.053	281.984	235.941	55.252	195.908	338.239	-14.723
1300	73.693	287.817	239.709	62.540	192.357	350.244	-14.073
1400	75.178	293.334	243.344	69.985	188.795	362.523	-13.526
1500	76.530	298.567	246.853	77.572	185.216	375.057	-13.061
1600	77.747	303.546	250.242	85.286	181.619	387.830	-12.661
1700	78.847	308.293	253.518	93.117	177.998	400.829	-12.316
1800	79.852	312.829	256.688	101.053	174.353	414.041	-12.015
1900	80.760	317.171	259.758	109.084	170.680	427.457	-11.752
2000	81.605	321.335	262.733	117.203	166.980	441.068	-11.520
2100	82.362	325.335	265.620	125.401	163.250	454.864	-11.314
2200	83.065	329.183	268.422	133.673	159.491	468.838	-11.132
2300	83.712	332.890	271.145	142.012	155.701	482.984	-10.969
2400	84.312	336.465	273.793	150.414	151.881	497.295	-10.823
2500	84.858	339.918	276.369	158.873	148.029	511.767	-10.693
2600	85.370	343.256	278.878	167.384	144.146	526.393	-10.575
2700	85.846	346.487	281.322	175.945	140.230	541.169	-10.470
2800	86.295	349.618	283.706	184.553	136.282	556.090	-10.374
2900	86.713	352.653	286.031	193.203	132.302	571.154	-10.288
3000	87.111	355.600	288.301	201.895	128.290	586.355	-10.209
3100	87.474	358.462	290.519	210.624	124.245	601.690	-10.138
3200	87.825	361.245	292.686	219.389	120.166	617.157	-10.074
3300	88.164	363.952	294.804	228.189	116.053	632.751	-10.016
3400	88.491	366.589	296.877	237.022	111.908	648.471	-9.963
3500	88.805	369.159	298.906	245.886	107.731	664.313	-9.914
3600	89.101	371.665	300.892	254.782	103.519	680.275	-9.871
3700	89.388	374.110	302.838	263.706	99.274	696.354	-9.831
3800	89.666	376.498	304.745	272.659	94.996	712.549	-9.795
3900	89.935	378.830	306.615	281.639	90.683	728.856	-9.762
4000	90.194	381.110	308.449	290.646	86.336	745.275	-9.732
4100	90.439	383.341	310.248	299.678	81.955	761.803	-9.705
4200	90.678	385.523	312.015	308.733	77.538	778.438	-9.681
4300	90.910	387.659	313.749	317.813	73.087	795.178	-9.660
4400	91.137	389.752	315.453	326.915	68.601	812.023	-9.640
4500	91.358	391.802	317.127	336.040	64.080	828.969	-9.622
4600	91.563	393.813	318.772	345.186	59.524	846.017	-9.607
4700	91.768	395.784	320.390	354.353	54.933	863.164	-9.593
4800	91.970	397.718	321.981	363.540	50.307	880.410	-9.581
4900	92.171	399.617	323.546	372.747	45.648	897.751	-9.570
5000	92.370	401.481	325.086	381.974	40.957	915.189	-9.561
5100	92.571	403.312	326.602	391.221	36.234	932.720	-9.553
5200	92.768	405.111	328.094	400.488	31.481	950.345	-9.546
5300	92.963	406.880	329.564	409.774	26.699	968.061	-9.541
5400	93.153	408.620	331.012	419.080	21.889	985.868	-9.536
5500	93.341	410.331	332.439	428.405	17.052	1003.763	-9.533
5600	93.525	412.014	333.845	437.748	12.189	1021.748	-9.530
5700	93.706	413.671	335.231	447.110	7.303	1039.819	-9.529
5800	93.883	415.302	336.597	456.489	2.393	1057.976	-9.528
5900	94.057	416.909	337.945	465.886	-2.537	1076.217	-9.528
6000	94.228	418.491	339.274	475.301	-7.488	1094.543	-9.529

PREVIOUS: March 1961 (1 atm) CURRENT: March 1961 (1 bar

Ethyne (C_2H_2) C_2H_2(g)

$\Delta_fH°(0\ K) = 60.986 \pm 0.29\ kJ\ mol^{-1}$

$S°(298.15\ K) = 219.330\ J\ K^{-1}\ mol^{-1}$ $\Delta_fH°(298.15\ K) = 52.467 \pm 0.29\ kJ\ mol^{-1}$

Vibrational Frequencies and Degeneracies

ν, cm^{-1}	ν, cm^{-1}	ν, cm^{-1}
3026.4(1)	3102.5(1)	3105.5(1)
1622.9(1)	1222.0(1)	826.0(1)
1342.2(1)	949.3(1)	2988.7(1)
1023.0(1)	943.0(1)	1443.5(1)

Ground State and Quantum Weight: 1 $\sigma = 4$

Point Group: D_{2h}

Bond Distances: H-C = 1.086 Å C-C = 1.337 Å

Bond Angles: H-C-H = 117° 22' H-C-C = 121° 19'

Product of the Moments of Inertia: $I_AI_BI_C = 5.4496 \times 10^{-117}\ g^3\ cm^6$

Enthalpy of Formation

The enthalpy of combustion of ethylene was measured by Rossini and Knowlton (1). From the value, $\Delta_rH°(298.15\ K) = -337.28\pm0.07\ kcal\ mol^{-1}$ for the reaction $C_2H_4(g) + 3\ O_2(g) = 2\ CO_2(g) + 2\ H_2O(\ell)$, the enthalpy of formation, $\Delta_fH°(298.15\ K)$, for $C_2H_4(g)$ was derived to be $12.54\pm0.07\ kcal\ mol^{-1}$. The values of $\Delta_fH°(298.15\ K)$ for $CO_2(g)$ and $H_2O(\ell)$ used for calculation were obtained from Rossini et al. (2).

Heat Capacity and Entropy

The molecular structure, bond distances and angles were obtained from Allen and Plyler (3). The vibrational frequencies were taken from Smith and Mills (4). Eight of the twelve frequencies were reported by Arnett and Crawford (5); Crawford et al. (6); Stoicheff (7); Feldman et al. (8); and Allen and Plyler (3). Herzberg (9) also reported the vibrational frequencies for ethylene as 3019.3 (1), 1623.3 (1), 1342.4 (1), 825.0 (1), 3272.3 (1), 1050 (1), 949.2 (1), 943 (1), 3105.5 (1), 995 (1), 298 .5 (1) and 1443.5 (1) which are very close to the values adopted. The infrared and Raman spectra of ethylene before 1945 have been summarized by Herzberg (9). The molecular structure of ethylene was also determined by Bartell and Bonham (10). Their results, r(C-H) = 1.085 Å, r(C-C) = 1.334 Å and H-C-H = 116 are in excellent agreement with the values adopted. The principal moments of inertia are $I_A = 0.5763 \times 10^{-39}$, $I_B = 2.8004 \times 10^{-39}$ and $I_C = 3.3767 \times 10^{-39}\ g\ cm^2$.

References

1. F. D. Rossini and J. W. Knowlton, J. Res. Nat. Bur. Stand. 19, 249 (1937).
2. U. S. Nat. Bur. Stand. Circ. 500, 1952.
3. H. C. Allen, Jr., and E. K. Plyler, J. Amer. Chem. Soc. 80, 2673 (1958).
4. W. L. Smith and I. M. Mills, J. Chem. Phys. 40, 2095 (1964).
5. R. L. Arnett and B. L. Crawford, J. Chem. Phys. 18, 118 (1950).
6. B. L. Crawford, J. E. Lancaster and R. Inskeep, J. Chem. Phys. 21, 678 (1953).
7. B. P. Stoicheff, J. Chem. Phys. 21, 755 (1953).
8. T. Feldman, J. Romanko and H. L. Welsh, Can. J. Phys. 34, 737 (1956).
9. G. Herzberg, "Infrared and Raman Spectra," D. Van Nostrand Company, Inc., New York, 1945.
10. L. S. Bartell and R. A. Bonham, J. Chem. Phys. 27, 1414 (1957).

Enthalpy Reference Temperature = T_r = 298.15 K Standard State Pressure = p° = 0.1 MPa

	J K⁻¹mol⁻¹			kJ mol⁻¹			
T/K	$C_p°$	S°	$-[G°-H°(T_r)]/T$	$H°-H°(T_r)$	$\Delta_fH°$	$\Delta_fG°$	Log K_f
0	0.	0.	INFINITE	-10.518	60.986	60.986	INFINITE
100	33.270	180.542	252.466	-7.192	58.194	60.476	-31.589
200	35.359	203.955	222.975	-3.804	55.542	63.749	-16.649
250	38.645	212.172	220.011	-1.960	54.002	65.976	-13.785
298.15	42.886	219.330	219.330	0.	52.467	68.421	-11.987
300	43.063	219.596	219.331	0.079	52.408	68.521	-11.930
350	48.013	226.602	219.873	2.355	50.844	71.330	-10.645
400	53.048	233.343	221.138	4.882	49.354	74.360	-9.710
450	57.907	239.874	222.858	7.657	47.951	77.571	-9.004
500	62.477	246.215	224.879	10.668	46.641	80.933	-8.455
600	70.663	258.348	229.456	17.335	44.294	88.017	-7.663
700	77.714	269.783	234.408	24.763	42.300	95.467	-7.124
800	83.840	280.570	239.511	32.847	40.637	103.180	-6.737
900	89.200	290.761	244.644	41.505	39.277	111.082	-6.447
1000	93.899	300.408	249.742	50.665	38.183	119.122	-6.222
1100	98.018	309.555	254.768	60.266	37.318	127.259	-6.043
1200	101.626	318.242	259.698	70.252	36.645	135.467	-5.897
1300	104.784	326.504	264.522	80.576	36.129	143.724	-5.775
1400	107.550	334.372	269.233	91.196	35.742	152.016	-5.672
1500	109.974	341.877	273.827	102.074	35.456	160.331	-5.583
1600	112.103	349.044	278.306	113.181	35.249	168.663	-5.506
1700	113.976	355.898	282.670	124.486	35.104	177.007	-5.439
1800	115.628	362.460	286.922	135.968	35.005	185.357	-5.379
1900	117.089	368.752	291.064	147.606	34.938	193.712	-5.326
2000	118.386	374.791	295.101	159.381	34.894	202.070	-5.278
2100	119.540	380.596	299.035	171.278	34.864	210.429	-5.234
2200	120.569	386.181	302.870	183.284	34.839	218.790	-5.195
2300	121.491	391.561	306.610	195.388	34.814	227.152	-5.159
2400	122.319	396.750	310.258	207.580	34.783	235.515	-5.126
2500	123.064	401.758	313.818	219.849	34.743	243.880	-5.096
2600	123.738	406.598	317.294	232.190	34.688	252.246	-5.068
2700	124.347	411.280	320.689	244.595	34.616	260.615	-5.042
2800	124.901	415.812	324.006	257.058	34.524	268.987	-5.018
2900	125.404	420.204	327.248	269.573	34.409	277.363	-4.996
3000	125.864	424.463	330.418	282.137	34.269	285.743	-4.975
3100	126.284	428.597	333.518	294.745	34.102	294.128	-4.956
3200	126.670	432.613	336.553	307.393	33.906	302.518	-4.938
3300	127.024	436.516	339.523	320.078	33.679	310.916	-4.921
3400	127.350	440.313	342.432	332.797	33.420	319.321	-4.906
3500	127.650	444.009	345.281	345.547	33.127	327.734	-4.891
3600	127.928	447.609	348.074	358.326	32.800	336.156	-4.877
3700	128.186	451.118	350.812	371.132	32.436	344.588	-4.865
3800	128.424	454.539	353.497	383.962	32.035	353.030	-4.853
3900	128.646	457.878	356.130	396.816	31.596	361.482	-4.842
4000	128.852	461.138	358.715	409.691	31.118	369.947	-4.831
4100	129.045	464.322	361.252	422.586	30.600	378.424	-4.821
4200	129.224	467.434	363.743	435.500	30.041	386.915	-4.812
4300	129.392	470.476	366.190	448.430	29.441	395.418	-4.803
4400	129.549	473.453	368.594	461.378	28.799	403.937	-4.795
4500	129.696	476.366	370.957	474.340	28.116	412.470	-4.788
4600	129.835	479.218	373.280	487.317	27.390	421.019	-4.781
4700	129.965	482.012	375.563	500.307	26.623	429.584	-4.774
4800	130.087	484.749	377.810	513.309	25.813	438.167	-4.768
4900	130.202	487.433	380.020	526.324	24.962	446.766	-4.763
5000	130.311	490.064	382.194	539.349	24.069	455.384	-4.757
5100	130.413	492.646	384.335	552.386	23.136	464.019	-4.753
5200	130.510	495.179	386.442	565.432	22.162	472.673	-4.748
5300	130.602	497.666	388.517	578.488	21.149	481.346	-4.744
5400	130.689	500.108	390.561	591.552	20.097	490.040	-4.740
5500	130.771	502.507	392.575	604.625	19.008	498.751	-4.737
5600	130.849	504.864	394.559	617.706	17.884	507.485	-4.734
5700	130.923	507.180	396.515	630.795	16.724	516.238	-4.731
5800	130.993	509.458	398.442	643.891	15.531	525.012	-4.728
5900	131.060	511.698	400.343	656.993	14.306	533.806	-4.726
6000	131.124	513.901	402.217	670.103	13.051	542.621	-4.724

PREVIOUS: September 1965 (1 atm) CURRENT: September 1965 (1 bar)

OXIRANE (C_2H_4O) IDEAL GAS M_r = 44.05316

$S°(298.15\ K)$ = 243.012 J K^{-1} mol^{-1}

$\Delta_f H°(0\ K)$ = -40.119 ± 0.63 kJ mol^{-1}
$\Delta_f H°(298.15\ K)$ = -52.635 ± 0.63 kJ mol^{-1}

Vibrational Frequencies and Degeneracies

ν, cm^{-1}	ν, cm^{-1}	ν, cm^{-1}
3005(1)	3063 (1)	1153(1)
1490(1)	[1345](1)	892(1)
1266(1)	807 (1)	3079(1)
1120(1)	3019 (1)	1143(1)
877(1)	1470 (1)	832(1)

Ground State Quantum Weight: 1 σ = 2
Point Group: C_{2v}
Bond Distances: C-H = 1.0802 Å C-O = 1.4363 Å C-C = 1.4728 Å
Bond Angles: H-C-H = 116° 51' H_2-C-C = 158° 5' H_2-C-O = 142° 45.6'
Product of the Moments of Inertia: $I_A I_B I_C = 7.42728 \times 10^{-116}\ g^3\ cm^6$

Enthalpy of Formation

The enthalpy of combustion of ethylene oxide (g) was determined by Pell and Pilcher (1) and Crog and Hunt (2) as -312.15±0.14 and -312.55±0.20 kcal mol^{-1}, respectively. The corresponding value of $\Delta_f H°(C_2H_4O$, g, 298.15 K) was evaluated to be -12.58±0.15 and -12.19±0.22 kcal mol^{-1}. The former value is adopted.

Heat Capacity and Entropy

The vibrational frequencies were taken from Lord and Nolin (3). The infrared and Raman spectra of ethylene oxide have also been examined by Linnett (4) and Thompson and Cave (5). The vibrational frequencies assigned were slightly different from the ones reported by Lord and Nolin (3). As a result of analogies between ethylene oxide and ethylene imine, small changes of Lord and Nolin's assignment have been made by Potts (6). The bond distances and angles were obtained from Cunningham et al. (7). The investigation of the structure of ethylene oxide by electron diffraction was reported by Ackermann and Mayer (8) and Igarashi (9). The principal moments of inertia are: $I_A = 3.2793 \times 10^{-39}$, $I_B = 3.8059 \times 10^{-39}$ and $I_C = 5.9511 \times 10^{-39}$ g cm^2.

The calculated heat capacities (307.2-371.2 K) are in good agreement with the experimentally measured ones reported by Kistiakowsky and Rice (10).

References

1. A. S. Pell and G. Pilcher, Trans. Faraday Soc. 61, 71 (1965).
2. R. S. Crog and H. Hunt, J. Phys. Chem. 46, 1162 (1942).
3. R. C. Lord and B. Nolin, J. Chem. Phys. 24, 656 (1956).
4. J. W. Linnett, J. Chem. Phys. 6, 692 (1938).
5. H. W. Thompson and W. T. Cave, Trans. Faraday Soc. 47, 946 (1951).
6. W. J. Potts, Spectrochim. Acta 21, 511 (1965).
7. G. L. Cunningham, A. W. Boyd, and W. D. Gwinn, J. Chem. Phys. 17, 211 (1949).
8. P. C. Ackermann and J. E. Mayer, J. Chem. Phys. 4, 377 (1936).
9. M. Igarashi, Bull. Chem. Soc. Japan 26, 330 (1953).
10. B. Kistiakowsky and W. W. Rice, J. Chem. Phys. 8, 618 (1940).

Oxirane (C_2H_4O) $C_2H_4O_1(g)$

Enthalpy Reference Temperature = T_r = 298.15 K Standard State Pressure = p° = 0.1 MPa

	J $K^{-1}mol^{-1}$			kJ mol^{-1}			
T/K	$C_p°$	S°	-[G°-H°(T_r)]/T	H°-H°(T_r)	$\Delta_f H°$	$\Delta_f G°$	Log K_f
0	0.	0.	INFINITE	-10.863	-40.119	-40.119	INFINITE
100	33.285	202.794	278.161	-7.537	-44.363	-35.641	18.617
200	36.701	226.467	246.985	-4.104	-48.426	-25.373	6.627
250	41.673	235.154	243.764	-2.153	-50.588	-19.360	4.045
298.15	47.901	243.012	243.012	0.	-52.635	-13.159	2.305
300	48.157	243.309	243.013	0.089	-52.711	-12.914	2.249
350	55.277	251.264	243.625	2 674	-54.705	-6.121	0.914
400	62.396	259.112	245.071	5.616	-56.527	0.946	-0.124
450	69.157	266.856	247.063	8.907	-58.172	8.230	-0.955
500	75.408	274.471	249.424	12.524	-59.648	15.689	-1.639
600	86.312	289.214	254.837	20.627	-62.139	31.000	-2.699
700	95.366	303.220	260.757	29.724	-64.090	46.685	-3.484
800	102.969	316.465	266.901	39.652	-65 578	62.616	-4.088
900	109.430	328.976	273.110	50.280	-66.671	78.710	-4.568
1000	114.963	340.800	279.293	61.507	-67.429	94 907	-4.957
1100	119.723	351.986	285.398	73.247	-67.909	111.166	-5.279
1200	123.832	362.584	291.392	85.430	-68.160	127.458	-5.548
1300	127.388	372.639	297.259	97.995	-68.226	143.763	-5.776
1400	130.472	382.195	302.987	110.892	-68.143	160.068	-5.972
1500	133.156	391.291	308.573	124.076	-67.944	176.362	-6.141
1600	135.498	399.961	314.016	137.511	-67.655	192.640	-6.289
1700	137.549	408.238	319.317	151.166	-67.297	208.898	-6.419
1800	139.350	416.152	324.478	165.013	-66.890	225.133	-6.533
1900	140.938	423.730	329.504	179.029	-66.447	241.344	-6.635
2000	142.342	430.996	334.398	193.194	-65.982	257.532	-6.726
2100	143.589	437.971	339.165	207.492	-65.505	273.696	-6.808
2200	144.699	444.677	343.810	221.908	-65.025	289.838	-6.882
2300	145.691	451.131	348.337	236.428	-64.549	305.957	-6.948
2400	146.580	457.351	352.750	251.042	-64.083	322.056	-7.009
2500	147.379	463.351	357.055	265.741	-63.632	338.136	-7.065
2600	148.101	469.146	361.255	280.516	-63.201	354.198	-7.116
2700	148.753	474.748	365.356	295.359	-62.793	370.243	-7.163
2800	149.344	480.168	369.360	310.264	-62.411	386.275	-7.206
2900	149.882	485.419	373.272	325.226	-62.059	402.293	-7.246
3000	150.373	490.508	377.095	340.239	-61.738	418.300	-7.283
3100	150.821	495.446	380.834	355.299	-61.451	434.296	-7.318
3200	151.232	500.241	384.491	370.402	-61.199	450.283	-7.350
3300	151.609	504.901	388.069	385.544	-60.984	466.264	-7.380
3400	151.956	509.432	391.572	400.723	-60.807	482.239	-7.409
3500	152.275	513.842	395.003	415.934	-60.670	498.208	-7.435
3600	152.571	518.135	398.364	431.177	-60.574	514.175	-7.460
3700	152.844	522.320	401.658	446.448	-60.519	530.139	-7.484
3800	153.098	526.399	404.887	461.745	-60.508	546.103	-7.507
3900	153.333	530.379	408.054	477.067	-60.540	562.067	-7.528
4000	153.553	534.264	411.161	492.411	-60.616	578.032	-7.548
4100	153.757	538.058	414.210	507.777	-60.738	593.999	-7.568
4200	153.947	541.765	417.203	523.162	-60.906	609.971	-7.586
4300	154.125	545.390	420.142	538.566	-61.120	625.947	-7.604
4400	154.292	548.935	423.029	553.987	-61.381	641.927	-7.621
4500	154.448	552.404	425.866	569.424	-61.688	657.915	-7.637
4600	154.595	555.801	428.654	584.876	-62.043	673.909	-7.652
4700	154.732	559.127	431.394	600.343	-62.446	689.913	-7.668
4800	154.862	562.386	434.089	615.822	-62.896	705.926	-7.682
4900	154.984	565.580	436.740	631.315	-63.393	721.948	-7.696
5000	155.099	568.712	439.349	646.819	-63.938	737.981	-7.710
5100	155.207	571.785	441.915	662.334	-64.530	754.025	-7.723
5200	155.310	574.800	444.442	677.860	-65.168	770.081	-7.736
5300	155.407	577.759	446.930	693.396	-65.852	786.150	-7.748
5400	155.499	580.665	449.379	708.941	-66.582	802.233	-7.760
5500	155.586	583.519	451.792	724.496	-67.357	818.328	-7.772
5600	155.669	586.323	454.170	740.058	-68.175	834.439	-7.783
5700	155.747	589.079	456.512	755.629	-69.037	850.565	-7.795
5800	155.821	591.788	458.822	771.208	-69.942	866.707	-7.806
5900	155.892	594.453	461.098	786.793	-70.887	882.862	-7.816
6000	155.960	597.073	463.342	802.386	-71.873	899.036	-7.827

PREVIOUS: September 1965 (1 atm) CURRENT September 1965 (1 bar)

Oxirane (C_2H_4O) $C_2H_4O_1(g)$

POTASSIUM CYANIDE $((KCN)_2)$ IDEAL GAS $M_r = 130.2320$

$S°(298.15\ K) = [373.129]\ J\ K^{-1}\ mol^{-1}$

$\Delta_f H°(0\ K) = [-8.2 \pm 42]\ kJ\ mol^{-1}$
$\Delta_f H°(298.15\ K) = [-8.4 \pm 42]\ kJ\ mol^{-1}$

Vibratonal Frequencies and Degeneracies

ν, cm^{-1}	ν, cm^{-1}	ν, cm^{-1}
[240](1)	[220](1)	[2158](1)
[130](1)	[225](1)	[2158](1)
[205](1)	[207](1)	[2158](1)
[130](1)	[207](1)	[2158](1)

Ground State Quantum Weight: [1] $\sigma = 4$
Point Group: $[D_{2h}]$
Bond Distances: K-C = [2.75] Å C-N = [1.16] Å
Bond Angles: N-C-K = [140]° C-K-C = [100]° K-C-K = [80]°
Product of the Moments of Inertia: $I_A I_B I_C = [2.948540 \times 10^{-112}]\ g^3\ cm^6$

Enthalpy of Formation

$\Delta_f H°((KCN)_2$, g, 298.15 K) was calculated based on an estimated enthalpy of dissociation, 40 kcal mol^{-1} for the reaction $(KCN)_2(g) \rightarrow 2\ KCN(g)$, which was obtained by comparison with that for $(NaCN)_2(g)$.

Heat Capacity and Entropy

The molecular structure and bond angles were assumed to be the same as those for KCN(cr) reported by Bijvoet and Lely [1]. The C-N bond distance was assumed to be the same as that for KCN(g), and the K-C bond distance was assumed to be 10% longer than that in KCN(g). The vibrational frequencies were estimated by comparison with those for $K_2F_2(g)$. The last six frequencies were adopted from the bending and asymmetric stretching frequencies for KCN(g). The three principal moments of inertia are: $I_A = 40.5737 \times 10^{-39}$, $I_B = 67.3413 \times 10^{-39}$ and $I_C = 107.9149 \times 10^{-39}\ g\ cm^2$.

Reference

1. J. M. Bijvoet and J. A. Lely, Rec. Trav. Chim. 59, 908 (1940).

Potassium Cyanide $((KCN)_2)$ $C_2K_2N_2(g)$

Enthalpy Reference Temperature = T_r = 298.15 K; Standard State Pressure = p° = 0.1 MPa

T/K	$C°_p$ (J K⁻¹mol⁻¹)	S° (J K⁻¹mol⁻¹)	$-[G°-H°(T_r)]/T$ (J K⁻¹mol⁻¹)	$H°-H°(T_r)$ (kJ mol⁻¹)	$\Delta_f H°$ (kJ mol⁻¹)	$\Delta_f G°$ (kJ mol⁻¹)	Log K_f
0	0.	0.	INFINITE	-24.786	-8.217	-8.217	INFINITE
100	77.547	266.448	465.521	-19.907	-9.825	-13.178	6.883
200	103.437	330.321	383.046	-10.545	-9.215	-16.652	4.349
250	107.785	353.916	374.932	-5.254	-8.720	-18.570	3.880
298.15	110.262	373.129	373.129	0.	-8.368	-20.502	3.592
300	110.337	373.812	373.132	0.204	-8.359	-20.578	3.583
350	112.029	390.955	374.482	5.766	-12.978	-22.435	3.348
400	113.310	406.001	377.500	11.400	-13.082	-23.779	3.105
450	114.405	419.411	381.425	17.094	-13.238	-25.107	2.914
500	115.421	431.518	385.839	22.839	-13.451	-26.415	2.760
600	117.361	452.734	395.269	34.479	-14.030	-28.957	2.521
700	119.200	470.965	404.811	46.308	-14.781	-31.388	2.342
800	120.890	486.995	414.102	58.314	-15.680	-33.700	2.200
900	122.393	501.322	423.011	70.480	-16.725	-35.892	2.083
1000	123.702	514.287	431.500	82.786	-17.931	-37.958	1.983
1100	124.828	526.131	439.573	95.214	-177.226	-30.675	1.457
1200	125.791	537.035	447.246	107.747	-176.702	-17.376	0.756
1300	126.614	547.137	454.546	120.368	-176.230	-4.118	0.165
1400	127.317	556.547	461.500	133.065	-175.804	9.105	-0.340
1500	127.920	565.352	468.133	145.828	-175.417	22.299	-0.777
1600	128.439	573.624	474.470	158.647	-175.069	35.468	-1.158
1700	128.888	581.425	480.534	171.514	-174.757	48.617	-1.494
1800	129.277	588.803	486.346	184.422	-174.481	61.749	-1.792
1900	129.616	595.802	491.924	197.367	-174.241	74.866	-2.058
2000	129.913	602.458	497.286	210.344	-174.041	87.971	-2.298
2100	130.175	608.803	502.447	223.349	-173.881	101.068	-2.514
2200	130.405	614.864	507.420	236.378	-173.766	114.158	-2.710
2300	130.610	620.666	512.218	249.429	-173.699	127.244	-2.890
2400	130.793	626.228	516.854	262.499	-173.684	140.328	-3.054
2500	130.956	631.571	521.336	275.587	-173.726	153.412	-3.205
2600	131.102	636.710	525.675	288.690	-173.831	166.499	-3.345
2700	131.234	641.660	529.880	301.807	-174.002	179.592	-3.474
2800	131.353	646.435	533.958	314.936	-174.248	192.692	-3.595
2900	131.460	651.046	537.916	328.077	-174.573	205.803	-3.707
3000	131.558	655.505	541.762	341.228	-174.984	218.926	-3.812
3100	131.647	659.820	545.501	354.388	-175.488	232.064	-3.910
3200	131.729	664.001	549.139	367.557	-176.094	245.221	-4.003
3300	131.803	668.056	552.682	380.734	-176.789	258.396	-4.090
3400	131.871	671.991	556.133	393.917	-177.609	271.596	-4.173
3500	131.934	675.815	559.498	407.108	-178.547	284.820	-4.251
3600	131.992	679.532	562.781	420.304	-179.619	298.074	-4.325
3700	132.046	683.150	565.986	433.506	-180.830	311.360	-4.396
3800	132.095	686.672	569.116	446.713	-182.188	324.680	-4.463
3900	132.141	690.104	572.174	459.925	-183.704	338.038	-4.528
4000	132.184	693.450	575.164	473.141	-185.374	351.436	-4.589
4100	132.224	696.714	578.089	486.362	-187.220	364.878	-4.649
4200	132.261	699.901	580.952	499.586	-189.258	378.368	-4.706
4300	132.295	703.013	583.754	512.814	-191.491	391.909	-4.761
4400	132.327	706.055	586.499	526.045	-193.373	405.450	-4.813
4500	132.358	709.029	589.189	539.279	-195.885	419.087	-4.865
4600	132.386	711.939	591.826	552.516	-198.594	432.783	-4.914
4700	132.413	714.786	594.412	565.756	-201.465	446.535	-4.963
4800	132.438	717.574	596.949	578.999	-204.323	460.327	-5.009
4900	132.461	720.305	599.439	592.244	-207.595	474.208	-5.055
5000	132.483	722.981	601.883	605.491	-211.075	488.157	-5.100
5100	132.504	725.605	604.283	618.740	-214.766	502.178	-5.143
5200	132.524	728.178	606.641	631.992	-218.669	516.273	-5.186
5300	132.542	730.703	608.958	645.245	-222.784	530.446	-5.228
5400	132.560	733.180	611.236	658.500	-227.113	544.698	-5.269
5500	132.577	735.613	613.475	671.757	-231.652	559.032	-5.309
5600	132.593	738.002	615.678	685.016	-236.402	573.451	-5.349
5700	132.608	740.349	617.844	698.276	-240.670	587.861	-5.387
5800	132.622	742.655	619.976	711.537	-245.745	602.440	-5.426
5900	132.635	744.922	622.075	724.800	-251.012	617.109	-5.463
6000	132.648	747.152	624.141	738.064	-256.468	631.869	-5.501

PREVIOUS: March 1966 (1 atm) CURRENT: March 1966 (1 bar)

LITHIUM CARBIDE (Li_2C_2) CRYSTAL M_r = 37.904

$S°(298.15\ K) = 58.6 \pm 4\ J\ K^{-1}\ mol^{-1}$

$\Delta_f H°(0\ K)$ = Unknown

$\Delta_f H°(298.15\ K) = -59.4 \pm 8\ kJ\ mol^{-1}$

Enthalpy of Formation

Data from the U. S. Nat. Bur. of Stand. (1)

Heat Capacity and Entropy

Data from the U. S. Nat. Bur. of Stand. (1)

Reference

1. G. T. Fursukawa, M. L. Reilly et al., U. S. Nat. Bur. Stand. Report 6928, Appendix B, 1960.

Lithium Carbide (Li_2C_2) $C_2Li_2(cr)$

Enthalpy Reference Temperature = T_r = 298.15 K Standard State Pressure = p° = 0.1 MPa

	J K⁻¹mol⁻¹			kJ mol⁻¹			
T/K	$C_p°$	S°	$-[G°-H°(T_r)]/T$	$H°-H°(T_r)$	$\Delta_f H°$	$\Delta_f G°$	Log K_f
0							
100							
200							
298.15	71.546	58.576	58.576	0.	-59.413	-66.494	11.649
300	71.965	59.020	58.577	0.133	-59.341	-66.538	11.585
400	87.404	82.023	61.608	8.166	-54.931	-69.576	9.086
500	95.060	102.405	67.777	17.314	-53.043	-73.493	7.678
600	99.705	120.169	75.062	27.064	-47.854	-78.066	6.796
700	102.926	135.791	82.644	37.203	-42.414	-83.528	6.233
800	105.353	149.698	90.173	47.620	-36.814	-89.782	5.862
900	107.361	162.225	97.494	58.258	-31.101	-96.745	5.615
1000	109.077	173.628	104.546	69.082	-25.287	-104.350	5.451
1100	110.583	184.095	111.308	80.066	-19.378	-112.542	5.344
1200	112.006	193.779	117.782	91.196	-13.375	-121.275	5.279
1300	113.345	202.797	123.979	102.464	-7.278	-130.513	5.244
1400	114.600	211.243	129.914	113.862	-1.086	-140.225	5.232
1500	115.813	219.191	135.603	125.383	5.203	-150.382	5.237
1600	116.985	226.703	141.064	137.022	11.588	-160.962	5.255
1700	118.156	233.831	146.313	148.780	-127.164	-164.769	5.063
1800	119.286	240.616	151.365	160.652	-119.829	-167.192	4.852
1900	120.416	247.096	156.234	172.637	-112.404	-170.025	4.674
2000	121.503	253.301	160.934	184.734	-104.891	-173.252	4.525

PREVIOUS: CURRENT: December 1960

Lithium Carbide (Li_2C_2) $C_2Li_2(cr)$

MAGNESIUM CARBIDE (MgC_2) CRYSTAL M_r = 48.327

S°(298.15 K) = 54.4 ± 8 J K^{-1} mol^{-1}

$\Delta_f H°$(0 K) = Unknown
$\Delta_f H°$(298.15 K) = 87.9 ± 21 kJ mol^{-1}

Enthalpy of Formation

Data from (1).

Heat Capacity and Entropy

Data from (1).

Reference

1. G. T. Fusukawa, M. L. Reilly et al., U. S. Nat. Bur. Stand. Report 6928, Appendix B, 1960.

Magnesium Carbide (MgC_2) C_2Mg_1(cr)

Enthalpy Reference Temperature = T_r = 298.15 K Standard State Pressure = p° = 0.1 MPa

T/K	$C_p°$ (J $K^{-1}mol^{-1}$)	S° (J $K^{-1}mol^{-1}$)	-[G°-H°(T_r)]/T (J $K^{-1}mol^{-1}$)	H°-H°(T_r) (kJ mol^{-1})	$\Delta_f H°$ (kJ mol^{-1})	$\Delta_f G°$ (kJ mol^{-1})	Log K_f
0							
100							
200							
298.15	56.233	54.392	54.392	0.	87.864	83.099	-14.559
300	56.442	54.740	54.393	0.104	87.906	83.070	-14.464
400	64.392	72.161	56.714	6.179	90.403	81.091	-10.589
500	68.408	86.989	61.326	12.832	93.063	78.455	-8.196
600	70.877	99.691	66.688	19.802	95.687	75.286	-6.554
700	72.634	110.753	72.210	26.980	98.221	71.684	-5.349
800	74.015	120.545	77.651	34.315	100.646	67.726	-4.422
900	75.145	129.329	82.914	41.774	102.952	63.472	-3.684
1000	76.149	137.300	87.960	49.340	96.539	59.679	-3.117
1100	77.027	144.599	92.782	56.999	98.573	55.895	-2.654
1200	77.864	151.337	97.384	64.744	100.636	51.925	-2.260
1300	78.701	157.603	101.778	72.572	102.735	47.780	-1.920
1400	79.454	163.463	105.977	80.480	-22.537	46.640	-1.740
1500	80.207	168.971	109.995	88.464	-19.007	51.459	-1.792
1600	80.919	174.170	113.845	96.520	-15.434	56.040	-1.830
1700	81.630	179.097	117.540	104.647	-11.818	60.397	-1.856
1800	82.341	183.783	121.091	112.846	-8.154	64.539	-1.873
1900	83.011	188.253	124.509	121.114	-4.444	68.477	-1.883
2000	83.722	192.529	127.804	129.450	-0.686	72.218	-1.886
2100	84.391	196.630	130.984	137.856	3.122	75.769	-1.885
2200	85.061	200.571	134.058	146.329	6.979	79.139	-1.879
2300	85.730	204.367	137.033	154.868	10.885	82.331	-1.870
2400	86.400	208.030	139.916	163.475	14.841	85.353	-1.858
2500	87.069	211.571	142.711	172.148	18.848	88.209	-1.843

PREVIOUS: CURRENT: December 1960

Magnesium Carbide (MgC_2) C_2Mg_1(cr)

CNC RADICAL (CNC) IDEAL GAS M_r = 38.0287

$S°(298.15\ K) = [230 \pm 8]\ J\ K^{-1}\ mol^{-1}$

$\Delta_f H°(0\ K) = [552 \pm 126]\ kJ\ mol^{-1}$
$\Delta_f H°(298.15\ K) = [556 \pm 126]\ kJ\ mol^{-1}$

Electronic Levels and Quantum Weights

ϵ_i, cm^{-1}	g_i
0	2
26.41	2
30338.53	4
34802.33	2

Vibrational Frequencies and Degeneracies

ν, cm^{-1}
[1200] (1)
321 (2)
[900] (1)

Point Group: $D_{\infty h}$ σ = 2
Bond Distances: C-N = 1.245 Å
Bond Angle: C-N-C = 180°
Rotational Constant: $B_0 = 0.452740\ cm^{-1}$

Enthalpy of Formation

Merer and Travis (1) state that the observed bond length in CNC(g) indicates double-bonded CN links. The value of the C = N dissociation energy is derived as 147 kcal from the enthalpy of combustion of n-butyl-isobutylidine amine determined by Coates and Sutton (2). However, due to possible extra binding energy from the unpaired electron, we assume here a slightly higher value of 160 kcal. This yields a $\Delta_f H°$(CNC, g, 298.15 K) = 133 kcal mol^{-1}. It is also possible to calculate another value by assuming that the ratio of the enthalpy of atomization of Si_2N to that of C_2N(g) is equal to the ratio of the dissociation energies of SiN(g) and CN(g); by this method $\Delta_f H°$(CNC, g, 298.15 K) = 144 kcal mol^{-1}.

Heat Capacity and Entropy

Merer and Travis (1) have observed the ultraviolet absorption sectrum of the CNC radical. They ave assigned the configuration, bond distance, and the electronic levels. In addition, they have derived ν_2; the remaining frequencies are estimated by comparison with the $^1\Pi$ state of C_3(g), which has been assigned by Weltner and McLeod (3).

References

1. A. J. Merer and D. N. Travis, Can. J. Phys. 44, 353 (1956).
2. G. E. Coates and L. E. Sutton, J. Chem. Soc. 1187 (1948).
3. W. Weltner and D. J. McLeod, J. Chem. Phys. 40, 1305 (1964); 45, 3096 (1966).

CNC Radical (C_2N) $C_2N_1(g)$

Enthalpy Reference Temperature = T_r = 298.15 K   Standard State Pressure = p° = 0.1 MPa

T/K	$C_p°$ (J K⁻¹mol⁻¹)	S° (J K⁻¹mol⁻¹)	$-[G°-H°(T_r)]/T$ (J K⁻¹mol⁻¹)	$H°-H°(T_r)$ (kJ mol⁻¹)	$\Delta_f H°$ (kJ mol⁻¹)	$\Delta_f G°$ (kJ mol⁻¹)	Log K_f
0	0.	0.	INFINITE	-11.080	551.829	551.829	INFINITE
100	32.963	188.237	267.905	-7.967	553.372	542.729	-283.493
200	40.683	213.673	234.935	-4.252	554.978	531.475	-138.807
250	43.425	223.053	231.645	-2.148	555.764	525.507	-109.799
298.15	45.751	230.905	230.905	0.	556.472	519.615	-91.034
300	45.835	231.188	230.906	0.085	556.498	519.386	-90.433
350	47.975	238.418	231.472	2.431	557.174	513.147	-76.583
400	49.849	244.949	232.755	4.878	557.787	506.814	-66.183
450	51.465	250.916	234.446	7.412	558.332	500.409	-58.086
500	52.844	256.412	236.371	10.020	558.807	493.947	-51.602
600	55.010	266.249	240.550	15.419	559.557	480.901	-41.866
700	56.572	274.852	244.849	21.002	560.073	467.748	-34.904
800	57.713	282.485	249.086	26.719	560.394	454.534	-29.678
900	58.563	289.334	253.184	32.535	560.552	441.291	-25.612
1000	59.208	295.539	257.114	38.425	560.576	428.037	-22.358
1100	59.706	301.206	260.868	44.372	560.486	414.786	-19.697
1200	60.097	306.419	264.450	50.363	560.300	401.549	-17.479
1300	60.410	311.242	267.866	56.389	560 031	388.330	-15.603
1400	60.663	315.728	271.126	62.443	559.689	375.134	-13.996
1500	60.871	319.921	274.241	68.520	559.283	361.966	-12.605
1600	61.043	323.855	277.220	74.616	558.820	348.826	-11.388
1700	61.187	327.560	280.074	80.728	558.305	335.717	-10.315
1800	61.309	331.061	282.810	86.853	557.742	322.639	-9.363
1900	61.414	334.379	285.438	92.989	557 135	309.594	-8.511
2000	61.503	337.532	287.964	99.135	556.488	296.583	-7.746
2100	61.580	340.534	290.397	105.289	555.802	283.604	-7.054
2200	61.648	343.400	292.741	111.450	555.080	270.659	-6.426
2300	61.707	346.142	295.004	117.618	554.324	257.748	-5.854
2400	61.759	348.770	297.190	123.792	553.536	244.870	-5.329
2500	61.805	351.292	299.304	129.970	552.716	232.025	-4.848
2600	61.846	353.716	301.350	136.153	551.865	219.214	-4.404
2700	61.883	356.051	303.333	142.339	550.985	206.437	-3.994
2800	61.916	358.302	305.256	148.529	550.077	193.692	-3.613
2900	61.946	360.476	307.123	154.722	549.141	180.981	-3.260
3000	61.973	362.576	308.937	160.918	548.178	168.302	-2.930
3100	61.998	364.609	310.700	167.117	547.188	155.656	-2.623
3200	62.020	366.577	312.416	173.318	546.173	143.042	-2.335
3300	62.041	368.486	314.086	179.521	545.132	130.460	-2.065
3400	62.061	370.339	315.713	185.726	544.065	117.910	-1.811
3500	62.079	372.138	317.300	191.933	542.974	105.392	-1 573
3600	62.096	373.887	318.848	198.141	541.859	92.905	-1.348
3700	62.113	375.588	320.358	204.352	540.719	80.450	-1.136
3800	62.129	377.245	321.834	210.564	539.556	68.026	-0.935
3900	62.145	378.859	323.275	216.778	538.369	55.632	-0.745
4000	62.160	380.433	324.685	222.993	537.159	43.270	-0.565
4100	62.176	381.968	326.063	229.210	535.926	30.938	-0.394
4200	62.191	383.466	327.412	235.428	534.670	18.636	-0.232
4300	62.207	384.930	328.733	241.648	533.392	6.365	-0.077
4400	62.224	386.360	330.026	247.870	532.092	-5.877	0.070
4500	62.241	387.759	331.294	254.093	530.769	-18.088	0.210
4600	62.259	389.127	332.536	260.318	529.425	-30.271	0.344
4700	62.277	390.466	333.754	266.545	528.059	-42.423	0.471
4800	62.297	391.777	334.950	272.773	526.671	-54.546	0.594
4900	62.318	393.062	336.123	279.004	525.262	-66.641	0.710
5000	62.339	394.321	337.274	285.237	523.833	-78.706	0.822
5100	62.362	395.556	338.405	291.472	522.382	-90.742	0.929
5200	62.387	396.767	339.515	297.709	520.911	-102.750	1.032
5300	62.412	397.956	340.607	303.949	519.419	-114.729	1.131
5400	62.439	399.123	341.680	310.192	517.907	-126.680	1.225
5500	62.468	400.269	342.735	316.437	516.376	-138.603	1.316
5600	62.498	401.395	343.772	322.686	514.824	-150.498	1.404
5700	62.530	402.501	344.793	328.937	513.253	-162.364	1.488
5800	62.563	403.589	345.797	335.192	511.662	-174.204	1.569
5900	62.598	404.659	346.786	341.450	510.051	-186.015	1.647
6000	62.634	405.711	347.759	347.711	508.421	-197.799	1.722

PREVIOUS: March 1967 (1 atm)   CURRENT: March 1967 (1 bar)

CNC Radical (C_2N) $C_2N_1(g)$

ETHANEDINITRILE $((CN)_2)$ IDEAL GAS M_r = 52.0354

$\Delta_fH°$(0 K) = 307.219 ± 1.80 kJ mol^{-1}

S°(298.15 K) = 241.565 J K^{-1} mol^{-1} $\Delta_fH°$(298.15 K) = 309.072 ± 1.80 kJ mol^{-1}

Vibrational Frequencies and Degeneracies

ν, cm^{-1}
2328.5(1)
850.6(1)
2149 (1)
507.2(2)
240 (2)

Ground State Quantum Weight: 1 σ = 2

Point Group: $D_{\infty h}$

Bond Distances: C–N = 1.157 Å C–C = 1.380 Å

Bond Angles: C–N–N–C = 180°

Rotational Constant: B_0 = 0.157545 ± 0.00015 cm^{-1}

Enthalpy of Formation

Knowlton and Prosen (1) report $\Delta_fH°$(298.15 K) = -261.94±0.43 kcal for the reaction $C_2N_2(g) + 2\ O_2(g) + 2\ CO_2(g) + N_2(g)$. When corrected for a change in the atomic weight of carbon, this becomes $\Delta_fH°$(298.15 K) = -261.95±0.43 kcal, whence $\Delta_fH°$(298.15 K) for $C_2N_2(g)$ is found to be 73.87±0.43 kcal mol^{-1}.

Heat Capacity and Entropy

Langseth and Moller (2) determined the vibrational frequencies by Raman spectroscopy. Moller and Stoicheff (3) determined B_0 by Raman spectroscopy and calculated the bond lengths listed above. These agree with lengths found from electron-diffraction measurements by Langseth and Moller (2) (CN distance = 1.15 Å, CC distance = 1.38 Å) and Pauling et al. (4) (CN distance = 1.16±0.02 A, CC distance = 1.37±0.02 Å). From measurements reported by Ruehrwein and Giauque (5), S°(298.15 K) is calculated by the 3rd law to be 57.63 cal K^{-1} mol^{-1}, in satisfactory agreement with the value calculated here.

References

1. J. W. Knowlton and E. J. Prosen, J. Res. Nat. Bur. Stand. 46, 489 (1951).
2. A. Langseth and C. K. Moller, Acta Chem. Scand. 4, 725 (1950).
3. C. K. Moller and B. P. Stoicheff, Can. J. Phys. 32, 625 (1954).
4. L. Pauling, H. D. Springall, and K. J. Palmer, J. Amer. Chem. Soc. 61, 927 (1939).
5. R. A. Ruehrwein and W. F. Giauque, J. Amer. Chem. Soc. 61, 2940 (1939).

Ethanedinitrile $((CN)_2)$ $C_2N_2(g)$

Enthalpy Reference Temperature = T_r = 298.15 K Standard State Pressure = p° = 0.1 MPa

T/K	$C_p°$	S°	$-[G°-H°(T_r)]/T$	$H°-H°(T_r)$	$\Delta_fH°$	$\Delta_fG°$	Log K_f
	J $K^{-1}mol^{-1}$			kJ mol^{-1}			
0	0.	0.	INFINITE	-12.626	307.219	307.219	INFINITE
100	36.399	191.121	286.323	-9.520	307.303	304.362	-158.982
200	48.910	220.447	246.531	-5.217	308.042	301.183	-78.661
250	53.388	231.863	242.482	-2.655	308.558	299.409	-62.558
298.15	56.746	241.565	241.565	0.	309.072	297.601	-52.138
300	56.859	241.917	241.566	0.105	309.092	297.529	-51.804
350	59.585	250.894	242.269	3.019	309.606	295.561	-44.110
400	61.799	259.000	243.862	6.055	310.079	293.522	-38.330
450	63.677	266.390	245.961	9.193	310.495	291.427	-33.828
500	65.332	273.186	248.348	12.419	310.850	289.288	-30.222
600	68.218	285.360	253.526	19.101	311.392	284.922	-24.805
700	70.704	296.067	258.853	26.050	311.752	280.479	-20.930
800	72.861	305.652	264.114	33.230	311.982	275.995	-18.021
900	74.717	314.344	269.220	40.612	312.117	271.487	-15.757
1000	76.302	322.301	274.136	48.165	312.184	266.969	-13.945
1100	77.650	329.638	278.852	55.864	312.199	262.446	-12.463
1200	78.795	336.445	283.371	63.688	312.171	257.924	-11.227
1300	79.769	342.791	287.701	71.618	312.109	253.406	-10.182
1400	80.598	348.734	291.850	79.637	312.015	248.894	-9.286
1500	81.308	354.320	295.831	87.733	311.895	244.389	-8.510
1600	81.918	359.587	299.652	95.895	311.748	239.893	-7.832
1700	82.445	364.569	303.326	104.114	311.577	235.407	-7.233
1800	82.901	369.295	306.861	112.382	311.382	230.932	-6.701
1900	83.299	373.788	310.266	120.692	311.165	226.469	-6.226
2000	83.648	378.070	313.550	129.040	310.925	222.017	-5.798
2100	83.954	382.159	316.720	137.421	310.663	217.578	-5.412
2200	84.225	386.071	319.784	145.830	310.379	213.152	-5.061
2300	84.465	389.820	322.748	154.265	310.073	208.740	-4.741
2400	84.679	393.419	325.619	162.722	309.746	204.341	-4.447
2500	84.870	396.880	328.400	171.199	309.397	199.956	-4.178
2600	85.041	400.212	331.099	179.695	309.026	195.586	-3.929
2700	85.195	403.425	333.718	188.207	308.634	191.230	-3.700
2800	85.335	406.525	336.263	196.734	308.220	186.889	-3.486
2900	85.461	409.522	338.738	205.274	307.785	182.563	-3.288
3000	85.576	412.421	341.146	213.826	307.328	178.253	-3.104
3100	85.680	415.229	343.491	222.388	306.850	173.958	-2.931
3200	85.775	417.951	345.776	230.961	306.350	169.680	-2.770
3300	85.863	420.592	348.003	239.543	305.828	165.417	-2.618
3400	85.943	423.156	350.176	248.134	305.285	161.170	-2.476
3500	86.016	425.649	352.297	256.732	304.720	156.939	-2.342
3600	86.084	428.073	354.368	265.337	304.134	152.725	-2.216
3700	86.147	430.432	356.392	273.948	303.526	148.528	-2.097
3800	86.205	432.730	358.371	282.566	302.896	144.347	-1.984
3900	86.259	434.970	360.306	291.189	302.244	140.183	-1.878
4000	86.309	437.155	362.200	299.818	301.571	136.036	-1.776
4100	86.356	439.287	364.055	308.451	300.876	131.906	-1.681
4200	86.399	441.368	365.871	317.089	300.159	127.794	-1.589
4300	86.439	443.402	367.650	325.731	299.420	123.699	-1.503
4400	86.477	445.389	369.395	334.376	298.659	119.621	-1.420
4500	86.513	447.333	371.105	343.026	297.877	115.561	-1.341
4600	86.546	449.235	372.783	351.679	297.073	111.518	-1.266
4700	86.577	451.096	374.429	360.335	296.247	107.493	-1.195
4800	86.606	452.919	376.046	368.994	295.398	103.486	-1.126
4900	86.634	454.705	377.633	377.656	294.528	99.497	-1.061
5000	86.660	456.456	379.192	386.321	293.635	95.525	-0.998
5100	86.684	458.172	380.724	394.988	292.720	91.572	-0.938
5200	86.707	459.856	382.229	403.658	291.783	87.637	-0.880
5300	86.729	461.508	383.710	412.329	290.824	83.720	-0.825
5400	86.750	463.129	385.165	421.003	289.842	79.822	-0.772
5500	86.769	464.721	386.597	429.679	288.837	75.942	-0.721
5600	86.788	466.285	388.006	438.357	287.811	72.080	-0.672
5700	86.805	467.821	389.393	447.037	286.761	68.237	-0.625
5800	86.822	469.331	390.759	455.718	285.688	64.413	-0.580
5900	86.838	470.815	392.103	464.401	284.592	60.607	-0.537
6000	86.853	472.275	393.427	473.086	283.472	56.820	-0.495

PREVIOUS: March 1961 (1 atm) CURRENT: March 1961 (1 bar)

Ethanedinitrile $((CN)_2)$ $C_2N_2(g)$

SODIUM CYANIDE ((NaCN)$_2$) IDEAL GAS M_r = 98.01494

$\Delta_fH°(0\ K)$ = [-8.3 ± 13] kJ mol^{-1}

S°(298.15 K) = [347.102] J K^{-1} mol^{-1} $\Delta_fH°(298.15\ K)$ = [-8.8 ± 13] kJ mol^{-1}

Vibrational Frequencies and Degeneracies

ν, cm^{-1}	ν, cm^{-1}	ν, cm^{-1}
[295](1)	[250](1)	[2176](1)
[175](1)	[290](1)	[2176](1)
[250](1)	[239](1)	[2176](1)
[170](1)	[239](1)	[2176](1)

Ground State Quantum Weight: 1 σ = [4]

Point Group: [D_{2h}]

Bond Distances: N-C = [1.16] Å C-Na = [2.19] Å

Bond Angles: C-Na-C = [105]° Na-C-Na = [75]° N-C-Na = [142.5]°

Product of the Moments of Inertia: $I_AI_BI_C$ = [4.483815 x 10^{-113}] g^3 cm^6

Enthalpy of Formation

The mass spectra of vapors from samples of NaCN have been observed by Porter (1) at temperatures around 1000 K. The results indicate that the compound evaporates as NaCN(g) and $Na_2(CN)_2$(g). A comparison of relative ion currents produced by electron bombardment of NaCN vapors effusing from single-and double-oven-type Knudsen cells, yielded information on the partial pressures of monomer and dimer. Based on the partial pressures for NaCN(g) and $(NaCN)_2$(g) reported at temperatures, 903-1049 K, the vapor pressures of $(NaCN)_2$(g) over NaCN(ℓ), 1078.2-1626.2 K, were calculated from the total vapor pressure measurements reported by Ingold (2). By the 2nd and 3rd law methods, the values of $\Delta_{vap}H°(298.15\ K)$ (ℓ → dimer) were evaluated to be 38.89 and 38.21 kcal mol^{-1}, respectively. Using the 3rd law value, the enthalpy of formation for $(NaCN)_2$(g) was evaluated.

Heat Capacity and Entropy

The molecular structure, bond distances and C-Na-C bond angle were estimated by Porter (1). The vibrational frequencies were estimated by comparison with those for NaCl(g). The last six frequencies were the bending and asymmetric stretching frequencies for NaCN(g). The principal moments of inertia are: I_A = 13.5707 x 10^{-39}, I_B = 51.0944 x 10^{-39} and I_C = 64.6652 x 10^{-39} g cm^2.

References

1. R. F. Porter, J. Chem. Phys. 35, 318 (1961).
2. C. K. Ingold, J. Chem. Soc. 123, 885 (1923).

Sodium Cyanide ((NaCN)$_2$) $C_2N_2Na_2$(g)

Enthalpy Reference Temperature = T_r = 298.15 K Standard State Pressure = p° = 0.1 MPa

T/K	$C_p°$ (J K^{-1}mol^{-1})	S°	-[G°-H°(T_r)]/T	H°-H°(T_r) (kJ mol^{-1})	$\Delta_fH°$	$\Delta_fG°$	Log K_f
0	0.	0.	INFINITE	-23.178	-8.299	-8.299	INFINITE
100	67.677	247.172	435.395	-18.822	-9.644	-13.427	7.013
200	98.488	305.757	356.722	-10.193	-9.489	-17.132	4.474
250	104.306	328.419	348.858	-5.110	-9.102	-19.087	3.988
298.15	107.686	347.102	347.102	0.	-8.786	-21.041	3.686
300	107.789	347.768	347.104	0.199	-8.777	-21.117	3.677
350	110.089	364.568	348.426	5.650	-8.631	-23.189	3.461
400	111.781	379.383	351.387	11.198	-13.984	-24.854	3.246
450	113.163	392.631	355.247	16.823	-14.207	-26.200	3.041
500	114.386	404.618	359.594	22.512	-14.466	-27.519	2.875
600	116.596	425.672	368.900	34.063	-15.087	-30.073	2.618
700	118.601	443.798	378.335	45.824	-15.818	-32.514	2.426
800	120.401	459.755	387.535	57.776	-16.636	-34.844	2.275
900	121.985	474.030	396.366	69.897	-17.535	-37.067	2.151
1000	123.355	486.955	404.789	82.166	-18.527	-39.185	2.047
1100	124.529	498.769	412.803	94.562	-19.633	-41.199	1.956
1200	125.531	509.648	420.426	107.066	-214.401	-38.226	1.664
1300	126.385	519.731	427.682	119.663	-213.953	-23.563	0.947
1400	127.115	529.124	434.596	132.339	-213.547	-8.933	0.333
1500	127.740	537.916	441.194	145.083	-213.178	5.669	-0.197
1600	128.278	546.178	447.500	157.884	-212.842	20.248	-0.661
1700	128.743	553.969	453.536	170.736	-212.538	34.807	-1.069
1800	129.146	561.339	459.322	183.631	-212.264	49.348	-1.432
1900	129.497	568.332	464.877	196.563	-212.018	63.875	-1.756
2000	129.805	574.982	470.218	209.529	-211.801	78.390	-2.047
2100	130.075	581.322	475.358	222.523	-211.611	92.895	-2.311
2200	130.314	587.379	480.314	235.543	-211.449	107.391	-2.550
2300	130.526	593.176	485.096	248.585	-211.316	121.881	-2.768
2400	130.715	598.735	489.716	261.647	-211.212	136.366	-2.968
2500	130.884	604.075	494.184	274.727	-211.139	150.846	-3.152
2600	131.035	609.211	498.510	287.823	-211.099	165.325	-3.321
2700	131.171	614.159	502.702	300.934	-211.093	179.803	-3.478
2800	131.294	618.932	506.768	314.057	-211.124	194.281	-3.624
2900	131.406	623.541	510.716	327.192	-211.193	208.760	-3.760
3000	131.507	627.998	514.552	340.338	-211.303	223.243	-3.887
3100	131.599	632.311	518.281	353.493	-211.458	237.731	-4.006
3200	131.683	636.491	521.910	366.658	-211.660	252.224	-4.117
3300	131.761	640.544	525.444	379.830	-211.912	266.724	-4.222
3400	131.831	644.478	528.887	393.010	-212.218	281.232	-4.321
3500	131.896	648.301	532.245	406.196	-212.581	295.751	-4.414
3600	131.956	652.017	535.521	419.389	-213.006	310.280	-4.502
3700	132.012	655.634	538.718	432.587	-213.497	324.823	-4.586
3800	132.063	659.155	541.841	445.791	-214.057	339.379	-4.665
3900	132.111	662.586	544.894	459.000	-214.692	353.951	-4.741
4000	132.155	665.931	547.878	472.213	-215.406	368.541	-4.813
4100	132.196	669.195	550.797	485.430	-216.204	383.149	-4.881
4200	132.234	672.381	553.654	498.652	-217.087	397.778	-4.947
4300	132.270	675.493	556.452	511.877	-218.069	412.429	-5.010
4400	132.303	678.534	559.192	525.106	-219.066	427.097	-5.070
4500	132.334	681.508	561.877	538.338	-220.208	441.794	-5.128
4600	132.364	684.417	564.510	551.573	-221.471	456.519	-5.184
4700	132.391	687.264	567.091	564.810	-222.845	471.272	-5.238
4800	132.417	690.051	569.624	578.051	-224.335	486.056	-5.289
4900	132.441	692.782	572.110	591.294	-225.947	500.872	-5.339
5000	132.464	695.458	574.550	604.539	-227.670	515.722	-5.388
5100	132.486	698.081	576.946	617.786	-229.538	530.608	-5.435
5200	132.506	700.654	579.301	631.036	-231.543	545.532	-5.480
5300	132.526	703.178	581.614	644.288	-233.691	560.496	-5.524
5400	132.544	705.655	583.888	657.541	-235.987	575.502	-5.567
5500	132.561	708.088	586.125	670.796	-238.436	590.552	-5.609
5600	132.577	710.476	588.324	684.053	-241.043	605.647	-5.649
5700	132.593	712.823	590.488	697.312	-243.813	620.791	-5.689
5800	132.608	715.129	592.617	710.572	-246.750	635.985	-5.728
5900	132.622	717.396	594.712	723.833	-249.859	651.231	-5.766
6000	132.635	719.625	596.776	737.096	-252.828	666.495	-5.802

PREVIOUS: March 1966 (1 atm) CURRENT: March 1966 (1 bar)

$S°(298.15\ K) = 233.16\ J\ K^{-1}\ mol^{-1}$

$\Delta_f H°(0\ K) = 282.4 \pm 63\ kJ\ mol^{-1}$
$\Delta_f H°(298.15\ K) = 286.6 \pm 63\ kJ\ mol^{-1}$

Electronic Levels and Quantum Weights

ε_i, cm^{-1}	g_i
0	3
[4050]	[2]
[6450]	[1]

Vibrational Frequencies and Degeneracies

ν, cm^{-1}
1074 (1)
381 (2)
1978 (1)

Point Group: $C_{\infty h}$ σ = 1
Bond Distance: C–C = [1.160] Å
Bond Angle: C–C–O = 180°
Rotational Constant: B_0 = [0.408483] cm^{-1}

Enthalpy of Formation

The adopted enthalpy of formation, $\Delta_f H°(C_2O,\ g,\ 298.15\ K) = 68.5\ kcal\ mol^{-1}$, was calculated from the enthalpy of reaction $\Delta_r H°(298.15\ K) = 65.5\ kcal\ mol^{-1}$ for $C_3O_2(g) = C_2O(g) + CO(g)$ using all JANAF functions (dated Sept. 30, 1965). The value of $\Delta_r H°$ was obtained from the 3rd law calculation of the equilibrium constant (Kp = 3.3 x 10^{-6} atm at 1060 K) which was determined in the studies of decomposition rate of $C_3O_2(g)$ by Palmer and Cross (1).

Jacox, Milligan, Moll, and Thompson (2) have observed the photodissociation of C_2O near 5000 Å which corresponds to a minimum of 57 kcal mol^{-1} for $C_2O(g) \rightarrow C(g) + CO(g)$. This yields $\Delta_f H°(C_2O,\ g,\ 298.15\ K) \leq 87\ kcal\ mol^{-1}$ which was not adopted because the irradiation was not monochromatic and thus possibly contained all wavelengths below 5000 Å.

Heat Capacity and Entropy

Jacox et al. (2) have observed three vibrational fundamentals in the infrared spectrum by matrix isolation, and suggested a linear structure with $^3\Sigma$ ground state for the CCO radical (g). They also estimated the bond distances C–C = 1.160 Å and C–O = 1.279 Å based on the structural analogy between the carbon suboxide and the CCO radical. Bayes (3) has predicted that the ^{1}Z and $^1\Sigma$ states lie respectively 0.5 and 0.8 eV above the ground state. These two elctronic states have been included in the tabulation.

References

1. H. B. Palmer and W. D Cross, Carbon 3, 475 (1966).
2. M. E. Jacox, D. E. Milligan, N. G. Moll, and W. E. Thompson, J. Chem. Phys. 43, 3745 (1965).
3. K. D. Bayes, J. Amer. Chem. Soc. 85, 1730 (1963).

Enthalpy Reference Temperature = T_r = 298.15 K — Standard State Pressure = p° = 0.1 MPa

T/K	$C_p°$ (J K⁻¹mol⁻¹)	S°	−[G°−H°(T_r)]/T	H°−H°(T_r) (kJ mol⁻¹)	$\Delta_f H°$	$\Delta_f G°$	Log K_f
0	0.	0.	INFINITE	-10.473	282.575	282.575	INFINITE
100	31.198	192.768	268.031	-7.526	283.950	273.529	-142.877
200	38.530	216.773	236.873	-4.020	285.348	262.574	-68.577
250	41.079	225.656	233.764	-2.027	286.020	256.802	-53.656
298.15	43.064	233.066	233.066	0.	286.604	251.121	-43.995
300	43.133	233.333	233.067	0.080	286.625	250.900	-43.686
350	44.880	240.116	233.599	2.281	287.146	244.904	-36.550
400	46.416	246.211	234.801	4.564	287.578	238.839	-31.189
450	47.795	251.759	236.381	6.920	287.919	232.725	-27.014
500	49.044	256.861	238.178	9.342	288.173	226.578	-23.670
600	51.227	266.001	242.071	14.358	288.454	214.228	-18.650
700	53.064	274.040	246.075	19.576	288.498	201.851	-15.062
800	54.626	281.230	250.028	24.962	288.374	189.479	-12.372
900	55.971	287.744	253.862	30.494	288.133	177.131	-10.280
1000	57.142	293.703	257.553	36.150	287.813	164.814	-8.609
1100	58.173	299.199	261.092	41.917	287.438	152.532	-7.243
1200	59.086	304.300	264.483	47.781	287.025	140.286	-6.106
1300	59.896	309.062	267.731	53.731	286.585	128.075	-5.146
1400	60.617	313.528	270.844	59.757	286.125	115.899	-4.324
1500	61.258	317.732	273.831	65.852	285.650	103.756	-3.613
1600	61.825	321.704	276.700	72.006	285.161	91.646	-2.992
1700	62.325	325.468	279.459	78.214	284.659	79.567	-2.445
1800	62.764	329.043	282.115	84.469	284.143	67.517	-1.959
1900	63.148	332.447	284.676	90.765	283.611	55.497	-1.526
2000	63.481	335.695	287.146	97.097	283.063	43.505	-1.136
2100	63.769	338.799	289.532	103.460	282.495	31.541	-0.785
2200	64.016	341.771	291.840	109.850	281.908	19.605	-0.465
2300	64.226	344.622	294.073	116.262	281.298	7.695	-0.175
2400	64.402	347.359	296.237	122.694	280.663	-4.187	0.091
2500	64.550	349.991	298.335	129.142	280.004	-16.043	0.335
2600	64.671	352.525	300.370	135.603	279.317	-27.871	0.560
2700	64.769	354.968	302.348	142.075	278.602	-39.672	0.768
2800	64.847	357.325	304.269	148.556	277.858	-51.447	0.960
2900	64.907	359.602	306.138	155.044	277.084	-63.194	1.138
3000	64.952	361.803	307.957	161.537	276.280	-74.914	1.304
3100	64.983	363.933	309.729	168.034	275.444	-86.607	1.459
3200	65.002	365.997	311.455	174.533	274.576	-98.273	1.604
3300	65.012	367.997	313.138	181.034	273.676	-109.910	1.740
3400	65.012	369.938	314.780	187.535	272.745	-121.520	1.867
3500	65.005	371.822	316.383	194.036	271.780	-133.102	1.986
3600	64.991	373.653	317.949	200.536	270.784	-144.656	2.099
3700	64.972	375.434	319.479	207.034	269.755	-156.182	2.205
3800	64.948	377.166	320.974	213.530	268.693	-167.680	2.305
3900	64.921	378.853	322.437	220.024	267.600	-179.149	2.399
4000	64.890	380.496	323.868	226.514	266.474	-190 590	2.489
4100	64.856	382.098	325.268	233.002	265.315	-202.002	2.574
4200	64.821	383.661	326.640	239.486	264 125	-213 386	2.654
4300	64.784	385.185	327.984	245.966	262.902	-224.741	2.730
4400	64.745	386.674	329.301	252.442	261.648	-236.067	2.802
4500	64.705	388.129	330.592	258.915	260.362	-247.364	2.871
4600	64.665	389.551	331.859	265.383	259.043	-258.633	2.937
4700	64.624	390.941	333.101	271.848	257.693	-269.872	2.999
4800	64.583	392.301	334.320	278.308	256.311	-281.082	3.059
4900	64.542	393.632	335.517	284.764	254.897	-292.263	3.116
5000	64.500	394.936	336.692	291.216	253.451	-303.415	3.170
5100	64.459	396.213	337.847	297.664	251.973	-314.537	3.222
5200	64.418	397.464	338.981	304.108	250.462	-325.631	3.271
5300	64.378	398.690	340.097	310.548	248.919	-336.695	3.318
5400	64.338	399.893	341.193	316.984	247.342	-347.730	3.364
5500	64.299	401.074	342.271	323.416	245.733	-358.735	3.407
5600	64.260	402.232	343.331	329.844	244.090	-369.710	3.449
5700	64.222	403.369	344.375	336.268	242.414	-380.656	3.488
5800	64.184	404.486	345.401	342.688	240.703	-391.572	3.526
5900	64.147	405.582	346.412	349.105	238.959	-402.459	3.563
6000	64.111	406.660	347.407	355.517	237.179	-413.315	3.598

PREVIOUS: September 1966 (1 atm) CURRENT: September 1966 (1 bar)

SILICON CARBIDE (SiC_2) IDEAL GAS M_r = 52.1075

$S°(298.15\ K) = 236.731\ J\ K^{-1}\ mol^{-1}$

$\Delta_f H°(0\ K) = 609.3 \pm 29\ kJ\ mol^{-1}$
$\Delta_f H°(298.15\ K) = 615.0 \pm 29\ kJ\ mol^{-1}$

Electronic Levels and Quantum Weights

ε_i, cm^{-1}	g_i
0	1
[13000]	[6]
20085	2

Vibrational Frequencies and Degeneracies

ν, cm^{-1}
853 (1)
300 (2)
1742 (1)

Point Group: $C_{\infty v}$ σ = 1
Bond Distances: C-C = [128] Å C-Si = [1.75] Å
Bond Angle: C-C-Si = 180°
Rotational Constant: B_0 = [0.201171] cm^{-1}

Enthalpy of Formation

The selected value is an average based on the equilibrium data summarized below. Drowart and co-workers have used the Knudsen effusion-mass spectrometric technique to determine the vapor equilibria over the systems SiC-graphite (1), SiC-silison (2) and boron-carbon-silicon (3). 3rd law analysis of the partial pressures of C_2Si and Si lead to the concordant values 146.7, 149.0 and 146.0 kcal mol^{-1}. Although the drifts all suggest that the entropy of C_2Si may be about 3.5 less than tabulated, this is of doubtful significance because the uncertainty bands are so large. The low condensation coefficient suggested by Drowart (2) may also be involved. Rates of free evaporation reported by Voronin (4) lead to apparent pressures which are smaller by a factor of 1/30, thus confirming the existence of a low evaporation coefficient. The adopted value of $\Delta_f H°(298.15\ K) = 147 \pm 7$ kcal mol^{-1} includes allowance for the unlikely possibility that the entropy may be in error by up to 3.5 cal K^{-1} mol^{-1}.

Source	Method	Reaction*	T/K	Data Points	$\Delta_r H°(298.15\ K)$ kcal mol^{-1} 2nd Law	3rd Law	Drift cal $K^{-1}mol^{-1}$	$\Delta_f H°(298.15\ K)$ kcal mol^{-1}
Drowart	Mass Spec.	A	2149-2316	7	31±9	38.98	3.5±4.1	146.7
Drowart	Mass Spec.	B	1935-2160	4	188±10	194.8	3.3±4.9	149.0
Verhaegen	Mass Spec.	A	2166-2344	5	30±16	38.31	3.9±7.2	146.0

*Reactions: (A) 2 C(graph) + Si(g) = C_2Si(g) (B) 2 SiC(α) = Si(ℓ) + C_2Si(g)

Heat Capacity and Entropy

The known molecular constants are from the assignments of Weltner and McLeod (5). These authors observed the infrared spectrum, emission spectrum and visible absorption spectrum of C_2Si in matrix isolation. The matrix data allowed more certain assignment of the A-X band system of the gas as observed by Kleman (6) and McKellar (7). Weltner concluded that this system probably involves transitions between a $^1\Sigma^+$ ground state and a $^1\Pi$ excited state. The two stretching frequencies were observed in the matrix infrared spectrum, whereas the bending mode, 300 cm^{-1}, was derived solely from the analysis of the A-X band system. The analysis presumed a linear unsymmetrical structure for the molecule.

A $^3\pi$ excited state is assumed to lie 7000 cm^{-1} below the $^1\pi$ state, which presumably arises from the same molecular orbital configuration. The effect of the estimated level is to increase the entropy by about 0.2 and 1.8 cal K^{-1} mol^{-1} at 3000 and 6000 K. Bond distances are estimated by comparison with those in Si_2, C_2 and C_3.

References

1. J. Drowart, G. DeMaria and M. G. Inghram, J. Chem. Phys. 29, 1015 (1958).
2. J. Drowart and G. DeMaria, pp. 16-23 in "Silicon Carbide," Edited by J. R. O'Connor and J. Smiltens, Pergamon Press, New York, 1960.
3. G. Verhaegen, F. E. Stafford and J. Drowart, J. Chem. Phys. 40, 1622 (1964).
4. N. I. Voronin, N. L. Makarova and B. F. Yudin, Issled. v Obl. Khim. Silikatov 1 Okislov, Akad. Nauk SSSR, Sb. Statei 203-8, 1965.
5. W. Weltner, Jr., and D. McLeod, Jr., J. Chem. Phys. 41, 235 (1964).
6. B. Kleman, Astrophys. J. 123, 162 (1956).
7. A. McKellar, J. Roy. Astron. Soc. Canada 41, 147 (1947).

Silicon Carbide (SiC_2) $C_2Si_1(g)$

Enthalpy Reference Temperature = T_r = 298.15 K Standard State Pressure = p° = 0.1 MPa

T/K	$C_p°$ (J $K^{-1}mol^{-1}$)	S°	$-[G°-H°(T_r)]/T$	$H°-H°(T_r)$ (kJ mol^{-1})	$\Delta_f H°$	$\Delta_f G°$	Log K_f
0	0.	0.	INFINITE	-11.055	609.313	609.313	INFINITE
100	33.354	193.558	274.049	-8.049	611.934	593.151	-309.831
200	41.224	219.393	240.775	-4.276	613.875	573.562	-149.799
250	43.686	228.867	237.472	-2.151	614.559	563.400	-117.716
298.15	45.611	236.731	236.731	0.	615.048	553.501	-96.971
300	45.678	237.014	236.732	0.084	615.064	553.119	-96.307
350	47.364	244.185	237.295	2.412	615.414	542.765	-81.003
400	48.832	250.607	238.564	4.817	615.630	532.370	-69.520
450	50.132	256.435	240.231	7.292	615.724	521.956	-60.587
500	51.292	261.779	242.122	9.828	615.709	511.537	-53.440
600	53.260	271.311	246.212	15.059	615.408	490.725	-42.721
700	54.831	279.643	250.405	20.467	614.822	469.988	-35.071
800	56.080	287.050	254.531	26.015	614.019	449.350	-29.340
900	57.074	293.715	258.521	31.674	613.047	428.823	-24.888
1000	57.869	299.771	262.348	37.423	611.939	408.412	-21.333
1100	58.510	305.317	266.006	43.243	610.715	388.118	-18.430
1200	59.031	310.431	269.497	49.121	609.392	367.939	-16.016
1300	59.462	315.174	272.831	55.046	607.979	347.875	-13.978
1400	59.823	319.594	276.015	61.011	606.482	327.922	-12.235
1500	60.133	323.732	279.060	67.009	604.907	308.080	-10.728
1600	60.409	327.622	281.974	73.036	603.258	288.345	-9.413
1700	60.662	331.292	284.769	79.090	551.363	269.162	-8.270
1800	60.907	334.766	287.451	85.169	549.808	252.606	-7.330
1900	61.153	338.066	290.028	91.272	548.234	236.138	-6.492
2000	61.410	341.209	292.509	97.400	546.643	219.753	-5.739
2100	61.685	344.212	294.901	103.554	545.041	203.448	-5.060
2200	61.985	347.088	297.208	109.737	543.431	187.219	-4.445
2300	62.314	349.851	299.437	115.952	541.819	171.064	-3.885
2400	62.676	352.511	301.593	122.201	540.209	154.979	-3.373
2500	63.073	355.077	303.682	128.488	538.607	138.960	-2.903
2600	63.504	357.559	305.706	134.817	537.016	123.006	-2.471
2700	63.971	359.964	307.672	141.191	535.441	107.112	-2.072
2800	64.471	362.300	309.581	147.612	533.887	91.277	-1.703
2900	65.001	364.571	311.438	154.086	532.358	75.496	-1.360
3000	65.558	366.784	313.246	160.613	530.857	59.768	-1.041
3100	66.140	368.943	315.008	167.198	529.387	44.090	-0.743
3200	66.740	371.053	316.727	173.842	527.951	28.458	-0.465
3300	67.355	373.116	318.405	180.547	526.551	12.870	-0.204
3400	67.980	375.136	320.044	187.313	525.189	-2.675	0.041
3500	68.609	377.115	321.646	194.143	523.866	-18.181	0.271
3600	69.239	379.057	323.214	201.035	138.430	-23.189	0.336
3700	69.864	380.963	324.749	207.990	137.600	-27.667	0.391
3800	70.480	382.834	326.253	215.008	136.806	-32.123	0.442
3900	71.083	384.673	327.727	222.086	136.050	-36.559	0.490
4000	71.669	386.480	329.174	229.224	135.329	-40.975	0.535
4100	72.234	388.256	330.593	236.419	134.642	-45.374	0.578
4200	72.775	390.004	331.987	243.670	133.988	-49.757	0.619
4300	73.290	391.722	333.356	250.973	133.365	-54.124	0.657
4400	73.776	393.413	334.702	258.327	132.769	-58.478	0.694
4500	74.232	395.076	336.025	265.727	132.199	-62.818	0.729
4600	74.656	396.712	337.327	273.172	131.651	-67.145	0.762
4700	75.047	398.322	338.607	280.657	131.122	-71.461	0.794
4800	75.404	399.905	339.868	288.180	130.610	-75.766	0.825
4900	75.728	401.464	341.109	295.737	130.111	-80.060	0.853
5000	76.018	402.997	342.332	303.325	129.621	-84.345	0.881
5100	76.274	404.504	343.536	310.939	129.138	-88.619	0.908
5200	76.498	405.988	344.723	318.578	128.657	-92.884	0.933
5300	76.689	407.447	345.892	326.238	128.177	-97.140	0.957
5400	76.850	408.882	347.046	333.915	127.692	-101.387	0.981
5500	76.980	410.293	348.183	341.607	127.202	-105.625	1.003
5600	77.082	411.681	349.304	349.310	126.702	-109.853	1.025
5700	77.156	413.046	350.411	357.022	126.188	-114.073	1.045
5800	77.204	414.389	351.502	364.740	125.659	-118.283	1.065
5900	77.228	415.709	352.579	372.462	125.112	-122.485	1.084
6000	77.229	417.007	353.642	380.185	124.544	-126.677	1.103

PREVIOUS: March 1967 (1 atm) CURRENT: March 1967 (1 bar)

Silicon Carbide (SiC_2) $C_2Si_1(g)$

CARBON (C_3) IDEAL GAS M_r = 36.033

$S°(298.15\ K) = 237.2 \pm 13\ J\ K^{-1}\ mol^{-1}$

$\Delta_f H°(0\ K) = 811 \pm 17\ kJ\ mol^{-1}$
$\Delta_f H°(298.15\ K) = 820 \pm 17\ kJ\ mol^{-1}$

Vibrational Frequencies and Degeneracies

ν, cm^{-1}
1225 (1)
63.5 (2)
2040 (1)

Ground State Quantum Weight: [1] $\sigma = 2$
Point Group: $D_{\infty h}$
Bond Distances: C-C = 1.277 Å
Bond Angle: C-C-C = 180°
Rotational Constant: $B_0 = 0.430334\ cm^{-1}$

Enthalpy of Formation

Drowart et al. (1) have measured mass spectrometrically the ion intensity ratio C_3^+/C^+ as a function of temperature from 1800 to 2700 K. We derive from a 2nd law analysis of their data $\Delta_f H°(2200\ K) = 185.2 \pm 3.3$ kcal, and $S°(2200\ K) = 74.9 \pm 2.8$ cal $K^{-1}\ mol^{-1}$ by calculation of the absolute pressure from the vapor pressure of monatomic carbon and a relative ionization cross-section ratio of 2.3. The entropy uncertainty includes 1.3 cal $K^{-1}\ mol^{-1}$ for uncertainty in the absolute pressure. Several earlier mass-spectrometric determinations (2, 3, 4) are in reasonable agreement. The absolute magnitude of the pressure is also confirmed by the work of Thorn and Winslow (5) who measured the total pressure of all species over graphite as 1.2 to 4.8 x 10^{-7} atm at 2400 K. Subtracting the JANAF values for C and C_2 we obtain C_3 = 5.4 x 10^{-8} to 2.2 x 10^{-7} atm, which compares with Drowart's value of 3.3 x 10^{-8} to 1.3 x 10^{-7} atm. The 2nd law enthalpy of formation is reported (5) to be $\Delta_f H°(2400\ K) = 184.4$ but the range is only 80 K. It does, however, confirm the magnitude of the mass-spectrometric values for both enthalpy and entropy. We have adopted a enthalpy of formation within the uncertainty limits of Dowart's measurements which with the adopted functions yields pressures in the 2000-3000 K range consistent with the measurements. We were not able to generate functions which agree exactly with the 2nd law entropies, since even direct summation over only the observed levels of the bending frequency gives an entropy which is already too large.

Heat Capacity and Entropy

The stretching frequencies are from Merer (6), Weltner et al. (7) and Weltner and McLeod (8). The bending frequency and its harmonics are from Gausset et al. (9) and Merer (6). The rotational constant and bond distance are from Gausset et al. (9).

As indicated briefly above, the calculated entropy of C_3 gas is higher than the experimental value even when a summation is performed only over the observed first six levels of the bending frequency. Treatment of the vibration as a harmonic oscillator or as a anharmonic oscillator gives values considerably higher. Strauss and Thiele (10) have also calculated functions based on a quartic potential function whlich still yields values several units higher than the experiments. In order to reduce the entropy to the approximate range of the measurements we have made the assumption that the potential function is of an unusual form above the sixth vibrational level. We assume that the large amplitude of the bending vibration causes a drastic change in the function, possibly due to the increasing repulsion between end carbons, as the amplitude increases. Thus, we assume a very steeply rising potential above the sixth vibrational level. We have chosen to represent this part of the potential by a harmonic oscillator of 650 cm^{-1}. The choice of 650 cm^{-1} is arbitrary, but is not unreasonable for a bending vibration, and yields functions which approximate the experimental data.

In order to explain the observed spectrum of C_3O_2, a molecule with a bending frequency of 63 cm^{-1}, Smith and Barrett (11) have proposed that potential maxima occur in the bending potential away from linearity. Thus, the adoption of an unusual potential function for C_3 bending may be justifiable. We do not claim that our functions are necessarily correct but only that they include all known facts and satisfy the experimental equilibrium observations. Further investigation of the carbon vapor pressure and of the higher vibrational levels of C_3 is desirable.

References

1. J. Drowart, R. P. Burns, G. DeMaria, and M. G. Inghram, J. Chem. Phys. 31, 1131 (1959).
2. W. A. Chupka and M. G. Inghram, J. Phys. Chem. 59, 100 (1955).
3. W. A. Chupka and M. G. Inghram, J. Chem. Phys. 21, 371 and 1313, (1953).
4. R. E. Honig, J. Chem. Phys. 22, 126 (1954).
5. R. J. Thorn and G. H. Winslow, J. Chem. Phys. 26, 186 (1957).
6. A. J. Merer, Can. J. Phys. 45, 4103 (1967).
7. W. Weltner, P. N. Walsh and C. L. Angell, J. Chem. Phys. 40, 1299 (1964).
8. W. Weltner and D. McLeod, J. Chem. Phys. 40, 1305 (1964).
9. L. Gausset, G. Harzberg, A. Lagerqvist, and B. Rosen, Astrophys. J. 142, 45 (1965).
10. H. L. Strauss and E. Thield, J. Chem. Phys. 46, 2473 (1967).
11. W. H. Smith and J. J. Barrett, J. Chem. Phys. 51, 1475 (1969).

Carbon (C_3) $C_3(g)$

Enthalpy Reference Temperature = T_r = 298.15 K Standard State Pressure = p° = 0.1 MPa

T/K	$C_p°$ (J K⁻¹mol⁻¹)	S° (J K⁻¹mol⁻¹)	$-[G°-H°(T_r)]/T$ (J K⁻¹mol⁻¹)	$H°-H°(T_r)$ (kJ mol⁻¹)	$\Delta_f H°$ (kJ mol⁻¹)	$\Delta_f G°$ (kJ mol⁻¹)	Log K_f
0	0.	0.	INFINITE	-11.761	811.456	811.456	INFINITE
100	42.459	193.021	272.308	-7.929	815.110	796.093	-415.836
200	39.861	221.795	240.748	-3.791	818.268	775.758	-202.607
298.15	37.740	237.246	237.246	0.	820.064	754.464	-132.179
300	37.719	237.479	237.247	0.070	820.086	754.057	-131.293
400	37.497	248.274	238.720	3.821	820.770	731.916	-95.578
500	38.380	256.727	241.504	7.612	820.580	709.709	-74.143
600	39.706	263.841	244.649	11.515	819.749	687.604	-59.861
700	41.125	270.069	247.844	15.557	818.472	665.676	-49.673
800	42.480	275.650	250.977	19.738	816.891	643.953	-42.046
900	43.714	280.726	254.005	24.049	815.098	622.441	-36.126
1000	44.815	285.390	256.913	28.477	813.156	601.138	-31.400
1100	45.786	289.708	259.701	33.008	811.105	580.035	-27.544
1200	46.647	293.729	262.371	37.630	808.974	559.122	-24.338
1300	47.413	297.494	264.929	42.334	806.781	538.389	-21.633
1400	48.095	301.033	267.383	47.110	804.538	517.828	-19.320
1500	48.710	304.372	269.739	51.951	802.255	497.428	-17.322
1600	49.267	307.534	272.003	56.850	799.940	477.181	-15.578
1700	49.773	310.536	274.182	61.802	797.595	457.081	-14.044
1800	50.233	313.394	276.282	66.803	795.226	437.119	-12.685
1900	50.660	316.122	278.307	71.848	792.834	417.289	-11.472
2000	51.053	318.731	280.264	76.934	790.421	397.586	-10.384
2100	51.417	321.230	282.155	82.057	787.989	378.004	-9.402
2200	51.760	323.630	283.986	87.216	785.538	358.538	-8.513
2300	52.081	325.938	285.761	92.408	783.069	339.184	-7.703
2400	52.384	328.161	287.481	97.632	780.584	319.938	-6.963
2500	52.667	330.305	289.152	102.885	778.081	300.796	-6.285
2600	52.936	332.376	290.775	108.165	775.562	281.754	-5.661
2700	53.189	334.379	292.353	113.471	773.025	262.809	-5.084
2800	53.430	336.318	293.888	118.802	770.472	243.959	-4.551
2900	53.659	338.197	295.384	124.157	767.902	225.200	-4.056
3000	53.877	340.019	296.842	129.534	765.315	206.531	-3.596
3100	54.084	341.789	298.263	134.932	762.711	187.947	-3.167
3200	54.283	343.510	299.650	140.350	760.089	169.448	-2.766
3300	54.476	345.183	301.005	145.788	757.450	151.031	-2.391
3400	54.660	346.812	302.328	151.245	754.793	132.695	-2.039
3500	54.834	348.399	303.622	156.720	752.118	114.437	-1.708
3600	55.003	349.946	304.887	162.212	749.425	96.256	-1.397
3700	55.164	351.455	306.126	167.720	746.713	78.149	-1.103
3800	55.321	352.929	307.338	173.244	743.982	60.116	-0.826
3900	55.472	354.368	308.526	178.784	741.233	42.156	-0.565
4000	55.618	355.774	309.689	184.339	738.464	24.266	-0.317
4100	55.757	357.149	310.830	189.907	735.676	6.445	-0.082
4200	55.893	358.494	311.949	195.490	732.868	-11.307	0.141
4300	56.024	359.811	313.047	201.086	730.040	-28.992	0.352
4400	56.151	361.100	314.124	206.695	727.192	-46.611	0.553
4500	56.275	362.364	315.182	212.316	724.324	-64.165	0.745
4600	56.392	363.602	316.222	217.949	721.434	-81.655	0.927
4700	56.506	364.816	317.243	223.594	718.524	-99.082	1.101
4800	56.618	366.007	318.246	229.250	715.593	-116.447	1.267
4900	56.726	367.175	319.233	234.918	712.641	-133.751	1.426
5000	56.831	368.322	320.203	240.596	709.667	-150.994	1.577
5100	56.932	369.449	321.158	246.284	706.672	-168.178	1.722
5200	57.031	370.555	322.097	251.982	703.654	-185.302	1.861
5300	57.127	371.642	323.022	257.690	700.614	-202.368	1.994
5400	57.221	372.711	323.932	263.407	697.552	-219.377	2.122
5500	57.312	373.762	324.828	269.134	694.468	-236.329	2.244
5600	57.402	374.795	325.712	274.870	691.361	-253.225	2.362
5700	57.488	375.812	326.582	280.614	688.232	-270.064	2.475
5800	57.573	376.813	327.439	286.367	685.080	-286.849	2.583
5900	57.655	377.798	328.284	292.129	681.904	-303.579	2.688
6000	57.735	378.767	329.118	297.898	678.706	-320.256	2.788

PREVIOUS: December 1969 (1 atm) CURRENT: December 1969 (1 bar)

Carbon (C_3) $C_3(g)$

ALUMINUM CARBIDE (Al_4C_3) CRYSTAL M_r = 143.95916

S°(298.15 K) = 88.969 ± 1.26 J K^{-1} mol^{-1}

T_{fus} = [2500] K

$\Delta_fH°$(0 K) = -210.84 ± 6.7 kJ mol^{-1}

$\Delta_fH°$(298.15 K) = -215.69 ± 6.7 kJ mol^{-1}

$\Delta_{fus}H°$ = Unknown

Enthalpy of Formation

The $\Delta_fH°$(298.15 K) = -51.55±1.6 kcal mol^{-1} is the average value of $\Delta_fH°$(298.15 K) = -49.7±1.2 kcal mol^{-1} by King and Armstrong (1) and that of $\Delta_fH°$(298.15 K) = -53.4±2.0 kcal mol^{-1} by Mah (2). Both measured the enthalpy of reaction Al_4C_3(cr) + $6O_2$(g) + $2Al_2O_3$(cr, α) + $3CO_2$(g) and used their $\Delta_rH°$(298.15 K) and the $\Delta_fH°$(298.15 K) of Al_2O_3(cr) and CO_2(g) to calculate the $\Delta_rH°$(298.15 K) of Al_4C_3(cr). The $\Delta_rH°$(298.15 K) measured by King and Armstrong was -1033.3±1.1 kcal mol^{-1}, the $\Delta_rH°$(298.15 K) measured by Mah (2) was -1029.6±1.9 kcal mol^{-1}.

Heat Capacity and Entropy

The low temperature heat capacities, 18-390 K, were measured by Saba and Furukawa (3). The high temperature heat capacities, 273-1173 K, were measured by Victor et al. (4). These two sets of heat capacity data were joined smoothly and extrapolated to 2000 K by Saba and Furukawa (3). The values of $C_p°$ above 2000 K were estimated by extrapolation graphically. S°(298.15 K) was reported by Saba and Furukawa (3), using S°(extrap. 20 K) = 0.013 cal K^{-1} mol^{-1}.

Fusion Data

T_{fus} was estimated by Ruff and Jellinek (5).

References

1. R. C. King and G. T. Armstrong, U. S. Nat. Bur. Stand. Report 8504, 1964.
2. A. D. Mah, U. S. Bur. Mines Report 6415, 1964.
3. W. G. Saba and G. T. Furukawa, U. S. Nat. Bur. Stand. Report 7587, 1962.
4. A. C. Victor, W. R. Thurber, and T. B. Douglas, U. S. Nat. Bur. Stand. Report 7437, 1962.
5. O. Ruff and E. Jellinek, Z. Anorg. Allg. Chem. 97, 312 (1916).

Aluminum Carbide (Al_4C_3) C_3Al_4(cr)

Enthalpy Reference Temperature = T_r = 298.15 K; Standard State Pressure = p° = 0.1 MPa

	J $K^{-1}mol^{-1}$			kJ mol^{-1}			
T/K	$C_p°$	S°	-[G°-H°(T_r)]/T	H°-H°(T_r)	$\Delta_fH°$	$\Delta_fG°$	Log K_f
0	0.	0.	INFINITE	-16.468	-210.843	-210.843	INFINITE
100	29.702	11.749	167.728	-15.598	-212.086	-210.181	109.787
200	82.098	49.200	98.655	-9.891	-214.566	-207.241	54.126
298.15	116.784	88.969	88.969	0.	-215.685	-203.356	35.627
300	117.298	89.693	88.971	0.217	-215.695	-203.279	35.394
400	138.507	126.645	93.865	13.112	-215.901	-199.095	25.999
500	150.892	158.981	103.729	27.626	-215.898	-194.894	20.360
600	159.201	187.262	115.345	43.150	-216.050	-190.683	16.600
700	165.180	212.270	127.440	59.381	-216.527	-186.423	13.911
800	169.707	234.633	139.466	76.134	-217.461	-182.065	11.888
900	173.264	254.834	151.181	93.288	-218.975	-177.557	10.305
1000	176.142	273.243	162.480	110.762	-263.532	-169.804	8.870
1100	178.527	290.146	173.328	128.499	-265.077	-160.356	7.615
1200	180.540	305.768	183.722	146.455	-266.575	-150.769	6.563
1300	182.268	320.288	193.675	164.597	-268.030	-141.059	5.668
1400	183.770	333.852	203.209	182.900	-269.446	-131.239	4.897
1500	185.096	346.577	212.347	201.345	-270.825	-121.318	4.225
1600	186.276	358.561	221.114	219.914	-272.171	-111.307	3.634
1700	187.334	369.886	229.536	238.596	-273.486	-101.213	3.110
1800	188.297	380.621	237.634	257.378	-274.774	-91.042	2.642
1900	189.171	390.826	245.430	276.252	-276.037	-80.800	2.221
2000	189.979	400.550	252.945	295.210	-277.278	-70.492	1.841
2100	190.631	409.835	260.196	314.241	-278.504	-60.123	1.495
2200	191.251	418.718	267.202	333.335	-279.719	-49.695	1.180
2300	191.836	427.232	273.976	352.490	-280.926	-39.212	0.891
2400	192.389	435.408	280.533	371.701	-282.124	-28.677	0.624
2500	192.907	443.273	286.886	390.966	-283.314	-18.093	0.378
2600	193.393	450.848	293.047	410.282	-284.499	-7.460	0.150
2700	193.845	458.155	299.028	429.644	-285.679	3.217	-0.062
2800	194.263	465.213	304.838	449.050	-1462.462	17.810	-0.332
2900	194.648	472.037	310.486	468.495	-1459.259	70.620	-1.272
3000	194.999	478.641	315.982	487.978	-1456.059	123.320	-2.147

PREVIOUS: June 1963 CURRENT: September 1965

Aluminum Carbide (Al_4C_3) C_3Al_4(cr)

CHROMIUM CARBIDE (Cr_7C_3) CRYSTAL M_r = 400.005

$\Delta_fH°(0\ K)$ = -163.51 ± 16.7 kJ mol^{-1} 1

S°(298.15 K) = 200.983 ± 1.26 J K^{-1} mol^{-1}

$\Delta_fH°(298.15\ K)$ = -160.67 ± 16.7 kJ mol^{-1}

Enthalpy of Formation

Mah (1) determined the heat of combustion of Cr_7C_3(cr) using oxygen bomb calorimetry. She reported $\Delta_cH°(298.15\ K)$ = -1193.0±0.9 kcal mol^{-1} based on the reaction Cr_7C_3(cr) + 33/4 O_2(g) = 7/2 Cr_2O_3(cr) + 3 CO_2(g). Using JANAF auxiliary data (2), we calculate $\Delta_fH°(298.15\ K)$ = -38.4 kcal mol^{-1} for Cr_7C_3(cr). Mah (1) reported that the six combustion runs gave 99.90 to 99.84% completion of combustion. Corrections for incomplete combustion, formation of small amounts of CO(g), and impurities were made by Mah (1). These corrections amount to no more than 1.04% of the Δ_cE value.

The pressures of CO(g) in equilibrium with a Cr_2O_3-Cr_7C_3-$Cr_{23}C_6$ mixture have been measured by Kelley et al. (3) and Kulkarni and Worrell (4). Kelley et al. (3) also examined the Cr_2O_3-Cr_3C_2-Cr_7C_3 system. Kulkarni and Worrell (4) measured the CO pressure using a torsion effusion technique while Kelley et al. (3) measured the pressure manometrically. The results of our analyses are tabulated in the following table. In all cases we have assumed unit activity for the condensed species.

Source	T/K	Data Points	Reaction	$\Delta_rH°(298.15\ K)$, kcal mol^{-1} 2nd Law	3rd Law	Drift cal K^{-1} mol^{-1}	$\Delta_fH°298.15\ K)$ kcal mol^{-1}
(3)	1307-1497	9	A	994.44	970.74	-16.9±4.3	-41.53
(4)	1111-1250	11[b]	B	220.48	218.99	- 1.3±0.6	-35.18
(3)	1505-1723	15[c]	B	211.34	220.89	5.8±2.0	-35.81

[a]Reactions: (A) 5 Cr_2O_3(cr) + 27 Cr_3C_2(cr) = 13 Cr_7C_3(cr) + 15 CO(g) (B) Cr_2O_3(cr) + 3 Cr_7C_3(cr) = $Cr_{23}C_6$(cr) + 3 CO(g).
[b]1234.6, 1250.0 K points deleted due to statistical test; effective range is now 1111-1220 K.
[c]1524 K point deleted due to statistical test.

We adopt $\Delta_fH°(298.15\ K)$ = -38.4±4.0 kcal mol^{-1} for Cr_7C_3(cr) based on the combustion data of Mah (1). The equilibrium studies by Kulkarni and Worrell (4) and Kelley et al. (3) are in reasonable agreement with the adopted value. Equilibrium studies involving Cr_7C_3(cr) by Hancock and Pidgeon (5), Alekseev and Shvartsman (6), and Bolgar et al. (7) are not considered (2).

Heat Capacity and Entropy

The heat capacity (54-295 K, 29 pts.) and enthalpy (480-1580 K, 15 pts.) have been measured by Kelley et al. (3). Using the Debye function suggested (3), we calculate S°(50 K) = 1.679 cal K^{-1} mol^{-1} and H°(50;0) = 62.66 cal K^{-1} mol^{-1}. The adopted heat capacity curve is based solely on the above mentioned study with the constraint that the heat capacity values derived from the enthalpy data join smoothly with the low temperature heat capacity values near 298.15 K. The observed $C_p°$ values deviate from the adopted values within ±0.2% for T < 298.15 K. The enthalpy data agrees within ±0.35% with the tabulated values except for the 1263.4 K data (1.66% or ~1.0 kcal mol^{-1} too low) and the two points at 1545.1 and 1578.1 K which were not included in the polynomial curve fit (3). The polynomial is used to extrapolate $C_p°$ values above 1500 K.

Phase Data

Storms (8) has summarized much of the phase data for the chromium carbides and reported a hexagonal structure for Cr_7C_3. The homogeneity range for Cr_7C_3 is not known (8).

Decomposition Data

The tentative phase diagram for the Cr-Cr_3C_2 system as proposed by Storms (8) indicated an incongruent melting at 2053 K.

References

1. A. D. Mah, U. S. Bur. Mines RI 7217, 1969.
2. JANAF Thermochemical Tables: Cr_2O_3(cr) and Cr_3C_2(cr), 12-31-73; CO_2(g), 9-30-65.
3. K. K. Kelley, F. S. Boericke, G. E. Moore, E. H. Huffman, and W. M. Bangert, U. S. Bur. Mines Tech. Paper 662, 1949.
4. A. D. Kulkarni and W. L. Worrell, Met. Trans. 3, 2363 (1972).
5. H. A. Hancock and L. M. Pidgeon, Trans. Met. Soc. AIME 227, 608 (1963).
6. V. I. Alekseev and L. A. Shvartsman, Fiz. Metallov i Metalloved 2, 545 (1961); Fiz. Khim. Osnovy Met. Protsessov. Komis. po Fiz. Khim. Osmovam Proiz. Stali. sb statei, 1964, pp. 414-21.
7. A. S. Bolgar, V. V. Fesenko, and G. P. Gordinko, Poroshk. Met. 38, 100 (1966).
8. E. K. Storms, "The Refractory Carbides," Academic Press, New York, 1967.

Chromium Carbide (Cr_7C_3) C_3Cr_7(cr)

Enthalpy Reference Temperature = T_r = 298.15 K; Standard State Pressure = p° = 0.1 MPa

T/K	$C_p°$ (J $K^{-1}mol^{-1}$)	S°	-[G°-H°(T_r)]/T	H°-H°(T_r) (kJ mol^{-1})	$\Delta_fH°$	$\Delta_fG°$	Log K_f
0	0.	0.	INFINITE	-34.397	-163.508	-163.508	INFINITE
100	80.414	39.481	355.684	-31.620	-163.089	-163.752	85.535
200	167.414	125.392	218.991	-18.720	-162.255	-164.668	43.007
298.15	209.748	200.983	200.983	0.	-160.666	-166.161	29.111
300	210.372	202.282	200.987	0.389	-160.628	-166.195	28.937
400	236.714	266.701	209.591	22.844	-159.033	-168.182	21.962
500	252.873	321.361	226.622	47.369	-156.663	-170.748	17.838
600	263.504	368.438	246.427	73.206	-154.596	-173.765	15.128
700	271.956	409.702	266.865	99.986	-152.839	-177.103	13.216
800	279.608	446.519	287.061	127.566	-151.313	-180.676	11.797
900	287.006	479.880	306.660	155.897	-150.046	-184.425	10.704
1000	294.369	510.499	325.534	184.966	-149.159	-188.297	9.836
1100	301.779	538.903	343.655	214.773	-148.794	-192.233	9.128
1200	309.269	565.481	361.045	245.324	-149.015	-196.177	8.539
1300	316.842	590.535	377.743	276.629	-149.909	-200.076	8.039
1400	324.494	614.295	393.798	308.695	-151.542	-203.879	7.607
1500	332.214	636.945	409.259	341.530	-153.940	-207.538	7.227
1600	339.992	658.634	424.171	375.140	-157.128	-211.012	6.889
1700	347.824	679.481	438.580	409.531	-161.123	-214.261	6.583
1800	355.703	699.585	452.525	444.707	-165.938	-217.252	6.304
1900	363.619	719.028	466.043	480.672	-171.585	-219.953	6.047
2000	371.568	737.882	479.166	517.431	-178.075	-222.334	5.807
2100	379.552	756.204	491.924	554.987	-185.417	-224.370	5.581
2200	387.543	774.045	504.344	593.342	-328.970	-221.450	5.258
2300	395.530	791.448	516.450	632.496	-325.008	-216.649	4.920
2400	403.526	808.451	528.264	672.449	-320.295	-212.036	4.615
2500	411.517	825.085	539.805	713.201	-314.828	-207.635	4.338

PREVIOUS: CURRENT: December 1973

Chromium Carbide (Cr_7C_3) C_3Cr_7(cr)

MAGNESIUM CARBIDE (Mg_2C_3) CRYSTAL M_r = 84.643

$S°(298.15\ K) = 100.41 \pm 13\ J\ K^{-1}\ mol^{-1}$

$\Delta_f H°(0\ K)$ = Unknown

$\Delta_f H°(298.15\ K) = 79.5 \pm 33\ kJ\ mol^{-1}$

Enthalpy of Formation

Data from Furukawa et al. (1).

Heat Capacity and Entropy

Data from Furukawa et al. (1).

References

1. G. T. Furukawa, M. L. Reilly et al., U. S. Nat. Bur. Stand. Report 6928, Appendix B, 1960.

Magnesium Carbide (Mg_2C_3) C_3Mg_2(cr)

Enthalpy Reference Temperature = T_r = 298.15 K Standard State Pressure = p° = 0.1 MPa

	J K⁻¹mol⁻¹			kJ mol⁻¹			
T/K	$C_p°$	S°	$-[G°-H°(T_r)]/T$	$H°-H°(T_r)$	$\Delta_f H°$	$\Delta_f G°$	Log K_f
0							
100							
200							
298.15	93.763	100.416	100.416	0.	79.496	74.173	-12.995
300	94.140	100.997	100.418	0.174	79.530	74.140	-12.909
400	107.361	130.046	104.288	10.303	81.481	72.052	-9.409
500	114.056	154.770	111.979	21.395	83.260	69.481	-7.259
600	118.198	175.949	120.919	33.018	84.614	66.591	-5.797
700	121.127	194.398	130.127	44.990	85.523	63.509	-4.739
800	123.386	210.724	139.200	57.219	86.012	60.326	-3.939
900	125.269	225.368	147.975	69.653	86.106	57.106	-3.314
1000	126.943	238.654	156.389	82.265	68.636	55.320	-2.890
1100	128.449	250.825	164.428	95.036	67.964	54.021	-2.565
1200	129.830	262.061	172.102	107.950	67.262	52.784	-2.298
1300	131.168	272.506	179.428	121.001	66.554	51.606	-2.074
1400	132.424	282.272	186.429	134.180	-188.965	56.817	-2.120
1500	133.679	291.452	193.128	147.486	-186.940	74.303	-2.587
1600	134.892	300.118	199.547	160.915	-184.884	91.652	-2.992
1700	136.064	308.331	205.706	174.463	-182.790	108.871	-3.345
1800	137.235	316.142	211.626	188.128	-180.652	125.967	-3.655
1900	138.365	323.592	217.325	201.907	-178.466	142.942	-3.930
2000	139.536	330.719	222.817	215.803	-176.227	159.800	-4.174
2100	140.666	337.554	228.119	229.813	-173.931	176.545	-4.391
2200	141.754	344.123	233.244	243.934	-171.579	193.180	-4.587
2300	142.884	350.449	238.203	258.166	-169.167	209.706	-4.763
2400	144.013	356.554	243.008	272.511	-166.692	226.126	-4.921
2500	145.101	362.455	247.669	286.967	-164.155	242.441	-5.066

PREVIOUS: CURRENT: December 1960

Magnesium Carbide (Mg_2C_3) C_3Mg_2(cr)

CARBON SUBOXIDE (C_3O_2) IDEAL GAS M_r = 68.0318

$S°(298.15\ K) = 276.07 \pm 0.8\ J\ K^{-1}\ mol^{-1}$

$\Delta_fH°(0\ K) = -96.819 \pm 1.84\ kJ\ mol^{-1}$
$\Delta_fH°(298.15\ K) = -93.638 \pm 1.84\ kJ\ mol^{-1}$

Vibrational Frequencies and Degeneracies

ν, cm^{-1}	ν, cm^{-1}	ν, cm^{-1}
2200(1)	2258(1)	550(2)
830(1)	1573(1)	63(2)
	577(2)	

Ground State Quantum Weight: 1 σ = 2
Point Group: $D_{\infty h}$
Bond Distances: C–C = 1.28 Å C–O = 1.16 Å
Bond Angles: C–C–C = 180° O–C–C = 180°
Rotational Constant: $B_0 = 0.073337\ cm^{-1}$

Enthalpy of Formation

Kybett et al. (1) have determined the enthalpy of combustion of the liquid form under its own pressure using the sample from the heat capacity determinations (2).

The average of four determinations gave $\Delta_fH°(\ell,\ 298.15\ K) = -28.03 \pm 0.24\ kcal\ mol^{-1}$; the authors report -29.03 due to an arithmetic error. The sample was 99.98 mole percent pure. McDougall et al. (2) measured $\Delta_{vap}H°(230\ K) = 6.421 \pm 0.012\ kcal\ mol^{-1}$ from which we calculate $\Delta_{vap}H°(298.15\ K) = 5.65 \pm 0.2\ kcal\ mol^{-1}$. Thus the enthalpy of formation of the gas is found to be $22.38 \pm 0.44\ kcal\ mol^{-1}$.

Heat Capacity and Entropy

There has been considerable controversy as to the configuration of this molecule and the assignment of its vibrational frequencies. However, the evidence seems fairly conclusive for the linearity of the molecule from the work of Lafferty et al. (3) and Miller et al. (4). The rotational constant adopted here is that found by Lafferty et al. (3) and the vibrational frequencies are those reported by Miller et al. (4), except for ν_7 which is taken from Miller et al. (5) and Borgers (7). The low frequency Π_u mode is confirmed by the work of McDougall et al. (2), who measured the low temperature entropy of C_3O_2(cr) and (ℓ) and also the enthalpy of vaporization. From a comparison of the entropy calculated statistically and that measured, they conclude $\nu_7 = 63.2\ cm^{-1}$. The values adopted are confirmed by 3rd law analysis of the vapor pressure data of McDougall et al. (5) which indicates a enthalpy of vaporization in fair agreement with their calorimetric value. The drift in the 3rd law analysis indicates that the entropy of the solid is too low or that of the gas is too high by 1 cal $K^{-1}\ mol^{-1}$. Alternatively as systematic error in the pressure measurements could have caused this deviation.

The bond distances are taken from the electron diffraction experiments of Livingston et al. (6) and are in good accord with the adopted rotational constant.

References

1. B. D. Kybett, G. K. Johnson, C. K. Barker and J. L. Margrave, J. Phys. Chem. 69, 3603 (1965).
2. L. A. McDougall and J. E. Kilpatrick, J. Chem. Phys. 42, 2311 (1965).
3. W. J. Lafferty, A. G. Maki and E. K. Plyler, J. Chem. Phys. 40, 224 (1964).
4. F. A. Miller and W. G. Fateley, Spectrochim. Acta 20, 253 (1964).
5. F. A. Miller, D. H. Lemmon and R. E. Witkowski, Spectrochim. Acta 21, 1709-16 (1965).
6. R. L. Livingston and C. N. R. Rao, J. Amer. Chem. Soc. 81, 285 (1959).
7. T. R. Borgers, Ph.D. Thesis, University of California, Berkeley, 1961.

Carbon Suboxide (C_3O_2) C_3O_2(g)

Enthalpy Reference Temperature = T_r = 298.15 K Standard State Pressure = p° = 0.1 MPa

	J K⁻¹mol⁻¹			kJ mol⁻¹			
T/K	$C_p°$	S°	$-[G°-H°(T_r)]/T$	$H°-H°(T_r)$	$\Delta_fH°$	$\Delta_fG°$	Log K_f
0	0.	0.	INFINITE	-15.018	-96.819	-96.819	INFINITE
100	45.309	217.230	328.013	-11.078	-95.963	-100.070	52.271
200	56.055	251.539	281.870	-6.066	-94.842	-104.603	27.320
250	62.055	264.706	277.147	-3.110	-94.230	-107.114	22.380
298.15	66.989	276.071	276.071	0.	-93.638	-109.649	19.210
300	67.161	276.486	276.072	0.124	-93.616	-109.749	19.109
350	71.401	287.167	276.906	3.591	-93.038	-112.484	16.787
400	74.976	296.941	278.808	7.253	-92.526	-115.298	15.056
450	78.069	305.955	281.330	11.081	-92.100	-118.171	13.717
500	80.806	314.324	284.216	15.054	-91.764	-121.087	12.650
600	85.504	329.486	290.524	23.377	-91.335	-126.996	11.056
700	89.406	342.969	297.071	32.129	-91.157	-132.957	9.921
800	92.664	355.126	303.580	41.237	-91.148	-138.931	9.071
900	95.380	366.202	309.932	50.643	-91.251	-144.898	8.410
1000	97.642	376.372	316.075	60.298	-91.428	-150.850	7.880
1100	99.527	385.770	321.989	70.159	-91.657	-156.781	7.445
1200	101.103	394.499	327.672	80.193	-91.926	-162.690	7.082
1300	102.426	402.646	333.129	90.371	-92.228	-168.575	6.773
1400	103.542	410.278	338.370	100.671	-92.560	-174.436	6.508
1500	104.490	417.455	343.406	111.074	-92.921	-180.272	6.278
1600	105.298	424.225	348.247	121.565	-93.313	-186.083	6.075
1700	105.993	430.630	352.907	132.130	-93.736	-191.868	5.895
1800	106.592	436.706	357.395	142.760	-94.192	-197.628	5.735
1900	107.112	442.483	361.722	153.446	-94.683	-203.361	5.591
2000	107.566	447.989	365.899	164.180	-95.209	-209.067	5.460
2100	107.964	453.247	369.934	174.957	-95.774	-214.746	5.342
2200	108.314	458.278	373.837	185.771	-96.378	-220.397	5.233
2300	108.625	463.100	377.613	196.619	-97.022	-226.020	5.133
2400	108.900	467.729	381.272	207.495	-97.708	-231.615	5.041
2500	109.146	472.179	384.820	218.398	-98.435	-237.179	4.956
2600	109.367	476.465	388.263	229.324	-99.205	-242.714	4.876
2700	109.565	480.596	391.607	240.270	-100.018	-248.218	4.802
2800	109.744	484.584	394.857	251.236	-100.875	-253.691	4.733
2900	109.905	488.438	398.018	262.218	-101.774	-259.134	4.668
3000	110.052	492.166	401.094	273.216	-102.717	-264.544	4.606
3100	110.185	495.777	404.090	284.228	-103.703	-269.921	4.548
3200	110.307	499.277	407.011	295.253	-104.733	-275.268	4.493
3300	110.419	502.673	409.858	306.290	-105.805	-280.581	4.441
3400	110.521	505.971	412.637	317.337	-106.919	-285.860	4.392
3500	110.615	509.176	415.349	328.393	-108.076	-291.106	4.345
3600	110.701	512.294	417.999	339.459	-109.274	-296.319	4.299
3700	110.781	515.328	420.589	350.533	-110.514	-301.498	4.256
3800	110.855	518.283	423.121	361.615	-111.796	-306.642	4.215
3900	110.924	521.163	425.598	372.704	-113.118	-311.752	4.175
4000	110.987	523.973	428.023	383.800	-114 481	-316.828	4.137
4100	111.046	526.714	430.396	394.902	-115.885	-321.870	4.101
4200	111 102	529.391	432.722	406.009	-117.329	-326.876	4.065
4300	111.153	532.005	435.000	417.122	-118.814	-331.848	4.031
4400	111.201	534.561	437.234	428.240	-120.339	-336.785	3.998
4500	111.246	537.061	439.425	439.362	-121.905	-341.687	3.966
4600	111.288	539.506	441.574	450.489	-123.511	-346.554	3.935
4700	111.328	541.900	443.683	461.620	-125.157	-351.384	3.905
4800	111.365	544.244	445.754	472.754	-126.845	-356.180	3.876
4900	111.400	546.541	447.788	483.892	-128.574	-360.939	3.848
5000	111.433	548.792	449.785	495.034	-130.345	-365.663	3.820
5100	111.464	550.999	451.748	506.179	-132.158	-370.351	3.793
5200	111.493	553.164	453.678	517.327	-134.015	-375.005	3.767
5300	111.521	555.288	455.575	528.477	-135.914	-379.620	3.741
5400	111.547	557.372	457.441	539.631	-137.859	-384.201	3.716
5500	111.572	559.419	459.276	550.787	-139.848	-388.745	3.692
5600	111.595	561.430	461.083	561.945	-141.883	-393.251	3.668
5700	111.618	563.405	462.861	573.106	-143 966	-397.722	3.645
5800	111.639	565.347	464.611	584.269	-146.096	-402.155	3.622
5900	111.659	567.255	466.334	595.434	-148.275	-406.552	3.599
6000	111.678	569.132	468.032	606.600	-150.504	-410.911	3.577

PREVIOUS: June 1968 (1 atm) CURRENT: June 1968 (1 bar)

Carbon Suboxide (C_3O_2) C_3O_2(g)

CARBON (C_4) IDEAL GAS M_r = 48.044

$S°(298.15 K) = 228.3 \pm 16.7 J K^{-1} mol^{-1}$

$\Delta_f H°(0 K) = 964 \pm 33 kJ mol^{-1}$

$\Delta_f H°(298.15 K) = 971 \pm 33 kJ mol^{-1}$

Vibrational Frequencies and Degeneracies

ν, cm^{-1}
[2150] (1)
[950] (1)
[1700] (1)
[550] (2)
[550] (2)

Ground State Configuration: [$^1\Sigma$] σ = 2

Point Group: [$D_{\infty h}$]

Bond Distances: C–C = [1.28] Å

Bond Angle: C–C–C = [180°]

Rotational Constant: B_0 = [0.171328] cm^{-1}

Enthalpy of Formation

The enthalpy of formation is obtained from the mass-spectrometric results of Drowart et al. (1) who obtained $\Delta_f H°(2400 K) = 226 \pm 7 kcal mol^{-1}$. This was obtained from a 2nd law analysis of the ion intensity ratio C_4^+/C^+ over a temperature range of 2200–2700 K. the absolute pressure, obtained from a knowledge of the monatomic carbon vapor pressure and an estimated ionization cross-section ratio of 4.96, yields a 2nd law entropy at 2400 K of $90.5 \pm 3 cal K^{-1} mol^{-1}$. The uncertainty in the ionization cross-section ratio is probably no more than a factor of two which gives an overall entropy uncertainty of $\pm$ 4.5 cal $K^{-1} mol^{-1}$. The enthalpy of formation at temperatures other than 2400 K depend on the functions adopted, for a discussion of these see below.

Heat Capacity and Entropy

The molecular properties of C_4 are all estimated. We have chosen a set of functions which is consistent with the mass-spectrometric observations of Drowart et al. (1). The thermodynamic functions of C_4 were first estimated by Pitzer and Clementi (2) but the basis for this estimate has completely changed. Originally they were calculated by analogy with C_3, C_3O_2 and CO_2, but the discovery of very low bending frequencies in C_3 and C_3O_2 complicated the picture by raising the calculated entropies significantly. Sandborn (3) has correlated the bending force constants in several related triatomics and has estimated the vibrational frequencies in C_4 and C_5. These vibrational frequencies yield high entropies outside the limits of the 2nd law determination. Since these frequencies are estimates based on a basically triatomic correlation we have chosen to adopt two bending frequencies of 550 cm^{-1} rather than the 200 cm^{-1} and 440 cm^{-1} estimated by Sandborn. In addition we have assumed that the ground state arising from a π^2 term is $^1\Sigma$ and not $^3\Sigma$ as assumed by Pitzer and Clementi. The basis for both these changes is to make the functions conform to the only available measured data. Use of the present table will yield values for the vapor pressure of C_4 in excellent agreement with the measurements in the 2000–3000 K range. Outside this range they provide only a general indication of the variation since they may not correspond to the real situation.

The stretching frequencies are from Sandborn (3) and the bond length is estimated from C_3.

References

1. J. Drowart, R. P. Burns, G. DeMaria and M. G. Inghram, J. Chem. Phys. 31, 1131 (1959).
2. K. S. Pitzer and E. Clementi, J. Am. Chem. Soc. 81, 4477 (1959).
3. R. H. Sandborn, J. Chem. Phys. 49, 4219 (1968).

Carbon (C_4) $C_4(g)$

Enthalpy Reference Temperature = T_r = 298.15 K   Standard State Pressure = $p°$ = 0.1 MPa

T/K	$C_p°$	S°	$-[G°-H°(T_r)]/T$	$H°-H°(T_r)$	$\Delta_f H°$	$\Delta_f G°$	Log K_f
	J $K^{-1}mol^{-1}$			kJ mol^{-1}			
0	0.	0.	INFINITE	-10.791	964.101	964.101	INFINITE
100	29.865	187.031	265.752	-7.872	966.782	948.459	-495.424
200	39.876	210.373	232.603	-4.446	968.902	929.293	-242.706
250	45.537	219.893	229.123	-2.308	969.858	919.279	-192.073
298.15	50.176	228.322	228.322	0.	970.688	909.460	-159.334
300	50.338	228.633	228.323	0.093	970.718	909.080	-158.285
350	54.319	236.701	228.951	2.712	971.454	898.747	-134.131
400	57.657	244.179	230.393	5.514	972.048	888.318	-116.002
450	60.512	251.139	232.316	8.470	972.491	877.823	-101.895
500	62.998	257.646	234.527	11.559	972.786	867.287	-90.605
600	67.134	269.512	239.389	18.074	972.988	846.161	-73.665
700	70.424	280.117	244.463	24.958	972.781	825.034	-61.565
800	73.067	289.700	249.528	32.137	972.276	803.959	-52.493
900	75.201	298.434	254.484	39.554	971.555	782.961	-45.442
1000	76.933	306.450	259.285	47.164	970.672	762.052	-39.806
1100	78.348	313.851	263.914	54.930	969.663	741.238	-35.198
1200	79.511	320.719	268.365	62.825	968.553	720.520	-31.363
1300	80.475	327.123	272.641	70.826	967.358	699.899	-28.122
1400	81.279	333.117	276.749	78.915	966.088	679.372	-25.348
1500	81.955	338.748	280.697	87.078	964.753	658.939	-22.946
1600	82.528	344.056	284.492	95.303	963.358	638.596	-20.848
1700	83.017	349.075	288.145	103.580	961.907	618.343	-18.999
1800	83.437	353.832	291.663	111.904	960.404	598.176	-17.359
1900	83.800	358.353	295.055	120.266	958.850	578.095	-15.893
2000	84.115	362.660	298.329	128.662	957.248	558.096	-14.576
2100	84.390	366.770	301.491	137.088	955.599	538.179	-13.386
2200	84.632	370.702	304.548	145.539	953.904	518.341	-12.307
2300	84.846	374.469	307.507	154.013	952.164	498.582	-11.323
2400	85.035	378.084	310.373	162.507	950.379	478.899	-10.423
2500	85.204	381.559	313.151	171.019	948.551	459.291	-9.596
2600	85.355	384.903	315.847	179.548	946.679	439.758	-8.835
2700	85.490	388.127	318.464	188.090	944.765	420.297	-8.131
2800	85.612	391.239	321.008	196.645	942.808	400.908	-7.479
2900	85.722	394.245	323.482	205.212	940.809	381.590	-6.873
3000	85.822	397.153	325.890	213.789	938.768	362.341	-6.309
3100	85.913	399.968	328.234	222.376	936.684	343.161	-5.782
3200	85.996	402.697	330.519	230.971	934.560	324.049	-5.290
3300	86.071	405.345	332.746	239.575	932.393	305.004	-4.828
3400	86.140	407.915	334.919	248.186	930.186	286.025	-4.394
3500	86.204	410.413	337.041	256.803	927.937	267.112	-3.986
3600	86.263	412.842	339.113	265.426	925.646	248.263	-3.602
3700	86.317	415.207	341.138	274.055	923.315	229.479	-3.240
3800	86.367	417.509	343.117	282.689	920.943	210.758	-2.897
3900	86.413	419.753	345.054	291.328	918.530	192.101	-2.573
4000	86.456	421.942	346.949	299.972	916.075	173.505	-2.266
4100	86.496	424.077	348.804	308.620	913.580	154.972	-1.974
4200	86.533	426.162	350.621	317.271	911.044	136.500	-1.698
4300	86.568	428.198	352.402	325.926	908.468	118.089	-1.434
4400	86.600	430.189	354.147	334.584	905.850	99.738	-1.184
4500	86.631	432.135	355.858	343.246	903.192	81.447	-0.945
4600	86.659	434.040	357.537	351.911	900.493	63.216	-0.718
4700	86.686	435.904	359.185	360.578	897.754	45.044	-0.501
4800	86.711	437.729	360.802	369.248	894.974	26.930	-0.293
4900	86.734	439.517	362.391	377.920	892.154	8.875	-0.095
5000	86.756	441.270	363.951	386.594	889.293	-9.122	0.095
5100	86.777	442.988	365.484	395.271	886.391	-27.062	0.277
5200	86.797	444.673	366.990	403.950	883.449	-44.944	0.451
5300	86.816	446.327	368.472	412.630	880.466	-62.769	0.619
5400	86.833	447.950	369.929	421.313	877.443	-80.538	0.779
5500	86.850	449.543	371.362	429.997	874.379	-98.250	0.933
5600	86.866	451.108	372.772	438.683	871.275	-115.906	1.081
5700	86.881	452.646	374.160	447.370	868.130	-133.507	1.223
5800	86.895	454.157	375.526	456.059	864.945	-151.052	1.360
5900	86.903	455.642	376.871	464.749	861.719	-168.541	1.492
6000	86.921	457.103	378.196	473.441	858.453	-185.976	1.619

PREVIOUS: December 1969 (1 atm)   CURRENT: December 1969 (1 bar)

Carbon (C_4) $C_4(g)$

TETRAMETHYLSILANE ($Si(CH_3)_4$) IDEAL GAS M_r = 88.22478

$S°(298.15\ K) = 361.189\ J\ K^{-1}\ mol^{-1}$

$\Delta_f H°(0\ K)$ = Unknown

$\Delta_f H°(298.15\ K) = [-286.604]\ kJ\ mol^{-1}$

Enthalpy of Formation

Source unknown.

Heat Capacity and Entropy

$C°_p$ taken from Janz et al. (1).

Reference

1. G. J. Janz, Y. Mikawa, and F. Behnke, Rensseler Polytechnic Institute, personal communication, March 16, 1960.

Tetramethylsilane ($Si(CH_3)_4$) $C_4H_{12}Si_1(g)$

Enthalpy Reference Temperature = T_r = 298.15 K Standard State Pressure = p° = 0.1 MPa

	J K⁻¹mol⁻¹			kJ mol⁻¹			
T/K	$C°_p$	S°	$-[G°-H°(T_r)]/T$	$H°-H°(T_r)$	$\Delta_f H°$	$\Delta_f G°$	Log K_f
0							
100							
200							
298.15	138.574	361.189	361.189	0.	-286.604	-148.063	25.940
300	139.076	362.047	361.191	0.257	-286.768	-147.202	25.630
400	167.695	406.038	366.996	15.617	-295.056	-99.399	12.980
500	193.761	446.324	378.886	33.719	-302.074	-49.646	5.186
600	216.271	483.685	393.276	54.245	-307.811	1.399	-0.122
700	235.852	518.525	408.708	76.871	-312.352	53.310	-3.978
800	253.006	551.160	424.497	101.331	-315.801	105.794	-6.908
900	268.153	581.853	440.292	127.405	-318.275	158.652	-9.208
1000	281.416	610.807	455.910	154.897	-319.909	211.739	-11.060
1100	293.047	638.185	471.247	183.632	-320.839	264.954	-12.582
1200	303.256	664.131	486.249	213.457	-321.187	318.227	-13.852
1300	312.210	688.765	500.888	244.241	-321.064	371.510	-14.927
1400	319.992	712.196	515.152	275.862	-320.568	424.768	-15.848
1500	326.352	734.499	529.037	308.192	-319.809	477.981	-16.645

PREVIOUS: December 1960 (1 atm) CURRENT: December 1960 (1 bar)

Tetramethylsilane ($Si(CH_3)_4$) $C_4H_{12}Si_1(g)$

2-BUTYNEDINITRILE (C_4N_2) IDEAL GAS M_r = 76.0574

$S°(298.15\ K) = 290.11 \pm 0.8\ J\ K^{-1}\ mol^{-1}$

$\Delta_fH°(0\ K) = 528.60 \pm 0.8\ kJ\ mol^{-1}$

$\Delta_fH°(298.15\ K) = 533.46 \pm 0.8\ kJ\ mol^{-1}$

Vibrational Frequencies and Degeneracies

ν, cm^{-1}	ν, cm^{-1}	ν, cm^{-1}
2290(1)	2241(1)	263(2)
2119(1)	1154(1)	472(2)
692(1)	504(2)	107(2)

Ground State Quantum Weight: 1 σ = 2

Point Group: $D_{\infty h}$

Bond Distances: C≡N = 1.14 Å C–C = 1.37 Å C≡C = 1.19 Å

Rotational Constant: $B_0 = 0.045397\ cm^{-1}$

Enthalpy of Formation

Armstrong and Marantz (1), measured the enthalpy of combustion of the liquid for which they obtain a $\Delta_fH°$(298.15 K) of 120.6 kcal mol^{-1} using 94.05 kcal mol^{-1} for $\Delta_fH°(CO_2$, g, 298.15 K). They calculate the $\Delta_fH°(C_4N_2$, g, 298.15 K) with a $\Delta_{vap}H$ taken from Saggiomo (2).

Heat Capacity and Entropy

Bond lengths determined by Miller and Hannan (3) from X-ray diffraction measurements on the solid. The vibrational frequencies were observed in the infrared and Raman by Miller et al. (4).

References

1. G. T. Armstrong and S. Marantz, J. Phys. Chem. 64, 1776 (1960).
2. A. J. Saggiomo, J. Org. Chem. 22, 1171 (1957).
3. F. A. Miller and R. B. Hannan Jr., J. Chem. Phys. 21, 110 (1953).
4. F. A. Miller, R. B. Hannan Jr., and L. R. Cousins, J. Chem. Phys. 23, 2127 (1955).

2-Butynedinitrile (C_4N_2) C_4N_2(g)

Enthalpy Reference Temperature = T_r = 298.15 K Standard State Pressure = p° = 0.1 MPa

T/K	$C_p°$ (J K⁻¹mol⁻¹)	S° (J K⁻¹mol⁻¹)	−[G°−H°(T_r)]/T (J K⁻¹mol⁻¹)	H°−H°(T_r) (kJ mol⁻¹)	$\Delta_fH°$ (kJ mol⁻¹)	$\Delta_fG°$ (kJ mol⁻¹)	Log K_f
0	0.	0.	INFINITE	-17.734	528.600	528.600	INFINITE
100	49.991	216.793	356.363	-13.957	529.237	523.919	-273.667
200	71.983	258.541	297.572	-7.806	531.171	517.925	-135.268
250	79.961	275.500	291.495	-3.999	532.341	514.480	-107.495
298.15	85.888	290.112	290.112	0.	533.460	510.937	-89.514
300	86.088	290.644	290.114	0.159	533.502	510.797	-88.938
350	90.899	304.290	291.181	4.588	534.590	506.926	-75.655
400	94.823	316.692	293.606	9.234	535.569	502.906	-65.673
450	98.158	328.058	296.811	14.061	536.417	498.770	-57.896
500	101.091	338.554	300.467	19.043	537.132	494.548	-51.665
600	106.147	357.446	308.425	29.413	538.205	485.923	-42.303
700	110.417	374.138	316.643	40.247	538.905	477.150	-35.605
800	114.050	389.126	324.782	51.475	539.340	468.295	-30.577
900	117.130	402.742	332.699	63.038	539.588	459.398	-26.663
1000	119.730	415.222	340.336	74.885	539.703	450.481	-23.531
1100	121.921	426.739	347.674	86.971	539.716	441.557	-20.968
1200	123.768	437.429	354.714	99.258	539.649	432.636	-18.832
1300	125.329	447.399	361.464	111.715	539.516	423.724	-17.025
1400	126.654	456.737	367.940	124.316	539.325	414.823	-15.477
1500	127.784	465.514	374.155	137.039	539.082	405.938	-14.136
1600	128.751	473.793	380.126	149.867	538.791	397.071	-12.963
1700	129.584	481.624	385.868	162.785	538.455	388.224	-11.929
1800	130.305	489.052	391.396	175.780	538.075	379.398	-11.010
1900	130.933	496.114	396.723	188.843	537.651	370.594	-10.188
2000	131.481	502.844	401.862	201.964	537.186	361.813	-9.450
2100	131.963	509.271	406.825	215.137	536.679	353.057	-8.782
2200	132.388	515.420	411.623	228.355	536.130	344.325	-8.175
2300	132.764	521.314	416.265	241.613	535.541	335.620	-7.622
2400	133.099	526.971	420.760	254.906	534.910	326.941	-7.116
2500	133.398	532.411	425.118	268.231	534.239	318.290	-6.650
2600	133.667	537.648	429.346	281.585	533.526	309.666	-6.221
2700	133.908	542.697	433.451	294.964	532.772	301.070	-5.825
2800	134.126	547.571	437.441	308.366	531.978	292.503	-5.457
2900	134.323	552.281	441.320	321.788	531.142	283.964	-5.115
3000	134.502	556.838	445.095	335.230	530.265	275.456	-4.796
3100	134.665	561.251	448.771	348.688	529.348	266.977	-4.499
3200	134.813	565.529	452.353	362.162	528.389	258.529	-4.220
3300	134.950	569.679	455.846	375.650	527.388	250.111	-3.959
3400	135.075	573.710	459.254	389.152	526.347	241.724	-3.714
3500	135.189	577.627	462.580	402.665	525.264	233.369	-3.483
3600	135.295	581.437	465.829	416.189	524.140	225.045	-3.265
3700	135.393	585.145	469.004	429.724	522.975	216.752	-3.060
3800	135.484	588.757	472.108	443.268	521.768	208.492	-2.866
3900	135.567	592.278	475.144	456.820	520.519	200.264	-2.682
4000	135.645	595.711	478.116	470.381	519.230	192.069	-2.508
4100	135.718	599.061	481.025	483.949	517.898	183.906	-2.343
4200	135.785	602.333	483.874	497.524	516.525	175.777	-2.186
4300	135.848	605.528	486.667	511.106	515.110	167.680	-2.037
4400	135.907	608.652	489.404	524.694	513.654	159.617	-1.895
4500	135.962	611.707	492.088	538.287	512.155	151.588	-1.760
4600	136.014	614.696	494.721	551.886	510.615	143.592	-1.631
4700	136.062	617.621	497.305	565.490	509.034	135.630	-1.507
4800	136.108	620.487	499.841	579.098	507.410	127.703	-1.390
4900	136.151	623.293	502.332	592.711	505.744	119.809	-1.277
5000	136.191	626.044	504.779	606.328	504.036	111.950	-1.170
5100	136.229	628.742	507.183	619.949	502.285	104.126	-1.066
5200	136.265	631.387	509.546	633.574	500.493	96.336	-0.968
5300	136.299	633.983	511.870	647.202	498.658	88.581	-0.873
5400	136.331	636.531	514.155	660.834	496.781	80.862	-0.782
5500	136.361	639.033	516.403	674.468	494.861	73.177	-0.695
5600	136.390	641.490	518.614	688.106	492.899	65.528	-0.611
5700	136.418	643.905	520.791	701.746	490.894	57.914	-0.531
5800	136.444	646.278	522.935	715.389	488.846	50.335	-0.453
5900	136.468	648.610	525.045	729.035	486.754	42.793	-0.379
6000	136.492	650.904	527.124	742.683	484.619	35.286	-0.307

PREVIOUS: March 1961 (1 atm) CURRENT: March 1961 (1 bar)

2-Butynedinitrile (C_4N_2) C_4N_2(g)

NICKEL CARBONYL ($Ni(CO)_4$) LIQUID M_r = 170.7416

$S°(298.15\ K) = [319.56 \pm 12.6]\ J\ K^{-1}\ mol^{-1}$ $\Delta_f H°(298.15\ K) = -631.78 \pm 8.4\ kJ\ mol^{-1}$

$T_{fus} = 253.86 \pm 0.1\ K$ $\Delta_{fus}H° = 13.832 \pm 0.021\ kJ\ mol^{-1}$

Enthalpy of Formation

The adopted enthalpy of formation is a weighted average based on enthalpy of combustion data of Smagina and Ormont (1) and Fischer et al. (2). The results of Smagina and Ormont (1) have been recalculated along the lines indicated by Sykes (3), using recent values of $\Delta_f H°$(NiO, cr) (4) and other auxiliary data (5). Their corrected value of $\Delta_f H°(Ni(CO)_4$, ℓ, 298.15 K) is -150.3±3.0 kcal mol^{-1} (average of 3 determinations using Pt fuse). Fischer et al. [2] made two sets of determinations yielding the following values of $\Delta_f H°(Ni(CO)_4$, ℓ, 298.15 K); -150.3±0.5 kcal mol^{-1} (average of 5 determinations using benzoic acid fuse) and -152.3±0.5 kcal mol^{-1} (average of 4 determinations using Pt fuse). Fischer et al. (2) mention that the difference in their two sets of values may be due to hydration effects but a more likely explanation is higher uncertainties in the percentage combustion, as suggested by Sykes (3). Therefore, we adopt an uncertainty of ±2.0 kcal mol^{-1}.

Heat Capacity and Entropy

We adopt the smooth values of $C_p°$ (90-270 K) reported by Spice et al. (6). Values above 270 K are based on a linear extrapolation of liquid $C_p°$ values from 255-270 K. The $C_p°$ data reported by Duncan and Murray (7) were calculated from gas phase $C_p°$ data via a poorly defined procedure and show large deviations from the adopted $C_p°$.

The value of S°(298.15 K) is difficult to assess because of the lack of $C_p°$ data below 90 K. The adopted value of S°(298.15 K) is chosen so as to get the best agreement between second and third law analysis of the consistent vapor pressure data of Suginuma and Satozaki (8), Walsh (9), and Baev (10). This treatment results in a value of S°(cr, 90 K) = 23.95 cal K^{-1} mol^{-1}. Spice et al. (6) had estimated a value of S°(cr, 90 K) = 20.6 cal K^{-1} mol^{-1} based on comparisons (at 90 and 120 K) with a "standard" heat capacity curve.

Fusion Data

The adopted values of T_{fus} and $\Delta_{fus}H°$ are those reported by Spice et al. (6). The value of $\Delta_{fus}H°$ is the mean of three calorimetrically determined values. Walsh (9) reported a value of T_{fus} approximately 2 K higher than the adopted value and also reported that T_{fus} is very sensitive to purity of the $Ni(CO)_4$. However, we find the purity determinations of Spice et al. (6) more than compelling than those of Walsh (9).

Vaporization Data

The normal boiling point at 1 atm, T_{vap} = 315.4±0.2 K, is based on the vapor pressure measurements of Suginuma and Satozaki (8), Walsh (9) and Baev (10); fugacity corrections have not been made. See $Ni(CO)_4$(g) table (5) for analysis of $\Delta_{vap}H°$.

References

1. E. I. Smagina and B. F. Ormont, J. Gen. Chem. (USSR) 25, 207 (1955).
2. A. K. Fischer, F. A. Cotton, and G. Wilkinson, J. Amer. Chem. Soc. 79, 2044 (1957).
3. K. W. Sykes, J. Chem. Soc., 2053 (1958).
4. A. D. Mah and L. B. Pankratz, U. S. Bur. Mines Bull. 668, 1976.
5. JANAF Thermochemical Tables: CO_2(g), 9-30-65; $Ni(CO)_4$(g), 3-31-78.
6. J. E. Spice, L. A. K. Staveley and G. A. Harrow, J. Chem. Soc., 100 (1955).
7. A. B. F. Duncan and J. W. Murray, J. Chem. Phys. 2, 636 (1934).
8. B. Suginuma and K. Satozaki, Bull. Inst. Phys. Chem. Research (Tokyo) 21, 432 (1942).
9. K. A. Walsh, U. S. Atomic Energy Commission Report LA-1649, 1953.
10. A. K. Baev, Obshch. Prikl. Khim. 2, 146 (1970).

Nickel Carbonyl ($Ni(CO)_4$) $C_4Ni_1O_4$(l)

Enthalpy Reference Temperature = T_r = 298.15 K Standard State Pressure = p° = 0.1 MPa

	J K⁻¹mol⁻¹			kJ mol⁻¹			
T/K	$C_p°$	S°	$-[G°-H°(T_r)]/T$	$H°-H°(T_r)$	$\Delta_f H°$	$\Delta_f G°$	Log K_f
0							
100	98.031	145.316	484.659	-33.934	-645.917	-624.661	326.289
170.000	138.490	207.499	358.113	-25.604	GLASS <--> LIQUID		
170.000	196.230	207.499	358.113	-25.604	TRANSITION		
200	197.903	239.525	337.985	-19.692	-640.692	-604.698	157.931
298.15	203.342	319.560	319.560	0.	-631.784	-588.980	103.187
300	203.426	320.818	319.564	0.376	-631.628	-588.715	102.504
400	208.949	380.097	327.609	20.995	-623.770	-575.626	75.169
500	214.472	427.315	342.983	42.166	-616.991	-564.395	58.962
600	219.995	466.906	360.424	63.889	-611.178	-554.437	48.268
700	225.518	501.234	378.141	86.165	-605.825	-545.398	40.698

PREVIOUS: CURRENT: March 1978

Nickel Carbonyl ($Ni(CO)_4$) $C_4Ni_1O_4$(l)

NICKEL CARBONYL ($Ni(CO)_4$) IDEAL GAS M_r = 170.7316

$\Delta_fH°$(0 K) = -605.71 ± 10.5 kJ mol^{-1}

S°(298.15 K) = 415.51 ± 8.4 J K^{-1} mol^{-1} $\Delta_fH°$(298.15 K) = -601.58 ± 10.5 kJ mol^{-1}

Vibrational Frequencies, Degeneracies, and Anharmonicities

ν, cm^{-1}	ν, cm^{-1}	x_{ij}, cm^{-1}
2132.4(1)	458.9(3)	x_{11} = -2.6
370.8(1)	423.1(3)	x_{15} = -11.0
380 (2)	79 (3)	x_{55} = -5.0
62 (2)	300 (3)	g_{55} = -4.5
2057.8(3)		All others = 0

Ground State Quantum Weight: 1 σ = 12

Point Group: T_d

Bond Distances: Ni–C = 1.82±0.03 Å C–O = 1.15±0.02 Å

Bond Angles: C–Ni–C = 109.4712° Ni–C–O = 180°

Product of the Moments of Inertia: $I_AI_BI_C$ = 5.14154 x 10^{-112} g^3 cm^6

Enthalpy of Formation

The value of $\Delta_fH°$(g, 298.15 K) is obtained from $\Delta_fH°$(ℓ, 298.15 K) (1) by addition of the enthalpy of vaporization, $\Delta_{vap}H°$(298.15 K) = 7.22±1.5 kcal mol^{-1}. The adopted value of the enthalpy of vaporization is selected from our 3rd law analysis of the vaporization data given below and is based primarily on the data of Suginuma and Satozaki (2), Baev (3) and Walsh (4). The uncertainty has been increased above that indicated in the table below to account for uncertainties in the free energy functions of the liquid and gas. Fugacity corrections have not been applied but they are expected to be within the stated uncertainty. The enthalpy of vaporization at the boiling point, $\Delta_{vap}H°$, is the difference between $\Delta_fH°$(g) and $\Delta_fH°$(ℓ) at the boiling point.

Data for the equilibrium Ni(cr) + 4CO(g) = $Ni(CO)_4$(g) have been reviewed by Kipnis (7). The data show considerable scatter and this has been attributed to differences in chemical potential of nickel (sheet or powder) used in the reaction (7, 8). Values of $\Delta_fH°$(298.15 K) derived from these studies are less negative than the adopted value by several kcal mol^{-1} and are less reliable due to uncertainty in the chemical potential of nickel.

Source	Data Points	T, K	$\Delta_{vap}H°$(298.15 K), kcal mol^{-1} 2nd Law	3rd Law	Drift cal K^{-1} mol^{-1}	$\Delta_fH°$(298.15 K) kcal mol^{-1}
Suginuma (2)	20	273-298	7.209±0.024	7.217±0.005	0.026±0.084	143.78±2.5
Suginuma (2)	18	273-307	7.285±0.015	7.224±0.005	-0.208±0.052	143.78±2.5
Baev (3)	33*	285-312	7.146±0.013	7.216±0.006	0.231±0.043	143.78±2.5
Walsh (4)	5	273-308	7.233±0.031	7.222±0.005	-0.037±0.105	143.78±2.5
Anderson (5)	9	253-316	6.933±0.025	7.209±0.043	0.961±0.086	143.79±2.5
Mittasch (6)	8	275-313	7.284±0.075	7.264±0.016	-0.068±0.255	143.74±2.5

*Point at 281.94 K rejected by statistical test.

Heat Capacity and Entropy

The adopted vibrational data, including anharmonicity, are from Jones et al. (9). The largest uncertainty in these assignments is in the frequency of the MCO bending vibration, ν_3. Jones et al. (9) chose 380 cm^{-1} based on a study of the spectra of substituted carbonyls by Bigorgne and Bouquet (10). Ross et al. (11) summarized values of S°(298.15 K) (96.2 to 99.8 cal K^{-1} mol^{-1}) obtained from various frequency assignments; most of this variation is due to alternate assignments for ν_3.

The adopted molecular structure data are from the electron diffraction study of Brockway and Cross (12) which conclusively indicates a tetrahedral structure. Moment of inertia calculations based on the adopted bond distances and angles yield $I_A=I_B=I_C$ = 8.01120 x 10^{-38} g cm^2.

The thermodynamic functions, including the effects of available vibrational anharmonicities, are calculated using the computer program of McDowell (13). Compared to a harmonic oscillator calculation using the same data the present results are identical at 298.15 K, while at 6000 K the entropy is 0.333 cal K^{-1} mol^{-1} higher.

References

1. JANAF Thermochemical Tables: $Ni(CO)_4$(ℓ), 3-31-78.
2. B. Suginuma and K. Satozaki, Bull. Inst. Phys. Chem. Research (Tokyo) 21, 432 (1942).
3. A. K. Baev, Obshch Prikl. Khim. 2, 146 (1970).
4. K. A. Walsh, U. S. Atomic Energy Commission Report LA-1649, 1953.
5. J. E. Anderson, J. Chem. Soc. 133, 1653 (1930).
6. A. Mittasch, Z. Physik. Chem. 40, 1 (1902).
7. A. Ya. Kipnis and N. F. Mikhailova, Zh. Prikl. Khim. (Leningrad) 45, 1450 (1972); A. Ya. Kipnis, Russ. J. Phys. Chem. 7, 775 (1962).
8. J. E. Spice, L. A. K. Staveley and G. A. Harrow, J. Chem. Soc., 100 (1955).
9. L. H. Jones, R. S. McDowell and M. Goldblatt, J. Chem. Phys. 48, 2663 (1968).
10. M. Bigorgne and G. Bouquet, Compt. Rend. 264, 1485 (1967).
11. L. W. Ross, F. H. Haynie and R. F. Hochman, J. Chem. Eng. Data 9, 339 (1964).
12. L. O. Brockway and P. C. Cross, J. Chem. Phys. 3, 828 (1935).
13. R. E. McDowell, QCPE #269, Quantum Chemistry Program Exchange, Indiana University (1974).

Nickel Carbonyl ($Ni(CO)_4$) $C_4Ni_1O_4$(g)

Enthalpy Reference Temperature = T_r = 298.15 K Standard State Pressure = p° = 0.1 MPa

T/K	$C_p°$ (J $K^{-1}mol^{-1}$)	S°	-[G°-H°(T_r)]/T	H°-H°(T_r) (kJ mol^{-1})	$\Delta_fH°$	$\Delta_fG°$	Log K_f
0	0.	0.	INFINITE	-30.495	-605.714	-605.714	INFINITE
100	84.479	286.831	532.198	-24.537	-606.311	-599.206	312.993
200	127.972	359.892	428.595	-13.741	-604.533	-592.611	154.774
298.15	149.289	415.507	415.507	0.	-601.576	-587.378	102.906
300	149.557	416.431	415.510	0.276	-601.519	-587.290	102.256
400	160.377	461.081	421.524	15.823	-598.734	-582.983	76.130
500	167.490	497.670	433.204	32.233	-596.716	-579.297	60.519
600	173.167	528.724	446.600	49.274	-595.585	-575.934	50.140
700	178.000	555.790	460.306	66.839	-594.943	-572.706	42.736
800	182.138	579.836	473.772	84.851	-594.376	-569.571	37.189
900	185.652	601.497	486.781	103.245	-594.059	-566.492	32.878
1000	188.623	621.216	499.253	121.963	-593.948	-563.436	29.431
1100	191.125	639.314	511.174	140.954	-594.007	-560.383	26.610
1200	193.238	656.037	522.558	160.175	-594.219	-557.319	24.259
1300	195.025	671.577	533.431	179.590	-594.575	-554.230	22.269
1400	196.543	686.087	543.822	199.171	-595.076	-551.109	20.562
1500	197.836	699.692	553.765	218.891	-595.721	-547.947	19.081
1600	198.949	712.496	563.289	238.731	-596.514	-544.737	17.784
1700	199.902	724.587	572.425	258.675	-597.462	-541.473	16.637
1800	200.748	736.037	581.199	278.708	-615.719	-537.434	15.596
1900	201.472	746.911	589.637	298.820	-616.893	-533.053	14.655
2000	202.121	757.262	597.762	319.000	-618.127	-528.608	13.806
2100	202.683	767.137	605.594	339.241	-619.424	-524.101	13.036
2200	203.192	776.578	613.153	359.535	-620.783	-519.529	12.335
2300	203.641	785.620	620.456	379.877	-622.208	-514.895	11.694
2400	204.049	794.296	627.520	400.262	-623.699	-510.199	11.104
2500	204.412	802.633	634.359	420.685	-625.257	-505.437	10.561
2600	204.744	810.657	640.986	441.143	-626.882	-500.613	10.057
2700	205.042	818.390	647.415	461.633	-628.575	-495.724	9.590
2800	205.317	825.852	653.655	482.151	-630.336	-490.771	9.155
2900	205.566	833.061	659.718	502.695	-632.164	-485.756	8.749
3000	205.798	840.034	665.613	523.264	-634.059	-480.674	8.369
3100	206.013	846.785	671.349	543.854	-636.021	-475.528	8.013
3200	206.213	853.329	676.934	564.466	-1014.883	-465.132	7.592
3300	206.396	859.678	682.376	585.096	-1015.322	-447.946	7.090
3400	206.568	865.842	687.682	605.745	-1015.825	-430.744	6.618
3500	206.730	871.832	692.858	626.409	-1016.390	-413.529	6.172
3600	206.882	877.658	697.911	647.090	-1017.019	-396.295	5.750
3700	207.024	883.328	702.846	667.786	-1017.711	-379.044	5.351
3800	207.158	888.851	707.668	688.495	-1018.466	-361.771	4.973
3900	207.284	894.234	712.383	709.217	-1019.284	-344.479	4.614
4000	207.405	899.483	716.995	729.951	-1020.166	-327.166	4.272
4100	207.521	904.606	721.509	750.698	-1021.112	-309.829	3.947
4200	207.631	909.608	725.928	771.455	-1022.122	-292.469	3.637
4300	207.736	914.495	730.257	792.224	-1023.197	-275.083	3.342
4400	207.836	919.272	734.499	813.002	-1024.338	-257.673	3.059
4500	207.932	923.944	738.657	833.791	-1025.546	-240.236	2.789
4600	208.024	928.515	742.735	854.589	-1026.822	-222.772	2.530
4700	208.113	932.989	746.735	875.395	-1028.167	-205.277	2.281
4800	208.200	937.372	750.661	896.211	-1029.583	-187.754	2.043
4900	208.285	941.666	754.516	917.035	-1031.071	-170.199	1.814
5000	208.367	945.874	758.301	937.868	-1032.634	-152.615	1.594
5100	208.446	950.001	762.019	958.709	-1034.273	-134.997	1.383
5200	208.522	954.050	765.673	979.557	-1035.991	-117.350	1.179
5300	208.596	958.022	769.265	1000.413	-1037.789	-99.665	0.982
5400	208.669	961.922	772.797	1021.276	-1039.673	-81.948	0.793
5500	208.738	965.752	776.271	1042.147	-1041.642	-64.195	0.610
5600	208.807	969.514	779.688	1063.024	-1043.702	-46.404	0.433
5700	208.874	973.210	783.051	1083.908	-1045.855	-28.577	0.262
5800	208.941	976.843	786.361	1104.799	-1048.103	-10.709	0.096
5900	209.006	980.416	789.620	1125.696	-1050.452	7.194	-0.064
6000	209.070	983.929	792.829	1146.600	-1052.904	25.143	-0.219

PREVIOUS: March 1978 (1 atm) CURRENT: March 1978 (1 bar)

Nickel Carbonyl ($Ni(CO)_4$) $C_4Ni_1O_4$(g)

$S°(298.15\ K) = 242.0 \pm 16.7\ J\ K^{-1}\ mol^{-1}$

$\Delta_f H°(0\ K) = 972 \pm 25\ kJ\ mol^{-1}$

$\Delta_f H°(298.15\ K) = 979 \pm 25\ kJ\ mol^{-1}$

Vibrational Frequencies and Degeneracies

ν, cm^{-1}	ν, cm^{-1}
[1920] (1)	[550] (2)
[770] (1)	[550] (2)
[2210] (1)	[550] (2)
[1430] (1)	

Ground State Configuration: $[^1\Sigma]$ $\sigma = 2$

Point Group: $[D_{\infty h}]$

Bond Distances: C–C = [1.28] Å

Bond Angle: C–C–C = [180°]

Rotational Constant: B_0 = [0.085664] cm^{-1}

Enthalpy of Formation

The enthalpy of formation is obtained from the mass-spectrometric results of Drowart et al. (1) who obtained $\Delta_f H°(2400\ K) = 230 \pm 5\ kcal\ mol^{-1}$. This was obtained from a 2nd law analysis of the ion intensity ration C_5^+/C^+ over a temperature range of 2200–2700 K. The absolute pressure, obtained from a knowledge of the monatomic carbon vapor pressure and an estimated ionization cross-section ratio of 6.3, yields a 2nd law entropy at 2400 K of $103.7 \pm 2\ cal\ K^{-1}\ mol^{-1}$. The uncertainty in the ionization cross-section ratio is probably no more than a factor of two which gives an overall entropy uncertainty of $\pm 3.5\ cal\ K^{-1}\ mol^{-1}$. The enthalpy of formation at temperatures other than 2400 K depend on the functions adopted, for a discussion of these see below.

Heat Capacity and Entropy

The molecular properties of C_5 are all estimated. We have chosen a set of functions which is consistent with the mass-spectrometric observations of Drowart et al. (1). The thermodynamic functions of C_5 were first estimated by Pitzer and Clementi (2) but the basis for this estimate has completely changed. Originally they were calculated by analogy with C_3, C_3O_2 and CO_2, but the discovery of very low bending frequencies in C_3 and C_3O_2 complicated the picture by raising the calculated entropies significantly. Sandborn (3) has correlated the bending force constants in several related triatomics and has estimated the vibrational frequencies in C_4 and C_5. These vibrational frequencies yield high entropies outside the limits of the 2nd law determination. Since these frequencies are estimates, based on a basically triatomic correlation we have chosen to adopt three doubly degenerate bending frequencies of 550 cm^{-1} in place of the 130 cm^{-1}, 330 cm^{-1} and 520 cm^{-1} frequencies estimated by Sandborn (3). The basis for this change is to make the functions conform to the only available measured data. Use of the present tables will yield values for the vapor pressure of C_4 is excellent agreement with the measurements in the 2000–3000 K range. Outside this range they provide only a general indication of the variation since they may not correspond to the real situation.

The stretching frequencies are from Sandborn (3) and the bond length is estimated equal to that in C_3.

References

1. J. Drowart, R. P. Burns, G. DeMaria, M. G. Inghram, J. Chem. Phys. 31, 1131 (1959).
2. K. S. Pitzer and E. Clementi, J. Am. Chem. Soc. 81, 4477 (1959).
3. R. H. Sandborn, J. Chem. Phys. 49, 4219 (1968).

Enthalpy Reference Temperature = T_r = 298.15 K Standard State Pressure = p° = 0.1 MPa

T/K	$C_p°$ (J K⁻¹mol⁻¹)	S° (J K⁻¹mol⁻¹)	$-[G°-H°(T_r)]/T$ (J K⁻¹mol⁻¹)	$H°-H°(T_r)$ (kJ mol⁻¹)	$\Delta_f H°$ (kJ mol⁻¹)	$\Delta_f G°$ (kJ mol⁻¹)	Log K_f
0	0.	0.	INFINITE	-11.914	972.398	972.398	INFINITE
100	30.260	195.633	285.525	-8.989	975.024	955.937	-499.330
200	45.669	220.670	247.105	-5.287	977.094	936.042	-244.469
250	54.253	231.807	242.941	-2.784	978.119	925.660	-193.406
298.15	61.135	241.972	241.972	0.	979.056	915.470	-160.387
300	61.372	242.350	241.973	0.113	979.090	915.075	-159.329
350	67.147	252.260	242.742	3.331	979.954	904.336	-134.965
400	71.899	261.546	244.519	6.811	980.674	893.482	-116.677
450	75.903	270.252	246.899	10.509	981.231	882.548	-102.444
500	79.347	278.432	249.648	14.392	981.622	871.561	-91.051
600	85.013	293.421	255.718	22.622	981.961	849.508	-73.956
700	89.478	306.874	262.081	31.355	981.829	827.437	-61.744
800	93.049	319.064	268.454	40.488	981.357	805.410	-52.588
900	95.926	330.195	274.705	49.942	980.638	783.458	-45.471
1000	98.259	340.427	280.772	59.655	979 736	761.595	-39.782
1100	100.164	349.885	286.631	69.579	978.691	739.831	-35.132
1200	101.731	358.670	292.273	79.676	977.532	718.167	-31.261
1300	103.029	366.865	297.699	89.916	976.277	696.603	-27.990
1400	104.113	374.541	302.916	100.275	974.938	675.140	-25.190
1500	105.025	381.756	307.934	110.733	973.524	653.774	-22.766
1600	105.798	388.560	312.763	121.276	972.041	632.506	-20.649
1700	106.457	394.994	317.412	131.889	970.494	611.332	-18.784
1800	107.023	401.096	321.893	142.564	968.885	590.251	-17.129
1900	107.513	406.895	326.216	153.291	967.218	569.261	-15.650
2000	107.938	412.421	330.389	164.064	965.493	548.360	-14.322
2100	108.310	417.697	334.422	174.877	963.713	527.547	-13.122
2200	108.637	422.743	338.323	185.725	961.877	506.820	-12.033
2300	108.925	427.579	342.099	196.603	959.988	486.177	-11.041
2400	109.181	432.220	345.758	207.509	958.045	465.618	-10.134
2500	109.409	436.682	349.306	218.439	956.049	445.142	-9.301
2600	109.613	440.977	352.750	229.390	954.001	424.745	-8.533
2700	109.796	445.117	356.095	240.360	951.900	404.429	-7.824
2800	109.960	449.113	359.346	251.348	949.748	384.192	-7.167
2900	110.109	452.974	362.508	262.352	947.544	364.032	-6.557
3000	110.244	456.710	365.586	273.370	945.289	343.949	-5.989
3100	110.367	460.327	368.584	284.400	942.982	323.942	-5.458
3200	110.479	463.832	371.506	295.443	940.624	304.011	-4.962
3300	110.581	467.234	374.356	306.496	938.215	284.154	-4.498
3400	110.675	470.536	377.137	317.559	935.755	264.371	-4.062
3500	110.761	473.746	379.851	328.631	933.244	244.660	-3.651
3600	110.840	476.867	382.503	339.711	930.682	225.023	-3.265
3700	110.913	479.905	385.094	350.798	928.069	205.457	-2.901
3800	110.981	482.864	387.629	361.893	925.406	185.962	-2.556
3900	111.043	485.747	390.108	372.994	922.692	166.539	-2.231
4000	111.101	488.559	392.534	384.101	919.927	147.185	-1.922
4100	111.155	491.303	394.910	395.214	917.111	127.901	-1.629
4200	111.206	493.983	397.237	406.332	914.245	108.687	-1.352
4300	111.252	496.600	399.517	417.455	911.328	89.541	-1.088
4400	111.296	499.158	401.753	428.583	908.361	70.464	-0.837
4500	111.337	501.660	403.945	439.714	905.343	51.455	-0.597
4600	111.376	504.107	406.096	450.850	902.275	32.513	-0.369
4700	111.412	506.503	408.207	461.990	899.156	13.639	-0.152
4800	111.446	508.849	410.279	473.132	895.987	-5.168	0.056
4900	111.477	511.147	412.315	484.279	892.767	-23.909	0.255
5000	111.507	513.399	414.314	495.428	889.496	-42.584	0.445
5100	111.536	515.608	416.278	506.580	886.176	-61.192	0.627
5200	111.562	517.774	418.209	517.735	882.804	-79.735	0.801
5300	111.587	519.899	420.108	528.892	879.383	-98.213	0.968
5400	111.611	521.985	421.976	540.052	875.910	-116.626	1.128
5500	111.634	524.033	423.813	551.215	872.388	-134.974	1.282
5600	111.655	526.045	425.620	562.379	868.815	-153.257	1.430
5700	111.676	528.021	427.399	573.546	865.191	-171.477	1.571
5800	111.695	529.964	429.151	584.714	861.517	-189.632	1.708
5900	111.713	531.873	430.876	595.885	857.793	-207.723	1.839
6000	111.730	533.751	432.575	607.057	854.018	-225.751	1.965

PREVIOUS: December 1969 (1 atm) CURRENT: December 1969 (1 bar)

IRON CARBONYL ($Fe(CO)_5$) LIQUID M_r = 195.8990

$S°(298.15\ K) = 337.08 \pm 4\ J\ K^{-1}\ mol^{-1}$

$T_{fus} = 253.1 \pm 0.1\ K$

$\Delta_f H°(298.15\ K) = -766.09 \pm 7.1\ kJ\ mol^{-1}$

$\Delta_{fus}H° = 13.226 \pm 0.013\ kJ\ mol^{-1}$

Enthalpy of Formation

The adopted enthalpy of formation is based on the enthalpy of combustion measurements of Cotton et al. (1). The average energy of combustion reported by these authors is $-386.86 \pm 1.7\ kcal\ mol^{-1}$ which yields a value of $\Delta_c H°(298.15\ K) = -385.82 \pm 1.7\ kcal\ mol^{-1}$. Using auxiliary JANAF data (2) we calculate the adopted value of $\Delta_f H°(298.15\ K) = -183.1 \pm 1.7\ kcal\ mol^{-1}$. Earlier combustion work by Roth (3) and Mittasch (4) appears less reliable due to incomplete characterization of products.

Heat Capacity and Entropy

The adopted heat capacity is derived by merging the $C°_p$ data of Leadbetter and Spice (5) (22 - 287 K) with $C°_p$ data derived from the enthalpy measurements of Welty et al. (6) (302 - 387 K). Leadbetter and Spice report a small anomaly in $C°_p$(cr) near 230 K which they attribute to the presence of a small amount of impurity; therefore we have smoothed through this anomaly. The heat capacity (5) and enthalpy data (6) of the liquid merge smoothly and yield a constant $C°_p = 55.876\ cal\ K^{-1}\ mol^{-1}$ from the melting point to 387 K. We assume $C°_p$ remains constant at this value above this temperature.

The adopted value of S°(298.15 K) is obtained by the appropriate integration of $C°_p$ data assuming $S°(25\ K) = 3.49 \pm 1.0\ cal\ K^{-1}\ mol^{-1}$. S°(25 K) is obtained by using a single Debye function with 18 degrees of freedom and θ_D = 155 K as suggested by Leadbetter and Spice (5). The uncertainty in this procedure is rather large since a single Debye function fits the observed $C°_p$ data quite poorly.

Fusion Data

The adopted values of T_{fus} and $\Delta_{fus}H°$ are those reported by Leadbetter and Spice (5). The value of T_{fus} agrees well with that listed by Brynestad (7). For $\Delta_{fus}H°$ the only other value is $\Delta_{fus}H° = 3250 \pm 40\ cal\ K^{-1}\ mol^{-1}$ reported by Mittasch (4). This is much less precise than the adopted value which is an average of two determinations.

Vaporization Data

The adopted boiling point, T_{vap}(1 atm) = ~ 377 K, is based on the vapor pressure data of Trautz and Badstubner (8) and Baev (9); fugacity corrections have not been made. See the enthalpy of formation section of the $Fe(CO)_5$(g) table (2) for details of the selected value of $\Delta_{vap}H°$.

References

1. F. A. Cotton, A. K. Fischer and G. Wilkinson, J. Amer. Chem. Soc. 81, 800 (1959).
2. JANAF Thermochemical Tables: Fe_2O_3(cr), 6-30-65; CO_2(g), 9-20-65; $Fe(CO)_5$(g), 3-31-78.
3. W. A. Roth, Angew. Chem. 42, 981 (1929).
4. A. Mittasch, Angew. Chem. 41, 827 (1928).
5. A. J. Leadbetter and J. E. Spice, Can. J. Chem. 37, 1923 (1959).
6. J. R. Welty, C. E. Wicks and H. O. Boren, U. S. Bur. Mines RI 6155, 1963.
7. J. Brynestad, ORNL/TM-5499, 1977.
8. M. Trautz and W. Badstubner, Z. ELectrokhem. 35, 799 (1929).
9. A. K. Baev, Obshch. Prikl. Khim., #2, 146 (1970).

Iron Carbonyl ($Fe(CO)_5$) $C_5Fe_1O_5$(l)

Enthalpy Reference Temperature = T_r = 298.15 K Standard State Pressure = p° = 0.1 MPa

T/K	$C°_p$ (J K⁻¹mol⁻¹)	S° (J K⁻¹mol⁻¹)	$-[G°-H°(T_r)]/T$ (J K⁻¹mol⁻¹)	$H°-H°(T_r)$ (kJ mol⁻¹)	$\Delta_f H°$ (kJ mol⁻¹)	$\Delta_f G°$ (kJ mol⁻¹)	Log K_f
0	0.	0.	INFINITE	-52.934	-787.554	-787.554	INFINITE
100	105.240	113.401	580.650	-46.725	-789.327	-756.258	395.029
200	164.163	204.889	370.766	-33.175	-786.457	-724.021	189.095
253.100	191.020	246.527	340.393	-23.758	CRYSTAL <--> LIQUID TRANSITION		
253.100	233.785	298.781	340.393	-10.532			
298.15	233.785	337.078	337.078	0.	-766.090	-696.975	122.107
300	233.785	338.524	337.083	0.432	-765.919	-696.547	121.280
400	233.785	405.780	346.253	23.811	-757.710	-674.716	88.109
500	233.785	457.948	363.569	47.190	-751.465	-654.729	68.399
600	233.785	500.572	382.958	70.568	-746.962	-635.831	55.354
700	233.785	536.610	402.401	93.947	-743.911	-617.569	46.084

PREVIOUS: CURRENT: March 1978

Iron Carbonyl ($Fe(CO)_5$) $C_5Fe_1O_5$(l)

IRON CARBONYL ($Fe(CO)_5$) IDEAL GAS M_r = 195.8990

$\Delta_f H°(0\ K) = -729.52 \pm 7.1\ kJ\ mol^{-1}$

$S°(298.15\ K) = 439.3 \pm 12.6\ J\ K^{-1}\ mol^{-1}$ $\Delta_f H°(298.15\ K) = -727.85 \pm 7.1\ kJ\ mol^{-1}$

Vibrational Frequencies, Degeneracies, and Anharmonicities

ν, cm^{-1}	ν, cm^{-1}	ν, cm^{-1}	x_{ij}, cm^{-1}	x_{ij}, cm^{-1}
2120.7 (1)	618.8 (1)	474.3 (2)	$x_{1,1}$ = -2.2	$x_{1,6}$ = -11.9
2041.7 (1)	429.0 (1)	104.9 (2)	$x_{2,10}$ = -6.9	$x_{1,10}$ = -7.2
442.8 (1)	100 (1)	74.3 (2)	$x_{6,6}$ = -7.7	$x_{2,6}$ = -14.1
413.4 (1)	2013.3 (2)	488 (2)	$x_{10,10}$ = -5.6	$x_{2,2}$ = $x_{6,10}$ = -2.0
383 (1)	645.0 (2)	375 (2)	$x_{1,2}$ = -9.3	all others = 0
2034.0 (1)	542.5 (2)	97.3 (2)		

Point Group: D_{3h} σ = 6

Bond Distances: Fe–C$_{eq}$ = 1.833 ± 0.004 Å Fe–C$_{ax}$ = 1.806 ± 0.005 Å C–O = 1.145 ± 0.003 Å

Bond Angles: Fe–C–O = 180° C$_{ax}$–Fe–C$_{ax}$ = 180° C$_{eq}$–Fe–C$_{eq}$ = 120° C$_{ax}$–Fe–C$_{eq}$ = 90°

Product of the Moments of Inertia: $I_A I_B I_C = 9.94833 \times 10^{-112}\ g^3\ cm^6$

Enthalpy of Formation

The adopted value of $\Delta_f H°$(g, 298.15 K) is obtained from $\Delta_f H°$(ℓ, 298.15 K) (1) by addition of the enthalpy of vaporization, $\Delta_{vap}H°(298.15\ K) = 9.14\pm1.2\ kcal\ mol^{-1}$. The adopted value of the enthalpy of vaporization is selected from our 3rd law analysis of the vaporization data below and is based primarily on the data of Trautz and Badstubner (2) and Baev (3). The uncertainty has been increased above that indicated in the table below to account for uncertainties in the free functions of the liquid and gas. Fugacity corrections have not been made but they are expected to be well within the stated uncertainty. The enthalpy of vaporization at the boiling point, $\Delta_{vap}H°$, is calculated from the difference between $\Delta_f H°$(g) and $\Delta_f H°$(ℓ) at the boiling point.

Source	Data Points	T/K	$\Delta_{vap}H°$(298.15 K), kcal mol^{-1} 2nd Law	3rd Law	Drift cal K^{-1} mol^{-1}	$\Delta_f H°$(298.15 K) kcal mol^{-1}
Trautz (2)	69	320-377	9.727±0.004	9.138±0.49	-1.688±0.011	173.96±1.7
Baev (3)	28	266-353	9.215±0.041	9.141±0.036	-0.224±0.125	173.96±1.7
Leadbetter (4)	16	293-306	9.754±0.095	9.219±0.017	-1.780±0.317	173.88±1.7
Gilbert (5)	12	254-306	9.207±0.106	9.241±0.042	0.121±0.381	173.86±1.7
Eyber (6)	13	258-293	9.505±0.014	9.247±0.025	-0.936±0.052	173.85±1.7
Dewar (7)	7	266-351	7.195±0.550	8.977±0.440	5.915±1.816	174.12±1.8

Heat Capacity and Entropy

The selected vibrational data, including anharmonicity, are from Jones et al. (8). The largest uncertainty in these assignments is for ν_9; Jones estimates it to lie at 100±15 cm^{-1}. The uncertainty in this assignment coupled with unknown anharmonicities for the low frequency vibrational modes lead to rather large uncertainties in the thermal functions. Following the arguments of Jones we estimate an uncertainty at ±3.0 cal K^{-1} mol^{-1} in S°(298.15 K).

The adopted molecular structure data are from the electron diffraction data of Beagley et al. (9) which clearly indicate that the Fe-C axial bond is shorter than the Fe-C equatorial bond. Moment of inertia calculations based on the adopted bond distances and angles yield $I_A = 9.078946 \times 10^{-38}\ g\ cm^2$ and $I_B=I_C= 1.046785 \times 10^{-37}\ g\ cm^2$. The electron diffraction study of Almenningen et al. (10) is in excellent agreement with the adopted data although a slightly smaller difference between the axial and equatorial Fe-C bond lengths is indicated. This would lead to an insignificant difference in S°(298.15 K).

The thermodynamic functions, including the effects of available vibrational anharmonicities, are calculated using the computer program of McDowell (11). Compared to a harmonic oscillator calculation using the same data the present results are identical at 298.15 K, while at 6000 K the entropy is 0.407 cal K^{-1} mol^{-1} higher.

References

1. JANAF Thermochemical Tables: $Fe(CO)_5$(ℓ), 3-31-78.
2. M. Trautz and W. Badstubner, Z. Electrokhem. 35, 799 (1929).
3. A. K. Baev, Obshch. Prikl. Khim. #2, 146 (1970).
4. A. J. Leadbetter and J. E. Spice, Can. J. Chem. 37, 1923 (1959).
5. A. G. Gilbert and K. G. P. Sulzmann, J. Electrochem. Soc. 121, 832 (1974).
6. G. Eyber, Z. Phys. Chem. 144A, 1 (1929).
7. J. Dewar and H. O. Jones, Proc. Roy. Soc. (London), 76, 558 (1905).
8. L. H. Jones, R. S. McDowell, M. Goldblatt and B. I. Swanson, J. Chem. Phys. 57, 2050 (1972).
9. B. Beagley, D. W. J. Cruickshank, P. M. Pinder, A. G. Robiette and G. M. Sheldrick, Acta Cryst. B25, 737 (1969).
10. A. Almenningen, A. Haaland and K. Wahl, Acta Chem. Scand. 23, 2245 (1969).
11. R. E. McDowell, QCPE #269, Quantum Chemistry Program Exchange, Indiana University, 1974.

Iron Carbonyl ($Fe(CO)_5$) $C_5Fe_1O_5(g)$

Enthalpy Reference Temperature = T_r = 298.15 K Standard State Pressure = p° = 0.1 MPa

T/K	$C_p°$ (J K^{-1}mol^{-1})	S°	$-[G°-H°(T_r)]/T$	$H°-H°(T_r)$ (kJ mol^{-1})	$\Delta_f H°$	$\Delta_f G°$	Log K_f
0	0.	0.	INFINITE	-33.145	-729.524	-729.524	INFINITE
100	90.843	299.071	568.282	-26.921	-731.282	-716.779	374.407
200	139.336	377.190	454.024	-15.367	-730.407	-702.431	183.456
298.15	170.703	439.286	439.286	0.	-727.849	-689.207	120.746
300	171.134	440.344	439.290	0.316	-727.794	-688.968	119.960
400	189.027	492.243	446.244	18.399	-724.880	-676.471	88.338
500	200.830	535.762	459.914	37.924	-722.489	-664.660	69.437
600	209.820	573.204	475.747	58.474	-720.815	-653.263	56.872
700	217.105	606.113	492.067	79.832	-719.784	-642.094	47.914
800	223.115	635.508	508.193	101.852	-719.325	-631.034	41.202
900	228.088	662.084	523.839	124.420	-719.472	-619.996	35.984
1000	232.209	686.335	538.894	147.442	-720.469	-608.900	31.806
1100	235.633	708.633	553.325	170.839	-722.455	-597.633	28.379
1200	238.490	729.261	567.137	194.549	-723.964	-586.239	25.518
1300	240.887	748.448	580.355	218.521	-723.888	-574.766	23.094
1400	242.910	766.375	593.009	242.713	-723.955	-563.293	21.017
1500	244.628	783.194	605.133	267.092	-724.162	-551.811	19.216
1600	246.097	799.030	616.761	291.630	-724.511	-540.311	17.639
1700	247.362	813.989	627.927	316.305	-725.955	-528.765	16.247
1800	248.458	828.159	638.661	341.097	-726.933	-517.139	15.007
1900	249.413	841.619	648.992	365.991	-742.160	-504.751	13.877
2000	250.250	854.434	658.946	390.975	-743.682	-492.217	12.855
2100	250.989	866.662	668.548	416.038	-745.279	-479.605	11.930
2200	251.644	878.353	677.821	441.170	-746.953	-466.913	11.086
2300	252.227	889.552	686.785	466.364	-748.706	-454.146	10.314
2400	252.751	900.298	695.459	491.614	-750.540	-441.301	9.605
2500	253.222	910.626	703.861	516.913	-752.456	-428.377	8.950
2600	253.647	920.566	712.005	542.256	-754.454	-415.374	8.345
2700	254.032	930.146	719.908	567.640	-756.537	-402.293	7.783
2800	254.387	939.391	727.583	593.062	-758.702	-389.133	7.259
2900	254.708	948.323	735.041	618.517	-760.951	-375.896	6.771
3000	255.006	956.963	742.296	644.003	-763.284	-362.578	6.313
3100	255.279	965.329	749.356	669.517	-765.698	-349.180	5.884
3200	255.534	973.438	756.233	695.058	-1116.495	-328.282	5.359
3300	255.768	981.305	762.934	720.623	-1117.171	-303.641	4.806
3400	255.988	988.944	769.470	746.211	-1117.960	-278.976	4.286
3500	256.193	996.367	775.847	771.820	-1118.861	-254.288	3.795
3600	256.386	1003.587	782.073	797.449	-1119.874	-229.571	3.331
3700	256.566	1010.614	788.156	823.097	-1120.999	-204.827	2.892
3800	256.737	1017.459	794.100	848.762	-1122.235	-180.047	2.475
3900	256.898	1024.130	799.913	874.444	-1123.583	-155.234	2.079
4000	257.051	1030.636	805.600	900.141	-1125.043	-130.388	1.703
4100	257.196	1036.985	811.167	925.854	-1126.615	-105.502	1.344
4200	257.335	1043.184	816.617	951.580	-1128.299	-80.577	1.002
4300	257.466	1049.241	821.957	977.320	-1130.098	-55.609	0.676
4400	257.593	1055.161	827.190	1003.073	-1132.010	-30.601	0.363
4500	257.714	1060.952	832.321	1028.839	-1134.038	-5.547	0.064
4600	257.831	1066.617	837.353	1054.616	-1136.182	19.553	-0.222
4700	257.942	1072.163	842.290	1080.405	-1138.445	44.703	-0.497
4800	258.050	1077.595	847.136	1106.204	-1140.828	69.902	-0.761
4900	258.155	1082.917	851.894	1132.015	-1143.332	95.152	-1.014
5000	258.256	1088.133	856.566	1157.835	-1145.961	120.454	-1.258
5100	258.353	1093.249	861.157	1183.666	-1148.717	145.811	-1.493
5200	258.449	1098.266	865.669	1209.506	-1151.602	171.219	-1.720
5300	258.541	1103.190	870.104	1235.355	-1154.619	196.688	-1.938
5400	258.631	1108.024	874.465	1261.214	-1157.773	222.213	-2.149
5500	258.719	1112.770	878.755	1287.081	-1161.066	247.797	-2.353
5600	258.804	1117.433	882.976	1312.958	-1164.504	273.445	-2.551
5700	258.888	1122.014	887.129	1338.842	-1168.089	299.153	-2.741
5800	258.970	1126.517	891.218	1364.735	-1171.827	324.929	-2.926
5900	259.050	1130.945	895.244	1390.636	-1175.723	350.764	-3.105
6000	259.129	1135.299	899.209	1416.545	-1179.782	376.673	-3.279

PREVIOUS: March 1978 (1 atm) CURRENT: March 1978 (1 bar)

Iron Carbonyl ($Fe(CO)_5$) $C_5Fe_1O_5(g)$

CHROMIUM CARBIDE ($Cr_{23}C_6$) CRYSTAL M_r = 1267.974

$\Delta_fH°(0\ K) = -333.57 \pm 41.9\ kJ\ mol^{-1}$

$S°(298.15\ K) = 612.044 \pm 4.2\ J\ K^{-1}\ mol^{-1}$ $\Delta_fH°(298.15\ K) = -328.44 \pm 41.9\ kJ\ mol^{-1}$

Enthalpy of Formation

Mah (1) determined the enthalpy of combustion of $Cr_{23}C_6$(cr) using oxygen bomb calorimetry. Using auxiliary JANAF data (2) and the value $\Delta_cH°(303.15\ K) = -3604.2\pm1.5\ kcal\ mol^{-1}$ reported by Mah (1) for the combustion of $Cr_{23}C_6$, we calculate $\Delta_fH°(298.15\ K) = -78.5\ kcal\ mol^{-1}$ for $Cr_{23}C_6$(cr). The combustion equation is $Cr_{23}C_6(cr) + 93/4\ O_2(g) = 23/2\ Cr_2O_3(cr) + 6\ CO_2(g)$. The factor (23/2) makes the Δ_fH value for $Cr_{23}C_6$ extremely sensitive to the corresponding value for Cr_2O_3(cr). Of the nine combustion runs, Mah (1) reported complete combustion in three runs and 99.40 to 99.85% completion in the remaining six. Corrections for incomplete combustion, formation of small amounts of CO(g) and impurities were made by Mah (1). These corrections amount to no more than 0.7% of the $\Delta_cE°$ value.

The pressures of CO(g) in equilibrium with a $Cr_{23}C_6$-Cr_2O_3-Cr mixture have been measured by Kelley et al. (3) and Kulkarni and Worrell (4). Our analysis of their data is tabulated below. Kulkarni and Worrell (4) measured the CO pressure using a torsion effusion technique while Kelley et al. (3) measured the pressure manometrically. In all cases unit activity is assumed for the condensed species.

Source	T/K	Comments[a]	Reaction[c]	$\Delta_fH°(298.15\ K)$, kcal mol^{-1} 2nd Law	3rd Law	Drift cal K^{-1} mol^{-1}	$\Delta_fH°(298.15\ K)$ kcal mol^{-1}
(4)	1170-1307	1/4 mm	A(15 pts.)	466.37	462.43	-3.2±2.5	-78.53
(4)	1158-1286	1/3 mm	A(8 pts.)	450.14	462.31	9.9±7.2	-78.41
(4)	1156-1283	1/2 mm	A(9 pts.)	451.46	467.72	13.3±3.3	-83.82
(3)	1571-1772		A(18 pts.)[b]	459.47	467.19	4.6±4.0	-83.29

[a]effusion orifice diameter
[b]1725, 1775 K points rejected due to statistical test.
[c]Reaction: (A)$Cr_{23}C_6$ (cr) + 2 Cr_2O_3(cr) = 27Cr(cr)

We adopt $\Delta_fH°(298.15\ K) = -78.5\pm10.0\ kcal\ mol^{-1}$ for $Cr_{23}C_6$(cr) based on the combustion data of Mah (1), with excellent support in the equilibrium studies by Kulkarni and Worrell (4) for orifice diameters of 1/4 and 1/3 mm. Kulkarni and Worrell (4) concluded that equilibrium was not obtained in the cell with the 1/2 mm orifice in order to provide an explanation for the low pressures observed. Equilibrium studies involving $Cr_{23}C_6$(cr) by Hancock and Pidgeon (5), Alekseev and Shvartsman (6), and Bolgar et al. (7) are not considered (2).

Heat Capacity and Entropy

The heat capacity (55-295 K, 29 pts.) and enthalpy (467-1697 K, 26 pts.) have been measured by Kelley et al. (3). When the research was performed, Kelley et al. (3) considered the sample to be Cr_4C. All the data are corrected to the composition $Cr_{23}C_6$, assuming that the sample actually contained excess Cr. The contribution of the assumed free Cr was subtracted from the reported heat capacity and enthalpy values using JANAF data for Cr (2). These corrections amount to 0.6 cal K^{-1} mol^{-1} at 55 K and 5.0 cal K^{-1} mol^{-1} at 295 K for the heat capacity data and are in a range of 1.00-1.85 kcal mol^{-1} for the enthalpy data. We adopt $S°(50\ K) = 4.861\ cal\ K^{-1}\ mol^{-1}$ and $H°(50;0) = 181.6\ cal\ mol^{-1}$. The enthalpy data are fit to a six term polynomial with a constraint to join smoothly in the 298.15 K region with the enthalpy derived from the heat capacity data. There is considerable scatter in the enthalpy data. Deviations of the observed data from the tabulated data are within ± 0.6% except for the 486.5 K data point which is approximately 1% (500 cal mol^{-1}) too high. The enthalpy polynomial is used to extrapolate the results to 2000 K.

Phase Data

Storms (8) has summarized the phase information in the Cr-C system and reported that $Cr_{23}C_6$ has a complex face-centered-cubic structure. The range of homogeneity for $Cr_{23}C_6$ is unknown (8).

Decomposition Data

The tentative phase diagram of the Cr-Cr_3C_2 system proposed by Storms (8) indicated that $Cr_{23}C_6$ has an incongruent melting point at 1793 K.

References

1. A. D. Mah, U. S. Bur. Mines RI 7217, 1969.
2. JANAF Thermochemical Tables: Cr_2O_3(cr) and Cr_3C_2(cr), 12-31-73; CO_2(g), 9-30-65; Cr(cr), 6-30-73.
3. K. K. Kelley, F. S. Boericke, G. E. Moore, E. H. Huffman, and W. M. Bangert, U. S. Bur. Mines Tech. Paper 662, 1949.
4. A. D. Kulkarni and W. L. Worrell, Met. Trans. 3, 2363 (1972).
5. H. A. Hancock and L. M. Pidgeon, Trans. Met. Soc. AIME 227, 608 (1963).
6. V. I. Alekseev and L. A. Shvartsman, Fiz. Metallov i Metalloved 2, 545 (1961); Fiz. Khim. Osnovy Met. Protsessov, Komis, po Fiz. Khim. Osmovam Proiz. Stali. sb statei, 414-21, 1964.
7. A. S. Bolgar, V. V. Fesenko, and G. P. Gordinko, Poroshk. Met. 38, 100 (1966).
8. E. K. Storms, "The Refractory Carbides," Academic Press, New York, 1967.

Chromium Carbide ($Cr_{23}C_6$) C_6Cr_{23}(cr)

Enthalpy Reference Temperature = T_r = 298.15 K Standard State Pressure = p° = 0.1 MPa

T/K	$C_p°$ (J $K^{-1}mol^{-1}$)	S°	$-[G°-H°(T_r)]/T$	$H°-H°(T_r)$ (kJ mol^{-1})	$\Delta_fH°$	$\Delta_fG°$	Log K_f
0	0.	0.	INFINITE	-104.755	-333.575	-333.575	INFINITE
100	250.316	117.835	1081.808	-96.397	-332.734	-334.089	174.510
200	509.447	383.581	666.329	-56.550	-331.274	-335.894	87.726
298.15	628.152	612.044	612.044	0.	-328.444	-338.693	59.338
300	629.692	615.935	612.056	1.164	-328.373	-338.757	58.983
400	704.376	808.005	637.736	68.108	-326.027	-342.211	44.688
500	753.538	970.840	688.496	141.172	-320.639	-346.880	36.238
600	785.751	1111.290	747.532	218.255	-315.571	-352.615	30.698
700	808.114	1234.112	808.454	297.960	-311.244	-359.142	26.799
800	829.072	1343.386	868.614	379.818	-307.585	-366.242	23.913
900	850.247	1442.252	926.944	463.778	-304.702	-373.756	21.692
1000	872.142	1532.961	983.070	549.890	-302.997	-381.534	19.929
1100	894.857	1617.144	1036.931	638.234	-302.930	-389.406	18.491
1200	918.329	1696.008	1088.602	728.888	-304.701	-397.203	17.290
1300	942.471	1770.464	1138.215	821.923	-308.594	-404.770	16.264
1400	967.178	1841.209	1185.923	917.402	-314.817	-411.950	15.370
1500	992.361	1908.794	1231.877	1015.375	-323.452	-418.601	14.577
1600	1017.946	1973.655	1276.225	1115.888	-334.571	-424.596	13.862
1700	1043.862	2036.143	1319.099	1218.976	-348.224	-429.817	13.207
1800	1070.058	2096.549	1360.622	1324.670	-364.452	-434.162	12.599
1900	1096.488	2155.112	1400.903	1432.996	-383.294	-437.533	12.029
2000	1123.115	2212.030	1440.043	1543.974	-404.778	-439.843	11.488

PREVIOUS: CURRENT: December 1973

Chromium Carbide ($Cr_{23}C_6$) C_6Cr_{23}(cr)

CALCIUM (Ca)

REFERENCE STATE

A_r = 40.08

0 to 716	K	crystal, alpha
716 to 1115	K	crystal, beta
1115 to 1773.658	K	liquid
above 1773.658	K	ideal monatomic gas

Refer to the individual tables for details.

Calcium (Ca) Ca_1(ref)

Enthalpy Reference Temperature = T_r = 298.15 K Standard State Pressure = p° = 0.1 MPa

T/K	C_p° (J K⁻¹mol⁻¹)	S° (J K⁻¹mol⁻¹)	−[G°−H°(T_r)]/T (J K⁻¹mol⁻¹)	H°−H°(T_r) (kJ mol⁻¹)	$\Delta_f H^\circ$ (kJ mol⁻¹)	$\Delta_f G^\circ$ (kJ mol⁻¹)	Log K_f
0	0.	0.	INFINITE	-5.736	0.	0.	0.
100	19.504	15.967	63.525	-4.756	0.	0.	0.
200	24.542	31.489	43.924	-2.487	0.	0.	0.
298.15	25.929	41.588	41.588	0.	0.	0.	0.
300	25.946	41.749	41.589	0.048	0.	0.	0.
400	26.868	49.322	42.615	2.683	0.	0.	0.
500	28.487	55.487	44.590	5.448	0.	0.	0.
600	30.382	60.846	46.861	8.391	0.	0.	0.
700	32.406	65.680	49.210	11.529	0.	0.	0.
716.000	32.737	66.416	49.586	12.051		ALPHA <--> BETA	
716.000	31.503	67.715	49.586	12.981		TRANSITION	
800	33.818	71.335	51.681	15.723	0.	0.	0.
900	36.713	75.484	54.097	19.249	0.	0.	0.
1000	39.706	79.507	56.437	23.069	0.	0.	0.
1100	42.763	83.434	58.714	27.192	0.	0.	0.
1115.000	43.226	84.017	59.050	27.837		BETA <--> LIQUID	
1115.000	35.000	91.676	59.050	36.377		TRANSITION	
1200	35.000	94.247	61.453	39.352	0.	0.	0.
1300	35.000	97.049	64.085	42.852	0.	0.	0.
1400	35.000	99.642	66.533	46.352	0.	0.	0.
1500	35.000	102.057	68.822	49.852	0.	0.	0.
1600	35.000	104.316	70.971	53.352	0.	0.	0.
1700	35.000	106.438	72.995	56.852	0.	0.	0.
1773.658	35.000	107.922	74.415	59.430		LIQUID <--> IDEAL GAS	
1773.658	20.837	191.956	74.415	208.477		FUGACITY = 1 bar	
1800	20.845	192.263	76.138	209.026	0.	0.	0.
1900	20.888	193.391	82.279	211.112	0.	0.	0.
2000	20.953	194.464	87.862	213.204	0.	0.	0.
2100	21.046	195.489	92.963	215.304	0.	0.	0.
2200	21.173	196.470	97.646	217.415	0.	0.	0.
2300	21.343	197.415	101.963	219.540	0.	0.	0.
2400	21.561	198.328	105.959	221.685	0.	0.	0.
2500	21.836	199.214	109.672	223.854	0.	0.	0.
2600	22.173	200.076	113.132	226.054	0.	0.	0.
2700	22.578	200.921	116.368	228.291	0.	0.	0.
2800	23.057	201.750	119.403	230.572	0.	0.	0.
2900	23.614	202.569	122.257	232.905	0.	0.	0.
3000	24.251	203.380	124.947	235.298	0.	0.	0.
3100	24.970	204.186	127.490	237.758	0.	0.	0.
3200	25.772	204.992	129.900	240.295	0.	0.	0.
3300	26.655	205.798	132.187	242.915	0.	0.	0.
3400	27.618	206.608	134.364	245.628	0.	0.	0.
3500	28.657	207.423	136.440	248.441	0.	0.	0.
3600	29.767	208.246	138.423	251.362	0.	0.	0.
3700	30.944	209.077	140.321	254.397	0.	0.	0.
3800	32.180	209.919	142.142	257.553	0.	0.	0.
3900	33.469	210.771	143.891	260.835	0.	0.	0.
4000	34.801	211.635	145.573	264.248	0.	0.	0.
4100	36.169	212.511	147.195	267.796	0.	0.	0.
4200	37.564	213.400	148.761	271.482	0.	0.	0.
4300	38.977	214.300	150.275	275.309	0.	0.	0.
4400	40.389	215.212	151.740	279.275	0.	0.	0.
4500	41.755	216.131	153.161	283.367	0.	0.	0.
4600	43.135	217.063	154.540	287.607	0.	0.	0.
4700	44.507	218.005	155.880	291.990	0.	0.	0.
4800	45.851	218.956	157.184	296.508	0.	0.	0.
4900	47.160	219.915	158.454	301.159	0.	0.	0.
5000	48.426	220.881	159.693	305.938	0.	0.	0.
5200	50.805	222.827	162.084	315.861	0.	0.	0.
5400	52.964	224.785	164.370	326.242	0.	0.	0.
5600	54.857	226.745	166.562	337.021	0.	0.	0.
5800	56.500	228.699	168.671	348.161	0.	0.	0.
6000	57.878	230.639	170.705	359.603	0.	0.	0.

PREVIOUS: December 1968 (1 atm) CURRENT: September 1983 (1 bar)

Calcium (Ca) Ca_1(ref)

CALCIUM, ALPHA (Ca) CRYSTAL A_r = 40.08

Pearson Notation: cF4

$S°(298.15\ K) = 41.588\ J\ K^{-1}\ mol^{-1}$

$T_{trs}\ (\alpha \rightarrow \beta) = 716 \pm 3\ K$

$\Delta_f H°(0\ K) = 0\ kJ\ mol^{-1}$

$\Delta_f H°(298.15\ K) = 0\ kJ\ mol^{-1}$

$\Delta_{trs} H° = 0.930 \pm 0.08\ kJ\ mol^{-1}$

Enthalpy of Formation

Zero by definition.

Heat Capacity and Entropy

The thermal functions given here are those recommended by CODATA (1). The low temperature heat capacity of α-CA(cr) has been measured many times: Griffil et al. (1.8 - 4.2 K, 2), Agarwal and Betterton (1.1 - 4.2 K, 3), Roberts (1.5 - 20 K, 4), Clusius and Vaughen (10-200 K, 5), Gunther (22-62 K, 6), and Eastman and Rodebush (67 - 293 K, 7). The adopted heat capacity values are derived assuming that the heat capacity curve should lie below and be similar to the shape of the α-Sr(cr) curve, but remain higher than the Mg curve. The experimental data would imply a crossing of the curves below 300 K which does not appear reasonable. The adopted curve is in reasonable agreement with Griffil et al. (2), Roberts (4), and Clusius and Vaughen (5).

The high temperature heat capacities for the α and β phasis are estimates based on the experimental enthalpy studies in Zalensinskii and Zulinski (8), Jauch (9), and Eastman et al. (10) with the two constraints that the values mesh smoothly with the low temperature results and show a similarity with the Mg and Sr results.

Transition Data

See Ca(β, cr) tables for details.

Sublimation Data

See the ideal gas table for details.

References

1. H. J. White, Jr., Chairman, CODATA Task Group on Chemical Thermodynamic Tables, report to be published, 1985.
2. M. Griffil, R. W. Vest, and J. F. Smith, J. Chem. Phys. 27, 1267 (1957).
3. K. L. Agarwal and J. O. Betterton, J. Low Temp. Phys. 17, 509 (1974).
4. L. M. Roberts, Proc. Phys. soc. (London) 70B, 738 (1957).
5. K. Clusius and J. V. Vaughen, J. Am. Chem. Soc. 52, 4686 (1930).
6. P. Gunther, Ann. Physik. 51, 828 (1916).
7. E. D. Eastman and W. H. Rodebush, J. Am. Chem. Soc. 40, 489 (1918).
8. E. Zalensinskii and R. Zulinskii, Bull. Intern. Acad. Polon. 1928A, 479 (1928).
9. R. Jauch, Diplomarbeit, Techn. Hochschule, Stuttgart, 1946.
10. E. D. Eastman, A, M. Williams and T. F. Young, J. Am. Chem. Soc. 46, 1178 (1924).

Calcium, Alpha (Ca) $Ca_1(cr)$

Enthalpy Reference Temperature = T_r = 298.15 K; Standard State Pressure = p° = 0.1 MPa

T/K	$C_p°$ (J K⁻¹mol⁻¹)	S° (J K⁻¹mol⁻¹)	$-[G°-H°(T_r)]/T$ (J K⁻¹mol⁻¹)	$H°-H°(T_r)$ (kJ mol⁻¹)	$\Delta_f H°$ (kJ mol⁻¹)	$\Delta_f G°$ (kJ mol⁻¹)	Log K_f
0	0.	0.	INFINITE	-5.736	0.	0.	0.
100	19.504	15.967	63.525	-4.756	0.	0.	0.
200	24.542	31.489	43.924	-2.487	0.	0.	0.
250	25.411	37.065	42.012	-1.237	0.	0.	0.
298.15	25.929	41.588	41.588	0.	0.	0.	0.
300	25.946	41.749	41.589	0.048	0.	0.	0.
350	26.320	45.774	41.906	1.354	0.	0.	0.
400	26.868	49.322	42.615	2.683	0.	0.	0.
450	27.643	52.531	43.541	4.045	0.	0.	0.
500	28.487	55.487	44.590	5.448	0.	0.	0.
600	30.382	60.846	46.861	8.391	0.	0.	0.
700	32.406	65.680	49.210	11.529	0.	0.	0.
716.000	32.737	66.416	49.586	12.051	-------- ALPHA <--> BETA -------		
800	34.613	70.150	51.551	14.879	-0.844	0.104	-0.007
900	36.819	74.354	53.853	18.451	-0.798	0.219	-0.013
1000	39.026	78.348	56.105	22.243	-0.826	0.333	-0.017
1100	41.232	82.171	58.302	26.256	-0.936	0.453	-0.022

PREVIOUS: December 1968 CURRENT: September 1983

Calcium, Alpha (Ca) $Ca_1(cr)$

CALCIUM, BETA (β-Ca) CRYSTAL A_r = 40.08

Pearson Notation: cI4

$S°(298.15 K) = [43.070] J K^{-1} mol^{-1}$

$T_{trs} (\alpha \to \beta) = 716 \pm 3 K$

$T_{fus(\beta \to \ell)} = 1115 \pm 2 K$

$\Delta_f H°(298.15 K) = [1.056] kJ mol^{-1}$

$\Delta_{trs}H° = 0.930 \pm 0.08 kJ mol^{-1}$

$\Delta_{fus}H° = 8.540 \pm 0.10 kJ mol^{-1}$

Enthalpy of Formation

The enthalpy of formation of Ca(β, cr) is calculated from that of the Ca(α, cr) by adding the enthalpy of transition and the difference in enthalpy, H°(716 K) - H°(298.15 K), between Ca(α, cr) and Ca(β, cr).

Heat Capacity and Entropy

The high temperatures heat capacities for the α and β phases are estimates based on the experimental enthalpy studies in Zalinsinski and Aulinski(1), Jauch (2), and Eastman et al. (3) with the two constraints that the values mesh smoothly with the low temperature results and show a similarity with the Mg and Sr results.

Transition and Fusion Data

According to Pearson (4), calcium exists in two crystalline modifications. The low temperature phase (α) has a face - centered - cubic (cF4) structure, as does α-Sr(cr). The high temperature phase (β) has a body - centered - cubic (cI2) structure, as does β - Sr(cr). In comparison, magnesium has a hexagonal - close - packed structure. The β-Ca(cr) region of stability is 716 ± 3 K to 1115 ± 2 K (5).

The enthalpies of transition and fusion are the values determined by Chiotti et al. (6) using an adiabatic calorimeter.

Sublimation Data

See the ideal gas table for details.

References

1. E. Zalensinskii and R. Zulinskii, Bull. Intern. Acad. Polon. 1928 A 479 (1928).
2. R. Jauch, Diplomarbeit, Techn. Hochochule, Stuttgart, 1946.
3. E. D. Eastman, A. M. Williams and T. F. Young, J. Am. Chem. Soc. 46, 1178 (1924).
4. W. B. Pearson, "A Handbook of Lattice Spacings and Structures of Metals and Alloys, " Pergamon Press, London, 1967.
5. F. X. Kayser and S. D. Soderquest, J. Phys. Chem. Solids 28, 2343 (1967).
6. P. Chiotti, G. J. Gartner, et al., J. Chem. Eng. Data 11, 571 (1966).

Calcium, Beta (Ca) $Ca_1(cr)$

Enthalpy Reference Temperature = T_r = 298.15 K Standard State Pressure = p° = 0.1 MPa

T/K	$C_p°$ (J K⁻¹mol⁻¹)	S° (J K⁻¹mol⁻¹)	$-[G°-H°(T_r)]/T$ (J K⁻¹mol⁻¹)	$H°-H°(T_r)$ (kJ mol⁻¹)	$\Delta_f H°$ (kJ mol⁻¹)	$\Delta_f G°$ (kJ mol⁻¹)	Log K_f
0							
100							
200							
250							
298.15	25.951	43.070	43.070	0.	1.056	0.614	-0.108
300	25.971	43.231	43.071	0.048	1.056	0.611	-0.106
350	26.345	47.260	43.388	1.355	1.057	0.537	-0.080
400	26.894	50.812	44.098	2.685	1.058	0.463	-0.060
450	27.708	54.026	45.025	4.051	1.061	0.388	-0.045
500	28.514	56.988	46.075	5.456	1.064	0.313	-0.033
600	30.099	62.336	48.349	8.392	1.057	0.163	-0.014
700	31.062	67.008	50.688	11.424	0.951	0.021	-0.002
716.000	31.503	67.715	51.061	11.925	-------- ALPHA <--> BETA -------		
800	33.818	71.335	53.001	14.668	0.	0.	0.
900	36.713	75.484	55.270	18.193	0.	0.	0.
1000	39.706	79.507	57.493	22.014	0.	0.	0.
1100	42.763	83.434	59.674	26.137	0.	0.	0.
1115.000	43.226	84.017	59.997	26.782	------- BETA <--> LIQUID -------		
1200	45.850	87.287	61.815	30.567	-7.729	0.622	-0.027
1300	48.949	91.080	63.920	35.307	-6.489	1.270	-0.051
1400	52.048	94.818	65.994	40.354	-4.943	1.811	-0.068
1500	55.566	98.527	68.039	45.731	-3.066	2.230	-0.078

PREVIOUS: December 1968 CURRENT: September 1983

Calcium, Beta (Ca) $Ca_1(cr)$

CALCIUM (Ca) LIQUID A_r = 40.08

$S°(298.15\ K) = [45.510]\ J\ K^{-1}\ mol^{-1}$ $\Delta_f H°(298.15\ K) = [7.788]\ kJ\ mol^{-1}$

$T_{fus}\ (\beta \rightarrow \ell) = 1115 \pm 2\ K$ $\Delta_{fus}H° = 8.540 \pm 0.10\ kJ\ mol^{-1}$

Enthalpy of Formation

The enthalpy of formation of the liquid is calculated from that of Ca(β, cr) by adding the enthalpy of fusion and the difference in enthalpy, H°(1115 K) - H°(298.15 K), between Ca(β, cr) and the liquid.

Heat Capacity and Entropy

The heat capacity of the liquid is chosen to be a value intermediate between the adopted values for Mg(ℓ) and Sr(ℓ).

Fusion Data

See the β-Ca(cr) table for details.

Vaporization Data

T_{vap} (1 bar) is calculated as the temperature at which $\Delta_r G° = 0$ for the process Ca(ℓ) = Ca(g). $\Delta_{vap}H°$ is calculated as the difference in the enthalpies of formation between the ideal gas and liquid at T_{vap}.

See the ideal gas table for details.

Calcium (Ca) $Ca_1(l)$

Enthalpy Reference Temperature = T_r = 298.15 K Standard State Pressure = p° = 0.1 MPa

T/K	$C_p°$ (J K⁻¹mol⁻¹)	S° (J K⁻¹mol⁻¹)	$-[G°-H°(T_r)]/T$ (J K⁻¹mol⁻¹)	$H°-H°(T_r)$ (kJ mol⁻¹)	$\Delta_f H°$ (kJ mol⁻¹)	$\Delta_f G°$ (kJ mol⁻¹)	Log K_f
0							
100							
200							
250							
298.15	35.000	45.510	45.510	0.	7.788	6.618	-1.159
300	35.000	45.727	45.511	0.065	7.804	6.611	-1.151
350	35.000	51.122	45.937	1.815	8.248	6.377	-0.952
400	35.000	55.796	46.884	3.565	8.670	6.080	-0.794
450	35.000	59.918	48.107	5.315	9.057	5.733	-0.665
500	35.000	63.606	49.476	7.065	9.404	5.345	-0.558
600	35.000	69.987	52.379	10.565	9.962	4.477	-0.390
700	35.000	75.382	55.290	14.065	10.323	3.532	-0.264
800	35.000	80.056	58.100	17.565	9.629	2.652	-0.173
900	35.000	84.178	60.773	21.065	9.603	1.779	-0.103
1000	35.000	87.866	63.301	24.565	9.283	0.924	-0.048
1100	35.000	91.202	65.688	28.065	8.660	0.116	-0.005
1115.000	35.000	91.676	66.035	28.590	------- BETA <--> LIQUID -------		
1200	35.000	94.247	67.943	31.565	0.	0.	0.
1300	35.000	97.049	70.076	35.065	0.	0.	0.
1400	35.000	99.642	72.096	38.565	0.	0.	0.
1500	35.000	102.057	74.014	42.065	0.	0.	0.
1600	35.000	104.316	75.838	45.565	0.	0.	0.
1700	35.000	106.438	77.576	49.065	0.	0.	0.
1773.658	35.000	107.922	78.806	51.643	------- FUGACITY = 1 bar -------		
1800	35.000	108.438	79.236	52.565	-148.674	2.211	-0.064
1900	35.000	110.331	80.823	56.065	-147.260	10.555	-0.290
2000	35.000	112.126	82.344	59.565	-145.852	18.824	-0.492
2100	35.000	113.834	83.803	63.065	-144.452	27.024	-0.672
2200	35.000	115.462	85.205	66.565	-143.062	35.157	-0.835
2300	35.000	117.018	86.555	70.065	-141.688	43.227	-0.982
2400	35.000	118.507	87.855	73.565	-140.333	51.237	-1.115
2500	35.000	119.936	89.110	77.065	-139.002	59.192	-1.237

PREVIOUS: December 1968 CURRENT: September 1983

Calcium (Ca) $Ca_1(l)$

CALCIUM (Ca) ALPHA-BETA-LIQUID A_r = 40.08

0 to 716	K	crystal, alpha
716 to 1115	K	crystal, beta
above 1115	K	liquid

Refer to the individual tables for details.

Calcium (Ca) $Ca_1(cr,l)$

Enthalpy Reference Temperature = T_r = 298.15 K Standard State Pressure = $p°$ = 0.1 MPa

	J K⁻¹mol⁻¹			kJ mol⁻¹			
T/K	$C_p°$	S°	$-[G°-H°(T_r)]/T$	$H°-H°(T_r)$	$\Delta_f H°$	$\Delta_f G°$	Log K_f
0	0.	0.	INFINITE	-5.736	0.	0.	0.
100	19.504	15.967	63.525	-4.756	0.	0.	0.
200	24.542	31.489	43.924	-2.487	0.	0.	0.
250	25.411	37.065	42.012	-1.237	0.	0.	0.
298.15	25.929	41.588	41.588	0.	0.	0.	0.
300	25.946	41.749	41.589	0.048	0.	0.	0.
350	26.320	45.774	41.906	1.354	0.	0.	0.
400	26.868	49.322	42.615	2.683	0.	0.	0.
450	27.643	52.531	43.541	4.045	0.	0.	0.
500	28.487	55.487	44.590	5.448	0.	0.	0.
600	30.382	60.846	46.861	8.391	0.	0.	0.
700	32.406	65.680	49.210	11.529	0.	0.	0.
716.000	32.737	66.416	49.586	12.051	ALPHA <--> BETA		
716.000	31.503	67.715	49.586	12.981	TRANSITION		
800	33.818	71.335	51.681	15.723	0.	0.	0.
900	36.713	75.484	54.097	19.249	0.	0.	0.
1000	39.706	79.507	56.437	23.069	0.	0.	0.
1100	42.763	83.434	58.714	27.192	0.	0.	0.
1115.000	43.226	84.017	59.050	27.837	BETA <--> LIQUID		
1115.000	35.000	91.676	59.050	36.377	TRANSITION		
1200	35.000	94.247	61.453	39.352	0.	0.	0.
1300	35.000	97.049	64.085	42.852	0.	0.	0.
1400	35.000	99.642	66.533	46.352	0.	0.	0.
1500	35.000	102.057	68.822	49.852	0.	0.	0.
1600	35.000	104.316	70.971	53.352	0.	0.	0.
1700	35.000	106.438	72.995	56.852	0.	0.	0.
1773.658	35.000	107.922	74.415	59.430	------- FUGACITY = 1 bar -------		
1800	35.000	108.438	74.909	60.352	-148.674	2.211	-0.064
1900	35.000	110.331	76.724	63.852	-147.260	10.555	-0.290
2000	35.000	112.126	78.450	67.352	-145.852	18.824	-0.492
2100	35.000	113.834	80.094	70.852	-144.452	27.024	-0.672
2200	35.000	115.462	81.665	74.352	-143.062	35.157	-0.835
2300	35.000	117.018	83.169	77.852	-141.688	43.227	-0.982
2400	35.000	118.507	84.610	81.352	-140.332	51.237	-1.115
2500	35.000	119.936	85.995	84.852	-139.002	59.192	-1.237

PREVIOUS: CURRENT: September 1983

Calcium (Ca) $Ca_1(cr,l)$

CALCIUM (Ca) IDEAL GAS A_r = 40.08

IP (Ca, g) = 49305.96 ± 0.08

S°(298.15 K) = 154.886 ± 0.020 J K^{-1} mol^{-1}

$\Delta_f H°$(0 K) = 177.3 ± 0.8 kJ mol^{-1}

$\Delta_f H°$(298.15 K) = 177.8 ± 0.8 kJ mol^{-1}

Electronic Level and Quantum Weight

State	ε_i, cm^{-1}	g_i
$^1S_0$	0.00	1
$^3P_0$	15157.901	1
$^3P_1$	15210.063	3
$^3P_2$	15315.943	5

Enthalpy of Formation

The adopted enthalpy of formation value for Ca(g), $\Delta_f H°$(298.15 K) = 177.8±0.8 kJ mol^{-1}, is that recommended by CODATA (1). This value was calculated, according to CODATA, from the vapor pressure measurements by Bogoslovskii et al. (773-1023 K, 2), Smith (844-965 K, 3), Bohdansky and Schins (1313-2060 K, 4), Schins et al. (1558-2115 K, 5). The study by DeMaria and Piacente (1126-1300 K, 6), and others cited in Hultgren et al. (7) were also considered.

Heat Capacity and Entropy

The information on electronic energy levels and quantum weights, given by Sugar and Corliss (8), is incomplete because many theoretically predicted levels have not been observed. Our calculations indicate that any reasonable method of filling in these missing levels (for n < 80) and cutting off the summation in the partition function (9) has no effect on the thermodynamic functions to 4000 K. This is a result of the high energy of all levels other than the ground state and the levels listed above; the next excited state is approximately 20335 cm^{-1} above the ground state. Although we list only a few levels, all levels listed by Sugar and Corliss (8) and the estimated levels (for n < 80) are considered in the calculation. The reported uncertainty in S°(298.15 K) is due to uncertainties in the relative atomic mass and the fundamental constants. Extension of these calculations above 6000 K may require consideration of additional excited states and use of different fill and cutoff procedures (9).

The thermal functions at 298.15 K differ from the CODATA recommendations (1) for two reasons. First, the entropy differs by 0.1094 J K^{-1} mol^{-1} because this table uses a reference pressure of 1 bar, whereas CODATA recommendations are based on 1 atm. Second, small changes, ~0.002 J K^{-1} mol^{-1}, arise due to the use of slightly different values for the fundamental constants.

References

1. J. D. Cox, Chairman, CODATA Task Force on Key Values for Thermodynamics, J. Chem. Thermodynamics 10, 903 (1978); CODATA Bulletin No. 22, 1977.
2. S. S. Bogoslovskii, V. K. Kulifiev, et al., Izvest. Vyss. Ucheb. Zaveden., Tsvet. Met. 12, 52 (1969).
3. J. F. Smith, Symposium on Thermodynamics of Nuclear Materials, IAEA, Vienna, 1962.
4. J. Bohdansky and H. E. J. Schins, J. Phys. Chem. 71, 215 (1967).
5. H. E. Schins, R. W. M. van Wijk, and B. Dorpema, Z. Metallk. 62, 330 (1971).
6. G. A. Maria and V. Piacente, J. Chem. Thermodynamics 6, 1 (1974).
7. R. Hultgren, P. D. Desai, et al., "Selected Values of the Thermodynamic Properties of the Elements", American Society for Metals, Metals Park, Ohio, 1973.
8. J. Sugar and C. Corliss, J. Phys. Chem. Ref. Data 8, 865 (1979).
9. J. R. Downey, Jr., The Dow Chemical Company, Report AFOSR-TR-78-0960, Contract No. F44620-75-1-0048, 1978.

Calcium (Ca) $Ca_1(g)$

Enthalpy Reference Temperature = T_r = 298.15 K Standard State Pressure = p° = 0.1 MPa

T/K	$C_p°$ (J $K^{-1}mol^{-1}$)	S° (J $K^{-1}mol^{-1}$)	$-[G°-H°(T_r)]/T$ (J $K^{-1}mol^{-1}$)	$H°-H°(T_r)$ (kJ mol^{-1})	$\Delta_f H°$ (kJ mol^{-1})	$\Delta_f G°$ (kJ mol^{-1})	Log K_f
0	0.	0.	INFINITE	-6.197	177.339	177.339	INFINITE
100	20.786	132.179	173.366	-4.119	178.437	166.816	-87.136
200	20.786	146.586	156.787	-2.040	178.247	155.227	-40.541
250	20.786	151.225	155.228	-1.001	178.036	149.496	-31.236
298.15	20.786	154.886	154.886	0.	177.800	144.020	-25.232
300	20.786	155.014	154.886	0.038	177.790	143.811	-25.040
350	20.786	158.219	155.139	1.078	177.524	138.168	-20.620
400	20.786	160.994	155.701	2.117	177.234	132.565	-17.311
450	20.786	163.442	156.428	3.156	176.911	127.001	-14.742
500	20.786	165.632	157.241	4.196	176.547	121.474	-12.690
600	20.786	169.422	158.965	6.274	175.684	110.538	-9.623
700	20.786	172.626	160.694	8.353	174.623	99.761	-7.444
800	20.786	175.402	162.363	10.431	172.508	89.255	-5.828
900	20.786	177.850	163.950	12.510	171.061	78.932	-4.581
1000	20.786	180.040	165.452	14.589	169.319	68.786	-3.593
1100	20.786	182.021	166.869	16.667	167.275	58.829	-2.794
1200	20.786	183.830	168.208	18.746	157.193	49.694	-2.163
1300	20.787	185.494	169.475	20.825	155.772	40.793	-1.639
1400	20.789	187.034	170.675	22.903	154.351	32.002	-1.194
1500	20.793	188.469	171.814	24.982	152.930	23.313	-0.812
1600	20.802	189.811	172.897	27.062	151.510	14.718	-0.480
1700	20.818	191.072	173.929	29.143	150.091	6.212	-0.191
1773.658	20.837	191.956	174.660	30.677	------- FUGACITY = 1 bar -------		
1800	20.845	192.263	174.915	31.226	0.	0.	0.
1900	20.888	193.391	175.858	33.313	0.	0.	0.
2000	20.953	194.464	176.762	35.405	0.	0.	0.
2100	21.046	195.489	177.629	37.504	0.	0.	0.
2200	21.173	196.470	178.464	39.615	0.	0.	0.
2300	21.343	197.415	179.267	41.740	0.	0.	0.
2400	21.561	198.328	180.043	43.885	0.	0.	0.
2500	21.836	199.214	180.792	46.054	0.	0.	0.
2600	22.173	200.076	181.517	48.254	0.	0.	0.
2700	22.578	200.921	182.220	50.491	0.	0.	0.
2800	23.057	201.750	182.903	52.773	0.	0.	0.
2900	23.614	202.569	183.567	55.105	0.	0.	0.
3000	24.251	203.380	184.214	57.498	0.	0.	0.
3100	24.970	204.186	184.845	59.958	0.	0.	0.
3200	25.772	204.992	185.462	62.495	0.	0.	0.
3300	26.655	205.798	186.066	65.115	0.	0.	0.
3400	27.618	206.608	186.658	67.828	0.	0.	0.
3500	28.657	207.423	187.240	70.641	0.	0.	0.
3600	29.767	208.246	187.812	73.562	0.	0.	0.
3700	30.944	209.077	188.375	76.597	0.	0.	0.
3800	32.180	209.919	188.931	79.753	0.	0.	0.
3900	33.469	210.771	189.480	83.035	0.	0.	0.
4000	34.801	211.635	190.023	86.448	0.	0.	0.
4100	36.169	212.511	190.561	89.996	0.	0.	0.
4200	37.564	213.400	191.094	93.682	0.	0.	0.
4300	38.977	214.300	191.623	97.509	0.	0.	0.
4400	40.389	215.212	192.149	101.475	0.	0.	0.
4500	41.755	216.131	192.672	105.568	0.	0.	0.
4600	43.135	217.063	193.192	109.808	0.	0.	0.
4700	44.507	218.005	193.710	114.190	0.	0.	0.
4800	45.851	218.956	194.226	118.708	0.	0.	0.
4900	47.160	219.915	194.740	123.359	0.	0.	0.
5000	48.426	220.881	195.253	128.138	0.	0.	0.
5100	49.646	221.852	195.765	133.042	0.	0.	0.
5200	50.805	222.827	196.276	138.061	0.	0.	0.
5300	51.914	223.805	196.786	143.198	0.	0.	0.
5400	52.964	224.785	197.296	148.442	0.	0.	0.
5500	53.952	225.766	197.805	153.789	0.	0.	0.
5600	54.857	226.745	198.312	159.221	0.	0.	0.
5700	55.712	227.723	198.820	164.750	0.	0.	0.
5800	56.500	228.699	199.327	170.361	0.	0.	0.
5900	57.222	229.671	199.833	176.047	0.	0.	0.
6000	57.878	230.639	200.338	181.803	0.	0.	0.

PREVIOUS: December 1968 (1 atm) CURRENT: September 1983 (1 bar)

Calcium (Ca) $Ca_1(g)$

CALCIUM, ION (Ca^+) IDEAL GAS M_r = 40.07945

Calcium, Ion (Ca^+) $Ca_1^+(g)$

$IP(Ca^+, g) = 95751.87 \pm 0.03\ cm^{-1}$

$S°(298.15\ K) = 160.649 \pm 0.02\ J\ K^{-1}\ mol^{-1}$

$\Delta_fH°(0\ K) = 767.17 \pm 0.2\ kJ\ mol^{-1}$

$\Delta_fH°(298.15\ K) = [773.828]\ kJ\ mol^{-1}$

Electronic Levels and Quantum Weights

State	ε_i, cm^{-1}	g_i
$^2S_{1/2}$	0	2
$^2D_{3/2}$	13650.19	4
$^2D_{5/2}$	13710.88	6
$^2P_{1/2}$	25191.51	2
$^2P_{3/2}$	25414.40	4
IP	95751.87	

Enthalpy of Formation

$\Delta_fH°(Ca^+, g, 0\ K)$ is calculated from $\Delta_fH°(Ca, g, 0\ K)$ (1) using the spectroscopic value of IP(Ca) = 49305.96 ± 0.08 cm^{-1} (589.830 ± 0.001 kJ mol^{-1}) from Sugar and Corliss (2). The ionization limit is converted from cm^{-1} to kJ mol^{-1} using the factor, 1 cm^{-1} = 0.01196266 kJ cm^{-1}, which is derived from the 1973 CODATA fundamental constants (3). Rosenstock et al. (4) and Levin and Lias (5) have summarized additional ionization and appearance potential data.

$\Delta_fH°(Ca^+, g, 298.15\ K)$ is calculated from $\Delta_fH°(Ca, g, 0\ K)$ by using IP(Ca) with JANAF (1) enthalpies, H°(0 K) − H°(298.15 K), for Ca(g), $Ca^+(g)$, and $e^-(g)$. $\Delta_fH°(Ca \rightarrow Ca^+ + e^-, 298.15\ K)$ differs from a room temperature threshold energy due to inclusion of these enthalpies and to threshold effects discussed by Rosenstock et al. (4). $\Delta_fH°(298.15\ K)$ should be changed by −6.197 kJ mol^{-1} if it is to be used in the ion convention that excludes the enthalpy of the electron.

Heat Capacity and Entropy

The information on electronic energy levels and quantum weights, given by Sugar and Corliss (2), is incomplete because many theoretically predicted levels have not been observed. Our calculations indicate that any reasonable method of filling in these missing levels and cutting off the summation in the partition function (6) has no effect on the thermodynamic functions to 6000 K. This is a result of the high energy of all levels other than the ground state and the first four excited states; the next excited state is approximately 52167 cm^{-1} above the ground state. Since inclusion of these higher excited states have little effect on the thermodynamic functions (to 6000 K), we list only the ground state and the first four excited states. The reported uncertainty in S°(298.15 K) is due to uncertainties in the relative ionic mass and the fundamental constants. Extension of these calculations above 6000 K may require consideration of the excited states and use of different fill and cutoff procedures (6).

References

1. JANAF Thermochemical Tables: Ca(g), 9-30-83; e^-(g), 3-31-82.
2. J. Sugar and C. Corliss, J. Phys. Chem. Ref. Data 8, 865 (1979).
3. E. R. Cohen and B. N. Taylor, J. Phys. Chem. Ref. Data 2, 663 (1973).
4. H. M. Rosenstock, K. Draxl et al., J. Phys. Chem. Ref. Data 6, Supp. 1 (1977).
5. R. D. Levin and S. G. Lias, U. S. Nat. Bur. Stand., NSRDS-NBS-71, 1982.
6. J. R. Downey, Jr., The Dow Chemical Company, AFOSR-TRS-78-0960, Contract No. F44620-75-1-0048, 1978.

Enthalpy Reference Temperature = T_r = 298.15 K; Standard State Pressure = p° = 0.1 MPa

T/K	$C_p°$ (J K⁻¹mol⁻¹)	S° (J K⁻¹mol⁻¹)	−[G°−H°(T_r)]/T (J K⁻¹mol⁻¹)	H°−H°(T_r) (kJ mol⁻¹)	$\Delta_fH°$ (kJ mol⁻¹)	$\Delta_fG°$ (kJ mol⁻¹)	Log K_f
0	0.	0.	INFINITE	-6.197	767.169		
100	20.786	137.942	179.129	-4.119			
200	20.786	152.349	162.550	-2.040			
250	20.786	156.988	160.991	-1.001			
298.15	20.786	160.649	160.649	0.	773.828	732.075	-128.257
300	20.786	160.777	160.649	0.038	773.857	731.816	-127.420
350	20.786	163.981	160.902	1.078	774.629	724.748	-108.163
400	20.786	166.757	161.464	2.117	775.379	717.570	-93.705
450	20.786	169.205	162.191	3.156	776.095	710.301	-82.450
500	20.786	171.395	163.004	4.196	776.771	702.954	-73.437
600	20.786	175.185	164.728	6.274	777.986	688.073	-59.902
700	20.786	178.389	166.457	8.353	779.004	673.004	-50.220
800	20.786	181.165	168.126	10.431	778.967	657.908	-42.957
900	20.786	183.613	169.713	12.510	779.599	642.734	-37.303
1000	20.786	185.803	171.214	14.589	779.936	627.506	-32.778
1100	20.786	187.784	172.632	16.667	779.970	612.259	-29.074
1200	20.787	189.593	173.971	18.746	771.967	597.645	-26.015
1300	20.789	191.257	175.238	20.825	772.625	583.091	-23.429
1400	20.792	192.798	176.438	22.904	773.282	568.487	-21.211
1500	20.800	194.232	177.577	24.983	773.940	553.836	-19.286
1600	20.814	195.575	178.660	27.064	774.600	539.141	-17.601
1700	20.838	196.838	179.693	29.147	775.261	524.405	-16.113
1800	20.874	198.030	180.679	31.232	627.251	511.840	-14.853
1900	20.927	199.160	181.622	33.322	629.333	505.372	-13.894
2000	20.999	200.235	182.526	35.418	631.416	498.793	-13.027
2100	21.095	201.262	183.394	37.523	633.500	492.111	-12.241
2200	21.218	202.246	184.228	39.638	635.583	485.329	-11.523
2300	21.369	203.192	185.032	41.767	637.665	478.453	-10.866
2400	21.550	204.105	185.808	43.913	639.745	471.486	-10.262
2500	21.764	204.989	186.558	46.078	641.819	464.433	-9.704
2600	22.009	205.847	187.283	48.267	643.886	457.297	-9.187
2700	22.286	206.683	187.986	50.481	645.943	450.081	-8.707
2800	22.594	207.499	188.669	52.725	647.984	442.789	-8.260
2900	22.931	208.298	189.332	55.001	650.005	435.425	-7.843
3000	23.296	209.081	189.977	57.312	652.003	427.992	-7.452
3100	23.686	209.851	190.606	59.661	653.970	420.492	-7.085
3200	24.098	210.610	191.219	62.050	655.901	412.930	-6.740
3300	24.529	211.358	191.818	64.481	657.790	405.308	-6.415
3400	24.977	212.097	192.404	66.956	659.631	397.629	-6.109
3500	25.438	212.827	192.977	69.477	661.417	389.897	-5.819
3600	25.908	213.551	193.538	72.044	663.142	382.115	-5.544
3700	26.385	214.267	194.089	74.659	664.800	374.285	-5.284
3800	26.865	214.977	194.629	77.321	666.386	366.412	-5.037
3900	27.345	215.681	195.160	80.031	667.893	358.498	-4.802
4000	27.822	216.379	195.682	82.790	669.317	350.547	-4.578
4100	28.294	217.072	196.195	85.596	670.653	342.561	-4.364
4200	28.757	217.759	196.700	88.448	671.898	334.543	-4.161
4300	29.210	218.441	197.198	91.347	673.048	326.497	-3.966
4400	29.651	219.118	197.689	94.290	674.104	318.425	-3.780
4500	30.077	219.789	198.172	97.276	675.077	310.330	-3.602
4600	30.487	220.455	198.649	100.305	675.943	302.214	-3.432
4700	30.880	221.115	199.120	103.373	676.708	294.081	-3.268
4800	31.254	221.769	199.585	106.480	677.376	285.933	-3.112
4900	31.608	222.417	200.045	109.623	677.946	277.772	-2.961
5000	31.942	223.059	200.499	112.801	678.424	269.600	-2.816
5100	32.255	223.694	200.947	116.011	678.808	261.419	-2.677
5200	32.546	224.324	201.391	119.251	679.108	253.232	-2.544
5300	32.817	224.946	201.829	122.520	679.318	245.040	-2.415
5400	33.066	225.562	202.263	125.814	679.447	236.845	-2.291
5500	33.293	226.171	202.692	129.132	679.497	228.648	-2.172
5600	33.500	226.773	203.117	132.472	679.483	220.450	-2.056
5700	33.685	227.367	203.537	135.831	679.392	212.253	-1.945
5800	33.851	227.954	203.953	139.208	679.237	204.059	-1.838
5900	33.997	228.534	204.365	142.601	679.021	195.868	-1.734
6000	34.124	229.107	204.772	146.007	678.751	187.681	-1.634

PREVIOUS: December 1970 (1 atm) CURRENT: September 1983 (1 bar)

Calcium, Ion (Ca^+) $Ca_1^+(g)$

CALCIUM CHLORIDE (CaCl) IDEAL GAS $M_r = 75.533$

$S°(298.15\ K) = 241.56 \pm 0.8\ J\ K^{-1}\ mol^{-1}$

$\Delta_f H°(0\ K) = -103.9 \pm 13\ kJ\ mol^{-1}$
$\Delta_f H°(298.15\ K) = -104.6 \pm 13\ kJ\ mol^{-1}$

Electronic Levels and Quantum Weights

State	ε_i, cm^{-1}	g_i
$X^2\Sigma^+$	0	2
$A^2\Pi$	16093.3	2
	16162.8	2
$B^2\Sigma$	16850.6	2
$C^2\Pi$	26498.9	2
	26574.7	2
$D[^2\Sigma]$	31107.8	2
$E[^2\Sigma]$	34266.4	2
$F^2\Pi$	35675	2
	35700	2
$G^2\Delta$	36705	2
	36710	2

$\omega_e = 369.8\ cm^{-1}$ $\omega_e x_e = 1.31\ cm^{-1}$ $\sigma = 1$
$B_e = 0.1516\ cm^{-1}$ $\alpha_e = [0.0007]\ cm^{-1}$ $r_e = 2.439$ Å

Enthalpy of Formation

Potter (1) investigated mass spectrometrically the gaseous equilibria among Ca, CaCl, and $CaCl_2$. Ion intensities were measured 2.5 eV above the ionization threshold. Using the equilibrium constants $K = I(CaCl^+)^2/I(Ca^+)I(CaCl_2^+)$ which we calculate from the reported ion intensities for the reaction $Ca(g) + CaCl_2(g) \rightarrow 2\ CaCl(g)$ in the temperature range 1252 - 1557 K, we obtain $\Delta_r H°(298.15\ K) = 15.81\ kcal\ mol^{-1}$ by the 3rd law method. The 3rd law drift is $-2.8\pm3.5\ cal\ K^{-1}\ mol^{-1}$. The enthalpy of formation is derived as $\Delta_f H°(CaCl, g, 298.15\ K) = -27.0\ kcal\ mol^{-1}$.

Zmbov (7) determined by mass spectrometry the enthalpy of the above reaction as $24.4\pm1.3\ kcal\ mol^{-1}$ with a drift of $12.5\pm9.4\ cal\ K^{-1}\ mol^{-1}$. This yields the enthalpy of formation, $\Delta_f H°(CaCl, g, 298.15\ K) = -22.7\ kcal\ mol^{-1}$.

Hildenbrand (2) also determined the equilibrium constants for the reaction (a) $Ca(g) + CaCl_2(g) \rightarrow 2\ CaCl(g)$ and the reaction (b) $Ca(g) + AlCl(g) \rightarrow CaCl(g) + Al(g)$ by mass spectrometry in the temperature range 1219-1353 K. 3rd law analysis of the reported equilibrium constants gives $\Delta_r H°(298.15\ K) = 21.08\ kcal\ mol^{-1}$ and drift = $-1.4\pm1.8\ cal\ K^{-1}\ mol^{-1}$ for reaction (a); and $\Delta_r H°(298.15\ K) = 24.20\ kcal\ mol^{-1}$ and drift = $0.5\pm1.3\ cal\ K^{-1}\ mol^{-1}$ for reaction (b). From 3rd law $\Delta_r H°(298.15\ K)$ of (a) and (b), we derived $\Delta_f H°(CaCl, g, 298.15\ K) = -24.4$ and $-23.3\ kcal\ mol^{-1}$, respectively, using all JANAF functions. Hildenbrand has also derived a corrected $D_0^0(Ca\text{-}Cl) = 95.0\ kcal\ mol^{-1}$ from a linear Birge-Sponer extrapolation of the ground state vibrational levels with correction for its ionicity. This corrected $D_0^0(Ca\text{-}Cl)$ yields the enthalpy of formation, $\Delta_f H°(CaCl, g, 298.15\ K) = -23.9\ kcal\ mol^{-1}$.

Ryabova and Gurvich (3) reported $D_0^0(Ca\text{-}Cl) = 102\pm6\ kcal\ mol^{-1}$ from flame spectra studies. This leads to $\Delta_f H°(CaCl, g, 298.15\ K) = -30.9\ kcal\ mol^{-1}$.

Hildenbrand (6) determined the dissociation energy of $CaCl_2(g)$ by electron impact as $\Delta_{at}H°\ (CaCl\text{-}Cl) = 5.04\pm0.02\ kcal\ mol^{-1}$ which, combined with JANAF $\Delta_f H°(298.15\ K)$ of $CaCl_2(g)$ and Cl(g), gives $\Delta_f H°(CaCl, g, 298.15\ K) = -23.1\ kcal\ mol^{-1}$.

A weighted average, $\Delta_f H°(CaCl, g, 298.15\ K) = -25 \pm 3\ kcal\ mol^{-1} (-104.6 \pm 13\ kJ\ mol^{-1})$ is adopted in the tabulation.

Heat Capacity and Entropy

The vibrational constants (ω_e and $\omega_e x_e$), electronic levels and ground state configuration were taken from Herzberg (4). Morgan and Barrow (5) made rotational analysis of the $C^2\Pi\text{-}X^2\Sigma$ system and reported $B_0 = 0.1516\ cm^{-1}$ and $r_e = 2.439$ Å for the ground state. The value of α_e is calculated from the Morse potential function. The F and G states were reported by Schutte (8).

References

1. N. D. Potter, Aeronutronic Div., Philco-Ford Corp., Newport Beach, Calif., personal communication, Dec. 4, 1969.
2. D. L. Hildenbrand, "Dissociation Energies and Chemical Bonding in the Alkaline-Earth Chlorides from Mass Spectrometric Studies," J. Chem. Phys., 52, 5751 (1970).
3. V. G. Ryabova and L. V. Gurvich, Teplofiz. Vysokikh Temp. Akad. Nauk SSSR 3, 652 (1965).
4. G. Herzberg, "Spectra of Diatomic Molecules," D. Van Nostrand Co., Inc., New York, 1950.
5. E. Morgan and R. F. Barrow, Nature 185, 754 (1960).
6. D. L. Hildenbrand, CPIA Publication No. 189, Vol. 1, page 86, Aug., 1969.
7. K. F. Zmbov, Chem. Phys. Lett. 4, 191 (1969).
8. M. Schutte, Naturwiss. 40, 528 (1953).

Calcium Chloride (CaCl) $Ca_1Cl_1(g)$

Enthalpy Reference Temperature = T_r = 298.15 K; Standard State Pressure = $p°$ = 0.1 MPa

T/K	$C_p°$ (J K⁻¹mol⁻¹)	S° (J K⁻¹mol⁻¹)	$-[G°-H°(T_r)]/T$ (J K⁻¹mol⁻¹)	$H°-H°(T_r)$ (kJ mol⁻¹)	$\Delta_f H°$ (kJ mol⁻¹)	$\Delta_f G°$ (kJ mol⁻¹)	Log K_f
0	0.	0.	INFINITE	-9.589	-103.862	-103.862	INFINITE
100	30.307	205.406	271.974	-6.657	-103.367	-112.860	58.952
200	33.959	227.636	244.777	-3.428	-103.925	-122.158	31.905
250	35.001	235.334	242.143	-1.702	-104.259	-126.679	26.468
298.15	35.665	241.559	241.559	0.	-104.600	-130.966	22.945
300	35.685	241.780	241.560	0.066	-104.613	-131.129	22.832
350	36.151	247.318	241.996	1.863	-104.982	-135.520	20.225
400	36.481	252.168	242.971	3.679	-105.370	-139.856	18.263
450	36.724	256.480	244.237	5.509	-105.791	-144.142	16.732
500	36.910	260.359	245.658	7.350	-106.250	-148.379	15.501
600	37.174	267.114	248.688	11.055	-107.303	-156.709	13.643
700	37.356	272.858	251.741	14.783	-108.551	-164.847	12.301
800	37.493	277.856	254.699	18.525	-110.852	-172.689	11.275
900	37.602	282.278	257.523	22.280	-112.483	-180.324	10.466
1000	37.693	286.245	260.200	26.045	-114.407	-187.761	9.808
1100	37.773	289.841	262.734	29.818	-116.632	-194.992	9.259
1200	37.846	293.131	265.132	33.599	-126.891	-201.384	8.766
1300	37.913	296.163	267.404	37.387	-128.489	-207.526	8.339
1400	37.976	298.975	269.560	41.182	-130.086	-213.546	7.968
1500	38.037	301.597	271.609	44.983	-131.681	-219.452	7.642
1600	38.095	304.054	273.561	48.789	-133.274	-225.252	7.354
1700	38.154	306.365	275.423	52.602	-134.866	-230.951	7.096
1800	38.213	308.548	277.203	56.420	-285.130	-234.347	6.801
1900	38.273	310.616	278.908	60.244	-285.306	-231.521	6.365
2000	38.336	312.580	280.543	64.075	-285.486	-228.685	5.973
2100	38.404	314.452	282.113	67.912	-285.673	-225.841	5.617
2200	38.477	316.241	283.624	71.756	-285.870	-222.987	5.294
2300	38.558	317.953	285.080	75.607	-286.081	-220.124	4.999
2400	38.646	319.596	286.484	79.467	-286.310	-217.251	4.728
2500	38.745	321.175	287.840	83.337	-286.562	-214.369	4.479
2600	38.854	322.697	289.152	87.217	-286.843	-211.476	4.249
2700	38.975	324.165	290.422	91.108	-287.159	-208.571	4.035
2800	39.110	325.585	291.652	95.012	-287.515	-205.654	3.837
2900	39.257	326.960	292.846	98.930	-287.919	-202.723	3.651
3000	39.418	328.294	294.006	102.864	-288.377	-199.778	3.478
3100	39.594	329.589	295.133	106.815	-288.895	-196.816	3.316
3200	39.784	330.849	296.229	110.783	-289.481	-193.837	3.164
3300	39.988	332.076	297.297	114.772	-290.140	-190.838	3.021
3400	40.207	333.273	298.338	118.782	-290.878	-187.818	2.885
3500	40.439	334.442	299.353	122.814	-291.702	-184.775	2.758
3600	40.685	335.585	300.343	126.870	-292.616	-181.707	2.637
3700	40.944	336.703	301.311	130.951	-293.625	-178.613	2.522
3800	41.215	337.799	302.257	135.059	-294.732	-175.490	2.412
3900	41.498	338.873	303.182	139.195	-295.941	-172.336	2.308
4000	41.792	339.927	304.087	143.359	-297.255	-169.150	2.209
4100	42.096	340.963	304.974	147.553	-298.674	-165.930	2.114
4200	42.409	341.981	305.843	151.779	-300.200	-162.674	2.023
4300	42.731	342.983	306.695	156.035	-301.833	-159.381	1.936
4400	43.060	343.969	307.531	160.325	-303.568	-156.048	1.853
4500	43.396	344.940	308.352	164.648	-305.394	-152.676	1.772
4600	43.738	345.898	309.158	169.004	-307.326	-149.262	1.695
4700	44.086	346.842	309.949	173.396	-309.360	-145.804	1.620
4800	44.438	347.774	310.728	177.822	-311.487	-142.302	1.549
4900	44.793	348.694	311.493	182.283	-313.703	-138.754	1.479
5000	45.151	349.602	312.246	186.780	-316.001	-135.161	1.412
5100	45.511	350.500	312.988	191.314	-318.378	-131.520	1.347
5200	45.873	351.387	313.717	195.883	-320.823	-127.833	1.284
5300	46.235	352.264	314.436	200.488	-323.336	-124.098	1.223
5400	46.597	353.132	315.145	205.130	-325.909	-120.315	1.164
5500	46.958	353.990	315.844	209.807	-328.535	-116.483	1.106
5600	47.317	354.840	316.532	214.521	-331.196	-112.605	1.050
5700	47.675	355.680	317.212	219.271	-333.905	-108.677	0.996
5800	48.029	356.513	317.882	224.056	-336.646	-104.702	0.943
5900	48.381	357.337	318.544	228.877	-339.413	-100.679	0.891
6000	48.729	358.153	319.197	233.732	-342.199	-96.610	0.841

PREVIOUS: June 1970 (1 atm) CURRENT: June 1970 (1 bar)

Calcium Chloride (CaCl) $Ca_1Cl_1(g)$

CALCIUM CHLORIDE ($CaCl_2$) CRYSTAL M_r = 110.986

$S°(298.15\ K) = 104.6 \pm 4\ J\ K^{-1}\ mol^{-1}$

$T_{fus} = 1045\ K$

$\Delta_f H°(0\ K) = -796.3 \pm 1.3\ kJ\ mol^{-1}$

$\Delta_f H°(298.15\ K) = -795.8 \pm 1.3\ kJ\ mol^{-1}$

$\Delta_{fus} H° = 28.543 \pm 0.8\ kJ\ mol^{-1}$

Enthalpy of Formation

Parker (1) has selected the standard enthalpy of formation at 298.15 K as -190.2 kcal mol^{-1} (-795.797 kJ mol^{-1}) which is adopted in the tabulation. The enthalpy of formation was derived from two independent paths. The first path involves $\Delta_f H°(Ca^{++}$, aq, ∞, 298.15 K) = -129.74 kcal mol^{-1} (1), $\Delta_f H°(Cl^-$, aq, ∞, 298.15 K) = -39.952 kcal mol^{-1} (2) and the enthaply of solution of $CaCl_2$(cr) in water (3, 4, 5, 6, 7). The second path involves the enthalpies of reaction of Ca(cr), CaO(cr) and $CaCl_2$(cr) in aqueous HCl solution (4, 8-20).

Heat Capacity and Entropy

Kelley and Moore (21) measured the low temperature heat capacities in the temperature range from 52.6 to 295.1 K and made an extrapolation to 0 K which yielded an entropy of 4.59±0.4 cal K^{-1} mol^{-1} at 50.12 K. We have adopted the measured heat capacities, but made our own extrapolation to 0 K by comparison with the measured heat capacities of $BaCl_2$(cr) (22) and $SrCl_2$(cr) (23) from 6 to 50 K. The extrapolation gives S°(50 K) = 2.45±1 cal K^{-1} mol^{-1}, which is used in the table. Moore (24) measured high temperature enthalpy data from 403.7 to 1667 K by drop calorimetry. The low temperature heat capacities and the heat capacities from the high temperature enthalpy data are smoothly joined at 298.15 K. The deviations of the observed enthalpies from the adopted values are about 2% near 400-500 K and only a few tenths percent at higher temperatures. Dworkin and Bredig (25) determined drop-calorimetrically H°(1045 K) -H°(298.15 K) = 14.3 kcal mol^{-1}, which is in very good agreement with the adopted value of 14.2 kcal mol^{-1}.

Fusion Data

The adopted melting point 1045 K was determined by Tokareva (26). Dworkin and Bredig (25) determined the enthalpy of fusion $\Delta_{fus}H°$(1045 K) = 6.78±0.1 kcal mol^{-1} by drop calorimetry. Chiotti et al. (27) also measured the enthalpy of fusion $\Delta_{fus}H°$(1055 K) = 6.79±0.2 kcal mol^{-1} in an adiabatic calorimeter. The adopted enthalpy of fusion, 6.822 kcal mol^{-1} (28.543 kJ mol^{-1}), is calculated from the difference between the observed relative enthalpy of the liquid and the adopted value for the crystal at the melting point.

Sublimation Data

The enthalpy of sublimation is calculated from the enthalpies of formation of $CaCl_2$(cr) and $CaCl_2$(g).

References

1. V. B. Parker, U. S. Nat. Bur. Stand., personal communication, Jan. 7, 1969.
2. U. S. Nat. Bur. Stand. Tech. Note 270-3, 1968.
3. Pickering, J. Chem. Soc. 53, 875 (1888).
4. Samoilov, Izvest. Akad. Nauk SSSR Otdel. Khim. Nauk 1952, 626.
5. Ditte, Compt. Rend. 85, 1103 (1877).
6. Baud, Ann. Chim. Phys. 1, 8 (1904).
7. Berthelot and Ilosvay, Ann. Chim. Phys. 29, 295 (1883).
8. Kapustinski and Samoilov, Izvest. Akad. Nauk SSSR Otdel Khim. Nauk 1950, 367.
9. P. Ehrlich, K. Peik and E. Koch, Z. Anorg. Allg. Chem. 324, 113 (1963).
10. Vedeneev, Zh. Fiz. Khim. 26, 1808 (1952).
11. Young, J. Amer. Chem. Soc. 66, 773 (1944).
12. Torgeson and Shomate, J. Amer. Chem. Soc. 69, 2103 (1947).
13. Taylor and Wells, J. Res. Nat. Bur. Stand. 21, 133 (1938).
14. Wells and Taylor, J. Res. Nat. Bur. Stand. 19, 215 (1937).
15. Backstrom, J. Amer. Chem. Soc. 51, 2678 (1929).
16. Thorvaldson, Brown, and Peaker, J. Amer. Chem. Soc. 51, 2678 (1929).
17. Guntz and Benoit, Ann. Chim. 20, 5 (1923).
18. Biltz and Wagner, Z. Anorg. Chem. 134, 1 (1924).
19. Geiseler and Buchner, Z. Anorg. Allg. Chem. 343, 286 (1966).
20. Kocherov, Gertman and Geld, Zh. Neorg. Khim. 4, 1106 (1959).
21. K. K. Kelley and G. E. Moore, J. Amer. Chem. Soc. 65, 1264 (1943).
22. R. M. Goodman and E. F. Westrum, Jr., J. Chem. Eng. Data 11, 294 (1966).
23. D. F. Smith, T. E. Gardner, B. B. Letson and A. R. Taylor, Jr., U. S. Bur. Mines RI 6316, 1963.
24. G. E. Moore, J. Amer. Chem. Soc. 65, 1700 (1943).
25. A. S. Dworkin and M. A. Bredig, J. Phys. Chem. 67, 697 (1963).
26. M. V. Tokareva, Zh. Neorg. Khim. 2, 1591 (1957).
27. P. Chiotti, G. J. Gartner, E. R. Stevens and Y. Saito, J. Chem. Eng. Data 11, 571 (1966).

Calcium Chloride ($CaCl_2$) Ca_1Cl_2(cr)

Enthalpy Reference Temperature = T_r = 298.15 K; Standard State Pressure = p° = 0.1 MPa

	J $K^{-1}mol^{-1}$			kJ mol^{-1}			
T/K	$C_p°$	S°	$-[G°-H°(T_r)]/T$	$H°-H°(T_r)$	$\Delta_f H°$	$\Delta_f G°$	Log K_f
0	0.	0.	INFINITE	-15.391	-796.270	-796.270	INFINITE
100	48.812	35.596	164.875	-12.928	-797.700	-780.762	407.828
200	67.357	76.550	111.119	-6.914	-796.992	-764.012	199.539
298.15	72.856	104.602	104.602	0.	-795.797	-748.073	131.059
300	72.927	105.053	104.603	0.135	-795.773	-747.777	130.200
400	75.647	126.448	107.502	7.578	-794.434	-731.979	95.587
500	77.153	143.502	113.054	15.224	-793.125	-716.519	74.854
600	78.199	157.663	119.342	22.992	-791.931	-701.312	61.055
700	79.370	169.801	125.704	30.868	-790.866	-686.294	51.212
800	80.919	180.495	131.897	38.878	-790.750	-671.318	43.833
900	83.094	190.145	137.841	47.074	-789.800	-656.447	38.099
1000	85.772	199.033	143.521	55.512	-788.919	-641.678	33.518
1045.000	87.090	202.837	145.994	59.401	------ CRYSTAL <--> LIQUID -----		
1100	88.701	207.345	148.950	64.236	-788.069	-626.995	29.774
1200	91.630	215.189	154.146	73.252	-794.974	-611.767	26.630
1300	94.558	222.639	159.130	82.561	-792.937	-596.580	23.971
1400	97.487	229.754	163.922	92.164	-790.616	-581.561	21.698
1500	100.416	236.579	168.540	102.059	-788.012	-566.717	19.735
1600	103.345	243.154	172.999	112.247	-785.124	-552.057	18.023
1700	106.274	249.507	177.314	122.728	-781.951	-537.586	16.518
1800	109.202	255.664	181.497	133.502	-927.169	-521.099	15.122
1900	112.131	261.647	185.558	144.568	-922.016	-498.679	13.710
2000	115.060	267.473	189.509	155.928	-916.586	-476.537	12.446

PREVIOUS: CURRENT: June 1970

Calcium Chloride ($CaCl_2$) Ca_1Cl_2(cr)

CALCIUM CHLORIDE ($CaCl_2$) LIQUID M_r = 110.986

$S°(298.15\ K)$ = [123.868] J K^{-1} mol^{-1} $\Delta_f H°(298.15\ K)$ = [-774.094] kJ mol^{-1}

T_{fus} = 1045 K $\Delta_{fus}H°$ = 28.543 ± 0.8 kJ mol^{-1}

Enthalpy of Formation

The enthalpy of formation is calculated from that of the crystal by adding the enthalpy of melting and the difference in enthalpy, H°(1045 K)-H°(298.15 K), between the crystal and the liquid.

Heat Capacity and Entropy

The constant heat capacity above the assumed glass transition at 700 K is derived from high temperature enthalpy data in a drop calorimeter from the melting point up to 1667 K by Moore (1). Below the glass transition point, the heat capacity is assumed to be the same as that of the crystal. Dworkin and Bredig (2) reported $C_p°(\ell)$ = 23.6 cal K^{-1} mol^{-1} from enthalpy measurements in a short temperature range by drop calorimetry.

S°(298.15 K) is calculated in a manner analogous to that used for the enthalpy of formation.

Fusion Data

See the crystal table for details.

Vaporization Data

The boiling point, T_{vap}, is calculated as the temperature at which $\Delta_r G$ = 0 for $CaCl_2(\ell) \rightarrow CaCl_2(g)$ at one bar. The enthalpy of vaporization is the difference in $\Delta_f H°$ at T_{vap} between liquid and gas.

References

1. G. E. Moore, J. Amer. Chem. Soc. 65, 1700 (1943).
2. A. S. Dworkin and M. A. Bredig, J. Phys. Chem. 67, 697 (1963).

Calcium Chloride ($CaCl_2$) $Ca_1Cl_2(l)$

Enthalpy Reference Temperature = T_r = 298.15 K; Standard State Pressure = p° = 0.1 MPa

T/K	$C_p°$ (J K⁻¹mol⁻¹)	S° (J K⁻¹mol⁻¹)	-[G°-H°(T_r)]/T (J K⁻¹mol⁻¹)	H°-H°(T_r) (kJ mol⁻¹)	$\Delta_f H°$ (kJ mol⁻¹)	$\Delta_f G°$ (kJ mol⁻¹)	Log K_f
0							
100							
200							
298.15	72.856	123.868	123.868	0.	-774.094	-732.115	128.264
300	72.927	124.319	123.869	0.135	-774.070	-731.855	127.427
400	75.647	145.714	126.768	7.578	-772.732	-717.983	93.759
500	77.153	162.768	132.320	15.224	-771.423	-704.449	73.593
600	78.199	176.929	138.608	22.992	-770.229	-691.169	60.172
700	79.370	189.067	144.970	30.868	-769.164	-678.078	50.599
700.000	79.370	189.067	144.970	30.868	GLASS <--> LIQUID TRANSITION		
700.000	102.533	189.067	144.970	30.868			
800	102.533	202.759	151.357	41.121	-766.804	-665.184	43.432
900	102.533	214.835	157.752	51.375	-763.797	-652.665	37.880
1000	102.533	225.638	164.010	61.628	-761.101	-640.465	33.454
1045.000	102.533	230.151	166.762	66.242	------ CRYSTAL <--> LIQUID -----		
1100	102.533	235.411	170.064	71.881	-758.720	-628.519	29.846
1200	102.533	244.332	175.887	82.135	-764.389	-616.153	26.820
1300	102.533	252.539	181.472	92.388	-761.408	-603.921	24.266
1400	102.533	260.138	186.822	102.641	-758.436	-591.918	22.085
1500	102.533	267.212	191.949	112.895	-755.474	-580.128	20.202
1600	102.533	273.829	196.862	123.148	-752.520	-568.534	18.561
1700	102.533	280.045	201.574	133.401	-749.576	-557.126	17.118
1800	102.533	285.906	206.098	143.655	-895.314	-543.679	15.777
1900	102.533	291.449	210.445	153.908	-890.974	-524.262	14.413
2000	102.533	296.709	214.628	164.161	-886.650	-505.073	13.191
2100	102.533	301.711	218.657	174.415	-882.345	-486.100	12.091
2200	102.533	306.481	222.541	184.668	-878.063	-467.332	11.096
2300	102.533	311.039	226.291	194.921	-873.809	-448.758	10.192
2400	102.533	315.403	229.913	205.174	-869.590	-430.368	9.367
2500	102.533	319.588	233.417	215.428	-865.410	-412.153	8.611
2600	102.533	323.610	236.809	225.681	-861.279	-394.104	7.918
2700	102.533	327.479	240.096	235.934	-857.203	-376.214	7.278
2800	102.533	331.208	243.284	246.188	-853.189	-358.473	6.687
2900	102.533	334.806	246.378	256.441	-849.247	-340.874	6.140
3000	102.533	338.282	249.384	266.694	-845.384	-323.410	5.631

PREVIOUS: CURRENT: June 1970

Calcium Chloride ($CaCl_2$) $Ca_1Cl_2(l)$

CALCIUM CHLORIDE ($CaCl_2$) CRYSTAL–LIQUID M_r = 110.986

0 to 1045 K crystal
above 1045 K liquid

Refer to the individual tables for details.

Calcium Chloride ($CaCl_2$) Ca_1Cl_2(cr,l)

Enthalpy Reference Temperature = T_r = 298.15 K Standard State Pressure = p° = 0.1 MPa

T/K	C_p° (J K⁻¹mol⁻¹)	S° (J K⁻¹mol⁻¹)	−[G°−H°(T_r)]/T (J K⁻¹mol⁻¹)	H°−H°(T_r) (kJ mol⁻¹)	$\Delta_f H^\circ$ (kJ mol⁻¹)	$\Delta_f G^\circ$ (kJ mol⁻¹)	Log K_f
0	0.	0.	INFINITE	-15.391	-796.270	-796.270	INFINITE
100	48.812	35.596	164.875	-12.928	-797.700	-780.762	407.828
200	67.357	76.550	111.119	-6.914	-796.992	-764.012	199.539
298.15	72.856	104.602	104.602	0.	-795.797	-748.073	131.059
300	72.927	105.053	104.603	0.135	-795.773	-747.777	130.200
400	75.647	126.448	107.502	7.578	-794.434	-731.979	95.587
500	77.153	143.502	113.054	15.224	-793.125	-716.519	74.854
600	78.199	157.663	119.342	22.992	-791.931	-701.312	61.055
700	79.370	169.801	125.704	30.868	-790.866	-686.294	51.212
800	80.919	180.495	131.897	38.878	-790.750	-671.318	43.833
900	83.094	190.145	137.841	47.074	-789.800	-656.447	38.099
1000	85.772	199.033	143.521	55.512	-788.919	-641.678	33.518
1045.000	87.090	202.837	145.994	59.401	CRYSTAL <--> LIQUID		
1045.000	102.533	230.151	145.994	87.944	TRANSITION		
1100	102.533	235.411	150.335	93.584	-758.720	-628.519	29.846
1200	102.533	244.332	157.801	103.837	-764.389	-616.153	26.820
1300	102.533	252.539	164.777	114.090	-761.408	-603.921	24.266
1400	102.533	260.138	171.321	124.344	-758.436	-591.918	22.085
1500	102.533	267.212	177.480	134.597	-755.474	-580.128	20.202
1600	102.533	273.829	183.298	144.850	-752.520	-568.534	18.561
1700	102.533	280.045	188.808	155.104	-749.576	-557.126	17.118
1800	102.533	285.906	194.041	165.357	-895.314	-543.679	15.777
1900	102.533	291.449	199.023	175.610	-890.974	-524.262	14.413
2000	102.533	296.709	203.777	185.864	-886.650	-505.073	13.191
2100	102.533	301.711	208.322	196.117	-882.345	-486.100	12.091
2200	102.533	306.481	212.677	206.370	-878.063	-467.332	11.096
2300	102.533	311.039	216.855	216.623	-873.809	-448.758	10.192
2400	102.533	315.403	220.871	226.877	-869.590	-430.368	9.367
2500	102.533	319.588	224.736	237.130	-865.410	-412.153	8.611
2600	102.533	323.610	228.462	247.383	-861.279	-394.104	7.918
2700	102.533	327.479	232.058	257.637	-857.202	-376.214	7.278
2800	102.533	331.208	235.533	267.890	-853.189	-358.473	6.687
2900	102.533	334.806	238.895	278.143	-849.247	-340.874	6.140
3000	102.533	338.282	242.150	288.397	-845.384	-323.410	5.631

PREVIOUS: CURRENT: June 1970

Calcium Chloride ($CaCl_2$) Ca_1Cl_2(cr,l)

CALCIUM CHLORIDE ($CaCl_2$) IDEAL GAS M_r = 110.986

$S°(298.15 K) = 290.3 \pm 8 J K^{-1} mol^{-1}$

$\Delta_fH°(0 K) = -471.7 \pm 4 kJ mol^{-1}$
$\Delta_fH°(298.15 K) = -471.5 \pm 4 kJ mol^{-1}$

Vibrational Frequencies and Degeneracies

ν, cm^{-1}
[217](1)
64 (2)
402 (1)

Ground State Quantum Weight: 1 σ = 2
Point Group: $D_{\infty h}$
Bond Distance: Ca-Cl - 2.51 Å
Bond Angle: Cl-Ca-Cl = 180°
Rotational Constant: $B_0 = 0.037737 cm^{-1}$

Enthalpy of Formation

2nd and 3rd law analyses of vapor pressure data are given below. The selected 3rd law enthalpy of vaporization is taken from Novikov's measurement (1) as $\Delta_{vap}H°(298.15 K) = 72.30 kcal mol^{-1}$ which is combined with $\Delta_fH°(298.15 K)$ of the liquid to give $\Delta_fH°(CaCl_2, g, 298.15 K) = -112.7\pm1 kcal mol^{-1}$ ($-471.5\pm4 kJ mol^{-1}$).

3rd law analyses indicate zero dirft for Novikov's data, positive and negative drifts for Hildenbrand's (2) and Bautista's (3) data, respectively. Their enthalpies are in good agreement within ±1 kcal. Wartenberg (4) reported there was no boiling at 1819 K at 62 mm Hg pressure, and his value, $\Delta_{vap}H°(298.15 K) = 71.1\pm0.7 kcal mol^{-1}$, serves as a guide for the minimum enthalpy of vaporization.

Source	Method	T/K	Data Points	$\Delta_{vap}H°(298.15 K)$, $kcal mol^{-1}$ 2nd Law	3rd Law	Drift $cal K^{-1} mol^{-1}$
Bautista and Margrave (3)	Langmuir	952-993	7	75.25	73.39	-1.7±6.9
				(80.34)*	(78.57)*	
Hildenbrand and Potter (2)	Torsion Effusion $0.0096 cm^2$ Orifice	1142-1228	12	64.99	71.24	5.3±0.4
Hildenbrand and Potter (2)	Torsion Effusion $0.015 cm^2$ Orifice	1111-1281	17	68.71	71.24	2.0±0.2
Hildenbrand and Potter (2)	Torsion Effusion Combined	1111-1281	29	68.91	71.24	2.4±0.3
Dewing (5)	Knudsen Effusion	1069-1225	4	59.37	71.03	10.2±1.6
Wartenberg and Bosse (4)	Boiling Point	1819	1	---	>71.10	---
Novikov and Gavryuchenkov (1)	Boiling Point	1591-1701	5	72.25	72.30	0.0±1.2
Lukashenko and Reutova (10)	Langmuir	973-1023	6	78.0	69.01	-9.0
	Knudsen Effusion	1073-1273	6	61.8	69.78	+8.0

*Value in parentheses is the enthalpy of sublimation.

Heat Capacity and Entropy

The linear structure and vibrational frequencies were determined by White and co-workers (6) in the matrix-infrared-spectrometric studies. The linear structure is consistent with experimental error with the electric-quadrupole-deflection studies (7), and electron-diffraction data (8). Hayes (9) has qualitatively rationalized the linear structure through molecular orbital theory. The bond length was measured by Akishin and Spiridonov (8).

References

1. G. I. Novikov and F. G. Gavryuchenkov, Zh. Neorg. Khim. 9, 260 (1964).
2. D. L. Hildenbrand and N. D. Potter, J. Phys. Chem. 67, 2231 (1963).
3. R. G. Bautista and J. L. Margrave, J. Phys. Chem. 67, 2411 (1963).
4. H. V. Wartenberg and O. Bosse, Z. Elektrochem. 28, 384 (1922).
5. E. W. Dewing, Aluminum Laboratories, quoted in ref. (2).
6. Unpublished work cited by D. L. Hildenbrand, "Dissociation Energies and Chemical Bonding in the Alkaline-Earth Chlorides from Mass Spectrometric Studies," J. Chem. Phys. 52, 5751 (1970).
7. L. Wharton, R. A. Berg and W. Klemperer, J. Chem. Phys. 39, 2023 (1963); 39, 2299 (1963).
8. P. A. Akishin and V. P. Spiridonov, Kristallografiya 2, 475 (1957).
9. E. F. Yates, J. Phys. Chem. 70, 3740 (1966).
10. E. E. Lukashenko and G. A. Reutova, Zhur. Fiz. Khim. 44, 600 (1970).

Calcium Chloride ($CaCl_2$) $Ca_1Cl_2(g)$

Enthalpy Reference Temperature = T_r = 298.15 K Standard State Pressure = p° = 0.1 MPa

T/K	$C_p°$ ($J K^{-1}mol^{-1}$)	S°	$-[G°-H°(T_r)]/T$	$H°-H°(T_r)$ ($kJ mol^{-1}$)	$\Delta_fH°$	$\Delta_fG°$	Log K_f
0	0.	0.	INFINITE	-15.118	-471.738	-471.738	INFINITE
100	49.371	230.265	340.874	-11.061	-471.573	-474.102	247.646
200	56.575	267.082	295.620	-5.708	-471.526	-476.652	124.489
250	58.277	279.904	291.236	-2.833	-471.520	-477.934	99.859
298.15	59.325	290.265	290.265	0.	-471.537	-479.169	83.948
300	59.357	290.632	290.266	0.110	-471.538	-479.216	83.439
350	60.073	299.839	290.991	3.097	-471.576	-480.493	71.710
400	60.566	307.895	292.611	6.113	-471.639	-481.763	62.912
450	60.919	315.050	294.714	9.151	-471.740	-483.023	56.068
500	61.178	321.482	297.075	12.204	-471.885	-484.269	50.591
600	61.525	332.670	302.102	18.340	-472.323	-486.708	42.372
700	61.740	342.171	307.165	24.504	-472.970	-489.057	36.494
800	61.882	350.425	312.068	30.686	-474.682	-491.195	32.072
900	61.980	357.720	316.743	36.879	-475.734	-493.199	28.625
1000	62.051	364.254	321.173	43.081	-477.090	-495.070	25.860
1100	62.104	370.171	325.363	49.289	-478.755	-496.790	23.591
1200	62.144	375.577	329.325	55.502	-488.464	-497.722	21.665
1300	62.175	380.552	333.077	61.718	-489.521	-498.451	20.028
1400	62.200	385.161	336.635	67.936	-490.584	-499.098	18.622
1500	62.221	389.453	340.014	74.158	-491.653	-499.669	17.400
1600	62.237	393.469	343.231	80.380	-492.730	-500.168	16.329
1700	62.251	397.242	346.298	86.605	-493.814	-500.600	15.382
1800	62.262	400.801	349.228	92.831	-643.580	-498.756	14.474
1900	62.272	404.168	352.032	99.057	-643.267	-490.719	13.491
2000	62.281	407.362	354.719	105.285	-642.969	-482.698	12.607
2100	62.288	410.401	357.299	111.513	-642.688	-474.692	11.807
2200	62.294	413.299	359.779	117.742	-642.431	-466.698	11.081
2300	62.299	416.068	362.167	123.972	-642.201	-458.715	10.418
2400	62.304	418.719	364.468	130.202	-642.004	-450.742	9.810
2500	62.308	421.263	366.690	136.433	-641.847	-442.776	9.251
2600	62.312	423.707	368.836	142.664	-641.738	-434.816	8.736
2700	62.315	426.058	370.912	148.895	-641.684	-426.858	8.258
2800	62.318	428.325	372.922	155.127	-641.692	-418.902	7.815
2900	62.321	430.512	374.871	161.359	-641.771	-410.944	7.402
3000	62.324	432.624	376.761	167.591	-641.929	-402.982	7.017
3100	62.326	434.668	378.596	173.824	-642.175	-395.013	6.656
3200	62.328	436.647	380.379	180.056	-642.514	-387.035	6.318
3300	62.330	438.565	382.113	186.289	-642.956	-379.045	6.000
3400	62.331	440.426	383.801	192.522	-643.506	-371.039	5.700
3500	62.333	442.232	385.445	198.756	-644.172	-363.016	5.418
3600	62.334	443.988	387.047	204.989	-644.958	-354.972	5.151
3700	62.335	445.696	388.609	211.222	-645.869	-346.905	4.897
3800	62.337	447.359	390.133	217.456	-646.910	-338.811	4.657
3900	62.338	448.978	391.622	223.690	-648.084	-330.688	4.429
4000	62.339	450.556	393.075	229.923	-649.393	-322.534	4.212
4100	62.340	452.095	394.496	236.157	-650.838	-314.344	4.005
4200	62.340	453.598	395.885	242.391	-652.420	-306.119	3.807
4300	62.341	455.065	397.245	248.625	-654.138	-297.853	3.618
4400	62.342	456.498	398.575	254.860	-655.989	-289.547	3.437
4500	62.343	457.899	399.878	261.094	-657.958	-281.198	3.264
4600	62.343	459.269	401.154	267.328	-660.063	-272.802	3.098
4700	62.344	460.610	402.405	273.563	-662.296	-264.360	2.938
4800	62.345	461.922	403.631	279.797	-664.650	-255.868	2.784
4900	62.345	463.208	404.834	286.031	-667.119	-247.327	2.637
5000	62.346	464.467	406.014	292.266	-669.696	-238.734	2.494
5100	62.346	465.702	407.173	298.501	-672.377	-230.088	2.357
5200	62.347	466.913	408.310	304.735	-675.151	-221.389	2.224
5300	62.347	468.100	409.427	310.970	-678.018	-212.636	2.096
5400	62.347	469.266	410.524	317.205	-680.968	-203.827	1.972
5500	62.348	470.410	411.603	323.439	-683.993	-194.963	1.852
5600	62.348	471.533	412.663	329.674	-687.077	-186.045	1.735
5700	62.348	472.637	413.705	335.909	-690.230	-177.070	1.623
5800	62.349	473.721	414.731	342.144	-693.437	-168.040	1.513
5900	62.349	474.787	415.740	348.379	-696.690	-158.953	1.407
6000	62.349	475.835	416.732	354.614	-699.982	-149.812	1.304

PREVIOUS: June 1970 (1 atm) CURRENT: June 1970 (1 bar)

Calcium Chloride ($CaCl_2$) $Ca_1Cl_2(g)$

CALCIUM FLUORIDE (CaF) IDEAL GAS M_r = 59.078403

$S°(298.15\ K) = 229.65 \pm 0.4\ J\ K^{-1}\ mol^{-1}$

$\Delta_f H°(0\ K) = -271 \pm 8\ kJ\ mol^{-1}$
$\Delta_f H°(298.15\ K) = -272 \pm 8\ kJ\ mol^{-1}$

Electronic Levels and Quantum Weights

State	ε_i, cm^{-1}	g_i
$X^2\Sigma^+$	0	2
$A^2\Pi$	16482	2
	16557	2
$B^2\Sigma$	18844	2
$C^2\Pi$	30256	2
	30285	2
$D^2\Sigma$	30772	2
$E^2\Sigma$	34135	2
$F^2\Pi$	37548	4

$\omega_e = 587.1\ cm^{-1}$ $\omega_e x_e = 2.74\ cm^{-1}$ $\sigma = 1$
$B_e = 0.3225\ cm^{-1}$ $\alpha_e = [0.0020]\ cm^{-1}$ $r_e = 2.01$ Å

Enthalpy of Formation

Hastie and Margrave (1) have recently reviewed the dissociation energy of CaF and conclude that it is 127.5±2.5 kcal mol^{-1}. The analysis below considers basically the same data except that all values are JANAF.

Hildenbrand and Murad (2) in a mass spectrometric determination of the equilibria among BF, BF_3, Ca, CaF and CaF_2 reported equilibrium constants for the reaction

(A) $Ca(g) + CaF_2(g) \rightarrow 2CaF(g)$.

Blue et al. (3) have also examined several equilibria mass-spectrometrically and report equilibrium constants for reaction

(B) $Ca(g) + CaF_2(cr) \rightarrow 2CaF(g)$ and reaction

(C) $Ca(g) + AlF(g) \rightarrow Al(g) + CaF(g)$.

Ryabova and Gurvich (4) give a enthalpy of dissociation from flame studies for the reaction

(D) $CaF(g) \rightarrow Ca(g) + F(g)$.

The remaining reactions considered by Hastie and Margrave (1) were not included since they involved subtracting reactions not performed simultaneously, or involve species such as SiF_2 whose enthalpy of formation is dependent on that of CaF and CaF_2.

Source	T/K	Reaction	Data Points	$\Delta_r H°$(298.15 K), kcal mol^{-1} 2nd law	3rd law	Drift cal K^{-1} mol^{-1}	$\Delta_f H°$(298.15 K) kcal mol^{-1}
Hildenbrand (2)	1583-1734	A	15	11.5±6.8	14.70±1.27	2.2±4.1	-65.5±1.8
Blue (3)	1271-1351	B	9	146.9±8.0	121.08±1.65	-19.4±6.2	-64.5±1.7
Blue (3)	1271-1351	C	10*	34.6±3.5	32.87±0.47	-1.4±2.7	-65.7±3.2
Ryobova (4)		D		135±7			-73.3±7
Hildenbrand (5)	1423-1443	A	2		32.97		-65.6±3.2

*1 point rejected due to failure of a statistical test

The auxiliary enthalpies of formation used were, in kcal mol^{-1}, AlF(g) = -63.4, Al(g) = 78, $CaF_2(g)$ = -187.5, $CaF_2(cr)$ = -293.0 and Ca(g) = 42.85.

We adopt a median value of $\Delta_f H°$(CaF, g, 298.15 K) = -65±2 kcal mol^{-1} (-271.960±8 kJ mol^{-1}) which corresponds to an enthalpy of dissociation of 126.7±2 kcal mol^{-1}.

Heat Capacity and Entropy

Herzberg (6) gives the vibrational constants and electronic levels. Harvey (7) has reported a value for B_0 which corresponds to a bond length of 2.02 Å. The Morse potential function was used to calculate α_e.

References

1. J. W. Hastie and J. L. Margrave, J. Chem. Eng. Data 13, 428 (1968).
2. D. L. Hildenbrand and E. Murad, J. Chem. Phys. 43, 1400 (1965).
3. G. D. Blue, J. W. Green, R. G. Bautista and J. L. Margrave, J. Phys. Chem. 67, 877 (1963).
4. V. G. Ryabova and L. V. Gurvich, Teplofiz. Vys. Temp. 2, 834 (1964).
5. D. L. Hildenbrand and E. Murad, J. Chem. Phys. 44, 1524 (1966).
6. G. Herzberg, "Spectra of Diatomic Molecules," D. Van Nostrand Co., Inc. New York, 1950.
7. A. Harvey, Proc. Roy. Soc. (London) 133, 336 (1931).

Calcium Fluoride (CaF) $Ca_1F_1(g)$

Enthalpy Reference Temperature = T_r = 298.15 K Standard State Pressure = p° = 0.1 MPa

T/K	$C_p°$ (J K^{-1}mol^{-1})	S°	-[G°-H°(T_r)]/T	H°-H°(T_r) (kJ mol^{-1})	$\Delta_f H°$	$\Delta_f G°$	Log K_f
0	0.	0.	INFINITE	-9.130	-270.942	-270.942	INFINITE
100	29.245	195.821	258.018	-6.220	-270.465	-279.932	146.221
200	31.424	216.653	232.661	-3.201	-271.180	-289.148	75.518
250	32.678	223.804	230.196	-1.598	-271.578	-293.593	61.343
298.15	33.661	229.647	229.647	0.	-271.960	-297.799	52.173
300	33.695	229.855	229.648	0.062	-271.975	-297.959	51.879
350	34.477	235.111	230.061	1.767	-272.370	-302.259	45.110
400	35.075	239.756	230.988	3.507	-272.775	-306.501	40.025
450	35.534	243.914	232.197	5.273	-273.205	-310.692	36.064
500	35.893	247.678	233.560	7.059	-273.671	-314.832	32.890
600	36.409	254.271	236.478	10.676	-274.733	-322.968	28.117
700	36.755	259.911	239.432	14.335	-275.989	-330.910	24.693
800	37.002	264.836	242.307	18.023	-278.299	-338.555	22.105
900	37.188	269.205	245.057	21.733	-279.942	-345.991	20.081
1000	37.335	273.131	247.672	25.460	-281.881	-353.229	18.451
1100	37.454	276.695	250.151	29.199	-284.124	-360.258	17.107
1200	37.556	279.959	252.500	32.950	-294.406	-366.446	15.951
1300	37.644	282.968	254.730	36.710	-296.031	-372.384	14.963
1400	37.722	285.761	256.848	40.478	-297.658	-378.196	14.111
1500	37.793	288.366	258.863	44.254	-299.287	-383.892	13.368
1600	37.860	290.807	260.784	48.037	-300.919	-389.479	12.715
1700	37.922	293.104	262.619	51.826	-302.552	-394.964	12.136
1800	37.983	295.274	264.373	55.621	-452.861	-398.142	11.554
1900	38.043	297.329	266.054	59.423	-453.082	-395.096	10.862
2000	38.103	299.282	267.667	63.230	-453.307	-392.038	10.239
2100	38.165	301.142	269.217	67.043	-453.537	-388.969	9.675
2200	38.230	302.919	270.709	70.863	-453.772	-385.889	9.162
2300	38.300	304.620	272.147	74.689	-454.016	-382.798	8.694
2400	38.375	306.252	273.534	78.523	-454.268	-379.697	8.264
2500	38.457	307.820	274.874	82.364	-454.533	-376.584	7.868
2600	38.548	309.330	276.170	86.215	-454.814	-373.460	7.503
2700	38.647	310.787	277.426	90.074	-455.115	-370.326	7.164
2800	38.757	312.194	278.642	93.944	-455.440	-367.180	6.850
2900	38.877	313.556	279.823	97.826	-455.794	-364.021	6.557
3000	39.009	314.876	280.970	101.720	-456.183	-360.850	6.283
3100	39.153	316.158	282.084	105.628	-456.612	-357.666	6.027
3200	39.309	317.403	283.168	109.551	-457.089	-354.466	5.786
3300	39.476	318.615	284.224	113.490	-457.618	-351.251	5.560
3400	39.657	319.797	285.253	117.447	-458.207	-348.019	5.347
3500	39.849	320.949	286.257	121.422	-458.860	-344.769	5.145
3600	40.053	322.074	287.236	125.417	-459.585	-341.499	4.955
3700	40.269	323.175	288.193	129.433	-460.387	-338.208	4.775
3800	40.496	324.251	289.127	133.471	-461.269	-334.895	4.603
3900	40.734	325.306	290.042	137.533	-462.238	-331.556	4.441
4000	40.982	326.341	290.936	141.619	-463.297	-328.192	4.286
4100	41.239	327.356	291.812	145.730	-464.448	-324.801	4.138
4200	41.506	328.353	292.670	149.867	-465.695	-321.380	3.997
4300	41.781	329.333	293.512	154.031	-467.038	-317.928	3.862
4400	42.063	330.296	294.337	158.223	-468.476	-314.444	3.733
4500	42.352	331.245	295.146	162.444	-469.996	-310.927	3.609
4600	42.648	332.179	295.941	166.694	-471.619	-307.375	3.490
4700	42.949	333.099	296.722	170.974	-473.338	-303.786	3.376
4800	43.254	334.007	297.489	175.284	-475.149	-300.160	3.266
4900	43.564	334.902	298.244	179.625	-477.046	-296.495	3.161
5000	43.878	335.785	298.986	183.997	-479.026	-292.790	3.059
5100	44.194	336.657	299.716	188.400	-481.086	-289.045	2.960
5200	44.512	337.518	300.435	192.835	-483.215	-285.259	2.865
5300	44.831	338.369	301.142	197.303	-485.417	-281.432	2.774
5400	45.152	339.210	301.840	201.802	-487.682	-277.562	2.685
5500	45.473	340.042	302.527	206.333	-490.004	-273.649	2.599
5600	45.793	340.864	303.204	210.896	-492.368	-269.695	2.516
5700	46.112	341.677	303.872	215.492	-494.785	-265.698	2.435
5800	46.430	342.482	304.531	220.119	-497.242	-261.657	2.356
5900	46.747	343.278	305.181	224.778	-499.731	-257.574	2.280
6000	47.060	344.067	305.822	229.468	-502.247	-253.448	2.206

PREVIOUS: December 1968 (1 atm) CURRENT: December 1968 (1 bar)

Calcium Fluoride (CaF) $Ca_1F_1(g)$

CALCIUM FLUORIDE (CaF_2) CRYSTAL M_r = 78.076806

$S°(298.15\ K) = 68.572 \pm 0.33\ J\ K^{-1}\ mol^{-1}$
$T_{trs} = 1424 \pm 20\ K\ (\alpha\text{-}\beta)$
$T_{fus} = 1691 \pm 5\ K$

$\Delta_f H°(0\ K) = -1223.0 \pm 6.3\ kJ\ mol^{-1}$
$\Delta_f H°(298.15\ K) = -1225.9 \pm 6.3\ kJ\ mol^{-1}$
$\Delta_{trs} H° = 4.770 \pm 0.4\ kJ\ mol^{-1}$
$\Delta_{fus} H° = 29.706 \pm 0.4\ kJ\ mol^{-1}$

Enthalpy of Formation

Torgeson and Sahama (1) measured the enthalpy of solution of CaO in aqueous 20.1% HF as -54.96±0.02 kcal mol^{-1}. Using $\Delta_f H°$(CaO, cr, 298.15 K) = -151.79 kcal mol^{-1} (2), $\Delta_f H°$(H_2O, ℓ, 298.15 K) = -68.317 kcal mol^{-1} (3) and $\Delta_f H°$(HF·4.4H_2O, 298.15 K) = -76.60 kcal mol^{-1}, we obtain $\Delta_f H°$(CaF_2, cr, 298.15 K) = -291.6±0.6 kcal mol^{-1}.

Smyshlyaev and Edeleva (5) report the solubility product of CaF_2 as 2.7±0.27 x 10^{-11} in dilute HCl, which yields ΔG°(298.15 K) = 14.42 kcal mol^{-1} for the reaction CaF_2(cr) → Ca^{++}(aq, ∞) + $2F^-$(aq, ∞). Using $\Delta_f G°$(Ca^{++},aq, ∞, 298.15 K) = -132.30 kcal mol^{-1} (2) and $\Delta_f G°$(F^-, aq, ∞, 298.15 K) = -66.95 kcal mol^{-1} (4) we obtain $\Delta_f G°$(CaF_2, cr, 298.15 K) = -280.64±0.6 kcal mol^{-1}, which, using the adopted functions, yields $\Delta_f H°$(CaF_2, cr, 298.15 K) = -293.1±0.6 kcal mol^{-1}.

Vecher and Vecher (17) have measured the emf of a high temperature solid state cell and report $\Delta_r G°$(1200) = -17.6±0.2 kcal mol^{-1} for the reaction CaO(cr) + MgF_2(cr) → CaF_2(cr) + MgO(cr). This reduces to $\Delta_r H°$(298.15 K) = -17.6 kcal mol^{-1}, using JANAF values for MgO and MgF_2 and CaO from reference 2, yields $\Delta_f H°$(CaF_2, cr, 298.15 K) = -294.4±1.3 kcal mol^{-1}.

Kohlrauch (18) reported the solubility of CaF_2 and fluorspar, which yield solubility products of 3.6 x 10^{-11} and 2.8 x 10^{-11}. From these we obtain, as above, $\Delta_f H°$(CaF_2, cr, 298.15 K) = -292.9 and -293.1 kcal mol^{-1}.

Guntz (19) made measurements on the neutralization of $Ca(OH)_2$(aq) and HF(aq) which yield $\Delta_f H°$(CaF_2, cr, 298.15 K) = -293.1 kcal mol^{-1}, according to Parker (2).

A median value of $\Delta_f H°$(CaF_2, cr, 298.15 K) = -293±1.5 kcal mol^{-1} (-1225.912±6.3 kJ mol^{-1}) is adopted.

Heat Capacity and Entropy

Huffman and Norwood (6) have measured the low temperature heat capacity in the range 3.6 to 30 K, Eucken and Schwers (7) in the range 17 to 86 K and Todd (8) from 54 to 296.5 K. We have fitted a smooth polynomial curve through the data of references 6 and 8 and this yields S°(298.15 K) = 16.38±0.08 cal K^{-1} mol^{-1} based on S°(3.6 K) = 0.0002 cal K^{-1} mol^{-1}. The values of reference 7 deviate considerably at the lowest temperatures but are in reasonable agreement at higher temperatures.

The high temperature enthalpy of calcium fluoride has been reported by Naylor (9) to 1789 K, Krestnikov and Karetnikov (10) to 1273 K, and Lyashenko (11) to 1490 K. All the data are in approximate agreement and the more extensive results of Naylor on a very pure sample are adopted. The heat capacities, below 1424 K, were derived from a polynomial fit of the enthalpy data of the form $H = aT^2 + bT + c + d/T + e/T^2$.

Transition Data

The temperature of transition is that reported by Naylor (9), the large uncertainty has been assigned since at 1424 K the enthalpies indicate complete conversion to β-CaF_2, and the next lower point is at 1402 K. Evidence in support of a transition in this region comes from Lyashenko (11) whose plot indicates a break between 1273 and 1413 K. In addition, $SrCl_2$ which has the fluorite structure has a transition near the melting point. The enthalpy of transition is that reported by Naylor (9).

Fusion Data

McCreary (12) reports the melting point of a 99.1% pure sample as 1687±5 K. Porter and Brown (13) report 1675±5 K as the melting point of 99.8% pure CaF_2 and Rogers et al. (14) obtained 1684 K as the melting point. The adopted melting point, 1691±5 K, for a sample of natural fluorite of high purity, was reported by Naylor (9). The enthalpy of melting, 7100 kcal mol^{-1} (29.706 kJ mol^{-1}), is that measured by Naylor (9). Three determinations from high temperature phase diagrams give 7250±350 cal mol^{-1} (15), 5500 cal mol^{-1} (16) and 9800 cal mol^{-1} (14).

Sublimation Data

See the ideal gas table for details.

References

1. D. R. Torgeson and T. G. Sahama, J. Amer. Chem. Soc. 70, 2156 (1948).
2. V. B. Parker, Nat. Bur. Stand., personal communication, Jan. 1969.
3. U. S. Nat. Bur. Stand. Tech. Note 270-3, 1968.
4. Consistent with the JANAF HF(g) value.
5. S. I. Smyshlaev and N. P. Edeleva, Izv. Vysshykh Uchebn. Zavedenii, Khim. i Khim. Tekhol. 5, 871 (1962).
6. D. R. Huffman and M. H. Norwood, Phys. Rev. 117, 709 (1960).
7. A. Eucken and F. Schwers, Ber. Deut. Physik. Ges. 15, 578 (1913).
8. S. S. Todd, J. Amer. Chem. Soc. 71, 4115 (1949).
9. B. F. Naylor, J. Amer. Chem. Soc. 67, 150 (1945).
10. A. N. Krestnikov and G. A. Karetnikov, Legkie Metal. 3, 29 (1934) and 4, 16 (1935).
11. V. S. Lyashenko, Metallurg. 10, 85 (1935).
12. W. J. McCreary, J. Amer. Chem. Soc. 77, 2113 (1955).
13. B. Porter and E. A. Brown, J. Amer. Ceram. Soc. 45, 49 (1962).
14. P. S. Rogers, J. W. Tomlinson and F. D. Richardson, Met. Soc. Conf. 8, 909 (1961).
15. F. Delbove, Compt. Rend. 252, 2192 (1961).
16. G. Petit and A. Cremieu, Compt. Rend. 243, 360 (1956).
17. D. V. Vecher and A. A. Vecher, Zhur. Fiz. Khim. 41, 2103 (1967).
18. F. Kohlrauch, Z. Physik, Chem. 64, 129 (1908).
19. A. Guntz, Ann. Chim. Phys. 3, 5 (1884).

Calcium Fluoride (CaF_2) Ca_1F_2(cr)

Enthalpy Reference Temperature = T_r = 298.15 K   Standard State Pressure = p° = 0.1 MPa

T/K	$C_p°$ (J $K^{-1}mol^{-1}$)	S° (J $K^{-1}mol^{-1}$)	$-[G°-H°(T_r)]/T$ (J $K^{-1}mol^{-1}$)	$H°-H°(T_r)$ (kJ mol^{-1})	$\Delta_f H°$ (kJ mol^{-1})	$\Delta_f G°$ (kJ mol^{-1})	Log K_f
0	0.	0.	INFINITE	-11.655	-1223.006	-1223.006	INFINITE
100	28.619	13.347	120.374	-10.703	-1225.941	-1208.642	631.330
200	56.769	43.555	74.450	-6.179	-1226.615	-1190.898	311.031
298.15	68.588	68.572	68.572	0.	-1225.912	-1173.496	205.592
300	68.772	68.997	68.574	0.127	-1225.891	-1173.171	204.267
400	73.864	89.630	71.353	7.311	-1224.561	-1155.791	150.931
500	76.220	106.375	76.736	14.819	-1223.184	-1138.759	118.965
600	78.517	120.471	82.880	22.554	-1221.865	-1121.999	97.679
700	81.090	132.764	89.146	30.532	-1220.577	-1105.456	82.490
800	83.918	143.774	95.298	38.781	-1220.132	-1088.995	71.104
900	86.952	153.832	101.251	47.323	-1218.769	-1072.684	62.257
1000	90.128	163.157	106.981	56.177	-1217.427	-1056.525	55.187
1100	93.412	171.901	112.489	65.353	-1216.093	-1040.499	49.409
1200	96.780	180.172	117.788	74.862	-1222.490	-1023.973	44.572
1300	100.207	188.054	122.892	84.711	-1219.910	-1007.532	40.483
1400	103.680	195.607	127.818	94.905	-1217.007	-991.302	36.986
1424.000	104.520	197.376	128.975	97.403	ALPHA <--> BETA TRANSITION		
1424.000	122.884	200.726	128.975	102.173			
1500	123.679	207.136	132.774	111.542	-1207.680	-975.581	33.973
1600	124.725	215.151	137.675	123.963	-1202.588	-960.273	31.350
1691.000	125.677	222.077	142.032	135.356	BETA <--> LIQUID		
1700	125.771	222.744	142.457	136.487	-1197.409	-945.287	29.045
1800	126.817	229.962	147.120	149.117	-1340.813	-928.397	26.941
1900	127.863	236.847	151.663	161.851	-1334.037	-905.669	24.899
2000	128.909	243.432	156.088	174.689	-1327.172	-883.301	23.069
2100	129.955	249.747	160.398	187.633	-1320.215	-861.278	21.423
2200	131.001	255.817	164.598	200.680	-1313.168	-839.589	19.934
2300	132.047	261.663	168.692	213.833	-1306.029	-818.222	18.582
2400	133.093	267.305	172.684	227.090	-1298.800	-797.167	17.350
2500	134.139	272.759	176.579	240.451	-1291.482	-776.416	16.222
2600	135.185	278.041	180.380	253.918	-1284.078	-755.959	15.187
2700	136.231	283.162	184.092	267.488	-1276.591	-735.789	14.235

PREVIOUS: CURRENT: December 1968

Calcium Fluoride (CaF_2) Ca_1F_2(cr)

CALCIUM FLUORIDE (CaF_2) LIQUID M_r = 78.076806

S°(298.15 K) = [92.573] J K^{-1} mol^{-1} $\Delta_f H°$(298.15 K) = [-1186.067] kJ mol^{-1}

T_{fus} = 1691 ± 5 K $\Delta_{fus}H°$ = 29.706 ± 0.4 kJ mol^{-1}

Enthalpy of Formation

$\Delta_f H°$(CaF_2, ℓ, 298.15 K) is calculated from that of the crystal by adding $\Delta_{fus}H°$ and the difference in enthalpy, H°(1691 K) - H°(298.15 K), between the crystal and liquid.

Heat Capacity and Entropy

The heat capacity is derived from the enthalpy measurements of Naylor (1). The heat capacity is constant in the real liquid range but at 1000 K a glass transition is assumed below which it follows that of the crystal. The entropy is calculated in a manner analogous to that used for the enthalpy of formation

Fusion Data

See the crystal table for details.

Vaporization data

The boiling point, T_{vap}, is calculated as the temperature at which the fugacity of CaF_2(g) is one bar for the process CaF_2(ℓ) = CaF_2(g). The enthalpy of vaporization is the difference in $\Delta_f H°$ at T_{vap} between the liquid and gas.

Reference

1. B. F. Naylor, J. Amer. Chem. Soc. 67, 150 (1945).

Calcium Fluoride (CaF_2) Ca_1F_2(l)

Enthalpy Reference Temperature = T_r = 298.15 K Standard State Pressure = p° = 0.1 MPa

T/K	$C_p°$ (J $K^{-1}mol^{-1}$)	S° (J $K^{-1}mol^{-1}$)	-[G°-H°(T_r)]/T (J $K^{-1}mol^{-1}$)	H°-H°(T_r) (kJ mol^{-1})	$\Delta_f H°$ (kJ mol^{-1})	$\Delta_f G°$ (kJ mol^{-1})	Log K_f
0							
100							
200							
298.15	68.588	92.573	92.573	0.	-1186.067	-1140.806	199.864
300	68.772	92.998	92.574	0.127	-1186.045	-1140.525	198.583
400	73.864	113.631	95.353	7.311	-1184.716	-1125.546	146.981
500	76.220	130.375	100.736	14.819	-1183.338	-1110.914	116.056
600	78.517	144.471	106.881	22.554	-1182.019	-1096.553	95.463
700	81.090	156.764	113.147	30.532	-1180.732	-1082.411	80.771
800	83.918	167.775	119.298	38.781	-1180.286	-1068.349	69.756
900	86.952	177.833	125.251	47.323	-1178.924	-1054.439	61.198
1000	90.128	187.157	130.981	56.176	-1177.581	-1040.680	54.360
1000.000	90.128	187.157	130.981	56.176	GLASS <--> LIQUID		
1000.000	99.914	187.157	130.981	56.176	TRANSITION		
1100	99.914	196.680	136.528	66.168	-1175.433	-1027.096	48.773
1200	99.914	205.374	141.908	76.159	-1181.347	-1013.072	44.098
1300	99.914	213.371	147.101	86.150	-1178.625	-999.159	40.147
1400	99.914	220.775	152.103	96.142	-1175.924	-985.456	36.768
1500	99.914	227.669	156.913	106.133	-1173.243	-971.944	33.846
1600	99.914	234.117	161.539	116.125	-1170.581	-958.611	31.295
1691.000	99.914	239.644	165.595	125.217	------- BETA <--> LIQUID -------		
1700	99.914	240.174	165.988	126.116	-1167.935	-945.444	29.050
1800	99.914	245.885	170.270	136.107	-1313.977	-930.222	26.994
1900	99.914	251.287	174.393	146.099	-1309.944	-909.012	24.990
2000	99.914	256.412	178.367	156.090	-1305.925	-888.014	23.193
2100	99.914	261.287	182.201	166.082	-1301.921	-867.217	21.571
2200	99.914	265.935	185.902	176.073	-1297.929	-846.611	20.101
2300	99.914	270.376	189.479	186.064	-1293.952	-826.185	18.763
2400	99.914	274.629	192.939	196.056	-1289.988	-805.933	17.541
2500	99.914	278.707	196.289	206.047	-1286.041	-785.845	16.419
2600	99.914	282.626	199.534	216.038	-1282.112	-765.915	15.387
2700	99.914	286.397	202.682	226.030	-1278.204	-746.135	14.435
2800	99.914	290.031	205.737	236.021	-1274.321	-726.500	13.553
2900	99.914	293.537	208.705	246.013	-1270.468	-707.003	12.735
3000	99.914	296.924	211.589	256.004	-1266.651	-687.639	11.973
3100	99.914	300.200	214.395	265.995	-1262.874	-668.401	11.262
3200	99.914	303.372	217.126	275.987	-1259.145	-649.284	10.598
3300	99.914	306.447	219.787	285.978	-1255.470	-630.282	9.977
3400	99.914	309.429	222.380	295.970	-1251.856	-611.392	9.393
3500	99.914	312.326	224.908	305.961	-1248.310	-592.606	8.844
3600	99.914	315.140	227.376	315.952	-1244.837	-573.921	8.327
3700	99.914	317.878	229.785	325.944	-1241.446	-555.331	7.840
3800	99.914	320.542	232.138	335.935	-1238.141	-536.833	7.379
3900	99.914	323.138	234.439	345.927	-1234.928	-518.419	6.943
4000	99.914	325.667	236.688	355.918	-1231.812	-500.087	6.530

PREVIOUS. CURRENT: December 1968

Calcium Fluoride (CaF_2) Ca_1F_2(l)

CALCIUM FLUORIDE (CaF_2) ALPHA-BETA-LIQUID M_r = 78.076806

0 to 1424 K		crystal, alpha
1424 to 1691 K		crystal, beta
above 1691 K		liquid

Refer to the individual tables for details.

Calcium Fluoride (CaF_2) Ca_1F_2(cr,l)

Enthalpy Reference Temperature = T_r = 298.15 K Standard State Pressure = p° = 0.1 MPa

T/K	C_p° (J K⁻¹mol⁻¹)	S° (J K⁻¹mol⁻¹)	$-[G^\circ-H^\circ(T_r)]/T$ (J K⁻¹mol⁻¹)	$H^\circ-H^\circ(T_r)$ (kJ mol⁻¹)	Δ_fH° (kJ mol⁻¹)	Δ_fG° (kJ mol⁻¹)	Log K_f
0	0.	0.	INFINITE	-11.655	-1223.006	-1223.006	INFINITE
100	28.619	13.347	120.374	-10.703	-1225.941	-1208.642	631.330
200	56.769	43.555	74.450	-6.179	-1226.615	-1190.898	311.031
298.15	68.588	68.572	68.572	0.	-1225.912	-1173.496	205.592
300	68.772	68.997	68.574	0.127	-1225.891	-1173.171	204.267
400	73.864	89.630	71.353	7.311	-1224.561	-1155.791	150.931
500	76.220	106.375	76.736	14.819	-1223.184	-1138.759	118.965
600	78.517	120.471	82.880	22.554	-1221.865	-1121.999	97.679
700	81.090	132.764	89.146	30.532	-1220.577	-1105.456	82.490
800	83.918	143.774	95.298	38.781	-1220.132	-1088.995	71.104
900	86.952	153.832	101.251	47.323	-1218.769	-1072.684	62.257
1000	90.128	163.157	106.981	56.177	-1217.427	-1056.525	55.187
1100	93.412	171.901	112.489	65.353	-1216.093	-1040.499	49.409
1200	96.780	180.172	117.788	74.862	-1222.490	-1023.973	44.572
1300	100.207	188.054	122.892	84.711	-1219.910	-1007.532	40.483
1400	103.680	195.607	127.818	94.905	-1217.007	-991.302	36.986
1424.000	104.520	197.376	128.975	97.403		ALPHA <--> BETA	
1424.000	122.884	200.726	128.975	102.173		TRANSITION	
1500	123.679	207.136	132.774	111.542	-1207.680	-975.581	33.973
1600	124.725	215.151	137.675	123.963	-1202.588	-960.273	31.350
1691.000	125.677	222.077	142.032	135.356		BETA <--> LIQUID	
1691.000	99.914	239.644	142.032	165.062		TRANSITION	
1700	99.914	240.174	142.550	165.961	-1167.935	-945.444	29.050
1800	99.914	245.885	148.134	175.953	-1313.977	-930.222	26.994
1900	99.914	251.287	153.422	185.944	-1309.944	-909.012	24.990
2000	99.914	256.412	158.444	195.936	-1305.925	-888.014	23.193
2100	99.914	261.287	163.227	205.927	-1301.921	-867.217	21.571
2200	99.914	265.935	167.790	215.918	-1297.929	-846.611	20.101
2300	99.914	270.376	172.155	225.910	-1293.952	-826.185	18.763
2400	99.914	274.629	176.337	235.901	-1289.988	-805.933	17.541
2500	99.914	278.707	180.350	245.893	-1286.041	-785.845	16.419
2600	99.914	282.626	184.209	255.884	-1282.112	-765.915	15.387
2700	99.914	286.397	187.925	265.875	-1278.204	-746.135	14.435
2800	99.914	290.031	191.507	275.867	-1274.321	-726.500	13.553
2900	99.914	293.537	194.965	285.858	-1270.468	-707.003	12.735
3000	99.914	296.924	198.307	295.850	-1266.651	-687.639	11.973
3100	99.914	300.200	201.542	305.841	-1262.874	-668.401	11.262
3200	99.914	303.372	204.675	315.832	-1259.145	-649.284	10.598
3300	99.914	306.447	207.712	325.824	-1255.470	-630.282	9.977
3400	99.914	309.429	210.660	335.815	-1251.856	-611.392	9.393
3500	99.914	312.326	213.524	345.807	-1248.310	-592.606	8.844
3600	99.914	315.140	216.308	355.798	-1244.837	-573.921	8.327
3700	99.914	317.878	219.016	365.789	-1241.446	-555.331	7.840
3800	99.914	320.542	221.653	375.781	-1238.141	-536.833	7.379
3900	99.914	323.138	224.222	385.772	-1234.928	-518.419	6.943
4000	99.914	325.667	226.726	395.763	-1231.812	-500.087	6.530

PREVIOUS: CURRENT: December 1968

Calcium Fluoride (CaF_2) Ca_1F_2(cr,l)

$S°(298.15\ K) = 273.80 \pm 2.1\ J\ K^{-1}\ mol^{-1}$

$\Delta_f H°(0\ K) = -782.6 \pm 8\ kJ\ mol^{-1}$
$\Delta_f H°(298.15\ K) = -784.5 \pm 8\ kJ\ mol^{-1}$

Vibrational Frequencies and Degeneracies

ν, cm^{-1}
484 (1)
163 (1)
554 (1)

Ground State Quantum Weight: 1 $\sigma = 2$
Point Group: C_{2v}
Bond Distance: Ca-F = 2.1 Å
Bond Angle: F-Ca-F = 135 ± 7°
Product of the Moments of Inertia: $I_A I_B I_C = 1.283897 \times 10^{-114}\ g^3\ cm^6$

Enthalpy of Formation

The enthalpy of formation is obtained from that of the crystal by adding $\Delta_{sub}H°(298.15\ K)$. The vapor pressure data of five authors are analyzed by 2nd and 3rd law methods to give $\Delta_{sub}H°(298.15\ K)$ as listed below. Note that the drifts are small and both positive and negative confirming the correctness of the functions used.

Reference	T/K	Data Points	$\Delta_{sub}H°(298.15\ K)$, kcal mol^{-1} 2nd Law	3rd Law	Drift cal K^{-1} mol^{-1}
Freeman (1)	1463-1668	9	106.2±3.7	105.66±0.9	-0.2±2.3
Blue et al. (2)	1246-1498	6	99.6±2.5	106.00±1.0	4.1±2.0
Schulz and Searcy (3)	1421-1689	34	102.4±0.3	104.00±0.2	1.0±0.2
Pottie (4)	1823	1		106.4	
Ruff and LeBoucher (5)	2086-2208	7	106.7±4.8	105.22±0.5	-0.7±2.2
Blue et al. (2)*	1242-1669	19	104.7±0.5		

*This set was not analyzed by the 3rd law since the absolute pressures were adjusted by the authors to match their weight loss values.

A median value of 105.5±0.5 kcal mol^{-1} (441.412±2.1 kJ mol^{-1}) is adopted for the enthalpy of sublimation which yields $\Delta_f H°(CaF_2, g, 298.15\ K) = -187.5 \pm 2$ kcal mol^{-1} (-784.5±8 kJ mol^{-1}).

Heat Capacity and Entropy

The bond length was measured by Aksihin and Spiridonov (6), but the diffraction patterns were interpreted as indicating a linear molecule. The vibrational frequencies and bond angle are those reported by Calder (7) using matrix isolation spectroscopy and isotopically enriched materials.

The principal moments of inertia are: $I_A = 2.0918 \times 10^{-39}$, $I_B = 23.7505 \times 10^{-39}$, and $I_C = 25.8423 \times 10^{-39}\ g\ cm^2$.

References

1. R. D. Freeman, Tech. Doc. Rept. ASD TDR 63-754, Part II, Oklahoma State University, 1965.
2. G. D. Blue, J. W. Green, R. G. Bautista and J. L. Margrave, J. Phys. Chem. 67, 877 (1963).
3. D. A. Schulz and A. W. Searcy, J. Phys. Chem. 67, 103 (1963).
4. R. W. Pottie, quoted in Reference 1.
5. O. Ruff and L. LeBoucher, Z. Anorg. Chem. 219, 376 (1934).
6. P. A. Akishin and V. P. Spiridonov, Kristallografiya, 2, 475 (1957).
7. G. V. Calder, Nat. Bur. Stand. Rept. 9389, July 1966.

Enthalpy Reference Temperature = T_r = 298.15 K Standard State Pressure = p° = 0.1 MPa

T/K	$C_p°$ (J $K^{-1}mol^{-1}$)	S°	$-[G°-H°(T_r)]/T$	$H°-H°(T_r)$ (kJ mol^{-1})	$\Delta_f H°$	$\Delta_f G°$	Log K_f
0	0.	0.	INFINITE	-12.658	-782.596	-782.596	INFINITE
100	39.182	224.769	315.945	-9.118	-782.944	-786.787	410.975
200	46.529	254.242	278.356	-4.823	-783.847	-790.267	206.397
250	49.301	264.937	274.632	-2.424	-784.200	-791.830	165.444
298.15	51.267	273.798	273.798	0.	-784.500	-793.272	138.978
300	51.331	274.115	273.799	0.095	-784.511	-793.326	138.130
350	52.798	282.144	274.430	2.700	-784.801	-794.772	118.613
400	53.870	289.268	275.848	5.368	-785.092	-796.177	103.970
450	54.666	295.661	277.700	8.082	-785.407	-797.544	92.577
500	55.269	301.454	279.791	10.832	-785.760	-798.874	83.458
600	56.100	311.610	284.271	16.403	-786.604	-801.421	69.770
700	56.627	320.300	288.812	22.042	-787.656	-803.811	59.981
800	56.981	327.886	293.233	27.723	-789.778	-805.930	52.622
900	57.228	334.613	297.464	33.434	-791.246	-807.864	46.887
1000	57.408	340.652	301.486	39.167	-793.024	-809.618	42.290
1100	57.543	346.131	305.299	44.914	-795.119	-811.179	38.520
1200	57.646	351.142	308.914	50.674	-805.266	-811.912	35.342
1300	57.727	355.759	312.342	56.443	-806.766	-812.405	32.643
1400	57.791	360.040	315.598	62.219	-808.281	-812.782	30.325
1500	57.843	364.029	318.695	68.001	-809.809	-813.051	28.313
1600	57.886	367.763	321.646	73.787	-811.352	-813.216	26.549
1700	57.922	371.274	324.463	79.578	-812.907	-813.285	24.989
1800	57.952	374.585	327.157	85.371	-963.146	-811.051	23.536
1900	57.977	377.719	329.736	91.168	-963.308	-802.598	22.065
2000	57.999	380.694	332.211	96.967	-963.482	-794.134	20.741
2100	58.017	383.524	334.587	102.768	-963.668	-785.663	19.542
2200	58.034	386.223	336.873	108.570	-963.866	-777.182	18.453
2300	58.048	388.803	339.076	114.374	-964.075	-768.691	17.458
2400	58.060	391.274	341.199	120.180	-964.298	-760.192	16.545
2500	58.071	393.645	343.250	125.986	-964.536	-751.682	15.706
2600	58.081	395.922	345.232	131.794	-964.790	-743.163	14.930
2700	58.090	398.115	347.151	137.602	-965.065	-734.634	14.212
2800	58.097	400.227	349.009	143.412	-965.364	-726.094	13.545
2900	58.104	402.266	350.810	149.222	-965.693	-717.543	12.924
3000	58.111	404.236	352.559	155.032	-966.056	-708.980	12.344
3100	58.116	406.142	354.256	160.844	-966.459	-700.404	11.802
3200	58.122	407.987	355.907	166.656	-966.910	-691.815	11.293
3300	58.126	409.775	357.512	172.468	-967.414	-683.210	10.814
3400	58.131	411.511	359.075	178.281	-967.978	-674.590	10.364
3500	58.135	413.196	360.597	184.094	-968.610	-665.952	9.939
3600	58.138	414.833	362.081	189.908	-969.315	-657.295	9.537
3700	58.142	416.426	363.529	195.722	-970.101	-648.617	9.157
3800	58.145	417.977	364.941	201.536	-970.973	-639.917	8.796
3900	58.147	419.487	366.321	207.351	-971.937	-631.192	8.454
4000	58.150	420.960	367.668	213.166	-972.998	-622.442	8.128
4100	58.153	422.396	368.986	218.981	-974.159	-613.664	7.818
4200	58.155	423.797	370.274	224.796	-975.425	-604.856	7.522
4300	58.157	425.165	371.535	230.612	-976.797	-596.017	7.240
4400	58.159	426.502	372.769	236.428	-978.276	-587.145	6.970
4500	58.161	427.809	373.977	242.243	-979.849	-578.239	6.712
4600	58.162	429.088	375.162	248.060	-981.538	-569.297	6.465
4700	58.164	430.339	376.322	253.876	-983.338	-560.315	6.227
4800	58.166	431.563	377.461	259.692	-985.244	-551.295	5.999
4900	58.167	432.762	378.577	265.509	-987.253	-542.234	5.780
5000	58.168	433.938	379.672	271.326	-989.361	-533.131	5.570
5100	58.170	435.090	380.748	277.143	-991.567	-523.984	5.367
5200	58.171	436.219	381.804	282.960	-993.860	-514.794	5.171
5300	58.172	437.327	382.841	288.777	-996.244	-505.558	4.983
5400	58.173	438.414	383.860	294.594	-998.711	-496.277	4.801
5500	58.174	439.482	384.862	300.412	-1001.254	-486.950	4.625
5600	58.175	440.530	385.846	306.229	-1003.860	-477.576	4.455
5700	58.176	441.560	386.815	312.046	-1006.538	-468.155	4.290
5800	58.177	442.572	387.767	317.864	-1009.276	-458.686	4.131
5900	58.177	443.566	388.705	323.682	-1012.068	-449.169	3.977
6000	58.178	444.544	389.627	329.500	-1014.907	-439.604	3.827

PREVIOUS: December 1968 (1 atm) CURRENT: December 1968 (1 bar)

CALCIUM HYDROXIDE (CaOH) IDEAL GAS M_r = 57.08734

$\Delta_f H°(0\ K)$ = -190.0 ± 21 kJ mol^{-1}

S°(298.15 K) = [235.47 ± 8] J K^{-1} mol^{-1} $\Delta_f H°(298.15\ K)$ = -193.9 ± 21 kJ mol^{-1}

Electronic Levels and Quantum Weights		Vibrational Frequencies and Degeneracies
ε_i, cm^{-1}	g_i	ν, cm^{-1}
0	[2]	[587](1)
[16050]	[2]	[466](2)
[16610]	[2]	[3650](1)
[18050]	[2]	

Point Group: [$C_{\infty v}$] σ = 1
Bond Distance: Ca–O = [2.03] Å O–H = [0.96] Å
Bond Angle: Ca–O–H = [180]°
Rotational Constant: B_0 = [0.318811] cm^{-1}

Enthalpy of Formation

The adopted $\Delta_f H°$(CaOH, g, 0 K) = -45.41±5 kcal mol^{-1}(-190.0±21 kJ mol^{-1}) is based on an assessment of D_0^0 values derived from flame spectra of CaOH, SrOH, and BaOH. Cotton and Jenkins (1) found both the monohydroxides and dihydroxides of the alkaline earths to be present in significant amounts in fuel-rich hydrogen-oxygen-nitrogen flames. They determined equilibrium constants for the reactions M(g) + H_2O(g) = MOH(g) + H(g) and M(g) + 2 H_2O(g) = $M(OH)_2$(g) + 2 H(g) and derived D_0^0 values. In earlier work, Ryabova and Gurvich (2) had considered CaOH to be the dominant compound, and Sugden and Schofield (3) had interpreted $Ca(OH)_2$ as dominant. Cotton and Jenkins (1) have recalculated the work of these last two investigations, considering both CaOH and $Ca(OH)_2$ to be present. Ryabova et al. (4) and Kalff and Alkemade (5) have made additional measurements.

Reference	As Published	As Recalculated by Cotton and Jenkins (1)	As Corrected for Current JANAF Auxiliary Data (6)
	D_0^0, kcal mol^{-1}	CaOH(g) = Ca(g) + OH(g)	
Ryabova and Gurvich (2)	100±8	99	
Sugden and Schofield (3)		102	
Cotton and Jenkins (1)	104±5		105.5
Ryabova et al. (4)	94±3		97.5[a]
Kalff and Alkemade (5)	102.4		

[a]An approximate correction of +3.5 kcal mol^{-1} is made, 2 kcal mol^{-1} assumed from the Cotton and Jekins' type calculation for the presence of both CaOH and $Ca(OH)_2$ and 1.5 kcal mol^{-1} indicated by the recalculation of Cotton and Jenkins' work (1) using auxiliary data from the current JANAF Tables (6).

The data analyses for BaOH(g) and $Ba(OH)_2$(g) indicate that flame-spectral data tend to give high dissociation energies. For $Ca(OH)_2$(g), the lowest value of D_0^0(HO-Ca-OH) was adopted (6). Similarly, D_0^0(Ca-OH) = 97.5 kcal mol^{-1} is adopted.

The ratio of the dissociation energies of the alkaline earth monohalides to those of the corresponding dihalides range from 0.40 to 0.51 with the ratio for the calcium fluorides being 0.47 (6). The similarity between the halides and hydroxides has been established (7-10). The ratio of the adopted values for the dissociation energies of CaOH(g) and $Ca(OH)_2$(g) is 0.47 where D_0^0 of the dihydroxide is defined by the reaction $Ca(OH)_2$ = Ca(g) + 2(OH)(g) and is 205.6 kcal mol^{-1}.

Heat Capacity and Entropy

The analogy between gaseous monohydroxides and monohalides, particularly the monofluorides, has been recognized (7-10). The molecular configuration is assumed to be linear in accordance with the prediction of Walsh (11) and the evidence that gaseous alkali metal hydroxides are linear (12-14). The ground state is assumed to be $^2\Sigma^+$ by analogy with CaF and CaCl (6). The electronic levels are estimated from the band spectra observed by James and Sugden (15), Gaydon (16), Zhitkevich et al. (17) and van der Hurk et al. (18), and the comparison with CaF and CaCl (6).

The Ca-O bond distance is estimated to be slightly larger, 0.02 Å, than the Ca-F distance (6) after noting the close similarity in bond distance of the alkali metal fluorides and hydroxides. The O-H bond distance is that in water (6).

The Ca-O stretching frequency, 587 cm^{-1}, is estimated to be the same as the CaF stretching frequency (6, 10). The O-H stretching frequency, 3650 cm^{-1}, is estimated from the alkali hydroxide series. The bending frequency, 466 cm^{-1}, is estimated by assuming that the ratio of the bending force constant to the stretching force constant is 0.022, which is the average ratio found in the alkali metal monohydroxides by Acquista and Abramowitz (10, 14).

The entropy in the present table is lower by 0.20 cal K^{-1} mol^{-1} at 298.15 K and 0.25 cal K^{-1} mol^{-1} at 1000 K than that proposed by Jackson (10); the data relevant to the calculations are nearly the same.

References

1. D. H. Cotton and D. R. Jenkins, Trans. Faraday Soc. 64, 2988 (1968).
2. V. B. Ryabova and L. V. Gurvich, Teplofiz. Vysokikh Temperatur, Akad. Nauk SSSR 3, 318 (1965).
3. T. M. Sugden and K. Schofield, Trans. Faraday Soc. 62, 566 (1966).
4. V. G. Ryabova, A. N. Khitrov, and L. V. Gurvich, Teplofiz. Vys. Temp. 10, 744 (1972).
5. P. J. Kalff and C. Th. J. Alkemade, J. Chem. Phys. 59, 2572 (1973).
6. JANAF Thermochemical Tables: Ca(g), 12-31-68; H_2O(g), 3-31-61; H(g), 6-30-74; CaCl(g), 6-30-70; CaF(g), 12-31-68; BaOH(g), $Ba(OH)_2$(g), $Ca(OH)_2$(g), 12-31-75.
7. R. C. Schoonmaker and R. F. Porter, J. Chem. Phys. 31, 830 (1959).
8. J. Berkowitz, D. J. Meschi, and W. A. Chupka, J. Chem. Phys. 33, 533 (1960).
9. F. E. Stafford and J. Berkowitz, J. Chem. Phsy. 40, 2963 (1964).
10. D. D. Jackson, Lawrence Livermore Laboratory, Univ. of Calif., Report UCRL-51137, Contract No. W-7405-Eng-48, Dec. 8, 1971.
11. A. D. Walsh, J. Chem. Soc. 1953, 2288 (1953).
12. N. Acquista, S. Abramowitz, and R. Lide, J. Chem. Phys. 49, 780 (1968).
13. R. L. Kuczkowski and R. Lide, J. Chem. Phys. 44, 3131 (1966).
14. N. Acquista and S. Abramowitz, U. S. Nat. Bur. Stand. Report 9905, July, 1968.
15. C. G. James and T. M. Sugden, Nature 175, 333 (1955).
16. A. G. Gaydon, Proc. Roy. Soc. (London) A231, 437 (1955).
17. V. F. Zhitkevich, A. I. Lyutyi, N. A. Nesterko, V. S. Rossikhin, and I. L. Tsikora, Optika i Spektroskopiya 14, 35 (1963).
18. J. Van der Hurk, Tj. Hollander, and C. Th. J. Alkemade, J. Quant. Spectrosc. Radiat. Transfer 13, 273 (1973).

Calcium Hydroxide (CaOH) $Ca_1H_1O_1$(g)

Enthalpy Reference Temperature = T_r = 298.15 K Standard State Pressure = p° = 0.1 MPa

	J K^{-1}mol^{-1}			kJ mol^{-1}			
T/K	$C_p°$	S°	-[G°-H°(T_r)]/T	H°-H°(T_r)	$\Delta_f H°$	$\Delta_f G°$	Log K_f
0	0.	0.	INFINITE	-10.429	-189.995	-189.995	INFINITE
100	30.146	195.611	270.666	-7.506	-191.004	-195.266	101.997
200	38.363	218.867	239.376	-4.102	-192.671	-198.857	51.936
250	42.001	227.836	236.191	-2.089	-193.335	-200.324	41.855
298.15	44.625	235.470	235.470	0.	-193.877	-201.620	35.323
300	44.711	235.746	235.471	0.083	-193.897	-201.668	35.113
350	46.687	242.795	236.023	2.370	-194.378	-202.924	30.285
400	48.139	249.129	237.272	4.743	-194.810	-204.115	26.655
450	49.226	254.864	238.913	7.178	-195.226	-205.253	23.825
500	50.062	260.096	240.774	9.661	-195.648	-206.345	21.557
600	51.270	269.337	244.785	14.732	-196.564	-208.400	18.143
700	52.148	277.309	248.875	19.904	-197.626	-210.291	15.692
800	52.879	284.321	252.876	25.156	-199.713	-211.914	13.837
900	53.544	290.589	256.724	30.478	-201.107	-213.359	12.383
1000	54.171	296.263	260.399	35.864	-202.774	-214.633	11.211
1100	54.769	301.454	263.898	41.311	-204.724	-215.727	10.244
1200	55.336	306.244	267.230	46.817	-214.692	-216.008	9.403
1300	55.868	310.695	270.404	52.377	-215.983	-216.065	8.682
1400	56.364	314.853	273.432	57.989	-217.260	-216.024	8.060
1500	56.823	318.758	276.325	63.649	-218.525	-215.891	7.518
1600	57.244	322.439	279.093	69.353	-219.781	-215.675	7.041
1700	57.631	325.921	281.747	75.097	-221.029	-215.380	6.618
1800	57.985	329.225	284.293	80.878	-370.947	-212.801	6.175
1900	58.311	332.369	286.741	86.693	-370.774	-204.020	5.609
2000	58.610	335.368	289.098	92.539	-370.606	-195.247	5.099
2100	58.888	338.234	291.370	98.414	-370.446	-186.483	4.639
2200	59.147	340.980	293.563	104.316	-370.299	-177.727	4.220
2300	59.390	343.614	295.683	110.243	-370.168	-168.976	3.838
2400	59.622	346.147	297.733	116.194	-370.059	-160.232	3.487
2500	59.845	348.585	299.719	122.167	-369.978	-151.490	3.165
2600	60.062	350.937	301.644	128.162	-369.929	-142.752	2.868
2700	60.276	353.208	303.512	134.179	-369.920	-134.015	2.593
2800	60.488	355.404	305.326	140.218	-369.956	-125.277	2.337
2900	60.701	357.530	307.090	146.277	-370.045	-116.537	2.099
3000	60.915	359.591	308.805	152.358	-370.194	-107.793	1.877
3100	61.132	361.592	310.476	158.460	-370.410	-99.043	1.669
3200	61.354	363.537	312.104	164.584	-370.700	-90.285	1.474
3300	61.580	365.428	313.691	170.731	-371.071	-81.516	1.290
3400	61.812	367.270	315.240	176.901	-371.531	-72.735	1.117
3500	62.049	369.065	316.753	183.094	-372.085	-63.939	0.954
3600	62.291	370.816	318.230	189.310	-372.741	-55.126	0.800
3700	62.538	372.526	319.675	195.552	-373.504	-46.294	0.654
3800	62.790	374.197	321.087	201.818	-374.380	-37.438	0.515
3900	63.047	375.832	322.470	208.110	-375.373	-28.558	0.382
4000	63.307	377.431	323.824	214.428	-376.487	-19.652	0.257
4100	63.571	378.998	325.151	220.772	-377.726	-10.716	0.137
4200	63.837	380.533	326.451	227.142	-379.092	-1.748	0.022
4300	64.105	382.038	327.727	233.539	-380.588	7.254	-0.088
4400	64.375	383.515	328.978	239.963	-382.210	16.291	-0.193
4500	64.644	384.965	330.206	246.414	-383.947	25.366	-0.294
4600	64.913	386.388	331.412	252.892	-385.820	34.482	-0.392
4700	65.181	387.787	332.597	259.397	-387.822	43.641	-0.485
4800	65.446	389.162	333.761	265.928	-389.948	52.843	-0.575
4900	65.709	390.515	334.905	272.486	-392.195	62.091	-0.662
5000	65.969	391.845	336.031	279.070	-394.558	71.386	-0.746
5100	66.224	393.154	337.138	285.679	-397.035	80.729	-0.827
5200	66.475	394.442	338.228	292.314	-399.616	90.121	-0.905
5300	66.720	395.710	339.300	298.974	-402.303	99.565	-0.981
5400	66.960	396.960	340.356	305.658	-405.089	109.060	-1.055
5500	67.193	398.191	341.397	312.366	-407.966	118.608	-1.126
5600	67.419	399.403	342.422	319.097	-410.921	128.208	-1.196
5700	67.639	400.599	343.432	325.850	-413.965	137.862	-1.263
5800	67.851	401.777	344.428	332.624	-417.083	147.571	-1.329
5900	68.055	402.938	345.410	339.419	-420.270	157.332	-1.393
6000	68.251	404.084	346.378	346.235	-423.521	167.150	-1.455

PREVIOUS: December 1975 (1 atm) CURRENT: December 1975 (1 bar)

Calcium Hydroxide (CaOH) $Ca_1H_1O_1$(g)

CALCIUM HYDROXIDE, ION ($CaOH^+$) IDEAL GAS M_r = 57.086791

$S°(298.15\ K) = [229.90 \pm 8.4]\ J\ K^{-1}\ mol^{-1}$

$\Delta_f H°(0\ K) = 366.733 \pm 63\ kJ\ mol^{-1}$

$\Delta_f H°(298.15\ K) = [369.084 \pm 63]\ kJ\ mol^{-1}$

Vibrational Frequencies and Degeneracies

ν, cm^{-1}
[580](1)
[460](2)
[3650](1)

Ground State Quantum Weight = [1] σ = 1

Point Group: [$C_{\infty v}$]

Bond Distances: Ca-O = [2.03] Å O-H = [0.96] Å

Bond Angle: Ca-O-H = [180°]

Rotational Constant: B_0 = [0.318830] cm^{-1}

Enthalpy of Formation

The ionization potential of CaOH(g) was deduced by Kelly and Padley (1) to be 5.9±0.1 eV. These authors quantitatively examined the total positive ion concentrations produced from Ca aqueous salt additives in fuel rich, premixed $H_2 + O_2 + N_2$ flames. Using current JANAF auxiliary data (2), we recalculate the ionization potential to be 5.79 eV.

Jensen (3) determined the enthalpy of reaction $\Delta_r H°(0\ K) = 35 \pm 10\ kcal\ mol^{-1}$ for $Ca(g) + OH(g) = CaOH^+(g) + e^-$ in atmospheric pressure $H_2 + O_2 + N_2$ flames using the microwave cavity resonance method. This value was calculated assuming a bent molecule for $CaOH^+$; the value is not significantly changed, within the uncertainty, by the change in the configuration. Using auxiliary data (2), we derive an ionization potential of 5.75 eV, which is in good agreement with the value derived from the data of Kelly and Padley (1).

We adopt an ionization potential of 5.77 eV (133.06 kcal mol^{-1}) which is average of the above two studies (1, 3). This leads to $\Delta_f H°(0\ K) = 87.651\ kcal\ mol^{-1}$ and $\Delta_f H°(298.15\ K) = 88.213\ kcal\ mol^{-1}$ (369.08 kJ mol^{-1}) for $CaOH^+(g)$. We assign an uncertainty of ±15 kcal mol^{-1}.

For comparison, the appearance potential of CaF(g) has been reported as 5.8±0.3 eV (4), 5.5±0.3 eV (4) and 6.0±0.5 eV (5). These values are all very similar to the ionization potential adopted here for CaOH(g). In addition, the ionization potential for Ca(g) is 6.11 eV (2).

Heat Capacity and Entropy

The molecular configuration is assumed to be linear, since experimental evidence indicates that the gaseous alkali metal hydroxides are linear (2, 6, 7, 8). In addition, Walsh (9) had predicted that triatomic molecules (containing hydrogen) with ten or less valence electrons ($CaOH^+$ has eight valence electrons) will be linear in their ground state. The molecule $CaOH^+$ is isoelectronic with KOH.

The bond dissociation energy for $CaOH^+$ (D_0^0 =105.4 kcal mol^{-1}, 2) for the process $CaOH^+(g) = Ca^+(g) + OH(g)$ is fairly close to that for CaOH (D_0^0 = 97.5 kcal mol^{-1}, 2). This suggests a similar bonding in these two molecules. Thus, bond distances are assumed to be the same as those adopted for CaOH(g, 2). The ground state quantum weight is assumed to be the same as that of KOH(g, 2).

References

1. R. Kelly and P. J. Padley, Trans. Faraday Soc. 67, 1384 (1971).
2. JANAF Thermochemical Tables: e^-, 3-31-65; $H_2(g)$ and $H_2O(g)$, 3-31-61; Ca(g), 12-31-68; OH(g) and $Ca^+(g)$, 12-31-70; H(g), 6-30-74, CaOH(g) and $Ca(OH)_2(g)$, 12-31-75.
3. D. E. Jensen, Combustion and Flame 12, 261 (1968).
4. G. D. Blue, J. W. Green, R. G. Bautista, and J. L. Margrave, J. Phys. Chem. 67, 877 (1963).
5. D. L. Hildenbrand and E. Murad, J. Chem. Phys. 43, 1400 (1965).
6. N. Acquista, S. Abramowitz, and D. R. Lide, J. Chem. Phys. 49, 780 (1968).
7. R. L. Kuczkowski and D. R. Lide, J. Chem. Phys. 44 3131 (1966).
8. N. Acquista and S. Abramowitz, Nat. Bur. Stands. Report 9905, July 1968.
9. A. D. Walsh, J. Chem. Soc. 1953, 2288 (1953).

Calcium Hydroxide, Ion ($CaOH^+$) $Ca_1H_1O_1^+(g)$

Enthalpy Reference Temperature = T_r = 298.15 K Standard State Pressure = p° = 0.1 MPa

T/K	$C_p°$ (J K⁻¹mol⁻¹)	S° (J K⁻¹mol⁻¹)	−[G°−H°(T_r)]/T (J K⁻¹mol⁻¹)	H°−H°(T_r) (kJ mol⁻¹)	$\Delta_f H°$ (kJ mol⁻¹)	$\Delta_f G°$ (kJ mol⁻¹)	Log K_f
0	0.	0.	INFINITE	-10.466	366.733		
100	30.213	189.861	265.270	-7.541			
200	38.571	213.223	233.828	-4.121			
250	42.199	222.238	230.629	-2.098			
298.15	44.799	229.905	229.905	0.	369.084	356.746	-62.501
300	44.884	230.182	229.906	0.083	369.104	356.670	-62.102
350	46.834	237.255	230.460	2.378	369.670	354.554	-52.914
400	48.263	243.607	231.713	4.758	370.284	352.353	-46.013
450	49.331	249.356	233.359	7.199	370.913	350.073	-40.635
500	50.151	254.598	235.225	9.687	371.535	347.724	-36.327
600	51.336	263.854	239.245	14.765	372.705	342.850	-29.848
700	52.198	271.835	243.344	19.943	373.727	337.791	-25.206
800	52.919	278.853	247.353	25.200	373.724	332.702	-21.723
900	53.576	285.124	251.207	30.525	374.412	327.530	-19.009
1000	54.198	290.801	254.887	35.914	374.826	322.296	-16.835
1100	54.791	295.995	258.392	41.364	374.957	317.034	-15.055
1200	55.355	300.787	261.727	46.871	367.070	312.396	-13.598
1300	55.884	305.239	264.905	52.433	367.859	307.807	-12.368
1400	56.378	309.398	267.936	58.047	368.662	303.158	-11.311
1500	56.834	313.304	270.832	63.708	369.477	298.451	-10.393
1600	57.253	316.985	273.603	69.412	370.301	293.689	-9.588
1700	57.637	320.468	276.258	75.157	371.132	288.875	-8.876
1800	57.987	323.772	278.807	80.939	223.294	286.223	-8.306
1900	58.305	326.916	281.257	86.753	225.544	289.658	-7.963
2000	58.595	329.914	283.615	92.599	227.790	292.974	-7.652
2100	58.859	332.780	285.889	98.472	230.026	296.178	-7.367
2200	59.099	335.524	288.083	104.370	232.249	299.277	-7.106
2300	59.318	338.156	290.203	110.291	234.452	302.274	-6.865
2400	59.518	340.684	292.254	116.233	236.631	305.175	-6.642
2500	59.700	343.118	294.240	122.194	238.779	307.987	-6.435
2600	59.866	345.463	296.166	128.172	240.889	310.714	-6.242
2700	60.018	347.725	298.033	134.166	242.954	313.360	-6.062
2800	60.158	349.910	299.847	140.175	244.967	315.931	-5.894
2900	60.286	352.023	301.610	146.198	246.919	318.430	-5.736
3000	60.404	354.069	303.325	152.232	248.803	320.864	-5.587
3100	60.512	356.052	304.994	158.278	250.609	323.236	-5.446
3200	60.613	357.974	306.620	164.334	252.330	325.551	-5.314
3300	60.705	359.841	308.204	170.400	253.956	327.814	-5.189
3400	60.791	361.654	309.750	176.475	255.481	330.029	-5.070
3500	60.871	363.418	311.258	182.558	256.895	332.201	-4.958
3600	60.944	365.134	312.731	188.649	258.192	334.334	-4.851
3700	61.013	366.804	314.170	194.747	259.364	336.432	-4.750
3800	61.077	368.432	315.577	200.852	260.405	338.501	-4.653
3900	61.136	370.020	316.952	206.962	261.310	340.544	-4.561
4000	61.192	371.568	318.299	213.079	262.073	342.566	-4.473
4100	61.244	373.080	319.616	219.201	262.690	344.570	-4.390
4200	61.293	374.556	320.907	225.327	263.159	346.561	-4.310
4300	61.339	375.999	322.171	231.459	263.477	348.543	-4.234
4400	61.381	377.410	323.411	237.595	263.645	350.519	-4.161
4500	61.422	378.790	324.626	243.735	263.676	352.491	-4.092
4600	61.460	380.140	325.818	249.879	263.548	354.465	-4.025
4700	61.495	381.462	326.988	256.027	263.267	356.445	-3.961
4800	61.529	382.757	328.137	262.178	262.840	358.432	-3.901
4900	61.561	384.026	329.264	268.333	262.268	360.429	-3.842
5000	61 591	385.270	330.372	274.490	261.557	362.440	-3.786
5100	61.619	386.490	331.461	280.651	260.710	364.465	-3.733
5200	61.646	387.687	332.530	286.814	259.736	366.508	-3.682
5300	61.671	388.861	333.582	292.980	258.633	368.572	-3.633
5400	61.695	390.014	334.617	299.148	257.410	370.658	-3.585
5500	61.718	391.147	335.634	305.319	256.074	372.767	-3.540
5600	61.740	392.259	336.635	311.492	254.640	374.901	-3.497
5700	61.761	393.352	337.621	317.667	253.098	377.062	-3.455
5800	61.780	394.426	338.591	323.844	251.460	379.251	-3.416
5900	61.799	395.482	339.546	330.023	249.735	381.468	-3.377
6000	61.817	396.521	340.487	336.204	247.929	383.717	-3.341

PREVIOUS: December 1975 (1 atm) CURRENT: December 1975 (1 bar)

Calcium Hydroxide, Ion ($CaOH^+$) $Ca_1H_1O_1^+(g)$

CALCIUM HYDROXIDE ($Ca(OH)_2$) CRYSTAL M_r = 74.09468

Calcium Hydroxide ($Ca(OH)_2$) $Ca_1H_2O_2(cr)$

$S°(298.15\ K) = 83.387 \pm 0.4\ J\ K^{-1}\ mol^{-1}$

$\Delta_fH°(0\ K) = -977.36 \pm 1.3\ kJ\ mol^{-1}$

$\Delta_fH°(298.15\ K) = -986.09 \pm 1.3\ kJ\ mol^{-1}$

Enthalpy of Formation

Taylor and Wells (1) measured enthalpies of solution of $Ca(OH)_2$(cr) and CaO(cr) in dilute HCl and obtained $\Delta_rH°(298.15\ K)$ = $-15.58\pm0.1\ kcal\ mol^{-1}$ for CaO(cr) + H_2O(ℓ) = $Ca(OH)_2$(cr) which leads to $\Delta_fH°(Ca(OH)_2$, cr, 298.15 K) = $-235.68\pm0.3\ kcal\ mol^{-1}$ using $\Delta_fH°$(CaO, cr, 298.15 K) = $-151.79\pm0.21\ kcal\ mol^{-1}$ (2) and $\Delta_fH°(H_2O$, ℓ, 298.15 K) = $-68.315\ kcal\ mol^{-1}$ (3). This value, $-235.68\pm0.3\ kcal\ mol^{-1}$ ($-986.09\ kJ\ mol^{-1}$) is adopted in the tabulation. They also measured directly the enthalpy of hydration of CaO to $Ca(OH)_2$ and found $\Delta_rH°(298.15\ K) = -15.43\pm0.1\ kcal\ mol^{-1}$ which leads to $\Delta_fH°(Ca(OH)_2$, cr, 298.15 K) = $-235.53\ kcal\ mol^{-1}$. Both measurements are in very good agreement. Literature $\Delta_rH°$ data (4, 5, 6, 7, 8) determined by these two methods were within the limit of $-15.40\pm0.3\ kcal\ mol^{-1}$ which is in good agreement with the value adopted.

JANAF analyses of dissociation pressure data (9, 10, 11) for $Ca(OH)_2$(cr) → CaO(cr) + H_2O(g) are listed below. The data of Halstead and Moore (9) and of Tamaru and Shiomi (10) are in good agreement, but the pressures reported by Johnston (11) are too low due to failure to reach equilibrium. The enthalpy of formation derived from 3rd law Δ_rH of Halstead and Moore (9) or Tamaru and Shiomi (10) is in good agreement with the value adopted. However, the decomposition of $Ca(OH)_2$ may yield non-standard state CaO in the final product which was shown in a similar decomposition of $Mg(OH)_2$. See $Mg(OH)_2$ table (2) for details.

$Ca(OH)_2(cr) = CaO(cr) + H_2O(g)$

Source	Method	T/K	Data Points	$\Delta_rH°$(298.15 K), kcal mol^{-1} 2nd law	3rd law	Drift cal K^{-1} mol^{-1}	$\Delta_fH°$(298.15 K) kcal mol^{-1}
Halstead and Moore (9)	Static	635-776.5	14	25.52	25.75±0.15	0.2±0.4	-235.34
Tamaru and Shiomi (10)	Static	694-776.5	8	26.48	25.90±0.07	-0.8±0.3	-235.49
Johnston (11)	Static	663-804	7	26.92	26.76±0.24	-0.5±1.0	-236.35

Heat Capacity and Entropy

The low temperature heat capacities up to 300 K are taken from the adiabatic calorimeter measurements (19-330 K) of Hatton et al. (12). Above 300 K, the heat capacities are based on the heat conduction calorimeter measurements (310-670 K) of Kobayashi (13) joined smoothly at 300 K with the low temperature heat capacities (12) and on a graphical comparison of the Cp vs T curve adopted for $Mg(OH)_2$(cr)(2). The entropy, S°(298.15 K) = $19.93\pm0.1\ cal\ K^{-1}\ mol^{-1}$ ($83.387\pm0.4\ J\ K^{-1}\ mol^{-1}$), is derived from the adopted low temperature heat capacities, based on a T^3 extrapolation to obtain S° = $0.070\ cal\ K^{-1}\ mol^{-1}$ at 20 K (12).

Decomposition Data

T_{dcm} is calculated as the temperature at which $\Delta_rG° = 0$ for the reaction $Ca(OH)_2(cr) = CaO(cr) + H_2O(g)$. Auxiliary data are from the JANAF Tables (2).

Under a pressure of 1000 bars, Wyllie and Tuttle (14) found that $Ca(OH)_2$ melts congruently at 1108 K.

References

1. K. Taylor and L. S. Wells, J. Res. Nat. Bur. Stand. 21, 133 (1938).
2. JANAF Thermochemical Tables: CaO(cr), 6-30-76; $Mg(OH)_2$(cr), 12-31-75; H_2O(g), 3-31-61.
3. U. S. Nat. Bur. Stand. Tech. Note 270-3, 1968.
4. J. Thomsen, Thermochemische Untersuchungen 3, 247 (1883).
5. M. Berthelot, Ann. Chim. Phys. 4, 531 (1875).
6. W. A. Roth and P. Chall, Z. Elektrochem. 34, 185 (1928).
7. T. Thorvaldson, W. G. Brown and C. R. Peaker, J. Amer. Chem. Soc. 51, 2687 (1929); 52, 80 (1930); 52, 910 (1930).
8. H. E. Schwietz and E. Hey, Z. Anorg. Chem. 217, 396 (1934).
9. P. E. Halstead and A. E. Moore, J. Chem. Soc. 1957, 3873 (1957).
10. S. Tamuru and K. Shiomi, Z. Phys. Chem. 161, 421 (1932).
11. J. Johnston, Z. Physik. Chem. 62, 330 (1908).
12. W. E. Hatton, D. L. Hildenbrand, G. C. Sinke and D. R. Stull, J. Amer. Chem. Soc. 81, 5028 (1959).
13. K. Kobayashi, Science Repts. Tohoku Univ., First Ser. 34, 153 (1950); Chem. Abst. 46, 332 (1952).
14. P. J. Wyllie and O. F. Tuttle, J. Petrol. 1, 1 (1960).

Enthalpy Reference Temperature = T_r = 298.15 K; Standard State Pressure = p° = 0.1 MPa

T/K	$C_p°$ (J $K^{-1}mol^{-1}$)	S° (J $K^{-1}mol^{-1}$)	$-[G°-H°(T_r)]/T$ (J $K^{-1}mol^{-1}$)	$H°-H°(T_r)$ (kJ mol^{-1})	$\Delta_fH°$ (kJ mol^{-1})	$\Delta_fG°$ (kJ mol^{-1})	Log K_f
0	0.	0.	INFINITE	-14.160	-977.358	-977.358	INFINITE
100	32.539	17.226	147.109	-12.988	-983.071	-955.793	499.255
200	68.576	52.183	90.862	-7.736	-985.692	-927.251	242.173
298.15	87.487	83.387	83.387	0.	-986.085	-898.421	157.400
300	87.780	83.929	83.389	0.162	-986.079	-897.877	156.334
400	98.408	110.820	86.983	9.535	-985.218	-868.582	113.425
500	103.931	133.431	94.075	19.678	-983.822	-839.579	87.710
600	107.445	152.696	102.280	30.249	-982.281	-810.874	70.593
700	110.709	169.508	110.708	41.160	-980.702	-782.431	58.386
800	113.512	184.479	119.010	52.375	-979.971	-754.111	49.238
900	116.022	197.998	127.048	63.855	-978.396	-725.974	42.134
1000	117.989	210.328	134.768	75.560	-976.978	-698.005	36.460

PREVIOUS: December 1971 CURRENT: December 1975

Calcium Hydroxide ($Ca(OH)_2$) $Ca_1H_2O_2(cr)$

CALCIUM HYDROXIDE ($Ca(OH)_2$) IDEAL GAS $M_r = 74.09468$

$\Delta_f H°(0\ K) = -603.17 \pm 37.7\ kJ\ mol^{-1}$

$S°(298.15\ K) = [285.6 \pm 8.4]\ J\ K^{-1}\ mol^{-1}$ $\Delta_f H°(298.15\ K) = -610.76 \pm 37.7\ kJ\ mol^{-1}$

Vibrational Frequencies and Degeneracies

ν, cm^{-1}
[484](1)
[163](1)
[554](1)
[3650](2)
[466](4)

Ground State Quantum Weight = [1] $\sigma = 2$

Point Group: [C_{2v}]

Bond Distances: Ca-O = [2.12] Å O-H = [0.96] Å

Bond Angles: O-Ca-O = [135°] Ca-O-H = [180°]

Product of the Moments of Inertia: $I_A I_B I_C = [1261.6507 \times 10^{-117}]\ g^3\ cm^6$

Enthalpy of Formation

Dissociation energies, D_0^0, for the reaction $Ca(OH)_2(g) = Ca(g) + 2\ OH(g)$ have been derived from flame-spectral measurements (1-3). Ryabova and Gurvich (1) believed the dominant reaction to be $Ca(g) + H_2O(g) = CaOH(g) + H(g)$, but they also considered the possibility that $Ca(g) + 2\ H_2O(g) = Ca(OH)_2(g) + 2\ H(g)$ was the dominant reaction and derived $D_0^0 = 200\pm20\ kcal\ mol^{-1}$. Sugden and Schofield (2) considered the dihydroxide to be the dominant product and derived $D_0^0 = 217\pm12\ kcal\ mol^{-1}$. Cotton and Jenkins (3) found both CaOH and $Ca(OH)_2$ to be present in significant amounts in fuel-rich hydrogen-oxygen-nitrogen flames and derived D_0^0 = 203.8±5 kcal mol^{-1}. Cotton and Jenkins (3) recalculated the work of Ryabova and Gurvich (1) and Sugden and Schofield (2) considering both CaOH and $Ca(OH)_2$ to be present and obtained the recalculated D_0^0 values of 199 and 201 kcal mol^{-1}, respectively.

A 3rd law analysis of the experimental equilibrium constants tabulated by Cotton and Jenkins (3) using current JANAF auxiliary data (4) leads to $D_0^0 = 210.4\ kcal\ mol^{-1}$ which is 6.6 kcal mol^{-1} higher than the 203.8 kcal mol^{-1} derived by Cotton and Jenkins (3). Applying this difference to the data of Ryabova and Gurvich (1) and Sugden and Schofield (2) as recalculated by Cotton and Jenkins (3) gives D_0^0 = 205.6 and 207.6 kcal mol^{-1}, respectively.

For $Ba(OH)_2(g)$ (4), the corrected dissociation energy of Ryabova and Gurvich (4), $D_0^0 = 208.8\ kcal\ mol^{-1}$, is in better agreement with the "adopted" value of 209.6 kcal mol^{-1}, based on good Knudsen-cell mass-spectrometric measurements, than are the corrected dissociation energies of Sugden and Schofield or Cotton and Jenkins (4). We adopt $D_0^0 = 205.6\ kcal\ mol^{-1}$ for the dissociation of $Ca(OH)_2$ from which is calculated $\Delta_f H°(0\ K) = -144.16\pm9\ kcal\ mol^{-1}$ ($-603.165\pm37.7\ kJ\ mol^{-1}$).

The enthalpy of dissociaton listed by Jackson (8) leads to $\Delta_f H°(Ca(OH)_2$, g, 298.15 K) = $-142.65\ kcal\ mol^{-1}$. Another recent compilation (9) lists $\Delta_f H°(298.15\ K) = -130\ kcal\ mol^{-1}$.

Heat Capacity and Entropy

The analogy between gaseous mono- and dihydroxides and gaseous mono- and dihalides, particularly the mono- and difluorides, has been recognized (5-8). The O-Ca-O bond angle is assumed to be the same as the F-Ca-F bond angle (4); the Ca-O-H bond angle is considered to be linear as in CaOH (4). The Ca-O bond distance is estimated to be slightly larger, 0.02Å, than the Ca-F bond distance in CaF_2 (4) after noting the close similarity in the bond distance of alkali metal fluorides and hydroxides. The O-H bond distance is taken to be the same as in water (4).

The vibrational frequencies are estimated to be the same as in CaF_2 (4) (O-Ca-O symmetrical and asymmetrical stretch, and bend) and as in CaOH (4), (O-H stretch and Ca-O-H bend). The principal moments of inertia are $I_A = 2.1629 \times 10^{-39}\ g\ cm^2$, $I_B = 23.0947 \times 10^{-39}$, and $I_C = 25.2576 \times 10^{-39}$.

Jackson (8) has used a different molecular configuration and different vibrational frequencies to estimate S°(298.15 K) = 68.530 cal $K^{-1}\ mol^{-1}$(1 atm). We assign an uncertainty of ±2.0 cal $K^{-1}\ mol^{-1}$ to the adopted entropy.

References

1. V. G. Ryabova and L. V. Gurvich, Teplofiz. Vysokikh Temperatur, Akad. Nauk SSSR 3, 318 (1965).
2. T. M. Sugden and K. Schofield, Trans. Faraday Soc. 62, 566 (1966),
3. D. H. Cotton and D. R. Jenkins, Trans. Faraday Soc. 64, 2988 (1968).
4. JANAF Thermochemical Tables: Ca(g), CaF_2(g), 12-31-68; HO(g), 12-31-70; H_2O(g), 3-31-61; H(g), 6-30-74; CaOH(g), $Ca(OH)_2$(g), $Ba(OH)_2$(g), 12-31-75.
5. R. C. Schoonmaker and R. F. Porter, J. Chem. Phys. 31, 830 (1959).
6. J. Berkowitz, D. J. Meschi, and W. A. Chupka, J. Chem. Phys. 33, 533 (1960).
7. F. E. Stafford and J. Berkowitz, J. Chem. Phys. 40, 2963 (1964).
8. D. D. Jackson, Lawrence Livermore Laboratory, Univ. of Calif., Report UCRL-51137, Contract No. W-7405-Eng-48, Dec. 8, 1971.
9. U. S. Natl. Bur. Stand. Tech. Note 270-6, 1971.

Calcium Hydroxide ($Ca(OH)_2$) $Ca_1H_2O_2(g)$

Enthalpy Reference Temperature = T_r = 298.15 K Standard State Pressure = p° = 0.1 MPa

T/K	$C_p°$ (J K⁻¹mol⁻¹)	S° (J K⁻¹mol⁻¹)	$-[G°-H°(T_r)]/T$ (J K⁻¹mol⁻¹)	$H°-H°(T_r)$ (kJ mol⁻¹)	$\Delta_f H°$ (kJ mol⁻¹)	$\Delta_f G°$ (kJ mol⁻¹)	Log K_f
0	0.	0.	INFINITE	-15.288	-603.165	-603.165	INFINITE
100	41.018	224.358	341.567	-11.721	-606.482	-599.918	313.365
200	60.577	258.745	291.969	-6.645	-609.279	-592.151	154.654
250	68.164	273.122	286.787	-3.416	-610.155	-587.762	122.806
298.15	73.450	285.605	285.605	0.	-610.764	-583.390	102.208
300	73.620	286.060	285.606	0.136	-610.783	-583.221	101.548
350	77.512	297.717	286.518	3.920	-611.231	-578.589	86.350
400	80.329	308.261	288.588	7.869	-611.562	-573.903	74.944
450	82.418	317.849	291.315	11.941	-611.831	-569.179	66.069
500	84.012	326.619	294.413	16.103	-612.075	-564.426	58.965
600	86.304	342.152	301.108	24.626	-612.583	-554.850	48.304
700	87.972	355.587	307.953	33.343	-613.197	-545.181	40.682
800	89.371	367.427	314.662	42.212	-614.812	-535.310	34.952
900	90.655	378.028	321.124	51.214	-615.716	-525.321	30.489
1000	91.876	387.643	327.302	60.341	-616.875	-515.217	26.912
1100	93.045	396.455	333.194	69.587	-618.300	-504.984	23.980
1200	94.158	404.599	338.809	78.948	-627.726	-493.988	21.503
1300	95.206	412.178	344.165	88.417	-628.461	-482.813	19.400
1400	96.185	419.270	349.279	97.987	-629.168	-471.583	17.595
1500	97.090	425.937	354.169	107.651	-629.853	-460.303	16.029
1600	97.922	432.230	358.853	117.403	-630.521	-448.978	14.658
1700	98.684	438.190	363.346	127.233	-631.175	-437.611	13.446
1800	99.380	443.850	367.663	137.137	-780.494	-423.996	12.304
1900	100.014	449.241	371.816	147.107	-779.723	-404.211	11.113
2000	100.591	454.386	375.817	157.138	-778.957	-384.467	10.041
2100	101.116	459.306	379.676	167.224	-778.201	-364.761	9.073
2200	101.594	464.022	383.403	177.360	-777.463	-345.090	8.193
2300	102.030	468.547	387.008	187.541	-776.749	-325.454	7.391
2400	102.427	472.898	390.496	197.764	-776.065	-305.847	6.657
2500	102.789	477.087	393.877	208.025	-775.418	-286.268	5.981
2600	103.120	481.125	397.155	218.321	-774.816	-266.714	5.358
2700	103.424	485.022	400.338	228.649	-774.267	-247.182	4.782
2800	103.702	488.789	403.430	239.005	-773.779	-227.669	4.247
2900	103.957	492.432	406.436	249.388	-773.360	-208.174	3.750
3000	104.192	495.961	409.362	259.796	-773.020	-188.690	3.285
3100	104.408	499.381	412.211	270.226	-772.765	-169.217	2.851
3200	104.608	502.699	414.987	280.677	-772.606	-149.751	2.444
3300	104.793	505.921	417.694	291.147	-772.551	-130.288	2.062
3400	104.963	509.052	420.335	301.635	-772.608	-110.824	1.703
3500	105.122	512.096	422.914	312.139	-772.786	-91.358	1.363
3600	105.269	515.060	425.432	322.659	-773.091	-71.885	1.043
3700	105.406	517.946	427.894	333.193	-773.531	-52.402	0.740
3800	105.533	520.759	430.301	343.740	-774.112	-32.903	0.452
3900	105.652	523.502	432.656	354.299	-774.840	-13.388	0.179
4000	105.763	526.178	434.960	364.870	-775.721	6.146	-0.080
4100	105.866	528.791	437.217	375.451	-776.757	25.706	-0.327
4200	105.964	531.343	439.428	386.043	-777.952	45.292	-0.563
4300	106.055	533.837	441.595	396.644	-779.309	64.910	-0.788
4400	106.140	536.277	443.719	407.254	-780.826	84.559	-1.004
4500	106.220	538.663	445.802	417.872	-782.492	104.244	-1.210
4600	106.296	540.998	447.847	428.498	-784.327	123.968	-1.408
4700	106.367	543.285	449.853	439.131	-786.326	143.735	-1.597
4800	106.434	545.525	451.823	449.771	-788.482	163.547	-1.780
4900	106.497	547.720	453.758	460.417	-790.794	183.405	-1.955
5000	106.557	549.873	455.658	471.070	-793.256	203.311	-2.124
5100	106.614	551.983	457.527	481.729	-795.867	223.268	-2.287
5200	106.667	554.054	459.363	492.393	-798.615	243.277	-2.444
5300	106.718	556.086	461.169	503.062	-801.504	263.341	-2.595
5400	106.766	558.081	462.945	513.736	-804.524	283.460	-2.742
5500	106.811	560.041	464.693	524.415	-807.669	303.636	-2.884
5600	106.855	561.966	466.413	535.098	-810.925	323.871	-3.021
5700	106.896	563.858	468.106	545.786	-814.302	344.165	-3.154
5800	106.935	565.717	469.773	556.478	-817.785	364.520	-3.283
5900	106.972	567.545	471.414	567.173	-821.368	384.933	-3.408
6000	107.008	569.344	473.032	577.872	-825.045	405.411	-3.529

PREVIOUS: December 1975 (1 atm) CURRENT: December 1975 (1 bar)

Calcium Hydroxide ($Ca(OH)_2$) $Ca_1H_2O_2(g)$

CALCIUM IODIDE (CaI) IDEAL GAS $M_r = 75.533$

$S°(298.15\ K) = 261.31 \pm 0.2\ J\ K^{-1}\ mol^{-1}$

$\Delta_f H°(0\ K) = -2.72 \pm 84\ kJ\ mol^{-1}$

$\Delta_f H°(298.15\ K) = -5.04 \pm 84\ kJ\ mol^{-1}$

Electronic Levels and Quantum Weights

State	ε_i, cm^{-1}	g_i
$X[^2\Sigma^+]$	0	2
$A[^2\Pi]$	15585.1	2
	15645.9	2
$B[^2\Sigma^+]$	15711.2	2
$C[^2\Pi]$	23314.0	2
	23741.9	2
$D[^2\Sigma^+]$	31062.0	2

$\omega_e = 238.3\ cm^{-1}$ $\omega_e x_e = 0.61\ cm^{-1}$ $\sigma = 1$

$B_e = [0.0667]\ cm^{-1}$ $\alpha_e = [0.00023]\ cm^{-1}$ $r_e = [2.88]$ Å

Enthalpy of Formation

The adopted $\Delta_f H°(0\ K) = -0.65 \pm 20\ kcal\ mol^{-1}$ is calculated from the $D_0^0 = 3 \pm 1$ eV($69 \pm 23\ kcal\ mol^{-1}$) selected by Gaydon (1). A linear Birge-Sponer extrapolation of ω_e and $\omega_e x_e$ data (2) with a correction for the ionic character of the molecule as described by Hildenbrand (3) gave $D_0^0 = 66.5\ kcal\ mol^{-1}$. Other D_0^0 values are 63.9 kcal mol^{-1} (4) as a lower bound from a consideration of ionic bonding forces and 64 kcal mol^{-1} (5) derived as a lower bound from a consideration of energy conservation and measured reaction threshold relative kinetic energy from crossed molecular beam experiments. Of the four ratios, $D_0^0(CaI)/\Delta_{at}H°(CaI_2)$, $D_0^0(CaI) = 69\ kcal\ mol^{-1}$ gives 0.445 which is closest to 0.46 found for a series of mono- and difluorides (6) and for other alkaline earth halide systems (8). $\Delta_f H°(298.15\ K) = -1.21 \pm 20\ kcal\ mol^{-1}$ is calculated from the adopted $\Delta_f H°(0\ K) = -0.65 \pm 20\ kcal\ mol^{-1}$.

Heat Capacity and Entropy

The ground state vibrational constants are from the compilation of Rosen (2). The value of $r_e = 2.88$ Å is assumed the same as the bond distance in CaI_2 (7). B_e is calculated from the adopted r_e. α_e is calculted assuming a Morse potential function.

The electronic levels and their probable designation are those given by Rosen (2).

References

1. A. G. Gaydon, "Dissociation Energies and Spectra of Diatomic Molecules," Third Ed., Chapman and Hall, London, 1968.
2. B. Rosen, Ed., "Spectroscopic Data Relative to Diatomic Molecules," Pergamon Press, New York, 1970.
3. D. L. Hildenbrand in "Advances in High Temperature Chemistry," Vol. 1, L. Eyring, Ed., Academic Press, New York, 1967, pp. 193-215.
4. K. S. Krasnov and N. V. Karaseva, Optics and Spectrosc. 19, 14 (1965).
5. C. A. Mims, S. Lin, and R. R. Herm, J. Chem. Phys. 57, 3099 (1972).
6. G. D. Blue, J. W. Green, T. C. Ehlert, and J. L. Margrave, Nature, 199, 804 (1963).
7. P. A. Akishin, V. P. Spiridonov, G. A. Sobolev, and V. A. Naumov, Zhur. Fiz. Khim. 32, 58 (1958).
8. JANAF Thermochemical Tables.

Calcium Iodide (CaI) $Ca_1I_1(g)$

Enthalpy Reference Temperature = T_r = 298.15 K						Standard State Pressure = $p°$ = 0.1 MPa	
	J K^{-1}mol^{-1}			kJ mol^{-1}			
T/K	$C_p°$	$S°$	$-[G°-H°(T_r)]/T$	$H°-H°(T_r)$	$\Delta_f H°$	$\Delta_f G°$	Log K_f
0	0.	0.	INFINITE	-10.010	-2.719	-2.719	INFINITE
100	32.513	223.091	293.124	-7.003	-2.218	-19.881	10.385
200	35.706	246.842	264.634	-3.558	-3.508	-37.083	9.685
250	36.301	254.880	261.907	-1.757	-4.271	-45.390	9.484
298.15	36.646	261.306	261.306	0.	-5.045	-53.240	9.327
300	36.656	261.532	261.306	0.068	-5.075	-53.538	9.322
350	36.886	267.201	261.753	1.907	-5.948	-61.548	9.186
400	37.044	272.138	262.750	3.755	-14.842	-69.156	9.031
450	37.159	276.508	264.040	5.610	-16.366	-75.854	8.805
500	37.247	280.428	265.486	7.471	-37.990	-80.481	8.408
600	37.373	287.230	268.560	11.202	-39.078	-88.880	7.738
700	37.461	292.998	271.650	14.944	-40.359	-97.081	7.244
800	37.529	298.005	274.638	18.694	-42.692	-104.982	6.855
900	37.586	302.429	277.485	22.449	-44.357	-112.671	6.539
1000	37.635	306.392	280.181	26.211	-46.318	-120.159	6.276
1100	37.679	309.981	282.729	29.976	-48.583	-127.436	6.051
1200	37.720	313.261	285.139	33.746	-58.889	-133.871	5.827
1300	37.758	316.282	287.420	37.520	-60.543	-140.052	5.627
1400	37.795	319.081	289.583	41.298	-62.209	-146.106	5.451
1500	37.832	321.690	291.637	45.079	-63.891	-152.040	5.294
1600	37.868	324.133	293.593	48.864	-65.595	-157.861	5.154
1700	37.906	326.430	295.457	52.653	-67.324	-163.575	5.026
1800	37.946	328.597	297.239	56.445	-217.755	-166.975	4.845
1900	37.989	330.650	298.944	60.242	-218.130	-164.144	4.513
2000	38.037	332.600	300.578	64.043	-218.541	-161.292	4.213
2100	38.092	334.457	302.148	67.850	-218.988	-158.419	3.940
2200	38.155	336.231	303.657	71.662	-219.472	-155.523	3.693
2300	38.227	337.928	305.110	75.481	-219.991	-152.605	3.466
2400	38.310	339.557	306.512	79.308	-220.544	-149.663	3.257
2500	38.406	341.123	307.865	83.143	-221.130	-146.698	3.065
2600	38.516	342.631	309.173	86.989	-221.747	-143.709	2.887
2700	38.640	344.087	310.440	90.847	-222.395	-140.695	2.722
2800	38.779	345.495	311.667	94.718	-223.073	-137.657	2.568
2900	38.935	346.858	312.857	98.603	-223.783	-134.594	2.424
3000	39.107	348.181	314.012	102.505	-224.525	-131.506	2.290
3100	39.295	349.466	315.135	106.425	-225.302	-128.392	2.163
3200	39.499	350.717	316.228	110.365	-226.117	-125.253	2.045
3300	39.720	351.936	317.292	114.326	-226.973	-122.088	1.932
3400	39.955	353.125	318.328	118.309	-227.875	-118.897	1.827
3500	40.206	354.287	319.339	122.317	-228.827	-115.678	1.726
3600	40.471	355.423	320.325	126.351	-229.832	-112.431	1.631
3700	40.749	356.536	321.289	130.412	-230.896	-109.155	1.541
3800	41.039	357.626	322.231	134.501	-232.024	-105.850	1.455
3900	41.341	358.696	323.152	138.620	-233.218	-102.514	1.373
4000	41.652	359.747	324.054	142.770	-234.483	-99.147	1.295
4100	41.972	360.779	324.937	146.951	-235.822	-95.747	1.220
4200	42.299	361.794	325.803	151.164	-237.237	-92.313	1.148
4300	42.633	362.794	326.652	155.411	-238.731	-88.845	1.079
4400	42.971	363.778	327.484	159.691	-240.303	-85.342	1.013
4500	43.314	364.747	328.301	164.005	-241.940	-81.803	0.950
4600	43.659	365.703	329.104	168.354	-243.663	-78.226	0.888
4700	44.005	366.645	329.893	172.737	-245.468	-74.610	0.829
4800	44.351	367.576	330.668	177.155	-247.349	-70.955	0.772
4900	44.697	368.494	331.431	181.607	-249.304	-67.260	0.717
5000	45.040	369.400	332.181	186.094	-251.330	-63.524	0.664
5100	45.381	370.295	332.920	190.615	-253.424	-59.747	0.612
5200	45.717	371.180	333.647	195.170	-255.578	-55.929	0.562
5300	46.049	372.054	334.364	199.759	-257.795	-52.069	0.513
5400	46.375	372.918	335.069	204.380	-260.067	-48.166	0.466
5500	46.694	373.771	335.765	209.033	-262.390	-44.221	0.420
5600	47.006	374.616	336.452	213.718	-264.748	-40.233	0.375
5700	47.310	375.450	337.129	218.434	-267.155	-36.203	0.332
5800	47.606	376.276	337.796	223.180	-269.598	-32.130	0.289
5900	47.894	377.092	338.456	227.955	-272.070	-28.014	0.248
6000	48.171	377.899	339.106	232.759	-274.567	-23.857	0.208

PREVIOUS: June 1974 (1 atm) CURRENT: June 1974 (1 bar)

CALCIUM IODIDE (CaI_2) CRYSTAL M_r = 293.8890

$S°(298.15\ K) = 145.29 \pm 0.21\ J\ K^{-1}\ mol^{-1}$

$T_{fus} = 1052 \pm 2\ K$

$\Delta_f H°(0\ K) = -535.8 \pm 2.1\ kJ\ mol^{-1}$

$\Delta_f H°(298.15\ K) = -536.8 \pm 2.1\ kJ\ mol^{-1}$

$\Delta_{fus} H° = 41.84 \pm 0.84\ kJ\ mol^{-1}$

Enthalpy of Formation

Ehrlich, Peik and Koch (1) derived $\Delta_f H°(CaI_2$, cr, 298.15 K) = -128.1 ± 0.4 kcal mol^{-1} from enthalpy of solution measurements of Ca(cr) and CaI_2(cr) in 0.1 N HI. An auxiliary HI enthalpy of solution value, -13.22 kcal mol^{-1}, calculated from data in referenece (2) was used in the derivation. This auxiliary value will be changed by incorporation of the accepted CODATA value of $\Delta_f H°(I^-$, aq, std. state, 298.15 K) = -13.60 kcal mol^{-1} (3) in the $\Delta_f H°$(HI, aq solution, 298.15 K) table, so that the above derived enthalpy of formation of CaI_2(cr) will be approximately -128.6 kcal mol^{-1}.

Combining $\Delta_f H°(Ca^{+2}$, aq, std. state, 298.15 K) = -129.74 kcal mol^{-1}, selected by Parker (4, 5) with the CODATA value for I^-(aq, std. state) (3) gives $\Delta_f H°(CaI_2$, aq, std. state, 298.15 K) = -156.94 kcal mol^{-1}. Further combination with the enthalpy of solution, -28.62 kcal mol^{-1} (5, 6) gives $\Delta_f H°(CaI_2$, cr, 298.15 K) = -128.32 kcal mol^{-1}. This enthalpy of solution value was based on several sets of data, one of which was that of Ehrlich, et al. (1, 6).

We adopt -128.3 ± 0.5 kcal mol^{-1} (-536.807 kJ mol^{-1}) because of the correlation of $\Delta_f H°(Ca^{+2}$, aq, std. state, 298.15 K) with the evaluation of data for several calcium compounds (4).

Heat Capacity and Entropy

Paukov, Khriplovich, and Korotkikh (7) have measured $C_p°$ (13.10 - 309.18 K). Our T^3 extrapolation is in agreement with their values of S°(13 K) = 0.516 cal K^{-1} mol^{-1} and H°(13 K) - H°(0 K) = 4.99 cal K^{-1} mol^{-1}. The low temperature heat capacity joins smoothly with a linear extrapolation from $C_p°$ = 18.31 cal K^{-1} mol^{-1} at 270 K to $C_p°$ = 22.0 cal K^{-1} mol^{-1} at the adopted T_{fus} = 1052 K. The linearly extrapolated heat capacity at 300 K is only 0.3% higher than the Paukov et al. (7) value and S°(298.15 K) = 34.72 cal $K^{-1}mol^{-1}$ is in agreement with Paukov et al. (7). Dworkin and Bredig (8) determined the heat capacity of the crystal to be 23.2 cal K^{-1} mol^{-1} (± 5%) near the melting point. By using the lower limit, $C_p°$ = 22.0 cal K^{-1} mol^{-1} of Dworkin and Bredig's value (8), H°(1052 K) - H°(298.15 K) is brought down to 15.24 kcal mol^{-1}, about 0.7% higher than Dworkin and Bredig's drop calorimeter measurement of 15.1 (± 0.2%) kcal mol^{-1} (8).

Fusion Data

Dworkin and Bredig (8) measured $\Delta_{fus}H°$ = 10.00 kcal mol^{-1} (± 2%) at T_{fus} = 1052 K ($S°_{fus}$ = 9.5 cal K^{-1} mol^{-1}) by drop calorimetry. Emons and Loeffelholz (9) determined $\Delta_{fus}H°$ = 6.65 kcal mol^{-1} (± 5%) and T_{fus} = 1053 K ($\Delta_{fus}S°$ = 6.3 cal K^{-1} mol^{-1}) by high temperature cryoscopy. The two pair of investigators are in much better agreement with other salts (9). An entropy of melting criteria for choosing one of these values is not conclusive because the entropies of melting of CdI_2, $CdCl_2$, and $CdBr_2$ which have similar layer crystal structure range from 7.5 to 9.5 cal K^{-1} mol^{-1} (10). Hutchison (11) found T_{fus} = 1018 K, noting that only approximate melting points were obtained.

Because drop calorimetry is a more direct measure of the enthalpy of melting, we adopt $\Delta_{fus}H°$ = 10.00 ± 0.20 kcal mol^{-1} at T_{fus} = 1052 ± 2 K.

Although unconfirmed by direct experimental evidence, the discrepancy in enthalpy of melting and the adjustments required in joining low temperature and high temperature heat capacities might be explained by the existence of a solid state transition in CaI_2 similar to the transitions in $BaCl_2$, $SrBr_2$, or $SrCl_2$ (8).

Sublimation Data

$\Delta_{sub}H°$(298.15 K) is calculated as the difference between the enthalpies of formation at 298.15 K of the ideal gas and crystal states.

References

1. P. Ehrlich, K. Peik, and E. Koch, Z. Anorg. Allgem. Chem. 324, 113 (1963).
2. U. S. Nat. Bur. Stand. Circ. 500, 1952.
3. ICSU-CODATA Task Group, J. Chem. Thermodynamics 4, 331 (1972).
4. V. B. Parker, personal communication, U. S. Nat. Bur. Stand., Jan. 1969.
5. U. S. Nat. Bur. Stand. Tech. Note 270-6, 1971.
6. V. B. Parker, personal communication, U. S. Nat. Bur. Stand., May 1974.
7. I. E. Paukov, L. M. Khriplovich, and A. M. Korotkikh, Russ. J. Phys. Chem. 43, 429 (1969).
8. A. S. Dworkin and M. A. Bredig, J. Phys. Chem. 67, 697 (1963).
9. H. H. Emons and B. Loeffelholz, Wiss. Z. Hochsch. Chem. Leuna-Merseburg 6, 261 (1964).
10. O. Kubaschewski, I. Ll. Evans, and C. B. Alcock, "Metallurgical Thermochemistry," Fourth Ed., Pergamon Press, New York, 1967.
11. J. F. Hutchison, U. S. At. Energy Comm. IS-T-50, 1965.

Calcium Iodide (CaI_2) Ca_1I_2(cr)

Enthalpy Reference Temperature = T_r = 298.15 K Standard State Pressure = p° = 0.1 MPa

T/K	$C_p°$	S°	$-[G°-H°(T_r)]/T$	$H°-H°(T_r)$	$\Delta_f H°$	$\Delta_f G°$	Log K_f
	J $K^{-1}mol^{-1}$			kJ mol^{-1}			
0	0.	0.	INFINITE	-17.965	-535.838	-535.838	INFINITE
100	61.840	67.860	210.957	-14.310	-536.213	-535.303	279.614
200	73.304	115.131	152.260	-7.426	-536.530	-534.267	139.536
298.15	77.157	145.287	145.287	0.	-536.807	-533.097	93.396
300	77.195	145.764	145.288	0.143	-536.813	-533.074	92.816
400	79.161	168.244	148.340	7.962	-553.267	-531.211	69.369
500	81.170	186.124	154.168	15.978	-596.213	-521.572	54.488
600	83.136	201.097	160.775	24.194	-594.694	-506.786	44.120
700	85.103	214.060	167.481	32.605	-593.188	-492.254	36.732
800	87.069	225.553	174.035	41.214	-592.553	-477.833	31.199
900	89.036	235.921	180.344	50.019	-591.064	-463.583	26.906
1000	91.002	245.403	186.383	59.021	-589.684	-449.494	23.479
1052.000	92.048	250.043	189.416	63.780	------ CRYSTAL <--> LIQUID -----		
1100	93.000	254.171	192.152	68.221	-588.422	-435.537	20.682
1200	94.977	262.348	197.664	77.620	-595.016	-421.067	18.329
1300	96.943	270.028	202.938	87.216	-592.776	-406.661	16.340
1400	98.910	277.284	207.992	97.009	-590.370	-392.433	14.642
1500	100.876	284.175	212.843	106.998	-587.808	-378.384	13.176

PREVIOUS: CURRENT: June 1974

Calcium Iodide (CaI_2) Ca_1I_2(cr)

CALCIUM IODIDE (CaI_2) LIQUID M_r = 293.8890

$S°(298.15\ K) = [178.942]\ J\ K^{-1}\ mol^{-1}$

$T_{fus} = 1052 \pm 2\ K$

$\Delta_f H°(298.15\ K) = [-500.170]\ kJ\ mol^{-1}$

$\Delta_{fus}H° = 41.84 \pm 0.84\ kJ\ mol^{-1}$

Enthalpy of Formation

$\Delta_f H°(CaI_2, \ell, 298.15\ K)$ is calculated from the sum of the enthalpy of formation of the crystal, the enthalpy of melting, and the difference in enthalpy, H°(1052 K) -H°(298.15 K), between the crystal and the liquid.

Heat Capacity and Entropy

The liquid heat capacity near the melting point, 24.7 cal K^{-1} mol^{-1}, was derived by Dworkin and Bredig (1) from drop calorimetric measurements. This value is adopted and is assumed constant from 700 K to 2500 K. A glass transition is assumed at 700 K below which the heat capacity is that of the crystal.

$S°(CaI_2, \ell, 298.15\ K)$ is calculated from the crystal entropy in a manner similar to that used for the enthalpy of formation calculation.

Vaporization Data

The temperature at which $\Delta_r G° = 0$ for the reaction $CaI_2(\ell) = CaI_2(g)$ at one bar is T_{vap}. Peterson and Hutchison (2) have extrapolated Knudsen effusion cell measurements in the 1076 - 1294 K range to obtain a normal boiling point of 1831 K (at 1 atm).

$\Delta_{vap}H°$ is the calculated difference between the enthalpies of formation of the ideal gas and the liquid at T_{vap}. See the ideal gas table for details.

References

1. A. S. Dworkin and M. A. Bredig, J. Phys. Chem. 67, 697 (1963).
2. D. T. Peterson and J. F. Hutchison, J. Chem. Eng. Data 15, 320 (1970).

Calcium Iodide (CaI_2) $Ca_1I_2(l)$

Enthalpy Reference Temperature = T_r = 298.15 K Standard State Pressure = p° = 0.1 MPa

T/K	$C_p°$ (J K⁻¹mol⁻¹)	S° (J K⁻¹mol⁻¹)	$-[G°-H°(T_r)]/T$ (J K⁻¹mol⁻¹)	$H°-H°(T_r)$ (kJ mol⁻¹)	$\Delta_f H°$ (kJ mol⁻¹)	$\Delta_f G°$ (kJ mol⁻¹)	Log K_f
0							
100							
200							
298.15	77.157	178.942	178.942	0.	-500.170	-506.494	88.736
300	77.195	179.419	178.943	0.143	-500.176	-506.534	88.195
400	79.161	201.899	181.995	7.962	-516.630	-508.036	66.343
500	81.170	219.779	187.823	15.978	-559.576	-501.762	52.419
600	83.136	234.752	194.430	24.194	-558.057	-490.342	42.688
700	85.103	247.715	201.136	32.605	-556.551	-479.176	35.757
700.000	85.103	247.715	201.136	32.605	GLASS <--> LIQUID TRANSITION		
700.000	103.345	247.715	201.136	32.605			
800	103.345	261.515	207.840	42.940	-554.190	-468.240	30.573
900	103.345	273.687	214.494	53.274	-551.171	-457.680	26.563
1000	103.345	284.576	220.967	63.609	-548.459	-447.441	23.372
1052.000	103.345	289.815	224.242	68.983	------ CRYSTAL <--> LIQUID -----		
1100	103.345	294.426	227.204	73.943	-546.063	-437.458	20.773
1200	103.345	303.418	233.186	84.278	-551.721	-427.056	18.589
1300	103.345	311.690	238.911	94.612	-548.742	-416.788	16.747
1400	103.345	319.349	244.387	104.947	-545.795	-406.749	15.176
1500	103.345	326.479	249.624	115.281	-542.888	-396.919	13.822
1600	103.345	333.148	254.638	125.616	-540.030	-387.281	12 643
1700	103.345	339.414	259.443	135.950	-537.232	-377.820	11.609
1800	103.345	345.321	264.051	146.285	-683.172	-366.311	10.630
1900	103.345	350.908	268.477	156.619	-679.094	-348.819	9.590
2000	103.345	356.209	272.732	166.954	-675.092	-331.540	8.659
2100	103.345	361.251	276.828	177.288	-671.165	-314.459	7.822
2200	103.345	366.059	280.776	187.623	-667.312	-297.564	7.065
2300	103.345	370.653	284.584	197.957	-663.528	-280.842	6.378
2400	103.345	375.051	288.263	208.292	-659.808	-264.284	5.752
2500	103.345	379.270	291.819	218.626	-656.147	-247.879	5.179

PREVIOUS: CURRENT: June 1974

Calcium Iodide (CaI_2) $Ca_1I_2(l)$

CALCIUM IODIDE (CaI_2) CRYSTAL–LIQUID M_r = 293.8890

0 to 1052 K crystal

above 1052 K liquid

Refer to the individual tables for details.

Calcium Iodide (CaI_2) Ca_1I_2(cr,l)

Enthalpy Reference Temperature = T_r = 298.15 K; Standard State Pressure = $p°$ = 0.1 MPa

T/K	$C_p°$ (J K⁻¹mol⁻¹)	$S°$ (J K⁻¹mol⁻¹)	$-[G°-H°(T_r)]/T$ (J K⁻¹mol⁻¹)	$H°-H°(T_r)$ (kJ mol⁻¹)	$\Delta_fH°$ (kJ mol⁻¹)	$\Delta_fG°$ (kJ mol⁻¹)	Log K_f
0	0.	0.	INFINITE	-17.965	-535.838	-535.838	INFINITE
100	61.840	67.860	210.957	-14.310	-536.213	-535.303	279.614
200	73.304	115.131	152.260	-7.426	-536.530	-534.267	139.536
298.15	77.157	145.287	145.287	0.	-536.807	-533.097	93.396
300	77.195	145.764	145.288	0.143	-536.813	-533.074	92.816
400	79.161	168.244	148.340	7.962	-553.267	-531.211	69.369
500	81.170	186.124	154.168	15.978	-596.213	-521.572	54.488
600	83.136	201.097	160.775	24.194	-594.694	-506.786	44.120
700	85.103	214.060	167.481	32.605	-593.188	-492.254	36.732
800	87.069	225.553	174.035	41.214	-592.553	-477.833	31.199
900	89.036	235.921	180.344	50.019	-591.064	-463.583	26.906
1000	91.002	245.403	186.383	59.021	-589.684	-449.494	23.479
1052.000	92.048	250.043	189.416	63.780	CRYSTAL <--> LIQUID	TRANSITION	
1052.000	103.345	289.815	189.416	105.620			
1100	103.345	294.426	193.898	110.580	-546.063	-437.458	20.773
1200	103.345	303.418	202.655	120.915	-551.721	-427.056	18.589
1300	103.345	311.690	210.729	131.249	-548.742	-416.788	16.747
1400	103.345	319.349	218.217	141.584	-545.795	-406.749	15.176
1500	103.345	326.479	225.200	151.918	-542.888	-396.919	13.822
1600	103.345	333.148	231.740	162.253	-540.030	-387.281	12.643
1700	103.345	339.414	237.892	172.587	-537.232	-377.820	11.609
1800	103.345	345.321	243.697	182.922	-683.172	-366.311	10.630
1900	103.345	350.908	249.194	193.256	-679.094	-348.819	9.590
2000	103.345	356.209	254.414	203.591	-675.092	-331.540	8.659
2100	103.345	361.251	259.382	213.925	-671.165	-314.459	7.822
2200	103.345	366.059	264.123	224.260	-667.312	-297.564	7.065
2300	103.345	370.653	268.655	234.594	-663.528	-280.842	6.378
2400	103.345	375.051	272.997	244.929	-659.808	-264.284	5.752
2500	103.345	379.270	277.164	255.263	-656.147	-247.879	5.179

PREVIOUS: CURRENT: June 1974

Calcium Iodide (CaI_2) Ca_1I_2(cr,l)

CALCIUM IODIDE (CaI_2) IDEAL GAS M_r = 293.8890

$S°(298.15\ K) = [327.57 \pm 8.4]\ J\ K^{-1}\ mol^{-1}$

$\Delta_f H°(0\ K) = -255.2 \pm 17\ kJ\ mol^{-1}$

$\Delta_f H°(298.15\ K) = -258.2 \pm 17\ kJ\ mol^{-1}$

Vibrational Frequencies and Degeneracies

$\nu,\ cm^{-1}$
[117](1)
[45](2)
[316](1)

Ground State Quantum Weight = [1] $\sigma = 2$

Point Group: $D_{\infty h}$

Bond Distance: Ca-I = 2.88 ± 0.03 Å

Bond Angle: I-Ca-I = [180°]

Rotational Constant: $B_0 = 0.008008\ cm^{-1}$

Enthalpy of Formation

Peterson and Hutchison (1) used a weight loss Knudsen effusion technique to observe vapor pressures of the liquid at 18 temperatures ranging from 1076 K to 1294 K. Our 2nd and 3rd law analyses of these data, after rejection of one point due to failure of a statistical test, yield $\Delta_{vap}H°(298.15\ K)$ (2nd Law) = 60.83 ± 2.04 kcal mol^{-1} and $\Delta_{vap}H°(298.15\ K)$ (3rd Law) = 57.84 ± 0.98 kcal mol^{-1} with a drift of -2.6 ± 1.7 cal $K^{-1}\ mol^{-1}$. We adopt $\Delta_{vap}H°(298.15\ K)$ = 57.8 ± 3 kcal mol^{-1} and combine that with the adopted enthalpy of formation of the liquid to obtain $\Delta_f H°(CaI_2$, g, 298.15 K) = -61.7 ± 4 kcal mol^{-1} (-258.153 ± 17 kJ mol^{-1}).

Heat Capacity and Entropy

The bond distance, as determined from an electron diffraction study, is taken from Akishin et al. (3) who also judged that the bond angle was 180° ± 10-20°. We have assumed a linear structure.

To estimate the vibrational frequencies, the stretching force constants for gaseous CaF, CaF_2, CaCl, and $CaCl_2$ are calculated from the vibrational frequencies (4) using a valence force model. The trend in the ratio of k(monohalide)/k_1(dihalide) indicates that k(CaI)/$k_1(CaI_2)$ = 1 is a reasonable approximation, an approximation which has been used by Brewer et al. (5). The stretching force constant for CaI(g) is calculated from the ground state vibrational frequency given by Rosen (6). The bending force constant is assumed to be 0.01 times the stretching force constant (5). The adopted vibrational frequencies are calculated from the estimated force constants. Other estimates for ν_1, ν_2, and ν_3 are 121, 77, 327 (7) and 118, 45, 321 (5).

We assign an uncertainty of ±2 cal $K^{-1}\ mol^{-1}$ to the entropy to allow for error in bond angle and vibrational frequency estimates.

References

1. J. F. Hutchison, U. S. At. Energy Comm. IS-T-50, 1965.
2. D. T. Peterson and J. F. Hutchison, J. Chem. Eng. Data 15, 320 (1970).
3. P. A. Akishin, V. P. Spiridonov and G. A. Sobolev, Zhur. Fiz. Khim. 31, 648 (1957).
4. JANAF Thermochemical Tables: CaF(g), CaF_2(g), 12-31-68; CaCl(g), $CaCl_2$(g), 6-30-70.
5. L. Brewer, G. R. Somayajulu, and E. Brackett, Chem. Rev. 63, 111 (1963).
6. B. Rosen, Ed., "Spectroscopic Data Relative to Diatomic Molecules," Pergamon Press, New York, 1970.
7. K. S. Krasnov and V. I. Svettsov, Izv. Vysshikh Uchebn. Zavedenii, Khim. i Khim. Tekhnol. 6, 167 (1963).

Calcium Iodide (CaI_2) $Ca_1I_2(g)$

Enthalpy Reference Temperature = T_r = 298.15 K; Standard State Pressure = p° = 0.1 MPa

T/K	$C_p°$	S°	$-[G°-H°(T_r)]/T$	$H°-H°(T_r)$	$\Delta_f H°$	$\Delta_f G°$	Log K_f
	J $K^{-1}mol^{-1}$			kJ mol^{-1}			
0	0.	0.	INFINITE	-16.011	-255.230	-255.230	INFINITE
100	53.623	264.518	380.113	-11.559	-254.808	-273.564	142.895
200	58.924	303.667	333.072	-5.881	-256.331	-291.775	76.204
250	60.001	316.942	328.564	-2.905	-257.233	-300.533	62.793
298.15	60.636	327.569	327.569	0.	-258.153	-308.790	54.099
300	60.655	327.944	327.570	0.112	-258.189	-309.104	53.820
350	61.076	337.327	328.310	3.156	-259.263	-317.508	47.385
400	61.360	345.503	329.959	6.217	-276.357	-325.205	42.467
450	61.561	352.742	332.096	9.291	-278.680	-331.171	38.441
500	61.707	359.236	334.491	12.373	-321.164	-333.079	34.796
600	61.901	370.505	339.582	18.554	-321.679	-335.416	29.201
700	62.020	380.057	344.699	24.750	-322.389	-337.653	25.196
800	62.098	388.344	349.648	30.957	-324.156	-339.669	22.178
900	62.152	395.661	354.362	37.169	-325.259	-341.545	19.823
1000	62.191	402.212	358.825	43.386	-326.664	-343.282	17.931
1100	62.220	408.141	363.043	49.607	-328.382	-344.863	16.376
1200	62.242	413.556	367.030	55.830	-338.151	-345.652	15.046
1300	62.259	418.538	370.803	62.055	-339.282	-346.231	13.912
1400	62.272	423.153	374.380	68.282	-340.443	-346.722	12.936
1500	62.283	427.449	377.776	74.510	-341.642	-347.129	12.088
1600	62.292	431.469	381.008	80.738	-342.890	-347.454	11.343
1700	62.300	435.246	384.088	86.968	-344.197	-347.700	10.684
1800	62.306	438.807	387.030	93.198	-494.241	-345.656	10.031
1900	62.311	442.176	389.845	99.429	-494.267	-337.401	9.276
2000	62.316	445.372	392.542	105.661	-494.367	-329.143	8.596
2100	62.320	448.413	395.131	111.892	-494.543	-320.877	7.981
2200	62.323	451.312	397.619	118.125	-494.792	-312.602	7.422
2300	62.326	454.083	400.014	124.357	-495.111	-304.314	6.911
2400	62.329	456.735	402.323	130.590	-495.493	-296.010	6.442
2500	62.331	459.280	404.550	136.823	-495.933	-287.690	6.011
2600	62.333	461.724	406.703	143.056	-496.426	-279.350	5.612
2700	62.335	464.077	408.784	149.289	-496.966	-270.991	5.243
2800	62.337	466.344	410.800	155.523	-497.550	-262.611	4.899
2900	62.338	468.531	412.753	161.757	-498.174	-254.209	4.579
3000	62.339	470.645	414.648	167.991	-498.836	-245.786	4.280
3100	62.341	472.689	416.487	174.225	-499.535	-237.339	3.999
3200	62.342	474.668	418.275	180.459	-500.274	-228.870	3.736
3300	62.343	476.586	420.013	186.693	-501.054	-220.376	3.488
3400	62.343	478.448	421.704	192.927	-501.877	-211.859	3.255
3500	62.344	480.255	423.351	199.162	-502.748	-203.316	3.034
3600	62.345	482.011	424.957	205.396	-503.672	-194.748	2.826
3700	62.346	483.719	426.522	211.631	-504.653	-186.153	2.628
3800	62.346	485.382	428.049	217.865	-505.695	-177.531	2.440
3900	62.347	487.001	429.540	224.100	-506.805	-168.881	2.262
4000	62.348	488.580	430.996	230.335	-507.986	-160.202	2.092
4100	62.348	490.119	432.420	236.569	-509.243	-151.492	1.930
4200	62.349	491.622	433.811	242.804	-510.580	-142.750	1.775
4300	62.349	493.089	435.173	249.039	-512.000	-133.976	1.627
4400	62.349	494.522	436.506	255.274	-513.502	-125.168	1.486
4500	62.350	495.924	437.810	261.509	-515.076	-116.325	1.350
4600	62.350	497.294	439.089	267.744	-516.745	-107.446	1.220
4700	62.350	498.635	440.341	273.979	-518.504	-98.530	1.095
4800	62.351	499.948	441.570	280.214	-520.349	-89.574	0.975
4900	62.351	501.233	442.774	286.449	-522.279	-80.580	0.859
5000	62.351	502.493	443.956	292.684	-524.290	-71.546	0.747
5100	62.352	503.728	445.116	298.919	-526.381	-62.470	0.640
5200	62.352	504.938	446.255	305.155	-528.544	-53.353	0.536
5300	62.352	506.126	447.373	311.390	-530.783	-44.194	0.436
5400	62.352	507.291	448.472	317.625	-533.090	-34.991	0.338
5500	62.352	508.436	449.552	323.860	-535.461	-25.746	0.245
5600	62.353	509.559	450.613	330.096	-537.881	-16.457	0.154
5700	62.353	510.663	451.657	336.331	-540.362	-7.124	0.065
5800	62.353	511.747	452.684	342.566	-542.892	2.253	-0.020
5900	62.353	512.813	453.694	348.801	-545.465	11.674	-0.103
6000	62.353	513.861	454.688	355.037	-548.075	21.139	-0.184

PREVIOUS: June 1974 (1 atm) CURRENT: June 1974 (1 bar)

Calcium Iodide (CaI_2) $Ca_1I_2(g)$

CALCIUM OXIDE (CaO) CRYSTAL $M_r = 56.0794$

$S°(298.15\ K) = 38.212 \pm 0.13\ J\ K^{-1}\ mol^{-1}$
$T_{fus} = 3200 \pm 50\ K$

$\Delta_f H°(0\ K) = -631.76 \pm 0.88\ kJ\ mol^{-1}$
$\Delta_f H°(298.15\ K) = -635.09 \pm 0.88\ kJ\ mol^{-1}$
$\Delta_{fus} H° = [79.5]\ kJ\ mol^{-1}$

Enthalpy of Formation

Huber and Holley (1) determined the enthalpy of combustion of calcium metal in a bomb calorimeter and derived the enthalpy of formation of calcium oxide (cr) as -151.9 kcal mol^{-1} which is adopted in the tabulation. The adopted value is in good agreement with the value, -151.9 kcal mol^{-1} (2) derived from solution calorimetry.

Heat Capacity and Entropy

Gmelin (3) measured low temperature Cp data from 4 to 300 K in an adiabatic calorimeter. We use his smoothed Cp values to derive $S°(298.15\ K) = 9.133 \pm 0.03$ cal K^{-1} mol^{-1} based on S° = 0.0001 cal K^{-1} mol^{-1} at 4 K. Lander (4) determined high temperature enthalpy data from 563.6 to 1176.4 K by drop calorimetry. The low temperature Cp and high temperature enthalpy data are smoothly joined at 298.15 K by a polynomial curve fitting method. The deviations of the observed enthalpies from the adopted values are about 0.2-1%, except the enthalpy value at 753 K (2.0%). Heat capacities above 1200 K are extrapolated from the adopted Cp functions. The extrapolated Cp at the melting point (2887 K), 14.8 J K mol^{-1}, is in reasonable agreement with the value 2 x 7.25 cal K^{-1} g-$atom^{-1}$ suggested by Kubaschewski (5).

Combination of the earlier low temperature Cp measurements of Nernst and Schwers (28-90 K) (6) and Parks and Kelley (87-293 K) (7) yields $S°(298.15\ K) = 9.5 \pm 0.2$ cal K^{-1} mol^{-1}, based on S°(298.15 K) = 0.04 cal K^{-1} mol^{-1} (8). These Cp measurements are less accurate than those of Gmelin (3), and are not adopted in the tabulation.

Fischer and Ertmer (9) determined high temperature enthalpy data by drop calorimetry in the temperature range from 0° to 1716°C. The accuracy was claimed to be approximately ±4%. We have not adopted their enthalpy data in the tabulation since the heat capacities which we derive from their data are always less than those of MgO (10) when the temperature is above 1000 K. The deviations between their enthalpy data and the adopted values are approximately 1.8% at 693 K, 3.3% at 1283 K and 5.3% at 1989 K.

Fusion Data

Schneider (11) adjusted earlier melting points (12-15) to IPTS-48, yielding values ranging from 2565° to 2630°C. Recent studies by Foex (16, 17), using solar energy to melt the center of a rotating crucible of oxide, gave values higher by about 300°C. Apparently the earlier values were depressed by formation of tungstates with the tungsten supports of the sample. Unfortunately, Foex reported three values: 2910°C (16, 1968), 2950°C(16, 1968) and 2900°C (17, 1969). We adopt an intermediate value of 2927°C = 3200 K and arbitrarily assign an uncertainty of 50 K.

The enthalpy of melting is assumed to be 19 kcal mol^{-1} which is calculated from the estimated $\Delta_{fus}S°$ = 6 cal K^{-1} mol^{-1} at the melting point. The latter is estimated to be the same as that of MgO (10).

References

1. E. J. Huber, Jr. and C. E. Holley, Jr., J. Phys. Chem. 60, 498 (1956).
2. U. S. Nat. Bur. Stand. Circ. 500, 1952.
3. E. Gmelin, Z. Naturforsch. 24a, 1794 (1969).
4. J. J. Lander, J. Amer. Chem. Soc. 73, 5794 (1951).
5. O. Kubaschewski, E. Ll. Evans and C. B. Alcock, "Metallurgical Thermochemistry," 4th Ed., Pergamon Press, London, 1967, p. 206.
6. W. Nernst and F. Schwers, "Untersuchungen uber die spezifischen Warme bei tiefen Temperaturen," Sigzb. Konig. Preuss. Akad. Wiss., 1914, p. 355.
7. G. S. Parks and K. K. Kelley, J. Phys. Chem. 30, 47 (1926).
8. K. K. Kelley and E. G. King, U. S. Bur. Mines Bull. 592, 1961.
9. W. A. Fischer and W. Ertmer, Archiv fur das Eisenhuttenwesen, 37, 275 (1966).
10. JANAF Thermochemical Tables: MgO(c), 12-31-65.
11. S. J. Schneider, "Compilation of the Melting Points of the Metal Oxides," NBS Monograph 68, 1963.
12. C. W. Kanolt, J. Wash. Acad. Sci. 3, 315 (1913); Z. Anorg. Chem. 85, 1411 (1914).
13. Ya. I. Ol'shanskii, Dokl. Akad. Nauk SSSR 59 , 1105 (1948).
14. R. C. Doman, J. B. Barr, N. R. McNally, and A. M. Alper, Bull. Amer. Ceram. Soc. 41, 584 (1962).
15. E. E. Schumacher, J. Amer. Chem. Soc. 48, 396 (1926).
16. M. Foex, Elec. MGD, Proc. Symp. 5, 3139 (1968) (Chem. Abstr. 70, 108848s, 1969); Solar Energy 9, 61 (1965).
17. J. P. Traverse and M. Foex, High Temp.-High Press. 1, 409 (1969).

Calcium Oxide (CaO) $Ca_1O_1(cr)$

Enthalpy Reference Temperature = T_r = 298.15 K Standard State Pressure = p° = 0.1 MPa

T/K	$C_p°$ (J K⁻¹mol⁻¹)	S° (J K⁻¹mol⁻¹)	$-[G°-H°(T_r)]/T$ (J K⁻¹mol⁻¹)	$H°-H°(T_r)$ (kJ mol⁻¹)	$\Delta_f H°$ (kJ mol⁻¹)	$\Delta_f G°$ (kJ mol⁻¹)	Log K_f
0	0.	0.	INFINITE	-6.749	-631.760	-631.760	INFINITE
100	14.715	6.222	69.166	-6.294	-633.739	-624.099	325.996
200	33.677	23.041	41.825	-3.757	-634.925	-613.887	160.331
298.15	42.120	38.212	38.212	0.	-635.089	-603.501	105.731
300	42.242	38.473	38.213	0.078	-635.086	-603.305	105.045
400	46.626	51.300	39.932	4.547	-634.738	-592.755	77.406
500	48.982	61.980	43.305	9.338	-634.242	-582.316	60.834
600	50.480	71.052	47.193	14.315	-633.787	-571.975	49.795
700	51.555	78.917	51.175	19.419	-633.449	-561.701	41.915
800	52.400	85.858	55.085	24.618	-634.112	-551.362	36.000
900	53.112	92.072	58.856	29.894	-634.064	-541.024	31.400
1000	53.735	97.700	62.463	35.237	-634.273	-530.677	27.720
1100	54.300	102.849	65.904	40.639	-634.749	-520.297	24.707
1200	54.831	107.596	69.183	46.096	-643.226	-509.239	22.167
1300	55.329	112.005	72.309	51.604	-643.009	-498.082	20.013
1400	55.810	116.123	75.293	57.161	-642.759	-486.943	18.168
1500	56.275	119.989	78.146	62.766	-642.475	-475.823	16.570
1600	56.727	123.636	80.876	68.416	-642.159	-464.723	15.172
1700	57.174	127.088	83.494	74.111	-641.810	-453.644	13.939
1800	57.609	130.369	86.007	79.850	-790.102	-440.375	12.779
1900	58.045	133.495	88.425	85.633	-788.275	-420.996	11.574
2000	58.471	136.483	90.754	91.459	-786.423	-401.713	10.492
2100	58.894	139.346	93.000	97.327	-784.547	-382.523	9.515
2200	59.317	142.096	95.170	103.238	-782.651	-363.423	8.629
2300	59.735	144.742	97.268	109.190	-780.739	-344.411	7.822
2400	60.149	147.293	99.299	115.184	-778.816	-325.481	7.084
2500	60.563	149.757	101.269	121.220	-776.887	-306.632	6.407
2600	60.978	152.140	103.180	127.297	-774.958	-287.860	5.783
2700	61.388	154.449	105.036	133.415	-773.036	-269.161	5.207
2800	61.798	156.689	106.841	139.575	-771.126	-250.534	4.674
2900	62.208	158.865	108.597	145.775	-769.238	-231.976	4.178
3000	62.618	160.981	110.308	152.016	-767.378	-213.481	3.717
3100	63.024	163.040	111.976	158.298	-765.554	-195.048	3.287
3200	63.434	165.048	113.604	164.621	-763.774	-176.673	2.884
3200.000	63.434	165.048	113.604	164.621	------ CRYSTAL <--> LIQUID -----		
3300	63.839	167.006	115.192	170.985	-762.047	-158.353	2.507
3400	64.245	168.918	116.744	177.389	-760.379	-140.084	2.152
3500	64.651	170.786	118.262	183.834	-758.779	-121.864	1.819
3600	65.057	172.613	119.746	190.319	-757.254	-103.688	1.504

PREVIOUS: June 1971 CURRENT: June 1973

Calcium Oxide (CaO) $Ca_1O_1(cr)$

CALCIUM OXIDE (CaO) LIQUID M_r = 56.0794

$S°(298.15\ K) = [62.321]\ J\ K^{-1}\ mol^{-1}$

$T_{fus} = 3200 \pm 50\ K$

$\Delta_f H°(298.15\ K) = [-557.335]\ kJ\ mol^{-1}$

$\Delta_{fus}H° = [79.5]\ kJ\ mol^{-1}$

Enthalpy of Formation

$\Delta_f H°(\ell,\ 298.15\ K)$ is calculated from $\Delta_f H°(cr,\ 298.15\ K)$ by adding $\Delta_{fus}H$ and the difference in enthalpy, H°(3200 K) - H°(298.15 K), between the crystal and liquid.

Heat Capacity and Entropy

A glass transition is assumed at 2100 K. Heat capacities of the liquid below 2100 K are assumed to be the same as those of the crystal. Above 2100 K the heat capacity is assumed to be constant at 7.50 cal K^{-1} g-atom^{-1}.

The entropy at 298.15 K is calculated in a manner analogous to that used for the enthalpy of formation.

Fusion Data

See the crystal table for details.

Decomposition

CaO does not vaporize simply to CaO(g); extensive dissociation of CaO(g) to Ca(g) was found mass-spectrometrically (1).

Reference

1. I. Pelchowitch, Phillips Research Repts. 9, 42 (1954).

Calcium Oxide (CaO) $Ca_1O_1(l)$

Enthalpy Reference Temperature = T_r = 298.15 K Standard State Pressure = p° = 0.1 MPa

T/K	$C_p°$	S°	$-[G°-H°(T_r)]/T$	$H°-H°(T_r)$	$\Delta_f H°$	$\Delta_f G°$	Log K_f
	J K⁻¹mol⁻¹			kJ mol⁻¹			
0							
100							
200							
298.15	42.120	62.321	62.321	0.	-557.335	-532.935	93.368
300	42.242	62.582	62.322	0.078	-557.332	-532.783	92.766
400	46.626	75.409	64.041	4.547	-556.984	-524.644	68.511
500	48.982	86.089	67.413	9.338	-556.488	-516.616	53.971
600	50.480	95.160	71.301	14.315	-556.033	-508.686	44.285
700	51.555	103.026	75.284	19.419	-555.695	-500.823	37.372
800	52.400	109.966	79.194	24.618	-556.358	-492.895	32.183
900	53.112	116.180	82.964	29.894	-556.310	-484.968	28.147
1000	53.735	121.809	86.572	35.237	-556.519	-477.032	24.918
1100	54.300	126.957	90.012	40.639	-556.995	-469.062	22.274
1200	54.831	131.705	93.291	46.096	-565.472	-460.415	20.041
1300	55.329	136.114	96.418	51.604	-565.255	-451.669	18.148
1400	55.810	140.232	99.402	57.161	-565.005	-442.941	16.526
1500	56.275	144.098	102.254	62.766	-564.721	-434.232	15.121
1600	56.727	147.744	104.984	68.416	-564.405	-425.543	13.893
1700	57.174	151.197	107.602	74.111	-564.056	-416.874	12.809
1800	57.609	154.477	110.116	79.850	-712.348	-406.017	11.782
1900	58.045	157.604	112.534	85.633	-710.521	-389.048	10.696
2000	58.471	160.592	114.862	91.459	-708.669	-372.176	9.720
2100	58.894	163.455	117.109	97.327	-706.793	-355.397	8.840
2100.000	58.894	163.455	117.109	97.327	GLASS <--> LIQUID		
2100.000	62.760	163.455	117.109	97.327	TRANSITION		
2200	62.760	166.374	119.282	103.603	-704.531	-338.717	8.042
2300	62.760	169.164	121.391	109.879	-702.296	-322.140	7.316
2400	62.760	171.835	123.437	116.155	-700.092	-305.659	6.652
2500	62.760	174.397	125.425	122.431	-697.922	-289.269	6.044
2600	62.760	176.859	127.356	128.707	-695.794	-272.965	5.484
2700	62.760	179.227	129.234	134.983	-693.714	-256.741	4.967
2800	62.760	181.510	131.060	141.259	-691.688	-240.594	4.488
2900	62.760	183.712	132.838	147.535	-689.724	-224.519	4.044
3000	62.760	185.840	134.569	153.811	-687.829	-208.510	3.630
3100	62.760	187.898	136.257	160.087	-686.011	-192.562	3.245
3200	62.760	189.890	137.902	166.363	-684.278	-176.673	2.884
3200.000	62.760	189.890	137.902	166.363	------ CRYSTAL <--> LIQUID -----		
3300	62.760	191.821	139.507	172.639	-682.638	-160.836	2.546
3400	62.760	193.695	141.073	178.915	-681.099	-145.047	2.228
3500	62.760	195.514	142.603	185.191	-679.668	-129.302	1.930
3600	62.760	197.282	144.097	191.467	-678.352	-113.596	1.648
3700	62.760	199.002	145.558	197.743	-677.158	-97.925	1.382
3800	62.760	200.676	146.986	204.019	-676.092	-82.285	1.131
3900	62.760	202.306	148.384	210.295	-675.160	-66.670	0.893
4000	62.760	203.895	149.752	216.571	-674.364	-51.079	0.667
4100	62.760	205.444	151.091	222.847	-673.711	-35.505	0.452
4200	62.760	206.957	152.404	229.123	-673.202	-19.946	0.248
4300	62.760	208.434	153.690	235.399	-672.840	-4.396	0.053
4400	62.760	209.876	154.950	241.675	-672.622	11.146	-0.132
4500	62.760	211.287	156.187	247.951	-672.538	26.683	-0.310

PREVIOUS: June 1971 CURRENT: June 1973

Calcium Oxide (CaO) $Ca_1O_1(l)$

CALCIUM OXIDE (CaO) CRYSTAL-LIQUID M_r = 56.0794

0 to 3200 K crystal
above 3200 K liquid

Refer to the individual tables for details.

Calcium Oxide (CaO) $Ca_1O_1(cr,l)$

Enthalpy Reference Temperature = T_r = 298.15 K   Standard State Pressure = p° = 0.1 MPa

T/K	C_p° (J K⁻¹mol⁻¹)	S° (J K⁻¹mol⁻¹)	$-[G^\circ - H^\circ(T_r)]/T$ (J K⁻¹mol⁻¹)	$H^\circ - H^\circ(T_r)$ (kJ mol⁻¹)	$\Delta_f H^\circ$ (kJ mol⁻¹)	$\Delta_f G^\circ$ (kJ mol⁻¹)	Log K_f
0	0.	0.	INFINITE	-6.749	-631.760	-631.760	INFINITE
100	14.715	6.222	69.166	-6.294	-633.739	-624.099	325.996
200	33.677	23.041	41.825	-3.757	-634.925	-613.887	160.331
298.15	42.120	38.212	38.212	0.	-635.089	-603.501	105.731
300	42.242	38.473	38.213	0.078	-635.086	-603.305	105.045
400	46.626	51.300	39.932	4.547	-634.738	-592.755	77.406
500	48.982	61.980	43.305	9.338	-634.242	-582.316	60.834
600	50.480	71.052	47.193	14.315	-633.787	-571.975	49.795
700	51.555	78.917	51.175	19.419	-633.449	-561.701	41.915
800	52.400	85.858	55.085	24.618	-634.112	-551.362	36.000
900	53.112	92.072	58.856	29.894	-634.064	-541.024	31.400
1000	53.735	97.700	62.463	35.237	-634.273	-530.677	27.720
1100	54.300	102.849	65.904	40.639	-634.749	-520.297	24.707
1200	54.831	107.596	69.183	46.096	-643.226	-509.239	22.167
1300	55.329	112.005	72.309	51.604	-643.009	-498.082	20.013
1400	55.810	116.123	75.293	57.161	-642.759	-486.943	18.168
1500	56.275	119.989	78.146	62.766	-642.475	-475.823	16.570
1600	56.727	123.636	80.876	68.416	-642.159	-464.723	15.172
1700	57.174	127.088	83.494	74.111	-641.810	-453.644	13.939
1800	57.609	130.369	86.007	79.850	-790.102	-440.375	12.779
1900	58.045	133.495	88.425	85.633	-788.275	-420.996	11.574
2000	58.471	136.483	90.754	91.459	-786.423	-401.713	10.492
2100	58.894	139.346	93.000	97.327	-784.547	-382.523	9.515
2200	59.317	142.096	95.170	103.238	-782.651	-363.423	8.629
2300	59.735	144.742	97.268	109.190	-780.739	-344.411	7.822
2400	60.149	147.293	99.299	115.184	-778.816	-325.481	7.084
2500	60.563	149.757	101.269	121.220	-776.887	-306.632	6.407
2600	60.978	152.140	103.180	127.297	-774.958	-287.860	5.783
2700	61.388	154.449	105.036	133.415	-773.036	-269.161	5.207
2800	61.798	156.689	106.841	139.575	-771.126	-250.534	4.674
2900	62.208	158.865	108.597	145.775	-769.238	-231.976	4.178
3000	62.618	160.981	110.308	152.016	-767.378	-213.481	3.717
3100	63.024	163.040	111.976	158.298	-765.554	-195.048	3.287
3200	63.434	165.048	113.604	164.621	-763.774	-176.673	2.884
3200.000	63.434	189.890	113.604	244.117	CRYSTAL <--> LIQUID		
3200.000	62.760	189.890	113.604	244.117	TRANSITION		
3300	62.760	191.821	115.945	250.393	-682.638	-160.836	2.546
3400	62.760	193.695	118.204	256.669	-681.099	-145.047	2.228
3500	62.760	195.514	120.387	262.945	-679.668	-129.302	1.930
3600	62.760	197.282	122.499	269.221	-678.352	-113.596	1.648
3700	62.760	199.002	124.543	275.497	-677.158	-97.925	1.382
3800	62.760	200.676	126.525	281.773	-676.092	-82.285	1.131
3900	62.760	202.306	128.447	288.049	-675.160	-66.670	0.893
4000	62.760	203.895	130.313	294.325	-674.364	-51.079	0.667
4100	62.760	205.444	132.127	300.601	-673.711	-35.505	0.452
4200	62.760	206.957	133.891	306.877	-673.202	-19.946	0.248
4300	62.760	208.434	135.607	313.153	-672.840	-4.396	0.053
4400	62.760	209.876	137.279	319.429	-672.622	11.146	-0.132
4500	62.760	211.287	138.908	325.705	-672.538	26.684	-0.310

PREVIOUS: CURRENT: June 1973

Calcium Oxide (CaO) $Ca_1O_1(cr,l)$

CALCIUM OXIDE (CaO) IDEAL GAS M_r = 56.0794

$\Delta_f H°(0\ K)$ = 45.1 ± 21 kJ mol^{-1}

$S°(298.15\ K)$ = 219.717 ± 0.42 J K^{-1} mol^{-1} $\Delta_f H°(298.15\ K)$ = 43.9 ± 21 kJ mol^{-1}

Electronic and Molecular Constants (σ = 1)

Source	State	e_i, cm^{-1}	g_i	r_e, Å	B_e, cm^{-1}	α_e, cm^{-1}	ω_e, cm^{-1}	$\omega_e x_e$, cm^{-1}
(1-3)	$X^1\Sigma^+$	0.0	1	1.822	0.44452	0.00338	733.4	5.28
(1)	$a\ ^3\Pi$	8225.	6	2.097	0.3353	0.0015	556.2	3.30
(1)	$A'^1\Pi$	8612.	2	2.092	0.3372	0.0021	545.7	2.54
(1, 4, 5)	$^3\Sigma^+$	[10000]	3	[1.906]	[0.4059]	[0.0014]	[719]	[2.1]
(1-3)	$A^1\Sigma^+$	11549.	1	1.906	0.40592	0.00137	718.9	2.11
(5)	$^3\Sigma^+$	[21000]	3	[2.00]	[0.369]	[0.003]	[560]	[4.]
(5)	$^3\Delta$	[21000]	6	[2.00]	[0.369]	[0.003]	[560]	[4.]
(5)	$^1\Delta$	[23000]	2	[2.00]	[0.369]	[0.003]	[560]	[4.]
(5)	$^3\Sigma^-$	[23000]	3	[2.00]	[0.369]	[0.003]	[560]	[4.]
(5)	$^1\Sigma^-$	[23000]	1	[2.00]	[0.369]	[0.003]	[560]	[4.]
(6)	$C^1\Sigma^+$	28772.	1	1.989	0.3731	0.0032	560.9	4.0
(5)	$^3\Pi$	[24000]	6	[1.95]	[0.3882]	[0.0055]	[581]	[3.3]
(7)	$B^1\Pi$	25913.	2	1.950	0.3882	0.0055	581.0	[3.3]

Enthalpy of Formation

We adopt D_0^0 = 91 ± 5 and $\Delta_f H°(298.15\ K)$ = 10.5 ± 5 kcal mol^{-1} based on mass-spectrometric data (8, 9) for four reactions analyzed below. We give "<" or ">" for ΔH° and D_0^0 values which may be biased due to our auxiliary data. JANAF differences in $\Delta_f H°(298.15\ K)$ for WO_3-WO_2 and MoO_3-MoO_2 (5) may be biased by up to -3.5 and +4.1 kcal mol^{-1}, respectively, leading to bias of the opposite sign in D_0^0 values derived from reactions B and C. Adjustments for this possible bias would improve the agreement in D_0^0 but leave the mean value almost unchanged. Kalff (10) used spectrometry of CO-N_2O flames to derive D_0^0 = 86.5 ± 4.6 kcal mol^{-1}. This value becomes ~89.5 kcal mol^{-1} when adjusted to be consistent with JANAF Gibbs energy functions. Our adopted D_0^0 = 91 ± 5 is similar to that of Rosen (6) and is comparable with other selected values (8, 11), considering the difference in CaO functions.

The controversy over D_0^0 of alkaline earth oxides has been reviewed in detail (8, 11-13). Uncertainty in the electronic partition function of CaO due to triplet states now is much reduced (1). Schofields's criticism (12) of flame studies is supported by new evidence for importance of hydroxides (10, 14, 15); thus, we dismiss higher flame values for D_0^0 (12). Valid criticism (8) also causes us to dismiss higher D_0^0 values derived from vaporization data. A linear Birge-Sponer extrapolation yields a very low estimate for D_0^0, just as it does for the alkali halides (13), unless $X^1\Sigma$ fails to dissociate to an excited state atom as predicted (17, 1).

Source	Method	Reaction[a]	T/K	Data Points	δS cal K^{-1}mol^{-1}	$\Delta_r H°(298.15\ K)$, kcal mol^{-1} 2nd law	3rd law	kcal/mol $\Delta_f H°(298.15\ K)$	D_0^0
(8)	Knudsen mass spec.	A	2168-2410	4	-2.3±9.7	24±22	29.4±3	12.7	88.8
	Knudsen mass spec.	B	2328-2334	2	67	209	51.9±3	>6.5	<95.0
	Knudsen mass spec.	C	2393-2410	3	12±83	85±100	55.8±3	≤15.5	<85.9
(9)	Knudsen mass spec.	D	2180-2386	3	1±31	31±72	27.9±6	5.6	95.8[b]

[a]Reactions: A) $Ca(g) + O_2(g) = CaO(g) + O(g)$; C) $Ca(g) + MoO_3(g) = CaO(g) + MoO_2(g)$;
B) $Ca(g) + WO_3(g) = CaO(g) + WO_2(g)$; D) $Ca(g) + SO(g) = CaO(g) + S(g)$.

[b]Ion intensities and D_0^0 reported (9) at 2180 K are discrepant by ~8 kcal mol^{-1}; we use ion intensities.

Heat Capacity and Entropy

Electronic levels (T_o) and vibrational-rotational constants of the observed states are from (1, 2, 3, 6, 7). The long-sought $a^3\Pi$ and $A'^1\Pi$ states of CaO, SrO and BaO were characterized by Field (1) using a new method for assignment of perturbations. This study resolved the long-standing controversy over low-lying electronic levels and confirmed that $X^1\Sigma^+$ is the ground state. We estimate the other potentially low-lying state ($^3\Sigma^+$) at 10000 cm^{-1} by assuming that it lies 1500 ± 1200 cm^{-1} (1, 4) below the isoconfigurational order of MgO(5). Rotational analysis of a band near 5470 Å was attributed (16) to CaO, but we question this assignment. Our thermodynamic functions are calculated using first-order anharmonic corrections to Q_r^i and Q_v^i in the partition function $Q = Q_t \sum_i Q_r^i Q_v^i g_i \exp(-c_2\epsilon_i/T)$.

References

1. R. W. Field, J. Chem. Phys. 60, 2400 (1974); R. W. Field, G. A. Capelle & C. R. Jones, J. Mol. Spectrosc. 54, 156 (1975).
2. L. Brewer and R. Hauge, J. Mol. Spectrosc. 25, 330 (1968).
3. H. Johansen, Ph.D. Thesis, Univ. California, Berkeley, UCRL-19649, 1970.
4. K. D. Carlson, K. Kaiser, C. Moser and A. C. WAhl, J. Chem. Phys. 52, 4678 (1970).
5. JANAF Thermochemical TAbles: CaO(g), MgO(g), BeO(g) 12-31-74; SrO(g), BaO(g) 6-30-74; WO_2(g), WO_3(g) 9-30-66; MoO_2(g), MoO_3(g) 6-30-68.
6. B. Rosen, "Spectroscopic Data Relative to Diatomics," Pergamon Press, New York, 1970, pp. 112-3.
7. A. Lagerqvist, Arkiv. Fysik 8, 83 (1954).
8. J. Drowart, G. Exsteen and G. Verhaegen, Trans. Faraday Soc. 60, 1290 (1964).
9. R. Colin, P. Goldfinger and M. Jeunehomme, Trans. Faraday Soc. 60, 306 (1964).
10. P. J. Kalff and C. Th. J. Alkemade, J. Chem. Phys. 59, 2572 (1973); 60, 1698 (1974).
11. L. Brewer and G. Rosenblatt, Advan. High Temp. Chem. 2, 1 (1969).
12. K. Schofield, Chem. Rev. 67, 707 (1967).
13. A. G. Gaydon, "Dissociation Energies and Spectra of Diatomic Moleculess," 3rd ed., Chapman and Hall Ltd. London, 1968, pp. 236-244, 109.
14. J. van der Hurk, Tj. Hollander and C. Th. J. Alkemade, J. Quant. Spectrosc. Radiat. Transfer 13, 273 (1973).
15. V. G. Ryabova, Teplofiz. Vys. Temp. 10, 744 (1972); 9, 290, 755 (1971).
16. G. A. Volynets, G. V. Kovalenok and V. A. Sokolov, Opt. Spektrosk. 36, 1034 (1974).
17. N. J. Stag and W. G. Richards, Mol. Phys. 27, 787 (1974).

Calcium Oxide (CaO) $Ca_1O_1(g)$

Enthalpy Reference Temperature = T_r = 298.15 K Standard State Pressure = p° = 0.1 MPa

T/K	$C_p°$ (J K^{-1}mol^{-1})	S°	$-[G°-H°(T_r)]/T$	$H°-H°(T_r)$ (kJ mol^{-1})	$\Delta_f H°$	$\Delta_f G°$	Log K_f
0	0.	0.	INFINITE	-8.953	45.057	45.057	INFINITE
100	29.136	186.730	247.175	-6.044	45.533	37.122	-19.390
200	30.381	207.194	222.620	-3.085	44.768	28.975	-7.568
298.15	32.455	219.717	219.717	0.	43.932	21.405	-3.750
300	32.491	219.918	219.717	0.060	43.917	21.266	-3.703
400	34.089	229.499	221.013	3.394	43.131	13.835	-1.807
500	35.139	237.226	223.508	6.859	42.301	6.604	-0.690
600	35.833	243.699	226.348	10.410	41.330	-0.447	0.039
700	36.313	249.260	229.233	14.019	40.172	-7.321	0.546
800	36.663	254.133	232.047	17.669	37.960	-13.911	0.908
900	36.947	258.468	234.747	21.349	36.412	-20.305	1.178
1000	37.218	262.375	237.317	25.058	34.569	-26.510	1.385
1100	37.532	265.936	239.760	28.794	32.428	-32.517	1.544
1200	37.947	269.219	242.079	32.567	22.266	-37.694	1.641
1300	38.518	272.278	244.286	36.389	20.797	-42.630	1.713
1400	39.291	275.159	246.389	40.278	19.378	-47.456	1.771
1500	40.298	277.903	248.399	44.255	18.035	-52.182	1.817
1600	41.551	280.542	250.326	48.345	16.792	-56.822	1.855
1700	43.046	283.105	252.179	52.573	15.674	-61.388	1.886
1800	44.755	285.613	253.967	56.962	-133.969	-63.682	1.848
1900	46.639	288.082	255.698	61.530	-133.357	-59.793	1.644
2000	48.642	290.525	257.378	66.293	-132.567	-55.941	1.461
2100	50.703	292.948	259.015	71.260	-131.592	-52.133	1.297
2200	52.757	295.354	260.612	76.434	-130.433	-48.375	1.149
2300	54.742	297.744	262.174	81.810	-129.098	-44.674	1.015
2400	56.603	300.113	263.706	87.378	-127.601	-41.036	0.893
2500	58.293	302.459	265.209	93.125	-125.962	-37.462	0.783
2600	59.778	304.775	266.686	99.030	-124.204	-33.956	0.682
2700	61.036	307.055	268.139	105.072	-122.357	-30.520	0.590
2800	62.057	309.294	269.569	111.229	-120.451	-27.153	0.507
2900	62.844	311.486	270.977	117.476	-118.515	-23.856	0.430
3000	63.404	313.627	272.363	123.790	-116.582	-20.624	0.359
3100	63.755	315.712	273.728	130.150	-114.681	-17.457	0.294
3200	63.918	317.739	275.072	136.535	-112.839	-14.350	0.234
3300	63.917	319.706	276.395	142.928	-111.082	-11.300	0.179
3400	63.776	321.613	277.697	149.314	-109.433	-8.301	0.128
3500	63.520	323.458	278.978	155.679	-107.913	-5.350	0.080
3600	63.172	325.243	280.239	162.015	-106.538	-2.439	0.035
3700	62.754	326.968	281.478	168.311	-105.323	0.435	-0.006
3800	62.283	328.635	282.698	174.564	-104.281	3.280	-0.045
3900	61.777	330.247	283.896	180.767	-103.420	6.099	-0.082
4000	61.250	331.804	285.075	186.918	-102.750	8.898	-0.116
4100	60.713	333.310	286.233	193.017	-102.274	11.683	-0.149
4200	60.175	334.767	287.371	199.061	-101.997	14.458	-0.180
4300	59.644	336.176	288.490	205.052	-101.920	17.230	-0.209
4400	59.125	337.542	289.589	210.990	-102.040	20.001	-0.237
4500	58.624	338.865	290.670	216.877	-102.344	22.777	-0.264
4600	58.142	340.148	291.731	222.715	-102.851	25.562	-0.290
4700	57.683	341.393	292.775	228.507	-103.554	28.361	-0.315
4800	57.246	342.603	293.800	234.253	-104.443	31.177	-0.339
4900	56.834	343.779	294.808	239.957	-105.514	34.014	-0.363
5000	56.446	344.923	295.799	245.620	-106.760	36.873	-0.385
5100	56.081	346.037	296.773	251.246	-108.175	39.760	-0.407
5200	55.739	347.123	297.731	256.837	-109.748	42.675	-0.429
5300	55.418	348.182	298.673	262.395	-111.478	45.623	-0.450
5400	55.119	349.215	299.600	267.922	-113.355	48.604	-0.470
5500	54.840	350.224	300.511	273.419	-115.371	51.621	-0.490
5600	54.580	351.209	301.408	278.890	-117.507	54.676	-0.510
5700	54.337	352.173	302.290	284.336	-119.775	57.771	-0.529
5800	54.110	353.116	303.158	289.758	-122.158	60.907	-0.549
5900	53.899	354.040	304.013	295.159	-124.648	64.083	-0.567
6000	53.702	354.944	304.854	300.539	-127.237	67.304	-0.586

PREVIOUS: December 1974 (1 atm) CURRENT: December 1974 (1 bar)

Calcium Oxide (CaO) $Ca_1O_1(g)$

$S°(298.15\ K) = 56.6 \pm 1.3\ J\ K^{-1}\ mol^{-1}$

T_{fus} = [2798] K

$\Delta_f H°(0\ K) = -472.1 \pm 2.9\ kJ\ mol^{-1}$

$\Delta_f H°(298.15\ K) = -473.2 \pm 2.9\ kJ\ mol^{-1}$

Enthalpy of Formation

Sabatier (1), Mourlot (2), and von Wartenberg (3), all measured the enthalpy of solution of CaS in hydrochloric acid solution. Sabatier determined $\Delta_r H°(238.8) = -26.3\pm1.0$ kcal mol^{-1} while Mourlot determined $\Delta_r H°(291) = -26.7\pm1.0$ kcal mol^{-1} for CaS(cr) + 2HCl(100 H_2O, aq) + $CaCl_2$(100 H_2O,aq) + H_2S(aq). We derive $\Delta_f H°$(CaS, cr, 298.15 K) = -112.7±2.0 kcal mol^{-1} and $\Delta_f H°$(CaS, cr, 298.15 K) = -112.3±2.0 kcal mol^{-1} respectively, based on the following auxiliary data: $\Delta_f H°$(HCl·100 H_2O, aq, 298.15 K) = -39.657 kcal mol^{-1} (4), $\Delta_f H°$(H_2S, aq, 298.15 K) = -9.5 kcal mol^{-1} (4), and $\Delta_f H°$($CaCl_2$·100 H_2O, aq, 298.15 K) = -208.849 kcal mol^{-1} (5). Von Wartenberg measured $\Delta_r H(293.15) = -20.6\pm0.4$ kcal mol^{-1} for CaS(cr) + 2HCl(30 H_2O, aq) + $CaCl_2$(30 H_2O, aq) + H_2S(g). We derive $\Delta_f H°$(CaS, cr, 298.15 K) = -113.9±0.6 kcal mol^{-1} based on the following auxiliary data: $\Delta_f H°$(HCl·30 H_2O, aq, 298.15 K) = -39.357 kcal mol^{-1} (4), $\Delta_f H°$(H_2S, g, 298.15 K) = -4.90 kcal mol^{-1} (4, 18), and $\Delta_f H°$($CaCl_2$·30 H_2O, aq, 298.15 K) = -208.33 kcal mol^{-1} (5).

Source	Method	Reaction[a]	T/K	Data Points	δS cal $K^{-1}mol^{-1}$	$\Delta_r H$(298.15 K), kcal mol^{-1} 2nd law	3rd law	$\Delta_f H°$(298.15 K) kcal mol^{-1}
(6)	Vap. Press.	A	1400-1700	Equation	-3.94	112.3	118.39±1.0	-88.9±30
(7)	Equilibrium	B	1019-1704	15	1.06±0.1	15.6±0.1	14.14±0.5	-113.01±0.8
(8)	Equilibrium	B	1031-1698	12	1.01±0.2	15.4±0.3	14.02±0.5	-112.89±0.8
(9)	Equilibrium	B	1173-1373	Equation	0.18	14.3	14.04±1.0	-112.91±1.0
(9)	Equilibrium	C	926-1026	Equation	0.03	-23.02	-23.06±1.0	-113.31±2.0
(10)	Equilibrium	D	1124-1348	6	5.5±0.6	-51.2±0.8	-58.0±1.0	-111.98±2.0
(11)	Equilibrium	D	1173-1393	10	7.3±3.2	-76.3±4.2	-66.8±1.6	-103.2±4
(11)	Equilibrium	E	1173-1393	10	-0.95±5	254.06±6.5	255.2±1.9	-117.9±4
(12)	Equilibrium	C	1400-1650	7	-2.8±0.3	17.3±0.5	22.3±0.5	-114.04±0.6
(13)	Mass Spec.	F	1709	1			212.5±10	-103.4±10
(14)	Mass Spec.	G	1849-2155	7	-11.8±2.0	-1.4±4.0	22.0±2.0	-116.3±8

[a]Reactions:
A) CaS(c) = CaS(g)
B) CaS(c) + H_2O(g) = CaO(c) + H_2S(g)
C) CaS(c) + 0.5O_2(g) = CaO(c) + 0.5S_2(g)
D) CaS(c) + 2SO_2(g) = $CaSO_4$(c) + S_2(g)
E) CaS(c) + 3$CaSO_4$(c) = 4CaO(c) + 4SO_2(g)
F) CaS(c) + CaS(c) = Ca(g) + S(g)
G) CaS(c) + 3S(g) = Ca(g) + 2S_2(g)

The calculated 3rd-law $\Delta_r H°$(298.15 K) may have an uncertainty of 0.5 kcal mol^{-1} since the JANAF Gibbs energy functions are partially based on the estimated Cp° data (above 300 K). The results of the majority of the equilibrium studies, (7, 8, 9, 10, 11, 12), are in very good agreement with the enthalpy of solution (1, 2, 3) studies within combined experimental errors. The mass spectrometric studies (13, 14) are not as reliable in condensed phase equilibrium as they are in gas phase equilibrium (18). A weighted average, $\Delta_f H°$(CaS, cr, 298.15 K) = -113.1±0.7 kcal mol^{-1} is adopted in the tabulation.

Heat Capacity and Entropy

Anderson (15) measured the low temperature heat capacities of CaS(cr) from 58.1 - 294.9 K in an isothermal calorimeter. We have smoothed his Cp° data by a polynomial curve fitting technique and obtain S°(298.15 K) = 13.5±0.3 cal K^{-1} mol^{-1} based on S°(60 K) = 1.192 cal K^{-1} mol^{-1}. The value of S°(60 K) is calculated from the combination of Debye and Einstein functions (θ_D = 284 and θ_E= 369). The Cp° values above 300 K are estimated by graphical extrapolation combined with a variant of method B of Kubaschewski et al. (16).

Fusion Data

The melting point of CaS(cr) is given as 2525°C (17).

References

1. M. Sabatier, Ann. Chim. Phys. 22, 5 (1881).
2. M. A. Mourlot, Ann. Chim. Phys. 17, 510 (1899).
3. H. von Wartenberg, Z. Anorg. Chem. 252, 136 (1943).
4. U. S. Nat. Bur. Stand. Tech. Note 270-3, 1968.
5. U. S. Nat. Bur. Stand. Tech. Note 270-6, 1971.
6. B. P. Nikonov and N. G. Otmakhova, Zhur. Fiz. Khim. 35, 1494 (1961).
7. N. J. Filipovska and H. B. Bell, Bull. Soc. Chim. Beograd 40, 499 (1975).
8. T. Rosenquist, J. Metals 3, 538 (1951).
9. T. Uno, Tetsu-to-Hagane 37, 14 (1951); data taken from Chem. Abstr. 46, 2382 (1952).
10. R. Schenck and F. Hammerschmidt, Z. Anorg. Chem. 210, 305 (1933).
11. J. Zawadzki, Z. Anorg. Allgem. Chem. 205, 180 (1932).
12. L. F.. Sander and G. W. Healy, Trans. Met. Soc. AIME 242, 1039 (1968).
13. J. Berkowitz and J. R. Marquart, J. Chem. Phys. 39, 275 (1963).
14. R. Colin, P. Goldfinger, and M. Jeunehomme, Trans. Faraday Soc. 60, 306 (1964).
15. C. T. Anderson, J. Amer. Chem. Soc. 53, 476 (1931).
16. O. Kubaschewski, E. L. Evans, and C. B. Alcock, "Mettalurgical Thermochemistry," Pergamon Press, Oxford, 1967.
17. R. Juza and K. Bluzen, Z. Physik. Chem. (Frankfurt) 17, 82 (1958).
18. JANAF Thermochemical Tables: CaS(g) 9-30-77; H_2S(g) 6-30-77.

Enthalpy Reference Temperature = T_r = 298.15 K — Standard State Pressure = p° = 0.1 MPa

T/K	$C_p°$ (J $K^{-1}mol^{-1}$)	S°	-[G°-H°(T_r)]/T	H°-H°(T_r) (kJ mol^{-1})	$\Delta_f H°$	$\Delta_f G°$	Log K_f
0	0.	0.	INFINITE	-8.999	-472.061	-472.061	INFINITE
100	25.945	14.477	94.617	-8.014	-472.746	-471.345	246.206
200	42.204	38.610	60.808	-4.440	-473.083	-469.779	122.694
298.15	47.442	56.601	56.601	0.	-473.210	-468.129	82.014
300	47.447	56.895	56.602	0.088	-473.213	-468.098	81.503
400	49.204	70.796	58.486	4.924	-475.608	-466.281	60.890
500	50.501	81.919	62.096	9.911	-477.314	-463.765	48.449
600	51.547	91.223	66.196	15.016	-478.737	-460.916	40.126
700	52.342	99.230	70.357	20.211	-480.027	-457.844	34.165
800	53.011	106.265	74.415	25.480	-482.164	-454.483	29.675
900	53.555	112.540	78.308	30.808	-536.618	-449.864	26.109
1000	54.099	118.211	82.020	36.192	-536.911	-440.212	22.994
1100	54.586	123.391	85.548	41.626	-537.471	-430.517	20.444
1200	55.020	128.159	88.903	47.107	-546.037	-420.137	18.288
1300	55.391	132.578	92.095	52.628	-545.919	-409.650	16.460
1400	55.731	136.695	95.135	58.184	-545.780	-399.173	14.893
1500	55.987	140.549	98.036	63.769	-545.629	-388.706	13.536
1600	56.317	144.172	100.807	69.384	-545.462	-378.250	12.349
1700	56.761	147.600	103.460	75.038	-545.271	-367.805	11.301
1800	57.195	150.857	106.003	80.736	-693.723	-355.161	10.306
1900	57.629	153.961	108.446	86.477	-692.057	-336.397	9.248
2000	58.032	156.927	110.797	92.261	-690.367	-317.722	8.298
2100	58.382	159.767	113.062	98.081	-688.659	-299.131	7.440
2200	58.743	162.491	115.247	103.938	-686.936	-280.623	6.663
2300	59.115	165.111	117.358	109.830	-685.201	-262.193	5.955
2400	59.496	167.634	119.401	115.761	-683.456	-243.839	5.307
2500	59.899	170.071	121.379	121.731	-681.704	-225.557	4.713
2600	60.291	172.428	123.298	127.740	-679.950	-207.346	4.166
2700	60.668	174.711	125.160	133.788	-678.201	-189.202	3.660
2800	61.045	176.924	126.969	139.874	-676.464	-171.123	3.192
2900	61.421	179.073	128.729	145.997	-674.747	-153.105	2.758
3000	61.798	181.161	130.442	152.158	-673.058	-135.147	2.353

PREVIOUS: December 1971 CURRENT: September 1977

CALCIUM SULFIDE (CaS) IDEAL GAS M_r = 72.14

$\Delta_f H°(0\ K) = -124.39 \pm 8.4\ kJ\ mol^{-1}$

$S°(298.15\ K) = 232.589 \pm 0.42\ J\ K^{-1}\ mol^{-1}$ $\Delta_f H°(298.15\ K) = -123.60 \pm 8.4\ kJ\ mol^{-1}$

Electronic and Molecular Constants (σ = 1)

Source	State	ε_i, cm^{-1}	g_i	r_e, Å	B_e, cm^{-1}	α_e, cm^{-1}	ω_e, cm^{-1}	$\omega_e x_e$, cm^{-1}
(3)	$X^1\Sigma^+$	0.0	1	2.3178	0.17667	0.000837	462.23	1.76
(4)	$^3\Pi$	[6800.]	6	[2.63]	[0.137]	[0.0006]	[330]	[1.5]
(4)	$^1\Pi$	[7200.]	2	[2.63]	[0.137]	[0.0006]	[330]	[1.5]
(3, 4)	$^3\Sigma^+$	[14000.]	3	[2.3864]	[0.16666]	[0.000605]	[409.04]	[0.818]
(3)	$A^1\Sigma^+$	15194.44	1	2.3864	0.16666	0.000605	409.04	0.818

Enthalpy of Formation

The adopted value for the enthalpy of formation, $\Delta_f H°$(CaS, g, 298.15 K) = 29.54±2 kcal mol^{-1}, is based on the Knudsen mass spectrometric studies analyzed below. The two independent studies are in very good agreement and we have adopted the average value. Our value is intermediate between $\Delta_f H°$(298.15 K) = 32 kcal mol^{-1} obtained by NBS (5) and $\Delta_f H°$(298.15 K) = 27.9±5 kcal mol^{-1} obtained by Mills (6) in a recent critical compilation. Using auxiliary JANAF data (4) and a recommended value for $D_0^0(S_2$, g) = 100.69±01 kcal mol^{-1} (4), we calculate D_0^0(CaS, g) = 78.7±2 kcal mol^{-1}.

Source	Method	Reaction[a]	T/K	Data Points	δS cal K^{-1}mol^{-1}	$\Delta_r H°$(298.15 K), kcal mol^{-1} 2nd law	3rd law	$\Delta_f H°$(298.15 K) kcal mol^{-1}	D_0^0
(1)	Mass Spec	A	2058-2319	9	2.83±3.5	28.32±7.5	22.32±7.5	29.72±1.7	78.6±2
(2)	Mass Spec	A	1709-1962	2	0.22	22.37	21.97±2	29.37±2	78.9±2

[a]Reaction: A) Ca(g) + S_2(g) = CaS(g) + (g)

Heat Capacity and Entropy

Electronic levels (T_{oo}) and vibrational-rotational constants for the observed states are from the optical study of Blues and Barrow (3). Other low-lying electronic states and their vibrational-rotational constants are estimated in isoconfigurational groups by analogy with CaO (4) and from trends observed in the known states of the other alkaline-earth oxides and sulfides (4). Uncertainty in the energy and constants for the estimated states may contribute as much as 2-3 cal K^{-1} mol^{-1} to S° at 3000 K. The molecular constants have been corrected to the natural isotopic abundances. The thermodynamic functions are calculated using first order anharmonic corrections to Q_r^i and Q_v^i in the partition function $Q = Q_t \sum_i Q_r^i Q_v^i g_i \exp(-c_2\varepsilon_i/T)$.

References

1. R. Colin, P. Goldfinger, and M. Jeunehomme, Trans. Faraday Soc. 60, 306 (1964).
2. J. R. Marquart and J. Berkowitz, J. Chem. Phys. 39, 283 (1963).
3. R. C. Blues and R. F. Barrow, Trans. Faraday Soc. 65, 646 (1969).
4. JANAF Thermochemical Tables: Ca(g), 12-31-68; S(g), 6-30-71; S_2(g), 12-31-65; BaO(g), SrO(g), 6-30-74; BeO(g), MgO(g), CaO(g), 12-31-74; BeS(g), MgS(g), SrS(g), BaS(g), 9-30-77.
5. U. S. Nat. Bur. Stand. Tech. Note 270-6, 1971.
6. K. C. Mills, "Thermodynamic Data for Inorganic Sulphides, Selenides, and Tellurides," Butterworths, London, 1974.

Calcium Sulfide (CaS) Ca_1S_1(g)

Enthalpy Reference Temperature = T_r = 298.15 K Standard State Pressure = p° = 0.1 MPa

T/K	$C_p°$ (J K^{-1}mol^{-1})	S°	$-[G°-H°(T_r)]/T$	$H°-H°(T_r)$ (kJ mol^{-1})	$\Delta_f H°$	$\Delta_f G°$	Log K_f
0	0.	0.	INFINITE	-9.358	124.385	124.385	INFINITE
100	29.604	197.623	262.034	-6.441	125.632	108.719	-56.789
200	32.744	219.086	235.716	-3.326	124.837	92.045	-24.040
298.15	34.803	232.589	232.589	0.	123.595	76.206	-13.351
300	34.830	232.804	232.590	0.064	123.570	75.912	-13.217
400	35.897	242.987	233.971	3.606	119.880	60.331	-7.878
500	36.487	251.066	236.610	7.228	116.808	45.785	-4.783
600	36.850	257.753	239.592	10.896	113.949	31.852	-2.773
700	37.106	263.454	242.604	14.595	111.162	18.389	-1.372
800	37.336	268.423	245.527	18.317	107.479	5.432	-0.355
900	37.624	272.837	248.321	22.064	51.444	-6.070	0.352
1000	38.061	276.822	250.975	25.847	49.550	-12.362	0.646
1100	38.732	280.478	253.493	29.684	47.393	-18.450	0.876
1200	39.703	283.888	255.885	33.603	37.265	-23.710	1.032
1300	41.007	287.115	258.164	37.636	35.895	-28.734	1.155
1400	42.641	290.212	260.344	41.815	34.657	-33.659	1.256
1500	44.564	293.218	262.435	46.174	33.581	-38.500	1.341
1600	46.705	296.161	264.452	50.736	32.695	-43.275	1.413
1700	48.971	299.060	266.402	55.519	32.015	-48.001	1.475
1800	51.260	301.924	268.296	60.530	-117.123	-50.483	1.465
1900	53.474	304.756	270.141	65.768	-115.961	-46.811	1.287
2000	55.523	307.552	271.942	71.220	-114.603	-43.206	1.128
2100	57.336	310.306	273.703	76.865	-113.070	-39.673	0.987
2200	58.863	313.009	275.429	82.677	-111.390	-36.217	0.860
2300	60.079	315.654	277.120	88.627	-109.598	-32.840	0.746
2400	60.977	318.231	278.780	94.683	-107.728	-29.542	0.643
2500	61.569	320.733	280.408	100.812	-105.816	-26.324	0.550
2600	61.878	323.155	282.006	106.987	-103.897	-23.182	0.466
2700	61.936	325.492	283.573	113.180	-102.004	-20.114	0.389
2800	61.781	327.742	285.111	119.367	-100.165	-17.115	0.319
2900	61.449	329.905	286.619	125.530	-98.409	-14.180	0.255
3000	60.977	331.980	288.096	131.652	-96.758	-11.304	0.197
3100	60.399	333.971	289.544	137.722	-95.232	-8.481	0.143
3200	59.743	335.878	290.963	143.729	-93.848	-5.705	0.093
3300	59.036	337.706	292.352	149.669	-92.622	-2.970	0.047
3400	58.298	339.457	293.711	155.536	-91.564	-0.269	0.004
3500	57.547	341.136	295.043	161.328	-90.685	2.403	-0.036
3600	56.796	342.747	296.345	167.045	-89.993	5.052	-0.073
3700	56.057	344.293	297.621	172.687	-89.494	7.685	-0.108
3800	55.336	345.778	298.868	178.257	-89.193	10.307	-0.142
3900	54.639	347.207	300.090	183.755	-89.094	12.923	-0.173
4000	53.972	348.581	301.285	189.186	-89.198	15.540	-0.203
4100	53.335	349.906	302.455	194.551	-89.508	18.162	-0.231
4200	52.730	351.184	303.600	199.854	-90.023	20.794	-0.259
4300	52.159	352.418	304.721	205.098	-90.742	23.441	-0.285
4400	51.620	353.611	305.819	210.287	-91.660	26.106	-0.310
4500	51.114	354.765	306.894	215.423	-92.763	28.794	-0.334
4600	50.639	355.884	307.946	220.511	-94.069	31.508	-0.358
4700	50.194	356.968	308.978	225.552	-95.568	34.254	-0.381
4800	49.778	358.020	309.989	230.550	-97.253	37.034	-0.403
4900	49.390	359.042	310.979	235.509	-99.117	39.851	-0.425
5000	49.027	360.037	311.951	240.429	-101.153	42.707	-0.446
5100	48.689	361.004	312.903	245.315	-103.356	45.606	-0.467
5200	48.374	361.946	313.837	250.168	-105.714	48.549	-0.488
5300	48.080	362.865	314.754	254.990	-108.226	51.539	-0.508
5400	47.806	363.761	315.653	259.784	-110.882	54.578	-0.528
5500	47.552	364.636	316.536	264.552	-113.673	57.668	-0.548
5600	47.315	365.491	317.402	269.295	-116.582	60.809	-0.567
5700	47.095	366.326	318.253	274.016	-119.617	64.003	-0.587
5800	46.890	367.143	319.089	278.715	-122.764	67.252	-0.606
5900	46.699	367.943	319.910	283.394	-126.014	70.556	-0.625
6000	46.522	368.727	320.718	288.055	-129.358	73.915	-0.643

PREVIOUS: September 1977 (1 atm) CURRENT: September 1977 (1 bar)

CALCIUM (Ca_2) IDEAL GAS M_r = 80.16

D_0^0 = 12.740 ± 0.010 kJ mol^{-1} (natural abundance)
$\Delta_f H$(0 K) = 341.92 ± 0.80 kJ mol^{-1}
S°(298.15 K) = 257.07 ± 0.5 J K^{-1} mol^{-1}
$\Delta_f H$(298.15 K) = 340.72 ± 0.8 kJ mol^{-1}

Spectroscopic Constants for $^{40}Ca_2$ in cm^{-1}

State	T_e	D°	ω_e	$\omega_e x_e$	B_e	$10^3\alpha_e$	$10^6 D_e$	r_e, Å
$^1\Sigma_g^+$	0	1063.1	65.046	1.0815[a]	0.046086	0.6966[c]	0.090731[e]	3.807
$^1\Sigma_u^+$	18964	22912	136.61	0.71797[b]	0.058178	0.2970[d]	0.043361[f]	4.277

(a) $+3.1093 \times 10^{-3}$ $(v + 1/2)^3$ + 7.0080×10^{-6} $(v + 1/2)^4$ − 2.4380×10^{-8} $(v + 1/2)^5$
(b) -2.8198×10^{-3} $(v + 1/2)^3$ − 1.4773×10^{-6} $(v + 1/2)^4$ + 1.692×10^{-6} $(v + 1/2)^5$ -5.545×10^{-9} $(v + 1/2)^6$
(c) -1.619×10^{-5} $(v + 1/2)^2$ + 1.60×10^{-6} $(v + 1/2)^3$
(d) -3.202×10^{-6} $(v + 1/2)^2$ − 6.58×10^{-8} $(v + 1/2)^3$
(e) $+1.330336 \times 10^{-9}$ $(v + 1/2)$ + 7.192454×10^{-10} $(v + 1/2)^2$
(f) $+3.283167 \times 10^{-10}$ $(v + 1/2)$ + 4.881575×10^{-4} $(v + 1/2)^2$

Enthalpy of Formation

The ground state dissociation energy for $^{40}Ca_2$ was determined by Vidal (1) to be D_e = 1095.4 ± 0.5 cm^{-1} which gives D_0^0 = 1063.1 cm^{-1} using the spectroscopic data above. The adopted value for natural abundance Ca_2 was obtained from the $^{40}Ca_2$ value by correcting for the presence of the minor isotopic species (14). Using this value and the recommended value for $\Delta_f H$°(Ca, g, 0 K) = 177.33 kJ mol^{-1} as deduced from CODATA (2), we calculate the adopted value of $\Delta_f H$°(Ca_2, g, 0 K) = 341.92 kJ mol^{-1}. Correcting this value to 298.15 K using the current value of H°(298.15 K) − H°(0 K) = 11.1895 kJ mol^{-1} for Ca_2(g) and the CODATA value for Ca(cr), we obtain $\Delta_f H$°(Ca_2, g, 298.15 K) = 340.72 kJ mol^{-1}. Note that Brewer and Winn (3) calculated a value of D_0^0 (Ca_2, natural abundance) = 12.47 ± 1.7 kJ mol^{-1} based on some earlier spectroscopic work of Balfour and Whitlock (4). Gurvich (9) adopted a value of $\Delta_f H$°(0 K) = 342.266 ± 2.6 kJ mol^{-1} which is also based on the earlier study of (4).

Heat Capacity and Entropy

The thermal functions are calculated using a direct summation technique, analogous to the alkali dimers (13). Included in the calculation are the ground state ($^1\Sigma_g^+$) and the $^1\Sigma_u^+$ excited state at T_e = 18964 cm^{-1}. Spectroscopic constants for both states are taken from a detailed experimental investigation by Vidal (1) and are considered to be the most accurate. Presumably, other excited states, lower in energy than the $^1\Sigma_u^+$ state, exist as in the case of Mg_2(g) (5), however no known experimental or theoretical constants have been published. The effect in the heat capacity of these missing states would probably be to cause the second Cp maximum to occur at lower temperatures, similiar to Mg_2(g) (5). As with other diatomics possessing relatively shallow ground states, Ca_2(g) is another example where the traditional integration methods of calculating thermal functions fail.

The rotational levels are extrapolated to high J values according to the method of Khachkuruzov (6) who proposed a simpler form of Wooley's method (7). For the ground state only, we included the so-called quasibound rotational levels above the dissociation limit (8) with J_{lim} = 209 and v_{max} = 34 as determined by Gurvich (9). Instead of performing separate calculations on all 21 isotopic species (made up of all possible dimeric combinations of the six isotopes), we adjust the spectroscopic constants of $^{40}Ca_2$ to a hypothetical natural abundance molecule using a standard reduced mass scaling routine (10). Thermal functions calculated in this manner should be trivially different from the exact procedure of performing the calculation on each isotope and combining the results according to the natural abundances of each species. The adopted value of S°(298.15 K) is 0.39 J K^{-1} mol^{-1} smaller than that calculated by (9). It should also be noted that the minor controversy in the literature over the correct assignment of the ground state (11) has apparently been satisfactorily resolved with the currently used data (12).

References

1. C. R. Vidal, J. Chem. Phys., 72, 1864 (1980).
2. J. D. Cox, J. Chem. Thermodynamics 10, 903 (1977).
3. L. Brewer and J. S. Winn, Faraday Symp. Chem. Soc. (14), 126 (1980).
4. W. J. Balfour and R. F. Whilock, Canad. J. Phys. 53, 472 (1975).
5. See for example JANAF Thermochemical Tables, Mg_2(g), 1983 and R. O. Jones, J. Chem. Phys. 71, 1300 (1979).
6. G. A. Khachkuruzov, Opt. Spectrosk. 30, 455 (1971).
7. H. W. Wooley, R. B. Scott, and F. G. Brickwedde, J. Res. Nat. Bur. Stand. 41, 379 (1948).
8. JANAF Thermochemical Tables: F_2(r), 6-30-82.
9. L. V. Gurvich, I. V. Veits, et al., "Thermodynamic Properties of Individual Substances", Volume 3, 3rd ed., Nauka, Moscow, 1983.
10. G. Herzberg, "Spectra of Diatomic Molecules", 2nd Ed., D. Van Nostrand Co, Inc., New York, 1950.
11. J. C. Wyss, J. Chem. Phys 71, 2949 (1979).
12. M. E. Rosenkrantz, M. Krauss, and W. J. Stevens, Chem. Phys. Lett. 89, 4 (1982).
13. JANAF Thermochemical Tables: Li_2(g), 12-31-83.
14. N. E. Holden, R. L. Martin, and I. L. Barnes, Pure Appl. Chem. 55, 1119 (1983).

Calcium (Ca_2) Ca_2(g)

Enthalpy Reference Temperature = T_r = 298.15 K Standard State Pressure = p° = 0.1 MPa

T/K	J K^{-1}mol^{-1} C_p°	S°	−[G°−H°(T_r)]/T	kJ mol^{-1} H°−H°(T_r)	$\Delta_f H$°	$\Delta_f G$°	Log K_f
0	0.	0.	INFINITE	-11.189	341.920	341.920	INFINITE
100	38.200	214.556	291.717	-7.716	343.432	325.170	-169.851
200	39.844	241.757	260.663	-3.781	342.830	307.074	-80.199
250	38.647	250.545	257.797	-1.813	342.297	298.193	-62.304
298.15	36.577	257.183	257.183	0.	341.637	289.757	-50.764
300	36.490	257.409	257.184	0.068	341.608	289.435	-50.395
350	34.120	262.855	257.619	1.833	340.761	280.804	-41.908
400	31.951	267.266	258.558	3.483	339.754	272.305	-35.559
450	30.117	270.920	259.735	5.033	338.579	263.943	-30.638
500	28.619	274.013	261.012	6.500	337.240	255.721	-26.715
600	26.434	279.023	263.615	9.245	334.100	239.702	-20.868
700	25.000	282.983	266.109	11.812	330.390	224.254	-16.734
800	24.030	286.254	268.428	14.260	324.450	209.584	-13.684
900	23.352	289.043	270.567	16.628	319.766	195.500	-11.347
1000	22.862	291.476	272.539	18.937	314.435	181.972	-9.505
1100	22.498	293.637	274.361	21.204	308.456	169.010	-8.026
1200	22.221	295.583	276.050	23.439	286.371	157.865	-6.872
1300	22.005	297.352	277.621	25.650	281.582	147.350	-5.921
1400	21.835	298.977	279.089	27.842	276.774	137.205	-5.119
1500	21.697	300.478	280.466	30.018	271.950	127.404	-4.437
1600	21.585	301.875	281.761	32.182	267.114	117.926	-3.850
1700	21.492	303.180	282.983	34.336	262.268	108.750	-3.341
1800	21.415	304.407	284.139	36.481	-39.934	104.281	-3.026
1900	21.350	305.563	285.237	38.619	-41.969	112.348	-3.089
2000	21.295	306.656	286.281	40.752	-44.021	120.523	-3.148
2100	21.248	307.694	287.276	42.879	-46.093	128.801	-3.204
2200	21.209	308.682	288.227	45.002	-48.191	137.179	-3.257
2300	21.177	309.624	289.137	47.121	-50.323	145.652	-3.308
2400	21.150	310.525	290.009	49.237	-52.496	154.219	-3.356
2500	21.130	311.388	290.847	51.351	-54.721	162.878	-3.403
2600	21.115	312.216	291.653	53.463	-57.009	171.627	-3.448
2700	21.106	313.013	292.430	55.574	-59.371	180.465	-3.491
2800	21.103	313.780	293.178	57.685	-61.823	189.392	-3.533
2900	21.105	314.521	293.902	59.795	-64.379	198.409	-3.574
3000	21.115	315.236	294.601	61.906	-67.053	207.516	-3.613
3100	21.131	315.929	295.278	64.018	-69.861	216.714	-3.652
3200	21.154	316.600	295.934	66.132	-72.820	226.006	-3.689
3300	21.184	317.252	296.570	68.249	-75.944	235.392	-3.726
3400	21.222	317.884	297.188	70.369	-79.250	244.875	-3.762
3500	21.267	318.500	297.788	72.494	-82.752	254.459	-3.798
3600	21.320	319.100	298.371	74.623	-86.464	264.146	-3.833
3700	21.380	319.685	298.940	76.758	-90.399	273.938	-3.867
3800	21.449	320.256	299.493	78.899	-94.569	283.841	-3.902
3900	21.525	320.814	300.033	81.048	-98.985	293.856	-3.936
4000	21.609	321.360	300.559	83.205	-103.654	303.987	-3.970
4100	21.700	321.895	301.073	85.370	-108.585	314.239	-4.003
4200	21.799	322.419	301.575	87.545	-113.783	324.614	-4.037
4300	21.904	322.933	302.066	89.730	-119.252	335.116	-4.071
4400	22.017	323.438	302.546	91.926	-124.987	345.748	-4.105
4500	22.135	323.934	303.016	94.133	-130.965	356.512	-4.138
4600	22.260	324.422	303.476	96.353	-137.225	367.412	-4.172
4700	22.391	324.902	303.926	98.586	-143.757	378.453	-4.206
4800	22.527	325.375	304.368	100.831	-150.548	389.635	-4.240
4900	22.667	325.841	304.802	103.091	-157.590	400.961	-4.274
5000	22.812	326.300	305.227	105.365	-164.874	412.434	-4.309
5100	22.961	326.753	305.645	107.654	-172.393	424.054	-4.343
5200	23.113	327.201	306.055	109.957	-180.128	435.823	-4.378
5300	23.269	327.642	306.458	112.277	-188.082	447.744	-4.413
5400	23.427	328.079	306.855	114.611	-196.236	459.817	-4.448
5500	23.586	328.510	307.244	116.962	-204.578	472.043	-4.483
5600	23.748	328.937	307.628	119.329	-213.076	484.420	-4.518
5700	23.910	329.358	308.005	121.712	-221.751	496.952	-4.554
5800	24.073	329.776	308.377	124.111	-230.574	509.637	-4.590
5900	24.237	330.189	308.743	126.526	-239.532	522.476	-4.626
6000	24.400	330.597	309.104	128.958	-248.611	535.468	-4.662

PREVIOUS: December 1974 (1 atm) CURRENT: September 1983 (1 bar)

Calcium (Ca_2) Ca_2(g)

CHLORINE (Cl) IDEAL GAS M_r = 35.453

IP(Cl, g) = 104591.0 ± 0.3 cm^{-1}
$S°$(298.15 K) = 165.189 ± 0.004 J K^{-1} mol^{-1}

$\Delta_f H°$(0 K) = 119.621 ± 0.006 kJ mol^{-1}
$\Delta_f H°$(298.15 K) = 121.302 ± 0.008 kJ mol^{-1}

Electronic Levels and Quantum Weights

State	ϵ_i, cm^{-1}	g_i
$^2P°_{3/2}$	0.0	4
$^2P°_{1/2}$	882.36	2

Enthalpy of Formation

The adopted value for the enthalpy of formation of Cl(g) is calculated from the dissociation energy of gaseous chlorine, D_0^0 (natural abundance) = 19999.10 ± 1.0 cm^{-1} (239.242 ± 0.012 kJ mol^{-1}). This value was calculated from the dissociation energy of $^{35}Cl_2$(g), D_0^0 = 19997.28 ± 0.10 cm^{-1}, obtained from the extensive spectroscopic measurements of Douglas and Hoy (1). CODATA (2) used a slightly different value (D_0^0 = 19997.25 ± 0.3 cm^{-1} for $^{35}Cl_2$), as obtained by LeRoy and Bernstein (3) from the measurements of Douglas et al. (4), Brewer and Winn (5), Huber and Herzberg (6), and Gurvich et al. (7) have adopted one or the other of these two cited values. Wagman et al. (16) adopted a dissociation energy which is 63 cm^{-1} greater than our adopted value. Detailed discussions on the spectroscopic information are given by LeRoy (8), Coxon (9) and Douglas and Hoy (1).

Heat Capacity and Entropy

Information on the electronic energy levels and quantum weight is taken from Moore (10, 11). Only two levels are included in the calculation. All other levels (observed and predicted) lie above 71954 cm^{-1}. Our calculations indicate that any reasonable method of filling in the missing levels and cutting off the summation in the partition function (12) has no effect on the thermodynamic properties to 6000 K. This is undoubtedly a result of the high energy of these levels. Therefore, we list only the ground state and the first excited state. Extension to higher temperatures may require consideration of the higher excited states and utilization of different fill and cutoff procedures.

The thermodynamic functions at 298.15 K are in agreement with recent CODATA recommendations (2) except for two minor differences. First, the entropy differs by 0.1094 J K^{-1} mol^{-1} because this table uses a standard-state pressure of 1 bar, whereas the CODATA recommendations are based on 1 atm. Second, entropy differences of the order of 0.001-0.004 J K^{-1} mol^{-1} for the monatomic halogens arise due to the use of slightly different values for R; this tables uses R = 8.31441 J K^{-1} mol^{-1}. Considering these minor changes, this table agrees within the estimated uncertainty with those by Hultgren et al. (13), Gurvich et al. (17), and Wagman et al. (16). The estimated uncertainty is due to uncertainties in the relative atomic mass and fundamental constants which are based on the 1981 scale (14) and the 1973 values (15), respectively.

Referenes

1. A. E. Douglas and A. R. Hoy, Can. J. Phys. 53, 1965 (1975).
2. J. D. Cox, chairman, ICSU-CODATA Task Group on Key Values for Thermodynamics, J. Chem. Thermodynamics 10, 903 (1978).
3. R. J. LeRoy and R. B. Berstein, Chem. Phys. Lett. 5, 42 (1970); J. Chem. Phys. 52, 3869 (1970); J. Mol. Spectrosc. 37, 109 (1971).
4. A. E. Douglas, C. C. Moeller and B. P. Stoicheff, Can. J. Phys. 41, 1174 (1963).
5. L. Brewer and J. S. Winn, Faraday Symp. Chem. Soc. (GB), No. 14, 126 (1980).
6. K. P. Huber and G. Herzberg, "Constants of Diatomic Molecules", Van Nostrand Reinhold Company, New York, 1979.
7. L. V. Gurvich, I. V. Veits et al., "Thermodynamic Properties of Individual Substances", 3rd ed., Vol. 1, Nauka, Moscow, 1978.
8. R. J. LeRoy in "Molecular Spectroscopy", Volume 1, R. F. Barrow, D. A. Long and D. J. Miller, Senior Reporters, The Chemical Society, London, 1973, Chapter 3.
9. J. A. Coxon in "Molecular Spectroscpy", Volume 1, R. F. Barrow, D. A. Long, and D. J. Miller, Senior Reporters, The Chemical Society, London, 1973, Chapter 4.
10. C. E. Moore, U. S. Nat. Bur. Stand., NSRDS-NBS-35, Volume I, 1970 [Reprint of NBS Circular 467, Volume 1, 1949].
11. C. E. Moore, U. S. Nat. Bur. Stand., NSRDS-NBS-34, 1970.
12. J. R. Downey, Jr., The Dow Chemical Company, AFOSR-TR-78-0960, Contract No. F44620-75-1-0048, 1978.
13. R. Hultgren, P. D. Desai et al., "Selected Values of the Thermodynamic Properties of the Elements", American Society for Metals, Metals Park, Ohio, 1973.
14. N. E. Holden and R. L. Martin, Pure Appl. Chem. 55, 1101 (1983).
15. E. R. Cohen and B. N. Taylor, J. Phys. Chem. Ref. Data 2, 663 (1973).
16. D. D. Wagman, W. H. Evans et al., J. Phys. Chem. Ref. Data 11, Supp. 2 (1982).

Chlorine (Cl) Cl_1(g)

Enthalpy Reference Temperature = T_r = 298.15 K Standard State Pressure = p° = 0.1 MPa

T/K	$C_p°$	$S°$	$-[G°-H°(T_r)]/T$	$H°-H°(T_r)$	$\Delta_f H°$	$\Delta_f G°$	Log K_f
	J $K^{-1}mol^{-1}$			kJ mol^{-1}			
0	0.	0.	INFINITE	-6.272	119.621	119.621	INFINITE
100	20.788	142.175	184.104	-4.193	120.244	115.476	-60.319
200	21.079	156.636	167.161	-2.105	120.813	110.482	-28.855
250	21.450	161.378	165.547	-1.042	121.066	107.869	-22.538
298.15	21.838	165.189	165.189	0.	121.302	105.306	-18.449
300	21.852	165.325	165.190	0.040	121.311	105.207	-18.318
350	22.202	168.720	165.457	1.142	121.553	102.504	-15.298
400	22.467	171.703	166.055	2.259	121.795	99.766	-13.028
450	22.644	174.361	166.833	3.387	122.035	96.998	-11.259
500	22.744	176.752	167.708	4.522	122.272	94.203	-9.841
600	22.781	180.905	169.571	6.800	122.734	88.546	-7.709
700	22.692	184.411	171.448	9.074	123.172	82.813	-6.180
800	22.549	187.432	173.261	11.337	123.585	77.019	-5.029
900	22.389	190.079	174.986	13.584	123.972	71.174	-4.131
1000	22.233	192.430	176.615	15.815	124.334	65.288	-3.410
1100	22.089	194.542	178.150	18.031	124.675	59.367	-2.819
1200	21.959	196.458	179.597	20.233	124.997	53.416	-2.325
1300	21.843	198.211	180.963	22.423	125.300	47.438	-1.906
1400	21.742	199.826	182.253	24.602	125.589	41.438	-1.546
1500	21.652	201.323	183.475	26.772	125.863	35.417	-1.233
1600	21.573	202.718	184.635	28.933	126.124	29.379	-0.959
1700	21.504	204.024	185.737	31.087	126.374	23.325	-0.717
1800	21.443	205.251	186.788	33.234	126.612	17.256	-0.501
1900	21.389	206.409	187.790	35.375	126.840	11.175	-0.307
2000	21.341	207.505	188.749	37.512	127.058	5.081	-0.133
2100	21.298	208.545	189.667	39.644	127.265	-1.023	0.025
2200	21.260	209.535	190.548	41.772	127.463	-7.136	0.169
2300	21.226	210.479	191.394	43.896	127.650	-13.259	0.301
2400	21.195	211.382	192.208	46.017	127.827	-19.389	0.422
2500	21.167	212.246	192.992	48.135	127.992	-25.526	0.533
2600	21.142	213.076	193.749	50.250	128.147	-31.670	0.636
2700	21.119	213.874	194.480	52.364	128.290	-37.820	0.732
2800	21.099	214.641	195.186	54.474	128.421	-43.974	0.820
2900	21.080	215.381	195.870	56.583	128.541	-50.133	0.903
3000	21.063	216.096	196.532	58.690	128.649	-56.297	0.980
3100	21.047	216.786	197.174	60.796	128.746	-62.463	1.052
3200	21.032	217.454	197.798	62.900	128.832	-68.633	1.120
3300	21.019	218.101	198.403	65.002	128.908	-74.804	1.184
3400	21.007	218.728	198.992	67.104	128.974	-80.979	1.244
3500	20.995	219.337	199.564	69.204	129.031	-87.155	1.301
3600	20.985	219.928	200.122	71.303	129.081	-93.332	1.354
3700	20.975	220.503	200.665	73.401	129.124	-99.511	1.405
3800	20.966	221.062	201.195	75.498	129.161	-105.691	1.453
3900	20.958	221.607	201.711	77.594	129.195	-111.871	1.498
4000	20.950	222.137	202.215	79.690	129.226	-118.053	1.542
4100	20.943	222.655	202.707	81.784	129.255	-124.235	1.583
4200	20.936	223.159	203.188	83.878	129.284	-130.418	1.622
4300	20.929	223.652	203.658	85.971	129.314	-136.602	1.659
4400	20.923	224.133	204.118	88.064	129.347	-142.786	1.695
4500	20.918	224.603	204.568	90.156	129.384	-148.972	1.729
4600	20.912	225.063	205.009	92.247	129.426	-155.158	1.762
4700	20.907	225.512	205.440	94.338	129.474	-161.345	1.793
4800	20.903	225.952	205.863	96.429	129.530	-167.533	1.823
4900	20.898	226.383	206.278	98.519	129.594	-173.723	1.852
5000	20.894	226.806	206.684	100.609	129.667	-179.914	1.880
5100	20.890	227.219	207.082	102.698	129.750	-186.106	1.906
5200	20.886	227.625	207.474	104.787	129.844	-192.300	1.932
5300	20.883	228.023	207.858	106.875	129.950	-198.496	1.956
5400	20.879	228.413	208.235	108.963	130.068	-204.694	1.980
5500	20.876	228.796	208.605	111.051	130.199	-210.895	2.003
5600	20.873	229.172	208.969	113.138	130.343	-217.098	2.025
5700	20.870	229.542	209.327	115.226	130.501	-223.304	2.046
5800	20.867	229.905	209.678	117.312	130.673	-229.512	2.067
5900	20.865	230.261	210.024	119.399	130.859	-235.724	2.087
6000	20.862	230.612	210.364	121.485	131.059	-241.939	2.106

PREVIOUS: June 1972 (1 atm) CURRENT: June 1982 (1 bar)

Chlorine (Cl) Cl_1(g)

CHLORINE, ION (Cl^+) IDEAL GAS M_r = 35.45245

IP(Cl^+, g) = 192070 ± 10 cm^{-1}

S°(298.15 K) = 167.556 ± 0.005 J K^{-1} mol^{-1}

$\Delta_f H°$(0 K) = 1370.807 ± 0.01 kJ mol^{-1}

$\Delta_f H°$(298.15 K) = [1378.801] kJ mol^{-1}

Electronic Levels and Quantum Weights

State	ϵ_i, cm^{-1}	g_i
$^3P_2$	0.0	5
$^3P_1$	696.1	3
$^3P_0$	996.4	1
$^1D_2$	11653.5	5
$^1S_0$	27877.8	1

Enthalpy of Formation

$\Delta_f H°$(Cl^+, g, 0 K) is calculated from $\Delta_f H°$(Cl, g, 0 K) (1) using the spectroscopic value of IP(Cl) = 104591.0±0.3 cm^{-1} (1251.187±0.004 kJ mol^{-1}) from Moore (2). The ionization limit is converted from cm^{-1} to kJ mol^{-1} using the factor, 1 cm^{-1} = 0.01196266 kJ mol^{-1}, which is derived from the 1973 CODATA fundamental constants (3). Rosenstock et al. (4) and Levin and Lias (5) have summarized additional ionization potential and appearance potential data.

$\Delta_f H°$(Cl^+, g, 298.15 K) is calculated from $\Delta_f H°$(Cl, g, 0 K) by using IP(Cl) with JANAF (1) enthalpies, H°(0 K)-H°(298.15 K), for Cl(g), Cl^+(g), and e^-(g). $\Delta_r H°$(Cl → Cl^+ + e^-, 298.15 K) differs from a room temperature threshold energy due to inclusion of these enthalpies and to threshold effects discussed by Rosenstock et al. (4). $\Delta_f H°$(298.15 K) should be changed by -6.197 kJ mol^{-1} if it is to be used in the ion convention that excludes the enthalpy of the electron.

Heat Capacity and Entropy

The information on electronic energy levels and quantum weights, given by Moore (2, 6), is incomplete because many theoretically predicted levels have not been observed. Our calculations indicate that any reasonable method of filling in these missing levels and cutting off the summation in the partition function (7) has no effect on the thermodynamic functions to 6000 K. This is a result of the high energy of all levels other than the ground state and the lowest four excited states; the next lowest level is approximately 93366 cm^{-1} above the ground state. Since inclusion of these upper levels has no effect on the thermodynamic functions (to 6000 K), we list only the ground state and four excited states, with the energy of these states taken from recent study by Moore (2). The reported uncertainty is S°(298.15 K) is due to uncertainties in the relative ionic mass, the fundamental constants, and the position of the four lowest excited states. Extension of these calculations above 6000 K may require consideration of the higher excited states and use of different fill and cutoff procedures (7).

References

1. JANAF Thermochemical Tables: Cl(g), 3-31-82; e^-(r), 3-31-82.
2. C. E. Moore, U. S. Nat. Bur. Stand., NSRDS-NBS-34, 1970.
3. E. R. Cohen and B. N. Taylor, J. Phys. Chem. Ref. Data 2, 663 (1973).
4. H. M. Rosenstock, K. Draxl et al., J. Phys. Chem. Ref. Data 6, Supp. 1 (1977).
5. R. D. Levin and S. G. Lias, U. S. Nat. Bur. Stand., NSRDS-NBS-71, 1982.
6. C. E. Moore, U. S. Nat. Bur. Stand., NSRDS-NBS-35, Volume I, 1970 [reprint of NBS Circular 467, Volume I, 1949].
7. J. R. Downey, Jr., The Dow Chemical Company, AFOSR-TR-78-0960, Contract No. F44620-75-1-0048, 1978.

Chlorine, Ion (Cl^+) Cl_1^+(g)

Enthalpy Reference Temperature = T_r = 298.15 K Standard State Pressure = p° = 0.1 MPa

T/K	$C_p°$ (J K⁻¹mol⁻¹)	S°	-[G°-H°(T_r)]/T	H°-H°(T_r) (kJ mol⁻¹)	$\Delta_f H°$	$\Delta_f G°$	Log K_f
0	0.	0.	INFINITE	-6.386	1370.807		
100	20.809	144.033	187.107	-4.307			
200	21.681	158.648	169.618	-2.194			
250	22.383	163.562	167.931	-1.092			
298.15	22.958	167.556	167.556	0.	1378.801	1355.845	-237.538
300	22.977	167.698	167.556	0.042	1378.850	1355.702	-236.049
350	23.388	171.273	167.838	1.202	1380.190	1351.738	-201.736
400	23.621	174.413	168.467	2.378	1381.530	1347.582	-175.976
450	23.713	177.202	169.286	3.562	1382.865	1343.258	-155.921
500	23.706	179.700	170.205	4.748	1384.192	1338.787	-139.862
600	23.519	184.008	172.158	7.111	1386.817	1329.458	-115.740
700	23.239	187.613	174.115	9.449	1389.398	1319.694	-98.477
800	22.949	190.697	176.000	11.758	1391.936	1309.562	-85.506
900	22.681	193.385	177.785	14.039	1394.436	1299.115	-75.399
1000	22.446	195.762	179.467	16.295	1396.902	1288.391	-67.299
1100	22.244	197.891	181.046	18.530	1399.340	1277.422	-60.660
1200	22.073	199.819	182.532	20.745	1401.753	1266.232	-55.118
1300	21.928	201.580	183.930	22.945	1404.146	1254.841	-50.420
1400	21.806	203.201	185.250	25.132	1406.520	1243.267	-46.387
1500	21.704	204.702	186.497	27.307	1408.878	1231.523	-42.885
1600	21.620	206.100	187.679	29.473	1411.223	1219.623	-39.817
1700	21.552	207.408	188.801	31.631	1413.556	1207.577	-37.104
1800	21.498	208.639	189.870	33.784	1415.878	1195.393	-34.689
1900	21.458	209.800	190.888	35.932	1418.191	1183.081	-32.525
2000	21.430	210.900	191.862	38.076	1420.495	1170.647	-30.574
2100	21.414	211.945	192.793	40.218	1422.792	1158.098	-28.806
2200	21.409	212.941	193.687	42.359	1425.081	1145.440	-27.196
2300	21.413	213.893	194.545	44.500	1427.363	1132.678	-25.724
2400	21.426	214.804	195.370	46.642	1429.639	1119.817	-24.372
2500	21.447	215.679	196.165	48.786	1431.909	1106.861	-23.127
2600	21.475	216.521	196.932	50.932	1434.173	1093.814	-21.975
2700	21.510	217.332	197.672	53.081	1436.431	1080.681	-20.907
2800	21.551	218.115	198.389	55.234	1438.683	1067.464	-19.914
2900	21.596	218.872	199.082	57.391	1440.930	1054.166	-18.988
3000	21.645	219.605	199.754	59.553	1443.171	1040.791	-18.122
3100	21.698	220.316	200.406	61.720	1445.409	1027.342	-17.311
3200	21.753	221.005	201.039	63.893	1447.642	1013.820	-16.549
3300	21.811	221.676	201.654	66.071	1449.872	1000.229	-15.832
3400	21.869	222.327	202.252	68.255	1452.099	986.569	-15.157
3500	21.929	222.962	202.835	70.445	1454.325	972.845	-14.519
3600	21.989	223.581	203.403	72.641	1456.550	959.057	-13.916
3700	22.049	224.184	203.956	74.843	1458.775	945.207	-13.344
3800	22.109	224.773	204.497	77.051	1461.002	931.297	-12.802
3900	22.168	225.348	205.024	79.265	1463.232	917.328	-12.286
4000	22.227	225.910	205.539	81.484	1465.466	903.303	-11.796
4100	22.284	226.460	206.043	83.710	1467.705	889.221	-11.329
4200	22.339	226.997	206.535	85.941	1469.950	875.084	-10.883
4300	22.393	227.524	207.017	88.178	1472.202	860.894	-10.458
4400	22.445	228.039	207.489	90.420	1474.463	846.652	-10.051
4500	22.496	228.544	207.951	92.667	1476.733	832.358	-9.662
4600	22.544	229.039	208.404	94.919	1479.014	818.013	-9.289
4700	22.591	229.524	208.849	97.175	1481.307	803.619	-8.931
4800	22.635	230.000	209.284	99.437	1483.612	789.175	-8.588
4900	22.677	230.467	209.712	101.702	1485.930	774.684	-8.258
5000	22.717	230.926	210.132	103.972	1488.262	760.145	-7.941
5100	22.755	231.376	210.544	106.246	1490 608	745.560	-7.636
5200	22.791	231.818	210.949	108.523	1492.970	730.928	-7.342
5300	22.825	232.253	211.347	110.804	1495.346	716.250	-7.059
5400	22.857	232.680	211.738	113.088	1497.739	701.528	-6.786
5500	22.887	233.100	212.122	115.375	1500.148	686.761	-6.522
5600	22.915	233.512	212.501	117.665	1502.573	671.951	-6.268
5700	22.941	233.918	212.873	119.958	1505.015	657.096	-6.022
5800	22.965	234.317	213.239	122.254	1507.474	642.199	-5.784
5900	22.987	234.710	213.600	124.551	1509.950	627.260	-5.553
6000	23.008	235.097	213.955	126.851	1512.442	612.277	-5.330

PREVIOUS: June 1965 (1 atm) CURRENT: June 1982 (1 bar)

Chlorine, Ion (Cl^+) Cl_1^+(g)

CHLORINE, ION (Cl^-) IDEAL GAS M_r = 35.45355

EA(Cl, g) = 3.617 ± 0.003 eV
S°(298.15 K) = 153.356 ± 0.005 J K^{-1} mol^{-1}

$\Delta_f H°$(0 K) = -229.36 ± 0.4 kJ mol^{-1}
$\Delta_f H°$(298.15 K) = [-233.954] kJ mol^{-1}

Electronic Level and Quantum Weight

State	ϵ_i, cm^{-1}	g_i
$^1S_0$	0.0	1

Enthalpy of Formation

$\Delta_f H°(Cl^-$, g, 0 K) is calculated from $\Delta_f H°$(Cl, g, 0 K) (1) using the adopted electron affinity of EA(Cl) = 3.617±0.003 eV (348.985±0.289 kJ mol^{-1}). This value, recommended by Hotop and Lineberger (10), is based on laser optogalvanic spectroscopic and plasma emission studies (5, 6). Additional information on Cl^-(g) may be obtained in the critical discussions of Hotop and Lineberger (2, 10), Rosenstock et al. (3) and Massey (4).

$\Delta_f H°(Cl^-$, g, 298.15 K) is obtained from $\Delta_f H°$(Cl, g, 0 K) by using EA(Cl) with JANAF (1) enthalpies, H°(0 K) - H°(298.15 K), for Cl^-(g), Cl(g), and e^-(g). $\Delta_r H(Cl^- \rightarrow Cl + e^-$, 298.15 K) differs from a room-temperature threshold energy due to inclusion of these enthalpies and to threshold effects discussed by Rosenstock et al. (3). $\Delta_f H°$(298.15 K) should be changed by +6.197 kJ mol^{-1} if it is to be used in the ion convention that excludes the enthalpy of the electron.

Heat Capacity and Entropy

The ground state configuration for Cl^-(g) is given by Hotop and Lineberger (2, 10), Rosenstock et al. (3) and Massey (4). Lacking any experimental evidence as to the stability of any excited states, we assume that no stable excited states exist.

The entropy at all temperatures (0-6000 K) agrees within 0.001 J K^{-1} mol^{-1} with the values of Gurvich et al. (8) except for one minor difference. The entropy differs by 0.1094 J K^{-1} mol^{-1} because this table uses a standard-state pressure of 1 bar, whereas the tabulation of Gurvich et al. (8) is based on 1 atm. The estimated uncertainty is due to uncertainties in the relative ionic mass and the fundamental constants which are based on the 1981 scale (9) and the 1973 values (7), respectively.

References

1. JANAF Thermochemical Tables: Cl(g), 6-30-82; e^-(g), 3-31-82.
2. H. Hotop and W. C. Lineberger, J. Phys. Chem. Ref. Data 4, 539 (1975).
3. H. M. Rosenstock, K. Draxl et al., J. Phys. Chem. Ref. Data 6, Supp. 1 (1977).
4. H. S. W. Massey, "Negative Ions," 3rd ed., Cambridge University Press, Cambridge, 1976.
5. I. S. McDermid and C. R. Webster, J. de Physique 44, C7-461 (1983).
6. G. Muck and H.-P. Popp, Z. Naturforsch. 23a, 1213 (1968).
7. E. R. Cohen and B. N. Taylor, J. Phys. Chem. Ref. Data 2, 663 (1973).
8. L. V. Gurvich, I. V. Veits et al., "Thermodynamic Properties of Individual Substances", 3rd ed., Vol. 1, Nauka, Moscow, 1978.
9. N. E. Holden and R. L. Martin, Pure Appl. Chem. 55, 1101 (1983).
10. H. Hotop and W. C. Lineberger, J. Phys. Chem. Ref. Data, to be published, 1985.

Chlorine, Ion (Cl^-) Cl_1^-(g)

Enthalpy Reference Temperature = T_r = 298.15 K Standard State Pressure = p° = 0.1 MPa

T/K	$C_p°$ (J $K^{-1}mol^{-1}$)	S° (J $K^{-1}mol^{-1}$)	-[G°-H°(T_r)]/T (J $K^{-1}mol^{-1}$)	H°-H°(T_r) (kJ mol^{-1})	$\Delta_f H°$ (kJ mol^{-1})	$\Delta_f G°$ (kJ mol^{-1})	Log K_f
0	0.	0.	INFINITE	-6.197	-229.364		
100	20.786	130.649	171.836	-4.119			
200	20.786	145.057	155.257	-2.040			
250	20.786	149.695	153.698	-1.001			
298.15	20.786	153.356	153.356	0.	-233.954	-240.167	42.076
300	20.786	153.485	153.356	0.038	-233.986	-240.205	41.823
350	20.786	156.689	153.610	1.078	-234.845	-241.174	35.993
400	20.786	159.464	154.172	2.117	-235.721	-242.019	31.604
450	20.786	161.913	154.899	3.156	-236.609	-242.753	28.178
500	20.786	164.103	155.711	4.196	-237.506	-243.388	25.427
600	20.786	167.892	157.435	6.274	-239.322	-244.394	21.276
700	20.786	171.097	159.164	8.353	-241.158	-245.094	18.289
800	20.786	173.872	160.833	10.431	-243.008	-245.530	16.031
900	20.786	176.320	162.420	12.510	-244.868	-245.734	14.262
1000	20.786	178.510	163.922	14.589	-246.736	-245.730	12.836
1100	20.786	180.492	165.340	16.667	-248.612	-245.539	11.660
1200	20.786	182.300	166.679	18.746	-250.492	-245.177	10.672
1300	20.786	183.964	167.945	20.824	-252.379	-244.657	9.830
1400	20.786	185.504	169.145	22.903	-254.270	-243.992	9.103
1500	20.786	186.938	170.284	24.982	-256.165	-243.192	8.469
1600	20.786	188.280	171.367	27.060	-258.065	-242.265	7.909
1700	20.786	189.540	172.400	29.139	-259.969	-241.220	7.412
1800	20.786	190.728	173.385	31.217	-261.878	-240.062	6.966
1900	20.786	191.852	174.328	33.296	-263.792	-238.797	6.565
2000	20.786	192.918	175.231	35.375	-265.710	-237.432	6.201
2100	20.786	193.932	176.097	37.453	-267.635	-235.971	5.869
2200	20.786	194.899	176.930	39.532	-269.565	-234.418	5.566
2300	20.786	195.823	177.732	41.610	-271.502	-232.777	5.287
2400	20.786	196.708	178.504	43.689	-273.446	-231.052	5.029
2500	20.786	197.557	179.249	45.768	-275.399	-229.245	4.790
2600	20.786	198.372	179.969	47.846	-277.360	-227.360	4.568
2700	20.786	199.156	180.666	49.925	-279.330	-225.400	4.361
2800	20.786	199.912	181.339	52.004	-281.309	-223.366	4.167
2900	20.786	200.642	181.993	54.082	-283.298	-221.262	3.985
3000	20.786	201.346	182.626	56.161	-285.297	-219.089	3.815
3100	20.786	202.028	183.241	58.239	-287.306	-216.849	3.654
3200	20.786	202.688	183.838	60.318	-289.324	-214.544	3.502
3300	20.786	203.327	184.419	62.397	-291.351	-212.176	3.358
3400	20.786	203.948	184.985	64.475	-293.386	-209.746	3.222
3500	20.786	204.550	185.535	66.554	-295.429	-207.256	3.093
3600	20.786	205.136	186.071	68.632	-297.478	-204.708	2.970
3700	20.786	205.706	186.594	70.711	-299.533	-202.103	2.853
3800	20.786	206.260	187.105	72.790	-301.593	-199.442	2.742
3900	20.786	206.800	187.603	74.868	-303.655	-196.727	2.635
4000	20.786	207.326	188.089	76.947	-305.720	-193.959	2.533
4100	20.786	207.839	188.565	79.025	-307.785	-191.139	2.435
4200	20.786	208.340	189.030	81.104	-309.850	-188.269	2.341
4300	20.786	208.829	189.485	83.183	-311.913	-185.350	2.252
4400	20.786	209.307	189.930	85.261	-313.973	-182.383	2.165
4500	20.786	209.774	190.365	87.340	-316.028	-179.369	2.082
4600	20.786	210.231	190.792	89.418	-318.077	-176.309	2.002
4700	20.786	210.678	191.211	91.497	-320.120	-173.206	1.925
4800	20.786	211.116	191.621	93.576	-322.155	-170.058	1.851
4900	20.786	211.544	192.023	95.654	-324.181	-166.868	1.779
5000	20.786	211.964	192.418	97.733	-326.198	-163.638	1.710
5100	20.786	212.376	192.805	99.811	-328.204	-160.366	1.642
5200	20.786	212.780	193.185	101.890	-330.198	-157.056	1.578
5300	20.786	213.175	193.559	103.969	-332.181	-153.707	1.515
5400	20.786	213.564	193.926	106.047	-334.151	-150.322	1.454
5500	20.786	213.945	194.286	108.126	-336.108	-146.899	1.395
5600	20.786	214.320	194.641	110.204	-338.051	-143.441	1.338
5700	20.786	214.688	194.989	112.283	-339.981	-139.949	1.282
5800	20.786	215.049	195.332	114.362	-341.896	-136.423	1.229
5900	20.786	215.405	195.669	116.440	-343.796	-132.864	1.176
6000	20.786	215.754	196.001	118.519	-345.682	-129.273	1.125

PREVIOUS: June 1965 (1 atm) CURRENT: June 1982 (1 bar)

Chlorine, Ion (Cl^-) Cl_1^-(g)

$S°(298.15\ K) = 245.68 \pm 6.3\ J\ K^{-1}\ mol^{-1}$

$\Delta_fH°(0\ K) = 192.8 \pm 12.6\ kJ\ mol^{-1}$

$\Delta_fH°(298.15\ K) = 192.9 \pm 12.6\ kJ\ mol^{-1}$

Electronic Levels and Quantum Weights

State	ε_i, cm^{-1}	g_i	ε_i, cm^{-1}	g_i	ε_i, cm^{-1}	g_i	ε_i, cm^{-1}	g_i
$^3\Sigma$	0	[3]	19868.1	[3]	21335.1	[3]	22182.5	[3]
	[11000]	[3]	19945.8	[3]	21640.4	[3]	22402.7	[3]
	17484.1	[3]	20969.1	[3]	22014.8	[3]	22966.2	[3]
	19826.3	[3]	21263.9	[3]	22072.3	[3]	[25000]	[3]

$\omega_e = 421.2\ cm^{-1}$ $\omega_ex_e = 0.74\ cm^{-1}$ $B_e = [0.17434]\ cm^{-1}$ $\alpha_e = [0.000459]\ cm^{-1}$ $r_e = [2.09]$ Å $\sigma = 1$

Enthalpy of Formation

Kulkarni and Dadape (1) have studied the reaction of $CoCl_2(g)$ with Co(cr) by a transpiration method. They used highly purified argon as the carrier gas and condensed the gaseous products on a mullite cold finger. Chemical and x-ray analyses of the condensate showed the presence of metallic cobalt. The transport of the Co metal was ascribed to the disproportionation of CoCl(g) according to the reaction 2 CoCl(g) = Co(cr) + $CoCl_2(g)$. Partial pressures for CoCl(g) and $CoCl_2(g)$ were reported for the temperature range 1395-1673 K. No consideration was given to possible dimerization of the dichloride. JANAF data (2) for $CoCl_2(g)$ and $Co_2Cl_4(g)$ are used to correct their $CoCl_2$ pressures for dimerization. The corrected pressures are 0.5% less than those of Kulkarni and Dadape (1) at 1395 K, and only 0.1% less at 1673 K. Hildenbrand (3) has also studied the disproportionation reaction of CoCl(g) by mass spectrometry. The monochloride was produced in the effusion cell (4) by vaporizing AgCl in the presence of metallic cobalt. The Ag^+ and $AgCl^+$ signals were also measured and used to derive equilibrium data for the process Ag(g + CoCl(g) = AgCl(g) + Co(g).

Source	Reaction	Data Points	T/K	$\Delta_rH°$(298.15 K), kcal mol^{-1} 2nd Law	3rd Law	Drift cal K^{-1} mol^{-1}	$\Delta_fH°$(298.15 K) kcal mol^{-1}	$\Delta_{at}H°$(298.15 K) kcal mol^{-1}
Kulkarni and Dadape (1)	A	9	1395-1673	83.3	79.65±1.8	-2.4±3.5	28.6±2.0	101.9
Hildenbrand (3)	B	7	1461-1605	20.0	14.19±0.53	-3.8±1.5	46.6±1.5	83.9
Hildenbrand (3)	C	7	1461-1605	6.21	9.87±0.35	2.4±1.1	45.6±1.5[a]	84.9

Reactions: (A) $CoCl_2(g)$ + Co(cr) = 2 CoCl(g) (B) $CoCl_2(g)$ + Co(g) = 2 CoCl(g) (C) Ag(g) + CoCl(g) = AgCl(g) + Co(g)

[a] $\Delta_rH°$(298.15 K)(3rd law) combined with D°(AgCl, g, 298.15 K) = 75.0±0.5 kcal mol^{-1} (4).

The two sets of measurements of Hildenbrand (3) lead to consistent values for $\Delta_fH°$, while the results of Kulkarni and Dadape (1) suggest a value roughly 20 kcal mol^{-1} lower. In the same paper, Kulkarni and Dadape (1) reported vapor pressures for $CoCl_2(\ell)$ which were also determined by their flow technique. These vapor pressures have been shown to be incompatible with JANAF data (see $CoCl_2(g)$ table). Furthermore, their value for the bond dissociation energy (D°(298.15 K) = 101.9 kcal mol^{-1}) of CoCl appears to be too high by comparison with similar data for other transition metal monochlorides. Values (5) of D° for the monochlorides of Cr, Mn, Ni, and Cu show a random variation of from 83.0 to 88.6 kcal mol^{-1}, suggesting a value for CoCl near 85 kcal mol^{-1}. The results of Hildenbrand (3) are in agreement with these predictions. It is further noted that Hildenbrand (4) has used a similar technique to study the monochlorides of Ca, Sr, and Ba. The results of these studies are quite consistent with JANAF heat of formation data (2). Thus, it appears that the method used by Hildenbrand (3, 4) is reliable, and we adopt an average value, $\Delta_fH°$(298.15 K) = 46.1 kcal mol^{-1}), of his two results. The uncertainty is estimated to be ± 3.0 kcal mol^{-1}. The adopted values for the ground state vibrational constants are judged to be of insufficient accuracy to warrant a Birge-Sponer calculation for D°(Co-Cl).

Heat Capacity and Entropy

The bond length is estimated as somewhat less than that for $CoCl_2$ (2). B_e is calculated from the value for r_e. The value for α_e is obtained from the other constants assuming a Morse potential function.

Several band systems have been observed in the electronic spectrum (6-9) of CoCl. Vibrational constants for the upper and lower states of the various systems have been summarized by Rosen (5). However, since it is not certain which of the observed electronic transitions involve the ground state, no assignments were made. Also, confusion exists as to whether some of the systems belong to triplet-triplet transitions (7-9), an electronic triplet (8), or whether they are all separate systems (9). We adopt the vibrational constants for the ground state suggested by Herzberg (10). Comparison of these values with those listed by Rosen (5) for the lower states of other systems for CoCl indicate that the adopted ω_e value may be uncertain by as much as ±5 cm^{-1}. Krishnamurty (8) has shown that the ground state electronic configuration for CoCl is either $^5\Sigma$ or $^3\Sigma$. We prefer a triplet ground state, since the electronic configuration for Co^{+1} is ^3F (11). Other transition metal monochlorides (5) and the corresponding unipositive metal ions (11) have the same ground state multiplicities. If the ground state is a quintet, then the value of S°(298.15 K) is increased by 1 cal K^{-1} mol^{-1}. The adopted electronic levels in the range 17,000-23,000 cm^{-1} are taken from the compilation of Rosen (5). Each level is assumed to be a separate system and is assigned a quantum weight of 3. Two triplet levels are also included at 11,000 and 25,000 cm^{-1}, since Rao and Rao (9) are believed to have observed two components of a triplet-triplet transition near 14,000 cm^{-1}. By analogy with similar data for other transition metal monochlorides (5), the lower triplet level is not expected to lie below 10,000 cm^{-1}. The adopted upper levels and assumed quantum weights contribute 0.31 cal K^{-1} mol^{-1} to the entropy of CoCl at 4000 K.

References

1. M. P. Kulkarni and V. V. Dadape, High Temp. Sci. 3, 277 (1971).
2. JANAF Thermochemical Tables: $CoCl_2(g)$, $Co_2Cl_4(g)$, 12-31-73; Co(g), 9-30-67; CaCl(g), 6-30-70; BaCl(g), SrCl(g), 12-31-72.
3. D. L. Hildenbrand, Stanford Research Institute, personal communication, January, 1973.
4. D. L. Hildenbrand, J. Chem. Phys. 52, 5751 (1970).
5. B. Rosen, Ed., "Spectroscopic Data Relative to Diatomic Molecules," Pergamon press, New York, 1970.
6. P. Mesnage, Compt. Rend. 200, 2072 (1935).
7. K. R. More, Phys. Rev. 54, 122 (1938).
8. V. G. Krishnamurty, Ind. J. Phys. 26, 177 (1952).
9. S. V. Krishna Rao and P. T. Rao, Ind. J. Phys. 35, 556 (1961).
10. G. Herzberg, "Spectra of Diatomic Molecules," 2nd ed., D. Van Nostrand Co., Inc., New York, 1963.
11. C. E. Moore, U. S. Nat. Bur. Stand. Circ. 467, 1952.

Enthalpy Reference Temperature = T_r = 298.15 K Standard State Pressure = p° = 0.1 MPa

T/K	$C_p°$	S°	$-[G°-H°(T_r)]/T$	$H°-H°(T_r)$	$\Delta_fH°$	$\Delta_fG°$	Log K_f
	J $K^{-1}mol^{-1}$			kJ mol^{-1}			
0	0.	0.	INFINITE	-9.447	192.797	192.797	INFINITE
100	29.838	210.257	275.505	-6.525	193.742	182.934	-95.555
200	33.228	232.008	248.842	-3.367	193.471	172.183	-44.970
250	34.377	239.554	246.253	-1.675	193.198	166.892	-34.870
298.15	35.139	245.679	245.679	0.	192.882	161.853	-28.356
300	35.163	245.896	245.679	0.065	192.870	161.661	-28.148
350	35.708	251.360	246.110	1.838	192.520	156.487	-23.354
400	36.097	256.155	247.072	3.633	192.135	151.365	-19.766
450	36.383	260.424	248.323	5.446	191.711	146.293	-16.981
500	36.599	264.269	249.728	7.270	191.249	141.271	-14.759
600	36.899	270.970	252.726	10.946	190.215	131.371	-11.437
700	37.097	276.674	255.750	14.647	189.043	121.654	-9.078
800	37.237	281.637	258.682	18.364	187.310	112.177	-7.324
900	37.342	286.030	261.482	22.093	185.835	102.872	-5.971
1000	37.426	289.969	264.137	25.832	184.133	93.743	-4.897
1100	37.495	293.539	266.650	29.578	182.168	84.796	-4.027
1200	37.556	296.804	269.029	33.330	179.904	76.042	-3.310
1300	37.611	299.812	271.283	37.089	177.193	67.493	-2.712
1400	37.664	302.602	273.421	40.853	173.908	59.175	-2.208
1500	37.718	305.202	275.454	44.622	171.635	51.061	-1.778
1600	37.775	307.638	277.390	48.396	169.616	43.090	-1.407
1700	37.837	309.930	279.238	52.177	167.697	35.241	-1.083
1800	37.906	312.094	281.003	55.964	149.513	27.798	-0.807
1900	37.984	314.146	282.694	59.758	147.344	21.095	-0.580
2000	38.073	316.097	284.316	63.561	145.178	14.507	-0.379
2100	38.175	317.957	285.874	67.373	143.016	8.026	-0.200
2200	38.289	319.735	287.373	71.196	140.858	1.649	-0.039
2300	38.419	321.440	288.817	75.032	138.706	-4.631	0.105
2400	38.566	323.078	290.211	78.881	136.561	-10.817	0.235
2500	38.730	324.656	291.557	82.745	134.423	-16.914	0.353
2600	38.914	326.178	292.860	86.627	132.294	-22.925	0.461
2700	39.119	327.651	294.121	90.529	130.175	-28.855	0.558
2800	39.346	329.077	295.344	94.452	128.069	-34.706	0.647
2900	39.596	330.462	296.532	98.399	125.977	-40.483	0.729
3000	39.872	331.809	297.685	102.372	123.901	-46.187	0.804
3100	40.174	333.121	298.807	106.374	121.844	-51.823	0.873
3200	40.503	334.402	299.900	110.408	-255.004	-55.620	0.908
3300	40.859	335.654	300.964	114.476	-255.587	-49.380	0.782
3400	41.244	336.879	302.003	118.580	-256.142	-43.123	0.662
3500	41.656	338.081	303.016	122.725	-256.669	-36.850	0.550
3600	42.097	339.260	304.007	126.913	-257.164	-30.562	0.443
3700	42.565	340.420	304.975	131.146	-257.625	-24.261	0.343
3800	43.060	341.562	305.923	135.427	-258.048	-17.948	0.247
3900	43.580	342.687	306.851	139.758	-258.433	-11.625	0.156
4000	44.124	343.797	307.761	144.143	-258.775	-5.292	0.069
4100	44.690	344.893	308.653	148.584	-259.073	1.049	-0.013
4200	45.275	345.977	309.529	153.082	-259.326	7.397	-0.092
4300	45.878	347.050	310.389	157.639	-259.530	13.750	-0.167
4400	46.496	348.111	311.235	162.258	-259.684	20.107	-0.239
4500	47.127	349.163	312.066	166.939	-259.787	26.467	-0.307
4600	47.766	350.206	312.884	171.684	-259.838	32.829	-0.373
4700	48.412	351.240	313.689	176.493	-259.835	39.191	-0.436
4800	49.061	352.266	314.482	181.366	-259.780	45.553	-0.496
4900	49.711	353.285	315.263	186.305	-259.671	51.913	-0.553
5000	50.357	354.295	316.034	191.308	-259.509	58.270	-0.609
5100	50.998	355.299	316.794	196.376	-259.294	64.624	-0.662
5200	51.630	356.295	317.544	201.508	-259.028	70.973	-0.713
5300	52.250	357.285	318.284	206.702	-258.712	77.316	-0.762
5400	52.856	358.267	319.016	211.957	-258.347	83.653	-0.809
5500	53.446	359.242	319.738	217.272	-257.937	89.982	-0.855
5600	54.017	360.211	320.452	222.646	-257.482	96.304	-0.898
5700	54.566	361.172	321.158	228.075	-256.987	102.617	-0.940
5800	55.093	362.125	321.856	233.558	-256.453	108.922	-0.981
5900	55.595	363.071	322.547	239.093	-255.839	115.213	-1.020
6000	56.071	364.010	323.230	244.676	-255.226	121.497	-1.058

PREVIOUS: December 1973 (1 atm) CURRENT: December 1973 (1 bar)

$S°(298.15\ K) = 101.182 \pm 0.021\ J\ K^{-1}\ mol^{-1}$
$T_{trs}(\alpha \to \beta) = 743\ K$
$T_{fus} = 918\ K$

$\Delta_f H°(0\ K) = -442.98 \pm 0.8\ kJ\ mol^{-1}$
$\Delta_f H°(298.15\ K) = -442.83 \pm 0.8\ kJ\ mol^{-1}$
$\Delta_{trs} H° = 3.77\ kJ\ mol^{-1}$
$\Delta_{fus} H° = 15.9\ kJ\ mol^{-1}$

Enthalpy of Formation

The heat of solution ($\Delta_{sln}H°$) of CsCl(cr) in water has been measured by Forcrand (1, 2), Haigh (3), Samoilov (4, 6), and Rodnikova (5). Based on their results, Parker (7) derived the corresponding $\Delta_{sln}H°$(aq, ∞, 298.15) values as 4.30, 4.32, 4.08, 4.25, 4.18 and 4.18 kcal mol^{-1}. Borob'ev (8) determined the same quantity and reported $\Delta_{sln}H°$(298.15 K) (CsCl, aq, ∞, H_2O) = 4.20±0.04 kcal mol^{-1}. Using this value, 4.20 kcal mol^{-1}, and $\Delta_f H°$(298.15 K) = -61.69 (9) and -39.952 kcal mol^{-1} (10) for Cs^+(aq, ∞) and Cl^-(aq, ∞, 298.15), respectively, we evaluate $\Delta_f H°$(CsCl, cr, 298.15 K) = -105.84±0.2 kcal mol^{-1}, which is adopted.

Heat Capacity and Entropy

Taylor (11) measured the low temperature heat capacities, 7.19 - 299.38 K, with an adiabatic calorimeter; the sample purity was >99.95%. Employing these data we derive the value S°(298.15 K) = 24.183 cal K^{-1} mol^{-1} based on S°(7.19 K) = 0.031 cal K^{-1} mol^{-1}.

The high temperature enthalpies, 385.2 - 904.9 K, were measured by Kaylor (21, 12, 13), using a Bunsen ice calorimeter and a CsCl sample of 99.8% purity. The heat capacities derived from the reported enthalpy data at temperatures 385.2 - 740.5 K appear too low (less than the adopted $C_p°$'s systematically by 0.9 cal K^{-1} mol^{-1} to join smoothly with the low temperature data at 298.15 K. The $C_p°$ values derived from the measured enthalpy data in the temperature range 760 - 920 K increase rapidly from 14.46 to 16.3 cal K^{-1} mol^{-1}. In order to rationalize this situation, we adopt the high temperature heat capacities, 298 - 743 K, for CsCl(α) obtained by linear extrapolation of the low temperature data. The $C_p°$ values for CsCl(β) are adjusted such that the rate of increase in heat capacities is linear and less rapid than the original one, and the total enthalpy remains essentially unchanged. The discrepancy between the high and low temperature $C_p°$ for CsCl(α) may be caused by the incomplete β → α phase conversion involved. In order to resolve this discrepancy, independent enthalpy measurements using a high purity sample with composition well characterized before and after each drop experiment seem necessary.

Transition Data

The temperature of transition (in K) has been reported by many investigators as 718 (14), 752 (15), 745 (16), 742 (21), 742.5 (12), 743 (20), and 733 (17), using different methods and samples of different purity. The value adopted is 743 K.

The heat of transition (in kcal mol^{-1}) was reported to be 1.8 (14), 1.55 (18), 1.4 (19), 0.80 (21), 0.58 (12), 0.90 (13), and 1.1 (17). A median value 0.90 kcal mol^{-1} is tentatively adopted.

Fusion Data

See the liquid table for details.

Sublimation Data

$\Delta_{sub}H°$(298.15 K) is calculated as the difference between $\Delta_f H°$(298.15 K) for CsCl(g) and CsCl(cr).

References

1. R. de Forcrand, Compt. Rend. 143, 98 (1906); 1 point, concentration m = 0.25 (19°C) where m is molality.
2. R. de Forcrand, Ann. Chim. Phys. [8] 24, 256 (1911); Compt. Rend. 152, 27 (1911); 1 point, m = 0.50 (15°C).
3. F. L. Haigh, J. Amer. Chem. Soc. 34, 1137 (1912); 3 points, m = 0.14 (21°C).
4. O. Y. Samoilov, Izv. Akad. Nauk SSSR, Otd. Khim. Nauk 1415 (1956); 1 point, m = 0.05 (25°C).
5. M. N. Rodnikova, Zh. Neorg. Khim. 3, 2417 (1958); 6 points, m = 0.05 (15, 35°C).
6. O. Y. Samoilov and M. N. Buslaeva, Stroenie Veshchestva i Spektroskopiya, Akad. Nauk SSSR, 102 (1960); 3 points, m = 0.06 (15, 35°C).
7. V. B. Parker, NSRDS-NBS 2, 1965.
8. A. F. Borob'ev, N. A. Ibragim, and S. M. Skuratov, Russ. J. Inorg. Chem. 11, 13 (1966).
9. S. R. Gunn, J. Phys. Chem. 71, 1386 (1967).
10. U. S. Nat. Bur. Stand. Tech. Note 270-3, 1968.
11. A. R. Taylor, Jr., T. E. Gardner, and D. F. Smith, U. S. Bur. Mines RI 6157, 1963.
12. C. E. Kaylor, G. E. Walden, and D. F. Smith, J. Phys. Chem. 64, 276 (1960).
13. D. F. Smith, C. E. Kaylor, G. E. Walden, A. R. Taylor, Jr., and J. B. Gayle, U. S. Bur. Mines RI 5832, 1961.
14. G. Wagner and L. Lippert, Z. Phys. Chem. B31, 263 (1936).
15. J. W. Menary, A. R. Ubbelonde, and I. Woodward, Proc. Roy. Soc. (London), A208, 158 (1951).
16. L. J. Wood, W. Seconda, and C. H. McBride, J. Amer. Chem. Soc. 80, 307 (1958).
17. A. J. Majumdar and R. Roy, J. Inorg. Nucl. Chem. 27, 1961 (1965).
18. J. Krogh-Moe, J. Amer. Chem. Soc. 82, 2399 (1960).
19. J. Krogh-Moe, J. Amer. Chem. Soc. 82, 6196 (1960).
20. J. W. Johnson, P. A. Agron, and M. A. Bredig, J. Amer. Chem. Soc. 77, 2734 (1955).
21. C. E. Kaylor, Ph. D. Thesis, University of Alabama, 1958.

Enthalpy Reference Temperature = T_r = 298.15 K Standard State Pressure = p° = 0.1 MPa

T/K	$C_p°$ (J K⁻¹mol⁻¹)	S° (J K⁻¹mol⁻¹)	-[G°-H°(T_r)]/T (J K⁻¹mol⁻¹)	H°-H°(T_r) (kJ mol⁻¹)	$\Delta_f H°$ (kJ mol⁻¹)	$\Delta_f G°$ (kJ mol⁻¹)	Log K_f
0	9.146	3.540	INFINITE	-12.452	-442.979	-442.979	INFINITE
100	44.091	47.798	145.996	-9.820	-443.952	-433.786	226.587
200	50.128	80.726	105.914	-5.038	-443.369	-423.827	110.692
298.15	52.442	101.182	101.182	0.	-442.835	-414.360	72.594
300	52.472	101.506	101.183	0.097	-442.829	-414.183	72.116
400	54.685	116.906	103.267	5.455	-444.485	-404.011	52.758
500	56.902	129.347	107.277	11.035	-443.823	-393.965	41.157
600	59.099	139.916	111.858	16.835	-442.946	-384.072	33.436
700	61.296	149.191	116.541	22.855	-441.858	-374.343	27.934
743.000	62.237	152.873	118.538	25.511	ALPHA <--> BETA TRANSITION		
743.000	63.388	157.941	118.538	29.276			
800	63.680	162.637	121.515	32.898	-436.758	-365.075	23.837
900	64.174	170.166	126.510	39.290	-435.321	-356.200	20.673
918.000	64.266	171.438	127.379	40.446	------- BETA <--> LIQUID -------		
1000	64.685	176.954	131.221	45.733	-500.972	-343.789	17.958
1100	65.187	183.142	135.664	52.227	-498.433	-328.193	15.585
1200	65.689	188.836	139.860	58.771	-495.849	-312.830	13.617
1300	66.191	194.114	143.833	65.365	-493.220	-297.685	11.961
1400	66.693	199.037	147.602	72.009	-490.548	-282.744	10.549
1500	67.195	203.656	151.187	78.703	-487.831	-267.995	9.332
1600	67.697	208.008	154.604	85.447	-485.073	-253.429	8.274
1700	68.241	212.129	157.867	92.245	-482.272	-239.037	7.345
1800	68.743	216.044	160.991	99.094	-479.431	-224.811	6.524
1900	69.245	219.774	163.988	105.993	-476.556	-210.743	5.794
2000	69.747	223.338	166.867	112.943	-473.651	-196.828	5.141

PREVIOUS: CURRENT: June 1968

CESIUM CHLORIDE (CsCl) LIQUID M_r = 168.3584

$S°(298.15 K) = [101.708] J K^{-1} mol^{-1}$

T_{fus} = 918 K

$\Delta_f H°(298.15 K) = [-434.468] kJ mol^{-1}$

$\Delta_{fus}H° = 15.9 \pm 4.2 kJ mol^{-1}$

Enthaply of Formation

$\Delta_f H°$(CsCl, ℓ, 298.15 K) is calculated from $\Delta_f H°$(CsCl, cr, 298.15 K) by adding $\Delta_{fus}H°$ and the difference in enthalpy, H°(918 K) -H°(298.15 K), between the crystal and liquid.

Heat Capacity and Entropy

The enthalpies of CsCl(ℓ) at temperatures 923.6 - 1168.0 K were measured with a Bunsen ice calorimeter by Kaylor (1, 2). Dworkin (3) derived the heat capacity of CsCl(ℓ) as 18.5 cal K^{-1} mol^{-1} from enthalpy data at temperatures 918 - 980 K. This value is adopted since it is in agreement with the enthalpy values reported by Kaylor et al. (1, 2). The heat capacity at temperatures above 1168 and below 923.6 K is assumed to be 18.5 cal K^{-1} mol^{-1}. The entropy is calculated in a manner analogous to that used for the enthalpy of formation.

Fusion Data

The melting temperature (in K) of CsCl(cr) has been determined by many investigators as 904 ± 3 (4), 919 (5), 912 (6), 911 (7), 918 (8), 916 (9), and 919 (13). The cause of the discrepancies is probably due to the difference in purity of the samples. The melting temperature adopted is 918 K.

The heat of melting (in kcal mol^{-1}) was reported to be 4.84 (14), 4.97 (15), 4.58 (10), 4.96 (11), and 4.79 (2), which were obtained calorimetrically. Using a modification of Kelley's method of obtaining heat of melting from freezing point data on binary systems involving CsCl (16), we derive the value of $\Delta_{fus}H°$ as 3.6 - 4.0 kcal mol^{-1}, based on the following systems: CsCl-NaCl (5), CsCl-CuCl (6), CsCl-AgCl (6), CsCl-LiCl (8) and CsCl-$Cs_3(PO_3)_3$ (12). The heat of melting for CsCl(cr) is tentatively adopted as 3.8 kcal mol^{-1}. The $\Delta_{fus}H°$ values obtained calorimetrically are not used due to the uncertainties of the state of the sample before and after the enthalpy measurements.

Vaporization Data

T_{vap} = 1597.3 K is the temperature at which the calculated total pressures of CsCl(g) and Cs_2Cl_2(g) equal one atmosphere. The vapor composition at T_{vap} is derived as CsCl 80.4% and Cs_2Cl_2 19.6%. The boiling point of CsCl(ℓ) has been determined by several investigators as 1562 (17), 1572 (18) and 1576 (19). The heat required to vaporize one mole of liquid to the above vapor mixture at T_{vap} is $\Delta_{vap}H°$.

References

1. C. E. Kaylor, G. E. Walden, and D. F. Smith, J. Phys. Chem. 64, 276 (1960).
2. D. F. Smith, C. E. Kaylor, G. E. Walden, A. R. Taylor, Jr., and J. B. Gayle, U. S. Bur. Mines RI 5832, 1961.
3. A. S. Dworkin, Oak Ridge National Laboratory, Oak Ridge, Tennessee, personal communication, December 1, 1964.
4. T. Carnelley and Carleton-Williams, J. Chem. Soc. 37, 125 (1880).
5. S. Zemczuzny and F. Rambach, Z. Anorg. Chem. 65, 415 (1910).
6. C. Sandonnini, Atti Acc. Linc. Ser. 5, 21, II, 77 (1912).
7. E. Korreng, Z. Anorg. Chem. 91, 194 (1915).
8. T. W. Richards and W. B. Meldrum, J. Amer. Chem. Soc. 39, 1816 (1917).
9. B. F. Markov and I. D. Panchenko, Russ. J. Gen. Chem. 25, 1987 (1955).
10. C. E. Kaylor, Ph. D. Thesis, University of Alabama, 1958.
11. C. E. Kaylor, G. E. Walden, and D. F. Smith, J. Phys. Chem. 64, 276 (1960).
12. G. A. Bukhalova and I. V. Mardirosova, Russ. J. Inorg. Chem. 12, 1158 (1967).
13. G. A. Bukhalova and D. V. Sementsova, Russ. J. Inorg. Chem. 10, 1027 (1965).
14. A. S. Dworkin and M. A. Bredig, J. Phys. Chem. 64, 269 (1960).
15. Determined by D. F. Smith, University of Alabama; quoted by Dworkin (14).
16. K. K. Kelley, U. S. Bur. Mines Bull. 393, 1936.
17. O. Ruff and S. Mugdan, Z. Anorg. Allgem. Chem. 117, 147 (1921).
18. I. G. Murgulescu and L. Topor, Rev. Roum. Chim. 12, 1077 (1967).
19. H. von Wartenberg and H. Schulz, Z. Elektrochem. 27, 568 (1921).

Cesium Chloride (CsCl) $Cl_1Cs_1(l)$

Enthalpy Reference Temperature = T_r = 298.15 K					Standard State Pressure = p° = 0.1 MPa		
	J $K^{-1}mol^{-1}$			kJ mol^{-1}			
T/K	$C_p°$	S°	$-[G°-H°(T_r)]/T$	$H°-H°(T_r)$	$\Delta_f H°$	$\Delta_f G°$	Log K_f
0							
100							
200							
298.15	77.404	101.708	101.708	0.	-434.468	-406.150	71.156
300	77.404	102.187	101.710	0.143	-434.416	-405.975	70.686
400	77.404	124.455	104.746	7.884	-433.690	-396.236	51.743
500	77.404	141.727	110.479	15.624	-430.867	-387.199	40.450
600	77.404	155.840	116.899	23.364	-428.050	-378.731	32.971
700	77.404	167.771	123.336	31.105	-425.242	-370.733	27.664
800	77.404	178.107	129.551	38.845	-422.445	-363.137	23.710
900	77.404	187.224	135.462	46.586	-419.659	-355.891	20.655
918.000	77.404	188.757	136.492	47.979	------- BETA <--> LIQUID -------		
1000	77.404	195.379	141.053	54.326	-484.013	-345.255	18.034
1100	77.404	202.757	146.333	62.066	-480.226	-331.563	15.745
1200	77.404	209.492	151.320	69.807	-476.446	-318.215	13.852
1300	77.404	215.688	156.036	77.547	-472.671	-305.182	12.262
1400	77.404	221.424	160.504	85.288	-468.902	-292.439	10.911
1500	77.404	226.764	164.745	93.028	-465.140	-279.966	9.749
1600	77.404	231.760	168.779	100.768	-461.386	-267.744	8.741
1700	77.404	236.452	172.624	108.509	-457.641	-255.756	7.858
1800	77.404	240.877	176.294	116.249	-453.909	-243.988	7.080
1900	77.404	245.062	179.804	123.990	-450.193	-232.427	6.390
2000	77.404	249.032	183.167	131.730	-446.498	-221.062	5.774
2100	77.404	252.808	186.394	139.470	-442.827	-209.880	5.220
2200	77.404	256.409	189.495	147.211	-439.186	-198.873	4.722
2300	77.404	259.850	192.480	154.951	-435.581	-188.030	4.270
2400	77.404	263.144	195.356	162.692	-432.019	-177.344	3.860
2500	77.404	266.304	198.131	170.432	-428.504	-166.805	3.485
2600	77.404	269.340	200.812	178.172	-425.045	-156.405	3.142
2700	77.404	272.261	203.405	185.913	-421.648	-146.138	2.827
2800	77.404	275.076	205.914	193.653	-418.318	-135.995	2.537
2900	77.404	277.792	208.346	201.394	-415.065	-125.970	2.269
3000	77.404	280.416	210.705	209.134	-411.893	-116.055	2.021

PREVIOUS: CURRENT: June 1968

Cesium Chloride (CsCl) $Cl_1Cs_1(l)$

CESIUM CHLORIDE (CsCl) ALPHA-BETA-LIQUID M_r = 168.3584

0 to 743 K crystal, alpha
743 to 918 K crystal, beta
above 918 K liquid

Refer to the individual tables for details.

Cesium Chloride (CsCl) $Cl_1Cs_1(cr,l)$

Enthalpy Reference Temperature = T_r = 298.15 K Standard State Pressure = $p°$ = 0.1 MPa

	J K⁻¹mol⁻¹			kJ mol⁻¹			
T/K	$C_p°$	S°	$-[G°-H°(T_r)]/T$	$H°-H°(T_r)$	$\Delta_f H°$	$\Delta_f G°$	Log K_f
0	9.146	3.540	INFINITE	-12.452	-442.979	-442.979	INFINITE
100	44.091	47.798	145.996	-9.820	-443.952	-433.786	226.587
200	50.128	80.726	105.914	-5.038	-443.369	-423.827	110.692
298.15	52.442	101.182	101.182	0.	-442.835	-414.360	72.594
300	52.472	101.506	101.183	0.097	-442.829	-414.183	72.116
400	54.685	116.906	103.267	5.455	-444.485	-404.011	52.758
500	56.902	129.347	107.277	11.035	-443.823	-393.965	41.157
600	59.099	139.916	111.858	16.835	-442.946	-384.072	33.436
700	61.296	149.191	116.541	22.855	-441.858	-374.343	27.934
743.000	62.237	152.873	118.538	25.511	ALPHA <--> BETA		
743.000	63.388	157.941	118.538	29.276	TRANSITION		
800	63.680	162.637	121.515	32.898	-436.758	-365.075	23.837
900	64.174	170.166	126.510	39.290	-435.321	-356.200	20.673
918.000	64.266	171.438	127.379	40.446	BETA <--> LIQUID		
918.000	77.404	188.757	127.379	56.345	TRANSITION		
1000	77.404	195.379	132.687	62.692	-484.013	-345.255	18.034
1100	77.404	202.757	138.727	70.433	-480.226	-331.563	15.745
1200	77.404	209.492	144.347	78.173	-476.446	-318.215	13.852
1300	77.404	215.688	149.600	85.914	-472.671	-305.182	12.262
1400	77.404	221.424	154.528	93.654	-468.902	-292.439	10.911
1500	77.404	226.764	159.168	101.394	-465.140	-279.966	9.749
1600	77.404	231.760	163.550	109.135	-461.386	-267.744	8.741
1700	77.404	236.452	167.702	116.875	-457.641	-255.756	7.858
1800	77.404	240.877	171.646	124.616	-453.909	-243.988	7.080
1900	77.404	245.062	175.400	132.356	-450.193	-232.427	6.390
2000	77.404	249.032	178.984	140.096	-446.498	-221.062	5.774
2100	77.404	252.808	182.410	147.837	-442.827	-209.880	5.220
2200	77.404	256.409	185.692	155.577	-439.186	-198.873	4.722
2300	77.404	259.850	188.842	163.318	-435.581	-188.030	4.270
2400	77.404	263.144	191.870	171.058	-432.019	-177.344	3.860
2500	77.404	266.304	194.785	178.798	-428.504	-166.805	3.485
2600	77.404	269.340	197.594	186.539	-425.045	-156.405	3.142
2700	77.404	272.261	200.306	194.279	-421.648	-146.138	2.827
2800	77.404	275.076	202.926	202.020	-418.318	-135.995	2.537
2900	77.404	277.792	205.461	209.760	-415.065	-125.970	2.269
3000	77.404	280.416	207.916	217.500	-411.893	-116.055	2.021

PREVIOUS: CURRENT: June 1968

Cesium Chloride (CsCl) $Cl_1Cs_1(cr,l)$

CESIUM CHLORIDE (CsCl) IDEAL GAS M_r = 168.3584

$\Delta_f H°(0\ K) = -237.98 \pm 4.2\ kJ\ mol^{-1}$
$S°(298.15\ K) = 256.070 \pm 0.021\ J\ K^{-1}\ mol^{-1}$
$\Delta_f H°(298.15\ K) = -240.16 \pm 4.2\ kJ\ mol^{-1}$

Electronic Level and Quantum Weight

State	ε_i, cm^{-1}	g_i
$^1\Sigma^+$	0	1

$\omega_e = 213.08\ cm^{-1}$ $\omega_e x_e = 0.733\ cm^{-1}$ $\sigma = 1$
$B_e = 0.07134\ cm^{-1}$ $\alpha_e = 0.000332\ cm^{-1}$ $r_e = 2.906$ Å

Enthalpy of Formation

The total pressures of CsCl(g) and Cs_2Cl_2(g) over CsCl(cr) and CsCl(ℓ) in the temperature range 605-1577 K have been determined by manometric, transpiration and Knudsen effusion methods by many investigators. Based on the reported results, the partial pressures of CsCl(g) and Cs_2Cl_2(g) are calculated using the derived Gibbs energy functions for CsCl(cr), CsCl(ℓ), CsCl(g) and Cs_2Cl_2(g) and the enthalpies of sublimation and vaporization which are adjusted so that the sum of the calculated partial pressures of CsCl(g) and Cs_2Cl_2(g) approaches the measured total pressures, and the evaluated 2nd and 3rd law $\Delta_r H°$(298.15 K) values are in reasonable agreement. The results obtained are presented in the table below. The value of $\Delta_f H°$(CsCl, g, 298.15 K) adopted is -57.4±1 kcal mol^{-1} (-240.162±4.2 kJ mol^{-1}).

Gaydon (10) reported D_0^0(Cs-Cl) = 4.6±0.2 eV or 106.08±4.61 kcal mol^{-1}. Using this D_0^0 value and $\Delta_f H°$(0 K) = 18.683 and 28.52 kcal mol^{-1} for Cs(g) and Cl(g), we evaluate $\Delta_f H°$(298.15 K) = -59.4±4.6 kcal mol^{-1} for CsCl(g), which is in fair agreement with the adopted value.

Source	Reaction*	Method	T/K	Data Points	$\Delta_r H°$(298.15 K), kcal mol^{-1} 2nd Law	3rd Law	Drift cal K^{-1} mol^{-1}	$\Delta_f H°$(298.15 K) kcal mol^{-1}
1. Niwa (1938)	1	A	793-893	11	46.81±0.35	49.92	3.7±0.4	-55.92
2. Cogin (1948)	1	A	690-891	28	47.80±0.31	47.88	0.1±0.4	-57.96
3. Treadwell (1953)	1	B	784-905	10	47.08±0.26	49.17	2.5±0.3	-56.67
4. Nesmeyanov (1960)	1	A	605-851	16	45.26±0.38	48.74	4.7±0.5	-57.10
5. Wartenberg (1921)	2	C	1335-1577	9	49.36±0.50	46.27	-2.1±0.3	-57.57
6. Ruff (1921)	2	C	1259-1568	7	51.61±1.08	46.17	-3.8±0.8	-57.67
7. Fiock (1926)	2	C	1098-1293	9	47.57±0.08	46.52	-0.9±0.1	-57.32
8. Kangro (1938)	2	B	1133-1263	3	48.73±2.09	47.00	-1.4±1.7	-56.84
9. Schrier (1961)	2	C	1165-1387	10	47.46±0.08	46.36	-0.9±0.1	-57.48
10. Murgulescu (1967)	1	A	1174-1353	10	47.97±0.15	46.54	-1.1±0.1	-57.30

*1 CsCl(cr) = CsCl(g); 2 CsCl(ℓ) = CsCl(g); A = Knudsen effusion; B = transpiration; and C = manometric.

Heat Capacity and Entropy

Clouser (11) studied the pure rotational spectra of CsCl(g) in the 0.96-3 mm range of the microwave region with the molecular-beam spectrometer, and derived the values of B_e, α_e, ω_e, $\omega_e x_e$ and r_e, which are adopted and corrected to the average isotopic species.

The Cs-Cl bond distance was measured as 3.06±0.03 Å by the electron diffraction method by Maxwell (12), which is not used. Similar values of B_e and α_e were reported by Honig (13) and Rice (14), determined from microwave and infrared spectra, respectively. Barrow (15) gave ω_e = 240 cm^{-1}($CsCl^{35}$) which is singificantly different from the adopted value 214.22 cm^{-1} for $CsCl^{35}$ and the value 209±6 cm^{-1} reported by Rice (14).

References

1. K. Niwa, J. Fac. Sci., Hokkaido Univ., Ser. III 2, 201 (1938).
2. G. E. Cogin and G. E. Kimball, J. Chem. Phys. 16, 1035 (1948).
3. W. D. Treadwell and W. Werner, Helv. Chim. Acta 36, 1436 (1953).
4. A. N. Nesmeyanov and L. A. Sazonov, Russ. J. Inorg. Chem. 5, 248 (1960).
5. H. von Wartenberg and H. Schulz, Z. Elektrochem. 27, 568 (1921).
6. O. Ruff and S. Mugdan, Z. Anorg. Allg. Chem. 117, 147 (1921).
7. E. F. Fiock and W. H. Rodebush, J. Amer. Chem. Soc. 48, 2522 (1926).
8. W. Kangro and H. W. Wieking, Z. Physik, Chem. (Leipzig) A183, 199 (1938).
9. E. E. Schrier, Ph. D. Thesis, Rensselaer Polytechnic Institute, 1961.
10. A. G. Gaydon, "Dissociation Energies," Chapman and Hall Ltd., London, 1953.
11. P. L. Clouser and W. Gordy, Phys. Rev. 134, A863 (1964).
12. L. R. Maxwell, S. B. Hendricks and V. M. Mosley, Phys. Rev. 52, 968 (1937).
13. A. Honig, M. Mandel, M. L. Stitch and C. H. Townes, Phys. Rev. 96, 629 (1954).
14. S. A. Rice and W. Klemperer, J. Chem. Phys. 27, 573 (1957).
15. R. F. Barrow and A. D. Caunt, Proc. Roy. Soc. (London) A219, 120 (1953).
16. I. G. Murgulescu and L. Topor, Rev. Roum. Chim. 12, 1077 (1967).

Cesium Chloride (CsCl) $Cl_1Cs_1(g)$

Enthalpy Reference Temperature = T_r = 298.15 K Standard State Pressure = p° = 0.1 MPa

T/K	$C_p°$ (J K^{-1}mol^{-1})	S°	$-[G°-H°(T_r)]/T$	$H°-H°(T_r)$ (kJ mol^{-1})	$\Delta_f H°$	$\Delta_f G°$	Log K_f
0	0.	0.	INFINITE	-10.130	-237.985	-237.985	INFINITE
100	33.194	217.326	288.246	-7.092	-238.551	-245.338	128.151
200	36.120	241.467	259.429	-3.593	-239.251	-251.857	65.778
250	36.643	249.588	256.677	-1.772	-239.660	-254.963	53.272
298.15	36.952	256.070	256.070	0.	-240.162	-257.867	45.177
300	36.961	256.299	256.071	0.068	-240.184	-257.977	44.918
350	37.175	262.014	256.522	1.922	-242.885	-260.558	38.886
400	37.330	266.988	257.526	3.785	-243.482	-263.042	34.350
450	37.450	271.392	258.827	5.655	-244.071	-265.451	30.813
500	37.548	275.343	260.284	7.530	-244.655	-267.795	27.976
600	37.702	282.203	263.383	11.292	-245.815	-272.314	23.707
700	37.827	288.025	266.498	15.069	-246.971	-276.640	20.643
800	37.935	293.083	269.512	18.857	-248.126	-280.799	18.334
900	38.034	297.557	272.384	22.656	-249.282	-284.814	16.530
1000	38.126	301.569	275.105	26.464	-317.569	-285.000	14.887
1100	38.215	305.207	277.679	30.281	-317.706	-281.737	13.379
1200	38.301	308.536	280.114	34.107	-317.840	-278.461	12.121
1300	38.386	311.605	282.420	37.941	-317.971	-275.174	11.057
1400	38.468	314.453	284.607	41.784	-318.100	-271.877	10.144
1500	38.550	317.110	286.686	45.635	-318.227	-268.571	9.352
1600	38.631	319.600	288.667	49.494	-318.354	-265.257	8.660
1700	38.712	321.945	290.556	53.361	-318.483	-261.934	8.048
1800	38.792	324.160	292.362	57.236	-318.616	-258.604	7.504
1900	38.872	326.259	294.091	61.119	-318.757	-255.266	7.018
2000	38.951	328.255	295.750	65.010	-318.911	-251.921	6.579
2100	39.030	330.157	297.343	68.909	-319.081	-248.567	6.183
2200	39.109	331.975	298.876	72.816	-319.274	-245.205	5.822
2300	39.188	333.715	300.354	76.731	-319.495	-241.833	5.492
2400	39.266	335.384	301.779	80.654	-319.750	-238.452	5.190
2500	39.345	336.989	303.155	84.585	-320.045	-235.058	4.911
2600	39.423	338.534	304.486	88.523	-320.388	-231.652	4.654
2700	39.502	340.023	305.775	92.469	-320.785	-228.232	4.415
2800	39.580	341.461	307.024	96.423	-321.242	-224.796	4.194
2900	39.658	342.851	308.236	100.385	-321.766	-221.342	3.987
3000	39.736	344.197	309.412	104.355	-322.365	-217.869	3.793
3100	39.814	345.501	310.555	108.333	-323.045	-214.375	3.612
3200	39.892	346.767	311.667	112.318	-323.812	-210.858	3.442
3300	39.970	347.995	312.750	116.311	-324.632	-207.318	3.282
3400	40.048	349.190	313.804	120.312	-325.574	-203.749	3.130
3500	40.126	350.352	314.831	124.321	-326.616	-200.151	2.987
3600	40.204	351.483	315.834	128.337	-327.763	-196.522	2.851
3700	40.282	352.586	316.812	132.361	-329.004	-192.861	2.723
3800	40.360	353.661	317.768	136.394	-330.355	-189.164	2.600
3900	40.438	354.710	318.702	140.433	-331.805	-185.432	2.484
4000	40.516	355.735	319.615	144.481	-333.386	-181.659	2.372
4100	40.593	356.737	320.508	148.537	-335.082	-177.845	2.266
4200	40.671	357.716	321.382	152.600	-336.895	-173.988	2.164
4300	40.749	358.674	322.239	156.671	-338.825	-170.087	2.066
4400	40.827	359.611	323.077	160.750	-340.577	-166.173	1.973
4500	40.905	360.530	323.899	164.836	-342.127	-162.250	1.883
4600	40.982	361.430	324.706	168.931	-344.225	-158.230	1.797
4700	41.060	362.312	325.496	173.033	-346.410	-154.163	1.713
4800	41.138	363.177	326.272	177.143	-348.679	-150.049	1.633
4900	41.216	364.026	327.034	181.260	-351.028	-145.887	1.555
5000	41.293	364.860	327.782	185.386	-353.449	-141.676	1.480
5100	41.371	365.678	328.518	189.519	-355.939	-137.416	1.407
5200	41.449	366.482	329.240	193.660	-357.960	-133.184	1.338
5300	41.526	367.272	329.950	197.809	-360.500	-128.837	1.270
5400	41.604	368.049	330.648	201.965	-362.809	-124.485	1.204
5500	41.682	368.813	331.335	206.130	-365.246	-120.073	1.140
5600	41.760	369.565	332.011	210.302	-367.853	-115.592	1.078
5700	41.837	370.305	332.677	214.481	-370.484	-111.064	1.018
5800	41.915	371.033	333.332	218.669	-373.133	-106.489	0.959
5900	41.993	371.751	333.977	222.864	-375.792	-101.869	0.902
6000	42.070	372.457	334.612	227.068	-378.457	-97.204	0.846

PREVIOUS: June 1968 (1 atm) CURRENT: June 1968 (1 bar)

Cesium Chloride (CsCl) $Cl_1Cs_1(g)$

$S°(298.15 K) = 87.03 \pm 4.2\ J\ K^{-1}\ mol^{-1}$
T_{fus} = 703 K

$\Delta_f H°(0\ K)$ = Unknown
$\Delta_f H°(298.15\ K) = -138.07 \pm 1.7\ kJ\ mol^{-1}$
$\Delta_{fus} H° = 10.230 \pm 2.1\ kJ\ mol^{-1}$

Enthalpy of Formation

There have been three calorimetric determinations of the enthalpy of formation. Thomsen (1) reported -32.875 kcal mol^{-1}. Bethelot (2) reported -35.6 kcal mol^{-1} and Wartenberg and Werth (3) reported -32.1±0.4 kcal mol^{-1}. In addition, values of the enthalpy of formation have been derived from several sets of equilibrium data. Noyes and Chow (4) from a study of cell potentials derived a $\Delta_r H°$ = 7.19 kcal mol^{-1} for the reaction Cu(cr) + HCl(∞,aq) → CuCl(cr) + 1/2 H_2(g) which yields $\Delta_f H°$(CuCl, 298.15 K) = -32.76 kcal mol^{-1}. Several investigators have studied the reaction 2 CuCl(cr) + H_2(g) → 2 Cu(cr) + 2 HCl(g) and its reverse reaction. These include Bagdasarian (5) who reported equilibrium constants which result in 2nd and 3rd law heats of reaction of 24.3 and 23.2 kcal mol^{-1}; Kapustinsky (6) who obtained similarly 20.1 and 21.8 kcal mol^{-1}; Shchukarev and Oranskaya (7) who obtained 25.4 and 21.5 kcal mol^{-1} for the 2nd and 3rd law heat of reaction. These yield values for $\Delta_f H°$(CuCl, 298.15 K) ranging from -32.1 to -34.8. Watanabe (8) from cell studies reports $\Delta_f H°$(CuCl, 298.15 K) = -34.6 kcal mol^{-1}. A value of $\Delta_f H°$(298.15 K) can also be obtained from $\Delta_f H°(CuCl_3$, g) which is fixed independently and known with good precision. Using the 3rd law heat of sublimation of CuCl(cr) to trimer determined by Magee (9) [see trimer table], we obtain $\Delta_f H°$(CuCl, cr, 298.15 K) = -33.01±0.3 kcal mol^{-1}. A weighted average of -33.0±0.4 kcal mol^{-1} (-138.072±1.7 kJ mol^{-1}) was adopted.

Heat Capacity and Entropy

The entropy of CuCl was obtained from the several pieces of equilibrium data reported above and the adopted $\Delta_f H°$(298.15 K). A weighted average of 20.8±1 cal $K^{-1}\ mol^{-1}$ was adopted for S°(CuCl, cr, 298.15 K). The enthalpy and heat capacity above 298.15 K have been reported by Krestovnikov and Karetnikov (10). However, these values were not adopted since they do not give reasonable heat capacities, and other data by these workers are not in agreement with established values. Heat capacities equal to those of AgCl, as given by Kelley (11), were adopted here above 298.15 K.

Fusion Data

The temperature and enthalpy of fusion were those selected by Kelley (12) from phase studies. The enthalpy of fusion was reported by Krestovnikov and Karetnikov (10) as 2.54 kcal mol^{-1}.

Sublimation Data

The enthalpy of sublimation was calculated from the adopted enthalpies of formation at 298.15 K for the reaction x CuCl(cr) → $(CuCl)_x$(g) where x = 1 or 3. See the individual ideal gas tables for details.

References

1. Thomsen, "Thermochemische Unterschungen," Barth, Leipzig (1882-1886).
2. Berthelot, Ann. Chim. Phys. 20, 504 (1880).
3. H. V. Wartenberg and H. Werth, Z. Physik. Chem. A151, 109 (1930).
4. A. A. Noyes and M. Chow, J. Amer. Chem. Soc. 40, 739 (1918).
5. A. B. Bagdasarian, Trans. Am. Electrochem. Soc. 51, 449 (1927).
6. A. F. Kapustinsky, J. Amer. Chem. Soc. 58, 460 (1936).
7. S. A. Shchukarev and M. A. Oranskaya, Z. Obshchei. Khim. 24, 1926 (1954).
8. M. Watanabe, Bull. Inst. Phys. Chem. Res. 9, 94 (1929).
9. D. W. Magee, Ph. D. Thesis, Ohio State University (1955).
10. A. N. Krestovnikov and G. A. Karetnikov, Jour. Gen. Chem. (USSR) 6, 955 (1936).
11. K. K. Kelley, U. S. Bur. Mines Bull. 584, 1960.
12. K. K. Kelley, U. S. Bur. Mines Bull. 393, 1936.

Enthalpy Reference Temperature = T_r = 298.15 K   Standard State Pressure = p° = 0.1 MPa

T/K	$C_p°$ (J $K^{-1}mol^{-1}$)	S° (J $K^{-1}mol^{-1}$)	$-[G°-H°(T_r)]/T$ (J $K^{-1}mol^{-1}$)	$H°-H°(T_r)$ (kJ mol^{-1})	$\Delta_f H°$ (kJ mol^{-1})	$\Delta_f G°$ (kJ mol^{-1})	Log K_f
0							
100							
200							
298.15	48.534	87.027	87.027	0.	-138.072	-120.876	21.177
300	48.771	87.328	87.028	0.090	-138.059	-120.769	21.028
400	56.902	102.675	89.065	5.444	-136.933	-115.157	15.038
500	59.831	115.711	93.128	11.291	-135.437	-109.882	11.479
600	61.505	126.773	97.838	17.361	-133.803	-104.924	9.134
700	62.760	136.354	102.671	23.578	-132.097	-100.244	7.480
703.000	62.791	136.623	102.815	23.767	------ CRYSTAL <--> LIQUID -----		
800	63.680	144.795	107.419	29.901	-130.348	-95.813	6.256
900	64.388	152.337	111.999	36.305	-128.580	-91.602	5.316
1000	65.019	159.155	116.379	42.776	-126.812	-87.588	4.575
1100	65.531	165.376	120.554	49.304	-125.064	-83.751	3.977
1200	66.011	171.099	124.531	55.881	-123.363	-80.071	3.485
1300	66.459	176.400	128.320	62.505	-121.753	-76.529	3.075
1400	66.874	181.341	131.932	69.172	-133.393	-72.700	2.712
1500	67.258	185.968	135.382	75.879	-131.866	-68.418	2.383

PREVIOUS.   CURRENT: March 1966

COPPER CHLORIDE (CuCl) LIQUID $M_r = 98.999$

$S°(298.15\ K) = [93.752]\ J\ K^{-1}\ mol^{-1}$ $\Delta_f H°(298.15\ K) = [-131.178]\ kJ\ mol^{-1}$

$T_{fus} = 703°K$ $\Delta_{fus}H° = 10.230\ kJ\ mol^{-1}$

Enthalpy of Formation

$\Delta_f H°(CuCl,\ \ell,\ 298.15\ K)$ is calculated from that of the crystal by adding $\Delta_{fus}H°$ and the difference in enthalpy, H°(703 K) - H°(298.15 K), between the crystal and liquid.

Heat Capacity and Entropy

The heat capacity was estimated to be constant at 8 cal $K^{-1}\ mol^{-1}$. The entropy was calculated in a manner analogous to that of the heat of formation.

Boiling Data

The boiling point and heat of vaporization were calculated from the respective liquid, monomer and trimer gas tables. The boiling point, $T_{vap} = 1485$ K, was taken as the point where the total pressure reached 1 atm. At this point the partial pressures are 0.021 atm of monomer and 0.979 atm of trimer.

Copper Chloride (CuCl) $Cl_1Cu_1(l)$

Enthalpy Reference Temperature = T_r = 298.15 K Standard State Pressure = p° = 0.1 MPa

T/K	$C_p°$ (J K⁻¹mol⁻¹)	S° (J K⁻¹mol⁻¹)	$-[G°-H°(T_r)]/T$ (J K⁻¹mol⁻¹)	$H°-H°(T_r)$ (kJ mol⁻¹)	$\Delta_f H°$ (kJ mol⁻¹)	$\Delta_f G°$ (kJ mol⁻¹)	Log K_f
0							
100							
200							
298.15	66.944	93.752	93.752	0.	-131.178	-115.987	20.320
300	66.944	94.166	93.754	0.124	-131.131	-115.893	20.179
400	66.944	113.425	96.379	6.818	-128.665	-111.189	14.520
500	66.944	128.363	101.338	13.513	-126.322	-107.093	11.188
600	66.944	140.569	106.890	20.207	-124.064	-103.461	9.007
700	66.944	150.888	112.457	26.901	-121.879	-100.201	7.477
703.000	66.944	151.174	112.622	27.102	------ CRYSTAL <--> LIQUID -----		
800	66.944	159.827	117.832	33.596	-119.759	-97.249	6.350
900	66.944	167.712	122.945	40.290	-117.701	-94.560	5.488
1000	66.944	174.765	127.781	46.985	-115.709	-92.096	4.811
1100	66.944	181.146	132.347	53.679	-113.795	-89.828	4.266
1200	66.944	186.971	136.659	60.373	-111.976	-87.731	3.819
1300	66.944	192.329	140.738	67.068	-110.296	-85.779	3.447
1400	66.944	197.290	144.603	73.762	-121.908	-83.545	3.117
1500	66.944	201.909	148.271	80.457	-120.394	-80.857	2.816
1600	66.944	206.229	151.760	87.151	-118.884	-78.271	2.555
1700	66.944	210.288	155.084	93.845	-117.378	-75.779	2.328
1800	66.944	214.114	158.259	100.540	-115.877	-73.375	2.129
1900	66.944	217.734	161.294	107.234	-114.381	-71.055	1.953
2000	66.944	221.167	164.203	113.929	-112.889	-68.813	1.797
2100	66.944	224.433	166.994	120.623	-111.404	-66.646	1.658
2200	66.944	227.548	169.676	127.317	-109.924	-64.549	1.533
2300	66.944	230.524	172.257	134.012	-108.451	-62.519	1.420
2400	66.944	233.373	174.745	140.706	-106.986	-60.554	1.318
2500	66.944	236.105	177.145	147.401	-105.528	-58.649	1.225
2600	66.944	238.731	179.464	154.095	-104.079	-56.803	1.141
2700	66.944	241.257	181.706	160.789	-102.639	-55.012	1.064
2800	66.944	243.692	183.876	167.484	-101.209	-53.274	0.994
2900	66.944	246.041	185.980	174.178	-399.992	-45.591	0.821
3000	66.944	248.311	188.020	180.873	-397.786	-33.408	0.582

PREVIOUS: CURRENT: March 1966

Copper Chloride (CuCl) $Cl_1Cu_1(l)$

COPPER CHLORIDE (CuCl) CRYSTAL–LIQUID M_r = 98.999

0 to 703 K crystal
above 703 liquid

Refer to the individual tables for details.

Copper Chloride (CuCl) $Cl_1Cu_1(cr,l)$

Enthalpy Reference Temperature = T_r = 298.15 K Standard State Pressure = p° = 0.1 MPa

	J K⁻¹mol⁻¹			kJ mol⁻¹			
T/K	C_p°	S°	$-[G^\circ-H^\circ(T_r)]/T$	$H^\circ-H^\circ(T_r)$	Δ_fH°	Δ_fG°	Log K_f
0							
100							
200							
298.15	48.534	87.027	87.027	0.	-138.072	-120.876	21.177
300	48.771	87.328	87.028	0.090	-138.059	-120.769	21.028
400	56.902	102.675	89.065	5.444	-136.933	-115.157	15.038
500	59.831	115.711	93.128	11.291	-135.437	-109.882	11.479
600	61.505	126.773	97.838	17.361	-133.803	-104.924	9.134
700	62.760	136.354	102.671	23.578	-132.097	-100.244	7.480
703.000	62.791	136.623	102.815	23.767	CRYSTAL <--> LIQUID		
703.000	66.944	151.174	102.815	33.996	TRANSITION		
800	66.944	159.827	109.215	40.490	-119.759	-97.249	6.350
900	66.944	167.712	115.285	47.184	-117.701	-94.560	5.488
1000	66.944	174.765	120.886	53.879	-115.709	-92.096	4.811
1100	66.944	181.146	126.079	60.573	-113.795	-89.828	4.266
1200	66.944	186.971	130.914	67.268	-111.976	-87.731	3.819
1300	66.944	192.329	135.435	73.962	-110.296	-85.779	3.447
1400	66.944	197.290	139.678	80.656	-121.908	-83.545	3.117
1500	66.944	201.909	143.675	87.351	-120.394	-80.857	2.816
1600	66.944	206.229	147.451	94.045	-118.884	-78.271	2.555
1700	66.944	210.288	151.029	100.740	-117.378	-75.779	2.328
1800	66.944	214.114	154.428	107.434	-115.877	-73.375	2.129
1900	66.944	217.734	157.666	114.128	-114.381	-71.055	1.953
2000	66.944	221.167	160.756	120.823	-112.889	-68.813	1.797
2100	66.944	224.433	163.711	127.517	-111.404	-66.646	1.658
2200	66.944	227.548	166.542	134.212	-109.924	-64.549	1.533
2300	66.944	230.524	169.260	140.906	-108.451	-62.519	1.420
2400	66.944	233.373	171.872	147.600	-106.986	-60.554	1.318
2500	66.944	236.105	174.387	154.295	-105.528	-58.649	1.225
2600	66.944	238.731	176.812	160.989	-104.079	-56.803	1.141
2700	66.944	241.257	179.152	167.684	-102.639	-55.012	1.064
2800	66.944	243.692	181.414	174.378	-101.209	-53.274	0.994
2900	66.944	246.041	183.602	181.072	-399.992	-45.591	0.821
3000	66.944	248.311	185.722	187.767	-397.786	-33.408	0.582

PREVIOUS: CURRENT: March 1966

COPPER CHLORIDE (CuCl) IDEAL GAS $M_r = 98.999$

$S°(298.15\ K) = 237.207\ J\ K^{-1}\ mol^{-1}$

$\Delta_f H°(0\ K) = 91.21 \pm 1.7\ kJ\ mol^{-1}$
$\Delta_f H°(298.15\ K) = 91.09 \pm 1.7\ kJ\ mol^{-1}$

Electronic Level and Multiplicity

State	ε_i, cm^{-1}	g_i
$^1\Sigma^+$	0	1

$\omega_e = 414.9\ cm^{-1}$ $\omega_e x_e = 1.577\ cm^{-1}$ $\sigma = 1$
$B_e = 0.1761\ cm^{-1}$ $\alpha_e = 0.001\ cm^{-1}$ $r_e = 2.050$ Å

Enthalpy of Formation

The enthalpy of formation was obtained from the equilibrium data of Brewer and Lofgren (1). They studied the reaction x Cu(cr) + x HCl(g) → Cu_2Cl_2(g) + x/2 H_2(g) by measuring the amount of CuCl formed when various ratios of HCl:H_2 were passed over heated copper. Brewer and Lofgren analyzed the data by a least squares fitting technique and deduced partial pressures of monomer and trimer. The monomer pressures were subjected to 2nd and 3rd law analysis and gave $\Delta_r H°(298.15\ K) = 46.85 \pm 3.2$ and $43.83 \pm 0.4\ kcal\ mol^{-1}$, respectively. The 3rd law value yields $\Delta_f H°(CuCl,\ g,\ 298.15\ K) = 21.77 \pm 0.4\ kcal\ mol^{-1}$, which was adopted.

Heat Capacity and Entropy

The molecular constants were corrected for normal isotopic abundance of both copper and chlorine and were calculated from the values for $^{65}Cu^{35}Cl$ given by Asundi, Rao and Brody (2). The rotational constants were in good agreement with those reported by Lagerqvist and Lazarova-Girsamov (3) for $Cu^{65}Cl^{35}$.

References

1. L. Brewer and N. L. Lofgren, J. Am. Chem. Soc. 72, 3038 (1950).
2. R. K. Asundi, P. R. Rao and J. K. Brody, Nature 192, 444 (1961).
3. A. Lagerqvist and V. Lazarova-Girsamov, Naturwissenschaften 48, 68 (1961).

Copper Chloride (CuCl) $Cl_1Cu_1(g)$

Enthalpy Reference Temperature = T_r = 298.15 K Standard State Pressure = p° = 0.1 MPa

	J K⁻¹mol⁻¹			kJ mol⁻¹			
T/K	$C_p°$	S°	$-[G°-H°(T_r)]/T$	$H°-H°(T_r)$	$\Delta_f H°$	$\Delta_f G°$	Log K_f
0	0.	0.	INFINITE	-9.471	91.212	91.212	INFINITE
100	29.899	201.663	267.142	-6.548	92.010	82.298	-42.988
200	33.356	223.487	240.381	-3.379	91.647	72.692	-18.985
250	34.503	231.061	237.784	-1.681	91.373	67.985	-14.205
298.15	35.261	237.207	237.207	0.	91.086	63.506	-11.126
300	35.285	237.425	237.208	0.065	91.074	63.335	-11.028
350	35.830	242.908	237.639	1.844	90.757	58.736	-8.766
400	36.220	247.719	238.605	3.646	90.426	54.184	-7.076
450	36.509	252.003	239.860	5.464	90.081	49.675	-5.766
500	36.730	255.861	241.270	7.295	89.724	45.204	-4.722
600	37.044	262.587	244.279	10.985	88.978	36.369	-3.166
700	37.257	268.315	247.314	14.701	88.184	27.663	-2.064
800	37.415	273.300	250.257	18.435	87.344	19.074	-1.245
900	37.538	277.715	253.067	22.183	86.455	10.594	-0.615
1000	37.641	281.675	255.733	25.942	85.512	2.215	-0.116
1100	37.729	285.267	258.258	29.710	84.500	-6.067	0.288
1200	37.807	288.553	260.647	33.487	83.401	-14.252	0.620
1300	37.879	291.582	262.912	37.272	82.171	-22.341	0.898
1400	37.945	294.392	265.061	41.063	67.656	-29.923	1.116
1500	38.008	297.012	267.105	44.860	66.274	-36.845	1.283
1600	38.068	299.467	269.052	48.664	64.893	-43.674	1.426
1700	38.126	301.777	270.909	52.474	63.514	-50.418	1.549
1800	38.182	303.957	272.685	56.289	62.136	-57.080	1.656
1900	38.236	306.023	274.386	60.110	60.759	-63.665	1.750
2000	38.290	307.986	276.017	63.937	59.382	-70.178	1.833
2100	38.342	309.855	277.585	67.768	58.005	-76.622	1.906
2200	38.394	311.640	279.092	71.605	56.627	-83.001	1.971
2300	38.445	313.348	280.545	75.447	55.247	-89.317	2.028
2400	38.496	314.985	281.946	79.294	53.866	-95.573	2.080
2500	38.546	316.558	283.299	83.146	52.481	-101.771	2.126
2600	38.595	318.070	284.608	87.003	51.093	-107.913	2.168
2700	38.645	319.528	285.874	90.865	49.700	-114.003	2.206
2800	38.694	320.934	287.101	94.732	48.303	-120.040	2.239
2900	38.743	322.293	288.292	98.604	-253.302	-120.032	2.162
3000	38.792	323.607	289.447	102.481	-253.914	-115.426	2.010
3100	38.840	324.880	290.570	106.362	-254.581	-110.799	1.867
3200	38.889	326.114	291.661	110.249	-255.304	-106.150	1.733
3300	38.937	327.311	292.723	114.140	-256.081	-101.476	1.606
3400	38.985	328.474	293.758	118.036	-256.913	-96.779	1.487
3500	39.033	329.605	294.766	121.937	-257.798	-92.057	1.374
3600	39.081	330.705	295.749	125.843	-258.734	-87.308	1.267
3700	39.129	331.777	296.708	129.753	-259.719	-82.533	1.165
3800	39.176	332.821	297.645	133.668	-260.750	-77.730	1.068
3900	39.224	333.839	298.560	137.588	-261.825	-72.900	0.976
4000	39.272	334.833	299.455	141.513	-262.939	-68.042	0.889
4100	39.319	335.803	300.329	145.443	-264.089	-63.155	0.805
4200	39.367	336.751	301.185	149.377	-265.272	-58.240	0.724
4300	39.414	337.678	302.023	153.316	-266.484	-53.296	0.647
4400	39.462	338.585	302.844	157.260	-267.721	-48.324	0.574
4500	39.509	339.472	303.648	161.209	-268.980	-43.324	0.503
4600	39.556	340.341	304.436	165.162	-270.256	-38.295	0.435
4700	39.604	341.192	305.209	169.120	-271.547	-33.239	0.369
4800	39.651	342.027	305.968	173.083	-272.850	-28.155	0.306
4900	39.698	342.845	306.712	177.050	-274.160	-23.043	0.246
5000	39.745	343.647	307.443	181.022	-275.476	-17.905	0.187
5100	39.793	344.435	308.160	184.999	-276.795	-12.740	0.130
5200	39.840	345.208	308.865	188.981	-278.114	-7.550	0.076
5300	39.887	345.967	309.558	192.967	-279.431	-2.334	0.023
5400	39.934	346.713	310.239	196.958	-280.744	2.906	-0.028
5500	39.981	347.446	310.909	200.954	-282.053	8.171	-0.078
5600	40.028	348.167	311.568	204.954	-283.354	13.460	-0.126
5700	40.075	348.876	312.216	208.959	-284.647	18.771	-0.172
5800	40.122	349.573	312.855	212.969	-285.932	24.106	-0.217
5900	40.169	350.260	313.483	216.984	-287.208	29.462	-0.261
6000	40.216	350.935	314.101	221.003	-288.473	34.840	-0.303

PREVIOUS: March 1966 (1 atm) CURRENT: March 1966 (1 bar)

Copper Chloride (CuCl) $Cl_1Cu_1(g)$

HYDROCHLORIC ACID-d (DCl) IDEAL GAS M_r = 37.467102

$S°(298.15\ K) = 192.741\ J\ K^{-1}\ mol^{-1}$

$\Delta_f H°(0\ K) = -93.131 \pm 0.21\ kJ\ mol^{-1}$
$\Delta_f H°(298.15\ K) = -93.345 \pm 0.21\ kJ\ mol^{-1}$

Electronic States and Molecular Constants (σ = 1)

State	ε_i, cm^{-1}	g_i	ω_e, cm^{-1}	$\omega_e x_e$, cm^{-1}	B_e, cm^{-1}	α_e, cm^{-1}	r_e, Å
$X^1\Sigma^+$	0	1	2144.0	26.90	5.4444	0.1121	1.2746
$B^1\Pi$	75160	2	2199.0	[26.9]	5.1793	[0.1121]	
$V^1\Sigma^+$	76520	1	684.6	[26.9]	1.555	[0.1121]	2.43
$C^1\Pi$	77525	2	2114.1	[26.9]	4.9605	[0.1121]	

Enthalpy of Formation (Provisional-evaluation incomplete)

The enthalpy of formation was calculated from the selected value of $\Delta_f H°$(298.15 K) of HCl(g) given by JANAF (1), the appropriate thermal functions (see H_2, D_2, DCl tables, and HCl (1)) and zero point energies. The zero point energies of H_2(g) and D_2(g) are taken from Herzberg and Monfile (2). The zero point energies of HCl(g) and DCl(g), including the Dunham correction, were calculated from the molecular constants taken from Rosen (3).

Heat Capacity and Entropy

The vibrational and rotational constants at the respective electronic levels were taken from Rosen (3) and were adjusted to Cl^{35} = 75.4% and Cl^{37} = 24.6%.

The National Bureau of Standards prepared this table (4) by critical analysis of data existing in 1972. Using molecular constants and $\Delta_f H°$ selected by NBS (4), we recalculate the table in terms of 1973 fundamental constants (5), 1975 atomic weights (6), and current JANAF reference states for the elements.

References

1. JANAF Thermochemical Tables, 2nd ed., NSRDS-NBS 37, June, 1971.
2. G. Herzberg and A. Monfils, J. Mol. Spectrosc. 5, 482 (1960).
3. B. Rosen, "Spectroscopic Data Relative to Diatomic Molecules," Pergamon Press, Oxford, 1970.
4. S. Abramowitz et al., U. S. Natl. Bur. Stand. Rept. 10904, 239, 1972.
5. CODATA Task Group on Fundamental Constants, CODATA Bulletin 11, December, 1973.
6. IUPAC Commission on Atomic Weights, Pure and Applied Chem. 47, 75 (1976); 37, 589 (1974).

Hydrochloric Acid-D (DCl) $Cl_1D_1(g)$

Enthalpy Reference Temperature = T_r = 298.15 K Standard State Pressure = p° = 0.1 MPa

T/K	$C_p°$ (J K⁻¹mol⁻¹)	S° (J K⁻¹mol⁻¹)	$-[G°-H°(T_r)]/T$ (J K⁻¹mol⁻¹)	$H°-H°(T_r)$ (kJ mol⁻¹)	$\Delta_f H°$ (kJ mol⁻¹)	$\Delta_f G°$ (kJ mol⁻¹)	Log K_f
0	0.	0.	INFINITE	-8.661	-93.131	-93.131	INFINITE
100	29.113	160.924	218.642	-5.772	-93.073	-94.076	49.140
200	29.124	181.107	195.407	-2.860	-93.157	-95.051	24.825
250	29.136	187.606	193.221	-1.404	-93.240	-95.516	19.957
298.15	29.170	192.741	192.741	0.	-93.345	-95.945	16.809
300	29.172	192.921	192.741	0.054	-93.349	-95.961	16.708
350	29.256	197.424	193.097	1.514	-93.478	-96.387	14.385
400	29.406	201.339	193.887	2.981	-93.618	-96.793	12.640
450	29.627	204.815	194.912	4.456	-93.763	-97.181	11.280
500	29.911	207.951	196.062	5.944	-93.905	-97.553	10.191
600	30.610	213.464	198.515	8.969	-94.170	-98.257	8.554
700	31.383	218.240	200.999	12.069	-94.397	-98.920	7.381
800	32.142	222.481	203.424	15.245	-94.582	-99.553	6.500
900	32.843	226.308	205.757	18.495	-94.732	-100.164	5.813
1000	33.467	229.801	207.990	21.811	-94.851	-100.762	5.263
1100	34.015	233.017	210.121	25.186	-94.948	-101.348	4.813
1200	34.491	235.998	212.154	28.612	-95.027	-101.926	4.437
1300	34.905	238.775	214.096	32.082	-95.094	-102.498	4.118
1400	35.266	241.375	215.953	35.591	-95.152	-103.065	3.845
1500	35.580	243.820	217.730	39.134	-95.205	-103.629	3.609
1600	35.857	246.125	219.433	42.706	-95.254	-104.189	3.401
1700	36.101	248.306	221.068	46.304	-95.301	-104.746	3.218
1800	36.318	250.376	222.639	49.925	-95.347	-105.300	3.056
1900	36.512	252.345	224.152	53.567	-95.395	-105.852	2.910
2000	36.687	254.222	225.608	57.227	-95.445	-106.401	2.779
2100	36.845	256.016	227.014	60.904	-95.498	-106.947	2.660
2200	36.989	257.733	228.372	64.596	-95.556	-107.491	2.552
2300	37.121	259.380	229.684	68.301	-95.619	-108.032	2.453
2400	37.242	260.963	230.955	72.019	-95.689	-108.570	2.363
2500	37.355	262.486	232.186	75.749	-95.766	-109.105	2.280
2600	37.460	263.953	233.380	79.490	-95.852	-109.637	2.203
2700	37.558	265.368	234.538	83.241	-95.946	-110.166	2.131
2800	37.650	266.736	235.664	87.002	-96.050	-110.691	2.065
2900	37.736	268.059	236.758	90.771	-96.164	-111.211	2.003
3000	37.819	269.339	237.823	94.549	-96.289	-111.728	1.945
3100	37.897	270.581	238.860	98.334	-96.424	-112.241	1.891
3200	37.971	271.785	239.870	102.128	-96.569	-112.749	1.840
3300	38.043	272.955	240.855	105.929	-96.724	-113.252	1.793
3400	38.111	274.091	241.816	109.736	-96.889	-113.750	1.748
3500	38.177	275.197	242.754	113.551	-97.063	-114.244	1.705
3600	38.241	276.273	243.670	117.372	-97.245	-114.732	1.665
3700	38.302	277.322	244.565	121.199	-97.434	-115.215	1.627
3800	38.362	278.344	245.441	125.032	-97.630	-115.693	1.590
3900	38.420	279.341	246.298	128.871	-97.831	-116.166	1.556
4000	38.477	280.315	247.136	132.716	-98.036	-116.633	1.523
4100	38.532	281.266	247.957	136.567	-98.245	-117.096	1.492
4200	38.586	282.195	248.761	140.422	-98.454	-117.553	1.462
4300	38.638	283.103	249.549	144.284	-98.665	-118.005	1.433
4400	38.690	283.992	250.322	148.150	-98.874	-118.452	1.406
4500	38.741	284.862	251.080	152.022	-99.082	-118.895	1.380
4600	38.790	285.714	251.823	155.898	-99.286	-119.333	1.355
4700	38.839	286.549	252.553	159.780	-99.486	-119.767	1.331
4800	38.888	287.367	253.270	163.666	-99.681	-120.196	1.308
4900	38.935	288.170	253.974	167.557	-99.869	-120.622	1.286
5000	38.982	288.957	254.666	171.453	-100.049	-121.043	1.265
5100	39.028	289.729	255.346	175.353	-100.221	-121.462	1.244
5200	39.074	290.487	256.015	179.259	-100.383	-121.876	1.224
5300	39.119	291.232	256.672	183.168	-100.535	-122.288	1.205
5400	39.164	291.964	257.319	187.082	-100.675	-122.697	1.187
5500	39.208	292.683	257.955	191.001	-100.803	-123.104	1.169
5600	39.252	293.390	258.582	194.924	-100.918	-123.508	1.152
5700	39.296	294.085	259.198	198.851	-101.020	-123.911	1.136
5800	39.339	294.769	259.806	202.783	-101.107	-124.312	1.120
5900	39.382	295.441	260.404	206.719	-101.180	-124.711	1.104
6000	39.424	296.104	260.994	210.660	-101.236	-125.110	1.089

PREVIOUS: June 1977 (1 atm) CURRENT: June 1977 (1 bar)

Hydrochloric Acid-D (DCl) $Cl_1D_1(g)$

HYPOCHLOROUS ACID-d (DOCl) IDEAL GAS M_r = 53.466502

$S°(298.15\ K) = 240.25 \pm 0.42\ J\ K^{-1}\ mol^{-1}$

$\Delta_f H°(0\ K) = -75.34 \pm 2.1\ kJ\ mol^{-1}$

$\Delta_f H°(298.15\ K) = -78.24 \pm 2.1\ kJ\ mol^{-1}$

Electronic Levels and Quantum Weights

State	ε_i, cm^{-1}	g_i
$^1A'$	0	1
$^3A''$	[22000]	3

Vibrational Frequencies and Degeneracies

ν, cm^{-1}
2666(1)
911(1)
728(1)

Point Group: C_s $\sigma = 1$

Bond Distances: O-Cl = 1.689 ± 0.003 Å O-D = 0.960 ± 0.004 Å

Bond Angle: D-O-Cl = 102.5 ± 0.5°

Product of the Moments of Inertia: $I_A I_B I_C = 9.0222 \times 10^{-117}\ g^3\ cm^6$

Enthalpy of Formation

$\Delta_f H°$(DOCl, g, 298.15 K) is calculated from the JANAF selection (1) for $\Delta_f H°$(298.15 K) of HOCl(g), the JANAF thermal functions (1), and the zero point energies of H_2(g), D_2(g), HOCl(g), and DOCl(g). The zero point energies of H_2(g) and D_2(g) are those given by Herzberg and Monfils (2). The zero-point energies of HOCl(g) and DOCl(g) are estimated from vibrational frequencies as listed by JANAF (1) using the first-order relation, zero-point energy = $(\omega_1 + \omega_2 + \omega_3)/2$.

Heat Capacity and Entropy

The molecular constants B_0 and C_0 are from the microwave work of Mirri et al. (3). The constant A_0 is calculated from the molecular structure given by Mirri et al. (3) and is somewhat uncertain. The rotational constants $A_0 = 11.105\ cm^{-1}$, $B_0 = 0.474852\ cm^{-1}$, and $C_0 = 0.454208\ cm^{-1}$ have been adjusted for the natural isotopic abundance of Cl. The vibrational frequencies are from the infrared spectroscopic work of Ashby (4) and that of Schwager and Arkell (5) and have not been corrected for isotopic abundances.

References

1. JANAF Thermochemical Tables: HOCl(g), 3-31-79.
2. G. Herzberg and A. Monfils, J. Mol. Spectrosc. 5, 482 (1960).
3. A. M. Mirri, F. Scappini, and G. Cazzoli, J. Mol. Spectrosc. 38, 218 (1971).
4. R. A. Ashby, J. Mol. Spectrosc. 23, 439 (1967); 40, 639 (1971).
5. I. Schwager and A. Arkell, J. Amer. Chem. Soc. 89, 6006 (1967).

Hypochlorous Acid-D (DOCl) $Cl_1D_1O_1(g)$

Enthalpy Reference Temperature = T_r = 298.15 K Standard State Pressure = p° = 0.1 MPa

	J K^{-1}mol^{-1}			kJ mol^{-1}			
T/K	$C_p°$	S°	$-[G°-H°(T_r)]/T$	$H°-H°(T_r)$	$\Delta_f H°$	$\Delta_f G°$	Log K_f
0	0.	0.	INFINITE	-10.319	-75.344	-75.344	INFINITE
100	33.286	202.209	272.143	-6.993	-76.301	-72.767	38.010
200	34.993	225.626	243.654	-3.606	-77.365	-68.814	17.972
250	36.758	233.619	240.871	-1.813	-77.840	-66.621	13.920
298.15	38.540	240.246	240.246	0.	-78.241	-64.422	11.287
300	38.607	240.485	240.247	0.071	-78.255	-64.337	11.202
350	40.306	246.566	240.724	2.045	-78.609	-61.988	9.251
400	41.785	252.048	241.802	4.098	-78.909	-59.593	7.782
450	43.057	257.045	243.222	6.220	-79.166	-57.162	6.635
500	44.157	261.639	244.837	8.401	-79.386	-54.705	5.715
600	45.987	269.858	248.339	12.912	-79.746	-49.734	4.330
700	47.488	277.063	251.938	17.588	-80.023	-44.709	3.336
800	48.767	283.490	255.487	22.402	-80.240	-39.649	2.589
900	49.872	289.299	258.927	27.335	-80.408	-34.564	2.006
1000	50.830	294.604	262.233	32.371	-80.538	-29.463	1.539
1100	51.659	299.489	265.401	37.497	-80.639	-24.350	1.156
1200	52.376	304.015	268.432	42.700	-80.716	-19.230	0.837
1300	52.997	308.233	271.334	47.969	-80.775	-14.103	0.567
1400	53.534	312.180	274.112	53.296	-80.822	-8.973	0.335
1500	54.000	315.890	276.774	58.673	-80.860	-3.840	0.134
1600	54.406	319.388	279.330	64.094	-80.894	1.296	-0.042
1700	54.760	322.698	281.784	69.553	-80.927	6.434	-0.198
1800	55.070	325.837	284.145	75.045	-80.961	11.574	-0.336
1900	55.342	328.821	286.419	80.566	-80.999	16.716	-0.460
2000	55.583	331.666	288.610	86.112	-81.044	21.860	-0.571
2100	55.796	334.384	290.726	91.681	-81.097	27.006	-0.672
2200	55.986	336.984	292.770	97.270	-81.162	32.156	-0.763
2300	56.156	339.476	294.747	102.878	-81.239	37.308	-0.847
2400	56.310	341.869	296.661	108.501	-81.330	42.464	-0.924
2500	56.449	344.171	298.515	114.139	-81.436	47.624	-0.995
2600	56.577	346.387	300.314	119.791	-81.559	52.789	-1.061
2700	56.696	348.525	302.060	125.454	-81.699	57.959	-1.121
2800	56.807	350.589	303.757	131.129	-81.857	63.135	-1.178
2900	56.912	352.584	305.406	136.815	-82.034	68.316	-1.230
3000	57.013	354.515	307.011	142.512	-82.228	73.503	-1.280
3100	57.111	356.386	308.574	148.218	-82.440	78.698	-1.326
3200	57.207	358.201	310.097	153.934	-82.670	83.899	-1.370
3300	57.303	359.963	311.581	159.659	-82.916	89.108	-1.410
3400	57.400	361.675	313.030	165.395	-83.177	94.325	-1.449
3500	57.499	363.340	314.443	171.139	-83.453	99.550	-1.486
3600	57.600	364.962	315.824	176.894	-83.740	104.783	-1.520
3700	57.704	366.541	317.174	182.660	-84.039	110.023	-1.553
3800	57.812	368.081	318.493	188.435	-84.346	115.273	-1.585
3900	57.924	369.585	319.784	194.222	-84.661	120.530	-1.614
4000	58.041	371.052	321.047	200.020	-84.981	125.795	-1.643
4100	58.162	372.487	322.285	205.830	-85.303	131.069	-1.670
4200	58.289	373.890	323.497	211.653	-85.627	136.350	-1.696
4300	58.421	375.263	324.685	217.488	-85.950	141.639	-1.721
4400	58.559	376.608	325.849	223.337	-86.270	146.935	-1.744
4500	58.701	377.926	326.992	229.200	-86.585	152.239	-1.767
4600	58.849	379.217	328.113	235.078	-86.893	157.549	-1.789
4700	59.002	380.485	329.214	240.970	-87.194	162.867	-1.810
4800	59.160	381.728	330.295	246.878	-87.484	168.190	-1.830
4900	59.322	382.950	331.358	252.802	-87.763	173.520	-1.850
5000	59.488	384.150	332.401	258.743	-88.029	178.855	-1.868
5100	59.659	385.330	333.428	264.700	-88.281	184.195	-1.887
5200	59.833	386.490	334.437	270.675	-88.518	189.540	-1.904
5300	60.011	387.631	335.430	276.667	-88.739	194.890	-1.921
5400	60.191	388.755	336.407	282.677	-88.942	200.243	-1.937
5500	60.375	389.861	337.369	288.705	-89.128	205.600	-1.953
5600	60.560	390.950	338.316	294.752	-89.295	210.960	-1.968
5700	60.747	392.024	339.249	300.817	-89.444	216.323	-1.982
5800	60.935	393.082	340.168	306.901	-89.572	221.689	-1.997
5900	61.124	394.125	341.074	313.004	-89.681	227.056	-2.010
6000	61.314	395.154	341.966	319.126	-89.771	232.425	-2.023

PREVIOUS: March 1979 (1 atm) CURRENT: March 1979 (1 bar)

Hypochlorous Acid-D (DOCl) $Cl_1D_1O_1(g)$

CHLORINE FLUORIDE (ClF) IDEAL GAS M_r = 54.451403

$\Delta_f H°(0\ K)$ = -50.197 ± 0.42 kJ mol^{-1}

S°(298.15 K) = 217.938 J K^{-1} mol^{-1} $\Delta_f H°(298.15\ K)$ = -50.292 ± 0.42 kJ mol^{-1}

Electronic States and Molecular Constants (σ = 1)

State	ε_i, cm^{-1}	g_i	ω_e, cm^{-1}	$\omega_e x_e$, cm^{-1}	B_e, cm^{-1}	α_e, cm^{-1}	r_e, Å
$X\ ^1\Sigma^+$	0	1	784.49	6.201	0.51409	0.004329	1.6281
$A\ ^3\Pi_0$	18721	6	312.74	2.207	0.37026	0.01390	1.92

Enthalpy of Formation

The selected value, $\Delta_f H°$(ClF, 298.15 K) = -12.02±0.1 kcal mol^{-1}, is based on spectroscopic data. Three studies of the visible absorption bands of ClF, by Wahrahftig (1), by Schmitz and Schumacher (2), and by Stricker (3) agree in indicating a band convergence limit at 21,514±10 cm^{-1}. Assignment of this limit to Cl($^2P_{1/2}$) + F($^2P_{3/2}$) is supported by the dissociative ionization threshold for ClF reported by Dibeler et al. (4) whose ion-pair threshold, when reassigned to a hot band, corroborates this choice. From this assignment it follows that D_0^0(ClF) = 58.98±0.02 kcal mol^{-1}.

Combining this value with $D_0^0(F_2)$ = 36.7±0.2 kcal mol^{-1}, reported by Berkowitz et al. (5), and with $D_0^0(Cl_2)$ = 57.177±0.006 kcal mol^{-1}, given by Douglas et al. (6), one obtains $\Delta_f H°$(ClF, 0 K) = -12.0 ± 0.1 kcal mol^{-1} (-50.200 ± 0.42 kJ mol^{-1}).

Calorimetric values for $\Delta_f H°$(298.15 K), which bracket the selected value, are as follows: -11.6 kcal mol^{-1}, Wicke (7); -11.7 kcal mol^{-1}, Wicke and Friz (8); -14.34 and -15.0 kcal mol^{-1}, Schmitz and Schumacher (9); and -14.4 ± 0.8 kcal mol^{-1}, Nuttall and Armstrong (10).

Heat Capacity and Entropy

The vibrational and rotational constants of the respective electronic levels are taken from Rosen (11) and are adjusted to Cl^{35} = 75.4% and Cl^{37} = 24.6%. The value of α_e listed for the ground state by Rosen (11) has been divided by 10 to agree with the original data (12). The National Bureau of Standards prepared this table (13) by critical analysis of data existing is 1972 but used the incorrect value (11) for α_e of the ground state. Using the correct molecular constants and $\Delta_f H°$ selected by NBS (13), we recalculate the table in terms of 1973 fundamental constants (14), 1975 atomic weights (15), and current JANAF reference states for the elements. The thermodyanmic functions of this table may be biased at temperatures above about 3000 K due to neglect of rotational cut-off.

References

1. A. L. Wahrhaftig, J. Chem. Phys. 10, 248 (1942).
2. H. Schmitz and H. J. Schumacher, Z. Naturforsch. 29, 359 (1947).
3. W. Stricker, Deutsche Luft-und Raumfahrt Forschungsbericht, 66-04 (1966).
4. V. H. Dibeler, J. A. Walker, and K. E. McCulloh, J. Chem. Phys. 53, 4414 (1970).
5. J. Berkowtiz, W. A. Chupka, P. M. Guyon, J. H. Holloway, and R. Spohr, J. Chem. Phys. 54, 5165 (1971).
6. A. E. Douglas, C. K. Moller, and B. P. Stoicheff, Can. J. Phys. 41, 1174 (1963).
7. E. Wicke, Nach. Akad. Wiss. Gottingen Math-Phys. Klasse p. 89 (1946).
8. E. Wicke and H. Friz, Z. ELectrochem. 57, 9 (1953).
9. H. Schmitz and H. J. Schumacher, Z. Naturforsch. 29, 362 (1947).
10. R. L. Nuttall and G. T. Armstrong, U. S. Nat. Bur. Stand. Report 10326, 1970.
11. B. Rosen, "Spectroscopic Data Relative to Diatomic Molecules," Pergamon Press, Oxford, 1970.
12. D. A. Gilbert, A. Roberts, and P. A. Griswold, Phys. Rev. 76, 1723 (1949).
13. S. Abramowitz et al., U. S. Nat. Bur. Stand. Report 10904, 239, 1972.
14. CODATA Task Group on Fundamental Constants, CODATA Bulletin 11, December, 1973.
15. IUPAC Commission on Atomic Weights, Pure and Applied Chem. 47, 75 (1976).

Chlorine Fluoride (ClF) $Cl_1F_1(g)$

Enthalpy Reference Temperature = T_r = 298.15 K Standard State Pressure = p° = 0.1 MPa

	J K^{-1}mol^{-1}			kJ mol^{-1}			
T/K	$C_p°$	S°	-[G°-H°(T_r)]/T	H°-H°(T_r)	$\Delta_f H°$	$\Delta_f G°$	Log K_f
0	0.	0.	INFINITE	-8.908	-50.197	-50.197	INFINITE
100	29.124	185.156	245.153	-6.000	-50.198	-50.745	26.506
200	30.126	205.546	220.809	-3.052	-50.234	-51.282	13.393
298.15	32.082	217.938	217.938	0.	-50.292	-51.784	9.072
300	32.118	218.136	217.939	0.059	-50.293	-51.793	9.018
400	33.751	227.613	219.220	3.358	-50.339	-52.285	6.828
500	34.866	235.273	221.689	6.792	-50.373	-52.768	5.513
600	35.619	241.700	224.503	10.319	-50.399	-53.244	4.635
700	36.143	247.233	227.364	13.908	-50.421	-53.717	4.008
800	36.522	252.085	230.157	17.542	-50.442	-54.186	3.538
900	36.807	256.404	232.838	21.210	-50.462	-54.653	3.172
1000	37.030	260.294	235.392	24.902	-50.483	-55.117	2.879
1100	37.210	263.832	237.819	28.614	-50.506	-55.580	2.639
1200	37.360	267.076	240.124	32.343	-50.531	-56.040	2.439
1300	37.487	270.072	242.314	36.085	-50.559	-56.498	2.270
1400	37.598	272.854	244.397	39.840	-50.591	-56.953	2.125
1500	37.697	275.451	246.382	43.605	-50.627	-57.407	1.999
1600	37.788	277.887	248.275	47.379	-50.667	-57.857	1.889
1700	37.873	280.181	250.085	51.162	-50.711	-58.305	1.792
1800	37.956	282.348	251.818	54.953	-50.758	-58.751	1.705
1900	38.040	284.402	253.479	58.753	-50.808	-59.193	1.627
2000	38.131	286.356	255.075	62.562	-50.858	-59.633	1.557
2100	38.233	288.219	256.609	66.380	-50.908	-60.071	1.494
2200	38.354	290.000	258.087	70.209	-50.954	-60.506	1.437
2300	38.500	291.708	259.512	74.051	-50.993	-60.939	1.384
2400	38.682	293.350	260.888	77.910	-51.020	-61.371	1.336
2500	38.908	294.934	262.218	81.789	-51.031	-61.802	1.291
2600	39.189	296.465	263.506	85.694	-51.018	-62.233	1.250
2700	39.535	297.950	264.754	89.629	-50.976	-62.665	1.212
2800	39.959	299.395	265.966	93.603	-50.895	-63.100	1.177
2900	40.470	300.806	267.143	97.624	-50.767	-63.538	1.144
3000	41.079	302.188	268.288	101.701	-50.580	-63.981	1.114
3100	41.796	303.547	269.404	105.843	-50.323	-64.432	1.086
3200	42.629	304.886	270.491	110.064	-49.983	-64.892	1.059
3300	43.587	306.212	271.554	114.373	-49.548	-65.364	1.035
3400	44.674	307.529	272.593	118.785	-49.004	-65.852	1.012
3500	45.894	308.842	273.610	123.312	-48.335	-66.356	0.990
3600	47.251	310.153	274.606	127.969	-47.528	-66.882	0.970
3700	48.742	311.468	275.585	132.767	-46.567	-67.433	0.952
3800	50.365	312.789	276.547	137.721	-45.437	-68.012	0.935
3900	52.115	314.120	277.493	142.844	-44.125	-68.622	0.919
4000	53.983	315.462	278.425	148.148	-42.617	-69.269	0.905
4100	55.960	316.819	279.345	153.645	-40.900	-69.956	0.891
4200	58.032	318.193	280.254	159.343	-38.964	-70.688	0.879
4300	60.184	319.583	281.152	165.254	-36.797	-71.469	0.868
4400	62.399	320.992	282.042	171.382	-34.392	-72.302	0.858
4500	64.658	322.420	282.923	177.735	-31.743	-73.193	0.850
4600	66.941	323.866	283.797	184.315	-28.845	-74.146	0.842
4700	69.228	325.330	284.665	191.123	-25.697	-75.164	0.835
4800	71.496	326.811	285.528	198.160	-22.297	-76.252	0.830
4900	73.725	328.308	286.386	205.421	-18.649	-77.413	0.825
5000	75.895	329.820	287.239	212.903	-14.757	-78.652	0.822
5100	77.985	331.344	288.089	220.597	-10.628	-79.970	0.819
5200	79.978	332.877	288.936	228.497	-6.269	-81.372	0.817
5300	81.858	334.419	289.779	236.589	-1.691	-82.860	0.817
5400	83.612	335.965	290.620	244.864	3.094	-84.436	0.817
5500	85.226	337.515	291.459	253.307	8.073	-86.102	0.818
5600	86.694	339.064	292.295	261.904	13.232	-87.861	0.820
5700	88.007	340.610	293.129	270.641	18.555	-89.713	0.822
5800	89.163	342.151	293.961	279.500	24.027	-91.660	0.825
5900	90.160	343.684	294.791	288.468	29.633	-93.702	0.830
6000	90.999	345.206	295.618	297.527	35.355	-95.841	0.834

PREVIOUS: June 1977 (1 atm) CURRENT: June 1977 (1 bar)

Chlorine Fluoride (ClF) $Cl_1F_1(g)$

LITHIUM CHLORIDEFLUORIDE (Li_2ClF) IDEAL GAS M_r = 68.333403

$\Delta_fH°(0\ K) = [-749]\ kJ\ mol^{-1}$

$S°(298.15\ K) = 268.04\ J\ K^{-1}\ mol^{-1}$ $\Delta_fH°(298.15\ K) = [-754]\ kJ\ mol^{-1}$

Enthalpy of Formation

The enthalpy of formation is from NBS (1).

Heat Capacity and Entropy

Data is from Koppang et. al (2).

References

1. U.S. Nat. Bur. Stand. Report 6297, (1959).
2. R. R. Koppang, C. M. Sherwood, and G. S. Bahn, "Some Provisional Tables of Species Thermodynamic Properties", Marquardt Corp., Van Nuys, California., October, 1959.

Lithium Chloride Fluoride (Li_2ClF) $Cl_1F_1Li_2(g)$

Enthalpy Reference Temperature = T_r = 298.15 K; Standard State Pressure = p° = 0.1 MPa

T/K	$C_p°$ (J K⁻¹mol⁻¹)	S° (J K⁻¹mol⁻¹)	$-[G°-H°(T_r)]/T$ (J K⁻¹mol⁻¹)	$H°-H°(T_r)$ (kJ mol⁻¹)	$\Delta_fH°$ (kJ mol⁻¹)	$\Delta_fG°$ (kJ mol⁻¹)	Log K_f
0	0.	0.	INFINITE	-13.422	-749.132	-749.132	INFINITE
100	35.723	215.376	315.709	-10.033	-749.642	-751.742	392.670
200	51.877	245.279	273.438	-5.632	-751.907	-752.971	196.656
298.15	62.839	268.040	268.040	0.	-753.957	-753.043	131.930
300	62.986	268.429	268.041	0.116	-753.992	-753.038	131.116
400	69.601	287.518	270.599	6.768	-755.887	-752.436	98.258
500	73.601	303.506	275.626	13.940	-764.049	-750.717	78.427
600	76.128	317.161	281.439	21.433	-766.074	-747.856	65.107
700	77.801	329.029	287.408	29.134	-767.836	-744.677	55.568
800	78.948	339.496	293.278	36.974	-769.442	-741.258	48.399
900	79.772	348.845	298.942	44.912	-770.974	-737.642	42.812
1000	80.383	357.282	304.361	52.921	-772.451	-733.859	38.333
1100	80.759	364.961	309.526	60.978	-773.891	-729.930	34.661
1200	81.096	372.003	314.443	69.071	-775.303	-725.871	31.596
1300	81.396	378.506	319.124	77.196	-776.694	-721.695	28.998
1400	81.657	384.548	323.584	85.349	-778.059	-717.413	26.767
1500	81.881	390.190	327.839	93.526	-779.398	-713.034	24.830
1600	82.014	395.478	331.903	101.721	-780.716	-708.567	23.132
1700	82.136	400.454	335.790	109.929	-1072.490	-689.665	21.191
1800	82.245	405.152	339.514	118.148	-1072.271	-667.153	19.360
1900	82.343	409.601	343.087	126.378	-1072.054	-644.652	17.723
2000	82.429	413.827	346.519	134.616	-1071.842	-622.163	16.249
2100	82.488	417.851	349.821	142.862	-1071.637	-599.684	14.916
2200	82.543	421.689	353.001	151.114	-1071.441	-577.215	13.705
2300	82.593	425.359	356.068	159.371	-1071.255	-554.754	12.599
2400	82.639	428.876	359.029	167.632	-1071.082	-532.301	11.585
2500	82.680	432.250	361.891	175.898	-1070.924	-509.855	10.653
2600	82.712	435.493	364.660	184.168	-1070.782	-487.415	9.792
2700	82.741	438.615	367.341	192.440	-1070.661	-464.980	8.996
2800	82.769	441.625	369.941	200.716	-1070.560	-442.550	8.256
2900	82.794	444.530	372.463	208.994	-1070.485	-420.122	7.567
3000	82.818	447.337	374.912	217.275	-1070.435	-397.697	6.925
3100	82.838	450.053	377.293	225.558	-1070.415	-375.273	6.323
3200	82.857	452.683	379.608	233.842	-1070.427	-352.849	5.760
3300	82.874	455.233	381.861	242.129	-1070.471	-330.424	5.230
3400	82.891	457.708	384.056	250.417	-1070.551	-307.997	4.732
3500	82.906	460.111	386.194	258.707	-1070.669	-285.567	4.262
3600	82.919	462.446	388.280	266.998	-1070.825	-263.133	3.818
3700	82.932	464.718	390.316	275.291	-1071.024	-240.695	3.398
3800	82.944	466.930	392.303	283.585	-1071.265	-218.251	3.000
3900	82.954	469.085	394.244	291.880	-1071.550	-195.799	2.622
4000	82.965	471.185	396.141	300.176	-1071.882	-173.340	2.264
4100	82.973	473.234	397.997	308.472	-1072.263	-150.872	1.922
4200	82.981	475.234	399.812	316.770	-1072.694	-128.394	1.597
4300	82.988	477.186	401.589	325.069	-1073.177	-105.905	1.286
4400	82.995	479.094	403.329	333.368	-1073.673	-83.407	0.990
4500	83.002	480.959	405.033	341.668	-1074.247	-60.895	0.707
4600	83.009	482.784	406.704	349.968	-1074.881	-38.369	0.436
4700	83.015	484.569	408.341	358.269	-1075.573	-15.829	0.176
4800	83.021	486.317	409.948	366.571	-1076.325	6.726	-0.073
4900	83.027	488.029	411.524	374.874	-1077.140	29.298	-0.312
5000	83.031	489.706	413.071	383.176	-1078.019	51.887	-0.542
5100	83.035	491.350	414.590	391.480	-1078.954	74.494	-0.763
5200	83.039	492.963	416.081	399.784	-1079.969	97.120	-0.976
5300	83.042	494.545	417.547	408.088	-1081.056	119.767	-1.180
5400	83.045	496.097	418.987	416.392	-1082.220	142.435	-1.378
5500	83.048	497.621	420.403	424.697	-1083.463	165.125	-1.568
5600	83.051	499.117	421.795	433.002	-1084.788	187.838	-1.752
5700	83.054	500.587	423.165	441.307	-1086.199	210.576	-1.930
5800	83.056	502.032	424.512	449.612	-1087.698	233.339	-2.101
5900	83.059	503.451	425.838	457.918	-1089.289	256.129	-2.268
6000	83.061	504.847	427.143	466.224	-1090.974	278.947	-2.428

PREVIOUS: December 1960 (1 atm) CURRENT: December 1960 (1 bar)

MAGNESIUM CHLORIDE FLUORIDE (MgClF) IDEAL GAS M_r = 78.756403

$S°(298.15\ K) = [260.227]\ J\ K^{-1}\ mol^{-1}$

$\Delta_f H°(0\ K) = [-567.4 \pm 21]\ kJ\ mol^{-1}$
$\Delta_f H°(298.15\ K) = [-569.0 \pm 21]\ kJ\ mol^{-1}$

Vibrational Frequencies and Degeneracies

ν, cm^{-1}
[719](1)
[205](1)
[453](1)

Ground State Quantum Weight: [1] σ = [1]
Point Group: C_1
Bond Distances: Mg-Cl = [2.18] Å Mg-F = [1.77] Å
Bond Angle: Cl-Mg-F = [180]°
Product of the Moments of Inertia: $I_A I_B I_C = [8.66593 \times 10^{-116}]\ g^3\ cm^6$

Enthalpy of Formation

The enthalpy of formation is based on an estimated $\Delta_r H°(298.15\ K) = 0$ for the reaction $MgCl_2(g) + MgF_2(g) \rightarrow 2MgClF(g)$, using $\Delta_f H°(298.15\ K) = -95.85$ and $-176.2\ kcal\ mol^{-1}$ for $MgCl_2(g)$ and $MgF_2(g)$, respectively.

Heat Capacity and Entropy

The bond angle is taken to be the value we have adopted for $MgCl_2(g)$. The Mg-Cl and Mg-F bond distances are assumed to be the same as those in $MgCl_2(g)$ and $MgF_2(g)$. The vibrational frequencies are the averages of those for $MgCl_2(g)$ and $MgF_2(g)$. The principal moments of inertia are: $I_A = 0.0766 \times 10^{-39}$, $I_B = 33.5997 \times 10^{-39}$ and $I_C = 33.6763 \times 10^{-39}\ g\ cm^2$.

Magnesium Chloride Fluoride (MgClF) $Cl_1F_1Mg_1(g)$

Enthalpy Reference Temperature = T_r = 298.15 K Standard State Pressure = p° = 0.1 MPa

T/K	$C_p°$ (J K⁻¹mol⁻¹)	S° (J K⁻¹mol⁻¹)	$-[G°-H°(T_r)]/T$ (J K⁻¹mol⁻¹)	$H°-H°(T_r)$ (kJ mol⁻¹)	$\Delta_f H°$ (kJ mol⁻¹)	$\Delta_f G°$ (kJ mol⁻¹)	Log K_f
0	0.	0.	INFINITE	-12.330	-567.353	-567.353	INFINITE
100	38.027	212.576	301.178	-8.860	-567.435	-569.773	297.619
200	45.162	241.231	264.656	-4.685	-568.250	-571.806	149.340
250	47.875	251.612	261.038	-2.356	-568.653	-572.648	119.648
298.15	49.911	260.227	260.227	0.	-569.024	-573.383	100.454
300	49.979	260.535	260.227	0.092	-569.038	-573.410	99.840
350	51.581	268.366	260.842	2.633	-569.413	-574.109	85.681
400	52.798	275.337	262.226	5.244	-569.786	-574.755	75.055
450	53.732	281.612	264.037	7.908	-570.164	-575.353	66.785
500	54.455	287.312	266.084	10.614	-570.551	-575.909	60.165
600	55.478	297.338	270.480	16.114	-571.371	-576.905	50.224
700	56.142	305.943	274.946	21.698	-572.272	-577.757	43.113
800	56.595	313.471	279.301	27.336	-573.276	-578.474	37.770
900	56.915	320.156	283.476	33.012	-574.406	-579.057	33.608
1000	57.149	326.166	287.450	38.716	-584.271	-578.795	30.233
1100	57.326	331.621	291.221	44.440	-585.713	-578.177	27.455
1200	57.461	336.615	294.799	50.180	-587.158	-577.428	25.135
1300	57.568	341.219	298.195	55.931	-588.608	-576.559	23.166
1400	57.653	345.489	301.422	61.693	-717.473	-572.410	21.357
1500	57.723	349.469	304.494	67.461	-717.584	-562.044	19.572
1600	57.780	353.196	307.423	73.237	-717.701	-551.671	18.010
1700	57.827	356.700	310.220	79.017	-717.827	-541.291	16.632
1800	57.867	360.007	312.895	84.802	-717.959	-530.902	15.406
1900	57.901	363.136	315.457	90.590	-718.099	-520.507	14.310
2000	57.930	366.107	317.916	96.382	-718.246	-510.103	13.323
2100	57.955	368.934	320.279	102.176	-718.398	-499.692	12.429
2200	57.976	371.631	322.552	107.973	-718.556	-489.274	11.617
2300	57.995	374.208	324.742	113.771	-718.718	-478.849	10.875
2400	58.012	376.677	326.855	119.572	-718.885	-468.416	10.195
2500	58.027	379.045	328.896	125.374	-719.054	-457.976	9.569
2600	58.040	381.321	330.869	131.177	-719.227	-447.530	8.991
2700	58.051	383.512	332.778	136.981	-719.401	-437.077	8.456
2800	58.062	385.623	334.628	142.787	-719.578	-426.617	7.959
2900	58.071	387.661	336.422	148.594	-719.756	-416.151	7.496
3000	58.080	389.630	338.163	154.401	-719.937	-405.679	7.063
3100	58.087	391.534	339.854	160.210	-720.119	-395.200	6.659
3200	58.094	393.379	341.498	166.019	-720.304	-384.716	6.280
3300	58.101	395.166	343.097	171.829	-720.491	-374.226	5.924
3400	58.106	396.901	344.654	177.639	-720.681	-363.731	5.588
3500	58.112	398.585	346.171	183.450	-720.876	-353.229	5.272
3600	58.117	400.223	347.650	189.261	-721.074	-342.722	4.973
3700	58.121	401.815	349.093	195.073	-721.278	-332.210	4.690
3800	58.125	403.365	350.500	200.885	-721.489	-321.692	4.422
3900	58.129	404.875	351.875	206.698	-721.706	-311.168	4.168
4000	58.133	406.347	353.219	212.511	-721.932	-300.638	3.926
4100	58.136	407.782	354.532	218.325	-722.168	-290.103	3.696
4200	58.139	409.183	355.817	224.138	-722.414	-279.562	3.477
4300	58.142	410.551	357.074	229.952	-722.672	-269.015	3.268
4400	58.144	411.888	358.305	235.767	-722.943	-258.461	3.068
4500	58.147	413.195	359.510	241.581	-723.229	-247.902	2.878
4600	58.149	414.473	360.691	247.396	-723.530	-237.336	2.695
4700	58.151	415.723	361.848	253.211	-723.849	-226.763	2.520
4800	58.153	416.948	362.984	259.026	-724.186	-216.183	2.353
4900	58.155	418.147	364.097	264.842	-724.544	-205.596	2.192
5000	58.157	419.322	365.190	270.657	-724.923	-195.001	2.037
5100	58.159	420.473	366.263	276.473	-725.325	-184.399	1.889
5200	58.160	421.603	367.316	282.289	-725.752	-173.788	1.746
5300	58.162	422.710	368.351	288.105	-726.203	-163.170	1.608
5400	58.163	423.798	369.368	293.922	-726.683	-152.542	1.476
5500	58.165	424.865	370.367	299.738	-727.190	-141.905	1.348
5600	58.166	425.913	371.350	305.554	-727.728	-131.259	1.224
5700	58.167	426.942	372.316	311.371	-728.297	-120.603	1.105
5800	58.168	427.954	373.267	317.188	-728.899	-109.936	0.990
5900	58.169	428.948	374.202	323.005	-729.536	-99.259	0.879
6000	58.170	429.926	375.122	328.822	-730.207	-88.571	0.771

PREVIOUS: March 1966 (1 atm) CURRENT: March 1966 (1 bar)

Magnesium Chloride Fluoride (MgClF) $Cl_1F_1Mg_1(g)$

SULFURYL CHLORIDE FLUORIDE (SO_2ClF) IDEAL GAS M_r = 118.510203

$S°(298.15 K) = 302.849 \pm 1.7 J K^{-1} mol^{-1}$

$\Delta_f H°(0 K) = [-549.1 \pm 21] kJ mol^{-1}$
$\Delta_f H°(298.15 K) = [-556.5 \pm 21] kJ mol^{-1}$

Vibrational Frequencies and Degeneracies

ν, cm^{-1}	ν, cm^{-1}	ν, cm^{-1}
1228(1)	510(1)	1467(1)
824(1)	423(1)	474(1)
629(1)	300(1)	308(1)

Ground State Quantum Weight: [1] σ = 1
Point Group: C_s
Bond Distances: S-O = [1.404] Å S-F = [1.530] Å S-Cl = [2.011] Å
Bond Angles: O-S-O = [123.5]° Cl-S-F = [98.06]°
Product of the Moments of Inertia: $I_AI_BI_C$ = [1.40403 x 10^{-113}] $g^3 cm^6$

Enthalpy of Formation

$\Delta_f H°$ is estimated as the mean of the $\Delta_f H°$ values (1) for $SO_2F_2(g)$ and $SO_2Cl_2(g)$. This corresponds to use of the mean bond energies, D(S-F) = 74 and D(S-Cl) = 36 kcal mol^{-1}, from SO_2F_2 and SO_2Cl_2, respectively. The resulting values of $\Delta_f G°$ are consistent with those methods of preparation for which thermodynamic tests are possible. Insufficient data preclude tests for preparation of SO_2ClF from AgF_2, CoF_3, MnF_3 and SbF_3. In other cases the tests are not stringent, since the calculated values of $\Delta_r G°$ are large and negative for the observed reactions.

Heat Capacity and Entropy

The point group and symmetry number are consistent with analysis of the spectral data, but the detailed structure is estimated from known coordinates of SO_2Cl in SO_2Cl_2 (1, 2) and SF in SO_2F_2 (1, 3). This estimate should be quite reliable since the SO_2 coordinates are almost identical in SO_2Cl_2 and SO_2F_2. The resulting structure has an O-S-Cl angle of 107.7°, an O-S-F angle of 108.4° and pricipal moments of inertia of $I_A = 16.4743 \times 10^{-39}$, $I_B = 29.0842 \times 10^{-39}$ and $I_C = 29.3031 \times 10^{-39}$ g cm^2.

Vibrational frequencies are based on infrared spectra of the vapor (4) and confirmed by Raman spectra of the liquid (5). Birchall and Gillespie (4) assigned δ(FSCl) to a weak Raman band at 195 cm^{-1}, but force-constant calculations based on SO_2Cl_2 and SO_2F_2 predict a value near 300 cm^{-1} (6, 7). It appears that the twist(FSCl) and δ(FSCl) are nearly accidentally degenerate, so we assign them arbitrarily to the Raman band at 308 cm^{-1} and the infrared band at 300 cm^{-1}. The adopted assignments are those of Toyuki and Shimizu (6) and Pfeiffer (7). All adopted values lie between the analogous modes for SO_2Cl_2 and SO_2F_2. δ(FSCl) = 195 cm^{-1} is unlikely since it would lie below δ(ClSCl) = 209 cm^{-1} in SO_2Cl_2; also the intensity of such bending modes should be strong not weak (7).

References

1. JANAF Thermochemical Tables: SO2F2(g) and SO2Cl2 (g), 6-30-71.
2. I. Hargittai, Acta Chim. (Budapest) 60, 231 (1969).
3. D. R. Lide, D. E. Mann and R. M. Fristrom, J. Chem. Phys. 26, 734 (1957).
4. T. Birchall and R. J. Gillespie, Spectrochim. Acta 22, 681 (1966).
5. R. J. Gillespie and E. A. Robinson, Spectrochim. Acta 18, 1473 (1962).
6. H. Toyuki and K. Shimizu, Bull. Chem. Soc. Japan 39, 2364 (1966).
7. M. Pfeiffer, Z. Phys. Chem. (Leipzig) 240, 380 (1969).

Sulfuryl Chloride Fluoride (SO_2ClF) $Cl_1F_1O_2S_1(g)$

Enthalpy Reference Temperature = T_r = 298.15 K Standard State Pressure = p° = 0.1 MPa

	J $K^{-1}mol^{-1}$			kJ mol^{-1}			
T/K	$C_p°$	S°	$-[G°-H°(T_r)]/T$	$H°-H°(T_r)$	$\Delta_f H°$	$\Delta_f G°$	Log K_f
0	0.	0.	INFINITE	-14.701	-549.075	-549.075	INFINITE
100	38.907	244.096	356.685	-11.259	-552.137	-539.994	282.064
200	57.944	276.982	308.997	-6.403	-554.817	-526.728	137.567
250	65.602	290.764	303.995	-3.308	-555.760	-519.591	108.563
298.15	71.592	302.849	302.849	0.	-556.472	-512.558	89.798
300	71.799	303.292	302.850	0.133	-556.496	-512.285	89.197
350	76.869	314.753	303.744	3.853	-557.072	-504.869	75.348
400	81.066	325.300	305.788	7.805	-559.737	-497.292	64.940
450	84.571	335.057	308.505	11.949	-560.757	-489.433	56.812
500	87.514	344.124	311.619	16.253	-561.744	-481.452	50.297
600	92.099	360.508	318.431	25.246	-563.048	-465.260	40.505
700	95.414	374.968	325.496	34.630	-563.877	-448.892	33.497
800	97.855	387.876	332.501	44.300	-564.410	-432.426	28.235
900	99.686	399.512	339.311	54.181	-617.878	-414.825	24.076
1000	101.086	410.091	345.868	64.223	-616.868	-392.318	20.493
1100	102.174	419.779	352.153	74.388	-615.818	-369.913	17.566
1200	103.034	428.707	358.165	84.650	-614.746	-347.605	15.131
1300	103.724	436.983	363.914	94.989	-613.663	-325.387	13.074
1400	104.284	444.690	369.412	105.391	-612.580	-303.253	11.315
1500	104.745	451.902	374.673	115.843	-611.503	-281.196	9.792
1600	105.129	458.674	379.714	126.337	-610.438	-259.210	8.462
1700	105.451	465.058	384.548	136.866	-609.390	-237.291	7.291
1800	105.724	471.093	389.190	147.426	-608.362	-215.433	6.252
1900	105.957	476.816	393.652	158.010	-607.356	-193.631	5.323
2000	106.157	482.256	397.948	168.616	-606.373	-171.881	4.489
2100	106.331	487.439	402.087	179.240	-605.414	-150.180	3.736
2200	106.483	492.390	406.080	189.881	-604.479	-128.524	3.052
2300	106.616	497.126	409.936	200.536	-603.568	-106.910	2.428
2400	106.733	501.666	413.664	211.204	-602.680	-85.336	1.857
2500	106.837	506.025	417.272	221.883	-601.815	-63.798	1.333
2600	106.929	510.217	420.767	232.571	-600.971	-42.294	0.850
2700	107.012	514.254	424.155	243.268	-600.146	-20.822	0.403
2800	107.086	518.147	427.443	253.973	-599.340	0.620	-0.012
2900	107.153	521.906	430.636	264.685	-598.551	22.033	-0.397
3000	107.213	525.540	433.739	275.403	-597.778	43.419	-0.756
3100	107.268	529.056	436.757	286.127	-597.019	64.780	-1.092
3200	107.317	532.463	439.695	296.857	-596.272	86.116	-1.406
3300	107.363	535.766	442.557	307.591	-595.536	107.429	-1.700
3400	107.404	538.972	445.345	318.329	-594.809	128.720	-1.978
3500	107.442	542.086	448.065	329.071	-594.089	149.990	-2.238
3600	107.477	545.113	450.719	339.817	-593.376	171.240	-2.485
3700	107.509	548.058	453.310	350.567	-592.666	192.469	-2.717
3800	107.539	550.925	455.841	361.319	-591.960	213.680	-2.937
3900	107.567	553.719	458.315	372.074	-591.255	234.872	-3.146
4000	107.592	556.443	460.735	382.832	-590.550	256.045	-3.344
4100	107.616	559.100	463.102	393.593	-589.844	277.201	-3.532
4200	107.638	561.693	465.418	404.355	-589.136	298.340	-3.710
4300	107.658	564.226	467.687	415.120	-588.425	319.463	-3.881
4400	107.677	566.702	469.909	425.887	-587.709	340.568	-4.043
4500	107.695	569.122	472.087	436.656	-586.989	361.656	-4.198
4600	107.712	571.489	474.222	447.426	-586.264	382.729	-4.346
4700	107.728	573.806	476.317	458.198	-585.533	403.786	-4.488
4800	107.742	576.074	478.371	468.972	-584.797	424.828	-4.623
4900	107.756	578.295	480.388	479.747	-584.054	445.855	-4.753
5000	107.769	580.473	482.368	490.523	-583.306	466.866	-4.877
5100	107.782	582.607	484.313	501.300	-582.552	487.862	-4.997
5200	107.793	584.700	486.223	512.079	-581.793	508.842	-5.111
5300	107.804	586.753	488.101	522.859	-581.030	529.810	-5.222
5400	107.814	588.768	489.946	533.640	-580.262	550.761	-5.328
5500	107.824	590.747	491.761	544.422	-579.492	571.698	-5.430
5600	107.834	592.690	493.546	555.205	-578.718	592.623	-5.528
5700	107.842	594.598	495.302	565.989	-577.944	613.532	-5.622
5800	107.851	596.474	497.030	576.773	-577.169	634.430	-5.714
5900	107.859	598.318	498.732	587.559	-576.395	655.311	-5.802
6000	107.866	600.131	500.406	598.345	-575.623	676.182	-5.887

PREVIOUS: June 1971 (1 atm) CURRENT. June 1971 (1 bar)

Sulfuryl Chloride Fluoride (SO_2ClF) $Cl_1F_1O_2S_1(g)$

PERCHLORYL FLUORIDE (ClO_3F) IDEAL GAS M_r = 102.449603

$S°(298.15 K) = 278.984 \pm 0.08 J K^{-1} mol^{-1}$

$\Delta_f H°(0 K) = -12.69 \pm 2.9 kJ mol^{-1}$
$\Delta_f H°(298.15 K) = -21.42 \pm 2.9 kJ mol^{-1}$

Vibrational Levels and Degeneracies

ν, cm^{-1}	ν, cm^{-1}
1061(1)	1315(2)
715(1)	589(2)
549(1)	405(2)

Ground State Quantum Weight: 1 σ = 3
Point Group: C_{3v}
Bond Distances: F-Cl = [1.63] Å O-Cl = [1.46] Å
Bond Angles: F-Cl-O = [95° 10'] O-Cl-O = [119° 12']
Product of the Moments of Inertia: $I_A I_B I_C = [4.29248 \times 10^{-114}] g^3 cm^6$

Enthalpy of Formation

Neugebauer and Margrave (1) measured the enthalpy of reaction of perchloryl fluoride and hydrogen.

$$ClO_3F(g) + 4H_2(g) = HF(aq., 0.6M) + HCl(aq, 0.6M) + 3H_2O(\ell)$$

The necessary, additional thermal data for the calculation of the enthalpy of formation was obtained by them from (2). Dibeler et al. (3) obtain a value of -5.3±4 kcal mol^{-1} from the observed appearance potential of Cl^+.

Heat Capacity and Entropy

The assignment of Lide and Mann (3), based upon the infrared spectra of the gas is supported by the Raman measurements of Powell and Lippincott (4).

The value of the spectroscopic constant B was obtained by Madden and Benedict (5) from a high resolution infrared spectrograph. The Cl-F bond distance was estimated assuming a single covalent bond with no ionic character as evidenced by the low dipole moment obtained by Maryott and Kreidler (6). Electron spin resonance results are explained by Brownstein (7), on the basis of a low dipole moment. The Cl-O distance was estimated to be that found in the Chlorate ion, Pauling (8). The resulting values of the bond angles and I_C are consistent with the measured I_B and the assumed bond lengths. The principal moments of inertia are: $I_A = I_B = [15.9597 \times 10^{-39}]$ and $I_C = [16.8522 \times 10^{-39}] g cm^2$.

A calorimetric entropy of 60.17 cal $K^{-1} mol^{-1}$ at 226.48 K was obtained by Koehler and Giauque (9) for the ideal gas. If the lattice doesn't discriminate between F and O atoms then at 0 K a residual entropy of 2.75 cal $K^{-1} mol^{-1}$ exists and the third law entropy would be 62.92 cal $K^{-1} mol^{-1}$. This value may be compared with 62.64 obtained statistically. The difference 0.3 cal $K^{-1} mol^{-1}$ is within the range found for other molecules and explained on the basis of some discrimination in the lattice, see Koehler and Giauque, (9) for references.

References

1. C. A. Neugebauer and J. L. Margrave, J. Amer. Chem. Soc. 79, 1338 (1957).
2. U. S. Nat. Bur. Stand. Circ. 500.
3. V. H. Dibeler, R. M. Reese and D. E. Mann, J. Chem. Phys. 27, 176 (1957).
4. D. R. Lide and D. E. Mann, J. Chem. Phys. 25, 1128 (1956).
5. F. X. Powell and E. R. Lippincott, J. Chem. Phys. 32, 1883 (1960).
6. R. P. Madden and W. S. Benedict from a high resolution infrared spectrograph.
7. A. A. Maryott and S. J. Kreidler, J. Chem. Phys. 27, 1221 (1957).
8. S. Brownstein, Can. J. Chem. 38, 1597 (1960).
9. L. Pauling, "The Nature of the Chemical Bond', 3rd ed. Cornell Univ. Press, Ithaca, 1960.
10. J. K. Koehler and W. F. Giauque, J. Amer. Chem. Soc. 80, 2659 (1958).

Perchloryl Fluoride (ClO_3F) $Cl_1F_1O_3(g)$

Enthalpy Reference Temperature = T_r = 298.15 K Standard State Pressure = p° = 0.1 MPa

T/K	$C_p°$ (J $K^{-1}mol^{-1}$)	S° (J $K^{-1}mol^{-1}$)	$-[G°-H°(T_r)]/T$ (J $K^{-1}mol^{-1}$)	$H°-H°(T_r)$ (kJ mol^{-1})	$\Delta_f H°$ (kJ mol^{-1})	$\Delta_f G°$ (kJ mol^{-1})	Log K_f
0	0.	0.	INFINITE	-13.299	-12.693	-12.693	INFINITE
100	35.404	227.211	326.596	-9.939	-16.599	4.645	-2.426
200	50.513	256.003	284.485	-5.696	-19.707	27.199	-7.104
250	58.338	268.132	280.018	-2.971	-20.729	39.051	-8.159
298.15	64.924	278.984	278.984	0.	-21.422	50.632	-8.871
300	65.158	279.387	278.986	0.120	-21.444	51.079	-8.894
350	70.991	289.881	279.801	3.528	-21.905	63.206	-9.433
400	75.945	299.693	281.681	7.205	-22.161	75.384	-9.844
450	80.137	308.887	284.199	11.110	-22.253	87.584	-10.167
500	83.677	317.519	287.104	15.208	-22.215	99.788	-10.425
600	89.199	333.291	293.513	23.867	-21.848	124.159	-10.809
700	93.178	347.356	300.219	32.996	-21.212	148.446	-11.077
800	96.092	359.998	306.915	42.467	-20.401	172.629	-11.272
900	98.267	371.448	313.459	52.190	-19.473	196.703	-11.416
1000	99.922	381.891	319.788	62.103	-18.466	220.669	-11.527
1100	101.204	391.477	325.875	72.162	-17.406	244.532	-11.612
1200	102.214	400.328	331.716	82.335	-16.310	268.297	-11.679
1300	103.022	408.543	337.313	92.598	-15.193	291.969	-11.731
1400	103.677	416.202	342.678	102.934	-14.063	315.554	-11.773
1500	104.215	423.374	347.821	113.330	-12.930	339.058	-11.807
1600	104.662	430.115	352.756	123.774	-11.801	362.488	-11.834
1700	105.036	436.471	357.495	134.260	-10.680	385.846	-11.856
1800	105.353	442.484	362.051	144.779	-9.573	409.139	-11.873
1900	105.624	448.188	366.436	155.329	-8.482	432.371	-11.887
2000	105.857	453.612	370.660	165.903	-7.411	455.546	-11.898
2100	106.058	458.781	374.734	176.499	-6.361	478.668	-11.906
2200	106.234	463.719	378.668	187.114	-5.333	501.741	-11.913
2300	106.388	468.445	382.469	197.745	-4.329	524.767	-11.918
2400	106.524	472.976	386.146	208.391	-3.349	547.749	-11.921
2500	106.644	477.327	389.707	219.049	-2.393	570.692	-11.924
2600	106.751	481.512	393.158	229.719	-1.459	593.597	-11.926
2700	106.846	485.542	396.506	240.399	-0.548	616.467	-11.926
2800	106.932	489.430	399.755	251.088	0.341	639.303	-11.926
2900	107.009	493.183	402.913	261.785	1.210	662.107	-11.926
3000	107.079	496.812	405.982	272.489	2.059	684.883	-11.925
3100	107.142	500.324	408.969	283.201	2.891	707.630	-11.923
3200	107.199	503.727	411.878	293.918	3.706	730.350	-11.922
3300	107.252	507.026	414.711	304.640	4.507	753.045	-11.920
3400	107.300	510.229	417.474	315.368	5.296	775.716	-11.917
3500	107.344	513.340	420.169	326.100	6.074	798.364	-11.915
3600	107.384	516.364	422.799	336.836	6.843	820.990	-11.912
3700	107.421	519.307	425.368	347.577	7.604	843.594	-11.909
3800	107.455	522.172	427.878	358.321	8.361	866.180	-11.906
3900	107.487	524.964	430.331	369.068	9.114	888.745	-11.903
4000	107.516	527.686	432.731	379.818	9.865	911.289	-11.900
4100	107.544	530.341	435.080	390.571	10.615	933.815	-11.897
4200	107.569	532.933	437.379	401.327	11.367	956.323	-11.894
4300	107.593	535.464	439.631	412.085	12.121	978.814	-11.890
4400	107.615	537.938	441.837	422.845	12.879	1001.286	-11.887
4500	107.636	540.357	443.999	433.608	13.641	1023.740	-11.883
4600	107.655	542.723	446.120	444.372	14.408	1046.178	-11.880
4700	107.673	545.038	448.200	455.139	15.180	1068.600	-11.876
4800	107.690	547.305	450.241	465.907	15.959	1091.005	-11.873
4900	107.706	549.526	452.245	476.677	16.744	1113.395	-11.869
5000	107.721	551.702	454.212	487.448	17.534	1135.767	-11.865
5100	107.735	553.835	456.145	498.221	18.330	1158.124	-11.862
5200	107.748	555.927	458.044	508.995	19.132	1180.464	-11.858
5300	107.761	557.980	459.910	519.770	19.938	1202.791	-11.854
5400	107.773	559.994	461.745	530.547	20.747	1225.100	-11.850
5500	107.784	561.972	463.549	541.325	21.560	1247.395	-11.847
5600	107.795	563.914	465.324	552.104	22.373	1269.677	-11.843
5700	107.805	565.822	467.070	562.884	23.187	1291.942	-11.839
5800	107.815	567.697	468.789	573.665	23.999	1314.195	-11.836
5900	107.824	569.540	470.481	584.447	24.808	1336.431	-11.832
6000	107.832	571.352	472.148	595.230	25.612	1358.656	-11.828

PREVIOUS: March 1961 (1 atm) CURRENT: March 1961 (1 bar)

PHOSPHORYL CHLORIDE FLUORIDE ($POClF_2$) IDEAL GAS M_r = 120.422966

$S°(298.15\ K) = 301.689\ J\ K^{-1}\ mol^{-1}$

$\Delta_f H°(0\ K) = [-961.8]\ kJ\ mol^{-1}$
$\Delta_f H°(298.15\ K) = [-970.7]\ kJ\ mol^{-1}$

Vibrational Frequencies and Degeneracies

ν, cm^{-1}	ν, cm^{-1}	ν, cm^{-1}
274(1)	424(2)	948(2)
410(1)	618(1)	1358(1)
	895(1)	

Ground State Quantum Weight: [1] $\sigma = 1$

Point Group: C_s

Bond Distances: P-Cl = 2.01 ± 0.04 Å P-F = 1.51 ± 0.03 Å P-O = 1.55 ± 0.03 Å

Bond Angles: F-P-F = 103° F-P-Cl = 103°

Product of the Moments of Inertia: $I_A I_B I_C = 1.90249 \times 10^{-113}\ g^3\ cm^6$

Enthaply of Formation

The $\Delta_f H°(298.15\ K)$ of $POClF_2$ was estimated to be -232 kcal mol^{-1} from the $\Delta_f H°(298.15\ K)$ of POF_3 by assuming bond energies of 120 kcal for D(P-F) and 80 kcal for D(P-Cl). The D(P-F) and D(P-Cl) values were taken from Neale and Williams (1) and Neale et al. (2).

Essentially the same $\Delta_f H°(298.15\ K)$ may be obtained by assuming a $\Delta_r H° = 0$ for the following reaction: $POCl_3(g) + POF_3(g) \rightarrow POCl_2F(g) + POF_2Cl(g)$.

Heat Capacity and Entropy

The molecular constants were determined by Brockway and Beach (3) and by Williams et al. (4). Brockway and Beach (3) used electron diffraction. Williams et al. (4) measured microwave spectra. Brockway and Beach (3) assumed that the F-P-F and F-P-Cl angles were equal in POF_2Cl and $POFCl_2$. An analysis of Brockway and Beach's data by Williams et al. (4) indicates that the angles in POF_2Cl, $POFCl_2$, $POCl_3$, and POF_3 should be 103 rather than the reported 106±3. The principal moments of inertia are: $I_A = 18.7062 \times 10^{-39}\ g\ cm^2$, $I_B = 31.3539 \times 10^{-39}\ g\ cm^2$, and $I_C = 32.4373 \times 10^{39}\ g\ cm^2$.
The Raman spectra and vibrational frequencies for POF_2Cl were reported by Delwaulle and Francois (5). The frequencies were reassigned by Delwaulle and Francois (6).

References

1. E. Neale and L. T. D. Williams, J. Chem. Soc. 2485 (1955) Part I.
2. E. Neale, L. T. D. Williams and V. T. Moores, J. Chem. Soc. 422 (1956) Part II.
3. L. O. Brockway and J. Y. Beach, J. Amer. Chem. Soc. 60, 1836 (1938).
4. Q. Williams, J. Sheridan, and W. Gordy, J. Chem. Phys. 20, 164 (1952).
5. M. L. Delwaulle and F. Francois, Compt. Rend. 222, 550 (1946).
6. M. L. Delwaulle and F. Francois, J. Chim. Phys. 46, 87 (1949).

Phosphoryl Chloride Fluoride ($POClF_2$) $Cl_1F_2O_1P_1(g)$

Enthalpy Reference Temperature = T_r = 298.15 K Standard State Pressure = p° = 0.1 MPa

	J $K^{-1}mol^{-1}$			kJ mol^{-1}			
T/K	$C_p°$	S°	$-[G°-H°(T_r)]/T$	$H°-H°(T_r)$	$\Delta_f H°$	$\Delta_f G°$	Log K_f
0	0.	0.	INFINITE	-14.198	-961.769	-961.769	INFINITE
100	38.154	245.398	353.098	-10.770	-965.044	-952.662	497.620
200	55.039	277.016	307.567	-6.110	-968.551	-939.053	245.256
250	62.542	290.122	302.789	-3.167	-969.739	-931.537	194.634
298.15	68.838	301.689	301.689	0.	-970.688	-924.090	161.897
300	69.062	302.116	301.691	0.128	-970.721	-923.801	160.848
350	74.631	313.192	302.553	3.724	-972.247	-915.844	136.682
400	79.327	323.473	304.533	7.576	-972.964	-907.735	118.538
450	83.259	333.051	307.176	11.644	-973.527	-899.546	104.417
500	86.541	341.998	310.216	15.891	-973.963	-891.302	93.114
600	91.583	358.249	316.895	24.812	-974.544	-874.709	76.150
700	95.154	372.650	323.851	34.159	-974.844	-858.043	64.028
800	97.736	385.534	330.771	43.810	-974.953	-841.348	54.934
900	99.646	397.161	337.512	53.684	-974.929	-824.648	47.861
1000	101.088	407.738	344.014	63.724	-974.811	-807.955	42.203
1100	102.200	417.427	350.254	73.891	-974.626	-791.278	37.575
1200	103.073	426.358	356.228	84.156	-1037.967	-773.542	33.671
1300	103.768	434.637	361.945	94.499	-1036.928	-751.548	30.198
1400	104.331	442.348	367.416	104.905	-1035.873	-729.635	27.223
1500	104.792	449.563	372.655	115.362	-1034.810	-707.798	24.648
1600	105.174	456.338	377.675	125.861	-1033.744	-686.032	22.397
1700	105.495	462.725	382.492	136.395	-1032.679	-664.333	20.412
1800	105.765	468.762	387.119	146.958	-1031.618	-642.696	18.651
1900	105.996	474.487	391.568	157.547	-1030.564	-621.118	17.076
2000	106.194	479.929	395.851	168.157	-1029.516	-599.596	15.660
2100	106.366	485.115	399.979	178.785	-1028.476	-578.125	14.380
2200	106.516	490.066	403.962	189.429	-1027.441	-556.704	13.218
2300	106.646	494.804	407.810	200.087	-1026.411	-535.330	12.158
2400	106.762	499.345	411.530	210.758	-1025.383	-514.001	11.187
2500	106.864	503.706	415.130	221.439	-1024.357	-492.714	10.295
2600	106.955	507.899	418.618	232.130	-1023.330	-471.469	9.472
2700	107.036	511.937	422.000	242.830	-1022.300	-450.263	8.711
2800	107.108	515.831	425.282	253.537	-1021.264	-429.095	8.005
2900	107.174	519.591	428.469	264.251	-1020.221	-407.965	7.348
3000	107.233	523.225	431.568	274.972	-1019.168	-386.871	6.736
3100	107.287	526.742	434.581	285.698	-1018.103	-365.811	6.164
3200	107.335	530.149	437.515	296.429	-1017.024	-344.787	5.628
3300	107.380	533.453	440.372	307.165	-1015.930	-323.797	5.125
3400	107.420	536.659	443.157	317.905	-1014.817	-302.839	4.653
3500	107.458	539.773	445.874	328.649	-1013.685	-281.916	4.207
3600	107.492	542.801	448.524	339.396	-1012.532	-261.024	3.787
3700	107.523	545.746	451.112	350.147	-1011.357	-240.165	3.391
3800	107.552	548.614	453.640	360.901	-1010.157	-219 338	3.015
3900	107.579	551.408	456.112	371.657	-1008.931	-198.542	2.659
4000	107.604	554.132	458.528	382.416	-1007.679	-177.779	2.322
4100	107.627	556.790	460.893	393.178	-1006.398	-157.048	2.001
4200	107.649	559.384	463.207	403.942	-1005.089	-136.347	1.696
4300	107.669	561.917	465.473	414.708	-1003.750	-115.679	1.405
4400	107.688	564.392	467.693	425.476	-1002.381	-95.042	1.128
4500	107.705	566.812	469.869	436.245	-1000.980	-74.436	0.864
4600	107.721	569.180	472.002	447.017	-999.548	-53.863	0.612
4700	107.737	571.497	474.095	457.789	-998.084	-33.320	0.370
4800	107.751	573.765	476.148	468.564	-996.588	-12.808	0.139
4900	107.765	575.987	478.163	479.340	-995.060	7.672	-0.082
5000	107.777	578.164	480.141	490.117	-993.500	28.120	-0.294
5100	107.789	580.299	482.084	500.895	-991.909	48.536	-0.497
5200	107.801	582.392	483.993	511.675	-990.286	68.921	-0.692
5300	107.811	584.445	485.869	522.455	-988.632	89.275	-0.880
5400	107.821	586.461	487.713	533.237	-986.948	109.597	-1.060
5500	107.831	588.439	489.527	544.019	-985 234	129.887	-1.234
5600	107.840	590.382	491.310	554.803	-983.491	150.147	-1.401
5700	107.849	592.291	493.065	565.587	-981.720	170.374	-1.561
5800	107.857	594.167	494.792	576.373	-979.922	190.571	-1.716
5900	107.864	596.011	496.492	587.159	-978.098	210.736	-1.866
6000	107.872	597.824	498.166	597.946	-976.248	230.870	-2.010

PREVIOUS: March 1963 (1 atm) CURRENT: March 1963 (1 bar)

Phosphoryl Chloride Fluoride ($POClF_2$) $Cl_1F_2O_1P_1(g)$

CHLORINE FLUORIDE (ClF_3) IDEAL GAS M_r = 92.448209

$S°(298.15\ K) = 281.600\ J\ K^{-1}\ mol^{-1}$

$\Delta_f H°(0\ K) = -154.69 \pm 2.9\ kJ\ mol^{-1}$
$\Delta_f H°(298.15\ K) = -158.87 \pm 2.9\ kJ\ mol^{-1}$

Vibrational Frequencies and Degeneracies

ν, cm^{-1}	ν, cm^{-1}
752 (1)	703 (1)
528 (1)	434 (1)
326 (1)	364 (1)

Ground State Quantum Weight: 1 $\sigma = 2$
Point Group: C_{2v}
Bond Distances: $Cl\text{-}F_1 = Cl\text{-}F_3 = 1.698$ Å $Cl\text{-}F_2 = 1.598$ Å
Bond Angles: $F_1\text{-}Cl\text{-}F_2 = F_2\text{-}Cl\text{-}F_3 = 87° 29'$ $F_3\text{-}Cl\text{-}F_1 = 185° 02'$
Product of Moments of Inertia: $I_A I_B I_C = 2.69336 \times 10^{-114}\ g^3\ cm^6$

Enthalpy of Formation

The $\Delta_f H°(298.15\ K) = -37.97\ kcal\ mol^{-1}$ was calculated from $\Delta_r H°(298.15\ K) = 25.83\ kcal\ mol^{-1}$ obtained by third law analysis of the equilibrium data of Schmitz and Schumacher (1) for the reaction $ClF_3(g) = ClF(g) + F_2(g)$. Similar values are obtained from the equilibrium data of Schafer and Wicke (2). Calorimetric values scatter widely, as summarized below; these were given little weight.

Source	Method	Reaction	Observation $kcal\ mol^{-1}$	Drift $cal\ K^{-1}\ mol^{-1}$	$\Delta_f H°(298.15\ K)$ $kcal\ mol^{-1}$
1a	$K_p°$(523-623 K)	$ClF_3(g) \rightarrow ClF(g) + F_2(g)$	$\Delta_r H°(298.15\ K)$ = 25.83	0.8 ± 0.3	-37.97
2a	$K_p°$(733 K)	$ClF_3(g) \rightarrow ClF(g) + F_2(g)$	$\Delta_r H°(298.15\ K)$ = 25.77	---	-37.91
2b	$K_p°$(570-680 K)	$ClF_3(g) \rightarrow ClF(g) + F_2(g)$	$\Delta_r H°(298.15\ K)$ = 26.12	6 ± 3	-38.26
1b	Calorimetric	$3\ NaCl + ClF_3 \rightarrow 3\ NaF + 2\ Cl_2$	$\Delta_r H°(291\ K)$ = -76.5		-40.02
1c	Calorimetric	$NaCl + 0.5\ F_2 = NaF + 0.5\ Cl_2$	$\Delta_r H°(291\ K)$ = -39.5		-42.0*
3	Calorimetric	$3\ NaCl + ClF_3 \rightarrow 3\ NaF + 2\ Cl_2$	$\Delta_r H°(298.15\ K)$ = -86.8		-29.7
4	Calorimetric	$0.5\ F_2 + NaCl \rightarrow NaF + 0.5\ Cl_2$	$\Delta_r H°(293\ K)$ = -39.3		-31.1**
5	Flow Reaction	$0.5\ Cl_2(g) + 1.5\ F_2(g) = ClF_3(g)$	$\Delta_r H°(473\ K)$ = -26.4		-26.4

* from combination of 1b and 1c **from combination of 3 and 4

Heat Capacity and Entropy

The structural constants were obtained from the microwave study of Smith (3). The planar distorted "T" structure is similar to that derived from X-ray studies of the solid by Burbank and Bensey (4). Fundamental frequencies are taken from the infrared and Raman spectra of Claassen et al. (5). These data confirm the "T" structure as opposed to the pyramidal structure used in earlier spectral analyses.

Claassen et al. (5) calculate thermodynamic functions virtually identical with the JANAF values. These authors also correct the third law entropy reported by Grisard et al. (6) from measurements of heat capacity (14-285 K) and vapor pressure (226-303 K). The entropy at the normal boiling point becomes $S°(284.91\ K) = 67.04\ kcal\ mol^{-1}$ when a non-ideality correction consistent with the dimer-monomer vapor equilibrium is applied. The corresponding statistical entropy, $S°(284.91\ K) = 66.59\ kcal\ mol^{-1}$, is consistent with the third law value within the probable error in the entropy of the liquid and the entrophy of vaporization. The principal moments of inertia are: $I_A = 6.1123 \times 10^{-39}$, $I_B = 18.1568 \times 10^{-39}$ and $I_C = 24.2691 \times 10^{-39}\ g\ cm^2$.

References

1. H. Schmitz and H. J. Schumacher, Z. Naturforsch. 2a, 362 (1947).
2a. K. Schafer and E. Wicke, Z. Elektrochem. 52, 205 (1948); Hot wire.
2b. K. Schafer and E. Wicke, Z. Elektrochem. 52, 205 (1948); From the Graph.
3. H. V. Wartenberg, Z. Anorg. Allg. Chem. 258, 356 (1949).
4. H. V. Wartenberg and O. Fitzner, A. Anorg. Allg. Chem. 151, 313 (1926).
5. C. F. Swinehart, Harshaw Chemical Co., personal communication, March, 1963.
3. D. F. Smith, J. Chem. Phys. 21, 609 (1953).
4. R. D. Burbank and F. N. Bensey, J. Chem. Phys. 21, 602 (1953).
5. H. H. Claassen, B. Weinstock and J. G. Malm, J. Chem. Phys. 28, 285 (1958).
6. J. W. Grisard, H. A. Bernhardt and G. D. Oliver, J. Amer. Chem. Soc. 73, 5725 (1951).

Chlorine Fluoride (ClF_3) $Cl_1F_3(g)$

Enthalpy Reference Temperature = T_r = 298.15 K Standard State Pressure = p° = 0.1 MPa

T/K	$C_p°$ (J K⁻¹mol⁻¹)	S° (J K⁻¹mol⁻¹)	$-[G°-H°(T_r)]/T$ (J K⁻¹mol⁻¹)	$H°-H°(T_r)$ (kJ mol⁻¹)	$\Delta_f H°$ (kJ mol⁻¹)	$\Delta_f G°$ (kJ mol⁻¹)	Log K_f
0	0.	0.	INFINITE	-13.648	-154.687	-154.687	INFINITE
100	37.121	227.838	330.328	-10.249	-157.104	-144.882	75.679
200	52.643	258.307	287.119	-5.762	-158.530	-131.999	34.475
250	59.050	270.771	282.625	-2.964	-158.794	-125.331	26.187
298.15	63.845	281.600	281.600	0.	-158.866	-118.877	20.827
300	64.005	281.995	281.601	0.118	-158.867	-118.629	20.655
350	67.760	292.157	282.395	3.417	-158.811	-111.926	16.704
400	70.605	301.399	284.202	6.879	-158.669	-105.237	13.743
450	72.781	309.847	286.589	10.466	-158.471	-98.569	11.442
500	74.465	317.606	289.308	14.149	-158.234	-91.926	9.603
600	76.839	331.409	295.203	21.723	-157.686	-78.715	6.853
700	78.380	343.377	301.249	29.489	-157.083	-65.600	4.895
800	79.428	353.916	307.187	37.383	-156.453	-52.574	3.433
900	80.170	363.317	312.911	45.365	-155.813	-39.627	2.300
1000	80.713	371.793	318.382	53.410	-155.171	-26.752	1.397
1100	81.121	379.506	323.594	61.503	-154.533	-13.941	0.662
1200	81.435	386.578	328.552	69.631	-153.904	-1.188	0.052
1300	81.681	393.106	333.270	77.788	-153.288	11.513	-0.463
1400	81.879	399.167	337.763	85.966	-152.687	24.168	-0.902
1500	82.039	404.822	342.047	94.162	-152.102	36.780	-1.281
1600	82.171	410.121	346.138	102.373	-151.534	49.353	-1.611
1700	82.280	415.106	350.050	110.596	-150.984	61.892	-1.902
1800	82.372	419.812	353.796	118.828	-150.449	74.398	-2.159
1900	82.451	424.267	357.389	127.070	-149.930	86.875	-2.388
2000	82.517	428.498	360.839	135.318	-149.421	99.326	-2.594
2100	82.575	432.526	364.158	143.573	-148.922	111.751	-2.780
2200	82.625	436.368	367.353	151.833	-148.426	124.152	-2.948
2300	82.669	440.042	370.434	160.098	-147.931	136.531	-3.101
2400	82.708	443.561	373.409	168.367	-147.431	148.888	-3.240
2500	82.742	446.938	376.283	176.639	-146.923	161.224	-3.369
2600	82.772	450.184	379.063	184.915	-146.402	173.540	-3.486
2700	82.799	453.308	381.755	193.193	-145.863	185.835	-3.595
2800	82.823	456.320	384.365	201.474	-145.302	198.110	-3.696
2900	82.844	459.227	386.897	209.758	-144.717	210.365	-3.789
3000	82.864	462.036	389.355	218.043	-144.102	222.599	-3.876
3100	82.882	464.753	391.743	226.331	-143.455	234.811	-3.957
3200	82.898	467.385	394.066	234.620	-142.773	247.003	-4.032
3300	82.912	469.936	396.327	242.910	-142.053	259.172	-4.102
3400	82.926	472.411	398.528	251.202	-141.293	271.319	-4.168
3500	82.938	474.815	400.674	259.495	-140.490	283.443	-4.230
3600	82.949	477.152	402.766	267.790	-139.643	295.544	-4.288
3700	82.960	479.425	404.807	276.085	-138.750	307.620	-4.343
3800	82.969	481.637	406.800	284.381	-137.809	319.671	-4.394
3900	82.978	483.792	408.747	292.679	-136.819	331.698	-4.443
4000	82.986	485.893	410.649	300.977	-135.778	343.698	-4.488
4100	82.994	487.943	412.509	309.276	-134.687	355.671	-4.531
4200	83.001	489.943	414.329	317.576	-133.544	367.618	-4.572
4300	83.007	491.896	416.111	325.876	-132.349	379.536	-4.610
4400	83.014	493.804	417.855	334.177	-131.101	391.426	-4.647
4500	83.019	495.670	419.563	342.479	-129.799	403.287	-4.681
4600	83.025	497.495	421.238	350.781	-128.444	415.119	-4.714
4700	83.030	499.280	422.879	359.084	-127.035	426.920	-4.745
4800	83.034	501.028	424.489	367.387	-125.573	438.691	-4.774
4900	83.039	502.740	426.069	375.691	-124.058	450.431	-4.802
5000	83.043	504.418	427.619	383.995	-122.489	462.140	-4.828
5100	83.047	506.063	429.141	392.299	-120.868	473.816	-4.853
5200	83.051	507.675	430.636	400.604	-119.195	485.461	-4.877
5300	83.054	509.257	432.104	408.909	-117.469	497.072	-4.899
5400	83.057	510.810	433.548	417.215	-115.693	508.651	-4.920
5500	83.060	512.334	434.966	425.521	-113.866	520.196	-4.940
5600	83.063	513.830	436.361	433.827	-111.988	531.708	-4.960
5700	83.066	515.301	437.733	442.134	-110.062	543.184	-4.978
5800	83.069	516.745	439.083	450.440	-108.087	554.628	-4.995
5900	83.071	518.165	440.412	458.747	-106.065	566.037	-5.011
6000	83.074	519.562	441.719	467.055	-103.996	577.411	-5.027

PREVIOUS: September 1965 (1 atm) CURRENT: September 1965 (1 bar)

CHLOROTRIFLUOROSILANE ($SiClF_3$) IDEAL GAS M_r = 120.533709

$S°(298.15\ K) = [308.68 \pm 16.7]\ J\ K^{-1}\ mol^{-1}$

$\Delta_f H°(0\ K) = [-1314 \pm 63]\ kJ\ mol^{-1}$

$\Delta_f H°(298.15\ K) = [-1318 \pm 63]\ kJ\ mol^{-1}$

Vibrational Frequencies and Degeneraciests

ν, cm^{-1}
[870](1)
[550](1)
[275](1)
[900](2)
[400](2)
[200](2)

Ground State Quantum Weight: 1 $\sigma = 3$

Point Group: C_{3v}

Product of the Moments of Inertia: $I_A I_B I_C = 22.869 \times 10^{-114}\ g^3\ cm^6$

Enthalpy of Formation

This value is that given by Henderson and Scheffee (1).

Heat Capacity and Entropy

All data is taken from Henderson and Scheffee (1). The principal moments of inertia are: $I_A = [19.9 \times 10^{-39}]\ g\ cm^2$, $I_B = [33.9 \times 10^{-39}]\ g\ cm^2$ and $I_C = [33.9 \times 10^{-39}]\ g\ cm^2$.

Reference

1. C. B. Henderson and R. S. Scheffee, "Survey of Thermochemical Data", Atlantic Research Corp., Alexandria, Va., January, 1960.

Chlorotrifluorosilane ($SiClF_3$) $Cl_1F_3Si_1(g)$

Enthalpy Reference Temperature = T_r = 298.15 K   Standard State Pressure = p° = 0.1 MPa

T/K	$C_p°$ (J K⁻¹mol⁻¹)	S° (J K⁻¹mol⁻¹)	$-[G°-H°(T_r)]/T$ (J K⁻¹mol⁻¹)	$H°-H°(T_r)$ (kJ mol⁻¹)	$\Delta_f H°$ (kJ mol⁻¹)	$\Delta_f G°$ (kJ mol⁻¹)	Log K_f
0	0.	0.	INFINITE	-16.612	-1313.526	-1313.526	INFINITE
100	46.502	240.797	369.839	-12.904	-1315.901	-1304.592	681.449
200	66.344	279.593	315.556	-7.193	-1317.280	-1292.674	337.612
250	73.615	295.204	309.955	-3.688	-1317.689	-1286.473	268.794
298.15	79.372	308.679	308.679	0.	-1317.960	-1280.434	224.327
300	79.571	309.171	308.681	0.147	-1317.968	-1280.201	222.903
350	84.392	321.812	309.668	4.250	-1318.142	-1273.891	190.118
400	88.261	333.343	311.918	8.570	-1318.230	-1267.563	165.527
450	91.361	343.925	314.894	13.064	-1318.250	-1261.227	146.399
500	93.851	353.684	318.292	17.696	-1318.216	-1254.893	131.098
600	97.504	371.141	325.680	27.276	-1318.039	-1242.243	108.147
700	99.969	386.368	333.286	37.158	-1317.768	-1229.631	91.756
800	101.689	399.836	340.779	47.246	-1317.454	-1217.061	79.466
900	102.929	411.889	348.022	57.480	-1317.123	-1204.531	69.909
1000	103.847	422.783	354.962	67.821	-1316.796	-1192.039	62.266
1100	104.544	432.715	361.586	78.242	-1316.486	-1179.578	56.014
1200	105.085	441.836	367.899	88.724	-1316.202	-1167.145	50.804
1300	105.512	450.264	373.914	99.255	-1315.952	-1154.734	46.398
1400	105.855	458.097	379.651	109.824	-1315.742	-1142.341	42.621
1500	106.134	465.410	385.127	120.424	-1315.578	-1129.961	39.349
1600	106.365	472.267	390.361	131.049	-1315.461	-1117.590	36.486
1700	106.557	478.721	395.371	141.695	-1365.572	-1104.779	33.946
1800	106.719	484.817	400.172	152.359	-1365.326	-1089.445	31.615
1900	106.857	490.590	404.781	163.038	-1365.089	-1074.125	29.530
2000	106.975	496.074	409.209	173.730	-1364.857	-1058.817	27.653
2100	107.077	501.296	413.471	184.433	-1364.628	-1043.521	25.956
2200	107.165	506.280	417.577	195.145	-1364.401	-1028.235	24.413
2300	107.243	511.045	421.538	205.865	-1364.169	-1012.960	23.005
2400	107.311	515.611	425.364	216.593	-1363.930	-997.695	21.714
2500	107.371	519.993	429.062	227.327	-1363.680	-982.441	20.527
2600	107.424	524.205	432.641	238.067	-1363.414	-967.196	19.431
2700	107.472	528.260	436.107	248.812	-1363.129	-951.962	18.417
2800	107.515	532.169	439.469	259.561	-1362.820	-936.740	17.475
2900	107.553	535.943	442.731	270.315	-1362.483	-921.528	16.599
3000	107.588	539.590	445.899	281.072	-1362.117	-906.329	15.781
3100	107.620	543.118	448.978	291.832	-1361.716	-891.143	15.016
3200	107.648	546.535	451.974	302.596	-1361.279	-875.970	14.299
3300	107.674	549.848	454.890	313.362	-1360.803	-860.811	13.626
3400	107.698	553.063	457.730	324.131	-1360.286	-845.668	12.992
3500	107.720	556.185	460.499	334.902	-1359.725	-830.540	12.395
3600	107.740	559.220	463.199	345.675	-1743.271	-804.968	11.680
3700	107.758	562.172	465.834	356.449	-1742.203	-778.918	10.996
3800	107.775	565.046	468.408	367.226	-1741.089	-752.899	10.349
3900	107.791	567.846	470.922	378.005	-1739.927	-726.909	9.736
4000	107.806	570.575	473.379	388.784	-1738.715	-700.949	9.153
4100	107.819	573.237	475.782	399.566	-1737.452	-675.021	8.600
4200	107.832	575.836	478.134	410.348	-1736.138	-649.123	8.073
4300	107.843	578.373	480.435	421.132	-1734.770	-623.258	7.571
4400	107.854	580.852	482.689	431.917	-1733.349	-597.425	7.092
4500	107.865	583.276	484.898	442.703	-1731.874	-571.625	6.635
4600	107.874	585.647	487.062	453.490	-1730.345	-545.859	6.198
4700	107.883	587.967	489.185	464.278	-1728.762	-520.126	5.781
4800	107.891	590.239	491.266	475.066	-1727.125	-494.427	5.380
4900	107.899	592.463	493.309	485.856	-1725.434	-468.764	4.997
5000	107.907	594.643	495.314	496.646	-1723.688	-443.135	4.629
5100	107.914	596.780	497.283	507.437	-1721.890	-417.542	4.277
5200	107.920	598.876	499.216	518.229	-1720.038	-391.984	3.938
5300	107.927	600.931	501.116	529.021	-1718.134	-366.463	3.612
5400	107.932	602.949	502.983	539.814	-1716.179	-340.978	3.298
5500	107.938	604.929	504.819	550.608	-1714.172	-315.530	2.997
5600	107.943	606.874	506.624	561.402	-1712.116	-290.118	2.706
5700	107.948	608.785	508.400	572.196	-1710.010	-264.745	2.426
5800	107.953	610.662	510.147	582.992	-1707.858	-239.408	2.156
5900	107.958	612.508	511.866	593.787	-1705.658	-214.108	1.896
6000	107.962	614.322	513.559	604.583	-1703.412	-188.847	1.644

PREVIOUS: December 1960 (1 atm)   CURRENT: December 1960 (1 bar)

CHLORINE FLUORIDE (ClF_5) IDEAL GAS M_r = 130.445015

$S°(298.15\ K) = 310.739\ J\ K^{-1}\ mol^{-1}$

$\Delta_fH°(0\ K) = [-229.8 \pm 63]\ kJ\ mol^{-1}$
$\Delta_fH°(298.15\ K) = [-238.5 \pm 63]\ kJ\ mol^{-1}$

Vibrational Frequencies and Degeneracies

ν, cm^{-1}	ν, cm^{-1}	ν, cm^{-1}
712(1)	488(1)	732 (2)
541(1)	346(1)	[440](2)
486(1)	375(1)	302 (2)

Ground State Quantum Weight: [1] $\sigma = 4$
Point Group: C_{4v}
Bond Distances: $(Cl-F)_{basal} = 1.72$ Å $(Cl-F)_{axial} = 1.62$ Å
Bond Angle: F-Cl-F = 90°
Product of the Moments of Inertia: $I_AI_BI_C = 2.47347 \times 10^{-113}\ g^3\ cm^6$

Enthalply of Formation

The heat of formation was estimated by assuming that the ratio [(A·B·E·(ClF) − A·B·E·(ClF_3))]/[A·B·E·(ClF_3) − A·B·E·(ClF_5)] was equal to [A·B·E·(BrF) − A·B·E·(BrF_3)]/[A·B·E·(BrF_3) − A·B·E·(BrF_5)], where A·B·E· represents average bond energy. All values were taken from these tables. A·B·E·(ClF_5) was thus calculated to be 36 kcal mol^{-1} which leads to $\Delta_fH°$ = −57±15 kcal mol^{-1}.

Heat Capacity and Entropy

Begun et al. (1) have reported the infrared spectra of the gas and the Raman spectra of the liquid. Their assignments have been adopted here, including one frequency ν_8 obtained from a valence force-field treatment. The values are consistent with those of IF_5(g) and BrF_5(g). Smith (2) had previously reported three infrared bands. The structure of the molecule has not been determined but the spectra can only be interpreted assuming C_{4v} symmetry, this is in accord with IF_5 and BrF_5(g). Begun (1) assumed a square pyramidal structure with the chlorine in the base of the pyramid. There is some evidence that the chlorine may not lie in the basal plane but this would not appreciably affect the calculations. The structure adopted was that given by Begun (1). The principal moments of inertia are $I_A = 37.3328 \times 10^{-39}$ and $I_B = I_C = 25.7400 \times 10^{-39}\ g\ cm^2$.

References

1. G. M. Begun, W. H. Fletcher, and D. F. Smith, J. Chem. Phys. 42, 2236 (1965).
2. D. F. Smith, Science 141, 1039 (1963).

Chlorine Fluoride (ClF_5) $Cl_1F_5(g)$

Enthalpy Reference Temperature = T_r = 298.15 K Standard State Pressure = p° = 0.1 MPa

T/K	$C_p°$ (J K⁻¹mol⁻¹)	S° (J K⁻¹mol⁻¹)	−[G°−H°(T_r)]/T (J K⁻¹mol⁻¹)	H°−H°(T_r) (kJ mol⁻¹)	$\Delta_fH°$ (kJ mol⁻¹)	$\Delta_fG°$ (kJ mol⁻¹)	Log K_f
0	0.	0.	INFINITE	-17.930	-229.765	-229.765	INFINITE
100	42.206	236.844	381.117	-14.427	-234.986	-206.628	107.931
200	75.005	276.255	319.001	-8.549	-237.949	-176.878	46.196
250	87.822	294.433	312.290	-4.464	-238.430	-161.544	33.753
298.15	97.165	310.739	310.739	0.	-238.488	-146.725	25.706
300	97.474	311.341	310.741	0.180	-238.484	-146.156	25.448
350	104.652	326.933	311.956	5.242	-238.254	-130.783	19.518
400	110.018	341.275	314.737	10.615	-237.832	-115.457	15.077
450	114.082	354.478	318.429	16.222	-237.281	-100.192	11.630
500	117.206	366.667	322.651	22.008	-236.639	-84.993	8.879
600	121.580	388.453	331.847	33.964	-235.183	-54.798	4.771
700	124.400	407.422	341.318	46.273	-233.589	-24.858	1.855
800	126.310	424.165	350.648	58.814	-231.922	4.847	-0.317
900	127.658	439.125	359.663	71.516	-230.215	34.341	-1.993
1000	128.641	452.628	368.295	84.333	-228.491	63.644	-3.324
1100	129.380	464.925	376.529	97.236	-226.763	92.774	-4.405
1200	129.948	476.208	384.372	110.204	-225.041	121.747	-5.299
1300	130.394	486.628	391.842	123.222	-223.332	150.576	-6.050
1400	130.751	496.305	398.962	136.280	-221.642	179.275	-6.689
1500	131.040	505.336	405.756	149.370	-219.974	207.853	-7.238
1600	131.277	513.801	412.247	162.486	-218.329	236.321	-7.715
1700	131.475	521.766	418.457	175.624	-216.709	264.687	-8.133
1800	131.641	529.285	424.408	188.780	-215.111	292.959	-8.501
1900	131.782	536.407	430.117	201.951	-213.533	321.141	-8.829
2000	131.902	543.169	435.602	215.135	-211.970	349.242	-9.121
2100	132.006	549.607	440.878	228.331	-210.417	377.264	-9.384
2200	132.097	555.750	445.961	241.536	-208.865	405.212	-9.621
2300	132.176	561.624	450.863	254.750	-207.310	433.091	-9.836
2400	132.245	567.251	455.596	267.971	-205.741	460.900	-10.031
2500	132.306	572.651	460.171	281.199	-204.152	488.645	-10.210
2600	132.361	577.841	464.598	294.432	-202.535	516.325	-10.373
2700	132.409	582.837	468.885	307.671	-200.883	543.942	-10.523
2800	132.452	587.653	473.041	320.914	-199.188	571.497	-10.661
2900	132.491	592.302	477.074	334.161	-197.444	598.991	-10.789
3000	132.527	596.794	480.990	347.412	-195.645	626.424	-10.907
3100	132.558	601.140	484.796	360.666	-193.786	653.795	-11.016
3200	132.587	605.349	488.498	373.923	-191.862	681.106	-11.118
3300	132.614	609.430	492.101	387.184	-189.868	708.356	-11.212
3400	132.638	613.389	495.611	400.446	-187.802	735.544	-11.300
3500	132.660	617.234	499.031	413.711	-185.659	762.670	-11.382
3600	132.680	620.972	502.367	426.978	-183.437	789.735	-11.459
3700	132.699	624.607	505.622	440.247	-181.135	816.736	-11.530
3800	132.716	628.146	508.800	453.518	-178.751	843.673	-11.597
3900	132.732	631.594	511.904	466.790	-176.282	870.547	-11.660
4000	132.747	634.955	514.939	480.064	-173.728	897.357	-11.718
4100	132.760	638.233	517.906	493.339	-171.089	924.101	-11.773
4200	132.773	641.432	520.809	506.616	-168.364	950.781	-11.825
4300	132.785	644.556	523.651	519.894	-165.552	977.393	-11.873
4400	132.796	647.609	526.433	533.173	-162.655	1003.940	-11.918
4500	132.806	650.593	529.159	546.453	-159.671	1030.421	-11.961
4600	132.816	653.512	531.831	559.734	-156.603	1056.833	-12.001
4700	132.825	656.369	534.451	573.016	-153.449	1083.178	-12.038
4800	132.833	659.165	537.020	586.299	-150.211	1109.456	-12.073
4900	132.841	661.904	539.541	599.583	-146.891	1135.665	-12.106
5000	132.849	664.588	542.015	612.867	-143.488	1161.804	-12.137
5100	132.856	667.219	544.444	626.153	-140.004	1187.876	-12.166
5200	132.862	669.799	546.830	639.438	-136.441	1213.879	-12.194
5300	132.869	672.330	549.174	652.725	-132.799	1239.812	-12.219
5400	132.875	674.813	551.478	666.012	-129.080	1265.675	-12.243
5500	132.880	677.252	553.743	679.300	-125.285	1291.468	-12.265
5600	132.885	679.646	555.970	692.588	-121.417	1317.194	-12.286
5700	132.891	681.998	558.160	705.877	-117.475	1342.847	-12.306
5800	132.895	684.309	560.315	719.166	-113.462	1368.432	-12.324
5900	132.900	686.581	562.436	732.456	-109.380	1393.948	-12.341
6000	132.904	688.815	564.524	745.746	-105.229	1419.392	-12.357

PREVIOUS: June 1966 (1 atm) CURRENT: June 1966 (1 bar)

Chlorine Fluoride (ClF_5) $Cl_1F_5(g)$

SULFUR CHLORIDE FLUORIDE ($SClF_5$) IDEAL GAS M_r = 162.505015

$S°(298.15\ K) = 319.931 \pm 1.3\ J\ K^{-1}\ mol^{-1}$

$\Delta_f H°(0\ K) = -1026.3 \pm 10.5\ kJ\ mol^{-1}$
$\Delta_f H°(298.15\ K) = -1038.9 \pm 10.5\ kJ\ mol^{-1}$

Vibrational Frequencies and Degeneracies

ν, cm^{-1}	ν, cm^{-1}	ν, cm^{-1}
855(1)	625 (1)	579(2)
707(1)	[332](1)	441(2)
602(1)	505 (1)	287(2)
402(1)	909 (2)	

Ground State Quantum Weight: [1] $\sigma = 4$

Point Group: C_{4v}

Bond Distances: S-F* = 1.566 ± 0.003 Å S-F** = 1.588 ±0.009 Å

S-Cl = 2.047 ± 0.003 Å

Bond Angles: Cl-S-F* = 90.7 ± 0.2° F*-S-F* = 90°

(* - equatorial ** - axial)

Product of the Moments of Inertia: $I_A I_B I_C = 6.65952 \times 10^{-113}\ g^3\ cm^6$

Enthalpy of Formation

Leach and Roberts (1) have reported calorimetric measurements on the heat of hydrolysis of liquid sulfur chloride pentafluoride in aqueous caustic. Re-evaluation of their data using more recent auxiliary heats of formation (2) for aqueous NaOH, Na_2SO_4, NaCl, and NaF leads to $\Delta_f H°(ClSF_4, \ell, 298.15\ K) = -252.9 \pm 2.0\ kcal\ mol^{-1}$. The adopted value for the gas is obtained from that for the liquid by addition of the heat of vaporization at 298.15 K. We estimate a value for $\Delta_{vap}H°(298.15\ K)$ equal to 4.6±0.5 kcal mol^{-1} from $\Delta_{vap}H° = 5.19$ kcal mol^{-1} reported by Griffiths (3) from an analysis of vapor pressure data. We assume that the reported $\Delta_{vap}H°$ value refers to a boiling temperature of 251.85 K, and we estimate the $\Delta_p C°$ of vaporization as -12 cal $K^{-1}\ mol^{-1}$. Earier vapor pressure measurements (4) lead to a lower value (3.9 kcal mol^{-1}) for $\Delta_{vap}H°(298.15\ K)$; however, we believe the earlier vapor pressure data are less reliable than the measurements of Griffiths (3).

Electron-impact studies by Harland and Thynne (5) on SF_6 and $ClSF_5$ have provided information on the appearance potentials (AP) for the formation of SF_5^+ and SF_5^- from these neutral molecules. Our analysis of their AP data (5) leads to $\Delta_f H°$ values which are more than 15 kcal mol^{-1} different than the adopted result.

The heat of atomization, $\Delta_a H°(0\ K)$, and S-Cl bond dissociation energy are calculated to be 431.8±4.5 kcal mol^{-1} and 59.6±6.0 kcal mol^{-1}, respectively. The latter value appears reasonable when compared with the mean values for the S-Cl bond strength in SCl_2 (63.8 kcal mol^{-1}).

Heat Capacity and Entropy

The structural parameters are taken from a study by Marsden and Bartell (6) who established the C_{4v} structure of $ClSF_5$ by a simultaneous least-squares analysis of electron diffraction and microwave spectroscopic data. The microwave data were based on the measurements of Kewley et al. (7). We use r_g distances to calculate the principal moments of inertia of $I_A = 30.9468 \times 10^{-39}$ and $I_B = I_C = 46.3888 \times 10^{-39}\ g\ cm^2$.

The vibrational spectrum of $ClSF_5$ has been studied using various spectroscopic techniques (3, 8, 9). The observed vibrational frequencies show good agreement, but conflicting assignments have been reported. Griffiths (3) has argued that the fundamentals assigned as $\nu_6 = 396\ cm^{-1}$ and $\nu_{11} = 270\ cm^{-1}$ by Cross et al. (8) should be reversed. A recent Raman study (9) of $ClSF_5$ isolated in an argon matrix has identified the band at 396 cm^{-1} as the Cl component of the S-Cl stretching frequency (ν_4). A comparison of the vibrational spectra of $ClSF_5$ and $ClSeF_5$ (10) and results from two force field studies (6, 11) suggest that the low-frequency SF_5 bending mode (ν_{11}) should lie in the range 270-290 cm^{-1}. We adopt $\nu_{11} = 287\ cm^{-1}$ which was observed in the gas phase infrared spectra by Griffiths (3). As suggested by others (10, 11) we believe that the SF_2 out-of-plane deformation (ν_6) has not been resolved, and we use an estimated value (332 cm^{-1}) calculated from a general valency-force field (11). An independent force-field calculation (6) supports this value ($\nu_6 = 317\ cm^{-1}$). All other fundamentals are rounded values taken from the gas phase infrared and liquid phase Raman study of Griffiths (3).

References

1. H. F. Leach and H. L. Roberts, J. Chem. Soc. 4693 (1960).
2. V. B. Parker, D. D. Wagman and D. Garvin, U. S. Nat. Bur. Stand. Interim Report 75-968, 1976.
3. J. E. Griffiths, Spectrochim. Acta 23A, 2145 (1967).
4. H. L. Roberts and N. H. Ray, J. Chem. Soc., 665 (1960).
5. P. Harland and J. C. J. Thynne, J. Phys. Chem. 73, 4031 (1969).
6. C. J. Marsden and L. S. Bartell, Inorg. Chem. 15, 3004 (1976).
7. R. Kewley, K. S. R. Murty, and T. M. Sugden, Trans. Faraday Soc. 58, 1284 (1962).
8. L. H. Cross, H. L. Roberts, P. Goggin, and L. A. Woodward, Trans. Faraday Soc. 56, 945 (1960).
9. R. R. Smardzewski, R. E. Noftle, and W. B. Fox, J. Mol. Spectrosc. 62, 449 (1976).
10. K. O. Christie, C. J. Schack, and E. C. Curtis, Inorg. Chem. 11, 583 (1972).
11. W. V. F. Brooks, M. Eshaque, C. Lau, and J. Passmore, Can. J. Chem. 54, 817 (1976).

Sulfur Chloride Fluoride ($SClF_5$) $Cl_1F_5S_1(g)$

Enthalpy Reference Temperature = T_r = 298.15 K Standard State Pressure = p° = 0.1 MPa

T/K	$C_p°$ (J $K^{-1}mol^{-1}$)	S°	$-[G°-H°(T_r)]/T$	$H°-H°(T_r)$ (kJ mol^{-1})	$\Delta_f H°$	$\Delta_f G°$	Log K_f
0	0.	0.	INFINITE	-18.458	-1026.280	-1026.280	INFINITE
100	42.459	243.875	393.298	-14.942	-1032.178	-1003.270	524.055
200	77.049	283.704	328.677	-8.995	-1036.714	-972.405	253.966
250	92.376	302.598	321.585	-4.747	-1038.051	-956.161	199.779
298.15	104.342	319.931	319.931	0.	-1038.887	-940.307	164.738
300	104.749	320.577	319.933	0.193	-1038.912	-939.696	163.616
350	114.465	337.485	321.246	5.684	-1039.419	-923.115	137.767
400	122.040	353.285	324.274	11.604	-1041.881	-906.393	118.363
450	127.964	368.015	328.326	17.860	-1042.606	-889.421	103.241
500	132.635	381.749	332.989	24.380	-1043.234	-872.363	91.135
600	139.350	406.569	343.230	38.003	-1043.695	-838.132	72.966
700	143.792	428.406	353.870	52.175	-1043.585	-803.874	59.986
800	146.853	447.818	364.423	66.716	-1043.129	-769.658	50.254
900	149.039	465.248	374.674	81.516	-1095.580	-734.432	42.625
1000	150.648	481.038	384.534	96.505	-1093.542	-694.413	36.272
1100	151.864	495.456	393.972	111.633	-1091.459	-654.601	31.084
1200	152.804	508.712	402.989	126.868	-1089.357	-614.979	26.769
1300	153.545	520.973	411.599	142.187	-1087.250	-575.533	23.125
1400	154.138	532.375	419.823	157.573	-1085.150	-536.249	20.008
1500	154.621	543.026	427.685	173.011	-1083.066	-497.115	17.311
1600	155.019	553.018	435.210	188.494	-1081.004	-458.119	14.956
1700	155.350	562.427	442.419	204.013	-1078.965	-419.251	12.882
1800	155.628	571.314	449.335	219.562	-1076.951	-380.502	11.042
1900	155.865	579.735	455.979	235.137	-1074.958	-341.866	9.399
2000	156.068	587.735	462.368	250.734	-1072.984	-303.333	7.922
2100	156.243	595.354	468.521	266.350	-1071.023	-264.899	6.589
2200	156.395	602.626	474.453	281.982	-1069.067	-226.558	5.379
2300	156.528	609.581	480.178	297.628	-1067.111	-188.305	4.277
2400	156.645	616.246	485.709	313.287	-1065.145	-150.138	3.268
2500	156.748	622.642	491.060	328.957	-1063.163	-112.054	2.341
2600	156.840	628.792	496.240	344.636	-1061.156	-74.049	1.488
2700	156.922	634.713	501.259	360.324	-1059.116	-36.122	0.699
2800	156.995	640.421	506.128	376.020	-1057.036	1.729	-0.032
2900	157.061	645.931	510.854	391.723	-1054.910	39.505	-0.712
3000	157.120	651.257	515.446	407.432	-1052.732	77.206	-1.344
3100	157.174	656.410	519.911	423.147	-1050.496	114.833	-1.935
3200	157.223	661.401	524.255	438.867	-1048.196	152.388	-2.487
3300	157.268	666.239	528.484	454.591	-1045.830	189.870	-3.005
3400	157.308	670.935	532.605	470.320	-1043.393	227.279	-3.492
3500	157.346	675.495	536.623	486.053	-1040.883	264.616	-3.949
3600	157.380	679.928	540.542	501.789	-1038.297	301.880	-4.380
3700	157.412	684.241	544.368	517.529	-1035.633	339.070	-4.787
3800	157.441	688.439	548.104	533.272	-1032.889	376.187	-5.171
3900	157.468	692.529	551.755	549.017	-1030.064	413.232	-5.535
4000	157.492	696.516	555.325	564.765	-1027.158	450.203	-5.879
4100	157.516	700.405	558.816	580.515	-1024.170	487.100	-6.206
4200	157.537	704.201	562.233	596.268	-1021.100	523.924	-6.516
4300	157.557	707.908	565.578	612.023	-1017.948	560.672	-6.811
4400	157.576	711.531	568.854	627.780	-1014.715	597.347	-7.091
4500	157.593	715.072	572.064	643.538	-1011.400	633.948	-7.359
4600	157.610	718.536	575.210	659.298	-1008.005	670.474	-7.613
4700	157.625	721.926	578.296	675.060	-1004.530	706.925	-7.857
4800	157.639	725.245	581.323	690.823	-1000.977	743.302	-8.089
4900	157.653	728.495	584.294	706.588	-997.347	779.604	-8.311
5000	157.665	731.680	587.210	722.354	-993.640	815.830	-8.523
5100	157.677	734.803	590.073	738.121	-989.859	851.982	-8.726
5200	157.689	737.864	592.886	753.889	-986.005	888.060	-8.921
5300	157.699	740.868	595.650	769.658	-982.079	924.062	-9.107
5400	157.709	743.816	598.366	785.429	-978.082	959.990	-9.286
5500	157.719	746.710	601.037	801.200	-974.016	995.842	-9.458
5600	157.728	749.552	603.664	816.973	-969.883	1031.621	-9.623
5700	157.736	752.344	606.248	832.746	-965.685	1067.324	-9.781
5800	157.744	755.087	608.791	848.520	-961.422	1102.954	-9.933
5900	157.752	757.784	611.293	864.295	-957.097	1138.510	-10.080
6000	157.759	760.435	613.757	880.070	-952.710	1173.991	-10.221

PREVIOUS: December 1977 (1 atm) CURRENT: December 1977 (1 bar)

Sulfur Chloride Fluoride ($SClF_5$) $Cl_1F_5S_1(g)$

IRON CHLORIDE (FeCl) IDEAL GAS M_r = 91.300

$S°(298.15\ K) = [257.573]\ J\ K^{-1}\ mol^{-1}$

$\Delta_f H°(0\ K) = 249.8 \pm 84\ kJ\ mol^{-1}$
$\Delta_f H°(298.15\ K) = 251.0 \pm 84\ kJ\ mol^{-1}$

Electronic Levels and Quantum Weights

State	ε_i, cm^{-1}	g_i
$^6\Sigma$	0	6
	[200]	[8]
	[1000]	[8]
	[4000]	[8]

$\omega_e = 404.92\ cm^{-1}$ $\omega_e x_e = 1.19\ cm^{-1}$ $\sigma = 1$
$B_e = [0.17795]\ cm^{-1}$ $\alpha_e = [0.00075]\ cm^{-1}$ $r_e = [2.09]$

Enthalpy of Formation

The dissociation energy of FeCl(g) was reported as 3±2 cal K^{-1} mol^{-1} (69±46 kcal mol^{-1}) by Gaydon (1). The corresponding $\Delta_f H°$(FeCl, g, 298.15 K) was evaluated to be 56±46 kcal mol^{-1}. By comparing D_0^0(Fe-O) = 101 kcal mol^{-1} in FeO(g) with the coordinate bond energies obtained in organic complexes and iron carbonyl halides of E(Fe-O) = 59.4 and E(Fe-Cl) = 37.3 kcal mol^{-1}, reported by Jones et al. (2), the D_0^0(Fe-Cl) in FeCl(g) was estimated as 101 x (37.3/59.4) = 63.3 kcal mol^{-1}, yielding $\Delta_f H°$(FeCl, g, 298.15 K) = 64 kcal mol^{-1}. The adopted value of $\Delta_f H°$(298.15 K) for FeCl(g) is 60±20 kcal mol^{-1} (251.040±84 kJ mol^{-1}).

Heat Capacity and Entropy

The ground state configuration, ω_e and $\omega_e x_e$ (corrected to the average isotopic species) were obtained from Herzberg (3). The Fe-Cl bond distance was assumed to be the same as that in $FeCl_2$(g). Hence the values of B_e and α_e were calculated. The electronic levels and quantum weights were estimated by comparison with the electronic levels in Fe^+(g) reported by Moore (4), using the reasoning suggested by Hougen et al. (5). The total 30 for the quantum weight was obtained from the ground multiplet of Fe^+ and was split arbitrarily. The entropy values may be in error by a few entropy units due to the uncertainties in the low lying electronic levels.

References

1. A. G. Gaydon, "Dissociation Energies," Chapman and Hall Ltd., London, 1953.
2. M. M. Jones, B. J. Yow and W. R. May, Inorg. Chem. 1, 166 (1962).
3. G. Herzberg, "Spectra of Diatomic Molecules," D. Van Nostrand Company, Inc., New York, 1950.
4. C. E. Moore, "Atomic Energy Levels," U.S. Nat. Bur. Stand. Circ. 467, Vol. II, 1952.
5. J. T. Hougen, G. E. Leroi and T. C. James, J. Chem. Phys. 34, 1670 (1961).

Iron Chloride (FeCl) $Cl_1Fe_1(g)$

Enthalpy Reference Temperature = T_r = 298.15 K Standard State Pressure = p° = 0.1 MPa

	J $K^{-1}mol^{-1}$			kJ mol^{-1}			
T/K	$C_p°$	S°	$-[G°-H°(T_r)]/T$	$H°-H°(T_r)$	$\Delta_f H°$	$\Delta_f G°$	Log K_f
0	0.	0.	INFINITE	-10.377	249.760	249.760	INFINITE
100	34.442	217.748	290.602	-7.285	250.973	239.255	-124.974
200	36.906	242.585	261.028	-3.689	251.282	227.351	-59.378
250	37.603	250.896	258.199	-1.826	251.198	221.376	-46.254
298.15	38.223	257.573	257.573	0.	251.040	215.646	-37.780
300	38.245	257.810	257.574	0.071	251.033	215.426	-37.509
350	38.782	263.747	258.041	1.997	250.814	209.508	-31.267
400	39.189	268.954	259.087	3.947	250.546	203.625	-26.591
450	39.472	273.587	260.445	5.914	250.227	197.779	-22.958
500	39.654	277.756	261.971	7.892	249.853	191.971	-20.055
600	39.813	285.003	265.224	11.867	248.926	180.477	-15.712
700	39.833	291.143	268.499	15.850	247.747	169.159	-12.623
800	39.813	296.460	271.669	19.833	246.265	158.029	-10.318
900	39.801	301.149	274.690	23.813	244.357	147.109	-8.538
1000	39.810	305.343	277.549	27.794	241.722	136.435	-7.127
1100	39.840	309.138	280.251	31.776	238.187	126.091	-5.988
1200	39.885	312.606	282.805	35.762	235.202	116.013	-5.050
1300	39.939	315.801	285.222	39.753	233.863	106.135	-4.265
1400	39.997	318.763	287.513	43.750	232.442	96.362	-3.595
1500	40.052	321.524	289.689	47.753	230.938	86.694	-3.019
1600	40.104	324.111	291.761	51.760	229.352	77.129	-2.518
1700	40.150	326.544	293.736	55.773	226.732	67.685	-2.080
1800	40.189	328.840	295.623	59.790	224.644	58.389	-1.694
1900	40.222	331.014	297.429	63.811	208.373	49.913	-1.372
2000	40.248	333.077	299.160	67.834	205.875	41.638	-1.087
2100	40.269	335.042	300.822	71.860	203.374	33.488	-0.833
2200	40.284	336.915	302.421	75.888	200.869	25.457	-0.604
2300	40.295	338.706	303.960	79.917	198.359	17.539	-0.398
2400	40.303	340.421	305.444	83.947	195.842	9.732	-0.212
2500	40.307	342.067	306.876	87.977	193.318	2.029	-0.042
2600	40.309	343.648	308.260	92.008	190.785	-5.573	0.112
2700	40.310	345.169	309.599	96.039	188.244	-13.077	0.253
2800	40.309	346.635	310.896	100.070	185.693	-20.486	0.382
2900	40.307	348.049	312.152	104.101	183.132	-27.805	0.501
3000	40.304	349.416	313.372	108.131	180.561	-35.035	0.610
3100	40.302	350.737	314.556	112.162	177.981	-42.179	0.711
3200	40.299	352.017	315.707	116.192	-172.910	-41.815	0.683
3300	40.297	353.257	316.826	120.222	-173.609	-37.708	0.597
3400	40.295	354.460	317.915	124.251	-174.349	-33.579	0.516
3500	40.293	355.628	318.976	128.280	-175.130	-29.427	0.439
3600	40.292	356.763	320.010	132.310	-175.952	-25.253	0.366
3700	40.293	357.867	321.018	136.339	-176.815	-21.055	0.297
3800	40.293	358.941	322.002	140.368	-177.717	-16.833	0.231
3900	40.295	359.988	322.963	144.398	-178.659	-12.587	0.169
4000	40.298	361.008	323.901	148.427	-179.641	-8.316	0.109
4100	40.302	362.003	324.819	152.457	-180.661	-4.021	0.051
4200	40.306	362.975	325.716	156.488	-181.719	0.300	-0.004
4300	40.312	363.923	326.593	160.519	-182.816	4.647	-0.056
4400	40.319	364.850	327.452	164.550	-183.950	9.019	-0.107
4500	40.326	365.756	328.293	168.582	-185.121	13.418	-0.156
4600	40.335	366.642	329.117	172.616	-186.330	17.844	-0.203
4700	40.345	367.510	329.925	176.649	-187.575	22.295	-0.248
4800	40.355	368.360	330.717	180.684	-188.858	26.774	-0.291
4900	40.367	369.192	331.494	184.721	-190.177	31.280	-0.333
5000	40.379	370.007	332.256	188.758	-191.534	35.813	-0.374
5100	40.392	370.807	333.004	192.796	-192.929	40.374	-0.414
5200	40.406	371.592	333.738	196.836	-194.361	44.963	-0.452
5300	40.421	372.361	334.460	200.878	-195.833	49.579	-0.489
5400	40.436	373.117	335.169	204.920	-197.345	54.223	-0.525
5500	40.453	373.859	335.866	208.965	-198.897	58.896	-0.559
5600	40.470	374.588	336.551	213.011	-200.492	63.598	-0.593
5700	40.488	375.305	337.224	217.059	-202.132	68.328	-0.626
5800	40.506	376.009	337.887	221.109	-203.817	73.088	-0.658
5900	40.525	376.702	338.539	225.160	-205.550	77.877	-0.689
6000	40.545	377.383	339.181	229.214	-207.333	82.695	-0.720

PREVIOUS: June 1965 (1 atm) CURRENT: June 1965 (1 bar)

Iron Chloride (FeCl) $Cl_1Fe_1(g)$

HYDROGEN CHLORIDE (HCl) IDEAL GAS $M_r = 36.46094$

$S°(298.15\ K) = 186.901\ J\ K^{-1}\ mol^{-1}$

$\Delta_f H°(0\ K) = -92.127 \pm 0.21\ kJ\ mol^{-1}$
$\Delta_f H°(298.15\ K) = -92.312 \pm 0.21\ kJ\ mol^{-1}$

Electronic Level and Quantum Weight

State	ε_i, cm^{-1}	g_i
${}^1\Sigma$	0	1

$\omega_e = 2889.59\ cm^{-1}$ $\omega_e x_e = 52.06\ cm^{-1}$ $\sigma = 1$
$B_e = 10.5884\ cm^{-1}$ $\alpha_e = 0.3037\ cm^{-1}$ $r_e = 1.2746\ Å$

Enthalpy of Formation

The selected value, Rossini's flow calorimetric measurement (1) of the direct combination of the elements, is essentially the same as that in the revised version (2) of NBS Circular 500. JANAF analyses of the more recent measurements are summarized below, while the earlier measurements have been reviewed by Rossini (1).

Experimenters	Date	Method	$\Delta_f H°(298.15\ K)$, kcal mol^{-1}	References
Johnson, Ambrose	1963	Comparison of HCl, H_2SO_4 and SO_2 by solution calorimetry	-22.14±0.12	2,3,4,5,6,7,8,9,10
Lachet, et al.	1949-52	Catalytic combination in flow calorimeter	-22.10±0.12	11,12
Roth, Richter	1934	Direct combination in bomb calorimeter	-21.90±0.05	13
von Wartenberg, Hanisch	1932	Direct combination in flow calorimeter	-21.90±0.01	14
Rossini	1932	Direct combination in flow calorimeter	-22.063±0.012	1
- - -	1922-63	$\Delta_f G°$ from e.m.f. and $\Delta_f S°$ from statistical mechanics	-22.02±0.03	15,16,17,18,19,20
Lewis	1906	Equilibria for reaction HCl(g) + 1/4 O_2(g) = 1/2 H_2O(g) + 1/2 Cl_2(g)	-22.30±0.44	21

The direct combination values of Roth-Richter (13) and von Wartenberg-Hanisch (14) are 0.7% less negative than that of Rossini; however, all of the other methods favor the more negative value. The Roth-Richter value may be affected by errors in the measurement of the extent of reaction by HCl titration (9, 22), since a set of experiments based on determination of the amount of H_2 gave a value of -22.02 kcal mol^{-1}.

Rossini's value is confirmed by a variety of independent methods. The solution calorimetry (3), though it is related through a complex reaction scheme, is based on species which are now quite well known. Emf measurements (15, 16, 17, 18, 19, 20) give $\Delta_f G°(298.15\ K) = -22.733$; assuming an uncertainty of 1.0 mv. in E°, this leads to $\Delta_f H°(298.15\ K) = -22.02\pm0.03$ when combined with the statistical entropies. The equilibria of Lewis (625-692 K) give $\Delta_f H°(298.15\ K) = -22.20\pm0.44$ (2nd law) and -22.30±0.04 (3rd law).

Heat Capacity and Entropy

Spectroscopic constants for HCl^{35} and HCl^{37} were selected from the work of Plyler et al. (23) and Haeusler and Barchewitz (24). The values were adjusted to 75.53% HCl^{35}.

References

1. R. D. Rossini, J. Res. Nat. Bur. Stand. 9, 679 (1932).
2. D. D. Wagman, U. S. Nat. Bur. Stand., personal communication, June 26, 1964.
3. W. H. Johnson and J. R. Ambrose, J. Res. Nat. Bur. Stand. 67A, 427 (1963).
4. J. R. Eckman and F. D. Rossini, ibid., 3, 597 (1929).
5. W. H. Johnson and S. Sunner, Acta Chem. Scand. 17, 1917 (1963).
6. M. Mansson and S. Sunner, ibid., 17, 723 (1963).
7. W. D. Good, J. L. Lacina and J. P. McCullough, J. Amer. Chem. Soc. 82, 5589 (1960).
8. C. E. Vanderzee and J. A. Swanson, J. Phys. Chem. 67, 2603 (1963).
9. C. E. Vanderzee and J. D. Nutter, ibid., 2521 (1963).
10. S. R. Gunn and L. G. Green, J. Chem. Eng. Data 8, 180 (1963).
11. J. R. Lacher, J. J. McKinley, C. M. Snow, L. Michel, G. Nelson and J. D. Park, J. Amer. Chem. Soc. 71, 1331 (1949).
12. J. R. Lacher, R. E. Scruby and J. D. Park, ibid., 74, 5292 (1952).
13. W. A. Roth and H. Richter, Z. Physik. Chem. A170, 123 (1934).
14. H. V. Wartenberg and K. Hanisch, ibid., A161, 463 (1932); A164, 144 (1933).
15. R. G. Bates and V. E. Bower, J. Res. Nat. Bur. Stand. 53A, 283 (1954).
16. H. S. Harned and T. R. Paxton, J. Phys. Chem. 57, 531 (1953).
17. J. G. Aston and F. L. Gittler, J. Amer. Chem. Soc. 77, 3173 (1955).
18. R. H. Gerke, ibid., 44, 1684 (1922).
19. N. Kameyana, H. Yamamoto and S. Oka, Proc. Imp. Acad (Japan) 3, 41 (1927); J. Soc. Chem. Ind. Japan 29, 679 (1926); from Chem. Abs. 21, 694 (1927).
20. S. R. Gupta, G. J. Hills and D. J. G. Ives, Trans. Faraday Soc. 59, 1874 (1963).
21. G. N. Lewis, J. Amer. Chem. Soc. 28, 1380 (1906).
22. W. A. Roth and Z. Elektrochem. 50, 107 (1944).
23. E. K. Plyler and E. D. Tidwell, Z. Elektrochem. 64, 717 (1960).
24. C. Haeusler and P. Barchewitz, Compt. Rend. 246, 3040 (1958).

Hydrogen Chloride (HCl) $Cl_1H_1(g)$

Enthalpy Reference Temperature = T_r = 298.15 K Standard State Pressure = p° = 0.1 MPa

T/K	$C_p°$ (J $K^{-1}mol^{-1}$)	S° (J $K^{-1}mol^{-1}$)	$-[G°-H°(T_r)]/T$ (J $K^{-1}mol^{-1}$)	$H°-H°(T_r)$ (kJ mol^{-1})	$\Delta_f H°$ (kJ mol^{-1})	$\Delta_f G°$ (kJ mol^{-1})	Log K_f
0	0.	0.	INFINITE	-8.640	-92.127	-92.127	INFINITE
100	29.116	155.087	212.797	-5.771	-92.214	-93.236	48.702
200	29.124	175.270	189.566	-2.859	-92.168	-94.284	24.625
250	29.130	181.770	187.381	-1.403	-92.219	-94.809	19.809
298.15	29.136	186.901	186.901	0.	-92.312	-95.300	16.696
300	29.137	187.081	186.902	0.054	-92.316	-95.318	16.596
350	29.149	191.574	187.256	1.511	-92.443	-95.809	14.299
400	29.175	195.467	188.045	2.969	-92.589	-96.280	12.573
450	29.223	198.906	189.064	4.429	-92.747	-96.732	11.228
500	29.304	201.989	190.205	5.892	-92.913	-97.166	10.151
600	29.575	207.353	192.629	8.835	-93.251	-97.985	8.530
700	29.985	211.942	195.068	11.811	-93.579	-98.747	7.369
800	30.494	215.978	197.434	14.835	-93.881	-99.465	6.494
900	31.055	219.602	199.699	17.912	-94.152	-100.146	5.812
1000	31.628	222.903	201.857	21.046	-94.388	-100.799	5.265
1100	32.186	225.944	203.910	24.237	-94.591	-101.430	4.817
1200	32.713	228.768	205.866	27.482	-94.766	-102.044	4.442
1300	33.203	231.406	207.730	30.779	-94.916	-102.644	4.124
1400	33.651	233.883	209.510	34.122	-95.046	-103.234	3.852
1500	34.060	236.219	211.214	37.508	-95.160	-103.815	3.615
1600	34.431	238.429	212.846	40.932	-95.261	-104.388	3.408
1700	34.767	240.527	214.413	44.393	-95.351	-104.956	3.225
1800	35.071	242.523	215.920	47.885	-95.435	-105.519	3.062
1900	35.348	244.426	217.371	51.406	-95.514	-106.077	2.916
2000	35.600	246.246	218.769	54.953	-95.590	-106.631	2.785
2100	35.830	247.989	220.119	58.525	-95.665	-107.181	2.666
2200	36.040	249.660	221.424	62.119	-95.742	-107.727	2.558
2300	36.233	251.267	222.687	65.732	-95.820	-108.270	2.459
2400	36.411	252.812	223.910	69.365	-95.903	-108.810	2.368
2500	36.575	254.302	225.097	73.014	-95.991	-109.346	2.285
2600	36.728	255.740	226.248	76.679	-96.086	-109.878	2.207
2700	36.870	257.129	227.366	80.359	-96.188	-110.407	2.136
2800	37.002	258.472	228.453	84.053	-96.298	-110.932	2.069
2900	37.127	259.772	229.511	87.760	-96.418	-111.452	2.007
3000	37.243	261.033	230.540	91.478	-96.547	-111.968	1.950
3100	37.354	262.256	231.544	95.208	-96.685	-112.480	1.895
3200	37.458	263.444	232.522	98.949	-96.834	-112.987	1.844
3300	37.557	264.598	233.477	102.699	-96.992	-113.490	1.796
3400	37.651	265.720	234.409	106.460	-97.159	-113.987	1.751
3500	37.740	266.813	235.319	110.229	-97.334	-114.480	1.709
3600	37.826	267.878	236.209	114.008	-97.518	-114.967	1.668
3700	37.908	268.915	237.079	117.795	-97.708	-115.449	1.630
3800	37.987	269.927	237.930	121.589	-97.905	-115.926	1.594
3900	38.063	270.915	238.763	125.392	-98.107	-116.397	1.559
4000	38.137	271.879	239.579	129.202	-98.313	-116.864	1.526
4100	38.208	272.822	240.378	133.019	-98.521	-117.325	1.495
4200	38.276	273.744	241.162	136.843	-98.731	-117.781	1.465
4300	38.343	274.645	241.930	140.674	-98.942	-118.232	1.436
4400	38.408	275.527	242.684	144.512	-99.152	-118.678	1.409
4500	38.471	276.391	243.423	148.356	-99.360	-119.120	1.383
4600	38.532	277.237	244.149	152.206	-99.564	-119.557	1.358
4700	38.592	278.067	244.862	156.062	-99.764	-119.989	1.334
4800	38.651	278.880	245.562	159.924	-99.959	-120.417	1.310
4900	38.708	279.677	246.250	163.792	-100.147	-120.842	1.288
5000	38.764	280.460	246.927	167.666	-100.328	-121.262	1.267
5100	38.819	281.228	247.592	171.545	-100.499	-121.679	1.246
5200	38.873	281.982	248.246	175.430	-100.662	-122.093	1.226
5300	38.927	282.723	248.889	179.320	-100.814	-122.503	1.207
5400	38.979	283.451	249.523	183.215	-100.954	-122.911	1.189
5500	39.030	284.167	250.146	187.116	-101.083	-123.317	1.171
5600	39.081	284.871	250.760	191.021	-101.198	-123.720	1.154
5700	39.131	285.563	251.364	194.932	-101.300	-124.121	1.137
5800	39.181	286.244	251.960	198.847	-101.387	-124.520	1.121
5900	39.230	286.914	252.547	202.768	-101.460	-124.919	1.106
6000	39.278	287.574	253.125	206.693	-101.517	-125.316	1.091

PREVIOUS: September 1964 (1 atm) CURRENT: September 1964 (1 bar)

Hydrogen Chloride (HCl) $Cl_1H_1(g)$

HYPOCHLOROROUS ACID (HOCl) IDEAL GAS M_r = 52.46034

$S°(298.15\ K) = 236.504 \pm 0.42\ J\ K^{-1}\ mol^{-1}$

$\Delta_f H°(0\ K) = -71.5 \pm 2.1\ kJ\ mol^{-1}$

$\Delta_f H°(298.15\ K) = -74.5 \pm 2.1\ kJ\ mol^{-1}$

Electronic Levels and Quantum Weights

State	ϵ_i, cm^{-1}	g_i
$^1A'$	0	1
$^3A''$	[22000]	3

Vibrational Frequencies and Degeneracies

ν, cm^{-1}
3609.5(1)
1239.4(1)
725 (1)

Point Group: C_s $\sigma = 1$

Bond Distances: O-Cl = 1.689 ± 0.003 Å O-H = 0.960 ± 0.005 Å

Bond Angle: H-O-Cl = 102.5 ± 0.5°

Product of the Moments of Inertia: $4.2740 \times 10^{-117}\ g^3\ cm^6$

Enthalpy of Formation

Our adopted value of $\Delta_f H°$(HOCl, g, 298.15 K) is based on the ultraviolet photometric equilibrium measurements of Knauth et al. (1) who determined K_{333} = 0.132±0.008 for the reaction $Cl_2O(g) + H_2O(g) = 2\ HOCl(g)$. Combining this result with JANAF auxiliary data (2), we obtain $\Delta_r H°$(298.15 K) = 2.66±0.10 kcal mol^{-1}. When this result is combined with $\Delta_f H°(H_2O$, g, 298.15 K) = -57.795±0.01 kcal mol^{-1} (3) and $\Delta_f H°(Cl_2O$, g, 298.15 K) = 19.46±0.4 kcal mol^{-1} from (4), we obtain $\Delta_f H°$(HOCl, g, 298.15 K) = -17.8±0.5 kcal mol^{-1}. This is in agreement with $\Delta_f H°$(HOCl, g, 298.15 K) = -18.0 kcal mol^{-1} derived from similar equilibrium measurements of Molina and Molina (5). Niki et al. (6) also measured the gas phase equilibrium constant for the same reaction using an infrared method and obtained K(295 K) = 0.068±0.01, from which we derive $\Delta_r H°$(298.15 K) = 2.75±0.2 kcal mol^{-1}, and $\Delta_f H°$(HOCl, g, 298.15 K) = -17.8±0.6 kcal mol^{-1} using the same auxiliary data as above. Hirsch et al. (7) have calculated a value of $\Delta_f H°$(HOCl, g, 298.15 K) = -19.2±4 kcal mol^{-1} from ab initio results.

Connick and Chia (8) determined $\Delta_r H°$(298.15 K) = 5.06±0.13 kcal mol^{-1} for the reaction $Cl_2(aq) + H_2O(\ell) = H^+(aq) + Cl^-(aq)$ + HOCl(aq). The auxiliary data (3) $\Delta_f H°(H_2O$, ℓ, 298.15 K) = -68.315 kcal mol^{-1}, $\Delta_f H°(Cl^-$, aq, 298.15 K) = -39.933 kcal mol^{-1} and $\Delta_f H°(Cl_2$, aq, 298.15 K) = -5.6 kcal mol^{-1} (9) give $\Delta_f H°$(HOCl, aq, 298.15 K) = -28.92±0.2 kcal mol^{-1}. Similarly Morris (10) determined $\Delta_r G°$(298.15 K) = 10.282 kcal mol^{-1} for the reaction HOCl(aq) = $H^+(aq) + OCl^-(aq)$. Using entropies from (9) we obtain $\Delta_r H°$(298.15 K) = 3.13±0.5 kcal mol^{-1}, which when combined with auxiliary data from (9), leads to $\Delta_f H°$(HOCl, aq, 298.15 K) = -28.73±0.5 kcal mol^{-1}. Combining the former value of $\Delta_f H°$(HOCl, aq, 298.15 K) with our adopted value for $\Delta_f H°$(HOCl, g, 298.15 K), we obtain $\Delta_r H°$(298.15 K) = -11.8±0.6 kcal mol^{-1} for the heat of solution at infinite dilution of HOCl(g).

Heat Capacity and Entropy

The molecular structure given above is taken from the microwave work of Mirri et al. (11). The rotational constants used to determine the moments of inertia are taken from the high resolution infrared spectrum of Wells et al. (12) for the ν_1 vibrational band and are in reasonable agreement with rotational constants calculated from the above structure (the product of the moments of inertia calculated is 2% lower than our adopted value of $I_A I_B I_C$). Rotational constants are adjusted for the natural isotopic abundance of Cl and are A_0 = 20.463 cm^{-1}, B_0 = 0.50214 cm^{-1}, and C_0 = 0.48910 cm^{-1}.

Vibrational frequencies are taken from Wells et al. (12 (ν_1)) and from Ashby (13 (ν_2, ν_3)). Values of ν are adjusted for the natural isotopic abundance of chlorine. The vibrations are expected to be anharmonic (11) but no anharmonicity constants are available. The lowest-lying excited state has been calculated to lie at 22,000 cm^{-1} (14). Omission of this state reduces $C_p°$ by 0.02 cal K^{-1} mol^{-1} and S° by 0.002 cal K^{-1} mol^{-1} at 3000 K.

References

1. H. D. Knauth, H. Alberti, and H. Clausen, J. Phys. Chem. 83, 1604 (1979).
2. JANAF Thermochemical Tables: $H_2O(g)$, 3-31-61; $Cl_2O(g)$, 12-31-65.
3. ICSU-CODATA Task Group on Key Values for Thermodynamics, J. Chem. Thermodynamics 10, 903 (1978).
4. R. Alqasmi, H. D. Knauth, and D. Rohlack, Ber. Bunsenges, Phys. Chem. 82, 217 (1978).
5. L. T. Molina and M. J. Molina, J. Phys. Chem. 82, 2410 (1978).
6. H. Niki, P. D. Maker, C. M. Savage, and L. P. Breitenbach, Chem. Phys. Letters 66, 325 (1979).
7. G. Hirsch, P. J. Bruna, S. D. Peyerimhoff, and R. J. Buenker, Chem. Phys. Lett. 52, 442 (1977).
8. R. E. Connick and Y. Chia, J. Amer. Chem. Soc. 81, 1280 (1959).
9. U. S. Nat. Bur. Stand. Tech. Note 270-3, 1968.
10. R. L. Jaffe and S. R. Langhoff, J. Chem. Phys. 68, 1638 (1978).
11. A. M. Mirri, F. Scappini, and G. Cazzoli, J. Mol. Spectrosc. 38, 218 (1971).
12. J. S. Wells, R. L. Sams, and W. J. Lafferty, J. Mol. Spectrosc. 77, 349 (1979).
13. R. A. Ashby, J. Mol. Spectrosc. 40, 639 (1971).
14. R. L. Jaffe and S. R. Langhoff, J. Chem. Phys. 68, 1638 (1978).

Hypochlorous Acid (HOCl) $Cl_1H_1O_1(g)$

Enthalpy Reference Temperature = T_r = 298.15 K Standard State Pressure = p° = 0.1 MPa

T/K	$C_p°$ (J K^{-1}mol^{-1})	S° (J K^{-1}mol^{-1})	$-[G°-H°(T_r)]/T$ (J K^{-1}mol^{-1})	$H°-H°(T_r)$ (kJ mol^{-1})	$\Delta_f H°$ (kJ mol^{-1})	$\Delta_f G°$ (kJ mol^{-1})	Log K_f
0	0.	0.	INFINITE	-10.224	-71.533	-71.533	INFINITE
100	33.284	198.866	267.844	-6.898	-72.615	-69.350	36.225
200	34.588	222.209	239.823	-3.523	-73.561	-65.717	17.164
250	35.898	230.063	237.110	-1.762	-74.037	-63.701	13.310
298.15	37.285	236.504	236.504	0.	-74.475	-61.670	10.804
300	37.339	236.735	236.505	0.069	-74.491	-61.591	10.724
350	38.759	242.599	236.965	1.972	-74.911	-59.407	8.866
400	40.077	247.862	238.004	3.943	-75.291	-57.165	7.465
450	41.260	252.652	239.369	5.977	-75.634	-54.879	6.370
500	42.307	257.054	240.920	8.067	-75.943	-52.556	5.490
600	44.045	264.928	244.281	12.388	-76.483	-47.826	4.164
700	45.437	271.825	247.733	16.864	-76.939	-43.013	3.210
800	46.604	277.971	251.136	21.468	-77.330	-38.139	2.490
900	47.625	283.520	254.431	26.180	-77.667	-33.219	1.928
1000	48.538	288.586	257.597	30.989	-77.960	-28.265	1.476
1100	49.366	293.252	260.629	35.885	-78.213	-23.283	1.106
1200	50.118	297.580	263.530	40.860	-78.433	-18.279	0.796
1300	50.800	301.619	266.306	45.907	-78.624	-13.258	0.533
1400	51.418	305.406	268.965	51.018	-78.792	-8.224	0.307
1500	51.977	308.973	271.514	56.188	-78.942	-3.178	0.111
1600	52.481	312.344	273.962	61.411	-79.078	1.878	-0.061
1700	52.936	315.540	276.314	66.683	-79.204	6.941	-0.213
1800	53.346	318.577	278.579	71.997	-79.323	12.012	-0.349
1900	53.716	321.471	280.761	77.351	-79.439	17.089	-0.470
2000	54.049	324.235	282.866	82.739	-79.556	22.173	-0.579
2100	54.351	326.880	284.899	88.159	-79.675	27.262	-0.678
2200	54.625	329.415	286.865	93.608	-79.800	32.357	-0.768
2300	54.874	331.848	288.769	99.084	-79.933	37.458	-0.851
2400	55.101	334.189	290.613	104.583	-80.075	42.565	-0.926
2500	55.310	336.442	292.401	110.103	-80.229	47.678	-0.996
2600	55.502	338.615	294.137	115.644	-80.397	52.798	-1.061
2700	55.679	340.714	295.823	121.203	-80.578	57.924	-1.121
2800	55.846	342.741	297.463	126.779	-80.775	63.058	-1.176
2900	56.002	344.704	299.058	132.372	-80.987	68.198	-1.228
3000	56.151	346.605	300.612	137.980	-81.216	73.346	-1.277
3100	56.294	348.449	302.125	143.602	-81.460	78.502	-1.323
3200	56.432	350.238	303.601	149.238	-81.719	83.666	-1.366
3300	56.567	351.977	305.041	154.888	-81.993	88.839	-1.406
3400	56.699	353.667	306.446	160.551	-82.281	94.020	-1.444
3500	56.832	355.313	307.819	166.228	-82.582	99.209	-1.481
3600	56.964	356.916	309.161	171.918	-82.894	104.408	-1.515
3700	57.098	358.478	310.472	177.621	-83.215	109.615	-1.547
3800	57.233	360.003	311.756	183 337	-83.544	114.831	-1.578
3900	57.371	361.491	313.012	189.068	-83.879	120.056	-1.608
4000	57.512	362.945	314.242	194.812	-84.219	125.289	-1.636
4100	57.656	364.367	315.448	200.570	-84.561	130.532	-1.663
4200	57.805	365.758	316.629	206.343	-84.903	135.782	-1.689
4300	57.957	367.120	317.788	212.131	-85.243	141.040	-1.713
4400	58.113	368.455	318.924	217.935	-85.580	146.306	-1.737
4500	58.273	369.762	320.039	223.754	-85.911	151.580	-1.759
4600	58.438	371.045	321.134	229.589	-86.236	156.861	-1.781
4700	58.607	372.303	322.209	235.442	-86.551	162.149	-1.802
4800	58.779	373.539	323.266	241.311	-86.856	167.444	-1.822
4900	58.956	374.753	324.304	247.198	-87.149	172.745	-1.841
5000	59.136	375.946	325.325	253.102	-87.429	178.052	-1.860
5100	59.319	377.119	326.329	259.025	-87.695	183.365	-1.878
5200	59.505	378.272	327.317	264.966	-87.945	188.682	-1.895
5300	59.695	379.408	328.289	270.926	-88.178	194.004	-1.912
5400	59.886	380.525	329.246	276.905	-88.394	199.330	-1.928
5500	60.079	381.626	330.189	282.903	-88.592	204.660	-1.944
5600	60.274	382.710	331.117	288.921	-88.771	209.994	-1.959
5700	60.471	383.779	332.032	294.958	-88.931	215.330	-1.973
5800	60.668	384.832	332.933	301.015	-89.071	220.670	-1.987
5900	60.866	385.871	333.821	307.092	-89.191	226.010	-2.001
6000	61.063	386.895	334.697	313.188	-89.291	231.354	-2.014

PREVIOUS: March 1979 (1 atm) CURRENT: March 1979 (1 bar)

Hypochlorous Acid (HOCl) $Cl_1H_1O_1(g)$

$S°(298.15 K) = 250.761 \pm 0.21 J K^{-1} mol^{-1}$

$\Delta_fH°(0 K) = [-132.77 \pm 8] kJ mol^{-1}$
$\Delta_fH°(298.15 K) = [-141.84 \pm 8] kJ mol^{-1}$

Vibrational Frequencies and Degeneracies

ν, cm^{-1}	ν, cm^{-1}
2201(1)	2211(2)
945(1)	952(2)
551(1)	663(2)

Ground State Quantum Weight: [1] σ = 3

Point Group: C_{3v}

Bond Distances: Si-H = 1.486 ± 0.008 Å Si-Cl = 2.049 Å

Bond Angles: H-Si-H = 110.42° ± 0.5° H-Si-Cl = 108.5° ± 0.5°

Product of the Moments of Inertia: $I_AI_BI_C = 1.60170 \times 10^{-115} g^3 cm^6$

Enthalpy of Formation

$\Delta_fH°$(298.15 K) is estimated from values for $SiCl_4$, $SiHCl_3$ and SiH_4 (1). Data for $SiHCl_3$ (1) suggest that $\Delta_fH°$ shows minor deviations from linearity in the chlorosilanes. We assume that $\Delta_fH°$ has a cubic variation with a constant third difference of -1.5 kcal mol^{-1}. $\Delta_fH°$ values of the chloromethanes yield an almost constant third difference of about -2.5 kcal mol^{-1} (2, 3, 1). This approximation corresponds to Allen's bond additivity scheme involving near-neighbor interactions taken two and three at a time (2). our adopted $\Delta_fH°$ is 0.4 kcal mol^{-1} more negative than that estimated by linear interpolation between $SiCl_4$ and SiH_4.

Heat Capacity and Entropy

The molecular structure is based on microwave data of Kewley et al. (4) for isotopic forms of the four monohalosilanes. The authors concluded that the silyl group is effectively the same in all four halides. We adopt effective (r_o) structural parameters rather than substitutional (r_s) parameters. These are in reasonable agreement with previous structures obtained by various methods (5). The principal moments of inertia are $I_A = 0.9972 \times 10^{-39}$ and $I_B = I_C = 12.6739 \times 10^{-39} g cm^2$.

Shimanouchi (6) selected vibrational frequencies from gas-phase infrared spectra of SiH_3X (X=F, Cl, Br) observed by Newman et al. (7). These authors' assignments for SiH_3F and SiH_3Br have been modified slightly by new data (1) obtained at higher resolution. Unpublished gas-phase frequencies of SiH_3X from the thesis of Robiette were quoted by Ball et al. (8). These are more consistent with the new data for SiH_3F and SiH_3Br, so we adopt them for SiH_3Cl. Monfils also obtianed partially resolved spectra of SiH_3Cl; his final paper (9) is not available to us.

We neglect excited states and assume the electronic ground state to be $^1A_1$ by analogy with CH_3Cl (10). Electronic absorption bands of SiH_3Cl were observed at ≥67000 cm^{-1} (11) and the absorption maximum in the lowest transition was estimated to be at ~62000 cm^{-1} (12). Contributions of these high-lying states would be negligible.

References

1. JANAF Thermochemical Tables: Cl_3HSi(g), 12-31-76; FH_3Si(g), H_4Si(g) 6-30-76; CH_3Cl(g) 6-30-72; Cl_4Si(g), 12-31-70; CCl_4(g), CH_2Cl_2(g), $CHCl_3$(g), 12-31-68; CH_4(g), 3-31-61.
2. S. A. Kudchadker and A. P. Kudchadker, J. Phys. Chem. Ref. Data 4, 457 (1975).
3. A. S. Rodgers, J. Chao, R. C. Wilhoit and B. J. Zwolinski, idid., 3, 17 (1974).
4. R. Kewley, P. M. McKinney and A. G. Robiette, J. Mol. Spectrosc. 34, 390 (1970).
5. L. E. Sutton, "Interatomic Distances," Spec. Publ. No. 11, 1958; "Interatomic Distances Supplement," Spec. Publ. No. 18, The Chemical Society, London, 1965.
6. T. Shimanouchi, NSRDS-NBS 39, 29, 1972.
7. C. Newman, J. K. O'Loane, S. R. Polo and M. K. Wilson, J. Chem. Phys. 25, 855 (1956).
8. D. F. Ball, M. J. Buttler and D. C. McKean, Spectrochim. Acta 21, 451 (1965).
9. A. Monfils, Acad. Roy. Belg., Classe Sci. Mem. 29, No. 5, 52 pp. (1956).
10. G. Herzberg, "Electronic Spectra of Polyatomic Molecules," D. Van Nostrand Co., Inc., New York, 1966.
11. S. Bell and A. D. Walsh, Trans. Faraday Soc. 62, 3005 (1966).
12. G. C. Causley, J. B. Clark and B. R. Russell, Chem. Phys. Lett. 38, 602 (1976).

Enthalpy Reference Temperature = T_r = 298.15 K Standard State Pressure = p° = 0.1 MPa

T/K	$C_p°$ (J $K^{-1}mol^{-1}$)	S°	$-[G°-H°(T_r)]/T$	$H°-H°(T_r)$ (kJ mol^{-1})	$\Delta_fH°$	$\Delta_fG°$	Log K_f
0	0.	0.	INFINITE	-11.441	-132.769	-132.769	INFINITE
100	33.561	207.803	288.915	-8.111	-135.660	-131.498	68.688
200	40.360	232.644	255.088	-4.489	-138.776	-126.152	32.948
250	45.846	242.233	251.574	-2.335	-140.376	-122.810	25.660
298.15	51.098	250.761	250.761	0.	-141.838	-119.292	20.900
300	51.292	251.078	250.762	0.095	-141.891	-119.152	20.746
350	56.258	259.365	251.406	2.786	-143.268	-115.252	17.200
400	60.660	267.171	252.893	5.711	-144.490	-111.164	14.517
450	64.557	274.545	254.893	8.843	-145.561	-106.932	12.412
500	68.034	281.530	257.210	12.160	-146.489	-102.589	10.717
600	74.008	294.480	262.361	19.272	-147.963	-93.663	8.154
700	78.961	306.272	267.804	26.928	-148.998	-84.525	6.307
800	83.092	317.093	273.298	35.036	-149.677	-75.265	4.914
900	86.538	327.085	278.726	43.523	-150.074	-65.937	3.827
1000	89.411	336.356	284.031	52.325	-150.257	-56.577	2.955
1100	91.809	344.994	289.185	61.390	-150.282	-47.206	2.242
1200	93.817	353.071	294.176	70.674	-150.195	-37.839	1.647
1300	95.504	360.649	299.001	80.142	-150.034	-28.482	1.144
1400	96.929	367.780	303.662	89.766	-149.830	-19.140	0.714
1500	98.140	374.510	308.163	99.521	-149.606	-9.812	0.342
1600	99.174	380.878	312.510	109.388	-149.382	-0.500	0.016
1700	100.062	386.917	316.711	119.351	-199.349	9.245	-0.284
1800	100.829	392.659	320.772	129.396	-198.933	21.504	-0.624
1900	101.496	398.129	324.701	139.513	-198.508	33.738	-0.928
2000	102.077	403.350	328.504	149.693	-198.082	45.951	-1.200
2100	102.587	408.343	332.188	159.926	-197.660	58.142	-1.446
2200	103.037	413.126	335.759	170.208	-197.247	70.313	-1.669
2300	103.435	417.715	339.223	180.532	-196.846	82.466	-1.873
2400	103.789	422.125	342.586	190.894	-196.460	94.602	-2.059
2500	104.105	426.368	345.853	201.289	-196.092	106.722	-2.230
2600	104.388	430.457	349.029	211.714	-195.745	118.828	-2.387
2700	104.642	434.402	352.118	222.165	-195.419	130.920	-2.533
2800	104.872	438.211	355.125	232.641	-195.116	143.001	-2.668
2900	105.079	441.895	358.054	243.139	-194.838	155.072	-2.793
3000	105.268	445.461	360.909	253.656	-194.585	167.133	-2.910
3100	105.439	448.915	363.692	264.192	-194.357	179.187	-3.019
3200	105.596	452.265	366.408	274.744	-194.156	191.233	-3.122
3300	105.739	455.517	369.059	285.311	-193.981	203.274	-3.218
3400	105.870	458.675	371.649	295.891	-193.831	215.310	-3.308
3500	105.991	461.746	374.179	306.484	-193.708	227.341	-3.393
3600	106.102	464.734	376.653	317.089	-577.761	249.831	-3.625
3700	106.205	467.642	379.073	327.705	-577.273	272.813	-3.851
3800	106.300	470.476	381.442	338.330	-576.811	295.782	-4.066
3900	106.388	473.238	383.760	348.964	-576.373	318.739	-4.269
4000	106.470	475.933	386.031	359.607	-575.958	341.685	-4.462
4100	106.546	478.563	388.256	370.258	-575.564	364.621	-4.645
4200	106.617	481.131	390.437	380.916	-575.191	387.548	-4.820
4300	106.683	483.641	392.575	391.581	-574.836	410.466	-4.986
4400	106.745	486.094	394.673	402.253	-574.499	433.376	-5.145
4500	106.803	488.493	396.731	412.930	-574.177	456.279	-5.296
4600	106.857	490.841	398.751	423.613	-573.870	479.174	-5.441
4700	106.908	493.140	400.735	434.302	-573.576	502.063	-5.580
4800	106.955	495.391	402.684	444.995	-573.293	524.947	-5.713
4900	107.000	497.597	404.599	455.693	-573.021	547.824	-5.840
5000	107.043	499.759	406.480	466.395	-572.758	570.696	-5.962
5100	107.083	501.879	408.330	477.101	-572.502	593.562	-6.079
5200	107.120	503.959	410.149	487.811	-572.252	616.424	-6.192
5300	107.156	506.000	411.938	498.525	-572.007	639.280	-6.300
5400	107.190	508.003	413.699	509.242	-571.766	662.133	-6.405
5500	107.222	509.970	415.432	519.963	-571.526	684.980	-6.505
5600	107.252	511.902	417.137	530.687	-571.288	707.824	-6.602
5700	107.280	513.801	418.816	541.413	-571.049	730.663	-6.696
5800	107.308	515.667	420.470	552.143	-570.809	753.498	-6.786
5900	107.334	517.502	422.099	562.875	-570.567	776.328	-6.873
6000	107.358	519.306	423.704	573.609	-570.319	799.155	-6.957

PREVIOUS: December 1976 (1 atm) CURRENT: December 1976 (1 bar)

AMMONIUM CHLORIDE (NH_4Cl) CRYSTAL M_r = 53.49146

$S°(298.15\ K) = 94.86 \pm 4.2\ J\ K^{-1}\ mol^{-1}$
T_{trs2} = 242.6 K
T_{trs1} = 457.7 K
T_{fus} = 793.2 K
T_{dcm} = 612 K

$\Delta_f H°(0\ K) = -311.391 \pm 0.84\ kJ\ mol^{-1}$
$\Delta_f H°(298.15\ K) = -314.553 \pm 0.84\ kJ\ mol^{-1}$
$\Delta_{trs2} = 1.046\ kJ\ mol^{-1}$
$\Delta_{trs1} = 3.950\ kJ\ mol^{-1}$

Enthalpy of Formation

The equilibrium pressures for the reaction $NH_4Cl(cr) = NH_3(g) + HCl(g)$, at 320-636 K, have been measured by many investigators. Smits and de Lange (2), using the densi-tensimeter, determined the vapor pressure of $NH_4Cl(cr)$ and the density of the saturated vapor simultaneously. The results obtained indicate that within the range 527-626 K the saturated vapor completely dissociates into gaseous NH_3 and HCl. Based on this conclusion, the reported vapor pressure data were employed to evaluate the enthalpy of dissociation of $NH_4Cl(cr)$ by both the 2nd and 3rd law methods. The results are presented as follows.

Source	T/K	$\Delta_r H°(298.15\ K)$, kcal mol^{-1} 2nd law	3rd law	Drift cal K^{-1} mol^{-1}
(1)	521.6-619.1	43.18±0.08	42.11	-2.0±0.2
(2)	527.9-625.6	42.16±0.22	42.14	-0.2±0.4
(3)	525.4-615.0	42.02±0.06	42.20	0.2±0.1
(4)	471.3-557.0	42.43±0.09	42.08	-0.3±0.2
(5)	319.0-352.7	43.83±0.38	42.17	-4.9±1.1
(6)	503.4-611.4	41.27±0.38	42.12	1.3±0.8

The value of $\Delta_r H°(298.15\ K)$ adopted is 42.15±0.10 kcal mol^{-1}, yielding $\Delta_f H°(NH_4Cl,\ cr,\ 298.15\ K) = -75.8\pm0.20$ kcal mol^{-1}. The corresponding $\Delta_r H°(298.15\ K)$ value evaluated from solution data, selected by Parker (7) is 42.06 kcal mol^{-1}.

Heat Capacity and Entropy

The heat capacities (20.1-523.2 K) of $NH_4Cl(cr)$ have been measured by the following investigators: Ewald 138-301 K (8); Simon 20.1-290.8 K (9); Klinhardt 323.2-523.2 K (10); Simon et al. 204.5-275.8 K (11); Ziegler and Messer 107.2-320.2 K(12); Extermann and Weigle 228.2-248.2 (13) and Popov and Galchenko 373.2-523.2 K (14). These $C_p°$ data were plotted and a smooth $C_p°$ curve was drawn and adopted. The $C_p°$ values above 523 K were obtained by graphical extrapolation of the previous $C_p°$ curve. S°(298.15 K) was derived from the adopted low temperature data, based on $S°(20.1\ K) = 0.114\ cal\ K^{-1}\ mol^{-1}$.

Transition Data

There are two transitions reported in the literature. The low temperature transition data (T_{trs2} and $\Delta_{trs2}H°$) were derived based on the heat capacity (107-320 K) data reported by Ziegler and Messer (12), in which the other T_{trs2} values given by previous investigators were reviewed. The high temperature transition data (T_{trs1} and $\Delta_{trs1}H°$) were taken from Klinkhardt (10). However, $\Delta_{trs1}H° = 1030$ and $\Delta H°(463.5\ K) = 1059.1\pm8.0$ cal mol^{-1} were reported by Scheffer (13), and Popov and Galchenko (14), respectively. Markowitz and Boryta (6) determined T_{trs1} to be 459.5±0.5 K.

Fusion Data

T_{fus} was obtained from Rassow (15)

Decomposition Data

The value of T_{dcm} was calculated as the temperature at which the vapor pressure of the decomposition products (NH_3 and HCl) equals 1 atm. By differential thermal analysis, T_{dcm} for $NH_4Cl(cr)$ was determined to be 611.4 K by Markowitz and Boryta (6).

References
1. A. Smith and R. P. Calvert, J. Amer. Chem. Soc. 36, 1363 (1914), using isoteniscope.
2. A. Smits and W. D. Lange, J. Chem. Soc. 2944 (1928), using densi-tensimeter.
3. H. Braune and S. Knoke, Z. Phys. Chem. 135, 49 (1928), Spiral Manometer.
4. W. H. Rodebush and J. C. Michalek, J. Amer. Chem. Soc. 51, 748 (1929), glass diaphragm manometer.
5. H. Wagner and K. Neumann, Z. Phys. Chem. 28, 51 (1961), torsion-effusion method.
6. M. M. Markowitz and D. A. Boryta, J. Phys. Chem. 66, 1477 (1962), differential thermal analysis.
7. V. B. Parker, NSRDS-NBS 2, 1965.
8. R. Ewald, Ann. Physik, 44, 1213 (1914).
9. F. Simon, ibid., 68, 241 (1922).
10. H. Klinkhardt, ibid., 84, 167 (1927).
11. F. Simon, C. V. Simon and M. Ruhemann, Z. Phys. Chem. 129, 339 (1927).
12. W. T. Ziegler and C. E. Messer, J. Amer. Chem. Soc., 63, 2694 (1941).
13. R. Extermann and J. Weigle, Helv. Phys. Acta, 15, 455 (1942).
14. M. M. Popov and G. L. Galchenko, J. Gen. Chem., USSR, 21, 2489 (1951).
15. H. Rassow, Z. Anorg. Chem. 114, 117 (1920).

Ammonium Chloride (NH_4Cl) $Cl_1H_4N_1(cr)$

Enthalpy Reference Temperature = T_r = 298.15 K — Standard State Pressure = p° = 0.1 MPa

T/K	$C_p°$	S°	$-[G°-H°(T_r)]/T$	$H°-H°(T_r)$	$\Delta_f H°$	$\Delta_f G°$	Log K_f
	J $K^{-1}mol^{-1}$			kJ mol^{-1}			
0	0.	0.	INFINITE	-22.698	-311.391	-311.391	INFINITE
100	37.656	24.552	163.795	-13.924	-311.523	-276.392	144.372
200	67.111	59.752	103.140	-8.678	-314.638	-239.828	62.637
298.15	86.441	94.860	94.860	0.	-314.553	-203.092	35.581
300	86.944	95.396	94.861	0.160	-314.558	-202.400	35.241
400	102.968	122.642	98.467	9.670	-314.054	-165.049	21.553
457.700	109.956	136.945	102.428	15.798		II <--> III	
457.700	85.898	145.537	102.428	19.731		TRANSITION	
500	90.500	153.330	106.407	23.462	-309.363	-128.299	13.403
600	101.671	170.818	115.701	33.071	-308.920	-92.110	8.019
700	112.801	187.327	124.764	43.795	-307.429	-56.079	4.185
800	123.888	203.113	133.577	55.629	-304.904	-20.332	1.328
900	134.976	218.360	142.155	68.585	-301.346	15.037	-0.873
1000	144.557	233.090	150.517	82.573	-296.853	49.958	-2.610
1100	152.716	247.259	158.672	97.446	-291.583	84.389	-4.007
1200	159.829	260.879	166.625	113.104	-285.636	118.311	-5.150
1300	163.176	273.774	174.376	129.217	-279.347	151.720	-6.096
1400	171.000	286.172	181.921	145.952	-272.549	184.628	-6.889
1500	175.728	298.149	189.273	163.314	-265.232	217.030	-7.558

PREVIOUS: December 1960 CURRENT: September 1965

Ammonium Chloride (NH_4Cl) $Cl_1H_4N_1(cr)$

AMMONIUM PERCHLORATE (NH_4ClO_4) CRYSTAL M_r = 117.48906

$S°(298.15\ K) = 184.180 \pm 1.26\ J\ K^{-1}\ mol^{-1}$
$T_{trs} = 513.15\ K$
$T_{dcm} = [513]\ K$

$\Delta_f H°(0\ K) = -277.78 \pm 0.17\ kJ\ mol^{-1}$
$\Delta_f H°(298.15\ K) = -295.77 \pm 0.17\ kJ\ mol^{-1}$
$\Delta_{trs} H° = 9.6 \pm 0.8\ kJ\ mol^{-1}$

Enthalpy of Formation

Gilliland and Johnson (1) and Birky and Hepler (2) have derived the enthalpy of formation for NH_4ClO_4(cr) from solution-calorimetric measurements. Their values are -70.74±0.32 and -70.63 kcal mol^{-1}, respectively. The mean of these is adopted here.

Heat Capacity and Entropy

Justice and Westrum (3) have measured the heat capacity from 5 to 350 K. The $C_p°$ values above 350 K were extrapolated. From the partial molar entropies of NH_4^+(aq) and ClO_4^-(aq) and the entropy of solution, Birky and Hepler (2) calculated a value of 43.6 cal K^{-1} mol^{-1} for S°(298.15 K). The fair agreement of this value with that derived from $C_p°$ suggests that there is no residual entropy at 0 K, as there would be if the NH_4^+ ion had a random orientation.

Transition and Decomposition Data

T_{trs} and $\Delta_{trs}H°$ were obtained (under a pressure of NH_3) by Markowitz (4). T_{dcm} was estimated from the report by Markowitz (4).

References

1. A. A. Gilliland and W. H. Johnson, J. Res. Nat. Bur. Stand. 65A, 67 (1961).
2. M. M. Birky and L. G. Hepler, J. Phys. Chem. 64, 686 (1960).
3. B. H. Justice and E. F. Westrum, Univ. of Michigan, Ann Arbor, Mich., personal communication, April 1961.
4. M. M. Markowitz, Foote Mineral Company, Exton, Pennsylvania, personal communication, September 20, 1962.

Ammonium Perchlorate (NH_4ClO_4) $Cl_1H_4N_1O_4$(cr)

Enthalpy Reference Temperature = T_r = 298.15 K Standard State Pressure = p° = 0.1 MPa

	J K⁻¹mol⁻¹			kJ mol⁻¹			
T/K	$C_p°$	S°	$-[G°-H°(T_r)]/T$	$H°-H°(T_r)$	$\Delta_f H°$	$\Delta_f G°$	Log K_f
0	0.	0.	INFINITE	-25.238	-277.778	-277.778	INFINITE
100	79.622	72.894	283.140	-21.025	-288.280	-223.321	116.651
200	107.236	137.419	195.200	-11.556	-292.995	-156.325	40.828
298.15	128.072	184.180	184.180	0.	-295.767	-88.607	15.524
300	128.449	184.973	184.182	0.237	-295.804	-87.322	15.204
400	148.699	224.699	189.462	14.095	-296.894	-17.615	2.300
500	168.950	260.052	200.101	29.975	-296.232	52.165	-5.450
513.150	171.628	264.472	201.694	32.215		I <--> II	
513.150	179.996	283.226	201.694	41.838		TRANSITION	
600	197.568	312.708	215.652	58.234	-283.459	119.958	-10.443
700	217.819	344.684	231.823	79.003	-278.431	186.822	-13.941
800	238.070	375.092	247.845	101.797	-271.621	252.842	-16.509
900	258.456	404.312	263.616	126.626	-263.001	317.902	-18.451
1000	278.571	432.586	279.107	153.480	-252.567	381.904	-19.949
1100	299.569	460.142	294.317	182.408	-240.258	444.771	-21.120
1200	317.984	487.016	309.260	213.307	-226.169	506.437	-22.045
1300	333.816	513.111	323.943	245.919	-210.548	566.864	-22.777
1400	347.066	538.350	338.362	279.984	-193.645	626.039	-23.358
1500	357.732	562.674	352.510	315.246	-175.712	683.966	-23.818

PREVIOUS: September 1961 CURRENT: December 1962

Ammonium Perchlorate (NH_4ClO_4) $Cl_1H_4N_1O_4$(cr)

MERCURY CHLORIDE (HgCl) IDEAL GAS M_r = 236.043

$\Delta_f H°(0\ K) = 82.54 \pm 9.6\ kJ\ mol^{-1}$

$S°(298.15\ K) = 260.005\ J\ K^{-1}\ mol^{-1}$ $\Delta_f H°(298.15\ K) = 78.45 \pm 9.6\ kJ\ mol^{-1}$

Electronic Level and Quantum Weight

State	ε_i, cm^{-1}	g_i
$^2\Sigma^+$	0	2

$\omega_e = 290.94\ cm^{-1}$ $\omega_e x_e = 1.5844\ cm^{-1}$ $\sigma = 1$

$B_e = [0.1121]\ cm^{-1}$ $\alpha_e = [0.0007]\ cm^{-1}$ $r_e = 2.23$ Å

Enthalpy of Formation

Gaydon (1) gives 23±2.3 kcal mol^{-1} for the heat of dissociation into gaseous atoms.

Heat Capacity and Entropy

Vibrational constants from Wieland (2) were weighted according to their natural isotopic abundances. Rotational constants were estimated using equation III.(3) Bond length was given by Herzberg (3).

References

1. A. G. Gaydon, "Dissociation Energies and Spectra of Diatomic Molecules," 2nd ed., Chapman and Hall, London 1953.
2. K. Wieland, Zeits. f. Elektrochemie 64, 761 (1960).
3. G. Herzberg, "Spectra of Diatomic Molecules," equation III, 123 p. 108, Van Nostrand, New York 1950.

Mercury Chloride (HgCl) $Cl_1Hg_1(g)$

Enthalpy Reference Temperature = T_r = 298.15 K Standard State Pressure = p° = 0.1 MPa

	J $K^{-1}mol^{-1}$			kJ mol^{-1}			
T/K	$C_p°$	S°	$-[G°-H°(T_r)]/T$	$H°-H°(T_r)$	$\Delta_f H°$	$\Delta_f G°$	Log K_f
0	0.	0.	INFINITE	-9.846	82.537	82.537	INFINITE
100	31.476	222.563	291.359	-6.880	82.342	73.258	-38.266
200	35.116	245.701	263.302	-3.520	81.598	64.458	-16.835
250	35.926	253.632	260.602	-1.743	78.869	60.376	-12.615
298.15	36.420	260.005	260.005	0.	78.450	56.853	-9.960
300	36.435	260.230	260.005	0.067	78.434	56.719	-9.876
350	36.778	265.874	260.450	1.898	78.016	53.133	-7.930
400	37.023	270.802	261.443	3.744	77.610	49.606	-6.478
450	37.207	275.173	262.730	5.599	77.211	46.130	-5.355
500	37.353	279.101	264.174	7.464	76.817	42.697	-4.461
600	37.574	285.932	267.248	11.211	76.034	35.947	-3.129
700	37.740	291.737	270.342	14.977	16.490	35.899	-2.679
800	37.878	296.786	273.339	18.758	16.342	38.682	-2.526
900	37.998	301.254	276.197	22.552	16.198	41.483	-2.408
1000	38.107	305.263	278.907	26.357	16.056	44.300	-2.314
1100	38.209	308.900	281.471	30.173	15.918	47.132	-2.238
1200	38.306	312.229	283.897	33.998	15.784	49.975	-2.175
1300	38.399	315.299	286.196	37.834	15.655	52.830	-2.123
1400	38.490	318.148	288.378	41.678	15.530	55.694	-2.078
1500	38.579	320.806	290.452	45.532	15.409	58.567	-2.039
1600	38.667	323.299	292.428	49.394	15.293	61.448	-2.006
1700	38.753	325.646	294.314	53.265	15.181	64.336	-1.977
1800	38.839	327.863	296.116	57.145	15.073	67.231	-1.951
1900	38.924	329.966	297.843	61.033	14.969	70.131	-1.928
2000	39.009	331.964	299.500	64.929	14.869	73.037	-1.908
2100	39.093	333.870	301.091	68.834	14.771	75.948	-1.889
2200	39.176	335.690	302.623	72.748	14.675	78.863	-1.872
2300	39.260	337.433	304.099	76.670	14.581	81.783	-1.857
2400	39.343	339.106	305.523	80.600	14.489	84.707	-1.844
2500	39.425	340.714	306.898	84.538	14.396	87.635	-1.831
2600	39.508	342.262	308.229	88.485	14.303	90.566	-1.819
2700	39.591	343.754	309.517	92.440	14.209	93.501	-1.809
2800	39.673	345.196	310.766	96.403	14.115	96.439	-1.799
2900	39.755	346.589	311.977	100.374	14.018	99.381	-1.790
3000	39.837	347.938	313.154	104.354	13.920	102.327	-1.782
3100	39.919	349.246	314.297	108.342	13.821	105.275	-1.774
3200	40.001	350.515	315.409	112.338	13.720	108.227	-1.767
3300	40.083	351.747	316.492	116.342	13.619	111.182	-1.760
3400	40.165	352.945	317.546	120.355	13.518	114.140	-1.754
3500	40.247	354.110	318.574	124.375	13.417	117.101	-1.748
3600	40.329	355.245	319.577	128.404	13.317	120.064	-1.742
3700	40.410	356.351	320.556	132.441	13.221	123.031	-1.737
3800	40.492	357.430	321.512	136.486	13.127	126.000	-1.732
3900	40.574	358.483	322.447	140.539	13.039	128.972	-1.727
4000	40.655	359.511	323.361	144.601	12.956	131.945	-1.723
4100	40.737	360.516	324.255	148.670	12.881	134.921	-1.719
4200	40.818	361.498	325.130	152.748	12.815	137.898	-1.715
4300	40.900	362.460	325.987	156.834	12.758	140.877	-1.711
4400	40.981	363.401	326.826	160.928	12.712	143.857	-1.708
4500	41.063	364.323	327.650	165.030	12.678	146.838	-1.704
4600	41.144	365.226	328.457	169.141	12.657	149.820	-1.701
4700	41.226	366.112	329.248	173.259	12.650	152.802	-1.698
4800	41.307	366.981	330.026	177.386	12.658	155.784	-1.695
4900	41.389	367.833	330.788	181.521	12.682	158.765	-1.692
5000	41.470	368.670	331.538	185.664	12.723	161.746	-1.690
5100	41.551	369.492	332.274	189.815	12.781	164.726	-1.687
5200	41.633	370.300	332.997	193.974	12.856	167.705	-1.685
5300	41.714	371.094	333.709	198.141	12.950	170.682	-1.682
5400	41.796	371.874	334.408	202.317	13.062	173.657	-1.680
5500	41.877	372.642	335.097	206.500	13.192	176.630	-1.677
5600	41.958	373.397	335.774	210.692	13.341	179.600	-1.675
5700	42.040	374.141	336.440	214.892	13.509	182.567	-1.673
5800	42.121	374.873	337.097	219.100	13.695	185.532	-1.671
5900	42.202	375.593	337.743	223.316	13.900	188.493	-1.669
6000	42.284	376.303	338.380	227.541	14.122	191.450	-1.667

PREVIOUS: December 1961 (1 atm) CURRENT: December 1961 (1 bar)

Mercury Chloride (HgCl) $Cl_1Hg_1(g)$

IODINE CHLORIDE (ICl) CRYSTAL M_r = 162.3575

$S°(298.15\ K) = 97.927\ J\ K^{-1}\ mol^{-1}$

$T_{fus} = 300.53\ K$

$\Delta_f H°(0\ K) = -36.395 \pm 0.105\ kJ\ mol^{-1}$

$\Delta_f H°(298.15\ K) = -35.422 \pm 0.105\ kJ\ mol^{-1}$

$\Delta_{fus} H° = 11.602 \pm 0.008\ kJ\ mol^{-1}$

Enthalpy of Formation

The enthalpy of formation is that adopted by Calder and Giauque (1) from an analysis of the equilibrium data of McMorris and Yost (2) and Beeson and Yost (3). This yielded a enthalpy of formation of ICl(g), from which the heat of formation of ICl(cr) was calculated by using the heat of vaporization determined by Calder and Giauque (1).

Heat Capacity and Entropy

The heat capacity has been measured by Calder and Giauque (1) from 17-322 K. The entropy was calculated based on S°(15 K) = 0.592 cal K^{-1} mol^{-1}. Above 300 K the heat capacity data was extrapolated smoothly.

Fusion Data

All data were measured by Calder and Giauque (1).

References

1. G. V. Calder and W. F. Giauque, J. Phys. Chem. 69, 2443 (1965).
2. J. McMorris and D. M. Yost, J. Amer. Chem. Soc. 54, 2247 (1932).
3. C. M. Beeson and D. M. Yost, J. Chem. Phys. 7, 44 (1939).

Iodine Chloride (ICl) $Cl_1I_1(cr)$

Enthalpy Reference Temperature = T_r = 298.15 K Standard State Pressure = p° = 0.1 MPa

T/K	$C_p°$ (J K⁻¹mol⁻¹)	S° (J K⁻¹mol⁻¹)	$-[G°-H°(T_r)]/T$ (J K⁻¹mol⁻¹)	$H°-H°(T_r)$ (kJ mol⁻¹)	$\Delta_f H°$ (kJ mol⁻¹)	$\Delta_f G°$ (kJ mol⁻¹)	Log K_f
0	0.	0.	INFINITE	-12.163	-36.395	-36.395	INFINITE
100	40.246	45.693	143.013	-9.732	-36.945	-29.015	15.156
200	49.539	77.011	102.805	-5.159	-36.357	-21.267	5.554
298.15	55.229	97.927	97.927	0.	-35.422	-14.049	2.461
300	55.325	98.268	97.928	0.102	-35.401	-13.917	2.423
300.530	55.589	98.366	97.928	0.131	------ CRYSTAL <--> LIQUID -----		
400	60.195	114.866	100.158	5.883	-42.174	-6.656	0.869
500	64.379	128.756	104.525	12.116	-61.826	4.083	-0.427
600	68.145	140.833	109.591	18.745	-58.889	16.995	-1.480
700	71.489	151.594	114.836	25.731	-55.624	29.389	-2.193
800	74.412	161.335	120.049	33.029	-52.065	41.294	-2.696
900	76.914	170.249	125.139	40.599	-48.250	52.737	-3.061
1000	78.994	178.464	130.066	48.398	-44.220	63.743	-3.330

PREVIOUS: September 1963 CURRENT: September 1965

Iodine Chloride (ICl) $Cl_1I_1(cr)$

IODINE MONOCHLORIDE (ICl) LIQUID M_r = 162.3575

$S°(298.15\ K) = [136.167]\ J\ K^{-1}\ mol^{-1}$

$T_{fus} = 300.53\ K$

$\Delta_f H°(298.15\ K) = [-23.929]\ kJ\ mol^{-1}$

$\Delta_{fus} H° = 11.602 \pm 0.008\ kJ\ mol^{-1}$

Enthalpy of Formation

The enthalpy of formation is calculated from that of the crystal by adding $\Delta_{fus}H°$ and the difference in enthalpy, H°(300.53 K) - H°(298.15 K), between the crystal and liquid.

Heat Capacity and Entropy

The heat capacity from 300.53 to 317.76 K has been measured by Calder and Giauque (1). Above 318 K it was extrapolated in a smooth manner. The entropy is calculated in a manner analogous to that used for the enthalpy of formation.

Reference

1. G. V. Calder and W. F. Giauque, J. Phys. Chem. 69, 2443 (1965).

Iodine Chloride (ICl) $Cl_1I_1(l)$

Enthalpy Reference Temperature = T_r = 298.15 K Standard State Pressure = p° = 0.1 MPa

T/K	$C_p°$ (J K⁻¹mol⁻¹)	S° (J K⁻¹mol⁻¹)	$-[G°-H°(T_r)]/T$ (J K⁻¹mol⁻¹)	$H°-H°(T_r)$ (kJ mol⁻¹)	$\Delta_f H°$ (kJ mol⁻¹)	$\Delta_f G°$ (kJ mol⁻¹)	Log K_f
0							
100							
200							
298.15	102.759	136.167	136.167	0.	-23.929	-13.958	2.445
300	102.851	136.802	136.168	0.190	-23.821	-13.896	2.420
300.530	103.856	136.972	136.170	0.241	------ CRYSTAL <--> LIQUID -----		
400	98.324	165.893	140.170	10.289	-26.275	-11.168	1.458
500	94.140	187.384	147.559	19.913	-42.536	-5.941	0.621
600	89.956	204.178	155.649	29.117	-37.025	0.853	-0.074
700	85.772	217.730	163.582	37.904	-31.958	6.759	-0.504
800	81.588	228.911	171.071	46.272	-27.329	11.969	-0.781
900	77.404	238.279	178.033	54.221	-23.135	16.625	-0.965
1000	73.220	246.217	184.465	61.753	-19.373	20.837	-1.088

PREVIOUS: September 1963 CURRENT: September 1965

Iodine Chloride (ICl) $Cl_1I_1(l)$

IODINE CHLORIDE (ICl) CRYSTAL–LIQUID M_r = 162.3575

0 to 300.53 K crystal
above 300.53 K liquid

Refer to the individual tables for details.

Iodine Chloride (ICl) $Cl_1I_1(cr,l)$

Enthalpy Reference Temperature = T_r = 298.15 K Standard State Pressure = $p°$ = 0.1 MPa

T/K	$C_p°$ (J K⁻¹mol⁻¹)	$S°$ (J K⁻¹mol⁻¹)	$-[G°-H°(T_r)]/T$ (J K⁻¹mol⁻¹)	$H°-H°(T_r)$ (kJ mol⁻¹)	$\Delta_fH°$ (kJ mol⁻¹)	$\Delta_fG°$ (kJ mol⁻¹)	Log K_f
0	0.	0.	INFINITE	-12.163	-36.395	-36.395	INFINITE
100	40.246	45.693	143.013	-9.732	-36.945	-29.015	15.156
200	49.539	77.011	102.805	-5.159	-36.357	-21.267	5.554
298.15	55.229	97.927	97.927	0.	-35.422	-14.049	2.461
300	55.325	98.268	97.928	0.102	-35.401	-13.917	2.423
300.530	55.589	98.366	97.928	0.131	CRYSTAL <--> LIQUID		
300.530	103.856	136.972	97.928	11.734	TRANSITION		
400	98.324	165.893	111.438	21.782	-26.275	-11.168	1.458
500	94.140	187.384	124.573	31.405	-42.536	-5.941	0.621
600	89.956	204.178	136.494	40.610	-37.025	0.853	-0.074
700	85.772	217.730	147.164	49.397	-31.958	6.759	-0.504
800	81.588	228.911	156.705	57.765	-27.329	11.969	-0.781
900	77.404	238.279	165.263	65.714	-23.135	16.625	-0.965
1000	73.220	246.217	172.972	73.245	-19.373	20.837	-1.088

PREVIOUS: CURRENT: September 1965

Iodine Chloride (ICl) $Cl_1I_1(cr,l)$

IODINE CHLORIDE (ICl) IDEAL GAS M_r = 162.3575

$S°(298.15\ K) = 247.567\ J\ K^{-1}\ mol^{-1}$

$\Delta_f H°(0\ K) = 19.141 \pm 0.105\ kJ\ mol^{-1}$
$\Delta_f H°(298.15\ K) = 17.506 \pm 0.105\ kJ\ mol^{-1}$

Electronic Levels and Quantum Weights

State	ε_i, cm^{-1}	g_i
$^1\Sigma^+$	0	1

$\omega_e = 382.18\ cm^{-1}$ $\omega_e x_e = 1.450\ cm^{-1}$ $\sigma = 1$
$B_e = 0.11298\ cm^{-1}$ $\alpha_e = 0.0005275\ cm^{-1}$ $r_e = 2.327$ Å

Enthalpy of Formation

The dissociation energy from spectroscopic measurements of ICl(g) has been summarized by Evans et al. (1) from which they adopt $D_0^0 = 49.64\ kcal\ mol^{-1}$. McMorris and Yost (2) and Beeson and Yost (3) studied the equilibria NOCl = NO + $0.5Cl_2$ and NO + ICl = NOCl + $0.5I_2(g)$ from which the equilibrium constant of ICl = 1/2 Cl_2 + 1/2 $I_2(g)$ can be obtained. Calder and Giauque (4) have used 3rd law analysis to obtain $D_0^0 = 49.578 \pm 0.025\ kcal\ mol^{-1}$. This value was adopted yielding $\Delta_f H°(298.15\ K) = 4.184 \pm 0.025\ kcal\ mol^{-1}$ ($17.506 \pm 0.105\ kJ\ mol^{-1}$).

Heat Capacity and Entropy

The molecular and vibrational constants were obtained from Herzberg (5). The values were then adjusted for the normal isotopic abundance of chlorine.

References

1. W. H. Evans, T. R. Munson and D. D. Wagman, J. Res. Nat. Bur. Stand. 55, 147 (1955).
2. J. McMorris and D. M. Yost, J. Amer. Chem. Soc. 54, 2247 (1932).
3. C. M. Beeson and D. M. Yost, J. Chem. Phys. 7, 44 (1939).
4. G. V. Calder and W. F. Giauque, J. Phys. Chem. 69, 2443 (1965).
5. G. Herzberg, "Spectra of Diatomic Molecules", D. Van Nostrand Co., New York 1950.

Iodine Chloride (ICl) $Cl_1I_1(g)$

Enthalpy Reference Temperature = T_r = 298.15 K Standard State Pressure = p° = 0.1 MPa

T/K	$C_p°$ (J K⁻¹mol⁻¹)	S° (J K⁻¹mol⁻¹)	$-[G°-H°(T_r)]/T$ (J K⁻¹mol⁻¹)	$H°-H°(T_r)$ (kJ mol⁻¹)	$\Delta_f H°$ (kJ mol⁻¹)	$\Delta_f G°$ (kJ mol⁻¹)	Log K_f
0	0.	0.	INFINITE	-9.555	19.141	19.141	INFINITE
100	30.179	211.592	277.844	-6.625	19.089	10.430	-5.448
200	33.787	233.701	250.772	-3.414	18.315	2.067	-0.540
250	34.859	241.364	248.149	-1.696	17.910	-1.948	0.407
298.15	35.548	247.567	247.567	0.	17.506	-5.737	1.005
300	35.570	247.787	247.568	0.066	17.490	-5.881	1.024
350	36.056	253.309	248.003	1.857	17.016	-9.740	1.454
400	36.401	258.147	248.975	3.669	8.539	-13.255	1.731
450	36.656	262.450	250.237	5.496	7.461	-15.914	1.847
500	36.849	266.323	251.655	7.334	-13.680	-16.554	1.729
600	37.125	273.067	254.678	11.033	-13.674	-17.130	1.491
700	37.313	278.805	257.725	14.756	-13.671	-17.706	1.321
800	37.452	283.797	260.679	18.494	-13.672	-18.283	1.194
900	37.562	288.215	263.498	22.245	-13.676	-18.859	1.095
1000	37.654	292.177	266.171	26.006	-13.685	-19.434	1.015
1100	37.734	295.770	268.701	29.776	-13.698	-20.009	0.950
1200	37.806	299.056	271.096	33.553	-13.718	-20.582	0.896
1300	37.871	302.085	273.364	37.336	-13.748	-21.152	0.850
1400	37.932	304.894	275.517	41.127	-13.792	-21.720	0.810
1500	37.991	307.513	277.564	44.923	-13.855	-22.285	0.776
1600	38.046	309.966	279.513	48.725	-13.942	-22.844	0.746
1700	38.100	312.274	281.373	52.532	-14.057	-23.397	0.719
1800	38.152	314.454	283.151	56.345	-14.204	-23.942	0.695
1900	38.203	316.518	284.853	60.162	-14.384	-24.478	0.673
2000	38.254	318.479	286.486	63.985	-14.600	-25.004	0.653
2100	38.303	320.346	288.054	67.813	-14.851	-25.518	0.635
2200	38.352	322.129	289.563	71.646	-15.134	-26.020	0.618
2300	38.400	323.835	291.016	75.483	-15.446	-26.508	0.602
2400	38.448	325.470	292.418	79.326	-15.783	-26.982	0.587
2500	38.495	327.041	293.772	83.173	-16.140	-27.441	0.573
2600	38.542	328.552	295.081	87.025	-16.513	-27.886	0.560
2700	38.589	330.007	296.347	90.881	-16.894	-28.316	0.548
2800	38.635	331.411	297.575	94.743	-17.281	-28.732	0.536
2900	38.682	332.768	298.765	98.608	-17.666	-29.134	0.525
3000	38.728	334.080	299.920	102.479	-18.046	-29.523	0.514
3100	38.774	335.351	301.043	106.354	-18.416	-29.899	0.504
3200	38.820	336.582	302.134	110.234	-18.773	-30.264	0.494
3300	38.866	337.778	303.196	114.118	-19.112	-30.618	0.485
3400	38.911	338.939	304.231	118.007	-19.431	-30.962	0.476
3500	38.957	340.067	305.239	121.900	-19.727	-31.296	0.467
3600	39.002	341.165	306.221	125.798	-19.997	-31.623	0.459
3700	39.048	342.234	307.180	129.701	-20.239	-31.943	0.451
3800	39.093	343.276	308.117	133.608	-20.453	-32.256	0.443
3900	39.138	344.292	309.031	137.519	-20.635	-32.564	0.436
4000	39.183	345.284	309.925	141.435	-20.785	-32.868	0.429
4100	39.229	346.252	310.799	145.356	-20.901	-33.169	0.423
4200	39.274	347.198	311.655	149.281	-20.984	-33.467	0.416
4300	39.319	348.123	312.492	153.211	-21.031	-33.763	0.410
4400	39.364	349.027	313.312	157.145	-21.042	-34.059	0.404
4500	39.409	349.912	314.116	161.083	-21.017	-34.355	0.399
4600	39.454	350.779	314.903	165.027	-20.955	-34.652	0.393
4700	39.499	351.628	315.676	168.974	-20.856	-34.951	0.388
4800	39.544	352.460	316.433	172.926	-20.720	-35.252	0.384
4900	39.589	353.276	317.177	176.883	-20.546	-35.557	0.379
5000	39.634	354.076	317.907	180.844	-20.335	-35.866	0.375
5100	39.678	354.861	318.624	184.810	-20.087	-36.178	0.371
5200	39.723	355.632	319.328	188.780	-19.801	-36.497	0.367
5300	39.768	356.389	320.020	192.754	-19.478	-36.821	0.363
5400	39.813	357.133	320.701	196.733	-19.118	-37.151	0.359
5500	39.858	357.864	321.370	200.717	-18.721	-37.489	0.356
5600	39.903	358.582	322.028	204.705	-18.288	-37.834	0.353
5700	39.947	359.289	322.676	208.698	-17.819	-38.187	0.350
5800	39.992	359.984	323.313	212.694	-17.314	-38.549	0.347
5900	40.037	360.668	323.940	216.696	-16.774	-38.919	0.345
6000	40.082	361.342	324.558	220.702	-16.199	-39.300	0.342

PREVIOUS: September 1965 (1 atm) CURRENT: September 1965 (1 bar)

Iodine Chloride (ICl) $Cl_1I_1(g)$

POTASSIUM CHLORIDE (KCl) CRYSTAL M_r = 74.5513

$S°(298.15\ K) = 82.55 \pm 0.17\ J\ K^{-1}\ mol^{-1}$

$T_{fus} = 1044\ K$

$\Delta_f H°(0\ K) = -436.38 \pm 0.25\ kJ\ mol^{-1}$

$\Delta_f H°(298.15\ K) = -436.68 \pm 0.25\ kJ\ mol^{-1}$

$\Delta_{fus} H° = 26.284\ kJ\ mol^{-1}$

Enthalpy of Formation

The enthalpy of formation was calculated from the enthalpy of hydrolysis of K(cr), the enthalpies of solution of HCl(g) and KCl(cr), and the standard enthalpy of neutralization of a strong acid and strong base. Extensive measurements are available for all these quantities. Values used in the calculation, in addition to the enthalpy of formation for HCl(g), are summarized below.

Source	Reaction	$\Delta_r H°(298.15\ K)$, kcal mol^{-1}
(1, 2, 3, 4)	$K(cr) + H_2O(\ell) \rightarrow KOH(\infty H_2O) + 1/2\ H_2(g)$	-46.966 ± 0.05
(5, 2, 6)	$HCl(g) \rightarrow HCl(\infty H_2O)$	-17.88 ± 0.015
(7)	$KCl(\infty H_2O) \rightarrow KCl(cr)$	-4.115 ± 0.010
(7)	$KOH(\infty H_2O) + HCl(\infty H_2O) \rightarrow KCl(\infty H_2O) + H_2O(\ell)$	-13.345 ± 0.025

Heat Capacity and Entropy

Low temperature heat capacities were obtained from the data (2.8-275 K) of Berg and Morrison (8). Previous measurements scatter around the selected values and are in good agreement. These include Strelkov et al. (12-298 K, 9), Clusius et al. (10-273 K, 10) and Southard and Nelson (17-285 K, 11). For other less extensive data see the review by Kelley and King (12). The entropy is based on the extrapolation $S°(2.8\ K) = 0.0006\ cal\ K^{-1}\ mol^{-1}$.

High temperature values are based on enthalpy data (373-1038 K) from Douglas (13). These values join smoothly with the low temperature values and are in excellent agreement with the enthalpy data (673-933 K) of Skuratov and Lapushkin (14) and the adiabatic heat capacities (334-721 K) of Mustajoki (15).

Fusion Data

The melting point and enthalpy of melting are from Douglas (13). Values of T_{fus} = 1043 K and 1045 K have been reported by Aukrust et al. (16) and Johnson and Bredig (17), respectively. The enthalpy of melting is confirmed by $\Delta_{fus}H° = 6.34\ kcal\ mol^{-1}$ obtained from enthalpy data by Dworkin and Bredig (18) and $\Delta_{fus}H°$ = 6.4 and 6.5 kcal mol^{-1} based on enthalpy and phase diagram studies of Aukrust et al. (16), respectively.

References

1. S. R. Gunn and L. G. Green, J. Am. Chem. Soc. 80, 4782 (1957).
2. S. R. Gunn, UCRL-7992, Livermore, Calif., July 1964.
3. C. E. Messer, L. G. Fasolino and C. E. Thalmayer, J. Amer. Chem. Soc. 77, 4524 (1955).
4. E. E. Ketchen and W. E. Wallace, J. Amer. Chem. Soc. 76, 4736 (1954).
5. C. E. Vanderzee and J. D. Nutter, J. Phys. Chem. 67, 2521 (1963).
6. S. R. Gunn and L. G. Green, J. Chem. Eng. Data 8, 180 (1963).
7. V. B. Parker, NSRDS-NBS 2, 1965.
8. W. T. Berg and J. A. Morrison, Proc. Roy. Soc. (London) A242, 467 (1957).
9. P. G. Strelkov, E. S. Itskevich, V. N. Kostryukov, and G. G. Mirskaya, Zhur. Fiz. Khim. 28 645 (1954).
10. K. Clusius, J. Goldman, and A. Perlick, Z. Naturforsch, 4a, 429 (1940).
11. J. C. Southard and R. A. Nelson, J. Am. Chem. Soc. 55, 4865 (1933).
12. K. K. Kelley and E. G. King, U.S. Bur. Mines Bull. 592, 1961.
13. T. B. Douglas, personal communication, U. S. Nat. Bur. Stand., May 23, 1961.
14. S. M. Skuratov and S. A. Lapushkin, J. Gen. Chem. (USSR) 21, 2485 (1951).
15. A. Mustajoki, Ann. Acad. Scient. Fennicae, Series A, I, 98, 7 (1951).
16. E. Aukrust, B. Bjorge, H. Flood, and T. Forland, Ann. New York Acad. Sci. 79, 830 (1960).
17. J. W. Johnson and M. A. Bredig, J. Phys. Chem. 62, 604 (1958).
18. A. S. Dworkin and M. A. Bredig, J. Phys. Chem. 64, 269 (1960).

Potassium Chloride (KCl) $Cl_1K_1(cr)$

Enthalpy Reference Temperature = T_r = 298.15 K Standard State Pressure = p° = 0.1 MPa

T/K	$C_p°$	S°	$-[G°-H°(T_r)]/T$	$H°-H°(T_r)$	$\Delta_f H°$	$\Delta_f G°$	Log K_f
	J $K^{-1}mol^{-1}$			kJ mol^{-1}			
0	0.	0.	INFINITE	-11.368	-436.379	-436.379	INFINITE
100	39.200	31.773	125.871	-9.410	-437.610	-427.776	223.447
200	48.438	62.630	87.169	-4.908	-437.221	-418.052	109.184
298.15	51.287	82.554	82.554	0.	-436.684	-408.761	71.613
300	51.329	82.872	82.555	0.095	-436.675	-408.587	71.141
400	53.137	97.895	84.592	5.321	-438.662	-398.837	52.083
500	54.752	109.927	88.494	10.717	-438.160	-388.935	40.632
600	56.317	120.048	92.931	16.270	-437.462	-379.152	33.008
700	58.003	128.852	97.447	21.984	-436.580	-369.501	27.573
800	60.078	136.727	101.873	27.883	-435.508	-359.989	23.505
900	62.911	143.957	106.152	34.024	-434.210	-350.625	20.350
1000	66.969	150.782	110.276	40.505	-432.610	-341.420	17.834
1044.000	69.287	153.713	112.045	43.502	------ CRYSTAL <--> LIQUID -----		
1100	71.965	157.406	114.261	47.460	-509.549	-327.785	15.565
1200	76.149	163.853	118.126	54.872	-506.096	-311.410	13.555
1300	79.496	170.086	121.885	62.662	-502.272	-295.339	11.867
1400	82.006	176.075	125.543	70.744	-498.159	-279.574	10.431
1500	83.680	181.794	129.104	79.035	-493.843	-264.111	9.197

PREVIOUS: December 1961 CURRENT: March 1966

Potassium Chloride (KCl) $Cl_1K_1(cr)$

POTASSIUM CHLORIDE (KCl) LIQUID M_r = 74.5513

$S°$(298.15 K) =[86.657] J K^{-1} mol^{-1} $\Delta_fH°$(298.15 K) = [-421.791]kJ mol^{-1}

$T_{fus}H°$ = 1044 K $\Delta_{fus}H°$ = 26.284 kJ mol^{-1}

Enthalpy of Formation

The enthalpy of formation was calculated from that of the crystal by adding $\Delta_{fus}H°$ and the difference in enthalpy, H°(1044 K) - H°(298.15 K), between the crystal and liquid.

Heat Capacity and Entropy

The constant heat capacity, 17.59 cal K^{-1} mol^{-1}, was derived from enthalpy data (1053-1173 K) of Douglas (1). This value was extrapolated above and below the experimental range. A lower value of 16.7 cal K^{-1} mol^{-1} was derived by Dworkin (2) from enthalpy data (1050-1110 K) obtained in an enthalpy of melting study. The entropy of the liquid is calculated in a manner analogous to that used for enthalpy of formation.

Fusion Data

The melting point and enthalpy of melting are from Douglas (1). Values of T_{fus} = 1043 K and 1045 K have been reported by Aukrust et al. (3) and Johnson and Bredig (4), respectively. The enthalpy of melting is confirmed by $\Delta_{fus}H°$ = 6.34 kcal mol^{-1} obtained from enthalpy data by Dworkin and Bredig (5) and $\Delta_{fus}H°$ = 6.4 and 6.5 kcal mol^{-1} based on enthalpy and phase diagram studies of Aukrust et al. (3), respectively.

Vaporization Data

T_{vap}(equilibrium vapor) = ~1710 K is the temperature resulting in unit total pressure based on the calculated partial pressures of KCl(g) and K_2Cl_2(g). This value may be compared with boiling points of 1688 and 1690 K observed respectively by Ruff and Mugdan (6) and von Wartenburg and Albrecht (7). $\Delta_{vap}H°$ (equilibrium vapor) at the boiling point is calculated as the enthalpy of vaporization of one mole of liquid to vapor containing 22.3 mole percent of dimer. According to the selected functions, the amount of dimer at lower temperatures increases gradually to about 32 mole percent at the melting point then decreases rapidly over the crystal. T_{vap}(monomer) = ~1750 K is taken as the temperature at which the calculated Gibbs energy change is zero for KCl(ℓ) → KCl(g), while $\Delta_{vap}H°$ (monomer) is the corresponding enthalpy of vaporization.

Data on vapor pressures and dimer-monomer equilibrium are reviewed on the tables for KCl(g) and K_2Cl_2(g), respectively.

References

1. T. B. Douglas, personal communication, U. S. Nat. Bur. Stand., May 23, 1961.
2. A. S. Dworkin, personal communication, Oak Ridge National Laboratory, Dec. 1, 1964.
3. E. Aukrust, B. Bjorge, H. Flood, and T. Forland, Ann. New York Acad. Sci. 79, 830 (1960).
4. J. W. Johnson and M. A. Bredig, J. Phys. Chem. 62, 604 (1958).
5. A. S. Dworkin and M. A. Bredig, J. Phys. Chem. 64, 269 (1960).
6. O. Ruff and S. Mugdan, Z. anorg. allgem. Chem. 117, 147 (1921).
7. H. von Wartenburg and P. Albrecht, Z. Elektrochem. 27, 162 (1921).

Potassium Chloride (KCl) Cl_1K_1(l)

Enthalpy Reference Temperature = T_r = 298.15 K Standard State Pressure = p° = 0.1 MPa

T/K	$C_p°$ (J $K^{-1}mol^{-1}$)	S° (J $K^{-1}mol^{-1}$)	$-[G°-H°(T_r)]/T$ (J $K^{-1}mol^{-1}$)	$H°-H°(T_r)$ (kJ mol^{-1})	$\Delta_fH°$ (kJ mol^{-1})	$\Delta_fG°$ (kJ mol^{-1})	Log K_f
0							
100							
200							
298.15	73.597	86.657	86.657	0.	-421.791	-395.090	69.218
300	73.597	87.112	86.658	0.136	-421.740	-394.925	68.763
400	73.597	108.285	89.545	7.496	-421.594	-385.925	50.397
500	73.597	124.707	94.996	14.855	-419.128	-377.293	39.415
600	73.597	138.126	101.100	22.215	-416.624	-369.160	32.138
700	73.597	149.471	107.221	29.575	-414.096	-361.449	26.972
800	73.597	159.298	113.130	36.934	-411.563	-354.101	23.120
900	73.597	167.966	118.751	44.294	-409.046	-347.070	20.143
1000	73.597	175.721	124.067	51.654	-406.568	-340.317	17.776
1044.000	73.597	178.890	126.311	54.892	------ CRYSTAL <--> LIQUID -----		
1100	73.597	182.735	129.087	59.013	-483.102	-329.200	15.632
1200	73.597	189.139	133.828	66.373	-479.702	-315.359	13.727
1300	73.597	195.030	138.312	73.733	-476.307	-301.801	12.127
1400	73.597	200.484	142.561	81.092	-472.917	-288.505	10.764
1500	73.597	205.561	146.593	88.452	-469.533	-275.451	9.592
1600	73.597	210.311	150.429	95.812	-466.154	-262.622	8.574
1700	73.597	214.773	154.084	103.171	-462.781	-250.005	7.682
1800	73.597	218.980	157.574	110.531	-459.415	-237.586	6.895
1900	73.597	222.959	160.911	117.891	-456.058	-225.354	6.195
2000	73.597	226.734	164.109	125.250	-452.711	-213.298	5.571

PREVIOUS: December 1961 CURRENT: March 1966

Potassium Chloride (KCl) Cl_1K_1(l)

POTASSIUM CHLORIDE (KCl) CRYSTAL–LIQUID M_r = 74.5513 Potassium Chloride (KCl) Cl_1K_1(cr,l)

0 to	1044 K	crystal
above	1044 K	liquid

Refer to the individual tables for details.

Enthalpy Reference Temperature = T_r = 298.15 K Standard State Pressure = p° = 0.1 MPa

	J K⁻¹mol⁻¹			kJ mol⁻¹			
T/K	C_p°	S°	$-[G^\circ-H^\circ(T_r)]/T$	$H^\circ-H^\circ(T_r)$	Δ_fH°	Δ_fG°	Log K_f
0	0.	0.	INFINITE	-11.368	-436.379	-436.379	INFINITE
100	39.200	31.773	125.871	-9.410	-437.610	-427.776	223.447
200	48.438	62.630	87.169	-4.908	-437.221	-418.052	109.184
298.15	51.287	82.554	82.554	0.	-436.684	-408.761	71.613
300	51.329	82.872	82.555	0.095	-436.675	-408.587	71.141
400	53.137	97.895	84.592	5.321	-438.662	-398.837	52.083
500	54.752	109.927	88.494	10.717	-438.160	-388.935	40.632
600	56.317	120.048	92.931	16.270	-437.462	-379.152	33.008
700	58.003	128.852	97.447	21.984	-436.580	-369.501	27.573
800	60.078	136.727	101.873	27.883	-435.508	-359.989	23.505
900	62.911	143.957	106.152	34.024	-434.210	-350.625	20.350
1000	66.969	150.782	110.276	40.505	-432.610	-341.420	17.834
1044.000	69.287	153.713	112.045	43.502	CRYSTAL <--> LIQUID		
1044.000	73.597	178.890	112.045	69.786	TRANSITION		
1100	73.597	182.735	115.547	73.907	-483.102	-329.200	15.632
1200	73.597	189.139	121.417	81.267	-479.702	-315.359	13.727
1300	73.597	195.030	126.856	88.626	-476.307	-301.801	12.127
1400	73.597	200.484	131.922	95.986	-472.917	-288.505	10.764
1500	73.597	205.561	136.664	103.346	-469.533	-275.451	9.592
1600	73.597	210.311	141.121	110.705	-466.154	-262.622	8.574
1700	73.597	214.773	145.323	118.065	-462.781	-250.005	7.682
1800	73.597	218.980	149.299	125.425	-459.415	-237.586	6.895
1900	73.597	222.959	153.072	132.784	-456.058	-225.354	6.195
2000	73.597	226.734	156.662	140.144	-452.711	-213.298	5.571

PREVIOUS: CURRENT: March 1966

Potassium Chloride (KCl) Cl_1K_1(cr,l)

POTASSIUM CHLORIDE (KCl) IDEAL GAS M_r = 74.5513

$\Delta_f H°(0\ K)$ = -212.89 ± 0.4 kJ mol^{-1}

$S°(298.15\ K)$ = 239.087 J K^{-1} mol^{-1} $\Delta_f H°(298.15\ K)$ = -214.68 ± 0.4 kJ mol^{-1}

Electronic Level and Quantum Weight

State	ε_i, cm^{-1}	g_i
$^1\Sigma$	0	1

ω_e = 278.6 cm^{-1} $\omega_e x_e$ = 1.157 cm^{-1} σ = 1

B_e = 0.127536 cm^{-1} α_e = 0.0007799 r_e = 2.6667 Å

Enthalpy of Formation

The enthalpy of formation was calculated from that of the liquid based on $\Delta_{vap}H°(298.15\ K)$ = 49.5 kcal mol^{-1} as selected from vapor pressure analyses summarized below. Partial pressures for the monomer were obtained from data of the first four sources by combination of total vapor pressures with PVT data or with transpiration data. Mass spectrometric values (5) are based on temperature variation of ion intensity. These same five studies were used in selection of $\Delta_r H°(298.15\ K)$ = 45.0 kcal mol^{-1} for the dimer-monomer equilibrium (see K_2Cl_2).

The resulting tables were tested by comparison with total pressure data over the range 819-1691 K. Dimer-monomer ratios calculated from the tables were used to convert observed total pressures to the monomer pressures which are analyzed below (6-11). The results are consistent except near the boiling point where the observed pressures (10, 11) are 6-18 percent larger than those calculated. Knudsen effusion and transpiration studies over the crystal were not analyzed since they are in essential agreement with the torsion effusion data of source 6.

Source	Method	Data Points	T/K	$\Delta_{vap}H°(298.15\ K)$, kcal mol^{-1} 2nd law	3rd law	Drift cal K^{-1} mol^{-1}
(1)	PVT + Vap. Press. (a)	15	1310-1404	51.04±0.40	49.67	-1.0±0.4
(2)	Trans. + Vap. Press.	10	1253-1352	50.75±0.80	49.49	-1.0±0.6
(3)	Trans. + Vap. Press. (b)	5	1261-1372	50.26±0.77	49.28	-0.8±0.6
(4)	Trans. + Vap. Press.	7	1250-1473	49.67±0.66	49.46	-0.1±0.5
(5)	Mass Spectrometry	-	774-1016	49.6 ±1.8	--	--
(6)	Total Vap. Press.	137	819-945	49.5	49.57	0.06
(7)	Total Vap. Press.	9	847-936	49.0	49.95	1.0±0.6
(8)	Total Vap. Press.	9	1179-1378	50.3	49.44	-0.8±0.2
(9)	Total Vap. Press.	9	1170-1466	51.0	49.42	-1.2±0.2
(10)	Total Vap. Press.	7	1393-1688	53.6	49.48	-2.7±0.4
(11)	Total Vap. Press.	16	1389-1691	50.2	49.22	-0.8±0.5

(a) Total vapor pressure (2, 3, 8).   (b) 1235 K was omitted.

Heat Capacity and Entropy

Rotational and vibrational constants are based on the microwave studies of Clouser and Gordy (12). Earlier microwave data of Honig et al. (13) and electric resonance data of Lee et al. (14) are in excellent agreement. Infrared spectra of Rice and Klemperer (15) yield θ_e = 261±6 cm^{-1} which is in good agreement. The constants were adjusted to a natural isotopic composition of 24.4 percent ^{37}Cl and 6.8 percent ^{41}K.

References

1. S. Datz, ORNL-2933, Oak Ridge National Laboratory, 1960.
2. E. E. Schrier, Ph.D. Thesis, Rensselaer Polytechnic Inst., Troy, New York, June 1961.
3. C. Beusman, ORNL-2323, Oak Ridge National Laboratory, 1957.
4. J. L. Barton and H. Bloom, J. Phys. Chem. 63, 1785 (1959); 60, 1413 (1956).
5. T. A. Milne and H. M. Klein, AFOSR TN 60-512, Menlo Park, California, 1960.
6. A. C. P. Pugh and R. F. Barrow, Trans. Faraday Soc. 54, 671 (1958).
7. V. Dietz, J. Chem. Phys. 4, 575 (1936).
8. E. F. Fiock and W. H. Rodebush, J. Am. Chem. Soc. 48, 2522 (1926).
9. G. I. Novikov and O. G. Polyachenok, Russ. J. Inorg. Chem. 6, 996 (1961).
10. O. Ruff and S. Mugdan, Z. anorg. allgem. Chem. 117, 149 (1921).
11. H. von Wartenburg and P. Albrecht, Z. elektrochem. 27, 162 (1921).
12. P. L. Clouser and W. Gordy, Phys. Rev. 134, A863 (1964).
13. A. Honig, M. Mandel, M. L. Stitch, and C. H. Townes, Phys. Rev. 96, 629 (1954).
14. C. A. Lee, B. P. Fabricand, R. O. Carlson, and I. I. Rabi, Phys. Rev. 91, 1395 (1953).
15. S. A. Rice and W. Klemperer, J. Chem. Phys. 27, 573 (1957).

Potassium Chloride (KCl) $Cl_1K_1(g)$

Enthalpy Reference Temperature = T_r = 298.15 K   Standard State Pressure = p° = 0.1 MPa

T/K	$C_p°$ (J K^{-1}mol^{-1})	S°	$-[G°-H°(T_r)]/T$	$H°-H°(T_r)$ (kJ mol^{-1})	$\Delta_f H°$	$\Delta_f G°$	Log K_f
0	0.	0.	INFINITE	-9.885	-212.893	-212.893	INFINITE
100	31.699	201.456	270.564	-6.911	-213.108	-220.243	115.043
200	35.270	224.739	242.394	-3.531	-213.841	-227.094	59.311
250	36.034	232.698	239.686	-1.747	-214.241	-230.362	48.131
298.15	36.496	239.087	239.087	0.	-214.681	-233.428	40.896
300	36.510	239.313	239.088	0.068	-214.700	-233.544	40.664
350	36.829	244.967	239.534	1.902	-217.616	-236.547	35.303
400	37.056	249.900	240.528	3.749	-218.231	-239.209	31.237
450	37.228	254.275	241.817	5.606	-218.826	-241.795	28.067
500	37.362	258.205	243.263	7.471	-219.403	-244.316	25.524
600	37.567	265.036	246.339	11.218	-220.511	-249.194	21.694
700	37.721	270.839	249.435	14.983	-221.578	-253.890	18.945
800	37.848	275.884	252.432	18.761	-222.626	-258.434	16.874
900	37.960	280.349	255.291	22.552	-223.679	-262.847	15.255
1000	38.061	284.354	258.000	26.353	-224.759	-267.141	13.954
1100	38.156	287.986	260.564	30.164	-304.842	-266.715	12.665
1200	38.246	291.309	262.989	33.984	-304.981	-263.243	11.459
1300	38.333	294.374	265.287	37.813	-305.117	-259.759	10.437
1400	38.417	297.218	267.468	41.651	-305.250	-256.265	9.561
1500	38.500	299.871	269.541	45.496	-305.379	-252.762	8.802
1600	38.582	302.359	271.515	49.351	-305.505	-249.250	8.137
1700	38.662	304.700	273.399	53.213	-305.630	-245.730	7.550
1800	38.742	306.912	275.200	57.083	-305.753	-242.203	7.029
1900	38.821	309.009	276.924	60.961	-305.878	-238.669	6.561
2000	38.900	311.002	278.579	64.847	-306.005	-235.129	6.141
2100	38.978	312.902	280.168	68.741	-306.136	-231.582	5.760
2200	39.056	314.717	281.698	72.643	-306.276	-228.028	5.414
2300	39.134	316.455	283.172	76.552	-306.425	-224.468	5.098
2400	39.211	318.122	284.593	80.470	-306.587	-220.902	4.808
2500	39.288	319.725	285.967	84.395	-306.766	-217.328	4.541
2600	39.365	321.267	287.295	88.327	-306.965	-213.746	4.294
2700	39.442	322.754	288.581	92.268	-307.187	-210.157	4.066
2800	39.519	324.190	289.827	96.216	-307.436	-206.559	3.853
2900	39.595	325.578	291.036	100.171	-307.714	-202.951	3.656
3000	39.672	326.922	292.210	104.135	-308.027	-199.333	3.471
3100	39.748	328.224	293.351	108.106	-308.376	-195.704	3.298
3200	39.825	329.487	294.461	112.084	-308.766	-192.064	3.135
3300	39.901	330.714	295.541	116.071	-309.190	-188.411	2.982
3400	39.977	331.906	296.593	120.065	-309.667	-184.744	2.838
3500	40.053	333.066	297.618	124.066	-310.191	-181.062	2.702
3600	40.130	334.195	298.619	128.075	-310.770	-177.365	2.573
3700	40.206	335.296	299.595	132.092	-311.404	-173.651	2.452
3800	40.282	336.369	300.549	136.116	-312.099	-169.918	2.336
3900	40.358	337.416	301.481	140.148	-312.856	-166.167	2.226
4000	40.434	338.439	302.392	144.188	-313.673	-162.396	2.121
4100	40.510	339.438	303.283	148.235	-314.560	-158.604	2.021
4200	40.586	340.415	304.156	152.290	-315.524	-154.788	1.925
4300	40.662	341.371	305.010	156.352	-316.565	-150.949	1.834
4400	40.738	342.307	305.847	160.422	-317.408	-147.112	1.746
4500	40.814	343.223	306.668	164.500	-318.543	-143.229	1.663
4600	40.889	344.121	307.472	168.585	-319.751	-139.320	1.582
4700	40.965	345.001	308.261	172.678	-321.016	-135.386	1.505
4800	41.041	345.865	309.036	176.778	-322.247	-131.436	1.430
4900	41.117	346.712	309.796	180.886	-323.658	-127.446	1.359
5000	41.193	347.543	310.543	185.001	-325.144	-123.427	1.289
5100	41.269	348.360	311.276	189.125	-326.705	-119.377	1.223
5200	41.344	349.162	311.997	193.255	-328.343	-115.296	1.158
5300	41.420	349.950	312.706	197.393	-330.056	-111.183	1.096
5400	41.496	350.725	313.403	201.539	-331.844	-107.037	1.035
5500	41.572	351.487	314.088	205.693	-333.706	-102.857	0.977
5600	41.648	352.237	314.763	209.854	-335.640	-98.642	0.920
5700	41.723	352.975	315.427	214.022	-337.300	-94.440	0.865
5800	41.799	353.701	316.080	218.198	-339.330	-90.162	0.812
5900	41.875	354.416	316.724	222.382	-341.422	-85.848	0.760
6000	41.951	355.120	317.358	226.573	-343.575	-81.498	0.710

PREVIOUS: March 1966 (1 atm)   CURRENT: March 1966 (1 bar)

Potassium Chloride (KCl) $Cl_1K_1(g)$

POTASSIUM PERCHLORATE ($KClO_4$) CRYSTAL M_r = 138.5489

$S°(298.15 K) = 151.04 \pm 0.8 J K^{-1} mol^{-1}$

$T_{trs} = 572.7 K$

$T_{fus} = 798 K$

$\Delta_f H°(0 K) = -422.1 \pm 1.0 kJ mol^{-1}$

$\Delta_f H°(298.15 K) = -430.1 \pm 4.0 kJ mol^{-1}$

$\Delta_{trs} H° = 13.77 kJ mol^{-1}$

$\Delta_{fus} H°$ = Unknown

Enthalpy of Formation

The enthalpy of decomposition of $KClO_4$(cr) to KCl(cr) and O_2(g) was measured by Johnson and Gilliland (1), yielding $\Delta_f H°(298.15 K) = -103.22 \pm 0.15 kcal mol^{-1}$. Hofmann and Martin (2) burned a mixture of paraffin and potassium perchlorate and obtained $\Delta_f E(298.15 K) = -1.73 kcal mol^{-1}$ for the constant volume process corresponding to the reaction: $KClO_4$(cr) = KCl(cr) + 2 O_2(g). Correction of their value for $\Delta(pV)$ gives $\Delta_r H°(298.15 K) = -0.55 kcal mol^{-1}$ which agrees reasonably well to the corresponding value, $-0.96 \pm 0.10 kcal mol^{-1}$, reported by Johnson and Gilliland (1). However, for the same reaction, $\Delta_r H°(298.15 K) = -2.55 \pm 0.18 kcal mol^{-1}$ was determined by Skuratov et al. (3). They measured this quantity by two different methods which gave concordant results. The value of $\Delta_f H°(298.15 K)$ reported was $-101.9 \pm 0.2 kcal mol^{-1}$. The value used here is the average of these four results.

Berthelot and Vieille (4) determined the difference between the enthalpy of explosion of ammonium picrate and that of an ammonium picrate-potassium perchlorate mixture, and also the difference between the enthalpy of explosion of potassium picrate and that of a potassium pricrate-potassium perchlorate mixture. They obtained $\Delta_r H° = 7.5 kcal mol^{-1}$ for the reaction $KClO_4$(cr) = KCl(cr) + 2 O_2(g). From that the value of $\Delta_f H°(298.15 K)$ was calculated to be $-111.9 kcal mol^{-1}$. This value was not used.

Heat Capacity and Entropy

The low temperature heat capacities (12.53-298.07 K) were measured by Latimer and Ahlberg (5). The heat capacities above 298.15 K were estimated by comparison with those of $NaClO_4$(cr). S°(298.15 K) was reported by Kelley et al. (6), based on the low temperature heat capacities measured by Latimer and Ahlberg (5), using $S°(12.59 K, extrap) = 0.14 cal K^{-1} mol^{-1}$.

Transition Data

T_{trs} and $\Delta_{trs}H°$ were obtained from the U. S. National Bureau of Standards (7).

Fusion Data

T_{fus} was taken from the National Research Council (8).

References

1. W. H. Johnson and A. A. Gilliland, J. Res. Natl. Bur. Std. 65A, 63 (1961).
2. K. A. Hofmann and P. H. Martin, Sitzber. preuss. Akad. Wiss. Physik. math. Kl. 448 (1932).
3. S. M. Skuratov, A. F. Vorob'ev and N. M. Privalova, Russ. J. Inorg. Chem. 7, 343 (1962).
4. M. Berthelot and P. Vieille, Ann. chim. et phys. [5] 27, 225 (1882).
5. W. M. Latimer and J. E. Ahlberg, J. Am. Chem. Soc. 52, 549 (1930).
6. K. K. Kelley and E. G. King, U. S. Bur. Mines Bull. 592, 1961.
7. U. S. Nat. Bur. Stand. Circ. 500, 1952.
8. Bulletin of the National Research Council, "Data on Chemicals for Ceramic Use", No. 118, June, 1949.

Potassium Perchlorate ($KClO_4$) $Cl_1K_1O_4$(cr)

Enthalpy Reference Temperature = T_r = 298.15 K — Standard State Pressure = p° = 0.1 MPa

T/K	$C_p°$ (J K⁻¹mol⁻¹)	S° (J K⁻¹mol⁻¹)	$-[G°-H°(T_r)]/T$ (J K⁻¹mol⁻¹)	$H°-H°(T_r)$ (kJ mol⁻¹)	$\Delta_f H°$ (kJ mol⁻¹)	$\Delta_f G°$ (kJ mol⁻¹)	Log K_f
0	0.	0.	INFINITE	-21.071	-422.147	-422.147	INFINITE
100	64.936	59.404	235.342	-17.594	-427.667	-385.936	201.592
200	88.408	112.278	161.419	-9.828	-429.837	-343.204	89.636
298.15	110.207	151.042	151.042	0.	-430.115	-300.282	52.608
300	112.968	151.733	151.045	0.206	-430.104	-299.477	52.144
400	138.490	187.967	155.809	12.863	-430.602	-255.709	33.392
500	153.971	220.606	165.568	27.519	-426.958	-212.379	22.187
572.700	162.758	242.115	173.939	39.045		I <--> II	
572.700	162.661	266.151	173.939	52.810		TRANSITION	
600	165.268	273.786	178.309	57.286	-408.365	-170.557	14.848
700	174.316	299.954	193.849	74.273	-402.719	-131.358	9.802
800	182.422	323.770	208.622	92.118	-396.375	-93.018	6.073
900	189.483	345.671	222.649	110.719	-389.427	-55.510	3.222
1000	195.811	365.969	235.979	129.990	-381.961	-18.803	0.982
1100	201.195	384.887	248.666	149.843	-453.021	21.744	-1.033
1200	206.271	402.613	260.764	170.219	-443.702	64.496	-2.807
1300	211.041	419.314	272.324	191.087	-433.965	106.454	-4.277
1400	215.504	435.119	283.392	212.417	-423.832	147.647	-5.509
1500	219.660	450.131	294.012	234.178	-413.329	188.102	-6.550

PREVIOUS: CURRENT: June 1963

Potassium Perchlorate ($KClO_4$) $Cl_1K_1O_4$(cr)

LITHIUM CHLORIDE (LiCl) CRYSTAL M_r = 42.394

$\Delta_f H°(0\ K) = -408.359 \pm 1.142\ J\ mol^{-1}$
$S°(298.15\ K) = 59.300 \pm 0.063\ J\ K^{-1}\ mol^{-1}$ $\Delta_f H°(298.15\ K) = -408.266 \pm 1.142\ kJ\ mol^{-1}$
$T_{fus} = 883 \pm 2\ K$ $\Delta_{fus}H° = 19.832 \pm 0.42\ kJ\ mol^{-1}$

Enthalpy of Formation

The enthalpy of formation was obtained as follows:

	$\Delta_f H°$(298.15 K), kcal mol^{-1}	Source
(1) $LiOH(aq, 100\ H_2O) + HCl(aq, 100\ H_2O) \rightarrow LiCl(aq, 201\ H_2O)$	-13.704 ± 0.014	(1)
(2) $Li(cr) + (aq, 1001\ H_2O)(\ell) \rightarrow LiOH\cdot(aq, 1000\ H_2O) + 1/2\ H_2(g)$	-53.142 ± 0.019	(2)
(3) $LiOH\cdot(aq, 1000\ H_2O) \rightarrow LiOH\cdot(aq, 100\ H_2O) + (aq, 900\ H_2O)(\ell)$	0.153 ± 0.040	(3)
(4) $1/2\ H_2(g) + 1/2\ Cl_2(g) \rightarrow HCl(g)$	-21.97 ± 0.09	(4)
(5) $HCl(g) + (aq, 100\ H_2O)(\ell) \rightarrow HCl\cdot(aq, 100\ H_2O)$	-17.575 ± 0.100	(5, 6)
(6) $LiCl(aq, 201\ H_2O) \rightarrow LiCl(cr) + (aq, 201\ H_2O)(\ell)$	8.660 ± 0.010	(7)
(1)+(2)+(3)+(4)+(5)+(6) gives Li(cr) + 1/2 Cl (g) → LiCl(cr)	-97.578 ± 0.273	

A direct calorimetric determination by Siemonsen and Siemonsen (8) gave -94.8 ± 0.6 kcal mol^{-1}. However, the solution route was thought to be inherently more accurate and was adopted.

Heat Capacity and Entropy

The heat capacity between 15 and 325 K has been measured by Shirley (9) and by Oetting et al. (10). The two sets of data are in substantial agreement and were plotted together and used to evalute the entropy at 298. The extrapolation to 0 K was done assuming a T^3 law and yielded S°(15 K) = 0.019 cal K^{-1} mol^{-1}. Above 298, Douglas et al. (11) have made enthalpy measurements to 1200 K. Rodigina et al. (12) also made enthalpy determinations to 1100 K. These two sets of measurements were averaged and joined smoothly to the low temperature values.

Fusion and Transition Data

The melting point has been reported as follows: 880 K, Douglas et al. (11); 883 K, Rodigina et al. (12); 883 K, Haendler et al. (13); 879 K, Flood et al. (14); 879 K, von Wartenberg and Schulz (15). The highest value of 883 K was taken as representing the most pure samples. At this temperature the enthalpy equations of Douglas et al. yield a heat of fusion of 4.718 kcal mol^{-1}, those of Rodigina et al., 4.669 kcal mol^{-1}. Dworkin and Bredig (16) report a value of 4.76 ± 0.10 kcal mol^{-1} and quote 4.83 kcal mol^{-1} from a private communication from D. E. Smith. The average value $\Delta_{fus}H°$ = 4.74 ± 0.10 kcal mol^{-1} was adopted.

A transition at 838 K has been reported by Kislova and Bergman (17) but is not supported by the enthalpy measurements quoted above.

References

1. T. W. Richards and A. W. Rowe, J. Amer. Chem. Soc. 44, 684 (1922).
2. S. R. Gunn and L. G. Green, J. Amer. Chem. Soc. 80, 4782 (1958).
3. T. W. Richards and A. W. Rowek J. Amer. Chem. Soc. 43, 770 (1921).
4. JANAF Thermochemical Tables: H2(g), 3-31-61; Cl2(g), 3-31-61; HCl(g), 3-31-61
5. C. M. Slansky, J. Amer. Chem. Soc. 62, 2430 (1940).
6. J. M. Sturtevan, J. Amer. Chem. Soc. 62, 3265 (1940).
7. E. Lange and F. Durr, Z. Phys. Chem. 121, 361 (1926).
8. H. Siemonsen and U. Siemonsen, Z. Electrochem. 56, 643 (1952).
9. D. A. Shirley, J. Amer. Chem. Soc. 82, 3841 (1960).
10. F. L. Oetting, W. E. Hatton and G. C. Sinke, unpublished work, Thermal Laboratory, Dow Chemical Company.
11. T. B. Douglas, J. L. Dever and A. W. Harman, U. S. Nat. Bur. Stand. Rept. 6297, 1959
12. E. N. Rodigina, K. Z. Gomelski, and V. F. Luginina, Zhur. Neorg. Khim. 4, 975 (1959).
13. H. M. Haendler, P. S. Sennett and C. M. Wheeler, J. Electrochem. Soc. 106, 264 (1959).
14. H. Flood, O. Fykse and S. Urnes, Z. Electrochem. 59, 364 (1955).
15. H. von Wartenberg and H. Schulz, Z. Electrochem. 27, 568 (1921).
16. A. S. Dworkin and M. A. Bredig, J. Phys. Chem. 64, 269 (1960).
17. Kislova and Bergman, Russ. J. Inorg. Chem. 5, 1210 (1960).

Lithium Chloride (LiCl) $Cl_1Li_1(cr)$

Enthalpy Reference Temperature = T_r = 298.15 K Standard State Pressure = p° = 0.1 MPa

T/K	$C_p°$ (J $K^{-1}mol^{-1}$)	S°	$-[G°-H°(T_r)]/T$	$H°-H°(T_r)$ (kJ mol^{-1})	$\Delta_f H°$	$\Delta_f G°$	Log K_f
0	0.	0.	INFINITE	-9.305	-408.359	-408.359	INFINITE
100	36.313	13.585	98.646	-8.506	-409.511	-400.685	209.296
200	43.350	41.045	63.555	-4.502	-408.867	-392.114	102.410
298.15	48.028	59.300	59.300	0.	-408.266	-384.019	67.279
300	48.095	59.597	59.301	0.089	-408.254	-383.869	66.838
400	50.969	73.853	61.225	5.051	-407.628	-375.837	49.079
500	53.342	85.484	64.948	10.268	-410.130	-367.643	38.407
600	55.593	95.409	69.217	15.715	-409.482	-359.202	31.271
700	57.781	104.144	73.595	21.384	-408.574	-350.891	26.184
800	59.936	112.000	77.912	27.270	-407.434	-342.726	22.378
883.000	61.702	118.003	81.402	32.318	------ CRYSTAL <--> LIQUID -----		
900	62.061	119.183	82.105	33.370	-406.085	-334.716	19.426
1000	64.170	125.831	86.149	39.682	-404.528	-326.868	17.074
1100	66.266	132.045	90.042	46.204	-402.762	-319.185	15.157
1200	68.358	137.901	93.788	52.935	-400.788	-311.673	13.567
1300	70.452	143.455	97.397	59.876	-398.606	-304.334	12.228
1400	72.534	148.752	100.877	67.025	-396.213	-297.170	11.088
1500	74.598	153.827	104.239	74.381	-393.610	-290.186	10.105
1600	76.663	158.707	107.492	81.945	-390.796	-283.381	9.251
1700	78.731	163.417	110.644	89.714	-533.008	-269.583	8.283
1800	80.797	167.976	113.703	97.691	-529.021	-254.202	7.377
1900	82.863	172.399	116.676	105.874	-524.833	-239.047	6.572
2000	84.927	176.702	119.571	114.263	-520.446	-224.119	5.853

PREVIOUS: CURRENT: June 1962

Lithium Chloride (LiCl) $Cl_1Li_1(cr)$

LITHIUM CHLORIDE (LiCl) LIQUID M_r = 42.394

$S°(298.15\ K) = [78.435]\ J\ K^{-1}\ mol^{-1}$ $\Delta_f H°(298.15\ K) = [-390.755]\ kJ\ mol^{-1}$

$T_{fus} = 883 \pm 2\ K$ $\Delta_{fus}H° = 19.832 \pm 0.42\ kJ\ mol^{-1}$

Enthalpy of Formation

$\Delta_f H°$(LiCl, ℓ, 298.15 K) is calculated from that of the crystal by adding $\Delta_{fus}H°$ and the difference in enthalpy, H°(883 K) - H°(298.15 K), between the crystal and liquid.

Heat Capacity and Entropy

The heat capacity data from the melting point to 1100 K and 1200 K has been reported by Rodigina et al. (1) and Douglas et al. (2). The equation given by the former authors for the enthalpy of the liquid does not agree with their measurements and has been replaced by $H°(T;293.15) = 18.000T - 1.5 \times 10^{-3}T^2 - 2279\ cal\ mol^{-1}$, which fits to ±0.3%. The two sets of results were averaged and the equations were assumed to hold up to the normal boiling point. The entropy is calculated in a manner analogous to that used for the enthalpy of formation.

Fusion Data

See the crystal table for details.

Vaporization Data

The equilibrium boiling point, T_{vap} = 1656 K, was taken from von Wartenburg and Schultz (3). The boiling point of the monomeric gas is defined as the temperature at which it reaches 1 bar pressure. This was obtained from the Gibbs energy crossover between liquid and monomeric gas.

References

1. E. N. Rodigina, K. Z. Gomelski and V. F. Luginina, Zhur. Neorg. Khim. 4, 975 (1959).
2. T. B. Douglas, J. L. Dever, and A. W. Harman, U.S. Nat. Bur. Stand. Report 6297, 1959.
3. H. von Wartenburg and H. Schultz, Z. Electrochem. 27, 568 (1921).

Lithium Chloride (LiCl) $Cl_1Li_1(l)$

Enthalpy Reference Temperature = T_r = 298.15 K Standard State Pressure = p° = 0.1 MPa

T/K	$C_p°$ (J K⁻¹mol⁻¹)	S° (J K⁻¹mol⁻¹)	$-[G°-H°(T_r)]/T$ (J K⁻¹mol⁻¹)	$H°-H°(T_r)$ (kJ mol⁻¹)	$\Delta_f H°$ (kJ mol⁻¹)	$\Delta_f G°$ (kJ mol⁻¹)	Log K_f
0							
100							
200							
298.15	48.028	78.435	78.435	0.	-390.755	-372.214	65.210
300	48.095	78.732	78.436	0.089	-390.743	-372.099	64.788
400	50.969	92.989	80.360	5.051	-390.117	-365.980	47.792
500	53.342	104.621	84.084	10.269	-392.618	-359.700	37.578
588.000	55.239	113.421	87.830	15.048	GLASS <--> LIQUID		
588.000	67.807	113.421	87.830	15.048	TRANSITION		
600	67.693	114.790	88.355	15.861	-391.825	-353.174	30.747
700	66.747	125.154	92.893	22.583	-389.864	-346.888	25.885
800	65.802	134.005	97.492	29.210	-387.983	-340.878	22.257
883.000	65.015	140.463	101.233	34.639	------ CRYSTAL <--> LIQUID -----		
900	64.856	141.701	101.986	35.743	-386.201	-335.098	19.449
1000	63.911	148.485	106.304	42.182	-384.517	-329.511	17.212
1100	62.965	154.532	110.418	48.525	-382.929	-324.088	15.390
1200	62.015	159.970	114.325	54.774	-381.437	-318.806	13.877
1300	61.070	164.897	118.028	60.929	-380.041	-313.644	12.602
1400	60.124	169.388	121.539	66.988	-378.739	-308.586	11.513
1500	59.165	173.503	124.868	72.953	-377.527	-303.618	10.573
1600	58.216	177.291	128.028	78.822	-376.408	-298.727	9.752
1700	57.284	180.793	131.030	84.597	-520.614	-286.728	8.810
1800	56.346	184.040	133.886	90.278	-518.922	-273.020	7.923
1900	55.403	187.062	136.606	95.866	-517.330	-259.402	7.131
2000	54.455	189.879	139 200	101.359	-515.839	-245.866	6.421

PREVIOUS: CURRENT: June 1962

Lithium Chloride (LiCl) $Cl_1Li_1(l)$

LITHIUM CHLORIDE (LiCl) CRYSTAL–LIQUID M_r = 42.394

0 to 883 K crystal
above 883 K liquid

Refer to the individual tables for details.

Lithium Chloride (LiCl) Cl_1Li_1(cr,l)

Enthalpy Reference Temperature = T_r = 298.15 K Standard State Pressure = $p°$ = 0.1 MPa

T/K	$C_p°$ (J K⁻¹mol⁻¹)	$S°$ (J K⁻¹mol⁻¹)	$-[G°-H°(T_r)]/T$ (J K⁻¹mol⁻¹)	$H°-H°(T_r)$ (kJ mol⁻¹)	$\Delta_f H°$ (kJ mol⁻¹)	$\Delta_f G°$ (kJ mol⁻¹)	Log K_f
0	0.	0.	INFINITE	-9.305	-408.359	-408.359	INFINITE
100	36.313	13.585	98.646	-8.506	-409.511	-400.685	209.296
200	43.350	41.045	63.555	-4.502	-408.867	-392.114	102.410
298.15	48.028	59.300	59.300	0.	-408.266	-384.019	67.279
300	48.095	59.597	59.301	0.089	-408.254	-383.869	66.838
400	50.969	73.853	61.225	5.051	-407.628	-375.837	49.079
500	53.342	85.484	64.948	10.268	-410.130	-367.643	38.407
600	55.593	95.409	69.217	15.715	-409.482	-359.202	31.271
700	57.781	104.144	73.595	21.384	-408.574	-350.891	26.184
800	59.936	112.000	77.912	27.270	-407.434	-342.726	22.378
883.000	61.702	118.003	81.402	32.318	CRYSTAL <--> LIQUID		
883.000	65.015	140.463	81.402	52.150	TRANSITION		
900	64.856	141.701	82.529	53.254	-386.201	-335.098	19.449
1000	63.911	148.485	88.792	59.693	-384.517	-329.511	17.212
1100	62.965	154.532	94.499	66.036	-382.929	-324.088	15.390
1200	62.015	159.970	99.732	72.285	-381.437	-318.806	13.877
1300	61.070	164.897	104.558	78.440	-380.041	-313.644	12.602
1400	60.124	169.388	109.031	84.499	-378.739	-308.586	11.513
1500	59.165	173.503	113.194	90.464	-377.527	-303.618	10.573
1600	58.216	177.291	117.083	96.333	-376.408	-298.727	9.752
1700	57.284	180.793	120.729	102.108	-520.614	-286.728	8.810
1800	56.346	184.040	124.157	107.789	-518.922	-273.020	7.923
1900	55.403	187.062	127.390	113.377	-517.330	-259.402	7.131
2000	54.455	189.879	130.444	118.870	-515.839	-245.866	6.421

PREVIOUS: CURRENT: June 1962

Lithium Chloride (LiCl) Cl_1Li_1(cr,l)

$S°(298.15\ K) = 212.919\ J\ K^{-1}\ mol^{-1}$

$\Delta_fH°(0\ K) = -195.57\ kJ\ mol^{-1}$
$\Delta_fH°(298.15\ K) = -195.72 \pm 12.6\ kJ\ mol^{-1}$

Electronic Level and Quantum Weight

State	ε_i, cm^{-1}	g_i
$^1\Sigma$	0.0	1

$\omega_e = 641.1 \pm 3.0\ cm^{-1}$ $\omega_ex_e = 4.2 \pm 0.3\ cm^{-1}$ $\sigma = 1$
$B_e = 0.7053\ cm^{-1}$ $\alpha_e = 0.0067\ cm^{-1}$ $r_e = 2.022$ Å

Enthalpy of Formation

Klemperer et al. (1) quote a private communication from W. H. Evans at the National Bureau of Standards giving $D_0^0 = 4.35$ cal $K^{-1}\ mol^{-1}$ which corresponds to $\Delta_fH°(0\ K) = -45.3 \pm 2.4\ kcal\ mol^{-1}$. A 3rd law analysis of the vapor pressure data of Nesmeyanov and Sazonov (2) in conjunction with the relative concentration of monomer, dimer, and trimer at 870 K from the work of Miller and Kusch (3) leads to a monomer enthalpy of sublimation at 298.15 K of 52.56 kcal mol^{-1}, this gives $\Delta_fH°(298.15\ K) = -45.02\ kcal\ mol^{-1}$. These values lead, however, to partial vapor pressures of the monomer which are less than half of the total vapor pressure near the boiling point. From the data of Miller and Kusch (3) it would be expected that the monomer would be the predominant species near the boiling point, a fact also substantiated by the work of Beusman (4), on the average molecular weight of the vapor. Thus it was decided to change the enthalpy of sublimation so as to give the correct normal boiling point and vapor pressures in the region 1400-1600 K and to approximate the relative concentrations of monomer, dimer, and trimer as found by Miller and Kusch at 870 K. A monomer heat of sublimation at 298.15 K of 50.8 kcal mol^{-1} was chosen, which gives $\Delta_fH°(298.15\ K) = -46.778\ kcal\ mol^{-1}$. The normal boiling point was taken as 1656 K as determined by von Wartenburg and Schulz (5) and the boiling point of LiCl(g) was found as 1701 K from the Gibbs energy crossover between liquid and gas.

Heat Capacity and Entropy

The spectroscopic constants were taken from Klemperer et al. (1). The value of α_e was calculated by the value of a_1 given by the Rittner potential function and the equation given by Klemperer et al. (1). The value of r_e was taken from (6).

References

1. W. Klemperer, W. G. Norris, A. Buchler and A. G. Emslie, J. Chem. Phys. 33, 1534 (1960).
2. An. N. Nesmeyanov and L. A. Sazonov, Zhur. Neorg. Khim. 4, 231 (1960).
3. R. C. Miller and P. Kusch, J. Chem. Phys. 25, 860 (1956); 27, 981 (1957).
4. C. Beusman, AEC Report ORNL-2323, 1957.
5. H. Jon Wartenburg and H. Schulz, Z. Electrochem 27, 568 (1921).
6. T. B. Douglas, J. L. Dever, and A. W. Harman, U.S. Nat. Bur. Stand. Report 6297, 1959.

Enthalpy Reference Temperature = T_r = 298.15 K Standard State Pressure = p° = 0.1 MPa

	J K⁻¹mol⁻¹			kJ mol⁻¹			
T/K	$C_p°$	S°	$-[G°-H°(T_r)]/T$	$H°-H°(T_r)$	$\Delta_fH°$	$\Delta_fG°$	Log K_f
0	0.	0.	INFINITE	-9.061	-195.567	-195.567	INFINITE
100	29.194	179.425	240.943	-6.152	-194.609	-202.367	105.706
200	31.011	200.100	215.893	-3.159	-194.976	-210.035	54.855
250	32.231	207.153	213.461	-1.577	-195.333	-213.759	44.663
298.15	33.245	212.919	212.919	0.	-195.719	-217.274	38.065
300	33.280	213.125	212.920	0.062	-195.735	-217.408	37.854
350	34.119	218.321	213.328	1.747	-196.171	-220.986	32.980
400	34.778	222.922	214.246	3.470	-196.661	-224.498	29.316
450	35.295	227.049	215.443	5.223	-197.220	-227.945	26.459
500	35.705	230.790	216.793	6.998	-200.852	-231.019	24.134
600	36.304	237.356	219.688	10.601	-202.049	-236.938	20.627
700	36.715	242.985	222.624	14.253	-203.158	-242.664	18.108
800	37.014	247.908	225.483	17.940	-204.217	-248.235	16.208
900	37.242	252.281	228.222	21.653	-205.255	-253.675	14.723
1000	37.425	256.215	230.828	25.387	-206.276	-259.000	13.529
1100	37.577	259.789	233.301	29.137	-207.282	-264.223	12.547
1200	37.707	263.065	235.647	32.901	-208.274	-269.356	11.725
1300	37.821	266.087	237.874	36.678	-209.256	-274.406	11.026
1400	37.924	268.894	239.990	40.465	-210.226	-279.381	10.424
1500	38.019	271.514	242.005	44.262	-211.182	-284.288	9.900
1600	38.107	273.970	243.927	48.069	-212.124	-289.131	9.439
1700	38.189	276.283	245.763	51.884	-358.291	-286.739	8.810
1800	38.268	278.468	247.520	55.706	-358.458	-282.525	8.199
1900	38.344	280.539	249.204	59.537	-358.622	-278.302	7.651
2000	38.417	282.508	250.820	63.375	-358.787	-274.071	7.158
2100	38.487	284.384	252.374	67.220	-358.952	-269.831	6.712
2200	38.556	286.176	253.870	71.073	-359.121	-265.583	6.306
2300	38.624	287.891	255.312	74.932	-359.294	-261.327	5.935
2400	38.690	289.537	256.704	78.797	-359.474	-257.064	5.595
2500	38.755	291.117	258.049	82.670	-359.663	-252.793	5.282
2600	38.820	292.639	259.351	86.548	-359.863	-248.515	4.993
2700	38.883	294.105	260.611	90.433	-360.077	-244.228	4.725
2800	38.946	295.520	261.833	94.325	-360.306	-239.933	4.476
2900	39.009	296.888	263.018	98.223	-360.552	-235.630	4.244
3000	39.071	298.211	264.169	102.127	-360.818	-231.318	4.028
3100	39.132	299.493	265.288	106.037	-361.105	-226.996	3.825
3200	39.193	300.737	266.376	109.953	-361.414	-222.665	3.635
3300	39.254	301.944	267.436	113.875	-361.747	-218.324	3.456
3400	39.314	303.116	268.468	117.804	-362.104	-213.973	3.287
3500	39.375	304.257	269.475	121.738	-362.487	-209.611	3.128
3600	39.435	305.367	270.456	125.679	-362.895	-205.237	2.978
3700	39.494	306.448	271.414	129.625	-363.331	-200.852	2.836
3800	39.554	307.502	272.350	133.578	-363.793	-196.454	2.700
3900	39.613	308.531	273.265	137.536	-364.282	-192.044	2.572
4000	39.673	309.534	274.159	141.500	-364.798	-187.621	2.450
4100	39.732	310.515	275.034	145.471	-365.342	-183.185	2.334
4200	39.791	311.473	275.890	149.447	-365.914	-178.735	2.223
4300	39.850	312.410	276.729	153.429	-366.514	-174.272	2.117
4400	39.909	313.327	277.550	157.417	-367.121	-169.795	2.016
4500	39.967	314.224	278.355	161.410	-367.768	-165.304	1.919
4600	40.026	315.103	279.144	165.410	-368.444	-160.797	1.826
4700	40.084	315.965	279.919	169.416	-369.148	-156.276	1.737
4800	40.143	316.809	280.678	173.427	-369.880	-151.739	1.651
4900	40.201	317.637	281.424	177.444	-370.641	-147.186	1.569
5000	40.260	318.450	282.157	181.467	-371.431	-142.618	1.490
5100	40.318	319.248	282.876	185.496	-372.244	-138.034	1.414
5200	40.376	320.031	283.583	189.531	-373.093	-133.434	1.340
5300	40.434	320.801	284.278	193.571	-373.974	-128.816	1.270
5400	40.492	321.557	284.962	197.618	-374.887	-124.182	1.201
5500	40.550	322.301	285.634	201.670	-375.833	-119.531	1.135
5600	40.608	323.032	286.295	205.728	-376.814	-114.862	1.071
5700	40.666	323.751	286.946	209.791	-377.831	-110.175	1.010
5800	40.724	324.459	287.587	213.861	-378.886	-105.470	0.950
5900	40.782	325.156	288.217	217.936	-379.978	-100.747	0.892
6000	40.840	325.842	288.839	222.018	-381.111	-96.005	0.836

PREVIOUS: June 1962 (1 atm) CURRENT: June 1962 (1 bar)

LITHIUM HYPOCHLORITE (LiOCl) IDEAL GAS M_r = 58.3934

$\Delta_f H°(0\ K) = -11.5 \pm 84\ kJ\ mol^{-1}$

$S°(298.15\ K) = [256.417]\ J\ K^{-1}\ mol^{-1}$ $\Delta_f H°(298.15\ K) = [-14.2 \pm 84]\ kJ\ mol^{-1}$

Vibrational Levels and Degeneracies

ν, cm^{-1}
[1150](1)
[500](1)
[700](1)

Ground State Quantum Weight: [1] σ = 1

Point Group: [Cs]

Bond Distances: Li–O = 1.82Å O–Cl = 1.73Å

Bond Angles: Li-O-Cl = 115°

Product of the Principal Moments of Inertia: $I_A I_B I_C = 1.83923 \times 10^{-115}\ g^3\ cm^6$

Enthalpy of Formation

Data from Hildenbrand. (1)

Heat Capacity and Entropy

Data from Hildenbrand. (1) The principal moments of inertia are: $I_A = 1.4098 \times 10^{-39}$, $I_B = 10.7387 \times 10^{-39}$, and $I_C = 12.1485 \times 10^{-39}\ g\ cm^2$.

Reference

1. D. L. Hildenbrand, "Tables of Thermodynamic Data for Performance Calculations, "Aeronutronic Division of Ford Motor Company, Newport Beach, Calif., September, 1959.

Lithium Hypochlorite (LiOCl) $Cl_1Li_1O_1(g)$

Enthalpy Reference Temperature = T_r = 298.15 K Standard State Pressure = p° = 0.1 MPa

T/K	$C_p°$ (J K⁻¹mol⁻¹)	S° (J K⁻¹mol⁻¹)	$-[G°-H°(T_r)]/T$ (J K⁻¹mol⁻¹)	$H°-H°(T_r)$ (kJ mol⁻¹)	$\Delta_f H°$ (kJ mol⁻¹)	$\Delta_f G°$ (kJ mol⁻¹)	Log K_f
0	0.	0.	INFINITE	-10.854	-11.525	-11.525	INFINITE
100	33.617	215.894	291.124	-7.523	-11.598	-14.337	7.489
200	37.908	240.313	260.189	-3.975	-12.866	-16.618	4.340
250	40.585	249.062	257.112	-2.012	-13.569	-17.475	3.651
298.15	42.961	256.417	256.417	0.	-14.226	-18.167	3.183
300	43.046	256.683	256.418	0.080	-14.250	-18.191	3.167
350	45.197	263.485	256.950	2.287	-14.903	-18.796	2.805
400	47.028	269.643	258.158	4.594	-15.557	-19.308	2.521
450	48.566	275.273	259.751	6.985	-16.236	-19.736	2.291
500	49.853	280.459	261.566	9.446	-19.953	-19.780	2.066
600	51.826	289.733	265.506	14.536	-21.242	-19.622	1.708
700	53.219	297.833	269.558	19.792	-22.374	-19.260	1.437
800	54.223	305.008	273.550	25.167	-23.414	-18.744	1.224
900	54.963	311.440	277.409	30.628	-24.407	-18.100	1.051
1000	55.520	317.261	281.107	36.153	-25.367	-17.348	0.906
1100	55.948	322.573	284.639	41.728	-26.303	-16.500	0.784
1200	56.284	327.456	288.006	47.340	-27.222	-15.569	0.678
1300	56.551	331.973	291.217	52.982	-28.130	-14.560	0.585
1400	56.767	336.172	294.280	58.649	-29.027	-13.483	0.503
1500	56.944	340.094	297.205	64.334	-29.915	-12.342	0.430
1600	57.090	343.774	300.002	70.036	-30.796	-11.141	0.364
1700	57.213	347.239	302.679	75.752	-176.908	-2.710	0.083
1800	57.316	350.512	305.247	81.478	-177.029	7.540	-0.219
1900	57.404	353.614	307.711	87.214	-177.158	17.798	-0.489
2000	57.480	356.560	310.081	92.959	-177.297	28.062	-0.733
2100	57.545	359.366	312.361	98.710	-177.449	38.334	-0.953
2200	57.602	362.044	314.559	104.467	-177.617	48.613	-1 154
2300	57.652	364.606	316.680	110.230	-177.801	58.900	-1.338
2400	57.696	367.061	318.728	115.998	-178.006	69.196	-1.506
2500	57.735	369.417	320.709	121.769	-178.233	79.501	-1.661
2600	57.770	371.682	322.626	127.545	-178.485	89.815	-1.804
2700	57.801	373.863	324.484	133.323	-178.764	100.139	-1.937
2800	57.828	375.965	326.285	139.105	-179.072	110.475	-2.061
2900	57.853	377.995	328.033	144.889	-179.411	120.821	-2.176
3000	57.876	379.957	329.732	150.675	-179.783	131.180	-2.284
3100	57.896	381.855	331.383	156.464	-180.189	141.553	-2.385
3200	57.915	383.693	332.989	162.254	-180.631	151.938	-2.480
3300	57.932	385.476	334.552	168.047	-181.109	162.338	-2.570
3400	57.947	387.205	336.076	173.841	-181.624	172.753	-2.654
3500	57.961	388.885	337.561	179.636	-182.178	183.184	-2.734
3600	57.974	390.518	339.009	185.433	-182.770	193.631	-2.810
3700	57.986	392.107	340.423	191.231	-183.401	204.095	-2.881
3800	57.997	393.653	341.803	197.030	-184.070	214.578	-2.950
3900	58.008	395.160	343.152	202.830	-184.779	225.078	-3.015
4000	58.017	396.629	344.471	208.632	-185.526	235.596	-3.077
4100	58.026	398.062	345.761	214.434	-186.313	246.134	-3.136
4200	58.034	399.460	347.023	220.237	-187.138	256.691	-3.192
4300	58.042	400.826	348.258	226.041	-188.003	267.269	-3.247
4400	58.049	402.160	349.468	231.845	-188.886	277.865	-3.299
4500	58.056	403.465	350.653	237.650	-189.821	288.483	-3.349
4600	58.062	404.741	351.815	243.456	-190.796	299.123	-3.397
4700	58.068	405.989	352.955	249.263	-191.810	309.784	-3.443
4800	58.073	407.212	354.072	255.070	-192.864	320.467	-3.487
4900	58.078	408.409	355.169	260.877	-193.958	331.173	-3.530
5000	58.083	409.583	356.246	266.685	-195.093	341.901	-3.572
5100	58.088	410.733	357.303	272.494	-196.265	352.653	-3.612
5200	58.092	411.861	358.341	278.303	-197.484	363.427	-3.651
5300	58.096	412.968	359.362	284.112	-198.746	374.226	-3.688
5400	58.100	414.054	360.364	289.922	-200.055	385.049	-3.725
5500	58.103	415.120	361.350	295.732	-201.411	395.896	-3.760
5600	58.107	416.167	362.320	301.543	-202.815	406.770	-3.794
5700	58.110	417.195	363.274	307.354	-204.269	417.668	-3.827
5800	58.113	418.206	364.212	313.165	-205.776	428.592	-3.860
5900	58.116	419.199	365.136	318.976	-207.336	439.542	-3.891
6000	58.119	420.176	366.045	324.788	-208.952	450.520	-3.922

PREVIOUS: December 1960 (1 atm) CURRENT: December 1960 (1 bar)

Lithium Hypochlorite (LiOCl) $Cl_1Li_1O_1(g)$

LITHIUM PERCHLORATE ($LiClO_4$) CRYSTAL M_r = 106.3916

$\Delta_f H°$(0 K) = Unknown
$\Delta_f H°$(298.15 K) = -380.7 ± 2.9 kJ mol^{-1}
$\Delta_{fus} H°$ = [29.29] kJ mol^{-1}

S°(298.15 K) = [126 ± 4.2] J K^{-1} mol^{-1}
T_{fus} = 509 K

Enthalpy of Formation

The heats of reaction ($\Delta_r H°$) involving $LiClO_4$(cr) has been determined by several investigators as follows:

Source	Reaction	$\Delta_r H°$(298.15 K), kcal mol^{-1}	$\Delta_f H°$(298.15 K) kcal mol^{-1}
Gilliland and Johnson (1)	$LiClO_4$(cr) + KCl(cr) → LiCl(cr) + $KClO_4$(cr)	-5.86±0.06	-90.34
Markowitz, et al. (2)	LiOH(225 H_2O) + $HClO_4$(225 H_2O) → $LiClO_4$ (451 H_2O) + H_2O(ℓ)	-13.54±0.09	-91.7
Birky and Hepler (3)	$LiClO_4$(cr) → Li^+(aq) + ClO_4^-(aq)	-6.31±0.13	-91.11

The adopted value, $\Delta_f H°$(298.15 K) = -91.0 kcal mol^{-1}, for $LiClO_4$(cr) is the weighted average of the three values listed.

Heat Capacity and Entropy

Heat capacities and S°(298.15 K) were estimated by comparison with those of $KClO_4$(cr).

Fusion Data

T_{fus} was obtained from Richards and Willard (4). The value of $\Delta_{fus} H°$(509 K) was estimated by comparison with the value, $\Delta_{fus} H°$(528 K) = 5.4 kcal mol^{-1}, for $NaClO_3$(cr) reported by Kelley (5).

References

1. A. A. Gilliland and W. H. Johnson, J. Res. Nat. Bur. Stand. 65A, 67 (1961).
2. M. M. Markowitz, R. F. Harris and H. Stewart, Jr., J. Phys. Chem. 63, 1325 (1959).
3. M. M. Birky and L. G. Hepler, J. Phys. Chem. 64, 686 (1960).
4. T. W. Richards and H. H. Willard, J. Amer. Chem. Soc., 32, 4 (1910).
5. K. K. Kelley, U. S. Bur. Mines Bull. 584, 1960.

Lithium Perchlorate ($LiClO_4$) $Cl_1Li_1O_4$(cr)

Enthalpy Reference Temperature = T_r = 298.15 K; Standard State Pressure = p° = 0.1 MPa

T/K	$C_p°$ (J $K^{-1}mol^{-1}$)	S° (J $K^{-1}mol^{-1}$)	$-[G°-H°(T_r)]/T$ (J $K^{-1}mol^{-1}$)	$H°-H°(T_r)$ (kJ mol^{-1})	$\Delta_f H°$ (kJ mol^{-1})	$\Delta_f G°$ (kJ mol^{-1})	Log K_f
0							
100							
200							
298.15	105.018	125.520	125.520	0.	-380.744	-253.911	44.484
300	105.437	126.171	125.522	0.195	-380.735	-253.125	44.073
400	129.704	160.077	129.980	12.039	-379.169	-210.770	27.524
500	143.511	190.659	139.119	25.770	-379.274	-168.682	17.622
509.000	144.340	193.227	140.053	27.065	------ CRYSTAL <--> LIQUID -----		
600	152.716	217.656	150.004	40.591	-375.571	-126.899	11.048
700	160.038	241.758	161.421	56.236	-371.197	-85.791	6.402
800	166.523	263.561	172.847	72.571	-366.282	-45.349	2.961
900	172.067	283.501	184.050	89.506	-360.909	-5.550	0.322
1000	176.983	301.890	194.926	106.963	-355.130	33.628	-1.757
1100	180.972	318.945	205.435	124.861	-349.006	72.210	-3.429
1200	184.933	334.862	215.564	143.157	-342.565	110.221	-4.798
1300	188.866	349.820	225.322	161.847	-335.799	147.680	-5.934
1400	192.771	363.959	234.724	180.929	-328.701	184.607	-6.888
1500	196.648	377.392	243.791	200.400	-321.266	221.014	-7.696

PREVIOUS: June 1963 CURRENT: March 1964

Lithium Perchlorate ($LiClO_4$) $Cl_1Li_1O_4$(cr)

LITHIUM PERCHLORATE ($LiClO_4$) LIQUID M_r = 106.3916

$S°(298.15\ K)$ = [164.611] $J\ K^{-1}\ mol^{-1}$

T_{fus} = 509 K

$\Delta_f H°(298.15\ K)$ = [-358.355] $kJ\ mol^{-1}$

$\Delta_{fus}H°$ = [29.3] $kJ\ mol^{-1}$

Enthalpy of Formation

$\Delta_f H°(LiClO_4, \ell, 298.15\ K)$ is calculated from $\Delta_f H°(LiClO_4, cr, 298.15\ K)$ by adding $\Delta_{fus}H°$ and the difference in enthalpy, $H°(509\ K) - H°(298.15\ K)$, between the crystal and liquid.

Heat Capacity and Entropy

$C_p°$ for $LiClO_4(\ell)$ was estimated by comparison with that for $NaClO_3(\ell)$. The entropy is calculated in a manner analogous to that used for the enthalpy of formation.

Fusion Data

The melting temperature (T_{fus}) was obtained from Richards and Willard (1). The value of $\Delta_{fus}H°(509\ K)$ was estimated by comparison with the value, $\Delta_{fus}H°(528\ K)$ = 5.4 $kcal\ mol^{-1}$, for $NaClO_3(cr)$ reported by Kelley (2).

References

1. T. W. Richards and H. H. Willard, J. Amer. Chem. Soc. 32, 4 (1910).
2. K. K. Kelley, U. S. Bur. Mines Bull. 584, 1960.

Lithium Perchlorate ($LiClO_4$) $Cl_1Li_1O_4(l)$

Enthalpy Reference Temperature = T_r = 298.15 K Standard State Pressure = p° = 0.1 MPa

T/K	$C_p°$ (J K⁻¹mol⁻¹)	S° (J K⁻¹mol⁻¹)	$-[G°-H°(T_r)]/T$ (J K⁻¹mol⁻¹)	$H°-H°(T_r)$ (kJ mol⁻¹)	$\Delta_f H°$ (kJ mol⁻¹)	$\Delta_f G°$ (kJ mol⁻¹)	Log K_f
0							
100							
200							
298.15	161.084	164.611	164.611	0.	-358.355	-243.178	42.604
300	161.084	165.608	164.614	0.298	-358.243	-242.463	42.217
400	161.084	211.949	170.933	16.406	-352.412	-204.763	26.739
500	161.084	247.894	182.864	32.515	-350.140	-168.166	17.568
509.000	161.084	250.767	184.039	33.965	------ CRYSTAL <--> LIQUID -----		
600	161.084	277.263	196.224	48.623	-345.150	-132.242	11.513
700	161.084	302.094	209.620	64.732	-340.312	-97.142	7.249
800	161.084	323.604	222.554	80.840	-335.624	-62.725	4.096
900	161.084	342.577	234.856	96.948	-331.077	-28.887	1.677
1000	161.084	359.549	246.492	113.057	-326.647	4.451	-0.232
1100	161.084	374.901	257.479	129.165	-322.313	37.351	-1.774
1200	161.084	388.918	267.856	145.274	-318.059	69.859	-3.041
1300	161.084	401.811	277.671	161.382	-313.876	102.015	-4.099
1400	161.084	413.749	286.970	177.490	-309.751	133.852	-4.994
1500	161.084	424.862	295.797	193.599	-305.678	165.395	-5.760

PREVIOUS: December 1960 CURRENT: March 1964

Lithium Perchlorate ($LiClO_4$) $Cl_1Li_1O_4(l)$

LITHIUM PERCHLORATE ($LiClO_4$)

CRYSTAL-LIQUID

M_r = 106.3916

0 to 509 K crystal
above 509 K liquid

Refer to the individual tables for details.

Lithium Perchlorate ($LiClO_4$) $Cl_1Li_1O_4$(cr,l)

Enthalpy Reference Temperature = T_r = 298.15 K Standard State Pressure = p° = 0.1 MPa

T/K	C_p° (J K⁻¹mol⁻¹)	S° (J K⁻¹mol⁻¹)	$-[G^\circ-H^\circ(T_r)]/T$ (J K⁻¹mol⁻¹)	$H^\circ-H^\circ(T_r)$ (kJ mol⁻¹)	Δ_fH° (kJ mol⁻¹)	Δ_fG° (kJ mol⁻¹)	Log K_f
0							
100							
200							
298.15	105.018	125.520	125.520	0.	-380.744	-253.911	44.484
300	105.437	126.171	125.522	0.195	-380.735	-253.125	44.073
400	129.704	160.077	129.980	12.039	-379.169	-210.770	27.524
500	143.511	190.659	139.119	25.770	-379.274	-168.682	17.622
509.000	144.340	193.227	140.053	27.065	CRYSTAL <--> LIQUID		
509.000	161.084	250.767	140.053	56.353	TRANSITION		
600	161.084	277.263	158.909	71.012	-345.150	-132.242	11.513
700	161.084	302.094	177.636	87.120	-340.312	-97.142	7.249
800	161.084	323.604	194.568	103.229	-335.624	-62.725	4.096
900	161.084	342.577	209.980	119.337	-331.077	-28.887	1.677
1000	161.084	359.549	224.103	135.446	-326.647	4.451	-0.232
1100	161.084	374.901	237.125	151.554	-322.313	37.351	-1.774
1200	161.084	388.918	249.199	167.662	-318.059	69.859	-3.041
1300	161.084	401.811	260.449	183.771	-313.876	102.015	-4.099
1400	161.084	413.749	270.978	199.879	-309.751	133.852	-4.994
1500	161.084	424.862	280.871	215.988	-305.678	165.395	-5.760

PREVIOUS: CURRENT: March 1964

Lithium Perchlorate ($LiClO_4$) $Cl_1Li_1O_4$(cr,l)

MAGNESIUM CHLORIDE (MgCl) IDEAL GAS $M_r = 59.758$

$S°(298.15\ K) = 233.413\ J\ K^{-1}\ mol^{-1}$

$\Delta_f H°(0\ K) = -43.3 \pm 42\ kJ\ mol^{-1}$
$\Delta_f H°(298.15\ K) = -43.5 \pm 42\ kJ\ mol^{-1}$

Electronic Level and Quantum Weight

State	ε_i, cm^{-1}	g_i
$^2\Sigma$	0	2

$\omega_e = 462.4\ cm^{-1}$ $\omega_e x_e = 2.02\ cm^{-1}$ $\sigma = 1$
$B_e = [0.2417]\ cm^{-1}$ $\alpha_e = [0.0014]\ cm^{-1}$ $r_e = 2.199$ Å

Enthalpy of Formation

The enthalpy of formation is calculated from $D_0^0 = 3.2$ eV reported by Herzberg (1). $D_0^0 = 2.7 \pm 0.7$ eV reported by Gaydon (2), yields $\Delta_f H°$(MgCl, g, 298.15 K) = 1.0±16 kcal mol^{-1}.

Heat Capacity and Entropy

Spectroscopic constants were reported by Herzberg (1) and adjusted for natural isotopic abundance. The bond length was reported by Morgan and Barrow (3). B_e and α_e were estimated by the method suggested by Herzberg (1).

References

1. G. Herzberg, "Spectra of Diatomic Molecules", 2nd ed., D. Van Nostrand Co., Inc., New York, 1950.
2. A. G. Gaydon, "Dissociation Energies and Spectra of Diatomic Molecules", 2nd ed., Chapman and Hall, London 1953.
3. E. Morgan and R. F. Barrow, Nature 192, 1182, (1961).

Magnesium Chloride (MgCl) $Cl_1Mg_1(g)$

Enthalpy Reference Temperature = T_r = 298.15 K Standard State Pressure = $p°$ = 0.1 MPa

T/K	$C_p°$ (J K⁻¹mol⁻¹)	S° (J K⁻¹mol⁻¹)	$-[G°-H°(T_r)]/T$ (J K⁻¹mol⁻¹)	$H°-H°(T_r)$ (kJ mol⁻¹)	$\Delta_f H°$ (kJ mol⁻¹)	$\Delta_f G°$ (kJ mol⁻¹)	Log K_f
0	0.	0.	INFINITE	-9.361	-43.287	-43.287	INFINITE
100	29.610	198.431	262.874	-6.444	-42.467	-51.910	27.115
200	32.761	219.902	236.542	-3.328	-42.877	-61.232	15.992
250	33.983	227.351	233.982	-1.658	-43.187	-65.786	13.745
298.15	34.828	233.413	233.413	0.	-43.514	-70.109	12.283
300	34.855	233.628	233.413	0.064	-43.527	-70.274	12.236
350	35.477	239.051	233.840	1.824	-43.889	-74.704	11.149
400	35.930	243.819	234.795	3.609	-44.272	-79.080	10.327
450	36.268	248.071	236.039	5.415	-44.675	-83.407	9.682
500	36.528	251.907	237.437	7.235	-45.098	-87.688	9.161
600	36.898	258.602	240.422	10.908	-46.009	-96.122	8.368
700	37.148	264.309	243.437	14.611	-47.014	-104.396	7.790
800	37.330	269.282	246.364	18.335	-48.128	-112.519	7.347
900	37.472	273.687	249.159	22.075	-49.366	-120.495	6.993
1000	37.587	277.642	251.813	25.828	-59.338	-127.613	6.666
1100	37.685	281.229	254.327	29.592	-60.880	-134.365	6.380
1200	37.772	284.512	256.707	33.365	-62.419	-140.978	6.137
1300	37.849	287.538	258.964	37.146	-63.955	-147.462	5.925
1400	37.921	290.346	261.107	40.935	-192.897	-150.661	5.621
1500	37.988	292.964	263.144	44.730	-193.076	-147.638	5.141
1600	38.051	295.418	265.085	48.532	-193.252	-144.603	4.721
1700	38.112	297.727	266.938	52.340	-193.427	-141.557	4.350
1800	38.170	299.907	268.710	56.154	-193.601	-138.501	4.019
1900	38.227	301.972	270.407	59.974	-193.773	-135.435	3.723
2000	38.282	303.934	272.034	63.800	-193.945	-132.361	3.457
2100	38.336	305.803	273.598	67.631	-194.117	-129.277	3.216
2200	38.390	307.588	275.103	71.467	-194.291	-126.185	2.996
2300	38.442	309.296	276.553	75.308	-194.466	-123.086	2.795
2400	38.494	310.933	277.951	79.155	-194.644	-119.979	2.611
2500	38.545	312.505	279.302	83.007	-194.826	-116.864	2.442
2600	38.596	314.018	280.609	86.864	-195.014	-113.741	2.285
2700	38.647	315.476	281.873	90.727	-195.208	-110.612	2.140
2800	38.697	316.882	283.098	94.594	-195.409	-107.475	2.005
2900	38.747	318.241	284.287	98.466	-195.619	-104.331	1.879
3000	38.796	319.555	285.441	102.343	-195.839	-101.179	1.762
3100	38.846	320.828	286.562	106.225	-196.071	-98.020	1.652
3200	38.895	322.062	287.652	110.112	-196.314	-94.854	1.548
3300	38.944	323.260	288.713	114.004	-196.571	-91.679	1.451
3400	38.992	324.423	289.746	117.901	-196.843	-88.496	1.360
3500	39.041	325.554	290.753	121.803	-197.131	-85.306	1.273
3600	39.090	326.655	291.735	125.709	-197.435	-82.106	1.191
3700	39.138	327.726	292.694	129.621	-197.758	-78.898	1.114
3800	39.186	328.771	293.629	133.537	-198.099	-75.681	1.040
3900	39.235	329.789	294.544	137.458	-198.460	-72.455	0.970
4000	39.283	330.783	295.437	141.384	-198.842	-69.219	0.904
4100	39.331	331.754	296.311	145.314	-199.246	-65.974	0.841
4200	39.379	332.702	297.166	149.250	-199.673	-62.718	0.780
4300	39.427	333.629	298.004	153.190	-200.124	-59.452	0.722
4400	39.475	334.536	298.824	157.135	-200.600	-56.175	0.667
4500	39.523	335.424	299.627	161.085	-201.102	-52.887	0.614
4600	39.571	336.293	300.415	165.040	-201.631	-49.588	0.563
4700	39.618	337.145	301.187	168.999	-202.188	-46.277	0.514
4800	39.666	337.979	301.945	172.964	-202.774	-42.953	0.467
4900	39.714	338.797	302.689	176.933	-203.391	-39.617	0.422
5000	39.762	339.600	303.419	180.906	-204.039	-36.269	0.379
5100	39.809	340.388	304.136	184.885	-204.719	-32.906	0.337
5200	39.857	341.162	304.841	188.868	-205.433	-29.531	0.297
5300	39.904	341.921	305.533	192.856	-206.180	-26.141	0.258
5400	39.952	342.668	306.214	196.849	-206.963	-22.737	0.220
5500	40.000	343.401	306.884	200.847	-207.783	-19.317	0.183
5600	40.047	344.122	307.542	204.849	-208.639	-15.883	0.148
5700	40.095	344.832	308.190	208.856	-209.535	-12.433	0.114
5800	40.142	345.529	308.828	212.868	-210.469	-8.967	0.081
5900	40.190	346.216	309.456	216.884	-211.444	-5.485	0.049
6000	40.237	346.892	310.074	220.906	-212.461	-1.986	0.017

PREVIOUS: March 1966 (1 atm) CURRENT: March 1966 (1 bar)

Magnesium Chloride (MgCl) $Cl_1Mg_1(g)$

MAGNESIUM CHLORIDE, ION ($MgCl^+$) IDEAL GAS M_r = 59.75745

$S°(298.15\ K) = [228.6 \pm 8]\ J\ K^{-1}\ mol^{-1}$

$\Delta_f H°(0\ K) = 646.6 \pm 84\ kJ\ mol^{-1}$
$\Delta_f H°(298.15\ K) = [652.704] \pm 84\ kJ\ mol^{-1}$

Electronic Levels and Quantum Weights

State	ε_i, cm^{-1}	g_i
$^1\Sigma^+$	0	1
$^3\Pi$	[9000]	6
$^1\Pi$	[11000]	2
$^3\Sigma^-$	[20000]	3
$^1\Delta$	[25000]	2
$^1\Sigma^+$	[30000]	1

$\omega_e = [400]\ cm^{-1}$ $\omega_e x_e = [2]\ cm^{-1}$ $\sigma = 1$
$B_e = [0.2415]\ cm^{-1}$ $\alpha_e = [0.00164]\ cm^{-1}$ $r_e = [2.2]$ Å

Enthalpy of Formation

The ionization potential of MgCl(g) was reported to be 7.2, 7.5 ± 0.5 and 7 ± 1 eV by Krasnov (1), Hildenbrand (9), and Beckett (2), respectively. The corresponding heats of formation at 298 K for $MgCl^+(g)$ are derived as 156, 163, and 151 kcal mol^{-1}, using $\Delta_f H°(MgCl, g, 0\ K) = -10.34$ kcal mol^{-1} from the JANAF MgCl(g) table (March 31, 1966). The value of $\Delta_f H°(298.15\ K)$ for $MgCl^+(g)$ is tentatively selected as 156±20 kcal mol^{-1} (652.704±84 kJ mol^{-1}).

Heat Capacity and Entropy

The ground state configuration is assumed to be the same as NaCl(g) which is esoelectronic with $MgCl^+(g)$. The six electronic states are obtained from Fougere (3). The corresponding electronic levels are estimated by comparison with the isoelectronic molecules BeS (4), BeO (5, 6), and MgO (5, 7). The bond distance is estimated by the method suggested by Krasnov (8). The values of ω_e and $\omega_e x_e$ are estimated by comparison with those for MgCl(g) and NaCl(g). B_e and α_e are calculated by the method recommended by Herzberg (5).

References

1. K. S. Krasnov, Teplofiz. Vysokikh Temperatur., Akad. Nauk SSSR 3, 927 (1965).
2. C. W Beckett and E. C. Cassidy, U. S. Nat. Bur. Stand. Rept. 8628, 1965.
3. P. F. Fougere and R. K. Nesbet, J. Chem. Phys. 44, 285 (1965).
4. G. Verhaegen and W. G. Richards, Proc. Phys. Soc. (London) 90, 579 (1967).
5. G. Herzberg, "Spectra of Diatomic Molecules," D. Van Nostrand Co., Inc., New York, 1950.
6. G. Verhaegen and W. G. Richards, J. Chem. Phys. 45, 1828 (1966).
7. L. Brewer and S. Trajmar, J. Chem. Phys. 36, 1585 (1962).
8. K. S. Krasnov, Zh. Strukt. Khim. 1, 209 (1960).
9. D. L. Hildenbrand, McDonnell Douglas Corporation, personal communication, July 24, 1968.

Magnesium Chloride, Ion ($MgCl^+$) $Cl_1Mg_1^+(g)$

Enthalpy Reference Temperature = T_r = 298.15 K Standard State Pressure = p° = 0.1 MPa

T/K	$C_p°$ (J K⁻¹mol⁻¹)	S° (J K⁻¹mol⁻¹)	$-[G°-H°(T_r)]/T$ (J K⁻¹mol⁻¹)	$H°-H°(T_r)$ (kJ mol⁻¹)	$\Delta_f H°$ (kJ mol⁻¹)	$\Delta_f G°$ (kJ mol⁻¹)	Log K_f
0	0.	0.	INFINITE	-9.516	646.579		
100	30.033	192.784	258.681	-6.590			
200	33.588	214.752	231.749	-3.399			
250	34.711	222.375	229.136	-1.690			
298.15	35.446	228.556	228.556	0.	652.704	621.302	-108.849
300	35.470	228.775	228.557	0.066	652.731	621.107	-108.144
350	35.996	234.285	228.991	1.853	653.436	615.780	-91.900
400	36.374	239.117	229.961	3.663	654.116	610.354	-79.704
450	36.655	243.419	231.221	5.489	654.773	604.844	-70.208
500	36.872	247.292	232.638	7.327	655.407	599.262	-62.604
600	37.186	254.045	235.659	11.031	656.606	587.919	-51.183
700	37.405	259.794	238.707	14.761	657.707	576.382	-43.010
800	37.572	264.800	241.662	18.510	658.697	564.696	-36.871
900	37.712	269.234	244.484	22.275	659.561	552.892	-32.089
1000	37.841	273.214	247.161	26.052	651.693	541.712	-28.296
1100	37.977	276.827	249.697	29.843	652.256	530.687	-25.200
1200	38.140	280.138	252.097	33.649	652.828	519.610	-22.618
1300	38.347	283.198	254.373	37.473	653.414	508.485	-20.431
1400	38.618	286.050	256.535	41.320	526.609	500.481	-18.673
1500	38.970	288.726	258.593	45.199	528.592	498.546	-17.361
1600	39.412	291.254	260.556	49.117	530.611	496.478	-16.208
1700	39.954	293.659	262.433	53.085	532.674	494.281	-15.187
1800	40.595	295.961	264.232	57.111	534.792	491.962	-14.276
1900	41.333	298.175	265.961	61.207	536.973	489.523	-13.458
2000	42.158	300.316	267.625	65.381	539.228	486.968	-12.718
2100	43.059	302.394	269.232	69.641	541.564	484.298	-12.046
2200	44.020	304.419	270.785	73.995	543.987	481.515	-11.433
2300	45.025	306.398	272.291	78.447	546.500	478.619	-10.870
2400	46.056	308.336	273.752	83.001	549.108	475.612	-10.351
2500	47.095	310.237	275.174	87.658	551.810	472.495	-9.872
2600	48.124	312.104	276.559	92.419	554.605	469.267	-9.428
2700	49.128	313.940	277.909	97.282	557.490	465.931	-9.014
2800	50.093	315.744	279.228	102.244	560.462	462.485	-8.628
2900	51.007	317.518	280.518	107.299	563.514	458.932	-8.266
3000	51.859	319.262	281.781	112.443	566.639	455.273	-7.927
3100	52.644	320.975	283.017	117.669	569.830	451.509	-7.608
3200	53.354	322.658	284.230	122.969	573.078	447.640	-7.307
3300	53.989	324.309	285.419	128.337	576.376	443.669	-7.023
3400	54.546	325.930	286.587	133.764	579.713	439.598	-6.754
3500	55.026	327.518	287.734	139.244	583.081	435.428	-6.498
3600	55.431	329.074	288.861	144.767	586.473	431.161	-6.256
3700	55.765	330.597	289.968	150.327	589.878	426.800	-6.025
3800	56.030	332.088	291.057	155.918	593.289	422.347	-5.806
3900	56.233	333.546	292.128	161.531	596.700	417.804	-5.596
4000	56.376	334.972	293.181	167.162	600.101	413.174	-5.395
4100	56.467	336.365	294.218	172.805	603.488	408.458	-5.204
4200	56.510	337.727	295.238	178.454	606.853	403.661	-5.020
4300	56.510	339.056	296.241	184.105	610.191	398.784	-4.844
4400	56.472	340.355	297.229	189.755	613.498	393.829	-4.675
4500	56.401	341.624	298.202	195.399	616.769	388.799	-4.513
4600	56.302	342.862	299.159	201.034	619.999	383.698	-4.357
4700	56.179	344.072	300.102	206.658	623.186	378.526	-4.207
4800	56.036	345.253	301.030	212.269	626.325	373.287	-4.062
4900	55.876	346.407	301.945	217.865	629.413	367.984	-3.923
5000	55.702	347.534	302.845	223.444	632.449	362.617	-3.788
5100	55.518	348.635	303.732	229.005	635.430	357.191	-3.658
5200	55.326	349.711	304.606	234.547	638.354	351.707	-3.533
5300	55.128	350.763	305.467	240.070	641.220	346.167	-3.412
5400	54.925	351.792	306.315	245.573	644.025	340.574	-3.294
5500	54.721	352.798	307.151	251.055	646.769	334.929	-3.181
5600	54.516	353.782	307.975	256.517	649.451	329.234	-3.071
5700	54.311	354.745	308.788	261.958	652.068	323.493	-2.964
5800	54.108	355.688	309.588	267.379	654.621	317.706	-2.861
5900	53.907	356.611	310.377	272.780	657.109	311.876	-2.761
6000	53.710	357.516	311.155	278.161	659.531	306.004	-2.664

PREVIOUS: June 1968 (1 atm) CURRENT: June 1968 (1 bar)

Magnesium Chloride, Ion ($MgCl^+$) $Cl_1Mg_1^+(g)$

NITROSYL CHLORIDE (ONCl) IDEAL GAS M_r = 65.4591

$S°(298.15\ K) = 261.680 \pm 0.17\ J\ K^{-1}\ mol^{-1}$

$\Delta_f H°(0\ K) = 53.622 \pm 0.42\ kJ\ mol^{-1}$
$\Delta_f H°(298.15\ K) = 51.714 \pm 0.42\ kJ\ mol^{-1}$

Vibrational Frequencies, Degeneracies and Anharmonicity Constants

ν, cm^{-1}	X_{ii}, cm^{-1}	X_{ij}, cm^{-1}
1799.7(1)	X_{11} = -17.8	X_{12} = 0.0
595.6(1)	X_{22} = - 2.6	X_{23} = -4.3
330.9(1)	X_{33} = - 1.0	X_{13} = -0.6

Ground State Quantum Weight: 1 $\sigma = 1$
Point Group: C_s
Bond Distances: O-N = 1.143 ± 0.012 Å N-Cl = 1.976 ± 0.005 Å
Bond Angle: O-N-Cl = 113.0° ± 0.7°
Rotational Constants: $A_0 = 2.8492\ cm^{-1}$ $B_0 = 0.19034\ cm^{-1}$ $C_0 = 0.17842\ cm^{-1}$

Enthalply of Formation

$\Delta_f H°(298.15\ K)$ is calculated from $\Delta_r H° = 18.43 \pm 0.06\ kcal\ mol^{-1}$ derived by 3rd law analysis of equilibrium data (1) for 2ONCl(g) + 2NO(g) + Cl_2(g). 2nd law analysis yields $\Delta_r H° = 18.39 \pm 0.07$ and an entropy deviation of $-0.09 \pm 0.16\ cal\ K^{-1}\ mol^{-1}$ from the 3rd law $\Delta_r S°$. Data analyzed cover a temperature range of 372.7 to 491.3 K and consist of a combination of three runs reported by Beeson and Yost (1). Combination is justified by the mutual consistency of the three runs, which involved excess NO, excess Cl_2, and roughly the stoichiometric ratio of NO and Cl_2, respectively. Earlier data reviewed by Bichowsky and Rossini (2) are considered less reliable.

Heat Capacity and Entropy

We list structural parameters which approximate the adopted rotational constants (3) obtained from microwave spectra of three isotopic molecules. These structural parameters are minor modifications of those given by Millen and Pannell (3). Changes are well within the ranges specified by the authors but give improved agreement with rotational constant A_0. Literature values include (O-N) = 1.139±0.012 (3), 1.14±0.02 (4, 7) and 1.17±0.06 Å(5, 6) (N-Cl) = 1.975±0.005 (3), 1.95±0.01 (4), 1.95±0.06 (5, 6) and 1.97±0.01 Å (7, 6); and angle (O-N-Cl) = 113.33°±0.67° (3), 116°±2° (4), 114°±3° (5, 6) and 113°±2°(7, 6). The earlier values (4, 5, 7) resulted from electron-diffraction data (4) and two separate analyses (5, 7) of older microwave data (6). Rotational constants are adjusted for the natural isotopic abundance of Cl. A_0 is relatively uncertain.

Observed ν values and anharmonicity constants are from the gas-phase infrared study of Jones et al. (8), who reviewed earlier spectral data. Values of ν are adjusted for the natural isotopic abundance of chlorine. The anharmonicity constants differ from, and supersede, earlier values (9) used by Gordon (10) in calculating thermodynamic functions for ONCl.

New functions have been calculated through the courtesy of B. J. McBride (11), who used the above vibrational and rotational constants with first and second order corrections (method NRRA02) in the NASA-Lewis computer program (12). We adopt these functions directly for 0 to 298.15 K and reproduce the functions at higher temperatures by integration of $C_p°$.

References

1. C. M. Beeson and D. Yost, J. Chem. Phys. 7, (1939).
2. F. R. Bichowsky and F. D. Rossini, "The Thermochemistry of the Chemical Substances," p. 215, Reinhold Publishing Corp., New York, 1936.
3. D. J. Millen and J. Pannell, J. Chem. Soc. 1322 (1961).
4. J. A. A. Ketelaar and K. J. Palmer, J. Amer. Chem. Soc. 59, 2629 (1937).
5. J. D. Rogers and D. Williams, J. Chem. Phys. 34, 2195 (1961).
6. J. D. Rogers, W. J. Pietenpol and D. Williams, Phys. Rev. 83, 431 (1951).
7. D. H. Whiffen, in L. E. Sutton (editor), "Interatomic Distances," Special Publication No. 11, The Chemical Society, London, 1958.
8. L. H. Jones, R. R. Ryan and L. B. Asprey, J. Chem. Phys. 49, 581 (1968).
9. L. Landau and W. H. Fletcher, J. Mol. Spectrosc. 4, 276 (1960).
10. J. S. Gordon, J. Chem. Eng. Data 7, 82 (1962).
11. B. J. McBride, NASA Lewis Research Center, Cleveland, Ohio, personal communication, Jan. 5, 1973.
12. B. J. McBride and S. Gordon, NASA-TN-D-4097, 135 pp, Aug., 1967; Scientific and Technical Aerospace Reports 5, p. 3608, N67-35192 (1967); CA. 68, 63364 (1968).

Nitrosyl Chloride (ONCl) $Cl_1N_1O_1(g)$

Enthalpy Reference Temperature = T_r = 298.15 K Standard State Pressure = p° = 0.1 MPa

T/K	$C_p°$ (J K⁻¹mol⁻¹)	S° (J K⁻¹mol⁻¹)	-[G°-H°(T_r)]/T (J K⁻¹mol⁻¹)	H°-H°(T_r) (kJ mol⁻¹)	$\Delta_f H°$ (kJ mol⁻¹)	$\Delta_f G°$ (kJ mol⁻¹)	Log K_f
0	0.	0.	INFINITE	-11.360	53.622	53.622	INFINITE
100	35.016	218.558	298.546	-7.999	52.623	56.874	-29.708
200	40.756	244.629	265.649	-4.204	51.988	61.406	-16.038
298.15	44.589	261.680	261.680	0.	51.714	66.096	-11.580
300	44.643	261.956	261.681	0.083	51.711	66.185	-11.524
400	47.149	275.162	263.465	4.679	51.629	71.027	-9.275
500	49.079	285.898	266.910	9.494	51.658	75.875	-7.927
600	50.687	294.992	270.852	14.484	51.761	80.709	-7.026
700	52.037	302.910	274.878	19.622	51.915	85.522	-6.382
800	53.166	309.934	278.829	24.884	52.104	90.311	-5.897
900	54.106	316.252	282.642	30.249	52.317	95.074	-5.518
1000	54.892	321.995	286.295	35.700	52.549	99.812	-5.214
1100	55.554	327.258	289.783	41.223	52.794	104.527	-4.964
1200	56.117	332.117	293.111	46.807	53.048	109.219	-4.754
1300	56.603	336.628	296.287	52.444	53.310	113.889	-4.576
1400	57.026	340.839	299.320	58.126	53.578	118.539	-4.423
1500	57.399	344.786	302.221	63.847	53.849	123.170	-4.289
1600	57.733	348.501	304.999	69.604	54.123	127.782	-4.172
1700	58.035	352.011	307.662	75.393	54.399	132.377	-4.067
1800	58.311	355.336	310.219	81.210	54.675	136.956	-3.974
1900	58.566	358.495	312.677	87.054	54.951	141.520	-3.891
2000	58.804	361.506	315.044	92.923	55.225	146.069	-3.815
2100	59.027	364.380	317.325	98.815	55.497	150.605	-3.746
2200	59.240	367.131	319.527	104.728	55.766	155.128	-3.683
2300	59.443	369.769	321.655	110.662	56.031	159.638	-3.625
2400	59.637	372.303	323.713	116.616	56.292	164.137	-3.572
2500	59.826	374.741	325.705	122.589	56.547	168.625	-3.523
2600	60.009	377.091	327.637	128.581	56.797	173.103	-3.478
2700	60.188	379.359	329.511	134.591	57.040	177.572	-3.435
2800	60.363	381.551	331.330	140.619	57.277	182.032	-3.396
2900	60.536	383.673	333.099	146.664	57.508	186.483	-3.359
3000	60.706	385.728	334.819	152.726	57.733	190.927	-3.324
3100	60.875	387.721	336.494	158.805	57.952	195.363	-3.292
3200	61.042	389.656	338.125	164.901	58.167	199.792	-3.261
3300	61.208	391.537	339.715	171.013	58.378	204.214	-3.232
3400	61.374	393.367	341.266	177.142	58.586	208.631	-3.205
3500	61.539	395.148	342.780	183.288	58.792	213.041	-3.179
3600	61.704	396.884	344.259	189.450	58.997	217.445	-3.155
3700	61.870	398.577	345.705	195.629	59.204	221.843	-3.132
3800	62.035	400.229	347.118	201.824	59.414	226.236	-3.110
3900	62.201	401.843	348.500	208.036	59.627	230.624	-3.089
4000	62.368	403.420	349.854	214.264	59.847	235.005	-3.069
4100	62.534	404.962	351.179	220.509	60.074	239.382	-3.050
4200	62.702	406.471	352.478	226.771	60.310	243.752	-3.032
4300	62.871	407.948	353.751	233.050	60.556	248.117	-3.014
4400	63.041	409.395	354.999	239.345	60.815	252.476	-2.997
4500	63.212	410.814	356.223	245.658	61.088	256.828	-2.981
4600	63.384	412.205	357.425	251.988	61.375	261.175	-2.966
4700	63.557	413.570	358.605	258.335	61.678	265.515	-2.951
4800	63.731	414.910	359.765	264.699	61.999	269.849	-2.937
4900	63.907	416.226	360.903	271.081	62.338	274.176	-2.923
5000	64.084	417.519	362.023	277.481	62.696	278.495	-2.909
5100	64.262	418.790	363.123	283.898	63.073	282.808	-2.897
5200	64.442	420.039	364.206	290.333	63.471	287.112	-2.884
5300	64.623	421.269	365.271	296.786	63.891	291.409	-2.872
5400	64.806	422.478	366.319	303.258	64.332	295.698	-2.860
5500	64.990	423.669	367.351	309.748	64.794	299.978	-2.849
5600	65.176	424.842	368.367	316.256	65.279	304.250	-2.838
5700	65.363	425.997	369.368	322.783	65.785	308.513	-2.827
5800	65.552	427.135	370.355	329.329	66.313	312.767	-2.817
5900	65.743	428.258	371.326	335.893	66.863	317.011	-2.807
6000	65.935	429.364	372.285	342.477	67.434	321.246	-2.797

PREVIOUS: December 1972 (1 atm) CURRENT: December 1972 (1 bar)

Nitrosyl Chloride (ONCl) $Cl_1N_1O_1(g)$

NITRYL CHLORIDE (NO_2Cl) IDEAL GAS M_r = 81.4585

Nitryl Chloride (NO_2Cl) $Cl_1N_1O_2(g)$

$S°$(298.15 K) = 272.187 J K^{-1} mol^{-1}

$\Delta_fH°$(0 K) = 17.54 ± 1.7 kJ mol^{-1}
$\Delta_fH°$(298.15 K) = 12.13 ± 1.7 kJ mol^{-1}

Vibrational Frequencies and Degeneracies

ν, cm^{-1}	ν, cm^{-1}
1293(1)	1685(1)
794(1)	411(1)
367(1)	651(1)

Ground State Quantum Weight: 1 σ = 2
Point Group: C_{2v}
Bond Distances: N-Cl = 1.83 Å N-O = 1.21 Å
Bond Angles: O-N-O = 129.5° O-N-Cl = 115.25°
Product of the Moments of Inertia: $I_AI_BI_C$ = 2.36673 x 10^{-114} g^3 cm^6

Enthalpy of Formation

The selected enthalpy of formation was obtained from the recalculation of the heat of reaction $ClNO_2 + NO = ClNO + NO_2$ which has been measured calorimetrically as -9.27 kcal mol^{-1} by Ray and Ogg (1). Not only the thermal effects of $2NO_2 = N_2O_4$ but also $NO + NO_2 = N_2O_3$ (a minor one) have been considered in calculation. Using all JANAF auxiliary data for ClNO, NO, NO_2, N_2O_3 and N_2O_4, the recalculation gives $\Delta_fH°$(298.15 K) = 2.90±0.4 kcal mol^{-1} for NO_2Cl(g). For the reaction $ClNO + N_2O_5 = ClNO_2 + 2NO_2$ also measured by Ray and Ogg (1), the recalculation gives $\Delta_fH°$(NO_2Cl, g, 298.15 K) = 2.56±0.4 kcal mol^{-1}.

Martin and Koehlein (2) have determined kinetically the heat of reaction, $\Delta_rH°$(273.15 K) = 3.10 kcal mol^{-1} for the reaction $NO_2 + NOCl = NO + NO_2Cl$ which gives $\Delta_fH°$(298.15 K) = 1.90±1 kcal mol^{-1} for NO_2Cl(g). Cordes and Johnston (3) have also determined kinetically the $\Delta_rH°$(0 K) = 29.5 kcal mol^{-1} for the reaction $NO_2Cl = NO_2 + Cl$ which gives $\Delta_fH°$(298.15 K) = 6.3±0.3 kcal mol^{-1}.

Heat Capacity and Entropy

The selected bond distances and bond angles were obtained from microwave spectrum studies by Clayton et al. (4). Millen and Sinnott (5), reported bond distance N-Cl = 1.840 Å, N-O = 1.202 Å and angle ONO = 130° 35' from the microwave measurement. These two are in good agreement.

The selected vibrational frequencies were obtained from infrared and Raman spectrum measurements by Ryason and Wilson (6). However, the assignment of the fundamental frequencies has been revised by Dodd et al. (7) and Morino and Tanaka (8). Morino and Tanaka's assignment was adopted. The principal moments of inertia are I_A = 6.3641 x 10^{-39}, I_B = 16.3632 x 10^{-39} and I_C = 22.7272 x 10^{-39} g cm^2.

References

1. J. Ray and R. Ogg, Jr., J. Chem. Phys. 31, 168 (1959).
2. H. Martin and E. Koehlein, Z. Physik. Chem. 17, 375 (1958).
3. H. Cordes and H. Johnston, J. Amer. Chem. Soc. 76, 4264 (1954).
4. L. Clayton, Q. Williams and T. L. Weatherly, J. Chem. Phys. 30, 1328 (1959).
5. D. J. Millen and K. M. Sinnott, J. Chem. Soc. 350 (1958).
6. R. Ryason and M. K. Wilson, J. Chem. Phys. 22, 2000 (1954).
7. R. E. Dodd, J. A. Rolfe and L. A. Woodward, Trans. Faraday Soc. 52, 145 (1956).
8. Y. Morino and T. Tanaka, J. Mol. Spectry. 16, 179 (1965).

Enthalpy Reference Temperature = T_r = 298.15 K Standard State Pressure = p° = 0.1 MPa

T/K	$C_p°$ (J $K^{-1}mol^{-1}$)	S° (J $K^{-1}mol^{-1}$)	$-[G°-H°(T_r)]/T$ (J $K^{-1}mol^{-1}$)	$H°-H°(T_r)$ (kJ mol^{-1})	$\Delta_fH°$ (kJ mol^{-1})	$\Delta_fG°$ (kJ mol^{-1})	Log K_f
0	0.	0.	INFINITE	-12.199	17.544	17.544	INFINITE
100	35.314	225.280	313.645	-8.837	15.094	27.338	-14.280
200	44.951	252.629	276.802	-4.835	13.211	40.377	-10.545
250	49.431	263.152	273.042	-2.473	12.578	47.246	-9.871
298.15	53.187	272.187	272.187	0.	12.134	53.965	-9.454
300	53.321	272.517	272.188	0.099	12.119	54.225	-9.441
350	56.696	280.995	272.850	2.851	11.807	61.269	-9.144
400	59.639	288.763	274.360	5.761	11.617	68.349	-8.925
450	62.214	295.939	276.364	8.809	11.527	75.447	-8.758
500	64.467	302.613	278.659	11.977	11.519	82.551	-8.624
600	68.161	314.709	283.681	18.617	11.691	96.744	-8.422
700	70.985	325.438	288.894	25.580	12.043	110.894	-8.275
800	73.153	335.064	294.074	32.792	12.513	124.985	-8.161
900	74.831	343.781	299.121	40.195	13.062	139.011	-8.068
1000	76.145	351.736	303.990	47.746	13.663	152.974	-7.991
1100	77.185	359.044	308.668	55.415	14.299	166.874	-7.924
1200	78.020	365.798	313.151	63.176	14.956	180.716	-7.866
1300	78.697	372.070	317.445	71.013	15.627	194.502	-7.815
1400	79.252	377.923	321.558	78.912	16.304	208.236	-7.769
1500	79.712	383.407	325.500	86.860	16.982	221.922	-7.728
1600	80.097	388.564	329.282	94.851	17.657	235.563	-7.690
1700	80.422	393.430	332.914	102.878	18.324	249.161	-7.656
1800	80.699	398.035	336.405	110.934	18.981	262.721	-7.624
1900	80.936	402.405	339.764	119.016	19.626	276.244	-7.594
2000	81.141	406.561	343.001	127.120	20.254	289.734	-7.567
2100	81.319	410.525	346.123	135.244	20.865	303.193	-7.542
2200	81.475	414.311	349.137	143.384	21.457	316.624	-7.518
2300	81.612	417.936	352.050	151.538	22.027	330.027	-7.495
2400	81.732	421.412	354.868	159.705	22.574	343.406	-7.474
2500	81.840	424.751	357.597	167.884	23.097	356.764	-7.454
2600	81.935	427.963	360.242	176.073	23.596	370.101	-7.435
2700	82.021	431.056	362.808	184.271	24.068	383.419	-7.418
2800	82.097	434.041	365.299	192.477	24.515	396.720	-7.401
2900	82.167	436.923	367.719	200.690	24.935	410.005	-7.385
3000	82.229	439.710	370.073	208.910	25.329	423.277	-7.370
3100	82.286	442.407	372.363	217.136	25.698	436.536	-7.356
3200	82.338	445.020	374.593	225.367	26.041	449.783	-7.342
3300	82.385	447.554	376.766	233.603	26.360	463.019	-7.329
3400	82.428	450.015	378.884	241.844	26.656	476.247	-7.317
3500	82.468	452.405	380.951	250.089	26.929	489.466	-7.305
3600	82.505	454.728	382.968	258.337	27.182	502.678	-7.294
3700	82.538	456.989	384.938	266.589	27.415	515.883	-7.283
3800	82.569	459.191	386.863	274.845	27.630	529.083	-7.273
3900	82.598	461.336	388.745	283.103	27.829	542.276	-7.263
4000	82.625	463.428	390.586	291.364	28.014	555.464	-7.254
4100	82.649	465.468	392.388	299.628	28.185	568.649	-7.245
4200	82.672	467.460	394.152	307.894	28.345	581.829	-7.236
4300	82.694	469.406	395.879	316.162	28.494	595.006	-7.228
4400	82.714	471.307	397.572	324.433	28.635	608.178	-7.220
4500	82.733	473.166	399.231	332.705	28.768	621.347	-7.212
4600	82.750	474.984	400.859	340.979	28.894	634.514	-7.205
4700	82.766	476.764	402.455	349.255	29.015	647.679	-7.198
4800	82.782	478.507	404.021	357.533	29.131	660.841	-7.191
4900	82.796	480.214	405.559	365.811	29.243	674.001	-7.185
5000	82.810	481.887	407.068	374.092	29.352	687.158	-7.179
5100	82.823	483.527	408.552	382.373	29.456	700.313	-7.173
5200	82.835	485.135	410.009	390.656	29.558	713.465	-7.167
5300	82.847	486.713	411.441	398.940	29.657	726.617	-7.161
5400	82.857	488.262	412.850	407.226	29.752	739.766	-7.156
5500	82.868	489.782	414.235	415.512	29.844	752.913	-7.151
5600	82.877	491.276	415.597	423.799	29.932	766.060	-7.146
5700	82.887	492.743	416.938	432.087	30.015	779.204	-7.141
5800	82.895	494.184	418.257	440.376	30.093	792.348	-7.136
5900	82.904	495.601	419.556	448.666	30.164	805.488	-7.131
6000	82.912	496.995	420.835	456.957	30.228	818.629	-7.127

PREVIOUS. December 1965 (1 atm) CURRENT: December 1965 (1 bar)

Nitryl Chloride (NO_2Cl) $Cl_1N_1O_2(g)$

SODIUM CHLORIDE (NaCl) CRYSTAL M_r = 58.44277

$S°(298.15\ K) = 72.115 \pm 0.21\ J\ K^{-1}\ mol^{-1}$
$T_{fus} = 1073.8 \pm 1.0\ K$

$\Delta_f H°(0\ K) = -410.693 \pm 0.34\ kJ\ mol^{-1}$
$\Delta_f H°(298.15\ K) = -411.120 \pm 0.34\ kJ\ mol^{-1}$
$\Delta_{fus} H° = 28.158 \pm 0.17\ kJ\ mol^{-1}$

Enthalpy of Formation

The enthalpy of formation is based on $\Delta_f H°(298.15\ K) = -57.39$ and $-39.95\ kcal\ mol^{-1}$ for $Na^+(aq,\infty)$ and $Cl^-(aq,\infty)$, respectively, and on the enthalpy of solution, $\Delta_{sln} H°(298.15\ K) = 0.922\ kcal\ mol^{-1}$ for NaCl. These values were reported in a personal communication from Wagman (1).

Heat Capacity and Entropy

The low temperature values are based on the heat capacities (3-266 K) reported by Barron et al. (2). The earlier work of Clusius et al., (3) results in a slightly higher entropy, $S°(298.15\ K) = 17.33\ cal\ K^{-1}\ mol^{-1}$, but a lower heat capacity at room temperature, $C_p°(298.15\ K) = 11.9\ cal\ K^{-1}\ mol^{-1}$. The high temperature values have been obtained by re-analysis of the enthalpies (672 - 1279 K) of Dawson et al. (4), using the new value for $C_p°(298.15\ K)$. The resulting tables agree within $\pm 0.1\ kcal\ mol^{-1}$ with the enthalpies (812 - 1058 K) of Roth and Bertram (5). The enthalpies (540 - 1037 K) of Magnus (6) agree well at the lower temperatures but are about $0.3\ kcal\ mol^{-1}$ smaller near the melting point.

Fusion Data

The melting temperature was taken from Dawson et al. (4) and the heat of melting was obtained by re-analysis of their crystal and liquid enthalpies. Values of $T_{fus} = 1073$ K and $\Delta_{fus} H° = 6.69 \pm 0.06\ kcal\ mol^{-1}$ were reported by Dworkin and Bredig (7).

Sublimation Data

The vapor in equilibrium with NaCl contains both monomeric and polymeric forms (for details, see the tables for NaCl gases).

References

1. D. D. Wagman, U. S. Nat. Bur. Stand., personal communication, June, 1964.
2. T. H. K. Barron, A. J. Leadbetter, and J. A. Morrison, Proc. Roy. Soc. (London) A279 (1376), 62 (1964).
3. K. Clusius, J. Goldman, and A. Perlick, Z. Naturforsch. 4a, 424 (1949).
4. R. Dawson, E. B. Brackett and T. E. Brackett, J. Phys. Chem. 67, 1669 (1963).
5. W. A. Roth and W. Bertram, Z. Elektrochem. 35, 38 (1929).
6. A. Magnus, Physik. Zeitschr. 14, 5 (1913).
7. A. S. Dworkin and M. A. Bredig, J. Phys. Chem. 64, 269 (1960).

Sodium Chloride (NaCl) $Cl_1Na_1(cr)$

Enthalpy Reference Temperature = T_r = 298.15 K   Standard State Pressure = p° = 0.1 MPa

T/K	$C_p°$	$S°$	$-[G°-H°(T_r)]/T$	$H°-H°(T_r)$	$\Delta_f H°$	$\Delta_f G°$	Log K_f
	J K⁻¹mol⁻¹			kJ mol⁻¹			
0	0.	0.	INFINITE	-10.611	-410.693	-410.693	INFINITE
100	34.932	23.723	114.140	-9.042	-411.920	-402.460	210.224
200	46.873	52.605	76.643	-4.807	-411.660	-393.046	102.653
298.15	50.509	72.115	72.115	0.	-411.120	-384.024	67.279
300	50.543	72.428	72.116	0.093	-411.110	-383.856	66.835
400	52.350	87.226	74.122	5.242	-413.318	-374.636	48.922
500	53.940	99.080	77.966	10.557	-412.890	-365.010	38.132
600	55.476	109.050	82.337	16.028	-412.252	-355.491	30.948
700	57.204	117.727	86.786	21.659	-411.408	-346.095	25.826
800	59.312	125.498	91.147	27.481	-410.346	-336.834	21.993
900	61.869	132.627	95.365	33.536	-409.038	-327.721	19.020
1000	64.865	139.296	99.428	39.868	-407.461	-318.768	16.651
1073.800	67.371	144.002	102.330	44.747	------ CRYSTAL <--> LIQUID -----		
1100	68.325	145.637	103.343	46.524	-405.588	-309.986	14.720
1200	71.965	151.742	107.123	53.542	-500.162	-298.950	13.013
1300	75.145	157.633	110.783	60.905	-496.764	-282.318	11.344
1400	77.530	163.293	114.333	68.544	-493.094	-265.958	9.923
1500	79.245	168.704	117.779	76.388	-489.224	-249.868	8.701

PREVIOUS: December 1960   CURRENT: September 1964

Sodium Chloride (NaCl) $Cl_1Na_1(cr)$

SODIUM CHLORIDE (NaCl) LIQUID M_r = 58.44277

$S°(298.15\ K) = [95.055]\ J\ K^{-1}\ mol^{-1}$

$T_{fus} = 1073.8 \pm 1.0\ K$

$\Delta_f H°(298.15\ K) = [-385.923]\ kJ\ mol^{-1}$

$\Delta_{fus} H° = 28.158 \pm 0.17\ kJ\ mol^{-1}$

Enthalpy of Formation

$\Delta_f H°$(NaCl, ℓ, 298.15 K) is calculated from that of the crystal by adding $\Delta_{fus}H°$ and the difference in enthalpy, H°(1073.8 K) - H°(298.15 K), between the crystal and liquid.

Heat Capacity and Entropy

These were obtained by analysis of the enthalpies (1076 - 1279 K) of Dawson et al. (1). The heat capacity was extrapolated to a constant value of 16.0 at the higher temperatures. A glass transition was assumed at 800 K and below this temperature $C_p°$ was taken to that of the crystal.

S°(298.15 K) is calculated in a manner similar to that used for the enthalpy of formation.

Fusion Data

The melting temperature was taken from Dawson et al. (1) and the heat of melting was obtained by reanalysis of their crystal and liquid enthalpies. Values of T_{fus} = 1073 K and $\Delta_{fus}H°$ = 6.69±0.06 kcal mol^{-1} were reported by Dworkin and Bredig (2).

Sublimation Data

The vapor in equilibrium with NaCl contains both monomeric and polymric forms (for details see the tables for NaCl gases).

References

1. R. Dawson, E. B. Brackett and T. E. Brackett, J. Phys. Chem. 67, 1669 (1963).
2. A. S. Dworkin and M. A. Bredig, J. Phys. Chem. 64, 269 (1960).

Sodium Chloride (NaCl) $Cl_1Na_1(l)$

Enthalpy Reference Temperature = T_r = 298.15 K   Standard State Pressure = p° = 0.1 MPa

T/K	$C_p°$ (J K⁻¹mol⁻¹)	S° (J K⁻¹mol⁻¹)	-[G°-H°(T_r)]/T (J K⁻¹mol⁻¹)	H°-H°(T_r) (kJ mol⁻¹)	$\Delta_f H°$ (kJ mol⁻¹)	$\Delta_f G°$ (kJ mol⁻¹)	Log K_f
0							
100							
200							
298.15	50.509	95.055	95.055	0.	-385.923	-365.667	64.063
300	50.543	95.367	95.055	0.093	-385.913	-365.541	63.646
400	52.350	110.165	97.061	5.242	-388.121	-358.615	46.830
500	53.940	122.019	100.905	10.557	-387.693	-351.283	36.698
600	55.476	131.989	105.276	16.028	-387.055	-344.058	29.953
700	57.204	140.666	109.725	21.659	-386.212	-336.955	25.144
800	59.312	148.437	114.086	27.481	-385.149	-329.988	21.546
800.000	59.312	148.437	114.086	27.481	GLASS <--> LIQUID	TRANSITION	
800.000	76.400	148.437	114.086	27.481			
900	74.852	157.354	118.410	35.050	-382.327	-323.264	18.762
1000	72.509	165.127	122.702	42.425	-379.707	-316.845	16.550
1073.800	70.668	170.225	125.796	47.708	------ CRYSTAL <-->	LIQUID -----	
1100	70.082	171.921	126.874	49.552	-377.364	-310.674	14.753
1200	68.325	177.937	130.883	56.464	-472.043	-302.265	13.157
1300	67.446	183.367	134.715	63.248	-469.224	-288.232	11.581
1400	67.070	188.351	138.371	69.972	-466.469	-274.413	10.238
1500	66.944	192.973	141.858	76.672	-463.744	-260.790	9.082
1600	66.944	197.293	145.190	83.366	-461.028	-247.349	8.075
1700	66.944	201.352	148.375	90.061	-458.317	-234.077	7.192
1800	66.944	205.178	151.426	96.755	-455.610	-220.964	6.412
1900	66.944	208.798	154.351	103.449	-452.909	-208.002	5.718
2000	66.944	212.232	157.160	110.144	-450.213	-195.182	5.098
2100	66.944	215.498	159.861	116.838	-447.524	-182.497	4.539
2200	66.944	218.612	162.461	123.533	-444.842	-169.939	4.035
2300	66.944	221.588	164.967	130.227	-442.169	-157.503	3.577
2400	66.944	224.437	167.386	136.921	-439.506	-145.184	3.160
2500	66.944	227.170	169.723	143.616	-436.855	-132.975	2.778

PREVIOUS: December 1960   CURRENT: September 1964

Sodium Chloride (NaCl) $Cl_1Na_1(l)$

SODIUM CHLORIDE (NaCl) CRYSTAL-LIQUID M_r = 58.44277

0 to 1073.8 K crystal
above 1073.8 K liquid

Refer to the individual tables for details.

Sodium Chloride (NaCl) Cl_1Na_1(cr,l)

Enthalpy Reference Temperature = T_r = 298.15 K Standard State Pressure = p° = 0.1 MPa

T/K	C_p° (J K⁻¹mol⁻¹)	S° (J K⁻¹mol⁻¹)	$-[G^\circ - H^\circ(T_r)]/T$ (J K⁻¹mol⁻¹)	$H^\circ - H^\circ(T_r)$ (kJ mol⁻¹)	$\Delta_f H^\circ$ (kJ mol⁻¹)	$\Delta_f G^\circ$ (kJ mol⁻¹)	Log K_f
0	0.	0.	INFINITE	-10.611	-410.693	-410.693	INFINITE
100	34.932	23.723	114.140	-9.042	-411.920	-402.460	210.224
200	46.873	52.605	76.643	-4.807	-411.660	-393.046	102.653
298.15	50.509	72.115	72.115	0.	-411.120	-384.024	67.279
300	50.543	72.428	72.116	0.093	-411.110	-383.856	66.835
400	52.350	87.226	74.122	5.242	-413.318	-374.636	48.922
500	53.940	99.080	77.966	10.557	-412.890	-365.010	38.132
600	55.476	109.050	82.337	16.028	-412.252	-355.491	30.948
700	57.204	117.727	86.786	21.659	-411.408	-346.095	25.826
800	59.312	125.498	91.147	27.481	-410.346	-336.834	21.993
900	61.869	132.627	95.365	33.536	-409.038	-327.721	19.020
1000	64.865	139.296	99.428	39.868	-407.461	-318.768	16.651
1073.800	67.371	144.002	102.330	44.747	CRYSTAL <--> LIQUID	TRANSITION	
1073.800	70.668	170.225	102.330	72.905			
1100	70.082	171.921	103.968	74.748	-377.364	-310.674	14.753
1200	68.325	177.937	109.886	81.661	-472.043	-302.265	13.157
1300	67.446	183.367	115.333	88.445	-469.224	-288.232	11.581
1400	67.070	188.351	120.373	95.169	-466.469	-274.413	10.238
1500	66.944	192.973	125.061	101.869	-463.744	-260.790	9.082
1600	66.944	197.293	129.442	108.563	-461.028	-247.349	8.075
1700	66.944	201.352	133.553	115.257	-458.317	-234.077	7.192
1800	66.944	205.178	137.427	121.952	-455.610	-220.964	6.412
1900	66.944	208.798	141.089	128.646	-452.909	-208.002	5.718
2000	66.944	212.232	144.561	135.341	-450.213	-195.182	5.098
2100	66.944	215.498	147.862	142.035	-447.524	-182.497	4.539
2200	66.944	218.612	151.008	148.729	-444.842	-169.939	4.035
2300	66.944	221.588	154.012	155.424	-442.169	-157.503	3.577
2400	66.944	224.437	156.888	162.118	-439.506	-145.184	3.160
2500	66.944	227.170	159.645	168.813	-436.855	-132.975	2.778

PREVIOUS: CURRENT: September 1964

Sodium Chloride (NaCl) Cl_1Na_1(cr,l)

$S°(298.15\ K) = 229.793\ J\ K^{-1}\ mol^{-1}$

$\Delta_f H°(0\ K) = -180.00 \pm 2.1\ kJ\ mol^{-1}$
$\Delta_f H°(298.15\ K) = -181.42 \pm 2.1\ kJ\ mol^{-1}$

Electronic Levels and Quantum Weights

State	ε_i, cm^{-1}	g_i
$^1\Sigma^+$	0	1

$\omega_e = 363.62\ cm^{-1}$ $\omega_e x_e = 1.72\ cm^{-1}$ $\sigma = 1$
$B_e = 0.21691\ cm^{-1}$ $\alpha_e = 0.001598\ cm^{-1}$ $r_e = 2.3606$ Å

Enthalpy of Formation

The enthalpy of formation is based on a enthalpy of sublimation $\Delta_{sub}H°(298.15\ K) = 54.9\ kcal\ mol^{-1}$ obtained from the more consistent sets of vapor pressure data by correction for the presence of dimer. Observed vapor pressures were converted to monomer pressures by use of dimer-monomer ratios calculated from JANAF Gibbs energy functions and $\Delta_r H°(298.15\ K) = 48.6\ kcal\ mol^{-1}$ for the reaction $(NaCl)_2(g) \rightarrow 2\ NaCl(g)$. Higher polymeric species were neglected. Analyses of the resulting monomer pressures are summarized below.

Source	T/K	$\Delta_{sub}H°(298.15\ K)$, $kcal\ mol^{-1}$ 2nd law	3rd law
Nesmeyanov & Sazonov (1)	743 - 948*	54.35	54.94
Zimm & Mayer (2)	760 - 897*	55.15	54.91
Mayer & Wintner (3)	890 - 944*	49.6	55.37
Niwa (4)	883 - 973*	52.6	55.03
Barton & Bloom (5)	1340 - 1540**	56.67	54.68
Flock & Rodebush (6)	1250 - 1428**	55.65	54.98
Wartenberg & Albrecht (7)	1429 - 1720**	60.0	54.45
Ruff & Mugden (8)	1433 - 1723**	59.5	54.44

*Knudsen effusion measurements over crystal **total pressures over liquid

Heat Capacity and Entropy

The molecular constants are based on molecular beam studies of the rotational spectra by Clouser and Gordy (9) and on microwave studies by Honig et al. (10). The infrared spectra of Rice and Klemperer (11) give a frequency in close agreement ($366\ cm^{-1}$).

References

1. A. N. Nesmeyanov and L. A. Sazanov, Zhur. Neorg. Khim. 2, 946 (1957).
2. B. H. Zimm and J. E. Mayer, J. Chem. Phys. 12, 362 (1944).
3. J. E. Mayer and J. H. Wintner, J. Chem. Phys. 6, 301 (1938).
4. K. Niwa, J. Fac. Sci., Hokkaido Univ., Ser. III, 2, 201 (1938).
5. J. L. Barton and H. Bloom, J. Phys. Chem. 60, 1413 (1956).
6. E. F. Flock and W. H. Rodebush, J. Amer. Chem. Soc. 48, 2522 (1926).
7. H. von Wartenberg and P. Albrecht, Z. Elektrochem. 27, 162 (1921).
8. O. Ruff and S. Mugdan, Z. Anorg. Allgem. Chem. 117, 147 (1921).
9. Clouser and Gordy, Bull. Am. Phys. Soc. 8, 326 (1963).
10. Honig, Mandel, Stitch and Townes, Phys. Rev. 96, 629 (1954).
11. Rice and Klemperer, J. Chem. Phys. 27, 573 (1957).

Enthalpy Reference Temperature = T_r = 298.15 K Standard State Pressure = p° = 0.1 MPa

T/K	$C°_p$ (J K⁻¹mol⁻¹)	S°	$-[G°-H°(T_r)]/T$	$H°-H°(T_r)$ (kJ mol⁻¹)	$\Delta_f H°$	$\Delta_f G°$	Log K_f
0	0.	0.	INFINITE	-9.616	-179.997	-179.997	INFINITE
100	30.394	193.507	260.320	-6.681	-179.858	-187.377	97.876
200	34.090	215.819	233.022	-3.440	-180.592	-194.620	50.829
250	35.125	223.546	230.379	-1.708	-180.993	-198.081	41.387
298.15	35.786	229.793	229.793	0.	-181.418	-201.334	35.273
300	35.806	230.014	229.794	0.066	-181.436	-201.458	35.077
350	36.272	235.571	230.231	1.869	-181.945	-204.756	30.558
400	36.604	240.437	231.209	3.691	-185.167	-207.769	27.132
450	36.852	244.764	232.479	5.528	-185.781	-210.557	24.441
500	37.044	248.657	233.906	7.375	-186.370	-213.279	22.281
600	37.324	255.437	236.946	11.095	-187.483	-218.555	19.027
700	37.524	261.206	240.010	14.838	-188.528	-223.650	16.689
800	37.680	266.228	242.980	18.598	-189.526	-228.599	14.926
900	37.809	270.673	245.815	22.373	-190.499	-233.424	13.548
1000	37.921	274.663	248.503	26.159	-191.468	-238.142	12.439
1100	38.022	278.282	251.049	29.957	-192.454	-242.761	11.528
1200	38.116	281.594	253.458	33.764	-290.239	-244.850	10.658
1300	38.205	284.649	255.741	37.580	-290.387	-241.062	9.686
1400	38.289	287.483	257.909	41.404	-290.532	-237.262	8.852
1500	38.371	290.128	259.969	45.237	-290.673	-233.452	8.130
1600	38.450	292.607	261.933	49.079	-290.811	-229.633	7.497
1700	38.528	294.940	263.806	52.927	-290.945	-225.805	6.938
1800	38.604	297.144	265.598	56.784	-291.076	-221.969	6.441
1900	38.679	299.234	267.313	60.648	-291.205	-218.127	5.997
2000	38.754	301.219	268.959	64.520	-291.332	-214.277	5.596
2100	38.828	303.112	270.541	68.399	-291.458	-210.421	5.234
2200	38.901	304.920	272.063	72.285	-291.585	-206.559	4.904
2300	38.973	306.651	273.529	76.179	-291.712	-202.692	4.603
2400	39.045	308,311	274.944	80.080	-291.843	-198.818	4.327
2500	39.117	309.906	276.311	83.988	-291.978	-194.940	4.073
2600	39.189	311.442	277.633	87.903	-292.118	-191.055	3.838
2700	39.260	312.922	278.913	91.826	-292.266	-187.165	3.621
2800	39.331	314.351	280.153	95.756	-292.422	-183.270	3.419
2900	39.402	315.733	281.356	99.692	-292.589	-179.369	3.231
3000	39.473	317.070	282.524	103.636	-292.768	-175.462	3.055
3100	39.544	318.365	283.660	107.587	-292.960	-171.548	2.891
3200	39.614	319.622	284.764	111.545	-293.167	-167.629	2.736
3300	39.685	320.842	285.839	115.510	-293.389	-163.702	2.591
3400	39.755	322.028	286.886	119.482	-293.629	-159.769	2.455
3500	39.825	323.181	287.907	123.461	-293.886	-155.828	2.326
3600	39.895	324.304	288.902	127.447	-294.162	-151.879	2.204
3700	39.965	325.398	289.874	131.440	-294.459	-147.923	2.088
3800	40.035	326.465	290.823	135.440	-294.777	-143.959	1.979
3900	40.105	327.506	291.750	139.447	-295.116	-139.985	1.875
4000	40.175	328.522	292.657	143.461	-295.480	-136.003	1.776
4100	40.245	329.515	293.544	147.482	-295.868	-132.011	1.682
4200	40.315	330.485	294.412	151.510	-296.280	-128.010	1.592
4300	40.385	331.435	295.262	155.545	-296.721	-123.998	1.506
4400	40.455	332.364	296.094	159.587	-297.148	-119.980	1.424
4500	40.524	333.274	296.911	163.636	-297.625	-115.949	1.346
4600	40.594	334.165	297.711	167.692	-298.139	-111.906	1.271
4700	40.664	335.039	298.496	171.755	-298.684	-107.851	1.199
4800	40.733	335.896	299.266	175.825	-299.260	-103.785	1.129
4900	40.803	336.737	300.022	179.901	-299.871	-99.706	1.063
5000	40.873	337.562	300.765	183.985	-300.509	-95.616	0.999
5100	40.942	338.372	301.494	188.076	-301.191	-91.511	0.937
5200	41.012	339.168	302.211	192.174	-301.912	-87.393	0.878
5300	41.082	339.949	302.916	196.278	-302.674	-83.260	0.821
5400	41.151	340.718	303.609	200.390	-303.479	-79.113	0.765
5500	41.221	341.474	304.290	204.509	-304.329	-74.950	0.712
5600	41.290	342.217	304.961	208.634	-305.227	-70.771	0.660
5700	41.360	342.948	305.621	212.767	-306.173	-66.576	0.610
5800	41.429	343.668	306.271	216.906	-307.170	-62.364	0.562
5900	41.499	344.377	306.911	221.052	-308.220	-58.135	0.515
6000	41.568	345.075	307.541	225.206	-309.167	-53.905	0.469

PREVIOUS: December 1964 (1 atm) CURRENT: December 1964 (1 bar)

SODIUM PERCHLORATE ($NaClO_4$) CRYSTAL M_r = 122.44037

$S°(298.15\ K)$ = [142.256] $J\ K^{-1}\ mol^{-1}$
T_{trs} = 581 K
T_{fus} = 755 K

$\Delta_f H°(0\ K)$ = Unknown
$\Delta_f H°(298.15\ K) = -382.752 \pm 0.92\ kJ\ mol^{-1}$
$\Delta_{trs}H°$ = [13.98] $kJ\ mol^{-1}$
$\Delta_{fus}H°$ = Unknown

Enthalpy of Formation

Gilliland and Johnson (1) measured the enthalpy of solution of $NaClO_4$(cr) and gave $\Delta_f H°(298.15\ K) = -92.48 \pm 0.22\ kcal\ mol^{-1}$ for $NaClO_4$(cr). The enthalpy of dilution of $NaClO_4$(cr) was measured by Vanderzee and Swanson. The data led to $\Delta_f H°(298.15\ K) = -91.464\ kcal\ mol^{-1}$ which agrees very well with the data reported by Gilliland and Johnson (1). The value of $\Delta_f H°(298.15\ K)$ was reported to be $-92.18\ kcal\ mol^{-1}$ (3), based upon the data prior to 1950. Vorob'ev, et al. (4) measured the enthalpy of decomposition of $NaClO_4$(cr) into NaCl(cr) and O_2(g) and gave $\Delta_f H°(298.15\ K) = -90.68 \pm 0.3\ kcal\ mol^{-1}$ for $NaClO_4$(cr). The value reported by Gilliland and Johnson (1) was used.

Heat Capacity and Entropy

Heat capacities and S°(298.15 K) wee estimated by comparison with those for $NaClO_3$(cr) and $KClO_4$(cr), respectively.

Transition Data

T_{trs} was reported in (3). $\Delta_{trs}H°$ was estimated from that for $KClO_4$(cr).

Fusion Data

T_{fus} was obtained from (5).

References

1. A. A. Gilliland and W. H. Johnson, J. Res. Nat. Bur. Stand. 65A, 67 (1961).
2. C. E. Vanderzee and J. A. Swanson, J. Phys. Chem. 67, 285 (1963).
3. U. S. Nat. Bur. Stand. Circ. 500, 1952.
4. A. F. Vorob'ev, et al., Proc. Acad. Sci. USSR, 135, 1439 (1960).
5. Bulletin of the National Research Council, No. 118, "Data on Chemicals for Ceramic Use," June 1949.

Sodium Perchlorate ($NaClO_4$) $Cl_1Na_1O_4$(cr)

Enthalpy Reference Temperature = T_r = 298.15 K Standard State Pressure = p° = 0.1 MPa

T/K	$C_p°$ (J K⁻¹mol⁻¹)	S° (J K⁻¹mol⁻¹)	$-[G°-H°(T_r)]/T$ (J K⁻¹mol⁻¹)	$H°-H°(T_r)$ (kJ mol⁻¹)	$\Delta_f H°$ (kJ mol⁻¹)	$\Delta_f G°$ (kJ mol⁻¹)	Log K_f
0							
100							
200							
298.15	111.294	142.256	142.256	0.	-382.752	-254.240	44.542
300	111.713	142.946	142.258	0.206	-382.738	-253.443	44.128
400	135.980	178.647	146.961	12.675	-383.569	-210.358	27.470
500	150.206	210.598	156.559	27.020	-380.228	-167.415	17.490
581.000	158.574	233.798	165.736	39.544	I <--> II	TRANSITION	
581.000	158.574	257.836	165.736	53.511			
600	160.247	262.966	168.734	56.539	-361.860	-125.708	10.944
700	168.511	288.302	184.037	72.986	-356.711	-86.747	6.473
800	175.728	311.286	198.528	90.207	-350.923	-48.569	3.171
900	181.586	332.327	212.242	108.076	-344.612	-11.149	0.647
1000	187.025	351.744	225.234	126.510	-337.858	25.544	-1.334
1100	191.962	369.803	237.565	145.461	-330.708	61.542	-2.922
1200	196.648	386.708	249.297	164.894	-419.964	99.312	-4.323
1300	201.083	402.625	260.485	184.782	-411.207	142.232	-5.715
1400	205.267	417.682	271.180	205.102	-402.083	184.465	-6.882
1500	209.200	431.979	281.428	225.828	-392.615	226.032	-7.871

PREVIOUS: CURRENT: June 1963

NICKEL CHLORIDE (NiCl) IDEAL GAS M_r = 94.143

$\Delta_f H°(0\ K)$ = 181.92 ± 4.2 kJ mol^{-1}

$S°(298.15\ K)$ = [251.9 ± 13] J K^{-1} mol^{-1} $\Delta_f H°(298.15\ K)$ = 182.00 ± 4.2 kJ mol^{-1}

Electronic Levels and Quantum Weights

ε_i, cm^{-1}	g_i	ε_i, cm^{-1}	g_i
0	[6]	[23108]	[6]
[1507]	[4]	[23796]	[4]
[8394]	[10]	[24788]	[4]
[9330]	[8]	[24836]	[2]
[10116]	[6]	[25036]	[6]
[10664]	[4]	[29071]	[4]
[13550]	[8]	[29593]	[2]
[14996]	[6]	[32500]	[10]
		[32524]	[8]

ω_e = [424] cm^{-1} $\omega_e x_e$ = [2] cm^{-1} σ = 1

B_e = [10.167] cm^{-1} α_e = [0.002] cm^{-1} r_e = [2.137] Å

Enthalpy of Formation

The adopted value of $\Delta_f H°(298.15\ K)$, is based on our third law analysis of unpublished mass spectrometry data (7 points, 1400-1540 K) of Hildenbrand (1) for the reaction Ni(g) + $NiCl_2$(g) = 2NiCl(g). Our analysis of this data yields 2nd and 3rd law values for $\Delta_f H°(298.15\ K)$ of 43.0±1.3 kcal mol^{-1} and 43.5±1.0 kcal mol^{-1}, respectively, with a drift of 0.64±1.17 cal K^{-1} mol^{-1}. These values are in good agreement with a value of $\Delta_f H°(298.15\ K)$ = 42.9±5.4 kcal mol^{-1} based on the dissociation energy of 88±5 kcal mol^{-1} reported by Bulewicz et al. (2) using flame photometry. Additional studies include an analysis by Rao and Dadape (3) of the equilibrium $NiCl_2$(g) + Ni(cr) + 2NiCl(g) by the flow method. Our analysis of their data (10 points, 1243-1298 K) using auxiliary JANAF data (4) yields the following values for $\Delta_f H°(298.15\ K)$; 36.9±1.6 kcal mol^{-1} (2nd law) and 28.5±0.8 kcal mol^{-1} (3rd law) with a very large drift of -12.85±2.42 cal K^{-1} mol^{-1}. The large drift probably indicates an experimental problem such as lack of equilibrium or the presence of side reactions. Other studies on the enthalpy of formation are summarized by Rao and Dadape (3) and are not considered here since the uncertainties in these studies are very large.

Heat Capacity and Entropy

The 18 observed band systems of NiCl(g) are adequately summarized by Suchard (5). All of these systems have been observed in emission only and the electronic states giving rise to these transitions have not been assigned. The only rotational analysis by Rao and Rao (6) indicates that the upper and lower states are $^2\Delta_{5/2}$ states but the lower state is not definitely identified with the ground state. In view of these difficulties we have chosen to use the electronic energy levels and quantum weights of Ni^+(g) (7) as an approximation to those of NiCl(g). This approximation generally gives upper limit values for the thermal functions; we estimate an uncertainty of ±3 cal K^{-1} mol^{-1} for S°(298.15 K).

Molecular constant data for the ground state are assigned assuming that the lower state of the G band system represents the ground state (5, 6). While this assignment is not certain, the molecular constant data for most of the observed systems are fairly similar (5) so no gross errors will be introduced if this assignment proves incorrect. On this basis the adopted values of B_e and ω_e are from Rao and Rao (6) and Rao et al. (8), respectively, while $\omega_e x_e$ and α_e are our best estimates.

References

1. D. L. Hildenbrand, SRI International, personal communication, May 3, 1978.
2. E. M. Bulewicz, L. F. Phillips and T. M. Sugden, Trans. Faraday Soc. 57, 921 (1961).
3. B. S. Rao and V. V. Dadape, High Temp. Sci. 3, 1 (1971).
4. JANAF Thermochemical Tables: Ni(cr), 12-31-76; $NiCl_2$(g), 9-30-77.
5. S. N. Suchard, "Spectroscopic Data," Vol. I, IFI/Plenum, New York, 1975.
6. N. V. K. Rao and P. T. Rao, Curr. Sci. 38, 589 (1969).
7. A. G. Shenstone, J. Res. Nat. Bur. Stand. 74A, 801 (1970).
8. S. V. K. Rao, S. Paddi Reddy, and P. T. Rao, Z. Physik 166, 261 (1962).

Nickel Chloride (NiCl) $Cl_1Ni_1(g)$

Enthalpy Reference Temperature = T_r = 298.15 K Standard State Pressure = p° = 0.1 MPa

T/K	$C_p°$ (J K^{-1}mol^{-1})	S° (J K^{-1}mol^{-1})	-[G°-H°(T_r)]/T (J K^{-1}mol^{-1})	H°-H°(T_r) (kJ mol^{-1})	$\Delta_f H°$ (kJ mol^{-1})	$\Delta_f G°$ (kJ mol^{-1})	Log K_f
0	0.	0.	INFINITE	-9.465	181.915	181.915	INFINITE
100	29.843	216.385	281.814	-6.543	182.874	171.431	-89.546
200	33.290	238.153	255.063	-3.382	182.626	160.032	-41.796
250	34.524	245.720	252.461	-1.685	182.340	154.415	-32.263
298.15	35.443	251.883	251.883	0.	182.004	149.067	-26.116
300	35.475	252.102	251.883	0.066	181.990	148.862	-25.919
350	36.268	257.632	252.318	1.860	181.592	143.372	-21.397
400	36.964	262.521	253.294	3.691	181.152	137.941	-18.013
450	37.586	266.911	254.567	5.555	180.676	132.568	-15.388
500	38.136	270.900	256.004	7.448	180.157	127.250	-13.294
600	39.023	277.936	259.089	11.308	178.923	116.780	-10.167
700	39.645	284.001	262.224	15.244	177.700	106.530	-7.949
800	40.049	289.324	265.286	19.230	176.750	96.427	-6.296
900	40.291	294.056	268.225	23.248	175.778	86.445	-5.017
1000	40.425	298.309	271.024	27.285	174.757	76.573	-4.000
1100	40.496	302.166	273.683	31.331	173.671	66.807	-3.172
1200	40.540	305.691	276.205	35.383	172.512	57.142	-2.487
1300	40.583	308.938	278.600	39.439	171.273	47.578	-1.912
1400	40.646	311.947	280.876	43.500	169.948	38.112	-1.422
1500	40.744	314.755	283.042	47.570	168.536	28.744	-1.001
1600	40.885	317.389	285.107	51.651	167.038	19.473	-0.636
1700	41.076	319.873	287.080	55.748	165.451	10.298	-0.316
1800	41.317	322.227	288.967	59.868	146.630	1.934	-0.056
1900	41.608	324.469	290.777	64.013	144.972	-6.059	0.167
2000	41.945	326.611	292.516	68.191	143.339	-13.966	0.365
2100	42.324	328.667	294.189	72.404	141.737	-21.791	0.542
2200	42.739	330.645	295.801	76.657	140.168	-29.542	0.701
2300	43.183	332.555	297.358	80.952	138.636	-37.221	0.845
2400	43.650	334.402	298.863	85.294	137.142	-44.835	0.976
2500	44.134	336.194	300.321	89.683	135.687	-52.387	1.095
2600	44.627	337.934	301.734	94.121	134.273	-59.882	1.203
2700	45.124	339.628	303.106	98.609	132.900	-67.324	1.302
2800	45.619	341.278	304.440	103.146	131.566	-74.715	1.394
2900	46.107	342.887	305.738	107.732	130.272	-82.059	1.478
3000	46.584	344.458	307.003	112.367	129.017	-89.359	1.556
3100	47.047	345.994	308.236	117.049	127.799	-96.618	1.628
3200	47.491	347.494	309.439	121.776	-250.218	-98.650	1.610
3300	47.916	348.962	310.615	126.546	-249.712	-93.921	1.487
3400	48.318	350.399	311.764	131.358	-249.172	-89.209	1.371
3500	48.698	351.805	312.888	136.209	-248.601	-84.512	1.261
3600	49.053	353.182	313.988	141.097	-248.000	-79.832	1.158
3700	49.384	354.530	315.066	146.019	-247.371	-75.169	1.061
3800	49.690	355.851	316.122	150.973	-246.717	-70.524	0.969
3900	49.973	357.146	317.157	155.956	-246.041	-65.896	0.883
4000	50.232	358.414	318.173	160.967	-245.342	-61.286	0.800
4100	50.468	359.658	319.169	166.002	-244.625	-56.693	0.722
4200	50.683	360.876	320.148	171.059	-243.890	-52.118	0.648
4300	50.877	362.071	321.109	176.138	-243.139	-47.561	0.578
4400	51.051	363.243	322.053	181.234	-242.374	-43.022	0.511
4500	51.207	364.392	322.982	186.347	-241.597	-38.500	0.447
4600	51.346	365.519	323.894	191.475	-240.809	-33.995	0.386
4700	51.469	366.625	324.792	196.616	-240.011	-29.508	0.328
4800	51.578	367.710	325 674	201.768	-239.205	-25.037	0.272
4900	51.672	368.774	326.543	206.931	-238.393	-20.584	0.219
5000	51.755	369.819	327.398	212.103	-237.576	-16.147	0.169
5100	51.826	370.844	328.240	217.282	-236.755	-11.727	0.120
5200	51.887	371.851	329.069	222.467	-235.933	-7.323	0.074
5300	51.939	372.840	329.886	227.659	-235.110	-2.934	0.029
5400	51.983	373.811	330.690	232.855	-234.287	1.439	-0.014
5500	52.019	374.766	331.483	238.055	-233.468	5.797	-0.055
5600	52.048	375.703	332.264	243.258	-232.652	10.140	-0.095
5700	52.072	376.625	333.034	248.465	-231.842	14.468	-0.133
5800	52.090	377.530	333.794	253.673	-231.040	18.782	-0.169
5900	52.104	378.421	334.543	258.882	-230.246	23.083	-0.204
6000	52.114	379.297	335.281	264.093	-229.463	27.370	-0.238

PREVIOUS: September 1977 (1 atm) CURRENT: September 1977 (1 bar)

Nickel Chloride (NiCl) $Cl_1Ni_1(g)$

CHLORINE OXIDE (ClO) IDEAL GAS M_r = 51.4524

$S°(298.15\ K) = 226.648\ J\ K^{-1}\ mol^{-1}$

$\Delta_fH°(0\ K) = 101.30 \pm 2.1\ kJ\ mol^{-1}$
$\Delta_fH°(298.15\ K) = 101.22 \pm 2.1\ kJ\ mol^{-1}$

Electronic Level and Quantum Weight

State	ε_i, cm^{-1}	g_i
$^2\Pi_{3/2}$	0	4

$\omega_e = 866 \pm 26\ cm^{-1}$ $\omega_e x_e = 7.5\ cm^{-1}$ $\sigma = 1$
$B_e = 0.643(3)$ $\alpha_e = [0.0069]\ cm^{-1}$ $D_e = 2.(2) \times 10^{-6}\ cm^{-1}$ $r_e = 1.546$ Å

Enthalpy of Formation

Durie and Ramsay (1) and Herzberg and Ramsay (2) report $D_0^0 = 22152 \pm 10\ cm^{-1}$ and $D_0^0 = 22141\ cm^{-1}$ respectively, and these values have been averaged.

Heat Capacity and Entropy

The spectroscopic constants are for the natural mixture of isotopes and are based on measurements by Durie and Ramsay (1) and by Porter (3). α_e has been estimated by the method given by Herzberg (4).

References

1. R. A. Durie and D. A. Ramsay, Can. J. Phys. 36, 35 (1958).
2. G. Herzberg and D. A. Ramsay, Disc. Faraday Soc. 9, 80 (1950).
3. G. Porter, Disc. Faraday Soc. 9, 60 (1950).
4. G. Herzberg, "Spectra of Diatomic Molecules", 2nd ed., D. Van Nostrand Company, Inc., New York, p. 108, 1950.

Chlorine Oxide (ClO) $Cl_1O_1(g)$

Enthalpy Reference Temperature = T_r = 298.15 K   Standard State Pressure = p° = 0.1 MPa

T/K	$C_p°$ (J K⁻¹mol⁻¹)	S° (J K⁻¹mol⁻¹)	$-[G°-H°(T_r)]/T$ (J K⁻¹mol⁻¹)	$H°-H°(T_r)$ (kJ mol⁻¹)	$\Delta_fH°$ (kJ mol⁻¹)	$\Delta_fG°$ (kJ mol⁻¹)	Log K_f
0	0.	0.	INFINITE	-8.850	101.301	101.301	INFINITE
100	29.115	194.124	253.544	-5.942	101.301	100.004	-52.237
200	29.811	214.434	229.475	-3.008	101.261	98.719	-25.783
250	30.651	221.173	227.162	-1.497	101.233	98.087	-20.494
298.15	31.546	226.648	226.648	0.	101.219	97.482	-17.078
300	31.580	226.843	226.649	0.058	101.219	97.459	-16.969
350	32.458	231.778	227.036	1.660	101.223	96.832	-14.451
400	33.228	236.164	227.908	3.302	101.243	96.204	-12.563
450	33.881	240.117	229.049	4.981	101.274	95.572	-11.094
500	34.428	243.716	230.338	6.689	101.314	94.936	-9.918
600	35.267	250.071	233.111	10.176	101.405	93.653	-8.153
700	35.862	255.555	235.935	13.734	101.500	92.353	-6.891
800	36.297	260.374	238.695	17.343	101.591	91.040	-5.944
900	36.627	264.669	241.346	20.990	101.675	89.716	-5.207
1000	36.884	268.541	243.875	24.666	101.752	88.383	-4.617
1100	37.092	272.067	246.280	28.365	101.821	87.043	-4.133
1200	37.264	275.302	248.566	32.083	101.884	85.696	-3.730
1300	37.409	278.291	250.739	35.817	101.940	84.345	-3.389
1400	37.536	281.068	252.807	39.564	101.990	82.990	-3.096
1500	37.647	283.661	254.779	43.324	102.033	81.631	-2.843
1600	37.747	286.094	256.661	47.093	102.069	80.270	-2.621
1700	37.839	288.385	258.460	50.873	102.098	78.906	-2.424
1800	37.923	290.551	260.183	54.661	102.120	77.541	-2.250
1900	38.001	292.603	261.836	58.457	102.133	76.175	-2.094
2000	38.075	294.554	263.424	62.261	102.137	74.809	-1.954
2100	38.145	296.414	264.951	66.072	102.131	73.443	-1.827
2200	38.212	298.190	266.421	69.890	102.114	72.077	-1.711
2300	38.277	299.890	267.840	73.715	102.086	70.712	-1.606
2400	38.339	301.520	269.209	77.545	102.046	69.349	-1.509
2500	38.399	303.086	270.533	81.382	101.993	67.988	-1.421
2600	38.458	304.594	271.815	85.225	101.927	66.629	-1.339
2700	38.516	306.046	273.056	89.074	101.847	65.273	-1.263
2800	38.572	307.448	274.259	92.928	101.753	63.920	-1.192
2900	38.627	308.802	275.427	96.788	101.645	62.570	-1.127
3000	38.682	310.113	276.562	100.654	101.523	61.225	-1.066
3100	38.735	311.382	277.664	104.525	101.388	59.884	-1.009
3200	38.788	312.613	278.737	108.401	101.239	58.547	-0.956
3300	38.840	313.807	279.782	112.282	101.078	57.216	-0.906
3400	38.892	314.967	280.800	116.169	100.905	55.889	-0.859
3500	38.944	316.095	281.792	120.061	100.723	54.568	-0.814
3600	38.994	317.193	282.761	123.957	100.530	53.252	-0.773
3700	39.045	318.262	283.706	127.859	100.330	51.941	-0.733
3800	39.095	319.304	284.629	131.766	100.124	50.636	-0.696
3900	39.145	320.320	285.531	135.678	99.912	49.337	-0.661
4000	39.195	321.312	286.413	139.595	99.696	48.043	-0.627
4100	39.244	322.280	287.276	143.517	99.479	46.754	-0.596
4200	39.293	323.227	288.121	147.444	99.260	45.471	-0.566
4300	39.342	324.152	288.948	151.376	99.042	44.193	-0.537
4400	39.391	325.057	289.759	155.313	98.826	42.919	-0.510
4500	39.439	325.943	290.553	159.254	98.614	41.651	-0.483
4600	39.488	326.810	291.332	163.201	98.405	40.387	-0.459
4700	39.536	327.660	292.096	167.152	98.202	39.128	-0.435
4800	39.584	328.493	292.845	171.108	98.006	37.873	-0.412
4900	39.632	329.309	293.581	175.069	97.817	36.623	-0.390
5000	39.680	330.111	294.304	179.034	97.635	35.376	-0.370
5100	39.728	330.897	295.014	183.005	97.463	34.133	-0.350
5200	39.776	331.669	295.711	186.980	97.299	32.892	-0.330
5300	39.823	332.427	296.397	190.960	97.145	31.655	-0.312
5400	39.871	333.172	297.071	194.945	97.001	30.421	-0.294
5500	39.918	333.904	297.734	198.934	96.866	29.189	-0.277
5600	39.966	334.623	298.386	202.928	96.741	27.960	-0.261
5700	40.013	335.331	299.028	206.927	96.626	26.732	-0.245
5800	40.061	336.027	299.660	210.931	96.521	25.508	-0.230
5900	40.108	336.713	300.282	214.939	96.425	24.283	-0.215
6000	40.155	337.387	300.895	218.952	96.338	23.062	-0.201

PREVIOUS: June 1961 (1 atm)   CURRENT: June 1961 (1 bar)

Chlorine Oxide (ClO) $Cl_1O_1(g)$

TITANIUM CHLORIDE OXIDE (OTiCl) IDEAL GAS M_r = 99.3324

$S°(298.15\ K) = [263.661]\ J\ K^{-1}\ mol^{-1}$

$\Delta_f H°(0\ K) = [-242.642]\ kJ\ mol^{-1}$
$\Delta_f H°(298.15\ K) = [-244.262]\ kJ\ mol^{-1}$

Vibrational Frequencies and Degeneracies

ν, cm^{-1}
[880](1)
[230](2)
[490](1)

Ground State Quantum Weight: [2] $\sigma = 1$
Point Group: $[C_{\infty v}]$
Bond Distances: O-Ti = [1.62] Å Ti-Cl = [2.17] Å
Bond Angle: O-Ti-Cl = [180]°
Rotational Constant: B_0 = [0.092252] cm^{-1}

Enthalpy of Formation

$\Delta_f H°$(TiOCl, g, 298.15 K) is calculated from an estimated value of $\Delta_f H°$((TiOCl, g, 0 K) = -58 kcal mol^{-1} (1).

Heat Capacity and Entropy

Molecular constants were estimated by Gordon (1).

Reference

1. J. S. Gordon, Astrosystems, Inc., Caldwell Twp., N. J., personal communication, January 10, 1963.

Titanium Chloride Oxide (OTiCl) $Cl_1O_1Ti_1(g)$

Enthalpy Reference Temperature = T_r = 298.15 K Standard State Pressure = p° = 0.1 MPa

T/K	$C°_p$ (J K⁻¹mol⁻¹)	S° (J K⁻¹mol⁻¹)	$-[G°-H°(T_r)]/T$ (J K⁻¹mol⁻¹)	$H°-H°(T_r)$ (kJ mol⁻¹)	$\Delta_f H°$ (kJ mol⁻¹)	$\Delta_f G°$ (kJ mol⁻¹)	Log K_f
0	0.	0.	INFINITE	-12.142	-242.642	-242.642	INFINITE
100	36.633	215.420	305.606	-9.019	-242.987	-245.587	128.282
200	46.229	244.103	268.229	-4.825	-243.685	-247.916	64.749
250	49.321	254.766	264.499	-2.433	-243.995	-248.937	52.012
298.15	51.654	263.661	263.661	0.	-244.262	-249.864	43.775
300	51.733	263.980	263.662	0.096	-244.272	-249.898	43.511
350	53.622	272.103	264.299	2.731	-244.521	-250.816	37.432
400	55.104	279.364	265.736	5.451	-244.750	-251.700	32.869
450	56.272	285.924	267.621	8.237	-244.969	-252.555	29.316
500	57.200	291.903	269.754	11.074	-245.182	-253.387	26.471
600	58.546	302.459	274.349	16.866	-245.612	-254.987	22.199
700	59.444	311.556	279.029	22.769	-246.061	-256.515	19.141
800	60.067	319.537	283.604	28.746	-246.526	-257.977	16.844
900	60.514	326.639	287.998	34.776	-247.050	-259.377	15.054
1000	60.844	333.032	292.187	40.845	-247.702	-260.714	13.618
1100	61.094	338.844	296.169	46.942	-248.549	-261.976	12.440
1200	61.288	344.168	299.950	53.062	-253.579	-263.032	11.449
1300	61.441	349.080	303.543	59.199	-254.101	-263.799	10.600
1400	61.563	353.638	306.960	65.349	-254.707	-264.522	9.869
1500	61.663	357.889	310.215	71.510	-255.412	-265.200	9.235
1600	61.745	361.871	313.321	77.681	-256.230	-265.826	8.678
1700	61.814	365.617	316.288	83.859	-257.174	-266.398	8.185
1800	61.871	369.152	319.127	90.043	-258.260	-266.910	7.746
1900	61.920	372.498	321.849	96.233	-259.502	-267.357	7.350
2000	61.962	375.675	324.462	102.427	-275.600	-267.281	6.981
2100	61.999	378.699	326.973	108.625	-277.943	-266.808	6.636
2200	62.030	381.584	329.390	114.827	-280.299	-266.222	6.321
2300	62.058	384.342	331.720	121.031	-282.671	-265.529	6.030
2400	62.082	386.984	333.968	127.238	-285.059	-264.733	5.762
2500	62.103	389.519	336.140	133.447	-287.463	-263.837	5.513
2600	62.123	391.955	338.240	139.659	-289.884	-262.844	5.281
2700	62.139	394.300	340.273	145.872	-292.324	-261.757	5.064
2800	62.155	396.560	342.243	152.087	-294.781	-260.580	4.861
2900	62.168	398.741	344.154	158.303	-297.256	-259.316	4.671
3000	62.181	400.849	346.009	164.520	-299.750	-257.965	4.492
3100	62.192	402.888	347.811	170.739	-302.262	-256.530	4.323
3200	62.202	404.863	349.563	176.959	-304.790	-255.015	4.163
3300	62.211	406.777	351.268	183.179	-307.336	-253.421	4.011
3400	62.220	408.634	352.928	189.401	-309.897	-251.748	3.868
3500	62.228	410.438	354.546	195.623	-312.473	-250.001	3.731
3600	62.235	412.191	356.123	201.846	-315.063	-248.179	3.601
3700	62.241	413.896	357.661	208.070	-726.763	-238.497	3.367
3800	62.247	415.556	359.163	214.295	-728.142	-225.282	3.097
3900	62.253	417.173	360.630	220.520	-729.586	-212.030	2.840
4000	62.258	418.749	362.063	226.745	-731.091	-198.741	2.595
4100	62.263	420.287	363.465	232.971	-732.657	-185.412	2.362
4200	62.267	421.787	364.835	239.198	-734.278	-172.046	2.140
4300	62.271	423.252	366.177	245.425	-735.952	-158.639	1.927
4400	62.275	424.684	367.490	251.652	-737.677	-145.194	1.724
4500	62.279	426.084	368.777	257.880	-739.448	-131.709	1.529
4600	62.282	427.452	370.038	264.108	-741.263	-118.183	1.342
4700	62.286	428.792	371.274	270.336	-743.119	-104.618	1.163
4800	62.289	430.103	372.486	276.565	-745.013	-91.013	0.990
4900	62.291	431.388	373.675	282.794	-746.942	-77.368	0.825
5000	62.294	432.646	374.842	289.023	-748.903	-63.683	0.665
5100	62.296	433.880	375.987	295.253	-750.893	-49.959	0.512
5200	62.299	435.089	377.112	301.482	-752.911	-36.196	0.364
5300	62.301	436.276	378.217	307.712	-754.954	-22.393	0.221
5400	62.303	437.441	379.303	313.943	-757.019	-8.552	0.083
5500	62.305	438.584	380.371	320.173	-759.105	5.327	-0.051
5600	62.307	439.707	381.420	326.404	-761.210	19.245	-0.180
5700	62.309	440.809	382.453	332.634	-763.331	33.201	-0.304
5800	62.310	441.893	383.468	338.865	-765.468	47.194	-0.425
5900	62.312	442.958	384.467	345.096	-767.619	61.223	-0.542
6000	62.314	444.006	385.451	351.328	-769.783	75.290	-0.655

PREVIOUS: September 1963 (1 atm) CURRENT: September 1963 (1 bar)

Titanium Chloride Oxide (OTiCl) $Cl_1O_1Ti_1(g)$

CHLORINE OXIDE (ClO_2) IDEAL GAS M_r = 67.4518

$S°(298.15\ K) = 257.229\ J\ K^{-1}\ mol^{-1}$

$\Delta_f H°(0\ K) = 107.07 \pm 6.3\ kJ\ mol^{-1}$
$\Delta_f H°(298.15\ K) = 104.60 \pm 6.3\ kJ\ mol^{-1}$

Electronic Level and Quantum Weight

ε_i, cm^{-1}	g_i
0	2

Vibrational Levels and Degeneracies

ν, cm^{-1}
945.3(1)
447.4(1)
1109 (1)

Point Group: C_{2v} σ = 2
No ε_1 available.
Rotational Constants: $A_{000} = 1.6006\ cm^{-1}$ $B_{000} = 0.33283\ cm^{-1}$ $C_{000} = 0.27553\ cm^{-1}$
$X_{11} = -4.4\ cm^{-1}$ $X_{33} = -2.0\ cm^{-1}$ $X_{23} = -13\ cm^{-1}$
$X_{22} = 0$ $X_{12} = -3.0\ cm^{-1}$ $X_{31} = -14.4\ cm^{-1}$
$g_{22} = 0$

Enthalpy of Formation

Evans et al. (1) have evaluated the available data and chosen the above value.

Heat Capacity and Entropy

Gordon (2) has calculated the thermodynamic functions from 298.15 to 6000 K by the method of Pennington and Kobe (3) which takes vibration-rotation interaction and anharmonicity into account. Gordon's data are from Coon and Ortiz (4). The functions below 298.15 K have been calculated for a rigidly rotating harmonic oscillator.

References

1. W. H. Evans, T. R. Munson, and D. D. Wagman, J. Res. Nat. Bur. Stand. 55, 147 (1955).
2. J. S. Gordon, personal communication, February, 1961.
3. R. E. Pennington and K. A. Kobe, J. Chem. Phys. 22, 1442 (1954).
4. J. B. Coon and E. Ortiz, J. Mol. Spectrosc. 1, 81 (1957).

Chlorine Oxide (ClO_2) $Cl_1O_2(g)$

Enthalpy Reference Temperature = T_r = 298.15 K Standard State Pressure = p° = 0.1 MPa

T/K	$C_p°$ (J K⁻¹mol⁻¹)	S° (J K⁻¹mol⁻¹)	$-[G°-H°(T_r)]/T$ (J K⁻¹mol⁻¹)	$H°-H°(T_r)$ (kJ mol⁻¹)	$\Delta_f H°$ (kJ mol⁻¹)	$\Delta_f G°$ (kJ mol⁻¹)	Log K_f
0	0.	0.	INFINITE	-10.803	107.071	107.071	INFINITE
100	33.828	216.682	291.366	-7.468	106.045	111.158	-58.063
200	37.643	241.204	260.743	-3.908	105.175	116.628	-30.460
298.15	41.995	257.229	257.229	0.	104.600	122.327	-21.431
300	42.078	257.489	257.230	0.078	104.592	122.437	-21.318
400	46.070	270.163	258.931	4.493	104.301	128.437	-16.772
500	49.137	280.788	262.268	9.260	104.224	134.484	-14.049
600	51.375	289.954	266.136	14.291	104.279	140.532	-12.234
700	53.007	298.001	270.125	19.513	104.411	146.564	-10.937
800	54.212	305.161	274.065	24.877	104.587	152.575	-9.962
900	55.128	311.601	277.884	30.346	104.791	158.561	-9.203
1000	55.840	317.447	281.553	35.895	105.010	164.524	-8.594
1100	56.321	322.792	285.062	41.503	105.234	170.465	-8.095
1200	56.762	327.712	288.414	47.158	105.459	176.385	-7.678
1300	57.162	332.272	291.615	52.854	105.686	182.286	-7.324
1400	57.521	336.521	294.672	58.589	105.916	188.170	-7.021
1500	57.840	340.501	297.596	64.357	106.148	194.037	-6.757
1600	58.062	344.241	300.396	70.152	106.376	199.889	-6.526
1700	58.272	347.767	303.080	75.969	106.596	205.727	-6.321
1800	58.468	351.104	305.656	81.806	106.809	211.551	-6.139
1900	58.652	354.270	308.132	87.662	107.012	217.365	-5.976
2000	58.823	357.283	310.515	93.536	107.205	223.168	-5.829
2100	58.965	360.156	312.811	99.426	107.384	228.962	-5.695
2200	59.103	362.902	315.025	105.329	107.549	234.748	-5.574
2300	59.235	365.532	317.165	111.246	107.698	240.526	-5.463
2400	59.362	368.056	319.233	117.176	107.831	246.297	-5.361
2500	59.484	370.482	321.235	123.118	107.946	252.064	-5.267
2600	59.596	372.817	323.174	129.072	108.043	257.827	-5.180
2700	59.705	375.068	325.055	135.037	108.121	263.587	-5.099
2800	59.811	377.242	326.880	141.013	108.179	269.344	-5.025
2900	59.915	379.342	328.653	146.999	108.219	275.098	-4.955
3000	60.015	381.375	330.377	152.996	108.239	280.853	-4.890
3100	60.110	383.345	332.054	159.002	108.241	286.607	-4.829
3200	60.203	385.255	333.686	165.018	108.225	292.360	-4.772
3300	60.295	387.108	335.277	171.043	108.192	298.115	-4.719
3400	60.386	388.910	336.828	177.077	108.143	303.871	-4.668
3500	60.476	390.662	338.342	183.120	108.080	309.628	-4.621
3600	60.563	392.366	339.819	189.172	108.003	315.388	-4.576
3700	60.650	394.027	341.261	195.233	107.915	321.150	-4.534
3800	60.737	395.646	342.671	201.302	107.816	326.915	-4.494
3900	60.822	397.224	344.050	207.380	107.709	332.682	-4.456
4000	60.906	398.765	345.399	213.466	107.595	338.452	-4.420
4100	60.989	400.270	346.719	219.561	107.476	344.225	-4.385
4200	61.072	401.741	348.011	225.664	107.353	350.000	-4.353
4300	61.154	403.179	349.278	231.775	107.228	355.780	-4.322
4400	61.236	404.586	350.519	237.895	107.102	361.561	-4.292
4500	61.317	405.963	351.736	244.023	106.977	367.345	-4.264
4600	61.397	407.311	352.929	250.158	106.852	373.132	-4.237
4700	61.477	408.633	354.100	256.302	106.731	378.922	-4.211
4800	61.556	409.928	355.250	262.453	106.612	384.715	-4.187
4900	61.635	411.198	356.379	268.613	106.498	390.511	-4.163
5000	61.714	412.444	357.488	274.781	106.388	396.308	-4.140
5100	61.792	413.667	358.577	280.956	106.283	402.108	-4.118
5200	61.871	414.867	359.648	287.139	106.183	407.908	-4.097
5300	61.948	416.047	360.701	293.330	106.089	413.712	-4.077
5400	62.026	417.205	361.737	299.529	105.999	419.517	-4.058
5500	62.103	418.344	362.756	305.735	105.914	425.323	-4.039
5600	62.180	419.464	363.759	311.949	105.834	431.132	-4.021
5700	62.256	420.565	364.746	318.171	105.757	436.941	-4.004
5800	62.333	421.648	365.717	324.401	105.684	442.753	-3.987
5900	62.408	422.715	366.674	330.638	105.613	448.564	-3.971
6000	62.484	423.764	367.617	336.882	105.544	454.378	-3.956

PREVIOUS: March 1961 (1 atm) CURRENT: March 1961 (1 bar)

Chlorine Oxide (ClO_2) $Cl_1O_2(g)$

PHOSPHORUS CHLORIDE (PCl) IDEAL GAS M_r = 66.42676

$S°(298.15\ K) = [237.879]\ J\ K^{-1}\ mol^{-1}$

$\Delta_f H°(0\ K) = [129.6]\ kJ\ mol^{-1}$
$\Delta_f H°(298.15\ K) = [129.0]\ kJ\ mol^{-1}$

Electronic Level and Quantum Weight

State	ε_i, cm^{-1}	g_i
$[^3\Sigma]$	0	3

$\omega_e = [472]\ cm^{-1}$ $\omega_e x_e = [1.89]\ cm^{-1}$ $\sigma = 1$
$B_e = [0.2443]\ cm^{-1}$ $\alpha_e = [0.00098]\ cm^{-1}$ $r_e = [2.04]$ Å

Enthalpy of Formation

The selected $\Delta_f H°(298.15\ K)$ was calculated by assuming that the second and third bonds in $PCl_3(g)$ are of equal strength, and that the first bond has a strength 20% greater than the other two. These assumptions wee used by Henderson and Scheffee (1). The selected $\Delta_f H°(298.15\ K)$ was calculated using these bond strengths and auxiliary data from JANAF Tables.

A $\Delta_f H°(298.15\ K) = 40\ kcal\ mol^{-1}$ was estimated by O'Brien and Perrin (2).

Sandoval et al. (3) have calculated a $\Delta_f H°(298.15\ K)$ for PCl from an appearance potential. They selected the process $PCl_3 + Cl^+ + e^- + PCl + Cl$ to explain this potential rather than the process $PCl_3 + Cl^+ + e^- + P + Cl_2$. The reason for their selection was a calculated $\Delta_f H°+(Cl)$, 10 $kcal\ mol^{-1}$ lower than the literature value (See Nat. Bur. Stand., Circ 500). Use of the JANAF $\Delta_f H°(298.15\ K)$ for PCl_3 eliminates this difference. Therefore, the process of the reported appearance potential would apparently be the second process which does not involve the neutral PCl radial.

Heat Capacity and Entropy

The molecular constants were estimated by Gordon (4).

References

1. C. B. Henderson and R. S. Scheffee, Survey of Thermochemical Data, Atlantic Research Corporation, Alexandria, Virginia, January 1960.
2. C. J. O'Brien and J. R. Perrin, "Estimation of the Heats of Formation of Gaseous Combustion Product Molecules," The Marquardt Corporation and J. Perrine, Olin Mathieson Chemical Corp. WSS/CI Paper 59-4, October, 1959.
3. A. A. Sandoval, H. C. Moser, and R. W. Kiser, J. Phys. Chem. 67, 124, (1963).
4. J. S. Gordon, Astrosystems International, Livingston, New Jersey, personal communication, April 25, 1963.

Phosphorus Chloride (PCl) $Cl_1P_1(g)$

Enthalpy Reference Temperature = T_r = 298.15 K Standard State Pressure = p° = 0.1 MPa

T/K	$C_p°$ (J K⁻¹mol⁻¹)	S° (J K⁻¹mol⁻¹)	$-[G°-H°(T_r)]/T$ (J K⁻¹mol⁻¹)	$H°-H°(T_r)$ (kJ mol⁻¹)	$\Delta_f H°$ (kJ mol⁻¹)	$\Delta_f G°$ (kJ mol⁻¹)	Log K_f
0	0.	0.	INFINITE	-9.338	129.612	129.612	INFINITE
100	29.561	203.014	267.234	-6.422	130.185	121.103	-63.258
200	32.631	224.417	240.997	-3.316	129.509	112.047	-29.264
250	33.861	231.837	238.446	-1.652	129.272	107.708	-22.504
298.15	34.718	237.879	237.879	0.	129.000	103.579	-18.147
300	34.745	238.094	237.879	0.064	128.989	103.421	-18.007
350	35.378	243.500	238.305	1.818	127.948	99.256	-14.813
400	35.840	248.256	239.258	3.599	127.537	95.186	-12.430
450	36.185	252.498	240.497	5.400	127.133	91.166	-10.582
500	36.449	256.324	241.892	7.216	126.736	87.191	-9.109
600	36.822	263.005	244.870	10.881	125.952	79.356	-6.909
700	37.073	268.702	247.878	14.577	125.179	71.651	-5.347
800	37.253	273.664	250.797	18.294	124.413	64.057	-4.182
900	37.391	278.060	253.587	22.026	123.653	56.558	-3.283
1000	37.502	282.006	256.235	25.771	122.897	49.144	-2.567
1100	37.595	285.585	258.743	29.526	122.144	41.805	-1.985
1200	37.675	288.859	261.118	33.289	57.822	35.613	-1.550
1300	37.747	291.878	263.370	37.061	57.849	33.761	-1.357
1400	37.812	294.678	265.507	40.839	57.874	31.907	-1.190
1500	37.873	297.288	267.540	44.623	57.896	30.051	-1.046
1600	37.930	299.735	269.476	48.413	57.916	28.194	-0.920
1700	37.983	302.036	271.325	52.209	57.934	26.336	-0.809
1800	38.035	304.208	273.092	56.010	57.949	24.477	-0.710
1900	38.085	306.266	274.784	59.816	57.963	22.617	-0.622
2000	38.133	308.221	276.407	63.627	57.974	20.757	-0.542
2100	38.181	310.082	277.967	67.442	57.982	18.896	-0.470
2200	38.227	311.860	279.468	71.263	57.987	17.034	-0.404
2300	38.272	313.560	280.913	75.088	57.987	15.173	-0.345
2400	38.317	315.190	282.308	78.917	57.983	13.311	-0.290
2500	38.361	316.755	283.654	82.751	57.974	11.450	-0.239
2600	38.405	318.260	284.957	86.589	57.958	9.589	-0.193
2700	38.448	319.710	286.217	90.432	57.937	7.730	-0.150
2800	38.491	321.109	287.438	94.279	57.909	5.870	-0.110
2900	38.534	322.461	288.623	98.130	57.874	4.013	-0.072
3000	38.576	323.768	289.773	101.986	57.831	2.156	-0.038
3100	38.618	325.034	290.890	105.845	57.782	0.301	-0.005
3200	38.660	326.260	291.976	109.709	57.727	-1.552	0.025
3300	38.702	327.451	293.033	113.577	57.665	-3.404	0.054
3400	38.744	328.607	294.063	117.450	57.598	-5.254	0.081
3500	38.785	329.730	295.066	121.326	57.526	-7.101	0.106
3600	38.826	330.823	296.044	125.207	57.450	-8.946	0.130
3700	38.868	331.888	296.998	129.091	57.371	-10.790	0.152
3800	38.909	332.925	297.930	132.980	57.291	-12.631	0.174
3900	38.950	333.936	298.840	136.873	57.211	-14.470	0.194
4000	38.991	334.923	299.730	140.770	57.131	-16.307	0.213
4100	39.031	335.886	300.600	144.671	57.053	-18.142	0.231
4200	39.072	336.827	301.452	148.576	56.979	-19.975	0.248
4300	39.113	337.747	302.285	152.486	56.910	-21.806	0.265
4400	39.154	338.647	303.101	156.399	56.846	-23.636	0.281
4500	39.194	339.527	303.901	160.316	56.790	-25.465	0.296
4600	39.235	340.389	304.685	164.238	56.743	-27.292	0.310
4700	39.275	341.233	305.454	168.163	56.705	-29.119	0.324
4800	39.316	342.060	306.208	172.093	56.677	-30.944	0.337
4900	39.356	342.871	306.948	176.026	56.662	-32.769	0.349
5000	39.397	343.667	307.674	179.964	56.659	-34.595	0.361
5100	39.437	344.447	308.387	183.906	56.669	-36.420	0.373
5200	39.477	345.214	309.088	187.851	56.694	-38.245	0.384
5300	39.518	345.966	309.777	191.801	56.733	-40.071	0.395
5400	39.558	346.705	310.454	195.755	56.788	-41.898	0.405
5500	39.598	347.431	311.120	199.713	56.858	-43.726	0.415
5600	39.639	348.145	311.775	203.675	56.946	-45.556	0.425
5700	39.679	348.847	312.419	207.641	57.050	-47.387	0.434
5800	39.719	349.537	313.053	211.610	57.171	-49.220	0.443
5900	39.759	350.217	313.677	215.584	57.309	-51.056	0.452
6000	39.799	350.885	314.292	219.562	57.465	-52.894	0.460

PREVIOUS: December 1963 (1 atm) CURRENT: December 1963 (1 bar)

Phosphorus Chloride (PCl) $Cl_1P_1(g)$

LEAD CHLORIDE (PbCl) IDEAL GAS $M_r = 242.653$

$S°(298.15\ K) = 259.61 \pm 0.21\ J\ K^{-1}\ mol^{-1}$

$\Delta_fH°(0\ K) = 16.74 \pm 50.2\ kJ\ mol^{-1}$
$\Delta_fH°(298.15\ K) = 15.06 \pm 50.2\ kJ\ mol^{-1}$

Electronic Levels and Quantum Weights

State	ε_i, cm^{-1}	g_i
$X\ ^2\Pi_{1/2}$	0	2
$^2\Pi_{3/2}$	8274	2
$A[^2\Sigma]$	21866.9	[2]
$B\ ^2\Sigma^+$	34937.5	2

$\omega_e = 302.5\ cm^{-1}$ $\omega_ex_e = 0.88\ cm^{-1}$ $\sigma = 1$
$B_e = 0.1172\ cm^{-1}$ $\alpha_e = [0.00047]\ cm^{-1}$ $r_e = 2.180$ Å

Enthalpy of Formation

$\Delta_fH°(0\ K)$ is obtained from $D_0^0(Pb\text{-}Cl) = 3.1\pm0.5$ eV (71.5 kcal mol^{-1}) which is the recommended value given by Gaydon (1). D_0^0 is combined with $\Delta_fH°$(Pb, g, 0 K) = 46.91±0.13 kcal mol^{-1} and $\Delta_fH°$(Cl, g, 0 K) = 28.587±0.002 kcal mol^{-1} (2) to give $\Delta_fH°$(PbCl, g, 0 K) = 4.0±12.0 kcal mol^{-1}. $\Delta_fH°$ at 298.15 K is 3.6±12.0 kcal mol^{-1}. The recommended value for D_0^0 is based mainly on results ($(D_0^0(LBX) = 3.2$ eV) of a linear Birge-Sponer extrapolation of the observed vibrational levels (v, 0-18) (3) for the ground state of PbCl. This same value for $D_0^0(LBX)$ is obtained from the results of two other spectrometric studies (4, 5), while the results of another similar study (6) lead to $D_0^0(LBX) = 2.7$ eV. The values of $D_0^0(LBX)$ are increased by only 0.1 eV when corrected for the ionic character of PbCl by application of the Hildenbrand formula (7). Other spectroscopic values for $D_0^0(LBS)$ based on linear Birge-Sponer extrapolations include 2.2 eV ($A^2\Sigma$, or $^2\Delta$) (5) and 2.4 eV ($B^2\Sigma$) (6). The dissociation products are assumed to be $Pb(^1D) + Cl(^2P)(A^2\Sigma$, or $^2\Delta)$ and $Pb(^3P) + Cl(^2P)\ (B^2\Sigma)$. Also, Wieland and Newburgh (8) first reported $D_0^0 = 3.12$ eV from observations of predissociation, but later work (9) gave 3.05, or possibly 3.40 eV. It is to be noted that our $\Delta_fH°$ values for PbCl and $PbCl_2$ (2) indicate no significant differences in the binding of these two molecules. The difference $D_0^0(PbCl\text{-}Cl)-D_0^0(Pb\text{-}Cl)$ is equal to only 2.2 kcal mol^{-1}.

Heat Capacity and Entropy

The vibrational constants, electronic states (except for $A^2\Sigma$), and levels are taken from the compilation of Rosen (10) for $Pb^{35}Cl$. The vibrational constants are adjusted for the natural isotopic abundances of chlorine. It is not clear whether the first excited state for PbCl is $^2\Sigma$ or $^2\Delta$. By analogy with that for PbF (11), we assume the first excited state to be $^2\Sigma$. However, if this state is $^2\Delta$, the entropy is increased by only 0.006 cal K^{-1} mol^{-1} at 4000 K and 0.05 cal K^{-1} mol^{-1} at 6000 K. The rotational constant is taken from the results of a rotational analysis of two bands of the A-X system for $^{208}Pb^{35}Cl$ by Singh and Singh (12). The rotational constant is adjusted for the natural isotopic abundances of lead and chlorine. The value for r_e is obtained from the adopted B_e value. The results of an earlier rotational analysis of the same bands by Rao and Rao (13) are in reasonable agreement with our adopted values. The value for α_e is estimated from the other constants assuming a Morse potential function.

References

1. A. G. Gaydon, "Dissociation Energies and Spectra of Diatomic Molecules," 3rd ed., Chapman and Hall Ltd., London 1968.
2. JANAF Thermochemical Tables: Pb(g), 3-31-62; Cl(g), 6-30-72; $PbCl_2$(g), 6-30-73.
3. G. D. Rochester, Proc. Roy. Soc. A153, 407 (1936).
4. F. Morgan, Phys. Rev. 49, 47 (1936).
5. G. Pannetier and DP. Deschamps, Bull. Soc. Chim. France 1965, 2933 (1965).
6. H. Cordes and F. Gehrke, Z. Physik Chem. 51, 281 (1966).
7. D. L. Hildenbrand in "Advances in High Temperature Chemistry," Vol. 1, L. Eyring, Ed., Academic Press, New York, 193-215, 1967.
8. K. Wieland and R. Newburgh, Helv. Phys. Acta 22, 590 (1949).
9. K. Wieland and R. Newburgh, Helv. Phys. Acta 25, 87 (1952).
10. B. Rosen, Ed., "Spectroscopic Data Relative to Diatomic Molecules," Pergamon Press, New York, 1970.
11. R. F. Barrow, D. Butler, J. W. C. Johns, and J. L. Powell, Proc. Phys. Soc. (London) 73, 317 (1959).
12. O. N. Singh and I. S. Singh, Curr. Sci. 37, 282 (1968).
13. V. S. Rao and P. T. Rao, Z. Physik 181, 58 (1964).

Lead Chloride (PbCl) $Cl_1Pb_1(g)$

Enthalpy Reference Temperature = T_r = 298.15 K; Standard State Pressure = $p°$ = 0.1 MPa

T/K	$C_p°$ (J K⁻¹mol⁻¹)	$S°$ (J K⁻¹mol⁻¹)	$-[G°-H°(T_r)]/T$ (J K⁻¹mol⁻¹)	$H°-H°(T_r)$ (kJ mol⁻¹)	$\Delta_fH°$ (kJ mol⁻¹)	$\Delta_fG°$ (kJ mol⁻¹)	Log K_f
0	0.	0.	INFINITE	-9.793	16.738	16.738	INFINITE
100	31.235	222.428	290.774	-6.835	16.474	7.365	-3.847
200	34.890	245.391	262.893	-3.500	15.765	-1.462	0.382
250	35.727	253.274	260.207	-1.733	15.418	-5.729	1.197
298.15	36.234	259.613	259.613	0.	15.062	-9.770	1.712
300	36.250	259.837	259.614	0.067	15.048	-9.924	1.728
350	36.597	265.453	260.056	1.889	14.657	-14.055	2.098
400	36.841	270.356	261.044	3.725	14.242	-18.129	2.367
450	37.021	274.707	262.325	5.572	13.805	-22.150	2.571
500	37.158	278.614	263.762	7.426	13.344	-26.120	2.729
600	37.358	285.408	266.820	11.153	12.366	-33.922	2.953
700	37.500	291.178	269.898	14.896	6.451	-40.757	3.041
800	37.611	296.193	272.878	18.652	5.340	-47.424	3.096
900	37.705	300.629	275.720	22.418	4.260	-53.955	3.131
1000	37.790	304.606	278.413	26.193	3.212	-60.367	3.153
1100	37.873	308.211	280.961	29.976	2.196	-66.675	3.166
1200	37.961	311.510	283.371	33.767	1.212	-72.892	3.173
1300	38.057	314.553	285.654	37.568	0.257	-79.029	3.175
1400	38.166	317.377	287.820	41.379	-0.686	-85.092	3.175
1500	38.289	320.014	289.880	45.202	-1.623	-91.089	3.172
1600	38.429	322.490	291.841	49.038	-2.560	-97.022	3.167
1700	38.585	324.824	293.713	52.888	-3.499	-102.898	3.162
1800	38.756	327.034	295.504	56.755	-4.443	-108.717	3.155
1900	38.941	329.135	297.219	60.640	-5.391	-114.485	3.147
2000	39.138	331.137	298.865	64.544	-6.346	-120.202	3.139
2100	39.345	333.051	300.448	68.468	-184.528	-118.756	2.954
2200	39.560	334.887	301.972	72.413	-185.106	-115.610	2.745
2300	39.780	336.650	303.441	76.380	-185.757	-112.437	2.554
2400	40.002	338.348	304.861	80.369	-186.484	-109.234	2.377
2500	40.225	339.985	306.233	84.380	-187.288	-105.999	2.215
2600	40.448	341.567	307.562	88.414	-188.169	-102.730	2.064
2700	40.668	343.098	308.850	92.470	-189.124	-99.426	1.924
2800	40.884	344.581	310.100	96.548	-190.152	-96.086	1.792
2900	41.095	346.019	311.313	100.647	-191.249	-92.707	1.670
3000	41.300	347.416	312.494	104.766	-192.408	-89.289	1.555
3100	41.499	348.773	313.642	108.906	-193.626	-85.832	1.446
3200	41.692	350.094	314.761	113.066	-194.896	-82.335	1.344
3300	41.877	351.380	315.851	117.244	-196.210	-78.797	1.247
3400	42.055	352.633	316.915	121.441	-197.563	-75.219	1.156
3500	42.226	353.854	317.953	125.655	-198.947	-71.600	1.069
3600	42.389	355.046	318.966	129.886	-200.356	-67.942	0.986
3700	42.546	356.209	319.957	134.133	-201.781	-64.244	0.907
3800	42.696	357.346	320.926	138.395	-203.218	-60.508	0.832
3900	42.840	358.457	321.875	142.672	-204.660	-56.734	0.760
4000	42.978	359.543	322.803	146.963	-206.100	-52.922	0.691
4100	43.110	360.606	323.712	151.267	-207.533	-49.075	0.625
4200	43.236	361.647	324.603	155.585	-208.954	-45.193	0.562
4300	43.358	362.665	325.476	159.914	-210.359	-41.277	0.501
4400	43.474	363.664	326.333	164.256	-211.742	-37.329	0.443
4500	43.586	364.642	327.173	168.609	-213.101	-33.350	0.387
4600	43.695	365.601	327.998	172.973	-214.433	-29.340	0.333
4700	43.799	366.542	328.808	177.348	-215.733	-25.303	0.281
4800	43.900	367.465	329.604	181.733	-217.001	-21.238	0.231
4900	43.997	368.371	330.386	186.128	-218.233	-17.146	0.183
5000	44.092	369.261	331.155	190.532	-219.428	-13.030	0.136
5100	44.184	370.135	331.910	194.946	-220.585	-8.891	0.091
5200	44.274	370.994	332.654	199.369	-221.703	-4.729	0.048
5300	44.361	371.838	333.385	203.801	-222.781	-0.546	0.005
5400	44.446	372.668	334.105	208.241	-223.819	3.657	-0.035
5500	44.529	373.484	334.814	212.690	-224.816	7.879	-0.075
5600	44.611	374.287	335.511	217.147	-225.773	12.118	-0.113
5700	44.691	375.078	336.198	221.612	-226.689	16.374	-0.150
5800	44.769	375.856	336.876	226.085	-227.566	20.646	-0.186
5900	44.846	376.622	337.543	230.566	-228.403	24.933	-0.221
6000	44.922	377.376	338.200	235.054	-229.202	29.233	-0.254

PREVIOUS: June 1973 (1 atm) CURRENT: June 1973 (1 bar)

Lead Chloride (PbCl) $Cl_1Pb_1(g)$

LEAD CHLORIDE, ION ($PbCl^+$) IDEAL GAS M_r = 242.65245

$\Delta_f H°(0\ K)$ = 741.0 ± 50.2 kJ mol^{-1}
$S°(298.15\ K)$ = [255.15 ± 8.4] J K^{-1} mol^{-1} $\Delta_f H°(298.15\ K)$ = [745.589] kJ mol^{-1}

Electronic Levels and Quantum Weights

State	ε_i, cm^{-1}	g_i
$^1\Sigma^+$	0	1
$^3\Pi$	[30000]	6

ω_e = [283] cm^{-1} $\omega_e x_e$ = [0.82] cm^{-1} σ = 1
B_e = [0.1062] cm^{-1} α_e = [0.00043] cm^{-1} r_e = [2.29] Å

Enthalpy of Formation

The $PbCl^+$ ion has been observed in several mass spectrometric studies (1-4) of the vapors in equilibrium with both solid and liquid $PbCl_2$. Electric deflection experiments of mass spectrometrically detected $PbCl_2$ beams by Buchler et al. (5) have shown neutral $PbCl_2$ to be the precursor of the ion. Reported values for the AP of $PbCl^+$ are 11.7 (1) and 10.7 cal K^{-1} mol^{-1} (2, 3, 4). The higher value of AP was obtained by the method of linear extrapolation; thus, it is judged to be somewhat less reliable. Also, the lower value was obtained with more sensitive equipment. Therefore, we adopt AP = 10.7±0.5 cal K^{-1} mol^{-1} for the process $PbCl_2(g) + e^- \rightarrow PbCl^+(g) + Cl(g) + 2e^-$, and we calculate $\Delta_f H°(PbCl^+, g, 0\ K)$ = 177.1±12.0 kcal mol^{-1} by combining the value for AP with $\Delta_f H°(PbCl_2, g, 0\ K)$ = -41.1±0.3 kcal mol^{-1} and $\Delta_f H°(Cl, g, 0\ K)$ = 28.587±0.002 kcal mol^{-1} (6). $\Delta_f H°$ at 298.15 K is 178.2 kcal mol^{-1}, and the corresponding IP for PbCl is 7.5±1.0 cal K^{-1} mol^{-1}. Our adopted $\Delta_f H°$ is considered a maximum value, since unknown kinetic energy factors could have been involved in the dissociative ionization of $PbCl_2$.

Heat Capacity and Entropy

The dissociation energy of $PbCl^+$ (D_0^0 = 69.5 kcal mol^{-1}) is slightly less than that for PbCl (D_0^0 = 71.5 kcal mol^{-1}). This implies that somewhat weaker binding occurs in the ion relative to the neutral molecule. Thus, one might expect $r_e(PbCl^+, g) > r_e(PbCl, g)$. We assume an increase in r_e of 5% and obtain $r_e(PbCl^+)$ = 2.29 Å with $r_e(PbCl)$ = 2.180 Å (6). The value of B_e is obtained from the estimated value for r_e. Values for ω_e and $\omega_e x_e$ are estimated by a procedure similar to that given for SrF^+ (6). The constant term (d_{ij}) used in Badger's rule (7) is based on molecular data for PbCl, PbS, CsCl, and BaCl (6). The value of α_e is obtained from the other constants assuming a Morse potential function.

$PbCl^+$ has 10 valence electrons and is isoelectronic with the Group III A monochlorides. By analogy with the electronic states and levels for these isoelectronic molecules (6, 8), we estimate the ground state configuration to be $^1\Sigma$ with a triplet level lying at 30,000 cm^{-1}. The uncertainty in the latter value is believed to be no greater than 5000 cm^{-1}. If the triplet level is decreased to 25,000 cm^{-1}, the entropy increases only 0.01 cal K^{-1} mol^{-1} at 4000 K and 1.14 cal K^{-1} mol^{-1} at 6000 K. The enthalpy at 0 K is -2.357 kcal mol^{-1}.

References

1. H. Bloom and J. W. Hastie, Aust. J. Chem. 19, 1003 (1966).
2. J. W. Hastie, H. Bloom and J. D. Morrison, J. Chem. Phys. 47, 1580 (1967).
3. H. Bloom and J. W. Hastie, J. Chem. Phys. 49, 2230 (1968).
4. J. W. Hastie and D. L. Swingler, High Temp. Sci. 1, 46 (1969).
5. A. Buchler, J. L. Stauffer and W. Klemperer, J. Amer. Chem. Soc. 86, 4544 (1964).
6. JANAF Thermochemical Tables: $PbCl_2(g)$, PbCl(g), and PbS(g), 6-20-73; Cl(g), 6-20-72; $SrF^+(g)$ and BaCl(g), 12-31-72; CSCl(g), 6-30-68; AlCl(g), 6-30-70.
7. R. M. Badger, J. Chem. Phys. 2, 128 (1934).
8. G. Herzberg, "Spectra of Diatomic Molecules," 2nd ed., D. Van Nostrand Co., Inc., New York, 1963.

Lead Chloride, Ion ($PbCl^+$) $Cl_1Pb_1^+(g)$

Enthalpy Reference Temperature = T_r = 298.15 K Standard State Pressure = p° = 0.1 MPa

T/K	$C_p°$	S°	$-[G°-H°(T_r)]/T$	$H°-H°(T_r)$	$\Delta_f H°$	$\Delta_f G°$	Log K_f
	J $K^{-1}mol^{-1}$			kJ mol^{-1}			
0	0.	0.	INFINITE	-9.860	741.001		
100	31.590	217.637	286.530	-6.889			
200	35.163	240.840	258.445	-3.521			
250	35.932	248.776	255.744	-1.742			
298.15	36.394	255.148	255.148	0.	745.589	715.833	-125.411
300	36.408	255.373	255.148	0.067	745.614	715.648	-124.605
350	36.723	261.010	255.593	1.896	746.269	710.602	-106.052
400	36.945	265.929	256.584	3.738	746.899	705.463	-92.124
450	37.108	270.290	257.869	5.590	747.505	700.247	-81.283
500	37.235	274.207	259.310	7.448	748.088	694.965	-72.602
600	37.420	281.013	262.377	11.182	749.195	684.235	-59.568
700	37.554	286.792	265.463	14.931	745.365	674.124	-50.304
800	37.660	291.814	268.449	18.691	746.337	663.881	-43.347
900	37.749	296.255	271.297	22.462	747.341	653.514	-37.929
1000	37.829	300.236	273.995	26.241	748.376	643.033	-33.589
1100	37.901	303.845	276.547	30.028	749.441	632.448	-30.032
1200	37.969	307.146	278.962	33.821	750.538	621.764	-27.065
1300	38.033	310.188	281.248	37.621	751.660	610.987	-24.550
1400	38.095	313.008	283.417	41.428	752.792	600.124	-22.391
1500	38.156	315.639	285.479	45.240	753.923	589.179	-20.517
1600	38.214	318.103	287.441	49.059	755.048	578.160	-18.875
1700	38.272	320.422	289.314	52.883	756.161	567.070	-17.424
1800	38.329	322.611	291.104	56.713	757.259	555.915	-16.132
1900	38.386	324.685	292.817	60.549	758.340	544.700	-14.975
2000	38.442	326.655	294.460	64.390	759.401	533.428	-13.932
2100	38.497	328.532	296.038	68.237	583.221	529.219	-13.164
2200	38.552	330.324	297.556	72.090	584.629	526.614	-12.503
2300	38.607	332.039	299.018	75.948	585.947	523.947	-11.899
2400	38.662	333.683	300.429	79.811	587.173	521.225	-11.344
2500	38.716	335.263	301.791	83.680	588.305	518.453	-10.832
2600	38.771	336.782	303.107	87.554	589.344	515.638	-10.359
2700	38.825	338.246	304.382	91.434	590.291	512.785	-9.920
2800	38.880	339.659	305.617	95.319	591.149	509.899	-9.512
2900	38.935	341.025	306.814	99.210	591.923	506.983	-9.132
3000	38.991	342.346	307.977	103.106	592.619	504.042	-8.776
3100	39.047	343.625	309.106	107.008	593.242	501.079	-8.443
3200	39.105	344.866	310.204	110.916	593.799	498.097	-8.131
3300	39.163	346.070	311.273	114.829	594.298	495.099	-7.837
3400	39.224	347.240	312.314	118.749	594.746	492.086	-7.560
3500	39.286	348.378	313.328	122.674	595.152	489.060	-7.299
3600	39.350	349.485	314.317	126.606	595.523	486.024	-7.052
3700	39.417	350.564	315.282	130.544	595.867	482.977	-6.818
3800	39.487	351.617	316.225	134.489	596.192	479.922	-6.597
3900	39.561	352.643	317.145	138.442	596.505	476.858	-6.387
4000	39.638	353.646	318.045	142.402	596.812	473.786	-6.187
4100	39.720	354.626	318.926	146.370	597.121	470.707	-5.997
4200	39.806	355.584	319.787	150.346	597.437	467.620	-5.816
4300	39.897	356.521	320.631	154.331	597.767	464.525	-5.643
4400	39.994	357.440	321.457	158.325	598.114	461.422	-5.478
4500	40.097	358.340	322.266	162.330	598.485	458.311	-5.320
4600	40.206	359.222	323.060	166.345	598.884	455.192	-5.169
4700	40.321	360.088	323.839	170.371	599.313	452.064	-5.024
4800	40.444	360.938	324.603	174.409	599.778	448.926	-4.885
4900	40.573	361.773	325.353	178.460	600.280	445.778	-4.752
5000	40.710	362.594	326.090	182.524	600.823	442.619	-4.624
5100	40.854	363.402	326.813	186.602	601.409	439.450	-4.501
5200	41.006	364.197	327.525	190.695	602.040	436.268	-4.382
5300	41.165	364.979	328.224	194.804	602.717	433.074	-4.268
5400	41.333	365.750	328.912	198.929	603.443	429.866	-4.158
5500	41.508	366.510	329.588	203.071	604.217	426.645	-4.052
5600	41.691	367.260	330.255	207.230	605.042	423.409	-3.949
5700	41.882	368.000	330.910	211.409	605.917	420.158	-3.850
5800	42.080	368.730	331.556	215.607	606.844	416.891	-3.755
5900	42.286	369.451	332.192	219.825	607.823	413.607	-3.662
6000	42.500	370.163	332.819	224.064	608.854	410.307	-3.572

PREVIOUS: June 1973 (1 atm) CURRENT: June 1973 (1 bar)

Lead Chloride, Ion ($PbCl^+$) $Cl_1Pb_1^+(g)$

SULFUR CHLORIDE (SCl) IDEAL GAS M_r = 67.513

$S°(298.15 K) = [237.32 \pm 2.5] J K^{-1} mol^{-1}$

$\Delta_f H°(0 K) = [155.65 \pm 16.7] kJ mol^{-1}$
$\Delta_f H°(298.15 K) = [156.46 \pm 16.7] kJ mol^{-1}$

Electronic Levels and Quantum Weights

ε_i, cm^{-1}	g_i
0	[2]
[400]	[2]
[25000]	[4]

$\omega_e = [536] cm^{-1}$ $\omega_e x_e = [2.08] cm^{-1}$ $\sigma = 1$
$B_e = [0.24060] cm^{-1}$ $\alpha_e = [0.00126] cm^{-1}$ $r_e = [2.04]$ Å

Enthalpy of Formation

The adopted results are $\Delta_f H°(SCl, g, 298.15 K) = 37.2 \pm 4.0 kcal mol^{-1}$ and $D_0^0(S-Cl) = 57.1 \pm 4.0 kcal mol^{-1}$. We calculate $\Delta_f H°$ from an estimated value for the primary bond dissociation energy of SCl_2 of $69.7 \pm 3.0 kcal mol^{-1}$, using auxiliary heat of formation data from JANAF (1). The value of $D_0^0(SCl-Cl)$ is obtained from the relationship $D_0^0(SCl-Cl)/\Delta_{at}H°(SCl_2) = 0.55 \pm 0.03$ with $\Delta_a H°(SCl_2) = 126.8 \pm 0.9 kcal mol^{-1}$ (1). The value of the ratio is taken to be the mean of three values calculated (1) for $SF_2(0.53 \pm 0.05)$, $SiCl_2(0.57 \pm 0.01)$, and $SiF_2(0.54 \pm 0.03)$. Converting $\Delta_f H°$ to 298.15 K, we obtain $\Delta_f H°(SCl, g, 298.15 K) = 37.4 kcal mol^{-1}$ which is in reasonable agreement with the previous estimate made by Mills (35±5, 2), McBride et al. (32.0, 3), and Benson (36.5±2, 4).

Perona et al. (5) have measured the hydrogen chloride infrared emission from reactions involving H and D atoms with SCl_2. From the observed highest vibrational level of HCl, they estimated an upper limit to $D_0^0(SCl-Cl)$ of $\leq 58 kcal mol^{-1}$. From this result, we calculate $\Delta_f H°(SCl, g, 0 K) \leq 25.5 kcal mol^{-1}$ and $D_0^0(SCl) \geq 68.8 kcal mol^{-1}$. We note that the results from this study predict that $D_0^0(S-Cl) > D_0^0(SCl-Cl)$ which is inconsistent with established trends in the bond dissociation energies (1) for other mono- and dihalides, such as the silicon chlorides, silicon fluorides, and sulfur fluorides. Factors which can influence the D_0^0 values estimated from infrared chemiluminescence studies have been discussed by Perona et al. (5). It seems likely that their estimate of D° may be too low due to the formation of some HCl in higher vibrational levels by an energy exchange mechanism rather than what is actually produced by the H/SCl_2 chemical reaction. For these reasons, we believe that the $\Delta_f H°(0 K)$ value ($19 \pm 3 kcal mol^{-1}$ recently reported by Takacs (6) based primarily on his analysis of the chemiluminscence work of Perona et al. (5) is suspect. The results adopted here predict $D_0^0(SCl-Cl) > D_0^0(S-Cl)$ by roughly 13 $kcal mol^{-1}$ which seems reasonable by comparison with bond energy data (1) for other halide systems.

Heat Capacity and Entropy

We estimate the S-Cl bond length to lie between those for SCl_2 and S_2Cl_2 (1). The value of B_e is calculated from the adopted r_e. We use Badger's rule (7) to estimate a value for ω_e. The equation is written as $\omega_e^2 = 3.159 \times 10^6/\mu(r_e-d_{ij})^3$, and we use molecular data (1) for SiCl to determine the constant $d_{ij} = 1.172$. The value of x_e is estimated from the expression $x_e\mu^{1/2} = 0.01595$ calculated from data for SiCl (1). By analogy with SiCl, SiF, and SF, we expect the ground state electronic configuration for SCl to be $^2\pi$. The ground state splitting (400 cm^{-1}) and double pi state at 25000 cm^{-1} are estimated from those observed for SF (1).

McBride et al. (3) have previously estimated thermal functions over an extended temperature range (0-6000 K), using molecular data similar to ours. Their entropies are consistently higher than our values by 0.6 $cal K^{-1} mol^{-1}$ at 298.15 K and 0.3 $cal K^{-1} mol^{-1}$ at 4000 K. The difference at 298.15 K arises primarily from the electronic contribution caused by treating the ground state as a single level (3) with a quantum weight of 4. We believe that the uncertainty in S°(298.15 K) should not exceed ±0.6 $cal K^{-1} mol^{-1}$. Their estimate of $\omega_e x_e = 5.85 cm^{-1}$ (3), obtained assuming a linear Birge-Sponer relation, seems rather high in comparison with similar data (1) for SiCl, SiF, and SF.

References

1. JANAF Thermochemical Tables: SCl_2(g), 6-30-78; SF(g) and SF_2(g), 6-30-76; SiF_2(g) and $SiCl_2$(g), 12-31-77; S_2Cl_2(g), 6-30-78; SiF(g) and SiCl(g), 12-31-76; Cl(g), 6-30-72; F(g), 9-30-65.
2. K. C. Mills, "Thermodynamic Data for Inorganic Sulphides, Selenides, and Tellurides," Butterworths, London, 1974.
3. B. J. McBride, S. Heinel, J. G. Ehlers, and S. Gordon, Nat. Aero. Space Admin., NASA SP 3001, 1963.
4. S. W. Benson, Chem. Rev. 78, 23 (1978).
5. M. J. Perona, D. W. Setser, and R. J. Johnson, J. Chem. Phys. 52, 6384 (1970).
6. G. A. Takacs, J. Chem. Eng. Data 23, 174 (1978).
7. R. M. Badger, J. Chem. Phys. 2, 128 (1934).

Sulfur Chloride (SCl) $Cl_1S_1(g)$

Enthalpy Reference Temperature = T_r = 298.15 K Standard State Pressure = p° = 0.1 MPa

T/K	$C_p°$	S°	$-[G°-H°(T_r)]/T$	$H°-H°(T_r)$	$\Delta_f H°$	$\Delta_f G°$	Log K_f
	J $K^{-1}mol^{-1}$			kJ mol^{-1}			
0	0.	0.	INFINITE	-9.819	155.648	155.648	INFINITE
100	30.207	200.110	269.024	-6.891	156.430	147.122	-76.848
200	35.382	222.704	240.708	-3.601	156.560	137.743	-35.975
250	36.817	230.770	237.939	-1.792	156.540	133.040	-27.797
298.15	37.541	237.324	237.324	0.	156.465	128.520	-22.516
300	37.560	237.556	237.325	0.069	156.461	128.346	-22.347
350	37.916	243.376	237.783	1.958	156.323	123.671	-18.457
400	38.073	248.451	238.807	3.858	153.917	119.106	-15.554
450	38.130	252.939	240.132	5.763	153.009	114.800	-13.326
500	38.140	256.958	241.618	7.670	152.016	110.610	-11.555
600	38.110	263.909	244.771	11.483	150.428	102.484	-8.922
700	38.068	269.781	247.935	15.292	149.054	94.604	-7.059
800	38.037	274.862	250.991	19.097	147.798	86.912	-5.675
900	38.018	279.341	253.897	22.899	93.484	80.454	-4.669
1000	38.010	283.346	256.645	26.701	93.561	79.002	-4.127
1100	38.011	286.969	259.240	30.502	93.615	77.543	-3.682
1200	38.019	290.276	261.690	34.303	93.649	76.080	-3.312
1300	38.032	293.320	264.008	38.106	93.662	74.616	-2.998
1400	38.050	296.139	266.203	41.910	93.658	73.151	-2.729
1500	38.070	298.765	268.288	45.716	93.635	71.687	-2.496
1600	38.093	301.223	270.270	49.524	93.595	70.225	-2.293
1700	38.118	303.533	272.159	53.335	93.538	68.766	-2.113
1800	38.145	305.712	273.963	57.148	93.466	67.310	-1.953
1900	38.173	307.775	275.689	60.964	93.379	65.860	-1.811
2000	38.202	309.734	277.343	64.782	93.278	64.414	-1.682
2100	38.232	311.599	278.930	68.604	93.163	62.973	-1.566
2200	38.263	313.378	280.456	72.429	93.035	61.539	-1.461
2300	38.295	315.080	281.924	76.257	92.893	60.110	-1.365
2400	38.327	316.710	283.340	80.088	92.739	58.688	-1.277
2500	38.360	318.275	284.706	83.922	92.573	57.273	-1.197
2600	38.394	319.780	286.027	87.760	92.394	55.864	-1.122
2700	38.428	321.230	287.304	91.601	92.203	54.463	-1.054
2800	38.463	322.628	288.541	95.446	92.000	53.069	-0.990
2900	38.498	323.979	289.739	99.294	91.785	51.682	-0.931
3000	38.535	325.284	290.903	103.145	91.559	50.303	-0.876
3100	38.572	326.549	292.032	107.001	91.323	48.932	-0.824
3200	38.610	327.774	293.130	110.860	91.076	47.568	-0.776
3300	38.649	328.962	294.198	114.723	90.820	46.213	-0.731
3400	38.689	330.117	295.238	118.589	90.556	44.865	-0.689
3500	38.731	331.239	296.250	122.460	90.284	43.525	-0.650
3600	38.774	332.331	297.237	126.336	90.005	42.193	-0.612
3700	38.819	333.394	298.200	130.215	89.721	40.869	-0.577
3800	38.865	334.429	299.140	134.100	89.433	39.552	-0.544
3900	38.913	335.440	300.058	137.988	89.142	38.243	-0.512
4000	38.962	336.425	300.955	141.882	88.849	36.942	-0.482
4100	39.014	337.388	301.832	145.781	88.556	35.648	-0.454
4200	39.067	338.329	302.690	149.685	88.264	34.361	-0.427
4300	39.122	339.249	303.529	153.594	87.975	33.081	-0.402
4400	39.179	340.149	304.351	157.509	87.689	31.808	-0.378
4500	39.239	341.030	305.157	161.430	87.407	30.541	-0.355
4600	39.300	341.893	305.946	165.357	87.132	29.280	-0.332
4700	39.363	342.739	306.720	169.290	86.863	28.025	-0.311
4800	39.428	343.568	307.479	173.230	86.603	26.776	-0.291
4900	39.495	344.382	308.224	177.176	86.352	25.533	-0.272
5000	39.564	345.181	308.955	181.129	86.110	24.294	-0.254
5100	39.635	345.965	309.673	185.089	85.880	23.060	-0.236
5200	39.708	346.735	310.378	189.056	85.661	21.830	-0.219
5300	39.783	347.492	311.071	193.031	85.455	20.604	-0.203
5400	39.859	348.236	311.753	197.013	85.262	19.383	-0.187
5500	39.937	348.969	312.423	201.003	85.082	18.165	-0.173
5600	40.017	349.689	313.082	205.000	84.916	16.949	-0.158
5700	40.098	350.398	313.730	209.006	84.765	15.737	-0.144
5800	40.181	351.096	314.368	213.020	84.629	14.527	-0.131
5900	40.265	351.784	314.997	217.042	84.509	13.320	-0.118
6000	40.350	352.461	315.616	221.073	84.404	12.114	-0.105

PREVIOUS: June 1978 (1 atm) CURRENT: June 1978 (1 bar)

Sulfur Chloride (SCl) $Cl_1S_1(g)$

SULFUR CHLORIDE, ION (SCl^+) IDEAL GAS M_r = 67.51245

$S°(298.15\ K)$ = [236.63 ± 2.5] J K^{-1} mol^{-1}

$\Delta_fH°(0\ K)$ = 1055.66 ± 2.9 kJ mol^{-1}
$\Delta_fH°(298.15\ K)$ = [1062.037] kJ mol^{-1}

Electronic Levels and Quantum Weights

State	ε_i, cm^{-1}	g_i
[$X^3\Sigma$]	0	[3]
[$a^1\Delta$]	[7000]	[2]
[$b^1\Sigma$]	[12000]	[1]
[$A^3\Pi_r$]	[25000]	[6]

ω_e = [553] cm^{-1} ω_ex_e = [2.19] cm^{-1} σ = 1
B_e = [0.26333] cm^{-1} α_e = [0.00142] cm^{-1} r_e = [1.95] Å

Enthalpy of Formation

Blais and Cottin (1) have measured the appearance potential (AP) of $SCl^+(g)$ from $SCl_2(g)$ as 12.5±0.2 eV using the electron impact technique. Subsequent work by Hartmann et al. (2) using the same method led to AP(SCl^+/SCl_2) = 12.2±0.1 eV which confirms the earlier results. We assume that the fragment ions produced in the dissociative ionization process $e^- + SCl_2(g) = SCl^+(g) + Cl(g) + 2\ e^-$ are formed with no excitation energy or excess kinetic energy, and we calculate the adopted value of $\Delta_fH°(0\ K)$ = 252.3±5.0 kcal mol^{-1} by combining the mean value (284.795 kcal mol^{-1}) of these two AP's with $\Delta_fH°$ data (3) for Cl(g) and $SCl_2(g)$.

Blais and Cottin (1) have also reported the electron impact AP for $SCl^+(g)$ from $S_2Cl_2(g)$ as 13.6±0.2 eV. Assuming that the ionization process is $e^- + S_2Cl_2(g) = SCl^+(g) + SCl(g) + 2e^-$, their AP value gives $\Delta_fH°(SCl^+$, g, 0 K) = 272.8 kcal mol^{-1} which is about 20 kcal mol^{-1} higher than the adopted value. Comparison of JANAF IPs (3) for SF (10.09±0.1 eV), SF_2 (10.29±0.3 eV), and SCl_2 (9.46±0.02 eV) suggests that the ionization value for SCl should lie near 9.3 eV. Our adopted data for SCl(g) and $SCl^+(g)$ give IP(SCl) = 9.33±0.4 eV. The higher $\Delta_fH°$ value of 272.8 kcal mol^{-1} corresponds to IP(SCl) = 10.2 ev which appears to be too high by comparison with the difference in the IPs for SF and SF_2 and the highly reliable IP value for SCl_2(3).

$\Delta_fH°(298.15\ K)$ is obtained from the adopted $\Delta_fH°(0\ K)$ value by using JANAF enthalpies H°(0 K) - H°(298.15 K) (3) for S(α, cr), $Cl_2(g)$, and $e^-(g)$. $\Delta_rH°(298.15\ K)$ determined for the direct ionization $SCl = SCl^+ + e^-$ from our tables differs from a room-temperature threshold energy due to inclusion of the enthalpies of all the species in this ionization process and to threshold effects discussed by Rosenstock et al. (4). $\Delta_fH°(298.15\ K)$ should be changed by -1.481 kcal mol^{-1} if it is to be used in the ion convention which excludes the enthalpy of the electron. The S-Cl bond strength to form $S^+(g)$ and Cl(g) by dissociation is calculated to be 80.8±5.0 kcal mol^{-1}.

Heat Capacity and Entropy

The value of r_e has been selected such that when it is substituted into the equation for Badger's rule (5) written as ω_e^2 = 3.159 x $10^6/\mu(r_e-d_{ij})^3$, a value of ω_e(553 cm^{-1}) is obtained which agrees with that (552.3 cm^{-1}) determined for $P^{35}Cl$ by Coxon and Wickramaaratchi (6). The value of d_{ij} is taken equal to 1.10 based on the molecular data for SiCl(3) and $P^{35}Cl$ (6). This procedure gives r_e = 1.95 Å which is only slightly lower than the S-Cl bond lengths (3) measured for SCl_2(2.015 Å) and S_2Cl_2(2.057 Å). We calculate B_e from the estimated value of r_e. Using the expression $x_e\mu^{1/2}$ = 0.01623, x_3 is determined to be 0.00396 which gives ω_ex_e = 2.19 cm^{-1} when multiplied by the estimated value of ω_e. The value of the constant in the above expression is based on the values determined for SiCl and PCl. The value of α_e is determined from the other estimated molecular constants assuming a Morse potential function.

Examination of the electronic configurations and levels for diatomic halides having 12 valence electrons, such as PF (3) PCl (6), NCl, and NF (7) suggests that the ground state for SCl^+ is most likely $^3\Sigma$, and several low-lying excited states are expected. We estimate the relative term value for the $a^1\Delta$ state from that observed for PF (7090.41 cm^{-1}). Relative term values for the $b^1\Sigma$ and $A^3\Pi_r$ states are estimated by analogy with those observed for PCl (6). As discussed by Coxon and Wickramaaratchi (6) for PCl, other singlet ($^1\Pi$) and triplet ($^3\Pi_i$ and $^3\Sigma$) states may be expected to lie in the range 20000 - 35000 cm^{-1}, similar to those observed for the isoelectronic molecules O_2, S_2, and SO (3). Inclusion of estimated term values for these additional levels changes our entropy by only about 0.05 cal K^{-1} mol^{-1} at 4500 K, and we prefer to omit these levels at this time. The adopted electronic entropy contribution from the excited states amounts to 0.01 cal K^{-1} mol^{-1} at 1500 K, and this contribution increases to 0.51 cal K^{-1} mol^{-1} at 4500 K.

References

1. J. C. Blais and M. Cottin, C. R. Acad. Sci. Paris, Series C, 276, 919 (1973).
2. O. R. Hartmann, K. H. Lebert, and H. U. Chun, Z. Phys. Chem. 92, 311 (1974).
3. JANAF Thermochemical Tables: Cl(g), 6-30-72; SCl_2, $SCl_2^+(g)$ and $S_2Cl_2(g)$, 6-30-78; $SF^+(g)$, 6-30-76; $SF_2^+(g)$ and SiCl(g), 12-31-76; PF(g) and SO(g), 6-30-77; $O_2(g)$ and $e^-(g)$, 3-31-77; $S_2(g)$ and S(α, cr), 9-30-77; $Cl_2(g)$, 9-30-65.
4. H. M. Rosenstock et al., J. Phys. Chem. Ref. Data 6, Suppl. 1, 1977; pp. I-28 and I-35.
5. R. M. Badger, J. Chem. Phys. 2, 128 (1934).
6. J. A. Coxon and M. A. Wickramaaratchi, J. Mol. Spectrosc. 68, 372 (1977).
7. S. N. Suchard, "Spectroscopic Data," Volume 1, Part B, IFI/Plenum, New York, 1975.

Sulfur Chloride, Ion (SCl^+) $Cl_1S_1^+(g)$

Enthalpy Reference Temperature = T_r = 298.15 K Standard State Pressure = p° = 0.1 MPa

T/K	$C_p°$ (J $K^{-1}mol^{-1}$)	S° (J $K^{-1}mol^{-1}$)	-[G°-H°(T_r)]/T (J $K^{-1}mol^{-1}$)	H°-H°(T_r) (kJ mol^{-1})	$\Delta_fH°$ (kJ mol^{-1})	$\Delta_fG°$ (kJ mol^{-1})	Log K_f
0	0.	0.	INFINITE	-9.183	1055.659		
100	29.303	202.547	265.263	-6.272			
200	31.737	223.512	239.676	-3.233			
250	33.002	230.735	237.187	-1.613			
298.15	33.958	236.633	236.633	0.	1062.037	1028.043	-180.109
300	33.989	236.844	236.634	0.063	1062.065	1027.832	-178.961
350	34.732	242.142	237.051	1.782	1062.798	1022.068	-152.535
400	35.291	246.818	237.985	3.533	1061.282	1016.290	-132.714
450	35.716	251.000	239.203	5.309	1061.283	1010.656	-117.314
500	36.046	254.781	240.575	7.103	1061.217	1005.037	-104.996
600	36.515	261.397	243.509	10.733	1061.524	993.779	-86.516
700	36.829	267.051	246.478	14.401	1062.088	982.446	-73.311
800	37.055	271.985	249.365	18.096	1062.801	971.021	-63.401
900	37.229	276.360	252.126	21.810	1010.477	960.582	-55.751
1000	37.375	280.290	254.749	25.541	1012.561	954.926	-49.880
1100	37.508	283.858	257.236	29.285	1014.638	949.062	-45.067
1200	37.639	287.127	259.592	33.042	1016.706	943.009	-41.048
1300	37.774	290.145	261.828	36.813	1018.766	936.784	-37.640
1400	37.917	292.950	263.952	40.597	1020.820	930.400	-34.714
1500	38.070	295.571	265.973	44.397	1022.870	923.870	-32.172
1600	38.234	298.033	267.901	48.212	1024.915	917.204	-29.944
1700	38.407	300.356	269.743	52.044	1026.959	910.409	-27.973
1800	38.589	302.557	271.505	55.893	1029.002	903.494	-26.219
1900	38.777	304.648	273.195	59.762	1031.046	896.466	-24.646
2000	38.970	306.642	274.818	63.649	1033.091	889.330	-23.227
2100	39.165	308.548	276.379	67.556	1035.140	882.091	-21.941
2200	39.362	310.375	277.883	71.482	1037.192	874.755	-20.769
2300	39.557	312.129	279.334	75.428	1039.248	867.326	-19.698
2400	39.751	313.816	280.736	79.394	1041.307	859.807	-18.713
2500	39.941	315.443	282.092	83.378	1043.369	852.202	-17.806
2600	40.127	317.013	283.405	87.382	1045.434	844.515	-16.967
2700	40.308	318.531	284.678	91.403	1047.502	836.748	-16.188
2800	40.483	320.000	285.913	95.443	1049.573	828.904	-15.463
2900	40.652	321.424	287.113	99 500	1051.646	820.986	-14.788
3000	40.815	322.805	288.280	103.573	1053.720	812.997	-14.156
3100	40.972	324.145	289.416	107.663	1055.796	804.939	-13.563
3200	41.122	325.449	290.521	111.767	1057.874	796.814	-13.007
3300	41.267	326.716	291.599	115.887	1059.953	788.623	-12.483
3400	41.406	327.950	292.650	120.021	1062.034	780.370	-11.989
3500	41.539	329.152	293.676	124.168	1064.117	772.055	-11.522
3600	41.667	330.324	294.678	128.328	1066.202	763.681	-11.081
3700	41.790	331.468	295.657	132.501	1068.290	755.249	-10.662
3800	41.909	332.584	296.614	136.686	1070.381	746.761	-10.265
3900	42.023	333.674	297.550	140.883	1072.477	738.217	-9.887
4000	42.134	334.739	298.467	145.090	1074.577	729.620	-9.528
4100	42.242	335.781	299.364	149.309	1076.682	720.970	-9.185
4200	42.346	336.800	300.243	153.539	1078.794	712.268	-8.858
4300	42.448	337.798	301.105	157.778	1080.914	703.516	-8.546
4400	42.548	338.775	301.950	162.028	1083.041	694.715	-8.247
4500	42.645	339.732	302.779	166.288	1085.177	685.865	-7.961
4600	42.741	340.670	303.593	170.557	1087.322	676.968	-7.687
4700	42.836	341.591	304.391	174.836	1089.478	668.024	-7.424
4800	42.929	342.493	305.176	179.124	1091.645	659.034	-7.172
4900	43.021	343.380	305.947	183.422	1093.824	649.999	-6.929
5000	43.112	344.250	306.704	187.728	1096.015	640.919	-6.696
5100	43.202	345.104	307.448	192.044	1098.219	631.795	-6.471
5200	43.291	345.944	308.181	196.369	1100.436	622.628	-6.254
5300	43.380	346.769	308.901	200.702	1102.668	613.418	-6.046
5400	43.468	347.581	309.610	205.045	1104.913	604.166	-5.844
5500	43.556	348.379	310.308	209.396	1107.173	594.872	-5.650
5600	43.643	349.165	310.994	213.756	1109.449	585.537	-5.462
5700	43.730	349.938	311.671	218.124	1111.739	576.161	-5.280
5800	43.816	350.700	312.337	222.502	1114.045	566.745	-5.104
5900	43.902	351.449	312.994	226.888	1116.367	557.289	-4.934
6000	43.987	352.188	313.641	231.282	1118.704	547.793	-4.769

PREVIOUS: September 1978 (1 atm) CURRENT: September 1978 (1 bar)

Sulfur Chloride, Ion (SCl^+) $Cl_1S_1^+(g)$

SULFUR CHLORIDE (S_2Cl) IDEAL GAS M_r = 99.573

Sulfur Chloride (S_2Cl) $Cl_1S_2(g)$

$S°$(298.15 K) = [291.7 ± 2.1] J K^{-1} mol^{-1}

$\Delta_fH°$(0 K) = 79.5 ± 8.4 kJ mol^{-1}
$\Delta_fH°$(298.15 K) = 78.6 ± 8.4 kJ mol^{-1}

Electronic Levels and Quantum Weights

State	ε_i, cm^{-1}	g_i
$[^2B_1]$	0	[2]
$[^2A_1]$	23000	[2]
$[^2A_1]$	26000	[2]
$[^2A_2]$	30000	[2]

Vibrational Frequencies and Degeneracies

ν, cm^{-1}
[500](1)
[200](1)
[550](1)

Point Group: $[C_s]$ σ = [1]
Bond Distances: S–Cl = [2.00] Å
S–S = [1.90] Å
Bond Angle: S–S–Cl = 105°
Product of Moments of Inertia: $I_AI_BI_C$ = $[4.34028 \times 10^{-114}]$ g^3 cm^6

Enthalpy of Formation

Perona et al. (1) have studied the infrared emission from vibrationally excited HCl and DCl formed in the reaction of H and D atoms on S_2Cl_2. From their observation that the highest HCl vibrational level occupied was ν = 7, they obtained an upper limit estimate of the first bond dissociation energy for S_2Cl_2 of D_0^O (S_2Cl–Cl) $\lesssim$ 53 kcal mol^{-1}. Combining this value with $\Delta_fH°$ data for gaseous Cl and S_2Cl_2 (2), we calculate that $\Delta_fH°(S_2Cl, g) \gtrsim$ 20.8 kcal mol^{-1}. A recent independent analysis of these thermochemical data for S_2Cl_2/S_2Cl was made by Takacs (3) who reported $\Delta_fH°(S_2Cl$, g, 0 K) = 17 ± 2 kcal mol^{-1} based on slightly different auxiliary Δ_fH datum for S_2Cl_2 (-7.7 kcal mol^{-1}) than what (-3.6 kcal mol^{-1}) was used here. In the same paper, Perona et al. (1) also used the infrared chemiluminescence technique to measure bond dissociation energies for $Cl_2$and the first S–Cl bond in SCl_2. While this technique leads to results in accord with the known thermochemistry of Cl_2/Cl (2), it appears to give biased results in the case of SCl_2 (see SCl(g) table for discussion). Factors which can adversely influence the D_0^O values determined from infrared chemiluminescence studies have been discussed by Perona et al. (1).

In a recent review of the thermochemistry of sulfur-containing species, Benson (4) has shown that in R_2S_2 compounds with R equal to H, alkyl, or aryl the first R–S bond dissociation energy exceeds the second one by 9.5 ± 1.0 kcal mol^{-1}. Assuming this same trend holds for R=Cl, we obtain $\Delta_rH°$ = 9.5 kcal mol^{-1} for $S_2Cl_2(g) + S_2(g) = 2\ S_2Cl(g)$ which leads to $\Delta_fH°(S_2Cl$, g, 0 K) = 18.3±1.0 kcal mol^{-1} by combining the value of $\Delta_rH°$ with JANAF $\Delta_fH°$ values for S_2Cl_2 and S_2 (2).

We choose to adopt a $\Delta_fH°$ value of 19.0±2.0 kcal mol^{-1} which is a weighted average of the two results derived above. The values of $\Delta_fH°$ at 298.15 K and of $\Delta_{at}H°$(0 K) are 18.8±2.0 kcal mol^{-1} and 140.9 ± 2.0 kcal mol^{-1}, respectively. Using JANAF data (2), we find bond dissociation energies in the S_2Cl radical are $D_0^O(S_2$–Cl) = 40.2 kcal mol^{-1} and D_0^O(S–SCl) = 83.9 kcal mol^{-1}.

Heat Capacity and Entropy

S_2Cl has 19 valence electrons, and according to the Walsh correlation diagram (5), the radical should have a bent structure with a bond angle somewhate greater than that (102°, 2) found for the 20 valence electron molecule SCl_2. We adopt a Cl–S–S angle of 105° which is intermediate between the measured angles for SCl_2 and S_2Cl_2(2). CNDO and INDO calculations (6) on S_2Cl as a function of bond angle show an energy minimum at 100°, supporting our adopted value. The S–Cl and S–S bond distances are estimated from experimentally measured structural data (2) for S_2Cl_2, SCl_2, and S_2. The principal moments of inertia are: I_A = 5.0613×10^{-39}, $I_B = 26.8624 \times 10^{-39}$, and $I_C = 31.9237 \times 10^{-39}$ g cm^2.

Wight and Andrews (7) have observed an absorption band system at 21930-26178 cm^{-1} during mercury are photolysis of Ar/S_2Cl_2 matrices and alkali metal S_2Cl_2 matrix reactions. They assigned this band system to S_2Cl and suggested that the S–Cl stretching frequency was ≥ 490 cm^{-1} based on the observed vibrational spacing in the absorption spectrum. We estimate ν_1 = 500 cm^{-1} based on the S–Cl stretching frequencies for S_2Cl_2 (466 and 461 cm^{-1}, 2), SCl_2(528 cm^{-1}, 2), and SCl_2^+ (530±30 cm^{-1}, 8). The ν_2 bending and ν_3 S–S stretching frequencies are estimated by similar comparisons to that for ν_1. The value of ν_3 is chosen to lie between those (2, 9) observed for S_2Cl_2 (546 cm^{-1}) and S_3(590 cm^{-1}).

An absorption band system has been observed at 22000-26000 cm^{-1} when gaseous S_2Cl_2 is flash photolyzed (10, 11), when Ar/S_2Cl_2 matrices are mercury arc photolyzed during deposition (7), and when S_2Cl_2 is reacted with alkali metal atoms in argon matrices (7). Wight and Andrews (7), along with McGrath (10), have assigned this spectrum to the S_2Cl radical while Donovan et al. (11) believe that portions of the spectrum are due to SCl and S_3. In addition, Donovan et al. (11) observed during flash photolysis of S_2Cl_2 a number of transient bands in the region 30221-34200 cm^{-1} which may be due to S_2Cl. By analogy with the known electronic levels for the isoelectronic radical cations SCl_2^+, OCl_2^+, and OF_2^+ (12), we predict the existence of three excited states for S_2Cl which should give rise to transitions from the ground state in the 20000-30000 cm^{-1} region. No attempt is made here to make definite assignments for the observed absorption spectra (7, 10, and 11) assigned to S_2Cl since this will require the results of higher resolution experiments. We prefer to predict the electronic configurations for the ground and excited states from the Walsh correlation diagram (5) and to estimate their relative term values based on the vertical ionization potentials (8, 12) observed in the photoelectron spectra of SCl_2, OCl_2, and OF_2. A detailed description of the factors considered in making these estimates can be found on the SF_2^+ gas table (2).

Additional support for the existence of the S_2Cl radical is provided by the matrix photolysis work of Herring et al. (6).

References

1. M. J. Perona, D. W. Setser, and R. J. Johnson, J. Chem. Phys. 52, 6384 (1970).
2. JANAF Thermochemical Tables: Cl(g), 6-30-72; S_2Cl_2(g), SCl(g), and SCl_2(g), 6-30-78; H(g), 2-31-77; HCl(g), 9-30-64; S_2(g), 9-30-77; SF_2(g), 12-31-76.
3. G. A. Takacs, J. Chem. Eng. Data, 23, 174 (1978).
4. S. W. Benson, Chem. Rev. 78, 23 (1978).
5. A. D. Walsh, J. Chem. Soc. 1953, 2266 (1953).
6. F. G. Herring, C. A. McDowell, and J. C. Tait, J. Chem. Phys. 57, 4564 (1972).
7. C. A. Wight and L. Andrews, J. Mol. Spectrosc. 72, 243 (1978).
8. B. Solouki, P. Rosmus, and H. Bock, Chem. Phys. Lett. 26, 20 (1974).
9. B. Meyer, Chem. Rev. 76, 367 (1976).
10. W. O. McGrath, J. Chem. Phys. 33, 297 (1960).
11. R. J. Donovan, D. Husain, and P. T. Jackson, Trans. Faraday Soc. 64, 1798 (1968).
12. A. B. Cornford, D. C. Frost, F. G. Herring, and C. A. McDowell, J. Chem. Phys. 55, 2820 (1971).

Enthalpy Reference Temperature = T_r = 298.15 K Standard State Pressure = p° = 0.1 MPa

	J $K^{-1}mol^{-1}$			kJ mol^{-1}			
T/K	$C_p°$	S°	$-[G°-H°(T_r)]/T$	$H°-H°(T_r)$	$\Delta_fH°$	$\Delta_fG°$	Log K_f
0	0.	0.	INFINITE	-12.474	79.499	79.499	INFINITE
100	38.122	243.386	333.376	-8.999	80.139	67.755	-35.391
200	45.979	272.300	296.209	-4.782	79.553	55.544	-14.507
250	48.894	282.889	292.515	-2.406	79.080	49.595	-10.362
298.15	50.954	291.686	291.686	0.	78.559	43.963	-7.702
300	51.021	292.002	291.687	0.094	78.538	43.748	-7.617
350	52.557	299.988	292.315	2.686	77.937	37.997	-5.671
400	53.677	307.083	293.726	5.343	72.857	32.511	-4.245
450	54.510	313.456	295.570	8.049	70.824	27.571	-3.200
500	55.140	319.233	297.652	10.791	68.663	22.886	-2.391
600	56.008	329.370	302.117	16.352	65.238	14.065	-1.225
700	56.558	338.048	306.645	21.982	62.339	5.771	-0.431
800	56.927	345.626	311.054	27.657	59.742	-2.131	0.139
900	57.186	352.346	315.276	33.364	-48.925	-7.557	0.439
1000	57.374	358.382	319.290	39.092	-48.777	-2.968	0.155
1100	57.514	363.857	323.096	44.837	-48.650	1.606	-0.076
1200	57.622	368.866	326.705	50.594	-48.548	6.170	-0.269
1300	57.706	373.482	330.128	56.360	-48.473	10.727	-0.431
1400	57.773	377.761	333.379	62.135	-48.426	15.279	-0.570
1500	57.828	381.749	336.472	67.915	-48.408	19.829	-0.690
1600	57.872	385.482	339.420	73.700	-48.419	24.378	-0.796
1700	57.910	388.992	342.234	79.489	-48.460	28.929	-0.889
1800	57.941	392.303	344.924	85.281	-48.529	33.483	-0.972
1900	57.967	395.436	347.501	91.077	-48.62t	38.042	-1.046
2000	57.990	398.410	349.973	96.875	-48.749	42.606	-1.113
2100	58.010	401.240	352.347	102.675	-48.898	47.177	-1.173
2200	58.027	403.939	354.631	108.477	-49.072	51.756	-1.229
2300	58.042	406.519	356.832	114.280	-49.270	56.344	-1.280
2400	58.056	408.989	358.954	120.085	-49.490	60.940	-1.326
2500	58.069	411.360	361.003	125.891	-49.734	65.547	-1.370
2600	58.081	413.637	362.984	131.699	-49.999	70.163	-1.410
2700	58.092	415.830	364.901	137.507	-50.284	74.790	-1.447
2800	58.104	417.942	366.758	143.317	-50.590	79.428	-1.482
2900	58.116	419.982	368.558	149.128	-50.915	84.077	-1.514
3000	58.129	421.952	370.305	154.941	-51.259	88.738	-1.545
3100	58.143	423.858	372.002	160.754	-51.621	93.410	-1.574
3200	58.158	425.704	373.652	166.569	-51.999	98.095	-1.601
3300	58.176	427.494	375.256	172.386	-52.393	102.791	-1.627
3400	58.195	429.231	376.818	178.204	-52.802	107.500	-1.652
3500	58.217	430.919	378.340	184.025	-53.225	112.221	-1.675
3600	58.242	432.559	379.823	189.848	-53.661	116.954	-1.697
3700	58.270	434.155	381.270	195.673	-54.107	121.699	-1.718
3800	58.301	435.709	382.683	201.502	-54.564	126.457	-1.738
3900	58.337	437.224	384.062	207.334	-55.029	131.227	-1.758
4000	58.376	438.702	385.409	213.169	-55.501	136.008	-1.776
4100	58.419	440.144	386.727	219.009	-55.980	140.802	-1.794
4200	58.467	441.552	388.016	224.854	-56.463	145.607	-1.811
4300	58.520	442.928	389.277	230.703	-56.949	150.424	-1.827
4400	58.577	444.274	390.511	236.558	-57.437	155.253	-1.843
4500	58.639	445.592	391.721	242.418	-57.925	160.092	-1.858
4600	58.706	446.881	392.906	248.286	-58.413	164.942	-1.873
4700	58.778	448.144	394.068	254.160	-58.899	169.803	-1.887
4800	58.854	449.383	395.207	260.041	-59.383	174.674	-1.901
4900	58.936	450.597	396.325	265.931	-59.862	179.555	-1.914
5000	59.022	451.789	397.423	271.829	-60.336	184.446	-1.927
5100	59.114	452.958	398.500	277.735	-60.804	189.346	-1.939
5200	59.209	454.107	399.559	283.652	-61.265	194.256	-1.951
5300	59.310	455.236	400.599	289.577	-61.718	199.174	-1.963
5400	59.415	456.345	401.621	295.514	-62.163	204.101	-1.974
5500	59.524	457.437	402.626	301.460	-62.598	209.036	-1.985
5600	59.637	458.510	403.614	307.418	-63.023	213.978	-1.996
5700	59.754	459.567	404.586	313.388	-63.438	218.928	-2.006
5800	59.874	460.607	405.543	319.369	-63.841	223.886	-2.016
5900	59.998	461.632	406.485	325.363	-64.233	228.850	-2.026
6000	60.126	462.641	407.413	331.369	-64.612	233.820	-2.036

PREVIOUS: June 1978 (1 atm) CURRENT: June 1978 (1 bar)

Sulfur Chloride (S_2Cl) $Cl_1S_2(g)$

$S°(298.15\ K) = 237.833 \pm 0.21\ J\ K^{-1}\ mol^{-1}$

$\Delta_f H°(0\ K) = 196.25 \pm 6.7\ kJ\ mol^{-1}$

$\Delta_f H°(298.15\ K) = 198.32 \pm 6.7\ kJ\ mol^{-1}$

Electronic States and Molecular Constants (σ = 1)

State	ε_i, cm^{-1}	g_i	ω_e, cm^{-1}	$\omega_e x_e$, cm^{-1}	B_e, cm^{-1}	α_e, cm^{-1}	r_e, Å	Source
X $^2\Pi_{1/2}$	0	2	533.5	2.15	0.2539	0.00154	2.063	(3)
$^2\Pi_{3/2}$	207.0	2	533.5	2.15	0.2539	0.00154	2.063	(3)
A $^2\Sigma^+$	23010	2	295.2	0.72	0.1970	0.0007	2.352	(3)
B $^2\Sigma^+$	34107	2	703.8	3.9	0.2769	0.0020	1.984	(6, 7)
B' $^2\Delta$	35618	4	509.1	5.6	0.2598	0.00240	2.035	(10)
C $^2\Pi$	41165	4	671.5	2.2	0.2865	0.0009	1.942	(12, 13)
D $^2\Pi$	44943	4	656.8	3.8	[0.2539]	[0.00154]	[2.063]	(15)

Enthalpy of Formation

The adopted value is based upon a 3rd law analysis of effusion mass spectrometric data by Farber (1) who obtained $\Delta_r H°(298.15\ K) = 27.1\ kcal\ mol^{-1}$ for $Si(g) + SiCl_2(g) = 2\ SiCl(g)$. With auxiliary JANAF data (2) this yields $\Delta_f H°(298.15\ K) = 47.4 \pm 1.6\ kcal\ mol^{-1}$. The error estimate is somewhat higher than that given by Farber and should more accurately reflect the uncertainties in relative cross-section values. A value of 88.3±3.0 kcal mol^{-1} is obtained for D_0^0 using auxiliary JANAF data (2). A linear Birge-Sponer extrapolation of the ground state vibrational data (3) yields D_0^0 = 93.8 kcal mol^{-1} in good agreement with the adopted value. Similar extrapolations for other electronic states yield higher values; Gaydon (4) obtained 104±12 kcal mol^{-1} from the B' $^2\Delta$ state. The value of D_0^0 may be compared to the average (per bond) enthalpies of atomization of 94.7, 96, and 101.9 kcal mol^{-1} for $SiCl_4$, $SiCl_3$, and $SiCl_2$, respectively (2).

Heat Capacity and Entropy

Vibrational and rotational constants and splitting for the ground state are taken from the analysis of the A-X system by Singhal and Verma (3). These are in excellent agreement with those of several other investigators (5-10, 12, 13, 15). Spectroscopic information for the A $^2\Sigma$ state is also taken from Singhal and Verma (3) and yields different molecular constants than an earlier study by Sanii and Verma (5). Molecular constants for the B $^2\Sigma$ state are taken from Ovcharenko (6) and Rai (7) and show good agreement with those of Ovcharenko (8) and Mishra (9). The molecular constants for the B' $^2\Delta$ state are taken from Verma (10) and are in good agreement with the less precise measurements by Ovcharenko (11). The vibrational properties of the C $^2\pi$ state were taken from Jevons (12) while the rotational constants are from Pandley (13). The latter differ slightly from those of Ovcharenko (14). The spin splittings of 2.727 cm^{-1} (10) and 10-12 cm^{-1} (13, 14) for the B' and C states, respectively, have been ignored in our calculations. The vibrational constants for the D $^2\Pi$ state were taken from Oldershaw (15); since no rotational study is available, we have used the ground state B_e and α_e as first approximations in our calculations. This introduces a nearly negligible error in the thermal functions. Oldershaw (15) has observed two additional sets of high energy levels but these are not included due to uncertainties in their states. It should be recognized that the $^4\Sigma^-$ state observed for SiF and SiI (2) has not been discovered for SiCl; by analogy with SiF and SiI it is expected to lie at 26,000±3,000 cm^{-1}. The stated uncertainty in the entropy should account for this level as well as the splittings mentioned above. The ground state is treated as two distinct levels due to the splitting of this state as expressed by the spin coupling constant (A = 207.2 cm^{-1}). This leads to slightly biased results at low temperatures (below 500 K); the stated uncertainty in S°(298.15 K) should account for this also. All molecular constant data has been corrected to reflect the natural isotopic abundances of Si and Cl.

References

1. M. Farber, Space Sciences Inc., personal communication, Oct. 4, 1976.
2. JANAF Thermochemical Tables: Si(g), 3-31-67; $SiCl_2(g)$, 12-31-70; Cl(g), 6-30-72; $SiCl_3(g)$, 12-31-69; $SiCl_4(g)$, 12-31-70; SiF(g), SiI(g), 12-31-76.
3. S. R. Singhal and R. D. Verma, Can. J. Phys. 49, 407 (1971).
4. A. G. Gaydon, "Dissociation Energies," Chapman and Hall Ltd., London, 1968.
5. N. Sanii and R. D. Verma, Can. J. Phys. 43, 960 (1965).
6. I. E. Ovcharenko and Y. Y. Kuzyakov, Opt. Spectrosc. 13, 362 (1962).
7. S. B Rai, J. Singh, K. N. Upadhya, and P. K. Rai, J. Phys. B, 7, 415 (1974).
8. I. E. Ovcharenko, L. N. Tunitskii, and V. I. Yakutin, Opt. Spectrosc. 8, 393 (1960).
9. R. K. Mishra and B. N. Khanna, Curr. Sci. 38, 361 (1969).
10. R. D. Verma, Can. J. Phys. 42, 2345 (1964).
11. I. E. Ovcharenko, Y. Y, Kuzyakov, and V. M. Tatevskii, Opt. Spectrosc. 14, 6 (1963).
12. W. Jevons, Proc. Phys. Soc. (London), 48, 563 (1936).
13. R. K. Pandey, K. N. Upadhya, and K. P. R. Nair, Indian J. Pure Appl. Phys. 9, 36 (1971).
14. I. E. Ovcharenko and Y. Y. Kuzyakov, Opt. Spectrosc. 20, 14 (1966).
15. G. A. Oldershaw and K. Robinson, J. Mol. Spectrosc. 38, 306 (1971).

Enthalpy Reference Temperature = T_r = 298.15 K — Standard State Pressure = p° = 0.1 MPa

T/K	$C_p°$ (J $K^{-1}mol^{-1}$)	S° (J $K^{-1}mol^{-1}$)	$-[G°-H°(T_r)]/T$ (J $K^{-1}mol^{-1}$)	$H°-H°(T_r)$ (kJ mol^{-1})	$\Delta_f H°$ (kJ mol^{-1})	$\Delta_f G°$ (kJ mol^{-1})	Log K_f
0	0.	0.	INFINITE	-9.885	196.245	196.245	INFINITE
100	32.744	200.342	268.867	-6.853	197.555	187.355	-97.864
200	34.712	223.757	241.075	-3.463	198.247	176.825	-46.182
298.15	35.780	237.833	237.833	0.	198.322	166.278	-29.131
300	35.795	238.054	237.833	0.066	198.319	166.080	-28.917
400	36.442	248.448	239.246	3.681	198.078	155.364	-20.288
500	36.834	256.626	241.933	7.346	197.679	144.729	-15.120
600	37.088	263.365	244.960	11.043	197.184	134.184	-11.682
700	37.266	269.097	248.009	14.761	196.618	123.728	-9.233
800	37.399	274.082	250.963	18.495	195.993	113.357	-7.401
900	37.504	278.493	253.782	22.240	195.316	103.067	-5.982
1000	37.591	282.449	256.454	25.995	194.592	92.856	-4.850
1100	37.667	286.035	258.983	29.758	193.824	82.719	-3.928
1200	37.734	289.316	261.376	33.528	193.014	72.654	-3.163
1300	37.795	292.339	263.643	37.304	192.164	62.658	-2.518
1400	37.851	295.142	265.794	41.087	191.273	52.729	-1.967
1500	37.905	297.755	267.838	44.875	190.341	42.865	-1.493
1600	37.956	300.203	269.786	48.668	189.369	33.065	-1.079
1700	38.004	302.505	271.643	52.466	138.177	23.774	-0.730
1800	38.052	304.679	273.419	56.269	137.352	17.068	-0.495
1900	38.098	306.738	275.119	60.076	136.526	10.408	-0.286
2000	38.143	308.693	276.749	63.888	135.700	3.792	-0.099
2100	38.187	310.555	278.315	67.705	134.872	-2.783	0.069
2200	38.231	312.332	279.821	71.526	134.043	-9.319	0.221
2300	38.275	314.033	281.272	75.351	133.212	-15.817	0.359
2400	38.318	315.663	282.671	79.180	132.378	-22.278	0.485
2500	38.362	317.228	284.022	83.014	131.540	-28.705	0.600
2600	38.405	318.733	285.328	86.853	130.697	-35.098	0.705
2700	38.449	320.183	286.593	90.695	129.851	-41.459	0.802
2800	38.493	321.583	287.817	94.543	128.999	-47.788	0.892
2900	38.539	322.934	289.005	98.394	128.141	-54.087	0.974
3000	38.585	324.241	290.158	102.250	127.279	-60.356	1.051
3100	38.632	325.507	291.278	106.111	126.412	-66.596	1.122
3200	38.681	326.735	292.367	109.977	125.540	-72.808	1.188
3300	38.732	327.926	293.427	113.847	124.664	-78.993	1.250
3400	38.784	329.083	294.458	117.723	123.785	-85.151	1.308
3500	38.839	330.208	295.464	121.604	122.904	-91.284	1.362
3600	38.896	331.303	296.444	125.491	-262.130	-86.930	1.261
3700	38.955	332.369	297.401	129.384	-262.599	-82.057	1.158
3800	39.017	333.409	298.335	133.282	-263.067	-77.171	1.061
3900	39.082	334.423	299.247	137.187	-263.533	-72.273	0.968
4000	39.151	335.414	300.139	141.099	-263.996	-67.363	0.880
4100	39.222	336.381	301.011	145.017	-264.453	-62.442	0.796
4200	39.296	337.327	301.865	14F.943	-264.903	-57.509	0.715
4300	39.374	338.253	302.700	152.877	-265.344	-52.566	0.639
4400	39.456	339.159	303.518	156.818	-265.773	-47.613	0.565
4500	39.541	340.047	304.320	160.768	-266.190	-42.650	0.495
4600	39.630	340.917	305.106	164.727	-266.592	-37.678	0.428
4700	39.722	341.770	305.877	168.694	-266.978	-32.697	0.363
4800	39.819	342.607	306.634	172.671	-267.346	-27.709	0.302
4900	39.919	343.429	307.377	176.658	-267.695	-22.712	0.242
5000	40.023	344.237	308.106	180.655	-268.024	-17.710	0.185
5100	40.131	345.030	308.822	184.663	-268.331	-12.700	0.130
5200	40.242	345.811	309.526	188.681	-268 616	-7.685	0.077
5300	40.358	346.578	310.218	192.711	-268.877	-2.665	0.026
5400	40.477	347.334	310.898	196.753	-269.114	2.361	-0.023
5500	40.600	348.078	311.567	200.807	-269.325	7.390	-0.070
5600	40.726	348.810	312.226	204.873	-269.511	12.423	-0.116
5700	40.856	349.532	312.874	208.952	-269.670	17.459	-0.160
5800	40.990	350.244	313.512	213.045	-269.804	22.497	-0.203
5900	41.128	350.946	314.141	217.150	-269.910	27.538	-0.244
6000	41.268	351.638	314.760	221.270	-269.988	32.580	-0.284

PREVIOUS: December 1976 (1 atm) CURRENT: December 1976 (1 bar)

STRONTIUM CHLORIDE (SrCl) IDEAL GAS M_r = 123.073

$S°(298.15\ K) = 252.284 \pm 0.42\ J\ K^{-1}\ mol^{-1}$

$\Delta_f H°(0\ K) = -122.49 \pm 8.4\ kJ\ mol^{-1}$
$\Delta_f H°(298.15\ K) = -123.85 \pm 8.4\ kJ\ mol^{-1}$

Electronic Levels and Quantum Weights

State	ε_i, cm^{-1}	g_i	State	ε_i, cm^{-1}	g_i
$X\ ^2\Sigma^+$	0.0	2	$D\ ^2\Sigma$	28822.9	2
$A\ ^2\Pi$	14818.3	2	$E\ ^2\Sigma$	32201.8	2
	15112.8	2	$F\ ^2\Pi$	32909.4	2
$B\ ^2\Sigma$	15779.5	2		32978.6	2
$C\ ^2\Pi$	25244.7	2	$G\ ^2\Delta$	34086.7	2
	25399.9	2		34112.8	2
			$H\ ^2\Sigma$	34256.7	2

$\omega_e = 300.94\ cm^{-1}$ $\omega_e x_e = 0.97\ cm^{-1}$ $\sigma = 1$
$B_e = [0.10271 \pm 0.003]\ cm^{-1}$ $\alpha_e = [0.00043]\ cm^{-1}$ $r_e = [2.55 \pm 0.04]$ Å

Enthalpy of Formation

$\Delta_f H°$ is a weighted average of results from three gas-phase equilibrium reactions studied mass spectrometrically by Hildenbrand (1) and Potter et al. (2). Analyses of these data are summarized below. The overall range of values is less than 4 kcal mol^{-1}, suggesting the approximate uncertainty in $\Delta_f H°$. From the adopted $\Delta_f H°$ we calculate D_0^0 = 98.2±2 kcal mol^{-1}. Independent methods give 98±4 (3), ≥ 99 (4), 95.6±1.6 (5), 97±6 (6) and 88.1 kcal mol^{-1} (7).

Electron-impact data (3) gave $D_0^0(ClSr\text{-}Cl)$ = 112.8±3.3 and we use $\Delta_{at}H°(SrCl_2)$ = 210.8±3 kcal mol^{-1} to derive D_0^0 = 98±4 kcal mol^{-1}. Chemiluminescence from crossed beams of Sr atoms with Cl_2 gave an approximate lower limit of 99 kcal mol^{-1} (4). Flame studies gave D_0^0 = 95.6±1.6 kcal mol^{-1} (5) and $D_0^0(ClSr\text{-}Cl)$ = 114±5 kcal mol^{-1} (6); we derive D_0^0 = 97±6 kcal mol^{-1} from the latter value using $\Delta_{at}H°(SrCl_2)$ = 210.8±3 kcal mol^{-1}. Ryabova et al. (6, 5) considered the flames to involve $SrCl_2$, SrOHCl and SrOH as well as SrCl. Schofield and Sugden (8) inferred that the flames involved mainly $SrCl_2$ rather than SrCl as assumed in earlier flame studies. Gaydon (9) discounted a value of D_0^0 derived from limited spectroscopic data on the ground state. Novikov and Gurvich (7) published additional emission spectra and derived D_0^0 = 88.1 kcal mol^{-1} by parabolic extrapolation. This value should be considered as approximate in light of Gaydon's precautionary notes, especially since the observed levels extend only to about 0.3 of the dissociation limit. Hildenbrand (1) found that ionicity corrections bring the Birge-Sponer extrapolation into reasonable agreement with D_0^0.

Source	Reaction	T/K	$\Delta_r H°$(298.15 K), kcal mol^{-1} 2nd law	3rd law	δS cal K^{-1} mol^{-1}	$\Delta_f H°$(298.15 K) kcal mol^{-1}	D_0^0
(1) Hildenbrand (1970)	A	1305-1416	16.9±4.2	14.93±0.6	1.5±3.1	-29.49±1.0	98.3
(2) Potter (1970)	A	1235-1540	13.6±1.1	13.97±0.5	-0.3±0.8	-29.97±1.0	98.8
(1) Hildenbrand (1970)	B	1305-1416	19.7±4.5	22.65±0.7	-2.2±3.3	-28.45±2.0	97.3
(2) Potter (1970)	B	1235-1540	7.0±4.7	18.79±2.6	-8.6±3.4	-32.31±3.2	101.1
(1) Hildenbrand (1970)	C	1661		2.07±2		-30.72±3.6	99.6

Reactions: (A) $Sr(g) + SrCl_2(g) = 2SrCl(g)$; (B) $Sr(g) + AlCl(g) = SrCl(g) + Al(g)$; (C) $Ca(g) + SrCl(g) = CaCl(g) + Sr(g)$.

Heat Capacity and Entropy

Electronic levels and vibrational constants are based on the review of Rosen (10) as modified by the additional data of Novikov and Gurvich (7). Vibrational constants are adjusted for natural isotopic abundances. B_e is calculated from r_e which is estimated by comparison of the bond lengths of MX_2(M=Mg, Ca, Sr, Ba and X=F, Cl) with those of MgCl, CaCl and MF(M=Mg, Ca, Sr, Ba). The value of α_e is estimated from the Morse potential function.

References

1. D. L. Hildenbrand, J. Chem. Phys. 52, 5751 (1970).
2. N. D. Potter, M. H. Boyer, F. Ju, D. L. Hildenbrand, E. Murad and W. F. Hall, AD-715567, 17-26, 1970; Aeronutronic Div., Philco Corp., Report U-4859, AFOSR 70-2311 TR, August 31, 1970.
3. D. L. Hildenbrand, Int. J. Mass Spectrom. Ion Phys. 4, 75 (1970).
4. C. D. Jonah and R. N. Zare, Chem. Phys. Letters 9, 65 (1971).
5. L. V. Gurvich, V. G. Ryabova, A. N. Khitrov and E. M. Starovoitov, Teplofiz. Vys. Temp. 9, 290 (1971).
6. V. G. Ryabova, L. V. Gurvich and A. N. Khitrov, Teplofiz. Vys. Temp. 9, 755 (1971).
7. M. M. Novikov and L. V. Gurvich, Optics and Spectroscopy 19, 76 (1965).
8. K. Schofield and T. M. Sugden, Trans. Faraday Soc. 67, 1054 (1971).
9. A. G. Gaydon, "Dissociation Energies and Spectra of Diatomic Molecules," 3rd Edition, Chapman and Hall, Ltd., London, 1968.
10. B. Rosen, Ed., "Spectroscopic Data Relative to Diatomic Molecules," Pergamon Press, New York, 1970.

Strontium Chloride (SrCl) $Cl_1Sr_1(g)$

Enthalpy Reference Temperature = T_r = 298.15 K Standard State Pressure = p° = 0.1 MPa

T/K	$C_p°$ (J $K^{-1}mol^{-1}$)	S° (J $K^{-1}mol^{-1}$)	$-[G°-H°(T_r)]/T$ (J $K^{-1}mol^{-1}$)	$H°-H°(T_r)$ (kJ mol^{-1})	$\Delta_f H°$ (kJ mol^{-1})	$\Delta_f G°$ (kJ mol^{-1})	Log K_f
0	0.	0.	INFINITE	-9.799	-122.487	-122.487	INFINITE
100	31.264	215.070	283.467	-6.840	-122.483	-131.734	68.811
200	34.916	238.054	255.566	-3.503	-123.153	-140.725	36.754
250	35.747	245.942	252.879	-1.734	-123.496	-145.084	30.314
298.15	36.252	252.284	252.284	0.	-123.846	-149.204	26.140
300	36.268	252.509	252.285	0.067	-123.860	-149.362	26.006
350	36.613	258.127	252.728	1.890	-124.250	-153.582	22.921
400	36.856	263.033	253.716	3.727	-124.665	-157.744	20.599
450	37.035	267.385	254.998	5.574	-125.107	-161.852	18.787
500	37.173	271.294	256.435	7.430	-125.575	-165.910	17.333
600	37.374	278.091	259.495	11.158	-126.593	-173.884	15.138
700	37.517	283.863	262.574	14.902	-127.718	-181.678	13.557
800	37.628	288.880	265.555	18.660	-128.950	-189.303	12.360
900	37.721	293.318	268.398	22.427	-130.917	-196.696	11.416
1000	37.803	297.296	271.093	26.204	-132.043	-203.945	10.653
1100	37.877	300.903	273.641	29.988	-141.070	-210.715	10.006
1200	37.945	304.201	276.052	33.779	-143.106	-216.957	9.444
1300	38.010	307.241	278.336	37.577	-145.141	-223.028	8.961
1400	38.073	310.060	280.503	41.381	-147.174	-228.944	8.542
1500	38.135	312.689	282.562	45.191	-149.205	-234.713	8.173
1600	38.196	315.152	284.523	49.008	-151.234	-240.348	7.847
1700	38.259	317.470	286.393	52.831	-290.177	-244.676	7.518
1800	38.324	319.659	288.181	56.660	-290.342	-241.994	7.022
1900	38.394	321.733	289.893	60.496	-290.511	-239.304	6.579
2000	38.469	323.704	291 534	64.339	-290.687	-236.604	6.179
2100	38.552	325.583	293.111	68.190	-290.871	-233.895	5.818
2200	38.644	327.378	294.629	72.049	-291.070	-231.178	5.489
2300	38.746	329.098	296.090	75.919	-291.288	-228.450	5.188
2400	38.861	330.750	297.500	79.799	-291.532	-225.713	4.913
2500	38.990	332.339	298.862	83.691	-291.807	-222.965	4.659
2600	39.133	333.871	300.179	87.597	-292.122	-220.206	4.424
2700	39.291	335.350	301.455	91.518	-292.484	-217.433	4.206
2800	39.466	336.782	302.691	95.456	-292.902	-214.646	4.004
2900	39.656	338.171	303.890	99.412	-293.384	-211.843	3.816
3000	39.863	339.518	305.056	103.388	-293.939	-209.022	3.639
3100	40.087	340.829	306.189	107.385	-294.576	-206.181	3.474
3200	40.327	342.106	307.291	111.406	-295.302	-203.318	3.319
3300	40.582	343.350	308.365	115.451	-296.124	-200.431	3.173
3400	40.852	344.566	309.412	119.523	-297.051	-197.518	3.034
3500	41.137	345.754	310.433	123.622	-298.088	-194.575	2.904
3600	41.435	346.917	311.431	127.751	-299.240	-191.602	2.780
3700	41.746	348.057	312.405	131.910	-300.511	-188.594	2.662
3800	42.069	349.174	313.358	136.100	-301.905	-185.551	2.551
3900	42.403	350.271	314.291	140.324	-303.423	-182.470	2.444
4000	42.746	351.349	315.204	144.581	-305.065	-179.348	2.342
4100	43.099	352.409	316.098	148.873	-306.832	-176.183	2.245
4200	43.460	353.452	316.975	153.201	-308.717	-172.974	2.151
4300	43.827	354.479	317.836	157.565	-310.725	-169.719	2.062
4400	44.201	355.491	318.680	161.967	-312.848	-166.415	1.976
4500	44.580	356.488	319.509	166.406	-315.082	-163.063	1.893
4600	44.963	357.472	320.324	170.883	-317.355	-159.664	1.813
4700	45.349	358.443	321.124	175.398	-319 773	-156.210	1.736
4800	45.737	359.402	321.912	179.953	-322.277	-152.704	1.662
4900	46.128	360.349	322.687	184.546	-324.860	-149.144	1.590
5000	46.518	361.285	323.449	189.178	-327.508	-145.532	1.520
5100	46.909	362.210	324.200	193.850	-330.199	-141.868	1.453
5200	47.299	363.125	324.940	198.560	-332.960	-138.148	1.388
5300	47.688	364.029	325.669	203.310	-335.763	-134.375	1.324
5400	48.074	364.924	326.388	208.098	-338.600	-130.549	1.263
5500	48.458	365.810	327.096	212.924	-341.461	-126.669	1.203
5600	48.838	366.686	327.796	217.789	-344.339	-122.738	1.145
5700	49.214	367.554	328.485	222.692	-347.223	-118.755	1.088
5800	49.585	368.413	329.167	227.632	-350.107	-114.722	1.033
5900	49.952	369.264	329.839	232.609	-352.773	-110.663	0.980
6000	50.312	370.107	330.503	237.622	-355.602	-106.536	0.927

PREVIOUS: December 1972 (1 atm) CURRENT: December 1972 (1 bar)

Strontium Chloride (SrCl) $Cl_1Sr_1(g)$

TITANIUM CHLORIDE (TiCl) IDEAL GAS M_r = 83.333

$S°(298.15\ K) = 249.214 \pm 8.4\ J\ K^{-1}\ mol^{-1}$

$\Delta_f H°(0\ K) = [154.13 \pm 41.8]\ kJ\ mol^{-1}$
$\Delta_f H°(298.15\ K) = [154.39 \pm 41.8]\ kJ\ mol^{-1}$

Electronic Levels and Quantum Weights

State	ϵ_i, cm^{-1}	g_i
$[^4\Sigma]$	0	[4]
	[1000]	[4]
	[2000]	[4]
	[4000]	[4]
	[6000]	[4]
	[8000]	[4]
	[10000]	[4]

$\omega_e = 379.7\ cm^{-1}$ $\omega_e x_e = 3.41\ cm^{-1}$ $\sigma = 1$
$B_e = [0.146]\ cm^{-1}$ $\alpha_e = [0.0013]\ cm^{-1}$ $r_e = [2.3]$ Å

Enthalpy of Formation

The enthalpy of formation, $\Delta_f H°(298.15\ K)$, of TiCl(g) is calculated from the dissociation energy which is estimated as 105 kcal mol^{-1}. This estimate is obtained from the relation $D(TiCl_4) < D_0^0(TiCl) < D(TiCl_3)$, where D represents the average energy per bond, which is valid for the titanium fluorides. The dissociation energy of TiF(g) from which the relation is derived was estimated relative to that of $TiF_2(g)$ by Zmbov and Margrave (1).

Extrapolations of the vibrational levels reported by Shenyavskaya (2), Rao (3) and Mace and Packer (4) yield values of the dissociation energy which are very low with respect to bond energy relations.

Heat Capacity and Entropy

The vibrational frequency, ω_e, and anharmonic vibrational term, $\omega_e x_e$, were reported by Shenyavskaya et al. (2). The interatomic distance is estimated from Guggenheimer's relation (5). B_e is calculated from r_e. The ground state term and electronic levels are estimated from the ground state multiplet of Ti^+ (6). α_e is estimated from the Morse potential function.

References

1. K. F. Zmbov and J. L. Margrave, J. Phys. Chem. 71, 2893 (1967).
2. E. A. Shenyaskaya, Yu. Ya. Kuzyakov, and V. M. Tatevskii, Opt. Spectrosc. (USSR) 12, 197 (1962).
3. V. R. Rao, Indian J. Phys. 23, 535 (1949).
4. K. R. Mace and A. H. Packer, Phys. Rev. 52, 1150 (1937).
5. K. M. Guggenheimer, Proc. Phys. Soc. (London) 58, 456 (1946).
6. C. E. Moore, U. S. Nat. Bur. Stand. Circ. 467, 1949.

Titanium Chloride (TiCl) $Cl_1Ti_1(g)$

Enthalpy Reference Temperature = T_r = 298.15 K Standard State Pressure = p° = 0.1 MPa

T/K	$C_p°$	S°	$-[G°-H°(T_r)]/T$	$H°-H°(T_r)$	$\Delta_f H°$	$\Delta_f G°$	Log K_f
	J $K^{-1}mol^{-1}$			kJ mol^{-1}			
0	0.	0.	INFINITE	-9.680	154.130	154.130	INFINITE
100	30.251	212.708	280.196	-6.749	155.045	144.050	-75.244
200	34.252	234.946	252.532	-3.517	154.840	133.093	-34.760
250	35.893	242.769	249.820	-1.763	154.622	127.681	-26.677
298.15	37.297	249.214	249.214	0.	154.390	122.513	-21.464
300	37.348	249.445	249.215	0.069	154.380	122.315	-21.297
350	38.622	255.300	249.674	1.969	154.134	116.990	-17.460
400	39.709	260.530	250.710	3.928	153.891	111.701	-14.587
450	40.614	265.261	252.068	5.937	153.654	106.441	-12.355
500	41.351	269.580	253.607	7.987	153.424	101.208	-10.573
600	42.402	277.220	256.922	12.179	152.974	90.807	-7.905
700	43.031	283.808	260.303	16.453	152.525	80.481	-6.006
800	43.387	289.579	263.610	20.776	152.073	70.220	-4.585
900	43.581	294.702	266.785	25.125	151.571	60.018	-3.483
1000	43.688	299.300	269.811	29.489	150.945	49.877	-2.605
1100	43.754	303.467	272.684	33.861	150.128	39.808	-1.890
1200	43.804	307.276	275.410	38.239	145.130	29.942	-1.303
1300	43.853	310.784	277.998	42.622	144.646	20.362	-0.818
1400	43.907	314.036	280.457	47.010	144.084	10.823	-0.404
1500	43.969	317.067	282.798	51.404	143.432	1.326	-0.046
1600	44.040	319.907	285.030	55.804	142.678	-8.124	0.265
1700	44.118	322.580	287.161	60.212	141.809	-17.523	0.538
1800	44.201	325.104	289.199	64.628	140.813	-26.867	0.780
1900	44.288	327.496	291.152	69.052	139.675	-36.152	0.994
2000	44.376	329.770	293.027	73.486	123.698	-44.920	1.173
2100	44.465	331.937	294.829	77.928	121.492	-53.297	1.326
2200	44.553	334.007	296.563	82.379	119.289	-61.568	1.462
2300	44.638	335.990	298.234	86.838	117.087	-69.740	1.584
2400	44.721	337.891	299.847	91.306	114.887	-77.816	1.694
2500	44.799	339.719	301.406	95.782	112.687	-85.800	1.793
2600	44.874	341.477	302.913	100.266	110.486	-93.696	1.882
2700	44.943	343.172	304.373	104.757	108.283	-101.507	1.964
2800	45.009	344.808	305.788	109.254	106.078	-109.237	2.038
2900	45.070	346.388	307.161	113.758	103.869	-116.887	2.105
3000	45.126	347.917	308.494	118.268	101.656	-124.462	2.167
3100	45.179	349.398	309.790	122.783	99.439	-131.963	2.224
3200	45.227	350.833	311.050	127.304	97.218	-139.392	2.275
3300	45.272	352.225	312.277	131.829	94.992	-146.751	2.323
3400	45.313	353.577	313.472	136.358	92.762	-154.044	2.367
3500	45.352	354.891	314.637	140.891	90.529	-161.270	2.407
3600	45.388	356.169	315.773	145.428	88.293	-168.433	2.444
3700	45.421	357.413	316.881	149.969	-323.044	-167.745	2.368
3800	45.453	358.625	317.964	154.512	-324.049	-163.534	2.248
3900	45.482	359.806	319.022	159.059	-325.110	-159.297	2.134
4000	45.511	360.958	320.056	163.609	-326.224	-155.031	2.024
4100	45.538	362.082	321.067	168.161	-327.388	-150.737	1.920
4200	45.564	363.180	322.057	172.716	-328.600	-146.413	1.821
4300	45.590	364.252	323.026	177.274	-329.857	-142.061	1.726
4400	45.615	365.301	323.975	181.834	-331.156	-137.679	1.634
4500	45.640	366.326	324.904	186.397	-332.493	-133.267	1.547
4600	45.664	367.329	325.816	190.962	-333.866	-128.824	1.463
4700	45.689	368.312	326.710	195.530	-335.271	-124.352	1.382
4800	45.714	369.274	327.586	200.100	-336.706	-119.849	1.304
4900	45.740	370.217	328.447	204.673	-338.167	-115.316	1.229
5000	45.766	371.141	329.292	209.248	-339.652	-110.753	1.157
5100	45.792	372.048	330.121	213.826	-341.157	-106.160	1.087
5200	45.819	372.937	330.936	218.406	-342.680	-101.538	1.020
5300	45.847	373.810	331.737	222.990	-344.218	-96.886	0.955
5400	45.876	374.667	332.524	227.576	-345.768	-92.204	0.892
5500	45.905	375.509	333.298	232.165	-347.328	-87.494	0.831
5600	45.936	376.337	334.059	236.757	-348.896	-82.756	0.772
5700	45.967	377.150	334.808	241.352	-350.468	-77.989	0.715
5800	46.000	377.950	335.545	245.951	-352.044	-73.195	0.659
5900	46.033	378.737	336.270	250.552	-353.621	-68.374	0.605
6000	46.068	379.511	336.984	255.157	-355.197	-63.526	0.553

PREVIOUS: December 1968 (1 atm) CURRENT: December 1968 (1 bar)

Titanium Chloride (TiCl) $Cl_1Ti_1(g)$

TUNGSTEN CHLORIDE (WCl) IDEAL GAS M_r = 219.303

$S°(298.15 K) = [261.895] J K^{-1} mol^{-1}$

$\Delta_f H°(0 K) = [553.74 \pm 41.8] kJ mol^{-1}$
$\Delta_f H°(298.15 K) = [553.54 \pm 41.8] kJ mol^{-1}$

Electronic Levels and Quantum Weights

State	ε_i, cm^{-1}	g_i
$[^2\Delta]$	0	[4]
	[5000]	[2]
	[15000]	[20]

ω_e = [455.4] cm^{-1} $\omega_e x_e$ = [1.46] cm^{-1} σ = 1
B_e = [0.11105] cm^{-1} α_e = [0.00042] cm^{-1} r_e = [2.26] Å

Enthalpy of Formation

The enthalpy of formation, $\Delta_f H°$(WCl, g, 298.15 K) = 132.3 kcal mol^{-1}, is calculated from the bond dissociation energy, D_0^0(W-Cl, 298.15 K) = 100±10 kcal mol^{-1}. This estimated $\Delta_{at}H°$(298.15 K) is obtained from the average bond dissociation energy for WCl_x(g) + W(g) + xCl(g) where x = 2,4,5,6, using all JANAF values.

Heat Capacity and Entropy

The bond distance is estimated to be the same as that in WCl_6(g). This distance is then used with Guggenheimer's relation for polar molecules to calculate the fundamental vibrational frequency ω_e (1). The anharmonicity correction x_e is estimated roughly by assuming $x_e = \omega_e/4(D° + 0.5\ \omega_e) = 0.0032$. The rotational constant B_e is calculated from the estimated bond distance. The value of α_e is calculated from the Morse potential function.

The ground state configuration, low lying electronic levels and their quantum weights are assumed to be the same as those estimated for WF(g) (2).

References

1. K. M. Guggenheimer, Proc. Phys. Soc. (London) 58, 456 (1946).
2. JANAF Thermochemical Table; WF(g), 3-31-67.

Enthalpy Reference Temperature = T_r = 298.15 K Standard State Pressure = p° = 0.1 MPa

T/K	$C_p°$ (J $K^{-1}mol^{-1}$)	S° (J $K^{-1}mol^{-1}$)	$-[G°-H°(T_r)]/T$ (J $K^{-1}mol^{-1}$)	$H°-H°(T_r)$ (kJ mol^{-1})	$\Delta_f H°$ (kJ mol^{-1})	$\Delta_f G°$ (kJ mol^{-1})	Log K_f
0	0.	0.	INFINITE	-9.370	553.736	553.736	INFINITE
100	29.633	226.867	291.391	-6.452	554.546	542.271	-283.253
200	32.811	248.370	265.027	-3.331	554.141	530.118	-138.453
250	34.017	255.829	262.465	-1.659	553.847	524.145	-109.514
298.15	34.843	261.895	261.895	0.	553.543	518.452	-90.831
300	34.869	262.111	261.896	0.064	553.531	518.234	-90.233
350	35.472	267.534	262.323	1.824	553.207	512.377	-76.468
400	35.908	272.300	263.278	3.609	552.877	506.567	-66.151
450	36.231	276.549	264.521	5.413	552.541	500.798	-58.131
500	36.478	280.380	265.918	7.231	552.199	495.067	-51.719
600	36.826	287.064	268.902	10.897	551.492	483.707	-42.110
700	37.066	292.760	271.913	14.593	550.750	472.467	-35.256
800	37.256	297.722	274.836	18.309	549.972	461.336	-30.122
900	37.426	302.120	277.628	22.043	549.157	450.306	-26.135
1000	37.595	306.072	280.278	25.794	548.306	439.367	-22.950
1100	37.771	309.663	282.788	29.562	547.420	428.516	-20.349
1200	37.955	312.958	285.167	33.349	546.501	417.747	-18.184
1300	38.146	316.003	287.423	37.154	545.550	407.056	-16.356
1400	38.341	318.837	289.567	40.978	544.567	396.439	-14.791
1500	38.537	321.489	291.608	44.822	543.553	385.894	-13.438
1600	38.733	323.982	293.554	48.685	542.507	375.417	-12.256
1700	38.926	326.336	295.414	52.568	541.428	365.007	-11.215
1800	39.116	328.567	297.194	56.470	540.317	354.661	-10.292
1900	39.305	330.687	298.902	60.391	539.172	344.378	-9.468
2000	39.492	332.707	300.542	64.331	537.992	334.156	-8.727
2100	39.682	334.639	302.120	68.290	536.777	323.994	-8.059
2200	39.875	336.489	303.640	72.268	535.525	313.891	-7.453
2300	40.074	338.266	305.107	76.265	534.237	303.845	-6.901
2400	40.281	339.976	306.525	80.283	532.912	293.856	-6.396
2500	40.499	341.625	307.896	84.322	531.550	283.923	-5.932
2600	40.729	343.218	309.224	88.383	530.155	274.046	-5.506
2700	40.972	344.759	310.512	92.468	528.700	264.223	-5.112
2800	41.228	346.254	311.762	96.578	527.152	254.455	-4.747
2900	41.499	347.705	312.976	100.714	525.473	244.745	-4.408
3000	41.783	349.117	314.158	104.878	523.636	235.096	-4.093
3100	42.081	350.492	315.308	109.071	521.601	225.511	-3.800
3200	42.390	351.833	316.428	113.295	519.336	215.995	-3.526
3300	42.711	353.142	317.521	117.550	516.819	206.554	-3.269
3400	43.041	354.422	318.588	121.837	514.015	197.193	-3.030
3500	43.378	355.674	319.629	126.158	510.863	187.920	-2.805
3600	43.722	356.901	320.648	130.513	507.267	178.743	-2.593
3700	44.069	358.104	321.644	134.902	468.364	169.863	-2.398
3800	44.419	359.284	322.619	139.327	467.172	161.811	-2.224
3900	44.768	360.442	323.574	143.786	466.013	153.791	-2.060
4000	45.116	361.580	324.510	148.280	464.886	145.800	-1.904
4100	45.460	362.698	325.428	152.809	463.793	137.836	-1.756
4200	45.798	363.798	326.328	157.372	462.735	129.899	-1.616
4300	46.130	364.879	327.212	161.969	461.712	121.986	-1.482
4400	46.453	365.944	328.080	166.598	460.725	114.097	-1.355
4500	46.767	366.991	328.933	171.259	459.775	106.230	-1.233
4600	47.069	368.022	329.772	175.951	458.861	98.384	-1.117
4700	47.360	369.038	330.597	180.672	457.983	90.557	-1.006
4800	47.637	370.038	331.408	185.422	457.141	82.748	-0.900
4900	47.901	371.023	332.206	190.199	456.336	74.957	-0.799
5000	48.151	371.993	332.993	195.002	455.566	67.181	-0.702
5100	48.387	372.949	333.767	199.829	454.831	59.421	-0.609
5200	48.607	373.891	334.529	204.679	454.129	51.675	-0.519
5300	48.813	374.818	335.281	209.550	453.462	43.942	-0.433
5400	49.003	375.733	336.021	214.441	452.826	36.221	-0.350
5500	49.179	376.633	336.752	219.350	452.222	28.512	-0.271
5600	49.339	377.521	337.472	224.276	451.648	20.813	-0.194
5700	49.485	378.396	338.182	229.218	451.104	13.124	-0.120
5800	49.616	379.257	338.883	234.173	450.587	5.445	-0.049
5900	49.733	380.107	339.574	239.140	450.098	-2.226	0.020
6000	49.837	380.943	340.257	244.119	449.634	-9.888	0.086

PREVIOUS: June 1967 (1 atm) CURRENT: June 1967 (1 bar)

ZIRCONIUM CHLORIDE (ZrCl) IDEAL GAS M_r = 126.673

$\Delta_f H°(0\ K)$ = 206.11 ± 20.9 kJ mol^{-1}

S°(298.15 K) = [254.309 ± 8.4] J K^{-1} mol^{-1} $\Delta_f H°(298.15\ K)$ = 205.43 ± 20.9 kJ mol^{-1}

Electronic Levels and Quantum Weights

state	ε_i, cm^{-1}	g_i
[$^4\Sigma$]	0	[4]
	[3000]	[4]
	[6000]	[4]
	[12000]	[4]
	[18000]	[4]
	[24000]	[4]
	[30000]	[4]

ω_e = [437] cm^{-1} $\omega_e x_e$ = [1.1] cm^{-1} σ = 1

B_e = [0.1248] cm^{-1} α_e = [0.00042] cm^{-1} r_e = [2.3] Å

Enthalpy of Formation

Potter (1) investigated mass spectrometrically the gaseous equilibria among ZrCl, $ZrCl_2$, $ZrCl_3$ and $ZrCl_4$. Ion intensities were measured 2.5 eV above the ionization threshold. Using the reported equilibrium constants for the reaction $ZrCl_3(g)$ + ZrCl(g) = 2 $ZrCl_2(g)$ in the temperature range 1978-2274 K, 2nd and 3rd law analyses give the enthalpies of reaction at 298 K as -10.36 and -12.83 kcal mol^{-1}, respectively. The 3rd law drift is 1.4 ± 2.5 cal K^{-1} mol^{-1}. Based on the 3rd law $\Delta_r H°(298.15\ K)$ and $\Delta_f H°(ZrCl_2$, g, 298.15 K) = -44.5 kcal mol^{-1} (2) and $\Delta_f H°(ZrCl_3$, g, 298.15 K) = -125.3 kcal mol^{-1} (3), we obtain $\Delta_f H°$(ZrCl, g, 298.15 K) = 49.1 kcal mol^{-1} which is adopted in the tabulation.

Farber et al. (4) also studied mass spectrometrically the reaction Zr(cr) + Cl(g) + ZnCl(g). The reported a second law heat of reaction $\Delta_r H°(298.15\ K)$ = 8.7 ± 1.4 kcal mol^{-1}. Using the equilibrium constants K = I(ZrCl)/I(Cl) which we calculated from their reported ion intensities of run 3 in the temperature range 1667-1896 K, we obtain $\Delta_r H°(298.15\ K)$ = 20.3 kcal mol^{-1} by the 3rd law method. The 3rd law drift is 6.1 ± 2.4 cal K^{-1} mol^{-1}. The enthalpy of formation is derived as $\Delta_f H°$(ZrCl, g, 298.15 K) = 49.2 kcal mol^{-1}.

Heat Capacity and Entropy

The bond distance is estimated to be the same as that in $ZrCl_4(g)$. The bond distance is then used with Guggenheimer's relation (5) for polar molecules to calculate the fundamental vibrational frequency ω_e. The anharmonicity correction x_e is estimated roughly by assuming $x_e = \omega_e/4(D° + 0.5\ \omega_e)$ = 0.0025. The rotational constant B_e is calculated from the estimated bond distance. The value of α_e is calculated from the Morse potential function.

The ground state configuration is taken from the ground state multiplet of Zr^+ reported by Moore (6). The electronic levels and the quantum weights are estimated to be the same as those of ZrF(g).

References

1. N. D. Potter, Aeronutronic Div., Philco-Ford Corp., Newport Beach, Calif., personal communication, dated Dec. 4, 1969.
2. JANAF Thermochemical Tables, $ZrCl_2(g)$ and $ZrCl_3(g)$, 12-31-69.
3. M. Farber, M. A. Frisch, G. Grenier and H. C. Ko, Space Sciences, Inc., Monrovia, Calif., Report No. AFRPL-TR-67-244, Contract F04611-67-C-0010, Nov., 1967.
4. K. M. Guggenheimer, Proc. Phys. Soc. (London) 58, 456 (1946).
5. C. E. Moore, U. S. Nat. Bur. Stand. Circ. 467, Vol. II, 1952.

Zirconium Chloride (ZrCl) $Cl_1Zr_1(g)$

Enthalpy Reference Temperature = T_r = 298.15 K Standard State Pressure = p° = 0.1 MPa

T/K	$C_p°$ (J K^{-1}mol^{-1})	S°	-[G°-H°(T_r)]/T	H°-H°(T_r) (kJ mol^{-1})	$\Delta_f H°$	$\Delta_f G°$	Log K_f
0	0.	0.	INFINITE	-9.411	206.112	206.112	INFINITE
100	29.736	219.074	283.981	-6.491	206.689	195.638	-102.191
200	33.033	240.706	257.458	-3.350	206.118	184.783	-48.260
250	34.211	248.212	254.882	-1.667	205.775	179.488	-37.502
298.15	35.004	254.309	254.309	0.	205.434	174.456	-30.564
300	35.030	254.526	254.310	0.065	205.421	174.264	-30.342
350	35.607	259.972	254.739	1.831	205.058	169.100	-25.237
400	36.032	264.755	255.698	3.623	204.685	163.989	-21.415
450	36.367	269.019	256.946	5.433	204.302	158.924	-18 447
500	36.655	272.866	258.348	7.259	203.910	153.903	-16.078
600	37.190	279.596	261.345	10.951	203.095	143.978	-12.534
700	37.744	285.370	264.374	14.698	202.239	134.192	-10.014
800	38.336	290.449	267.322	18.501	201.343	124.532	-8.131
900	38.947	294.999	270.149	22.365	200.400	114.986	-6.674
1000	39.553	299.134	272.844	26.291	199.401	105.549	-5.513
1100	40.131	302.932	275.409	30.275	198.339	96.215	-4.569
1200	40.667	306.447	277.851	34.315	193.496	87.202	-3.796
1300	41.155	309.721	280.178	38.407	192.833	78.371	-3.149
1400	41.590	312.788	282.399	42.545	192.169	69.591	-2.596
1500	41.972	315.670	284.522	46.723	191.489	60.859	-2.119
1600	42.302	318.390	286.554	50.937	190.776	52.173	-1.703
1700	42.586	320.963	288.503	55.182	190.016	43.533	-1.338
1800	42.825	323.404	290.375	59.453	189.192	34.940	-1.014
1900	43.025	325.725	292.175	63.746	188.291	26.395	-0.726
2000	43.191	327.937	293.908	68.057	187.296	17.899	-0.467
2100	43.327	330.047	295.579	72.383	186.195	9.456	-0.235
2200	43.438	332.065	297.192	76.721	163.650	1.814	-0.043
2300	43.526	333.998	298.751	81.070	161.877	-5.503	0.125
2400	43.597	335.852	300.258	85.426	160.105	-12.742	0.277
2500	43.653	337.633	301.718	89.789	158.331	-19.908	0.416
2600	43.696	339.346	303.132	94.156	156.554	-27.002	0.542
2700	43.729	340.996	304.504	9°.528	154.771	-34.028	0.658
2800	43.755	342.587	305.836	102.902	152.982	-40.988	0.765
2900	43.775	344.123	307.130	107.278	151.185	-47.884	0.862
3000	43.790	345.607	308.388	111.657	149.381	-54.718	0.953
3100	43.801	347.043	309.612	116.036	147.568	-61.491	1.036
3200	43.811	348.434	310.804	120.417	145.746	-68.206	1.113
3300	43.819	349.782	311.964	124.798	143.917	-74.863	1.185
3400	43.825	351.090	313.096	129.180	142.080	-81.466	1.252
3500	43.832	352.361	314.200	133.563	140.236	-88.014	1.314
3600	43.839	353.596	315.277	137.947	138.386	-94.509	1.371
3700	43.846	354.797	316.329	142.331	136.531	-100.953	1.425
3800	43.854	355.966	317.357	146.716	134.673	-107.346	1.476
3900	43.863	357.105	318.361	151.102	132.812	-113.691	1.523
4000	43.873	358.216	319.344	155.489	130.950	-119.988	1.567
4100	43.884	359.300	320.305	159.877	129.088	-126.238	1.608
4200	43.896	360.357	321.246	164.266	127.229	-132.443	1.647
4300	43.909	361.390	322.168	168.656	125.372	-138.604	1.684
4400	43.924	362.400	323.071	173.047	123.520	-144.722	1.718
4500	43.940	363.387	323.956	177.441	121.674	-150.797	1.750
4600	43.957	364.353	324.824	181.835	119.835	-156.832	1.781
4700	43.975	365.299	325.675	186.232	118.005	-162.827	1.810
4800	43.994	366.225	326.510	190.630	-444.752	-157.165	1.710
4900	44.014	367.132	327.330	195.031	-446.213	-151,159	1.611
5000	44.035	368.021	328.135	199.433	-447.680	-145.122	1.516
5100	44.057	368.894	328.925	203.838	-449.150	-139.057	1.424
5200	44.079	369.749	329.702	208.245	-450.621	-132 962	1.336
5300	44.102	370.589	330.466	212.654	-452.090	-126.839	1.250
5400	44.126	371.414	331.217	217.065	-453.556	-120.689	1.167
5500	44.150	372.224	331.955	221.479	-455.016	-114.511	1.088
5600	44.175	373.019	332.681	225.895	-456.470	-108.307	1.010
5700	44.200	373.801	333.396	230.314	-457.915	-102.077	0.935
5800	44.226	374.570	334.099	234.735	-459.349	-95.821	0.863
5900	44.251	375.327	334.791	239.159	-460.772	-89.541	0.793
6000	44.277	376.071	335.473	243.586	-462.182	-83.238	0.725

PREVIOUS. December 1969 (1 atm) CURRENT: December 1969 (1 bar)

Zirconium Chloride (ZrCl) $Cl_1Zr_1(g)$

CHLORINE (Cl_2) REFERENCE STATE M_r = 70.906

0 to 6000 K Ideal Diatomic Gas

D_0^0 = 239.242 ± 0.012 kJ mol^{-1} (natural abundance)
S°(298.15 K) = 223.079 ± 0.040 J K^{-1} mol^{-1}

$\Delta_f H°$(0 K) = 0 kJ mol^{-1}
$\Delta_f H°$(298.15 K) = 0 kJ mol^{-1}

Spectroscopic Data for $^{35}Cl_2$ in cm^{-1}(a)

State	T_e	D°(f)	ω_e	$\omega_e x_e$(c)	B_e	$10^3 \alpha_e$(d)	$10^7 D_e$	r_e, Å	Reference
$^1\Sigma_g^+$	0	19997.28	559.751	2.6943	0.24415	1.5163	1.9195	1.9879	(1(r_e), 2)
$^3\Pi_{0^+}$	17832.8	20879.64(b)	246.7	2.6747	0.16254	2.022	2.365(e)	2.4354	(1(r_e), 3)

(a) Constants for ground state applicable to v ≤ 40 only.
(b) The $^3\Pi_{0^+}$ state dissociates to Cl($^2P_{3/2}$) + Cl($^2P_{1/2}$).
(c) $-3.32527\times10^{-3}(v + 1/2)^3 - 2.27337\times10^{-4}(v + 1/2)^4$ for $^1\Sigma_g^+$; $-0.23826(v + 1/2)^3 + 9.4555\times10^{-3}(v + 1/2)^4$ for $^3\Pi_o^+$
(d) $-3.9078\times10^{-6}(v + 1/2)^2 + 7.0811\times10^{-8}(v + 1/2)^3$ for $^1\Sigma_g^+$; $-1.102\times10^{-4}(v + 1/2)^2 + 9.2\times10^{-7}(v + 1/2)^3$ for $^3\Pi_o^+$
(e) $+2.25\times10^{-8}(v + 1/2) - 1.5\times10^{-9}(v + 1/2)^2$
(f) The dissociation energy of each state is relative to the lowest level of the ground state.

Enthalpy of Formation

Zero by definition. Refer to the monatomic chlorine gas table for a discussion of the dissociation energy.

Heat Capacity and Entropy

The thermal functions are calculated using a direct summation technique. Included in the calculation are the ground state plus the $^3\Pi_{0^+}$ excited state at T_e = 17833 cm^{-1}. To account, in an approximate way, for the presence of the $^3\Pi_{0^-}$, $^3\Pi_{1u}$, and $^3\Pi_{2u}$ excited states (all near T_e = 17000 cm^{-1}) the rotational states of the $^3\Pi_{0^+}$ state are artificially weighted by a factor of six.

For the ground state only, the calculation is performed including the so-called quasi-bound rotational levels above the dissociation limit (4,5). For a given vibrational quantum number, v, the summation over rotational states is carried out up to a rotational quantum number, J, given by $J = J_{lim}(1-v/v_{max})$, where J_{lim} and v_{max} were determined by Gurvich et al. (3) to be 418 and 61, respectively. We adopt the $^{35}Cl_2$ ground state spectroscopic constants and dissociation energy of Douglas and Hoy (1) who gave Dunham type coefficients which fitted the spectroscopic data for v ≤ 40. For 41 ≤ v ≤ 60, we use the exact G_v, B_v, and D_v data as given by Douglas and Hoy (1).

For the $^3\Pi_{0^+}$ state, we use the spectroscopic constants given in Gurvich et al. (3) as deduced from Clyne and Coxon (6). The excited states contribute significantly to the heat capacity above 3000 K but the effect on the Gibbs free energy function is considerably smaller (amounting to only 0.4 J K^{-1} mol^{-1} at 6000 K). The rotational levels are extrapolated to high J values according to the method of Khachkuruzov (7) who proposed a simpler form of Woolley's method (8).

The spectroscopic constants of the hypothetical natural abundance species are determined from the $^{35}Cl_2$ data using a standard reduced mass scaling routine. Our adopted value for S°(298.15 K) differs by 0.003 J K^{-1} mol^{-1} from values recommended by CODATA (11) and Gurvich et al. (3) and by 0.007 J K^{-1} mol^{-1} from the value recommended by Wagman et al. (9). For this comparison, the values reported by CODATA (11) and Gurvich et al. (3) are increased by 0.1094 J K^{-1} mol^{-1} to account for a change in standard-state pressure from 1 atm to 1 bar.

Phase Data

Hultgren et al. (10) had recommended a triple point of 172.16 K (0.01374 atm) and a boiling point of 239.10 K (1 atm). These values should prove reasonably accurate, although they have not been evaluated by the present authors, and are furnished for the convenience of the reader. As a result of the low values, the reference state for chlorine is chosen to be the ideal diatomic gas at all temperatures.

References

1. A. E. Douglas and A. R. Hoy, Can. J. Phys. 53, 1965 (1975).
2. K. P. Huber and G. Herzberg, "Constants of Diatomic Molecules," Van Nostrand-Reinhold Co., New York, 1979.
3. L. V. Gurvich, I. V. Veits et al., "Thermodynamic Properties of Individual Substances," 3rd ed., Vol. I, Nauka, Moscow, 1978.
4. F. M. Mies and P. S. Julienne, J. Chem. Phys. 77, 6162 (1982).
5. Reference (3), Part 1, pp. 38-40.
6. M. A. Clyne and J. A. Coxon, J. Mol. Spectrosc. 33, 381 (1970).
7. G. A. Khachkuruzov, Opt. Spectrosc. 30, 455 (1971).
8. H. W. Woolley, R. B. Scott and F. G. Brickwedde, J. Res. Nat. Bur. Stand. 41, 379 (1948).
9. D. D. Wagman, W. H. Evans et al., J. Phys. Che. Ref. Data 11, Supp. 2 (1982).
10. R. Hultgren, P. D. Desai et al., "Selected Values of the Thermodynamic Properties of the Elements", American Society for Metals, Metals Park, Ohio, 1973.
11. J. D. Cox, chairman, ICSU-CODATA Task Group on Key Values for Thermodynamics, J. Chem. Thermodynamics 10, 903 (1978).

Chlorine (Cl_2) Cl_2(ref)

Enthalpy Reference Temperature = T_r = 298.15 K Standard State Pressure = p° = 0.1 MPa

T/K	$C_p°$ (J K^{-1}mol^{-1})	S° (J K^{-1}mol^{-1})	$-[G°-H°(T_r)]/T$ (J K^{-1}mol^{-1})	$H°-H°(T_r)$ (kJ mol^{-1})	$\Delta_f H°$ (kJ mol^{-1})	$\Delta_f G°$ (kJ mol^{-1})	Log K_f
0	0.	0.	INFINITE	-9.181	0.	0.	0.
100	29.299	189.007	251.696	-6.269	0.	0.	0.
200	31.720	209.964	226.120	-3.231	0.	0.	0.
250	32.988	217.183	223.633	-1.613	0.	0.	0.
298.15	33.949	223.079	223.079	0.	0.	0.	0.
300	33.981	223.289	223.080	0.063	0.	0.	0.
350	34.731	228.586	223.496	1.782	0.	0.	0.
400	35.296	233.263	224.431	3.533	0.	0.	0.
450	35.728	237.446	225.648	5.309	0.	0.	0.
500	36.064	241.228	227.020	7.104	0.	0.	0.
600	36.546	247.849	229.956	10.736	0.	0.	0.
700	36.872	253.509	232.926	14.408	0.	0.	0.
800	37.109	258.449	235.814	18.108	0.	0.	0.
900	37.291	262.831	238.577	21.828	0.	0.	0.
1000	37.438	266.767	241.203	25.565	0.	0.	0.
1100	37.562	270.342	243.692	29.315	0.	0.	0.
1200	37.672	273.615	246.051	33.077	0.	0.	0.
1300	37.771	276.634	248.289	36.849	0.	0.	0.
1400	37.864	279.437	250.415	40.631	0.	0.	0.
1500	37.954	282.052	252.438	44.422	0.	0.	0.
1600	38.042	284.504	254.366	48.221	0.	0.	0.
1700	38.131	286.813	256.207	52.030	0.	0.	0.
1800	38.223	288.995	257.969	55.848	0.	0.	0.
1900	38.322	291.065	259.657	59.675	0.	0.	0.
2000	38.428	293.033	261.277	63.512	0.	0.	0.
2100	38.544	294.911	262.834	67.361	0.	0.	0.
2200	38.672	296.707	264.333	71.222	0.	0.	0.
2300	38.813	298.429	265.778	75.096	0.	0.	0.
2400	38.966	300.084	267.173	78.985	0.	0.	0.
2500	39.132	301.678	268.522	82.889	0.	0.	0.
2600	39.308	303.216	269.827	86.811	0.	0.	0.
2700	39.493	304.703	271.091	90.751	0.	0.	0.
2800	39.685	306.143	272.317	94.710	0.	0.	0.
2900	39.880	307.539	273.508	98.688	0.	0.	0.
3000	40.075	308.894	274.665	102.686	0.	0.	0.
3100	40.267	310.211	275.791	106.703	0.	0.	0.
3200	40.451	311.492	276.886	110.739	0.	0.	0.
3300	40.624	312.740	277.954	114.793	0.	0.	0.
3400	40.783	313.955	278.995	118.864	0.	0.	0.
3500	40.925	315.139	280.011	122.949	0.	0.	0.
3600	41.048	316.294	281.003	127.048	0.	0.	0.
3700	41.149	317.420	281.972	131.158	0.	0.	0.
3800	41.226	318.518	282.919	135.277	0.	0.	0.
3900	41.278	319.590	283.846	139.402	0.	0.	0.
4000	41.305	320.636	284.753	143.532	0.	0.	0.
4100	41.306	321.656	285.640	147.663	0.	0.	0.
4200	41.280	322.651	286.510	151.792	0.	0.	0.
4300	41.229	323.621	287.361	155.918	0.	0.	0.
4400	41.152	324.568	288.196	160.037	0.	0.	0.
4500	41.051	325.492	289.015	164.147	0.	0.	0.
4600	40.927	326.393	289.818	168.246	0.	0.	0.
4700	40.781	327.272	290.605	172.332	0.	0.	0.
4800	40.615	328.129	291.378	176.402	0.	0.	0.
4900	40.429	328.964	292.137	180.454	0.	0.	0.
5000	40.226	329.779	292.882	184.487	0.	0.	0.
5100	40.007	330.573	293.613	188.499	0.	0.	0.
5200	39.774	331.348	294.331	192.488	0.	0.	0.
5300	39.528	332.103	295.037	196.454	0.	0.	0.
5400	39.271	332.840	295.730	200.394	0.	0.	0.
5500	39.004	333.558	296.411	204.307	0.	0.	0.
5600	38.729	334.258	297.081	208.194	0.	0.	0.
5700	38.447	334.941	297.739	212.053	0.	0.	0.
5800	38.159	335.608	298.386	215.883	0.	0.	0.
5900	37.867	336.257	299.023	219.685	0.	0.	0.
6000	37.571	336.891	299.649	223.456	0.	0.	0.

PREVIOUS: September 1965 (1 atm) CURRENT: June 1982 (1 bar)

Chlorine (Cl_2) Cl_2(ref)

COBALT CHLORIDE ($CoCl_2$) CRYSTAL M_r = 129.8392

$S°(298.15\ K) = 109.265 \pm 0.21\ J\ K^{-1}\ mol^{-1}$

$T_{fus} = 1013 \pm 2\ K$

$\Delta_f H°(0\ K) = -314.39 \pm 1.7\ kJ\ mol^{-1}$

$\Delta_f H°(298.15\ K) = -312.54 \pm 1.7\ kJ\ mol^{-1}$

$\Delta_{fus} H° = [44.77 \pm 6.3]\ kJ\ mol^{-1}$

Enthalpy of Formation

The adopted value for $\Delta_f H°(298.15\ K)$ of $-74.7\ kcal\ mol^{-1}$ is that given in NBS Technical Note 270-4 (1). This value is based mainly on results derived from heat of solution measurements on Co(cr) (2) and $CoCl_2$(cr) (3) in aqueous HCl, $\Delta_f H°(CoCl_2$, cr, 298.15 K) = $-74.9\pm1.0\ kcal\ mol^{-1}$, and from a 3rd law analysis of several sets of equilibrium data (4-11). Results of earlier heat of solution measurements on Co(cr) (12, 13) in aqueous HCl are in reasonable agreement ($\pm0.4\ kcal\ mol^{-1}$) with those used in our calculations (2), while the results of Thomsen (14) are some 2 kcal mol^{-1} more negative. Results of a 2nd and 3rd law analysis of the equilibrium data with JANAF functions are tabulated below. Auxiliary $\Delta_f H°$ data used in the analysis are (in kcal mol^{-1}): $\Delta_f H°$(HCl, g, 298.15 K) = -22.063 ± 0.05 and $\Delta_f H°(FeCl_2$, cr, 298.15 K) = -81.7 ± 0.1 (15).

Source	Reaction*	Method	Data Points	T/K	$\Delta_r H°$(298.15 K), kcal mol^{-1} 2nd law	3rd law	Drift cal K^{-1} mol^{-1}	$\Delta_f H°$(298.15 K) kcal mol^{-1}
Crut (4)	A	Static	1	718		23.4		67.5
Jellinek and Uloth (5)	A	Dynamic	3	673-873	36.5	30.46±1.59	-7.8±1.7	74.6±1.7
Sano (6)	A	Static	7	759-831	32.8	30.66±0.16	-2.8±0.4	74.8±0.3
Partington and Towndrow (7)	A	Static	5	673-773	32.9	30.91±0.30	-2.7±1.5	75.0±0.4
Shchukarev et al. (8)	A		10	573-973	28.8	29.91±0.39	1.4±0.2	74.0±0.5
Egan (9)	B	emf	2	673;723	-74.1	-75.09±0.10	-1.5	75.1±0.1
Hamby and Scott (10)	B	emf	1	973		-73.09		73.1
Sheldon and Gee (11)	C	emf	Equation	490-760	-6.4	-7.10±0.22	-1.1±0.02	74.6±0.3

*Reactions: (A) $CoCl_2(cr) + H_2(g) = Co(cr) + 2\ HCl(g)$ (B) $Co(cr) + Cl_2(g) = CoCl_2(cr)$ (C) $Fe(cr) + CoCl_2(cr) = FeCl_2(cr) + Co(cr)$

Three (5-7) of the five sets of equilibrium data on the H_2 reduction of $CoCl_2$(cr) yield $\Delta_f H°$ values which are in good agreement ($\pm0.3\ kcal\ mol^{-1}$) with the value ($-74.9\ kcal\ mol^{-1}$) obtained by solution calorimetry (2, 3). Also, the results obtained from our analysis of the recent emf study of reaction (C) provide further confirmation for the adopted value of $\Delta_f H°$.

Heat Capacity and Entropy

$C_p°$ data below 300 K are based on the low temperature measurements of Chisholm and Stout (16) (11-300 K) and Kostryukova (17) (1.8-4 K). The $C_p°$ data of Kostryakova (17) below 3 K involve bias due to the presence of trace amounts of water in the $CoCl_2$ sample. His results (17) contain a small anomaly near 2.8 K which presumably arises from the $CoCl_2 \cdot 6\ H_2O$ anti-ferromagnetic transition at 2.3 K (18). The effect of the hexahydrate on the $C_p°$ data above 3 K is judged to be negligible. The $C_p°$ data above 3 K are joined graphically with those of Chisholm and Stout at 11 K (16). The latter data set contains a lambda peak at 24.71 K. $C_p°$ near the maximum is 4.1 cal K^{-1} mol^{-1}, and the enthalpy associated with the transition is on the order of a few calories. The existence of this anomaly has been confirmed from heat capacity measurements by Shalyt (19)(30-75 K, T_{trs} = 24.9 K) and Trapeznikowa et al. (20) (13-130 K, T_{trs} = 24.90 K). This anomaly is apparently associated with the cooperative ordering of the magnetic moments of the cobaltous ions. The latter two sets of $C_p°$ data are judged to be less reliable than that of Chisholm and Stout (16) and are not used in the calculation of S°(298.15 K). S°(298.15 K) is obtained from integration of the adopted $C_p°$'s and is based on S°(3 K) = 0.003 cal K^{-1} mol^{-1}.

$C_p°$ data above 300 K are estimated from Cp-T curves for $FeCl_2$ (15) and $MnCl_2$ (21). The results of the only high temperature enthalpy study (22) reported for $CoCl_2$(cr) have been analyzed by Kelley (21). However, a comparison of these $C_p°$'s with those for the isostructural chlorides $FeCl_2$ (15) and $MnCl_2$ (21) indicate that these values are probably too high by 3.5% at 400 K and 27.3% at 1000 K. Other comparisons (21, 15) exist which indicate that it is not unusual for the $C_p°$ data of these Russian workers to be in error by several percent.

Fusion Data

See the liquid table for details

Sublimation Data

See the ideal gas table for details.

References

1. U. S. Nat. Bur. Stand. Tech. Note 270-4, 1969.
2. G. Rienacker, J. W. Birckenstaedt, M. Schubert-Birckenstaedt, and G. Techel, Z. Anorg. Chem. 289, 29 (1957).
3. W. Hieber and A. Woerner, Z. Elektrochem. 40, 256 (1934).
4. M. G. Crut, Bull. Soc. Chim. 35, 550 (1924).
5. K. Jellinek and R. Uloth, Z. Physik. Chem. 119, 161 (1926).
6. K. Sano, J. Chem. Soc. Japan 58, 370 (1937).
7. J. R. Partington and R. P. Towndrow, Trans. Faraday Soc. 35, 553 (1939).
8. S. A. Shchukarev, T. A. Tolmacheva, and M. A. Orauskaya, Zhur. Obshchei Khim. 24, 2093 (1954).
9. J. J. Egan, Brookhaven National Laboratory, Report BNL-9343, July, 1965.
10. D. C. Hamby and A. B. Scott, J. Electrochem. Soc. 117, 319 (1970).
11. R. A. J. Shelton and R. Gee, University of Manchester, personal communication, January, 1973.
12. M. Schubert-Birckenstaedt, Z. Anorg. Chem. 276, 227 (1954).
13. W. Biltz, Z. Anorg. Chem. 134, 25 (1924).
14. J. Thomsen, "Thermochemische Untersuchungen," Vol. III, Barth, Leipzig, 1883.
15. JANAF Thermochemical Tables: HCl(g), 9-30-64; $FeCl_2$(cr), 12-31-70; NaF(cr), 12-31-68; BaF_2(cr), 12-31-72; MgF_2(cr), 3-31-66; Na_3AlF_6(cr), 12-31-68.
16. R. C. Chisholm and J. W. Stout, J. Chem. Phys. 36, 972 (1962).
17. M. O. Kostryukova, Soviet Physics JETP, 28, 236 (1969).
18. W. K. Robinson and A. S. Friedberg, Phys. Rev. 117, 402 (1960).
19. S. S. Shalyt, J. Exptl. Theoret. Phys. (USSR) 8, 518 (1938).
20. O. Trapeznikowa, L. Schnbnikov, and G. Miljutin, Physik. Z. Sowjetunion 9, 237 (1936).
21. K. K. Kelley, U. S. Bur. Mines Bull. 584, 1960.
22. A. N. Krestovnikov and G. A. Karetnikov, J. Gen. Chem. (USSR) 7, 6 (1937).

Cobalt Chloride ($CoCl_2$) Cl_2Co_1(cr)

Enthalpy Reference Temperature = T_r = 298.15 K Standard State Pressure = p° = 0.1 MPa

T/K	$C_p°$ (J $K^{-1}mol^{-1}$)	S°	$-[G°-H°(T_r)]/T$	$H°-H°(T_r)$ (kJ mol^{-1})	$\Delta_f H°$	$\Delta_f G°$	Log K_f
0	0.	0.	INFINITE	-15.801	-314.394	-314.394	INFINITE
100	48.158	36.945	173.312	-13.637	-315.663	-299.689	156.541
200	71.337	79.256	116.256	-7.400	-314.374	-284.115	74.203
298.15	78.492	109.266	109.266	0.	-312.545	-269.647	47.241
300	78.576	109.751	109.267	0.145	-312.508	-269.381	46.903
400	81.714	132.821	112.391	8.172	-310.520	-255.304	33.339
500	83.387	151.243	118.382	16.431	-308.570	-241.727	25.253
600	84.600	166.552	125.170	24.829	-306.698	-228.536	19.896
700	85.939	179.699	132.042	33.360	-304.875	-215.653	16.092
800	86.818	191.234	138.735	41.999	-303.536	-202.967	13.252
900	87.529	201.501	145.149	50.717	-301.883	-190.496	11.056
1000	88.241	210.762	151.255	59.507	-300.401	-178.201	9.308
1013.000	88.316	211.902	152.026	60.655	------ CRYSTAL <--> LIQUID -----		
1100	88.826	219.200	157.054	68.361	-299.134	-166.045	7.885
1200	89.370	226.952	162.560	77.271	-298.121	-153.993	6.703
1300	89.872	234.126	167.793	86.233	-297.514	-142.010	5.706
1400	90.374	240.804	172.772	95.245	-297.441	-130.053	4.852
1500	90.876	247.057	177.518	104.308	-296.317	-118.134	4.114

PREVIOUS: CURRENT: June 1973

Cobalt Chloride ($CoCl_2$) Cl_2Co_1(cr)

COBALT CHLORIDE ($CoCl_2$) LIQUID M_r = 129.8392

S°(298.15 K) = [-149.021] J K^{-1} mol^{-1} $\Delta_f H°$(298.15 K) = [-271.514] kJ mol^{-1}

T_{fus} = 1013 ± 2 K $\Delta_{fus}H°$ = [44.8 ± 6.3] kJ mol^{-1}

Enthalpy of Formation

$\Delta_f H°$ is calculated from that of the crystal by addition of the estimated value for $\Delta_{fus}H°$ and the difference in enthalpy, H°(1013 K) - H°(298.15 K), between the crystal and liquid. Independent values of $\Delta_f H°$(ℓ) (or $\Delta_f H°$(cr)) can be obtained from a 2nd and 3rd law analysis of two sets (1, 2) of emf data for the formation cell of molten $CoCl_2$. Results of our analysis of these data are tabulated below.

Source	Data Points	T/K	$\Delta_r H°$(298.15 K), kcal mol^{-1} 2nd Law	3rd Law	Drift cal K^{-1} mol^{-1}	$\Delta_f H°$(ℓ) kcal mol^{-1}	$\Delta_f H°$(cr) kcal mol^{-1}
Devoto and Guzzi (1)	4	1023-1173	-75.8	-65.77±1.2	9.2±1.8	-65.8	-75.6
Hamby and Scott (2)	2	1073-1173	-65.9	-63.67±0.3	2.0	-63.7	-73.5

Even though both data sets contain few points and show significant positive drift, we note that the average $\Delta_f H°$(cr) value (-74.6 kcal mol^{-1}) obtained from these results is in excellent agreement with our adopted value (-74.7 kcal mol^{-1}).

Heat Capacity and Entropy

$C_p°$ is assumed constant at 23.70 cal K^{-1} mol^{-1} in the temperature range 700-2000 K and is estimated from liquid heat capacities (3) for $FeCl_2$, $MgCl_2$, $CaCl_2$ and $NiCl_2$ (4). A glass transition is assumed at 700 K below which $C_p°$ is that of the crystal. S°(298.15 K) is calculated in a manner analogous to that used for the enthalpy of formation.

Fusion Data

T_{fus} is that obtained by Schafer and Krehl (5) from cooling curve measurements. Other reported values for T_{fus} are 997 K (6) and 1008 K (7). Our adopted value for T_{fus} is that selected by Brewer et al. (8), while Kelley (9) chose T_{fus} = 1000 K. $\Delta_{fus}H°$ is calculated from $\Delta_{fus}H° = \Delta_{fus}S° \times T_{fus}$ with $\Delta_{fus}S°$ = 3.51±0.5 cal K^{-1} g-$atom^{-1}$. $\Delta_{fus}S°$ is estimated from similar data (3, 4) for the isostructural chlorides $FeCl_2$, $MgCl_2$, $MnCl_2$, $NiCl_2$, and $CdCl_2$. Kelley (9) has analyzed freezing point depression data for several binary systems which involve $CoCl_2$. Unfortunately, the values of $\Delta_{fus}H°$ derived by Kelley (9) are very discrepant and probably of no use in defining $\Delta_{fus}H°$. Reported values fall in the range 6.71-8.75 kcal mol^{-1}, and Kelley (9) recommended $\Delta_{fus}H°$ = 7.39 kcal mol^{-1}. This value is 3.31 kcal mol^{-1} less than our estimated value and is also inconsistent with $\Delta_f H°$ values derived for the crystal and liquid from equilibrium data. See Enthalpy of Formation Section for $CoCl_2$(cr, ℓ). The discrepancies which arise in Kelley's analysis of the freezing point data may be associated with the ability of the divalent cobalt ion to form complex ions.

Vaporization Data

See the ideal gas table for details.

References

1. G. Devoto and A. Guzzi, Gazz. Chim. Ital. 59, 591 (1929).
2. D. C. Hamby and A. B. Scott, J. Electrochem. Soc. 117, 319 (1970).
3. JANAF Thermochemical Tables: $FeCl_2$(ℓ), 12-31-70; $MgCl_2$(ℓ), 12-31-65; $CaCl_2$(ℓ), 6-30-70.
4. K. K. Kelley, U. S. Bur. Mines Bull. 584, 1960.
5. H. Schafer and K. Krehl, Z. Anorg. Chem. 268, 25 (1952).
6. A. Ferrari and A. Baroni, Atti Accad. Liucei, 7, 848 (1928); CA 23 7448 (1929).
7. H. Basett and W. L. Badwell, J. Chem. Soc. 1931, 2479, 2486 (1931).
8. L. Brewer, G. R. Somayajulu and E. Brackett, Chem. Rev. 63, 111 (1963).
9. K. K. Kelley, U. S. Bur. Mines Bull. 393, 1936.

Cobalt Chloride ($CoCl_2$) Cl_2Co_1(l)

Enthalpy Reference Temperature = T_r = 298.15 K Standard State Pressure = p° = 0.1 MPa

T/K	$C_p°$ (J $K^{-1}mol^{-1}$)	S°	-[G°-H°(T_r)]/T	H°-H°(T_r) (kJ mol^{-1})	$\Delta_f H°$	$\Delta_f G°$	Log K_f
0							
100							
200							
298.15	78.492	149.021	149.021	0.	-271.514	-240.469	42.129
300	78.576	149.507	149.023	0.145	-271.477	-240.276	41.836
400	81.714	172.577	152.147	8.172	-269.489	-230.175	30.058
500	83.387	190.999	158.137	16.431	-267.539	-220.574	23.043
600	84.600	206.307	164.925	24.829	-265.667	-211.358	18.400
700	85.941	219.448	171.797	33.355	-263.849	-202.451	15.107
700.000	85.941	219.448	171.797	33.355	GLASS <--> LIQUID		
700.000	99.161	219.448	171.797	33.355	TRANSITION		
800	99.161	232.689	178.600	43.271	-261.233	-193.827	12.656
900	99.161	244.368	185.271	53.187	-258.381	-185.575	10.770
1000	99.161	254.816	191.712	63.103	-255.774	-177.628	9.278
1013.000	99.161	256.097	192.530	64.392	CRYSTAL <--> LIQUID		
1100	99.161	264.267	197.885	73.019	-253.444	-169.928	8.069
1200	99.161	272.895	203.782	82.936	-251.426	-162.428	7.070
1300	99.161	280.832	209.408	92.852	-249.865	-155.078	6.231
1400	99.161	288.181	214.775	102.768	-248.888	-147.827	5.515
1500	99.161	295.022	219.899	112.684	-246.910	-140.675	4.899
1600	99.161	301.422	224.797	122.600	-244.688	-133.664	4.364
1700	99.161	307.433	229.483	132.516	-242.375	-126.796	3.896
1800	99.161	313.101	233.972	142.432	-256.339	-119.770	3.476
1900	99.161	318.462	238.279	152.348	-254.300	-112.238	3.086
2000	99.161	323.549	242.417	162.264	-252.271	-104.814	2.737

PREVIOUS: CURRENT: June 1973

Cobalt Chloride ($CoCl_2$) Cl_2Co_1(l)

COBALT CHLORIDE ($CoCl_2$) CRYSTAL-LIQUID M_r = 129.8392

0 to 1013 K crystal
above 1013 K liquid

Refer to the individual tables for details.

Cobalt Chloride ($CoCl_2$) Cl_2Co_1(cr,l)

Enthalpy Reference Temperature = T_r = 298.15 K Standard State Pressure = p° = 0.1 MPa

	J K⁻¹mol⁻¹			kJ mol⁻¹			
T/K	C_p°	S°	-[G°-H°(T_r)]/T	H°-H°(T_r)	$\Delta_f H^\circ$	$\Delta_f G^\circ$	Log K_f
0	0.	0.	INFINITE	-15.801	-314.394	-314.394	INFINITE
100	48.158	36.945	173.312	-13.637	-315.663	-299.689	156.541
200	71.337	79.256	116.256	-7.400	-314.374	-284.115	74.203
298.15	78.492	109.266	109.266	0.	-312.545	-269.647	47.241
300	78.576	109.751	109.267	0.145	-312.508	-269.381	46.903
400	81.714	132.821	112.391	8.172	-310.520	-255.304	33.339
500	83.387	151.243	118.382	16.431	-308.570	-241.727	25.253
600	84.600	166.552	125.170	24.829	-306.698	-228.536	19.896
700	85.939	179.699	132.042	33.360	-304.875	-215.653	16.092
800	86.818	191.234	138.735	41.999	-303.536	-202.967	13.252
900	87.529	201.501	145.149	50.717	-301.883	-190.496	11.056
1000	88.241	210.762	151.255	59.507	-300.401	-178.201	9.308
1013.000	88.316	211.902	152.026	60.655	CRYSTAL <--> LIQUID		
1013.000	99.161	256.097	152.026	105.423	TRANSITION		
1100	99.161	264.267	160.585	114.050	-253.444	-169.928	8.069
1200	99.161	272.895	169.589	123.967	-251.426	-162.428	7.070
1300	99.161	280.832	177.845	133.883	-249.865	-155.078	6.231
1400	99.161	288.181	185.467	143.799	-248.888	-147.827	5.515
1500	99.161	295.022	192.545	153.715	-246.910	-140.675	4.899
1600	99.161	301.422	199.152	163.631	-244.688	-133.664	4.364
1700	99.161	307.433	205.347	173.547	-242.375	-126.796	3.896
1800	99.161	313.101	211.177	183.463	-256.339	-119.770	3.476
1900	99.161	318.462	216.684	193.379	-254.300	-112.239	3.086
2000	99.161	323.549	221.901	203.295	-252.271	-104.814	2.737

PREVIOUS: CURRENT: June 1973

Cobalt Chloride ($CoCl_2$) Cl_2Co_1(cr,l)

COBALT DICHLORIDE ($CoCl_2$) IDEAL GAS $M_r = 129.8392$

$\Delta_fH°(0\ K) = -95.0 \pm 8\ kJ\ mol^{-1}$

$S°(298.15\ K) = [298.51 \pm 8.4]\ J\ K^{-1}\ mol^{-1}$ $\Delta_fH°(298.15\ K) = -93.7 \pm 8\ kJ\ mol^{-1}$

Electronic Levels and Quantum Weights

State	ε_i, cm^{-1}	g_i	State	ε_i, cm^{-1}	g_i
$^4\Sigma_g^-$	0	[4]	$^4\Sigma_g^-$	[10425]	[4]
$^4\Phi_g$	[210]	[8]	$^4\Delta_g$	[14197]	[8]
$^4\Pi_g$	[2408]	[8]	$^4\Pi_{g°}$	[21270]	[8]

Vibrational Frequencies and Degeneracies

ν, cm^{-1}
[325] (1)
94.5 (2)
493 (1)

Point Group: $D_{\infty h}$

Bond Distance: Co-Cl = 2.117 Å Bond Angle: Cl-Co-Cl = 180° Rotational constant: $B_0 = 0.053048\ cm^{-1}$ $\sigma = 2$

Enthalpy of Formation

$\Delta_fH°$ is calculated from $\Delta_{sub}H° = 52.3 \pm 1.7\ kcal\ mol^{-1}$ by addition of $\Delta_fH°(CoCl_2,\ cr,\ 298.15\ K) = -74.7 \pm 0.3\ kcal\ mol^{-1}$ (1). The adopted value for $\Delta_{sub}H°$ is based on results of a 2nd and 3rd law analysis of extensive sublimation data reported by Hill et al. (2). These studies included both Knudsen-effusion and torsion-effusion measurements which were performed in platinum-10% rhodium effusion cells. Orifice diameters of the cells varied from 0.0518 cm to 0.1645 cm. The extent of dimerization (~1-2%) was also determined from mass spectrometric measurements, and the measured vapor pressures were corrected to represent equilibrium pressure of the monomer. Results of our analysis of sixty-four vapor pressure points measured in five different cells are given below. Derived Δ_fH values show no significant trends with orifice size; thus, a mean value of the five results is adopted. Also included in this table is a value for Δ_fH which was obtained from an analysis of equilibrium data (3) for $CoO(cr) + Cl_2(g) = CoCl_2(g) + 1/2\ O_2(g)$. Orlov (3) has not reported sufficient information to allow a small correction to be made for dimerization; even so, his results still give a value for $\Delta_fH°(-22.1\ kcal\ mol^{-1})$ which agrees fairly well with that adopted.

Source	Method	T/K	Data Points	$\Delta_rH°(298.15\ K)$, $kcal\ mol^{-1}$ 2nd law	3rd law	Drift cal K^{-1} mol^{-1}	$\Delta_fH°$ (298.15 K) kcal mol^{-1}
(2)	Knudsen-Cell A	680-800	13		52.35±0.13	0.4±0.5	22.4±0.4
(2)	Knudsen-Cell B	697-814	14(a)	51.9	52.29±0.10	0.5±0.4	22.4±0.4
(2)	Knudsen-Cell C	715-810	10	52.9	52.14±0.16	-0.9±0.9	22.6±0.5
(2)	Torsion-Cell A	759-834	10	56.7	52.38±0.52	-5.4±3.2	22.3±0.8
(2)	Torsion-Cell B	745-824	15	55.0	52.30±0.17	-3.4±0.4	22.4±0.5
(3)	Transpiration	1323-1523	Equation	16.6	34.77±2.1	12.7	22.1±2.2(b)

(a)Two points rejected due to failure of a statistical test. (b)$\Delta_fH°(CoO,\ cr,\ 298.15\ K) = -56.82 \pm 0.10\ kcal\ mol^{-1}$ (1).

Other vapor pressures for $CoCl_2$(cr, ℓ) have been measured by static (4) and transpiration (5, 6) methods. In all three studies the presence of a small amount of dimer in the saturated vapor was not considered. JANAF pressures based on (a) selected values of the heats of sublimation and vaporization for the monomer and dimer, (b) the mass spectrometric value of $2\Delta_{sub}H(monomer) - \Delta_{sub}H(dimer) = -38.99\ kcal\ mol^{-1}$ (see Co_2Cl_4(g) table) at 298.15 K, and (c) adopted entropies for the condensed and gaseous species are compared with the observed total pressures in the table given below.

Source	Method	Process	T/K	Mole Fraction Dimer Calc. Range	P_{calc}/P_{obs} Range
(4)	Static	Sublimation	906-997	0.051.0.086	0.073-0.156
(4)	Static	Vaporization	1022-1256	0.082-0.092	0.198-0.816
(5)	Transpiration	Vaporization	1015-1144	0.090-0.092	0.531-0.743
(6)	Transpiration	Sublimation	925-1010	0.057-0.091	0.875-1.096
(6)	Transpiration	Vaporization	1019-1073	0.092	0.825-0.883

Our predictions agree reasonably well with the transpiration data of Schafer and Krehl (6), while the static measurements of Maier (4) show significant deviations from our pressures, particularly at lower temperatures. We note that JANAF analysis (1) of Maier's vapor pressures for $BaCl_2$, $AlCl_3$, and $FeCl_2$ indicate the existence of temperature dependent errors in his measurements; thus, the $CoCl_2$ results are probably not reliable. The transpiration data of Kulkarni and Dadape (5) show similar discrepancies.

Heat Capacity and Entropy

Measurements of the infrared (7-9) and electronic (10-12) spectra of both matrix-isolated and gaseous $CoCl_2$ have been interpreted on the basis of a linear structure of $D_{\infty h}$ symmetry. This structure is adopted, and the Co-Cl bond length is taken to be equal to the value determined by Tremmel et al. (13). The antisymmetric stretching frequency (ν_3) is that observed in the infrared absorption spectra of gaseous CoC_2 by Leroi et al. (7). This value has been confirmed from results of two infrared matrix isolation studies (8, 9). The doubly degenerate bending frequency (ν_2) was observed by Thompson and Carlson (8) in the infrared spectra of $CoCl_2$ isolated in an argon matrix. The infrared-inactive symmetric stretching frequency (ν_1) is estimated by comparison with similar data for $FeCl_2$ (1) and $NiCl_2$ (14). Some additional support for this value is provided by calculations of DeKock and Gruen (11). Assuming linear model, they calculated $\nu_1 = 332\ cm^{-1}$ from $\nu_3 = 493\ cm^{-1}$. Also, Clifton and Gruen (15) observed a vibrational spacing of 332 cm^{-1} in the fluorescence spectrum of argon matrix-isolated $CoCl_2$ which may be ν_1. A similar assignment for $NiCl_2$ was recently proposed (14) based on observations of vibrational spacings in its fluorescence spectrum.

The electronic spectra of $CoCl_2$ have been extensively investigated both in the gas phase (10, 11, 16) and in various matrices (9, 12, 15, 17). Unfortunately, most of the results are conflicting and difficult to interpret. It is not certain whether the ground state configuration is $^4\Phi$, or $^4\Sigma$. Theroretical arguments based on a ligand-field model (12, 18) favor the former, but recently Lever and Hollebone (19) interpreted the electronic spectra of $CoCl_2$ in terms of an "orbital angular over-lap" model and preferred the latter ground state configuration. Other problems associated with band assignments in the $CoCl_2$ spectra have been dealt with by Smith (18). The adopted states and levels comprise one of two sets of assignments proposed by Lever and Hollebone (19). The alternate functions derived from their second set of assignments (19) suggest that the adopted entropies may be biased by as much as +2 cal K^{-1} mol^{-1} at 298 K and +0.2 cal K^{-1} mol^{-1} at 4000 K. Functions based on the assignments of Smith (18) agree with those adopted to better than 1 cal K^{-1} mol^{-1} at all temperatures.

References - References are listed on the Co_2Cl_4(g) table.

Cobalt Chloride ($CoCl_2$) Cl_2Co_1(g)

Enthalpy Reference Temperature = T_r = 298.15 K Standard State Pressure = p° = 0.1 MPa

T/K	$C_p°$	S°	$-[G°-H°(T_r)]/T$	$H°-H°(T_r)$	$\Delta_fH°$	$\Delta_fG°$	Log K_f
	J $K^{-1}mol^{-1}$			kJ mol^{-1}			
0	0.	0.	INFINITE	-15.271	-95.041	-95.041	INFINITE
100	51.604	236.844	350.003	-11.316	-94.519	-98.535	51.469
200	57.705	275.056	303.902	-5.769	-93.920	-102.821	26.854
250	58.855	288.064	299.478	-2.853	-93.778	-105.064	21.952
298.15	59.623	298.500	298.500	0.	-93.722	-107.244	18.789
300	59.648	298.869	298.501	0.110	-93.720	-107.328	18.687
350	60.221	308.109	299.230	3.108	-93.704	-109.598	16.357
400	60.657	316.180	300.855	6.130	-93.738	-111.866	14.608
450	61.011	323.346	302.963	9.172	-93.821	-114.128	13.248
500	61.318	329.790	305.329	12.230	-93.947	-116.378	12.158
600	61.872	341.019	310.369	18.390	-94.314	-120.832	10.519
700	62.395	350.596	315.448	24.604	-94.808	-125.214	9.344
800	62.891	358.961	320.375	30.868	-95.843	-129.456	8.453
900	63.344	366.395	325.083	37.181	-96.596	-133.614	7.755
1000	63.737	373.090	329.555	43.535	-97.550	-137.678	7.192
1100	64.064	379.181	333.793	49.926	-98.746	-141.635	6.726
1200	64.324	384.767	337.812	56.346	-100.223	-145.471	6.332
1300	64.522	389.924	341.625	62.789	-102.135	-149.168	5.994
1400	64.666	394.711	345.247	69.249	-104.615	-152.696	5.697
1500	64.765	399.176	348.696	75.720	-106.081	-156.077	5.435
1600	64.828	403.358	351.983	82.200	-107.295	-159.369	5.203
1700	64.862	407.289	355.121	88.685	-108.414	-162.589	4.996
1800	64.875	410.997	358.124	95.172	-125 807	-165.450	4.801
1900	64.873	414.505	361.000	101.660	-127.196	-167.615	4.608
2000	64.860	417.832	363.759	108.146	-128.597	-169.706	4.432
2100	64.842	420.996	366.410	114.631	-130.011	-171.727	4.271
2200	64.821	424.012	368.960	121.115	-131.438	-173.680	4.124
2300	64.799	426.893	371.417	127.595	-132.882	-175.568	3.987
2400	64.779	429.650	373.786	134.074	-134.342	-177.392	3.861
2500	64.762	432.294	376.074	140.551	-135.820	-179.156	3.743
2600	64.748	434.834	378.285	147.027	-137.316	-180.860	3.634
2700	64.738	437.277	380.425	153.501	-138.832	-182.506	3.531
2800	64.733	439.632	382.498	159.975	-140.367	-184.095	3.434
2900	64.733	441.903	384.507	166.448	-141.923	-185.630	3.344
3000	64.738	444.098	386.457	172.921	-143.497	-187.110	3.258
3100	64.747	446.221	388.351	179.396	-145.090	-188.538	3.177
3200	64.761	448.277	390.192	185.871	-521.514	-188.141	3.071
3300	64.778	450.270	391.982	192.348	-521.715	-177.720	2.813
3400	64.800	452.204	393.725	198.827	-521.932	-167.292	2.570
3500	64.825	454.083	395.423	205.308	-522.165	-156.858	2.341
3600	64.853	455.909	397.078	211.792	-522.413	-146.418	2.124
3700	64.884	457.686	398.692	218.279	-522.674	-135.970	1.920
3800	64.917	459.417	400.268	224.769	-522.949	-125.515	1.725
3900	64.952	461.104	401.806	231.262	-523.234	-115.052	1.541
4000	64.989	462.749	403.309	237.759	-523.529	-104.582	1.366
4100	65.028	464.354	404.778	244.260	-523.832	-94.105	1.199
4200	65.068	465.922	406.216	250.765	-524.143	-83.620	1.040
4300	65.108	467.453	407.622	257.274	-524.458	-73.127	0.888
4400	65.149	468.950	408.999	263.786	-524.778	-62.628	0.743
4500	65.191	470.415	410.347	270.303	-525.100	-52.121	0.605
4600	65.232	471.848	411.669	276.825	-525.424	-41.606	0.472
4700	65.274	473.251	412.964	283.350	-525.748	-31.085	0.345
4800	65.315	474.626	414.235	289.879	-526.072	-20.557	0.224
4900	65.356	475.973	415.481	296.413	-526.394	-10.022	0.107
5000	65.396	477.294	416.704	302.950	-526.714	0.519	-0.005
5100	65.435	478.589	417.905	309.492	-527.032	11.067	-0.113
5200	65.474	479.860	419.084	316.037	-527.346	21.621	-0.217
5300	65.511	481.108	420.243	322.587	-527.658	32.181	-0.317
5400	65.548	482.333	421.381	329.140	-527.966	42.747	-0.413
5500	65.583	483.536	422.500	335.696	-528.271	53.319	-0.506
5600	65.617	484.718	423.601	342.256	-528.573	63.896	-0.596
5700	65.650	485.880	424.683	348.820	-528.873	74.478	-0.683
5800	65.681	487.022	425.748	355.386	-529.170	85.066	-0.766
5900	65.711	488.145	426.796	361.956	-529.422	95.656	-0.847
6000	65.740	489.249	427.828	368.528	-529.706	106.252	-0.925

PREVIOUS: December 1974 (1 atm) CURRENT: December 1974 (1 bar)

Cobalt Chloride ($CoCl_2$) Cl_2Co_1(g)

CESIUM CHLORIDE ((CsCl)$_2$) IDEAL GAS M_r = 336.7168

$S°(298.15 K) = [383.445 \pm 8.4]\ J\ K^{-1}\ mol^{-1}$

$\Delta_f H°(0\ K) = -656.08 \pm 10.5\ kJ\ mol^{-1}$

$\Delta_f H°(298.15\ K) = -659.82 \pm 10.5\ kJ\ mol^{-1}$

Vibrational Frequencies and Degeneracies

ν, cm^{-1}	ν, cm^{-1}
[140](1)	[76](1)
[67](1)	[151](1)
[141](1)	[156](1)

Ground State Quantum Weight: [1] σ = [4]

Point Group: [D_{2h}]

Bond Distance: Cs-Cl = [3.18] Å

Bond Angle: Cl-Cs-Cl = [93.2]°

Product of the Moments of Inertia: $I_A I_B I_C = [3.62360 \times 10^{-111}]\ g^3\ cm^6$

Enthalpy of Formation

Datz (1) studied the molecular association equilibrium in cesium chloride vapors by measuring the temperature dependence of the molecular weight of gaseous CsCl. The molecular weights were determined by measurement of the absolute pressure exerted by a known weight of completely vaporized salt contained in an isothermal fused silica bulb of known volume. The pressure sensing element was a fused silica manometer containing Au(ℓ). Based on the reported results, we derive the equilibrium constants for the reaction $Cs_2Cl_2(g) = 2CsCl(g)$ in the temperature range 1224 - 1418 K. The enthalpy change of this reaction is evaluated by the 2nd and 3rd law method as 39.3±1.1 and 42.9±0.6 kcal mol^{-1}, respectively. The drift in the 3rd law value is equivalent to an entropy error of 2.7±0.8 cal K^{-1} mol^{-1}. Using the 3rd law $\Delta_r H°$ and $\Delta_f H°$(CsCl, g, 298.15 K) = -57.4 kcal mol^{-1}, we obtain $\Delta_f H°$(298.15 K) = -157.7±2.5 kcal mol^{-1} (-659.817±10.5 kJ mol^{-1}) for $Cs_2Cl_2(g)$, which is adopted.

Schrier (2) determined the vapor pressures of molten cesium chloride at temperatures between 1165 and 1387 K by the boiling point or manometric method. The vapor analysis was performed using the transpiration method. An inert gas, N_2, was passed over the CsCl sample at a given temperature. The salt vapor entrained by the carrier gas was collected, weighed and analyzed. The data obtained were combined with the boiling point measurements to yield the average molecular weight of the vapor. Association into dimer was postulated. The mole fraction in the vapor and the partial pressure of CsCl(g) and $Cs_2Cl_2(g)$ were calculated. The equilibrium constants, 1209 - 1350 K, for dissociation of the Cs_2Cl_2 dimer were derived. Using the reported equilibrium constants, we evaluate the enthalpy change $\Delta_r H°$(298.15 K) of the reaction $Cs_2Cl_2(g) = 2CsCl(g)$ to be 34.1±2.9 and 43.2±0.9 kcal mol^{-1} by the 2nd and 3rd law methods. Based on the 3rd law $\Delta_r H°$(298.15 K) value and $\Delta_f H°$(CsCl, g, 298.15 K) = -57.4 kcal mol^{-1}, we obtain $\Delta_f H°$(298.15 K) = -158.0 kcal mol^{-1} for $Cs_2Cl_2(g)$, which is in agreement with the adopted value.

Miller and Kusch (3) made an analysis of the velocity distribution of the CsCl molecules which escape through an ideal aperture from an isothermal enclosure. They found that there was no definite evidence that any species other than diatomic molecules were present in the beams in measurable amounts, in other words, the dimerization must be small, e.g. <1.5 per cent. However, according to our calculation the concentration of dimer is 15 - 34 per cent in the temperature range 600 - 900 K, and 33 -20 per cent at the temperatures 1000 - 1600 K.

Heat Capacity and Entropy

The molecular structure is assumed to be planar, having D_{2h} symmetry. Based on the bond distances of Cs-Cs and Cl-Cl estimated by Berkowitz (4), the Cs-Cl bond distance and Cl-Cs-Cl bond angle are calculated. The vibrational frequencies were calculated by Berkowitz (5), on the basis of the potential function for an ionic model. The principal moments of inertia are: $I_A = 62.8570 \times 10^{-39}$, $I_B = 210.7205 \times 10^{-39}$, and $I_C = 273.5775 \times 10^{-39}\ g\ cm^2$.

References

1. S. Datz, Oak Ridge National Laboratory, ORNL-2933, Oak Ridge, Tennessee, May 1960.
2. E. E. Schrier, Ph. D. Thesis, Rensselaer Polytechnic Institute, 1961.
3. R. C. Miller and P. Kusch, J. Chem. Phys. 25, 860 (1956).
4. J. Berkowitz, J. Chem. Phys. 29, 1386 (1958).
5. J. Berkowitz, J. Chem. Phys. 32, 1519 (1960).

Cesium Chloride ((CsCl)$_2$) $Cl_2Cs_2(g)$

Enthalpy Reference Temperature = T_r = 298.15 K Standard State Pressure = p° = 0.1 MPa

T/K	$C_p°$ (J K⁻¹mol⁻¹)	S°	-[G°-H°(T_r)]/T	H°-H°(T_r) (kJ mol⁻¹)	$\Delta_f H°$	$\Delta_f G°$	Log K_f
0	0.	0.	INFINITE	-20.880	-656.082	-656.082	INFINITE
100	71.695	298.145	454.495	-15.635	-658.046	-657.970	343.688
200	79.843	351.168	390.868	-7.940	-658.750	-657.609	171.750
250	80.993	369.121	384.785	-3.916	-659.184	-657.277	137.330
298.15	81.617	383.445	383.445	0.	-659.817	-656.856	115.079
300	81.635	383.950	383.446	0.151	-659.848	-656.838	114.366
350	82.029	396.566	384.442	4.243	-664.864	-655.599	97.843
400	82.287	407.537	386.658	8.352	-665.676	-654.219	85.432
450	82.465	417.240	389.527	12.471	-666.474	-652.739	75.768
500	82.593	425.935	392.741	16.597	-667.266	-651.171	68.027
600	82.760	441.010	399.567	24.866	-668.843	-647.804	56.396
700	82.862	453.776	406.423	33.147	-670.427	-644.172	48.069
800	82.928	464.845	413.049	41.437	-672.023	-640.312	41.808
900	82.973	474.615	419.357	49.732	-673.637	-636.252	36.927
1000	83.005	483.359	425.328	58.031	-809.528	-624.612	32.626
1100	83.029	491.271	430.969	66.333	-809.133	-606.139	28.783
1200	83.048	498.497	436.299	74.637	-808.749	-587.702	25.582
1300	83.062	505.145	441.343	82.942	-808.375	-569.297	22.875
1400	83.073	511.301	446.123	91.249	-808.011	-550.920	20.555
1500	83.082	517.032	450.661	99.557	-807.659	-532.569	18.546
1600	83.090	522.395	454.979	107.865	-807.323	-514.240	16.788
1700	83.096	527.432	459.094	116.175	-807.006	-495.933	15.238
1800	83.101	532.182	463.024	124.485	-806.713	-477.643	13.861
1900	83.106	536.675	466.783	132.795	-806.452	-459.368	12.629
2000	83.109	540.938	470.385	141.106	-806.230	-441.107	11.521
2100	83.113	544.993	473.842	149.417	-806.058	-422.855	10.518
2200	83.115	548.860	477.165	157.728	-805.946	-404.610	9.607
2300	83.118	552.554	480.363	166.040	-805.906	-386.369	8.775
2400	83.120	556.092	483.445	174.352	-805.949	-368.128	8.012
2500	83.122	559.485	486.419	182.664	-806.089	-349.883	7.310
2600	83.124	562.745	489.293	190.976	-806.339	-331.630	6.663
2700	83.125	565.882	492.072	199.289	-806.713	-313.365	6.062
2800	83.126	568.905	494.762	207.601	-807.223	-295.084	5.505
2900	83.128	571.822	497.369	215.914	-807.883	-276.782	4.985
3000	83.129	574.640	499.898	224.227	-808.708	-258.455	4.500
3100	83.130	577.366	502.353	232.540	-809.709	-240.097	4.046
3200	83.131	580.006	504.739	240.853	-810.901	-221.704	3.619
3300	83.131	582.564	507.059	249.166	-812.213	-203.277	3.218
3400	83.132	585.045	509.316	257.479	-813.787	-184.801	2.839
3500	83.133	587.455	511.515	265.792	-815.575	-166.276	2.482
3600	83.133	589.797	513.657	274.105	-817.588	-147.697	2.143
3700	83.134	592.075	515.745	282.419	-819.807	-129.061	1.822
3800	83.134	594.292	517.783	290.732	-822.259	-110.362	1.517
3900	83.135	596.451	519.773	299.046	-824.924	-91.597	1.227
4000	83.135	598.556	521.716	307.359	-827.868	-72.757	0.950
4100	83.136	600.609	523.616	315.673	-831.059	-53.840	0.686
4200	83.136	602.612	525.473	323.986	-834.497	-34.843	0.433
4300	83.137	604.569	527.290	332.300	-838.185	-15.760	0.191
4400	83.137	606.480	529.068	340.614	-841 534	3.344	-0.040
4500	83.137	608.348	530.809	348.927	-844.494	22.462	-0.261
4600	83.138	610.175	532.514	357.241	-848.564	41.772	-0.474
4700	83.138	611.963	534.186	365.555	-852.825	61.172	-0.680
4800	83.138	613.714	535.824	373.869	-857.269	80.666	-0.878
4900	83.138	615.428	537.432	382.183	-861.887	100.254	-1.069
5000	83.139	617.108	539.008	390.496	-866.667	119.937	-1.253
5100	83.139	618.754	540.556	398.810	-871.600	139.718	-1.431
5200	83.139	620.368	542.075	407.124	-875.608	159.442	-1.602
5300	83.139	621.952	543.568	415.438	-880.674	179.395	-1.768
5400	83.139	623.506	545.034	423.752	-885.289	199.358	-1.928
5500	83.140	625.032	546.474	432.066	-890.178	219.442	-2.084
5600	83.140	626.530	547.890	440.380	-895.424	239.664	-2.235
5700	83.140	628.001	549.283	448.694	-900.731	259.981	-2.382
5800	83.140	629.447	550.653	457.008	-906.089	280.391	-2.525
5900	83.140	630.868	552.000	465.322	-911.485	300.893	-2.664
6000	83.140	632.266	553.326	473.636	-916.906	321.487	-2.799

PREVIOUS: June 1968 (1 atm) CURRENT: June 1968 (1 bar)

Cesium Chloride ((CsCl)$_2$) $Cl_2Cs_2(g)$

COPPER CHLORIDE ($CuCl_2$) CRYSTAL M_r = 134.4520

$S°(298.15\ K) = 108.085\ J\ K^{-1}\ mol^{-1}$

$\Delta_f H°(0\ K) = -206.65 \pm 6.3\ kJ\ mol^{-1}$
$\Delta_f H°(298.15\ K) = -205.85 \pm 6.3\ kJ\ mol^{-1}$

Enthalpy of Formation

Shchukarev and Oranskaya have reported equilibrium constants for the dissociation $CuCl_2(cr) \rightarrow CuCl(cr/\ell) + 0.5\ Cl_2(g)$, which are in general agreement with earlier reports by Ephraim (2) and Tarasenkov et al. (3). These data were subjected to third law analysis which revealed a significant trend of 13±0.5 cal $K^{-1}\ mol^{-1}$, indicating that equilibrium conditions were not reached. However an average value of $\Delta_r H°(298.15\ K) = 16.5 \pm 1.9\ kcal\ mol^{-1}$ yields a heat of formation of $CuCl_2(cr) = -49.5 \pm 2.5\ kcal\ mol^{-1}$ which is in good agreement with the adopted value taken from (4).

Heat Capacity and Entropy

The low temperature heat capacity has been measured from 11-300 K by Stout and Chisholm (5). The integration of this data from $S°(10\ K) = 0.111\ cal\ K^{-1}\ mol^{-1}$ gave $S°(298.15\ K) = 25.833\ cal\ K^{-1}\ mol^{-1}$. The heat capacity above 298.15 K was obtained by smooth extrapolation of the above data, since the heat capacities of Krestovnikov and Karetnikov (6) were not compatible with the low temperature data.

Decomposition Data

The temperature of decomposition, T_{dcm} = 766 K, was estimated from the table for $CuCl_2(cr)$, $CuCl(\ell)$ and $Cl_2(g)$ so that the pressure of Cl_2 is 1 bar.

References

1. S. A. Shchukarev and M. A. Oranskaya, Zhur. Obshchei Khim. 24, 1926 (1954).
2. F. Ephraim, Ber. 50, 1069 (1917).
3. D. I. Taresenkov, A. I. Grigorovich and A. V. Boroslovskaya, Collected Works of the Central State Institute for Non-Ferrous Metallurgy, Part 1, 1, 69 (1929).
4. U. S. Nat. Bur. Stand. Circ. 500, 1952.
5. J. W. Stout and R. C. Chisholm, J. Chem. Phys. 36, 979 (1962).
6. A. N. Krestovnikov and G. A. Karetnikov, Journ. Gen. Chem. (USSR) 6, 955 (1936).

Copper Chloride ($CuCl_2$) $Cl_2Cu_1(cr)$

Enthalpy Reference Temperature = T_r = 298.15 K — Standard State Pressure = p° = 0.1 MPa

T/K	$C_p°$	S°	$-[G°-H°(T_r)]/T$	$H°-H°(T_r)$	$\Delta_f H°$	$\Delta_f G°$	Log K_f
	J K⁻¹mol⁻¹			kJ mol⁻¹			
0	0.	0.	INFINITE	-14.983	-206.648	-206.648	INFINITE
100	46.777	41.020	167.042	-12.602	-207.848	-192.046	100.314
200	65.647	80.525	114.499	-6.795	-207.091	-176.458	46.086
298.15	71.881	108.085	108.085	0.	-205.853	-161.680	28.326
300	71.923	108.530	108.087	0.133	-205.828	-161.405	28.103
400	75.061	129.678	110.949	7.492	-204.433	-146.806	19.171
500	77.278	146.679	116.450	15.115	-202.947	-132.569	13.849
600	78.785	160.906	122.706	22.920	-201.393	-118.638	10.328
700	79.956	173.142	129.057	30.860	-199.800	-104.971	7.833
800	80.885	183.881	135.253	38.903	-198.181	-91.535	5.977
900	81.655	193.453	141.197	47.030	-196.550	-78.302	4.545
1000	82.370	202.093	146.862	55.232	-194.919	-65.251	3.408
1100	83.001	209.974	152.246	63.501	-193.306	-52.362	2.486
1200	83.607	217.222	157.363	71.831	-191.732	-39.619	1.725
1300	84.190	223.937	162.229	80.221	-190.242	-27.004	1.085
1400	84.750	230.197	166.863	88.668	-201.992	-14.093	0.526
1500	85.287	236.063	171.282	97.170	-200.566	-0.721	0.025

PREVIOUS: CURRENT: March 1966

Copper Chloride ($CuCl_2$) $Cl_2Cu_1(cr)$

PHOSPHORYL CHLORIDE FLUORIDE ($POFCl_2$) IDEAL GAS M_r = 136.877563

$S°(298.15 K) = 320.381 J K^{-1} mol^{-1}$

$\Delta_f H°(0 K) = [-758.8] kJ mol^{-1}$

$\Delta_f H°(298.15 K) = [-765.7] kJ mol^{-1}$

Vibrational Frequencies and Degeneracies

ν, cm^{-1}	ν, cm^{-1}	ν, cm^{-1}
207 (1)	372 (1)	620 (1)
254 (1)	386 (1)	894 (1)
330 (1)	547 (1)	1331 (1)

Ground State Multiplicity = [1] $\sigma = 1$

Point Group= C_s

Bond Distances: P-Cl = 1.99 ± 0.04 Å P-F = 1.50 ± 0.03 Å P-O = 1.54 ± 0.03 Å

Bond Angle: Cl-P-Cl = F-P-Cl = 103° O-P-Cl = O-P-F = 115.35°

Product of the Moments of Inertia: $I_A I_B I_C = 4.70925 \times 10^{-113} g^3 cm^6$

Enthalpy of Formation

The $\Delta_f H°(298.15 K)$ of $POFCL_2$ was estimated to be -183 kcal mol^{-1} from the $\Delta_f H°(298.15 K)$ of $POCl_3$ by assuming bond energies of 120 kcal mol^{-1} for D(P-F) and 80 kcal mol^{-1} for D(P-Cl). The D(P-F) and D(P-Cl) values were taken from Neale and Williams (1) and Neale et al. (2).

Essentially the same $\Delta_f H°(298.15 K)$ may be obtained by assuming a $\Delta_r H° = 0$ for the following reaction:

$$POCl_3(g) + POF_3(g) \rightarrow POCl_2F(g) + POF_2Cl(g)$$

Heat Capacity and Entropy

The molecular constants were determined by Brockway and Beach (3) and by Williams et al. (4). Brockway and Beach used electron diffraction. Williams et al. (4) measured microwave spectra. Brockway and Beach (3) assumed that the F-P-F and F-P-Cl angles were equal in POF_2Cl and $POFCl_2$. An analysis of their data by Williams et al. (4) indicates that the angles in POF_2Cl, $POFCl_2$, and $POCl_3$, and POF_3 should be 103° rather than the reported 106° ± 3°. The principal moments of inertia are: $I_A = 26.9230 \times 10^{-39}$, $I_B = 37.9561 \times 10^{-39}$, and $I_C = 46.0837 \times 10^{-39}$.

The Raman spectra and vibrational frequencies for $POFCl_2$ were reported by Delwaulle and Francois (5). These frequency assignments are summarized by Delwaulle and Francois (6).

References

1. E. Neale and L. T. D. Williams, J. Chem. Soc., 2485, Part I (1955).
2. E. Neale, L. T. D. Williams, and V. T. Moores, J. Chem. Soc. Part II, 422 (1956).
3. L. O. Brockway and J. Y. Beach, J. Amer. Chem. Soc. 60, 1836 (1938).
4. Q. Williams, J. Sheridan and W. Gordy, J. Chem. Phys. 20, 164 (1952).
5. M. L. Delwaulle and F. Francois, Compt. Rend. 222, 550 (1946).
6. M. L. Delwaulle and F. Francois, J. Chim. Phys. 46, 87 (1949).

Phosphoryl Chloride Fluoride ($POCl_2F$) $Cl_2F_1O_1P_1(g)$

Enthalpy Reference Temperature = T_r = 298.15 K Standard State Pressure = p° = 0.1 MPa

T/K	$C_p°$ (J K⁻¹mol⁻¹)	S° (J K⁻¹mol⁻¹)	$-[G°-H°(T_r)]/T$ (J K⁻¹mol⁻¹)	$H°-H°(T_r)$ (kJ mol⁻¹)	$\Delta_f H°$ (kJ mol⁻¹)	$\Delta_f G°$ (kJ mol⁻¹)	Log K_f
0	0.	0.	INFINITE	-16.431	-758.808	-758.808	INFINITE
100	44.533	253.224	381.368	-12.814	-761.896	-749.365	391.428
200	66.285	291.266	327.264	-7.199	-764.503	-735.924	192.204
250	73.746	306.896	321.658	-3.690	-765.184	-728.696	152.253
298.15	79.319	320.381	320.381	0.	-765.672	-721.622	126.425
300	79.508	320.873	320.383	0.147	-765.688	-721.349	125.598
350	84.039	333.482	321.368	4.240	-766.782	-713.857	106.537
400	87.659	344.949	323.610	8.536	-767.117	-706.272	92.230
450	90.586	355.449	326.573	12.994	-767.342	-698.651	81.097
500	92.977	365.121	329.950	17.585	-767.483	-691.011	72.189
600	96.579	382.412	337.288	27.074	-767.575	-675.704	58.825
700	99.094	397.499	344.835	36.865	-767.492	-660.397	49.279
800	100.901	410.856	352.269	46.870	-767.293	-645.110	42.121
900	102.232	422.821	359.455	57.030	-767.016	-629.853	36.556
1000	103.235	433.646	366.341	67.305	-766.685	-614.630	32.105
1100	104.007	443.523	372.915	77.669	-766.319	-599.442	28.465
1200	104.613	452.600	379.183	88.101	-829.501	-583.210	25.386
1300	105.095	460.993	385.157	98.587	-828.321	-562.734	22.611
1400	105.485	468.797	390.856	109.117	-827.137	-542.349	20.235
1500	105.805	476.086	396.298	119.682	-825.956	-522.048	18.179
1600	106.070	482.923	401.500	130.276	-824.780	-501.826	16.383
1700	106.292	489.360	406.481	140.895	-823.612	-481.677	14.800
1800	106.479	495.441	411.256	151.533	-822.455	-461.597	13.395
1900	106.639	501.202	415.840	162.189	-821.311	-441.580	12.140
2000	106.776	506.676	420.246	172.860	-820.180	-421.624	11.012
2100	106.895	511.888	424.486	183.544	-819.065	-401.723	9.992
2200	106.999	516.864	428.573	194.239	-817.965	-381.875	9.067
2300	107.089	521.622	432.516	204.943	-816.881	-362.078	8.223
2400	107.169	526.181	436.325	215.656	-815.815	-342.327	7.451
2500	107.240	530.558	440.007	226.377	-814.764	-322.620	6.741
2600	107.303	534.765	443.571	237.104	-813.731	-302.954	6.086
2700	107.359	538.816	447.024	247.837	-812.714	-283.328	5.481
2800	107.409	542.721	450.373	258.576	-811.713	-263.740	4.920
2900	107.455	546.491	453.622	269.319	-810.727	-244.187	4.398
3000	107.496	550.135	456.779	280.067	-809.755	-224.668	3.912
3100	107.533	553.660	459.848	290.818	-808.796	-205.180	3.457
3200	107.567	557.075	462.833	301.573	-807.848	-185.725	3.032
3300	107.597	560.385	465.739	312.331	-806.910	-166.298	2.632
3400	107.626	563.598	468.570	323.093	-805.979	-146.899	2.257
3500	107.651	566.718	471.330	333.856	-805.055	-127.528	1.903
3600	107.675	569.751	474.022	344.623	-804.133	-108.183	1.570
3700	107.697	572.701	476.649	355.391	-803.212	-88.864	1.255
3800	107.717	575.574	479.215	366.162	-802.289	-69.569	0.956
3900	107.736	578.372	481.722	376.935	-801.362	-50.299	0.674
4000	107.753	581.100	484.172	387.709	-800.428	-31.053	0.406
4100	107.769	583.761	486.569	398.485	-799.485	-11.830	0.151
4200	107.784	586.358	488.914	409.263	-798.529	7.369	-0.092
4300	107.798	588.894	491.210	420.042	-797.559	26.546	-0.322
4400	107.811	591.372	493.458	430.822	-796.572	45.700	-0.543
4500	107.823	593.795	495.661	441.604	-795.566	64.831	-0.753
4600	107.834	596.165	497.820	452.387	-794.540	83.939	-0.953
4700	107.845	598.484	499.938	463.171	-793.491	103.026	-1.145
4800	107.855	600.755	502.014	473.956	-792.417	122.089	-1.329
4900	107.864	602.979	504.052	484.742	-791.318	141.130	-1.504
5000	107.873	605.158	506.053	495.528	-790.192	160.148	-1.673
5100	107.881	607.295	508.017	506.316	-789.038	179.144	-1.835
5200	107.889	609.389	509.946	517.105	-787.855	198.116	-1.990
5300	107.896	611.445	511.842	527.894	-786.642	217.066	-2.139
5400	107.903	613.462	513.705	538.684	-785.400	235.992	-2.283
5500	107.910	615.442	515.537	549.475	-784.128	254.895	-2.421
5600	107.916	617.386	517.338	560.266	-782.825	273.775	-2.554
5700	107.922	619.296	519.110	571.058	-781.493	292.630	-2.682
5800	107.928	621.173	520.854	581.850	-780.130	311.463	-2.805
5900	107.933	623.018	522.570	592.643	-778.738	330.271	-2.924
6000	107.938	624.832	524.259	603.437	-777.318	349.056	-3.039

PREVIOUS: March 1963 (1 atm) CURRENT: March 1963 (1 bar)

Phosphoryl Chloride Fluoride ($POCl_2F$) $Cl_2F_1O_1P_1(g)$

IRON CHLORIDE ($FeCl_2$) CRYSTAL M_r = 126.753

$S°$(298.15 K) = 117.947 ± 3.3 J K^{-1} mol^{-1}

T_{fus} = 950 K

$\Delta_f H°$(0 K) = -344.419 ± 0.42 kJ mol^{-1}

$\Delta_f H°$(298.15 K) = -341.833 ± 0.42 kJ mol^{-1}

$\Delta_{fus} H°$ = 43.012 ± 0.21 kJ mol^{-1}

Enthalpy of Formation

Koehler and Coughlin (1) determined $\Delta_r H°$(298.15 K) = -4.06±0.05 kcal mol^{-1} for Fe(α) + 2 HCl(12.731 H_2O) = $FeCl_2$(cr) + H_2(g) by solution calorimetry. This leads to $\Delta_f H°$($FeCl_2$, cr, 298.15 K) = -81.7±0.1 kcal mol^{-1} which is adopted in the tabulation. The value $\Delta_f H°$(298.15 K) [HCl(12.731 H_2O)] = -38.812 kcal mol^{-1} (2) is used in the calculation.

Li and Gregory (3) also determined calorimetrically $\Delta_r H°$(298.15 K) = -19.5±0.2 kcal mol^{-1} for $FeCl_2$(cr) + Fe^{++}(aq) + 2 Cl^-(aq) in aqueous solution at concentrations near 10^{-3} molal. Using $\Delta_f H°$(Fe^{++}, aq, ∞ 298.15 K) = -21.3 kcal mol^{-1} (4) and $\Delta_f H°$(Cl^-, aq, ∞ 298.15 K) = -39.952 kcal mol^{-1} (2), we have obtained $\Delta_f H°$($FeCl_2$, cr, 298.15 K) = -81.7 kcal mol^{-1} which is in excellent agreement with the adopted value.

Kangro and Petersen (5) determined the equilibrium constants for the reaction $FeCl_2$(cr) + H_2(g) = Fe(cr) + 2 HCl(g) in the temperature range from 484-662°C. By use of 2nd and 3rd law analyses, the enthalpy of reaction is evaluated as 37.81 and 38.24 kcal mol^{-1}, respectively. Using the 3rd law $\Delta_r H°$(298.15 K), we obtain $\Delta_f H°$($FeCl_2$, cr, 298.15 K) = -82.36±0.5 kcal mol^{-1} which is in reasonable agreement with the value adopted.

Heat Capacity and Entropy

Kelley and Moore (6) measured the low temperature heat capacities in the temperature range from 53.2 - 295.0 K and made an extrapolation to 0 K which yielded an entropy of 5.19±0.8 cal K^{-1} mol^{-1} at 50.12 K. This value also includes the uncoupling energy of Fe^{++} (Rℓn5 = 3.2 cal K^{-1} mol^{-1}). We have adopted the measured heat capacities, but made a minor adjustment of the entropy at 53.2 K to 4.93 cal K^{-1} mol^{-1} which is still within the limit of uncertainty, but yields the entropy at 298.15 K as 28.19 cal K^{-1} mol^{-1}. The latter was found by E. F. Westrum (7).

Moore (8) measured high temperature enthalpy data from 670.5 to 941 K by drop calorimetry. The low temperature heat capacities and high temperature enthalpy data were smoothly joined at 298.15 K. The $C_p°$ values above 941 K were obtained by graphical extrapolation. Oetting and Gregory (9) measured high temperature heat capacities in the temperature range from 60 to 500 C in a constant heating adiabatic calorimeter. Their $C_p°$ values are in good agreement with those found by Moore (8).

Fusion Data

T(fus) and $\Delta_{fus} H°$ were taken from Moore (8).

Sublimation Data

$\Delta_{sub} H°$(298.15 K) for the monomer and dimer are calculated from the adopted enthalpies of formation of the crystal and the respective gaseous species.

References

1. M. F. Koehler and J. P. Coughlin, J. Phys. Chem. 63, 605 (1959).
2. U. S. Nat. Bur. Stand. Tech. Note 270-3, 1968.
3. J. C. M. Li and N. W. Gregory, J. Amer. Chem. Soc. 74, 4670 (1952).
4. U. S. Nat. Bur. Stand. Tech. Note 270-4, 1969.
5. W. Kangro and E. Petersen, Z. Anorg. Chem. 261, 157 (1950).
6. K. K. Kelley and G. E. Moore, J. Amer. Chem. Soc. 65, 1264 (1943).
7. E. F. Westrum, Jr., Dept. of Chem., Univ. of Michigan, personal communication to L. E. Wilson and N. W. Gregory as quoted in J. Phys. Chem. 62, 437 (1958).
8. G. E. Moore, J. Amer. Chem. Soc. 65, 1700 (1943).
9. F. L. Oetting and N. W. Gregory, J. Phys. Chem. 65, 138 (1961).

Iron Chloride ($FeCl_2$) Cl_2Fe_1(cr)

Enthalpy Reference Temperature = T_r = 298.15 K Standard State Pressure = p° = 0.1 MPa

	J $K^{-1}$$mol^{-1}$			kJ mol^{-1}			
T/K	$C_p°$	S°	$-[G°-H°(T_r)]/T$	$H°-H°(T_r)$	$\Delta_f H°$	$\Delta_f G°$	Log K_f
0	0.	0.	INFINITE	-16.273	-344.419	-344.419	INFINITE
100	50.895	44.782	181.905	-13.712	-345.192	-330.163	172.460
200	70.734	88.442	124.802	-7.272	-343.558	-315.664	82.443
298.15	76.664	117.947	117.947	0.	-341.833	-302.342	52.969
300	76.735	118.421	117.948	0.142	-341.800	-302.097	52.600
400	79.668	140.921	120.996	7.970	-340.070	-289.125	37.756
500	81.638	158.919	126.840	16.039	-338.424	-276.581	28.894
600	83.107	173.937	133.473	24.278	-336.904	-264.358	23.014
700	84.375	186.845	140.197	32.654	-335.526	-252.378	18.833
800	85.508	198.187	146.751	41.149	-334.345	-240.583	15.708
900	86.504	208.317	153.039	49.751	-333.492	-228.919	13.286
950.000	86.944	213.006	156.072	54.087	------ CRYSTAL <--> LIQUID -----		
1000	87.354	217.476	159.032	58.445	-333.282	-217.320	11.352
1100	88.013	225.834	164.730	67.214	-333.906	-205.678	9.767
1200	88.554	233.516	170.147	76.043	-333.928	-194.039	8.446
1300	88.962	240.621	175.298	84.920	-332.267	-182.449	7.331
1400	89.287	247.226	180.202	93.833	-330.663	-170.986	6.380
1500	89.507	253.394	184.879	102.773	-329.125	-159.634	5.559
1600	89.684	259.177	189.343	111.733	-327.659	-148.383	4.844
1700	89.817	264.618	193.613	120.708	-327.220	-137.202	4.216
1800	89.907	269.754	197.702	129.695	-326.248	-126.053	3.658
1900	89.953	274.617	201.623	138.688	-339.460	-114.254	3.141
2000	89.956	279.231	205.389	147.684	-338.904	-102.416	2.675

PREVIOUS: June 1965 CURRENT: December 1970

IRON CHLORIDE ($FeCl_2$) LIQUID M_r = 126.753

$S°(298.15\ K)$ = [139.876] J K^{-1} mol^{-1}

T_{fus} = 950 K

$\Delta_f H°(298.15\ K)$ = [-311.336] kJ mol^{-1}

$\Delta_{fus}H°$ = 43.012 ± 0.21 kJ mol^{-1}

Enthalpy of Formation

$\Delta_f H°(FeCl_2$, ℓ, 298.15 K) is calculated from $\Delta_f H°$(cr, 298.15 K) by adding $\Delta_{fus}H°$ and the difference in enthalply, H°(950 K) - H°(298.15 K), between the crystal and liquid.

Heat Capacity and Entropy

Moore (1) derived the constant heat capacity from enthalpy data measured in the temperature range from 950 to 1100 K in a drop calorimeter. The constant $C°_p$ is assumed for all the other temperatures. The entropy is calculated a manner similar to that used for the enthalpy of formation.

Fusion Data

T_{fus} and $\Delta_{fus}H°$ were taken from Moore (1).

Vaporization Data

The boiling point, T_{vap} = 1297 K, is the calculated temperature at which the sum of the partial pressures of $FeCl_2$(g) and Fe_2Cl_4(g) over $FeCl_2$(ℓ) reaches one atmosphere. At T_{vap}, the mole fraction of the dimer is 0.1635. The enthalpy of vaporization is calculated to be the difference between the enthalpies of formation at 1297 K for one mole of $FeCl_2$(ℓ) vaporizing to the equilibrium mixture of 0.719 moles of $FeCl_2$(g) and 0.1403 moles of Fe_2Cl_4(g).

Reference

1. G. E. Moore, J. Amer. Chem. Soc. 65, 1700 (1943).

Iron Chloride ($FeCl_2$) Cl_2Fe_1(l)

Enthalpy Reference Temperature = T_r = 298.15 K Standard State Pressure = p° = 0.1 MPa

T/K	$C°_p$ (J $K^{-1}mol^{-1}$)	S° (J $K^{-1}mol^{-1}$)	$-[G°-H°(T_r)]/T$ (J $K^{-1}mol^{-1}$)	$H°-H°(T_r)$ (kJ mol^{-1})	$\Delta_f H°$ (kJ mol^{-1})	$\Delta_f G°$ (kJ mol^{-1})	Log K_f
0							
100							
200							
298.15	102.173	139.876	139.876	0.	-311.336	-278.383	48.772
300	102.173	140.508	139.878	0.189	-311.256	-278.179	48.435
400	102.173	169.902	143.886	10.406	-307.137	-267.784	34.969
500	102.173	192.701	151.454	20.624	-303.343	-258.391	26.994
600	102.173	211.330	159.928	30.841	-299.844	-249.734	21.741
700	102.173	227.080	168.425	41.058	-296.624	-241.641	18.031
800	102.173	240.723	176.628	51.276	-293.721	-233.988	15.278
900	102.173	252.757	184.432	61.493	-291.253	-226.675	13.156
950.000	102.173	258.281	188.174	66.602	------ CRYSTAL <--> LIQUID -----		
1000	102.173	263.522	191.812	71.710	-289.520	-219.603	11.471
1100	102.173	273.260	198.781	81.928	-288.695	-212.637	10.097
1200	102.173	282.151	205.363	92.145	-287.330	-205.802	8.958
1300	102.173	290.329	211.589	102.362	-284.328	-199.130	8.001
1400	102.173	297.901	217.487	112.580	-281.420	-192.686	7.189
1500	102.173	304.950	223.085	122.797	-278.604	-186.447	6.493
1600	102.173	311.544	228.410	133.014	-275.881	-180.393	5.889
1700	102.173	317.738	233.484	143.232	-274.200	-174.487	5.361
1800	102.173	323.578	238.329	153.449	-271.997	-168.686	4.895
1900	102.173	329.103	242.962	163.666	-283.985	-162.302	4.462
2000	102.173	334.343	247.402	173.884	-282.207	-155.944	4.073

PREVIOUS: June 1965 CURRENT: December 1970

Iron Chloride ($FeCl_2$) Cl_2Fe_1(l)

IRON CHLORIDE ($FeCl_2$) CRYSTAL–LIQUID M_r = 126.753

0 to 950 K crystal
above 950 K liquid

Refer to the individual tables for details.

Iron Chloride ($FeCl_2$) Cl_2Fe_1(cr,l)

Enthalpy Reference Temperature = T_r = 298.15 K Standard State Pressure = $p°$ = 0.1 MPa

T/K	$C_p°$ (J K⁻¹mol⁻¹)	$S°$ (J K⁻¹mol⁻¹)	$-[G°-H°(T_r)]/T$ (J K⁻¹mol⁻¹)	$H°-H°(T_r)$ (kJ mol⁻¹)	$\Delta_fH°$ (kJ mol⁻¹)	$\Delta_fG°$ (kJ mol⁻¹)	Log K_f
0	0.	0.	INFINITE	-16.273	-344.419	-344.419	INFINITE
100	50.895	44.782	181.905	-13.712	-345.192	-330.163	172.460
200	70.734	88.442	124.802	-7.272	-343.558	-315.664	82.443
298.15	76.664	117.947	117.947	0.	-341.833	-302.342	52.969
300	76.735	118.421	117.948	0.142	-341.800	-302.097	52.600
400	79.668	140.921	120.996	7.970	-340.070	-289.125	37.756
500	81.638	158.919	126.840	16.039	-338.424	-276.581	28.894
600	83.107	173.937	133.473	24.278	-336.904	-264.358	23.014
700	84.375	186.845	140.197	32.654	-335.526	-252.378	18.833
800	85.508	198.187	146.751	41.149	-334.345	-240.583	15.708
900	86.504	208.317	153.039	49.751	-333.492	-228.919	13.286
950.000	86.944	213.006	156.072	54.087	CRYSTAL <--> LIQUID		
950.000	102.173	258.281	156.072	97.099	TRANSITION		
1000	102.173	263.522	161.315	102.207	-289.520	-219.603	11.471
1100	102.173	273.260	171.056	112.425	-288.695	-212.637	10.097
1200	102.173	282.151	179.949	122.642	-287.330	-205.802	8.958
1300	102.173	290.329	188.129	132.859	-284.328	-199.130	8.001
1400	102.173	297.901	195.703	143.077	-281.420	-192.686	7.189
1500	102.173	304.950	202.754	153.294	-278.604	-186.447	6.493
1600	102.173	311.544	209.350	163.511	-275.881	-180.393	5.889
1700	102.173	317.738	215.545	173.729	-274.200	-174.487	5.361
1800	102.173	323.578	221.386	183.946	-271.997	-168.686	4.895
1900	102.173	329.103	226.911	194.163	-283.985	-162.302	4.462
2000	102.173	334.343	232.153	204.381	-282.207	-155.944	4.073

PREVIOUS: CURRENT: December 1970

Iron Chloride ($FeCl_2$) Cl_2Fe_1(cr,l)

IRON DICHLORIDE ($FeCl_2$) IDEAL GAS $M_r = 126.753$

$\Delta_f H°(0\ K) = -141.59 \pm 2.1\ kJ\ mol^{-1}$

$S°(298.15\ K) = 299.28 \pm 4.2\ J\ K^{-1}\ mol^{-1}$ $\Delta_f H°(298.15\ K) = -141.00 \pm 2.1\ kJ\ mol^{-1}$

Electronic Levels and Quantum Weights

State	ε_i, cm^{-1}	g_i
$^5\Delta$	0	10
$^5\Pi$	4800	10
$^5\Sigma$	7140	5

Vibrational Frequencies and Degeneracies

ν, cm^{-1}
[327](1)
88 (2)
492 (1)

Point Group: $[D_{\infty h}]$ $\sigma = 2$

Bond Distance: Fe-Cl = [2.17] Å

Bond Angle: Cl-Fe-Cl = [180]°

Rotational Constant: $B_0 = [0.050489]\ cm^{-1}$

Enthalpy of Formation

Using the Gibbs energy functions of monomeric and dimeric $FeCl_2(g)$ with the adopted enthalpy of dimerization at 298.15 K, $\Delta_{dim}H° = -35.7\ kcal\ mol^{-1}$ (1), we have derived partial pressure of monomeric $FeCl_2$ from the reported pressure data (2-7). Schafer (4) determined Kp for the equilibrium reaction C assuming the dimer concentration was insignificant at the equilibrium conditions.

Source	Method	Reaction	T/K	Data Points	$\Delta_r H°(298.15\ K)$, kcal mol^{-1} 3rd law	2nd law	Drift cal K^{-1} mol^{-1}	$\Delta_f H°(298.15\ K)$ kcal mol^{-1}
(2)	Static	A	1012-1179	29	40.61±0.11	41.3	-0.6±0.2	-33.80
(3)	Static	A	972-1268	22	40.70±0.40	39.1	1.1±0.4	-33.71
(4)	Carrier gas	A	980-1106	6	40.71±0.27	41.7	-0.9±1.3	-33.70
(5)	Mass spectrometric	B	621-701	8[a]	47.74±0.11	45.7	3.1±0.3	-33.96
(6)	Torsion effusion	B	685-760	Eqn	43.38±0.11	47.0	1.9	-33.32
(7)	Torsion effusion	B	725-825	Eqn	48.28±0.17	49.9	-2.2	-33.42
(7)	Mass effusion (Mo cell)	B	725-825	Eqn	48.53±0.14	49.9	-1.7	-33.17
(7)	Mass effusion (Fe cell)	B	725-825	Eqn	48.37±0.12	49.5	-1.5	-33.33
(14)	Torsion effusion	B	643-715	28	48.28±0.08	49.1	-1.1±0.3	-33.42
(4)	Equilibrium	C	1205-1375	9	≥10.47±0.50	5.1	4.2±0.1	≥-33.66

Reaction: A) $FeCl_2(l) \rightarrow FeCl_2(g)$; B) $FeCl_2(cr) \rightarrow FeCl_2(g)$; C) $Fe(cr) + 2\ HCl(g) \rightarrow FeCl_2(g) + H_2(g)$.

[a] 1 point rejected due to failure of a statistical test.

The adopted value, $\Delta_f H°(FeCl_2,\ g,\ 298.15\ K) = -33.7\ kcal\ mol^{-1}$, is in very good agreement with those derived from the reactions A and C. The adopted value is also between the more discrepant results for reaction B derived from mass spectrometry and from effusion studies. This implies that the total pressures calculated from the monomer and dimer tables fit the observed vapor pressures over liquid much better than those over solid. For the liquid, the ratio of the calculated to the observed vapor pressure is about 1.0 ± 0.1 and for the solid, the ratio is about 1.3 ± 0.2

Heat Capacity and Entropy

The molecular configuration is assumed to be linear, since experimental evidence indicated that other transition metal dichlorides are linear (8). The bond distance is estimated by comparison with that of $Fe_2Cl_6(g)$ reported by Zasorin (9). The ground state ($^5\Delta$), low lying electronic levels and the quantum weights were obtained from the work of Dekock and Gruen (10).

The asymmetric stretching frequency was observed in the infrared spectra of the vapor at 492 cm^{-1} by Leroi (11) and at 494 cm^{-1} in Ar matrices by Thompson (12) and Frey (13). Normal coordinate analysis yields the stretching force constant $k_1 = 2.23$ md/Å which leads to the stretching vibrational frequency, $\nu_1 = 327\ cm^{-1}$, by a valence force field model. The bending frequency, $\nu_2 = 88\ cm^{-1}$ was observed in the matrix infrared spectra (12).

The drift indicated that the entropy of $FeCl_2(g)$ is made more consistent by introducing two low lying electronic levels and a relatively low bending frequency. The enthalpies derived by 2nd and 3rd law analyses are in good agreement.

References

1. JANAF Thermochemical Tables: $Fe_2Cl_4(g)$, 12-31-70.
2. C. Beusman, Oak Ridge Nat. Lab. ORNL-2323, June 1957.
3. C. G. Maier, U.S. Bur. Mines, Tech. Paper 360, 1929.
4. H. Schafer and K. Krehl, Z. Anorg. Chem. 268, 25 (1952); H. Schafer, L. Bayer, G. Breil, K. Etzel and K. Krehl, Z. Anorg. Chem. 278, 300 (1955); H. Schafer and G. Breil, Z. Anorg. Chem. 283, 304 (1956).
5. R. C. Schoomaker and R. F. Porter, J. Chem. Phys. 29, 116 (1958).
6. R. J. Sime and N. W. Gregory, J. Phys. Chem. 64, 86 (1960).
7. A. S. Kana'an, J. R. McCreary, D. E. Peterson and R. J. Thorn, High Temp. Sci. 1, 222 (1969).
8. A. Buchler, J. L. Stauffer and W. Klemperer, J. Chem. Phys. 40, 3471 (1964).
9. E. Z. Zasorin, N. G. Rambidi and P. A. Akishin, Zh. Strukt. Khim. 4, 910 (1963).
10. C. W. DeKock and D. M. Gruen, J. Chem. Phys. 44, 4387 (1966); 49, 4521 (1968).
11. G. E. Leroi, T. C. James, J. T. Hougen and W. Klemperer, J. Chem. Phys. 36, 2879 (1962).
12. K. R. Thompson and K. D. Carlson, J. Chem. Phys. 49, 4379 (1968).
13. R. A. Frey, R. D. Werder and H. H. Gunthard, J. Mol. Spectry. 35, 260 (1970).
14. D. L. Hildenbrand, McDonald Douglas Astronautics Company, personal communication, Dec. 23, 1970.

Iron Chloride ($FeCl_2$) $Cl_2Fe_1(g)$

Enthalpy Reference Temperature = T_r = 298.15 K Standard State Pressure = p° = 0.1 MPa

T/K	$C_p°$ (J $K^{-1}mol^{-1}$)	S° (J $K^{-1}mol^{-1}$)	$-[G°-H°(T_r)]/T$ (J $K^{-1}mol^{-1}$)	$H°-H°(T_r)$ (kJ mol^{-1})	$\Delta_f H°$ (kJ mol^{-1})	$\Delta_f G°$ (kJ mol^{-1})	Log K_f
0	0.	0.	INFINITE	-14.277	-141.590	-141.590	INFINITE
100	45.723	242.588	347.586	-10.500	-141.148	-145.899	76.210
200	53.738	277.015	304.447	-5.486	-140.941	-150.761	39.375
250	56.054	289.272	300.223	-2.738	-140.949	-153.217	32.013
298.15	57.570	299.283	299.283	0.	-141.001	-155.575	27.256
300	57.618	299.639	299.284	0.107	-141.004	-155.666	27.104
350	58.695	308.607	299.990	3.016	-141.099	-158.102	23.595
400	59.456	316.497	301.570	5.971	-141.237	-160.522	20.962
450	60.010	323.534	303.626	8.958	-141.423	-162.922	18.912
500	60.425	329.879	305.940	11.970	-141.662	-165.299	17.269
600	60.999	340.950	310.879	18.043	-142.307	-169.969	14.797
700	61.392	350.384	315.865	24.163	-143.184	-174.514	13.022
800	61.713	358.603	320.705	30.319	-144.343	-178.914	11.682
900	62.022	365.890	325.328	36.506	-145.905	-183.147	10.630
1000	62.346	372.441	329.717	42.724	-148.171	-187.173	9.777
1100	62.695	378.399	333.876	48.976	-151.312	-190.907	9.065
1200	63.067	383.870	337.817	55.263	-153.876	-194.412	8.463
1300	63.456	388.934	341.557	61.590	-154.766	-197.754	7.946
1400	63.852	393.651	345.111	67.955	-155.709	-201.026	7.500
1500	64.244	398.069	348.496	74.360	-156.706	-204.229	7.112
1600	64.625	402.228	351.726	80.803	-157.757	-207.363	6.770
1700	64.986	406.157	354.813	87.284	-159.813	-210.411	6.465
1800	65.322	409.881	357.770	93.800	-161.311	-213.344	6.191
1900	65.630	413.421	360.606	100.348	-176.969	-215.491	5.924
2000	65.907	416.795	363.332	106.925	-178.831	-217.470	5.680
2100	66.152	420.016	365.955	113.528	-180.679	-219.356	5.456
2200	66.364	423.099	368.483	120.154	-182.516	-221.155	5.251
2300	66.545	426.053	370.922	126.800	-184.347	-222.870	5.062
2400	66.696	428.888	373.279	133.462	-186.176	-224.506	4.886
2500	66.819	431.613	375.558	140.138	-188.007	-226.066	4.723
2600	66.915	434.236	377.765	146.825	-189.845	-227.552	4.572
2700	66.987	436.763	379.904	153.520	-191.692	-228.967	4.430
2800	67.038	439.200	381.978	160.222	-193.552	-230.313	4.297
2900	67.068	441.553	383.992	166.927	-195.427	-231.593	4.171
3000	67.081	443.827	385.949	173.635	-197.319	-232.808	4.054
3100	67.079	446.027	387.851	180.343	-199.231	-233.960	3.942
3200	67.063	448.156	389.703	187.050	-549.462	-227.626	3.716
3300	67.035	450.219	391.506	193.755	-549.512	-217.567	3.444
3400	66.997	452.220	393.262	200.457	-549.616	-207.507	3.188
3500	66.950	454.161	394.974	207.154	-549.772	-197.443	2.947
3600	66.896	456.047	396.645	213.846	-549.980	-187.373	2.719
3700	66.836	457.879	398.275	220.533	-550.240	-177.298	2.503
3800	66.771	459.660	399.867	227.213	-550.551	-167.214	2.299
3900	66.701	461.394	401.423	233.887	-550.912	-157.122	2.104
4000	66.628	463.082	402.943	240.554	-551.321	-147.019	1.920
4100	66.553	464.726	404.430	247.213	-551.778	-136.906	1.744
4200	66.476	466.329	405.885	253.864	-552.280	-126.781	1.577
4300	66.397	467.892	407.309	260.508	-552.826	-116.644	1.417
4400	66.318	469.418	408.703	267.143	-553.416	-106.494	1.264
4500	66.238	470.907	410.069	273.771	-554.047	-96.329	1.118
4600	66.158	472.362	411.407	280.391	-554.718	-86.150	0.978
4700	66.078	473.784	412.719	287.003	-555.429	-75.957	0.844
4800	65.998	475.174	414.006	293.607	-556.177	-65.747	0.715
4900	65.920	476.534	415.268	300.202	-556.963	-55.522	0.592
5000	65.842	477.865	416.507	306.791	-557.786	-45.280	0.473
5100	65.765	479.168	417.723	313.371	-558.644	-35.021	0.359
5200	65.690	480.445	418.917	319.944	-559.539	-24.746	0.249
5300	65.616	481.695	420.090	326.509	-560.469	-14.452	0.142
5400	65.543	482.921	421.242	333.067	-561.436	-4.141	0.040
5500	65.471	484.123	422.374	339.618	-562.439	6.188	-0.059
5600	65.401	485.302	423.488	346.161	-563.480	16.536	-0.154
5700	65.333	486.459	424.582	352.698	-564.560	26.903	-0.247
5800	65.266	487.595	425.659	359.228	-565.680	37.290	-0.336
5900	65.201	488.710	426.718	365.751	-566.842	47.696	-0.422
6000	65.137	489.805	427.760	372.268	-568.048	58.121	-0.506

PREVIOUS: December 1970 (1 atm) CURRENT: December 1970 (1 bar)

Iron Chloride ($FeCl_2$) $Cl_2Fe_1(g)$

DICHLOROSILANE (SiH_2Cl_2) IDEAL GAS M_r = 101.00738

$S°(298.15 K) = 286.734 \pm 0.34\ J\ K^{-1}\ mol^{-1}$

$\Delta_f H°(0\ K) = [-313.01 \pm 12.6]\ kJ\ mol^{-1}$

$\Delta_f H°(298.15\ K) = [-320.49 \pm 12.6]\ kJ\ mol^{-1}$

Vibrational Frequencies and Degeneracies

ν, cm^{-1}	ν, cm^{-1}
2224(1)	2237(1)
954(1)	602(1)
527(1)	876(1)
188(1)	590(1)
710(1)	

Ground State Quantum Weight = [1] σ = 2

Point Group: C_{2v}

Bond Distances: Si-H = 1.480 Å Si-Cl = 2.033 Å

Bond Angles: H-Si-H = 111.3° Cl-Si-Cl = 109.72°

Product of the Moments of Inertia: $I_A I_B I_C = 7.53113 \times 10^{-114}\ g^3\ cm^6$

Enthalpy of Formation

$\Delta_f H°$(298.15 K) is estimated from values for $SiCl_4$, $SiHCl_3$, and SiH_4(1). Data for $SiHCl_3$ (1) suggest that $\Delta_f H°$ shows minor deviations from linearity in the chlorosilanes. We assume that $\Delta_f H°$ has a cubic variation with a constant third difference of -1.5 kcal mol^{-1}. $\Delta_f H°$ values of the chloromethanes yield an almost constant third difference of about -2.5 kcal mol^{-1} (2, 3, 1). This approximation corresonds to Allen's bond additivity scheme involving near-neighbor interactions taken two and three at a time (2). Our adopted $\Delta_f H°$ is 1.5 kcal mol^{-1} more negative than that estimated by linear interpolation between $SiCl_4$ and SiH_4.

Heat Capacity and Entropy

The molecular structure is based on microwave data of Davis and Gerry (4) for three isotopic forms of dichlorosilane. Structural parameters are essentially substitutional (r_s) values. They are in reasonable agreement with an early electron diffraction study (5). The principal moments of inertia are $I_A = 5.9928 \times 10^{-39}$, $I_B = 33.0411 \times 10^{-39}$ and $I_C = 38.0343 \times 10^{-39}$ g cm^2.

Vibrational frequencies are those selected by Shimanouchi (6) from a gas-phase infrared study of SiH_2Cl_2, $SiDHCl_2$ and SiD_2Cl_2 done by Christensen and Nielsen (7). Assignments are based on band contours and isotopic rules and are supported by theoretical calculations (7, 4). Two frequencies, ν_4 = 188 and infrared inactive ν_5 = 710 cm^{-1}, are from liquid-phase Raman spectra (8) which also support the vibrational assignment. Christensen and Nielsen (7) found a weak infrared band at ~190 cm^{-1} but could not estimate a definite band contour.

We neglect excited states and assume the electronic ground state to be a singlet. A recent vacuum ultraviolet study of Si_2HCl_2 (9) is not available to us. We assume that there is little shift in electronic levels between SiH_2Cl_2 and SiH_3Cl (1), since there is little shift in absorption spectra between CH_2Cl_2 and CH_3Cl (10). This implies that contributions from the excited states in SiH_2Cl_2 should be unimportant, just as we believe them to be in SiH_3Cl (1).

References

1. JANAF Thermochemical Tables: $Cl_3HSi(g)$, 12-31-76; $H_4Si(g)$, 6-30-76; $Cl_4Si(g)$, 12-31-70.
2. S. A. Kudchadker and A. P. Kudchadker, J. Phys. Chem. Ref. Data 4, 457 (1975).
3. A. S. Rodgers, J. Chao, R. C. Wilhoit and B. J. Zwolinski, ibid., 3, 17 (1974).
4. R. W. Davis and M. C. L. Gerry, J. Mol. Spectrosc. 60, 117 (1976).
5. L. O. Brockway and I. E. Coop, Trans. Faraday Soc. 34, 1429 (1938).
6. T. Shimanouchi, J. Phys. Chem. Ref. Data 2, 134 (1973).
7. D. H. Christensen and O. F. Nielsen, J. Mol. Specrosc. 27, 499 (1968); 33, 425 (1970).
8. J. A. Hawkins and M. K. Wilson, J. Chem. Phys. 21, 360 and 1122 (1953).
9. G. C. Causley and B. R. Russell, J. Electron Spectrosc. Relat. Phenom. 8, 71 (1976).
10. B. R. Russell, L. O. Edwards and J. W. Raymonda, J. Amer. Chem. Soc. 95, 2129 (1973).

Dichlorosilane (SiH_2Cl_2) $Cl_2H_2Si_1(g)$

Enthalpy Reference Temperature = T_r = 298.15 K Standard State Pressure = p° = 0.1 MPa

T/K	$C_p°$	S°	$-[G°-H°(T_r)]/T$	$H°-H°(T_r)$	$\Delta_f H°$	$\Delta_f G°$	Log K_f
	J $K^{-1}mol^{-1}$			kJ mol^{-1}			
0	0.	0.	INFINITE	-13.385	-313.013	-313.013	INFINITE
100	38.443	234.601	333.516	-9.892	-315.697	-309.800	161.823
200	49.901	264.436	292.044	-5.522	-318.237	-302.916	79.114
250	56.488	276.287	287.727	-2.860	-319.441	-298.945	62.461
298.15	62.174	286.734	286.734	0.	-320.494	-294.900	51.665
300	62.375	287.119	286.735	0.115	-320.532	-294.741	51.319
350	67.358	297.120	287.514	3.362	-321.488	-290.365	43.335
400	71.521	306.394	289.301	6.837	-322.308	-285.861	37.330
450	75.028	315.027	291.686	10.503	-323.002	-281.263	32.648
500	78.029	323.091	294.427	14.332	-323.585	-276.593	28.895
600	82.940	337.770	300.453	22.390	-324.464	-267.106	23.254
700	86.822	350.858	306.735	30.886	-325.026	-257.498	19.215
800	89.965	362.663	313.001	39.730	-325.343	-247.827	16.181
900	92.536	373.413	319.125	48.860	-325.470	-238.128	13.821
1000	94.653	383.276	325.054	58.222	-325.459	-228.422	11.932
1100	96.406	392.382	330.766	67.778	-325.348	-218.723	10.386
1200	97.864	400.835	336.257	77.494	-325.171	-209.038	9.099
1300	99.084	408.718	341.531	87.343	-324.956	-199.368	8.011
1400	100.112	416.100	346.597	97.304	-324.723	-189.717	7.078
1500	100.983	423.037	351.464	107.360	-324.490	-180.081	6.271
1600	101.726	429.579	356.143	117.497	-324.271	-170.462	5.565
1700	102.363	435.765	360.647	127.702	-374.252	-160.408	4.929
1800	102.912	441.632	364.984	137.966	-373.859	-147.840	4.290
1900	103.389	447.209	369.166	148.282	-373.463	-135.294	3.719
2000	103.805	452.523	373.202	158.642	-373.070	-122.769	3.206
2100	104.169	457.597	377.101	169.041	-372.684	-110.264	2.743
2200	104.490	462.451	380.871	179.474	-372.310	-97.776	2.322
2300	104.775	467.102	384.520	189.938	-371.951	-85.305	1.937
2400	105.027	471.566	388.055	200.428	-371.611	-72.850	1.586
2500	105.252	475.858	391.481	210.942	-371.291	-60.408	1.262
2600	105.454	479.991	394.807	221.478	-370.995	-47.979	0.964
2700	105.635	483.974	398.036	232.032	-370.724	-35.560	0.688
2800	105.799	487.819	401.174	242.604	-370.480	-23.152	0.432
2900	105.947	491.534	404.226	253.192	-370.264	-10.751	0.194
3000	106.081	495.128	407.197	263.793	-370.077	1.643	-0.029
3100	106.203	498.608	410.090	274.408	-369.920	14.031	-0.236
3200	106.315	501.982	412.909	285.034	-369.791	26.414	-0.431
3300	106.417	505.255	415.658	295.670	-369.691	38.794	-0.614
3400	106.510	508.433	418.340	306.317	-369.619	51.171	-0.786
3500	106.596	511.522	420.958	316.972	-369.574	63.547	-0.948
3600	106.676	514.526	423.516	327.636	-753.706	86.382	-1.253
3700	106.749	517.450	426.015	338.307	-753.295	109.712	-1.549
3800	106.816	520.298	428.459	348.985	-752.907	133.031	-1.829
3900	106.879	523.073	430.850	359.670	-752.539	156.341	-2.094
4000	106.937	525.780	433.189	370.361	-752.190	179.641	-2.346
4100	106.991	528.421	435.480	381.057	-751.856	202.933	-2.585
4200	107.042	531.000	437.724	391.759	-751.534	226.216	-2.813
4300	107.089	533.519	439.922	402.466	-751.222	249.492	-3.031
4400	107.133	535.981	442.078	413.177	-750.917	272.761	-3.238
4500	107.174	538.389	444.191	423.892	-750.616	296.023	-3.436
4600	107.212	540.745	446.265	434.611	-750.317	319.278	-3.626
4700	107.249	543.052	448.300	445.334	-750.017	342.527	-3.807
4800	107.283	545.310	450.297	456.061	-749.714	365.770	-3.980
4900	107.314	547.522	452.259	466.791	-749.406	389.006	-4.147
5000	107.345	549.691	454.186	477.524	-749.091	412.235	-4.307
5100	107.373	551.817	456.079	488.260	-748.766	435.459	-4.460
5200	107.400	553.902	457.941	498.998	-748.430	458.676	-4.607
5300	107.425	555.948	459.771	509.740	-748.081	481.886	-4.749
5400	107.449	557.956	461.570	520.483	-747.718	505.090	-4.886
5500	107.472	559.928	463.341	531.229	-747.337	528.286	-5.017
5600	107.493	561.865	465.083	541.978	-746.940	551.476	-5.144
5700	107.514	563.767	466.798	552.728	-746.524	574.658	-5.266
5800	107.533	565.637	468.486	563.480	-746.088	597.833	-5.384
5900	107.552	567.476	470.148	574.235	-745.632	621.000	-5.498
6000	107.569	569.284	471.785	584.991	-745.152	644.159	-5.608

PREVIOUS: December 1976 (1 atm) CURRENT: December 1976 (1 bar)

Dichlorosilane (SiH_2Cl_2) $Cl_2H_2Si_1(g)$

MERCURIC CHLORIDE ($HgCl_2$) CRYSTAL M_r = 271.496

$S°(298.15 K) = [144.494 \pm 6.3] J K^{-1} mol^{-1}$ $\Delta_f H°(298.15 K) = -230.12 \pm 4.2 kJ mol^{-1}$

$T_{fus} = 550 K$ $\Delta_{fus}H = 19.414 \pm 0.21 kJ mol^{-1}$

Enthalpy of Formation

Taken from (1).

Heat Capacity and Entropy

$C_p°$ was determined from the enthalpy data of Topol and Ransom (2) and Ewald (3). Above 550 the curve was smoothly extrapolated. The entropy was estimated by adjusting the value to give the best fit of the melting, sublimation, and vaporization data.

Melting

T_{fus} was taken from (1). $\Delta_{fus}H$ was given by Topol and Ransom (2).

References

1. U. S. Nat. Bur. Stand. Circular 500, 1952.
2. L. E. Topol and L. D. Ransom, J. Phys. Chem. 64, 1339 (1960).
3. R. Ewald, Ann. Phys. Chem. (4) 44, 1213 (1914).

Mercury Chloride ($HgCl_2$) $Cl_2Hg_1(cr)$

Enthalpy Reference Temperature = T_r = 298.15 K Standard State Pressure = p° = 0.1 MPa

T/K	$C_p°$ (J K⁻¹mol⁻¹)	S° (J K⁻¹mol⁻¹)	$-[G°-H°(T_r)]/T$ (J K⁻¹mol⁻¹)	$H°-H°(T_r)$ (kJ mol⁻¹)	$\Delta_f H°$ (kJ mol⁻¹)	$\Delta_f G°$ (kJ mol⁻¹)	Log K_f
0							
100							
200							
298.15	73.906	144.494	144.494	0.	-230.120	-184.022	32.240
300	73.973	144.952	144.496	0.137	-230.098	-183.736	31.991
400	76.986	166.664	147.436	7.691	-228.779	-168.475	22.001
500	79.224	184.075	153.079	15.498	-227.271	-153.570	16.043
550.000	80.500	191.685	156.247	19.491	------ CRYSTAL <--> LIQUID -----		
600	81.672	198.742	159.498	23.546	-225.569	-138.986	12.100
700	83.948	211.509	166.035	31.832	-282.429	-118.132	8.815
800	85.688	222.830	172.441	40.312	-279.728	-94.844	6.193
900	87.646	233.037	178.615	48.979	-276.859	-71.903	4.173
1000	89.508	242.368	184.531	57.837	-273.816	-49.292	2.575
1100	91.287	250.983	190.185	66.878	-270.604	-26.994	1.282
1200	92.968	258.999	195.589	76.092	-267.231	-4.995	0.217
1300	94.563	266.504	200.759	85.469	-263.704	16.716	-0.672
1400	96.065	273.567	205.710	95.000	-260.034	38.150	-1.423
1500	97.571	280.246	210.458	104.682	-256.221	59.316	-2.066

PREVIOUS: CURRENT: March 1962

Mercury Chloride ($HgCl_2$) $Cl_2Hg_1(cr)$

MERCURIC CHLORIDE ($HgCl_2$) LIQUID M_r = 271.496

$S°(298.15\ K)$ = [164.471] J K^{-1} mol^{-1} $\Delta_fH°(298.15\ K)$ = [-213.222] kJ mol^{-1}

T_{fus} = 550 K $\Delta_{fus}H°$ = 19.414 ± .21 kJ mol^{-1}

Enthalpy of Formation

$\Delta_fH°(HgCl_2, \ell, 298.15\ K)$ is calculated from that of the crystal by adding $\Delta_{fus}H°$ and the difference in enthalpy, H°(550 K) - H°(298.15 K), between the crystal and liquid.

Heat Capacity and Entropy

The heat capacity was estimated to conform with the bromide and iodide, the value given by Topol and Ransom (1) was discounted. A glass type transition was assumed to be 367 K below which the heat capacity was equal to that of the crystal. The entropy was obtained from S°(cr), the entropy of fusion and the estimated heat capacity.

Fusion and Vaporization

T_{fus} and T_{vap} were taken from (2). $\Delta_{fus}H°$ was given by Topol and Ransom (1) and $\Delta_{vap}H$ was obtained from Prideaux (3) and Johnson (4); T_{vap} = 577 K and $\Delta_{vap}H°$ = 58.9 kJ mol^{-1}.

References

1. L. E. Topol and L. D. Ransom, J. Phys. Chem. 64, 1339 (1960).
2. U. S. Nat. Bur. Stand. Circular 500, 1952.
3. E. B. R. Prideaux, J. Chem. Soc. (London) 97, 2033 (1910).
4. F. M. G. Johnson, J. Amer. Chem. Soc. 33, 777 (1911).

Mercury Chloride ($HgCl_2$) $Cl_2Hg_1(l)$

Enthalpy Reference Temperature = T_r = 298.15 K Standard State Pressure = p° = 0.1 MPa

T/K	$C_p°$ (J $K^{-1}mol^{-1}$)	S° (J $K^{-1}mol^{-1}$)	$-[G°-H°(T_r)]/T$ (J $K^{-1}mol^{-1}$)	$H°-H°(T_r)$ (kJ mol^{-1})	$\Delta_fH°$ (kJ mol^{-1})	$\Delta_fG°$ (kJ mol^{-1})	Log K_f
0							
100							
200							
298.15	102.090	164.471	164.471	0.	-213.222	-173.080	30.323
300	102.090	165.103	164.473	0.189	-213.148	-172.831	30.093
400	102.090	194.472	168.477	10.398	-209.174	-159.994	20.893
500	102.090	217.253	176.039	20.607	-205.264	-148.152	15.477
550.000	102.090	226.983	180.235	25.711	------ CRYSTAL <--> LIQUID -----		
600	102.090	235.866	184.506	30.816	-201.401	-137.093	11.935
700	102.090	251.603	192.996	41.025	-256.338	-120.107	8.962
800	102.090	265.235	201.193	51.234	-251.907	-100.948	6.591
900	102.090	277.260	208.990	61.443	-247.497	-82.343	4.779
1000	102.090	288.016	216.364	71.652	-243.104	-64.228	3.355
1100	102.090	297.746	223.327	81.861	-238.723	-46.552	2.211
1200	102.090	306.629	229.904	92.070	-234.355	-29.275	1.274
1300	102.090	314.800	236.125	102.278	-229.997	-12.362	0.497
1400	102.090	322.366	242.018	112.487	-225.648	4.216	-0.157
1500	102.090	329.409	247.612	122.696	-221.309	20.484	-0.713

PREVIOUS: CURRENT: March 1962

Mercury Chloride ($HgCl_2$) $Cl_2Hg_1(l)$

MERCURY CHLORIDE ($HgCl_2$) CRYSTAL–LIQUID M_r = 271.496

0 to 550 K crystal
above 550 K liquid

Refer to the individual tables for details.

Mercury Chloride ($HgCl_2$) Cl_2Hg_1(cr,l)

Enthalpy Reference Temperature = T_r = 298.15 K Standard State Pressure = p° = 0.1 MPa

T/K	C_p°	S°	$-[G^\circ-H^\circ(T_r)]/T$	$H^\circ-H^\circ(T_r)$	Δ_fH°	Δ_fG°	Log K_f
	J K⁻¹mol⁻¹			kJ mol⁻¹			
0							
100							
200							
298.15	73.906	144.494	144.494	0.	-230.120	-184.022	32.240
300	73.973	144.952	144.496	0.137	-230.098	-183.736	31.991
400	76.986	166.664	147.436	7.691	-228.779	-168.475	22.001
500	79.224	184.075	153.079	15.498	-227.271	-153.570	16.043
550.000	80.500	191.685	156.247	19.491	CRYSTAL <--> LIQUID		
550.000	102.090	226.983	156.247	38.905	TRANSITION		
600	102.090	235.866	162.517	44.009	-205.106	-140.798	12.258
700	102.090	251.603	174.149	54.218	-260.043	-123.811	9.239
800	102.090	265.235	184.701	64.427	-255.612	-104.652	6.833
900	102.090	277.260	194.331	74.636	-251.202	-86.047	4.994
1000	102.090	288.016	203.171	84.845	-246.808	-67.932	3.548
1100	102.090	297.746	211.333	95.054	-242.428	-50.257	2.387
1200	102.090	306.629	218.910	105.263	-238.060	-32.980	1.436
1300	102.090	314.800	225.976	115.472	-233.701	-16.067	0.646
1400	102.090	322.366	232.594	125.681	-229.353	0.511	-0.019
1500	102.090	329.409	238.816	135.890	-225.014	16.779	-0.584

PREVIOUS: CURRENT: March 1962

Mercury Chloride ($HgCl_2$) Cl_2Hg_1(cr,l)

MERCURIC CHLORIDE ($HgCl_2$) IDEAL GAS M_r = 271.496

$S°(298.15\ K) = 294.793\ J\ K^{-1}\ mol^{-1}$

$\Delta_f H°(0\ K) = -142.31 \pm 6.3\ kJ\ mol^{-1}$

$\Delta_f H°(298.15\ K) = -146.29 \pm 6.3\ kJ\ mol^{-1}$

Vibrational Levels and Multiplicities

ν, cm^{-1}
360 (1)
70 (2)
413 (1)

Ground State Quantum Weight: 1 $\sigma = 2$

Point Group: $D_{\infty h}$

Bond Distance: Hg-Cl = 2.31 A

Bond Angle: O-Hg-Cl = 180°

Rotational Constant: $B_o = 0.044554\ cm^{-1}$

Enthalpy of Formation

The heat of formation of the crystal as given in (1) was combined with the 3rd law enthalpy of sublimation at 298.15 K from the data of Niwa and Zhibata (2), Ruf and Treadwell (3) and Johnson (4).

Heat Capacity and Entropy

The vibrational constants were given by Klemperer and Lindeman (5). The bond length is a weighted average of the values given by Braune and Knoke (6), Gregg et al. (7), Maxwell and Mosely (8), and Akishin et al. (9).

References

1. U. S. Nat. Bur. Stand. Circular 500, "Selected Values of Chemical Thermodynamic Proeprties,".
2. K. Niwa and Z. Shibata, J. Fac. Sci. Hokkaido Imp. Univ. Ser. III, 2, 183 (1938).
3. R. Ruf and W. D. Treadwell, Helv. Chim. Acta 37, 1941 (1954).
4. F. M. G. Johnson, J. Amer. Chem. Soc. 33, 777 (1911).
5. W. Klemperer and L. Lindeman, J. Chem. Phys. 25, 397 (1956).
6. H. Braune and S. Knoke, Naturwiss. 21, 349 (1933).
7. A. H. Gregg, G. C. Hampson, G. I. Jenkins, P. L. F. Jones and L. E. Sutton, Trans. Faraday Soc. 33, 852, 1937.
8. L. R. Maxwell and V. M. Mosley, Phys. Rev. 57, 21 (1940).
9. P. A. Akishin, V. P. Spiridinov and A. N. Khodchenkov, Zhur. Fiz. Khim. 33, 20 (1959).

Mercury Chloride ($HgCl_2$) $Cl_2Hg_1(g)$

Enthalpy Reference Temperature = T_r = 298.15 K Standard State Pressure = p° = 0.1 MPa

T/K	$C_p°$	S°	$-[G°-H°(T_r)]/T$	$H°-H°(T_r)$	$\Delta_f H°$	$\Delta_f G°$	Log K_f
	$J\ K^{-1}mol^{-1}$			$kJ\ mol^{-1}$			
0	0.	0.	INFINITE	-14.543	-142.313	-142.313	INFINITE
100	46.437	237.359	343.680	-10.632	-143.019	-144.133	75.287
200	54.452	272.269	300.012	-5.549	-143.559	-145.015	37.874
250	56.690	284.678	295.742	-2.766	-146.092	-145.198	30.338
298.15	58.110	294.793	294.793	0.	-146.294	-145.007	25.405
300	58.154	295.152	294.794	0.108	-146.301	-144.999	25.247
350	59.140	304.196	295.506	3.041	-146.475	-144.768	21.605
400	59.826	312.140	297.099	6.017	-146.627	-144.514	18.872
450	60.319	319.217	299.170	9.021	-146.766	-144.241	16.743
500	60.683	325.592	301.499	12.046	-146.896	-143.954	15.039
600	61.174	336.703	306.467	18.141	-147.147	-143.341	12.479
700	61.478	346.157	311.479	24.275	-206.159	-136.116	10.157
800	61.680	354.381	316.339	30.434	-205.779	-126.136	8.236
900	61.819	361.654	320.977	36.609	-205.403	-116.203	6.744
1000	61.920	368.173	325.376	42.796	-205.031	-106.312	5.553
1100	61.995	374.078	329.540	48.992	-204.663	-96.458	4.580
1200	62.053	379.475	333.479	55.195	-204.301	-86.637	3.771
1300	62.097	384.444	337.211	61.402	-203.945	-76.846	3.088
1400	62.133	389.047	340.751	67.614	-203.593	-67.082	2.503
1500	62.162	393.335	344.115	73.829	-203.248	-57.343	1.997
1600	62.185	397.347	347.318	80.046	-202.909	-47.628	1.555
1700	62.205	401.118	350.373	86.266	-202.577	-37.933	1.166
1800	62.221	404.674	353.292	92.487	-202.252	-28.257	0.820
1900	62.235	408.038	356.086	98.710	-201.935	-18.599	0.511
2000	62.247	411.231	358.764	104.934	-201.626	-8.958	0.234
2100	62.258	414.268	361.335	111.159	-201.328	0.668	-0.017
2200	62.267	417.165	363.808	117.385	-201.041	10.280	-0.244
2300	62.274	419.933	366.188	123.612	-200.767	19.879	-0.451
2400	62.281	422.583	368.483	129.840	-200.507	29.467	-0.641
2500	62.287	425.126	370.698	136.069	-200.262	39.044	-0.816
2600	62.292	427.569	372.839	142.298	-200.033	48.612	-0.977
2700	62.297	429.920	374.910	148.527	-199.822	58.171	-1.125
2800	62.301	432.185	376.915	154.757	-199.630	67.723	-1.263
2900	62.305	434.372	378.859	160.987	-199.457	77.268	-1.392
3000	62.309	436.484	380.745	167.218	-199.302	86.807	-1.511
3100	62.312	438.527	382.576	173.449	-199.167	96.342	-1.623
3200	62.315	440.506	384.355	179.680	-199.050	105.873	-1.728
3300	62.317	442.423	386.086	185.912	-198.951	115.400	-1.827
3400	62.320	444.284	387.771	192.144	-198.868	124.924	-1.919
3500	62.322	446.090	389.411	198.376	-198.801	134.447	-2.007
3600	62.324	447.846	391.010	204.608	-198.746	143.967	-2.089
3700	62.326	449.553	392.569	210.841	-198.702	153.487	-2.167
3800	62.327	451.216	394.091	217.073	-198.667	163.005	-2.241
3900	62.329	452.835	395.577	223.306	-198.639	172.522	-2.311
4000	62.330	454.413	397.028	229.539	-198.615	182.039	-2.377
4100	62.332	455.952	398.446	235.772	-198.592	191.555	-2.440
4200	62.333	457.454	399.833	242.006	-198.568	201.070	-2.501
4300	62.334	458.921	401.191	248.239	-198.540	210.585	-2.558
4400	62.335	460.354	402.519	254.472	-198.506	220.099	-2.613
4500	62.336	461.754	403.820	260.706	-198.464	229.612	-2.665
4600	62.337	463.125	405.094	266.940	-198.411	239.125	-2.715
4700	62.338	464.465	406.343	273.173	-198.346	248.636	-2.763
4800	62.339	465.778	407.568	279.407	-198.265	258.145	-2.809
4900	62.340	467.063	408.769	285.641	-198.168	267.653	-2.853
5000	62.340	468.322	409.947	291.875	-198.053	277.158	-2.895
5100	62.341	469.557	411.104	298.109	-197.918	286.661	-2.936
5200	62.342	470.767	412.240	304.343	-197.762	296.161	-2.975
5300	62.342	471.955	413.355	310.577	-197.584	305.658	-3.012
5400	62.343	473.120	414.451	316.812	-197.384	315.151	-3.048
5500	62.343	474.264	415.529	323.046	-197.160	324.641	-3.083
5600	62.344	475.388	416.588	329.280	-196.911	334.126	-3.117
5700	62.344	476.491	417.629	335.515	-196.639	343.606	-3.149
5800	62.345	477.575	418.653	341.749	-196.341	353.082	-3.180
5900	62.345	478.641	419.661	347.984	-196.019	362.552	-3.210
6000	62.346	479.689	420.653	354.218	-195.672	372.016	-3.239

PREVIOUS: March 1962 (1 atm) CURRENT: March 1962 (1 bar)

Mercury Chloride ($HgCl_2$) $Cl_2Hg_1(g)$

MERCUROUS CHLORIDE (Hg_2Cl_2) CRYSTAL M_r = 472.086

$\Delta_f H°(0\ K) = -260.537 \pm 0.21\ kJ\ mol^{-1}$

S°(298.15 K) = 192.535 J K^{-1} mol^{-1} $\Delta_f H°(298.15\ K) = -264.927 \pm 0.21\ kJ\ mol^{-1}$

Enthalpy of Formation

The adopted enthalpy of formation is based on the average Gibbs energy of formation, -50.315 kcal mol^{-1}, as measured by Gerke (1) and Hills and Ives (2), and the entropy of formation.

Heat Capacity and Entropy

The data of Pollitzer (3) from 22-199 K was treated by Kelley (4). The heat capacities listed by Kelley were raised by 0.06 to match the entropy data of Gerke (1). Data above room temperature were obtained by smooth extrapolation. A mean value of the entropy of formation of -46.614 cal K^{-1} mol^{-1} was adopted from the cell measurements of Gerke (1).

Sublimation

The vapor pressure over mercurous chloride reaches 1 atmosphere at 655 K according to Smith and Menzies (5). At this point the vapor is dissociated into Hg(g) and $HgCl_2$(g).

References

1. R. H. Gerke, J. Amer. Chem. Soc. 44, 1684 (1922).
2. G. J. Hills and D. J. G. Ives, J. Chem. Soc. (London) 318 (1951).
3. F. Pollitzer, Electroch. 19, 513 (1913).
4. K. K. Kelley, U. S. Bur. Mines Bull. 592 (1960).
5. A. Smith and A. W. C. Menzies, J. Amer. Chem. Soc. 32, 1541 (1910).

Mercury Chloride (Hg_2Cl_2) Cl_2Hg_2(cr)

Enthalpy Reference Temperature = T_r = 298.15 K						Standard State Pressure = p° = 0.1 MPa	
	J $K^{-1}mol^{-1}$			kJ mol^{-1}			
T/K	$C_p°$	S°	$-[G°-H°(T_r)]/T$	$H°-H°(T_r)$	$\Delta_f H°$	$\Delta_f G°$	Log K_f
0	0.	0.	INFINITE	-23.476	-260.537	-260.537	INFINITE
100	79.245	92.186	278.165	-18.598	-261.981	-244.855	127.899
200	95.479	153.076	201.673	-9.719	-261.310	-227.924	59.528
298.15	101.968	192.535	192.535	0.	-264.927	-210.485	36.876
300	102.048	193.166	192.537	0.189	-264.904	-210.147	36.590
400	106.023	223.084	196.589	10.598	-263.497	-192.096	25.085
500	109.328	247.106	204.367	21.370	-261.751	-174.442	18.224
600	112.131	267.293	213.217	32.446	-259.735	-157.165	13.682
700	114.558	284.762	222.217	43.781	-375.019	-127.101	9.484
800	116.817	300.210	231.019	55.352	-371.305	-91.936	6.003
900	118.826	314.086	239.491	67.136	-367.399	-57.247	3.323
1000	120.667	326.703	247.591	79.112	-363.317	-23.003	1.202
1100	122.340	338.283	255.316	91.264	-359.072	10.824	-0.514
1200	123.846	348.994	262.682	103.574	-354.680	44.258	-1.926
1300	125.169	358.960	269.709	116.027	-350.158	77.320	-3.107
1400	126.311	368.279	276.421	128.602	-345.521	110.030	-4.105
1500	127.277	377.028	282.839	141.283	-340.788	142.405	-4.959

PREVIOUS: CURRENT: December 1961

Mercury Chloride (Hg_2Cl_2) Cl_2Hg_2(cr)

POTASSIUM CHLORIDE $((KCl)_2)$ — IDEAL GAS — M_r = 149.1026

$S°(298.15\ K)$ = [352.864] $J\ K^{-1}\ mol^{-1}$

$\Delta_f H°(0\ K)$ = -614.43 ± 4.2 $kJ\ mol^{-1}$
$\Delta_f H°(298.15\ K)$ = -617.64 ± 4.2 $kJ\ mol^{-1}$

Vibrational Frequencies and Degeneracies

ν, cm^{-1}	ν, cm^{-1}
[219](1)	[104](1)
[108](1)	[184](1)
[180](1)	[100](1)

Ground State Quantum Weight: 1 — σ = 4
Point Group: D_{2h}
Bond Distance: Cl-K = [2.932] Å
Bond Angle: Cl-K-Cl = [100.8]°
Product of the Moments of Inertia: $I_A I_B I_C$ = [2.87407 x 10^{-112}] $g^3\ cm^6$

Enthalpy of Formation

The enthalpy of formation is calculated from that of KCl(g) using $\Delta_r H°(298.15\ K)$ = 45.0 $kcal\ mol^{-1}$ for $K_2Cl_2(g) \rightarrow 2\ KCl(g)$ as selected from data analyses summarized below. The three experimental techniques include PVT measurements (1), combination of total vapor pressure and transpiration data (2,3,4), and mass spectrometric studies of the variation of ion intensities with temperature (5). The results are reasonably consistent, considering the uncertainties inherent in the problem.

The resulting tables were tested by comparison with other data. Calculated total vapor pressures are in satisfactory agreement with the observed values over the range 820-1500 K; however, near the boiling point they are lower by 6-18 percent [see KCl(g)]. Knudsen effusion data and transpiration data over KCl(cr) have been reviewed by Pugh and Barrow (6). These data suggest negligible dimer formation when combined with the total vapor pressures of Pugh and Barrow (6) but significant dimer formation when combined with the lower (20-30 percent) total pressures of Dietz (7). Calculated dimer formation (22 mole percent at 900 K) agrees with the latter combination but calculated total pressures agree with those of Pugh and Barrow (6). Velocity distribution studies of Miller and Kusch (8) suggest about 10 mole percent of dimer at 900 K. Thus, while minor inconsistencies do exist, it is not apparent that they exceed the combined experimental errors.

Source	Method	Points	T/K	$\Delta_r H°(298.15\ K)$, $kcal\ mol^{-1}$ 2nd law	3rd law	Drift $cal\ K^{-1}\ mol^{-1}$
(1)	PVT data	15	1310-1404	46.70±2.1	46.01	-0.5±1.6
(2)	Trans. + Vap. Press.	10	1253-1352	46.24±3.5	44.98	-1.0±2.7
(3)	Trans. + Vap. Press.	5	1261-1372	43.0 ±4.5	44.16	0.7±3.4
(4)	Trans. + Vap. Press.	7	1250-1473	42.54±2.9	45.21	2.2±2.3
(5)	Mass spectrometry	-	774-1016	41.9±3	--	--

Heat Capacity and Entropy

The structure and vibrational frequencies are those calculated from an ionic model by Berkowitz (9). The sixth fundamental was arbitrarily lowered from 206 to 100 cm^{-1} since the resulting entropy increase (1.4 $cal\ K^{-1}\ mol^{-1}$ at 1300 K.) improved the overall agreement with the equilibrium data. Based on electron diffraction studies of the monomer-dimer vapor, Akishin and Rambidi (10) have derived the bond distance and angle to be 2.81 Å and 98°. Interpretation of the diffraction data, however, is complicated by the presence of only about 30 mole percent of dimer at the temperature of measurement. Adoption of this dimer structure would reduce the entropy by 0.23 $cal\ K^{-1}\ mol^{-1}$. Principal moments of inertia for the Berkowitz structure are I_A = 45.3554 x 10^{-39}, I_B = 60.0935 x 10^{-39}, and I_C = 105.4488 x 10^{-39} $g\ cm^2$.

Reference

1. S. Datz, ORNL-2933, Oak Ridge National Laboratory, 1960.
2. E. E. Schrier, Ph.D. Thesis, Rensselaer Polytechnic Inst., Troy, N.Y., June 1961.
3. C. Beusman, ORNL-2323, Oak Ridge National Laboratory, 1957.
4. J. L. Barton and H. Bloom, J. Phys, Chem. 63, 1785 (1959); 60, 1413 (1956).
5. T. A. Milne and H. M. Klein, AFOSR TN 60-512, Menlo Park, Calif., 1960.
6. A. C. P. Pugh and R. F. Barrow, Trans. Faraday Soc. 54, 671 (1958).
7. V. Dietz, J. Chem. Phys. 4, 575 (1936).
8. R. C. Miller and P. Kusch, J. Chem. Phys. 27, 981 (1957); 25, 860 (1956).
9. J. Berkowitz, J. Chem. Phys. 32, 1519 (1960); 29, 1358 (1958).
10. P. A. Akishin and N. G. Rambidi, Zhur. Neorg. Khim. 4, 718 (1959).

Potassium Chloride $((KCl)_2)$ — $Cl_2K_2(g)$

Enthalpy Reference Temperature = T_r = 298.15 K — Standard State Pressure = p° = 0.1 MPa

T/K	$C_p°$ ($J\ K^{-1}mol^{-1}$)	S°	$-[G°-H°(T_r)]/T$	$H°-H°(T_r)$ ($kJ\ mol^{-1}$)	$\Delta_f H°$	$\Delta_f G°$	Log K_f
0	0.	0.	INFINITE	-20.133	-614.430	-614.430	INFINITE
100	67.552	269.905	422.606	-15.270	-615.944	-616.913	322.242
200	78.301	321.028	360.199	-7.834	-616.734	-617.550	161.287
250	79.953	338.697	354.191	-3.874	-617.141	-617.708	129.063
298.15	80.865	352.864	352.864	0.	-617.642	-617.774	108.232
300	80.892	353.364	352.866	0.150	-617.665	-617.775	107.564
350	81.473	365.881	353.853	4.210	-623.105	-617.548	92.164
400	81.857	376.787	356.053	8.294	-623.947	-616.696	80.532
450	82.122	386.444	358.903	12.394	-624.751	-615.741	71.473
500	82.314	395.107	362.098	16.505	-625.523	-614.699	64.217
600	82.565	410.139	368.889	24.750	-626.989	-612.395	53.314
700	82.718	422.879	375.715	33.014	-628.388	-609.851	45.508
800	82.817	433.931	382.317	41.292	-629.764	-607.109	39.640
900	82.885	443.690	388.604	49.577	-631.164	-604.194	35.066
1000	82.934	452.425	394.557	57.868	-632.636	-601.119	31.399
1100	82.971	460.332	400.183	66.163	-792.128	-588.672	27.954
1200	82.998	467.552	405.501	74.462	-791.749	-570.192	24.820
1300	83.020	474.197	410.533	82.763	-791.377	-551.745	22.169
1400	83.037	480.350	415.303	91.066	-791.014	-533.325	19.899
1500	83.051	486.079	419.832	99.370	-790.660	-514.931	17.931
1600	83.062	491.439	424.142	107.676	-790.316	-496.560	16.211
1700	83.071	496.475	428.250	115.982	-789.982	-478.211	14.694
1800	83.079	501.224	432.174	124.290	-789.663	-459.881	13.345
1900	83.086	505.716	435.927	132.598	-789.360	-441.568	12.140
2000	83.092	509.978	439.524	140.907	-789.077	-423.270	11.055
2100	83.096	514.032	442.976	149.217	-788.819	-404.986	10.073
2200	83.101	517.898	446.295	157.526	-788.590	-386.714	9.182
2300	83.104	521.592	449.489	165.837	-788.398	-368.452	8.368
2400	83.108	525.129	452.567	174.147	-788.247	-350.197	7.622
2500	83.110	528.521	455.538	182.458	-788.144	-331.947	6.936
2600	83.113	531.781	458.408	190.769	-788.095	-313.700	6.302
2700	83.115	534.918	461.184	199.081	-788.109	-295.454	5.716
2800	83.117	537.941	463.872	207.392	-788.190	-277.206	5.171
2900	83.119	540.857	466.477	215.704	-788.347	-258.954	4.664
3000	83.121	543.675	469.003	224.016	-788.587	-240.695	4.191
3100	83.122	546.401	471.456	232.328	-788.915	-222.427	3.748
3200	83.124	549.040	473.840	240.641	-789.340	-204.147	3.332
3300	83.125	551.598	476.157	248.953	-789.849	-185.853	2.942
3400	83.126	554.079	478.413	257.266	-790.478	-167.541	2.574
3500	83.127	556.489	480.609	265.578	-791.216	-149.209	2.227
3600	83.128	558.831	482.750	273.891	-792.079	-130.854	1.899
3700	83.129	561.108	484.837	282.204	-793.069	-112.474	1.588
3800	83.130	563.325	486.873	290.517	-794.194	-94.065	1.293
3900	83.130	565.484	488.861	298.830	-795.459	-75.624	1.013
4000	83.131	567.589	490.803	307.143	-796.859	-57.150	0.746
4100	83.132	569.642	492.701	315.456	-798.415	-38.639	0.492
4200	83.132	571.645	494.557	323.769	-800.139	-20.087	0.250
4300	83.133	573.601	496.373	332.082	-802.032	-1.493	0.018
4400	83.133	575.512	498.150	340.396	-803.545	17.094	-0.203
4500	83.134	577.381	499.890	348.709	-805.656	35.768	-0.415
4600	83.134	579.208	501.594	357.022	-807.931	54.491	-0.619
4700	83.135	580.996	503.265	365.336	-810.331	73.261	-0.814
4800	83.135	582.746	504.902	373.649	-812.681	92.060	-1.002
4900	83.135	584.460	506.509	381.963	-815.404	110.937	-1.183
5000	83.136	586.140	508.085	390.276	-818.294	129.871	-1.357
5100	83.136	587.786	509.631	398.590	-821.350	148.864	-1.525
5200	83.136	589.400	511.150	406.904	-824.573	167.919	-1.687
5300	83.137	590.984	512.641	415.217	-827.962	187.038	-1.843
5400	83.137	592.538	514.106	423.531	-831.516	206.222	-1.995
5500	83.137	594.064	515.546	431.845	-835.233	225.473	-2.141
5600	83.137	595.562	516.962	440.158	-839.109	244.794	-2.283
5700	83.138	597.033	518.354	448.472	-842.452	264.089	-2.420
5800	83.138	598.479	519.723	456.786	-846.551	283.538	-2.554
5900	83.138	599.900	521.070	465.100	-850.789	303.058	-2.683
6000	83.138	601.298	522.395	473.413	-855.163	322.651	-2.809

PREVIOUS: March 1966 (1 atm) — CURRENT: March 1966 (1 bar)

Potassium Chloride $((KCl)_2)$ — $Cl_2K_2(g)$

LITHIUM CHLORIDE $((LiCl)_2)$ IDEAL GAS $M_r = 84.788$

$S°(298.15\ K) = [288.767]\ J\ K^{-1}\ mol^{-1}$

$\Delta_f H°(0\ K) = -595.61 \pm 12.6\ kJ\ mol^{-1}$
$\Delta_f H°(298.15\ K) = -598.55 \pm 12.6\ kJ\ mol^{-1}$

Vibrational Levels and Degeneracies

ν, cm^{-1}	ν, cm^{-1}
[330](1)	[356](1)
[296](1)	335 (1)
[385](1)	460 (1)

Ground State Quantum Weight: 1 $\sigma = 4$

Point Group: D_{2h}

Bond Distance: Li-Cl = 2.23 Å Li-Li = 2.62 Å

Bond Angle: Cl-Li-Cl = 108 ± 4°

Product of the Moments of Inertia $I_A I_B I_C = 6.41793 \times 10^{-114}\ g^3\ cm^6$

Enthalpy of Formation

An enthalpy of sublimation at 298.15 K of 52.1 kcal mol^{-1} was chosen to fit the vapor pressure data and relative concentrations of monomer, dimer and trimer as described in the table for LiCl(g). This value is in reasonable agreement with the value of 53.5 kcal mol^{-1} from a 3rd law analysis of the vapor pressure data of Nesmeyanov and Sazonov (1) and the relative concentrations of monomer, dimer, and trimer from the work of Miller and Kusch (2). Also a value of 54.1 kcal mol^{-1} was found by Milne and Klein (3) from mass spectrometer studies.

Heat Capacity and Entropy

Klemperer and Norris (4) observed two bands in the infrared spectrum and made tentative assignments. The remaining four frequencies were calculated by Berkowitz (5) from a reasonable model. The bond lengths and angles were determined from the electron diffraction experiments of Bauer et al. (6). The principal moments of inertia are: $I_A = 3.9605 \times 10^{-39}$, $I_B = 38.3234 \times 10^{-39}$, and $I_C = 42.2839 \times 10^{-39}\ g\ cm^2$.

References

1. An. N. Nesmeyanov and L. A. Sazonov, Zhur. Neorg. Khim. 4, 231 (1960).
2. R. C. Miller and P. Kusch, J. Chem. Phys. 25, 860 (1956).
3. T. A. Milne and H. M. Klein, J. Chem. Phys. 33, 1628 (1960).
4. W. Klemperer and W. G. Norris, J. Chem. Phys. 34, 1071 (1961).
5. J. Berkowitz, J. Chem. Phys. 32, 1519 (1960).
6. S. H. Bauer, T. Ino and R. F. Porter, J. Chem. Phys. 33, 685 (1960).

Lithium Chloride $((LiCl)_2)$ $Cl_2Li_2(g)$

Enthalpy Reference Temperature = T_r = 298.15 K Standard State Pressure = p° = 0.1 MPa

	J K⁻¹mol⁻¹			kJ mol⁻¹			
T/K	$C_p°$	S°	$-[G°-H°(T_r)]/T$	$H°-H°(T_r)$	$\Delta_f H°$	$\Delta_f G°$	Log K_f
0	0.	0.	INFINITE	-15.486	-595.607	-595.607	INFINITE
100	41.503	225.773	345.688	-11.991	-596.014	-598.223	312.479
200	62.653	261.707	295.121	-6.683	-597.427	-599.845	156.663
250	68.505	276.364	289.937	-3.393	-598.012	-600.380	125.443
298.15	72.195	288.767	288.767	0.	-598.546	-600.788	105.256
300	72.309	289.214	288.769	0.134	-598.567	-600.802	104.609
350	74.860	300.565	289.659	3.817	-599.128	-601.130	89.714
400	76.631	310.683	291.666	7.607	-599.765	-601.374	78.531
450	77.902	319.786	294.294	11.472	-600.522	-601.531	69.824
500	78.841	328.045	297.262	15.391	-607.417	-600.983	62.784
600	80.103	342.541	303.634	23.344	-609.063	-599.538	52.194
700	80.885	354.952	310.100	31.396	-610.533	-597.832	44.611
800	81.403	365.789	316.398	39.512	-611.909	-595.923	38.910
900	81.761	375.398	322.430	47.672	-613.252	-593.844	34.466
1000	82.020	384.027	328.166	55.861	-614.572	-591.617	30.903
1100	82.213	391.854	333.605	64.073	-615.872	-589.258	27.982
1200	82.360	399.014	338.762	72.302	-617.157	-586.782	25.542
1300	82.475	405.611	343.654	80.544	-618.431	-584.199	23.473
1400	82.567	411.726	348.300	88.797	-619.693	-581.519	21.697
1500	82.640	417.426	352.721	97.057	-620.939	-578.748	20.154
1600	82.701	422.761	356.933	105.324	-622.170	-575.895	18.801
1700	82.751	427.776	360.955	113.597	-913.861	-558.614	17.164
1800	82.794	432.507	364.800	121.874	-913.562	-537.725	15.604
1900	82.829	436.985	368.482	130.155	-913.271	-516.854	14.209
2000	82.860	441.234	372.014	138.440	-912.992	-495.997	12.954
2100	82.886	445.278	375.408	146.727	-912.726	-475.154	11.819
2200	82.909	449.134	378.672	155.017	-912.477	-454.323	10.787
2300	82.929	452.820	381.816	163.309	-912.250	-433.503	9.845
2400	82.947	456.350	384.849	171.603	-912.047	-412.692	8.962
2500	82.962	459.736	387.777	179.898	-911.874	-391.889	8.1[illegible]8
2600	82.976	462.990	390.607	188.195	-911.735	-371.093	7.455
2700	82.988	466.122	393.347	196.493	-911.635	-350.301	6.777
2800	82.999	469.140	396.000	204.793	-911.576	-329.512	6.147
2900	83.009	472.053	398.573	213.093	-911.565	-308.724	5.561
3000	83.018	474.867	401.069	221.394	-911.603	-287.936	5.013
3100	83.026	477.590	403.494	229.697	-911.695	-267.146	4.501
3200	83.033	480.226	405.851	238.000	-911.843	-246.352	4.021
3300	83.039	482.781	408.143	246.303	-912.049	-225.552	3.570
3400	83.046	485.260	410.375	254.607	-912.316	-204.745	3.146
3500	83.051	487.667	412.549	262.912	-912.646	-183.930	2.745
3600	83.056	490.007	414.669	271.218	-913.039	-163.104	2.367
3700	83.061	492.283	416.736	279.523	-913.496	-142.266	2.008
3800	83.065	494.498	418.753	287.830	-914.019	-121.416	1.669
3900	83.069	496.655	420.723	296.137	-914.607	-100.550	1.347
4000	83.073	498.759	422.648	304.444	-915.262	-79.668	1.040
4100	83.076	500.810	424.529	312.751	-915.983	-58.770	0.749
4200	83.079	502.812	426.369	321.059	-916.771	-37.853	0.471
4300	83.082	504.767	428.170	329.367	-917.627	-16.916	0.205
4400	83.085	506.677	429.933	337.675	-918.509	4.038	-0.048
4500	83.088	508.544	431.659	345.984	-919.481	25.015	-0.290
4600	83.090	510.370	433.350	354.293	-920.524	46.016	-0.523
4700	83.092	512.157	435.008	362.602	-921.634	67.039	-0.745
4800	83.095	513.907	436.634	370.911	-922.811	88.087	-0.959
4900	83.097	515.620	438.228	379.221	-924.057	109.160	-1.164
5000	83.098	517.299	439.793	387.531	-925.373	130.259	-1.361
5100	83.100	518.944	441.329	395.841	-926.748	151.385	-1.550
5200	83.102	520.558	442.837	404.151	-928.206	172.538	-1.733
5300	83.104	522.141	444.318	412.461	-929.737	193.721	-1.909
5400	83.105	523.695	445.774	420.771	-931.345	214.934	-2.079
5500	83.106	525.219	447.205	429.082	-933.032	236.176	-2.243
5600	83.108	526.717	448.611	437.393	-934.799	257.450	-2.401
5700	83.109	528.188	449.994	445.704	-936.650	278.757	-2.555
5800	83.110	529.633	451.355	454.015	-938.587	300.097	-2.703
5900	83.111	531.054	452.694	462.326	-940.612	321.471	-2.846
6000	83.112	532.451	454.011	470.637	-942.727	342.879	-2.985

PREVIOUS. June 1962 (1 atm) CURRENT: June 1962 (1 bar)

Lithium Chloride $((LiCl)_2)$ $Cl_2Li_2(g)$

MAGNESIUM CHLORIDE ($MgCl_2$) CRYSTAL M_r = 95.211

$S°(298.15\ K) = 89.629\ J\ K^{-1}\ mol^{-1}$

$T_{fus} = 987\ K$

$\Delta_f H°(0\ K) = -641.196 \pm 0.46\ kJ\ mol^{-1}$

$\Delta_f H°(298.15\ K) = -641.616 \pm 0.46\ kJ\ mol^{-1}$

$\Delta_{fus} H° = 43.095 \pm 0.21\ kJ\ mol^{-1}$

Enthalpy of Formation

The adopted $\Delta_f H°(298.15\ K) = -153.35 \pm 0.11\ kcal\ mol^{-1}$ is from enthalpy of solution measurements reported by Shomate and Huffman (1). Shomate and Huffman's enthalpy of solution of Mg in 1 M HCL ($\Delta_f H = -111.322 \pm 0.041\ kcal\ mol^{-1}$) has been substantiated for calibration purposes by Gunn and Cunningham (2) ($\Delta_f H = -111.3\ kcal\ mol^{-1}$) and by Westrum and Eyring (3) ($\Delta_f H = -111.27\ kcal\ mol^{-1}$). Auxiliary enthalpy of dilution and enthalpy of formation data for HCl were taken from Parker (4) and from Wagman et al. (5).

Heat Capacity and Entropy

High temperature heat content data by Moore (6) were joined by Shomate correlation with low temperature heat capacity data reported by Kelley and Moore. $C_p°(cr)$ above T_{fus} is a linear extrapolation from 700 K. $S°(298.15\ K)$ is derived from the low temperature data and is based on $S°(53.6\ K) = 2.006\ cal\ K^{-1}\ mol^{-1}$.

Fusion Data

$\Delta_{fus} H° = 10.30 \pm 0.05\ kcal\ mol^{-1}$ is taken from enthalpy measurements of Moore (6). T_{fus} = 987 K is from (8).

Sublimation Data

The value of $\Delta_{sub} H°(298.15\ K)$ was derived by 2nd and 3rd law analyses of vapor pressure data. See the ideal gas table for details.

References

1. C. H. Shomate and E. H. Huffman, J. Amer. Chem. Soc. 65, 1625 (1943).
2. S. R. Gunn and B. B. Cunningham, J. Amer. Chem. Soc. 79, 1563 (1957).
3. E. F. Westrum Jr. and L. Eyring, J. Amer. Chem. Soc. 74, 2045 (1952).
4. V. B. Parker, NSRDS-NBS 2, April, 1965.
5. D. D. Wagman et al., U. S. Nat. Bur. Stand. Technical Note 270-1, 1965.
6. U. S. Nat. Bur. Stand. Circ. 500, 1952.

Magnesium Chloride ($MgCl_2$) $Cl_2Mg_1(cr)$

Enthalpy Reference Temperature = T_r = 298.15 K — Standard State Pressure = p° = 0.1 MPa

T/K	$C_p°$ (J K⁻¹mol⁻¹)	S° (J K⁻¹mol⁻¹)	$-[G°-H°(T_r)]/T$ (J K⁻¹mol⁻¹)	$H°-H°(T_r)$ (kJ mol⁻¹)	$\Delta_f H°$ (kJ mol⁻¹)	$\Delta_f G°$ (kJ mol⁻¹)	Log K_f
0	0.	0.	INFINITE	-13.759	-641.196	-641.196	INFINITE
100	40.294	25.956	146.771	-12.081	-643.073	-625.817	326.893
200	63.575	62.664	95.923	-6.652	-642.688	-608.599	158.950
298.15	71.379	89.629	89.629	0.	-641.616	-592.087	103.731
300	71.496	90.071	89.630	0.132	-641.593	-591.780	103.038
400	75.709	111.276	92.493	7.513	-640.237	-575.376	75.136
500	78.145	128.451	98.022	15.215	-638.773	-559.328	58.433
600	79.873	142.857	104.326	23.119	-637.269	-543.580	47.323
700	81.253	155.274	110.737	31.176	-635.756	-528.085	39.406
800	82.550	166.209	117.001	39.366	-634.254	-512.806	33.483
900	83.847	176.007	123.022	47.686	-632.773	-497.714	28.887
987.000	84.977	183.795	128.040	55.030	------ CRYSTAL <--> LIQUID -----		
1000	85.144	184.908	128.773	56.135	-639.916	-482.073	25.181
1100	86.417	193.083	134.252	64.714	-638.519	-466.356	22.145
1200	87.662	200.656	139.474	73.418	-637.007	-450.771	19.622
1300	88.880	207.721	144.455	82.245	-635.383	-435.316	17.491
1400	90.071	214.351	149.213	91.193	-761.057	-416.823	15.552
1500	91.235	220.605	153.766	100.258	-757.861	-392.346	13.663
1600	92.372	226.530	158.131	109.439	-754.559	-368.086	12.017
1700	93.482	232.163	162.321	118.732	-751.153	-344.035	10.571
1800	94.565	237.538	166.352	128.135	-747.647	-320.187	9.292
1900	95.621	242.679	170.235	137.644	-744.043	-296.537	8.152
2000	96.650	247.610	173.981	147.258	-740.346	-273.080	7.132

PREVIOUS: December 1960 CURRENT: December 1965

Magnesium Chloride ($MgCl_2$) $Cl_2Mg_1(cr)$

MAGNESIUM CHLORIDE ($MgCl_2$) LIQUID M_r = 95.211

$S°(298.15\ K) = [129.480]\ J\ K^{-1}\ mol^{-1}$

$T_{fus} = 987\ K$

$\Delta_f H°(298.15\ K) = [-601.577]\ kJ\ mol^{-1}$

$\Delta_{fus} H° = 43.095 \pm 0.21\ kJ\ mol^{-1}$

Enthalpy of Formation

$\Delta_f H°(MgCl_2, \ell, 298.15\ K)$ is calculated from $\Delta_f H°(MgCl_2, cr, 298.15\ K)$ by adding $\Delta_{fus}H°$ and the difference in enthalpy, H°(987 K) - H°(298.15 K), between the crystal and liquid.

Heat Capacity and Entropy

A constant $C_p°(\ell) = 22.0\ cal\ K^{-1}\ mol^{-1}$ over the temperature range 1006-1428 K is from the high temperature enthalpy data of Moore (1). This constant value was assumed to hold from an assumed glass transition of 660 K to 3000 K. $C_p°(\ell)$ below 660 K is taken to be that of the crystal. The entropy is calculated in a manner analogous to that used for the enthalpy of formation.

Fusion Data

See the crystal table for details.

Vaporization Data

T_{vap} is calculated as the temperature at which the Gibbs energy change of the reaction $MgCl_2(\ell) = MgCl_2(g)$ is zero. The difference between $\Delta_f H°$ for $MgCl_2(\ell)$ and $MgCl_2(g)$ at T_{vap} is $\Delta_{vap}H°$.

Reference

1. G. E. Moore, J. Amer. Chem. Soc. 65, 1700 (1943).

Magnesium Chloride ($MgCl_2$) $Cl_2Mg_1(l)$

Enthalpy Reference Temperature = T_r = 298.15 K; Standard State Pressure = p° = 0.1 MPa

T/K	$C_p°$ (J K⁻¹mol⁻¹)	S° (J K⁻¹mol⁻¹)	$-[G°-H°(T_r)]/T$ (J K⁻¹mol⁻¹)	$H°-H°(T_r)$ (kJ mol⁻¹)	$\Delta_f H°$ (kJ mol⁻¹)	$\Delta_f G°$ (kJ mol⁻¹)	Log K_f
0							
100							
200							
298.15	71.379	129.480	129.480	0.	-601.577	-563.929	98.798
300	71.496	129.922	129.482	0.132	-601.553	-563.696	98.148
400	75.709	151.128	132.344	7.513	-600.197	-551.276	71.989
500	78.145	168.303	137.873	15.215	-598.733	-539.214	56.331
600	79.873	182.554	144.164	23.034	-597.314	-527.443	45.918
660.000	85.617	190.408	148.013	27.981	GLASS <--> LIQUID TRANSITION		
660.000	92.048	190.415	148.013	27.985			
700	92.048	195.831	150.592	31.667	-595.225	-515.943	38.500
800	92.048	208.122	157.032	40.872	-592.708	-504.791	32.959
900	92.048	218.964	163.323	50.077	-590.342	-493.945	28.668
987.000	92.048	227.458	168.608	58.085	CRYSTAL <--> LIQUID		
1000	92.048	228.662	169.381	59.282	-596.730	-482.642	25.211
1100	92.048	237.435	175.175	68.486	-594.706	-471.331	22.382
1200	92.048	245.444	180.702	77.691	-592.694	-460.204	20.032
1300	92.048	252.812	185.969	86.896	-590.692	-449.244	18.051
1400	92.048	259.634	190.990	96.101	-716.109	-435.271	16.240
1500	92.048	265.984	195.781	105.306	-712.774	-415.328	14.463
1600	92.048	271.925	200.356	114.510	-709.448	-395.607	12.915
1700	92.048	277.505	204.732	123.715	-706.130	-376.093	11.556
1800	92.048	282.767	208.922	132.920	-702.822	-356.775	10.353
1900	92.048	287.744	212.941	142.125	-699.523	-337.640	9.282
2000	92.048	292.465	216.800	151.330	-696.234	-318.679	8.323
2100	92.048	296.956	220.511	160.534	-692.957	-299.881	7.459
2200	92.048	301.238	224.084	169.739	-689.692	-281.240	6.677
2300	92.048	305.330	227.528	178.944	-686.441	-262.747	5.967
2400	92.048	309.247	230.852	188.149	-683.206	-244.395	5.319
2500	92.048	313.005	234.063	197.354	-679.988	-226.177	4.726

PREVIOUS: December 1960 CURRENT: December 1965

Magnesium Chloride ($MgCl_2$) $Cl_2Mg_1(l)$

MAGNESIUM CHLORIDE ($MgCl_2$) CRYSTAL–LIQUID M_r = 95.211

0 to 987 K crystal
above 987 K liquid

Refer to the individual tables for details.

Magnesium Chloride ($MgCl_2$) Cl_2Mg_1(cr,l)

Enthalpy Reference Temperature = T_r = 298.15 K Standard State Pressure = p° = 0.1 MPa

T/K	C_p° (J K⁻¹mol⁻¹)	S° (J K⁻¹mol⁻¹)	$-[G^\circ-H^\circ(T_r)]/T$ (J K⁻¹mol⁻¹)	$H^\circ-H^\circ(T_r)$ (kJ mol⁻¹)	Δ_fH° (kJ mol⁻¹)	Δ_fG° (kJ mol⁻¹)	Log K_f
0	0.	0.	INFINITE	-13.759	-641.196	-641.196	INFINITE
100	40.294	25.956	146.771	-12.081	-643.073	-625.817	326.893
200	63.575	62.664	95.923	-6.652	-642.688	-608.599	158.950
298.15	71.379	89.629	89.629	0.	-641.616	-592.087	103.731
300	71.496	90.071	89.630	0.132	-641.593	-591.780	103.038
400	75.709	111.276	92.493	7.513	-640.237	-575.376	75.136
500	78.145	128.451	98.022	15.215	-638.773	-559.328	58.433
600	79.873	142.857	104.326	23.119	-637.269	-543.580	47.323
700	81.253	155.274	110.737	31.176	-635.756	-528.085	39.406
800	82.550	166.209	117.001	39.366	-634.254	-512.806	33.483
900	83.847	176.007	123.022	47.686	-632.773	-497.714	28.887
987.000	84.977	183.795	128.040	55.030	CRYSTAL <--> LIQUID		
987.000	92.048	227.458	128.040	98.125	TRANSITION		
1000	92.048	228.662	129.341	99.321	-596.730	-482.642	25.211
1100	92.048	237.435	138.775	108.526	-594.706	-471.331	22.382
1200	92.048	245.444	147.335	117.731	-592.694	-460.204	20.032
1300	92.048	252.812	155.169	126.936	-590.692	-449.244	18.051
1400	92.048	259.634	162.390	136.141	-716.109	-435.271	16.240
1500	92.048	265.984	169.087	145.345	-712.774	-415.328	14.463
1600	92.048	271.925	175.331	154.550	-709.448	-395.607	12.915
1700	92.048	277.505	181.179	163.755	-706.130	-376.093	11.556
1800	92.048	282.767	186.678	172.960	-702.822	-356.775	10.353
1900	92.048	287.744	191.867	182.165	-699.523	-337.640	9.282
2000	92.048	292.465	196.780	191.369	-696.234	-318.679	8.323
2100	92.048	296.956	201.444	200.574	-692.957	-299.881	7.459
2200	92.048	301.238	205.884	209.779	-689.692	-281.240	6.677
2300	92.048	305.330	210.119	218.984	-686.441	-262.747	5.967
2400	92.048	309.247	214.169	228.189	-683.206	-244.395	5.319
2500	92.048	313.005	218.048	237.393	-679.988	-226.177	4.726

PREVIOUS: CURRENT: December 1965

Magnesium Chloride ($MgCl_2$) Cl_2Mg_1(cr,l)

MAGNESIUM CHLORIDE ($MgCl_2$) IDEAL GAS M_r = 95.211

$\Delta_f H°(0\ K) = -392.57 \pm 2.1\ kJ\ mol^{-1}$

$S°(298.15\ K) = 277.027 \pm 2.1\ J\ K^{-1}\ mol^{-1}$ $\Delta_f H°(298.15\ K) = -392.46 \pm 2.1\ kJ\ mol^{-1}$

Vibrational Frequencies and Degeneracies

ν, cm^{-1}
590 (1)
88 (2)
[273](1)

Ground State Quantum Weight: 1 $\sigma = 2$

Point Group: $D_{\infty h}$

Bond Distance: Mg–Cl = 2.18 Å

Bond Angle: Cl–Mg–Cl = 180°

Rotational Constant: $B_0 = 0.050027\ cm^{-1}$

Enthalpy of Formation

Analysis of vapor pressure data on $MgCl_2$ is clouded by conflicting evidence for the amount of dimer in the vapor. Berkowitz and Marquart (1) by mass spectrometer studies found about 2% dimer at 920 K and derived $\Delta_{dim}H = -39.2\ kcal\ mol^{-1}$ and $\Delta_{dim}S = -29.9\ cal\ K^{-1}\ mol^{-1}$. Schrier and Clark (2) from gas transpiration data calculated about 30% dimer in the saturated vapor at 1300 K with $\Delta_{dim}H = -32.0\ kcal\ mol^{-1}$ and $\Delta_{dim}S = -18.8\ cal\ K^{-1}\ mol^{-1}$. Thermodynamic functions for the gaseous monomer adopted here are clearly not compatible with large amounts of dimer and it is assumed for the present that the mass spectrometer results are correct.

Several vapor pressure studies have been published and third law analyses are given below. Berkowitz and Marquart (1) gave a single point while the remaining publications gave only equations but not individual points. Total pressures calculated from the equations were corrected for dimer.

Source	Method	T/K	$\Delta_{sub}H°(298.15\ K)$ kcal mol^{-1} Third Law
Berkowitz and Marquart (1)	Mass spectrometer	920	58.10
Schrier and Clark (2)	Boiling Point	1208–1413	59.55–59.68
Hildenbrand et al. (3)	Torsion-effusion	802–985	59.55–59.19
Hildenbrand et al. (4)	Torsion-effusion	800–970	59.53–59.38
Fischer et al. (5)	"Bell" method (total vapor pressures)	1136–1435	59.30–59.56

The selected value for $\Delta_{sub}H°(298.15\ K)$ is 59.55 kcal mol^{-1} which is combined with $\Delta_f H°(298.15\ K)$ of the crystal to give $\Delta_f H°(g,\ 298.15\ K) = -93.80\ kcal\ mol^{-1}$.

Heat Capacity and Entropy

White, Mann, and co-workers (6) by matrix infrared spectroscopy determined the molecule to be linear with the asymmetric stretching frequency at 590 cm^{-1} and the doubly degenerate bending frequency at 88 cm^{-1}. The symmetric stretching frequency was estimated as 273 cm^{-1}. The asymmetric stretching was observed by Buchler and Klemperer (7) at 597 cm^{-1} and by Randall et al. (8) at 598 cm^{-1}, both by infrared spectroscopy of the vapor at high temperature. Electric deflection experiments by Klemperer and co-workers as summarized by Hayes (9) and the electron diffraction data of Akishin and Spiridonov (10) agree with the linear structure within their error limits of about 20. The Mg-Cl bond length was determined as 2.18 Å by Akishin and Spiridonov (10).

References

1. J. Berkowitz and J. R. Marquart, J. Chem. Phys. 37, 1853 (1962).
2. E. E. Schrier and H. M. Clark, J. Phys. Chem. 67, 1259 (1963).
3. D. L. Hildenbrand, W. F. Hall, F. Ju, and N. D. Potter, J. Chem. Phys. 40, 2882 (1964).
4. D. L. Hildenbrand et al., Aeronutronic Report U-3183, June, 1965.
5. W. Fischer, T. Petzel, and S. Lauter, Z. Anorg. Allg. Chem. 333, 226 (1964).
6. Unpublished work cited by D. L. Hildenbrand, J. Chem. Phys., in press.
7. A. Buchler and W. Klemperer, J. Chem. Phys. 29, 121 (1958).
8. S. P. Randall, F. T. Greene, and J. L. Margrave, J. Phys. Chem. 63, 758 (1959).
9. E. F. Hayes, J. Phys. Chem. 70, 3740 (1966).
10. P. A. Akishin and V. P. Spiridonov, Kristallografiya 2, 475 (1957).

Magnesium Chloride ($MgCl_2$) $Cl_2Mg_1(g)$

Enthalpy Reference Temperature = T_r = 298.15 K Standard State Pressure = p° = 0.1 MPa

T/K	$C_p°$ (J $K^{-1}mol^{-1}$)	S° (J $K^{-1}mol^{-1}$)	$-[G°-H°(T_r)]/T$ (J $K^{-1}mol^{-1}$)	$H°-H°(T_r)$ (kJ mol^{-1})	$\Delta_f H°$ (kJ mol^{-1})	$\Delta_f G°$ (kJ mol^{-1})	Log K_f
0	0.	0.	INFINITE	-14.285	-392.565	-392.565	INFINITE
100	46.426	220.299	325.112	-10.481	-392.316	-394.494	206.063
200	53.484	254.921	282.150	-5.446	-392.325	-396.688	103.604
250	55.615	267.097	277.959	-2.715	-392.384	-397.772	83.110
298.15	57.097	277.027	277.027	0.	-392.459	-398.803	69.869
300	57.145	277.380	277.028	0.106	-392.462	-398.842	69.445
350	58.249	286.277	277.728	2.992	-392.557	-399.898	59.682
400	59.057	294.110	279.296	5.926	-392.668	-400.940	52.357
450	59.658	301.102	281.337	8.894	-392.796	-401.966	46.659
500	60.116	307.413	283.635	11.889	-392.942	-402.977	42.099
600	60.747	318.434	288.542	17.935	-393.296	-404.953	35.254
700	61.150	327.830	293.500	24.031	-393.744	-406.862	30.360
800	61.420	336.014	298.314	30.160	-394.302	-408.699	26.685
900	61.610	343.260	302.913	36.312	-394.989	-410.459	23.822
1000	61.748	349.759	307.278	42.481	-404.413	-411.422	21.490
1100	61.851	355.649	311.412	48.661	-405.414	-412.074	19.568
1200	61.931	361.035	315.326	54.850	-406.418	-412.636	17.962
1300	61.993	365.994	319.035	61.046	-407.424	-413.113	16.599
1400	62.042	370.590	322.556	67.248	-535.845	-410.346	15.310
1500	62.082	374.872	325.902	73.455	-535.508	-401.393	13.978
1600	62.115	378.880	329.090	79.665	-535.176	-392.463	12.813
1700	62.143	382.647	332.130	85.878	-534.850	-383.553	11.785
1800	62.166	386.199	335.036	92.093	-534.531	-374.663	10.872
1900	62.185	389.561	337.819	98.311	-534.220	-365.790	10.056
2000	62.202	392.751	340.486	104.530	-533.916	-356.933	9.322
2100	62.217	395.786	343.048	110.751	-533.623	-348.091	8.658
2200	62.229	398.681	345.511	116.973	-533.341	-339.263	8.055
2300	62.240	401.447	347.884	123.197	-533.071	-330.447	7.505
2400	62.250	404.096	350.171	129.421	-532.816	-321.643	7.000
2500	62.258	406.638	352.379	135.647	-532.577	-312.849	6.537
2600	62.266	409.080	354.513	141.873	-532.357	-304.064	6.109
2700	62.272	411.430	356.578	148.100	-532.156	-295.288	5.713
2800	62.278	413.695	358.578	154.327	-531.976	-286.518	5.345
2900	62.284	415.880	360.516	160.555	-531.820	-277.755	5.003
3000	62.288	417.992	362.397	166.784	-531.687	-268.996	4.684
3100	62.293	420.034	364.224	173.013	-531.580	-260.242	4.385
3200	62.297	422.012	365.999	179.242	-531.499	-251.490	4.105
3300	62.301	423.929	367.725	185.472	-531.445	-242.741	3.842
3400	62.304	425.789	369.406	191.703	-531.419	-233.993	3.595
3500	62.307	427.595	371.043	197.933	-531.421	-225.245	3.362
3600	62.310	429.350	372.638	204.164	-531.450	-216.496	3.141
3700	62.312	431.058	374.194	210.395	-531.508	-207.747	2.933
3800	62.315	432.719	375.712	216.626	-531.593	-198.996	2.735
3900	62.317	434.338	377.195	222.858	-531.707	-190.242	2.548
4000	62.319	435.916	378.643	229.090	-531.847	-181.484	2.370
4100	62.321	437.455	380.059	235.322	-532.015	-172.723	2.201
4200	62.323	438.956	381.444	241.554	-532.211	-163.958	2.039
4300	62.324	440.423	382.798	247.786	-532.432	-155.187	1.885
4400	62.326	441.856	384.124	254.019	-532.681	-146.411	1.738
4500	62.327	443.256	385.423	260.251	-532.955	-137.630	1.598
4600	62.328	444.626	386.695	266.484	-533.255	-128.841	1.463
4700	62.330	445.967	387.942	272.717	-533.582	-120.046	1.334
4800	62.331	447.279	389.164	278.950	-533.935	-111.244	1.211
4900	62.332	448.564	390.364	285.183	-534.313	-102.434	1.092
5000	62.333	449.823	391.540	291.416	-534.718	-93.616	0.978
5100	62.334	451.058	392.695	297.650	-535.149	-84.790	0.868
5200	62.335	452.268	393.829	303.883	-535.608	-75.955	0.763
5300	62.336	453.456	394.943	310.117	-536.092	-67.111	0.661
5400	62.337	454.621	396.037	316.350	-536.605	-58.257	0.564
5500	62.337	455.765	397.113	322.584	-537.144	-49.394	0.469
5600	62.338	456.888	398.170	328.818	-537.713	-40.521	0.378
5700	62.339	457.991	399.210	335.052	-538.311	-31.637	0.290
5800	62.339	459.075	400.233	341.286	-538.939	-22.743	0.205
5900	62.340	460.141	401.239	347.520	-539.597	-13.837	0.123
6000	62.341	461.189	402.230	353.754	-540.287	-4.920	0.043

PREVIOUS: December 1969 (1 atm) CURRENT: December 1969 (1 bar)

Magnesium Chloride ($MgCl_2$) $Cl_2Mg_1(g)$

MOLYBDENUM CHLORIDE OXIDE (MoO_2Cl_2) IDEAL GAS M_r = 198.8448

$\Delta_fH°(0\ K) = -628.99 \pm 14.6\ kJ\ mol^{-1}$

$S°(298.15\ K) = 337.787 \pm 5.44\ J\ K^{-1}\ mol^{-1}$ $\Delta_fH°(298.15\ K) = -633.04 \pm 14.6\ kJ\ mol^{-1}$

Vibrational Frequencies and Degeneracies

ν, cm^{-1}	ν, cm^{-1}	ν, cm^{-1}
972 (1)	134 (1)	156 (1)
453 (1)	233 (1)	437 (1)
347 (1)	996 (1)	287 (1)

Ground State Quantum Weight: 1 σ = 2

Point Group: C_{2v}

Bond Distances: Mo–O = 1.75 ± 0.10 Å Mo–Cl = 2.28 ± 0.03 Å

Bond Angles: O–Mo–O = 109.5° Cl–Mo–Cl = 113°

O–Mo–O plane perpendicular to Cl–Mo–Cl plane

Product of Moments of Inertia: $I_AI_BI_C = 1.09994 \times 10^{-112}\ g^3\ cm^6$

Enthalpy of Formation

Graham and Hepler (1) determined calorimetrically $\Delta_rH° = -65.3 \pm 0.5\ kcal\ mol^{-1}$ for the reaction:

$$MoO_2Cl_2(cr) + 4\ NaOH(\infty H_2O) + Na_2MoO_4(\infty H_2O) + 2\ NaCl(\infty H_2O) + 2\ H_2O(\ell).$$

Combining this result with the following enthalpy of formation data; $\Delta_fH°(Na_2MoO_4, \infty H_2O) = -353\pm0.4\ kcal\ mol^{-1}$ (2); $\Delta_fH°(NaCl, \infty H_2O) = -97.33\pm0.09\ kcal\ mol^{-1}$ (3, 4); $\Delta_fH°(NaOH, \infty H_2O) = -112.46\pm0.1\ kcal\ mol^{-1}$ (3, 4); $\Delta_fH°(H_2O, \ell) = -68.315\ kcal\ mol^{-1}$ (5); we derive $\Delta_fH°(MoO_2Cl_2, cr) = -169.6\pm1.0\ kcal\ mol^{-1}$.

Shchukarev et al. (6) measured the heats of solution of MoO_3(cr), MoO_2Cl_2(cr) and NaCl(cr) in NaOH(aq) and derived $\Delta_rH° = -45.56\pm0.5\ kcal\ mol^{-1}$ for the reaction $MoO_2Cl_2(cr) + 2\ NaOH{\cdot}77\ H_2O + MoO_3(cr) + 2\ NaCl(cr) + H_2O(\ell)$. Using JANAF auxiliary data (3, 4), we obtain $\Delta_fH°(MoO_2Cl_2, cr, 298.15\ K) = -172.86\pm1.0\ kcal\ mol^{-1}$.

The two investigations differ only in the heat of solution of MoO_2Cl_2(cr), both samples were better than 99.9% pure, and multiple determinations on two different samples were employed in each case. We adopt an average of the two values, -171.2±1.7 kcal mol^{-1} for $\Delta_fH°(MoO_2Cl_2$, cr, 298.15 K). Shchukarev et al. (8) reported an equation for the vapor pressure over the solid from which we have calcualted a 2nd law $\Delta_{sub}H°(298.15\ K) = 19.9\ kcal\ mol^{-1}$ with an estimated uncertainty of ±2 kcal mol^{-1}, a ΔCp correction of -4.1 cal K^{-1} mol^{-1} was used (3). Thus, our adopted value is $\Delta_fH°(MoO_2Cl_2, g, 298.15\ K) = -151.3\pm3.5\ kcal\ mol^{-1}$.

Shchukarev et al. (7) reported equilibrium constants for the reaction $MoO_2Cl_2(g) + 0.5\ O_2(g) = MoO_3(cr) + Cl_2(g)$. Shchukarev et al. (8) and Hultgren and Brewer (9) reported equilibrium data for the reaction $MoO_3(cr) + 2\ HCl(g) = MoO_2Cl_2(g) + H_2O(g)$. 2nd and 3rd law analysis of these equilibrium data give the following results:

Source	Data Points	$\Delta_rH°(298.15\ K)$, kcal mol^{-1} 2nd law	3rd law	Drift cal K^{-1} mol^{-1}	$-\Delta_fH°(298.15\ K)$ kcal mol^{-1}
(7)	7	-15.6	-30.11±2.0	-15.0	148.0 ± 3.0
(8)	Eqn	12.1	13.8 ±1.0	2.95	150.6 ± 1.2
(9) Set I	5	27.2	15.2 ±1.7	-21.37	149.2 ± 2.0
Set II	5	26.5	15.4 ±3.2	-21.75	149.4 ± 3.5

These equilibrium data are in agreement with the adopted enthalpy of formation but are suspect because of the significant drifts and because the range of validity of the equation (8) is estimated.

Heat Capacity and Entropy

The molecular structural data are from the compilation of Sutton (10). Infrared data (11, 12, 13) for vapor phase MoO_2Cl_2 have indicated that the molecule possesses a distorted tetrahedral structure by reason of the similarity of its spectra to that of CrO_2Cl_2. Principal moments of inertia are $I_A = 32.2143 \times 10^{-39}$, $I_B = 53.4145 \times 10^{-39}$, and $I_C = 63.9238 \times 10^{-39}\ g\ cm^2$.

The vibrational frequencies 972, 996, 453, and 437 have been observed in the infrared vapor phase spectra of MoO_2Cl_2 by Barraclough and Stals (11). These frequencies were assigned to the symmetric and asymmetric Mo–O and Mo–Cl stretches, respectively. Similar infrared data for these four frequencies have been reported by Iorns and Stafford (12) and Ward and Stafford (13). The rest of the frequencies are from the infrared and Raman studies of crystalline MoO_2Cl_2 by Adams and Churchill (14). Assignments are made by comparison with data reported for $CrO_2Cl_2(\ell)$ by Miller et al. (15).

References

1. R. L. Graham and L. G. Hepler, J. Phys. Chem. 63, 723 (1959).
2. R. L. Graham and L. G. Hepler, J. Amer. Chem. Soc. 79, 4045 (1957). Value based upon JANAF values for MoO_3 and NaOH.
3. JANAF Thermochemical Tables: NaCl(cr), 9-30-64; NaOH(cr), 3-31-66; MoO2Cl2(cr), 3-31-67.
4. V. B. Parker, NSRDS-NBS 2, 1965.
5. U. S. Nat. Bur. Stand. Tech. Note 270-3, 1968.
6. S. A. Shchukarev, I. V. Vasil'kova and B. N. Sharupin, Vestnik Leningrad Univ. 14, No. 4, Ser. Fiz. Khim. No. 1, 73 (1959).
7. S. A. Shchukarev, I. V. Vasil'kova and B. N. Sharupin, Vestnik Leningrad Univ. 16, No. 4, Ser. Fiz. Khim. No. 1, 87 (1961).
8. S. A. Shchukarev, I. V. Vasil'kova and B. N. Sharupin, Vestnik Leningrad Univ. 16, No. 22, Ser. Fiz. Khim. No. 4, 130 (1961).
9. N. Hultgren and L. Brewer, J. Phys. Chem. 60, 947 (1956).
10. L. E. Sutton, "Tables of Interatomic Distances and Configuration in Molecules and Ions", Special Publication No. 11, The Chemical Society, London, 1958.
11. C. G. Barraclough and J. Stals, Aust. J. Chem. 19, 741 (1966).
12. T. V. Iorns and F. E. Stafford, J. Amer. Chem. Soc. 88, 4819 (1966).
13. B. G. Ward and F. E. Stafford, J. Amer. Chem. 7, 2569 (1968).
14. D. M. Adams and R. G. Churchill, J. Chem. Soc., 2310 (1968).
15. F. A. Miller, G. L. Carlson, and W. B. White, Spectrochim. Acta, 1959, 709 (1959).

Molybdenum Chloride Oxide (MoO_2Cl_2) $Cl_2Mo_1O_2(g)$

Enthalpy Reference Temperature = T_r = 298.15 K Standard State Pressure = p° = 0.1 MPa

T/K	$C_p°$ (J $K^{-1}mol^{-1}$)	S°	$-[G°-H°(T_r)]/T$	$H°-H°(T_r)$ (kJ mol^{-1})	$\Delta_fH°$	$\Delta_fG°$	Log K_f
0	0.	0.	INFINITE	-18.400	-628.990	-628.990	INFINITE
100	53.401	261.666	405.432	-14.377	-631.275	-620.507	324.120
200	74.277	305.899	345.270	-7.874	-632.563	-609.158	159.096
250	80.563	323.183	339.165	-3.996	-632.879	-603.266	126.046
298.15	85.220	337.787	337.787	0.	-633.039	-597.546	104.688
300	85.378	338.315	337.789	0.158	-633.043	-597.326	104.004
350	89.169	351.772	338.843	4.525	-633.085	-591.368	88.257
400	92.191	363.883	341.228	9.062	-633.036	-585.411	76.447
450	94.615	374.887	344.366	13.734	-632.921	-579.465	67.262
500	96.572	384.960	347.929	18.516	-632.761	-573.533	59.917
600	99.467	402.841	355.629	28.327	-632.357	-561.724	48.902
700	101.440	418.331	363.505	38.378	-631.905	-549.987	41.041
800	102.829	431.973	371.228	48.596	-631.446	-538.316	35.148
900	103.835	444.145	378.666	58.932	-631.002	-526.702	30.569
1000	104.585	455.126	385.772	69.354	-630.591	-515.136	26.908
1100	105.156	465.122	392.538	79.843	-630.224	-503.608	23.914
1200	105.600	474.291	398.974	90.381	-629.915	-492.112	21.421
1300	105.952	482.758	405.097	100.960	-629.673	-480.638	19.312
1400	106.235	490.621	410.929	111.569	-629.508	-469.181	17.505
1500	106.465	497.959	416.489	122.205	-629.430	-457.732	15.940
1600	106.656	504.836	421.798	132.861	-629.448	-446.286	14.570
1700	106.815	511.307	426.875	143.535	-629.570	-434.835	13.361
1800	106.949	517.416	431.737	154.223	-629.806	-423.374	12.286
1900	107.064	523.202	436.400	164.924	-630.167	-411.896	11.324
2000	107.161	528.696	440.878	175.635	-630.661	-400.396	10.457
2100	107.246	533.926	445.186	186.356	-631.302	-388.868	9.673
2200	107.320	538.917	449.334	197.084	-632.098	-377.305	8.958
2300	107.384	543.689	453.333	207.820	-633.062	-365.703	8.305
2400	107.441	548.261	457.194	218.561	-634.206	-354.055	7.706
2500	107.491	552.648	460.925	229.307	-635.542	-342.355	7.153
2600	107.535	556.864	464.534	240.059	-637.094	-330.598	6.642
2700	107.575	560.924	468.029	250.814	-638.903	-318.776	6.167
2800	107.610	564.837	471.417	261.574	-641.032	-306.881	5.725
2900	107.643	568.613	474.704	272.336	-679.446	-294.855	5.311
3000	107.671	572.263	477.896	283.102	-680.420	-281.577	4.903
3100	107.698	575.794	480.997	293.871	-681.430	-268.265	4.520
3200	107.721	579.214	484.013	304.642	-682.475	-254.921	4.161
3300	107.743	582.529	486.949	315.415	-683.552	-241.543	3.823
3400	107.763	585.745	489.807	326.190	-684.660	-228.132	3.505
3500	107.781	588.870	492.593	336.967	-685.798	-214.689	3.204
3600	107.798	591.906	495.310	347.746	-686.963	-201.212	2.920
3700	107.813	594.860	497.961	358.527	-688.152	-187.703	2.650
3800	107.827	597.735	500.549	369.309	-689.363	-174.160	2.394
3900	107.840	600.536	503.077	380.092	-690.593	-160.586	2.151
4000	107.852	603.267	505.547	390.877	-691.839	-146.981	1.919
4100	107.864	605.930	507.963	401.663	-693.098	-133.344	1.699
4200	107.874	608.529	510.327	412.450	-694.367	-119.676	1.488
4300	107.884	611.068	512.640	423.237	-695.644	-105.977	1.287
4400	107.893	613.548	514.906	434.026	-696.926	-92.250	1.095
4500	107.902	615.973	517.125	444.816	-698.210	-78.493	0.911
4600	107.910	618.345	519.300	455.607	-699.495	-64.708	0.735
4700	107.917	620.665	521.432	466.398	-700.777	-50.894	0.566
4800	107.924	622.937	523.523	477.190	-702.056	-37.053	0.403
4900	107.931	625.163	525.574	487.983	-703.329	-23.185	0.247
5000	107.937	627.343	527.588	498.776	-1295.980	-3.561	0.037
5100	107.943	629.481	529.565	509.570	-1298.917	22.317	-0.229
5200	107.948	631.577	531.507	520.365	-1301.991	48.252	-0.485
5300	107.953	633.633	533.414	531.160	-1305.199	74.250	-0.732
5400	107.958	635.651	535.289	541.955	-1308.540	100.308	-0.970
5500	107.963	637.632	537.132	552.751	-1312.010	126.429	-1.201
5600	107.967	639.577	538.944	563.548	-1315.607	152.615	-1.424
5700	107.971	641.488	540.726	574.345	-1319.328	178.866	-1.639
5800	107.975	643.366	542.480	585.142	-1323.171	205.184	-1.848
5900	107.979	645.212	544.205	595.940	-1327.131	231.567	-2.050
6000	107.983	647.027	545.904	606.738	-1331.210	258.020	-2.246

PREVIOUS: June 1970 (1 atm) CURRENT: June 1970 (1 bar)

Molybdenum Chloride Oxide (MoO_2Cl_2) $Cl_2Mo_1O_2(g)$

SODIUM CHLORIDE $((NaCl)_2)$ IDEAL GAS M_r = 116.88554

$S°(298.15\ K) = [325.432]\ J\ K^{-1}\ mol^{-1}$

$\Delta_f H°(0\ K) = -562.47 \pm 8.4\ kJ\ mol^{-1}$
$\Delta_f H°(298.15\ K) = -566.10 \pm 8.4\ kJ\ mol^{-1}$

Vibrational Frequencies and Degeneracies

ν, cm^{-1}	ν, cm^{-1}
[265](1)	[155](1)
[159](1)	[222](1)
[226](1)	[260](1)

Ground State Quantum Weight = 1 $\sigma = 4$
Point Group: D_{2h}
Bond Distance: Cl-Na = [2.624] Å
Bond Angle: Cl-Na-Cl = [108.4]°
Product of the Moments of Inertia: $I_A I_B I_C = [6.84195 \times 10^{-113}]\ g^3\ cm^6$

Enthalpy of Formation

An analysis is given below of the more extensive measurements relating to the dimer dissociation, $(NaCl)_2(g) \rightarrow 2\ NaCl(g)$. The mass spectrometric studies suggest that higher polymeric species may be neglected. Velocity distribution analyses are consistent with this procedure. When reduced to 298.15 K, the observed heats of dissociation of the dimer show a range of more than 7 kcal mol^{-1}, reflecting the considerable uncertainty involved in such measurements. A value of $\Delta_r H°(298.15\ K) = 48.6$ kcal mol^{-1} was selected for the dissociation; combining this with $\Delta_{sub}H°(298.15\ K) = 54.9$ kcal mol^{-1} for sublimation to the monomer leads to $\Delta_{sub}H°(298.15\ K) = 61.2$ kcal mol^{-1} for sublimation to the dimer.

Source	Method	T/K	$\Delta_r H°(T)$ kcal mol^{-1}	$\Delta_r H°(298.15\ K)$ kcal mol^{-1} 2nd Law	3rd Law
(1)	Molecular weight from vapor pressures	1350	51.0	52.8	52.0
(2)	Molecular weight from P-V-T measurements	1337	50.6	52.4	54.1
(3)	Double oven effusion with mass spectrometer	1002	47.4	48.6	--
(4)	Knudsen effusion with mass spectrometer	870	47.3	48.3	--
(5)	Knudsen effusion with mass spectrometer	930	44.5	45.6	--
(6)	Velocity distribution analysis	940	49.8	50.9	--
(7)	Velocity distribution analysis	980	46.6	47.8	--

Heat Capacity and Entropy

The structure and vibrational frequencies are those calculated from an ionic model by Berkowitz (8). Electron diffraction studies by Akishin and Rambidi (9) gave a similar bond angle (108°) and a slightly shorter bond distance (2.50 Å); however, interpretation of the diffraction pattern was complicated by uncertainty in the monomer-dimer ratio for the vapor. Principal moments of inertia for the Berkowitz structure are $I_A = 17.9886 \times 10^{-39}$, $I_B = 53.3306 \times 10^{-39}$ and $I_C = 71.3192 \times 10^{-39}\ g\ cm^2$.

References

1. J. L. Barton and H. Bloom, J. Phys. Chem. 63, 1785 (1959).
2. S. Datz, W. T. Smith and E. H. Taylor, J. Chem. Phys. 34, 558 (1961).
3. P. A. Akishin, L. N. Gorokhov and L. N. Sidorov, Zhur. Fiz. Khim. 33, 2822 (1959); Vestn. Mosk. Univ. Ser. Mat. Mekh. Astron. Fiz. Khim. 1959, No. 6, 194 (1959).
4. L. N. Gorokhov, Yu. S. Khodeev, P. A. Akishin, Zhur. Neorg. Khim. 3, 2597 (1958).
5. T. A. Milne and H. M. Klein, J. Chem. Phys. 33, 1628 (1960).
6. M. Eisenstadt, V. S. Rao and G. M. Rothberg, J. Chem. Phys. 30, 604 (1959).
7. R. C. Miller and P. Kusch, J. Chem. Phys. 27, 981 (1957); 25, 860 (1956).
8. J. Berkowitz, J. Chem. Phys. 32, 1519 (1960), 29, 1386 (1958).
9. P. A. Akishin and Rambidi, Z. Physik. Chem. (Leipzig) 213, 111 (1960).

Sodium Chloride $((NaCl)_2)$ $Cl_2Na_2(g)$

Enthalpy Reference Temperature = T_r = 298.15 K   Standard State Pressure = p° = 0.1 MPa

T/K	$C_p°$ (J $K^{-1}mol^{-1}$)	S°	$-[G°-H°(T_r)]/T$	$H°-H°(T_r)$ (kJ mol^{-1})	$\Delta_f H°$	$\Delta_f G°$	Log K_f
0	0.	0.	INFINITE	-18.446	-562.467	-562.467	INFINITE
100	57.464	248.550	391.674	-14.312	-563.924	-565.116	295.186
200	74.137	294.802	332.524	-7.544	-565.105	-565.794	147.770
250	77.102	311.696	326.720	-3.756	-565.585	-565.911	118.241
298.15	78.786	325.432	325.432	0.	-566.095	-565.929	99.148
300	78.836	325.919	325.433	0.146	-566.116	-565.928	98.537
350	79.928	338.160	326.397	4.117	-566.770	-565.847	84.448
400	80.656	348.883	328.552	8.133	-572.843	-565.251	73.814
450	81.163	358.414	331.350	12.179	-573.698	-564.250	65.496
500	81.531	366.986	334.492	16.247	-574.503	-563.156	58.833
600	82.016	381.897	341.187	24.426	-575.988	-560.745	48.817
700	82.312	394.564	347.930	32.644	-577.346	-558.096	41.646
800	82.505	405.569	354.462	40.885	-578.623	-555.258	36.255
900	82.639	415.295	360.692	49.143	-579.860	-552.263	32.053
1000	82.734	424.007	366.595	57.412	-581.102	-549.130	28.684
1100	82.805	431.896	372.179	65.689	-582.392	-545.871	25.921
1200	82.859	439.103	377.460	73.972	-777.291	-537.611	23.402
1300	82.901	445.737	382.460	82.260	-776.933	-517.653	20.800
1400	82.934	451.882	387.202	90.552	-776.580	-497.721	18.570
1500	82.961	457.605	391.707	98.847	-776.233	-477.815	16.639
1600	82.983	462.960	395.995	107.144	-775.893	-457.931	14.950
1700	83.002	467.991	400.083	115.443	-775.560	-438.069	13.460
1800	83.017	472.736	403.989	123.744	-775.235	-418.226	12.137
1900	83.030	477.225	407.726	132.047	-774.919	-398.401	10.953
2000	83.041	481.484	411.309	140.350	-774.613	-378.592	9.888
2100	83.051	485.536	414.748	148.655	-774.319	-358.799	8.925
2200	83.059	489.399	418.054	156.960	-774.039	-339.018	8.049
2300	83.066	493.092	421.237	165.267	-773.775	-319.251	7.250
2400	83.073	496.627	424.305	173.574	-773.531	-299.494	6.518
2500	83.078	500.018	427.266	181.881	-773.309	-279.747	5.845
2600	83.083	503.277	430.127	190.189	-773.113	-260.009	5.224
2700	83.088	506.413	432.895	198.498	-772.944	-240.277	4.648
2800	83.092	509.434	435.575	206.807	-772.808	-220.551	4.114
2900	83.095	512.350	438.172	215.116	-772.706	-200.830	3.617
3000	83.098	515.167	440.692	223.426	-772.641	-181.111	3.153
3100	83.101	517.892	443.139	231.736	-772.617	-161.394	2.719
3200	83.104	520.531	445.516	240.046	-772.637	-141.677	2.313
3300	83.106	523.088	447.828	248.356	-772.700	-121.959	1.930
3400	83.108	525.569	450.078	256.667	-772.812	-102.238	1.571
3500	83.110	527.978	452.270	264.978	-772.974	-82.513	1.231
3600	83.112	530.319	454.406	273.289	-773.187	-62.782	0.911
3700	83.114	532.597	456.488	281.601	-773.455	-43.045	0.608
3800	83.116	534.813	458.520	289.912	-773.779	-23.301	0.320
3900	83.117	536.972	460.504	298.224	-774.162	-3.546	0.047
4000	83.118	539.076	462.443	306.536	-774.605	16.219	-0.212
4100	83.120	541.129	464.337	314.847	-775.111	35 996	-0.459
4200	83.121	543.132	466.189	323.159	-775.679	55.785	-0.694
4300	83.122	545.088	468.001	331.472	-776.319	75.589	-0.918
4400	83.123	546.999	469.775	339.784	-776.945	95.402	-1.133
4500	83.124	548.867	471.512	348.096	-777.685	115.235	-1.338
4600	83.125	550.694	473.214	356.409	-778.512	135.087	-1.534
4700	83.125	552.481	474.881	364.721	-779.415	154.957	-1.722
4800	83.126	554.231	476.516	373.034	-780.395	174.847	-1.903
4900	83.127	555.945	478.120	381.346	-781.457	194.759	-2.076
5000	83.128	557.625	479.693	389.659	-782.589	214.692	-2.243
5100	83.128	559.271	481.237	397.972	-783.821	234.650	-2.403
5200	83.129	560.885	482.754	406.285	-785.145	254.632	-2.558
5300	83.129	562.469	484.243	414.598	-786.566	274.642	-2.707
5400	83.130	564.022	485.706	422.911	-788.086	294.678	-2.850
5500	83.130	565.548	487.144	431.224	-789.711	314.745	-2.989
5600	83.131	567.046	488.557	439.537	-791.444	334.841	-3.123
5700	83.131	568.517	489.947	447.850	-793.288	354.969	-3.253
5800	83.132	569.963	491.314	456.163	-795.249	375.131	-3.378
5900	83.132	571.384	492.659	464.476	-797.328	395.328	-3.500
6000	83.133	572.781	493.983	472.789	-799.214	415.524	-3.617

PREVIOUS: December 1964 (1 atm)   CURRENT: December 1964 (1 bar)

Sodium Chloride $((NaCl)_2)$ $Cl_2Na_2(g)$

NICKEL CHLORIDE ($NiCl_2$) CRYSTAL M_r = 129.596

$S°(298.15\ K) = 98.157 \pm 0.21\ J\ K^{-1}\ mol^{-1}$

$T_{fus} = 1304 \pm 4\ K,\ p = 2.5\ atm$

$\Delta_f H°(0\ K) = -305.389 \pm 0.21\ kJ\ mol^{-1}$

$\Delta_f H°(298.15\ K) = -304.930 \pm 0.21\ kJ\ mol^{-1}$

$\Delta_{fus} H° = 77.170 \pm 0.84\ kJ\ mol^{-1}$

Enthalpy of Formation

$\Delta_f H°(298.15\ K)$ of -72.88 ± 0.5 kcal mol^{-1} is based on our 3rd law analysis of equilibrium data for the reaction $NiCl_2(cr) + H_2(g) \rightarrow Ni(cr) + 2HCl(g)$ of Busey and Giauque (1). This value is selected from the studies shown below mainly because of the attention to purity of the sample, assurance of equilibrium, and correction for diffusion of hydrogen through pyrex by these authors. The results of Busey and Giauque (1) show excellent agreement between 2nd and 3rd law values with a consequent low drift. Our analysis of the other studies is given below and the large drifts for some of the studies indicate a lack of equilibrium which is attained very slowly in this system; this is certainly the case for Shchukarev et al. (2) and possibly also for (3, 4, 5). The emf results of Egan (6) and Gee and Shelton (7) are about 0.5 kcal mol^{-1} more negative than the adopted value and we increase the uncertainty in the event that these results prove to be more accurate.

Source	Reaction[a]	Method	Data Points	T/K	$\Delta_r H°(298.15\ K)$, kcal mol^{-1} 2nd Law	3rd Law	Drift cal K^{-1} mol^{-1}	$\Delta_f H°(298.15\ K)$ kcal mol^{-1}
Giauque and Busey (1)	A	static	15	630 - 738	28.86±0.05	28.75±0.02	-0.16±0.08	72.88±0.1
Sano (3)	A	static	5	661 - 792	29.96±0.38	29.26±0.13	-0.96±0.52	73.39±0.1
Egan (6)	B	emf	2	673, 723	-69.73	-73.32±0.36	-5.13	73.32±0.4
Shchukarev et al. (2)	A	circulation	7	573 - 823	26.52±0.13	28.90±0.58	3.35±0.19	73.03±0.6
Jellinek and Uloth (4)	A	dynamic	3	573 - 723	31.50±3.09	29.81±0.84	-2.64±4.80	73.94±0.8
Berger and Crut (5)	A	static	16	583 - 718	31.84±0.44	29.57±0.46	-3.52±0.67	73.70±0.5
Gee and Shelton (7)	C	emf	Equation	470 - 750	-0.97	-1.20±0.08	-0.39	73.50±0.3
Gee and Shelton (7)	D	emf	Equation	530 - 800	-7.35	-8.34±0.31	-1.51	73.36±0.3

[a]Reactions:
A) $NiCl_2(cr) + H_2(g) \rightarrow Ni(cr) + 2HCl(g)$
B) $Ni(cr) + Cl_2(g) \rightarrow NiCl_2(cr)$
C) $Co(cr) + NiCl_2(g) \rightarrow CoCl_2(cr) + Ni(cr)$
D) $Fe(cr) + NiCl_2(g) \rightarrow FeCl_2(cr) + Ni(cr)$

Heat Capacity and Entropy

Heat capacity data below 300 K are based on the measurements of Kostryokova (8) (2-30 K) and Busey and Giauque (9) (14.14-336.36 K). The two sets of data are joined smoothly via a least squares procedure; Kostryokova's values (8) are higher than those of Busey and Giaugue (9) by 5.3% at 15 K and 0.3% at 30 K. The results of Busey and Giauque (9) indicate a sharp lambda peak at 52.35 K which is apparently associated with the cooperative ordering of the magnetic moments of the nickel ions. $C°_p$ near the maximum is 6.74 cal K^{-1} mol^{-1} and the enthalpy associated with this transition is of the order of a few calories. The only other measurements in this temperature region are by Trapeznikowa et al. (10) (13-129 K) who report three peaks in $C°_p$ at 49.55 K, 57-58 K, and 60-61 K. We put no weight on these results since the authors apparently suffered from an impure sample among other experimental problems (9); their results range from 4 to 13% higher than those of Busey and Giauque over the temperature range 40-130 K. Earlier results by Kostryokova (11) below 16 K are in good agreement with the adopted $C°_p$ values. S°(298.15 K) is obtained from integration of the adopted $C°_p$'s and is based on $C°_p(2\ K) = 0.0055$.

$C°_p$ data above 300 K are based on the high temperature enthalpy data of Coughlin (12) (376-1282 K). These data are corrected to IPTS 68 (13) and then joined smoothly with the low temperature $C°_p$ data of Busey and Giauque (9) to yield the adopted $C°_p$ values. Data above 1300 K are extrapolated. The high temperature enthalpy data of Krestovnikov and Karetinkov (14) range from 2 to 11% higher than those of Coughlin and are believed to be of lower accuracy.

Fusion Data

See the liquid table for details.

Sublimation Data

T_{sub} = ~1245 K is calculated as the temperature at which $\Delta_r G = 0$ for $NiCl_2(cr) \rightarrow NiCl_2(g)$ (15). See the enthalpy of formation section of the $NiCl_2(g)$ table (15) for the derivation of the adopted enthalpy of sublimation.

References

1. R. H. Busey and W. F. Giauque, J. Amer. Chem. Soc. 75, 1791 (1953).
2. S. A. Shchukarev, T. A. Tolmacheva, and M. A. Oranskaya, J. Gen. Chem. (USSR) 24, 2063 (1954).
3. K. Sano, J. Chem. Soc. Japan 58, 376 (1937).
4. K. Jellinek and R. Uloth, Z. Physik. Chem. 119, 161 (1926).
5. E. Berger and G. Crut, Compt. Rend. 173, 977 (1921).
6. J. J. Egan, U. S. Atomic Energy Commission Rept. BNL9343, 1965.
7. R. Gee and R. A. J. Shelton, Trans. Inst. Min. Metall., Sect. C 85, C208 (1976).
8. M. O. Kostryokova, Soviet Physics JETP, 69, 941 (1975).
9. R. H. Busey and W. F. Giauque, J. Amer. Chem. Soc. 74, 4443 (1952).
10. O. N. Trapeznikowa, L. V. Shubnikov, and G. Miliutin, Physik Z. Sowjetunion 9, 237 (1936).
11. M. O. Kostryokova, Soviet Physics, JETP 28, 236 (1969). M. O. Kostryokova and O. A. Zarubina, Soviet Physics, JETP Letters 7, 10 (1968).
12. J. P. Coughlin, J. Amer. Chem. Soc. 73, 5314 (1951).
13. T. B. Douglas, J. Res. Nat. Bur. Stand. 73A, 451 (1969).
14. A. N. Krestovnikov and G. A. Karetnikov, J. Gen. Chem. (USSR) 6, 955 (1936).
15. JANAF Thermochemical Tables: $NiCl_2(g)$, 9-30-77.

Nickel Chloride ($NiCl_2$) $Cl_2Ni_1(cr)$

Enthalpy Reference Temperature = T_r = 298.15 K Standard State Pressure = p° = 0.1 MPa

T/K	$C°_p$ (J $K^{-1}mol^{-1}$)	S°	$-[G°-H°(T_r)]/T$	$H°-H°(T_r)$ (kJ mol^{-1})	$\Delta_f H°$	$\Delta_f G°$	Log K_f
0	0.	0.	INFINITE	-14.426	-305.389	-305.389	INFINITE
52.350	28.361	11.674	278.918	-13.990	Cp LAMBDA MAXIMUM	TRANSITION	
52.350	29.122	11.674	278.918	-13.990			
100	43.878	32.394	156.473	-12.408	-306.791	-290.384	151.681
200	64.890	70.800	104.534	-6.747	-306.057	-274.184	71.610
298.15	71.660	98.157	98.157	0.	-304.930	-258.779	45.337
300	71.747	98.601	98.159	0.133	-304.908	-258.492	45.008
400	76.289	119.910	101.033	7.551	-303.688	-243.202	31.759
500	78.898	137.244	106.597	15.324	-302.454	-228.225	23.842
600	79.881	151.724	112.945	23.267	-301.421	-213.481	18.585
700	80.313	164.068	119.389	31.275	-300.407	-198.895	14.842
800	80.900	174.824	125.660	39.331	-299.137	-184.480	12.045
900	82.067	184.412	131.664	47.473	-297.846	-170.225	9.880
1000	84.072	193.153	137.382	55.771	-296.473	-156.117	8.155
1100	87.077	201.298	142.826	64.320	-294.932	-142.155	6.750
1200	91.186	209.043	148.024	73.223	-293.120	-128.343	5.587
1300	96.469	216.542	153.007	82.596	-290.929	-114.698	4.609
1304.000	96.705	216.839	153.202	82.982	------ CRYSTAL <--> LIQUID -----		
1400	102.973	223.922	157.809	92.558	-288.244	-101.239	3.777
1500	110.733	231.284	162.462	103.232	-284.946	-87.992	3.064
1600	119.771	238.712	166.995	114.747	-280.911	-74.989	2.448
1700	130.107	246.276	171.435	127.230	-276.016	-62.264	1.913
1800	141.752	254.036	175.808	140.812	-287.283	-49.140	1.426

PREVIOUS: CURRENT: September 1977

Nickel Chloride ($NiCl_2$) $Cl_2Ni_1(cr)$

NICKEL CHLORIDE ($NiCl_2$) LIQUID M_r = 129.596

$S°(298.15\ K) = [151.810]\ J\ K^{-1}\ mol^{-1}$ $\Delta_f H°(298.15) = [-233.516]\ kJ\ mol^{-1}$

$T_{fus} = 1304 \pm 4\ K, p \cong 2.5\ atm$ $\Delta_{fus}H° = 77.170 \pm 0.84\ kJ\ mol^{-1}$

Enthalpy of Formation

$\Delta_f H°(NiCl_2$, l, 298.15 K) is calculated from that of the crystal by addition of $\Delta_{fus}H°$ and the difference in enthalpy, H°(1304 K) - H°(298.15 K), between the crystal and liquid (1).

Heat Capacity and Entropy

The adopted values of $C_p°$ are obtained from our analysis of the high temperature enthalpy data of Coughlin (2) (1305-1337 K) after correction to IPTS 68 (3). The constant $C_p°$ value obtained from these data is assumed to be valid above an assumed glass transition temperature at 880 K. The assumption of a constant $C_p°$ over such a broad range based on data over a very short range leads to a rather large uncertainty in the thermal functions. Below the glass transition at 880 K, $C_p°$ is assumed to be that of the crystal.

S°(298.15 K) is calculated in a manner analogous to that used for the enthalpy of formation.

Fusion Data

The temperature and heat of melting are obtained from our analysis of Coughlin's (2) measurements after correction to IPTS 68 (3). He observed pre-melting beginning at 1288 K and continuing to 1305 K where the enthalpy change again becomes regular. We adopt a melting point near the upper end of this range, T_{fus} = 1304±4 K. The adopted enthalpy of melting is 18.444±0.2 kcal mol^{-1} at T_{fus}.

Phase Data

$NiCl_2(\ell)$ will not exist as a stable phase at 1 bar pressure according to calculations from our tables (1). We calculate sublimation at ~1245 K in good agreement with the measured value of 1241±3 K at 730 torr (4). The adopted melting point of 1304±4 K is from Coughlin's work which was conducted at 2-3.5 atm in the presence of inert gas. These pressures are high enough to increase the sublimation point above the melting point and a relatively short liquid range will be observed. Information on the enthalpy of sublimation is in the enthalpy of formation section of the ideal gas table (1).

References

1. JANAF Thermochemical Tables: $NiCl_2$(cr), $NiCl_2$(g), 9-30-77.
2. J. P. Coughlin, J. Amer. Chem. Soc. 73, 5314 (1951).
3. T. B. Douglas, J. Res. Nat. Bur. Stand. 73A, 451 (1969).
4. H. Schaefer, Z. Anorg. Allg. Chem. 278, 300 (1955).

Nickel Chloride ($NiCl_2$) $Cl_2Ni_1(l)$

Enthalpy Reference Temperature = T_r = 298.15 K Standard State Pressure = p° = 0.1 MPa

T/K	$C_p°$	S°	$-[G°-H°(T_r)]/T$	$H°-H°(T_r)$	$\Delta_f H°$	$\Delta_f G°$	Log K_f
	J $K^{-1}mol^{-1}$			kJ mol^{-1}			
0							
100							
200							
298.15	71.659	151.810	151.810	0.	-233.517	-203.362	35.628
300	71.747	152.253	151.811	0.133	-233.495	-203.175	35.376
400	76.289	173.562	154.685	7.551	-232.275	-193.250	25.236
500	78.898	190.896	160.249	15.324	-231.041	-183.638	19.184
600	79.881	205.376	166.597	23.267	-230.007	-174.259	15.171
700	80.313	217.720	173.041	31.275	-228.993	-165.038	12.315
800	80.900	228.476	179.312	39.331	-227.724	-155.989	10.185
880.000	81.773	236.225	184.139	45.836	GLASS <--> LIQUID TRANSITION		
880.000	101.186	236.225	184.139	45.836			
900	101.186	238.499	185.322	47.860	-226.045	-147.103	8.538
1000	101.186	249.160	191.182	57.978	-222.853	-138.504	7.235
1100	101.186	258.804	196.898	68.097	-219.741	-130.220	6.184
1200	101.186	267.608	202.429	78.215	-216.714	-122.216	5.320
1300	101.186	275.708	207.758	88.334	-213.778	-114.461	4.599
1304.000	101.186	276.018	207.967	88.739	------ CRYSTAL <--> LIQUID -----		
1400	101.186	283.206	212.883	98.453	-210.936	-106.929	3.990
1500	101.186	290.187	217.807	108.571	-208.194	-99.596	3.468
1600	101.186	296.718	222.537	118.690	-205.555	-92.443	3.018
1700	101.186	302.852	227.082	128.808	-203.024	-85.451	2.626
1800	101.186	308.636	231.454	138.927	-217.755	-77.891	2.260

PREVIOUS: CURRENT September 1977

Nickel Chloride ($NiCl_2$) $Cl_2Ni_1(l)$

NICKEL CHLORIDE ($NiCl_2$) CRYSTAL-LIQUID M_r = 129.596

0 to 1304 K crystal
above 1304 K liquid

Refer to the individual tables for details.

Nickel Chloride ($NiCl_2$) $Cl_2Ni_1(cr,l)$

Enthalpy Reference Temperature = T_r = 298.15 K Standard State Pressure = $p°$ = 0.1 MPa

T/K	$C_p°$ (J K⁻¹mol⁻¹)	S° (J K⁻¹mol⁻¹)	$-[G°-H°(T_r)]/T$ (J K⁻¹mol⁻¹)	$H°-H°(T_r)$ (kJ mol⁻¹)	$\Delta_f H°$ (kJ mol⁻¹)	$\Delta_f G°$ (kJ mol⁻¹)	Log K_f
0	0.	0.	INFINITE	-14.426	-305.389	-305.389	INFINITE
52.350	28.361	11.674	278.918	-13.990	Cp LAMBDA MAXIMUM TRANSITION		
52.350	29.122	11.674	278.918	-13.990			
100	43.878	32.394	156.473	-12.408	-306.791	-290.384	151.681
200	64.890	70.800	104.534	-6.747	-306.057	-274.184	71.610
298.15	71.660	98.157	98.157	0.	-304.930	-258.779	45.337
300	71.747	98.601	98.159	0.133	-304.908	-258.492	45.008
400	76.289	119.910	101.033	7.551	-303.688	-243.202	31.759
500	78.898	137.244	106.597	15.324	-302.454	-228.225	23.842
600	79.881	151.724	112.945	23.267	-301.421	-213.481	18.585
700	80.313	164.068	119.389	31.275	-300.407	-198.895	14.842
800	80.900	174.824	125.660	39.331	-299.137	-184.480	12.045
900	82.067	184.412	131.664	47.473	-297.846	-170.225	9.880
1000	84.072	193.153	137.382	55.771	-296.473	-156.117	8.155
1100	87.077	201.298	142.826	64.320	-294.932	-142.155	6.750
1200	91.186	209.043	148.024	73.223	-293.120	-128.343	5.587
1300	96.469	216.542	153.007	82.596	-290.929	-114.698	4.609
1304.000	96.705	216.839	153.202	82.982	CRYSTAL <--> LIQUID TRANSITION		
1304.000	101.186	276.018	153.202	160.152			
1400	101.186	283.206	161.873	169.866	-210.936	-106.929	3.990
1500	101.186	290.187	170.198	179.984	-208.194	-99.596	3.468
1600	101.186	296.718	177.903	190.103	-205.555	-92.443	3.018
1700	101.186	302.852	185.075	200.222	-203.024	-85.451	2.626
1800	101.186	308.636	191.780	210.340	-217.755	-77.891	2.260

PREVIOUS: CURRENT: September 1977

Nickel Chloride ($NiCl_2$) $Cl_2Ni_1(cr,l)$

$S°(298.15\ K) = 294.233 \pm 8.4\ J\ K^{-1}\ mol^{-1}$

$\Delta_f H°(0\ K) = -74.17 \pm 0.25\ kJ\ mol^{-1}$
$\Delta_f H°(298.15\ K) = -73.932 \pm 0.25\ kJ\ mol^{-1}$

Electronic Levels and Quantum Weights

State	ε_i, cm^{-1}	g_i	State	ε_i, cm^{-1}	g_i
$^3\Pi g$	0	6	$^1\Delta g$	15466	2
$^3\Sigma g^-$	[879]	3	$^1\Delta g$	18236	2
$3\Delta g$	2454	6	$^3\Sigma g^-$	21701	3
$^3\Phi g$	5853	6	$^1\Pi g$	[26740]	2
$^1\Sigma g^+$	10601	1	$^1\Sigma g^+$	[27234]	1
$^1\Pi g$	11727	2	$^1\Gamma g$	[28825]	2
$^3\Pi g$	12984	6	$^1\Sigma g^+$	[55433]	1
$^1\Phi g$	14717	2			

Vibrational Frequencies and Degeneracies

ν, cm^{-1}
360 (1)
87 (2)
523 (1)

Point Group: $D_{\infty h}$ σ = [2]

Bond Distance: Ni-Cl = [2.09] Å

Bond Angle: Cl-Ni-Cl = 180°

Rotational Constant: B_0 = [0.054428] cm^{-1}

Enthalpy of Formation

The adopted value of $\Delta_f H°(298.15\ K)$ is calculated from $\Delta_{sub}H°(298.15\ K) = 55.21\pm0.3\ kcal\ mol^{-1}$ by addition of $\Delta_f H°(NiCl_2$, cr, 298.15 K) (1). The adopted value is an average derived from our 3rd law analysis of the sublimation data of Ratkovskii et al. (2), McCreary and Thorn (3), Schaefer et al. (4), and Hildenbrand (5). The data of Maier (6) are rejected because there is an apparent temperature dependent error at the high temperature end of his data.

Source	Method	T/K	Data Points	$\Delta_r H°(298.15\ K)$, kcal mol^{-1} 2nd Law	3rd Law	Drift cal K^{-1} mol^{-1}	$\Delta_f H°(298.15\ K)$ kcal mol^{-1}
Ratkovskii et al. (2)	Knudsen	973 - 1203	Equation	57.14	55.36±0.29	-1.6	-17.52±0.8
McCreary and Thorn (3)	Torsional Recoil	823 - 881	Equation	55.68	55.01±0.03	-0.80	-17.87±0.5
McCreary and Thorn (3)	Mass Effusion	823 - 881	Equation	54.85	55.08±0.01	0.27	-17.80±0.5
Schaefer et al. (4)	Transpiration	973 - 1056	10	55.25±1.50	55.62±0.23	0.36±1.47	-17.26±0.7
Maier (6)	Static (pure sample)	1047 - 1236	18	54.00±0.89	55.39±0.42	1.19±0.75	-17.49±0.7
Hildenbrand (5)	Torsion Effusion	706 - 771	19	55.14±0.34	54.99±0.08	-0.19±0.46	-17.89±0.5

Schoonmaker et al. (7) observed no dimers in their Knudsen-mass spectroscopic study while Schaefer and Binnewies (8) observed dimerization to the extent of 0.15% at 796 K by a similar technique. We have not corrected for this small dimerization effect. Of more importance may be the observation of significant quantities of $SiCl_4(g)$ and $Si_2Cl_6(g)$ by Schoonmaker et al. (7) indicating reaction between $NiCl_2(g)$ and the quartz liner in their experiment. Schaefer and Binnewies (8) make no mention of Si_xCl_y species even though their Knudsen cell contains a quartz crucible. Therefore we ignore this effect while recognizing the need for further work in the characterization of the vapor over $NiCl_2$.

Heat Capacity and Entropy

The adopted gas phase values of $\nu_2 = 87\ cm^{-1}$ and $\nu_3 = 523\ cm^{-1}$ are from the matrix isolation work of Thompson and Carlson (9), we have added 2 cm^{-1} to the observed values to account for the matrix effect. The symmetric stretching frequency, ν_1, has not been observed in the infrared spectrum and this is one of the major indications that the molecule is linear. The adopted value of ν_1 (360 cm^{-1}) is from the observed vibrational spacing of the fluorescence spectrum by Gruen et al. (10) and is in good agreement with the value of 351 cm^{-1} calculated by Milligan et al. (11).

The rotational constant is calculated assuming the molecule is linear with the bond distance estimated to be 2.09 Å in accordance with Brewer et al. (12). Further support for a linear structure comes from the molecular beam work of Buchler et al. (13) which indicates a zero dipole moment for a number of transition metal difluorides including NiF_2.

The ground state of the linear $NiCl_2$ molecule is established as $^3\Pi g$ from numerous studies on the Laporte-forbidden d-d transitions (14 - 16) as well as a study of the Laporte-allowed charge-transfer transitions (17). The very complex electronic spectra for both the gas phase and matrix isolated $NiCl_2$ have been studied by numerous investigators and interpreted using crystal field or ligand field theory. Our adopted values for the excited states are from Lever and Hollebone (14) who used an "orbital angular overlap" model to fit the spin-allowed and spin-forbidden d-d transition spectra. These assignments are in good agreement with those of DeKock and Gruen (16, 18) and Smith (15) but differ from those of Hougen et al. (19).

References

1. JANAF Thermochemical Tables: $NiCl_2$(cr), 9-30-77.
2. I. A. Ratkovskii et al., Izv. Vyssh. Uchebn. Zaved. Khim. Khim. Tekhnol. 19, 407 (1976).
3. J. R. McCreary and R. J. Thorn, J. Chem. Phys. 48, 3290 (1968).
4. H. Schaefer, L. Bayer, G. Breil, K. Etzel, and K. Krehl, Z. Anorg. Allg. Chem. 278, 300 (1955).
5. D. L. Hildenbrand, SRI International, personal communication, May 3, 1978.
6. C. G. Maier, U. S. Bur. Mines Tech. Paper 360, 1, 1925.
7. R. C. Schoonmaker, A. H. Friedman, and R. F. Porter, J. Chem. Phys. 31, 1586 (1959).
8. H. Schaefer and M. Binnewies, Z. Anorg. Allg. Chem. 410, 251 (1974).
9. K. R. Thompson and K. D. Carlson, J. Chem. Phys. 49, 4379 (1968).
10. D. M. Gruen, J. R. Clifton, and C. W. Dekock, J. Chem. Phys. 42, 902 (1964).
11. D. E. Milligan, M. E. Jacox, and J. D. McKinley, J. Chem. Phys. 42, 902 (1964).
12. L. Brewer, G. R. Somayajulu, and E. Brackett, Chem. Rev. 63, 111 (1963).
13. A. Buchler, J. L. Stauffer, and W. Klemperer, J. Chem. Phys. 40, 3471 (1964).
14. A. B. P. Lever and B. R. Hollebone, Inorg. Chem. 11, 2183 (1972).
15. D. W. Smith, Inorg. Chimica Acta 5, 231 (1971).
16. C. W. DeKock and D. M. Gruen, J. Chem. Phys. 44, 4387 (1966).
17. C. W. DeKock and D. M. Gruen, J. Chem. Phys. 49, 4521 (1968).
18. C. W. DeKock and D. M. Gruen, J. Chem. Phys. 46, 1096 (1967).
19. J. T. Hougen, G. E. Leroi, and T. C. James, J. Chem. Phys. 34, 1670 (1961).

Enthalpy Reference Temperature = T_r = 298.15 K   Standard State Pressure = p° = 0.1 MPa

T/K	$C_p°$ (J $K^{-1}mol^{-1}$)	S°	$-[G°-H°(T_r)]/T$	$H°-H°(T_r)$ (kJ mol^{-1})	$\Delta_f H°$	$\Delta_f G°$	Log K_f
0	0.	0.	INFINITE	-14.206	-74.171	-74.171	INFINITE
100	45.245	237.999	342.346	-10.435	-73.819	-77.973	40.729
200	53.251	271.980	299.409	-5.486	-73.797	-82.160	21.458
250	56.051	284.178	295.178	-2.750	-73.854	-84.245	17.602
298.15	58.077	294.233	294.233	0.	-73.931	-86.240	15.109
300	58.144	294.592	294.234	0.108	-73.935	-86.316	15.029
350	59.697	303.677	294.948	3.055	-74.039	-88.372	13.189
400	60.859	311.728	296.552	6.071	-74.170	-90.411	11.806
450	61.749	318.950	298.646	9.137	-74.331	-92.432	10.729
500	62.451	325.494	301.009	12.243	-74.536	-94.432	9.865
600	63.502	336.979	306.073	18.543	-75.146	-98.359	8.563
700	64.265	346.828	311.208	24.933	-75.750	-102.170	7.624
800	64.839	355.448	316.211	31.390	-76.080	-105.922	6.916
900	65.270	363.111	321.004	37.897	-76.423	-109.632	6.363
1000	65.590	370.006	325.565	44.440	-76.805	-113.302	5.918
1100	65.824	376.269	329.894	51.012	-77.241	-116.931	5.553
1200	65.993	382.004	334.001	57.603	-77.741	-120.518	5.246
1300	66.114	387.291	337.900	64.209	-78.317	-124.060	4.985
1400	66.200	392.194	341.605	70.825	-78.979	-127.554	4.759
1500	66.260	396.763	345.132	77.448	-79.732	-130.998	4.562
1600	66.300	401.041	348.494	84.076	-80.583	-134.388	4.387
1700	66.327	405.061	351.704	90.707	-81.540	-137.723	4.232
1800	66.344	408.853	354.775	97.341	-99.755	-140.282	4.071
1900	66.354	412.440	357.716	103.976	-100.838	-142.504	3.918
2000	66.360	415.844	360.538	110.612	-101.931	-144.669	3.778
2100	66.363	419.082	363.250	117.248	-103.035	-146.779	3.651
2200	66.365	422.169	365.858	123.884	-104.150	-148.836	3.534
2300	66.367	425.119	368.371	130.521	-105.279	-150.842	3.426
2400	66.370	427.944	370.795	137.158	-106.422	-152.798	3.326
2500	66.374	430.653	373.135	143.795	-107.581	-154.707	3.232
2600	66.380	433.257	375.398	150.433	-108.756	-156.569	3.146
2700	66.389	435.762	377.587	157.071	-109.949	-158.385	3.064
2800	66.400	438.177	379.709	163.710	-111.159	-160.157	2.988
2900	66.413	440.507	381.765	170.351	-112.388	-161.885	2.916
3000	66.428	442.759	383.761	176.993	-113.635	-163.571	2.848
3100	66.445	444.937	385.699	183.637	-114.900	-165.215	2.784
3200	66.465	447.047	387.584	190.282	-493.016	-161.629	2.638
3300	66.486	449.092	389.417	196.930	-492.660	-151.278	2.395
3400	66.509	451.078	391.201	203.580	-492.318	-140.939	2.165
3500	66.532	453.006	392.940	210.232	-491.988	-130.609	1.949
3600	66.557	454.880	394.634	216.886	-491.670	-120.288	1.745
3700	66.583	456.704	396.287	223.543	-491.361	-109.976	1.553
3800	66.609	458.480	397.901	230.203	-491.061	-99.673	1.370
3900	66.635	460.211	399.476	236.865	-490.768	-89.377	1.197
4000	66.661	461.898	401.016	243.530	-490.481	-79.088	1.033
4100	66.687	463.545	402.521	250.197	-490.196	-68.807	0.877
4200	66.712	465.152	403.993	256.867	-489.914	-58.533	0.728
4300	66.737	466.722	405.434	263.539	-489.632	-48.265	0.586
4400	66.762	468.257	406.844	270.214	-489.348	-38.004	0.451
4500	66.785	469.757	408.226	276.892	-489.062	-27.750	0.322
4600	66.808	471.225	409.579	283.571	-488.771	-17.501	0.199
4700	66.829	472.662	410.906	290.253	-488.475	-7.260	0.081
4800	66.850	474.069	412.208	296.937	-488.173	2.976	-0.032
4900	66.869	475.448	413.484	303.623	-487.863	13.205	-0.141
5000	66.887	476.799	414.737	310.311	-487.546	23.428	-0.245
5100	66.904	478.124	415.967	317.000	-487.221	33.644	-0.345
5200	66.919	479.423	417.175	323.692	-486.888	43.854	-0.441
5300	66.933	480.698	418.361	330.384	-486.546	54.057	-0.533
5400	66.945	481.949	419.527	337.078	-486.196	64.254	-0.622
5500	66.956	483.178	420.674	343.773	-485.838	74.444	-0.707
5600	66.966	484.384	421.800	350.469	-485.474	84.628	-0.789
5700	66.975	485.570	422.909	357.166	-485.102	94.805	-0.869
5800	66.982	486.734	423.999	363.864	-484.725	104.976	-0.945
5900	66.987	487.880	425.072	370.563	-484.343	115.139	-1.019
6000	66.992	489.005	426.129	377.262	-483.959	125.297	-1.091

PREVIOUS: September 1977 (1 atm)   CURRENT: September 1977 (1 bar)

CHLORINE OXIDE (Cl_2O) IDEAL GAS M_r = 86.9054

$\Delta_fH°$(0 K) = 89.69 ± 6.7 kJ mol^{-1}

$S°$(298.15 K) = 267.959 J K^{-1} mol^{-1} $\Delta_fH°$(298.15 K) = 87.86 ± 6.7 kJ mol^{-1}

Vibrational Frequencies and Degeneracies

ν, cm^{-1}
640(1)
300(1)
686(1)

Ground State Quantum Weight: 1 σ = 2

Point Group: C_{2v}

Bond Distance: Cl-O = 1.701 ± 0.02 Å

Bond Angle: Cl-O-Cl = 110.8 ± 1°

Product of the Moments of Inertia: $I_AI_BI_C$ = 1.17198 x 10^{-114} g^3 cm^6

Enthalpy of Formation

Kustodina et al. (1) derived the enthalpy of formation of Cl_2O dissolved in CCl_4 from solution calorimetry at 278, 283, 298 and 308 K. The enthalpy of formation in CCl_4 was interpolated to 285 K. Yost and Felt (2) measured the partial vapor pressures of Cl_2O above its CCl_4 solution at 273 and 298 K which yield the heat of solution -6.59 kcal mol^{-1} at 285 K. The $\Delta_fH°$(298.15 K) = 21.0 kcal mol^{-1} was calculated from these two values. Wallace and Goodeve (3) obtained the heat of explosion of Cl_2O(g): their data give $\Delta_fH°$(Cl_2O, g, 298.15 K) = 21.4 kcal mol^{-1} which is in very good agreement with the above. Also Gunther and Wekua (4) measured the enthalpy of explosion; their data give $\Delta_fH°$(Cl_2O, g, 298.15 K) = 24.7 kcal mol^{-1}.

Source	Reaction	T/K	$\Delta_rH°$(T) kcal mol^{-1}	$\Delta_fH°$(298.15 K) kcal mol^{-1}
(1)	Cl_2(g) + 1/2 O_2(g) → Cl_2O(in CCl_4)	285	14.38	-
(2)	Cl_2O(in CCl_4) → Cl_2O(g)	285	6.59	21.0
(3)	Cl_2O(g) → Cl_2(g) + 1/2 O_2(g)	298	-21.4	21.4
(4)	Cl_2O(g) → Cl_2(g) + 1/2 O_2(g)	298	-24.7	24.7

Heat Capacity and Entropy

The vibrational frequencies were obtained from Rochkind and Pimentel (5). The infrared spectrum of Cl_2O(g) has been investigated previously beginning with the work of Bailey and Cassie (6). Hedberg (7) proposed stretching frequencies at 688 cm^{-1} (ν_1) and 969 cm^{-1} (ν_3) but left the bending frequency unknown. Previous evidence concerning the bending mode (ν_2) (9) was shown by Hedberg (7) to be spurious. Rochkind and Pimentel (5) present infrared studies of Cl_2O gas and condensed phases which reveal the bending frequency and require reassignment of the stretching fundamentals. The bond distance and bond angle were obtained from Dunitz and Hedberg (8).

The principal moments of inertia are: I_A = 2.0224x 10^{-39}, I_B = 23.0829 x 10^{-39} and I_C = 25.1033 x 10^{-39} g cm^2.

References

1. V. A. Kustodina, K. P. Mishchenko and I. E. Flis, Zh. Prikl. Khim. 35, 1374 (1962).
2. D. M. Yost and R. C. Felt, J. Amer. Chem. Soc. 56, 68 (1934).
3. J. I. Wallace and C. F. Goodeve, Trans. Faraday Soc. 27, 648 (1931).
4. P. Gunther and K. Wekua, Z. Physik Chem. 154, 193 (1931).
5. M. M. Rochkind and G. C. Pimentel, J. Chem. Phys. 42, 1361 (1965).
6. C. R. Bailey and A. B. D. Cassie, Proc. Roy. Soc. (London) A142, 192 (1933).
7. K. Hedberg, J. Chem. Phys. 19, 509 (1951).
8. J. D. Dunitz and K. Hedberg, J. Amer. Chem. Soc. 72, 3108 (1950).
9. G. B. B. M. Sutherland and W. B. Penny, Proc. Roy. Soc. (London) A156, 678 (1936).

Chlorine Oxide (Cl_2O) Cl_2O_1(g)

Enthalpy Reference Temperature = T_r = 298.15 K Standard State Pressure = p° = 0.1 MPa

T/K	$C°_p$ (J $K^{-1}mol^{-1}$)	S° (J $K^{-1}mol^{-1}$)	-[G°-H°(T_r)]/T (J $K^{-1}mol^{-1}$)	H°-H°(T_r) (kJ mol^{-1})	$\Delta_fH°$ (kJ mol^{-1})	$\Delta_fG°$ (kJ mol^{-1})	Log K_f
0	0.	0.	INFINITE	-11.695	89.691	89.691	INFINITE
100	35.494	223.344	306.542	-8.320	88.702	93.934	-49.066
200	42.255	249.979	272.170	-4.438	88.091	99.436	-25.970
250	45.365	259.752	268.733	-2.245	87.936	102.293	-21.373
298.15	47.810	267.959	267.959	0.	87.864	105.065	-18.407
300	47.893	268.255	267.960	0.089	87.862	105.172	-18.312
350	49.863	275.793	268.551	2.535	87.852	108.059	-16.127
400	51.378	282.554	269.886	5.067	87.886	110.943	-14.488
450	52.548	288.676	271.639	7.667	87.950	113.822	-13.212
500	53.458	294.262	273.626	10.318	88.035	116.692	-12.191
600	54.749	304.131	277.910	15.733	88.239	122.405	-10.656
700	55.590	312.638	282.277	21.253	88.459	128.082	-9.558
800	56.163	320.101	286.548	26.842	88.681	133.727	-8.731
900	56.569	326.741	290.652	32.480	88.895	139.345	-8.087
1000	56.866	332.717	294.565	38.152	89.100	144.939	-7.571
1100	57.090	338.148	298.284	43.851	89.294	150.514	-7.147
1200	57.262	343.123	301.816	49.569	89.475	156.071	-6.794
1300	57.398	347.712	305.172	55.302	89.645	161.614	-6.494
1400	57.506	351.970	308.365	61.047	89.802	167.144	-6.236
1500	57.594	355.940	311.406	66.802	89.945	172.663	-6.013
1600	57.666	359.660	314.306	72.565	90.075	178.174	-5.817
1700	57.726	363.158	317.078	78.335	90.190	183.676	-5.644
1800	57.777	366.459	319.731	84.110	90.290	189.172	-5.490
1900	57.820	369.584	322.273	89.890	90.373	194.664	-5.352
2000	57.857	372.550	324.713	95.674	90.438	200.151	-5.227
2100	57.888	375.374	327.059	101.461	90.484	205.635	-5.115
2200	57.916	378.068	329.317	107.252	90.510	211.118	-5.013
2300	57.940	380.643	331.493	113.044	90.513	216.600	-4.919
2400	57.961	383.109	333.593	118.840	90.493	222.082	-4.833
2500	57.980	385.476	335.621	124.637	90.447	227.566	-4.755
2600	57.996	387.750	337.582	130.435	90.376	233.052	-4.682
2700	58.011	389.939	339.481	136.236	90.278	238.542	-4.615
2800	58.024	392.049	341.321	142.038	90.152	244.035	-4.553
2900	58.036	394.085	343.106	147.841	89.998	249.533	-4.495
3000	58.047	396.053	344.838	153.645	89.816	255.038	-4.441
3100	58.057	397.956	346.521	159.450	89.606	260.549	-4.390
3200	58.066	399.800	348.157	165.256	89.369	266.066	-4.343
3300	58.074	401.587	349.749	171.063	89.107	271.592	-4.299
3400	58.081	403.321	351.300	176.871	88.820	277.126	-4.258
3500	58.088	405.004	352.810	182.679	88.511	282.669	-4.219
3600	58.094	406.641	354.283	188.488	88.182	288.221	-4.182
3700	58.100	408.233	355.720	194.298	87.835	293.782	-4.147
3800	58.105	409.782	357.122	200.108	87.472	299.354	-4.115
3900	58.110	411.291	358.492	205.919	87.096	304.935	-4.084
4000	58.114	412.763	359.830	211.730	86.710	310.525	-4.055
4100	58.118	414.198	361.139	217.542	86.316	316.125	-4.027
4200	58.122	415.598	362.419	223.354	85.918	321.735	-4.001
4300	58.126	416.966	363.671	229.166	85.518	327.355	-3.977
4400	58.129	418.302	364.898	234.979	85.119	332.983	-3.953
4500	58.132	419.609	366.099	240.792	84.722	338.621	-3.931
4600	58.135	420.886	367.276	246.605	84.332	344.267	-3.909
4700	58.138	422.137	368.430	252.419	83.948	349.922	-3.889
4800	58.141	423.361	369.562	258.233	83.575	355.585	-3.870
4900	58.143	424.559	370.672	264.047	83.213	361.256	-3.851
5000	58.145	425.734	371.762	269.862	82.864	366.934	-3.833
5100	58.147	426.886	372.831	275.676	82.529	372.619	-3.816
5200	58.149	428.015	373.882	281.491	82.211	378.309	-3.800
5300	58.151	429.122	374.914	287.306	81.909	384.007	-3.785
5400	58.153	430.209	375.928	293.121	81.625	389.709	-3.770
5500	58.155	431.276	376.924	298.937	81.360	395.417	-3.755
5600	58.157	432.324	377.904	304.752	81.113	401.130	-3.742
5700	58.158	433.354	378.868	310.568	80.885	406.846	-3.728
5800	58.160	434.365	379.816	316.384	80.677	412.567	-3.716
5900	58.161	435.359	380.749	322.200	80.488	418.290	-3.703
6000	58.162	436.337	381.668	328.016	80.319	424.017	-3.691

PREVIOUS: December 1965 (1 atm) CURRENT: December 1965 (1 bar)

Chlorine Oxide (Cl_2O) Cl_2O_1(g)

TITANIUM CHLORIDE OXIDE ($TiOCl_2$) IDEAL GAS M_r = 134.7854

$S°(298.15\ K) = [321.001]\ J\ K^{-1}\ mol^{-1}$

$\Delta_f H°(0\ K) = [-543.909]\ kJ\ mol^{-1}$
$\Delta_f H°(298.15\ K) = [-545.552]\ kJ\ mol^{-1}$

Vibrational Frequencies and Degeneracies

ν, cm^{-1}	ν, cm^{-1}
[455](1)	[250](1)
[849](1)	[250](1)
[120](1)	[180](1)

Ground State Quantum Weights: [1] $\sigma = 2$

Point Group: [C_{2v}]

Bond Distances: O-Ti = [1.62] Å Ti-Cl = [2.20] Å

Bond Angle: O-Ti-Cl = [120]°

Product of the Moments of Inertia: $I_A I_B I_C = [4.63214 \times 10^{-113}]\ g^3\ cm^6$

Enthalpy of Formation

$\Delta_f H°(0\ K)$ was estimated as -130 kcal mol^{-1} by Gordon (1). $\Delta_f H°(298.15\ K)$ was then calculated.

Heat Capacity and Entropy

Molecular constants were estimated by Gordon (1). Principal moments of inertia are: $I_A = 17.8783 \times 10^{-39}\ g\ cm^2$, $I_B = 42.7411 \times 10^{-39}\ g\ cm^2$, and $I_C = 60.6194 \times 10^{-39}\ g\ cm^2$.

Reference

1. J. S. Gordon, Astrosystems, Inc., Caldwell Twp., N. J., personal communication, January 10, 1963.

Titanium Chloride Oxide ($TiOCl_2$) $Cl_2O_1Ti_1(g)$

Enthalpy Reference Temperature = T_r = 298.15 K Standard State Pressure = p° = 0.1 MPa

T/K	$C_p°$ (J $K^{-1}mol^{-1}$)	S° (J $K^{-1}mol^{-1}$)	$-[G°-H°(T_r)]/T$ (J $K^{-1}mol^{-1}$)	$H°-H°(T_r)$ (kJ mol^{-1})	$\Delta_f H°$ (kJ mol^{-1})	$\Delta_f G°$ (kJ mol^{-1})	Log K_f
0	0.	0.	INFINITE	-16.710	-543.909	-543.909	INFINITE
100	51.425	252.861	380.084	-12.722	-544.846	-541.740	282.976
200	65.445	293.526	327.402	-6.775	-545.309	-538.428	140.623
250	69.250	308.562	322.172	-3.403	-545.448	-536.691	112.135
298.15	71.952	321.001	321.001	0.	-545.552	-534 994	93.729
300	72.041	321.447	321.003	0.133	-545.555	-534.929	93.139
350	74.138	332.717	321.888	3.790	-545.643	-533.150	79.568
400	75.739	342.726	323.879	7.539	-545.719	-531.360	69.389
450	76.976	351.721	326.481	11.358	-545.792	-529.561	61.470
500	77.945	359.883	329.420	15.232	-545.867	-527.754	55.134
600	79.330	374.226	335.725	23.101	-546.035	-524.116	45.628
700	80.243	386.528	342.125	31.082	-546.241	-520.447	38.836
800	80.870	397.286	348.362	39.140	-546.476	-516.747	33.740
900	81.318	406.839	354.338	47.250	-546.780	-513.013	29.775
1000	81.647	415.424	360.025	55.399	-547.219	-509.240	26.600
1100	81.896	423.218	365.421	63.577	-547.862	-505.412	24.000
1200	82.088	430.353	370.539	71.777	-552.693	-501.397	21.825
1300	82.240	436.930	375.396	79.993	-553.020	-497.110	19.974
1400	82.361	443.029	380.012	88.224	-553.438	-492.794	18.386
1500	82.459	448.715	384.405	96.465	-553.958	-488.445	17.009
1600	82.541	454.039	388.592	104.715	-554.596	-484.057	15.803
1700	82.608	459.045	392.591	112.973	-555.366	-479.626	14.737
1800	82.665	463.769	396.415	121.236	-556.281	-475.145	13.788
1900	82.714	468.239	400.079	129.505	-557.357	-470.609	12.938
2000	82.755	472.483	403.594	137.779	-573.294	-465.558	12.159
2100	82.791	476.522	406.971	146.056	-575.482	-460.118	11.445
2200	82.822	480.374	410.221	154.337	-577.690	-454.572	10.793
2300	82.849	484.056	413.351	162.620	-579.919	-448.926	10.195
2400	82.873	487.582	416.371	170.907	-582.172	-443.183	9.646
2500	82.894	490.966	419.288	179.195	-584.450	-437.345	9.138
2600	82.913	494.217	422.108	187.485	-586.753	-431.415	8.667
2700	82.929	497.347	424.837	195.777	-589.084	-425.396	8.230
2800	82.944	500.363	427.481	204.071	-591.441	-419.291	7.822
2900	82.958	503.274	430.044	212.366	-593.827	-413.101	7.441
3000	82.970	506.087	432.532	220.663	-596.241	-406.828	7.084
3100	82.981	508.807	434.949	228.960	-598.682	-400.473	6.748
3200	82.991	511.442	437.299	237.259	-601.150	-394.041	6.432
3300	83.000	513.996	439.584	245.558	-603.643	-387.530	6.134
3400	83.008	516.474	441.810	253.859	-606.161	-380.943	5.852
3500	83.016	518.880	443.977	262.160	-608.701	-374.283	5.586
3600	83.023	521.219	446.091	270.462	-611.261	-367.549	5.333
3700	83.029	523.494	448.152	278.764	-1022.938	-352.955	4.983
3800	83.035	525.708	450.164	287.068	-1024.297	-334.829	4.603
3900	83.041	527.865	452.129	295.371	-1025.725	-316.667	4.241
4000	83.046	529.968	454.049	303.676	-1027.217	-298.467	3.898
4100	83.051	532.018	455.925	311.981	-1028.768	-280.229	3.570
4200	83.055	534.020	457.761	320.286	-1030.375	-261.953	3.258
4300	83.059	535.974	459.557	328.592	-1032.034	-243.637	2.960
4400	83.063	537.883	461.316	336.898	-1033.739	-225.283	2.674
4500	83.066	539.750	463.038	345.204	-1035.487	-206.890	2.402
4600	83.070	541.576	464.726	353.511	-1037.273	-188.457	2.140
4700	83.073	543.362	466.380	361.818	-1039.093	-169.985	1.889
4800	83.076	545.111	468.002	370.126	-1040.943	-151.474	1.648
4900	83.079	546.824	469.593	378.433	-1042.819	-132.923	1.417
5000	83.081	548.503	471.155	386.741	-1044.718	-114.335	1.194
5100	83.084	550.148	472.687	395.049	-1046.636	-95.708	0.980
5200	83.086	551.761	474.193	403.358	-1048.570	-77.044	0.774
5300	83.088	553.344	475.671	411.667	-1050.516	-58.342	0.575
5400	83.090	554.897	477.124	419.976	-1052.473	-39.603	0.383
5500	83.092	556.422	478.552	428.285	-1054.437	-20.829	0.198
5600	83.094	557.919	479.956	436.594	-1056.406	-2.018	0.019
5700	83.096	559.390	481.337	444.903	-1058.379	16.828	-0.154
5800	83.097	560.835	482.695	453.213	-1060.352	35.709	-0.322
5900	83.099	562.256	484.031	461.523	-1062.325	54.623	-0.484
6000	83.100	563.652	485.347	469.833	-1064.296	73.571	-0.640

PREVIOUS: September 1963 (1 atm) CURRENT: September 1963 (1 bar)

Titanium Chloride Oxide ($TiOCl_2$) $Cl_2O_1Ti_1(g)$

SULFURYL CHLORIDE (SO_2Cl_2) IDEAL GAS M_r = 134.9648

$\Delta_fH°(0\ K)$ = -348.56 ± 2.1 kJ mol^{-1}

$S°(298.15\ K)$ = 311.095 ± 0.4 J K^{-1} mol^{-1} $\Delta_fH°(298.15\ K)$ = -354.80 ± 2.1 kJ mol^{-1}

Vibrational Frequencies and Degeneracies (ω, cm^{-1}, all singly degenerate)

1205	577	406	209	208	1434	388	586	363

Ground State Quantum Weight: [1] σ = 2

Point Group: C_{2v}

Bond Distances: S–O = 1.404 ± 0.004 Å S–Cl = 2.011 ± 0.005 Å

Bond Angles: O–S–O = 123.5 ± 0.8° Cl–S–Cl = 100.0 ± 0.7°

Product of the Moments of Inertia: $I_AI_BI_C$ = 3.80579 x 10^{-113} g^3 cm^6

Enthalpy of Formation

The adopted $\Delta_fH°$ = -84.8 ± 0.5 kcal mol^{-1} is based on Kp for dissociation to SO_2 and Cl_2 (see below). The relatively thorough study of Arii (2) gave similar values of Kp near the pressure of SO_2Cl_2 and at reduced pressures. The entropy discrepancy is large (δS = -11 cal K^{-1} mol^{-1}), but this may be due to the small temperature range and to uncertainties in absorption on the activated charcoal catalyst. Absorption uncertainties probably exist in all of the Kp studies, but the high-temperature data of Trautz (3) have an additional uncertainty; i.e., Kp is very sensitive to small systematic errors in the data since the SO_2Cl_2 is almost completely decomposed. Exclusion of data (3) above 416 K yields a 3rd law $\Delta_rH°$ which agrees satisfactorily with Arii (2). Data of Londergan (1) yield a 3rd law $\Delta_rH°$ which is smaller by about 0.4 kcal mol^{-1}; these data yield almost no entropy discrepancy, but the experimental method (1) appears to be less satisfactory.

The adopted $\Delta_fH°$ is consistent with old calorimetric data of Thomsen (5) for hydrolysis of the liquid but inconsistent with recent data of Neale and Williams (4), on which the selections of NBS 270-3 (6) are probably based. Consistency tests require Δ_vH; thus, we note that in the absence of a catalyst $SO_2Cl_2(g)$ persists to well above the boiling point even though it is unstable with respect to SO_2 and Cl_2. Evidence for this conclusion includes studies of catalysts (3, 7), vapor densities (8), boiling point and vapor pressures (9, 10, 7). The boiling point (69.5 ± 0.3°C) and $\Delta_{vap}S$ (20.8 cal K^{-1} mol^{-1}) are almost exactly those predicted by comparison with SO_2ClF (7.3°C and 20.6 cal K^{-1} mol^{-1}) and SO_2F_2 (-55.3°C and 21.1 cal K^{-1} mol^{-1}), which have normal vapors. Vapor pressure data (9) for SO_2Cl_2 yield Δ_vH = 7.1 kcal mol^{-1} at 69.5°C and $\Delta_vH°$ = 7.8 kcal mol^{-1} at 25°C. Trautz et al. (10) used a condensation method to measure $\Delta_{vap}H$ = 6.67 kcal mol^{-1} near Tb, but we dismiss this value since the resulting $\Delta_{vap}S$ (19.5 cal K^{-1} mol^{-1}) is too small for normal vapor and much too small for dissociating vapor.

Source	Data	Reaction[a]	T/K	$\Delta_rH°$(298.15 K), kcal mol^{-1} 2nd law	3rd law	δS, cal $K^{-1}mol^{-1}$	$\Delta_fH°$(298.15 K), kcal mol[b] gas	liquid
(1)	Kp	A	340-373	13.08±0.05	13.41±0.3	-0.9±0.1	-84.36	(-92.1)
(2)	Kp	A	303-323	10.41±0.13	13.84±0.2	-10.9±0.4	-84.79	(-92.6)
(3)	Kp	A	375-464	6.33±0.51	14.2 ±1.1	-19.2±1.2	-85.2	(-93.0)
			375-416	---	13.91±0.4	---	-84.86	(-92.7)
(4)	Δ_rH	B	295-296	---	-63.4 ±0.5[c]	---	(-86.1)	-93.9
(5)	Δ_rH	B	291-293	---	-63.6 ±0.5			
(6)	(4?)	B?	298?		-63.1	---	-87.0	-94.2

[a] A) $SO_2Cl_2(g) \rightarrow SO_2(g) + Cl_2(g)$; B) $SO_2Cl_2(\ell) + 2\ H_2O(\ell) \rightarrow [H_2SO_4 + 2HCl](nH_2O)$

[b] Values in parentheses calculated from $\Delta_{vap}H°$ = 7.8 kcal mol^{-1} at 25°C.

[c] Calorimetric Δ_rH adjusted to 25°C.

Heat Capacity and Entropy

Bond lengths and bond angles are those derived from new electron-diffraction data by Hargittai (11). These differ significantly from early results of Palmer (12). Microwave spectra were reported by Abbar (13), but structural parameters were not derived from the data. The adopted structure has an O-S-Cl angle of 107.7° and principal moments of inertia of I_A = 24.0520 x 10^{-39}, I_B = 36.0706 x 10^{39} and I_C = 43.8672 x 10^{39} g cm^2.

The vibrational assignment is from force-constant calculations of Toyuki and Shimizu (14). Pfeiffer's assignment (15) is only slightly different and results in negligible change in the thermodynamic functions. Spectral data were summarized by Gillespie and Robinson (16) in their study of the liquid-phase Raman spectra. Only those frequencies above 500 cm^{-1} are based on gas-phase spectra. Trautz et al. (10) found Cp = 15.4 ± 0.3 J K^{-1} mol^{-1} by calorimetry at 43 mmHg and 16 to 99°C; this Cp is too low, like $\Delta_{vap}H$ (10), and the error is almost 20%.

References

1. M. C. Londergan, Ph.D. Thesis, Iowa State College, 1943; Iowa State College J. Science 17, 95 (1942).
2. K. Arii, Bull. Inst. Phys.-Chem. Research 10, 256 (1931).
3. M. Trautz, Z. Elektrochem. 14, 534 (1908).
4. E. Neale and L. T. B. Williams, J. Chem. Soc. 1954, 2156 (1954).
5. J. Thomsen, "Thermochemische Untersuchungen", Vol. II, pp. 310-2, Barth, Leipzig, 1882.
6. U. S. Nat. Bur. Stand. Tech. Note 270-3, 1968.
7. R. H. McKee and C. M. Salls, Ind. Eng. Chem. 16, 279, 351 (1924).
8. K. Heumann and P. Kochlin, Berichte 16, 602 (1883).
9. V. M. Vdovenko and T. V. Kovaleva, J. Appl. Chem. USSR 31, 81 (1958); Zhur. Priklad. Khim. 31, 89 (1958).
10. M. Trautz, E. Baisch and A. von Dechend, Z. Elektrochem. 14, 271, 275 (1908).
11. I. Hargittai, Acta. Chim. (Budapest) 60, 231 (1969); 57, 403 (1968).
12. K. J. Palmer, J. Amer. Chem. Soc. 60, 2360 (1938).
13. C. Abbar, compt. Rend. 261, 365, 5047 (1965); 257, 2022 (1963).
14. H. Toyuki and K. Shimizu, Bull. Chem. Soc. Japan 39, 2364 (1966).
15. M. Pfeiffer, Z. Physik. Chem. (Leipzig) 240, 380 (1969).
16. R. J. Gillespie and E. A. Robinson, Can. J. Chem. 39, 2171 (1961).

Sulfuryl Chloride (SO_2Cl_2) $Cl_2O_2S_1(g)$

Enthalpy Reference Temperature = T_r = 298.15 K Standard State Pressure = p° = 0.1 MPa

T/K	$C_p°$ (J $K^{-1}mol^{-1}$)	S°	-[G°-H°(T_r)]/T	H°-H°(T_r) (kJ mol^{-1})	$\Delta_fH°$	$\Delta_fG°$	Log K_f
0	0.	0.	INFINITE	-16.028	-348.555	-348.555	INFINITE
100	43.179	245.860	370.394	-12.453	-351.487	-338.589	176.861
200	64.505	282.771	317.789	-7.004	-353.628	-324.765	84.820
250	71.750	297.982	312.336	-3.589	-354.309	-317.466	66.331
298.15	77.095	311.095	311.095	0.	-354.803	-310.323	54.367
300	77.276	311.573	311.097	0.143	-354.820	-310.047	53.984
350	81.651	323.826	312.055	4.120	-355.204	-302.553	45.154
400	85.216	334.969	314.233	8.294	-357.706	-294.923	38.513
450	88.175	345.182	317.112	12.631	-358.588	-287.030	33.318
500	90.655	354.604	320.396	17.104	-359.455	-279.030	29.150
600	94.520	371.494	327.538	26.373	-360.562	-262.831	22.881
700	97.321	386.286	334.896	35.973	-361.236	-246.485	18.393
800	99.387	399.423	342.156	45.813	-361.643	-230.062	15.021
900	100.940	411.223	349.186	55.833	-415.006	-212.518	12.334
1000	102.128	421.922	355.933	65.989	-413.905	-190.078	9.929
1100	103.052	431.701	362.383	76.250	-412.775	-167.749	7.966
1200	103.783	440.700	368.539	86.593	-411.629	-145.525	6.335
1300	104.370	449.031	374.414	97.002	-410.478	-123.396	4.958
1400	104.847	456.784	380.024	107.463	-409.330	-101.356	3.782
1500	105.240	464.031	385.386	117.968	-408.190	-79.398	2.765
1600	105.566	470.834	390.516	128.509	-407.065	-57.515	1.878
1700	105.840	477.242	395.431	139.080	-405.957	-35.702	1.097
1800	106.073	483.299	400.146	149.676	-404.871	-13.954	0.405
1900	106.271	489.039	404.675	160.293	-403.810	7.735	-0.213
2000	106.442	494.495	409.030	170.929	-402.775	29.368	-0.767
2100	106.590	499.692	413.225	181.581	-401.770	50.950	-1.267
2200	106.719	504.653	417.269	192.246	-400.796	72.486	-1.721
2300	106.833	509.400	421.172	202.924	-399.855	93.977	-2.134
2400	106.933	513.949	424.944	213.613	-398.949	115.428	-2.512
2500	107.021	518.316	428.592	224.310	-398.080	136.843	-2.859
2600	107.100	522.515	432.124	235.016	-397.247	158.223	-3.179
2700	107.170	526.558	435.547	245.730	-396.453	179.573	-3.474
2800	107.233	530.457	438.867	256.450	-395.697	200.893	-3.748
2900	107.290	534.221	442.091	267.176	-394.980	222.186	-4.002
3000	107.342	537.859	445.223	277.908	-394.302	243.457	-4.239
3100	107.388	541.379	448.268	288.645	-393.662	264.705	-4.460
3200	107.431	544.790	451.232	299.386	-393.058	285.932	-4.667
3300	107.469	548.096	454.117	310.131	-392.490	307.142	-4.862
3400	107.505	551.305	456.929	320.879	-391.956	328.335	-5.044
3500	107.537	554.422	459.670	331.632	-391.454	349.512	-5.216
3600	107.567	557.451	462.344	342.387	-390.981	370.676	-5.378
3700	107.594	560.399	464.955	353.145	-390.535	391.827	-5.532
3800	107.620	563.269	467.504	363.906	-390.115	412.967	-5.677
3900	107.643	566.065	469.996	374.669	-389.716	434.095	-5.814
4000	107.665	568.790	472.432	385.434	-389.338	455.214	-5.944
4100	107.685	571.449	474.814	396.202	-388.976	476.323	-6.068
4200	107.704	574.044	477.146	406.971	-388.628	497.423	-6.186
4300	107.721	576.579	479.429	417.742	-388.293	518.516	-6.299
4400	107.738	579.055	481.665	428.515	-387.966	539.601	-6.406
4500	107.753	581.477	483.857	439.290	-387.647	560.678	-6.508
4600	107.767	583.845	486.005	450.066	-387.333	581.748	-6.606
4700	107.781	586.163	488.111	460.843	-387.023	602.812	-6.699
4800	107.793	588.432	490.178	471.622	-386.714	623.869	-6.789
4900	107.805	590.655	492.205	482.402	-386.406	644.920	-6.875
5000	107.816	592.833	494.196	493.183	-386.096	665.964	-6.957
5100	107.827	594.968	496.151	503.965	-385.784	687.003	-7 036
5200	107.836	597.062	498.072	514.748	-385.470	708.034	-7.112
5300	107.846	599.116	499.959	525.532	-385.152	729.060	-7.185
5400	107.855	601.132	501.814	536.317	-384.831	750.080	-7.256
5500	107.863	603.111	503.638	547.103	-384.506	771.093	-7.323
5600	107.871	605.055	505.432	557.890	-384.177	792.102	-7.388
5700	107.878	606.964	507.196	568.678	-383.845	813.104	-7.451
5800	107.885	608.840	508.933	579.466	-383.509	834.101	-7.512
5900	107.892	610.685	510.642	590.255	-383.171	855.089	-7.570
6000	107.899	612.498	512.324	601.044	-382.832	876.075	-7.627

PREVIOUS: June 1971 (1 atm) CURRENT: June 1971 (1 bar)

Sulfuryl Chloride (SO_2Cl_2) $Cl_2O_2S_1(g)$

TUNGSTEN CHLORIDE OXIDE (WO_2Cl_2) CRYSTAL M_r = 286.7548

$\Delta_fH°(0\ K)$ = Unknown

S°(298.15 K) = [200.832] J K^{-1} mol^{-1} $\Delta_fH°(298.15\ K) = -780.32 \pm 5.9$ kJ mol^{-1}

Enthalpy of Formation

Shchukarev et al. (1) measured calorimetrically the enthalpies of reaction (1) and (2) at 298.15 K at -67.3±0.2 and -13.6±0.11 kcal mol^{-1}, respectively. The reactions are given as follows: (1) WO_2Cl_2(cr) + 4 NaOH(aq, 16.3 H_2O) + Na_2WO_4(aq, 80000 H_2O) + 2 NaCl(aq, 40000 H_2O) + 2 H_2O(ℓ) and (2) H_2WO_4(cr) + 2 NaOH(aq, 16.3 H_2O) + Na_2WO_4(aq, 20000 H_2O) + 2 H_2O(ℓ).

Based on these data and the enthalpies of dilution for NaOH(aq), NaCl(aq) and Na_2WO_4(aq), we obtain $\Delta_rH°(298.15\ K)$ = 53.77±1.0 kcal mol^{-1} for WO_2Cl_2(cr) + 2 OH^-(aq, ∞) + H_2WO_4(cr) + 2 Cl^-(aq,∞). This leads to $\Delta_fH°(WO_2Cl_2$, cr, 298.15 K) = -186.5 kcal mol^{-1} using $\Delta_fH°(H_2WO_4$, cr, 298.15 K) = -270.5±0.4 kcal mol^{-1}.

Heat Capacity and Entropy

$C_p°(300)$ = 25.0 cal K^{-1} mol^{-1} is estimated using Kopp's rule. Heat capacities at higher temperature are estimated from those of WO_2(cr), WO_3(cr), WCl_6(cr). The entropy S°(298.15 K) = 48.0 cal K^{-1} mol^{-1} is calculated from ΔS°(598.6 K) = 32.27±1.3 cal K^{-1} mol^{-1} for WO_2Cl_2(cr) + WO_2Cl_2(g). The value of ΔS°(598.6 K) is obtained from the 2nd law analysis of the vapor pressure data given by Shchukarev and Suvorov (4).

Decomposition Data

Shchukarev and Suvorov (4) found that WO_2Cl_2(cr) does not melt but decomposes. The saturated vapor over WO_2Cl_2(cr) consists mainly of $WOCl_4$(g) which will disproportionate to form WO_2Cl_2(g) and WCl_6(g). The estimated T_{dcm} = 642 K is derived by interpolation to one atmosphere in the total pressures of $WOCl_4$(g), WO_2Cl_2(g) reported by Shchukarev and Suvorov (4).

References

1. S. A. Shchukarev, I. V. Vasil'kova and G. I. Novikov, Zh. Neorg. Khim. 3, 2642 (1958).
2. V. B. Parker, U. S. Nat. Bur. Stand., NSRDS-NBS 2, 1965.
3. We have assumed $\Delta_{dil}H°$ = 0 for Na_2WO_4(aq, 20000 H_2O) + Na_2WO_4(aq, 80000 H_2O).
4. S. A. Shchukarev and A. V. Suvorov, Vestnik Leningrad. Univ. 16, No. 4, Ser. Fiz. i Khim., No. 1, 87 (1961). Earlier data reported by Shchukarev et al., Zh. Neorg. Khim. 3, 2630 (1958); 5, 1650 (1960), have been revised and they are not used here.
5. JANAF Thermochemical Tables: H_2WO_4(cr), 3-31-67.

Tungsten Chloride Oxide (WO_2Cl_2) $Cl_2O_2W_1$(cr)

Enthalpy Reference Temperature = T_r = 298.15 K Standard State Pressure = p° = 0.1 MPa

T/K	$C_p°$ (J $K^{-1}mol^{-1}$)	S° (J $K^{-1}mol^{-1}$)	$-[G°-H°(T_r)]/T$ (J $K^{-1}mol^{-1}$)	$H°-H°(T_r)$ (kJ mol^{-1})	$\Delta_fH°$ (kJ mol^{-1})	$\Delta_fG°$ (kJ mol^{-1})	Log K_f
0							
100							
200							
298.15	104.408	200.832	200.832	0.	-780.316	-702.781	123.124
300	104.600	201.478	200.834	0.193	-780.285	-702.300	122.281
400	115.060	233.003	205.062	11.176	-778.207	-676.597	88.354
500	125.520	259.809	213.391	23.209	-775.319	-651.511	68.063
600	135.562	283.597	223.145	36.272	-771.605	-627.085	54.593
700	144.557	305.202	233.347	50.298	-767.106	-603.344	45.022
800	151.042	324.947	243.581	65.092	-761.993	-580.292	37.889
900	155.854	343.031	253.641	80.451	-756.449	-557.909	32.380
1000	158.992	359.629	263.422	96.207	-750.625	-536.159	28.006

PREVIOUS: September 1962 CURRENT: March 1967

Tungsten Chloride Oxide (WO_2Cl_2) $Cl_2O_2W_1$(cr)

TUNGSTEN CHLORIDE OXIDE (WO_2Cl_2) IDEAL GAS M_r = 286.7548

$\Delta_fH°(0\ K)$ = -668.2 ± 25 kJ mol^{-1}

$S°(298.15\ K)$ = [353.927] J K^{-1} mol^{-1} $\Delta_fH°(298.15\ K)$ = -671.5 ± 25 kJ mol^{-1}

Vibrational Frequencies and Degeneracies

ν, cm^{-1}	ν, cm^{-1}	ν, cm^{-1}
984 (1)	[400](1)	[150](1)
972 (1)	[300](1)	[100](1)
[400](1)	[300](1)	[100](1)

Ground State Quantum Weight: 1 σ = [2]

Point Group: [C_{2v}]

Bond Distances: W-Cl = [2.26] Å W-O = [1.81] Å

Bond Angles: Cl-W-Cl = [113]° O-W-O = [109.47]° O-W-Cl = [90]°

Product of the Moments of Inertia: $I_AI_BI_C$ = [1.16512 x 10^{-112}] g^3 cm^6

Enthalpy of Formation

$\Delta_fH°(WO_2Cl_2$, g, 298.15 K) = -160.5 kcal mol^{-1} is obtained from $\Delta_rH°$(298.15 K) = 26.0 kcal mol^{-1} for $WO_2Cl_2(cr) \rightarrow WO_2Cl_2(g)$. The latter is calculated by the 3rd law method from the partial pressures of $WO_2Cl_2(g)$ given by Shchukarev and Suvorov (6). 2nd law analysis gives $\Delta_rH°$(598.6 K) = 24.1±0.8 kcal mol^{-1} $\Delta_rH°$(298.15 K) = 26.0 kcal mol^{-1}. These reported partial vapor pressures of $WO_2Cl_2(g)$ were derived by the optical tensimetric method from total pressures over $WO_2Cl_2(cr)$ considering the species $WOCl_4(g)$, $WO_2Cl_2(g)$ and $WCl_6(g)$. Since the decomposition of $WO_2Cl_2(cr)$ was very complicated, we tentatively assign 6 kcal mol^{-1} uncertainty to the adopted enthalpy of formation.

Shchukarev and Suvorov (6) have also derived a log Kp equation for the disproportionation process, 2 $WOCl_4(g) = WO_2Cl_2(g) + WCl_6(g)$. 3rd law analysis gives $\Delta_rH°$(298.15 K) = 10.3 kcal mol^{-1} for the above process. This leads to $\Delta_fH°(WO_2Cl_2$, g, 298.15 K) = -145.7±16 kcal mol^{-1}. The 2nd law $\Delta_rH°$(590 K) is 11.0 kcal mol^{-1} $\Delta_rH°$(298.15 K) = 10.1 kcal mol^{-1}. The drift in the 3rd law analysis is 0.3 cal K^{-1} mol^{-1}.

Heat Capacity and Entropy

The molecular configuration is assumed to be a distorted tetrahedron of C_{2v} symmetry, similar to that of $MoO_2Cl_2(g)$ (1), and $CrO_2Cl_2(g)$ (2). The bond distances W-O and W-Cl are estimated to be the same as those in $WO_3(g)$ and $WCl_6(g)$, respectively. The bond angles are assumed to be the same as those in $MoO_2Cl_2(g)$ (1). The principal moments of inertia are: I_A = 33.9761 x 10^{-39}, I_B = 53.4238 x 10^{-39} and I_C = 64.1894 x 10^{-39} g cm^2.

The vibrational frequencies, 984 cm^{-1} and 972 cm^{-1}, were observed in the infrared spectra Barraclough and Stals (3). They tentatively assigned these two as the symmetric and asymmetric W-O stretching frequencies. The rest of the frequencies are estimated by comparison with those observed in the infrared and Raman spectra for $CrO_2Cl_2(g)$ (4, 5). They are not listed in point group order.

References

1. L. E. Sutton, "Tables of Interatomic Distances and Configuration in Molecules and Ions," Special Publication No. 11, The Chemical Society, London, 1958.
2. K. J. Palmer, J. Amer. Chem. Soc. 60, 2360 (1938).
3. C. G. Barraclough and J. Stals, Aust. J. Chem. 19, 741 (1966).
4. H. Stammreich, K. Kawai and Y. Tavares, Spectrochim. Acta 1959, 438 (1959).
5. F. A. Miller, G. L. Carlson and W. B. White, Spectrochim. Acta 1959, 709 (1959).
6. S. A. Shchukarev and A. V. Suvorov, Vestnik Leningrad. Univ. 16, No. 4, Ser. Fiz. i Khim., No. 1, 87 (1961). Earlier data reported by Shchukarev et al., Zh. Neorg. Khim. 3, 2630 (1958); 5, 1650 (1960), have been revised and they are not used here.

Tungsten Chloride Oxide (WO_2Cl_2) $Cl_2O_2W_1(g)$

Enthalpy Reference Temperature = T_r = 298.15 K Standard State Pressure = p° = 0.1 MPa

T/K	$C_p°$ (J $K^{-1}mol^{-1}$)	S° (J $K^{-1}mol^{-1}$)	-[G°-H°(T_r)]/T (J $K^{-1}mol^{-1}$)	H°-H°(T_r) (kJ mol^{-1})	$\Delta_fH°$ (kJ mol^{-1})	$\Delta_fG°$ (kJ mol^{-1})	Log K_f
0	0.	0.	INFINITE	-19.504	-668.198	-668.198	INFINITE
100	59.017	273.602	424.320	-15.072	-670.235	-660.403	344.959
200	77.605	320.992	361.629	-8.127	-671.247	-650.101	169.789
250	83.093	338.927	355.341	-4.104	-671.457	-644.787	134.721
298.15	87.199	353.927	353.927	0.	-671.532	-639.642	112.063
300	87.339	354.467	353.929	0.161	-671.533	-639.445	111.337
350	90.723	368.194	355.006	4.616	-671.497	-634.098	94.634
400	93.447	380.493	357.436	9.223	-671.376	-628.762	82.108
450	95.648	391.631	360.627	13.952	-671.192	-623.446	72.368
500	97.435	401.804	364.243	18.781	-670.963	-618.153	64.578
600	100.093	419.820	372.043	28.666	-670.427	-607.640	52.900
700	101.913	435.394	380.006	38.772	-669.848	-597.221	44.565
800	103.198	449.092	387.803	49.031	-669.271	-586.886	38.320
900	104.131	461.303	395.304	59.400	-668.717	-576.621	33.466
1000	104.827	472.312	402.463	69.849	-668.199	-566.417	29.587
1100	105.358	482.329	409.275	80.360	-667.727	-556.261	26.415
1200	105.771	491.515	415.751	90.917	-667.305	-546.147	23.773
1300	106.098	499.995	421.909	101.511	-666.937	-536.066	21.539
1400	106.361	507.867	427.771	112.134	-666.625	-526.011	19.626
1500	106.576	515.213	433.359	122.781	-666.372	-515.977	17.968
1600	106.753	522.097	438.692	133.448	-666.182	-505.957	16.518
1700	106.902	528.574	443.791	144.131	-666.057	-495.947	15.239
1800	107.027	534.688	448.672	154.828	-665.998	-485.943	14.102
1900	107.133	540.477	453.353	165.536	-666.009	-475.940	13.084
2000	107.224	545.975	457.848	176.254	-666.092	-465.934	12.169
2100	107.303	551.208	462.170	186.980	-666.249	-455.923	11.340
2200	107.371	556.201	466.331	197.714	-666.484	-445.901	10.587
2300	107.431	560.976	470.343	208.454	-666.797	-435.869	9.899
2400	107.484	565.549	474.216	219.200	-667.191	-425 821	9.268
2500	107.531	569.938	477.957	229.951	-667.668	-415.754	8.687
2600	107.572	574.156	481.577	240.706	-668.227	-405.666	8.150
2700	107.609	578.217	485.081	251.465	-668.894	-395.555	7.652
2800	107.643	582.131	488.478	262.228	-669.707	-385.417	7.190
2900	107.672	585.909	491.773	272.994	-670.703	-375.248	6.759
3000	107.699	589.559	494.972	283.762	-671.911	-365.039	6.356
3100	107.724	593.091	498.080	294.533	-673.373	-354.786	5.978
3200	107.746	596.512	501.103	305.307	-675.119	-344.483	5.623
3300	107.766	599.827	504.045	316.083	-677.174	-334.120	5.289
3400	107.785	603.045	506.909	326.860	-679.570	-323.689	4.973
3500	107.802	606.169	509.701	337.639	-682.371	-313.182	4.674
3600	107.817	609.207	512.423	348.420	-685.669	-302.589	4.390
3700	107.832	612.161	515.079	359.203	-724.329	-291.705	4.118
3800	107.845	615.037	517.672	369.987	-725.329	-279.999	3.849
3900	107.857	617.838	520.204	380.772	-726.348	-268.266	3.593
4000	107.868	620.569	522.679	391.558	-727.383	-256.507	3.350
4100	107.879	623.233	525.100	402.345	-728.431	-244.722	3.118
4200	107.889	625.832	527.467	413.134	-729.490	-232.912	2.897
4300	107.898	628.371	529.784	423.923	-730.556	-221.076	2.686
4400	107.906	630.852	532.053	434.713	-731.628	-209.215	2.484
4500	107.914	633.277	534.276	445.504	-732.701	-197.331	2.291
4600	107.922	635.649	536.454	456.296	-733.775	-185.422	2.106
4700	107.928	637.970	538.589	467.089	-734.847	-173.490	1.928
4800	107.935	640.242	540.683	477.882	-735.916	-161.534	1.758
4900	107.941	642.468	542.738	488.676	-736.978	-149.556	1.594
5000	107.947	644.649	544.755	499.470	-738.034	-137.558	1.437
5100	107.952	646.786	546.734	510.265	-739.082	-125.537	1.286
5200	107.957	648.882	548.679	521.060	-740.120	-113.498	1.140
5300	107.962	650.939	550.589	531.856	-741.149	-101.437	1.000
5400	107.967	652.957	552.466	542.653	-742.167	-89.358	0.864
5500	107.971	654.938	554.311	553.450	-743.175	-77.260	0.734
5600	107.975	656.884	556.125	564.247	-744.172	-65.142	0.608
5700	107.979	658.795	557.910	575.045	-745.158	-53.008	0.486
5800	107.983	660.673	559.665	585.843	-746.134	-40.856	0.368
5900	107.986	662.519	561.393	596.641	-747.101	-28.690	0.254
6000	107.990	664.334	563.094	607.440	-748.058	-16.504	0.144

PREVIOUS: March 1967 (1 atm) CURRENT: March 1967 (1 bar)

Tungsten Chloride Oxide (WO_2Cl_2) $Cl_2O_2W_1(g)$

$\Delta_f H°$(0 K) = Unknown
$\Delta_f H°$(298.15 K) = -359.4 ± 0.8 kJ mol^{-1}
$\Delta_{fus}H°$ = 21.88 ± 0.84 kJ mol^{-1}

S°(298.15 K) = 135.98 ± 2.1 J K^{-1} mol^{-1}
T_{fus} = 774 K

Enthalpy of Formation

$\Delta_f H°$ of -85.9 ± 0.2 kcal mol^{-1} is obtained from a 3rd law analysis of six sets of electromotive force data (3, 4, 6, 7) for three different reactions, namely, (A) Pb(cr) + 2 AgCl(cr) = 2 Ag(cr) + $PbCl_2$(cr), (B) Pb(cr) + Hg_2Cl_2(cr) = 2 Hg(ℓ) + $PbCl_2$(cr), and (C) Pb(ℓ) + Cl_2(g) = $PbCl_2$(cr). Seven other sets of similar data (1, 2, 5) give $\Delta_f H°$ values which are within ±0.2 kcal mol^{-1} of adopted value, while one set of emf data (7) for (D) 2CuCl(cr) + Pb(cr) = $PbCl_2$(cr) + 2 Cu(cr) gives a $\Delta_f H°$ value of -85.4 kcal mol^{-1}. A $\Delta_f H°$ value can be obtained also from a 3rd law analysis of equilibrium data (8) for (E) $PbCl_2$(cr) + H_2(g) = Pb(ℓ) + 2 HCl(g). Auxiliary enthalpies of formation used in our analysis are $\Delta_f H°$(298.15 K), kcal mol^{-1}: AgCl, cr, = -30.370 (9), CuCl, cr = -33.0 ± 0.4 (10), Pb, ℓ = 1.025 (10), Hg_2Cl_2, cr = -63.319 ± 0.05 (10) and HCl, g = -22.063 ± 0.05 (10). We assign an uncertainty of ±0.2 kcal mol^{-1} to $\Delta_f H°$ to reflect the estimated error in the adopted value for S°(298.15 K).

Source	Reaction	Data Points	T/K	$-\Delta_r H°$(298.15 K), kcal mol^{-1} 2nd law	3rd law	Drift cal K^{-1} mol^{-1}	$\Delta_f H°$(298.15 K) kcal mol^{-1}
(1)	A	5	273-298	24.5	25.06±0.04	-1.8±0.3	85.8±0.1
	B	7	273-298	22.0	22.34±0.03	-1.3±0.3	85.7±0.1
(2) Set I	A	5	273-294	24.8	25.26±0.03	-1.7±0.1	86.0±0.1
Set II	A	5	273-294	24.9	25.25±0.03	-1.4±0.2	86.0±0.1
Set I	B	5	273-298	22.2	22.36±0.02	-0.7±0.1	85.7±0.1
Set II	B	5	273-296	22.2	22.36±0.02	-0.7±0.3	85.7±0.1
(3)	A	5	278-318	25.1	25.17±0.03	-0.1±0.5	85.9±0.1
Set I	B	5	278-318	22.7	22.54±0.04	0.4±0.7	85.9±0.1
Set II	B	5	288-308	22.7	25.54±0.04	0.5±0.7	85.9±0.1
(4)	A	3	288-308	25.2	25.17±0.02	0.2±0.9	85.9±0.1
(5)	C	Eqn	673-771	82.1	86.87±0.55	-6.6	85.8±0.6
(6)	C	2	723-773	86.8	86.98±0.02	-0.2	85.9±0.1
(7)	D	Eqn	363-553	19.8	19.44±0.13	0.9	85.4±1.0
	A	Eqn	293-573	24.9	25.20±0.80	-0.7	85.9±0.9
(8)	E	3	633-773	23.3	33.7 ±2.4	14.7±7.8	76.8±2.5

Heat Capacity and Entropy

$C_p°$ data below 298.15 K have been reported by Nerst (15.6-87.7 K, 11), Eucken (106.5-290.7 K, 12), and Bronsted (283 K, 13). We combine these results with $C_p°$ = 18.52 cal K^{-1} mol^{-1} (14) at 330 K and obtain S°(298.15 K) = 33.7 cal K^{-1} mol^{-1} based on S°(15.6 K) = 0.9 cal K^{-1} mol^{-1}. Kelley and King (15) analyzed the same $C_p°$'s and found S°(298.15 K) = 34.0 ± 1.0 cal K^{-1} mol^{-1}. Data obtained from cell measurements on both the crystal and liquid allow independent calculations of S°(298.15 K). Thirteen of the 23 sets of emf data htat we have analyzed are more compatible with a value for S°(298.15 K) which is at least one cal K^{-1} mol^{-1} less than that determined from the low temperature data. Since the cell measurements are judged to be more reliable, we adopt S°(298.15 K) = 32.5 ± 0.5 cal K^{-1} mol^{-1} and omit the $C_p°$ data below 298.25 K.

$C_p°$'s above 300 K are based on the smoothed relative enthalpies of Alim et al. (16). Their measurements were made in an isothermal diphenyl ether calorimeter over the temperature range 400 to 950 K. The $PbCl_2$ was reported to have a purity of 99.96%, and samples of this material were sealed in platinum capsules for the drop experiments. Their enthalpies are fitted to a polynomial which represents the data to within ±0.4%. The calculated $C_p°$'s above 300 K are joined graphically with those of Eucken (12) at 298.15 K by comparison with the $C_p°$-T curve for $BaCl_2$(cr) (10). Our adopted enthalpies are in reasonable agreement with those of Kelley (17). His results are based on the enthalpy measurements of Krestovnikov and Karetnikov (18), and most of the points agree with ours to better than a percent but show a maximum deviation of -1.3% at 500 K. Other enthalpy data that have been recently reported are those of Bloom and Tricklebank (693-773 K, 19) and Thompson and Flengas (577-745 K, 20). Both of these studies show rather poor agreement with our adopted results. Deviations from our values are -4.2% at 673 K to -5.0% at 773 K for (19), _0.1% at 680 K and -6.2% at 745 K for (20). Both drop investigations were conducted with quartz capsules which might have been slowly attacked by the $PbCl_2$ at the higher temperatures. The earlier enthalpy measurements of Goodwin and Kalmus (21) are considered less reliable and not included in our analysis.

Modestova (22) has reported the existence of a transition for $PbCl_2$ at 695 K based on differential thermal analysis and x-ray studies. This transition has not been observed by other workers, and we prefer not to include it. However, if it does exist, we speculate that the enthalpy associated with it is quite small; thus, our functions would be only slightly altered.

Fusion Data

T_{fus} = 774 K is from the enthalpy study of Alim et al. (16). Other values reported for T_{fus} based on enthalpy measurements are 771 (21), 774 (18), 774 (19) and 772 (20). DTA measurements (24) gave T_{fus} equal to 774 K, while cryoscopic studies (23) yielded 768 K. $\Delta_{fus}H°$ is calculated to be 5.23 kcal mol^{-1} from the adopted enthalpies for the crystal and liquid at T_{fus}. Published values (all in kcal mol^{-1}) include 5.15 (21), 5.70 (23), 2.98 (18), 5.3(19), 5.56 (20), and 5.03 (24).

Sublimation Data

See the ideal gas table for details.

References: References continued on the liquid table.
1. P. Gunther, Z. Elektrochem. 23, 197 (1917).
2. C. Krahmer, Z. Elektrochem. 26, 97 (1920).
3. R. H. Gerke, J. Amer. Chem. Soc. 44, 1684 (1922).
4. W. Jahn-Held and K. Jellinek, Z. Elektrochem. 42, 401 (1936).
5. B. F. Markov, Yu. K. Delimarskii and I. D. Panchenko, Zhur. Fiz. Khim. 28, 1987 (1954).
6. Yu. K. Delimarskii and Yu. G. Roms, Ukr. Khim. Zh. 29, 781 (1963).
7. W. G. Bugden and R. A. J. Sheldon, Inst. Mining Met., Trans., Sect. C 79, C215 (1970).
8. N. Parravano and G. Malzuori, Gazz. Chim. Ital. 56, 3 (1926).
9. U. S. Natl. Bur. Stand. Tech. Note 270-4, 1969.
10. JANAF Thermochemical Tables: CuCl(cr), 3-31-66; Pb(ℓ), 3-31-62; Hg_2Cl_2(cr), 12-31-61; HCl(g), 9-30-64; $BaCl_2$(cr), 12-31-72.

Enthalpy Reference Temperature = T_r = 298.15 K — Standard State Pressure = p° = 0.1 MPa

T/K	$C_p°$ (J $K^{-1}mol^{-1}$)	S°	-[G°-H°(T_r)]/T	H°-H°(T_r) (kJ mol^{-1})	$\Delta_f H°$	$\Delta_f G°$	Log K_f
0							
100							
200							
298.15	77.073	135.980	135.980	0.	-359.406	-314.121	55.033
300	77.170	136.457	135.981	0.143	-359.375	-313.840	54.644
400	80.094	159.059	139.043	8.006	-357.711	-298.911	39.034
500	82.985	177.242	144.922	16.160	-355.942	-284.413	29.712
600	85.881	192.628	151.622	24.603	-354.020	-270.285	23.530
700	88.772	206.083	158.460	33.336	-356.781	-255.694	19.080
774.000	90.915	215.110	163.451	39.984	------ CRYSTAL <--> LIQUID -----		
800	91.667	218.126	165.179	42.358	-354.476	-241.407	15.762
900	94.558	229.090	171.680	51.669	-351.870	-227.427	13.200
1000	97.454	239.203	177.933	61.270	-348.961	-213.753	11.165
1100	100.345	248.626	183.936	71.160	-345.746	-200.385	9.516
1200	103.240	257.482	189.699	81.339	-342.222	-187.324	8.154
1300	106.131	265.860	195.238	91.808	-338.397	-174.569	7.014
1400	109.027	273.831	200.570	102.566	-334.283	-162.119	6.049
1500	111.918	281.451	205.710	113.613	-329.891	-149.973	5.223

PREVIOUS: March 1962 CURRENT: June 1973

LEAD CHLORIDE ($PbCl_2$) LIQUID M_r = 278.106

S°(298.15 K) = [153.396] J K^{-1} mol^{-1}

T_{fus} = 774 K

$\Delta_f H°$(298.15 K) = [-344.251] kJ mol^{-1}

$\Delta_{fus}H°$ = 21.88 ± 0.84 kJ mol^{-1}

Enthalpy of Formation

$\Delta_f H°(PbCl_2$, ℓ, 298.15 K) is calculated from that of the crystal by addition of $\Delta_{fus}H°$ and the difference in enthalpy, H°(774 K)-H°(298.15 K), between the crystal and liquid. Independent values of $\Delta_f H°$(ℓ) (or $\Delta_f H°$(cr)) can be obtained from a 2nd and 3rd analysis of nine sets of emf data (1-9) for the formation cell of molten $PbCl_2$ ((A) Pb(ℓ) + Cl_2(g) = $PbCl_2$(ℓ)) and three sets of equilibrium data (10, 11) for (B) $PbCl_2$(ℓ) + H_2(g) = Pb(ℓ) + 2 HCl(g). Auxiliary enthalpies of formation used in the analysis are $\Delta_f H°$(Pb, ℓ, 298.15 K) = 1.025 kcal mol^{-1} (12) and $\Delta_f H°$(HCl, g, 298.15 K) = -22.063±0.05 kcal mol^{-1} (12). Five of the nine sets of emf data yield values for $\Delta_f H°$(cr) which are in good agreement (±0.2 kcal mol^{-1}) with the adopted value. Also, the results of the extensive emf measurements of Warner and Seifert (8) by the electronic commutator method are quite compatible with our adopted functions.

Source	Reaction	Data Points	T/K	$-\Delta_r H°$, kcal mol^{-1} 2nd law	3rd law	Drift cal $K^{-1}mol^{-1}$	$-\Delta_f H°$(298.15 K) kcal mol^{-1} (ℓ)	(cr)
Lorenz and Fox (1)	A	14	776-942	86.9	82.45±0.58	5.1±0.5	81.43	85.05
Hildenbrand and Ruhle (2)	A	7	774-880	88.9	82.95±0.62	7.2±0.3	81.93	85.55
Lorenz and Velde (3)	A	47[a]	826-1016	87.1	83.18±0.52	4.4±0.2	82.16	85.78
Wachter and Hildenbrand (4)	A	18	772-855	84.9	83.31±0.11	2.0±0.3	82.29	85.91
Holub et al. (5)	A	3	843-923	86.0	83.08±0.27	3.3±0.2	82.06	85.68
Lantratov and Alabyshev (6)	A	47[b]	814-954	82.5	83.69±0.15	1.4±0.1	82.67	86.29
Delimarskii and Roms (7)	A	2	8231, 873	83.5	83.56		82.54	86.16
Warner and Seifert (8)	A	Equation	775-1139	83.4	83.32±0.03	0.1	82.30	85.92
Hagemark and Hengstenberg (9)	A	Equation	773-893	84.6	83.51±0.15	1.3±0.1	82.49	86.11
Bagdasarian (10) Set I	B	10	876-978	40.0	38.89±0.36	-1.2±1.4	81.99	85.61
Set II	B	21	871-1220	44.2	38.08±1.26	-5.8±0.3	81.18	84.80
Jellinek and Rudat (11)	B	5	833-1133	37.8	38.62±0.51	0.8±1.1	81.72	85.34

(3)[a] and (1)[b] points rejected due to failure of a statistical test.

Heat Capacity and Entropy

$C_p°$ is assumed constant at 26.65 cal K^{-1} mol^{-1} over the temperature range 500-2000 K and is based on the smoothed relative enthalpies (774-950 K) of Alim et al. (13). A glass transition is assumed at 500 K below which $C_p°$ is that of the crystal. The smoothed enthalpies of Kelley (14) (800-1000 K) deviate by +4.0% from our values, while those of Thompson and Flengas (15) (813-945 K) deviate by ±1%. The enthalpies of Bloom and Tricklebank (16) are consistently lower than our values by -3.1% at 823 K, and the deviation increases to -12.6% at 1073 K. Warner and Seifert (8) have determined $C_p°$ data for $PbCl_2$(ℓ) from their emf measurements. They reported values of (all in cal K^{-1} mol^{-1}) 25.7 (800 K), 26.2 (900 K), 26.5 (1000 K) and 26.7 (1100 K). Bizouard and Pauty (17) measured $C_p°$ for the liquid as 24.7 cal K^{-1} mol^{-1}. S°(298.15 K) is calculated in a manner analogous to that used for the enthalpy of formation.

Vaporization Data

T_{vap} is the temperature at which $\Delta_r G°$ = 0 for the process $PbCl_2$(ℓ) → $PbCl_2$(g). $\Delta_{vap}H°$ is calculated as the difference between the enthalpies of formation of the gas and liquid at T_{vap}.

References - PbCl2(ℓ)
1. R. Lorenz and M. G. Fox, Z. Physik. Chem. 63, 109 (1908).
2. J. H. Hildenbrand and G. C. Ruhle, J. Amer. Chem. Soc. 49, 722 (1927).
3. R. Lorenz and H. Velde, Z. Anorg. Chem. 183, 81 (1929).
4. A. Wachter and J. H. Hildenbrand, J. Amer. Chem. Soc. 52, 4655 (1930).
5. L. Holub, F. Neubert, and F. Sauerwald, Z. Physik. Chem. 174, 161 (1935).
6. M. F. Lantratov and A. F. Alabyshev, J. Appl. Chem. (USSR) 26, 235 (1953).
7. Yu. K. Delimarskii and Yu. G. Roms, Ukr. Khim. Zh. 29, 781 (1963).
8. T. B. Warner and R. L. Seifert, J. Phys. Chem. 69, 1034 (1965).
9. K. Hagemark and D. Hengstenberg, J. Chem. Eng. Data 11, 596 (1966).
10. A. B. Bagdasarian, Trans. Amer. Electrochem. Soc. 51, 449 (1927).
11. K. Jellinek and A. Rudat, Z. Physik. Chem. 143, 244 (1929).
12. JANAF Thermochemical Tables: Pb(ℓ), 3-31-62; HCl(g), 9-30-64.
13. A. S. Alim, P. Kurtz, and W. D. Van Vorst, J. Chem. Eng. Data 13, 24 (1968).
14. K. K. Kelley, U. S. Bur. Mines Bull. 584, 1960.
15. W. T. Thompson and S. N. Flengas, Can. J. Chem. 49, 1550 (1971).
16. H. Bloom and S. B. Tricklebank, Aust. J. Chem. 19, 187 (1966).
17. M. Bizouard and F. Pauty, Compt. rend. 252, 514 (1961).

References - $PbCl_2$(cr) - continued
11. W. Nerst, Ann. Physik 36, 395 (1911).
12. A. Eucken, Physik. Z. 10, 586 (1910).
13. J. N. Bronsted, Z. Elektrochem. 18, 714 (1912).
14. A. Magnus, Ann. Physik. 31, 597 (1910).
15. K. K. Kelley and E. G. King, U. S. Bur. Mines Bull. 592, 1961.
16. A. S. Alim, P. Kurtz and W. D. Van Vorst, J. Chem. Eng. Data 13, 24 (1968).
17. K. K. Kelley, U. S. Bur. Mines Bull. 584, 1960.
18. A. N. Krestovnikov and G. A. Karetnikov, J. Gen. Chem. (USSR) 6, 955 (1936).
19. H. Bloom and S. B. Tricklebank, Aust. J. Chem. 19, 187 (1966).
20. W. T. Thompson and S. N. Flengas, Can. J. Chem. 49, 1550 (1971).
21. H. M. Goodwin and H. T. Kalmus, Phys. Rev. 28, 1 (1960).
22. T. P. Modestova, Russ. J. Inorg. Chem. 5, 804 (1960).
23. M. Blanc and G. Petit, Compt. rend. 248, 1305 (1959).
24. L. A. Raznitskii, V. A. Kholler, and S. E. Filippova, Russ. J. Phys. Chem. 44, 299 (1970).

Lead Chloride ($PbCl_2$) Cl_2Pb_1(l)

Enthalpy Reference Temperature = T_r = 298.15 K Standard State Pressure = p° = 0.1 MPa

T/K	$C_p°$ (J $K^{-1}mol^{-1}$)	S°	$-[G°-H°(T_r)]/T$	$H°-H°(T_r)$ (kJ mol^{-1})	$\Delta_f H°$	$\Delta_f G°$	Log K_f
0							
100							
200							
298.15	77.073	153.396	153.396	0.	-344.251	-304.160	53.288
300	77.170	153.873	153.398	0.143	-344.221	-303.911	52.916
400	80.094	176.475	156.460	8.006	-342.557	-290.723	37.965
500	82.985	194.659	162.338	16.160	-340.787	-277.967	29.039
500.000	82.985	194.659	162.338	16.160	GLASS <--> LIQUID TRANSITION		
500.000	111.504	194.659	162.338	16.160			
600	111.504	214.988	169.470	27.311	-336.158	-265.840	23.143
700	111.504	232.177	177.232	38.461	-336.502	-253.680	18.930
774.000	111.504	243.382	183.030	46.712	------ CRYSTAL <--> LIQUID -----		
800	111.504	247.066	185.051	49.611	-332.068	-242.151	15.811
900	111.504	260.199	192.686	60.762	-327.623	-231.178	13.417
1000	111.504	271.947	200.035	71.912	-323.164	-220.701	11.528
1100	111.504	282.575	207.063	83.063	-318.689	-210.671	10.004
1200	111.504	292.277	213.766	94.213	-314.194	-201.050	8.751
1300	111.504	301.202	220.153	105.363	-309.687	-191.804	7.707
1400	111.504	309.465	226.241	116.514	-305.181	-182.905	6.824
1500	111.504	317.158	232.049	127.664	-300.686	-174.328	6.071
1600	111.504	324.354	237.595	138.814	-296.207	-166.050	5.421
1700	111.504	331.114	242.900	149.965	-291.751	-158.052	4.856
1800	111.504	337.488	247.979	161.115	-287.320	-150.315	4.362
1900	111.504	343.516	252.850	172.265	-282.917	-142.824	3.927
2000	111.504	349.236	257.528	183.416	-278.544	-135.564	3.541

PREVIOUS: March 1962 CURRENT: June 1973

Lead Chloride ($PbCl_2$) Cl_2Pb_1(l)

LEAD CHLORIDE ($PbCl_2$) CRYSTAL-LIQUID M_r = 278.106 Lead Chloride ($PbCl_2$) $Cl_2Pb_1(cr,l)$

0 to 774 K crystal
above 774 K liquid

Refer to the individual tables for details.

Enthalpy Reference Temperature = T_r = 298.15 K; Standard State Pressure = $p°$ = 0.1 MPa

T/K	$C_p°$ (J K⁻¹mol⁻¹)	$S°$ (J K⁻¹mol⁻¹)	$-[G°-H°(T_r)]/T$ (J K⁻¹mol⁻¹)	$H°-H°(T_r)$ (kJ mol⁻¹)	$\Delta_fH°$ (kJ mol⁻¹)	$\Delta_fG°$ (kJ mol⁻¹)	Log K_f
0							
100							
200							
298.15	77.073	135.980	135.980	0.	-359.406	-314.121	55.033
300	77.170	136.457	135.981	0.143	-359.375	-313.840	54.644
400	80.094	159.059	139.043	8.006	-357.711	-298.911	39.034
500	82.985	177.242	144.922	16.160	-355.942	-284.413	29.712
600	85.881	192.628	151.622	24.603	-354.020	-270.285	23.530
700	88.772	206.083	158.460	33.336	-356.781	-255.694	19.080
774.000	90.915	215.110	163.451	39.984	CRYSTAL <--> LIQUID		
774.000	111.504	243.382	163.451	61.867	TRANSITION		
800	111.504	247.066	166.109	64.766	-332.068	-242.151	15.811
900	111.504	260.199	175.848	75.916	-327.623	-231.178	13.417
1000	111.504	271.947	184.881	87.067	-323.164	-220.701	11.528
1100	111.504	282.575	193.286	98.217	-318.689	-210.671	10.004
1200	111.504	292.277	201.137	109.367	-314.194	-201.050	8.751
1300	111.504	301.202	208.496	120.518	-309.687	-191.804	7.707
1400	111.504	309.465	215.416	131.668	-305.181	-182.905	6.824
1500	111.504	317.158	221.946	142.818	-300.686	-174.328	6.071
1600	111.504	324.354	228.124	153.969	-296.207	-166.050	5.421
1700	111.504	331.114	233.985	165.119	-291.751	-158.052	4.856
1800	111.504	337.488	239.560	176.269	-287.320	-150.315	4.362
1900	111.504	343.516	244.874	187.420	-282.917	-142.824	3.927
2000	111.504	349.236	249.951	198.570	-278.544	-135.564	3.541

PREVIOUS: CURRENT: June 1973

Lead Chloride ($PbCl_2$) $Cl_2Pb_1(cr,l)$

LEAD CHLORIDE ($PbCl_2$) IDEAL GAS M_r = 278.106

$\Delta_f H°(0\ K) = -172.03 \pm 1.3\ kJ\ mol^{-1}$

S°(298.15 K) = 317.205 ± 2.9 J K^{-1} mol^{-1} $\Delta_f H°(298.15\ K) = -174.05 \pm 1.3\ kJ\ mol^{-1}$

Point Group: C_{2v} σ = 2

Bond Distance: Pb-Cl = 2.46 ± 0.02 Å

Bond Angle: Cl-Pb-Cl = 96 + 3°

Product of Moments of Inertia: $I_AI_BI_C$ = 5.90383 x 10^{-113} g^3 cm^6

Electronic Levels and Quantum Weights

State	ε_i, cm^{-1}	g_i
$^1A_1$	0	1
$^3B_1$	[22000]	3

Vibrational Frequencies and Degeneracies

ν_i, cm^{-1}
329 (1)
80 (1)
307 (1)

Enthalpy of Formation

$\Delta_f H°$ of -41.6 ± 0.3 kcal mol^{-1} is obtained from $\Delta_f H°(cr)$ = -85.9 ± 0.2 kcal mol^{-1} by addition of $\Delta_{sub}H°(298.15\ K)$. The adopted value for $\Delta_{sub}H°(298.15\ K)$ is 44.3 ± 0.1 kcal mol^{-1} and is based on a 3rd law analysis of six sets of vapor pressure data (4, 9, 11, 14, 15) for both the crystal and liquid. Results of our analysis of a total of eighteen sets of Vp data are given below. Seven (1, 2, 3, 7, 8, 10, 12) of the remaining twelve sets yield $\Delta_f H°(g)$ values which are within 0.3 kcal mol^{-1} of the adopted value. Mass spectrometric studies (16) of the equilibrium vapors over the condensed phases indicated that the predominant species is the monomer.

Source	Reaction*	T/K	Data Points	$\Delta_{sub}H°$(or v), kcal mol^{-1} 2nd law	3rd law	Drift cal K^{-1} mol^{-1}	$\Delta_f H°(PbCl_2$, g, 298.15 K) kcal mol^{-1}
(1)	A	673-748	4	47.6	45.41±0.21	-3.2±0.1	40.5±0.4
	B	773-1223	19	42.5	40.93±0.49	-1.6±0.2	41.3±0.7
(2)	B	1043-1227	7	37.6	40.62±0.42	2.6±0.8	41.7±0.6
(3)	B	955-1266	12[a]	39.8	40.75±0.25	0.8±0.3	41.5±0.4
(4)	A	708-766	5	37.1	44.34±0.53	9.8±3.2	41.6±0.7
	B	783-865	7	44.3	40.69±0.30	-4.4±0.7	41.6±0.5
(5)	B	1033-1043	2	36.1	40.19±0.06	3.9	42.1±0.3
(6)	A	653-733	7	49.0	45.13±0.31	-5.6±0.2	40.8±0.5
(7)	B	953-1215	9[a]	43.6	40.94±0.69	-2.5±0.9	41.3±0.9
(8)	B	865-1272	18	38.9	40.60±0.50	1.6±0.5	41.7±0.7
(9)	B	950-1250	Eqn	40.9	40.67±0.04	-0.2	41.6±0.3
(10)	B	931-1154	4	41.0	40.74±0.24	-0.2±0.8	41.5±0.4
(11)	B	923	1		40.70		41.6±0.3
(12)	B	821-1259	56[b]	39.9	40.73±0.24	0.8±0.1	41.5±0.4
(13)	A	708-758	3	45.1	42.93±0.37	-3.0±6.8	43.0±0.6
	B	793-893	3	31.3	40.24±1.14	10.6±3.8	41.6±0.3
(14)	B	914-1249	Eqn	40.2	40.69±0.12	0.5	41.6±0.3
(15)	B	1071	1		40.63		41.6±0.3

*A) $PbCl_2(cr) = PbCl_2(g)$ B) $PbCl_2(\ell) = PbCl_2(g)$. (1)[a] and (2)[b] points rejected due to failure of a statistical test

Heat Capacity and Entropy

The bond length is taken from the electron diffraction study of Lister and Sutton (17). Electric deflection experiments by Buchler et al. (18) showed that $PbCl_2$ possessed a permanent dipole moment; thus, the molecule must have a bent structure. Infrared studies (19) of matrix isolated $PbCl_2$ have led to a value for the bond angle of 96 ± 3° which we adopt. This value is based on the observed Cl isotopic splitting of the antisymmetrical stretching frequency (ν_3). The principal moments of inertia are: $I_A = 23.7690 \times 10^{-39}$, $I_B = 39.3510 \times 10^{-39}$, and $I_C = 63.1200 \times 10^{-39}$ g cm^2.

The vibrational frequencies are from the matrix infrared studies of Hastie et al. (19). Both of the stretching frequencies (ν_1 and ν_3) were observed in matrices of N_2, Ar, and Ne. The adopted values are gas-phase frequencies which were estimated by Hastie et al. (19) from their matrix data. Other matrix infrared (20) and matrix Raman (21) studies have led to stretching frequencies which deviate by less than 30 cm^{-1} from our adopted values. Also, ν_1 is in agreement with the value of 314 cm^{-1} which was observed by Beattie and Perry (22) in a gas-phase Raman study at 1000°C. The bending frequency (ν_2) is somewhat more uncertain that the stretching frequencies. Hastie et al. (19) observed a relatively weak band at 80 cm^{-1} when $PbCl_2$ was trapped in a neon matrix and assigned this band to ν_2. Gas-phase (22) and matrix (21) Raman studies have indicated that ν_2 lies much closer to 100 cm^{-1}. The gas-phase Raman spectrum (22) was obtained in the presence of an excess of chlorine to suppress decomposition and subsequent resonance fluorescence. We prefer the lower value (80 cm^{-1}) but assign an uncertainty of ±20 cm^{-1} which corresponds to 0.5 cal K^{-1} mol^{-1} in S°(298.15 K).

The ground state is assumed to be $^1A_1$ based on an analogy with those for other Group IV A dichlorides (23). The upper $^1B_1$ level at 31000 cm^{-1} was observed by Hastie et al. (23). Also included is a triplet level ($^3B_1$) at 22000 cm^{-1}. The energy separation for the $^1B_1$-$^3B_1$ states (9000 cm^{-1}) is assumed to be similar to that observed for other Group IV A dichlorides (23).

References

1. E. D. Eastman and L. H. Duschak, U. S. Bur. Mines Tech. Paper 225, 73 (1919).
2. H. V. Wartenberg and O. Boase, Z. Electrochem. 28, 384 (1922).
3. C. G. Maier, U. S. Bur. Mines Tech. Paper 360, 28 (1925).
4. F. Volmer, Physik. Z. 30, 590 (1929).
5. B. Greiner and K. Jellinek, Z. Physik. Chem. A165, 97 (1933).
6. K. Niwa, M. Sato, and M. Yasiyama, J. Chem. Soc. Japan 60, 918 (1939).
7. D. N. Tarasenkov and P. A. Kozhmyakov, J. Gen. Chem. (USSR) 5, 830 (1935).
8. W. Jahn-Held and K. Jellinek, Z. Elektrochem. 43, 491 (1937).
9. J. L. Barton and H. Bloom, J. Phys. Chem. 60, 1413 (1956).
10. G. I. Novikov and O. G. Polyachenok, Russ. J. Inorg. Chem. 6, 996 (1961).
11. H. Bloom and J. W. Hastie, J. Phys. Chem. 72, 2361 (1968).
12. L. C. George and J. W. Jensen, U. S. Bur. Mines RI 6421, 1964.
13. I. I. Naryshkin, V. I. Glazov, and M. U. Klots, Tr. Leningrad. Politekh. Inst. No. 304, 31 (1970).
14. L. C. George, R. M. Doerr, and A. Visnapuu, J. Chem. Eng. Data 14, 23 (1969).
15. H. Bloom, S. R. Richards, and D. J. Williams, Aust. J. Chem. 25, 2077 (1972).
16. H. Bloom and J. W. Hastie, Aust. J. Chem. 19, 1003 (1966).
17. M. W. Lister and L. E. Sutton, Trans. Faraday Soc. 37, 406 (1941).
18. A. Buchler, J. L. Stauffer,, and W. Klemperer, J. Am. Chem. Soc. 86, 4544 (1964).
19. J. W. Hastie, R. H. Hauge, and J. L. Margrave, High Temp. Sci. 3, 56 (1971).
20. L. Andrews and D. L. Frederick, J. Am. Chem. Soc. 92, 775 (1970).
21. G. A. Ozin and A. Vander Voet, J. Chem. Phys. 56, 4768 (1970).
22. I. R. Beattie and R. O. Perry, J. Chem. Soc. A1970, 2429 (1970).
23. J. W. Hastie, R. H. Hauge and J. L. Margrave, J. Mol. Spectroscopy 29, 152 (1969).

Lead Chloride ($PbCl_2$) $Cl_2Pb_1(g)$

Enthalpy Reference Temperature = T_r = 298.15 K Standard State Pressure = p° = 0.1 MPa

T/K	$C_p°$ (J $K^{-1}mol^{-1}$)	S°	$-[G°-H°(T_r)]/T$	$H°-H°(T_r)$ (kJ mol^{-1})	$\Delta_f H°$	$\Delta_f G°$	Log K_f
0	0.	0.	INFINITE	-14.038	-172.033	-172.033	INFINITE
100	44.386	262.096	363.997	-10.190	-172.863	-176.489	92.189
200	52.288	295.698	322.179	-5.296	-173.532	-179.824	46.965
250	54.098	307.577	318.109	-2.633	-173.793	-181.367	37.895
298.15	55.184	317.205	317.205	0.	-174.054	-182.802	32.026
300	55.218	317.547	317.207	0.102	-174.065	-182.857	31.838
350	55.946	326.117	317.882	2.882	-174.357	-184.299	27.505
400	56.442	333.622	319.390	5.693	-174.673	-185.698	24.250
450	56.793	340.291	321.349	8.524	-175.014	-187.056	21.713
500	57.050	346.289	323.548	11.371	-175.380	-188.375	19.679
600	57.392	356.724	328.233	17.094	-176.178	-190.900	16.619
700	57.602	365.587	332.952	22.845	-181.921	-192.487	14.364
800	57.740	373.289	337.524	28.612	-182.870	-193.932	12.662
900	57.836	380.095	341.883	34.391	-183.797	-195.259	11.333
1000	57.904	386.193	346.014	40.178	-184.701	-196.483	10.263
1100	57.955	391.714	349.922	45.972	-185.583	-197.619	9.384
1200	57.994	396.759	353.618	51.769	-186.441	-198.675	8.648
1300	58.025	401.402	357.117	57.570	-187.283	-199.660	8.022
1400	58.049	405.703	360.436	63.374	-188.124	-200.581	7.484
1500	58.068	409.709	363.589	69.180	-188.973	-201.441	7.015
1600	58.084	413.457	366.590	74.987	-189.838	-202.244	6.603
1700	58.098	416.978	369.451	80.797	-190.723	-202.992	6.237
1800	58.109	420.300	372.185	86.607	-191.631	-203.688	5.911
1900	58.119	423.442	374.800	92.418	-192.567	-204.333	5.617
2000	58.127	426.423	377.308	98.231	-193.532	-204.927	5.352
2100	58.135	429.259	379.715	104.044	-371.749	-198.357	4.934
2200	58.142	431.964	382.029	109.857	-372.389	-190.086	4.513
2300	58.149	434.548	384.256	115.672	-373.130	-181.783	4.128
2400	58.157	437.023	386.404	121.487	-373.975	-173.446	3.775
2500	58.166	439.398	388.476	127.303	-374.927	-165.071	3.449
2600	58.176	441.679	390.479	133.121	-375.985	-156.657	3.147
2700	58.188	443.875	392.416	138.939	-377.148	-148.199	2.867
2800	58.202	445.991	394.292	144.758	-378.414	-139.697	2.606
2900	58.220	448.034	396.110	150.579	-379.777	-131.147	2.362
3000	58.241	450.008	397.874	156.402	-381.232	-122.549	2.134
3100	58.266	451.918	399.587	162.228	-382.773	-113.901	1.919
3200	58.297	453.769	401.251	168.056	-384.392	-105.203	1.717
3300	58.332	455.563	402.870	173.887	-386.081	-96.452	1.527
3400	58.373	457.305	404.445	179.722	-387.830	-87.649	1.347
3500	58.421	458.998	405.980	185.562	-389.632	-78.794	1.176
3600	58.476	460.644	407.476	191.407	-391.476	-69.887	1.014
3700	58.537	462.247	408.934	197.257	-393.353	-60.928	0.860
3800	58.606	463.809	410.358	203.114	-395.254	-51.919	0.714
3900	58.683	465.333	411.748	208.979	-397.171	-42.858	0.574
4000	58.767	466.819	413.107	214.851	-399.094	-33.749	0.441
4100	58.859	468.272	414.434	220.732	-401.016	-24.591	0.313
4200	58.960	469.691	415.733	226.623	-402.928	-15.387	0.191
4300	59.068	471.080	417.004	232.525	-404.824	-6.138	0.075
4400	59.184	472.439	418.249	238.437	-406.697	3.156	-0.037
4500	59.308	473.770	419.468	244.362	-408.539	12.491	-0.145
4600	59.440	475.075	420.662	250.299	-410.347	21.868	-0.248
4700	59.579	476.355	421.834	256.250	-412.114	31.283	-0.348
4800	59.725	477.611	422.983	262.215	-413.836	40.735	-0.443
4900	59.878	478.844	424.110	268.195	-415.509	50.223	-0.535
5000	60.037	480.055	425.217	274.191	-417.130	59.744	-0.624
5100	60.203	481.246	426.304	280.203	-418.695	69.297	-0.710
5200	60.374	482.417	427.372	286.232	-420.201	78.880	-0.792
5300	60.551	483.568	428.421	292.278	-421.647	88.492	-0.872
5400	60.732	484.702	429.453	298.342	-423.031	98.130	-0.949
5500	60.918	485.818	430.468	304.424	-424.352	107.794	-1.024
5600	61.108	486.917	431.466	310.526	-425.608	117.480	-1.096
5700	61.301	488.000	432.448	316.646	-426.798	127.189	-1.166
5800	61.497	489.068	433.415	322.786	-427.924	136.918	-1.233
5900	61.696	490.121	434.368	328.946	-428.982	146.666	-1.298
6000	61.896	491.160	435.306	335.125	-429.976	156.431	-1.362

PREVIOUS: June 1973 (1 atm) CURRENT: June 1973 (1 bar)

Lead Chloride ($PbCl_2$) $Cl_2Pb_1(g)$

LEAD CHLORIDE, ION ($PbCl_2^+$) IDEAL GAS M_r = 278.10545

$\Delta_fH°(0\ K) = 812.26 \pm 50.2\ kJ\ mol^{-1}$

$S°(298.15\ K) = [318.57 \pm 8.4]\ J\ K^{-1}\ mol^{-1}$ $\Delta_fH°(298.15\ K) = [816.298]\ kJ\ mol^{-1}$

Electronic Levels and Quantum Weights

State	ε_i, cm^{-1}	g_i
$^2A_1$	0	[2]
$^2B_1$	[20000]	[2]

Vibrational Frequencies and Degeneracies

ν, cm^{-1}
[300](1)
[100](1)
[350](1)

Point Group: $[C_{2v}]$ $\sigma = 2$

Bond Distance: Pb-Cl = [2.30] Å

Bond Angle: Cl-Pb-Cl = [116]°

Product of the Moments of Inertia: $I_AI_BI_C = [3.37567 \times 10^{-113}]\ g^3\ cm^6$

Enthalpy of Formation

The appearance potential for $PbCl_2^+$ from $PbCl_2(g)$ has been determined spectrometrically as (in eV) 11.2±0.2 (1), 10.3±0.2 (2) and 10.3±0.5 (3). Hastie et al. (2) preferred the lower AP for $PbCl_2^+$, since this value was obtained with more sensitive equipment. Subsequent measurements by Hastie and Swingler (4) confirm the lower result. They reported an ionization efficiency curve for $PbCl_2^+$ from which we obtain AP = 10.0 eV by the linear extrapolation method. Also, molecular orbital calculations by Hastie and Margrave (5) indicated IP($PbCl_2$,g) = 10.3 eV. We adopt a mean value from (2, 3, and 4) of 10.2±0.5 eV (235.2 kcal mol^{-1}) for the process $PbCl_2(g) + e^- \rightarrow PbCl_2^+(g) + 2e^-$ which gives $\Delta_fH°(PbCl_2^+, g, 0\ K) = 194.1 \pm 12.0\ kcal\ mol^{-1}$ when AP is combined with $\Delta_fH°(PbCl_2, g, 0\ K) = -41.1 \pm 0.3\ kcal\ mol^{-1}$ (6). $\Delta_fH°(298\ K) = 195.1\ kcal\ mol^{-1}$.

Heat Capacity and Entropy

The correlation diagram of Walsh (7) predicts a bent configuration for $PbCl_2^+$, since the molecule has seventeen valence electrons. Also, Walsh (7) predicted that the apex angle for a non-hydride AB_2 molecule would decrease markedly in going from a 17- to an 18- electron molecule. We assume that the bond angle in $PbCl_2^+$ is 20° larger than that for $PbCl_2$ (6). The adopted value (116°) is consistent with JANAF bond angles for the isoelectronic molecules BCl_2(112°) and $AlCl_2$(120°) (6). The bond length is assumed to lie between that for PbCl and $PbCl_2$ (6). The principal moments of inertia are: $I_A = 13.0315 \times 10^{-39}$, $I_B = 44.7956 \times 10^{-39}$ and $I_C = 57.8271 \times 10^{-39}\ g\ cm^2$.

The vibrational frequencies are estimated by comparison with those for BCl_2, CCl_2, $SiCl_2$, and $PbCl_2$ (6). The ground state is assumed to be $^2A_1$ by analogy with that for the isoelectronic molecule NO_2 (8). MO calculations by Hastie and Margrave (5) for the Group III A dihalides predicted an upper $^2B_1$ state with the $^2A_1$-$^2B_1$ separation in the energy range 1-2 eV. We have included this state and estimated the $^2A_1$-$^2B_1$ separation (20,000 cm^{-1}) to be somewhat greater than that for AlO_2 (15,000 cm^{-1}) (6), BCl_2 (11,000 cm^{-1}) (6) and NO_2 (<15,000 cm^{-1}) (8).

References

1. H. Bloom and J. W. Hastie, Aust. J. Chem. 19, 1003 (1966).
2. J. W. Hastie, H. Bloom, and J. D. Morrison, J. Chem. Phys. 47, 1580 (1967).
3. H. Bloom and J. W. Hastie, J. Chem. Phys. 49, 2230 (1968).
4. J. W. Hastie and D. L. Swingler, High Temp. Sci. 1, 46 (1969).
5. J. W. Hastie and J. L. Margrave, J. Phys. Chem. 73, 1105 (1969).
6. JANAF Thermochemical Tables: $PbCl_2$(g) and PbCl(g), 6-30-73; BCl_2(g) and $AlCl_2$(g), 6-30-72; CCl_2(g), 12-31-68; $SiCl_2$(g), 12-31-70.
7. A. D. Walsh, J. Chem. Soc. 1953, 2266 (1953).
8. G. Herzberg, "Electronic Spectra of Polyatomic Molecules," D. Van Nostrand Co., Inc., New York, 1966.

Lead Chloride, Ion ($PbCl_2^+$) $Cl_2Pb_1^+(g)$

Enthalpy Reference Temperature = T_r = 298.15 K Standard State Pressure = p° = 0.1 MPa

	J K⁻¹mol⁻¹			kJ mol⁻¹			
T/K	$C_p°$	S°	$-[G°-H°(T_r)]/T$	$H°-H°(T_r)$	$\Delta_fH°$	$\Delta_fG°$	Log K_f
0	0.	0.	INFINITE	-13.896	812.264		
100	43.788	263.871	365.095	-10.122			
200	51.970	297.156	323.520	-5.273			
250	53.863	308.973	319.465	-2.623			
298.15	55.006	318.566	318.566	0.	816.298	800.890	-140.313
300	55.041	318.906	318.567	0.102	816.326	800.794	-139.431
350	55.809	327.452	319.240	2.874	817.066	798.147	-119.117
400	56.334	334.941	320.744	5.679	817.783	795.395	-103.868
450	56.705	341.599	322.698	8.505	818.476	792.555	-91.997
500	56.978	347.588	324.893	11.348	819.145	789.638	-82.493
600	57.340	358.011	329.569	17.065	820.420	783.616	-68.220
700	57.564	366.868	334.281	22.811	816.751	778.185	-58.069
800	57.711	374.565	338.846	28.575	817.877	772.599	-50.446
900	57.812	381.369	343.200	34.352	819.026	766.870	-44.508
1000	57.885	387.464	347.327	40.137	820.198	761.012	-39.751
1100	57.939	392.983	351.231	45.928	821.394	755.036	-35.854
1200	57.981	398.027	354.923	51.724	822.613	748.950	-32.601
1300	58.013	402.669	358.420	57.524	823.848	742.762	-29.845
1400	58.039	406.969	361.736	63.327	825.085	736.478	-27.478
1500	58.060	410.974	364.886	69.132	826.313	730.106	-25.425
1600	58.077	414.722	367.885	74.938	827.527	723.652	-23.625
1700	58.091	418.243	370.745	80.747	828.719	717.123	-22.035
1800	58.103	421.564	373.477	86.557	829.889	710.525	-20.619
1900	58.113	424.706	376.091	92.367	831.031	703.862	-19.351
2000	58.122	427.687	378.597	98.179	832.144	697.140	-18.207
2100	58.130	430.523	381.003	103.992	656.005	697.479	-17.349
2200	58.138	433.227	383.316	109.805	657.443	699.419	-16.606
2300	58.145	435.811	385.542	115.619	658.781	701.297	-15.927
2400	58.153	438.286	387.689	121.434	660.014	703.119	-15.303
2500	58.161	440.660	389.760	127.250	661.140	704.891	-14.728
2600	58.170	442.942	391.762	133.067	662.160	706.621	-14.196
2700	58.179	445.137	393.699	138.884	663.075	708.313	-13.703
2800	58.190	447.253	395.574	144.702	663.887	709.973	-13.245
2900	58.203	449.295	397.391	150.522	664.601	711.607	-12.817
3000	58.218	451.269	399.154	156.343	665.222	713.217	-12.418
3100	58.234	453.178	400.867	162.166	665.757	714.807	-12.044
3200	58.253	455.027	402.530	167.990	666.213	716.382	-11.694
3300	58.275	456.820	404.148	173.816	666.598	717.944	-11.364
3400	58.299	458.560	405.723	179.645	666.920	719.495	-11.054
3500	58.326	460.250	407.257	185.476	667.189	721.037	-10.761
3600	58.356	461.894	408.752	191.310	667.413	722.572	-10.484
3700	58.388	463.493	410.210	197.148	667.601	724.102	-10.222
3800	58.424	465.051	411.633	202.988	667.762	725.627	-9.974
3900	58.462	466.569	413.022	208.832	667.904	727.147	-9.739
4000	58.504	468.050	414.379	214.681	668.035	728.665	-9.515
4100	58.548	469.495	415.706	220.533	668.163	730.179	-9.303
4200	58.595	470.906	417.004	226.390	668.295	731.690	-9.100
4300	58.644	472.285	418.273	232.252	668.439	733.198	-8.907
4400	58.697	473.634	419.516	238.119	668.600	734.702	-8.722
4500	58.751	474.954	420.734	243.992	668.783	736.202	-8.546
4600	58.808	476.246	421.926	249.870	668.995	737.699	-8.377
4700	58.867	477.511	423.096	255.753	669.239	739.189	-8.215
4800	58.928	478.751	424.242	261.643	669.520	740.675	-8.060
4900	58.990	479.967	425.367	267.539	669.841	742.154	-7.911
5000	59.055	481.159	426.471	273.441	670.206	743.626	-7.769
5100	59.120	482.329	427.555	279.350	670.616	745.090	-7.631
5200	59.187	483.478	428.619	285.265	671.075	746.546	-7.499
5300	59.255	484.606	429.665	291.187	671.583	747.993	-7.372
5400	59.324	485.714	430.693	297.116	672.143	749.429	-7.249
5500	59.393	486.803	431.703	303.052	672.755	750.855	-7.131
5600	59.464	487.874	432.697	308.995	673.419	752.269	-7.017
5700	59.534	488.927	433.674	314.945	674.136	753.671	-6.907
5800	59.605	489.963	434.636	320.902	674.907	755.060	-6.800
5900	59.676	490.983	435.582	326.866	675.731	756.434	-6.697
6000	59.746	491.987	436.514	332.837	676.608	757.795	-6.597

PREVIOUS: June 1973 (1 atm) CURRENT: June 1973 (1 bar)

Lead Chloride, Ion ($PbCl_2^+$) $Cl_2Pb_1^+(g)$

SULFUR CHLORIDE (SCl_2) LIQUID M_r = 102.966

$S°(298.15\ K) = 183.678 \pm 4.2\ J\ K^{-1}\ mol^{-1}$ $\Delta_f H°(298.15\ K) = -49.79 \pm 2.1\ kJ\ mol^{-1}$

Enthalpy of Formation

Trautz and Hoffmann (1) have measured calorimetrically the enthalpy of solution of gaseous chlorine in liquid dichlorodisulfane at 298.15 K. They reported $\Delta_{sln}H° = -9.8 \pm 0.4$ kcal mol^{-1} for a final state consisting of SCl_2 dissolved in an excess of S_2Cl_2. Combining their measured $\Delta_{sln}H$ value with $\Delta_f H°(S_2Cl_2$, ℓ, 298.15 K) = -13.9 ± 0.5 kcal mol^{-1} (2), we obtain the adopted result of $\Delta_f H°(SCl_2$, ℓ, 298.15 K) = -11.9 ± 0.5 kcal mol^{-1}. In arriving at this value, no enthalpy corrections are applied for dissolution of the SCl_2 in the excess S_2Cl_2 and for the temperature difference since we expect both corrections to be small and they tend to cancel. Our adopted value agrees with the selected values of NBS (-12 kcal mol^{-1}, 3) and Mills (-11.8 kcal mol^{-1}, 4).

Heat Capacity and Entropy

The heat capacity of liquid SCl_2 is estimated as 21.75 cal K^{-1} mol^{-1} using the value of 7.25 cal K^{-1} g-atom^{-1} recommended by Kubaschewski et al. (5). This estimated value of $C_p°$ corresponds to a specific heat of 0.21 cal g K^{-1} which agrees with the experimentally measured value for S_2Cl_2 of 0.22 cal g K^{-1} (2). We assume that $C_p°$ is constant in the temperature range 298-800 K. The value of S°(298.15 K) is selected as 43.9 ± 1.0 cal K^{-1} mol^{-1} which minimizes differences between the second and third law entropies of vaporization determined from vapor pressure data (6) on a pure stabilized sample of SCl_2. Further details on the results of these analyses can be found in the heat of formation section of the SCl_2(g) table (2).

Vaporization Data

T_{vap} is the temperature at which the Gibbs energy of vaporization becomes zero. The value of $\Delta_{vap}H°$ is calculated as the difference between the $\Delta_f H°$ values for the gas and liquid at T_{vap}. The experimental value of T_{vap} determined from vapor pressure measurements is 332.7 K (6).

References

1. M. Trautz and A. Hoffmann, Z. Elektrochem. 35, 118 (1929).
2. JANAF Thermochemical Tables: S_2Cl_2(ℓ), 6-30-78; SCl_2(g), 6-30-78.
3. U. S. Nat. Bur. Stand. Tech. Note 270-3, 1968.
4. K. C. Mills, "Thermodynamic Data for Inorganic Sulphides, Selenides, and Tellurides," Butterworths, London, 1974.
5. O. Kubaschewski, E. LL. Evans, and C. B. Alcock, "Metallurgical Thermochemistry," 4th ed., Pergamon Press, New York, 1967.
6. R. J. Rosser and F. R. Whitt, J. Appl. Chem. 10, 229 (1960).

Sulfur Chloride (SCl_2) $Cl_2S_1(l)$

Enthalpy Reference Temperature = T_r = 298.15 K Standard State Pressure = p° = 0.1 MPa

	J $K^{-1}mol^{-1}$			kJ mol^{-1}			
T/K	$C_p°$	S°	$-[G°-H°(T_r)]/T$	$H°-H°(T_r)$	$\Delta_f H°$	$\Delta_f G°$	Log K_f
0							
100							
200							
298.15	91.002	183.678	183.678	0.	-49.790	-28.485	4.990
300	91.002	184.241	183.679	0.168	-49.726	-28.353	4.937
400	91.002	210.420	187.249	9.269	-48.693	-21.639	2.826
500	91.002	230.727	193.989	18.369	-47.092	-15.075	1.575
600	91.002	247.318	201.537	27.469	-45.209	-8.843	0.770
700	91.002	261.346	209.105	36.569	-43.127	-2.945	0.220
800	91.002	273.498	216.411	45.669	-40.938	2.647	-0.173

PREVIOUS: CURRENT: June 1978

Sulfur Chloride (SCl_2) $Cl_2S_1(l)$

$S°(298.15\ K) = 281.628 \pm 0.13\ J\ K^{-1}\ mol^{-1}$

$\Delta_fH°(0\ K) = -16.43 \pm 3.3\ kJ\ mol^{-1}$
$\Delta_fH°(298.15\ K) = -17.57 \pm 3.3\ kJ\ mol^{-1}$

Electronic Levels and Quantum Weights

State	ε_i, cm^{-1}	g_i
$^1A_1$	0	1
$^1A_2$	25810	1
$^1B_2$	29762	1

Vibrational Frequencies and Degeneracies

ν, cm^{-1}
528 (1)
205 (1)
525 (1)

Point Group: C_{2v} σ = 2
Bond Distance: S-Cl = 2.015 ± 0.001 Å
Bond Angle: Cl-S-Cl = 102.73 ± 0.05°
Product of the Moments of Inertia:
$I_AI_BI_C = 5.91980 \times 10^{-114}\ g^3\ cm^6$

Enthalpy of Formation

Rosser and Whitt (1) have reported the results of a vapor pressure study of SCl_2. Measurements were conducted on a sample of SCl_2 stabilized with PCl_3 to suppress decomposition. Using JANAF free energy functions (2), we analyze their vapor pressure points (1) lying above 298 K by the second and third law methods. In the analysis we select $S°(SCl_2, \ell, 298.15\ K) = 43.9\ cal\ K^{-1}\ mol^{-1}$ which minimizes the difference between the 2nd and 3rd law entropies of vaporization. Our results are $\Delta_{vap}H°$ (2nd law) = $\Delta_{vap}H°$ (3rd law) = 7.75±0.02 kcal mol^{-1} which leads to the adopted heat of formation of $\Delta_fH°(SCl_2$, g, 298.15 K) = -4.2±0.8 kcal mol^{-1} when combined with $\Delta_fH°(SCl_2$, ℓ, 298.15 K) = -11.9±0.5 kcal mol^{-1} (2). The adopted value of $\Delta_fH°$ is rounded to the nearest 0.1 kcal mol^{-1}. The overall uncertainty is estimated as ±0.8 kcal mol^{-1} based on contributions from the uncertainty in $\Delta_fH°$ of the liquid (±0.5 kcal mol^{-1}) and from the uncertainty in the entropy (±1.0 cal $K^{-1}\ mol^{-1}$) of the liquid (2) which amounts to roughly 0.3 kcal mol^{-1} in $\Delta_fH°$.

Barton and Yost (3) have studied the dissociation of $S_2Cl_2(g)$ at various temperatures. Their measured dissociation data above 760 K have been reanalyzed by Yost and Russell (4) who have determined equilibrium constants for the reactions (a) $S_2(g) + Cl_2(g) = S_2Cl_2(g)$ and (b) $0.5S_2(g) + Cl_2(g) = SCl_2(g)$ by combining the dissociation pressures with molecular data and an estimated value for the $\Delta_fH°$ of SCl_2. Second and third law analyses of the equilibrium data for reaction (b) give $\Delta_rH°$(298.15 K) (2nd law) = $\Delta_rH°$(298.15 K) (3rd law) = -20.69±0.01 kcal mol^{-1} which corresponds to $\Delta_fH°(SCl_2$, g, 298.15 K) = -5.3 kcal mol^{-1} using the recent JANAF (2) value of $\Delta_fH°$ for $S_2(g)$. This $\Delta_fH°$ value for $SCl_2(g)$ is some 1.1 kcal mol^{-1} more negative than that obtained from an analysis of vapor pressure data described above. We find a similar discrepancy in the equilibrium data for reaction (a) where the derived $\Delta_fH°$ value of $S_2Cl_2(g)$ is 3 kcal mol^{-1} more negative than that obtained from an analysis of vapor pressure data. (See $S_2Cl_2(g)$ table for details). We conclude that the equilibrium constants reported by Yost and Russell (4) are biased, and we give no weight to the $\Delta_fH°$ values derived from their data.

Values of $\Delta_fH°$ previously selected by Colwell (5) and Mills (6) have been based primarily on the equilibrium data of Yost and Russell (4). The value selected by NBS (7) is -4.7 kcal mol^{-1}. Our adopted results give an enthalpy of atomization and mean S-Cl bond dissociation energy of 126.8±0.9 kcal mol^{-1} and 63.4 kcal mol^{-1}, respectively.

Heat Capacity and Entropy

Early electron diffraction studies (8, 9) and more recent microwave measurements (10, 11) have shown that the SCl_2 molecule has C_{2v} symmetry. We adopt structural parameters from the microwave work of Davis and Gerry (11). The bond length and angle refer to an average structure of $^{32}S^{35}Cl_2$ in the (000) vibrational state obtained from a quadratic potential function based solely on microwave data (11). This average structure is only slightly different from the effective structure determined by Murray et al. (10) but quite different from the molecular structure derived from electron diffraction measurements (8, 9) which are considered less accurate. The symmetrical stretch (ν_1) and bending (ν_2) frequencies are those observed by Frankiss and Harrison (12) in the Raman spectra of the vapor at 335 K. The asymmetrical stretching frequency (νn_3) was not detected in this recent Raman study (12) and still remains relatively uncertain (±10 cm^{-1}). We use $\nu_3 = 525\ cm^{-1}$ based on the infrared measurements of Savoie and Tremblay (13), as suggested by Frankiss and Harrison (12). The electronic levels and configurations are taken from the CNDO/2 MO calculations and electronic absorption spectral measurements of Colton and Rabalais (14). The principal moments of inertia are: $I_A = 5.8026 \times 10^{-38}$, $I_B = 29.1706 \times 10^{-38}$, and $I_C = 34.9733 \times 10^{-39}\ g\ cm^{-2}$.

Our ideal gas thermal functions are essentially an extension of those previously published by Frankiss and Harrison (12) who also reviewed data published prior to their paper. All earlier functions require revision since they are based on liquid state vibrational frequencies and the molecular structure determined by electron diffraction.

References
1. R. J. Rosser and F. R. Whitt, J. Appl. Chem. 10, 229 (1960).
2. JANAF Thermochemical Tables: $SCl_2(\ell)$, 6-30-78; $S_2(g)$, 9-30-77.
3. R. C. Barton and D. M. Yost, J. Amer. Chem. Soc. 57, 307 (1935).
4. D. M. Yost and H. Russell, Jr., "Systematic Inorganic Chemistry," Prentice-Hall, Inc., New York, 1944.
5. K. H. Colwell, M. Sc. Thesis, Leeds, 1963, reported in H. Mackle and P. A. G. O'Hare, Trans. Faraday Soc. 60, 666 (1964).
6. K. C. Mills, "Thermodynamic Data for Inorganic Sulphides, Selenides, and Tellurides," Butterworths, London, 1974.
7. U.S. Nat. Bur. Stand. Tech. Note 270-3, 1968.
8. D. P. Stevenson and J. Y. Beach, J. Amer. Chem. Soc. 60, 2872 (1938).
9. Y. Morino, Y. Minata, T. Ito, and J. Nakamura, J. Phys. Soc. Japan, 17, 37 (1962).
10. J. T. Murray, W. A. Little, Jr., Q. Williams, and T. L. Weatherly, J. Chem. Phys. 65, 985 (1976).
11. R. W. Davis and M. C. L. Gerry, J. Mol. Spectrosc. 65, 455 (1977).
12. S. G. Frankiss and D. J. Harrison, Spectrochim. Acta, 31A, 161 (1975).
13. R. Savoie and J. Tremblay, Can. J. Spectrosc. 17, 73 (1972).
14. R. J. Colton and J. W. Rabalais, J. Electron Spectrosc. and Related Phenomena, 3, 345 (1974).

Enthalpy Reference Temperature = T_r = 298.15 K Standard State Pressure = p° = 0.1 MPa

T/K	$C_p°$ (J $K^{-1}mol^{-1}$)	S° (J $K^{-1}mol^{-1}$)	$-[G°-H°(T_r)]/T$ (J $K^{-1}mol^{-1}$)	$H°-H°(T_r)$ (kJ mol^{-1})	$\Delta_fH°$ (kJ mol^{-1})	$\Delta_fG°$ (kJ mol^{-1})	Log K_f
0	0.	0.	INFINITE	-12.445	-16.425	-16.425	INFINITE
100	37.966	233.457	323.231	-8.977	-16.559	-19.752	10.317
200	45.873	262.274	286.144	-4.774	-17.035	-22.770	5.947
250	48.815	272.842	282.455	-2.403	-17.303	-24.173	5.051
298.15	50.895	281.628	281.628	0.	-17.573	-25.472	4.463
300	50.962	281.943	281.629	0.094	-17.583	-25.521	4.444
350	52.512	289.922	282.255	2.683	-17.879	-26.820	4.003
400	53.643	297.011	283.665	5.338	-20.406	-27.989	3.655
450	54.482	303.380	285.508	8.043	-21.404	-28.885	3.353
500	55.117	309.155	287.588	10.783	-22.461	-29.658	3.098
600	55.992	319.288	292.051	16.342	-24.119	-30.935	2.693
700	56.547	327.964	296.576	21.971	-25.508	-31.958	2.385
800	56.919	335.540	300.983	27.646	-26.745	-32.794	2.141
900	57.179	342.260	305.203	33.351	-81.016	-32.400	1.880
1000	57.368	348.295	309.216	39.079	-80.881	-27.005	1.411
1100	57.509	353.770	313.021	44.823	-80.759	-21.624	1.027
1200	57.618	358.778	316.629	50.580	-80.651	-16.253	0.707
1300	57.703	363.394	320.051	56.346	-80.559	-10.890	0.438
1400	57.770	367.673	323.301	62.120	-80.486	-5.534	0.206
1500	57.825	371.660	326.394	67.900	-80.430	-0.182	0.006
1600	57.870	375.394	329.341	73.685	-80.393	5.166	-0.169
1700	57.908	378.903	332.154	79.473	-80.375	10.513	-0.323
1800	57.939	382.214	334.844	85.266	-80.377	15.860	-0.460
1900	57.966	385.347	337.421	91.061	-80.398	21.207	-0.583
2000	57.988	388.321	339.892	96.859	-80.440	26.555	-0.694
2100	58.008	391.151	342.266	102.659	-80.501	31.906	-0.794
2200	58.025	393.850	344.550	108.460	-80.582	37.261	-0.885
2300	58.040	396.430	346.750	114.264	-80.685	42.620	-0.968
2400	58.053	398.900	348.872	120.068	-80.810	47.983	-1.044
2500	58.065	401.270	350.920	125.874	-80.958	53.352	-1.115
2600	58.076	403.548	352.901	131.681	-81.128	58.728	-1.180
2700	58.086	405.740	354.818	137.490	-81.322	64.111	-1.240
2800	58.095	407.852	356.674	143.299	-81.540	69.501	-1.297
2900	58.104	409.891	358.474	149.109	-81.782	74.900	-1.349
3000	58.113	411.861	360.221	154.919	-82.047	80.307	-1.398
3100	58.121	413.767	361.918	160.731	-82.336	85.724	-1.444
3200	58.130	415.612	363.567	166.544	-82.647	91.150	-1.488
3300	58.139	417.401	365.172	172.357	-82.979	96.586	-1.529
3400	58.149	419.137	366.733	178.171	-83.332	102.033	-1.568
3500	58.160	420.823	368.255	183.987	-83.702	107.490	-1.604
3600	58.172	422.461	369.738	189.804	-84.089	112.958	-1.639
3700	58.186	424.055	371.185	195.621	-84.490	118.437	-1.672
3800	58.200	425.607	372.596	201.441	-84.902	123.927	-1.703
3900	58.217	427.119	373.975	207.262	-85.324	129.428	-1.733
4000	58.235	428.593	375.322	213.084	-85 752	134.940	-1.762
4100	58.256	430.032	376.639	218.909	-86.185	140.463	-1.790
4200	58.279	431.436	377.927	224.735	-86.619	145.996	-1.816
4300	58.304	432.807	379.188	230.565	-87.052	151.539	-1.841
4400	58.331	434.148	380.421	236.396	-87.481	157.093	-1.865
4500	58.362	435.459	381.630	242.231	-87.904	162.656	-1.888
4600	58.395	436.742	382.814	248.069	-88.318	168.229	-1.910
4700	58.430	437.998	383.975	253.910	-88.721	173.810	-1.932
4800	58.469	439.229	385.113	259.755	-89.111	179.400	-1.952
4900	58.511	440.435	386.230	265.604	-89.485	184.998	-1.972
5000	58.555	441.618	387.326	271.457	-89.843	190.603	-1.991
5100	58.603	442.778	388.402	277.315	-90.181	196.216	-2.010
5200	58.653	443.916	389.459	283.178	-90.499	201.834	-2.027
5300	58.707	445.034	390.497	289.046	-90.794	207.459	-2.045
5400	58.764	446.132	391.517	294.919	-91.066	213.089	-2.061
5500	58.823	447.210	392.520	300.799	-91.313	218.724	-2.077
5600	58.885	448.271	393.506	306.684	-91.535	224.363	-2.093
5700	58.950	449.314	394.476	312.576	-91.729	230.006	-2.108
5800	59.018	450.340	395.430	318.474	-91.896	235.652	-2.122
5900	59.089	451.349	396.369	324.380	-92.034	241.300	-2.136
6000	59.162	452.343	397.294	330.292	-92.143	246.951	-2.150

PREVIOUS: June 1978 (1 atm) CURRENT: June 1978 (1 bar)

SULFUR CHLORIDE, ION (SCl_2^+) IDEAL GAS M_r = 102.96545

Sulfur Chloride, Ion (SCl_2^+) $Cl_2S_1^+(g)$

S°(298.15 K) = [287.32 ± 4.2] J K^{-1} mol^{-1}

$\Delta_f H°$(0 K) = 896.43 ± 5.4 kJ mol^{-1}
$\Delta_f H°$(298.15 K) = [901.485] kJ mol^{-1}

Electronic Levels and Quantum Weights

State	ε_i, cm^{-1}	g_i
$^2B_1$	0	2
$^2A_1$	20325	2
$^2A_1$	[21000]	2
$^2A_2$	22422	2
$^2A_2$	34198	2

Vibrational Frequencies and Degeneracies

ν, cm^{-1}
530±30(1)
[200] (1)
[530] (1)

Point Group: [C_{2v}] σ = 2
Bond Distance: S-Cl = [2.00] Å
Bond Angle: Cl-S-Cl = [105]°
Product of the Moments of Inertia: $I_A I_B I_C$ = [5.65102 x 10^{-114}] g^3 cm^6

Enthalpy of Formation

$\Delta_f H°$(0 K) is calculated from IP = 9.46±0.02 eV by addition of the JANAF $\Delta_f H°$(0 K) value for SCl_2 (1). The selected value of IP is calculated as the adiabatic IP from the first band system observed in the HeI photoelectron spectrum of SCl_2 by Solouki et al. (2). The adiabatic IP is 0.21 eV less than the vertical value determined by photoelectron spectroscopy (2) and 0.24-0.74 eV less than the vertical IP's measured by electron impact (3, 4).

$\Delta_f H°$(298.15 K) is obtained from the adopted $\Delta_f H°$(0 K) value by using JANAF enthalpies (1) for S(α, cr), Cl_2(g), and e^-(g). $\Delta_r H°$(298.15 K) determined for the direct ionization SCl_2 = SCl_2^+ + e^- from our tables differs from a room-temperature threshold energy due to inclusion of the enthalpies of all the species in this ionization process and to threshold effects discussed by Rosenstock et al. (5). $\Delta_f H°$(298.15 K) should be changed by -1.481 kcal mol^{-1} if it is to be used in the ion convention which excludes the enthalpy of the electron.

Heat Capacity and Entropy

We perform bond energy calculations which show that the S-Cl primary bond dissociation energy for SCl_2^+ (66.6 kcal mol^{-1}) is quite similar to that for the neutral dichloride (69.7 kcal mol^{-1}), indicating that the removal of the electron during ionization should produce minimal structural changes. According to the Walsh diagram (6) for AB_2 molecules, the most loosely bound electron in SCl_2 occupies a $4b_1$ orbital which is S-Cl antibonding and Cl-Cl bonding. Thus, we would expect a slight increase in the bond angle and a decrease in the S-Cl bond distance upon removing this electron to form SCl_2^+. We estimate the Cl-S-Cl bond angle as 105° (102.7° for SCl_2) and the S-Cl bond distance as 2.00 Å (2.015 Å for SCl_2. Based on this estimated structure, the principal moments of inertia are: I_A = 5.4345 x 10^{-39}, I_B = 29.6437 x 10^{-39}, and I_C = 35.0782 x 10^{-39} g cm.

The photoelectron spectrum of SCl_2 has been measured by Solouki et al. (2) who reported observing a vibrational spacing of 530±30 cm^{-1} in the first band system. They assigned this progression to the S-Cl symmetric stretching frequency ν_1. By analogy with the observed vibrational frequencies for SCl_2 (1), the difference between ν_1 and ν_3 is expected to be small. We adopt $\nu_1 = \nu_3$ = 530 cm^{-1}, and we estimate the bending frequency to be 5 cm^{-1} lower than that for SCl_2 (205 cm^{-1}, 1).

The electronic configurations of the ground and excited states are predicted from Walsh's empirical diagram (6) which correlates well with the observed photoelectron spectra and calculated orbital energies for the related dihalide molecules SCl_2(2), OCl_2, and OF_2 (7). We use relative term values for three of the four predicted states calculated from the vertical ionization energies reported by Solouki et al. (2). We include a second $^2A_1$ state at 21000 cm^{-1} based on the fact that this state has been observed in the PE spectrum of OCl_2 (7), lying 2258 cm^{-1} above the first $^2A_1$ state. Inclusion of this state is also supported by the results of orbital energy calculations (2) on SCl_2. The enthalpy between 0 K and 298.15 K is -2.9764 kcal mol^{-1}.

References

1. JANAF Thermochemical Tables: SCl_2(g), 6-30-78; S(α, cr), 9-30-77; e^-(g), 3-31-77; Cl_2(g), 9-30-65.
2. B. Solouki, P. Rosmus, and H. Bock, Chem. Phys. Lett. 26, 20 (1974).
3. O. R. Hartmann, K. H. Lebert, and H. U. Chun, Z. Phys. Chem. 92, 311 (1974).
4. J. C. Blais and M. Cottin, C. R. Acad. Sci. Paris, Ser. C, 276, 919 (1973).
5. H. M. Rosenstock et al., "Energetics of Gaseous Ions," J. Phys. Chem. Ref. Data, 6, Suppl. 1, 1977; pp. I-28 and I-35.
6. A. D. Walsh, J. Chem. Soc. 1953, 2266 (1953).
7. A. B. Cornford, D. C. Frost, F. G. Herring, and C. A. McDowell, J. Chem. Phys. 55, 2820 (1971).

Enthalpy Reference Temperature = T_r = 298.15 K Standard State Pressure = p° = 0.1 MPa

T/K	$C_p°$	S°	$-[G°-H°(T_r)]/T$	$H°-H°(T_r)$	$\Delta_f H°$	$\Delta_f G°$	Log K_f
	J $K^{-1}mol^{-1}$			kJ mol^{-1}			
0	0.	0.	INFINITE	-12.453	896.427		
100	38.080	239.130	328.914	-8.978			
200	45.858	267.979	291.835	-4.771			
250	48.783	278.542	288.148	-2.402			
298.15	50.860	287.322	287.322	0.	901.485	885.633	-155.159
300	50.928	287.636	287.323	0.094	901.512	885.535	-154.185
350	52.479	295.610	287.949	2.681	902.254	882.813	-131.753
400	53.613	302.695	289.357	5.335	900.765	880.074	-114.926
450	54.456	309.061	291.199	8.038	900.806	877.476	-101.855
500	55.095	314.833	293.278	10.777	900.787	874.888	-91.399
600	55.975	324.962	297.738	16.334	901.205	869.675	-75.712
700	56.533	333.636	302.262	21.962	901.892	864.369	-64.500
800	56.908	341.211	306.667	27.635	902.733	858.952	-56.084
900	57.170	347.930	310.886	33.340	850.540	854.505	-49.594
1000	57.361	353.964	314.897	39.067	852.753	854.827	-44.652
1100	57.503	359.438	318.701	44.810	854.953	854.928	-40.597
1200	57.613	364.446	322.307	50.566	857.139	854.829	-37.210
1300	57.698	369.061	325.729	56.332	859.308	854.548	-34.336
1400	57.767	373.340	328.979	62.105	861.461	854.101	-31.867
1500	57.822	377.327	332.070	67.885	863.594	853.501	-29.722
1600	57.867	381.060	335.017	73.669	865.709	852.759	-27.840
1700	57.905	384.570	337.829	79.458	867.805	851.885	-26.175
1800	57.937	387.880	340.519	85.250	869.882	850.889	-24.692
1900	57.964	391.013	343.095	91.045	871.939	849.777	-23.362
2000	57.988	393.987	345.566	96.843	873.977	848.558	-22.162
2100	58.009	396.817	347.939	102.643	875.994	847.237	-21.074
2200	58.028	399.516	350.223	108.445	877.991	845.821	-20.082
2300	58.047	402.096	352.423	114.249	879.968	844.314	-19.175
2400	58.065	404.567	354.544	120.054	881.922	842.722	-18.341
2500	58.083	406.937	356.593	125.862	883.855	841.049	-17.573
2600	58.103	409.216	358.573	131.671	885.765	839.299	-16.862
2700	58.124	411.409	360.490	137.482	887.653	837.476	-16.202
2800	58.148	413.523	362.346	143.296	889.518	835.583	-15.588
2900	58.176	415.564	364.146	149.112	891.361	833.625	-15.015
3000	58.208	417.537	365.893	154.931	893.183	831.603	-14.479
3100	58.244	419.446	367.590	160.754	894.983	829.520	-13.977
3200	58.286	421.296	369.240	166.580	896.765	827.380	-13.506
3300	58.334	423.090	370.845	172.411	898.528	825.184	-13.062
3400	58.388	424.833	372.407	178.247	900.276	822.935	-12.643
3500	58.449	426.526	373.929	184.089	902.011	820.635	-12.247
3600	58.517	428.174	375.413	189.937	903.734	818.286	-11.873
3700	58.592	429.778	376.861	195.792	905.450	815.889	-11.518
3800	58.676	431.341	378.274	201.656	907.160	813.445	-11.182
3900	58.767	432.867	379.655	207.528	908.868	810.957	-10.862
4000	58.866	434.356	381.003	213.409	910.577	808.425	-10.557
4100	58.973	435.811	382.323	219.301	912.291	805.850	-10.267
4200	59.088	437.233	383.613	225.204	914.011	803.233	-9.990
4300	59.211	438.625	384.876	231.119	915.743	800.575	-9.725
4400	59.342	439.988	386.113	237.047	917.488	797.876	-9.472
4500	59.479	441.323	387.326	242.988	919.250	795.138	-9.230
4600	59.624	442.632	388.514	248.943	921.032	792.361	-8.998
4700	59.776	443.916	389.679	254.913	922.836	789.544	-8.775
4800	59.934	445.176	390.822	260.898	924.666	786.688	-8.561
4900	60.098	446.413	391.944	266.900	926.522	783.795	-8.355
5000	60.267	447.629	393.045	272.918	928.408	780.863	-8.158
5100	60.442	448.824	394.127	278.953	930.326	777.893	-7.967
5200	60.622	450.000	395.191	285.007	932.277	774.885	-7.784
5300	60.805	451.156	396.236	291.078	934.264	771.839	-7.607
5400	60.993	452.294	397.263	297.168	936.287	768.756	-7.436
5500	61.183	453.415	398.274	303.277	938.348	765.635	-7.271
5600	61.377	454.519	399.269	309.404	940.448	762.475	-7.112
5700	61.573	455.607	400.248	315.552	942.588	759.278	-6.958
5800	61.770	456.680	401.211	321.719	944.768	756.044	-6.809
5900	61.969	457.738	402.160	327.906	946.990	752.771	-6.665
6000	62.169	458.781	403.095	334.113	949.254	749.460	-6.525

PREVIOUS: June 1978 (1 atm) CURRENT June 1978 (1 bar)

Sulfur Chloride, Ion (SCl_2^+) $Cl_2S_1^+(g)$

SULFUR CHLORIDE (ClSSCl) LIQUID $M_r = 135.026$

$S°(298.15\ K) = 223.844 \pm 4.2\ J\ K^{-1}\ mol^{-1}$ $\Delta_fH°(298.15\ K) = -58.158 \pm 2.1\ kJ\ mol^{-1}$

Enthalpy of Formation

Feher and Strater (1) have measured calorimetrically the heats of chlorination of sulfur and liquid dichlorodisulfane at 298.15 K. Recalculation of their calorimetric data using 1975 atomic weights (2) and a $\Delta C_p°$ of 10.80 cal K^{-1} mol^{-1} for the formation reactions $2S(rh) + Cl_2(g) \rightarrow S_2Cl_2(l)$ leads to the adopted $\Delta_fH°$ of -13.9 ± 0.5 kcal mol^{-1}. (-58.158 ± 2.1 kJ mol^{-1}).

Other calorimetric values for $\Delta_fH°$ which were published prior to 1930 have been recently summarized by Mills (3). Two of these four values, which range from -14.3 to -17.6 kcal mol^{-1}, agree with the adopted $\Delta_fH°$ value within the combined experimental error. Our selected value agrees with that from Mills (3) but is some 0.3 kcal mol^{-1} less negative than the value (-14.2 kcal mol^{-1}) recommended by NBS (4).

Heat Capacity and Entropy

$C_p°$ has been measured calorimetrically as 29.706 cal K^{-1} mol^{-1} by Trautz and Rick (5) at 295.15 K and by Ogier (6) as the mean $C_p°$ value over the temperature range 285-343 K. We adopt this value and assume $C_p°$ is constant from 298.15 K to 1000 K. The value of S°(298.15 K) is selected as 53.5 ± 1.0 cal K^{-1} mol^{-1} which minimizes differences between the second and third law entropies of vaporization determined from two independent sets (7, 8) of vapor pressure data. Further details on the results of these analyses can be found in the heat of formation section of the S_2Cl_2 gas phase table (9).

Vaporization Data

T_{vap} is the temperature at which the Gibbs energy of vaporization becomes zero. The value of $\Delta_{vap}H°$ is calculated as the difference between the $\Delta_fH°$ values for the gas and liquid at T_{vap}. Experimental values of T_{vap} determined from vapor pressure measurements include 411 K (7) and 408.9 K (8).

References - $S_2Cl_2(l)$

1. F. Feher and F. Strater, Z. Anorg. Allg. Chem. 322, 172 (1963).
2. IUPAC Commission on Atomic Weights, Pure and Appl. Chem. 47, 75 (1976).
3. K. C. Mills, "Thermodynamic Data For Inorganic Sulphides, Selenides, and Tellurides," Butterworths, London, 1974.
4. U.S. Nat. Bur. Stand. Tech. Note 270-3, 1968.
5. M. Trautz and A. Rick, Z. Elektrochem. 35, 127 (1929).
6. J. Ogier, and C. R. Hebd, Seanc. Acad. Sci., Paris, 92, 922 (1881).
7. E. H. Harvey and H. A. Schuette, J. Amer. Chem. Soc. 48, 2065 (1926).
8. M. Trautz, A. Rick, and H. Acker, Z. Elektrochem. 35, 122 (1929).
9. JANAF Thermochemical Tables: $S_2Cl_2(g)$, 6-30-78.

References - $S_2Cl_2(g)$

1. E. H. Harvey and H. A. Schuette, J. Amer. Chem. Soc. 48, 2065 (1926).
2. M. Trautz, A. Rick, and H. Acker, Z. Elektrochem. 35, 122 (1929).
3. R. C. Barton and D. M. Yost, J. Amer. Chem. Soc. 57, 307 (1935).
4. D. M. Yost and H. Russell, Jr., "Systematic Inorganic Chemistry," Prentice-Hall, Inc., New York, 1944.
5. JANAF Thermochemical Tables: $S_2Cl_2(\ell)$, 6-30-78.
6. F. G. Bodewig and J. A. Plambeck, J. Elektrochem. Soc. 116, 607 (1969).
7. K. C. Mills, "Thermodynamic Data for Inorganic Sulphides, Selenides, and Tellurides," Butterworths, London, 1974.
8. U.S. Nat. Bur. Stand. Tech. Note 270-3, 1968.
9. S. G. Frankiss and D. J. Harrison, Spectrochim. Acta, 31A, 161 (1975).
10. S. G. Frankiss, J. Mol. Struct. 2, 271 (1968).
11. B. Beagley, G. H. Eckersley, D. P. Brown, and D. Tomlison, Trans. Faraday Soc. 65, 2300 (1969).
12. R. J. Colton and J. W. Rabalais, J. Electron Spectrosc. Related Phenomena, 3, 345 (1974).
13. P. N. Noble and L. Mei, J. Chem. Phys. 61, 541 (1974).

Sulfur Chloride (ClSSCl) $Cl_2S_2(l)$

Enthalpy Reference Temperature = T_r = 298.15 K Standard State Pressure = p° = 0.1 MPa

	J K⁻¹mol⁻¹			kJ mol⁻¹			
T/K	$C_p°$	S°	$-[G°-H°(T_r)]/T$	$H°-H°(T_r)$	$\Delta_fH°$	$\Delta_fG°$	Log K_f
0							
100							
200							
298.15	124.290	223.844	223.844	0.	-58.158	-39.271	6.880
300	124.290	224.613	223.846	0.230	-58.075	-39.154	6.817
400	124.290	260.369	228.721	12.659	-58.310	-33.318	4.351
500	124.290	288.103	237.927	25.088	-57.308	-27.213	2.843
600	124.290	310.764	248.236	37.517	-55.681	-21.336	1.857
700	124.290	329.923	258.572	49.946	-53.617	-15.771	1.177
800	124.290	346.520	268.551	62.375	-51.310	-10.519	0.687
900	124.290	361.159	278.044	74.804	-155.115	-3.405	0.198
1000	124.290	374.254	287.022	87.233	-150.135	13.185	-0.689

PREVIOUS. March 1978 CURRENT: June 1978

Sulfur Chloride (ClSSCl) $Cl_2S_2(l)$

SULFUR CHLORIDE (ClSSCl) IDEAL GAS M_r = 135.026

$S°(298.15 K) = 327.21 \pm 0.4 J K^{-1} mol^{-1}$

$\Delta_f H°(0 K) = -15.2 \pm 4 kJ mol^{-1}$
$\Delta_f H°(298.15 K) = -16.7 \pm 4 kJ mol^{-1}$

Vibrational Frequencies and Degeneracies

ν, cm^{-1}	ν, cm^{-1}
546 (1)	Internal Rotation
466 (1)	461 (1)
202 (1)	244 (1)

Ground State Quantum Weight: 1 σ(overall) = 2
Point Group: C_2
Bond Distances: S-Cl = 2.057 ± 0.002 Å S-S = 1.931 ± 0.005 Å
Bond Angles: S-S-Cl = 108.2 ± 0.3° Dihedral Angle = 84.8 ± 1.3°
Product of Moments of Inertia: $I_A I_B I_C = 6.36049 \times 10^{-113} g^3 cm^6$
Reduced Moment of Inertia: $I_{red} = 5.487 \times 10^{-39} g cm^2$
Potential Function for Internal Rotation: $V = 11.99 - 5.60(1-\cos 2\phi) - 0.70(1-\cos 3\phi) kcal mol^{-1}$

Enthalpy of Formation

Harvey and Schuette (1) and Trautz et al. (2) have used a static method to measure vapor pressures for liquid S_2Cl_2 at various temperatures below the normal boiling point. Barton and Yost (3) measured the extent of dissociation for gaseous S_2Cl_2 above 760 K. These dissociation data have been reanalyzed by Yost and Russell (4) who reported equilibrium constants for the reaction $S_2(g) + Cl_2(g) = S_2Cl_2(g)$. We analyze these data by the 2nd and 3rd law methods with the results presented below:

Source	Reaction	Data Points	T/K	$\Delta_r H°(298.15 K)$, kcal mol^{-1} 2nd law	3rd law	Drift cal $K^{-1} mol^{-1}$	$\Delta_f H°(298.15 K)$ kcal mol^{-1}
(1)	A	12[a]	304-411	9.46	9.89±0.07	1.1±0.2	4.01
(2)	A	Set 1 20[b]	374-410	10.0	9.93±0.01	-0.1±0.2	3.97
		Set 2 39[c]	329-374	10.15	9.95±0.02	-0.6±0.1	3.95
(2)	A	22	336-405	10.23	9.98±0.04	-0.8±0.1	3.92
(4)	B	Equation	760-1100	-35.4	-38.1 ±0.9	-3.1±7	7.4

Reaction: (A) $S_2Cl_2(l) = S_2Cl_2(g)$ (B) $S_2(g) + Cl_2(g) = S_2Cl_2(g)$.

[a] Points below 298.15 K not analyzed; 2 points rejected due to failure of a statistical test.
[b] Two points rejected due to failure of a statistical test.
[c] Three points rejected.

We adopt $\Delta_f H°(S_2Cl_2, g, 298.15 K) = -4.0 kcal mol^{-1}$ which is the rounded mean value determined from the four 3rd law enthalpies of vaporization. We estimate the uncertainty in the $\Delta_f H°$ value as ±1.0 kcal mol^{-1} which includes contributions from the uncertainties in the values of $\Delta_f H°(298.15 K)$ and S°(298.15 K) for the liquid (5). $\Delta_f H°(0 K)$ is -3.6 kcal mol^{-1} and the enthalpy of atomization is 192.1±1.0 kcal mol^{-1}.

Bodewig and Plambeck (6) determined $\Delta_r G = -12.4 kcal mol^{-1}$ for $2 S(\ell) + 2 Cl^- = S_2Cl_2(g) + 2 e^-$ in fused LiCl-KCL eutectic at 693 K. This datum has been analyzed by Mills (7) who reported $\Delta_f H°(298.15 K)$ value of -5.8±2 kcal mol^{-1}. Our selected $\Delta_f H°$ value is some 0.4 kcal mol^{-1} less negative than that recommended by NBS (8) and is only 0.1 kcal mol^{-1} less negative than the value adopted by Mills (7).

Heat Capacity and Entropy

The heat capacity and entropy values from 298.15 K to 1500 K are those calculated by Frankiss and Harrison (9) who used gas-phase infrared ($\nu_1 = 546 cm^{-1}$, 10) and Raman (9) frequencies and the most recent structural data determined in an electron diffraction study by Beagley et al (11). Contributions to the thermal functions for internal rotation were determined by integration of the potential function from ϕ = 0° to 2π. Terms in the potential function were established from the torsional vibration ($\nu_4 = 92 cm^{-1}$) and the dihedral angle (84.8°, 11) of the two rotamers. Thermal functions are calculated below 298.15 K by treating the torsion as a harmonic oscillator and including its contribution in the vibrational partition function. Justification for this treatment lies in the fact that values of the ratio V_o/RT become extremely large (>20) at these lower temperatures. We add R$\ln$ 2 to the calculated entropies below 298.15 K to account for the optical activity of the skew rotamer. The absorption spectra (12) reported for S_2Cl_2 vapor indicate no low-lying electronic states which would contribute to the thermal functions below 1500 K. The principal moments of inertia are: $I_A = 15.2331 \times 10^{-39}$, $I_B = 60.6645 \times 10^{-39}$, and $I_C = 68.8282 \times 10^{-39} g cm^2$.

Thermodynamic functions previously published for S_2Cl_2 have been reviewed by Frankiss and Harrison (9) who concluded that all previous calculations had treated incorrectly the internal rotation of the SCl top. Noble and Mei (14) have discussed the vibrational spectral data on S_2Cl_2 published prior to 1974.

References

See the liquid table for details.

Sulfur Chloride (ClSSCl) $Cl_2S_2(g)$

Enthalpy Reference Temperature = T_r = 298.15 K Standard State Pressure = p° = 0.1 MPa

T/K	$C_p°$ (J $K^{-1}mol^{-1}$)	S°	$-[G°-H°(T_r)]/T$	$H°-H°(T_r)$ (kJ mol^{-1})	$\Delta_f H°$	$\Delta_f G°$	Log K_f
0	0.	0.	INFINITE	-16.479	-15.210	-15.210	INFINITE
100	49.149	260.367	385.447	-12.508	-15.531	-20.162	10.532
200	64.455	299.547	333.379	-6.766	-16.111	-24.572	6.418
298.15	72.768	327.215	327.215	0.	-16.736	-28.669	5.023
300	72.881	327.665	327.216	0.135	-16.748	-28.743	5.005
400	77.032	349.244	330.130	7.646	-21.901	-32.460	4.239
500	79.350	366.698	335.754	15.472	-25.503	-34.705	3.626
600	80.818	381.301	342.162	23.484	-28.292	-36.269	3.158
700	81.839	393.840	348.670	31.619	-30.523	-37.417	2.792
800	82.588	404.819	355.016	39.842	-32.422	-38.270	2.499
900	83.140	414.580	361.102	48.130	-140.368	-36.736	2.132
1000	83.529	423.361	366.897	56.464	-139.482	-25.269	1.320
1100	83.780	431.335	372.398	64.831	-138.609	-13.890	0.660
1200	83.931	438.632	377.618	73.217	-137.758	-2.589	0.113
1300	83.990	445.353	382.573	81.614	-136.939	8.641	-0.347
1400	83.969	451.576	387.282	90.012	-136.158	19.810	-0.739
1500	83.902	457.368	391.764	98.406	-135.422	30.925	-1.077

PREVIOUS: June 1978 (1 atm) CURRENT: June 1978 (1 bar)

Sulfur Chloride (ClSSCl) $Cl_2S_2(g)$

$\Delta_fH°(0\ K) = -168.74 \pm 3.3\ kJ\ mol^{-1}$

$S°(298.15\ K) = [281.333 \pm 0.8]\ J\ K^{-1}\ mol^{-1}$ $\Delta_fH°(298.15\ K) = -168.62 \pm 3.3\ kJ\ mol^{-1}$

Electronic Levels and Quantum Weights

State	ε_i, cm^{-1}	g_i
[$^1A_1$]	0	[1]
[$^3B_1$]	[23000]	[3]
	[28300]	[1]
	[29950]	[1]

Vibrational Frequencies and Degeneracies

ν, cm^{-1}
514.0(1)
202.2(1)
502.7(1)

Point Group: C_{2v} $\sigma = 2$

Bond Distance: Si-Cl = [2.03] Å

Bond Angle: Cl-Si-Cl = 105° ± 3°

Product of the Moments of Inertia: $I_AI_BI_C = [5.55300 \times 10^{-114}]\ g^3\ cm^6$

Enthalpy of Formation

The adopted $\Delta_fH°(SiCl_2$, g, 298.15 K) = -40.3 kcal mol^{-1} is the mean value derived from four studies (2, 3, 7, 8). JANAF analyses of all the equilibrium data are summarized below. The $\Delta_fH°$(298.15 K) values are calculated from the third law $\Delta_rH°$(298.15 K) values using auxiliary JANAF data (1). Four sets of data (2, 3, 7, 8) yield average values of $\Delta_fH°$ (from third law) which agree within 0.4 kcal mol^{-1} and values of $\Delta_rS°$(second law) which are in reasonable agreement with the JANAF functions. Two of these equilibrium studies used flow techniques, one used mass spectrometry, while the fourth used a static technique. The adopted $\Delta_fH°$(298.15 K) value corresponds to $\Delta_{at}H°$(0 K) = 204.2 kcal mol^{-1} and an average bond energy of 102.1 kcal mol^{-1}. This latter value compares with 94.7 kcal mol^{-1} for $SiCl_4(g)$, implying that the average bond energy in $SiCl_2(g)$ is 7% larger.

Two other flow studies (4, 5) yield large entropy errors, suggesting that there are temperature-dependent errors in $K°_p$. The reported equilibrium constants are reasonable at the lower temperatures but deviate progressively at the higher temperatures. In contrast, the static study of Ishino (6) gave $K°_p$ values which differ by a factor of five from the adopted ones. These data yield a consistent entropy but $\Delta_rH°$ (third law) is lower by 5 kcal mol^{-1}.

Source	Method	Data Points	T/K	Reaction*	$\Delta_rH°$(298.15 K), kcal mol^{-1} 2nd Law	3rd Law	Drift cal K^{-1} mol^{-1}	$\Delta_fH°$(298.15 K) kcal mol^{-1}
Schafer et al. (2)	flow	Equation	1273-1473	A	80.56	77.93	-1.9	-40.2
Teichmann (3)	flow	Equation	1223-1573	A	80.83	78.06	-2.0	-40.2
Chechentsev (4)	flow	3	1473-1573	A	131.78	75.96	-37±6	-41.2
Antipin (5)	flow	5	1273-1673	A	113.99	75.00	-26±3	-41.7
Ishino et al. (6)	static	27	1448-1573	A	70.24	72.97	1.8±4	-42.7
Schafer (7)	static	Equation	1400-1593	A	79.78	78.38	-0.9	-40.0
Farber (8)	mass spec	10	1593-1792	B	-31.62	-30.95	0.4±2.0	-40.8

*Reactions (A) $Si(cr) + SiCl_4(g) = 2\ SiCl_2(g)$ (B) $Si(g) + SiCl_4(g) = 2\ SiCl_2(g)$

Heat Capacity and Entropy

Maass, et al. (9) observed the infrared spectra of $SiCl_2$ in argon, neon, and nitrogen matrices. They observed and assigned all three fundamental frequencies. The bond angle was calculated to be 105±3° from the isotopic splitting. The bond length is assumed to be the same as in SiH_2Cl_2 (1). The earlier work of Milligan and Jacox (10) is consistent with these results. The principal moments of inertia are: $I_A = 5.1016 \times 10^{-39}$, $I_B = 30.5397 \times 10^{-39}$, $I_C = 35.6413 \times 10^{-39}\ g\ cm^2$.

The ground state is assumed to be $^1A_1$ as suggested by Huckel-molecular-orbital calculation (12). Singlet excited states (11, 13) are included near 28000 and 30000 cm^{-1}, but the values are placed in brackets to emphasize the uncertainty in the analysis. Also tentatively adopted is the suggestion by Hastie et al. (13) that the continuous spectrum near 23000 cm^{-1} is due to a triplet excited state.

References

1. JANAF Thermochemical Tables: $SiCl_4(g)$, 12-31-70; $SiH_2Cl_2(g)$, 12-31-76; Si(g), 3-31-67; Cl(g), 6-30-72.
2. H. Schafer, H. Bruderreck and B. Mörcher, Z. Anorg. Allg. Chem. 352, 122 (1967).
3. R. Teichmann and E. Wolf, Z. Anorg. Allg. Chem. 347, 145 (1966).
4. V. N. Chechentsev, L. A. Firsanova and O. K. Fedorchuk, Izv. Vyssh. Ucheb. Zaved., Tsvetn. Met. 8(4), 97 (1965).
5. P. F. Antipin and V. V. Sergeev, Russ. J. Appl. Chem. 27, 737 (1954).
6. T. Ishino, A. Matsumoto and S. Yamagishi, Kogyo Kagaku Zasshi 68, 262 (1965).
7. H. Schafer and J. Nickl, Z. Anorg. Allg. Chem. 274, 250 (1953).
8. M. Farber and R. D. Srivastava.
9. G. Maass, R. H. Hauge, and J. L. Margrave, Z. Anorg. Allg. Chem. 392, 295 (1972).
10. D. E. Milligan and M. E. Jacox, J. Chem. Phys. 49, 1938 (1968).
11. R. K. Asundi, M. Karim and R. Samuel, Proc. Phys. Soc. 50, 581 (1936).
12. J. W. Hastie and J. L. Margrave, J. Phys. Chem. 73, 1105 (1969).
13. J. W. Hastie, R. H. Hauge and J. L. Margrave, J. Mol. Spectrosc. 29, 152 (1969).

Enthalpy Reference Temperature = T_r = 298.15 K Standard State Pressure = p° = 0.1 MPa

T/K	$C°_p$ (J $K^{-1}mol^{-1}$)	S°	$-[G°-H°(T_r)]/T$	$H°-H°(T_r)$ (kJ mol^{-1})	$\Delta_fH°$	$\Delta_fG°$	Log K_f
0	0.	0.	INFINITE	-12.524	-168.740	-168.740	INFINITE
100	38.144	232.778	323.282	-9.050	-168.445	-172.439	90.073
200	46.298	261.819	285.885	-4.813	-168.424	-176.462	46.087
250	49.217	272.481	282.167	-2.421	-168.501	-178.464	37.288
298.15	51.248	281.333	281.333	0.	-168.615	-180.373	31.601
300	51.313	281.651	281.334	0.095	-168.620	-180.446	31.418
350	52.810	289.679	281.965	2.700	-168.769	-182.405	27.222
400	53.893	296.806	283.383	5.369	-168.938	-184.342	24.073
450	54.693	303.202	285.236	8.085	-169.122	-186.256	21.620
500	55.297	308.997	287.327	10.835	-169.321	-188.150	19.656
600	56.125	319.158	291.809	16.410	-169.754	-191.875	16.704
700	56.648	327.852	296.352	22.050	-170.234	-195.525	14.590
800	56.998	335.441	300.774	27.733	-170.759	-199.103	13.000
900	57.243	342.169	305.006	33.446	-171.328	-202.612	11.759
1000	57.421	348.210	309.030	39.180	-171.942	-206.056	10.763
1100	57.553	353.689	312.845	44.929	-172.599	-209.435	9.945
1200	57.655	358.701	316.460	50.689	-173.299	-212.754	9.261
1300	57.734	363.320	319.889	56.459	-174.042	-216.012	8.679
1400	57.798	367.600	323.146	62.236	-174.830	-219.211	8.179
1500	57.849	371.590	326.244	68.018	-175.663	-222.352	7.743
1600	57.891	375.325	329.196	73.805	-176.541	-225.436	7.360
1700	57.926	378.836	332.014	79.596	-227.644	-228.017	7.006
1800	57.956	382.147	334.708	85.390	-228.387	-228.018	6.617
1900	57.981	385.282	337.288	91.187	-229.137	-227.977	6.268
2000	58.003	388.256	339.763	96.987	-229.895	-227.896	5.952
2100	58.021	391.087	342.140	102.788	-230.662	-227.778	5.666
2200	58.038	393.786	344.427	108.591	-231.439	-227.622	5.404
2300	58.053	396.366	346.629	114.395	-232.228	-227.431	5.165
2400	58.068	398.837	348.754	120.201	-233.031	-227.205	4.945
2500	58.081	401.208	350.805	126.009	-233.847	-226.946	4.742
2600	58.095	403.486	352.787	131.818	-234.680	-226.653	4.554
2700	58.110	405.679	354.706	137.628	-235.529	-226.329	4.379
2800	58.126	407.793	356.564	143.440	-236.396	-225.972	4.216
2900	58.144	409.833	358.366	149.253	-237.281	-225.584	4.063
3000	58.164	411.804	360.115	155.069	-238.183	-225.166	3.920
3100	58.188	413.712	361.813	160.886	-239.102	-224.717	3.786
3200	58.215	415.560	363.464	166.706	-240.037	-224.238	3.660
3300	58.247	417.352	365.070	172.529	-240.987	-223.729	3.541
3400	58.284	419.091	366.633	178.356	-241.951	-223.192	3.429
3500	58.326	420.781	368.156	184.186	-242.926	-222.626	3.323
3600	58.374	422.425	369.641	190.021	-628.061	-211.571	3.070
3700	58.429	424.025	371.089	195.861	-628.637	-199.994	2.823
3800	58.492	425.584	372.503	201.707	-629.217	-188.402	2.590
3900	58.561	427.104	373.884	207.560	-629.799	-176.794	2.368
4000	58.639	428.588	375.233	213.420	-630.378	-165.171	2.157
4100	58.725	430.037	376.552	219.288	-630.951	-153.533	1.956
4200	58.819	431.453	377.842	225.165	-631.514	-141.882	1.765
4300	58.922	432.838	379.105	231.052	-632.064	-130.218	1.582
4400	59.033	434.194	380.342	236.950	-632.597	-118.541	1.407
4500	59.154	435.522	381.553	242.859	-633.109	-106.852	1.240
4600	59.283	436.824	382.741	248.781	-633.598	-95.152	1.080
4700	59.421	438.100	383.905	254.716	-634.059	-83.441	0.927
4800	59.567	439.353	385.047	260.665	-634.490	-71.721	0.780
4900	59.722	440.582	386.168	266.630	-634.887	-59.993	0.640
5000	59.885	441.791	387.269	272.610	-635.250	-48.257	0.504
5100	60.055	442.978	388.349	278.607	-635.573	-36.514	0.374
5200	60.234	444.146	389.411	284.621	-635.857	-24.765	0.249
5300	60.420	445.295	390.455	290.654	-636.098	-13.010	0.128
5400	60.612	446.426	391.481	296.705	-636.295	-1.252	0.012
5500	60.811	447.540	392.490	302.776	-636.446	10.510	-0.100
5600	61.016	448.638	393.483	308.868	-636.550	22.273	-0.208
5700	61.227	449.720	394.460	314.980	-636.606	34.039	-0.312
5800	61.442	450.786	395.422	321.113	-636.613	45.805	-0.413
5900	61.663	451.838	396.369	327.268	-636.571	57.570	-0.510
6000	61.887	452.877	397.302	333.446	-636.477	69.334	-0.604

PREVIOUS: December 1977 (1 atm) CURRENT: December 1977 (1 bar)

STRONTIUM CHLORIDE ($SrCl_2$) CRYSTAL (ALPHA-BETA) M_r = 158.526

$S°(298.15\ K) = 114.809 \pm 0.17\ J\ K^{-1}\ mol^{-1}$

$T_{trs} = 1000 \pm 10\ K$

$T_{fus} = 1147 \pm 1\ K$

$\Delta_f H°(0\ K) = -829.34 \pm 2.5\ kJ\ mol^{-1}$

$\Delta_f H°(298.15\ K) = -828.85 \pm 2.5\ kJ\ mol^{-1}$

$\Delta_{trs} H° = [6.008 \pm 1.3]\ kJ\ mol^{-1}$

$\Delta_{fus} H° = 16.221 \pm 0.63\ kJ\ mol^{-1}$

Enthalpy of Formation

$\Delta_f H°$ is from (1) and was selected by Parker in a manner analogous to that described (2) for $\Delta_f H°(BaCl_2, cr)$. The adopted value is essentially that derived by Ehrlich et al (3) from their measurements of $\Delta_{sol} H°$ of Sr(cr) and $SrCl_2$(cr) in HCl(aq, 550 H_2O). Earlier solution data for Sr (4) and $SrCl_2$ (5) are in good agreement.

Parker (2) rejected $\Delta_f H° = -210\ kcal\ mol^{-1}$ obtained by Siemonsen (6) from direct chlorination and $\Delta_f H° = -201.0\ kcal\ mol^{-1}$ which would result from SrO(cr) based on Mah's direction combustion (7). Parker noted that these more negative values are inconsistent with data for $SrCl_2$(cr, aq) and related compounds. They are also less consistent with equilibrium data discussed on the tables for $SrCl_2$(ℓ, g). Impurity effects are of concern in all these studies involving Sr(cr), but the evidence favors the solution calorimetry.

Heat Capacity and Entropy

$C_p°$ is based on adiabatic calorimetry (7.4-300 K) of Smith et al. (8) with omission of the point at 11.22 K which appears to be about 30% low. S° is calculated from $C_p°$ using $S°(7\ K) = 0.012\ cal\ K^{-1}\ mol^{-1}$. $C_p°$ above 300 K is from constrained fitting of enthalpy data (418-936 K for α-phase and 1034-1134 K for β-phase) from Dworkin and Bredig (9).

The transition near 1000 K is gradual and may be second order rather than first order as we have treated it. Gradual transitions also are observed for BaF_2, SrF_2 and perhaps CaF_2, all of which have α-phases of the fluorite structure. Reasonable $C_p°$ curves are obtained by omitting from the curve fits those enthalpy data from about 50 below to 30 above T_{trs}. The resulting Gibbs-energy functions agree well with those obtained (9) by second-order analysis.

Deviations of the fitted data (9) from the adopted functions for α-phase are -0.5 to +0.3% except for +2.0% at 417.6 K. Data of Efremova and Matizen (10) deviate by +0.6 to +2.2% (805-959 K), while the equation of Janz et al. (11) agrees almost exactly near 850 K but deviates by -4.8% at 600 K. The β-phase deviations are -0.4 to +0.3% (9), 0.0 to -1.0% (10, 1047-1142 K) and -2 to 0% (11, 1050-1147 K).

Transition Data

T_{trs} = 1000 K is arbitrarily selected as the peak of the gradual transition. Reported values include 1003 (9), 1002 (11) and 990 K (10). $\Delta_{trs} H°$ is calculated as the difference in the enthalpy curve fits (cf. above) at T_{trs}. $\Delta_{trs} H°$ is assigned an uncertainty of 0.3 $kcal\ mol^{-1}$ as a first-order transition, but the result is placed in brackets to emphasize that the transition may be a second order. Reported values of $\Delta_{trs} H°$ include 1.65 (9) and 0.65 (11) $kcal\ mol^{-1}$.

Fusion Data

See the liquid table for details.

Sublimation Data

See the ideal gas table for details.

References

1. U. S. Nat. Bur. Stand. Tech. Note 270-6, 1971.
2. V. B. Parker, U. S. Nat. Bur. Stand. Report 10074, 164, 1969.
3. P. Ehrlich, K. Peik and E. Koch, Z. Anorg. Allg. Chem. 324, 113 (1963).
4. A. Guntz and F. Benoit, Ann. Chim. (Paris) [9] 20, 5 (1923).
5. O. Ya. Samoilov, Izv. Akad. Nauk SSSR, Otdel. Khim. Nauk 627 (1952); 337 (1950).
6. H. Siemonsen, Z. Elektrochem. 55, 327 (1951).
7. A. D. Mah, U. S. Bur. Mines, RI-6171, 1963.
8. D. F. Smith, T. E. Gardner, B. B. Letson and A. R. Taylor, U. S. Bur. Mines RI-6316, 1963.
9. A. S. Dworkin and M. A. Bredig, J. Chem. Eng. Data 8, 416 (1963); J. Phys. Chem. 67, 697 (1963).
10. R. I. Efremova and E. V. Matizen, Izv. Sib. Otd. Akad. Nauk SSSR, Ser. Khim. Nauk, No. 1, 3 1970.
11. G. J. Janz, F. J. Kelly and J. L. Perano, Trans. Faraday Soc. 59, 2718 (1963).

Strontium Chloride ($SrCl_2$) $Cl_2Sr_1(cr)$

Enthalpy Reference Temperature = T_r = 298.15 K — Standard State Pressure = p° = 0.1 MPa

T/K	$C_p°$	S°	$-[G°-H°(T_r)]/T$	$H°-H°(T_r)$	$\Delta_f H°$	$\Delta_f G°$	Log K_f
	J K⁻¹mol⁻¹			kJ mol⁻¹			
0	0.	0.	INFINITE	-16.242	-829.344	-829.344	INFINITE
100	54.296	41.459	178.054	-13.659	-831.172	-813.612	424.988
200	70.735	85.571	121.583	-7.202	-830.241	-796.320	207.978
298.15	75.588	114.809	114.809	0.	-828.850	-779.965	136.647
300	75.659	115.277	114.810	0.140	-828.823	-779.661	135.751
400	78.889	137.512	117.820	7.877	-827.285	-763.503	99.703
500	81.321	155.385	123.603	15.891	-825.670	-747.743	78.116
600	83.659	170.415	130.185	24.138	-823.985	-732.315	63.754
700	86.580	183.518	136.887	32.642	-822.187	-717.177	53.516
800	90.793	195.336	143.465	41.497	-820.171	-702.309	45.856
900	96.977	206.364	149.846	50.866	-818.397	-687.644	39.910
1000	105.826	217.013	156.032	60.981	-815.052	-673.287	35.169
1000.010	105.827	217.014	156.033	60.982	ALPHA <--> BETA	TRANSITION	
1000.010	123.010	223.022	156.033	66.990			
1100	123.010	234.745	162.663	79.290	-811.429	-659.613	31.322
1147.000	123.010	239.892	165.723	85.071	BETA <--> LIQUID		
1200	123.010	245.448	169.123	91.591	-806.836	-646.014	28.120
1300	123.010	255.294	175.378	103.892	-802.254	-632.798	25.426
1400	123.010	264.410	181.415	116.193	-797.681	-619.935	23.130
1500	123.010	272.897	187.235	128.494	-793.117	-607.398	21.151
1600	123.010	280.836	192.839	140.795	-788.562	-595.166	19.430
1700	123.010	288.293	198.237	153.096	-920.931	-582.038	17.884
1800	123.010	295.324	203.437	165.397	-914.533	-562.288	16.317
1900	123.010	301.975	208.450	177.698	-908.151	-542.893	14.925
2000	123.010	308.285	213.285	189.999	-901.787	-523.833	13.681

PREVIOUS: CURRENT: December 1972

Strontium Chloride ($SrCl_2$) $Cl_2Sr_1(cr)$

STRONTIUM CHLORIDE ($SrCl_2$) LIQUID M_r = 158.526

$S°$(298.15 K) = [135.820] J K^{-1} mol^{-1} $\Delta_f H°$(298.15 K) = [-805.352] kJ mol^{-1}

T_{fus} = 1147 ± 1 K $\Delta_{fus}H°$ = 16.221 ± 0.63 kJ mol^{-1}

Enthalpy of Formation

$\Delta_f H°$ is calculated from that of the crystal by addition of $\Delta_{fus}H°$ and the difference in enthalpy, H°(1147 K)-H°(298.15 K), between the crystal and liquid. An independent value of $\Delta_f H°$ for liquid (and crystal) may be derived from equilibrium data (1) for Ca(ℓ) + $SrCl_2$(ℓ) → Sr(ℓ) + $CaCl_2$(ℓ) at 1273 K. Our third-law analysis yields $\Delta_r H°$(298.15 K) = 6.5±1.3 kcal mol^{-1} or $\Delta_f H°$(ℓ) = -191.5±1.5 kcal mol^{-1} and $\Delta_f H°$(cr) = -197.2±1.5 kcal mol^{-1}. This confirms the adopted values as opposed to the more negative values (cf. $SrCl_2$, cr).

Heat Capacity and Entropy

The constant $C_p°$ of 25.0 cal K^{-1} mol^{-1} is a compromise chosen by comparison with $BaCl_2$ and $CaCl_2$. Our fits of the enthalpy data yield 26.8 (2) and 23.5 (3) cal K^{-1} mol^{-1}; the adopted value is roughly their mean. The value 26.7 cal K^{-1} mol^{-1} was also reported (4). Deviations from our adopted functions are ±0.3% (2, 1154-1204 K), +1.1 to -1.7% (3, 1152-1797 K) and -1.1 to -0.2% (4, equation, 1148-1287 K). $C_p°$ is taken equal to that of the crystal from 298.15 to 800 K, the assumed glass transition. S°(298.15 K) is calculated in a manner analogous to that used for the enthalpy of formation.

Fusion Data

T_{fus} is selected from 1146 (2), 1147±2 (3) and 1148 K (4) reported in the enthalpy studies. $\Delta_{fus}H°$ is obtained from the difference in the adopted enthalpy fits for liquid and β-phases. Published values include 3.85 (2, 3), 3.88 (2), 3.65±0.17 (4) and 3.73 kcal mol^{-1} (5), the last value obtained by dynamic differential calorimetry.

Vaporization Data

T_{vap} is calculated as the temperature at which $\Delta_r G°$ = 0 for $SrCl_2$(ℓ) → $SrCl_2$(g). $\Delta_{vap}H°$ is calculated as the corresponding $\Delta_r H°$.

References

1. H. Feschotte-Ostertag, Ann. Chim. (Paris) 6, 791 (1961).
2. A. S. Dworkin and M. A. Bredig, J. Chem. Eng. Data 8, 416 (1963); J. Phys. Chem. 67, 697 (1963).
3. R. I. Efremova and E. V. Matizen, Izv. Sib. Otd. Akad. Nauk SSSR, Ser. Khim. Nauk, No. 1, 3, 1970.
4. G. J. Janz, F. J. Kelly and J. L. Perano, Trans. Faraday Soc. 59, 2718 (1963).
5. H. H. Emons and B. Loeffelholz, Wiss. Z. Tech. Hochsch. Chem., Leuna-Merseburg 6, 261 (1964).

Strontium Chloride ($SrCl_2$) Cl_2Sr_1(l)

Enthalpy Reference Temperature = T_r = 298.15 K Standard State Pressure = p° = 0.1 MPa

T/K	$C_p°$ (J K⁻¹mol⁻¹)	S° (J K⁻¹mol⁻¹)	-[G°-H°(T_r)]/T (J K⁻¹mol⁻¹)	H°-H°(T_r) (kJ mol⁻¹)	$\Delta_f H°$ (kJ mol⁻¹)	$\Delta_f G°$ (kJ mol⁻¹)	Log K_f
0							
100							
200							
298.15	75.588	135.820	135.820	0.	-805.351	-762.730	133.627
300	75.659	136.288	135.822	0.140	-805.323	-762.465	132.757
400	78.889	158.524	138.832	7.877	-803.786	-748.408	97.732
500	81.321	176.397	144.615	15.891	-802.170	-734.749	76.759
600	83.659	191.426	151.196	24.138	-800.486	-721.422	62.805
700	86.580	204.529	157.898	32.642	-798.687	-708.385	52.860
800	90.793	216.348	164.476	41.497	-796.671	-695.619	45.419
800.010	90.793	216.349	164.477	41.498	GLASS <--> LIQUID TRANSITION		
800.010	104.600	216.349	164.477	41.498			
900	104.600	228.668	170.938	51.957	-793.806	-683.126	39.648
1000	104.600	239.688	177.271	62.417	-790.116	-671.027	35.051
1100	104.600	249.658	183.406	72.877	-794.343	-658.930	31.290
1147.000	104.600	254.034	186.211	77.793	BETA <--> LIQUID		
1200	104.600	258.759	189.312	83.337	-791.591	-646.742	28.152
1300	104.600	267.132	194.980	93.797	-788.849	-634.782	25.506
1400	104.600	274.883	200.414	104.257	-786.117	-623.033	23.246
1500	104.600	282.100	205.622	114.717	-783.394	-611.480	21.294
1600	104.600	288.851	210.615	125.177	-780.680	-600.107	19.591
1700	104.600	295.192	215.406	135.637	-914.890	-587.725	18.059
1800	104.600	301.171	220.006	146.097	-910.333	-568.611	16.501
1900	104.600	306.826	224.428	156.557	-905.792	-549.751	15.114
2000	104.600	312.191	228.683	167.017	-901.269	-531.129	13.872
2100	104.600	317.295	232.782	177.477	-896.769	-512.732	12.754
2200	104.600	322.161	236.735	187.937	-892.298	-494.550	11.742
2300	104.600	326.811	240.551	198.397	-887.862	-476.570	10.823
2400	104.600	331.262	244.239	208.857	-883.470	-458.782	9.985
2500	104.600	335.532	247.805	219.317	-879.131	-441.176	9.218
2600	104.600	339.635	251.259	229.777	-874.852	-423.742	8.513
2700	104.600	343.582	254.606	240.237	-870.645	-406.472	7.864
2800	104.600	347.386	257.852	250.697	-866.521	-389.356	7.264
2900	104.600	351.057	261.003	261.157	-862.488	-372.386	6.707
3000	104.600	354.603	264.064	271.617	-858.558	-355.553	6.191

PREVIOUS: CURRENT: December 1972

Strontium Chloride ($SrCl_2$) Cl_2Sr_1(l)

STRONTIUM CHLORIDE ($SrCl_2$) ALPHA-BETA-LIQUID M_r = 158.526

0 to 1000 K crystal, alpha
1000 to 1147 K crystal, beta
above 1147 K liquid

Refer to the individual tables for details.

Strontium Chloride ($SrCl_2$) Cl_2Sr_1(cr,l)

Enthalpy Reference Temperature = T_r = 298.15 K Standard State Pressure = p° = 0.1 MPa

T/K	C_p° (J K⁻¹mol⁻¹)	S° (J K⁻¹mol⁻¹)	−[G°−H°(T_r)]/T (J K⁻¹mol⁻¹)	H°−H°(T_r) (kJ mol⁻¹)	$\Delta_f H^\circ$ (kJ mol⁻¹)	$\Delta_f G^\circ$ (kJ mol⁻¹)	Log K_f
0	0.	0.	INFINITE	0.	-813.101	-813.101	INFINITE
100	65.775	38.171	178.944	-14.077	-831.590	-813.701	425.034
200	71.271	85.512	121.591	-7.216	-830.254	-796.322	207.978
298.15	75.588	114.809	114.809	0.	-828.850	-779.965	136.647
300	75.659	115.277	114.810	0.140	-828.823	-779.661	135.751
400	78.889	137.512	117.820	7.877	-827.285	-763.503	99.703
500	81.321	155.385	123.603	15.891	-825.670	-747.743	78.116
600	83.659	170.415	130.185	24.138	-823.985	-732.315	63.754
700	86.580	183.518	136.887	32.642	-822.187	-717.177	53.516
800	90.793	195.336	143.465	41.497	-820.171	-702.309	45.856
900	96.977	206.364	149.846	50.866	-818.397	-687.644	39.910
1000	105.826	217.013	156.032	60.981	-815.052	-673.287	35.169
1000.010	105.827	217.014	156.033	60.982	ALPHA <--> BETA		
1000.010	123.010	223.022	156.033	66.990	TRANSITION		
1100	123.010	234.745	162.663	79.290	-811.429	-659.613	31.322
1147.000	123.010	239.892	165.723	85.071	BETA <--> LIQUID		
1147.000	104.600	254.034	165.723	101.293	TRANSITION		
1200	104.600	258.759	169.729	106.836	-791.591	-646.742	28.152
1300	104.600	267.132	176.903	117.296	-788.849	-634.782	25.506
1400	104.600	274.883	183.629	127.756	-786.117	-623.033	23.246
1500	104.600	282.100	189.956	138.216	-783.394	-611.480	21.294
1600	104.600	288.851	195.928	148.676	-780.680	-600.107	19.591
1700	104.600	295.192	201.582	159.136	-914.890	-587.725	18.059
1800	104.600	301.171	206.950	169.596	-910.333	-568.611	16.501
1900	104.600	306.826	212.060	180.056	-905.792	-549.751	15.114
2000	104.600	312.191	216.933	190.516	-901.269	-531.129	13.872
2100	104.600	317.295	221.592	200.976	-896.769	-512.732	12.754
2200	104.600	322.161	226.053	211.436	-892.298	-494.550	11.742
2300	104.600	326.811	230.334	221.896	-887.862	-476.570	10.823
2400	104.600	331.262	234.447	232.356	-883.470	-458.782	9.985
2500	104.600	335.532	238.406	242.816	-879.131	-441.176	9.218
2600	104.600	339.635	242.221	253.276	-874.852	-423.742	8.513
2700	104.600	343.582	245.902	263.736	-870.645	-406.472	7.864
2800	104.600	347.386	249.459	274.196	-866.521	-389.356	7.264
2900	104.600	351.057	252.900	284.656	-862.488	-372.386	6.707
3000	104.600	354.603	256.231	295.116	-858.558	-355.553	6.191

PREVIOUS: CURRENT: December 1972

Strontium Chloride ($SrCl_2$) Cl_2Sr_1(cr,l)

STRONTIUM CHLORIDE ($SrCl_2$) IDEAL GAS M_r = 158.526

$S°(298.15\ K) = 316.340 \pm 5.0\ J\ K^{-1}\ mol^{-1}$

$\Delta_fH°(0\ K) = -471.91 \pm 6.3\ kJ\ mol^{-1}$
$\Delta_fH°(298.15\ K) = -473.21 \pm 6.3\ kJ\ mol^{-1}$

Vibrational Frequencies and Degeneracies

ν, cm^{-1}	ν, cm^{-1}	ν, cm^{-1}
270 (1)	44 (1)	300 (1)

Ground State Quantum Weight: 1 $\sigma = 2$

Point Group: C_{2v}

Bond Distance: Sr-Cl = 2.67 ± 0.03 Å

Bond Angle: Cl-Sr-Cl = 120° ± 20°

Product of the Moments of Inertia: $I_AI_BI_C = 5.44364 \times 10^{-113}\ g^3\ cm^6$

Enthalpy of Formation

$\Delta_fH°(298.15\ K)$ is calculated from $\Delta_fH°(cr)$ by addition of $\Delta_{sub}H° = 85.0 \pm 1.5$ kcal mol^{-1} derived by third law analysis of the vapor-pressure equation of Hildenbrand et al. (1, 2). The equation (1) summarized five torsion-effusion runs (102 total points) and is reasonably consistent with three boiling pressures in the mmHg range measured by Novikov (3). Effusion studies (4-7) yield lower values of $\Delta_{sub}H°$, but that of Loehman et al. (6) is consistent with the adopted value. Langmuir data (6) suggest that the sublimation coefficient is 0.3±0.2. Equilibria derived from flame studies (8, 9) and mass-spectrometric effusion (10) confirm the adopted $\Delta_fH°$ via independent paths (cf. below). Mass spectra (11) at 1220 K showed that the saturated vapor consists predominantly of $SrCl_2$ with considerably less than 1% of dimer.

Source	Method	T/K	$\Delta_{sub}H°$(298.15 K), kcal mol^{-1} 2nd law	3rd law	Drift cal K^{-1} mol^{-1}	$\Delta_fH°$(298.15 K) kcal mol^{-1}
(1, 2) Hildenbrand (1965)	Torsion effusion[b]	1220-1425	84.0	85.00	-0.8	-113.1
(3) Novikov (1964)	Boiling Point	1620-1710	75.8±9.4	85.52±0.7	-5.8±5.6	-112.6
(4) Novikov (1970)	Total effusion[b]	1137-1422	75.2	82.3	-5.1	
(5) Naryshkin (1968)	Effusion[b]	893-1141	75.6	78.0	-2.3	
(6) Loehman (1965)	Effusion	1195-1307	82.9±3.9	84.2±0.7	-1.0±3.1	-113.9
(6) Loehman (1965)	Langmuir[c]	968-1119	78.7±3.6	86.3±1.5	-7.4±3.5	-111.8
(7) Van Westenburg (1964)	Effusion	1175-1287	80.3±0.5	82.8±0.2	-2.0±0.4	
(8) Schofield (1971)	Flame Kp[d]	~2134-2350	-9.1±3.7[d]	-1.6±0.8[d]	-3.3±1.6	-110.7±2
(9) Ryabova (1971)	Flame Kp	1813-2209	9.0±25[d]	-2.0±7.2[d]	5.5±12	-111.1±7
(10) Hildenbrand (1970)	Mass spec. Kp[e]	1661		3.7[e]		-112.7±2

[b] Pressures obtained from an equation.
[c] Assuming a sublimation coefficient of 1.0; agreement with (1) requires $\alpha = 0.5$.
[d] $\Delta_rH°$ for $Sr(g) + 2\ HCl(g) = SrCl_2(g) + 2\ H(g)$. [e] $\Delta_rH°$ for $Sr(g) + CaCl_2(g) = SrCl_2(g) + Ca(g)$.

Heat Capacity and Entropy

Vibrations ν_2 and ν_3 are the values quoted by Hildenbrand (10) and derived from IR spectra of molecules in matrix isolation (12). The bending mode was observed (13). Hastie et al. (14) failed to observe ν_2 but estimated a higher value of 60±15 cm^{-1}. Agreement is better for ν_1 = 269 (10, 13) or 274 cm^{-1} (14) and for ν_3 = 300±7 (15, gas), 300 (10, 13) or 307 cm^{-1} (14). We adopt a rounded value of 270 cm^{-1} for ν_1.

Bond length is from gas-phase electron diffraction (16) which also gave the bond angle 180° ± 30°. Later studies (17) of electric deflection of molecular beams revealed a permanent electric dipole moment, leading to the conclusion that $SrCl_2$ is bent. The matrix spectra (10, 12) were first interpreted to give an angle of 155° but reanalysis (13) gave the angle 112°, calculated from the three observed fundamentals. An angle of 130° ± 8° was derived by Hastie et al. (14) from the Cl-isotope splitting of ν_3. We adopt an intermediate value of 120° ± 20°. S° would change by +0.1 cal K^{-1} mol^{-1} for 112° and -1.3 cal K^{-1} mol^{-1} for 155°. The singlet ground state is taken by analogy with $BaCl_2(g)$. Principal moments of inertia are $I_A = 11.5986 \times 10^{-39}$, $I_B = 62.9539 \times 10^{-39}$, and $I_C = 74.5525 \times 10^{-39}\ g\ cm^2$.

References

1. D. L. Hildenbrand, Stanford Res. Inst., personal communication, 1973; D. L. Hildenbrand, E. Murad, N. D. Potter, L. P. Theard and W. F. Hall, Aeronutronic Division, Philco Corp., Report No. U-3183 Contract AF 49 (838)-1397, June 30, 1965.
2. D. L. Hildenbrand, Proc. 2nd Meeting, ICRPG Working Group on Thermochemistry, CPIA Publ. No. 54U, pp. 79-80, Aug., 1964.
3. G. I. Novikov and F. G. Gavryuchenkov, Russ. J. Inorg. Chem. 9, 260 (1964).
4. G. I. Novikov, I. A. Rat'kovskii and L. Ya. Kris'ko, Dokl. Akad. Nauk Beloruss. SSR 14, 918 (1970).
5. I. I. Naryshkin, V. I. Glazov and V. M. Kharlamov, J. Appl. Chem. USSR 41, 1260 (1968).
6. R. E. Loehman, R. A. Kent and J. L. Margrave, J. Chem. Eng. Data 10, 297 (1965).
7. J. A. Van Westenburg, Ph.D. Thesis, Iowa State Univ., 1964; Diss. Abstr. 25, 1129 (1964).
8. K. Schofield and T. M. Sugden, Trans. Faraday Soc. 67, 1054 (1971).
9. V. G. Ryabova, L. V. Gurvich, and A. N. Khitrov, Teplofiz. Vys. Temp. 9, 755 (1971).
10. D. L. Hildenbrand, J. Chem. Phys. 52, 5751 (1970).
11. D. L. Hildenbrand, Int. J. Mass Spectrom. Ion Phys. 4, 75 (1970).
12. D. White, Univ. of Pennsylvania, personal communication, 1972.
13. S. Hemple, U. S. Army Foreign Science & Technology Center, Charlottesville, Va., personal communication, 1973.
14. J. W. Hastie, R. H. Hauge and J. L. Margrave, High Temp. Sci. 3, 56 (1971).
15. V. I. Baikov, Optics and Spectroscopy 27, 502 (1969).
16. P. A. Akishin, V. P. Spiridonov, G. A. Solobev and V. A. Naumov, Zh. Fiz. Khim. 32, 58 (1958); Kristallografiya 2, 475 (1957).
17. L. Wharton, R. A. Berg and W. Klemperer, J. Chem. Phys. 39, 2023 (1963).

Strontium Chloride ($SrCl_2$) $Cl_2Sr_1(g)$

Enthalpy Reference Temperature = T_r = 298.15 K; Standard State Pressure = p° = 0.1 MPa

T/K	$C_p°$ (J $K^{-1}mol^{-1}$)	S°	$-[G°-H°(T_r)]/T$	$H°-H°(T_r)$ (kJ mol^{-1})	$\Delta_fH°$	$\Delta_fG°$	Log K_f
0	0.	0.	INFINITE	-14.449	-471.910	-471.910	INFINITE
100	46.112	259.888	363.999	-10.411	-472.284	-476.567	248.933
200	53.351	294.507	321.380	-5.375	-472.773	-480.640	125.530
250	54.877	306.591	317.253	-2.666	-472.985	-482.588	100.831
298.15	55.775	316.339	316.339	0.	-473.210	-484.411	84.867
300	55.802	316.685	316.341	0.103	-473.220	-484.480	84.356
350	56.397	325.334	317.022	2.909	-473.485	-486.337	72.582
400	56.798	332.893	318.544	5.740	-473.782	-488.153	63.746
450	57.080	339.600	320.518	8.587	-474.112	-489.930	56.870
500	57.286	345.625	322.732	11.446	-474.475	-491.668	51.364
600	57.559	356.096	327.446	17.190	-475.293	-495.032	43.096
700	57.727	364.982	332.190	22.955	-476.234	-498.249	37.180
800	57.836	372.698	336.782	28.733	-477.295	-501.323	32.733
900	57.912	379.515	341.158	34.521	-479.102	-504.185	29.262
1000	57.966	385.620	345.305	40.315	-480.078	-506.920	26.479
1100	58.007	391.146	349.225	46.114	-488.965	-509.191	24.179
1200	58.038	396.195	352.932	51.916	-490.871	-510.945	22.241
1300	58.062	400.842	356.441	57.721	-492.785	-512.541	20.594
1400	58.081	405.145	359.768	63.528	-494.706	-513.988	19.177
1500	58.096	409.153	362.928	69.337	-496.634	-515.298	17.944
1600	58.109	412.903	365.936	75.147	-498.570	-516.480	16.861
1700	58.119	416.426	368.803	80.959	-637.427	-516.360	15.866
1800	58.128	419.748	371.542	86.771	-637.519	-509.236	14.778
1900	58.136	422.891	374.162	92.584	-637.624	-502.106	13.804
2000	58.142	425.873	376.674	98.398	-637.747	-494.970	12.927
2100	58.147	428.710	379.085	104.213	-637.893	-487.828	12.134
2200	58.152	431.415	381.403	110.028	-638.067	-480.678	11.413
2300	58.156	434.000	383.634	115.843	-638.276	-473 520	10.754
2400	58.160	436.475	385.784	121.659	-638.528	-466.351	10.150
2500	58.163	438.850	387.860	127.475	-638.832	-459.171	9.594
2600	58.166	441.131	389.865	133.292	-639.197	-451.978	9.080
2700	58.168	443.326	391.805	139.108	-639.634	-444.769	8.605
2800	58.171	445.442	393.683	144.925	-640.152	-437.542	8.162
2900	58.173	447.483	395.503	150.742	-640.762	-430.296	7.750
3000	58.175	449.455	397.269	156.560	-641.474	-423.026	7.366
3100	58.176	451.363	398.983	162.377	-642.299	-415.732	7.005
3200	58.178	453.210	400.649	168.195	-643.246	-408.408	6.667
3300	58.179	455.000	402.269	174.013	-644.323	-401.053	6.348
3400	58.180	456.737	403.845	179.831	-645.539	-393.664	6.048
3500	58.182	458.423	405.381	185.649	-646.900	-386.236	5.764
3600	58.183	460.062	406.877	191.467	-648.411	-378.767	5.496
3700	58.184	461.657	408.336	197.285	-650.078	-371.255	5.241
3800	58.185	463.208	409.760	203.104	-651.904	-363.695	4.999
3900	58.185	464.720	411.150	208.922	-653.889	-356.085	4.769
4000	58.186	466.193	412.508	214.741	-656.035	-348.422	4.550
4100	58.187	467.630	413.835	220.560	-658.341	-340.703	4.341
4200	58.187	469.032	415.132	226.378	-660.800	-332.927	4.141
4300	58.188	470.401	416.402	232.197	-663.416	-325.090	3.949
4400	58.189	471.739	417.644	238.016	-666.182	-317.190	3.766
4500	58.189	473.046	418.861	243.835	-669.091	-309.226	3.589
4600	58.190	474.325	420.053	249.654	-672.072	-301.201	3.420
4700	58.190	475.577	421.221	255.473	-675.229	-293.105	3.257
4800	58.191	476.802	422.366	261.292	-678.503	-284.940	3.101
4900	58.191	478.002	423.489	267.111	-681.886	-276.706	2.950
5000	58.191	479.177	424.591	272.930	-685.364	-268.402	2.804
5100	58.192	480.330	425.673	278.749	-688.913	-260.030	2.663
5200	58.192	481.460	426.735	284.568	-692.560	-251.585	2.527
5300	58.192	482.568	427.778	290.388	-696.276	-243.069	2.396
5400	58.193	483.656	428.803	296.207	-700.052	-234.483	2.268
5500	58.193	484.724	429.810	302.026	-703.877	-225.826	2.145
5600	58.193	485.772	430.800	307.845	-707.743	-217.099	2.025
5700	58.194	486.802	431.773	313.665	-711.641	-208.303	1.909
5800	58.194	487.814	432.731	319.484	-715.560	-199.438	1.796
5900	58.194	488.809	433.673	325.304	-719.284	-190.530	1.687
6000	58.194	489.787	434.600	331.123	-723.193	-181.536	1.580

PREVIOUS: December 1972 (1 atm) CURRENT: December 1972 (1 bar)

Strontium Chloride ($SrCl_2$) $Cl_2Sr_1(g)$

TITANIUM CHLORIDE ($TiCl_2$) CRYSTAL M_r = 118.786

$S°(298.15\ K) = 87.345 \pm 4.2\ J\ K^{-1}\ mol^{-1}$

$\Delta_f H°(0\ K) = -514.76 \pm 16.7\ kJ\ mol^{-1}$

$\Delta_f H°(298.15\ K) = -515.47 \pm 16.7\ kJ\ mol^{-1}$

Enthalpy of Formation

The enthalpy of formation of $TiCl_2$(cr) has been measured calorimetrically. The chosen value of $\Delta_f H°$(298.15 K) is the average of two such determinations. Values of $\Delta_f H°$(298.15 K) are also calculated from the heats of reaction for the processes (A) $2TiCl_3(cr) = TiCl_2(cr) + TiCl_4(g)$, (B) $3TiCl_2(cr) = 2TiCl_3(g) + Ti(cr)$, and (C) $2TiCl_2(cr) = TiCl_4(g) + Ti(cr)$. These enthalpies of reaction are combined with auxiliary JANAF values for the heats of formation of $TiCl_3$(cr), $TiCl_4$(g), $TiCl_3$(cr) and Ti(cr). The following table summarizes the data used to calculate $\Delta_f H°$(298.15 K).

Source	Reaction	Method	Data Points	T/K	$\Delta_r H°$(298.15 K) 2nd law	3rd law	Drift cal K^{-1} mol^{-1}	$\Delta_f H°$(298.15 K) kcal mol^{-1}
Clifton and MacWood (1)		Calorimetric						-123.4, -123.8
Krieve et al. (2)		Calorimetric						-122.4
Farber and Darnell (3)	A	Effusion	11	615-732	36.0±1.4	38.9	4.2±2.1	-123.7
Sanderson and MacWood (4)	A	Knudsen	22*	593-720	35.6±0.8	39.6	6.1±1.3	-123.0
Sanderson and MacWood (4)	A	Static	15**	679-821	38.5±0.5	35.9	-3.4±0.7	-126.7
Farber and Darnell (5)	B	Effusion	13	798-893	94.4±1.9	112.6	21.4±2.2	-123.1
Farber and Darnell (5)	C	Effusion	11	794-894	51.5±1.4	63.0	13.6±1.7	-122.7
Hartman and Rinds (6)	A	Effusion	20	815-974	38.6±0.5	35.7	-3.3±0.5	-126.9
Skinner and Ruehrwein (7)	A	Manometric	5	802-928	37.6±0.5	36.0	-1.9±0.6	-126.9

*Four points rejected due to failure of a statistical test.

**Two points rejected due to failure of a statistical test.

Heat Capacity and Entropy

The heat capacity fo $TiCl_2$(cr) has been measured by Kim and Stout (8) over the range 6 to 300 K. They calculated the adopted value of S°(298.15 K) from their measurements based on S°(6 K) = 0.017 cal K^{-1} mol^{-1}. The heat capacity above 300 K is estimated by graphical extrapolation. The large uncertainty on S°(298.15 K) is due to the possibiility of sizeable electronic contributions to the heat capacity below 6 K which would account for the rather low value of S°(298.15 K) reported by Kim and Stout (8).

Sublimation Data

The enthalpy of sublimation is calculated from the enthalpies of formation of $TiCl_2$(cr) and $TiCl_2$(g). The temperature of sublimation is taken as the point at which $\Delta_r G°$ = 0 for the process $TiCl_2(cr) = TiCl_2(g)$. These values are hypothetical since $TiCl_2$ would be decomposed to $TiCl_3$(g) and $TiCl_4$(g) below this temperature.

References

1. D. G. Clifton and G. E. MacWood, J. Phys. Chem. 60, 311 (1956).
2. W. F. Krieve, S. P. Vango and D. M. Mason, J. Chem. Phys. 25, 519 (1956).
3. M. Farber and A. J. Darnell, J. Phys. Chem. 59, 156 (1955).
4. B. S. Sanderson and G. E. MacWood, J. Phys. Chem. 60, 316 (1956).
5. M. Farber and A. J. Darnell, J. Chem. Phys. 25, 526 (1956).
6. H. Hartman and G. Rinck, Z. Physik, Chem. 11, 213 (1957).
7. G. B. Skinner and R. A. Ruehrwein, J. Phys. Chem. 59, 113 (1955).
8. S. Kim and J. W. Stout, University of Chicago, personal communication, 1967.

Titanium Chloride ($TiCl_2$) Cl_2Ti_1(cr)

Enthalpy Reference Temperature = T_r = 298.15 K; Standard State Pressure = p° = 0.1 MPa

T/K	$C_p°$ (J $K^{-1}mol^{-1}$)	S°	$-[G°-H°(T_r)]/T$	$H°-H°(T_r)$ (kJ mol^{-1})	$\Delta_f H°$	$\Delta_f G°$	Log K_f
0	0.	0.	INFINITE	-13.301	-514.759	-514.759	INFINITE
100	37.777	25.966	142.867	-11.690	-516.621	-499.490	260.907
200	61.505	61.095	93.500	-6.481	-516.366	-482.347	125.976
298.15	69.831	87.345	87.345	0.	-515.469	-465.829	81.611
300	69.899	87.777	87.347	0.129	-515.449	-465.521	81.054
400	73.392	108.374	90.131	7.297	-514.365	-449.040	58.639
500	76.484	125.090	95.502	14.794	-513.179	-432.843	45.219
600	78.375	139.202	101.640	22.537	-511.894	-416.896	36.294
700	80.266	151.426	107.898	30.469	-510.522	-401.170	29.936
800	82.157	162.267	114.029	38.590	-509.025	-385.649	25.180
900	84.048	172.053	119.941	46.901	-507.426	-370.322	21.493
1000	85.939	181.006	125.606	55.400	-505.785	-355.176	18.552
1100	87.831	189.286	131.024	64.088	-504.161	-340.194	16.154
1200	89.722	197.009	136.204	72.966	-506.540	-325.239	14.157
1300	91.613	204.265	141.163	82.033	-504.226	-310.223	12.465
1400	93.504	211.124	145.918	91.289	-501.811	-295.390	11.021
1500	95.395	217.640	150.484	100.734	-499.307	-280.732	9.776
1600	97.286	223.857	154.877	110.368	-496.728	-266.245	8.692
1700	99.178	229.811	159.111	120.191	-494.086	-251.920	7.741
1800	101.069	235.534	163.199	130.203	-491.395	-237.752	6.899
1900	102.960	241.049	167.152	140.405	-488.668	-223.735	6.151
2000	104.851	246.378	170.980	150.795	-500.607	-209.409	5.469

PREVIOUS: June 1968 CURRENT: December 1968

Titanium Chloride ($TiCl_2$) Cl_2Ti_1(cr)

TITANIUM CHLORIDE ($TiCl_2$) IDEAL GAS M_r = 118.786

$\Delta_f H°(0\ K) = -236.77 \pm 12.6\ kJ\ mol^{-1}$

$S°(298.15\ K) = [278.338 \pm 12.6]\ J\ K^{-1}\ mol^{-1}$ $\Delta_f H°(298.15\ K) = -237.23 \pm 12.6\ kJ\ mol^{-1}$

Electronic Levels and Quantum Weights

ε_i, cm^{-1}	g_i
0	[3]
[7000]	[6]
[17000]	[6]
[22000]	[15]

Vibrational Frequencies and Degeneracies

ν, cm^{-1}
[279](1)
[196](2)
[439](1)

Ground State Quantum Weight: 3 σ = 2

Point Group: [$D_{\infty h}$]

Bond Distance: Ti-Cl = [2.3] Å

Bond Angle: Cl-Ti-Cl = [180]°

Rotational Constant: B_0 = [0.044943] cm^{-1}

Enthalpy of Formation

Gross and Levi (1) have measured the equilibrium $2TiCl_3(g) + Ti(cr) \rightarrow 3TiCl_2(g)$ at 1723 K and obtain $\Delta_r G°(1723) = 21.7$ kcal mol^{-1} from four measurements. This yields $\Delta_f H°(TiCl_2,\ g,\ 298.15\ K) = -56.7 \pm 3$ kcal mol^{-1}. They discuss the measurements of Farber and Darnell (2), on the vapor pressure of $TiCl_2(cr)$, and show that the results are probably in error due to disproportionation of $TiCl_3(g)$ on condensation. In the work of Skinner and Ruehrwein (3) on the reaction $2NaCl(g) + Ti(cr) = 2Na(g) + TiCl_2(g)$, Gross and Levi indicate that the basic assumption, of conversion to $TiCl_2$ only, is erroneous. In the 3rd law analysis of both these data sets significant drifts are obtained which tends to confirm the above arguments.

Heat Capacity and Entropy

The interatomic distances are estimated from those of $TiCl_2(cr)$ reported by Benzinger and Rundle (4). The vibrational frequencies are estimated from a valence force field model.

The electronic levels are assumed to correspond to the inverted states of $NiCl_2(g)$ (5). The linear configuration is adopted, since experimental evidence indicates that other transition metal dihalides are linear (6, 7, 8, 9).

References

1. P. Gross and D. L. Levi, 16e Cong. Intern. Chim. Pure et Appl. 628, (1958).
2. M. Farber and A. J. Darnell, J. Chem. Phys. 25, 526 (1956).
3. G. B. Skinner and R. A. Ruehrwein, J. Phys. Chem. 59, 113 (1965).
4. N. C. Benzinger and R. E. Rundle, Acta Cryst. 1, 274 (1949).
5. C. W. DeKock and D. M. Gruen, J. Chem. Phys. 44, 4387 (1966).
6. R. A. Berg and O. Sinanoglu, J. Chem. Phys. 32, 1082 (1960).
7. J. T. Hougen, G. E. Leroi, and T. C. James, J. Chem. Phys. 34, 1670 (1961).
8. G. E. Leroi, T. C. James, J. T. Hougen and W. Klemperer, J. Chem. Phys. 36, 2879 (1962).
9. A. Buchler, J. L. Stauffer and W. Klemperer, J. Chem. Phys. 40, 3471 (1964).

Titanium Chloride ($TiCl_2$) $Cl_2Ti_1(g)$

Enthalpy Reference Temperature = T_r = 298.15 K Standard State Pressure = p° = 0.1 MPa

T/K	$C_p°$ (J $K^{-1}mol^{-1}$)	S° (J $K^{-1}mol^{-1}$)	$-[G°-H°(T_r)]/T$ (J $K^{-1}mol^{-1}$)	$H°-H°(T_r)$ (kJ mol^{-1})	$\Delta_f H°$ (kJ mol^{-1})	$\Delta_f G°$ (kJ mol^{-1})	Log K_f
0	0.	0.	INFINITE	-13.545	-236.767	-236.767	INFINITE
100	41.124	223.342	326.011	-10.267	-236.961	-239.569	125.138
200	53.079	256.188	283.484	-5.459	-237.108	-242.108	63.232
250	55.821	268.352	279.276	-2.731	-237.163	-243.351	50.846
298.15	57.503	278.338	278.338	0.	-237.233	-244.537	42.842
300	57.555	278.694	278.339	0.106	-237.236	-244.583	42.586
350	58.701	287.658	279.044	3.015	-237.333	-245.800	36.684
400	59.490	295.551	280.624	5.971	-237.455	-247.001	32.255
450	60.054	302.592	282.681	8.960	-237.600	-248.186	28.809
500	60.469	308.942	284.995	11.974	-237.764	-249.354	26.050
600	61.025	320.020	289.936	18.051	-238.144	-251.637	21.907
700	61.371	329.455	294.924	24.172	-238.583	-253.852	18.943
800	61.605	337.666	299.765	30.321	-239.058	-256.001	16.715
900	61.782	344.933	304.387	36.491	-239.600	-258.087	14.979
1000	61.938	351.450	308.773	42.677	-240.272	-260.107	13.587
1100	62.098	357.361	312.926	48.878	-241.135	-262.050	12.444
1200	62.280	362.772	316.858	55.097	-246.173	-263.787	11.482
1300	62.496	367.765	320.584	61.336	-246.687	-265.234	10.657
1400	62.750	372.406	324.122	67.598	-247.266	-266.639	9.948
1500	63.044	376.745	327.487	73.887	-247.918	-268.001	9.333
1600	63.373	380.824	330.694	80.208	-248.652	-269.316	8.792
1700	63.732	384.677	333.757	86.563	-249.478	-270.583	8.314
1800	64.115	388.330	336.689	92.955	-250.407	-271.798	7.887
1900	64.513	391.807	339.499	99.386	-251.451	-272.959	7.504
2000	64.919	395.127	342.198	105.858	-267.309	-273.608	7.146
2100	65.326	398.304	344.794	112.370	-269.369	-273.872	6.812
2200	65.728	401.352	347.296	118.923	-271.400	-274.038	6.506
2300	66.120	404.282	349.711	125.515	-273.406	-274.113	6.225
2400	66.499	407.105	352.044	132.146	-275.387	-274.102	5.966
2500	66.861	409.827	354.301	138.814	-277.348	-274.008	5.725
2600	67.206	412.456	356.487	145.518	-279.290	-273.836	5.501
2700	67.531	414.998	358.608	152.255	-281.217	-273.589	5.293
2800	67.837	417.460	360.666	159.023	-283.131	-273.272	5.098
2900	68.125	419.845	362.665	165.822	-285.034	-272.886	4.915
3000	68.394	422.159	364.610	172.648	-286.930	-272.435	4.744
3100	68.647	424.406	366.503	179.500	-288.819	-271.920	4.582
3200	68.884	426.590	368.347	186.377	-290.701	-271.345	4.429
3300	69.107	428.713	370.144	193.276	-292.579	-270.711	4.285
3400	69.317	430.779	371.897	200.198	-294.452	-270.020	4.148
3500	69.516	432.791	373.608	207.139	-296.320	-269.275	4.019
3600	69.705	434.752	375.280	214.101	-298.181	-268.475	3.895
3700	69.884	436.664	376.913	221.080	-709.134	-259.836	3.668
3800	70.056	438.530	378.510	228.077	-709.745	-247.685	3.405
3900	70.220	440.352	380.073	235.091	-710.402	-235.517	3.154
4000	70.378	442.132	381.602	242.121	-711.100	-223.332	2.916
4100	70.531	443.872	383.100	249.167	-711.837	-211.129	2.690
4200	70.679	445.573	384.567	256.227	-712.608	-198.907	2.474
4300	70.822	447.238	386.005	263.302	-713.410	-186.667	2.268
4400	70.960	448.868	387.415	270.391	-714.240	-174.407	2.070
4500	71.094	450.464	388.799	277.494	-715.092	-162.129	1.882
4600	71.224	452.028	390.156	284.610	-715.964	-149.831	1.701
4700	71.350	453.561	391.489	291.739	-716.851	-137.514	1.528
4800	71.472	455.065	392.798	298.880	-717.750	-125.179	1.362
4900	71.590	456.539	394.084	306.033	-718.657	-112.824	1.203
5000	71.703	457.987	395.347	313.198	-719.568	-100.451	1.049
5100	71.812	459.408	396.590	320.373	-720.482	-88.059	0.902
5200	71.916	460.803	397.811	327.560	-721.393	-75.650	0.760
5300	72.016	462.174	399.013	334.757	-722.300	-63.223	0.623
5400	72.111	463.521	400.195	341.963	-723.200	-50.779	0.491
5500	72.201	464.845	401.358	349.179	-724.091	-38.319	0.364
5600	72.286	466.147	402.504	356.403	-724.969	-25.842	0.241
5700	72.365	467.427	403.631	363.635	-725.834	-13.350	0.122
5800	72.439	468.686	404.742	370.876	-726.683	-0.843	0.008
5900	72.508	469.925	405.837	378.123	-727.515	11.679	-0.103
6000	72.571	471.144	406.915	385.377	-728.327	24.214	-0.211

PREVIOUS: December 1968 (1 atm) CURRENT: December 1968 (1 bar)

Titanium Chloride ($TiCl_2$) $Cl_2Ti_1(g)$

TUNGSTEN CHLORIDE (WCl_2) CRYSTAL M_r = 254.756

$\Delta_fH°(0\ K)$ = Unknown

$S°(298.15\ K) = [130.54]\ J\ K^{-1}\ mol^{-1}$ $\Delta_fH°(298.15\ K) = -257.32 \pm 12.6\ kJ\ mol^{-1}$

Enthalpy of Formation

The enthalpy of formation, $\Delta_fH°(WCl_2,\ cr,\ 298.15\ K) = -61.5\pm3\ kcal\ mol^{-1}$, is derived from the heat of combustion, $\Delta_cH° = -140\pm3\ kcal\ mol^{-1}$ for $WCl_2(cr) + 3/2\ O_2(g) \rightarrow WO_3(cr) + Cl_2(g)$, reported by Shchukarev et al. (1). The value of $\Delta_cH°$ was determined calorimetrically, but no further information about the combustion process was given in the paper.

Heat Capacity and Entropy

$C_p°(298.15\ K) = 6.2\ cal\ K^{-1}\ g\text{-}atom^{-1}$ and $C_p°(900\ K) = 7.25\ cal\ K^{-1}\ g\text{-}atom^{-1}$ are estimated using the method described by Kubaschewski and Evans (2). Between 298 and 900, which is near the decomposition temperature, the heat capacity is obtained by linear interpolation.

The entropy, $S°(298.15\ K) = 31.2\ cal\ K^{-1}\ mol^{-1}$, is estimated by the method of Latimer (3). Shchukarev et al. (1), have estimated $\Delta_rS° = -30\ cal\ K^{-1}\ mol^{-1}$ for $W(cr) + Cl_2(g) = WCl_2(cr)$ which leads to $S°(WCl_2,\ cr,\ 298.15\ K) = 31.1\ cal\ K^{-1}\ mol^{-1}$, using JANAF auxiliary data.

Decomposition Data

The decomposition temperature, T_{dcm} = 862 K, is obtained by extrapolation to one atmosphere total pressure of a log P versus 1/T fit of the decomposition pressure data reported by Shchukarev et al. (4). Decomposition pressure data for Cl:W ratios less than two suggest that the decomposition reaction is not simple.

References

1. S. A. Shchukarev, G. I. Novikov, I. V. Vasil'kova, A. V. Suvorov, N. V. Andreeva, B. N. Sharupin, and A. K. Baev, Zhur. Neorg. Khim. 5, 1650 (1960).
2. O. Kubaschewski and E. Ll. Evans, "Metallurgical Thermochemistry," Pergamon Press, New York, 1958.
3. W. M. Latimer, "Oxidation Potentials," 2nd Ed., Prentice-Hall, Inc., New York, 1952.
4. S. A. Shchukarev, G. I. Novikov, and N. V. Andreeva, Vestnik Leningrad. Univ. 14, No. 10, Ser. Fiz. i Khim., No. 2, 78 (1959).

Tungsten Chloride (WCl_2) $Cl_2W_1(cr)$

Enthalpy Reference Temperature = T_r = 298.15 K Standard State Pressure = p° = 0.1 MPa

T/K	$C_p°$ (J K⁻¹mol⁻¹)	S° (J K⁻¹mol⁻¹)	$-[G°-H°(T_r)]/T$ (J K⁻¹mol⁻¹)	$H°-H°(T_r)$ (kJ mol⁻¹)	$\Delta_fH°$ (kJ mol⁻¹)	$\Delta_fG°$ (kJ mol⁻¹)	Log K_f
0							
100							
200							
298.15	77.822	130.541	130.541	0.	-257.316	-219.988	38.541
300	77.864	131.022	130.542	0.144	-257.280	-219.757	38.263
400	80.052	153.721	133.622	8.040	-255.318	-207.544	27.102
500	82.241	171.817	139.511	16.153	-253.290	-195.833	20.459
600	84.429	187.006	146.194	24.487	-251.145	-184.541	16 066
700	86.617	200.185	152.986	33.039	-248.866	-173.619	12.956
800	88.805	211.893	159.631	41.810	-246.440	-163.033	10.645
900	91.002	222.480	166.035	50.801	-243.859	-152.760	8.866
1000	93.190	232.181	172.171	60.010	-241.119	-142.784	7.458
1100	95.378	241.166	178.040	69.439	-238.220	-133.089	6.320
1200	97.567	249.559	183.654	79.086	-235.159	-123.666	5.383
1300	99.755	257.455	189.030	88.952	-231.935	-114.504	4.601
1400	101.943	264.927	194.187	99.037	-228.548	-105.597	3.940
1500	104.131	272.035	199.142	109.341	-224.998	-96.938	3.376
1600	106.320	278.826	203.911	119.863	-221.285	-88.521	2.890
1700	108.508	285.337	208.511	130.605	-217.409	-80.341	2.469
1800	110.696	291.601	212.954	141.565	-213.371	-72.393	2.101
1900	112.884	297.645	217.253	152.744	-209.172	-64.675	1.778
2000	115.073	303.491	221.420	164.142	-204.812	-57.182	1.493

PREVIOUS: September 1962 CURRENT: December 1966

Tungsten Chloride (WCl_2) $Cl_2W_1(cr)$

TUNGSTEN CHLORIDE (WCl_2) IDEAL GAS M_r = 254.756

$S°(298.15\ K) = [309.416]\ J\ K^{-1}\ mol^{-1}$

$\Delta_f H°(0\ K) = [-13.0 \pm 105]\ kJ\ mol^{-1}$
$\Delta_f H°(298.15\ K) = [-12.6 \pm 105]\ kJ\ mol^{-1}$

Electronic Levels and Quantum Weights

ε_i, cm^{-1}	g_i
0	[5]
[5600]	[10]
[8900]	[10]

Vibrational Frequencies and Degeneracies

ν, cm^{-1}
[360](1)
[60](2)
[423](1)

Ground State Quantum Weight: [5] σ = [2]
Point Group: [$D_{\infty h}$]
Bond Distance: W-Cl = [2.26] Å
Bond Angle: Cl-W-Cl = [180]°
Rotational Constant: B_0 = [0.046548] cm^{-1}

Enthalpy of Formation

The value of $\Delta_f H°(WCl_2$, g, 298.15 K) = -3±25 kcal mol^{-1} is calculated from the heat of formation of the crystal plus the heat of sublimation, $\Delta_{sub}H°$(298.15 K) = 58.5±25 kcal mol^{-1}. The latter is estimated from the heats of sublimation of $TiCl_2$, $CrCl_2$, $MnCl_2$, $FeCl_2$, $CoCl_2$, $NiCl_2$ given by Brewer et al. (1). The uncertainty covers the difference between the heats of sublimation among the above compounds and also the possible error in transference from the first row transition metal dichlorides to tungsten dichloride.

Heat Capacity and Entropy

The molecular configuration of WCl_2(g) is assumed to be linear. The bond distance is estimated to be the same as that in WCl_6(g). The ground state configuration ($^5\Sigma$), the low lying electronic levels and their quantum weights are assumed to be the same as those for $CrCl_2$(g), observed by DeKock and Gruen (3).

The vibrational frequencies are calculated from a valence force field model. The stretching force constant, K_1, is estimated to be the same as that of $HgCl_2$(g), 2.7 millidyne/Å. The bending force constant is assumed to be one hundredth of the stretching force constant.

References

1. L. Brewer, G. R. Somayajulu and E. Brackett, Chem. Rev. 63, 111 (1963).
2. S. A. Shchukarev, G. I. Novikov, I. V. Vasil'kova, A. V. Suvorov, N. V. Andreeva, B. N. Sharupin, and A. K. Baev, Zh. Neorg. Khim. 5, 1650 (1960).
3. C. W. DeKock and D. M. Gruen, J. Chem. Phys. 44, 4387 (1966).

Tungsten Chloride (WCl_2) Cl_2W_1(g)

Enthalpy Reference Temperature = T_r = 298.15 K Standard State Pressure = p° = 0.1 MPa

T/K	$C_p°$ (J K⁻¹mol⁻¹)	S° (J K⁻¹mol⁻¹)	-[G°-H°(T_r)]/T (J K⁻¹mol⁻¹)	H°-H°(T_r) (kJ mol⁻¹)	$\Delta_f H°$ (kJ mol⁻¹)	$\Delta_f G°$ (kJ mol⁻¹)	Log K_f
0	0.	0.	INFINITE	-14.626	-13.024	-13.024	INFINITE
100	46.708	251.950	358.279	-10.633	-12.595	-17.928	9.365
200	54.413	286.912	314.630	-5.544	-12.551	-23.286	6.082
250	56.638	299.310	310.364	-2.764	-12.547	-25.971	5.426
298.15	58.060	309.416	309.416	0.	-12.552	-28.556	5.003
300	58.105	309.775	309.417	0.107	-12.552	-28.655	4.989
350	59.097	318.811	310.128	3.039	-12.564	-31.338	4.677
400	59.789	326.750	311.720	6.012	-12.581	-34.019	4.442
450	60.288	333.823	313.790	9.015	-12.607	-36.697	4.260
500	60.657	340.195	316.117	12.039	-12.640	-39.372	4.113
600	61.159	351.302	321.082	18.132	-12.737	-44.710	3.892
700	61.485	360.756	326.092	24.265	-12.877	-50.029	3.733
800	61.740	368.983	330.950	30.426	-13.060	-55.324	3.612
900	61.984	376.269	335.588	36.612	-13.283	-60.594	3.517
1000	62.261	382.813	339.989	42.824	-13.541	-65.838	3.439
1100	62.595	388.763	344.157	49.067	-13.828	-71.054	3.374
1200	62.996	394.226	348.104	55.346	-14.136	-76.242	3.319
1300	63.465	399.286	351.849	61.668	-14.455	-81.405	3.271
1400	63.997	404.009	355.408	68.041	-14.781	-86.543	3.229
1500	64.579	408.443	358.797	74.469	-15.106	-91.657	3.192
1600	65.198	412.631	362.032	80.958	-15.427	-96.750	3.159
1700	65.838	416.603	365.126	87.509	-15.741	-101.823	3.129
1800	66.486	420.384	368.092	94.126	-16.047	-106.878	3.102
1900	67.126	423.996	370.940	100.806	-16.346	-111.916	3.077
2000	67.748	427.455	373.680	107.550	-16.640	-116.939	3.054
2100	68.339	430.775	376.320	114.355	-16.934	-121.946	3.033
2200	68.892	433.967	378.869	121.217	-17.232	-126.940	3.014
2300	69.401	437.041	381.331	128.132	-17.539	-131.920	2.996
2400	69.860	440.004	383.715	135.095	-17.863	-136.886	2.979
2500	70.268	442.865	386.024	142.102	-18.209	-141.838	2.964
2600	70.622	445.628	388.263	149.147	-18.582	-146.776	2.949
2700	70.924	448.299	390.438	156.225	-19.013	-151.699	2.935
2800	71.174	450.883	392.551	163.330	-19.545	-156.604	2.921
2900	71.374	453.384	394.605	170.458	-20.223	-161.487	2.909
3000	71.527	455.806	396.605	177.603	-21.077	-166.344	2.896
3100	71.637	458.154	398.553	184.762	-22.155	-171.170	2.884
3200	71.707	460.429	400.451	191.930	-23.494	-175.956	2.872
3300	71.740	462.636	402.302	199.102	-25.120	-180.696	2.860
3400	71.740	464.778	404.109	206.276	-27.072	-185.382	2.848
3500	71.711	466.857	405.872	213.449	-29.415	-190.005	2.836
3600	71.656	468.877	407.594	220.618	-32.247	-194.555	2.823
3700	71.578	470.839	409.277	227.780	-70.433	-198.827	2.807
3800	71.481	472.747	410.922	234.933	-70.955	-202.290	2.781
3900	71.367	474.602	412.531	242.075	-71.495	-205 739	2.756
4000	71.238	476.407	414.106	249.206	-72.050	-209.174	2.732
4100	71.098	478.165	415.647	256.323	-72.620	-212.595	2.708
4200	70.948	479.876	417.156	263.425	-73.204	-216.002	2.686
4300	70.791	481.544	418.634	270.512	-73.799	-219.395	2.665
4400	70.627	483.169	420.082	277.583	-74.404	-222.774	2.645
4500	70.459	484.755	421.502	284.637	-75.016	-226.140	2.625
4600	70.288	486.301	422.894	291.675	-75.634	-229.491	2.606
4700	70.114	487.811	424.259	298.695	-76.256	-232.829	2.588
4800	69.940	489.286	425.599	305.697	-76.880	-236.154	2.570
4900	69.765	490.726	426.913	312.683	-77.503	-239.465	2.553
5000	69.591	492.134	428.203	319.651	-78.125	-242.765	2.536
5100	69.419	493.510	429.471	326.601	-78.742	-246.051	2.520
5200	69.248	494.856	430.715	333.534	-79.355	-249.326	2.505
5300	69.079	496.174	431.938	340.451	-79.960	-252.589	2.489
5400	68.912	497.463	433.139	347.350	-80.557	-255.840	2.475
5500	68.749	498.726	434.320	354.233	-81.144	-259.081	2.461
5600	68.589	499.964	435.482	361.100	-81.720	-262.311	2.447
5700	68.432	501.176	436.623	367.951	-82.285	-265.531	2.433
5800	68.278	502.365	437.747	374.787	-82.836	-268.741	2.420
5900	68.128	503.531	438.852	381.607	-83.373	-271.941	2.408
6000	67.982	504.675	439.939	388.412	-83.896	-275.133	2.395

PREVIOUS: December 1966 (1 atm) CURRENT: December 1966 (1 bar)

Tungsten Chloride (WCl_2) Cl_2W_1(g)

ZIRCONIUM CHLORIDE ($ZrCl_2$) CRYSTAL M_r = 162.126

$S°$(298.15 K) = [110.039 ± 12.6] J K^{-1} mol^{-1}

T_{fus} = [1000] K

$\Delta_f H°$(0 K) = Unknown

$\Delta_f H°$(298.15 K) = [-430.95 ± 41.8] kJ mol^{-1}

$\Delta_{fus} H°$ = [26.778 ± 12.6] kJ mol^{-1}

Enthalpy of Formation

The value of $\Delta_f H°$($ZrCl_2$, cr, 298.15 K) = -103.0 kcal mol^{-1} is calculated from the estimated $\Delta_{sub} H°$(298.15 K) = 58.5 kcal mol^{-1} for $ZrCl_2$(cr) → $ZrCl_2$(g), using $\Delta_f H°$($ZrCl_2$, g, 298.15 K) = -44.5 kcal mol^{-1} (1). The estimated $\Delta_{sub} H°$(298.15 K) is obtained as an average of the heats of sublimation at 298.15 K for $TiCl_2$, VCl_2, $CrCl_2$, $MnCl_2$, $FeCl_2$, $CoCl_2$, $NiCl_2$ given by Brewer, et al (2).

Heat Capacity and Entropy

The heat capacities of $ZrCl_2$(cr) are estimated based on the assumption that $C_p°$($ZrCl_2$, cr) = 1/2[$C_p°$($ZrCl_4$, cr) + $C_p°$(Zr, cr)]. The entropy, $S°$(298.15 K), is also estimated in a manner similar to that of the heat capacity.

Fusion Data

The melting point was estimated to be 1000 K by Brewer (3) and the heat of melting was also estimated by Glassner (4).

References

1. JANAF Thermochemical Tables: $ZrCl_2$(g), 12-31-69.
2. L. Brewer, G. R. Somayajulu and E. Brackett, Chem. Rev. 63, 111 (1963).
3. L. Brewer, "The Chemistry and Metallurgy of Miscellaneous Materials: Thermodynamics," L. L. Quill, Ed., McGraw-Hill New York, 1949.
4. A. Glassner, "The Thermochemical Properties of the Oxides, Fluorides and Chlorides to 2500 K," ANL Report 5750, Argonne National Laboratory, 1957.

Zirconium Chloride ($ZrCl_2$) Cl_2Zr_1(cr)

Enthalpy Reference Temperature = T_r = 298.15 K Standard State Pressure = p° = 0.1 MPa

T/K	$C_p°$	$S°$	$-[G°-H°(T_r)]/T$	$H°-H°(T_r)$	$\Delta_f H°$	$\Delta_f G°$	Log K_f
	J $K^{-1}mol^{-1}$			kJ mol^{-1}			
0							
100							
200							
298.15	72.592	110.039	110.039	0.	-430.952	-385.660	67.566
300	72.710	110.489	110.041	0.134	-430.927	-385.379	67.101
400	76.015	131.884	112.935	7.580	-429.511	-370.407	48.370
500	78.241	149.094	118.502	15.296	-427.991	-355.805	37.171
600	80.040	163.520	124.834	23.211	-426.399	-341.516	29.732
700	81.714	175.987	131.271	31.301	-424.748	-327.498	24.438
800	83.128	186.990	137.562	39.543	-423.055	-313.721	20.484
900	84.542	196.863	143.611	47.927	-421.339	-300.157	17.421
1000	85.923	205.843	149.392	56.450	-419.608	-286.785	14.980
1000.000	85.923	205.843	149.392	56.450	------ CRYSTAL <--> LIQUID -----		
1100	86.692	214.073	154.904	65.086	-417.894	-273.586	12.992
1200	87.446	221.649	160.155	73.793	-419.951	-260.319	11.331
1300	88.199	228.680	165.159	82.578	-417.807	-247.103	9.929
1400	88.617	235.232	169.933	91.420	-415.657	-234.053	8.733
1500	88.910	241.357	174.492	100.297	-413.534	-221.155	7.701

PREVIOUS: June 1962 CURRENT: December 1969

ZIRCONIUM CHLORIDE ($ZrCl_2$) LIQUID M_r = 162.126

$S°(298.15 K) = [122.493] J K^{-1} mol^{-1}$

$T_{fus} = [1000] K$

$\Delta_f H°(298.15 K) = [-411.594] kJ mol^{-1}$

$\Delta_{fus} H° = [26.778 \pm 12.6] kJ mol^{-1}$

Enthalpy of Formation

$\Delta_f H°(ZrCl_2, \ell, 298.15 K)$ is calculated from $\Delta_f H°(ZrCl_2, cr, 298.15 K)$ by adding $\Delta_{fus}H°$ and the difference in enthalpy, H°(1000 K)-H°(298.15 K), between the crystal and liquid.

Heat Capacity and Entropy

The heat capacities are assumed to be constant at 7.25 cal K^{-1} g-atom^{-1}. The entropy is calculated in a manner analogous to that used for the enthalpy of formation.

Fusion Data

The melting point was estimated to be 1000 K by Brewer (1), and the heat of melting was also estimated by Glassner (2).

Vaporization Data

T_{vap} is the temperature at which the Gibbs energy change approaches zero for the process $ZrCl_2(\ell) \rightarrow ZrCl_2(g)$. The difference between $\Delta_f H°$ for $ZrCl_2(\ell)$ and $ZrCl_2(g)$ at T_{vap} is $\Delta_{vap}H°$.

References

1. L. Brewer, "The Chemistry and Metallurgy of Miscellaneous Materials: Thermodynamics," L L. Quill, Ed., McGraw-Hill Book Company, New York, 1949.
2. A. Glassner, "The Thermochemical Properties of the Oxides, Fluorides and Chlorides to 2500 K," ANL Report 5750, Argonne National Laboratory, 1957.

Zirconium Chloride ($ZrCl_2$) $Cl_2Zr_1(l)$

Enthalpy Reference Temperature = T_r = 298.15 K Standard State Pressure = p° = 0.1 MPa

T/K	$C_p°$	S°	$-[G°-H°(T_r)]/T$	$H°-H°(T_r)$	$\Delta_f H°$	$\Delta_f G°$	Log K_f
	J K⁻¹mol⁻¹			kJ mol⁻¹			
0							
100							
200							
298.15	91.002	122.493	122.493	0.	-411.594	-370.015	64.825
300	91.002	123.056	122.495	0.168	-411.535	-369.758	64.381
400	91.002	149.236	126.064	9.269	-408.464	-356.300	46.528
500	91.002	169.542	132.805	18.369	-405.560	-343.598	35.895
600	91.002	186.134	140.352	27.469	-402.783	-331.468	28.857
700	91.002	200.162	147.920	36.569	-400.121	-319.794	23.863
800	91.002	212.314	155.227	45.669	-397.571	-308.495	20.143
900	91.002	223.032	162.177	54.770	-395.138	-297.507	17.267
1000	91.002	232.620	168.750	63.870	-392.831	-286.785	14.980
1000.000	91.002	232.620	168.750	63.870	------ CRYSTAL <--> LIQUID -----		
1100	91.002	241.294	174.957	72.970	-390.652	-276.286	13.120
1200	91.002	249.212	180.820	82.070	-392.315	-265.759	11.568
1300	91.002	256.496	186.365	91.170	-389.856	-255.313	10.259
1400	91.002	263.240	191.618	100.271	-387.448	-245.054	9.143
1500	91.002	269.518	196.604	109.371	-385.103	-234.966	8.182
1600	91.002	275.391	201.347	118.471	-382.829	-225.031	7.347
1700	91.002	280.908	205.866	127.571	-380.638	-215.236	6.613
1800	91.002	286.110	210.181	136.671	-378.541	-205.568	5.965
1900	91.002	291.030	214.308	145.772	-376.549	-196.013	5.389
2000	91.002	295.698	218.262	154.872	-374.673	-186.560	4.872

PREVIOUS: June 1962 CURRENT: December 1969

Zirconium Chloride ($ZrCl_2$) $Cl_2Zr_1(l)$

ZIRCONIUM CHLORIDE ($ZrCl_2$)

CRYSTAL–LIQUID

M_r = 162.126

0 to 1000 K crystal
above 1000 K liquid

Refer to the individual tables for details.

Zirconium Chloride ($ZrCl_2$) Cl_2Zr_1(cr,l)

Enthalpy Reference Temperature = T_r = 298.15 K Standard State Pressure = p° = 0.1 MPa

	J K⁻¹mol⁻¹			kJ mol⁻¹			
T/K	C_p°	S°	$-[G°-H°(T_r)]/T$	$H°-H°(T_r)$	$\Delta_f H°$	$\Delta_f G°$	Log K_f
0							
100							
200							
298.15	72.592	110.039	110.039	0.	-430.952	-385.660	67.566
300	72.710	110.489	110.041	0.134	-430.927	-385.379	67.101
400	76.015	131.884	112.935	7.580	-429.511	-370.407	48.370
500	78.241	149.094	118.502	15.296	-427.991	-355.805	37.171
600	80.040	163.520	124.834	23.211	-426.399	-341.516	29.732
700	81.714	175.987	131.271	31.301	-424.748	-327.498	24.438
800	83.128	186.990	137.562	39.543	-423.055	-313.721	20.484
900	84.542	196.863	143.611	47.927	-421.339	-300 157	17.421
1000	85.923	205.843	149.392	56.450	-419.608	-286.785	14.980
1000.000	85.923	232.620	149.392	83.228	CRYSTAL <--> LIQUID		
1000.000	91.002	232.620	149.392	83.228	TRANSITION		
1100	91.002	241.294	157.359	92.328	-390.652	-276.286	13.120
1200	91.002	249.212	164.688	101.428	-392.315	-265.759	11.568
1300	91.002	256.496	171.474	110.528	-389.856	-255.313	10.259
1400	91.002	263.240	177.791	119.629	-387.448	-245.054	9.143
1500	91.002	269.518	183.699	128.729	-385.103	-234.966	8.182
1600	91.002	275.391	189.248	137.829	-382.829	-225.031	7.347
1700	91.002	280.908	194.479	146.929	-380.638	-215.236	6.613
1800	91.002	286.110	199.427	156.029	-378.541	-205.568	5.965
1900	91.002	291.030	204.120	165.130	-376.549	-196.013	5.389
2000	91.002	295.698	208.583	174.230	-374.673	-186.560	4.872

PREVIOUS: CURRENT: December 1969

Zirconium Chloride ($ZrCl_2$) Cl_2Zr_1(cr,l)

ZIRCONIUM CHLORIDE ($ZrCl_2$) IDEAL GAS M_r = 162.126

$\Delta_fH°(0\ K) = -185.75 \pm 20.9\ kJ\ mol^{-1}$

$S°(298.15\ K) = [292.558 \pm 12.6]\ J\ K^{-1}\ mol^{-1}$ $\Delta_fH°(298.15\ K) = -186.19 \pm 20.9\ kJ\ mol^{-1}$

Electronic Levels and Quantum Weights

ε_i, cm^{-1}	g_i
0	[3]
[7000]	[6]
[17000]	[6]
[22000]	[15]

Vibrational Frequencies and Degeneracies

ν, cm^{-1}
[346] (1)
[92] (2)
[461] (1)

Ground State Quantum Weight = [3] σ = [2]

Point Group = [$D_{\infty h}$]

Bond Distance: Zr-Cl = [2.3] Å

Bond Angle: Cl-Zr-Cl = [180]°

Rotational Constant: B_0 = [0.044943] cm^{-1}

Enthalpy of formation

Potter (1) investigated mass spectrometrically the gaseous equilibria among ZrCl, $ZrCl_2$, $ZrCl_3$ and $ZrCl_4$. Ion intensities were measured 2.5 cal K^{-1} mol^{-1} above the ionization threshold. Using the reported equilibrium constants for the reaction $ZrCl_4(g) + ZrCl_2(g) \rightarrow 2\ ZrCl_3(g)$ in the temperature range 1673-2110 K, second and third law analyses give the heat of reaction at 298 K as -15.8 and +1.7 kcal mol^{-1}, respectively. The third law drift is 8.9±3.7 cal K^{-1} mol^{-1}. Based on the third law heat of reaction and $\Delta_fH°(ZrCl_4$, g, 298.15 K) = -207.77 kcal mol^{-1} (2) and $\Delta_fH°(ZrCl_3$, g, 298.15 K) = -125.3 kcal mol^{-1} (2), we obtain $\Delta_fH°(ZrCl_2$, g, 298.15 K) = -44.5 kcal mol^{-1} (-186.188±20.9 kJ mol^{-1}) which is adopted in the tabulation.

Farber et al. (3) also studied mass spectrometrically the reaction (a) $Zr(cr) + ZrCl_4(g) \rightarrow 2\ ZrCl_2(g)$ and the reaction (b) $Zr(cr) + 2\ Cl(g) \rightarrow ZrCl_2(g)$. They reported second law heats of reaction as $\Delta_rH°(298.15\ K)$ = 28.8±1.8 kcal mol^{-1} for reaction (a); and -141.1±2.2 kcal mol^{-1} for reaction (b). Third law analysis of the equilibrium constants which were calculated from their ion intensities of run 3 (the only available values) gives $\Delta_fH°(ZrCl_2$, g, 298.15 K) = -43.4 kcal mol^{-1}. See JANAF $ZrCl_3(g)$ table (2) for details.

Heat Capacity and Entropy

The molecular configuration is assumed to be linear, since experimental evidence indicates that other transition metal dichlorides are linear (4, 5, 6, 7, 8, 9) even though some fluorides are bent (10). The bond distance is assumed to be the same as that of $ZrCl_4(g)$. The electronic levels and quantum weights are estimated to be the same as those of $TiCl_2(g)$ (2).

The vibrational frequencies are calculated from a valence force field model. The stretching force constant is estimated to be 2.5 millidyne/Å and the bending force constant 0.05 millidyne/Å.

References

1. N. D. Potter, Aeronutronic Div., Philco-Ford Corp., Newport Beach, Calif., personal communication, Dec. 4, 1969.
2. JANAF Thermochemical Tables: $ZrCl_4(g)$ and $ZrCl_3(g)$, 12-31-69; $TiCl_2(g)$, 12-31-68.
3. M. Farber, M. A. Frisch, G. Grenier and H. C. Ko, Space Sciences, Inc., Monrovia, Calif., Rep. No. AFRPL-TR-67-244, Contract F04611-67-C-0010, Nov., 1967.
4. R. A. Berg and O. Sinanoglu, J. Chem. Phys. 32, 1082 (1960).
5. J. T. Hougen, G. E. Leroi and T. C. James, J. Chem. Phys. 34, 1670 (1961).
6. G. E. Leroi, T. C. James, J. T. Hougen and W. Klemperer, J. Chem. Phys. 36, 2879 (1962).
7. A. Buchler, J. L. Stauffer and W. Klemperer, J. Chem. Phys. 40, 3471 (1964).
8. M. E. Jacox and D. E. Milligan, J. Chem. Phys. 51, 4143 (1969).
9. K. R. Thompson and K. D. Carlson, J. Chem. Phys. 49, 4379 (1968).
10. J. W. Hastie, R. H. Hauge and J. L. Margrave, J. Chem. Phys. 51, 2648 (1969).

Zirconium Chloride ($ZrCl_2$) $Cl_2Zr_1(g)$

Enthalpy Reference Temperature = T_r = 298.15 K Standard State Pressure = p° = 0.1 MPa

T/K	$C_p°$ (J $K^{-1}mol^{-1}$)	S°	-[G°-H°(T_r)]/T	H°-H°(T_r) (kJ mol^{-1})	$\Delta_fH°$	$\Delta_fG°$	Log K_f
0	0.	0.	INFINITE	-14.243	-185.753	-185.753	INFINITE
100	45.422	235.909	340.900	-10.499	-185.807	-189.092	98.771
200	53.791	270.254	297.731	-5.495	-186.034	-192.282	50.219
250	56.153	282.529	293.500	-2.743	-186.116	-193.834	40.499
298.15	57.676	292.558	292.558	0.	-186.188	-195.314	34.218
300	57.724	292.915	292.559	0.107	-186.191	-195.371	34.017
350	58.793	301.898	293.266	3.021	-186.266	-196.895	29.385
400	59.543	309.801	294.849	5.981	-186.346	-198.408	25.909
450	60.085	316.847	296.909	8.972	-186.436	-199.911	23.205
500	60.488	323.199	299.226	11.987	-186.536	-201.402	21.040
600	61.033	334.280	304.171	18.065	-186.781	-204.354	17.791
700	61.374	343.716	309.163	24.187	-187.098	-207.259	15.466
800	61.606	351.927	314.007	30.336	-187.498	-210.112	13.719
900	61.782	359.194	318.631	36.506	-187.996	-212.910	12.357
1000	61.937	365.711	323.019	42.692	-188.603	-215.647	11.264
1100	62.097	371.622	327.173	48.894	-189.322	-218.318	10.367
1200	62.280	377.033	331.106	55.112	-193.867	-220.696	9.607
1300	62.495	382.026	334.833	61.351	-194.270	-222.916	8.957
1400	62.749	386.666	338.372	67.613	-194.700	-225.103	8.399
1500	63.043	391.005	341.737	73.902	-195.166	-227.259	7.914
1600	63.372	395.084	344.945	80.222	-195.672	-229.382	7.489
1700	63.732	398.937	348.009	86.577	-196.226	-231.473	7.112
1800	64.114	402.590	350.941	92.970	-196.837	-233.529	6.777
1900	64.512	406.068	353.751	99.401	-197.514	-235.549	6.476
2000	64.918	409.387	356.451	105.872	-198.267	-237.532	6.204
2100	65.325	412.564	359.048	112.384	-199.106	-239.475	5.957
2200	65.727	415.612	361.550	118.937	-221.368	-240.630	5.713
2300	66.120	418.543	363.965	125.530	-222.833	-241.472	5.484
2400	66.498	421.365	366.298	132.161	-224.275	-242.252	5.272
2500	66.861	424.087	368.555	138.829	-225.696	-242.972	5.077
2600	67.205	426.716	370.742	145.532	-227.098	-243.635	4.895
2700	67.531	429.258	372.863	152.269	-228.485	-244.244	4.725
2800	67.837	431.720	374.921	159.038	-229.860	-244.803	4.567
2900	68.125	434.106	376.921	165.836	-231.224	-245.312	4.419
3000	68.394	436.420	378.866	172.662	-232.579	-245.775	4.279
3100	68.647	438.666	380.759	179.514	-233.928	-246.193	4.148
3200	68.884	440.850	382.603	186.391	-235.272	-246.567	4.025
3300	69.107	442.973	384.400	193.291	-236.610	-246.899	3.908
3400	69.317	445.039	386.153	200.212	-237.943	-247.191	3.798
3500	69.516	447.051	387.865	207.154	-239.271	-247.444	3.693
3600	69.704	449.012	389.536	214.115	-240.593	-247.658	3.593
3700	69.884	450.925	391.169	221.094	-241.907	-247.837	3.499
3800	70.056	452.791	392.767	228.091	-243.213	-247.979	3.409
3900	70.220	454.612	394.329	235.105	-244.509	-248.088	3.323
4000	70.378	456.392	395.858	242.135	-245.792	-248.163	3.241
4100	70.531	458.132	397.356	249.181	-247.061	-248.207	3.162
4200	70.679	459.833	398.824	256.241	-248.314	-248.220	3.087
4300	70.821	461.498	400.262	263.316	-249.549	-248.203	3.015
4400	70.960	463.128	401.672	270.405	-250.763	-248.157	2.946
4500	71.094	464.724	403.056	277.508	-251.955	-248.085	2.880
4600	71.224	466.288	404.413	284.624	-253.122	-247.986	2.816
4700	71.350	467.821	405.746	291.753	-254.263	-247.862	2.755
4800	71.472	469.325	407.055	298.894	-816.312	-236.096	2.569
4900	71.590	470.800	408.341	306.047	-817.047	-224.001	2.388
5000	71.703	472.247	409.605	313.212	-817.768	-211.891	2.214
5100	71.812	473.668	410.847	320.387	-818.473	-199.766	2.046
5200	71.916	475.063	412.069	327.574	-819.158	-187.628	1.885
5300	72.016	476.434	413.270	334.770	-819.822	-175.477	1.729
5400	72.111	477.781	414.452	341.977	-820.463	-163.313	1.580
5500	72.201	479.105	415.616	349.192	-821.079	-151.138	1.435
5600	72.285	480.407	416.761	356.417	-821.668	-138.952	1.296
5700	72.365	481.687	417.889	363.649	-822.228	-126.756	1.162
5800	72.439	482.946	419.000	370.890	-822.759	-114.549	1.032
5900	72.508	484.185	420.094	378.137	-823.259	-102.335	0.906
6000	72.571	485.404	421.173	385.391	-823.728	-90.112	0.784

PREVIOUS: December 1969 (1 atm) CURRENT: December 1969 (1 bar)

Zirconium Chloride ($ZrCl_2$) $Cl_2Zr_1(g)$

COBALT CHLORIDE ($CoCl_3$) IDEAL GAS M_r = 165.2922

$\Delta_f H°(0\ K) = -161.78 \pm 10.5\ kJ\ mol^{-1}$

$S°(298.15\ K) = [334.209 \pm 8.4]\ J\ K^{-1}\ mol^{-1}$ $\Delta_f H°(298.15\ K) = -163.59 \pm 10.5\ kJ\ mol^{-1}$

Electronic Levels and Quantum Weights

ε_i, cm^{-1}	g_i
0	[5]
[500]	[5]
[1000]	[5]
[5000]	[5]
[10000]	[5]

Vibrational Frequencies and Degeneracies

ν, cm^{-1}
[500] (1)
[250] (1)
[550] (2)
[150] (2)

Ground State Quantum Weight = [5] $\sigma = 3$

Point Group: [C_{3v}]

Bond Distance: Co-Cl = [2.09] Å

Bond Angle: Cl-Co-Cl = [100]°

Product of the Moments of Inertia: $I_A I_B I_C = [7.9528 \times 10^{-113}]\ g^3\ cm^6$

Enthalpy of Formation

Schafer and Krehl (1) studied the reaction of chlorine with $CoCl_2$ by a transpiration method. They ascribed the enhanced volatility of $CoCl_2$(cr,ℓ) in the presence of Cl_2 to the formation of gaseous $CoCl_3$ according to the reaction $Cl_2(g) + 2\ CoCl_2(g) = 2\ CoCl_3(g)$. Partial pressures for the three species involved in the reaction were reported for the temperature range 918-1073 K. No consideration was given to possible dimerization of either the di- or trichloride. Subsequent measurements by Schafer and Breil (2) have confirmed the increased volatility of $CoCl_2$(ℓ) in the presence of Cl_2. However, the latter study was conducted at only one temperature (1056 K). JANAF data (3) for $CoCl_2$(g) and Co_2Cl_4(g) are used to correct their pressures (1, 2) for dimerization of the dichloride. Our corrected $CoCl_2$ pressures are 5% less than those reported by Schafer and Krehl (1) at 918 K and 11.4% less at 1073 K. We tentatively assume that dimerization of the trichloride is negligible over the temperature range investigated and treat the equilibrium pressures by the second and third law methods. Results of the analysis are: $\Delta_r H°$(298.15 K) (2nd Law) = -32.9 and $\Delta_r H°$(298.15 K) (3rd Law) = -33.36±2.0 kcal mol^{-1}. The third law drift is -0.5±2.9 cal K^{-1} mol^{-1}. The measurements of Schafer and Breil (2) at 1056 K give $\Delta_r H°$(298.15 K) (3rd Law) = -33.31 kcal mol^{-1}. Combining the average value for the third law heats of reaction with $\Delta_f H°(CoCl_2$, g, 298.15 K) = -22.4±2.0 kcal mol^{-1} (3), we obtain $\Delta_f H°(CoCl_3$, g, 298.15 K) = -39.1±2.5 kcal mol^{-1}. This value is adopted and corresponds to a bond dissociation energy of $D_0^0(CoCl_2-Cl$, 298.15 K) = 45.7 kcal mol^{-1}. A comparison of this value with that for the monochloride (D_0^0(Co-Cl, 0 K) = 84.4 kcal mol^{-1}) and dichloride (D_0^0(CoCl-Cl) = 97.5 kcal mol^{-1}) clearly reveals the instability of the trichloride molecule.

Heat Capacity and Entropy

The structure of CoCl is not known; however, it can be limited to two possibilities. The molecule has either a pyramidal structure of C_{3v} symmetry, or a planar structure of D_{3h} symmetry. By analogy with the geometries for PCl_3 (3), $SiCl_3$ (3) and TiF_3 (3, 4), we adopt the pyramidal structure. The bond length is taken equal to that for CoCl (3). The Cl-Co-Cl bond angle is estimated as 100°. The principal moments of inertia are: $I_A = I_B = 36.1659 \times 10^{-39}$, and $I_C = 60.3628 \times 10^{-39}\ g\ cm^2$.

The vibrational frequencies are estimated by comparison with those for PCl_3 (3), $SiCl_3$ (3) and TiF_3 (4). The ground state quantum weight is taken equal to that for Co^{+3} (5). The electronic levels are estimated from those for Fe^{+2} (5), since few levels have been observed in the spectrum of Co^{+3}. The Fe^{+2} ion is isoelectronic with Co^{+3}. The adopted upper levels and quantum weights contribute 0.65 cal K^{-1} mol^{-1} at 298.15 K and 2.4 cal K^{-1} mol^{-1} at 3000 K to the entropy of $CoCl_3$.

References

1. H. Schafer and K. Krehl, Z. Anorg. Chem. 268, 25 (1952).
2. H. Schafer and G. Breil, Z. Anorg. Chem. 283, 304 (1956).
3. JANAF Thermochemical Tables: $CoCl_2$(g), CoCl(g), and Co_2Cl_4(g), 12-31-73; PCl_3(g), 6-30-70; $SiCl_3$(g), 12-31-69; TiF_3(g), 6-30-68.
4. J. W. Hastie, R. H. Hauge, and J. L. Margrave, J. Chem. Phys. 51, 2648 (1969).
5. C. E. Moore, U. S. Nat. Bur. Stand. Circ. 467, 1952.

Cobalt Chloride ($CoCl_3$) Cl_3Co_1(g)

Enthalpy Reference Temperature = T_r = 298.15 K Standard State Pressure = p° = 0.1 MPa

T/K	$C_p°$ (J $K^{-1}mol^{-1}$)	S° (J $K^{-1}mol^{-1}$)	$-[G°-H°(T_r)]/T$ (J $K^{-1}mol^{-1}$)	$H°-H°(T_r)$ (kJ mol^{-1})	$\Delta_f H°$ (kJ mol^{-1})	$\Delta_f G°$ (kJ mol^{-1})	Log K_f
0	0.	0.	INFINITE	-16.731	-161.783	-161.783	INFINITE
100	48.842	266.224	394.607	-12.838	-162.780	-160.283	83.723
200	66.145	305.685	340.909	-7.045	-163.452	-157.483	41.130
250	72.162	321.128	335.443	-3.579	-163.570	-155.974	32.589
298.15	76.238	334.209	334.209	0.	-163.594	-154.508	27.069
300	76.366	334.681	334.210	0.141	-163.593	-154.451	26.892
350	79.198	346.680	335.152	4.035	-163.540	-152.931	22.824
400	81.067	357.387	337.274	8.045	-163.463	-151.421	19.774
450	82.281	367.010	340.053	12.131	-163.389	-149.920	17.402
500	83.059	375.723	343.191	16.266	-163.337	-148.427	15.506
600	83.853	390.948	349.918	24.618	-163.327	-145.448	12.662
700	84.136	403.899	356.728	33.020	-163.469	-142.459	10.630
800	84.213	415.140	363.342	41.438	-164.200	-139.376	9.100
900	84.219	425.060	369.660	49.860	-164.703	-136.246	7.907
1000	84.212	433.933	375.651	58.282	-165.459	-133.046	6.950
1100	84.214	441.959	381.320	66.703	-166.499	-129.757	6.162
1200	84.233	449.287	386.683	75.125	-167.855	-126.359	5.500
1300	84.268	456.031	391.762	83.550	-169.671	-122.831	4.935
1400	84.317	462.277	396.578	91.979	-172.073	-119.141	4.445
1500	84.375	468.097	401.154	100.414	-173.472	-115.309	4.015
1600	84.440	473.544	405.510	108.854	-174.624	-111.393	3.637
1700	84.509	478.665	409.664	117.302	-175.685	-107.409	3.300
1800	84.580	483.498	413.633	125.756	-193.019	-103.068	2.991
1900	84.651	488.073	417.432	134.218	-194.348	-98.035	2.695
2000	84.721	492.416	421.073	142.686	-195.686	-92.931	2.427
2100	84.789	496.552	424.570	151.162	-197.033	-87.760	2.183
2200	84.854	500.498	427.932	159.644	-198.392	-82.525	1.959
2300	84.917	504.271	431.170	168.133	-199.765	-77.228	1.754
2400	84.975	507.886	434 291	176.627	-201.154	-71.870	1.564
2500	85.031	511.356	437.305	185.128	-202.561	-66.455	1.388
2600	85.082	514.692	440.218	193.633	-203.988	-60.982	1.225
2700	85.129	517.904	443.036	202.144	-205.438	-55.454	1.073
2800	85.173	521.001	445.765	210.659	-206.911	-49.873	0.930
2900	85.212	523.990	448.412	219.178	-208.409	-44.238	0.797
3000	85.248	526.880	450.979	227.701	-209.933	-38.551	0.671
3100	85.280	529.676	453.473	236.228	-211.482	-32.813	0.553
3200	85.308	532.384	455.897	244.757	-587.870	-25.251	0.412
3300	85.333	535.009	458.255	253.289	-588.043	-7.667	0.121
3400	85.354	537.557	460.550	261.824	-588.240	9.923	-0.152
3500	85.372	540.031	462.786	270.360	-588.460	27.519	-0.411
3600	85.387	542.436	464.965	278.898	-588.704	45.122	-0.655
3700	85.399	544.776	467.090	287.437	-588.968	62.732	-0.886
3800	85.408	547.054	469.165	295.978	-589.251	80.349	-1.104
3900	85.415	549.272	471.191	304.519	-589.551	97.974	-1.312
4000	85.419	551.435	473.170	313.060	-589.866	115.607	-1.510
4100	85.420	553.544	475.104	321.602	-590.194	133.248	-1.698
4200	85.420	555.603	476.997	330.144	-590.532	150.897	-1.877
4300	85.417	557.612	478.848	338.686	-590.877	168.554	-2.048
4400	85.413	559.576	480.661	347.228	-591.228	186.220	-2.211
4500	85.406	561.496	482.436	355.769	-591.581	203.893	-2.367
4600	85.399	563.373	484.175	364.309	-591.936	221.574	-2.516
4700	85.389	565.209	485.880	372.848	-592.288	239.263	-2.659
4800	85.379	567.007	487.551	381.387	-592.638	256.959	-2.796
4900	85.367	568.767	489.191	389.924	-592.983	274.663	-2.928
5000	85.353	570.492	490.799	398.460	-593.321	292.373	-3.054
5100	85.339	572.182	492.379	406.995	-593.651	310.090	-3.176
5200	85.324	573.839	493.929	415.528	-593.973	327.814	-3.293
5300	85.308	575.464	495.452	424.060	-594.284	345.543	-3.406
5400	85.292	577.058	496.949	432.590	-594.585	363.279	-3.514
5500	85.274	578.623	498.420	441.118	-594.876	381.020	-3.619
5600	85.256	580.159	499.866	449.644	-595.154	398.766	-3.720
5700	85.238	581.668	501.288	458.169	-595.422	416.517	-3.817
5800	85.219	583.150	502.686	466.692	-595.679	434.273	-3.911
5900	85.199	584.607	504.063	475.213	-595.880	452.029	-4.002
6000	85.179	586.039	505.417	483.732	-596.104	469.792	-4.090

PREVIOUS: December 1973 (1 atm) CURRENT: December 1973 (1 bar)

Cobalt Chloride ($CoCl_3$) Cl_3Co_1(g)

COPPER CHLORIDE ((CuCl)$_3$) IDEAL GAS M_r = 296.997

$\Delta_f H°(0\ K) = -258.50 \pm 2.1\ kJ\ mol^{-1}$

S°(298.15 K) = [429.526] $J\ K^{-1}\ mol^{-1}$ $\Delta_f H°(298.15\ K) = -258.57 \pm 2.1\ kJ\ mol^{-1}$

Vibrational Frequencies and Degeneracies

ν, cm^{-1}	ν, cm^{-1}
[100](1)	350 (2)
[150](1)	[300](2)
[250](1)	[60](2)
[200](1)	[80](2)

Ground State Quantum Weight: [1]

Point Group: D_{3h} σ = 6

Bond Distance: Cu-Cl = 2.16 Å

Bond Angles: Cl-Cu-Cl = 150° Cu-Cl-Cu = 90°

Product of the Moments of Inertia: $I_A I_B I_C = 2.02938 \times 10^{-111}\ g^3\ cm^6$

Enthalpy of Formation

The heat of sublimation was determined by second and third law analysis of the following vapor pressure data. Magee (1) measured both vapor pressure by torsion effusion and vapor density in the range 533-644 K and found only trimer present. The second and third law analyses gave $\Delta_{sub}H°(298.15\ K) = 37.37\pm0.04$ and $37.24\pm0.02\ kcal\ mol^{-1}$ with a drift of $-0.25\pm0.07\ cal\ K^{-1}\ mol^{-1}$. Shelton (2), using Knudsen effusion in the range 547-657 K, obtained $\Delta_{sub}H°(298.15\ K) = 40.35\pm1.4$ and $37.42\pm0.4\ kcal\ mol^{-1}$ with a drift of $-4.9\ cal\ K^{-1}\ mol^{-1}$. Fedorov and Shakhova (3), using transport methods, obtained vapor pressure over the liquid from 751-1057 K. Using second and third law analysis these gave $\Delta_{sub}H°(298.15\ K) = 38.6\pm1.0$ and $39.27\pm0.5\ kcal\ mol^{-1}$.

Brewer and Lofgren (4) obtained equilibrium constants for the reaction 3 Cu(cr) + 3 HCl(g) → Cu_3Cl_3(g) + 1.5 H_2(g). Second and third law analysis of these constants gave $\Delta_r H°(298.15\ K) = 3.9\pm0.6$ and $4.39\pm0.3\ kcal\ mol^{-1}$ which leads to $\Delta_f H° = -61.8\pm0.5\ kcal\ mol^{-1}$ ($-258.571\pm2.1\ kJ\ mol^{-1}$). This was the value adopted since it is dependent only on HCl(g) and does not involve three times the uncertainty of CuCl(cr) as do the sublimation experiments, which yield values in good agreement but of larger uncertainty.

Heat Capacity and Entropy

The structure of the trimer has been investigated by Wong and Schomaker (5). They conclude that the molecule is planar with D_{3h} symmetry and a Cu-Cl bond length of 2.16 Å. They also state that a very large amplitude of the symetric in-plane bending appears plausible from the diffraction pattern. The infra-red spectrum has been investigated by Klemperer et al. (6). They observed one strong absorption with a maximumm at 350 cm^{-1}; they attribute this to bond stretching motion of type E'. They also suggest that the broad absorption may involve another mode of approximately equal frequency.

In order to estimate the remaining frequencies we assumed that the other infra-red active E' mode would be close to the observed 350 cm^{-1} and this was estimated as 300 cm^{-1}. The in-plane bending is the third E' model and since this has a very large amplitude it was estimated as 60 cm^{-1}. By analogy with other D_{3h} rings the A_2' mode was assumed to be in the same region as the E' stretches and was taken to be 250 cm^{-1}. The 2A_1' stretches were estimated as 100 and 150 cm^{-1}, while the out of plane A_2" was taken as 200 cm^{-1} and the E" as 80 cm^{-1}. These values gave calculated entropies in excellent agreement with the equilibrium data and thus support the estimates.

The principal moments of inertia are: $I_A = I_B = 100.4872 \times 10^{-39}\ g\ cm^2$, $I_C = 200.9745 \times 10^{-39}\ g\ cm^2$.

References

1. D. W. Magee, Ph. D. Thesis, Ohio State University (1955).
2. R. A. J. Shelton, Trans. Faraday Soc. 57, 2113 (1961).
3. P. I. Fedorov and M. N. Shakhova, Izv. Vyssahikh Uchebn. Zavendenii, Khim. i Khim. Tekhnol. 4, 550 (1961).
4. L. Brewer and N. L. Lofgren, J. Amer. Chem. Soc. 72, 3038 (1950).
5. C. H. Wong and V. Schomaker, J. Phys. Chem. 61, 358 (1957).
6. W. Klemperer, S. A. Rice and R. S. Berry, J. Amer. Chem. Soc. 79, 1810 (1957).

Copper Chloride ((CuCl)$_3$) Cl_3Cu_3(g)

Enthalpy Reference Temperature = T_r = 298.15 K Standard State Pressure = p° = 0.1 MPa

T/K	$C_p°$	S°	$-[G°-H°(T_r)]/T$	$H°-H°(T_r)$	$\Delta_f H°$	$\Delta_f G°$	Log K_f
	J $K^{-1}mol^{-1}$			kJ mol^{-1}			
0	0.	0.	INFINITE	-28.723	-258.503	-258.503	INFINITE
100	91.029	308.854	533.649	-22.480	-258.633	-258.157	134.848
200	116.232	381.296	440.710	-11.883	-258.633	-257.665	67.295
250	121.471	407.846	431.562	-5.929	-258.595	-257.427	53.786
298.15	124.570	429.526	429.526	0.	-258.571	-257.204	45.061
300	124.664	430.297	429.529	0.231	-258.571	-257.196	44.782
350	126.724	449.680	431.055	6.519	-258.570	-256.967	38.350
400	128.120	466.698	434.468	12.892	-258.596	-256.737	33.526
450	129.105	481.849	438.907	19.324	-258.655	-256.501	29.774
500	129.823	495.490	443.894	25.798	-258.743	-256.257	26.771
600	130.778	519.252	454.531	38.832	-259.017	-255.737	22.264
700	131.365	539.459	465.257	51.942	-259.438	-255.159	19.040
800	131.749	557.027	475.654	65.099	-260.002	-254.511	16.618
900	132.015	572.561	485.575	78.288	-260.724	-253.783	14.729
1000	132.206	586.481	494.982	91.499	-261.619	-252.965	13.214
1100	132.348	599.088	503.882	104.727	-262.731	-252.048	11.969
1200	132.457	610.609	512.303	117.968	-264.119	-251.018	10.927
1300	132.541	621.215	520.278	131.218	-265.912	-249.857	10.039
1400	132.608	631.040	527.843	144.475	-307.574	-247.320	9.228
1500	132.662	640.190	535.031	157.739	-309.850	-242.937	8.460
1600	132.707	648.754	541.874	171.008	-312.135	-238.402	7.783
1700	132.744	656.800	548.400	184.280	-314.428	-233.723	7.181
1800	132.775	664.389	554.635	197.556	-316.732	-228.909	6.643
1900	132.801	671.568	560.602	210.835	-319.047	-223.967	6.157
2000	132.823	678.380	566.322	224.116	-321.376	-218.903	5.717
2100	132.842	684.861	571.814	237.399	-323.718	-213.722	5.316
2200	132.859	691.042	577.094	250.684	-326.078	-208.428	4.949
2300	132.874	696.948	582.178	263.971	-328.456	-203.028	4.611
2400	132.886	702.603	587.078	277.259	-330.854	-197.523	4.299
2500	132.898	708.028	591.809	290.548	-333.275	-191.918	4.010
2600	132.908	713.240	596.379	303.839	-335.721	-186.215	3.741
2700	132.917	718.257	600.801	317.130	-338.193	-180.418	3.490
2800	132.925	723.091	605.083	330.422	-340.693	-174.529	3.256
2900	132.932	727.755	609.233	343.715	-1243.833	-150.563	2.712
3000	132.938	732.262	613.259	357.008	-1244.005	-112.861	1.965
3100	132.944	736.621	617.169	370.302	-1244.357	-75.151	1.266
3200	132.949	740.842	620.968	383.597	-1244.888	-37.427	0.611
3300	132.954	744.933	624.663	396.892	-1245.599	0.317	-0.005
3400	132.959	748.902	628.259	410.188	-1246.487	38.085	-0.585
3500	132.963	752.756	631.761	423.484	-1247.548	75.881	-1.132
3600	132.966	756.502	635.174	436.780	-1248.778	113.711	-1.650
3700	132.970	760.145	638.503	450.077	-1250.168	151.576	-2.140
3800	132.973	763.691	641.751	463.374	-1251.710	189.481	-2.605
3900	132.976	767.146	644.922	476.672	-1253.396	227.429	-3.046
4000	132.979	770.512	648.020	489.969	-1255.215	265.422	-3.466
4100	132.981	773.796	651.048	503.267	-1257.157	303.462	-3.866
4200	132.983	777.000	654.009	516.566	-1259.211	341.550	-4.248
4300	132.986	780.130	656.905	529.864	-1261.365	379.689	-4.612
4400	132.988	783.187	659.741	543.163	-1263.609	417.879	-4.961
4500	132.989	786.176	662.517	556.462	-1265.931	456.120	-5.295
4600	132.991	789.098	665.237	569.761	-1268.322	494.415	-5.614
4700	132.993	791.959	667.903	583.060	-1270.770	532.761	-5.921
4800	132.994	794.759	670.517	596.359	-1273.265	571.160	-6.215
4900	132.996	797.501	673.081	609.659	-1275.800	609.612	-6.499
5000	132.997	800.188	675.596	622.958	-1278.365	648.115	-6.771
5100	132.999	802.821	678.065	636.258	-1280.952	686.671	-7.033
5200	133.000	805.404	680.489	649.558	-1283.554	725.277	-7.285
5300	133.001	807.937	682.870	662.858	-1286.164	763.933	-7.529
5400	133.002	810.424	685.209	676.158	-1288.777	802.639	-7.764
5500	133.003	812.864	687.508	689.459	-1291.389	841.393	-7.991
5600	133.004	815.261	689.768	702.759	-1293.994	880.194	-8.210
5700	133.005	817.615	691.990	716.059	-1296.590	919.042	-8.422
5800	133.006	819.928	694.176	729.360	-1299.173	957.936	-8.627
5900	133.007	822.202	696.327	742.661	-1301.743	996.873	-8.826
6000	133.007	824.437	698.443	755.961	-1304.296	1035.854	-9.018

PREVIOUS: March 1966 (1 atm) CURRENT: March 1966 (1 bar)

Copper Chloride ((CuCl)$_3$) Cl_3Cu_3(g)

TRICHLOROFLUOROSILANE ($SiCl_3F$) IDEAL GAS M_r = 153.442903

$S°(298.15\ K) = [336.055 \pm 16.7]\ J\ K^{-1}mol^{-1}$

$\Delta_f H°(0\ K) = -838.80 \pm 62.8\ kJ\ mol^{-1}$

$\Delta_f H°(298.15\ K) = -840.98 \pm 62.8\ kJ\ mol^{-1}$

Vibrational Levels and Quantum Weights

ν, cm^{-1}
[925](1)
[350](1)
[500](2)
[270](2)
[200](2)

Ground State Quantum Weight: 1 σ = 3

Point Group: C_{3v}

Product of the Moments of Inertia: $I_A I_B I_C = [142,900 \times 10^{-117}]\ g^3\ cm^6$

Enthalpy of Formation

All data are from Henderson and Scheffee (1).

Heat Capacity and Entropy

All data are from Henderson and Scheffee (1).

Reference

1. C. B. Henderson and R. S. Scheffee, "Survey of Thermochemical Data", Atlantic Research Corp., Alexandria, Va., January, 1960.

Trichlorofluorosilane ($SiCl_3F$) $Cl_3F_1Si_1(g)$

Enthalpy Reference Temperature = T_r = 298.15 K; Standard State Pressure = p° = 0.1 MPa

T/K	$C_p°$ (J K⁻¹mol⁻¹)	S° (J K⁻¹mol⁻¹)	$-[G°-H°(T_r)]/T$ (J K⁻¹mol⁻¹)	$H°-H°(T_r)$ (kJ mol⁻¹)	$\Delta_f H°$ (kJ mol⁻¹)	$\Delta_f G°$ (kJ mol⁻¹)	Log K_f
0	0.	0.	INFINITE	-19.214	-838.796	-838.796	INFINITE
100	55.092	255.734	407.697	-15.196	-840.866	-829.187	433.123
200	78.839	302.234	343.986	-8.350	-841.220	-817.279	213.451
250	85.487	320.584	337.514	-4.233	-841.131	-811.303	169.513
298.15	90.081	336.055	336.055	0.	-840.984	-805.570	141.133
300	90.231	336.613	336.057	0.167	-840.977	-805.351	140.224
350	93.713	350.797	337.169	4.770	-840.782	-799.428	119.308
400	96.325	363.489	339.680	9.524	-840.557	-793.535	103.625
450	98.321	374.955	342.973	14.392	-840.310	-787.672	91.431
500	99.873	385.398	346.701	19.348	-840.050	-781.837	81.678
600	102.076	403.816	354.726	29.454	-839.505	-770.246	67.056
700	103.521	419.667	362.898	39.739	-838.952	-758.746	56.618
800	104.511	433.559	370.880	50.143	-838.410	-747.325	48.795
900	105.217	445.912	378.543	60.631	-837.892	-735.971	42.715
1000	105.735	457.026	385.845	71.180	-837.404	-724.673	37.853
1100	106.126	467.122	392.782	81.774	-836.951	-713.423	33.878
1200	106.429	476.370	399.368	92.403	-836.537	-702.211	30.566
1300	106.667	484.899	405.623	103.058	-836.165	-691.033	27.766
1400	106.857	492.811	411.572	113.734	-835.839	-679.881	25.367
1500	107.012	500.189	417.237	124.428	-835.561	-668.751	23.288
1600	107.140	507.099	422.639	135.136	-835.333	-657.638	21.470
1700	107.246	513.598	427.801	145.855	-885.335	-646.091	19.852
1800	107.335	519.730	432.739	156.585	-884.981	-632.028	18.341
1900	107.411	525.536	437.472	167.322	-884.640	-617.985	16.990
2000	107.476	531.047	442.014	178.066	-884.312	-603.959	15.774
2100	107.532	536.292	446.379	188.817	-883.997	-589.949	14.674
2200	107.581	541.296	450.581	199.573	-883.697	-575.954	13.675
2300	107.624	546.079	454.630	210.333	-883.412	-561.972	12.763
2400	107.661	550.660	458.536	221.097	-883.142	-548.002	11.927
2500	107.694	555.056	462.310	231.865	-882.888	-534.043	11.158
2600	107.724	559.280	465.959	242.636	-882.651	-520.094	10.449
2700	107.750	563.346	469.491	253.410	-882.430	-506.154	9.792
2800	107.774	567.265	472.913	264.186	-882.225	-492.222	9.183
2900	107.795	571.048	476.232	274.964	-882.037	-478.296	8.615
3000	107.814	574.702	479.454	285.745	-881.863	-464.377	8.086
3100	107.831	578.238	482.584	296.527	-881.704	-450.464	7.590
3200	107.847	581.662	485.627	307.311	-881.556	-436.555	7.126
3300	107.861	584.980	488.588	318.096	-881.419	-422.651	6.690
3400	107.874	588.201	491.470	328.883	-881.290	-408.751	6.280
3500	107.886	591.328	494.279	339.671	-881.166	-394.855	5.893
3600	107.897	594.367	497.017	350.460	-1265.195	-370.501	5.376
3700	107.907	597.324	499.688	361.251	-1264.658	-345.656	4.880
3800	107.916	600.201	502.296	372.042	-1264.118	-320.825	4.410
3900	107.925	603.005	504.842	382.834	-1263.571	-296.009	3.965
4000	107.933	605.737	507.331	393.627	-1263.013	-271.206	3.542
4100	107.940	608.403	509.763	404.420	-1262.440	-246.418	3.139
4200	107.947	611.004	512.143	415.215	-1261.849	-221.644	2.757
4300	107.954	613.544	514.472	426.010	-1261.234	-196.885	2.392
4400	107.960	616.026	516.752	436.806	-1260.593	-172.140	2.044
4500	107.965	618.452	518.985	447.602	-1259.922	-147.410	1.711
4600	107.971	620.825	521.173	458.399	-1259.217	-122.695	1.393
4700	107.975	623.147	523.318	469.196	-1258.476	-97.996	1.089
4800	107.980	625.420	525.422	479.994	-1257.695	-73.313	0.798
4900	107.984	627.647	527.485	490.792	-1256.873	-48.647	0.519
5000	107.988	629.828	529.510	501.591	-1256.006	-23.998	0.251
5100	107.992	631.967	531.498	512.390	-1255.093	0.633	-0.006
5200	107.996	634.064	533.451	523.189	-1254.131	25.246	-0.254
5300	107.999	636.121	535.369	533.989	-1253.120	49.840	-0.491
5400	108.003	638.140	537.253	544.789	-1252.059	74.414	-0.720
5500	108.006	640.122	539.105	555.589	-1250.944	98.968	-0.940
5600	108.008	642.068	540.927	566.390	-1249.778	123.502	-1.152
5700	108.011	643.980	542.718	577.191	-1248.558	148.014	-1.356
5800	108.014	645.858	544.480	587.992	-1247.284	172.504	-1.554
5900	108.016	647.705	546.214	598.794	-1245.957	196.972	-1.744
6000	108.019	649.520	547.921	609.596	-1244.576	221.416	-1.928

PREVIOUS: December 1960 (1 atm) CURRENT: December 1960 (1 bar)

Trichlorofluorosilane ($SiCl_3F$) $Cl_3F_1Si_1(g)$

IRON CHLORIDE ($FeCl_3$) CRYSTAL M_r = 162.206

$S°(298.15\ K) = 142.336\ J\ K^{-1}\ mol^{-1}$

$T_{fus} = 577\ K$

$\Delta_f H°(0\ K) = -400.833 \pm 0.84\ kJ\ mol^{-1}$

$\Delta_f H°(298.15\ K) = -399.405 \pm 0.84\ kJ\ mol^{-1}$

$\Delta_{fus} H° = 43.095\ kJ\ mol^{-1}$

Enthalpy of Formation

The enthalpy change, $\Delta_r H°(298.15\ K) = -24.49\pm0.07\ kcal\ mol^{-1}$, for the reaction Fe(cr) + 3(HCl·12.731 H_2O)(ℓ) + 1/2(H_2O_2·12.580 H_2O)(ℓ) = $FeCl_3$(cr) + H_2(g) + 45.483 H_2O(ℓ) was determined by Koehler and Coughlin (1). This leads to $\Delta_f H°$(298.15 K) = -95.46±0.20 kcal mol^{-1} for $FeCl_3$(cr), using $\Delta_f H°$(298.15 K) = -38.9±0.05 and -68.32±0.01 kcal mol^{-1} for hydrochloric acid and hydrogen peroxide solution, respectively, obtained from Rossini et al. (2). From $\Delta_r H°$(298.15 K) = -38.0±0.2 kcal mol^{-1} for the reaction $FeCl_3$(cr) = Fe^{+++}(aq) reported by Li and Gregory (3), the value of $\Delta_f H°$($FeCl_3$, cr, 298.15 K) was calculated as -93.26±0.2 kcal mol^{-1}. The former $\Delta_f H°$(298.15 K) value is adopted.

The equilibrium pressures of the reaction 2 $FeCl_2$(cr) + Cl_2(g) = 2$FeCl_3$(cr), in the temperature range of 435.7-482.2 K, were measured by Schafer and Oehler (4). Using the reported partial pressures for Cl_2(g), the enthalpy change $\Delta_r H°$(298.15 K) of the reaction was evaluated as -27.49±0.44 and -27.21 kcal mol^{-1} by the 2nd and 3rd law methods, respectively. Based on the 3rd law value for $\Delta_r H°$(298.15 K), the value of $\Delta_f H°$($FeCl_3$, cr, 298.15 K) was derived to be -95.36±0.12 kcal mol^{-1} which is in good agreement with the adopted value.

Heat Capacity and Entropy

The low temperature (51-298.16 K) heat capacities and high temperature (339.6-560.9 K) heat content of $FeCl_3$(cr) were determined by Todd and Coughlin (5). The low temperature $C°_p$ and the derived high temperature $C°_p$ were plotted. The two $C°_p$ curves were joined smoothly at 298.15 K). The $C°_p$ values above 560.9 K were obtained by graphical extrapolation. The value of S°(298.15 K) was derived from the low temperature data reported by Todd and Coughlin (5), using S°(51 K) = 6.29 cal K^{-1} mol^{-1}. The value of S°(51 K) was selected to make the second and third law values of $\Delta_r H°$, derived from the equilibrium data reported by Schafer and Oehler, (4), in good agreement. The difference in S°(51 K) between the selected value, 6.29, and the reported value, 4.50 cal K^{-1} mol^{-1}, by Todd and Coughlin (5), is assumed to be the magnetic entropy remaining at 51 K which was unaccounted for in the report.

Fusion Data

T_{fus} and $\Delta_{fus}H°$ were obtained from Todd and Coughlin (5).

Sublimation Data

The difference between $\Delta_f H°$(298.15 K) for Fe_2Cl_6(g) and 2$FeCl_3$(cr) is the enthalpy of sublimation at 298.15 K. The former is obtained by the 2nd and 3rd law analyses of related equilibrium data. See Fe_2Cl_6(g) table for details.

References

1. M. F. Koehler and J. P. Coughlin, J. Phys. Chem. 63, 605 (1959).
2. U. S. Nat. Bur. Stand. Circ. 500, 1952.
3. J. C. M. Li and N. W. Gregory, J. Amer. Chem. Soc. 74, 4670 (1952).
4. H. Schafer and E. Oehler, Z. Anorg. Alleg. Chem. 271, 206 (1953).
5. S. S. Todd and J. P. Coughlin, J. Amer. Chem. Soc. 73, 4184 (1951).

Iron Chloride ($FeCl_3$) $Cl_3Fe_1(cr)$

Enthalpy Reference Temperature = T_r = 298.15 K Standard State Pressure = p° = 0.1 MPa

	J $K^{-1}mol^{-1}$			kJ mol^{-1}			
T/K	$C°_p$	S°	$-[G°-H°(T_r)]/T$	$H°-H°(T_r)$	$\Delta_f H°$	$\Delta_f G°$	Log K_f
0	0.	0.	INFINITE	-19.706	-400.833	-400.833	INFINITE
100	58.254	54.924	220.004	-16.508	-402.425	-378.960	197.948
200	86.219	105.867	150.854	-8.997	-401.240	-355.834	92.934
298.15	96.650	142.336	142.336	0.	-399.405	-333.930	58.503
300	96.860	142.935	142.338	0.179	-399.366	-333.524	58.072
400	106.692	172.135	146.251	10.354	-397.025	-311.913	40.732
500	119.872	197.194	153.973	21.611	-393.977	-290.964	30.397
577.000	121.754	214.368	160.911	30.845	------ CRYSTAL <--> LIQUID -----		
600	123.428	219.165	163.053	33.667	-390.455	-270.691	23.566
700	128.658	238.611	172.484	46.289	-386.667	-251.027	18.732
800	131.796	256.012	181.857	59.324	-382.796	-231.914	15.142
900	133.470	271.644	190.980	72.598	-379.131	-213.278	12.378
1000	133.888	285.731	199.763	85.968	-376.113	-195.021	10.187
1100	133.888	298.491	208.167	99.357	-373.992	-177.000	8.405
1200	133.888	310.141	216.187	112.745	-371.336	-159.229	6.931
1300	133.888	320.858	223.832	126.134	-367.049	-141.727	5.695
1400	133.888	330.780	231.121	139.523	-362.860	-124.553	4.647
1500	133.888	340.017	238.076	152.912	-358.769	-107.674	3.750

PREVIOUS: CURRENT: June 1965

Iron Chloride ($FeCl_3$) $Cl_3Fe_1(cr)$

IRON CHLORIDE ($FeCl_3$) LIQUID M_r = 162.206

$S°(298.15\ K) = [200.657]\ J\ K^{-1}\ mol^{-1}$

$T_{fus} = 577\ K$

$\Delta_f H°(298.15\ K) = [-362.815]\ kJ\ mol^{-1}$

$\Delta_{fus}H° = 43.095\ kJ\ mol^{-1}$

Enthalpy of Formation

$\Delta_f H°(FeCl_3, \ell, 298.15\ K)$ is calculated from $\Delta_f H°(FeCl_3, cr, 298.15\ K)$ by adding $\Delta_{fus}H°$ and the difference in enthalpy, $H°(577\ K) - H°(298.15\ K)$, between the crystal and liquid.

Heat Capacity and Entropy

The heat capacity for $FeCl_3(\ell)$ was taken from Todd and Coughlin (1) and assumed as constant between 298.15 and 1500 K. $S°(298.15\ K)$ is calculated in a manner analogous to that used for the enthalpy of formation.

Fusion Data

T_{fus} and $\Delta_{fus}H°$ were reported by Todd and Coughlin (1).

Vaporization Data

The boiling point (T_{vap}) is determined as the temperature at which the Gibbs energy change ($\Delta_r G°$) of the reaction $2FeCl_3(\ell) = Fe_2Cl_6(g)$ is zero. The corresponding enthalpy change at T_{vap} is the enthalpy vaporization.

References

1. S. S. Todd and J. P. Coughlin, J. Amer. Chem. Soc. 73, 4184 (1951).

Iron Chloride ($FeCl_3$) $Cl_3Fe_1(l)$

Enthalpy Reference Temperature = T_r = 298.15 K Standard State Pressure = p° = 0.1 MPa

T/K	$C_p°$ (J K⁻¹mol⁻¹)	S°	$-[G°-H°(T_r)]/T$	$H°-H°(T_r)$ (kJ mol⁻¹)	$\Delta_f H°$	$\Delta_f G°$	Log K_f
0							
100							
200							
298.15	133.888	200.657	200.657	0.	-362.815	-314.728	55.139
300	133.888	201.485	200.660	0.248	-362.708	-314.430	54.747
400	133.888	240.002	205.911	13.636	-357.152	-299.187	39.070
500	133.888	269.879	215.828	27.025	-351.973	-285.302	29.805
577.000	133.888	289.056	224.351	37.335	------ CRYSTAL <--> LIQUID -----		
600	133.888	294.289	226.932	40.414	-347.118	-272.429	23.717
700	133.888	314.928	238.067	53.803	-342.563	-260.345	19.427
800	133.888	332.806	248.817	67.192	-338.338	-248.892	16.251
900	133.888	348.576	259.042	80.580	-334.558	-237.944	13.810
1000	133.888	362.683	268.713	93.969	-331.522	-227.382	11.877
1100	133.888	375.444	277.845	107.358	-329.401	-217.057	10.307
1200	133.888	387.093	286.471	120.747	-326.745	-206.980	9.010
1300	133.888	397.810	294.629	134.136	-322.458	-197.174	7.923
1400	133.888	407.732	302.358	147.524	-318.269	-187.695	7.003
1500	133.888	416.970	309.694	160.913	-314.177	-178.511	6.216

PREVIOUS: CURRENT: June 1965

Iron Chloride ($FeCl_3$) $Cl_3Fe_1(l)$

IRON CHLORIDE ($FeCl_3$) CRYSTAL-LIQUID M_r = 162.206

0 to 577 K crystal
above 577 K liquid

Refer to the individual tables for details.

Enthalpy Reference Temperature = T_r = 298.15 K   Standard State Pressure = p° = 0.1 MPa

T/K	C_p° (J K⁻¹mol⁻¹)	S° (J K⁻¹mol⁻¹)	−[G°−H°(T_r)]/T (J K⁻¹mol⁻¹)	H°−H°(T_r) (kJ mol⁻¹)	$\Delta_f H^\circ$ (kJ mol⁻¹)	$\Delta_f G^\circ$ (kJ mol⁻¹)	Log K_f
0	0.	0.	INFINITE	-19.706	-400.833	-400.833	INFINITE
100	58.254	54.924	220.004	-16.508	-402.425	-378.960	197.948
200	86.219	105.867	150.854	-8.997	-401.240	-355.834	92.934
298.15	96.650	142.336	142.336	0.	-399.405	-333.930	58.503
300	96.860	142.935	142.338	0.179	-399.366	-333.524	58.072
400	106.692	172.135	146.251	10.354	-397.025	-311.913	40.732
500	119.872	197.194	153.973	21.611	-393.977	-290.964	30.397
577.000	121.754	214.368	160.911	30.845	CRYSTAL <--> LIQUID		
577.000	133.888	289.056	160.911	73.940	TRANSITION		
600	133.888	294.289	165.924	77.019	-347.103	-272.414	23.716
700	133.888	314.928	185.774	90.408	-342.548	-260.330	19.426
800	133.888	332.806	203.060	103.797	-338.323	-248.877	16.250
900	133.888	348.576	218.370	117.186	-334.543	-237.929	13.809
1000	133.888	362.683	232.108	130.574	-331.507	-227.367	11.876
1100	133.888	375.444	244.568	143.963	-329.386	-217.041	10.306
1200	133.888	387.093	255.967	157.352	-326.730	-206.965	9.009
1300	133.888	397.810	266.471	170.741	-322.443	-197.159	7.922
1400	133.888	407.732	276.211	184.130	-318.254	-187.679	7.002
1500	133.888	416.970	285.291	197.518	-314.162	-178.496	6.216

PREVIOUS: CURRENT: June 1965

IRON CHLORIDE ($FeCl_3$) IDEAL GAS M_r = 162.206

$\Delta_f H°(0\ K) = -253.07 \pm 5.0\ kJ\ mol^{-1}$

$S°(298.15\ K) = [344.206]\ J\ K^{-1}\ mol^{-1}$ $\Delta_f H°(298.15\ K) = -253.13 \pm 5.0\ kJ\ mol^{-1}$

Vibrational Frequencies and Degeneracies

ν, cm^{-1}
[350](1)
[160](1)
[310](2)
[130](2)

Ground State Quantum Weight: [6] σ = [6]

Point Group: D_{3h}

Bond Distance: Fe-Cl = [2.17] Å

Bond Angle: Cl-Fe-Cl = [120]°

Product of the Moments of Inertia: $I_A I_B I_C = [1.43810 \times 10^{-112}]g^3\ cm^6$

Enthalpy of Formation

The equilibrium pressures for the reaction $Fe_2Cl_6(g) = 2FeCl_3(g)$ were determined by Kangro and Bernstorff (1) and Schafer (2), respectively. Using their vapor pressure data the corresponding values of $\Delta_r H°(298.15\ K)$ for that reaction were evaluated by both the 2nd and 3rd law methods. Based on the 3rd law values for $\Delta_r H°(298.15\ K)$, the heats of formation $\Delta_f H°(298.15\ K)$ for $FeCl_3(g)$ were derived. The results obtained are presented as follows.

Source	T/K	$\Delta_r H°(298.15\ K)$, kcal mol^{-1} 2nd law	3rd law	$\Delta_f H°(298.15\ K)$ kcal mol^{-1}
Kangro and Bernstorf (1)	760.2-1014.2	34.99±0.44	35.73	-60.34±0.30
Schafer (2)	673.2-1073.2	34.81±0.02	34.93	-60.74±0.20

The adopted value of $\Delta_f H°(FeCl_3,\ g,\ 298.15\ K)$ is -60.5 ± 1.2 kcal mol^{-1}.

Heat Capacity and Entropy

The molecular structure was assumed to be the same as that of $AlCl_3(g)$ due to the similarities in structure between $Fe_2Cl_6(g)$ and $Al_2Cl_6(g)$. The Fe-Cl bond distance was estimated by comparison with those for $Fe_2Cl_6(g)$ reported by Zasorin et al. (3) and Hassel and Viervoll (4). Both the ground state quantum weight and vibrational frequencies were estimated so that the second and third law values of $\Delta_r H°(298.15\ K)$, derived from the vapor pressure data, for the reaction $Fe_2Cl_6(g) = 2FeCl_3(g)$ are in reasonable agreement. The principal moments of inertia are: $I_A = I_B = 41.5834 \times 10^{-39}$ and $I_C = 83.1667 \times 10^{-39}\ g\ cm^2$.

References

1. W. Kangro and H. Bernstorff, Z. Anorg. Allg. Chem. 263, 316 (1950).
2. H. Schafer, Z. Anorg. Allg. Chem. 259, 53 (1949).
3. E. Z. Zasorin, N. G. Rambidi and P. A. Akishin, Zh. Strukt. Khim., 4, 910 (1963).
4. O. Hassel and H. Viervoll, Acta Chem. Scand. 1, 149 (1947).

Iron Chloride ($FeCl_3$) $Cl_3Fe_1(g)$

Enthalpy Reference Temperature = T_r = 298.15 K — Standard State Pressure = p° = 0.1 MPa

T/K	$C_p°$ (J $K^{-1}mol^{-1}$)	S°	$-[G°-H°(T_r)]/T$	$H°-H°(T_r)$ (kJ mol^{-1})	$\Delta_f H°$	$\Delta_f G°$	Log K_f
0	0.	0.	INFINITE	-18.214	-253.068	-253.068	INFINITE
100	56.528	269.115	409.062	-13.995	-253.639	-251.593	131.419
200	72.371	314.146	351.179	-7.406	-253.377	-249.627	65.196
250	75.716	330.687	345.475	-3.697	-253.233	-248.707	51.964
298.15	77.702	344.206	344.206	0.	-253.132	-247.845	43.421
300	77.762	344.687	344.208	0.144	-253.129	-247.812	43.148
350	79.085	356.780	345.160	4.067	-253.070	-246.931	36.852
400	79.981	367.403	347.290	8.045	-253.060	-246.055	32.132
450	80.615	376.862	350.060	12.061	-253.106	-245.178	28.459
500	81.077	385.381	353.173	16.104	-253.211	-244.292	25.521
600	81.692	400.222	359.814	24.245	-253.604	-242.475	21.109
700	82.070	412.846	366.510	32.435	-254.248	-240.573	17.952
800	82.318	423.822	373.003	40.655	-255.192	-238.559	15.576
900	82.489	433.528	379.199	48.896	-256.560	-236.403	13.720
1000	82.613	442.226	385.075	57.151	-258.657	-234.060	12.226
1100	82.704	450.104	390.634	65.417	-261.659	-231.441	10.990
1200	82.774	457.303	395.894	73.691	-264.118	-228.605	9.951
1300	82.828	463.931	400.876	81.971	-264.939	-225.613	9.065
1400	82.872	470.071	405.602	90.257	-265.854	-222.554	8.304
1500	82.907	475.790	410.093	98.546	-266.862	-219.426	7.641
1600	82.935	481.141	414.368	106.838	-267.964	-216.228	7.059
1700	82.959	486.170	418.445	115.132	-270.111	-212.940	6.543
1800	82.979	490.912	422.341	123.429	-271.737	-209.531	6.080
1900	82.996	495.399	426.069	131.728	-287.557	-205.326	5.645
2000	83.010	499.657	429.643	140.028	-289.615	-200.945	5.248
2100	83.023	503.707	433.074	148.330	-291.689	-196.461	4.887
2200	83.033	507.570	436.373	156.633	-293.779	-191.877	4.556
2300	83.043	511.261	439.549	164.937	-295.889	-187.198	4.251
2400	83.051	514.795	442.611	173.241	-298.020	-182.427	3.970
2500	83.058	518.186	445.567	181.547	-300.174	-177.566	3.710
2600	83.065	521.444	448.423	189.853	-302.353	-172.619	3.468
2700	83.071	524.579	451.186	198.160	-304.559	-167.587	3.242
2800	83.076	527.600	453.861	206.467	-306.792	-162.474	3.031
2900	83.080	530.515	456.455	214.775	-309.054	-157.280	2.833
3000	83.085	533.332	458.971	223.083	-311.345	-152.007	2.647
3100	83.088	536.056	461.414	231.392	-313.664	-146.658	2.471
3200	83.092	538.694	463.788	239.701	-664.312	-133.809	2.184
3300	83.095	541.251	466.096	248.010	-664.785	-117.224	1.855
3400	83.098	543.732	468.343	256.320	-665.316	-100.623	1.546
3500	83.100	546.140	470.532	264.630	-665.902	-84.006	1.254
3600	83.103	548.482	472.665	272.940	-666.542	-67.372	0.978
3700	83.105	550.758	474.745	281.250	-667.233	-50.718	0.716
3800	83.107	552.975	476.775	289.561	-667.974	-34.046	0.468
3900	83.109	555.134	478.756	297.872	-668.760	-17.354	0.232
4000	83.111	557.238	480.692	306.183	-669.589	-0.640	0.008
4100	83.112	559.290	482.584	314.494	-670.459	16.094	-0.205
4200	83.114	561.293	484.434	322.805	-671.366	32.850	-0.409
4300	83.115	563.248	486.245	331.117	-672.308	49.628	-0.603
4400	83.116	565.159	488.017	339.428	-673.281	66.429	-0.789
4500	83.118	567.027	489.752	347.740	-674.283	83.251	-0.966
4600	83.119	568.854	491.451	356.052	-675.312	100.097	-1.137
4700	83.120	570.642	493.117	364.364	-676.365	116.964	-1.300
4800	83.121	572.392	494.751	372.676	-677.441	133.855	-1.457
4900	83.122	574.105	496.353	380.988	-678.537	150.769	-1.607
5000	83.123	575.785	497.925	389.300	-679.651	167.704	-1.752
5100	83.123	577.431	499.468	397.612	-680.784	184.663	-1.891
5200	83.124	579.045	500.982	405.925	-681.933	201.643	-2.026
5300	83.125	580.628	502.470	414.237	-683.099	218.646	-2.155
5400	83.126	582.182	503.932	422.550	-684.281	235.672	-2.280
5500	83.126	583.707	505.369	430.862	-685.479	252.719	-2.400
5600	83.127	585.205	506.781	439.175	-686.695	269.788	-2.516
5700	83.128	586.676	508.170	447.488	-687.928	286.879	-2.629
5800	83.128	588.122	509.536	455.800	-689.180	303.992	-2.738
5900	83.129	589.543	510.880	464.113	-690.453	321.126	-2.843
6000	83.129	590.940	512.203	472.426	-691.749	338.282	-2.945

PREVIOUS: June 1965 (1 atm) CURRENT: June 1965 (1 bar)

Iron Chloride ($FeCl_3$) $Cl_3Fe_1(g)$

$S°(298.15 K) = 313.717 \pm 0.4 J K^{-1} mol^{-1}$

$\Delta_f H°(0 K) = -491.15 \pm 4.2 kJ mol^{-1}$
$\Delta_f H°(298.15 K) = -496.22 \pm 4.2 kJ mol^{-1}$

Vibrational Frequencies and Degeneracies

ν, cm^{-1}	ν, cm^{-1}
2261(1)	811(2)
499(1)	600(2)
254(1)	176(2)

Ground State Quantum Weight: [1] $\sigma = 3$
Point Group: C_{3v}
Bond Distances: Si-H = 1.4655 Å Si-Cl = 2.0118 Å
Bond Angles: H-Si-Cl = 108.32° Cl-Si-Cl = 110.60°
Product of the Moments of Inertia: $I_A I_B I_C = 7.59150 \times 10^{-113} g^3 cm^6$

Enthalpy of Formation

$\Delta_f H°(298.15 K)$ is calculated from that of $SiCl_4(g)$ (1) using the difference $\Delta(\Delta_f H°) = 39.8 kcal mol^{-1}$ taken from the analysis of Hunt and Sirtl (2). The authors derived $\Delta_f H°(SiHCl_3, g, 298.15 K) = -116.9\pm0.7 kcal mol^{-1}$ from six previously reported equilibrium studies. These were at temperatures (773-1200 K) and compositions where the dominant equilibrium was $SiHCl_3 + HCl = SiCl_4 + H_2$. Selected data (2) were either close to or extrapolated to infinite residence time. Equilibria were reduced to $\Delta_f H°(298.15 K)$ using auxiliary data (2) which are similar to JANAF values (1), except for $\Delta_f H°(SiCl_4, g, 298.15 K)$. Since this is now 1.7 kcal mol^{-1} more negative, we change $\Delta_f H°(SiHCl_3)$ accordingly. Our adopted value is 1.8 kcal mol^{-1} more negative than that estimated by linear interpolation between $SiCl_4(g)$ and $SiH_4(g)$. (1)

We dismiss two calorimetric values due to probable bias. These include $\Delta_f H°(\ell) = -112 kcal mol^{-1}$ from oxygen-bomb calorimetry (3) and $\Delta_f H°(g) = -117.4\pm1.7 kcal mol^{-1}$ from solution calorimetry (4) in dilute aqueous NaOH. Revised auxiliary data (1, 5) change these values to $\Delta_f H°(g) = -115$ (3) and $-124 kcal mol^{-1}$. Wagman et al. (5) adopted $-122.6 kcal mol^{-1}$, presumably based on Wolf (4). The combustion result is suspect because the chemical reaction was determined only qualitatively and included corrosion reactions of the bomb (3). In our approximate revision of the solution result, we eliminate $Na_2SiO_3(cr)$ by taking the difference between the overall reactions (4) for $SiCl_4(\ell)$ and $SiHCl_3(g)$. This yields the difference $\Delta(\Delta_f H°) = \Delta_f H°(SiHCl_3, g) - \Delta_f H°(SiCl_4, g) = 34.6$ compared to 39.8 kcal mol^{-1} (2). The solution result is suspect due to the possibility that the two hydrolyses did not proceed to equivalent degrees. Solution results (4) for $SiCl_4$, $SiBr_4(\ell)$ and $SiI_4(cr)$ also have a negative bias compared to our adopted values (1).

Heat Capacity and Entropy

The molecular structure is based on microwave data of Mitzlaff et al. (6) for twleve isotopic forms of $SiHCl_3$ and $SiDCl_3$. Structural parameters are substitutional (r_s) values. They are in reasonable agreement with previous data (6) determined by various methods. The principal moments of inertia are $I_A = I_B = 34.3279 \times 10^{-39}$ and $I_C = 64.4220 \times 10^{-39} g cm^2$.

Vibrational frequencies are those selected by Shimanouchi (7) from gas-phase infrared spectra of Buerger and Ruoff (8). Assignments are based on band contours and isotopic splittings and are consistent with recent liquid-phase Raman spectra (9).

We neglect excited states and assume the electronic ground state to be a singlet. We expect that contributions from excited states should be unimportant, as discussed on the tables for SiH_2Cl_2, SiH_3Cl and SiH_3Br (1).

References

1. JANAF Thermochemical Tables: $Cl_4Si(g)$, 12-31-70; $Cl_2H_2Si(g)$, $ClH_3Si(g)$, $BrH_3Si(g)$, $Br_4Si(\ell)$, 12-31-76; $I_4Si(cr)$, $H_4Si(g)$, 6-30-76; O_2Si(quartz), 6-30-67; ClH(g), 9-30-64; HNaO(cr), 12-31-70; ClNa(cr), 9-30-64.
2. L. P. Hunt and E. Sirtl, J. Electrochem. Soc. 119, 1741 (1972); 121, 919 (1974).
3. H. Reuther, Chem. Tech. (Berlin) 2, 331 (1950).
4. E. Wolf, Z. Anorg. Allgem. Chem. 313, 228 (1961); 319, 168 (1962); Z. Chem. 7, 283 (1967).
5. U.S. Nat. Bur. Stand. Tech. Note 270-3, 1968.
6. M. Mitzlaff, R. Holm and H. Hartmann, Z. Naturforsch. A22, 1415 and 288 (1967); Theor. Chim. Acta 12, 330 (1969).
7. T. Shimanouchi, J. Phys. Chem. Ref. Data 1, 205 (1972).
8. H. Buerger and A. Ruoff, Spectrochim. Acta A26, 1449 (1970).
9. V. S. Dernova et al., Dokl. Akad. Nauk SSSR 222, 147 (1975).

Enthalpy Reference Temperature = T_r = 298.15 K Standard State Pressure = p° = 0.1 MPa

	J $K^{-1}mol^{-1}$			kJ mol^{-1}			
T/K	$C_p°$	S°	$-[G°-H°(T_r)]/T$	$H°-H°(T_r)$	$\Delta_f H°$	$\Delta_f G°$	Log K_f
0	0.	0.	INFINITE	-16.151	-491.151	-491.151	INFINITE
100	46.854	248.357	372.097	-12.374	-493.507	-484.572	253.115
200	63.216	286.041	320.258	-6.844	-495.059	-475.003	124.058
250	70.039	300.901	314.931	-3.507	-495.699	-469.914	98.183
298.15	75.455	313.717	313.717	0.	-496.222	-464.899	81.448
300	75.641	314.185	313.719	0.140	-496.241	-464.704	80.912
350	80.107	326.194	314.657	4.038	-496.680	-459.412	68.563
400	83.657	337.131	316.793	8.135	-497.025	-454.063	59.295
450	86.512	347.155	319.617	12.392	-497.286	-448.677	52.081
500	88.851	356.395	322.839	16.778	-497.478	-443.265	46.308
600	92.465	372.932	329.843	25.853	-497.691	-432.399	37.644
700	95.144	387.396	337.053	35.240	-497.729	-421.511	31.454
800	97.212	400.241	344.164	44.862	-497.642	-410.628	26.811
900	98.848	411.789	351.047	54.668	-497.466	-399.761	23.202
1000	100.164	422.275	357.654	64.621	-497.231	-388.917	20.315
1100	101.235	431.873	363.971	74.693	-496.960	-378.099	17.954
1200	102.115	440.721	370.003	84.862	-496.671	-367.306	15.988
1300	102.845	448.924	375.762	95.111	-496.381	-356.537	14.326
1400	103.455	456.569	381.264	105.427	-496.103	-345.790	12.902
1500	103.970	463.724	386.525	115.799	-495.845	-335.063	11.668
1600	104.406	470.449	391.562	126.218	-495.617	-324.352	10.589
1700	104.779	476.790	396.391	136.678	-545.602	-313.207	9.624
1800	105.100	482.788	401.026	147.172	-545.221	-299.548	8.693
1900	105.378	488.478	405.480	157.696	-544.843	-285.909	7.860
2000	105.620	493.890	409.766	168.247	-544.474	-272.291	7.112
2100	105.832	499.048	413.896	178.819	-544.116	-258.691	6.435
2200	106.018	503.976	417.879	189.412	-543.773	-245.107	5.820
2300	106.182	508.692	421.726	200.022	-543.450	-231.539	5.258
2400	106.328	513.214	425.444	210.648	-543.147	-217.984	4.744
2500	106.458	517.558	429.043	221.287	-542.870	-204.441	4.272
2600	106.575	521.735	432.528	231.939	-542.619	-190.909	3.835
2700	106.679	525.759	435.907	242.602	-542.398	-177.386	3.432
2800	106.774	529.641	439.186	253.275	-542.208	-163.871	3.057
2900	106.859	533.389	442.370	263.956	-542.050	-150.361	2.708
3000	106.936	537.013	445.464	274.646	-541.925	-136.857	2.383
3100	107.007	540.521	448.475	285.343	-541.834	-123.357	2.079
3200	107.071	543.919	451.404	296.047	-541.774	-109.859	1.793
3300	107.129	547.215	454.258	306.757	-541.746	-96.361	1.525
3400	107.183	550.414	457.039	317.473	-541.748	-82.865	1.273
3500	107.233	553.521	459.752	328.194	-541.778	-69.368	1.035
3600	107.278	556.543	462.399	338.919	-925.984	-45.409	0.659
3700	107.320	559.483	464.983	349.649	-925.647	-20.953	0.296
3800	107.359	562.345	467.508	360.383	-925.331	3.493	-0.048
3900	107.395	565.135	469.975	371.121	-925.032	27.932	-0.374
4000	107.428	567.854	472.388	381.862	-924.746	52.364	-0.684
4100	107.459	570.507	474.749	392.607	-924.468	76.788	-0.978
4200	107.488	573.097	477.060	403.354	-924.196	101.206	-1.259
4300	107.515	575.627	479.323	414.104	-923.924	125.617	-1.526
4400	107.541	578.099	481.540	424.857	-923.649	150.021	-1.781
4500	107.564	580.516	483.713	435.612	-923.367	174.420	-2.025
4600	107.586	582.880	485.843	446.370	-923.074	198.812	-2.258
4700	107.607	585.194	487.932	457.130	-922.767	223.197	-2.481
4800	107.627	587.460	489.982	467.891	-922.442	247.576	-2.694
4900	107.645	589.679	491.994	478.655	-922.097	271.948	-2.899
5000	107.662	591.854	493.970	489.420	-921.728	296.312	-3.096
5100	107.678	593.986	495.910	500.187	-921.333	320.670	-3.284
5200	107.694	596.077	497.816	510.956	-920.909	345.018	-3.466
5300	107.708	598.129	499.690	521.726	-920.455	369.359	-3.640
5400	107.722	600.142	501.531	532.497	-919.968	393.691	-3.808
5500	107.735	602.119	503.342	543.270	-919.446	418.013	-3.970
5600	107.747	604.060	505.124	554.044	-918.888	442.325	-4.126
5700	107.759	605.967	506.876	564.820	-918.293	466.627	-4.276
5800	107.770	607.841	508.601	575.596	-917.661	490.919	-4.421
5900	107.781	609.684	510.298	586.374	-916.989	515.199	-4.561
6000	107.791	611.495	511.970	597.152	-916.276	539.467	-4.696

PREVIOUS: December 1976 (1 atm) CURRENT: December 1976 (1 bar)

LITHIUM CHLORIDE $((LiCl)_3)$ IDEAL GAS M_r = 127.182

$S°(298.15\ K) = [335.747]\ J\ K^{-1}\ mol^{-1}$

$\Delta_f H°(0\ K) = -997.47\ kJ\ mol^{-1}$

$\Delta_f H°(298.15\ K) = -1004.72 \pm 20.9\ kJ\ mol^{-1}$

Vibrational Levels and Degeneracies

ν, cm^{-1}
[650](6)
[220](6)

Point Group: D_{3h} $\sigma = 6$

Bond Distance: Li-Cl = [2.2] Å

Bond Angle: Li-Cl-Li = [120]°

Product of the Moment of Inertia: $I_A I_B I_C = 2.67006 \times 10^{-112}\ g^3\ cm^6$

Enthalpy of Formation

The heat of sublimation at 298.15 K was chosen so as to give the correct ratio of monomer, dimer and trimer at 870 K. The value used was 52.6 kcal mol^{-1} which compares very well with the value 53.6 kcal mol^{-1} obtained from a third law analysis of the data of Nesmeyanov and Sazonov (1) in conjunction with the relative concentrations of monomer, dimer, and trimer at 870 K from the work of Miller and Kusch. (2).

Heat Capacity and Entropy

The postulated structure, bond lengths and frequencies were all taken from Hildenbrand et al. (3). The principal moments of inertia are: $I_A = I_B = 51.1090 \times 10^{-39}$ and $I_C = 102.2179 \times 10^{-39}\ g\ cm^2$.

References

1. An. N. Nesmeyanov and L. A. Sazonov, Zhur. Neorg. Khim. 4, 231 (1960).
2. R. C. Miller and P. Kusch, J. Chem. Phys. 25, 860 (1950).
3. D. L. Hildenbrand, A. M. Saul, P. R. Basford, S. E. Stephanon, L. Larson and A. Uyehara, Aeronutronic Tenth Quarterly Report on Contract Nord 17980 Report C-898 (1960).

Lithium Chloride $((LiCl)_3)$ $Cl_3Li_3(g)$

Enthalpy Reference Temperature = T_r = 298.15 K Standard State Pressure = p° = 0.1 MPa

	J $K^{-1}mol^{-1}$			kJ mol^{-1}			
T/K	$C_p°$	S°	$-[G°-H°(T_r)]/T$	$H°-H°(T_r)$	$\Delta_f H°$	$\Delta_f G°$	Log K_f
0	0.	0.	INFINITE	-20.383	-997.466	-997.466	INFINITE
100	56.629	250.069	413.649	-16.358	-999.293	-993.747	519.080
200	84.270	298.513	344.567	-9.211	-1002.228	-987.045	257.790
250	94.366	318.444	337.386	-4.735	-1003.565	-983.092	205.406
298.15	102.005	335.747	335.747	0.	-1004.721	-979.042	171.524
300	102.261	336.379	335.749	0.189	-1004.763	-978.883	170.439
350	108.285	352.616	337.017	5.459	-1005.859	-974.482	145.433
400	112.864	367.388	339.904	10.993	-1006.965	-969.925	126.659
450	116.369	380.893	343.719	16.728	-1008.164	-965.224	112.040
500	119.083	393.300	348.066	22.617	-1018.497	-959.462	100.234
600	122.907	415.376	357.491	34.731	-1020.781	-947.432	82.481
700	125.387	434.521	367.158	47.154	-1022.641	-935.055	69.775
800	127.072	451.381	376.653	59.782	-1024.251	-922.431	60.229
900	128.263	466.421	385.807	72.552	-1025.735	-909.613	52.793
1000	129.134	479.982	394.558	85.424	-1027.126	-896.636	46.835
1100	129.788	492.322	402.893	98.372	-1028.447	-883.522	41.955
1200	130.292	503.638	410.823	111.377	-1029.713	-870.291	37 883
1300	130.688	514.083	418.370	124.427	-1030.938	-856.956	34.433
1400	131.004	523.780	425.557	137.512	-1032.124	-843.528	31.472
1500	131.261	532.827	432.410	150.626	-1033.269	-830.017	28.904
1600	131.472	541.305	438.954	163.763	-1034.380	-816.430	26.654
1700	131.647	549.281	445.211	176.919	-1471.169	-781.247	24.005
1800	131.795	556.810	451.204	190.091	-1469.964	-740.698	21.494
1900	131.920	563.940	456.952	203.277	-1468.764	-700.216	19.250
2000	132.027	570.709	462.472	216.475	-1467.574	-659.797	17.232
2100	132.120	577.153	467.780	229.682	-1466.399	-619.437	15.408
2200	132.200	583.301	472.893	242.898	-1465.244	-579.133	13.750
2300	132.270	589.179	477.822	256.122	-1464.117	-538.880	12.238
2400	132.332	594.810	482.580	269.352	-1463.024	-498.676	10.853
2500	132.386	600.213	487.178	282.588	-1461.972	-458.517	9.580
2600	132.435	605.406	491.626	295.829	-1460.968	-418.398	8.406
2700	132.478	610.405	495.933	309.075	-1460.019	-378.318	7.319
2800	132.516	615.224	500.108	322.324	-1459.131	-338.271	6.311
2900	132.551	619.875	504.158	335.578	-1458.310	-298.255	5.372
3000	132.582	624.369	508.091	348.835	-1457.563	-258.266	4.497
3100	132.611	628.717	511.912	362.094	-1456.894	-218.301	3.678
3200	132.636	632.927	515.628	375.357	-1456.308	-178.356	2.911
3300	132.660	637.009	519.245	388.621	-1455.808	-138.428	2.191
3400	132.681	640.970	522.767	401.888	-1455.399	-98.514	1.513
3500	132.701	644.816	526.200	415.158	-1455.081	-58.610	0.875
3600	132.719	648.555	529.547	428.429	-1454.858	-18.714	0.272
3700	132.735	652.191	532.812	441.701	-1454.730	21.177	-0.299
3800	132.751	655.731	536.001	454.976	-1454.699	61.065	-0.839
3900	132.765	659.180	539.115	468.251	-1454.766	100.954	-1.352
4000	132.778	662.541	542.159	481.529	-1454.931	140.847	-1.839
4100	132.790	665.820	545.135	494.807	-1455.196	180.744	-2.303
4200	132.801	669.020	548.047	508.086	-1455.560	220.649	-2.744
4300	132.812	672.145	550.897	521.367	-1456.025	260.564	-3.165
4400	132.822	675.198	553.687	534.649	-1456.528	300.486	-3.567
4500	132.831	678.183	556.421	547.931	-1457.167	340.425	-3.952
4600	132.839	681.103	559.100	561.215	-1457.912	380.380	-4.319
4700	132.847	683.960	561.726	574.499	-1458.756	420.351	-4.672
4800	132.855	686.757	564.302	587.784	-1459.701	460.342	-5.010
4900	132.862	689.496	566.829	601.070	-1460.748	500.354	-5.334
5000	132.869	692.181	569.309	614.357	-1461.900	540.387	-5.645
5100	132.875	694.812	571.744	627.644	-1463.141	580.444	-5.945
5200	132.881	697.392	574.136	640.932	-1464.504	620.527	-6.233
5300	132.886	699.923	576.485	654.220	-1465.979	660.638	-6.511
5400	132.892	702.407	578.794	667.509	-1467.567	700.778	-6.779
5500	132.897	704.846	581.064	680.799	-1469.274	740.948	-7.037
5600	132.901	707.240	583.296	694.088	-1471.101	781.150	-7.286
5700	132.906	709.593	585.491	707.379	-1473.053	821.386	-7.527
5800	132.910	711.904	587.651	720.670	-1475.134	861.657	-7.760
5900	132.914	714.176	589.776	733.961	-1477.346	901.965	-7.985
6000	132.918	716.410	591.868	747.253	-1479.695	942.312	-8.204

PREVIOUS: June 1962 (1 atm) CURRENT: June 1962 (1 bar)

Lithium Chloride $((LiCl)_3)$ $Cl_3Li_3(g)$

PHOSPHORYL CHLORIDE ($OPCl_3$) IDEAL GAS M_r = 153.33216

$S°(298.15\ K) = 325.454\ J\ K^{-1}\ mol^{-1}$

$\Delta_fH°(0\ K) = -554.2 \pm 1.7\ kJ\ mol^{-1}$
$\Delta_fH°(298.15\ K) = -559.8 \pm 1.7\ kJ\ mol^{-1}$

Vibrational Levels and Multiplicities

ν, cm^{-1}	ν, cm^{-1}	ν, cm^{-1}
1290(1)	486(1)	337(1)
581(2)	267(2)	193(2)

Point Group: C_{3v} $\sigma = 3$
Bond Distances: P-O = 1.45 ± 0.03 Å P-Cl = 1.99 ± .02 Å
Bond Angle: Cl-P-Cl = 103.6 ± 2°
Product of the Moments of Inertia: $I_AI_BI_C = 1.01841 \times 10^{-112}\ g^3\ cm^6$

Enthalpy of Formation

The $\Delta_fH°(298.15\ K)$ for $POCl_3(g)$ was derived from the $\Delta_fH°(298.15\ K)$ of $POCl_3(\ell)$ and the $\Delta_{vap}H°(298.15\ K)$. The $\Delta_fH°(298.15\ K)$ of the $POCl_3(\ell)$ was calculated from the heat of hydrolysis observed by Charnley and Skinner (1) for the reaction:

$$POCl_3(\ell) + (n+3)\ H_2O(\ell) \rightarrow [H_3PO_4 + 3\ HCl]\ in\ n\ H_2O(\ell)$$

The auxiliary thermochemical data, on the enthalpies of formation of water and aqueous solutions of HCl and of H_3PO_4, were taken from the (2), and were adjusted for consistency with JANAF Tables. The heat of vaporization was measured and corrected to 298.15 K by Ott and Giauque (3).

Heat Capacity and Entropy

The molecular constants are from the microwave spectral data of Williams et al. (4). In order to determine the structure Williams et al. (4) assumed the P-Cl distance equal to 1.99 Å. This was the smallest value allowed (2.02±0.03 Å) by the electron diffraction work of Brockway and Beach (5).

Ott and Giauque (3) have given the moments of inertia obtained from a weighted average of the iotopic species. $I_A \neq I_B$ due to isotopic loss of symmetry. The S°(298.15 K) calculated from calorimetric data by Ott and Giauque (3) was 77.75 cal K^{-1} mol^{-1} and the S°(298.15 K), calculated from the spectral data was 77.77 cal K^{-1} mol^{-1}.

The principal moments of inertia are: $I_A = I_B = 42.0514 \times 10^{-39}$ and $I_C = 57.5920 \times 10^{-39}\ g\ cm^2$.

References

1. T. Charnley and H. A. Skinner, J. Chem. Soc. 450, (1953).
2. U. S. Nat. Bur. Stand. Circ. 500, 1952.
3. J. B. Ott and W. F. Giauque, J. Amer. Chem. Soc. 82, 1308 (1960).
4. Q. Williams, J. Sheridan and W. Gordy, J. Chem. Phys. 20, 164 (1952).
5. L. O. Brockway and J. Y. Beach, J. Amer. Chem. Soc. 60, 1836 (1938).

Phosphoryl Chloride ($OPCl_3$) $Cl_3O_1P_1(g)$

Enthalpy Reference Temperature = T_r = 298.15 K Standard State Pressure = p° = 0.1 MPa

T/K	$C_p°$ (J K⁻¹mol⁻¹)	S° (J K⁻¹mol⁻¹)	$-[G°-H°(T_r)]/T$ (J K⁻¹mol⁻¹)	$H°-H°(T_r)$ (kJ mol⁻¹)	$\Delta_fH°$ (kJ mol⁻¹)	$\Delta_fG°$ (kJ mol⁻¹)	Log K_f
0	0.	0.	INFINITE	-17.880	-554.226	-554.226	INFINITE
100	50.035	251.381	392.051	-14.067	-557.121	-543.474	283.881
200	72.927	293.857	332.888	-7.806	-559.135	-529.144	138.198
250	79.947	310.927	326.826	-3.975	-559.553	-521.594	108.981
298.15	84.917	325.454	325.454	0.	-559.819	-514.257	90.096
300	85.082	325.980	325.456	0.157	-559.828	-513.975	89.491
350	88.953	339.399	326.507	4.512	-560.724	-506.246	75.553
400	91.957	351.481	328.886	9.038	-560.889	-498.451	65.091
450	94.338	362.455	332.016	13.698	-560.969	-490.641	56.952
500	96.257	372.497	335.569	18.464	-560.982	-482.825	50.440
600	99.115	390.317	343.246	28.242	-560.865	-467.202	40.674
700	101.093	405.753	351.098	38.258	-560.616	-451.610	33.700
800	102.506	419.349	358.797	48.442	-560.283	-436.060	28.472
900	103.544	431.485	366.211	58.747	-559.894	-420.555	24.408
1000	104.325	442.437	373.295	69.142	-559.467	-405.095	21.160
1100	104.925	452.410	380.041	79.606	-559.016	-389.680	18.504
1200	105.395	461.560	386.458	90.123	-622.120	-373.228	16.246
1300	105.770	470.012	392.564	100.682	-620.869	-352.537	14.165
1400	106.072	477.862	398.380	111.275	-619.619	-331.943	12.385
1500	106.320	485.189	403.926	121.895	-618.373	-311.439	10.845
1600	106.526	492.057	409.221	132.537	-617.133	-291.017	9.501
1700	106.697	498.521	414.286	143.199	-615.904	-270.672	8.317
1800	106.843	504.624	419.137	153.876	-614.688	-250.400	7.266
1900	106.967	510.404	423.790	164.567	-613.487	-230 194	6.328
2000	107.073	515.893	428.259	175.269	-612.303	-210.052	5.486
2100	107.165	521.119	432.557	185.981	-611.140	-189.968	4.725
2200	107.245	526.107	436.697	196.701	-610.000	-169.938	4.035
2300	107.315	530.876	440.689	207.429	-608.886	-149.961	3.406
2400	107.377	535.444	444.542	218.164	-607.800	-130.031	2.830
2500	107.432	539.829	448.267	228.905	-606.745	-110.146	2.301
2600	107.481	544.043	451.870	239.650	-605.723	-90.302	1.814
2700	107.524	548.100	455.359	250.400	-604.736	-70.497	1.364
2800	107.563	552.011	458.742	261.155	-603.784	-50.728	0.946
2900	107.598	555.787	462.023	271.913	-602.869	-30.992	0.558
3000	107.630	559.435	465.210	282.674	-601.992	-11.287	0.197
3100	107.658	562.965	468.307	293.439	-601.151	8.389	-0.141
3200	107.685	566.383	471.319	304.206	-600.346	28.038	-0.458
3300	107.708	569.697	474.250	314.976	-599.576	47.663	-0.754
3400	107.730	572.913	477.105	325.748	-598.838	67.266	-1.033
3500	107.750	576.036	479.887	336.522	-598.130	86.846	-1.296
3600	107.768	579.071	482.600	347.298	-597.449	106.408	-1.544
3700	107.785	582.024	485.247	358.075	-596.791	125.950	-1.778
3800	107.801	584.899	487.832	368.855	-596.154	145.475	-2.000
3900	107.815	587.699	490.357	379.635	-595.533	164.984	-2.210
4000	107.829	590.429	492.825	390.418	-594.925	184.476	-2.409
4100	107.841	593.092	495.238	401.201	-594.325	203.954	-2.598
4200	107.852	595.691	497.599	411.986	-593.730	223.417	-2.779
4300	107.863	598.229	499.910	422.772	-593.135	242.866	-2.950
4400	107.873	600.709	502.173	433.558	-592.538	262.301	-3.114
4500	107.883	603.133	504.389	444.346	-591.933	281.722	-3.270
4600	107.891	605.504	506.562	455.135	-591.317	301.130	-3.419
4700	107.900	607.825	508.692	465.924	-590.688	320.524	-3.562
4800	107.907	610.096	510.781	476.715	-590.042	339.905	-3.699
4900	107.915	612.321	512.830	487.506	-589.376	359.272	-3.830
5000	107.921	614.502	514.842	498.298	-588.689	378.625	-3.955
5100	107.928	616.639	516.817	509.090	-587.977	397.965	-4.076
5200	107.934	618.735	518.757	519.883	-587.238	417.289	-4.192
5300	107.940	620.791	520.663	530.677	-586.471	436.600	-4.303
5400	107.945	622.808	522.536	541.471	-585.675	455.896	-4.410
5500	107.950	624.789	524.377	552.266	-584.849	475.176	-4.513
5600	107.955	626.734	526.188	563.061	-583.990	494.442	-4.612
5700	107.959	628.645	527.968	573.857	-583.100	513.692	-4.707
5800	107.964	630.523	529.720	584.653	-582.177	532.926	-4.800
5900	107.968	632.368	531.445	595.450	-581.220	552.143	-4.888
6000	107.972	634.183	533.142	606.247	-580.231	571.344	-4.974

PREVIOUS: June 1962 (1 atm) CURRENT: June 1962 (1 bar)

Phosphoryl Chloride ($OPCl_3$) $Cl_3O_1P_1(g)$

PHOSPHORUS CHLORIDE (PCl_3) IDEAL GAS M_r = 137.33276

$\Delta_fH°(0\ K) = -285.50 \pm 5.4\ kJ\ mol^{-1}$

$S°(298.15\ K) = 311.682 \pm .42\ J\ K^{-1}\ mol^{-1}$ $\Delta_fH°(298.15\ K) = -288.70 \pm 5.4\ kJ\ mol^{-1}$

Vibrational Frequencies and Degeneracies

ν, cm^{-1}	ν, cm^{-1}
510 (1)	507 (2)
259 (1)	187 (2)

Ground State Quantum Weight = 1 σ = 3

Point Group: C_{3v}

Bond Distance: P-Cl = 2.039 ± 0.0014 Å

Bond Angle: Cl-P-Cl = 100.27 ± 0.09°

Product of the Moments of Inertia: $I_AI_BI_C = 6.05196 \times 10^{-113}\ g^3\ cm^6$

Enthalpy of Formation

Neale and Williams (1) measured the heat of hydrolysis of liquid phosphorus trichloride and reported a $\Delta_rH° = -67.5\pm0.3$ kcal mol^{-1} for the following reaction:

$$PCl_3(\ell) + (X_2 + 3)\ H_2O \rightarrow H_3PO_3(aq) + 3\ HCl(aq)$$

with X_2 in range 4500 to 7500. Combining this result with enthalpy of formation data for aqueous H_3PO_3, -229.4 kcal mol^{-1} (2, 5, 16), H_2O (2) and aqueous HCl (2), we derive $\Delta_fH°(PCl_3, \ell, 0\ K) = -76.7$ kcal mol^{-1}. Charnley and Skinner (3) made an independent investigation of the hydrolysis of liquid phosphorus trichloride and obtained enthalpy data in very good agreement with Neale and Williams' results. Their results lead to $\Delta_fH°(PCl_3, \ell, 0\ K) = -76.5$ kcal mol^{-1} when combined with heat of formation data (2, 5, 16).

Neale and Williams (4) also investigated the hydrolysis of liquid phosphorus trichloride in aqueous bromine solution with the formation of phosphoric acid rather than phosphorus acid which results when hydrolysis occurs in water. They reported a $\Delta_rH°$ = -137.9 kcal mol^{-1} for the following reaction:

$$PCl_3(\ell) + Br_2(aq) + (X_2 + 4)H_2O \rightarrow H_3PO_4(aq) + 3\ HCl(aq) + 2\ HBr(aq)$$

with X_2 in range 3500 to 6300. Combining this result with the following enthalpy of formation data: $\Delta_fH°(H_3PO_4 \cdot 1\ H_2O)$ = -306.4±0.5 kcal mol^{-1} (5), Δ H°(Br, aq) = -0.2 $kcal^{-1}$ mol (6), HCl(aq) and HBr(aq) (2), along with dilution data for $H_3PO_4(aq)$ (2), we derive $\Delta_fH°(PCl_3, 298.15\ K) = -77.0$ kcal mol^{-1}.

The enthalpy of formation from white phosphorus of gaseous PCl_3, $\Delta_fH°(PCl_3, g, 298.15\ K) = -69.0\pm1.1$ kcal mol^{-1}, is calculated from $\Delta_fH°(PCl_3, \ell, 298.15\ K) = -76.7\pm1.0$ kcal mol^{-1}, an average value of the three results given above, and the enthalpy of vaporization of PCl_3. $\Delta_{vap}H°(298.15\ K) = 7.7\pm0.1$ kcal mol^{-1} is determined from a 2nd law analysis of the combined vapor pressure data of Nisel'son and Seryakov (7) and Regnault (8). The smoothed vapor pressure data of Regnault as reported by Stull (9) were used.

An independent value for the heat of formation can be obtained from the work of Duus and Mykytiuk (17) on the reaction 1.5 $CaF_2 + PCl_3(g) + 1.5\ CaCl_2 + PF_3(g)$. The authors reported $\Delta_fH°(630\ K) = -2.97$ kcal mol^{-1}. From a re-examination of their technique we believe that this heat evolution is likely to be too small, perhaps by 100%. The analytical results indicate that the reaction is 99% complete or better; thus, we can use this with JANAF entropies to calculate $\Delta_rH°(298.15\ K) \leq 3.55$ kcal mol^{-1}. Finch et al. (18) have reported $\Delta_fH°(CaF_2, cr, 0\ K) - \Delta_fH°(CaCl_2, cr, 0\ K) = -100.87\pm0.02$ kcal mol^{-1}; thus, we obtain $\Delta_fH°(PCl_3, g, 298.15\ K) > -70.1\pm2.0$ kcal mol^{-1} from $\Delta_fH°(PF_3, g, 298.15\ K) = -224.9\pm0.9$ kcal mol^{-1}. This value is not adopted because of the lack of calorimeter calibration, but it offers an attractive possibility for further study.

Heat Capacity and Entropy

The molecular structure and bond length given above for PCl_3 are from the electron diffraction studies of Hedberg and Iwasahi (11). The adopted values are in good agreement within experimental error with the following structural data reported by Kisliuk and Townes (12) from microwave studies: r_{P-Cl} = 2.043±0.003 Å, the angle Cl-P-Cl = 100°6'±20'. The vibrational frequencies are from the recent infrared and Raman studies of Frankiss and Miller (13) and are in excellent agreement with those of Muller et al. (14) and Davis and Oetjen (15). The principal moments of inertia are: $I_A = I_B = 32.3918 \times 10^{-39}$ and $I_C = 57.6799 \times 10^{-39}$ g cm^2.

References

1. E. Neale and L. T. D. Williams, J. Chem. Soc. 1952, 4535.
2. U. S. Nat. Bur. Stand. Tech. Note 270-3, 1968.
3. T. Charnley and H. A. Skinner, J. Chem. Soc. 1953, 450.
4. E. Neale and L. T. D. Williams, J. Chem. Soc. 1954, 2156.
5. R. J. Irving and H. McKerrell, Trans. Faraday Soc. 63, 2582 (1967).
6. C. H. Wu, M. M. Birky, and L. G. Hepler, J. Phys. Chem. 67, 1202 (1963).
7. L. A. Nisel'son and G. V. Seryakov, Russ. J. Inorg. Chem. (English Transl.) 5, 548 (1960).
8. H. V. Regnault, Mem. Paris 26, 339 (1862).
9. D. R. Stull, Ind. Eng. Chem. 39, 517 (1947).
10. JANAF Thermochemical Tables: P(red,V), 6-30-61.
11. K. Hedberg and M. Iwasaki, J. Chem. Phys. 36, 589 (1962).
12. P. Kisliuk and C. H. Townes, J. Chem. Phys. 18, 1109 (1950).
13. S. G. Frankiss and F. A. Miller, Spectrochimica Acta 21, 1235 (1965).
14. A. Muller, E. Niecke, B. Krebs, and O. Glemser, Z. Naturforsch. 23b, 588 (1968).
15. P. W. Davis and R. A. Oetjen, J. Mol. Spectroscopy 2, 253 (1958).
16. A. Finch, P. J. Gardner, K. S. Hussain and K. K. Sen Gupta, Chem. Commun. 872 (1968).
17. H. C. Duus and D. P. Mykytiuk, J. Chem. Eng. Data 9, 585 (1964).
18. A. Finch, P. J. Gardner and C. J. Steadman, Can. J. Chem. 46, 3447 (1968).

Phosphorus Chloride (PCl_3) $Cl_3P_1(g)$

Enthalpy Reference Temperature = T_r = 298.15 K; Standard State Pressure = p° = 0.1 MPa

T/K	$C_p°$ (J K⁻¹mol⁻¹)	S° (J K⁻¹mol⁻¹)	-[G°-H°(T_r)]/T (J K⁻¹mol⁻¹)	H°-H°(T_r) (kJ mol⁻¹)	$\Delta_fH°$ (kJ mol⁻¹)	$\Delta_fG°$ (kJ mol⁻¹)	Log K_f
0	0.	0.	INFINITE	-15.932	-285.497	-285.497	INFINITE
100	46.469	246.997	368.917	-12.192	-287.012	-281.592	147.088
200	62.949	284.738	317.995	-6.651	-288.291	-275.824	72.038
250	68.098	299.372	312.842	-3.368	-288.527	-272.679	56.973
298.15	71.581	311.682	311.682	0.	-288.696	-269.610	47.235
300	71.693	312.125	311.683	0.133	-288.702	-269.492	46.923
350	74.221	323.377	312.566	3.784	-289.564	-266.207	39.729
400	76.036	333.413	314.556	7.543	-289.749	-262.857	34.326
450	77.369	342.450	317.162	11.380	-289.892	-259.487	30.120
500	78.370	350.656	320.107	15.274	-290.007	-256.102	26.755
600	79.738	365.075	326.433	23.185	-290.177	-249.304	21.704
700	80.599	377.436	332.858	31.205	-290.297	-242.482	18.094
800	81.175	388.239	339.119	39.296	-290.388	-235.645	15.386
900	81.576	397.824	345.120	47.434	-290.463	-228.798	13.279
1000	81.867	406.435	350.828	55.607	-290.528	-221.943	11.593
1100	82.085	414.249	356.244	63.805	-290.587	-215.081	10.213
1200	82.251	421.398	361.380	72.022	-354.218	-207.136	9.016
1300	82.381	427.987	366.253	80.254	-353.502	-194.908	7.832
1400	82.485	434.096	370.884	88.498	-352.794	-182.735	6.818
1500	82.569	439.790	375.290	96.751	-352.094	-170.613	5.941
1600	82.638	445.121	379.489	105.011	-351.404	-158.537	5.176
1700	82.696	450.133	383.499	113.278	-350.723	-146.503	4.502
1800	82.744	454.861	387.333	121.550	-350.054	-134.510	3.903
1900	82.784	459.336	391.006	129.826	-349.397	-122.553	3.369
2000	82.819	463.583	394.530	138.107	-348.754	-110.631	2.889
2100	82.849	467.625	397.915	146.390	-348.127	-98.740	2.456
2200	82.875	471.479	401.172	154.676	-347.517	-86.879	2.063
2300	82.898	475.164	404.310	162.965	-346.927	-75.045	1.704
2400	82.918	478.692	407.336	171.256	-346.359	-63.236	1.376
2500	82.936	482.078	410.258	179.549	-345.814	-51.451	1.075
2600	82.951	485.331	413.084	187.843	-345.295	-39.687	0.797
2700	82.965	488.462	415.818	196.139	-344.803	-27.942	0.541
2800	82.978	491.479	418.466	204.436	-344.340	-16.215	0.302
2900	82.989	494.391	421.035	212.734	-343.907	-4.504	0.081
3000	82.999	497.205	423.527	221.034	-343.503	7.193	-0.125
3100	83.008	499.927	425.948	229.334	-343.128	18.877	-0.318
3200	83.017	502.562	428.301	237.635	-342.782	30.548	-0.499
3300	83.024	505.117	430.590	245.938	-342.464	42.210	-0.668
3400	83.031	507.595	432.819	254.240	-342.171	53.862	-0.827
3500	83.038	510.002	434.990	262.544	-341.902	65.506	-0.978
3600	83.043	512.342	437.106	270.848	-341.653	77.143	-1.119
3700	83.049	514.617	439.170	279.152	-341.422	88.773	-1.253
3800	83.054	516.832	441.185	287.458	-341.204	100.397	-1.380
3900	83.058	518.989	443.153	295.763	-340.998	112.015	-1.500
4000	83.063	521.092	445.075	304.069	-340.798	123.629	-1.614
4100	83.066	523.143	446.954	312.376	-340.601	135.237	-1.723
4200	83.070	525.145	448.792	320.682	-340.403	146.840	-1.826
4300	83.073	527.100	450.591	328.990	-340.200	158.439	-1.925
4400	83.077	529.010	452.351	337.297	-339.988	170.033	-2.019
4500	83.080	530.877	454.076	345.605	-339.764	181.622	-2.108
4600	83.082	532.703	455.765	353.913	-339.524	193.206	-2.194
4700	83.085	534.489	457.421	362.221	-339.265	204.784	-2.276
4800	83.087	536.239	459.045	370.530	-338.983	216.356	-2.354
4900	83.090	537.952	460.638	378.839	-338.676	227.923	-2.430
5000	83.092	539.631	462.201	387.148	-338.341	239.482	-2.502
5100	83.094	541.276	463.735	395.457	-337.975	251.036	-2.571
5200	83.096	542.890	465.242	403.767	-337.575	262.581	-2.638
5300	83.098	544.472	466.722	412.076	-337.141	274.118	-2.702
5400	83.099	546.026	468.176	420.386	-336.671	285.647	-2.763
5500	83.101	547.551	469.606	428.696	-336.161	297.167	-2.822
5600	83.102	549.048	471.011	437.007	-335.613	308.677	-2.879
5700	83.104	550.519	472.393	445.317	-335.023	320.177	-2.934
5800	83.105	551.964	473.753	453.627	-334.392	331.667	-2.987
5900	83.107	553.385	475.090	461.938	-333.718	343.145	-3.038
6000	83.108	554.782	476.407	470.249	-333.001	354.610	-3.087

PREVIOUS: June 1970 (1 atm) CURRENT: June 1970 (1 bar)

Phosphorus Chloride (PCl_3) $Cl_3P_1(g)$

THIOPHOSPHORYL CHLORIDE ($SPCl_3$) IDEAL GAS M_r = 169.39276

$S°(298.15\ K) = 337.347\ J\ K^{-1}\ mol^{-1}$

$\Delta_f H°(0\ K) = [-376.2]\ kJ\ mol^{-1}$
$\Delta_f H°(298.15\ K) = [-380.0]\ kJ\ mol^{-1}$

Vibrational Frequencies and Degeneracies

ν, cm^{-1}	ν, cm^{-1}	ν, cm^{-1}
169(2)	[247](2)	247(1)
431(1)	540 (2)	751(1)

Ground State Quantum Weight: 1 $\sigma = 3$
Point Group: C_{3v}
Bond Distances: P-S = 1.85 ± 0.02 Å P-Cl = 2.02 ± 0.01 Å
Bond Angle: Cl-P-Cl = 100.5° ± 1°
Product of the Moments of Inertia: $I_A I_B I_C = 2.06885 \times 10^{-112}\ g^3\ cm^6$

Enthalpy of Formation

$\Delta_f H°(298.15\ K)$ was estimated by Henderson and Scheffee (1). Henderson and Scheffee use white (α) phosphorus as a reference state.

Heat Capacity and Entropy

Ziomek and Piotrowski (2) have chosen the most probable values for the wave numbers from the infrared spectral data of Cilento et al. (3) and the Raman spectral data of Simon and Schulze (4), Thatte (5), Gerding and Westrick (6) and Delwaulle and Francois (7). Ziomek and Piotrowski (2) found their normal coordinate treatments gave the listed wave numbers as fundamentals and lent support for the 247 cm^{-1} band missing from the Raman spectral data. The structure (bond distances and angles) was determined from microwave data by Williams et al. (8). The principal moments of inertia are: $I_A = 56.7997 \times 10^{-39}$, $I_B = I_C = 60.3519 \times 10^{-39}\ g\ cm^2$.

1. C. B. Henderson and R. S. Scheffee, "Survey of Thermochemical Data," Atlantic Research Corporation, Alexandria, Virginia, January 1960.
2. J. S. Ziomek and E. A. Piotrowski, J. Chem. Phys. 34, 1087 (1961).
3. G. Cilento, D. A. Ramsey, and R. N. Jones, J. Amer. Chem. Soc. 71, 2753 (1949).
4. A. Simon and G. Schulze, Naturwissenschaften 25, 669 (1937).
5. V. N. Thatte, Nature 138, 468 (1936).
6. H. Gerding and R. Westrick, Rec. Trav. Chim. 61, 842 (1942).
7. M. Delwaulle and F. Francois, Compt. Rend. 220, 817 (1945).
8. Q. Williams, J. Sheridan, and W. Gordy, J. Amer. Chem. Phys. 20, 164 (1952).

Thiophosphoryl Chloride ($SPCl_3$) $Cl_3P_1S_1(g)$

Enthalpy Reference Temperature = T_r = 298.15 K — Standard State Pressure = p° = 0.1 MPa

T/K	$C_p°$ (J K⁻¹mol⁻¹)	S°	$-[G°-H°(T_r)]/T$	$H°-H°(T_r)$ (kJ mol⁻¹)	$\Delta_f H°$	$\Delta_f G°$	Log K_f
0	0.	0.	INFINITE	-19.001	-376.202	-376.202	INFINITE
100	54.294	258.141	408.225	-15.008	-378.154	-372.596	194.624
200	77.707	303.800	345.230	-8.286	-379.893	-366.512	95.723
250	84.855	321.951	338.800	-4.212	-380.359	-363.110	75.868
298.15	89.847	337.347	337.347	0.	-380.744	-359.753	63.027
300	90.009	337.903	337.348	0.166	-380.758	-359.622	62.616
350	93.755	352.074	338.459	4.765	-381.839	-356.000	53.130
400	96.516	364.783	340.969	9.526	-384.453	-352.193	45.992
450	98.589	376.276	344.264	14.406	-385.479	-348.107	40.407
500	100.173	386.749	347.996	19.376	-386.520	-343.896	35.927
600	102.377	405.223	356.036	29.512	-388.049	-335.219	29.183
700	103.791	421.118	364.225	39.826	-389.223	-326.318	24.350
800	104.745	435.044	372.225	50.255	-390.187	-317.263	20.715
900	105.418	447.422	379.905	60.765	-444.146	-307.016	17.819
1000	105.908	458.556	387.223	71.333	-443.673	-291.805	15.242
1100	106.275	468.668	394.174	81.943	-443.192	-276.641	13.137
1200	106.558	477.928	400.773	92.585	-506.284	-260.443	11.337
1300	106.779	486.466	407.041	103.253	-505.035	-240.007	9.644
1400	106.956	494.386	413.000	113.940	-503.802	-219.667	8.196
1500	107.100	501.770	418.675	124.643	-502.585	-199 414	6.944
1600	107.218	508.686	424.087	135.359	-501.387	-179.242	5.852
1700	107.316	515.189	429.257	146.086	-500.210	-159.144	4.890
1800	107.398	521.326	434.203	156.821	-499.053	-139.115	4.037
1900	107.468	527.134	438.942	167.565	-497.919	-119.149	3.276
2000	107.528	532.648	443.491	178.315	-496.807	-99.243	2.592
2100	107.580	537.896	447.862	189.070	-495.721	-79.391	1.975
2200	107.625	542.902	452.069	199.830	-494.659	-59.591	1.415
2300	107.664	547.687	456.123	210.595	-493.626	-39.838	0.905
2400	107.698	552.269	460.035	221.363	-492.621	-20.130	0.438
2500	107.728	556.666	463.813	232.134	-491.646	-0.463	0.010
2600	107.755	560.892	467.466	242.909	-490.703	19.165	-0.385
2700	107.779	564.959	471.002	253.685	-489.792	38.759	-0.750
2800	107.801	568.879	474.428	264.464	-488.915	58.318	-1.088
2900	107.820	572.663	477.750	275.245	-488.072	77.847	-1.402
3000	107.838	576.318	480.975	286.028	-487.264	97.348	-1.695
3100	107.853	579.854	484.108	296.813	-486.488	116.822	-1.968
3200	107.868	583.279	487.154	307.599	-485.745	136.271	-2.224
3300	107.881	586.598	490.118	318.386	-485.034	155.699	-2.465
3400	107.893	589.819	493.003	329.175	-484.351	175.104	-2.690
3500	107.904	592.947	495.814	339.965	-483.696	194.491	-2.903
3600	107.914	595.987	498.555	350.756	-483.065	213.858	-3.103
3700	107.923	598.944	501.228	361.548	-482.455	233.209	-3.292
3800	107.931	601.822	503.837	372.340	-481.863	252.543	-3.471
3900	107.939	604.625	506.386	383.134	-481.285	271.862	-3.641
4000	107.947	607.358	508.876	393.928	-480.719	291.166	-3.802
4100	107.953	610.024	511.311	404.723	-480.159	310.457	-3.955
4200	107.960	612.625	513.692	415.519	-479.604	329.733	-4.101
4300	107.966	615.166	516.023	426.315	-479.048	348.997	-4.239
4400	107.971	617.648	518.304	437.112	-478.489	368.247	-4.372
4500	107.976	620.074	520.539	447.909	-477.922	387.484	-4.498
4600	107.981	622.448	522.729	458.707	-477.345	406.710	-4.618
4700	107.985	624.770	524.875	469.506	-476.755	425.922	-4.734
4800	107.990	627.043	526.980	480.304	-476.148	445.121	-4.844
4900	107.993	629.270	529.045	491.103	-475.522	464.308	-4.950
5000	107.997	631.452	531.071	501.903	-474.873	483.481	-5.051
5100	108.001	633.591	533.061	512.703	-474.201	502.642	-5.148
5200	108.004	635.688	535.014	523.503	-473.503	521.789	-5.241
5300	108.007	637.745	536.933	534.304	-472.776	540.922	-5.331
5400	108.010	639.764	538.819	545.104	-472.020	560.042	-5.417
5500	108.013	641.746	540.672	555.906	-471.232	579.147	-5.500
5600	108.015	643.692	542.495	566.707	-470.412	598.237	-5.580
5700	108.018	645.604	544.287	577.509	-469.558	617.313	-5.657
5800	108.020	647.483	546.050	588.311	-468.670	636.373	-5.731
5900	108.023	649.329	547.785	599.113	-467.747	655.418	-5.803
6000	108.025	651.145	549.492	609.915	-466.788	674.446	-5.872

PREVIOUS: September 1962 (1 atm) CURRENT: September 1962 (1 bar)

Thiophosphoryl Chloride ($SPCl_3$) $Cl_3P_1S_1(g)$

TRICHLOROSILYL ($SiCl_3$) IDEAL GAS M_r = 134.4445

Trichlorosilyl ($SiCl_3$) $Cl_3Si_1(g)$

$S°(298.15 K) = [-318.189 \pm 4.2] J K^{-1} mol^{-1}$

$\Delta_f H°(0 K) = -389.25 \pm 16.7 kJ mol^{-1}$
$\Delta_f H°(298.15 K) = -390.37 \pm 16.7 kJ mol^{-1}$

Vibrational Frequencies and Degeneracies

ν, cm^{-1}
470 (1)
[254](1)
582 (2)
[176](2)

Ground State Quantum Weight = [2] σ = 3
Point Group: C_{3v}
Bond Distance: Si-Cl = [2.0118] Å
Bond Angle: Cl-Si-Cl = 110.9°
Product of the Moments of Inertia: $I_A I_B I_C = [7.36634 \times 10^{-113}] g^3 cm^6$

Enthalpy of Formation

The adopted value for the heat of formation of $SiCl_3(g)$ is based on the mass spectrometric study of the silicon subchlorides by Farber and Srivastava (1). Our 2nd and 3rd law analysis of ten data points (1593-1792 K) for the equilibrium $SiCl_4(g)$ + $SiCl_2(g)$ = 2 $SiCl_3(g)$ yield $\Delta_r H°$(298.15 K)(2nd law) = 6.4±1.1 kcal mol^{-1} and $\Delta_r H°$(298.15 K)(3rd law) = 12.0±0.5 kcal mol^{-1} with a drift of 3.3±0.6 cal K^{-1} mol^{-1}. Using the 3rd law heat of reaction and auxiliary data (2), we calculate and adopt $\Delta_f H°(SiCl_3$, g, 298.15 K) = -93.3±2 kcal mol^{-1}. This leads to $\Delta_a H°$(0 K) = 285.5±2 kcal mol^{-1} or an average bond energy of 95.2 kcal mol^{-1}; a value which is within 0.5 kcal mol^{-1} of the average bond energy in $SiCl_4(g)$.

Many less definitive studies leading to the heat of formation of $SiCl_3(g)$ are reported in the literature. These studies yield values which are 5-10 kcal mol^{-1} less negative than our adopted value. These works include the relative reaction rates of $SiCl_3$ by Kerr et al. (3), decomposition studies of $SiCl_4$ by adiabatic compression by Vurzel et al. (4), interpretation of electron impact studies on several molecules by Steele et al. (5), examination of infrared emission of a reaction involving $SiCl_3(g)$ by Cadman et al. (6), silane kinetic studies by Walsh and Wells (7), and negative ion studies by Wang et al. (10) and Pabst et al. (11).

Heat Capacity and Entropy

Jacox and Milligan (8) studied the formation of $SiCl_3$ by photolysis of $HSiCl_3$ in matrices of argon, nitrogen and CO. The two stretching fundamentals were assigned from the infrared spectra. Isotopic splittings were found to be consistent with an angle of 72±5° between the threefold axis and each of the Si-Cl bonds. This yields a bond angle of 110.9° which is similar to that of $HSiCl_3$. Non-planarity of $SiCl_3$ is consistent with the electron-spin-resonance studies of Roncin (9).

We adopt the results of Jacox (8) and estimate the bond length and two bending vibrations by analogy with $HSiCl_3$ (2). The ground state quantum weight is taken as two, but excited electronic levels are neglected. Jacox (8) observed unstructured absorptions corresponding to excitation energies of 30000-34500 cm^{-1} and 41000-46000 cm^{-1}. The principal moments of inertia are: $I_A = I_B = 33.7539 \times 10^{-39}$ and $I_C = 64.6554 \times 10^{-39}$ g cm^2.

References

1. M. Farber and R. D. Srivastava, Spaces Sciences, Inc., personal communication, Oct. 4, 1976.
2. JANAF Thermochemical Tables: $SiCl_4$(g), 12-31-70; $SiCl_2$(g), 9-30-77; $SiHCl_3$(g), 12-31-76.
3. J. A. Kerr, B. J. A. Smith A. F. Trotman-Dickenson, and J. C. Young, J. Chem. Soc. A1968, 510 (1968).
4. F. B. Vurzel, L. S. Polak, and V. A. Shchipachev, Kinet. Katal. 7, 1068 (1966).
5. W. C. Steele, L. D. Nichols, and F. G. A. Stone, J. Amer. Chem. Soc. 84, 4441 (1962).
6. P. Cadman, G. M. Tilsley, and A. F. Trotman-Dickenson, J. Chem. Soc. Faraday I 69, 914 (1973).
7. R. Walsh and J. M. Wells, J. Chem. Soc. Faraday I 72, 1212 (1976).
8. M. E. Jacox and D. E. Milligan, J. Chem. Phys. 49, 3130 (1968).
9. J. Roncin, Mol. Cryst. 3, 117 (1967).
10. J. Ling-Fai Wang, J. L. Margrave, and J. L. Franklin, J. Chem. Phys. 61, 1357 (1974).
11. R. E. Pabst, D. Perry, J. L. Margrave, and J. L. Franklin, Proc. of the 23rd Annual Conference on Mass Spectroscopy, p. 153, 1975.

Enthalpy Reference Temperature = T_r = 298.15 K — Standard State Pressure = p° = 0.1 MPa

T/K	$C_p°$	S°	$-[G°-H°(T_r)]/T$	$H°-H°(T_r)$	$\Delta_f H°$	$\Delta_f G°$	Log K_f
	J $K^{-1}mol^{-1}$			kJ mol^{-1}			
0	0.	0.	INFINITE	-15.868	-389.246	-389.246	INFINITE
100	46.997	253.930	374.811	-12.088	-390.100	-386.759	202.022
200	62.205	291.584	324.423	-6.568	-390.315	-383.310	100.110
250	67.220	306.032	319.335	-3.326	-390.351	-381.554	79.721
298.15	70.733	318.189	318.189	0.	-390.367	-379.858	66.549
300	70.847	318.627	318.190	0.131	-390.368	-379.792	66.128
350	73.464	329.755	319.063	3.742	-390.369	-378.030	56.418
400	75.376	339.696	321.032	7.466	-390.359	-376.267	49.135
450	76.798	348.660	323.612	11.272	-390.342	-374.507	43.472
500	77.877	356.810	326.531	15.140	-390.320	-372.748	38.941
600	79.365	371.151	332.805	23.008	-390.276	-369.238	32.145
700	80.311	383.461	339.183	30.995	-390.245	-365.735	27.291
800	80.946	394.230	345.405	39.060	-390.238	-362.234	23.651
900	81.392	403.791	351.371	47.178	-390.263	-358.732	20.820
1000	81.715	412.384	357.050	55.334	-390.322	-355.226	18.555
1100	81.958	420.184	362.440	63.518	-390.419	-351.712	16.701
1200	82.143	427.324	367.554	71.724	-390.555	-348.188	15.156
1300	82.289	433.905	372.408	79.946	-390.732	-344.650	13.848
1400	82.405	440.007	377.021	88.181	-390.953	-341.097	12.727
1500	82.499	445.696	381.412	96.426	-391.218	-337.527	11.754
1600	82.576	451.023	385.598	104.680	-391.530	-333.938	10.902
1700	82.640	456.031	389.595	112.941	-442.066	-329.881	10.136
1800	82.694	460.756	393.419	121.208	-442.246	-323.277	9.381
1900	82.740	465.229	397.082	129.479	-442.434	-316.662	8.706
2000	82.779	469.474	400.596	137.755	-442.634	-310.038	8.097
2100	82.813	473.513	403.973	146.035	-442.847	-303.403	7.547
2200	82.842	477.366	407.222	154.318	-443.075	-296.757	7.046
2300	82.868	481.050	410.352	162.603	-443.320	-290.101	6.588
2400	82.890	484.577	413.372	170.891	-443.585	-283.434	6.169
2500	82.910	487.961	416.288	179.181	-443.872	-276.755	5.782
2600	82.928	491.213	419.108	187.473	-444.182	-270.064	5.426
2700	82.943	494.343	421.837	195.767	-444.518	-263.361	5.095
2800	82.957	497.360	424.481	204.062	-444.881	-256.645	4.788
2900	82.970	500.271	427.044	212.358	-445.272	-249.915	4.501
3000	82.981	503.084	429.532	220.656	-445.691	-243.172	4.234
3100	82.992	505.805	431.949	228.954	-446.137	-236.414	3.984
3200	83.001	508.440	434.298	237.254	-446.611	-229.641	3.749
3300	83.009	510.994	436.584	245.555	-447.111	-222.853	3.527
3400	83.017	513.473	438.809	253.856	-447.635	-216.050	3.319
3500	83.024	515.879	440.977	262.158	-448.181	-209.231	3.123
3600	83.031	518.218	443.090	270.461	-832.898	-191.935	2.785
3700	83.037	520.493	445.152	278.764	-833.065	-174.128	2.458
3800	83.043	522.708	447.164	287.068	-833.247	-156.317	2.149
3900	83.048	524.865	449.128	295.373	-833.439	-138.500	1.855
4000	83.052	526.968	451.048	303.678	-833.638	-120.678	1.576
4100	83.057	529.018	452.925	311.983	-833.839	-102.852	1.310
4200	83.061	531.020	454.761	320.289	-834.038	-85.021	1.057
4300	83.065	532.974	456.557	328.595	-834.232	-67.185	0.816
4400	83.068	534.884	458.315	336.902	-834.415	-49.344	0.586
4500	83.072	536.751	460.038	345.209	-834.585	-31.500	0.366
4600	83.075	538.577	461.725	353.516	-834.737	-13.651	0.155
4700	83.078	540.363	463.380	361.824	-834.869	4.199	-0.047
4800	83.080	542.113	465.002	370.132	-834.976	22.053	-0.240
4900	83.083	543.826	466.593	378.440	-835.056	39.909	-0.425
5000	83.085	545.504	468.154	386.748	-835.107	57.766	-0.603
5100	83.088	547.149	469.687	395.057	-835.125	75.624	-0.775
5200	83.090	548.763	471.193	403.366	-835.108	93.481	-0.939
5300	83.092	550.346	472.671	411.675	-835.056	111.338	-1.097
5400	83.094	551.899	474.124	419.984	-834.965	129.194	-1.250
5500	83.096	553.424	475.552	428.294	-834.834	147.048	-1.397
5600	83.097	554.921	476.956	436.603	-834.664	164.898	-1.538
5700	83.099	556.392	478.337	444.913	-834.451	182.746	-1.675
5800	83.100	557.837	479.695	453.223	-834.197	200.590	-1.807
5900	83.102	559.257	481.031	461.533	-833.900	218.428	-1.934
6000	83.103	560.654	482.347	469.843	-833.560	236.261	-2.057

PREVIOUS: December 1977 (1 atm) CURRENT: December 1977 (1 bar)

Trichlorosilyl ($SiCl_3$) $Cl_3Si_1(g)$

$S°(298.15\ K) = 139.750 \pm 1.3\ J\ K^{-1}\ mol^{-1}$

$T_{trs} = 220.1\ K$

$\Delta_f H°(0\ K) = -724.06 \pm 4.2\ kJ\ mol^{-1}$

$\Delta_f H°(298.15\ K) = -721.74 \pm 4.2\ kJ\ mol^{-1}$

$\Delta_{trs} H° = 0\ kJ\ mol^{-1}$

Enthalpy of Formation

The enthalpy of formation, $\Delta_f H°(298.15\ K)$, of $TiCl_3(cr)$ has been measured by several investigators. The results of these measurements are as follows.

Source	Method	$\Delta_f H°(298.15\ K)$, kcal mol^{-1}
Schaffer et al. (1954) (1)	Calorimetric, $TiCl_4 + Hg = TiCl_3 + 1/2\ Hg_2Cl_2$	-171.9, -172.4
Clifton and MacWood (1956) (2)	Solution calorimetric	-172.4
Johnson et al. (1956) (3)	Calorimetric, $TiCl_4(\ell) + HI(g) = TiCl_3(cr) + HCl(g) + 1/2\ I_2$	-172.5
Krieve et al. (1956) (5)	Solution calorimetric	-171.3
Krieve and Mason (1956) (5)	Equilibrium, $TiCl_3(cr) + HCl(g) = TiCl_4(g) + 1/2\ H_2(g)$	-172.9
Altman et al. (1956) (6)	Equilibrium, $Ti(cr) + 3\ TiCl_4(\ell) = 4\ TiCl_3(cr)$	-169.4, -171.0
Head (1960) (7)	Equilibrium, $2\ TiCl_4(\ell) + H_2(g) = 2\ TiCl_3(cr) + 2\ HCl(g)$ $TiCl_4(\ell) + Hg(\ell) = 1/2\ Hg_2Cl_2(cr) + TiCl_3(cr)$ $TiCl_4(\ell) + Ag(cr) = AgCl(cr) + TiCl_3(cr)$	-172.3

The chosen value of -172.5 ± 1.0 kcal mol^{-1} is that reported by Johnson et al. (3).

Heat Capacity and Entropy

The heat capacity and entropy of $TiCl_3(cr)$ has been measured over the temperature range 54 to 1000 K by King et al. (8). The value of $S°(298.15\ K)$ is calculated from these data based on $S°(51\ K) = 4.29$ cal $K^{-1}\ mol^{-1}$.

Transition Data

A second order transition at 220.1 K was observed by King et al. (8). The heat capacity at this temperature is in excess of 146.8 cal $K^{-1}\ mol^{-1}$. King et al. (8) measured the value of $H°(240\ K) - H°(200\ K)$ as 1.213 kcal mol^{-1}.

Sublimation Data

The enthalpy of sublimation is calculated from the enthalpies of formation of $TiCl_3(cr)$ and $TiCl_3(g)$ at the sublimation temperature. The sublimation temperature is taken as the point at which $\Delta_r G° = 0$ for the process $TiCl_3(cr) = TiCl_3(g)$.

References

1. H. Schaffer, G. Breil and G. Pfeffer, Z. Anorg. Chem. 276, 325 (1954).
2. D. G. Clifton and G. E. MacWood, J. Phys. Chem. 60, 309 (1956).
3. W. H. Johnson, A. A. Gilliland and E. T. Prosen, J. Res. Nat. Bur. Stand. 64A, 515 (1956).
4. W. F. Krieve, S. P. Vango and D. M. Mason, J. Chem. Phys. 25, 59 (1956).
5. W. F. Krieve and D. M. Mason, J. Chem. Phys. 25, 524 (1956).
6. D. Altman, M. Farber and D. M. Mason, J. Chem. Phys. 25, 531 (1956).
7. R. B. Head, Aust. J. Chem. 13, 332 (1960).
8. U. S. Bur. Mines RI 5799, 1961.

Enthalpy Reference Temperature = T_r = 298.15 K — Standard State Pressure = p° = 0.1 MPa

T/K	$C_p°$ (J $K^{-1}mol^{-1}$)	S° (J $K^{-1}mol^{-1}$)	$-[G°-H°(T_r)]/T$ (J $K^{-1}mol^{-1}$)	$H°-H°(T_r)$ (kJ mol^{-1})	$\Delta_f H°$ (kJ mol^{-1})	$\Delta_f G°$ (kJ mol^{-1})	Log K_f
0	0.	0.	INFINITE	-20.920	-724.059	-724.059	INFINITE
100	54.438	44.350	227.777	-18.343	-726.410	-701.668	366.514
200	101.926	95.504	149.101	-10.719	-725.260	-677.126	176.847
298.15	97.161	139.750	139.750	0.	-721.740	-654.469	114.660
300	97.195	140.351	139.752	0.180	-721.701	-654.052	113.881
400	98.959	168.557	143.585	9.989	-719.711	-631.807	82.506
500	100.577	190.816	150.882	19.967	-717.830	-610.050	63.731
600	102.048	209.285	159.120	30.099	-715.971	-588.668	51.248
700	103.248	225.106	167.443	40.364	-714.102	-567.598	42.355
800	104.426	238.970	175.535	50.748	-712.193	-546.799	35.702
900	105.583	251.337	183.283	61.249	-710.263	-526.241	30.542
1000	106.704	262.519	190.656	71.863	-708.375	-505.896	26.425
1100	107.818	272.742	197.661	82.589	-706.589	-485.735	23.066
1200	108.926	282.171	204.315	93.426	-708.889	-465.613	20.268
1300	110.029	290.933	210.645	104.374	-706.581	-445.433	17.898
1400	111.125	299.127	216.676	115.432	-704.254	-425.432	15.873
1500	112.215	306.831	222.432	126.599	-701.924	-405.597	14.124

PREVIOUS: March 1964 CURRENT: June 1968

TITANIUM CHLORIDE ($TiCl_3$) IDEAL GAS M_r = 154.239

$\Delta_f H°(0\ K) = -536.09 \pm 6.3\ kJ\ mol^{-1}$

$S°(298.15\ K) = [316.895 \pm 4.2]\ J\ K^{-1}\ mol^{-1}$ $\Delta_f H°(298.15\ K) = -539.32 \pm 6.3\ kJ\ mol^{-1}$

Electronic Levels and Quantum Weights

ε_i, cm^{-1}	g_i	ε_i, cm^{-1}	g_i
0	[2]	[4000]	[2]
[600]	[2]	[10000]	[2]
[1500]	[2]		

Vibrational Frequencies and Degeneracies

ν, cm^{-1}	ν, cm^{-1}
[530](1)	[590](2)
[275](1)	[260](2)

Point Group: [C_{3v}] σ = 3

Bond Distance: Ti-Cl = [2.3] Å

Bond Angle: Cl-Ti-Cl = [100]°

Products of the Moments of Inertia: $I_A I_B I_C = [1.34297 \times 10^{-112}]\ g^3\ cm^6$

Enthalpy of Formation

The enthalpy of formation of $TiCl_3(g)$ is calculated from the heats of reaction for the processes, (A) $3TiCl_4(g) + Ti(cr) = 4TiCl_3(g)$, and (B) $TiCl_3(cr) = TiCl_3(g)$, combined with auxiliary JANAF heats of formation for $TiCl_4(g)$, $TiCl_3(cr)$ and Ti(cr). Equilibrium measurements of reaction (A) have been reported by Gross and Levi (1). Vapor pressure measurements for reaction (B) have been reported by Sanderson and MacWood (2) and by Farber and Darnell (3). Second and third law analyses of these data yield the following results.

Source	Reaction	Data Points	T/K	$\Delta_r H°$(298.15 K), kcal mol^{-1} 2nd Law	3rd Law	Drift cal K^{-1} mol^{-1}	$\Delta_f H°$(298.15 K) k cal mol^{-1}
Gross and Levi (1)	A	3	1223		31.6±2.0		-128.5
Sanderson and MacWood (2)	B	9*	628-823	42.8±1.1	42.8	-0.1±1.5	-129.3
Farber and Darnell (3)	B	8	636-723	45.3±2.1	45.8	0.8±3.1	-126.2

*One point rejected due to failure of statistical test.

The chosen value of $\Delta_f H°$(298.15 K) is -128.9 kcal mol^{-1}. This value is the average of the first two determinations.

Heat Capacity and Entropy

The interatomic distance is estimated from that of the solid, given as 2.46 Å by Natta et al. (4), and from comparisons of bond distances of crystal and gas phase tetrahalides of titanium. The pyramidal bond angle is estimated assuming $TiCl_3(g)$ similar to the group V trihalides. The principal moments of inertia are: $I_A = I_B = 42.8614 \times 10^{-39}$ and $I_C = 73.1025 \times 10^{-39}\ g\ cm^2$. The vibrational frequencies are estimated from valence force field predictions and comparisons with group V trihalides. The electronic levels are estimated from the levels of Ti^{+3} (5).

References

1. P. Gross and D. L. Levi, Congres International de Chemie Pure et Appliquies, 16e Congres, Paris, 1957; Memories presents a la Section de Chemie Minerale, 627 (1958).
2. B. S. Sanderson and G. E. MacWood, J. Phys. Chem. 60, 314 (1956).
3. M. Farber and A. J. Darnell, J. Phys. Chem. 59, 156 (1955).
4. G. Natta, P. Corradini and G. Allegia, J. Polym. Sci. 51, 399 (1961).
5. C. E. Moore, U. S. Nat. Bur. Stand. Circ. 467, 1949.

Titanium Chloride ($TiCl_3$) $Cl_3Ti_1(g)$

Enthalpy Reference Temperature = T_r = 298.15 K Standard State Pressure = p° = 0.1 MPa

T/K	$C_p°$ (J $K^{-1}mol^{-1}$)	S°	-[G°-H°(T_r)]/T	H°-H°(T_r) (kJ mol^{-1})	$\Delta_f H°$	$\Delta_f G°$	Log K_f
0	0.	0.	INFINITE	-15.375	-536.091	-536.091	INFINITE
100	42.233	254.787	373.049	-11.826	-537.471	-533.773	278.814
200	61.090	290.122	323.216	-6.619	-538.737	-529.527	138.298
250	67.847	304.518	318.065	-3.387	-539.097	-527.179	110.148
298.15	72.567	316.895	316.895	0.	-539.318	-524.862	91.954
300	72.719	317.344	316 896	0.134	-539.324	-524.773	91.371
350	76.177	328.828	317.795	3.862	-539.462	-522.335	77.954
400	78.645	339.171	319.831	7.736	-539.541	-519.883	67.890
450	80.423	348.542	322.509	11.715	-539.584	-517.423	60.061
500	81.717	357.086	325.546	15.770	-539.604	-514.959	53.797
600	83.362	372.145	332.092	24.032	-539.616	-510.029	44.402
700	84.264	385.070	338.759	32.418	-539.626	-505.097	37.691
800	84.774	396.358	345.268	40.872	-539.646	-500.163	32.657
900	85.073	406.362	351.511	49.366	-539.724	-495.224	28.742
1000	85.258	415.336	357.453	57.883	-539.933	-490.270	25.609
1100	85.378	423.468	363.090	66.415	-540.341	-485.285	23.044
1200	85.457	430.900	368.436	74.957	-544.936	-480.135	20.900
1300	85.509	437.743	373.507	83.506	-545.026	-474.732	19.075
1400	85.543	444.081	378.325	92.059	-545.205	-469.318	17.510
1500	85.563	449.984	382.908	100.614	-545.486	-463.889	16.154
1600	85.572	455.506	387.274	109.171	-545.884	-458.437	14.966
1700	85.575	460.694	391.442	117.728	-546.412	-452.956	13.918
1800	85.571	465.585	395.427	126.285	-547.085	-447.439	12.984
1900	85.564	470.212	399.242	134.842	-547.917	-441.882	12.148
2000	85.555	474.600	402.901	143.398	-563.609	-435.822	11.383
2100	85.544	478.774	406.415	151.953	-565.551	-429.385	10.680
2200	85.533	482.753	409.796	160.507	-567.512	-422.855	10.040
2300	85.521	486.555	413.051	169.060	-569 494	-416.236	9.453
2400	85.509	490.195	416.190	177.611	-571.499	-409.529	8.913
2500	85.497	493.685	419.221	186.162	-573.530	-402.739	8.415
2600	85.486	497.038	422.149	194.711	-575.587	-395.867	7.953
2700	85.474	500.264	424.983	203.259	-577.673	-388.915	7.524
2800	85.463	503.373	427.728	211.806	-579.788	-381.885	7.124
2900	85.453	506.371	430.388	220.351	-581.934	-374.779	6.751
3000	85.442	509.268	432.969	228.896	-584.109	-367.599	6.400
3100	85.431	512.070	435.476	237.440	-586.315	-360.346	6.072
3200	85.420	514.782	437.912	245.982	-588.550	-353.021	5.762
3300	85.409	517.410	440.282	254.524	-590.813	-345.626	5.471
3400	85.397	519.960	442.588	263.064	-593.103	-338.162	5.195
3500	85.385	522.435	444.834	271.603	-595.416	-330.630	4.934
3600	85.373	524.840	447.023	280.141	-597.750	-323.031	4.687
3700	85.360	527.179	449.158	288.678	-1009.200	-307.580	4.342
3800	85.347	529.455	451.241	297.213	-1010.333	-288.602	3.967
3900	85.333	531.672	453.275	305.747	-1011.531	-269.594	3.611
4000	85.319	533.832	455.262	314.280	-1012.792	-250.554	3.272
4100	85.304	535.939	457.205	322.811	-1014.108	-231.481	2.949
4200	85.289	537.994	459.104	331.341	-1015.475	-212.376	2.641
4300	85.274	540.001	460.962	339.869	-1016.887	-193.238	2.347
4400	85.257	541.961	462.781	348.395	-1018.339	-174.067	2.066
4500	85.241	543.877	464.561	356.920	-1019.824	-154.863	1.798
4600	85.224	545.750	466.306	365.444	-1021.338	-135.624	1.540
4700	85.206	547.583	468.016	373.965	-1022.875	-116.354	1.293
4800	85.188	549.377	469.692	382.485	-1024.431	-97.049	1.056
4900	85.170	551.133	471.337	391.003	-1025.999	-77.712	0.828
5000	85.152	552.854	472.950	399.519	-1027.576	-58.344	0.610
5100	85.133	554.540	474.533	408.033	-1029.156	-38.943	0.399
5200	85.114	556.193	476.088	416.545	-1030.737	-19.512	0.196
5300	85.095	557.814	477.614	425.056	-1032.313	-0.051	0.
5400	85.075	559.404	479.114	433.564	-1033.880	19.441	-0.188
5500	85.056	560.965	480.588	442.071	-1035 437	38.961	-0.370
5600	85.036	562.497	482.037	450.575	-1036.979	58.510	-0.546
5700	85.016	564.002	483.462	459.078	-1038.503	78.085	-0.716
5800	84.996	565.481	484.864	467.579	-1040.007	97.688	-0.880
5900	84.976	566.933	486.242	476.077	-1041.488	117.316	-1.039
6000	84.956	568.361	487.599	484.574	-1042.944	136.969	-1.192

PREVIOUS: December 1968 (1 atm) CURRENT: December 1968 (1 bar)

Titanium Chloride ($TiCl_3$) $Cl_3Ti_1(g)$

ZIRCONIUM CHLORIDE ($ZrCl_3$) CRYSTAL M_r = 197.579

$S°(298.15\ K) = [145.812 \pm 12.6]\ J\ K^{-1}\ mol^{-1}$

$\Delta_f H°(0\ K)$ = Unknown

$\Delta_f H°(298.15\ K) = [-714.21 \pm 62.8]\ kJ\ mol^{-1}$

Enthalpy of Formation

Turnbull and Watts (1) determined the equilibrium pressures for the disproportionation of $ZrCl_3$(cr) in the temperature range from 613 K to 723 K by a modified dew point technique. 2nd and 3rd law analyses of their vapor pressure-temperature equation give the same enthalpy of reaction $\Delta_r H°(298.15\ K) = 30.7\ kcal\ mol^{-1}$ for $2ZrCl_3(cr) \rightarrow ZrCl_2(cr) + ZrCl_4(g)$. The adopted value, $\Delta_f H°(ZrCl_3$, cr, 298.15 K) = $-170.7\ kcal\ mol^{-1}$, is calculated from the enthalpy of reaction at 298.15 K, and $\Delta_f H°(ZrCl_4$, g, 298.15 K) = $-207.77\ kcal\ mol^{-1}$ (2) and $\Delta_f H°(ZrCl_2$, cr, 298.15 K) = $-103\ kcal\ mol^{-1}$ (3). The assigned uncertainty, ±15 kcal mol^{-1} is mainly due to the uncertainty in the enthalpy of formation of $ZrCl_2$(cr), and also the dew point technique used.

Heat Capacity and Entropy

The heat capacities of $ZrCl_3$(cr) are estimated from the $C°_p$ values of $ZrCl_4$(cr) by deduction of one Cl atom $C°_p$ contributions which are calculated as $1/4[C°_p(ZrCl_4, cr) - C°_p(Zr, cr)]$. The S°(298.15 K) is calculated as 34.85 cal $K^{-1}\ mol^{-1}$ based on the assumption that $S°(ZrCl_3$, cr, 298.15 K) = $S°(ZrCl_4$, cr, 298.15 K) − $1/4[S°(ZrCl_4$, cr, 298.15 K) − S°(Zr, cr, 298.15 K)].

Sublimation Data

The enthalpy of sublimation is calculated from the difference in the enthalpies of formation of gas and crystal at 298.15 K. The sublimation temperature is taken as the point at which $\Delta_r G = 0$ for $ZrCl_3(cr) = ZrCl_3(g)$.

References

1. A. G. Turnbull and J. A. Watts, Aust. J. Chem. 16, 947 (1963).
2. JANAF Thermochemical Table: $ZrCl_4$(g), 12-31-69.
3. JANAF Thermochemical Table: $ZrCl_2$(cr), 12-31-69.

Zirconium Chloride ($ZrCl_3$) Cl_3Zr_1(cr)

Enthalpy Reference Temperature = T_r = 298.15 K Standard State Pressure = p° = 0.1 MPa

T/K	$C°_p$ (J K⁻¹mol⁻¹)	S° (J K⁻¹mol⁻¹)	−[G°−H°(T_r)]/T (J K⁻¹mol⁻¹)	H°−H°(T_r) (kJ mol⁻¹)	$\Delta_f H°$ (kJ mol⁻¹)	$\Delta_f G°$ (kJ mol⁻¹)	Log K_f
0							
100							
200							
298.15	96.190	145.812	145.812	0.	-714.209	-646.327	113.234
300	96.311	146.408	145.814	0.178	-714.172	-645.906	112.462
400	100.705	174.755	149.649	10.042	-712.072	-623.463	81.416
500	103.470	197.536	157.023	20.257	-709.839	-601.567	62.845
600	105.583	216.593	165.406	30.712	-707.523	-580.129	50 505
700	107.378	233.005	173.918	41.361	-705.149	-559.083	41.719
800	109.052	247.454	182.224	52.184	-702.725	-538.382	35.153
900	110.596	260.388	190.203	63.166	-700.270	-517.986	30.063
1000	112.081	272.121	197.817	74.304	-697.794	-497.865	26.006
1100	113.177	282.855	205.067	85.567	-695.328	-477.992	22.698
1200	114.223	292.748	211.967	96.937	-696.602	-458.120	19.941
1300	115.269	301.932	218.538	108.412	-693.654	-438.366	17.614
1400	116.315	310.514	224.805	119.992	-690.657	-418.840	15.627
1500	117.152	318.567	230.790	131.666	-687.633	-399.530	13.913
1600	117.989	326.155	236.515	143.424	-684.602	-380.422	12.420
1700	118.700	333.329	242.001	155.258	-681.581	-361.504	11.108
1800	119.453	340.137	247.266	167.168	-678.583	-342.762	9.947
1900	119.872	346.607	252.325	179.135	-675.638	-324.186	8.912
2000	120.290	352.766	257.195	191.143	-672.773	-305.763	7.986

PREVIOUS: June 1964 CURRENT: December 1969

Zirconium Chloride ($ZrCl_3$) Cl_3Zr_1(cr)

ZIRCONIUM CHLORIDE ($ZrCl_3$) IDEAL GAS M_r = 197.579

$S°(298.15\ K) = [339.340 \pm 12.6]\ J\ K^{-1}\ mol^{-1}$

$\Delta_f H°(0\ K) = -522.60 \pm 20.9\ kJ\ mol^{-1}$

$\Delta_f H°(298.15\ K) = -524.26 \pm 20.9\ kJ\ mol^{-1}$

Electronic Levels and Quantum Weights

ε_i, cm^{-1}	g_i
0	[2]
[1800]	[2]
[4500]	[2]
[12000]	[2]
[30000]	[2]

Vibrational Frequencies and Degeneracies

ν, cm^{-1}	ν, cm^{-1}
[490](1)	[333](2)
[185](1)	[131](2)

Point Group: $[C_{3v}]$ σ = [3]

Bond Distance: Zr-Cl = [2.3] Å

Bond Angle: Cl-Zr-Cl = [100]°

Product of the Moments of Inertia: $I_A I_B I_C = [1.54256 \times 10^{-112}]\ g^3\ cm^6$

Enthalpy of Formation

Potter (1) investigated mass spectrometrically the equilibrium $CaCl(g) + ZrCl_4(g) \rightarrow CaCl_2(g) + ZrCl_3(g)$ in the temperature range 1163-1157 K. Ion intensities were measured 2.5 eV above threshold and they were used in the calculation of equilibrium constants. Using the reported equilibrium constants, the enthalpies of reaction at 298 K are evaluated by the 3rd law and 2nd law methods as -3.69 and -16.67 kcal mol^{-1}, respectively. The 3rd law drift is 9.31±1.25 cal K^{-1} mol^{-1}. Using the 3rd law $\Delta_r H°(298.15\ K) = -3.69$ kcal mol^{-1}, $\Delta_f H°(ZrCl_4, g, 298.15\ K) = -207.77$ kcal mol^{-1} (2), $\Delta_f H°(CaCl, g, 298.15\ K) = -28.8$ kcal mol^{-1} (3) and $\Delta_f H°(CaCl_2, g, 298.15\ K) = -115.0$ kcal mol^{-1} (4), we obtain $\Delta_f H°(ZrCl_3, g, 298.15\ K) = -125.3$ kcal mol^{-1}, which is adopted in the tabulation.

Farber et al. (5) also studied mass spectrometrically the reaction $Zr(cr) + 3\ ZrCl_4(g) \rightarrow 4\ ZrCl_3(g)$ in the temperature range from 1579 K to 1729 K. They reported a second law heat of reaction as $\Delta_r H°(298.15\ K) = 59.2 \pm 1.6$ kcal mol^{-1}. Since the attainment of equilibrium was questionable in their studies, the heat of reaction obtained from the third law method is generally more reliable. Using their reported ion intensities of run 3 (temperature range from 1667 K to 1896 K, five points), we have calculated the equilibrium constants $K_1 = (I_{ZrCl_2})^2/(I_{ZrCl_3})(I_{ZrCl})$ for reaction (1) $ZrCl_3(g) + ZrCl(g) \rightarrow 2\ ZrCl_2(g)$; and $K_2 = (I_{ZrCl_3})^2/(I_{ZrCl_4})(I_{ZrCl_2})$ for reaction (2) $ZrCl_4(g) + ZrCl_2 \rightarrow 2\ ZrCl_3(g)$. Third law analyses of the equilibrium constants give $\Delta_r H°(298.15\ K) = -11.0$ kcal mol^{-1} for reaction (1) and its drift 4.1±1.7 cal K^{-1} mol^{-1}; and $\Delta_r H°(298.15\ K) = 1.4$ kcal mol^{-1} for reaction (2) and its drift 11.1±1.2 cal K^{-1} mol^{-1}. Using the 3rd law heats of reaction, and $\Delta_f H°(ZrCl, g, 298.15\ K) = 49.1$ kcal mol^{-1} (2) and $\Delta_f H°(ZrCl_4, g, 298.15\ K) = -207.77$ kcal mol^{-1} (2), we obtain $\Delta_f H°(ZrCl_3, g, 298.15\ K) = -124.9$ kcal mol^{-1} and $\Delta_f H°(ZrCl_2, g, 298.15\ K) = -43.4$ kcal mol^{-1} which are in good agreement with the values adopted.

Heat Capacity and Entropy

The molecular structure is assumed to be a pyramid similar to the group VA trichlorides. The bond distance is assumed to be the same as that of $ZrCl_4(g)$. The principal moments of inertia are: $I_A = I_B = 45.9362 \times 10^{-39}$ and $I_C = 73.1025 \times 10^{-39}$ g cm^2. The electronic levels are estimated as three times higher than those of $TiCl_3(g)$, based on the ratio of the first splitting interval between Ti(IV) and Zr(IV), given by Moore (6). The quantum weights are assumed to be the same as those of $TiCl_3(g)$.

Vibrational frequencies are calculated from the estimated force constants by the valence force method (7). The force constants are estimated from those of PCl_3, $AsCl_3$ and $SbCl_3$ listed by Herzberg (7).

References

1. N. D. Potter, Aeronutronic Div., Philco-Ford Corp., Newport Beach, Calif., private communication, dated Dec. 4, 1969.
2. JANAF Thermochemical Table: ZrCl(g) and $ZrCl_4(g)$, 12-31-69.
3. The enthalpy of formation and the free energy functions of CaCl(g) are preliminary values calculated by the Physicochemical Laboratory, The Dow Chemical Company, Midland, Mich.
4. The enthalpy of formation and the free energy functions of $CaCl_2(g)$ are preliminary values calculated by the Physicochemical Studies, The Dow Chemical Company, Midland, Mich.
5. M. Farber, M. A. Frisch, G. Grenier and H. C. Ko, Space Sciences, Inc., Monrovia, Calif., Rep. No. AFRPL-TR-67-244, Contract F04611-67-C-0010, Nov., 1967.
6. C. E. Moore, U.S. Nat. Bur. Stand. Circ. 467, 1949.
7. G. Herzberg, "Infrared and Raman Spectra," D. Van Nostrand Company, Inc., New York, 1945.

Zirconium Chloride ($ZrCl_3$) $Cl_3Zr_1(g)$

Enthalpy Reference Temperature = T_r = 298.15 K Standard State Pressure = p° = 0.1 MPa

T/K	$C_p°$ (J $K^{-1}mol^{-1}$)	S° (J $K^{-1}mol^{-1}$)	$-[G°-H°(T_r)]/T$ (J $K^{-1}mol^{-1}$)	$H°-H°(T_r)$ (kJ mol^{-1})	$\Delta_f H°$ (kJ mol^{-1})	$\Delta_f G°$ (kJ mol^{-1})	Log K_f
0	0.	0.	INFINITE	-17.615	-522.602	-522.602	INFINITE
100	54.064	267.133	402.036	-13.490	-523.731	-520.688	271.979
200	69.657	310.172	346.126	-7.191	-524.181	-517.416	135.135
250	73.530	326.163	340.579	-3.604	-524.239	-515.717	107.753
298.15	76.025	339.340	339.340	0.	-524.255	-514.074	90.064
300	76.104	339.811	339.342	0.141	-524.255	-514.011	89.497
350	77.945	351.688	340.276	3.994	-524.251	-512.304	76.457
400	79.367	362.192	342.371	7.928	-524.232	-510.598	66.677
450	80.536	371.610	345.106	11.927	-524.203	-508.895	59.071
500	81.533	380.148	348.190	15.979	-524.163	-507.197	52.986
600	83.156	395.163	354.801	24.217	-524.064	-503.812	43.861
700	84.387	408.079	361.511	32.597	-523.959	-500.445	37.344
800	85.308	419.410	368.055	41.084	-523.871	-497.093	32.457
900	85.986	429.499	374.332	49.651	-523.833	-493.748	28.656
1000	86.482	438.586	380.310	58.275	-523.869	-490.404	25.616
1100	86.842	446.846	385.989	66.943	-523.998	-487.053	23.128
1200	87.101	454.414	391.380	75.641	-527.945	-483.463	21.045
1300	87.281	461.394	396.501	84.360	-527.752	-479.764	19.277
1400	87.402	467.867	401.370	93.095	-527.601	-476.079	17.763
1500	87.475	473.899	406.007	101.839	-527.507	-472.402	16.451
1600	87.511	479.546	410.428	110.589	-527.483	-468.730	15.302
1700	87.518	484.852	414.652	119.340	-527.546	-465.056	14.289
1800	87.502	489.854	418.692	128.091	-527 706	-461.376	13.389
1900	87.470	494.584	422.563	136.840	-527.979	-457.684	12.583
2000	87.426	499.070	426.277	145.585	-528.377	-453.975	11.857
2100	87.372	503.334	429.846	154.325	-528.913	-450.242	11.199
2200	87.313	507.397	433.279	163.059	-550.923	-445.735	10.583
2300	87.251	511.277	436.587	171.787	-552.190	-440.925	10.014
2400	87.187	514.989	439.777	180.509	-553.486	-436.060	9.491
2500	87.123	518.547	442.857	189.225	-554.811	-431.140	9.008
2600	87.060	521.963	445.834	197.934	-556.169	-426.166	8.562
2700	86.999	525.247	448.715	206.637	-557.560	-421.140	8.147
2800	86.940	528.410	451.505	215.334	-558.986	-416.061	7.762
2900	86.883	531.460	454.210	224.025	-560.446	-410.931	7.402
3000	86.829	534.404	456.834	232.711	-561.941	-405 750	7.065
3100	86.778	537.251	459.383	241.391	-563.470	-400.519	6.749
3200	86.729	540.005	461.859	250.066	-565.033	-395.238	6.452
3300	86.683	542.673	464.268	258.737	-566.627	-389.907	6.172
3400	86.639	545.260	466.612	267.403	-568.251	-384.527	5.908
3500	86.597	547.771	468.895	276.065	-569.901	-379.100	5.658
3600	86.558	550.210	471.120	284.723	-571.576	-373.624	5.421
3700	86.520	552.581	473.290	293.376	-573.271	-368.103	5.197
3800	86.484	554.888	475.407	302.027	-574.983	-362.535	4.983
3900	86.450	557.134	477.474	310.673	-576.709	-356.922	4.780
4000	86.418	559.322	479.493	319.317	-578.443	-351.264	4.587
4100	86.386	561.456	481.466	327.957	-580.183	-345.563	4.403
4200	86.356	563.537	483.396	336.594	-581.924	-339.819	4.226
4300	86.327	565.569	485.283	345.228	-583.663	-334.034	4.058
4400	86.300	567.553	487.130	353.860	-585.394	-328.209	3.896
4500	86.273	569.492	488.939	362.488	-587.115	-322.345	3.742
4600	86.247	571.388	490.711	371.114	-588.822	-316.442	3.593
4700	86.222	573.243	492.447	379.738	-590.511	-310.502	3.451
4800	86.197	575.058	494.150	388.359	-1153.115	-292.910	3.188
4900	86.174	576.835	495.819	396.977	-1154.411	-274.975	2.931
5000	86.151	578.576	497.457	405.593	-1155.697	-257.015	2 685
5100	86.129	580.281	499.064	414.207	-1156.969	-239.028	2.448
5200	86.107	581.954	500.642	422.819	-1158.224	-221.017	2.220
5300	86.087	583.594	502.192	431.429	-1159.458	-202.982	2.001
5400	86.066	585.203	503.714	440.037	-1160.667	-184.924	1.789
5500	86.047	586.782	505.210	448.642	-1161.850	-166.844	1.585
5600	86.028	588.332	506.681	457.246	-1163.003	-148.742	1.387
5700	86.009	589.854	508.127	465.848	-1164.123	-130.621	1.197
5800	85.991	591.350	509.549	474.448	-1165.210	-112.479	1.013
5900	85.974	592.820	510.948	483.046	-1166.260	-94.320	0.835
6000	85.957	594.265	512.324	491.643	-1167.271	-76.143	0.663

PREVIOUS: December 1969 (1 atm) CURRENT: December 1969 (1 bar)

Zirconium Chloride ($ZrCl_3$) $Cl_3Zr_1(g)$

COBALT CHLORIDE ($(CoCl_2)_2$) IDEAL GAS M_r = 259.6784

$S°(298.15\ K) = [450.400 \pm 16.7]\ J\ K^{-1}\ mol^{-1}$

$\Delta_fH°(0\ K) = -352.91 \pm 29.3\ kJ\ mol^{-1}$
$\Delta_fH°(298.15\ K) = -350.62 \pm 29.3\ kJ\ mol^{-1}$

Electronic Levels and Quantum Weights

ϵ_i, cm^{-1}	g_i
0	[4]
[210]	[8]
[2804]	[8]
[10425]	[4]
[14197]	[8]
[21270]	[8]

Vibrational Frequencies and Degeneracies

ν, cm^{-1}	ν, cm^{-1}	ν, cm^{-1}
[50](1)	[125](1)	323 (1)
[75](1)	[150](1)	[143] (1)
[100](1)	[175](1)	289 (1)
[110](1)	[200](1)	432.5(1)

Point Group: $[D_{2h}]$ σ = 4
Bond Distances: Co-Cl = [2.09] Å Co-Cl'(bridge) = [2.2] Å
Bond Angles: Co-Cl'(bridge)-Co = [90]° Cl'(bridge)-Co-Cl = [135]° Cl'(bridge)-Co-Cl'(bridge) = [90]°
Product of Moments of Inertia: $I_AI_BI_C = [1.34964 \times 10^{-111}]\ g^3\ cm^6$

Enthalpy of Formation

Mass spectrometric studies (1, 2) have indicated the presence of small amounts of dimer in the saturated vapor over $CoCl_2$(cr). The existence of the dimeric species was inferred from the ion current intensities observed for $Co_2Cl_4^+$. This ion is believed to be produced by simple ionization of the dimer molecules. Schoonmaker et al. (1) have combined their mass spectral data with results of effusion vapor pressure measurements to give partial pressures for the monomer and dimer at 817 K. These results are analyzed by the third law method with JANAF free energy functions to give ΔH(dimerization, 298.15 K) = -38.99 kcal mol^{-1} for the process 2 $CoCl_2$(g) = Co_2Cl_4(g). This result is combined with $\Delta_fH°(CoCl_2$, g, 298.15 K) = -22.4±2.0 kcal mol^{-1} (3) to give $\Delta_fH°(Co_2Cl_4$, g, 298.15 K) = -83.8 kcal mol^{-1} (350.619±29.3 kJ mol^{-1}) which we adopt. The uncertainty assigned to $\Delta_fH°$ is estimated to be ±7.0 kcal mol^{-1} and arises mainly from uncertainties in the values of S°(298.15 K) for dimer and $\Delta_fH°$ for the monomer.

Heat Capacity and Entropy

A bridge-bond structure of D_{2h} symmetry is assigned to Co_2Cl_4. A similar structure has been proposed for Be_2Cl_4(3) and other dimeric transition-metal dihalides (3-5). The Co_2Cl_2 ring system is assumed to be square and planar. The Cl-Co-Cl'(bridge) bond angle is estimated as 135°. The outer Co-Cl bond lengths are estimated from that for $CoCl_2$ (3), while the ring Co-Cl'(bridge) bond lengths are estimated to be somewhat larger. The principal moments of inertia are: $I_A = 28.4941 \times 10^{-39}$, $I_B = 203.8550 \times 10^{-39}$, and $I_C = 232.3490 \times 10^{-39}$ g cm^2.

The three observed vibrational frequencies (ν_9, ν_{11}, and ν_{12}) are taken from the matrix-infrared study of Thompson and Carlson (5). The rest of the frequencies except for ν_{10} are estimated by comparison with similar data for other dimeric molecules (3). The value for ν_{10} (143 cm^{-1}) was obtained by Thompson and Carlson (5) from a simplified normal coordinate analysis. Also, Leroi et al. (4) obseved a band at 422±10 cm^{-1} in the infrared spectrum of gaseous $CoCl_2$ near 1000 K. The band was assigned to an outer-bond stretching frequency of the dimer. This interpretation agrees with our adopted results. The electronic levels and quantum weights are arbitrarily taken equal to those for the monomer (3).

References - $(CoCl_2)_2$

1. R. C. Schoonmaker, A. H. Friedman and R. F. Porter, J. Chem. Phys. 31, 1586 (1959).
2. S. D. Hill, C. A. Cleland, A. Adams, A. Landsberg and F. E. Block, J. Chem. Eng. Data 14, 84 (1969).
3. JANAF Thermochemical Tables: $CoCl_2$(g), 12-31-73; Be_2Cl_4(g), 6-30-65; Fe_2Cl_4(g), 12-31-70; Mg_2Cl_4(g), 12-31-69.
4. G. E. Leroi, T. C. Jones, J. T. Hougen and W. Klemperer, J. Chem. Phys. 36, 2879 (1962).
5. K. R. Thompson and K. D. Carlson, J. Chem. Phys. 49, 4379 (1968).

Table References - $CoCl_2$(g)

1. JANAF Thermochemical Tables: $CoCl_2$(cr), 6-30-73; CoO(cr) and $FeCl_2$(g), 12-31-70; $BaCl_2$(g), 12-31-72; $AlCl_3$(g), 6-30-70.
2. S. D. Hill, C. A. Cleland, A. Adams, A. Landsberg and F. E. Block, J. Chem. Eng. Data 14, 84 (1969).
3. A. K. Orlov, Russ. J. Phys. Chem. 45, 1352 (1971).
4. C. G. Maier, U. S. Bur. Mines, Tech. Paper 360, 1929.
5. M. P. Kulkarni and V. V. Dadape, High Temp. Sci. 3, 277 (1971).
6. H. Schafer and K. Krehl, Z. Anorg. Chem. 268, 25 (1952).
7. G. E. Leroi, T. C. James, J. T. Hougen, and W. Klemperer, J. Chem. Phys. 36, 2879 (1962).
8. K. R. Thompson and K. D. Carlson, J. Chem. Phys. 49, 4379 (1968).
9. M. E. Jacox and D. E. Milligan, J. Chem. Phys. 51, 4143 (1969).
10. J. T. Hougen, G. E. Leroi and T. C. James, J. Chem. Phys. 34, 1670 (1961).
11. C. W. DeKock and D. M. Gruen, J. Chem. Phys. 44, 4387 (1966).
12. C. W. DeKock and D. M. Gruen, J. Chem. Phys. 49, 4521 (1968).
13. J. Tremmel, A. A. Ivanov, Gy. Schultz and I. Hargittai, Chem. Phys. Letters 23, 533 (1973).
14. D. M. Gruen, J. R. Clifton and C. W. DeKock, J. Chem. Phys. 48, 2394 (1968).
15. J. R. Clifton and D. M. Gruen, Applied Spectroscopy 24, 53 (1970).
16. A. Trutia and M. Musa, Spectrochimica Acta 23, 1165 (1967).
17. M. Musa, Rev. Rommaine Chim. 15, 705 (1970).
18. D. W. Smith, Inorganica Chim. Acta 5, 231 (1971).
19. A. B. P. Lever and B. R. Hollebone, Inorg. Chem. 11, 2183 (1972).

Cobalt Chloride ($(CoCl_2)_2$) Cl_4Co_2(g)

Enthalpy Reference Temperature = T_r = 298.15 K Standard State Pressure = p° = 0.1 MPa

	J K⁻¹mol⁻¹			kJ mol⁻¹			
T/K	$C_p°$	S°	$-[G°-H°(T_r)]/T$	$H°-H°(T_r)$	$\Delta_fH°$	$\Delta_fG°$	Log K_f
0	0.	0.	INFINITE	-30.198	-352.913	-352.913	INFINITE
100	100.830	322.221	559.182	-23.696	-353.278	-346.163	180.817
200	121.951	400.502	461.920	-12.283	-351.761	-339.640	88.705
250	125.421	428.124	452.488	-6.091	-351.117	-336.687	70.347
298.15	127.432	450.400	450.400	0.	-350.619	-333.955	58.507
300	127.493	451.189	450.403	0.236	-350.601	-333.851	58.129
350	128.835	470.949	451.960	6.646	-350.154	-331.096	49.413
400	129.766	488.217	455.436	13.113	-349.800	-328.399	42.885
450	130.455	503.543	459.946	19.619	-349.543	-325.740	37.811
500	131.001	517.317	465.006	26.156	-349.376	-323.106	33.755
600	131.862	541.281	475.780	39.301	-349.283	-317.865	27.673
700	132.569	561.662	486.629	52.523	-349.477	-312.617	23.328
800	133.183	579.405	497.141	65.811	-350.788	-307.199	20.058
900	133.716	595.123	507.171	79.157	-351.571	-301.707	17.511
1000	134.167	609.236	516.684	92.552	-352.794	-296.106	15.467
1100	134.536	622.041	525.689	105.988	-354.531	-290.358	13.788
1200	134.829	633.760	534.213	119.457	-356.857	-284.426	12.381
1300	135.052	644.562	542.291	132.951	-360.073	-278.267	11.181
1400	135.216	654.576	549.958	146.465	-364.438	-271.817	10.142
1500	135.330	663.909	557.248	159.993	-366.787	-265.114	9.232
1600	135.406	672.646	564.190	173.530	-368.637	-258.274	8.432
1700	135.451	680.856	570.814	187.073	-370.301	-251.325	7.722
1800	135.473	688.599	577.144	200.619	-404.514	-243.691	7.072
1900	135.478	695.924	583.205	214.167	-406.721	-234.696	6.452
2000	135.472	702.873	589.016	227.714	-408.948	-225.585	5.892
2100	135.460	709.483	594.596	241.261	-411.199	-216.362	5.382
2200	135.443	715.784	599.963	254.806	-413.476	-207.030	4.916
2300	135.426	721.804	605.130	268.350	-415.781	-197.595	4.488
2400	135.410	727.567	610.113	281.891	-418.117	-188.059	4.093
2500	135.396	733.095	614.922	295.432	-420.486	-178.424	3.728
2600	135.385	738.405	619.570	308.971	-422.891	-168.695	3.389
2700	135.378	743.514	624.067	322.509	-425.333	-158.871	3.074
2800	135.375	748.438	628.421	336.046	-427.814	-148.957	2.779
2900	135.377	753.188	632.642	349.584	-430.333	-138.954	2.503
3000	135.384	757.778	636.737	363.122	-432.891	-128.863	2.244
3100	135.394	762.217	640.714	376.661	-435.487	-118.686	2.000
3200	135.410	766.516	644.578	390.201	-1187.745	-104.879	1.712
3300	135.429	770.683	648.337	403.743	-1187.558	-71.042	1.125
3400	135.451	774.726	651.995	417.287	-1187.407	-37.211	0.572
3500	135.478	778.653	655.558	430.833	-1187.289	-3.383	0.050
3600	135.507	782.470	659.030	444.382	-1187.203	30.441	-0.442
3700	135.539	786.183	662.417	457.935	-1187.148	64.264	-0.907
3800	135.573	789.798	665.722	471.490	-1187.120	98.085	-1.348
3900	135.609	793.320	668.948	485.049	-1187.119	131.906	-1.767
4000	135.647	796.754	672.101	498.612	-1187.140	165.728	-2.164
4100	135.686	800.104	675.182	512.179	-1187.182	199.550	-2.542
4200	135.726	803.374	678.196	525.749	-1187.242	233.374	-2.902
4300	135.768	806.568	681.144	539.324	-1187.316	267.199	-3.246
4400	135.809	809.690	684.030	552.903	-1187.402	301.026	-3.574
4500	135.851	812.742	686.857	566.486	-1187.498	334.854	-3.887
4600	135.893	815.729	689.626	580.073	-1187.600	368.686	-4.187
4700	135.935	818.652	692.340	593.665	-1187.708	402.519	-4.473
4800	135.977	821.514	695.002	607.260	-1187.818	436.355	-4.749
4900	136.018	824.318	697.612	620.860	-1187.930	470.193	-5.012
5000	136.059	827.067	700.174	634.464	-1188.041	504.033	-5.266
5100	136.098	829.761	702.688	648.072	-1188.152	537.876	-5.509
5200	136.137	832.404	705.158	661.683	-1188.260	571.720	-5.743
5300	136.175	834.998	707.583	675.299	-1188.366	605.567	-5.968
5400	136.212	837.544	709.966	688.918	-1188.468	639.416	-6.185
5500	136.247	840.043	712.309	702.541	-1188.568	673.267	-6.394
5600	136.282	842.499	714.612	716.168	-1188.666	707.119	-6.596
5700	136.315	844.911	716.876	729.798	-1188.762	740.973	-6.790
5800	136.346	847.282	719.104	743.431	-1188.858	774.830	-6.978
5900	136.377	849.613	721.297	757.067	-1188.864	808.680	-7.160
6000	136.406	851.906	723.455	770.706	-1188.939	842.537	-7.335

PREVIOUS: December 1973 (1 atm) CURRENT: December 1973 (1 bar)

Cobalt Chloride ($(CoCl_2)_2$) Cl_4Co_2(g)

IRON CHLORIDE $((FeCl_2)_2)$ IDEAL GAS $M_r = 253.506$

$S°(298.15\ K) = [464.504 \pm 12.6]\ J\ K^{-1}\ mol^{-1}$

$\Delta_f H°(0\ K) = -433.84 \pm 4.2\ kJ\ mol^{-1}$
$\Delta_f H°(298.15\ K) = -431.37 \pm 4.2\ kJ\ mol^{-1}$

Electronic Levels and Quantum Weights

ε_i, cm^{-1}	g_i
0	[10]
[4600]	[10]
[7140]	[5]

Vibrational Frequencies and Degeneracies

ν, cm^{-1}	ν, cm^{-1}	ν, cm^{-1}	ν, cm^{-1}
[30](1)	[90](1)	[150](1)	249(1)
[50](1)	110 (1)	[180](1)	325(1)
[80](1)	[125](1)	[200](1)	438(1)

Ground State Quantum Weight: [10] Point Group: $[D_{2h}]$ σ = [4]

Bond Distances: Fe-Cl = [2.17] Å Fe-Cl(bridge) = [2.30] Å

Bond Angles: Fe-Cl(bridge)-Fe = [90]° Cl(bridge)-Fe-Cl(bridge) = [90]°
Cl-Fe-Cl(bridge) = [135]°

Product of the Moments of Inertia: $I_A I_B I_C = [1.70248 \times 10^{-111}]\ g^3\ cm^6$

Enthalpy of Formation

Beusman (1) reported the partial pressures of the monomer and dimer in the temperature range from 980-1167 K using both a static method to determine the total pressures of these two species and a transpiration method to determine the ratio of monomer to dimer. Schoonmaker and Porter (2) determined mass spectrometrically the partial pressures of the monomer and dimer in the temperature range from 621-658 K. Based on the combination of these two sets of partial pressure data, we have derived a smoothed equation log Kp(atm) = -7549/T + 6.506 for $Fe_2Cl_4(g) = 2\ FeCl_2(g)$. 2nd law analysis of the equilibrium constants gives the enthalpy and the entropy of the dimerization at 298 K as $\Delta_d H° = -35.7\ kcal\ mol^{-1}$ and $\Delta_d S° = -32.0\ cal\ K^{-1}\ mol^{-1}$ which lead to the adopted $\Delta_f H°(Fe_2Cl_4,\ g,\ 298.15\ K) = -103.1\ kcal\ mol^{-1}$ ($-431.370 \pm 4.2\ kJ\ mol^{-1}$) and $S°(Fe_2Cl_4,\ g,\ 298.15\ K) = 111.0\ cal\ K^{-1}\ mol^{-1}$, using $\Delta_f H°(FeCl_2,\ g,\ 298.15\ K) = -33.7\ kcal\ mol^{-1}$ and $S°(FeCl_2,\ g,\ 298.15\ K) = 71.50\ cal\ K^{-1}\ mol^{-1}$ (3).

The equilibrium constants derived from the two different experimental techniques are reasonably consistent considering the uncertainties inherent in the problems. Combination of these two sets of data leads to reliable 2nd law heat and entropy since the temperature ranges are different by 400 K.

Heat Capacity and Entropy

A planar structure with D_{2h} symmetry was suggested by Leroi et al. (4), Thompson and Carlson (5). Two iron and two chlorine atoms form a square. The outer Fe-Cl distance is estimated to be the same as in $FeCl_2(g)$ and the square Fe-Cl' distance is assumed to be slightly larger. The Cl-Fe-Cl(bridge) angle is estimated to be 135°. The principal moments of inertia are: $I_A = 31.1433 \times 10^{-39}$, $I_B = 218.7538 \times 10^{-39}\ g\ cm^2$ and $I_C = I_A + I_B$. The electronic levels and quantum weights were arbitrarily estimated to be the same as in $FeCl_2(g)$.

There are only four vibrational frequencies observed in the matrix isolated infrared spectrum, $\nu_{10} = 110\ cm^{-1}$ and $\nu_{12} = 436.2\ cm^{-1}$ (5), $\nu_9 = 325\ cm^{-1}$, $\nu_{11} = 249\ cm^{-1}$ and $\nu_{12} = 438\ cm^{-1}$ (6). The other eight fundamental vibrational frequencies were estimated by comparison with other dimer molecules and also adjusted so that they give $S°(Fe_2Cl_4,\ g,\ 298.15\ K) = 111.0\ cal\ K^{-1}\ mol^{-1}$ which has been discussed in the earlier Enthalpy of Formation part. The order of the frequencies listed above is arbitrary and not related to their species types.

References

1. C. Beusman, Oak Ridge National Lab. ORNL-2323, June 1957.
2. R. C. Schoonmaker and R. F. Porter, J. Chem. Phys. 29, 116 (1958).
3. JANAF Thermochemical Table: $FeCl_2(g)$, 12-31-70.
4. G. E. Leroi, T. C. James, J. T. Hougen and W. Klemperer, J. Chem. Phys. 36, 2879 (1962).
5. K. R. Thompson and K. D. Carlson, J. Chem. Phys. 49, 4379 (1968).
6. R. A. Frey, R. D. Werder and H. H. Gunthard, J. Mol. Spectrosc. 35, 260 (1970).

Iron Chloride $((FeCl_2)_2)$ $Cl_4Fe_2(g)$

Enthalpy Reference Temperature = T_r = 298.15 K Standard State Pressure = p° = 0.1 MPa

T/K	$C°_p$ (J K⁻¹mol⁻¹)	S° (J K⁻¹mol⁻¹)	$-[G°-H°(T_r)]/T$ (J K⁻¹mol⁻¹)	$H°-H°(T_r)$ (kJ mol⁻¹)	$\Delta_f H°$ (kJ mol⁻¹)	$\Delta_f G°$ (kJ mol⁻¹)	Log K_f
0	0.	0.	INFINITE	-29.849	-433.845	-433.845	INFINITE
100	97.710	339.737	570.950	-23.121	-433.786	-428.745	223.953
200	119.006	415.490	475.843	-12.071	-432.348	-424.281	110.811
250	123.307	442.549	466.563	-6.004	-431.794	-422.331	88.241
298.15	125.875	464.503	464.503	0.	-431.370	-420.548	73.678
300	125.953	465.282	464.506	0.233	-431.356	-420.481	73.212
350	127.673	484.836	466.046	6.576	-431.022	-418.697	62.487
400	128.846	501.966	469.488	12.991	-430.794	-416.953	54.448
450	129.679	517.193	473.958	19.455	-430.676	-415.231	48.199
500	130.290	530.889	478.978	25.955	-430.677	-413.516	43.200
600	131.116	554.723	489.674	39.029	-431.040	-410.058	35.699
700	131.659	574.977	500.450	52.169	-431.895	-406.500	30.333
800	132.077	592.586	510.890	65.357	-433.336	-402.782	26.299
900	132.451	608.164	520.849	78.583	-435.607	-398.837	23.148
1000	132.822	622.138	530.292	91.847	-439.312	-394.572	20.610
1100	133.205	634.816	539.226	105.148	-444.796	-389.804	18.510
1200	133.604	646.423	547.683	118.489	-449.159	-384.650	16.743
1300	134.013	657.133	555.695	131.869	-450.210	-379.233	15.238
1400	134.424	667.080	563.300	145.291	-451.406	-373.729	13.944
1500	134.830	676.368	570.532	158.754	-452.747	-368.135	12.820
1600	135.221	685.082	577.422	172.257	-454.232	-362.447	11.833
1700	135.591	693.291	583.998	185.797	-457.765	-356.623	10.958
1800	135.935	701.051	590.288	199.374	-460.217	-350.604	10.174
1900	136.249	708.409	596.313	212.983	-491.018	-343.040	9.431
2000	136.531	715.405	602.094	226.623	-494.258	-335.168	8.754
2100	136.780	722.073	607.650	240.288	-497.494	-327.134	8.137
2200	136.996	728.441	612.997	253.977	-500.732	-318.945	7.573
2300	137.181	734.535	618.149	267.687	-503.976	-310.610	7.054
2400	137.335	740.376	623.121	281.413	-507.232	-302.133	6.576
2500	137.460	745.985	627.924	295.153	-510.506	-293.520	6.133
2600	137.559	751.379	632.570	308.904	-513.804	-284.775	5.721
2700	137.633	756.572	637.067	322.664	-517.129	-275.903	5.338
2800	137.685	761.578	641.425	336.430	-520.485	-266.908	4 979
2900	137.718	766.410	645.652	350.200	-523.876	-257.792	4.643
3000	137.732	771.079	649.755	363.973	-527.304	-248.558	4.328
3100	137.731	775.596	653.742	377.746	-530.770	-239.209	4.031
3200	137.716	779.968	657.619	391.518	-1230.874	-214.901	3.508
3300	137.690	784.206	661.391	405.289	-1230.615	-183.156	2.899
3400	137.653	788.315	665.064	419.056	-1230.458	-151.417	2.326
3500	137.607	792.305	668.642	432.819	-1230.402	-119.682	1.786
3600	137.554	796.181	672.132	446.577	-1230.445	-87.946	1.276
3700	137.494	799.949	675.535	460.329	-1230.586	-56.209	0.794
3800	137.430	803.615	678.858	474.076	-1230.823	-24.466	0.336
3900	137.361	807.184	682.103	487.815	-1231.152	7.284	-0.098
4000	137.289	810.660	685.273	501.548	-1231.571	39.045	-0.510
4100	137.214	814.050	688.373	515.273	-1232.076	70.816	-0.902
4200	137.137	817.355	691.405	528.990	-1232.666	102.601	-1.276
4300	137.059	820.581	694.372	542.700	-1233.337	134.401	-1.633
4400	136.980	823.731	697.276	556.402	-1234.085	166.217	-1.973
4500	136.901	826.809	700.120	570.096	-1234.909	198.051	-2.299
4600	136.821	829.817	702.907	583.782	-1235.805	229.905	-2.611
4700	136.741	832.758	705.639	597.460	-1236.771	261.778	-2.909
4800	136.662	835.636	708.317	611.131	-1237.806	293.673	-3.196
4900	136.584	838.453	710.945	624.793	-1238.907	325.590	-3.471
5000	136.507	841.212	713.522	638.448	-1240.074	357.530	-3.735
5100	136.430	843.914	716.053	652.094	-1241.305	389.494	-3.989
5200	136.355	846.563	718.537	665.734	-1242.600	421.483	-4.234
5300	136.281	849.160	720.977	679.365	-1243.960	453.497	-4.469
5400	136.209	851.706	723.375	692.990	-1245.384	485.538	-4.697
5500	136.137	854.205	725.731	706.607	-1246.875	517.606	-4.916
5600	136.068	856.657	728.047	720.217	-1248.434	549.702	-5.127
5700	135.999	859.065	730.324	733.821	-1250.064	581.825	-5.332
5800	135.933	861.430	732.565	747.417	-1251.767	613.979	-5.529
5900	135.868	863.753	734.768	761.007	-1253.547	646.162	-5.721
6000	135.804	866.036	736.937	774.591	-1255.409	678.375	-5.906

PREVIOUS: December 1970 (1 atm) CURRENT: December 1970 (1 bar)

Iron Chloride $((FeCl_2)_2)$ $Cl_4Fe_2(g)$

MAGNESIUM CHLORIDE ($(MgCl_2)_2$) IDEAL GAS M_r = 190.618

$S°(298.15\ K) = [418.810 \pm 20.9]\ J\ K^{-1}\ mol^{-1}$

$\Delta_f H°(0\ K) = -954.41 \pm 29.3\ kJ\ mol^{-1}$
$\Delta_f H°(298.15\ K) = -954.37 \pm 29.3\ kJ\ mol^{-1}$

Vibrational Frequencies and Degeneracies

ν, cm^{-1}	ν, cm^{-1}	ν, cm^{-1}
[500](1)	[240](1)	[130](2)
[400](1)	[200](1)	[100](2)
295 (1)	[150](1)	[80](2)

Ground State Quantum Weight: 1 $\sigma = 4$
Point Group: [D_{2h}]
Bond Distance: Mg-Cl = [2.18] Å Mg-Cl'(ring) = [2.3] Å
Bond Angles: Cl'-Mg-Cl'(in ring) = [90°] Cl-Mg-Cl' = [135°]
Product of the Moments of Inertia: $I_A I_B I_C = 1.3313 \times 10^{-111}\ g^3\ cm^6$

Enthalpy of Formation

References relative to the monomer-dimer ratio in magnesium chloride vapor are given in the discussion for monomeric ideal gas. The selected best value for the enthalpy of dimerization at 900 K is -39.2 kcal mol^{-1}. Combined with enthalpy of formation data for the monomer and the heat capacity of the dimer, this yields $\Delta_f H°(298.15\ K) = -228.10\ kcal\ mol^{-1}$.

Heat Capacity and Entropy

A planar structure with D_{2h} symmetry was assumed, with the outer Mg-Cl distance the same as in $MgCl_2$ and the ring Mg-Cl distance slightly larger. The ring was taken to be square, and the Cl-Mg-Cl' angle 135°. This structure gives individual moments of inertia of $I_A = 31.12 \times 10^{-39}\ g\ cm^2$, $I_B = 191.86 \times 10^{-39}\ g\ cm^2$ and $I_C = I_A + I_B$.

A reasonable set of frequencies was estimated by comparison with $MgCl_2$ monomer and other dimer molecules. The single observed frequency is from Buchler and Klemperer (1) who erroneously ascribed it to the bending mode of the monomer. The estimated values were adjusted until the entropy of dimerization at 900 K was -29.9 kcal mol^{-1}, as selected in the discussion for ideal monomeric gas. The calculated thermodynamic properties of monomer and dimer give total vapor pressures in good agreement with experiment and yield a boiling point for magnesium chloride of 1624 K at one atmosphere pressure.

References

1. A. Buchler and W. Klemperer, J. Chem. Phys. 29, 121 (1958).

Magnesium Chloride ($(MgCl_2)_2$) $Cl_4Mg_2(g)$

Enthalpy Reference Temperature = T_r = 298.15 K Standard State Pressure = p° = 0.1 MPa

T/K	$C°_p$ (J K⁻¹mol⁻¹)	S° (J K⁻¹mol⁻¹)	$-[G°-H°(T_r)]/T$ (J K⁻¹mol⁻¹)	$H°-H°(T_r)$ (kJ mol⁻¹)	$\Delta_f H°$ (kJ mol⁻¹)	$\Delta_f G°$ (kJ mol⁻¹)	Log K_f
0	0.	0.	INFINITE	-28.396	-954.409	-954.409	INFINITE
100	91.553	298.917	522.238	-22.332	-955.452	-945.642	493.953
200	115.181	371.122	430.020	-11.780	-954.989	-935.971	244.451
250	120.397	397.431	420.950	-5.880	-954.669	-931.253	194.575
298.15	123.611	418.931	418.931	0.	-954.370	-926.771	162.366
300	123.710	419.696	418.933	0.229	-954.359	-926.600	161.335
350	125.913	438.942	420.448	6.473	-954.077	-921.996	137.600
400	127.437	455.861	423.839	12.809	-953.830	-917.430	119.804
450	128.529	470.937	428.250	19.209	-953.622	-912.893	105.966
500	129.335	484.523	433.209	25.657	-953.457	-908.377	94.897
600	130.418	508.208	443.792	38.649	-953.264	-899.382	78.298
700	131.090	528.366	454.470	51.727	-953.274	-890.404	66.443
800	131.534	545.902	464.827	64.860	-953.517	-881.409	57.550
900	131.842	561.413	474.713	78.029	-954.025	-872.368	50.631
1000	132.065	575.316	484.091	91.225	-972.015	-861.830	45.017
1100	132.230	587.911	492.965	104.441	-973.162	-850.756	40.399
1200	132.357	599.422	501.364	117.670	-974.317	-839.577	36.546
1300	132.456	610.021	509.320	130.911	-975.483	-828.302	33.282
1400	132.534	619.840	516.868	144.161	-1231.477	-810.602	30.244
1500	132.598	628.986	524.041	157.417	-1229.960	-780.592	27.183
1600	132.650	637.545	530.870	170.680	-1228.454	-750.683	24.507
1700	132.693	645.588	537.384	183.947	-1226.961	-720.869	22.150
1800	132.729	653.174	543.608	197.218	-1225.482	-691.141	20.056
1900	132.760	660.351	549.566	210.493	-1224.020	-661.496	18.186
2000	132.786	667.161	555.276	223.770	-1222.575	-631.927	16.504
2100	132.809	673.641	560.760	237.050	-1221.150	-602.429	14.985
2200	132.829	679.819	566.032	250.332	-1219.748	-572.999	13.605
2300	132.846	685.724	571.109	263.615	-1218.372	-543.633	12.346
2400	132.861	691.379	576.003	276.901	-1217.026	-514.325	11.194
2500	132.874	696.802	580.727	290.188	-1215.713	-485.073	10.135
2600	132.886	702.014	585.293	303.476	-1214.435	-455.873	9.159
2700	132.896	707.029	589.709	316.765	-1213.198	-426.720	8.255
2800	132.906	711.863	593.986	330.055	-1212.004	-397.614	7.418
2900	132.914	716.527	598.132	343.346	-1210.856	-368.549	6.638
3000	132.922	721.033	602.154	356.638	-1209.756	-339.523	5.912
3100	132.929	725.391	606.059	369.930	-1208.708	-310.532	5.232
3200	132.935	729.612	609.855	383.223	-1207.712	-281.575	4.596
3300	132.941	733.703	613.546	396.517	-1206.770	-252.648	3.999
3400	132.946	737.671	617.139	409.811	-1205.884	-223.749	3.437
3500	132.951	741.525	620.638	423.106	-1205.053	-194.875	2.908
3600	132.955	745.271	624.048	436.401	-1204.279	-166.023	2.409
3700	132.959	748.913	627.374	449.697	-1203.560	-137.193	1.937
3800	132.963	752.459	630.619	462.993	-1202.898	-108.382	1.490
3900	132.966	755.913	633.788	476.290	-1202.291	-79.586	1.066
4000	132.969	759.280	636.883	489.586	-1201.740	-50.806	0.663
4100	132.972	762.563	639.908	502.884	-1201.243	-22.039	0.281
4200	132.975	765.767	642.867	516.181	-1200.800	6.717	-0.084
4300	132.978	768.896	645.762	529.479	-1200.411	35.462	-0.431
4400	132.980	771.953	648.595	542.776	-1200.074	64.200	-0.762
4500	132.982	774.942	651.370	556.075	-1199.790	92.930	-1.079
4600	132.984	777.865	654.088	569.373	-1199.558	121.655	-1.381
4700	132.986	780.725	656.752	582.671	-1199.379	150.374	-1.671
4800	132.988	783.525	659.364	595.970	-1199.251	179.091	-1.949
4900	132.990	786.267	661.926	609.269	-1199.176	207.806	-2.215
5000	132.991	788.953	664.440	622.568	-1199.153	236.519	-2.471
5100	132.993	791.587	666.907	635.867	-1199.183	265.233	-2.717
5200	132.994	794.170	669.330	649.167	-1199.267	293.947	-2.953
5300	132.996	796.703	671.709	662.466	-1199.403	322.664	-3.180
5400	132.997	799.189	674.047	675.766	-1199.596	351.384	-3.399
5500	132.998	801.629	676.345	689.066	-1199.843	380.108	-3.610
5600	132.999	804.026	678.603	702.365	-1200.148	408.837	-3.813
5700	133.000	806.380	680.824	715.665	-1200.512	437.572	-4.010
5800	133.001	808.693	683.009	728.965	-1200.935	466.314	-4.200
5900	133.002	810.966	685.159	742.266	-1201.420	495.064	-4.383
6000	133.003	813.202	687.274	755.566	-1201.967	523.822	-4.560

PREVIOUS: December 1969 (1 atm) CURRENT: December 1969 (1 bar)

Magnesium Chloride ($(MgCl_2)_2$) $Cl_4Mg_2(g)$

MOLYBDENUM CHLORIDE ($MoCl_4$) CRYSTAL M_r = 237.752

$\Delta_f H°(0\ K)$ = Unknown

$S°(298.15\ K) = [223.84 \pm 12.6]\ J\ K^{-1}\ mol^{-1}$ $\Delta_f H°(298.15\ K) = -477.0 \pm 8\ kJ\ mol^{-1}$

T_{fus} = 590 K $\Delta_{fus}H° = [16.736 \pm 8]\ kJ\ mol^{-1}$

Enthalpy of Formation

The adopted $\Delta_f H°(298.15\ K) = 114\ kcal\ mol^{-1}$ was determined by Shchukarev et al. (1) by solution calorimetry. The authors measured the enthalpy of solution of $MoCl_4$ in the aqueous solvent $FeCl_3$ + HCl and obtained the enthalpy of formation from a comparison with the enthalpy of solution of $MoCl_5$ measured in the same solvent.

The decomposition pressures of $2MoCl_5(cr) = 2MoCl_4(cr) + Cl_2(g)$ (80 - 150°C) were measured manometrically by Saeki and Matsuzaki (2). From the reported data we evaluate the enthalpy change for the above reaction by 3rd law analysis. Based on $\Delta_f H°(MoCl_5, cr, 298.15\ K) = -126\ kcal\ mol^{-1}$ and the 3rd law $\Delta_r H°(298.15\ K) = 24.132\ kcal\ mol^{-1}$, we obtain $\Delta_f H°(MoCl_4, cr, 298.15\ K) = -113.94\ kcal\ mol^{-1}$, which is in very good agreement with the value reported by Shchukarev et al. (1).

Heat Capacity and Entropy

The heat capacity is estimated by comparison with that of WCl_4(cr). The estimated $S°(298.15\ K) = 53.5\ cal\ K^{-1}\ mol^{-1}$ is based on the assumption that the entropy difference between $MoCl_4$(cr) and $MoCl_5$(cr) is about the same as that between WCl_4(cr) and WCl_5(cr). The entropy obtained from the second law analysis of the data of Saeki and Matsuzaki (2) agrees with the adopted one within the error of the vapor pressure data.

Fusion Data

See the liquid table for details.

References
1. S. A. Shchukarev, G. I. Novikov, I. V. Vasil'kova, A. V. Suvorov, N. V. Andreeva, B. N. Sharupin and A. K. Baev, Russ. J. Inorg. Chem. 5, 802 (1960).
2. Y. Saeki and R. Matsuzaki, Denki Kagaku 33, 155-8 (1965).

Molybdenum Chloride ($MoCl_4$) $Cl_4Mo_1(cr)$

Enthalpy Reference Temperature = T_r = 298.15 K Standard State Pressure = p° = 0.1 MPa

T/K	$C_p°$ (J K⁻¹mol⁻¹)	S° (J K⁻¹mol⁻¹)	$-[G°-H°(T_r)]/T$ (J K⁻¹mol⁻¹)	$H°-H°(T_r)$ (kJ mol⁻¹)	$\Delta_f H°$ (kJ mol⁻¹)	$\Delta_f G°$ (kJ mol⁻¹)	Log K_f
0							
100							
200							
298.15	129.704	223.844	223.844	0.	-476.976	-402.164	70.458
300	129.704	224.646	223.846	0.240	-476.906	-401.700	69.942
400	135.143	262.705	228.999	13.482	-473.059	-377.207	49.258
500	140.582	293.445	238.908	27.269	-468.964	-353.711	36.952
590.000	145.478	317.108	249.072	40.141	------ CRYSTAL <--> LIQUID -----		
600	146.022	319.557	250.226	41.599	-464.515	-331.072	28.822
700	151.461	342.470	261.799	56.469	-459.661	-309.210	23.074
800	157.318	363.082	273.192	71.912	-454.339	-288.074	18.809
900	162.758	381.937	284.242	87.926	-448.531	-267.635	15.533
1000	166.942	399.306	294.891	104.415	-442.329	-247.865	12.947
1100	170.707	415.397	305.124	121.301	-435.806	-228.731	10.862
1200	174.054	430.400	314.945	138.546	-429.003	-210.205	9.150
1300	176.565	444.434	324.372	156.080	-421.994	-192.255	7.725
1400	178.657	457.595	333.423	173.841	-414.847	-174.850	6.524
1500	180.749	469.993	342.118	191.811	-407.583	-157.960	5.501

PREVIOUS: CURRENT: December 1968

Molybdenum Chloride ($MoCl_4$) $Cl_4Mo_1(cr)$

MOLYBDENUM CHLORIDE ($MoCl_4$) LIQUID M_r = 237.752

$S°(298.15\ K) = [245.525]\ J\ K^{-1}\ mol^{-1}$

T_{fus} = 590 K

$\Delta_f H°(298.15\ K) = [-462.837]\ kJ\ mol^{-1}$

$\Delta_{fus}H° = [16.736 \pm 8]\ kJ\ mol^{-1}$

Enthalpy of Formation

$\Delta_f H°(MoCl_4, \ell, 298.15\ K)$ is calculated from $\Delta_f H°(MoCl_4, cr, 298.15\ K)$ by adding $\Delta_{fus}H°$ and the difference in enthalpy, H°(590 K) - H°(298.15 K), between the crystal and liquid.

Heat Capacity and Entropy

The heat capacities are assumed to be constant at 7 cal K^{-1} g-atom. The entropy is calculated in a manner analogous to that used for the enthalpy of formation.

Fusion Data

T_{fus} is taken from Gutmann (1). $\Delta_{fus}H°$ is estimated such that the derived $\Delta_f G°$ for $MoCl_4(\ell)$ yields a reasonable boiling point which is expected to be between 600 and 700 K from comparisons with related compounds.

Vaporization Data

T_{vap} is the temperature at which the Gibbs energy change ($\Delta_r G°$) of the following process $MoCl_4(\ell) = MoCl_4(g)$ is zero. The difference between $\Delta_f H°$ for $MoCl_4(\ell)$ and $MoCl_4(g)$ at T_{vap} is the enthalpy of vaporization.

Reference

1. V. Gutmann, "Halogen Chemistry," Vol. 3, Academic Press, New York, 1967.

Molybdenum Chloride ($MoCl_4$) $Cl_4Mo_1(l)$

Enthalpy Reference Temperature = T_r = 298.15 K; Standard State Pressure = p° = 0.1 MPa

T/K	$C_p°$ (J K⁻¹mol⁻¹)	S° (J K⁻¹mol⁻¹)	−[G°−H°(T_r)]/T (J K⁻¹mol⁻¹)	H°−H°(T_r) (kJ mol⁻¹)	$\Delta_f H°$ (kJ mol⁻¹)	$\Delta_f G°$ (kJ mol⁻¹)	Log K_f
0							
100							
200							
298.15	146.440	245.525	245.525	0.	-462.837	-394.490	69.113
300	146.440	246.431	245.528	0.271	-462.736	-394.066	68.613
400	146.440	288.559	251.271	14.915	-457.488	-371.977	48.575
500	146.440	321.236	262.118	29.559	-452.535	-351.178	36.687
590.000	146.440	345.474	273.036	42.739	------ CRYSTAL <--> LIQUID -----		
600	146.440	347.935	274.264	44.203	-447.772	-331.356	28.847
700	146.440	370.509	286.442	58.847	-443.144	-312.321	23.306
800	146.440	390.063	298.200	73.491	-438.621	-293.941	19.192
900	146.440	407.311	309.384	88.135	-434.184	-276.124	16.026
1000	146.440	422.740	319.961	102.779	-429.826	-258.796	13.518
1100	146.440	436.698	329.950	117.423	-425.545	-241.901	11.487
1200	146.440	449.440	339.384	132.067	-421.343	-225.392	9.811
1300	146.440	461.161	348.306	146.711	-417.224	-209.230	8.407
1400	146.440	472.013	356.760	161.355	-413.194	-193.383	7.215
1500	146.440	482.117	364.784	175.999	-409.257	-177.820	6.192

PREVIOUS: CURRENT: December 1968

Molybdenum Chloride ($MoCl_4$) $Cl_4Mo_1(l)$

MOLYBDENUM CHLORIDE ($MoCl_4$) CRYSTAL–LIQUID M_r = 237.752

0 to 590 K crystal
above 590 K liquid

Refer to the individual tables for details.

Molybdenum Chloride ($MoCl_4$) Cl_4Mo_1(cr,l)

Enthalpy Reference Temperature = T_r = 298.15 K Standard State Pressure = p° = 0.1 MPa

	J K⁻¹mol⁻¹			kJ mol⁻¹			
T/K	C_p°	S°	$-[G^\circ-H^\circ(T_r)]/T$	$H^\circ-H^\circ(T_r)$	Δ_fH°	Δ_fG°	Log K_f
0							
100							
200							
298.15	129.704	223.844	223.844	0.	-476.976	-402.164	70.458
300	129.704	224.646	223.846	0.240	-476.906	-401.700	69.942
400	135.143	262.705	228.999	13.482	-473.059	-377.207	49.258
500	140.582	293.445	238.908	27.269	-468.964	-353.711	36.952
590.000	145.478	317.108	249.072	40.141	CRYSTAL <--> LIQUID		
590.000	146.440	345.474	249.072	56.877	TRANSITION		
600	146.440	347.935	250.699	58.342	-447.772	-331.356	28.847
700	146.440	370.509	266.244	72.986	-443.144	-312.321	23.306
800	146.440	390.063	280.526	87.630	-438.621	-293.941	19.192
900	146.440	407.311	293.674	102.274	-434.184	-276.124	16.026
1000	146.440	422.740	305.823	116.918	-429.826	-258.796	13.518
1100	146.440	436.698	317.096	131.562	-425.545	-241.901	11.487
1200	146.440	449.440	327.601	146.206	-421.343	-225.392	9.811
1300	146.440	461.161	337.430	160.850	-417.224	-209.230	8.407
1400	146.440	472.013	346.661	175.494	-413.194	-193.383	7.215
1500	146.440	482.117	355.358	190.138	-409.257	-177.820	6.192

PREVIOUS: CURRENT: December 1968

Molybdenum Chloride ($MoCl_4$) Cl_4Mo_1(cr,l)

MOLYBDENUM CHLORIDE ($MoCl_4$) IDEAL GAS $M_r = 237.752$

$S°(298.15 K) = [371.941] J K^{-1} mol^{-1}$

$\Delta_f H°(0 K) = [-388.91 \pm 12.6] kJ mol^{-1}$

$\Delta_f H°(298.15 K) = [-384.93 \pm 12.6] kJ mol^{-1}$

Electronic Levels and Quantum Weights

ε_i, cm^{-1}	g_i
0	[3]
[10000]	[3]
[30000]	[3]

Vibrational Frequencies and Degeneracies

ν, cm^{-1}	ν, cm^{-1}
[390](1)	[410](3)
[120](2)	[120](3)

Point Group: T_d σ = [12]

Bond Distance: Mo-Cl = [2.27] Å

Bond Angle: Cl-Mo-Cl = [109.4712]°

Product of the Moments of Inertia: $I_A I_B I_C = [5.29405 \times 10^{-112}] g^3 cm^6$

Enthalpy of Formation

Shchukarev et al. (1), derived the value $\Delta_f H°(MoCl_4, g, 298.15 K) = -89 kcal mol^{-1}$ from the decomposition pressures of $2MoCl_3(cr) = MoCl_2(cr) + MoCl_4(g)$, 480-600°C, measured by a transpiration method. However, this value is adjusted to be -92±3 $kcal mol^{-1}$, in order to obtain a reasonable boiling point (see the liquid table for details).

Heat Capacity and Entropy

Vibrational frequencies and quantum weights are estimated by comparison with those of $WCl_4(g)$. The molecular configuration is assumed to be tetrahedral similar to those of $WCl_4(g)$, $SiCl_4(g)$, $TiCl_4(g)$ and $PbCl_4(g)$. The bond distance is estimated to be the same as that of $MoCl_5(g)$. The principal moments of inertia are $I_A = I_B = I_C = 80.8964 \times 10^{-39} g cm^2$. The low lying electronic levels are estimated to be 0, 10000, and 30000, all triplets, by comparison with those of $WCl_4(g)$.

Reference

1. S. A. Shchukarev, G. I. Novikov, I. V. Vasil'kova, A. V. Suvorov, N. V. Andreeva, B. N. Sharupin and A. K. Baev, Zh. Neorg. Khim. 5, 1650-4 (1960).

Molybdenum Chloride ($MoCl_4$) $Cl_4Mo_1(g)$

Enthalpy Reference Temperature = T_r = 298.15 K — Standard State Pressure = p° = 0.1 MPa

	J $K^{-1}mol^{-1}$			kJ mol^{-1}			
T/K	$C_p°$	S°	$-[G°-H°(T_r)]/T$	$H°-H°(T_r)$	$\Delta_f H°$	$\Delta_f G°$	Log K_f
0	0.	0.	INFINITE	-26.931	-388.912	-388.912	INFINITE
100	69.235	279.286	452.572	-17.329	-385.625	-375.050	195.906
200	89.478	334.366	380.690	-9.265	-385.479	-364.471	95.190
250	94.767	354.944	373.540	-4.649	-385.219	-359.248	75.061
298.15	98.109	371.941	371.941	0.	-384.928	-354.272	62.067
300	98.213	372.548	371.943	0.182	-384.916	-354.081	61.651
350	100.529	387.873	373.148	5.154	-384.595	-348.967	52.081
400	102.140	401.408	375.851	10.223	-384.271	-343.900	44.909
450	103.298	413.509	379.375	15.360	-383.950	-338.873	39.335
500	104.154	424.439	383.344	20.548	-383.637	-333.881	34.880
600	105.306	443.539	391.830	31.025	-383.040	-323.987	28.206
700	106.021	459.830	400.409	41.594	-382.488	-314.189	23.445
800	106.494	474.020	408.743	52.222	-381.981	-304.467	19.880
900	106.822	486.583	416.707	62.888	-381.521	-294.806	17.110
1000	107.060	497.851	424.268	73.583	-381.112	-285.193	14.897
1100	107.239	508.064	431.429	84.298	-380.760	-275.619	13.088
1200	107.380	517.401	438.210	95.030	-380.471	-266.074	11.582
1300	107.497	526.001	444.636	105.774	-380.252	-256.550	10.308
1400	107.600	533.971	450.736	116.529	-380.111	-247.040	9.217
1500	107.696	541.398	456.536	127.294	-380.053	-237.538	8.272
1600	107.791	548.352	462.059	138.068	-380.085	-228.036	7.445
1700	107.889	554.889	467.329	148.852	-380.214	-218.530	6.715
1800	107.992	561.059	472.367	159.646	-380.447	-209.013	6.065
1900	108.103	566.901	477.190	170.451	-380.791	-199.480	5.484
2000	108.222	572.449	481.815	181.267	-381.256	-189.926	4.960
2100	108.348	577.732	486.258	192.095	-381.851	-180.345	4.486
2200	108.482	582.775	490.531	202.937	-382.587	-170.733	4.054
2300	108.623	587.601	494.648	213.792	-383.474	-161.084	3.658
2400	108.770	592.227	498.618	224.661	-384.526	-151.393	3.295
2500	108.921	596.670	502.452	235.546	-385.754	-141.655	2.960
2600	109.075	600.945	506.158	246.446	-387.184	-131.863	2.649
2700	109.231	605.064	509.745	257.361	-388.856	-122.012	2.360
2800	109.387	609.040	513.221	268.292	-390.834	-112.094	2.091
2900	109.543	612.881	516.592	279.238	-429.084	-102.049	1.838
3000	109.698	616.597	519.864	290.201	-429.883	-90.759	1.580
3100	109.849	620.197	523.042	301.178	-430.706	-79.441	1.339
3200	109.997	623.687	526.133	312.170	-431.551	-68.097	1.112
3300	110.140	627.074	529.141	323.177	-432.417	-56.725	0.898
3400	110.279	630.364	532.070	334.198	-433.303	-45.327	0.696
3500	110.412	633.562	534.924	345.233	-434.205	-33.903	0.506
3600	110.540	636.674	537.708	356.280	-435.121	-22.453	0.326
3700	110.661	639.705	540.424	367.340	-436.046	-10.977	0.155
3800	110.776	642.657	543.075	378.412	-436.978	0.523	-0.007
3900	110.885	645.536	545.666	389.495	-437.911	12.049	-0.161
4000	110.987	648.345	548.198	400.589	-438.842	23.598	-0.308
4100	111.084	651.087	550.674	411.693	-439.766	35.171	-0.448
4200	111.174	653.765	553.097	422.806	-440.677	46.766	-0.582
4300	111.258	656.382	555.468	433.927	-441.572	58.382	-0.709
4400	111.336	658.940	557.791	445.057	-442.447	70.019	-0.831
4500	111.409	661.443	560.067	456.194	-443.296	81.675	-0.948
4600	111.476	663.893	562.297	467.339	-444.115	93.351	-1.060
4700	111.538	666.291	564.484	478.489	-444.902	105.043	-1.167
4800	111.595	668.640	566.630	489.646	-445.650	116.752	-1.271
4900	111.648	670.941	568.735	500.808	-446.358	128.476	-1.370
5000	111.696	673.197	570.802	511.976	-1038.408	145.945	-1.525
5100	111.741	675.410	572.832	523.147	-1040.704	169.655	-1.738
5200	111.781	677.580	574.825	534.324	-1043.098	193.411	-1.943
5300	111.818	679.709	576.784	545.504	-1045.583	217.213	-2.141
5400	111.851	681.800	578.710	556.687	-1048.158	241.064	-2.332
5500	111.882	683.852	580.603	567.874	-1050.817	264.963	-2.516
5600	111.909	685.869	582.464	579.063	-1053.556	288.911	-2.695
5700	111.934	687.850	584.296	590.255	-1056.372	312.908	-2.867
5800	111.956	689.797	586.098	601.450	-1059.260	336.956	-3.035
5900	111.977	691.711	587.872	612.647	-1062.215	361.053	-3.197
6000	111.995	693.593	589.618	623.845	-1065.238	385.201	-3.353

PREVIOUS: December 1968 (1 atm) CURRENT: December 1968 (1 bar)

Molybdenum Chloride ($MoCl_4$) $Cl_4Mo_1(g)$

TUNGSTEN CHLORIDE OXIDE ($OWCl_4$) CRYSTAL M_r = 351.6614

$S°(298.15 K) = [172.8 \pm 16.7]\ J\ K^{-1}\ mol^{-1}$

T_{fus} = 484 K

$\Delta_f H°(0 K)$ = Unknown

$\Delta_f H°(298.15 K) = -671.11 \pm 8.4\ kJ\ mol^{-1}$

$\Delta_{fus} H° = 45.380 \pm 12.6\ kJ\ mol^{-1}$

Enthalpy of Formation

Shchukarev, et al. (1) measured calorimetrically the enthalpies of reaction (1) and (2) at 298.15 K as -131.5±0.3 and -13.6±0.11 kcal mol^{-1}, respectively. The reactions are given as follows:

(1) $WOCL_4(cr) + 6\ NaOH(aq,\ 16.3\ H_2O) \rightarrow Na_2WO_4(aq,\ 190000\ H_2O) + 4\ NaCl(aq,\ 50000\ H_2O) + 3\ H_2O(\ell)$

(2) $H_2WO_4(cr) + 2\ NaOH(aq,\ 16.3\ H_2O) \rightarrow Na_2WO_4(aq,\ 20000\ H_2O) + 2\ H_2O(\ell)$

Based on these data and the enthalpies of dilution for NaOH(aq)[2], and $Na_2WO_4(aq)$[3], we obtain $\Delta_r H°(298.15 K) = -118.0 \pm 1.6$ kcal mol^{-1} for $WOCl_4(cr) + 4OH^-(aq, \infty) + H_2WO_4(cr) + 4\ Cl^-(aq, \infty) + H_2O(\ell)$. This leads to $\Delta_r H°(WOCl_4, cr, 298.15 K) = 160.4 \pm 2$ kcal mol^{-1}, using $\Delta_f H°(H_2WO_4, cr, 298.15 K) = -270.5 \pm 0.4$ kcal mol[5].

Heat Capacity and Entropy

$C_p°(300 K) = 35.0\ cal\ K^{-1}\ mol^{-1}$ is estimated using Kopp's rule. Since the melting point of $WOCl_4(cr)$ is approximately 500 K, which is relatively low for oxygen to reach its maximum contribution, we estimate $C_p°(500 K) = 40.0\ cal\ K^{-1}\ mol^{-1}$ based on the values for $WO_2(cr)$, $WO_3(cr)$ and $WCl_6(cr)$.

The entropy, $S°(298.15 K) = 41.3\ cal\ K^{-1}\ mol^{-1}$, is calculated from $\Delta S°(457 K) = 44.5\ cal\ K^{-1}\ mol^{-1}$ for $WOCl_4(cr) \rightarrow WOCl_4(g)$ obtained from the 2nd law analysis of the vapor pressure equation given by Shchukarev and Suvorov (4). See $WOCl_4(g)$ table (Mar. 31, 1967) for details.

It is surprising that this entropy is smaller than the corresponding value for $WOF_4(cr)$. See $WOF_4(cr)$ table (Mar. 31, 1967) for details.

Fusion Data

T_{fus} is calculated as the temperature at which the Gibbs energies of formation of crystal and liquid are equal. The difference between the enthalpies of formation of crystal and liquid at the melting point is $\Delta_{fus}H°$. This $\Delta_{fus}H°$ is derived from 2nd law analyses of the vapor pressure equation determined by Shchukarev and Suvorov (4). See $WOCl_4(\ell, g)$ tables (Mar. 31, 1967) for details.

References

1. S. A. Shchukarev, I. V. Vasil'kova and G. I. Novikov, Zh. Neorg. Khim. 3, 2642 (1958).
2. V. B. Parker, "Thermal Properties of Aqueous Uni-univalent Electrolytes," NSRDS-NBS 2, U. S. Nat. Bur. Stand., 1965.
3. We have assumed ΔH° dil = 0±0.5 kcal mol^{-1} for $Na_2WO_4(aq,\ 20000\ H_2O) \rightarrow Na_2WO_4(aq,\ 190000\ H_2O)$.
4. S. A. Shchukarev and A. V. Suvorov, Vestnik Leningrad. Univ. 16, No. 4, Ser. Fiz. i Khim., No. 1, 87 (1961). Earlier data reported by Shchukarev et al., Zh. Neorg. Khim. 1, 357 (1956); 3, 2630 (1958); 5, 1650 (1960); have been revised and they are not used here.
5. JANAF Thermochemical Tables: $H_2WO_4(cr)$, 3-31-67.

Tungsten Chloride Oxide ($OWCl_4$) $Cl_4O_1W_1(cr)$

Enthalpy Reference Temperature = T_r = 298.15 K   Standard State Pressure = p° = 0.1 MPa

T/K	$C_p°$ (J K⁻¹mol⁻¹)	S° (J K⁻¹mol⁻¹)	$-[G°-H°(T_r)]/T$ (J K⁻¹mol⁻¹)	$H°-H°(T_r)$ (kJ mol⁻¹)	$\Delta_f H°$ (kJ mol⁻¹)	$\Delta_f G°$ (kJ mol⁻¹)	Log K_f
0							
100							
200							
298.15	146.248	172.799	172.799	0.	-671.114	-549.292	96.234
300	146.440	173.704	172.802	0.271	-671.041	-548.536	95.509
400	156.900	217.265	178.671	15.438	-666.763	-508.327	66.381
484.000	165.686	247.984	188.095	28.986	------ CRYSTAL <--> LIQUID -----		
500	167.360	253.400	190.099	31.651	-661.736	-469.283	49.026
600	177.820	284.879	203.325	48.932	-655.856	-431.331	37.551
700	185.560	312.904	217.015	67.122	-649.238	-394.425	29.432
800	190.790	338.049	230.600	85.959	-642.114	-358.504	23.408
900	193.719	360.694	243.818	105.188	-634.717	-323.495	18.775
1000	196.230	381.238	256.549	124.689	-627.155	-289.319	15.112

PREVIOUS: September 1962   CURRENT: March 1967

Tungsten Chloride Oxide ($OWCl_4$) $Cl_4O_1W_1(cr)$

TUNGSTEN CHLORIDE OXIDE ($OWCl_4$) LIQUID M_r = 341.6614

$S°(298.15 K) = [253.565] J K^{-1} mol^{-1}$ $\Delta_f H°(298.15 K) = [-630.573] kJ mol^{-1}$

T_{fus} = 484 K $\Delta_{fus}H° = 45.380 \pm 12.6 kJ mol^{-1}$

Enthalpy of Formation

$\Delta_f H°(WOCl_4, \ell, 298.15 K)$, is calculated from that of the gas by subtracting the enthalpy of vaporization, $\Delta_{vap}H°(298.15 K) = 13.643 kcal mol^{-1}$. The latter is obtained by the 3rd law method from the vapor pressure equation given by Shchukarev and Surorov (1). 2nd law analysis gives $\Delta_{vap}H°(501.9 K) = 10.32 kcal mol^{-1}$ [($\Delta_{vap}H°(298.15 K) = 13.643 kcal mol^{-1}$] and the boiling point T_{vap} = 493 K. Earlier data reported by Shchukarev et al. (2) have been revised and they are not used here.

Heat Capacity and Entropy

The heat capacity is assumed to be constant at 7.25 cal K^{-1} g-atom^{-1} as suggested by Kubaschewski and Evans (3).

The entropy, $S°(298.15 K) = 60.6 cal K^{-1} mol^{-1}$, is calculated from $\Delta S°(501.9 K) = 20.9 cal K^{-1} mol^{-1}$ for $WOCl_4(\ell)$ + $WOCl_4(g)$ obtained from the 2nd law analysis of the vapor pressure equation given by Shchukarev and Suvorov (1).

Fusion Data

See the crystal table for details.

Vaporization Data

T_{vap} is calculated as the temperature at which the Gibbs energies of formation of liquid and gas are equal. The difference between the enthalpies of formation of liquid and gas at the boiling point is $\Delta_{vap}H°$.

References

1. S. A. Shchukarev and A. V. Surorov, Vestn. Leningrad. Univ. 16, No. 4, Ser. Fiz. i Khim., No. 1, 87 (1961).
2. A. Shchukarev et al., Zh. Neorg. Khim. 1, 357 (1956); 3, 2630 (1958); and 5, 1650 (1960).
3. O. Kubaschewski and E. Ll. Evans, "Metallurgical Thermochemistry", Pergamon Press, New York, 1958.

Tungsten Chloride Oxide ($OWCl_4$) $Cl_4O_1W_1(l)$

Enthalpy Reference Temperature = T_r = 298.15 K; Standard State Pressure = p° = 0.1 MPa

T/K	$C_p°$ (J K⁻¹mol⁻¹)	S° (J K⁻¹mol⁻¹)	$-[G°-H°(T_r)]/T$ (J K⁻¹mol⁻¹)	$H°-H°(T_r)$ (kJ mol⁻¹)	$\Delta_f H°$ (kJ mol⁻¹)	$\Delta_f G°$ (kJ mol⁻¹)	Log K_f
0							
100							
200							
298.15	182.004	253.565	253.565	0.	-630.573	-532.832	93.350
300	182.004	254.691	253.569	0.337	-630.434	-532.226	92.669
400	182.004	307.050	260.708	18.537	-623.123	-500.601	65.372
484.000	182.004	341.744	271.857	33.825	------ CRYSTAL <--> LIQUID -----		
500	182.004	347.663	274.188	36.738	-616.109	-470.788	49.183
600	182.004	380.847	289.283	54.938	-609.310	-442.365	38.511
700	182.004	408.903	304.419	73.138	-602.682	-415.067	30.973
800	182.004	433.206	319.032	91.339	-596.194	-388.709	25.380
900	182.004	454.643	332.933	109.539	-589.826	-363.157	21.077
1000	182.004	473.819	346.079	127.740	-583.563	-338.309	17.671

PREVIOUS: CURRENT: March 1967

Tungsten Chloride Oxide ($OWCl_4$) $Cl_4O_1W_1(l)$

TUNGSTEN CHLORIDE OXIDE ($OWCl_4$) CRYSTAL–LIQUID M_r = 351.6614

0 to 484 K crystal
above 484 K liquid

Refer to the individual tables for details.

Tungsten Chloride Oxide ($OWCl_4$) $Cl_4O_1W_1$(cr,l)

Enthalpy Reference Temperature = T_r = 298.15 K Standard State Pressure = $p°$ = 0.1 MPa

T/K	$C_p°$ (J K⁻¹mol⁻¹)	$S°$ (J K⁻¹mol⁻¹)	$-[G°-H°(T_r)]/T$ (J K⁻¹mol⁻¹)	$H°-H°(T_r)$ (kJ mol⁻¹)	$\Delta_fH°$ (kJ mol⁻¹)	$\Delta_fG°$ (kJ mol⁻¹)	Log K_f
0							
100							
200							
298.15	146.248	172.799	172.799	0.	-671.114	-549.292	96.234
300	146.440	173.704	172.802	0.271	-671.041	-548.536	95.509
400	156.900	217.265	178.671	15.438	-666.763	-508.327	66.381
484.000	165.686	247.984	188.095	28.986	CRYSTAL <--> LIQUID		
484.000	182.004	341.744	188.095	74.366	TRANSITION		
500	182.004	347.663	193.107	77.278	-616.109	-470.788	49.183
600	182.004	380.847	221.716	95.478	-609.310	-442.365	38.511
700	182.004	408.903	246.504	113.679	-602.682	-415.067	30.973
800	182.004	433.206	268.357	131.879	-596.194	-388.709	25.380
900	182.004	454.643	287.888	150.080	-589.826	-363.158	21.077
1000	182.004	473.819	305.539	168.280	-583.563	-338.309	17.671

PREVIOUS: CURRENT: March 1967

Tungsten Chloride Oxide ($OWCl_4$) $Cl_4O_1W_1$(cr,l)

TUNGSTEN CHLORIDE OXIDE ($OWCl_4$) IDEAL GAS M_r = 341.6614

$\Delta_f H°(0\ K) = -568.09 \pm 20.9\ kJ\ mol^{-1}$

$S°(298.15\ K) = [377.079]\ J\ K^{-1}\ mol^{-1}$ $\Delta_f H°(298.15\ K) = -573.49 \pm 20.9\ kJ\ mol^{-1}$

Vibrational Frequencies and Degeneracies

ν, cm^{-1}	ν, cm^{-1}	ν, cm^{-1}
[1000](1)	[350](1)	[200](1)
[1000](1)	[350](1)	[200](1)
[480](1)	[300](1)	[100](1)
[480](1)	[300](1)	[100](1)

Ground State Quantum Weight = 1 σ = [2]

Point Group: [C_{2v}]

Bond Distances: W-Cl = [2.26] Å W-O = [1.81] Å

Bond Angles: Cl*-W-O* = Cl*-W-Cl* = [120]°

Cl*-W-Cl** = Cl**-W-O* = [90]°

Cl**-W-Cl** = [180]°

*Equatorial **Axial

Product of the Moments of Inertia: $I_A I_B I_C = [5.87434 \times 10^{-112}]\ g^3\ cm^6$

Enthalpy of Formation

The enthalpy formation, $\Delta_f H°(WOCl_4$, g, 298.15 K) = -137.0 kcal mol^{-1}, is calculated from $\Delta_r H°$(298.15 K) = 23.4 kcal mol^{-1} for $WOCl_4$(cr) → $WOCl_4$(g). The latter is obtained by the 3rd law method from the vapor pressure equation given by Shchukarev and Suvorov (1). Second law analysis gives $\Delta_r H°$(457.2 K) = 21.9 kcal mol^{-1}(($\Delta_r H°$(298.15 K) = 23.4 kcal mol^{-1}).

*Earlier data reported by Shchukarev et al. (2) have been revised and they are not used here.

Heat Capacity and Entropy

The molecular configuration is assumed to be a trigonal bipyramid similar to that of SOF_4(g), reported by Goggin et al. (3). The bond distances W-O and W-Cl are estimated to be the same as those in WO_3(g) and WCl_6(g), respectively. The principal moments of inertia are: $I_A = 67.5706 \times 10^{-39}$, $I_B = 82.6053 \times 10^{-39}$ and $I_C = 105.2432 \times 10^{-39}\ g\ cm^2$.

All vibrational frequencies are estimated by comparison with those observed in the infrared and Raman spectra for SOF_4(g), (3), $MoOCl_4$(g) (4), WO_2Cl_2(g) (5), CrO_2Cl_2(g) (6, 7), WO_3(g) (8), and WCl_6(g) (8). The frequencies are not listed in point group order.

References

1. S. A. Shchukarev and A. V. Suvorov, Vestnik Leningrad Univ. 16, No. 4, Ser. Fiz. i Khim., No. 1, 87 (1961).
2. S. A. Shchukarev et al., Zh. Neorg. Khim. 1, 357 (1956), 3, 2630 (1956), 5, 1650 (1960).
3. P. L. Goggin, H. L. Roberts and L. A. Woodward, Trans. Faraday Soc. 57, 1877 (1961).
4. T. V. Iorns and F. E. Stafford, J. Amer. Chem. Soc. 88, 4819 (1966).
5. C. G. Barraclough and J. Stals, Aust. J. Chem. 19, 741 (1966).
6. F. A. Miller, G. L. Carlson and W. B. White, Spectrochim. Acta 1959, 709 (1959).
7. H. Stammreich, K. Kawai and Y. Tavares, Spectrochim. Acta 1959, 438 (1959).
8. JANAF Thermochemical Tables: WO_3(g), 9-30-66; WCl_6(g), 12-31-66.

Tungsten Chloride Oxide ($OWCl_4$) $Cl_4O_1W_1$(g)

Enthalpy Reference Temperature = T_r = 298.15 K Standard State Pressure = p° = 0.1 MPa

T/K	$C_p°$ (J $K^{-1}mol^{-1}$)	S°	-[G°-H°(T_r)]/T	H°-H°(T_r) (kJ mol^{-1})	$\Delta_f H°$	$\Delta_f G°$	Log K_f
0	0.	0.	INFINITE	-22.271	-568.087	-568.087	INFINITE
100	63.815	283.395	460.920	-17.752	-571.497	-552.408	288.549
200	92.057	337.421	386.393	-9.794	-573.077	-532.573	139.094
250	100.261	358.894	378.796	-4.975	-573.382	-522.407	109.151
298.15	106.138	377.079	377.079	0.	-573.493	-512.577	89.801
300	106.335	377.736	377.081	0.197	-573.494	-512.199	89.182
350	110.987	394.492	378.394	5.635	-573.456	-501.984	74.917
400	114.621	409.559	381.364	11.278	-573.301	-491.783	64.220
450	117.495	423.232	385.268	17.084	-573.058	-481.607	55.904
500	119.794	435.735	389.698	23.018	-572.747	-471.462	49.253
600	123.162	457.895	399.265	35.178	-571.990	-451.274	39.287
700	125.440	477.062	409.041	47.615	-571.125	-431.222	32.178
800	127.036	493.922	418.619	60.243	-570.209	-411.298	26.855
900	128.188	508.955	427.836	73.007	-569.277	-391.490	22.721
1000	129.045	522.508	436.637	85.871	-568.352	-371.786	19.420
1100	129.697	534.839	445.012	98.809	-567.447	-352.173	16.723
1200	130.203	546.147	452.975	111.805	-566.573	-332.642	14.480
1300	130.604	556.585	460.549	124.846	-565.738	-313.181	12.584
1400	130.926	566.276	467.759	137.924	-564.948	-293.784	10.961
1500	131.188	575.318	474.632	151.030	-564.207	-274.441	9.557
1600	131.405	583.792	481.192	164.160	-563.520	-255.146	8.330
1700	131.586	591.764	487.464	177.309	-562.890	-235.893	7.248
1800	131.739	599.289	493.469	190.476	-562.321	-216.674	6.288
1900	131.868	606.416	499.228	203.656	-561.817	-197.486	5.429
2000	131.980	613.183	504.758	216.849	-561.382	-178.322	4.657
2100	132.076	619.624	510.076	230.052	-561.019	-159.178	3.959
2200	132.159	625.770	515.196	243.264	-560.731	-140.050	3.325
2300	132.232	631.647	520.132	256.483	-560.524	-120.933	2.746
2400	132.297	637.276	524.897	269.710	-560.400	-101.824	2.216
2500	132.353	642.678	529.501	282.942	-560.363	-82.717	1.728
2600	132.404	647.870	533.954	296.180	-560.412	-63.611	1.278
2700	132.449	652.868	538.266	309.423	-560.578	-44.500	0.861
2800	132.489	657.685	542.446	322.670	-560.896	-25.380	0.473
2900	132.526	662.335	546.500	335.921	-561.407	-6.247	0.113
3000	132.559	666.828	550.437	349.175	-562.138	12.909	-0.225
3100	132.588	671.175	554.262	362.432	-563.133	32.093	-0.541
3200	132.615	675.385	557.981	375.693	-564.422	51.313	-0.838
3300	132.640	679.467	561.601	388.955	-566.028	70.579	-1.117
3400	132.662	683.427	565.126	402.221	-567.983	89.898	-1.381
3500	132.683	687.273	568.562	415.488	-570.349	109.281	-1.631
3600	132.702	691.011	571.911	428.757	-573.218	128.738	-1.868
3700	132.719	694.647	575.180	442.028	-611.453	148.474	-2.096
3800	132.736	698.186	578.370	455.301	-612.028	169.020	-2.323
3900	132.750	701.634	581.487	468.575	-612.622	189.582	-2.539
4000	132.764	704.996	584.533	481.851	-613.230	210.159	-2.744
4100	132.777	708.274	587.511	495.128	-613.845	230.751	-2.940
4200	132.789	711.474	590.425	508.406	-614.462	251.359	-3.126
4300	132.800	714.598	593.276	521.686	-615.077	271.981	-3.304
4400	132.810	717.652	596.068	534.966	-615.685	292.617	-3.474
4500	132.820	720.636	598.803	548.248	-616.279	313.267	-3.636
4600	132.829	723.556	601.484	561.530	-616.857	333.930	-3.792
4700	132.837	726.412	604.112	574.814	-617.412	354.605	-3.941
4800	132.845	729.209	606.689	588.098	-617.942	375.292	-4.084
4900	132.853	731.948	609.217	601.383	-618.442	395.990	-4.221
5000	132.860	734.632	611.699	614.668	-618.909	416.697	-4.353
5100	132.866	737.263	614.135	627.955	-619.340	437.414	-4.480
5200	132.873	739.844	616.528	641.241	-619.732	458.138	-4.602
5300	132.879	742.375	618.879	654.529	-620.083	478.870	-4.720
5400	132.884	744.858	621.189	667.817	-620.390	499.608	-4.833
5500	132.889	747.297	623.459	681.106	-620.653	520.351	-4.942
5600	132.894	749.691	625.692	694.395	-620.869	541.099	-5.047
5700	132.899	752.043	627.888	707.685	-621.038	561.849	-5.149
5800	132.904	754.355	630.049	720.975	-621.159	582.604	-5.247
5900	132.908	756.627	632.175	734.265	-621.231	603.358	-5.342
6000	132.912	758.861	634.268	747.556	-621.254	624.114	-5.433

PREVIOUS: March 1967 (1 atm) CURRENT: March 1967 (1 bar)

Tungsten Chloride Oxide ($OWCl_4$) $Cl_4O_1W_1$(g)

LEAD CHLORIDE ($PbCl_4$) IDEAL GAS M_r = 349.012

$\Delta_f H°(0\ K) = [-540.3 \pm 84]\ kJ\ mol^{-1}$

$S°(298.15\ K) = [381.659 \pm 12.6]\ J\ K^{-1}\ mol^{-1}$ $\Delta_f H°(298.15\ K) = [-552.4 \pm 84]\ kJ\ mol^{-1}$

Vibrational Frequencies and Degeneracies

ν, cm^{-1}
331 (1)
90 (2)
352 (3)
103 (3)

Ground State Quantum Weight: [1] σ = 12

Point Group: T_d

Bond Distance: Pb-Cl = 2.43 Å

Bond Angle: Cl-Pb-Cl = [109.4712°]

Product of the Moments of Inertia: $I_A I_B I_C = [7.96655 \times 10^{-112}]\ g^3\ cm^6$

Enthalpy of Formation

The adopted $\Delta_f H°(0\ K) = -131.6 \pm 20\ kcal\ mol^{-1}$ is derived from the dissociation energy for the process $PbCl_4(g) = Pb(g) + 4Cl(g)$. D° = 12.70 eV is estimated from an intercomparison of the lead mono-, di-, and tetra- halides (1). The adopted D° value for $PbCl_4(g)$ is greater than the D° value for $PbCl_2(g)$ by a factor of 2.02. The auxiliary values $\Delta_f H°(Pb, g, 0\ K) = 46.91\ kcal\ mol^{-1}$ (1) and $\Delta_f H°(Cl, g, 0\ K) = 28.587\ kcal\ mol^{-1}$ (1) are used. $\Delta_f H°(PbCl_4, g, 298.15\ K) = -132.0 \pm 20\ kcal\ mol^{-1}$ is calculated from $\Delta_f H°(0\ K)$.

Heat Capacity and Entropy

The bond distance is given by Lister and Sutton (2). The vibrational frequencies are those determined by Clark nd Hunter (3) from Raman spectra of $PbCl_4$ in n-pentane solution. Clark and Hunter (3) also observed an electronic transition at 35,250 cm^{-1} which they claim is the lowest allowed electronic transition of the molecule. The present calculations assume a singlet electronic gound state.

The principal moments of inertia are: $I_A = I_B = I_C = 92.7022 \times 10^{-39}\ g\ cm^2$.

References

1. JANAF Thermochemical Tables: PbF(g), $PbF_2(g)$, $PBF_4(g)$, PbBr(g), $PbBr_2(g)$, PbI(g), and $PbI_2(g)$, 12-31-73; PbCl(g) and $PbCl_2(g)$, 6-30-73; Pb(g), 3-31-62; Cl(g), 6-30-72.
2. M. W. Lister and L. E. Sutton, Trans. Faraday Soc. 37, 393 (1941).
3. R. J. H. Clark and B. K. Hunter, J. Mol. Struct. 9, 354 (1971).

Lead Chloride ($PbCl_4$) $Cl_4Pb_1(g)$

Enthalpy Reference Temperature = T_r = 298.15 K Standard State Pressure = p° = 0.1 MPa

T/K	$C_p°$ (J K⁻¹mol⁻¹)	S° (J K⁻¹mol⁻¹)	$-[G°-H°(T_r)]/T$ (J K⁻¹mol⁻¹)	$H°-H°(T_r)$ (kJ mol⁻¹)	$\Delta_f H°$ (kJ mol⁻¹)	$\Delta_f G°$ (kJ mol⁻¹)	Log K_f
0	0.	0.	INFINITE	-23.449	-540.279	-540.279	INFINITE
100	74.317	284.403	465.502	-18.110	-544.070	-528.296	275.953
200	93.443	342.811	390.673	-9.572	-548.403	-511.046	133.471
250	97.857	364.177	383.300	-4.781	-550.403	-501.476	104.778
298.15	100.536	381.659	381.659	0.	-552.403	-491.873	86.174
300	100.618	382.281	381.661	0.186	-552.482	-491.497	85.577
350	102.425	397.936	382.893	5.265	-555.354	-481.092	71.799
400	103.662	411.699	385.651	10.419	-557.633	-470.328	61.419
450	104.540	423.962	389.239	15.625	-559.944	-459.277	53.311
500	105.184	435.011	393.273	20.869	-562.286	-447.966	46.799
600	106.044	454.271	401.881	31.434	-567.065	-424.656	36.970
700	106.573	470.661	410.565	42.067	-576.831	-399.736	29.829
800	106.922	484.916	418.987	52.743	-581.849	-374.095	24.426
900	107.163	497.525	427.027	63.448	-586.892	-347.823	20.187
1000	107.337	508.825	434.651	74.174	-591.961	-320.989	16.767
1100	107.466	519.061	441.867	84.914	-597.056	-293.645	13.944
1200	107.564	528.417	448.695	95.666	-665.752	-264.758	11.525
1300	107.641	537.029	455.163	106.426	-670.134	-231.164	9.288
1400	107.702	545.009	461.299	117.193	-674.572	-197.231	7.359
1500	107.752	552.441	467.130	127.966	-679.075	-162.979	5.675
1600	107.792	559.397	472.682	138.743	-683.650	-128.424	4.193
1700	107.826	565.933	477.977	149.524	-688.302	-93.580	2.875
1800	107.854	572.097	483.036	160.308	-693.036	-58.460	1.696
1900	107.878	577.929	487.879	171.095	-697.855	-23.075	0.634
2000	107.898	583.462	492.521	181.884	-702.763	12.566	-0.328
2100	107.916	588.727	496.978	192.675	-884.984	55.570	-1.382
2200	107.931	593.748	501.263	203.467	-889.691	100.468	-2.385
2300	107.944	598.546	505.389	214.261	-894.562	145.584	-3.306
2400	107.956	603.140	509.367	225.056	-899.604	190.915	-4.155
2500	107.966	607.547	513.207	235.852	-904.820	236.460	-4.941
2600	107.975	611.782	516.917	246.649	-910.208	282.217	-5.670
2700	107.983	615.857	520.507	257.447	-915.795	328.185	-6.349
2800	107.991	619.784	523.983	268.245	-921.614	374.364	-6.984
2900	107.997	623.574	527.352	279.045	-927.701	420.756	-7.579
3000	108.003	627.236	530.621	289.845	-934.078	467.362	-8.137
3100	108.008	630.777	533.795	300.645	-940.781	514.186	-8.664
3200	108.013	634.206	536.879	311.447	-947.835	561.232	-9.161
3300	108.018	637.530	539.879	322.248	-955.255	608.504	-9.632
3400	108.022	640.755	542.799	333.050	-963.068	656.007	-10.078
3500	108.025	643.886	545.643	343.852	-971.329	703.746	-10.503
3600	108.029	646.929	548.414	354.655	-980.122	751.728	-10.907
3700	108.032	649.889	551.117	365.458	-1024.303	800.154	-11.296
3800	108.035	652.770	553.754	376.262	-1030.843	849.551	-11.678
3900	108.037	655.577	556.329	387.065	-1037.413	899.121	-12.042
4000	108.040	658.312	558.845	397.869	-1044.002	948.859	-12.391
4100	108.042	660.980	561.303	408.673	-1050.602	998.763	-12.724
4200	108.044	663.583	563.708	419.477	-1057.201	1048.827	-13.044
4300	108.046	666.126	566.060	430.282	-1063.792	1099.049	-13.351
4400	108.048	668.610	568.363	441.087	-1070.365	1149.423	-13.645
4500	108.050	671.038	570.617	451.892	-1076.912	1199.946	-13.929
4600	108.051	673.413	572.826	462.697	-1083.427	1250.616	-14.201
4700	108.053	675.736	574.991	473.502	-1089.902	1301.425	-14.464
4800	108.054	678.011	577.114	484.307	-1096.333	1352.373	-14.717
4900	108.056	680.239	579.196	495.113	-1102.712	1403.454	-14.961
5000	108.057	682.422	581.239	505.918	-1109.036	1454.663	-15.197
5100	108.058	684.562	583.244	516.724	-1115.301	1506.000	-15.425
5200	108.059	686.660	585.212	527.530	-1121.503	1557.458	-15.645
5300	108.060	688.719	587.146	538.336	-1127.639	1609.036	-15.858
5400	108.061	690.739	589.046	549.142	-1133.706	1660.728	-16.064
5500	108.062	692.722	590.913	559.948	-1139.703	1712.533	-16.264
5600	108.063	694.669	592.748	570.755	-1145.628	1764.445	-16.458
5700	108.064	696.581	594.553	581.561	-1151.481	1816.462	-16.646
5800	108.065	698.461	596.328	592.367	-1157.260	1868.583	-16.828
5900	108.066	700.308	598.075	603.174	-1162.964	1920.801	-17.005
6000	108.066	702.124	599.794	613.980	-1168.595	1973.115	-17.177

PREVIOUS: December 1973 (1 atm) CURRENT: December 1973 (1 bar)

Lead Chloride ($PbCl_4$) $Cl_4Pb_1(g)$

$\Delta_fH°(0\ K) = -660.57 \pm 1.3\ kJ\ mol^{-1}$

$S°(298.15\ K) = 330.945 \pm 0.21\ J\ K^{-1}$ $\Delta_fH°(298.15\ K) = -662.75 \pm 1.3\ kJ\ mol^{-1}$

Vibrational Frequencies and Degeneracies

ν, cm^{-1}	ν, cm^{-1}
425 ± 1 (1)	620 ± 1 (3)
149 ± 2 (2)	220 ± 4 (3)

Ground State Quantum Weight: 1 σ = 12

Point Group: T_d

Bond Distance: Si-Cl = 2.0173 ± 0.0034 Å

Bond Angle: Cl-Si-Cl = 109.4712°

Product of the Moments of Inertia: $I_AI_BI_C = 2.60769 \times 10^{-112}\ g^3\ cm^6$

Enthalpy of Formation

Gross et al. (1) measured the enthalpy of formation of $SiCl_4(\ell)$ by the direct chlorination of silicon in a bomb calorimeter. Corrections for the difference in enthalpy content between liquid and gaseous chlorine at 25°C and 1 atm led to the value of $\Delta_fH°(SiCl_4, \ell, 298.15\ K) = -165.49 \pm 0.16\ kcal\ mol^{-1}$. Combination of $\Delta_fH°(\ell, 298.15\ K)$ with $\Delta_{vap}H°(298.15\ K) = 7.09\ kcal\ mol^{-1}$ gives $\Delta_fH°(SiCl_4, g, 298.15\ K) = -158.4 \pm 0.3\ kcal\ mol^{-1}$ which is adopted in the tabulation.

Schafer and Heine (2) measured both enthalpy of solution of Si(cr) and $SiCl_4(\ell)$ in HF solution containing AgF which gave $\Delta_rH°(298.15\ K) = 42.7 \pm 0.4\ kcal\ mol^{-1}$ for the overall reaction $SiCl_4(\ell) + 4Ag(cr) = Si(cr) + 4AgCl(cr)$ and led to $\Delta_fH°(SiCl_4, \ell, 298.15\ K) = -164.18\ kcal\ mol^{-1}$ in agreement with the value adopted, using $\Delta_fH°(AgCl, cr, 298.15\ K) = -30.37\ kcal\ mol^{-1}$ (3).

Wolf (4) determined the heats of solution of $SiCl_4(\ell)$ and $Na_2SiO_3.8H_2O(cr)$ in NaOH solution and derived $\Delta_rH°(298.15\ K) = -129.3\ kcal\ mol^{-1}$ for $SiCl_4(\ell) + 13NaOH(27.8\ H_2O) = Na_2SiO_3(cr) + 4NaCl + 3H_2O(\ell) + 7NaOH(52.14\ H_2O)$ which leads to $\Delta_fH°(SiCl_4, \ell, 298.15\ K) = -166.7 \pm 2\ kcal\ mol^{-1}$, using all JANAF auxiliary data and heats of dilution of NaOH from Parker (5). However, Wolf's data are suspect due to uncertainties in how he converted his data from Na_2SiO_3(calorimetric solution) to $Na_2SiO_3(cr)$. Beezer and Mortimer (6) measured the heat of solution $\Delta_rH°(298.15\ K) = -97.0\ kcal\ mol^{-1}$ for $SiCl_4(\ell) + 185\ HF(4.335\ H_2O) = 4HCl(11\ H_2O) + [H_2SiF_6, 179\ HF, 758\ H_2O]$ and combined with those of Good (7) for $[H_2SiF_6, 41\ HF, 174\ H_2O]$ solution by neglecting the unknown heats of dilution between the two calorimetric solutions and obtained $\Delta_fH°(\ell, 298.15\ K) = -169.9\ kcal\ mol^{-1}$ which is in poor agreement with the value adopted. In both the measurements of Woof and Beezer, the final state of the calorimetric solution is not well defined or not well known, thus it leads to large uncertainties in their results.

Other calorimetric studies involving aqueous hydrolysis of $SiCl_4(\ell)$ to form SiO_2(colloidal solution) have been reviewed by Gross et al. (1). Since the SiO_2(colloidal solution) is an ill-defined state, the heat of formation of $SiCl_4(\ell)$ derived from hydrolysis data is quite unreliable.

Heat Capacity and Entropy

Morino and Murata (8) measured the adopted bond distance by electron diffraction. Their value was confirmed by Ryan and Hedberg (9). Morino et al. (10) observed the adopted vibrational frequencies from the infrared and Raman spectra from the vapor. Infrared data of Smith (11) confirm that ν_3 is about 12 cm^{-1} larger in the vapor than in the liquid. The principal moments of inertia are: $I_A = I_B = I_C = 63.8879 \times 10^{-39}\ g\ cm^2$.

Vaporization Data

Vapor pressure data ranging from 20° below T_{fus}(supercooled liquid) to 10° above T_{vap} are in reasonable agreement as summarized below. The analysis is based on fugacities calculated from vapor pressures by the approximation of Pitzer (19) using T_c = 506 K (20), P_c = 37.1 atm (21) and ω = 0.26. At 298.15 K the calculated f/P is 0.982 compared with 0.984 observed by Wood (22). The analysis also presumes a linear variation of the liquid phase Cp based on 33.3 $calK^{-1}\ mol^{-1}$ at 210 K and 34.2 at 300 K. The latter value is from data of Vold (17) since earlier data (77-294 K) of Latimer (18) appear to have large positive bias below T_{fus}. The Cp at 210 K assumes Latimer's value to have about 1 percent bias by analogy with his data for $CCl_4(\ell)$.

Source	Data Points	T/K	$\Delta_{vap}S°(298.15\ K)$, cal $K^{-1}\ mol^{-1}$	$\Delta_{vap}H°(298.15\ K)$, kcal mol^{-1}
Combined (14), (15), (22)	19	275-341	21.45 ± 0.04	7.09 ± 0.01
(12)	Equation	183-204	21.13	7.19
(13)	6	303-323	20.85 ± 0.13	6.89 ± 0.04
(14)	12	298-341	21.48 ± 0.09	7.10 ± 0.03
(15)	6	275-330	21.40 ± 0.05	7.08 ± 0.01
(16)	13	273-333	21.55 ± 0.08	7.13 ± 0.03

References

1. P. Gross, C. Hayman and S. Mwroka, Trans. Faraday Soc. 65, 2856 (1969.
2. H. Schafer and H. Heine, Z. Anorg. Allg. Chem. 332 25, (1964).
3. U. S. Nat. Bur. Stand. Tech. Note 270-3, 1968.
4. E. Wolf, Z. Anorg. Chem. 313, 228 (1961).
5. V. B. Parker, NSRDS-NBS 2, 1965.
6. A. E. Beezer and C. T. Mortimer, J. Chem. Soc. 1964, 2727 (1964).
7. W. D. Good, J. L. Lacina, B. L. DePrater and J. P. McCullough, J. Phys. Chem. 68, 579 (1964).
8. Y. Morino and Y. Marate, Bull. Chem. Soc. Japan 38, 104 (1965). r_g = 2.0193 appears to be a misprint.
9. R. R. Ryan and K. Hedberg, J. Chem. Phys. 50, 4986 (1969).
10. Y. Morino, Y. Murata, T. Ito and J. Nakamura, J. Phys. Soc. Japan 17, Suppl. B-II, 37 (1962).
11. A. L. Smith, J. Chem. Phys. 21, 1997 (1953).
12. P. Balk and D. Dong, J. Phys. Chem. 68, 960 (1964).
13. W. R. King and L. N. Canjar, J. Chem. Eng. Data 7, 351 (1962).
14. L. A. Nisel'son and G. V. Seryakov, Russ. J. Inorg. Chem. 5, 548 (1960).
15. A. C. Jenkins and G. F. Chambers, Ind. Eng. Chem. 46, 2367 (1954).
16. K. Kearby, J. Amer. Chem. Soc. 58, 374 (1936).
17. R. D. Vold, J. Amer. Chem. Soc. 59, 1515 (1937).
18. W. M. Latimer, J. Amer. Chem. Soc. 44, 90 (1922).
19. G. N. Lewis, M. Randall, K. S. Pitzer and L. Brewer, "Thermodynamics", 2nd ed., pp. 605-29, McGraw-Hill Book Co., Inc., New York, 1961.
20. K. A. Kobe and R. E. Lynn, Chem. Revs. 52, 117 (1953).
21. W. Menzer, Naturwissenschafen 45, 126 (1958).
22. S. E. Wood, J. Amer. Chem. Soc. 59, 1510 (1937).

Enthalpy Reference Temperature = T_r = 298.15 K; Standard State Pressure = p° = 0.1 MPa

	J $K^{-1}mol^{-1}$			kJ mol^{-1}			
T/K	$C_p°$	S°	$-[G°-H°(T_r)]/T$	$H°-H°(T_r)$	$\Delta_fH°$	$\Delta_fG°$	Log K_f
0	0.	0.	INFINITE	-19.404	-660.570	-660.570	INFINITE
100	57.203	250.191	402.620	-15.243	-662.499	-649.333	339.177
200	78.605	297.154	338.877	-8.344	-662.854	-635.967	166.097
250	85.418	315.465	332.406	-4.235	-662.833	-629.246	131.474
298.15	90.261	330.945	330.945	0.	-662.746	-622.784	109.109
300	90.420	331.504	330.947	0.167	-662.741	-622.536	108.393
350	94.081	345.732	332.062	4.784	-662.596	-615.846	91.910
400	96.788	358.480	334.582	9.559	-662.410	-609.179	79.551
450	98.820	370.003	337.888	14.452	-662.194	-602.538	69.941
500	100.372	380.499	341.632	19.433	-661.957	-595.922	62.256
600	102.529	399.005	349.693	29.587	-661.443	-582.763	50.734
700	103.910	414.921	357.901	39.914	-660.908	-569.692	42.511
800	104.841	428.861	365.918	50.354	-660.376	-556.697	36.349
900	105.495	441.249	373.613	60.873	-659.860	-543.769	31.560
1000	105.972	452.390	380.943	71.447	-659.370	-530.896	27.731
1100	106.329	462.508	387.905	82.063	-658.910	-518.071	24.601
1200	106.604	471.772	394.513	92.710	-658.485	-505.287	21.995
1300	106.819	480.314	400.789	103.382	-658.099	-492.536	19.790
1400	106.991	488.236	406.756	114.073	-657.754	-479.814	17.902
1500	107.130	495.623	412.437	124.779	-657.454	-467.114	16.266
1600	107.245	502.541	417.854	135.498	-657.200	-454.434	14.836
1700	107.340	509.045	423.029	146.227	-707.173	-441.320	13.560
1800	107.420	515.183	427.980	156.966	-706.790	-425.693	12.353
1900	107.488	520.993	432.724	167.711	-706.419	-410.087	11.274
2000	107.546	526.508	437.276	178.463	-706.061	-394.500	10.303
2100	107.596	531.756	441.651	189.220	-705.721	-378.930	9.425
2200	107.639	536.762	445.862	199.982	-705.400	-363.376	8.628
2300	107.677	541.548	449.919	210.747	-705.102	-347.837	7.900
2400	107.710	546.131	453.833	221.517	-704.830	-332.309	7.233
2500	107.740	550.529	457.613	232.289	-704.587	-316.793	6.619
2600	107.766	554.755	461.269	243.065	-704.375	-301.285	6.053
2700	107.789	558.823	464.807	253.842	-704.197	-285.785	5.529
2800	107.810	562.743	468.235	264.622	-704.054	-270.291	5.042
2900	107.829	566.527	471.560	275.404	-703.948	-254.802	4.589
3000	107.845	570.182	474.786	286.188	-703.880	-239.315	4.167
3100	107.861	573.719	477.921	296.973	-703.848	-223.830	3.772
3200	107.875	577.144	480.969	307.760	-703.853	-208.346	3.401
3300	107.887	580.463	483.933	318.548	-703.892	-192.860	3.053
3400	107.899	583.684	486.820	329.337	-703.963	-177.374	2.725
3500	107.909	586.812	489.633	340.128	-704.064	-161.885	2.416
3600	107.919	589.852	492.374	350.919	-1088.341	-135.931	1.972
3700	107.928	592.809	495.049	361.712	-1088.075	-109.479	1.546
3800	107.936	595.687	497.660	372.505	-1087.827	-83.034	1.141
3900	107.944	598.491	500.209	383.299	-1087.592	-56.595	0.758
4000	107.951	601.224	502.701	394.094	-1087.366	-30.162	0.394
4100	107.958	603.890	505.136	404.889	-1087.143	-3.735	0.048
4200	107.964	606.491	507.519	415.685	-1086.917	22.687	-0.282
4300	107.969	609.032	509.850	426.482	-1086.683	49.104	-0.596
4400	107.975	611.514	512.133	437.279	-1086.435	75.514	-0.896
4500	107.980	613.941	514.368	448.077	-1086.169	101.919	-1.183
4600	107.984	616.314	516.559	458.875	-1085.880	128.318	-1.457
4700	107.989	618.636	518.706	469.674	-1085.564	154.710	-1.719
4800	107.993	620.910	520.812	480.473	-1085.215	181.096	-1.971
4900	107.996	623.137	522.877	491.272	-1084.830	207.473	-2.212
5000	108.000	625.319	524.904	502.072	-1084.405	233.842	-2.443
5100	108.003	627.457	526.894	512.872	-1083.938	260.203	-2.665
5200	108.007	629.555	528.848	523.673	-1083.424	286.553	-2.878
5300	108.010	631.612	530.768	534.473	-1082.862	312.894	-3.084
5400	108.012	633.631	532.654	545.275	-1082.250	339.223	-3.281
5500	108.015	635.613	534.508	556.076	-1081.584	365.541	-3.472
5600	108.018	637.559	536.331	566.878	-1080.865	391.846	-3.655
5700	108.020	639.471	538.124	577.679	-1080.090	418.137	-3.832
5800	108.022	641.350	539.887	588.482	-1079.259	444.415	-4.002
5900	108.025	643.196	541.623	599.284	-1078.370	470.678	-4.167
6000	108.027	645.012	543.331	610.087	-1077.423	496.924	-4.326

PREVIOUS: December 1970 (1 atm) CURRENT: December 1970 (1 bar)

TITANIUM TETRACHLORIDE ($TiCl_4$) CRYSTAL M_r = 189.712

$S°(298.15\ K) = 209.069 \pm 0.4\ J\ K^{-1}\ mol^{-1}$

$T_{fus} = 249.05\ K$

$\Delta_f H°(0\ K) = [-818.798]\ kJ\ mol^{-1}$

$\Delta_f H°(298.15\ K) = [-815.026]\ kJ\ mol^{-1}$

$\Delta_{fus} H° = 9.966\ kJ\ mol^{-1}$

Heat of Formation

The enthalpy of formation, $\Delta_f H°(298.15\ K)$, of $TiCl_4(cr)$ is calculated from the corresponding quantity for $TiCl_4(\ell)$ and the values of H°(249.05 K) - H°(298.15 K) for both crystal and liquid.

Heat Capacity and Entropy

The heat capacity of $TiCl_4(cr)$ has been reported by Latimer (1) and by Furukawa (2). The data from the latter investigation are adopted. These data lead to a value of 44.5 cal $K^{-1}\ mol^{-1}$ for S°(249.05 K). This compares with Latimer's value (1) of 43.8 cal $K^{-1}\ mol^{-1}$ for the same quantity. The heat capacity of the solid is extrapolated to a value of 18R at 600 K, based on the known heat capacities of $TiBr_4(cr)$ and $TiI_4(cr)$. The hypothetical entropy at 298.15 K is base don the integration of the extrapolated curve and the value at the melting point.

Melting Data

The triple point is reported to be 249.05 K, with a measured enthalpy of fusion of 2.382 kcal mol^{-1}, by Furukawa (2). Latimer (1) reported values of 248.0 K and 2.233 kcal mol^{-1} for these quantities. The former data are adopted.

References

1. W. M. Latimer, J. Amer. Chem. Soc. 44, 90 (1922).
2. G. T. Furukawa, personal communication, U. S. Nat. Bur. Stand., 1964.

Titanium Chloride ($TiCl_4$) $Cl_4Ti_1(cr)$

Enthalpy Reference Temperature = T_r = 298.15 K   Standard State Pressure = p° = 0.1 MPa

T/K	$C_p°$	S°	$-[G°-H°(T_r)]/T$	$H°-H°(T_r)$	$\Delta_f H°$	$\Delta_f G°$	Log K_f
	J K⁻¹mol⁻¹			kJ mol⁻¹			
0	0.	0.	INFINITE	-26.963	-818.798	-818.798	INFINITE
100	84.480	92.582	312.283	-21.970	-820.189	-790.820	413.082
200	112.380	160.341	220.513	-12.034	-818.245	-762.083	199.036
249.046	124.710	186.262	211.264	-6.227	------ CRYSTAL <--> LIQUID -----		
298.15	129.460	209.069	209.069	0.	-815.026	-735.167	128.798
300	129.704	209.871	209.072	0.240	-814.959	-734.672	127.918
400	139.540	248.622	214.288	13.733	-811.019	-708.488	92.519
500	145.600	280.480	224.436	28.022	-806.613	-683.358	71.390

PREVIOUS: March 1964   CURRENT: December 1967

TITANIUM CHLORIDE ($TiCl_4$) LIQUID M_r = 189.692

$S°(298.15 K) = 221.93 \pm 0.4 J K^{-1} mol^{-1}$ $\Delta_f H°(298.15 K) = -804.2 \pm 4.2 kJ mol^{-1}$

$T_{fus} = 249.05 K$ $\Delta_{fus}H° = 9.966 kJ mol^{-1}$

Enthalpy of Formation

The enthalpy of formation, $\Delta_f H°(TiCl_4, \ell, 298.15 K)$, is calculated from that of $TiCl_4$(g) and the value of $\Delta_{vap}H°(298.15 K)$. The value of $\Delta_{vap}H°(298.15 K)$ is determined from 2nd and 3rd law analyses of fugacity as follows.

Source	Method	T/K	Data Points	$\Delta_{vap}H°(298.15 K)$ 2nd law	3rd law	Drift cal $K^{-1} mol^{-1}$	$\Delta_f H°(298.15 K)$ kcal mol^{-1}
Pike and Foster (1)	Manometric	363-415	18*	9.82±0.02	9.83	0.0±0.1	-192.2
Seryakov et al. (2)	Manometric	358-412	8	9.79±0.03	9.84	0.1±0.1	-192.2
Schafer et al. (3)	Isoteniscope	313-358	17*	9.86±0.02	9.82	-0.1±0.0	-192.2
Schafer et al. (3)	Isoteniscope	313-357	16**	9.80±0.01	9.82	0.0±0.0	-192.2
Schafer et al. (3)	Isoteniscope	313-357	16	9.91±0.02	9.83	-0.2±0.1	-192.2
Schafer et al. (3)	Isoteniscope	312-359	17	9.90±0.02	9.83	-0.2±0.1	-192.2
Weed (4)	Manometric	298-319	***	9.79±0.01	9.82	0.1±0.0	-192.2
Weed (4)	Manometric	298-319	***	9.81±0.01	9.82	0.0±0.0	-192.2
Weed (4)	Spectra	250-302	***	9.69±0.01	9.80	0.4±0.0	-192.2

* Two Points rejected to failure of a statistical test

** Three points rejected due to failure of a statistical test

*** Selected points from extensive data

3rd law analyses of the several vapor pressure sets result in drifts which are proportional to the mean temperatures. Conversion of these data to fugacity eliminates this temperature dependence and the drifts are then scattered randomly about a constant value. This constant value is made to be essentially zero by methods described on the $TiCl_4$(g) table. The conversion to fugacity was made using Pitzer's method for normal fluids (5). The calculation was carried out using the value of 45.7 atm. for the critical pressure, P_c, reported by Minzer (6) and estimated values of 643±15 K for the critical temperature, T_c, and 0.23±0.03 for the acentric constant, ω. Schafer et al. (7) reported a calorimetric determination of $\Delta_{vap}H°(298.15 K)$. Their value of 9.9±0.02 kcal mol^{-1} was used to check the reliability of their equipment, assuming that the vapor pressure data (3) gave a more accurate determination of $\Delta_{vap}H°(298.15 K)$. The adopted value of $\Delta_{vap}H°(298.15 K)$ is 9.82±0.02 kcal mol^{-1}.

Heat Capacity and Entropy

The entropy and heat capacity of $TiCl_4(\ell)$ has been measured by Furukawa (8). His reported values are adopted.

Fusion Data

See the crystal table for details.

Vaporization Data

T_{vap} is taken as the temperature at which Kp = 1 for the reaction $TiCl_4(\ell) = TiCl_4(g)$. The vapor pressure data are discussed above.

References

1. F. P. Pike and C. T. Foster, Jr., J. Chem. Eng. Data 4, 305 (1959).
2. G. V. Seryakov, S. A. Vaks, L. S. Sidorina, Zhur. Obshch. Khim. 30, 2130 (1960).
3. H. Schafer and F. Zeppenick, Z. Anorg. Chem. 272, 274 (1953).
4. H. C. Weed, Ph.D. Thesis, The Ohio State University, 1957.
5. G. N. Lewis and M. Randall, "Thermodynamics," McGraw-Hill Book Co., Inc., New York, 1961.
6. W. Minzer, Naturwissenschaften 45, 126 (1958).
7. H. Schafer, G. Breil and G. Pfeffer, Z. Anorg. Chem. 276, 325 (1954).
8. G. T. Furukawa, U. S. Nat. Bur. Stand., personal communication, 1964.

Titanium Chloride ($TiCl_4$) Cl_4Ti_1(l)

Enthalpy Reference Temperature = T_r = 298.15 K Standard State Pressure = p° = 0.1 MPa

T/K	$C_p°$ (J $K^{-1}mol^{-1}$)	S°	$-[G°-H°(T_r)]/T$	$H°-H°(T_r)$ (kJ mol^{-1})	$\Delta_f H°$	$\Delta_f G°$	Log K_f
0							
100							
200	144.700	164.061	235.188	-14.226	-809.575	-754.157	196.965
249.046	144.910	195.826	224.422	-7.122	------ CRYSTAL <--> LIQUID -----		
298.15	145.205	221.925	221.925	0.	-804.165	-728.139	127.567
300	145.220	222.824	221.928	0.269	-804.069	-727.668	126.698
400	146.170	264.741	227.637	14.842	-799.049	-702.966	91.798
500	147.080	297.456	238.447	29.504	-794.269	-679.502	70.987
600	147.985	324.352	250.589	44.258	-789.606	-656.987	57.196
700	148.892	347.232	262.801	59.101	-784.994	-635.249	47.403
800	149.800	367.173	274.628	74.036	-780.383	-614.172	40.101
900	150.709	384.869	285.912	89.061	-775.789	-593.672	34.456
1000	151.620	400.795	296.617	104.178	-771.267	-573.680	29.966

PREVIOUS: March 1964 CURRENT: December 1967

Titanium Chloride ($TiCl_4$) Cl_4Ti_1(l)

TITANIUM CHLORIDE ($TiCl_4$) CRYSTAL–LIQUID M_r = 189.692

0 to 249.05 K crystal
above 249.05 K liquid

Refer to the individual tables for details.

Titanium Chloride ($TiCl_4$) Cl_4Ti_1(cr,l)

Enthalpy Reference Temperature = T_r = 298.15 K Standard State Pressure = p° = 0.1 MPa

T/K	C_p° (J K⁻¹mol⁻¹)	S° (J K⁻¹mol⁻¹)	$-[G^\circ-H^\circ(T_r)]/T$ (J K⁻¹mol⁻¹)	$H^\circ-H^\circ(T_r)$ (kJ mol⁻¹)	Δ_fH° (kJ mol⁻¹)	Δ_fG° (kJ mol⁻¹)	Log K_f
0	0.	0.	INFINITE	-30.240	-811.213	-811.213	INFINITE
100	84.480	92.582	345.052	-25.247	-812.605	-783.235	409.120
200	112.380	160.341	236.897	-15.311	-810.661	-754.498	197.055
249.046	124.710	186.262	224.422	-9.504	CRYSTAL <--> LIQUID TRANSITION		
249.046	144.910	195.826	224.422	-7.122			
298.15	145.205	221.925	221.925	0.	-804.165	-728.139	127.567
300	145.220	222.824	221.928	0.269	-804.069	-727.668	126.698
400	146.170	264.741	227.637	14.842	-799.049	-702.966	91.798
500	147.080	297.456	238.447	29.504	-794.269	-679.502	70.987
600	147.985	324.352	250.589	44.258	-789.606	-656.988	57.196
700	148.892	347.232	262.801	59.101	-784.994	-635.249	47.403
800	149.800	367.173	274.628	74.036	-780.383	-614.172	40.101
900	150.709	384.869	285.912	89.061	-775.789	-593.672	34.456
1000	151.620	400.795	296.617	104.178	-771.267	-573.680	29.966

PREVIOUS: CURRENT: December 1967

Titanium Chloride ($TiCl_4$) Cl_4Ti_1(cr,l)

TITANIUM CHLORIDE ($TiCl_4$) IDEAL GAS M_r = 189.692

$S°(298.15\ K) = 354.889 \pm 2.9\ J\ K^{-1}\ mol^{-1}$

$\Delta_f H°(0\ K) = -761.58 \pm 3.8\ kJ\ mol^{-1}$
$\Delta_f H°(298.15\ K) = -763.16 \pm 3.8\ kJ\ mol^{-1}$

Vibrational Frequences and Degeneracies

ν, cm^{-1}	ν, cm^{-1}
368 (1)	498.5(3)
111 (2)	131 (3)

Ground State Quantum Weight: 1 $\sigma = 12$
Point Group: T_d
Bond Distance: Ti-Cl = 2.185 Å
Bond Angle: Cl-Ti-Cl = 109.4712°
Product of the Moments of Inertia: $I_A I_B I_C = 4.21057 \times 10^{-112}\ g^3\ cm^6$

Enthalpy of Formation

$\Delta_f H°(298.15\ K)$ of $TiCl_4(g)$ and $TiCl_4(\ell)$ has been measured by several investigators. The results of these measurements are as follows.

Source	Method	Reaction	$\Delta_f H°(298.15\ K)$, kcal mol^{-1} $TiCl_4(\ell)$	$TiCl_4(g)$
Johnson et al. (1)	Calorimetric	$Ti(cr) + 2\ Cl_2(g) = TiCl_4(g)$	(-192.2)	-182.4±0.7
Farber and Darnell (2)	Equilibrium	$TiO_2(cr) + 4\ HCl(g) = TiCl_4(g) + 2\ H_2O(g)$	(-192.7)	-182.9±0.5
Skinner and Ruehrwein (3)	Calorimetric	$Ti(cr) + xCl_2(g) = [TiCl_4 + (x-2)Cl_2]$ soln	-190.3±3.0	(-180.5)
Gross et al. (4)	Calorimetric	$Ti(cr) + xCl_2(\ell) = [TiCl_4 + (x-2)Cl_2]$ soln	-191.5±0.3	(-181.7)
Krieve et al. (5)	Calorimetric	$Ti(cr) + xCl_2(g) = [TiCl_4 + (x-2)Cl_2]$ soln	-190.0±0.4	(-180.2)
Thomsen (6)		$TiCl_4(\ell) + 2\ H_2O(\ell) = 4\ HCl_{(0.002\ m)} + TiO_2(cr)$	-194.5	(-184.7)

The chosen value of $\Delta_f H°(298.15\ K)$ is that reported by Johnson et al. (1). This investigation has the advantage of being independent of the enthalpy of vaporization of chlorine and any heats of solution in deriving the value of the enthalpy of formation.

Heat Capacity and Entropy

The adopted value for the interatomic distance is that reported by Kimura et al. (7). The tetrahedral structure was established by the Raman work of Bhagavantam (8). The vibrational frequencies determined from the infrared and Raman spectra of $TiCl_4$ by Hawkins and Carpenter (9) are adjusted downward 8 cm^{-1} for ν_2 and ν_4 so that the enthalpies of vaporization determined by both 2nd and 3rd law methods are in agreement. See $TiCl_4(\ell)$ table for details. The principal moments of inertia are: $I_A = I_B = I_C = 74.9515 \times 20^{-39}\ g\ cm^2$.

References

1. W. H. Johnson, R. A. Nelson and E. J. Prosen, J. Res. Nat. Bur. Stand. 62A, 49 (1959).
2. M. Farber and M. Darnell, J. Chem. Phys. 23, 1460 (1955).
3. G. B. Skinner and R. A. Ruehrwein, J. Phys. Chem. 59, 113 (1955).
4. P. Gross, C. Hayman and D. L. Levi, Trans. Faraday Soc. 53, 1601 (1957).
5. W. F. Krieve, S. P. Vango and D. M. Mason, J. Chem. Phys. 25, 519 (1956).
6. J. Thomsen, Thermochemische Untersuchungen, I. (Verlag, J. A. Barth, Leipzig, 1882).
7. M. Kimura, K. Kimura, M. Aoki and S. Shibata, Bull, Chem. Soc. Japan 29, 95 (1956).
8. S. Bhagavantam, Indian J. Phys. 7, 79 (1932).
9. N. J. Hawkins and D. R. Carpenter, J. Chem. Phys. 23, 1700 (1955).

Titanium Chloride ($TiCl_4$) $Cl_4Ti_1(g)$

Enthalpy Reference Temperature = T_r = 298.15 K; Standard State Pressure = $p°$ = 0.1 MPa

T/K	$C_p°$ (J $K^{-1}mol^{-1}$)	S°	$-[G°-H°(T_r)]/T$	$H°-H°(T_r)$ (kJ mol^{-1})	$\Delta_f H°$	$\Delta_f G°$	Log K_f
0	0.	0.	INFINITE	-21.613	-761.583	-761.583	INFINITE
100	67.436	265.427	432.828	-16.740	-763.094	-751.010	392.287
200	86.118	318.509	363.378	-8.974	-763.321	-738.791	192.953
250	91.800	338.374	356.444	-4.517	-763.265	-732.664	153.082
298.15	95.614	354.889	354.889	0.	-763.162	-726.779	127.329
300	95.735	355.481	354.891	0.177	-763.157	-726.553	126.504
350	98.486	370.458	356.068	5.037	-763.022	-720.463	107.523
400	100.452	383.744	358.712	10.013	-762.875	-714.393	93.290
450	101.892	395.663	362.167	15.073	-762.724	-708.342	82.222
500	102.970	406.457	366.065	20.196	-762.574	-702.307	73.370
600	104.441	425.371	374.417	30.572	-762.288	-690.281	60.094
700	105.366	441.546	382.880	41.066	-762.026	-678.301	50.615
800	105.982	455.658	391.114	51.635	-761.781	-666.358	43.509
900	106.412	468.167	398.994	62.256	-761.591	-654.442	37.983
1000	106.723	479.396	406.482	72.914	-761.528	-642.542	33.563
1100	106.956	489.579	413.581	83.598	-761.659	-630.638	29.947
1200	107.134	498.893	420.307	94.303	-765.972	-618.594	26.927
1300	107.273	507.474	426.687	105.024	-765.777	-606.321	24.362
1400	107.384	515.428	432.745	115.757	-765.666	-594.060	22.165
1500	107.474	522.840	438.507	126.500	-765.655	-581.804	20.260
1600	107.547	529.779	443.997	137.251	-765.758	-569.544	18.594
1700	107.609	536.301	449.237	148.009	-765.990	-557.274	17.123
1800	107.660	542.453	454.246	158.773	-766.366	-544.987	15.815
1900	107.703	548.275	459.043	169.541	-766.900	-532.674	14.644
2000	107.741	553.801	463.644	180.313	-782.294	-519.876	13.578
2100	107.773	559.058	468.063	191.089	-783.940	-506.714	12.604
2200	107.801	564.072	472.314	201.867	-785.606	-493.474	11.717
2300	107.825	568.865	476.409	212.649	-787.297	-480.157	10.905
2400	107.846	573.454	480.357	223.432	-789.015	-466.767	10.159
2500	107.865	577.857	484.170	234.218	-790.762	-453.304	9.471
2600	107.882	582.088	487.855	245.005	-792.542	-439.771	8.835
2700	107.897	586.160	491.421	255.794	-794.357	-426.168	8.245
2800	107.910	590.084	494.875	266.584	-796.208	-412.498	7.695
2900	107.922	593.871	498.224	277.376	-798.097	-398.761	7.182
3000	107.933	597.530	501.474	288.169	-800.024	-384.958	6.703
3100	107.943	601.069	504.630	298.963	-801.988	-371.090	6.253
3200	107.951	604.496	507.697	309.757	-803.989	-357.158	5.830
3300	107.960	607.818	510.681	320.553	-806.025	-343.163	5.432
3400	107.967	611.041	513.586	331.349	-808.093	-329.106	5.056
3500	107.974	614.171	516.415	342.146	-810.191	-314.988	4.701
3600	107.980	617.213	519.173	352.944	-812.315	-300.809	4.365
3700	107.986	620.171	521.863	363.742	-1223.559	-278.783	3.936
3800	107.991	623.051	524.488	374.541	-1224.487	-253.236	3.481
3900	107.996	625.857	527.051	385.340	-1225.483	-227.664	3.049
4000	108.000	628.591	529.556	396.140	-1226.541	-202.066	2.639
4100	108.004	631.258	532.004	406.940	-1227.654	-176.440	2.248
4200	108.008	633.860	534.398	417.741	-1228.815	-150.787	1.875
4300	108.012	636.402	536.741	428.542	-1230.017	-125.105	1.520
4400	108.015	638.885	539.034	439.343	-1231.253	-99.395	1.180
4500	108.019	641.312	541.280	450.145	-1232.517	-73.657	0.855
4600	108.021	643.687	543.481	460.947	-1233.802	-47.890	0.544
4700	108.024	646.010	545.638	471.749	-1235.101	-22.096	0.246
4800	108.027	648.284	547.752	482.552	-1236.408	3.726	-0.041
4900	108.029	650.512	549.827	493.355	-1237.718	29.576	-0.315
5000	108.032	652.694	551.863	504.158	-1239.024	55.452	-0.579
5100	108.034	654.833	553.861	514.961	-1240.322	81.355	-0.833
5200	108.036	656.931	555.823	525.765	-1241.605	107.283	-1.078
5300	108.038	658.989	557.750	536.568	-1242.871	133.235	-1.313
5400	108.040	661.009	559.643	547.372	-1244.113	159.211	-1.540
5500	108.041	662.991	561.504	558.176	-1245.329	185.210	-1.759
5600	108.043	664.938	563.334	568.980	-1246.515	211 230	-1.970
5700	108.044	666.850	565.134	579.785	-1247.666	237.272	-2.174
5800	108.046	668.729	566.904	590.589	-1248.781	263.333	-2.372
5900	108.047	670.576	568.645	601.394	-1249.857	289.413	-2.562
6000	108.049	672.392	570.359	612.199	-1250.891	315.511	-2.747

PREVIOUS: December 1967 (1 atm) CURRENT: December 1967 (1 bar)

Titanium Chloride ($TiCl_4$) $Cl_4Ti_1(g)$

TUNGSTEN CHLORIDE (WCl_4) CRYSTAL M_r = 325.662

$\Delta_f H°(0\ K)$ = Unknown

$S°(298.15\ K) = [198.322]\ J\ K^{-1}\ mol^{-1}$

$\Delta_f H°(298.15\ K) = -443.09 \pm 62.8\ kJ\ mol^{-1}$

Enthalpy of Formation

$\Delta_f H°(WCl_4$, cr, 298.15 K) = -105.9 kcal mol^{-1} is calculated from $\Delta_r H°(298.15\ K)$ = 58.9 kcal mol^{-1} for 3 $WCl_4(cr) = WCl_2(cr) + 2\ WCl_5(g)$. The value of $\Delta_r H°(298.15\ K)$ is calculated by the 3rd law method from decomposition pressure data in the temperature range from 354°C to 436°C reported by Shchukarev et al. (1). The 2nd law, $\Delta_r H°(298.15\ K) = 73.6 \pm 7.9$ kcal mol^{-1} and the 3rd law drift is -24±12 cal $K^{-1}\ mol^{-1}$ which would correspond to a total entropy discrepancy of -8±4 cal $K^{-1}\ mol^{-1}$ per mole of WCl_4. This entropy discrepancy is probably within the combined uncertainty of the data and the estimated entropies for the three species. Shchukarev et al. (2) have shown that decomposition pressures are essentially independent of the composition in the solid phase until WCl_4(cr) is almost completely decomposed to WCl_2(cr). This is consistent with the assumption of solid phase activities of unity which we have used in the equilibrium analysis.

Heat Capacity and Entropy

$C_p°(298.15\ K)$ = 6.2 cal K^{-1} g-atom and $C_p°(700\ K)$ = 7.25 cal K^{-1} g-atom are estimated using the method described by Kubaschewski and Evans (3). Between 298 and 700 K, which is near the decomposition temperature, the heat capacity is obtained by linear interportion.

The entropy of WCl_4(cr), $S°(298.15\ K)$ = 47.4 cal $K^{-1}\ mol^{-1}$, is estimated by the method of Latimer (4).

Decomposition Data

T_{dcm} = 771 K (f = 1 atm) is calculated as the temperature at which the Gibbs energy change of the reaction 3 $WCl_4(cr) = WCl_2(cr) + 2\ WCl_5(g)$ approaches zero. We have also derived an approximate value of decomposition temperature, T_{dcm} = 744 K, by extrapolation to one atmosphere total pressure of a log P versus 1/T fit of the decomposition pressure data reported by Shchukarev et al. (5). The difference between the two decomposition temperatures is related to the 2nd law entropy and the adopted entropy.

References

1. S. A. Shchukarev, G. I. Novikov and N. V. Andreeva, Vestnik Leningrad. Univ. 14, No. 4, Ser. Fiz. I Khim., No. 1, 120 (1959).
2. S. A. Shchukarev et al., Vestnik Leningrad. Univ. 14, No. 10, Ser. Fiz I Khim. No. 2, 78 (1959).
3. O. Kubaschewski and E. Ll. Evans, "Metallurgical Thermochemistry," Pergamon Press, New York, 1958.
4. W. M. Latimer, "Oxidation Potentials," 2nd ed., Prentice-Hall, Inc., New York, 1952.
5. S. A. Shchukarev et al., Vestnik Leningrad. Univ. 14, No. 4, Ser. Fiz. I Khim., No. 1, 120 (1959).

Tungsten Chloride (WCl_4) Cl_4W_1(cr)

Enthalpy Reference Temperature = T_r = 298.15 K Standard State Pressure = p° = 0.1 MPa

T/K	$C_p°$ (J K⁻¹mol⁻¹)	S° (J K⁻¹mol⁻¹)	$-[G°-H°(T_r)]/T$ (J K⁻¹mol⁻¹)	$H°-H°(T_r)$ (kJ mol⁻¹)	$\Delta_f H°$ (kJ mol⁻¹)	$\Delta_f G°$ (kJ mol⁻¹)	Log K_f
0							
100							
200							
298.15	129.704	198.322	198.322	0.	-443.086	-359.456	62.975
300	129.830	199.124	198.324	0.240	-443.016	-358.937	62.496
400	135.290	237.222	203.482	13.496	-439.164	-331.485	43.288
500	140.750	267.997	213.401	27.298	-435.019	-305.038	31.867
600	146.210	294.142	224.732	41.646	-430.492	-279.460	24.329
700	151.670	317.090	236.319	56.540	-425.543	-254.673	19.004
800	157.130	337.700	247.724	71.981	-420.147	-230.626	15.058
900	162.507	356.526	258.782	87.970	-414.287	-207.283	12.030
1000	167.025	373.888	269.436	104.452	-408.012	-184.616	9.643
1100	170.875	389.994	279.672	121.353	-401.390	-162.594	7.721
1200	173.929	404.997	289.498	138.598	-394.493	-141.188	6.146
1300	176.439	419.019	298.928	156.119	-387.387	-120.366	4.836
1400	178.657	432.178	307.981	173.876	-380.110	-100.098	3.735
1500	180.623	444.572	316.678	191.841	-372.689	-80.355	2.798
1600	182.422	456.287	325.040	209.995	-365.145	-61.111	1.995
1700	184.096	467.397	333.090	228.321	-357.493	-42.343	1.301
1800	185.770	477.967	340.848	246.815	-349.738	-24.028	0.697
1900	187.276	488.052	348.332	265.468	-341.892	-6.147	0.169
2000	188.698	497.695	355.561	284.268	-333.969	11.319	-0.296

PREVIOUS: September 1962 CURRENT: December 1966

Tungsten Chloride (WCl_4) Cl_4W_1(cr)

TUNGSTEN CHLORIDE (WCl_4) IDEAL GAS M_r = 325.662

$\Delta_f H°(0\ K) = -335.3 \pm 33\ kJ\ mol^{-1}$

$S°(298.15\ K) = 379.285\ J\ K^{-1}\ mol^{-1}$ $\Delta_f H°(298.15\ K) = -336.0 \pm 33\ kJ\ mol^{-1}$

Electronic Levels and Quantum Weights

ε_i, cm^{-1}	g_i
0	[3]
[10000]	[3]
[30000]	[3]

Vibrational Frequencies and Degeneracies

ν, cm^{-1}	ν, cm^{-1}
[380](1)	[390](3)
[110](2)	[115](3)

Point Group: [T_d] σ = [12]

Bond Distance: W-Cl = [2.26] Å

Bond Angle: Cl-W-Cl = [109.4712]°

Product of the Moments of Inertia: $I_A I_B I_C = [5.15566 \times 10^{-112}]\ g^3\ cm^6$

Enthalpy of Formation

$\Delta_f H°(WCl_4, g, 298.15\ K) = -80.3\ kcal\ mol^{-1}$ is calculated from $\Delta_r H° = 37.7\ kcal\ mol^{-1}$ for $WCl_6(g) + WCl_4(g) + Cl_2(g)$. The value of $\Delta_r H°(298.15\ K)$ is calculated by the 3rd law method from the equation for log Kp given by Shchukarev and Suvorov (1). The 2nd law $\Delta_r H°(298.15\ K)$ is 43.6 kcal mol^{-1} and the 3rd law drift is -9.6 cal $K^{-1}\ mol^{-1}$.

Heat Capacity and Entropy

The molecular configuration is assumed to be tetrahedral similar to those of $SiCl_4$, $TiCl_4$, VCl_4, $GeCl_4$, $ZrCl_4$, $SnCl_4$, $PbCl_4$. The bond distance is estimated to be the same as that in $WCl_6(g)$. The principal moments of inertia are: $I_A = I_B = I_C = 80.1853 \times 10^{-39}\ g\ cm^2$.

All vibrational frequencies are estimated from those of $SiCl_4$, $TiCl_4$, VCl_4, $GeCl_4$, $SnCl_4$ and $PbCl_4$ given by Nakamoto (2).

If it is considered that the metal atoms are effectively ionized, then the W^{4+} will have 2 d electrons in a field of chloride ions. The electronic splitting would be roughly the inverse of that in $NiCl_2(g)$, as given by DeKock and Gruen (3), where Ni has a d^8 configuration. By analogy with the splitting in $NiCl_2$, the low lying electronic levels are estimated to be 10000 cm^{-1} and 30000 cm^{-1}, both triplets.

References

1. S. A. Shchukarev and A. V. Suvorov, Vestnik Leningrad. Univ. 16, No. 4, Ser. Fiz. 1 Khim., No. 1, 87 (1961).
2. Kazuo Nakamoto, "Infrared Spectra of Inorganic and Coordination Compounds," John Wiley & Sons, Inc., New York, 1963.
3. D. W. DeKock and D. M. Gruen, J. Chem. Phys. 44, 4387 (1966).

Tungsten Chloride (WCl_4) $Cl_4W_1(g)$

Enthalpy Reference Temperature = T_r = 298.15 K Standard State Pressure = p° = 0.1 MPa

T/K	$C_p°$ (J K⁻¹mol⁻¹)	S° (J K⁻¹mol⁻¹)	-[G°-H°(T_r)]/T (J K⁻¹mol⁻¹)	H°-H°(T_r) (kJ mol⁻¹)	$\Delta_f H°$ (kJ mol⁻¹)	$\Delta_f G°$ (kJ mol⁻¹)	Log K_f
0	0.	0.	INFINITE	-22.616	-335.256	-335.256	INFINITE
100	70.716	285.257	460.891	-17.563	-336.680	-326.443	170.516
200	90.681	341.319	388.116	-9.359	-336.559	-316.182	82.578
250	95.721	362.138	380.897	-4.690	-336.284	-311.119	65.005
298.15	98.865	379.285	379.285	0.	-335.975	-306.300	53.662
300	98.963	379.897	379.287	0.183	-335.963	-306.115	53.299
350	101.124	395.325	380.501	5.189	-335.619	-301.168	44.947
400	102.620	408.932	383.221	10.284	-335.265	-296.270	38.689
450	103.691	421.084	386.765	15.444	-334.910	-291.417	33.827
500	104.481	432.052	390.754	20.649	-334.557	-286.604	29.941
600	105.540	451.203	399.279	31.154	-333.874	-277.078	24.122
700	106.197	467.525	407.891	41.744	-333.229	-267.664	19.973
800	106.630	481.736	416.253	52.386	-332.630	-258.338	16.868
900	106.931	494.314	424.241	63.065	-332.081	-249.085	14.457
1000	107.148	505.592	431.822	73.770	-331.584	-239.891	12.531
1100	107.312	515.812	439.000	84.493	-331.140	-230.744	10.957
1200	107.439	525.155	445.796	95.231	-330.750	-221.634	9.647
1300	107.544	533.759	452.236	105.980	-330.415	-212.555	8.541
1400	107.635	541.732	458.347	116.739	-330.136	-203.500	7.593
1500	107.719	549.161	464.156	127.507	-329.913	-194.463	6.772
1600	107.801	556.116	469.689	138.283	-329.746	-185.438	6.054
1700	107.885	562.654	474.967	149.067	-329.636	-176.423	5.421
1800	107.974	568.823	480.011	159.860	-329.583	-167.412	4.858
1900	108.069	574.663	484.841	170.662	-329.588	-158.403	4.355
2000	108.173	580.209	489.472	181.474	-329.651	-149.392	3.902
2100	108.284	585.489	493.919	192.297	-329.776	-140.376	3.492
2200	108.403	590.529	498.197	203.131	-329.962	-131.353	3.119
2300	108.530	595.351	502.317	213.978	-330.212	-122.320	2.778
2400	108.663	599.973	506.290	224.838	-330.529	-113.275	2.465
2500	108.802	604.411	510.127	235.711	-330.913	-104.215	2.177
2600	108.945	608.681	513.836	246.598	-331.365	-95.138	1.911
2700	109.091	612.796	517.425	257.500	-331.913	-86.042	1.665
2800	109.239	616.766	520.903	268.416	-332.593	-76.924	1.435
2900	109.388	620.602	524.275	279.348	-333.444	-67.778	1.221
3000	109.537	624.313	527.548	290.294	-334.495	-58.600	1.020
3100	109.684	627.907	530.728	301.255	-335.789	-49.383	0.832
3200	109.830	631.391	533.819	312.231	-337.355	-40.120	0.655
3300	109.972	634.773	536.827	323.221	-339.218	-30.803	0.488
3400	110.110	638.058	539.757	334.225	-341.410	-21.425	0.329
3500	110.245	641.252	542.611	345.243	-343.994	-11.978	0.179
3600	110.374	644.360	545.395	356.274	-347.062	-2.449	0.036
3700	110.499	647.385	548.110	367.318	-385.476	7.363	-0.104
3800	110.618	650.334	550.762	378.373	-386.215	17.990	-0.247
3900	110.732	653.209	553.352	389.441	-386.955	28.636	-0.384
4000	110.840	656.013	555.884	400.520	-387.691	39.302	-0.513
4100	110.942	658.752	558.359	411.609	-388.420	49.986	-0.637
4200	111.039	661.426	560.782	422.708	-389.136	60.688	-0.755
4300	111.130	664.040	563.153	433.816	-389.835	71.406	-0.867
4400	111.215	666.596	565.475	444.934	-390.513	82.141	-0.975
4500	111.295	669.096	567.750	456.059	-391.165	92.890	-1.078
4600	111.370	671.543	569.980	467.193	-391.786	103.654	-1.177
4700	111.440	673.939	572.166	478.333	-392.373	114.430	-1.272
4800	111.505	676.286	574.311	489.480	-392.922	125.219	-1.363
4900	111.566	678.586	576.416	500.634	-393.430	136.020	-1.450
5000	111.622	680.840	578.482	511.793	-393.892	146.829	-1.534
5100	111.674	683.051	580.510	522.958	-394.308	157.648	-1.615
5200	111.722	685.220	582.503	534.128	-394.673	168.474	-1.692
5300	111.766	687.349	584.461	545.302	-394.985	179.307	-1.767
5400	111.807	689.438	586.386	556.481	-395.243	190.145	-1.839
5500	111.844	691.490	588.279	567.664	-395.444	200.988	-1.909
5600	111.879	693.506	590.140	578.850	-395.588	211.833	-1.976
5700	111.910	695.436	591.971	590.039	-395.672	222.681	-2.041
5800	111.939	697.433	593.772	601.232	-395.697	233.530	-2.103
5900	111.966	699.347	595.545	612.427	-395.661	244.378	-2.164
6000	111.990	701.229	597.291	623.625	-395.563	255.225	-2.222

PREVIOUS December 1966 (1 atm) CURRENT: December 1966 (1 bar)

Tungsten Chloride (WCl_4) $Cl_4W_1(g)$

ZIRCONIUM CHLORIDE ($ZrCl_4$) CRYSTAL M_r = 233.032

$S°(298.15 K) = 181.42 \pm 0.7 J K^{-1} mol^{-1}$

$T_{fus} = 710 \pm 2 J K^{-1} mol^{-1}$

$\Delta_f H°(0 K) = -981.59 \pm 1.7 kJ mol^{-1}$

$\Delta_f H°(298.15 K) = -980.52 \pm 1.7 kJ mol^{-1}$

$\Delta_{fus} H° = 50.2 \pm 12.6 kJ mol^{-1}$

Enthalpy of Formation

Gal'chenko et al. (1) determined enthalpy of formation of $ZrCl_4$(cr) by the direct chlorination of highly pure Zr metal in a bomb calorimeter. Complete chlorination of the metal was accomplished as indicated by the absence of unreacted metal and of lower chlorides. They reported $\Delta_f H°(ZrCl_4$, cr, 298.15 K) = -234.17±0.28 kcal mol^{-1}.

Gal'chenko et al. (2) had also earlier determined the enthalpy of formation of $ZrCl_4$(cr) by a similar technique and reported -234.35±2.1 kcal mol^{-1}. Gross et al. (6) measured calorimetrically the chlorination of Zr(cr) with liquid chlorine. His results led to $\Delta_f H°$(298.15 K) = -234.7 ± 0.4 kcal mol^{-1} for $ZrCl_4$(cr). We adopt $\Delta_f H°$(298.15 K) = -234.35 ± 0.4 kcal mol^{-1} which is a weighted mean of these three studies. This adopted value is identical to that suggested by NBS (4).

Less reliable studies leading to $\Delta_f H°$(298.15 K) values have been reported by Siemonsen and Siemonsen (7), and Beck (8). Equilibrium studies by Hildenbrand et al. (3) and Morozov and Korshunov (5) suggest $\Delta_f H°$(298.15 K) values 1-3 kcal mol^{-1} less negative than our adopted value.

Heat Capacity and Entropy

Todd (9) measured the low temperature heat capacities from 52.6 to 296.7 K, and made an extrapolation to 0 K which yielded an entropy of 8.12 cal K^{-1} mol^{-1} at 51 K. We adopt the measured heat capacities but make our own extrapolation to 0 K, based on the ratio of the measured heat capacities of ZrF_4 (10), TiF_4 (11) and $TiCl_4$ (12) from 6 to 50 K. This extrapolation gives S°(50 K) = 6.758±0.7 cal K^{-1} mol^{-1} which is adopted. Coughlin and King (13) measured high temperature enthalpy data from 335.9 to 566.8 K by drop calorimetry. Their data are smoothly joined with Todd's low temperature heat capacities.

Fusion Data

The melting point, T_{fus} = 710±2 K, has been observed by Rahlfs and Fischer (14), Palko et al. (15), and Denisova et al. (16, 17). Nisel'son (18) observed a melting temperature of 708.7 K. Heat of melting data may be extracted from Palko et al. (15) and Denisova et al. (16, 17). A weighted average $\Delta_{fus}H°$ = 12±3 kcal mol^{-1} is adopted for the heat of melting.

Sublimation Data

Several sets of sublimation studies are summarized below, using a 2nd and 3rd analysis.

Source	Method	Data Points	T/K	$\Delta_{sub}H°$(298.15 K), kcal mol^{-1} 2nd law	3rd law	Drift cal k^{-1} mol^{-1}
(14)	static	11	535-607	27.36±0.43	26.31±0.13	-1.9±0.8
(15)	static (diaphragm)	50[a]	503-575	25.75±0.20	26.39±0.17	1.1±0.3
	static (cap. bridge)	36[b]	580-689	25.87±0.28	26.40±0.16	0.8±0.4
(19)	glass Bourdon gage	eqn	503-603	26.19	26.33	0.3
(16)	static	17[c]	625-709	25.51±0.34	26.42±0.11	1.3±0.5
(20)	mass spec.	eqn	373-448	27.14	26.66	-1.2
(21)	--	eqn	376-493	26.59	26.44	-0.4

(a)3, (b)2, (c)1 points rejected due to failure of a statistical test.

A mean of the 3rd law values, $\Delta_{sub}H°$(298.15 K) = 26.42±0.12 kcal mol^{-1}, is adopted. The sublimation temperature, T_{sub}, is obtained from the Gibbs energy crossover between the crystal and gas. Since T_{sub} is lower than T_{fus}, the liquid phase is thermodynamically unstable at normal conditions.

In analyzing the vapor pressures for the sublimation studies, corrections were made for non-ideality by means of the equation $\Delta G°/T = -R\ln p - Bp/T$. The Bertholet equation of state and the critical constants T_c = 776.65 K and p_c = 57.4 atm as determined by Nisel'son and Sokolova (22) are used to calculate B.

References

1. G. L. Gal'chenko, D. A. Gedakyan, and B. I. Timofeev, Russ. J. Inorg. Chem. 13, 159 (1968).
2. G. L. Gal'chenko, D. A. Gedakyan, B. I. Timofeev, and S. M. Skuratov, Dokl. Akad. Nauk SSSR 161, 1081 (1965).
3. D. L. Hildenbrand, L. P. Theard, W. F. Hall, F. Ju, F. S. La Viola, and N. D. Potter, Aeronutronic Div., Philco-Ford Corp., Newport, Calif., Pub. No. U-1734, NOw 61-0905-C, June, 1962.
4. U. S. Nat. Bur. Stand. Tech. Note 270-5, 1971.
5. I. S. Morozov and B. G. Korshunov, Khim. Redkikh Elementov, Akad. Nauk SSSR, Inst. Obshchei i Neorg. Khim., No. 2, 102 (1955); Chem. Abstr. 50, 3136c (1956).
6. P. Gross, C. Hayman, and D. L. Levi, Trans. Faraday Soc. 53, 1285 (1957).
7. H. Siemonsen and U. Siemonsen, Z. Elektrochem. 56, 643 (1952).
8. G. Beck, Z. anorg. allgem. Chem. 174, 31 (1928).
9. S. S. Todd, J. Amer. Chem. Soc. 72, 2914 (1950).
10. E. F. Westrum, Jr., J. Chem. Eng. Data 10, 140 (1965).
11. R. D. Euler and E. F. Westrum, Jr., J. Phys. Chem. 65, 132 (1961).
12. G. T. Furukawa, U. S. Nat. Bur. Stand., personal communication, 1964.
13. J. P. Coughlin and E. G. King, J. Amer. Chem. Soc. 72, 2262 (1950).
14. O. Rahlfs and W. Fischer, Z. anorg. allgem. Chem. 211, 349 (1933).
15. A. A. Palko, A. D. Ryon, and D. W. Kuhn, J. Phys. Chem. 62, 319 (1958).
16. N. D. Denisova, E. K. Safronov, and O. N. Bystrova, Russ. J. Inorg. Chem. 11, 1171 (1966).
17. N. D. Denisova, E. K. Safranov, A. I. Pustil'nik, and O. N. Bystrova, Russ. J. Phys. Chem. 41, 30 (1967).
18. L. A. Nisel'son, Russ. J. Inorg. Chem. 7, 354 (1962).
19. K. Funaki and K. Uchimura, Denki Kagaku 33, 167 (1965); Chem. Abstr. 63, 3688c (1965).
20. Yu. S. Khodeev and K. I. Tuirel'nikov, Vest. Nosk. Univ., Khim 14, 611 (1973).
21. E. K. Safronov, L. N. Shelst, and S. M. Selifonov, Nauck. Tr., Gos. Nauch-Issled. Proekt. Inst. Redkometal. Prom. 20, 127 (1968).
22. L. A. Nisel'son and T. D. Sokolova, Russ. J. Inorg. Chem. 7, 1382 (1962).

Zirconium Chloride ($ZrCl_4$) Cl_4Zr_1(cr)

Enthalpy Reference Temperature = T_r = 298.15 K; Standard State Pressure = p° = 0.1 MPa

T/K	$C_p°$ (J $K^{-1}mol^{-1}$)	S°	$-[G°-H°(T_r)]/T$	$H°-H°(T_r)$ (kJ mol^{-1})	$\Delta_f H°$	$\Delta_f G°$	Log K_f
0	0.	0.	INFINITE	-24.924	-981.585	-981.585	INFINITE
100	77.990	70.509	278.997	-20.849	-984.220	-952.064	497.307
200	108.282	135.750	192.067	-11.263	-982.903	-920.258	240.347
298.15	119.788	181.418	181.418	0.	-980.520	-889.999	155.924
300	119.913	182.160	181.421	0.222	-980.471	-889.438	154.865
400	125.394	217.491	186.201	12.516	-977.676	-859.509	112.240
500	128.700	245.848	195.387	25.230	-974.729	-830.306	86.741
600	131.127	269.533	205.825	38.225	-971.690	-801.705	69.795
700	133.135	289.899	216.415	51.439	-968.586	-773.619	57.728
800	134.976	307.799	226.741	64.846	-965.428	-745.981	48.708
900	136.649	323.794	236.652	78.428	-962.235	-718.742	41.715
1000	138.239	338.274	246.101	92.173	-959.019	-691.859	36.139

PREVIOUS: December 1969 CURRENT: June 1975

Zirconium Chloride ($ZrCl_4$) Cl_4Zr_1(cr)

ZIRCONIUM CHLORIDE ($ZrCl_4$) IDEAL GAS M_r = 233.032

$\Delta_fH°(0\ K) = -868.68 \pm 2.1\ kJ\ mol^{-1}$

$S°(298.15\ K) = 367.72 \pm 0.4\ J\ K^{-1}\ mol^{-1}$ $\Delta_fH°(298.15\ K) = -869.98 \pm 2.1\ kJ\ mol^{-1}$

Vibrational Frequencies and Degeneracies

ν, cm^{-1}	ν, cm^{-1}
377(1)	418(3)
98(2)	113(3)

Ground State Quantum Weight: [1] σ = 12

Point Group: T_d

Bond Distance: 2.32 ± 0.02 Å

Bond Angle: Cl-Zr-Cl = 109.4712°

Product of the Moments of Inertia: $I_AI_BI_C = 6.03340 \times 10^{-112}\ g^3\ cm^6$

Enthalpy of Formation

The enthalpy of formation of gaseous $ZrCl_4$, $\Delta_fH°(298.15\ K) = -207.76 \pm 0.5\ kcal\ mol^{-1}$ is calculated from the enthalpies of formation and sublimation of the crystal (1).

Hildenbrand et al. (14) determined the equilibrium constants by the transpiration method for the reaction: $ZrO_2(cr) + 4\ HCl(g) = ZrCl_4(g) + 2\ H_2O(g)$. 2nd and 3rd law analyses for their data are given below. The enthalpy of formation derived from this study is in fair agreement with the value adopted.

Series	Data Points	T/K	$\Delta_rH°(298.15\ K)$, kcal mol^{-1} 2nd law	3rd law	Drift cal K^{-1} mol^{-1}	$\Delta_fH°(298.15\ K)$ kcal mol^{-1}
I	10	1171-1373	28.94±0.44	29.16±0.15	0.17±0.34	-205.80
II	9	1169-1374	31.36±0.60	29.65±0.28	-1.19±0.47	-205.11

Heat Capacity and Entropy

The adopted vibrational frequencies are from the work of Clark et al. (2, 3), who recorded the Raman spectra of $ZrCl_4$ in the vapor phase (380-420°C). These studies indicated that $ZrCl_4$ is a tetrahedral monomer in the vapor phase. Shimanouchi (4), in his compilation of molecular vibrational frequencies, adopted the values of Clark et al. (2, 3) for $ZrCl_4(g)$. Pontrelli (5) also observed in the Raman spectra all four fundamental vibrational frequencies (376, 99, 418, and 110 cm^{-1}). These values are in excellent agreement with those of Clark et al. (2, 3).

Other experimental studies involving the vibrational frequencies are as follows. Wilmshurst (6) observed one fundamental vibrational frequency (421 cm^{-1}) in the infrared spectrum of $ZrCl_4$ vapor, while Buchler et al. (7) found the same fundamental mode at 423 cm^{-1}. Delwaulle and Francois (8) observed the symmetric stretching vibration (383 cm^{-1}) from the Raman spectra of $ZrCl_4$ in PCl_5 and $POCl_3$. Bobovich (9) reported the symmetric stretching vibration (315 cm^{-1}) from a partially polarized band in his Raman spectrometric studies.

Spiridonov et al. (10) measured the bond distance Zr-Cl = 2.32±0.02 Å in the vapor phase by electron diffraction. The molecule was also found to be a regular tetrahedron. The same bond distance was also reported by Lister and Sutton (11) and Kimura et al. (12). Rahlfs and Fischer (13), through vapor density measurements, had earlier concluded that $ZrCl_4$ was monomeric in the vapor phase. The principal moments of inertia are $I_A=I_B=I_C=84.4995 \times 10^{-19}\ g\ cm^2$.

Much literature has been published on the inter-relationships between force constants and vibrational frequencies. Since the majority of these articles are based on estimated frequencies, they will not be further discussed or referenced. The same situation is true for thermodynamic tabulations of $ZrCl_4(g)$. One exception is that Clark et al. (2) calculated thermodynamic properties based on their experimental vibrational frequencies. Their tabulation is identical to ours in the range 100-1000 K.

References

1. JANAF Thermochemical Tables: $ZrCl_4$(cr), 6-30-75; ZrO_2(cr), 12-31-65; HCl(g), 9-30-64; H_2O(g), 3-31-61.
2. R. J. H. Clark, B. K. Hunter, and D. M. Rippon, Chem. Ind. (London) 1971; Inorg. Chem. 11, 56 (1971).
3. R. J. H. Clark and D. M. Rippon, J. Mol. Spectrosc. 44, 479 (1972); Advanc. Raman Spectrosc. 1, 493 (1972).
4. T. Shimanouchi, J. Phys. Chem. Ref. Data 3, 269 (1974).
5. G. N. Pontrelli, Diss. Abstr. 23, 4130 (1963).
6. J. K. Wilmshurst, J. Mol. Spectrosc. 5, 343 (1960).
7. A. Buchler, J. B. Berkowitz-Mattuck, and D. H. Dugre, J. Chem. Phys. 34, 2202 (1961).
8. M. L. Dewaulle and F. Francois, Bull. Soc. Chim. 13, 205 (1946); J. Phys. Radium 7, 15 (1946).
9. Ya. S. Bobovich, Optics and Spectrosc. 11, 85 (1961).
10. V. P. Spiridonov, P. A. Akiskin, and V. I. Tsirel'nikov, Zh. Strukt. Khim. 3, 329 (1962).
11. M. W. Lister and L. E. Sutton, Trans. Faraday Soc. 37, 393 (1941).
12. M. Kimura, R. Kimura, M. Aoki, and S. Shibata, Bull. Chem. Soc. Japan 29, 95 (1956).
13. O. Rabitz and W. Fischer, Z. anorg. allgem. Chem. 211, 349 (1933).
14. D. L. Hildenbrand, L. P. Theard, W. F. Hall, I. Ju, F. S. La Viola, and N. D. Potter, Aeronutronic Div., Philco-Ford Corp., Newport, Calif., Pub. No. U-1734, NOw 61-0905-C, June, 1962.

Zirconium Chloride ($ZrCl_4$) $Cl_4Zr_1(g)$

Enthalpy Reference Temperature = T_r = 298.15 K; Standard State Pressure = p° = 0.1 MPa

T/K	$C_p°$ (J $K^{-1}mol^{-1}$)	S°	$-[G°-H°(T_r)]/T$	$H°-H°(T_r)$ (kJ mol^{-1})	$\Delta_fH°$	$\Delta_fG°$	Log K_f
0	0.	0.	INFINITE	-22.561	-868.681	-868.681	INFINITE
100	70.794	274.385	448.659	-17.427	-870.258	-858.489	448.429
200	89.827	330.054	376.481	-9.285	-870.384	-846.599	221.109
250	94.964	350.691	369.317	-4.656	-870.209	-840.671	175.649
298.15	98.234	367.716	367.716	0.	-869.979	-835.003	146.289
300	98.336	368.324	367.718	0.182	-869.970	-834.786	145.349
350	100.611	383.665	368.925	5.159	-869.701	-828.943	123.713
400	102.199	397.209	371.630	10.232	-869.419	-823.139	107.491
450	103.342	409.316	375.157	15.372	-869.136	-817.371	94.878
500	104.188	420.250	379.128	20.561	-868.857	-811.635	84.791
600	105.328	439.355	387.620	31.041	-868.332	-800.240	69.667
700	106.036	455.649	396.203	41.612	-867.872	-788.929	58.871
800	106.505	469.840	404.540	52.241	-867.492	-777.679	50.777
900	106.831	482.405	412.507	62.908	-867.213	-766.470	44.485
1000	107.066	493.673	420.070	73.604	-867.047	-755.286	39.452
1100	107.241	503.866	427.232	84.320	-867.003	-744.113	35.335
1200	107.374	513.224	434.015	95.051	-870.797	-732.717	31.894
1300	107.479	521.822	440.443	105.793	-870.467	-721.224	28.979
1400	107.562	529.791	446.544	116.546	-870.189	-709.755	26.481
1500	107.629	537.214	452.344	127.305	-869.975	-698.303	24.317
1600	107.684	544.162	457.868	138.071	-869.835	-686.863	22.424
1700	107.730	550.692	463.138	148.842	-869.783	-675.430	20.753
1800	107.768	556.851	468.175	159.617	-869.829	-663.996	19.269
1900	107.801	562.678	472.996	170.395	-869.985	-652.558	17.940
2000	107.829	568.208	477.620	181.177	-870.265	-641.108	16.744
2100	107.853	573.470	482.060	191.961	-870.682	-629.640	15.661
2200	107.874	578.488	486.330	202.747	-892.570	-617.403	14.659
2300	107.892	583.283	490.442	213.536	-893.714	-604.870	13.737
2400	107.908	587.876	494.407	224.326	-894.886	-592.287	12.891
2500	107.922	592.281	498.234	235.117	-896.088	-579.654	12.111
2600	107.934	596.514	501.933	245.910	-897.323	-566.972	11.391
2700	107.945	600.588	505.512	256.704	-898.593	-554.242	10.722
2800	107.955	604.513	508.978	267.499	-899.900	-541.465	10.101
2900	107.964	608.302	512.338	278.295	-901.244	-528.640	9.522
3000	107.972	611.962	515.598	289.092	-902.627	-515.768	8.980
3100	107.979	615.503	518.764	299.889	-904.048	-502.850	8.473
3200	107.986	618.931	521.841	310.688	-905.505	-489.885	7.997
3300	107.992	622.254	524.834	321.487	-906.998	-476.873	7.548
3400	107.998	625.478	527.747	332.286	-908.524	-463.817	7.126
3500	108.003	628.609	530.584	343.086	-910.079	-450.715	6.727
3600	108.007	631.651	533.349	353.887	-911.660	-437.567	6.349
3700	108.012	634.611	536.046	364.687	-913.263	-424.376	5.991
3800	108.016	637.491	538.678	375.489	-914.883	-411.142	5.652
3900	108.019	640.297	541.248	386.291	-916.517	-397.864	5.329
4000	108.023	643.032	543.759	397.093	-918.157	-384.544	5.022
4100	108.026	645.699	546.213	407.895	-919.800	-371.183	4.729
4200	108.029	648.302	548.612	418.698	-921.441	-357.783	4.450
4300	108.031	650.844	550.960	429.501	-923.073	-344.343	4.183
4400	108.034	653.328	553.259	440.304	-924.692	-330.865	3.928
4500	108.036	655.756	555.510	451.108	-926.294	-317.351	3.684
4600	108.038	658.130	557.715	461.911	-927.872	-303.801	3.450
4700	108.040	660.454	559.876	472.715	-929.423	-290.219	3.225
4800	108.042	662.728	561.995	483.519	-1491.880	-264.985	2.884
4900	108.044	664.956	564.074	494.324	-1493.016	-239.413	2.552
5000	108.046	667.139	566.113	505.128	-1494.130	-213.818	2.234
5100	108.047	669.279	568.115	515.933	-1495.217	-188.200	1.928
5200	108.049	671.377	570.081	526.738	-1496.274	-162.563	1.633
5300	108.050	673.435	572.012	537.543	-1497.295	-136.905	1.349
5400	108.052	675.455	573.909	548.348	-1498.277	-111.227	1.076
5500	108.053	677.437	575.773	559.153	-1499.217	-85.532	0.812
5600	108.054	679.384	577.606	569.958	-1500.111	-59.821	0.558
5700	108.055	681.297	579.408	580.764	-1500.958	-34.094	0.312
5800	108.056	683.176	581.181	591.569	-1501.754	-8.352	0.075
5900	108.058	685.023	582.926	602.375	-1502.497	17.402	-0.154
6000	108.058	686.839	584.643	613.181	-1503.185	43.169	-0.376

PREVIOUS: June 1975 (1 atm) CURRENT: June 1975 (1 bar)

Zirconium Chloride ($ZrCl_4$) $Cl_4Zr_1(g)$

MOLYBDENUM CHLORIDE ($MoCl_5$) CRYSTAL M_r = 273.205

$S°(298.15\ K)$ = [238.488 ± 12.6] J K^{-1} mol^{-1}

T_{fus} = 467 K

$\Delta_f H°(0\ K)$ = Unknown

$\Delta_f H°(298.15\ K)$ = -527.18 ± 8.4 kJ mol^{-1}

$\Delta_{fus} H°$[18.8 ± 6.3] kJ mol^{-1}

Enthalpy of Formation

The enthalpy of formation is obtained from Shchukarev et al. (1). The authors derived the $\Delta_f H°(MoCl_5$, cr, 298.15 K) = -126 kcal mol^{-1} by measuring the enthalpy of solution of $MoCl_5$ in NaOH(aq), which was compared with the enthalpies of solution of MoO_3 and NaCl measured in the same solvent.

Heat Capacity and Entropy

The heat capacity is estimated by comparison with that of WCl_5(cr). S°(298.15 K) is estimated as 57 cal K^{-1} mol^{-1} from the vapor pressure data for $MoCl^5$(cr) → $MoCl^5$(g) measured by Shchukarev et al. (2).

Fusion Data

See the liquid table for details.

References

1. S. A. Shchukarev, G. I. Novikov, I. V. Vasil'kova, A. V. Suvorov, N. V. Andreeva, B. N. Sharupin, and A. K. Baev, Zh. Neorg. Khim. 5, 1650 (1960).
2. S. A. Shchukarev, I. V. Vasil'kova and B. N. Sharupin, J. Gen. Chem. USSR 26, 2335 (1956).

Molybdenum Chloride ($MoCl_5$) Cl_5Mo_1(cr)

Enthalpy Reference Temperature = T_r = 298.15 K; Standard State Pressure = p° = 0.1 MPa

	J $K^{-1}mol^{-1}$			kJ mol^{-1}			
T/K	$C_p°$	S°	$-[G°-H°(T_r)]/T$	$H°-H°(T_r)$	$\Delta_f H°$	$\Delta_f G°$	Log K_f
0							
100							
200							
298.15	155.645	238.488	238.488	0.	-527.184	-423.483	74.192
300	155.728	239.451	238.491	0.288	-527.097	-422.840	73.623
400	167.360	285.854	244.739	16.446	-522.070	-388.825	50.775
467.000	174.929	312.345	252.573	27.913	------ CRYSTAL <--> LIQUID -----		
500	178.657	324.413	256.920	33.747	-516.246	-356.170	37.209
600	189.954	358.022	271.026	52.198	-509.492	-324.773	28.274
700	198.740	387.996	285.634	71.654	-501.888	-294.578	21.982
800	205.016	414.962	300.143	91.855	-493.657	-265.517	17.337
900	209.618	439.388	314.279	112.598	-484.982	-237.517	13.785
1000	212.966	461.657	327.920	133.737	-475.997	-210.500	10.995

PREVIOUS: CURRENT: December 1968

Molybdenum Chloride ($MoCl_5$) Cl_5Mo_1(cr)

MOLYBDENUM CHLORIDE ($MoCl_5$) LIQUID M_r = 273.205

$S°(298.15\ K) = [273.807]\ J\ K^{-1}\ mol^{-1}$

$T_{fus} = 467\ K$

$\Delta_f H°(298.15\ K) = [-510.114]\ kJ\ mol^{-1}$

$\Delta_{fus} H° = [18.828 \pm 6.3]\ kJ\ mol^{-1}$

Enthalpy of Formation

$\Delta_f H°(MoCl_5, \ell, 298.15\ K)$ is calculated from $\Delta_f H°(MoCl_5, cr, 298.15\ K)$ by adding $\Delta_{fus} H°$ and the difference in enthalpy, $H°(467\ K) - H°(298.15\ K)$, between the crystal and liquid.

Heat Capacity and Entropy

The heat capacity is assumed constant and estimated as 42 cal K^{-1} mol^{-1} or 7 cal K^{-1} g-atom^{-1}. The entropy is calculated in a manner analogous to that used for the enthalpy of formation.

Fusion Data

T_{fus} is taken from Shchukarev et al. (1). $\Delta_{fus} H°$ is estimated by comparison with that of WCl_5(cr).

Vaporization Data

T_{vap} is the temperature at which the difference between the Gibbs energy of formation for $MoCl_5(\ell)$ and $MoCl_5(g)$ is zero. The difference between $\Delta_f H°(MoCl_5, \ell)$ and $\Delta_f H°(MoCl_5, g)$ at T_{vap} is $\Delta_{vap} H°$.

Reference

1. S. A. Shchukarev, G. I. Novikov, A. V. Suvorov and V. K. Moksimov, Russ. J. Inorg. Chem. (English Transl.) 4, 935 (1959).

Molybdenum Chloride ($MoCl_5$) $Cl_5Mo_1(l)$

Enthalpy Reference Temperature = T_r = 298.15 K — Standard State Pressure = p° = 0.1 MPa

T/K	$C_p°$	S°	$-[G°-H°(T_r)]/T$	$H°-H°(T_r)$	$\Delta_f H°$	$\Delta_f G°$	Log K_f
	J K⁻¹mol⁻¹			kJ mol⁻¹			
0							
100							
200							
298.15	175.728	273.807	273.807	0.	-510.114	-416.944	73.047
300	175.728	274.894	273.811	0.325	-509.991	-416.366	72.496
400	175.728	325.448	280.703	17.898	-503.548	-386.141	50.425
467.000	175.728	352.662	289.125	29.672	------ CRYSTAL <--> LIQUID -----		
500	175.728	364.661	293.719	35.471	-497.452	-357.500	37.348
600	175.728	396.700	308.294	53.043	-491.576	-330.064	28.735
700	175.728	423.788	322.908	70.616	-485.856	-303.600	22.655
800	175.728	447.254	337.017	88.189	-480.253	-277.947	18.148
900	175.728	467.951	350.438	105.762	-474.748	-252.990	14.683
1000	175.728	486.466	363.131	123.335	-469.329	-228.642	11.943
1100	175.728	503.215	375.117	140.907	-463.995	-204.832	9.727
1200	175.728	518.505	386.438	158.480	-458.745	-181.504	7.901

PREVIOUS: CURRENT: December 1968

Molybdenum Chloride ($MoCl_5$) $Cl_5Mo_1(l)$

MOLYBDENUM CHLORIDE ($MoCl_5$) CRYSTAL-LIQUID M_r = 273.205

0 to 467 K crystal
above 467 K liquid

Refer to the individual tables for details.

Molybdenum Chloride ($MoCl_5$) Cl_5Mo_1(cr,l)

Enthalpy Reference Temperature = T_r = 298.15 K Standard State Pressure = p° = 0.1 MPa

T/K	C_p° (J K⁻¹mol⁻¹)	S° (J K⁻¹mol⁻¹)	−[G°−H°(T$_r$)]/T (J K⁻¹mol⁻¹)	H°−H°(T$_r$) (kJ mol⁻¹)	$\Delta_f H^\circ$ (kJ mol⁻¹)	$\Delta_f G^\circ$ (kJ mol⁻¹)	Log K_f
0							
100							
200							
298.15	155.645	238.488	238.488	0.	-527.184	-423.483	74.192
300	155.728	239.451	238.491	0.288	-527.097	-422.840	73.623
400	167.360	285.854	244.739	16.446	-522.070	-388.825	50.775
467.000	174.929	312.345	252.573	27.913	CRYSTAL <--> LIQUID	TRANSITION	
467.000	175.728	352.662	252.573	46.741			
500	175.728	364.661	259.580	52.540	-497.452	-357.500	37.348
600	175.728	396.700	279.844	70.113	-491.576	-330.064	28.735
700	175.728	423.788	298.522	87.686	-485.856	-303.600	22.655
800	175.728	447.254	315.680	105.259	-480.253	-277.947	18.148
900	175.728	467.951	331.472	122.832	-474.748	-252.990	14.683
1000	175.728	486.466	346.062	140.404	-469.329	-228.642	11.943
1100	175.728	503.215	359.599	157.977	-463.995	-204.832	9.727
1200	175.728	518.505	372.213	175.550	-458.745	-181.504	7.901

PREVIOUS: CURRENT: December 1968

Molybdenum Chloride ($MoCl_5$) Cl_5Mo_1(cr,l)

MOLYBDENUM CHLORIDE ($MoCl_5$) IDEAL GAS M_r = 273.205

$\Delta_f H°(0\ K) = -445.94 \pm 4.2$ kJ mol^{-1}

$S°(298.15\ K) = [397.774 \pm 12.6]$ J K^{-1} mol^{-1} $\Delta_f H°(298.15\ K) = -447.69 \pm 4.2$ kJ mol^{-1}

Electronic Levels and Quantum Weights

ε_i, cm^{-1}	g_i
0	[2]
[7000]	[2]
[14000]	[2]

Vibrational Frequencies and Degeneracies

ν, cm^{-1}	ν, cm^{-1}
[490](1)	[350](2)
[400](1)	[170](2)
[410](1)	[100](2)
[150](1)	[190](2)

Point Group: D_{3h} σ = 6

Bond Distance: Mo-Cl = 2.27 ± 0.02 Å

Bond Angles: Cl*-Mo-Cl* = 120° Cl*-Mo-Cl** = 90° Cl**-Mo-Cl** = 180°

*Equatorial **Axial

Product of the Moments of Inertia: $I_A I_B I_C = 1.02598 \times 10^{-111}$ g^3 cm^6

Enthalpy of Formation

The adopted $\Delta_f H°(298.15\ K)$ = 107.0 kcal mol^{-1} is obtained from the vapor pressure data for $MoCl_5(cr) = MoCl_5(g)$ measured by Shchukarev et al. (1) and Saeki and Matsuzaki (2).

Source	Method	Reaction	T/K	$\Delta_r H°(298.15\ K)$, kcal mol^{-1} 2nd law	3rd law	Drift cal K^{-1} mol^{-1}	$\Delta_f H°(298.15\ K)$ kcal mol^{-1}
Shchukarev	Transpiration	$MoCl_5(cr) = MoCl_5(g)$	343 - 435	19.04 ± 0.45	19.18	0.3	-106.82
Saeki	Manometric	$MoCl_5(cr) = MoCl_5(g)$	298 - 474	16.79 ± 0.06	18.94	5.6	-107.06

*Calculation based on third law $\Delta_r H°(298.15\ K)$

Heat Capacity and Entropy

The vibrational frequencies are estimated by comparison with those of $NbCl_5$ obtained from infrared and Raman spectra by Carlson (3). Structure, bond distance and bond angle are taken from Ewens et al. (4). The principal moments of inertia are: $I_A = 91.0085 \times 10^{-39}$ and $I_B = I_C = 106.1766 \times 10^{-39}$ g cm^2. The electronic levels are estimated to be 0, 7000 and 14000 cm^{-1}, all doublets, by comparison with those of $WCl_5(g)$.

References

1. S. A. Shchukarev, I. V. Vasil'kova and B. N. Sharupin, J. Gen. Chem. USSR 26, 2335-9 (1956).
2. Y. Saeki and R. Matsuzaki, Denki Kagaku 33, 155-8 (1965).
3. G. L. Carlson, Spectrochim. Acta 19, 1291 (1963).
4. R. V. G. Ewens and M. W. Lister, Trans. Faraday Soc. 34, 1358-62 (1938).

Molybdenum Chloride ($MoCl_5$) $Cl_5Mo_1(g)$

Enthalpy Reference Temperature = T_r = 298.15 K Standard State Pressure = p° = 0.1 MPa

T/K	$C_p°$ (J K^{-1}mol^{-1})	S°	$-[G°-H°(T_r)]/T$	$H°-H°(T_r)$ (kJ mol^{-1})	$\Delta_f H°$	$\Delta_f G°$	Log K_f
0	0.	0.	INFINITE	-25.793	-445.945	-445.945	INFINITE
100	77.392	287.812	494.800	-20.699	-448.621	-429.448	224.321
200	107.562	352.265	408.393	-11.226	-448.584	-410.160	107.123
250	114.900	377.117	399.718	-5.650	-448.174	-400.598	83.700
298.15	119.482	397.774	397.774	0.	-447.688	-391.478	68.585
300	119.624	398.514	397.777	0.221	-447.668	-391.129	68.102
350	122.781	417.207	399.246	6.286	-447.114	-381.750	56.973
400	124.971	433.753	402.545	12.483	-446.537	-372.451	48.637
450	126.543	448.568	406.850	18.773	-445.952	-363.226	42.162
500	127.704	461.964	411.702	25.131	-445.366	-354.065	36.989
600	129.265	485.397	422.087	37.986	-444.208	-335.914	29.244
700	130.234	505.402	432.595	50.964	-443.082	-317.955	23.726
800	130.878	522.837	442.809	64.022	-441.994	-300.154	19.598
900	131.332	538.279	452.575	77.134	-440.950	-282.487	16.395
1000	131.674	552.135	461.850	90.285	-439.953	-264.934	13.839
1100	131.950	564.698	470.638	103.466	-439.010	-247.479	11.752
1200	132.192	576.190	478.962	116.674	-438.125	-230.106	10.016
1300	132.417	586.780	486.854	129.904	-437.307	-212.805	8.551
1400	132.639	596.601	494.346	143.157	-436.558	-195.564	7.297
1500	132.862	605.760	501.472	156.432	-435.886	-178.374	6.212
1600	133.092	614.342	508.261	169.730	-435.294	-161.227	5.264
1700	133.328	622.418	514.741	183.051	-434.790	-144.113	4.428
1800	133.570	630.045	520.937	196.395	-434.381	-127.027	3.686
1900	133.817	637.274	526.871	209.765	-434.074	-109.961	3.023
2000	134.067	644.144	532.565	223.159	-433.880	-92.908	2.427
2100	134.317	650.691	538.035	236.578	-433.808	-75.861	1.887
2200	134.566	656.945	543.299	250.022	-433.872	-58.815	1.396
2300	134.811	662.933	548.371	263.491	-434.083	-41.763	0.948
2400	135.053	668.675	553.265	276.984	-434.455	-24.699	0.538
2500	135.288	674.193	557.992	290.501	-435.003	-7.615	0.159
2600	135.516	679.504	562.565	304.042	-435.753	9.495	-0.191
2700	135.736	684.622	566.991	317.604	-436.748	26.638	-0.515
2800	135.947	689.562	571.281	331.189	-438.052	43.823	-0.818
2900	136.150	694.337	575.442	344.794	-475.633	61.111	-1.101
3000	136.342	698.956	579.483	358.418	-475.769	79.622	-1.386
3100	136.525	703.429	583.409	372.062	-475.934	98.137	-1.654
3200	136.697	707.766	587.228	385.723	-476.128	116.659	-1.904
3300	136.860	711.975	590.945	399.401	-476.350	135.187	-2.140
3400	137.012	716.063	594.565	413.094	-476.598	153.722	-2.362
3500	137.155	720.037	598.093	426.803	-476.870	172.265	-2.571
3600	137.287	723.903	601.535	440.525	-477.160	190.816	-2.769
3700	137.410	727.666	604.893	454.260	-477.466	209.375	-2.956
3800	137.523	731.332	608.172	468.007	-477.782	227.942	-3.133
3900	137.627	734.905	611.376	481.764	-478.104	246.518	-3.302
4000	137.722	738.391	614.508	495.532	-478.425	265.102	-3.462
4100	137.808	741.793	617.571	509.308	-478.741	283.694	-3.614
4200	137.885	745.115	620.569	523.093	-479.046	302.294	-3.760
4300	137.955	748.360	623.503	536.885	-479.333	320.901	-3.898
4400	138.017	751.532	626.377	550.684	-479.599	339.514	-4.031
4500	138.071	754.634	629.193	564.488	-479.836	358.132	-4.157
4600	138.118	757.670	631.953	578.298	-480.040	376.757	-4.278
4700	138.159	760.641	634.659	592.112	-480.205	395.384	-4.394
4800	138.193	763.550	637.314	605.929	-480.328	414.015	-4.505
4900	138.221	766.399	639.920	619.750	-480.404	432.649	-4.612
5000	138.243	769.192	642.477	633.573	-1071.814	457.013	-4.774
5100	138.259	771.930	644.989	647.398	-1073.463	487.607	-4.994
5200	138.271	774.615	647.456	661.225	-1075.201	518.232	-5.206
5300	138.278	777.248	649.880	675.052	-1077.021	548.892	-5.410
5400	138.280	779.833	652.263	688.880	-1078.922	579.588	-5.606
5500	138.277	782.371	654.605	702.708	-1080.896	610.319	-5.796
5600	138.271	784.862	656.909	716.536	-1082.941	641.087	-5.980
5700	138.261	787.309	659.176	730.362	-1085.051	671.891	-6.157
5800	138.248	789.714	661.406	744.188	-1087.224	702.734	-6.329
5900	138.231	792.077	663.600	758.012	-1089.452	733.614	-6.495
6000	138.212	794.400	665.761	771.834	-1091.738	764.532	-6.656

PREVIOUS: December 1968 (1 atm) CURRENT: December 1968 (1 bar)

Molybdenum Chloride ($MoCl_5$) $Cl_5Mo_1(g)$

NIOBIUM CHLORIDE ($NbCl_5$) CRYSTAL M_r = 270.1714

S°(298.15 K) = [214.05 ± 4.2] J K^{-1} mol^{-1}

T_{fus} = 478.9 ± 1.5 K

$\Delta_f H°$(0 K) = Unknown

$\Delta_f H°$(298.15 K) = -797.47 ± 4.2 kJ mol

$\Delta_{fus} H°$ = 33.878 ± 0.96 kJ mol^{-1}

Enthalpy of Formation

The adopted value for the enthalpy of formation of $NbCl_5$(cr) is $\Delta_f H°$(298.15 K) = -190.6 ± 1.0 kcal mol^{-1} and is based on the study by Gross et al. (2). This value was determined from measurements of the enthalpy evolved in the reaction Nb(cr) + 5/2 Cl_2(ℓ) = $NbCl_5$(cr). Schafer and Kahlenberg (1) also determined the enthalpy formation of $NbCl_5$(cr) via calorimetric measurement of the heats of solution of Nb(cr) and $NbCl_5$(cr) in hydrofluoric acid. Their procedure consisted of five steps; four were determined experimentally and one was based on literature data. Using their data and auxiliary results (7), we calculate $\Delta_f H°$(298.15 K) = -190.0 ± 1.0 kcal mol^{-1} for $NbCl_5$(cr). This value is in good agreement with our adopted value. Shchukarev et al. (3) determined a $\Delta_f H°$(298.15 K) value from a heat of hydrolysis for $NbCl_5$(cr). A recalculation of this data, using current auxiliary $\Delta_f H°$(298.15 K) data for Nb_2O_5(cr) (10), H_2O(ℓ) (7), and HCl(g) (10), yields $\Delta_f H°$(298.15 K) = -192.8 ± 0.7 kcal mol^{-1}.

Three compilations on Nb species suggested a similar or closely related $\Delta_f H°$(298.15 K) value for $NbCl_5$(cr) (6, 8, 9): in particular, the NBS Technical Note 270 Series suggested -190.6 kcal mol^{-1} (9).

Heat Capacity and Entropy

Keneshea et al. (5) measured the saturation enthalpy increments above 298.15 K for the condensed phases of $NbCl_5$ in a drop calorimeter up to the critical point (804 ± 3 K). A figure presented by Keneshea et al. (5) indicated roughly 30 data points, the lowest occurring at approximately 360 K. The differences between the saturation and standard enthalpy increments for the crystal phase are negligible, so that the heat capacity values which we adopt are those which are derived from the reported enthalpy equation, H°(T) - H°(298.15 K) = [-10.53 + 3.535 x 10^{-2}T] ± 0.07 kcal mol^{-1}. This equation is reported to apply to the temerature region 298.15 - 478.9 K.

Schafer and Kahlenberg (1) estimated the heat capacity of $NbCl_5$(cr) to be given by $C_p°$ = 38.0 - 3 x $10^5 T^{-2}$ cal K^{-1} mol^{-1}. For the crystal range 298.15 - 478.9 K, these estimated $C_p°$ values vary from 34.63 to 36.69 cal K^{-1} mol^{-1}. These values are in fair agreement with the adopted values and were estimated by comparison with $ZrCl_4$(cr) and $HfCl_4$(cr). Amosov (4) also estimated the heat capacity of $NbCl_5$ based on Neumann and Koppe's rule, $C_p°$ = 26.71 + 35.2 x 10^{-3}T cal K^{-1} mol^{-1}. These latter values vary between 37.20 and 43.57 cal K^{-1} mol^{-1} over the region 298.15 - 478.9 K and differ significantly from those $C_p°$ values derived from the experimental data of Keneshea et al. (5).

As there is no low temperature heat capacity data reported in the literature, the entropy of $NbCl_5$(cr) at 298.15 K is calculated from the equation $\Delta_{sub}S°$(298.15 K) = S°(g, 298.15 K) - S°(cr, 298.15 K); S°(g, 298.15 K) = 96.56 cal K^{-1} mol^{-1} as given in the $NbCl_5$(g) table (10) and $\Delta_{sub}S°$(298.15 K) = 45.40 cal K^{-1} mol^{-1} as obtained Keneshea et al. (5) based on their evaluation of the available vapor pressure data by a modified Σ approach.

Fusion Data

See the liquid table for details.

Sublimation Data

The enthalpy of sublimation, $\Delta_{sub}H°$(298.15 K) is the difference between the $\Delta_f H°$(298.15 K) values for $NbCl_5$(g) and $NbCl_5$(cr). For sublimation studies are summarized in the $NbCl_5$(ℓ) table.

References

1. H. Schafer and F. kahlenberg, Z. Anorg. Allg. Chem. 305, 291 (1960).
2. P. Gross, C. Hayman, D. L. Levi and G. L. Wilson, Trans. Faraday Soc. 56, 318 (1960).
3. S. A. Shchukarev, M. A. Oranskaya and T. S. Shemyakina, Russ. J. Inorg. Chem. 5, 1036 (1960).
4. V. M. Amosov, Izv. Vysshikh Uchebn. Zavedenii, Tsvetn. Met. 6, 103 (1963).
5. F. J. Keneshea, D. Cubicciotti, G. Withers and H. Eding, J. Phys. Chem. 72, 1272 (1968).
6. V. I. Lavrentev and Ya. I. Gerassimov, At. Energy Rev. 1968, 7 (1968).
7. U. S. Nat. Bur. Stand. Tech. Note 270-3, 1968.
8. J. O. Hill, I. G. Worsley and L. G. Hepler, Chem. Rev. 71, 127 (1971).
9. U. S. Nat. Bur. Stand. Tech. Note 270-5, 1971.
10. JANAF Thermochemical Tables: Nb_2O_5(cr), 12-31-72; HCl(g), 9-3-64' $NbCl_5$(g), 12-31-74.

Niobium Chloride ($NbCl_5$) Cl_5Nb_1(cr)

Enthalpy Reference Temperature = T_r = 298.15 K Standard State Pressure = p° = 0.1 MPa

	J $K^{-1}mol^{-1}$			kJ mol^{-1}			
T/K	$C_p°$	S°	-[G°-H°(T_r)]/T	H°-H°(T_r)	$\Delta_f H°$	$\Delta_f G°$	Log K_f
0							
100							
200							
298.15	147.904	214.053	214.053	0.	-797.470	-684.141	119.859
300	147.904	214.968	214.056	0.274	-797.400	-683.438	118.997
400	147.904	257.518	219.858	15.064	-793.792	-646.006	84.360
478.900	147.904	284.145	228.321	26.734	------ CRYSTAL <--> LIQUID -----		
500	147.904	290.522	230.813	29.854	-790.494	-609.446	63.668
600	147.904	317.488	243.080	44.645	-787.397	-573.530	49.930
700	147.904	340.287	255.380	59.435	-784.442	-538.121	40.155

PREVIOUS: CURRENT: December 1974

Niobium Chloride ($NbCl_5$) Cl_5Nb_1(cr)

NIOBIUM CHLORIDE ($NbCl_5$) LIQUID M_r = 270.1714

$S°(298.15 K) = [260.970] J K^{-1} mol^{-1}$

$T_{fus} = 478.9 \pm 1.5$ K

$\Delta_f H°(298.15 K) = [-773.354] kJ mol^{-1}$

$\Delta_{fus} H° = 33.878 \pm 0.96$ kJ mol^{-1}

Enthalpy of Formation

$\Delta_f H°(NbCl_5, \ell, 298.15 K)$ is calculated from that of $NbCl_5$(cr) by adding $\Delta_{fus}H°$ and the difference in enthalpy, H°(478.9 K) - H°(298.15 K), between the crystal and liquid.

Heat Capacity and Entropy

The liquid phase heat capacity values are derived from the enthalpy equation reported by Keneshea et al. (5). The equation is used for the region 478.9 - 600 K, i.e. that region in which the saturation heat capacity and the heat capacity at constant pressure are essentially the same in value. This equation is used also to extrapolate to 700 K and to an assumed glass transition temperature at T_{gls} = 350 K. Below T_{gls} the heat capacity values are those of the crystal. S°(298.15 K) is calculated in a manner analogous to that used for $\Delta_f H°$(298.15 K).

Fusion Data

The adopted value of T_{fus} is that reported by Keneshea et al. (5). This value, 205.7°C or 478.9 K, was obtained by visual observation on four different samples in evacuated capsules. Other values for T_{fus} reported in the literature are: 205.3±1.5 C° (6), 204.8°C (4), 206.8±0.3°C (3), 204.5°C (2) and 204.7°C (1). Earlier determinations of the melting point have been summarized by Schafer and Pietruck (1) and Meyer et al. (3).

The adopted heat of fusion, $\Delta_{fus}H° = 8.097 \pm 0.23$ kcal mol^{-1}, is calculated as the difference at 478.9 K in the enthalpy equations for $NbCl_5(\ell)$ and $NbCl_5$(cr) as reported by Keneshea et al. (5). Meyer et al. (3), from an examination of the $NbCl_5$-$NbOCl_3$ system, reported a value of 8.30±0.4 kcal mol^{-1} for $\Delta_{fus}H°$. Voitovich et al. (4) studied the $NbCl_5$-S_2Cl_2 system and calculated $\Delta_{fus}H°$ = 8.15 kcal mol^{-1}. These two latter values are in excellent agreement with our adopted value.

Johnson and Cubicciotti (6), in their study of the orthobaric densities of $NbCl_5$, discussed the possible molecular behavior of $NbCl_5$. The explanation served to explain the large entropy of melting, $\Delta_{fus}S°$ = 16.91 kcal mol^{-1}. Based on currently available information, they suggested that between the melting point and about 650 K the substance changes from a dimeric solid to a monomeric liquid. This change was suggested to manifest itself in the unusually large volume change (6) (27±2%) and entropy of fusion and in the inverse curvature of the liquid density curve just above the melting point. For additional references, refer to the $NbCl_5$(g) table.

Vaporization Data

T_{vap} is calculated as that temperature for which $\Delta_r G° = 0$ for the process $NbCl_5(\ell) = NbCl_5(g)$. $\Delta_{vap}H°$ is calculated as the difference between the $\Delta_f H°$ values for $NbCl_5$(g) and $NbCl_5(\ell)$ at T_{vap}. Seven vaporization studies are summarized in the $NbCl_5$(g) table. Sheka et al. (2) studied the vapor-liquid equilibrium in the $NbCl_5$-$POCl_3$ system and measured a boiling point (at 760 mm) of 521.7 K for pure $NbCl_5$. The vaporization studies discussed in the $NbCl_5$(g) table which included the boiling point yielded normal boiling points (p = 760 mm) in the range 519.2 - 520.6 K.

References - $NbCl_5(\ell)$

1. H. Schafer and C. Pietruck, Z. Anorg. Allg. Chem. 267, 174 (1951).
2. I. A. Sheka, B. A. Voitovich and L. A. Nisel'son, Russ. J. Inorg. Chem. 4, 813 (1959).
3. G. Meyer, J. F. Oosterom, and W. J. Van Oeversen, Rec. Trav. Chim. 80, 502 (1961).
4. B. A. Voitovich, A. S. Barabanova, and N. Kh. Tumanova, Russ. J. Inorg. Chem. 6, 1286 (1961).
5. F. J. Keneshea, D. Cubicciotti, G. Withers, and H. Eding, J. Phys. Chem. 72, 1272 (1963).
6. J. W. Johnson, and D. Cubicciotti, High Temp. Sci. 2, 9 (1970).

References - $NbCl_5$(g)

1. M. A. Opykhtina and N. A. Fleischer, Zh. Obshch. Khim. 7, 2016 (1937).
2. D. N. Tarasenkov and A. V. Komandin, Zh. Obshch. Khim. 10, 1319 (1940).
3. K. M. Alexander and F. Fairbrother, J. Chem. Soc. (London) 1949, S223 (1949).
4. J. B. Ainscough, R. J. W. Holt, and F. W. Trowse, J. Chem. Soc. (London) 1957, 1034 (1957).
5. W. 't Hart and G. Meyer, Rec. Trav. Chim. 83, 1233 (1964).
6. L. A. Nisel'son, A. I. Pustil'nik, and T. D. Sokolova, Russ. J. Inorg. Chem. 9, 574 (1964).
7. H. Schafer and W. Polert, Z. Anorg. Allg. Chem. 353, 78 (1967).
8. L. A. Reznitskii, Russ. J. Phys. Chem. 41, 737 (1967).
9. J. W. Johnson and D. Cubicciotti, High Temp. Sci. 2, 9 (1970).
10. J. W. Johnson, W. J. Silva, and D. Cubicciotti, High Temp. Sci. 2, 20 (1970).
11. JANAF Thermochemical Tables: Nb(cr), 12-31-73; Cl_2(g), 9-30-65.
12. H. Skinner and L. Sutton, Trans. Faraday Soc. 36, 668 (1940).
13. A. Zalkin and D. E. Sands, Acta Crystallog. 11, 615 (1958); 12, 723 (1959).
14. J. Gaunt and J. B. Ainscough, Spectrochim. Acta 10, 52 (1957).
15. G. L. Carlson, Spectrochim. Acta 19A, 1291 (1963).
16. D. L. Kepert and R. S. Nyholm, J. Chem. Soc. (London) 1965, 2871 (1965).
17. R. A. Walton and B. J. Brisdon, Spectrochim. Acta 28A, 2489 (1967).
18. R. D. Werder, R. A. Frey, and H. H. Gunthard, J. Chem. Phys. 47, 4159 (1967).
19. V. P. Spiridonov and G. V. Romanov, Vestn. Mosk. Univ., Khim. 23, 10 (1968).
20. V. P. Spiridonov and G. V. Romanov, Vestn. Mosk. Univ., Khim. 24, 65 (1969).
21. I. R. Beattie and G. A. Ozin, J. Chem. Soc. (London) A1969, 1691 (1969).
22. S. P. So, J. Mol. Struct. 16, 311 (1973).

Niobium Chloride ($NbCl_5$) Cl_5Nb_1(l)

Enthalpy Reference Temperature = T_r = 298.15 K; Standard State Pressure = p° = 0.1 MPa

	J K⁻¹mol⁻¹			kJ mol⁻¹			
T/K	$C_p°$	S°	-[G°-H°(T_r)]/T	H°-H°(T_r)	$\Delta_f H°$	$\Delta_f G°$	Log K_f
0							
100							
200							
298.15	147.904	260.970	260.970	0.	-773.354	-674.012	118.084
300	147.904	261.885	260.973	0.274	-773.283	-673.396	117.249
350	147.904	284.684	262.773	7.669	-771.430	-656.897	98.036
350.000	147.904	284.684	262.773	7.669	------- GLASS <--> LIQUID ------		
350.000	228.292	284.684	262.773	7.669	TRANSITION		
400	224.848	314.949	267.451	18.999	-765.740	-640.926	83.696
478.900	218.448	354.886	278.680	36.495	------ CRYSTAL <--> LIQUID -----		
500	216.564	364.264	282.096	41.084	-755.148	-610.971	63.828
600	206.397	402.878	299.132	62.248	-745.678	-583.044	50.759
700	194.430	433.819	316.241	82.304	-737.456	-556.607	41.535
800	180.611	458.900	332.560	101.072	-730.635	-531.254	34.687

PREVIOUS: CURRENT: December 1974

Niobium Chloride ($NbCl_5$) Cl_5Nb_1(l)

NIOBIUM CHLORIDE ($NbCl_5$)

CRYSTAL-LIQUID

M_r = 270.1714

0 to 478.9 K crystal
above 478.9 K liquid

Refer to the individual tables for details.

Niobium Chloride ($NbCl_5$) Cl_5Nb_1(cr,l)

Enthalpy Reference Temperature = T_r = 298.15 K   Standard State Pressure = $p°$ = 0.1 MPa

T/K	$C_p°$	$S°$	$-[G°-H°(T_r)]/T$	$H°-H°(T_r)$	$\Delta_fH°$	$\Delta_fG°$	Log K_f
	J K⁻¹mol⁻¹			kJ mol⁻¹			
0							
100							
200							
298.15	147.904	214.053	214.053	0.	-797.470	-684.141	119.859
300	147.904	214.968	214.056	0.274	-797.400	-683.438	118.997
400	147.904	257.518	219.858	15.064	-793.792	-646.006	84.360
478.900	147.904	284.145	228.321	26.734	CRYSTAL <--> LIQUID TRANSITION		
478.900	218.448	354.886	228.321	60.612			
500	216.564	364.264	233.862	65.201	-755.148	-610.971863.828	
600	206.397	402.878	258.937	86.364	-745.678	-583.044	50.759
700	194.430	433.819	281.789	106.421	-737.456	-556.607	41.535
800	180.611	458.900	302.414	125.189	-730.635	-531.254	34.687

PREVIOUS: CURRENT: December 1974

Niobium Chloride ($NbCl_5$) Cl_5Nb_1(cr,l)

NIOBIUM CHLORIDE ($NbCl_5$) IDEAL GAS M_r = 270.1714

$\Delta_f H°(0\ K) = -701.48 \pm 8.4\ kJ\ mol^{-1}$

$S°(298.15\ K) = 404.111 \pm 3.14\ J\ K^{-1}\ mol^{-1}$ $\Delta_f H°(298.15\ K) = -703.33 \pm 8.4\ kJ\ mol^{-1}$

Vibrational Frequencies and Degeneracies

ν, cm^{-1}	ν, cm^{-1}
394.0 (1)	[492.4](2)
317.0 (1)	179.9 (2)
[414.1](1)	54.1 (2)
[143.5](1)	148.0 (2)

Ground State Quantum Weight: [1] σ = 6

Point Group: D_{3h}

Bond Distance: Nb-Cl = 2.28 ± 0.02 Å

Bond Angles: Cl*-Nb-Cl* = 120° Cl*-Nb-Cl** = 90° Cl**-Nb-Cl** = 180°

(* = equatorial ** = axial)

Product of the Moments of Inertia: $I_A I_B I_C = 1.05340 \times 10^{-111}\ g^3\ cm^6$

Enthalpy of Formation

The vapor pressures over $NbCl_5$(cr) and $NbCl_5$(ℓ) have been measured by seven investigators. The reported vapor pressures are corrected for vapor non-ideality by means of the equation $\Delta G°/T = -R\ln p - Bp/T$. The Berthelot equation of state and critical constants T_c = 803.5±2 K and p_c = 48.2 atm. reported by Johnson et al. (9, 10) are used to calculate B. The corrected vapor pressures are used to calculate $\Delta_r H°$(298.15 K) by both 2nd and 3rd law methods. By means of comparison, Nisel'son et al. (6) measured the orthobaric densities of $NbCl_5$(ℓ), as did Johnson et al. (9), and reported T_c = 807 K and p_c = 46 atm.

Source	Method	T/K	Data Points	$\Delta_r H°$(298.15 K) kcal mol^{-1} 2nd law	3rd law	Drift cal $K^{-1} mol^{-1}$
$NbCl_5(cr) = NbCl_5(g)$						
Opykhtina and Fleischer (1)	transpiration	373-Tm	Eqn	20.74	22.28	3.6
Tarasenkov and Komandin (2)	static	446-Tm	Eqn	19.20	22.41	7.0
Alexander and Fairbrother (3)	static	402-Tm	7*	22.02±0.30	22.69±0.09	1.5±0.7
Schafer and Polert (7)	flow	351-384	25	22.71±0.10	22.52±0.03	-0.5±0.3
$NbCl_5(ℓ) = NbCl_5(g)$						
Opykhtina and Fleisher (1)	transpiration	Tm-503	Eqn	16.42	16.55	0.3
Tarasenkov and Komandin (2)	static	Tm-506	Eqn	18.41	16.73	-3.4
	boiling	504-520	Eqn	15.08	16.70	3.2
Alexander and Fairbrother (3)	static	Tm-528	13	16.98±0.05	16.91±0.01	-0.2±0.1
Ainscough et al. (4)	boiling	516-530	Eqn	16.65	16.76	0.2
Hart and Meyer (5)	static	544-594	32*	16.45±0.08	16.79±0.03	0.6±0.1
Johnson et al. (10)	boiling	503-594	10	16.42±0.04	16.76±0.04	0.6±0.1

*One point is rejected due to statistical test.

The adopted value for $\Delta_f H°$(298.15 K) of $NbCl_5$(g) is -168.1±2.0 kcal mol^{-1} (-703.330±8.4 kJ mol^{-1}). This value is based on a rounded mean of the 3rd law results for both the sublimation and vaporization studies and the appropriate $\Delta_f H°$(298.15 K) value for the crystal or liquid.

Reznitskii (8) measured the enthalpy of chlorination for Nb(cr) and reported $\Delta_f H°$(680 K) = -169.1±1 kcal mol^{-1} for the reaction Nb(cr) + 5/2 Cl_2(g) = $NbCl_5$(g). Using auxiliary data for Nb(cr) and Cl_2(g) (11), we calculated $\Delta_f H°$(298.15 K) = -170.1±1.0 kcal mol^{-1} for $NbCl_5$(g).

Heat Capacity and Entropy

Monomeric $NbCl_5$(g) was shown by Spiridonov and Romanov (19, 20), using electron diffraction techniques, to have a trigonal bipyramidal structure of D_{3h} symmetry: all the Nb-Cl bond lengths being equal within experimental uncertainty, 2.28±0.2 Å. Skinner and Sutton (12) earlier used electron diffraction techniques and had suggested the same structure although a square pyramidal structure was consistent with the experimental results. We adopt the results of Spiridonov and Romanov (19, 20). The principal moments of inertia are: $I_A = 91.8121 \times 10^{-39}$ and $I_B = I_C = 107.1141 \times 10^{-39}\ g\ cm^2$.

A normal coordinate treatment of $NbCl_5$(g) in the Urey-Bradley force fields was performed by So (22) using the reported vibrational frequencies of Beattie and Ozin (21). This work by So (22) was intended to check the correctness of the reported fundamental frequencies and predict those frequencies which had not been observed (ω_3, ω_4, ω_5). The calculations of So (22) suggested that the ω_5 value reported for $NbCl_5$(g) was incorrect. Beattie and Ozin (21) had recorded the gas phase Raman spectra of $NbCl_5$, $NbBr_5$, $TaCl_5$ and $TaBr_5$ and has assigned ω_5 only in the case of $NbCl_5$. It was noted that this appeared only very weakly in the spectra. We adopt the results of So (22) which support the work of Beattie and Ozin (21) except for the ω_5 assignment.

There are many references in the literature to spectroscopic observations (IR and Raman) of $NbCl_5$ in the solid state, in organic solvents, and in matrices. Many problems arise in any attempt in relating the observed spectra. Niobium pentachloride was shown to be dimeric in the solid phase (13). There are also indications that $NbCl_5$ is dimeric in solution (15, 16, 17).

On the basis of these frequencies and the adopted structure, we calculate S°(298.15 K) = 96.56±0.75 cal K^{-1} mol^{-1}, as was also calculated by So (22). This compares to a value of 90.29 cal K^{-1} mol^{-1} calculated by Gaunt and Ainscough (14) and 95.52 cal K^{-1} mol^{-1} by Werder et al. (18).

References

References are listed on the liquid table.

Niobium Chloride ($NbCl_5$) Cl_5Nb_1(g)

Enthalpy Reference Temperature = T_r = 298.15 K Standard State Pressure = p° = 0.1 MPa

T/K	$C_p°$ (J $K^{-1}mol^{-1}$)	S°	-[G°-H°(T_r)]/T	H°-H°(T_r) (kJ mol^{-1})	$\Delta_f H°$	$\Delta_f G°$	Log K_f
0	0.	0.	INFINITE	-26.343	-701.481	-701.481	INFINITE
100	80.548	293.472	501.092	-20.762	-703.961	-684.800	357.703
200	107.409	358.746	414.693	-11.189	-704.080	-665.472	173.803
250	114.494	383.531	406.048	-5.629	-703.751	-655.855	137.033
298.15	119.040	404.111	404.111	0.	-703.330	-646.667	113.293
300	119.183	404.848	404.114	0.220	-703.313	-646.315	112.534
350	122.372	423.475	405.577	6.264	-702.811	-636.855	95.045
400	124.611	439.970	408.865	12.442	-702.274	-627.469	81.939
450	126.230	454.745	413.157	18.715	-701.717	-618.152	71.753
500	127.434	468.110	417.994	25.058	-701.151	-608.897	63.611
600	129.061	491.501	428.351	37.890	-700.012	-590.553	51.412
700	130.076	511.477	438.834	50.850	-698.887	-572.399	42.713
800	130.749	528.893	449.027	63.893	-697.791	-554.404	36.199
900	131.217	544.322	458.774	76.993	-696.731	-536.545	31.140
1000	131.556	558.166	468.033	90.133	-695.713	-518.802	27.099
1100	131.808	570.716	476.806	103.301	-694.740	-501.158	23.798
1200	132.001	582.194	485.117	116.492	-693.814	-483.601	21.051
1300	132.151	592.766	492.996	129.700	-692.935	-466.119	18.729
1400	132.271	602.564	500.477	142.921	-692.106	-448.703	16.741
1500	132.368	611.693	507.591	156.154	-691.333	-431.345	15.021
1600	132.448	620.238	514.367	169.395	-690.624	-414.036	13.517
1700	132.514	628.270	520.833	182.643	-689.989	-396.769	12.191
1800	132.570	635.846	527.014	195.897	-689.434	-379.537	11.014
1900	132.617	643.015	532.933	209.156	-688.969	-362.334	9.961
2000	132.657	649.818	538.608	222.420	-688.602	-345.152	9.014
2100	132.691	656.292	544.059	235.688	-688.341	-327.987	8.158
2200	132.721	662.465	549.302	248.958	-688.194	-310.831	7.380
2300	132.748	668.365	554.352	262.232	-688.174	-293.679	6.670
2400	132.771	674.016	559.221	275.508	-688.294	-276.525	6.018
2500	132.791	679.436	563.922	288.786	-688.573	-259.363	5.419
2600	132.809	684.645	568.465	302.066	-689.038	-242.186	4.866
2700	132.825	689.657	572.862	315.347	-689.729	-224.986	4.353
2800	132.839	694.488	577.120	328.631	-717.062	-207.273	3.867
2900	132.852	699.150	581.248	341.915	-717.071	-189.066	3.405
3000	132.864	703.654	585.253	355.201	-717.126	-170.858	2.975
3100	132.875	708.010	589.143	368.488	-717.230	-152.648	2.572
3200	132.884	712.229	592.924	381.776	-717.379	-134.434	2.194
3300	132.893	716.318	596.602	395.065	-717.572	-116.213	1.839
3400	132.901	720.286	600.182	408.355	-717.805	-97.987	1.505
3500	132.908	724.138	603.668	421.645	-718.076	-79.753	1.190
3600	132.915	727.883	607.067	434.936	-718.379	-61.511	0.892
3700	132.921	731.524	610.382	448.228	-718.710	-43.260	0.611
3800	132.927	735.069	613.617	461.520	-719.062	-25.000	0.344
3900	132.932	738.522	616.775	474.813	-719.430	-6.731	0.090
4000	132.937	741.[illegible]88	619.861	488.107	-719.807	11.549	-0.151
4100	132.941	745.170	622.878	501.401	-720.187	29.837	-0.380
4200	132.946	748.374	625.828	514.695	-720.564	48.135	-0.599
4300	132.949	751.502	628.714	527.990	-720.930	66.442	-0.807
4400	132.953	754.559	631.540	541.285	-721.280	84.757	-1.006
4500	132.956	757.547	634.307	554.580	-721.608	103.079	-1.197
4600	132.960	760.469	637.018	567.876	-721.907	121.410	-1.379
4700	132.963	763.328	639.675	581.172	-722.173	139.745	-1.553
4800	132.965	766.128	642.280	594.469	-722.398	158.086	-1.720
4900	132.968	768.869	644.836	607.765	-722.580	176.432	-1.881
5000	132.971	771.556	647.343	621.062	-722.712	194.780	-2.035
5100	132.973	774.189	649.805	634.359	-722.792	213.131	-2.183
5200	132.975	776.771	652.222	647.657	-1413.486	240.792	-2.419
5300	132.977	779.304	654.596	660.954	-1413.781	272.608	-2.687
5400	132.979	781.790	656.928	674.252	-1413.986	304.425	-2.945
5500	132.981	784.230	659.221	687.550	-1414.198	336.250	-3.193
5600	132.983	786.626	661.474	700.848	-1414.370	368.077	-3.433
5700	132.984	788.980	663.691	714.147	-1414.501	399.908	-3.665
5800	132.986	791.292	665.871	727.445	-1414.589	431.741	-3.888
5900	132.987	793.566	668.016	740.744	-1414.633	463.574	-4.104
6000	132.989	795.801	670.127	754.043	-1414.633	495.408	-4.313

PREVIOUS: December 1974 (1 atm) CURRENT: December 1974 (1 bar)

Niobium Chloride ($NbCl_5$) Cl_5Nb_1(g)

PHOSPHORUS CHLORIDE (PCl_5) IDEAL GAS M_r = 208.23876

$S°(298.15\ K)$ = 364.288 J K^{-1} mol^{-1}

$\Delta_fH°(0\ K)$ = -354.72 ± 4.2 kJ mol^{-1}
$\Delta_fH°(298.15\ K)$ = -360.18 ± 4.2 kJ mol^{-1}

Vibrational Frequencies and Degeneracies

ν, cm^{-1}	ν, cm^{-1}	ν, cm^{-1}
394 (1)	394(1)	465(1)
[176](1)	592(2)	335(2)
100 (2)	280(2)	

Ground State Quantum Weight: 1 σ = 6
Point Group D_{3h}
Bond Distances: $P-Cl^{*}$ = 2.04Å $P-Cl^{**}$ = 2.19 Å
Bond Angles: $Cl^{*}-P-Cl^{*}$ = 120° $Cl^{**}-P-Cl^{**}$ = 180° $Cl^{*}-P-Cl^{**}$ = 90
(* = equatorial ** = axial)
Product of the Moments of Inertia: $I_AI_BI_C = 6.3873 \times 10^{-112}\ g^3\ cm^6$

Enthalpy of Formation

$\Delta_fH°(298.15\ K)$ was calculated from the $\Delta_rH°(298.15\ K)$ for the reaction: $PCl_3(g) + Cl_2(g) \rightarrow PCl_5(g)$. A third law calculation using experimental equilibrium constants, and free energy functions from JANAF Tables gave a $\Delta_rH°(298.15\ K)$ = -20.884 kcal mol^{-1}. This $\Delta_rH°$ is the average of (1) $\Delta_rH°(298.15\ K)$ = -20.766±0.26 kcal mol^{-1} and (2) $\Delta_rH°(298.15\ K)$ = -21.001 ± 0.24 kcal mol^{-1}. The first, $\Delta_rH°(1)$, was obtained from eleven experimental equilibrium constants measured by Holland (1) as corrected by Nernst (2). The second, $\Delta_rH°[2]$, was obtained from thirteen experimental equilibrium constants measured by Fischer and Jubermann (3).

For the same reaction, $PCl_3(g) + Cl_2(g) = PCl_5(g)$, a 3rd law calculation by Stevenson and Yost (4) gave a $\Delta H°(0\ K)$ = -21.320±0.107 kcal mol^{-1}. With a new assignment of fundamental frequencies Wilmshurst and Bernstein (5) reported a $\Delta E°(0\ K)$ = -20.525± 0.16 kcal mol^{-1}.

The $\Delta_fH°(298.15\ K)$ values corresponding to these heats of reaction are:

$\Delta_fH°(298.15\ K)$	Source
-82.9 kcal mol^{-1}	(4)
-82.1 kcal mol^{-1}	(5)
-81.9 kcal mol^{-1}	Calculated using Gibbs energy functions from JANAF Tables

Heat Capacity and Entropy

The fundamental vibrational frequencies were assigned by Wilmshurst and Bernstein (5) from their infrared and Raman spectral data, except the frequency, 176 cm^{-1}, which they calculated. Electron diffraction data of Rouault (6) and Sargent and Schomaker (7), show the free PCl_5 molecule to be a trigonal bipyramid of D_{3h} symmetry. The bond lengths were calculated from force constants and Badger's rule by Wilmshurst and Bernstein (5). Their values are in good agreement with the electron diffraction bond lengths of Rouault (6). The principal moments of inertia are $I_A = 73.5006 \times 10^{-39}$ and $I_B = I_C = 93.2215 \times 10^{-39}$ g cm^2.

References

1. C. Holland, Z. Electrochem 18, 234 (1912).
2. W. Nernst, Z. Electrochem. 22, 37 (1916).
3. W. Fischer and O. Jubermann, Z. Anorg. Allg. Chim. 235, 337 (1938).
4. D. P. Stevenson and D. M. Yost, J. Chem. Phys. 9, 403 (1941).
5. J. K. Wilmshurst and H J. Bernstein, J. Chem. Phys. 27, 661 (1957).
6. M. Rouault, Ann. Phys. 14, 78(1940).
7. Sargen and Schomaker, quoted by Stevenson and Yost (4).

Phosphorus Chloride (PCl_5) $Cl_5P_1(g)$

Enthalpy Reference Temperature = T_r = 298.15 K Standard State Pressure = p° = 0.1 MPa

T/K	$C_p°$	S°	$-[G°-H°(T_r)]/T$	$H°-H°(T_r)$	$\Delta_fH°$	$\Delta_fG°$	Log K_f
	J $K^{-1}mol^{-1}$			kJ mol^{-1}			
0	0.	0.	INFINITE	-22.852	-354.723	-354.723	INFINITE
100	63.002	267.936	451.647	-18.371	-358.410	-336.183	175.604
200	95.704	322.664	374.087	-10.285	-360.181	-313.307	81.827
250	105.426	345.132	366.097	-5.241	-360.276	-301.572	63.010
298.15	111.890	364.288	364.288	0.	-360.184	-290.271	50.854
300	112.096	364.981	364.290	0.207	-360.178	-289.838	50.465
350	116.747	382.631	365.673	5.935	-360.682	-278.059	41.498
400	120.070	398.449	368.799	11.860	-360.452	-266.270	34.771
450	122.505	412.739	372.900	17.927	-360.141	-254.515	29.543
500	124.332	425.745	377.544	24.100	-359.772	-242.799	25.365
600	126.825	448.652	387.538	36.668	-358.917	-219.482	19.108
700	128.396	468.329	397.708	49.435	-357.963	-196.317	14.649
800	129.443	485.547	407.634	62.330	-356.949	-173.293	11.315
900	130.175	500.838	417.156	75.313	-355.900	-150.399	8.729
1000	130.705	514.582	426.223	88.359	-354.829	-127.623	6.666
1100	131.101	527.059	434.832	101.450	-353.745	-104.955	4.984
1200	131.404	538.480	443.000	114.576	-416.229	-81.307	3.539
1300	131.641	549.007	450.755	127.728	-414.364	-53.473	2.149
1400	131.830	558.770	458.126	140.902	-412.508	-25.781	0.962
1500	131.983	567.871	465.142	154.093	-410.661	1.777	-0.062
1600	132.109	576.393	471.832	167.298	-408.826	29.213	-0.954
1700	132.213	584.405	478.220	180.514	-407.005	56.535	-1.737
1800	132.301	591.965	484.332	193.740	-405.199	83.750	-2.430
1900	132.375	599.120	490.186	206.974	-403.412	110.865	-3.048
2000	132.439	605.912	495.804	220.215	-401.646	137.886	-3.601
2100	132.494	612.375	501.203	233.462	-399.904	164.820	-4.100
2200	132.541	618.540	506.397	246.713	-398.190	191.671	-4.551
2300	132.583	624.432	511.402	259.970	-396.506	218.445	-4.961
2400	132.619	630.076	516.230	273.230	-394.857	245.146	-5.335
2500	132.651	635.490	520.893	286.493	-393.247	271.780	-5.679
2600	132.680	640.693	525.401	299.760	-391.677	298.350	-5.994
2700	132.705	645.701	529.764	313.029	-390.152	324.861	-6.285
2800	132.728	650.528	533.992	326.301	-388.674	351 315	-6.554
2900	132.748	655.186	538.091	339.575	-387.243	377.718	-6.803
3000	132.767	659.686	542.070	352.850	-385.860	404.072	-7.036
3100	132.783	664.040	545.934	366.128	-384.526	430.381	-7.252
3200	132.799	668.256	549.691	379.407	-383.238	456.648	-7.454
3300	132.812	672.343	553.346	392.688	-381.995	482.875	-7.643
3400	132.825	676.308	556.905	405.969	-380.793	509.065	-7.821
3500	132.837	680.158	560.372	419.253	-379.630	535.220	-7.988
3600	132.847	683.900	563.751	432.537	-378.500	561.343	-8.145
3700	132.857	687.540	567.048	445.822	-377.398	587.435	-8.293
3800	132.866	691.084	570.266	459.108	-376.319	613.496	-8.433
3900	132.874	694.535	573.408	472.395	-375.256	639.530	-8.566
4000	132.882	697.899	576.478	485.683	-374.204	665.537	-8.691
4100	132.889	701.180	579.480	498.972	-373.155	691.518	-8.810
4200	132.896	704.383	582.416	512.261	-372.104	717.473	-8.923
4300	132.902	707.510	585.289	525.551	-371.045	743.403	-9.031
4400	132.908	710.565	588.102	538.841	-369.969	769.308	-9.133
4500	132.913	713.552	590.856	552.132	-368.872	795.188	-9.230
4600	132.918	716.474	593.555	565.424	-367.748	821.044	-9.323
4700	132.923	719.332	596.201	578.716	-366.591	846.874	-9.412
4800	132.927	722.131	598.796	592.008	-365.395	872.680	-9.497
4900	132.931	724.872	601.341	605.301	-364.156	898.461	-9.578
5000	132.935	727.557	603.838	618.595	-362.869	924.215	-9.655
5100	132.939	730.190	606.290	631.888	-361.531	949.944	-9.729
5200	132.943	732.771	608.698	645.182	-360.136	975.645	-9.800
5300	132.946	735.304	611.063	658.477	-358.682	1001.320	-9.869
5400	132.949	737.789	613.387	671.772	-357.167	1026.966	-9.934
5500	132.952	740.228	615.671	685.067	-355.586	1052.584	-9.997
5600	132.955	742.624	617.916	698.362	-353.939	1078.172	-10.057
5700	132.957	744.977	620.125	711.658	-352.223	1103.730	-10.115
5800	132.960	747.290	622.298	724.953	-350.437	1129.259	-10.170
5900	132.962	749.562	624.435	738.250	-348.579	1154.754	-10.223
6000	132.964	751.797	626.540	751.546	-346.648	1180.218	-10.275

PREVIOUS: September 1962 (1 atm) CURRENT: September 1962 (1 bar)

Phosphorus Chloride (PCl_5) $Cl_5P_1(g)$

TANTALUM CHLORIDE ($TaCl_5$) CRYSTAL M_r = 358.2129

$S°(298.15\ K) = [221.75 \pm 6.3]\ J\ K^{-1}\ mol^{-1}$

$T_{fus} = 489.7 \pm 1.5\ K$

$\Delta_f H°(0\ K)$ = Unknown

$\Delta_f H°(298.15\ K) = -858.98 \pm 4.2\ kJ\ mol^{-1}$

$\Delta_{fus} H° = 35.146 \pm 2.1\ kJ\ mol^{-1}$

Enthalpy of Formation

Gross et al. (3) measured the heat evolved in the reaction Ta(cr) + 5/2 Cl_2(ℓ) = $TaCl_5$(cr). Using their data for 6 runs we calculate $\Delta_f H°(298.15\ K) = -205.6\ kcal\ mol^{-1}$ for $TaCl_5$(cr). Gal'chenko et al. (4, 5) also determined the heat of reaction for the direct chlorination of Ta(15 runs involving 2 samples). They reported $\Delta_f H°(298.15\ K) = -205.05 \pm 0.18\ kcal\ mol^{-1}$. We adopt an intermediate value, $\Delta_f H°(298.15\ K) = -205.3 \pm 1.0\ kcal\ mol^{-1}$ ($-858.975 \pm 4.2\ kJ\ mol^{-1}$), based on these two studies (3, 4, 5).

Schafer and Kahlenberg (1) determined the heat of formation of $TaCl_5$(cr) via calorimetric measurements of the heats of solution of Ta(cr) and $TaCl_5$(cr) in hydrofluoric acid. Their procedure involved eight steps and resulted in $\Delta_f H°(298.15\ K) = -205.0 \pm 0.3\ kcal\ mol^{-1}$. This is in good agreement with our adopted value. In the case of $NbCl_5$(cr) [$TaCl_5$(cr)] the $\Delta_f H°$(298.15 K) value derived from the data of Schafer and Kahlenberg (1, 2) is 0.6 kcal mol^{-1} [0.3 kcal mol^{-1}] more positive than that derived from the data of Gross et al. (3).

Heat Capacity and Entropy

The heat capacity for $TaCl_5$(cr) is assumed to be the same as that for $NbCl_5$(cr) (6). We adopt $S°(298.15\ K) = 53.0 \pm 1.5\ cal\ K^{-1}\ mol^{-1}$ in order to reproduce the sublimation and vaporization data. Schafer and Kahlenberg (1, 2) assumed a similar relationship based on a comparison with the corresponding values for the halides of Zr, Hf, and U.

Fusion Data

See the liquid table for details.

Sublimation Data

The enthalpy of sublimation is discussed in the $TaCl_5$(g) table (6). Seven sublimation studies are analyzed and lead to a 3rd law result of $\Delta_{sub} H°(298.15\ K) = 22.5\ kcal\ mol^{-1}$.

References

1. H. Schafer and F. Kahlenberg, Z. Anorg. Allg. Chem. 294, 243 (1958).
2. H. Schafer and F. Kahlenberg, Z. Anorg. Allg. Chem. 305, 291 (1960).
3. P. Gross, C. Hayman, D. L. Levi, and G. L. Wilson, Trans. Faraday Soc. 56, 318 (1960).
4. G. L. Gal'chenko, D. A. Gedakyan, B. I. Timofiev, S. M. Skuratov, T. I. Serebryakova, and G. V. Samsonov, Dokl. Akad. Nauk. SSSR 170, 132 (1966).
5. G. L. Gal'chenko, D. A. Gedakyan, B. I. Timofiev, Russ. J. Inorg. Chem. 13, 159 (1968).
6. JANAF Thermochemical Tables: $NbCl_5$(cr) and $TaCl_5$(g), 12-31-74.

Tantalum Chloride ($TaCl_5$) Cl_5Ta_1(cr)

Enthalpy Reference Temperature = T_r = 298.15 K Standard State Pressure = p° = 0.1 MPa

T/K	$C_p°$	S°	$-[G°-H°(T_r)]/T$	$H°-H°(T_r)$	$\Delta_f H°$	$\Delta_f G°$	Log K_f
	J $K^{-1}mol^{-1}$			kJ mol^{-1}			
0							
100							
200							
298.15	147.904	221.752	221.752	0.	-858.975	-746.448	130.775
300	147.904	222.667	221.755	0.274	-858.905	-745.750	129.847
400	147.904	265.216	227.556	15.064	-855.349	-708.578	92.531
489.700	147.904	295.142	237.288	28.331	------ CRYSTAL <--> LIQUID -----		
500	147.904	298.220	238.511	29.854	-852.096	-672.267	70.231
600	147.904	325.186	250.778	44.645	-849.046	-636.591	55.420
700	147.904	347.986	263.078	59.435	-846.139	-601.413	44.878

PREVIOUS: CURRENT: December 1974

Tantalum Chloride ($TaCl_5$) Cl_5Ta_1(cr)

TANTALUM CHLORIDE ($TaCl_5$) LIQUID M_r = 358.2129

$S°(298.15\ K)$ = [268.135] $J\ K^{-1}\ mol^{-1}$

T_{fus} = 489.7 ± 1.5 K

$\Delta_f H°(298.15\ K)$ = [-834.347] $kJ\ mol^{-1}$

$\Delta_{fus}H°$ = 35.146 ± 2.1 $kJ\ mol^{-1}$

Enthalpy of Formation

$\Delta_f H°(TaCl_5$, ℓ, 298.15K) is calculated from that of $TaCl_5$(cr) by adding $\Delta_{fus}H°$ and the difference in enthalpy, H°(489.7 K) - H°(298.15 K), between the crystal and liquid.

Heat Capacity and Entropy

The liquid phase heat capacity is assumed to be identical to that for $NbCl_5$(ℓ) (4). As in the case of $NbCl_5$(ℓ), there is a glass transition assumed at 350 K. S°(298.15 K) is calculated in a manner analogous to that used for $\Delta_f H°$(298.15 K)

Fusion Data

We adopt T_{fus} = 478.9± 1.5 K (216.5°C) based on the work by Schafer and Pietruck (216.5°C, 1), Sheka et al. (216.5°C, 2), and Voitovich et al. (216.7°C, 3). Earlier reported melting points range between 204°C and 221°C and have been summarized by Schafer and Pietruck (1).

The adopted heat of fusion, $\Delta_{fus}H°$ = 8.4±0.5 kcal mol^{-1}, is based on the analysis of the S_2Cl_2-$TaCl_5$ system by Voitovich et al. (3). A similar study on the S_2Cl_2-$NbCl_5$ system (3) yielded a heat of melting for $NbCl_5$ which was within 0.063 kcal mol^{-1} of the result determined by drop calorimetry (4). The discussion of the molecular behavior of $NbCl_5$(ℓ) (4) is assumed to also apply to $TaCl_5$: that is, between the melting point and ~650 K, the substance changes from a dimeric solid to a monomeric liquid.

Vaporization Data

T_{vap} is calculated as that temperature for which $\Delta_r G°$ = 0 for the process $TaCl_5$(ℓ) = $TaCl_5$(g). $\Delta_{vap}H°$ is calculated as the difference between the $\Delta_f H°$ values for $TaCl_5$(g) and $TaCl_5$(ℓ) at T_{vap}. Three vaporization studies are summarized in the $TaCl_5$(g) table (4). Sheka et al. (2) studied the vapor-liquid equilibrium in the $TaCl_5$-$POCl_3$ system and measured a boiling point (at 760 mm) of 507.3 K for pure $TaCl_5$. The vaporization studies discussed in the $TaCl_5$(g) table which included the boiling point yielded normal boiling points (p = 760 mm) in the range 506.0 - 506.9 K.

References

1. H. Schafer and C. Pietruck, Z. Anorg. Allg. Chem. 267, 174 (1951).
2. I. A. Sheka, B. A. Voitovich, and L. A. Nisel'son, Russ. J. Inorg. Chem. 4, 813 (1959).
3. B. A. Voitovich, A. S. Barabanova, and N. Kh. Tumanova, Russ. J. Inorg. Chem. 6, 1286 (1961).
4. JANAF Thermochemical Tables: $NbCl_5$(ℓ), $NbCl_5$(g) and $TaCl_5$(g), 12-31-74.

Tantalum Chloride ($TaCl_5$) Cl_5Ta_1(l)

Enthalpy Reference Temperature = T_r = 298.15 K Standard State Pressure = p° = 0.1 MPa

T/K	$C_p°$ (J K⁻¹mol⁻¹)	S° (J K⁻¹mol⁻¹)	$-[G°-H°(T_r)]/T$ (J K⁻¹mol⁻¹)	$H°-H°(T_r)$ (kJ mol⁻¹)	$\Delta_f H°$ (kJ mol⁻¹)	$\Delta_f G°$ (kJ mol⁻¹)	Log K_f
0							
100							
200							
298.15	147.904	268.135	268.135	0.	-834.347	-735.649	128.883
300	147.904	269.050	268.138	0.274	-834.277	-735.037	127.981
350	147.904	291.849	269.938	7.669	-832.452	-718.643	107.252
350.000	147.904	291.849	269.938	7.669	------- GLASS <--> LIQUID	TRANSITION ------	
350.000	228.280	291.849	269.938	7.669			
400	224.848	322.113	274.616	18.999	-826.786	-702.774	91.773
489.700	217.482	366.911	287.580	38.849	------ CRYSTAL <-->	LIQUID -----	
500	216.564	371.429	289.261	41.084	-816.238	-673.014	70.309
600	206.397	410.042	306.297	62.247	-806.816	-645.274	56.176
700	194.430	440.983	323.406	82.304	-798.643	-619.014	46.191
800	180.611	466.064	339.725	101.072	-791.858	-593.834	38.773

PREVIOUS: CURRENT: December 1974

Tantalum Chloride ($TaCl_5$) Cl_5Ta_1(l)

TANTALUM CHLORIDE ($TaCl_5$) CRYSTAL–LIQUID M_r = 358.2129 Tantalum Chloride ($TaCl_5$) Cl_5Ta_1(cr,l)

0 to 489.7 K crystal
above 489.7 K liquid

Refer to the individual tables for details.

Enthalpy Reference Temperature = T_r = 298.15 K — Standard State Pressure = p° = 0.1 MPa

T/K	C_p° (J K⁻¹mol⁻¹)	S° (J K⁻¹mol⁻¹)	$-[G°-H°(T_r)]/T$ (J K⁻¹mol⁻¹)	$H°-H°(T_r)$ (kJ mol⁻¹)	$\Delta_f H°$ (kJ mol⁻¹)	$\Delta_f G°$ (kJ mol⁻¹)	Log K_f
0							
100							
200							
298.15	147.904	221.752	221.752	0.	-858.975	-746.448	130.775
300	147.904	222.667	221.755	0.274	-858.905	-745.750	129.847
400	147.904	265.216	227.556	15.064	-855.349	-708.578	92.531
489.700	147.904	295.142	237.288	28.331	CRYSTAL <--> LIQUID	TRANSITION	
489.700	217.482	366.911	237.288	63.477			
500	216.564	371.429	240.005	65.712	-816.238	-673.014	70.309
600	206.397	410.042	265.250	86.875	-806.816	-645.274	56.176
700	194.430	440.983	288.223	106.932	-798.643	-619.014	46.191
800	180.611	466.064	308.940	125.700	-791.858	-593.834	38.773

PREVIOUS: CURRENT: December 1974

Tantalum Chloride ($TaCl_5$) Cl_5Ta_1(cr,l)

TANTALUM CHLORIDE ($TaCl_5$) IDEAL GAS M_r = 358.2129

$\Delta_fH°$(0 K) = -763.07 ± 8.4 kJ mol^{-1}

S°(298.15 K) = 412.973 ± 3.14 J K^{-1} mol^{-1} $\Delta_fH°$(298.15 K) = -764.84 ± 8.4 kJ mol^{-1}

Vibrational Frequencies and Degeneracies

ν, cm^{-1}	ν, cm^{-1}
406.0 (1)	[458.0](2)
324.0 (1)	180.9 (2)
[377.2](1)	53.9 (2)
[112.1](1)	127.0 (2)

Ground State Quantum Weight: [1] σ = 6

Point Group: D_{3h}

Bond Distance: Ta-Cl = 2.27 ± 0.02 Å

Bond Angles: Cl*-Ta-Cl* = 120° Cl*-Ta-Cl** = 90° Cl**-Ta-Cl* = 180°

(* = equatorial ** = axial)

Product of the Moments of Inertia: $I_AI_BI_C = 1.02598 \times 10^{-111}$ g^3 cm^6

Enthalpy of Formation

The vapor pressures over $TaCl_5$(cr) and $TaCl_5$(ℓ) have been measured by eight investigators. The reported vapor pressures are corrected for vapor non-ideality by means of the equation ΔG°/T = Rlnp-Bp/T. The Berthelot equation of state and critical constants T_c = 767 K and P_c = 43 atm. reported by Nisel'son et al. (6) are used to calculate B. The corrected vapor pressures are used to calculate $\Delta_rH°$(298.15 K) by both 2nd and 3rd law methods.

Source	Method	T/K	Data Points	$\Delta_rH°$(298.15 K) kcal mol^{-1} 2nd law	3rd law	Drift cal $K^{-1}mol^{-1}$
$TaCl_5$(cr) = $TaCl_5$(g)						
Opykhtina and Fleischer (1)	transpiration	393-Tm	Eqn	23.12	22.46	-1.5
Tarasenkov and Komandin (2)	static	301-475	Eqn	20.63	22.26	4.3
Alexander and Fairbrother (3)	static	412-Tm	10*	22.94±0.09	22.76±0.03	-0.4±0.2
Shchukarev and Kurbanov (5)	membrane	425-Tm	Eqn	23.46	22.67	-1.7
Schafer and Polert (7)	flow	350-385	17	22.75±0.06	22.53±0.02	-0.6±0.2
Saeki et al. (8)	static	363-Tm	Eqn	22.31	22.53	0.6
Brink and Stevenson (9)	transpiration	431-473	28	21.80±0.20	22.59±0.08	1.8±0.4
$TaCl_5$(ℓ) = $TaCl_5$(g)						
Alexander and Fairbrother (3)	static	Tm-511	6*	17.17±0.13	16.79±0.01	-0.8±0.3
Ainscough et al. (4)	boiling point	502-513	Eqn	16.73	16.62	-0.2
Saeki et al. (8)	static	490-505	Eqn	17.17	16.66	-1.0

*One point rejected due to a statistical test.

The adopted value for $\Delta_fH°$(298.15 K) of $TaCl_5$(g) is -182.8±2.0 kcal mol^{-1}. This value is calculated using the rounded mean 3rd law result for the seven sublimation studies, $\Delta_{sub}H°$(298.15 K) = 22.5 kcal mol^{-1}. The rounded mean 3rd law result for the three vaporization studies leads to a $\Delta_fH°$(298.15 K) value for the gas which is 0.1 kcal mol^{-1} more positive.

Heat Capacity and Entropy

$TaCl_5$(g) was shown by Spiridonov and Romanov (15, 16), using electron diffraction techniques, to have a trigonal bipyramidal structure of D_{3h} symmetry: all the Ta-Cl bond lengths being equal within experimental uncertainty, Ta-Cl = 2.27±0.02 Å. Skinner and Sutton (10) earlier used electron diffraction techniques and had suggested the same structure although a square pyramidal structure was consistent with their experimental results. We adopt the results of Spiridonov and Romanov (15, 16). The principal moments of inertia are: $I_A = 91.0085 \times 10^{-39}$ and $I_B = I_C = 106.1766 \times 10^{-39}$ g cm^2.

A normal coordinate treatment of $TaCl_5$(g) in the Urey-Bradley force fields was performed by So (18) using the reported vibrational frequencies of Beattie and Ozin (17). This work by So (18) was intended to check the correctness of the reported fundamental frequencies and predict those frequencies which had not been observed (ν_3, ν_4, ν_5). Beattie and Ozin (17) had recorded the gas phase Raman spectra of $NbCl_5$, $NbBr_5$, $TaCl_5$, and $TaBr_5$. We adopt the results of So (18) which support the work of Beattie and Ozin (17).

There are many references in the literature to spectroscopic observations (IR and Raman) of $TaCl_5$ in the solid state and in organic solvents. Many problems arise in any attempt in relating the observed spectra. Tantalum pentachloride was shown to be dimeric in the solid phase (11), while there are indications that it is also dimeric in solution (12, 13, 14).

References

1. M. A. Opykhtina and N. A. Fleischer, Zh. Obshch. Khim. 7, 2016 (1937).
2. D. N. Taransenkov and A. V. Komandin, Zh. Obshch. Khim. 10, 1319 (1940).
3. K. M. Alexander and F. Fairbrother, J. Chem. Soc. (London) 1949, S223 (1949).
4. J. B. Ainscough, R. J. W. Holt, and F. W. Trowse, J. Chem. Soc. (London) 1957, 1034 (1957).
5. S. A. Shchukarev and A. R. Kurbanov, Vestn. Leningr. Univ. 17, No. 10, Ser. Fiz. i Khim. No. 2, 144 (1962).
6. L. A. Nisel'son, A. I. Prestil'nik, and T. D. Sokolova, Russ. J. Inorg. Chem. 9, 574 (1964).
7. H. Schafer and W. Polert, Z. Anorg. Allg. Chem. 353, 78 (1967).
8. Y. Saeki, R. Matsuzaki, M. Yanai, and K. Funaki, Kogyo Kagaku Zasshi 71, 350 (1968).
9. J. M. Brink and F. D. Stevenson, J. Chem. Eng. Data 17, 143 (1972).
10. H. Skinner and L. Sutton, Trans. Faraday Soc. 36, 668 (1940).
11. A. Zalkin and D. E. Sands, Acta Crystallog. 11, 615 (1958); 12, 723 (1959).
12. G. L. Carlson, Spectrochim. Acta 19A, 1291 (1963).
13. D. L. Kepert and R. S. Nyholm, J. Chem. Soc. (London) 1965, 2871 (1965).
14. R. A. Walton and B. J. Brisdon, Spectrochim. Acta. 28A, 2489 (1967).
15. V. P. Spiridonov and G. V. Romanov, Vest. Mosk. Univ., Khim. 23, 10 (1968).
16. V. P. Spiridonov and G. V. Romanov, Vest. Mosk. Univ., Khim. 24, 65 (1969).
17. I. R. Beattie and G. A. Ozin, J. Chem. Soc. (London) A 1969, 1691 (1969).
18. S. P. So, J. Mol. Struct. 16, 311 (1973).

Tantalum Chloride ($TaCl_5$) Cl_5Ta_1(g)

Enthalpy Reference Temperature = T_r = 298.15 K Standard State Pressure = p° = 0.1 MPa

T/K	$C_p°$ (J $K^{-1}mol^{-1}$)	S°	-[G°-H°(T_r)]/T	H°-H°(T_r) (kJ mol^{-1})	$\Delta_fH°$	$\Delta_fG°$	Log K_f
0	0.	0.	INFINITE	-26.869	-763.071	-763.071	INFINITE
100	82.867	300.387	511.324	-21.094	-765.570	-746.743	390.059
200	109.077	367.054	423.672	-11.324	-765.647	-727.758	190.071
250	115.849	392.176	414.929	-5.688	-765.283	-718.325	150.086
298.15	120.137	412.973	412.973	0.	-764.835	-709.321	124.270
300	120.271	413.716	412.975	0.222	-764.817	-708.976	123.444
350	123.250	432.494	414.451	6.315	-764.294	-699.710	104.426
400	125.326	449.095	417.764	12.532	-763.741	-690.522	90.173
450	126.821	463.948	422.085	18.838	-763.173	-681.403	79.095
500	127.928	477.370	426.953	25.208	-762.602	-672.348	70.240
600	129.419	500.837	437.367	38.082	-761.469	-654.404	56.971
700	130.346	520.862	447.900	51.073	-760.361	-636.648	47.507
800	130.960	538.310	458.134	64.141	-759.277	-619.049	40.420
900	131.386	553.761	467.917	77.259	-758.215	-601.584	34.915
1000	131.693	567.620	477.206	90.414	-757.182	-584.237	30.517
1100	131.922	580.183	486.006	103.595	-756.186	-566.991	26.924
1200	132.097	591.670	494.339	116.796	-755.237	-549.834	23.934
1300	132.234	602.249	502.239	130.013	-754.334	-532.753	21.406
1400	132.343	612.052	509.737	143.242	-753.470	-515.741	19.243
1500	132.431	621.186	516.866	156.481	-752.635	-498.790	17.369
1600	132.503	629.736	523.656	169.728	-751.825	-481.894	15.732
1700	132.563	637.770	530.134	182.981	-751.048	-465.047	14.289
1800	132.613	645.349	536.327	196.240	-750.323	-448.245	13.008
1900	132.656	652.520	542.255	209.504	-749.666	-431.481	11.862
2000	132.692	659.325	547.940	222.771	-749.085	-414.750	10.832
2100	132.723	665.800	553.399	236.042	-748.581	-398.045	9.901
2200	132.751	671.975	558.650	249.316	-748.157	-381.363	9.055
2300	132.774	677.877	563.706	262.592	-747.820	-364.699	8.283
2400	132.795	683.528	568.582	275.870	-747.577	-348.047	7.575
2500	132.814	688.949	573.289	289.151	-747.439	-331.403	6.924
2600	132.830	694.159	577.838	302.433	-747.418	-314.763	6.324
2700	132.845	699.172	582.240	315.717	-747.526	-298.120	5.767
2800	132.858	704.004	586.503	329.002	-747.779	-281.472	5.251
2900	132.869	708.666	590.635	342.288	-748.192	-264.811	4.770
3000	132.880	713.171	594.645	355.576	-748.782	-248.133	4.320
3100	132.889	717.528	598.539	368.864	-749.569	-231.432	3.900
3200	132.898	721.747	602.324	382.154	-750.570	-214.703	3.505
3300	132.906	725.837	606.005	395.444	-788.263	-197.468	3.126
3400	132.913	729.804	609.588	408.735	-789.332	-179.549	2.758
3500	132.920	733.657	613.078	422.026	-790.439	-161.598	2.412
3600	132.926	737.402	616.480	435.319	-791.578	-143.615	2.084
3700	132.931	741.044	619.798	448.612	-792.744	-125.600	1.773
3800	132.937	744.589	623.035	461.905	-793.932	-107.554	1.478
3900	132.941	748.042	626.196	475.199	-795.135	-89.476	1.198
4000	132.946	751.408	629.285	488.493	-796.349	-71.366	0.932
4100	132.950	754.691	632.304	501.788	-797.565	-53.226	0.678
4200	132.954	757.895	635.256	515.083	-798.777	-35.057	0.436
4300	132.957	761.023	638.144	528.379	-799.980	-16.859	0.205
4400	132.960	764.080	640.972	541.675	-801.166	1.367	-0.016
4500	132.963	767.068	643.741	554.971	-802.330	19.619	-0.228
4600	132.966	769.990	646.454	568.267	-803.465	37.898	-0.430
4700	132.969	772.850	649.113	581.564	-804.566	56.199	-0.625
4800	132.972	775.649	651.720	594.861	-805.628	74.525	-0.811
4900	132.974	778.391	654.277	608.158	-806.646	92.873	-0.990
5000	132.976	781.078	656.786	621.456	-807.615	111.239	-1.162
5100	132.978	783.711	659.249	634.754	-808.531	129.626	-1.328
5200	132.980	786.293	661.668	648.052	-809.390	148.029	-1.487
5300	132.982	788.826	664.043	661.350	-810.188	166.449	-1.640
5400	132.984	791.312	666.377	674.648	-810.924	184.883	-1.788
5500	132.986	793.752	668.671	687.946	-811.594	203.331	-1.931
5600	132.987	796.148	670.926	701.245	-812.196	221.789	-2.069
5700	132.989	798.502	673.144	714.544	-812.729	240.258	-2.202
5800	132.990	800.815	675.325	727.843	-1551.944	261.537	-2.355
5900	132.992	803.088	677.471	741.142	-1552.546	292.809	-2.592
6000	132.993	805.324	679.584	754.441	-1553.118	324.091	-2.821

PREVIOUS: December 1974 (1 atm) CURRENT: December 1974 (1 bar)

Tantalum Chloride ($TaCl_5$) Cl_5Ta_1(g)

TUNGSTEN CHLORIDE (WCl_5) CRYSTAL M_r = 361.115

$S°(298.15\ K) = [217.6]\ J\ K^{-1}\ mol^{-1}$

$T_{fus} = 526\ K$

$\Delta_f H°(0\ K)$ = Unknown

$\Delta_f H°(298.15\ K) = -512.96 \pm 41.8\ kJ\ mol^{-1}$

$\Delta_{fus} H° = 20.573 \pm 4.2\ kJ\ mol^{-1}$

Enthalpy of Formation

The enthalpy of formation, $\Delta_f H°(WCl_5, cr, 298.15\ K) = -122.6\ kcal\ mol^{-1}$, is calculated from that of the gas less the heat of sublimation, $\Delta_{sub} H°(298.15\ K) = 24.0\ kcal\ mol^{-1}$. The latter is calculated by the 3rd law method from partial pressures of WCl_5(g) derived from the total pressure data* of Shchukarev et al. (1). Dimer pressures, calculated from the same authors dimer-monomer dissociation data, are subtracted from the total pressures in order to obtain the monomer pressures. Analyses of the monomer pressures over both crystal and liquid give similar results and the average values for $\Delta_{sub} H°(298.15\ K)$ are 20.6 (2nd law) and 24.0 $kcal\ mol^{-1}$ (3rd law) with a drift of about 7 $cal\ K^{-1}\ mol^{-1}$.

*Earlier data reported by Schukarev et al. (2), have been revised and are not used here.

Heat Capacity and Entropy

$C°_p(298) = 6.2\ cal\ K^{-1}\ mol^{-1}$ and $C°_p(526) = 7.25\ cal\ K^{-1}\ mol^{-1}$ are estimated using the method described by Kubaschewski and Evans (3). Between 298.15 and 528, which is the melting point, the heat capacity is obtained by linear interpolation.

The entropy, $S°(298.15\ K) = 52\ cal\ K^{-1}\ mol^{-1}$, is listed by comparison with those of WCl_2(cr), WCl_4(cr) and WCl_6(cr).

Fusion Data

The melting point was determined from vapor pressure data by Shchukarev et al. (1). The heat of melting is the difference at the melting point between the 2nd law enthalpies of vaporization and sublimation derived from total vapor pressure data of Schchukarev et al. (1).

Sublimation Data

See the ideal gas table for details.

References

1. S. A. Shchukarev, G. I. Novikov and N. V. Andreeva, Vestnik Leningrad. Univ. 14, No. 4 Ser. Fiz. i Khim., No. 1, 120 (1959).
2. S. A. Shchukarev et al., Zhur. Neorg. Khim. 1, 357 (1956); 3, 2650 (1958).
3. O. Kubaschewski and E. LI. Evans, "Metallurgical Thermochemistry," Pergamon Press, New York, 1958.

Tungsten Chloride (WCl_5) Cl_5W_1(cr)

Enthalpy Reference Temperature = T_r = 298.15 K   Standard State Pressure = p° = 0.1 MPa

T/K	$C°_p$ (J K⁻¹mol⁻¹)	S° (J K⁻¹mol⁻¹)	$-[G°-H°(T_r)]/T$ (J K⁻¹mol⁻¹)	$H°-H°(T_r)$ (kJ mol⁻¹)	$\Delta_f H°$ (kJ mol⁻¹)	$\Delta_f G°$ (kJ mol⁻¹)	Log K_f
0							
100							
200							
298.15	155.645	217.568	217.568	0.	-512.958	-401.811	70.396
300	155.875	218.531	217.571	0.288	-512.872	-401.122	69.842
400	167.435	264.957	223.823	16.454	-507.845	-364.608	47.613
500	178.996	303.562	236.011	33.775	-501.966	-329.460	34.418
526.000	182.004	312.711	239.577	38.468	------ CRYSTAL <--> LIQUID -----		
600	189.954	337.194	250.129	52.239	-495.140	-295.585	25.733
700	198.531	367.151	264.746	71.684	-487.476	-262.921	19.619
800	205.016	394.106	279.260	91.877	-479.177	-231.402	15.109
900	209.618	418.534	293.399	112.621	-470.422	-200.952	11.663
1000	212.966	440.806	307.043	133.763	-461.356	-171.494	8.958
1100	215.811	461.240	320.145	155.205	-452.069	-142.956	6.788
1200	217.903	480.113	332.699	176.896	-442.606	-115.271	5.018
1300	219.325	497.614	344.720	198.762	-433.042	-88.381	3.551
1400	220.288	513.905	356.230	220.745	-423.429	-62.229	2.322
1500	220.915	529.126	367.254	242.808	-413.806	-36.765	1.280

PREVIOUS: December 1962   CURRENT: December 1966

Tungsten Chloride (WCl_5) Cl_5W_1(cr)

TUNGSTEN PENTACHLORIDE (WCl_5) LIQUID M_r = 361.115

$S°(298.15 K) = [248.500] J K^{-1} mol^{-1}$ $\Delta_f H°(298.15 K) = -495.386 kJ mol^{-1}$

T_{fus} = 526 K $\Delta_{fus}H° = 20.573 \pm 4.2 kJ mol^{-1}$

Heat of Formation

The heat of formation $\Delta_f H°(WCl_5, \ell, 298.15 K) = -118.4 kcal mol^{-1}$, is calculated from that of WCl_5(cr) by adding the heat of melting and the difference between H°(526;298.15) for the crystal and liquid.

Heat Capacity and Entropy

The heat capacity is assumed to be constant at 7.25 cal $K^{-1} mol^{-1}$.

The entropy of $WCl_5(\ell)$, $S°(298.15 K) = 59.393 cal K^{-1} mol^{-1}$ is calculated in a manner analogous to that of the heat of formation.

Melting Data

See the crystal table for details.

Vaporization Data

The boiling point (f = 1 atm), 561.4 K, is calculated as the temperature at which the Gibbs energy of formation for both $WCl_5(\ell)$ and $WCl_5(g)$ are equal. The difference in the enthalpies of formation of $WCl_5(\ell)$ and $WCl_5(g)$ at the boiling point is the enthalpy of vaporization. Shchukarev et al. (1) revised their previous data (2) and derived the boiling point as 561.2 K from their vapor pressure measurements.

References

1. S. A. Shchukarev, G. I. Novikov and N. V. Andreeva, Vestnik Leningrad. Univ. 14, No. 4, Ser. Fiz. i Khim., No. 1, 120 (1959).
2. S. A. Shchukarev and G. I. Novikov, Zhur. Neorg. Khim. 1, 357 (1956).

Tungsten Chloride (WCl_5) $Cl_5W_1(l)$

Enthalpy Reference Temperature = T_r = 298.15 K Standard State Pressure = p° = 0.1 MPa

T/K	$C_p°$ (J K⁻¹mol⁻¹)	S°	$-[G°-H°(T_r)]/T$	$H°-H°(T_r)$ (kJ mol⁻¹)	$\Delta_f H°$	$\Delta_f G°$	Log K_f
0							
100							
200							
298.15	182.004	248.498	248.498	0.	-495.387	-393.462	68.933
300	182.004	249.624	248.502	0.337	-495.252	-392.830	68.398
400	182.004	301.983	255.641	18.537	-488.190	-359.764	46.980
500	182.004	342.596	269.121	36.738	-481.432	-328.444	34.312
526.000	182.004	351.823	272.983	41.470	------ CRYSTAL <--> LIQUID -----		
600	182.004	375.780	284.216	54.938	-474.870	-298.465	25.984
700	182.004	403.836	299.352	73.138	-468.450	-269.574	20.116
800	182.004	428.139	313.966	91.339	-462.144	-241.595	15.775
900	182.004	449.576	327.866	109.539	-455.933	-214.400	12.443
1000	182.004	468.752	341.012	127.740	-449.808	-187.892	9.814
1100	182.004	486.099	353.426	145.940	-443.762	-161.994	7.692
1200	182.004	501.935	365.152	164.140	-437.791	-136.643	5.948
1300	182.004	516.503	376.241	182.341	-431.891	-111.787	4.492
1400	182.004	529.991	386.748	200.541	-426.061	-87.382	3.260
1500	182.004	542.548	396.721	218.742	-420.301	-63.392	2.208

PREVIOUS: December 1962 CURRENT: December 1966

Tungsten Chloride (WCl_5) $Cl_5W_1(l)$

TUNGSTEN CHLORIDE (WCl_5) CRYSTAL–LIQUID M_r = 361.115

0 to 526 K crystal
above 526 K liquid

Refer to the individual tables for details.

Tungsten Chloride (WCl_5) Cl_5W_1(cr,l)

Enthalpy Reference Temperature = T_r = 298.15 K Standard State Pressure = p° = 0.1 MPa

T/K	C_p° (J K⁻¹mol⁻¹)	S° (J K⁻¹mol⁻¹)	$-[G^\circ-H^\circ(T_r)]/T$ (J K⁻¹mol⁻¹)	$H^\circ-H^\circ(T_r)$ (kJ mol⁻¹)	Δ_fH° (kJ mol⁻¹)	Δ_fG° (kJ mol⁻¹)	Log K_f
0							
100							
200							
298.15	155.645	217.568	217.568	0.	-512.958	-401.811	70.396
300	155.875	218.531	217.571	0.288	-512.872	-401.122	69.842
400	167.435	264.957	223.823	16.454	-507.845	-364.608	47.613
500	178.996	303.562	236.011	33.775	-501.966	-329.460	34.418
526.000	182.004	312.711	239.577	38.468	CRYSTAL <--> LIQUID TRANSITION		
526.000	182.004	351.823	239.577	59.041			
600	182.004	375.780	254.931	72.509	-474.870	-298.465	25.984
700	182.004	403.836	274.250	90.710	-468.450	-269.574	20.116
800	182.004	428.139	292.001	108.910	-462.144	-241.595	15.775
900	182.004	449.576	308.342	127.111	-455.933	-214.400	12.443
1000	182.004	468.752	323.441	145.311	-449.808	-187.892	9.814
1100	182.004	486.099	337.452	163.511	-443.762	-161.994	7.692
1200	182.004	501.935	350.509	181.712	-437.791	-136.643	5.948
1300	182.004	516.503	362.725	199.912	-431.891	-111.787	4.492
1400	182.004	529.991	374.197	218.113	-426.061	-87.382	3.260
1500	182.004	542.548	385.006	236.313	-420.301	-63.392	2.208

PREVIOUS: CURRENT: December 1966

Tungsten Chloride (WCl_5) Cl_5W_1(cr,l)

TUNGSTEN PENTACHLORIDE (WCl_5) IDEAL GAS M_r = 361.116

$S°(298.15\ K)$ = [405.693] $J\ K^{-1}\ mol^{-1}$

$\Delta_f H°(0\ K) = 410.9 \pm 33\ kJ\ mol^{-1}$
$\Delta_f H°(298.15\ K) = -412.5 \pm 33\ kJ\ mol^{-1}$

Electronic Levels and Quantum Weights

ε_i, cm^{-1}	g_i
0	[2]
[7000]	[2]
[14000]	[2]

Vibrational Frequencies and Degeneracies

ν, cm^{-1}	ν, cm^{-1}
[480](1)	[350](2)
[400](1)	[150](2)
[380](1)	[100](2)
[130](1)	[180](2)

Ground State Quantum Weight: [2] σ = [6]
Point Group: [D_{3h}]
Bond Distance: W-Cl = [2.26] Å
Bond Angles: Cl*-W-Cl* = [120]° Cl*-W-Cl** = [90]° Cl**-W-Cl** = [180]°
Product of the Moments of Inertia: $I_A I_B I_C = [9.98 \times 10^{-112}]\ g^3\ cm^6$

Enthalpy of Formation

The heat of formation, $\Delta_f H°(WCl_5$, g, 298.15 K) = -98.6 kcal mol^{-1}, is calculated from $\Delta_r H°(298.15\ K)$ = 19.4 kcal mol^{-1} for $WCl_6(g) + WCl_5(g) + 1/2\ Cl_2(g)$. The value of $\Delta_r H°(298.15\ K)$ is calculated by the third law method from the equation for log Kp given by Shchukarev (1). The second law $\Delta_r H°(298.15\ K)$ is 24.6 kcal mol^{-1} and the third law drift is -8.5 cal $K^{-1}\ mol^{-1}$.

Heat Capacity and Entropy

The molecular configuration is assumed to be a trigonal bipyramid similar to that of $MoCl_5$ determined by electron diffraction by Ewens and Lister (2). The bond distance is estimated to be the same as that in $WCl_6(g)$. The three principal moments of inertia are: $I_A = 90.2084 \times 10^{-39}$ and $I_B = I_C = 105.2432 \times 10^{-39}\ g\ cm^2$.

All vibrational frequencies are estimated from those of $SbCl_5$, PCl_5, $NbCl_5$ and $TaCl_5$ obtained from infrared and Raman spectra by Carlson (3).

If it is considered that the metal atoms are effectively ionized, then the W^{5+} will have a single d electron in a field of chloride ions. The electronic splitting would be roughly the inverse of that in $CuCl_2(g)$ as given by DeKock and Gruen (4), where Cu has a d^9 configuration. By analogy with the splitting in $CuCl_2$, the low lying electronic levels are estimated to be 7000 cm^{-1}, both doublets.

References
1. S. A. Shchukarev and A. V. Shvorov, Vestnik Leningrad. Univ. 16, No. 4, Ser. Fiz. i Khim., No. 1, 87 (1961).
2. R. V. G. Ewens and M. W. Lister, Trans. Faraday Soc. 34, 1358 (1938).
3. G. L. Carlson, Spectrochim. Acta 19, 1291 (1963).
4. C. W. DeDock and D. M. Gruen, J. Chem. Phys. 44, 4387 (1966).

Tungsten Chloride (WCl_5) $Cl_5W_1(g)$

Enthalpy Reference Temperature = T_r = 298.15 K Standard State Pressure = p° = 0.1 MPa

	J $K^{-1}mol^{-1}$			kJ mol^{-1}			
T/K	$C°_p$	S°	$-[G°-H°(T_r)]/T$	$H°-H°(T_r)$	$\Delta_f H°$	$\Delta_f G°$	Log K_f
0	0.	0.	INFINITE	-26.248	-410.866	-410.866	INFINITE
100	79.886	293.996	503.816	-20.982	-413.531	-394.718	206.179
200	108.885	359.776	416.397	-11.324	-413.475	-375.794	98.147
250	115.882	384.884	407.651	-5.692	-413.047	-366.420	76.559
298.15	120.235	405.693	405.693	0.	-412.542	-357.485	62.630
300	120.370	406.438	405.696	0.223	-412.522	-357.143	62.184
350	123.362	425.232	407.173	6.321	-411.945	-347.959	51.930
400	125.434	441.848	410.489	12.544	-411.340	-338.859	44.250
450	126.918	456.712	414.814	18.854	-410.721	-329.836	38.286
500	128.014	470.144	419.686	25.229	-410.097	-320.882	33.522
600	129.486	493.625	430.108	38.110	-408.853	-303.156	26.392
700	130.400	513.659	440.648	51.108	-407.636	-285.637	21.314
800	131.006	531.114	450.889	64.180	-406.458	-268.289	17.517
900	131.434	546.570	460.678	77.303	-405.325	-251.086	14.573
1000	131.757	560.436	469.972	90.463	-404.240	-234.008	12.223
1100	132.019	573.006	478.777	103.652	-403.205	-217.035	10.306
1200	132.250	584.503	487.115	116.866	-402.221	-200.154	8.712
1300	132.467	595.098	495.019	130.102	-401.285	-183.353	7.367
1400	132.681	604.922	502.523	143.359	-400.399	-166.623	6.217
1500	132.900	614.084	509.658	156.638	-399.560	-149.954	5.222
1600	133.125	622.668	516.456	169.940	-398.768	-133.340	4.353
1700	133.357	630.746	522.944	183.264	-398.022	-116.774	3.588
1800	133.596	638.375	529.146	196.611	-397.323	-100.250	2.909
1900	133.841	645.605	535.087	209.983	-396.671	-83.764	2.303
2000	134.088	652.476	540.786	223.380	-396.070	-67.311	1.758
2100	134.336	659.024	546.262	236.801	-395.520	-50.887	1.266
2200	134.583	665.279	551.531	250.247	-395.024	-34.488	0.819
2300	134.827	671.267	556.607	263.717	-394.588	-18.110	0.411
2400	135.067	677.010	561.505	277.212	-394.214	-1.750	0.038
2500	135.301	682.529	566.237	290.730	-393.906	14.596	-0.305
2600	135.528	687.840	570.812	304.272	-393.664	30.932	-0.621
2700	135.748	692.959	575.242	317.836	-393.520	47.260	-0.914
2800	135.958	697.900	579.535	331.421	-393.510	63.584	-1.186
2900	136.160	702.674	583.699	345.027	-393.676	79.911	-1.439
3000	136.352	707.293	587.742	358.653	-394.047	96.247	-1.676
3100	136.534	711.767	591.671	372.297	-394.666	112.600	-1.897
3200	136.706	716.105	595.493	385.959	-395.563	128.976	-2.105
3300	136.868	720.314	599.212	399.638	-396.764	145.387	-2.301
3400	137.020	724.402	602.834	413.332	-398.302	161.837	-2.486
3500	137.162	728.376	606.364	427.042	-400.237	178.339	-2.662
3600	137.294	732.242	609.807	440.764	-402.663	194.903	-2.828
3700	137.416	736.005	613.167	454.500	-440.440	211.733	-2.989
3800	137.529	739.671	616.448	468.247	-440.547	229.360	-3.153
3900	137.632	743.245	619.654	482.005	-440.658	246.991	-3.308
4000	137.727	746.731	622.788	495.774	-440.770	264.624	-3.456
4100	137.813	750.133	625.852	509.551	-440.876	282.261	-3.596
4200	137.890	753.455	628.851	523.336	-440.971	299.899	-3.730
4300	137.960	756.700	631.787	537.128	-441.049	317.540	-3.857
4400	138.021	759.873	634.662	550.927	-441.105	335.182	-3.979
4500	138.075	762.975	637.479	564.732	-441.132	352.825	-4.095
4600	138.122	766.010	640.240	578.542	-441.127	370.469	-4.207
4700	138.163	768.981	642.948	592.357	-441.083	388.112	-4.313
4800	138.196	771.890	645.604	606.175	-440.996	405.753	-4.415
4900	138.224	774.740	648.210	619.996	-440.862	423.393	-4.513
5000	138.246	777.533	650.769	633.819	-440.677	441.028	-4.607
5100	138.263	780.271	653.281	647.645	-440.438	458.661	-4.698
5200	138.274	782.956	655.749	661.472	-440.140	476.287	-4.784
5300	138.281	785.590	658.175	675.299	-439.782	493.907	-4.868
5400	138.283	788.174	660.558	689.128	-439.360	511.520	-4.948
5500	138.280	790.712	662.902	702.956	-438.873	529.126	-5.025
5600	138.274	793.203	665.206	716.783	-438.318	546.720	-5.100
5700	138.264	795.650	667.473	730.610	-437.695	564.305	-5.171
5800	138.251	798.055	669.704	744.436	-437.001	581.878	-5.240
5900	138.234	800.418	671.899	758.260	-436.237	599.438	-5.307
6000	138.214	802.741	674.061	772.083	-435.401	616.984	-5.371

PREVIOUS: December 1966 (1 atm) CURRENT: December 1966 (1 bar)

Tungsten Chloride (WCl_5) $Cl_5W_1(g)$

IRON CHLORIDE $((FeCl_3)_2)$ IDEAL GAS M_r = 324.412

$\Delta_f H°(0\ K) = -658.27 \pm 8.4\ kJ\ mol^{-1}$

$S°(298.15\ K) = [536.937]\ J\ K^{-1}\ mol^{-1}$ $\Delta_f H°(298.15\ K) = -654.38 \pm 8.4\ kJ\ mol^{-1}$

Vibrational Frequencies and Degeneracies

ν, cm^{-1}	ν, cm^{-1}	ν, cm^{-1}
[300](1)	[110](1)	[230](1)
[200](1)	[350](1)	[85](1)
[120](1)	[100](1)	[70](1)
[95](1)	[18](1)	[260](1)
[35](1)	[310](1)	[110](1)
[250](1)	[110](1)	[90](1)

Ground State Quantum Weight: [1] σ = 4

Point Group: D_{2h}

Bond Distances: Fe-Cl = 2.11 ± 0.03 Å Fe-Cl_{bridge} = 2.28 ± 0.03 Å

Bond Angles: Cl-Fe-Cl = 128 ± 3° Cl_{bridge}-Fe-Cl_{bridge} = 92 ± 3°

Product of the Moments of Inertia: $I_A I_B I_C = 7.05018 \times 10^{-111}\ g^3\ cm^6$

Enthalpy of Formation

The vaporization of $FeCl_3(cr)$ is complex. It involves three possible chemical reactions, i.e. (1) $2FeCl_3(cr) = Fe_2Cl_6(g)$; (2) $FeCl_3(cr) = FeCl_3(g)$; and (3) $2FeCl_3(cr) = 2FeCl_2(cr) + Cl_2(g)$. Although these equilibria have been studied previously by many investigators, a review of their reported data reveals inconsistencies and errors. At room temperature the main vapor species present above $FeCl_3(cr)$ is $Fe_2Cl_6(g)$, a dimer of $FeCl_3(g)$. As temperature increases, the dimer gradually decomposes into monomers ($FeCl_3$, g). Simultaneously, part of $FeCl_3(cr)$ also decomposes into $FeCl_2(cr)$ and $Cl_2(g)$. Therefore in order to obtain the partial pressure of $Fe_2Cl(g)$, the partial pressures of other vapor species, e.g. $Cl_2(g)$ and $FeCl_3(g)$, must be known and deducted from the observed total pressures. However, only few investigators made such corrections. The total pressures above $FeCl_3(cr)$, at 500-600 K, have been determined by Maier (1) and Johnstone et al. (2), respectively. The partial pressures of $Fe_2Cl_6(g)$ were evaluated by the subtraction of the calculated equilibrium pressures of $Cl_2(g)$, obtained from reaction (3), from the reported total pressures. The partial pressures of $FeCl_3(g)$ were found to be insignificant at these temperatures. Using the partial pressures obtained for $Fe_2Cl_6(g)$, the enthalpy change $\Delta_r H°(298.15\ K)$ of reaction (1) was derived by both the second and third law methods. Then the value of $\Delta_f H°(Fe_2Cl_6$, g, 298.15 K) was calculated. The results are given in this table.

Wilson and Gregory (3) have investigated the vaporization and thermal decomposition equilibria of $FeCl_3(cr)$ by gas saturation flow and diaphragm gage techniques. Using the partial pressures for $Fe_2Cl_6(g)$, calculated from an equation, the values of $\Delta_r H°(298.15\ K)$ and $\Delta_f H°(298.15\ K)$ were also evaluated and listed in the table. Based on the total pressures measured by Stirnemann (4) and Sano (5), the equations for the pressures of $Fe_2Cl_6(g)$ and $Cl_2(g)$ in equilibrium with $FeCl_3(cr)$ and $FeCl_2(cr)$ as a function of temperature were reported by Wilson and Gregory (3). The corresponding values of $\Delta_r H°(298.15\ K)$ and $\Delta_f H°(298.15\ K)$ were hence calculated. A brief summary and discussion of results of some other previous investigators were also given in the same report.

Source	T/K	$\Delta_r H°(298.15\ K)$, kcal mol^{-1} 2nd law	3rd law	$\Delta_f H°(298.15\ K)$* kcal mol^{-1}
Maier (1)	489.6-591.5	32.18±0.40	34.67	-156.25±0.50
Johnstone, et al. (2)	505.2-562.2	34.81±0.49	34.50	-156.42±0.50
Wilson and Gregory (3)	463.4-577.2	34.46	34.45	-156.47±0.40
Stirnemann (4)	526.2-574.2	35.04	34.49	-156.43±0.50
Sano (5)	513.2-569.2	30.67	34.76	-156.16±0.50

The value of $\Delta_f H°(298.15\ K)$ for $Fe_2Cl_6(g)$ is selected as $-156.4\pm2.0\ kcal\ mol^{-1}$.

Heat Capacity and Entropy

The molecular structure, bond distances and angles were obtained from Zasorin et al. (6). The vibrational frequencies were estimated by comparison with those for $Al_2Cl_6(g)$ such that the values of ΔH° derived from the equilibrium data by the 2nd and 3rd law methods are in reasonable agreement. The principal moments of inertia are: $I_A = 116.3664 \times 10^{-39}$, $I_B = 194.7424 \times 10^{-39}$ and $I_C = 311.1088 \times 10^{-37}\ g\ cm^2$.

References

1. C. G. Maier, U. S. Bur. Mines Tech. Paper 360, 1925.
2. H. F. Johnstone, H. C. Weingartner and W. E. Winsche, J. Amer. Chem. Soc. 64, 241 (1942).
3. L. E. Wilson and N. W. Gregory, J. Phys. Chem. 62, 433 (1958).
4. E. Stirnemann, Neues Jahrb. Mineral., Geol. Palaont. 52A, 334 (1925).
5. K. Sano, J. Chem. Soc. Japan 59, 1073 (1938).
6. E. Z. Zasorin, N. G. Rambidi and P. A. Akishin, Zh. Strukt. Khim. 4, 910 (1963).

Iron Chloride $((FeCl_3)_2)$ $Cl_6Fe_2(g)$

Enthalpy Reference Temperature = T_r = 298.15 K Standard State Pressure = p° = 0.1 MPa

T/K	$C_p°$ (J $K^{-1}mol^{-1}$)	S°	$-[G°-H°(T_r)]/T$	$H°-H°(T_r)$ (kJ mol^{-1})	$\Delta_f H°$	$\Delta_f G°$	Log K_f
0	0.	0.	INFINITE	-40.447	-658.269	-658.269	INFINITE
100	132.329	365.442	683.671	-31.823	-659.226	-637.855	333.181
200	164.177	469.299	552.585	-16.657	-656.711	-617.413	161.252
250	170.184	506.640	539.778	-8.285	-655.470	-607.734	126.979
298.15	173.663	536.937	536.937	0.	-654.378	-598.641	104.879
300	173.768	538.011	536.940	0.321	-654.338	-598.295	104.172
350	176.052	564.982	539.065	9.071	-653.317	-589.037	87.909
400	177.587	588.597	543.811	17.914	-652.411	-579.917	75.729
450	178.665	609.579	549.975	26.822	-651.626	-570.904	66.269
500	179.449	628.446	556.894	35.776	-650.968	-561.971	58.709
600	180.486	661.263	571.635	53.777	-650.035	-544.268	47.383
700	181.121	689.137	586.480	71.860	-649.620	-526.680	39.301
800	181.537	713.351	600.858	89.994	-649.813	-509.113	33.242
900	181.824	734.751	614.569	108.163	-650.863	-491.473	28.524
1000	182.030	753.919	627.563	126.356	-653.374	-473.648	24.741
1100	182.183	771.276	639.851	144.567	-657.699	-455.437	21.627
1200	182.300	787.133	651.473	162.792	-660.940	-436.946	19.020
1300	182.391	801.729	662.477	181.026	-660.909	-418.282	16.807
1400	182.463	815.248	672.913	199.269	-661.066	-399.613	14.910
1500	182.521	827.839	682.826	217.519	-661.411	-380.927	13.265
1600	182.569	839.620	692.262	235.773	-661.945	-362.212	11.825
1700	182.609	850.689	701.259	254.032	-664.568	-343.420	10.552
1800	182.642	861.128	709.853	272.295	-666.151	-324.484	9.416
1900	182.670	871.003	718.077	290.560	-696.123	-304.051	8.359
2000	182.694	880.374	725.960	308.828	-698.572	-283.353	7.400
2100	182.715	889.288	733.527	327.099	-701.052	-262.531	6.530
2200	182.733	897.788	740.801	345.371	-703.566	-241.590	5.736
2300	182.748	905.911	747.805	363.645	-706.120	-220.534	5.008
2400	182.762	913.689	754.556	381.921	-708.715	-199.367	4.339
2500	182.774	921.150	761.071	400.198	-711.358	-178.090	3.721
2600	182.785	928.319	767.367	418.476	-714.050	-156.706	3.148
2700	182.795	935.218	773.457	436.755	-716.796	-135.217	2.616
2800	182.803	941.866	779.353	455.035	-719.598	-113.626	2.120
2900	182.811	948.281	785.068	473.315	-722.457	-91.934	1.656
3000	182.818	954.478	790.613	491.597	-725.373	-70.143	1.221
3100	182.824	960.473	795.996	509.879	-728.347	-48.253	0.813
3200	182.830	966.278	801.227	528.162	-1427.977	-11.418	0.186
3300	182.835	971.904	806.314	546.445	-1427.259	32.838	-0.520
3400	182.840	977.362	811.265	564.729	-1426.656	77.074	-1.184
3500	182.844	982.662	816.087	583.013	-1426.164	121.294	-1.810
3600	182.848	987.813	820.786	601.297	-1425.780	165.501	-2.401
3700	182.852	992.823	825.368	619.582	-1425.499	209.700	-2.960
3800	182.855	997.699	829.839	637.868	-1425.315	253.891	-3.490
3900	182.858	1002.449	834.205	656.153	-1425.223	298.079	-3.992
4000	182.861	1007.079	838.469	674.439	-1425.218	342.267	-4.470
4100	182.864	1011.594	842.636	692.726	-1425.293	386.455	-4.923
4200	182.866	1016.001	846.712	711.012	-1425.444	430.645	-5.356
4300	182.869	1020.303	850.699	729.299	-1425.663	474.840	-5.768
4400	182.871	1024.508	854.602	747.586	-1425.946	519.041	-6.162
4500	182.873	1028.617	858.423	765.873	-1426.287	563.249	-6.538
4600	182.875	1032.637	862.167	784.160	-1426.681	607.466	-6.898
4700	182.877	1036.570	865.836	802.448	-1427.123	651.690	-7.243
4800	182.878	1040.420	869.433	820.736	-1427.610	695.926	-7.573
4900	182.880	1044.191	872.961	839.024	-1428.138	740.172	-7.890
5000	182.881	1047.885	876.423	857.312	-1428.704	784.428	-8.195
5100	182.883	1051.507	879.821	875.600	-1429.306	828.697	-8.488
5200	182.884	1055.058	883.156	893.888	-1429.941	872.977	-8.769
5300	182.885	1058.542	886.433	912.177	-1430.609	917.270	-9.040
5400	182.886	1061.960	889.652	930.465	-1431.310	961.577	-9.301
5500	182.887	1065.316	892.815	948.754	-1432.043	1005.897	-9.553
5600	182.889	1068.611	895.925	967.043	-1432.810	1050.229	-9.796
5700	182.890	1071.848	898.983	985.332	-1433.613	1094.576	-10.031
5800	182.890	1075.029	901.991	1003.621	-1434.454	1138.938	-10.257
5900	182.891	1078.156	904.951	1021.910	-1435.337	1183.314	-10.476
6000	182.892	1081.230	907.863	1040.199	-1436.265	1227.705	-10.688

PREVIOUS: June 1965 (1 atm) CURRENT: June 1965 (1 bar)

Iron Chloride $((FeCl_3)_2)$ $Cl_6Fe_2(g)$

MOLYBDENUM CHLORIDE ($MoCl_6$) CRYSTAL M_r = 308.658

$\Delta_fH°$(0 K) = Unknown

S°(298.15 K) = [255.2 ± 16.7] J K^{-1} mol^{-1}

$\Delta_fH°$(298.15 K) = [-523 ± 42] kJ mol^{-1}

Enthalpy of Formation

Novikov and Galitskii (1), assumed that the equilibrium $MoCl_6$(cr) = $MoCl_5$(g) + 1/2 Cl_2(g) exists. They then calculated values for the constant of the above equilibrium and estimated the thermodynamic characteristics, $\Delta_rH°$(T) = 4 kcal mol^{-1} and Δ_rST) = 8 cal K^{-1} mol^{-1}, of the equilibrium $MoCl_6$(cr) = $MoCl_5$(cr) + 1/2 Cl_2(g), from which they obtained the $\Delta_fH°$($MoCl_6$, cr, 298.15 K) = -125 kcal mol^{-1}.

Heat Capacity and Entropy

The adopted heat capacity is the same as that of WCl_6(α, cr), since the heat capacities of W and Mo are almost the same. The entropy is estimated from that of $MoCl_5$(cr) by the addition of 4 cal K^{-1} mol^{-1} due to one extra chlorine atom in the molecule.

Vaporization Data

The sublimation point is calculated from the Gibbs energy crossover between crystal and gas. However, it should be noted that $MoCl_6$(cr) is unstable with respect to $MoCl_5$(cr) at all temperatures and thus sublimation could occur only under an atmosphere of Cl_2.

Reference

1. G. I. Novikov and N. V. Galitskii, Zh. Neorg. Khim. 10, 576-8 (1965).

Molybdenum Chloride ($MoCl_6$) Cl_6Mo_1(cr)

Enthalpy Reference Temperature = T_r = 298.15 K — Standard State Pressure = p° = 0.1 MPa

T/K	$C_p°$ (J K⁻¹mol⁻¹)	S° (J K⁻¹mol⁻¹)	$-[G°-H°(T_r)]/T$ (J K⁻¹mol⁻¹)	$H°-H°(T_r)$ (kJ mol⁻¹)	$\Delta_fH°$ (kJ mol⁻¹)	$\Delta_fG°$ (kJ mol⁻¹)	Log K_f
0							
100							
200							
298.15	175.310	255.224	255.224	0.	-523.000	-391.033	68.507
300	175.728	256.310	255.227	0.325	-522.908	-390.215	67.942
400	192.464	309.156	262.320	18.734	-517.364	-346.787	45.286
500	209.200	353.955	276.271	38.842	-510.519	-304.907	31.853
600	223.007	393.425	292.572	60.512	-502.362	-264.531	23.029
700	229.702	428.347	309.526	83.175	-493.387	-225.594	16.834
800	233.049	459.266	326.350	106.333	-484.049	-187.973	12.273
900	233.886	486.768	342.675	129.684	-474.626	-151.529	8.795
1000	234.304	511.432	358.339	153.093	-465.239	-116.133	6.066
1100	234.722	533.783	373.288	176.544	-455.901	-81.675	3.878
1200	235.141	554.224	387.526	200.037	-446.612	-48.065	2.092
1300	235.559	573.063	401.083	223.574	-437.373	-15.228	0.612
1400	235.840	590.531	413.999	247.145	-428.198	16.900	-0.631
1500	235.978	606.808	426.317	270.737	-419.104	48.375	-1.685

PREVIOUS: CURRENT: December 1968

MOLYBDENUM CHLORIDE ($MoCl_6$) IDEAL GAS M_r = 308.658

$\Delta_f H°(0\ K)$ = [-438.25 ± 83.7] kJ mol^{-1}

S°(298.15 K) = [419.654 ± 16.7] J K^{-1} mol^{-1} $\Delta_f H°(298.15\ K)$ = [-439.32 ± 83.7] kJ mol^{-1}

Vibrational Frequencies and Degeneracies

ν, cm^{-1}	ν, cm^{-1}
[347](1)	[144](3)
[326](2)	[171](3)
[390](3)	[106](3)

Ground State Quantum Weight: [1] σ = 24

Point Group: O_h

Bond Distance: Mo-Cl = [2.26] Å

Bond Angle: Cl-Mo-Cl = [90]°

Product of the Moments of Inertia: $I_A I_B I_C$ = [1.74003 x 10^{-111}] g^3 cm^6

Enthalpy of Formation

Novikov and Galitskii (1), assumed that the equilibrium $MoCl_6(cr) = MoCl_5(g) + 1/2\ Cl_2(g)$ exists. They then calculated values for the constant of the above equilibrium and estimated the thermodynamic characteristics, $\Delta_r H°(T)$ = 2.5 kcal mol^{-1} and $\Delta_r S°$ = 10 cal K^{-1} mol^{-1} of the equilibrium $MoCl_6(g) = MoCl_5(g) + 1/2\ Cl_2(g)$, from which they obtained the $\Delta_f H°(MoCl_6$, g, 298.15 K) = -105 kcal mol^{-1}.

Heat Capacity and Entropy

The vibrational frequencies are estimated by comparison with those of MoF_6 and WF_6, given by Nakamoto (2), and WCl_6 given by Evans and Lo (3).

Structure, bond distance and bond angle are estimated to be the same as those of WCl_6. The principal moments of inertia are: $I_A = I_B = I_C$ = 120.2779 x 10^{-39} g cm^2.

References

1. G. I. Novikov and N. V. Galitskii, Zh. Neorg. Khim. 10, 576-82 (1965).
2. K. Nakamoto, "Infrared Spectra of Inorganic and Coordination Compounds," John Wiley and Sons, Inc., New York - London, 1963.
3. J. C. Evans and G. Y. S. Lo, J. Mol. Spectrosc. 26, 147 (1968).

Molybdenum Chloride ($MoCl_6$) $Cl_6Mo_1(g)$

Enthalpy Reference Temperature = T_r = 298.15 K Standard State Pressure = p° = 0.1 MPa

	J $K^{-1}mol^{-1}$			kJ mol^{-1}			
T/K	$C_p°$	S°	$-[G°-H°(T_r)]/T$	$H°-H°(T_r)$	$\Delta_f H°$	$\Delta_f G°$	Log K_f
0	0.	0.	INFINITE	-31.060	-438.253	-438.253	INFINITE
100	94.690	285.524	537.586	-25.206	-441.626	-412.774	215.611
200	131.116	364.480	432.505	-13.605	-440.980	-384.002	100.291
250	139.219	394.683	422.003	-6.830	-440.179	-369.847	77.275
298.15	144.125	419.654	419.654	0.	-439.320	-356.378	62.436
300	144.276	420.546	419.657	0.267	-439.286	-355.864	61.961
350	147.587	443.052	421.427	7.569	-438.354	-342.033	51.046
400	149.853	462.917	425.396	15.008	-437.410	-328.338	42.876
450	151.463	480.665	430.569	22.543	-436.468	-314.760	36.536
500	152.644	496.687	436.393	30.147	-435.533	-301.287	31.475
600	154.221	524.669	448.840	45.497	-433.696	-274.611	23.907
700	155.193	548.521	461.419	60.972	-431.910	-248.239	18.524
800	155.833	569.289	473.633	76.525	-430.177	-222.119	14.503
900	156.276	587.670	485.302	92.132	-428.498	-196.213	11.388
1000	156.595	604.153	496.377	107.776	-426.876	-170.492	8.906
1100	156.832	619.090	506.864	123.448	-425.317	-144.929	6.882
1200	157.013	632.744	516.794	139.141	-423.829	-119.506	5.202
1300	157.154	645.318	526.203	154.849	-422.418	-94.203	3.785
1400	157.266	656.968	535.132	170.570	-421.092	-69.006	2.575
1500	157.357	667.822	543.620	186.302	-419.859	-43.901	1.529
1600	157.431	677.980	551.704	202.041	-418.726	-18.874	0.616
1700	157.493	687.526	559.415	217.788	-417.700	6.084	-0.187
1800	157.545	696.529	566.785	233.539	-416.793	30.986	-0.899
1900	157.588	705.048	573.840	249.296	-416.012	55.841	-1.535
2000	157.626	713.133	580.604	265.057	-415.370	80.658	-2.107
2100	157.658	720.824	587.100	280.821	-414.878	105.446	-2.623
2200	157.686	728.159	593.346	296.588	-414.549	130.216	-3.092
2300	157.711	735.169	599.361	312.358	-414.396	154.973	-3.520
2400	157.732	741.882	605.161	328.130	-414.433	179.728	-3.912
2500	157.751	748.321	610.759	343.905	-414.677	204.489	-4.273
2600	157.768	754.508	616.170	359.681	-415.152	229.265	-4.606
2700	157.783	760.463	621.404	375.458	-415.902	254.064	-4.915
2800	157.796	766.201	626.474	391.237	-416.991	278.896	-5.203
2900	157.808	771.739	631.388	407.017	-454.386	303.823	-5.472
3000	157.819	777.089	636.156	422.799	-454.363	329.968	-5.745
3100	157.829	782.264	640.786	438.581	-454.398	356.113	-6.000
3200	157.838	787.275	645.286	454.364	-454.488	382.259	-6.240
3300	157.846	792.132	649.663	470.149	-454.631	408.410	-6.465
3400	157.853	796.844	653.923	485.934	-454.823	434.565	-6.676
3500	157.860	801.420	658.072	501.719	-455.060	460.727	-6.876
3600	157.866	805.867	662.116	517.505	-455.336	486.897	-7.065
3700	157.872	810.193	666.060	533.292	-455.645	513.074	-7.243
3800	157.877	814.403	669.908	549.080	-455.979	539.259	-7.413
3900	157.882	818.504	673.666	564.868	-456.333	565.455	-7.573
4000	157.887	822.501	677.337	580.656	-456.699	591.659	-7.726
4100	157.891	826.400	680.925	596.445	-457.068	617.873	-7.872
4200	157.895	830.205	684.435	612.234	-457.432	644.095	-8.010
4300	157.898	833.920	687.868	628.024	-457.785	670.326	-8.143
4400	157.902	837.550	691.229	643.814	-458.119	696.566	-8.269
4500	157.905	841.099	694.520	659.604	-458.425	722.811	-8.390
4600	157.908	844.569	697.744	675.395	-458.698	749.065	-8.506
4700	157.911	847.965	700.904	691.186	-458.929	775.322	-8.617
4800	157.913	851.290	704.003	706.977	-459.113	801.585	-8.723
4900	157.916	854.546	707.042	722.769	-459.245	827.852	-8.825
5000	157.918	857.736	710.024	738.560	-1050.702	859.850	-8.983
5100	157.920	860.864	712.951	754.352	-1052.391	898.079	-9.198
5200	157.922	863.930	715.825	770.144	-1054.158	936.340	-9.406
5300	157.924	866.938	718.648	785.937	-1055.996	974.636	-9.606
5400	157.926	869.890	721.422	801.729	-1057.901	1012.968	-9.799
5500	157.928	872.788	724.148	817.522	-1059.868	1051.336	-9.985
5600	157.929	875.634	726.827	833.315	-1061.891	1089.739	-10.165
5700	157.931	878.429	729.463	849.108	-1063.964	1128.179	-10.339
5800	157.932	881.176	732.055	864.901	-1066.084	1166.657	-10.507
5900	157.934	883.875	734.605	880.694	-1068.244	1205.170	-10.670
6000	157.935	886.530	737.115	896.488	-1070.444	1243.720	-10.828

PREVIOUS: December 1968 (1 atm) CURRENT: December 1968 (1 bar)

Molybdenum Chloride ($MoCl_6$) $Cl_6Mo_1(g)$

TUNGSTEN CHLORIDE, ALPHA (α-WCl_6) CRYSTAL M_r = 396.568

$S°(298.15 K) = [238.488] J K^{-1} mol^{-1}$
$T_{trs} (\alpha_1 \rightarrow \alpha_2) = 450 \pm 10 K$
$T_{trs} (\alpha_2 \rightarrow \beta) = 503 \pm 3 K$

$\Delta_f H°(0 K)$ = Unknown
$\Delta_f H°(298.15 K) = -593.71 \pm 25.1 kJ mol^{-1}$
$\Delta_{trs} H° (\alpha_1 \rightarrow \alpha_2) = [4.2] kJ mol^{-1}$
$\Delta_{trs} H° (\alpha_2 \rightarrow \beta) = 15.774 kJ mol^{-1}$

Enthalpy of Formation

Shchukarev et al. (1) have measured calorimetrically the heat of solution of WCl_6(cr) and also of H_2WO_4(cr) in a 12% solution of NaOH as -188.0±0.3 and -13.6±0.11 kcal mol^{-1}, respectively. Based on these data and a heat of dilution value for Na_2WO_4 which is essentially zero, we obtain $\Delta_r H°(298.15 K) = -174.4\pm0.3$ kcal mol^{-1} for $WCl_6(cr) + 6 NaOH(aq, 16.3 H_2O) \rightarrow H_2WO_4(cr) + 6 NaCl(aq, 22000 H_2O) + 2 H_2O(\ell)$. This reaction gives $\Delta_f H°(WCl_6$, cr, 298.15 K) = -141.9±6 kcal mol^{-1}, using a revised $\Delta_f H°(H_2WO_4$, cr, 298.15 K) = 270.5 kcal mol^{-1}, and other auxiliary data from JANAF Tables and from Parker (2). The authors have suggested a possible uncertainty of the order of 5 kcal mol^{-1} due to the slowness of the dissolution of WCl_6(cr).

Heat Capacity and Entropy

Welty (5) has measured the enthalpy changes for $WCl_6(\alpha$, cr) in the temperature range from 406 to 502.4 K by drop calorimetry. Few points were measured in the region of the $\alpha_1 \rightarrow \alpha_2$ transition and no attempt was made to investigate the phase present at the conclusion of the drop. Thus, these data appear to be inadequate to define the heat capacities of the α_1 and α_2 phase. The adopted heat capacities are estimated so that they are reasonably consistent with the enthalpy data.

The entropy, S°(298.15 K) = 57.0 cal K^{-1} mol^{-1}, is calculated from that of the gas using $\Delta S°$(298.15 K) = 43.155 cal K^{-1} mol^{-1} from the vaporization and sublimation data given in the WCl_6(g) table (December 31, 1966).

Transition Data

Transition temperatures and heats of transition have been reported by the following investigators, and their results are summarized as follows:

		$\alpha_1 \rightarrow \alpha_2$ Transition		$\alpha_2 \rightarrow \beta$ Transition	
Source	Method	T/K	$\Delta_{trs}H°$ kcal mol^{-1}	T/K	$\Delta_{trs}H°$ kcal mol^{-1}
Ketelaar et. al (3)	Vapor pressure	441	--	500	3.4
Stevenson (4)	Cooling curve	458	--	503	--
	Vapor pressure	--	--	503	1.9
Welty (5)	Drop calorimetric	--	--	504	3.4
Shchukarev et al. (6)	Vapor pressure	--	--	504	3.4
Dobortin	Approximate calorimetric	--	--	493 ± 10	5.5

The adopted $\Delta_f H°(450 K) (\alpha_1 \rightarrow \alpha_2)$ and $\Delta_f H°(503 K) (\alpha_2 \rightarrow \beta)$ are calculated as 1.0 and 3.77 kcal mol^{-1}, respectively, based on the adopted heat capacities and the enthalpy data. The enthalpies are assumed to refer to α_1 phase as the final state, although no experimental evidence for this is given.

Fusion Data

See the β-crystal table for details.

Sublimation Data

See the ideal gas table for details.

References

1. S. A. Shchukarev, I. V. Vasilkova and G. I. Novikov, Zh. Neorg. Khim. 3, 2642 (1958).
2. V. B. Parker, NSRDS-NBS 2, 1965.
3. J. A. A. Ketelaar, G. W. Oosterhout and P. B. Braun, Rec. Trav. Chim. 62, 597 (1943).
4. F. D. Stevenson, Ph.D. Thesis, Oregon State University (1962); U.S. Bur. Mines RI 6367, 1964.
5. J. R. Welty, Ph.D. Thesis, Oregon State University (1962).
6. S. A Shchukarev and A. V. Suvorov, Vestnik Leningrad. Univ. 16, No. 4, Ser. Fiz. i. Khim., No. 1, 87 (1961).
7. R. B. Dobrotin, A. V. Suvorov and Y. V. Kondrat'ev Vestnik Leningrad. Univ. 19, No. 4 Fiz. i. Khim. No. 1, 95 (1964).

Tungsten Chloride, Alpha (WCl_6) Cl_6W_1(cr)

Enthalpy Reference Temperature = T_r = 298.15 K Standard State Pressure = p° = 0.1 MPa

T/K	$C_p°$	S°	$-[G°-H°(T_r)]/T$	$H°-H°(T_r)$	$\Delta_f H°$	$\Delta_f G°$	Log K_f
	J K^{-1} mol^{-1}			kJ mol^{-1}			
0							
100							
200							
298.15	175.418	238.488	238.488	0.	-593.710	-455.544	79.809
300	175.728	239.574	238.491	0.325	-593.618	-454.687	79.168
400	192.464	292.420	245.584	18.734	-588.082	-409.178	53.433
450	200.832	315.572	252.090	28.567	-584.830	-387.006	44.923
450.000	200.832	324.870	252.090	32.751	-----------	I <--> II	----------
500	209.200	346.463	260.459	43.002	-577.043	-365.681	38.202
503.000	209.702	347.716	260.976	43.630	----------	II <--> III	---------
600	222.170	385.865	278.144	64.633	-568.866	-324.158	28.220
700	229.367	420.706	296.072	87.244	-559.872	-284.076	21.198
800	232.840	451.598	313.620	110.383	-550.476	-245.315	16.017
900	234.095	479.108	330.508	133.740	-540.969	-207.741	12.057
1000	234.304	503.792	346.624	157.168	-531.485	-171.225	8.944

PREVIOUS: December 1962 CURRENT: December 1966

Tungsten Chloride, Alpha (WCl_6) Cl_6W_1(cr)

TUNGSTEN CHLORIDE, BETA (β-WCl_6) CRYSTAL M_r = 396.568

$S°(298.15\ K) = [280.604]\ J\ K^{-1}\ mol^{-1}$
$T_{trs}\ (\alpha_2 \rightarrow \beta) = 503 \pm 3\ K$
$T_{fus} = 555 \pm 2\ K$

$\Delta_f H°(0\ K)$ = Unknown
$\Delta_f H°(298.15\ K) = [-572.877]\ kJ\ mol^{-1}$
$\Delta_{trs} H°(\alpha_2 \rightarrow \beta) = [15.774]\ kJ\ mol^{-1}$
$\Delta_{fus} H° = [6.694]\ kJ\ mol^{-1}$

Enthalpy of Formation

The enthalpy of formation, $\Delta_f H°(WCl_6,\ \beta,cr,\ 298.15\ K) = -136.921\ kcal\ mol^{-1}$, is calculated from that of $WCl_6(\alpha,cr)$ by adding the heat of $\alpha_2 \rightarrow \beta$ transition and the difference between H°(593;298.15) for $WCl_6(\alpha,cr)$ and $WCl_6(\beta,cr)$.

Heat Capacity and Entropy

Welty (1) has measured the enthalpy changes for $WCl_6(\beta,cr)$ in the temperature range from 508 to 553 K by drop calorimetry. Because of the short temperature range, poor distribution of pints and lack of identification of the phase present at the conclusion of each drop, we feel that the enthalpy data are insufficient to define the heat capacity accurately. The adopted heat capacities are estimated so that they are consistent with the enthalpy data within their probable uncertainty.

The entropy of $WCl_6(\beta,cr)$, $S°(298.15\ K) = 67.066\ cal\ K^{-1}\ mol^{-1}$ is calculated in a manner analogous to that of the heat of formation.

Transition Data

See the α-crystal table for details.

Fusion Data

Melting point and heat of melting have been reported by the following investigators, and their results are summarized as follows:

Source	Method	T_{fus}/K	$\Delta_{fus}H°$(kcal mol^{-1})
Ketelaar et al. (2)	Vapor pressure	557	2.3
Stevenson (3)	Cooling curve	555	-
	Vapor Pressure	555	1.5
Welty (1)	Drop calorimetric	556	1.6
Shchukarev et al. (4)	Vapor pressure	564	2.0
Dobrotin et al. (5)	Approximate calorimetric	545 ± 12	4.2

The adopted heat of melting, $\Delta_f H°(555\ K) = 1.6\ kcal\ mol^{-1}$, is calculated from the adopted heat capacities and the enthalpy data. The enthalpies are assumed to refer to α_1 phase as the final state, although no experimental evidence for this is available.

Sublimation Data

$\Delta_{sub}H°(298.15\ K)$ is calculated as the difference between $\Delta_f H°(298.15\ K)$ for $WCl_6(\beta,cr)$ and $WCl_6(g)$.

References

1. J. R. Welty, Ph.D. Thesis, Oregon State University (1962).
2. J. A. A. Ketelaar, G. W. Oosterhout and P. B. Braun, Rec. Trav. Chim. 62, 597 (1943).
3. T. D. Stevenson, Ph.D. Thesis, Oregon State University (1962); U. S. Bur. Mines RI 6367, 1964.
4. S. A. Shchukarev and A. V. Suvorov, Vestnik Leningrad. Univ. 16, No. 4, Ser. Fiz. i Khim., No. 1, 87 (1961).
5. R. B. Dobrotin, A. V. Suvorov and Y. V. Kondrat'ev, Vestnik Leningrad. Univ. 19, No. 4, Ser. Fiz. i Khim., No. 1, 95 (1964).

Tungsten Chloride, Beta (WCl_6) $Cl_6W_1(cr)$

Enthalpy Reference Temperature = T_r = 298.15 K; Standard State Pressure = p° = 0.1 MPa

T/K	$C_p°$ (J $K^{-1}mol^{-1}$)	S° (J $K^{-1}mol^{-1}$)	$-[G°-H°(T_r)]/T$ (J $K^{-1}mol^{-1}$)	$H°-H°(T_r)$ (kJ mol^{-1})	$\Delta_f H°$ (kJ mol^{-1})	$\Delta_f G°$ (kJ mol^{-1})	Log K_f
0							
100							
200							
298.15	188.280	280.604	280.604	0.	-572.877	-447.269	78.360
300	188.280	281.769	280.608	0.348	-572.763	-446.490	77.741
400	188.280	335.934	287.993	19.176	-566.808	-405.309	52.928
500	188.280	377.947	301.938	38.004	-561.208	-365.588	38.193
503.000	188.280	379.075	302.397	38.569	---------- II <--> III	---------	
555.000	188.280	397.597	310.463	48.360	-------- III <--> LIQUID	-------	
600	188.280	412.275	317.554	56.832	-555.834	-326.972	28.465
700	188.280	441.298	333.212	75.660	-550.623	-289.242	21.584
800	188.280	466.439	348.329	94.488	-545.539	-252.250	16.470
900	188.280	488.616	362.708	113.316	-540.561	-215.889	12.530
1000	188.280	508.453	376.308	132.144	-535.676	-180.077	9.406
1100	188.280	526.398	389.150	150.972	-530.878	-144.750	6.874
1200	188.280	542.780	401.280	169.800	-526.160	-109.857	4.782
1300	188.280	557.851	412.752	188.628	-521.518	-75.354	3.028
1400	188.280	571.804	423.621	207.456	-516.952	-41.205	1.537
1500	188.280	584.794	433.938	226.284	-512.460	-7.380	0.257

PREVIOUS: CURRENT: December 1966

Tungsten Chloride, Beta (WCl_6) $Cl_6W_1(cr)$

TUNGSTEN CHLORIDE (WCl_6) LIQUID M_r = 396.568

$S°(298.15\ K) = [284.868]\ J\ K^{-1}\ mol^{-1}$ $\Delta_f H°(298.15\ K) = [-569.405]\ kJ\ mol^{-1}$

$T_{fus} = 555 \pm 2\ K$ $\Delta_{fus}H° = 6.694\ kJ\ mol^{-1}$

Enthalpy of Formation

$\Delta_f H°(WCl_6, \ell, 298.15\ K)$ is calculated from that of $WCl_6(\beta, cr)$ by adding the enthalpy of fusion and the difference in enthalpy, H°(555 K)-H°(298.15 K), between the β- crystal and liquid.

Heat Capacity and Entropy

Welty (1), has measured the enthalpy changes for $WCl_6(\ell)$ in the temperature range from 560 to 603 K by drop calorimetry. Since the temperature range was very short and no attempt was made to investigate the phase present at the conclusion of the drop, we feel that the enthalpy data are insufficient to define the heat capacity accurately. The adopted heat capacities are estimated so that they are reasonably consistent with the enthalpy data.

S°(298.15 K) is calculated in a manner analogous to that used for the enthalpy of formation.

Fusion Data

See the β-crystal table for details.

Vaporization Data

T_{vap} is calculated as the temperature at which the Gibbs energies of formation for both $WCl_6(\ell)$ and $WCl_6(g)$ are equal. The difference in the enthalpies of formation of $WCl_6(\ell)$ and $WCl_6(g)$ at T_{vap} is the enthalpy of vaporization.

Ketelaar, et al. (2), have determined the boiling point and the heat of vaporization as 609.7 K and 15.24 kcal mol^{-1}, respectively, from vapor pressure measurements. Shchukarev and Suvorov (3, 4) revised their previous data and gave the boiling point and the heat of vaporization as 616.2 K and 14.7 kcal mol^{-1}, respectively, from their vapor pressure measurements.

See the ideal gas table for additional information.

References

1. J. R. Welty, Ph.D. Thesis, Oregon State University (1962).
2. J. A. A. Ketelaar, G. W. Oosterhout and P. B. Braun, Rec. Trav. Chim. 62, 597 (1943).
3. S. A. Shchukarev and A. V. Suvorov, Vestnik Leningrad. Univ. 16, No. 4, Ser. Fiz. i Khim., No. 1, 87 (1961).
4. S. A. Shchukarev and G. I. Novikov, Zh. Neorg. Khim. 1, 357 (1956).
5. S. A. Shchukarev, G. I. Novikov, A. V. Suvorov and A. K. Baev, Zh. Neorg. Khim. 3, 2630 (1958).

Tungsten Chloride (WCl_6) $Cl_6W_1(l)$

Enthalpy Reference Temperature = T_r = 298.15 K Standard State Pressure = p° = 0.1 MPa

	J K⁻¹mol⁻¹			kJ mol⁻¹			
T/K	$C_p°$	S°	$-[G°-H°(T_r)]/T$	$H°-H°(T_r)$	$\Delta_f H°$	$\Delta_f G°$	Log K_f
0							
100							
200							
298.15	200.832	284.868	284.868	0.	-569.405	-445.068	77.974
300	200.832	286.110	284.872	0.372	-569.267	-444.296	77.359
400	200.832	343.886	292.749	20.455	-562.057	-403.739	52.723
500	200.832	388.701	307.625	40.538	-555.202	-364.958	38.127
555.000	200.832	409.659	316.716	51.584	-------- III <--> LIQUID -------		
600	200.832	425.317	324.281	60.621	-548.573	-327.535	28.514
700	200.832	456.275	340.983	80.704	-542.106	-291.209	21.730
800	200.832	483.092	357.108	100.788	-535.767	-255.801	16.702
900	200.832	506.747	372.446	120.871	-529.533	-221.180	12.837
1000	200.832	527.907	386.953	140.954	-523.394	-187.249	9.781
1100	200.832	547.048	400.651	161.037	-517.340	-153.928	7.309
1200	200.832	564.523	413.589	181.120	-511.367	-121.155	5.274
1300	200.832	580.598	425.826	201.204	-505.470	-88.877	3.571
1400	200.832	595.481	437.419	221.287	-499.649	-57.050	2.129
1500	200.832	609.337	448.424	241.370	-493.901	-25.637	0.893

PREVIOUS: December 1962 CURRENT: December 1966

Tungsten Chloride (WCl_6) $Cl_6W_1(l)$

TUNGSTEN CHLORIDE (WCl_6)   ALPHA1-ALPHA2-BETA-LIQUID M_r = 396.568

0 to 450 K crystal, alpha 1
450 to 503 K crystal, alpha 2
503 to 555 K crystal, beta
above 555 K liquid

Refer to the individual tables for details.

Tungsten Chloride (WCl_6) Cl_6W_1(cr,l)

Enthalpy Reference Temperature = T_r = 298.15 K   Standard State Pressure = p° = 0.1 MPa

T/K	C_p° (J K⁻¹mol⁻¹)	S° (J K⁻¹mol⁻¹)	$-[G°-H°(T_r)]/T$ (J K⁻¹mol⁻¹)	$H°-H°(T_r)$ (kJ mol⁻¹)	$\Delta_fH°$ (kJ mol⁻¹)	$\Delta_fG°$ (kJ mol⁻¹)	Log K_f
0							
100							
200							
298.15	175.418	238.488	238.488	0.	-593.710	-455.544	79.809
300	175.728	239.574	238.491	0.325	-593.618	-454.687	79.168
400	192.464	292.420	245.584	18.734	-588.082	-409.178	53.433
450	200.832	315.572	252.090	28.567	-584.830	-387.006	44.923
450.000	200.832	324.870	252.090	32.751	----------	I <--> II	----------
500	209.200	346.463	260.459	43.002	-577.043	-365.681	38.202
503.000	209.702	347.716	260.976	43.630		II <--> III	
503.000	188.280	379.075	260.976	59.404		TRANSITION	
555.000	188.280	397.597	272.923	69.194		III <--> LIQUID	
555.000	200.832	409.659	272.923	75.889		TRANSITION	
600	200.832	425.317	283.773	84.926	-548.573	-327.535	28.514
700	200.832	456.275	306.262	105.009	-542.106	-291.209	21.730
800	200.832	483.092	326.727	125.092	-535.767	-255.801	16.702
900	200.832	506.747	345.441	145.176	-529.533	-221.180	12.837
1000	200.832	527.907	362.648	165.259	-523.394	-187.249	9.781
1100	200.832	547.048	378.555	185.342	-517.340	-153.928	7.309
1200	200.832	564.523	393.335	205.425	-511.367	-121.155	5.274
1300	200.832	580.598	407.130	225.508	-505.470	-88.877	3.571
1400	200.832	595.481	420.058	245.592	-499.649	-57.050	2.129
1500	200.832	609.337	432.221	265.675	-493.901	-25.637	0.893

PREVIOUS: CURRENT: December 1966

Tungsten Chloride (WCl_6) Cl_6W_1(cr,l)

TUNGSTEN CHLORIDE (WCl_6) IDEAL GAS M_r = 396.568

$S°(298.15\ K)$ = [419.166] J K^{-1} mol^{-1}

$\Delta_fH°(0\ K)$ = -491.99 ± 25.1 kJ mol^{-1}
$\Delta_fH°(298.15\ K)$ = -493.71 ± 25.1 kJ mol^{-1}

Vibrational Frequencies and Degeneracies

ν, cm^{-1}	ν, cm^{-1}
[360](1)	[140](3)
[340](2)	[180](3)
375 (3)	[120](3)

Ground State Quantum Weight = 1 σ = 24
Point Group: O_h
Bond Distance: W-Cl = 2.26 Å
Bond Angle: Cl-W-Cl = 90°
Product of the Moments of Inertia: $I_AI_BI_C = 1.74003 \times 10^{-111}\ g^3\ cm^6$

Enthalpy of Formation

The enthalpy of formation of WCl_6(g), $\Delta_fH°(298.15\ K)$ = -118.0 kcal mol^{-1}, is calculated from that of the crystal plus the adopted heat of sublimation, $\Delta_{sub}H°(298.15\ K)$ = 23.9 kcal mol^{-1} for WCl_6(α, cr) → WCl_6(g). The latter was obtained from third law analysis of the vapor pressure data (1, 2, 3, 4), using all JANAF functions (Dec. 31, 1966).

Source	Phase Studied	T/K	Data Points	$\Delta_{sub}H°(298.15\ K)$, kcal mol 2nd law	3rd law	Drift cal K^{-1} mol^{-1}	$\Delta_fH°(298.15\ K)$ kcal mol^{-1}
Stevenson (1)	α, cr	462 - 502	38	20.9 ± 0.5	23.74	6.2 ± 1.0	-118.16
	β*, cr	503 - 551	15	23.1 ± 0.2	23.89	1.5 ± 0.3	-118.01
	1*	557 - 598	13	24.2 ± 0.1	23.89	-0.6 ± 0.1	-118.01
Shchukarev (2) et al.	α, cr	**	--	22.0	23.80	4.0	-118.10
	β*, cr		--	23.8	23.98	0.4	-117.92
	1*		--	24.1	23.98	-0.2	-117.92
Ketelaar (3) et al.	α, cr	425 - 500	21	23.1 ± 0.1	23.94	1.7 ± 0.4	-117.96
	β*, cr	504 - 554	21	24.4 ± 0.1	23.91	-0.8 ± 0.1	-117.99
	1*	523 - 635	21	24.4 ± 0.1	23.85	-1.0 ± 0.1	-118.05
Vernon (4)	α, cr	298 - 423	4	16.1 ± 4.1	21.02	10.8 ± 12.8	-120.88

Vernon (4) has also determined the equilibrium constants in the temperature range from 1360 to 1600 K for the decomposition of WCl_6(g) to W(cr) and Cl_2(g) by reaction of the sample with a hot tungsten filament in a vacuum bulb. Using the reported equilibrium constants, the heat of formation of WCl_6(g) at 298 K is calculated by the second and third law method as -131.8 ± 2.9 kcal mol^{-1} and -135.2 kcal mol^{-1}, respectively. The third law drift is 2.4 ± 1.9 cal K^{-1} mol^{-1}. Unfortunately the reported equilibrium constants may be seriously in error since atomic chlorine is the main species at the pressure and temperature studied.

* Enthalpy changes are converted to $\Delta_{sub}H°(298.15\ K)$ for WCl_6(α, cr) → WCl_6(g).
** Vapor presssure equations for the crystal (α, β) and the liquid to gas are given.

Heat Capacity and Entropy

Ewens and Lister (5) have determined the octahedral structure for WCl_6(g) by electron diffraction and found the bond length to be 2.26 Å.

The vibrational frequency of ν_3, 375 cm^{-1}, was observed in the infrared spectra by Evans (6). The values of ν_1, ν_2, and ν_5 (360, 340 and 180 cm^{-1}, respectively) are estimated based on those of the aqueous ion $PtCl_6^{2-}$ reported by Woodward and Creighton (7). The values of ν_4 and ν_6 (140 and 120 cm^{-1}, respectively) are estimated by comparison with those of WF_6(g).

The principal moments of inertia are: $I_A = I_B = I_C = 120.2779 \times 10^{-39}\ g\ cm^2$.

References

1. F. D. Stevenson, Ph.D. Thesis, Oregon State University (1962); U.S. Bur. Mines RI. 6367, 1964.
2. S. A. Shchukarev and A. V. Suvorov, Vestnik Leningrad. Univ. 16, No. 4, Ser. Fiz. i Khim., No. 1, 87 (1961). Previous vapor pressure measurements, cited in Zh. Neorg. Khim. 1, 357 (1956) and 3, 2630 (1958), have been revised and are not used as reference here.
3. J. A. A. Ketelaar, G. W. Oosterhout and P. B. Braun, Rec. Trav. Chim. 62, 597 (1943).
4. A. A. Vernon, J. Amer. Chem. Soc. 59, 1832 (1937).
5. R. V. G. Ewens and M. W. Lister, Trans. Faraday Soc. 34, 1358 (1938).
6. J. C. Evans, The Dow Chemical Company, private communication, Dec. 1, 1966.

Tungsten Chloride (WCl_6) Cl_6W_1(g)

Enthalpy Reference Temperature = T_r = 298.15 K Standard State Pressure = p° = 0.1 MPa

T/K	$C_p°$ (J $K^{-1}mol^{-1}$)	S°	$-[G°-H°(T_r)]/T$	$H°-H°(T_r)$ (kJ mol^{-1})	$\Delta_fH°$	$\Delta_fG°$	Log K_f
0	0.	0.	INFINITE	-30.790	-491.987	-491.987	INFINITE
100	93.109	285.822	536.701	-25.088	-495.672	-466.591	243.722
200	130.658	364.112	431.993	-13.576	-495.281	-437.471	114.256
250	138.942	394.234	421.510	-6.819	-494.537	-423.101	88.402
298.15	143.941	419.166	419.166	0.	-493.712	-409.416	71.728
300	144.093	420.057	419.169	0.266	-493.679	-408.893	71.195
350	147.459	442.539	420.936	7.561	-492.765	-394.833	58.926
400	149.758	462.389	424.902	14.995	-491.824	-380.907	49.741
450	151.390	480.127	430.070	22.526	-490.873	-367.100	42.612
500	152.586	496.143	435.889	30.127	-489.921	-353.398	36.919
600	154.182	524.116	448.330	45.472	-488.029	-326.272	28.404
700	155.165	547.963	460.901	60.943	-486.175	-299.459	22.346
800	155.812	568.727	473.110	76.494	-484.368	-272.910	17.819
900	156.259	587.106	484.775	92.099	-482.613	-246.583	14.311
1000	156.581	603.587	495.846	107.741	-480.914	-220.449	11.515
1100	156.821	618.523	506.330	123.412	-479.273	-194.483	9.235
1200	157.003	632.176	516.257	139.104	-477.691	-168.663	7.342
1300	157.146	644.749	525.663	154.811	-476.170	-142.973	5.745
1400	157.259	656.399	534.591	170.532	-474.711	-117.398	4.380
1500	157.351	667.252	543.077	186.263	-473.316	-91.924	3.201
1600	157.426	677.410	551.159	202.002	-471.986	-66.542	2.172
1700	157.488	686.956	558.869	217.747	-470.723	-41.241	1.267
1800	157.541	695.959	566.237	233.499	-469.529	-16.012	0.465
1900	157.585	704.478	573.291	249.255	-468.406	9.152	-0.252
2000	157.623	712.562	580.054	265.016	-467.359	34.260	-0.895
2100	157.655	720.253	586.549	280 780	-466.391	59.317	-1.475
2200	157.683	727.588	592.794	296.547	-465.505	84.329	-2.002
2300	157.708	734.598	598.808	312.316	-464.706	109.303	-2.482
2400	157.730	741.311	604.607	328.088	-463.999	134.245	-2.922
2500	157.749	747.750	610.205	343.862	-463.388	159.159	-3.325
2600	157.766	753.937	615.615	359.638	-462.874	184.050	-3.698
2700	157.781	759.892	620.849	375.415	-462.486	208.924	-4.042
2800	157.794	765.630	625.918	391.194	-462.262	233.787	-4.361
2900	157.807	771.167	630.831	406.974	-462.243	258.644	-4.659
3000	157.818	776.517	635.599	422.755	-462.457	283.505	-4.936
3100	157.827	781.692	640.229	438.537	-462.947	308.378	-5.196
3200	157.836	786.703	644.728	454.321	-463.741	333.271	-5.440
3300	157.845	791.560	649.104	470.105	-464.864	358.195	-5.670
3400	157.852	796.273	653.364	485.890	-466.346	383.157	-5.886
3500	157.859	800.848	657.513	501.675	-468.248	408.169	-6.092
3600	157.865	805.296	661.556	517.461	-470.659	433.242	-6.286
3700	157.871	809.621	665.500	533.248	-508.441	458.581	-6.474
3800	157.876	813.831	669.348	549.036	-508.567	484.718	-6.663
3900	157.881	817.932	673.106	564.823	-508.711	510.859	-6.842
4000	157.886	821.929	676.776	580.612	-508.868	537.004	-7.013
4100	157.890	825.828	680.365	596.401	-509.027	563.153	-7.175
4200	157.894	829.633	683.873	612.190	-509.183	589.306	-7.329
4300	157.898	833.348	687.307	627.979	-509.327	615.462	-7.476
4400	157.901	836.978	690.667	643.769	-509.451	641.622	-7.617
4500	157.904	840.527	693.958	659.560	-509.548	667.783	-7.751
4600	157 907	843.997	697.182	675.350	-509.612	693.948	-7.880
4700	157.910	847.393	700.342	691.141	-509.634	720.111	-8.003
4800	157.913	850.718	703.440	706.932	-509.609	746.276	-8.121
4900	157.915	853.974	706.479	722.724	-509.531	772.440	-8.234
5000	157.917	857.164	709.461	738.515	-509.395	798.601	-8.343
5100	157.920	860.292	712.388	754.307	-509.195	824.759	-8.447
5200	157.922	863.358	715.262	770.099	-508.927	850.912	-8.548
5300	157.924	866.366	718.085	785.891	-508.586	877.060	-8.644
5400	157.926	869.318	720.858	801.684	-508.170	903.201	-8.737
5500	157.927	872.216	723.584	817.476	-507.676	929.333	-8.826
5600	157.929	875.062	726.264	833.269	-507.099	955.455	-8.912
5700	157.930	877.857	728.899	849.062	-506.439	981.567	-8.995
5800	157.932	880.604	731.491	864.855	-505.693	1007.666	-9.075
5900	157.933	883.303	734.041	880.649	-504.860	1033.751	-9.152
6000	157.935	885.958	736.551	896.442	-503.939	1059.821	-9.227

PREVIOUS: December 1966 (1 atm) CURRENT: December 1966 (1 bar)

Tungsten Chloride (WCl_6) Cl_6W_1(g)

TUNGSTEN CHLORIDE $((WCl_5)_2)$ IDEAL GAS M_r = 722.230

Tungsten Chloride $((WCl_5)_2)$ $Cl_{10}W_2(g)$

S°(298.15 K) = [713.581] J K^{-1} mol^{-1}

$\Delta_f H°(0 K)$ = -873.16 ± 41.8 kJ mol^{-1}
$\Delta_f H°(298.15 K)$ = -868.60 ± 41.8 kJ mol^{-1}

Vibrational Frequencies and Degeneracies

ν, cm^{-1}	ν, cm^{-1}	ν, cm^{-1}
[500](2)	[150](2)	[90](2)
[400](2)	[125](4)	[80](2)
[350](2)	[100](6)	[60](4)
[250](2)		[40](2)

Ground State Quantum Weight: [3] σ = [4]
Point Group: $[D_{2h}]$
Bond Distance: W-Cl = [2.26] Å
Bond Angle: Cl*-W-Cl* = [90]° W-Cl^*_{bridge}-W = [90]°
Cl**-W-Cl** = [180]° Cl*-W-Cl** = [90]°
* = Equatorial ** = Axial
Product of the Moments of Inertia: $I_A I_B I_C$ = [6.64026 x 10^{-110}] g^3 cm^6

Enthalpy of Formation

The enthalpy of formation, $\Delta_f H°(W_2Cl_{10}$, g, 298.15 K) = -207.6 kcal mol^{-1}, is calculated from $\Delta_r H°(298.15 K)$ = 10.4 kcal mol^{-1} for $W_2Cl_{10}(g)$ = 2 $WCl_5(g)$, using all JANAF functions (Dec. 31, 1966). The value of $\Delta_r H°(298.15 K)$ is obtained from third law analysis of the partial pressures of $WCl_5(g)$ and $W_2Cl_{10}(g)$ derived from PVT data determined by Shchukarev et al. (1). The second law $\Delta_r H°(298.15 K)$ is 10.4±0.3 kcal mol^{-1}.

Heat Capacity and Entropy

The molecular configuration is estimated by analogy with that of Mo_2Cl_{10} reported by Sands and Zalkin (2). Chlorine atoms are placed octahedrally around two tungsten atoms and these two octahedra share one common edge with six chlorine and two tungsten atoms in a plane. The bond length is estimated to be the same as that in $WCl_6(g)$. The principal moments of inertia are: I_A = 210.4863 x 10^{-39}, I_B = 546.8356 x 10^{-39}, and I_C = 576.9051 x 10^{-39} g cm^2.

All vibrational frequencies are estimated based on those of WCl_5, $(AlCl_3)_2$ and Cs_2Cl_2, such that S°(W_2Cl_{10}, g, 607 K) = 216.65 cal K^{-1} mol^{-1}. This entropy is derived from ΔS°(607 K) = 19.92±0.5 cal K^{-1} mol^{-1} for $W_2Cl_{10}(g)$ = $2WCl_5(g)$, obtained by 2nd law analysis of the data of Shchukarev et al. (1). The frequencies are not in point group order.

References

1. S. A. Shchukarev, G. I. Novikov and N. V. Andreeva, Vestnik Leningrad. Univ. 14, No. 4, Ser. Fiz. i Khim., No. 1, 120 (1959).
2. D. Z. Sands and A. Zalkin, Acta Cryst. 12, 723 (1959).

Enthalpy Reference Temperature = T_r = 298.15 K Standard State Pressure = p° = 0.1 MPa

T/K	$C_p°$ (J $K^{-1}mol^{-1}$)	S° (J $K^{-1}mol^{-1}$)	-[G°-H°(T_r)]/T (J $K^{-1}mol^{-1}$)	H°-H°(T_r) (kJ mol^{-1})	$\Delta_f H°$ (kJ mol^{-1})	$\Delta_f G°$ (kJ mol^{-1})	Log K_f
0	0.	0.	INFINITE	-60.408	-873.157	-873.157	INFINITE
100	200.630	455.551	934.652	-47.910	-876.522	-825.651	431.276
200	246.532	611.501	737.242	-25.148	-872.964	-775.991	202.668
250	256.975	667.725	717.886	-12.540	-870.764	-752.000	157.122
298.15	263.467	713.581	713.581	0.	-868.598	-729.322	127.774
300	263.668	715.212	713.586	0.488	-868.515	-728.459	126.836
350	268.139	756.214	716.814	13.790	-866.254	-705.295	105.259
400	271.243	792.234	724.036	27.279	-864.001	-682.454	89.119
450	273.470	824.317	733.429	40.900	-861.764	-659.895	76.599
500	275.117	853.219	743.987	54.616	-859.548	-637.585	66.608
600	277.333	903.592	766.512	82.248	-855.192	-593.603	51.678
700	278.710	946.454	789.232	110.055	-850.946	-550.342	41.067
800	279.620	983.734	811.266	137.975	-846.815	-507.682	33.148
900	280.252	1016.707	832.296	165.970	-842.799	-465.532	27.019
1000	280.708	1046.260	852.240	194.019	-838.901	-423.824	22.138
1100	281.047	1073.031	871.114	222.108	-835.121	-382.500	18.163
1200	281.307	1097.497	888.975	250.226	-831.461	-341.516	14.866
1300	281.509	1120.021	905.893	278.367	-827.921	-300.831	12.088
1400	281.671	1140.890	921.942	306.527	-824.503	-260.415	9.716
1500	281.801	1160.327	937.194	334.701	-821.209	-220.239	7.669
1600	281.908	1178.518	951.714	362.886	-818.042	-180.278	5.885
1700	281.997	1195.611	965.563	391.082	-815.003	-140.511	4.317
1800	282.071	1211.732	978.796	419.285	-812.097	-100.920	2.929
1900	282.135	1226.985	991.461	447.495	-809.327	-61.486	1.690
2000	282.188	1241.458	1003.602	475.712	-806.700	-22.195	0.580
2100	282.235	1255.227	1015.259	503.933	-804.221	16.969	-0.422
2200	282.275	1268.357	1026.467	532.158	-801.897	56.018	-1.330
2300	282.310	1280.906	1037.259	560.388	-799.736	94.965	-2.157
2400	282.341	1292.921	1047.663	588.620	-797.745	133.822	-2.913
2500	282.369	1304.448	1057.705	616.856	-795.930	172.599	-3.606
2600	282.393	1315.523	1067.410	645.094	-794.292	211.308	-4.245
2700	282.414	1326.181	1076.798	673.334	-792.890	249.958	-4.836
2800	282.434	1336.452	1085.889	701.577	-791.799	288.560	-5.383
2900	282.451	1346.363	1094.701	729.821	-791.100	327.131	-5.892
3000	282.467	1355.939	1103.250	758.067	-790.846	365.685	-6.367
3100	282.481	1365.201	1111.552	786.314	-791.125	404.239	-6.811
3200	282.494	1374.170	1119.619	814.563	-791.996	442.810	-7.228
3300	282.505	1382.863	1127.465	842.813	-793.506	481.421	-7.620
3400	282.516	1391.297	1135.101	871.064	-795.718	520.086	-7.990
3500	282.526	1399.486	1142.539	899.316	-798.755	558.828	-8.340
3600	282.535	1407.445	1149.787	927.569	-802.799	597.671	-8.672
3700	282.543	1415.187	1156.856	955.823	-877.571	637.023	-8.993
3800	282.551	1422.722	1163.754	984.078	-877.024	677.950	-9.319
3900	282.558	1430.061	1170.489	1012.333	-876.508	718.864	-9.628
4000	282.564	1437.215	1177.068	1040.589	-876.012	759.765	-9.922
4100	282.570	1444.192	1183.498	1068.846	-875.522	800.654	-10.200
4200	282.576	1451.002	1189.787	1097.103	-875.025	841.530	-10.466
4300	282.581	1457.651	1195.939	1125.361	-874.508	882.394	-10.719
4400	282.586	1464.147	1201.961	1153.619	-873.959	923.246	-10.960
4500	282.591	1470.498	1207.858	1181.878	-873.365	964.085	-11.191
4600	282.595	1476.709	1213.636	1210.138	-872.714	1004.911	-11.411
4700	282.599	1482.787	1219.298	1238.397	-871.995	1045.720	-11.622
4800	282.603	1488.736	1224.849	1266.657	-871.198	1086.514	-11.824
4900	282.606	1494.563	1230.294	1294.918	-870.312	1127.292	-12.017
5000	282.609	1500.273	1235.637	1323.179	-869.328	1168.048	-12.203
5100	282.613	1505.869	1240.881	1351.440	-868.239	1208.786	-12.380
5200	282.615	1511.357	1246.030	1379.701	-867.037	1249.500	-12.551
5300	282.618	1516.740	1251.087	1407.963	-865.714	1290.190	-12.716
5400	282.621	1522.023	1256.056	1436.225	-864.265	1330.854	-12.873
5500	282.623	1527.209	1260.939	1464.487	-862.684	1371.491	-13.025
5600	282.626	1532.302	1265.739	1492.749	-860.968	1412.096	-13.171
5700	282.628	1537.304	1270.460	1521.012	-859.112	1452.670	-13.312
5800	282.630	1542.219	1275.103	1549.275	-857.114	1493.212	-13.448
5900	282.632	1547.051	1279.671	1577.538	-854.970	1533.716	-13.578
6000	282.634	1551.801	1284.167	1605.801	-852.679	1574.181	-13.704

PREVIOUS: December 1966 (1 atm) CURRENT: December 1966 (1 bar)

Tungsten Chloride $((WCl_5)_2)$ $Cl_{10}W_2(g)$

COBALT (Co) REFERENCE STATE A_r = 58.9332

0 to 700.1	K	crystal, alpha
700.1 to 1768	K	crystal, beta
1768 to 3184.943	K	liquid
above 3184.943	K	ideal monatomic gas

Refer to the individual tables for details.

Cobalt (Co) Co_1(ref)

Enthalpy Reference Temperature = T_r = 298.15 K — Standard State Pressure = p° = 0.1 MPa

T/K	C_p° (J K⁻¹mol⁻¹)	S° (J K⁻¹mol⁻¹)	$-[G^\circ-H^\circ(T_r)]/T$ (J K⁻¹mol⁻¹)	$H^\circ-H^\circ(T_r)$ (kJ mol⁻¹)	Δ_fH° (kJ mol⁻¹)	Δ_fG° (kJ mol⁻¹)	Log K_f
0	0.	0.	INFINITE	-4.771	0.	0.	0.
100	13.909	7.681	50.176	-4.249	0.	0.	0.
200	22.226	20.588	32.287	-2.340	0.	0.	0.
298.15	24.802	30.067	30.067	0.	0.	0.	0.
300	24.833	30.221	30.068	0.046	0.	0.	0.
400	26.527	37.597	31.062	2.614	0.	0.	0.
500	28.200	43.700	32.996	5.352	0.	0.	0.
600	29.665	48.973	35.229	8.246	0.	0.	0.
700	31.045	53.650	37.533	11.282	0.	0.	0.
700.100	31.047	53.655	37.535	11.285		ALPHA <--> BETA	
700.100	30.583	54.300	37.535	11.737		TRANSITION	
800	32.426	58.497	39.894	14.883	0.	0.	0.
900	34.518	62.433	42.181	18.227	0.	0.	0.
1000	36.987	66.194	44.396	21.799	0.	0.	0.
1100	39.832	69.849	46.544	25.635	0.	0.	0.
1200	43.221	73.445	48.636	29.771	0.	0.	0.
1300	48.660	77.110	50.685	34.354	0.	0.	0.
1394.000	54.978	80.719	52.587	39.215		Cp LAMBDA MAXIMUM	
1394.000	54.999	80.719	52.587	39.215		TRANSITION	
1400	44.225	80.930	52.708	39.511	0.	0.	0.
1500	39.748	83.794	54.688	43.659	0.	0.	0.
1600	38.284	86.307	56.587	47.552	0.	0.	0.
1700	37.782	88.608	58.404	51.347	0.	0.	0.
1768.000	37.990	90.092	59.594	53.921		BETA <--> LIQUID	
1768.000	40.501	99.251	59.594	70.113		TRANSITION	
1800	40.501	99.977	60.305	71.409	0.	0.	0.
1900	40.501	102.167	62.452	75.459	0.	0.	0.
2000	40.501	104.245	64.490	79.509	0.	0.	0.
2100	40.501	106.221	66.430	83.560	0.	0.	0.
2200	40.501	108.105	68.282	87.610	0.	0.	0.
2300	40.501	109.905	70.053	91.660	0.	0.	0.
2400	40.501	111.629	71.750	95.710	0.	0.	0.
2500	40.501	113.282	73.378	99.760	0.	0.	0.
2600	40.501	114.871	74.944	103.810	0.	0.	0.
2700	40.501	116.399	76.451	107.860	0.	0.	0.
2800	40.501	117.872	77.904	111.910	0.	0.	0.
2900	40.501	119.293	79.307	115.961	0.	0.	0.
3000	40.501	120.666	80.663	120.011	0.	0.	0.
3100	40.501	121.994	81.975	124.061	0.	0.	0.
3184.943	40.501	123.089	83.057	127.501		LIQUID <--> IDEAL GAS	
3184.943	26.233	240.840	83.057	502.529		FUGACITY = 1 bar	
3200	26.233	240.963	83.800	502.924	0.	0.	0.
3300	26.245	241.771	88.574	505.548	0.	0.	0.
3400	26.269	242.555	93.092	508.174	0.	0.	0.
3500	26.305	243.317	97.373	510.802	0.	0.	0.
3600	26.354	244.058	101.437	513.435	0.	0.	0.
3700	26.416	244.781	105.302	516.073	0.	0.	0.
3800	26.491	245.487	108.982	518.719	0.	0.	0.
3900	26.579	246.176	112.491	521.372	0.	0.	0.
4000	26.679	246.850	115.841	524.035	0.	0.	0.
4200	26.919	248.157	122.111	529.394	0.	0.	0.
4400	27.209	249.416	127.869	534.806	0.	0.	0.
4600	27.546	250.633	133.180	540.281	0.	0.	0.
4800	27.930	251.813	138.099	545.827	0.	0.	0.
5000	28.358	252.962	142.671	551.456	0.	0.	0.
5200	28.829	254.083	146.934	557.174	0.	0.	0.
5400	29.342	255.181	150.923	562.990	0.	0.	0.
5600	29.896	256.258	154.666	568.913	0.	0.	0.
5800	30.492	257.317	158.187	574.951	0.	0.	0.
6000	31.003	258.351	161.508	581.056	0.	0.	0.

PREVIOUS: September 1967 (1 atm) — CURRENT: September 1967 (1 bar)

Cobalt (Co) Co_1(ref)

COBALT (Co) CRYSTAL $M_r = 58.9332$

Cobalt (Co) $Co_1(cr)$

$S°(298.15\ K) = 30.067 \pm 0.10\ J\ K^{-1}\ mol^{-1}$
$T_{trs}(\alpha \rightarrow \beta) = 700\ K$
$T_{fus} = 923 \pm 1\ K$

$\Delta_f H°(0\ K) = 0\ kJ\ mol^{-1}$
$\Delta_f H°(298.15\ K) = 0\ kJ\ mol^{-1}$
$\Delta_{tr} H° = 0.452\ kJ\ mol^{-1}$
$\Delta_{fus} H°(923\ K) = 16.192 \pm 0.25\ kJ\ mol^{-1}$

Enthalpy of Formation

Zero by definition.

Heat Capacity and Entropy

The heat capacity of Co(cr) is adopted from Hultgren et al. (1). Their selected values are based on the data from nineteen investigations. The entropy values are based on $S°(1\ K) = 0.0011\ cal\ K^{-1}\ mol^{-1}$. The heat capacity has a maximum value of 13.14 $cal\ K^{-1}\ mol^{-1}$ at the Curie point (1394 K).

Transition Data

The exact nature of the α(hcp) - β(fcc) transition is yet to be clearly elucidated. Three investigations (2, 3, 4) indicate that it is not a simple time- and temperature-dependent phenomenon. Crystalline cobalt exists as pure β-phase above 700 K whereas mixtures of alpha and beta cobalt commonly coexist in varying proportions below this temperature. The values of T_{trs} and $\Delta_{trs}H°$ are those selected by Hultgren et al. (1).

Fusion Data

The temperature and enthalpy of melting are those selected by Hultgren et al. (1).

Sublimation Data

The enthalpy of sublimation is calculated from the data of Edwards et al. (5). See the ideal gas table for details.

References

1. R. Hultgen, R. L. Orr and K. K. Kelley "Supplement to Selected Values of Thermodynamic Properties of Metal and Alloys, "University of California, Berkeley, Calif., (1966).
2. O. S. Edwards and H. Lipson, J. Inst. Metals 69, 177 (1943).
3. A. R. Rtoiano and J. L. Tokich, Trans. Am. Inst. Mining Met. Engrs. 175, 728 (1948).
4. P. R. Rao, Proc. Indian Acad. Sci. Sect. A, 61, 230 (1965).
5. J. W. Edwards, H. L. Johnston and W. E. Ditmars, J. Am. Chem. Soc. 73, 4729 (1951).

Enthalpy Reference Temperature = T_r = 298.15 K — Standard State Pressure = p° = 0.1 MPa

T/K	$C_p°$ (J K⁻¹mol⁻¹)	S° (J K⁻¹mol⁻¹)	$-[G°-H°(T_r)]/T$ (J K⁻¹mol⁻¹)	$H°-H°(T_r)$ (kJ mol⁻¹)	$\Delta_f H°$ (kJ mol⁻¹)	$\Delta_f G°$ (kJ mol⁻¹)	Log K_f
0	0.	0.	INFINITE	-4.771	0.	0.	0.
100	13.909	7.681	50.176	-4.249	0.	0.	0.
200	22.226	20.588	32.287	-2.340	0.	0.	0.
250	23.984	25.738	30.475	-1.184	0.	0.	0.
298.15	24.802	30.067	30.067	0.	0.	0.	0.
300	24.833	30.221	30.068	0.046	0.	0.	0.
350	25.682	34.113	30.373	1.309	0.	0.	0.
400	26.527	37.597	31.062	2.614	0.	0.	0.
450	27.390	40.772	31.967	3.962	0.	0.	0.
500	28.200	43.700	32.996	5.352	0.	0.	0.
600	29.665	48.973	35.229	8.246	0.	0.	0.
700	31.045	53.650	37.533	11.282	0.	0.	0.
700.100	31.047	53.655	37.535	11.285	ALPHA <--> BETA		
700.100	30.583	54.300	37.535	11.737	TRANSITION		
800	32.426	58.497	39.894	14.883	0.	0.	0.
900	34.518	62.433	42.181	18.227	0.	0.	0.
1000	36.987	66.194	44.396	21.799	0.	0.	0.
1100	39.832	69.849	46.544	25.635	0.	0.	0.
1200	43.221	73.445	48.636	29.771	0.	0.	0.
1300	48.660	77.110	50.685	34.354	0.	0.	0.
1394.000	54.978	80.719	52.587	39.215	Cp LAMBDA MAXIMUM		
1394.000	54.999	80.719	52.587	39.215	TRANSITION		
1400	44.225	80.930	52.708	39.511	0.	0.	0.
1500	39.748	83.794	54.688	43.659	0.	0.	0.
1600	38.284	86.307	56.587	47.552	0.	0.	0.
1700	37.782	88.608	58.404	51.347	0.	0.	0.
1768.000	37.990	90.092	59.594	53.921	BETA <--> LIQUID		
1800	37.656	90.776	60.142	55.141	-16.269	0.293	-0.009
1900	37.656	92.812	61.809	58.906	-16.553	1.221	-0.034
2000	37.656	94.744	63.408	62.672	-16.838	2.164	-0.057
2100	37.656	96.581	64.944	66.437	-17.122	3.121	-0.078
2200	37.656	98.333	66.422	70.203	-17.407	4.092	-0.097
2300	37.656	100.006	67.846	73.969	-17.691	5.076	-0.115
2400	37.656	101.609	69.220	77.734	-17.976	6.071	-0.132
2500	37.656	103.146	70.546	81.500	-18.260	7.079	-0.148

PREVIOUS: CURRENT: September 1967

Cobalt (Co) $Co_1(cr)$

COBALT (Co) LIQUID A_r = 58.9332

$S°(298.15\ K) = [41.000]\ J\ K^{-1}\ mol^{-1}$

$T_{fus} = 1768\ K$

$\Delta_f H°(298.15\ K) = [18.004]\ kJ\ mol^{-1}$

$\Delta_{fus} H° = 16.192 \pm 0.25\ kJ\ mol^{-1}$

Enthalpy of Formation

The enthalpy of formation of Co(ℓ) is calculated from that of the crystal by adding $\Delta_{fus}H°$ and the difference in enthalpy, H°(1768K) −H°(298.15 K), between the crystal and liquid.

Heat Capacity and Entropy

The heat capacity of Co(ℓ) selected by Hultgren et al. (1) is used at temperatures above 1789 K. The heat capacities at temperatures below the melting point are estimated by assuming a glass transition at 1100 K and heat capacities similar to those of the crystal below 1100 K.

Fusion Data

See the crystal table for details.

Vaporization Data

The boiling point is calculated from the adopted thermodynamic functions and the chosen enthalpy of sublimation at 298.15 K so that the Gibbs energy functions calculated by integration of the crystal liquid data and by statistical methods from the gas phase are equal at the boiling point.

Reference

1. R. Hultgren, R. L. Orr and K. K. Kelley, 'Supplement to Selected Values of Thermodynamic Properties of Metals and Alloys," University of California, Berkeley, Calif. (1966).

Cobalt (Co) $Co_1(l)$

Enthalpy Reference Temperature = T_r = 298.15 K Standard State Pressure = p° = 0.1 MPa

	J K⁻¹mol⁻¹			kJ mol⁻¹			
T/K	$C_p°$	S°	−[G°−H°(T_r)]/T	H°−H°(T_r)	$\Delta_f H°$	$\Delta_f G°$	Log K_f
0							
100							
200							
298.15	24.811	41.000	41.000	0.	18.004	14.745	-2.583
300	24.842	41.153	41.000	0.046	18.004	14.724	-2.564
350	25.686	45.046	41.306	1.309	18.004	14.178	-2.116
400	26.527	48.531	41.995	2.614	18.005	13.631	-1.780
450	27.390	51.706	42.900	3.963	18.005	13.084	-1.519
500	28.200	54.634	43.929	5.353	18.005	12.538	-1.310
600	29.665	59.909	46.162	8.248	18.006	11.444	-0.996
700	30.878	64.568	48.465	11.272	17.994	10.351	-0.772
800	32.426	68.786	50.745	14.433	17.555	9.323	-0.609
900	34.518	72.723	52.971	17.777	17.555	8.294	-0.481
1000	36.987	76.484	55.135	21.349	17.555	7.265	-0.379
1100	39.832	80.140	57.243	25.187	17.556	6.236	-0.296
1100.000	39.832	80.140	57.243	25.187	GLASS <--> LIQUID		
1100.000	39.832	80.140	57.243	25.187	TRANSITION		
1200	39.999	83.613	59.297	29.178	17.412	5.211	-0.227
1300	40.166	86.821	61.293	33.187	16.837	4.214	-0.169
1400	40.334	89.804	63.224	37.212	15.705	3.282	-0.122
1500	40.501	92.593	65.090	41.255	15.600	2.401	-0.084
1600	40.501	95.207	66.891	45.305	15.757	1.517	-0.050
1700	40.501	97.662	68.630	49.355	16.012	0.619	-0.019
1768.000	40.501	99.251	69.777	52.109	------- BETA <--> LIQUID -------		
1800	40.501	99.977	70.308	53.405	0.	0.	0.
1900	40.501	102.167	71.928	57.455	0.	0.	0.
2000	40.501	104.245	73.492	61.505	0.	0.	0.
2100	40.501	106.221	75.004	65.555	0.	0.	0.
2200	40.501	108.105	76.466	69.606	0.	0.	0.
2300	40.501	109.905	77.881	73.656	0.	0.	0.
2400	40.501	111.629	79.251	77.706	0.	0.	0.
2500	40.501	113.282	80.580	81.756	0.	0.	0.
2600	40.501	114.871	81.868	85.806	0.	0.	0.
2700	40.501	116.399	83.119	89.856	0.	0.	0.
2800	40.501	117.872	84.334	93.906	0.	0.	0.
2900	40.501	119.293	85.515	97.956	0.	0.	0.
3000	40.501	120.666	86.664	102.006	0.	0.	0.
3100	40.501	121.994	87.783	106.057	0.	0.	0.
3184.943	40.501	123.089	88.710	109.497	------- FUGACITY = 1 bar -------		
3200	40.501	123.280	88.872	110.107	-374.813	1.773	-0.029
3300	40.501	124.526	89.934	114.157	-373.387	13.519	-0.214
3400	40.501	125.736	90.969	118.207	-371.962	25.222	-0.387
3500	40.501	126.910	91.979	122.257	-370.541	36.883	-0.550
3600	40.501	128.051	92.965	126.307	-369.124	48.504	-0.704
3700	40.501	129.160	93.929	130.357	-367.712	60.085	-0.848
3800	40.501	130.240	94.870	134.407	-366.307	71.628	-0.985
3900	40.501	131.292	95.790	138.457	-364.910	83.135	-1.113
4000	40.501	132.318	96.691	142.508	-363.523	94.605	-1.235

PREVIOUS: CURRENT: September 1967

Cobalt (Co) $Co_1(l)$

COBALT (Co) ALPHA-BETA-LIQUID A_r = 58.9332

0 to 700.1 K crystal, alpha
700.1 to 1768 K crystal, beta
above 1768 K liquid

Refer to the individual tables for details.

Cobalt (Co) $Co_1(cr,l)$

Enthalpy Reference Temperature = T_r = 298.15 K Standard State Pressure = $p°$ = 0.1 MPa

	J K⁻¹mol⁻¹			kJ mol⁻¹			
T/K	C_p°	S°	$-[G^\circ-H^\circ(T_r)]/T$	$H^\circ-H^\circ(T_r)$	$\Delta_f H^\circ$	$\Delta_f G^\circ$	$\log K_f$
0	0.	0.	INFINITE	-4.771	0.	0.	0.
100	13.909	7.681	50.176	-4.249	0.	0.	0.
200	22.226	20.588	32.287	-2.340	0.	0.	0.
250	23.984	25.738	30.475	-1.184	0.	0.	0.
298.15	24.802	30.067	30.067	0.	0.	0.	0.
300	24.833	30.221	30.068	0.046	0.	0.	0.
350	25.682	34.113	30.373	1.309	0.	0.	0.
400	26.527	37.597	31.062	2.614	0.	0.	0.
450	27.390	40.772	31.967	3.962	0.	0.	0.
500	28.200	43.700	32.996	5.352	0.	0.	0.
600	29.665	48.973	35.229	8.246	0.	0.	0.
700	31.045	53.650	37.533	11.282	0.	0.	0.
700.100	31.047	53.655	37.535	11.285		ALPHA <--> BETA	
700.100	30.583	54.300	37.535	11.737		TRANSITION	
800	32.426	58.497	39.894	14.883	0.	0.	0.
900	34.518	62.433	42.181	18.227	0.	0.	0.
1000	36.987	66.194	44.396	21.799	0.	0.	0.
1100	39.832	69.849	46.544	25.635	0.	0.	0.
1200	43.221	73.445	48.636	29.771	0.	0.	0.
1300	48.660	77.111	50.685	34.354	0.	0.	0.
1394.000	54.978	80.719	52.587	39.215		Cp LAMBDA MAXIMUM	
1394.000	54.978	80.719	52.587	39.215		TRANSITION	
1400	44.225	80.930	52.708	39.511	0.	0.	0.
1500	39.748	83.794	54.688	43.659	0.	0.	0.
1600	38.284	86.307	56.587	47.552	0.	0.	0.
1700	37.782	88.608	58.404	51.347	0.	0.	0.
1768.000	37.990	90.092	59.594	53.921		BETA <--> LIQUID	
1768.000	40.501	99.251	59.594	70.113		TRANSITION	
1800	40.501	99.977	60.305	71.409	0.	0.	0.
1900	40.501	102.167	62.452	75.459	0.	0.	0.
2000	40.501	104.244	64.490	79.509	0.	0.	0.
2100	40.501	106.221	66.430	83.560	0.	0.	0.
2200	40.501	108.105	68.282	87.610	0.	0.	0.
2300	40.501	109.905	70.053	91.660	0.	0.	0.
2400	40.501	111.629	71.750	95.710	0.	0.	0.
2500	40.501	113.282	73.378	99.760	0.	0.	0.
2600	40.501	114.871	74.944	103.810	0.	0.	0.
2700	40.501	116.399	76.451	107.860	0.	0.	0.
2800	40.501	117.872	77.904	111.910	0.	0.	0.
2900	40.501	119.293	79.307	115.960	0.	0.	0.
3000	40.501	120.666	80.663	120.011	0.	0.	0.
3100	40.501	121.994	81.975	124.061	0.	0.	0.
3184.943	40.501	123.089	83.057	127.501	-------	FUGACITY = 1 bar	-------
3200	40.501	123.280	83.246	128.111	-374.813	1.773	-0.029
3300	40.501	124.526	84.478	132.161	-373.387	13.519	-0.214
3400	40.501	125.735	85.673	136.211	-371.963	25.222	-0.387
3500	40.501	126.910	86.835	140.261	-370.541	36.883	-0.550
3600	40.501	128.050	87.964	144.311	-369.124	48.504	-0.704
3700	40.501	129.160	89.063	148.361	-367.712	60.085	-0.848
3800	40.501	130.240	90.132	152.411	-366.307	71.629	-0.985
3900	40.501	131.292	91.174	156.462	-364.911	83.135	-1.113
4000	40.501	132.318	92.190	160.512	-363.523	94.606	-1.235

PREVIOUS: CURRENT: September 1967

Cobalt (Co) $Co_1(cr,l)$

COBALT (Co) IDEAL GAS $A_r = 58.9332$

IP(Co, g) = 63430 ± 50 cm^{-1}

S°(298.15 K) = 179.52 ± 0.01 J K^{-1} mol^{-1}

$\Delta_f H°(0\ K) = 425.1 \pm 2.1\ kJ\ mol^{-1}$

$\Delta_f H°(298.15\ K) = [426.7]\ kJ\ mol^{-1}$

Electronic Levels and Quantum Weights

State	ε_i, cm^{-1}	g_i
$^4F_{9/2}$	0	10
$^4F_{7/2}$	816.00	8
$^4F_{5/2}$	1406.84	6
$^4F_{3/2}$	1809.33	4
.	.	.
.	.	.
.	.	.
IP	63430	

Enthalpy of Formation

The enthalpy of formation is the enthalpy of sublimation at 298.15 K. The adopted value is that reported by Edwards et al. (1). Their value is used because their vapor pressure measurements for Fe (1) and Cu (2) have proved to be very reliable. The results obtained by Vintaikin and Tomash (3) are rejected because their vapor pressure data for Fe (4) resulted in a value for the enthalpy of sublimation which was higher than the accepted value [see JANAF Fe(g) Table for details]. The remaining vapor pressure data are rejected either because the drift was too large or because there were not enough points. Second and third law analyses of the available vapor pressure data gave the following results, with reaction A corresponding to Co(c) = Co(g) and reaction B corresponding to Co(ℓ) = Co(g).

Source	Reaction	Method	Data Points	T/K	$\Delta_r H°$(298.15 K), kcal mol^{-1} 2nd law	3rd law	Drift cal K^{-1} mol^{-1}	$\Delta_f H°$(298.15K) kcal mol^{-1}
Edwards (1)	A	Langmuir	9	1363-1522	101.0 ± 1.0	101.54	±0.4 ± 0.7	101.54
Kornev (5)	A	Knudsen	8*	1323-1523	98.8 ± 0.8	95.37	-2.4 ± 0.6	95.37
Dancy (6)	B	Langmuir	1	1823	--	90.51	--	94.82
Ruff(7)	B	Boiling Pt.	1	2648	--	99.35	--	103.66
Nesmeyanov (8)	A	Knudsen	3	1391-1540	82.4 ± 2.4	82.31	-0.1 ± 1.6	82.31
Nesmeyanov (9)	A	Knudsen	12*	1345-1541	91.7 ± 2.4	83.19	-5.9 ± 1.7	83.19
Vintaikin (3)	A	Knudsen	Eqn	1373-1523	102.9	103.34	+0.3	103.34

*One point rejected due to failure of a statistical test.

Heat Capacity and Entropy

The electronic levels are taken from Moore (10). All observed levels were used but only the first few levels are listed above.

Unobserved levels below 20,000 cm^{-1} are estimated.

References

1. J. W. Edwards, H. L. Johnston and W. E. Ditmars, J. Am. Chem. Soc. 73, 4729 (1951).
2. J. W. Edwards, H. L. Johnston and W. E. Ditmars, J. Am. Chem. Soc. 75, 2467 (1953).
3. E. Z. Vintaikin and Ya. Tomash, Russ. J. Phys. Chem. (English Transl.) 35, 1042 (1961).
4. E. Z. Vintaikin, Dokl. Akad. Nauk SSSR 117, 632 (1957).
5. Iu. V. Kornev and V. N. Golubkin Dokl. Akad. Nauk SSSR 99, 565 (1954).
6. T. E. Dancy, J. Iron Steel Inst. (London) 167, 160 (1951).
7. O. Ruff and F. Keilig, Z. Anorg. Chem. 129, 321 (1914).
8. An. N. Nesmeyanov and D. D. Man, Dokl. Akad. Nauk SSSR 123, 1064 (1959).
9. An. N. Nesmeyanov and D. D. Man, Izv. Akad. Nauk SSSR, Otd. Tekhn. Nauk, No. 1 (1960).
10. C. E. Moore, U. S. Nat. Bur. Stand. Circ, 467 (1949).

Cobalt (Co) $Co_1(g)$

Enthalpy Reference Temperature = T_r = 298.15 K Standard State Pressure = p° = 0.1 MPa

T/K	$C_p°$ (J $K^{-1}mol^{-1}$)	S°	$-[G°-H°(T_r)]/T$	$H°-H°(T_r)$ (kJ mol^{-1})	$\Delta_f H°$	$\Delta_f G°$	Log K_f
0	0.	0.	INFINITE	-6.360	425.087	425.087	INFINITE
100	20.793	156.132	198.942	-4.281	426.644	411.799	-215.102
200	21.452	170.670	181.570	-2.180	426.836	396.819	-103.639
250	22.214	175.536	179.892	-1.089	426.771	389.322	-81.344
298.15	23.024	179.518	179.518	0.	426.676	382.117	-66.945
300	23.054	179.660	179.518	0.043	426.673	381.841	-66.484
350	23.841	183.274	179.802	1.215	426.583	374.376	-55.873
400	24.508	186.503	180.441	2.425	426.487	366.924	-47.915
450	25.034	189.421	181.279	3.664	426.378	359.486	-41.728
500	25.425	192.080	182.229	4.926	426.250	352.060	-36.779
600	25.890	196.762	184.271	7.495	425.924	337.251	-29.360
700	26.093	200.771	186.349	10.095	425.489	322.504	-24.066
800	26.186	204.262	188.375	12.710	424.503	307.891	-20.103
900	26.251	207.350	190.315	15.331	423.781	293.356	-17.026
1000	26.320	210.119	192.159	17.960	422.837	278.913	-14.569
1100	26.399	212.631	193.908	20.596	421.637	264.576	-12.564
1200	26.480	214.932	195.565	23.240	420.145	250.361	-10.898
1300	26.558	217.054	197.138	25.892	418.214	236.287	-9.494
1400	26.624	219.025	198.632	28.551	415.716	222.383	-8.297
1500	26.673	220.864	200.053	31.216	414.233	208.628	-7.265
1600	26.705	222.586	201.408	33.885	413.009	194.962	-6.365
1700	26.719	224.206	202.702	36.556	411.885	181.369	-5.573
1800	26.715	225.733	203.939	39.228	394.495	168.135	-4.879
1900	26.697	227.177	205.125	41.899	393.116	155.597	-4.278
2000	26.666	228.546	206.262	44.567	391.734	143.132	-3.738
2100	26.626	229.846	207.354	47.232	390.348	130.736	-3.252
2200	26.579	231.083	208.405	49.892	388.958	118.406	-2.811
2300	26.528	232.264	209.417	52.547	387.564	106.139	-2.410
2400	26.475	233.391	210.392	55.198	386.164	93.933	-2.044
2500	26.423	234.471	211.334	57.842	384.758	81.786	-1.709
2600	26.375	235.507	212.244	60.482	383.348	69.695	-1.400
2700	26.331	236.501	213.124	63.118	381.933	57.658	-1.115
2800	26.293	237.458	213.976	65.749	380.514	45.674	-0.852
2900	26.264	238.380	214.802	68.377	379.092	33.740	-0.608
3000	26.243	239.270	215.603	71.002	377.667	21.856	-0.381
3100	26.233	240.130	216.380	73.626	376.241	10.019	-0.169
3184.943	26.233	240.840	217.023	75.854	------- FUGACITY = 1 bar -------		
3200	26.233	240.963	217.136	76.249	0.	0.	0.
3300	26.245	241.771	217.870	78.873	0.	0.	0.
3400	26.269	242.555	218.584	81.498	0.	0.	0.
3500	26.305	243.317	219.280	84.127	0.	0.	0.
3600	26.354	244.058	219.958	86.760	0.	0.	0.
3700	26.416	244.781	220.619	89.398	0.	0.	0.
3800	26.491	245.487	221.265	92.043	0.	0.	0.
3900	26.579	246.176	221.895	94.697	0.	0.	0.
4000	26.679	246.850	222.510	97.360	0.	0.	0.
4100	26.793	247.510	223.112	100.033	0.	0.	0.
4200	26.919	248.157	223.700	102.719	0.	0.	0.
4300	27.058	248.792	224.277	105.417	0.	0.	0.
4400	27.209	249.416	224.841	108.131	0.	0.	0.
4500	27.372	250.029	225.394	110.859	0.	0.	0.
4600	27.546	250.633	225.936	113.605	0.	0.	0.
4700	27.733	251.227	226.468	116.369	0.	0.	0.
4800	27.930	251.813	226.990	119.152	0.	0.	0.
4900	28.139	252.391	227.502	121.956	0.	0.	0.
5000	28.358	252.962	228.006	124.780	0.	0.	0.
5100	28.589	253.526	228.501	127.628	0.	0.	0.
5200	28.829	254.083	228.987	130.498	0.	0.	0.
5300	29.081	254.635	229.466	133.394	0.	0.	0.
5400	29.342	255.181	229.937	136.315	0.	0.	0.
5500	29.614	255.721	230.401	139.263	0.	0.	0.
5600	29.896	256.258	230.858	142.238	0.	0.	0.
5700	30.189	256.789	231.308	145.242	0.	0.	0.
5800	30.492	257.317	231.752	148.276	0.	0.	0.
5900	30.701	257.833	232.189	151.296	0.	0.	0.
6000	31.003	258.351	232.621	154.381	0.	0.	0.

PREVIOUS: September 1967 (1 atm) CURRENT: September 1967 (1 bar)

Cobalt (Co) $Co_1(g)$

COBALT, ION (Co^+) IDEAL GAS M_r = 58.93265

IP(Co^+, g) = 137795 ± 10 cm^{-1}

S°(298.15 K) = 178.346 ± 0.05 J K^{-1} mol^{-1}

$\Delta_f H°$(0 K) = 1183.88 ± 1.0 kJ mol^{-1}

$\Delta_f H°$(298.15 K) = [1191. 597] kJ mol^{-1}

Electronic Level and Quantum Weight

State	ϵ_i, cm^{-1}	g_i
$^3F_4$	0.0	9
$^3F_3$	950.51	7
$^3F_2$	1597.32	5
$^5F_5$	3350.58	9
.	.	.
.	.	.
.	.	.
$^5H_6$	125969.5	13
IP	137795	

Enthalpy of Formation

$\Delta_f H°(Co^+$, g, 0 K) is calculated from $\Delta_f H°$(Co, g, 0 K) (1) using the spectroscopic value of IP(Co) = 63430 ± 10 cm^{-1} (758.79 ± 0.12 kJ mol^{-1}) from Sugar and Corliss (2). The ionization limit is converted from cm^{-1} to kJ mol^{-1} using the factor, 1 cm^{-1} = 0.01196266 kJ cm^{-1}, which is derived from the 1973 CODATA fundamental constants (3). Rosenstock et al. (4) and Levin and Lias (5) have summarized additional ionization and appearance potential data.

$\Delta_f H°(Co^+$, g, 298.15 K) is calculated from $\Delta_f H°$(Co, g, 0 K) by using IP(Co) with JANAF (1) enthalpies, H°(0 K) - H°(298.15 K), for Co(g), Co^+(g), and e^-(g). $\Delta_f H°(Co \rightarrow Co^+ + e^-$, 298.15 K) differs from a room temperature threshold energy due to inclusion of these enthalpies and to threshold effects discussed by Rosenstock et al. (4). $\Delta_f H°$(298.15 K) should be changed by -6.197 kJ mol^{-1} if it is to be used in the ion convention that excludes the enthalpy of the electron.

Heat Capacity and Entropy

The information on electronic energy levels and quantum weights, given by Sugar and Corliss (2), is incomplete because many theoretically predicted levels have not been observed. Although we have listed only the ground, the first excited state, the highest observed excited state, and the ionization potential for Co^+(g), all levels listed by Sugar and Corliss (2), as well as estimated levels, are used in the calculation. The observed levels are too numerous to list completely. The calculations indicate that for Co^+(g), the thermodynamic functions are independent of the estimated missing levels (for n=3), the cut-off procedure, and the inclusion of n=7 levels up to 1000 K; the Gibbs energy function showing variations of 0.05% at this temperature. The reported uncertainty in S°(298.15 K) is due to uncertainties in the relative ionic mass, and the fundamental constants. Extension of these calculations above 6000 K may require consideration of the higher excited states (n>7), and use of different fill and cut-off procedures (6).

References

1. JANAF Thermochemical Tables: Co(g), 9-30-67; e^-(g), 3-31-82.
2. J. Sugar and C. Corliss, J. Phys. Chem. Ref. Data 10, 1097 (1981).
3. E. R. Cohen and B. N. Taylor, J. Phys. Chem. Ref. Data 2, 663 (1973).
4. H. M. Rosenstock, K. Draxl et al., J. Phys. Chem. Ref. Data 6, Supp. 1 (1977).
5. R. D. Levin and S. G. Lias, U. S. Natl. Bur. Stand., NSRDS-NBS-71, 1982.
6. J. R. Downey, Jr., The Dow Chemical Company, AFOSR-TR-78-0960, Contract No. F44620-75-1-0048, 1978.

Cobalt, Ion (Co^+) Co_1^+(g)

Enthalpy Reference Temperature = T_r = 298.15 K   Standard State Pressure = p° = 0.1 MPa

T/K	$C_p°$ (J K⁻¹mol⁻¹)	S° (J K⁻¹mol⁻¹)	-[G°-H°(T_r)]/T (J K⁻¹mol⁻¹)	H°-H°(T_r) (kJ mol⁻¹)	$\Delta_f H°$ (kJ mol⁻¹)	$\Delta_f G°$ (kJ mol⁻¹)	Log K_f
0	0.	0.	INFINITE	-6.291	1183.879		
100	20.787	155.255	197.384	-4.213			
200	21.116	169.718	180.341	-2.125			
250	21.635	174.482	178.709	-1.057			
298.15	22.272	178.346	178.346	0.	1191.597	1141.133	-199.922
300	22.297	178.484	178.346	0.041	1191.630	1140.819	-198.634
350	22.981	181.973	178.620	1.173	1192.539	1132.279	-168.983
400	23.606	185.083	179.237	2.338	1193.438	1123.609	-146.728
450	24.133	187.895	180.045	3.532	1194.323	1114.827	-129.406
500	24.559	190.461	180.961	4.750	1195.190	1105.947	-115.538
600	25.165	194.996	182.932	7.238	1196.863	1087.940	-94.714
700	25.585	198.908	184.941	9.777	1198.444	1069.660	-79.819
800	25.946	202.348	186.906	12.354	1199.499	1051.222	-68.638
900	26.307	205.425	188.796	14.966	1200.846	1032.605	-59.931
1000	26.681	208.216	190.601	17.615	1202.002	1013.848	-52.958
1100	27.067	210.777	192.320	20.303	1202.932	994.985	-47.248
1200	27.485	213.150	193.958	23.030	1203.602	976.049	-42.486
1300	28.002	215.369	195.521	25.803	1203.871	957.072	-38.456
1400	28.740	217.469	197.014	28.638	1203.626	938.094	-35.001
1500	29.883	219.488	198.445	31.564	1204.484	919.101	-32.006
1600	31.654	221.469	199.822	34.635	1205.740	900.036	-29.383
1700	34.287	223.462	201.154	37.924	1207.312	880.883	-27.066
1800	37.981	225.521	202.450	41.528	1192.933	861.922	-25.012
1900	42.853	227.700	203.721	45.560	1194.993	843.479	-23.189
2000	48.890	230.047	204.978	50.138	1197.599	824.913	-21.545
2100	55.916	232.599	206.232	55.371	1200.861	806.202	-20.053
2200	63.578	235.375	207.492	61.342	1204.861	787.318	-18.693
2300	71.383	238.374	208.769	68.091	1209.639	768.234	-17.447
2400	78.752	241.570	210.069	75.604	1215.180	748.927	-16.300
2500	85.120	244.919	211.395	83.808	1221.413	729.374	-15.239
2600	90.025	248.358	212.751	92.579	1228.212	709.559	-14.255
2700	93.179	251.820	214.134	101.755	1235.416	689.476	-13.339
2800	94.494	255.238	215.541	111.153	1242.843	669.120	-12.483
2900	94.073	258.552	216.967	120.595	1250.314	648.498	-11.681
3000	92.159	261.713	218.406	129.918	1257.665	627.620	-10.928
3100	89.071	264.687	219.852	138.988	1264.763	606.500	-10.219
3200	85.153	267.455	221.297	147.705	896.695	586.930	-9.581
3300	80.724	270.008	222.735	156.002	904.447	577.129	-9.135
3400	76.057	272.349	224.160	163.842	911.740	567.100	-8.712
3500	71.364	274.486	225.568	171.212	918.561	556.863	-8.311
3600	66.801	276.432	226.954	178.119	924.913	546.438	-7.929
3700	62.471	278.202	228.316	184.580	930.815	535.842	-7.565
3800	58.437	279.814	229.650	190.623	936.291	525.093	-7.218
3900	54.728	281.283	230.955	196.279	941.371	514.205	-6.887
4000	51.353	282.626	232.231	201.580	946.088	503.191	-6.571
4100	48.306	283.856	233.475	206.560	950.474	492.064	-6.269
4200	45.568	284.986	234.688	211.251	954.558	480.833	-5.980
4300	43.119	286.029	235.870	215.683	958.370	469.508	-5.703
4400	40.933	286.995	237.021	219.884	961.936	458.098	-5.438
4500	38.985	287.893	238.142	223.878	965.280	446.609	-5.184
4600	37.250	288.730	239.233	227.688	968.422	435.048	-4.940
4700	35.707	289.514	240.294	231.334	971.384	423.420	-4.706
4800	34.332	290.251	241.327	234.835	974.180	411.732	-4.481
4900	33.107	290.946	242.333	238.206	976.826	399.986	-4.264
5000	32.015	291.604	243.312	241.461	979.335	388.189	-4.055
5100	31.040	292.228	244.265	244.613	981.718	376.342	-3.855
5200	30.168	292.822	245.193	247.672	983.985	364.449	-3.661
5300	29.386	293.390	246.097	250.649	986.145	352.514	-3.474
5400	28.686	293.932	246.978	253.552	988.206	340.540	-3.294
5500	28.056	294.453	247.837	256.389	990.173	328.528	-3.120
5600	27.489	294.953	248.673	259.166	992.053	316.481	-2.952
5700	26.978	295.435	249.490	261.889	993.851	304.401	-2.790
5800	26.517	295.900	250.286	264.563	995.570	292.290	-2.632
5900	26.099	296.350	251.063	267.193	997.259	280.147	-2.480
6000	25.721	296.785	251.821	269.784	998.843	267.979	-2.333

PREVIOUS: September 1967 (1 atm)   CURRENT: June 1984 (1 bar)

Cobalt, Ion (Co^+) Co_1^+(g)

COBALT, ION (Co^-) IDEAL GAS M_r = 58.93375

Cobalt, Ion (Co^-) $Co_1^-(g)$

EA(Co, g) = 0.661 ± 0.010 eV

S°(298.15 K) = 178.412 ± 0.005 $J\ K^{-1}\ mol^{-1}$

$\Delta_f H°(0\ K)$ = 361.31 ± 4.7 $kJ\ mol^{-1}$

$\Delta_f H°(298.15\ K)$ = [356.649] $kJ\ mol^{-1}$

Electonic Levels and Quantum Weights

State	ε_i, cm^{-1}	g_i
$^3F_4$	0.0	9
$^3F_3$	910	7
$^3F_2$	1560	5

Enthalpy of Formation

$\Delta_f H°(Co^-, g, 0\ K)$ is calculated from $\Delta_f H°(Co, g\ 0\ K)$ (1) using the adopted electron affinity of EA(Co) = 0.661 ± 0.010 eV (63.776 ± 0.965 $kJ\ mol^{-1}$). This value, recommended by Hotop and Lineberger (2), is based on a laser photodetachment electron spectrometry study (3). Additional information on $Co^-(g)$ may be obtained in the critical discussions of Hotop and Lineberger (2, 4), Rosenstock et al. (5) and Massey (6).

$\Delta_f H°(Co^-, g, 298.15\ K)$ is obtained from $\Delta_f H°(Co, g, 0\ K)$ by using EA(Co) with JANAF (1) enthalpies, H°(0 K) - H°(298.15 K), for $Co^-(g)$, Co(g), and $e^-(g)$. $\Delta_r H°(Co^- \rightarrow Co + e^-, 298.15\ K)$ differs from a room-temperature threshold energy due to inclusion of these enthalpies and to threshold effects discussed by Rosenstock et al. (5). $\Delta_f H°(298.15\ K)$ should be changed by +6.197 $kJ\ mol^{-1}$ if it is to be used in the ion convention that excludes the enthalpy of the electron.

Heat Capacity and Entropy

The ground state and fine-structure separations for $Co^-(g)$ have been experimentally determined by Corderman et al. (3). Lacking any experimental evidence as to the existence of stable excited states, we assume no stable excited states exist.

References

1. JANAF Thermochemical Tables: Co(g), 9-30-67; $e^-(g)$, 3-31-82.
2. H. Hotop and W. C. Lineberger, J. Phys. Chem. Ref. Data, to be published, 1985.
3. R. R. Corderman, P. C. Engelking and W. C. Lineberger, J. Chem. Phys. 70, 4474 (1979).
4. H. Hotop and W. C. Lineberger, J. Phys. Chem. Ref. Data 4, 539 (1975).
5. H. M. Rosenstock, K. Draxl et al., J. Phys. Chem. Ref. Data 6, Supp. 1 (1977).
6. H. S. W. Massey, "Negative Ions", 3rd ed., Cambridge University Press, Cambridge, 1976.

Enthalpy Reference Temperature = T_r = 298.15 K   Standard State Pressure = p° = 0.1 MPa

	J K⁻¹mol⁻¹			kJ mol⁻¹			
T/K	$C_p°$	S°	$-[G°-H°(T_r)]/T$	$H°-H°(T_r)$	$\Delta_f H°$	$\Delta_f G°$	Log K_f
0	0.	0.	INFINITE	-6.307	361.311		
100	20.788	155.256	197.536	-4.228			
200	21.191	169.734	180.420	-2.137			
250	21.768	174.521	178.777	-1.064			
298.15	22.438	178.412	178.412	0.	356.649	318.675	-55.831
300	22.465	178.550	178.412	0.042	356.607	318.440	-55.445
350	23.154	182.066	178.688	1.182	355.445	312.170	-46.589
400	23.755	185.198	179.310	2.355	354.274	306.068	-39.968
450	24.233	188.025	180.123	3.556	353.087	300.114	-34.836
500	24.582	190.597	181.044	4.777	351.878	294.292	-30.745
600	24.937	195.117	183.024	7.256	349.385	283.007	-24.638
700	24.962	198.966	185.033	9.753	346.767	272.150	-20.308
800	24.794	202.290	186.987	12.242	343.577	261.739	-17.090
900	24.531	205.195	188.852	14.709	340.621	251.684	-14.607
1000	24.233	207.765	190.618	17.147	337.409	241.972	-12.639
1100	23.932	210.060	192.283	19.555	333.902	232.595	-11.045
1200	23.646	212.130	193.852	21.934	330.067	223.552	-9.731
1300	23.383	214.012	195.331	24.285	325.756	214.847	-8.633
1400	23.144	215.736	196.728	26.611	320.846	206.496	-7.704
1500	22.931	217.326	198.049	28.915	316.924	198.467	-6.911
1600	22.740	218.799	199.300	31.198	313.235	190.691	-6.225
1700	22.571	220.173	200.488	33.464	309.627	183.143	-5.627
1800	22.420	221.458	201.618	35.713	289.736	176.101	-5.110
1900	22.286	222.667	202.694	37.948	285.842	169.894	-4.671
2000	22.166	223.807	203.722	40.171	281.936	163.892	-4.280
2100	22.059	224.886	204.704	42.382	278.018	158.086	-3.932
2200	21.964	225.910	205.645	44.583	274.090	152.468	-3.620
2300	21.878	226.884	206.547	46.775	270.154	147.027	-3.339
2400	21.800	227.814	207.414	48.959	266.209	141.758	-3.085
2500	21.731	228.702	208.248	51.135	262.257	136.654	-2.855
2600	21.668	229.553	209.051	53.305	258.298	131.708	-2.646
2700	21.611	230.370	209.826	55.469	254.333	126.915	-2.455
2800	21.559	231.155	210.574	57.627	250.363	122.268	-2.281
2900	21.511	231.910	211.296	59.781	246.387	117.763	-2.121
3000	21.468	232.639	211.996	61.930	242.408	113.396	-1.974
3100	21.429	233.342	212.673	64.074	238.424	109.161	-1.839
3200	21.392	234.022	213.330	66.215	-140.377	106.828	-1.744
3300	21.359	234.680	213.967	68.353	-142.942	114.593	-1.814
3400	21.328	235.317	214.585	70.487	-145.512	122.435	-1.881
3500	21.300	235.935	215.187	72.619	-148.088	130.354	-1.945
3600	21.273	236.534	215.771	74.747	-150.671	138.346	-2.007
3700	21.249	237.117	216.340	76.873	-153.262	146.409	-2.067
3800	21.227	237.683	216.895	78.997	-155.862	154.544	-2.124
3900	21.206	238.234	217.435	81.119	-158.472	162.746	-2.180
4000	21.186	238.771	217.961	83.238	-161.094	171.016	-2.233
4100	21.168	239.294	218.475	85.356	-163.728	179.351	-2.285
4200	21.151	239.804	218.977	87.472	-166.376	187.751	-2.335
4300	21.135	240.301	219.467	89.586	-169.040	196.214	-2.384
4400	21.120	240.787	219.946	91.699	-171.719	204.739	-2.431
4500	21.106	241.262	220.415	93.810	-174.415	213.325	-2.476
4600	21.093	241.725	220.873	95.920	-177.129	221.972	-2.521
4700	21.080	242.179	221.322	98.029	-179.863	230.677	-2.564
4800	21.069	242.622	221.761	100.136	-182.617	239.441	-2.606
4900	21.058	243.057	222.191	102.243	-185.393	248.263	-2.647
5000	21.047	243.482	222.613	104.348	-188.191	257.141	-2.686
5100	21.038	243.899	223.026	106.452	-191.013	266.076	-2.725
5200	21.028	244.307	223.431	108.555	-193.859	275.066	-2.763
5300	21.020	244.708	223.829	110.658	-196.731	284.111	-2.800
5400	21.011	245.100	224.219	112.759	-199.629	293.211	-2.836
5500	21.004	245.486	224.602	114.860	-202.554	302.364	-2.872
5600	20.996	245.864	224.979	116.960	-205.508	311.571	-2.906
5700	20.989	246.236	225.348	119.059	-208.492	320.831	-2.940
5800	20.982	246.601	225.712	121.158	-211.506	330.144	-2.973
5900	20.976	246.959	226.069	123.256	-214.506	339.506	-3.006
6000	20.970	247.312	226.420	125.353	-217.573	348.921	-3.038

PREVIOUS: CURRENT: June 1984 (1 bar)

Cobalt, Ion (Co^-) $Co_1^-(g)$

COBALT FLUORIDE (CoF_2) CRYSTAL M_r = 96.930006

$S°(298.15\ K) = 82.026 \pm 0.42\ J\ K^{-1}\ mol^{-1}$

$T_{fus} = 1400 \pm 7.0\ K$

$\Delta_f H°(0\ K) = -670.4 \pm 4.2\ kJ\ mol^{-1}$

$\Delta_f H°(298.15\ K) = -671.53 \pm 4.2\ kJ\ mol^{-1}$

$\Delta_{fus} H° = 58.819 \pm 12.6\ kJ\ mol^{-1}$

Enthalpy of Formation

Heus and Egan (1) measured the emf of a solid state galvanic cell which involved the reactions: $Al(cr) + 3/2\ CoF_2(cr) + AlF_3(cr) + 3/2\ Co(cr)$, E(873 K) = 1.611 volts; $Mg(cr) + CoF_2(cr) + MgF_2(cr) + Co(cr)$, E(873 K) = 2.231 volts. Combining these results with auxiliary thermodynamic data (2) for the reactants and products, we derive $\Delta_f H°(CoF_2$, cr, 298.15 K) = -160.2 and -159.95 kcal mol^{-1}.

Several equilibrium studies (3, 4, 5) involving $CoF_2(cr)$ have been reported in the literature. 2nd and 3rd law analyses of these equilibrium data are summarized below, auxiliary data for CoO(cr) and $CoCl_2(cr)$ are from (6).

Source	Reaction	T/K	Data Points	$\Delta_r H°(298.15\ K)$, kcal mol^{-1} 2nd law	3rd law	Drift cal K^{-1} mol^{-1}	*$\Delta_f H°(298.15\ K)$ kcal mol^{-1}
(3)	$CoF_2(cr) + 2HCl(g) \rightleftarrows CoCl_2(cr) + 2HF(g)$	588-805	3	4.1	1.75±0.4	-1.7±0.1	-162.6±2.0
(4)	$CoF_2(cr) + H_2(g) \rightleftarrows Co(cr) + 2HF(g)$	673-873	3	35.2	31.5±1.0	-4.2±3.5	-161.8±2.0
(5)	$CoF_2(cr) + H_2O(g) \rightleftarrows CoO(cr) + 2HF(g)$	773-1023	5	30.2	31.1±0.2	0.9±0.3	-160.45±1.0

*3rd law values except for the first reaction.

We note that Heus and Egan (1), in the same paper, measured $\Delta_f H°(AlF_3$, cr, 298.15 K) = -360.2 kcal mol^{-1} which is in excellent agreement with the JANAF value (2). Domange (5) also measured similar equilibria for MgF_2 and CaF_2, which were very consistent with the JANAF enthalpies of formation (2). Jellinek and Rudat (4) reported similar equilibria for $PbF_2(cr)$, which show excellent consistency with the JANAF enthalpy of formation (2). Thus, it is apparent that these three sets are probably quite reliable, and we adopt $\Delta_f H°(CoF_2$, cr, 298.15 K) = -160.5±1 kcal mol^{-1} (-671.532±4.2 kJ mol^{-1}).

Heat Capacity and Entropy

Catalano and Stout (7) measured low temperature heat capacities in the temperature range 10 to 300 K. Heat capacity data above 300 K are calculated from the high temperature (468-1400 K) enthalpy data of Binford et al. (8). Both sets of data were smoothed by computer and joined at 298.15 K.

The value of S°(298.15 K) is obtained from the low temperature heat capacity data of Catalano and Stout (7) and is based on S°(10 K) = 0.013 cal K^{-1} mol^{-1} (0.05439 J K^{-1} mol^{-1}).

Fusion Data

See the liquid table for details.

Sublimation Data

Kana'an et al. (9) made sublimation studies of CoF_2 using Knudsen and Langmuir techniques over the temperature range 972-1241 K. 2nd and 3rd law analyses of these sublimation pressures are summarized below.

Method	T/K	Data Points	$\Delta_{sub} H°(298.15\ K)$ kcal mol^{-1} 2nd law	3rd law	Drift cal K^{-1} mol^{-1}
Knudsen	1058-1242	17	75.0	75.1±0.1	0.1±0.2
Langmuir	972-1032	8	72.8	75.5±0.2	2.6±1.4

The selected value is $\Delta_{sub} H°(298.15\ K) = 75.3 \pm 3.0$ kcal mol^{-1}(315.055 ± 12.6 kJ mol^{-1}).

References

1. R. J. Heus and J. J. Egan, Z. Phys. Chem. 49, 38 (1966).
2. JANAF Thermochemical Tables: $AlF_3(cr)$, 6-30-70; $MgF_2(cr)$, 3-31-66; $PbF_2(cr)$, 6-30-69; $CaF_2(cr)$, 12-31-68.
3. K. Jellinek and R. Koop, Z. Phys. Chem. A145, 305 (1929).
4. K. Jellinek and A. Rudat, Z. Anorg. Chem. 175, 281 (1928).
5. L. Domange, Ann. Chim. 7, 225 (1937).
6. U. S. Nat. Bur. Stand. Tech. Note 270-4, 1969.
7. E. Catalano and J. W. Stout, J. Chem. Phys. 23, 1803 (1955).
8. J. S. Binford, Jr., J. M. Strohmenger, and T. H. Hebert, J. Phys. Chem. 71, 2404 (1967).
9. A. S. Kana'an, G. Besenbruch, and J. L. Margrave, J. Inorg. Nucl. Chem. 28, 1035 (1966).

Cobalt Fluoride (CoF_2) $Co_1F_2(cr)$

Enthalpy Reference Temperature = T_r = 298.15 K; Standard State Pressure = p° = 0.1 MPa

T/K	$C_p°$ (J $K^{-1}mol^{-1}$)	S°	$-[G°-H°(T_r)]/T$	$H°-H°(T_r)$ (kJ mol^{-1})	$\Delta_f H°$	$\Delta_f G°$	Log K_f
0	0.	0.	INFINITE	-12.459	-670.395	-670.395	INFINITE
100	33.492	25.002	134.693	-10.969	-672.334	-657.029	343.197
200	57.584	56.746	87.999	-6.250	-672.454	-641.555	167.557
298.15	68.909	82.026	82.026	0.	-671.532	-626.562	109.771
300	69.041	82.453	82.028	0.128	-671.508	-626.283	109.046
400	75.714	103.350	84.831	7.408	-670.016	-611.424	79.844
500	78.927	120.625	90.315	15.155	-668.372	-596.965	62.365
600	80.789	135.193	96.613	23.148	-666.746	-582.837	50.741
700	82.015	147.744	103.041	31.292	-665.190	-568.977	42.458
800	82.902	158.756	109.331	39.540	-664.152	-555.271	36.255
900	83.592	168.562	115.378	47.866	-662.824	-541.742	31.442
1000	84.161	177.400	121.145	56.254	-661.698	-528.351	27.598
1100	84.646	185.444	126.631	64.695	-660.813	-515.062	24.458
1200	85.077	192.828	131.844	73.182	-660.208	-501.841	21.845
1300	85.475	199.654	136.801	81.710	-660.032	-488.654	19.634
1400	85.839	206.002	141.519	90.276	-660.415	-475.460	17.740
1400.000	85.839	206.002	141.519	90.276	------ CRYSTAL <--> LIQUID -----		
1500	86.186	211.936	146.018	98.877	-659.771	-462.269	16.098
1600	86.517	217.509	150.314	107.512	-658.858	-449.131	14.663
1700	86.831	222.764	154.423	116.180	-657.832	-436.054	13.398
1800	87.140	227.736	158.359	124.878	-673.055	-422.744	12.268
1900	87.441	232.455	162.135	133.607	-672.247	-408.860	11.240
2000	87.734	236.948	165.765	142.366	-671.420	-395.019	10.317

PREVIOUS: CURRENT: June 1970

Cobalt Fluoride (CoF_2) $Co_1F_2(cr)$

COBALT FLUORIDE (CoF_2) LIQUID M_r = 96.930006

$S°(298.15 K) = [86.238] J K^{-1} mol^{-1}$

$T_{fus} = 1400 \pm 7.0 K$

$\Delta_f H°(298.15 K) = [-637.691] kJ mol^{-1}$

$\Delta_{fus} H° = 58.819 \pm 12.6 kJ mol^{-1}$

Enthalpy of Formation

$\Delta_f H°(CoF_2, \ell, 298.15 K)$ is calculated from $\Delta_f H°(CoF_2, cr, 298.15 K)$ by adding the enthalpy of fusion and the difference in enthalpy, H°(1400 K) - H°(298.15), between the crystal and liquid.

Heat Capacity and Entropy

The heat capacity of liquid CoF_2 is estimated by comparison with those for $FeCl_2(\ell)$, $MgF_2(\ell)$, and $PbF_2(\ell)$ and is assumed constant in the temperature range 298 - 3000 K. Binford et al. (1) reported Cp = 30.3 cal $K^{-1} mol^{-1}$ for liquid CoF_2 at 1444 K. This value is based on only two enthalpy points near the melting point and appears high in comparison with heat capacities for other liquid metal dihalides. Therefore, the value is not adopted.

The entropy is calculated in a manner analogous to that used for the enthalpy of formation.

Fusion Data

T_{fus} = 1400 K is from the high temperature enthalpy studies of Binford et al. (1). $\Delta_{fus}H°$ is calculated from their enthalpy data. Premelting of the sample is assumed to be responsible for the rapid rise in enthalpy near the melting point. This enthalpy is included in $\Delta_{fus}H°$.

Vaporization Data

T_{vap} is the temperature at which the Gibbs energy change for the process $CoF_2(\ell) \rightarrow CoF_2(g)$ equals zero. The difference between the enthalpy of formation of the gas and liquid at T_{vap} is $\Delta_{vap}H°$.

Reference

1. J. S. Bindord, Jr., J. M. Strohmenger, and T. H. Hebert, J. Phys. Chem., 71, 2404 (1967).

Cobalt Fluoride (CoF_2) $Co_1F_2(l)$

Enthalpy Reference Temperature = T_r = 298.15 K Standard State Pressure = p° = 0.1 MPa

	J K⁻¹mol⁻¹			kJ mol⁻¹			
T/K	$C_p°$	S°	-[G°-H°(T_r)]/T	H°-H°(T_r)	$\Delta_f H°$	$\Delta_f G°$	Log K_f
0							
100							
200							
298.15	104.600	86.238	86.238	0.	-637.691	-593.977	104.062
300	104.600	86.885	86.240	0.194	-637.602	-593.706	103.373
400	104.600	116.976	90.343	10.654	-632.929	-579.787	75.712
500	104.600	140.317	98.090	21.114	-628.573	-567.012	59.235
600	104.600	159.388	106.765	31.574	-624.480	-555.088	48.325
700	104.600	175.512	115.464	42.034	-620.608	-543.832	40.581
800	104.600	189.480	123.863	52.494	-617.358	-533.055	34.805
900	104.600	201.800	131.851	62.954	-613.896	-522.728	30.338
1000	104.600	212.820	139.407	73.414	-610.698	-512.772	26.784
1100	104.600	222.790	146.541	83.874	-607.794	-503.123	23.891
1200	104.600	231.891	153.280	94.334	-605.216	-493.723	21.491
1300	104.600	240.264	159.653	104.794	-603.108	-484.522	19.468
1400	104.600	248.015	165.691	115.254	-601.596	-475.460	17.740
1400.000	104.600	248.015	165.691	115.254	------ CRYSTAL <--> LIQUID -----		
1500	104.600	255.232	171.423	125.714	-599.094	-466.535	16.246
1600	104.600	261.983	176.874	136.174	-596.356	-457.786	14.945
1700	104.600	268.324	182.069	146.634	-593.537	-449.212	13.803
1800	104.600	274.303	187.029	157.094	-606.999	-440.510	12.783
1900	104.600	279.958	191.772	167.554	-604.461	-431.330	11.858
2000	104.600	285.324	196.317	178.014	-601.932	-422.283	11.029
2100	104.600	290.427	200.678	188.474	-599.409	-413.362	10.282
2200	104.600	295.293	204.869	198.934	-596.889	-404.562	9.606
2300	104.600	299.943	208.902	209.394	-594.367	-395.876	8.991
2400	104.600	304.394	212.789	219.854	-591.840	-387.301	8.429
2500	104.600	308.664	216.539	230.314	-589.305	-378.830	7.915
2600	104.600	312.767	220.162	240.774	-586.757	-370.461	7.443
2700	104.600	316.714	223.665	251.234	-584.194	-362.191	7.007
2800	104.600	320.519	227.057	261.694	-581.612	-354.016	6.604
2900	104.600	324.189	230.343	272.154	-579.007	-345.933	6.231
3000	104.600	327.735	233.531	282.614	-576.379	-337.941	5.884

PREVIOUS: CURRENT: June 1970

Cobalt Fluoride (CoF_2) $Co_1F_2(l)$

COBALT FLUORIDE (CoF_2) CRYSTAL-LIQUID M_r = 96.930006 Cobalt Fluoride (CoF_2) Co_1F_2(cr,l)

0 to 1400 K crystal
above 1400 K liquid

Refer to the individual tables for details.

Enthalpy Reference Temperature = T_r = 298.15 K; Standard State Pressure = p° = 0.1 MPa

T/K	C_p° (J K⁻¹mol⁻¹)	S° (J K⁻¹mol⁻¹)	$-[G°-H°(T_r)]/T$ (J K⁻¹mol⁻¹)	$H°-H°(T_r)$ (kJ mol⁻¹)	$\Delta_fH°$ (kJ mol⁻¹)	$\Delta_fG°$ (kJ mol⁻¹)	Log K_f
0	0.	0.	INFINITE	-12.459	-670.395	-670.395	INFINITE
100	33.492	25.002	134.693	-10.969	-672.334	-657.029	343.197
200	57.584	56.746	87.999	-6.250	-672.454	-641.555	167.557
298.15	68.909	82.026	82.026	0.	-671.532	-626.562	109.771
300	69.041	82.453	82.028	0.128	-671.508	-626.283	109.046
400	75.714	103.350	84.831	7.408	-670.016	-611.424	79.844
500	78.927	120.625	90.315	15.155	-668.372	-596.965	62.365
600	80.789	135.193	96.613	23.148	-666.746	-582.837	50.741
700	82.015	147.744	103.041	31.292	-665.190	-568.977	42.458
800	82.902	158.756	109.331	39.540	-664.152	-555.271	36.255
900	83.592	168.562	115.378	47.866	-662.824	-541.742	31.442
1000	84.161	177.400	121.145	56.254	-661.698	-528.351	27.598
1100	84.646	185.444	126.631	64.695	-660.813	-515.062	24.458
1200	85.077	192.828	131.844	73.182	-660.208	-501.841	21.845
1300	85.475	199.654	136.801	81.710	-660.032	-488.654	19.634
1400	85.839	206.002	141.519	90.276	-660.415	-475.460	17.740
1400.000	104.600	248.015	141.519	149.094	CRYSTAL <--> LIQUID TRANSITION		
1400.000	104.600	248.015	141.519	149.094			
1500	104.600	255.232	148.862	159.554	-599.094	-466.535	16.246
1600	104.600	261.983	155.724	170.014	-596.356	-457.786	14.945
1700	104.600	268.324	162.163	180.474	-593.537	-449.212	13.803
1800	104.600	274.303	168.228	190.934	-606.999	-440.510	12.783
1900	104.600	279.958	173.961	201.394	-604.461	-431.330	11.858
2000	104.600	285.324	179.396	211.854	-601.932	-422.283	11.029
2100	104.600	290.427	184.563	222.314	-599.409	-413.363	10.282
2200	104.600	295.293	189.487	232.774	-596.889	-404.562	9.606
2300	104.600	299.943	194.189	243.234	-594.367	-395.876	8.991
2400	104.600	304.394	198.688	253.694	-591.840	-387.301	8.429
2500	104.600	308.664	203.003	264.154	-589.305	-378.830	7.915
2600	104.600	312.767	207.146	274.614	-586.757	-370.461	7.443
2700	104.600	316.714	211.131	285.074	-584.194	-362.191	7.007
2800	104.600	320.519	214.971	295.534	-581.612	-354.016	6.604
2900	104.600	324.189	218.674	305.994	-579.007	-345.933	6.231
3000	104.600	327.735	222.250	316.454	-576.379	-337.941	5.884

PREVIOUS: CURRENT: June 1970

Cobalt Fluoride (CoF_2) Co_1F_2(cr,l)

COBALT FLUORIDE (CoF_2) IDEAL GAS M_r = 96.930006

$\Delta_f H°$(0 K) = -355.82 ± 12.6 kJ mol^{-1}

S°(298.15 K) = 278.04 ± 12.6 J K^{-1} mol^{-1} $\Delta_f H°$(298.15 K) = -356.48 ± 12.6 kJ mol^{-1}

Electronic Levels and Quantum Weights

State	ε_i, cm^{-1}	g_i
$^4\Sigma$	0	4
	[300]	[4]
	[4000]	[4]
	[9000]	[4]

Vibrational Frequencies and Degenracies

ν, cm^{-1}
[600](1)
[151](1)
762 (1)

Point Group: C_{2v} σ = 2

Bond Distance: Co-F = [1.72] Å

Bond Angle: F-Co-F = [165]°

Product of Moments of Inertia: $I_A I_B I_C$ = 6.57815 x 10^{-116} g^3 cm^6

Enthalpy of Formation

$\Delta_f H°$(CoF_2, g, 298.15 K) = -85.2±3.0 kcal mol^{-1} (-355.477±12.6 kJ mol^{-1}), is calculated from the enthalpy of formation of the crystal, $\Delta_f H°$(CoF_2, cr, 298.15 K) = -160.5±4.0 kcal mol^{-1}, and the enthalpy of sublimation, $\Delta_{sub}H°$(298.15 K) = 75.3±3.0 kcal mol^{-1}. The sublimation value is determined from a 3rd law analysis of the vapor pressure data for CoF_2 reported by Kana'an et al. (1). See the crystal table for details.

Heat Capacity and Entropy

Buchler et al. (2) investigated the deflection of a molecular beam of CoF_2(g) by an electric field and found that within the sensitivity of the apparatus the molecule was nonpolar and consequently possessed a linear structure. Hastie et al. (3) recently investigated the infrared absorption spectra of several transition metal difluorides through the use of matrix isolation techniques. Isoptopic shift measurements for matrix isolated NiF_2, ZnF_2, and CuF_2 in neon and argon indicated that these species were slightly bent. Based upon these results, the authors estiamted the F-Co-F angle as 165°. This value is adopted here. The bond length was estimated by Brewer et al. (4). The principal moments of inertia are: I_A = 0.1934 x 10^{-39}, I_B = 18.3484 x 10^{-39}, and I_C = 18.5417 x 10^{-39} g cm^2.

Hastie et al. (3) isolated CoF_2 in neon and argon matrices and observed the ν_3 fundamental at 762 cm^{-1}. They also calculated ν_1 = 600 cm^{-1} by a valence force field method and estimated ν_2 as 151 cm^{-1} by applying the variation in the frequencies for the chlorides of Co, Ni, Cu, and Zn to the fluorides. The electronic levels and quantum weights are estimated by comparison with the electronic energy levels for Co^{++}(g) ion (5) and those for $CoCl_2$(g) observed by Hougen et al. (6).

References

1. A. S. Kana'an, G. Besenbruch, and J. L. Margrave, J. Inorg. Nucl. Chem., 28, 1035 (1966).
2. A. Buchler, J. L. Stauffer, and W. Klemperer, J. Chem. Phys., 40, 3471 (1964).
3. J. W. Hastie, R. H. Hauge, and J. L. Margrave, High Temperature Science, 1, 76 (1969).
4. L. Brewer, G. R. Somayajulu, and E. Brackett, Chem. Rev., 63, 111 (1963).
5. C. E. Moore, U. S. Nat. Bur. Stand. Circ. 467, Volume II, 1952.
6. J. T. Hougen, G. E. Leroi, and T. C. James, J. Chem. Phys., 34, 1670 (1961).

Cobalt Fluoride (CoF_2) Co_1F_2(g)

Enthalpy Reference Temperature = T_r = 298.15 K Standard State Pressure = p° = 0.1 MPa

T/K	$C_p°$ (J $K^{-1}mol^{-1}$)	S° (J $K^{-1}mol^{-1}$)	-[G°-H°(T_r)]/T (J $K^{-1}mol^{-1}$)	H°-H°(T_r) (kJ mol^{-1})	$\Delta_f H°$ (kJ mol^{-1})	$\Delta_f G°$ (kJ mol^{-1})	Log K_f
0	0.	0.	INFINITE	-12.940	-355.821	-355.821	INFINITE
100	41.091	227.574	320.907	-9.333	-355.643	-360.595	188.355
200	47.569	258.314	282.621	-4.861	-356.010	-365.424	95.439
250	49.634	269.160	278.877	-2.429	-356.236	-367.752	76.838
298.15	51.210	278.042	278.042	0.	-356.477	-369.949	64.813
300	51.263	278.359	278.043	0.095	-356.486	-370.032	64.428
350	52.536	286.361	278.672	2.691	-356.741	-372.270	55.558
400	53.528	293.444	280.084	5.344	-357.024	-374.470	48.901
450	54.305	299.795	281.928	8.040	-357.343	-376.632	43.718
500	54.920	305.550	284.007	10.772	-357.700	-378.756	39.568
600	55.822	315.648	288.462	16.312	-358.528	-382.892	33.334
700	56.459	324.303	292.979	21.927	-359.500	-386.878	28.869
800	56.956	331.876	297.377	27.599	-361.038	-390.652	25.507
900	57.377	338.609	301.591	33.316	-362.319	-394.279	22.883
1000	57.753	344.674	305.601	39.073	-363.824	-397.752	20.776
1100	58.095	350.195	309.408	44.865	-365.588	-401.062	19.045
1200	58.410	355.263	313.021	50.691	-367.644	-404.198	17.594
1300	58.699	359.950	316.453	56.546	-370.141	-407.147	16.359
1400	58.964	364.310	319.718	62.430	-373.205	-409.882	15.293
1500	59.205	368.387	322.828	68.338	-375.255	-412.428	14.362
1600	59.424	372.215	325.796	74.270	-377.045	-414.847	13.543
1700	59.622	375.823	328.634	80.223	-378.734	-417.157	12.818
1800	59.801	379.236	331.351	86.194	-396.684	-419.075	12.161
1900	59.963	382.474	333.957	92.182	-398.618	-420.267	11.554
2000	60.108	385.554	336.461	98.186	-400.545	-421.356	11.005
2100	60.239	388.489	338.869	104.203	-402.465	-422.350	10.505
2200	60.356	391.295	341.189	110.233	-404.375	-423.252	10.049
2300	60.461	393.980	343.426	116.274	-406.272	-424.067	9.631
2400	60.555	396.555	345.586	122.325	-408.154	-424.800	9.246
2500	60.638	399.029	347.675	128.385	-410.019	-425.455	8.889
2600	60.712	401.409	349.696	134.452	-411.864	-426.036	8.559
2700	60.777	403.701	351.654	140.527	-413.686	-426.547	8.252
2800	60.834	405.912	353.553	146.608	-415.483	-426.990	7.966
2900	60.882	408.048	355.395	152.693	-417.253	-427.370	7.698
3000	60.924	410.113	357.185	158.784	-418.994	-427.689	7.447
3100	60.959	412.111	358.925	164.878	-420.705	-427.951	7.211
3200	60.988	414.047	360.617	170.975	-797.197	-426.385	6.960
3300	61.011	415.924	362.265	177.075	-797.416	-414.793	6.566
3400	61.029	417.746	363.870	183.177	-797.604	-403.196	6.194
3500	61.041	419.515	365.435	189.281	-797.761	-391.593	5.844
3600	61.050	421.235	366.961	195.385	-797.888	-379.986	5.513
3700	61.054	422.907	368.450	201.491	-797.986	-368.377	5.201
3800	61.054	424.536	369.905	207.596	-798.056	-356.765	4.904
3900	61.051	426.121	371.326	213.701	-798.101	-345.151	4.623
4000	61.044	427.667	372.715	219.806	-798.121	-333.536	4.356
4100	61.035	429.174	374.074	225.910	-798.119	-321.922	4.101
4200	61.023	430.645	375.404	232.013	-798.096	-310.307	3.859
4300	61.009	432.081	376.705	238.115	-798.055	-298.694	3.628
4400	60.992	433.483	377.980	244.215	-797.996	-287.082	3.408
4500	60.973	434.853	379.228	250.313	-797.924	-275.471	3.198
4600	60.953	436.193	380.452	256.409	-797.838	-263.862	2.996
4700	60.931	437.504	381.652	262.504	-797.742	-252.254	2.803
4800	60.908	438.787	382.829	268.595	-797.637	-240.649	2.619
4900	60.883	440.042	383.984	274.685	-797.526	-229.047	2.442
5000	60.858	441.272	385.118	280.772	-797.410	-217.446	2.272
5100	60.831	442.477	386.230	286.857	-797.291	-205.848	2.108
5200	60.804	443.658	387.323	292.938	-797.171	-194.252	1.951
5300	60.776	444.816	388.397	299.017	-797.052	-182.659	1.800
5400	60.747	445.951	389.453	305.094	-796.936	-171.068	1.655
5500	60.718	447.066	390.490	311.167	-796.825	-159.479	1.515
5600	60.689	448.160	391.510	317.237	-796.721	-147.891	1.379
5700	60.659	449.234	392.513	323.305	-796.625	-136.306	1.249
5800	60.629	450.288	393.501	329.369	-796.539	-124.723	1.123
5900	60.599	451.324	394.472	335.430	-796.420	-113.143	1.002
6000	60.569	452.343	395.428	341.489	-796.348	-101.563	0.884

PREVIOUS: June 1970 (1 atm) CURRENT: June 1970 (1 bar)

Cobalt Fluoride (CoF_2) Co_1F_2(g)

COBALT FLUORIDE (CoF_3) CRYSTAL M_r = 115.928409

Cobalt Fluoride (CoF_3) $Co_1F_3(cr)$

$S°(298.15\ K)$ = [94.6 ± 12.6] J K^{-1} mol^{-1}

T_{fus} = [1200 ± 200] K

$\Delta_f H°(0\ K)$ = Unknown

$\Delta_f H°(298.15\ K)$ = -790.36 ± 12.6 kJ mol^{-1}

$\Delta_{fus} H°$ = Unknown

Enthalpy Reference Temperature = T_r = 298.15 K							Standard State Pressure = p° = 0.1 MPa
	J $K^{-1}mol^{-1}$			kJ mol^{-1}			
T/K	$C_p°$	S°	$-[G°-H°(T_r)]/T$	$H°-H°(T_r)$	$\Delta_f H°$	$\Delta_f G°$	Log K_f
0							
100							
200							
298.15	91.797	94.558	94.558	0.	-790.358	-718.893	125.947
300	92.471	95.128	94.560	0.170	-790.320	-718.450	125.093
400	97.470	122.481	98.255	9.691	-788.197	-694.811	90.733
500	99.701	144.498	105.375	19.561	-786.113	-671.708	70.173
600	100.454	162.742	113.461	29.569	-784.209	-649.010	56.501
700	101.219	178.285	121.638	39.653	-782.489	-626.614	46.759
800	101.981	191.849	129.584	49.812	-781.344	-604.393	39.463
900	102.839	203.909	137.185	60.051	-779.930	-582.361	33.799
1000	103.868	214.796	144.410	70.386	-778.703	-560.477	29.276
1100	105.027	224.750	151.268	80.830	-777.675	-538.706	25.581
1200	106.248	233.940	157.779	91.393	-776.867	-517.019	22.505
1300	107.587	242.497	163.970	102.084	-776.412	-495.386	19.905
1400	108.935	250.519	169.869	112.910	-776.429	-473.771	17.677
1500	110.320	258.082	175.500	123.873	-775.330	-452.188	15.747

PREVIOUS: CURRENT: December 1970

Enthalpy of Formation

Fowler et al. (1) determined calorimetrically the enthalpy of fluorination of crystalline CoF_2 at 473.15 K. For the reaction 2 $CoF_2(cr)$ + $F_2(g)$ → 2 $CoF_3(cr)$ they reported $\Delta_r H°(473.15\ K)$ = -52.0±3.0 kcal mol^{-1} which yields $\Delta_r H°(298.15\ K)$ = -52.4±3.0 kcal mol^{-1} based upon our functions for $CoF_2(cr)$ and $CoF_3(cr)$. Combining this result with the enthalpy of formation of $CoF_2(cr)$ (2), we derive $\Delta_f H°(CoF_3, cr, 298.15\ K)$ = -186.7±3.0 kcal mol^{-1}.

Jessup et al. (3) investigated calorimetrically this same fluorination reaction and reported $\Delta_r H°(523.15\ K)$ = -56.1±2.0 kcal mol^{-1}. This result leads to $\Delta_f H°(CoF_3, cr, 298.15\ K)$ = -188.8±2.0 kcal mol^{-1}. In the same investigation Jessup et al. (3) reported $\Delta_r H°(608\ K)$ = -74.0±4.0 kcal mol^{-1} for the reduction reaction 2 $CoF_3(cr)$ + $H_2(g)$ → 2 $CoF_2(cr)$ + 2 HF(g). Corrected to 298.15 K we calculate $\Delta_r H°(298.15\ K)$ = -73.3±4.0 kcal mol^{-1}. Combining this result with enthalpy of formation data for $CoF_2(cr)$(2). and HF(g) (4), we derive $\Delta_f H°(CoF_3, cr, 298.15\ K)$ = -189.0±4.0 kcal mol^{-1}. We note that the addition of these two results by Jessup et al. (3) leads to $\Delta_r H°(298.15\ K)$ = -129.9±6.0 kcal mol^{-1} for the reaction $H_2(g)$ + $F_2(g)$ → 2 HF(g) which is quite consistent with the result -130.28±0.4 kcal mol^{-1} calculated from the selected value for the enthalpy of formation of HF(g) (4). Thus, it is apparent that these two results are probably quite reliable, and we adopted $\Delta_f H°(CoF_3, cr, 298.15\ K)$ = -188.9±3.0 kcal mol^{-1} (-790.358±12.6 kJ mol^{-1}).

Jellinek and Koop (5) reported equilibrium constants for the reduction reaction given above in the temperature range 470-595 K. 2nd and 3rd law analyses of these equilibrium data give $\Delta_r H°(298.15\ K)$ of 31.7 and 29.3±0.6 kcal mol^{-1}, respectively; the drift is -4.7±1.5 cal K^{-1} mol^{-1}. These values are in considerable disagreement with the calorimetric value of -73.3±4.0 kcal mol^{-1} given by Jessup et al. [3], and no weight is given to these equilibrium data.

Heat Capacity and Entropy

Heat capacity data for $CoF_3(cr)$ are estimated by comparison with those for $FeF_2(cr)$, $FeF_3(cr)$, $TiF_3(cr)$, and $CoF_2(cr)$. $S°(298.15\ K)$ is estimated to be 22.6 cal K^{-1} mol^{-1} (94.5584 J K^{-1} mol^{-1}) from the ionic entropy contributions given by Kubaschewski, Evans, and Alcock (6). We note that this method leads to $S°(298.15\ K)$ = 19.4 cal K^{-1} mol^{-1} for $CoF_2(cr)$ which is in good agreement with the experimental value of 19.60±0.10 cal K^{-1} mol^{-1} (2).

Fusion Data

T_{fus} is estimated by comparison with those for other transition metal halides. We note that Ruff and Ascher [7] reported $CoF_3(cr)$ being thermally unstable around 900 K, while Stewart (8) recently showed that $CoF_2(cr)$ was formed by hydrolytic reactions and not by thermal decomposition. The derived $\Delta_f G°$ values for $CoF_3(cr)$ indicate that it is thermodynamically more stable than $CoF_2(cr)$ in agreement with the findings of Stewart (8). In the temperature range 298-900 K, the Gibbs energy change for the decomposition reaction 2 $CoF_3(cr)$ → 2 $CoF_2(cr)$ + $F_2(g)$ are in the range +10-20 kcal mol^{-1}.

References

1. R. D. Fowler, W. B. Burford III, J. M. Hamilton, Jr., R. G. Sweet, C. E. Weber, J. S. Kasper, and I. Litant, Ind. Eng. Chem., 39, 292 (1947).
2. JANAF Thermochemical Tables: $CoF_2(cr)$, 6-30-70.
3. R. S. Jessup, F. G. Brickwedde, and M. T. Wechsler, J. Res. Nat. Bur. Stand. 44, 457 (1950).
4. JANAF Thermochemical Tables: HF(g), 12-31-68.
5. K. Jellinek and R. Koop, Z. Physik. Chem., 145, 305 (1929).
6. O. Kubaschewski, E. LL. Evans, and C. B. Alcock, "Metallurgical Thermochemistry," Fourth Ed., Pergamon Press Ltd., London, 1967.
7. O. Ruff and E. Ascher, Z. Anorg. Chem., 183, 193 (1929).
8. D. F. Stewart, unpublished results quoted in R. Colton and J. H. Canterford, "Halides of the First Row Transition Metals," John Wiley and Sons Ltd., London, 1969.

Cobalt Fluoride (CoF_3) $Co_1F_3(cr)$

COBALT OXIDE (CoO) CRYSTAL M_r = 74.9326

$S°(298.15\ K) = 52.993 \pm 0.34\ J\ K^{-1}\ mol^{-1}$

$T_{fus} = 2103 \pm 10\ K$

$\Delta_f H°(0\ K) = -238.1 \pm 0.4\ kJ\ mol^{-1}$

$\Delta_f H°(298.15\ K) = -237.7 \pm 0.4\ kJ\ mol^{-1}$

$\Delta_{fus} H°$ = Unknown

Enthalpy of Formation

Boyle et al. (1) measured the enthalpy of combustion of metallic cobalt at 303.16 K. Their analyses of the combustion products indicated that some overoxidation of the metal had occurred. Correction of their results to correspond to CoO was made on the assumption that the excess oxygen was combined as Co_3O_4. Also the analyses indicated that the combustion products had attacked the silica-glass capsules, used to hold the metal samples, resulting in appreciable amounts of silicate formation. Corrections for these two side reactions amounted to three percent of the total measured energy of combustion. From these data, we derive $\Delta_f H°$(CoO, cr, 298.15 K) = -57.0±0.3 kcal mol^{-1}. Roth and Havekoss (2) also investigated calorimetrically the combustion of metallic cobalt and reported $\Delta_f H°$(CoO, cr, 298.15 K) = -57.2±0.2 kcal mol^{-1}.

2nd and 3rd law analyses of several equilibrium studies (3-12), emf measurements (13), and mass-spectrometric studies (14) involving CoO(cr) are discussed in the JANAF Thermochemical Tables (18), $\Delta_f H°$(CoO, cr, 298.15 K) = -56.82±0.10 kcal mol^{-1} (-237.735±0.4 kJ mol^{-1}) was adopted. A recent critical evaluation by Cyr et al. (19) recommended $\Delta_f H°$(CoO, cr, 298.15 K) = -237.5±0.4 kJ mol^{-1}, based on 2nd and 3rd law analyses. Since this value differs from the value previously adopted by JANAF by only 0.2 kJ mol^{-1} and includes ten new references in the evaluation, it is adopted for the current table.

Heat Capacity and Entropy

The adopted heat capacities for CoO(cr) in the temperature range 51-298 K are those of King (15). These data have a pronounced maximum at 287.3 K, at which point the heat capacity is greater than 17.6 cal K^{-1} mol^{-1}. Assayag and Bizette (16) also measured heat capacities for CoO(cr) between 180-400 K and reported a maximum at 298.7 K. Presumably this anomaly is of the magnetic type. Heat capacity data above 300 K are calculated from the high temperature (400-1800 K) enthalpy data of King and Christensen (17). Both sets of data were smoothed by computer and joined at 298.15 K.

The value of S°(298.15 K) is obtained from the low temperature heat capacity dataof King (15) and is based on S°(51 K) = 0.42 cal K^{-1} mol^{-1}.

Fusion Data

The adopted T_{fus} of 2103±10 K was recommended by IUPAC [20] based on the work of Coutures [21].

References

1. B. J. Boyle, E. G. King, and K. C. Conway, J. Amer. Chem. Soc., 76, 3835 (1954).
2. W. A. Roth and H. Havekoss, Z. Anorg. Chem. 195, 239 (1931).
3. G. Chaudron, Ann. Chim. 16, 221 (1921).
4. L. Wohler and O. Balz, Z. Elektrochem. 27, 406 (1921).
5. P. H. Emmett and J. F. Shultz, J. Amer. Chem. Soc. 51, 3249 (1929).
6. Z. Shibata and I. Mori, Z. Anorg. Chem. 212, 305 (1933).
7. O. J. Kleppa, Svensk. Kem. Tid. 55, 18 (1943).
8. J. Aubry and C. Gleitzer, Bull. Soc. Chim. France 2086 (1960).
9. V. F. Balakirev and G. I. Chufarov, Doklady Akad. Nauk SSSR 138, 112 (1961).
10. R. Schenck and H. Wesselkoch, Z. Anorg. Chem. 184, 39 (1929).
11. M. Watanabe, Bull. Inst. Phys. Chem. Res. (Tokyo) 9, 676 (1930); C. A. 25:3224.
12. P. H. Emmett and J. F. Shultz, J. Amer. Chem. Soc. 52, 1782 (1930).
13. K. Kiukkola and C. Wagner, J. Electrochem. Soc. 104, 379 (1957).
14. R. T. Grimley, R. P. Burns, and M. G. Inghram, J. Chem. Phys. 45, 4158 (1966).
15. E. G. King, J. Amer. Chem. Soc. 79, 2399 (1957).
16. G. Assayag and H. Bizette, Compt. Rend. 239, 238 (1954).
17. E. G. King and A. U. Christensen, Jr., J. Amer. Chem. Soc. 80, 1800 (1958).
18. JANAF Thermochemical Tables, 2nd ed., NSRDS-NBS 37, 1971.
19. J. P. Cyr, J. Dellacherie, D. Balesdent, J. Chem. Eng. Data 26, 319 (1981).
20. J. Hlavac, Collaborative Study Group, IUPAC Commission on High Temperatures and Refractory Materials, Private communication, February 26, 1979.
21. J. P. Coutures, Thesis (1971).

Cobalt Oxide (CoO) $Co_1O_1(cr)$

Enthalpy Reference Temperature = T_r = 298.15 K Standard State Pressure = p° = 0.1 MPa

T/K	$C_p°$	S°	$-[G°-H°(T_r)]/T$	$H°-H°(T_r)$	$\Delta_f H°$	$\Delta_f G°$	Log K_f
	J $K^{-1}mol^{-1}$			kJ mol^{-1}			
0	0.	0.	INFINITE	-9.434	-238.056	-238.056	INFINITE
100	19.038	9.109	97.044	-8.794	-239.390	-230.867	120.593
200	45.376	30.904	58.385	-5.496	-239.458	-222.172	58.025
298.15	55.279	52.993	52.993	0.	-237.735	-213.988	37.490
300	54.831	53.333	52.994	0.102	-237.706	-213.840	37.233
400	52.932	68.698	55.099	5.439	-236.422	-206.088	26.912
500	53.932	80.628	59.052	10.788	-235.342	-198.632	20.751
600	54.312	90.497	63.495	16.201	-234.402	-191.381	16.661
700	54.513	98.883	67.966	21.642	-233.625	-184.275	13.751
800	54.819	106.180	72.297	27.107	-233.428	-177.207	11.570
900	55.312	112.664	76.428	32.612	-232.970	-170.209	9.879
1000	56.011	118.526	80.349	38.176	-232.709	-163.251	8.527
1100	56.886	123.904	84.068	43.820	-232.656	-156.309	7.423
1200	57.915	128.897	87.598	49.559	-232.827	-149.363	6.502
1300	59.070	133.578	90.957	55.407	-233.353	-142.390	5.721
1400	60.329	138.001	94.160	61.377	-234.348	-135.357	5.050
1500	61.672	142.208	97.224	67.476	-234.217	-128.288	4.467
1600	63.086	146.233	100.163	73.713	-233.707	-121.241	3.958
1700	64.559	150.102	102.987	80.095	-232.966	-114.234	3.510
1800	66.078	153.835	105.709	86.627	-248.354	-106.981	3.105
1900	67.639	157.449	108.337	93.312	-247.588	-99.147	2.726
2000	69.233	160.959	110.881	100.156	-246.676	-91.357	2.386
2100	70.856	164.376	113.347	107.160	-245.615	-83.617	2.080
2200	72.505	167.710	115.743	114.328	-244.401	-75.930	1.803
2300	74.174	170.970	118.074	121.662	-243.033	-68.303	1.551
2400	75.860	174.162	120.344	129.163	-241.508	-60.738	1.322
2500	77.563	177.294	122.560	136.834	-239.825	-53.240	1.112
2600	79.283	180.369	124.724	144.676	-237.981	-45.813	0.920
2700	81.011	183.394	126.841	152.691	-235.975	-38.459	0.744
2800	82.747	186.371	128.914	160.879	-233.806	-31.183	0.582
2900	84.496	189.305	130.946	169.241	-231.473	-23.987	0.432
3000	86.253	192.199	132.940	177.778	-228.974	-16.874	0.294

PREVIOUS: CURRENT: December 1970

Cobalt Oxide (CoO) $Co_1O_1(cr)$

COBALT SULFATE ($CoSO_4$) CRYSTAL (ALPHA - BETA) M_r = 154.9908

$\Delta_f H°(0\ K) = -878.95 \pm 1.3\ kJ\ mol^{-1}$
$S°(298.15\ K) = 117.377 \pm 4.2\ J\ K^{-1}\ mol^{-1}$ $\Delta_f H°(298.15\ K) = -888.26 \pm 1.3\ kJ\ mol^{-1}$
$T_{trs}\ (\alpha \rightarrow \beta) = 964 \pm 5\ K$ $\Delta_{trs}H° = 2.134 \pm 0.4\ kJ\ mol^{-1}$

Enthalpy of Formation

Adami and King (1) determined calorimetrically the enthalpies of solution of crystalline CoO and α-$CoSO_4$ in hydrochloric acid solution. From these measurements they derived the enthalpy for the reaction CoO(cr) + $H_2SO_4 \cdot 7.068\ H_2O(\ell)$ → $CoSO_4$(cr) + $8.068\ H_2O\ (\ell)$ as $\Delta_r H°(298.15\ K) = -14.280 \pm 0.110\ kcal\ mol^{-1}$. Combining this result with enthalpies of formation for CoO (2) $H_2SO_4 \cdot 7.068\ H_2O$ (3), and H_2O (3), we derive $\Delta_f H°(\alpha$-$CoSO_4$, cr, 298.15 K) = $-212.3 \pm 0.3\ kcal\ mol^{-1}(-888.263 \pm 1.3\ kJ\ mol^{-1})$ which is adopted.

The total pressures of SO_3(g), SO_2(g), and O_2(g) over $CoSO_4$(cr) in the temperature range 1100-1300 K have been determined by manometric (4-7) and transpiration (8) methods. Previous investigators (4-6) analyzed these data solely in terms of two equilibria, namely (A) $CoSO_4$(cr) = CoO(cr) + SO_3(g) and (B) SO_3(g) = SO_2(g) + 0.5 O_2(g). However, Warner (10) has shown that $CoSO_4$ takes part in three successive, univariant, decomposition equilibria as the temperature is increased. Co_3O_4 is the decomposition product at low temperatures (<1150 K), a mixture of Co_3O_4 and CoO at intermediate tempertures, and CoO at high temperatures. Second and third law analyses of those points involving only equilibria (A) and (B) are given below. The appearance of Co_3O_4 as a decomposition product was determined by comparing the oxygen partial pressures that were calculated from the total pressure data with those for the decomposition reaction (C) Co_3O_4(cr) ⇄ 3 CoO(cr) + 0.5 O_2(g) (2). Espelund and Flengas (9) measured the Gibbs energy of reaction (A) in a high temperature solid electrolytic cell. Here, no Co_3O_4 was formed over the entire temperature range (827-1133 K), since the oxygen pressures in the cell were too low to oxidize the CoO. Results of an analysis of these data are also given below:

Source	Reaction	Method	Data Points	T/K	$\Delta_r H°$(298.15 K), kcal mol^{-1} 2nd Law	3rd Law	Drift cal K^{-1} mol^{-1}	$\Delta_f H°$(298.15 K) kcal mol^{-1}
Marchal (4)	A	Manometric	17	1143-1303	54.5	60.4±0.5	4.9 ±0.6	-211.8±0.8
Schenck and Raub (5)	A	Manometric	14	1099-1243	68.5	60.9±0.7	-6.3 ±1.1	-212.3±1.0
Warner (6)	A	Manometric	2	1164-1175	56.9	59.3±0.3	2.1	-210.8±0.3
Ingraham (7)	A	Manometric	9	1144-1210	59.1	59.1±0.1	0.04±0.91	-210.6±0.4
Alcock et al. (8)	D**	Transpiration	8	1131-1217	-191.1	-166.8±1.4	20.2 ±5.5	-211.2±1.5
Espelund and Flengas (9)	A	emf	28	827-1177	59.5	59.5±0.3	0.08±0.22	-211.0±0.6

** Reaction: (D) 2 CoO(cr) + 2 SO_2(g) + O_2(g) = 2 $CoSO_4$(cr)

These results indicate that the Gibbs energy functions and equilibrium data are consistent within experimental error.

Heat Capacity and Entropy

The low temperature (52-298 K) heat capacities are those reported by Weller (11). No anomalies appear in these data. Heat capacities in the temperature range 300-2000 K are estimated by comparison with those for $CuSO_4$ (2).

$S°(298.15\ K)$ is determined from the C_p data based on $S°(51\ K) = 1.56 + 2.75\ cal\ K^{-1}\ mol^{-1}$. The 1.56 is a lattice contribution, while the 2.75 is the entropy associated with the ordering of the Co^{+2} spin moments. In assigning the magnetic entropy, it is assumed that all of the contribution remains to be extracted below the minimum temperature (52 K) of the heat capacity measurements (11).

Transition Data

The temperature (T_{trs}) and the heat ($\Delta_{trs}H°$) of the α-β transition for $CoSO_4$ are from the differential thermal analysis studies of Ingraham and Marier (12). Other temperatures reported for the transition are 890 (6) and 903 K (13).

Decomposition Data

T_{dcm} is the temperature at which the Gibbs energy change for the process $CoSO_4$(cr) → CoO(cr) + SO_3(g) is zero.

References
1. L. H. Adami and E. G. King, U. S. Bur. Mines RI 6617, 1965.
2. JANAF Thermochemical Tables: CoO(cr), 2-31-70; $CuSO_4$(cr), 6-30-66; SO_3(g), 9-30-65; SO_2(g), 6-30-61; Co_3O_4(cr), 12-31-71.
3. U. S. Nat. Bur. Stand. Tech. Note 270-3, 1968.
4. G. Marchal, J. Chim. Phys. 22, 559 (1925).
5. R. Schenck and E. Raub, Z. Anorg. Chem. 178, 225 (1929).
6. J. S. Warner, Trans. Met. Soc. AIME, 221, 591 (1961).
7. T. R. Ingraham, Can. Met. Quart. 3, 221 (1964).
8. C. B. Alcock, K. Sudo, and S. Zador, Trans. Met. Soc. AIME, 233, 655 (1965).
9. A. W. Espelund and S. N. Flengas, Can. J. Chem. 49, 1545 (1971).
10. J. S. Warner, Ph.D. Thesis, Columbia University, 1964.
11. W. W. Weller, U. S. Bur. Mines RI 6669, 1965.
12. T. R. Ingraham and P. Marier, Can. Met. Quart. 4, 169 (1965).
13. C. Duval and C. Wadier, Anal. Chim. Acta, 23, 541 (1960).

Cobalt Sulfate ($CoSO_4$) $Co_1O_4S_1$(cr)

Enthalpy Reference Temperature = T_r = 298.15 K — Standard State Pressure = p° = 0.1 MPa

T/K	$C_p°$ (J K^{-1}mol^{-1})	S°	$-[G°-H°(T_r)]/T$	$H°-H°(T_r)$ (kJ mol^{-1})	$\Delta_f H°$	$\Delta_f G°$	Log K_f
0	0.	0.	INFINITE	-17.238	-878.952	-878.952	INFINITE
100	44.454	37.748	192.584	-15.484	-884.218	-851.311	444.679
200	80.281	80.789	126.133	-9.069	-887.177	-817.096	213.403
298.15	103.219	117.377	117.377	0.	-888.263	-782.408	137.075
300	103.638	118.017	117.379	0.191	-888.269	-781.751	136.115
400	119.286	150.056	121.642	11.366	-890.202	-746.171	97.440
500	131.545	178.049	130.182	23.933	-890.418	-710.133	74.187
600	140.666	202.877	140.269	37.565	-889.584	-674.139	58.689
700	147.277	225.081	150.827	51.977	-888.064	-638.344	47.634
800	152.047	245.074	161.380	66.955	-886.571	-602.736	39.355
900	155.394	263.190	171.702	82.339	-937.599	-566.301	32.867
964.000	156.792	273.916	178.135	92.332	ALPHA <--> BETA		
964.000	157.221	276.129	178.135	94.466	TRANSITION		
1000	158.113	281.910	181.767	100.142	-932.148	-525.304	27.439
1100	160.415	297.089	191.571	116.071	-928.946	-484.775	23.020
1200	162.465	311.137	200.956	132.217	-925.920	-444.531	19.350
1300	164.306	324.215	209.941	148.556	-923.232	-404.528	16.254
1400	165.979	336.453	218.545	165.071	-921.020	-364.713	13.608
1500	167.569	347.959	226.793	181.749	-917.705	-325.090	11.321
1600	169.075	358.822	234.708	198.582	-914.048	-285.700	9.327
1700	170.498	369.115	242.314	215.561	-910.211	-246.545	7.575
1800	171.921	378.901	249.633	232.683	-922.559	-207.326	6.016
1900	173.259	388.232	256.684	249.942	-918.819	-167.692	4.610
2000	174.598	397.153	263.486	267.335	-915.002	-128.257	3.350

PREVIOUS: CURRENT: December 1971

Cobalt Sulfate ($CoSO_4$) $Co_1O_4S_1$(cr)

COBALT OXIDE (Co_3O_4) CRYSTAL M_r = 240.7972

$\Delta_f H°(0\ K) = -896.3 \pm 4.0\ kJ\ mol^{-1}$

$S°(298.15\ K) = 114.29 \pm 4.2\ J\ K^{-1}\ mol^{-1}$ $\Delta_f H°(298.15\ K) = -910.0 \pm 4.0\ kJ\ mol^{-1}$

Enthalpy of Formation

Bugden and Pratt (1) measured Gibbs energies of reaction for (a) 2Cu(cr) + Co_3O_4(cr) + 3CoO(cr) + Cu_2O(cr) and (b) Ni(cr) + Co_3O_4(cr) + 3CoO(cr) + NiO(cr) from high temperature solid electrolytic cells. These data were subjected to 2nd and 3rd law analyses with results summarized in the following table:

Source	Reaction	Method	T/K	No. Of Points	$\Delta_r H°(298.15\ K)$ kcal mol^{-1} 2nd law	3rd law	Drift cal K^{-1} mol^{-1}	$\Delta_f H°(298.15K)$*
Bugden and Pratt (1)	a	emf	900-1150	Eqn	5.1	6.3±0.2	1.1	-217.5±0.8
	b	emf	850-1100	Eqn	-12.0	-10.6±0.2	1.3	-217.1±0.8
Foote and Smith (3)	c	Manometric	1073-1243	6	34.5	47.2±1.4	10.6±1.3	-217.7±1.5
Watanabe (4)	c	Manometric	1128-1233	4	38.1	47.9±0.7	8.3±0.8	-218.4±1.0
Chufarov et al. (5)	c	Manometric	973-1173	4	27.9	46.4±2.9	17.0±0.6	-216.9±3.0
Balakirev and Chufarov (6)	c	Manometric	923-1173	5	33.8	46.1±2.5	11.3±2.5	-216.6±3.0
Roiter and Paladine (7)	c	Manometric	1139-1220	5	36.4	47.1±0.6	9.0±2.0	-217.6±1.0
Ingraham (8) Set I	c	Manometric	1101-1159	6	41.4	46.8±0.2	4.5±0.4	-217.3±0.5
Set II	c	Manometric	1086-1219	19	35.6	46.9±0.8	9.7±0.2	-217.4±1.0
O'Bryan and Parravano (9)	c	X-ray & Wt. Loss	1073-1243	Eqn	40.3	47.3±0.8	5.9	-217.8±1.0

*3rd law values based on $\Delta_f H°$(CoO,cr, 298.15 K) = -56.82 ± 0.1 kcal mol^{-1}(2).

Both sets of emf data are much more compatible with the adopted functions than the dissociation pressure data (see discussion below). The results obtained with the Cu/Cu_2O reference couple are judged to be more reliable and lead to the adopted enthalpy of formation, $\Delta_f H°(Co_3O_4$,cr, 298.15 K) = -217.5±1.0 kcal mol^{-1}, when used in conjunction with JANAF (2) enthalpies of formation for CoO and Cu_2O.

Equilibrium oxygen pressures for the reaction (c) Co_3O_4(cr) = 3CoO(cr) + 0.5 O_2(g) have been determined by numerous investigators (3-9). 2nd and 3rd law analyses of these data are given in the above table. Within each set the 2nd and 3rd law $\Delta_r H°$(298.15 K) values are not in agreement, and all sets contain large positive drifts. Warner (10) has suggested that these discrepancies arise as a result of an error in the measured standard entropy (11) for Co_3O_4. However, once the magnetic contribution is included (see entropy write-up) the real error in this value is probably no greater than one cal K^{-1} mol^{-1}. The other possible source of error is lack of true equilibrium conditions. Recent thermogravimetric (9), X-ray (9), and kinetic (9, 12) studies of the Co_3O_4-CoO-O_2 system all point to an irreversible reaction. Therefore, no weight has been given to the dissociation pressure data in our evaluation.

Heat Capacity and Entropy

The low temperature heat capacities (54-296.3 K) are those reported by King (11). Heat capacities in the temperature range 300-1000 K are calculated from the high temperature enthalpy data of King and Christensen (13); a constrained curve fitting technique was used to join smoothly these data with the low temperature data at 298 K. No anomalies appear in either set of data.

Roth (14) observed a maximum in the magnetic susceptibility of Co_3O_4 at 40 K, while Mossbauer studies by Kundig et al. (15) indicated a Neel temperature of 33.0±1.0 K. With the assumption that Co_3O_4 is a normal 2-3 spinel (16), this transition can be associated with the antiferromagnetic ordering of the Co^{+2} ion spin moments. Therefore, the entropy is based on S°(51 K) = 1.36 + 2.75 = 4.11 cal K^{-1} mol^{-1}, where 1.36 is a lattice contribution and 2.75 is the magnetic entropy. In assigning the magnetic entropy, it is assumed that all of the contribution remains to be extracted below the minmum temperature (54 K) of the heat capacity measurements.

Decomposition Data

T_{dcm} is the temperature at which the Gibbs energy change for the process Co_3O_4(cr) + 3CoO(cr) + 0.5 O_2(g) is zero.

References

1. W. G. Bugden and J. N. Pratt, Inst. Mining Met., Trans., Sect. C., 79, 221 (1970).
2. JANAF Thermochemical Tables; CoO(cr), 12-31-70; Cu_2O(cr), 6-30-66.
3. H. W. Foote and E. K. Smith, J. Amer. Chem. Soc., 30, 1344 (1908).
4. M. Watanable, Bull. Inst. Phys. Chem. Research (Tokyo), 12, 255 (1933); CA 27 2617.
5. G. I. Chufarov, M. G. Zhuravleva, and E. P. Tatievskaya, Dokl. Akad. Nauk SSSR, 73, 1209 (1950).
6. V. F. Balakirev and G. I. Chufarov, Dokl. Akad. Nauk SSSR, 138, 112 (1961).
7. B. D. Roiter and A. E. Paladino, J. Amer. Soc., 45, 128 (1962).
8. T. R. Ingraham, Can. Met. Quart., 3, 221 (1964).
9. H. M. O'Bryan, Jr., and G. Parravano in "Reactivity of Solids", G. M. Schwab, Ed., Elsevier Publishing Co., Amsterdam, 1965.
10. J. S. Warner, Ph.D. Thesis, Columbia University, 1964.
11. E. G. King, J. Amer. Chem. Soc., 79, 2399 (1957).
12. A. R. Reti, P. L. T. Brian, and L. C. Hoagland, Ind. Eng. Chem., Process Design Develop., 5, 171 (1966).
13. E. G. King and A. U. Christensen, Jr., J. Amer. Chem. Soc., 80, 1800 (1958).
14. W. L. Roth, J. Phys. Chem. Solids, 25, 1 (1964).
15. W. Kundig, M. Kobelt, H. Appel, G. Constabaris, and R. H. Linduist, J. Phys. Chem. Solids, 30, 819 (1969)
16. P. Cossee, J. Inorg. Nucl. Chem., 8, 483 (1958).

Cobalt Oxide (Co_3O_4) Co_3O_4(cr)

Enthalpy Reference Temperature = T_r = 298.15 K; Standard State Pressure = p° = 0.1 MPa

T/K	$C_p°$ (J $K^{-1}mol^{-1}$)	S°	$-[G°-H°(T_r)]/T$	$H°-H°(T_r)$ (kJ mol^{-1})	$\Delta_f H°$	$\Delta_f G°$	Log K_f
0	0.	0.	INFINITE	-17.968	-896.309	-896.309	INFINITE
100	34.368	31.487	198.112	-16.663	-902.377	-868.560	453.689
200	87.595	72.039	124.547	-10.501	-907.767	-832.428	217.408
298.15	123.051	114.286	114.286	0.	-910.020	-794.871	139.258
300	123.512	115.049	114.289	0.228	-910.038	-794.157	138.275
400	142.666	153.657	119.423	13.694	-910.220	-755.468	98.654
500	152.850	186.590	129.647	28.471	-909.774	-716.825	74.886
600	162.808	215.325	141.580	44.247	-908.999	-678.302	59.051
700	173.628	241.224	153.992	61.062	-907.802	-639.941	47.753
800	185.272	265.161	166.410	79.001	-907.337	-601.601	39.280
900	197.552	287.688	178.645	98.139	-905.043	-563.516	32.706
1000	210.296	309.158	190.630	118.528	-902.293	-525.712	27.460
1100	223.384	329.814	202.350	140.210	-899.139	-488.203	23.183
1200	236.727	349.821	213.809	163.214	-895.639	-450.998	19.631
1300	250.262	369.303	225.024	187.562	-892.206	-414.087	16.638
1400	263.943	388.349	236.012	213.272	-889.197	-377.422	14.082
1500	277.742	407.029	246.793	240.355	-881.838	-341.108	11.878

PREVIOUS: December 1970 CURRENT: December 1971

Cobalt Oxide (Co_3O_4) Co_3O_4(cr)

Part II contains the continuation
of the *JANAF Thermochemical Tables*
on pp. 927–1856.